ERRATA
ASM Handbook- Volume 7

Thank you for your order of ASM Handbook, Volume 7, Powder Metal Technologies and Applications.

Please note an error in the byline for the article "History of Powder Metallurgy" on page 3. The byline should read:
"Revised by Donald G. White,
Metal Powder Industries Federation and APMI International"

The corrected byline in typeset form is given below for pasting over the current byline.

Revised by Donald G. White, Metal Powders Industries Federation and APMI International

Contents

ASM Handbook®

Volume 7
Powder Metal Technologies and Applications

Prepared under the direction of the
ASM International Handbook Committee

ASM INTERNATIONAL®
The Materials
Information Society

First printing, December 1998

Library of Congress Cataloging-in-Publication Data

ASM International

ASM handbook.
Vols. 1–2 have title: Metals handbook.
Includes bibliographical references and indexes.
Contents: v. 1. Properties and selection—irons,
steels, and high-performance alloys—v. 2. Properties
and selection—nonferrous alloys and special-purpose
materials—[etc.]—v. 7. Powder metal technologies and applications

1. Metals—Handbooks, manuals, etc. 2. Metal-work—
Handbooks, manuals, etc. I. ASM International. Handbook
Committee. II. Metals Handbook.
TA459.M43 1990 620.1'6 90-115
SAN 204-7586

ISBN 0-87170-387-4

ASM International®
Materials Park, OH 44073-0002

Printed in the United States of America

Foreword

In recognition of the ongoing development and growth of powder metallurgy (P/M) materials, methods, and applications, ASM International offers the new Volume 7 of *ASM Handbook*. *Powder Metal Technologies and Applications* is a completely revised and updated edition of *Powder Metallurgy,* Volume 7 of the 9th Edition *Metals Handbook,* published in 1984. This new volume provides comprehensive updates that reflect the continuing improvements in traditional P/M technologies as well as significant new coverage of emerging P/M materials and manufacturing methods.

The ASM Handbook Committee, the editors, the authors, and the reviewers have collaborated to produce a book that meets the high technical standards of the *ASM Handbook* series. In addition to in-depth articles on production, testing and characterization, and consolidation of powders, the new volume expands coverage on the performance of P/M materials, part shaping methods, secondary operations, and advanced areas of engineering research such as process modeling. This extensive coverage is designed to foster increased awareness of the current status and potential of P/M technologies. To all who contributed toward the completion of this task, we extend our sincere thanks.

Alton D. Romig, Jr.
President
ASM International

Michael J. DeHaemer
Managing Director
ASM International

Policy on Units of Measure

By a resolution of its Board of Trustees, ASM International has adopted the practice of publishing data in both metric and customary U.S. units of measure. In preparing this Handbook, the editors have attempted to present data in metric units based primarily on Système International d'Unités (SI), with secondary mention of the corresponding values in customary U.S. units. The decision to use SI as the primary system of units was based on the aforementioned resolution of the Board of Trustees and the widespread use of metric units throughout the world.

For the most part, numerical engineering data in the text and in tables are presented in SI-based units with the customary U.S. equivalents in parentheses (text) or adjoining columns (tables). For example, pressure, stress, and strength are shown both in SI units, which are pascals (Pa) with a suitable prefix, and in customary U.S. units, which are pounds per square inch (psi). To save space, large values of psi have been converted to kips per square inch (ksi), where 1 ksi = 1000 psi. The metric tonne ($kg \times 10^3$) has sometimes been shown in megagrams (Mg). Some strictly scientific data are presented in SI units only.

To clarify some illustrations, only one set of units is presented on artwork. References in the accompanying text to data in the illustrations are presented in both SI-based and customary U.S. units. On graphs and charts, grids corresponding to SI-based units usually appear along the left and bottom edges. Where appropriate, corresponding customary U.S. units appear along the top and right edges.

Data pertaining to a specification published by a specification-writing group may be given in only the units used in that specification or in dual units, depending on the nature of the data. For example, the typical yield strength of steel sheet made to a specification written in customary U.S. units would be presented in dual units, by the sheet thickness specified in that specification might be presented only in inches.

Data obtained according to standardized test methods for which the standard recommends a particular system of units are presented in the units of that system. Wherever feasible, equivalent units are also presented. Some statistical data may also be presented in only the original units used in the analysis.

Conversions and rounding have been done in accordance with ASTM Standard E 380, with attention given to the number of significant digits in the original data. For example, an annealing temperature of 1570 °F contains three significant digits. In this case, the equivalent temperature would be given as 855 °C; the exact conversion to 854.44 °C would not be appropriate. For an invariant physical phenomenon that occurs at a precise temperature (such as the melting of pure silver), it would be appropriate to report the temperature as 961.93 °C or 1763.5 °F. In some instances (especially in tables and data compilations), temperature values in °C and °F are alternatives rather than conversions.

The policy of units of measure in this Handbook contains several exceptions to strict conformance to ASTM E 380; in each instance, the exception has been made in an effort to improve the clarity of the Handbook. The most notable exception is the use of g/cm^3 rather than kg/m^3 as the unit of measure for density (mass per unit volume).

SI practice requires that only one virgule (diagonal) appear in units formed by combination of several basic units. Therefore, all of the units preceding the virgule are in the numerator and all units following the virgule are in the denominator of the expression; no parentheses are required to prevent ambiguity.

Preface

On behalf of the ASM Handbook Committee, it is a pleasure to introduce this fully revised and updated edition of Volume 7, *Powder Metal Technologies and Applications* as part of the *ASM Handbook* series. Since the first publication of Volume 7 in 1984 as part of the 9th Edition *Metals Handbook,* substantial new methods, technologies, and applications have occurred in powder metallurgy. These developments reflect the continuing growth of powder metallurgy (P/M) as a technology for net-shape fabrication, new materials, and innovative manufacturing processes and engineering practices.

Net-shape or near-net-shape fabrication is a key objective in many P/M applications. Many factors influence the economics and performance of P/M fabrication, and new methods and process improvements are constantly considered and developed. In this regard, the new Volume 7 provides completely updated information on several emerging technologies for powder shaping and consolidation. Examples include all new articles on powder injection molding, binder assisted extrusion, warm compaction, spray forming, powder extrusion, pneumatic isostatic forging, field activated sintering, cold sintering, and the consolidation of ultrafine and nanocrystalline materials. New articles also cover process modeling of injection molding, isostatic pressing, and rigid die compaction.

Traditional press-and-sinter fabrication and high-density consolidation remain the major topic areas in the new Volume 7. This coverage includes new articles in several practical areas such as resin impregnation, dimensional control, machining, welding, heat treatment, and metallography of P/M materials. The traditional processes of rigid die compaction and sintering are also covered extensively with several updated articles on major production factors such as tooling, die design, compressibility and compaction, sintering practices, and atmosphere control. An overview article, "Powder Shaping and Consolidation Technologies," also compares and summarizes the alternatives and factors that can influence the selection of a P/M manufacturing method. Coverage is also expanded on high-density consolidation and high-performance P/M materials such as powder forged steels.

Multiple articles on powder production and characterization methods have also been revised or updated in several key areas. The article on atomization is fully revised from the previous edition, and several new articles have been added to the Section "Powder Characterization and Testing." In particular, the new article by T. Allen, "Powder Sampling and Classification," is a key addition that provides essential information for accurate characterization of particle size distributions. The variability of sieve analysis is also covered in more detail in this new Volume.

The new Volume 7 also provides detailed performance and processing information on a wide range of advanced and conventional P/M materials. Ferrous P/M materials are covered in several separate articles, and more detailed information on corrosion, wear, fatigue, and mechanical properties are discussed in separate articles. New articles also provide information on several advanced materials such as aluminum-base composites and reactive-sintered intermetallics.

This extensive volume would not have been possible without the guidance of the section editors and the dedicated efforts of the contributing authors. I would also like to thank Erhard Klar for organizing the previous edition, which formed the core for the structure of the new edition.

Finally, special thanks are extended to ASM staff—particularly to project editor Steve Lampman—for their dedicated efforts in developing and producing this Volume.

Peter W. Lee
The Timken Company
Member, ASM Handbook Committee

Officers and Trustees of ASM International (1997–1998)

Members of the ASM Handbook Committee (1997–1998)

Previous Chairmen of the ASM Handbook Committee

Contributors and Reviewers

Stanley Abkowitz
Dynamet Technology

Samuel Allen
Massachusetts Institute of Technology

Terry Allen

David E. Alman
U.S. Department of Energy
Albany Research Center

Sundar Atre
Pennsylvania State University

Christopher Avallone
International Specialty Products

Satyajit Banerjee
Breed Technologies Inc.

J. Banhart
Fraunhofer Institute

Daniel Banyash
Dixon Ticonderoga Company

Tim Bell
DuPont Company

David Berry
OMG Americas

Ram Bhagat
Pennsylvania State University

Pat Bhave
Thermal Technology Inc.

Sherri Bingert
Los Alamos National Laboratory

Jack Bonsky
Advanced Manufacturing Center
Cleveland State University

Robert Burns
Cincinnati Incorporated

Donald Byrd
Wyman Gordon Forgings

John Carson
Jenike and Johanson Inc.

Francois Chagnon
Quebec Metal Powders

Tom Chirkot
Patterson-Kelley Company
Harsco Corporation

Stephen Claeys
Pyron Corporation

John Conway
Crucible Compaction Metals

Kevin Couchman
Sinter Metals Inc.
Pennsylvania Pressed Metals Division

F. Robert Dax
Concurrent Technologies Corporation

Amedeo deRege
Domfer Metal Powders

R. Doherty
Drexel University

Ian Donaldson
Presmet

Carl Dorsch
Latrobe Steel Company

John Dunkley
Atomising Systems Ltd.

William Eisen
Crucible Research

Mark Eisenmann
Moft Metallurgical Corporation

Victor Ettel
Inco Technical Services Limited

Daniel Eylon
University of Dayton

Zhigang Fang
Smith International

B. Lynn Ferguson
Deformation Control Technology Inc.

Howard Ferguson
Metal Powder Products Inc.

Richard Fields
National Institute of Standards and Technology

Gavin Freeman
Sherritt International Corporation

Sam Froes
University of Idaho

Randall German
Pennsylvania State University

Herbert Giesche
Alfred University
New York State College of Ceramics

Howard Glicksman
DuPont Company

Kinyon Gorton
Caterpillar Inc.

Mark Greenfield
Kennametal Inc.

Joanna Groza
University of California at Davis

E.Y. Gutmanas
Technion—Israel Institute of Technology

Richard Haber
Rutgers University

Jack A. Hammill, Jr.
Hoeganaes Corporation

Francis Hanejko
Hoeganaes Corporation

John Hebeisen
Bodycotte, IMT

Ralph Hershberger
UltraFine Powder Technology Inc.

Gregory Hildeman
Alcoa Technical Center

Craig Hudson
Sinter Metals Inc.

Ronald Iacocca
Pennsylvania State University

M.I. Jaffe

W. Brian James
Hoeganaes Corporation

John Johnson
Howmet Corporation

Brian Kaye
Laurentian University

Pat Kenkel
Burgess-Norton Manufacturing Company

Mark Kirschner
BOC Gases

Erhard Klar

Richard Knight
Drexel University

Walter Knopp
P/M Engineering & Consulting

John Kosco
Keystone Powdered Metal Company

Sriram Krishnaswami
MARC Analysis Research Corporation

David Krueger
BASF Corporation

Howard Kuhn
Concurrent Technologies Corporation

Chaman Lall
Sinter Metals Inc.

Larry Lane
Brush Wellman Inc.

Alan Lawley
Drexel University

Jai-Sung Lee
Hanyang University

Peter Lee
The Timken Company

Louis W. Lherbier
Dynamet Inc.

Deepak Madan
F.W. Winter Inc. & Company

Craig Madden
Madden Studios

Gary Maddock
Carpenter Technology Corporation

Dan Marantz
Flame Spray Industries Inc.

Alain Marcotte
U.S. Bronze Powders

James Marder
Brush Wellman

Millard S. Masteller
Carpenter Technology Corporation

Ian Masters
Sherritt International Corporation

Paul E. Matthews

Steve Miller
Nuclear Metals Inc.

Wojciech Misiolek
Lehigh University

John Moll
Crucible Research Center

In-Hyung Moon
Hanyang University

Ronald Mowry
C.I. Hayes Inc.

Kenneth Moyer
Magna Tech P/M Labs

Charles Muisener
Loctite Corporation

Zuhair Munir
University of California at Davis

Anil Nadkarni
OMG Americas

K.S. Narasimhan
Hoeganaes Corporation

Ralph Nelson
DuPont Company

Bernard North
Kennametal Inc.

W. Glen Northcutt
Lockheed Martin

Scott Nushart
ATM Corporation

James Oakes
Teledyne Advanced Materials

Barbara O'Neal
AVS Inc.

Lanny Pease III
Powder Tech Associates Inc.

Kenneth Pinnow

Michael Pohl
Horiba Laboratory Products

John Porter
Cincinnati Inc.

Peter Price

Tom Prucher
Burgess-Norton Manufacturing Company

David Pye
Pye Metallurgical Consulting Inc.

Thomas Reddoch
Ametek Inc.

John Reinshagen
Ametek Inc.

Melvin Renowden
Air Liquide America

Frank Rizzo
Crucible Compaction Metals

Prasan Samal
OMG Americas

Howard Sanderow
Management and Engineering Technologies

G. Sathyanarayanan
Lehigh University

Barbara Shaw
Pennsylvania State University

Haskell Sheinberg
Los Alamos National Laboratory

George Shturtz
Carbon City Products

John Simmons
B.I. Thortex

Ronald Smith
Drexel University

Richard Speaker
Air Liquide America

Robert Sprague
Consultant

Victor Straub
Keystone Carbon Company

C. Suryanarayana
Colorado School of Mines

Bruce Sutherland
Westaim Corporation

Rajiv Tandon
Phillips Origen Powder Metallurgy

Pierre Taubenblat
Promet Associates

Mark Thomason
Sinterite Furnace Division
Gasbarre Products Inc.

Juan Trasorras
Federal Mogul

Yves Trudel
Quebec Metal Powders Limited

John Tundermann
Inco Alloys International Inc.

Christian Turner
Hasbro Inc.

William Ullrich
AcuPowder Int.

Arvind Varma
University of Notre Dame

Jack T. Webster
Webster-Hoff Corporation

Bruce Weiner
Brookhaven Instruments

Greg West
National Sintered Alloys

Donald White
Metal Powder Industries Federation

George White
BOC Gases

Eric Whitney
Pennsylvania State University

Jeff Wolfe
Kennametal Inc.

John Wood
University of Nottingham

C. Fred Yolton
Crucible Materials

Antonios Zavaliangos
Drexel University

Robert Zimmerman
Arburg Inc.

Contents

Introduction

History of Powder Metallurgy

Revised by Charles White, Metal Powder Industries Federation and APMI International

POWDER METALLURGY has been called a lost art. Unlike clay and other ceramic materials, the art of molding and firing practical or decorative metallic objects was only occasionally applied during the early stages of recorded history. Sintering of metals was entirely forgotten during the succeeding centuries, only to be revived in Europe at the end of the 18th century, when various methods of platinum powder production were recorded (Table 1).

Metal powders such as gold, copper, and bronze, and many powdered oxides (particularly iron oxide and other oxides used as pigments), were used for decorative purposes in ceramics, as bases for paints and inks, and in cosmetics since the beginnings of recorded history. Powdered gold was used to illustrate some of the earliest manuscripts. It is not known how these powders were produced, but it is possible that some of the powders were obtained by granulation after the metal was melted. Low melting points and resistance to oxidation (tarnishing) favored such procedures, especially in the case of gold powder. The use of these powders for pigments and ornamental purposes is not true powder metallurgy, because the essential features of the modern art are the production of powder and its consolidation into a solid form by the application of pressure and heat at a temperature below the melting point of the major constituent.

Early man learned by chance that particles of metal could be joined together by hammering, resulting in a solid metallic structure. In time, man learned how to build furnaces and develop temperatures high enough to melt and cast metals and to form lower melting alloys, such as copper and tin to make bronze.

Earliest Developments

Long before furnaces were developed that could approach the melting point of metal, P/M principles were used. About 3000 B.C., the Egyptians used a "sponge iron" for making tools. In this early process, iron oxide was heated in a charcoal and crushed shell fire, which was intensified by air blasts from bellows to reduce the oxide to a spongy metallic iron. The resulting hot sponge iron was then hammered to weld the particles together. Final shapes were obtained by simple forging procedures. Although the product often contained large amounts of nonmetallic impurities, some remarkably solid and sound structures have been discovered (Ref 1).

W.D. Jones (Ref 2) wrote of a process modification developed by African tribes. After reduction, the sponge was broken into powder particles, washed, and sorted by hand to remove as much of the slag and gangue as possible. The powder was then either compacted or sintered into a porous material, which was subsequently forged. Another example of ancient reduction of iron oxide was carried out in the fabrication of the Delhi Pillar, which weighs 5.9 metric tons (6.5 tons).

These crude forms of powder metallurgy ultimately led to the development of one of the commercial methods for producing iron powder. By grinding the sponge iron into fine particles, and heating in hydrogen to remove oxides and anneal or soften the particles, this process is today a viable technique for producing high-quality iron powder.

Powder metallurgy practices were used by the Incas and their predecessors in making platinum before Columbus made his voyage to the "New World" in 1492. The technique used was based on the cementing action of a lower melting binder, a technique similar to the present practice of making sintered carbides.

The technique consisted of cementing platinum grains (separated from the ore by washing and selection) by the addition of an oxidation-resistant gold-silver alloy of a fairly low melting point to wet the grains, drawing them together by surface tension and forming a raw ingot suitable for further handling (Ref 3).

A color change from the yellow of the sintered material to the whitish platinum of the final metal was caused by diffusion during heating prior to working. Heating is thought to have been accomplished by means of charcoal fires fanned by blowpipes. Analyses of these alloys vary considerably. The platinum content ranged from 26 to 72%, and the gold content ranged from 16 to 64%. Silver additions were found to vary from 3 to 15%, and amounts of copper up to 4% were traced.

Powder Metallurgy of Platinum

The metallurgy of platinum, as practiced in the 18th and 19th centuries in Europe, is considered

Table 1 Major historical developments in powder metallurgy

Date	Development	Origin
3000 B.C.	"Sponge iron" for making tools	Egypt, Africa, India
A.D. 1200	Cementing platinum grains	South America (Incas)
1781	Fusible platinum-arsenic alloy	France, Germany
1790	Production of platinum-arsenic chemical vessels commercially	France
1822	Platinum powder formed into solid ingot	France
1826	High-temperature sintering of platinum powder compacts on a commercial basis	Russia
1829	Wollaston method of producing compact platinum from platinum sponge (basis of modern P/M technique)	England
1830	Sintering compacts of various metals	Europe
1859	Platinum fusion process	
1870	Patent for bearing materials made from metal powders (forerunner of self-lubricating bearings)	United States
1878–1900	Incandescent lamp filaments	United States
1915–1930	Cemented carbides	Germany
Early 1900s	Composite metals	United States
	Porous metals and metallic filters	United States
1920s	Self-lubricating bearings (used commercially)	United States
1940s	Iron powder technology	Central Europe
1950s and 1960s	P/M wrought and dispersion-strengthened products, including P/M forgings	United States
1970s	Hot isostatic pressing, P/M tool steels, and superplastic superalloys	United States
1980s	Rapid solidification and powder injection molding technology	United States
1990s	Intermetallics, metal-matrix composites, spray forming, nanoscale powders, and warm compaction	United States, England

to be one of the most important stages of development for modern powder metallurgy. For the first time, complete records were available that provided insight into the various methods of powder production and the processing of these powders into solid, useful implements.

Between 1750 and 1825, considerable attention was given to the manufacture of platinum. In 1755, Lewis (Ref 4) discovered that when a lead-platinum alloy was oxidized at high temperatures, a spongy, workable mass remained after lead oxide impurities had been volatilized. Scheffer (Ref 5) found that when platinum was heated with arsenic, the platinum showed signs of melting. This finding was confirmed in 1781 by Achard (Ref 6), who described the production of a fusible platinum-arsenic alloy, probably by forming the eutectic containing 87% Pt and melting at 600 °C (1110 °F). Achard formed solid platinum by hot hammering a sponge, welding the individual particles into a large solid. The sponge was obtained by high-temperature working of the platinum-arsenic alloy, which caused volatilization of the arsenic.

This procedure formed the basis for a method of producing platinum that was first used in about 1790 in commercially manufactured chemical vessels by Jannetty in Paris. Mercury was used later in a similar process by von Mussin-Puschkin (Ref 7). Other metals worked in this way include palladium, by using sulfur instead of arsenic, and iridium (using phosphorus). Ridolfi (Ref 8) made malleable platinum for chemical vessels using lead and sulfur.

In 1786, Rochon (Ref 9) successfully produced solid platinum without using arsenic by welding small pieces of scrap platinum. He produced malleable platinum by uniting purified platinum grains.

Knight (Ref 10) found that if chemically precipitated platinum powder was heated at high temperatures in a clay crucible, it softened and could be compressed and forged. Tilloch (Ref 11) put platinum powder into tubes made of rolled platinum sheet, which were then heated and forged to produce a compact mass. In 1813, Leithner (Ref 12) reported production of thin, malleable platinum sheets by drying out successive layers of powder suspended in turpentine and heating the resulting films at high temperatures without pressure.

In 1882, a French process was reported by Baruel (Ref 13), in which 14 kg (30 lb) of platinum powder was made into a solid ingot by a series of operations. Platinum was precipitated in powdered form, slightly compressed in a crucible, and heated to white heat. The powder was then put in a steel matrix and put under pressure with a screw coining press. The compact platinum was repeatedly reheated and re-pressed until a solid ingot was formed. The final heat treatments were made in a charcoal fire at lower temperatures. Because the platinum powder was placed in the steel die while hot, this process was based on the hot pressing technique.

In Russia in 1826, a high-temperature sintering operation was applied to previously compressed powder compacts on a commercial basis for the first time. This was in contrast to methods based on hot pressing. Sobolewskoy (Ref 14) described sifted platinum powder pressed into a cast iron cylinder that featured a steel punch actuated by a screw press. The resulting compacts were annealed for 1½ days at high temperature in a porcelain firing kiln. The final product was highly workable, especially if the platinum powder had been well washed and was of high purity. Annealing, however, caused a decrease in volume; a cylinder 100 mm (4 in.) in diameter and 19 mm (¾ in.) in height shrank 19 mm (¾ in.) and 6 mm (¼ in.) in these dimensions, respectively.

Another Russian method was reported by Marshall (Ref 15) in 1832. Platinum powder in a ring-shaped iron mold was pressed by a screw press, heated to a red heat, and re-pressed. After working in a rolling mill, the compacted discs were used as coins.

The Wollaston process of producing compact platinum from platinum sponge powder is generally considered the beginning of modern powder metallurgy. At least 16 years prior to his publication of 1829 (Ref 16), describing the manufacture of a product much superior to that of contemporary manufacturers, Wollaston devised the foundations for modern P/M technique. Wollaston was the first to realize all the difficulties connected with the production of solid platinum ingot from powdered metal, and thus concentrated on the preparation of the powder. He found that pressing the powder while wet into a hard cake (to be subsequently baked at red heat) was best done under considerable pressure. In addition, because available screw presses were not powerful enough, Wollaston developed a horizontal toggle press of the simple construction shown in Fig. 1. Wollaston used the following nine steps in the manufacture of compact platinum metal (Ref 17):

1. Precipitating ammonium-platinum-chloride from diluted solutions
2. Slowly decomposing the finely divided and carefully washed ammonium-platinum-chloride precipitate into loose sponge powder
3. Grinding this sponge powder without applying pressure to the powder particles, thus avoiding any burnishing of the particles and preserving all the surface energy of the particles
4. Sieving the sponge powder
5. Washing the sponge powder with water to remove all remnants of volatile salts
6. Separating fine particles from coarser particles through sedimentation (only the finest sponge particles were used)
7. Pressing the wet mass containing the finest platinum particles into a cylindrical cake
8. Drying the wet cake very slowly and then heating it to about 800 to 1000 °C (1475 to 1830 °F)
9. Forging the cake while it was still hot

By applying these steps, Wollaston succeeded in producing compact platinum, which when rolled into thin sheet was practically free of gas blisters. Crucibles made from this sheet were the best quality platinum implements of their time. Wollaston's process was used for more than a generation and became obsolete only with the advent of the platinum fusion procedure developed by Sainte-Claire Deville and Debray in 1859 (Ref 17). They succeeded in producing a powerful flame with illuminating gas and oxygen, the oxygen being manufactured from manganese dioxide. However, the fused metal which they produced was superior to Wollaston platinum in quality and homogeneity, and the fusion procedure was also less expensive and quicker than the Wollaston method. Fusion, therefore, was soon adopted by every platinum refinery. It

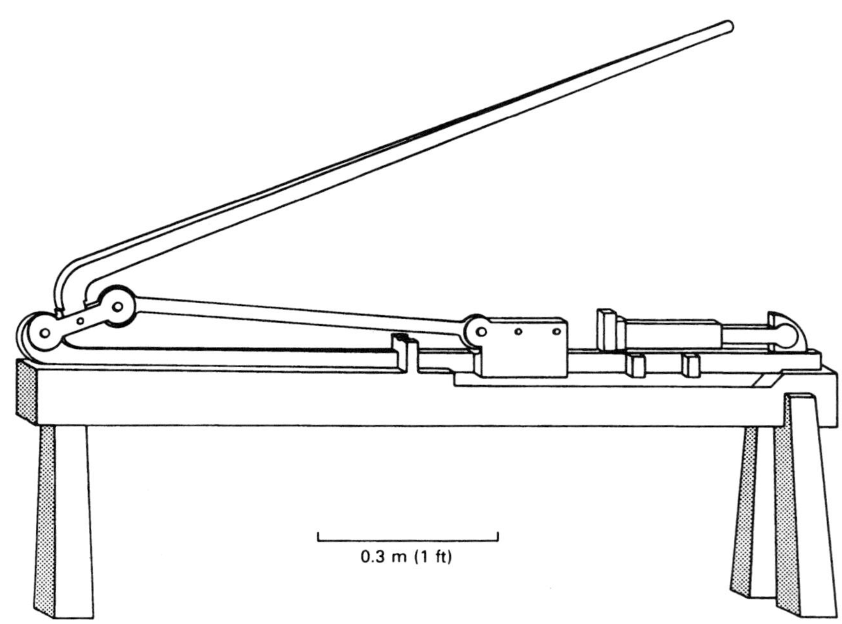

Fig. 1 Simple toggle press used by Wollaston for making platinum powder compacts

is still considered the superior method for manufacturing standard-quality platinum.

Further Developments

The use of P/M technology to form intricately shaped parts by pressing and sintering was introduced in the 19th century. In 1830, while determining the atomic weight of copper, Osann (Ref 18) found that the reduced metal could be sintered into a compact. Osann then developed a process for making impressions of coins from copper powder produced by the reduction of precipitated copper carbonate (Cu_2CO_3).

Osann found that reduction was best done at the lowest possible temperatures that could be used to produce a metal powder of the fineness known in platinum manufacture. High reduction temperatures resulted in granular masses that did not sinter well. Contamination of the powder by the atmosphere was eliminated by using the powder immediately after reduction or storing it in closed glass bottles. The powder was separated into three grades, determined by particle size, before use. To make an impression of a coin, fine powder was sprayed on the surface, followed by layers of coarser grades. The powder and a die were placed in a ring-shaped mold and compressed by the pressure of hammer blows on a punch or use of a knuckle press. Volume of the copper powder was reduced to one-sixth of the original powder during compression. Sintering was done at temperatures close to the melting point of copper, after the compacts were placed in airtight copper packets sealed with clay. A nondistorted 20% shrinkage occurred, but the sintered copper was harder and stronger than cast copper.

Osann also produced medals of silver, lead, and copper by the same procedure. Although he considered his process especially suitable as an alternative to the electrotype method of reproducing coins and medallions, Osann advocated its use as an initial production method for these articles. He believed powder metallurgy could be used for producing printing type and for making convex and concave mirrors by pressing on glass. Osann thought that measurement of the shrinkage of copper compacts could be used to calculate temperature, as the shrinkage of clay cylinders was used in the Wedgewood pyrometer.

Among the advancements in the P/M industry during the second half of the 19th century were Gwynn's attempts to develop bearing materials from metal powders. Patents issued to Gwynn in 1870 (Ref 19) were the forerunners of a series of developments in the area of self-lubricating bearings. Gwynn employed a mixture of 99 parts of powdered tin, prepared by rasping or filing, and 1 part of petroleum-still residue. The two constituents were stirred while being heated. A solid form of desired shape was then produced by subjecting the mixture to extreme pressure while enclosing it in a mold. The patent specifically states that journal boxes made by this method or lined with material thus produced would permit shafts to run at high speeds without using any other lubrication.

Commercial Developments

The first commercial application of powder metallurgy occurred when carbon, and later osmium, zirconium, vanadium, tantalum, and tungsten, was used for incandescent lamp filaments. Methods were developed from 1878 to 1898 for making carbon filaments by the extrusion and subsequent sintering of carbonaceous materials.

Osmium filaments were used for a short time from 1898 to 1900. Auer von Welsbach (Ref 20) described the production of filaments of osmium by chemical precipitation of the powder and formation of a mixture with sugar syrup, which served both as binder and, if osmium oxide powder was used instead of the metal, as reducing agent as well. The mixture was squirted through fine dies, and the resulting fine threads were subsequently fired in protective atmospheres to carburize and volatilize the binder, reduce the oxide, and sinter the metal particles into a coherent metallic wire for use as an electrical conductor.

The osmium electric lamp was soon succeeded by tantalum filament lights, which were used widely from 1903 to 1911. The general procedure (Ref 21) was similar to that used for osmium, with the exception that tantalum had to be purified by a vacuum treatment to become ductile. Similar techniques were used for the production of filaments from zirconium, vanadium, and tungsten; with tungsten, especially, extruded wires were bent into hairpin shapes before sintering to shape them for use as filaments. Because lack of ductility was the major shortcoming of these filaments, attempts were made to improve this property by the addition of a few percent of a lower-melting, ductile metal. Tungsten powder was mixed with 2 to 3% Ni, pressed into a compact, and sintered in hydrogen at a temperature slightly below the melting point of nickel. The resulting bars could be drawn, and nickel was removed from the final filaments by a vacuum heat treatment at a high temperature (Ref 22). Although this process was not commercially successful, it was an important step toward the industrial development of cemented carbides and composite materials.

Tungsten was soon recognized as the best material for lamp filaments. The problem, however, was to devise an economical procedure for producing these filaments in large quantities. A number of procedures to produce powdered tungsten had been worked out earlier. In 1783, the D'Elhujar brothers (Ref 23) first produced tungsten powder by heating a mixture of tungstic acid and powdered charcoal, cooling the mixture, and removing the small cake, which crumbled to a powder of globular particles. The purification of tungsten powder by boiling, scrubbing, and skimming to remove soluble salts, iron oxide, clay, and compounds of calcium and magnesium was reported by Polte (Ref 24).

Coolidge Process. At the beginning of the 20th century, Coolidge (Ref 25) made the important discovery that tungsten could be worked in a certain temperature range and would retain its ductility at room temperature. Few changes have been made over the years on the Coolidge procedure; it is still the standard method of producing incandescent lamp filaments. In this method, very fine tungsten oxide powder, WO_3, is reduced by hydrogen. The powder is pressed into compacts, which are presintered at 1200 °C (2190 °F) to strengthen them so that they can be clamped into contacts. They receive a final sintering treatment near 3000 °C (5430 °F) by passing a low-voltage, high-current density current through the compacts. During sintering, the compacts shrink and reach a density near 90% that of solid tungsten. The sintered compacts can be worked only at temperatures near 2000 °C (3630 °F). When heated to this temperature, they can be swaged into rounds. With increasing amounts of warm work, tungsten becomes more ductile, the swaging temperature can be progressively lowered, and the swaged bars can be drawn into fine wire at relatively low temperatures.

Other Refractory Metals. The procedures developed for the production of tungsten often were adaptable to the manufacture of molybdenum. Lederer (Ref 26) developed a method of making molybdenum using powdered molybdenum sulfide. The sulfide, mixed with amorphous sulfur and kneaded into a paste, was formed into a filament. When exposed to air, the filaments became strong enough to be placed in a furnace. Heating in hydrogen resulted in formation of hydrogen sulfide and sintering of the metal into solid filaments. A similar process was patented by Oberländer (Ref 27), who used molybdenum chloride and other halides as starting materials. When the chloride was treated with a reducing agent such as ether, a paste was obtained.

Tungsten, molybdenum, and tantalum are the three most important refractory metals used today in the lamp, aerospace, electronics, x-ray, and chemical industries. Other refractory metals of minor significance were developed by the P/M method in the early 1900s, notably niobium, thorium, and titanium. However, at the same time another development, originating in refractory metal processing, took form and rapidly grew to such importance that it far overshadows the parent field. Cemented carbides have become one of the greatest industrial developments of the century.

Cemented Carbides. Ordinary drawing dies were unsatisfactory for drawing tungsten wires and filaments. The need for a harder material to withstand greater wear became urgent. Because it was known that tungsten granules combined readily with carbon at high temperatures to give an extremely hard compound, this material was used as the basis for a very hard, durable tool material known as cemented carbide. The tungsten carbide particles, present in the form of finely divided, hard, strong particles, are bonded into a solid body with the aid of a metallic cementing agent. Early experiments with a number of metals established that this cementing agent had to possess the following properties to permit solidification of the hard metal body:

- Close chemical affinity for the carbide particles
- A relatively low melting point
- Limited ability to alloy with the carbide
- Great ductility (not to be impaired by the cementing operation)

Cobalt satisfied these requirements most closely. The early work was carried out mainly in Germany by Lohmann and Voigtländer (Ref 28) in 1914, by Liebmann and Laise (Ref 29) in 1917, and by Schröter (Ref 30) from 1923 to 1925. Krupp (Ref 31) perfected the process in 1927 and marketed the first product of commercial importance, "Widia." In 1928 this material was introduced to the United States, and the General Electric Company, which held the American patent rights, issued a number of licenses. The process entails carefully controlled powder manufacture, briquetting a mixture of carbide and metallic binder (usually 3 to 13% Co), and sintering in a protective atmosphere at a temperature high enough to allow fusion of the cobalt and partial alloying with the tungsten carbide. The molten matrix of cobalt and partly dissolved tungsten carbide forms a bond, holding the hard particles together and giving the metallic body sufficient toughness, ductility, and strength to permit its effective use as tool material.

Composite Metals. The next development in powder metallurgy was the production of composite metals used for heavy-duty contacts, electrodes, counterweights, and radium containers. All of these composite materials contain refractory metal particles, usually tungsten, and a cementing material with a lower melting point, present in various proportions. Copper, copper alloys, and silver are frequently used; cobalt, iron, and nickel are used less frequently. Some combinations also contain graphite. The first attempt to produce such materials was recorded in the patent of Viertel and Egly (Ref 32) issued shortly after 1900. The procedures used either were similar to those developed for the hard metals (Ref 33) or called for introduction of the binder in liquid form by dipping or infiltration. In 1916, Gebauer (Ref 34) developed such a procedure, which was developed further by Baumhauer (Ref 35) and Gillette (Ref 36) in 1924. Pfanstiehl (Ref 37) obtained patent protection in 1919 for a heavy metal, consisting of tungsten and a binder that contained copper and nickel.

Porous Metal Bearings and Filters. In addition to the development of refractory metals and their carbides, another important area of powder metallurgy that gained attention during the early 1900s was that of porous metal bearings. Special types of these porous bearings are referred to as self-lubricating.

The modern types of bearings, usually made of copper, tin, and graphite powders and impregnated with oil, were first developed in processes patented by Loewendahl (Ref 38) and Gilson (Ref 39 and 40). Gilson's material was a bronze structure, in which finely divided graphite inclusions were uniformly distributed. It was produced by mixing powdered copper and tin oxides with graphite, compressing the mixture, and heating it to a temperature at which the oxides were reduced by the graphite and the copper and tin could diffuse sufficiently to give a bronzelike structure. Excess graphite (up to 40 vol%) was uniformly distributed through this structure. The porosity was sufficient to allow for the introduction of at least 2% oil. The process was later improved by Boegehold and Williams (Ref 41), Claus (Ref 42), and many others, primarily by utilization of elemental metal powders rather than oxides.

Metallic filters were the next stage in the development of these porous metals, and patents date back as far as 1923 (Ref 43), when Claus patented a process and machine to mold porous bodies from granular powder.

Post-War Developments

Infiltration techniques, porous materials, iron powder cores for ratio tuning devices, P/M permanent magnets, and W-Cu-Ni heavy metal compositions were developed during the periods between 1900, World War I, and the late 1920s. At the beginning of World War II in Europe, iron powder technology began its advance to commercial viability. The most spectacular development of iron parts made by powder metallurgy was during World War II in central Europe, where paraffin-impregnated sintered iron driving bands for military projectiles were extensively used. German powder metallurgists found this technique effective as a substitute for scarce gilding metal, a copper-zinc alloy containing 5 to 10% Zn. Production reached a peak of 3175 metric tons (3500 tons) per month for this application.

The advent of mass production in the automotive industry made possible the use of iron and copper powders in large tonnages and spawned many of the technological advances of the modern P/M industry. The automobile has been the basis for most industrial applications of P/M, even in fields unrelated to the automotive industry. The first commercial application of a P/M product, the self-lubricating bearing, was used in an automobile in 1927. It was made from a combination of copper and tin powders to produce a porous bronze bearing capable of retaining oil within its pores by capillary attraction. At about the same time, self-lubricating bearings were introduced to the home appliance market as a refrigerator compressor component.

Through the 1940s and early 1950s, copper powder and the self-lubricating bearing were the principal products of powder metallurgy. Since then, iron powder and steel P/M mechanical components such as gears, cams, and other structural shapes have become dominant. While copper powder remains an important P/M material, consumed on the order of 21,000 metric tons (23,000 tons) per year, it is overshadowed by iron and iron-base powders with markets of 318,000 metric tons (350,000 tons) per year.

Since the end of World War II, and especially with the advent of aerospace and nuclear technology, developments have been widespread with regard to the powder metallurgy of refractory and reactive metals such as tungsten, molybdenum, niobium, titanium, and tantalum and of nuclear metals such as beryllium, uranium, zirconium, and thorium.

All of the refractory metals are recovered from their ores, processed, and formed using P/M techniques. With the reactive metals, powder metallurgy is often used to achieve higher purity or to combine them with other metals or nonmetallics to achieve special properties. Nuclear power plants use fuel elements often made by dispersing uranium oxide in a metal powder (aluminum, for example) matrix. The control rods and neutron shielding may use boron powder in a matrix of nickel, copper, iron, or aluminum. Tungsten combined with nickel and copper powders is used widely as a shielding component in applications where intricate configuration involving machining is required, such as in cobalt-60 containers. In aerospace, beryllium and titanium are used extensively. Rocket skirts, cones, and heat shields are often formed from niobium. Molybdenum is widely used in missile and rocket engine components. Nozzles for rockets used in orbiting space vehicles often are made from tungsten via the P/M process in order to maintain critical dimensional tolerances.

The 1950s and 1960s witnessed the emergence of P/M wrought products. These are fully dense metal systems that began as powders. Hot isostatically pressed superalloys, P/M forgings, P/M tool steels, roll compacted strip, and dispersion-strengthened copper are all examples. Each of these processes and materials is covered in separate articles in this Volume.

The commercialization of powder-based high-performance material emerged as a major breakthrough in metalworking technology in the 1970s by opening up new markets through superior performance, coupled with the cost effectiveness of material conservation and longer operational life.

Recent Developments

In the late 1970s, the experimental programs involving P/M wrought products began spilling over into the commercial industrial sector, principally in the form of P/M tool steels and P/M forgings. With the advent of P/M forgings, no longer were properties compromised by density. Fully dense components capable of combining the alloying flexibility and the net and near-net design features of powder metallurgy were very marketable. The later 1970s and early 1980s witnessed a significant metallurgical breakthrough in the recognition of P/M techniques for eliminating segregation and ensuring a fully homogeneous, fine-grained, pore-free, high-alloy structure. Categorized as P/M wrought metals, they led to the perfection of extremely high-purity metal powders and improved consolidation techniques such as hot isostatic pressing (HIP). The 1980s also saw the commercialization of ultrarapid solidification and injection molding technology. Both of these developments are also covered in separate articles in this Volume.

Commercial powder metallurgy now spans the density spectrum from highly porous metal filters through self-lubricating bearings and P/M parts with controlled density to fully dense P/M wrought metal systems. The P/M parts and products industry in North America has estimated sales of more than $3 billion. It comprises 150 companies that make conventional P/M parts and products from iron- and copper-base powders and about 50 companies that make specialty P/M products such as superalloys, tool steels, porous products, friction materials, strip for electronic applications, high-strength permanent magnets, magnetic powder cores and ferrites, tungsten carbide cutting tools and wear parts, rapid solidification rate (RSR) products, and metal injection molded parts and tool steels. Powder metallurgy is international in scope with growing industries in all of the major industrialized countries. The value of U.S. metal powder shipments (including paste and flake) was $1.854 billion in 1995. Annual worldwide metal powder production exceeds 1 million tons.

Trends and new developments include:

- Improved manufacturing processes such as HIP, P/M forging, metal injection molding (MIM), and direct powder rolling through increased scientific investigation of P/M technology by government, academic, and industrial research and development programs
- Fully dense P/M products for improved strength properties and quality in automobiles, diesel and turbine engines, aircraft parts, and industrial cutting and forming tools
- Commercialization of technologies such as MIM, rapid solidification, P/M forging, spray forming, high-temperature vacuum sintering, warm compacting, and both cold and hot isostatic pressing
- The use of P/M hot-forged connecting rods in automobiles and a P/M camshaft for four- and eight-cylinder automobile engines. The use of P/M composite camshafts in automotive engines and main bearing caps

A review of major historical developments in powder metallurgy is presented in Table 1.

Powder Metallurgy Literature

A number of literary works are worthy of mention in connection with the background of powder metallurgy. One of the earliest works of significance was *Principles of Powder Metallurgy* by W.D. Jones, published in 1937 in England (Ref 44). It was updated in 1960 and published as *Fundamental Principles of Powder Metallurgy* (Ref 45). The first Russian publication was by Bal'shin (Ref 46) and appeared in 1938; the first comprehensive text in German, *Pulvermetallurgie und Sinterwerkstoffe*, was published by R. Kieffer and W. Hotop in 1943 (Ref 47). In the United States, the first publication was by H.H. Hausner in 1947 (Ref 48), followed closely by P. Schwarzkopf (Ref 49). Two years later, the first of four volumes of a treatise on powder metallurgy, a major work by C.G. Goetzel (Ref 50), was published. Some current "Selected References" on powder metallurgy science and technology are listed at the end of this article.

Powder Metallurgy Trade Associations

The advancement of powder metallurgy from a laboratory curiosity to an industrial technology has been influenced greatly by various professional societies and the P/M trade association, whose annual technical conference proceedings chronicle the maturing of the technology. In 1944, an organization called the Metal Powder Association was founded by a group of metal powder producers in the United States. It was reorganized in 1958 as the Metal Powder Industries Federation, a trade association whose representation embraced the commercial and technological interests of the total metal powder producing and consuming industries. International in scope, the Federation consists of the following autonomous associations, which together represent the primary elements of the P/M and particulate materials industries:

- *Powder Metallurgy Parts Association:* Members are companies that manufacture P/M parts for sale on the open market.
- *Metal Powder Producers Association:* Members are producers of metal powders in any form for any use.
- *Powder Metallurgy Equipment Association:* Members are manufacturers of P/M processing equipment and supplies, including compacting presses, sintering furnaces, belts, tools and dies, and atmospheres.
- *Refractory Metals Association:* Members are manufacturers of powders or products from tungsten, molybdenum, tantalum, niobium, and cobalt.
- *Advanced Particulate Materials Association (APMA):* Members are companies that use P/M or other related processes to produce any of a wide variety of materials not covered by the other MPIF associations as well as companies that have proprietary P/M parts manufacturing facilities. It also includes emerging technologies that use the powders as precursors in manufacturing processes.
- *Metal Injection Molding Association (MIMA):* Members are international companies that use the metal or ceramic injection molding process to form parts.

MPIF also has both Overseas and Affiliate/Consultant classes of membership.

The Federation generates industry statistics, process and materials standards, industrial public relations and market development, government programs, research, and various educational programs and materials.

The technology's "professional" society is APMI International. As distinguished from the Federation, APMI members are individuals, not companies. Members are kept informed of developments in P/M technology through local section activities, conferences, and publications, including the *International Journal of Powder Metallurgy and Powder Technology.* It is the only professional society organized specifically to serve the powder metallurgist and the P/M industry.

Many of the major professional societies are also active in powder metallurgy, usually through committees working on standards, conferences, or publications. This includes the ASM International, the Metallurgical Society, SAE, the American Society for Testing and Materials, and the Society of Manufacturing Engineers.

REFERENCES

1. H.C.H. Carpenter and J.M. Robertson, The Metallography of Some Ancient Egyptian Implements, *J. Iron Steel Inst.*, Vol 121, 1930, p 417–448
2. W.D. Jones, *Fundamental Principles of Powder Metallurgy*, London, 1960, p 593
3. P. Bergsöe, The Metallurgy and Technology of Gold and Platinum Among the Pre-Columbian Indians, *Ing. Skrift. (A)*, Vol 44, 1937
4. W. Lewis, Experimental Examination of a White Metallic Substance Said to Be Found in the Gold Mines of Spanish West Indies, *Philos. Trans. R. Soc.*, Vol 48, 1755, p 638
5. H.T. Scheffer, *Handlingar*, Vol 13, 1752, p 269–275
6. K.F. Achard, *Nouveaux Mem. Acad. R. Sci.*, Vol 12, 1781, p 103–119
7. A. von Mussiin-Puschkin, *Allgem. J. Chem.*, Vol 4, 1800, p 411
8. C. Ridolfi, *Quart. J. Sci. Lit. Arts*, Vol 1, 1816, p 259–260 (From Giornale di scienza ed arti, Florence, 1816)
9. A. Rochon, *J. Phys. Chem. Arts*, Vol 47, 1798, p 3–15 (Rochon states that this was written in 1786 as part of his voyage to Madagascar)
10. R. Knight, A New and Expeditious Process for Rendering Platina Malleable, *Philos. Mag.*, Vol 6, 1800, p 1–3
11. A. Tilloch, A New Process of Rendering Platina Malleable, *Philos. Mag.*, Vol 21, 1805, p 175
12. Leithner, Letter quoted by A.F. Gehlen, *J. Chem. Phys.*, Vol 7, 1813, p 309, 514
13. M. Baruel, Process for Procuring Pure Platinum, Palladium, Rhodium, Iridium, and Osmium from the Ores of Platinum, *Quart. J. Sci. Lit. Arts*, Vol 12, 1822, p 246–262
14. P. Sobolewskoy, *Ann. Physik Chem.*, Vol 109, 1834, p 99
15. W. Marshall, An Account of the Russian Method of Rendering Platinum Malleable, *Philos. Mag.*, Vol 11 (No. II), 1832, p 321–323
16. W.H. Wollaston, On a Method of Rendering Platina Malleable (Bakerian Lecture for 1828), *Philos. Trans. R. Soc.*, Vol 119, 1829, p 1–8
17. J.S. Streicher, *Powder Metallurgy*, J. Wulff, Ed., American Society for Metals, 1942, p 16

18. G. Osann, *Ann. Physik Chem.*, Vol 128, 1841, p 406
19. U.S. Patents 101,863; 101,864; 101,866; and 101,867, 1870
20. U.S. Patent 976,526, 1910
21. U.S. Patents 899,875, 1908 and 912,246, 1909
22. C.R. Smith, *Powder Metallurgy*, J. Wulff, Ed., American Society for Metals, 1942, p 4
23. A.W. Deller, *Powder Metallurgy*, J. Wulff, Ed., American Society for Metals, 1942, p 582
24. U.S. Patent 735,293, 1903
25. U.S. Patent 963,872, 1910
26. U.S. Patent 1,079,777, 1913
27. U.S. Patent 1,208,629, 1916
28. German Patents 289,066, 1915; 292,583, 1916; 295,656, 1916; 295,726, 1916. Swiss Patents 91,932 and 93,496, 1919
29. U.S. Patents 1,343,976 and 1,343,977, 1920
30. German Patent 420,689, 1925. U.S. Patent 1,549,615, 1925
31. British Patents 278,955, 1927, and 279,376, 1928. Swiss Patent 129,647, 1929. U.S. Patent 1,757,846, 1930
32. U.S. Patent 842,730, 1907
33. U.S. Patents 1,418,081, 1922; 1,423,338, 1922; and 1,531,666, 1925
34. U.S. Patent 1,223,322, 1917
35. U.S. Patent 1,512,191, 1924
36. U.S. Patent 1,539,810, 1925
37. U.S. Patent 1,315,859, 1919
38. U.S. Patent 1,051,814, 1913
39. U.S. Patent 1,177,407, 1916
40. E.G. Gilson, *General Electric Rev.*, Vol 24, 1921, p 949–951
41. U.S. Patents, 1,642,347, 1927; 1,642,348, 1927; 1,642,349, 1927; and 1,766,865, 1930
42. U.S. Patent 1,648,722, 1927
43. U.S. Patent 1,607,389, 1926
44. W.D. Jones, *Principles of Powder Metallurgy*, Arnold, London, 1937
45. W.D. Jones, *Fundamental Principles of Powder Metallurgy*, Arnold, London, 1960
46. M.Y.J. Bal'shin, *Metal Ceramics*, Gonti, 1938 (in Russian)
47. R. Kieffer and W. Hotop, *Pulvermetallurgie und Sinterwerkstoffe*, Springer, 1943; Re-issue Springer, 1948
48. H.H. Hausner, *Powder Metallurgy*, Chemical Publishing Co., 1947
49. P. Schwarzkopf, *Powder Metallurgy*, Macmillan, 1947
50. C.G. Goetzel, *Treatise on Powder Metallurgy*, Vol 1–4, Interscience, 1949

SELECTED REFERENCES

- M.E. Fayed and L. Oteen, Ed., *Handbook of Science & Technology,* Chapman & Hall, 1997
- R.M. German, *Powder Metallurgy Science,* Metal Powder Industries Federation, 1994
- A. Lawley, *The Production of Metal Powders,* Metal Powder Industries Federation, 1992
- F. Thümmler and R. Oberacker, *An Introduction to Powder Metallurgy,* I. Jenkins and J.V. Wood, Ed., The Institute of Materials, 1993
- A.J. Yule and J.J. Dunkley, *Atomization of Melts for Powder Production and Spray Deposition,* Oxford Science Publications, 1994

Powder Metallurgy Methods and Design*

Howard I. Sanderow, Management & Engineering Technologies

THE POWDER METALLURGY (P/M) process is a near-net or net-shape manufacturing process that combines the features of shape-making technology for powder compaction with the development of final material and design properties (physical and mechanical) during subsequent densification or consolidation processes (e.g., sintering). It is critical to recognize this interrelationship at the outset of the design process because a subtle change in the manufacturing process can cause a significant change in material properties.

General P/M Design Considerations

To begin a design using powder processing, six key design considerations must be recognized. With the variety of powder processing schemes available, the selection of the appropriate method depends to a great extent on these design constraints.

Size. Due to the physical nature of the processes and the physical limits of commercial manufacturing equipment, product size has certain critical boundaries. For some powder processes, the product size is quite limited (such as metal injection molding, MIM), while for hot-isostatic pressing (HIP), size is not considered a serious constraint.

Shape Complexity. Powder metallurgy is a flexible process capable of producing complex shapes. The ability to develop complex shapes in powder processing is determined by the method used to consolidate the powders. Because a die or mold provides the container for the consolidation step, the ease of manufacture of the container and the ability to remove a green compact (unsintered) from the container, in most cases, determines the allowable shape complexity of a given part.

Tolerances. Control of dimensional tolerances, a demanding feature of all near-net or net-shape manufacturing processes, is a complex issue in powder processing. Tolerances are determined by such process parameters as powder characteristics, compaction parameters, and the sinter cycle. The amount of densification during sintering and the uniformity of that shrinkage controls dimensional tolerance in most P/M products. Due to the very small amount of size change during sintering of conventional press-and-sinter P/M parts, these products typically have the closest dimensional tolerances, as compared to HIP parts, which require the largest spread in tolerances.

Material Systems. Powder shape, size, and purity are important factors in the application of a powder processing technique. For some consolidation processes or steps, powders must be smooth, spherical particles, but for other processes a much more irregular powder shape is required. Nearly every material and alloy system is available in powder form. For some materials such as cemented carbides, copper-tungsten composites and the refractory metals (tungsten, molybdenum, tantalum, etc.), powder processing is the only commercially viable manufacturing process.

As an example, for "press and sinter" processing, an irregular powder shape and distribution of particle sizes are desired for adequate green strength and sinter response. Hot isostatic pressing requires spherical powders (gas atomized) for lowest impurities and good particle packing. The MIM process also prefers spherical particles, but very small particle size (10 to 20 μm) is needed to ensure proper rheology, homogeneous distribution in the plastic binder, and excellent sinter response.

Properties. The functional response of any product is determined by its physical or mechanical properties. In powder processing these properties are influenced directly by the product density, the raw material (powder), and the processing conditions (most often the sintering cycle). As P/M materials deviate from full density, the properties decrease (as shown for the tensile properties—and electrical conductivity—of pure copper in Fig. 1). The mechanical response for 4% Ni steels is found in Fig. 2.

Quantity and Cost. The economic feasibility of P/M processing is typically a function of the number of pieces being produced. For conventional press-and-sinter processing, production quantities of at least 1,000 to 10,000 pieces are desired in order to amortize the tooling investment. In contrast, isostatic processing can be feasible for much lower quantities, in some cases as small as 1 to 10 pieces. On a per pound basis, the approximate costs for steel P/M parts produced by various methods are roughly as follows:

Condition	Density range, g/cm^3	1997 selling price(a), $/lb
Pressed and sintered	6.0–7.1	2.45–2.70
Pressed, sintered, sized	6.0–7.1	2.90–3.20
Copper infiltrated	7.3–7.5	3.50–3.55
Warm formed	7.2–7.4	3.10–3.30
Double pressed and sintered	7.2–7.4	4.00–4.10
Metal injection molded	7.5–7.6	45.0–70.0
Hot forged	7.8	5.00–5.50
Double press and sinter + HIP	7.87	6.00–7.00

(a) These numbers are only averages; smaller parts are more expensive and larger parts less expensive per pound.

Powder Processing Techniques

In order to understand the design restrictions of each powder processing method, it is best to review these processes individually. The P/M manufacturing methods can be divided into two main categories: conventional press-and-sinter methods and full-density processes.

Conventional (Press-and-Sinter) Processes. The conventional press-and-sinter process technologies follow the steps outlined in Fig. 3. The

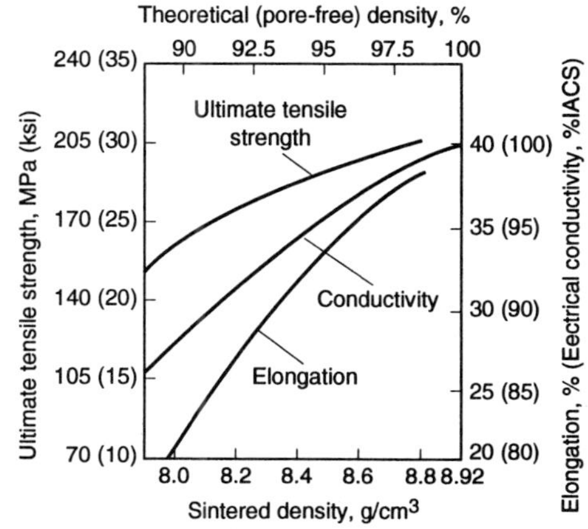

Fig. 1 Properties of pure copper. Source: Ref 1

*Adapted from article in *Materials Selection and Design*, Vol 20, *ASM Handbook*, 1997, p 745–753

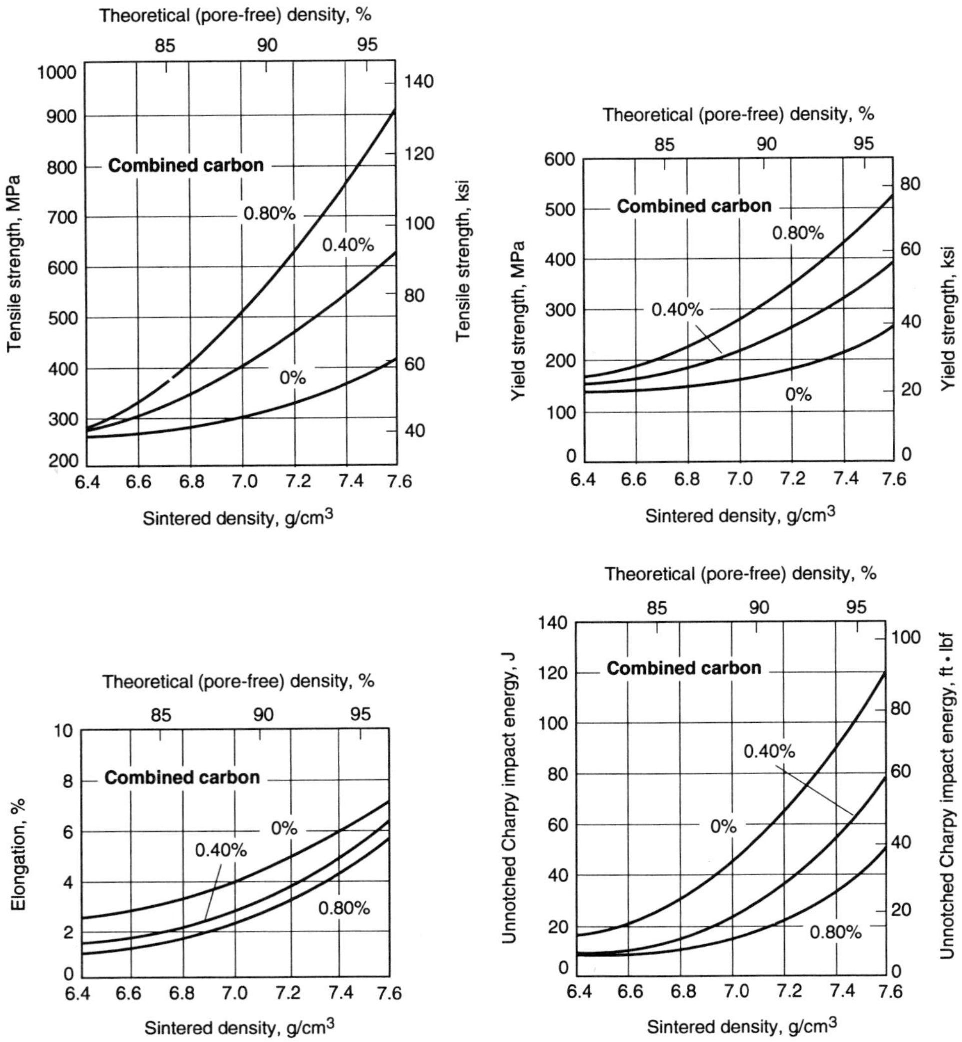

Fig. 2 Effects of density on mechanical properties of as-sintered 4% Ni steel. Source: Ref 2

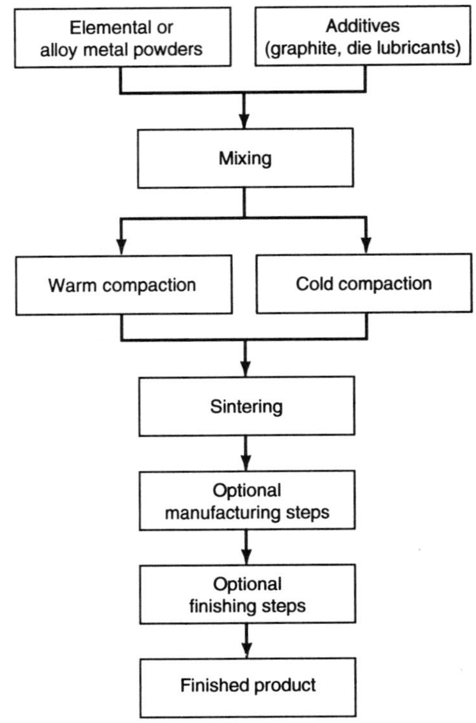

Fig. 3 General steps in the P/M process

various powder ingredients are selected to satisfy the process constraints and still meet the requirements of the end product. For example, in cold compaction irregularly shaped powders are used to ensure adequate green strength and structural integrity of the as-pressed product. Special solid lubricants are added to the powder blend to reduce friction between the powder particles and the tooling. If these lubricants might contaminate the metal powder particles, then an alternate consolidation method would be needed.

Because powder is compacted in hard tooling using a vertical compaction motion, the product size and shape are limited by the constraints of available press capacity, powder compressibility, and the density level required in the product. For most conventional P/M products these limitations have a maximum size of about 160 cm^2 (25 in.^{2}) compaction area, part thickness of about 75 mm (3 in.), and a weight of 2.2 kg (5 lb). However, parts as large as 200 mm (8 in.) diameter by 100 mm (4 in.) thick, weighing 14.5 kg (32 lb), have been produced on conventional equipment. Even parts 380 mm (15 in.) in diameter by 6 mm (¼ in.) thick have been produced by conventional P/M methods.

After compaction the green compact is sintered in a controlled-atmosphere furnace. Dimensional tolerance control is determined by the maximum temperature of the sintering cycle and the metallurgical changes that occur during sintering. If solid-state diffusion is the primary sintering mechanism, very little densification occurs, dimensional change is minimal, and tolerance control is very good. This practice is followed for most P/M steels where size change during sintering is held to less than 0.3%. In contrast, other alloy systems utilize liquid-phase formation as the primary sintering mechanism, causing a significant increase in density, large dimensional changes, and much lower tolerance control. Examples of these material systems include cemented carbides where dimensional changes of 6 to 8% are typical and tolerance control is in the range of ±0.25 mm (±0.010 in.). In addition to its effects on dimensional tolerance levels, the sintering step also plays a significant role in determining the final physical and mechanical properties of the product. Higher sintering temperatures and longer sintering times promote pore rounding and increase densification, thereby improving critical mechanical properties such as tensile strength, duc-

Table 1 Effect of sintering conditions on the mechanical properties of two P/M nickel steels

	MPIF FN0205(a)		MPIF FN0208(b)	
	Belt(c)	Vacuum(d)	Belt(c)	Vacuum(d)
Tensile strength, MPa (ksi)	380 (55)	552 (80)	448 (65)	758 (110)
Yield strength, MPa (ksi)	193 (28)	414 (60)	331 (48)	586 (85)
Elongation, %	4	7	2	4
Impact energy, J (ft · lbf)	19 (14)	38 (28)	11 (8)	33 (24)
Hardness, HRB	64	80	80	90
Sintered density, g/cm^3	(e)	7.32	(e)	7.30

All samples pressed to a green density of 7.2 g/cm^3. (a) 1–3% Ni, 0.3–0.6% C. (b) 1–3% Ni, 0.6–0.9% C. (c) Belt, 30 min at 1125 °C (2060 °F) in nitrogen/endo atmosphere. (d) Vacuum, 2 h at 1260 °C (2300 °F) with nitrogen backfill. (e) Density not reported but estimable by MPIF Standard 35. Source: Ref 3

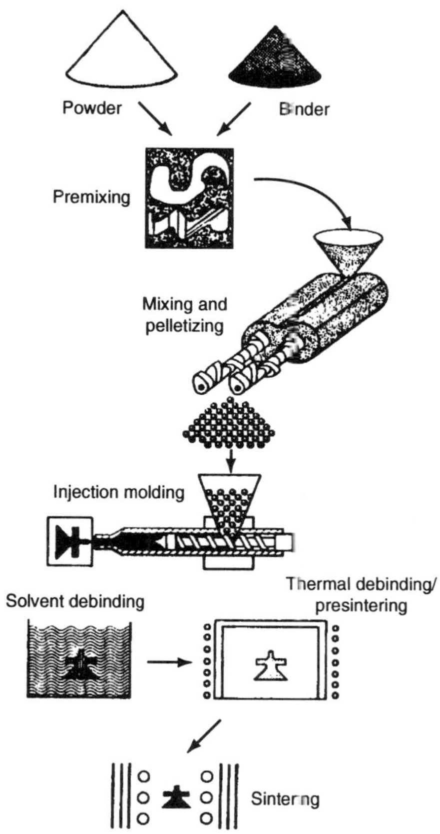

Fig. 4 Metal injection molding process. Source: Ref 7

Table 2 Effect of sintering conditions on the properties of magnetic P/M iron (0.45 wt% P)

Sintering conditions		Maximum magnetic	Coercive	Maximum	Tensile strength		Elongation,
Atmosphere(a)	Temperature, °C (°F)	induction (B_{max}), kG	force (H_c), Oe	permeability	MPa	ksi	%
10% H_2	1120 (2050)	13.2	2.3	2620	345	50	3
75% H_2	1120 (2050)	13.3	2.0	3220	355	52	7
100% H_2	1120 (2050)	13.4	1.7	3680	372	54	5
100% H_2	1200 (2200)	13.7	1.3	5710	400	58	14

(a) Balance N_2. Source: Ref 4

Table 3 Properties of powder forged steels

	Hardness,	Tensile strength		Yield strength		Elongation,	Impact toughness	
Alloy	HRC	MPa	ksi	MPa	ksi	%	J	ft · lbf
10C60	23	793	115	690	100	11	2.7	2
11C60	28	895	130	620	90	11	4	3
4620	28	965	140	895	130	24	81	60
	38	1310	190	1070	155	20	47	35
4640	38	1310	190	1070	155	17	34	25
	48	1585	230	1310	190	11	16	12
4660	38	1310	190	1070	155	15	27	20
	48	1585	230	1310	190	10	13.5	10

Source: Ref 6

tility, impact resistance, and fatigue limit (see Table 1). The sintering process is extremely important in determining the magnetic response of soft magnetic P/M alloys. As shown in Table 2 for the Fe-0.45 wt% P alloy, increasing the hydrogen content in the sintering atmosphere and raising the sintering temperature improved the maximum permeability more than 100%, the tensile strength more than 15%, and the ductility more than 300%. In a similar manner, the mechanical properties and corrosion resistance of P/M stainless steels are strongly dependent on the sintering process parameters (Ref 5).

Warm compaction is used to increase the green density and green strength of P/M steel parts. When combined with high-temperature sintering, this process can provide mechanical properties equivalent to double press-double sinter processing at a lower cost. Due to the much higher green strength, warm compacted parts can be machined in the green condition. This technique can also be used to produce insulated magnetic cores, a composite material suitable for high-frequency electromagnetic systems.

Full-Density Processes. The second group of powder process technologies are formulated specifically to yield a product as close to full density as possible. This contrasts significantly with the previous conventionally processed products where attainment of full density was not the primary goal. The full-density processes include powder forging (P/F), metal injection molding (MIM), hot isostatic pressing (HIP), roll compaction, hot pressing and extrusion.

Powder Forging. In P/F a preform is manufactured using conventional P/M process techniques and then hot formed in confined dies to cause sufficient material deformation so that nearly all the porosity is eliminated. Due to the high costs in developing the preform design and maintaining forging tools and automated production systems of the P/F process, it has been limited, in most commercial practices, to high-volume products such as automotive connecting rods and transmission components. The P/F process has been successful in developing mechanical properties in P/F steel comparable to wrought steels (see Table 3). This process successfully overcomes the mechanical property limitations imposed by the residual porosity in conventional P/M products.

Metal Injection Molding. The MIM process combines the structural benefits of metallic materials with the shape complexity of plastic injection molding technology. A uniform mixture of powder and binders is prepared and injected into a mold (see Fig. 4). The MIM powders are typically spherical in shape and much finer in particle size than those used for conventional cold-die compaction (MIM powder, 10 to 20 μm; conventional die-compaction powders, 50 to 150 μm). The binders are formulated specially to provide the proper rheological properties during injection molding as well as ease of binder removal after the molding step. Once the part is ejected from the mold, the binder material is removed using either solvent extraction or thermal processes (or both). After the debinding step the part is then sintered to complete the process. Due to the large amount of binder in the MIM starting material (up to 40% by volume), the MIM part undergoes a large reduction in size (as much as 20% linear shrinkage) during sintering. Dimensional tolerances, therefore, are not as good as in conventional die compaction and a straightening or coining step is sometimes needed.

Hot Isostatic Pressing. This fully dense process method is the least constrained technique. However, due to its very low production rate, costly equipment, and unique tool requirements, the HIP process is normally relegated to expensive materials such as tool steels, superalloys, titanium, and so forth. The process also requires high-purity powders (generally spherical in shape), and it is considered only a near-net-shape process. The powders are vibrated in place in a container, which is then evacuated and sealed. These metal or ceramic containers are placed in the HIP vessel, which applies an isostatic pressure (using a gaseous medium) and temperature to the container and the powder mass. This combination of heat and pressure on the container consolidates the powder to its final shape, as defined by the initial container configuration. The container must be removed from the HIP part after the process cycle, typically by machining or chemical etching.

Other Full-Density Methods. The remaining full-density consolidation methods are used infrequently in commercial practice and are limited to specialty materials. For example, roll compaction is used to form certain soft magnetic alloys, composite materials, and compositions unique to powder metallurgy. Hot pressing is used when the deformation characteristics of the base powder require high temperatures to achieve plastic flow and adequate consolidation. The powder extrusion process requires a container, and it is similar to roll compaction and limited to specialty materials not suitable for conventional extrusion methods, such as composites, titanium, and nuclear materials.

Table 4 Comparison of powder processing methods

Characteristic	Conventional	MIM	HIP	P/F
Size	Good	Fair	Excellent	Good
Shape complexity	Good	Excellent	Very good	Good
Density	Fair	Very good	Excellent	Excellent
Dimensional tolerance	Excellent	Good	Poor	Very good
Production rate	Excellent	Good	Poor	Excellent
Cost	Excellent	Good	Poor	Very good

Table 5 Application of powder processing methods

	Conventional die compaction	MIM	HIP	P/F
Material	Steel, stainless steel, brass, copper	Steel, stainless steel	Superalloys, titanium, stainless steel, tool steel	Steel
Production quantity	>5000	>5000	1–1000	>10,000
Size, lb	<5	<¼	5–5000	<5
Dimensional tolerance	±0.001 in./in.	±0.003 in./in.	±0.020 in./in.	±0.0015 in./in.
Mechanical properties	~80–90% wrought	~90–95% wrought	Greater than wrought	Equal to wrought
Price per pound	$0.50–5.00	$1–10	>$100	$1–5

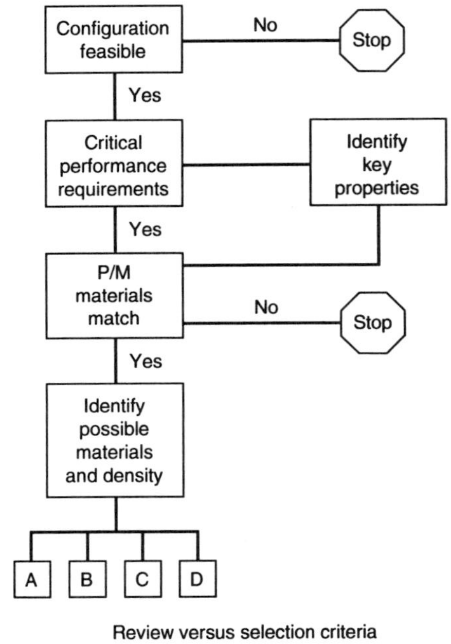

Fig. 5 Material selection method

Comparison of Powder Processing Methods

Effective application of powder processing methods requires a general comparison of the major design features, focusing on the similarities, differences, advantages, and disadvantages of each method. Table 4 provides a qualitative comparison, while Table 5 offers more specific design information. Characteristics for each processing method are summarized below.

Conventional die compaction:

- Widest range of most frequently used engineering materials, including iron, steel, stainless steel, brass, bronze, copper, and aluminum
- Most applicable to medium-to-high production volumes; small- to medium-size parts such as gears, sprockets, pulleys, cams, levers, and pressure plates (automotive, appliances, power tools, sporting equipment, office machines, and garden tractors are typical markets)
- Greatest density range, including high-porosity filters, self-lubricating bearings, and high-performance structural parts
- Limited physical and mechanical properties caused by residual porosity
- Most cost-competitive of the powder processes
- Wide range of applications from low- to high-stress applications

Powder forging:

- Potentially applicable to all engineering materials now hot forged, but actual applications currently limited to low-alloy steels
- Product applications limited to high-volume products such as automotive connecting rods and transmission components as well as power tool parts
- Mechanical properties equivalent to wrought steel
- Most cost-competitive of the full-density processes for medium-to-large parts

Metal injection molding:

- Limited range of materials, though most standard engineering alloys available as well as several specialty alloys
- Limited to relatively small, highly complex shaped products for medium-to-high production volumes
- Greatest range in shape complexity including high aspect ratios
- More costly than conventional die-compaction processes
- Superior physical and mechanical properties as compared to conventional process, due to higher density

Hot isostatic pressing:

- Materials limited only by the inherent cost of the process, therefore typically applied only to expensive materials
- Most suited for low-to-medium production volumes
- Competitive against large casting or forging products where substantial machining is needed to obtain the final product
- Much shape detail is machined after HIP processing; not normally a "net-shape" manufacturing process
- Physical and mechanical properties meet or exceed those of cast or wrought materials

Conventional Die Compaction

In order to use die-compaction technology effectively, the designer must be familiar with the limitations and design constraints related to product shape and special features. A shape or feature can be die compacted provided that (a) it can be ejected from the tooling and (b) the tools that form the feature have sufficient strength to withstand the repeated compaction loads. Due to the vertical closure of the tooling and the lack of tool motions perpendicular to the pressing direction, part removal from the tools controls many features. Examples of features that cannot be accommodated in die compaction, and therefore require secondary machining operations, include undercuts, reverse taper (larger on bottom than on top), annular grooves, and threads.

The specific design issues for the powder process technologies would require an extensive discussion well beyond the limits of this article. An excellent source for additional details is the manual on P/M product design prepared by the Metal Powder Industries Federation (MPIF) trade association (Ref 8). Additional information on tooling and part design is also included in the article "Powder Metallurgy Presses and Tooling" in this Volume.

Secondary Operations

A variety of secondary manufacturing and finishing operations may be required to complete the part, to improve properties, or to calibrate dimensional tolerances. Because die-compacted parts have residual porosity that may affect the response to these secondary operations, several guidelines are provided in this section.

Repressing. Die-compacted parts can be repressed in a second tool set in order to reduce the amount of porosity or reduce dimensional variation. By increasing the part density, critical physical and mechanical properties will be improved. This secondary densification step can be performed on the entire part or just specific features, for example, gear teeth. The sizing of P/M parts for improved dimensional tolerances is widely performed, especially for the manufacture of bearings.

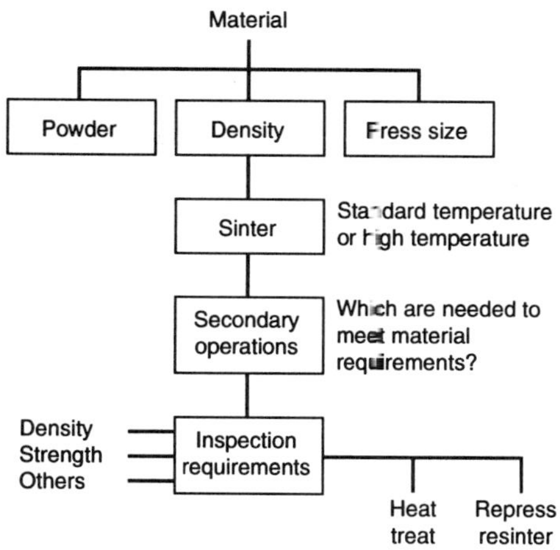

Fig. 6 How material selection affects process

Impregnation and Infiltration. Several procedures are used to fill the residual porosity in conventional P/M parts. Oil impregnation adds an internal lubricant to the product, which is useful in bearing and wear applications. The part can be resin impregnated to seal the pores for pressure tightness and improved machinability or to prevent the intrusion of undesired plating chemicals. Infiltration with a lower melting metal than the base material (such as copper-infiltrated steel or copper-infiltrated tungsten) is also used to seal porosity, but most often to increase mechanical properties or create unique composite structures.

Steam treating, also known as steam oxidizing, is a low-temperature (540 °C, or 1000 °F, 1 to 2 h) heat treatment process in which P/M steel parts are exposed to superheated steam. The process can be conducted in batch, pressurized furnaces, or continuous belt furnaces. The steam reacts with the iron surface converting it to an adherent, protective blue-grey iron oxide (Fe_3O_4). Because the steam can penetrate the porosity, the oxide layer can be 0.020 to 0.050 in. deep depending on the processing conditions. Steam treating enhances the product by:

- Increasing wear resistance
- Increasing surface hardness
- Improving corrosion resistance
- Increasing compressive yield strength
- Providing low-pressure leak tightness

Formation of this oxide layer unfortunately reduces tensile strength and ductility ~10 to 20%, depending on the material system and processing conditions.

Heat Treatment. Due to the presence of residual porosity in die-compacted P/M parts, heat-treat practices should utilize gaseous or noncorrosive liquid media (such as quench oil rather than water or fused salts). In addition, because the porosity allows for the penetration of gaseous media (i.e., carburizing gas), special steps must be considered when trying to develop a carburized case in P/M steels. The surface (or entire part) must be high enough in density to prevent the carburizing gas from penetrating the pore network. A density of 7.2 g/cm^3 or greater is desired. Copper infiltration can also be used to seal the part for a case-carburizing treatment. Heat treatment of P/M steels is most effective in improving mechanical properties when parts have a density greater than 7.0 g/cm^3.

Finishing. The variety of finishing operations—machining, plating, deburring, joining—must also consider the effects of residual porosity. Improvements in machinability have been achieved through resin impregnation (seals the porosity) and by adding machinability aids to the original powder blend. Surface finishing may also require resin impregnation if the part density is low enough to allow entrapment of plating or finishing chemicals. These liquids can cause internal corrosion of the P/M part if allowed to penetrate the open porosity network. For welding and brazing, precautions are necessary to ensure a sound joint (Ref 8).

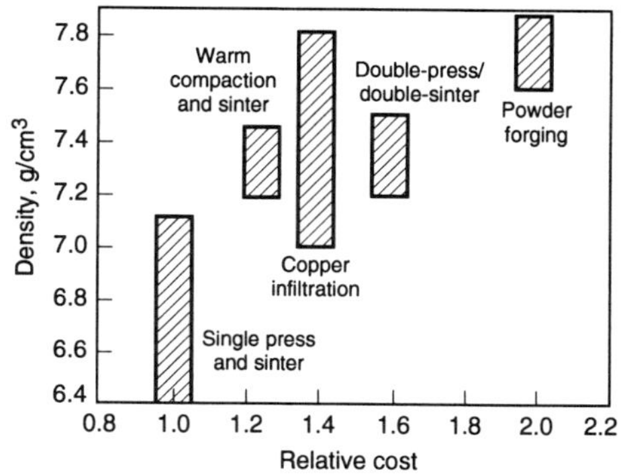

Fig. 7 Relative cost versus ferrous density of several P/M processes. Source: Hoeganaes Corporation

Material Selection

Once the die-compaction processing method has been selected the designer must consider the product configuration and material requirements. Previous sections have discussed the design constraints regarding shape and configuration. The next step in the procedure (see Fig. 5) is comparing the critical part performance requirements with the available P/M materials. Reference 5 serves as a very useful source of material property data. In many cases, more than one material can satisfy the design performance requirements. Therefore, the various production methods must be considered because these affect the ultimate cost of the product.

Figure 6 provides the second evaluation scheme. Starting from the material selected, this leads directly to the density needed in the critical section of the part. Knowing the desired density leads to a preliminary selection of the powder and press size necessary to reach the required density. From this point the remainder of the processing steps are considered. For example, is a conventional sintering process or a "high-temperature" process needed? Will a repressing-resintering process be necessary to achieve the density selected? What secondary processing treatments are needed, and how will the density affect these processes? Using a standard press-and-sinter process as a baseline, the cost of high-temperature sintering may add a 10 to 20% premium, while a double-press/double-sinter process could result in as much as a 40% premium (Fig. 7). Therefore, the designer must recognize that selecting a material and density level leads directly to a processing cycle and the cost associated with the process. A judicious review and selection of materials can ensure the most cost-effective product design.

Bearings

Due to the unique self-lubricating characteristics of P/M bearings, these products have additional design guidelines. After conventional die compaction and sintering, bearings are vacuum-oil impregnated to fill the interconnected porosity with oil. The design of P/M bearings and selection of the optimal alloy depend primarily on shaft velocity and bearing load. Other design factors include type of operation (stop-start versus continuous), shaft surface finish, type of lubricant, and conditions of uneven loading. The MPIF bearing standard (Ref 9) provides a systematic guide for these design issues:

- Recommended load as a function of shaft velocity
- Environmental factors that can reduce permissible loads
- Recommended press fits
- Running clearances
- Dimensional tolerances

The selection of a bearing alloy normally considers the following factors:

- Chemical compatibility with the lubricant
- Thermal conductivity to dissipate heat
- Mechanical properties to maintain structural integrity and withstand press-fit installation forces
- Antigalling with the shaft material
- Wear characteristics with the shaft material

Six commercial alloys dominate the self-lubricating bearing designs: bronze, diluted bronze, plain carbon steel, iron-copper alloys, copper steels, and iron-graphite. Graphite can be used as a solid lubricant and is added to the powder mix prior to compacting. More information on these materials can be found in Ref 9 and the article "Powder Metallurgy Bearings" in this Volume.

In some products the designer can incorporate bearing surfaces into the part, thereby eliminating the need for separately fabricated and installed bearings. The following design restrictions must be followed:

- Bearing areas should have a minimum of 10% interconnected porosity to provide sufficient oil capacity.

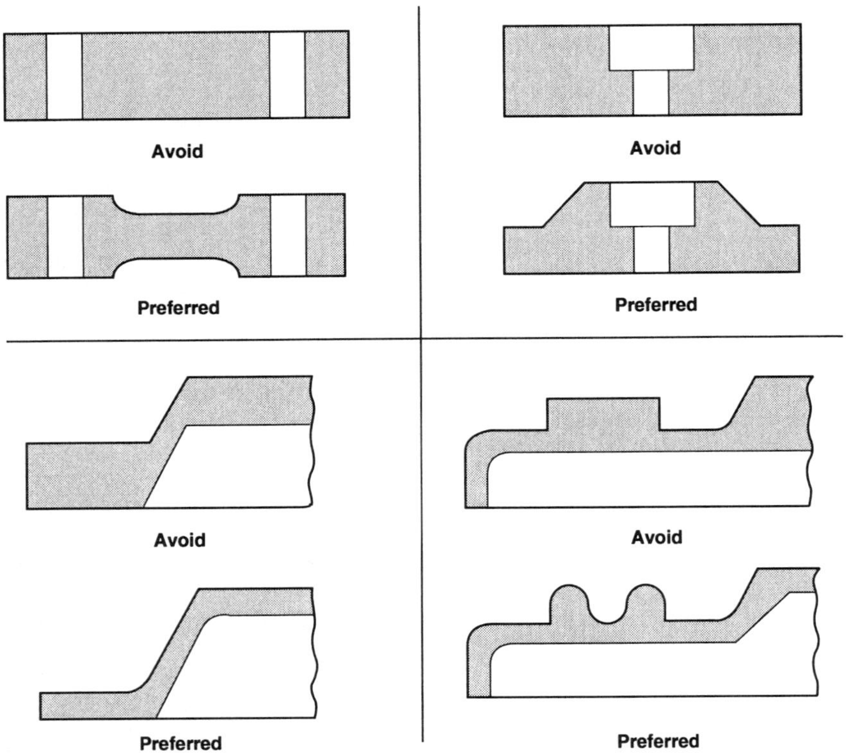

Fig. 8 Designs for maintaining uniform wall thickness in MIM parts. Source: Ref 8

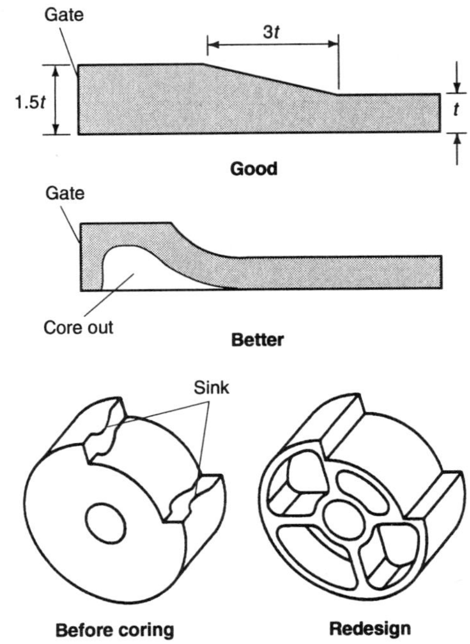

Fig. 9 Redesign of a MIM part to establish uniform wall thickness. Source: Ref 8

- Infiltration is not allowed in the bearing areas because the infiltration will fill the open pores.
- Machining on the bearing surfaces must be performed to avoid sealing or closing off the open porosity.

Metal Injection Molding

The MIM process is similar to plastic injection molding: (1) material is fed into the mold cavity through gates, which may leave a visible mark on the part; (2) the mold must open, leaving a parting line on the part; and (3) the part must be ejected from the mold cavity using ejector pins, which also may leave visible marks on the part. Location of the parting line is critical to the orientation of part features. To the extent possible, all features should be oriented perpendicular to the plane of the parting line to facilitate removing the part from the mold. Features not so oriented require additional tooling, such as side pulls.

The optimal location of gates is a design balance between product and processing requirements. In general, gates are located on the parting line, positioned to direct flow onto a core rod or cavity wall. Where wall thicknesses vary, gates are located so the material flows from thicker sections to thinner sections. Ejector pins push the molded part off of the mold cores. There must be enough pins to free the part off of the core without damaging or distorting the as-molded part. Typically, these pins are located near features requiring the highest ejection forces, such as cored holes. Because the pins leave permanent marks, they should be positioned on recessed features, surfaces that will be machined, or unused areas such as runners.

Injection-molded parts are normally placed on flat plates or shelves for thermal debinding and sintering. A geometric feature that allows the part to rest securely with no additional support is preferred. Long spans, cantilevers, or delicate resting points may require special fixtures or setters to minimize distortion during sintering. A postsintering straightening operation may also be required.

Shape and Features. The design features that can be produced by MIM are similar to those made by conventional plastic injection molding or die casting. Dimensions and proportions are different due to the small size of MIM parts and molding characteristics of the MIM materials.

To facilitate release and ejection of the molded part, draft is normally in the range of ½ to 2°. Draft should be uniform on all features of the part to facilitate mold fabrication and be near the upper range with more complex part features or multiple cores.

Where feasible, walls should be of uniform thickness throughout the part (Fig. 8). Variations in thickness can lead to distortion, internal stresses, voids, cracking, and sink marks. In addition, thickness variations cause nonuniform shrinkage during sintering, leading to greater size variation and looser tolerances. A thickness in the range of 1.25 to 6.35 mm (0.050 to 0.250 in.) is preferred. Common ways of modifying a part to make wall thickness more uniform or ease the transition from thick to thin are shown in Fig. 9. More detailed design tips are found in Ref 8.

Material Selection. The material systems currently in commercial use include ferrous alloys (low-alloy steel, stainless steels, soft magnetic alloys), nonferrous alloys (brass, bronze), tungsten carbide, pure nickel, electronic alloys (Invar, Kovar), and tungsten-copper composites. The physical and mechanical properties of several MIM engineering alloys (low-alloy steel and stainless steels) have been standardized by MPIF (Ref 10). After sintering, the residual porosity in MIM parts is very low and not interconnected. With densities typically in excess of 96% of theoretical density, the resultant mechanical properties are superior to conventional die-compacted P/M materials and more closely match the properties of investment castings in similar alloys. Therefore, the limitations discussed previously regarding the residual porosity in conventional die-compacted products do not apply to MIM parts. Secondary operations can be performed with no restrictions.

Powder Forging

The design issues in P/F are similar to the requirement of any precision, closed-die forging. The difference is the starting preform; in the case of P/F, the preform is a sintered powder metal part, typically 80 to 85% of theoretical density, with a shape similar to the final part configuration. By contrast, in a precision closed-die forging the preform is a wrought steel blank with very little shape detail. Preform design for P/F fabrication determines the extent of product shape detail required to meet the performance requirements of the finished P/F part. Preform design is a complex, iterative process currently modeled by computer simulation software programs to help reduce design time and development costs.

In the forging step, the P/M preform is removed from the reheat (or sintering) furnace, coated with a die lube, and forged in a heated, closed die operation. The forging process reduces the preform height and forces metal into the recesses of the closed die. This step also brings all features to their final tolerances and densities.

Configuration guidelines, typical of precision closed-die forged parts, also apply to P/F parts as follows:

- Radii on inside corners of the forging as large as possible to promote metal flow around corners in the tool and promote complete fill of all details.
- Radii of at least 1 mm (0.040 in.) on all outside corners of the forging to aid in material flow to define features.
- Shape of the forging should be such that, when placed in the die, the lateral forces will be balanced. Shapes that are symmetrical along a vertical plane, such as connecting rods and shapes that are axisymmetric (or nearly so), are preferred.
- Zero draft is possible on surfaces formed by the die and core rod, but not by the upper punch.
- Re-entrant angles (undercuts) cannot be forged.
- Axial tolerances—in the direction of forging—are driven by variations in the mass of metal in the preform. Lateral tolerances are driven by metal flow as the cavity fills. Typical axial tolerance of 0.25 to 0.5 mm (0.010 to 0.020 in.) are encountered, with diametric tolerances of 0.003 to 0.005 mm/mm of diameter.
- Concentricity of a P/F part is determined by the quality and density distribution in the preform. Concentricity is normally double that of the preform.

REFERENCES

1. F.V. Lenel, *Powder Metallurgy—Principles and Applications,* Metal Powder Industries Federation, 1980, p 426
2. L.F. Pease III and V.C. Potter, Mechanical Properties of P/M Materials, *Powder Metallurgy,* Vol 7, *ASM Handbook* (formerly *Metals Handbook,* 9th ed.), American Society for Metals, 1984, p 467
3. H.I. Sanderow, H. Rodrigues, and J.D. Ruhkamp, New High Strength 4100 Alloy P/M Steels, *Prog. Powder Metall.,* Vol 41, Metal Powder Industries Federation, 1985, p 283
4. D. Gay and H. Sanderow, The Effect of Sintering Conditions on the Magnetic and Mechanical Properties of Warm Compacted Fe-P P/M Steels, *Advances in Powder Metallurgy and Particulate Materials—1996,* Vol 6, Metal Powder Industries Federation, 1996, p 20–127
5. "Material Standards for P/M Structural Parts," Standard 35 1994 edition, Metal Powder Industries Federation, 1994
6. "Standard Specification for Powder Forged (P/F) Ferrous Structural Parts," B 848-94, *Annual Book of ASTM Standards,* American Society for Testing and Materials
7. R.M. German, *Powder Metallurgy Science,* 2nd ed., Metal Powder Industries Federation, 1994, p 193
8. *Powder Metallurgy Design Manual,* 2nd ed., Metal Powder Industries Federation, 1995
9. "Material Standards for P/M Self-Lubricating Bearings," Standard 35 1991–1992 edition, Metal Powder Industries Federation, 1991
10. "Material Standards for Metal Injection Molded Parts," Standard 35 1993–1994 edition, Metal Powder Industries Federation, 1993

Advances in Powder Metallurgy Applications

F.H. "Sam" Froes, Institute for Materials and Advanced Processes (IMAP), University of Idaho
John Hebeisen, Bodycote IMT, Inc.

POWDER METALLURGY parts can be broadly separated into three categories: those in which the P/M approach allows a lower-cost component to be produced, an intermediate category of cost-effective/high-performance parts, and those in which the P/M approach leads to a part with enhanced mechanical property characteristics (Ref 1–3). The first approach generally results in a product with lower mechanical properties than wrought product, the third is normally a higher-cost approach. The low-cost P/M product is the more traditional commercial approach, the second has been more recently established commercially, and high-performance parts represent a developing facet of P/M technology.

Many P/M parts are now used in a variety of industries, including automobiles, household appliances, yard and garden equipment, computers, fabric industry equipment, and orthodontic devices (Ref 4). The growth in North American metal powder shipments up to 1996 is shown in Table 1 (Ref 5).

In 1995, North American powder metal shipments totaled 437,774 tons, a 2% increase from the previous year. Iron and steel powder shipments amounted to 347,172 tons, up 2.7% from 1994. Parts applications for iron/steel powders accounted for a record-high 312,974 tons, an increase of 3.1%. However, it was the first time in four years that the parts market improved by a percentage of less than double digits. In the 1992 to 1994 period, the parts market for North American iron and steel powder producers, the biggest market of its kind in the world, grew at an average annual rate of 18.6%.

Automotive parts continue to be the leading application of P/M parts. The typical U.S. automobile contains about 14 kg (30 lb) of P/M parts with an increase expected in the next several years (Fig. 1). A slightly lower amount is used in Japanese automobiles (Table 2) (Ref 6), but automotive application of P/M parts is still the dominant use in Japan as well (Fig. 2). The growth of P/M parts in automobiles is due to increased use of P/M components in engines, transmissions, brakes, airbags, and other complex parts. Emerging automotive applications are also described at the end of this article.

Table 1 North American metal powder shipments

	Shipments, tons				
	1992	1993	1994	1995	1996
Iron and steel	246,300	287,550	337,850	347,172	350,603
Copper and copper base	20,000	22,400	23,100	23,216	22,891
Aluminum	29,700	29,500	43,700	37,044	34,179
Molybdenum	2,500(E)	2,500(E)	2,500(E)	2,500(E)	2,500(E)
Tungsten	1,450	1,900	1,450	1,445	1,000(E)
Tungsten carbide	4,500	5,200	6,200	10,846	11,200(E)
Nickel	9,900	9,600	10,000	10,476	11,600(E)
Tin	950	1,100	1,250	1,010	1,010
Total	**315,300**	**359,750**	**426,050**	**433,709**	**434,983**

E, estimate. Source: Ref 5

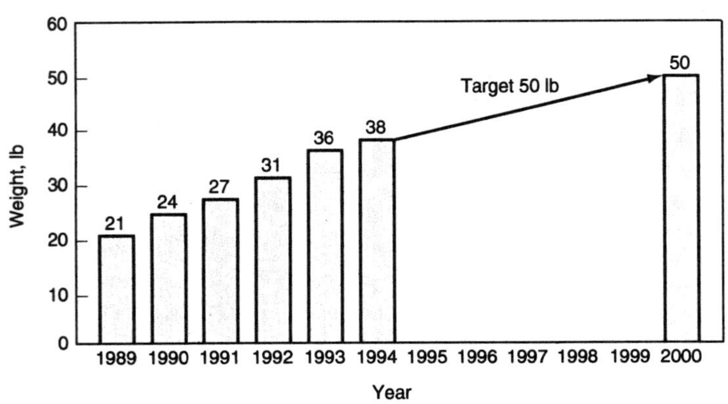

Fig. 1 Growth of powder metallurgy in Ford automobiles

Net Shape Capability

A major advantage of the P/M approach is the net-shape capability, particularly for high-strength materials. This has led to continuing developments in P/M technologies for parts production.

Table 2 Average weight of P/M components used in each Japanese car

Year	Cars, $\times 10^3$	PM Part, t	Weight kg/car	Weight lb/car
1980	11,175	33,923	3.03	6.68
1987	12,350	52,921	4.29	9.46
1988	12,819	60,046	4.68	10.3
1989	12,953	70,138	5.41	11.9
1990	13,592	75,459	5.55	12.2
1991	13,145	75,099	5.71	12.6

Source: Ref 6

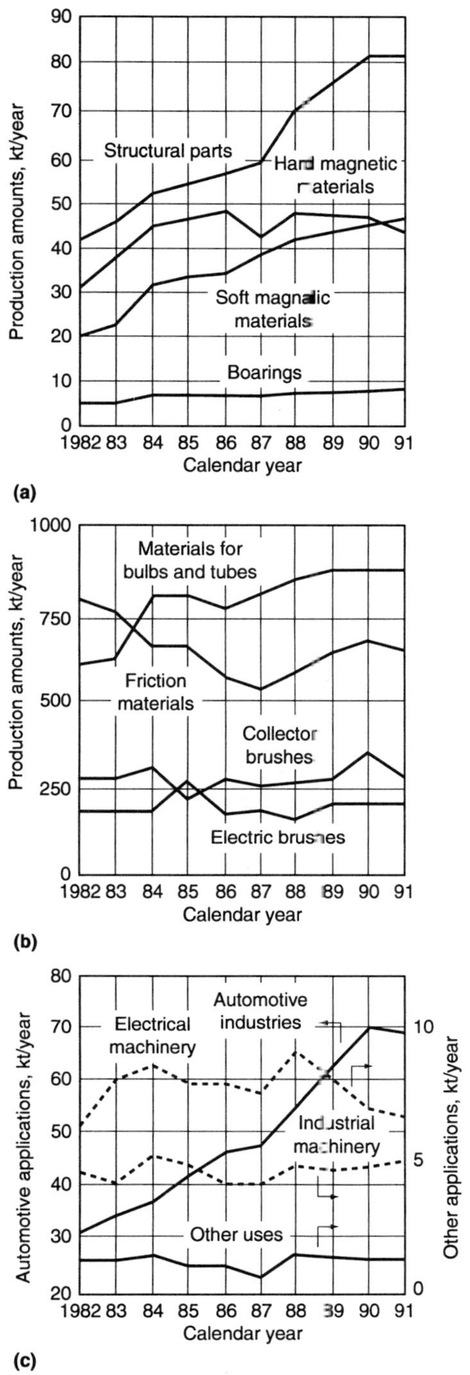

Fig. 2 Annual production amounts for P/M parts in Japan. Source: Ref 6

Warm Compaction. One recent development in P/M production is warm compaction, which allows the production of higher-density ferrous P/M parts via a single compaction process. The process utilizes heated tooling and powder during the compaction step. The powder and tooling are typically heated between 130 and 150 °C (260 and 300 °F). In order for the powder premix to perform at these temperatures, a proprietary lubricant system has been developed that provides lower die ejection forces than conventional lubricants. This lubricant system also incorporates a polymeric binder system to limit segregation and provide enhanced flow characteristics of the powder premix. By utilizing warm compaction technology, the green density of the consolidated part can be increased from 0.10 to 0.25 g/cm^3 over traditionally processed (single-pressed/single-sintered) materials. The green strength is typically increased between 50 and 100%.

Table 3 (Ref 7) compares the green properties of warm-compacted and cold-compacted P/M parts. This increase in green strength provides advantages such as a reduction in green chipping and cracking due to part handling prior to sintering and makes possible the crack-free compaction of complex multilevel parts. Additionally, the higher green strength provides an opportunity to machine the P/M part in the green state. This capability is critical in the use of high-performance alloy systems that achieve high hardness in the as-sintered state. Warm compaction also enables P/M fabricators to single press and single sinter ferrous P/M parts to densities as high as approximately 7.4 g/cm^3, which is considerably higher than cold-compacted-and-sintered P/M parts.

Warm compaction, although applicable to all ferrous material systems, produces the greatest benefits when coupled with high-performance compositions such as diffusion-alloyed steels or molybdenum-prealloyed steels. Achieving densities in excess of 7.25 g/cm^3 using these compositions results in mechanical properties that are comparable to steel forgings and ductile iron castings.

Table 3 Effect of processing on the green properties of ferrous compacts

Base material	Processing technique	Compaction pressure, MPa	Compaction pressure, tsi	Green density, g/cm^3	Green strength, MPa	Green strength, psi	Peak die ejection force, MPa	Peak die ejection force, tsi
Ancorsteel 85HP(a)	Warm compaction	415	30	7.14	23.2	3370	29.6	2.15
		550	40	7.31	25.4	3685	33.5	2.43
		700	50	7.37	24.7	3580	32.0	2.32
	Cold compaction	415	30	7.00	9.9	1430	37.2	2.70
		550	40	7.19	12.2	1770	50.7	3.68
		700	50	7.29	13.4	1950	53.8	3.90
Distaloy 4800A(b)	Warm compaction	415	30	7.07	28.3	4100	27.4	1.99
		550	40	7.29	30.6	4445	31.7	2.30
		700	50	7.36	31.1	4515	32.3	2.37
	Cold compaction	415	30	6.93	12.2	1770	37.2	2.70
		550	40	7.15	15.0	2170	48.5	3.52
		700	50	7.26	16.9	2450	52.0	3.77

(a) Ancorsteel 85HP is a prealloyed steel powder containing 2.0% Ni, 0.85% Mo, 0.4% graphite, and 0.6% lubricant. (b) Distaloy 4800A is a diffusion-alloyed steel powder containing 4% Ni, 1.5% Cu, 0.50% Mo, 0.5% graphite, and 0.6% lubricant. Source: Ref 7

Stainless Steels. Other developments in conventional P/M steels include new and improved stainless steels, with the goal of improving compressibility and corrosion resistance. Developing applications include automobile parts such as a solenoid spacer in an electronic fuel injector, sealing washers in the water pump, and brake components. They also include other applications such as bearing holders and pulleys in computers.

Metal injection molding (MIM), which is finding increased use in small parts manufacturing, uses fine powders (5–10 μm) that exhibit good sintering densification in combination with binders that hold the particles in place for transportation (Ref 8). The basic five steps involved are feedstock characterization, mixing rheology, injection molding, debinding, and sintering (see Fig. 4 in the article "Powder Metallurgy Methods and Design" in this Volume). With tolerances as low as 0.3%, sizing is generally not required. Densification to a usable article is accomplished in the sintering furnace. However, surrounding these seemingly simple steps there is a great deal of "know-how" that has been developed.

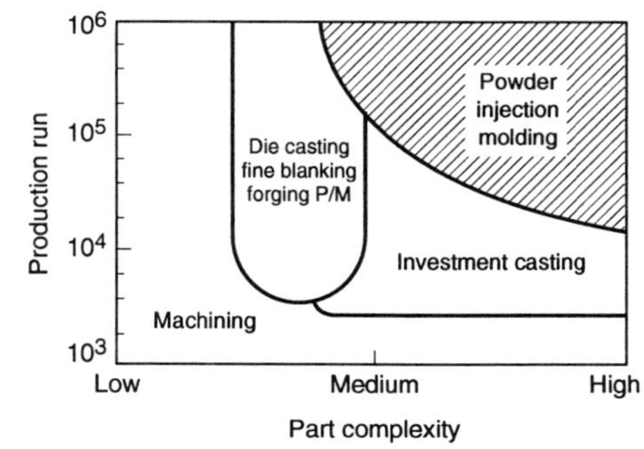

Fig. 3 Section of the part-size/production-run/part-complexity diagram where the MIM process is most effective. Source: Ref 8

Metal injection molding occupies a certain region in the part-size/production-run/part-complexity scenario (Fig. 3). Sizes are generally in the 1 to 200 g range, but parts as large as 1 kg have been made by MIM. The near-net-shape capabilities of the process can reduce machining costs to low levels, especially for long runs (≥10,000 parts). For small complex components, cost savings can be up to 80% compared to conventional approaches. A wide range of geometric options are possible including undercuts, tapered external surfaces, and crossholes. Short production runs can be cost effective, but because of die and powder costs, this only occurs with very complex shapes and/or hard materials.

Some MIM fabricators produce runs of 2000 to 5000 pieces, particularly on more expensive parts (Ref 9). They can do a short run such as

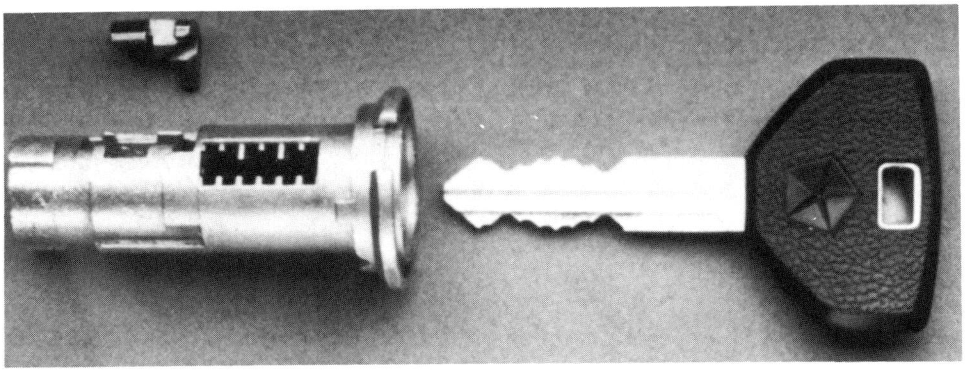

Fig. 4 MIM part (upper left) for an automobile ignition lock. The key forces the MIM part into contact with a security switch. Courtesy of SSI Technologies

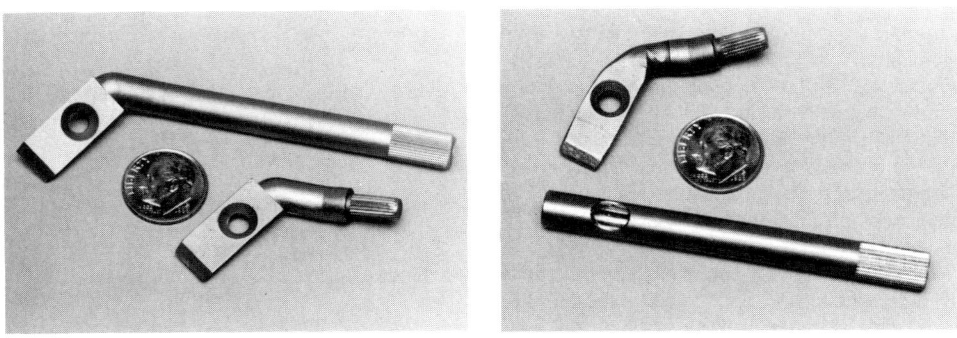

(a) (b)

Fig. 5 (a) Single-piece MIM part that replaced (b) a two-piece automobile turn signal lever assembly. The smaller MIM part in (a) was the first version, while the larger MIM part is the finished version that replaced the two-part assembly shown in (b). Courtesy of the Remington Arms Division of E.I. Du Pont de Nemours & Company, Inc.

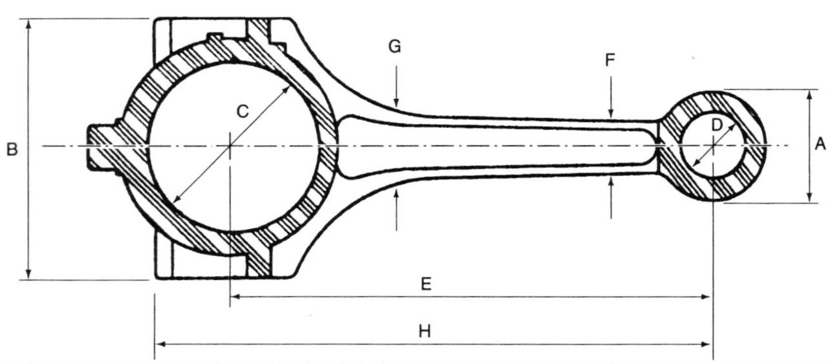

Dimension	Powder forge variation(a), %	Conventional forge variation(a), %
A: Pin end outside diameter	0.19	2.09
B: Bolt boss width	0.11	0.28
C: Crank bore diameter	0.14	0.31
D: Pin bore diameter	0.15	(b)
E: Bore-to-bore center distance	0.10	0.17
F: I-beam width at pin end	0.42	3.16
G: I-beam width at crack end	0.35	0.95
H: Bolt head seat location	0.06	1.11
I: Crank end thickness(c)	0.58	1.13
J: Pin end thickness(c)	0.45	0.48

(a) Variation evaluated as the ratio of the range of measurements to the mean dimension. (b) Not applicable. Pin end forged solid. (c) Not shown in schematic

Fig. 6 Dimensional control of P/F versus conventional forging. Source: Ref 10

these because there is no danger to the tooling at setup, while in conventional P/M there is more risk. The capital equipment (presses) for injection molding is more economical than that for large-scale P/M presses. Tool life is at least 300,000 pieces. These factors help offset the added short-term material cost.

Two of the first MIM production parts in automobiles were parts for an ignition lock (Fig. 4) and a single-part replacement (Fig. 5a) for a two-part turn signal lever assembly (Fig. 5b). Both have been in service since July 1988.

Figure 4 shows the entire ignition lock and the MIM subcomponent. As the key is inserted in the lock, the cam-shaped MIM part moves away and depresses an electrical switch, which is part of the security system. The initial design of the part was too small and complicated for the model shop to make, and it was prototyped from the MIM tooling.

The turn signal indicator lever is an example of the replacement of a two-piece assembly (Fig. 5b) with a single MIM part. The lower portion of Fig. 5(a) shows the first version of the MIM part, and the upper view shows the final 19.0 g (0.670 oz) MIM part that replaced the assembly. The MIM material is iron with 2% Ni, sintered and then case hardened. It replaced AISI 4037 and SAE 1018 case hardened. The MIM part succeeded because of its superior strength compared to the two-piece assembly. The core properties of the materials are 415 MPa (60 ksi) tensile strength and 15% elongation at 60 HRB.

Hot powder forging (P/F) continues to be attractive for the production of fully dense P/M parts for demanding applications (Ref 10–12). In this process a loose powder is blended to the desired composition and pressed to a forging preform having the general shape of the final component. Powder forging has proven to be competitive when the overall economics are improved through some combination of enhanced machining characteristics, mechanical properties, and dimensional or weight tolerances.

It was initially believed that P/F products would displace a wide variety of conventionally processed P/M parts, as well as a significant number of conventional forgings. However, the number of high-volume applications has been limited to bearing races, connecting rods, and ring gears.

One of the more dramatic applications in the mid-1980s was P/F connecting rods. Compared with conventional forging, P/F technology improved weight and dimensional control (Fig. 6) and reduced machining requirements (Fig. 7). In addition, P/F connecting rods are always forged in one piece, where the rod and cap are separated by "fracture splitting." Use of this technique eliminates several machining operations. In addition, the irregular, mating fracture surfaces (ductile failure mode) provide an intimate interlock between rod and cap. This virtually eliminates both "cap shift"—rotation of the cap relative to the rod—and lateral movement of the cap relative to the rod. Cap shift can lead to accelerated wearing of surfaces and, in extreme cases,

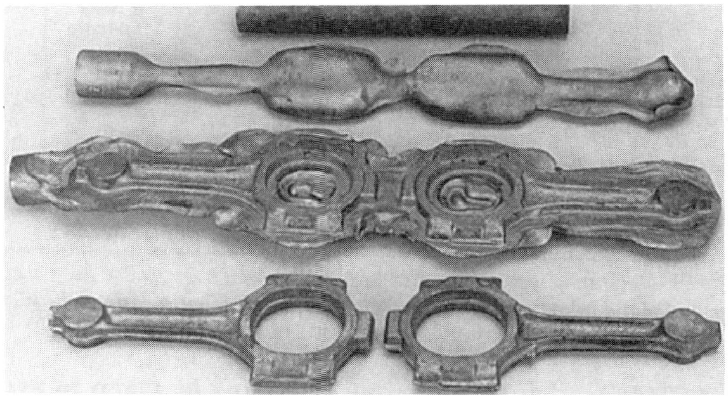

(a)

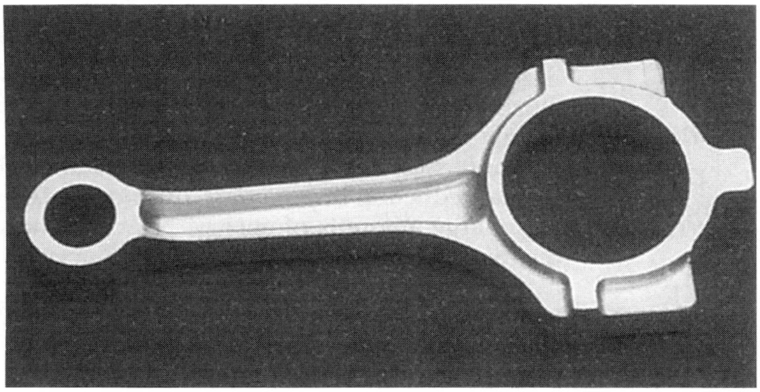

(b)

Fig. 7 Comparison of fabrication sequence for connecting rods. (a) Typical sequence for conventional forging of rods. The pin end is forged solid and requires machining steps that P/F eliminates. (b) Powder forged connecting rod for a Ford 4.6 L V-8 engine. Note its good surface finish, near-net shape, and small weight control pad at the crank end.

bearing seizure. Lateral movement can result in high shear stress on connecting rod bolts at high engine revolutions per minute.

Cost-Effective, High-Performance Parts

So-called "mid-range" parts are not generally considered exceptionally low in cost, but rather as parts for which a cost reduction and improved behavior can both be obtained. There are many parts in this category that are the subject of current production programs. They generally fall into three categories of parts: billet materials, P/M near-net shapes, and clad composite materials.

Billet Materials. A good example of billet materials are high-speed steel billets made by hot isostatic pressing (HIP) of gas-atomized powders. This material is current bill-of-material for about 8000 tons/year of high-speed steels manufactured in both the United States and Europe (Ref 13). It is typically competitively priced with cast and wrought material and has a strong advantage for the larger size products due to its fine, uniform carbide size. This gives it manufacturing advantages (especially grindability) and performance advantages over conventional cast and wrought material. Other widely used P/M billet materials include sputtering targets for applying the coatings used in data storage applications. Important materials are powders such as chromium, chromium alloys, and cobalt alloys. Important characteristics are 100% density, high purity, and uniformity of chemical composition, traits that are the trademark of HIP P/M materials.

Powder metallurgy near-net shapes have made their mark for production applications in several industries. Power turbines for instance have used rotors, rings, and disks (Ref 14) HIP densified from steel alloys such as 12% Cr (X 20 CrMo V 12 1) and 9% Cr 1% Mo (Grade 91). The P/M alloys are superior to their wrought counterparts in that the properties developed are isotropic and are equivalent to the maximum wrought properties in the forging direction. Powder metallurgy near-net shape manifolds and valve bodies have become standard production items in the North Sea oil fields in alloys such as austenitic and duplex (ferritic-austenitic) stainless steels (Ref 15). Manifold sections can be made as net shapes with no machining except for the mating surfaces on connecting flanges. Individual parts are then welded into sections for use in underwater applications. Also, P/M near-net shapes are finding acceptance for pump and valve components in hostile environments.

The advantage these parts have compared to closed-die forgings is that they have the same properties as the maximum forged properties in all directions (isotropic), but with inside detail that is not possible with forgings. Also, the delivery time is much shorter for first deliveries for the P/M parts due to much more simplified tooling.

HIP cladding of powder materials on solid substrates has become an important engineering technique in a number of applications involving improving corrosion, erosion, and wear properties. The advantage of the technique is that the P/M alloy can be engineered precisely for the environmental factors and then applied in sparing quantities in such a way that a metallurgical bond is created with a low-cost substrate that will withstand the rigors of the application. A good example is tooling for plastic extrusion. Corrosion resistance can be provided, for example, by selecting a gas-atomized nickel-chromium alloy as a base and blending in tungsten carbides for wear resistance. This powder can be HIP bonded to the inside surface of relatively inexpensive steel barrels to provide a duplex material capable of giving long life in extruder service.

Advanced Synthesis

Far-from-equilibrium synthesis leads to novel constitutional and microstructural effects and enhanced physical and mechanical properties (Ref 16). A number of these techniques are summarized in the sections that follow in Table 4 with emphasis on the amount of departure from equilibrium.

Table 4 Departure from equilibrium achieved in various processes

Process	Maximum departure from equilibrium, kJ/N_a
Solid-state quench	16
Quench from liquid (rapid solidification)	24
Mechanical alloying	30
Condensation from vapor	160

N_a, Avogadro's number of atoms

Rapid Solidification

Rapid solidification (RS) allows extension of solubility limits, production of novel phases, and more refined microstructures than ingot metallurgy (I/M) techniques. The greatly increased chemistry/microstructure "window" can lead to enhanced mechanical and physical properties.

Aluminum. Five families of alloys being explored using the RS approach are the high-strength corrosion-resistant alloys based on traditional 7000-series alloys, low-density aluminum-lithium alloys with increased lithium levels over those possible using the I/M approach, dispersoid-strengthened elevated-temperature alloys based on low-solubility/low-diffusion-rate additions such as the transitional metals (iron, molybdenum, nickel) and rare earth elements (cerium), wear-resistant high-silicon alloys, and recycled alloys (in which conventional ingot processing would lead to excessive segregation).

Magnesium. Cast magnesium alloys exhibit lower than desirable strength, ductility, and creep behavior, and the nonprotective oxide skin can lead to severe corrosion problems. High-strength, corrosion-resistant magnesium alloys containing rare earth additions (yttrium, neodymium, cerium) have been developed using

Table 5 Tensile properties of elevated-temperature P/M RS aluminum alloys at 315 °C (longitudinal)

Alloy	Ultimate tensile strength, MPa	Yield strength, MPa	Elongation, %
Al-Fe-Ce	270	225	7
Al-Fe-Mo	235	210	10
Al-Fe-V-Si	310	300	7
Al-Zr-Cr-Mn	235	215	...

Table 6 Room-temperature mechanical properties of MA aluminum alloys (longitudinal)

Process	Alloy	Ultimate tensile strength, MPa	Yield strength, MPa	Elongation, %	Fracture toughness (K_{Ic}) MPa$\sqrt{m}$	Density, g/cm^3
Mechanical alloying	Al-Li-Mg-O-C	510	450	10	45	2.55
Ingot metallurgy	7075-T73	505	435	13	32	2.81

Source: Ref 2, 3

RS. Mg-Al-Zn-*X* (*X* = Si, Y, Nd, or Ce) alloys can exhibit tensile strengths ranging between 450 and 510 MPa. These alloys have strength and ductility combinations equivalent to high-strength aluminum alloys and are five times more corrosion resistant than the most resistant conventional magnesium alloys.

Titanium. The major concentration on RS terminal alloys to date has been to enhance elevated-temperature capability beyond I/M alloy levels (i.e., >700 °C) through dispersion hardening (Ref 2, 3). The additions of erbium and other rare earth elements produce dispersoids that resist coarsening at least up to 800 °C; however, much further optimization is required particularly in increasing the volume fraction of second-phase particles. Alloys containing additions such as iron offer the potential for extremely high-strength levels (>1400 MPa ultimate tensile strength), with possible applications as replacement for steel in components such as landing gears.

Titanium Aluminide Intermetallics. Improved ductility can be achieved by disordering, grain refinement, and deoxidation of the matrix, while good elevated-temperature properties are achieved due to a dispersion of fine, thermodynamically stable, second-phase particles. In combination with the near-net-shape advantages offered by the P/M approach, RS may offer some advantages for the processing of γ alloys over the I/M approach; presently, the same cannot be said for RS processing of α_2-Ti_3Al alloys.

Spray Deposition

In spray deposition, finely divided molten metal droplets, produced by disintegration of a stream of molten metal using high-energy inert gases, impinge on a substrate before they completely solidify. This allows some of the characteristics of RS to be achieved in combination with a near-net-shape capability.

Dramatic enhancements in mechanical behavior can result because of the refinement in constituent particle size. An example of this is shown in Fig. 8 for 7075 aluminum. The spray deposition of titanium-base alloys also has been investigated only recently with much further development necessary.

Mechanical Alloying

Mechanical alloying (MA) is a process in which heavy working of powder particles results in intimate alloying by repeated welding and fracturing. This process has the same attributes as RS: extension of solubility limits, production of novel structures and refinement of the microstructure (down to the nanostructure range); additionally, it allows production of a dispersion of second-phase particles.

Aluminum. High-temperature alloys with titanium additions, low-density alloys with magnesium additions, and extra-low-density alloys with lithium additions have been developed through MA. The mechanical property combination that can be obtained in an MA aluminum-base alloy is shown in Table 6. The MA aluminum-lithium alloys exhibit minimal degradation of properties when stressed in the transverse direction and are characterized by excellent general corrosion resistance that is 100× better than that exhibited by the I/M alloy.

Magnesium. Mechanical alloying has been employed in the development of "supercorroding" alloys for submarine applications, such as a heat source in diver suits, and as a hydrogen gas generator.

Titanium. Mechanical alloying of titanium alloys has resulted in supersaturated solid solutions, metastable crystalline and glassy phases as well as nanometer-sized grain structures.

Titanium Aluminide Intermetallics. Work on MA of the titanium aluminides indicates that this is an interesting fabrication method for both the α_2 and γ families of alloys, resulting in the formation of surprisingly stable nanosized grains, which are stable even after compaction by HIP (Ref 17–19). In dispersoid-strengthened material in comparison to RS alloys, MA alloys exhibit finer grain structures and absence of dispersoid-free zones near grain boundaries; further dispersoids do not coarsen significantly up to very high temperatures; however further work is needed to optimize the product.

Plasma Processing

In plasma processing, solid feedstock is converted to the vapor phase at elevated temperatures due to the highly concentrated enthalpy. Chemical reactions occur, and a solid product nucleates and grows at a lower temperature. A number of studies have led to the formation of nanometer-sized metallic powders.

Vapor Deposition

Production of alloys directly from the vapor phase allows even greater flexibility in constitutional and microstructural development than RS and MA. Either monolithic material or alternate layers of two or more metals can be produced (Ref 21, 22).

Aluminum. The strength of a vapor deposited Al-7.5Cr-1.5Fe alloy is significantly higher than the strength of RS alloys up to at least 250 °C, a result of the fine microstructure produced by the vapor deposition technique. It is also possible to produce nanostructured layered structures that exhibit novel strength and modulus combinations (Ref 21, 22).

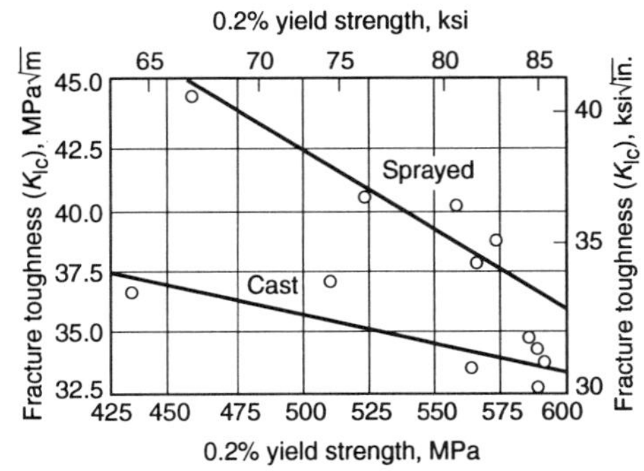

Fig. 8 Enhanced fracture toughness in spray-deposited Al-7075

Titanium. The electron beam vapor quenching technique has been used to produce alloys that are not possible by I/M or even RS. One example is the production of titanium-magnesium alloys. Magnesium boils below the melting point of titanium making production of a liquid alloy impossible by conventional methods. Using the electron beam vapor quenching technique up to 28 wt% Mg can be alloyed with titanium. This is a potentially attractive alloy because magnesium lowers the density of titanium in excess of 1% for each 1 wt% Mg added.

Thermohydrogen Processing

Thermohydrogen processing (THP) is a technique in which hydrogen is added to a metal or alloy to enhance processing and improve final mechanical properties (Ref 23–26). To date it has only been applied to any extent to titanium-base materials. The THP technique has been used to improve the processibility of alloys such as Ti-6Al-4V and to refine the microstructure with a concurrent enhancement in crack-initiation-related properties such as fatigue.

Preliminary studies have shown that the THP approach can also be used to refine the microstructure of Ti_3Al-type alloys, and the technique is being used on a pilot-plant basis at the All Union Institute for Light Alloys Moscow, Russia, to improve processibility. Because of the conventionally very low solubility for hydrogen in TiAl-base alloys, use of innovative techniques such as RS are required in combination with THP.

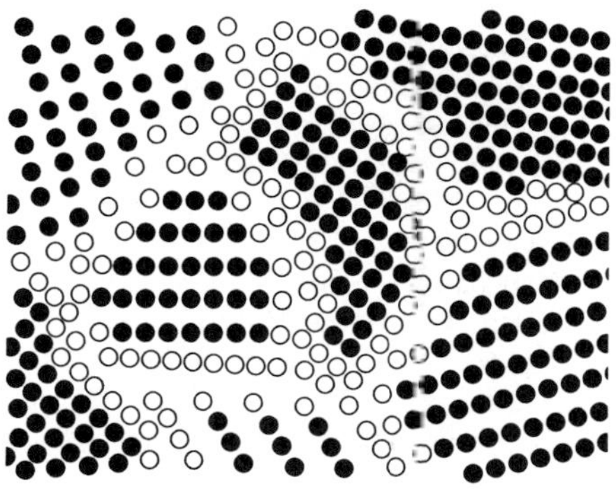

Fig. 9 Schematic of nanostructured material. Black circles represent atoms in normal lattice positions within the grains; white circles indicate atoms that are "relaxed" in grain-boundary regions. Source: Ref 27

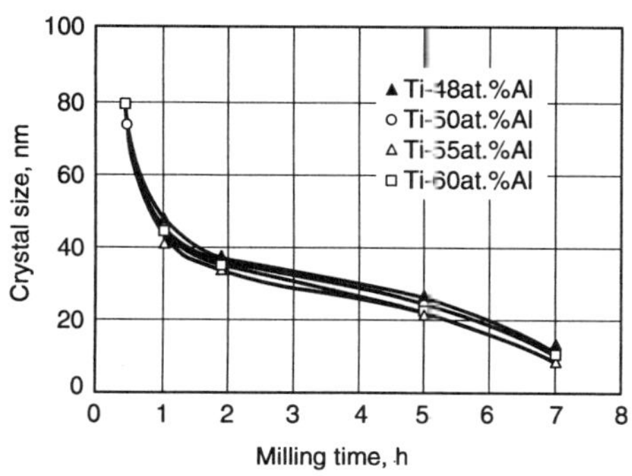

Fig. 10 Reduction in grain size with MA time determined from x-ray diffraction line broadening, for a number of TiAl-type alloys

Nanostructured Materials

Nanostructured materials are materials with at least one dimension in the nanometer range (1 nm = 10^{-9} m), generally ≤100 nm. Because of their novel combinations of mechanical, physical, and magnetic properties they have received considerable attention in the past few years (Ref 27–30). The nanostructures can be one-dimensional (layered), two-dimensional (fibrous), or three-dimensional (crystallites). However, the vast majority of work to date has been on the third type, which is generally produced using a P/M approach, and this article only considers this type of nanostructured material. A schematic representation of a nanostructured crystallite material is shown in Fig. 9 (Ref 27). The large fraction of atoms located in the grain boundary regions results in novel processing possibilities and enhanced combinations of mechanical, physical, and magnetic behavior compared to materials, with a more conventional grain size (>1 µm). Because of the "far-from-equilibrium" nature of nanostructured powder, compaction is a major concern, particularly retention of nanometer-sized (≤100 nm) grains.

Methods of production of nanostructured materials have concentrated on gas condensation methods, but chemical, mechanical (e.g. mechanical alloying) and plasma (see earlier) techniques have also been used (Ref 27–30). During the MA process the powder particle size and also the grain size are continuously refined, and very often nanometer-sized grains are produced (Fig. 10) (Ref 1, 2). Not only can nanostructured grains be mechanically alloyed, but these grains do not show excessive growth during the compaction cycle (Ref 1, 2).

Metal-Matrix Composites

Aluminum. AMC Ltd. has used a P/M approach involving mechanical attrition and HIP to achieve an exceptionally uniform dispersion of particles and consistent mechanical properties (Table 7) (Ref 1, 2).

Magnesium. A silicon carbide particulate reinforced magnesium alloy MMC has been developed for defense and automotive applications (Ref 1, 2). With a density of 2 g/cm^3 and a modulus of 63 GPa, thin-walled tube can be manufactured 18% lighter than an equivalent aluminum alloy tube. The composite also has good wear resistance and a reduced coefficient of thermal expansion compared with magnesium alloys.

Titanium. A recent program on fabrication of components using the blended elemental (BE) MMC titanium approach had as its main goal development of low-cost products that could compete for automotive applications while maintaining acceptable mechanical property levels (Ref 1, 2). Expensive extra-low chloride hydride-dehydride (HDH) titanium powder and HIP are out of consideration for automobile use because of the added cost. In a recent fabrication program, titanium sponge fines, master alloy, and boron (or boride) were blended together and given a MA (kneading) treatment, followed by cold pressing and sintering. The TiB formed was judged to be the optimal reinforcement because of lack of reaction with the matrix and good physical and mechanical characteristics. The MA treatment broke up large remnant NaCl particles (100–200 µm in diameter), resulting in elimination of large pores. Mechanical properties of Ti-6Al-4V with 10 vol% TiB_2 are shown in Fig. 11 in comparison with HDH materials. The fatigue behavior is enhanced, and wear resistance is excellent. Other titanium alloys have also been shown to exhibit attractive properties fabricated using this new approach (Ref 1, 2) with 10 vol% TiB.

Table 7 Typical mechanical properties for Al 2124 particle-reinforced MMC produced by the P/M route

SiC content, vol%	Elastic modulus (*E*), GPa	Ultimate tensile strength, MPa	0.2% yield strength, MPa	Elongation, %
17	100	600	400	8
25	115	700	460	7
35	135	540	450	2.5

Advanced Automotive Use

The use of P/M parts in automobiles can only increase due to the influence of the Partnership for Next Generation Vehicle (PNGV) program. The PNGV program is a "network" between the government, the car manufacturers (USCAR: Chrysler, Ford, and GM) and Suppliers/Universities/Others with Government funding currently at greater than $100 million per year. In summary, the goal of the PNGV program is to have all of the features of today's car, with three times the fuel efficiency, affordability, manufacturability, repairability, and recyclability (greater than 80% of the vehicle to be recyclable). Clearly, this fuel efficiency must come mainly from a vehicle weight reduction, with the body and chassis potentially showing the greater weight reductions.

In Japan, a number of automotive components have been targeted by the Japanese Titanium Society for possible titanium use. In this regard, a comparison of the manufacturing costs of low-cost titanium BE parts with wrought titanium and current steel parts is shown in Fig. 12. The material developed requires little machining and is cost competitive with current steel parts.

Use of spray-formed and extruded parts have proven to be cost effective for use in the rotors of the high-efficiency Mazda Miller cycle automotive engine (Ref 31–34). This engine results in a reduction in fuel consumption of about 20%. Future plans are to use sprayed aluminum-

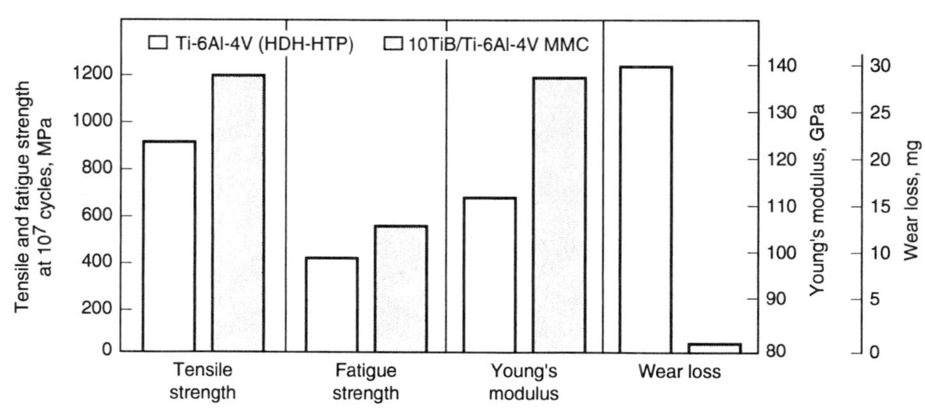

Fig. 11 Comparison of mechanical properties of Ti-6Al-4V + 10 vol% and HDH hot isostatically pressed Ti-6Al-4V

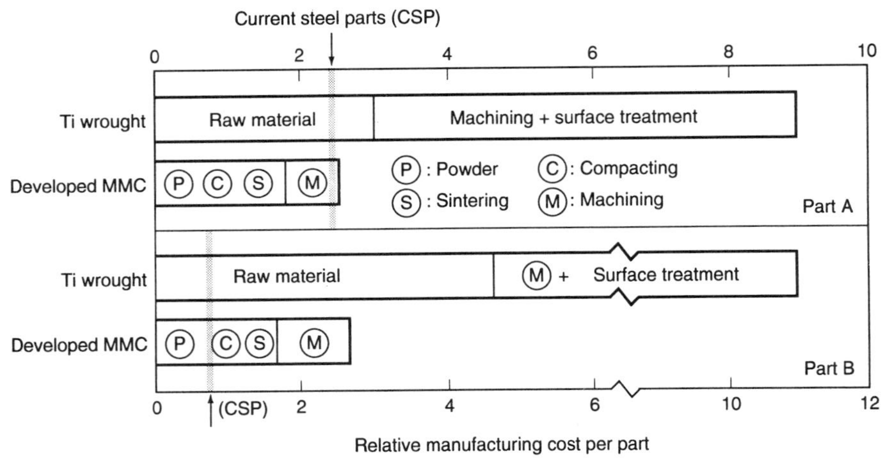

Fig. 12 Comparison of manufacturing costs of low-cost titanium BE composite parts with conventional wrought titanium and current steel parts

silicon in the connector rods, cylinders, and air-conditioning compressor of the Mazda EUNOS 800 (XEDOS 9), featuring a V-6 Miller cycle engine (Ref 31–34).

Aluminum alloy MMCs have been introduced in various automotive engine components, and applications are steadily increasing. Examples include Toyota diesel engine pistons selectively reinforced with 5% short alumina fiber (Ref 35), where the MMC leads to a weight reduction of 5 to 10%, wear reduced by four times, and seizure stress doubled; Honda cast aluminum cylinder blocks locally reinforced with 12% short alumina fiber and 9% C (Ref 36). Particulate-reinforced aluminum alloy MMCs, containing typically 15 to 30% SiC, are poised for automotive use, with promising applications including brake discs, drums, and calipers. The new Lotus Elise sports car has aluminum MMC discs for both front and rear brakes. Disc weight is reduced by about 50% compared with the alternative cast iron, and better thermal diffusivity gives increased resistance to overheating of the friction surfaces.

REFERENCES

1. F.H. Froes, C.M. Ward-Close, P.R. Taylor, and W.A. Baeslack, *P/M in Aerospace, Defense and Demanding Applications,* F.H. Froes, Ed., Metal Powder Industries Federation, 1995, p 3
2. F.H. Froes and P.D. Desai, "Recent Developments in Powder Metallurgy Processing Techniques," Report 5, MIAC, Purdue University, Oct 1994
3. F.H. Froes et al., *Light Materials for Transportation Systems,* N.J. Kim, Ed., Postech, Pohang, Korea, 1993, p 27
4. P.K. Johnson, *Industrial Heating Powder Metallurgy Supplement,* June 1996, p 4
5. www.mpif.org/indust.html, 10 Feb 1998, Metal Powder Industries Federation
6. Y. Morioka, Recent Trends in Powder Metallurgy Industry and Technology, *J. Jpn. Soc. Powder Powder Metall.,* Vol 4 (No. 8), Aug 1993, p 755–762
7. Product literature, Hoeganaes Corporation
8. P.J. Vervoort, R. Vetter, and J. Duszczyk, *Adv. Perform. Mat.,* Vol 3 (No. 2), 1995, p 121
9. L. Pease III, Ferrous Powder Metallurgy Materials, *Properties and Selection: Irons, Steels, and High-Performance Alloys,* Vol 1, *ASM Handbook,* ASM International, 1990, p 819
10. D.R. Bankovic and D.A. Yeager, *Adv. Mater. Process.,* Vol 146 (No. 2), 1994, p 26–30
11. H.A. Kuhn and B.L. Ferguson, *Powder Forging,* Metal Powder Industries Federation, 1990
12. D.R. Bankovic and A.J. Schlimm, *Near-Net-Shape Manufacturing,* P.W. Lee and B.L. Ferguson, Ed., ASM International, 1993, p 145
13. W.B. Eisen, P/M Tool and High Speed Steel—A Comprehensive Review, *Advanced Particulate Materials and Processes,* F.H. Froes and J.C. Hebeisen, Ed., Metal Powder Industries Federation, 1997
14. R. Ekbom, The Application of HIP P/M Technique for Gas and Steam Turbines, *Modern Developments in Powder Metallurgy,* Vol 18, P.U. Gummeson and D.A. Gustafson, Ed., Metal Powder Industries Federation, 1988
15. C.G. Hjorth and P. Ekstrom, *The HIP Process and Products,* Stainless Steel Europe, The Netherlands, June 1995
16. F.H. Froes et al., *Proceedings of International Conference on Novel Techniques in Synthesis and Processing of Advanced Materials,* J. Singh and S.M. Copley, Ed., ASM International, 1995, p 1
17. J. Hebeisen, P. Tylus, D. Zick, D.K. Mukhopadhyay, K. Brand, C. Suryanarayana, and F.H. Froes, presented at MPIF, ADDA Conference, *P/M in Aerospace, Defense and Demanding Applications,* F.H. Froes, Ed., Metal Powder Industries Federation, 1995, p 363
18. F.H. Froes, C. Suryanarayana, D.K. Mukhopadhyay, G. Korth, and J. Hebeisen, Compaction of Nanograined Gamma Titanium Aluminides, *1995 Int. Conf. and Exhibition on PM and Particulate Materials,* Metal Powder Industries Federation, 1995, p 227
19. L. Öveçöglu et al., Grain Growth Behavior of Nanograined Gamma TiAl Compacted by Hot Isostatic Pressing, *ASM International Conference HIP '96,* F.H. Froes, R.Widmer, and J. Hebeisen, Ed., ASM International, 1996, p 227
20. P.R. Taylor and S. Pirzada, *Adv. Perform. Mater.,* Vol 1 (No. 1), 1994, p 37
21. R.L. Bickerdike, D. Clark, J.N. Eastabrook, G. Hughes, W.N. Mair, P.G. Partridge, and H.C. Ranson, *Rapidly Solidified Materials,* P.W. Lee and R.S. Carbonara, Ed., ASM International, 1987, p 137
22. R.L. Bickerdike, D. Clark, G. Hughes, M.C. McConnell, W.M. Mair, P.G. Partridge, and B.W. Viney, *Rapidly Solidified Materials,* P.W. Lee and R.S. Carbonara, Ed., ASM International, 1987, p 145
23. F.H. Froes, D. Eylon, and C. Suryanarayana, *JOM,* March 1990, p 26
24. O.N. Senkov, J.J. Jones, and F.H. Froes, *JOM,* July 1996, p 42
25. F.H. Froes and O.N. Senkov, Thermohydrogen Processing of Titanium Alloys, *Int. Mater. Rev.,* 1998 (update planned for 1999)
26. F.H. Froes and C. Suryanarayana, Powder Processing of Titanium Alloys, *Reviews in Particulate Materials,* A. Bose, R. German, and A. Lawley, Ed., Metal Powder Industries Federation, 1993
27. H. Gleiter, *Prog. Mater. Sci.,* Vol 33, 1990, p 223
28. C. Suryanarayana and F.H. Froes, *Metall. Trans. A,* Vol 23, 1992, p 1071
29. R.W. Siegel, *Ann. Rev. Mater. Sci.,* Vol 21, 1991, p 559
30. F.H. Froes, C. Suryanarayana, G.H. Chen, A. Frefer, and D.K. Mukhopadhyay, *P/M in Aerospace, Defense and Demanding Applications,* F.H. Froes, Ed., Metal Powder Industries Federation, 1993, p 1
31. F.H. Froes and J. Storer, Ed., *Recent Advances in Titanium Metal Matrix Composites,* TMS, 1995
32. T. Saito, T. Furuta, and T. Yamaguchi, *Recent Advances in Titanium Metal Matrix Composites,* TMS, 1995, p 33
33. K. Shibue, presented at MPIF, ADDA Conference, *P/M in Aerospace, Defense and Demanding Applications,* F.H. Froes, Ed., Metal Powder Industries Federation, 1995, p 61
34. H. Sano, *Met. Powder Rep.,* Jan 1994, p 26
35. T.W. Clyne and P.J. Withers, *Metal Matrix Composites,* Cambridge University Press, 1993, p 459
36. *Adv. Mater. Process.,* June 1993, p 20

Process Modeling and Design

Robert Dax, R.J. Henry, T. McCabe, and P. Barrous-Antolin, Concurrent Technologies Corporation

CONCURRENT ENGINEERING methodologies drastically reduce or eliminate the need for trial and error methods in the design of process tooling or prototype components. Both the materials data and the analysis tools are readily available for product design activity. However, the process design activity has, in general, been hampered by a lack of readily available data on the mechanical behavior of materials under the conditions encountered during manufacturing. Furthermore, the mechanical behavior of metal powders is fundamentally different than that of metallic solids leading to the necessity to develop new process analysis tools tailored to this unique behavior. This article provides an update on process models used in the production of components by metal powder die compaction, powder injection molding, and hot isostatic pressing. More information on modeling of die compaction is in the article "Mechanical Behavior of the Metal Powders and Powder Compaction Modeling" in this Volume.

Metal Powder Compaction

In the die compaction process, shown schematically in Fig. 1, a die cavity of the desired geometric shape is filled with metal powder. Pressure, on the order of 350 to 700 MPa (50 to 100,000 psi), is applied by the axial movement of one or more punches. The pressure causes the metal powder particles to mechanically interlock and cold weld together into a porous mass of the approximate shape and dimensions desired for the final component. This as-pressed shape, commonly referred to as a green part, is then heated to an elevated temperature, on the order of 1120 to 1200 °C (2050 to 2200 °F), for 30 to 120 min to sinter the metal powder particles into a solid mass. There are several limiting factors in the use of die compaction to manufacture engineering components:

- Geometric shape is limited by the required axial movement of the punches.
- Friction between the powder particles and the tooling surfaces, combined with interparticle friction, hinders the uniform consolidation of the metal powder leading to density variations.
- Density variations in the green compacts lead to nonuniform size changes during sintering.
- Mechanical properties of the sintered components are a function of the local density variations.

However, the cost advantages of directly producing a near-net or netshape component combined with excellent part-to-part uniformity and reproducibility have made die compaction the process of choice for a wide variety of automotive, electronic, and consumer parts.

The key to successfully broadening the application of die compaction to the manufacture of engineering components is the development of a meaningful constitutive description of the deformation behavior of the powder during compaction. This description is complicated, requiring knowledge of the mechanical interaction of a large number of powder particles of varying sizes that are not uniformly distributed throughout the die and including the friction effects which create density variations during the compaction process. This effort is further complicated by the functional relationship between density and the elastic and plastic properties of the compact that, in turn, affects the frictional forces.

Given a constitutive equation describing the mechanical properties of the metal powder and a description of the frictional forces between the metal powder and the walls of the tooling, numerical solution techniques can be used to solve the mechanical equations of state that describe the compaction process. For example, finite element method codes, such as NIKE (Cray Research, Eagan, MN) or ABAQUS (National Supercomputer Centre, Linkoping, Sweden) have been used to produce solutions describing these density variations. The frictional forces between the metal powder and the walls of the tooling have been estimated using descriptors incorporated into these finite element programs. The material constitutive model is usually derived from experiments simulating the compaction process. To understand the limitations of these constitutive models, it is helpful to compare the deformation behavior of metal powder compacts with that of full density metals.

Constitutive Behavior

Constitutive material models provide a mathematical description of the deformation response of the material to an applied stress. In their simplest form, constitutive material models can be reduced to the equations describing stress -strain behavior in elastic loading. In their more complex form, constitutive material models include the effects of strain (strain hardening), strain rate, temperature, microstructure, anisotropy, and time (creep or viscoelastic deformation) on the stress-strain behavior. In practice, constitutive models for full density materials are generally configured to describe the stress state required under a given set of conditions to cause plastic flow. Ideally they can, furthermore, provide quantitative measures of the final dimensions of the material body after deformation and knowledge of the final mechanical properties and microstructure. Typically, these models describe the stress state by separating it into hydrostatic and deviatoric components, as shown in Fig. 2. The hydrostatic or spherical stress com-

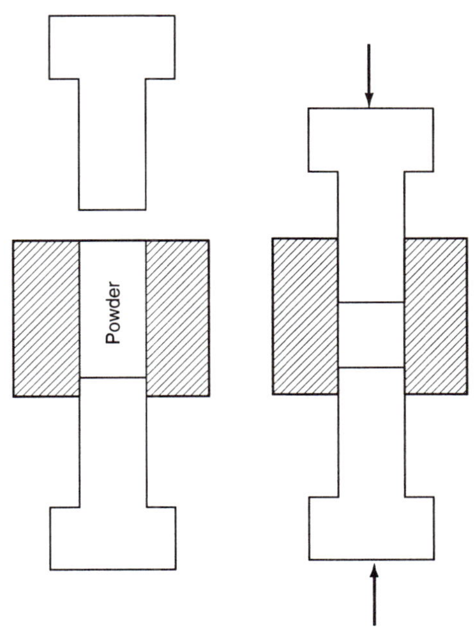

Fig. 1 Schematic of the die compaction process

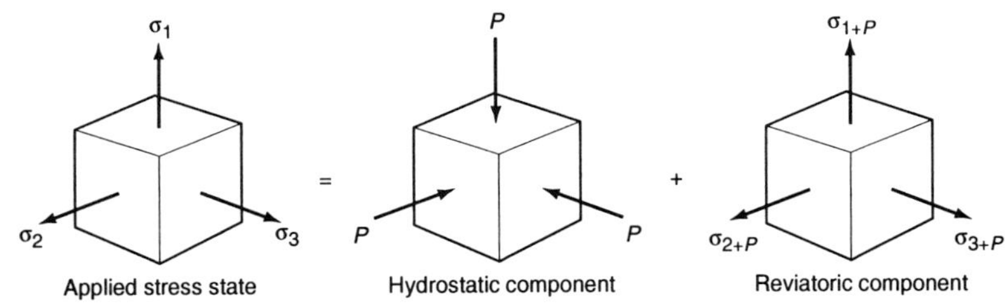

Fig. 2 Separation of generic stress state into components. Source: Ref 1

ponent is the first invariant of the stress tensor and is equal to the average stress and opposite in sign:

$$P = -1/3(s_1 + s_2 + s_3) = J_1 \quad \text{(Eq 1)}$$

The deviatoric or octahedral shear stress component is the second invariant of the stress tensor and represents the effective shear stress developed by the applied stresses:

$$J'_2 = 1/3[(s_1 - s_2)^2 + (s_2 - s_3)^2 + (s_3 - s_1)^2]^{1/2} \quad \text{(Eq 2)}$$

The hydrostatic stress is related to the volume change of a fully dense material; the deviatoric stress is related to the distortion or shape change of a fully dense material.

A familiar example of a constitutive model for a full density material is the von Mises yield criterion which predicts the onset of plastic deformation when the deviatoric stress reaches a set or constant value determined by the nature and condition of the material in question:

$$J'_2 = C_1 \quad \text{(Eq 3)}$$

A plot of this criterion in three-dimensional principle stress space, as shown in Fig. 3, produces a regular right cylinder which defines the yield surface. All values of principle stresses lying on (or outside) the surface of the cylinder produce plastic flow. This surface degenerates into the familiar von Mises yield ellipse for two-dimensional stress plots. If the material work hardens during plastic flow, the diameter of the cylinder expands, thus indicating higher values of stress are required to produce additional plastic flow.

The graphical depiction of the von Mises yield criterion can be simplified by plotting the von Mises criterion as a function of the deviatoric and hydrostatic stress components as shown in Fig. 4. Plastic flow occurs for all stress states located on (or above) the dotted line. If the material workhardens, the yield surface moves progressively to higher values of J'_2. Using J'_2-P space to plot the yield criterion clearly shows that yielding is determined by the deviatoric stress component (J'_2) and is not affected by the hydrostatic stress component (J_1).

The deformation response of metal powders to an applied stress is considerably more complicated than that of solid metals. This response may be categorized into two regimes (Ref 1): (a) an aggregate of loose particles during the early stages of consolidation, and (b) a porous body at the latter stages of compaction.

During the initial stages of compaction, the densification of aggregates of loose particles occurs by particle reorientation and sliding. As compaction continues, localized particle deformation occurs at the particle contact points. As compaction continues, particle yielding occurs on a broader basis leading. At this stage of compaction, the powder compact begins to deform as a porous body.

During densification, the compaction pressure increases as the resistance of the metal powder compact increases. At low compaction pressures, particle sliding is the operative mechanism and large increases in density per unit compaction pressure occur. Later in the compaction process, the particle rearrangement and packing associated with the sliding mechanism is inhibited and particle deformation begins to occur. Higher compaction pressures are required to continue densification due to the need to locally exceed the flow stress of the powder particles. Thus, the rate of densification per unit compaction pressure is reduced.

The constitutive behavior of metal powder compacts will be distinctly different from that of the parent metallic material. The presence of the pores in the structure allows the creation of localized shear stresses under hydrostatic loading leading to plastic flow. Thus, the yield surface will show dependence on both the deviatoric and the hydrostatic stress components. Figure 4 also presents a typical plot of the yield surface of a powder compact as a function of the pressure and deviatoric components of stress. Because the yield surface has become a segment of an ellipse, which is a function of both the hydrostatic and deviatoric components of stress, both shape change (distortion) and volume change (densification) will occur during yielding of powder materials. The Normality Flow Rule requires that the direction of the plastic strain increment be perpendicular to the yield surface. Thus, the plastic strain increment, $d\varepsilon$, can be viewed as two components, a volume strain increment, $d\varepsilon_v$, parallel to the direction of the pressure and a distortion strain increment, $d\varepsilon_d$, in the direction of the deviatoric stress component. Because the volume strain increment is in the same direction as the pressure, the volume of the compact is reduced by that increment.

As the metal powder compact hardens, the elliptical yield surface will move to higher levels of deviatoric and hydrostatic stress combinations. This hardening or increase in flow stress is due both to densification of the porous body and work hardening of the individual metal powder particles. Thus, during compaction, the yield surface of the metal powder compact in J'_2-P space can be represented by a family of ellipses as shown in Fig. 5. For points to the right of the line marked "Critical state line," the volume strain increment, $d\varepsilon_v$, will be in the same direction as the pressure and the volume of the compact is reduced by that increment. For points to the left of the same line, the volume strain increment is in the opposite direction to the pressure and the volume of the compact is increased by that increment. This strain produces a decrease in density called shear dilatancy and describes a stress state under which a metal powder compact

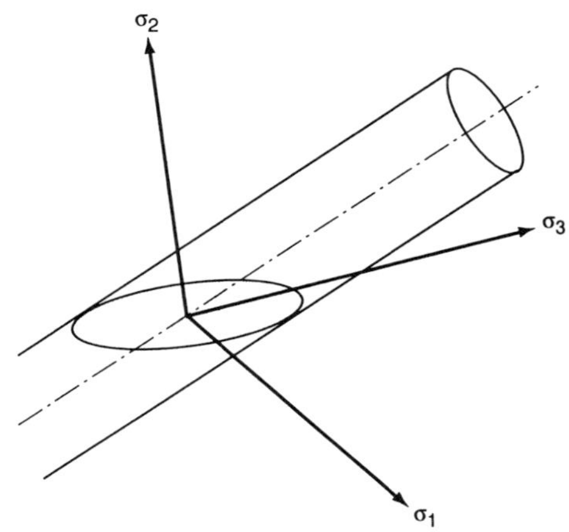

Fig. 3 von Mises yield surface in three-dimensional stress space. Source: Ref 1

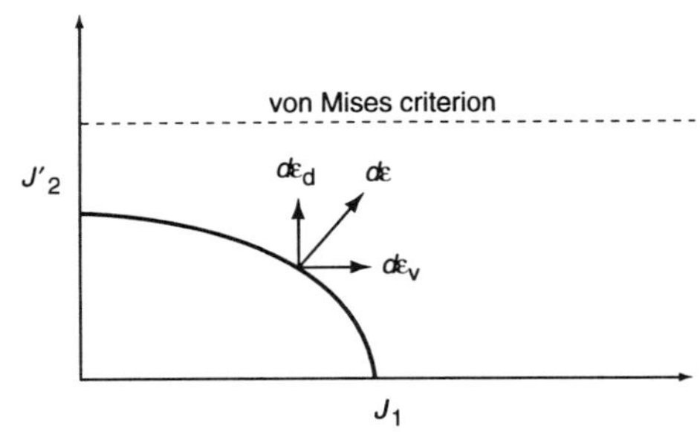

Fig. 4 Plot of yield criteria in J'_2-P space. Source: Ref 1

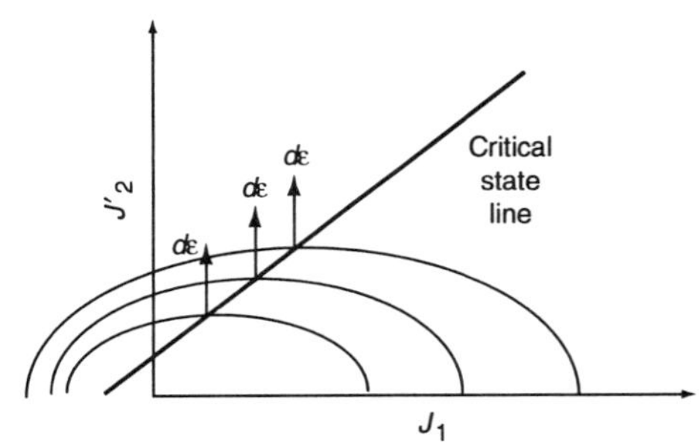

Fig. 5 Family of yield surfaces representing increasing densification. Source: Ref 1

will fail. The point on the yield surface that divides the dilatancy region from the densification region is called the *critical state*. As densification occurs, the locus of critical state values traces the path defined as the *critical state line* that represents the limiting values of J'_2 and J_1 for shear dilatancy.

The yield or flow stress (YS) of a metal powder during compaction can be related to the deviatoric (J'_2) and hydrostatic (J_1) components of the stress state (Ref 2, 3) as follows:

$$AJ_1^2 + BJ'_2 = C(YS)^2 \quad \text{(Eq 4)}$$

where the coefficients, A, B, and C, are functions of relative density and the degree of cold work that has been induced into the powder during compaction. This general equation describes an elliptical yield surface in J'_2-J_1 space as shown in Fig. 4 and 5. The application of this constitutive relationship in describing the deformation or compaction of a metal powder is dependent on incorporating experimentally verified flow stress data into a finite element code capable of handling the large strains associated with metal powder consolidation. Because these codes generally calculate stresses based on displacements or strains, experimentally verified data relating the elastic properties as a function of green density are needed to accurately predict the onset of plastic flow.

Materials Data Requirements

Traditionally, the deformation characteristics of metal powders were analyzed by determining the green density as a function of pressure by running a series of die compaction tests. Specialized techniques were required to minimize the effects of friction on green density. However, these tests were flawed by the inability to vary the state of stress under which compaction was occurring.

The Advanced Metalworking System, described elsewhere (Ref 1, 4), is an experimental apparatus which permits a variety of metalworking tests including direct, indirect, and hydrostatic extrusion, isostatic compaction, and triaxial compaction. It has been employed to evaluate the triaxial compaction of metal powders and provides independent control of pressure and the superimposed axial stress.

Triaxial compaction testing can be described schematically as shown in Fig. 6. The metal powder to be tested, either precompacted or loose fill, is placed in an elastomer sleeve which uses two solid metal anvils as closures. After the chamber has been filled with fluid and sealed, but prior to pressurization, the powder filled elastomer sleeve is moved in contact with the load cell (fixed platen assembly). An annular ram assembly is then advanced to pressurize the fluid environment in the chamber. The superimposed axial stress is then applied to the specimen by applying an axial force through advancement of the force stem portion of the annular ram

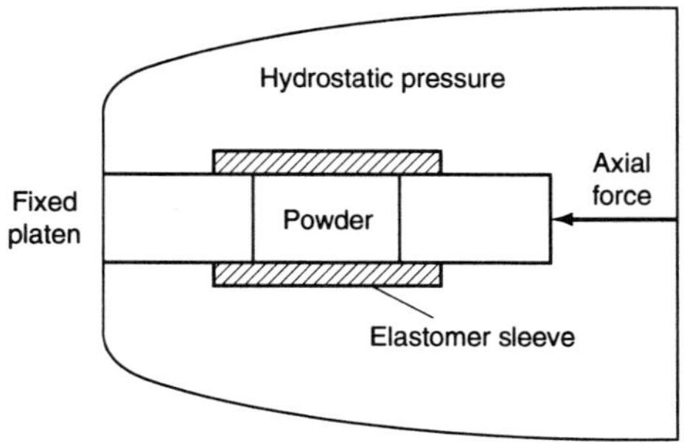

Fig. 6 Schematic representation of triaxial testing

assembly, thus, creating a triaxial state of stress. During this stage, the annular ram withdraws to keep the fluid volume constant and maintain a constant environmental pressure in the chamber. The hydrostatic chamber pressure is measured with the force stem, while hydrostatic pressure load plus the superimposed axial load is measured by the load cell pin. The positions of the main and secondary rams in the annular ram assembly are measured using LVDTs.

The data are usually analyzed by two methods. Figure 7 presents schematically the volumetric, axial, and radial strain measurements for a metal powder specimen plotted as a function of total axial stress (axial pressure component plus the applied axial stress). For this example, a hydrostatic pressure of 103 MPa (15 ksi) has been applied prior to applying the superimposed axial stress. Thus, the radial and axial strains are equal from zero stress to 103 MPa (15 ksi), and then deviate as the superimposed axial stress is increased. The radial strain is calculated from the measured values for axial and volumetric strains.

Given the initial density of the metal powder specimen, the strain data can be used to calculate the relative density of the specimen at any point during the test. Thus at any point during compaction testing, the density values can be related to the flow stress of the powder compact. By separating the state of stress into hydrostatic (pressure) and deviatoric (axial compression) components, these data can be used to generate isodensity yield surfaces. Typical data are shown in Fig. 8.

Mechanical measurement of elastic properties of green metal powder compacts has, in general, been unsuccessful. The strength of the green structure, while sufficient to allow handling in manufacture, is insufficient to permit the use of conventional tensile testing techniques. Similarly, the ductility of the green structure is insufficient to permit the use of compression testing techniques. Resonant frequency testing methods can be used to evaluate elastic properties of green powder specimens (Ref 5).

A transducer is used to induce vibrations in the powder specimen. A second transducer is used to measure the intensity of those vibrations as a function of frequency. The frequency at which the maximum intensity occurs is referred to as the resonant frequency and is related to the density and elastic modulus as follows:

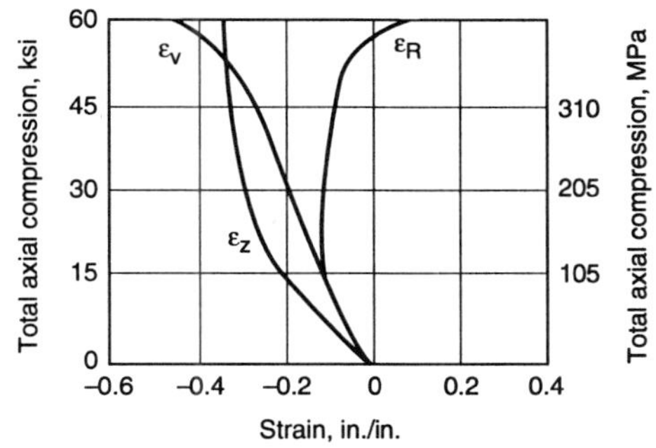

Fig. 7 Schematic strain data from triaxial testing of metallic powder

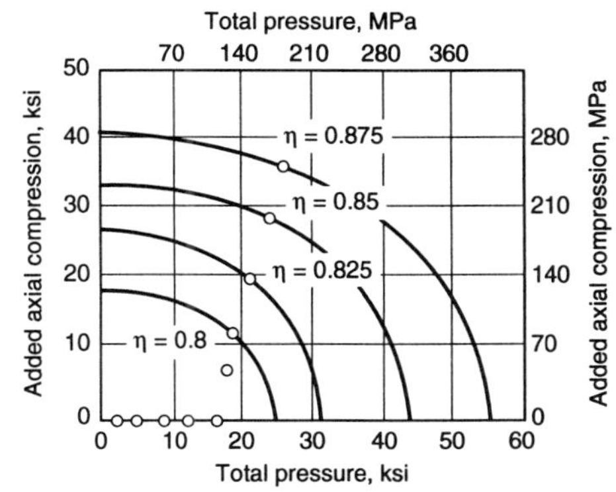

Fig. 8 Experimentally determined yield surface for metallic powder. Source: Ref 4

$$f_R = KE/r \quad \text{(Eq 5)}$$

where K is a factor that accounts for the geometric aspects of the specimen. Figure 9 represents data showing Young's modulus and shear modulus as a function of density for green specimens pressed from iron powders.

Die Compaction Process Simulation Model

The finite element method is commonly applied to metalworking deformation processes to predict the stress state present in the component during manufacture. This method involves dividing the component into very small, but finite, regions following a set procedure based on the rules of application of the particular finite element methodology being used. This technique allows replacing a set of differential equations describing the overall constitutive behavior of the material with an equivalent but approximate set of algebraic equations which describe the material behavior locally at a fixed nodal point. By writing algebraic equations for each nodal point and then using the computer to simultaneously solve those equations for all nodal points, a solution describing the local values of the characteristic of interest can be generated. This solution may provide a strain map, show temperature gradients, or areas of isostress.

These methods have been widely applied to evaluate the state of stress or the temperature

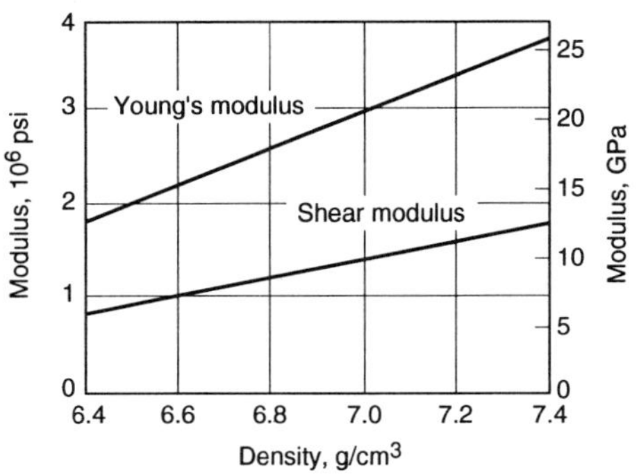

Fig. 9 Young's and Shear moduli as a function of density for green compacts

distribution in a component, such as a turbine blade, during its use. However, the application of this technique to processing is complicated by the requirement that some nodes may be required to slide past others, for example to describe the flow of metal along the surface of a forging die. Furthermore, this sliding may require incorporation of dynamic friction conditions. Both of these situations present serious programming problems to the finite element analyst.

The use of finite element techniques to describe metal powder die compaction is complicated by the large strains which occur during compaction—strains which exceed those that can be accounted for using traditional stress analysis finite element techniques. This distorts the geometric grid and deleteriously affects the predictive capabilities of the finite element system. Thus it becomes necessary to define a new nodal pattern after partial compaction has occurred. This remeshing is quite tedious and time consuming leading the necessity to develop techniques which automatically remesh. The application of finite element techniques to metal powder die compaction is further complicated by the nonuniform movement of the powder particles during compaction creating a variation in density throughout the compact. Both the load transfer characteristics within the metal powder compact and the nature of frictional behavior between the powder compact surface and the die wall are a function of density. These functional relationships with density are unknown and difficult to account for in the existing finite element methods.

Recent work has focused on incorporating the elasto-plastic constitutive equations for metal powder deformation into a finite element methodology. Figure 10 presents density gradient predictions for simple right cylinders of varying aspect ratios. Although the effects of friction have not yet been fully incorporated, the model predicts density variations consistent with the experimental results published in the literature.

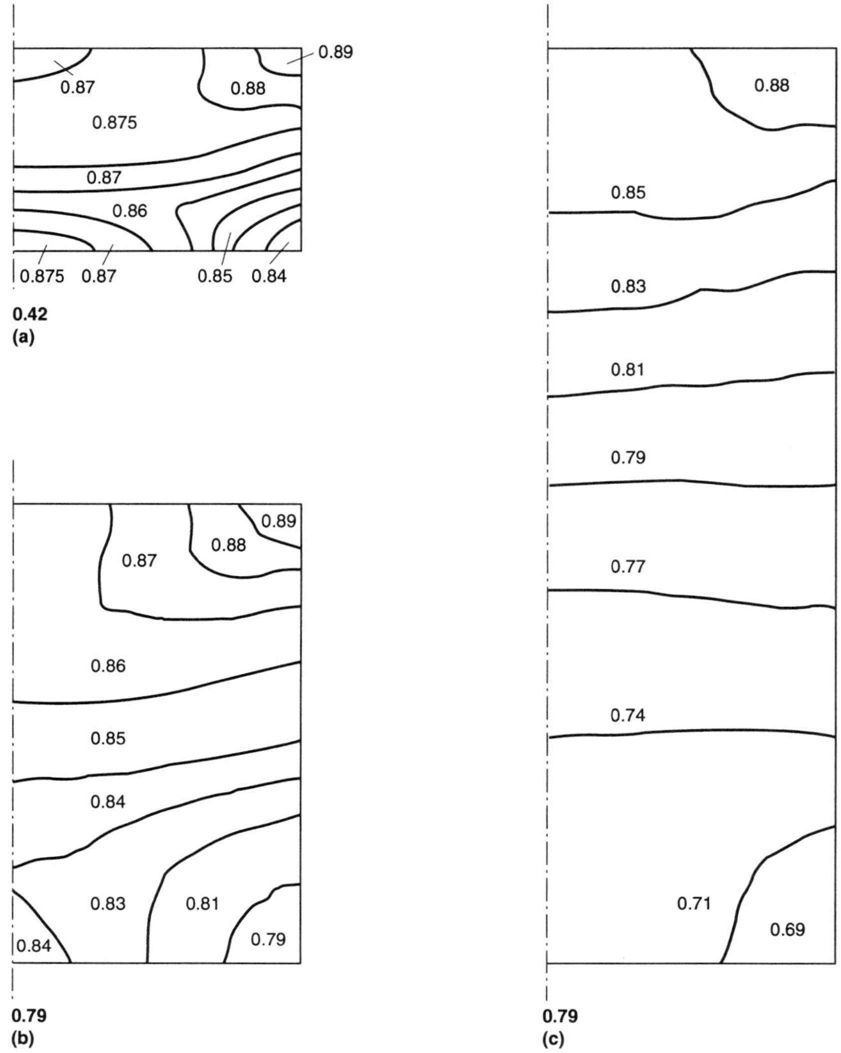

Fig. 10 Isodensity lines predicted by the die compaction model.

Powder Injection Molding

Powder Injection Molding (PIM) is a net-shape processing technique for making complex metal and ceramic components. In the PIM process, a metal or ceramics powder, mixed with a suitable carrier, is injected into a die cavity in a manner similar to plastic injection molding (Ref 6).

The process uses a feedstock that is composed of fine (typically <20 μm) powder particles blended with waxes and polymers. After injection, the green preform shape is ejected from the die cavity and then subjected to a debinding process to remove the polymeric material prior to sintering. This extraction of the binder can be accomplished by thermal or chemical means. The resulting porous shell (approximately 65 to 70% of theoretical density) is heated to 1200 to 1370 °C (2200 to 2500 °F) for 30 to 120 min to sinter the metal powder particles into a solid mass. The resulting component is about 30 to 35% smaller in volume than the molded component and typically 96 to 98% of theoretical density.

This process offers two significant advantages over die compaction. The geometric shape is not limited by axial punch movement, and the mechanical properties of the sintered component are generally higher, primarily due to a higher final density. However, powder injection tooling is generally more expensive than die compaction tooling and the size of parts that can be made is limited by the difficulty of removing the polymeric carrier without damaging the integrity of the green preformed component. Powder injection molding presents an economical route to mass produce small complex shaped parts (typically less than 50 g) that are difficult to form or machine by conventional methods.

In PIM, the mold filling phase is the most critical step in the manufacture of a part. Defects, such as sink marks, voids, weld lines, and density variations, can occur from poor tooling design or improper selection of processing parameters, such as injection temperature and pressure. Traditionally, the design of PIM process tooling, as well as the selection of processing parameters, has been made by expensive and time consuming trial and error procedures. An efficient way to reduce these costs and improve the design to manufacture cycle time is through the use of process simulation. Parametric studies of the injection variables can be conducted by computer simulation. The results of these studies provide guidance in process planning and reduce process development time.

The determination of a meaningful constitutive description of the flow behavior of the polymer/powder mixture during injection is a key to successfully broadening the application of powder injection molding to the manufacture of engineering components by using computer simulation as a tool. This description is complicated, requiring knowledge of two phase non-Newtonian flow that incorporates the effect of local shear rate on viscosity and the effect of viscous heating on the temperature distribution in the component.

These constitutive equations describing viscous flow can be used to solve the conservation of heat, mass, and momentum equations that describe fluid flow. For example, finite difference methods have been used to produce solutions describing the flow during the filling of the mold and the subsequent temperature gradients developed during cooling.

Constitutive Behavior. The mathematics of describing the constitutive behavior of the molten polymer-particulate mixture has been presented elsewhere (Ref 7). In essence, by combining the solution of the mathematical description of the flow with certain geometric constraints, the relationship between stress and displacement, the constitutive behavior, can be quantitatively described.

Thus, powder injection molding is viewed as a fluid flow problem where classical equations of state can be used to describe the behavior of the polymer-particulate mixture. The melt is assumed to be incompressible and the governing differential equations for the conservation of mass, momentum, and energy are known from first principles. These governing differential equations are solved by assuming an appropriate viscosity model (for polymer-powder particle mixtures, viscosity is assumed to be a function of shear rate and temperature) and incorporating appropriate values for key materials properties, such as density, specific heat, and thermal conductivity.

Two additional process characteristics are incorporated into this mathematical description—the volume fraction of the mold that is filled by the melt and the volume fraction of the molten material that has solidified. By tracking the volume fraction of the mold that has been filled at a given time, and relating that volume to the velocity vector for the flow system, the location of the melt front can be calculated at any time during fill. Similarly, by assuming that the solidification rate is proportional to the liquid fraction of the polymeric binder and proportional to the square of the undercooling of the melt, the solid fraction of solidified polymer can be calculated at any given time. Coupling this calculation with the thermal equilibrium requirements provides the basis for determining the location of the liquid/solid interface at any time throughout the die cavity.

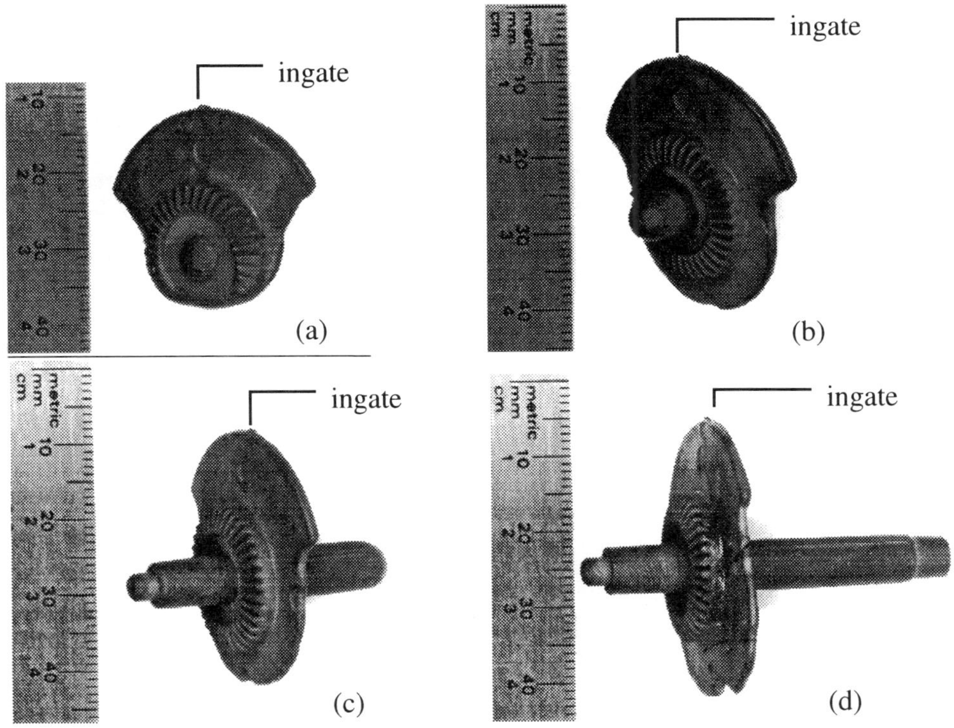

Fig. 11 Experimental results showing flow front location at different times during filling. Injection melt temperature is 138 °C and initial die temperature is 24 °C. (a) to (d) represent increasing time. Source: Ref 18

Powder Injection Molding Process Simulation. During the last decade, many process modeling techniques have been developed in material processing. Tremendous success has been achieved in the development and application of process modeling technology in molten metal operations. Both the finite difference method and the finite element method have been used successfully in casting processes that involve filling a sand or metal mold with molten metal to produce parts with complex three-dimensional (3-D) geometries (Ref 8–13). Considerable progress has also been made in modeling polymer and powder injection molding processes (Ref 14–17). Most of polymer processing and PIM models have been based on the generalized Hele-Shaw model for thin cavities. This may be adequate for modeling parts comprised of thin-wall sections, but may not be effective for modeling general parts with arbitrary complex 3-D geometries. Development of a general program capable of simulating the PIM process in arbitrary 3-D geometries has become increasingly imperative. This could be accomplished by using either the finite element method (FEM), the finite difference method (FDM), finite volume method, or a combination of them. Generally speaking, the FEM technique gives a more precise description of the geometry, and possibly more accurate results than FDM. However, FEM requires greater memory storage, more complicated programming, and longer computing time. Hence, the finite difference technique has been applied in this article.

The PIM simulation model is based on an existing casting simulation program which models the mold filling and solidification of general 3-D metal castings. This casting model was modified to include special features, such as viscous heating, associated with the PIM process. In essence, the geometric shape to be analyzed, the component, is divided into a large number of brick-type cells. Finite difference formulations are written for each equation included in the descriptive set incorporated into the constitutive behavior equations. These equations are then solved simultaneously leading to a coupled analysis of velocity, temperature, shear rate, and mass transfer (Ref 7).

The finite difference method is similar to the finite element method in that it provides a means for converting a partial differential equation into a set of algebraic equations. It differs from the finite element method in that the region of interest is divided into n discrete intervals or control volumes at which values of the differential equation are approximated using mathematical techniques such as a Taylor series expansion. Although some accuracy may be lost because stepped rectangular cells are used to represent sloped boundaries, the FDM technique allows the use of a large number of cells (elements) because of its simplicity in programming and efficiency in computing. For a given amount of computer memory, the FDM method can compute values for 20 to 50 times more cells than the FEM technique. As such, it is capable of describing small localized phenomena and large thermal or velocity gradients. Thus, it is especially suitable for modeling PIM and polymer processing where the thermal and momentum gradients are large due to low thermal conductivity combined with a rapid filling rate.

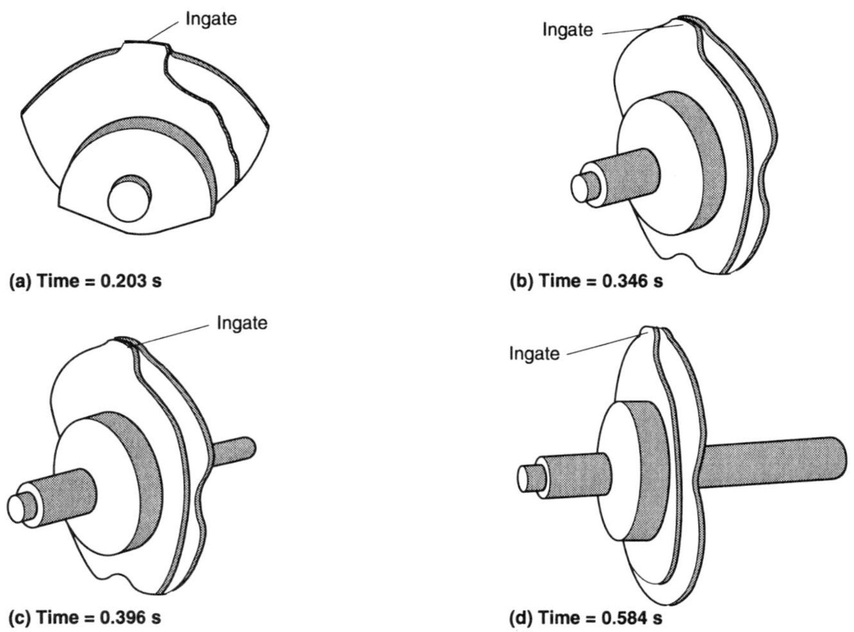

Fig. 12 Numerical simulation results showing flow front locations at different times during filling under similar conditions to Fig. 11. Injection melt temperature is 130 °C, initial die temperature is 25 °C and injection pressure is 3 MPa, with sufficient venting. Source: Ref 18

In the computing technique incorporated into the model, the complete partial differential equations describing the PIM processes are solved implicitly by the finite difference method. The technique takes into account the time-dependent non-Newtonial fluid flow behavior with free surfaces and viscous heating phenomena in arbitrary 3-D geometries. The complex 3-D domain is subdivided into small brick-type cells. Sloped surfaces are approximated in a zig-zag fashion. A uniform mesh is used throughout the model. The typical number of cells used for an industrial part is approximately 100,000 to 500,000. The discrete values of the velocity components are located at cell faces, while all other variables are located at the centers of the cell (Ref 8). Useful information, such as flow front locations, distribution of velocities, temperature, shear rates, and weld lines are obtained through the simulation. Figures 11 and 12 compare the flow fronts from an injected part (Fig. 11) and the numerical simulation of the injection process (Fig. 12).

There are several advantages of the computational and modeling techniques used:

- No term of the momentum and energy transport equations is neglected in the formulation. The partial differential equations are solved as they appear in their original forms, so that the model is capable of predicting any flow configuration, including jetting phenomena.
- The model can readily analyze general three-dimensional geometries.
- The finite difference method (FDM) is used.
- As a result of the simplicity in the formulation and the computation of this technique, the input of the geometry and the meshing of the part are much easier than with FEM techniques.

Several assumptions and simplifications are made to streamline this modeling technique. Variations in density and the heat capacity of feedstock and mold materials are assumed negligible in space and time. The solidification rate of the feedstock is assumed proportional to the square of the degree of undercooling and proportional to the liquid fraction of the binder material. Perfect contact is assumed between the feedstock and the mold during filling, so that heat transfer at the melt-mold interface is controlled by conduction. Heat transfer from the melt to the air in the mold cavity is assumed negligible since the filling time is short and the heat capacity of the air per unit volume is small.

Hot Isostatic Pressing

Hot isostatic pressing (HIP) modeling is based on a phenomenological constitutive relation for powder consolidation under high pressure and high temperature. The consolidation of metal powders via HIP can produce cost-effective, high-quality, and high-performance components. The advantages of HIP processing over casting include a uniform, fine grain structure without shrinkage porosity, inclusions, and segregation; a decrease in machining time due to making parts nearer to net shape and with a finer microstructure; and reduced residual stresses that can cause parts to undergo distortions during finish machining. In spite of these advantages, the cost-effective manufacture of parts with complex shapes via HIP has been hindered by difficulties in achieving suitable dimensional control.

Currently, the containers that hold the powder and define the component shape are designed by trial and error; this is very costly and time consuming. In each iteration, the container is redesigned and another part is HIPed to determine the effect of the design modification. Depending on part size and alloy, the cost can exceed $100,000 per trial (1998 dollars), and each iteration can take months to complete. To reduce costs associated with shape control by trial and error, it is necessary to accurately predict the final shape of the as-HIPed component during the design stage.

With computer models that accurately predict the as-HIPed shapes, simulations may replace actual HIP trials, thereby reducing the expense of container design and the lead times for new components. Using a computer model, a container design iteration could be measured in days or weeks rather than months, and the cost of actual HIP trials would be saved.

General Conditions of HIP Modeling. The HIP methods is essentially a thermomechanical process. In HIP, powder consolidation and heat transfer occur simultaneously. The physical mechanisms involved in the powder consolidation include plastic yielding, creep, and diffusion. The local stress and temperature in the powder compact govern the contribution of these mechanisms to the powder consolidation. Heat transfer (mainly heat conduction) in the powder compact is strongly influenced by the local relative density. Therefore, the analysis of HIP is a coupled thermomechanical analysis.

Modeling the HIP process requires the solution of a boundary value problem described by:

- Equation of motion
- Conservation of mass/continuity equation
- A set of constitutive equations
- A proper set of initial and boundary conditions

To fully specify the constitutive model of the powder and can material, a variety of experiments should be performed. The testing approach and data analysis is described in the next section. Prior to discussing the constitutive model, it is also important to distinguish between the microscopic (powder particle) deformation associated with individual particles and the macroscopic (powder aggregate) deformation of the powder aggregate idealized as a continuum. At the macroscopic level, field variables such as stress and strain are defined over a small representative material element containing enough particles to qualify for a macroscopic regime. Clearly, an exact solution of these field variables would require the solution of a boundary value problem involving the exact number, shape, and size of these particles. At the microscopic level, the field variables of each particle are obtained in some average sense. Thus, the particle quantities are defined as those obtained by averaging the actual particle values (obtained from the unified theories of dislocation driven plasticity for isotropic materials) over the macroscopic material element. These two modes of deformation, which constitute the basic philosophy of the proposed methodology, facilitate the tracking of internal state evolution of the material as a whole and the individual particles in some average sense. Details of the development of the viscoplastic constitutive model for the densification of metal powders during HIP have been reported previously (Ref 19, 20). Additional information is also contained in the article "Principles and Process Modeling of Higher Density Consolidation" in this Volume.

Constitutive Model. Power-law creep represents the flow behavior of the container material. The flow behavior of the powder is represented as a combination of the particle flow behavior and the powder aggregate flow behavior described by the following macroscopic, viscoplastic potential function or yield function, ϕ (Ref 19, 20). This function represents the evolution of yield loci as a function of relative density.

$$\phi = S^2 + bp^2 - cs^2 = 0 \qquad \text{(Eq 6)}$$

where S is the magnitude of the deviatoric stress in the powder aggregate, s is a measure of the deviatoric stress in the powder particle, p is the pressure, and the parameters b and c are functions of relative density to be determined from experiments.

Based on this constitutive model, the following material properties are required:

- Flow behavior of the particles
- Constitutive functions. $b(\rho)$ and $c(\rho)$ for the powder aggregate
- Specific heat of the powder as a function of temperature
- Thermal conductivity of the powder as a function of density

In addition, the accuracy of the model must be verified by some basic finite element simulations and empirical testing. The experimental tests are required to fully specify the constitutive model of the powder and can material. The experimental characterization of powder determines the flow behavior of particles and the constitutive functions of the aggregate.

Partial HIP Testing. The macroscopic solution of the boundary value problem involving powder aggregate under hydrostatic pressure can be obtained from the analytical expression given below and the power-law creep model for the particle flow behavior. The corresponding density evolution takes the form:

$$\dot{\rho} = A\exp\left(-\frac{Q}{RT}\right)\left(\frac{3}{2}\right)^{(n+1)/2} p^n \left(\frac{b(\rho)}{c(\rho)}\right)^{(n+1)/2} \rho^2 \qquad \text{(Eq 7)}$$

This differential equation can be numerically integrated to obtain a distribution of densities at different applied temperature and pressure cycles. As can be seen from Eq 7, the rate of change in density strongly depends on the temperature and pressure. Other important characteristics of Eq 7 are a threshold value of temperature below which the dependency of density rate on the pressure is weak and a range of pressure magnitude below which the rate of change in the density weakly depends on the temperature. The understanding of this interaction between the temperature and pressure dependence can be very important in avoiding HIP cycles that cause densification to occur too quickly, which could cause a densification front to form, or too slowly, exposing the material to elevated temperature longer and limiting productivity. However, the overall objective is to derive accurate material properties using a series of interrupted HIP cycles at different dwell times to produce density distributions ranging from a low relative density (about 75%) up to full density.

Effects of Container. There are several reasons for distortion of the final geometry, which can be directly related to the preparation of the container. Some of these reasons include incomplete filling and degassing of the powder, poor powder quality, and nonuniform can thickness including thick and thin regions near welds. Even if these problems are corrected and the powder filling operation is sound, inhomogeneous compaction rates will always prevail due to the uneven resistance offered by different sections of the container, such as at corners. The thickness of the container provides shielding of the gas pressure in the HIP unit. To properly account for this difference between the gas pressure and the effective pressure inside the powder, a numerical scheme based on the flow properties of the container at different temperatures and strain rates has been developed and incorporated into the material model. Details of this derivation have been reported previously (Ref 21). Unlike pressure, temperature shielding due to the container thickness was found to be negligible, and thus, was not considered. It should be emphasized that this shielding effect is more pronounced in bulky parts where temperature and pressure gradients can be significant.

Calibration and Experimental Verification. Calibration of a HIP material model for the powder aggregate involves the following steps:

1. Obtain partially HIPed compact bars with short HIP cycles not exceeding the rate of temperature or pressure application for actual HIP production cycles
2. Machine the partially HIPed bars into compression specimens
3. Measure the density of every compression specimen
4. Calculate the pressure correction factor (effective pressure applied to the powder aggregate as opposed to the gas pressure applied by the HIP vessel on the outer wall of the container) (Ref 21)
5. Calculate the relationship between relative density and HIP temperature and pressure
6. Perform compression tests on specimens with different densities at different strain rates and temperatures
7. Correct true stress-true strain curves to account for volume change during compression testing
8. Measure the density of compression specimens and obtain a relationship between axial stress and final density for each specimen
9. Using results from above, calculate the functions $b(\rho)$ and $c(\rho)$ that represent the weighting coefficients in the deformation macroscopic potential function (Ref 21)
10. Measure thermal conductivity and specific heat on partially HIPed and fully dense material over a range of relative densities and obtain a relationship between relative density and thermal conductivity for a temperature range of 25 °C to the HIP temperature

The following sections describe the partial HIP testing, isothermal compression tests, and thermophysical properties measurement used in the evaluation of the power-law creep parameters and the ratio of constitutive functions, $b(\rho)/c(\rho)$ for Alloy 625.

Compression Tests of Fully Dense Alloy 625. Tests were conducted on a servohydraulic testing system with a vacuum furnace. To avoid possible oxidation of the specimens during testing, a vacuum environment at 10^{-4} torr or lower was used. A computer should control the testing system with a data acquisition system employed to perform true strain rate tests under stroke control. Load and stroke data are required and converted to true stress and true strain by assuming constant volume of the specimen during deformation. A boron nitride treatment is applied to the grooved ends of the specimens and the compression platens. Experiments show that lubrication of both the specimen and the platens greatly reduced the friction and enabled uniform deformation during compression. The specimens are heated in a vacuum furnace to test temperatures ranging from 260 °C to 1149 °C. Prior to testing, the specimens are soaked at the testing temperature for three minutes to avoid a thermal gradient. Subsequently, the specimens were uniaxially compressed to a total true strain of 0.6 (60%). Immediately after the compression test, the specimens are quenched with forced helium gas to retain the deformed microstructure.

Compression Tests of Partially Dense Alloy 625. Compression testing is also required on partially dense materials in order to determine the thermomechanical behavior of the aggregate powder. Specimens with minimum initial relative density (75%) are compressed from a true strain of 0.2 (20%) to a true strain of 0.8 (80%). Three additional specimens with initial relative densities of approximately 83%, 90%, and 96% are used to complete the curve of axial stress versus density in the high-density regime. After the tests, specimens are immediately quenched with helium gas to retain the deformed microstructures.

Unlike fully dense material in which the volume of the specimen remains constant, the powder compacts densify during the compression testing. To account for this, conservation of mass is used in the calculation of true stress-true strain. From the converted true stress-true strain data, a relationship between uniaxial stress and relative density was obtained.

Thermophysical Tests. Four partially HIPed test specimens should be tested for thermophysical properties between room temperature and 1150 °C. Thermal diffusivity (α) values are measured using a technique such as laser flash. A differential scanning calorimeter (DSC) should be used to measure specific heats. Testing is conducted under vacuum to reduce oxidation of the test samples. Thermal conductivity (λ) values were calculated from the following:

$$\lambda = \alpha C_p \rho \quad \text{(Eq 8)}$$

where ρ is the material bulk density.

A range of relative densities of specimens, such as 0.72, 0.78, 0.90 and 1.0, are needed. Thermal conductivity results showed a strong dependency on the relative density mainly in the high temperature regime. It is also important to note that the conductivity values increase well above the transformation temperature.

The thermal conductivity of the aggregate powder, $\hat{\kappa}$ (ρ, T) is a function of both density and temperature. This function can be expressed as a product of two functions:

$$\hat{\kappa}(\rho, T) = \hat{R}(\rho)\hat{\kappa}_p(T) \quad \text{(Eq 9)}$$

where, $\hat{\kappa}_p$ is the thermal conductivity of the powder particle, and $\hat{R}$ is a function that depends only on the relative density.

Calibration of a Power-Law Creep Equation for Fully Dense Materials. A series of uniaxial isothermal compression tests are needed on fully dense material of the alloy powder. The test conditions, temperature, and strain rate should be chosen to be within the diffusion creep regime for the material.

The power-law creep is usually described in the following form:

$$\dot{\bar{\varepsilon}} = A \exp\left(-\frac{Q}{RT}\right)\bar{\sigma}^n \quad \text{(Eq 10)}$$

where, Q, R, and T are the activation energy, universal gas constant, and absolute temperature; and $\dot{\bar{\varepsilon}}$ and $\bar{\sigma}$ are the average strain rate and flow stress, respectively. A and n are the creep parameters.

There are several ways to estimate the activation energy. In one way, two compression tests are conducted. The first is at constant temperature with a rapid increase in strain rate, and the second is conducted at constant strain rate with a rapid decrease in temperature.

For the two tests with equal temperatures ($T_1 = T_2$), Equation 10 could be rearranged to obtain n as follows:

$$n = \ln(\dot{\varepsilon}_1 / \dot{\varepsilon}_2)/\ln(\sigma_1 / \sigma_2) \quad \text{(Eq 11)}$$

Similarly, for two tests with equal strain rates ($\dot{\varepsilon}_1 = \dot{\varepsilon}_2$), rearrangement of Eq 10 yields the following expression for Q:

$$Q = nR \ln(\sigma_1 / \sigma_2)/[T_1 - T_2 / T_1 T_2] \quad \text{(Eq 12)}$$

To fully describe the power-law creep behavior of alloy 625 (Eq 10), the only parameter left to be determined is A. However, for consistency, both A and n were obtained using the following procedure which took into account a wider range of strain rates typical of those found in the HIP process. Equation 10 was rearranged as follows:

$$\dot{\varepsilon} \exp(-Q/RT) = A\sigma^n$$

or

$$Z = A\sigma^n \quad \text{(Eq 13)}$$

where Z is the Zener-Hollomon parameter. Equation 13 can be plotted using the strain rate, temperature, and flow stress data from the compression tests of the fully dense Alloy 625. The data are then fitted to a power-law distribution, and thus the power-law constants A and n are identified.

REFERENCES

1. W.L. Otto, J. Trasorras, J.P. Bandstra, and A.R. Austen, Powder Constitutive Model Development from Triaxial Compaction Data, Advances in Powder Metallurgy, Vol 1, *Proc. of the 1990 Powder Metallurgy Conf. and Exhibition,* p 133–146
2. H.A. Kuhn and C. Downey, Deformation Characteristics and Plasticity Theory of Sintered Powder Materials, *Int. J. Powder Metall.,* Vol 7, 1971, p 15
3. S. Shima and M. Oyane, Plasticity Theory for Porous Metals, *Int. J. Mech. Sci.,* Vol 18, 1976, p 285
4. A.L. Gurson and R.A. Posteraro, Yield Function for Metal Powders for Use in the Numerical Simulation of Powder Compaction, *Model Based Design of Materials and Processes,* Russell et al., Ed., TMS, 1992
5. C.-Y. Yu and T.R. Prucher, "Measuring Young's Modulus and Shear Modulus—A Comparison of Dynamic and Mechanical Techniques," *1993 Advances in Powder Metallurgy and Particulate Materials,* American Powder Metallurgy Institute, 1993
6. R.M. German, *Powder Injection Molding,* Metal Powder Industries Federation, 1990
7. C.M. Wang et al., A Finite Difference Computing Model for 3-D Simulation of the Powder Injection Molding Process, *Modern. Developments in Powder Metallurgy,* American Powder Metallurgy Institute, 1992
8. C.M. Wang, "Computer Modeling of Fluid Flow and Heat Transfer in Metal Castings," Ph.D. Thesis, University of Pittsburgh, 1990
9. C.M. Wang, A.J. Paul, and R.A. Stoehr, "Modeling Foundry Castings for a Rational Process Design System," Proc. of TMS Symp. on Numerical Simulation of Casting in Automotive Applications, East Lansing, MI, May 1991
10. P.N. Hansen, Solidification Processes: Computer Simulation and Modeling, *Work Shop,* P.R. Sahm and P.N. Hansen, Ed., CIATF-Cairo, 1983
11. P.N. Hansen et al., Fast and Adequate Preprocessing for Simulation of Real Complex Shaped Castings, *Modeling of Casting, Welding and Advanced Solidification Processes—V,* M.R. Ozgu and K.W. Mahin, Ed., TMS, 1990
12. M. Lipinski, W. Schaefer, and S. Andersen, Modeling of Combined Heat and Fluid Flow for Determination of Filling Sequence for Real Complex Shaped Castings, *Modeling of Casting, Welding and Advanced Solidification Processes—V,* M. Rappax, M.R. Ozgu, and K.W. Mahin, Ed., 1990
13. R.A. Stoehr and C.M. Wang, Advances in Fluid Flow, Heat Transfer, and Solidification Modeling and Application to Actual Foundry Problems, *Modeling of Casting, Welding and Advanced Solidification Processes—V,* M. Rappax, M.R. Ozgu, and K.W. Mahin, Ed., TMS, 1990
14. H.P. Wang, E.W. Liang, and E.M. Perry, "FE-MAP-RTM: A Resin Transfer Molding Process Simulator," Conf. Proc. for the Advanced Composites Conf. and Exhibition (ACCE), Detroit, MI, 1991
15. K.K. Wang and V.W. Wang, Computer-Aided Mold Design and Manufacturing, *Injection Molding Fundamentals,* A.I. Isagev, Ed., Mercel Dekker, 1987
16. D. Lee et al., Analysis of Powder Injection Molding Process, *Advances in Powder Metallurgy,* Vol 3, Metal Powder Industries Federation, 1989, p 1–15
17. D. Lee et al., "Modeling and Testing of Powder Injection Molding Processes," Proc. of 1988 Int. Powder Metallurgy Conf. (Orlando, FL), 5–10 June 1988, p 417–429
18. C.M. Wang, K.E. Carr, and T.J. McCabe, Computer Simulation of the Powder Injection Molding Process, *1994 Advances in Powder Metallurgy and Particulate Materials,* American Powder Metallurgy Institute, 1994
19. J.R.L. Trasorras, M.E. Canga, and W. Eisen, Modeling the Hot Isostatic Pressing of Titanium Parts to Near-Net Shape, *1994 Advances in Powder Metallurgy and Particulate Materials,* American Powder Metallurgy Institute, 1994
20. R.E. Smelser et al., "On the Modeling of Near-Net Shape Hot Isostatic Pressing," *Mechanics in Material Processing and Manufacturing,* ASME, 1994
21. J.F. Zarzour, J.R.L. Trasorras, J. Xu, and J.J. Conway, "Experimental Calibration of a Constitutive Model for Hot Isostatic Pressing (HIP) of Metallic Powders," *1995 Advances in Powder Metallurgy and Particular Materials,* American Powder Metallurgy Institute, 1995

Metal Powder Production and Characterization

Introduction

Yves Trudel, Quebec Metal Powders Limited

METAL POWDER PRODUCTION techniques are used to manufacture a wide spectrum of metal powders designed to meet the requirements of a large variety of applications. Powders of virtually all metals can be produced. Various powder production processes allow precise control of the chemical composition and the physical characteristics of powders and allow tailoring of specific attributes for targeted applications. Development and technical innovation in metal powder production processes are constantly pursued to meet the quality, cost, and performance requirements of existing and emerging applications.

General Methods of Metal Powder Production. Metal powders are produced by mechanical methods or chemical methods. The commonly used methods include water and gas atomization, milling, mechanical alloying, electrolysis, and chemical methods including reduction of oxides.

Suitable methods for powder production depend on required production rates, powder properties, and the physical and chemical properties of the material. Chemical and electrolytic methods are useful for producing high-purity powders. Mechanical comminution (or milling) is the most widely used method of powder production for hard metals and oxides. Secondary milling of oxide-reduced, atomized, or electrolytic powders is also a very common and economical practice. Nonetheless, mechanical comminution can be a limited method of metal powder production due to the ductility of most metals (although hydriding is a technique that can enhance powder production by milling for some metals, see the article "Milling of Ductile and Brittle Materials" in this Section).

Atomization is perhaps the most versatile method that produces metal powders over a wide range of production rates (from 1 to 10^5 tons/yr) and a wide variety of powder sizes from 10 to 1000 μm (Ref 1). Typical particle sizes for atomized metal powders are summarized in Fig. 1, and more detailed coverage on the basic principles of atomization are detailed in the next article in this Section. Nonetheless, a growing number of P/M applications (Fig. 2) require submicron powder sizes that are produced by chemical or physical means. Ultrafine (i.e., submicron) powders cannot be effectively produced by grinding or conventional (gas and/or liquid) atomization. Some basic physical and chemical methods of ultrafine powder production are discussed in the article "Ultrafine and Nanophase Powders" in this Volume.

Another article also discusses powder blending and mixing techniques, including the recently developed binder-treatment blending used for metal powders. Binder-treated premixes (Fig. 3) lead to reduced dusting and segregation of alloy additions and enhanced powder flow and die filling. This results in improvements in density uniformity, weight variation, alloy homogeneity, and dimensional control of P/M parts.

Ferrous and Nonferrous Powders. Metal powders produced by mechanical and chemical methods are summarized in Fig. 4 and 5, respectively. Ferrous powders are of prime importance to the P/M industry because they account for more than 80% of all powder shipments (tonnage). The production and usage of iron and steel powders in North America has grown at a compounded rate of about 6% per year from the mid-1980s to the mid-1990s, reaching more than 350,000 tons in 1996. The production of iron and steel powders in Japan and Europe during the same year was 173,000 and 114,000 tons, respectively.

Most ferrous powders are currently produced by water atomization of liquid steel. This process, pioneered in the 1960s by the A.O. Smith Company, has since been adopted by the main iron and steel powders producers in North America, who have developed a wide spectrum of elemental and prealloyed powders designed for specific needs of P/M and powder forging appli-

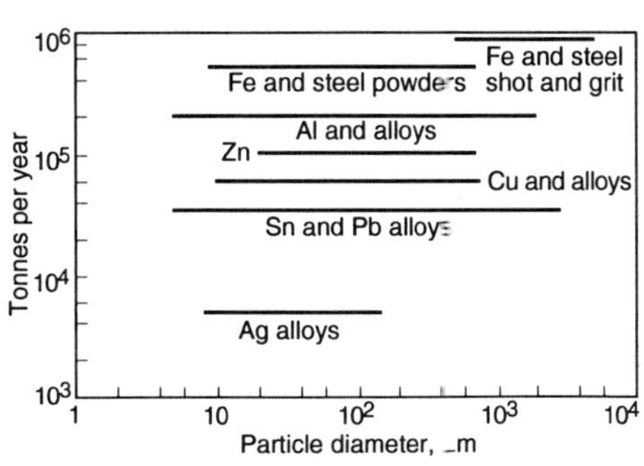

Fig. 1 Approximate ranges of useful particle size and world production of different melt-atomized alloys. Source: Ref 1

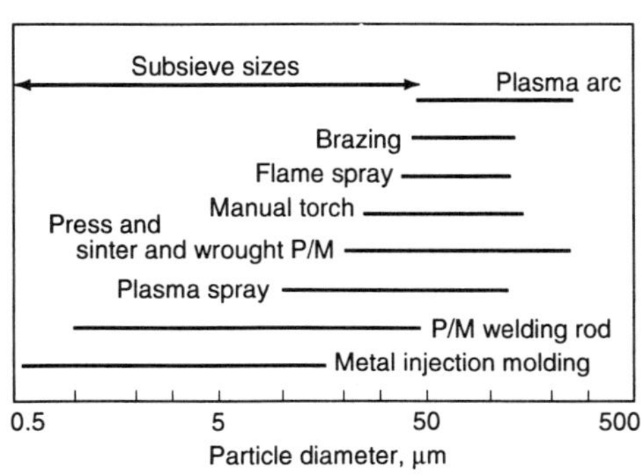

Fig. 2 Some ranges of useful particle size for different applications of metal powders. Source: Ref 2

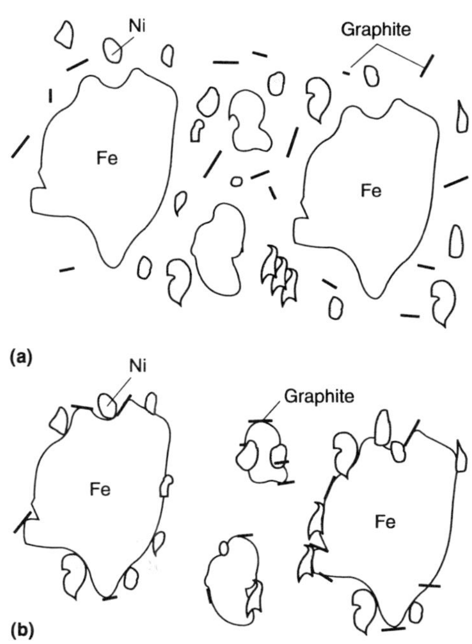

Fig. 3 Alloy distribution. (a) Regular premix. (b) Segregation-free premix with powder binder and blending

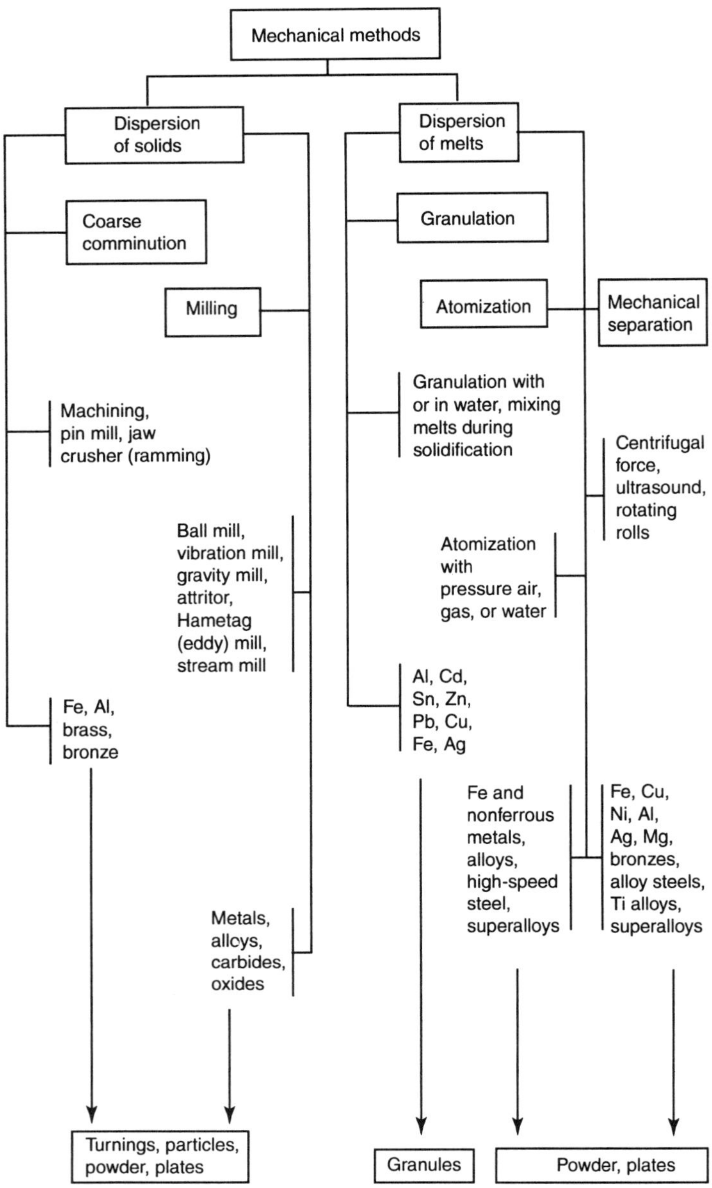

Fig. 4 Mechanical methods of powder production. Source: Ref 3

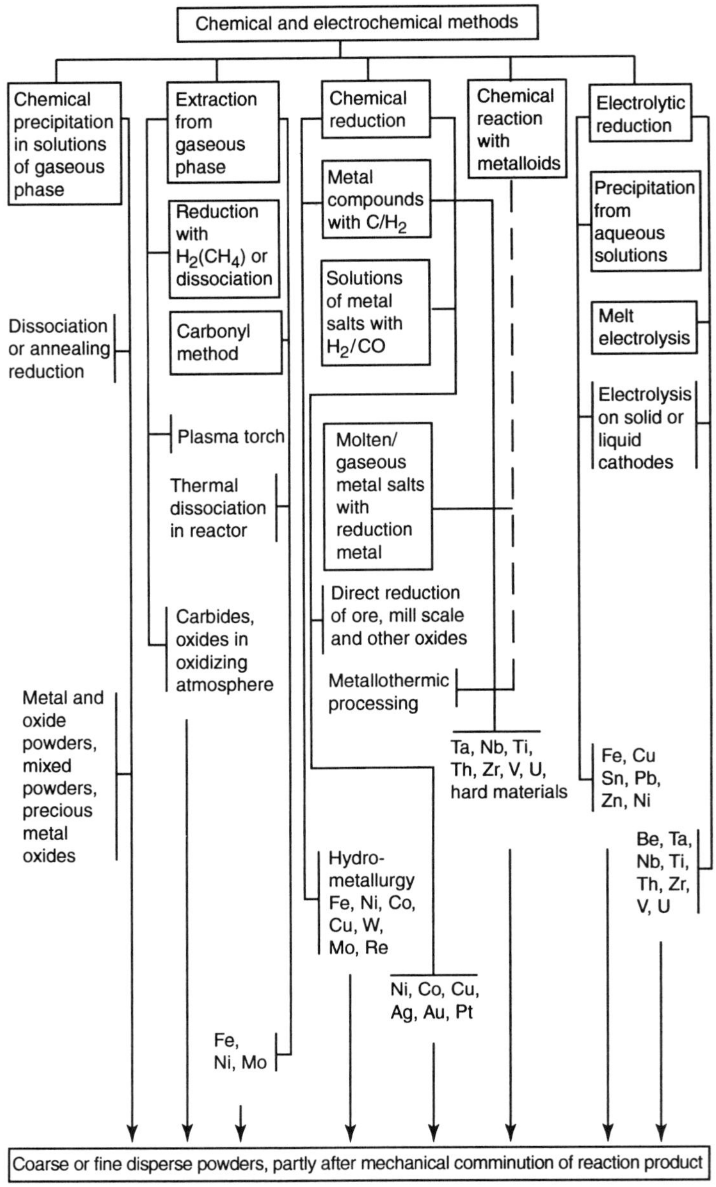

Fig. 5 Chemical and electrochemical methods of powder production. Source: Ref 3

cations. The techniques used to manufacture stainless steel powders, tool steel powders, and high-alloy materials are also described in the Section of this Volume devoted to the production of iron and steel powders.

The production of nonferrous powders is covered in this Section. Copper and copper-base powders, aluminum powders, nickel powders, and tungsten powders are among the many types of metal powders whose production techniques are described and discussed.

Powder Characterization and Testing. Reliable techniques for powder characterization and testing are required to evaluate the chemical and physical properties of metal powders, both as individual particles and in bulk form. This is important to ensure adequate control of powder production processes and to ensure that the required properties for parts manufacturing are met. This Section also features a series of articles on characterization and testing of powders.

The representativity of samples taken to test metal powders is key to the characterization process. The required sampling techniques that ensure such representativity are discussed. The various techniques for analyzing the chemical composition of metal powders are described, together with specific testing procedures that characterize the physical properties of powders. These include particle size distribution, particle shape, and surface area. Each of these parameters may have a significant effect on the physical and mechanical behavior of powders used for the manufacturing of P/M parts. Bulk properties of metal powders are also discussed in this Section.

A description of test methods for apparent density, a critical property for ensuring part-to-part weight consistency, and powder flow rate is included. Finally, an article describes test methods for determining the compacting behavior of metal powders. This includes compressibility, a key factor in reaching high part density, and green strength, which is a measurement of the mechanical strength of P/M parts in the as-pressed condition.

REFERENCES

1. A.J. Yule and J.J. Dunkley, *Atomization of Melts for Powder Production and Spray Deposition*, Oxford University Press, 1994
2. A. Lawley, *Atomization,* Metal Powder Industries Federation, 1991
3. Andrej Šalak, *Ferrous Powder Metallurgy,* Cambridge International Science Publishing, 1995

Atomization

John J. Dunkley, Atomising Systems Limited, Sheffield, England

ATOMIZATION is the dominant method for producing metal and prealloyed powders from aluminum, brass, iron, low-alloy steels, stainless steels, tool steels, superalloys, titanium alloys, and other alloys. Current atomization technology is the result of steady advances over the last 50 years since the first large-scale production of atomized iron powder during World War II. By the early 1980s atomized powders had accounted for more than 60% by weight of all metal powders produced in North America, with a higher percentage today due to the continued application of prealloyed powders and the diminishing importance of reduced iron powder. Currently, it is estimated that worldwide atomization capacity of metal powders exceeds 10^6 metric tons/year. Atomization has become the dominant mode of powder production because high production rates favor economy of scale and because prealloyed powders can only be produced by atomization.

Atomization is simply the breakup of a liquid into fine droplets. Any material available in liquid form can be atomized. In the case of high melting materials, the result is frozen droplets, that is, a powder. Typically, the size of atomized powders is smaller than 150 μm, although larger-sized powders can be produced (in which case atomization is referred to as "shotting" or "granulation"). Atomization is also synonymous with the term nebulization, which is applied to atomization of aerosols in the chemical/pharmaceutical industry.

The general types of atomization processes encompass a number of industrial and research methods. Industrial methods include:

- Two-fluid atomization, where a liquid metal is broken up into droplets by impingement of high-pressure jets of gas, water, or oil (Fig. 1a and b)
- Centrifugal atomization, where a liquid stream is dispersed into droplets by the centrifugal force of a rotating disk, cup, or electrode (Fig. 1c)
- Vacuum or soluble-gas atomization, where a molten metal is supersaturated with a gas that causes atomization of the metal in a vacuum (Fig. 1d)
- Ultrasonic atomization, where a liquid metal film is agitated by ultrasonic vibration (Fig. 1e)

Water, gas, centrifugal, ultrasonic, and soluble-gas atomization are all used in commercial production, but two-fluid atomization methods with gas (including air) or water account for more than 95% of atomization capacity worldwide (Ref 1).

This article describes the key process variables and production factors for the industrially applied methods of atomization, while briefly mentioning some other methods. It also briefly reviews the effect of atomization methods and process variables on key powder characteristics such as average particle size, particle size distribution or screen analysis, particle shape, chemical composition (including surface composition), and microstructure. These fundamental

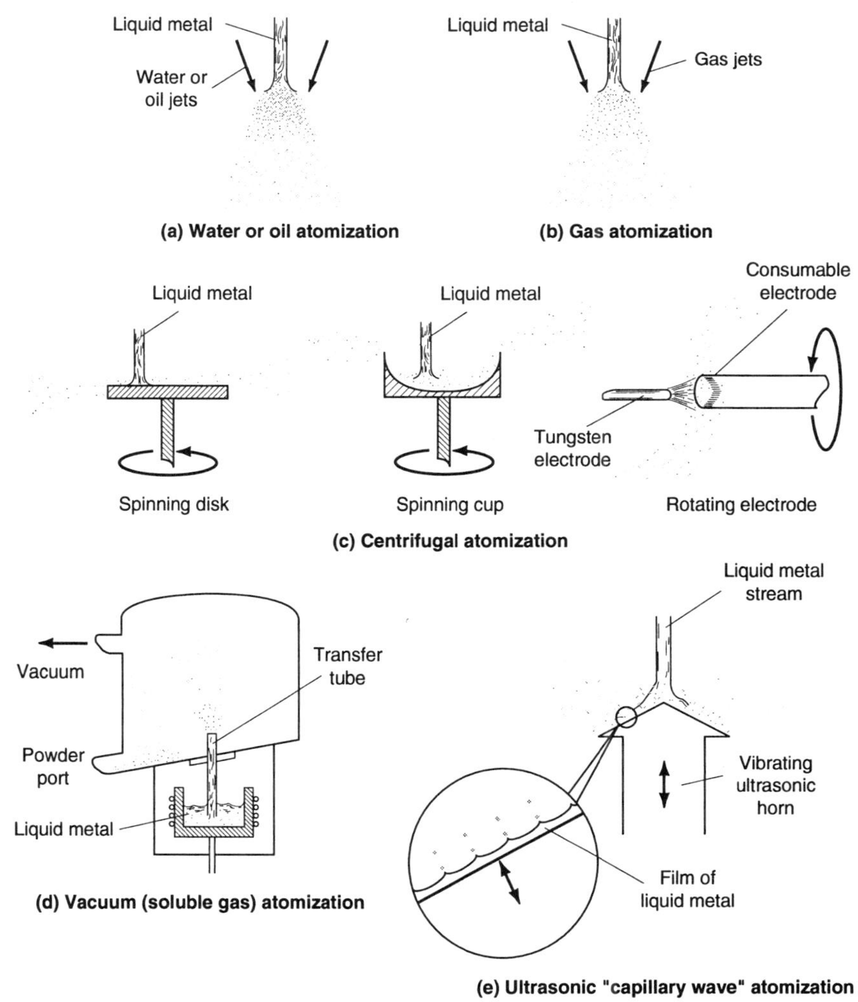

Fig. 1 Atomization process

properties are also discussed in more detail in separate articles in this Volume.

General Characteristics of Atomized Metal Powders

Particle Size. Atomized metal powders are generally found to follow log normal distributions before any screening is done. In other words, screen analysis data or particle size distributions form straight or nearly straight lines when plotted cumulatively on log-probability paper, as shown in Fig. 2. Two numbers—d_m median mass diameter and σ_g geometric standard deviation—define the entire size distribution. The geometric standard deviation represents the slope of the curve. It is a measure of the width of the distribution, or the spread of particle size about the median value.

Standard deviation is obtained as the ratio of the particle size diameters taken at 84.1 and 50.0% of the cumulative weight plot, respectively (Ref 2). A σ_g value of 1.0 signifies that all particles are the same size. Typical σ_g values of water- and gas-atomized metal powders may vary from about 1.7 to 2.3 under good conditions, while bad operation or setup can raise it to as high as 7. Centrifugal and ultrasonic atomization can achieve tighter distributions, as good as 1.3 in some cases (see Fig. 2, line 3 and 7). Vibrating orifice techniques produce almost monosized distributions with σ_g as low as 1.05.

A number of powder particle size distributions are shown in Fig. 2, including gas, water, centrifugal, and ultrasonic. It can be seen that median sizes can range from 10 μm or less in the case of ultrahigh pressure water atomization and some special gas atomizers, to well over 1 mm for very low pressure water granulation. In practical terms, very large particles are normally water quenched because of the huge equipment needed to allow them to freeze in gas. Thus, gas atomizers and most atomizers other than water atomizers are only used for powders finer than about 500 μm.

Particle Shape. The principal characteristics that set atomized powders apart from powder produced by other methods such as electrolysis, reduction, precipitation, and so forth, is their shape and density (Fig. 3). It is generally true to say that atomized powders are free of fine porosity and are relatively compact, with high packing densities and low surface areas, compared with these other types of powder. This means that they have good flow characteristics, good compressibilities, and lower sintering activities than other types.

Packing densities range from as high as 65% of solid for spherical powders such as inert gas, ultrasonic, and centrifugally atomized powders frozen in inert gas, to as low as 20% for some alloys prone to oxidation or very rapid freezing in water atomization.

Purity. The concept of purity covers several aspects. It can be defined as an absence of undesirable contaminants. These can be of three types: bulk dissolved impurities, surface impurities, and inclusions (both of nonmetallic particles and of other metals). It is a general characteristic of atomized powders that there are no impurities of the type found in chemically or electrolytically produced powders that can have chloride or sulfate surface contamination. However, dissolved impurities such as oxygen, sulfur, or carbon can be found, normally unchanged in concentration from the melt. Inclusions are also mainly determined by melt practice, except where sloppy handling allows contamination after manufacture, an ever-present problem for all powders. Surface oxidation is often thought of as the main purity index, but is not always as important as the other two types, especially where oxides are readily reduced in sintering. Obviously, inert atmosphere processes minimize contamination, while air atomization introduces the most impurities. Water atomization lies between these extremes.

Gas porosity is a special case of contamination. If porosity is found in any atomized powder, the first suspicion is always of gas solution in the melt, normally hydrogen, but possibly oxygen or steam. Hydrogen-bearing melt stock (e.g., cathode nickel, copper, cobalt) and dampness should

Table 1 Operating conditions and properties of various water atomized powders

	Powder properties(a)			Operating parameters(a)			
Alloy	Median size, μm	Apparent density(b), g/mL	Oxygen, ppm	Metal flow (M), kg/min	Water flow (W), L/min	*W/M* ratio	Water pressure, MPa
Fe-0.1% C	55	3.2	10,000(c)	150	950	6.3	11
Ag	21	3.4	...	18	100	5.5	50
Fe-0.1% Si	175	3.5	3,400	155	850	5.5	5.8
Fe-15% Si	35	3.6	900	110	250	2.27	25
Cu-5% Cr	24	3.18	...	27	180	6.7	20
316L	50	2.69	1,200	19	120	6.3	8.5
Co	29	3.28	...	27	180	6.7	20
Au-2% Cu	88	...	...	7	40	5.7	13.7
Ag-22Cu-3Zn	72	3.95	110	9	42	4.7	17.2
M2 tool steel	53	2.56	2,100	72	410	5.7	13.8
304L	67	2.63	1,600	65	370	5.7	10.2
Ni-B-Si	51	4.26	...	24	160	6.7	17.2

(a) All data from V-jet atomizer with 40–50° apex angle, nitrogen purged. (b) –250 μm. (c) Air purged

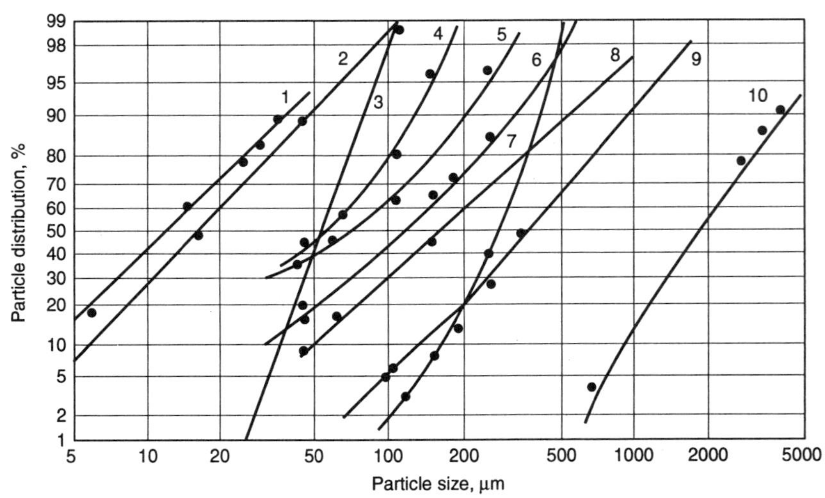

Curve No.	Powder	Mean mass diameter (d_m), μm	Geometric standard deviation (σ_g)
1	Copper, water atomized at 540 bar (by sedigraph)	12.5	2.4
2	Zinc, gas atomized at 20 bar	16	2.1
3	67Pb-37Sn ultrasonically atomized	64	1.31
4	M2 tool steel, water atomized at 140 bar	54	2.13
5	60Pb-40Sn solder, gas atomized at 8 bar	70	2.5
6	Bronze, air atomized at 5 bar	120	2.33
7	Zinc, centrifugally atomized	270	1.4
8	Zinc, water atomized at 55 bar	170	2.58
9	Copper shot, water atomized at 20 bar	280	2.1
10	Phosphor-copper shot, water atomized at 0.5 bar	1950	1.9

Fig. 2 Typical particle size distributions of atomized powders

be checked and melt deoxidation practice reviewed.

Composition and Microstructure. Atomization has the fundamental advantage over other methods of powder production: the composition is completely flexible, being limited only by melt miscibility or volatility considerations. Thus, tungsten-copper or iron-silver cannot be produced and zinc-titanium is tricky, but a vast range of alloys normally made by casting or forging can be made and even greatly extended due to the rapid freezing rate inherent in atomization. Such materials as brasses, stainless steels, superalloys, NiAl, and so forth, can readily be atomized.

The powder particle sizes resulting from atomization allow cooling rates many orders of magnitude above those in casting processes, ranging from 10^2 to 10^7 K/s. This is now known as rapid solidification (RS), but well before the invention of this terminology, such materials as CuPb bearing alloys and P/M high-speed-steel were being produced by atomization and P/M compaction methods. These alloys remain important to engineering, with outputs of thousands of tons per year, while massive research effort over the last 20 years has lead to a rather disappointing level of production of RS products such as amorphous metals and supersaturated solid-solution alloys. Some examples of recent applications that rely on fairly high solidification rates are hypereutectic aluminum-silicon alloys, used in Japan for high-temperature wear resistance in automotive applications, and certain "supermagnets" based on rare earths that have become a billion-dollar business worldwide. Only perhaps 1 to 3% of atomized powders are utilized because of microstructural considerations.

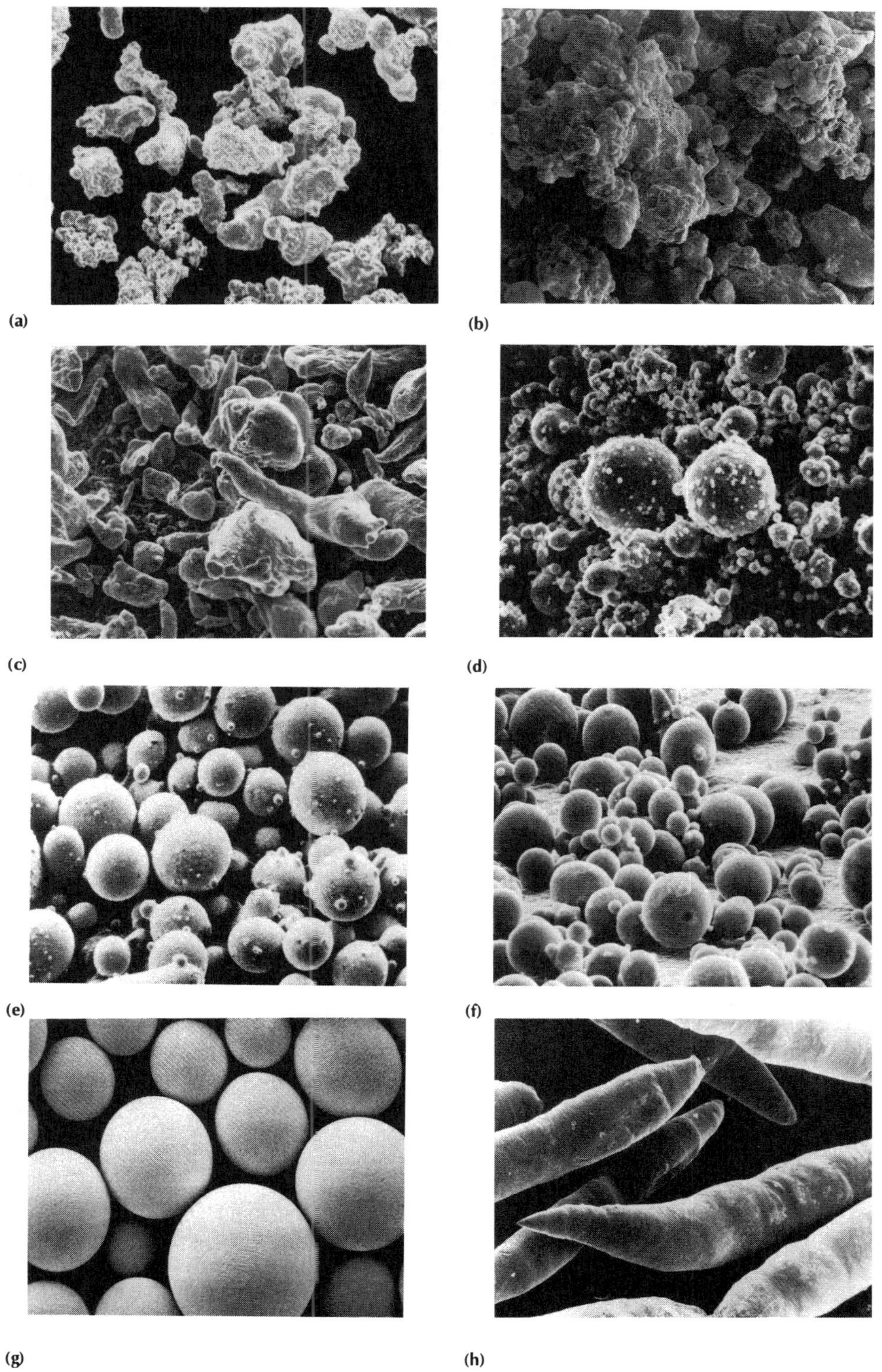

Fig. 3 Examples of atomized powders. (a) Water-atomized copper. (b) Water-atomized iron, apparent density 2.9 g/cm^3. (c) Air atomized aluminum. (d) Helium-atomized aluminum. (e) Nitrogen-atomized high-speed steel. (f) Vacuum-atomized IN-100 superalloy. (g) PREP-atomized René 95 superalloy (dendritic structure). (h) Rotating perforated cup-produced aluminum needles. Courtesy of Huppmann and Dalal

Water Atomization

In terms of tonnage, water atomization is now the preeminent mode of atomization for metal—especially ferrous metal—powders. The largest commercial application of water atomization is iron powder for press-and-sinter applications, and about 60 to 70% of the world's production of iron powder, which now totals some 600,000 to 700,000 metric tons/year, is produced by water atomization. Water atomization is also used for the commercial production of copper, copper alloys, nickel, nickel alloys, stainless steels, tool steels, and soft magnetic powders for pressing and sintering. Water-atomized nickel alloy powders are widely used in thermal spray coating and brazing. Water-atomized precious metals are used in dental amalgams, sintered electrical contacts, and brazing pastes.

As a general criterion, any metal or alloy that does not react violently with water can be water atomized, provided it can be melted and poured satisfactorily. However, it is found in practice that metals melting below about 500 °C give extremely irregular powders due to ultrarapid freezing, which is often undesirable. Thus, zinc is the lowest melting metal produced commercially in this way.

In general, water atomization is less expensive than the other methods of atomization because of the low cost of the medium (water), low energy use for pressurization compared with gas or air, and the very high productivity that can be achieved (up to 30 tons/h or about 500 kg/min). The primary limitations of water atomization are powder purity and particle shape, particularly with more reactive metals and alloys. These generally give irregular powders of (relatively) high oxygen content.

A schematic flow sheet of water atomization is shown in Fig. 4. The major components of a typical installation include a melting facility, an atomizing chamber, water pumping/recycling system, and powder dewatering and drying equipment. Melting of metals follows standard procedures. Air induction melting, arc melting, and fuel heating are suitable procedures.

Typically, the molten metal is poured, either directly or by means of a ladle or runner, into a tundish (Fig. 4), which is essentially a reservoir that supplies a uniform and controlled head of molten metal to the tundish nozzle. The nozzle, which is located at the base of the tundish, controls the shape and size of the metal stream and directs it through an atomizing nozzle system in which the metal stream is disintegrated into fine droplets by the high-velocity water jets. The slurry of powder and water is pumped to a first-stage dewatering device (e.g., cyclone, magnetic system, etc.), which often feeds a second-stage (e.g., vacuum filter) dewatering unit to reduce drying energy use.

Process Variables

Water atomization involves several variables for the distinct stages of melting, atomization, and particle solidification (Fig. 4b). Many of these variables are interrelated, and each atomization unit is unique to some degree with specific operating conditions. Typical operating conditions range as follows:

- Metal flow (single nozzle), 1 to 500 kg/min
- Water flow rate, 20 to 2000 L/min
- Water velocity (at exit), 10 to 500 m/s
- Water pressure (at exit), 5 to 150 MPa
- Metal superheat, 75 to 150 °C above melting point

Commercial water atomization units typically operate with water flow to metal flow ratios from 2-to-1 to 10-to-1. Operating conditions for specific powders and configurations are listed in Table 1.

In conventional commercial water atomization, the water pressure is usually in the range of 5 to 20 MPa (750 to 3000 psi) resulting in mass median particle sizes of 30 to 150 μm. More specifically, typical molding (100 mesh) powders are normally made using pressures of 7 to 13 MPa (1200 to 2000 psi). However, much higher water pressures (50 to 150 MPa) are used to produce finer powders with median particle size of 5 to 20 μm. This use of high-pressure atomization is driven by the demands of the injection molding and thermal spray industries that require such fine powders.

Water-Jet Nozzle. The nozzles that direct and control the atomizing medium are critical in influencing the efficiency of the disintegration of the liquid metal stream. Many of the nozzle designs are proprietary or described in patent literature (Ref 3). However, nozzle designs in most commercial water atomization units are variants of either V-jet nozzles (Fig. 5a) or annular cone-jet nozzles (Fig. 5b) that are concentric with the metal stream. Annular ring nozzles are utilized in some large-capacity plants making iron powder, but are rarely used elsewhere because they are less flexible and harder to make than designs based on "V" nozzle configurations with two or more discrete openings located symmetrically about the axis defined by the metal stream.

The liquid metal stream typically falls a certain distance (typically about 15 to 25 cm) before being hit by the water jets. A partial vacuum is often created above the impingement point of the water, and the gas/air entrained by the water streams can disrupt the liquid metal before it actually meets the water sprays (Ref 4). It should be noted that the water jet that hits the metal stream is typically only a few percent by volume of water droplets, and 98% or so gas.

Another class of water atomizers is used principally at low pressures and for very high flows. This is asymmetrical atomization (Fig. 6), where the melt is poured onto a horizontal jet of water. This is used on a large scale (~500 kg/min) for steel and iron shot production and a huge scale (~5000 kg/min) for the granulation of blast furnace slags. However, these are millimeter range products and are not discussed here in detail.

Mechanism of Water Atomization. Based on a review of models for various atomization methods (Ref 5), the basic mechanism in water atomization appears based on momentum transfer, where the metal stream is broken up under the impact from water droplets (Fig. 7). Droplet formation by impact rather than shear was first proposed by Grandzol and Tallmadge for water atomization (Ref 6, 7). This model reflects the dispersed nature of the water and an inverse proportionality between particle size and the normal

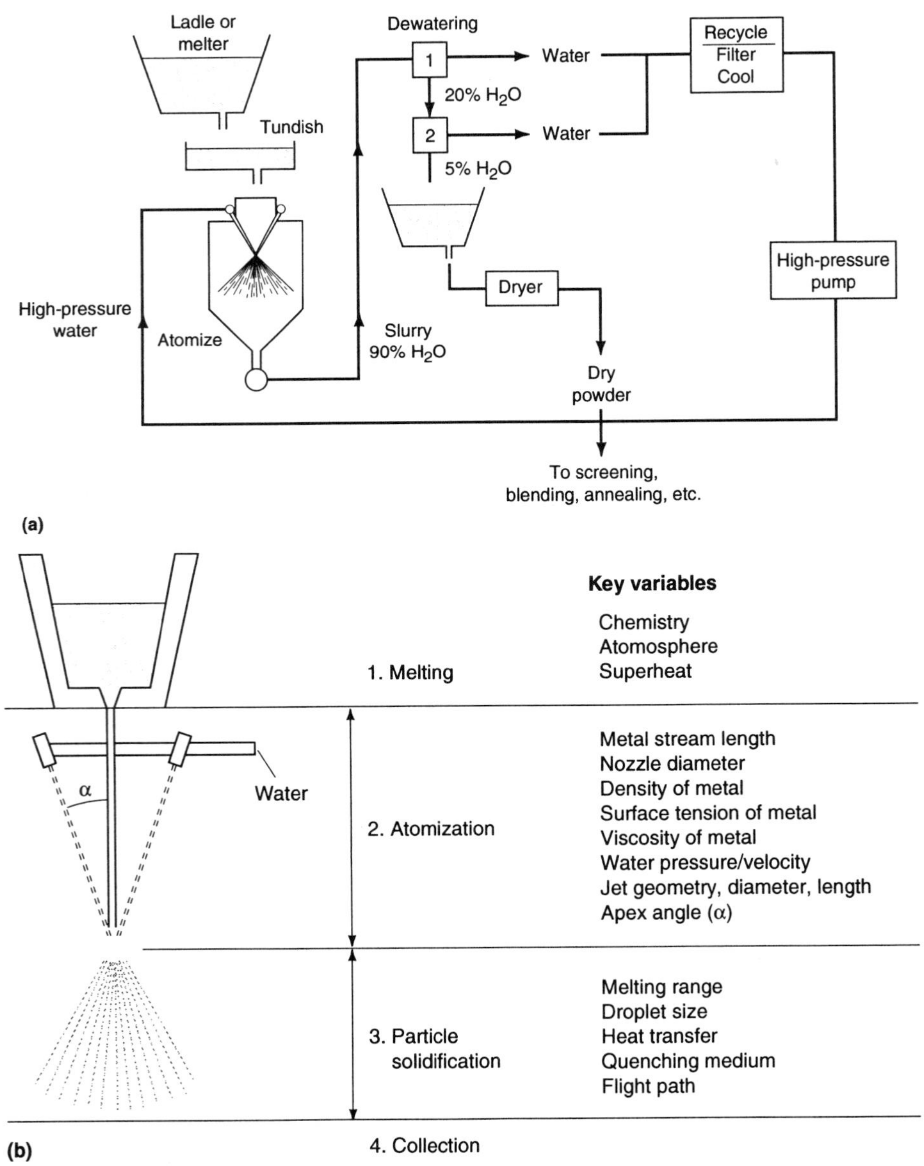

Fig. 4 Water atomization system. (a) Large-scale (1,000 to 100,000 tons/year) water atomizing system. (b) Key variables in the various stages of the water atomization process

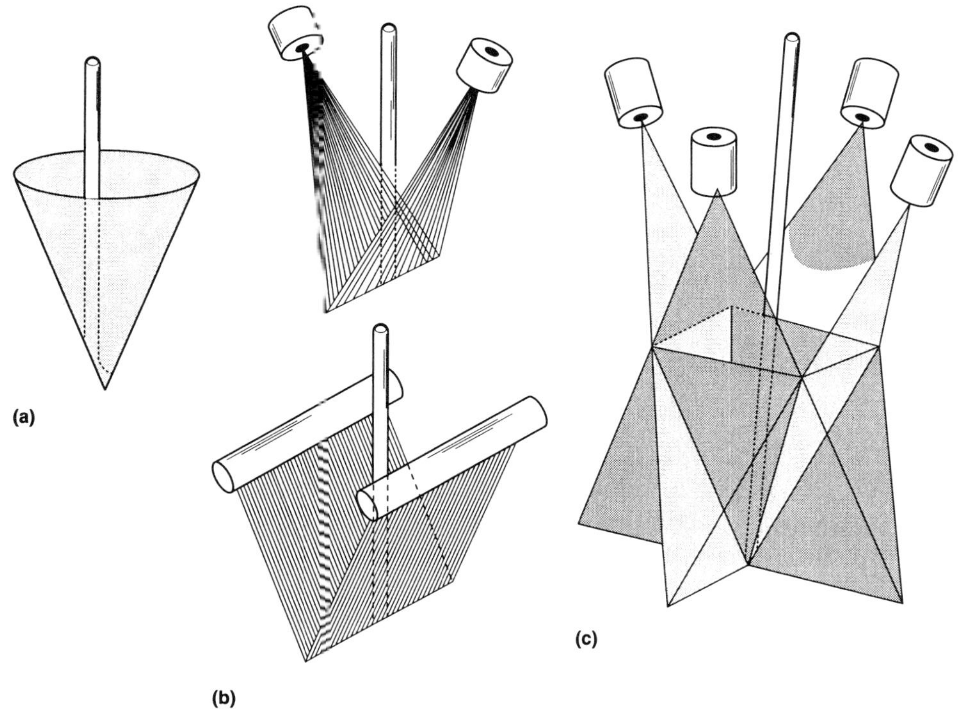

Fig. 5 Water-jet configuration. (a) Annular jet. (b) Open V-jets. (c) Closed V-jets

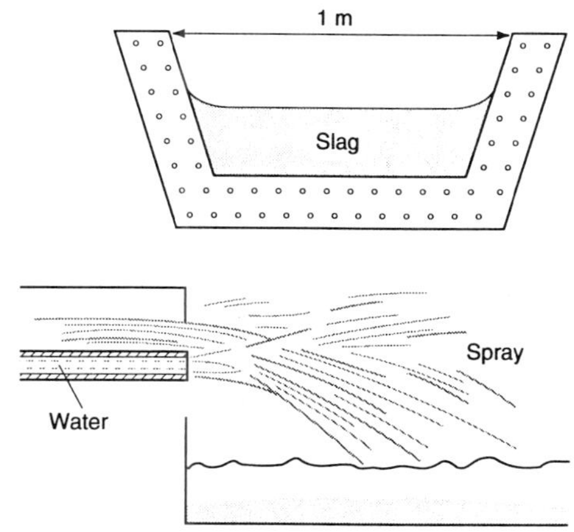

Fig. 6 Asymmetric water atomization of slag

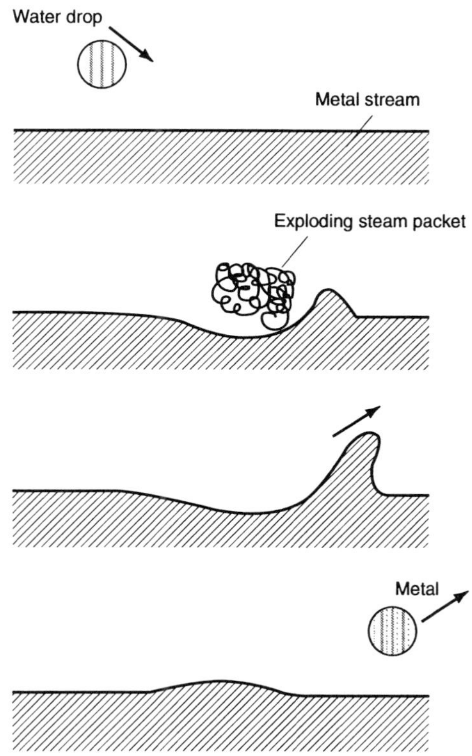

Fig. 7 Impact model of water atomization mechanism. Source: Ref 5

velocity component of the water (with respect to the metal stream axis). Specifically, Grandzol and Tallmadge showed that drop velocity (V_W) is related to the mass median particle size (d_m) such that:

$$d_m = \frac{A}{V_W} = \frac{14{,}900}{V_W}\left(\frac{1}{n}\right)^{1/3} \quad \text{(Eq 1)}$$

where A is a constant, d_m is in μm, V_W is in m/s, and n is the number of metal droplets formed from the impact of one water droplet. In this mathematical form, water flow rate, water pressure, momentum and energy, jet length, and metal flow rate are not primary parameters influencing d_m; they do, however, affect particle size by influencing water-jet velocities and the number of water droplets. Also, the number of metal droplets produced per water droplet can vary without altering the basic details of the model.

Collisions during atomization can also influence particle size, particle size distributions, and particle shape (Ref 8, 9). Collisions reduce the efficiency of atomization, and the probability of particle collisions increased with increasing metal flow rate. Thus, the maximum of average particle size and geometric standard deviation are achieved in the center of the atomizing stream, where metal concentration generally is highest.

Collisions also cause gas entrapment and microinhomogeneity from internal (primary) particle boundaries. Collisions between particles usually overlap with the earlier stages of atomization. These collisions are minimized or practically eliminated in centrifugal modes of atomization, in which traffic density of the particle decreases rapidly with distance.

Particle Size

In practice, average particle size is controlled by the pressure of the water jets. Higher water pressure directly influences water velocity (V_W), as well as changing the particle size of the water droplets in the jet. These further control particle size, but in a complex and poorly understood way. The relation between d_m and water pressure (P) is shown in Fig. 8 (Ref 10) for various metal powders.

Based on the straight log-log plots in Fig. 8, the relationship between P and d_m can be expressed as:

$$d_m = KP^{-m} \quad \text{(Eq 2)}$$

where K and m are constants for a given material, atomization unit, or nozzle configuration. Variations for two nozzle configurations are shown in Fig. 9 (Ref 11). When d_m is in μm, the parameter m is typically in the range 0.6 to 0.8 for water pressures from 0.1 to 50 MPa.

The value of K is also a function of the molten metal or alloy that is being atomized; it is actually a measure of the ease with which a given molten metal stream can be disintegrated. Values of K are almost one order of magnitude higher for aluminum and zinc than for copper and iron-base alloys, meaning that the latter materials require a much lower pressure for atomization. In practice, this means that for two alloys water atomized under identical operating conditions, d_m can vary by a factor of about 6. The value of K is probably determined by the physical and chemical properties of the molten alloy, in particular viscosity and surface tension. Modification of the surface tension by the atmosphere (steam) to allow the formation of oxides is also expected to be important in terms of disintegration of the molten metal stream. Density and thermal conductivity are considered to be of secondary importance in their influence on the magnitude of K.

Other Factors Influencing Particle Size. Besides the velocity of the water jet, particle size after atomization is also influenced by melt superheat temperatures and the apex angle (α) be-

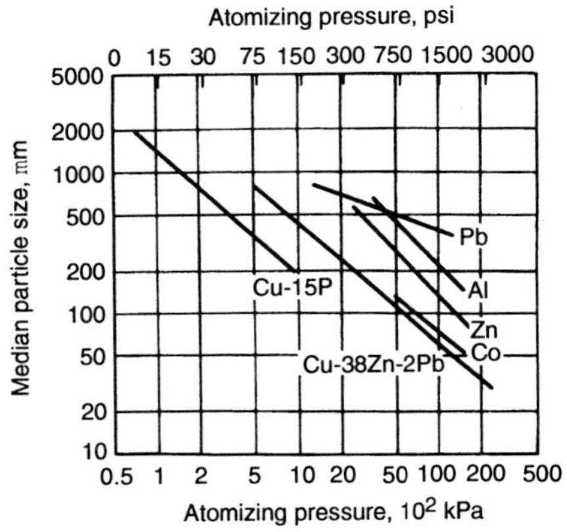

Fig. 8 Particle size of water-atomized metals as a function of atomizing pressure. Source: Ref 10

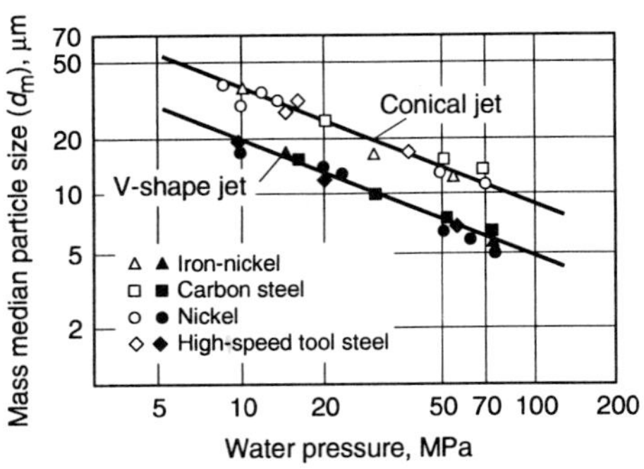

Fig. 9 Mass median particle size (d_m) versus water pressure for nozzle with V-shaped jets (solid symbols) and conical jets (open symbols). Source: Ref 11

tween the jets and metal stream. For water atomization, d_m and σ_g are not particularly sensitive to metal flow rate and the ratio of water flow to metal flow rates. Commercial water atomization units typically operate with water flow-to-metal flow ratios of 5-to-1 to 10-to-1. Experience with water atomization shows that a very low ratio of the water flow-to-metal flow rate (less than 3 to 1) will result in coarser powders. Increasing the water-to-metal ratio above about 6 to 1, however, does not have a significant effect on particle size (Ref 12).

Table 2 Representative properties of water-atomized powders

Metal alloy	Mass median particle size (d_m), µm	Standard deviation, σ_g	Apparent density, g/cm^3	Oxygen content, ppm
Ag	37	2.16	3.76	...
Ag-28Cu	32	2.3	3.37	285
Ag-25 Sn-5 Cu-1 Zn	35	2.40	3.11	260
Au	130	2.69	7.82	...
Au-20Ni	62	2.9	...	57
Bi	79	3.33	3.58	...
Cd	185	2.00	1.29	...
Cu	90	2.40	5.07	387
Cu	46	2.30	3.48	350
Cu-0.3 Mg	31	2.51	2.01	1105
Cu-30 Pb	77	2.30	4.75	...
Fe	75	2.13	3.43	4850
Fe	58	2.15	3.47	...
Fe-15 Si	32	1.8	3.37	...
Fe-45 Si	63	1.6	2.57	...
Ni	100	2.10	4.50	645
Ni-5 Al	34	2.29	2.85	710
Ni-Cr-B-Si	51	1.69	4.26	...
Pt-10 Rh	250	2.80	12.3	...
Sn	160	2.19	1.48	314
Sn	90	2.19	1.23	620
Zn	58	2.05	1.82	...
Stellite F	90	2.24	4.32	400
304 L	52	2.16	3.08	2000
316 L	48	2.33	3.12	2600
316 L	26	1.75	2.75	2000

Source: Ref 1

Effect of Superheat. Increasing the melt superheat reduces d_m in water atomization primarily through the effect of temperature on melt viscosity and surface tension. For zinc, an increase in superheat from 100 to 300 °C (180 to 540 °F) reduces d_m from 150 to 100 µm (Ref 10). Similarly, in a cobalt-base alloy, increasing the superheat by about 150 °C resulted in a decrease in d_m of about 13.5% (Ref 13). In addition to its effect on particle size, superheat is used as an operating variable to prevent freezing of the alloy in the exit nozzle of the tundish and to avoid premature freezing of droplets. However, it is very rare to use superheat to control particle size; its effect on particle shape is much more important.

Effect of Apex Angle. Particle size is influenced by the apex angle, as a larger angle increases the velocity component normal to the metal stream. Data reproduced in Fig. 10(a) show that d_m decreases with increasing angle between the axis of the water jet nozzle and the liquid stream. According to Ref 6, this dependence is of the form:

$$d_m \approx \frac{1}{V \sin \alpha} \qquad \text{(Eq 3)}$$

However, there is a practical limit to the magnitude of the apex angle. Above about 50°, flow of molten metal from the base of the tundish is impaired because of welled-up water, and the incidence of metal freeze-up in the exit nozzle is also found to increase at higher apex angles, especially at higher pressures (Fig. 10). To decrease attenuation in water pressure, a short water-jet length is preferred, but again there are practical limits. It is normally found that using

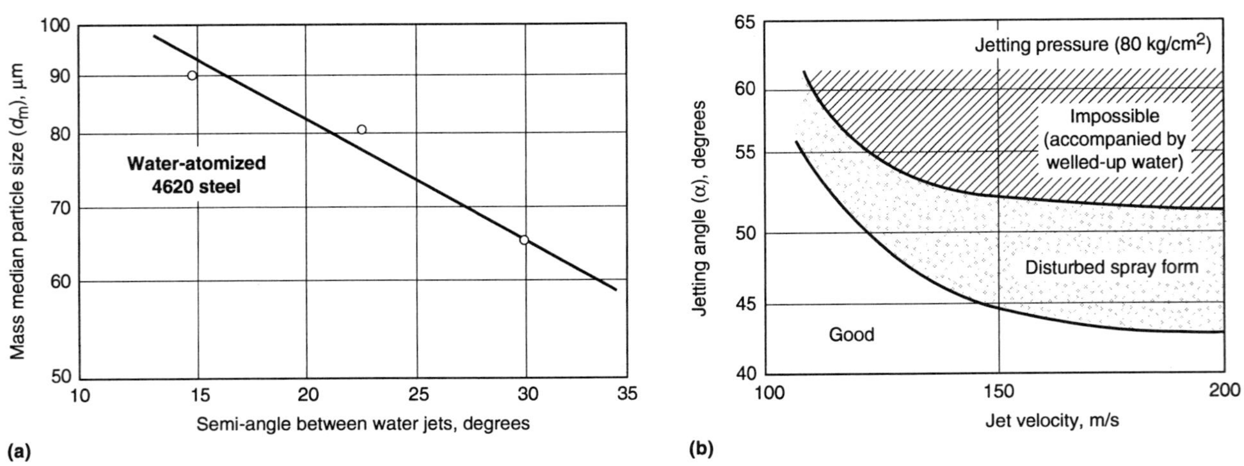

Fig. 10 Effect of jet angle on (a) mass median particle diameter and on (b) stability of water atomization at various water velocities. Source: Ref 1, 14

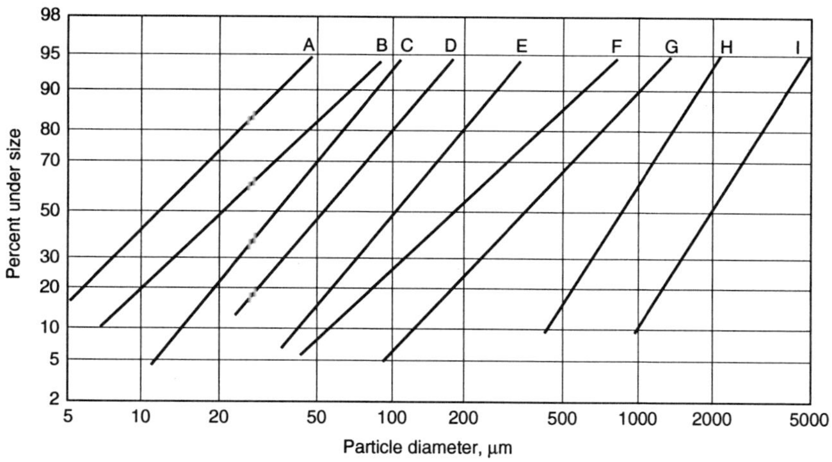

	Alloy and water pressure	σ_g
A	Copper atomized at 54 MPa	2.40
B	18Cr-10Ni-2.5Mo stainless steel atomized at 50 MPa	2.40
C	Fe-15%Si atomized at 20 MPa	1.70
D	M2 high-speed steel atomized at 14 MPa	2.13
E	Fe-45%Si for welding electrodes atomized at 4 MPa	1.70
F	Zinc for alkaline manganese batteries atomized at 5.5 MPa	2.34
G	Copper shot atomized at 2.0 MPa	2.30
H	15% phosphor-copper shot atomized at 0.15 MPa	1.76
I	15% phosphor-copper shot atomized at 0.05 MPa	1.75

Fig. 11 Typical water-atomized particle size distributions. Source: Ref 14

apex angle to control fineness leads to dangerously unstable conditions. It is far better to select a good stable setup with essentially zero rejection of melt and spray toward the jets and tundish and then provide the necessary pressure to achieve the desired particle size.

Particle Size Distribution

The factors affecting σ_g are somewhat debatable and poorly understood. The data shown in Fig. 11 (Ref 14) might lead to suggestions that very coarse distributions, toward the millimeter range, are tighter (lower σ_g). It is easily argued that extremely low pressures must give tighter distributions as liquid drops cannot be stable in the centimeter range. However, in the author's experience, there is little firm evidence linking σ_g to pressure where particle sizes are in the range from 10 to 200 μm. While it is quite easy to design atomizers with too high an apex angle allowing rejection of the coarse droplets (which can give poor σ_g values up to well over 3.0), it is extremely difficult, indeed practically impossible, to make a jet system that produces a uniformly low value of σ_g below 2.0 for a range of metals.

As a general characteristic of water atomization, d_m and σ_g are not particularly sensitive to metal flow rate and the ratio of water flow to metal flow rate. Figure 12 shows that atomization of Fe-15wt%Si powder at rates from 20 to 110 kg/min (stream diameters from 5 to 13 mm) gives very uniform results.

The main influence on σ_g, once a good and stable atomizing jet setup is ensured, appears to be melt chemistry. There is no satisfactory explanation of this yet published, but there is no doubt that alloys such as Ni-Cr-B-Si and Fe-Si(15–45 wt%) give consistently narrower distributions ($\sigma_g \sim 1.6$–1.8) than other metals such as copper, nickel, and iron. It can be speculated that this is related to the tendency of such alloys to spherical shape, in turn related to the characteristics of the oxide films formed on the particle surfaces. Those on "self-fluxing" alloys seem to be liquid below the melting point of the alloy and allow it to spherodize before freezing. As a particle of a given mass can appear on sieving to have various sizes depending on its shape, it may be that a spherical shape, which allows minimum variation in shape and thus reported size, favors low σ_g. However this rationale is largely speculation that requires verification.

Figures 13 and 14 (Ref 16) illustrate the effects of nozzle diameter and pouring temperature for water-atomized copper powders on σ_g and d_m. The large effect of the tank atmosphere—air versus nitrogen (Fig. 13) is interesting. In fact, it is known that if copper is atomized with a large dissolved oxygen content, the powder is much finer (about 30% reduction in d_m at constant pressure). Thus, the large difference between nitrogen and air atmosphere sizes may be due to this effect. Similarly, the chemistry may be directly influencing σ_g. The generally lower values of σ_g at lower metal flow rates reflect the importance of collisions in broadening the distributions.

The use of values of σ_g, in combination with median particle size according to Fig. 11, permits the estimation of powder yields as a function of atomizing pressure. Where particles are needed in a narrow size range, for example, −150+53 μm, the impact of σ_g on yields is very important (see Fig. 15).

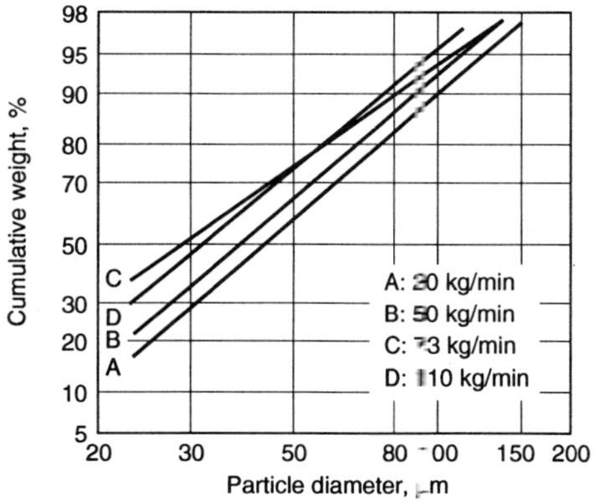

Fig. 12 Effect of metal flow rate on size of water-atomized Fe-15wt%Si powder. Source: Ref 15

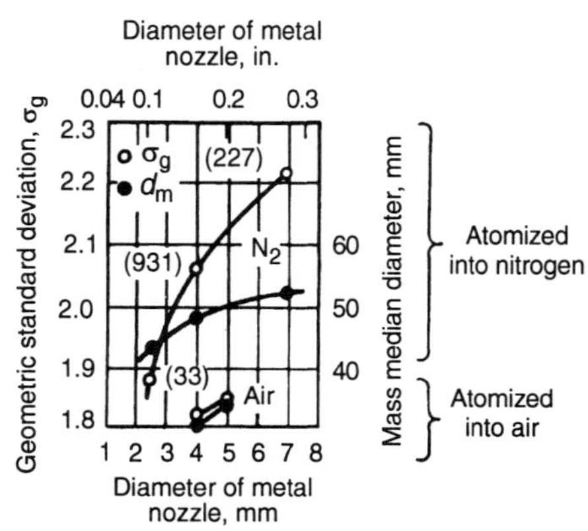

Fig. 13 Effect of metal-stream diameter on geometric distribution of water atomized copper powder. Pouring temperature 1200 °C (2190 °F); figures in parentheses show average metal flow rates in g/s. Source: Ref 15

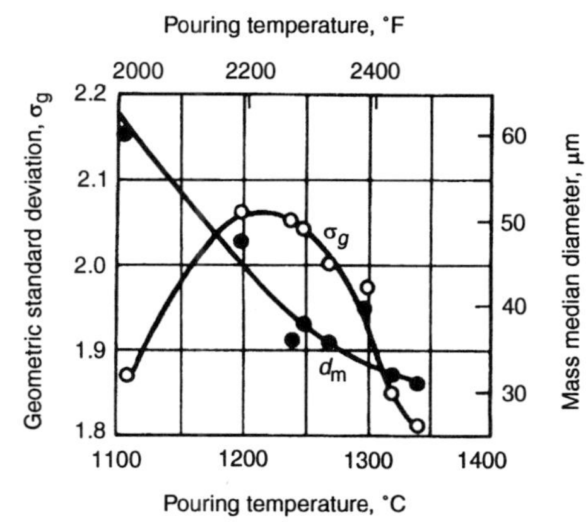

Fig. 14 Effect of pouring temperature on distribution of water-atomized copper powder. Metal stream diameter 4 mm (0.16 in.); water flow 320 L/min (85 gal/min); water pressure 13.2 MPa (1915 psi); atmosphere nitrogen. Source: Ref 11

Powder Characteristics

Water atomized powders generally are quite irregular in shape compared to gas atomized powders and have relatively higher surface oxygen contents (depending on the alloy reactivity). Gas atomized powders, on the other hand, generally are more spherical or rounded in shape and, if atomized by an inert gas, generally have lower oxygen (oxide) contents. There are, of course, exceptions in each type. For high-volume, low-cost production, water atomization generally is preferred over gas atomization, providing powder characteristics are compatible with the application. Typical characteristics of various water atomized powders are summarized in Table 2. Factors affecting powder characteristics are summarized in Table 3 (Ref 17).

Particle Shape. The irregular shape of water atomized powders is attributed by some authors (Ref 10, 18) to the relative rates of cooling and spheroidizing of liquid droplets. For example, if the spheroidization time (τ_{sph}) of the liquid droplet is short compared to its solidification time (τ_{sol}), particle shape tends to be spherical; if spheroidization time is long, particles tend to be irregular. Thus, water atomized powders tend to be more irregular because cooling rates for water atomization are one to two orders of magnitude larger than for gas (nitrogen or argon) atomization. The particle shape of water atomized powder also becomes more irregular if small amounts of impurities lower surface tension and thus increase spheroidization time.

Figure 16 (Ref 10) compares solidification times for 100 μm particles of different melting point metals with the apparent densities of water atomized powders as a percentage of solid. It can be seen that metals melting above about 500 °C have long freezing times. This is due to the steam film formed at such temperatures. By contrast, metals such as tin, lead, and zinc freeze very fast due to nucleate boiling heat transfer, unimpeded by steam films. However, a fusible alloy has so little thermal head that it cools slowly. The density data mirror the resulting curve. It should be noted that none of the metals showing high densities has a major affinity for oxygen and that such alloys as stainless steel show densities of only about 30%.

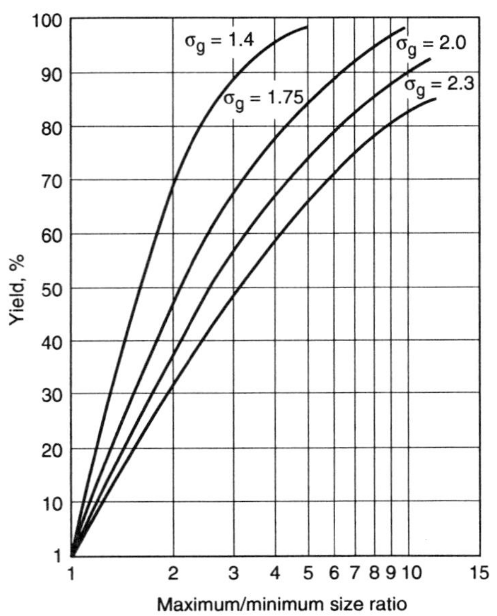

Fig. 15 Master yield curve showing yield between two sizes as a function of size ratio and standard deviation (σ_g)

Oxidation. During water atomization, most metals react with water or steam to form oxides, following a chemical reaction of the form: xMe + yH$_2$O = MexOy + yH$_2$. Oxidation generally increases with increasing superheat of the liquid metal and depends on particle size. Some materials exhibit a minimum if oxygen content is plotted versus particle size. Such a minimum may result from two opposing factors. With increasing particle size, cooling time, and exposure to oxidation increase, oxide layers become thicker. Conversely, with decreasing particle size, the specific surface area of a powder increases rapidly.

In addition to the amount of oxidation, the nature and composition of oxides and their spatial distribution within a particle can have a great effect on the usefulness and subsequent processing requirements of a powder. Surface analytical techniques have shown that the surfaces of water atomized alloys are typically highly enriched in the oxides of elements that have high negative free energies of oxide formation and high diffusivities at elevated temperature. Oxide films have thicknesses ranging from tens to thousands of nanometers.

This phenomenon is also true for gas atomized powders in which some oxidation has occurred during atomization. Typical oxygen contents of water atomized metal powder are:

Metal	Oxygen content, ppm
Silver	100
Gold-nickel	<100
Copper	<800
Copper-tin	<400
Nickel	<500
Iron	1,000–3,000
Cobalt	2,000–3,000
Iron-carbon	500
Aluminum	2,000
Fe-13Cr	1,000–3,000
Fe-45Si	2,000
Ni-Cr-B-Si	250–1,000
Fe-80Mn	30,000

Source: Ref 10

Within the group of alloys that contain elements with intermediate oxygen affinity (chromium, manganese, and silicon), silicon-

Table 3 Effect of water atomization parameters on powder properties

Parameter	Mean particle size	Width of particle size distribution	Particle irregularity	Particle oxygen content
Nozzle design				
Apex angle	–	0	+	+
Water jet length	+	0	–	–
Molten metal	0	0	0	0
Stream length	+	+	–	–
Atomizing liquid				
Pressure	–	–	+	+
Flow rate	–	–	+	0
Velocity	–	–	+	+
Viscosity	–	0	+	+
Molten metal				
Viscosity	+	0	–	–
Surface tension	+	0	–	–
Melting temperature range	0	0	+	0
Liquidus	+	+	–	0
Superheat	–	+	–	+
Metal flow rate	+	+	–	0
Metal stream diameter	+		–	0
				+
Atmosphere oxygen content				
Over melt	0	0	0	+
In atomizing tank	0	0	0	+
Tank parameters				
Flight path	0	0	0	+

Note: A plus sign indicates that there is some direct relationship between the powder property and the atomization parameter, while a minus sign signifies some inverse relationship. Zeroes (0) indicate the absence of a strong relationship or that data are conflicting. Source: Ref 17

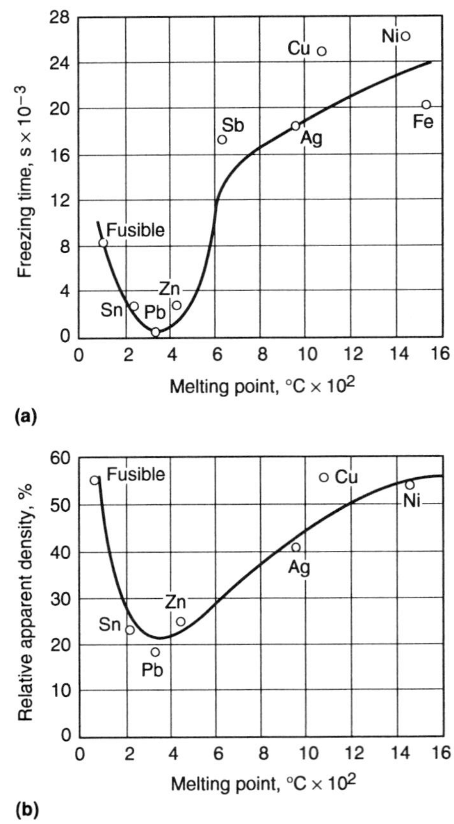

Fig. 16 Freezing time and relative density of powders as a function of melting point. (a) Freezing time of 100 μm particles versus melting point. (b) Relative apparent density of water-atomized powders versus melt points. Source: Ref 10

containing alloys generally produce low-oxygen-content powders, while manganese-containing alloys become heavily oxidized during water atomization. This tendency is attributed to the more protective nature of silicon oxide films compared to manganese oxide films. Figure 17 shows this effect for 304L stainless steel.

Oil Atomization

Atomization of metal has been done commercially with oil, which should avoid the problem of powder oxidation (Ref 1). Commercial application is limited to production of relatively high-carbon (0.4 wt% C) steel as an inoculant in continuous casting of steel. Oil-atomized powders resembled water-atomized powder with d_m typically at 70 μm and sometimes down to 30 μm. Oil-atomized powders have densities intermediate between water- and gas-atomized powders as the quench rate is slower and oxidation much less. However, carbon pickup occurs to an extent depending on metal temperature and carbide-forming tendency (Ref 19).

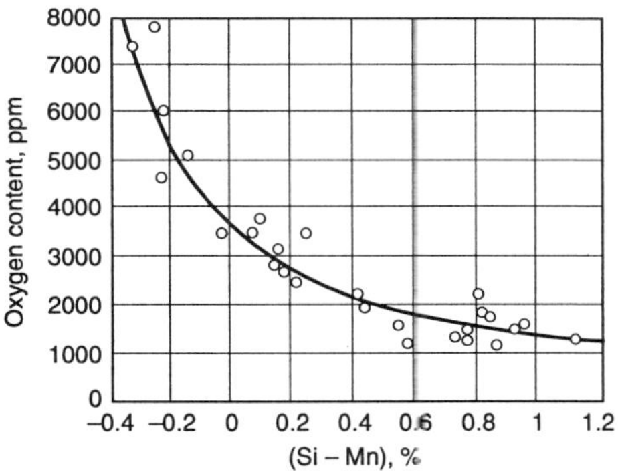

Fig. 17 Effect of silicon and manganese content on the oxygen content of 304L powders

Gas Atomization

Gas atomization is the process where the liquid metal is disrupted by a high-velocity gas such as air, nitrogen, argon, or helium. Atomization occurs by kinetic energy transfer from the atomizing medium to the metal.

Gas atomization differs from water atomization in many respects. Rather than being dominated by the pressure of the medium like water atomization, it is found that the gas-to-metal ratio is the dominant factor in controlling particle size, with median particle size being related inversely to the square root of the gas-to-metal ratio. The reasons for this fundamental difference are easily explained. In the case of gases, increases in pressure above 0.1 MPa (the pressure at which sonic velocity is reached) give only very small increments in gas velocity. In contrast, to reach sonic velocity (in air/nitrogen) with a water jet, a pressure of nearly 40 MPa is needed and the velocity increases uniformly as the square root of the pressure. Also, gas atomization takes place by the action of a continuum on another, while in water atomization a stream of droplets (in an entrained gas flow) acts on a continuum. The density of the water medium is about a thousand times higher than typical gases, giving much greater "punch" or short-range forces.

Gas-atomizing units also come in a much wider range of designs than water atomizers and are classified as either "confined" or "free-fall" nozzle configurations (Fig. 18). There is also a third type, "internal mixing," where the gas and metal are mixed together before expanding into the atomizing chamber (discussed in the following sections). Free-fall gas units are very similar in design to water-atomizing units. However, due to the rapid velocity decay as the gas moves away from the jet, in free-fall gas units it is very difficult to bring mean diameter of powder below 50 to 60 μm on iron-base material. High efficiency is thus difficult to obtain in free-fall systems, although special design and configuration of nozzle arrangements can produce relatively fine powder at reasonable gas-to-metal ratios for high-velocity oxyfuel thermal spray, plasma tungsten arc (PTA) welding, and hot isostatic pressing applications. As well as vertical designs, resembling water atomizing designs, there are a number of asymmetric horizontal designs where a vertical, inclined, or sometimes horizontal melt stream is atomized by essentially horizontal gas jets. These designs are widely used in zinc, aluminum, and copper alloy air atomizers.

Closed or "confined" nozzle designs enhance the yield of fine powder particles (~10 μm) by maximizing gas velocity and density on contact with the metal. However, although confined designs are more efficient, they can be prone to freezing of the molten metal at the end of the tundish nozzle, which rapidly blocks the nozzle. Also, the interaction of the gas stream with the nozzle tip can generate either suction or positive pressure, varying from suction that can triple metal flow rate to back pressure sufficient to stop it and blow gas back into the tundish. Thus, great care is needed in setting up close-coupled nozzles, and the closer the coupling, the greater the care (as well as the efficiency). Water-bench test techniques (Ref 20), where the melt is substituted by water as a model liquid, have allowed consistent hot performance to be achieved by cold testing. Tests of suction alone have proved poor predictors of hot performance because the gas stream is considerably affected by its interaction with the liquid being atomized.

Confined designs are of two types: conventional (Fig. 18) or ultrasonic (Fig. 19). The ultrasonic design uses the Hartman tube principle to apply high-frequency pulsation to the gas stream, with gas exit velocities reported to be Mach 2 to 2.5 and the major pulsation frequency at about 100,000 Hz.

In many confined designs, the circulation created by the gas flowing down the side of the tundish nozzle causes the molten metal to flow across the face of the ceramic nozzle to its edge, where it is sheared by the flowing gas (Ref 21–23). These nozzles are referred to as "prefilming" and are quite widely used (see Fig. 20).

Worldwide annual tonnage of inert gas-atomized powder is much less than that of water-atomized powders, probably amounting to no more than 50,000 tons/year. Metal feed rates are lower than in water atomization, and melt size is

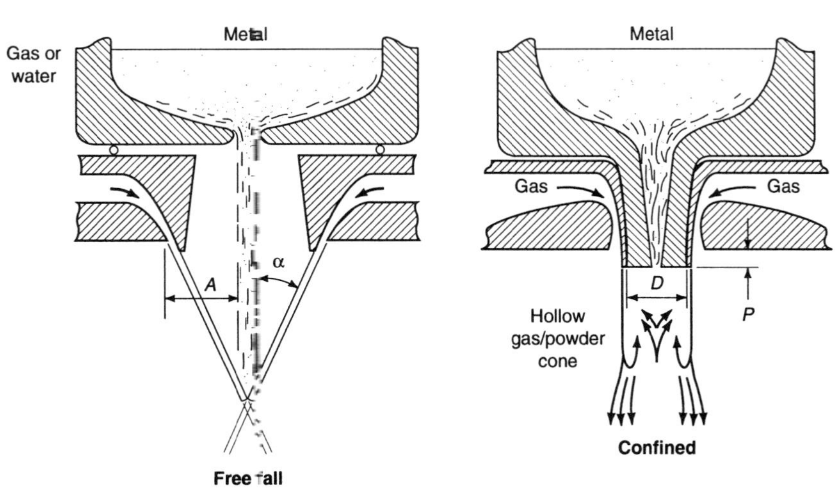

Fig. 18 Two-fluid atomization with (a) free-fall design (gas or water) and (b) continued nozzle design (gas only). Design characteristics: α, angle formed by free-falling molten metal and atomizing medium; A, distance between molten metal and nozzle; D, diameter of confined molten metal nozzle; P, protrusion length of metal nozzle

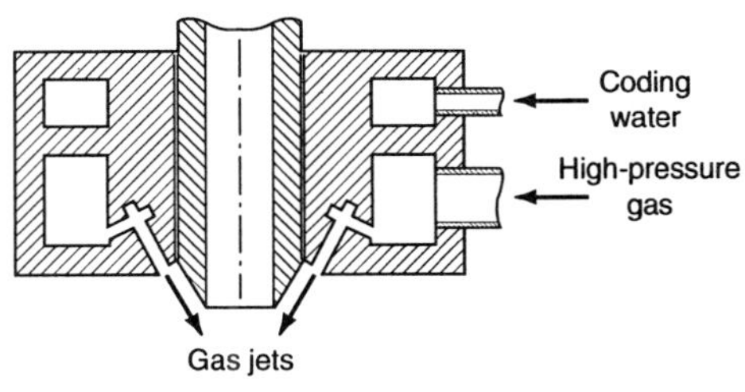

Fig. 19 Ultrasonic gas atomizer (U.S. patent 2,997,245)

smaller. However tonnage of air-atomized powders, especially zinc and aluminum, but also tin, lead, and copper alloys, probably exceeds 300,000 tons/year. Most of these air atomizers operate continuously for many hours or even days. Multinozzle units are often used to boost output on aluminum and zinc.

Process Variables

Conventional atomization pressures are typically in the range 0.5 to 4 MPa (70 to 600 psi), and gas velocities in the nozzles range from Mach 1 to 3. However, in free-fall atomizers, measured gas velocities in the impingement area normally have fallen to 50 to 150 m/s (for air/nitrogen). Typically, gas-atomized powder is spherical with a log normal size distribution. Mean particle size is usually in the range 10 to 300 μm with a standard deviation of about 2. Oxygen content is about 100 ppm. Powder cost is extremely sensitive to the economies of scale. Prealloyed ferrous, nonferrous, and specialty alloys are made by inert gas atomization.

For free-fall gas atomization, the key process variables are similar to those of water atomization (Fig. 4b). In a confined-nozzle design, the major parameters are the geometry of the tundish nozzle tip, the gas jet apex angle, and gas jet diameter (or width for an annular design), the number of jets, and the horizontal spacing between the jets and the tundish nozzle center line.

In conventional gas or air atomization, typical metal flow rates through single orifice nozzles range from about 1 to 90 kg/min (2 to 200 lb/min). Typical gas flow rate ranges from 1 to 50 m^3/min (40 to 1600 ft^3/min) at gas pressures in the range of 350 kPa to 4 MPa (50 to 600 psi). Effective gas velocities are very difficult to measure and depend on nozzle design, ranging from 20 m/s (66 ft/s) to supersonic velocities. The temperature differential between the melting point of the metal and the temperature at which the molten metal is atomized (superheat of the molten metal) is generally about 75 to 150 °C (135 to 270 °F). In gas atomization with argon or helium, the cost of gas consumption is significant, and a means of circulation to facilitate gas reuse is desirable in larger-scale facilities.

In practice, for a given gas nozzle design and size, average particle size is controlled by the pressure of the atomizing medium and the melt flow rate (controlled by nozzle diameter and nozzle suction). For all nozzles, the velocity of the gas usually "chokes" at sonic velocity (about 300 m/s, or 1000 ft/s, for nitrogen and argon) in the narrowest region of the nozzle if the upstream gas pressure is at least 1.9 times the external pressure.

Consequently, the amount of gas flow (A) depends on gas pressure, temperature, and nozzle area. For ideal conditions and zero velocity on the entrance side of the nozzle, gas flow is:

$$A = a\left(\frac{2}{k+1}\right)^{k+1/2\,(k-1)} \frac{p\sqrt{2g}}{\sqrt{RT}} \qquad \text{(Eq 4)}$$

where a is the cross section of gas nozzle at exit, k equals C_p/C_v, the ratio of specific heat at constant pressure and volume, p is the gas pressure in gas reservoir, T is the temperature in gas reservoir, R is the gas constant, and g is the acceleration due to gravity. For nitrogen, with $k = 1.4$:

$$A \approx 4 \times 10^6 \frac{ap}{\sqrt{T}} \text{ g/s} \qquad \text{(Eq 5)}$$

As a compressible fluid passes through a nozzle, a drop in pressure and a simultaneous increase in velocity result. If the pressure drops sufficiently, a point is reached where, in order to accommodate the increased volume due to expansion, the nozzle unit must diverge. Thus, nozzles for supersonic velocities must converge to a minimum section and diverge again.

Gas efficiency can be compared on the basis of how much powder surface is generated per unit volume of gas spent. This criterion accounts for higher gas consumption requirements when higher gas pressures are applied in producing finer powders. Confined- or "close-coupled-" nozzle designs give higher efficiencies at com-

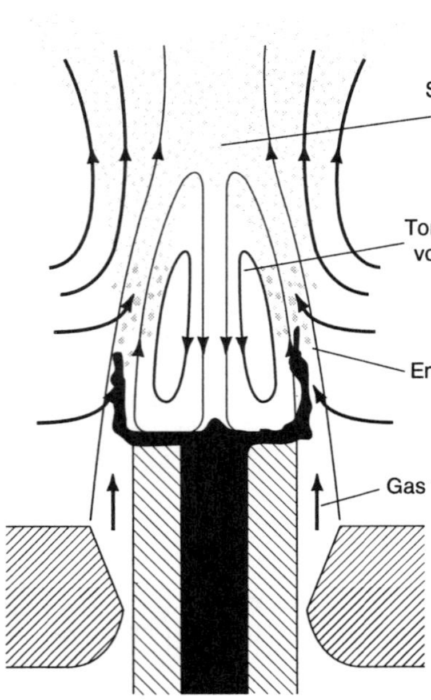

(a)

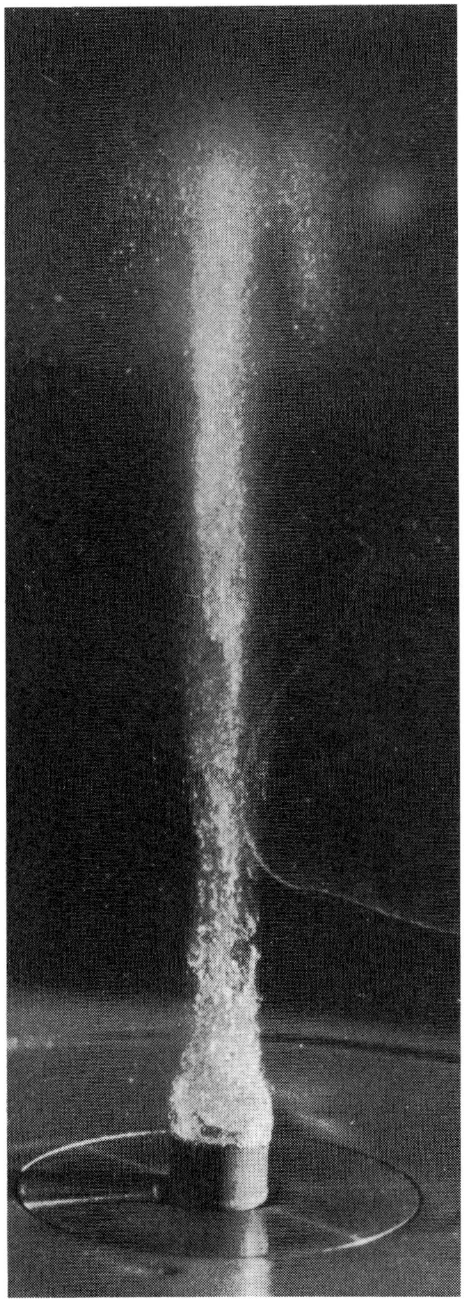

(b)

Fig. 20 Prefilming operation for gas atomization. (a) The prefilming operation of a closed nozzle. (b) The atomization of aluminum powder (25 μm). Source: Ref 22, 23

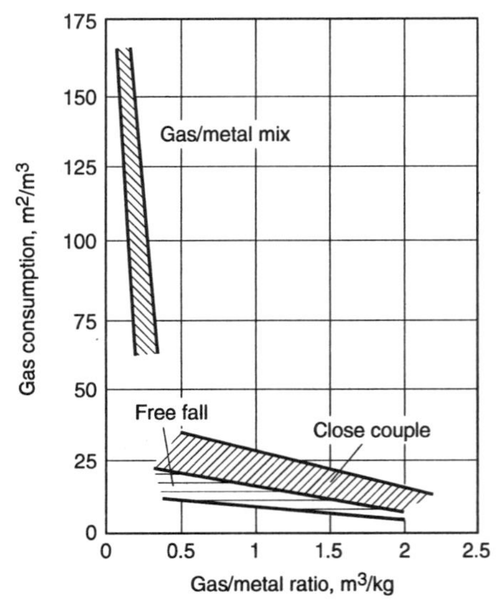

Fig. 21 Gas efficiency expressed as powder area produced per unit volume of gas consumption (m^2/m^3). Source: Ref 24

parable gas/metal ratios (Fig. 21) (Ref 24). A more simple method is to use the equation:

$$d_m = k/\sqrt{G/M} \quad \text{(Eq 6)}$$

Where k is a constant for the process and metal, and G/M is the gas/metal ratio, which is variously measured in kg/kg or cubic meters of gas per metal mass (m^3/kg). A typical plot of d_m versus G/M is shown in Fig. 22 with data from several published papers. For a given metal, the value of k can be taken as an index of efficiency in terms of fineness achieved per unit gas consumed (a lower value of k is a higher efficiency). If d_m is in microns (μm), k is the median size when using one cubic meter of gas per kilogram of metal, a not untypical level.

Typical values of k shown in Fig. 22 range from 40 to 60 μm for close-coupled gas nozzles.

Models of Gas Atomization

The basic conceptual mechanism of droplet formation in gas atomization is a three-step process first proposed by Dombrowski and Johns (Ref 25). Figure 23 shows this model for the disintegration of a liquid sheet by a high-velocity jet of gas. The initiation of sinuous waves, or other disturbances that rapidly increase in amplitude in stage I (see the following paragraphs), is followed by fragmentation, which forms ligaments in stage II. Breakdown of ligaments into droplets occurs in stage III. Thus, there are three stages involved in the atomization process: formation of wave with a particular wavelength λ and wave number k (i.e., $2\pi/\lambda$), removal of the ligament, and spheroidization of the ligament.

Bradley (Ref 26, 27) developed a detailed mathematical model of the first two stages and invoked Rayleigh instability for the final stage. In the first stage of the atomization process, it is necessary to determine the wave number k_{max} with the fastest growing amplitude. In the second stage of atomization, Bradley suggested that the ligament diameter D is related to the wavelength λ_{max} (i.e., $2\pi/k_{max}$) by a linear equation:

$$D = \varepsilon\lambda_{max} = 2\pi\varepsilon/k_{max} \quad \text{(Eq 7)}$$

where the parameter ε (of the order of 0.25) was determined in earlier studies on the air atomization of water.

In the third stage, the cylindrical ligament of diameter D breaks up by Rayleigh instability into a series of spherical droplets of diameter d such that:

$$d = D/0.53 \quad \text{(Eq 8)}$$

From the above two equations, the predicted droplet diameter d (and hence the powder particle diameter) is given by:

$$d = 11.86\varepsilon/k_{max} \quad \text{(Eq 9)}$$

Bradley (Ref 26) also gave a graphical method of evaluating the droplet diameter for Mach numbers (M) between 0.1 and 0.9. This was in terms of a dimensionless parameter L defined as:

$$L = k_{max}\gamma/\rho_g U_s^2 \quad \text{(Eq 10)}$$

where γ is the liquid/gas interfacial energy that opposes the growth of the wave, ρ_g is gas density, and U_s is the sonic velocity of the gas. The relation for L comes from a dimensional analysis that suggests a negligible effect of liquid metal viscosity on the wave development, at least in determining k_{max}. The resultant droplet diameter d is then given by:

$$d = 2.95\gamma/L\rho_g U_s^2 \quad \text{(Eq 11)}$$

For all practical purposes, L is a function of Mach numbers (M); a plot of the dimensionless quantity L against M is given in Fig. 24. This is a universal curve for the atomization of liquid metals. Knowing the value of M during gas atomization, L is obtained from Fig. 24. Substitution of this value of L in Eq 11 gives the predicted liquid droplet diameter D. Unfortunately, predicting or measuring M is practically impossible as it varies very rapidly in space and atomization does not take place instantaneously at a point.

Bradley's model has been shown to be accurate within a factor of two with some experimental data (Ref 1). However, there are some obvious limitations. As presently constructed, Bradley's model does not give any indication of the origin of the spread in particle sizes produced by gas atomization. This phenomenon clearly needs to be included in a comprehensive model of the process. There are also observations reported by See and Johnston (Ref 28) that need to be explained. One observation, revealed by high-speed photography, is that under impact of the gas stream a hollow "cone" of metal is formed that breaks up and then the whole process repeats itself. If the Bradley model is valid, then this cone has to be the fastest growing wave of the Bradley analysis. It is not immediately obvious why the process is cyclic rather than continuous. The work of Miller observing the process also shows marked instabilities on a 10 to 100 ms scale. (Ref 29).

Particle Size Prediction with Adjustable Parameters. Bradley's model does not require the use of any adjustable parameter because ε was determined from Bradley's earlier studies on the air atomization of water. However, another approach based on impact force and work (Ref 30) can be used to predict particle size from gas atomization. This method is based on the

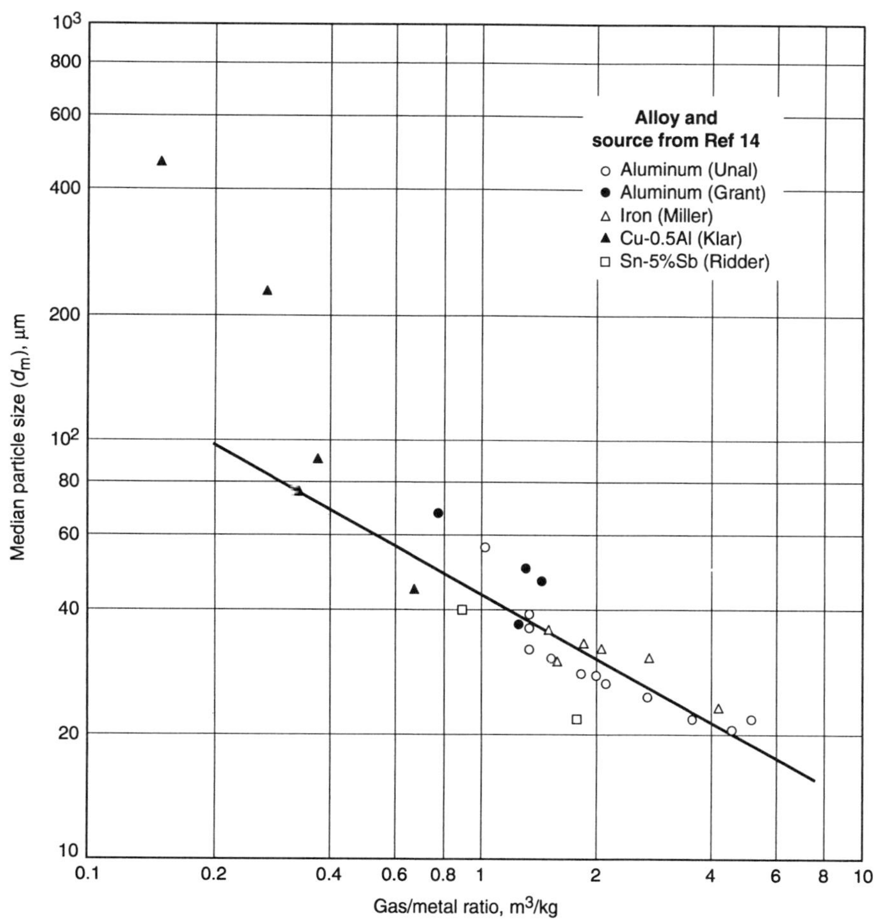

Fig. 22 Effect of gas-to-metal ratio on median particle size. Source: Ref 14

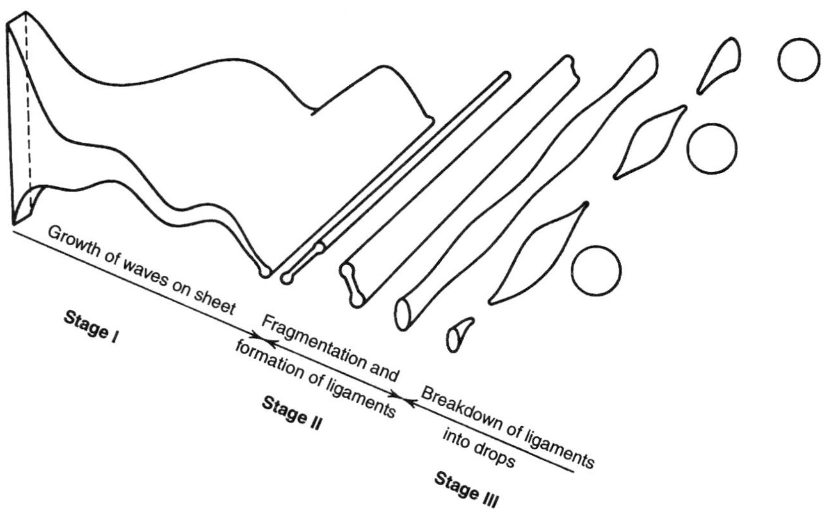

Fig. 23 Model for the disintegration of a liquid sheet by a high-velocity gas jet. Source: Ref 25

following expression (which was originally derived for centrifugal atomization, Ref 31):

$$d = \sqrt{\frac{6 \cdot \sigma}{\rho_m \cdot a}} \quad \text{(Eq 12)}$$

where σ is surface tension, ρ_m is the alloy density, and a is acceleration associated with the particular gas/metal interaction. From an analysis of the basic variables for inert-gas atomization (Ref 30), Eq 12 expands into the form:

$$d = \sqrt{\frac{A \cdot \sigma}{\rho_m \frac{M_g}{M_m} \cdot V_g(1 - B\eta D)}} \quad \text{(Eq 13)}$$

where η is the dynamic viscosity of the metal (in $N \cdot s/m^2$), V_g is the velocity of the gas leaving the nozzles, D is the diameter of the metal stream, and M_g and M_m are mass flow rates of the gas and metal, respectively.

Constants A and B in Eq 13 are system constants that need to be defined for the individual atomization system in order to enable use of the model. By plotting results from atomization runs with different values of the parameters the constants A and B can be estimated and the system thereby characterized. This model has been used in two atomization plants to further develop gas atomization processes, and it accurately predicts experimental results (Ref 30). It is noteworthy that it has the same functional relationship as Eq 6.

Gas-Atomized Powders

Particle Size and Size Distribution. Gas-atomized powders generally have a log normal size distribution, where the standard deviation σ is normally around 2.0 ± 0.3. Lower values are associated with very small metal flow rates, and thus, perhaps falsely, with fine particle sizes (which result from low metal flow rates and constant gas flow rates). Data from Unal (Ref 32) are shown in Fig. 25. Particle size is controlled by changes in gas-to-metal flow ratio.

By comparing the literature on gas-atomized metal powders and by using a relationship developed by Wigg (Ref 33), Lubanska (Ref 34) proposed the following empirical equation for the average particle size (d_m) of gas-atomized powder:

$$\frac{d_m}{d_{Met\,Str}} = K\left[\frac{v_m}{v_g}\frac{1}{W}\left(1 + \frac{M}{A}\right)\right]^{1/2} \quad \text{(Eq 14)}$$

where W is the Weber number ($W = \rho V^2 d_{Met\,Str}/\sigma$), d_m is the mass median particle diameter, $d_{Met\,Str}$ is the diameter of metal stream; v_m is the kinematic viscosity of liquid metal, v_g is the kinematic viscosity of atomizing medium, V is the velocity of atomizing medium, ρ is the density of the gas, σ is the surface tension of liquid metal, M is the mass flow rate of liquid metal, A is the mass flow rate of atomizing medium, and K is a constant.

The individual terms of this equation are dimensionless; the Weber number measures the ratio of inertial to surface tension forces. Within the explicit form for the Weber number, particle size decreases with decreasing surface tension of the liquid metal and with increasing velocity of the atomizing medium.

Data from Ref 34 for various metals are depicted graphically in Fig. 26. The log/log scale obscures an order of magnitude scatter band, and attempts to apply the 1968 Lubanska correlation to practical nozzle systems have been fruitless. Many of the parameters in it are there simply to provide dimensionless elegance and are based on no experimental data whatever. As in the case of Bradley's Eq 13, the inclusion of a velocity term, when the velocity field is extremely complex and variable, nullifies any attempt at prediction, as predicting the velocity is as difficult as predicting the particle size. Thus, more recent practical workers in the field have tended to leave detailed process modeling, even with high-powered computational fluid dynamic (CFD) techniques, to academics and to use either simple empirical correlations such as Eq 6 or advanced observational techniques to understand the process. The CFD techniques have been used to predict gas flow patterns with some success, but modeling the interaction of the gas flow with the metal, and the effect of the metal on the gas flow—involving both intensive momentum and heat transfer (the latter alters both the volume and physical properties of the gas)—has proved beyond the most powerful computers, and the

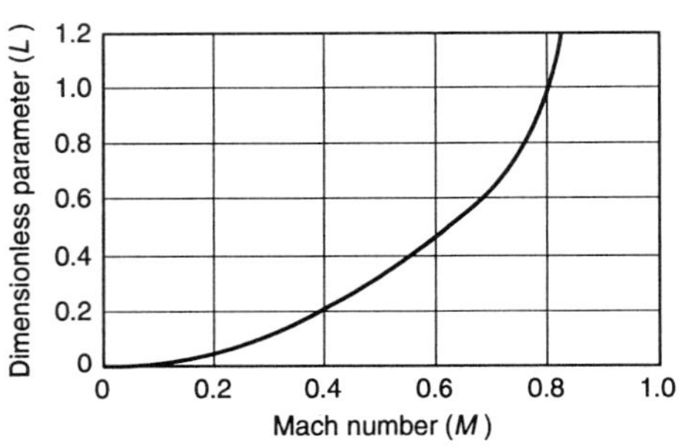

Fig. 24 Universal curve of L against M. Source: Ref 2

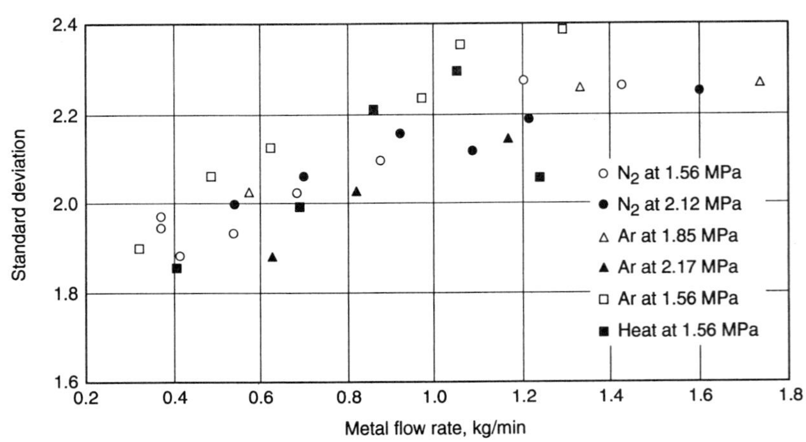

Fig. 25 Variation of geometric standard deviation (σ_g) with metal flow rates. Source: Ref 32

observed fact that atomization is essentially a nonsteady state process (Ref 29) makes it extremely doubtful that any satisfactory solution is possible in the foreseeable future.

Various other empirical equations are given in the selected references at the end of this article. Some of these empirical relationships can be rationalized in terms of physical models for droplet formation. The so-called maximum stability criterion, for instance, gives maximum drop size as a function of surface tension of the liquid metal (σ), gas density (ρ_g) and velocity (V) of the atomizing medium as:

$$d_{crit} \approx \frac{\sigma}{\rho_g V^2} \quad \text{(Eq 15)}$$

This equation applies to the breakup of ligaments (Ref 28). Secondary disintegration into particles smaller than d_{crit} occurs only if the dynamic pressure due to the gas stream velocity exceeds the restoring force of surface tension. Due primarily to the higher surface tensions of liquid metals, energy requirements and, therefore, gas pressures used for atomizing metals are generally higher than for other liquids—typically between 350 and 2750 kPa (50 and 400 psi).

Powder Cleanliness. Impurity control is of paramount importance for high-performance applications. Currently, inert gas atomization systems have been perfected so that the oxygen content of a powder does not significantly exceed that of the liquid metal prior to atomization. To minimize the presence of oxides in the molten metal prior to atomization, particularly in the case of alloys containing readily oxidizable elements, good melting and deoxidation practices are important. Precautions during transfer of the liquid metal from ladle to tundish and avoidance of turbulence inside the tundish are equally important. In the case of the most critical superalloys, recent workers have developed a bottom-pouring electroslag remelting furnace that discharges using a ceramic-free pouring nozzle to minimize all possibility of inclusion pickup (Ref 35).

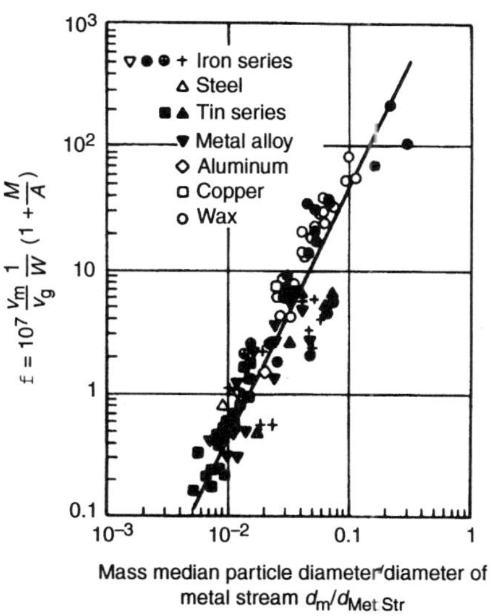

Fig. 26 Correlation of atomization data for liquid metals. See Eq 14 in text for line equation. Source: Ref 34

Gas-atomized metal powders may contain small amounts of the atomizing gas within individual particles, which can cause microporosity, especially in the case of argon. This gas probably originates from mechanical entrapment that occurs when particles collide during atomization. Porosity can also originate from hydrogen dissolved in the melt. In practical terms, porosity in atomized particles is found mostly in the coarser particles. Thus, trends to screen finer to reduce the maximum possible inclusion size have helped to reduce this problem. In the case of high-speed steels, severe porosity on heat treatment was eliminated by changing from inert, insoluble argon to reactive, soluble nitrogen, which merely incorporated itself into the carbides.

Particle Shape. The surface of (inert) gas-atomized powders is ideally smooth and spherical. However, in practice, especially when finer powders are produced, "satelliting" is seen (see Fig. 3e). This is the decoration of the larger particles by attached smaller particles. This is believed to be caused by the circulation of gas within the atomizing chamber that lifts finer particles back into the spray plume, where they collide with larger, still partly molten particles. It is found that this problem is much less severe when water quench sprays are used, which scrub out the fines and also that it gets much worse as finer median sizes are produced.

The spherical shape and generally smooth surfaces of gas-atomized powders limit the development of green strength in cold compaction. In addition, prealloying before atomization increases particle hardness and strength, which reduces compressibility. These intrinsic limitations in gas-atomized powders have stimulated development of elevated-temperature consolidation processes to achieve high density; examples include hot extrusion and hot isostatic pressing. However, spherical particles have ideal flow characteristics and are desirable for feeding to thermal spray processes.

The far larger volumes of air-atomized powders do not all suffer from the drawback of high sphericity, as can be seen in Fig. 3(c), which shows air-atomized aluminum powder with a characteristic shape caused by a strong oxide film. Zinc, aluminum, and copper-zinc (brasses) are similar to this and can be cold compacted. However, if the oxide-forming tendencies are minimal, for example in phosphorous deoxidized copper or bronze, good spheres can result, such as filter grades of 10% Sn bronze.

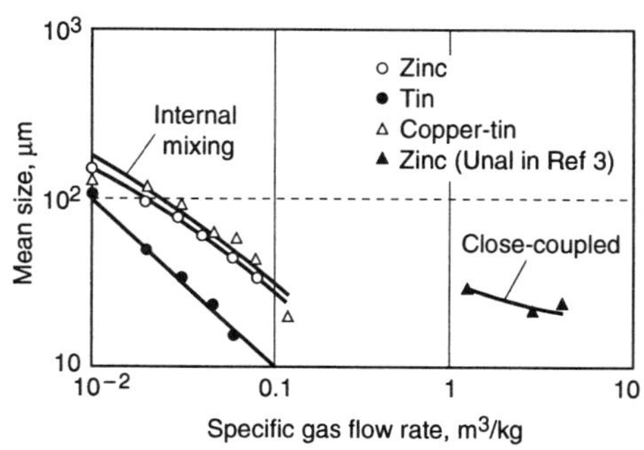

Fig. 27 Dependence of mean particle size on gas-to-metal ratio. Source: Ref 36

Other Gas Atomization Methods

Ultrasonic gas atomization (Fig. 1e and 19) is claimed to allow finer particles, but no published comparative work has clearly demonstrated significant advantage over similar close-coupled nozzles. No commercial operator is known to use the method, and the largest melter used has been in the 10 to 50 kg size range.

Internal Mixing Nozzles. Typical values of k in Eq 6 shown in Fig. 22 range from 40 to 60 μm for close-coupled gas nozzles. Far higher efficiencies are obtained in another type of nozzle, the internal mixing nozzle, where gas-to-metal ratios are of a totally different order of magnitude than in conventional gas atomization (Fig. 21). In an internal mixing atomizer, gas and metal are mixed under pressure, and expansion and atomization take place at the nozzle exit into the atomization chamber. Two such processes are used today for the production of superalloy powders on a commercial scale (Ref 24). The first of the two gas/mix processes is known as vacuum atomization (Fig. 1d), where dissolved gas in the liquid metal adds an extra atmosphere pressure difference in the beginning of the atomization to promote the expansion of the gas in solution in the metal. In addition to the dissolved gas, some gas is also added to the system to finally break up the liquid into fine droplets as the dissolved gas alone would not be sufficient to create the fine particles that can be achieved with this process.

The vacuum atomization process is only operated by one company and is highly proprietary. The lack of any other operators, even when the original patents have expired, may indicate that the rather complex plant may be too expensive to justify the gas savings.

In another type of internal mixing nozzle (Ref 36), very high efficiency was achieved (Fig. 27), but low outputs. The k value in Eq 6 was below 10. The fundamental drawback of the internal mixing concept is the engineering difficulty of arranging to pressurize the melt to the same pressure as the gas. There are also problems with erosion of the ceramic nozzles employed as the metal velocity, normally only 1 to 3m/s in a pouring nozzle, could be expected to rise to the same order of magnitude as the gas velocities, that is, perhaps to 100 m/s. Thus, applications are so far very limited indeed.

Centrifugal Atomization Methods

In centrifugal atomization, centrifugal force breaks up the liquid and throws off the molten metal as a spray of droplets that then solidify as powder particles. As shown in Fig. 1, there are several different types, which are discussed. In general, centrifugal atomization methods are far more energy efficient than gas and water atomization, where only about 1% of the jet energy is used in the disintegration of the metal stream (Ref 1). In contrast, the energy used in centrifugal methods is low as all the rotational work directly accelerates the metal droplets, instead of the atomizing medium, as in the case of two-fluid techniques. Centrifugal atomization also generally leads to a much narrower spread in particle sizes than does gas atomization (see Fig. 2), with σ_g as low as 1.2 to 1.4 in some cases.

As the process depends on the solution to the problem of finding a compatible material for the spinning disk or cup, the applications that have been, and are currently, used on a significant industrial scale are quite distinct. They are reviewed in order of ascending melting point of alloy.

Spinning Disk Atomization of Electronic Grade Solder. Solder powder for electronic applications has a very demanding specification; it must be perfectly spherical and satellite-free, it must be very low in oxygen content (~100 ppm), and it must have a very narrow size distribution. In 1997, more than half of the demand was for type III size grade that is nominally –45+25 μm, and often demanded as –40+30 μm. Attempts to make this product with inert gas atomization have now virtually ceased as yields are as little as 5% and the avoidance of satellites is very difficult.

In the United States, Japan, and Europe, there are many producers using spinning disk methods to make this product. This is possible because a steel disk is well wetted by normal Sn63Pb37 solder and is not eroded very fast. A disk with diameter of 40 to 100 mm running at speeds of 30,000 to 60,000 rpm can produce this material with good yields of 30 to 70% and at (gross) rates of 50 to 100 kg/h. The plant is only about 2 to 3 m in diameter and inert gas filled. Unfortunately, a small amount of ultrafine particles is produced as a secondary peak in the distribution, and some problems with satellites are found. Annual production, if the modest yields are taken into account, is probably currently several thousand tons per year.

Spinning-Cup Atomization of Zinc, Aluminum, and Magnesium. When used on higher-melting metals, it is difficult to run at the sort of speeds that are used on solder. However, there are markets for coarser powders of zinc (alkaline batteries), aluminum (chemical), and magnesium (flares) that have made this a significant process. In all cases, the cup is 100 to 200 mm diameter, running at moderately high speeds, from 3,000 to 10,000 rpm. The vessel size needed is very large, up to 12 m in diameter, but productivity can be very high. In the case of zinc for battery applications, which is commonly required as –600+100 μm material, air atomization may give yields of 70 to 80%. Using a 5 kW spinning cup, 98% can be achieved, at outputs of several tons per hour, and with no compressor costs (an equivalent air atomizer might use several hundred kilowatts of compressed air). However, large output is needed to justify the large-scaled plant needed. Many thousands of tons per year are made in this way.

Aluminum is also processed in this way, but because of the large size of plant needed to freeze coarse particles, it is done in open air, which produces needles (see Fig. 3d). In this case, a perforated steel or cast iron cup is used to make a series of streams of metal, which break up into needles due to the oxide film on them. Production is considerable, probably thousands of tons per year.

Magnesium powder for flares was made in several similar systems in the United States during the Vietnam war at rates of 1 ton/h. Excellent yields of a powder almost free of the very dangerous –100 μm fines were obtained. In this case, the fact that liquid magnesium does not attack iron simplified cup manufacture. Due to lack of current demand, this production process is seldom used.

The rapid solidification rate (RSR) process, which was first developed in the 1970s by Pratt and Whitney for making superalloy powders, is another form of centrifugal atomization. To overcome the problems of the material in handling high melting and aggressive alloys, the process employs a high-speed water-cooled rotating disk (20,000 to 30,000 rpm), which breaks up the molten metal stream. To enhance solidification rates, the resulting droplets are then hit by high-pressure helium gas as they leave the periphery of the rotating disk. The RSR powders are spherical, with an average particle diameter <100 μm. High cooling rates (>10^5 °C/s) are achieved in the small powder particles, and this leads to a high degree of compositional homogeneity and fine-scale microstructure. An early plant of this type is shown in Fig. 28.

The first RSR unit was operated in late 1975, producing IN-100 powder as a model, and patents were issued in 1978 (Ref 37) and 1982 (Ref 38). In addition to superalloys, the process has been used to produce specialty aluminum alloy, beryllium alloy, molybdenum, titanium alloy, and silicide powders. The RSR powders currently produced by Pratt and Whitney are nickel-base superalloys, steels, and aluminum alloys. These alloy types are induction melted in vacuum. In a more recent development of the RSR process, reactive metals (titanium, molybdenum) and silicides have been arc melted prior to atomization. For titanium alloys, melt sizes up to 45 kg have been atomized successfully.

For nickel-base alloys, representative operating conditions are disk rotation speed 2500 radians/s, melt flow rate 0.2 kg/s, gas (helium) velocity 170 m/s, gas mass flow rate 1 kg/s, helium backfill partial pressure 33 kPa, and melt superheat 100 °C (180 °F). A major consideration in the design of this rotating disk atomizer was the desire to achieve extremely high cooling rates in

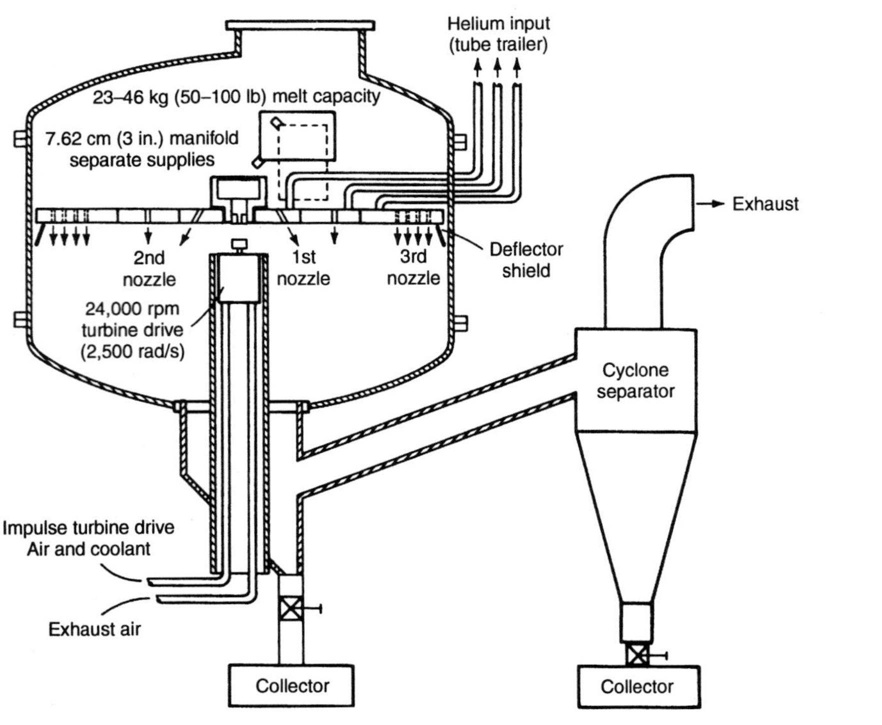

Fig. 28 Schematic of first-generation RSR machine. The second-generation machine incorporates closed-loop helium recirculation, higher atomization speeds, and a three-fold increase in melt capacity. The third-generation machine retains these features, but the melt capacity is increased to 900 kg (2000 lb).

the droplets by convective cooling in the helium gas. Depending on the diameter of the rotating disk, droplets are ejected at velocities in the range 40 to 110 m/s. Typically RSR droplets smaller than 100 μm diameter cool at rates of about 10^5 °C/s.

This technique produces notably narrow distributions and is used commercially to make plasma spraying powders where size ranges such as −150+53 or −100+38 μm may be needed. However, the uncontrolled geometry of the skull on the rotating disk, and the problems of out-of-balance forces, degrades the size distribution significantly compared with that achieved with lower-melting materials where the cup or disk has precise geometry. It also results in heavy maintenance costs on the spinning-cup assembly. As the melting process is normally batched, this is a major drawback. Lower-melting materials are easily supplied continuously, avoiding or minimizing startup and maintenance problems. As a result of these disadvantages, gas atomization has continued to dominate the production of higher-melting alloy powders. The use of huge volumes of costly helium is another drawback, and the benefits of the rapidly solidified microstructures have yet to find major commercial paybacks outside the aerospace sector. Thus, Pratt & Whitney remains the only operator in the 1990s, and the few attempts that have been made to develop similar devices elsewhere have not been productive. Production is probably less than 1000 tons/year.

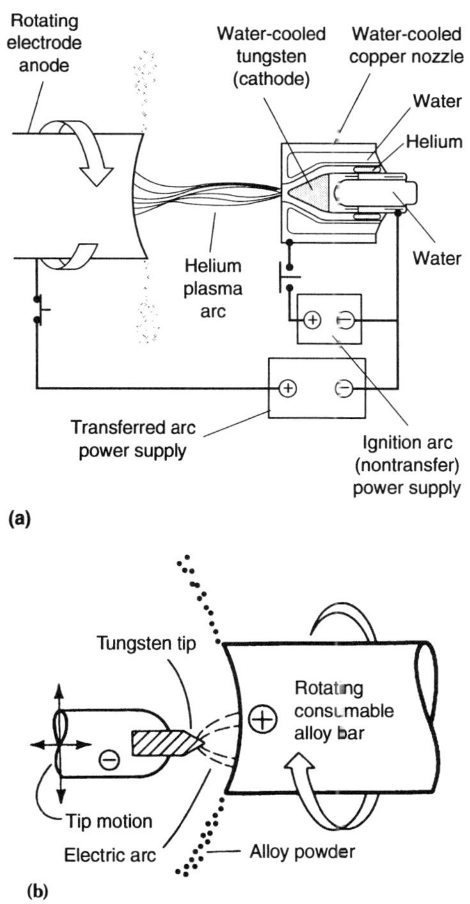

Fig. 29 Schematic of the REP. (a) Plasma arc rotating electrode process (PREP). (b) Tungsten tip rotating electrode process (REP)

Rotating Electrode Process

The rotating electrode process (REP) is a commercial form of centrifugal atomization that overcomes the material problem by spinning the metal before it is melted. Thus, the spinning element is a consumable electrode. Material in the form of a rod electrode is rotated at about 15,000 rpm while it is melted by an arc. The molten metal is ejected centrifugally in the form of molten-metal droplets that solidify before hitting the walls of the inert-gas-filled chamber. The REP concept was developed by Nuclear Metals, Inc., and resulted in the granting of patents (Ref 39).

Figure 29(a) shows the process carried out using a plasma torch process (PREP). The original method (REP) used a tungsten-tipped cathode as shown in Fig. 29(b). Use of a transferred arc helium plasma torch eliminates the tungsten contamination characteristic of the earlier design. This was a critical factor in titanium powder production.

The rotating electrode and plasma rotating electrode processes offer certain advantages over other powder-making processes. Titanium alloys are optimally produced by these methods, because the corrosive nature of molten titanium and the difficulty involved in containing this molten material are overcome. Generally, many high-duty materials to be made into powder benefit in that there is no liquid metal/container contact. This ensures that ceramic particles are not inadvertently added. When the addition of foreign particles from external sources is prevented, these processes provide a method for making powders to exact standards of cleanliness. However the bar stock must first be sufficiently clean itself.

As with other centrifugal methods, size distributions can be held within tighter ranges than is commonly achieved by gas atomization. A cumulative plot of weights passing through progressively finer screens versus screen size is shown in Fig. 30. The effect of rotation speed is illustrated, and the median particle size d_{50} is approximately defined by:

$$d_{50} = \frac{K}{\omega\sqrt{D}} \qquad \text{(Eq 16)}$$

where ω is the rotation rate, D is the diameter of the electrode, and K is a constant for a given alloy for a limited range of arc power. Hollow gas-filled particles, which cause thermally induced porosity in compacts, are not generated. Powders made by these techniques have demonstrated a high degree of sphericity along with good surface quality. Particles also pour readily into molds of complex shape. They give consistent packing at approximately 65% full density so that near-net shapes can be obtained by HIP. However, a major drawback is the mechanical limitations on rotational speed, which limit the minimum median particle size to about 50 to 150 μm (depending on alloy density, etc.). Also, the cost of making a high-quality bar of metal is very significant and productivity is low and energy consumption high compared with other techniques.

The process has only been commercially operated by one firm and is effectively a very "niche" one, with a few highly specialized industrial ap-

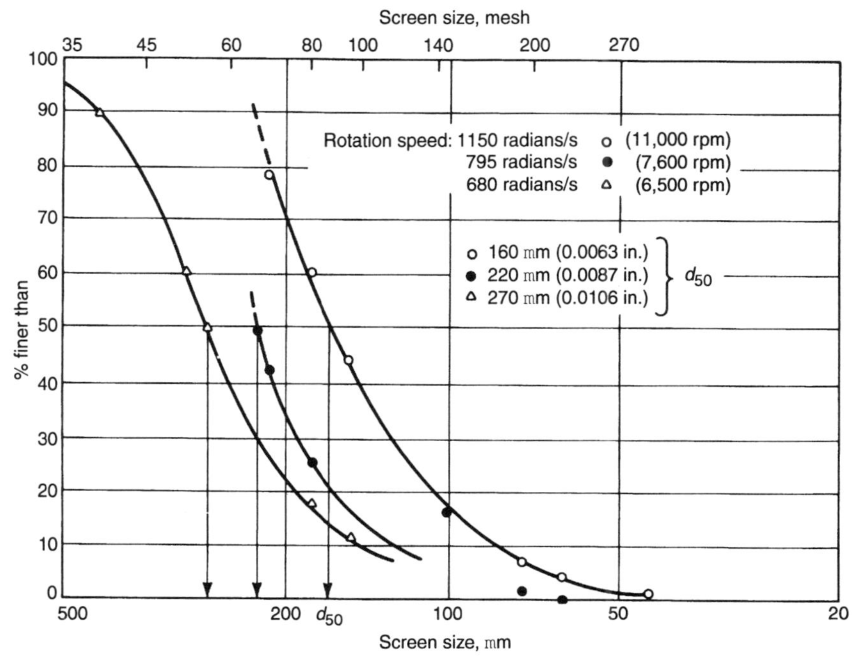

Fig. 30 Cumulative plot of REP particle size distributions. Weight percent of sample finer than a given screen size versus screen size. Electrode: C-1018 steel, 63.5 mm (2½ in.) diameter

plications and total annual output not exceeding 1000 tons/year.

Models of Centrifugal Atomization

Centrifugal atomization is more amenable to detailed analysis (see Ref 1, 40–42) than two fluid methods because the velocity of the particles at the edge of the disk/cup is fairly well defined, in contrast to the very poorly defined velocity term in two fluid atomization theory. As melt flow rate increases, the mechanism changes from the formation of drops to ligaments and then sheets. The analysis of centrifugal atomization developed by Champagne and Angers (Ref 40–42) allows for a quantitative prediction of mean particle diameter in the direct droplet formation mode. In its present form, the model gives no indication of the spread of droplet sizes.

By simple force balance between centrifugal force and surface tension forces, the following expression can be derived:

$$d = \sqrt{\frac{A \cdot \sigma}{\rho_m \omega^2 \cdot R}} \quad \text{(Eq 17)}$$

where ω is the angular velocity (rad/s) of the rotating disk (or the bar in REP) and R is the disk radius (m). A is a constant with value 6 in the elementary case.

From particle size distributions reported on the atomization of the Merl 76 (Ref 43), comparison is made between actual results and predicted results (Fig. 31). Mean particle sizes with the model are plotted as curve 1 in Fig. 31 for A = 5.25. Actual powder sizes are coarser than those predicted, which may be explained by the influence of viscosity particularly at higher speeds. In effect, the molten film is "slipping" on the disk surface, making its speed less than the peripheral speed of the disk. If a term that is counteracting the centrifugal acceleration and that includes both the viscosity and the speed at the rim of the disk is added to the basic model, the following relationship is suggested by Tornberg (Ref 24):

$$d = \sqrt{\frac{A \cdot \sigma}{\rho_m \cdot \omega^2 \cdot R\,(1 - (\omega \cdot R)^n \cdot \eta)}} \quad \text{(Eq 18)}$$

In this curve, the constant $A = 4.8$ and the exponent $n = 0.93$.

The reason why no significant effect of viscosity has been observed in REP atomization (Ref 42) can be explained by the relatively low speed at the rim of the small diameter rotating bar in REP and the very thin film due to the low melting rate of the process. The viscosity effect probably increases with increased disk velocity and thereby also metal velocity at the point where the metal leaves the disk to form a particle. The viscosity factor therefore includes both the radius and the angular speed of the disk. To reach high metal velocities in centrifugal atomization a large disk radius or a high speed of rotation is necessary. The mechanical limitations on radius and speed make it difficult to obtain the finest powder particles via centrifugal atomization. However, as Eq 17 shows, high density, low surface tension alloys are finer, and there are reports of powders as fine as 25 μm being made from dense alloys such as SnBi (Ref 1). This topic is discussed in Ref 44.

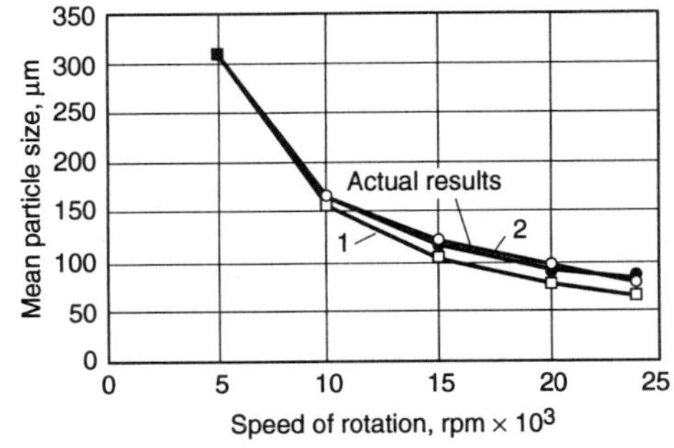

Fig. 31 Calculated and actual powder particle sizes of Merl 76 superalloy powder made by centrifugal atomization (with disk radius of 45 mm). Curve 1 is a plot of Eq 17, which does not account for viscosity effect. Curve 2 is a plot of Eq 18 with a viscosity factor. Source: Ref 30

Champagne and Angers (Ref 41, 42) have demonstrated that a bimodal distribution of particle size occurs due to a mechanism in which liquid drops separating from the electrode rim remain connected momentarily by a necked-down column of liquid metal that breaks into smaller drops when the major spheroid is freed. This effect of bimodal particle size distribution is shown in Fig. 32. Although smaller particles are not present in large quantities by weight, their absolute numbers may be significant. It is this effect, giving rise to 3 to 8 μm particles, which can then form satellites by collision with larger particles, that has given some problems to electronic solder producers.

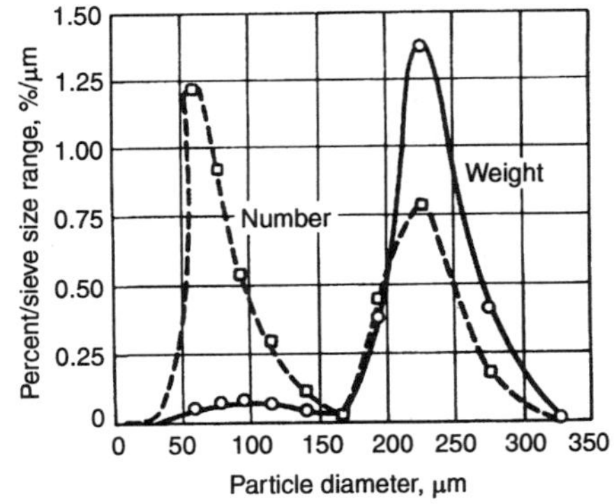

Fig. 32 REP bimodal particle size distribution attributed to satellite drop formation. Source: Ref 41, 42

Other Methods

Ultrasonic (Vibrational) Atomization. This very specialized technique was, until recently, merely a laboratory curiosity, despite being reported as early as the 1960s. The principle is that a liquid film, if vibrated sufficiently strongly, will form waves that will in turn break off to form droplets. This effect is observable at low frequencies and has been demonstrated on metals such as zinc using low kHz frequencies in the Ukraine. However, it has recently been developed by several leading solder producers to make electronic solder powder for paste production. As mentioned above, this is needed with very narrow particle size and perfectly spherical shape.

The major drawback of the process is that, similar to the spinning-cup process, it places high demands on materials of construction for the solid surface. It must resist melt attack, be wettable (attack and wetting are two sides of the same coin, so this is always a compromise), and have suitable acoustic properties. Therefore, it is not surprising that it has not found application on higher-melting materials. However, it can be made to work on solders, although with a low output, typically 20 to 40 kg/hour (for one horn). Frequencies used range from 20 to 80 kHz, and the particle size distribution is practically free of ultrafine particles. Standard deviation σ_g of 1.4 or better is achievable, giving yields similar to centrifugal techniques and far better than gas atomization. Equipment is compact (typical vessel diameter less than 1 m) and energy use small (~1 kW). Production by this method is probably around 1000 tons/year (gross).

Other Methods of Atomization. Besides the industrial types of atomization processes, there are also many atomization methods of potential scientific interest that have been reported in the literature. These methods include:

- *Roller atomization:* A mechanical process that feeds a stream of molten metal between rapidly rotating rolls and discharging it into flake, acicular, or irregular spherical particles. This has been found to be generally impractical due to accretions on the rolls and very variable heat transfer to them.
- *Vibrating electrode:* This mechanical approach produces high-purity powder by the vibration of a consumable electrode. The electrode, forming a resonant rod with a fixed and free end, is continuously moved between rollers toward a slowly rotating, water-cooled copper disk. Atomization occurs in an arc struck between the disk and vibrating electrode end. The size of spherical particles formed is controlled by changing the length of the resonant rod. Productivity is low, cost high, and particles coarse.
- *Melt drop (vibrating orifice) technique:* In this method, molten metal contained within a closed, pressurized crucible is subjected to vibratory oscillations. Forcing the metal through a nozzle at the bottom of the crucible and into a vacuum or inert gas chamber causes it to form a jet and undergo Rayleigh breakup into very uniform droplets. Values of σ_g of less than 1.05 have been reported. The tech-

nique has been demonstrated with aluminum, beryllium, copper, lead, mild steel, and some superalloys (Ref 45) and has recently been proposed for solder (Ref 46). The technique is in limited industrial use in the 1990s, making precision solder balls for electronics with sizes in the range 500 to 1000 μm. It is limited as production is related to the square of particle size and nozzle blockage is obviously a worry for holes in the 200 to 20 μm range.

- *Variants of centrifugal atomization:* Besides the more common methods of the rotating electrode and rotating disk or cup processes, other methods include centrifugal impact atomization, shot casting, laser-spin atomization, and the Durarc process (Ref 1). Few of these are commercially exploited, but new variants on old ideas are frequently reported.

REFERENCES

1. A. Lawley, *Atomization,* Metal Powder Industries Federation, 1992
2. T. Allen, *Particle Size Measurement,* 3rd ed., Chapman and Hall, 1981, p 136
3. J.J. Dunkley, *Powder Metall.,* Vol 29 (No. 1), 1986, p 10
4. Y. Seki, H. Takigawa, and N. Kawai, Effect of Atomization Variables on Powder Characteristics, *Met. Powder Rep.,* Jan 1990, p 38–40
5. A.J. Aller and A. Losada, Models in Metal Powder Atomization, *Powder Metall. Sci. Technol.,* Vol 2 (No. 2), 1991, p 13
6. R.J. Grandzol and J.A. Tallmadge, Effect of Jet Angle on Water Atomization, *Int. J. Powder Metall. Powder Technol.,* Vol 11 (No. 2), 1975, p 103–116
7. R.J. Grandzol and J.A. Tallmadge, Water Jet Atomization of Molten Steel, *Am. Inst. Chem. Engineers J.,* Vol 19 (No. 6), 1973, p 1149–1158
8. E. Klar and J. Fesko, On the Particle Shape of Atomized Metal Powders, *Prog. Powder Metall.,* Vol 37, 1981, p 47–66
9. T. Takeda and K. Minagawa, Effects of Atmosphere During Water Atomization on Properties of Copper Powders, *J. Jpn. Soc. Powder Metall.,* Vol 25 (No. 7), 1978, p 213–218
10. J.J. Dunkley, The Production of Metal Powders by Water Atomization, *Powder Metall. Int.,* Vol 10 (No. 1), 1978, p 38–41
11. Y. Seki, S. Okamoto, H. Takigawa, and N. Kawai, Effect of Atomization Variables on Powder Characteristics in the High-Pressure Water Atomization Process, *Met. Powder Rep.,* Vol 45 (No. 1), 1990, p 38
12. J.J. Dunkley and J.D. Palmer, Factors Affecting Particle Size of Atomized Metal Powders, *Powder Metall.,* Vol 29 (No. 4), 1986, p 287
13. S. Small and T.J. Bruce, The Comparison of Characteristics of Water and Inert Gas Atomized Powders, *Int. J. Powder Metall.,* Vol 4 (No. 3), 1968, p 7
14. J.J. Dunkley, *Atomization of Metal Powders in Powder Metallurgy: An Overview,* I. Jenkins and J. Wood, Ed., Institute of Metals, 1991, p 2–21
15. A.J. Yule and J.J. Dunkley, *Atomization of Melts,* Oxford University Press, 1994, p 64
16. G.C. Lewis et al., Atomization of Liquids in High Velocity Gas Streams, *Ind. Eng. Chem.,* Vol 40 (No. 1), 1948, p 67–74
17. J. Reinshagen and A. Neupaver, Principles of Atomization, *Physical Chemistry of Powder Metals Production and Processing,* W.M. Smith, Ed., TMS/AIME, 1989, p 16
18. O.S. Nichiporenko, Shaping of Powder Particles During the Atomization of a Melt by Water, *Sov. Powder Metall., Met. Ceram.,* Vol 15 (No. 9), 1976, p 665–669
19. J.J. Dunkley, An Assessment of Atomization with Hydrocarbons, *Met. Powder Rep.,* Feb 1992, p 20–23
20. S.A. Miller, S. Savkar, and R.S. Miller, A Systematic Approach to Designing Close-Coupled Gas Atomization Nozzles, *Advances in Powder Metallurgy 1994,* Vol 1, Metal Powder Industries Federation, 1994, p 95–104
21. J. Anderson and R. Figliola, Observations of Gas Atomization Process Dynamics, *Modern Developments in Powder Metallurgy,* Vol 20, Metal Powder Industries Federation, 1988, p 205–223
22. A. Unal, Flow Separation and Liquid Rundown in a Gas Atomization Process, *Metall. Trans. B.,* Vol 20B, Oct 1989, p 613–621
23. A. Unal, Influence of Nozzle Geometry in Gas Atomization of Rapidly Solidified Aluminum Powders, Dec 1987, p 1029–1039
24. C. Tornberg, Gas Efficiency in Different Atomization Systems, *Powder Production and Spray Forming,* Vol 1, *Advances in Powder Metallurgy and Particulate Materials 1992,* Metal Powder Industries Federation, 1992, p 127–135
25. N. Dombrowski and W.R. Johns, The Aerodynamic Instability and Disintegration of Viscous Liquid Sheets, *Chem Eng. Sci.,* Vol 18, 1963, p 203–214
26. D. Bradley, *J. Phys. D, Appl. Phys.,* Vol 6, 1973, p 1724
27. D. Bradley, *J. Phys. D, Appl. Phys.,* Vol 6, 1973, p 2267
28. J.B. See and G.H. Johnston, *Powder Technol.,* Vol 21, 1978, p 119
29. S.A. Miller and R.S. Miller, Real Time Visualization of Close-Coupled Gas Atomization, *Advances in Powder Metallurgy 1992,* Vol 1, Metal Powder Industries Federation, 1992, p 113–125
30. C. Tornberg, Particle Size Prediction in an Atomization System, *Powder Production and Spray Forming,* Vol 1, *Advances in Powder Metallurgy and Particulate Materials 1992,* Metal Powder Industries Federation, 1992, p 137–150
31. J.M. Wentzell, Particle Size Prediction from the Spinning Disk Atomizer, *Powder Metall. Int.,* Vol 18, 1986, p 16
32. A. Unal, Effect of Processing Variables on Particle Size in Gas Atomization of Rapidly Solidified Aluminium Powders, *Mater. Sci. Technol.,* Vol 3, Dec 1987, p 1029–1039
33. L.D. Wigg, Drop-Size Prediction for Twin-Fluid Atomizers, *J. Inst. Fuel,* Nov 1964, p 500–505
34. H. Lubanska, Correlation of Spray Ring Data for Gas Atomization of Liquid Metals, *J. Met.,* Vol 22 (No. 2), 1970, p 45–49
35. M.G. Benz et al., "Atomization of Electroslag Refined Metal," U.S. Patent 5,310,165, 1994
36. S.M. Sheikhaliev and J.J. Dunkley, A Novel Internal Mixing Gas Atomizer for Fine Powder Production, *Advances in Powder Metallurgy 1996,* Vol 1, Metal Powder Industries Federation, 1996, p 161–170
37. P. Holiday and R. Patterson, "Apparatus for Producing Metal Powder," U.S. Patent 4,078,873, 1978
38. P. Holiday and R. Patterson, "Method for Producing Metal Powder," U.S. Patent 4,343,750, 1982
39. U.S. Patents 3,099,041 and 3,802,816, 1974
40. B. Champagne and R. Angers, *Powder Metall. Int.,* Vol 16 (No. 3), 1984, p 125
41. B. Champagne and R. Angers, Size Distributions by the Rotating Electrode Process, *Modern Developments in Powder Metallurgy,* Vol 12, H.H. Hausner, H.W. Antes, and G.D. Smith, Ed., Metal Powders Industries Federation, 1981, p 83
42. B. Champagne and R. Angers, *Int. J. Powder Metall. Powder Technol.,* Vol 16 (No. 4), 1980, p 359
43. Superplastic PM Superalloys by Centrifugal Atomization, *Met. Powder Rep.,* Vol 43 (No. 10), 1988, p 688
44. R. Angers, Centrifugal Atomization, Influence of Process Parameters on Size Distribution, Advances in Powder Metallurgy and Particulate Materials, Vol 1, MPIF, 1992, p 79–88
45. F. Aldinger, E. Linck, N. Claussen, A Melt Drop Technique for the Production of High-Purity Metal Powders, *Mod. Dev. P/M,* Vol 9, 1977, p 141–151
46. P. Yim et al., Production and Characterization of Mono-Sized Sn38Pb Alloy Balls, *Int. J. Powder Metall. Powder Technol.,* Vol 32 (No. 2), 1996, p 155–164

SELECTED REFERENCES

- J.P. Cherre and A. Accary, A Study of Atomization Mechanisms through Centrifugal Atomization, *Proc. Int. Conf. PM,* London, Vol 2, 1990, p 11–15
- J.J. Dunkley, Evaluating the Performance of Atomizers, *Advances in Powder Metallurgy,* Vol 2, Metal Powders Industries Federation, 1989, p 1–14
- A. Lawley, *Atomization,* Metal Powder Industries Federation, 1992
- A.H. Lefebvre, *Atomization and Sprays,* Hemisphere, 1989

- S.P. Mehrotra, Mathematical Modeling of Gas Atomization Process for Metal Powder Production, *Powder Metall. Int.,* Vol 13 (No. 2), 1981, p 80–84; Vol 13 (No. 3), 1981, p 132–135
- H. Schmitt, Mathematical-Physical Considerations Regarding the Production of Metal Powders for Powder Metallurgy, *Powder Metall. Int.,* Vol 11 (No. 1), 1979, p 17–21; Vol 11 (No. 2), 1979, p 68–71
- J.B. See and G.H. Johnston, Interactions between Nitrogen Jets and Liquid Lead and Tin Streams, *Powder Technol.,* Vol 21 (No. 2), 1981, p 119–133
- S.L. Shinde and G.S. Tendolkar, Analyses of Atomization—A Review, *Powder Metall. Int.,* Vol 9 (No. 4), 1977, p 180–184
- A.J. Yule and J.J. Dunkley, *Atomization of Melts, Oxford University Press, 1994*

Milling of Brittle and Ductile Materials

Revised by Carl C. Koch, Materials Science and Engineering, North Carolina State University

MILLING OF MATERIALS, whether hard and brittle or soft and ductile, is of prime interest and of economic importance to the P/M industry. Mechanical comminution is the most widely used method of powder production for hard metals and oxide powders. Secondary milling of spongy cakes of oxide-reduced, atomized, or electrolytic powders is the most common milling process; hammer and rod mills are used for this type of milling. Depending on the degree to which the material is sintered, either primary particle size distribution is reestablished during milling or larger agglomerates are produced.

Mechanical comminution is restricted to relatively hard, brittle metals (electrolytic iron or bismuth, for example), some reactive metals such as beryllium and metal hydrides, ductile metals used for producing metal flakes, and chemically embrittled materials such as sensitized stainless steel.

Increasing interest in metal powder with particle sizes that are finer than the particle sizes of powders produced by atomization has reactivated interest in milling, particularly in solid-state alloying or high-energy milling. However, milling of metal powders has received minimal attention to date. Research conducted on ball milling of metals is primarily proprietary and empirical, and thus restricted to specialized P/M applications. Objectives of milling include:

- Particle size reduction (comminution or grinding)
- Particle size growth
- Shape change (flaking)
- Agglomeration
- Solid-state alloying (mechanical alloying)
- Solid-state blending (incomplete alloying)
- Modifying, changing, or altering properties of a material (density, flowability, or work hardening)
- Mixing or blending of two or more materials or mixed phases
- Nonequilibrium processing of metastable phases such as amorphous alloys, extended solid solutions, and nanocrystalline structures

The milling operation fractures, deforms (cold works), or cold welds the impacted particles. Milling also may produce polymorphic transformations (Ref 1), as is the case with lead oxide and stainless steel (Ref 2). The specific effect that milling has on a powder depends on the physical and chemical properties of the powder, the vacuum, gaseous, or liquid environment in which the operation is conducted, and milling conditions. Selection of the milling process is based on the desired result of the milling operation, the behavior of the powder under milling conditions (if known), the characteristics of the powder, and the physical and mechanical properties of the material.

During milling, four types of forces act on particulate material: impact, attrition, shear, and compression. Impact is the instantaneous striking of one object by another. Both objects may be moving or one may be stationary. Attrition is the production of wear debris or particles created by the rubbing action between two bodies. This type of milling force is preferred when the material is friable and exhibits minimal abrasiveness. Shear consists of cutting or cleaving of particles and usually is combined with other types of force. Shear contributes to fracturing by breaking particles into individual pieces with a minimum of fines. Compression is the slow application of compressive forces to a body (crushing or squeezing of particulate material). This type of milling action usually is associated with jaw crushers and the breaking of large agglomerates of hard, nonductile material.

Design and process improvement has been based primarily on empirical and semi-empirical data. Advancement has been made in understanding the nature of brittle fracture of single particles of relatively homogeneous materials, such as glass. Fundamental research on milling ductile metals is lacking, although much is known phenomenologically about the process. Comminution is not adequately understood as it applies to the detailed mechanism involved when particles are impacted by colliding mediums.

Fracture occurs in hard, brittle materials, with minimal particle deformation and agglomeration by welding. Particle deformation results from attempts to comminute or blend particulate materials. Deformation, cold welding, and fracturing occur in varying degrees with both hard and soft ductile materials. Although powders of <100 mesh (<150 μm) are most frequently milled, particles up to 6 mm (0.2 in.) and larger can be processed.

Principles of Milling—Phenomenological Description

The impact process is shown in Fig. 1. This model represents the moment of collision, at which particles are trapped between two colliding balls within a space occupied by a dense cloud, dispersion, or mass of powder particles. This phenomenon is typical in dry and wet milling operations that use colliding milling mediums such as tumbler, vibratory, and attrition ball mills.

The number of particles within the incremental volume subjected to impact compression stress may range from a single particle to several million, depending on the average particle size and the radius of the ball or curved surfaces contacting the trapped incremental volume. The

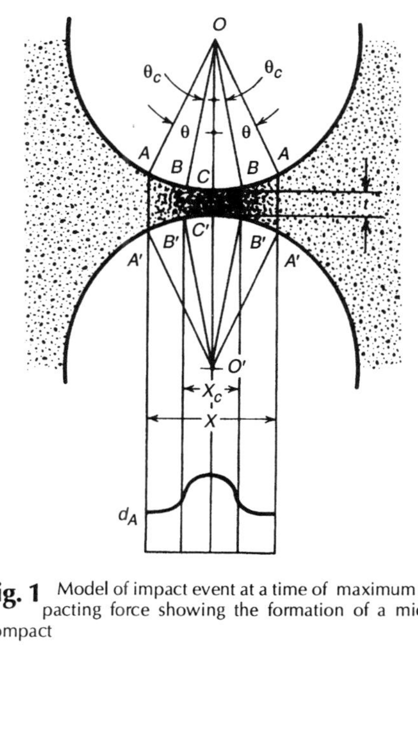

Fig. 1 Model of impact event at a time of maximum impacting force showing the formation of a microcompact

number of particles trapped between two balls on impact in a loose powder mass should increase as the ratio of the ball diameter (D) to the particle diameter (d) increases. Also, the probability of a particle being trapped and impacted between the two balls increases as powder fill (percentage of space between the balls filled with powder) is increased and apparent density of the agitated or fluidized powder increases.

At the high ratios of D/d encountered in milling of fine particles (<100 mesh, or <150 μm), the trapped volume consists of particles within the microvolume $BB'BB'$ (Fig. 1). Outside this incremental volume, the apparent density of the powder sharply decreases to the density of the agitated loose powder.

The process of trapping an incremental volume between two balls within a randomly agitated charge of balls and powder, assuming no rotation or transverse motion of the curved surfaces, is shown in Fig. 2. Between the stages shown in Fig. 2(a) and (c), a certain volume of agitated powder is trapped between two impacting balls. The amount of powder trapped and the size of the trapped volume depends on many factors, including particle size, apparent density of the particle cloud, the presence of a slurry and its concentration and viscosity, air or gas pressure within the mill, surface roughness of the balls, and velocity of the balls.

In a gaseous atmosphere prior to impact, the majority of particles is ejected as the balls approach each other, while the remaining fraction becomes trapped and compacted between the decelerating balls at the moment of collision (Fig. 2a-c). If impacting force is sufficient, the trapped incremental volume of powder is compacted to form an agglomerate particle or pellet (Fig. 2d), which is released (Fig. 2e) when elastic energy impels the balls apart. If bonding by welding or adhesion occurs between contacting surfaces of the particles and bonding forces are sufficient, the agglomerate does not break apart. Similarly, particles may bond to the surfaces of the balls, which may become coated with the powder, as shown in Fig. 3 and 4.

If the compressive impact forces exerted by the balls and particles are sufficient, particles fracture or compact within a critical zone of diameter X_c and thickness (t), within the boundaries of B-B' and B-B' (Fig. 1). The initial zone of compaction theoretically does not exceed $AA'AA'$. The incremental volume of compacted powder consequently consists of surviving original particles and fragments of fractured particles. Surviving particles include those particles protected by larger particles that underwent fracture and fine particles that absorbed and dissipated impact force without fracturing. Density of the trapped bed of particles varies from the maximum at the point of first contact between the two balls (C-C' in Fig. 1) to the density of the freely agitated powder outside the two A-A' boundaries.

As the balls decelerate during impaction because of the cushioning effect of the trapped powder, a radial displacement of powder particles away from the line of contact O-O' (Fig. 1) occurs in the direction of least resistance to particle flow. The greater the curvature of the balls, the greater the radial wedge effect acting on the trapped particles and the higher the velocity of ejection. Radial displacement of powder is resisted by interparticle friction and friction between particles and the ball surface. Such displacement is dependent on the roughness of the particle and ball surfaces and the degree of adhesive binding or cold welding between agglomerate and ball surfaces.

Comminution is similar to impact compacting of a radially unrestrained micropowder compact. Compaction begins with a powder mass that is characterized by relatively large spaces between particles compared to the particle size. The finished product is a mass of powder with comparatively small pores. The first stage of compaction

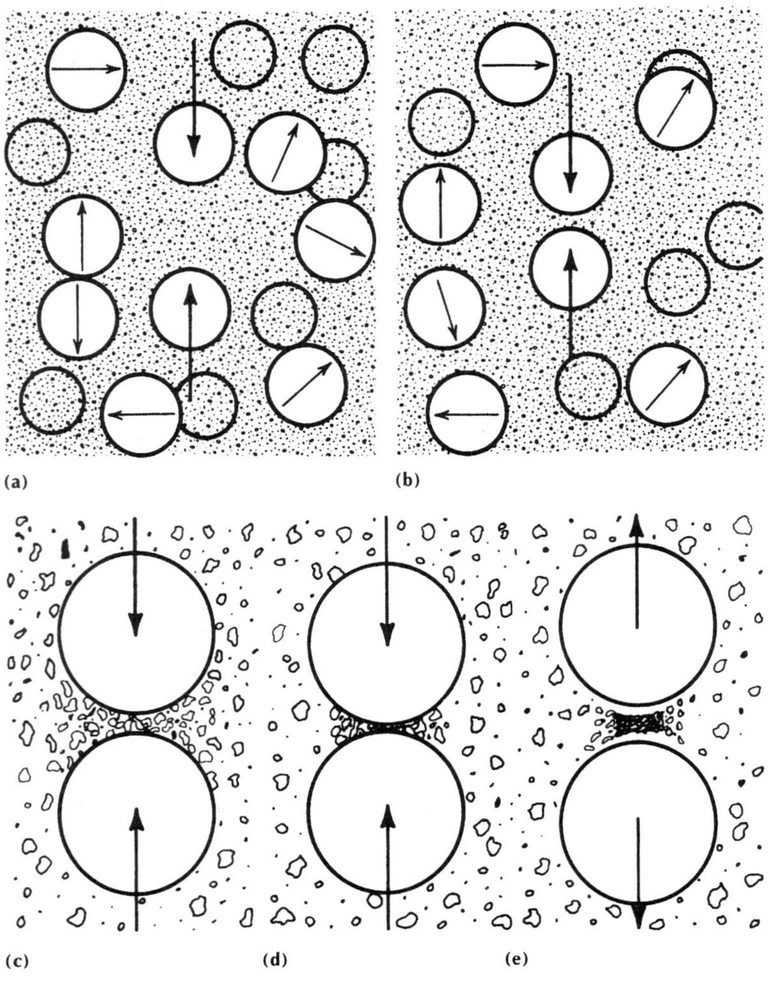

Fig. 2 Process of trapping an incremental volume of powder between two balls in a randomly agitated charge of balls and powder. (a) through (c) Trapping and compaction of particles. (d) Agglomeration. (e) Release of agglomerate by elastic energy

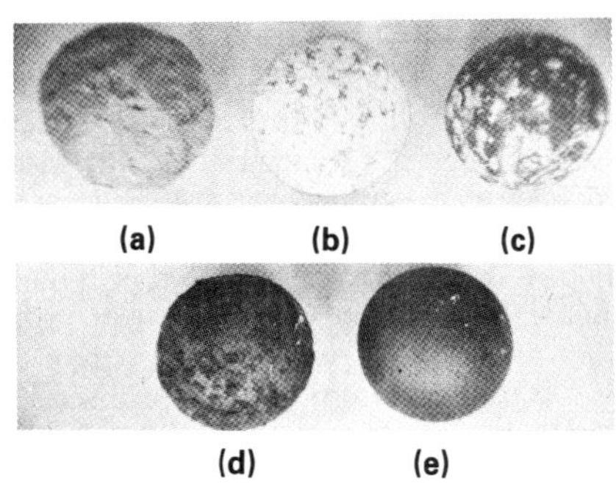

Fig. 3 Grinding balls coated with powder (a) Cement. (b) Marble. (c) Silica. (d) Silicon. (e) Corundum

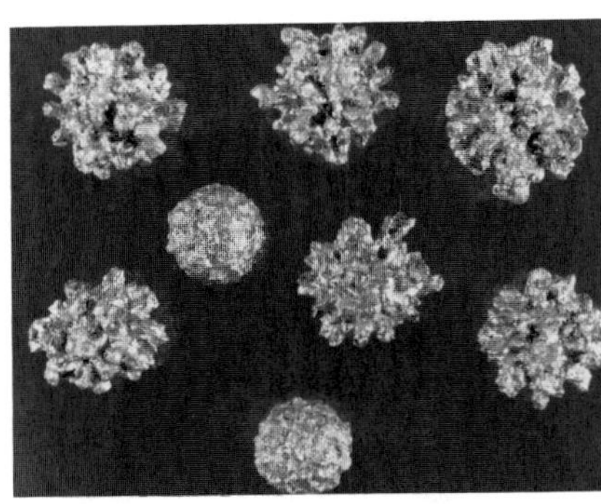

Fig. 4 Spikes of magnesium formed on 12.7-mm (½-in.) diam balls. After prolonged milling under nonreactive milling conditions in a vibratory ball mill

consists of rearrangement and restacking of particles. Particles slide past one another with a minimum of deformation and fracture, producing some fines—especially from brittle, irregularly shaped particles. During this stage, flowability is reduced, and void spaces are partially filled with fine particles.

Particle shape plays an important role during this stage. Spherical particles, having the greatest flowability and least friction are almost totally ejected from between colliding balls; pancake or flake particles, having the least flowability and greatest frictional resistance to flow, tend to be retained between ball surfaces. Particles with irregular surfaces also tend to mechanically bond together to form an agglomerate.

The second stage of compaction involves elastic and plastic deformation of particles. During this stage, cold welding may occur between metal particles. Elastic deformation plays a relatively minor role, whereas plastic deformation and cold welding have major effects, depending on the ductility of the materials and the milling environment. For hard, brittle ceramic and metallic materials, plastic deformation and welding have little effect on comminution. However, plastic deformation and cold welding drastically alter the mechanism of milling for ductile metals. Most metals work harden during the second stage of compaction, which increases deformation resistance.

Finally, the third stage of compaction, involving particle fracture, results in further deformation and/or fragmentation of the particles. Densification of brittle ceramic and metallic particulate materials in the microbed occurs when small pores are filled with fragmented particles. Metal powder and soft ceramic particles achieve higher densities when pores are filled by flow processes.

As shown in Fig. 1, the density of the powder bed reaches a maximum at the core of the zone of compaction, *C-C′*. The strength of the microcompact formed at this point depends on the strength of the bonds between particles, which in turn depends on factors such as conformance of the bonding surfaces, oxide films, surface cleanliness, temperature, and impact force.

The size of the microcompact depends on the above factors, as well as on particle shape, surface roughness of the balls and particles, and the size of the balls and particles. Thus, particle-to-particle bond strength ranges from very weak for ceramics to very strong for metals. For most ceramic materials, the microcompact disintegrates because of weak van der Waals bonds and because of the release of elastic strain in the particles when the balls rebound.

The microcompact ranges from very small for slightly ductile nonmetals, such as sodium chloride, to relatively large for highly ductile metal powders, depending on powder characteristics and the type of attractive forces associated with particle size. Particle size and surface area ranges associated with surface forces are:

Particle size or surface area	Surface force
50–10 μm	van der Waals and low level chemisorption forces
10–0.1 μm	Chemisorption forces predominate
200–500 m²/g	Chemisorption and valency forces
500–1000 m²/g	Valency forces

If chemical species that tend to chemisorb on the surface of the particles in a dry or gaseous environment can be eliminated or reduced, larger and stronger agglomerates tend to form. If the species in the environment promote adhesion between particle surfaces, agglomerates tend to be larger. In general, chemical reactions between fluid or wet environments and powder inhibit agglomerate formation and assist comminution, especially for metal powders. The less reactive the milling fluid, the greater the tendency for particles to agglomerate by welding. Moisture in the powder promotes packing of large masses of powder in ball mills. Drying the powder at 100 °C (212 °F) or above often alleviates or eliminates this problem, especially with fine aluminum alloy powders.

Brittle Fracture. The Griffith theory assumes that a brittle material has minute cracks dispersed within it and that fracture emanates from these cracks (Ref 3). The stress σ_c at which a crack propagates depends on the size of the crack. The general form of the fracture equation for modes of loading that lead to catastrophic failure is:

$$\sigma_c = A\left(\frac{\gamma E}{c(1-\mu^2)}\right)^{1/2} \qquad \text{(Eq 1)}$$

where A is a numerical constant that depends on crack geometry and its location, mode of loading, and dimensions of the particle (for example, A is $\sqrt{2/\pi}$ for the slit crack in an infinite sheet); c is the size of the crack, E is the modulus of elasticity, μ is Poisson's ratio, and γ is the surface energy of the milled substance. Thus, measured strength depends on the size of the crack, elastic properties, and surface energy. For brittle materials, γ is between 10^3 and 10^4 erg/cm^2.

The stress σ for fracture of a particle may be represented by:

$$\sigma = \sqrt{\frac{2Er}{L}} \qquad \text{(Eq 2)}$$

where L is the length of the crack, and r is the radius of the crack tip. When stress at the crack tip equals the strength of the cohesion between atoms, all new surface energy is supplied by the stress itself from elastic strain energy and corresponds roughly to the Griffith model.

Three types of flaws (facial, volume, or edge) have been identified by Gilvarry (Ref 4), with respect to location on the surface or interior of single particles. The initial fracture surface in a single particle usually starts in a facial flaw on the surface of the original particle. However, insufficient surface flaws exist to explain the large number of fragments formed in brittle fracture. When an internal flaw is activated, two surfaces in apposition are created, each of which contains flaws intersected by the propagating fracture. New fracture surfaces are produced in turn by fractures propagating from these volume flaws. Similarly, edges formed by intersecting fracture surfaces contain flaws distributed linearly along the edge (edge flaws), which on activation generate new fractures.

According to Gilvarry, the three types of flaws are distributed independently of each other, as a consequence of the randomness of fracture surface orientation and the fracturing stress. Activated flaws are distributed randomly, independent of any stress-producing fracture. More specifically, the position of a flaw within a domain is independent of the position of all other flaws. The probability of a subdomain containing a given number of flaws is independent of the number contained in any other subdomain. Single fractures are initiated by an external

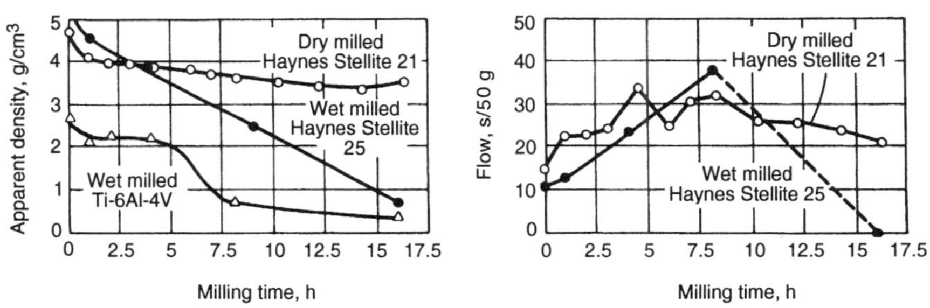

Fig. 5 Effect of vibratory milling time on apparent density and Hall flowability of cobalt-base and titanium-base powders

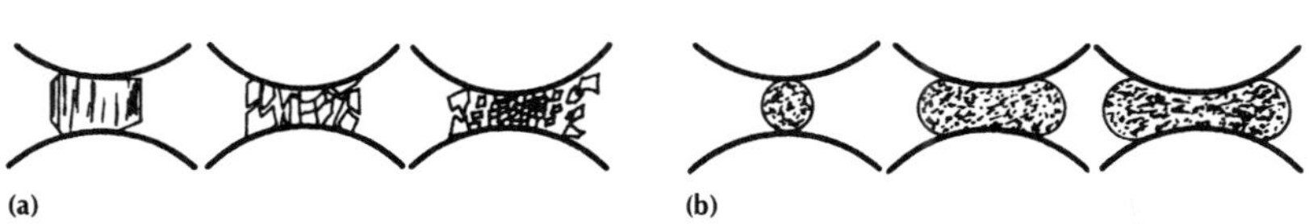

Fig. 6 Effect of impact. (a) Brittle single particle. (b) Ductile single spherical particle

stress system that is relieved with the propagation of the first facial flaws. After initial fracture, flaws are activated by stress waves that are generated by the application and withdrawal of the external stress. According to Gilvarry, flaws exhibit a Poisson-type distribution.

During a milling operation, the conditions for single fracture are satisfied only occasionally. The amount of time and energy required to produce fracture varies with the grinding process and environment. As particle size decreases, flaw concentration and size diminish to a point where edge flaws no longer remain in the particles activated by stress (Ref 4). At this point, fracture stops.

Ultrafine Grinding of Brittle and Hard Materials. The primary function of a mill is to sufficiently stress as many individual particles of the charge as possible, to induce fracture with a minimum expenditure of energy. During milling, a major portion of the energy input is expended on various internal processes that do not contribute to particle failure. As grinding proceeds into the ultrafine region, fracture mode varies, and conditions that were insignificant when particles were relatively large gradually become controlling factors.

Eventually, the milling process reaches an effectiveness limit where the physical and chemical effects associated with surfaces, edges, and corners multiply to inhibit comminution. Edge length per unit volume varies directly with the square of specific surface area, and the number of corners per unit volume varies as the cube of specific surface area. Thus, the probability that an individual particle will be stressed to failure becomes increasingly small.

If the particle deforms plastically, fracture is difficult to induce, and the probability that stress will occur is reduced further. As fragments decrease in size, the tendency to aggregate increases, and fracture resistance increases. Particle fineness approaches a limit as milling continues, and maximum energy is expended. According to Ref 5, the following major factors contribute to grind limit:

- Increasing resistance to fracture
- Increasing cohesion between particles, with decreasing particle size causing agglomeration
- Excessive clearance between impacting surfaces, which is minimized as ball diameter or radius of curvature decreases
- Coating of grinding medium by fine particles that cushion the microbed of particles from impact
- Surface roughness of grinding medium (highly polished, hard mediums that retain a minimum root mean square surface roughness during milling are most effective)
- Bridging of large particles to protect smaller particles in the microbed
- Increasing apparent viscosity as particle size decreases
- Decreasing internal friction of slurry as particle size decreases, thereby reducing production of finer particles by wear mechanism and probability of impacting particles

Generally, as batch comminution proceeds over an extended time, the mean applied stress needed for particle failure increases, while the magnitude of local stresses available to initiate fracture decreases.

Milling of Single Particles. Particles and powders are classified according to particle size, particle size distribution, particle shape, and surface texture. These powder characteristics determine flowability, apparent density, color, sinterability, compactibility, and the properties of sintered products (Fig. 5).

Particle characteristics determine the manner in which particles are deformed or fractured and their resistance to breakage. Thus, the fracture mechanics of single particles must be considered in the selection of milling equipment and milling conditions.

Figure 6 illustrates the crushing of single particles between impacting balls. This condition is encountered in the milling of relatively large particles or a fraction of a powder, as with large granules of a ceramic material, hard metals, or large atomized particles of a ductile metal powder. Under these conditions, many single-particle impact events occur. With ceramics and hard metals, large brittle particles are rapidly reduced to fine and ultrafine powders. The milling of a brittle single particle is shown in Fig. 6(a). During milling of ductile metals (Fig. 6b), individual particles do not fracture, but deform and thereby undergo shape change with little or no change in mass (Fig. 7). Large single particles of brittle materials shatter into fragments, while ductile single particles deform, first into a pancake shape and then into a flake shape.

Milling Parameters and Powder Characteristics

Grinding elements in ball mills travel at different velocities. Consequently, collision force, direction, and kinetic energy between two or more elements vary greatly within the ball charge. Frictional wear or rubbing forces act on the particles, as well as collision energy. These forces are derived from the rotational motion of the balls and movement of particles within the mill and contact zones of colliding balls.

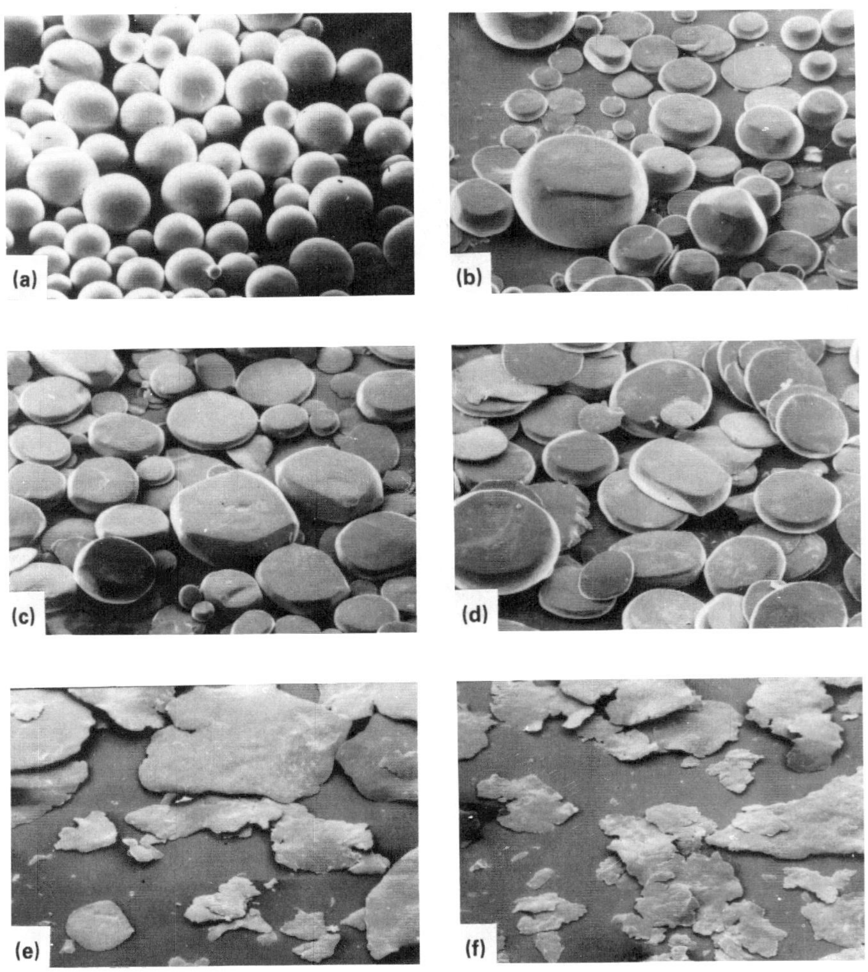

Fig. 7 Effect of milling time on particle shape change of spherical Ti-6Al-4V alloy particles. (a) As-received powder. (b) After 1 h. (c) After 2 h. (d) After 4 h. (e) After 8 h. (f) After 16 h

Surface roughness of the balls is a significant factor in milling—the rougher the surface, the greater the frictional energy expended, thus causing increased abrasive action. Frictional forces exert a wearing action between particles in contact with one another and reduce particle size by attrition, producing wear debris. Impact forces effectively reduce particle size in grinding applications involving coarser particle fractions or hard and/or brittle materials.

Impact forces are desirable for deformation processing of metal powders, such as the production of flake powders. Figure 8 shows the change in flake width of iron powder with milling time in a vibratory ball mill. Attrition grinding is most effective in fine grinding both hard and ductile materials and in mechanical alloying. In most ball mills, particles are subjected to both impact and attrition forces. When forces are highly energetic, welding may occur between powder particles, between the powder and the balls, and between powder particles and the walls of the chamber.

Force of impact is directly proportional to the mass of the milling medium. Consequently, processing rate and the forces acting on particles are a direct function of the effective diameter and mass of the milling bodies. Actual collision forces vary widely from theoretical values because of the complex motion of the medium and varying velocity, path lengths, and trajectories of individual milling bodies. Milling fluid viscosity also effects collision rate.

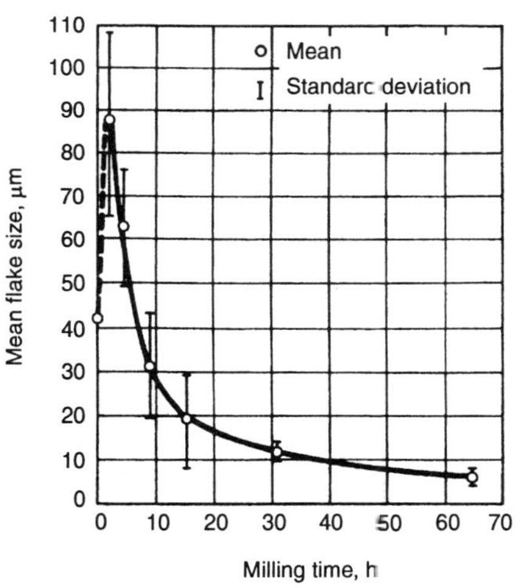

Fig. 8 Relationship between flake size and milling time for electrolytic iron milled in a Megapact vibratory ball mill. Source: Ref 6

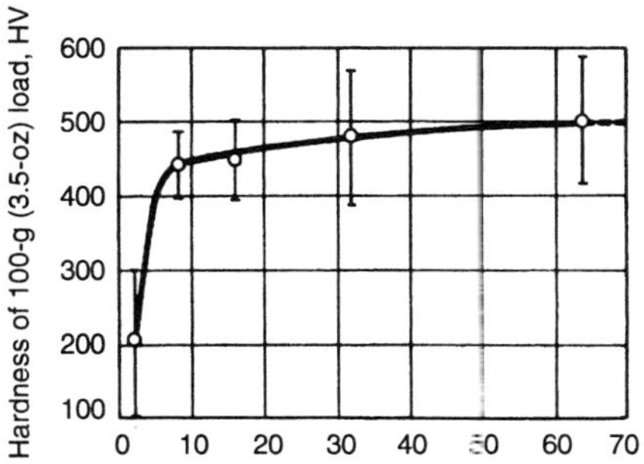

Fig. 9 Effect of milling time on microhardness of Nickel 123 powder

Generally, the size and density of the milling medium selected are determined by the deformation and fracture resistance for metals. For hard, brittle materials, fracture resistance is the only selection criterion. Large, dense grinding mediums are used to mill larger and stronger particulates, whereas smaller diameter mediums are used for finer grinding. For example, ceramic beads having a diameter of 1.6 mm (0.06 in.) and a density of 3.9 g/cm^3 may be used for reducing coarse, thin metal flakes to very fine flakes. Milling elements of such a small size cannot be used effectively in conventional tumbler ball mills. For fine grinding of tough, hard materials, tungsten carbide mediums are recommended.

Often, a compromise must be made in the selection of ball size and material. Although tungsten carbide balls are economical for use in small-scale experimental mills or for milling expensive materials, costs may be prohibitive for some large-scale milling operations. Some mediums, such as stainless steel grinding balls, may not be readily available in large quantities in certain sizes; thus, optimum milling conditions may not be satisfied.

The surface area and particle size of nonmetallic materials change continuously. Total surface area of the medium increases as particle size decreases. With metals, however, two additional factors must be considered in addition to comminution. Metal particles may agglomerate by cold welding at impact, thus consuming surface area and causing a shape change because of ductility and the ability to cold weld.

Thus, depending on the dominant process during milling (fracturing, welding, or microforging), a particle may (a) become smaller in size through fracturing, (b) grow in size through agglomeration by welding, or (c) change from an equiaxed shape to a platelet or flake-like particle by microforging. For metals, changes in surface area and particle size measurements do not provide a meaningful criterion for comparing the effect of changes in process parameters, competing milling processes, and equipment.

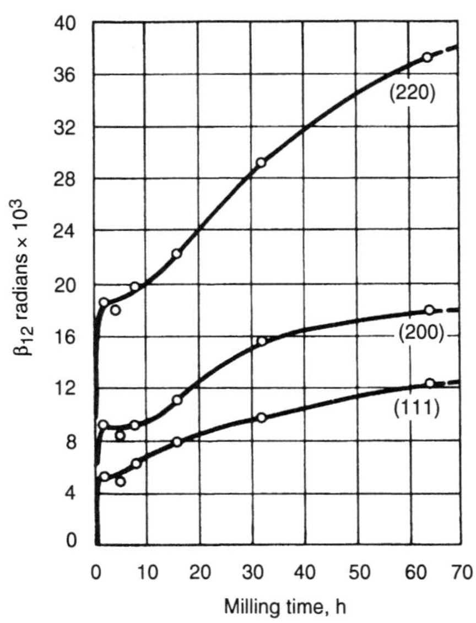

Fig. 10 Relationship between x-ray line broadening and vibratory milling time for Nickel 123 powder. Numbers in parentheses refer to coordinates of atomic planes in face-centered cubic structure

A more useful criterion for assessing the milling process is one that reflects the spectrum of structural and physical changes occurring in the metal and that is sensitive to differences in processing parameters. Microhardness measurement of individual particles large enough to accommodate the smallest Knoop or diamond pyramid hardness indentation provides a suitable measure of the effect of milling on metal powders, because hardness is a measure of cold work and internal defects produced by milling. As shown in Fig. 9, hardness measurement is most meaningful up to milling times that produce maximum levels of cold works.

X-ray line broadening is sensitive to both the amount of cold work and the refinement in crystalline structure that occurs with continued kneading and working of the metal well after saturation cold work. Changes in the deformation rate of a metal powder produced by a given set of milling parameters are shown by the relationship of x-ray line broadening to milling time (Fig. 10).

Energy Relationship

When a metal is plastically deformed by cold working, most of the mechanical energy of the deformation process is converted into heat. However, a small amount (approximately 5%) is stored in the metal, thereby raising its internal energy. The energy associated with permanent lattice strain or cold work is minimal for hard, brittle particles, but can be large for ductile materials. The energy expended to overcome the friction between particles is translated to heat and performs no useful work in milling.

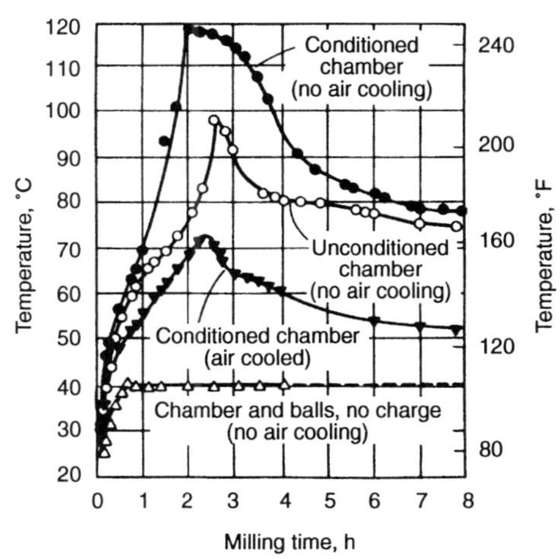

Fig. 11 Heating curve for 280 cm^3 (17 in.^{3}) stainless steel milling chamber during vibration milling of Fe-27Ni-16Cr

If the temperature of the powder rises above a certain point, the cold worked metal particles may undergo recovery and recrystallization. Heat is generated by particle deformation, and by elastic deformation of metal grinding balls and grinding chamber walls. Figure 11 illustrates typical temperature versus milling time curves for dry vibratory milled Fe-27Ni-16Cr alloy powder (Ref 7). Generally, the temperature rises during severe cold working, abruptly falls just before cold work attains saturation levels, and then slowly decreases after extended milling times. Water-jacketed milling chambers usually are required for large, high-energy vibratory and attrition mills that reach temperatures above 200 °C (390 °F).

Milling Processes

Changes in powder particle morphology (Fig. 12) that occur during milling of metal powders are produced by the following processes:

Process	Symbolic designation
Microforging	M
Fracture	F
Agglomeration	A
Welding (atomic bonding)	A_w
Mechanical interlocking	A_m
Autohesion (van der Waals forces)	A_a
Deagglomeration	D

Microforging. The initial and predominant process during milling is compression shape forming of ductile metal particles by impact from the milling medium, as shown in Fig. 6(b). Individual particles, or a cluster of particles, are impacted repeatedly by milling medium so that they deform (flatten) with little or no net change in individual mass.

Fracture. After a period of milling, individual particles deform to the extent that cracks initiate, propagate, and ultimately fracture the particles. Fissures, cracks, defects, and inclusions in particles facilitate fracturing. Particles formed from irregular or spongy particles contain fissures and cracks that facilitate compression, fatigue failure, and fragmentation, compared to the smooth, relatively nonporous spheroidal particles formed by atomization. Some friable metals (antimony, bismuth, silicon, manganese, and chromium, for example) and brittle intermetallic alloys (such as copper-aluminum, aluminum-magnesium, copper-tin, nickel-iron, nickel-zirconium, and phosphorus-copper) can be milled easily into powders.

Agglomeration of particles may occur by welding, mechanical interlocking of spongy or rough surfaces, or autohesion. Autohesion is the molecular interaction of particles among themselves, characterized by van der Waals forces. Deagglomeration is a process that breaks up agglomerates formed by autohesion without further disintegration of the individual powder particles.

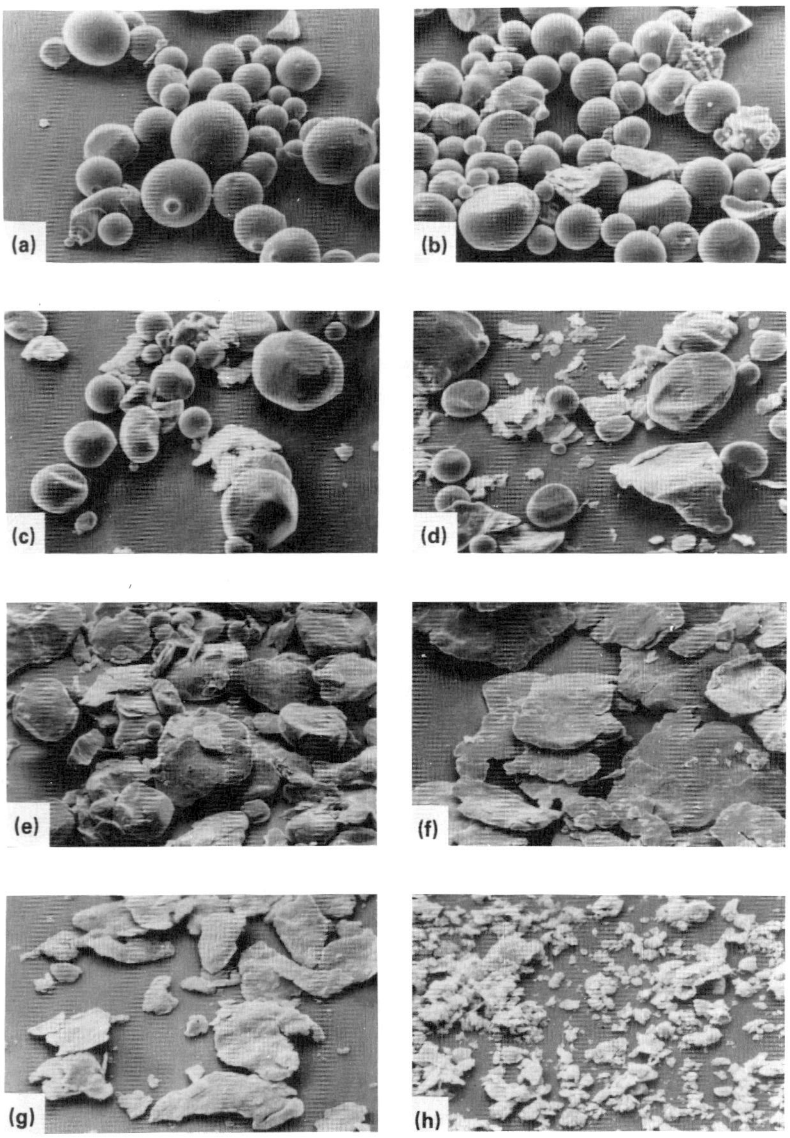

Fig. 12 Scanning electron micrographs of Haynes Stellite 21 cobalt-base alloy powder. Milled in ethyl alcohol with aluminum nitrate grinding aid. (a) As-received powder. (b) After 1 h. (c) After 2 h. (d) After 4 h. (e) After 8 h. (f) After 16 h. (g) After 32 h. (h) After 64 h

Mechanism of Milling—Phenomenology

One type of milling mechanism consists of an initial microforging stage in which particles are deformed in the absence of agglomeration by welding and fracture. Eventually, particles become so severely deformed and embrittled by cold work that they enter a secondary stage, during which the particle fracture by a fatigue failure mechanism and/or by the fragmentation of fragile flakes. Fragments generated by this mechanism may continue to reduce in size in the absence of strong agglomerating forces by alternate microforging and fracture. When fracture dominates and the role of microforging is negligible in the absence of welding agglomeration, flakes continue to become smaller until molecular interaction occurs.

As particles (particularly flakes) become finer, coupling forces tend to become greater, and agglomerates become stronger. Eventually, milling forces that deagglomerate the particles reach equilibrium with the coupling forces present, and an equilibrium agglomerate particle size results.

When cold welding between particles occurs, the initial process is microforging. This is followed by a secondary stage involving a closed loop process of fracture (F), microforging (M), and agglomeration by welding (A_w) as shown in Fig. 13.

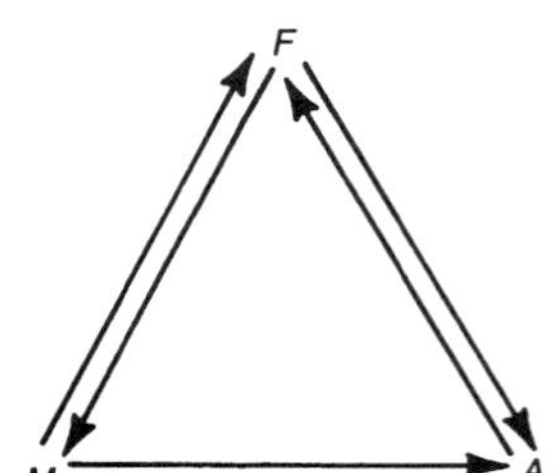

Fig. 13 A closed loop process of fracture (F), microforging (M), and agglomeration by welding (A_w)

Constant kneading, fracturing, and rewelding produce a composite particle that consists of fragmented original and composite particles. At later stages in the milling process, composite particles are composed entirely of fragmented composite particles. The degree of refinement and blending of the composite structure is a function of milling time, as is shown in Fig. 14 and 15 for wet-milled iron and dry-milled nickel powder, respectively.

After a period of time, cold welding and fracturing attain a steady-state equilibrium. Average particle size obtained at this stage depends on the relative ease with which agglomerates can be formed by welding, fatigue, and fracture strength of composite particles, and resistance of particles to deformation. Particle size distribution is narrow, because particles larger than average are reduced in size at the same rate that fragments smaller than average grow through accretion of smaller particles and wear debris produced by attrition.

This process is shown in Fig. 8 and 16. When a large number of fine particles are produced by attrition, the degree of refinement is very high. If particles are spongy, rough, or covered with asperities, microforging may be accompanied by agglomeration through mechanical interlocking.

Dispersion of Oxide Films, Hard Particles, and Wear Debris

The strengthening effect produced by ultrafine oxide particles through solid-state or mechanical alloying is an important factor affecting final equilibrium particle size and the time required to reach equilibrium particle size. Oxide particles, derived from oxide films originally present on the particle surface and/or from oxygen in the chamber environment, become entrapped between cold welded surfaces by kneading. This process is illustrated in Fig. 17 using a fluid medium. The strengthening effect of the dispersoid, cold working, and the composite structure greatly increase toughness, strength, and fracture resistance of individual particles. Under these conditions, particles assume a more equiaxed shape. More importantly, the degree of hardening may limit the amount of structural refinement if milling energy is insufficient.

Because the mechanical alloying effect depends on achieving the ultimate degree of refinement, sufficient energy is required for milling. More information on mechanical alloying of oxide dispersion-strengthened superalloys is in the article "Production of Nickel-Base Powders" in this Volume.

Milling Environment

Surface-active agents and lubricants generally are used to nullify the forces of welding (or autohesion) and thereby inhibit agglomeration. Surface-active agents adsorbed on a particle surface interfere with welding and lower the surface tension of solid material. Because the energy required for milling is equal to the product of new surface area generated times the surface tension, a reduction in surface tension results in the use of shorter milling times and/or finer powders.

Corrosion reactions between the metal and grinding fluid or environment in reactive grinding also facilitate comminution. The greater the corrosion resistance of a metal, the less effective are corrosive agents. For example, aluminum nitrate, an effective grinding aid for nickel milled in ethyl alcohol, is significantly less effective for either type 316L stainless steel or Haynes Stellite 21.

Milling of iron powder in water, a highly reactive environment for this metal, produces an ultrafine mixture of iron oxide and metal particles. Heptane is one of the least reactive fluid milling environments for metals. However, addition of a grinding aid such as cetyl alcohol may reduce agglomeration by welding. The effectiveness of a given additive depends on its reactivity with the metal being comminuted; the more reactive the metal, the more effective the additive.

Metal powders milled in organic or inorganic fluids retain small amounts of the fluid dispersed throughout each particle (Ref 8). Thus, hydrocarbons containing hydrogen and carbon and carbohydrates containing hydrogen, carbon, and oxygen are likely to introduce carbon and/or oxygen into the particle. The hydrogen subsequently escapes as a gas or is absorbed into the metal lattice on heating or sintering. For example, Fig. 18 shows the amounts of oxygen, carbon, and nitrogen picked up during milling of electrolytic iron in heptane. Hydrated aluminum oxide surface films on aluminum alloy powders produce blisters during subsequent rolling into sheet after solid-state blending, unless removed by heating at 370 to 480 °C (700 to 900 °F) in an inert atmosphere or in a flowing inert gas stream.

Some metals, such as aluminum, nickel, and copper, react with certain alcohols during milling to form complex metallo-organic compounds. For example, aluminum reacts with isopropyl alcohol. Other metals, such as titanium and zirconium, may react explosively with chlorinated fluids such as carbon tetrachloride. Chlorinated fluids should never be used with reactive metals.

According to Arias (Ref 9), two classifications of chemical activity exist between metal powders and the milling environment—reactive and nonreactive milling. In reactive milling, metal powders react extensively with the milling fluid during milling. The metal powder is comminuted to a particle size much smaller than the starting powder. Balls, mill chamber walls, and powder of the same metal do not weld together during milling.

During nonreactive milling, the metal powder either reacts mildly or not at all with the milling fluid during milling. Powder particles weld together to a great extent during ball milling. Additionally, balls, mill chamber walls, and powder of the same metal may weld together during milling. Welding of powders to balls and cham-

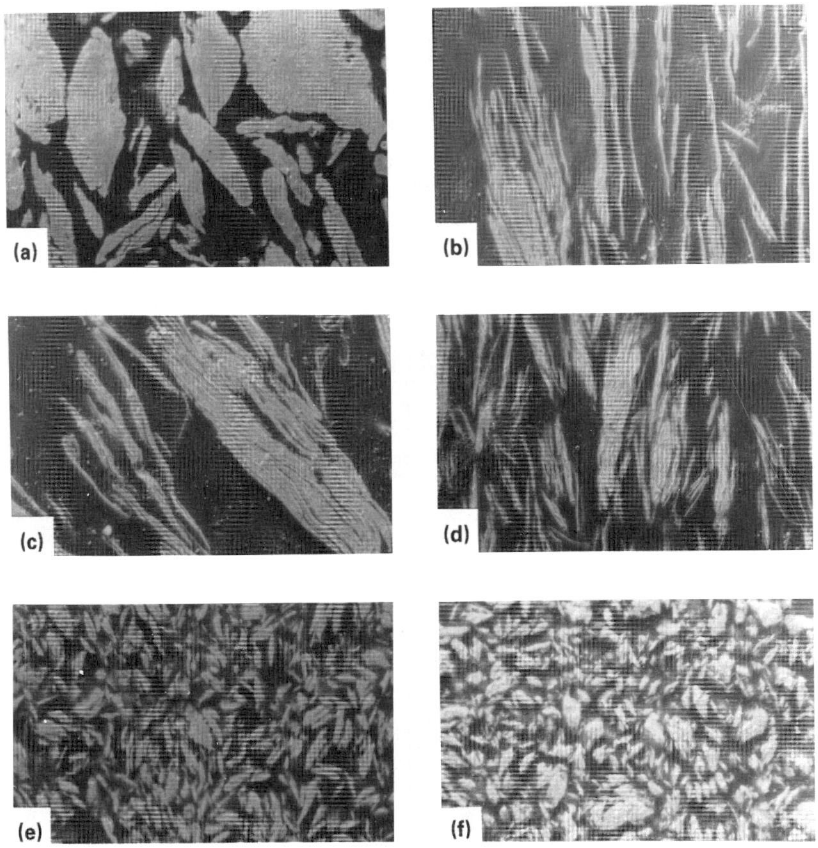

Fig. 14 Electrolytic iron showing transverse cross sections of particles. Milled in heptane. (a) As-received powder. (b) After 4 h. (c) After 8 h. (d) After 16 h. (e) After 32 h. (f) After 64 h

ber walls may occur even if they are made of different alloys.

Reactive and nonreactive milling may occur in wet or dry environments. Mechanical alloying consists of dry milling in a vacuum or an inert gas, or in a flowing gas that may be reactive. In mechanical alloying, the welding process may be controlled by monitoring the oxygen concentration in the grinding chamber (Ref 10).

Milling of metal powders produces external shape and textural changes and internal structural changes. The extent of these changes is determined largely by milling parameters, milling environment, and physical and chemical properties of the metal or alloy being milled. These changes in turn affect the physical properties of the metal being processed and the resulting consolidation products. Powder characteristics associated with milling include:

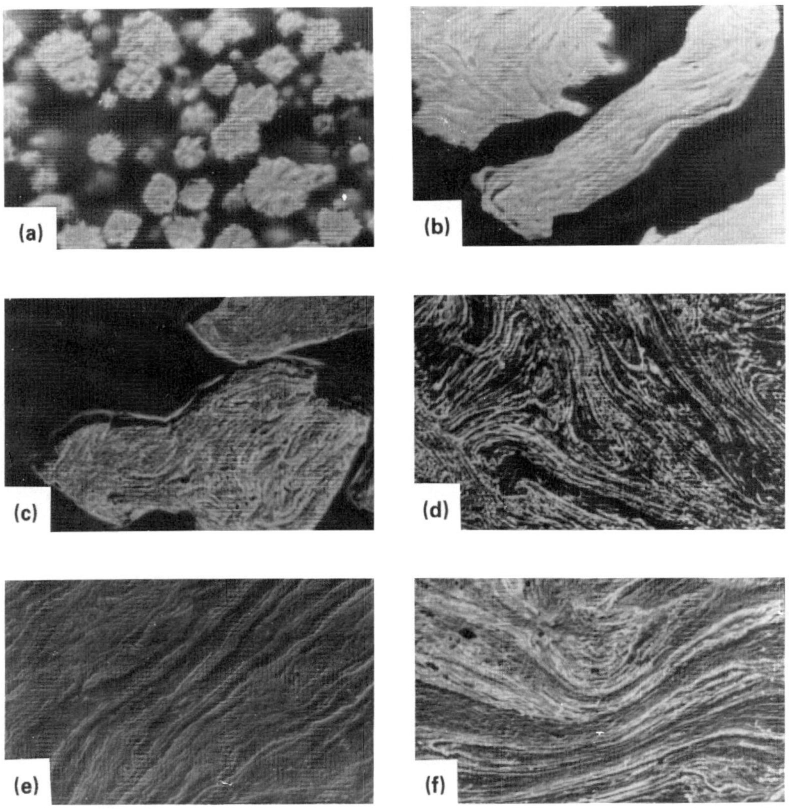

Fig. 15 Nickel 123 showing transverse cross sections of particles. Dry milled. (a) As-received powder. (b) After 4 h. (c) After 8 h. (d) After 16 h. (e) After 32 h. (f) After 64 h

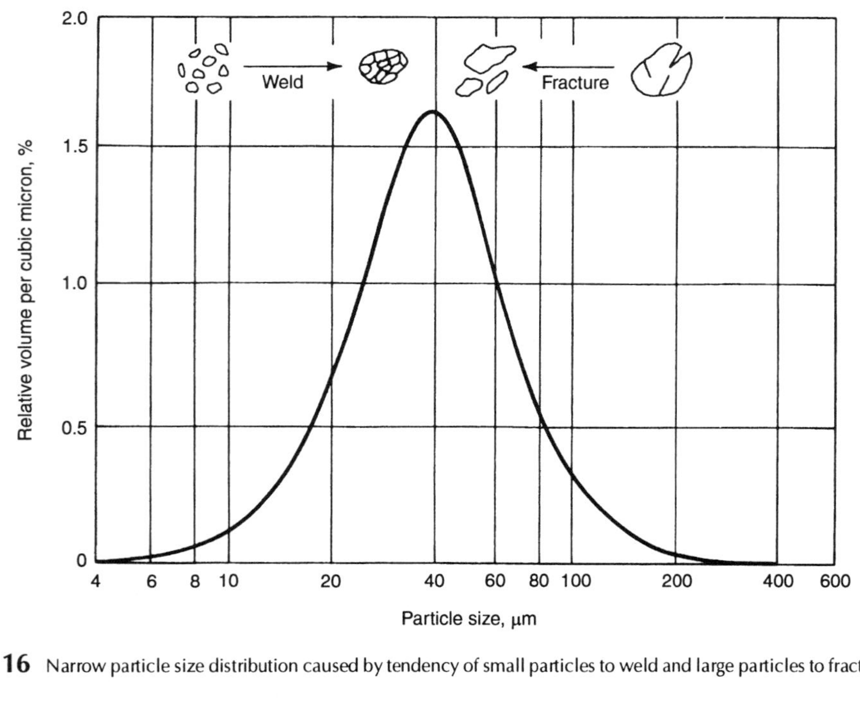

Fig. 16 Narrow particle size distribution caused by tendency of small particles to weld and large particles to fracture

External

- Shape
- Texture
- Reflectivity
- Particle size and particle size distribution

Internal

- Composite structure or composite agglomerates
- Cold work (stored energy)
- High dislocation density
- Improved homogeneity (alloys)
- Finer structure (laminar structure)
- Refined crystallite size, nanocrystalline grain size
- Chemical composition (incorporation of gases and fluids from the milling environment)

Physical

- Mechanical properties (resistance to deformation and hardness)
- Physical properties (resistivity)
- Chemical properties (corrosion)
- P/M properties (apparent density, flowability, sinterability, compressibility, and sintered density)

Knowledge of the variables that influence characteristics of a milled product can be used to improve product performance and material properties and to achieve innovative P/M applications not possible by conventional metallurgy. Some of these include silicon carbide whisker-reinforced composites, oxide-dispersion-strengthened alloys, and superconducting compounds such as Nb_3Sn.

Milling Equipment

During comminution, the milling machine stresses the maximum number of individual particles in a powder mass to initiate fracture with a minimum of energy. The motion of milling medium and charge varies between types of mills, with respect to movement and trajectories of individual balls, movement of the mass of balls, and the degree of energy applied to impact, shear, attrition, and compression forces acting on powder particles.

Tumbler Ball Mills

In the tumbler mill shown in Fig. 19, the motion of the balls or rods depends on the rotation speed of the chamber walls. Balls may roll, or "cascade," down the surface of the charge in a series of parallel layers or they may project, or "cataract," into space, falling on the balls and particles at the toe of the charge below. A detailed mathematical discussion of the trajectories followed by balls and particles is given in Ref 11.

Ball Mill Charge Parameters. The rate of milling of a powder is a function of the quantity of powder in the total volume between the balls and attains a maximum value when the powder fill is 100% (that is, the space between the balls is completely filled with powder) (Ref 11). Speed of the mill should be adjustable to ensure operation with the required proportions of balls for cataracting and cascading. This proportion is largely dependent on the speed of rotation and the coefficient of friction of the charge material. With increasing ball size, the rate of milling increases, attaining a maximum when balls reach sizes capable of crushing all particles in the feed powder.

If the slip of the charge against the mill chamber wall is assumed to be negligible, critical rotational speed of the mill may be calculated:

$$N_c = 76.6\sqrt{\frac{1}{D}}$$

where D is the mill diameter, in feet; and N_c is the critical speed of the mill, in revolutions per minute. These parameters are approximate and may not be valid for metal particles that tend to agglomerate by welding.

Solid-state blending can be achieved in tumbler ball mills using 6 to 25 mm (0.25 to 1 in.) diam balls, if mill diameter is approximately 1.5 to 1.8 m (5 to 6 ft) and other conditions that promote cold welding agglomeration are satisfied. For large-scale production of mechanically alloyed superalloys, tumbler mills are replacing attrition mills. Milling times in tumbler mills are longer to accomplish the same level of blending achieved in the attrition or vibratory mill, but overall productivity is substantially greater. Tumbler mills usually are used to comminute or flake metals, using a grinding aid or lubricant to prevent cold welding agglomeration and to minimize oxidation (Ref 12 and 13).

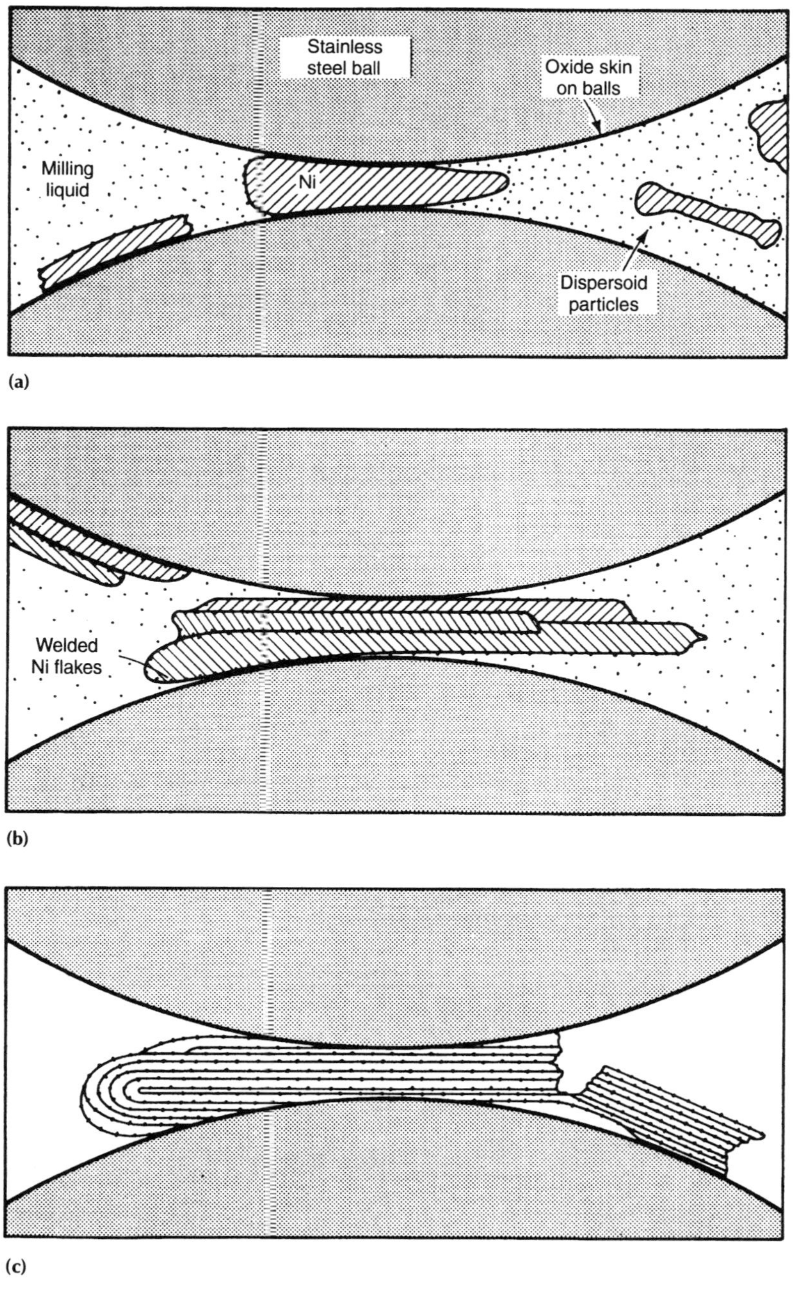

Fig. 17 Nonreactive milling process. (a) Initial milling: nickel flattened into flakes; dispersoid embedded in nickel particles. (b) Intermediate milling: nickel flake formation continues; flakes weld; embedding of dispersoid particles in nickel continues. (c) Final milling: multilayered nickel flakes fold and/or break and reweld; all dispersoid kneaded into milled powder. Source: Ref 8

Vibratory Ball Mills

Vibratory Tube Mill. In the laboratory vibratory tube mill shown in Fig. 20, oscillatory motion of the balls is complicated. The path of the balls and particles is determined by many factors, including vibrational speed, amplitude, curvature of the sides of the mill chamber wall, horizontal motion of the mill, and charge contact with the upper surface of the mill chamber. The balls gyrate around the chamber wall, sometimes traveling horizontally in a spiral path. Gyration in cylindrical chambers occasionally produces deformation and/or wear grooves in the inner walls of the chamber. These grooves may extend 360° or less, depending on whether the balls make contact with the upper surface of the chamber wall. Balls also revolve at different rates and directions, causing substantial shearing action, which is desirable in mixing and blending operations.

Impact forces acting on powders in a vibratory mill are a function of the rate of milling, amplitude of vibration, and mass of the milling medium. High-energy milling forces can be obtained by using high vibrational frequencies and

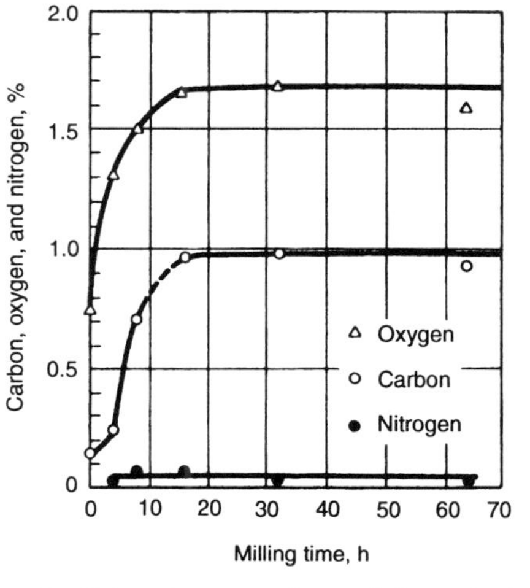

Fig. 18 Pickup of oxygen, carbon, and nitrogen in electrolytic iron. Milled in heptane in a Megapact vibratory mill. Source: Ref 7

small amplitudes of vibration. The mill shown in Fig. 20 operates at 3300 rpm, with a 2 mm (0.08 in.) amplitude, reaching a maximum acceleration rate of 12.2 *g*, where *g* is the gravitational acceleration at 9.81 m/s^2 (32.3 ft/s^2). Large production mills operate at relatively low vibrational frequencies and high amplitude (for example, 1000 to 1500 rpm and up to 12 mm, or 0.48 in.).

The vibratory ball mill is an excellent means of producing solid-state alloyed and dispersion-strengthened metals in amounts up to 4.5 kg (10 lb) or more, depending on the apparent density of the powder. Figure 21(a) and (b) show the microstructure of an aluminum-iron-cerium alloy and a transmission electron micrograph of oxide dispersion in the same alloy after solid-state alloying in a laboratory vibratory mill of the type shown in Fig. 20.

In large tube mills of the type shown in Fig. 22, vibratory motion of the medium decreases from the chamber walls to the center of the mill tube; consequently, the milling effect is less at the center of the mill than adjacent to the chamber walls. Efficient operation is obtained at a ball fill of 60 to 80% of the volume of the mill chamber for tubes about 500 mm (20 in.) in diameter, with 100% fill of the volume between the balls.

In vibratory mills, the grinding medium receives rapid impulses at a rate proportional to the vibrational frequency of the mill. Impact forces acting on the powder exceed shearing and friction forces. The entire charge slowly revolves counter-clockwise to the oscillatory vibrations, so that grinding and intense mixing occur simultaneously. Vibratory mills utilize smaller mediums because of higher impact forces, frequencies, and acceleration; thus, a higher specific surface is available for milling. The rate of processing in a vibratory mill is the following:

- Proportional to the density of balls (diameter constant)
- Proportional to the diameter of the balls (density constant)
- Proportional to the cube of the frequency of vibration
- Negligible for speeds less than 900 to 1000 rpm
- Proportional to the square root of ball diameter/mean particle diameter ratio
- Not significantly affected by chamber diameter
- Increased as amount of powder in mill decreases
- Greater with balls than cylinders or other shapes

Sweco vibratory mills (Fig. 23) are equipped with a grinding chamber in the form of a vertical cylinder with a solid center axis. A double-ended motor with eccentric weights rigidly attached to the bottom of the chamber generates high-frequency, three-dimensional vibrations. The chamber and motor assembly are mounted on a base and are supported by compression springs. Vibration of the grinding medium within the chamber creates the milling action. Vibrations are transmitted from the sides and base of the chamber to the grinding medium. Particles trapped within the medium are broken down by high-frequency impaction.

The medium is packed to provide near-maximum packing density. The packed mass slowly gyrates horizontally, rises near the outer wall of the chamber, and descends as it approaches the inner wall. This motion facilitates distribution of the charge in dry grinding and serves to maintain solids in suspension in wet milling. The Sweco mill is not widely used to mill metal powders. It is particularly ill suited for high-density metals, because high density causes particles to settle to the bottom and become caked.

Attrition Mills

Milling in an attrition mill (Fig. 24) is effected by the stirring action of an agitator that has a vertical rotating shaft with horizontal arms. This motion causes a differential movement between the balls and the material being milled, thus providing a substantially higher degree of surface contact than is achieved in tumbler or vibratory mills.

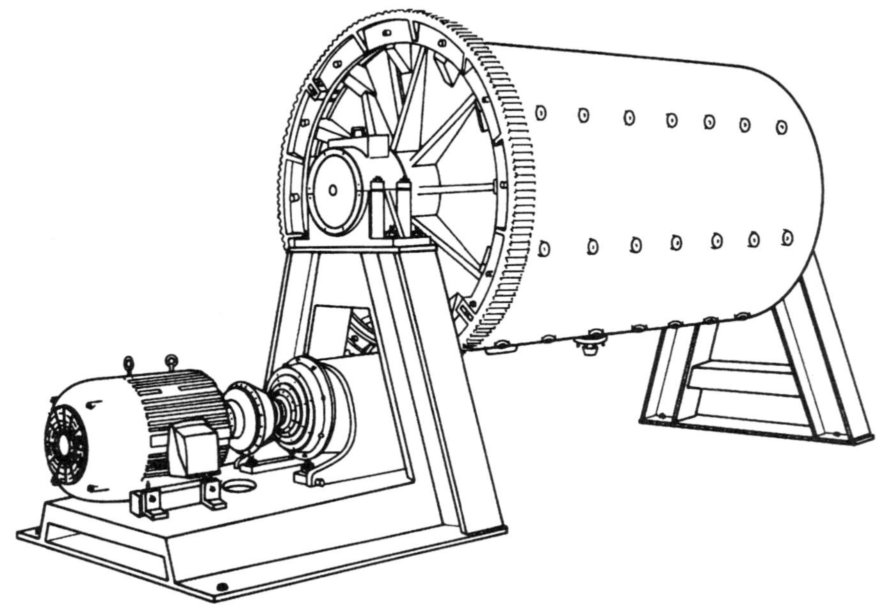

Fig. 19 Tumbler mill used for milling metal powders

Fig. 20 Megapact vibratory ball mill. Courtesy of Battelle Memorial Institute

Milling is accomplished by impact and shear forces. The rotating charge of balls and milling product form a vortex at the upper end of the stirring shaft, into which the milling product and balls are drawn. The milling product is impacted by balls traveling in various trajectories that collide within the dilated charge of medium and powder.

While ball mills use large mediums, normally 12.7 mm (0.5 in.) or larger, and run at low rotational speeds of 10 to 50 rpm, the attrition mill agitator rotates at speeds ranging from 60 rpm for production units to 300 rpm for laboratory units and uses mediums that range from 3 to 6 mm (0.125 to 0.25 in.). Power input to attrition mills is used to agitate the medium, not to rotate or vibrate the heavy containment vessel.

For hard materials such as ceramics, carbides, and hard metals, attrition mills are more efficient than conventional tumbler and vibratory ball mills. The main advantages of attrition mills for mixing and blending tungsten carbide/cobalt cutting tool powders include short milling time, production of fine particle size (submicron sized), and enhanced smearing of cobalt onto carbide particles. Attrition mills effectively mill metals in inert atmospheres, such as in solid-state or mechanical alloying processes, but are not cost effective. Product output is relatively low with attrition mills, compared to large tumbler and vibratory ball mills. Consequently, tumbler ball mills usually are used for production runs of over 135 to 180 kg (300 to 400 lb) per day.

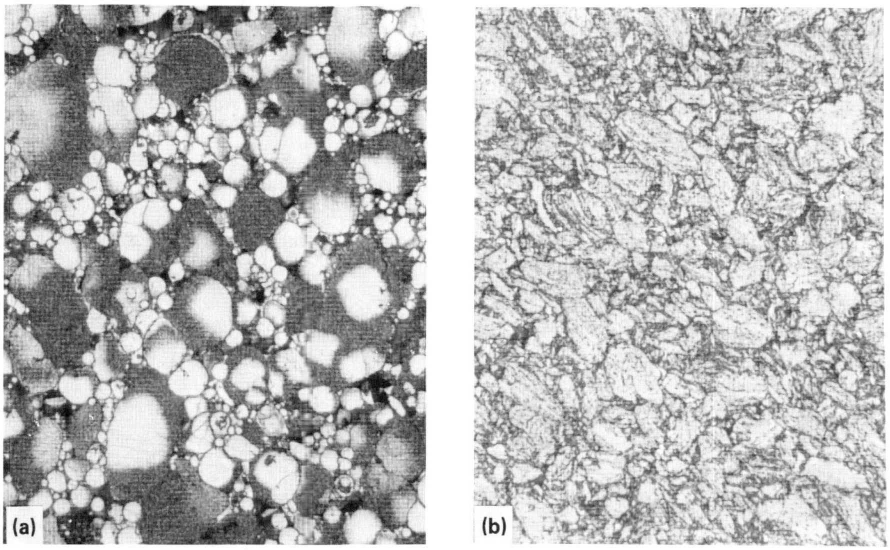

Fig. 21 Homogenization of Al-Fe-Ce alloy by high-energy milling. (a) Untreated rapidly solidified powder after hot pressing. (b) Hot pressed rapidly solidified powder after high-energy milling in a Megapact mill

Fig. 22 Pilot- and production-size vibratory mills

Mechanism of Attrition Mills. The central rotating shaft of an attrition mill, equipped with several horizontal arms, exerts sufficient stirring action to tumble the grinding medium randomly through the entire chamber volume, causing irregular movement by the following:

- Impact action on the medium
- Rotational force on the medium
- Tumbling force as medium fills the void left by the shaft

For fine grinding, both impact and shearing forces must be present. In attrition mills, impaction is caused by constant impinging of the grinding medium, due to irregular movement. Shearing action is produced by random movement of the balls in different rotational directions, which exerts shearing forces on the adjacent slurry. The strongest medium agitation occurs at a point located two-thirds of the way from the center. Grinding does not occur against the chamber walls, which serve as a container rather than a grinding surface. Minimal wear of chamber walls ensures long service life.

In attrition grinding, grinding time is related to medium diameter and agitator speeds, within given limits, as:

$$t = \frac{kd}{\sqrt{n}}$$

where t is grinding time required to reach a certain median particle size; k is a constant that varies with the slurry being processed and the type of medium and mill being used; d is diameter of the medium; and n is shaft movement, in revolutions per minute.

Attrition mills are classified as batch-, continuous-, or circulation-type mills. In the batch mill, material is fed into the jacketed chamber and is ground until the desired dispersion and particle size are achieved. Premixing is accomplished in the grinding chamber. Chamber walls are jacketed so that either hot or cold water can be circulated to control and maintain the temperature of the batch. Batch attrition mills can process high-density material, such as tungsten carbide, as well as viscous materials, and are suitable for dry grinding and for processing dispersion-strengthened metals.

Continuous attrition mills, best suited for large production output, consist of a tall, narrow, jacketed chamber through which a well-remixed slurry is pumped in at the bottom and discharged at the top. Grids located at the bottom and top retain the grinding medium.

The circulation grinding system is comprised of an attrition mill and a large holding tank, generally ten times the volume of the grinding unit. The circulation attrition mill is filled with medium and contains grids that restrain the medium while the slurry passes through. Attrition mills generally have high pumping rates. Usually, the contents of the holding tank are passed through the system at a rate of ten times per hour.

One advantage of the circulation attrition mill is that large quantities of material can be handled with a minimum investment in grinding medium and equipment. The slurry can be monitored continuously, and processing can be stopped when the desired particle size dispersion is achieved.

Coldstream Impact Process

The Coldstream process is a high-velocity process in which material is entrained in a gas stream and projected against a stationary target. After material has struck the target and shattered, it is removed from the impact chamber by suction. Material is then transported to a classifier, which allows oversized products to drop into a storage vessel for subsequent impact against the target. The finished product is separated by the classifier and transported to collection containers.

The process chamber is lined with tungsten carbide or other wear-resistant material. The supersonic nozzle and target generally are made of cemented tungsten carbide, but other wear-resistant materials or the product being comminuted can be used to make these components. Figure 25 shows a high-speed closeup of a raw material stream impacting a target and shattering.

Rapidly expanding gases exiting the nozzle create a strong cooling effect through adiabatic expansion. This effect is greater than the heat generated by pulverization. Therefore, the process operates at well below room temperature. A compressor capable of operation at 56 m^3/min (2000 ft^3/min) at 6.9 MPa (1000 psi) is used. Such a compressor can operate two or three Coldstream impact systems simultaneously.

The Coldstream impact process is used to pulverize hard, abrasive, relatively expensive materials such as tungsten carbide, tungsten alloys, molybdenum, tool steels, beryllium, and other alloys. The process rapidly converts a feedstock of particles 6 mesh or smaller to micron size. Process benefits include simplicity, low operating cost, low operating temperature (which prevents oxidation), and the ability to retain high purity and particle size control over the material being pulverized.

Many standard products are supplied as atomized powders. The higher cost of additional Coldstream processing is justified by production of a finer powder than is commercially available by straight atomization. Most atomized powders have a slight oxide film on the surface of each particle, which is particularly noticeable in coarse powders of nonstainless steel composition. This brittle oxide film is stripped from the metal and broken up into fine dust during Coldstream processing. Lower density and smaller particle size facilitate separation from the metal particles by air classification. Coldstream processing generally produces a relatively oxide-free powder.

High-Energy Milling

Vibratory, attrition, and large-diameter tumbler mills may be classified as high-energy ball mills and, as such, may be used effectively in solid-state or mechanical alloying processes. Figure 26 provides the capabilities of typical size reduction equipment for grinding a material of a given feed size to a desired product size, when the milling objective is comminution of a hard and/or brittle metal or ceramic material.

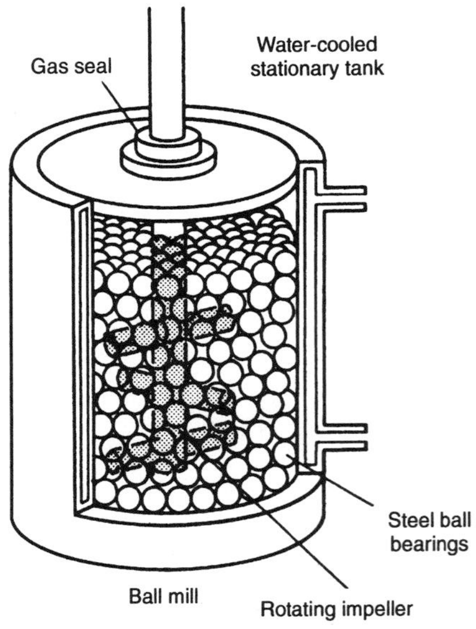

Fig. 24 Attrition ball mill

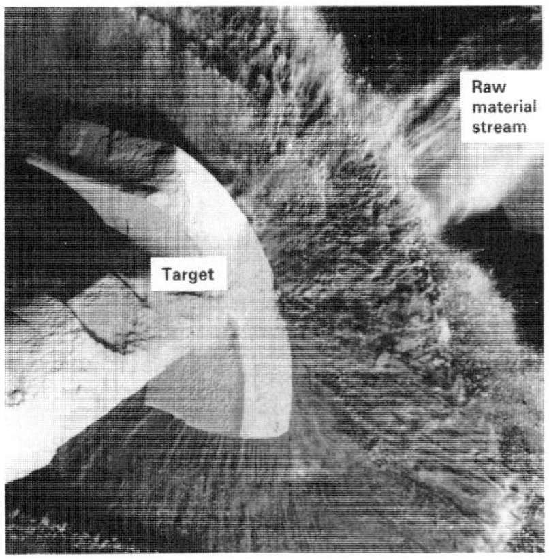

Fig. 25 Raw material stream impacting a target and shattering in Coldstream impact process

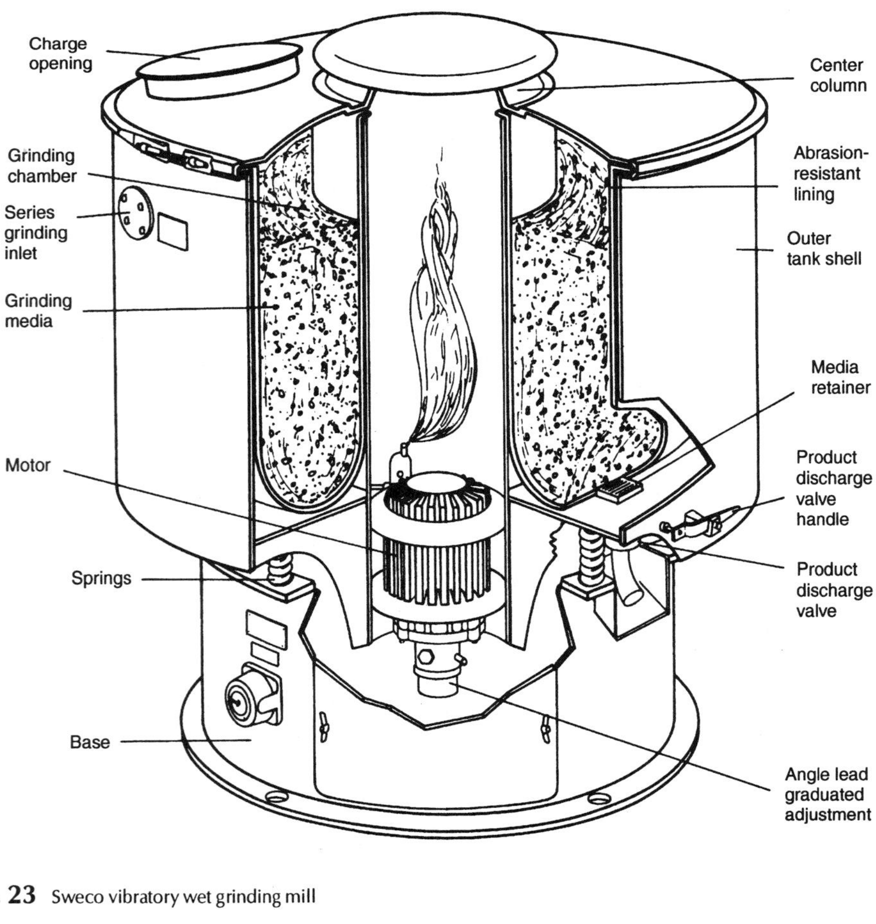

Fig. 23 Sweco vibratory wet grinding mill

The following applications are suited to high-energy milling:

- Flaking metal powders (Ref 12–14)
- Mechanical alloying (Ref 6, 7, 10, 15, 16)
- Dispersion strengthening (Ref 5, 8, 16, 17)
- Composite powders (Ref 18)
- Particle shape control
- Control of powder flowability
- Apparent density control
- Tungsten carbide and tungsten carbide/cobalt alloys
- P/M alloy blending (Ref 18)
- Alloying of immiscible alloying elements (Ref 10)
- Particle size control
- Ultrafine powders (Ref 5, 19)
- Activated sintering (Ref 7)

With the explosion of interest in the last 15 years to use mechanical milling/alloying for production of metastable materials such as amorphous alloys, extended solid solutions, quasicrystalline materials, metastable crystalline phases, and nanostructured materials, many laboratory-scale mills have been used for research. These mills typically use small quantities of powder, and in some cases, they are more energetic than the larger industrial mills. An often used mill used in research, especially in the United States, is Model 8000 shaker mill of Spex (Metuchen, N.J.). This high energy ball mill agitates the charge of powder and balls in three mutually perpendicular directions at approximately 1200 rpm. Ball velocities in such shaker mills are high, ~5 m/s. A given reaction, such as amorphization, can typically take place an order-of-magnitude faster in the Spex mill than in the usual vibratory or attritor mill.

Planetary ball mills also have been used often for research studies, particularly by European investigators. The vial containing the grinding media rotates about two separate parallel axes. A planetary arm of radius, R_p, is centered at a point about which it rotates at angular velocity, ω_p. The vial with radius R_v rotates (usually in the direction opposite to the planet arm) with angular velocity, ω_v. Typically, the commercially available Fritsch Pulverisette 5 (Germany) has been used. This mill also has been modified by some researchers for better control over the energy variation and milling temperature. There are several examples of specially designed mills for experimental research. These include the vibratory mill designed by Bakker and coworkers (Ref 20), which employs a single large steel ball (6 cm diameter) in a tungsten carbide lined steel vial which is attached to a vibrating frame. Calka and coworkers (Ref 21) designed an experimental ball mill, which can alter the milling action from predominantly impact to mainly shearing by changing the placement of strong permanent magnets, which are external to the vial. The recently developed Zoz horizontal attritor mill is produced for both laboratory and pilot plant powder quantities (Ref 22).

Ball Milling as a Nonequilibrium Processing Tool. The application of high energy ball milling that has stimulated the most research interest in recent years is its use as a nonequilibrium processing tool. It has been realized that mechanical alloying/milling can be used to synthesize metastable structures, in analogy to other nonequilibrium processing methods such as rapid solidification and physical vapor deposition. However, the precursor phase in the case of mechanical alloying/milling is typically a crystalline solid or solids rather than liquid or vapor. The thermodynamics and kinetic factors that govern metastable phase formation, therefore, can be very different. The variety of metastable structures that can be produced by ball milling are described in more detail in the article "Mechanical Alloying" in this Volume.

Modeling of the Milling Process. The qualitative descriptions of the physics/mechanics of the ball milling process discussed earlier in this article have been complemented by a number of attempts to quantify, that is, model, the process during the last ten years. McDermott (Ref 23) and Davis et al. (Ref 24) examined the mechanics of the Spex shaker mill. The model consists of a simplified version of the movements of the vial combined with the inferred ball movements based on classical mechanics and analytic geometry. The vial motion was recorded on videotape after slowing its apparent velocity by use of a high precision stroboscope. After digitizing these data, that is, ball displacements and velocities, a computer program performed calculations of a number of impacts for various ball and kinetic energy values. Martin and Gaffet (Ref 25) and Magini and Isonna (Ref 26) modeled the mechanics of ball milling in a planetary mill. A schematic of a planetary mill is given in Fig. 27. The Ω is the angular velocity of the ball mill plate, and ω is the angular velocity of the vials. The energy transferred per unit area scales with Ω^2, and the frequency of the occurrence of the impacts scales with ω. It has been shown that plots of Ω versus ω provide maps of regions where certain metastable structures can be formed as illustrated in Fig. 29. Courtney (Ref 27) performed a number of modeling studies of ball milling. These results, as well as the results of others, are summarized by Courtney (Ref 27). The modeling is divided into two broad categories, local and global. Local modeling considers the mechanical action that transpires when powder particles are trapped between the colliding milling balls. Global modeling considers the characteristics of specific milling devices and how these affect the local impact events in the ball-powder-ball collisions. Courtney summarizes the modeling efforts with the warning that

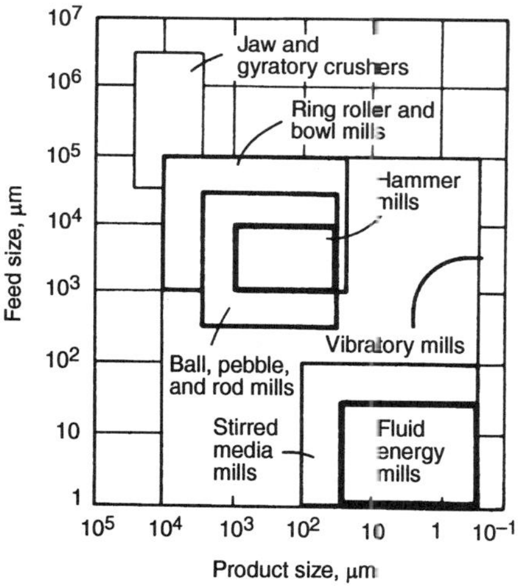

Fig. 26 Typical size capabilities of common classes of size reduction equipment

Fig. 27 Particles (+8 mesh) of copper powder hammer milled from oxide-reduced sinter cake. Courtesy of SCM Metal Products

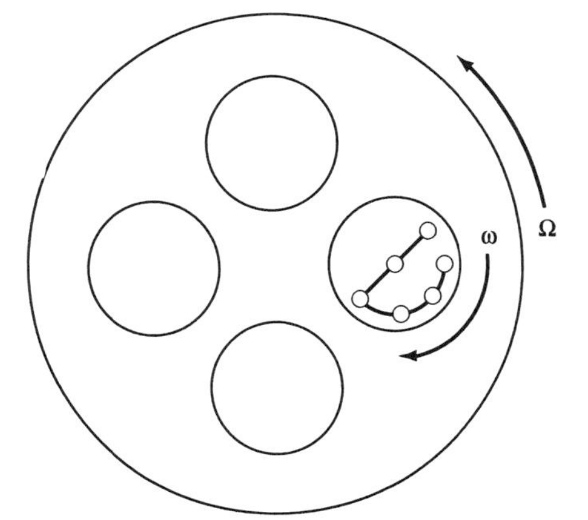

Fig. 28 Schematic drawing of a planetary ball mill. The Ω and ω are the angular velocities of the ball mill plate and of the vials, respectively.

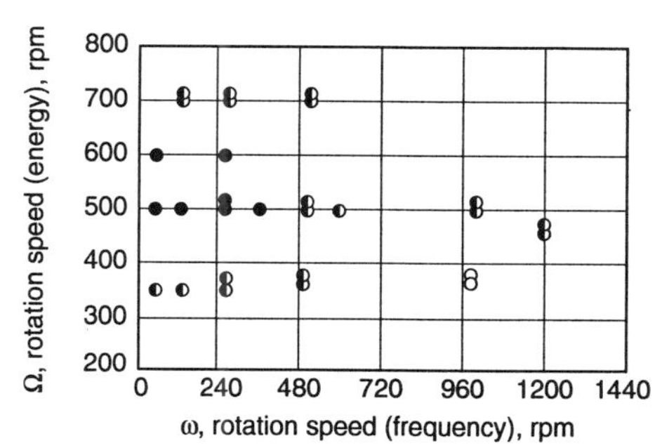

Fig. 29 Ball-milling parameter phase diagram for $Ni_{10}Zr_7$ (large circles) and $Ni_{11}Zr_9$ (small circles). Filled circle: homogeneous amorphous end product structure. Semifilled circle: mixture of amorphous and crystalline structures. Open circle: crystalline structure. Source: Ref 25

modeling of such a complex process can never be exact. However, at least partial success is achieved if the models identify the important material and process variables that determine the alloying kinetics, powder size, morphology, and so on. Indeed, such success has been achieved with some of the models.

Hammer and Rod Mills

Hammer and rod mills are used in the P/M industry to mill large quantities of sinter cakes, which are intermediate products in the production of metal powders by reduction of their oxides. If such powders are to be used in conventional P/M parts processing, it is essential that they possess good green strength and compressibility characteristics. For these reasons, the sinter cakes must be ground into powder, typically −80 mesh, while introducing minimal cold work and densification. Hammer and rod mills with built-in screens or other separation devices are best suited to achieve these properties. Figure 27 shows the blocky particle shape of a copper powder produced by hammer milling of a porous sinter cake.

ACKNOWLEDGMENT

The article was revised and updated from the following: W.E. Kuhn, I.L. Friedman, W. Summers, and A. Szegvari, Milling of Brittle and Ductile Materials, *Powder Metallurgy,* Vol 7, *ASM Handbook,* 1985.

REFERENCES

1. M. Senna and H. Kuno, Polymorphic Transformation of PbO by Isothermal Wet Ball-Milling, *J. Am. Ceram. Soc.,* Vol 54 (No. 5), May 1971
2. W.E. Kuhn and H. Lucky, Characterization of Ball-Milled Type 316-L Stainless Steel Powder, *Fine Particles,* W. Kuhn and J. Ehretsmann, Ed., The Electrochemical Society, Inc., 1974, p 95–121
3. A.A. Griffith, *Trans. Roy. Soc. Lond.,* Vol A221, 1921, p 163
4. J.J. Gilvarry and B.M. Borgstrom, Fracture and Comminution of Brittle Solids, *AIME Trans.,* Vol 220, 1961, p 380
5. C.C. Harris, On the Limit of Communition, *Trans. Soc. Mining. Eng.,* March 1967, p 17–30
6. W.E. Kuhn, High Energy Milling of Electrolytic Iron Powder, *Modern Developments in Powder Metallurgy,* Vol 12, Metal Powder Industries Federation, Princeton, NJ, 1980, p 195–218
7. A.N. Patel and W.E. Kuhn, Influence of High Energy Vibra-Milling on the Annealed and Sintered Structures and Properties of a P/M Austenitic Stainless Steel, *Modern Developments in Powder Metallurgy,* Vol 13, Metal Powder Industries Federation, Princeton, NJ, 1980, p 27–50
8. A. Arias, "Chemical Reactions of Metal Powders with Organic and Inorganic Liquids During Ball Milling," NASA TN D-8015, Sept 1975
9. A. Arias, "Oxide Dispersion Strengthened Nickel Produced by Nonreactive Milling," NASA TMX-3331, Jan 1976
10. J.S. Benjamin and T.E. Volin, The Mechanism of Mechanical Alloying, *Met. Trans.,* Vol 5, Aug 1974, p 1929
11. E. Rose and R.M.E. Sullivan, *A Treatise on the Internal Mechanics of Ball, Tube, and Rod Mills,* Chemical Publishing Co., New York, 1958
12. U.S. Patent 4,172,720, Oct 1979
13. British Patent 1,224,736, Jan 1969
14. U.S. Patent 4,115,107, Sept 1978
15. P.S. Gilman and W.D. Nix, The Structure and Properties of Aluminum Alloys Produced by Mechanical Alloying: Powder Processing and Resultant Powder Structures, *Met. Trans.,* Vol 12A, May 1981, p 813
16. B.N. Babich, et al., Production of a Dispersion-Strengthened Nickel-Tungsten-Aluminum Alloy by Mechanical Alloying, *Poroshkovaya Metall.,* Vol 235 (No. 7), July 1982, p 44–47
17. J.S. Benjamin and M.V. Bonford, Dispersion Strengthened Aluminum Made by Mechanical Alloying, *Met. Trans.,* Vol 8A, Aug 1977, p 1301
18. O. Webster, Effect of Lithium on the Mechanical Properties and Microstructure of SiC Whisker Reinforced Aluminum Alloys, *Met. Trans.,* Vol 13A, Aug 1982, p 1516
19. H.G. Domazer, New Milling Process for Fine Grinding of Powder Alloys in a Gas-Tight Attrition Mill, *Powder Metall. Int.,* Vol 10 (No. 3), 1978, p 150
20. A.W. Weeber, A.J.H. Wester, W.J. Haag, and H. Bakker, *Physica B,* Vol 145, 1987, p 349
21. A. Calka and A.P. Radlinski, *Mater. Sci. Eng. A,* Vol 134, 1991, p 1350
22. H. Zoz, *Mater. Sci. Forum,* Vol 419, 1995, p 179–181
23. B.T. McDermott, M.S. thesis, North Carolina State University, 1988
24. R.M. Davis, B.T. McDermott, and C.C. Koch, *Metall. Trans. A,* Vol 19, 1988, p 2867
25. G. Martin and E. Gaffet, *J. Phys. France (Colloques),* Vol 51, 1990, p C4-71
26. M. Magini and A. Isonna, *Mater. Trans., JIM,* Vol 36, 1995, p 123
27. T.H. Courtney, *Reviews in Particulate Materials 2,* A. Bose, R.M. German, and A. Lawley, Ed., Metal Powder Industries Federation, 1994, p 63

SELECTED REFERENCES

- C.C. Koch, Mechanical Milling and Alloying, *Processing of Metals and Alloys,* Vol 15, *Materials Science and Technology—A Comprehensive Treatment,* R.W. Cahn, Ed., VCH, 1991, p 193
- P.S. Gilman and J.S. Benjamin, *Ann. Rev. Mater. Sci.,* Vol 13, 1983, p 279
- T.H. Courtney, Modeling of Mechanical Milling and Mechanical Alloying, *Reviews in Particulate Materials 2,* A. Bose, R.M. German, and A. Lawley, Ed., Metal Powder Industries Federation, 1994, p 63
- C.C. Koch, Materials Synthesis by Mechanical Alloying, *Ann. Rev. Mater. Sci.,* Vol 19, 1984, p 121
- J.S. Benjamin, Mechanical Alloying, *Scientific American,* Vol 234, 1976, p 40

Chemical and Electrolytic Methods of Powder Production

Reviewed and revised by Pierre W. Taubenblat, PROMET Associates

CHEMICAL AND PHYSIOCHEMICAL METHODS of metal powder production allow great variations in powder properties. The wide variety of processing variables and production parameters currently available permit close control of particle size and shape. Powders made by reduction of oxides, precipitation from solution or from a gas, thermal decomposition, chemical embrittlement, hydride decomposition, and thermit reactions belong in this classification. The most widely used processes within this category include oxide reduction, precipitation from solution, and thermal decomposition.

Oxide Reduction

The production of iron, copper, tungsten, and molybdenum powders from their respective oxides are well-established commercial processes. Detailed process descriptions for these oxide-reduced powders can be found in the articles "Production of Iron Powders," "Production of Copper Powders," and "Production of Tungsten and Carbide Powders" in this Volume. On a smaller scale, oxide reduction is also used for production of cobalt and nickel powders.

Oxide-reduced powder grades of iron and copper compete with powder grades made by other processes. Oxide-reduced powders characteristically exhibit the presence of pores within each powder particle and thus are called sponge powders. This sponginess is controlled by the amount and size of the pores and accounts for the good compactibility (high green strength) and sinterability of such powders.

Processing conditions for oxide reduction, which are discussed in detail in Ref 1, are based on the generally known equilibria for the reduction reactions using hydrogen, carbon monoxide, and carbon as the reducing media, as shown in Fig. 1. The ratios of carbon monoxide to carbon dioxide, hydrogen to water, and the partial pressure of oxygen permit determination of the minimum ratios necessary to maintain reducing conditions at a given temperature and at a total gas pressure of 1 atm.

In practice, however, reduction temperatures are usually much higher than indicated by thermodynamic data. Final powder properties such as particle size, particle porosity, and hydrogen loss that critically determine performance properties (apparent density, powder flow, and compacting and sintering properties) primarily depend on purity and size of the starting material and the kinetics of the reduction process. The kinetics of the reduction process depend on composition and flow rate of the reducing gas, reduction temperature, temperature profile in the furnace, and bed depth of the oxide if reduction is performed in a stationary system.

Improved reduction rates are possible with nonstationary reduction in rotary kilns or fluidized bed reactors because of improved access of the reducing gas to the metal oxide particles. For many reducible oxides, however, it is difficult to control sintering and "freezing" of the fluidized bed. The former H-iron process (see the article "Production of Iron Powders" in this Volume), a fluidized bed reduction of iron oxide, used a high hydrogen pressure of approximately 3.5 MPa (500 psi) and a low reduction temperature of approximately 540 °C (1000 °F) to avoid freezing or caking of the bed. The low reduction temperature, however, caused the porous iron powder to be pyrophoric.

Process Variables. The various combinations of processing parameters used by powder manufacturers to produce numerous grades of powder are proprietary. The most important process variable is the reduction temperature. Typically, low reduction temperatures result in powders possessing fine pores, large specific surface areas, and high green strength. High reduction temperatures (>0.6 T_m) produce large intraparticle pores and small specific surface area powders that exhibit high compressibility. Extremely low reduction temperatures (<0.3 T_m) can readily produce pyrophoric powder. High temperatures can cause excessive sintering and agglomeration, which lead to difficulties with the breakup of the sinter cake.

With tungsten and molybdenum, oxide reduction is used partly for economic reasons, because the melting points of these metals are very high. Reduction processes, which use hydrogen as the reducing medium, are similar for both tungsten and molybdenum oxides. For molybdenum trioxide, however, reduction is performed in two stages to control particle size. Due to the high vapor presence of molybdenum trioxide, the first step is carried out at 600 to 700 °C (1110 to 1290 °F). The second step, the reduction of molybdenum dioxide to molybdenum monoxide, is carried out at around 900 to 1100 °C (1650 to 2000 °F).

In contrast to atomized powders where oxides often are enriched on the surface of a particle, oxide-reduced powders, at least when freshly reduced or stabilized against tarnishing, contain most of their residual oxides within the particles.

Precipitation from Solution

Production of metal powders by hydrometallurgical processing is based on leaching an ore or ore concentrate, followed by precipitating the metal from the leach solution. Although basic precipitation reactions have been known for more than 100 years, commercial use of this process did not flourish until the 1950s, as a result of increasing interest and work on lower-grade ores. Metal precipitation from solution can be accomplished directly by electrolysis, cementation, or chemical reduction. Indirect precipitation can be achieved by first precipitating a compound of the metal (hydroxide, such as carbonate, or oxalate, for example), followed by heating, decomposition, and reduction.

The most widely used commercial processes based on hydrometallurgy are copper cementation and the separation and precipitation of copper, nickel, and cobalt from salt solutions by reduction with hydrogen (Sheritt Gordon process). In the 1960s and 1970s, several pilot plants using hydrometallurgical processing (solvent extraction and reduction with sulfur dioxide and hydrogen) were operating temporarily in the United States for the production of high-purity copper powder.

In its simplest form, copper cementation recovers copper from acidic dump leach solutions as an impure powder precipitate. Due to the presence of significant amounts of iron and silicates, low apparent density, and high green strength, such copper powders find use in P/M

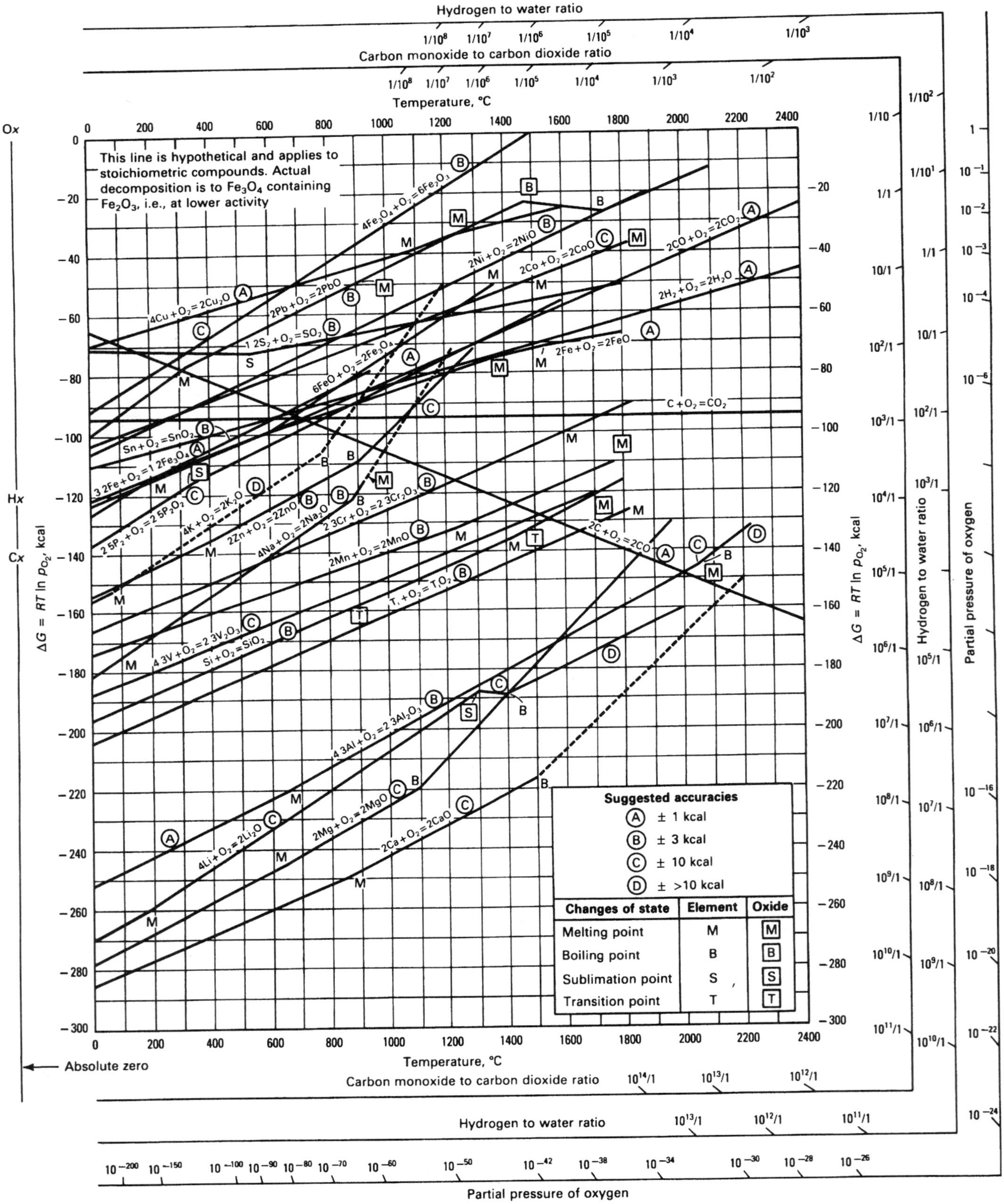

Fig. 1 Standard free energy of formation of metal oxides. To convert kcal to kJ, multiply kcal by 4.184. Source: Ref 1

friction composite components. They are not used in conventional structural parts.

Nickel powders are now produced in large quantities directly from their ores by precipitation. The powder has a purity of at least 99.8%, with the major impurities being cobalt, iron, and sulfur, due to their presence in the original nickel ore. Powder can be produced in a variety of size distributions and is quite uniform. Similar techniques have been developed for cobalt.

Processing Conditions. For divalent ions, these processes consist of precipitation from an aqueous source using hydrogen. The basic concept is that a metallic ion such as nickel, copper, or cobalt in the solution reacts with gas (hydrogen) by the following reaction:

$$M^{2+} + H_2 \rightarrow M^o + 2H^+$$

or if the solution is ammoniacal:

$$M^{2+} + 2NH_3 + H_2 \rightarrow M^o + 2NH_4^+$$

Generally, processing begins with leaching of ores and includes purification and separation stages prior to reduction. Reduction potential may be estimated by comparing the electrochemical potential of the metal ion and that of the hydrogen ions as a function of the partial pressure of hydrogen and pH of the solution. For reduction to occur, the hydrogen potential must be greater than the metal potential. Metal concentration has a minimal effect on electrochemical potential. These relationships are discussed in detail in Ref 2 and are shown in Fig. 2, which indicates that copper can be reduced in very acidic solutions. Higher pH values are required for nickel and cobalt. To obtain practical reaction rates, the actual process is carried out at elevated temperatures and pressures. For complete reduction, pH is increased by adding ammonia.

Separation of metals is based on differences in the stability of complexed metal ions (ammines) subjected to hydrolyzing and oxidizing reactions. Potential-pH diagrams with the thermodynamics of oxygen reduction and hydrogen oxidation provide the necessary information to predict separation potential within a given system. Figure 3, a potential-pH diagram for the copper-ammonia-water system, is discussed in detail in Ref 3.

The use of additives and control of nucleation, particle growth, and particle agglomeration allow the production of powders with a wide range of particle sizes, particle density, and particle shape with specific surface areas from less than 1 m^2/g to about 8 m^2/g.

Coprecipitation or successive precipitation of different metals from solution allows the production of alloyed and composite powders. Spray drying extends this capability to innumerable combinations. For more information, see the article "Spray Drying and Granulation" in this Volume.

Thermal Decomposition

Of the group of thermally decomposed powders, those produced by thermal decomposition of carbonyls are the most important. Both iron and nickel are produced by decomposition of the respective carbonyls.

Processing Conditions. Carbonyls are obtained by passing carbon monoxide over spongy metal at specific temperatures and pressures. Iron pentacarbonyl, $Fe(CO)_5$, is a liquid at room temperature, boiling at 103 °C (217 °F). Nickel tetracarbonyl, $Ni(CO)_4$, boils at 43 °C (109 °F). When the pressure is reduced to 1 atm and the temperature is raised correspondingly, both of these carbonyls decompose to re-form the metal and carbon monoxide. The latter is recycled to form more carbonyl and to continue the process. These reactions are expressed as follows:

$$Fe + 5CO \leftrightarrow Fe(CO)_5$$

$$Ni + 4CO \leftrightarrow Ni(CO)_4$$

Powder is produced by boiling the carbonyls in heated vessels at atmospheric pressure under conditions that allow the vapors to decompose within the heated space and not on the sides of the container. The powder is collected and sieved and may be milled, followed by an anneal in hydrogen. The chemical purity of the powders can be very high (over 99.5%), with the principal impurities being carbon, nitrogen, and oxygen. Particle size can be controlled very closely. Iron carbonyl powder is usually spherical in shape and very fine (less than 10 μm), while the nickel powder is usually quite irregular in shape, porous, and fine. Detailed descriptions of the iron and nickel carbonyl processes are given in the articles "Production of Nickel-Base Powders" and "Production of Iron Powders" in this Volume.

Other Chemical Methods

Precipitation from Salt Solutions. A number of metal powders are made by the precipitation of their soluble salts as insoluble hydroxide, carbonate, or oxalate. On heating, these compounds decompose into the respective metals or metal oxides and gaseous products. Examples include the production of uranium dioxide, platinum, selenium, tellurium, silver nickel, and silver cadmium oxide compounds through coprecipitation. Thorium dioxide dispersion-strengthened nickel, known as TD nickel, used to be made by chemical precipitation of a nickel salt solution containing colloidal thorium dioxide.

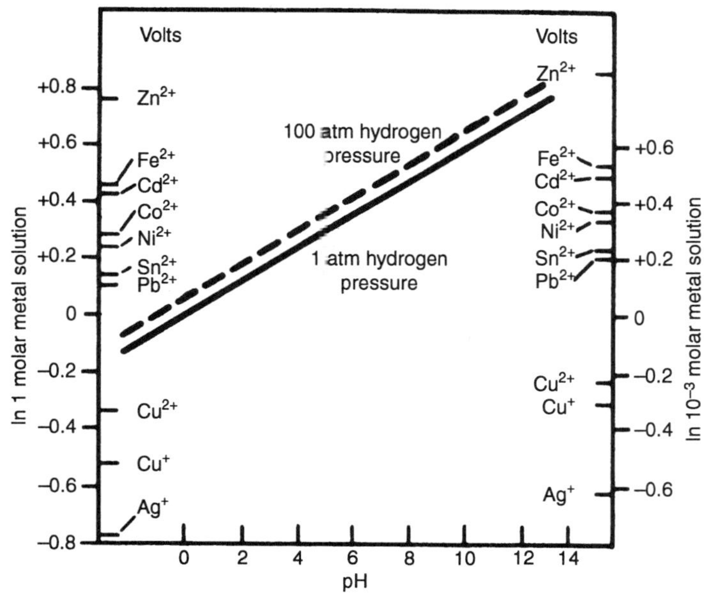

Fig. 2 Potential of 1 and 10^{-3} molar metal solutions and hydrogen potential at varying pH at 25 °C (75 °F). Source: Ref 2

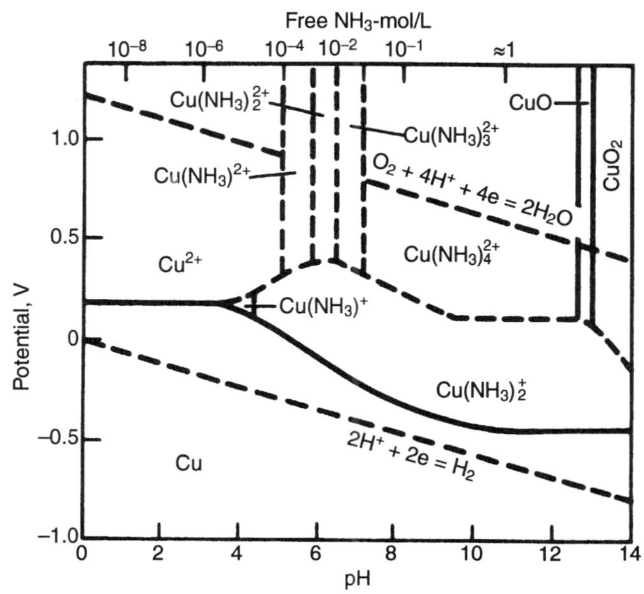

Fig. 3 Potential-pH diagram for the Cu-NH_3-H_2O system. Total NH_3 equals 1 mol/L. Source: Ref 3

Precipitation from a gas is used for certain reactive metals such as titanium and zirconium. In these cases, the gaseous metal chlorides are reduced with liquid sodium or magnesium (Kroll process). The reaction product is leached with dilute hydrogen chloride to remove sodium chloride and/or magnesium chloride. The result is a metallic sponge material that is subsequently pulverized.

Hydride Decomposition. For titanium and zirconium, the so-called hydride-dehydride process also may be used to make powder. The metals are heated in hydrogen in the form of chips or turnings. The brittle hydrides are milled into powder and then dehydrided under vacuum at elevated temperature. The resulting sinter cake is milled into powder.

Chemical embrittlement processes of past commercial importance include the carburization of iron, sulfurizing of 2-81 Permalloy, and the sensitizing (with carbon) of stainless steel (Ref 4). These reactions produce grain-boundary precipitates that make it possible to mill or dissolve the material into powder.

Thermit Reactions. Some difficult-to-reduce metals may be produced either by thermit reactions (the reduction of an oxide with a metallic powder that has a very high negative free energy of oxide formation) or by reduction with metal hydrides. Thermit reactions include the reduction of chromium trioxide with magnesium and the reduction of uranium dioxide with calcium or magnesium. Calcium hydride, by means of its nascent hydrogen, which forms at elevated temperature by decomposition, can reduce chromium trioxide and other difficult-to-reduce oxides.

An alumino-thermit method also is used to produce tungsten monocarbide powders. Metallic iron is added to the reaction charge in quantities to control the calculated reaction temperature within the range of 2410 to 2480 °C (4370–4500 °F). It has been found that this process can not be controlled to produce macrocrystalline tungsten carbide powders, which are very low in titanium, tantalum, and niobium content (Ref 5).

Electrodeposition

Electrodeposition of metals from aqueous solutions produces a variety of metal powders. This article discusses the methods used to produce powders electrolytically, and describes the types of metal powders made and their physical and chemical characteristics. Detailed information on the processing variables that permit close control of the chemical and physical properties of electrolytic powders can be found in the articles "Production of Iron Powders," "Production of Copper Powders," and "Production of Precious Metal Powders: Silver, Gold, Palladium, and Platinum" in this Volume.

There are two practical methods of obtaining powder by electrodeposition: (1) direct deposition of a loosely adhering powdery or spongy deposit that can easily be disintegrated mechanically into fine particles, and (2) deposition of a dense, smooth, brittle layer of refined metal that can be ground into powder. For example, copper and silver produce powdery or spongy cathode deposits. Conversely, iron and manganese produce coherent cathode deposits; because these deposits will be crushed and ground into powder, it is highly desirable that they be brittle. Brittleness of the cathode deposit can be achieved by proper control of the electrolytic cell conditions.

Currently, only iron, copper, and silver powders are produced commercially to any extent by electrodeposition.

Direct Deposition

Direct deposition of powder or sponge on the cathode is achieved by controlling the composition (concentration of metal and pH), temperature, and rate of circulation of the electrolyte; current density; size and type of anode and cathode and their distance from each other; quantity and type of addition agent; and removal of deposits at the cathode (brush-down interval). These and other variables are discussed elsewhere in this Volume.

The shape of electrolytically produced powder particles depends on the metal deposited and the operating conditions. Electrolytic copper deposited from a sulfate/sulfuric acid electrode forms dendritic (fernlike) particles, as shown in Fig. 4. Silver electrolytic powder deposited from a silver nitrate solution generally is coarse and regularly shaped (Fig. 5). Electrolytic iron, copper, and silver are the most widely used powders made by this technique, but tin, chromium, beryllium, antimony, cadmium, lead, and zinc powders also have been produced by electrodeposition.

Electrolytic copper powder has been produced commercially since the early 1900s. Modern practice, which utilizes a copper sulfate/sulfuric acid electrolyte, closely follows the original process developed in France.

Electrolytic copper production follows the same electrochemical principles that apply to copper refining. However, the conditions of electrodeposition are modified so that a powdery or spongy material is deposited on lead antimonial cathodes instead of the smooth, adherent deposit desired in electrorefining. Formation of powder deposits is facilitated by low copper ion concentration and high acid content in the electrolyte. High cathode current density and use of an electrolytically refined copper anode (instead of the impure cast copper anode used in refining operations) also contribute to powder formation.

Powder characteristics can be altered to a considerable extent by control of the operating conditions. Further size control can be achieved by use of addition agents such as colloids, which promote the formation of finer powders. The dendritic structure of copper particles is shown in Fig. 4. Details of the process variables and properties and applications of electrolytic copper powders are discussed in the article "Production of Copper Powders" in this Volume.

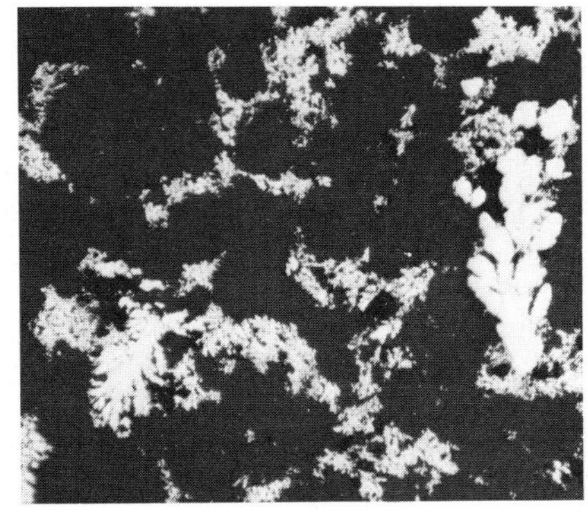

Fig. 4 Electrolytic copper powder, showing dendritic structure. 85×

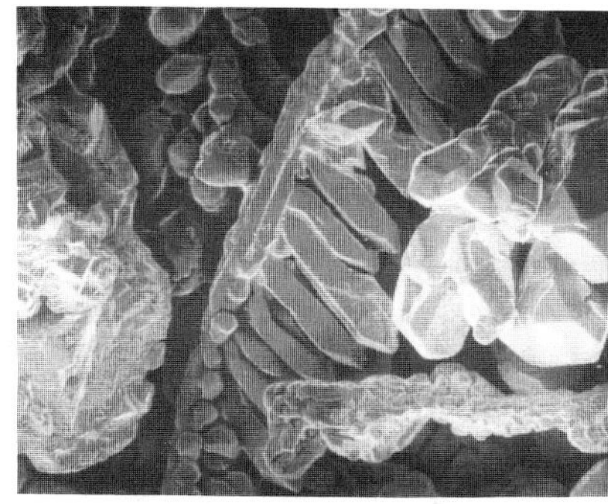

Fig. 5 Electrolytic silver powder. 60×

Electrolytic silver powder can be made by using a silver nitrate electrolyte and pure silver anodes. The powder particles deposited on the cathode are generally regularly shaped (see Fig. 5) and coarse, with particle sizes ranging from 40 μm to several millimeters, depending on the operating conditions and current densities. Details of the electrolytic and chemical processes are discussed in the article "Production of Precious Metal Powders" in this Volume.

Zinc Powder. The most recent metal powder to be produced electrolytically is zinc powder, which is made by using an electrolyte consisting of sodium hydroxide and sodium zincate. Zinc content is maintained at about 55 g/L to yield a deposit of suitable particle size. Current density is about 1076 A/m^2 (100 A/ft^2). Under these conditions, the product is a loosely adherent sponge that can be easily scraped or brushed off the cathode and processed into powder.

Brittle Cathode Process

Metals characterized by high electrolytic polarization yield coherent deposits that can be made brittle under controlled operating conditions. Metals of this type include iron, manganese, nickel, and cobalt, with only iron and manganese being produced in commercial quantities.

Iron. Although electrolytic iron powders were made experimentally in the late 19th century, the first commercial production occurred in 1904. The electrolyte, composed of ferrous and ammonium sulfates, was maintained in a slightly alka-

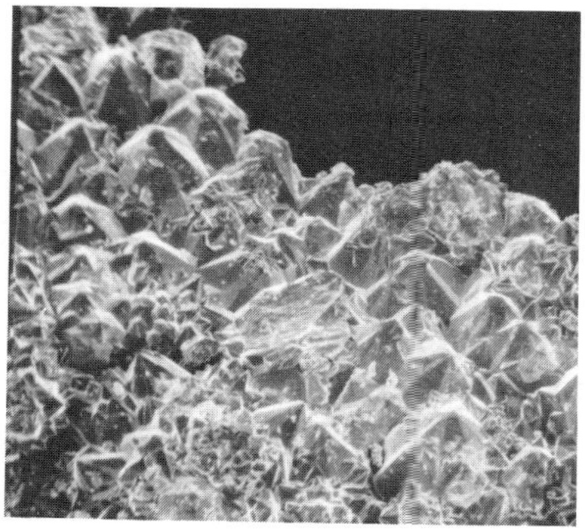

Fig. 6 Scanning electron micrograph of electrodeposited iron sheet, indicating dendritic grain structure. 25×

line condition. As a result, oxides were deposited along with the iron on the cathode, producing a brittle deposit that could be readily powdered. Steel anodes were used. The product was not particularly pure, but the application (electrical loading coils) did not require high purity. A later modification permitted the production of pure iron powder.

In modern practice, a chloride or sulfate electrolyte is used with soluble low-carbon steel or ingot iron anodes and stainless steel cathodes. Cathodes are stripped of deposited iron. The iron deposits are pulverized subsequently in ball or hammer mills, and the powder then is annealed in hydrogen to make it softer. The dendritic structure of the powder (see Fig. 6) is altered by crushing and ball milling to form irregularly shaped, flaky particles with a fairly rough surface. Details of this process can be found in the article "Production of Iron Powders" in this Volume.

Manganese powder is produced by electrodeposition from a purified manganese sulfate electrolyte using insoluble lead anodes and stainless steel cathodes. The brittle cathode deposit is readily removed. Because the powder is pyrophoric, it is ground into powder in ball or hammer mills under a protective atmosphere.

REFERENCES

1. F.D. Richardson and J.H.E. Jeffes, The Thermodynamics of Substances of Interest in Iron and Steelmaking from 0 °C to 2400 °C, *J. Iron Steel Inst.,* Vol 160, 1948, p 261
2. F.A. Schaufelberger, *J. Met.,* Vol 8 (No. 5), 1956, p 695–704
3. J. Halpern, *J. Met.,* Vol 9 (No. 2), 1957, p 280–289
4. F.V. Lenel, *Powder Metallurgy Principles and Applications,* Metal Powder Industries Federation, 1980, p 33, 44
5. C.J. Terry and J.D. Frank, "Macrocrystalline Tungsten Monocarbide Powder and Process for Producing," U.S. Patent 4834963, 16 Dec 1986

Ultrafine and Nanophase Powders

ULTRAFINE POWDERS are typically defined by particle size distributions in the submicron range (Fig. 1) with equivalent diameters of approximately ten to a few hundred nanometers. Ultrafine powders are not new, but only until recently have engineers and applied scientists realized their potential in producing nanocrystalline (or nanophase) materials, in which the distance between neighboring defects, grain boundaries, or interphase boundaries is in the order of interatomic spacing.

This article briefly reviews dedicated processes that have been developed to generate economical yields of fine and ultrafine metal powders. Fine metal powders have been commercially available for many years, and metal powders as fine as 3 µm are produced economically by atomization methods. Particle size distributions as low as 1 µm are possible with atomization, although precipitation and carbonyl vapor deposition processes are normally more economical methods near the ultrafine regime.

Ultrafine (submicron) powders can be produced either by physical or chemical methods. Neither mechanical grinding nor conventional (gas or liquid) atomization can produce powders extending into the submicron size ranges (the electrohydrodynamic atomization technique is a special case, of interest only for research production of milligram quantities). Mechanical alloying is of interest because it can produce powders, which in many cases have a nanocrystalline microstructure. However, it cannot produce ultrafine powders directly.

Fine Metal Powders

Ralph H. Hershberger, UltraFine Powder Technology Inc.

Fine metals have been commercially available for years by various methods. One of the earliest methods was the precipitation of elemental precious metal powders. Fine refractory metal powders are also produced by milling.

Elemental iron or nickel powders are also produced to fine particle distributions with the carbonyl process. On a tonnage basis, carbonyl iron still accounts for a large percentage of fine powders used for metal injection molding. Carbonyl iron is produced to very tight distributions (Table 1) with typical particle sizes ranging from 3 to 8 µm. Tables 2 and 3 summarize carbonyl iron grades from BASF, which is the world's largest supplier of carbonyl iron. More information on carbonyl iron is also contained in the article "Production of Iron Powders" in this Volume.

Some alloy parts are produced using elemental carbonyl iron and/or nickel as a base, but this method has obvious chemical limitations. In this regard, prealloyed powders from atomization offer more flexibility in alloy composition. So-called "residual products" met the initial demand for fine, prealloyed powders. Prior to 1980, the demands for fine alloy powders were satisfied by screening the fine residuals from a standard as-atomized distribution. Although this method avoided investment in technology development and dedicated equipment, it had several weaknesses. First, the supply of fine powders was constantly at risk. The available fine powders were a function of the production and sales of a coarse powder stream. This dependency jeopardized the supply continuity. Secondly, the powders frequently were not discrete particles and had satellites or were dog-boned shaped. Finally, if the products were sold on an as-yielded basis, the prices were prohibitively high and few markets could support production. Until manufacturers became convinced that a sustainable fine powder market existed, end users had to contend with these limitations. Ultimately, the limitations of residual fines prevented further growth.

Today a full range of fine alloy powders are produced by atomization methods (Table 4). Ametek, Specialty Metals Division in the United States, uses a specialized high-pressure, water-atomized system. Pacific Metals (PAMCO) of Japan uses high-pressure water and a recently announced combination water- and gas-atomization system (Ref 1). Mitsubishi employs a water-atomized process to create a feedstock, which is subsequently cold worked to create nodular shape powders. The primary commercial gas-atomization operations include UltraFine Powder Technology in the United States, Anval in Sweden, and Osprey Metals in Wales.

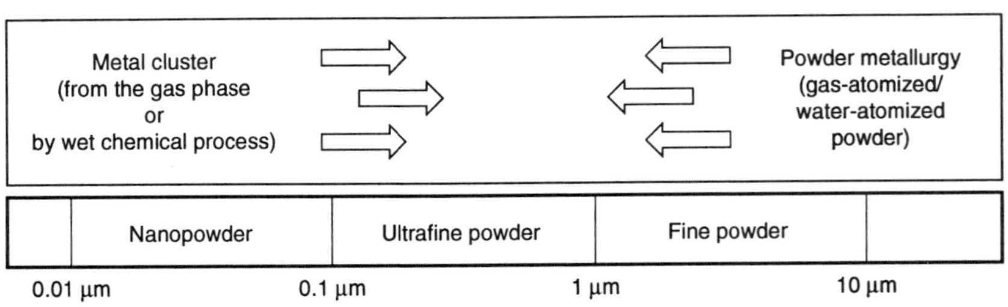

Fig. 1 Size range and powder production methods

Table 1 Typical composition and particle size distribution of carbonyl iron powder

BASF grade	Mean size, µm	Iron, wt%	Carbon(a), wt%	Nitrogen(a), wt%	Oxygen(a), wt%	Other	Particle size distribution 10% less than	50% less than	90% less than
Reduced standard powders									
CL	7–8	>99.5	0.05	0.01	0.2	...	5.0 µm	10.0 µm	25.0 µm
CM	5–6	>99.5	0.05	0.01	0.2	...	4.0 µm	9.0 µm	22.0 µm
CS	4–5	>99.5	0.05	0.01	0.2	...	4.0 µm	8.0 µm	18.0 µm
CN	5–6	>99.5	0.04	0.01	0.2	...	3.5 µm	8.0 µm	18.0 µm
CC	4–6	>99.5	0.05	0.01	0.3	0.1 wt% SiO_2 coating	3.0 µm	6.0 µm	11.0 µm
SU	1–1.5	99.4	0.1	0.01	0.5	...	0.70 µm	1.67 µm	3.43 µm
SM	1–1.6	99.4	0.1	0.01	0.5	...	0.81 µm	1.91 µm	3.66 µm
Iron-phosphide	3–5	88	0.3–0.7	0.1	0.4–0.6	9–10 wt% P	1.5 µm	4.0 µm	10.0 µm

(a) Maximum wt% unless a range is specified

Table 2 Carbonyl iron powders for powder metallurgy and injection molding

BASF grade	Mean size, µm	Iron, wt%	Carbon (max), wt%	Nitrogen (max), wt%	Oxygen (max), wt%	Characteristic properties
Reduced standard powders						
CL	7–8	>99.5	≤0.05	≤0.01	≤0.2	Soft, spherical powder
CM	5–6	>99.5	≤0.05	≤0.01	≤0.2	Soft, spherical powder
CS	4–5	>99.5	≤0.05	≤0.01	≤0.2	Soft, spherical powder
CN	5–6	>99.5	≤0.04	≤0.01	≤0.2	Soft, spherical powder
Unreduced standard powders for injection molding						
OM	4–5	>97.8	≤0.9	≤0.9	≤0.4	Unreduced, hard powder; agglomerates broken up by grinding
ON	4–5	>97.5	≤1.2	≤0.1	≤1.2	Unreduced, hard powder with low N content and higher O content
OS	4–5	>97.3	≤0.9	≤0.9	0.7% SiO_2	Unreduced, hard powder; SiO_2 coated
OX	3–4	>96.2	≤0.9	≤0.9	5% α-Fe_2O_3	More stable form in debinding with improved sinter properties
OX	3–4	>94.7	≤0.9	≤0.9	10% α-Fe_2O_3	With lower or higher α-Fe_2O_3 content on request

Table 3 Carbonyl iron powders for electronic and microwave applications

BASF grade(a)	Mean size, µm	Iron min, wt%	Carbon max, wt%	Nitrogen max, wt%	Oxygen max, wt%	Bulk density, g/cm³	Characteristic properties
For electronic parts							
EN	4–5	>97.4	<1.0	<1.0	<0.4	2.6	General-purpose product
EW	4–5	>97.3	<1.0	<1.0	<0.8	2.3	High ohmic resistance and high Q factor at 5–10 MHz
EQ	3–5	>97.2	<1.0	<1.0	<0.4	2.7	High Q factor at 4–15 MHz
ES	3–4	>97.4	<1.1	<1.1	<0.4	2.2	High Q factor at 30–100 MHz
SP	4–6	>99.5	<0.05	<0.01	<0.2	2.2	Ductile attainable toroidal µ = 75(b)
SQ	4–6	>99.5	<0.06	<0.01	<0.4	1.9	Specially insulated; high Q factor at 100–200 KHz; attainable toroidal µ: 68(b)
SL	7–8	>99.5	<0.05	<0.01	<0.2	2.6	attainable toroidal µ: 65(b)
SD	5–6	>99.5	<0.06	<0.01	<0.2	2.4	attainable toroidal µ: 60(b)
SB	5–6	>99.5	<0.1	<0.01	<0.2	2.6	attainable toroidal µ: 55(b)
HM	2.6–3.5	>96.0	<2.0	<2.0	<0.4	2.9	High permeability; high Q factor; wide particle size distribution
HL	2.8–3.5	>97.0	<1.0	<1.0	<0.5	2.7	General-purpose H product
HS	2.4–3.3	>97.0	<1.0	<1.0	<0.5	2.2	High Q factor, particularly at 10 MHz
HF	2.0–2.5	>97.0	<1.0	<1.0	<0.7	2.0	High Q factor, 30 to 100 MHz
HQ	1.6–1.9	>97.0	<1.0	<1.0	<1.0	1.9	High Q factor, 10 to 100 MHz
For microwave absorption							
EA	3–4	>97.3	<1.2	<1.2	<0.4	...	Narrow particle size distribution
EB	3–4	>97.3	<1.0	<1.0	<0.4	...	Somewhat wider particle size distribution

(a) E brands: mechanically hard, mean particle size 3–8 µm; H brands: mechanically hard, mean particle size 1.6–3.5 µm; S brands: mechanically soft, mean particle size 3–8 µm; in each case, the second letter indicates the properties, for example, L, large; S, small; Q, high quality; M, medium; P, good permeability; W, high resistance. (b) With 0.3% of binder and 1.5 GPa pressure

Table 4 Major commercial fine atomized powder producers, systems, and products

Country	Producer	Powder production system	Grades produced
Japan	Pacific Metals (PAMCO)	High-pressure water	Stainless, tool steels, nickel and cobalt alloys
	Mitsubishi	High-pressure water, shape modified	Stainless, tool steels, nickel and cobalt alloys
	Kobe	Water atomization	Stainless and tool steels
Sweden	Anval	Nitrogen gas atomization Plasma-heated tundish	Stainless steels
United Kingdom	Osprey	Proprietary gas atomization	Stainless, tool steels, low alloy, and cobalt alloys
United States	UltraFine Powder Technology	Proprietary gas atomization	Stainless, tool steel, low alloy steels, copper, nickel, and cobalt alloys
	Ametek	High-pressure water atomization	Stainless, tool steels, and nickel alloys

Atomization Methods

All atomization processes convert the kinetic energy of the atomizing media, into surface free energy to form powder particles. The greater the energy introduced to the system, the greater the energy available for particle formation. Because smaller particles have a higher surface-to-volume ratio than larger particles, the mean size for the as-atomized particle distribution is reduced. For water atomization, this relationship was described by Grandzol (Ref 2) to be:

$$d_m = A/V_w$$

wher d_m is mean particle diameter of the as-atomized distribution, V_w is water droplet velocity, and A is a constant.

Atomization is a notoriously inefficient energy conversion process. Conventional water atomizing systems are only ~5% efficient, and gas atomizing is even less efficient, converting <1% of the kinetic energy into surface free energy (Ref 3). Nonetheless, typical powder distributions for water-atomized products contain a significant amount of particles below 45 µm (see the article "Atomization" in this Volume).

In residual fines, particles below 45 µm must be separated. This step in fine powder production is commonly done by either screening or classification. As powders become finer, screening is less viable because the particles tend to blind the screen openings.

For particles in this size range, classification is the preferred method for size separation. Unfortunately, classification is inefficient, with recovery rates between 65 and 40% of the particle sizes present, depending on the cut point.

For fine powder to be produced, more energy must be introduced into the system per material atomized than is required for conventional atomization. The common measurement is a water/metal or gas/metal ratio. Three factors must be optimized: atomizing media mass, flow rate of the quenching media, and flow rate of the liquid metal. The energy available from kinetic energy is ($\frac{1}{2}mV^2$), which is linearly proportional to the mass but to the square of the media velocity. This suggests higher flow rates of atomizing media, that is, water or gas, which requires higher pressure pumping systems to deliver the quenching media to the molten stream. Because pumping systems have natural limitations, to transfer sufficient energy, the water/metal or gas/metal ratio must include a reduction in the metal flow rate.

Typical flow rates for a conventional water-atomization system are 150 lb/min (Ref 4). For conventional gas atomization the rate is closer to 40 lb/min, and in some fine powder systems, the rate drops to 10 lb/min. This places an additional strain on the refractory system because the melt must be held at temperature for a longer period of time to complete the atomization run.

Another key factor in the process is the atomizing jet design. Manufacturers consider this element of the system to be a key factor in the viability of their process, and designs are seldom released publicly except as part of licensing packages. Lawley (Ref 4) has published an excellent review of atomization jet designs. Most fine-powder atomization systems employ some type of a close-coupled system to provide real-time feedback. During gas atomization, other factors physically limit the production. Because gas is significantly less dense than water, much higher flow rates must be used to deliver enough gas momentum and kinetic energy through the atomizing nozzle to the metal stream. This places extreme demands on pumps, hoses, couplings, and other plumbing. These technical difficulties are accompanied by economic issues

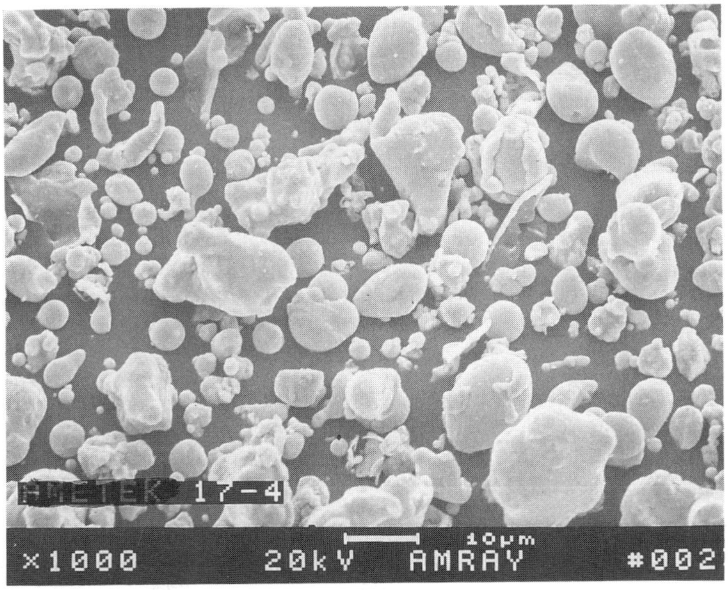

Fig. 2 Scanning electron microscopy of Ametek 17-4 PH stainless high-pressure, water-atomized powder particles. 1000×

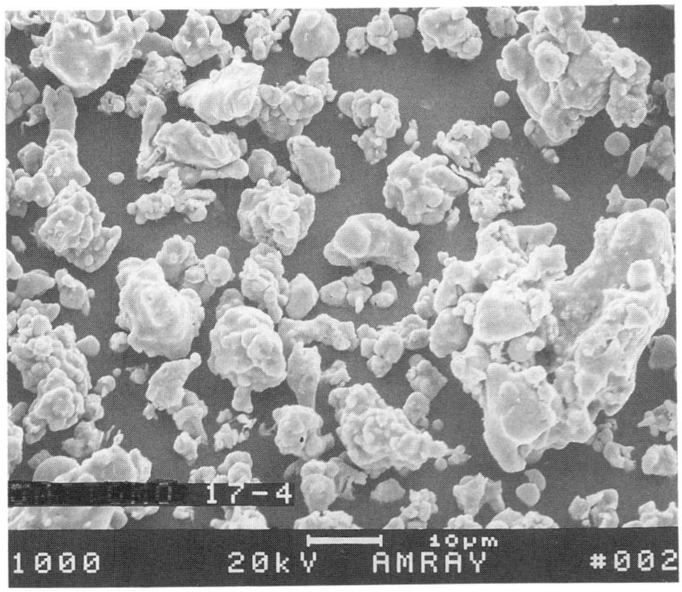

Fig. 3 Scanning electron microscopy of PAMCO 17-4 PH high-pressure, water-atomized powder particle. 1000×

Table 5 Comparative properties of gas- and water-atomized powders

Properties	Gas atomized standard	Water atomized
Tap density, g/cm^3	4.83	4.20
Pycnometer	7.685	7.649
Carbon, wt%	0.03	0.085
Oxygen, wt%	0.055	0.45
Distribution		
d_{100}	52.00 µm	44.00 µm
d_{90}	31.00 µm	22.00 µm
d_{50}	15.40 µm	11.50 µm
d_{10}	6.40 µm	4.60 µm

Source: Ref 5

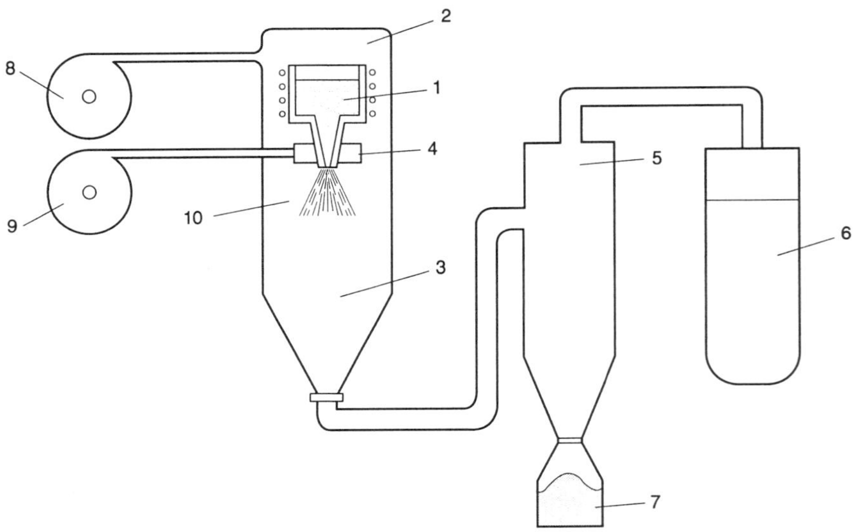

Fig. 4 Schematic of the ultrafine powder gas-atomization system. Raw materials are melted in a vacuum induction furnace (1, 2), and a lower pressurized chamber (3) receives powder from the atomization die (4). Powder is removed from the lower chamber into cyclones (5) and collection bins (6). Pressurizing and atomization gases are recycled (7, 8) and liquefied. At the beginning of a new atomization cycle, the gases are evaporated and reintroduced into the system (10).

associated with the scale up of these systems, which render many designs uneconomical for commercial use.

High-Pressure Water Atomization. Work performed by the National Research Institute in Japan has resulted in commercial water-atomizing systems, which have been adopted by Ametek and Pacific Metals. Typical properties for gas- and water-atomized products, as presented by the various processes, are shown in Table 5 (Ref 5). Figures 2 and 3 show the morphology of the same powders.

Gas Atomization. The first company to dedicate itself to fine powders was UltraFine Powder Technology. Their process (Fig. 4) employs two zone vacuum chambers, an atomizing chamber, and a gas recycle system. This allows for the upper and lower chambers to have independent pressures, which are key to the process.

Raw materials are melted in a conventional induction melt furnace inside an evacuated upper chamber. When the melt approaches the proper temperature, stopper rods are pulled from the bottom of the crucible, and the metal flows through the orifices and atomizing jets. Figure 5 shows a typical as-atomized distribution for copper and 17-4 PH stainless steel.

This system has the advantages of relatively large heats (up to 1270 kg); adaptability for argon, nitrogen, or helium atomizing gases; and an argon recovery/recycle system. It presently accommodates ferrous-, nickel-, cobalt-, and copper-base alloys. Figure 6 shows a typical 316L stainless particle.

The Osprey Process. Although Osprey process atomizing devices were originally developed for the production of spray-deposited preforms, they also can be used to produce high-purity powders in a wide controlled range of particle size distributions (Ref 6 and 7). Powders produced using Osprey atomizers include silver and gold brazing alloy powders, tin-lead soldering powders, aluminum-magnesium powders for explosives, copper-tin-lead bearing alloys, and nickel- and cobalt-base hard-facing alloys. In addition high-melting-point alloys such as nickel superalloys (Fig. 7), Alnico permanent magnet alloys, and stainless steel powders have been produced using Osprey atomizing equipment.

Alloys that have been processed by the Osprey forming process include stainless steels, high-speed steels, and nickel-base superalloys, al-

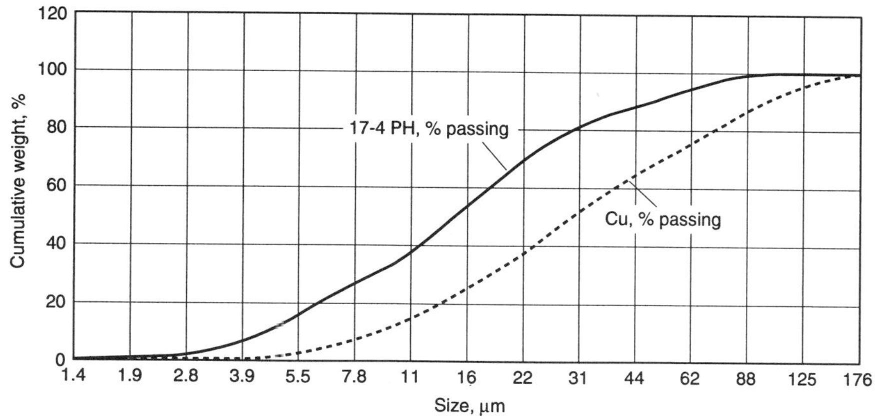

Fig. 5 As-atomized distribution for copper and 17-4 PH stainless

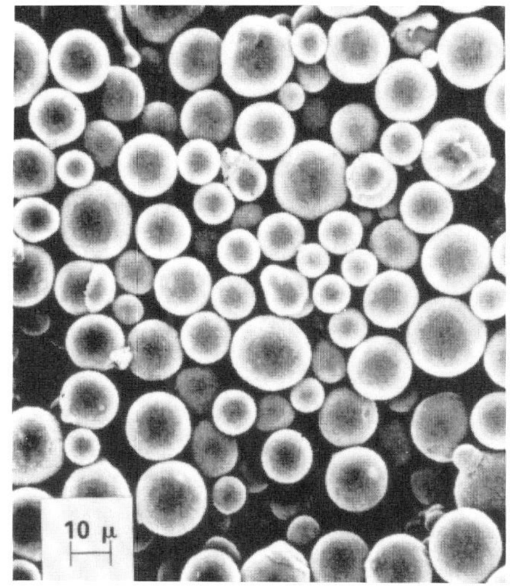

Fig. 7 Osprey atomized superalloy powder (MAR M002)

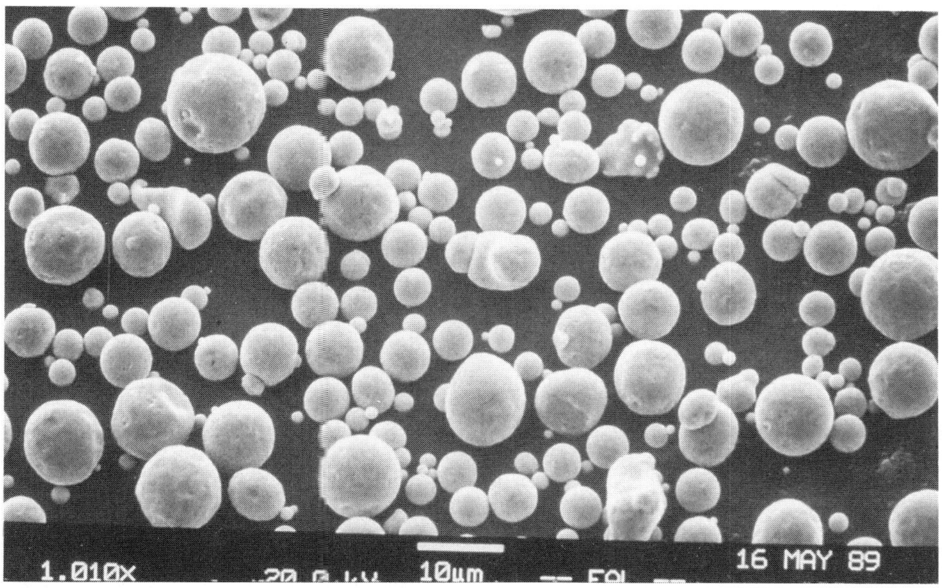

Fig. 6 Scanning electron microscopy of 316L stainless steel gas-atomized powder

Table 6 As-atomized diameters for various alloys produced by the Osprey system

Alloy	Mean particle diameter as-atomized, μm
D2	12.90
M2	15.8
Ti5	10.43
T42	12.57
FeSiB	14.21
Ni21Cr9Mo4Nb	10.49
FeCrNi	10.45
304L	13.9
316L	13.0
317L	13.85
17-4PH	13.59
310S	14.95
434	16.14

Source: Ref 8

though many materials appear to be compatible with the process. Alloy development has centered on high-alloy ferrous metals, Stellite alloys, superalloys, and composite materials. Because an inert atmosphere is maintained during spraying, oxygen levels similar to conventional ingot metallurgy products are attained, typically in the range of 20 to 40 ppm for superalloys. The high preform density ensures that no interconnected porosity is present in the preform, preventing internal oxidation during transfer of the material to subsequent consolidation and forming operations. (See the article "Spray Forming" for more information on the production of preforms by Osprey.)

In terms of atomization, Osprey systems have equipment common to other gas-atomization systems, for example, pressure chamber, furnace and tundish, atomization nozzles, and collection chambers. The key difference between the Osprey system and other high-pressure gas systems is that it involves atomization by ultrasonic agitation. The process produces an as-atomized distribution (Table 6) that falls between the distributions of a carbonyl process and high-pressure water atomization. Osprey is commercial for heats up to 250 kg across a wide range of alloys.

Anval Gas Atomization System. The Anval system is a modified close-coupled, dual-nozzle atomizing system. Melting is performed in a five metric ton, air melt, tandem furnace system. Atomization occurs after the metal has been poured from a plasma-heated tundish (PHT), which removes many of the oxides normally found in air melted products. The system produces large batches of metallurgically clean, spherical powders typical of a nitrogen gas-atomized system.

Powder Comparisons. Atomization methods are capable of producing powder distributions as fine as 1 μm, but they are effectively limited to ~3 μm for general production. Precipitation and carbonyl vapor deposition processes are normally more economical methods to create fine particles in the range.

The primary differences between water and gas-atomized fine powders are:

- Shape
- Purity level, that is, residual gas levels
- Production costs

Gases have a lower heat-carrying capacity than water so the metal stream does not quench as rapidly in a gas-atomized system. The residence time of a liquid droplet in the atomizing chamber is longer in a gas system than a water atomizer. This additional time permits the driving forces to minimize the surface free energy by forming a sphere. Although this effect is somewhat alloy dependent, fine gas-atomized powders are generally more spherical than water-atomized products. Because most of the commercial fine gas-atomized powder systems are vacuum or atmosphere controlled, the typical values for oxygen and nitrogen are lower than water-atomized powders. For applications that do not reduce the soluble gas levels in process, this is a critical difference. The third difference is production cost. On a momentum transfer basis, water is significantly less expensive than gas. Water systems do not require the special plumbing associated with gases and their subsequent recovery. In addition, the elimination of a vacuum cycle allows water-atomized systems to be configured to have a shorter cycle time than gas-atomized systems.

Typical vacuum-gas systems include a 1 to 2 h vacuum cycle, a 1 to 2 h melting time, and a 1 to 2 h atomization time. Cooling time for the atomized powder is additional. A typical 500 kg water-atomized heat can turn around in melt and atomize in 2 h. With dual tundishes, it can be run on a semi-continuous basis.

All of these reductions in equipment mean that the capital costs and resultant depreciation are lower for water-based systems compared with gas atomizers. An excellent treatment of these variables has been performed by Dunkley (Ref 9). The remaining factors of labor, yield, and overhead tend to be producer specific.

Applications

The physical and economic differences between the two types of powders result in clearly defined applications for each. Fine powders are used in both consolidated and unconsolidated forms. Markets include metal injection molding (MIM), electronic pastes and inks, thermal spray, additives, radar absorption coatings, and filtration.

Metal Injection Molding. The growth forecasts for MIM encouraged development of fine powder production processes. The first MIM parts were made from available powders generated by carbonyl (iron or nickel) processes. On a tonnage basis, carbonyl iron still accounts for a large percentage of the powders used by MIM. Nonetheless, both water- and gas-atomized powders are used extensively in MIM. Typically water-atomized powders are used to improve green strength and lower costs, while gas-atomized powders are preferred for moldability, lower oxide content, and tolerance control.

Electronics. Precious and nonprecious metals are used in the electronics industry for inks and pastes, conductive path, and capacitor terminations. Due to the extreme space limitations that these applications impose, finer distributions, (<10 μm) are required (Ref 10). Because atomization yields for all but a few elements in this range are extremely low, precipitation is a more economical production method, especially for precious metal formulations. Because copper has a low viscosity and atomizes easily, it is an exception, and it is available as either a precipitated or atomized product in these size ranges. The powders are blended with a polymer and applied to a substrate via precise silk screening techniques. Continued miniaturization is driving these applications into the submicron power range.

Thermal Spray. Thermal spray equipment has evolved to produce extremely dense metallic

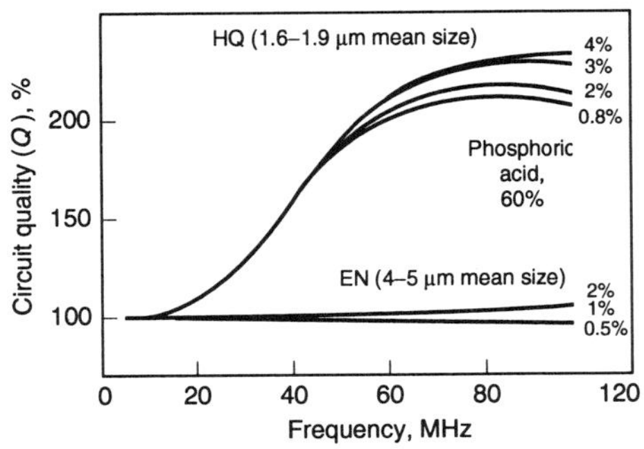

Fig. 8 Circuit quality with different levels of insulation on the primary carbonyl iron powders. See Table 3 for powder designations. Courtesy BASF

coatings. Gas-atomized powders have always been preferred over water-atomized powders for both fine and coarser distributions. The spray process favors spherical, free-flowing particles over irregular water-atomized particles.

Two types of equipment have been designed that require fine powders. Plasma equipment requires powders in the 5 to 45 μm range, and high velocity oxygen fuel (HVOF) equipment requires an even tighter range of 15 to 45 μm, with controlled intermediate ranges. Although HVOF has been enthusiastically embraced for carbide coatings (carbides had the most room for density improvement offered by HVOF), metals are a significant part of the HVOF supply mix. The market for fine powder from dedicated production processes is primarily an economic market. Metal alloys have always been available from fractions of coarse as-atomized distributions. The opportunity for dedicated fine powder systems is to supply these distributions in higher yields and lower overall costs.

Additives. Many industrial products are formulations of many particulate constituents. The most cost effective production method for these powders is normally grinding, but grinding has drawbacks. Grinding can lead to explosions, can introduce contaminants such as dirt and oxygen, and can generate a wider than desired particle size distribution. For these reasons, fine powders produced by the methods discussed in this chapter have found a niche in specialized applications where purity, size control, and distribution consistency are necessary.

Metal filled polymers require a consistent size distribution and shape to achieve desired properties. Conductive polymers need high-conductivity powders with low or controlled oxygen levels. In some cases, applications of fine powders are required to ensure an even loading in the polymer structure.

Grinding wheel formulations use fine-atomized powders because they disperse more uniformly than coarse distributions. The lower oxygen levels of gas-atomized products allows sintering at lower temperatures.

Radar Absorption. Although most of this work remains classified, fine carbonyl and prealloyed powders have been used in radar absorption coatings for stealth fighter planes. Soft magnetic compositions are used in fine sizes, which

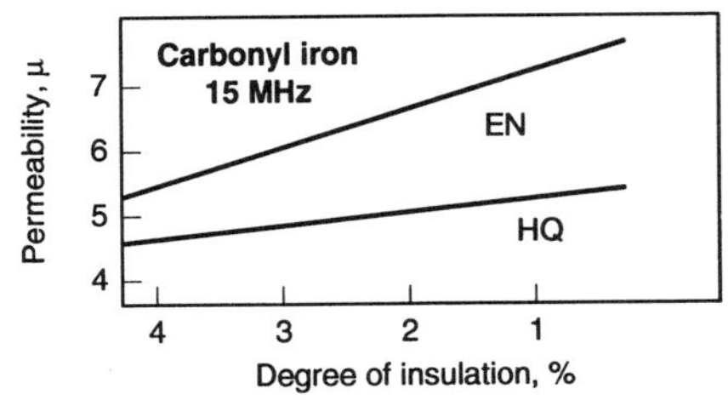

Fig. 9 Effect of powder insulation on permeability. Courtesy BASF

offer the best combination of absorption and weight reduction. Below the Curie point, carbonyl iron is cost effective. Above the Curie point, prealloyed materials are necessary to maintain performance. Table 3 lists two grades of microwave absorbing carbonyl powders.

Filtration. Metal filter media is used in very demanding and corrosive applications. New applications require removal of finer and finer particles. Clean room applications are the most demanding, and this places tremendous demands on filter-reduced pore size and control. Smaller particles can create finer pore sizes. Typical compositions are 316L and 304 stainless and Hastelloys. The preferred morphology is an irregular shape that provides green strength during the fabrication process, which favors water-atomized particles. However, in extremely corrosive environments, the purity of gas-atomized particles makes them a preferred material.

Magnetic Power Cores. Fine carbonyl iron powders with suitable insulator coatings improve the performance of electrical power cores. It is possible to achieve a substantial reduction in eddy current losses compared with cores produced from coarser powder. This reduction in eddy current losses is achieved when the insulation breaks conduction between the individual, primary powder particles.

Figure 8 shows circuit quality with two types of carbonyl iron powder as a function of the frequency and the degree of insulation. The insulation was carried out with a 60% phosphoric acid. The powders, insulated and bound with 5% Epikote, were compressed with a compaction pressure of 0.6 GPa to rod cores having a diameter of 9.2 mm and a length of 20 mm.

With an increasing degree of insulation, the quality at first increases, and then reaches a saturation value typical for the particular powder. The thickness of the insulating layer depends on the proportion of phosphoric acid. This depresses the content of the pure iron, and so the permeability falls (Fig. 9). This means that the gain in quality is detrimental to the permeability.

It is therefore necessary to find a compromise between sufficient additions of an insulating material for high quality without excessive reduction of permeability.

The aim of the insulation is to cover the powder particles with a layer that is as thin as possible, but completely surrounds the powder particle. The layer should also be thick enough to prevent damage during compaction under high pressure. Finally, the binding agent content also

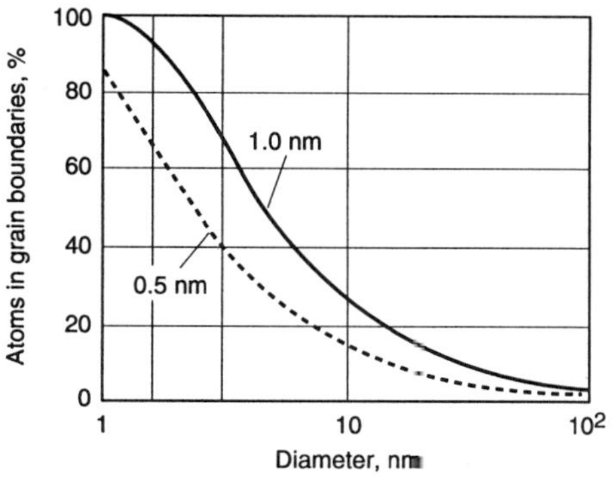

Fig. 10 Percentage of atoms in grain boundaries of a nanophase material as a function of grain diameter, assuming that the average grain boundary thickness ranges from 0.5 to 1.0 nm, that is, ~2 to 4 atomic planes wide. Source: Ref 11

has an effect on quality and permeability. A higher binding agent content improves quality within certain limits, but also reduces the permeability.

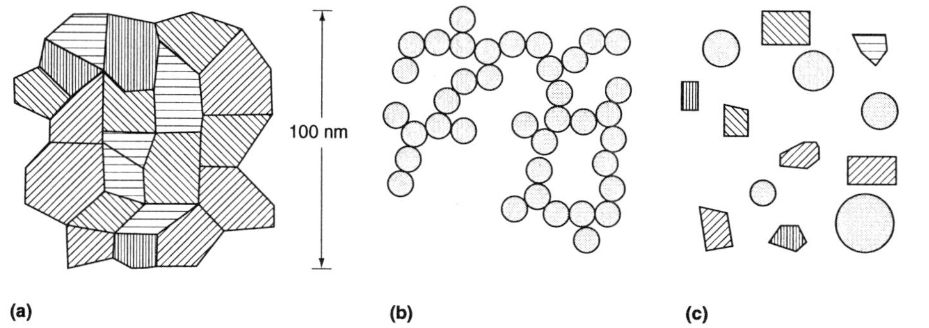

Fig. 11 Three different types of nanoscale materials: (a) nanocrystalline material, (b) agglomerated nanopowder, and (c) nanosuspension

Table 7 Comparison of the most frequently used gas phase methods for production of metal nanopowders

Method	Advantage	Disadvantage	Typical capacity
Flame reaction	Large output	Broad distribution of particle size, ionic impurities	>1 tonne/day
Plasma reaction	Large output	Broad distribution of particle size, ionic impurities	> 1 tonne/day
Chemical vapor reaction	Narrow distribution of particle size	Ionic impurities	200 kg/day
Inert gas condensation	No impurities	Small output	<10 kg/day

Metallic Nanopowders

Harald Eifert, Fraunhofer Resource Center—Delaware
Bernd Günther, Fraunhofer Institute IFAM

Today there are a number of established applications of ultrafine powders or nanocrystalline materials. At present, applications are essentially limited to magnetic recording media and nanocrystalline soft magnetic materials used in transformer cores (Ref 11, 12). However, the potential applications in advanced engineering and technology are numerous. Some of the applications already under development are:

- Electrically conductive inks or pastes
- Catalysts
- Sintering accelerators
- Microfilters
- Magnetic recording media
- Magnetic fluids
- High-strength/high-temperature construction materials
- Nanocomposites
- Drug carriers
- Radar absorbing coatings
- Wear-resistant coatings (tool materials)

Compared to ceramic processing, metal nanopowders are much more difficult to process and handle than powders in the micron size range. Nevertheless, different applications require metallic nanopowders and different production processes to serve this need. This Section provides an introduction in the different processes commercially used to produce metal nanopowders with specific emphasis on gas phase techniques for production of powders with a low-impurity level. Example applications describe the state-of-the-art methods in using metal nanopowders in industry.

The term nanopowder is usually used to describe powders with a particle diameter of <1 μm = 1000 nm (Fig. 1). Nanoscale powders have been used in the paint coating industry in tremendous volumes for decades—if made from carbon and metal oxides. Pigment soot, also named carbon black, is a carbon nanoscale powder with particle size 10 to 100 nm and is probably the most prominent example of this type of nanopowder. Nanoscaled silica powders are used as filler additives for a variety of organic suspensions in order to tailor their rheological behavior. The use of ultrafine carbide and nitride powders provides hardmetal tools with a remarkable increase in bending strength resulting in an extended time to failure.

The term nanopowder is also used to designate important changes in properties with the reduction of particle size. In this context the term is sometimes used for powders smaller than a certain threshold, where a dramatic change in properties occurs. Changes in properties can be caused, for example, by the increase in surface atoms of nanopowders. Figure 10 (Ref 13) shows the relative amount of surface atoms of powder particles as the function of the particle diameter.

Nanoscale materials are used for production of advanced material components in three different types, as shown in Fig. 11. A nanocrystal or nanocrystalline material is a dense material with grain sizes in the nanometer range. Nanopowder is often used as aggregated powder particles, either as a suspension or as a loosely packed layer. A nanosuspension, or nanodispersion, is usually the name for a suspension of single nanoparticles, either in a liquid or in a solid matrix (Ref 14).

Production Methods

The production techniques used for fine (micron scale) powders generally fail in the submicron (nanometer) range. Methods like gas or water atomization have a lower particle size limit of ~1 to 5 μm. Milling is used for particle size reduction of very brittle materials like tungsten carbide or ceramics, but fail for ductile materials like precious metals.

A variety of methods are available for the production of ultrafine powders with submicron particle size. Gas phase reactions, spray drying, or precipitation methods are used most frequently. Thousands of tons of ceramic nanopowders have been produced for decades basically via thermal decomposition from the gas phase (aerosol process). One of the basic principles is the inert gas condensation (IGC). With IGC, high-quality powders with low chemical impurities from precursor materials, and low amounts of oxides or nitrides from the production process can be produced. Other principles like flame reduction or plasma reduction use the decomposition and reduction of metal salts in a gas flame or plasma. The chemical vapor reaction (CVR) process uses the reaction of metal chlorides and hydrogen in a hot wall reactor.

Table 7 gives a comparison of the most commonly used four basic methods in gas phase production of metal nanopowders. Information from approximately twelve companies active in production of metal nanopowders for industrial applications is the source of the different production techniques in Table 7. Additional companies are active in producing powders by liquid-phase techniques or by ball milling (Table 8).

The method used most frequently for the production of dry metal nanopowders with very low impurity content is based on the condensation of a supersatured metal vapor in the presence of a nonreacting cooling gas such as helium. This method is described in more detail in the next section.

Inert Gas Condensation

The IGC method is performed by evaporating and condensing of the respective material in a vacuum cylinder filled with a low pressure of an inert gas (e.g., 10 mbar helium). Evaporation of the respective material can be completed by magnetron sputtering, electron beam, laser, or Joule heating of a crucible. The vapor nucleates homogeneously due to collisions with the cold inert gas atoms. The growth of particles within the convection gas flow occurs basically via coalescence. This type of growth process is intrinsic and leads to a Gaussian distribution of the resulting particle sizes on a logarithmic scale. The size of the primary particles that are agglomerated in a chain or network-like structure can be controlled via the type and pressure of the inert gas within approximately one to two orders of magnitude. Via thermophoretic effects, the particle fog condenses preferably on cooled substrates that are implemented into the container, for example, in the form of rotating stainless steel container filled with liquid nitrogen. The particles build up a flaky powder layer on the cooling wall and can be removed continuously by a scraper.

The fundamentals of the gas phase evaporation process are well understood and used for several process variations. Figure 12 shows the sketch of equipment used in a lab scale environment by Fraunhofer for production of precious metals nanopowders with a rate of ~1 to 2 kg/day.

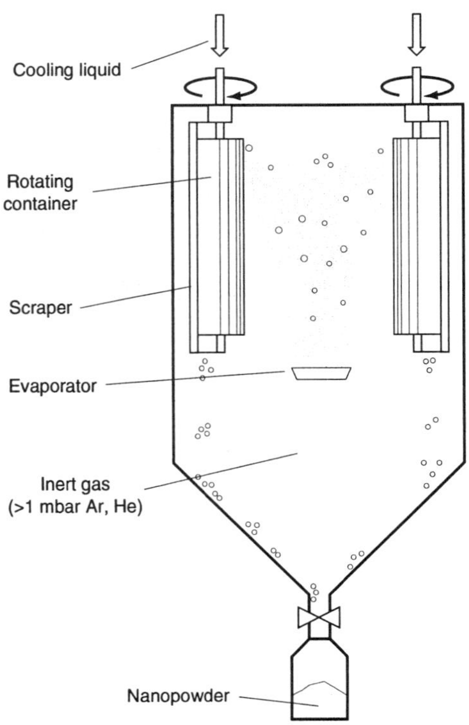

Fig. 12 Sketch of pilot scale equipment to produce metallic nanopowders with the inert gas condensation technique. Source: Ref 14

Table 8 Overview on production processes used in industry

Production process	Number of companies (different processes possible)	Main products
Gas phase processes		
Inert gas condensation	4	Precious metals, Ni, Fe, Pt, semiconductor powders
Chemical vapor condensation	2	Ta and other refractory metals, semiconductor powders
Precipitation	3	Precious metals
Plasma or flame enhanced chemical vapor condensation	3	Mainly ceramics, side products are precious metals and others
Spray condensation	3	Carbides
Chemical precipitation in the liquid phase	5–10	Precious metals
Milling processes	~6	Carbides

Process Variations in Inert Gas Condensation. Several variations of the process exist, for example, evaporation in a reactive gas instead of an inert gas and formation of the respective nitrides or oxides of an evaporated metal. Other variations of the process are using an enforced gas flow transporting the particles to a porous substrate, where the particles are separated from the carrier gas. Additionally to these variations, the powder produced in the gas phase can be deposited on different substrates depending on the planned application. Figure 13 shows the process variations possible for production of dry powder, dense pressed parts, dispersions, or filter applications.

Quality Affecting Factors in Production of Metallic Nanopowders. The main problems in production and processing are caused by the small particle size and the corresponding high specific surface area of nanopowders. As a main quality issue, the pickup of oxides or nitrides during processing or storage need to be pre-

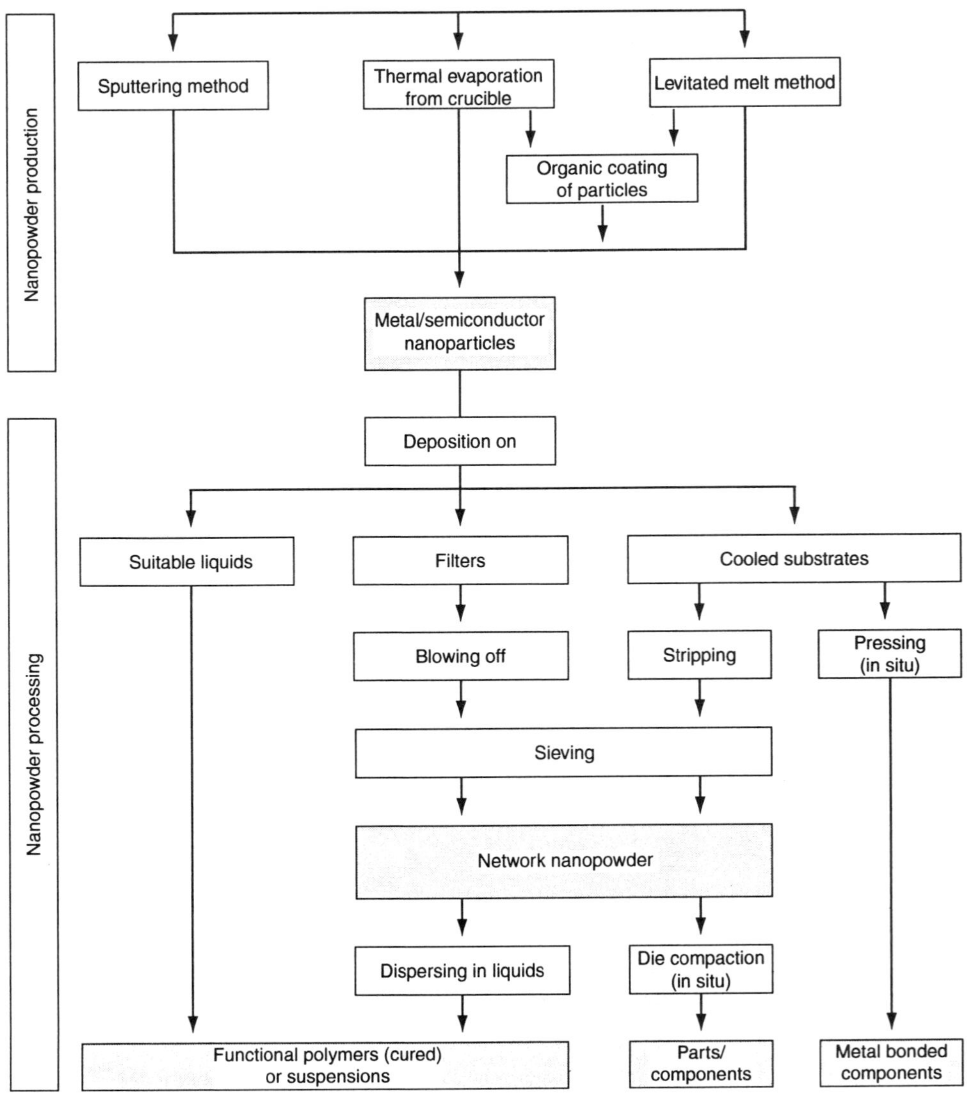

Fig. 13 Process variation of the inert gas condensation for production of different types of nanoscale materials

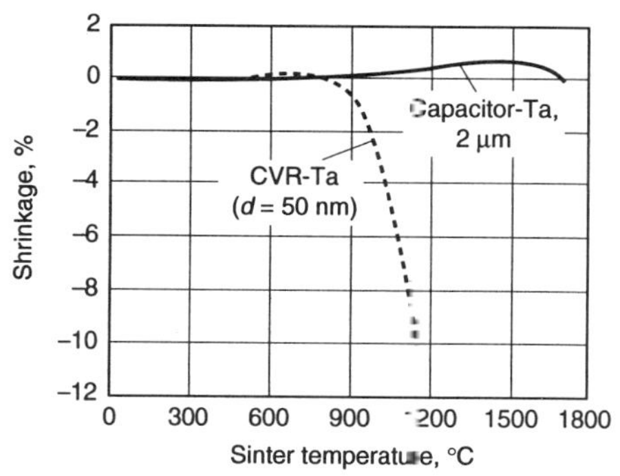

Fig. 14 Decrease in sinter temperature of a tantalum nanopowder with decreasing powder particle size. Source: Ref 16

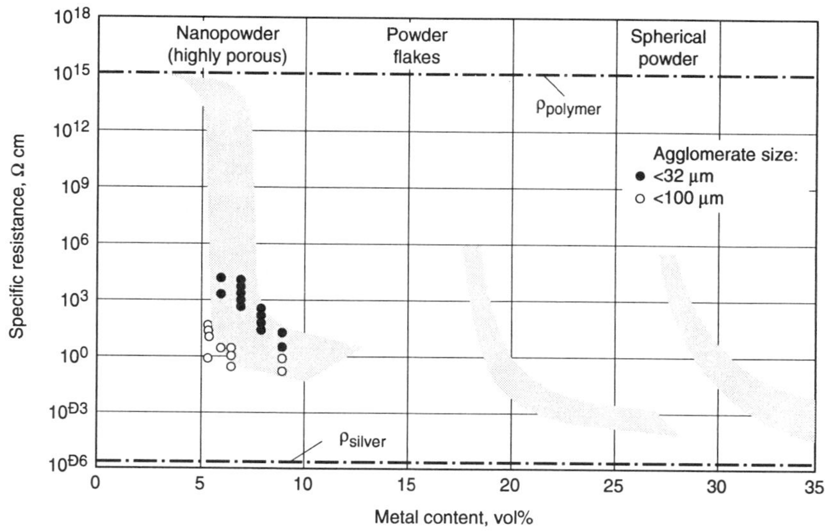

Fig. 15 Shift of the percolation threshold of electrical conductivity in filled polymers by usage of highly porous nanosized powders

vented. As nanopowders are often used or processed in suspension, the degree of agglomeration is also a quality issue due to the need of deagglomeration of the powders. Both factors, the pickup of gaseous impurities and the agglomeration, increase with decreasing particle size. For nanopowders the development of technologies that integrate production and processing steps is essential.

Applications

Because of the very fine grain sizes, nanocrystalline materials exhibit a variety of properties that are different and often considerably improved in comparison with those of conventional coarse-grained polycrystalline materials. These properties include increased strength/hardness, enhanced diffusivity, improved ductility/toughness, reduced density, reduced elastic modulus, higher electrical resistivity, increased specific heat, higher thermal expansion coefficient, lower thermal conductivity, and superior soft magnetic properties in comparison with conventional coarse-grained materials. All of these properties are being extensively investigated to explore possible applications, and the origin and consequences of some of these properties are discussed in Ref 15.

Dense parts and porous coatings from ceramic nanopowders have attracted interest for decades, and many low-cost processing routes are available. In particular, the use of oxide nanopowders in optics, electronic, and cosmetics, (ultraviolet, UV, protection) is established. Several potential applications of metallic nanopowders are investigated and reported. The following two examples are typical for the use of nanopowders for functional materials.

Applications Based on the High-Sinter Activity of Metallic Nanopowders. With regard to solid state sintering theory and experimental proof, it is well known that a decrease in particle size enhances sinter activity and leads to intense shrinkage of pressed powders at temperatures well below those attained at conventional powder size. This property is particularly useful in electronic packaging technology, where electrically conductive bonding materials play an essential role. The cold welding properties combined with ductility make metal alloy nanopowders suitable sinter aids for metal to metal bonding. The high sinter activity can be usefully applied in diffusion bonding of parts. The nanopowders act as an intermediate layer. It was shown that after the bonding process no welding interface is visible in metallographic cross section.

In particular, sintering of ultrafine refractory metals provides a means to lower the sintering temperature. Due to the fact that the performance of tantalum capacitors is closely related to the specific anode surface area, the use of nanosized tantalum powders at a reduced sintering temperature was investigated (Ref 16). Figure 14 shows the decrease in sintering temperature obtained by use of 50 nm powder for tantalum capacitors compared with standard quality 2 μm powder.

Nanopowder-Polymer Composites for Microelectronic Applications. Metallic filling powders in polymers play an important role for the realization of electrical conductive adhesives, radio frequency shielding polymers, or magnetic polymeric layers. In most applications, the use of high-aspect ratio fibers and flakes is advantageous.

A reduction of the relative amount of metal powder in an electrical conductive adhesive or polymer is possible by using porous nanoscale powders (Fig. 15). With this reduction of filler volume, adhesives with a better thermo-cycling behavior can be produced for applications in automotive microelectronics (Ref 17).

REFERENCES

1. T. Tingskog, "Fully Dense Powder Products," presented at the Gorham Advanced Materials, Inc. World Powder Metallurgy Markets—1997 Conference (Atlanta, GA), 1997
2. R.J. Grandzol and J.A. Tallmadge, Water Jet Atomization of Molten Steel, *AIChE J.*, Vol 19 (No. 6), 1973, p 1149
3. P.U. Gummeson, private communication, 1997
4. A. Lawley, *Atomization: The Production of Fine Metal Powders,* Metal Powder Industries Federation, 1992, p 19, 113–120
5. UltraFine Powder Technology, Internal Comparative Testing Results of Competitive Powders
6. R. Howells, "Cost Effective Production of Pre-Alloyed MIM Feed Stock," 1995 Int. Powder Injection Moulding Symp. (Pennsylvania), Pennsylvania State University, 1995
7. G.R. Dunstan et al., *Progress in Powder Metallurgy,* Metal Powder Industries Federation, 1981, p 23
8. R. Howells, "Cost Effective Production of Pre-Alloyed MIM Feed Stock," 1995 Int. Powder Injection Moulding Symp. (Pennsylvania), Pennsylvania State University, 1995
9. J.J. Dunkley, The Impact of System Design on Metal Powder Production Costs, *Int. J. Powder Metall.,* Vol 19 (No. 2), 1983, p 107
10. *Technic Gold Flake 55, 554, 560, and Silpowder 130, 188, 330, 333,* Technic Engineered Products Division, Woonsocket, RI
11. B. Günther, "Nano-Powders May Extend PM Part's Function," Metal Powder Report 02, 1995
12. R.W. Siegel, Creating Nanophase Materials, *Sci. Am.,* Dec 1996
13. R.W. Siegel, *Ann. Rev. Mater. Sci.,* Vol 21, 1991, p 550–578
14. H. Eifert and B. Günther, "New Products made from Ultrafine Metal Powders," Int. Conf. PM '94 (Paris), 1994
15. C. Suryanaeayana, Nanocrystalline Materials, *Int. Mater. Rev.,* Vol 40 (No. 2), 1995, p 41–64
16. H. Eifert, B. Günther, T. König, R.L. Meisel, and G. Winter, "CVR Production and Sinter Characteristics of Tantalum Nano-Powders," Int. Conf. on Tungsten, Refractory Metals and Alloys (Orlando, FL), 16–19 Nov 1997
17. S. Uotthous et al., Study of Isotropically Conductive Bondings Filled with Aggregates of Nano-sized Ag Particles, *IEEE Transactions on Components, Packaging and Manufacturing Technology Part A,* Vol 20, No. 1, 1997, p 15–20

Mechanical Alloying

C. Suryanarayana, Department of Metallurgical and Materials Engineering, Colorado School of Mines

Mechanical alloying is a simple and useful technique to synthesize both equilibrium and non-equilibrium phases of commercially useful materials starting from elemental powders. It is also an economical process with important technical advantages. One of the greatest advantages of mechanical alloying is in the synthesis of novel alloys that are not possible by any other technique, such as alloying of normally immiscible elements. This is because mechanical alloying is a completely solid-state processing technique and therefore limitations imposed by the phase diagrams do not apply here.

Mechanical alloying (MA) is normally a dry, high-energy ball milling technique and has been employed to produce a variety of commercially useful and scientifically interesting materials. This technique was developed during the late 1960s by John S. Benjamin and his colleagues at the Paul D. Merica Research Laboratory of the International Nickel Company (INCO) essentially to combine the advantages of precipitation hardening and oxide-dispersion strengthening in several nickel- and iron-base superalloys (Ref 1, 2).

The formation of an amorphous phase by mechanical grinding of an yttrium-cobalt intermetallic compound in 1981 (Ref 3) and in the nickel-niobium system by ball milling of blended elemental powder mixtures in 1983 (Ref 4) brought about the recognition that MA is a potential nonequilibrium processing technique. Since the mid-1980s, a number of investigations have been carried out to synthesize a variety of stable and metastable phases including supersaturated solid solutions, crystalline and quasicrystalline intermediate phases, and amorphous alloys. Efforts have been made since the early 1990s to understand the process fundamentals of MA. Additionally, it has been recognized that MA can be used to induce chemical (displacement) reactions in powder mixtures at room temperature or at least at much lower temperatures than normally required (Ref 5). This simple but effective processing technique has been applied to metals, ceramics, polymers, and composite materials. The important attributes of MA are:

- Production of fine dispersion of second phase particles
- Extension of solid solubility limits
- Refinement of grain sizes down to nanometer range
- Synthesis of novel crystalline and quasicrystalline phases
- Development of amorphous (glassy) phases
- Disordering of ordered intermetallics
- Possibility of alloying of difficult to alloy elements
- Inducement of chemical (displacement) reactions at low temperatures

The different facets of MA have been reviewed periodically (Ref 6–8) and the literature available up to 1994 has been collected together in an annotated bibliography (Ref 9). Additionally, a number of conferences are devoted to the science and technology of mechanically alloyed materials and the proceedings of the annual International Symposia on Metastable and Mechanically Alloyed and Nanocrystalline Materials (ISMANAM) are regularly published in *Materials Science Forum* published by Trans Tech Publications, Switzerland.

Two different terms are used in the literature to denote the processing of powder particles in high-energy ball mills. Mechanical alloying describes the process when mixtures of powders (of different metals or alloys/compounds) are milled together. Material transfer is involved in this process to obtain a homogeneous alloy. On the other hand, milling of uniform (often stoichiometric) composition powders, such as pure metals, intermetallics, or prealloyed powders, where material transfer is not required for homogenization, has been termed mechanical milling (MM). Mechanical milling of ordered intermetallics has been shown to lead to disordering or amorphization, and the nature of these phase transformations and the reasons for their occurrence have been reviewed (Ref 10). Some investigators have referred to MM as mechanical grinding (MG). Since "grinding" is normally considered an abrasive machining process that involves mainly shear stresses and chip formation, the term *milling* is preferred because it includes the more complex triaxial, perhaps partly hydrostatic, stress states that can occur during ball milling of powders (Ref 7). It should be realized, however, that MA is a generic term, and some investigators use this term to include mechanical milling/grinding as well. This article uses the term MA only to include both MA and MM.

The Process of Mechanical Alloying

Figure 1 is a schematic showing the path of raw materials using the MA process. The raw materials, the type of mill used, the process of consolidation, and the details of heat treatment differ depending on the type of product desired, but the processing route remains essentially the same. It is possible that some minor steps are either added or deleted in some special circumstances. The actual process of MA starts with mixing of powders in the right proportion and loading the powder into the mill along with the grinding medium (generally steel balls). This mix is then milled for the desired length of time until a steady state is reached. A steady state occurs when the composition of every powder particle is the same as the proportion of the elements in the starting powder mix. Sometimes the powder is milled to an intermediate state either to form metastable phases or to achieve certain desired properties. The milled powder is then consolidated into a bulk shape and heat treated to obtain the desired microstructure and properties.

Raw Materials

The raw materials used for MA are widely available commercially pure powders that have particle sizes from 1 to 200 μm. The powder particle size is not critical, except that it should be smaller than the grinding ball size, because the powder particle size decreases exponentially with time and reaches a small value after a few minutes of milling. These powders fall into the broad categories of pure metals, master alloys, prealloyed powders, and refractory compounds. The oxygen content of commercially pure metal powders ranges from 0.05 to 0.2 wt%. Therefore, if one is interested in studying phase formation, it becomes important to choose the proper purity of the powder. This is important because, more often than not, the nature and amount of impurities in the system decide the nature of the final phase formed and the chemical constitution of the alloy. Dispersion strength-

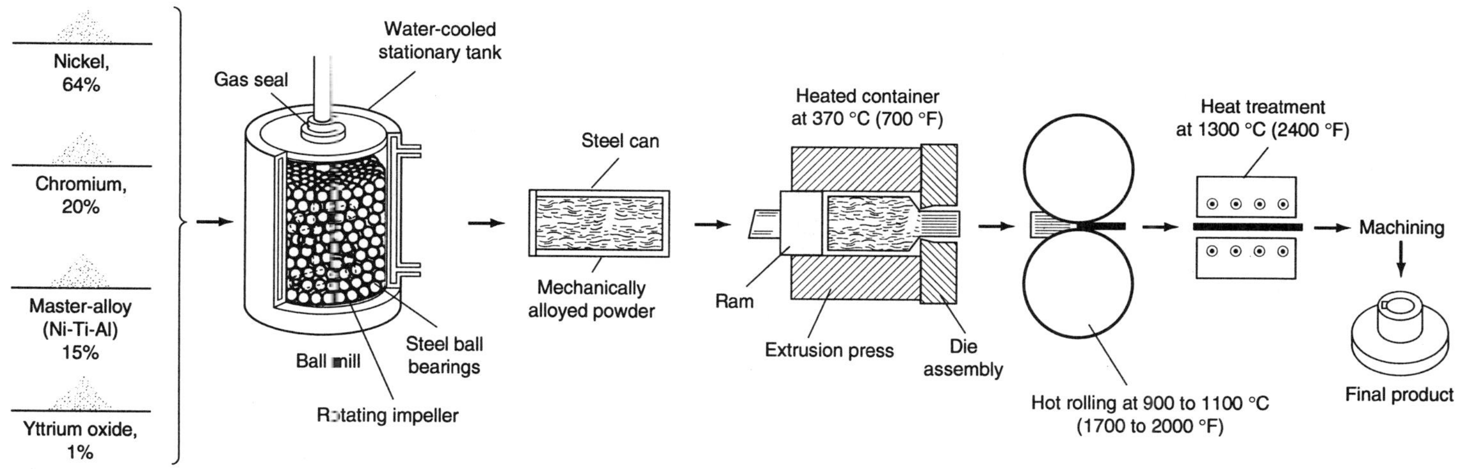

Fig. 1 Processing path in producing a product from powders by mechanical alloying. Source: Ref 1

ened materials usually contain additions of carbides, nitrides, and oxides. Oxides are the most common and these alloys are known as oxide-dispersion strengthened (ODS) materials. In the 1960s, the powder charge for MA consisted of at least 15 vol% of a ductile compressibly deformable metal powder to act as a host or a binder. However, in the 1990s, mixtures of fully brittle materials have been milled successfully resulting in alloy formation. Thus, the requirement of having a ductile metal powder during milling is no longer necessary. Accordingly, ductile-ductile, ductile-brittle, and brittle-brittle powder mixtures are milled to produce novel alloys. Mixtures of solid powder particles and liquids have also been milled (Ref 11).

Process Control Agents

A process control agent (PCA) is added to the powder mixture during milling, especially when the powder mix involves a substantial fraction of a ductile component. The PCAs are mostly organic compounds, which act as surface-active agents. The PCA adsorbs on the surface of the powder particles and minimizes cold welding between powder particles and thereby inhibits agglomeration. The surface-active agents adsorbed on particle surfaces interfere with cold welding and lower the surface tension of the solid material. Because the energy required for milling is equal to the product of new surface area generated times the surface tension, a reduction in surface tension results in the use of shorter milling times and/or finer powders.

A wide range of PCAs have been used in practice. These are mostly organic compounds used at a level of ~1 to 4 wt% of the total powder charge and include stearic acid [$CH_3(CH_2)_{16}COOH$], hexane (C_5H_{14}), oxalic acid [$(COOH)_2 \cdot 2H_2O$], methanol, ethanol, acetone, isopropyl alcohol, heptane, Nopcowax-22DSP, octane, toluene, trichlorotrifluoroethane, DDAA (diodecyl dimethyl ammonium acetate), silicone grease, etc. Graphite, alumina, aluminum nitrate, and sodium chloride have also been used as PCAs. The majority of these compounds decompose during milling and interact with the powder and form compounds, which get incorporated in the form of dispersions inside the powder particles during milling. Thus, hydrocarbons containing hydrogen and carbon and carbohydrates containing hydrogen, carbon, and oxygen are likely to introduce carbon and/or oxygen into the powder particles, resulting in the formation of carbides and oxides that are uniformly dispersed in the matrix. These are not necessarily harmful to the alloy system because they can contribute to dispersion strengthening of the material. The hydrogen subsequently escapes as a gas or is absorbed into the metal lattice on heating or sintering.

Minimization or avoidance of cold welding of powder particles among themselves and to the grinding tools can also be achieved by other means. For example, the presence of air in the milling container acts like a PCA and prevents cold welding of the powders. Milling of the powders at very low temperatures (e.g., in liquid nitrogen) referred to as cryomilling also has been shown to minimize welding, probably due to the increased brittleness of the powder particles at such low temperatures. Metal powders (with face-centered cubic structure) milled in a hydrogen atmosphere have been found to become brittle and not stick to themselves or the container.

Some metals, such as aluminum, nickel, and copper, react with certain alcohols during milling to form complex metallo-organic compounds. For example, aluminum reacts with isopropyl alcohol. Other metals such as titanium and zirconium can react explosively with chlorinated fluids such as carbon tetrachloride. Chlorinated fluids should never be used with reactive metals. Reactive metals such as titanium and zirconium, when milled in presence of air, can pick up substantial amounts of oxygen and nitrogen and cause phase changes, including formation of new phases (Ref 12).

Reactive milling (milling of metal powders in the presence of reactive solids/liquids/gases enabling a chemical reaction to take place) has also been employed to synthesize metal oxides, carbides, and nitrides (Ref 13–15). Thus, milling of titanium in a nitrogen atmosphere has produced titanium nitride; several other nitrides have also been produced in a similar way. Milling of tungsten with carbon (graphite) has produced tungsten carbide. Milling of aluminum with carbon (or by use of a PCA containing carbon) produced aluminum carbide (Al_4C_3), which gets dispersed inside the aluminum matrix and improves the performance of the alloy. The carbide in this case formed only partially during milling; subsequent heat treatment was required to complete the reaction of carbon with aluminum. However, in other cases, the reaction can take place entirely during milling, only after heat treatment, or in part during both the stages.

The choice of a PCA for milling depends on the nature of the powder being milled and the purity of the final product desired. Use of a PCA normally introduces some impurities into the powder (see the section "Powder Contamination" at the end of this article). Thus, to produce a high-purity alloy, use of a PCA should be avoided. It should also be realized that there is no universal PCA. One would have to decide on a PCA looking at the possible interaction between the metal powder being milled and the components in the PCA.

Types of Mills

A number of mills are used to produce mechanically alloyed powders. They differ in capacity, efficiency of milling, and additional arrangements for cooling, heating, etc.

Shaker mills, such as SPEX mills (SPEX CertPrep, Inc., Metuchen, NJ) (Fig. 2a), which mill about 10 g of the powder at a time, are most commonly used for laboratory investigations and for alloy screening purposes. The common variety of the mill has one vial containing the sample and grinding balls, secured in the clamp and swung energetically back and forth several thousand times a minute. The back-and-forth shaking motion is combined with lateral movements of the ends of the vial so that the vial appears to be tracing a figure-8 or infinity sign as it moves. With each swing of the vial, the balls impact against the sample and the end of

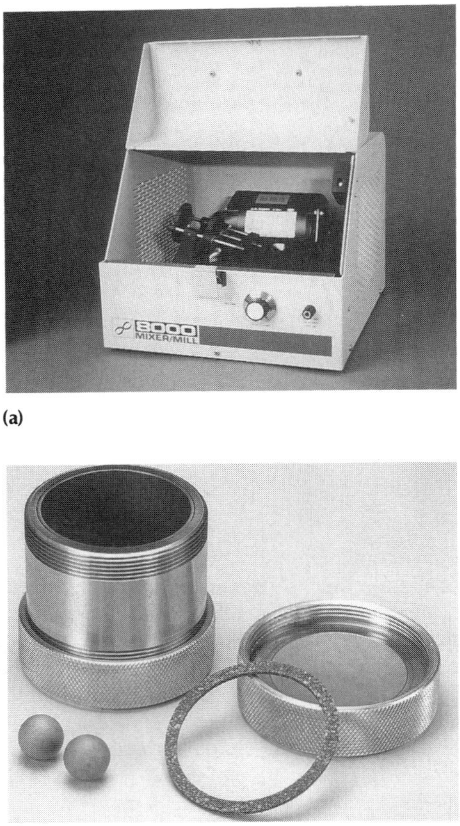

(a)

(b)

Fig. 2 (a) SPEX 8000 mixer/mill in the assembled condition. (b) Tungsten carbide vial set consisting of the vial, lid, gasket, and balls. Courtesy of SPEX CertiPrep, Metuchen, NJ

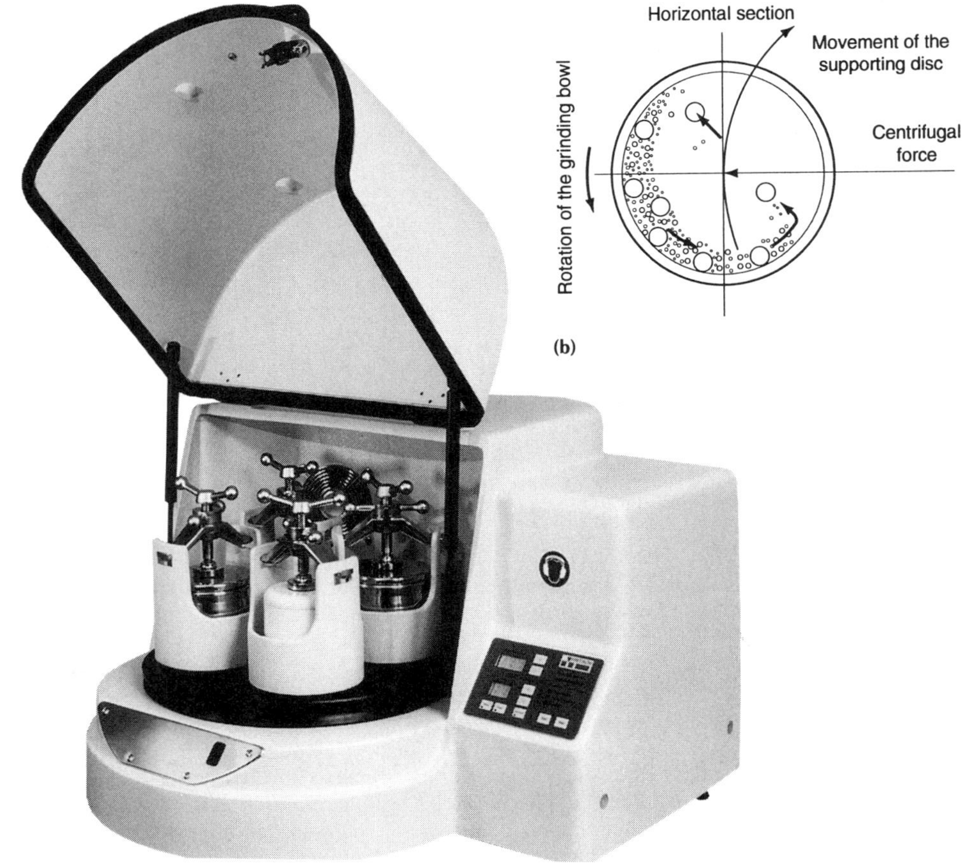

(a)

(b)

Fig. 3 (a) Fritsch Pulverisette P-5 four-station ball mill. (b) Schematic depicting the ball motion inside the ball mill. Courtesy of Gilson Company, Inc., Worthington, OH

the vial, both milling and mixing the sample. Because of the amplitude (~5 cm) and speed (~1200 rpm) of the clamp motion, the ball velocities are high (~5 m/s), and consequently the force of the ball's impact is unusually great. Therefore, these mills can be considered as high-energy variety. The most recent design of the mills allows for simultaneously milling the powder in two vials to increase the throughput. A variety of vial materials is available for the SPEX mills and these include hardened steel, alumina, tungsten carbide, zirconia, stainless steel, silicon nitride, agate, plastic, and methacrylate. A typical example of a tungsten carbide vial, gasket, and grinding balls for the SPEX mill is shown in Fig. 2b. Majority of the research on the fundamental aspects of MA has been carried out using some version of these SPEX mills.

The planetary ball mill (referred to as Pulverisette) is another popular mill for conducting MA experiments which allows a few hundred grams of the powder to be milled at a time (Fig. 3a). These are manufactured by Fritsch GmbH in Germany and marketed by Gilson Co., Inc. in the United States and Canada. The planetary ball mill owes its name to the planetlike movement of its vials. These are arranged on a rotating support disk, and a special drive mechanism causes them to rotate around their own axes. The centrifugal force produced by the vials rotating around their axes and by the rotating support disk act on the vial contents, consisting of material to be ground and the grinding balls. Because the vials and the supporting disk rotate in opposite directions, the centrifugal forces alternately act in like and opposite directions. This causes the grinding balls to run down the inside wall of the vial—the friction effect—followed by the material being ground and grinding balls lifting off and traveling freely through the inner chamber of the vial and colliding against the opposing inside wall—the impact effect (Fig. 3b).

Even though the disk and the vial rotation speeds could not be independently controlled in the earlier versions, it is possible to do so in the modern versions. In a single mill one can have either two (Pulverisette 5 or 7) or four (Pulverisette 5) milling stations. Recently, a single-station mill was also developed (Pulverisette 6). Grinding vials and balls are available in nine different materials—agate, silicon nitride, sintered corundum 1, sintered corundum 2, zirconia, chrome steel, Cr-Ni steel, tungsten carbide, and plastic polyamide. Even though the linear velocity of the balls in this type of mill is higher than that in the SPEX mills, the frequency of impact is much higher in the SPEX mills. Hence, in comparison to SPEX mills, Fritsch Pulverisette can be considered lower energy mills.

Attritors are the mills in which large quantities of powder (from a few pounds to 100 lb) can be milled at a time (Fig. 4a). Commercial attritors are available from Union Process, Akron, OH. The velocity of the grinding medium is much lower (~0.5 m/s) than in Fritsch or SPEX mills, and consequently the energy of the attritors is low. Attritors of different sizes and capacities are available. The grinding tanks or containers are available either in stainless steel or stainless steel coated inside with alumina, silicon carbide, silicon nitride, zirconia, rubber, and polyurethane. A variety of grinding media also is available—glass, flint stones, steatite ceramic, mullite, silicon carbide, silicon nitride, sialon, alumina, zirconium silicate, zirconia, stainless steel, carbon steel, chrome steel, and tungsten carbide.

The operation of an attritor is simple. The powder to be milled is placed in a stationary tank with the grinding media. This mixture is then agitated by a shaft with arms, rotating at a high speed of ~250 rpm (Fig. 4b). This causes the media to exert both shearing and impact forces on the material. The laboratory attritor works up to ten times faster than conventional jar mills.

Commercial mills for MA are much larger in size than the mills described above and can process several hundred pounds at a time. Mechanical alloying for commercial production is carried out in ball mills of up to 3000 lb capacity (Fig. 5). The milling time decreases with an increase in the energy of the mill. For example, a process that takes only a few minutes in the SPEX mill can take hours in an attritor and a few days in a commercial mill.

Several new designs of mills have been developed in recent years for specialized purposes. These include the rod mills, vibrating frame mills, and the equipment available from Dymatron, Inc., Cincinnati, OH; Super Misuni NEV-MA-8 from Nisshin Giken, Co., Ltd., Tokyo, Japan (with the ability to control the temperature of milling from very low temperatures by spraying liquid nitrogen and up to 300 °C by electrical heating); Uni-Ball-Mill from Australian Scientific Instruments, Canberra, Australia (it is possible to control the nature and magnitude of impact of the balls in this machine with the help of magnets); HEMill from M.B.N. srl, Rome, Italy; and Zoz Maschinenbau GmbH, Kreuztal, Germany. The Uni-Ball-Mill with the spatially adjustable high field strength magnet (on the left side) that allows for continuously variable mill kinetics is shown in Fig. 6. Some special equipment is also designed for specific laboratory applications. The operating details and special features of these mills can be found in the respective brochures available from the manufacturers.

Powder Handling

Mechanical alloying involves powders with very small sizes and these should be handled with caution and care. Because of the large surface area, they are highly reactive and can be pyrophoric and can cause health problems when inhaled. Precautions should be taken not to open the powder to atmosphere immediately after milling since this can lead to oxidation of the powders and in some situations they can even catch fire.

(a)

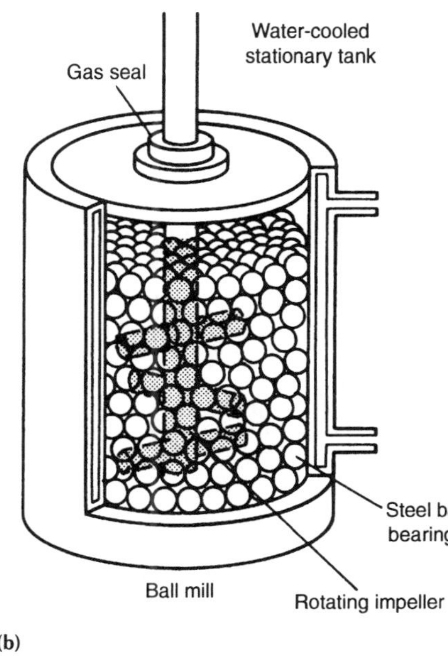

(b)

Fig. 4 (a) Model 1-S attritor. (b) Arrangement of rotating arms on a shaft in the attrition ball mill. Courtesy of Union Process, Akron, OH

Mechanism of Alloying

During high-energy milling, the powder particles are repeatedly flattened, fractured, and rewelded. Whenever two steel balls collide, some amount of powder is trapped between them. Typically, around 1000 particles with an aggregate weight of ~0.2 mg are trapped during each collision. The force of the impact plastically deforms the powder particles, creates new surfaces, and enables the particles to weld together, which leads to an increase in particle size. In the early stages of milling, the particles are soft (using either ductile-ductile or ductile-brittle material combination) and their tendency to weld together and form large particles is high. A broad range of particle size develops, with some as large as three times bigger than the starting particles. The composite particles at this stage have a characteristic layered structure consisting of various combinations of the starting constituents. With continued deformation, the particles become work hardened and fracture by a fatigue failure mechanism and/or by the fragmentation of fragile flakes. Fragments generated by this mechanism can continue to reduce in size in the absence of strong agglomerating forces. At this stage, the tendency to fracture predominates over cold welding. Due to the continued impact of grinding balls, the structure of the particles is steadily refined, but the particle size continues to be the same. Consequently, the interlayer spacing decreases and the number of layers in a particle increases. As previously mentioned, the rate of refinement of the internal structure (particle size, crystallite size, lamellar spacing, etc.) is roughly logarithmic with processing time and therefore the size of the starting particles is relatively unimportant. In a few minutes to an hour,

Fig. 5 Commercial production-size ball mills used for mechanical alloying. Courtesy of Inco Alloys International

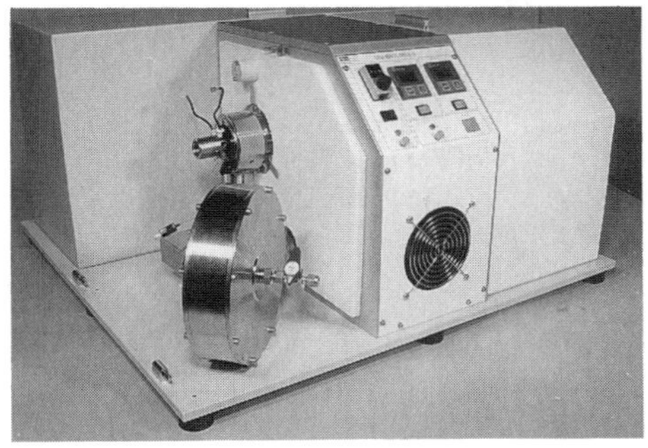

Fig. 6 Uni-Ball-Mill with the provision of a spatially adjustable high strength magnet. Courtesy of Australian Scientific Instruments, Fyshwick, Australia

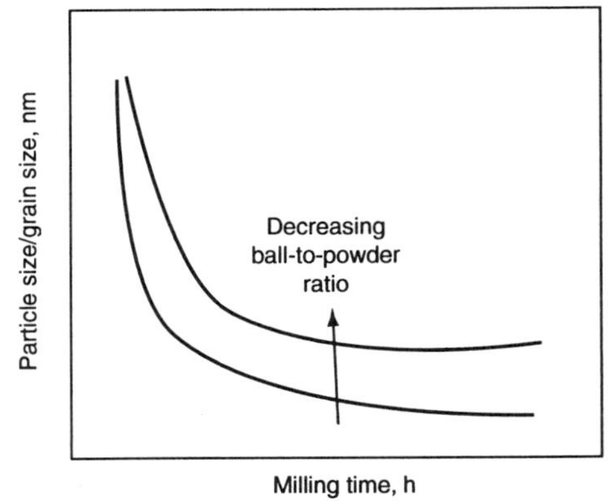

Fig. 7 Refinement of particle and grain sizes with milling time. Rate of refinement increases with higher ball-to-powder weight ratios.

the lamellar spacing usually becomes small and the crystallite (or grain) size is refined to nanometer (1 nm = 10^{-9} m or 10 Å) dimensions (Fig. 7). The ease with which this can be achieved is one reason why MA has been extensively employed to produce nanocrystalline materials (Ref 8, 9, 16).

After milling for a certain length of time, steady-state equilibrium is attained when a balance is achieved between the rate of welding, which tends to increase the average particle size, and the rate of fracturing, which tends to decrease the average composite particle size. At this stage, each particle contains substantially all of the starting ingredients in the proportion they were mixed together, and the particles reach saturation hardness due to the accumulation of strain energy. The particle size distribution at this stage is narrow, because particles larger than average are reduced in size at the same rate that fragments smaller than average grow through agglomeration of smaller particles (Fig. 8). The average particle size obtained at this stage depends on the relative ease with which agglomerates can be formed by cold welding, fatigue and fracture strength of composite particles, and resistance of particles to deformation.

From the foregoing explanation, it is clear that during MA, heavy deformation is introduced into the particles. This is manifested by the presence of a variety of crystal defects, such as dislocations, vacancies, stacking faults, and increased number of grain boundaries. The presence of this defect structure increases the diffusivity of solute elements into the matrix. Further, the refined microstructural features decrease the diffusion distances. Additionally, the slight rise in temperature during milling further aids the diffusion behavior, and consequently, true alloying takes place among the constituent elements. While this alloying generally takes place nominally at room temperature, sometimes it may be necessary to heat treat the mechanically alloyed powder for alloying to be achieved. This is particularly true when formation of intermetallics is desired.

Consolidation

Widespread application of mechanically alloyed powders requires production of the powder in tonnage quantities and also efficient methods of consolidating the powders into bulk shapes. All the consolidation methods generally used in powder metallurgy processes can also be used for mechanically alloyed powders. However, because the powder particles in the mechanically alloyed condition are smaller (typically a few microns, even though the grain size may be only a few nanometers) in size than those used in conventional powder metallurgy operations, some special precautions need to be taken to minimize their activity and high level of interparticle friction.

Conventional consolidation of powder to full density through processes such as hot extrusion and hot isostatic pressing normally requires use of high pressures and high temperatures for extended periods of time. Unfortunately, this results in loss of the benefits achieved due to the metastable effects or nanostructures obtained by MA. Therefore, novel and innovative methods of consolidating the mechanically alloyed powders are required.

Mechanically alloyed powders have been consolidated into useful bulk shapes by several processes. Because mechanically alloyed powders have a high hardness, cold compaction is not an option. Further, ODS alloys have been found not to densify during simple sintering. The most common method of consolidation is hot compaction followed by hot extrusion, or by direct hot extrusion at temperatures greater than half the melting point. This process could be used if one is not concerned with the loss of metastable effects, such as the crystallization of the amorphous phase in the powder. Some of the consolidation methods used for mechanically alloyed powders, both on laboratory and industrial scales, include hot isostatic pressing, powder rolling, Ceracon processing, plasma activated sintering, electrodischarge compaction, and explosive forming (shock methods). These operations are usually carried out at temperatures considerably lower than those used for conventional powders. For example, while conventional γ-

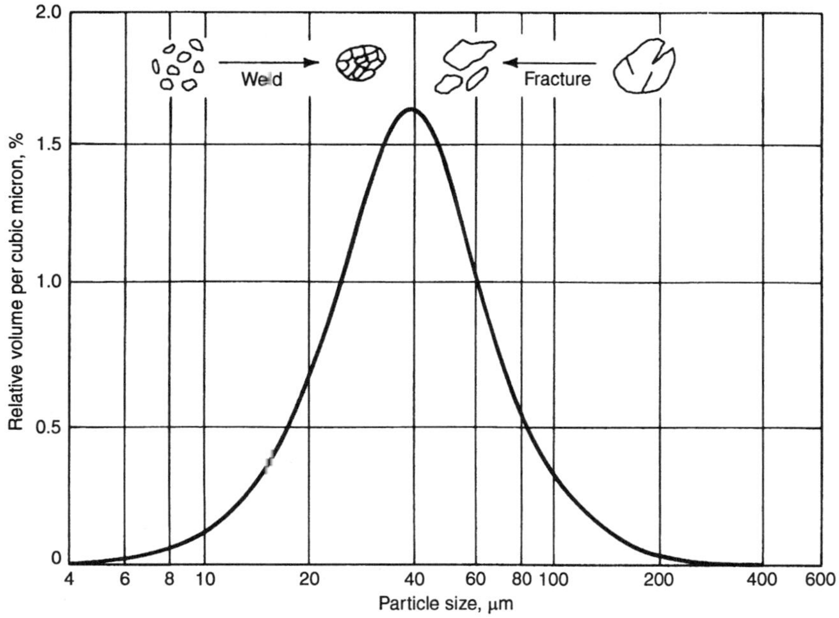

Fig. 8 Narrow particle size distribution caused by tendency of small particles to weld together and large particles to fracture under steady-state conditions

Table 1 Room temperature and elevated temperature mechanical properties of commercial ODS nickel- and iron-base superalloys (see Tables 2 and 3 for the chemical compositions of the alloys)

Alloy	Test temperature		0.2% Yield strength		Ultimate tensile strength		Elongation, %	Reduction in area, %
	°C	°F	MPa	ksi	MPa	ksi		
MA 6000	Room temperature		1220	177	1253	182	7.2	6.5
	871	1600	675	98	701	102	2.2	4.6
	982	1800	445	65	460	67	2.8	1.9
MA 754	Room temperature		586	85	965	140	21	33
	871	1600	214	31	248	36	31	58
	982	1800	169	25	190	28	18	34
MA956	Room temperature		517	75	655	95	20	35
	1000	1832	97	14	100	15	...	...
	1100	2192	69	10	72	11	12	30

TiAl powders are hot isostatically pressed at about 1100 °C, mechanically alloyed γ-TiAl powders could be consolidated to full density by hot isostatically pressing them at about 750 °C.

Oxide-Dispersion Strengthened Alloys

Use of MA to synthesize metastable phases is a recent development. The major use of MA has been in the development of commercial ODS nickel-, iron-, and aluminum-base alloys.

Mechanically alloyed materials are strong both at room and elevated temperatures (Table 1). The elevated temperature strength of these materials is derived from more than one mechanism. First, the uniform dispersion (with a spacing of ~100 nm) of very fine (5 to 50 nm) oxide particles (commonly used are Y_2O_3 (yttria), ThO_2 (thoria), and La_2O_3 (lanthana)), which are stable at high temperatures, inhibit dislocation motion in the metal matrix, and increase the resistance of the alloy to creep deformation. Another function of the dispersoid particles is to inhibit the recovery and recrystallization processes, which allow a very stable large grain size to be obtained. These large grains resist grain rotation during high temperature deformation. A stable large grain size can also be obtained by secondary recrystallization mechanisms. The very homogeneous distribution of alloying elements during MA gives both the solid-solution strengthened and precipitation-hardened alloys more stability at elevated temperatures and overall improvement in properties. Mechanically alloyed materials also have excellent oxidation and hot corrosion resistance. The increased resistance to oxidation-sulfidation attack is due to the homogeneous distribution of the alloying elements and the improved scale adherence due to the dispersoid itself (Ref 17, 18).

Commercial Processing of ODS Alloys

Commercial ODS alloys all use raw materials, such as elemental powders, crushed master alloy (intermetallic) powders, and prealloyed powders. Yttria powder lying within a narrow range of crystallite sizes (optimum strength of the alloy is achieved when the crystallite size is <30 nm) and metal powders are blended together to form a bulk preblend feedstock, the chemical composition of which is verified before actual processing starts. These ingredients are charged into large, horizontal axis ball mills containing alloy steel balls. Each ball mill is used for the production of one type of alloy. Cross contamination between different alloys is thus avoided, while pick-up from the through-hardened steel balls and the internal parts of the mill is prevented by the accumulation of a protective superalloy coating. During milling, an equilibrium powder particle size distribution is established and each individual particle acquires the alloying constituents in the correct proportions. The process is considered complete when the composite particles have a finely layered structure, not visible under an optical microscope.

The mechanically alloyed powder from the mill is then collected into stainless steel containers, and powder from several different containers (production runs) can be blended together at this stage. After a full chemical analysis, the blend is consolidated at elevated temperatures by either extrusion or hot isostatic pressing (HIP). In most cases extrusion is preferred because it is less expensive than HIP and contributes to the work necessary for secondary recrystallization. The powder is loaded into mild steel extrusion cans measuring up to 300 mm in diam and 900 mm in length and containing up to 280 kg powder. The powder is soaked for several hours in a furnace at the extrusion temperature, during which time homogenization of the powder and true alloying occur, if it has not already occurred during milling. Extrusion is carried out in commercial presses at temperatures, ratios, and speeds that are independent variables and material dependent. The ODS alloys have a very fine equiaxed submicron grain structure in the as-extruded condition. The mild steel from the extrusion cans is removed either by machining or by pickling. However, the consolidation process chosen depends on the properties, shape, and size required for the final mill product.

Once consolidated, the mechanically alloyed billets are worked to final shape, such as sheet, plate, bar, tube, etc., using normal hot working methods such as extrusion and hot rolling. Virtually all processing of the nickel-base alloy is performed hot, while cold working operations are possible with the iron-base alloys. This material is then subjected to a secondary recrystallization anneal at temperatures at or above 1300 °C to produce a coarse grain structure. The combination of oxide dispersion strengthening and coarse grain structure provides outstanding creep and stress rupture properties at elevated temperatures of >1000 °C.

Applications

The technique of MA has been shown to produce a variety of materials. The most important reason for the development of MA was the production of ODS materials in which fine particles of Y_2O_3 or ThO_2 were uniformly dispersed in a nickel- or iron-base alloy. In the mid-1980s, it was realized that MA is also capable of producing true alloys from elements that are not

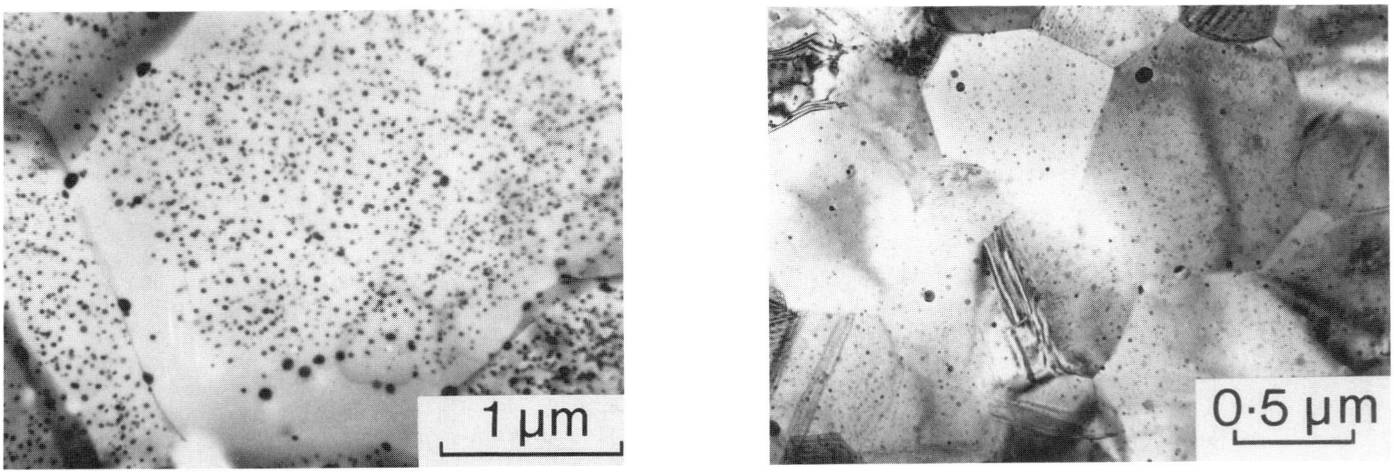

(a) (b)

Fig. 9 Transmission electron micrographs showing the difference in the matrix grain size, and size and distribution of dispersoids. (a) Rapidly solidified and hot isostatically pressed at 850 °C. (b) Mechanically alloyed and hot isostatically pressed at 1000 °C

either easy to form by conventional means or sometimes even impossible to prepare, e.g., elements which are immiscible under equilibrium conditions. Investigations have revealed that metastable phases, such as supersaturated solid solutions, non-equilibrium crystalline or quasicrystalline intermediate phases, and amorphous alloys, can be synthesized by MA (Ref 7–9). In addition, nanostructures with a grain size of a few nanometers, typically <100 nm, are produced. These metastable phases have interesting combinations of physical, chemical, mechanical, and magnetic properties and are being widely explored for potential applications. It is impossible to provide details of each of these developments, so only a very brief survey of the ODS alloys is presented here. Details of the structure and properties of metastable alloys are in Ref 7–10.

Ever since it was realized that MA could synthesize metastable phases, attempts have been made to compare the results of MA with those of rapid solidification, another non-equilibrium processing technique, even though the mechanisms by which metastable phases form by these two techniques are different. For example, solid solutions have been formed in the whole composition range in the copper-silver system by both techniques. A similar situation was obtained in several other systems by rapid solidification methods but not MA. Solid solutions have been obtained in the full composition range in the Cu-Fe, AlSb-InSb, and Cu-Co systems by MA but not by rapid solidification. Similarly, the levels of solid solubility achieved differ in various systems using the two techniques.

A similar situation exists with respect to the formation of metastable intermediate phases and amorphous alloys. Formation of an amorphous phase has been reported in the titanium-aluminum system by MA, which is not possible by rapid solidification methods. An important difference between these methods is that while amorphous phases form in the vicinity of eutectic compositions by rapid solidification, they form near the equiatomic composition by MA. Further, the glass formation range (the composition range over which glasses can form) is wider in the mechanically alloyed condition than in rapid solidification.

Formation of nanostructures has been reported in many alloy systems by mechanical alloying methods, but not by rapid solidification methods. But, devitrification of amorphous phases obtained by rapid solidification has led to the synthesis of nanostructure composites useful for magnetic applications (Ref 19).

Another important difference between mechanically alloyed and rapidly solidified alloys containing dispersoids appears to be in the size and distribution of the dispersoids. Figure 9 shows a pair of transmission electron micrographs comparing the matrix grain sizes, and size and distribution of dispersoids in HIP compacts of Ti_3Al-base alloys containing Er_2O_3 dispersoids. The rapidly solidified alloy was Ti_3Al to which 2 wt% Er was added and then hot isostatically pressed at 850 °C. The mechanically alloyed material was Ti-25Al-10Nb-3V-1Mo (at.%) to which 2 wt% Er was added and the alloy powder was hot isostatically pressed at 1000 °C. Even after HIP at a higher temperature, in comparison to the rapidly solidified alloy, the mechanically alloyed material showed a finer matrix grain size, more uniform distribution of the dispersoids, absence of large dispersoids at the grain boundaries, and absence of dispersoid-free zones near the grain boundaries.

The majority of these and other differences between materials processed by the two techniques could be traced to the fact that rapid solidification is a liquid-to-solid transformation while MA is a fully solid-state transformation.

The major industrial applications of mechanically alloyed materials have been in thermal processing, glass processing, energy production, aerospace, and other industries. These applications are based on the oxide-dispersion strengthening effect achieved in mechanically alloyed nickel-, iron-, and aluminum-base alloys.

Nickel-Base Alloys

Typical compositions of the commercially available mechanically alloyed nickel-base superalloys are presented in Table 2. The most significant advantage of oxide-dispersion strengthened superalloys is the increased stress rupture properties. Figure 10 compares the specific rupture strength (strength/density) for a 1000 h life as a function of temperature for several nickel-base superalloys used for turbine blade applications. Mar-M200 is a nickel-base alloy containing by weight percent 9.0 Cr, 5.0 Al, 2.0 Ti, 12.0 W, 10.0 Co, 1.0 Nb, and 1.8Hf, while PWA454 is a nickel-base alloy containing 10.0 Cr, 5.0 Al, 1.5 Ti, 12.0 Ta, 4.0 W, and 5.0 Co, and TD Ni is nickel containing 2.0 wt% ThO_2. It is clear from this figure that the MA 6000 alloy can maintain a given stress for a much longer time than a conventional alloy for similar vane applications. This is mainly due to the benefits of the combined strengthening modes in the mechanically alloyed material.

The ODS alloys continue to find applications in a wide variety of industries. The two production facilities of Inco Alloys International have a combined annual powder capacity approaching 300,000 kg. The yield of the final product varies greatly with the product form and size, but the final wrought capacity is over 200,000 kg.

The mechanically alloyed superalloys are considered mainly for three groups of applications—gas turbine vanes, turbine blades, and sheets for use in oxidizing/corrosive atmospheres. The largest use of MA 754 is as vanes and bands for aircraft gas turbine engines (Fig. 11). For applications requiring good resistance to thermal fatigue, such as gas turbine vanes, Inconel MA 754 is given a strong texture. The majority of the grains are aligned so that their <100> axes are parallel to the principal working direction and along the length of the bar. Such texture results in low modulus of elasticity (149 GPa or 21.6×10^3 ksi) in the longitudinal direction. The low modulus improves resistance to thermal fatigue by lowering stresses for given

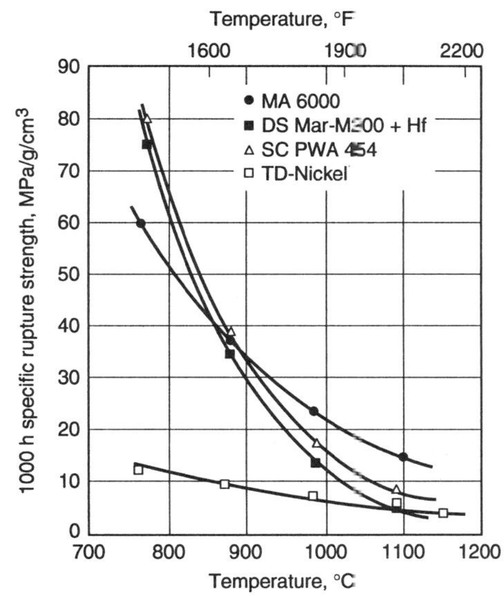

Fig. 10 Comparison of 1000 h specific rupture strength of INCONEL MA6000 with dispersion strengthened Mar-M200 + Hf, TD-Ni, and single-crystal PWA 454. Courtesy of Inco Alloys International

Fig. 11 Some typical gas turbine engine components made of INCONEL alloy MA 754. Courtesy of Inco Alloys International

Table 2 Nominal compositions (wt%) of mechanically alloyed nickel-base superalloys

Alloy	Ni	Cr	Al	Ti	Mo	W	Y_2O_3	Ta
INCONEL alloy MA 754	bal	20	0.3	0.5	-	...	0.6	...
INCONEL alloy MA 757	bal	16	4.0	0.5	...	...	0.6	...
INCONEL alloy MA 758	bal	30	0.3	0.5	...	...	0.6	...
INCONEL alloy MA 760	bal	20	6.0	...	2.0	3.5	0.95	...
INCONEL alloy MA 6000	bal	15	4.5	0.5	2.0	4.0	1.1	2.0
TMO-2(a)	bal	6	4.2	0.8	2.0	12.4	1.1	4.7

(a) This alloy additionally contains 9.7 wt% cobalt.

thermal strains. Components are fabricated from bar stock using state-of-the-art machining and brazing processes typical for conventional wrought superalloys. Alloy MA 758 is used in a number of industrial applications where its high chromium content makes it resistant to extremes of temperature and environment. The alloy is used in the glass industry for high temperature components requiring both elevated temperature strength and resistance to extremely corrosive molten glass. Alloy MA 758 is also used for internal combustion engine components, mainly in critical fuel injection parts. One novel industrial application of alloy MA 754 is a high-temperature atmosphere-circulation fan in a "floating" furnace design being commercialized in Japan. Large rounds are used for the hub and plate material is used for blades of the fan, which operates at temperatures >1100 °C.

New product forms of the commercial alloys continue to be developed. Large diameter, thin wall tubing of alloy MA 754 has been produced and evaluated for radiant tube applications, and alloy MA 758 has been used as tubing in heat exchangers and process equipment operating at very high temperatures.

The alloy MA 754 is used for brazed nozzle guide vane and band assemblies in U.S. military aero engines. The principal advantages of the alloy for these applications are thermal fatigue resistance, long-term creep strength, and high melting point.

The MA 6000 alloy is a more complex alloy developed as a blade material for advanced gas turbines. It is used for first and second stage turbine vanes and blades machined from solid bar. Unlike cast alloys, MA 6000 exhibits nearly flat rupture-life curves at high temperatures due to the combination of oxide-dispersion strengthening and high grain-to-width ratios (typically >10 to 1). Because of its composition, MA 6000 has excellent resistance to oxidation and sulfidation. The characteristics of this alloy allow blade cooling to be reduced or eliminated as the metal temperature can be increased by 100 °C or more in engines where the stresses are medium or low.

Iron-Base Alloys

Table 3 lists the chemical compositions of mechanically alloyed iron-base superalloys. These alloys combine the high-temperature strength and stability of oxide-dispersion strengthening with excellent resistance to oxidation, carburization, and hot corrosion. These alloys are suitable for use in gas turbine combustion chambers. Incoloy alloy MA 956 is particularly well suited for use in heat processing applications. For example, vacuum furnace fixtures made of MA 956 have shown excellent durability and are able to compete with wrought molybdenum, which is also used in these applications. In comparison to molybdenum, MA 956 is ~30% lower in density, providing weight savings and cost advantages. Further, since MA 956 has a lower vapor pressure than molybdenum, it will not coat the inside of the vacuum chamber or the parts being heat treated. Thus, MA 956 rods, flats, and sheets are used in numerous atmosphere and vacuum furnace applications including muffles, baskets, trays, and thermowells. Alloy MA 956 in tubing form has also been used for high temperature, severe service applications such as coke injection lance pipes in steel making. The alloy MA 956 is also being used in the glass processing industry because of its resistance to attack by molten glass. Because of this corrosion resistance, the alloy is being evaluated for applications such as firing-kiln rollers, muffle tubes, and furnace racks. Other applications include molten-glass resistance heaters, thermocouple protection tubes, glass-processing components used in nuclear waste disposal, and the bushings used to make single and multistrand fibers.

More recently, MA 957 has been evaluated for use as the fuel cladding in fast neutron, breeder reactors. Conventional austenitic alloys are unsuitable for this application due to the dimensional swelling phenomenon caused by the high neutron fluxes. The mechanically alloyed materials are also being evaluated for heat exchanger components in high-temperature gas-cooled reactors. Figure 12 shows some typical high-temperature applications of the alloy MA 956.

Aluminum-Base Alloys

The success of mechanically alloyed superalloys led to the development of dispersion-strengthened aluminum alloys. Table 4 lists the compositions of the mechanically alloyed dispersion-strengthened aluminum alloys. Because an aluminum oxide layer is always present either on the surface of the powder particles at the start of processing or during milling, its incorporation into the alloy contributes to significant improvements in the properties of the alloy. Further, since aluminum is a ductile metal, PCAs are added to assist in minimizing cold welding during processing. Aluminum carbides are formed during MA by the decomposition of the PCA. Both the oxide or carbide type dispersions are approximately 30 to 50 nm in size and stabilize the ultrafine grain size. This results in a 50% increase in strength, higher fracture toughness, and improved resistance to stress corrosion cracking and fatigue crack growth of the mechanically alloyed materials. IncoMAP alloy AL-9052 has a density 5% less than that of conventional age-hardenable aluminum alloy of comparable strengths such as 2024. With its combination of lightweight, high strength, and corrosion resistance, IncoMAP alloy AL-9052 is evaluated for aerospace applications where marine corrosion is also a factor.

Addition of lithium to mechanically alloyed aluminum alloys has produced an ultra lightweight alloy, IncoMAP AL-905XL. Its density is 8% lower and stiffness 10% greater than the age-hardenable conventional alloy 7075-T73 of comparable strength. The excellent combination of the properties makes this alloy very attractive for airframe applications. In particular, the freedom from age-hardening treatments makes it possible to produce forgings and heavy sections with homogeneous metallurgical structures.

Recently, high-strength aluminum-titanium alloys have been developed using MA by dispersing nanometer or submicron-sized Al_3Ti intermetallic particles (in addition to the Al_2O_3 and Al_4C_3 dispersoids from addition of PCAs) in an aluminum matrix. Similar approaches could be used to develop high strength alloys in other systems.

Magnesium-base Alloys

A useful application of the MA technique was in the production of supercorroding magnesium alloys that operate as short-circuited galvanic cells to corrode (react) rapidly and predictably with an electrolyte, such as seawater, to produce heat and hydrogen gas (Ref 20, 21). Such an alloy system is suitable as a heat source for warming deep-sea divers, as a gas generator to provide gas for buoyancy, or as a fuel in hydrogen engines or fuel cells. The corrosion rate of alloys can be maximized by providing (a) a short electrolyte path, (b) a large amount of exposed surface area, and (c) a strong bond (weld) between the cathode and the anode. It is also useful to provide a very low resistance path for external currents to flow through the corroding pairs. All these requirements can be met with MA processing. Consequently, magnesium-base alloys containing Fe, Cu, C, Cr, or Ti have been evaluated for such applications. The Mg-5 to 20at.% Fe alloy is ideal because of its extremely fast reaction rate, high power output, and the high percentage of theoretical completion of the actual reaction. For corrodable release links an alloy with a slower reaction rate, such as Mg-5at.% Ti is useful.

Table 3 Nominal compositions (wt%) of mechanically alloyed iron-base superalloys

Alloy	Fe	Cr	Al	Ti	Mo	Y_2O_3
INCOLOY alloy MA 956	bal	20	4.5	0.5	...	0.5
INCOLOY alloy MA 957	bal	14	...	1.0	0.3	0.25

Table 4 Nominal compositions (wt%) of mechanically alloyed dispersion-strengthened aluminum-base alloys

Alloy	Al	Mg	Li	C	O
IncoMAP alloy AL-9021	bal	1.5	...	1.1	0.8
IncoMAP alloy AL-9052	bal	4.0	...	1.1	0.6
IncoMAP alloy AL-905XL	bal	4.0	1.3	1.1	0.6

There have also been a number of investigations in recent years to examine use of MA to produce metal hydrides. This is because metal hydrides are materials for safe storage of hydrogen, and they can store hydrogen with a higher volume density than liquid hydrogen. However, these are sensitive to surface oxidation and hence can be a limiting factor in their commercial utilization. Nanocrystalline hydrides have a high density of defects and interfaces that could enhance diffusion; therefore, nanocrystalline intermetallics would not require activation treatments at high temperatures and pressures after exposure to air (Ref 22). In comparison to coarse-grained materials, mechanically alloyed nanocrystalline intermetallics exhibit a narrower absorption plateau and a lower plateau pressure. Their hydrogen storage behavior is typical of amorphous systems. Several magnesium-base and iron-base intermetallics are being evaluated for this application.

Displacement Reactions

Mechanical activation of solids to induce chemical reactions has been used for over 30 years (Ref 5). A resurgence in this activity occurred in 1989 when it was recognized that MA can be used as the basis of a chemical refining process (Ref 23) by demonstrating that the pure metal copper can be synthesized when CuO and calcium were ball-milled together at room temperature. Simultaneous reduction of CuO and ZnO by calcium has also been shown to result in the formation of β-brass. Metal synthesis directly from oxides or chlorides has now been extended to the synthesis of a number of metals (Zn, Ti, Zr, Ta, Gd, Er, Sm, V, W and some intermetallics, e.g. $SmCo_5$). A concise review on this subject is found in Ref 24.

Most of the reactions studied to date have been displacement reactions:

$$MO + R \rightarrow M + RO$$

where a metal oxide (MO) is reduced by a reductant (R) to the metal (M). Metal chlorides have also been reduced to metals this way.

The displacement reactions studied by MA are characterized by a large negative free energy change at room temperature and are therefore thermodynamically feasible at room temperature. However, commercial operations by pyrometallurgical techniques are conducted at elevated temperatures to overcome the kinetic barriers and achieve sufficiently high reaction rates. Mechanical alloying can provide the means to substantially increase the reaction kinetics of the displacement reactions because the repeated fracturing and welding of powder particles increases the area of contact between the particles and allows fresh surfaces to come into contact repeatedly. This allows the reaction to proceed without diffusion through the product layer. Additionally, the high defect densities induced during MA accelerate the diffusion process. As a consequence, these reactions will now occur at room temperature. If a reaction cannot occur at room temperature, the particle refinement and consequent reduction in diffusion distances (due to microstructural refinement) can at least significantly reduce the reaction temperatures.

Depending on the milling conditions, two entirely different reaction kinetics have been noted. The reaction can extend to a very small volume during each collision between the grinding media leading to a gradual transformation. Alternatively, if the reaction enthalpy is sufficiently high, a self-propagating combustion reaction can be initiated. Such reactions require a critical milling time for the combustion reaction to be ignited. If the temperature of the powder (or milling vial) is recorded during MA, a sudden increase in temperature at a critical time indicates the onset of the combustion event. Measurements of the ignition time provide a useful means of characterizing the structural and chemical evolution during MA.

The product of the displacement reactions normally consists of two phases—the metal (or a compound) and the oxide or chloride associated with the reductant. The removal of the unwanted reaction by-product can be difficult due to the high reactivity of the metal phase associated with nanocrystalline grain sizes and intermixing of the phases induced by the MA process. The by-product phase can be easily removed if the metal particles are embedded in a continuous matrix. Removal of the by-product is achieved by leaching the product mixture in a dilute acid or hot water, or by vacuum distillation. The use of carbon or hydrogen as the reductant produces

(a)

(b)

Fig. 12 (a) Furnace baskets made of INCOLOY alloy MA 956 operating in air at temperatures >1200 °C. (b) Removable hearth of vacuum heat treating furnace fabricated from 1.0 and 0.5 in. diam rod of INCOLOY alloy MA 956. The hearth legs are encased in alumina sleeves to prevent direct contact with the graphite heating elements. Courtesy of Inco Alloys International

gaseous CO_2 or water vapor as the byproduct and obviates the need for leaching/distillation.

Process parameters such as milling temperature, ball diameter, ball-to-powder weight ratio, use of a process control agent, and relative proportion of the reactant phases seem to play an important role on the nature and amount of product phase obtained by the displacement reactions and these need to be optimized for the best yields.

The mechanically driven displacement reaction offers a number of advantages over conventional metal processing techniques. First, it enables the reduction of a number of oxides and halides to pure metals at room temperature, thus effecting energy savings. Second, if a number of components are reduced simultaneously, it is possible to produce an alloy without first having to convert the oxides to pure metals and then to the desired alloy. Third, for powder metallurgy applications, it allows the direct formation of powder product without first having to manufacture the bulk alloy and then convert it to powder form. Thus, a number of high-temperature processes can be combined into one single room-temperature process with the potential for significant cost savings.

Powder Contamination

A major concern in the MA process is the impurities that get into the powder and contaminate it. The small size of the powder particles, availability of large surface area, and formation of fresh surfaces during milling all contribute to the contamination of the powder. Thus, it appears as though powder contamination is an inherent drawback of the technique, unless special measures are taken to avoid/minimize it.

As mentioned earlier, in the early stages of MA, the metal powder coats the surface of the grinding medium and the inner walls of the container. This was expected to have prevented contamination of the powder and so no attention was paid to the problem of powder contamination. However, when different results were reported by different groups of researchers on the same alloy system, it was recognized that contamination could be a problem. This problem appears to be ubiquitous and is now encountered in many investigations, especially when reactive metals such as titanium and zirconium are milled. The magnitude of contamination appears to depend on the time of milling, the intensity of milling, the atmosphere in which the powder is milled, and difference in strength/hardness of the powder, and the milling medium. Whereas 1 to 4 wt% iron has been found to be normally present in most of the powders milled with the steel grinding medium, amounts as large as 7 wt% (20 at.%) iron in a tungsten-carbon mixture milled for 310 h and 13 wt% (33 at.%) iron in pure tungsten milled for 50 h in a SPEX mill were also reported. These are very high levels of contamination. Similarly, large amounts of oxygen (up to 36.5 at.%) and nitrogen (up to 22.6 at.%) have also been reported to be present in aluminum-titanium powders milled for 400 h in a low-energy ball mill.

Contamination of metal powders can be traced to chemical purity of the starting powders, milling atmosphere, milling equipment, and the process control agents added to the powders. Contamination from chemical purity can be either substitutional or interstitial in nature, while contamination from the milling atmosphere is essentially interstitial in nature and that from milling equipment is mainly substitutional, even though carbon from the steel equipment can be an interstitial impurity. Contamination from the PCA leads to interstitial contamination. The presence of interstitial impurities such as oxygen and nitrogen is deleterious to reactive metals like titanium and zirconium, and therefore maximum impurity levels are generally specified for acceptable microstructural and mechanical properties. Substantial amounts of nitrogen and oxygen (the amount of nitrogen is much higher than oxygen) are picked up during the milling of titanium and zirconium alloys and the presence of these impurities leads to a change in the constitution of the alloys (Ref 25–27). For example, the formation of an amorphous phase in titanium-aluminum alloys and a crystalline phase with a face-centered cubic (fcc) structure in powders milled beyond the formation of an amorphous phase have been attributed to the presence of large quantities of nitrogen in these alloys (Ref 12).

Several attempts have been made in recent years to minimize the powder contamination during MA. One way of minimizing the contamination from the grinding medium and container is to use the same material for the container and grinding medium as the powder being milled. In this case also there will be wear of the grinding medium and this gets incorporated into the powder. Thus, even though there is no contamination, the chemistry of the final powder will be quite different from the starting powder; the metallic content (of the container and balls) would be higher than in the initial powder. This can be compensated for if we know how much of the metallic content is increased.

The above solution may be possible occasionally; but it is difficult in many cases due to the nonavailability of the special grinding medium and containers. The problem is becoming more and more complex because the technique of MA is being applied to a variety of materials, such as metals, alloys, ceramics, polymers, and composites, and it is impossible to get containers of all these types of materials. If a container of the same material to be milled is not available, then a thin adherent coating on the internal surface of the container with the material to be milled will minimize the contamination. If this is not possible, then a simple rule that could be followed to minimize contamination is that the container and grinding medium should be harder/stronger than the powder being milled.

The problem of milling atmosphere is serious and has been found to be the major cause of contamination in many cases. In fact, it has been observed that if the container is not properly sealed, the atmosphere surrounding the container, usually air (containing nitrogen and oxygen), leaks into the container and contaminates the powder. Thus, when reactive metals like titanium and zirconium are milled in improperly sealed containers, the powders are contaminated with nitrogen and oxygen. It has been reported that flushing with argon gas will not remove oxygen and nitrogen absorbed on the internal surfaces. Pickup of impurities during milling would reduce the pressure within the container allowing outside atmosphere to continuously leak into the container through an ineffective seal. In practice, it has been noted that if it is difficult to open the container lid due to the vacuum present inside, it is an indication that contamination of the powder is minimum.

Attempts have been made to improve the container seal integrity to prevent the outside atmosphere leaking inside. Use of high-purity argon (99.998%) atmosphere and improvements of the seal quality resulted in the processing of high-quality titanium alloy powder with as little as 100 ppm oxygen and 15 ppm nitrogen (Ref 26). This process, however, may not be economically viable and hence may not be feasible on an industrial scale.

It is also important to remember that cross contamination could occur if a container that was used earlier to mill some powder is used again to mill another powder without properly cleaning it.

The level of contamination can be different under different processing conditions and is dependent on the type of mill, intensity of milling, nature of the powder, nature of the grinding medium and container, ambient atmosphere, ball-to-powder weight ratio, seal integrity, and others. Claims have been made in the literature about the superiority of certain mills and practices over others, but systematic investigations on milling the same powder under identical conditions in different mills and evaluating the contamination levels have not been undertaken.

Modeling

From the previous description it is easy to realize that MA is a complex process. Like any other process, modeling of MA is carried out to identify the salient factors affecting the process and to establish process control instrumentation. By modeling the process effectively, it is possible to lower the number of actual experiments to be conducted to optimize the process and achieve a particular application.

The number of variables involved in the MA process is very large. For a particular alloy system, the variables include the type of mill, intensity of milling, type of milling media, ball-to-powder weight ratio, the atmosphere under which the powder is milled, purity of the powders, milling time, milling temperature, and nature and amount of the PCA used. These have a significant effect on the constitution of the powder. Even on a local scale, the nature of impacts between two balls, the frequency of impacts, and the amount of powder trapped between two balls during a collision could vary from point to point. Thus, modeling the MA process is a difficult task. In spite of this, some attempts have been made (Ref 28–30) and moderate success has been achieved in modeling the mechanics of the process. From the actual experiments conducted, attempts have been made to correlate the phases formed with the process parameters during milling. But, the ability to predict the final chemical constitution of the powder (type and description of phases formed) has not been achieved. It should be realized that due to the stochastic nature of the MA process, it is difficult to make absolute predictions.

REFERENCES

1. J.S. Benjamin, *Sci. Amer.,* Vol 234 (No. 5), 1976, p 40–48
2. J.S. Benjamin, *Met. Powder Rep.,* Vol 45, 1990, p 122–127
3. A.E. Yermakov, Y.Y. Yurchikov, and V.A. Barinov, *Phys. Met. Metallogr.,* Vol 52 (No. 6), 1981, p 50–58
4. C.C. Koch, O.B. Cavin, C.G. McKamey, and J.O. Scarbrough, *Appl. Phys. Lett.,* Vol 43, 1983, p 1017–1019
5. G. Heinicke, *Tribochemistry,* Akademie Verlag, Berlin, Germany, 1984
6. P.S. Gilman and J.S. Benjamin, *Ann. Rev. Mater. Sci.,* Vol 13, 1983, p 279–300
7. C.C. Koch, Mechanical Milling and Alloying, *Materials Science and Technology—A Comprehensive Treatment,* Vol 15, R.W. Cahn, Ed., VCH, 1991, p 193–245
8. C. Suryanarayana, *Metals and Mater.,* Vol 2, 1996, p 195–209
9. C. Suryanarayana, *Bibliography on Mechanical Alloying and Milling,* Cambridge International Science Publishing, 1995
10. H. Bakker, G.F. Zhou, and H. Yang, *Prog. Mater. Sci.,* Vol 39, 1995, p 159–241
11. E. Ivanov, *Mater. Sci. Forum,* Vol 88–90, 1992, p 475–480
12. C. Suryanarayana, *Intermetallics,* Vol 3, 1995, p 153–160
13. E.G. Avvakumov, *Mechanical Methods of Activation of Chemical Processes,* Nauka, Novosibirsk, Russia, 1986
14. A. Calka, J.J. Nikolov, and J.S. Williams, *Mater. Sci. Forum,* Vol 225–227, 1996, p 527–532
15. G. Jangg, in *New Materials by Mechanical Alloying Techniques,* E. Arzt and L. Schultz, Ed., Deutsche Gesellschaft für Metallkunde, Oberursel, Germany, 1989, p 39–52
16. C.C. Koch, *Nanostructured Mater.,* Vol 2, 1993, p 109–129; Vol 9, 1997, p 13–22
17. G.A.J. Hack, *Metals and Mater.,* Vol 3, 1987, p 457–462
18. J.J. Fischer, J.J. deBarbadillo, and M.J. Shaw, *Heat Treating,* Vol 23 (No. 5), 1991, p 15–16
19. C. Suryanarayana, *Int. Mater. Rev.,* Vol 40, 1995, p 41–64
20. S.A. Black, Report of Civil Eng. Lab. (Navy), Port Hueneme, CA, 1979
21. S.S. Sergev, S.A. Black, and J.F. Jenkins, U.S. Patent 4, 264, 362, 13 Aug 1979
22. P. Tessier, L. Zaluski, A. Zaluska, J.O. Ström-Olsen, and R. Schulz, *Mater. Sci. Forum,* Vol 225–227, 1996, p 869–874
23. G.B. Scahffer and P.G. McCormick, *Appl. Phys. Lett.,* Vol 55, 1989, p 45–46
24. P.G. McCormick, *Mater. Trans., JIM,* Vol 36, 1995, p 161–169
25. G.H. Chen, C. Suryanarayana, and F.H. Froes, *Metall. Mater. Trans. A,* Vol 26, 1995, p 1379–1387
26. P.S. Goodwin and C.M. Ward-Close, *Mater. Sci. Forum,* Vol 179–181, 1995, p 411–418
27. T. Klassen, M. Oehring, and R. Bormann, *J. Mater. Res.,* Vol 9, 1994, p 47–52
28. T.H. Courtney, *Rev. Part. Mater.,* Vol 2, 1994, p 63–116
29. M. Magini and A. Iasonna, *Mater. Trans., JIM,* Vol 36, 1995, p 123–133
30. M. Abdellaoui and E. Gaffet, *Acta Mater.,* Vol 44, 1996, p 725–734

Spray Drying and Granulation

GRANULATION is a term used to describe two different types of processes. In one definition, granulation is the intentional agglomeration of fine particles into larger clusters to improve certain powder properties. For example, bulk powders typically have a low bulk density, do not readily flow, are dusty, and have low thermal conductivity. When properly granulated, the same powder pours easily, exhibits a high and uniform bulk density, does not experience dusting losses, and more efficiently transfers thermal energy. For example, in the case of a powder used as a feed material for high speed presses, the granules should typically be greater than 50 μm (2 mils), but less than 1000 μm (40 mils) in diameter, have a spherical shape, and should not be strong enough to retain their identity in the compacted part.

Granulation is also defined as the production of metal particles by agitating molten metal. The most commonly used techniques are spray drying and spray granulation. Another method is water granulation. These three methods are discussed in this article. Other granulation methods include agitation and pressure technique for ceramics (Ref 1).

Spray Granulation

Spray granulation (Ref 2) forms granules by atomizing a liquid or a binder solution into a fluidized powder bed (Fig. 1). Fluidization is achieved by directing a heated gas, which is usually air, through a distributor at the bottom of the powder bed. The gas imparts a vigorous motion to the particles, which prevents the formation of large lumps. The binding liquid is usually sprayed into the powder bed with a two-fluid nozzle. Spray granulators can be designed to operate in either a batch or a continuous mode.

The formation of granules in a spray granulator occurs through the random nucleation of small seed agglomerates, followed by the growth of these seeds to the desired size. Growth occurs either by the layering of powder onto the seeds or by the agglomeration of seeds to form larger granules. Granule growth by seed agglomeration forms irregular shapes.

Granule size increases as the fraction of the bed exposed to the binding liquid is reduced and as the spray nozzle is adjusted to give coarser droplets (Ref 3). Increasing the intensity of agitation of the bed (with a higher gas velocity) decreases the size of the granules. There is an upper limit on granule size because of the tendency of the powder bed to defluidize. However, spray granulation can form larger granules than is usually possible by spray drying because of longer residence times.

Spray granulation is commonly used in the pharmaceutical industry to prepare feedstock for tablet presses. It is used less frequently in the metal powder industry. However, both spray drying and fluid bed conversion are scaleable technologies and together provide the means for producing bulk quantities of nanophase composite powders at low manufacturing cost. A few examples of fluidized bed granulation are described below.

Metal Powder Granulation in a Plasma-Spouted/Fluidized Bed. A direct current (DC) plasma-spouted/fluidized bed was applied to the granulation of spherical alloy grains from metal powder mixtures. From a mixture of iron powder (149 to 210 μm) and aluminum powder (74 to 88 μm and 125 to 149 μm), alloy grains from 1 to 5 mm in diameter were obtained. The grains exhibited a dense homogeneous core and a porous nonhomogeneous shell structure. The grains were quite spherical; the average aspect ratio of the grains was >0.85. The mass fraction of iron in the core section of product grains depended on the Al/Fe ratio of initial powder mixtures but was insensitive to the size of product grains and to the size of initial Al particles. Selective growth of seed grains was observed when sufficient seed grains were added to the starting powder mixture (Ref 4).

Contamination-Free Processing of Pyrophoric Rare Earths and Abrasive Ceramic Powders. The production of exactly limited particle fractions is an important process for rare earths (Sm-Co, Nd-Fe-B) and ceramic (oxides, carbides, nitrides, borides) products. Extremely fine products with tight upper particle size limits can be produced by a fluidized bed process (Ref 5). This allows practically wear- and contamination-free grinding of extremely hard products. The necessary grinding and classification can either be wet or dry, or a combination of both procedures (Ref 5).

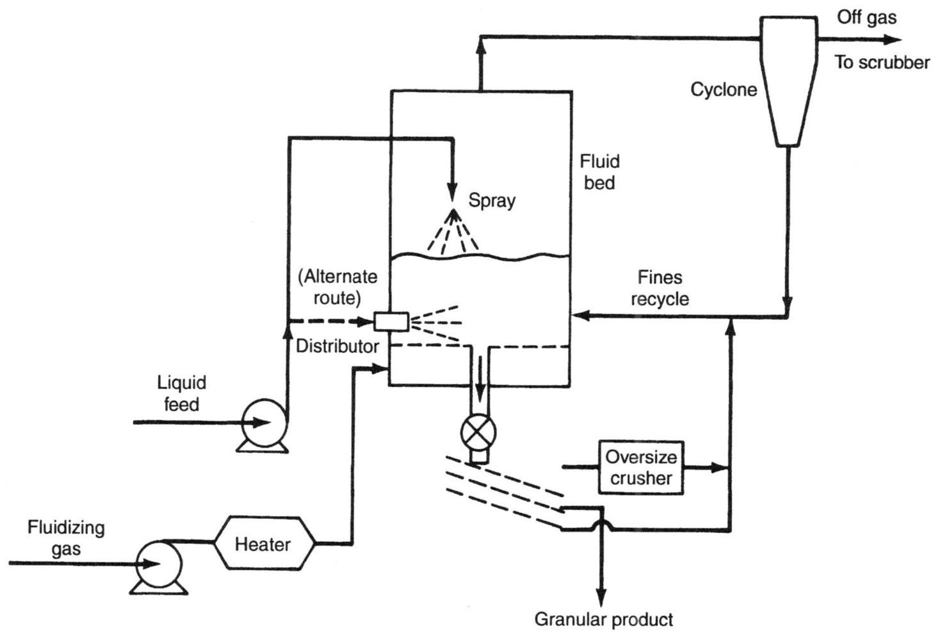

Fig. 1 Fluidized bed spray granulator. Source: Ref 3

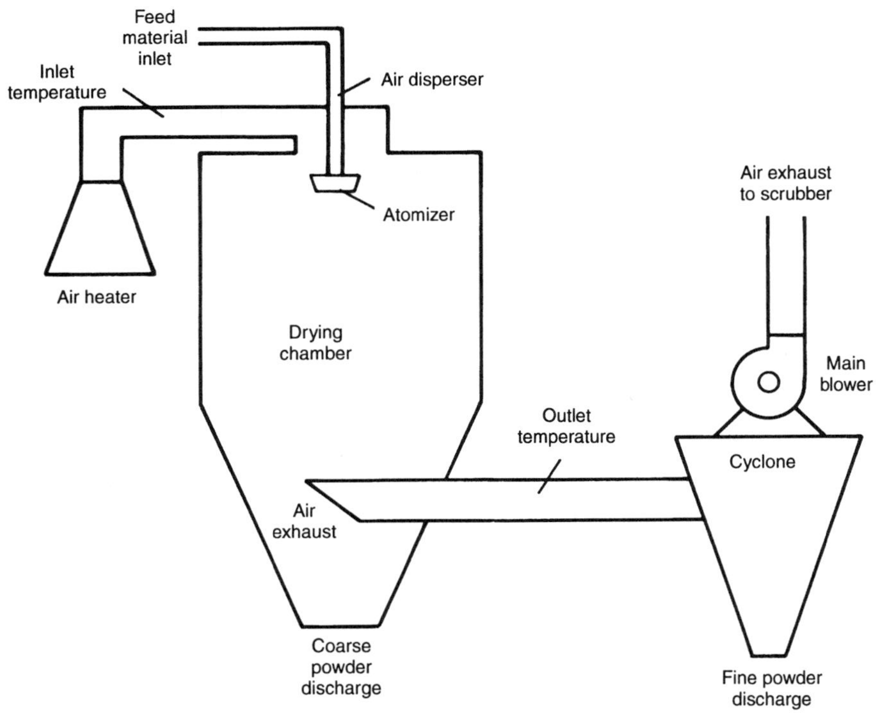

Fig. 2 Typical spray dryer arrangement

Water Granulation

Water granulation is a process for producing coarse metal particles by pouring molten metals through a screen into water or by agitating molten metals into droplets with subsequent water quenching. Granulation of liquid metal offers a simple technique for solidification of metals, but pouring of liquid metal into water has always been regarded as hazardous. Nonetheless, several plants for granulation of steel and different kinds of ferroalloys have shown that granulation can be done in a safe and reliable way (Ref 6).

For example, granulation of ferrosilicon and Si-metal has a risk of explosions when done at high flow rates (Ref 6). The explosion potential is considered to be due to the rapid formation of steam. Light metals like ferrosilicon and Si-metal have a comparatively low falling velocity in the water and thus result in a high power generation per unit volume of water. To decrease the power concentration, the heated water has to be efficiently replaced, or the heat has to be distributed to a big volume. It has been found that the replacement of the water from a small cooling zone in a big water tank is rather difficult, but metal can be efficiently distributed over a big water surface to decrease explosion potential (Ref 6).

Another trial of water granulation of ferrosilicon was performed by impingement of molten alloy stream against stationary ceramic plate or rotary cone followed by water quenching of the drops. The test stand permitted variations of nozzle-plate/cone and plate/cone-water distances and adjustment of the inclination of the plate. Quality and size of the product was affected chiefly by the plate/cone-water distance. Better quality was obtained with impingement against rotary cone. Results provided the basis for development of an industrial-scale installation for granulation of ferrosilicon (Ref 7).

Spray Drying

Spray drying is a powder-producing process in which a slurry of liquids and solids or a solution is atomized into droplets in a chamber through which heated gases, usually air, are passed. Figure 2 represents a typical spray dryer utilizing a disk atomizer with co-current air flow. Examples of commercially spray-dried powders are shown in Fig. 3. Spray drying is used widely in the pharmaceutical, chemical, and food industries. It is used to a lesser extent in the metals-related industries.

Spray drying offers several advantages over other powder-processing techniques, particularly in applications requiring agglomerates for subsequent pressing and sintering operations. Spray drying also is one of the most economical ways of drying slurries.

Spray drying is a continuous rather than a batch process that allows close control of agglomerate size, bulk density, and moisture content. Agglomerate shape is spherical, facilitating excellent flowability. Particles have very short exposure time. Residence time within the dryer may range from 2 to 20 s, depending on dryer

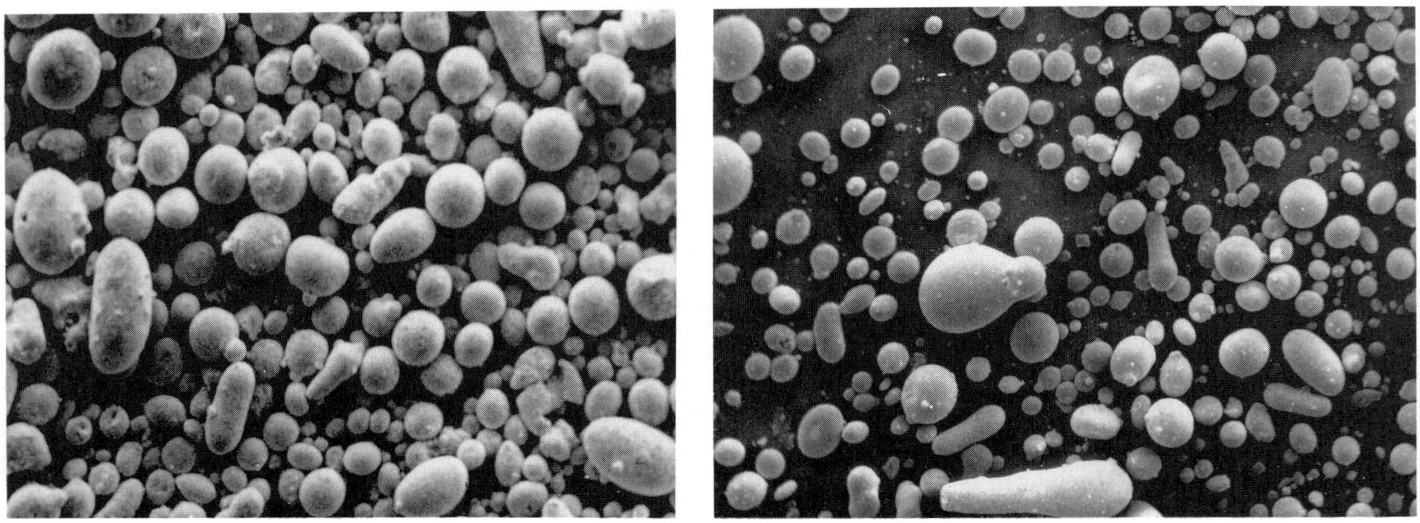

(a) (b)

Fig. 3 Micrographs of commercially spray-dried granules. (a) Ferrite, 75×. (b) Zirconia, 30×

size. Therefore, heat-sensitive materials can be spray dried. Lubricating additives can be added easily for die pressing applications,

Process Description

During spray drying, atomized droplets dry rapidly because of their high surface-to-volume ratio. Coarse dried solids fall to the bottom of the chamber and are continuously collected, typically through a rotary airlock valve. The finer solids, which are entrained in the gases exiting the chamber, are removed by and collected from a cyclone separator. Baghouse collectors or wet scrubbers can be used to clean ultrafine particles from the air exhausted from the cyclone. Baghouse collectors and electrostatic precipitators also can be substituted for cyclone collectors. An alternate spray dryer design includes the introduction of the air through ductwork above the cyclone exit.

Size of the drying chamber is dictated by (a) product characteristics, (b) product throughput, (c) atomization technique, (d) the properties of the slurry or solution, and (e) the evaporative capacity of the heating unit and associated fans. The evaporative capacity is a function of the allowable inlet and outlet temperatures of the dryer. These temperatures are, in turn, dictated by the heat sensitivity of the product. The interrelationship of these factors is shown in Fig. 4 (Ref 8). Generally, a spray drying facility is designed as an integral unit so that all component capacities are compatible.

Some special applications in the cemented carbide industry require the use of organic solvents, such as acetone, ethanol, hexane, and methanol. Because of the flammable nature of these materials, closed-system drying technology (see Ref 9–11) requires that the medium be spray dried using a nonoxidizing gas, such as nitrogen. By recycling the nitrogen that contains organic vapors, organic emissions are eliminated, and the solvent can be recycled.

Closed-cycle systems are gastight and operate slightly above atmospheric pressure. Most of the dried particles are collected from the dryer chamber; a small percentage of fines that pass through the chamber in the gas stream is collected in the cyclone.

The drying gas subsequently passes through a scrubber/condenser to recover the solvent and to recondition the nitrogen. Typically, gas temperature is controlled by a recirculating scrubber and heat exchangers that are cooled by water or brine. For some spray drying applications, a cooling system may be required.

Droplet-Air Mixing. The air flow pattern in the drying chamber controls the completeness of moisture removal from the droplet, the maximum temperature that the granules will experience, and the formation of wall deposits. Droplet-air mixing in spray dryers is determined by the location of the air disperser and the atomizing device and is classified according to the relative direction between the droplets and the air. These classifications are termed cocurrent, countercurrent, and mixed-flow mixing.

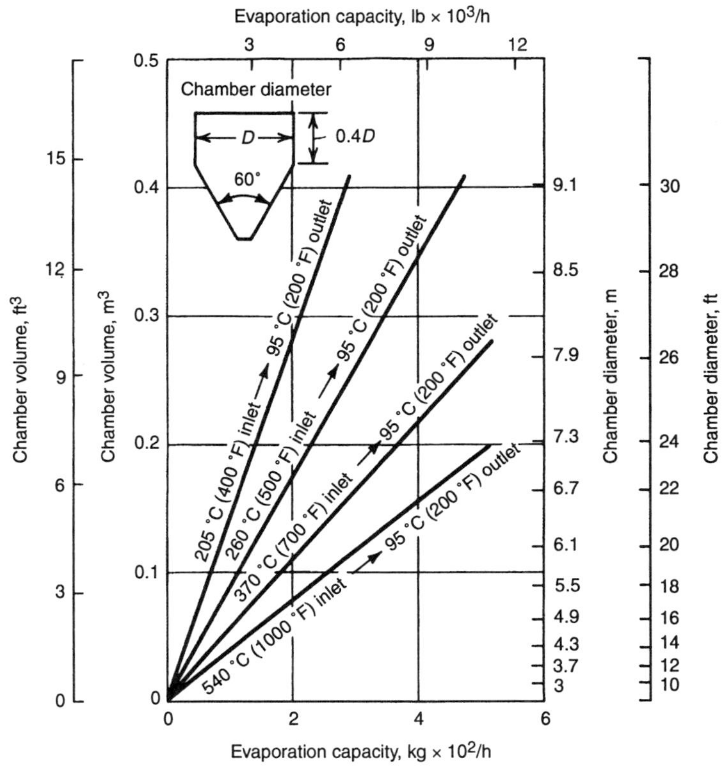

Fig. 4 Relationship between chamber volume and evaporation capacity of spray dryers. Data based on retention time of 20 s. Source: Ref 8

Cocurrent conditions exist when the atomizing device is located near the air disperser at the top of the dryer, as in Fig 5(a). The droplets are exposed to the hottest air immediately after formation, but high evaporation rates maintain low product temperatures. As the moisture content of the droplets decreases, they come into contact with cooler air and surface temperature does not increase appreciably. Cocurrent air flow is common in dryers equipped with rotary atomizers.

In cocurrent drying, the maximum temperature to which larger dried particles are exposed is approximately 10 °C (18 °F) less than the outlet temperature of the dryer. This is caused by the evaporative cooling effect that occurs when the liquid leaves the droplet while passing through the heated zone. Finer particles, which dry more quickly, experience temperatures approaching those of the inlet gas.

Countercurrent conditions occur when the atomizing device is placed at the top of the chamber and the air disperser is located at the bottom (Fig. 5b). Immediately after formation, the droplets contact cool humid air. However, as their moisture content decreases, they are exposed to increasingly hotter air. Because high internal temperatures can be realized in the granules, the organic binders should not be heat sensitive.

When the slurry is introduced from the bottom and gas from the top of the dryer, droplets dry both in their upward and downward trajectories. Therefore, narrower, shorter dryers can be used. However, if partially dried products cannot be exposed to the high temperatures at the top of the dryer, this mode of spraying may not be applicable.

Mixed-flow conditions are a combination of cocurrent and countercurrent air flow. This type is commonly utilized in fountain-type dryers, where a nozzle atomizer is located at the base of the drying chamber and the air disperser is placed at the top of the chamber, as in Fig. 5(c). Mixed-flow conditions are frequently used in combination with pneumatic nozzles in small laboratory dryers because the fountain-like spray pattern increases droplet trajectory and provides sufficient airborne time to dry large droplets in small chambers.

Control of Powder Properties

Spray drying can be used to remove water from a slurry or to create agglomerates of fine particles contained within the slurry. Agglomeration is achieved by using a binder (discussed below in this section).

Most of the applications of spray drying in the metals industry require the formation of free-flowing agglomerates. Many of these powders are used for producing pressed parts. Therefore, the agglomerate size distribution and bulk density are the two most important properties of spray-dried powders.

Agglomerate size distribution is a function of atomization conditions and the properties of the slurry. Generally, a lower solids content yields a finer average agglomerate size. The maximum attainable solids content varies with material, but usually can be increased by using deflocculating or suspending agents.

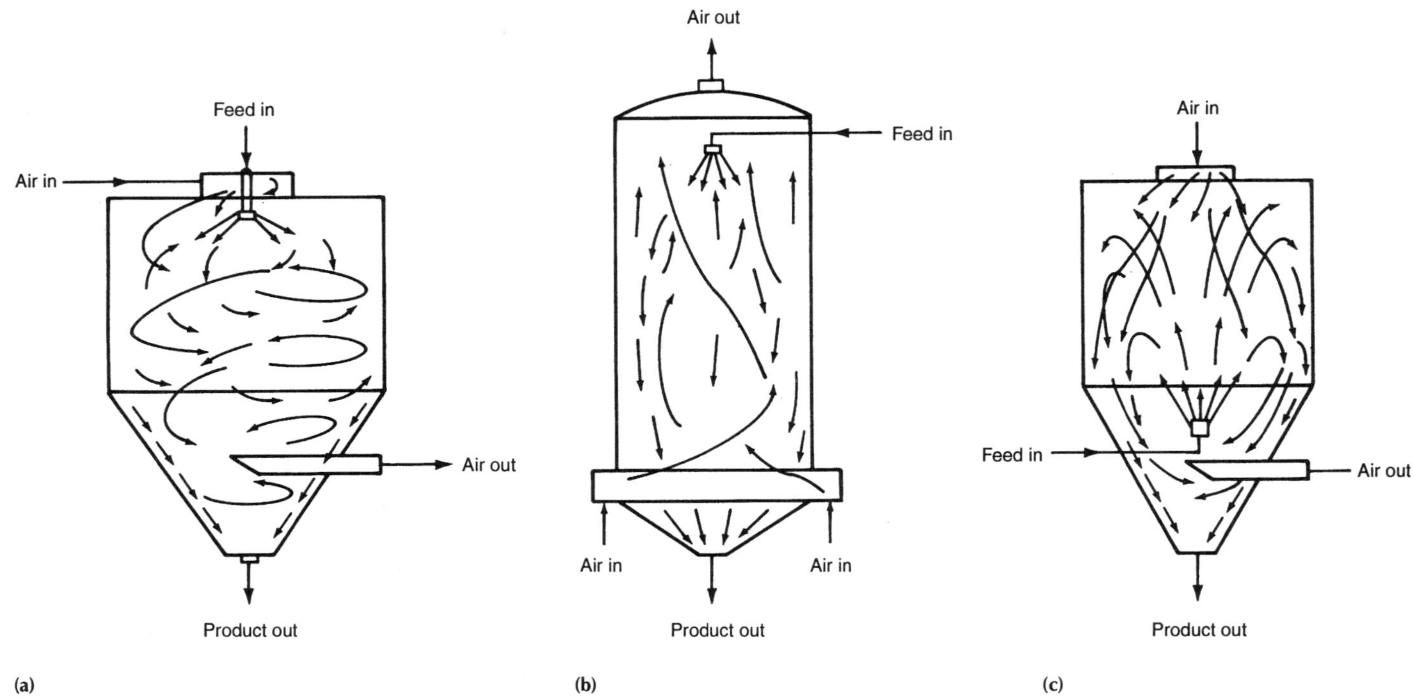

Fig. 5 Types of droplet-air mixing in spray dryers. (a) Cocurrent. (b) Countercurrent. (c) Mixed-flow. Source: Ref 12

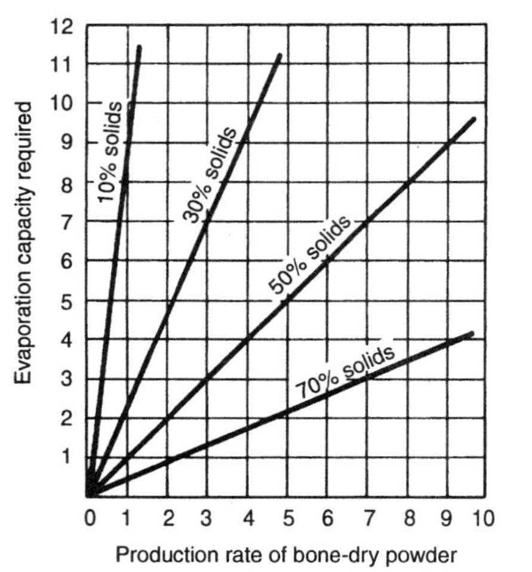

Fig. 6 Relationship of evaporation capacity of spray dryers to production rate and slurry solids concentration. Source: Ref 3

Bulk density is a function of the solids content of the slurry, inlet temperature, and slurry additives. Lower bulk densities generally are achieved from slurries with low solids contents. Also, excessive inlet temperatures can cause lower bulk densities. Rapid evaporation of the liquids causes the partially dried droplets to expand rapidly, thus decreasing density. Introduction of frothing agents may entrap air in the slurry, which also leads to lower bulk densities. Typically, frothing agents are not added in metallurgical applications.

Moisture content of a powder can be controlled by the inlet and outlet temperature of the spray dryer in conjunction with the slurry feed rate. Moisture levels below 0.1% are possible. For a given airflow and inlet temperature, outlet temperature is controlled by the rate of slurry feed and the solids content. With higher percentages of solids, less water must be evaporated, which leads to higher throughput of dry product. If a product can withstand higher inlet temperatures, throughput can be increased (see Fig. 6).

Binders for Agglomeration. Suitable binder materials must be homogeneously dispersable (preferably soluble) in the liquid used to form the slurry. When dry, binders must form a coating and/or adhere to the material being agglomerated. They must impart the required strength and crush resistance to the particle for subsequent handling. In addition to the liquid, solids, and binders used to formulate a slurry, various other additives may be necessary. The following are typical components of spray drying slurries:

Organic binders

- Polyvinyl alcohol
- Gum arabic
- Other natural gums
- Carboxy-methyl cellulose salts
- Polyvinyl acetate
- Methyl cellulose
- Ethyl cellulose
- Polyvinyl butyral dispersions
- Protein colloids
- Acrylic resin emulsions
- Ethylene oxide polymers
- Water-soluble phenolics
- Lignin sulfonates
- Propylene glycol alginates
- Flour
- Starches

Inorganic binders

- Sodium silicate
- Boric acid
- Borax
- Carbonates
- Nitrates
- Oxylates
- Oxychlorides

Plasticizers

- Glycerine
- Ethylene glycol
- Triethylene glycol
- Dibutyl phthalate
- Diglycerol
- Ethanolamines
- Propylene glycol
- Glycerol monochlorhydrin
- Polyoxyethylene aryl ether

Deflocculating agents

- Sodium hexametaphosphate
- Sodium molybdenate
- Tetrasodium pyrophosphate
- Ammonium citrate
- Ammonium oxalate
- Ammonium tartrate
- Ammonium chloride
- Monoethylamine

Wetting agents

- Synthetic detergents
- Alkylaryl sulfonates
- Alkylaryl sulfates
- Soaps

Suspending agents (high molecular weight)

- Sodium carboxymethyl cellulose
- Methyl cellulose
- Ethylene oxide polymers

Plasticizers may be used with binding materials that are hard or brittle and that tend to crack during drying. Suspending agents may be needed to prevent solids from settling within the slurry. Deflocculating agents aid in the formation of slurries by preventing the agglomeration of fine particles. Wetting agents also may be used to maintain solids in suspension. Some slurries have a tendency to foam during mixing. Antifoaming agents or defoamers may be used to control this action. Chemical activators also may be used as additives to aid in subsequent sintering or processing of powders.

Atomization Techniques and Agglomerate Size Distribution. Three standard techniques are used to atomize slurries for spray drying:

- Single-fluid nozzle atomization
- Centrifugal (rotating disk) atomization
- Two-fluid nozzle atomization

Table 1 gives the advantages and disadvantages of these techniques for atomization of slurries. Table 2 gives the relative agglomerate sizes produced by these techniques.

As shown in Table 2, the largest agglomerate sizes (600 μm) are achieved by the single-fluid nozzle. The centrifugal (rotating disk) atomizer yields agglomerate sizes up to 300 μm, and the two-fluid nozzle produces agglomerates only up to about 200 μm in size. Centrifugal atomization yields the narrowest agglomerate size distribution range, followed by single-fluid atomization and two-fluid atomization.

Centrifugal atomization entails the introduction of the slurry into a horizontally rotating disk that is equipped with vanes or holes through which the slurry exits. The slurry is atomized into fine droplets, the average size of which is a function of the design and peripheral speed of the wheel. Most wheel designs incorporate a wear-resistant material for extended life.

Single-fluid atomization is capable of producing the largest diameter agglomerates. However, because of the high pressures required to force the slurry through a single, small orifice to produce the desired droplet size, considerable downtime may be required to replace worn parts and to unplug the nozzle. Use of a nozzle in the fountain spray position extends the residence time of the droplets in the chamber. As a result, smaller dryers can be used for single-fluid atomization of materials that are not heat-sensitive.

Two-fluid atomization, which uses a pressurized air blast to break up a slurry stream into droplets, produces the widest range of agglomerate sizes of all three atomization techniques. However, it is a relatively easy technique for producing atomized droplets. Wear problems do not exist because high pressures are not employed. The orifices used are typically larger than those used for single-fluid atomization; therefore, plugging is not as severe a problem. This technique does, however, require the use and additional expense of pressurized air.

Table 1 Advantages and disadvantages of various spray drying atomization systems

Atomization system	Advantages	Disadvantages
Centrifugal (rotating disk)	High feed rates, less downtime, low-pressure pumps, never plugs	Larger diameter dryer required, coarse agglomerate size not obtainable
Single-fluid nozzle	Large agglomerate capability, smaller dryers can be used	Downtime due to part wear and plugging, high pressure pumps
Two-fluid nozzle	Less part wear	Broad agglomerate size distribution, compressed air needed

Table 2 Relative agglomerate sizes produced by various atomizing systems

Atomizing system	Mean agglomerate size, μm
Centrifugal (rotating disk)	
High speed	25–100
Medium speed	50–200
Low speed	100–300
Single-fluid nozzle	
High pressure	25–100
Medium pressure	50–200
Low pressure	100–300
Very low pressure	200–600
Two-fluid nozzle	
High pressure	10–50
Medium pressure	25–100
Low pressure	50–200

Source: Ref 12

Applications

Spray drying applications are most prevalent in the pharmaceutical, chemical, and food industries. There are, however, several areas in the metals industry that utilize spray drying techniques. These include production of cemented carbides, mineral processing, production of iron powders, production of oxide-dispersion-strengthened alloys, and production of powders used for thermal spraying applications.

Agglomeration by spray drying is a basic process that has unlimited possibilities in combining different materials to produce composite powders, which behave homogeneously in bulk. Investigations in producing custom-made powders have led to various multicomponent powders. The first step of this technique consists of the agglomeration of the starting powder, for example, metallic, oxide, and nonoxide hard materials. The second step includes plasma densification. Using the high energy of the plasma, materials with extremely high melting points can be melted.

Cemented Carbides (Tungsten Carbide/Cobalt). Historically, cemented carbides have been pressed from powders produced by various agglomeration techniques. With the advent of closed-cycle systems, production of cemented carbides by spray drying increased considerably in the 1980s. Closed-cycle spray drying is required for most cemented carbide powders because the binders that are used are soluble only in volatile organic fluids.

The nitrogen drying gas that is used in the spray drying of cemented carbides is heated to 75 to 100 °C (170 to 210 °F), depending on the milling liquid used. The solids content of the slurry varies from 75 to 80%. Viscosity of milled slurries is sometimes modified with stabilizers, such as stearic acid (0.3 to 0.5 wt%). Pressures for single-fluid nozzle atomization range from 590 to 1470 kPa (85 to 213 psi). Pressure is a function of the particular type powder, slurry viscosity, and binder content. Recent work related to spray drying for cemented carbides is described in Ref 13–19.

Multicomponent Oxide Powders for Plasma Spraying. The concept of multicomponent powders consists of two steps. The first step is the agglomeration of the starting powder by the spray drying process. The second step involves plasma densification.

During spray drying, the starting powder combination is suspended in a liquid (water or solvent) and simultaneously mixed with binder and possible other needed auxiliary agents, for example, wetting agents, defoamers, etc. The suspension is sprayed through a nozzle into a heated chamber. Using rapid vaporizing, the formed droplets are dried briefly and micropellets are formed from the starting powder due to the adherence of the binder. This procedure and the properties of the resulting micropellets have certain advantages:

- There are almost no limits to the combination of different materials.
- Particle sizes of the micropellets are reproducible, dependent upon the process parameters.
- The starting components are homogeneously distributed in the micropellets.
- Nearly ideal spherical shape produces powders with excellent flowability.
- Lower energy consumption compared to other melting or crushing methods
- Easier control on mean diameter and size range

However, a key disadvantage of many spray dried powders is the poor mechanical resistance of the coating due to particle breakage (Ref 20). Particle breakage results in poor transport during the plasma spray coating process, which thus makes spray dried powder difficult to use for thermal spray. Fused and crushed powders are still the most widely used powder types for plasma spraying because they give deposits with good thermophysical properties and mechanical resistance.

To apply the advantages of spray dried powders for plasma spray coatings, a second step of plasma densification is required (Ref 20–23). In this step, the micropellets from spray drying are brought into a high energy plasma. Depending on the powder, one can choose whether the process takes place under atmospheric, inert atmospheric, vacuum, or underwater plasma spraying conditions. This additional step has the following advantages.

During plasma densification, reactions between the components of the micropellets occur, which during further processing, such as plasma spraying, leads finally to a more stable microstructure in the product. Without the plasma treatment, the reaction time is not sufficient, so the resulting metastable microstructure can lead to certain disadvantages and to property changes during use of the material.

Plasma densification enables high cooling rates so that disadvantages in microstructure—brittle phases or the solutioning of important phases—will not occur. This is an important advantage in contrast to conventional melting processes.

The plasma densified powder obtained has a spherical, smooth surface, a high density, and a porosity approaching zero. Experience in this technique concerning the plasma densification of metals, metallic hard materials, and ceramics shows that the multicomponent powder conception is practicable for nearly all materials. To characterize the properties, plasma sprayed coatings of plasma densified powders were investigated. Results revealed that coatings of plasma densified powders show better resistance to wear than coatings of agglomerated powders.

ACKNOWLEDGMENT

This article was adapted from "Granulation and Spray Drying" by Stanley Lukasiewicz in *Ceramic and Glasses*, ASM International, 1991 and "Spray Drying of Metal Powders" by David Houck in *Powder Metallurgy*, Volume 7, *ASM Handbook*, 1984.

REFERENCES

1. S. Lukasiewicz, Granulation and Spray Drying, *Ceramics and Glasses*, Vol 4, *Engineered Materials Handbook*, ASM International, 1991, p 100–108
2. J.S. Reed, *Introduction to the Principles of Ceramic Processing*, John Wiley & Sons, 1988
3. C.E. Capes, Particle Size Enlargement, *Handbook of Powder Technology*, Vol 1, J.C. Williams and T. Allen, Ed., Elsevier Scientific, 1980
4. M. Tsukada, K. Goto, R.H. Yamamoto, and M. Horio, Metal Powder Granulation in a Plasma-Spouted/Fluidized Bed, *Powder Technol.*, Vol 82 (No. 3), Mar 1995, p 347–353
5. H. Prem and D.J. Eddington, Contamination-Free Processing of Pyrophoric Rare Earths and Abrasive Ceramic Powders, *Powder Handl. Process.*, Vol 1 (No. 1), Mar 1989, p 101–107
6. P.-A. Lundstrom and A. West, Granulation of Ferroalloys and Silicon-Metal, *52nd Electric Furnace Conf.*, (Nashville, TN), Iron and Steel Society/AIME, 13–16 Nov 1994, p 309–315
7. L. Bulkowski et al, Trials of Water Granulation of Ferrosilicon, *Pr. Inst. Metal. Zelaza*, Vol 47 (No. 3), 1995, p 23–27
8. R.H. Perry and C.H Chilton, Spray Dryers, *Chemical Engineers Handbook*, 5th ed., McGraw Hill, 1973, p 58–63
9. A.O. Jensen and K. Masters, "Spray Dryer for Producing Tungsten Carbide Products," Bulletin F-125, Niro Atomizer, Inc., Columbia, MD
10. "Closed Circuit Spray Drying Systems," Bulletin 1342, Anhydro, Inc., Attleboro Falls, MA
11. "Closed Cycle Systems," Bowen Engineering Bulletin, Bowen Engineering, Inc., Somerville, NJ
12. K. Masters, *Spray Drying*, 4th ed., John Wiley & Sons, 1985
13. L. Wu, Nanostructured Tungsten Carbide/Cobalt Alloys: Processing and Properties, *Diss. Abstr. Int.*, Vol 54 (No. 9), Mar 1994, p 196
14. L.E. McCandlish, B.H. Kear, and S.J. Bhatia, Spray Conversion Process for the Production of Nanophase Composite Powders, U.S. Patent No. 5,352,269, 9 July 1991
15. S. Danzglock, Vacuum Plasma Spraying, *Metalloberfläche*, Vol 45 (No. 10), Oct 1991, p 455–458
16. B.-K. Kim, H.-S. Chung, L.E. McCandlish, and B.H. Kear, Fluid Bed Synthesis of Nanophase WC/Co Composite Powders, *Novel Powder Processing*, Vol 7, *Advances in Powder Metallurgy and Particulate Materials*, Metal Powder Industries Federation, 1992, p 51–61
17. B. Kim, "Synthesis, Processing and Characterization of WC/Co Nanophase Composites," *Diss. Abstr. Int.*, Vol 52 (No. 3), Sept 1991, p 243
18. G. Nagarajan and K. Sadananda, Hard-Metal Powder Granulation for Spray Drying (Retroactive Coverage), *PMAI Newsl.*, Vol 13 (No. 1), 12–15 Dec 1986
19. E. Lugscheider, H. Eschnauer, A. Nisch, and Z. Li, Plasma Treated Multi-Component Powders on the Basis of Metallic Hard Materials, *12th Int. Plansee Seminar '89*, Vol 3, Metallwerk Plansee GmbH, Reutte, Austria, 1989, p 221–236
20. A. Denoirjean, A Vardelle, A. Grimaud, P. Fauchais, E. Lugscheider, I. Rass, H.L. Heijen, P. Chandler, R. McIntyre, and T. Cosack, Plasma Densification of Zirconia Powders: Optimization for Thermal Barrier Coatings in IC Engines and Gas Turbines (Brite Project P2280), *Thermal Spray: Int. Advances in Coatings Technology*, ASM International, 1992, p 975–982
21. Z. Bartnik, P. Bialucki, S. Kozerski, G. Clinton, K. Davies, P. Bork, B. Schrader, F. Guglielmi, and L. Pawlowski, Improvements in Manufacturing Technology of Wear Resistant Plasma Sprayed Cr_2O Coatings, *Thermal Spray: Int. Advances in Coatings Technology*, ASM International, 1992, p 983–993
22. P. Luo, P.R. Strutt, T.D. Xiao, Synthesis of Chromium Silicide—Silicon Carbide Composite Powders, *Mater. Sci. Eng. B*, Vol 17 (No. 1–3), 28 Feb 1993, p 126–130
23. E. Lugscheider, M. Loch, and H.G. Suk, Powder Technology—State of the Art, *Thermal Spray: Int. Advances in Coatings Technology*, ASM International, 1992, p 555–559

Rotating Electrode Process

Steven A. Miller, Starmet Corporation
Peter R. Roberts, American Superconductor

THE ROTATING ELECTRODE process is a method for producing metal powders where the end of a metal bar is melted while it is rotated about its longitudinal axis. Molten metal is centrifugally ejected and forms droplets that solidify to spherical powder particles. The basic process is shown in Fig. 1. The electrode can be melted by any energy source, but usually an electric arc or plasma is used and the process is then identified by the acronyms REP or PREP, respectively. In this article REP is used generically unless a distinction needs to be made between the two processes. The REP concept was first invented by Starmet Corporation (then Nuclear Metals, Inc.), and resulted in the granting of patents (Ref 1, 2). While Starmet remains the sole source of REP/PREP (REP and PREP are registered trademarks of the Starmet Corporation) powder, powders are also made in Russia, China, and Japan using similar rotating electrode technology.

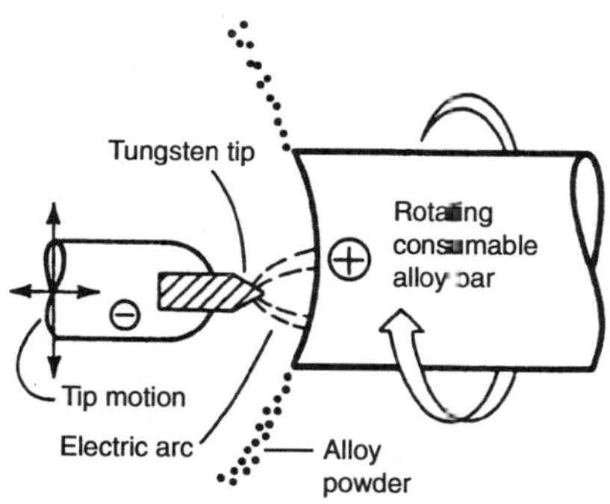

Fig. 1 Schematic of REP

The rotating electrode process has several inherent characteristics that make it uniquely suitable for the fabrication of specific alloy powders to provide manufacturing and product advantages. First, REP is a means of contactless melting and atomization to make powder with the highest level of cleanliness possible. This is a critical feature for reactive, high-melting-temperature alloys that are aggressively corrosive in their molten state and attack conventional ceramic crucibles. Such alloys are routinely atomized by REP without incurring contamination. Examples are titanium, zirconium, molybdenum, and vanadium alloys. Second, REP powder is almost perfectly spherical and practically satellite free. Because the atomized droplets are dispersed and move radially away from each other there is little opportunity for collisions between droplets and particles and the resulting coalescence of the two into irregularly shaped clusters. This single-particle nature of the powder spheres results in REP powder being very free flowing and having a high packing density, approximately 65%. Also, REP powder has both a tighter size distribution and a larger median size than can be produced by gas atomization. Finally, because REP atomization is produced by centrifugal forces rather than by aerodynamic drag, the powder is essentially porosity free when compared to gas atomized powder (Ref 3).

Fig. 2 Rotating electrode/plasma rotating electrode process machine for generation of specialty spherically shaped powders

The rotating electrode process has evolved to include equipment that consists of a vacuum/controlled-atmosphere tank 2440 mm (96 in.) in diameter by 300 mm long in which powder is produced. The tank dimensions are determined by the trajectory and solidification distance of the largest droplets of the molten spray. This is mounted with its circular plane section in the vertical position. Figure 2 shows a typical production setup. The consumable rotating electrode is introduced through a seal-and-bearing assembly; the long axis of the electrode is horizontal and centrally located in the tank and is made the anode of a direct-current power circuit. The permanent cathode may be a simple tungsten-tipped device provided with adequate cooling (REP) or a transferred arc plasma torch (PREP). This equipment projects through the other face of the tank to oppose the rotating electrode. Usually, melting is conducted under inert gas; the preferred medium is helium, which offers improved heat transfer properties and electric arc characteristics.

Equipment

There are two standard types of machines used in REP: short bar and long bar.

Short-Bar Apparatus. The short-bar machine (Fig. 3) accepts consumable anodes up to 89 mm (3½ in.) in diameter by 250 mm (10 in.) long. The anode is held in a collet in a precision spindle, the head of which projects into the tank through a rotating seal mechanism.

Usually at least 80% of the length of a short bar is converted to powder. Electrode stub removal and the introduction of new electrodes to the collet is performed manually through a glove port that is located in the front of the machine adjacent to the cathode or plasma torch. Short-bar methodology is appropriate for converting experimental quantities of material, alloys that are inherently brittle, and materials that have a low specific stiffness where an electrode of long aspect ratio is not practical.

Long-Bar Apparatus. Increased productivity and conversion efficiency are realized in the long-bar operation (Fig. 4), which is designed to consume 63.5 mm (2½ in.) diam electrodes that are up to 1830 mm (72 in.) long. Typically the assembly includes a precision spindle similar to that used in the short-bar machine. Instead of being fixed in position relative to the tank, the spindle is mounted on a table that also carries the drive motor and the electrical transfer brush mechanism. The table moves toward the tank from an extended starting position, feeding the electrode through the special seal-and-bearing assembly.

When the head of the spindle approaches the seal housing, the process is interrupted while the stub of the consumed electrode, now typically 230 to 250 mm (9 to 10 in.) long, is pushed through the seal by another long bar that is mounted in the spindle head when it is retracted to the starting position. Stubs produced in long-bar machines may be joined to new bars to obtain effectively 100% conversion to powder.

The rotational speed used is determined by the desired particle size. Standard speeds range between 314 rad/s (3000 rpm) and 1570 rad/s (15,000 rpm). Smaller diameter short-bars have been rotated up to 2620 rad/s (25,000 rpm). Because high-speed rotation rates are employed, electrodes must have precise dimensions; they must also be straight to keep mechanical out-of-balance forces to a minimum.

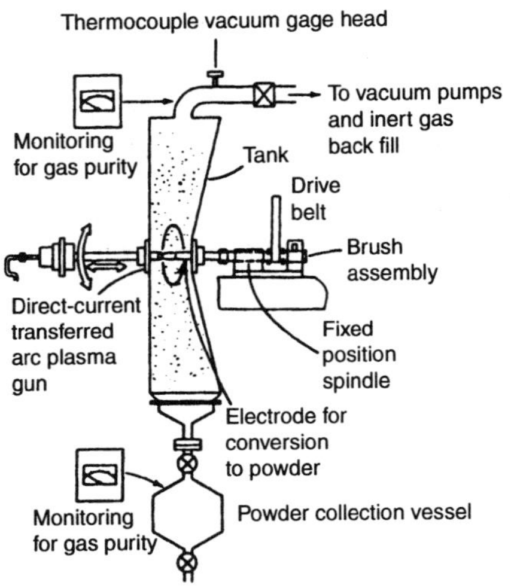

Fig. 3 Short-bar PREP

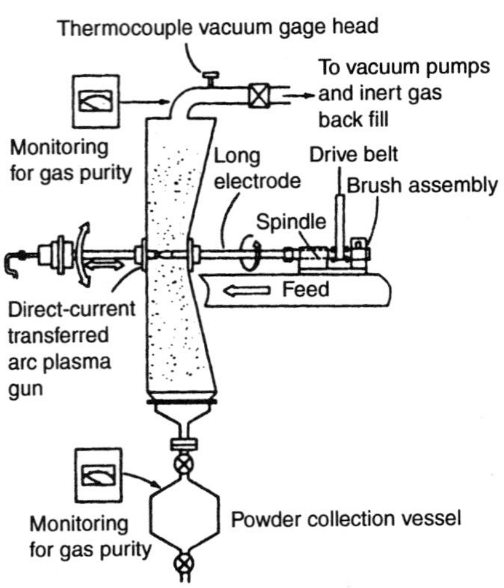

Fig. 4 Long-bar PREP

Particle Size Distribution

Accurately controlled and maintained rotation speed of the anode is necessary to obtain a desired range of particle size distribution. Molten droplet diameter of a given material is determined by the properties of the liquid metal, the centrifugal ejecting forces (related to rotation speed), and to a limited extent the aerodynamics of the droplet trajectory through the inert cover gas. The following equation predicting median droplet size is obtained from a force balance of the centrifugal forces acting on the molten surface tending to cause atomization with the surface tension forces resisting atomization:

$$d = \frac{0.000346k}{\omega}\sqrt{\frac{\gamma}{\rho D}} \qquad \text{(Eq 1)}$$

where d is the median droplet diameter (microns); ω is rotation rate (rad/s); γ and ρ are the surface tension (dynes/cm) and density (g/cm^3) of the alloy being atomized, respectively; D is the electrode diameter (cm); and k is an empirical constant principally determined by the method of droplet formation which is in turn controlled by the melting rate (Ref 4). For any particular alloy, the material properties are fixed and Eq 1 can be further simplified to:

$$d = \frac{K}{\omega\sqrt{D}} \qquad \text{(Eq 2)}$$

where K is a constant (over a limited melting rate range) that has been determined for many alloy systems (Ref 5).

Figure 5 shows the powder size distributions of 1018 steel and Ti-6Al-4V produced at different rotational speeds. Within the steel data there is a clear reduction in particle size produced with increasing rotational speeds as predicted by Eq 1 and 2. Equation 1 also predicts that Ti-6Al-4V, due to its decreased density (approximately 57%) and increased surface tension (approximately 107%) compared to 1018 steel, would have greater inherent resistance to centrifugal

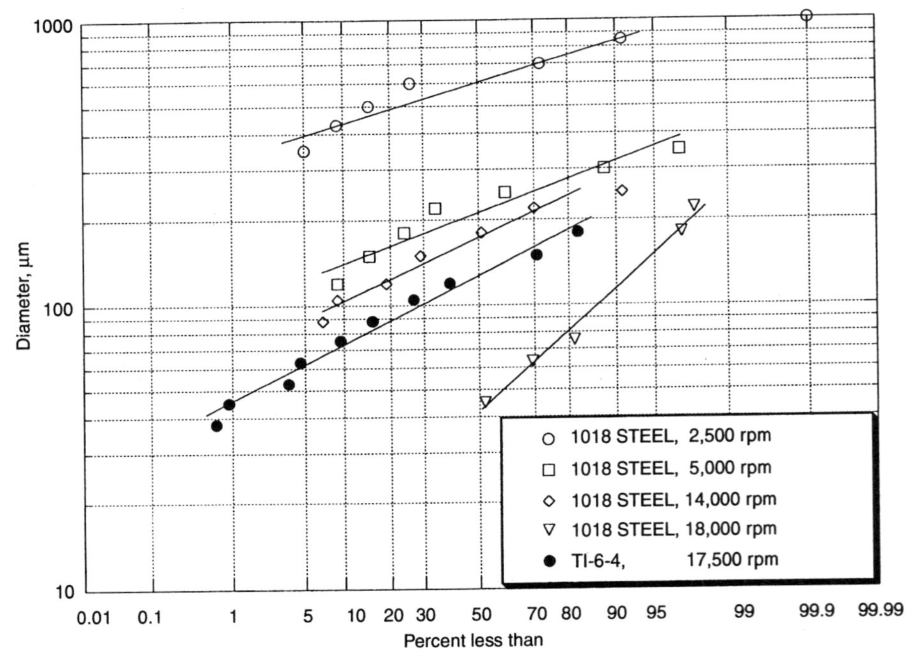

Fig. 5 Effect of rotational speed and alloy properties on particle size

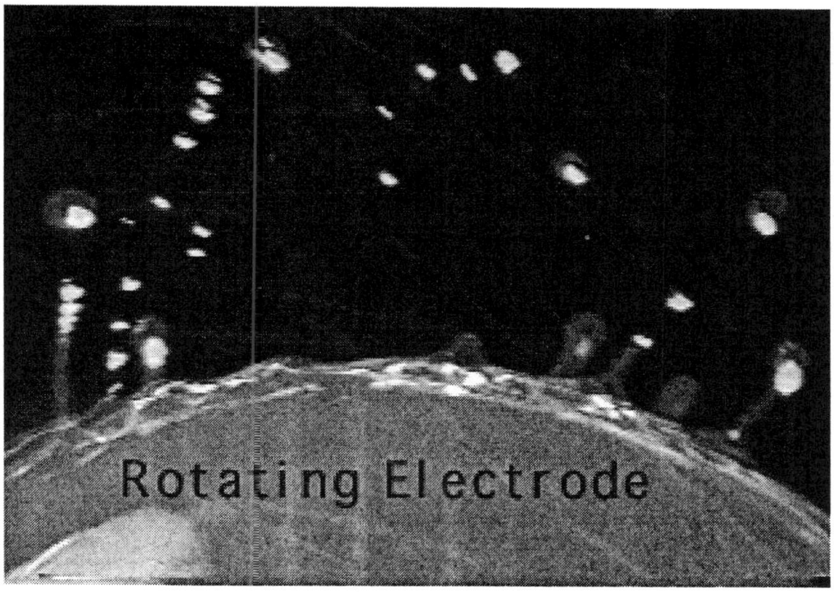

Fig. 6 High speed video of the droplet formation process. Instabilities formed in the liquid metal foil grow into ligaments, which ultimately separate into primary and secondary molten droplets

atomization. Again, the data of Fig. 5 confirm the trend of the prediction, although the magnitude of the shift in sizes due to the change in melt properties is actually greater than that predicted by Eq 1.

High-speed video images of the commercial REP process have shown the powder is formed by direct droplet production; that is, the liquid metal forms discrete droplets, with little or no ligament formation, as soon as it detaches from the bar surface (Fig. 6). Spheroidization of the liquid is nearly instantaneous; droplets tend to come off in clusters of stringers, and the droplet flight path is nearly tangential to the bar surface (Ref 6).

Champagne and Angers (Ref 7, 8) have demonstrated that a bimodal number distribution of particle size occurs due to this mechanism in which liquid drops separating from the electrode rim remain connected momentarily by a necked-down column of liquid metal that breaks into smaller drops when the major spheroid is freed. Bimodal particle size distribution is shown in Fig. 7. Although smaller particles are not present in large quantities by weight, their absolute numbers may be significant. This behavior is superimposed on distributions and is shown as a function of anode melting rate (Fig. 8) and rotation rate (Fig. 9).

The ordinate values of these curves are expressed in terms of the percentage of retained particles on a given sieve size divided by the size range between sieves, which eliminates the effect of this range on the shape of the curve. Additional information can be found in Ref 7 and 8.

Aerospace Applications

Spherical metal powders made by REP or by gas atomization are not well suited for cold pressing into green compacts. Therefore, spherical powders tend to be used in specialized applications where consolidation is achieved by hot isostatic pressing (HIP), or other high-temperature processing in which interparticle voids are more readily closed. These particles flow well into complex mold shapes and can be tapped to a reproducible density of packing to provide fully dense parts that closely approach the dimensions of the finished component (Ref 9).

Much research has been performed on near-net shape compaction for military airframe parts in titanium alloys. An example of a complex shape that can be made from titanium alloy powders pressed to near-net shape is shown in Fig. 10. Overseas research has resulted in as hot isostatically pressed PREP superalloy powder being used in both land- and air-based gas turbines (Ref 10, 11). This use of powders can offer both economic and metallurgical advantages. Near-net-shape technology ensures more efficient material utilization, which is important for high-cost materials such as titanium. Additionally, the metallurgical benefits include improved homogeneity and control of microstructure to achieve enhanced mechanical properties (Ref 12).

Because REP prevents contact of the melted alloy with any container material, it provides a decided advantage over other methods of making contamination-free spherical particles. This process generates powder with cleanliness and composition approximating that of the precursor electrode. Molten titanium is extremely aggressive and reacts with all container materials; consequently, REP is ideally suited for production of clean titanium alloy powders in commercial quantities. Typical characteristics for Ti-6Al-4V powder made by REP are shown in Table 1.

Tensile tests conducted on hot isostatically pressed Ti-6Al-4V compacted from samples of rotating electrode processed powder indicate excellent properties, as shown in Table 2. Subsequent improvements in the process, which included plasma torch melting, have provided plasma rotating electrode processed Ti-6Al-4V P/M compacts with fatigue properties that are comparable or superior to those for cast and wrought materials. More recent work has demonstrated that plasma rotating electrode processed Ti-6Al-4V compacts generally have a narrower fatigue distribution than other forms. This allows design engineers to work with minimum property values in advanced aerospace structures. Careful control of cleanliness is critical to obtain this type of performance.

The nonmetallic inclusions that occurred in early developmental work were associated with the initiation of fatigue failures in compacted pieces made from these materials. This problem has been eliminated by ensuring that the overall

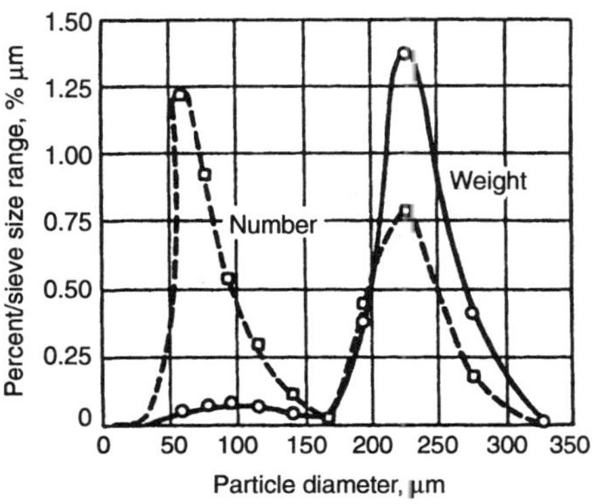

Fig. 7 Bimodal particle size distribution attributed to satellite drop formation

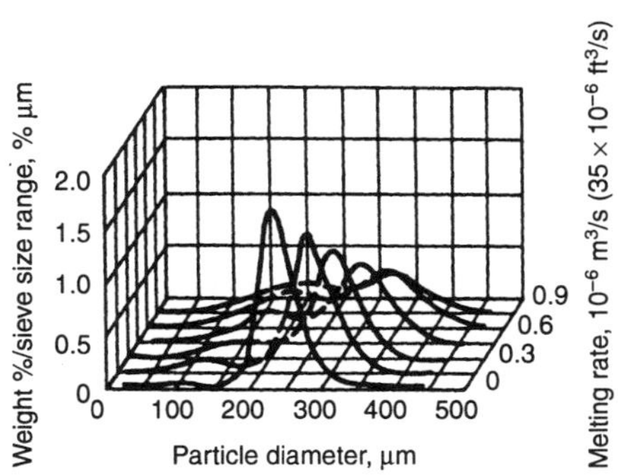

Fig. 8 Particle size distribution as a function of melting rate

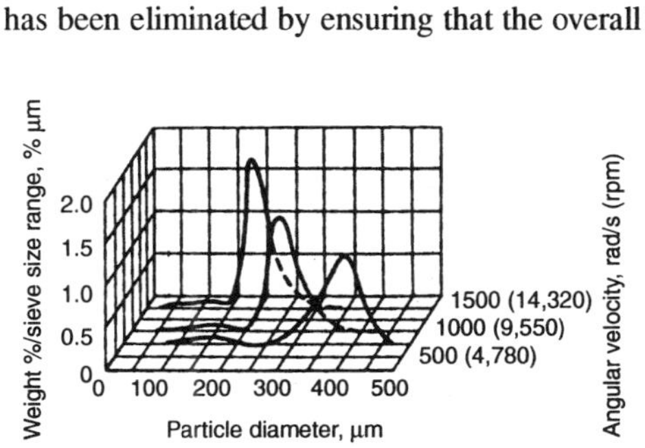

Fig. 9 Particle size distrubution as a function of rotation rate

Table 1 Typical Ti-6Al-4V REP powder compositions and properties

Element	Composition, wt%
Aluminum	5.50–6.75
Vanadium	3.50–4.50
Oxygen(a)	0.13–0.20
Oxygen(b)	0.05–0.13
Iron	0.30 max
Carbon	0.10 max
Nitrogen	0.05 max
Hydrogen	0.0125 max
Tungsten	<10 ppm
Other	0.4 max
Titanium	Bal
Screen analysis, μm	**% retained**
500	0
354	1
250	5.5
177	43
125	38
88	9.4
63	3
44	0.1
<44	0
Other properties	
Median particle size (d_m), μm	175
Particle size range, μm	50–500
Bulk density (60% theoretical), g/cm^3	2.65
Tap density (65% theoretical), g/cm^3	2.90
Flow rate, s/50 g	24–32
Surface rate, m^2/g	0.008

(a) Standard grade. (b) ELI grade (extra-low-interstitial, 0.13% oxygen max). Source: Ref 13

Table 2 Typical tensile properties of hot isostatically pressed Ti-6Al-4V rotating electrode processed powder

Orientation	Tensile strength		0.2% offset yield strength		Elongation	Reduction
	MPa	ksi	MPa	ksi	(4D), %	in area, %
L	938.4	136.1	850.8	123.4	20.0	37.0
	936.3	135.8	868.1	125.9	18.0	37.4
T	950.8	137.9	863.3	125.2	18.0	40.2
	936.3	135.8	848.8	123.1	18.0	35.6
S	932.9	135.3	843.3	122.3	23.0	42.2
AMS 4928-H	941.9	136.6	848.8	123.1	20.0	39.1
	896.4(a)	130(a)	827.4(a)	120(a)	10(a)	25(a)

Note: Consolidated material made by HIP at 950 °C (1750 °F) for 10 h at 100 MPa (15 ksi). Vacuum annealed for 10 h at 700 °C (1300 °F). Hydrogen after vacuum annealing equals 0.0057%. (a) Minimum. Source: Ref 13

powder-making and handling process is designed so that foreign particles are excluded. As atomized powder is removed from the plasma rotating electrode machine under inert gas cover in closed collection vessels and is subsequently passed through a processing tower located within a class 100 clean room. This assembly sieves and passes particles whose sizes lie within desired upper and lower limits, and extracts a representative sample from a given lot of powder prior to loading into appropriate containers. The entire powder processing tower can be evacuated and backfilled with inert gas. The overall system is shown schematically in Fig. 11.

Fig. 10 Gas turbine engine compressor rotor made from hot isostatically pressed plasma rotating electrode processed Ti-6Al-4V powder

Ti-6Al-4V powder is also being used for the production of continuously reinforced metal-matrix composites. In one approach the powder is mixed with organic binders and cast as a thin tape. The tapes are then alternatively laid up with layers of continuous silicon carbide fibers and hot pressed to form a sheet. An alternative technology involves the plasma spraying of the titanium alloy powder directly onto the fibers to produce reinforced foils.

Other Powder Applications

Although REP and PREP are ideally suited to the production of clean powders for subsequent consolidation to fully dense parts, there are other applications for spherical metal particles. Wide application has been found for low-carbon steel powder in photocopier and printer applications, where rotating electrode processed spherical powder acts as a carrier for carbon black used in the printing cycle. In this application, the required particle size distribution is narrow enough so that the ability to optimize yields in the desired size range has made REP an attractive method for powder generation.

Because of their excellent flow characteristics, REP titanium, nickel-base, and high-alloy powder are being used in laser welding, repair, and enhancement of aircraft turbine parts. The same alloy powders are also being employed in the development of new free-form fabrication techniques (see the section "Laser-Based Direct Fabrication" in the article "Powder Metallurgy Methods for Rapid Prototyping" in this Volume).

Coarse particles in narrow size ranges meet unique medical needs. Cobalt-chromium alloys and titanium alloys are vacuum sintered or plasma sprayed onto the stems of prosthetic devices to form porous coatings. The porous structures so generated promote part fixation by bone ingrowth. In addition, the finer size ranges are used for consolidation to near-net shape for various implanted prostheses.

Recent Process Development

Rotating electrode processed powder has been typically used for applications requiring a large median size (in excess of 100 μm). New product and manufacturing approaches require REP powder characteristics, but with maximum sizes

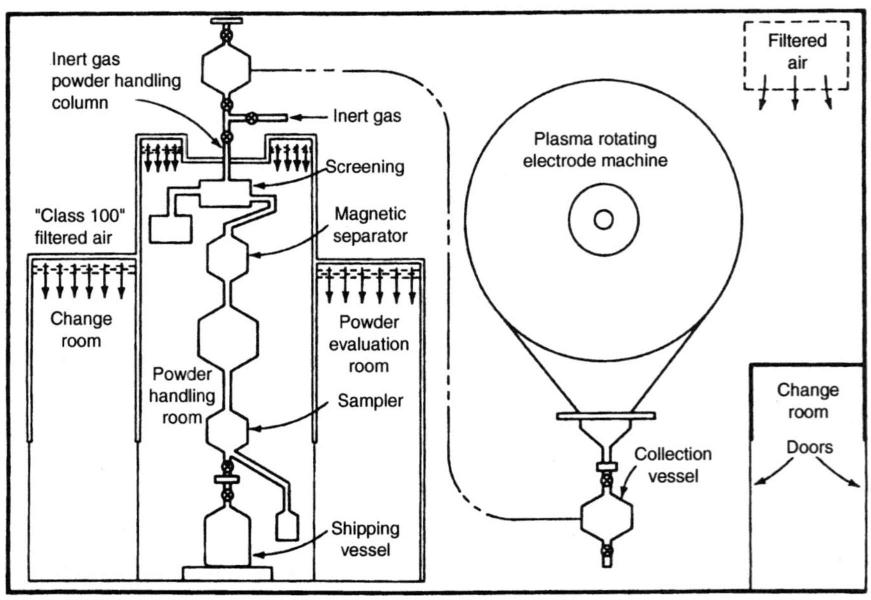

Fig. 11 Ultraclean powder generation and clean-room processing

being less than 100 μm (aerospace) and 44 μm (metal injection molding). In order to meet these needs two new approaches have been investigated to extend REP capability. Two modifications to the REP process have been developed for the production of finer powders.

The first, liquid argon quenching (LArQ) involves impact of the molten REP droplets with a high-velocity film of liquid argon. While balance issues limit rotational speeds to approximately 625 rad/s, for brittle alloys that cannot be rotated faster, LArQ produces almost a 5 to 1 decrease in median particle size (Ref 14).

The second, involves the close-coupled gas atomization of the molten REP-produced droplets. As the droplets formed by REP are being ejected from the electrode surface they are hit by a high-velocity gas jet. The aerodynamic drag imposed on the droplets causes further subdivision and atomization of the droplets. The atomization process is driven by aerodynamics, REP simply provides a means of injecting the liquid metal into the gas flow. This approach allows the melt and the gas flow to be close coupled for reactive alloys, maximizing the atomization efficiency. Close coupling has already been shown to provide substantial yield improvement compared to remote atomization for conventional alloys. Yields of Ti-6Al-4V powder are shown in Fig. 12, with large gains being made in the –44 and –100 μm yields (Ref 15).

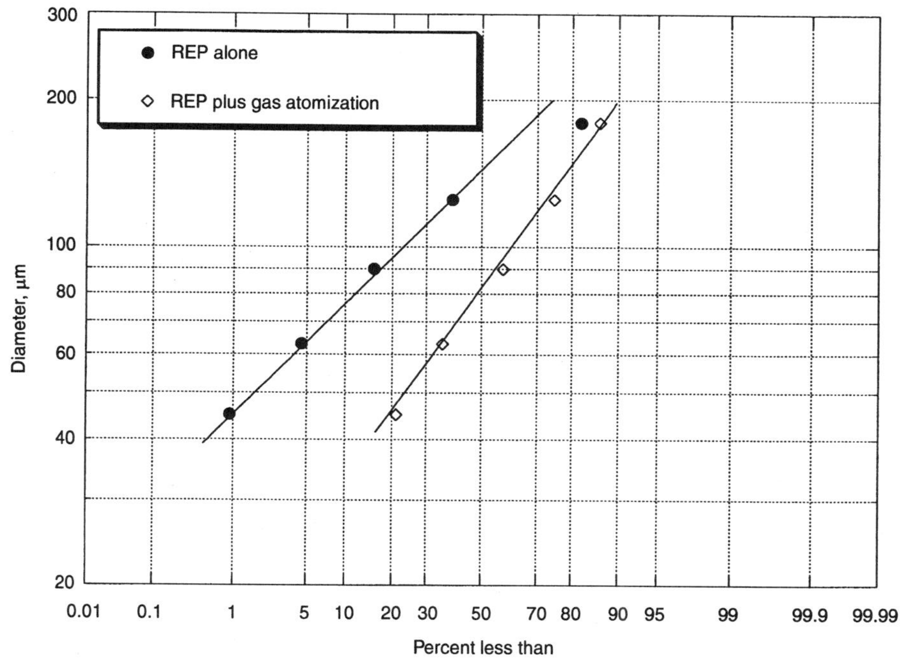

Fig. 12 Comparison of powder yields produced by REP alone and gas-assisted REP

REFERENCES

1. A.R. Kaufman, Method and Apparatus for Making Powder, U.S. Patent 3,099,041
2. A.R. Kaufman, Production of Pure, Spherical Powders, U.S. Patent 3,802,816
3. P.R. Roberts, The Production of PREP Titanium Powder, *1989 Advances in Powder Metallurgy,* Vol 3, Metal Powder Industries Federation, 1989, p 427–438
4. R. Angers et al., Inverted Disk Centrifugal Atomization of 2024, *Int. J. Powder Metall.,* Vol 30 (No. 4), 1994, p 429–434
5. P.R. Roberts, "Powders Made by the Rotating Electrode Process," Starmet Int. Report, 1986
6. S.A. Miller, Gas Enhanced Rotating Electrode Atomization, *Advanced Particulate Materials and Processes,* F.H. Froes and J.C. Hebeisen, Ed., Metal Powder Industries Federation, p 457–454
7. B. Champagne and R. Angers, Fabrication of Powders by the Rotating Electrode Process, *Int. J. Powder Metall. Powder Technol.,* Vol 16 (No. 4), 1980, p 359–367
8. B. Champagne and R. Angers, Size Distribution of Powders Atomized by the Rotating Electrode Process, *Modern Development in Powder Metallurgy,* Proc. 1980 International Powder Metallurgy Conference (Washington, D.C.), H. Hausner, H. Antes, and G. Smith, Ed., Metal Powder Industries Federation, 1981, p 83–104
9. C.F. Yolton et al., Powder Metallurgy of Titanium Aluminide Components, *Progress in Powder Metallurgy 1986,* Vol 42, Metal Powder Industries Federation, 1986, p 479–488
10. B.I. Boris et al., Prospect of HIP Application in Aerospace and Power Plant Industry, *Hot Isostatic Pressing,* F.H. Froes and J.C. Hebeisen, Ed., ASM International, 1996, p 39–49
11. H.X. Chen et al., Development and Application of HIP Technology in CISRI, *Advanced Particulate Materials and Processes,* F.H. Froes and J.C. Hebeisen, Ed., Metal Powder Industries Federation, p 251–263
12. D. Eylon et al., Improved Properties and Forgeability of Gamma Titanium Aluminide Alloys through Prealloyed Powder Metallurgy, *Proc. Plansee-Seminar 93,* H. Bildstein and R. Eck, Ed., 1993, p 552–563
13. V.C. Petersen and V.K. Chandhok, "Manufacturing Process for Hot Isostatic Pressing of Large Titanium PM Shapes," Contract No. F33615-77-C-5005, Interim Report No. AFML-IR-184-7T(1), Air Force Materials Laboratory, 1977
14. P.R. Roberts et al., Liquid Argon Quenching (LArQ) of PREP Generated Titanium Alloy Molten Metal Particles," *Sixth World Conference on Titanium,* P. Lacombe et al., Ed., 1989, p 661–665
15. S.A. Miller and N.F. Levoy, Rotating Electrode Atomization for the Production of Fine Powders from Reactive Metals, *1997 International Conf. on Powder Metallurgy and Particulate Materials,* to be published

Blending and Premixing of Metal Powders and Binders

Revised by Thomas Chirkot, Patterson-Kelley Co., Division of Harsco Corp.

BLENDING OF POWDERS is defined as the thorough intermingling of powders of the same nominal composition. The implication with blending is that the constituents in the vessel are virtually identical except for some minor physical characteristics. For example, one may be interested in blending several lots of powder that have slight variations in moisture or color. Premixing is the preparation of a uniform mixture of two or more components. The implication with mixing is that the constituents in the vessel are different. They may differ in chemical composition or may be physically dissimilar. Many properties of powders and sintered parts—such as powder flow, apparent density, ejection stress, delubrication behavior, dimensional change, and mechanical strength—are quite sensitive to even small changes in particle size distribution and to fluctuations in the concentrations of components within a powder mixture.

In many cases, the P/M parts producer is interested in relatively large, uniformly blended powder lots, because costly and time-consuming process adjustments, often of an empirical nature, must be made. Adjustments are often made from lot to lot, even when the various lots are within given specifications.

Lubricated, single-component metal powders comprise the most important two-component systems in powder metallurgy. Important three- and four-component systems include iron-carbon-lubricant, copper-tin-lubricant, iron-copper-carbon-lubricant, and copper-tin-carbon-lubricant. The widespread use of powder mixtures rather than prealloyed powders is attributable to both economic and technical factors: powder mixtures often are less expensive, have better compacting properties, and sometimes permit shorter sintering times.

Blending and Premixing Variables

The degree by which the formulation components differ governs the amount of energy needed to produce adequate results. From an economic perspective, it is important to understand the mechanisms of powder movement.

Mixing can be divided into two categories, random and ordered. Random mixing straddles the fine line between the definition of blending and mixing. In this category, the physical characteristics in the batch constituents are similar. This similarity allows the movement of the mixing vessel to be controlled by the laws of probability.

For these laws to be truly operative, the walls of the mixing vessel must have only a minimal influence on particle motion. A horizontal drum mixer is an example of a machine where wall interaction is simple. More complex vessel shapes can only be described by differential equations that characterize the system. Although significant investigation has gone into random mixing, the exercise is mostly academic because the conditions of true random behavior are generally not encountered in an industrial environment.

With ordered mixing, differences in the physical characteristics become important. In fact, these differences may have a favorable influence on mixing behavior. The ordering process is achieved by several factors acting individually or in concert. Mechanical ordering is caused by movement of the vessel. The ordering may be due to a tumbling-induced movement causing a splitting and merging of powder as the vessel rotates, or the movement may be augmented by baffles.

Properties of the materials also contribute to ordering. A rough-surfaced particle may trap smaller particles. Moisture within a particle may contribute to capillary attraction. On the molecular level, electrostatic and van der Waals attractions may be important.

Diffusion, convection, and shear are generally recognized as the most common mixing mechanisms. The diffusive mechanism is a particle phenomenon where vessel motion allows particles to redistribute themselves through contact with other particles or by dropping into a newly created void space.

The convective mechanism deals with groups of particles. Convective cells are induced within the mixer through agitation of the powder bed. Within a cell, the diffusive mechanism may also be operating. The shear mechanism creates slip planes with the bed and can cause substantial particle degradation.

These mechanisms exist on an energy continuum extending from the low-energy environment of diffusion to the high-energy environment of shear. Thus, the economic importance of defining the degree of uniformity of blending needed for each particular application is obvious.

Most types of mixing equipment achieve their highest degree of particle distribution rather quickly when dealing with dry, free-flowing, noncaking materials. Any extension of the mixing cycle usually results in a plus or minus variation from some acceptable mixing norm. Therefore, it is prudent to determine the minimum processing time needed to achieve the desired result and to terminate processing at that time.

Control over mixing time becomes more important when using higher-velocity mixing devices, in which particle fragmentation can introduce severe changes in particle behavior and a tendency toward segregation. Although the determination of minimum or optimum mixing is usually done by measuring specific technological powders properties as a function of time, a general method based on measuring the uniformity of a mixture follows.

The quality of a powder mixture improves as the number of contacts between the different components of the mixture increases. Figure 1 schematically illustrates various powder dispersions. A truly ordered particle arrangement (Fig. 1a) is a theoretical standard. The ordering may be ephemeral if no adhesional properties are operating to maintain order. The agglomerated (Fig. 1b) and the demixed or segregated (Fig. 1d) arrangements represent extreme and undesirable particle dispersions. From a practical point of view, the statistical distribution of the components (Fig. 1c) provides a reference mark against which the actual distribution can be measured.

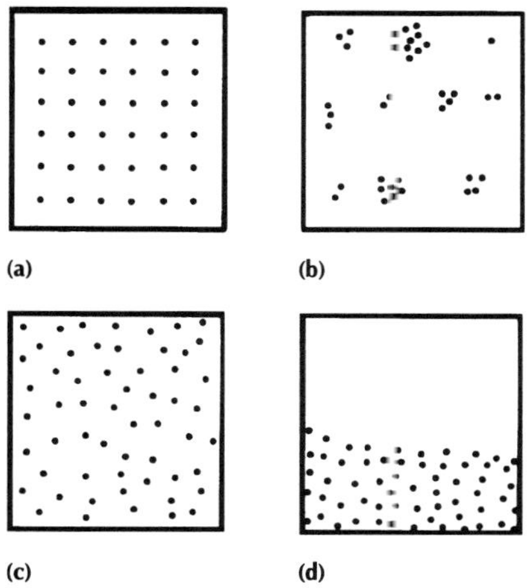

Fig. 1 Schematic representation of particle patterns in a powder mixture. (a) Ordered. (b) Agglomerated. (c) Statistical (random) distribution. (d) Demixed or segregated

This does not necessarily mean that a statistical distribution is best or even attainable for a given application.

The quality of a mixture can be described quantitatively by comparing the actual distribution of the components with that of a statistical mixture. An example of such a comparison, in terms of the number of contacts between identical particles in the binary iron-copper system, is shown in Fig. 2. In this sample the addition of a stabilizer minimizes segregation of components and helps produce a near-random (bottom curve) mixture.

Statistical Analysis of the Blend. Analytical determination of the standard deviation of a mixture is described as a statistical measure of how much the concentration of one constituent in a sample varies from the true concentration that characterizes the entire blend. For ideal conditions, the standard deviation for a random mixture can be calculated if the mechanism of mixing is known. Thus, formulas for the calculation of standard deviation for the random mixture for certain cases of diffusional (Ref 1), convective (Ref 2), and shear mixing (Ref 3) exist. A statistical mixture of a typical two-component P/M premix depends on the component mass ratio, the average component masses, the standard deviations of the component particle mass distributions, and the sample size (Ref 4). Using statistics provides a structured method of analysis, but also has drawbacks. The standard deviation is related to the number of particles in the sample, and therefore the standard deviation is sensitive to particle size. Also, standard deviation is a raw statistical concept that does not directly indicate when a proper mix has been achieved. The experimenter must decide the acceptance criteria for the samples. Further details on statistical techniques are found in Ref 5 and 6.

The quality of a mixture can be assessed in terms of its uniformity, either on a macroscale or a microscale. The results of structure/property relationships, often of an empirical nature, usually determine which degree of uniformity must be attained in a given application. When evaluation of the blend or premix is desired, thief probes using spaced compartments are used to obtain multiple samples throughout the cross section of the batch, whether in a mixer, drum, or other storage container. The thief provides a reasonable look at sections within the mixing vessel, but there are some problems with its use. Inserting the probe through the surface of the powder will possibly force some surface material down with it. Secondly, when the sample compartments are opened, better flowing material can preferentially fill the sample port and bias the result.

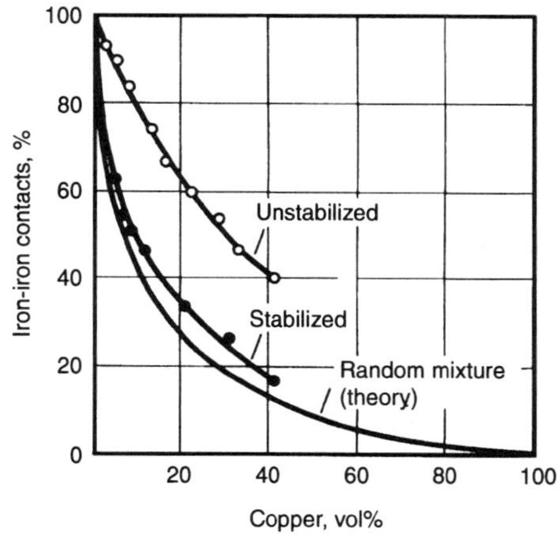

Fig. 2 Effect of stabilizer on iron-iron contact formation in binary iron-copper system. Lower curve represents theoretical random mixture.

Sample size should be related to the method of quality determination. Scale of scrutiny is a term used to define sampling criteria. If the mixed product is to be used in large quantities for a subsequent processing step, the scale of scrutiny can be a bag of the packaged product. If the mixed product is ultimately used for small parts, the scale of scrutiny can be a few grams. Ideally, the closer the sample size approaches the actual quantity used in analysis, the more accurate and consistent the results will be, because sample segregation will have minimal effect. Subdivisions of samples can yield nonrepresentative results, particularly when the materials are prone to segregate. To eliminate or reduce the effect of segregation related to particle size, reduction of large particle sizes by grinding or milling the sample is recommended. The number of samples taken should be determined by the application of the mix. Other factors such as analysis methods, analysis costs, and how the first few samples analyze to the expected assay ultimately dictate how many samples are required (Ref 7).

If the percentage of one component in a batch of blend or premix is X, and the percentage found in a spot sample is X_i, then the difference is $(X - X_i)$, which can be a plus or minus value. The average of all the X_i readings could approach the value of X if enough samples were checked; the average of all the plus or minus variations of $(X - X_i)$ would approach 0. To obtain a measure of the overall sample—the significance of all the spot samples—the statistical variance and standard deviation are computed. This involves averaging the squares of the individual sample errors, as illustrated in the simple example that follows. The following data represent analytical results of six spot samples (Ref 7) taken from a batch of mixed powder. The percentage of ingredient A in batch:

$$X = 40$$

Percentages measured in the spot samples:

$$\begin{array}{ll} X_1 = 41 & X_4 = 42 \\ X_2 = 45 & X_5 = 36 \\ X_3 = 38 & X_6 = 38 \end{array}$$

To compute variance $(X - X_1)^2$.

$$\begin{array}{r} (40-41)^2 = 1 \\ (40-45)^2 = 25 \\ (40-38)^2 = 4 \\ (40-42)^2 = 4 \\ (40-36)^2 = 16 \\ (40-38)^2 = 4 \\ \hline 54 \end{array}$$

$$\frac{54}{6} = 9 \quad \sqrt{9} = 3$$

The number 9 is the biased variance, and 3% is the biased standard deviation. These values quantitatively characterize the batch of mixed powder. In mathematical form, it can be expressed:

$$S^2 = \frac{\sum_{i=1}^{n} (X - X_i)^2}{n}$$

where S^2 is biased sample variance, S is biased standard deviation, i is spot sample number, X_i is spot sample measurement, X is actual content of ingredient, and n is number of spot samples.

To compensate for the increased uncertainty entailed using very few spot samples, the unbiased variance is used. The difference is that $n - 1$ is placed in the denominator instead of n. The expression then becomes:

$$S^2 = \frac{\sum_{i=1}^{n} (X - X_i)^2}{n - 1}$$

The unbiased sample variance for the foregoing example is calculated to be:

$$S^2 = \frac{54}{6 - 1} = 10.8$$

$$S = \sqrt{10.8} = 3.3$$

Thus, 3.3% is the unbiased standard deviation. It is a statistical measure of the probable variation, plus or minus, from 40% of ingredient A

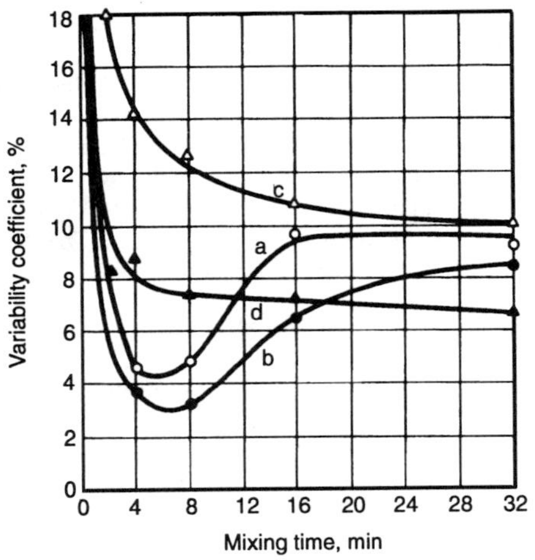

Fig. 3 Effect of particle size and shape of components of 90%Fe-10%Cu mixtures on degree of blending. Quality of blending improves as variability coefficient decreases. Particle size and shape for components: a, Cu, 200 to 300 μm; Fe, <63 μm of spherical particle shape; b, Cu, 200 to 315 μm; Fe, 100 to 200 μm of spherical particle shape; c, Cu, 200 to 315 μm; Fe, <63 μm of irregular particle shape; d, Cu, 200 to 315 μm; Fe, 100 to 200 μm of irregular particle shape

in the blend, as measured by the sampling operation.

For more information on sampling, see the article "Sampling and Classification of Metal Powders" in this Volume.

Effect of Powder Characteristics

The present state-of-the-art blending and mixing does not permit a calculation or prediction of blending results based on available powder component data. Part of the reason for this is that some of the forces and phenomena operating during blending are not measurable and their relationships to blending are not known.

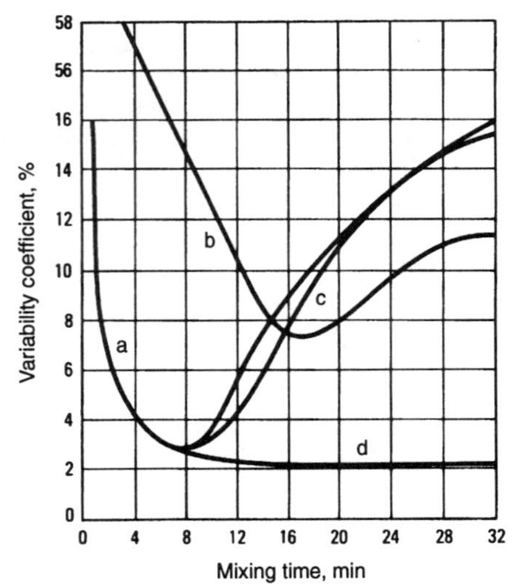

Fig. 4 Effect of stabilizer on degree of blending of 90%Fe-10%Al mixtures. Particle size for both components: 100 to 200 μm. Stabilizer: benzeneloleic acid. a, without stabilizer; b, with 1.0% stabilizer; c, 1.0% stabilizer added after 8 min of blending; d, 3% stabilizer added after 8 min of blending

Table 1 Particle size distribution of the copper powders listed in Table 2

Mesh size	wt%
+100	1
−100 to +150	5
−150 to +200	35
−200 to +250	10
−250 to +325	30
−325	19

There is a general agreement in the literature (Ref 7–9) that the primary reason for segregation is the different flow rates caused by powder properties, or velocity gradients created by mixing, handling, transporting, or movement of materials. Segregation mechanisms that interfere with random motion of particles operate in combination, and reduction of industrial problems depends on minimizing the more severe ones (Ref 7). Some of the variables affecting flow, and thus segregation, are:

- Particle shape, size, and size distribution
- Surface roughness
- Surface hardness
- Particle density
- Material density
- Batch volume
- Ratio of components
- Electrostatic phenomena
- Mass flow conditions

Examination of the variables causing segregation indicates that segregation is a material, rather than a mechanical, problem. Campbell and Bauer (Ref 8) point out that segregation also occurs in processing equipment. Therefore, the cure lies in control, adjustment, or change of the variables affecting the powder or its mass behavior.

For fine powders, particularly those able to hold static charges, electrostatic charge phenomena often are at the core of the blending problem. In such cases, charge dissipators (such as water) and surfactants should be investigated first. The effects of particle size, shape, density, and surface characteristics (Ref 10) should then be evaluated if the problem persists.

Particle Size. As a general rule, the greater the uniformity of particle size of the material to be blended or components to be premixed, the greater the possibility for consistent and uniform results. As the particle size range and distribution change, a change in blendability should be expected. Larger particles of the same or of different materials have a strong tendency to segregate during blending or premixing. Elimination of the larger particle sizes, if feasible, is beneficial. Slowing down the blender, or the use of vacuum blending (wherein the atmosphere that supports and fluidizes the fines is absent), or mechanical or thermal agglomeration (as in the so-called diffusion-bonded grades of iron-base powders) also are possible solutions to segregation problems. Spray drying of multicomponent mixtures is another solution.

Table 3 Apparent and tap densities of tungsten as affected by powder particle size

Particle size, μm	Specific surface, mm^2/g	Apparent density (d_a), g/cm^3	Tap density (d_f), g/cm^3	d_f/d_a
0.5	5480	2.58	4.32	1.68
3–4	1120	3.56	5.64	1.57
14–16	220	3.92	5.10	1.30

Table 2 Apparent and tap densities of three differently shaped copper powders of similar particle size distribution

Particle shape	Apparent density (d_a), g/cm^3	Tap density(a) (d_f), g/cm^3	d_f/d_a
Spherical	4.5	5.3	1.18
Irregular	2.3	3.1	1.35
Flake	0.4	0.7	1.74

(a) Vibrated in Synchron vibrator, at constant amplitude, for 10 min

Particle Shape. The blendability of powder is strongly influenced by particle shape. Figure 3 shows the variability coefficient—that is, the standard deviation divided by the average value of the measured property—of 90%Fe-10%Cu mixtures for different particle sizes and shapes. For spherical iron, the quality of mixing improves quickly (lower coefficient of variability), but results in demixing as mixing is continued. Consequently, blending should be stopped once a random or near-random distribution has been achieved. Demixing or overblending is often caused by the accumulation of electrostatic charges, which often can be dissipated by the addition of a small amount of water. Surfactants can improve the flow of materials. This is illustrated in Fig. 4, which shows the beneficial effect of oleic acid for 90%Fe-10%Al mixtures.

The net effect of both particle size and shape on blendability tends to be the result of a combination of surface conditions (smooth, rough, or jagged) and the basic particle shape.

Density. One of the long-standing problems of blending and premixing has been density differences of the powders being processed. Low-density materials tend to "float" and collect at the top of a mass, while high-density components tend to "sink" to the bottom of the mix.

Table 4 Apparent and tap densities of three types of iron powders

Particle size, μm	Type of iron powder, wt%: Carbonyl	Electrolytic	Reduced
0–10	2.0	6.6	1.8
10–15	59.0	9.2	17.0
15–30	36.0	53.0	41.0
30–44	3.0	31.2	40.2
Apparent density, g/cm^3	3.42	2.56	1.89
Tap density, g/cm^3	4.09	3.42	2.12
d_f / d_a	1.20	1.34	1.12

The situation is aggravated if the heavy particles are coarse and the light particles are fines. Size and density differences can balance out, on the other hand, if the fines are denser than the coarse components. This is true of many water-atomized metal powders, in which the coarse fractions have more irregular particle shape, that is, their effective density is lower. For mixtures where small particles tend to coat the larger ones, a good blend usually depends on the uniform size of the small particles and a narrow ratio of the two components (Ref 7).

Hausner (Ref 11) has compiled some of the variables affecting loose powder density, as follows:

- Material density
- Particle shape
- Particle density (particle porosity)
- Average particle size
- Particle size distribution
- Average particle shape
- Particle shape distribution
- Specific surface area of the powder
- Oxide films
- Additions, such as lubricants
- Medium surrounding the particles

Hausner has shown the change in apparent and tap densities with differently shaped particles (Tables 1 and 2) and reported on tests made by Kothari (Ref 12) reflecting the apparent and tapped densities as affected by powder particle size (Tables 3 and 4). He has also described the importance of friction conditions in a mass of powders and the effects of particle size and shape on friction, as well as apparent and tap densities.

For more information on particle size and size distribution, particle shape, and density, see the related articles in this Volume.

Equipment for Blending and Premixing

Many different types of machines are available for blending and premixing solids. The high density, abrasive nature, frictional characteristics, and tendency to segregate of metal powders, however, reduce the suitability of some machines. A blender suitable for metal powders should:

- Achieve a maximum blending in minimum time
- Have a gentle mixing action to avoid particle degradation
- Provide repeatability of blending

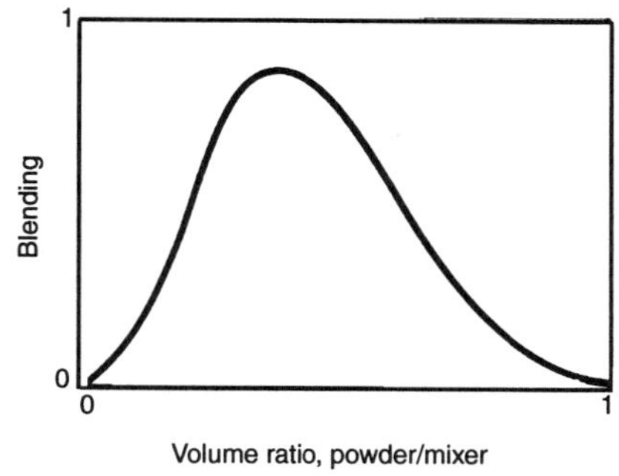

Fig. 5 Effect of powder fill in mixer on quality of mixing

- Provide for complete batch discharge
- Provide accessibility for ease of cleaning
- Have low power consumption
- Require low maintenance
- Provide dust-tight operation

The above list indicates that the use of high-shear paddle- or plough-type devices is inappropriate for metal powders, with few exceptions. Thus, the choice narrows down to tumble-type and low-shear agitated-type machines.

Tumble-type blenders meet the greatest number of the aforementioned criteria, making them the most desirable. Furthermore, scale-up from laboratory size tumblers is very precise. Drum-type, double cone, twin shell, and cross-flow, as well as other tumbler shapes, all produce some quality or degree of blending and also permit straightforward scale-up. Some of these have obvious shortcomings, such as lack of material movement along the axis of rotation, incomplete discharging, and lack of accessibility to the internal surfaces for cleaning. Figures 5 and 6 illustrate the effects of the ratio of total powder volume to mixer volume and the effect of baffle plates in cylindrical mixers on mixing quality.

Low-shear agitated-type blenders, using ribbons, slow-speed paddles, screw-type augers, and other means of moving components in a stationary vessel, are inferior to tumblers. Excessive power consumption resulting from driving the submerged mixing device through the mass of dense metal powders, grinding action caused by the relative motion between the moving and stationary parts, excessive wear of the agitator in an abrasive medium, and lack of precision mixing capability are the main drawbacks. Also, ribbon blender effectiveness does not scale up in direct proportion to laboratory models.

The ideal mixer has not yet been invented. Material variables have such an overwhelming influence on the results obtained from any mixing operation that it is mandatory to sample and

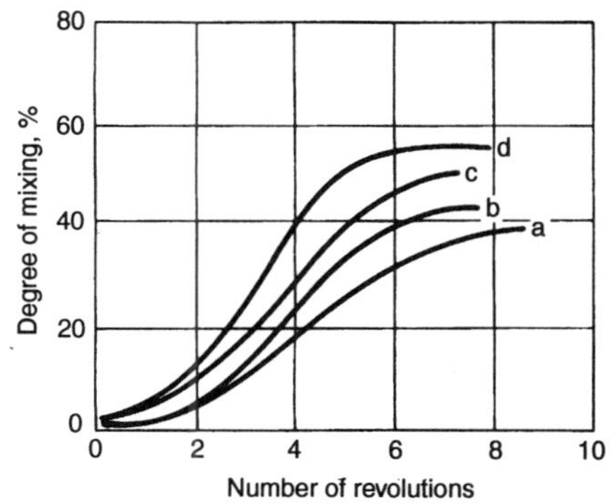

Fig. 6 Effect of baffles on quality of mixing in cylindrical mixer. Radius of cylindrical mixer: *R*, rotational speed: 25 rpm. a, no baffle plates; b, height of baffle plate: 0.66*R*; c, height of baffle plates: 0.5*R*; d, height of baffle plates: 0.33*R*

analyze to arrive at the best combination of materials and equipment for each application.

REFERENCES

1. R. Hogg, *Chem. Eng. Sci.,* Vol 21 (No. 11), 1966, p 1025
2. C.C. Harris, *Powder Technol.,* Elsevier Sequoia S-A, Vol 12 (No. 85), 1975
3. A. Brouthman, G.N. Wollan, and S.M. Feldman, *Chem. Metall. Eng.,* Vol 52 (No. 4), 1945, p 102
4. *Pulvermetallurgie Sinter und Verbundwerkstoffe,* W. Schatt, Ed., VEB Deutscher Verlag fur Grundstoffindustrie, Leipzig, 1979
5. W. Volk, Industrial Statistics, *Chem. Eng.,* March 1956
6. S.S. Weidenbraum, Mixing of Solids, *Advances in Chemical Engineering,* Academic Press, 1958
7. J.J. Fisher, Solid-Blending, *Chem. Eng.,* 1960, p 107–128
8. H. Campbell and W.C. Bauer, Cause and Cure of Demixing in Solids-Solids Mixers, *Chem. Eng.,* 12 Sept 1966, p 179–185
9. M.B. Donald and B. Roseman, *Chem. Eng.,* Vol 7, 1962, Part I, p 749, Part II, p 823, Part III, p 992
10. J.J. Fisher, Liquid-Solids Blending, *Chem. Eng.,* 5 Feb 1962
11. H.H. Hausner, Friction Conditions in a Mass of Metal Powder, *Int. J. Powder Metall.,* Vol 3 (No. 4), 1967, p 7–13
12. N.C. Kothari, Credit by H.H. Hausner in Friction Conditions in a Mass of Metal Powders, *Int. J. Powder Metall.,* Vol 3 (No. 4), 1967, p 7–13

Effect of Blending Techniques on Properties of Metal Powder Mixes

François Chagnon, Quebec Metal Powders Limited

THE PRODUCTION OF P/M PARTS generally requires admixing of various constituents into metal powders in order to promote part compaction and ejection from the die and obtain adequate strength through diffusion of these elements during the sintering process. Solid lubricants are generally used in a range of 0.5 to 1.5%, depending on the part density and size, to allow part ejection and maintain good surface finish. In the case of iron and steel powders, constituents such as graphite, copper, nickel, or ferroalloys are used to increase strength through diffusion during the sintering process. Finally, other elements such as manganese sulfide or boron nitride may also be admixed to improve machinability of P/M parts. These elements must be uniformly mixed to ensure part-to-part consistency during the compaction and sintering operations. Also, because of differences in particle shape, size, and density of these various powders and low specific gravity, fine particles are susceptible to demixing and dusting, which are a major cause of variations in chemical composition of mixes and hence of the dimensional change and mechanical properties of P/M parts.

Two blending techniques are generally used in the P/M industry to prepare press-ready mixes. The first technique consists of dry mixing the various constituents to get a homogeneous mixture. The second technique, binder treatment blending, makes use of binding agents to reduce segregation and dusting during powder handling.

Dry Mixing of Metal Powders

Dry mixing is the most common way to produce press-ready mixes. Metal powders and additives are charged into a blender, mixed until a homogeneous mixture is achieved, and discharged into containers. Adding elemental constituents to metal powders can affect the physical properties of the mix such as flow rate and apparent density. If the additives are very fine, they have a tendency to segregate and create dust during handling. It is a common practice in the P/M industry to adjust the physical properties of a mix by proper use of additives, particularly the lubricant. Figures 1 and 2 illustrate how the type of lubricant can affect apparent density and flow rate, respectively, of iron powder mixes. Stearate-base lubricants generally increase apparent density and improve flow rate of iron powder mixes. Mixes containing ethylene bisstearamide (EBS) wax exhibit lower apparent density and slower flow rates than mixes containing stearates. Polyethylene wax significantly lowers the apparent density and adversely affects the flow rate of iron mixes

Admixed elements, particularly the lubricant, can also affect the green properties, compressibility, and green strength of the P/M parts as illustrated in Fig. 3 and 4. Powdered EBS and polyethylene waxes in iron powder mixes reduce compressibility and improve green strength. Stearate lubricants have good lubricating properties and maintain good mix compressibility, but lower the green strength.

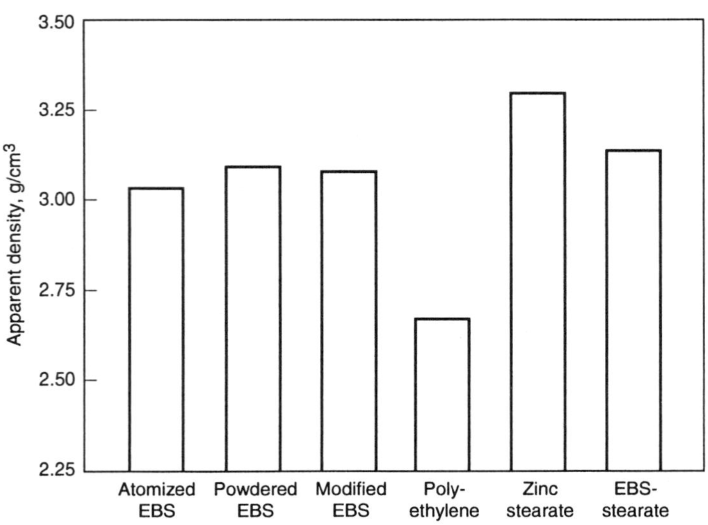

Fig. 1 Effect of lubricant types on apparent density of iron powder mixes (FC0205 mixes with 0.75% lubricant)

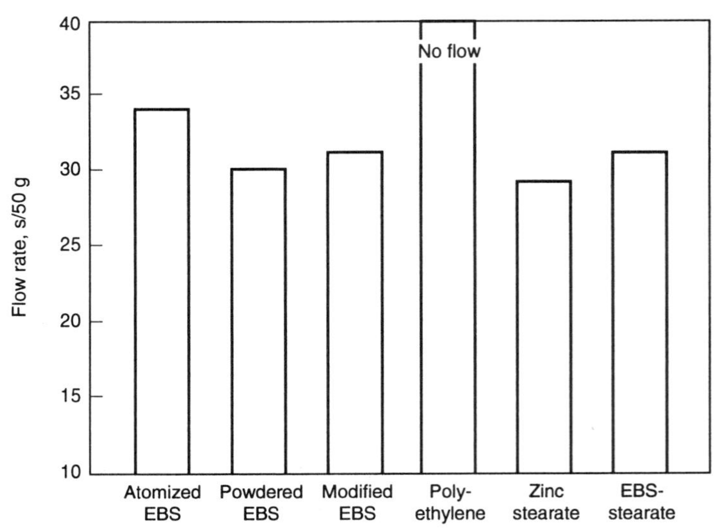

Fig. 2 Effect of lubricant types on flow rate of iron powder mixes (FC0205 mixes with 0.75% lubricant)

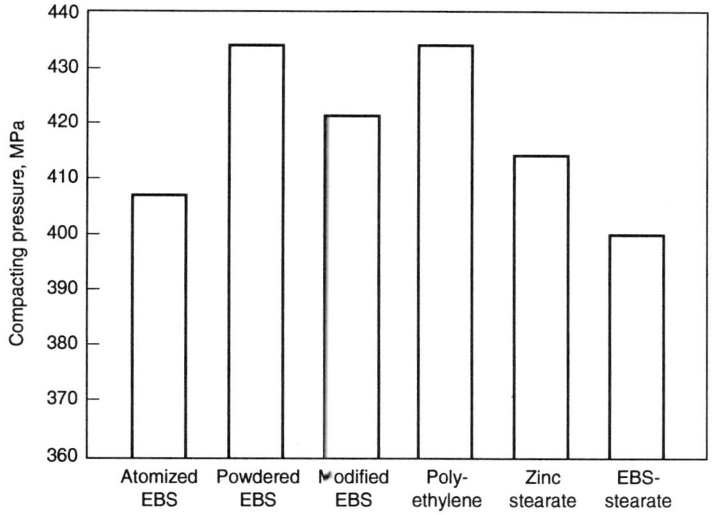

Fig. 3 Effect of lubricant types on compacting pressure of FC0205 specimens (mixes containing 0.75% lubricant)

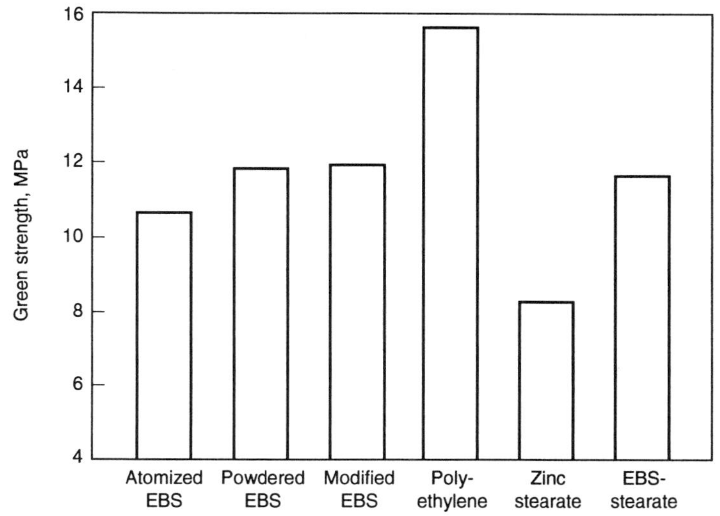

Fig. 4 Effect of lubricant types on green strength of FC0205 specimens (mixes containing 0.75% lubricant)

Binder Treatment of Metal Powder Mixes

This technique consists of binding the fine additives to the coarser ones by adding a small quantity of binding agent to the powder. This binder creates a thin film that covers the particles causing the fine additives to adhere to the base powder. Binders can be of two types: liquid or solid (Ref 1). In the case of liquid binder (Fig. 5a), a dry mix is produced and the liquid binder is subsequently added to the mixture. Further homogenization is required to ensure that the liquid is evenly dispersed throughout the mix. This operation can be done in conventional blender units commonly used in the P/M industry. Kerosene is a well-known liquid binder, but tall oil (Ref 2), types of glycol, glycerin, and polyvinyl alcohol (Ref 3) may also be used to prevent segregation.

As shown in Fig. 5(b), solid binders must be dissolved in an appropriate volatile solvent that acts as a carrier. The solution is then sprayed onto the mix. After homogenization, the solvent is extracted and the binder itself remains as a thin coating with the fine additives on the surface of powder particles. The solvent can be recovered in a condensation chamber and recycled. For iron powder mixes, many types of polymeric binders have been evaluated and proven their efficiency (Ref 4–8). Another derivative of liquid binder is described in Ref 9. Oleic acid is sprayed onto an iron-powder-base mix containing zinc stearate, graphite, and other additives. The blender is then heated above the melting point of the lubricant, 110 to 130 °C, and the liquid lubricant acts as a binder. The mixture is then cooled to room temperature and packaged.

Effect of Blending Techniques on the Physical and Green Properties of Metal Powder Mixes Dust Resistance. One attribute of binder-treated mixes is a better resistance to segregation and dusting. A test method to evaluate the efficiency of the binder agent is described in Ref 8. This is a measure of the resistance of the powder mix to dusting when fluidized by a stream of gas, generally air or nitrogen. Figure 6 illustrates a schematic of the apparatus used in this test method. An air stream is injected at a

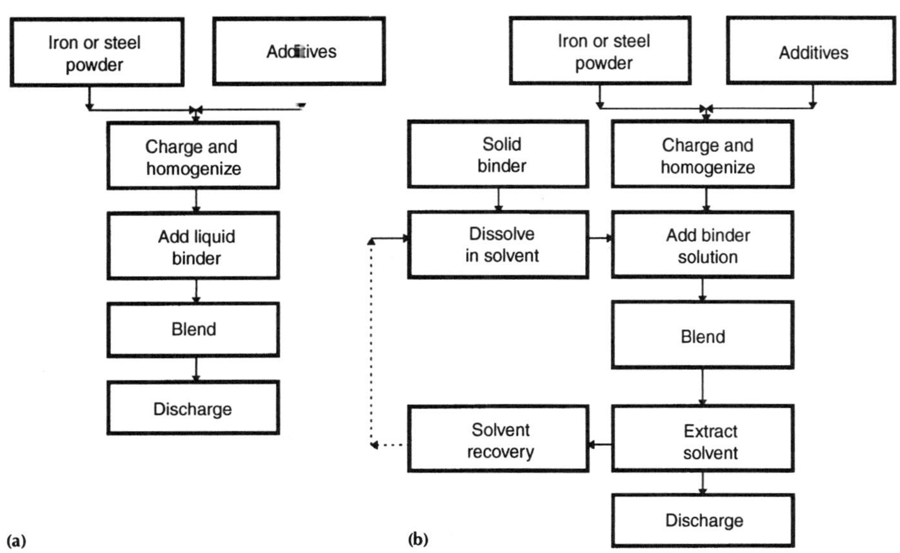

Fig. 5 Processing techniques of binder-treated mixes (Ref 1)

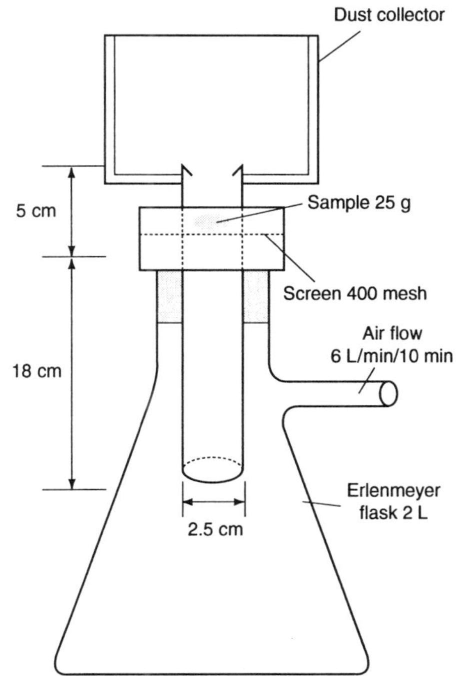

Fig. 6 Dust-resistance apparatus

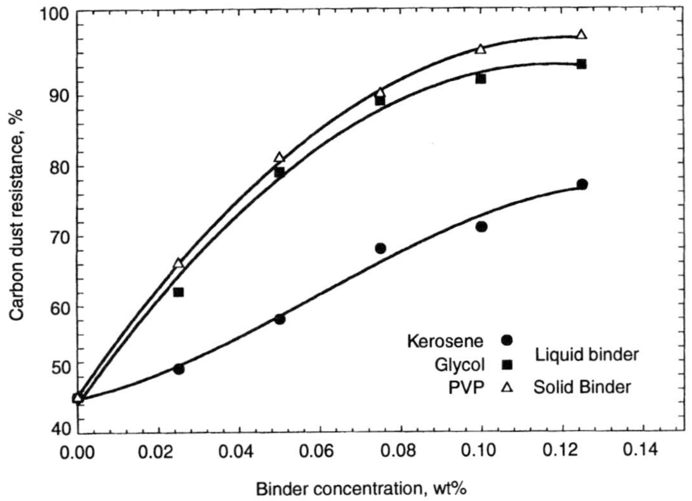

Fig. 7 Effect of binder concentration on carbon dust resistance of iron powder mixes (F0008 with 0.80% lubricant) (Ref 1)

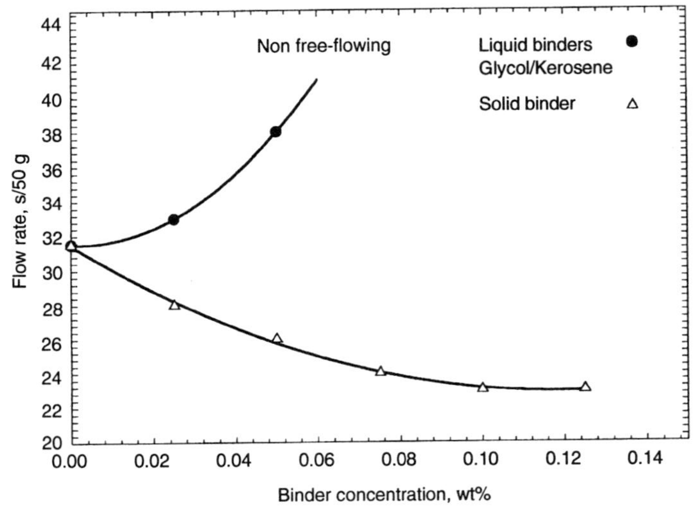

Fig. 8 Effect of binder concentration on flow rate of iron powder mixes (F0008 with 0.80% lubricant) (Ref 1)

constant flow rate of 6.0 L/min for 10 min through a 2.5 cm diam tube with a 400 mesh screen upon which the mix sample is placed. This causes the mix to bubble and the fine particles, such as graphite, to be entrained as a result of a large surface to volume ratio and low specific gravity. The graphite and other similar materials are then deposited in the dust collector. The dust resistance factor is the ratio of the chemical content of the element before and after the test.

Figure 7 illustrates the effect of binder concentration, solid or liquid, on the graphite and lubricant dust resistance of a F0008 mix containing 0.8% zinc stearate. Without the addition of a binder, the typical dust resistance of the mix is 45%. This means that only 45% of the graphite and the lubricant was retained in the test sample. The various binders improve the dust resistance up to a concentration of 0.10%. The kerosene has the lowest dust resistance factor, 70% for a 0.10% addition. Glycol, which is also a liquid binder, shows better results: 90% for an addition of 0.10%. A solid binder, such as polyvinyl pyrrolidone (PVD), shows the best results. Dust resistance of about 95% is reached for a 0.10% binder addition.

Flow rate is an important parameter in the production of metal powder mixes because it directly affects the quality and the productivity of compacting presses. As shown in Fig. 8, liquid binders deteriorate mix flow rate, potentially limiting their use. However, the use of a solid binder significantly improves the flow rate of powder mixes. This is an important attribute because it directly impacts on the productivity and the quality of the parts (Ref 10). A better flow rate results in increased press productivity because the die cavity can be filled faster than with a regular mix. Also, because of the superior flow characteristics, complex and thin-walled parts can be produced with better control of weight and density. An example of how the quality and productivity are enhanced by the use of binder-treated mixes is illustrated in Fig. 9. In this example, an index of 100% was assigned to the standard condition. The part weight (C_p) index in this condition was 1.3 for the regular mix and 2.4 for the binder-treated mix. The C_p index is a direct measurement of the scatter, which indicates that the binder-treated mix achieved a reduction of nearly 55% in weight variation. Another interesting point is that even with an increase in press productivity to 113%, the C_p index is nevertheless 25% higher than the regular mix processed at a lower production rate.

Green properties —mix compressibility and green strength—are also critical parameters in the production of P/M parts and can be affected by the blending technique. The film rigidity of the binder can be controlled by using a plasticizer (Ref 11), which significantly improves the green strength of P/M parts (Fig.10) and maintains good mix compressibility. For multilevel parts, the powder mix must adequately fill the die cavity with a constant quantity of powder for each press stroke. If the different sections of the die cavity are not well filled, the powder transfer during the compacting cycle can create density gradients and negatively affect the green strength. As shown in Fig. 11, laboratory results show a 25% green strength improvement, while for production parts the improvement is nearly 45%. This is caused by a lower density gradient throughout the different sections of the part.

Binder-Treated Mixes for High-Density Parts. Recent innovations (Ref 12–19) have adapted the binder-treatment technology to successfully heat and press metal powder mixes at temperatures up to 150 °C. This compaction technique can increase green densities by about 0.1 to 0.2 g/cm^3 as compared with parts compacted at room temperature. However, these mixes must show good physical characteristics, particularly flow rate and apparent density, over a wide range of temperatures. To withstand these temperatures, powder mixes are produced with a high-melting-point lubricant and binder treated to improve flow rate and stabilize apparent density. As shown in Fig. 12, binder-treated mixes must be used for these specific temperature con-

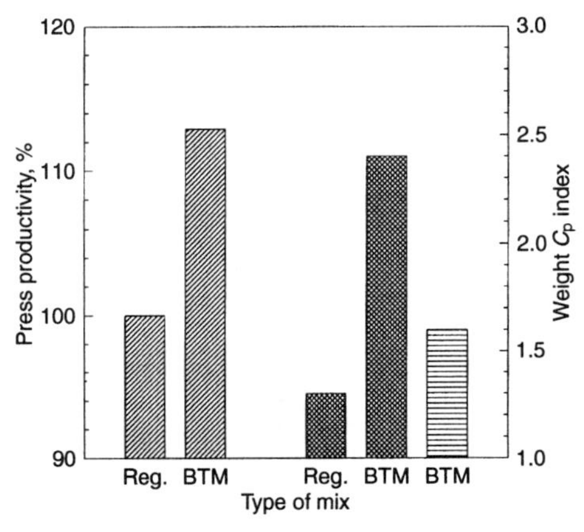

Fig. 9 Comparison of regular (reg) and binder-treated mixes (BTM) processed on a production press (FC0208 modified)

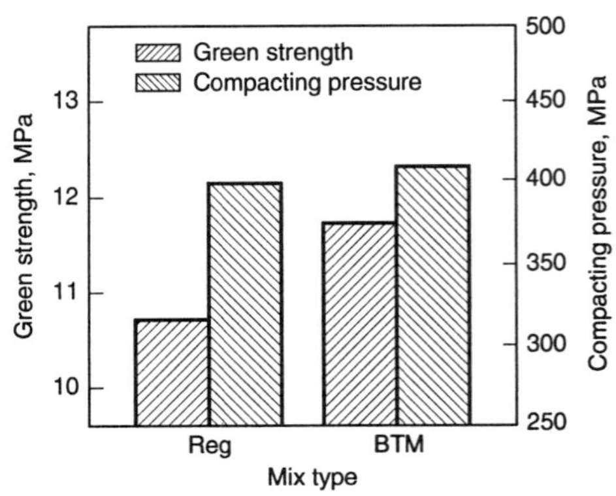

Fig. 10 Effect of binder treatment on green properties of laboratory specimens (FC0205 modified mixes)

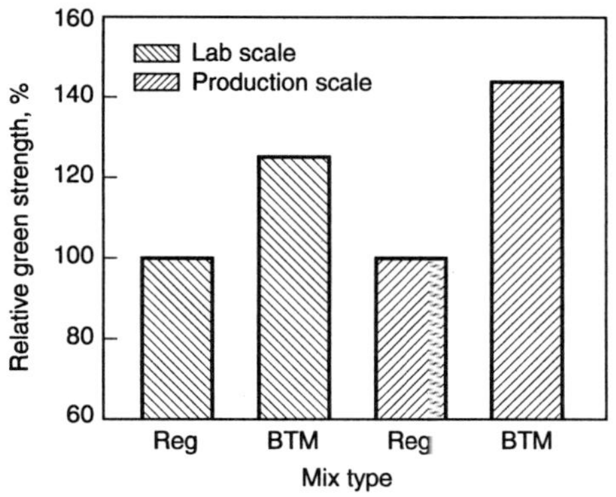

Fig. 11 Effect of binder treatment on relative green strength of laboratory specimens and production parts (FC0205 modified mixes)

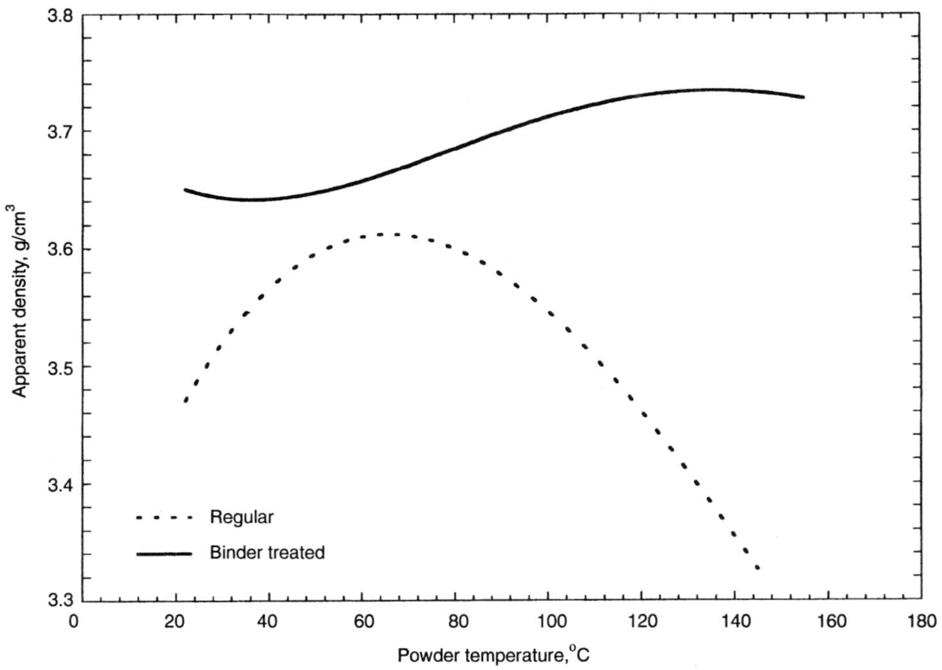

Fig. 12 Effect of powder temperature on apparent density of powder mixes in a die (FN0205 with 0.6% lubricant)

ditions because they show consistent apparent density at the working temperature. As previously mentioned, stable apparent density and good flow rate are key parameters to achieve constant die filling from stroke to stroke and hence minimize press adjustment and maintain good part-to-part consistency.

REFERENCES

1. F. Gosselin, M. Gagné, and Y. Trudel, Segregation-Free Blends: Processing Parameters and Product Properties, *International Conf. on Powder Metallurgy,* Vol 1, The Institute of Metals, 1990, p 297
2. U.F. Engström, Powder Mixture Containing Talloil Free of Segregation, U.S. Patent 4,676,831, 1987
3. U.F. Engström, Homogeneous Iron Based Powder Mixtures Free of Segregation, U.S. Patent 4,483,905, 1984
4. F.J. Semel and S. Luk, Iron-Based Powder Mixtures Containing Binder-Lubricant, U.S. Patent 5,298,055, 1994
5. F.J. Semel and S. Luk, Method for Preparing Binder-Treated Metallurgical Powders Containing an Organic Lubricant, U.S. Patent 5,256,185, 1993
6. F.J. Semel, Iron-Based Powder Mixtures, U.S. Patent 4,834,800, 1989
7. M. Murakami, H. Suzuki, H. Sakuma, T. Hayami, and J. Chosokabe, Powder Mixture for Powder Metallurgy and Binder Therefor, U.S. Patent 5,286,275, 1994
8. F. Gosselin, Segregation-Free Metallurgical Powder Blend Using Polyvinyl Pyrrolidone Binder, U.S. Patent 5,069,714, 1991
9. I. Sakuranda, R. Okabe, T. Omura, Y. Kiyota, and S. Takajo, Iron Base Powder Mixture and Method, U.S. Patent 5,135,566, 1992
10. S.H. Luk and J.A. Hamill, Dust and Segregation-Free Powders for Flexible P/M Processing, *Advances in Powder Metallurgy and Particulate Materials,* Vol 1, Metal Powder Industries Federation, 1993, p 153–168
11. C. Gélinas, F. Chagnon, Y. Trudel, and S. Pelletier, Optimizing Properties of Binder-Treated Ferrous Powder Premixes, *Advances in Powder Metallurgy and Particulate Materials,* Vol 3, Metal Powder Industries Federation, 1995, p 3-45 to 3-56
12. S.H. Luk, H.G. Rutz, and M.A. Lutz, Properties of High Density Ferrous P/M Materials; a Study of Various Processes, *Advances in Powder Metallurgy and Particulate Materials,* Vol 5, Metal Powder Industries Federation, 1994, p 135–153
13. F. Chagnon, C. Gélinas, and Y. Trudel, Development of High Density Materials for P/M Applications, *Advances in Powder Metallurgy and Particulate Materials,* Vol 3, Metal Powder Industries Federation, 1994, p 199–206
14. I. Donaldson and F.G. Hanejko, An Investigation into the Effects of Processing Methods on the Mechanical Characteristics of High Performance Ferrous P/M Materials, *Advances in Powder Metallurgy and Particulate Materials,* Vol 2, Metal Powder Industries Federation, 1995, p 5-51 to 5-67
15. H.G. Rutz, A.J. Rawlings, and T.M. Cimino, Advanced Properties of High Density Ferrous Powder Metallurgy Materials, *Advances in Powder Metallurgy and Particulate Materials,* Vol 3, Metal Powder Industries Federation, 1995, p 10-97 to 10-115
16. D. Barrow and F. Chagnon, New Binder-Treated Mixes for High Efficiency Compaction, *European Conference on Advanced Metallurgy Materials,* Birmingham, 1995, p 50–56
17. T.M. Cimino, A.J. Rawlings, and H.G. Rutz, Properties of Several ANCORDENSE Processed High Performance Materials, *Advances in Powder Metallurgy and Particulate Materials,* Vol 4, Metal Powder Industries Federation, 1996, p 13-337 to 13-352
18. S. St.-Laurent and F. Chagnon, Key Parameters for Warm Compaction of High Density Materials, *Advances in Powder Metallurgy and Particulate Materials,* Vol 2, Metal Powder Industries Federation, 1996, p 5-125 to 5-138
19. H.G. Rutz, J. Khanuja, and S. Kassam, Single Compaction to Achieve High Density in Ferrous P/M Materials in Automotive Applications, *Advances in Powder Metallurgy and Particulate Materials,* Vol 4, Metal Powder Industries Federation, 1996, p 13-363 to 13-374

Production of Iron Powder*

IRON POWDER represents the largest tonnage of raw materials used in P/M fabricating. Iron powder leads in commercial P/M applications because (1) it is inexpensive to produce compared to nonferrous metals; (2) it possesses superior properties, particularly strength, compared to other metals or nonmetals; (3) it has a favorable strength-to-weight-to-cost ratio; (4) it alloys readily, especially with carbon, and consequently possesses all of the attributes of the iron-carbon system (including heat treatability); and (5) it is available in sufficient quantity. Iron powder is not limited to the manufacture of P/M parts. About one-third of all iron powder produced is used for applications such as welding rods, flame cutting, food enrichment, and electronic, magnetic, and chemical applications.

Production of iron powder by the reduction of iron oxide, and in particular the reduction of iron ore by carbon, is the oldest method of producing iron powder. The Swedish sponge iron process, which was developed in Höganäs, Sweden in the early 1900s, was originally intended to produce metallic iron in sponge form as the raw material for steelmaking. A modified version of the process is still one of the important industrial methods of producing iron powder and is practiced in Sweden and in the United States. A similar adaptation also is used in the Soviet Union.

The Swedish sponge iron process is one of many developed for the direct reduction of iron ore to metallic iron at temperatures below the melting point of iron. However, metallic irons produced by most of these direct reduction processes are not suitable for use as iron powder in the production of P/M iron and steel products because they contain too high a percentage of impurities. Also, they do not have the required physical and mechanical properties, such as particle size and distribution, particle shape, low internal particle porosity, and lack of strain hardening.

Most powders currently used for P/M stainless steels, low-alloy steels, and tool steels are produced by atomization of the molten alloy (where melting points of these alloys are lower than that of pure iron). Water-atomized steel powders became available in the early 1960s as a direct result of the invention and commercialization of water-atomized iron powders made from low-carbon iron. Atomized iron and steel powders require the same type of melting and atomizing facilities and methods, but this article only discusses iron powder production by water atomization. Production of alloy steel powders is described in the article "Production of Steel Powders" in this Volume.

A significant portion of low-carbon irons for powder metallurgy are produced by water atomization. Nonetheless, this article also focuses on production of iron powder by reduction and other chemical processes below the melting point of iron. The methods discussed in this article are:

- The Höganäs process
- The Pyron process
- Carbonyl vapor metallurgy
- Electrolytic iron
- Fluidized-bed reduction
- Water-atomized iron powders

The focus is on iron powder production, although some of these processes are also adapted to produce alloy powders.

The Höganäs Process

The Höganäs process uses pure magnetite (Fe_3O_4) ore found in northern Sweden, which has an iron content of approximately 71.5% after beneficiation by grinding and magnetic separation. The impurities present in small amounts are not in solid solution in the oxide, but exist as discrete phases. The ore is of consistent quality and is available in sufficient quantities for continued usage.

Coke breeze or another carbon source that provides the reducing agent is required to produce sponge iron powder. Additionally, limestone is used to react with the sulfur contained in the coke and to prevent its inclusion in the iron powder as an impurity. The ore is prepared for reduction and drying in a rotary kiln, followed by magnetic separation. Coke and limestone are mixed in the proportion of 85% coke and 15% limestone. This mixture also is dried in a rotary kiln and then crushed to uniform size.

The ore and the coke-limestone mixture are then charged into ceramic tubes, which consist of silicon carbide. A pair of concentric steel charging tubes is lowered to the bottom of the ceramic tubes. The ore is fed between the steel tubes. The coke-limestone mixture is fed within the inner of the two concentric charging tubes and between the outer charging tube and the inner wall of the ceramic tube. The charging tubes are then withdrawn from the ceramic tube, leaving the ore and the reduction mixture in contact with one another, but not intermixed.

Charged ceramic tubes are placed on kiln cars that carry them into the kiln. Ceramic tubes are arranged in rows of six tubes by six tubes per kiln car. Cars are pushed into a long tunnel kiln where the reduction occurs. Every 53 min, one car is pushed into one end of the kiln, while another car is removed at the other end. The total time that a car is in the kiln is about 68 h. Gas burners heat 150 m (500 ft) of the tunnel to a temperature of 1260 °C (2300 °F); the remaining 19 m (62 ft) are cooled with circulating air. The tops of the kiln cars form the bottom of the heating chamber.

Within the hot zone of the tunnel kiln, several chemical reactions occur. Decomposition of the limestone generates carbon dioxide, which oxidizes the carbon in the coke to form carbon monoxide. Carbon monoxide reacts with magnetite to form additional carbon dioxide and ferrous oxide. The ferrous iron oxide is further reduced by the carbon monoxide to metallic iron. Metallic iron particles are sintered together to form sponge iron cake. Additionally, a reaction occurs between the sulfur contained in the coke and the limestone admixed to the coke.

The thermodynamics and the kinetics of the process by which iron oxide is reduced to metallic iron have been studied repeatedly, but it is difficult to apply the results of these studies to the conditions found in the Höganäs sponge iron process. When the material in the ceramic tubes has been cooled to room temperature at the end of the reduction process, 96% of the oxide has been reduced to iron. The iron contains 0.3% C. The reduced iron particles sinter together to form a hollow cylinder of sponge iron that has pores between and within the iron particles.

Figure 1 shows the cross section of +40 mesh ground magnetite powder. Figure 2 is a micrograph of the cross section of +40 mesh sponge iron powder obtained by reduction of the mag-

*Revised and updated by Stephen Claeys, Pyron Corporation; Christopher Avallone, International Specialty Products; Amédéo F. De Rege, Domfer Metal Powders Ltd; Yves Trudel, Quebec Metal Powders; and Prasan Samal, OMG Americas

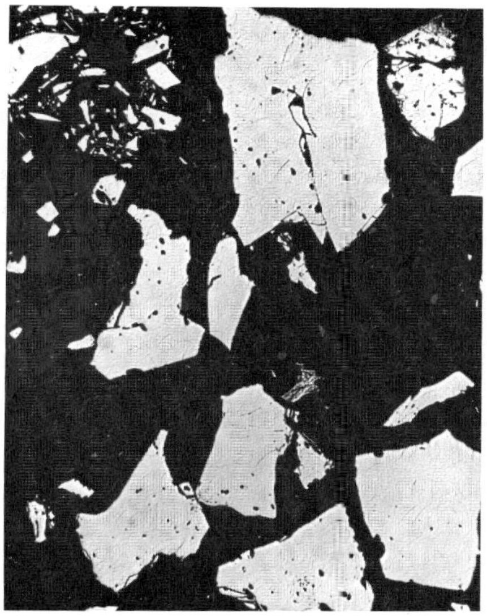

Fig. 1 Cross section of +40 mesh ground magnetite. 30×

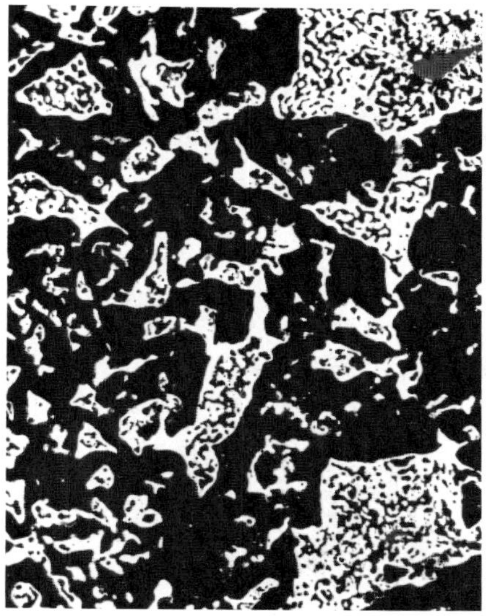

Fig. 2 Cross section of +40 mesh sponge iron obtained by reduction of ground magnetite. 125×

Fig. 3 Cross section of sponge iron particles. Etched with 2% nital. 180×

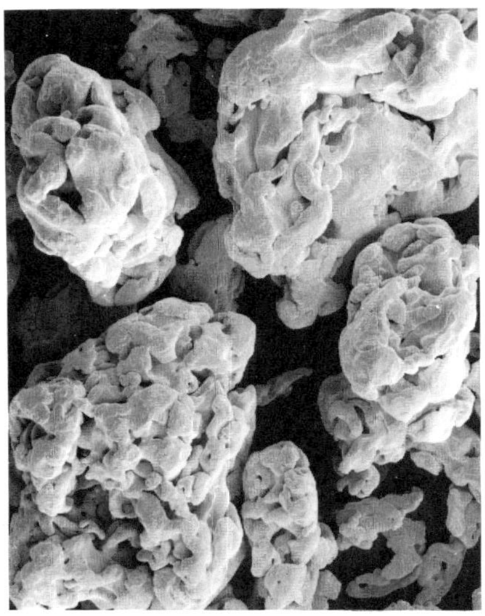

Fig. 4 Scanning electron micrograph of particles of annealed sponge iron powder. 180×

netite at 1250 °C (2280 °F). While the magnetite particles are dense, those of the reduced iron powder are highly porous. Figure 3 illustrates sponge iron particles etched with 2% nital after polishing. Residual unreduced oxide is evident in Fig. 3. Also, etching highlights the very small grain size of the iron produced by reduction of the oxide.

After removal from the kiln, cars are transferred to the discharge section. Cylindrical cakes of sponge iron are removed from the ceramic tubes and dropped into a tooth crusher, where they are broken into pieces approximately 25 mm (1 in.) in diameter. The emptied ceramic tubes are cleaned and transferred to the charging station, where they are reloaded for successive reduction runs.

The 25 mm (1 in.) diam sponge iron lumps undergo a series of grinding, magnetic separation, and screening steps to produce a powder of the desired particle size distribution and to remove nonmagnetic impurities, while preserving the porous nature of the powder particles that makes them well suited for P/M fabrication. Consequently, grinding is continued only until about 65% of the powder passes through a 100 mesh screen. For P/M purposes, however, a –100 mesh powder is generally required. Oversize powder is separated by screening and is used for coating welding electrodes and other non-P/M applications.

Only minor changes in chemical composition of the powder occur during grinding, magnetic separation, and screening. The resultant powder contains approximately 1% O and 0.3% C. However, the powder has been strain hardened considerably. Consequently, it is annealed in a belt furnace.

Powder is loaded onto a solid stainless steel belt that carries it into the furnace through a powder-lock arrangement. This mechanism seals the entrance of the furnace and prevents air from entering. The powder is heated to about 870 °C (1600 °F) in an atmosphere of dissociated ammonia. During annealing, the carbon in the powder is essentially removed, and the oxygen content is reduced from 1% from about 0.30%. At the same time, strain hardening introduced during grinding is relieved. During annealing, the powder is loosely sintered, but requires only light grinding and screening to produce a finished product ready for packing and shipping to P/M fabricators.

Reduction and annealing steps in the production of sponge iron are essentially continuous, but there are several holding areas between operations. These holding areas assist in producing a powder of uniform quality through an "equalizing" process, which consists of recirculating the powder by withdrawing it from the bottom of the containers in the holding areas and feeding it back at the top.

Figure 4 is a scanning electron micrograph of Hoeganaes Ancor MH-100 iron powder particles. Irregular shape of the particle surfaces and particle porosity are visible. Typical powder properties are:

Apparent density, g/cm^3	2.58
Composition, %	
Iron	98.2
Silicon dioxide	0.20
Carbon	0.01
Hydrogen loss	0.26
Sulfur	0.01
Phosphorus	0.01
Green density at compacting pressure of 414 MPa (30 tsi), g/cm^3	6.42
Green strength at compacting pressure of 414 MPa (30 tsi), MPa (psi)	16.5 (2400)
Screen analysis (U.S. mesh)	

Mesh No.	Percent
+80 mesh	Trace
–80/+100	1
–100/+150	18
–150/+200	26
–200/+250	9
–250/+325	24
–325	22

The Pyron Process

Revised by Stephen Claeys, Pryon Inc.

The Pyron process provides an alternative for the production of iron powders by the reduction of oxides. Instead of using ground iron ore as the raw material for reduced iron powders, the Pyron process uses mill scale taken from steel mills that produce plain carbon steel products, such as sheet, rod, wire, plate, and pipe. Mill scales containing alloying elements other than manganese

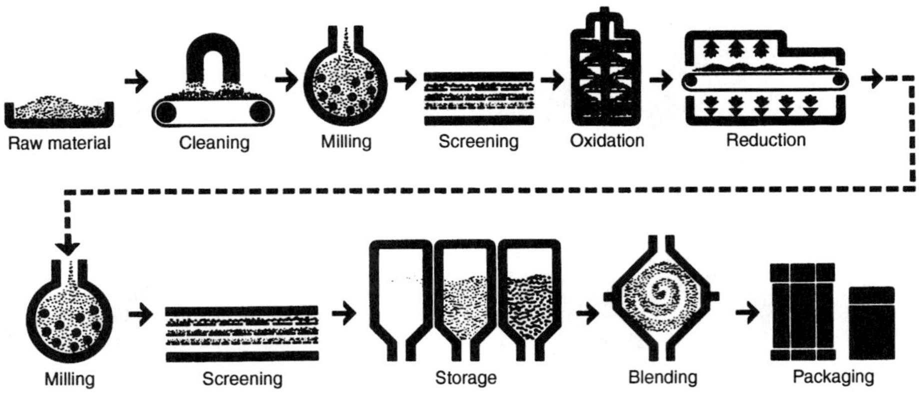

Fig. 5 Flowchart of Pyron iron powder process

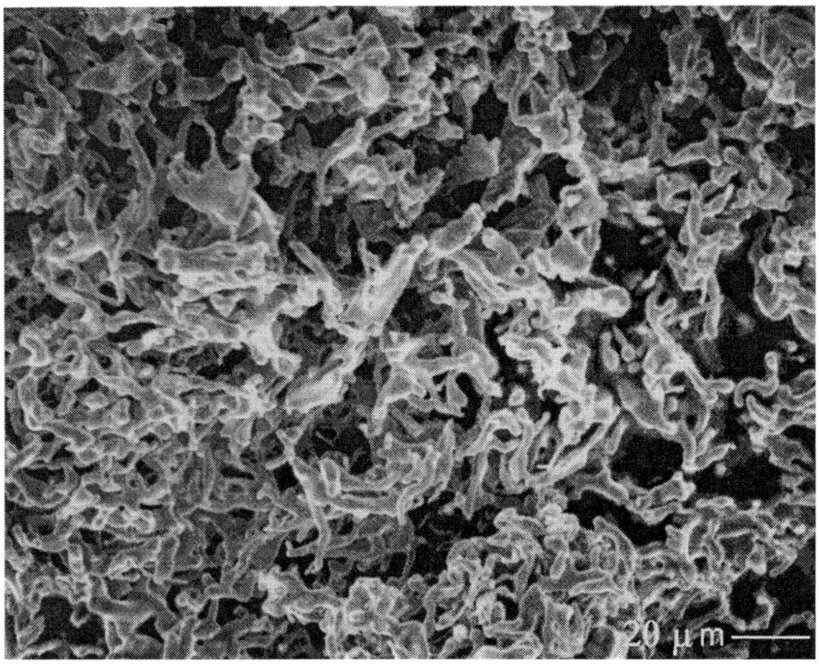

Fig. 6 Scanning electron micrograph of Pyron iron particles

Table 1 Physical properties of Pyron iron powders

Properties		P-100	LD-80	R-80	R-12	AC-325
Apparent density, g/cm^3		2.3–2.5	1.75–2.10	1.0–1.5	1.0–1.5	2.2–2.5
Hall flow rate, s/50 g		27–35	35–poor	Poor	Poor	Poor
Screen analysis, %						
Tyler mesh No.	Particle size, µm					
+20	850	...	...	...	2 max	...
+35	425	...	...	...	10–20	...
+60	250	...	...	...	20–30	...
+80	180	Trace	2 max	2 max	10–20	...
+100	150	2 max	1–12	1–12	5–15	Trace
+150	106	10–15	15–30	15–30	10–20	Trace
+200	75	15–25	15–30	15–30	5–15	0.2
+325	45	25–40	20–40	20–40	3–10	5 max
–325	<45	28–45	15–35	15–35	12 max	95 min

are not used. A flowchart of the Pyron process is shown in Fig. 5.

Process Conditions. Mill scales from different mills are mixed by layering in a bed that is large enough to supply the plant with a constant feed for up to 3 months. Scale is scalped as it is fed to the plant to remove large, undesirable objects. It is then magnetically cleaned to remove any sand, dirt, or other nonmagnetic material. The mill scale is then ground in a continuous ball milling operation to –100 mesh. Careful control of this operation ensures desired particle size distribution.

Oxidation at about 980 °C (1800 °F) converts the iron oxides that are present in the mill scale, FeO and Fe_3O_4, to ferric oxide (Fe_2O_3). This operation is critical to the Pyron process and is necessary to ensure the uniform properties of Pyron iron powders. Gas-fired, fluidized-bed roasters are used for this operation.

Hydrogen reduction is done in electric furnaces 37 m (120 ft) long. The prepared oxide is carried from the roasters through the reduction furnace on 183 cm (72 in.) wide belts. It is converted to iron:

$$Fe_2O_3 + 3H_2 \rightarrow 2Fe + 3H_2O$$

Hydrogen in this reaction is supplied by pipeline from a nearby chemical manufacturer. It circulates through the furnace in a totally enclosed system that removes water and returns the unspent hydrogen to the furnace.

Reduction is done at about 980 °C (1800 °F). Small adjustments in temperature and belt speed are needed to control quality. The resulting sinter cake is friable, and a simple milling operation converts it to iron powder. Particle size distribution is determined by the original grinding of the mill scale. From the milling operation, the powder is conveyed to large storage bins, where it is subsequently scalped and blended.

Pyron iron powder has fine porosity and a sponge-like microstructure (Fig. 6 and 7). The pores in the interior structure of Pyron iron powder particles (Fig. 7) are considerably finer than those in Swedish sponge iron powder particles, because in the latter the long-term reducing treatment at higher temperature coarsens the pores. Because of the fine pore structure, compacts from Pyron iron powder sinter faster than those from other commercial iron powders. Properties of several grades of Pyron iron powder are given in Table 1. Table 2 lists characteristics and uses of hydrogen-reduced iron powders produced by the Pyron process. Typical chemical compositions (wt%) of Pyron powders are:

Total iron, wt%	97.0–98.5
Carbon, wt%	0.01–0.05
Sulfur, wt%	0.005
Phosphorus, wt%	0.012
Manganese, wt%	0.40–0.65
Acid insolubles:	
AC-325	0.20–0.90
Other grades	0.20–0.45
Oxygen (typical), wt%	0.70–1.75

Carbonyl Iron Powders

Revised by Sotiri Papoulias,
International Specialty Products

Very fine iron powders, like nickel powders, can be produced by carbonyl decomposition (va-

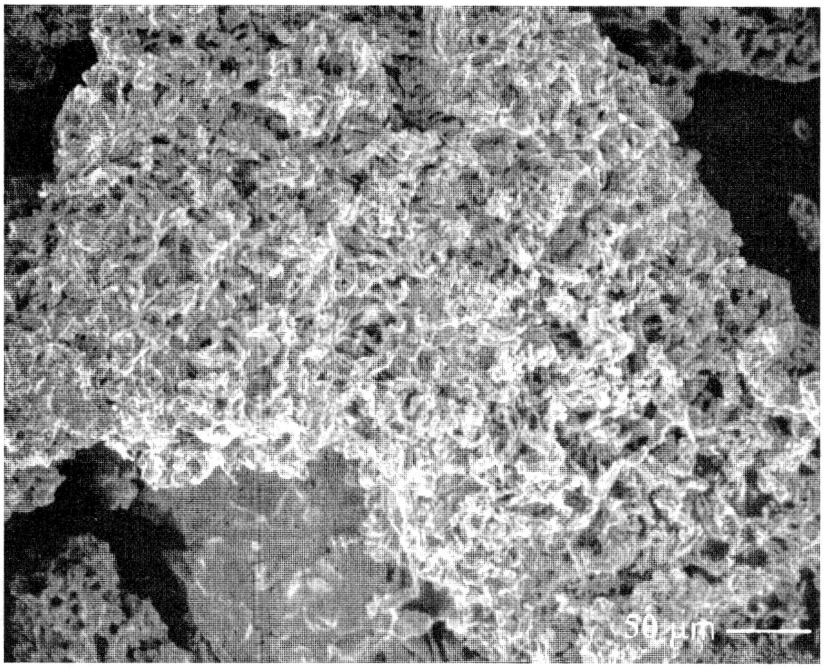

Fig. 7 Optical micrograph of cross section of Pyron iron

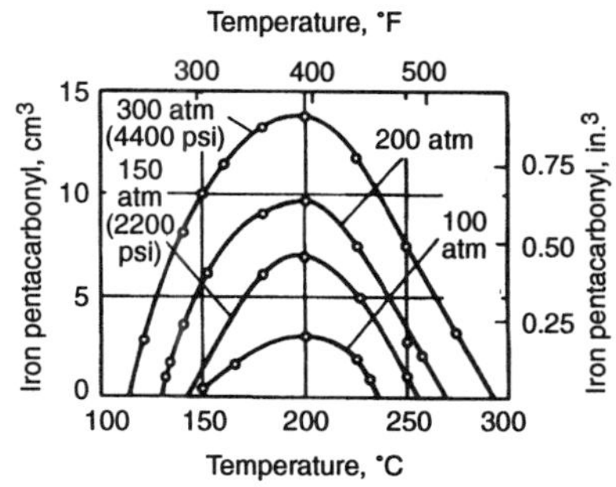

Fig. 8 Effect of system pressure of 100 to 300 atm (1470 to 4400 psi) and temperature on the formation of iron pentacarbonyl. Source: *J. Chem. Soc.*, Vol 121, 1922, p 29–32

por metallurgy) techniques. The first patents covering the carbonyl process were registered to Ludwig Mond in 1890. A brief review of early metal carbonyl research, with emphasis on his work, is presented in the article "Production of Nickel-Base Powders" in this Volume. In addition, some of the different types of metal carbonyls that can be formed and the formation and decomposition reactions of carbonyls used for the production of primary metal and powder products are described in that article in the discussion of nickel carbonyl powders. This section, however, deals only with the production of iron powders produced by the decomposition of iron pentacarbonyl, $Fe(CO)_5$.

Process Conditions. Iron pentacarbonyl, the raw material from which carbonyl iron powder is produced, is a liquid with a boiling point of 102.8 °C (217 °F). It is formed by passing carbon monoxide over a reduced sponge iron at relatively high pressures and at temperatures ranging from 170 to 200 °C (340 to 390 °F). The physical properties of iron pentacarbonyl are:

Formula	$Fe(CO)_5$
Color and state (room temperature)	Viscous yellow liquid
Molecular weight	195.9
Iron, %	28.51%
Melting point, °C (°F)	−21 (−5.8)
Boiling point, °C (°F)	102.8 (217)
Specific gravity (room temperature)	1.457
Heat of formation, kJ/g · mol (kcal/g · mol)	−964.0 (−230.2)

Typically, the feed stocks used are high-surface-area oxidized iron powders or iron turnings that are reduced in hydrogen or another suitable atmosphere prior to carbonylation. The presence of oxygen or oxides on the surface of the iron hinders the reaction, whereas the presence of catalysts improves the rate of formation.

The reaction is exothermic. To achieve commercial production rates, carbonylation is conducted at 130 to 180 atm (1900 to 2600 psi) and temperatures of 170 to 175 °C (340 to 350 °F). As shown in Fig. 8, iron pentacarbonyl formation increases with increasing carbon monoxide pressure. Increasing the temperature increases the rate of reaction. Increasing pressure prevents excessive decomposition of the carbonyl. At temperatures above 200 °C (390 °F), carbonyl yield is reduced rapidly by the increased conversion of carbon monoxide to carbon and carbon dioxide by the disproportionation reaction. Iron pentacarbonyl is then condensed and purified by distillation, whereupon it reverts to di-iron nonacarbonyl, $Fe_2(CO)_9$.

The rate of decomposition for di-iron nonacarbonyl is temperature dependent, with maximum rates achieved at about 200 to 250 °C (390 to 480 °F). At higher temperatures, the iron produced oxidizes in the carbon monoxide atmosphere, and high-carbon powder is formed.

Present Commercial Processes. In the United States, International Specialty Products (formerly GAF Corporation) currently produces high-purity iron powder using carbonyl decomposition technology. In Europe, BASF produces carbonyl iron powders using the same decomposition product, but employing scrap iron as source material. The GAF (ISP) process starts with high-purity iron. Iron powder, or sponge, which is initially treated under hydrogen to reduce surface oxides to metallic iron. This feedstock is then reacted with carbon monoxide under pressure at elevated temperature to form liquid iron pentacarbonyl. Subsequently, the carbonyl is vaporized and thermally decomposed to form "crude" carbonyl iron powder. The crude powder is refined by chemical processing and mechanically separated to yield various size grades of powder. The carbon content of the powder may reach a maximum of 0.8%, which is reduced to as low as 0.075% to satisfy specific requirements.

Table 2 Properties and uses of Pyron hydrogen-reduced iron powders

Property	Uses
High strength-to-weight ratio	Low-density, high-strength iron bearings and structural parts requiring a high percentage of interconnecting porosity for oil retention
High green strength	Low-density fragile parts; additives in blend of other iron powders to increase green strength. Friction material compounds
Low apparent density	Friction material compounds; additive in blends of other iron powders to reduce apparent density
Dimensional stability	Close tolerance, low- and medium-density Fe-Cu or Fe-Cu-C bearings and structural parts requiring good strength, machinability, and hardenability

Powder Properties. High-purity iron powders produced using carbonyl technology are typically spherical and are available in several grades that range from 2 to 9 μm Fisher subsieve size, with apparent densities of 1.2 to 3.2 g/cm³. The powder can be produced at or near atmospheric pressure. Particles characteristically exhibit an onionskin structure due to minute carbon deposits in alternate layers. The high purity of these powders makes them an excellent starting material for the production of magnetic cores and electronic components. This powder oxidizes readily in air and is packaged under an inert gas to facilitate storage.

These powders are essentially free of nonferrous metals and contain 0.3% O, 0.075 to 0.8% C (depending on the grade), and 0.05 to 0.9% N. Alloy powders also are available that nominally contain 92.5% Fe (minimum), 0.8% C, 0.3% O, and 6.0% N. Primary applications of these powders include:

- High-frequency cores for radio transmitters, telephones, televisions, direction finders, VHF and UHF circuitry, and radar equipment
- Magnetic-fluid clutch and brake systems
- Carbide and diamond cutting tools
- Chemicals
- Pharmaceuticals (iron supplements, multivitamins)
- Catalysts
- P/M materials and alloys
- Cermets
- Magnetic coatings and tapes
- Radar-absorbing materials
- Food enrichment and animal feed

More information on applications is given in the article "Specialty Applications of Metal Powders" in this Volume.

Electrolytic Iron Powder

Revised by Prasan Samal, OMG Americas

Iron powder made using electrolytic deposition has the highest level of purity among all types of commercially available iron powders. It possesses very irregular particle shape and consequently offers relatively large surface areas per unit mass (typically 0.1 to 1.0 m^2/g). As a result of its high purity and irregular particle shape, it exhibits excellent compressibility and green strength. Its high level of purity and large surface area also render it desirable for many chemical, catalyst, and food-additive applications. Despite these unique attributes, the usage of electrolytic iron powder is limited due to its high cost of manufacture. Before improvements in the atomization of molten metal permitted the production of iron powders with relatively low oxygen contents and with moderate levels of purity, electrolytic iron powder was used widely for the manufacture of conventional P/M parts.

For many years, the telephone industry used electrolytic iron powder for manufacture of magnetic cores. Magnetic core and P/M magnet applications make up a sizable portion of the total volume of usage of electrolytic powder currently. Other applications include dry toner carrier for plain-paper photocopiers, iron-enrichment additives in foods, exothermic food heaters and hand warmers, catalysts, and chemical reagents. Of all other commercially available iron powders, only the carbonyl iron powders offer similarly high level of purity. However, carbonyl iron powders contain larger amounts of carbon and nitrogen, and are somewhat higher in cost.

Process Conditions. Manufacture of a metal powder by the electrolytic process can be achieved by one of two methods: a loosely adhering, powdery deposit may be formed directly on a permanent cathode, or a smooth, dense layer of refined metal may be deposited on a cathode and subsequently milled to obtain powder. Electrochemical polarization characteristics of the metal being processed largely determine which process can be used. In metals with low electrochemical polarization properties, such as copper, silver, zinc, and cadmium, the deposition potential varies little with current density; thus, spongy or powdery deposits, which can be scraped off as a powder, can be formed easily on the cathode. In metals with high electrochemical polarization properties, such as iron, nickel, and cobalt, a large change in the cathode potential is required to cause a relatively small change in current density. Consequently, smooth, coherent deposits are obtained with these metals. Electrolytic iron is, therefore, deposited on the cathode as a dense, brittle, and relatively thick layer, which is pulverized for the manufacture of the powder.

Table 3 Chemical and physical properties of compacting-grade iron powder

Property	Electrolytic	Reduced	Atomized
Chemical analysis, %			
Total iron	99.61	98.80	99.15
Insolubles	0.02	0.10	0.17
Carbon	0.02	0.04	0.015
Hydrogen loss	0.29	0.30	0.16
Manganese	0.002	...	0.20
Sulfur	0.01	0.007	0.015
Phosphorus	0.002	0.010	0.01
Physical properties			
Apparent density, g/cm^3	2.31	2.40	3.00
Flow rate, s/50 g	38.2	30.0	24.5
Sieve analysis, %			
+100	0.5	0.1	2.0
−100/+150	13.1	7.0	17.0
−150/+200	22.6	22.0	28.0
−200/+325	29.4	17.0	22.0
−325	34.4	27.7	22.0
Compacting properties(a)			
Green density, g/cm^3	6.72	6.51	6.72
Green strength, MPa (psi)	19.7 (2800)	19.0 (2700)	8.4 (1200)

(a) At 414 MPa (60 ksi) with 1% zinc stearate

Two types of electrolytes are commonly used for the electrodeposition of iron: chloride and sulfate baths. Each has certain limitations and advantages over the other. Chloride baths can have higher concentrations of iron ions due to the greater solubility of ferrous chloride in water. High ionic concentration of the bath increases its conductivity, thus lowering power consumption. Chloride baths are, however, more corrosive than sulfate baths. The iron powder and chips produced in chloride baths contain small amounts of chlorides (about 0.10%), which are usually more objectionable than the residual sulfur found in products of sulfate baths.

Commercial Processes. Currently, OMG Americas (formerly SCM Metal Products) is the sole producer of electrolytic iron powder in the United States. The production process is based on a sulfate electrolyte with soluble anodes.

The electrolytic cells measure about 3 m (10 ft) long by 0.75 m (2.5 ft) wide by 0.75 m (2.5 ft) deep. The electrolyte consists of a mixture of ferrous and ammonium sulfates. Each cell contains 17 consumable anodes interspersed with 16 reusable cathodes. The iron from the anodes is deposited on the cathodes.

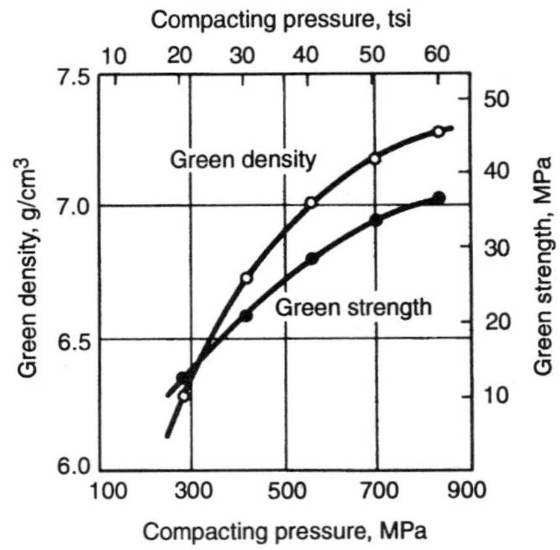

Fig. 9 Compacting properties of electrolytic OMG A-210 iron powder. Powder admixed with 0.5% zinc stearate for lubrication

Electrodeposited iron is stripped off the cathodes periodically, and the deposits are broken into chips, approximately 15 mm (⅝ in.) × 15 mm (⅝ in.) in size. Chips are ground in closed-type continuous ball mills or batch-type mills as determined by the fineness of powder required. Mills are purged with an inert atmosphere, such as nitrogen, to prevent excessive oxidation. Milled powder is classified using various types of screens and classifiers. Although some powder is used in the as-milled condition for applications such as food enrichment and chemical reagents, other applications require that the powder be annealed. Annealing softens the powder, increasing its compressibility, and modifies its magnetic characteristics. Annealing is usually carried out in hydrogen-containing atmospheres, which also reduces the oxygen content of the powder.

Blending to ensure uniformity is the final step in processing. Samples are taken from the blended powder for conducting various quality control tests, including chemical analyses and physical properties.

Powder Properties. Table 3 compares the properties of a compacting grade of electrolytic iron powder (annealed) with those of typical atomized and reduced-iron powders. Figure 9 graphically depicts the compacting properties of this powder.

The high compressibility and green strength of this powder make it very suitable for manufacture of intricate and fragile parts. Also, as a result of its intrinsic low hardness and freedom from refractory and oxide contaminants, it does not cause excessive die wear. Electrolytic iron powder also offers superior sintered properties, such as ductility and dimensional accuracy, achievable with reasonable compacting pressures. Figure 10 shows the relationship of sintered properties to green density. Table 4 lists the

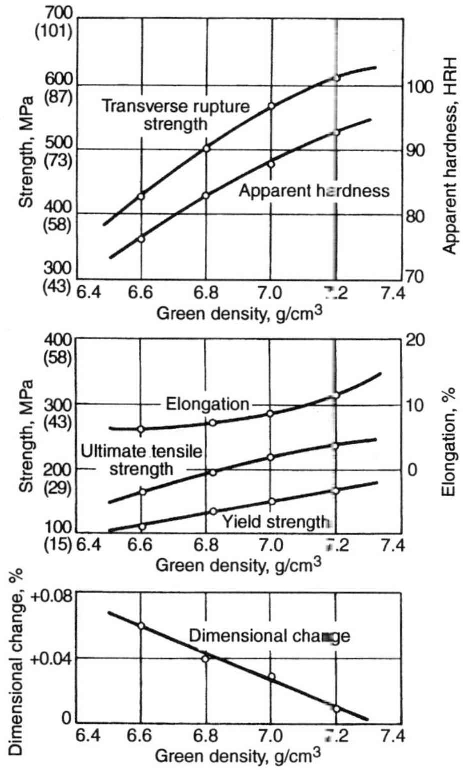

Fig. 10 Mechanical properties of pressed and sintered electrolytic iron powder compacts. Note: OMG A-210 powder was mixed with 0.5% zinc stearate, pressed into standard MPIF test bars, and sintered in dissociated ammonia atmosphere at 1120 °C (2050 °F) for 30 min.

Table 4 Properties of various grades of electrolytic iron powder

Property	Compacting powders		Ferromagnetic core powders				Photocopier powders,	Food enrichment
	A-210	A-220	A-233	A-234	A-251	M-908	A-277	powder, A-131
Chemical analysis, %								
Iron	99.61	99.25	...	...	...	...	99.50	98.30
Carbon	0.02	0.02	0.02	0.02	0.02	0.02	0.02	Arsenic, <0.0004
Hydrogen loss	0.29	0.65	0.35	0.65	0.15	0.20	0.40	Lead, <0.0010
								Mercury, <0.0002
Physical properties								
Apparent density, g/cm^3	2.31	2.30	2.31	1.75	2.30	2.25	2.50	2.02
Sieve analysis, %								
−100 mesh	0.5	...	...	...	...	...	...	20–40 μm, 25%
−100 + 150	13.1	...	<50 μm	<40 μm	−150 mesh	−100 mesh	4.0	10–20 μm, 35%
−150 + 200 mesh	22.6	0.8	...	...	...	...	41.0	0–10 μm, 40%
−200 + 325 mesh	29.4	14.1	...	...	...	...	51.0	...
−325 mesh	34.4	85.1	...	...	...	...	4.0	...
Green density(a), g/cm^3	6.72	...	...	...	...	...	...	...
Permeability	...	...	45	25	80	80	...	...
Q-value	...	...	Medium	Highest	Low	Low	...	...

(a) At 414 MPa (60 ksi). Source: OMG Americas

properties of various grades of electrolytic iron powders.

Applications. The high green strength, high compressibility, irregular particle shape, and high purity of electrolytic iron powder make it suitable for a number of P/M applications. It is ideal for high-density, close-tolerance applications. Secondary operations are readily performed on high-density components made from electrolytic iron powder. Because of the irregular particle shape, re-pressed parts made from electrolytic iron powder have interconnected porosity eliminated at a lower density level than with more regularly shaped atomized iron powders. Hence, the risks of entrapment of plating solutions or infiltration of brazing filler material are minimized.

In applications requiring case-hardening heat treatment, electrolytic iron powder offers a very uniform case depth due to high density and freedom from interconnected porosity. For optimal control of case hardening, a minimum density of 7.2 g/cm^3 is recommended. In full-dense applications, the low oxide inclusion content (acid insolubles) of electrolytic iron powder accounts for the superior dynamic properties of the parts, such as fatigue and impact strengths. The superior green strength of electrolytic iron powder also makes it ideal for use as the matrices for abrasive cutting wheels and diamond-bonded saw segments.

Annealed electrolytic iron powder is used in the manufacture of ferromagnetic cores for radios, televisions, and other audio devices. The common designs of ferromagnetic cores made from these powders are the E-type, C-type, and toroids.

Electrolytic iron powder is also used for making soft magnetic parts because of its high compressibility and freedom from impurities. For maximum magnetic properties, soft magnetic parts should be pressed to the highest practical density from a powder mix that contains a minimum amount of lubricant. To minimize carbon contamination, sintering is best carried out in dissociated ammonia, vacuum, or hydrogen. Other magnetic applications include magnetic strip produced by direct powder rolling and additives for Nd-Fe-B magnets.

Annealed electrolytic iron powder, screened to closely controlled particle size range, is used as a developer or carrier in toners for plain-paper copier machines. The irregular particle shape and magnetic quality of high-purity iron powder result in a magnetic brush of high loading characteristics that allows the toner to be uniformly transferred to the paper, thus providing high-contrast copies.

Unannealed, ultrafine (average particle size of about 10 μm) electrolytic iron powder is used for enrichment of food products. This grade of electrolytic iron powder meets or exceeds all specifications of the *Food Chemical Codex*. In addition to possessing the high purity required for this application, these powders exceed other iron powders, such as carbonyl and reduced-iron powders, in bioavailability because of their large specific surface area.

Other food-related applications include uses as desiccants for packaged food, to getter oxygen; and as self-contained food warmers especially for military and camping food packages. Similar disposable heating devices made using electrolytic iron are also sold as hand warmers for camping and other outdoor activities. Some electrolytic iron powder is used in the oxygen candles by the mining industry.

The high purity and large specific surface area of electrolytic iron powder make it an excellent candidate for many chemical reagent and catalyst applications.

Fluidized-Bed Reduction

Fluidized-bed processes currently are used extensively in industrial applications. Process variations include fluidized-bed combustion, gasification, heat treatment, and catalytic reactions. Attractive characteristics of fluidized-bed processes include high, uniform heat transfer between gas and solid and high reaction rates due to the large exposed surface area of the fluidized-bed solid material. Fluidized-bed oxidation of metals and reduction of metal oxides are possible, but defluidization caused by sticking, as well as the pyrophoricity of reduced powders, has restricted industrial usage.

This article describes the processes of fluidized-bed reduction of iron oxides that were used from 1958 to 1970. Oxide reduction, atomization, electrolytic deposition, and carbonyl processes have replaced the fluidized-bed process for current iron powder production.

Typically, in the fluidized-bed process a granular material is kept in fluid motion by using a fluidizing gas. In many cases, the fluidizing gas reacts chemically with the granular solid. Thus, in fluidized-bed oxidation, the fluidizing gas typically is air or oxygen. In fluidized-bed reduction, the fluidizing gas contains hydrogen or carbon monoxide, or a mixture of the two.

Principles of Fluidized Reduction Processes. Many of the principles of fluidized-bed reduction were developed as a by-product of research aimed at direct ironmaking, that is, bypassing the reduction of iron ores in a blast furnace and replacing this procedure with continuous fluidized-bed reduction of iron ore. Iron processed in this manner is referred to as direct-reduced iron. To a great extent, these principles also apply to the reduction of high-purity iron ore or iron oxide, where the reduced iron is used in metallurgical applications without further processing.

For direct-reduced iron production, ore feed in the form of pellets is preferred (Ref 2). Fully oxidized pellets, such as hematite (Fe_2O_3) pellets, should be used, because magnetite (Fe_3O_4) forms a relatively compact iron shell during reduction that strongly resists gas diffusion into the pellets, thus increasing reduction time (Ref 3). Pellets made from rich magnetite also disintegrate during reduction because of the high degree of swelling (Ref 4).

Sticking of the particles during reduction is perhaps the most serious problem encountered in fluidized-bed reduction of metal oxides, because it results in defluidization of the bed. The susceptibility of iron ore to sticking is a complex relationship between percent reduction, bed temperature, and gas velocity (Ref 5). By reducing low-gangue ore, Agarwal et al. (Ref 5) found that the susceptibility to defluidization is highest between about 620 and 730 °C (1150 and 1350 °F) for products reduced more than 90%. Below about 620 °C (1150 °F), the tendency to defluidize is minimal, but the resulting product is usually pyrophoric. At temperatures above 700 °C (1300 °F), the tendency to defluidize decreases.

Similar findings were observed by Bondarenko et al. (Ref 6) who established that the zone of fluidization expands by using larger pellets and increasing gas velocity and temperature from 550 to 900 °C (1020 to 1650 °F). Evidently, high temperature decreases surface activity of reduced particles, and sticking is decreased.

The extremely high surface area of direct-reduced iron contributes to rapid reoxidation (pyrophoric iron). The reoxidation of direct-reduced iron (the reaction of iron with oxygen and water) is a rapid process in which spontaneous ignition of iron powder may occur.

The tendency of direct-reduced iron to reoxidation depends on reduction temperature. The higher the temperature, the less the degree of reoxidation. Raw material composition also has an effect on the oxidation or degree of reactivity (Ref 7). According to Jensen (Ref 8), the presence of cementite (Fe_3C) has a passivating effect on oxidation at temperatures up to 150 °C (330 °F). At higher temperatures, this material oxidizes faster due to cementite decomposition. As discussed in Ref 8, high-temperature reoxidation may be prevented successfully by:

- Passivation during the manufacture of the product
- Keeping the product dry prior to use
- Avoiding contact with hot objects or other sources of thermal energy above about 200 °C (390 °F)

Advantages of reduction in a fluidized bed include:

- Effective interaction of gases and solids in the bed and high rates of heat and mass transfer between the solid and fluid allow high velocity of reactions.
- Fluidized-bed reduction is an isothermic process, which allows a narrow temperature range to be maintained.
- Gas recirculation is possible.
- Reduction in a fluidized bed can be converted into a fully automatic, continuous process.

Disadvantages include:

- Not all iron ore materials can be reduced in a fluidized bed because of the tendency of these particles to stick together.
- Particle size is limited from about 6.5 mm (0.25 in.) to a few microns, or about 0.001 in.
- Swelling of iron ore materials during reduction can result in disintegration and excessive loss of particles from the bed by the transport of material of the reaction zone (carryover).
- Pyrophoricity hinders storage and transport of reduced iron.
- A sharp temperature gradient exists near the bottom of the bed (Ref 9), which causes overheating of the grid.
- Countercurrent rates of fluid throughput are limited to the range over which the bed may be fluidized, the minimum fluidization velocity, or the carryover velocity.
- The hydrodynamics of fine materials is mismatched to the reduction process, because fluidization occurs at a gas velocity lower than that required for reduction.

Critical Fluidization Velocity. Critical velocity is the reduction gas velocity at which the fluidization of material is achieved without carryover from the reaction zone. References 10 to 18 discuss empirical expressions for the parameters of the critical fluidization velocity. The derivations consider particle diameter, density of the solid and fluid, viscosity, and voidage.

Nu-Iron Process. Developed by the United States Steel Company in 1950 (Ref 19), the Nu-iron process uses fine iron ore (1.65 mm, or 0.065 in.) in the fluidized-bed reactor at temperatures of 600 to 700 °C (1110 to 1290 °F), which eliminates sticking and pyrophoricity. The reducing gas (73% H and 16% CO) is obtained by catalytic conversion of natural gas with steam. Before entering the reactor, iron ore is preheated to 375 °C (705 °F) by the off gas; then it is preheated to 925 °C (1695 °F) in a special furnace. The sponge iron, reduced to 90 to 95%, is briquetted and used in electric steelmaking furnaces.

The H-iron process was developed by Hydrocarbon Research Company and Bethlehem Steel Corporation. The process consists of reducing fine iron ore or mill scale (0.04 to 0.8 mm, or 0.0016 to 0.032 in.) by high-pressure hydrogen at 35 atm (515 psi) at a temperature of 540 °C (1000 °F).

Hydrogen is obtained by the water-steam or oxygen re-forming of natural gas. High hydrogen pressure is used to increase reduction velocity and decrease reactor size. To increase hydrogen utilization, three- and four-zone reactors are used, with capacities of 45 to 90 metric tons/year (50 to 100 tons/year). The reduced iron powder is annealed at 650 to 870 °C (1200 to 1600 °F) to render the powder nonpyrophoric. Depending on the reduction degree, the resulting reduced iron can be used for P/M structural parts (90 to 95% reduction degree) or in electric steelmaking furnaces (75% reduction degree).

ONIA Process. Developed in France, the ONIA (Office National Industriel d'Azote) process differs from the H-iron process. A low pressure of 5 atm (73.5 psi) is used. The composition of the reducing gas is 85 to 87% H and 14 to 16% CO. To avoid sticking, the process uses a reduction gas velocity significantly greater than the critical fluidization velocity. The sponge iron is used as a melt stock in electric furnaces and as a molding-grade powder.

Commercial Viability. As discussed in Ref 20, all of the fluidized-bed processes have been limited commercially because of sticking, high degrees of pyrophoricity due to extremely high microporosity, unsatisfactory quality of iron powder for molding applications, or a combination of these. Sticking can be eliminated by using a "semifluidized" bed in a reactor with pull-out grids (Ref 21). In this process, iron ore pellets (2 to 6 mm, or 0.08 to 0.24 in.) are reduced by the products of catalytic conversion of natural gas (35% H, 18% CO, 40% N).

Pellets are reduced in a two-stage fluidized-bed process. At first, a reduction degree of r = 30% is reached. After this stage, the fluidized bed sticks, forming a sponge cake. The second reduction stage (r = 95 to 99%) is accomplished in a semifluidized pulsating bed. Gas velocity during the impulse fluidization substantially exceeds the critical fluidization velocity, provided that the cake breaks and that there is a uniform gas supply. After cooling by nitrogen, the reduced iron is not pyrophoric because of high reduction temperatures of 800 to 870 °C (1470 to 1600 °F).

Laboratory experiments with a drum-type rotary fluidized bed also have shown methods of eliminating sticking (Ref 22). The advantages of fluidized-bed technology can be realized by use of an "improved" process that combines the fluidized bed with a mechanical or hydraulic device for breaking the sticking material. The technical feasibility to expand and develop fluidized-bed reduction processes depends on:

- Possibility to increase the temperature level of reduction with a simultaneous elimination of sticking and pyrophoricity
- Possibility to increase gas utilization efficiency

- Elimination of temperature gradient between the top and bottom of the fluidized bed

Water Atomization

Atomization processes are widely used for making both iron and steel powders. Atomization consists of mechanical disintegration of molten metal stream into small droplets that are rapidly cooled by means of conductive or/and convective heat transfer. The disintegration of a liquid stream by the impingement of high-pressure jets of water, gas or oil is called water, gas or oil atomization, respectively. Other modified versions of atomization are centrifugal, vacuum, ultrasonic, rotating electrode process, plasma rotating electrode process, electron beam rotating process, and rotating disk atomization. These processes are discussed in more detail in the article "Atomization" in this Volume.

Atomization of molten iron is a flexible process that can be adapted to the production of low-carbon iron or alloy steel powders. Production of atomized iron powder is limited commercially to processes based on water atomization. Steel powders are also produced primarily by water atomization, although inert-gas atomization is used in production of some alloy steel powders (as discussed in the article "Production of Steel Powders" in this Volume).

This section focuses exclusively on water-atomized iron powders from the following four production facilities:

- Quebec Metal Powders process
- Domfer process
- Kobe/Kobelco process
- Kawasaki process

Production and properties of atomized low-alloy and high-alloy steel powders (by both water or gas atomization) are discussed in the article "Production of Steel Powders" in this Volume.

The QMP Iron Powder Process*

In the Quebec Metal Powders (QMP) process, freshly refined high-purity iron passes through horizontal water jets, which produce granulated iron that is processed further by drying, grinding, decarburization, milling, and blending (Fig. 11). Because of the nature of the manufacturing process, the QMP iron powder grades (ATOMET 25, 28, 29, and 30) are highly porous powders with high compressibility for low- and medium-density P/M parts production.

The QMP process is based on conversion of high-purity, high-carbon iron received from the nearby smelter plant of QIT-Fer et Titane Inc. The liquid iron, desulfurized and recarburized at QIT to QMP specifications, is delivered as molten iron with a typical as-received analysis of:

- Carbon: 3.5 wt%
- Manganese: 0.008 wt%
- Silicon: 0.04 wt%
- Sulfur: 0.008 wt%

Due to the nature of both the ore deposit and the smelting process, the QIT iron is characterized by a low content of residual elements and an exceptional consistency of chemical composition. The high purity of the iron make it a well-adapted starting material for the production of both steel and iron powders (Fig. 12). Steel powder QMP production is discussed in the article "Production of Steel Powders" in this Volume.

Granulation. On arrival at the QMP iron powder plant, the hot metal is poured into one of two 82 metric ton (90 ton) induction holding furnaces. After temperature adjustment, the molten iron is poured from the induction furnace into a large tundish. The bottom of the tundish is equipped with four refractory nozzles that act as metering orifices. The streams of liquid iron flowing from the nozzles are disintegrated with high-pressure horizontal water jets directed into a large quench tank. Air drawn through ports in the quench tank cause the coarse granulated iron to be partially oxidized. The wet slurry from the granulation chamber is pumped to a magnetic dewaterer that feeds a vacuum filter to remove most of the water. The wet powder is then dried in a gas-fired rotary kiln dryer. The particle size of the as-granulated iron is typically –8 mesh. The carbon content of the iron drops to approximately 3% during granulation due to the partial oxidation of the granules that have an average oxygen content of about 5%.

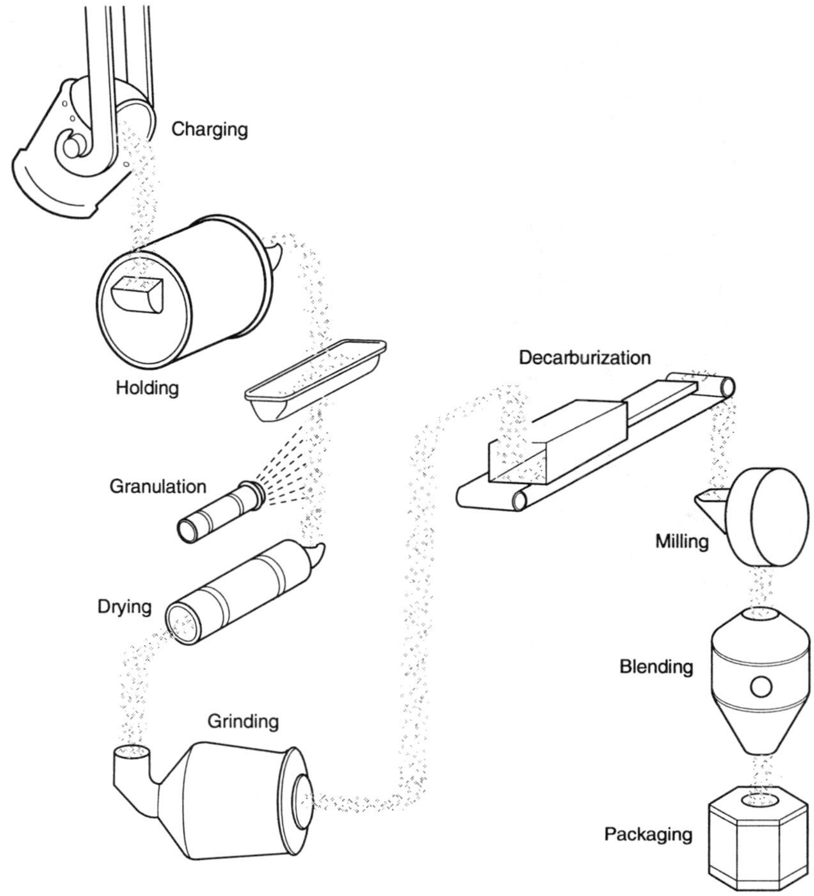

Fig. 11 Schematic flow diagram of the QMP iron powder manufacturing process

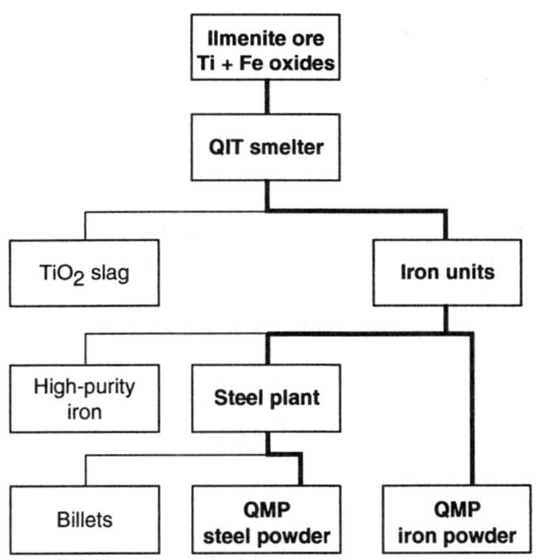

Fig. 12 Material flow through the QIT/QMP facilities for manufacturing iron and steel powders from ilmenite ore

*Contributed by Yves Trudel, Quebec Metal Powders Limited

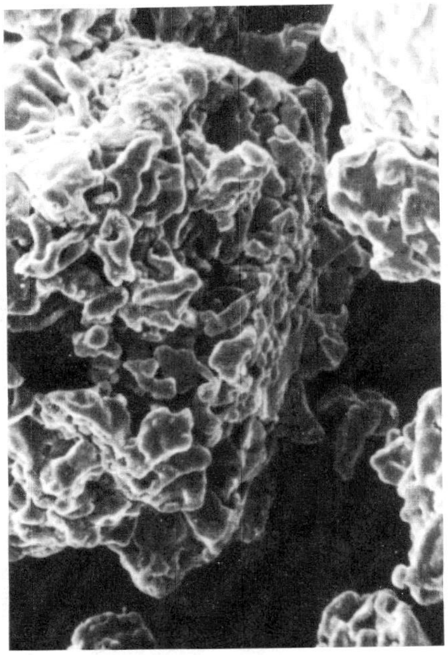

Fig. 13 Scanning electron micrograph picture of ATOMET 28 iron powder. 700×

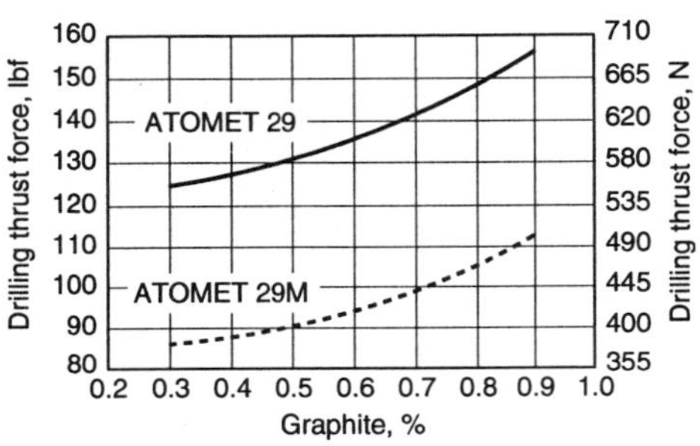

Fig. 14 Effect of boron nitride (ATOMET 29M) and carbon content on the thrust force measured when drilling sintered specimens

Table 5 Typical chemical and physical properties of QMP ATOMET iron powders

Property		QMP grade			
		ATOMET 25	ATOMET 28	ATOMET 29	ATOMET 30
Chemical analysis, wt%					
Carbon		0.05	0.05	0.05	0.01
Oxygen		0.18	0.17	0.17	0.14
Sulfur		0.005	0.005	0.005	0.005
Manganese		0.008	0.008	0.008	0.008
Iron		99+	99+	99+	99+
Apparent density, g/cm^3		2.52	2.83	2.93	2.93
Flow rate, s/50g		29	27	26	26
Sieve analysis, wt%					
U.S. Mesh	**μm**				
+70	+212	Trace	Trace	Trace	Trace
−70 +100	−212 +150	2	5	5	8
−100 +200	−150 +75	36	51	51	52
−200 +325	−75 +45	31	22	22	22
−325	−45	31	22	22	18

Table 6 Typical compaction and sintering properties of QMP ATOMET iron powders

Property	QMP Grade			
	ATOMET 25	ATOMET 28	ATOMET 29	ATOMET 30
Green density, g/cm^3	6.4	6.7	6.7	6.8
Compacting pressure(a), MPa (tsi)	360 (26.0)	455 (33.0)	455 (33.0)	450 (32.5)
Green strength(a), MPa (tsi)	13.1 (1900)	13.8 (2000)	13.8 (2000)	12.4 (1800)
Dimensional change from die size(b), %	−0.04	+0.17	+0.17	+0.25
Transverse rupture strength(b), MPa (ksi)	703 (102)	897 (130)	897 (130)	862 (125)
Tensile strength(b), MPa (ksi)	448 (69)	483 (70)	483 (70)	500 (72.5)

(a) ATOMET powder + 0.75% zinc stearate. (b) ATOMET powder + 0.9% C + 2.0% Cu + 0.75% zinc stearate sintered for 30 min at 1120 °C (2050 °F) in endogas

Ball Milling. Coarsely granulated iron is then ball milled to powder size. The grinding is done in a Hardinge air-swept mill. At the discharge of the mill, the powder is pneumatically conveyed to a cyclone and discharged on a series of screens that allow grinding to be adjusted to specific particle size ranges. The ball-milled powder is analyzed for its chemical composition and particle size distribution and transferred to decarburization.

Decarburization. The granulated and ball-milled powder is stored in dedicated bins according to the oxygen and carbon contents of the product. Powder from various bins is fed into a blender to prepare chemically controlled feed for the decarburization furnaces. Quantities are determined so as to produce a blend of powder with a uniform oxygen-to-carbon ratio close to 2. The blended powder is then transferred to decarburizing furnaces. The continuous-belt furnaces, equipped with 120 cm (48 in.) wide solid belts, are maintained under a reducing atmosphere of dissociated ammonia. The ball-milled powder is fed onto the belt to form a bed approximately 25 mm (1 in.) thick and heated to temperatures in the range of 980 to 1040 °C (1800 to 1900 °F).

The carbon and the oxygen contained in the powder then reacts to form CO and CO_2 as reaction products. Hydrogen from dissociated ammonia contributes to further reduce the oxygen content of the powder. After cooling under a protective atmosphere, the product emerges from the furnaces as a loosely sintered spongy cake of pure iron, containing typically 0.05% C and 0.17% O.

Milling. At the discharge of the furnaces, the iron powder cake falls into a sawtooth crusher. The crushed cake is further processed in a series of attrition mills. The mills pulverize the cake into a powder of controlled particle size distribution and apparent density.

Blending and Packaging. The iron powder plant is equipped with a 36,000 kg (80,000 lb) capacity product blender in which individual batches of milled powder are homogenized to prepare large lots of powder of constant properties. For P/M applications, press-ready mixes are also manufactured by blending the plain iron powder with selected additives such as lubricants, graphite, and copper powders. The QMP iron powder is then packaged to be shipped worldwide.

Each powder lot is sampled and submitted to a series of quality control tests to certify that its chemical composition, physical characteristics, compressibility, green strength, dimensional change, and sintered strength meet customer requirements. The QMP quality system has been registered to ISO 9001 since 1993 and to QS-9000 since 1995.

Iron Powder Characteristics. The QMP iron powder process, because of its flexibility, allows high-quality iron powders to be produced for all the major applications of ferrous powders. The main use of QMP's iron powders is the production of structural parts by the P/M technique.

Because of the nature of their manufacturing process, ATOMET iron powders are highly porous, as illustrated in Fig. 13. This spongy structure and the high degree of purity of the product confer to P/M grades a combination of compressibility and green strength that make these products well suited for low- and medium-density applications. Typical chemical and physical properties of QMP iron powders (ATOMET 25, 28, 29, and 30) are given in Table 5. The porous nature of ATOMET iron powder particles also promotes sintering and permits rapid carbon pickup. Table 6 shows typical compaction behavior and sintered properties of iron powder grades.

Machinable P/M Grade. A patented machinable P/M grade, ATOMET 29M, has been developed specifically for P/M parts requiring machining. This product uses hexagonal boron nitride as a machinability enhancer. The boron-nitride enhancement allows a reduced drilling thrust force of P/M materials, even at high car-

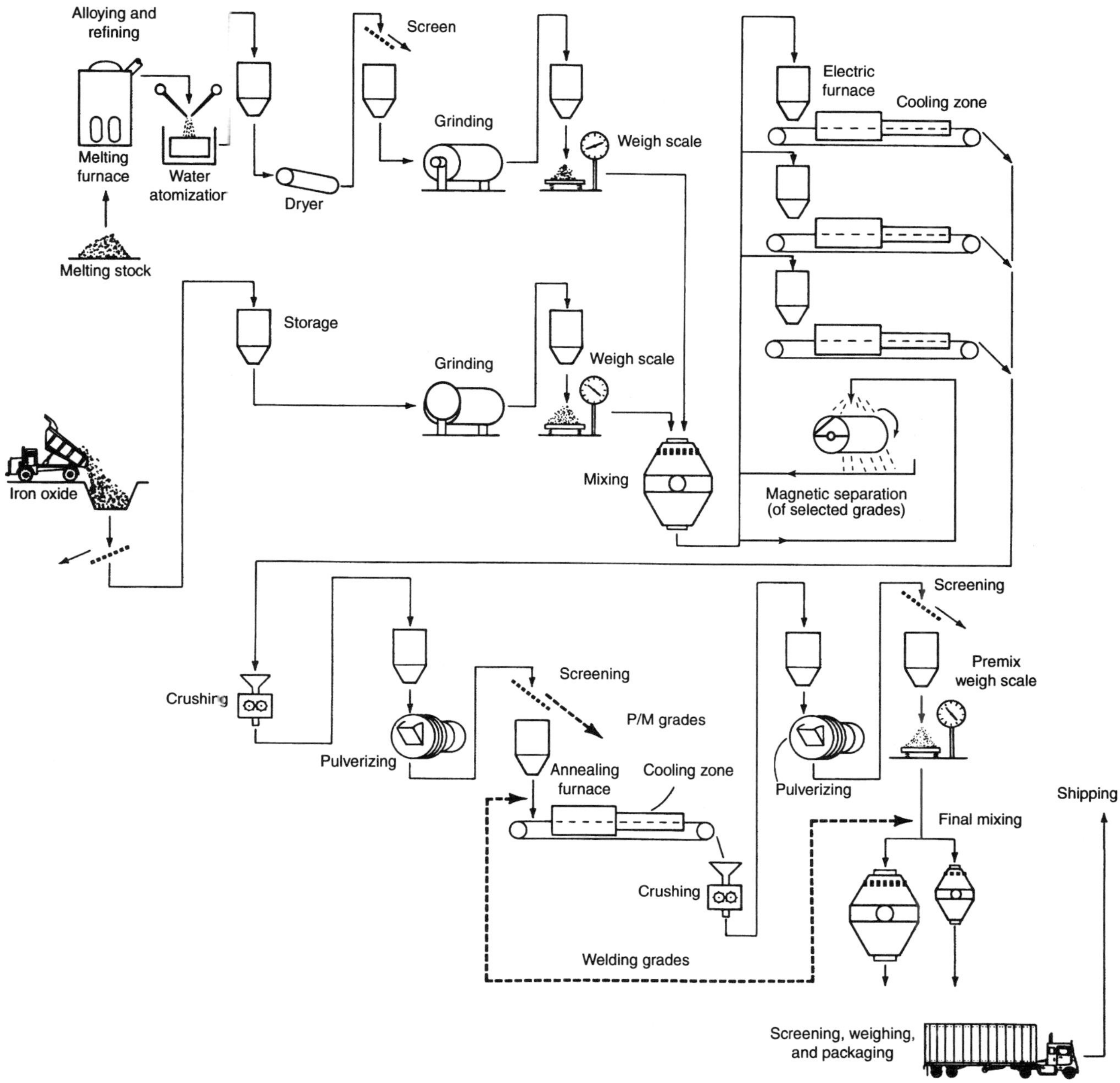

Fig. 15 Flowchart of the Domfer powder process

bon contents (Fig. 14), and also contributes to increased tool life.

Special Powder Applications. Iron powders produced at QMP can also be tailored to meet the needs of a variety of non-P/M applications such as welding electrode coating, aluminum alloying, flame cutting, and scarfing. Also, special grades are manufactured for magnetic particle inspection, food additives, pharmaceutical applications, and carrier materials for photocopying machines.

The Domfer Process**

The Domfer process of making iron and steel powder was developed in 1952 at Iberville, Quebec. It consists of water atomizing a high-carbon iron melt to a granular "shot," which is then ball milled to powder size. The ground shot, which contains carbon, is mixed with ground mill scale (iron oxide) that contains oxygen. The carbon and oxygen are chemically combined in a belt furnace at sintering temperatures to form carbon monoxide gas, thus leaving a pure iron cake. The cake is disintegrated to powder size and further refined. A flowchart of the Domfer powder process in its present form which evolved from the original invention is given in Fig. 15.

Melting and Refining. High-quality selected steel scrap is melted in a 12.7 metric ton (14 ton) coreless induction furnace with an addition of low-sulfur carbon to make a melt of approximately 3.5% C. For low-alloy steel powders, nickel and molybdenum also are added to the furnace.

Approximately 30% of the furnace charge is tapped into a ladle, deslagged, and desulfurized if necessary. Molten metal is poured through a tundish with bottom openings to control the size of the stream. The molten metal is disintegrated by high-pressure jets of water to a granular material approximately 0.8 to 3.2 mm (0.03 to 0.12 in.) in size. This shot is dewatered in a cyclone and dried in a gas-fired dryer.

Grinding. Because the high-carbon shot is very brittle, it can be ground to the required powder size in an air-swept ball mill. Air flow controls the size of particles that leave the mill,

**Contributed by Amédéo F. De Rege, Domfer Limited

Table 7 Typical properties for Domfer welding, flame-cutting, and friction grades

Property	Domfer grade							
	MP61(a)	MP62(a)	MP64F(b)	MP64C(b)	MP43(c)	MP31(d)	MP33(d)	MP61H(d)
Screen analysis, wt%								
U.S. Mesh No.								
+20	Nil						Nil	Nil
+30	1			Nil			<1	<1
+40	...	Nil		...	Nil		...	...
+50	27	...	Nil	<1	...	Nil	12	15
+60	...	Trace	...	...	Trace	...	...	...
+80	27	4	0.2	17	3	2	15	35
+100	7	6	0.2	16	5	4	7	12
+140	...	19	0.9	30	18	16	...	...
+200	22	24	9.2	21	23	26	30	20
+325	...	26	38.2	10	28	28	...	...
Pan(e)	16	21	51.3	6	23	24	36	18
Physical properties								
Apparent density, g/cm^3	2.86	2.60	3.03	3.50	2.57	2.55	2.21	3.20
Chemical analysis, wt%								
Carbon	0.05	0.045	4.0	4.0	0.10	0.07	0.03	0.83
Hydrogen loss	0.48	0.73	...	...	0.47	0.46	0.72	0.61
Total iron	98.4	98.4	95	95	98.3	98	98	98

(a) Welding electrodes grade. (b) Hardfacing applications grade. (c) Flame-cutting and scarfing grade. (d) Friction applications grade. (e) Fraction collected under the finer sieve

Table 8 Typical properties of Domfer P/M grade powders

Property	Domfer grade								
	MP30	MP32	MP35	MP37R	MP37HD	MP39HSR	MP52	MP55HD	MP56HD
Screen analysis, wt%									
U.S. Mesh No.									
+40	Nil	Nil	Nil	Nil	Nil	Nil	Nil	Nil	Nil
+60	Trace	Trace	Trace	Trace	Trace	Trace	Trace	Trace	Trace
+80	4	4	5	3	4	3	2	3	7
+100	6	6	8	6	7	5	7	6	7
+140	20	21	22	19	23	18	18	20	24
+200	22	23	25	28	26	24	24	25	25
+325	23	24	24	29	26	29	30	31	25
−325	25	22	16	15	14	21	19	15	12
Chemical analysis, wt%									
Carbon	0.018	0.008	0.006	0.008	0.012	0.540	0.004	0.008	0.012
Sulfur	0.026	0.025	0.024	0.236	0.243	0.025	0.016	0.021	0.021
Oxygen	0.43	0.36	0.32	0.31	0.31	0.36	0.26	0.30	0.33
Manganese	0.54	0.54	0.54	0.78	0.73	0.54	0.39	0.45	0.45
Molybdenum	...	...	...	...	...	...	0.71	0.75	1.10
Nickel	...	...	...	...	...	...	0.47	1.82	1.90
Physical properties									
Apparent density, g/cm^3	2.60	2.59	2.94	2.61	2.85	2.71	2.63	2.92	2.95
Hall flow rate, s/50 g	30	29	36	30	27	29	30	27	27
Mechanical properties(a)									
Compressibility(b), g/cm^3	6.56	6.69	6.76	6.66	6.72	6.35	6.69	6.60	6.65
Green density of following tests, g/cm^3	6.4	6.4	6.8	6.4	6.8	6.6	6.8	6.8	6.8
Green strength, MPa (psi)	9.8 (1420)	8.4 (1220)	7.9 (1150)	8.6 (1250)	9.0 (1300)	12.0 (1730)	11.7 (1700)	8.6 (1250)	8.5 (1230)
Transverse rupture strength(c), MPa (ksi)	725 (105)	740 (107)	1070 (155)	915 (133)	1080 (157)	1035 (150)	1000 (145)	945 (137)	980 (142)
Hardness(c), HRB	61	72	88	78	89	90	(4)	(4)	(4)
Dimensional change from die size(c), %	0.04	0.02	0.20	0.09	0.19	−0.34	0.08	0.24	0.25

(a) Of the base powder, admixed with 1% graphite, 2% copper and 0.75% lubricant. (b) Expressed as the green density obtained at 30 tsi compacting pressure. (c) Sintered at 1120 °C (2050 °F) for 30 min, under $N_2 + H_2$ atmosphere. (d) Hardness of these sinter-hardening grades depends on the cooling rate. It typically exceeds 20, 30, and 35 HRC for MP52, HP55HD, and MP56HD, respectively.

and the powder is recovered from a combination dropout and cyclone separator. Selected mill scale, which is a by-product of rolling steel, is similarly ground to powder. The ground shot and mill scale are blended to achieve a carbon-to-oxygen ratio that results in decarburization and deoxidation, thus leaving essentially pure iron.

Decarburizing. The mixed powder is purified by magnetic separation (for selected grades) and laid on a steel belt that passes through an electrically heated furnace. The reaction between the high-carbon iron and mill scale occurs at a high temperature. The carbon monoxide gas that is generated provides both a reducing atmosphere and a furnace seal. The resulting pure iron sinters to a cake that is cooled to room temperature before leaving the furnace. The cake is reduced to powder size in an attrition mill and is screened to the desired size. It is then blended in a double-cone blender to the required specifications.

In the first stage of processing, carbon and oxygen levels are reduced to the specified values. This powder is suitable for the manufacture of welding rods and for lancing, cutting, and scarfing operations as well as for friction applications. Table 7 lists several iron powders of welding and friction grades that are produced by the Domfer process.

Annealing. A pure, soft powder is required by the P/M parts industry to reduce die wear and to lower the pressing tonnage required to attain the desired density. For this purpose, powder from the first stage is blended and fed to a second belt furnace. The conditions employed produce a very soft cake that can be reduced to powder size without work hardening. For selected P/M grades, decarburizing and annealing are conducted in a single-pass operation with modified furnace conditions.

Annealing typically reduces carbon to less than 0.01% and oxygen to less than 0.35%. It also produces soft powder with good compressibility. Powder premixes containing lubricant, graphite, and copper are blended to customer specifications. Table 8 lists several grades of compactable Domfer iron powders.

Quality Assurance. The Domfer process is a multistep operation that incorporates testing and blending between the stages to permit a wide range of powder qualities. The process sequence allows close control of uniformity from batch to batch. For most users, especially P/M parts manufacturers, this uniformity is the most important requirement.

For welding rod applications, essential properties to be controlled include carbon, oxygen, and sulfur content, screen analysis, apparent density, and flow characteristics. For additional P/M parts applications, sample bars are pressed and sintered to measure:

- Compressibility
- Green strength at this density
- Dimensional change during sintering
- Strength and hardness of the sintered bar

Domfer quality system has been registered as conforming to the ISO-9002 Standards since September 1994. QS-9000 accreditation is sought for 1998.

Kobe/Kobelco Process

Kobe Steel Company and Kobelco Metal Powders produce water-atomized iron powder using a special V-jet water atomization process. Typical properties of four Kobelco water-atomized ferrous powders are summarized in Tables 9 and

Table 9 Typical chemical and sieve analysis of Kobe ferrous water-atomized powder grades

Property	Kobe grade 500MA	300MA	300MC	300ME
Chemical analysis, %				
Carbon	0.001	0.001	0.001	0.001
Silicon	0.01	0.01	<0.01	<0.01
Manganese	0.16	0.16	0.16	0.17
Phosphorus	0.004	0.004	0.004	0.004
Sulfur	0.009	0.009	0.009	0.009
Nickel	...	...	...	0.01
Chromium	...	...	...	0.02
Oxygen	0.14	0.13	0.09	0.13
Sieve analysis, %				
+60	0.2	...	0.2	...
−60/+100	14	5	16	5
−100/+325	61	68	66	68
−326	25	27	18	27

10. Grades 500MA, 300MA, and 300ME achieve green densities of 6.8 g/cm^3 at 415 MPa (60 ksi). Grade 300MC has the highest compressibility, achieving a 6.90 g/cm^3 and 9.6 MPa (1.4 ksi) green strength at 415 MPa (60 ksi) with 0.75% zinc stearate (Fig. 16). Kobelco has also produced a 300MS grade iron powder with lower compressibility and green density (6.7 g/cm^3 at 415 MPa, or 60 ksi) but improved cleanliness, that is, low inclusion level (size: 50 μm max and area ratio: 0.03% max), which finds successful applications in hot powder forging of connecting rods and other automotive parts.

Kobe Steel Company also produces water-atomized iron powder grades such as:

- 300M offering high compressibility (6.9 g/cm^3 at 440 MPa, or 64 ksi)
- 300MH offering slightly higher purity and compressibility level (7.0 g/cm^3 at the same compacting pressure)
- 300NH achieving 7.1 g/cm^3 green density after extended reduction/annealing treatment (Ref 24)

Kawasaki Process

Kawasaki Steel Corporation has produced water-atomized iron powder (KIP 304AS grade) with high compressibility (7.05 g/cm^3 green density at 1% zinc stearate pressed at 490 MPa), higher purity, improved compactibility, lower inclusion content, high green strength, and low ejection force. These property improvements are achieved by the stricter selection of scrap, more rigorous refining in the electric arc furnace, extended high-temperature first annealing, and the incorporation of a second annealing.

Figure 17(a) shows the new Kawasaki process in comparison to standard water-atomization processing (Fig. 17b). Scrap is selected more carefully than for the standard operation, and the melt is atomized by water applied at a 30% lower pressure and 25% lower volume than in the standard operation. The reduced volume of

Table 10 Typical properties of Kobe water-atomized ferrous powders

Property	Kobe grade 500MA	300MA	300MC	300ME
Powder properties				
Apparent density, g/cm^3	2.92	2.91	2.87	2.93
Flow rate, s	25.1	24.7	25.8	24.6
Green density(a), g/cm^3	6.82	6.82	6.90	6.82
Green strength, MPa (psi)	8.3 (1200)	8.3 (1200)	9.7 (1400)	7.9 (1150)
Sintered properties(b)				
Dimensional change(c),%	+0.21	+0.23	+0.33	+0.23
Transverse rupture strength(c), MPa (ksi)	965 (140)	979 (142)	1020 (148)	1000 (145)

(a) Green strength and density with 0.76% zinc stearate lubricant. (b) Sintered at 1120 °C (2048 °F) for 30 min in 75% H_2 + 25% N_2 gas. (c) 2% Cu and 0.90% graphite added with zinc stearate

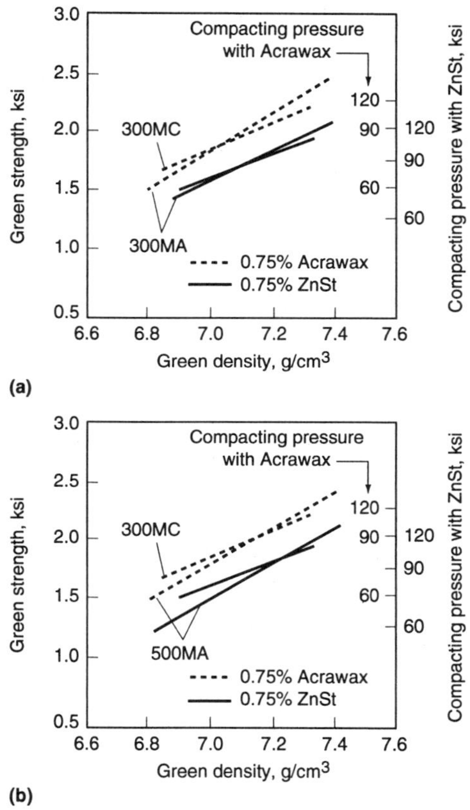

Fig. 16 Green strength and density comparison of high compressibility 300MC powder grade with (a) 300MA grade and (b) 500MA grade of Kobelco. ksi × 6.8947 = MPa

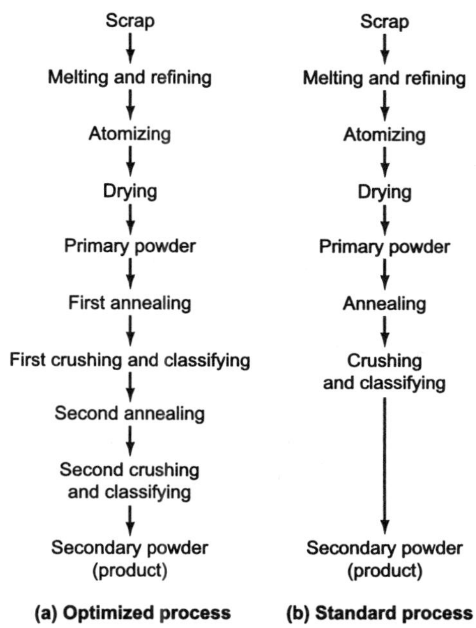

Fig. 17 Kawasaki manufacturing processes of atomized iron powders. (a) Optimized process of powders with high compressibility and green strength. (b) Standard process

atomizing water facilitates a higher degree of particle spheroidizing, which thus increases the apparent density of the primary powder. The lower water pressure results in an increase in average particle size to about 120 μm, which is almost twice the size of primary powders from the standard process.

The primary powder is annealed in hydrogen at a slightly higher temperature and a longer duration than the anneal in the standard process. After annealing, crushing and classification are executed. The first crushing in the new process is harder than that under standard conditions. In addition, a second anneal crush and classification sequence is carried out. The second crushing was softer, and the sieve size of the second classification was 44% larger than under standard conditions.

The purpose of this modified production process is to obtain powders that allow high compressibility, low ejection force, and high green strength during high-pressure pressing. High compressibility improves die life, while lower forces for ejection reduce the potential for cracking of more complex shapes during part compaction and transfer.

However, compressibility and high green strength require careful optimization of particle shape. Compressibility (or fill density) increases for spheroidized particles, while compact green strength depends on particle "entanglements" that are enhanced by more irregular shaped particles. Therefore, the compatibility of high compressibility and improved compactability (green strength) requires optimization of factors besides

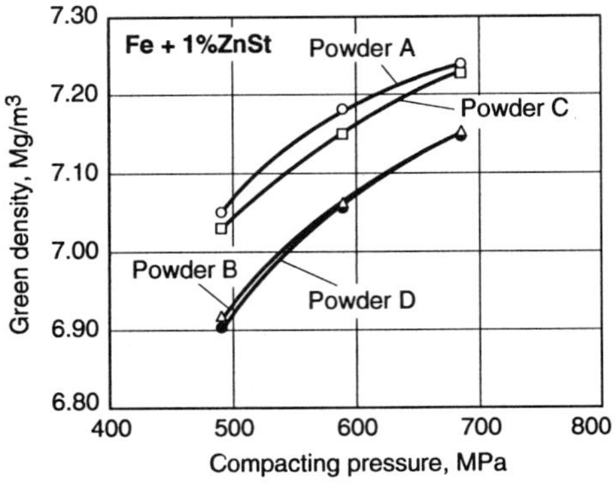

Fig. 18 Relationship between green density and compacting pressure for four iron powders in Table 11. Source: Ref 25

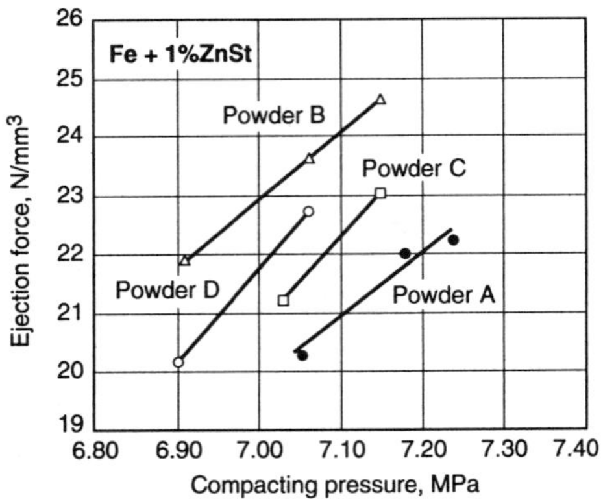

Fig. 19 Relationship between ejection force and green density for four iron powders in Table 11. Source: Ref 25

just particle shape. Green strength is enhanced by preventing excessive spheroidization, while compressibility is improved by other factors such as reductions in impurities and inclusions, reduced dislocation density, and greater coarseness in the particle size distribution. Accordingly, various manufacturing conditions are optimized in the Kawasaki process to obtain iron powder with a good combination of green strength and compressibility (Fig. 18) and low ejection force requirements (Fig. 19). The characteristics of powders A, B, C, and D are summarized in Table 11.

Table 11 Characteristics of secondary iron powder compared in Fig. 18 and 19

	Process conditions	Flow rate (s/50 g)	Particle size distribution, %						
			+180 μm	+150 μm	+106 μm	+75 μm	+63 μm	+45 μm	−45 μm
Powder A	Optimized (Fig. 17a)	23.5	5.2	14.2	24.7	24.0	8.2	13.5	10.2
Powder B	Standard (Fig. 17b)	25.4	0.1	7.3	19.3	26.7	10.0	12.5	24.1
Powder C	Commercial iron powder	21.6	1.5	8.4	19.4	26.2	9.6	16.4	18.5
Powder D	Commercial iron powder	25.1	0.1	4.0	15.3	22.9	9.5	14.0	34.2

REFERENCES

1. R.L. Mond and A.E. Wallis, Researches on the Metallic Carbonyls, *J. Chem. Soc.*, Vol 121, 1922, p 29–32
2. J.G. Stephenson, The Influence of Iron Ore Feed Characteristics on Direct Reduction, *16th Ann. Conf. Metall.*, CIM, 1977
3. J.K. Brimacoude and V. Venkateswaran, Literature Review of the Production and Use of Direct Reduced Iron, *DRI—Technology and Economics of Production and Use*, Iron and Steel Society of AIME, 1980, p 216
4. V. Bogdandy et al., *Archiv Eisenhüttenwes.*, Vol 34, 1963, p 401–409
5. J.C. Agarwal et al., Fluidized Bed Technology, *Chem. Eng. Prog. Symp.*, Vol 62 (No. 67), 1966, p 107
6. E.I. Bondarenko et al., Production of Sponge Iron in Fluidized Bed Reactor, *Visocotemp. Endoter. Proc. Metall.*, Moscow, 1968
7. W. Pietsch, "Storage, Shipping and Handling of Direct Reduced Iron," Society of Mechanical Engineers—American Institute of Mechanical Engineers Fall Meeting (Denver), 1976
8. H.B. Jensen and R.M. Smailer, The Handling, Storage, and Shipment of Direct Reduced Iron, *DRI—Technology and Economics of Production and Use*, Iron and Steel Society of AIME, 1980, p 96
9. A.G. Fane and C.Y. Wen, Fluidized-Bed Reactors, *Handbook of Multiphase Systems*, Hemisphere Publishing, 1982, p 8–104
10. P.N. Rowe, The Effect of Bubbles on Gas-Solids Contacting in Fluidized Beds, *Chem. Eng. Prog. Symp. Ser.*, Vol 58 (No. 38), 1962, p 42
11. J.R. Grace, Fluidized Bed Hydrodynamics, *Handbook of Multiphase Systems*, Hemisphere Publishing, 1982
12. O.M. Todes, *Methods and Processes of Chemical Technology*, 1956
13. H.U.A. Schenk, *Archiv. Eisenhüttenwes.*, Vol 33 (No. 4), 1962, p 211–216
14. I.V. Kogevnikov, *Bekoksovaya Metallurgia Galesa*, Metallurgia, 1970
15. D. Geldart and A.R. Abrahamson, The Effect of Fines on the Behavior of Gas Fluidized Beds of Small Particles, *Fluidization*, Plenum Press, 1980, p 453
16. L. Bogdandy and H. Engell, *Vosstanoulenie Gelesnich Rud*, Metallurgia, 1971, p 197
17. J. Yerushalmi, A.E. McIver, and A.M. Squires, "Preprint of GVC/AI Che-Joint Meeting," E-31, Munich, 1974
18. J. Yerushalmi, D.H. Turner, and A.M. Squires, The Fast Fluidized Bed, *Ind. Eng. Chem., Process Des. Dev.*, Vol 15 (No. 1), 1976, p 47
19. Direct Reduced Iron, *Technology and Economics of Production and Use*, Iron and Steel Society of AIME, 1980, p 90
20. S.I. Hulthen, Five Decades of Iron Powder Production, *J. Powder Metall. Powder Technol.*, Vol 17 (No. 2), 1981
21. USSR Patent No. 624723, 1978
22. T. Hagane, Countercurrent Reduction of Powder in a Drum-Type Rotary Fluidized Bed, *J. Iron Steel Inst. Jpn.*, Vol 66 (No. 13), 1980, p 1995–2003
23. *Met. Powder Rep.*, Vol 44, 1989, p 535
24. *Met. Powder Rep.*, Vol 145, 1990, p 470–473
25. S. Saito et al., Characteristics of High Compressibility Atomized Iron Powder with Improved Compactibility, *Advances in Powder Metallurgy*, Vol 5, Metal Powder Industries Federation, 1991, p 105–117

Production of Steel Powders

STEEL POWDERS can be classified into three general types (Fig. 1) depending on the alloying process. The prealloyed powders are produced by melting and subsequent atomization, so that powder particles are of similar alloy composition. In contrast, admixed powders are alloyed during sintering, which requires sufficient diffusion and mass transport during the sintering process for homogenous alloying. The bonded or diffusion alloyed powders are the third type; they are similar to admixed grades, except that the powder fines for alloying are bonded to the surface of iron particle. This step reduces segregation and thus improves the chemical homogeneity of the alloy part after sintering.

The elements commonly used in powder metallurgy for the alloying of steel are carbon, nickel, copper, and molybdenum. Prealloyed steel powders with nickel and/or molybdenum are common because they have a low affinity for oxygen and only minor effect on compressibility. In contrast, manganese has a much stronger affinity for oxygen (Fig. 2), which requires better atmosphere control or higher furnace temperatures during sintering (see the article "Production Sintering Practices" in this Volume). In addition, manganese (due to oxide formation and significant solid-solution hardening) degrades powder compressibility of prealloyed steel powders (see the article "Compressibility and Compactibility of Metal Powders" in this Volume). Therefore, even though manganese is a powerful hardenability agent in steels, it is not commonly used in excess of 0.5% in prealloyed powders.

In steel P/M parts made from prealloyed or diffusion alloyed powders, carbon is usually mixed as graphite to form steel during sintering. Carbon has a strong hardening effect that significantly reduces compressibility (Fig. 3), which is why even many prealloyed steel powders have low residual carbon content. Impurities such as sulfur, nitrogen, oxygen, and phosphorus also have a detrimental effect on green strength (see the article "Compressibility and Compactibility of Metal Powders" in this Volume).

This article briefly reviews the production methods and characteristics of low-alloy, tool steel, and stainless steel powders. Emphasis is placed on atomized prealloyed powders. Numerous steel powder blends also are produced, but it would be difficult to summarize the various kinds of standard and custom mixes developed by powder producers for specific customer requirements. In many respects, ferrous blends will be influenced by the characteristics of the base iron powder, such as size distribution, particle shape, and composition. Figure 4 summarizes the particle size distributions from various iron powder production methods. Further information is provided in the article "Production of Iron Powders" in this Volume.

Designations of Ferrous Powder Materials. Ferrous powders are commonly designated by specifications or standards such as MPIF standard 35 (Ref 3). A cross index of various P/M standards and specifications is listed in Appendix 1 of this Volume. Chemical compositions of ferrous P/M structural materials in MPIF standard 35 are listed in Table 1.

Ferrous Materials Produced from Elemental Powders. The standard MPIF designations for ferrous P/M materials manufactured from elemental iron powders include a prefix of one or more letters (the first of which is F to indicate an iron-base material), four numerals, and a suffix. The second letter in the prefix, which may be C for copper or N for nickel, identifies the principal alloying element (if one is specified); the percentage of that element is indicated by the first two digits. The third and fourth digits indicate the amount of carbon in the compacted and sintered part; the code designation 00 indicates less than 0.3%, 05 indicates 0.3 to 0.6%, and 08 indicates 0.6 to 0.9%. The suffix is used to indicate the minimum 0.2% yield strength of the as-sintered parts and the minimum ultimate tensile strength of heat treated materials in units of 6.894 MPa (1 ksi). The letters HT indicate heat treated. Ferrous materials produced from elemental powders include the iron and carbon steel, iron-copper and copper steels, and iron-nickel and nickel steels listed in Table 1.

Ferrous Materials Produced from Prealloyed Powders. In the case of prealloyed P/M low-alloy steels, the four-digit series described above is replaced with a designation derived from the American Iron and Steel Institute (AISI) coding

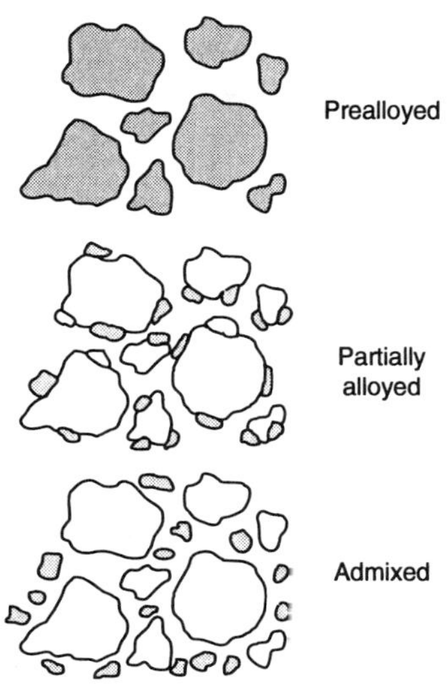

Fig. 1 Schematic main alloying methods

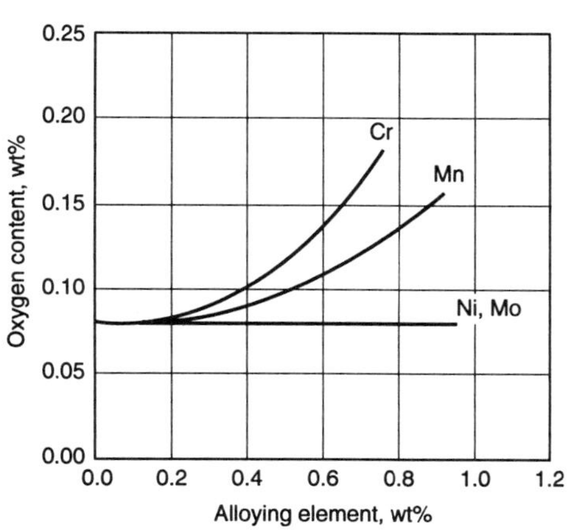

Fig. 2 Effect of alloying elements on residual oxygen content of prealloyed steel powders. Source: Ref 1

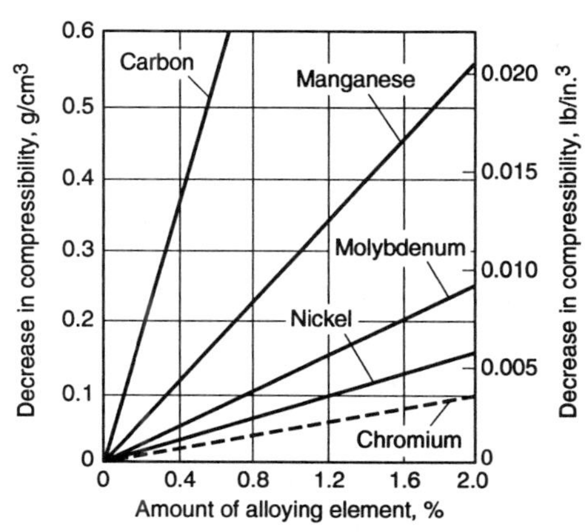

Fig. 3 Effect of alloy elements on compressibility

system. An example would be FL-4605-100HT, where the prefix FL indicates a prealloyed low-alloy ferrous material and the four digits (4605) indicate a nickel-molybdenum low-alloy steel.

When a prealloyed P/M steel powder is modified with elemental additions, a three or four alpha-numeric designator is used. For example, FLN-4205 indicates a prealloyed steel powder (FL) with an elemental nickel (N) addition. Compositions of prealloyed low-alloy P/M steels are given in Table 1.

Sinter-hardened materials are also produced from prealloyed low-alloy steel powders containing nickel, molybdenum, manganese, and/or chromium as the major alloying elements. They may also contain elemental additions of copper and/or nickel. Due to these compositions, these materials may be quench hardened during the cooling cycle following sintering, thereby eliminating a subsequent heat treating step. Compositions of sinter-hardened P/M steels are given in Table 1. They are generally used in applications where high strength and wear resistance are required.

Diffusion-alloyed materials are powders in which the alloying elements of molybdenum, nickel, and copper are added as finely divided elements or oxides to the iron powder. They are then coreduced with the iron powders at an annealing step, resulting in firm attachment and partial diffusion of the elements to the iron. This partial alloying increases hardenability compared to elemental mixtures; although for a given composition, hardenability of diffusion-alloyed material is not as high as a prealloyed material.

Diffusion-alloyed powders exhibit good compressibility. Diffusion bonding of the alloying elements also reduces the tendency toward powder segregation. These materials are also amenable to the warm compaction process. Compositions of diffusion-alloyed steels, which are indicated by the prefix FD in MPIF standard 35, are listed in Table 1.

Copper-Infiltrated Irons and Steels. Infiltration of ferrous P/M compacts with copper (10 to 20% Cu) increases their strength, hardness, fatigue resistance, and impact energy. Densities of ≥7.5 g/cm^3 are readily achieved. Compositions of infiltrated irons and steels, which are indicated by the prefix FX in MPIF standard 35, are listed in Table 1. Additional information is included in the article "Copper-Infiltrated Steel" in this Volume.

Table 1 Chemical compositions of ferrous P/M structural materials

MPIF designation	Composition range (min to max)(a), wt% Fe	C	Cu	Ni	Mo
P/M iron and carbon steel					
F-0000	97.7–100	0–0.3	...	...	...
F-0005	97.4–99.7	0.3–0.6	...	...	...
F-0008	97.1–99.4	0.6–0.9	...	...	...
P/M iron-copper and copper steel					
FC-0200	93.8–98.5	0.0–0.3	1.5–3.9	...	...
FC-0205	93.5–98.2	0.3–0.6	1.5–3.9	...	...
FC-0208	93.2–97.9	0.6–0.9	1.5–3.9	...	...
FC-0505	91.4–95.7	0.3–0.6	4.0–6.0	...	...
FC-0508	91.1–95.4	0.6–0.9	4.0–6.0	...	...
FC-0808	88.1–92.4	0.6–0.9	6.0–9.0	...	...
FC-1000	87.2–90.5	0.0–0.3	9.5–10.5	...	...
P/M iron-nickel and nickel-steel					
FN-0200	92.2–99.0	0.0–0.3	0.0–2.5	1.0–3.0	...
FN-0205	91.9–98.7	0.3–0.6	0.0–2.5	1.0–3.0	...
FN-0208	91.6–98.4	0.6–0.9	0.0–2.5	1.0–3.0	...
FN-0405	89.9–96.7	0.3–0.6	0.0–2.0	3.0–5.5	...
FN-0408	89.6–96.4	0.6–0.9	0.0–2.0	3.0–5.5	...
P/M low-alloy steel					
FL-4205	95.90–98.75	0.4–0.7	...	0.35–0.55	0.50–0.85
FL-4405	96.35–98.85	0.4–0.7	...	...	0.75–0.95
FL-4605	94.20–97.50	0.4–0.7	...	1.70–2.00	0.40–1.10
FLN-4205	93.95–97.76	0.4–0.7	...	1.35–2.50(b)	0.49–0.85
FLN2-4405	93.35–97.95	0.4–0.7	...	1.00–3.00	0.65–0.95
FLN4-4405	91.35–95.95	0.4–0.7	...	3.00–5.00	0.65–0.95
FLN6-4405	89.35–93.95	0.4–0.7	...	5.00–7.00	0.65–0.95
FLNC-4405	90.35–96.95	0.4–0.7	1.0–3.0	1.00–3.00	0.65–0.95
P/M sinter-hardened steel					
FLN2-4408	93.15–97.75	0.6–0.9	...	1.00–3.00	0.65–0.95
FLN4-4408	91.15–95.75	0.6–0.9	...	3.00–5.00	0.65–0.95
FLN6-4408	89.15–93.75	0.6–0.9	...	5.00–7.00	0.65–0.95
FLN-4608	91.00–93.41	0.6–0.9	...	3.6–5.0(c)	0.39–1.10
FLC-4608	91.00–96.41	0.6–0.9	1.0–3.0	1.60–2.00	0.39–1.10
FLC-4908	92.40–95.10	0.6–0.9	1.0–3.0	...	1.30–1.70
FLNC-4408	90.15–96.75	0.6–0.9	1.0–3.0	1.00–3.00	0.65–0.95
P/M diffusion-alloyed steel					
FD-0205	93.15–96.45	0.3–0.6	1.3–1.7	1.55–1.95	0.4–0.6
FD-0208	92.85–96.15	0.6–0.9	1.3–1.7	1.55–1.95	0.4–0.6
FD-0405	90.70–94.40	0.3–0.6	1.3–1.7	3.60–4.40	0.4–0.6
FD-0408	90.40–94.10	0.6–0.9	1.3–1.7	3.60–4.40	0.4–0.6
P/M copper-infiltrated iron and steel					
FX-1000	82.8–92.0	0.0–0.3	8.0–14.9	...	...
FX-1005	82.5–91.7	0.3–0.6	8.0–14.9	...	...
FX-1008	82.2–91.4	0.6–0.9	8.0–14.9	...	...
FX-2000	72.7–85.0	0.0–0.3	15.0–25.0	...	...
FX-2005	72.4–84.7	0.3–0.6	15.0–25.0	...	...
FX-2008	72.1–84.4	0.6–0.9	15.0–25.0	...	...

(a) MPIF standards require that the total amount of all other elements be ≤2.0%. (b) At least 1.0% Ni is admixed as elemental powder. (c) At least 2% Ni is admixed as elemental powder. Source: Ref 3

Steel Powders*

Fully prealloyed steel powders have been made by several manufacturers since the late 1960s. Because the alloying elements are homogeneously distributed throughout their powder particles, the overall compressibility is usually moderate-poor due to increased strength of the powder particles. This is a drawback when powders are pressed at room temperature as in traditional P/M process, but this characteristic becomes irrelevant for hot powder forging applications. That is why fully prealloyed powders are widely used in powder forging applications.

Both gas- and water-atomization processes are available to produce low-alloy steel, tool steel, 12% Cr steel, high-speed steel, and stainless steel powders. The method of production preference is determined to a large extent by the powder properties desired. Water-atomized powders generally are quite irregular in shape and have relatively high surface oxygen contents. Gas-atomized powders, on the other hand, generally are more spherical or rounded in shape and, if atomized by an inert gas, generally have lower oxygen (oxide) contents. There are, of course, exceptions in each type. For high-volume, low-cost production, water atomization generally is preferred over gas atomization, providing powder characteristics are compatible with the application. In spite of these important differences, the two processes are similar in many other respects.

Atomized ferrous powders must be softened by reducing annealing at 900 °C (1652 °F) in dry, purified hydrogen to facilitate pressing and to achieve the required green density and strength necessary for subsequent processing.

*This section is adapted in part from A.K. Sinha, Recent Developments in Iron and Steel Powder Production for High Performance P/M Components, *Production of Iron, Steel, and High Quality Product Mix,* ASM International, 1992, p 57–69.

Plain and Low-Alloy Steel Powders

In the 1970s, a host of low-alloy steel powders was introduced by Hoeganaes, notably iron-phosphorus alloyed steel powders for high-performance P/M parts and soft magnets as well as for a range of Astaloy prealloyed water-atomized steel powders for powder forging and wear-resistant applications. More recently, prealloyed low-alloy steel powders containing nickel and molybdenum have now captured a permanent place in the manufacture of high-strength structural P/M parts. Silicon and manganese as well as manganese and chromium are also considered to have great potential as inexpensive alloying elements in the production of high-strength P/M steel components.

Hoeganaes has produced several water-atomized steel powders and designated them Ancorsteel grades such as 1000, 1000B, 1000C, 4600V, 85HP, and 150HP for high-performance applications. Table 2 lists the typical analysis and properties of base powder Ancorsteel 1000, 1000B, and 1000C grades. Ancorsteel 4600V has been shown to have a combination of high tensile and fatigue strengths. Table 3 illustrates the typical properties of Ancorsteel base powders 85HP, 150HP, and 4600V. Controlled additions of copper, nickel, and graphite into these Ancorsteel prealloy powders produce high sintered density and ultimate tensile strength. Accelerated cooling from sintering temperature (or sinter hardening) can be used to develop high-strength martensitic structures.

Quebec Metal Powders (QMP) water-atomized low-alloy steel grades include ATOMET 1001, 1001 HP, 4001, 4201, 4401, 4601, and 4701. High chemical purity, exceptional cleanliness, good apparent density and flow rate, high hardenability, and superior compressibility make them ideal material for high-performance P/M and powder forging applications. Typical chemical compositions of plain steel and low-alloy steel powders produced at QMP are shown in Table 4. The physical properties of all grades of Atomet low-alloy steel powders are similar as a result of well-controlled processing conditions. The typical physical characteristics of 1001 HP and 1001 plain steel powders are given in Table 5. While prealloying is generally known to reduce the compressibility of ferrous powders, Atomet low-alloy powders exhibit high compressibility characteristics (Table 6) due to the high purity of the liquid steel feedstock and specially developed atomization and annealing techniques.

As in the case of plain steel powders, low-alloy steel powders can also be admixed with graphite and metallic additives such as nickel and copper powders using either regular or binder-treatment blending techniques. The combination of prealloyed and admixed elements allows a wide spectrum of strength and hardness to be achieved in the as-sintered or heat treated conditions. Due to its high content of alloying elements, Atomet 4701 is particularly well suited for sinter-hardening applications.

Mannesmann Demag has introduced water-atomized MSP4 prealloyed powder containing 4% Ni, 0.5% Mo, 0.01% C, and 0.15% O. Advantages of this powder are (a) its competitive price with heat treatable diffusion-alloyed steel powders based on Cr-Mn-Mo using vacuum annealing and (b) the use of traditional conveyor belt sintering at 1120 °C (2050 °F) together with rapid cooling to produce a homogeneous martensitic structure with close dimensional tolerance and without the need for sizing.

Mixing of 0.3 to 0.6% C as graphite and copper powder (up to 2%) with MSP4 compensates for shrinkage during sintering and volume increase during martensitic transformation. This powder is finding application as high-performance automotive cams and synchronizer hubs (Ref 5).

Table 2 Typical analysis and properties of Ancorsteel 1000 grades

Property	Ancorsteel 1000 grades		
	1000	1000B	1000C
Composition, %			
Carbon	<0.01	<0.01	<0.01
Oxygen	0.14	0.09	0.07
Nitrogen	0.002	0.001	0.001
Sulfur	0.018	0.009	0.007
Phosphorus	0.009	0.005	0.004
Silicon	<0.01	<0.01	<0.01
Manganese	0.20	0.10	0.07
Chromium	0.07	0.03	0.03
Copper	0.10	0.05	0.03
Nickel	0.08	0.05	0.04
Apparent density, g/cm^3	2.94	2.92	2.92
Flow rate, s/50 g	26	26	25
Sieve analysis, %			
+60	Trace	Trace	Trace
−60/+100	10	12	17
−100/+325	68	57	70
−325	22	21	13

Source: Ref 4

Table 3 Typical properties of prealloyed base powders Ancorsteel 85HP, 150HP, and 4600V

Property	Ancorsteel grades		
	85HP	150HP	4600V
Composition, wt%			
Nickel	...	...	1.82
Molybdenum	0.85	1.50	0.54
Manganese	0.14	0.14	0.17
Carbon	<0.01	<0.01	<0.01
Oxygen	0.07	0.07	0.16
Nitrogen	0.014	0.014	0.005
Sulfur	0.01	0.01	0.002
Apparent density, g/cm^3	2.90	2.90	2.96
Flow rate, s/50 g	24.0	24.0	23.8
Sieve analysis, %			
+100	10	10	11
−100/+325	70	70	68
−325	20	20	21

Source: Ref 4

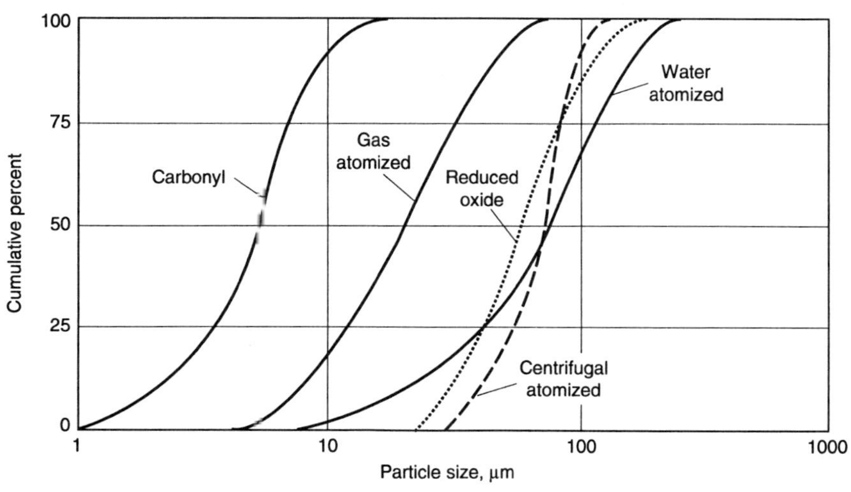

Fig. 4 Cumulative particle size distributions of some common iron powders. Source: Ref 2

Table 4 Typical chemical composition of Atomet steel powders

Atomet powder grade	Alloying elements, wt %				
	Mn	Ni	Mo	Cr	Other
Low-alloy grades					
4001	0.16	...	0.50	...	...
4201	0.28	0.45	0.60	...	...
4401	0.15	...	0.85	...	...
4601	0.20	1.80	0.55	...	...
4701	0.45	0.90	1.00	0.45	...
Plain steel grades					
1001 HP	0.06	...	...	...	(0.004 C, 0.06 O, 0.006 S)
1001	0.06	...	...	...	(0.003 C, 0.10 O, 0.008 S)

Table 5 Typical physical properties of Atomet plain steel powders

	1001	1001 HP
Apparent density, g/cm^3	2.92	2.92
Flow rate, s/50 g	26	25
Sieve analysis, U.S. mesh (microns), %		
+70/ (+212)	Trace	Trace
−70/+100 (−212/+150)	10	12
−100/+200 (−150/+75)	38	40
−200/+325 (−75/+45)	26	27
−325 (−45)	26	21

Kobelco ultrahigh-pressure water-atomized steel powders have high purity, consistent high quality, good compressibility, and compactibility. They are available for a wide range of industrial applications. Grades include 300MC, 300ME, 300MS, and 4600A (Table 7).

Kawasaki Steel Corporation has developed a grade of low-alloy steel powder, 4600ES, which is suitable for the production of high-density and high-strength P/M components through double-pressing/double-sintering and heat treatment operations. Grade 4100VS powder (Table 8) obtains green density of 7.15 g/cm^3 at compaction pressure of 690 MPa. This is attributed to the low oxygen, carbon, and nitrogen contents. By including a vacuum reduction annealing method, green strength is better than 6.91 g/cm^3 for an earlier grade of 4100. Table 8 lists the compositions of KIP alloy steel powders.

Tool Steel Powders

The basic problem with high-speed steels is their strong tendency for segregation, which results in a fairly coarse and nonuniform structure; this, in turn, affects the safety in heat treatment, makes the grinding operation more difficult, and limits the toughness. The use of proper P/M technique has solved this problem.

Tool steel powders are produced by gas- or water-atomization methods (as discussed in more detail in the section "Tool Steel Atomization" in this article). Gas (argon or nitrogen)-atomized tool steel powders have particles that are spherical in shape with high apparent densities (about 6 g/cm^3), which thus requires hot or cold isostatic pressing consolidation. Standard and proprietary compositions of gas-atomized tool steel powders are given in Table 9.

Water-atomized tool steel powders have irregularly shaped particles and are suitable for conventional die compaction and sintering to higher theoretical density. Typical compositions and powder properties of water-atomized high-speed steel powders are given in Table 10. Small (0.15 to 0.2%) amounts of graphite sometimes are added to the powder to adjust or increase carbon content of the sintered product.

Powder cleanliness is important for tool steel powders. Powder producers provide high standards of cleanliness through campaigning of heats, dedication of production equipment, use of powder cleaning equipment, and meticulous housekeeping. Producers include Powders Ltd., Davy Loewy, and Edgar Allen Tools Ltd. in the United Kingdom and CMI in the United States, who have developed high-speed steel (HSS) powder production technology based on water-atomization process. On the other hand, Uddeholm in Sweden, Crucible Inc. in the United States, and Kobe Steel in Japan have produced HSS powders by inert-gas-atomization process.

The Kobe atomization process involves high-purity, high-pressure gas atomization of molten steel from the tundish into a rapidly solidified powder with a fine homogeneous carbide or carbonitride structure, heating in a nitriding furnace, and blending in order to homogenize contained nitrogen. It is then transformed to a mild steel can, evacuated, and sealed (Fig. 5). This nitrogen-containing HSS powder contains 0.6 to 1.4% C, 0 to 1% N, 4% Cr, 6% Mo, 6% W, and 3.5% V.

The ASEA STORA process (ASP) for the high-speed and tool steel powders was introduced in the 1970s. This uses inert gas atomization of prealloyed melt to produce relatively coarse, essentially spherical metal powders. In the new horizontal design, the melt is disintegrated into very small drops by a gas jet. The droplets are flung horizontally where they cool and solidify very rapidly into spherical powder particles. The powder is then screened prior to leaving the atomization chamber and finally poured directly into a storage container.

A limiting factor for HSS material is the size of the fracture-initiating defect to which the strength level can be directly related. The smaller the defect the higher the strength according to the equation:

$$\sigma = \text{constant} \times \left(\frac{K_{Ic}}{\sqrt{d}}\right)$$

where K_{Ic} is the fracture toughness of the material and d is the size of the critical size defect. In the ASP-processed material, the harmful effect of large carbide is nonexistent (Ref 6).

Diffusion-Alloyed Powders

The demand for making high-strength, high-ductility P/M parts through single pressing and sintering has led to the development of several types of composite-type, partially prealloyed steel powder containing copper, nickel, and molybdenum. They exhibit high compressibility and are heat treatable for maximum properties. They represent a real breakthrough in iron-base sintered alloys (Ref 7).

The production of these powders involves the heat treatment of a mixture of iron powder and fine metal alloy powders in reducing atmosphere. Diffusion bonding of the alloying ele-

Table 6 Typical compressibility and green strength of Atomet low-alloy steel powders, mixed with 0.75% ZnSt

Property	Atomet grade 4001	4201	4401	4601	4701
Compacting pressure, at 6.8 g/cm^3, MPa (tsi)	420 (30.5)	440 (32.0)	420 (30.5)	500 (36.2)	540 (39.0)
Green strength, MPa (psi)	10.3 (1500)	11 (1600)	11 (1600)	10.8 (1560)	12.8 (1850)

Table 7 Kobelco standard steel powder grades, typical properties

Grade	Chemical analysis, % C	Si	Mn	P	S	Ni	Cr	O	Screen analysis, % +60	−60/+100	−100/+325	−325	Powder properties: Apparent density, g/cm^3	Flow rate, sec	Green density, g/cm^3
500MA	0.001	0.01	0.16	0.004	0.009	...	...	0.14	0.2	14	61	25	2.92	25.1	6.82
300MA	0.001	0.01	0.16	0.004	0.009	...	...	0.13	...	5	68	27	2.91	24.7	6.82
300MC	0.001	<0.01	0.16	0.004	0.009	...	...	0.09	0.2	16	66	18	2.87	25.8	6.90
300ME	0.001	<0.01	0.17	0.004	0.009	0.01	0.02	0.13	...	5	68	27	2.93	24.6	6.82

Table 8 Compositions of Kawasaki (KIP) prealloy and diffusion-alloyed powders

Powder	Chemical composition, % C	Si	Mn	P	S	Ni	Cu	Mo	Cr	N	O
Fully prealloyed											
4600 ES	0.001	0.01	0.08	0.017	0.007	1.12	0.42	0.23	...	0.001	0.09
4600 AS	0.003	0.01	0.08	0.009	0.004	1.45	0.53	0.47	...	0.001	0.12
4100 VS	0.02	0.04	0.83	0.021	0.015	...	...	0.29	1.05	0.001	0.10
Diffusion alloyed											
SIGMALOY 415S	0.004	0.01	0.05	0.005	0.005	4.31	1.60	1.48	...	0.001	0.11
SIGMALOY 2010	0.001	0.01	0.05	0.005	0.003	1.93	...	1.03	...	...	0.07

Source: Ref 4

ments into the iron powder particles takes place during this process (Ref 8).

Diffusion-alloyed steels were introduced in the late 1970s to bridge the gap between premix and prealloyed steels. The premixes were produced by mixing elemental alloying powders, such as nickel, copper, and molybdenum with high compressibility pure iron-base powders. Although the premixes were able to achieve high green density, the alloying powders were likely to segregate and not completely alloy during sintering, and the resultant mechanical properties were insufficient for high-strength sintered P/M parts. In contrast, the prealloyed steel powders produced a homogeneous structure, but were much less compressible than the premixed powders.

The diffusion-alloyed steel powders overcome the specific limitations of both premixed and prealloyed powders by combining the compressibility of a pure iron powder with the lack of segregation of a prealloyed powder. The diffusion-alloyed steel powders have been successfully used to produce high-strength sintered P/M products through single-press/single-sinter processing.

The advantages of diffusion-alloyed steel powders are (Ref 8):

- Maintenance of high inherent compressibility of the base iron
- Increased green strength
- Reduced risk of alloy segregation and dusting during transportation of the powder mix
- Homogeneous distribution of alloying elements
- Good sinterability
- Outstanding stability of this composite-type base powder

Diffusion-alloyed powders have also led to the elimination of double-pressing/double-sintering and sizing because of the high hardness obtained in the sintered condition (Ref 8).

Commercial Grades. Several different trademarks are used for diffusion-alloyed grades, and some of the more prevalent grades are briefly described below.

Distaloy grades are produced by Hoeganaes and Högänäs including Distaloy grades SA, AB, SE, SH, AE, AG, 4600A, and 4800A. The Distaloy grades are produced using either sponge iron or atomized Ancorsteel 1000 B powders and varying amounts of Cu, Ni-Cu, Ni-Mo-Cu, and Ni-Mo as alloying elements for bonding to the surface of the iron powder particles. Distaloy AG is based on the supercompressible iron powder (ASC 100.29) atomized iron powder with 8% Ni and 1% Mo additions. Mixed with 0.5% graphite and lubricant, Distaloy AG can produce parts (after single pressing at 600 MPa and single sintering at 1150 °C, or 2100 °F), with a sintered density of 7.32 g/cm^3, a tensile strength of more than 1000 MPa, elongation of 4%, and hardness of 320 HV20 without heat treatment.

Distaloy 4800A is a diffusion-bonded steel powder with 4% Ni, 0.5% Mo, and 1.5% Cu. Mixed with 2% Ni, 0.3 to 0.8% graphite, and 0.75% Acrawax using Ancorbond process, it can produce column tilt levers by double pressing to a density of 7.3 g/cm^3 with impact and hardness values exceeding the specified minimum (27.2 J and 68 HR15N, respectively). Carburizing treatment, followed by tempering at 232 °C (450 °F), imparts a wear-resistant surface. This P/M part only requires honing of the pivot hole to meet the specified tolerance.

Sigmaloy grades are diffusion-alloyed powders produced by Kawasaki. The composite-type alloyed Sigmaloy 415S and 2010 steel powders use water-atomized pure iron powder as the base material (see Table 8 for composition). Sigmaloy 2010 can achieve tensile strength of 1500 MPa, an impact toughness of 21 J, a rotating bend fatigue strength of 460 MPa and a contact fatigue strength of 2560 MPa after normal pressing, sintering, and case hardening. Tensile strength of 1920 MPa with an impact strength of 53 J is obtained after double-pressing/double-sintering, bright quenching, and tempering (Ref 9). The increased properties of sintered compacts made from Sigmaloy 2010 are attributed to the strain-induced martensitic transformation of

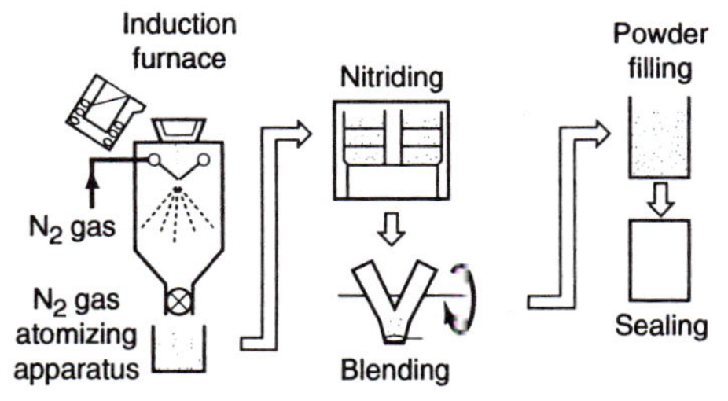

Fig. 5 Flowsheet of the Kobe high-speed steel powder manufacturing process

Table 9 Nominal compositions of gas-atomized tool steels

Alloy	Composition, %									
	C	Mn	Si	Cr	W	Mo	V	Co	S	Fe
M4	1.35	0.30	0.30	4.25	5.75	4.0	4.0	...	...	bal
M42	1.10	...	...	3.75	1.50	9.50	1.15	8.00	...	bal
T15	1.55	0.30	0.30	4.00	12.25	...	5.00	5.00	...	bal
CPM 76(a)	1.50	0.30	0.30	3.75	9.75	5.25	3.10	8.50	...	bal
CPM 10V(a)	2.45	0.50	0.90	5.25	0.30	1.30	9.75	...	0.07	bal
CPM T440V(a)	1.55	...	...	17.00	...	0.40	5.50	...	...	bal
CPM 15V(a)	3.40	0.50	0.90	5.25	...	1.30	14.50	...	0.07	bal
ASP23 (2023)(b)	1.27	0.30	0.30	4.20	6.40	5.0	3.10	...	...	bal
ASP30 (2030)(b)	1.27	0.30	0.30	4.20	6.40	5.0	3.10	8.50	...	bal
ASP60 (2060)(b)	2.30	0.30	0.40	4.00	6.50	7.00	6.50	10.50	...	bal

(a) Trademark of Crucible Service Centers. (b) Trademark of Soderfors Powder AB, Sweden. Grade designations in parentheses are made by electroslag heating.

Table 10 Composition and properties of water-atomized high-speed tool steels

Properties	Tool steel			
	M2	M3 type 2	M42	T15
Composition, %				
Carbon	0.85	1.20	1.10	1.60
Chromium	4.15	4.10	3.75	4.40
Tungsten	6.30	6.00	1.50	12.50
Molybdenum	5.00	5.00	9.50	...
Vanadium	1.85	3.00	1.15	...
Cobalt	...	...	8.00	5.00
Iron	bal	bal	bal	bal
Oxygen content, ppm	<1000	<1000	<1000	<1000
Physical properties				
Apparent density, g/cm^3	2.2	2.1	2.3	1.8
Tap density, g/cm^3	3.1	3.0	3.3	2.4
Flow rate, s/50 g	45	40	30	50
Sieve analysis (Tyler)				
+100 mesh (<150 μm)	...	...	...	...
−100/+150	13	13	13	13
−150/+200	22	22	22	22
−200/+325	30	30	30	30
−325 (<44 μm)	35	35	35	35
Green density(a), g/cm^3				
Pressed at 620 MPa (45 tsi)	6.2	6.0	6.0	6.15
Pressed at 830 MPa (60 tsi)	6.6	6.4	6.3	6.55
Green transverse-rupture strength(a), MPa (ksi)				
Pressed at 620 MPa (45 tsi)	23 (3.3)	24 (3.5)	21 (3.0)	43 (6.2)
Pressed at 830 MPa (60 tsi)	52 (7.5)	48 (7.0)	41 (6.0)	69 (10)

(a) Green properties determined using die wall lubrication

nickel-rich retained austenite in the sintered compact during tensile or fatigue tests. Sigmaloy 2010 and 415S steel powder find applications in automobiles as valve guides, shock absorber parts, oil pump rotors, timing pulleys, and so forth (Ref 8).

Atomet diffusion-bonded powders are produced by Quebec Metal Powders Ltd. The typical properties of Atomet diffusion-bonded powders are given in Table 11. Because of the multiphase structure of materials made from diffusion-bonded powders, these exhibit excellent dynamic properties. Like other diffusion-alloyed powders, they are well suited for high-density applications where impact and fatigue resistance are critical.

Mannesmann Demag's new diffusion-alloyed "master alloy" grade ULTRAPAC-20 Cu gives a segregation-free copper distribution. The copper content can be adjusted by mixing ULTRAPAC-20 Cu with plain iron powder (Ref 10).

Other Alloying Methods

Admixing. Press-ready mixes contain all the necessary alloying additions for powder users. To preserve premix uniformity against the natural tendency to demix during handling and subsequent processing, various proprietary mixing processes use patented binders to bond additives to the base powders. The benefits of premix processing with binders are:

- Improved (or uniform) flow rate and better die filling while retaining similar green strength
- Potential for improved productivity
- Reduced variability in the sintered properties
- Opportunities for new alloy development
- Utilization of fine particle additives
- Potential for achieving increased P/M part densities and weight control without the need to resort to double-pressing/double-sintering techniques

Hoeganaes QMP, and Kawasaki have developed these "segregation-free" premixed powders by bonding graphite and other alloying additives to the surface of iron powder.

Mechanical Alloying (MA). In the MA process, composite metallic (or ceramic) powders are produced by simultaneous and repeated sequences of extensive plastic deformation, cold welding, and fracturing of a mixture of metallic and alloying ingredient particles during a dry, high-energy (attrition, vibratory, or large-diameter tumbler) ball-milling process.

The mechanically alloyed powders are characterized by dense, intimate mixing of constituent metals on a fine scale, homogeneous with a grain-refined (submicron grain size) microstructure, extended solid solubility, and formation of nonequilibrium phases. These particles have irregular shape, suitable for high packing density during compaction, and, in addition, they are free of the interdendritic microsegregation and pores that are occasionally encountered. This results in a unique combination of high strength and corrosion resistance. Another advantage is the production of amorphous alloys where an extended range of compositions can be processed, which is not possible by rapid solidification.

Mechanical alloying is the most successful method for the production of high-temperature creep-resistant fine (submicron) oxide-dispersion-strengthened (ODS) iron-base superalloys, amorphous iron-titanium, and iron-tantalum alloy powders. Further information is contained in the article "Mechanical Alloying" in this Volume.

Rapid Solidification Process (RSP). In RSP, the local solidification time is reduced with increasing cooling rate. Typically, the cooling rates of conduction processes range between 10^6 and 10^8 °C/s (1.8×10^6 and 1.8×10^8 °F/s), whereas the convection processes may be limited to the range 10^4 and 10^6 °C/s (1.8×10^4 and 1.8×10^6 °F/s).

Important attributes of RSP are: increased homogeneity, highly refined microstructure and second-phase particle refinement, extended solid-solubility limit (i.e., alloying flexibility), and formation of unique nonequilibrium crystalline and noncrystalline (amorphous or glassy) metastable phases that have significant influence on the properties and structural engineering applications of alloys.

Rapidly solidified low-alloy steel powders with a fine homogeneous structure and dispersion of fine, stable sulfide and oxide inclusions such as MnS, VS, SiO_2, MgO, and Al_2O_3 have been produced for hot consolidation processing to high-strength and ultrahigh-strength P/M parts. This enhancement of mechanical properties is attributed to the retention of fine grain size during austenitizing at high temperatures because of effective pinning of grain boundaries by finely dispersed, stable inclusions. Production of stress-corrosion-resistant NiMoLa ultrahigh-strength steel via RSP is another landmark in the development of high-performance P/M parts (Ref 11). The composition of this steel is similar to 4340 NiCrMo steel where chromium is replaced by higher concentration of molybdenum (1.5% instead of 0.25%), and lanthanum (as $LaNi_5$) is introduced into the melt to balance phosphorus and sulfur to promote stable fine $LaPo_4$ and La_2O_2S inclusions (Ref 12).

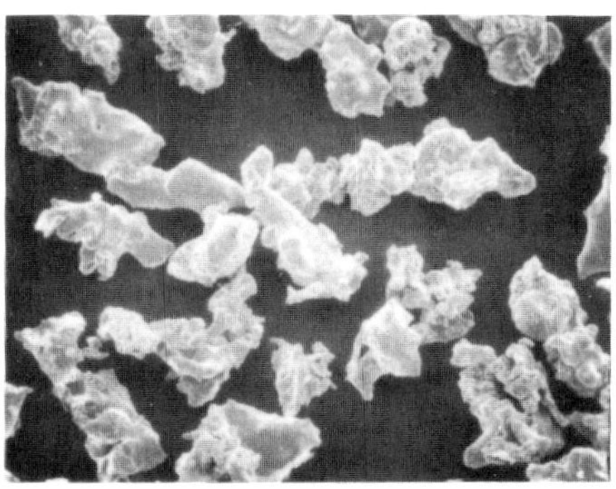

Fig. 6 Scanning electron micrograph of water-atomized 304L stainless steel. –100 mesh. 150×

Other potentials in RSP technology include the development of amorphous soft magnetic materials, amorphous ferromagnetic Fe-B-Si alloys for transformer applications, crystalline soft magnetic Fe-B-Si-Al alloy (Ref 13), and hard magnetic alloys based on crystalline Fe-Nd-B alloy (Ref 14).

Stainless Steel and High-Alloy Powders

E. Klar and P.K. Samal,
OMG AMERICAS

Most stainless steel and other high-alloy powders are made by water atomization or by inert gas and centrifugal atomization. Information on these processes can be found in the article "Atomization" in this Volume. Water-atomized powders generally are of irregular particle shape and widely used for cold compaction and sintering. The densities of the resulting sintered parts are typically below about 95% of theoretical. It is for this reason that dynamic mechanical properties as well as a number of physical properties are below those of their fully dense counterparts. For the same reason, such parts often can tolerate greater levels of interstitials, although the tendency here is toward lower levels of interstitials in both powders and sintered parts.

Table 11 Typical properties of Atomet diffusion-bonded powders

Property	Atomet grade	
	DB46	DB48
Chemical analysis, %		
Carbon	0.005	0.005
Oxygen	0.10	0.10
Sulfur	0.007	0.007
Manganese	0.16	0.16
Molybdenum	0.50	0.50
Copper	1.45	1.45
Nickel	1.80	4.00
Physical properties		
Apparent density, g/cm³	3.00	3.00
Flow rate, s/50 g	24	24
Sieve analysis, U.S. mesh (microns), %		
+70 (+212)	Trace	Trace
–70/+100 (–212/+150)	9	9
–100/+200 (–150/+75)	38	38
–200/+325 (–75/+45)	29	29
–325 (–45)	24	24
Compacting pressure at 7.0 g/cm³, MPa (tsi)	540 (39.0)	545 (39.5)
Green strength at 7.0 g/cm³, MPa (ksi)	11 (1.6)	11 (1.6)
Dimensional change from die size(a), %	+0.18	+0.01
Transverse rupture strength(a), MPa (ksi)	1275 (185)	1450 (210)
Ultimate tensile strength(a), MPa (ksi)	620 (90)	725 (105)

(a) Atomet DB powder + 0.6 graphite + 0.75 ZnSt, pressed to 7.0 g/cm³, sintered 25 min at 1120 °C in nitrogen-base atmosphere

Gas-atomized powders generally have spherical particle shape and require consolidation techniques such as hot isostatic pressing, extrusion, or metal injection molding. The properties of parts produced by these methods are generally equal and sometimes superior to their ingot metallurgy (I/M) counterparts. For these materials, the control of interstitials usually is very important. The superior properties derive from a more uniform and cleaner microstructure as a result of the rapid quenching inherent in the powder-making processes and from the careful control of contamination with ceramic and other materials during powder making and consolidation.

Three important classes of high-alloy powders are discussed in more detail in this section. They include stainless steel powders, tool steel powders, and superalloy powders. Aluminum alloy powders are discussed in a separate article in this Volume.

Stainless Steel Powders

Water-Atomized Stainless Steel Powders. Early experiments concentrated on producing stainless steel powders from elemental components and from alloy powders by sensitization embrittlement and grinding of stainless steel sheet. Water atomization became the established process in the 1950s for producing stainless steel powders for conventional die compaction and sintering. Over the years, slight modifications to the existing wrought compositions led to improved compacting properties. Recently, further modifications in composition, combined with processing precautions, have led to improvements in corrosion resistance.

Melting of virgin raw materials is performed in open air or vacuum induction furnaces. Low manganese concentrations (<0.3%) and deoxidation with ferrosilicon to achieve 0.7 to 1.0% Si in the alloy prevent excessive oxidation of the powder during water atomization in an inert gas (nitrogen)-purged atomization chamber. Typical water pressures for producing predominantly –80 mesh powder are about 14 MPa (2 ksi).

Compositions and powder properties of commercial grades of stainless steel powders are given in Table 12. These powders normally are used in the as-atomized condition, although martensitic and ferritic grades may be annealed to improve green strength and compressibility. Particle shape (Fig. 6) and other powder characteristics are controlled to produce powders with apparent densities ranging from 2.5 to 3.2 g/cm^3, adequate green strength (Fig. 7a), and good compressibility (Fig. 7b). In Fig. 7(a) and (b), the higher green densities are typically obtained for the lower alloyed grades. Also, low contents of interstitials (carbon, oxygen, nitrogen) improve green strength, compressibility, and corrosion resistance. For applications requiring high green strength, Acrawax or stearic acid, instead of lithium stearate, may be used as lubricant. However, the improvement of green strength comes at the expense of reduced compressibility.

To ensure that a stainless steel powder is free from contamination with iron, low-alloy steel, and other powders capable of producing galvanic corrosion, the copper sulfate or Ferroxyl test may be applied. Details of these tests are described in "Sintering of Stainless Steel" in this Volume (see the article "Production Sintering Practices").

Gas-Atomized Stainless Steel Powders. Gas (nitrogen and argon)-atomized stainless steel powders have particles that are spherical in shape (Fig. 8). These powders have high apparent densities of about 5 g/cm^3 and excellent flow rates. They require special methods of consolidation such as hot isostatic pressing, cold isostatic pressing, or extrusion. Oxygen contents

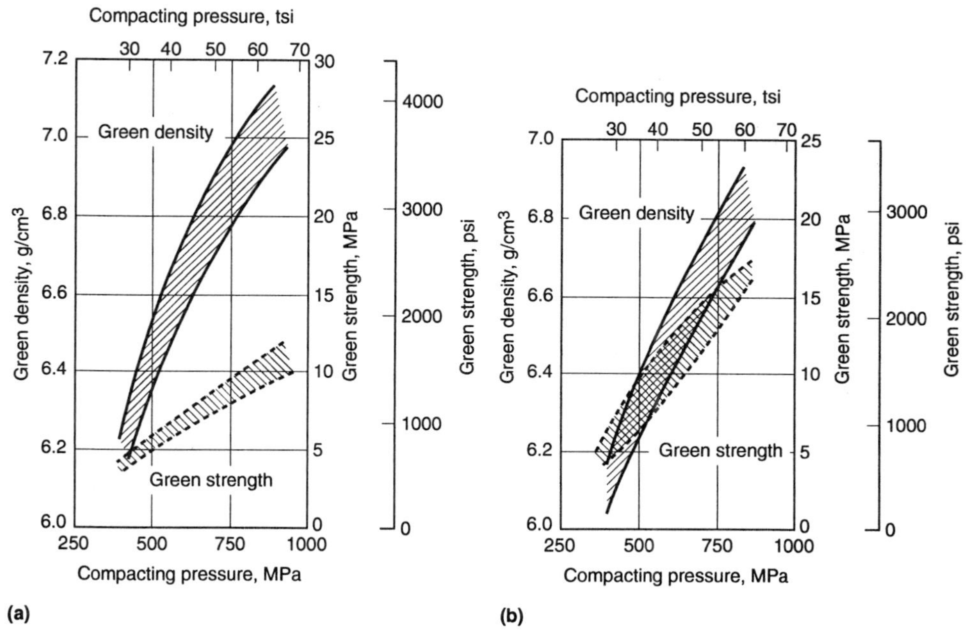

Fig. 7 Typical compressibility and green strength ranges of stainless steel powders. Die lubricant, 1% lithium stearate. (a) Austenitic grades. (b) Ferritic and martensitic grades

Table 12 Commercial P/M grades of water-atomized stainless steel powder

	Composition, wt%												Apparent density(a),	Flow rate(a),	Sieve analysis, %	
	Cr	Ni	Si	Mo	Cu	Sn	Mn	C	S	P	Fe	O (ppm)	N (ppm)	g/cm^3	s/50 g	+100 mesh	–325 mesh
Austenitic grades																	
303	17–18	12–13	0.5–0.8	...	...	...	0.3(b)	0.03(b)	0.1–0.3	0.03(b)	bal	1000–2500	200–500	3.0–3.2	24–28	3(b)	40–60
303LSC	17–18	12–13	0.5–0.8	...	2(c)	1(c)	0.3(b)	0.03(b)	0.1–0.3	0.03(b)	bal	1000–2500	200–500	2.8–3.0	26–30	1–3	30–40
304L	18–19	10–12	0.7–0.9	...	...	...	0.3(b)	0.03(b)	0.03(b)	0.03(b)	bal	1000–2500	200–500	2.5–2.8	28–32	1–4	35–45
ULTRA 304L	19(b)	11(c)	0.8(c)	...	0.8(c)	2(c)	0.2(c)	0.02(c)	0.01(c)	0.01(c)	bal			2.7(c)	30(c)	3(c)	40(c)
316L	16.5–17.5	13–14	0.7–0.9	2–2.5	...	...	0.3(b)	0.03(b)	0.03(b)	0.03(b)	bal	1000–2500	200–500	2.6–3.0	24–32	1–4	35–45
316LSC	16.5–17.5	13–14	0.7–0.9	2–2.5	2(c)	1(c)	0.3(b)	0.03(b)	0.03(b)	0.03(b)	bal	1000–2500	200–500	2.9(c)	25(c)	1(c)	42(c)
317L	19(c)	14(c)	0.8(c)	3(c)	...	...	0.3(b)	0.3(b)	0.3(b)	0.3(b)	bal	1000–2500	200–500	2.9(c)	27(c)	2(c)	44(c)
SS100	20(c)	17(c)	0.8(c)	5(c)	...	...	0.2(c)	0.2(c)	0.3(b)	0.2(c)	bal	1000–2500	200–500	2.9(c)	29(c)	2(c)	45(c)
Martensitic grades(d)																	
410L	12–13	...	0.7–0.9	...	...	...	0.3(b)	0.3(b)	0.3(b)	0.3(b)	bal	1500–2500	200–500	2.9(c)	27(c)		39(c)
Ferritic grades																	
430L	16–17	...	0.7–0.9	...	...	...	0.3(b)	0.3(b)	0.3(b)	0.3(b)	bal	1500–2500	200–500	2.8–3.0	27–30	1–3	40–50
434L	16–18	...	0.7–0.9	0.5–1.5	...	...	0.3(b)	0.3(b)	0.3(b)	0.3(b)	bal	1500–2500	200–500	2.8–3.0	27–30	1–3	40–50

(a) Determined on unlubricated powder. (b) Maximum. (c) Typical. (d) With C <0.03%, this grade is ferritic; graphite addition to the low-carbon grade powder renders it martensitic upon sintering.

are similar to those of wrought stainless steels. Much emphasis is being placed on avoiding cross-contamination and contamination with ceramic (slag) particles. The latter are controlled to very small levels of concentration as well as to very small sizes, typically less than 25 μm and preferably less than 10 μm. Such control is essential if fatigue life, toughness, and other properties are to be maximized and exceed those of I/M materials.

Commercially available grades include AISI grades 410, 440C, 446, 304L, 304, 347, 316L, 316, and 317. Chemical compositions are similar or identical to conventional wrought grades. Special compositional adjustments are unnecessary due to the absence of water during atomization and due to the special consolidation methods.

Tool Steel Atomization

Powder metallurgy tool steel development in the United States began in the 1970s. Success of P/M tool steels is based mainly on the uniform microstructure obtainable, compared to wrought and conventionally produced products. Superior uniformity of composition leads to excellent toughness and less distortion during heat treatment. Tool life is increased. Reduced grinding costs due to superior grindability in comparison to conventional tool steels is due to the finer and more uniformly distributed carbides.

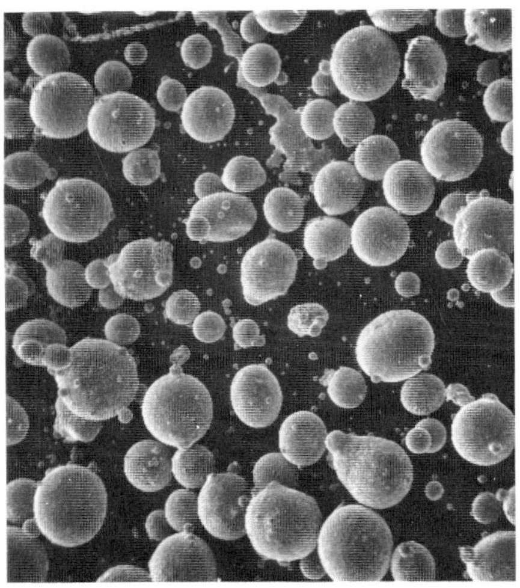

Fig. 8 Scanning electron micrograph of nitrogen-atomized 316L stainless steel. 65×

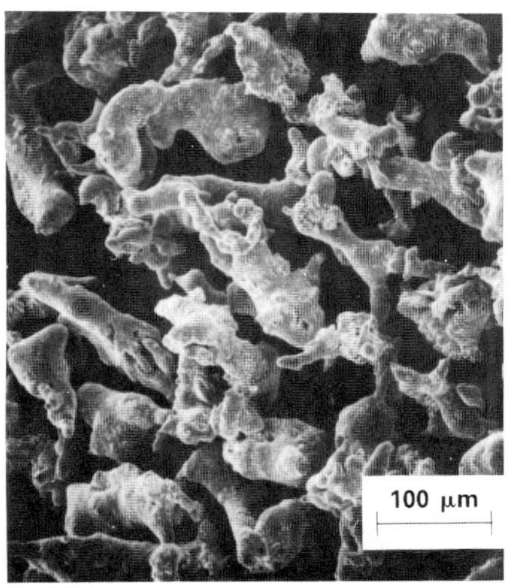

Fig. 9 Scanning electron micrograph of water atomized high-speed tool steel

As in the case of stainless steel powders, both water-atomized and inert-gas-atomized tool steel powders are available. They also differ in particle shape and chemical composition and require different consolidation techniques, as previously noted in the section "Tool Steel Powders."

Water-atomized tool steel powders are produced by induction melting of virgin raw materials and/or scrap. These powders have irregularly shaped particles (Fig. 9) and are suitable for conventional die compaction and sintering to high or theoretical density. Unlike stainless steel and low-alloy steel powders, water-atomized tool steel powders are vacuum deoxidized and vacuum annealed to reduce their oxygen contents and improve their compacting properties. Deoxidation occurs at elevated temperature as a solid-state reaction between the oxygen and carbon within individual particles, in which the oxygen content of the powder is lowered from about 1500 to 3000 ppm to below 1000 ppm.

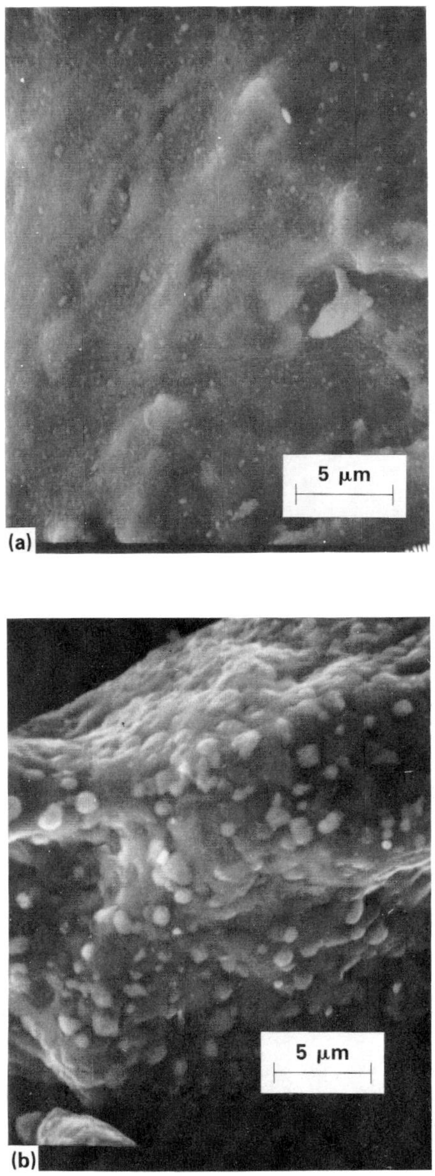

Fig. 10 Microstructures of water-atomized T15 tool steel powder. (a) As-atomized. (b) After annealing

Annealing lowers hardness from 700 to less than 300 HV. Further reduction of oxygen to less than 100 ppm occurs during sintering. Water-atomized tool steel powders can be pulverized for very fine particle size and accelerated sintering characteristics.

Commercially available water-atomized tool steel powders include several high-speed steels (M2, M3/2, M4, M35, M42-Si, T15, T42) and hot- and cold-work steels (D2, H13). Typical compositions and powder properties of some of these powders are given in Table 10. Small (0.15 to 0.2%) amounts of graphite sometimes are added to the powder to adjust or increase carbon content of the sintered product.

Water-atomized tool steel powders may also be added to low-alloy iron powders to enhance their wear resistance (Ref 15).

Figure 10 illustrates the microstructures of water-atomized T15 tool steel before and after annealing. Annealing depletes the carbon from the martensitic matrix by forming carbides.

Gas-atomized tool steel powders have spherical particle shape (Fig. 11) and are usually consolidated to full density by HIP, extrusion, or MIM. In the early years of this technology, emphasis was placed on small carbide size and uniform microstructure, as well as the capability to make higher-alloyed compositions using P/M techniques. In recent years, the control and minimization of oxide inclusions have been in the foreground of technical efforts. Improvements in melting, that is, electroslag heating, slag separation, and atomizing (Ref 16, 17) have led to still lower oxygen levels of less than 50 ppm, smaller inclusion size (the goal is <10 μm) and the re-

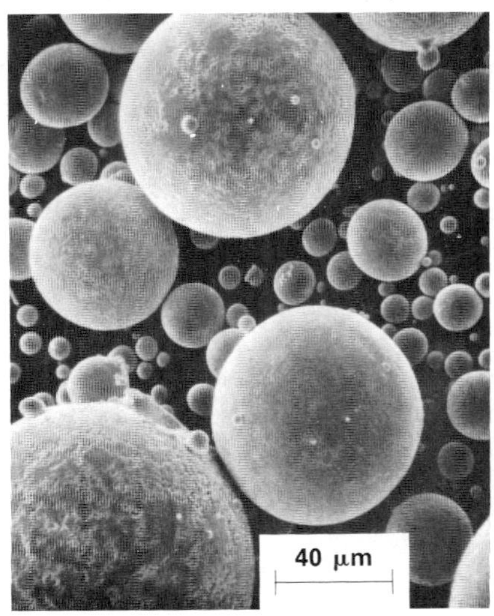

Fig. 11 Scanning electron micrograph of nitrogen-atomized T15 tool steel

Table 13 Compositions of several P/M superalloys

Alloy	Composition, %														
	C	Cr	Mo	W	Ta	Ti	Nb	Co	Al	Hf	Zr	B	Ni	Fe	V	Y_2O_3
IN-100	0.07	12.5	2.8	...	...	4.3	...	16.0	5.0	...	0.016	0.02	bal	...	0.9	...
René 88	0.05	16.0	4.0	4.0	...	3.75	0.7	13.0	2.2	...	0.05	0.015	bal	...	...	...
René 95	0.06	13.0	3.5	3.5	...	2.5	3.5	8.0	3.5	...	0.05	0.010	bal	...	...	...
MERL 76	0.02	12.4	3.2	...	...	4.3	1.4	18.5	5.0	0.4	0.06	0.02	bal	...	...	...
AF 115	0.05	10.5	2.8	6.0	...	3.9	1.7	15.0	3.8	2.0	...	...	bal	...	...	...
PA101	0.1	12.5	...	4.0	4.0	4.0	...	9.0	3.5	1.0	...	...	bal	...	...	...
Low-carbon Astroloy	0.04	15.0	5.0	...	...	3.5	...	17.0	4.0	...	0.4	0.025	bal	...	...	...
MA 754	0.05	20.0	...	...	...	0.5	...	...	0.3	...	...	...	bal	1.0	...	0.6
MA 956	...	20.0	...	...	...	0.5	...	...	4.5	...	...	...	...	bal	...	0.5
MA 6000	0.05	15.0	2.0	4.0	2.0	2.5	...	...	4.5	...	0.15	0.01	bal	...	...	1.1
Stellite 31	0.5	25.5	...	7.5	...	...	...	bal	...	...	...	...	10.5	2.0	...	...
Udimet Alloy 720	0.025	16.0	3.0	1.25	...	5.0	...	15.2	2.5	...	0.03	0.018	bal	...	...	...
N18	0.015	11.5	6.5	...	...	4.35	...	15.7	4.35	0.45	...	0.015	bal	...	...	...
KM4	0.030	12	4	...	...	4	2	18	4	...	0.030	0.030	bal	...	...	...

duction of other defects (e.g., argon porosity). Closed processing of the powder minimizes macroinclusions. The refined microstructure of the powder is retained in the consolidated products and accounts for marked improvements in toughness, bend strength, fatigue life, and other properties. Table 9 summarizes widely used grades of gas-atomized tool steels.

Superalloy Powders

Superalloy powders are typically made by vacuum induction melting and filtering of high-purity raw materials followed by inert atomization such as argon gas or centrifugal atomization. The presence in superalloys of elements with very high oxygen affinities (i.e., titanium, aluminum, hafnium, and zirconium) prohibits water atomization because the oxides formed during atomization (via $Me + H_2O \rightarrow MeO + H_2$) are nonreducible and in part responsible for the so-called prior particle boundaries (PPBs) in the consolidated products with their detrimental effects on mechanical properties. Even with inert atomized powders, PPBs can be present and must be dealt with. Furthermore, superalloys with large amounts of second-phase γ' are practically unworkable by conventional hot-working techniques for large parts and are therefore made by P/M techniques.

Cleanliness, minimization of the amount and size of ceramic inclusions (e.g., by screening the powder, Ref 18), and other defects (i.e., inert gas porosity, Ref 19) are even more important than is described in the previous section for high-performance P/M tool steels. In a recent paper (Ref 20), combining electroslag refining with atomizing and/or spray forming is described and shown to produce very clean materials. To facilitate the achievement of these demanding requirements, manufacturers of superalloy parts, unlike manufacturers in the conventional powder metals industry, often make their own powders.

The goal of combining sulfidation and oxidation resistance with ever-increasing elevated-temperature strength (e.g., in gas turbine engines) for this class of P/M materials has led, in recent years, to increasing efforts and use of ODS superalloys. The invention of mechanical alloying in 1966 by Benjamin (Ref 21) permits the introduction of a finely dispersed oxide into even very highly alloyed and reactive-element-containing alloys.

Mechanically alloyed powders are made by dry, high-energy milling of blends of metal powders, master alloys, and refractory compounds. The starting powders can be relatively coarse, for example, 50 to 100 μm average diameter. The dispersion of the oxides (typically yttria or alumina) after milling can have uniform interparticle spacings of less than 1 μm. The amount of oxide added to superalloys ranges from about 0.4 to 1.5 wt% (1 to 2.7 vol% for yttria).

Table 13 summarizes both conventional and yttria-dispersion-strengthened P/M superalloy compositions. For details on processing and properties of consolidated materials, see the article "Superalloys" in this Volume.

REFERENCES

1. Y. Trudel, M. Gagné, and F. Racicot, Design Criteria for the Manufacturing of Low Alloy Steel Powders, *Adv. Powder Metall.,* Vol 5, 1991, p 45–58
2. R. German, *Powder Metallurgy of Iron and Steel,* John Wiley & Sons, 1998
3. Standard 35, Metal Powder Industries Federation, 1997
4. A.K. Sinha, Recent Developments in Iron and Steel Powder Production for High Performance P/M Components, *Production of Iron, Steel, and High Quality Product Mix,* ASM International, 1992, p 57–69
5. *Met. Powder Rep.,* Vol 45 (No. 7), 1990, p 526–527
6. C. Tornberg and P. Billgren, *Met. Powder Rep.,* Vol 47 (No. 6), 1992, p 19–23
7. G.F. Bocchini, *Met. Powder Rep.,* Vol 47 (No. 2), 1992, p 48–53
8. P. Lindskog, *Met. Powder Rep.,* Vol 47 (No. 1), 1992, p 32–37
9. O. Furukimi, K. Yano, and S. Takajo, *Int. J. Powder Metall.,* Vol 27 (No. 4), 1991, p 331–337
10. *Met. Powder Rep.,* Vol 47 (No. 5), 1992, p 12–16
11. J.F. Watton, J.B. Olson, and M. Cohen, Innovations of Ultrahigh Strength Steel Technology, *Proc. 34th Army Sagamore Materials Conf.,* G.B. Olson, M. Azrin, and E.S. Wright, Ed., Plenum Press, 1990
12. G.B. Olson, Innovations of Ultrahigh Strength Steel Technology, *Proc. 34th Army Sagamore Materials Conf.,* G.B. Olson, M. Azrin, and E.S. Wright, Ed., Plenum Press, 1990
13. N.C. Koon and R. Hasegawa, *Rapidly Solidified Crystalline Alloys,* S.K. Das, B.H. Kear, and C.M. Adam, Ed., TMS, 1985, p 245–262
14. E.Y. Gutamanas, *Prog. Mater. Sci.,* Vol 34 (No. 4), 1991, p 261–366
15. N.C. Trilk, T. Davis, P. Woods, and P.R. Brewin, Mixed Powder Blends for the Best Performance in Wear, Rolling Contact, Mechanical Properties, and General Use, *Powder Metall. Partic. Mater.,* Vol 3 (No. 10), 1995, p 29–40
16. M. Hull, Powdermet Sweden AB: Hipping as an Innovative Manufacturing Route, *Powder Metall.,* Vol 39 (No. 3), 1996, p 174–176
17. M. Hull, Soderfors Powder AB: New Generation High Speed Steel Powders, *Powder Metall.,* Vol 39 (No. 3), 1996, p 176–178
18. D.R. Chang, D.D. Krueger, and R.A. Sprague, Superalloy Powder Processing, Properties, and Turbine Disc Applications, *Superalloys 1984,* TMS/Metallurgical Society AIME, 1984, p 245–273
19. E. Huron, R.L. Casey, M.F. Henry, and D.P. Mourer, The Influence of Alloy Chemistry and Powder Production Methods on Porosity in a P/M Nickel-Base Superalloy, *Superalloys 1996,* R.D. Kissinger et al., Ed., Minerals, Metals and Materials Society/AIME, 1996, p 667–676
20. M.G. Benz, W.T. Carter, F.G. Miller, and R.M. Forbes-Jones, Electroslag Refining as a Clean Liquid Metal Source for Atomization and Spray Forming of Superalloys, *Superalloys 1996,* R.D. Kissinger et al., Ed., Minerals, Metals and Materials Society/AIME, 1996, p 723–728
21. T.J. Benjamin, *Dispersion Strengthened Superalloys by Mechanical Alloying,* TMS/AIME, Vol 1, 1970, p 2943–2951

Production of Copper Powders

David F. Berry, OMG AMERICAS
Erhard Klar, Consultant

THE SELF-LUBRICATING BRONZE BEARING, invented in the 1920s, and its subsequent development brought about the commercial production of copper powders. The first large-scale production processes were oxide reduction and electrolytic deposition. Electrolytic copper powders enabled the development of copper-graphite electrical brushes and copper-base friction materials in the 1930s. Copper powder made by the cementation process was also used in copper-base friction materials for a period of time, but ceased in the 1970s. Other hydrometallurgical processes for copper powder production were developed in the 1950s and 1960s. Some powders were commercially used for a number of years. Further growth of copper powder production took place with the development, during World War II, of the ferrous structural parts industry, which uses copper additions for alloy strengthening. In the 1950s, production of copper and copper alloy powders by water atomization became commercially significant. By the early 1980s, competition from oxide-reduction and atomization processes caused commercial production of electrolytic copper powder to cease in the United States, but it is still commercially produced in Europe and Asia, where water atomization is the main competitive commercial process.

This article describes the fundamentals of the four processes used for the production of copper powders, the physical and mechanical properties of the powders produced, and the applications of the copper powders. More detailed information is found in the articles of this Volume devoted to the individual powder-producing methods.

Production of Copper Powder by the Reduction of Copper Oxide

Reduction of copper oxide is the oldest and (in the United States) still one of the main commercial processes for making copper powder, along with water atomization. In this process, particulate copper oxide is converted to copper at elevated temperature by reducing gases. The product, a sintered porous cake, is then milled to a powder. The starting material for reduction was originally copper millscale and cement copper. However, as demand for copper powder exceeded supply and higher purity was required, particulate copper of high purity (chopped scrap or atomized copper) was oxidized to form cuprous or cupric oxide or a mixture of both.

Melting of Copper

In recent years, the increasing emphasis on purity has supplanted both copper mill scale and cement copper as the starting materials for making oxide-reduced copper powders.

Copper melting is performed in both fuel-fired and induction-heated furnaces. Strong oxide formers, such as aluminum and silicon, are kept low; thus, the molten metal remains fluid and can be poured without difficulty. Aluminum and silicon oxides also make the powder less compressible and very abrasive. In addition, lead and tin cause problems in pouring the melt due to buildups and clogging in the furnaces and nozzles.

In some applications of copper powder (for example, metal-graphite brushes and friction parts), good electrical or thermal conductivity is important. For this reason, impurity levels must be kept low. The detrimental effect on electrical conductivity caused by some impurities that are present in solid solution is shown in Fig. 1. Thermal conductivity is similarly dependent on the impurity content.

Atomization and Shotting of Copper

Large-scale atomization of copper usually is conducted in air as a continuous process. Liquid copper may be atomized directly from a tube in the side of the furnace wall or through a tundish. Both air and water may be used as atomizing media, and horizontal atomization with high-pressure air into a rotating drum eliminates the need for drying of the powder. For details of atomization, including nozzle configurations and pressure relationships, see the article "Atomization" in this Volume.

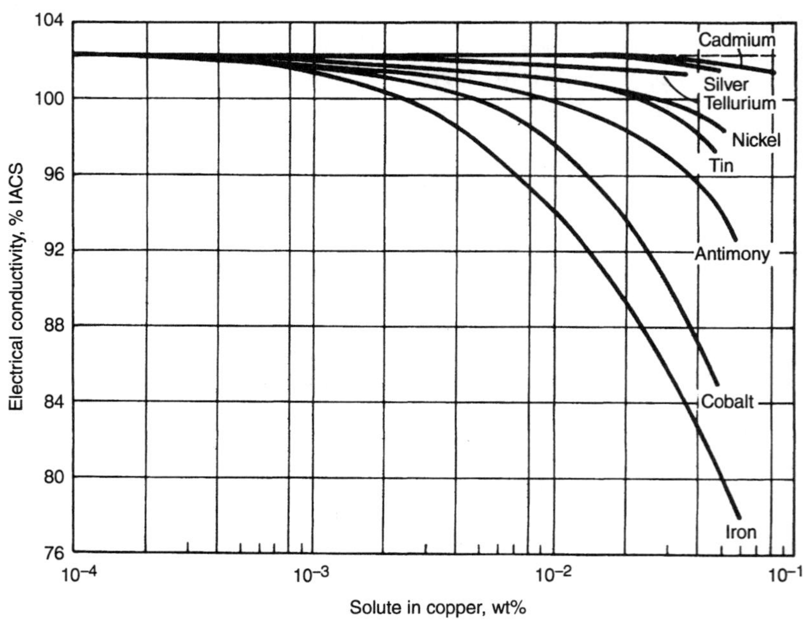

Fig. 1 Effect of impurities in solid solution on electrical conductivity of oxygen-free copper

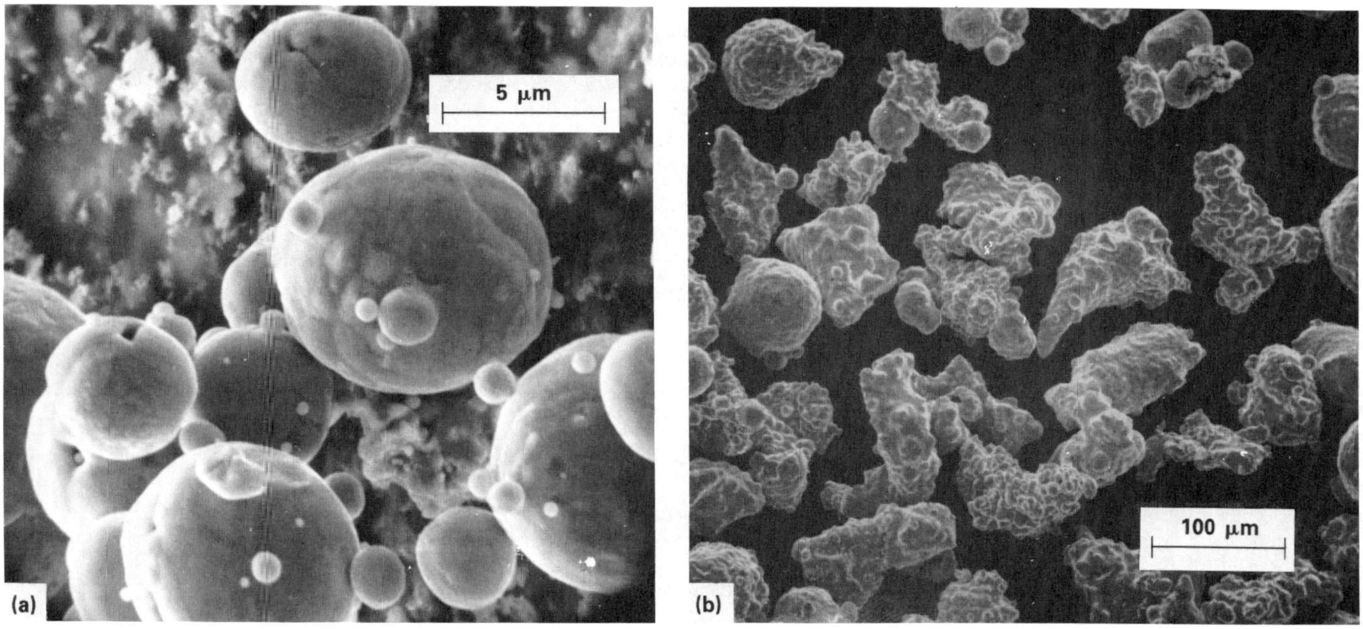

Fig. 2 Scanning electron micrographs of copper powder. (a) Air atomized. (b) Water atomized

Scanning electron micrographs of air-atomized and water-atomized copper powders are shown in Fig. 2. The water-atomized powder has a more irregular particle shape, and its oxide content is somewhat lower. At this process stage, however, particle shape is not very important. The coarse shot-type powder that is produced by low-pressure air or water atomization (Fig. 3) is the typical starting material for the pure copper oxide process.

Oxidation of Copper Powder

Oxidation of air-atomized, water-atomized, or shotted copper is performed to radically change the shape of the powder and thus enhance control over the various engineering properties of parts made from the powder. Completely oxidized and reduced powders with a wholly spongy (porous) structure and entirely solid powders, as typified by a gas-atomized powder, constitute the extremes of available copper powders. Partial oxidation results in intermediate structures.

The oxidation of copper is a well-known process. There are two copper oxides, the red cuprous oxide, Cu_2O, and the black cupric oxide, CuO. At high temperatures and for the so-called thick film range, oxidation follows the parabolic rate law, in which film thickness (y) increases with the square root of time ($y = \sqrt{Kpt + c}$). At low temperatures, linear, logarithmic, and cubic oxidation rates have been observed, depending on the history of the oxide. Free energies, heats of reaction, and rates of copper oxide formation are shown in Table 1.

In commercial practice, oxidation or roasting of copper powder normally is done in air at temperatures above 650 °C (1200 °F). Oxidation in rotary kilns or fluidized beds provides faster oxidation rates by increasing the contact area between powder and oxidizing gas. However, due to the strongly exothermic nature of the oxidation reactions, these operations are more difficult to control than roasting in a belt conveyor furnace.

Grinding of Copper Oxide. Both oxides of copper are brittle and easy to grind to −100 mesh powder. The oxide particles themselves are porous. Figure 4 shows oxidized shot before and after grinding.

Table 1 Free energies, heats, and rates of copper oxide formation

$2(Cu) + \frac{1}{2}(O_2) = \langle Cu_2O\rangle$ exothermic
$\Delta G = -41{,}166 - 1.27 \times 10^{-3}\, T \ln T + 3.7 \times 10^{-3} T^2 - 1.80 \times 10^{-7} T^3 + 27.881\, T$
$k = 957\, e{-37{,}100/RT}\ g^2cm^{-4}h^{-1}$

$\langle Cu\rangle + \frac{1}{2}(O_2) = \langle CuO\rangle$ exothermic
$\Delta G = -37{,}353 - 0.16\, T \ln T - 1.69 \times 10^{-3}\, T^2 - 9 \times 10^{-8} T^3 + 25.082T$
$\Delta H = -38{,}170 + 1.30\, T + 0.99 \times 10^{-3}\, T^2 + 0.57 \times 10^5\, T^{-1}$

$\langle Cu_2O\rangle + \frac{1}{2}(O_2) = 2\langle CuO\rangle$ exothermic
$\Delta G = -33{,}550 + 0.95\, T \ln T - 3.75 \times 10^{-3}\, T^2 + 22.340\, T$
$\Delta H = -35{,}710 + 3.28 \times T - 0.40 \times 10^{-3}\, T^2 - 0.20 \times 10^5\, T^{-1}$
$k = 0.0268\, e{-20{,}140/RT}\ g^2cm^{-4}h^{-1}$

ΔG is free energy; ΔH is heat; values of ΔG and ΔH are given in calories per gram mole; k is a rate constant mathematically derived; T is the absolute temperature, degrees kelvin; R is the absolute gas constant; ln is the natural logarithm (base e, where e = 2.7182).

Reduction of Copper Oxide

Reduction of particulate copper oxide is generally accomplished on a stainless steel belt in a continuous belt furnace. The depth of the oxide bed is approximately 25 mm (1 in.). The typical reduction temperature ranges from 425 to 650 °C (800 to 1200 °F). Reduction occurs gradually from top to bottom of the bed. The reducing atmosphere in the furnace generally flows countercurrent to the motion of the conveyor belt.

Reducing atmospheres may include hydrogen, dissociated ammonia, water-reformed natural gas, or other endothermic or exothermic gas mixtures. Because reduction of copper oxides with hydrogen or carbon monoxide is exothermic, a careful balancing of oxide particle size, reducing gas species, and reduction temperature is necessary to optimize the reduction rate and to control the pore structure. Hydrogen diffuses readily through solid copper and is a more effective reducing agent than carbon monoxide, particularly at low temperatures. At higher tempera-

Fig. 3 Air-shotted copper

tures, however, all reduction reactions involving either hydrogen or carbon monoxide proceed almost to completion. Free energies and heats of reaction for copper oxide reduction with hydrogen and carbon monoxide are shown in Table 2.

Control of Powder Properties. By manipulating the reduction process, the particle porosity, pore size, and particle size distribution of the finished powder can be controlled over a wide range. As with other metal oxides, low reduction temperatures generally produce particles having fine internal porosity and correspondingly high specific surface. High reduction temperatures produce particles containing large pores and low specific surface. High reduction temperatures generally result in more interparticle sintering and more complete reduction.

Post-Reduction Procedure. The reduced copper oxide emerges from the reduction furnace as a porous cake. It is broken into smaller pieces in a jaw crusher or similar equipment, followed by fine grinding in hammer mills. The amount of work hardening is moderate, and the resulting powders have good compressibility and green strength. A scanning electron micrograph of a typical copper powder is shown in Fig. 5.

Various combinations of the controlling parameters during melting, atomization, oxidation, reduction, and grinding permit the manufacture of powders with characteristics tailored to the requirements of all major applications.

The reduced and milled powder is screened and/or classified and, if necessary, blended and lubricated. These processing steps are carefully controlled to avoid losses or uncontrolled changes in characteristics such as apparent density, amount of fines, and powder flow. Some powder grades are treated with proprietary antioxidants to stabilize them against oxidation. Without this treatment, copper powders, particularly when exposed to moist air, generally tarnish, with attendant losses in green strength and other side effects. With increasing tarnish, the color of the powder changes from orange to purple to black. Simultaneously, oxygen content increases from a typical 0.1 or 0.2% to several tenths of a percent, and as much as approximately 1%. Copper powders with large specific surface areas are more sensitive to tarnishing.

Finished Powders

Finished lots of powder are subjected to a series of tests designed to ensure performance in the various applications. The single most important application for copper powder is for self-lubricating bronze bearings. In the United States, these bearings are made predominantly from elemental blends of copper and tin. Since approximately 1960, preblended and lubricated mixtures of 90% Cu and 10% Sn that have graded dimensional change characteristics during sintering have increased in use.

Table 3 lists various grades of copper powders produced by the oxide reduction process; apart from bronze bearings, applications include copper-base friction materials, electrical contacts and brushes, diamond abrasive wheels, and copper additions to iron mixes for structural parts (Ref 1). Noncompacting applications include copper fillers in plastics, catalysts, and many chemical applications. The product brochures and data sheets of the powder producers generally provide details on the characteristics and performance properties of their powders (including sintered properties), as well as recommendations for specific uses.

Table 2 Free energies, heats, and partial gas pressures for copper oxide reduction with hydrogen and carbon monoxide

$\langle Cu_2O\rangle + (H_2) \leftrightarrow 2\langle Cu\rangle + (H_2O)$ exothermic
$\Delta G = -16{,}260 + 2.21\, T \ln T + 1.28\times10^{-3} T^2 + 3.8\times10^{-7} T^3 - 24.768T$
$\Delta H_{298.1K} = -17{,}023$
Temperature, °C (°F): 450 (840), 900 (1650), 950 (1740), 1000 (1830), 1050 (1920)
P_{H_2}, torr(a): 0.0104, 0.0150, 0.0207, 0.0283

$\langle Cu_2O\rangle + (CO) \leftrightarrow 2\langle Cu\rangle + (CO_2)$ exothermic
$\Delta G = -27{,}380 + 1.47\, T\ln T - 1.4\times10^{-3} T^2 + 0.5\times10^{-6} T^3 - 7.01\, T$
$\Delta H = -27{,}380 - 1.47\, T + 1.4\times10^{-3} T^2 - 1.1\times10^{-6} T^3$
Temperature, °C (°F): 25 (77), 900 (1650), 1050 (1920), 1083 (1980)
P_{CO}, torr(a): 0.021, 0.068, 0.085

$2\langle CuO\rangle + (H_2) \leftrightarrow \langle Cu_2O\rangle + (H_2O)$ exothermic
$\Delta G = -24{,}000 - 0.01\, T\ln T + 5.4\times10^{-3} T^2 - 3.7\times10^{-7} T^3 + 22.896T$
$\Delta H_{298.1K} = -23{,}543$

$\langle CuO\rangle + (H_2) = (H_2O) + \langle Cu\rangle$ exothermic
$\Delta H_{290K} = -31{,}766$

$2\langle CuO\rangle + (CO) \leftrightarrow \langle Cu_2O\rangle + (CO_2)$ exothermic
$\Delta H = -33{,}300$

ΔG is free energy; ΔH is heat; values of ΔG and ΔH are given in calories per gram mole; P is pressure; T is the absolute temperature, degrees kelvin; and ln is the natural logarithm (base *e, where e = 2.7182). (a) Total pressure is 1 atm.*

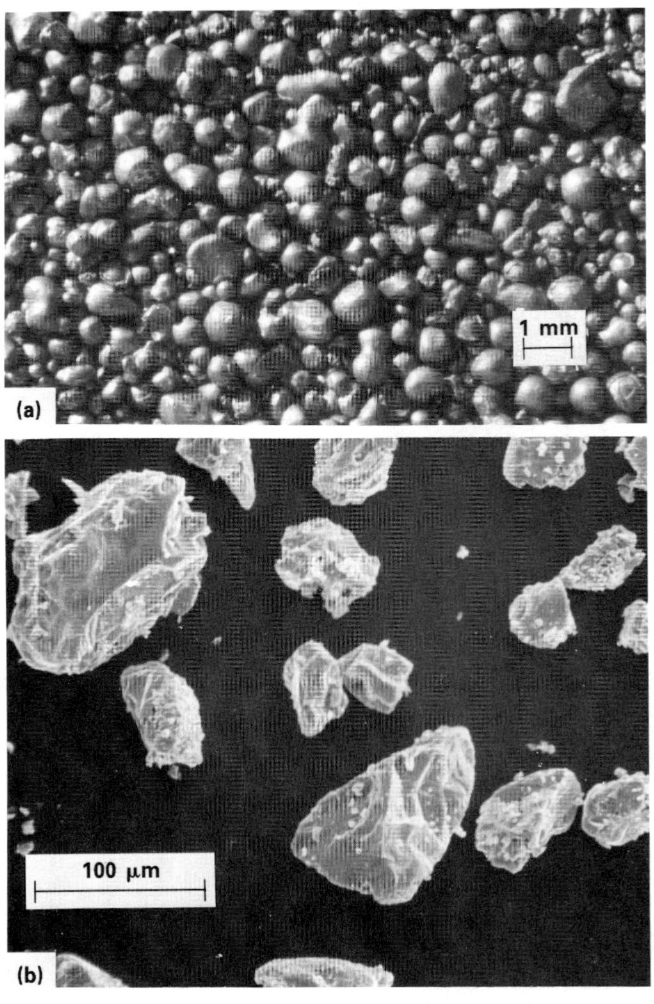

Fig. 4 Micrographs of copper oxide. (a) As-oxidized copper shot. (b) Scanning electron micrograph of copper shot after grinding

Production of Copper Powder by Electrolysis

Production of electrolytic copper powder, as shown in Fig. 6, follows the same electrochemical principles that apply to the electrorefining of copper. The conditions of deposition, however, are changed to yield a powdery or spongy deposit, instead of the strongly adherent product desired in electrorefining. The formation of powdery deposits is favored by low copper ion concentration and high acid content in the electrolyte, as well as by high cathode current density.

Although these conditions yield spongy deposits, control of additional variables is necessary to produce powders that meet commercial requirements. Other factors are quantity and type of addition agent, temperature and circulation rate of the electrolyte, size and type of anode and cathode, electrode spacing, and brush-down interval. For more detailed information on electrolytic production of metal powders, see the article "Electrodeposition of Metal Powders" in this Volume.

Effects of Electrolyte Composition

The composition of the electrolyte is a major factor in the production of powder. Copper concentration in the electrolyte must be sufficiently low to prevent adherent deposits. In the desired range, current efficiency rises with increasing copper ion concentration, as shown in Fig. 7, with a maximum of 92% in the range of 23 to 33 g/L (3 to 4.4 oz/gal) copper. Above approximately 33 g/L (4.4 oz/gal), current efficiency decreases, and a hard deposit is produced instead of powder. Apparent density and particle size also increase with a rise in copper ion concentration.

Acid Concentration. A high acid concentration favors the formation of powder. As shown in Fig. 8, current efficiency increases to a maximum at a concentration of 120 g/L (16 oz/gal) of sulfuric acid, then gradually falls as the acid concentration increases. Continued rise in acid concentration leads to passivity. Apparent density decreases with increasing acid concentration.

Addition Agents. The copper sulfate/sulfuric acid electrolyte occasionally is modified to alter powder characteristics. The addition of colloidal materials, such as glue or glucose, leads to the formation of fine powder deposits, possibly because the colloid retards the evolution of hydrogen at the cathode. Table 4 indicates the effects of a number of additions to the electrolyte.

The addition of surfactants is reported to yield a powder with controllable particle size at a current density of 215 A/m^2 (20 A/ft^2), in contrast with the 700 to 1100 A/m^2 (64 to 100 A/ft^2) normally used, which results in considerable reduction in power cost (Ref 2). Small quantities of copper chloride have been added to the electrolyte to increase the dendritic character of the powder particles and to increase the yield of fine powder due to the polarizing effect of the chloride ions (Ref 3). The addition of sodium sulfate reportedly reduces the cathode current density, and as the sulfate content is increased, the powder becomes finer (Ref 4). By contrast, replacement of the normal sulfuric acid electrolyte by a sulfamate electrolyte favors the formation of a coarse copper powder (Ref 5).

Effects of Operating Conditions

Alteration of operating conditions can affect process variables such as current efficiency and the formation and size of particles.

Current Density. High current density favors the formation of powder, but has only a minor effect on current efficiency. A marked decrease in particle size occurs with rising current density. For example, in an electrolyte containing 25 g/L (3.3 oz/gal) copper and 120 g/L (16 oz/gal) free sulfuric acid, increasing the current density from 600 to 1000 A/m^2 (55 to 90 A/ft^2) raised the quantity of powder passing through a 300 mesh sieve from 20 to 96%.

Temperature. Raising the operating temperature of the cell increases the current efficiency and reduces the cell voltage. However, cell operation is difficult at temperatures higher than 60 °C (140 °F), and powders obtained at high temperatures are coarser than those produced at lower temperatures. Generally, electrolytic cells

Fig. 5 Scanning electron micrograph of copper oxide reduced copper powder

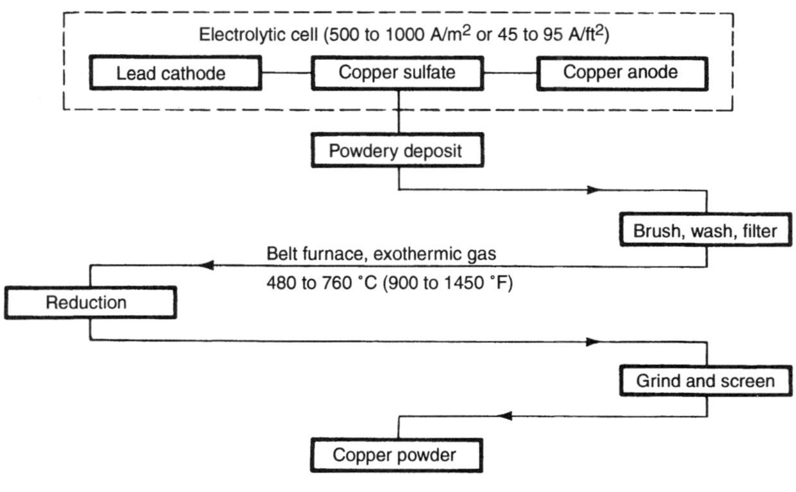

Fig. 6 Flowchart for production of electrolytic copper powder

are operated at temperatures between 25 and 60 °C (77 and 140 °F).

Powder Removal. The method used to remove the powder from the cathode has a significant effect on powder characteristics. Usually the powder is removed mechanically by brushing. The brush-down interval aids in control of the particle size of the deposit and, as shown in Fig. 9, the powder becomes coarser as the interval is increased from 15 to 60 min. Figure 10 indicates that the apparent density increases as the brush-down interval is extended. Frequent brush-down also limits variations in cathode current density. In another approach to powder removal, an organic extract is used in combination with sodium lauryl sulfate to discharge the powder automatically from the cathode without brush-down (Ref 6).

Powder Production

The major producers of electrolytic copper powder generally follow the procedures used in electrolytic refining. However, as has been mentioned, changes in operating conditions are required to obtain a deposit of powder. Typical conditions for the production of copper powder are illustrated in Fig. 6 and can be summarized as:

Condition	Quantity
Copper	5–15 g/L (0.6–2 oz/gal)
Sulfuric acid	150–175 g/L (20–23 oz/gal)
Temperature	25–60 °C (77–140 °F)
Anode current density	430–550 A/m^2 (39–50 A/ft^2)
Cathode current density	700–1100 A/m^2 (64–100 A/ft^2)
Cell potential	1.0–1.5 V

Generally, the anodes are electrolytically refined copper, and the cathodes are lead alloy sheet. In a typical installation, the cathodes are 61 by 86 by 0.95 cm (24 by 34 by 0.37 in.) in dimension (Ref 7). Both anodes and cathodes are short to allow enough space at the bottom of the tank for the collection of the powder. The electrodes are arranged parallel to one another in lead-lined, rubber-lined, or plastic tanks, typically 3.4 m long by 1.1 m wide by 1.2 m high (11 ft long by 3.6 ft wide by 3.9 ft high). Each cell contains 18 cathodes spaced at 16 cm (6.4 in.) intervals and 19 anodes that are hung alternately. To ensure uniform current density and to eliminate short circuits in the cells, the electrodes are inspected frequently for prevention of excessive nodule buildup.

Typically, the electrolyte is pumped to an elevated storage tank, from which it flows by gravity into the tops of the cells and out through the bottoms. Thus, circulation is from top to bottom of the tanks, a type of circulation that yields a more homogeneous powder than that obtained by bottom-to-top circulation. The overflow returns to basement storage tanks to be recirculated.

The copper deposits on the cathode in the form of dendritic particles. To prevent short circuits between anode and cathode and heavy accumulation of powder on the cathode, both of which would decrease the cathode current density, the deposit is removed periodically by brushing.

After operating the cell for several days, the power is turned off. Most of the electrolyte is drained from the cell, leaving enough solution to cover the powder. The anodes and cathodes are

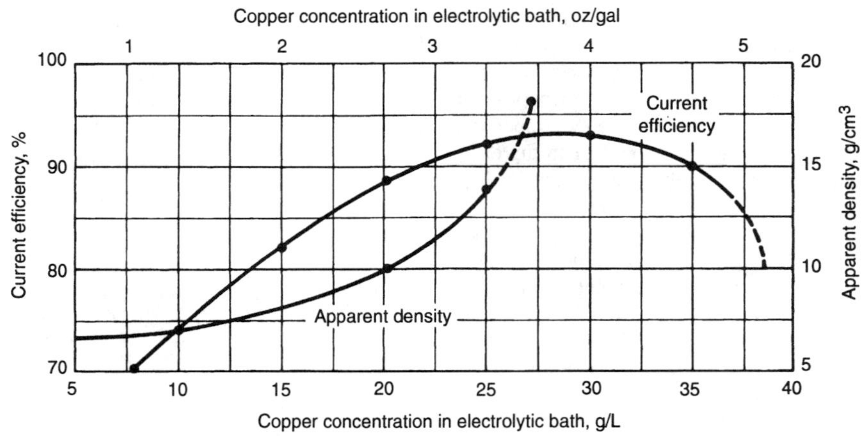

Fig. 7 Effect of copper concentration on current efficiency and apparent density. Source: Ref 1

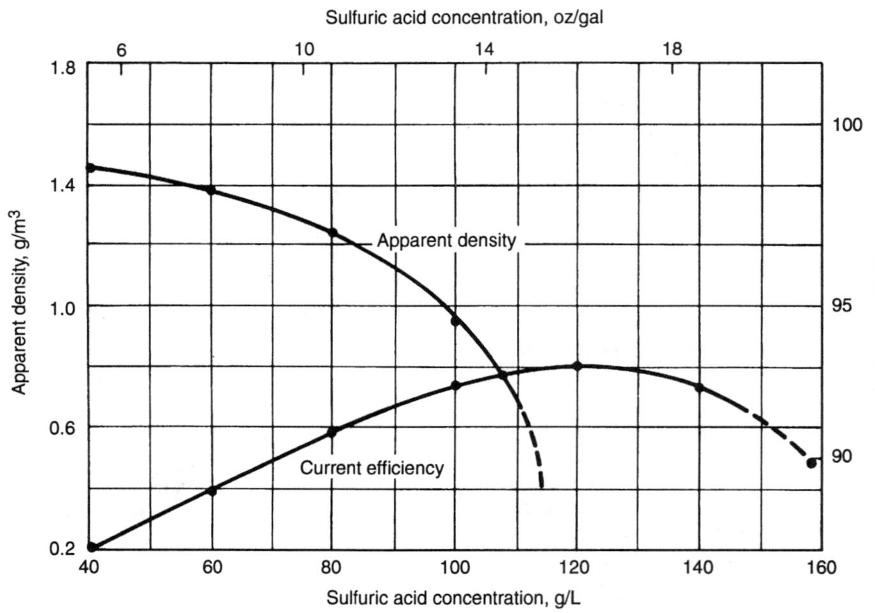

Fig. 8 Effect of acid concentration on current efficiency and apparent density. Source: Ref 1 and 14

Table 3 Typical properties of commercial grades of copper powders produced by oxide reduction

Chemical properties			Physical properties								Compacting properties at 165 MPa (12 tsi)(a)		
					Sieve analysis, % Tyler								
Copper, %	Hydrogen loss, %	Acid insoluble, %	Apparent density, g/cm^3	Hall flow, s/50 g	+60	+100	+150	+200	+325	−325	Density, g/cm^3	Strength(b), MPa	Strength(b), psi
99.8	0.13	0.06	2.91	26	0.4	39.7	46.6	13.3	...	...	...	...	...
99.8(c)	0.13	0.03	3.00	22	...	0.1	0.6	15.5	42.8	41.1	6.15	8.6	1240
99.8(c)	0.13	0.04	2.83	23	...	...	0.1	9.5	33.4	57.0	6.12	9.7	1400
99.8(c)	0.16	0.04	2.75	24	...	...	0.1	7.3	29.0	63.6	6.03	10.4	1500
99.7	0.18	0.06	2.51	...	...	...	...	0.5	7.0	92.6	6.04	...	...
99.7	0.21	0.06	2.31	...	...	...	...	...	1.2	98.5	...	...	...
99.6(d)	0.28	0.10	1.61	...	...	...	0.1	2.8	10.3	86.7	6.0	20.0	2900
99.6(d)	0.26	0.10	1.36	...	...	...	0.1	1.5	7.9	90.5	5.97	22.8	3300
99.5(d)	0.26	0.10	0.94	...	...	...	0.1	0.2	1.4	98.6	5.90	29.0	4200

(a) Measured with die wall lubricant only. (b) Transverse rupture strength. (c) For bronze self-lubricating bearings. (d) For friction materials and electrical brushes washed down and removed, the remaining electrolyte is drained from the cell, and the powder is removed.

Thorough washing of the powder is essential. All traces of the electrolyte must be removed to prevent the powder from becoming oxidized. In addition, any remaining sulfate damages the heating elements if an electric furnace is used for subsequent drying and treatment of the powder. Various methods are employed to wash the powder. Although centrifuging to remove the electrolyte and wash the powder yields a clean product, the particles are compacted, and production of a low-density powder by this procedure is difficult.

In another method, the powder is transferred into a large tank, and water is added to produce a slurry that is pumped into a filter. In the filter, the powder is dewatered, washed several times, and again dewatered. Because the wet powder oxidizes readily due to its finely divided state and active surface, addition of a stabilizer is desirable. Treatment with an aqueous solution of gelatin protects the powder from oxidation in the intervals between successive operations (Ref 8).

The addition of surface-active agents during washing or subsequent powder treatment also protects the powder from oxidation. Next, the powder must be subjected to a furnace operation to obtain an acceptable grade.

Furnace Operation

After thorough washing and filtering, the wet powder is ready for furnace processing. The furnace operation also alters certain properties, particularly particle size and shape, apparent density, and green strength. In a typical operation, the powder is transferred into the charge box of a mesh belt electric furnace (Ref 7). To prevent the powder from falling through the belt, a continuous sheet of high wet-strength paper is fed to the belt, and then the powder is transferred to the paper. A roller compresses the powder to improve heat transfer. As it enters the furnace, water is driven off and the paper burns, but not before the powder has sintered sufficiently to prevent it from falling through the belt.

The furnace atmosphere is produced in exothermic gas units in which natural gas and air are blended to yield an atmosphere containing 17% H_2, 12% CO, 4% CO_2, and the balance of nitrogen. The gas is refrigerated to lower the dew point to the range of –22 to –40 °C (–8 to –40 °F). The gas enters the furnace from the discharge end and, because it is refrigerated, aids in cooling the powder cake. The furnace operation dries the powder, alters the particle shape, reduces the oxides, and sinters the fines. The discharge temperature is sufficiently low to prevent reoxidation of the powder cake.

By varying the furnace temperature between 480 and 760 °C (900 and 1400 °F) and altering the time of exposure, considerable change can be made in the content of fines, apparent density, and dimensional characteristics. Upon completion of the furnace operation, the cake is broken and is ready for grinding.

Grinding and Finishing

Fine grinding is performed in high-speed, water-cooled hammer mills in which feed rate, mill speed, and screen openings under the mill can be varied to obtain the powder characteristics desired; thus, grinding is another operation in which powder properties can be changed. The powder leaving the hammer mills is fed to screens, where the oversize is separated and returned for additional grinding. The –100 mesh powder is classified in an air classifier, and the fines are transferred to the blending operation. Oversize material is returned for regrinding or is used as melting stock.

The products of the grinding and classifying operations range in apparent density from approximately 1 to 4 g/cm^3. They are stored in drums to which a drying agent such as silica gel or camphor is added to prevent further oxidation. To produce high volume finished powder that meets customer specifications, powder is selected from various batches in the proper proportions and mixed in a blender. The lot is sampled before being removed from the blender, and if required, adjustments of the particle size distribution are made before the powder is packed in shipping drums. Table 5 lists the physical characteristics of several representative blends.

Process Modifications

Although most electrolytic powder is produced as previously described, other methods have been used. To obtain a very low density product, one producer employs small-size cells

Table 4 Effect of addition agents on current efficiency and particle size

Test No.	Addition agent	Solution strength, %	Voltage, V	Current efficiency, %	Sieve analysis, % –200 mesh	Sieve analysis, % –300 mesh
1	...	...	1.0	95.9	74.6	55.0
2	Boric acid	0.5	1.0	95.2	...	100
3	Glucose	0.5	1.2	85.4	...	100
4	Glycerine	0.5	1.9	94.7	...	100
5	Glue	0.5	1.5	94.5	...	100

Source: Ref 1

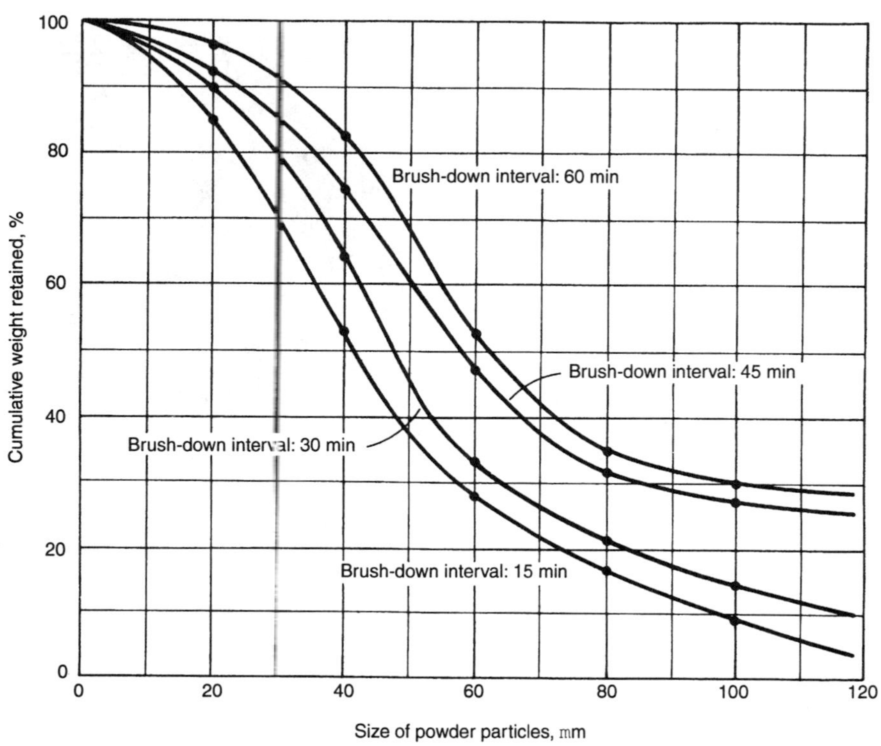

Fig. 9 Effect of brush-down interval on particle size. Source: Ref 1 and 14

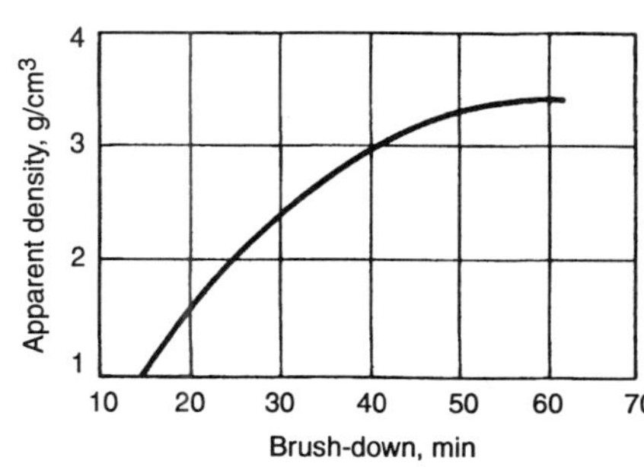

Fig. 10 Effect of brush-down interval on apparent density. Source: Ref 1 and 14

and handles the powder lightly in both the cell and the finishing operations. The resulting powder shape is fern-like and has an apparent density of 0.9 to 1.3 g/cm^3.

Another process for the production of copper powder uses a vertical rotating titanium cathode that is partially immersed in a copper sulfate/sulfuric acid electrolyte (Ref 10). In the operation, the powder is collected continuously above the liquid level by scraping the cathode, followed by continuous dewatering, washing, and drying steps.

Properties of Electrolytic Copper Powder

The properties of electrolytic copper powder are dependent on various characteristics of the operation and, therefore, can often be controlled by altering certain process variables.

Purity of powder prepared by the electrolytic process is high, with a copper content that usually exceeds 99.5% (Table 5). A measure of the oxygen content is obtained by exposing a sample of powder to hydrogen at an elevated temperature as specified in American Society for Testing and Materials standard ASTM E 159 or Metal Powder Industries Federation standard MPIF 02, and ISO 4491/2. Generally, the hydrogen loss ranges from 0.1 to 0.5%, depending on the apparent density and particle size distribution of the powder. Nitric acid insolubles are determined by ASTM E 194 or MPIF 08 standard procedures and generally are less than 0.05%.

Particle size distribution for copper powder is selected to meet the requirements of the application and can be varied over a wide range. As noted in Table 5, several combinations are produced, and these are only representative. For example, the –325 mesh fraction can be varied from 5 to 90%.

Apparent densities of powders are available from 1.0 to 4.0 g/cm^3. Densities that are somewhat lower and higher can be produced depending on process and electrolyte conditions.

Flow rate is shown in relationship to apparent density in Fig. 11. Generally, powders with densities of less than about 1.3 g/cm^3 do not flow, powders with apparent densities of 1.3 to 2.3 g/cm^3 have poor flow rates, and powders with higher apparent densities flow freely. At approximately 2.2 g/cm^3, which is the transition range, flow depends on the content of fine particles of the powder, because relatively fine powders have poor flowability and relatively coarse powders flow freely. The flow rates of the usual blends of electrolytic copper powder range from 25 to 40 s/50 g.

Green density is a function of the compacting pressure. As shown in Fig. 12, the green density for the blend used rises from 7.2 to 8.0 g/cm^3 as the compacting pressure is increased from 275 to 550 MPa (920 to 40 tsi).

Green strength increases with the compacting pressure, as indicated in Fig. 12. In this example, the green strength rises from less than 15 to 24 MPa (2200 to 3500 psi) as the compacting pressure is increased from 275 to 550 MPa (20 to 40 tsi).

Particle shape of electrolytic copper generally is dendritic when deposited on the cathode (Fig. 13). During subsequent operations, however, the dendrites are rounded somewhat.

Compacting pressure is an important variable to consider, because the pressing and sintering conditions have a marked influence on the properties of sintered compacts. If sound compacts are to be obtained, gases from the atmosphere, from the products of reduction, or from the lubricant must escape during sintering. When the compacting pressure is too high, the flow through interconnected pores may be obstructed, and the gases cannot escape. Compacting pressures no higher than 275 MPa (20 tsi) should be used in the production of large, thick parts from electrolytic copper powder, although higher pressures can be used for thin-walled parts (Ref 12).

Dimensional changes in a typical powder blend lubricated with lithium stearate are shown in relationship to sintering time in Fig. 14. The lubricated powder blend was pressed as indicated and sintered at 1000 °C (1830 °F) in an atmosphere of dissociated ammonia atmosphere. Relatively stable dimensional changes are achieved at normal sintering times.

Tensile strength and elongation of a typical powder blend are shown in Fig. 15 in relationship to compacting pressure and sintering time. The powder was lubricated with lithium stearate, pressed as indicated, and sintered at 1000 °C (1830 °F) in an atmosphere of dissociated ammonia. The graph demonstrates that good tensile properties can be obtained with short sintering times.

Electrical conductivity that is high can be achieved. However, a high-purity powder, a characteristic of powder produced by the electrolytic process, must be used. Figure 16 shows the relationship between electrical conductivity and sintered density; high conductivity can be achieved only in high-density compacts. Electrical conductivity can be increased by coining and resintering (Fig. 17).

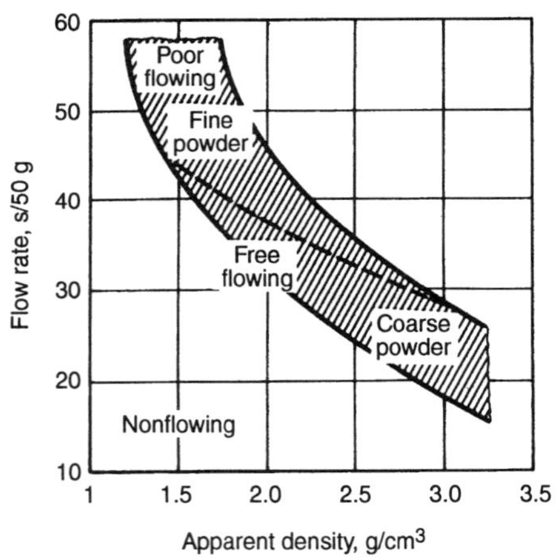

Fig. 11 Relationship between apparent density and flow rate. Source: Ref 11

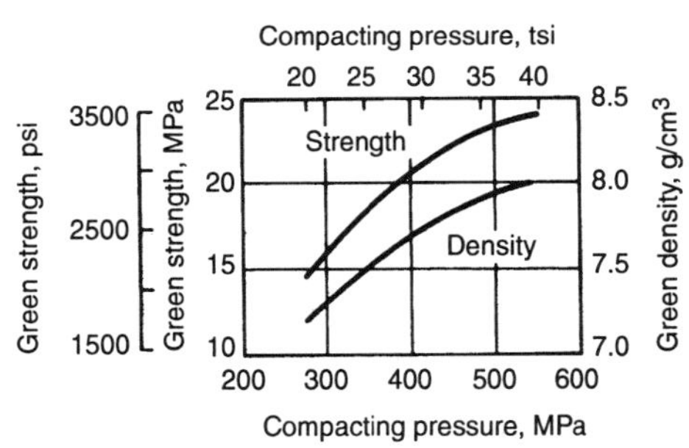

Fig. 12 Effect of compacting pressure on green strength and green density. Source: Ref 9

Fig. 13 Electrolytic copper powder showing dendritic structure. 85×. Source: Ref 9

Table 5 Properties of typical commercial grades of electrolytic copper powders

Chemical properties			Physical properties						
Copper, % min	Hydrogen loss, % max	Acid insoluble, % max	Apparent density, g/cm^3	Sieve analysis, % Tyler +60	+100	+150	+200	+325	–325
99.8	0.15	0.06	2.5–2.7	...	5 max	1–13	11–24	20–30	40–55
99.8	0.15	0.06	2.3–2.5	...	1 max	6 max	...	50–60	40–50
99.7	0.20	0.06	2.0–2.3	...	...	...	0.8	5–15	85–95
99.7	0.20	0.06	1.75–1.95	...	...	...	trace	5–15	85–95
99.7	0.20	0.06	1.25–1.45	...	...	...	trace	10 max	90 min
99.7	0.20	0.06	0.9–1.1	...	...	...	trace	10 max	90 min
99.7	0.20	0.06	0.65–0.75	...	...	...	trace	10 max	90 min

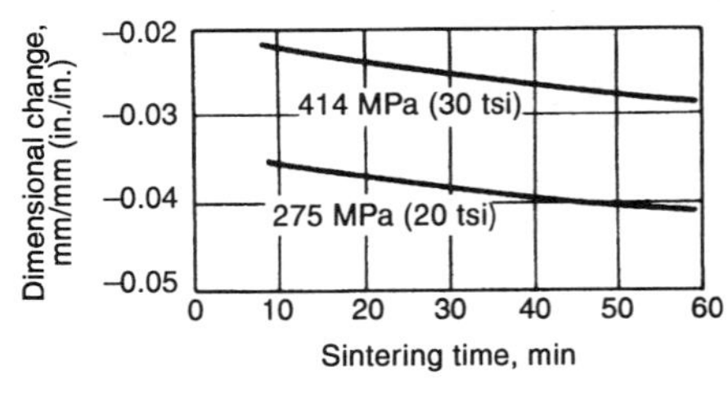

Fig. 14 Effect of sintering time on dimensional change. Source: Ref 9

Applications

Presently, electrolytic copper powder is produced in Germany, Italy, Russia, Japan, India, and Brazil where it is used in most copper powder applications (Ref 13). However, in the United States, since the early 1980s, when electrolytic copper powder production ceased, applications such as bronze bearings, copper-base friction materials and copper additions to iron mixes have, with a few exceptions, been converted to powders made by oxide reduction or water atomization. But electrolytic copper is still the powder of choice in many electrical and electronic applications, where high electrical and thermal conductivity is required. Using suitable fabricating procedures, parts can be produced with conductivities of 90% IACS and higher. Complex parts such as armature bearing blocks, contacts for circuit breakers, shading coils for contactors, heavy-duty contacts for circuit breakers, switch gear components for use in switch boxes with capacities up to 600 A, and components for 150 and 250 A fuse blowouts are in regular production. Heat sinks for diodes used in silicon rectifiers for the alternating current system in automobiles and electrode tools for electrical discharge machining also are produced from electrolytic copper powder.

Although recently-developed oxide-reduced grades with higher purity, lower apparent density, higher green strength (see Table 3) and higher surface area (up to 0.8 m^2/g), have found increased usage in the manufacture of electrical brushes, electrolytic copper is still the preferred powder, having a combination of purity, high electrical conductivity, high surface area and particle morphology (dendritic shape) which allow a large volume of graphite to be accommodated while maintaining a copper matrix with high conductivity and high strength.

Electrolytic copper powder is used with various nonmetallic materials to produce friction parts, such as brake bands or clutch disks. Premixes of iron-copper or iron-copper-carbon are employed in various automotive applications—cams, sprockets, gears, piston rings for small-bore engines—and similar service.

Production of Copper Powder by Atomization

Water atomization of copper—that is, the disintegration of a molten stream of high-quality copper with high-pressure water jets—produces copper powders of compacting-grade quality. The resulting dried powder then may be subjected to an elevated-temperature treatment that further modifies its characteristics and engineering properties.

Nearly spherical powders result from the atomization of liquid copper with inert gas or air. These powders are used in copper flake production and other specialty applications. Spherically shaped copper powders have insufficient green strength for use in conventional P/M. To render them compactible, their particle shape and morphology may be changed by deformation and/or oxidation and reduction, as described in the section "Production of Copper Powder by the Reduction of Copper Oxide" in this article. Particle shapes of gas- and water-atomized copper powder are shown in Fig. 18. The following section discusses details specific to the water atomization of copper and subsequent processing. For more information, see the article "Atomization" in this Volume.

Commercial Practice

Liquid copper is superheated to approximately 1150 to 1200 °C (2100 to 2200 °F), utilizing flow rates of 27 kg/min (60 lb/min) or more. Generally, to produce a predominantly −100 mesh powder, water pressures of 10 to 14 MPa (1500 to 2000 psi) are used. Atomization may be conducted in an air or inert (nitrogen gas) atmosphere.

During melting of the copper, impurity content must be controlled to achieve good fluidity and high conductivity (Fig. 1). The need for oxygen control at this stage of the process depends on subsequent processing and end use of the powder. In conventional refining of copper, poling and steam-generated hydrogen keep copper oxidation under control in accordance with the equilibrium curves shown in Fig. 19. High oxygen content tends to produce a more irregular powder, and subsequent reduction of the atomized powder further improves compactibility through agglomeration and pore generation. If the powder is used in the as-atomized condition, lower oxygen contents are generally preferred,

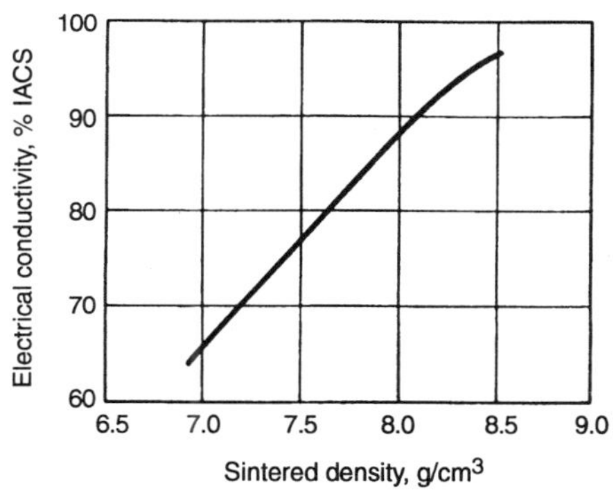

Fig. 16 Effect of sintered density on electrical conductivity. Source: Ref 9

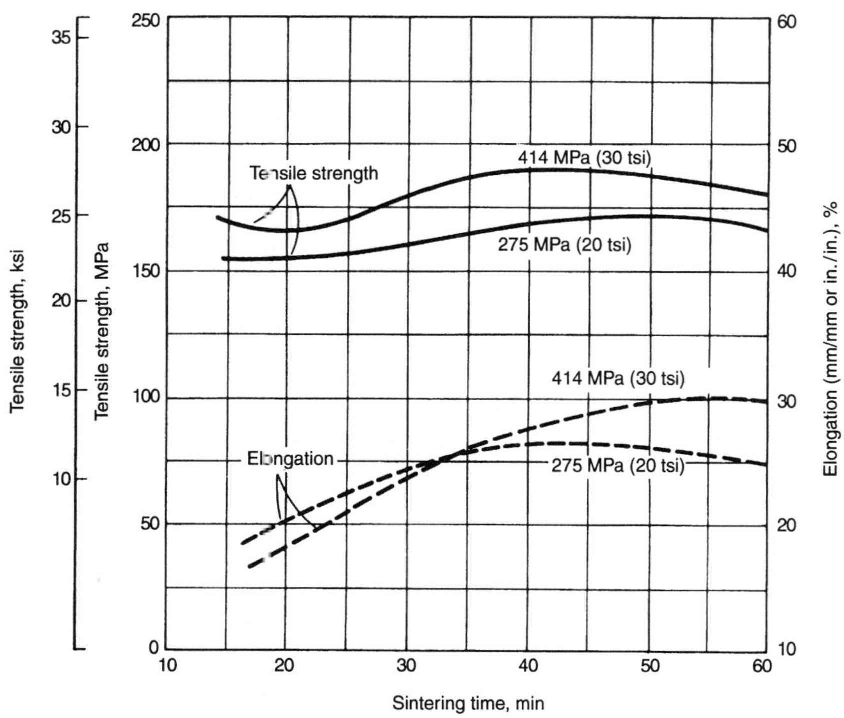

Fig. 15 Effect of compacting and sintering conditions on tensile properties. Source: Ref 9

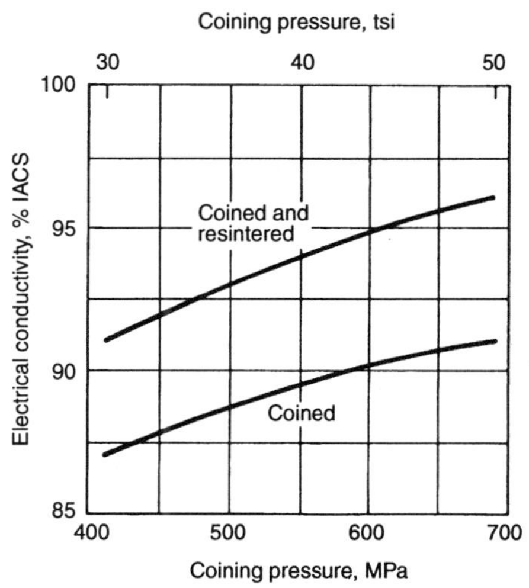

Fig. 17 Effect of coining and resintering on electrical conductivity. Source: Ref 9

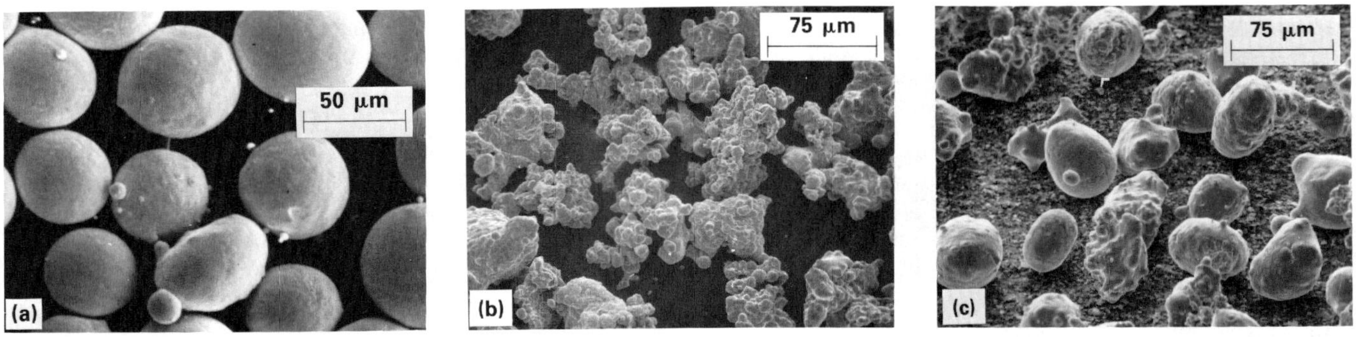

Fig. 18 Scanning electron micrographs of gas- and water-atomized copper powders. (a) Nitrogen atomized. (b) Water atomized, apparent density of 3.04 g/cm^3. (c) Water atomized, apparent density of 4.60 g/cm^3

because of the detrimental effects of oxygen in many applications.

The apparent densities of −100 mesh gas-atomized copper powders, as a consequence of their spherical particle shape, range between 4 and 5 g/cm^3. Oxygen picked up during atomization is present partly as surface oxide and partly as copper oxide throughout the bulk of a copper particle. Removal of oxygen requires reduction temperatures of approximately 700 °C (1290 °F) or higher. At these temperatures, considerable sintering occurs, which in turn requires substantial milling of the sinter cake.

During reduction, hydrogen readily diffuses through solid copper to react with oxygen and form steam. The large steam molecules, unable to diffuse through solid copper, force their way outward through grain boundaries—phenomenon known as hydrogen embrittlement of copper that manifests itself in the formation of blisters or cracks. Figure 20 illustrates the grain boundary widening of air-atomized copper particles due to this phenomenon. These defects improve both compactibility and sintering rate during liquid phase sintering of copper mixed with tin.

Alloying Additions

Some applications of copper powders require apparent densities lower than those attainable with water atomization of pure copper. These powders can be produced by addition of small amounts, up to 0.2%, of certain elements (for example, magnesium, calcium, titanium, and lithium) to the liquid copper prior to atomization (Fig. 21 and 22). These metals are believed to decrease the surface tension of copper and/or to form thin oxide films on the particle surface during atomization. Magnesium additions are most frequently used to produce compacting-grade copper powders for applications such as bronze bearings, filters, structural parts, and additives for iron powders. These powders may have apparent densities as low as 2 g/cm^3.

The addition of small amounts (0.1 to 0.3%) of phosphorus to the liquid copper, prior to atomization, allows the production of a powder that is very spherical and very low in oxygen. During atomization, even with air, the phosphorus oxidizes preferentially and forms protective gaseous phosphorus pentoxide (P_2O_5). Such powders have apparent densities up to approximately 5.5 g/cm^3. Spherical copper powders with closely controlled particle size ranges are used in applications such as thermal spray coatings, metal impregnated plastics, and heat exchangers. Irregular copper powders are used in compacting applications such as bronze mixes for self-lubricating bearings, additions to iron mixes, friction materials, electrical brushes, diamond cutting wheels, and electrical parts requiring high strength and electrical/thermal conductivity. Irregular copper powders are also used in copper brazing pastes and various chemical applications such as catalysts and in the production of copper compounds. Table 6 lists the properties of typical commercial atomized copper powders. The specific surface areas of these powders are from 0.02 m^2/g for coarse spherical gas atomized powder, to 0.2 m^2/g for fine water atomized powder.

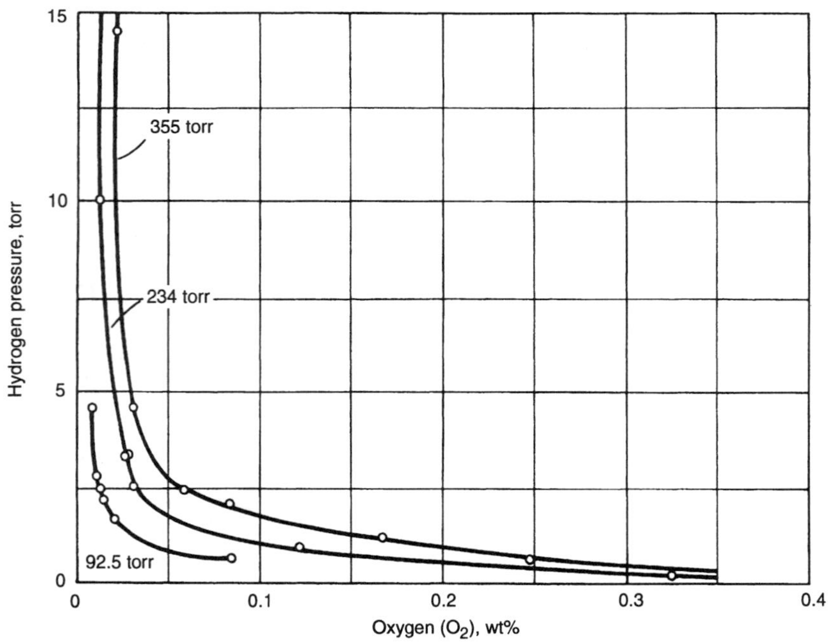

Fig. 19 Effect of partial pressures of hydrogen and steam on oxygen content of liquid copper at 1150 °C (2100 °F). Cu + $H_2 \leftrightarrow$ 2Cu + H_2O reaction at various water vapor pressures

Production of Copper Powder by Hydrometallurgical Processing

Hydrometallurgical methods may be used to produce a number of metal powders, including copper, cobalt, and nickel. The basic processing steps consist of preparing pregnant liquor by leaching ore or another suitable raw material,

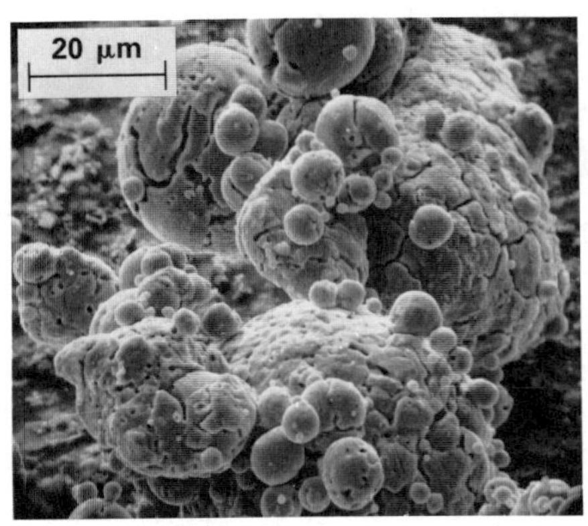

Fig. 20 Hydrogen-embrittled, air-atomized copper after reduction in hydrogen

Table 6 Properties of typical commercial grades of water and gas-atomized copper powders

Chemical properties			Physical properties							
				Sieve analysis, % Tyler						
Copper, % min	Hydrogen loss, % max	Acid insoluble, % max	Apparent density, g/cm^3	+60	+80	+100	+150	+200	+325	−325
99.0(a)	NA	NA	4.5–5.5	5 max	30–60	30–60	15 max	...	...	...
99.0(a)	NA	NA	4.5–5.5	...	2 max	20–50	50–75	10 max	trace	...
98.5(a)	0.7	NA	4.5–5.5	...	trace	0.2 max	5 max	2 max	bal	60–90
98.5(a)	0.7	NA	4.5–5.5	...	...	...	...	0.5 max	bal	95 min
99.3(b)	0.3	0.1	2.5–2.7	...	...	0.8 max	35 max	70 max	bal	5 max
99.3(b)	0.3	0.1	2.5–2.8	...	...	1 max	20 max	25 max	40 max	30–45
99.3(b)	0.3	0.1	2.5–2.8	...	...	0.5 max	10 max	20 max	bal	42–55
99.3(b)	0.3	0.1	2.8–3.0	...	...	trace	1 max	15 max	bal	55–65
99(c)	0.35	NA	2.1–2.4	...	...	5 max	15–25	10–20	15–35	20–40
99(c)	0.35	NA	2.3–2.6	...	...	1 max	10 max	5–20	15–30	60–70
99(b)	0.5	0.1	2.1–2.5	...	...	...	1 max	3 max	14 max	85 min

NA, not applicable. (a) Air atomized. (b) Water atomized/annealed. (c) Water atomized + Mg

followed by the precipitation of the metal from this solution. For copper, the most important precipitation methods are cementation, reduction with hydrogen or sulfur dioxide, and electrolysis. Use of several leach-precipitation steps or the inclusion of flotation, solvent extraction, or ion exchange improves the purity of the final material.

Despite several process development efforts in the 1950s and 1960s that led to radical improvements in purity and properties of hydrometallurgically produced copper powders, commercialization attempts in the United States have failed. Causes for this failure include the increasing cost of energy, coupled with the high energy requirements of some hydrometallurgical methods and the difficulty in economically producing powders with a wide range of properties suitable for various uses.

Cement copper, like other hydrometallurgically produced copper powders, has low apparent density and high specific surface area (1 m^2/g). The particles are spongy, because they are agglomerates of very small primary particles (Fig. 23). Although cement copper often has higher green strength characteristics than most other copper powders, its sintering activity, when used alone or in 90/10 bronze, is inferior because of the presence of finely divided, unreducible aluminum and silicon oxides. The primary use of cement copper is in composite friction material applications.

Leaching

The leaching of copper oxide and copper sulfide ores entails partial dissolution to cupric sulfate ($CuSO_4$) with sulfuric acid (H_2SO_4) plus iron sulfate ($FeSO_4$). The presence of pyrite (FeS_2) in many ore deposits, and its reaction with water and oxygen to form iron sulfate and sulfuric acid, represents an important source of acid. In dump leaching, the pH of the leach solution is maintained between 1.5 and 3.0, partly to protect bacteria that promote and accelerate the oxidation of pyrite and sulfidic copper minerals and also to avoid hydrolysis of iron salts.

The copper content of the pregnant solutions varies from less than 1 g/L (0.15 oz/gal) to several grams per liter (ounces per gallon); for vat leaching, it can be much higher. Other leaching methods include ammonia leaching, which is used for certain copper oxide ores, and the so-called leach-precipitation-flotation method, which is applied to mixed oxide sulfide ores.

Cementation

The precipitation of a metal from its solution by the addition of another less noble metal is known as cementation. The fundamental equation for copper recovery from copper-bearing pregnant liquors on iron is:

$$Fe + CuSO_4 = Cu + FeSO_4$$

In practice, the copper bearing solution is passed over scrap iron, such as detinned and shredded cans. Subsequent separation, washing, reduction, and pulverizing produces a copper powder that contains considerable amounts of iron and acid insolubles, such as alumina and silica. Contamination with gangue varies and depends on the nature of the pregnant liquor.

Significant reduction in iron and alumina levels is possible through the use of V-trough or inverted precipitators that allow faster precipitation rates and more efficient iron utilization. Ta-

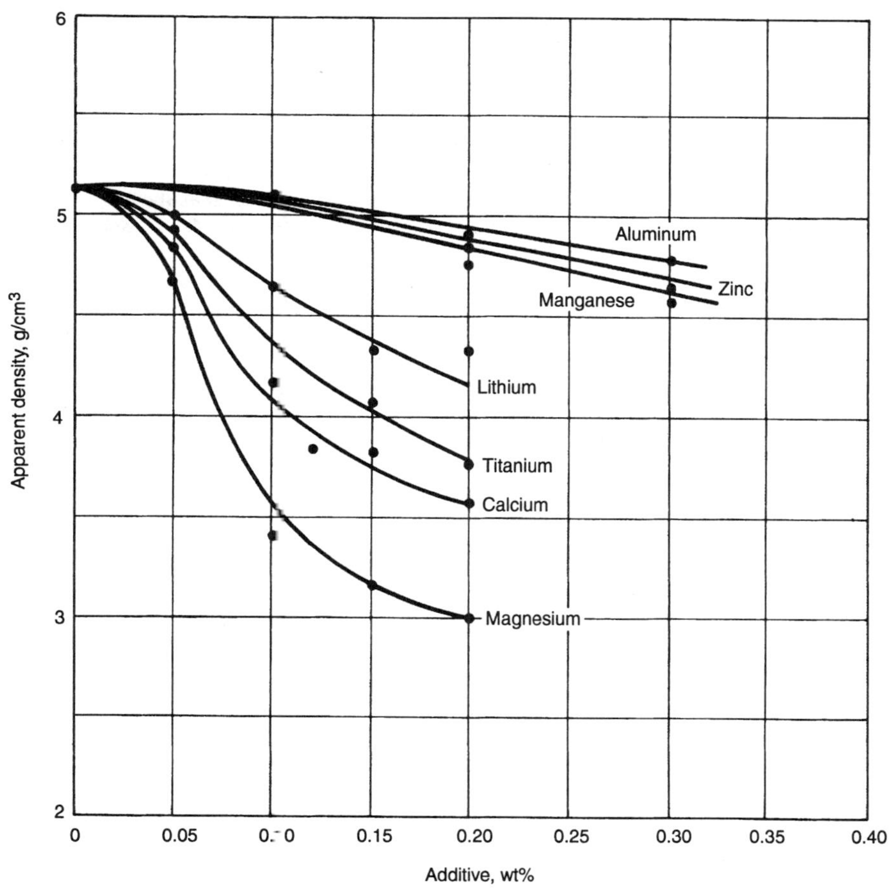

Fig. 21 Effect of additions to the molten copper on apparent density of atomized copper powder

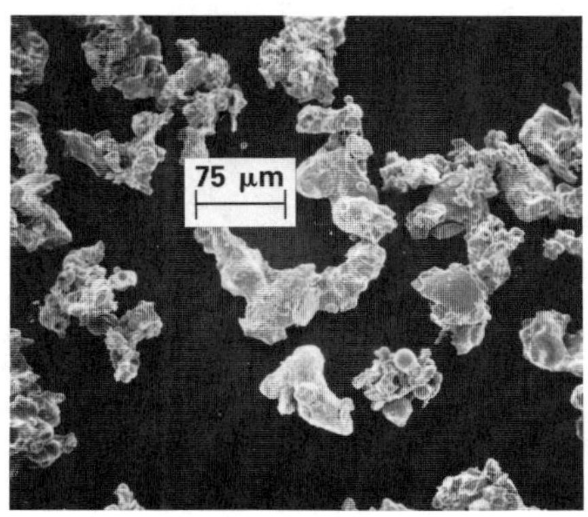

Fig. 22 Scanning electron micrograph of water-atomized copper containing 0.5% Li

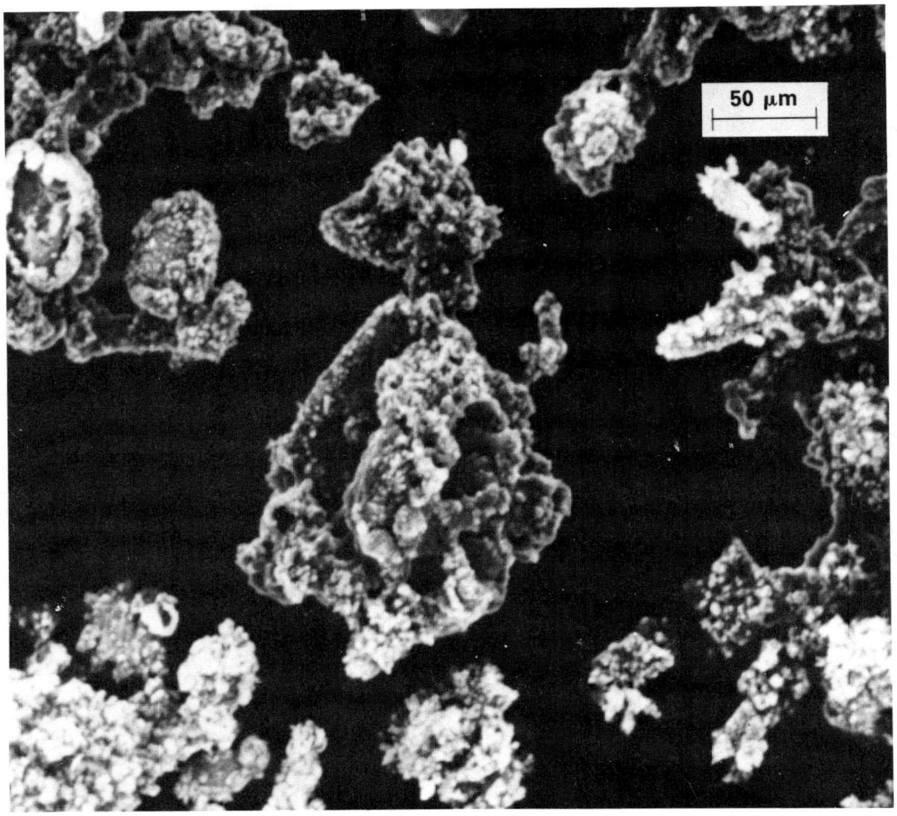

Fig. 23 Scanning electron micrograph of cement copper

Table 7 Chemical analyses of cement copper from various locations (dry basis)

Component	Composition, wt%, for location A	B	C	D
Total copper	75	83.0	87.4	85.0
Iron	6	2.4	0.7	10.0
Sulfur	1	0.5	...	1.1
Nitric acid insolubles	2	...	0.7	1.9
Hydrogen loss	16	...	...	...
Calcium oxide	...	0.08	...	...
Aluminum oxide (alumina)	...	1.2	0.5	...
Silicon dioxide	...	0.4	...	...
Lead	...	...	0.2	...
Oxygen	...	...	9.5	...

ble 7 shows chemical analyses of cement copper from different sources. Copper and iron are present, partly as oxides.

Additional Reduction Processes

Electrowinning may be used to recover copper from leaching solutions containing more than approximately 25 g/L (3.4 oz/gal) of copper (Harlan process). The electrolytic cells are equipped with insoluble lead-antimony anodes and 99% Ni cathodes. The copper powder does not adhere, but falls to the bottom of the cells. Temperature of the electrolyte is 60 °C (140 °F); cathode current density varies from 1350 to 2700 A/m^2 (125 to 250 A/ft^2). When the copper concentration falls below 15 g/L (2 oz/gal), the electrolyte is drained and used for ore leaching. Particle size of the powder prior to furnace processing is 1 to 25 μm, with excellent purity (>99.9%). The power requirements for electrowinning of copper are approximately ten times as large as those for electrorefining of copper with soluble anodes.

Copper may be concentrated from low-content leach solutions by solvent extraction, followed by stripping with dilute sulfuric acid into an aqueous solution and electrowinning. Carboxylic acid and hydroxylamine-based compounds have been found to be selective solvents of low water solubility, to have good stability, and to be compatible with inexpensive diluents. Direct powder precipitation with hydrogen or ammonia is an alternative to stripping the metal from the organic solvent into an aqueous solution.

Metals can be precipitated from their acid or basic solutions by reduction with hydrogen. Sulfuric acid, ammoniacal ammonium carbonate, and ammoniacal ammonium sulfate solutions have been used to produce copper powder by this method. Sulfuric acid leaching of a cement copper and hydrogen reduction of the filtered solution in an autoclave at 120 to 140 °C (250 to 280 °F) and 3 MPa (425 psi) is reported to produce a precipitate with a purity of about 100% Cu. Drying and furnace processing in a reducing atmosphere at 540 to 790 °C (1000 to 1450 °F) increases particle size, due to agglomeration of the very fine powder.

REFERENCES

1. E. Klar and D. Berry, Copper P/M Products, Vol 2, *Properties and Selection: Nonferrous Alloys and Pure Metals,* Vol 2, *ASM Handbook,* 9th ed., ASM International, 1979, p 392
2. S. Harper and A.A. Marks, Electrodeposition of Copper Powder with the Aid of Surfactants, *Copper Development Association/American Society for Metals Conf. on Copper,* Vol 3, Conference Paper No. 059/2, Copper Development Association, 1972
3. L.I. Gurevich and A.V. Pomosov, The Effect of Chloride on Electrodeposition of Powdered Copper Precipitates, *Sov. Powder Metall. Met. Ceram.,* Jan 1969, p 10–15
4. V.P. Artamonov and A.V. Pomosov, Effect of Foreign Electrolytes on the Production of Copper Powder by Contact Deposition, *Izv. V.U.Z. Tsvetn. Metall.,* No. 2, 1976, p 30–34 (in Russian); *Metall. Abstracts,* No. 54-0503, Nov 1976
5. S.L. Kotovskaya et al., Manufacture of Coarse Copper Powder from Sulfamate Electrolytes, *Sov. Powder Metall. Met. Ceram.,* Feb 1973, p 93–96
6. S.K. Singh and D.D. Akerkar, A Continuous Self-Regulating Method of Making Copper Powder by Electrolysis, *NML Tech. J.,* Vol 17, 1975, p 23–26
7. F. Willis and E.J. Clugston, Production of Electrolytic Copper Powder, *J. Electrochem. Soc.,* Vol 106, 1959, p 362–366
8. A.V. Pomosov, M.I. Numberg, and E.G. Krymakove, Protection of Copper Powder Against Corrosion during Manufacture and Storage, *Sov. Powder Metall. Met. Ceram.,* March 1976, p 175–177
9. Technical data from AMAX Metal Powders, AMAX Copper, Inc., 1968
10. D.L. Adamson and W.M. Toddenham, "Production of High Quality Electrolytic Copper Powder," American Institute of Mining, Metallurgical and Petroleum Engineers Annual Meeting (New York), 1–4 March 1971
11. E. Peissker, *Metal Powders,* Norddeutsche Affinerie, Aug 1974
12. P.W. Taubenblat, W.E. Smith, and C.E. Evans, Production of P/M Parts from Copper Powders, *Precis. Met.,* April 1972, p 41
13. D.E. Tyler and W.T. Black, Introduction to Copper and Copper Alloys, Vol 2, *Properties and Selection: Nonferrous Alloys and Special-Purpose Materials, ASM Handbook,* ASM International, 1990, p 216–240
14. D. Kumar and A.K. Gaur, Electrochemical Studies on Production of Electrolytic Copper Powders, *J. Electrochem. Soc. India,* July 1973, p 211–216

Production of Copper Alloy Powders

Thomas W. Pelletiers, II, United States Bronze Powders, Inc.
David F. Berry, OMG Americas

COMMERCIAL COPPER ALLOY POWDERS, including brasses, bronzes, and nickel silvers, are manufactured in a similar manner. Usually, the same integrated manufacturing facilities are used to complete the melting process, atomization, final screening, and blending of a lot or batch.

The powder-producing process is similar to a foundry operation in which high-purity virgin metals are charged to a melting furnace in preweighed batches to be processed at predetermined heating rates and times. High-purity raw materials are required, because melting produces minimal refinement (only a partial transfer of contaminants from the melt to the formed slag occurs). To ensure continuity, homogeneity, and uninterrupted atomization, the molten batch of alloy is transferred to a second furnace with a greater holding capacity than the primary furnace melting rate. Induction heating is preferred for at least one furnace to ensure the constant induced metal movement required for alloy homogeneity and to ensure uniform lead dispersion in lead-bearing alloys.

Atomization is achieved by particulation of a controlled, constant-flowing, molten stream emitted from the secondary furnace by medium-pressure dry air. A typical melting and atomizing sequence is shown in Fig. 1. Subsequent reduction of oxides is not required for standard P/M grades.

Air-cooled, atomized powder is collected and passed over a primary control screen (sieve) to remove oversize particles. Usually, these are remelted concurrently during the processing of each alloy. Finally, the screened alloy powder may be blended with dry organic lubricants,

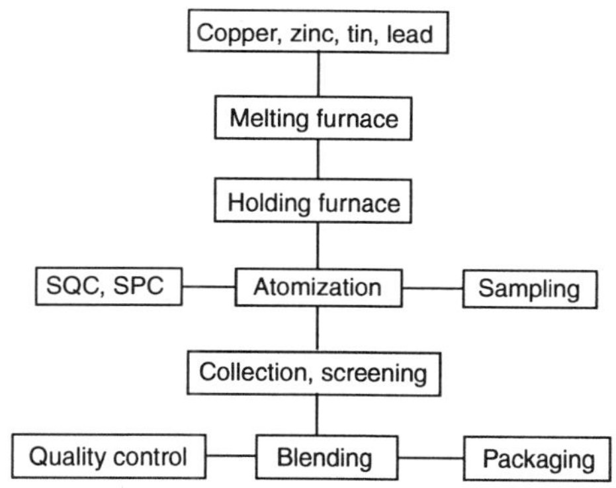

Fig. 1 Flowchart for copper alloy powder air atomization. SQC, statistical quality control; SPC, statistical process control

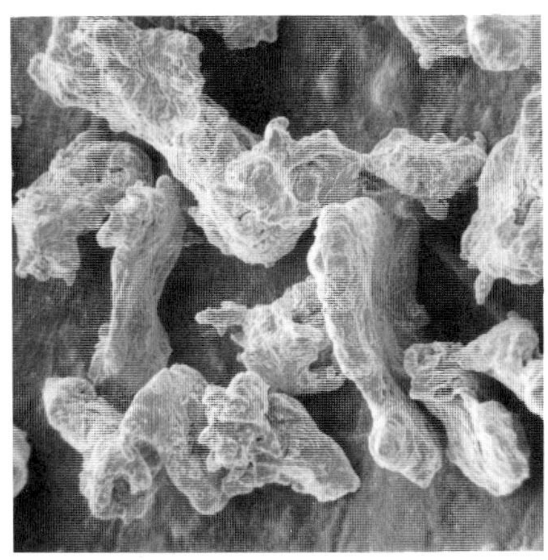

Fig. 2 Scanning electron micrograph of prealloyed, air-atomized brass (80%Cu-18%Zn-2%Pb). 165×

Table 1 Physical properties of typical brass, bronze, and nickel silver alloy compositions

Property	Brass(a)	Bronze(a)	Nickel Silver(a)(b)
Sieve analysis, %			
+100 mesh	2.0 max	2.0 max	2.0 max
−100 + 200	15–35	15–35	15–35
−200 + 325	15–35	15–35	15–35
−325	60 max	60 max	60 max
Physical properties			
Apparent density	3.0–3.2	3.3–3.5	3.0–3.2
Flow rate, s/50 g	24–26	...	...
Mechanical properties			
Compressibility(c) at 414 MPa (300 tsi), g/cm^3	7.6	7.4	7.6
Green strength(c) at 414 MPa (300 tsi), MPa (psi)	10–12 (1500–1700)	10–12 (1500–1700)	9.6–11 (1400–1600)

(a) Nominal mesh sizes; brass, −60 mesh; bronze, −60 mesh; nickel silver, −100 mesh. (b) Contains no lead. (c) Compressibility and green strength data of powders lubricated with 0.5% lithium stearate

Table 2 Properties of four grades of filter materials produced by loose powder sintering spherical powders

Particle size of spherical powder particles		Tensile strength		Recommended minimum filter thickness		Largest dimensions of particles retained, μm	Viscous permeability coefficient, m^2
Mesh range	Range in μm	MPa	ksi	mm	in.		
20–30	850–600	20–22	2.9–3.2	3.2	0.125	50–250	2.5×10^{-4}
30–40	600–425	25–28	3.6–4.1	2.4	0.095	25–50	1×10^{-4}
40–60	425–250	33–35	4.8–5.1	1.6	0.063	12–25	2.7×10^{-5}
80–120	180–125	33–35	4.8–5.1	1.6	0.063	2.5–12	9×10^{-6}

Source: Ref 1

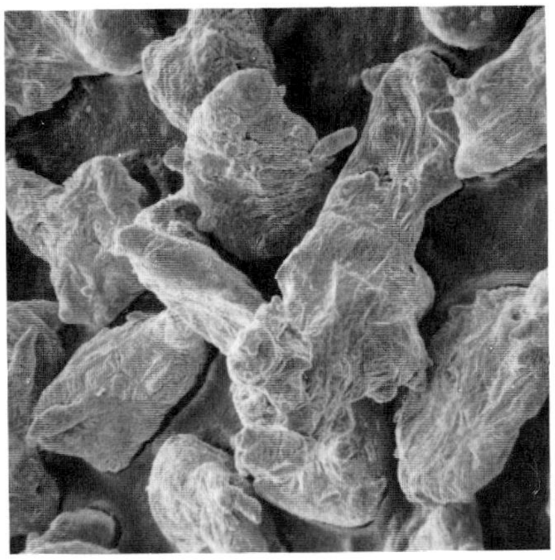

Fig. 3 Scanning electron micrograph of prealloyed, air-atomized bronze (89%Cu-9%Sn-2%Zn). 165×

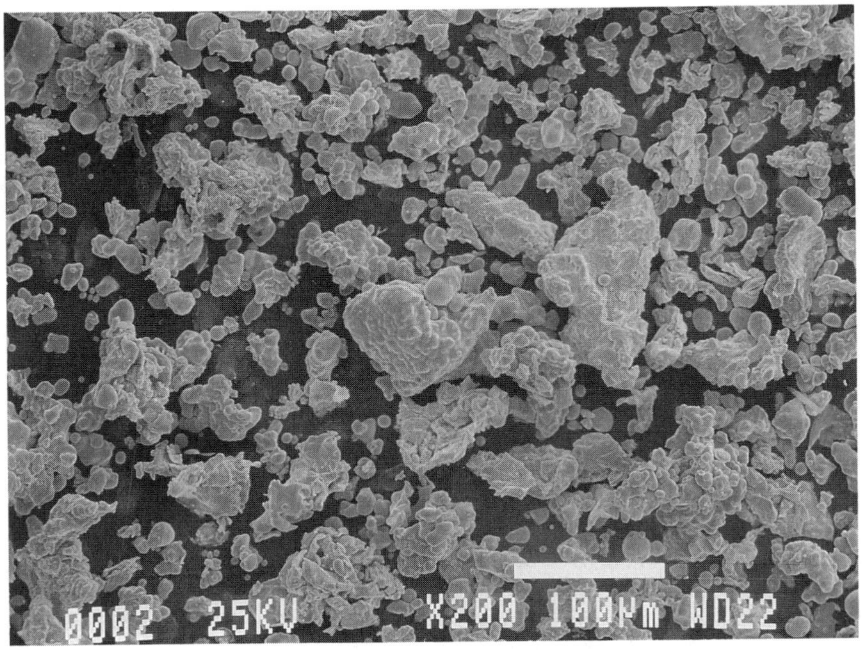

Fig. 4 Scanning electron micrograph of a typical prealloyed water-atomized bronze powder (90%Cu-10%Sn); apparent density 3.4 g/cm^3. 200×

such as lithium and zinc stearate, for use in the manufacture of P/M structural components.

Adjustment of powder properties (particle size distribution, apparent density, green strength, etc.) of each alloy is accomplished by controlled manipulation of the atomizing process conditions (e.g., atomizing air flow rate, molten metal temperature, nozzle configuration, etc.). Powder properties are maintained by periodically recording the atomizing parameters and by evaluating the properties of representative samples from the in-process atomized product.

A variety of copper-base alloys can be manufactured by the atomizing process; however, commercial P/M applications normally are confined to a rather narrow range of specific, single-phase (α) compositions.

Brasses

Brasses comprise the major portion of copper-base alloy powders used for parts fabrication, in comparison with prealloyed bronze and nickel silver materials. Typical copper-zinc brass powders contain zinc contents ranging from 10 to 30%. Lead may be added in small amounts (1 to 2%) to improve machinability of the sintered compact. A photomicrograph of an 80%Cu-18%Zn-2%Pb alloy powder is shown in Fig. 2. The melting temperatures of these alloys range from 1045 °C (1910 °F) for 90%Cu-10%Zn to 960 °C (1760 °F) for 70%Cu-30%Zn. As zinc content increases, melting temperature is lowered.

Additional superheat, or the temperature in excess of the alloy melting temperature, depends on heat losses within the manufacturing system and the physical property requirements of the atomized powders. Typical physical properties of brass alloy compositions are given in Table 1.

Bronzes

Prealloyed atomized bronze compositions are not used extensively as base powders for compacted parts fabrication because of their nodular particle form and high apparent density, both of which contribute to poor compacted green strength. Common prealloyed compositions are 90%Cu-10%Sn and 85%Cu-15%Sn, prepared in the same manner as brass powder, except that high-purity elemental copper and tin are used. A scanning electron micrograph of an 89%Cu-9%Sn-2%Zn alloy powder is shown in Fig. 3. Typical physical properties of a bronze alloy composition are given in Table 1.

Prealloyed bronze powders are also made commercially by water atomizing. Application is more extensive in Europe, where 90/10 prealloyed bronze powders are incorporated in bronze premixes for bearing manufacture. Low green strength due to high apparent density (3.2 to 3.6 g/cm^3) is overcome by incorporating lower apparent density copper powders (see previous section on copper powder manufacture) and choice of lubricants that have a less deleterious effect on green strength. Physical properties are similar to air-atomized powders, but particle morphology is different (Fig. 4). Powders contain 0.1 to 0.2% P to aid sintering.

Spherical 89/11 bronze powders are used to make filters. These are made by horizontal air atomizing and dry collection. The spherical shape is achieved by addition of small amounts of phosphorus, 0.2 to 0.45% (in the form of a

Fig. 5 Assorted filters made from P/M bronze. Courtesy of Arrow Pneumatics, Inc.

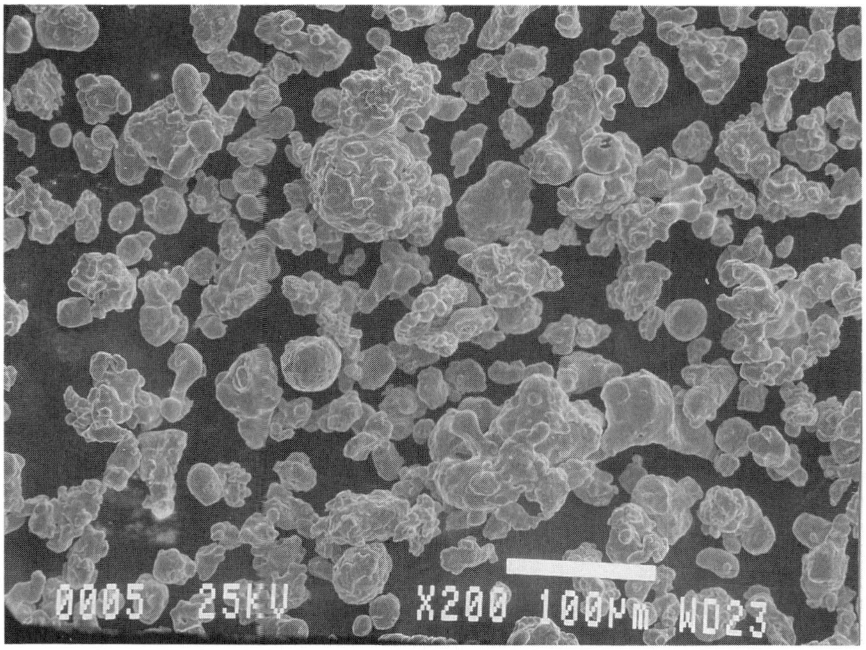

Fig. 6 Scanning electron micrograph of a typical diffusion-alloyed bronze powder (90%Cu-10%Sn); apparent density 2.6 g/cm^3. 200×

Cu/15% P alloy) to the molten bronze prior to atomizing. During air atomizing, surface oxidation of atomized molten particles of bronze and brass, which cause them to solidify in an irregular shape (see Fig. 3) is prevented. The oxygen in the air preferentially reacts with phosphorus to form phosphorus pentoxide (P_2O_5), which is volatile at atomizing temperatures.

The spherical powders are screened to produce a number of grades, each with a narrow particle size range. An assortment of filters made from bronze powders and properties of four grades of filters are shown in Fig. 5 and Table 2.

Microbearings. A more recent development has been prealloyed bronze powders for microbearings. These are very small bearings, often weighing less than 1 g, used in electronic equipment such as computers, audiocassette players, and videocassette recorders. Most powders used in this application are made by diffusion alloying tin into copper powder to produce a substantially alloyed powder with particles of uniform composition and a particle shape that gives high green strength (apparent density of 2.3 to 2.7 g/cm^3). As previously stated, water-atomized bronze powders have relatively high apparent density (3.2 to 3.6 g/cm^3) and are usually limited to application in high-density structural parts. Figure 6 shows the particle morphology of a typical diffusion-alloyed bronze powder.

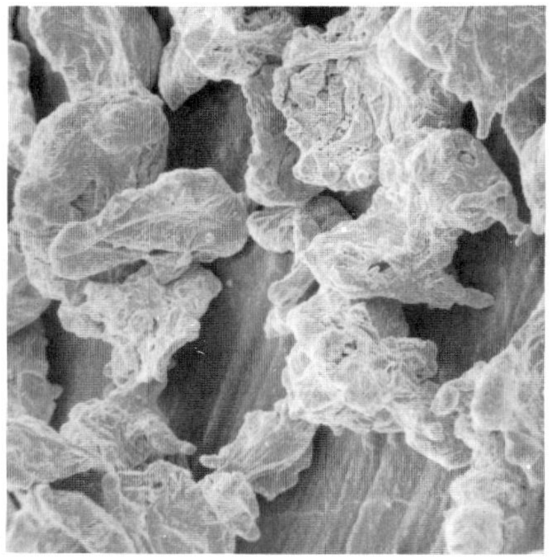

Fig. 7 Scanning electron micrograph of prealloyed, air-atomized nickel silver (63%Cu-18%Ni-17%Zn-2%Pb). 165×

Nickel Silvers

Only one nickel-silver-base alloy composition, 65%Cu-18%Ni-17%Zn, is commonly used in the P/M industry. This alloy can be modified by the addition of lead to improve machinability. Foundry practices are similar to those employed for brasses, except the melting temperature is in excess of 1093 °C (2000 °F). A micrograph of a 63%Cu-18%Ni-17%Zn-2%Pb alloy powder is shown in Fig. 7. Typical physical properties of a nickel silver alloy composition are given in Table 1.

REFERENCE

1. F.R. Lenel, *Powder Metallurgy Principles and Applications,* Metal Powder Industries, 1980

Production of Tin Powders

William J. Ullrich, ACuPowder International, LLC

TIN POWDERS are used extensively in the production of porous, self-lubricating bronze bearings and as constituents in soldering and brazing pastes and powders. Low-melting-point tin powders normally are produced by air atomization, although other methods of production such as chemical precipitation and electrodeposition have been used (Ref. 1). Tin powders also are used in powder metallurgy (P/M) structural parts, friction disks, clutches, brake linings, metal graphite brushes, diamond abrasive grinding wheels, bronze filters, peen plating, plasma arc spraying, chemical formulations, additives for rubber and plastics, chemical manufacturing, smokeless powder for pyrotechnics, and tin flake.

Melting

High-purity pig tin is melted in a gas-fired or electrically heated crucible (cast iron, clay, graphite, or ceramic). Melt temperature is controlled to maintain the desired degree of superheat above the low melting point of tin (232 °C, or 449 °F).

Atomization

Atomization results in a fine powder whose average particle size can be regulated over a wide range. Resultant powders are of high purity. Excessive oxidation, characteristic of other atomized products, does not occur because of the rapid chilling effect of the expanding gases released through the nozzle Thus, the oxygen content of atomized tin is normally below 0.2%. The thin film of oxide produced during atomization with steam or air is sufficient to inhibit further oxidation of the particles.

The molten tin can be atomized vertically (up or down) or horizontally, depending on the nozzle direction, to produce a powder (Fig. 1). There are two atomization techniques: annular nozzle and cross-jet atomization. In an annular nozzle, the gas stream aspirates liquid tin into the nozzle, where it is disintegrated into tiny droplets by the high-velocity gas stream. In cross-jet atomization, the gas is at right angles to the molten tin stream. This method usually produces coarser particles than annular nozzles.

To atomize uniformly fine tin powders, temperature, stream diameter, and flow rate of the molten tin and temperature, pressure, velocity, and angle of impingement of the atomizing gas must be well controlled. Pressures commonly employed to atomize tin powder range from 345 to 1725 kPa (50 to 250 psi). Generally, finer powders require the higher pressure; however, the exact pressure is related to the nozzle design used.

Compressed air is usually used as an atomizing medium. It is frequently preheated in a gas-fired heat exchanger to prevent solidification of tin within or around the orifice caused by the chilling effect of expanding air as it is released through the nozzle. Nozzles are designed to facilitate the atomization of several different particle size ranges, usually by changing orifice diameter and air pressure.

A blower (fan) at the end of the system pulls the atomized tin powder from the atomization chamber into a cyclone, where it is collected. The finest particles, which have not settled in the cyclone, are then retained in the cyclone filters. The tin powder collected in the cyclone is sieved to remove the oversized particles, most commonly +100, +200, or +325 mesh.

A uniform mixture of tin powder of the desired lot size is produced by tumbling in a blender. A sample of the blended tin is then analyzed for physical and chemical properties. The tin powder is packed in steel or waterproof fiber containers weighing up to 320 kg (700 lb).

Properties of Tin Powder

The high-purity virgin tin melt stock meets the requirements of ASTM grade:

Element	Composition (max), %
Tin	99.8(a)
Antimony	0.04
Arsenic	0.05
Lead	0.05
Copper	0.04
Bismuth	0.015
Iron	0.015
Sulfur	0.01

(a) Minimum (99.9% typical)

Impurities in the tin melt stock carry over to the powder and usually are measured by atomic absorption spectroscopy. The tin powder purity is directly related to the purity of the ingot melted.

Table 1 Commercially available tin powders

	Sample A	Sample B	Sample C	Sample D	Sample E	Sample F	Sample G	Sample H
Sieve analysis, %								
+100 mesh	48.0	0.2	0.5 max	0.3 max	0.3 max	...	...	...
+150 mesh	46.5	5.2	5 max	3 max	2 max	...	...	...
+200 mesh	4.0	11.3	4–12	3–8	5 max	0.1	...	...
+325 mesh	1.0	29.0	15–30	12–25	2–8	...	2 max	...
−325 mesh	0.5	53.6	65–75	70–85	90 min	96 min	96 min	98 min
Fisher subsieve size, average particle size, μm	...	...	12–18	10–15	8–11	8–10	7–9	1–3
Apparent density, g/cm³	3.35	3.90	3.7–4.2	3.7–4.2	3.3–3.8	3.0–3.5	3.0–3.5	1.3–2.0

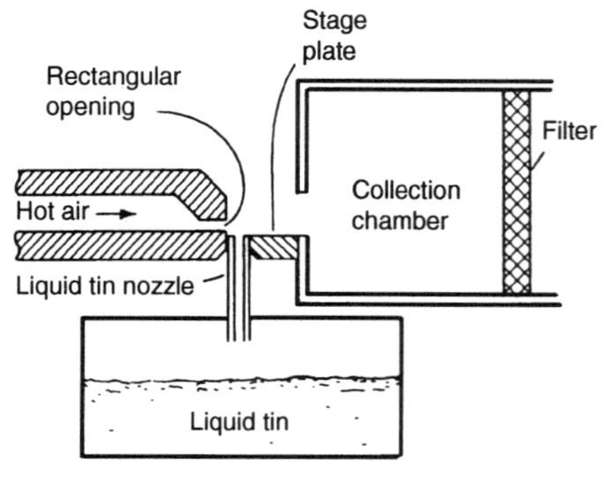

Fig. 1 Schematic of apparatus for atomizing tin powder

Fig. 2 Scanning electron micrograph of air-atomized tin powder. Magnification 500×

During atomization, a thin oxide film, which inhibits further oxidation of the particles, forms on the particle surfaces. Oxygen pickup commonly is measured according to MPIF 02, "Method for Determination of Hydrogen Loss of Metal Powder," published by the Metal Powder Industries Federation. A typical oxygen content of tin powder determined by this method is 0.05%.

Physical properties of tin powder depend on atomization conditions and sieving procedures. Coarser grades of tin powder are atomized from larger metal streams using lower air pressures. The finer grades of powder require higher pressures and small molten metal streams. Sieving is an independent means of controlling the size of the largest particles in the powder. Generally, 100-, 200-, or 325-mesh sieves are used to remove oversize particles.

Particle Size Distribution. Most grades of tin powder are classified by particle size. Methods of measuring particle size distribution include:

- Sieve analysis, as specified by ASTM B 214 and MPIF 5. Usually 100-, 150-, 200-, and 325-mesh sieves are used.
- Fisher subsieve size (FSSS), as specified by ASTM B 330 and MPIF 32, provides a quick measure of the average particle size of the subsieve (–325 mesh) fraction of fine tin powder.
- Other particle size analysis methods, such as the use of micromesh sieves, light scattering, sedimentation, and electrical and optical sensing zone instruments, are used to measure subsieve size particles.

Table 1 gives sieve analysis data and Fisher subsieve size diameters for several commercially available tin powders.

Apparent density varies with particle size and shape. It is measured as specified by ASTM B 212 and B 417 and MPIF 4 and 28. Apparent densities of some commercially available tin powders are given in Table 1.

Surface area usually is measured by inert gas adsorption, or Brunauer-Emmett-Teller (BET) techniques. The specific surface area of the tin powder varies between approximately 1.0 m^2/g for the finest grade to 0.1 m^2/g for the coarsest grade.

Particle shape and surface topography can be observed by light or scanning electron microscopy. Air-atomized tin particles are rounded to spherical in shape with relatively smooth surfaces. Figure 2 is a scanning electron micrograph of atomized tin powder particles displaying typical rounded shape.

REFERENCE

1. C.G. Goetzel, *Treatise on Powder Metallurgy,* Vol 1, Interscience, 1949, p 199–200

Production of Aluminum and Aluminum-Alloy Powder

A. Ünal, D.D. Leon, T.B. Gurganus, and G.J. Hildeman, Aluminum Company of America

ALUMINUM POWDERS and ALUMINUM alloys are produced almost exclusively by gas atomization. For most applications, the atomizing gas is compressed air; however, in some special cases, inert gases (helium, nitrogen, and argon) are used. Between 25 and 30 countries in the world are known to have production facilities for aluminum powder at an estimated total capacity of 200,000 tons per year (Ref 1). A large portion of this capacity is in North America where annual shipments are about 40,000 tons (Ref 2). Atomized aluminum powders are used in a variety of applications that include pyrotechnics, explosives, rocket fuel, thermite welding, aluminothermic reduction, chemical processes (as catalyst or reagent), additives for lightweight concrete, pharmaceuticals, and pigments for paints and printing inks. With few exceptions, the demand is for unalloyed powder of various standards of purity. Aluminum alloy P/M, historically a small percentage (about 1%) of the total powder market, has been enjoying a revival in recent years in automotive applications because of the need to reduce weight, lower emissions, and boost fuel economy. Viable P/M applications include engine cam caps and air conditioning compressor parts. Powder metallurgy offers competition to wrought aluminum castings and stampings because of its net shape advantage.

Historical Background

Aluminum powder was first used commercially as flakelike pigment products called aluminum bronze powder in the United States around 1900. It was made from aluminum sheet by the Bessemer dry stamping process developed fifty years earlier for the manufacture of gold powders. Stamping mills were both inefficient and dangerous to operate, because, unlike gold powders, aluminum powder forms an explosive mixture with air over a wide range of metal-to-air ratios (Ref 3). Several fatal fires and explosions marred the early days of the aluminum powder industry.

Two major breakthroughs developed by Hall (Ref 4) occurred in aluminum powder production in the late 1920s. First, the development of safer ball mill production processes that are still employed today for milling powders into flakes. Second, the introduction of the atomization process, which allowed the manufacture of the forerunners of today's atomized powders. Atomized aluminum powder was used initially only as raw material to produce aluminum flake pigments by ball milling. After World War II and the development of aluminized high explosives, a major market for atomized aluminum was created. The application of atomized powders broadened in postwar years from military explosives to commercial markets listed above.

To meet the individual needs of each of these markets, manufacturers now produce many grades of aluminum powders in several different general categories: granules, regular and coated atomized powders, spherical powders, high-purity powders, alloy powders, blended powders, and dedusted atomized powders. All are produced by the same basic gas atomization technique, but starting materials may differ and, as with flake powders, further processing may be required. Powders atomized in inert gas (nitrogen, argon, and helium), which have a spherical shape, are preferred for some applications. Water atomization of aluminum is not practiced because it is dangerous, due to the creation of hydrogen and the potential presence of rust in the equipment. The latter can lead to explosions through thermite reaction with the dry powder. Some manufacturers still use the dry stamping process for converting foil scrap into coarse particulate (granules), although it is now a very small percentage.

Gas Atomization

There are many methods that have been used to produce aluminum and aluminum alloy powders including gas and centrifugal atomization, ultrasonic or pulsed atomization, melt spinning with attrition, and mechanically alloying. Gas atomization is used for several applications.

In gas atomization, a high-velocity gas jet disintegrates molten aluminum into droplets that solidify to form the powder. In the Alcoa process (Fig. 1), the material flow during atomization is vertically upward (also known as "updraught"). Molten metal of the appropriate composition is supplied from a holding or melting furnace at the required temperature to the atomizing bay. The liquid is drawn from the bay through a liquid delivery tube into the atomizing nozzle. This is achieved by the aspirating effect (suction) caused at the nozzle end of the delivery tube by the flow of the high-pressure atomizing gas in the nozzle. When the liquid metal meets the high-velocity gas, it is broken up into droplets and sprayed as a jet. The droplets are quenched by the gaseous atmosphere in the chamber to solidify as powder particles. These particles, together with a substantial volume of cooling air, are then drawn through a chiller chamber into the collection system consisting of two sets of cyclones. After the cyclones, the powder is transported in an atmosphere of inert gas to the screens and pack-outs where they are packed under inert gas into drums and bins or onto railroad cars or trucks.

Because aluminum powder forms an explosive mixture with air over a wide range of metal-to-air ratios, aluminum powder plants have explosion hazards associated with them. The Alcoa process (Fig. 1) incorporates several safety features. The bottom of the chiller chamber is closed rather than open to prevent ingress of combustible dust and foreign matter into the atomizing zone. Cooling air entering the chiller chamber is filtered in order to remove foreign matter and ignition sources, such as rust particles. The chiller chamber walls are reinforced. Explosion vents are located in the chiller chamber, along the exhaust duct, and in other critical locations, such as the lids of the cyclone abatement chambers. These vents open up and minimize pressure buildup in the event of an explosion. To reduce explosion hazards between the atomizing nozzle and the cyclones, the airborne

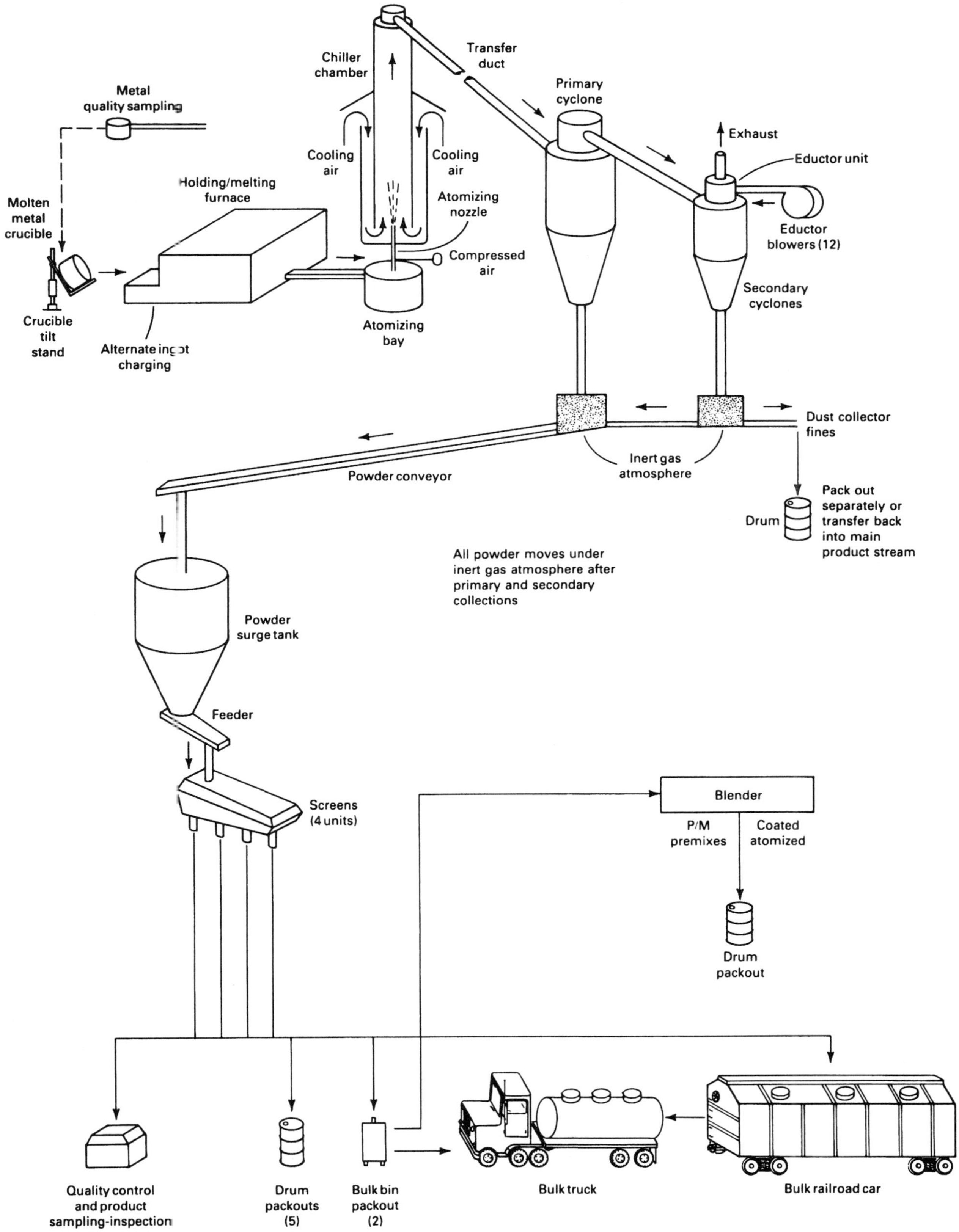

Fig. 1 Schematic of the Alcoa process for atomizing aluminum powder. Entire operation is under computer control. Powder is packed in drums or bins or is loaded for bulk shipment in trucks or railroad cars.

concentration of aluminum powder is kept to a level below which forms an explosive mixture (lower explosive limit, LEL). Airflow in the system is created by means of an eductor, so that powder does not have to travel through a rotating fan, which eliminates a significant static electricity ignition source. The whole plant is built in stainless steel and fully grounded to eliminate the danger of sparking and static buildup. The

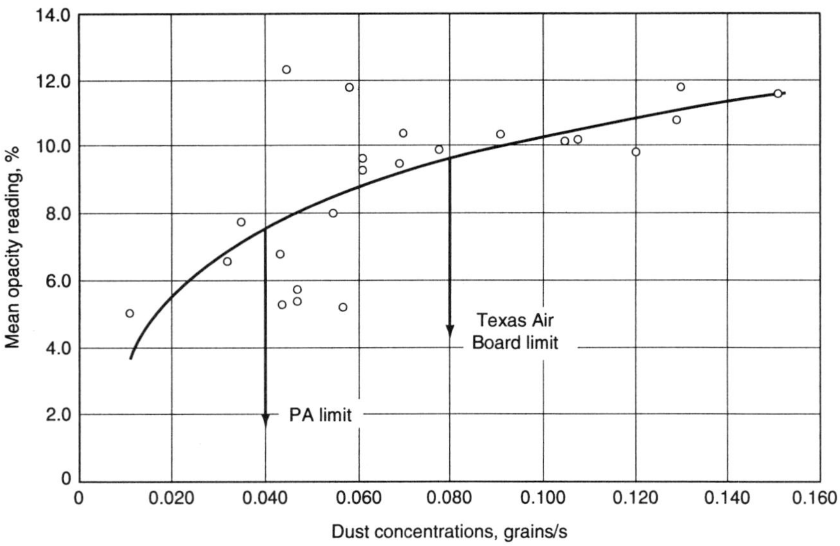

Fig. 2 Dust concentration and mean opacity readings measured in a pilot aluminum atomizer

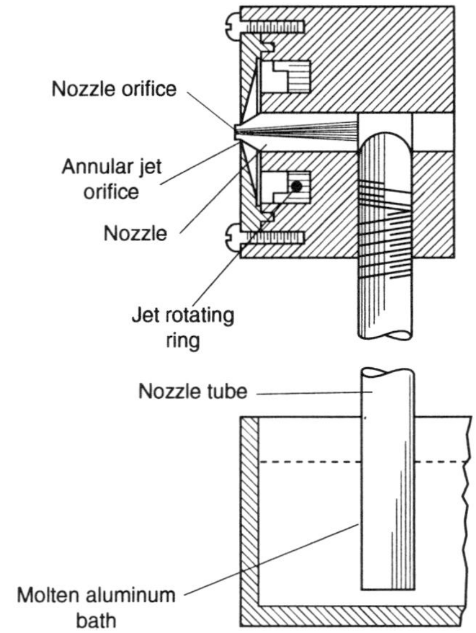

Fig. 3 Aspirating nozzle design developed by E.J. Hall in 1920s. Source: Ref 4

transport of powder after the cyclones is done in an atmosphere of inert gas, as noted above, to substantially reduce the explosion hazard. A certain level of oxygen is maintained in the inert gas atmosphere to ensure that the powder does not become pyrophoric.

Workers are removed entirely from areas containing the greatest hazards by substitution of computerized controls, remote TV surveillance, and automated operations. Conductive shoes are provided to workers in powder handling areas. Good housekeeping and training of operators are essential elements of safe operation in an aluminum powder plant. Regular audits of plant equipment and operating practices are carried out to ensure compliance. Areas of improvement identified in such audits and by risk assessment procedures are implemented on a timely basis.

The industry has recognized the hazards of aluminum powder. Manufacturers and users of aluminum powders cooperate in reporting and analyzing incidents through the auspices of The Aluminum Association. The Association holds safety workshops and supports research relating to safety funded by the members.

Control of particulate emissions in industrial plants is an important criterion. In the United States, these requirements vary from state to state. Pennsylvania has one of the strictest specifications limiting particulate emissions rate to 0.09 g/Nm^3 (0.04 grains/ft^3) of dry process air for aluminum powder plants. To meet these stringent requirements, manufacturers of aluminum powders often collect the powder in two stages of cycloning followed by a baghouse. Opacity of the plume from the plant is monitored and recorded. The upper limit for opacity is often specified at 10% max. These two specifications of emissions do not in general correspond to each other for aluminum powder plants (Fig. 2). This may be attributed to fine particles (<1 μm) causing a disproportionately high obscuration of the sunlight in the plume than coarser particles would cause.

Vertically upward aspirating processes, such as that operated by Alcoa, permit better control of the metal delivery rate to the nozzle and thus the production of the widest range of powder grades: fine, medium, and coarse. Aluminum powder atomizing can be accomplished also by directing the molten metal stream either horizontally or vertically downward. Systems that utilize metal-head pressure to induce molten metal flow through the nozzle often atomize either vertically downward or horizontally. Aspirating systems, on the other hand, usually prefer to atomize horizontally or vertically upward. Multiple nozzles are often used in commercial operation to achieve the required rates of production.

Atomizing Nozzle Technology. The first nozzle used for atomizing aluminum powders was invented by Hall in the late 1920s (Ref 4). In the Hall design, the atomizing gas (compressed air) is delivered through an annular orifice around the nozzle at a converging angle, as shown in Fig. 3. The gas flow creates a suction (aspiration) effect at the tip of the nozzle that is utilized to draw the molten metal into the nozzle. The amount of gas delivered by the nozzle is controlled by the size of the air gap, and pressure and temperature of the gas. Rate of metal flow and resultant powder particle size are influenced by the aspirating force, the nozzle metal orifice diameter, and the vertical distance between the nozzle and the molten metal level.

The Hall nozzle is a good example of the general class of confined nozzles, or close-coupled nozzles. This nozzle design aims to contact the liquid metal with the gas at or close to the gas jet exit plane, thus making use of the high jet velocities at that point. It leads to efficient breakup of the liquid, resulting in finer powders, and is the preferred method for aluminum and rapidly solidified fine powders of other metals. It can be operated vertically upwards or downwards or in the horizontal position. Careful design of the nozzle is required, however, to avoid the quenching of the metal by the gas jet, which can lead to premature solidification at the nozzle tip.

In the operation of the confined design nozzles, it has been observed that the molten metal presents itself as a thin film to the gas jet. This filming effect, which is beneficial for efficient breakup of the liquid, occurs from the gas flow pattern and toroidal recirculation vortexes above the nozzle (Fig. 4). The vortexes in this flow pattern cause the liquid to spread radially on the top of the delivery tube into a thin film. This is the preferred operating regime for producing fine powders. When the metal flow rate is too high, the filming effect is partially lost, and some of the metal is pulled into the low-velocity wake and does not film effectively (Ref 5, 6). Such operation leads to coarser powders and to the formation of undesirable flakes in the powder.

The toroidal recirculation vortexes of confined nozzles are caused by fluid entrainment on the inner boundaries of the annular jet (Ref 7). When a jet exits from an annular nozzle, a mixing zone is formed due to the shearing action of the jet on the surrounding fluid. As there is no supply of fluid in the central region to satisfy the entrainment requirements of the inside boundaries, the jet is forced to draw this fluid from the mainstream jet itself. By forming recirculation vortexes, the jet is able to feed itself with its own fluid and hence satisfy entrainment requirements. Associated with the vortexes is a zone of subatmospheric pressure at the tip of the nozzle, and the resulting radial pressure forces draw in the annular jet toward the axis. As the converging streamlines approach the axis, a region of stagnation is set up at some distance downstream from the axis. The high pressure at this point redirects the flow away from the axis, until finally the combined jet expands in the same way as a single round jet. Similar recirculation flows

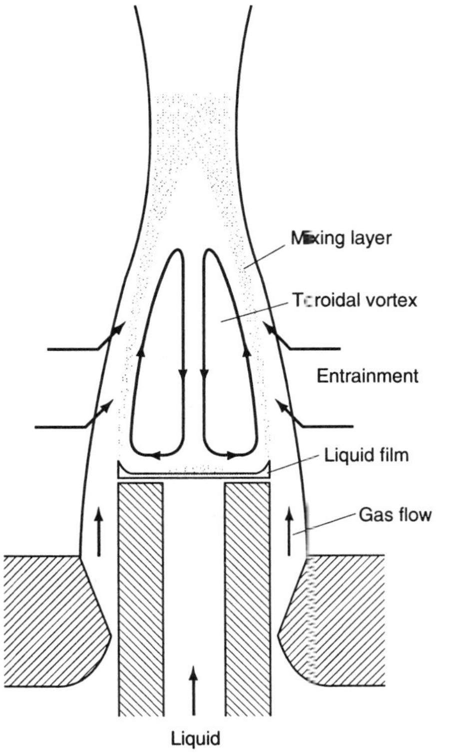

Fig. 4 Fluid flow pattern at the tip of a confined nozzle during atomization

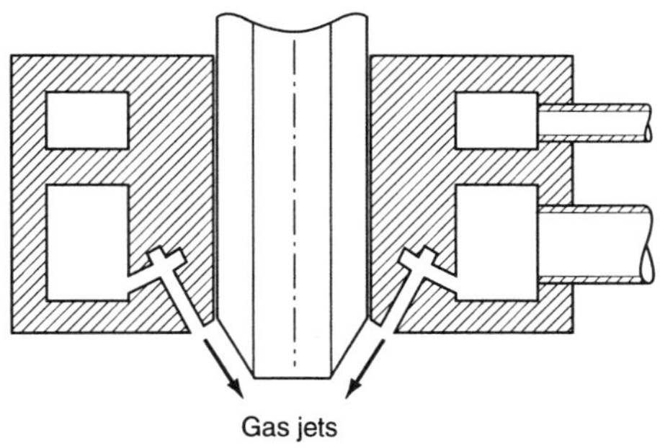

Fig. 5 Schematic of USGA nozzle design. Source: Ref 10

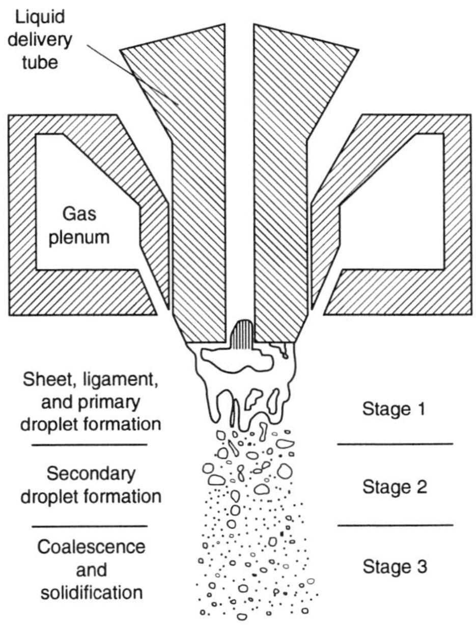

Fig. 6 Particle formation mechanism proposed for the NIST nozzle that operates vertically downwards. Source: Ref 15

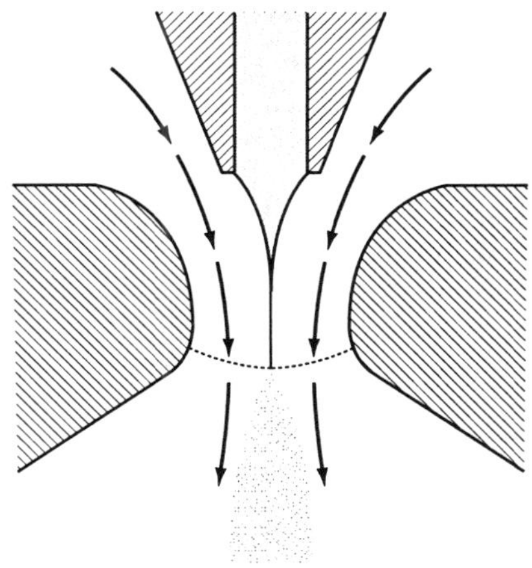

Fig. 7 Schematic of Nanoval nozzle design. Source: Ref 11

can also form on the outside due to entrainment on the outer boundary if the amount of fluid available is restricted, as in the case of a jet being blown into a closed container. It is important to avoid recirculating flows in the vicinity of the atomizing jets that can lead to multiple collisions between the spray and the powder particles and result in excessive satelliting.

The low pressure (suction, aspiration) created at the nozzle tip is a useful feature of the confined design. It is employed to draw the liquid metal into the nozzle from the crucible in updraught operation, as mentioned above. Additionally, the presence of suction at the point of contact between the gas and the liquid avoids the potentially hazardous blowback effects observed in free fall nozzles for certain jet geometries. The level of suction created by a nozzle is found to be extremely sensitive to nozzle geometry, atomizing gas, and operating pressure (Ref 8).

Nozzle designs used in the aluminum powder industry are proprietary and information on their performance is not available. Information has become available in recent years on several different confined nozzle designs developed for use with molten metals, some of which have been applied to aluminum. It is instructive to review some of the more important designs.

Ultrasonic gas atomization (USGA) has been promoted by Grant (Ref 9). This nozzle invented by Kohlswa Jernwerk of Sweden and patented in the United States in 1961 uses a modification of the original design with a multitude of discrete holes for gas delivery. Each gas jet contains a pair of resonating (Hartmann) cavities in its path, as indicated in Fig. 5. The original patent claims that ultrasonic vibrations are created in the gas jet that lead to the production of finer powders. Published reports of experimental work with metal powders have not supported that claim. Further, the modified design has been shown to lead to unnecessarily high frictional losses (Ref 12).

NIST Design. The nozzle developed at NIST is similar to the USGA design, but it does not contain the Hartmann cavities (Ref 13, 14). The gas jets are inclined to metal delivery axis at an included angle of 45 °C (Fig. 6). With such designs, too, under certain conditions, positive pressure rather than suction is created at the tip of the nozzle. The selection of operating conditions then becomes crucial. Recent work by Mates (Ref 6, 16) has shown that filming of the molten metal takes place also in this design. It is, however, sensitive to the operating parameters and is not always complete, and metal can exit the nozzle in the form of ligaments and sheet (Fig. 6) as proposed by NIST workers.

Nanoval Nozzle Design. Nanoval has recently introduced a nozzle design (Fig. 7) in which the molten metal is delivered to the throat of a converging-diverging nozzle in the form of a thin stream (Ref 10). It is thus conceptually similar to the free fall design; but because the metal stream is always kept very thin and it is delivered directly to the throat of the nozzle, this design is capable of producing very fine powders. Nanoval has produced stainless steel powders of 6 μm median diameter by this method in pilot quantities. No filming is likely to occur in this design, because it is necessarily restricted to very low rates of production and best suited possibly for precious metals. The manufacturer also claims substantial gas savings.

Free fall nozzle designs (Fig. 8) are also known to be used in the aluminum powder industry. In this design, the liquid metal issues in the form of a stream from a tundish and falls 50 to 200 mm by gravity. Then it is atomized by the gas directed at a point, either by means of discrete jets or an annular nozzle concentric with the metal stream. This method is easy to operate but highly inefficient and is not suitable for making fine powders.

Formation of Powder Particles. Recent studies on the operation of confined nozzles have shown that the liquid film obtained in these nozzles breaks up in two stages to form the powder particles. The first stage (primary breakup) occurs at the point of first contact between the liquid film and the gas jet along the edge of the delivery (Fig. 4). Liquid droplets formed at this stage can have up to 500 μm diameter under normal operating conditions with aluminum. These droplets undergo further disintegration in flight (secondary breakup), as is evident from the absence of such large particles in the resulting powders. Similar stages are described for the NIST design by Ridder et al. (Ref 13) for the breakup ligaments and the droplets formed.

Primary breakup is not well understood, but it is known from experiments that unsuitable nozzle designs or operating conditions can lead to incomplete breakup and/or formation of filaments at the nozzle tip, which may eventually become flakes in the powder. It is accepted that proper completion of the primary breakup stage is essential for the production of fine powders. The film-

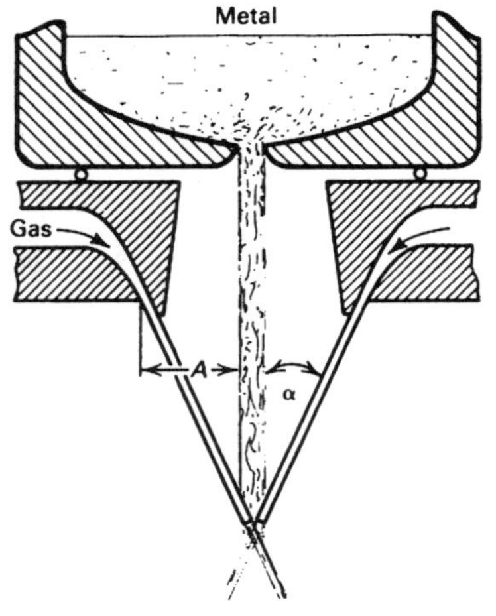

Fig. 8 Atomization design characteristics: α, angle formed by free falling molten metal and impinging gas; *A*, distance between molten metal and gas nozzle

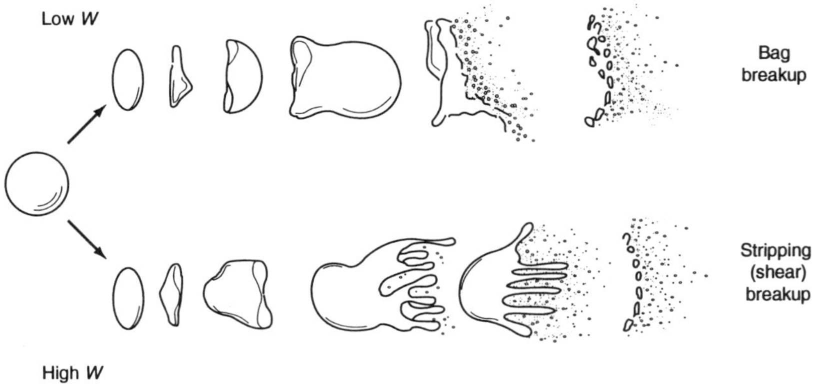

Fig. 9 Secondary breakup mechanisms proposed by Lane. Source: Ref 17

(a) (b)

Fig. 10 Metal film left in the tip of the nozzle in the rundown mode

ing mechanism facilitates the breakup. Also, swirl in the atomizing gas is likely to be helpful. Indeed, swirl is one of the important features of the nozzle design patented by Hall in 1928 for making metal powders. Most interestingly, Miller (Ref 11) has recently introduced nonaxisymmetric metal delivery tubes and has found them to be beneficial in the atomization of superalloys. He reported that they lead to the formation of multiple plumes at the tip of the nozzle, which would break the ligaments into smaller parts and thus enhance the efficiency of the primary breakup stage. Miller also notes that with asymmetric nozzles, the pinching of the plume is much reduced. He suggests that the resulting reduction in the level of droplet collision and coalescence would also contribute to the improved efficiency observed. It is important to note that the largest benefits were observed at the higher range of atomization rates (lower range of the gas/metal ratios) that correspond to coarse powders.

Secondary Breakup. The breakup of single droplets in a flowing gas has been studied using materials that are liquid at room temperature, such as water, oils, and glycerin. Lane (Ref 15), one of the earliest to publish work in the field, identified two main mechanisms of breakup as "bag" mechanism and "stripping" mechanism, shown in Fig. 9 applicable at low and high Weber numbers (*W*), respectively.

Experimental evidence for the occurrence of secondary breakup in metal powder production can be obtained from studies of the atomizing spray at very low metal flow rates. At such low rates, individual droplets become visible (Fig. 10) and show "tails" in the direction of the gas flow as suggested by Lane. Weber numbers obtained in metal atomization are generally very high, and Ünal (Ref 7) concluded on that basis that breakup occurs by stripping. In that mode, the droplet first deforms in such a way as to present a convex surface to the gas flow. The edges of this saucer-shaped droplet are then drawn out into a thin sheet and torn into ligaments, which are later broken into smaller droplets. These particles (daughter particles) form the fine size range of the powder. As droplets get progressively smaller through disintegration, a point comes at which surface tension forces, which are inversely proportional to diameter ($4\sigma/d$), become sufficiently great to resist the aerodynamic drag forces responsible for the breakup. At this point, secondary breakup ceases, and the droplets solidify to form the coarse range of powder particles. From earlier studies of the breakup of single droplets, Ünal (Ref 7) deduced that this occurs when the Weber number (*W*) takes on a value of 13. That is, for stable particles:

$$W = \rho_g u^2 d/\sigma = 13$$

This equation can be used to calculate the size (*d*) of the stable droplets:

$$d = 13\sigma/\rho_g u^2$$

A range of stable particle sizes is obtained because the gas decelerates and the particles accelerate as a function of distance from the nozzle. The lower limit for stable sizes is that obtained at the tip when the relative velocity is a maximum. This has been estimated at ~30 μm for aluminum and for the common atomizing gases, such as nitrogen, argon, and helium. The largest size found in the powder can be used to estimate the relative velocity at which secondary disintegration is complete. These estimates have indicated that the process is likely to be completed while the gas flow is still supersonic in a well designed nozzle.

It is important to appreciate that the presence of stable particles in the final product can apply only to such liquids as molten metals, which have high surface tension and low viscosity. For

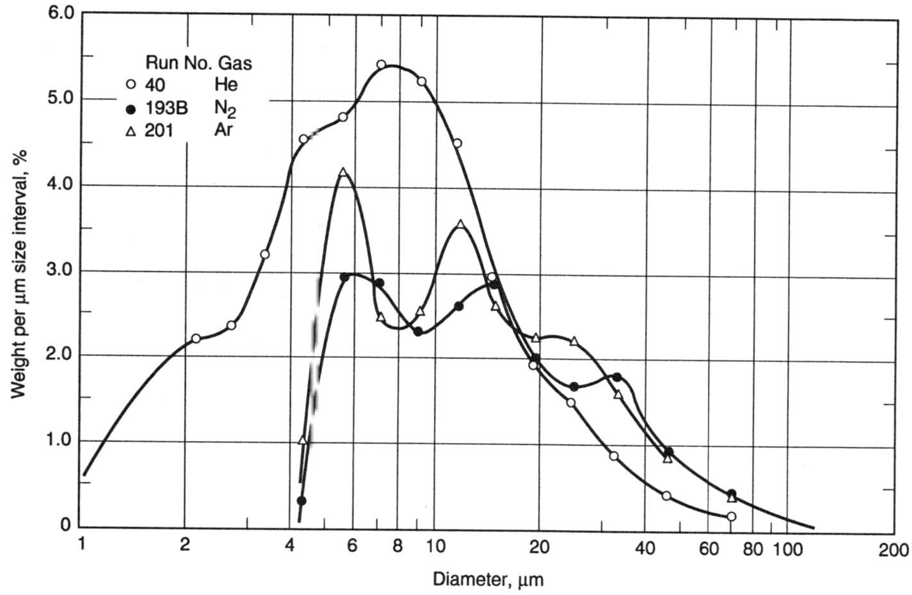

Fig. 11 Frequency distribution for AA 2014 powders made using helium, argon, and nitrogen. Source: Ref 20

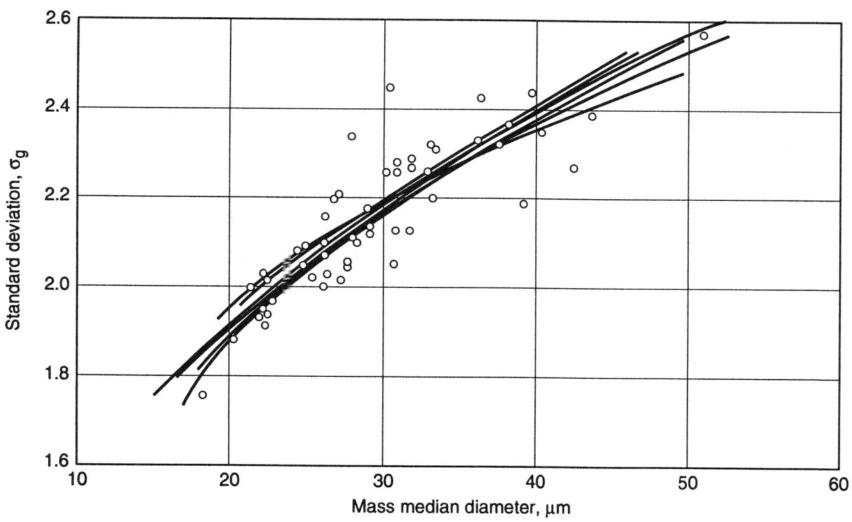

Fig. 12 Variation of geometric standard deviation with median particle size for as-atomized particles

other liquids with low surface tension, the stable sizes may be so small as to be comparable to that of the daughter particles, in which case they would be indistinguishable. Ünal et al. (Ref 19) found that to be true with wax powders atomized in air. These powders have a tighter size distribution represented by the Rosin-Ramler law as opposed to the log-normal law that generally applies to atomized metal powders.

Particle Size Distribution

When various mechanisms operate to produce a powder, these occurrences should appear as peaks in the frequency distribution plots. Nitrogen and argon atomized powders in Fig. 11 show three areas of peaks. The first, at ~5 μm, is considered to represent the daughter particles (Ref 21, 22). The third peak belongs to the stable sizes, which range from ~30 μm to 100 μm and greater. The intermediate peaks are taken to indicate considerable overlapping between the daughter particles of large droplets breaking up at lower relative velocity and stable sizes of initially small drops shattering at high relative velocity. The peaks are not so well defined in the case of helium atomized powder, showing complete merging.

The distribution of particle sizes in atomized metal powders is often found to comply approximately with the log-normal law (Ref 21, 22). The two mechanisms for the production of fine and coarse droplets described above are not compatible with this law, which is strictly unimodal. The wide range of sizes covered is its main advantage in representing satisfactorily the size distribution of gas-atomized metal powders.

Several mean diameters are used to describe the size of a powder. The most common is the mass median diameter, d_m, i.e., the 50% point on the cumulative weight versus particle size graph. Two other useful mean diameters are the Sauter mean diameter d_{vs} and the volume mean diameter d_{vm}, defined by:

$$d_{vs} = \Sigma x^3 dN / \Sigma x^2 dN = 100/(\Sigma d\Phi/x)$$
$$d_{vm} = \Sigma x^4 dN / \Sigma x^3 dN = \Sigma x d\Phi/100$$

where $1x$ = mean diameter in a size band, dN = number of particles in size band, $d\vartheta$ = weight percent in size band. Sauter mean diameter is the diameter of a sphere with the same surface area per unit volume as the powder and is sensitive to changes in the fine particle range of the powder. Volume mean diameter, on the other hand, is a moment mean, sensitive to changes in the coarse particle range. Both these two diameters are easily calculated from the sizing data and are routinely reported in sizing reports obtained in light scattering type instruments.

A single mean diameter is not sufficient to describe a powder; information is also required on the spread of sizes about the mean. This is measured by the geometric standard deviation of the log normal distribution and is calculated:

$$\sigma_g = (d_{84.1}/d_{15.9})^{1/2} = (d_{84.1}/d_{50}) = (d_{50}/d_{15.9})$$

The σ_g values for gas-atomized aluminum powders tend to increase as the mean particle size increases indicating that the distribution becomes less tight for coarse powders. The typical range of σ_g is from 1.8 to ~2.5 (Fig. 12).

In the case of log-normal distribution of sizes, all mean diameters are related to each other. Hence, if one diameter and the value of σ_g are known, all other mean diameters can be calculated. The relationship between the two common mean diameters, d_{vs} and d_{vm}, and the mass median diameter, d_m, is given by the following equations:

$$\ln d_{vm} = \ln d_m + 1/2\ln^2 \sigma_g$$
$$\ln d_{vs} = \ln d_m - 1/2\ln^2 \sigma_g$$
$$d_m^{\,2} = d_{vs}\, d_{vm}$$

These relationships are useful in checking consistency of the data and the applicability of the log-normal law to any size distribution obtained. Figure 12 can be used for quick estimates of the σ_g from the ratio of these mean diameters.

Control of Particle Size. By modifying process parameters and screen sizes, the particle size distribution of atomized powders can be adjusted to produce a wide range of products from aluminum granules (typically +200 mesh) to fine atomized powders (up to 99%, –325 mesh). Figure 13 shows particle size distribution curves for typical coarse, medium, and fine grades of atomized powders. Besides the wide range of grades available from direct atomizing, others are possible from blending screen fractions and air classification.

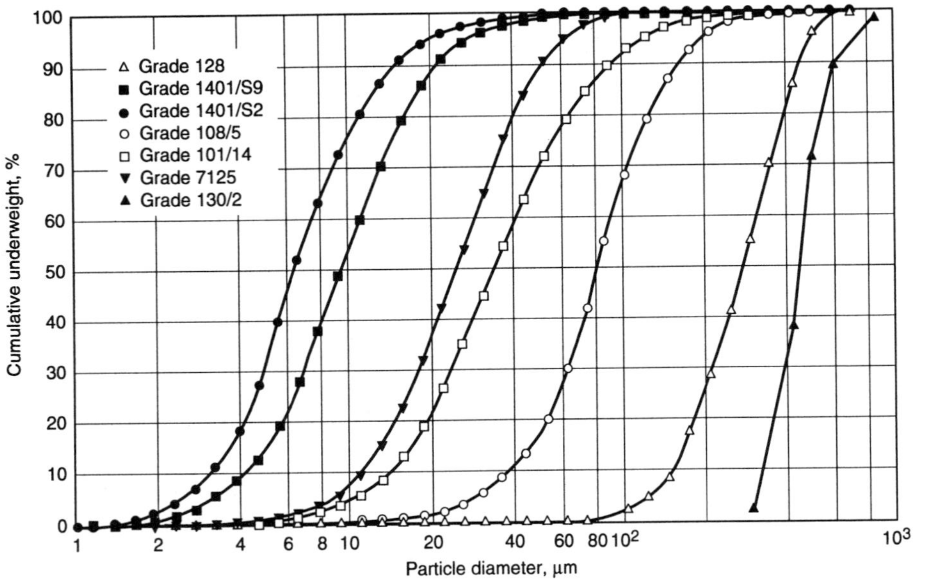

Fig. 13 Particle size distribution measured by Microtrac in some Alcoa powder grades (grade 130/2 is by sonic sieve method)

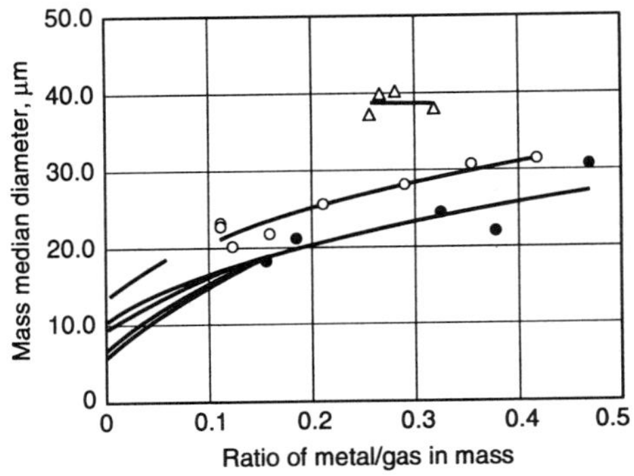

Fig. 14 Variation of mass median diameter for aluminum powders with the ratio of metal/gas

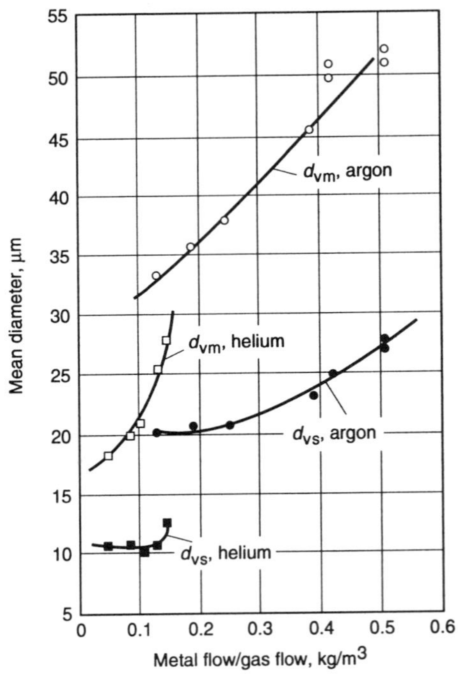

Fig. 15 Comparison of argon and helium gases on the basis of gas volume used per unit weight of powder produced

Important process parameters in atomizing pure metals include nozzle configuration, nature, pressure, and temperature of the atomizing gas and delivery rate, and temperature of the molten metal. For alloy powders, nature and content of the alloying elements also need to be taken into account. Special precautions are often needed in atomizing alloys that contain alloying elements that evaporate or oxidize readily.

The geometry of a nozzle controls the flow of the atomizing gas and is therefore of the utmost importance in any atomization process. Information on the influence of nozzle design is scanty and empirical, and it mostly originates from patent literature. A recent study on a confined design nozzle has amply illustrated the influence of nozzle configuration (Ref 5) on the particle size of the as-atomized aluminum powders. Some results from that study are shown in Fig. 14. To eliminate variation due to differences in the amount of gas delivered by different designs, the results were replotted against the ratio of gas used per unit metal atomized. It is noted that a good design can produce substantially finer powders for the same amount of gas used. In subsequent gas flow studies, good nozzle designs have been found to maintain the supersonic nature of the jet farther in the jet plume than poorer designs, thus allowing the secondary breakup process to take place in a high-velocity gas jet.

The influence of nature and pressure of the atomizing gas on particle size of the as-atomized powder are well established. Helium produces much finer powders than those produced by nitrogen and argon under similar conditions using the same nozzle (Fig. 15). Comparison on the basis of volume of gas used per unit weight of powder produced also shows helium to be superior to argon. Helium atomized powders as fine as 13.5 μm median diameter have been obtained using unheated gas. Under similar operating conditions, nitrogen atomized powders were ~23 μm and argon atomized powders were ~25 μm. This difference is attributed to the higher gas velocities attained in helium in supersonic flow. Similar benefits are expected from higher gas temperatures due to increased velocities. Further, heated gases will help reduce the danger of the premature freezing of the metal at the nozzle tip. With respect to gas pressure, it is known that a given nozzle, when all other parameters are constant, will produce finer powders as the gas pressure is increased. However, as the gas flow rate also increases linearly with pressure, the observed effect is similar to a more favorable gas/metal flow ratio, as well as that of increased pressure. When the results are compared on a gas/metal flow ratio basis, higher operating pressures are still beneficial.

During manufacture, particle size control in the as-atomized state is achieved by the selection of the rate of atomization (metal delivery rate). For any given nozzle configuration and operating parameters, the rates have to be slowed down in order to make finer powders. The median diameter is found to be approximately proportional to the square root of the rate of atomization. This may be more generally expressed as a function of the metal/gas flow rate ratio (Fig. 14). Superheat in the molten metal was found to have only a small effect on the powder size providing that there was sufficient heat in the liquid to prevent premature solidification. This is reasonable, as any change would come about primarily through changes in such properties of liquid metal as surface tension, density, and viscosity. These properties are only mildly dependent on temperature.

As metal flowrate is increased and the powder becomes coarser for any given atomizing conditions, in general, the spread in powder sizes (as measured by σ_g) also increases (Fig. 15). This spread has been taken as indirect evidence for the presence of stable particle sizes in the powder after the completion of secondary breakup (Ref 5, 7, 8, 21).

Particle Morphology

The morphology of atomized aluminum powders is influenced strongly by the amount of oxygen present in the gas phase during atomization. Particles atomized in an inert gas show spherical features, whereas air-atomized particles are of irregular shape (also known as "nodular" powders). This effect of oxygen on morphology has been attributed to the pinning effect of the oxide nuclei that form on the droplets before they have solidified (Ref 23). Many such

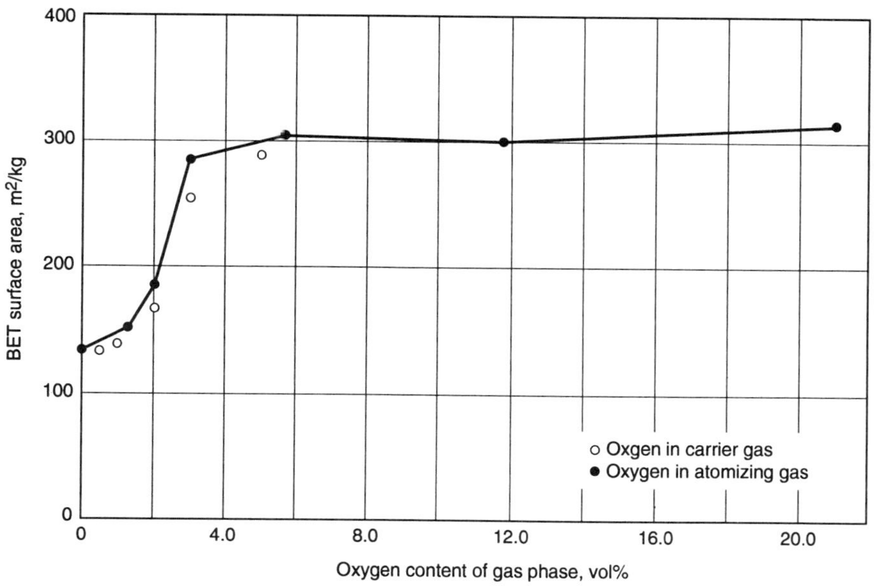

Fig. 16 Influence of oxygen content of the gaseous medium on BET surface area of aluminum powders

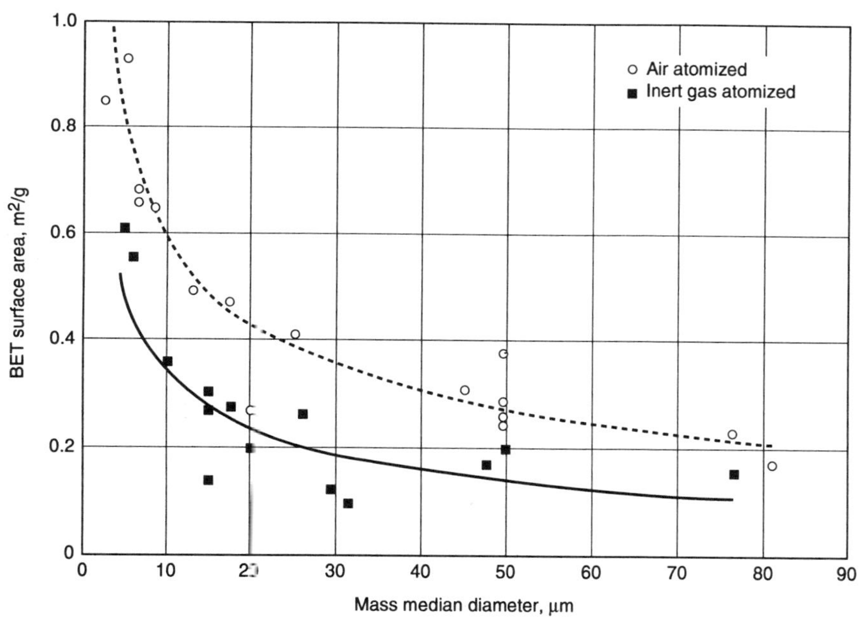

Fig. 17 BET surface area for aluminum powders atomized in air and in inert gas

nuclei form rapidly on a droplet when there is a high level of oxygen in the gas phase leading to effective pinning at those points. Whereas at low oxygen levels or in inert gas, the rate of oxidation would be slow and not cause pinning in which case surface tension forces would be able to maintain the spherical shape of the droplet until solidification. Tests with gaseous atmospheres of varying oxygen content have shown that the change in morphology takes place rather sharply at an oxygen level between 2 and 4%. Powders made in a gaseous atmosphere with 2% oxygen or less show essentially the same degree of sphericity as inert gas-atomized powders. When oxygen is above 4%, the deviation sphericity approaches those levels observed in air-atomized powders. The change in shape of the particles as a function of oxygen level in gas is attributed to the distortion of the droplets by oxide formation on the surface. Brunauer-Emmet-Teller (BET) surface area measurements indicate that the change in morphology takes place at an oxygen level of ~2% in the case of pure aluminum powders (Fig. 16).

Particle size and superheat are also known to influence the degree of sphericity of powder particles. Smaller particles tend to be closer to spherical than larger particles made under the same conditions.

Another feature of atomized aluminum powders, particularly important in milling applications, is the surface quality of the particles. Finer powders tend to have excellent surface quality. In others, cells can be seen.

Satellites form by warm/cold welding of smaller particles onto large particles during atomization. If the collision takes place before the larger particle has solidified, the satellite particle will be embedded.

Specific Surface Area. The specific surface area of aluminum powders is determined by the size distribution and the morphology of the particles. Powders atomized in air show a larger specific surface than inert gas-atomized powders due to the irregular shape. Figure 17 shows specific surface area of aluminum powders measured by the method of gas adsorption (BET). Fine air-atomized powders can have specific surface area values approaching $1 m^2/g$. Medium sized powders are typically in the 0.2 to 0.4 m^2/g. For spherical powders at any given mass median diameter, the specific surface area values are about one-half of that of air-atomized powders.

It is possible to estimate the surface area (S) of atomized powders from the distribution of particle sizes. This is related to the Sauter mean diameter d_{vs} that was defined above. For powder particles of perfect spheres, the relationship is:

$$S = \rho d_{vs}$$

Because atomized aluminum powders are not geometrically perfect spheres, they show a greater surface area. For inert gas-atomized powders, a factor of 1.3 leads to good estimates (Ref 21). A greater factor in the range of 2.6 to 3.5 is needed for the irregularly shaped air-atomized powders.

Surface Oxide Content

Aluminum reacts readily with moisture or free oxygen in the air during manufacture to form an oxide coating on the powder surface. The amount of surface oxide formed during atomization is strongly dependent on the oxygen level of the atomizing atmosphere and particle size of the powder produced. The oxygen content of fine powders atomized in air can approach 1% level by weight (Fig. 18). For coarser powders, 10.41% oxygen is more typical. A mean oxide thickness can be calculated from the oxygen content of the powder if the specific surface area has also been determined. It is then found that the thickness of oxide is, in fact, greater for coarser powders despite the lower overall oxygen content (Fig. 19). Finest powders under 10 μm show a mean oxide thickness of 14 to 15 nm, whereas with coarser powders, the thickness can reach 10 nm.

Spherical powders, atomized in an inert gas, contain substantially less oxygen compared to air-atomized powders. This is in part due to the smaller specific surface area of these powders. However, when compared to the mean oxide thickness, the oxide is actually between 2 and 5 nm thinner with relatively little dependence on the size of the powder.

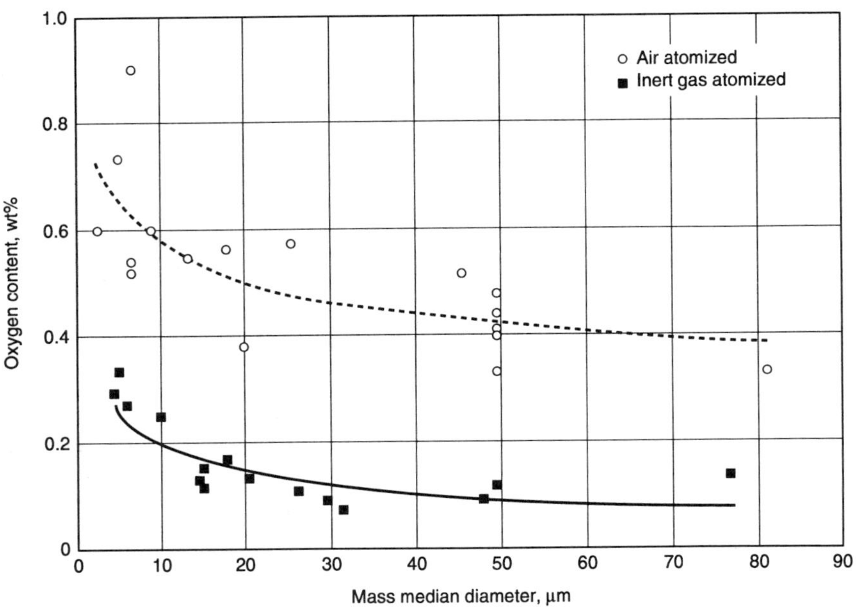

Fig. 18 Oxygen content for aluminum powders

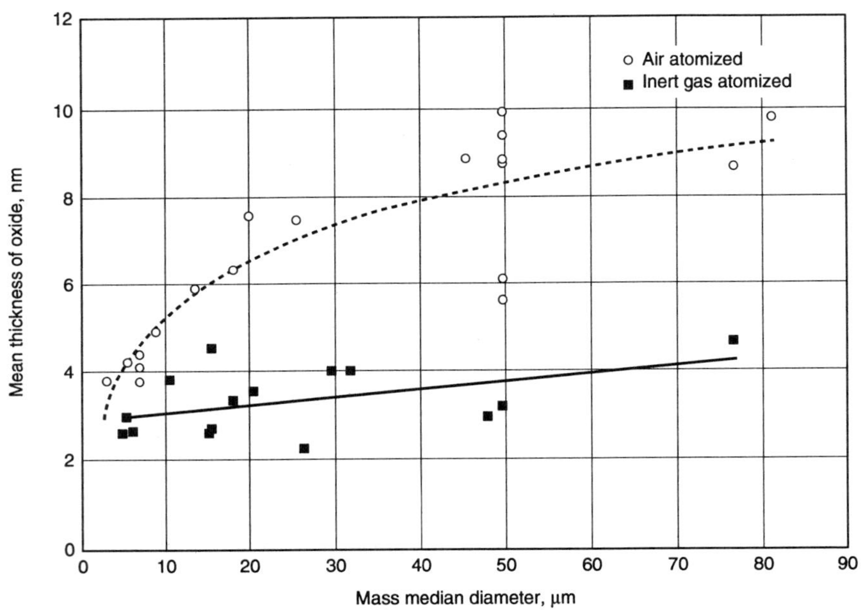

Fig. 19 Mean oxide thickness for aluminum powders

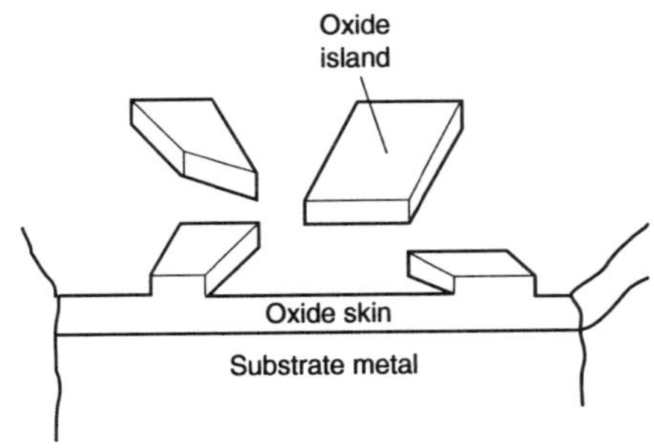

Fig. 20 Simplified model of nonuniform oxide layer consisting of thick islands on a thin skin

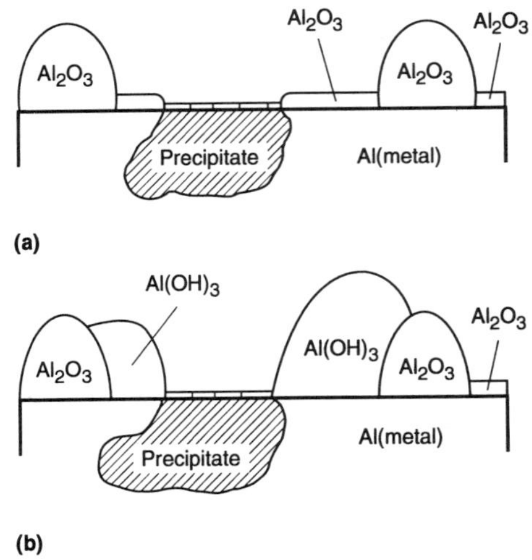

Fig. 21 Model of oxide formed on gas-atomized Al5Mn6Cr powder particle. (a) In the as-atomized state. (b) After exposure to a humid atmosphere. Source: Ref 24

While the mean oxide thickness is useful for comparison between different powders, it does not provide any information on the nature of the oxide skin. Indeed, recent observations have indicated that the oxide on aluminum powders is not a uniform thickness (Fig. 20). Most of the surface is covered with a thin oxide film with interspersed high hills up to 20 nm thickness (Fig. 21). The high hills represent the oxide that forms at discrete nucleation sites while the droplet is still in the molten state. As noted above, the distortion of the droplets by the nuclei is considered to be the cause of the irregular morphology of the air-atomized powders. The space between the mounts is covered by a thin skin that mostly forms in the solid state. A similar description has been offered for alloyed powders. The greater mean thickness observed in coarser air-atomized powders is consistent with this mechanism because they will have spent a longer period in the molten state exposed to oxygen.

Temperature of the metal would obviously have a bearing on the amount of oxide formed in powders. The results relate to material obtained in regular atomizing conditions with a superheat of 150 to 200 °C. If higher temperatures were to be employed, higher oxide levels would be expected.

The surface of the aluminum powder is hygroscopic and will react with moisture to form $Al(OH)_3$ by hydration of the oxide and corrosion of the metallic aluminum especially in fine powders. A minimum relative humidity of 60% is needed for hydration promoted by prolonged exposure to moisture and temperature. The presence of manganese and chromium has been found to sharply decrease the formation of $Al(OH)_3$ in alloyed powders.

Chemical and Physical Properties

The chemical compositions of unalloyed atomized aluminum powders are shown in Table 1. Iron and silicon are the major contaminants for both regular and high-purity powders. In many cases, atomizers have access to molten metal from a nearby aluminum smelter for regular purity metal grades. For high-purity powder, metal is typically bought in the form of purified ingots (sows) and prepared in furnaces that are suitably lined.

High-purity aluminum is resistant to attack by acids, but it dissolves in a mixture of nitric and hydrochloric acids. Solutions of alkali hydroxides rapidly attack aluminum, with evolution of hydrogen and formation of the corresponding soluble aluminate. Aluminum reacts vigorously with bromine and iodine to form aluminum hal-

Table 1 Typical chemical analyses of atomized aluminum powders

Type of powder	Composition, wt %			Other metallics	
	Aluminum	Iron	Silicon	Each	Total
Atomized powders					
Typical	99.7	...	...	...	...
Maximum	...	0.25(a)	0.15(a)	0.05	0.15
High-purity atomized powders					
Minimum	99.97	...	...	...	...
Typical	99.976	0.007	0.008	...	0.009

(a) Iron plus silicon, 0.30 wt% max

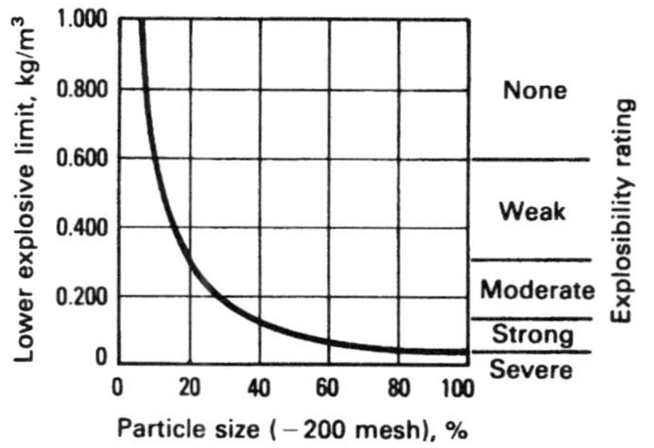

Fig. 22 Effect of aluminum powder particle size on explosibility. Shown by comparing lower explosion limit with the percentage of particles less than 200 mesh in size

Table 2 Typical physical properties of atomized aluminum powders

Wrought density (metal), kg/m^3	2700
Melting point, °C (°F)	660 (1220)
Boiling point, °C (°F)	2430 (4410)
Surface tension at 800 °C (1470 °F), N/m	0.865
Apparent density, kg/m^3	800–1300
Tap density, kg/m^3	1200–1500
Melting point of oxide, °C (°F)	2045 (3720)
Oxygen content, wt%	0.1–1.0

Source: Ref 5

Table 3 Relative explosibility of various dusts

Material	Lower explosive limit, oz/ft^3	Rate of pressure rise, psi
Aluminum	0.045	20,000
Magnesium	0.030	15,000
Polyethylene	0.020	7,500
Flour	0.050	3,700
Coal	0.050	2,000
Coffee	0.085	150

1 oz/ ft^3 = 1 kg/m^3. 1 psi/ s = 6.9 kPa/ s

ides. Dry, oxygen-free chlorine or hydrogen chloride gas also reacts with aluminum to form aluminum chloride. Finely divided aluminum can react violently when exposed to halogenated hydrocarbons, such as methylene chloride and carbon tetrachloride, especially if the operation is carried out under pressure.

Aluminum is stable in air because of its thin, natural oxide film. In finely divided powder form, however, aluminum is more chemically reactive and hydrates when exposed to moisture as described above. Powders also react with water to liberate hydrogen and form aluminum hydroxide and release heat in the process.

Physical Properties. The physical properties of aluminum and pure aluminum powders are listed in Table 2. The real density of aluminum powder approaches that of the base metal, but both apparent density and tap density vary as a function of particle size distribution.

Explosion Potential

Many materials, including metals, polymers, and composites, are potentially explosive in the powdered form. These dusts may be the product of a given process or an unwanted byproduct. Safe handling of potentially explosive powders requires recognition of the hazards, proper handling techniques, and awareness of explosion prevention methods and fire fighting techniques.

The typical explosion occurs due to the simultaneous occurrence of the following:

- A dust cloud
- An oxygen-containing agent
- An ignition source

Common fuels include flammable gases, degradation by products, plastics and other carbonaceous particles, vapors deposited during alloying, or other molten metal handling operations, and aluminum or other metal dust cloud or layer. Common oxidizers include oxygen containing polymers, salt oxides, metal oxides (like Fe_2O_3), carbon dioxide, water, and, of course, air. Typical igniters include hot surfaces, electric or static arcs, spontaneous combustion, pyrophoric materials, impact sparks, thermite reactions, and matches and lighters.

The probability of explosion for different materials is ranked by the United States Bureau of Mines as severe, strong, moderate, weak, and none. These ratings only describe the probability of an explosion occurring as a function of the minimum amount of airborne material required to initiate and sustain a catastrophic reaction. These ratings do not describe the force of the reaction.

Aluminum, magnesium, and lithium have severe ratings. They are aggressive reducing agents whose reactions occur with a drastic release of energy. On the other hand, copper, stainless steel, and nickel do not oxidize readily and therefore are low on the explosibility list. In Table 3, we compare these metal powders to nonmetallic materials.

The rate of pressure rise is the main property that separates metal powders like aluminum and magnesium from organic materials like plastics and flour. If corn flour can destroy a silo at 3,700 psi/s, imagine what a metal dust at 5 to 6 times that will do!

Explosibility is particle size dependent, as noted in Fig. 22. One of the first rules used to predict explosibility potential is to characterize the amount of material finer than 200 mesh (75 μm) in a sample. Two things become apparent. First, below 10% –200 mesh, aluminum powders have typically shown no explosion hazards. Second, above the 10% level, the minimum amount required drops exponentially to a strong hazard, above 40%, the explosion probability is at the severe level.

Preventive measures must be geared at controlling and/or eliminating as many of the legs of the triangle as possible: the fuel, the oxidizer, and/or the igniter. Table 4 shows several properties of aluminum that need to be taken into consideration when establishing preventive measures.

Prevent formation of dust clouds when handling explosive powders. Housekeeping is the most effective prevention method. Using plant air to blow away dirt is prohibited. Natural hair brushes should be used to clean up work areas. Metal dust pans or containers are preferred.

Eliminate/Control Ignition Sources. The most obvious sources include: hot surfaces, smoking, and electrical grounding of equipment. More rigorous items include: The need for special nonsparking hand tooling, dust-tight/explosion-proof electrical switches, connectors, and other equipment, selective use of conductive floors and attention to welding and torching operations.

In the event of a fire or explosion, some simple rules for metal dusts are:

- Know and recognize the hazard
- Evacuate and let the event run its course
- For small, localized fires, initiate alarm procedure first and then only use Class D fire extinguishers or dry inert granular material (for example, sand) to smother the fire.
- *Do not use water.* Water reacts with aluminum and many other finely divided metals to form hydrogen, which is an even worse problem, and water usually comes out of the hose under pressure. This in turn creates a dust cloud that can turn a simple metal fire into a dust explosion.

Your local fire department should be aware of the hazards around metal fires and prepared in proper fire fighting methods.

Selected References. More information on the topic of powder handling and the hazards involved for aluminum powders include:

Table 4 Explosibility characteristics of aluminum powders

Minimum concentration (lower explosive limit), kg/ m^3 (oz/ ft^3)	0.045 (0.045)
Minimum igniting energy, mJ (cal)	50 (0.012)
Ignition temperature of cloud, °C (°F)	650 (1200)
Ignition temperature of layer, °C (°F)	760 (1400)
Maximum explosion pressure (gage), kPa (psig)	503 (73)
Maximum rate of pressure rise, MPa (psi/s)	140 (20,000 +)

- "Recommendations for Storage and Handling of Aluminum Powders and Paste," Publication TR-2 by the Aluminum Association, Inc.
- "Explosibility of Metal Powders," Bureau of Mines Report 6516, United States Department of Interior, 1964
- Material safety data sheets

Applications

Aluminum powders have physical and metallurgical characteristics related to their method of manufacture that make them extremely important in a variety of applications. They can propel rockets, improve personal hygiene, make computers more reliable, refine exotic alloys, and make the family sedan or the newest Air Force fighter lighter in weight and more efficient while providing beauty and protection against sunlight and moisture. Powders, in the form of P/M parts for structural and nonstructural applications, also hold the key to some of the most exciting new developments in the aluminum future.

As powder, aluminum is used in blasting agents where the concentrated energy from the high heat of combustion allows explosive manufacturers to better match energy release to specific rock characteristics or mining conditions. This eliminates or minimizes over or under blasting and reduces total rock breakage costs. This same energy release characteristic makes powders an important element in the solid fuel rockets used for national defense and to launch space probes. Each launch of the Space Shuttle requires 350,000 pounds of aluminum powder. Energy release is also an important criteria in the metallurgical industry, where aluminum powder is used as a heat source as well as a reducing agent. Such applications include production of ferro-alloys like ferro-columbium, pure metal, such as chromium and nonferrous alloys like boron-chromium. Other exothermic applications include hot topping compounds, stress relieving, exothermic welding, and powder lancing.

Powder compounds derived from aluminum powders also find a wide variety of uses in the chemical and plastics industries. These uses range from highly selective catalysts through powerful reducing agents to inert fillers. Aluminum alkyls, produced from powders, are a group of compounds used as feedstock in the production of biodegradable detergents and plasticizers. Aluminum chlorhydroxide and aluminum glycinate are widely used constituents in deodorant and antiperspirant formulations. As a filler for epoxy resins, aluminum powders can reduce shrinkage, increase thermal conductivity by an order of magnitude and reduce thermal expansion as well as reduce permeability and swelling attributable to water, oxygen, or other penetrants.

Powders flattened into tiny flakes by steel balls in a rotating mill provide the metallic pigments for paints and coatings. Leafing pigments, processed to rise to the surface of paint, reflect heat in roof coatings and can protect nonaluminum structures, such as bridges and tanks, from the weather. Nonleafing pigments, processed to remain suspended in paint or coating, provide the metallic finish for cars, trucks, and other items, such as inks and football helmets. The ability to tailor the pigment size and shape to achieve variations in appearance gives product developers an extra degree of freedom in design.

In the electronics area, aluminum nitride P/M ceramics are excellent candidates for electronic packages and substrates, especially with multichip modules, because of thermal expansion and heat transfer characteristics. The use of improved thermal conductivity to yield a reduction in operating temperatures of a typical module by 10 °C (20 °F), will double the reliability of the circuit.

The most exciting application for aluminum powders is in the production of P/M parts for structural and nonstructural applications in the transportation and commercial areas. These press and sinter products, blends of aluminum and elemental alloy powders, are pressed into intricate configurations and sintered to yield net or near-net shapes. There are two basic classes of commercial press and sinter aluminum alloys: 601AB (Al-0.25Cu-0.6Si-1Mg) and 201AB (Al-4.4Cu-0.8Si-0.5Mg). Alloy 601AB displays moderate strength (tensile strength is 237 MPa, or 34.5 ksi) with excellent corrosion resistance, while alloy 201AB develops high mechanical properties (tensile strength is 331 MPa, or 48 ksi) in both the as-sintered and heat-treated conditions.

Parts produced from these materials have the exceptional properties associated with aluminum: light weight, corrosion resistance, high thermal and electrical conductivity, good machinability, and excellent response to a variety of finishing processes. They also offer competitive raw material prices on a cost per unit volume basis.

However, the primary driver for the application of P/M aluminum is the ability to produce complex net or near-net shape parts that need minimal or no machining. This capability makes P/M parts cost-competitive with many castings, extrusion, forgings, and machine-screw products. In addition, the P/M parts can be further processed by hot or cold forging to eliminate porosity and improve strength. Mechanical properties of such products compare favorably with those of conventional wrought alloys and are superior to those of typical press and sinter parts. The niche for P/M forgings is in the strength/cost gap between castings and conventional forgings, and in applications such as connecting rods, gears, and pistons.

Newer aluminum powder products take advantage of rapid solidification technology to produce prealloyed P/M materials with strength, toughness, fatigue, corrosion resistance, and elevated temperature performance not achievable with conventional wrought alloys. These powders can be blended with nonmetallic powders to produce P/M-based metal-matrix composites (P/M MMC) that have "next generation" properties in the areas of stiffness, fatigue, wear, and physical property control. Such new materials have been aimed at the aerospace market and have found uses in both new aircraft platform production and aging aircraft retrofit programs. The relatively limited applications coupled with high projected costs have resulted in reduced commercial interest by both producers and users. The technology, however, stands ready to solve the tough next generation product needs, and the ultimate progression of market use from aerospace to the commercial area could generate significant, high volume applications.

Aluminum powder products span an incredible range of uses from beautification to national defense and from the mundane to the highly sophisticated. Life without these products would not be the same, and they are the key to exciting new developments for the future.

REFERENCES

1. P.D. Liddiard, Aluminium Powder Metallurgy in Perspective, *Powder Metall.,* Vol 27 (No. 4), 1984, p 193–200
2. A.J. Yule and J.J. Dunkley, *Atomization of Melts for Powder Production and Spray Deposition,* Clarendon Press, Oxford, 1994
3. K.L. Cashdollar, Flammability of Metals and Other Elemental Dust Clouds, *Proc. Safety Progress,* Vol 13 (No. 3), 1994, p 1139–1145
4. E.J. Hall, Process for Disintegrating Metal, U.S. Patent 1,659,291, 14 February 1928
5. A. Ünal, Influence of Nozzle Geometry in Gas Atomization of Rapidly Solidified Aluminium Alloy, *Mater. Sci. Technol.,* Vol 4, 1988, p 909–915
6. S.P. Mates and G.S. Settles, High Speed Imaging of Liquid Metal Atomization by Two Different Close-Coupled Nozzles, *Advances in Powder Metallurgy and Particulate Materials,* Vol 1, T.M. Cadle and K.S. Narasimhan, Ed., Metal Powder Industries Federation/APMI, 1996, p 1–67 to 1–80
7. A. Ünal, Effect of Processing Variables on Particle Size in Gas Atomization of Rapidly Solidified Aluminium Powders, *Mater. Sci. Technol.,* Vol 3, 1987, p 1029–1039
8. A. Ünal, Gas Flow in Atomization Nozzles, *Proc. Int. Symp. on Physical Chemistry of Powder Metals Production and Processing* (St. Mary's, PA), W.M. Small, Ed., The Minerals, Metals and Materials Society, 1989

9. N.J. Grant, in *Industrial Materials Science and Engineering,* Marcel Dekker, L.E. Murr, Ed., 1984, p 243–272
10. L. Gerking, Powder from Metallic and Ceramic Melts by Laminar Gas Streams at Supersonic Speeds, *Powder Met. Int.,* Vol 25 (No. 2), 1992, p 59–65
11. S.A. Miller, L.A. Wojcik, R.S. Miller, D.P. Mourer, and R.W. Christensen, A Preliminary Investigation into the Use of Nonaxisymmetric Close-Coupled Nozzles, *Advances in Powder Metallurgy and Particulate Materials,* Vol 1, M. Phillips and J. Porter, Ed., Metal Powder Industries Federation/APMI, 1995, p 31–40
12. A. Ünal, Frictional Losses in Ultrasonic Gas Atomization Nozzles, *Powder Metall.,* Vol 33 (No. 4), 1990, p 327–333
13. S.D. Ridder, S.A. Osella, P.I. Espina, and F.S. Biancaniello, Intelligent Control of Particle Size Distribution during Gas Atomization, *Int. J. Powder Metall.,* Vol 28 (No. 2), p 133–147
14. S.D. Ridder and F.S. Biancaniello, *Mater. Sci. Eng.,* Vol 98, 1989, p 47–51
15. W.R. Lane, Shatter of Drops in Streams of Air, *Ind. Eng. Chem.,* Vol 43 (No. 6), 1951, p 1312–1317
16. S.P. Mates and G.S. Settles, A Flow Visualization Study of the Gas Dynamics of Liquid Metal Atomization Nozzles, *Advances in Powder Metallurgy and Particulate Materials,* Vol 1, M. Phillips and J. Porter, Ed., Metal Powder Industries Federation/APMI, 1995, p 15–29
17. A. Ünal, BET Surface Area and Morphology of Rapidly Solidified Aluminium Powders, *Powder Metall.,* Vol 32, 1989, p 31–34
18. A. Nylund and I. Olefjord, Degassing of USGA-atomized Al5Mn6Cr Powder after Exposure to a Humid Atmosphere, *Mater. Sci. Eng. A,* Vol 34, 1991, p 1225–1228
19. A. Ünal, M.J. Naylor, and H.B. McShane, Modelling of Metal Powder Production Using a Wax Atomiser, *Powder Metall.,* Vol 33 (No. 3), 1990, p 259–268
20. H. Förster and H.-C. Neubing, Bestimmung der Oxidfilmdicke auf der OberflächevonAluminiumpulver, *Aluminium,* Vol 70 (No. 7/8), 1994, p 447–450 (in German)
21. A. Ünal, Production of Rapidly Solidified Aluminium Alloy Powders by Gas Atomization and Their Applications, *Powder Metall.,* Vol 33 (No. 1), 1990, p 53–64
22. A. Ünal, Liquid Break-Up in Gas Atomization of Fine Aluminium Powders, *Metall. Trans. B,* Vol 20, 1989, p 61–69
23. S. Özbilen, A. Ünal, and T. Sheppard, Influence of Oxygen on Morphology and Oxide Content of Gas Atomized Aluminum Powders, *Physical Chemistry of Powder Metals Production and Processing,* W.M. Small, Ed., TMS, 1989, p 489–505
24. J.F. Flumerfelt and I.E. Anderson, High Purity Aluminum Powder Processed Using High Pressure Gas Atomization, *Advances in Powder Metallurgy and Particulate Materials,* Vol 1, T.M. Cadle and K.S. Narasimhan, Ed., Metal Powder Industries Federation/APMI, 1996, p 1–87 to 1–96

Production of Titanium Powder

J.H. Moll and C.F. Yolton, Crucible Research Center, Crucible Materials Corporation

TITANIUM is used for many aerospace and nonaerospace applications because of an attractive combination of low density, good mechanical properties, and good corrosion resistance. More recently, titanium aluminide alloys are being developed for applications requiring high-temperature strength. In general, titanium is expensive to produce and fabricate. With the advent of the titanium aluminide alloys, this is further complicated by segregation problems in these highly alloyed materials. Powder metallurgy (P/M) processing of titanium offers potential solutions to these problems. One area for cost reduction with P/M processing is the production of near-net shapes eliminating yield losses associated with ingot breakdown, forging, and machining. With highly alloyed materials such as the titanium aluminide alloys, processing from ingot requires repeated homogenizing heat treatments and hot working operations to minimize segregation. Powder metallurgy products are inherently homogeneous and do not require any special processing to eliminate segregation. Powder metallurgy processing also provides an opportunity to produce alloys and structures that are not possible through conventional ingot processing.

Titanium powder is produced by a variety of processes. These include chemical reduction, hydride/dehydride, gas atomization, plasma-rotating electrode, and mechanical alloying. This article will describe each of the processes, the characteristics of the resulting powder, and some end applications.

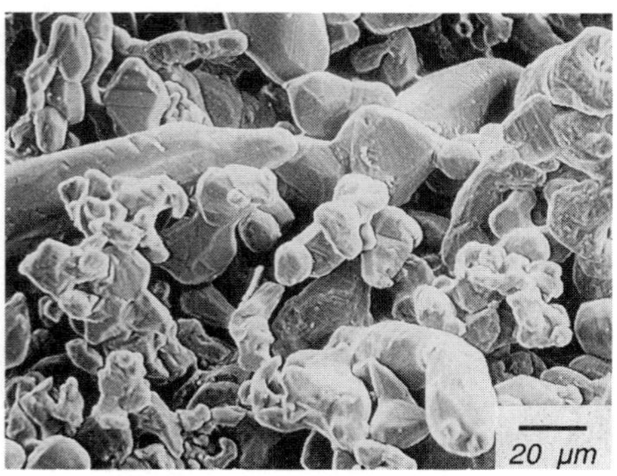

Fig. 1 Sponge fines made by the Hunter process

Chemical Reduction

A primary source of titanium powder is "sponge fines," so called because of their origin and appearance (Fig. 1). Sponge fines are small irregular particles of nominally pure titanium produced during the production of titanium sponge from titanium tetrachloride. The commercial production of titanium metal involves the chlorination of natural or synthetically produced rutile, TiO_2, in the presence of carbon (Ref 1). The resulting titanium tetrachloride is reduced to metallic titanium by the Kroll process using magnesium (Ref 2) or by the Hunter process using sodium (Ref 3). Most of the remnant chloride is removed by vacuum distillation or by water leaching. The final sponge fines typically contain 0.12 to 0.15 wt% Cl. The inherent chloride content of conventional sponge fines can lead to incomplete densification and limited fatigue properties. A typical screen analysis and chemical composition of sponge fines produced by sodium reduction are given in Tables 1 and 2, respectively. Table 3 gives tensile properties of Ti-6Al-4V compacts produced from sponge fines and aluminum-vanadium master alloy.

Commercially pure (CP) titanium is also being produced in Russia from titanium dioxide by reduction with calcium hydride (Ref 4) as follows:

$$TiO_2 + 2CaH_2 \rightarrow Ti + 2CaO + 2H_2$$

The reaction is performed at temperatures ranging from 1100 °C (2000 °F) to 1200 °C (2200 °F). Prealloyed powder is produced by reacting mixtures of oxides of different elements or master alloy powders. Because the process does not use titanium tetrachloride, the powders contain low levels of chlorine. Table 4 gives the composition of CP titanium and prealloyed Ti-6Al-4V

Table 1 Screen analysis of sodium reduced titanium sponge fines

Mesh size (U.S.)		Particle size, μm		Weight percent retained
+80		+177		0
−80	+100	−177	+149	0.1
−100	+140	−149	+105	11.2
−140	+200	−105	+74	32.9
−200	+230	−74	+64	5.0
−230	+325	−64	+45	23.3
−325		−45		17.5

Source: Ref 1

Table 2 Typical chemical analysis of Hunter process titanium sponge fines sinter compacted Ti-6Al-4V produced from these fines and Al-V master alloy powder

Element	Analysis, wt%	
	Sponge fines	Sintered compact
Al	...	6.2
V	...	4.1
O	0.13	0.24
N	0.03	0.016
H	0.07	0.002
C	0.02	0.02
Fe	0.02	0.18
Na	0.10	0.10
Cl	0.13	0.12

Source: Ref 1

Table 3 Typical tensile properties of Ti-6Al-4V compacts made from blended sponge fines and Al-V master alloy powder

Material	0.2% yield strength		Tensile strength		Elongation, %	Reduction of area, %
	MPa	ksi	MPa	ksi		
Cold isostatic press and hot isostatic press	827	120	917	133	13	26
Press and sinter	868	126	945	137	15	25
Wrought mill annealed	923	134	978	142	16	44
MIL-T-9047 (min)	827	120	896	130	10	25

Sponge fines from the Hunter process. Source: Ref 1

powders made by the process. The process also allows the direct production of titanium aluminides.

Powder particles made by the calcium hydride process (Fig. 2) have a spongy morphology and consist of a conglomeration of small particles fused together during the reaction process. The particles are readily sintered due to their irregular shape and large surface area. Figure 3 shows the particle size distribution for CP titanium. The average particle size is 41 μm. Fully dense articles with the Ti-6Al-4V composition have been made by synthesizing the composition from a blend of metal hydride reduced CP titanium and master alloy powders and consolidating by hot isostatic pressing. Table 5 shows mechanical properties of the fully dense products.

Sponge fines with <400 ppm chlorine can be produced by an electrolytic process (Ref 5). Reportedly, low chlorine sponge can also be produced by a reaction of TiO_2 with fluorine salts followed by reduction with aluminum in a zinc carrier medium (Ref 1, 6).

Figure 4 shows elemental powders are being used to manufacture a variety of corrosion resistant fasteners. A technology has also been developed to make porous titanium plates for a variety

Table 4 Chemical composition of powders made by the calcium hydride reduction method

Element	Commercially pure titanium	Ti-6Al-4V
Al	...	4.65
V	...	3.80
Fe	0.11	...
Ni	0.07	...
C	0.03	...
N	0.06	0.06
O	0.19	0.20
H	0.34	0.30
Ca	0.04	0.06
Si	0.05	...
Cl	0.004	0.003
Ti	bal	bal

Source: Ref 4

Table 5 Mechanical properties of Ti-6Al-4V articles synthesized from calcium hydride reduced Ti and Al-V master alloy powders

Material	Temperature 25 °C (76 °F)	205 °C (400 °F)	472 °C (800 °F)
Tension			
Ultimate strength, MPa (ksi)	966 (140)	749 (109)	594 (86)
Yield strength, MPa (ksi)	865 (125)	604 (88)	465 (67)
Elongation, %	13	27	29
Reduction of area, %	32	43	48
Young's modulus, GPa (10^6 psi)	117 (16.9)	101 (14.6)	91 (13.2)
Compression			
Yield strength, MPa (ksi)	934 (136)	639 (93)	472 (68)
Young's modulus, GPa (10^6 psi)	118 (17.1)	117 (17.0)	91 (13.2)
Shear			
Ultimate strength, MPa (ksi)	641 (93)	523 (76)	417 (60)
Bearing, e/D = 1.5(a)			
Ultimate strength, MPa (ksi)	1615 (234)	1242 (180)	993 (144)
Yield strength, MPa (ksi)	1374 (199)	999 (145)	773 (112)
Bearing, e/D = 2.0			
Ultimate strength, MPa (ksi)	2010 (292)	1571 (228)	1265 (183)
Yield strength, MPa (ksi)	1643 (238)	1222 (177)	916 (133)
Plane-strain fracture toughness			
K_{Ic}, MPa$\sqrt{m}$ (ksi$\sqrt{in.}$)	98 (87)(b)	...	...
Axial fatigue			
Unnotched, $R = 0.1$, $K_t = 1.0$			
10^3 cycles, MPa (ksi)	930 (135)	...	...
10^5 cycles, MPa (ksi)	700 (102)	...	...
10^7 cycles, MPa (ksi)	375 (54)	...	...
Notched, $R = 0.1$, $K_t = 1.0$			
10^3 cycles, MPa (ksi)	550 (80)	...	...
10^5 cycles, MPa (ksi)	375 (54)	...	...
10^7 cycles, MPa (ksi)	190 (28)	...	...
Creep			
0.2%, 100 h, MPa (ksi)	...	...	305 (44)
0.2%, 1000 h, MPa (ksi)	...	...	235 (34)
Rupture, 100 h, MPa (ksi)	...	...	585 (85)
Rupture, 1000 h, MPa (ksi)	...	...	490 (71)

(a) e/D is edge distance/pin diameter. (b) Circumferential specimen with crack propagation in the radial direction. Source: Ref 4

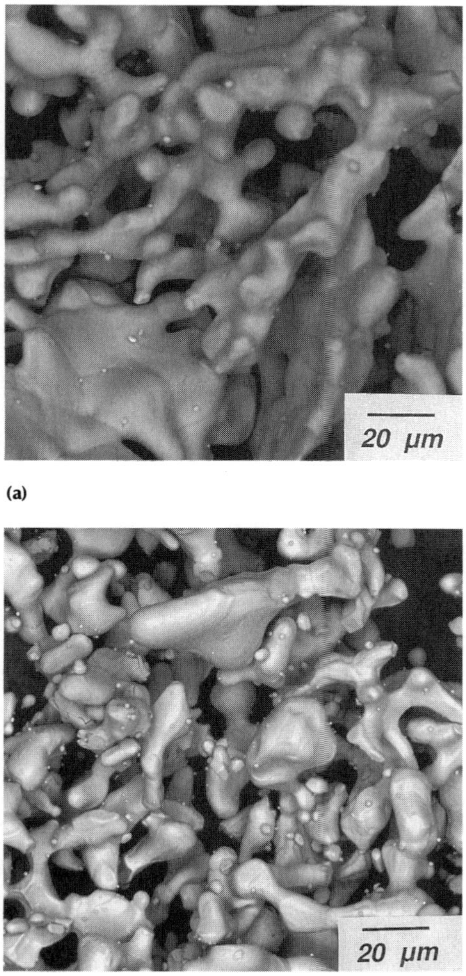

(b)

Fig. 2 Powders produced by the calcium hydride reduction method. (a) Commercially pure titanium. (b) Ti-6Al-4V

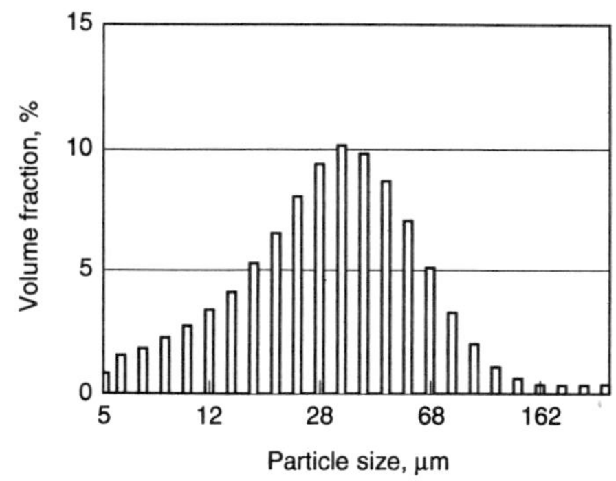

Fig. 3 Particle size distribution of commercially pure titanium powder made by the calcium hydride reduction method. Source: Ref 4

Fig. 4 Titanium parts made from powder made by the calcium hydride reduction process. Courtesy of AdMa International Inc.

of applications including supports for the electrolyte membrane in water electrolysis cell technology (Ref 4). Titanium powders have also been evaluated for use in automotive applications such as valve spring retainer caps and connecting rods. Reportedly, the parts demonstrated the necessary mechanical properties and provided significant weight savings (Ref 7).

Fully dense parts can be made from blended elemental powder using a cold plus hot isostatic process designated CHIP (Ref 8). The process, developed by Dynamet Technology, involves cold pressing of the blended powders in inexpensive elastomeric tooling under hydrostatic pressure. The "green" preform is then sintered to 95% minimum density. Full density is attained by hot isostatic pressing and the near-net shape is machined to final dimensions. Figure 5 shows a missile component made by the CHIP process.

Fig. 5 Sidewinder missile component fabricated by the CHIP process. The left part is the near-net machining preform; the right part is the finished machined component. Courtesy of Dynamet Technology

Hydride/Dehydride Process

The hydride/dehydride (HDH) process for making titanium powder is based on the reversible interaction of titanium and hydrogen. Titanium has a very high affinity for hydrogen and is easily hydrogenated by heating titanium in a hydrogen atmosphere. Hydrogenated titanium is very brittle and can be crushed to a fine powder. The hydrogen can then be simply removed by heating the powder in a dynamic vacuum. The minimum hydrogenation temperature for commercially pure titanium is about 400 °C (750 °F) at 0.007 MPa (1 psi) positive hydrogen pressure. The time required varies depending upon the cross-sectional area of the starting stock. Typically, machine turnings can be hydrogenated in

Fig. 6 Powder made by the hydride/dehydride process

Table 6 Composition of alloys synthesized from blends of hydride/dehydride commercially pure titanium and master alloy powders

Element	Alloy P662–1	Alloy P6/4–1
Al	5.65	5.09
Cu	0.73	<0.01
Fe	0.81	0.1
Sn	2.1	0.01
V	5.55	4.17
Nb	0.45	1.17
Mo	0.02	0.01
W	0.01	0.01
Cr	0.01	0.02
Si	0.05	0.02
C	0.02	0.02
H	0.0023	0.0015
N	0.0060	0.0060
O	0.1060	0.1310

Source: Ref 9

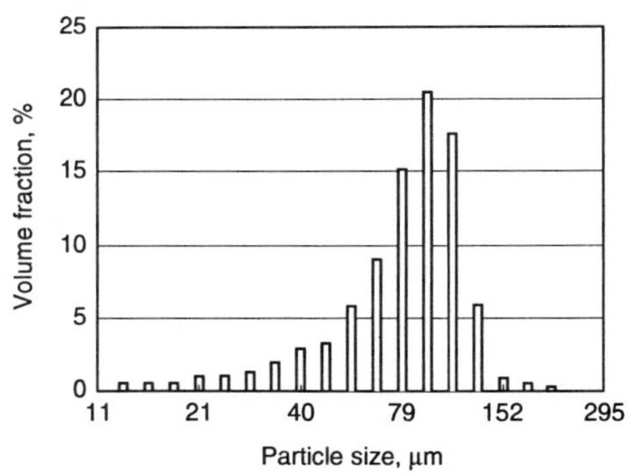

Fig. 7 Particle size distribution of a commercially pure powder made by the hydride/dehydride method. Source: Ref 9

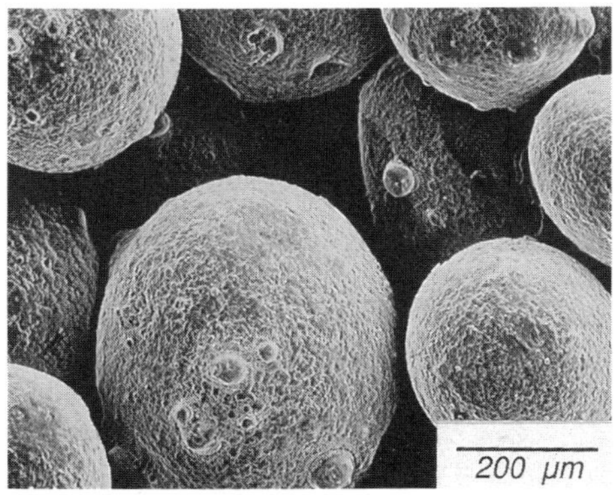

Fig. 8 Ti-48Al-2Cr-2Nb (at.%) titanium aluminide powder made by the titanium gas atomization process

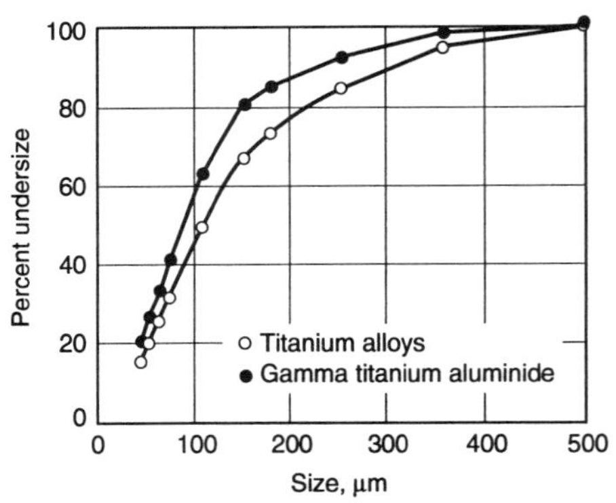

Fig. 9 Size distribution of powders made by the titanium gas atomization process

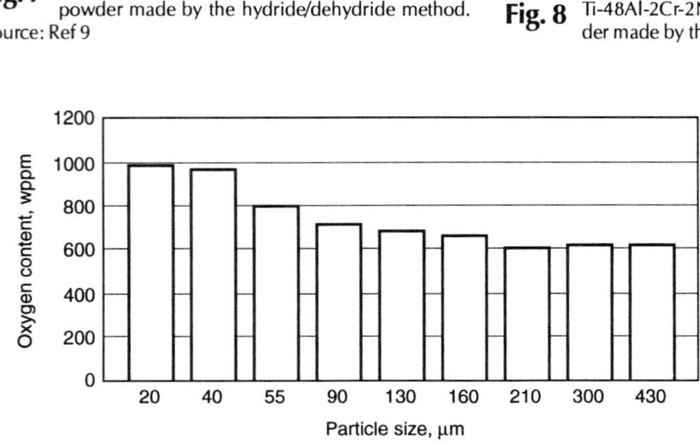

Fig. 10 Effect of particle size on oxygen content of gas atomized Ti-6Al-4V powder wppm, weight parts per million

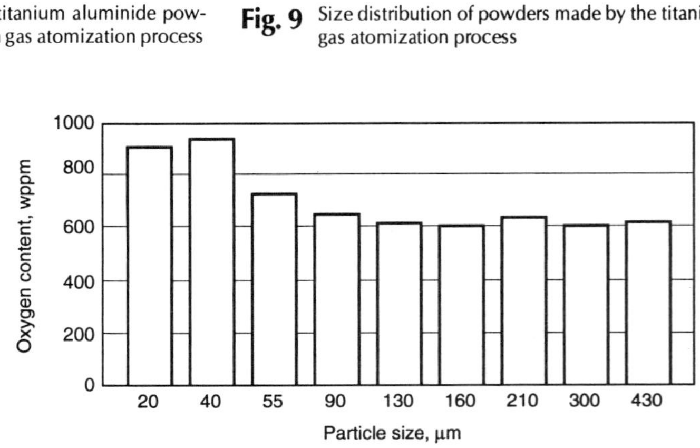

Fig. 11 Effect of particle size on oxygen content of gas atomized titanium aluminide powder, wppm, ppm by weight

four hours. Starting stock can be a variety of pure or prealloyed titanium materials such as ingot, billet, scrap, or machine turnings. Care must be taken to ensure the cleanliness of the starting material because any extraneous material may carry over into the resulting powder (Ref 9). The resulting powder, Fig. 6, is angular and is amenable to cold compaction and subsequent densification by sintering and/or hot isostatic pressing. Figure 7 shows a particle size distribution for commercially pure titanium powder produced by the HDH process. Hydride/dehydride commercially pure titanium powder has been used to make Ti-6Al-4V and Ti-6Al-6V-2Sn by blending with the appropriate master alloy, cold isostatic pressing, sintering, and hot isostatic processing (Ref 10). Table 6 gives the compositions of the synthesized compositions, and the room temperature tensile properties of these materials are given in Table 7.

Gas Atomization

A gas atomization process for titanium, designated TGA, was developed by Crucible Materials Corporation (Ref 11–13). The TGA process involves some unique characteristics such as:

- Vacuum-induction skull melting a charge in a water-cooled copper crucible
- Tapping the furnace through the bottom into a nozzle
- Induction heating of the metal nozzle

The resulting molten metal stream is atomized with inert gas. Starting stock for atomization can be virgin raw materials or prealloyed materials, such as ingot or bars. As the charge is melted in the crucible, a thin skull forms at the crucible wall so that the molten titanium is always contained in a solid skull of the same composition. After the charge is fully molten, it can be held in the molten state for an extended time to ensure complete homogeneity of the melt. The vigorous stirring action produced by the strong inductive field and the ability to hold the metal in the molten state for an extended period of time are particular advantages of the melting method.

To initiate the pour in the TGA process, power to a second induction coil causes localized melting at the center of the bottom of the skull. The liquid metal is then bottom poured through a metal nozzle and free falls through the atomization die, which contains a ring of high-pressure jets. The force of the atomization gas disintegrates the stream into tiny droplets, which rapidly solidify into powder particles as they fall through the cooling tower. The atomization gas carries the powder particles into a cyclone collector, and they are finally collected in a cooled canister.

The powder produced by the TGA process is spherical and free flowing (Fig. 8). Figure 9 shows the size distribution of various titanium and titanium aluminide alloys. Typical tap densities for gas atomized titanium alloys range from 60 to 70% of solid density. Flow rates, depending on the alloy, range from 25 to 35 s, as measured according to ASTM B 213. Interstitial element levels in gas atomized titanium and titanium aluminide alloys can be held to maximums of 800 ppm oxygen, 150 ppm carbon, and 100 ppm nitrogen. There is essentially no increase in carbon, nitrogen, or hydrogen over the melt stock used for atomization. The increase in oxygen content from the starting stock to –35 mesh powder is ~200 ppm. Figures 10 and 12 show the effect of powder particle size on oxygen content of Ti-6Al-4V and Ti-48Al-2Cr-2Nb (at.%), respectively. As can be noted, oxygen content increases with decreasing particle size due to increasing surface area. Nitrogen content is unaffected by particle size.

Table 7 Room temperature properties of hot isostatically pressed compacts of alloys synthesized from blends of hydride/dehydride commercially pure titanium and master alloy powders

Property	Alloy P662–1	Alloy P6/4–1
Yield strength, MPa (ksi)	920 (134)	868 (126)
Ultimate strength, MPa (ksi)	1040 (151)	965 (140)
Elongation, %	15	14
Reduction of area, %	33	31

Source: Ref 9

Fig. 13 Gamma titanium aluminide sheet produced from hot isostatically pressed gas atomized powder. Courtesy of Plansee

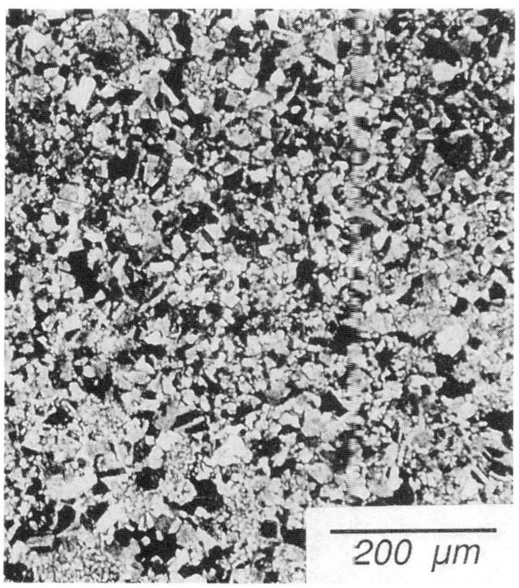

Fig. 12 Microstructure of hot isostatically pressed Ti-48Al-2Cr-2Nb (at.%) powder made by gas atomization

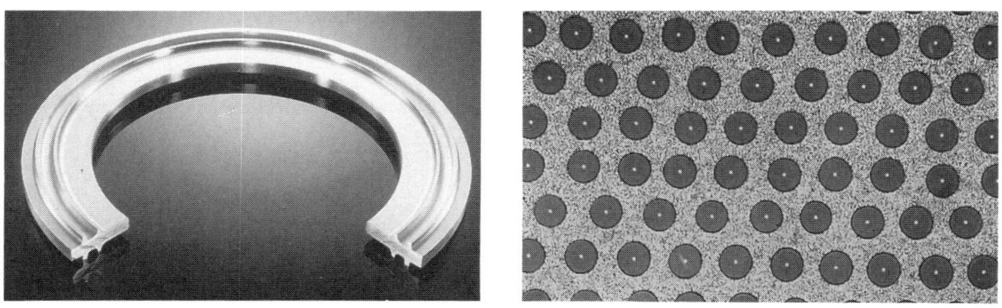

(a) (b)

Fig. 14 Ring containing a composite insert (a) and microstructure of the Ti-6Al-4V/SiC composite (b) made by tape casting using gas atomized powder. Courtesy of Atlantic Research Corporation

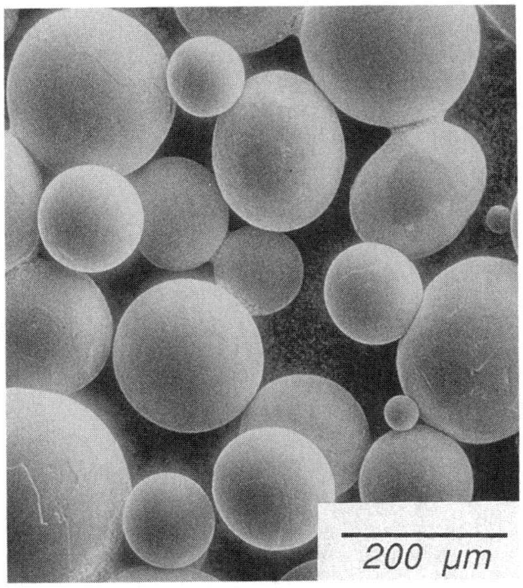

Fig. 15 Ti-6Al-4V powder made by the plasma-rotating electrode process

The TGA process has been used to produce a wide variety of materials such as CP titanium, conventional alpha-beta and beta alloys, and alpha two and gamma titanium aluminide alloys, as well as specialized compositions. Consolidation of the powder results in a highly uniform microstructure, as exemplified with a gamma titanium aluminide alloy in Fig. 12. Gas atomized gamma titanium aluminide powder is being considered for producing forging and machining preforms and near-net shapes for engine application. A process is also being developed to produce gamma titanium aluminide sheet from hot isostatically pressed plate (Ref 14); see Fig. 13. The P/M process has been reported to be more efficient and exhibit higher yields than sheet produced from ingot. It is also reported that the strength and ductility of the P/M sheet are comparable to sheet produced from ingot. Gamma titanium aluminide sheet is being evaluated for inner shrouds on the F-119 engine for transition into production engines for the Joint Strike Fighter and F-22A (Ref 15). Other potential applications for TGA powder include metal matrix composites made by induction-plasma deposition (Ref 16, 17) and tape casting (Ref 18). Titanium matrix composites are being considered for use in a variety of rotating and nonrotating parts in turbine engines to reduce weight and increase stiffness (Ref 19). Figure 14 shows the microstructure of a Ti-6Al-4V/SiC composite ring and its microstructure made by tape casting using gas atomized powder.

Sumitomo Sitix has also developed a gas atomization process for titanium (Ref 20, 21). In this process, a bar of titanium sponge or titanium alloy is lowered through an induction coil. As the bar melts, argon or helium gas jets break up the molten metal into droplets, which solidify as they are swirled within the atomization chamber. The resulting powder is fine with a mean particle size of ~70 μm. The process has been used to make CP titanium, Ti-6Al-4V, and Ti-36Al.

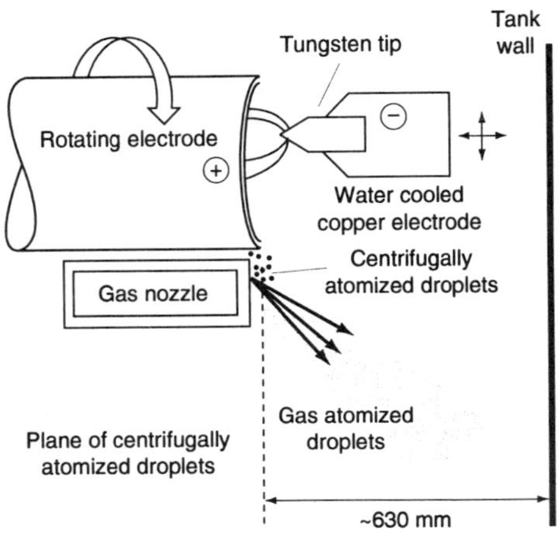

Fig. 16 Conceptual schematic of gas-assisted, rotating-electrode process. Source: Ref 23

Plasma-Rotating Electrode Process

The plasma-rotating electrode process (PREP) is a centrifugal atomization process for making titanium prealloyed powder developed by Starmet (Ref 22). In this process, a helium plasma is used to melt the end of a rapidly rotating bar and molten droplets are spun off and solidify in flight in a helium atmosphere (see the article "Rotating Electrode Process" in this Volume). Melting and atomization are conducted in a circular stainless steel tank ~2440 mm in diameter maintained at a positive pressure of helium. The electrodes are prealloyed bars nominally 60 to 65 mm in diameter and are spun at speeds up to 15,000 rpm.

The PREP powder (Fig. 15) is spherical, has good flow characteristics, and a tap density of ~65% of theoretical density. The size of the powder particles depends on the alloy, the electrode diameter, and the rotational speed at the time of atomization. Typical sizes for Ti-6Al-4V powders are between 100 and 300 μm with a d_{50} of ~175 μm (Ref 23).

Compared to gas atomized powder, PREP powder is relatively coarse. Various process modifications have been investigated to produce finer powder. The use of a large diameter disk electrode rather than a bar electrode has been reported to decrease the mean particle size by more than 40% and to reduce cost of input material (Ref 23). In another process modification, the metal droplets were quenched into liquid argon. The result was a major reduction in powder particle size (Ref 23). In yet another modification, gas atomization was coupled with the rotating-electrode process (REP), as shown in Fig. 16. The REP powder process is similar to PREP, except that a tungsten arc is used instead of a plasma to melt the end of the rotating electrode (Ref 22). As shown in Fig. 17, the yield of –44 μm titanium powder increased from less than 1% with regular REP to greater than 20% with gas atomization assisted REP.

An engine mount support (Fig. 18) made from PREP Ti-6Al-4V powder has been qualified for use on the F/A-18A Hornet fighter aircraft (Ref 25). Table 8 gives mechanical property data used to qualify the hot isostatically pressed near-net shape. The PREP powder is also used to produce a porous coating on hip prostheses for bone ingrowth (Fig. 19).

Mechanical Alloying

Mechanical alloying is a high-energy ball milling process, which can be used to produce materials with unique compositions and microstructure. The process was originally developed to produce nickel and iron-base oxide-strengthened alloys (Ref 26). The application of the process to titanium-base alloys is relatively recent and generally involves loading a mix of powders and a grinding medium (usually hardened steel balls) into a container under a protective argon atmosphere. A small amount of a process control agent is normally added to pre-

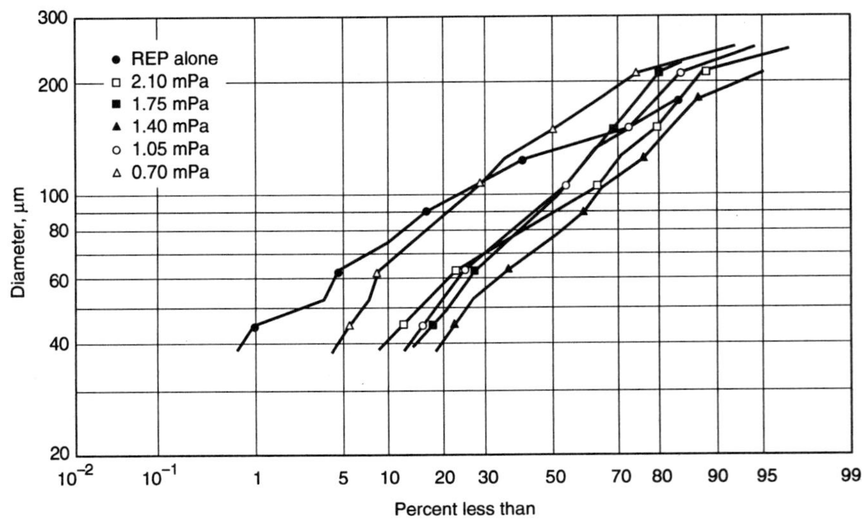

Fig. 17 Titanium powder size distributions produced by the rotating-electrode process (REP) assisted by gas atomization at various pressures. Source: Ref 23

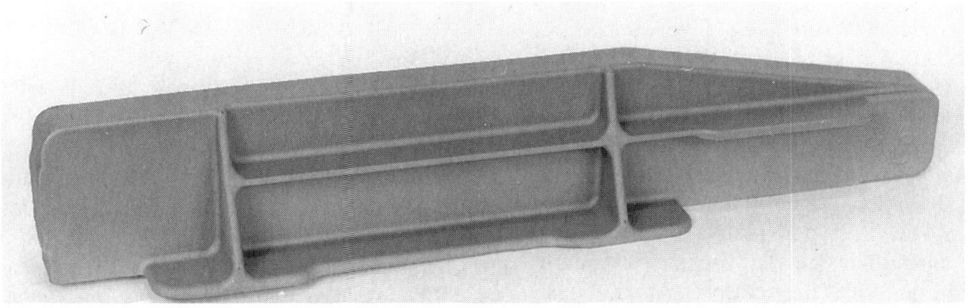

Fig. 18 Near-net engine mount support made by using PREP powder and hot isostatic pressing. Source: Ref 24

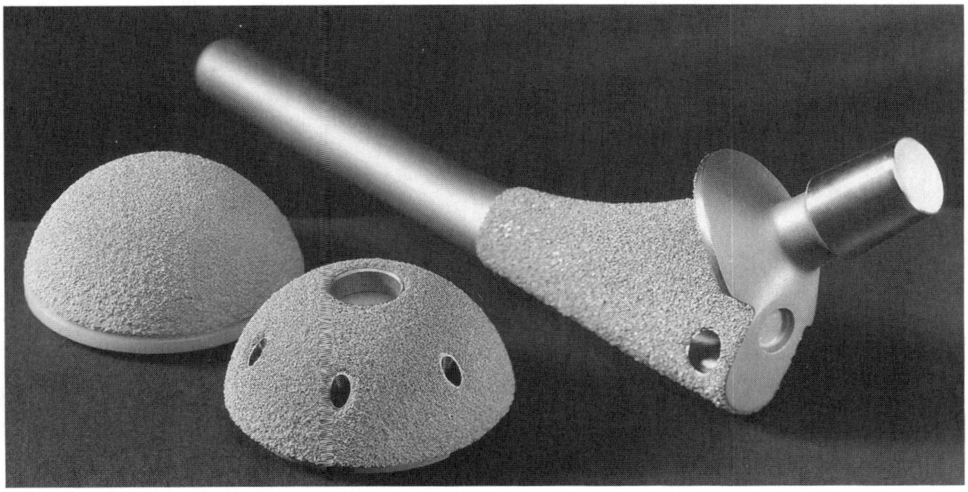

Fig. 19 Hip prostheses coated with PREP powder. Courtesy of Starmet

Table 8 Room temperature static mechanical properties of P/M Ti-6Al-4V engine mount supports in the mill annealed condition

Property	Minimum specification	Minimum result	Maximum result	Average result	Number of tests
0.2% yield strength, MPa (ksi)	827 (120)	861 (125)	916 (133)	888 (129)	34
Ultimate strength, MPa (ksi)	896 (130)	923 (134)	978 (142)	958 (139)	34
Elongation, %	10	12	16	14	16
Reduction of area, %	20	37	42	39	16
Compressive yield strength, MPa (ksi)	868 (126)	888 (129)	992 (144)	944 (137)	13
Ultimate shear strength, MPa (ksi)	523 (76)	613 (89)	640 (93)	627 (91)	8
Bearing yield strength, edge distance/pin diameter, $e/D = 1.5$, MPa (ksi)	1123 (163)	1260 (183)	1351 (191)	1316 (191)	8
Ultimate bearing strength, MPa (ksi)	1315 (191)	1557 (226)	1610 (238)	1598 (232)	8

Source: Ref 24

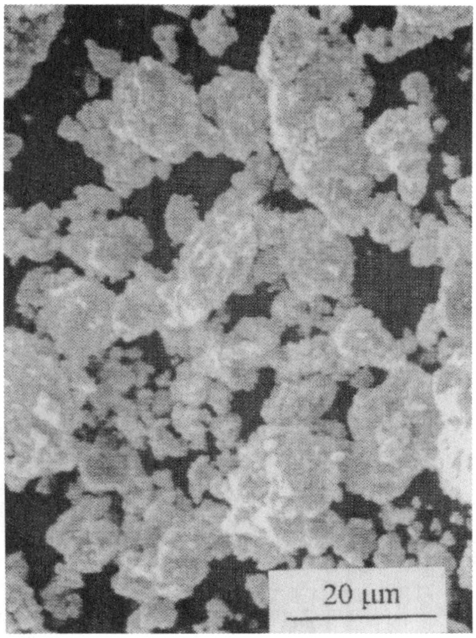

Fig. 20 Example of mechanically alloyed powder. Starting powders of titanium hydride and a master alloy were blended in the proportion to produce a mixture of 66 vol% TiAl and 34 vol% Ti_5Si_3. Source: Ref 32

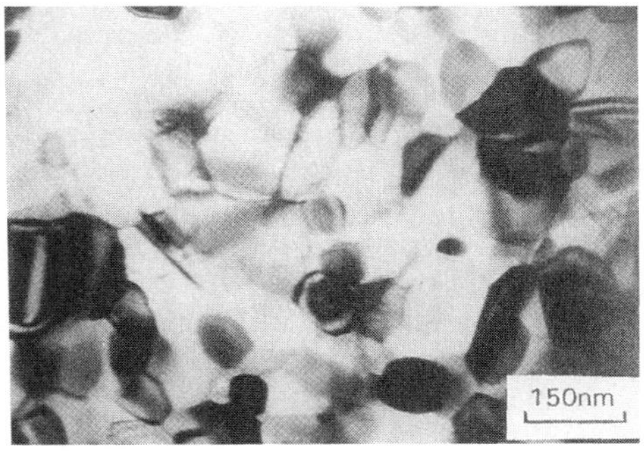

Fig. 21 Nanostructured grains in Ti-55Al (at.%) after mechanical alloying and hot isostatic pressing. Source: Ref 26

vent excessive cold welding of the powder particles during the milling operation. The material is then processed on a high-energy ball mill. Alloying occurs as the result of repeated welding, fracturing, and rewelding of the powder particles (Ref 27). Much of the work to date has involved the development of unique structural and microstructural characteristics such as supersaturated solid solutions, metastable crystalline phases, dispersion strengthened alloys, amorphous phases, and nanostructured materials. Extensive reviews of the application of mechanical alloying to titanium are contained in Ref 27 to 30. Mechanical alloying is also being studied as a technique for producing nanostructured intermetallic alloys (Ref 30, 31) and composites such as Ti-6Al-4V/SiC (Ref 32) and titanium aluminide/titanium silicide (Ref 33). Figure 20 shows an example of mechanically alloyed powder. Figure 21 shows an example of the extremely fine microstructure that can be attained using mechanically alloyed powder.

REFERENCES

1. F.H. Froes and D. Eylon, Powder Metallurgy of Titanium Alloys, *Int. Mater. Rev.,* Vol 35 (No. 3), 1990, p 162–182
2. W.J. Kroll, *J. Electrochem. Soc.,* Vol 78, 1940, p 35
3. M.A. Hunter, *J. Am. Chem. Soc.,* Vol 32, 1910, p 330
4. V. Moxson, O.N. Senkov, and F.H. Froes, Production and Characterization of Titanium Powder Products for Environmental, Medical and Other Applications, *Advanced Particulate Materials & Processes,* F.H. Froes and J.C. Hebeisen, Ed., American Powder Metallurgy Institute, 1997, p 387–394
5. M.V. Ginatta, G. Orsello, P. Perotti, and R. Berruti, A Newly Designed Plant for the Electrowinning of Titanium from Molten Salts, *Sixth World Conf. on Titanium,* P. Lacombe, R. Tricot, and G. Beranger, Ed., Société Française de Métallurgie, 1988, p 753–757
6. F.H. Hayes, H.B. Bomberger, F.H. Froes, L. Kaufman, and H.M. Burte, Advances in Titanium Extraction Metallurgy, *JOM,* Vol 36 (No. 6), 1984, p 70–75
7. E.S. Brosius, J.C. Malek, N.K. Petek, and M.J. Trzcinski, Blended Elemental Powder for Automotive Applications, *Int. Conf. on Titanium Products and Applications,* Titanium Development Association, Oct 1986

8. S. Abkowitz, P.F. Weihrauch, and H.L. Heussi, The Commercial Application of Low Cost Titanium Composites, *JOM,* Vol 47 (No. 8), 1995, p 40–41
9. F.H. Froes and D. Aeolian, Production of Titanium Powder, *Powder Metallurgy,* Vol 7, *Metals Handbook,* 9th ed., American Society for Metals, 1984, p 164–168
10. E.S.P. Payne, E.W. Timko, J.D. Hall, J. Hebeisen, O.N. Senkov, and F.H. Froes, Evaluation of a New Hydride-Dehydride (HDH) Titanium Powder, *Advanced Particulate Materials & Processes,* F.H. Froes and J.C. Hebeisen, Ed., American Powder Metallurgy Institute, 1997, p 357–361
11. U.S. Patents 4,544,404; 4,999,051; 5,084,091; and 5,213,610
12. J.H. Moll, C.F. Yolton, and B.J. McTiernan, P/M Processing of Titanium Aluminides, *Int. J. Powder Metall.,* Vol 26 (No. 2), April 1990, p 149–155
13. C.F. Yolton, Gas Atomized Titanium and Titanium Aluminide Alloys, *P/M in Aerospace and Defense Technologies,* Vol 1, F.H. Froes, compiler, Metal Powder Industries Federation, 1990, p 123–131
14. C.F. Yolton, U. Habel, and H. Clemens, Gamma Titanium Aluminide PM Sheet Production, *Advanced Particulate Materials & Processes,* F.H. Froes and J.C. Hebeisen, Ed., American Powder Metallurgy Institute, 1997, p 161–168
15. S. Kandebo, Caeser Targets Tech Transfer, *Aviat. Week Space Technol.,* 9 Feb 1998, p 89
16. D. Backman, Metal Matrix Composites and IPM: A Modeling Perspective, *JOM,* Vol 42 (No. 7), 1990, p 17–20
17. P.A. Siemers and J.J. Jackson, Ti_3Al/SCS-6 MMC Fabrication by Induction Plasma Deposition, *Titanium Aluminide Composites,* P.R. Smith et al., Ed., Report W1-TR-91-4020, Wright-Patterson AFB, 1991, p 233–251
18. J.T. Niemann and J.F. Edd, Fabrication of Titanium Aluminide Composite by Tape Casting, *Titanium Aluminide Composites,* P.R. Smith et al., Ed., Report W1-TR-91-4020, Wright-Patterson AFB, 1991, p 300–314
19. S.A. Singerman and J.J. Jackson, Titanium Metal Composites for Aerospace Applications, *Superalloys 1996,* R.D. Kissinger et al., Ed., TMS, 1996, p 579–586
20. T. Furukawa, Sumitomo Sitix in Titanium Powders, *Am. Met. Mark,* Vol 101 (No. 165), 26 Aug 1993, p 4
21. N. Arimoto, S. Mori, H. Shiraishi, and K. Yamasaki, Manufacturing Technique of Titanium Powders by Gas Atomizing, *Mater. Jpn.,* Vol 36 (No. 6), 1995, p 792–794 (in Japanese)
22. P. Lowenstein, Specialty Powders by the Rotating Electrode Process, *Prog. Powder Metall.,* Vol 37, 1982
23. W.T. Nachtrab, P.R. Roberts, and H.A. Newborn, Powder Metallurgy of Advanced Titanium Alloys, *Key Eng. Mater.,* Vol 77–78, 1993, p 115–140
24. S.A. Miller and N.F. Levoy, Rotating Electrode Atomization for the Production of Fine Powders from Reactive Metals, *Advances in Powder Metallurgy & Particulate Materials,* Metal Powder Industries Federation, 1997, p 5-23 to 5-30
25. A.A. Sheinker, G. Chanani, and J.W. Bohen, Evaluation and Application of Prealloyed Titanium P/M Parts for Airframe Structures, *Int. J. Powder Metall.,* Vol 23 (No. 3), 1987, p 171–176
26. J.S. Benjamin, Dispersion Strengthened Superalloys by Mechanical Alloying, *Metall. Trans.,* Vol 1, American Society for Metals, Oct 1970, p 2943–2951
27. F.H. Froes and C. Suryanarayana, Powder Processing of Titanium Alloys, *Rev. Part. Metals,* Vol 1, 1993, p 223–275
28. P.S. Goodwin and C.M. Ward-Close, Mechanical Alloying of Titanium Based Materials, *Mater. Sci. Forum,* Vol 235–238, 1997, p 53–58
29. C. Suryanarayana and F.H. Froes, Mechanical Alloying of Titanium-Base Alloys, *Adv. Mater.,* Vol 5 (No. 2), 1993, p 96–106
30. J. Ahn and K. Lee, Preparation of Ti-Base Intermetallic Compounds by Mechanical Alloying (Overview), *Mater. Trans., JIM,* Vol 36 (No. 2), 1995, p 297–304
31. O.N. Senkov, N. Srisukhumbowornchai, F.H. Froes, M. Öveçglu, E. Cerri, and E. Evangelista, Synthesis of TiAl-Based Intermetallics with Nanocrystalline Structure by Mechanical Alloying and Hot Isostatic Pressing: Microstructure Stability and Properties, *Advanced Particulate Materials & Processes,* F.H. Froes and J.C. Hebeisen, Ed., American Powder Metallurgy Institute, 1997, p 129–138
32. X. Chen, E.G. Baburaj, F.H. Froes, and A. Vassel, Ti-6Al-4V/SiC Composites by Mechanical Alloying and Hot Isostatic Pressing, *Advanced Particulate Materials & Processes,* F.H. Froes and J.C. Hebeisen, Ed., American Powder Metallurgy Institute, 1997, p 185–192
33. O.N. Senkov, F.H. Froes, and E.G. Baburaj, Synthesis of a Nanocrystalline Titanium Aluminide—Titanium Silicide Particulate Composite by Mechanical Alloying and Thermohydrogen Processing, *Advanced Particulate Materials & Processes,* F.H. Froes and J.C. Hebeisen, Ed., American Powder Metallurgy Institute, 1997, p 193–194

Production of Nickel-Base Powders

NICKEL-BASE POWDERS are produced by the decomposition of nickel carbonyl, by the reduction of aqueous solution of a nickel salt with hydrogen under pressure (the Sherritt process), by inert gas or water atomization, and by mechanical alloying. Of these, decomposition of nickel carbonyl and the Sherritt process, which produce metallic nickel from its ore, are the most important commercially. This article briefly reviews each of these methods.

Carbonyl Vapormetallurgy

John H. Tundermann,
Inco Alloys International, Inc.

The production of nickel powders by the decomposition of nickel carbonyl dates back to a process developed in 1889 by Ludwig Mond and his coworkers Carl Langer and Friedrich Quincke. According to Mond (Ref 1), the firm of Brunner, Mond, and Company was endeavoring to prepare chlorine from ammonium chloride, which in turn was obtained as a by-product of the ammonia-soda process, widely known as the Solvay process.

One of the obstacles to the successful production of chlorine was the unusually rapid corrosion of the nickel valves on the brick-lined tanks in which ammonium chloride was vaporized. Laboratory study of this corrosion proved that it resulted from the small amount of carbon monoxide used to sweep ammonia from the vaporizing tanks. Carbon monoxide was partially converted into carbon dioxide, and a black mixture of nickel and amorphous carbon was deposited. This observation led to a more elaborate study of the reaction of carbon monoxide with finely divided nickel.

Mond and coworkers found that four molecules of carbon monoxide at atmospheric pressure and at temperatures between 40 and 100 °C (105 and 212 °F) react with active nickel to form a colorless gas, nickel tetracarbonyl:

$$Ni + 4CO \overset{50\ °C}{\leftrightarrow} Ni(CO)_4$$

They further demonstrated that the reaction is readily reversible by heating the nickel tetracarbonyl to temperatures in the 150 to 300 °C (300 to 570 °F) range to yield pure nickel and carbon monoxide:

$$Ni(CO)_4 \overset{230\ °C}{\leftrightarrow} Ni + 4CO$$

Nickel that has been in contact with air does not react with carbon monoxide, which is why carbonyls were not discovered earlier.

Encouraged by their success in preparing nickel tetracarbonyl, Mond and coworkers immediately attempted to prepare carbonyls from all the metals at their disposal. However, it was Berthelot who first announced the formation of a volatile compound of carbon monoxide and iron, iron pentacarbonyl, in June 1891. Subsequent discoveries were made of many metal carbonyls, including cobalt, iron, molybdenum, and ruthenium carbonyls.

After preliminary laboratory trials, Mond built an experimental plant near Birmingham, England, and subsequently developed the Mond-Langer process for separating nickel from Canadian matte. By 1895. the plant successfully produced 1½ tons of nickel weekly from Canadian matte containing 40% Ni. Mond subsequently opened a refinery in Clydach, Wales (Mond Nickel Company), and during the 27 years the refinery operated, the plant processed more than 82,000 metric tons (90,000 tons) of nickel in the form of pellets that exhibited greater purity than nickel produced by any other commercial process at that time.

In this process, which in modified form is still used at the refinery in Clydach, nickel oxide produced from roasting nickel sulfide is reduced to nickel sponge by hydrogen; activated by sulfiding, and volatilized as carbonyl in an atmospheric reactor. Nickel carbonyl produced in this manner is decomposed directly in the powder and pellet units as part of a continuous process of refining.

Metal Carbonyls

Many metals form carbonyls; in fact, all of the metals of the first, second, and third transition metal series have been converted into one or more types of carbonyls. Additionally, several of the lanthanides and actinides have formed carbonyls. Nickel forms a single carbonyl, $Ni(CO)_4$, in its zero valent form:

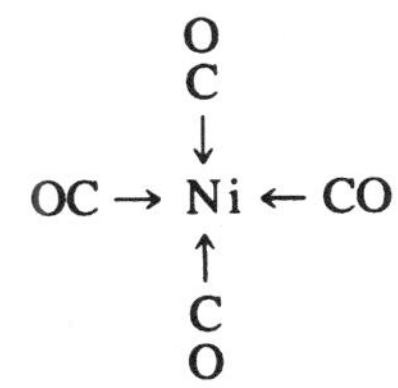

Nickel also forms a hydridocarbonyl, $H_2Ni_2(CO)_6$, where nickel has an oxidation number of −1. The carbon monoxide ligands in nickel carbonyl can be replaced by other ligands such as phosphines, phosphites, and certain unsaturated hydrocarbons where a high degree of electron density is capable of allowing "back bonding π," as well as the conventional σ bonding typical of donor ligands.

The iron penta compound, iron pentacarbonyl or $Fe(CO)_5$, rapidly condenses into the bimetallic species diiron nonacarbonyl $Fe_2(CO)_9$, which on heating further condenses into the trimetallic triiron dodecacarbonyl $Fe_3(CO)_{12}$. The nonacarbonyl compound exhibits two types of carbonyl bonding, the donor σ type and the bonding bridge type (π):

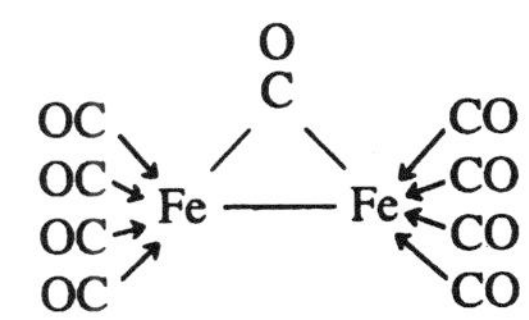

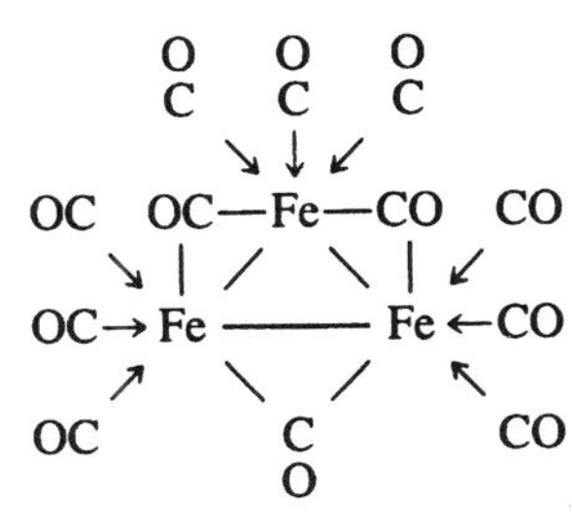

Chromium, molybdenum, and tungsten each form a single hexacarbonyl, a hexacoordinated octahedron. Cobalt forms a binuclear carbonyl, dicobalt octacarbonyl, $Co_2(CO)_8$:

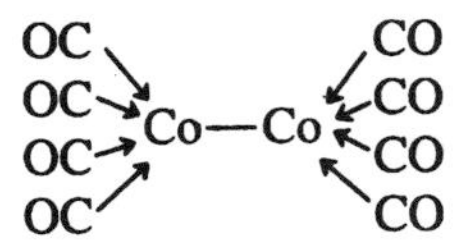

which condenses to the tetranuclear species, containing bridged cobalt atoms:

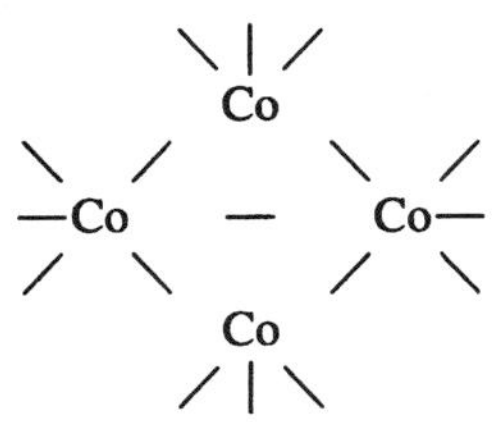

The condensing continues with the formation of what is known as a cluster carbonyl:

$$Co_8(CO)_{18}C^{-2}$$

Several types of carbonyl clusters exist; one of the largest is the tetracapped pentagonal prism:

$$Rh_{15}(CO)_{28}C_2^-$$

Some metals have avoided becoming "pure" carbonyls to date, but halo derivatives of copper, gold, platinum, and palladium have been synthesized:

$$Cu(CO)Cl \quad Pt(CO)_2Cl_2$$

$$Au(CO)Cl \quad Pd_2(CO)_2Cl_4$$

Heterocarbonyls also have been produced, including:

$$(CO)_5MnCo(CO)_4$$

$$[(CO)_5Re]_2Fe(CO)_4$$

$$Re_2Ru(CO)_{12}$$

$$Ru_2Os(CO)_{12}$$

Metal carbonyl carbides also exist, such as $Fe_5(CO)_{15}C$ and the dicarbide clusters $Ru_{10}C_2(CO)_{24}$, $Co_{13}C_2(CO)_{24}$, and $Rh_{15}C_2(CO)_{23}$. Another example of metal carbonyl is the dinitrogen derivative:

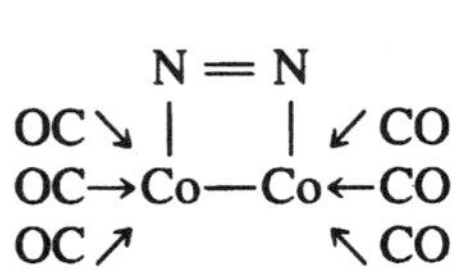

A complete list of all the metal carbonyls that can be formed is beyond the scope of this article. The above examples are provided to illustrate the diversity of the chemical forms of metal carbonyls.

Metal Carbonyl Formation and Decomposition

While the equation for metal carbonyls is generally written as follows, the mechanism is far more complex than the equation implies:

$$x\text{Me} + y\text{CO} \leftrightarrow \text{Me}_x(\text{CO})_y$$

At room temperature, carbonyls typically form liquids, vapors, and colored crystals. Carbonyls may be formed, volatilized, and condensed in a continuous process to separate them from the inert constituents of the feed. Some must be separated with elaborate techniques, while still others are intractable materials. They may be formed as liquids or as solutions in organic solvents and separated from inert solids by filtration. Crude carbonyls are separated from each other and residual impurities by distillation, sublimation, recrystallization, or selective solution of the carbonyl. Purified carbonyls are decomposed by heating into the carbon monoxide and pure metal. When a metal carbonyl can be separated from its impurities, subsequent decomposition generally results in high-purity metals. Exceptions to this include the formation of carbides.

Nickel tetracarbonyl was the first carbonyl discovered in the 1890s. Iron pentacarbonyl and cobalt octacarbonyl followed soon after; these remain the only carbonyls whose singular chemical properties have been applied commercially to extractive metallurgy, although it is speculated that ruthenium separation by carbonyl formation should ensue. In the formation of nickel tetracarbonyl, a mixture of freshly prepared nickel metal and nickel sulfide is heated in the presence of carbon monoxide.

Thermal decomposition reactions of commercial interest for cobalt, iron, and nickel carbonyls occur at temperatures about 200 °C (390 °F), where carbonyls can be handled in vapor form. Under these conditions, reaction kinetics permit acceptable powder production rates to be maintained. Carbonyls are heated rapidly to the desired decomposition temperature; at this temperature, nuclei form in the vapor to provide the required sites for metal deposition.

Decomposition products are the pure metal and carbon monoxide. During decomposition in commercial decomposers, the disproportionation reaction of carbon monoxide is catalyzed by the freshly formed metals. Iron is much more active as a catalyst than either nickel or cobalt, and therefore carbon contents in nickel carbonyl-derived powder are related to the trace iron content:

$$2CO \underset{Fe}{\rightarrow} CO_2 + C$$

Nickel Tetracarbonyl

The versatility of the carbonyl extraction of nickel is characterized not only by selectivity in formation, but also by the relative ease of separation and decomposition to high-purity metal. Under the conditions employed for reaction at atmospheric pressure, carbonyl-forming impurities in crude nickel metal do not enter the gas phase. Thus, extraction of nickel by its volatilization from crude metal as carbonyl vapor is a highly selective process.

For a given condition of surface activity, the rate of gaseous or liquid carbonyl formation is determined by the temperature of the reaction and increases with increasing partial pressure of carbon monoxide. Whereas the rate of carbonyl formation increases with increasing temperature, the equilibrium mole fraction of carbonyl formed decreases sharply with increasing temperature. At a given total pressure, there is an optimal temperature at which metal carbonyl will form, as shown in Fig. 1.

Kinetic data for the nickel tetracarbonyl reaction can be correlated by means of a modified first-order rate equation. The ultimate degree of metal conversion of "initial surface activity" has a definite value at a given temperature and pressure, which permits the derivation of a functional relationship between surface activity and metal conversion. Substituting this relationship and integrating the first-order rate equation yields a semiempirical model for correlating nickel tetracarbonyl formation rate data at a given temperature (Ref 3):

$$\ln \frac{a^o}{a^o - x} = K_o \left(P_{coi} - \sqrt[b]{\frac{P_{cai}}{K_e}} \right) t$$

where a^o is the initial activity of the surface, b is the number of moles of carbon monoxide per mole of carbonyl, K_o is the rate constant in grams per square centimeter per hour, K_e is the equilibrium constant, P_{cai} is the partial pressure of carbonyl at the interface in torr, P_{coi} is the partial pressure of carbon monoxide at the inter-

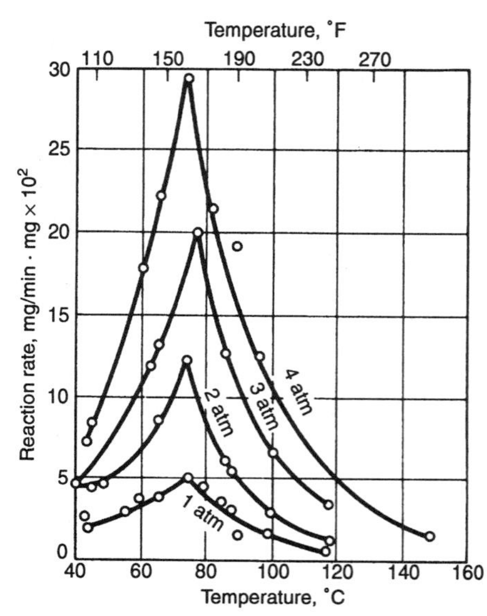

Fig. 1 Influence of system pressure and temperature on reaction rate in the formation of nickel tetracarbonyl. Source: Ref 2

face in torr, t is time in seconds, and x is the fraction of metal reacted.

From an engineering design point of view, conversion of nickel to a carbonyl at a fixed temperature depends primarily on the partial pressure of the carbon monoxide present. The effect of increased system pressure on the conversion of nickel to a carbonyl has been shown experimentally and demonstrated commercially. The volume change inherent in the carbonyl reaction—four carbon monoxide molecules combining in one molecule of nickel tetracarbonyl—also points to an increase in carbon monoxide pressure as a means of improving reaction kinetics. In addition to directly increasing the rate of reaction, increased pressure stabilizes the carbonyl, thus permitting reaction at higher temperatures for a further increase in rate. The nickel tetracarbonyl formation reaction is highly exothermic:

$$\mathrm{Ni} + 4\mathrm{CO} \rightarrow \mathrm{Ni(CO)_4}$$
$$\Delta_f H^{298} = +160.4\ \mathrm{kJ/g \cdot mol}\ (38.4\ \mathrm{kcal/g \cdot mol})$$

Accordingly, high conversion-rate commercial nickel carbonylation processes require large heat removal systems.

At room temperature, nickel tetracarbonyl is a colorless volatile liquid with a high vapor pressure. It boils at about 43 °C (109.4 °F) and begins to decompose at 60 °C (140 °F) or less, depending on conditions. It is only slightly soluble in water, but is highly miscible with many organic solvents. Some physical properties of nickel tetracarbonyl include:

Chemical formula	$Ni(CO)_4$
Color and state	Colorless liquid
Molecular weight	170.75
Nickel, %	34.37
Melting point	–25 °C (–13 °F)
Boiling point	43 °C (109.4 °F) at 101.1 kPa (751 mm Hg) 42.1 °C (107.8 °F) at atmospheric pressure
Specific gravity	1.36153 at 0 °C (32 °F), 1.27132 at 36 °C (97 °F)
Heat of formation ($\Delta_f H^{298}$)	–602.5 kJ/g · mol (–144.0 kcal/g · mol)

Nickel tetracarbonyl vapor is exceedingly toxic when inhaled; 50 ppb of the compound in the air is the maximum allowable concentration designated by the American Conference of Governmental and Industrial Hygienists. Nickel tetracarbonyl should be handled only by personnel who are familiar with its toxic properties.

The stability of metal carbonyls varies considerably. Nickel tetracarbonyl begins to liberate carbon monoxide and nickel at 0 °C (32 °F) in a vacuum, and the decomposition becomes rapid above 60 °C (140 °F) in an inert gas under atmospheric pressure. Thermal decomposition is endothermic and is markedly inhibited by the presence of carbon monoxide. The mechanism of formation of nickel powder by the thermal decomposition of nickel tetracarbonyl is complex. In the formation of powder particles from the gaseous phase, several simultaneously occurring processes exert an influence. These include the formation of complex nuclei, secondary crystallization of nickel on the surfaces of the finest particles, the interaction of particles in the crystallization process, and side reactions such as the decomposition of carbon monoxide. At a given temperature, the homogeneous part of the decomposition of nickel tetracarbonyl follows the rate equation (Ref 3):

$$r = \frac{K_o P_{ca}}{1 + K_g P_{co}}$$

where r is the decomposition rate in grams per cubic centimeter per hour, K_g is the adsorption constant of carbon monoxide on metal torr^{-1}, K_o is the rate constant in grams per square centimeter per hour, P_{ca} is the partial pressure of carbonyl in torr, and P_{co} is the partial pressure of carbon monoxide in torr. The decomposition rate for nickel tetracarbonyl is directly proportional to the partial pressure of the carbonyl and inversely proportional to the partial pressure of the liberated carbon monoxide.

Conditions that influence the formation of the self-nucleating nickel particles during carbonyl decomposition (such as process temperature and the concentration and rate of the nickel tetracarbonyl supply to the decomposer) can vary greatly, which affects the physical and technical properties of the powders produced. Additions to the nickel tetracarbonyl gas stream sent to the decomposer can alter the mechanism by which the powder is formed, as well as its resulting morphology. For example, hydroquinone vaporized with the carbonyl acts as a free radical trap, resulting in the formation of macroparticles of nickel metal. Nickel tetracarbonyl can be photochemically decomposed to form elemental nickel by light of wavelengths shorter than 390 nm (3900 Å).

The most important industrial application of nickel tetracarbonyl is for refining nickel, in which nickel pellets and powder products are formed. Gas-plated products and a wide range of metallic and nonmetallic (nickel-coated graphite) powders can be produced by forming coatings of nickel on various surfaces by decomposing the carbonyl under special conditions. Nickel tetracarbonyl is manufactured on a limited basis to serve the chemical industry. The carbonyl serves as a catalyst for organic synthesis, as well as providing a means for the manufacture of other organonickel compounds. It also is used for vapor plating on smooth surfaces (such as plastic and metal) and to form nickel molds for the glass industry.

Commercial Processes

Currently, Inco Ltd. produces high-purity nickel powders by the thermal decomposition of nickel tetracarbonyl. Gaseous nickel tetracarbonyl is formed by reacting carbon monoxide with nickel concentrates under controlled conditions; subsequent thermal decomposition of the gas permits recovery of the nickel as a fine metallic powder and nickel pellets. The process, which affords a high degree of purity with respect to metallic elements other than nickel, produces nickel powders with extremely uniform particle size and structure.

The original Mond-Langer process has been greatly improved during its continued use. Commercial nickel refineries employing the carbonyl process to produce nickel powder are located on the site of the original Mond facility in Clydach, Wales, and in Copper Cliff, a subdivision of Sudbury, Ontario.

The Clydach refinery began nickel output from Sudbury cupriferrous nickel matte in 1902. Current plant feed matte is granular nickel oxide that contains small amounts of copper, cobalt, and iron, as well as siliceous matter. This refinery still uses the basic concept of the Mond-Langer process and produces nickel products using atmospheric pressure carbonyl processing. The latest innovations involve replacement of the many small hearth-type reducers and volatilizers of the original works by large rotary kilns arranged in trains of two kilns with a smaller sulfiding unit. In the initial unit of each train, the nickel feed matte is reduced to metal by a countercurrent flow of preheated hydrogen. The reduced nickel matte is then activated by sulfiding and is volatilized as carbonyl in the final kiln by a countercurrent flow of carbon monoxide. The resulting nickel tetracarbonyl vapor is decomposed directly into pure nickel pellets and powders in reactors or decomposers.

The Copper Cliff Nickel Refinery began operations in 1973 with two newly developed techniques, including the use of top-blown rotary converters that take a 64 metric ton (70 ton) charge of feed materials. The other manufacturing method uses the Inco pressure carbonyl process for the recovery of nickel from a variety of nickel-bearing feed materials, including the removal of nickel from copper, cobalt, and precious metal concentrates. The refinery produces pellets and powder of 99.9+% nickel purity and a codecomposed iron-nickel powder.

Environmental controls at the plant include a totally enclosed process that constantly recycles all process reagents. The Copper Cliff refinery uses the same basic process chemistries, but forms nickel tetracarbonyl at high pressure, which is necessary to extract nickel in the presence of high copper concentrations.

The plant feed is a mixture of oxides and sulfides of nickel, copper, iron, and cobalt, other crude metallics, and partially processed precious metals and refinery residues that contain nickel. This mixture is charged along with coke into one of two 64 metric ton (70 ton) capacity, top-blown rotary converters, where it is melted and partly blown to slag silicates and some of the iron. The converter-treated hot material is subsequently granulated by pouring it through high-velocity water jets. At this point, the granules contain 65 to 75% Ni, 15 to 20% Cu, 2 to 3% Fe, and 3% S present as sulfides.

The metallic granules are batch-reacted with carbon monoxide at temperatures up to 180 °C (360 °F) and pressures up to 70 atm (1030 psi) in one of three 136 metric ton (150 ton) capacity rotating reactors. Nickel, and some iron, is extracted as a crude carbonyl vapor, while copper, cobalt, precious metals, and impurities are retained in the residue. Removal of iron is controlled to between 20 and 50% of the input iron level. Iron concentration in the granules rarely exceeds 4%. Cobalt in the feed does not form $Co_2(CO)_8$ or $Co_3(CO)_{12}$, because these compounds occur only when carbon monoxide pressures of 150 atm (2205 psi) are achieved.

Formation of nickel tetracarbonyl is quite exothermic; consequently, the reactors are water cooled. Extracted carbonyl vapor is liquefied and stored at atmospheric pressure. The carbon monoxide carrier gas also is recycled. The liquefied product stream, containing mutually soluble nickel and iron carbonyls, subsequently is pumped to distillation columns, where it is separated into nickel tetracarbonyl vapor and an iron-rich liquid carbonyl. Fractionating the crude carbonyl liquid is achieved easily, because the boiling points of the nickel and iron carbonyls are 43 °C (109.4 °F) and 102.8 °C (217 °F), respectively. The system produces nickel tetracarbonyl vapor of 99.998% purity, with the bottom liquor having nickel-to-iron ratios of up to 3 to 7. At the top of the column, nickel tetracarbonyl vapor is siphoned off and fed directly into powder decomposers. Somewhat below the top of the column, liquid nickel tetracarbonyl is taken off and either held in storage or used to feed the nickel pellet decomposers.

The powder decomposers are steel cylinders, the walls of which are heated with high-capacity electrical resistance heaters. Nickel carbonyl vapor is introduced into the top of the decomposer chamber slightly above atmospheric pressure, where it contacts the heated decomposer walls that are preset at a temperature between 250 and 350 °C (480 and 660 °F). The thermal shock decomposes nickel tetracarbonyl into nickel powder, with the release of carbon monoxide. The latter is recycled through filters to the main gas compressor to be recycled to the pressure carbonyl reactors, while the powder is collected, gas purged, stabilized with an oxide coating, and transferred to storage completely free from carbonyl and carbon monoxide.

The dangerous nature of the gaseous and liquid carbonyl compounds formed in the Inco processes is fully countered by established, effective, and practical safety measures. All product discharge systems are purged completely prior to powder and pellet packaging.

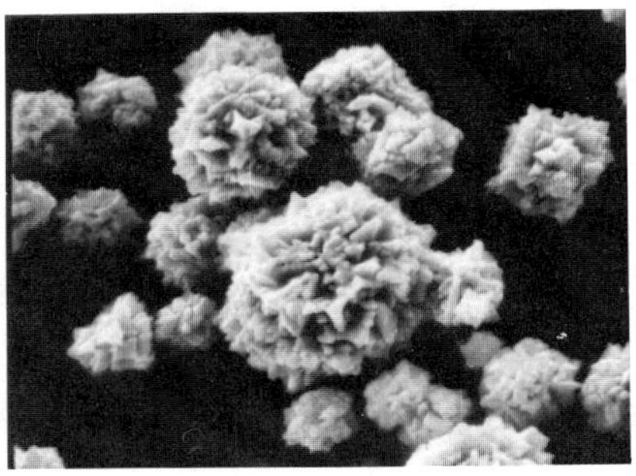

Fig. 2 Scanning electron micrograph of general-purpose nickel powder produced by carbonyl decomposition. 3000×

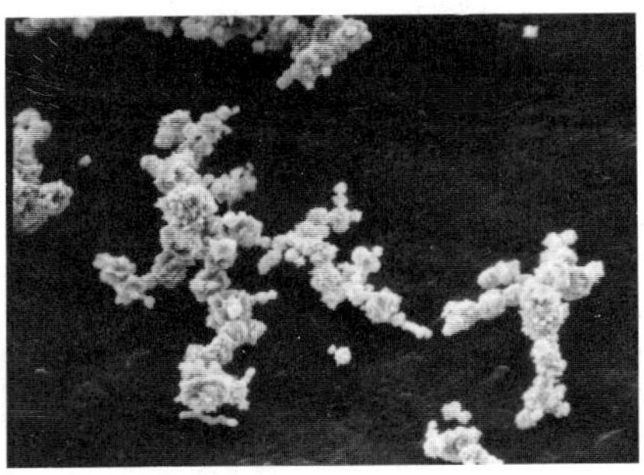

Fig. 3 Scanning electron micrograph of the filamentary-type nickel powder produced by carbonyl decomposition. 1000×

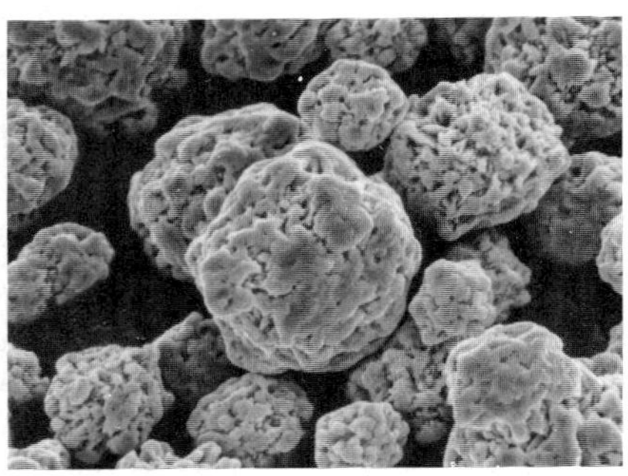

Fig. 4 Scanning electron micrograph of high-density (fine) nickel powder produced by carbonyl decomposition. 1000×

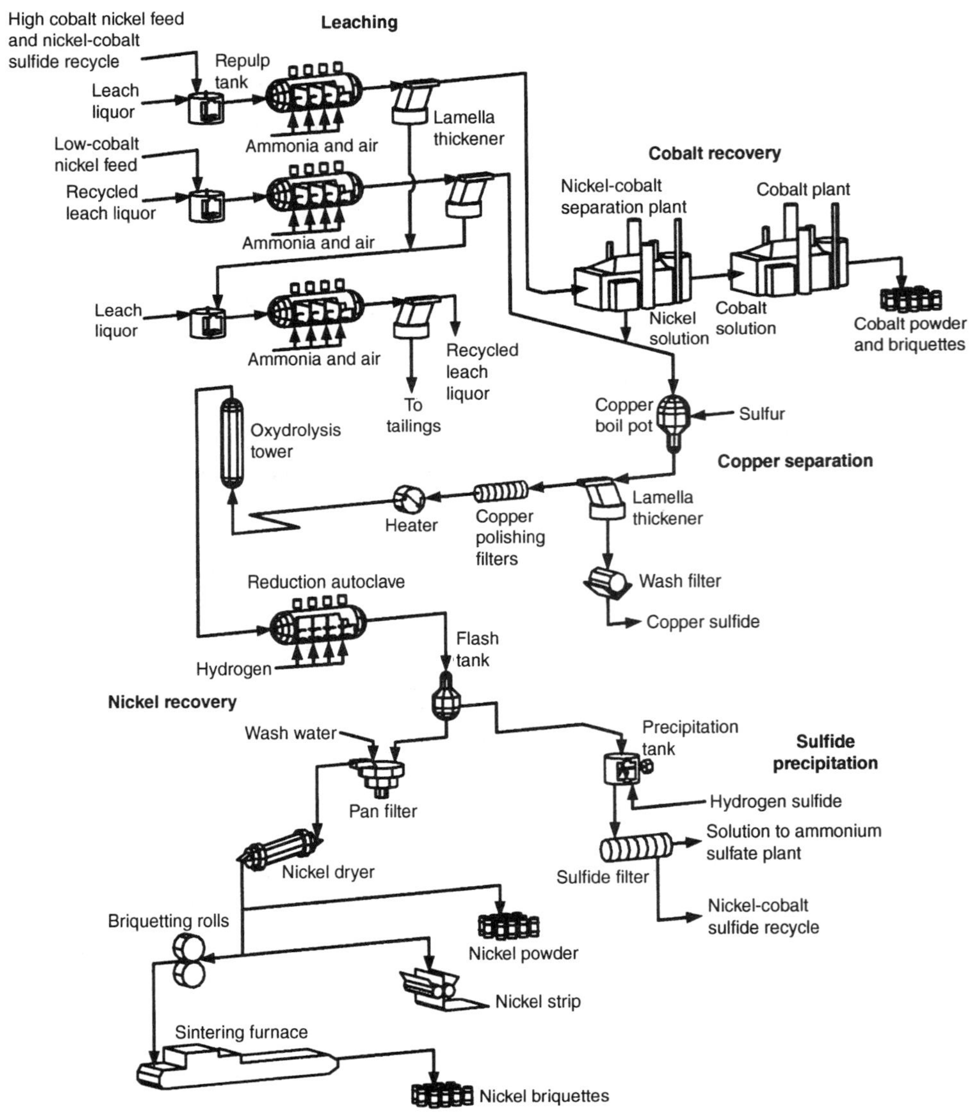

Fig. 5 Flowsheet of the Sherritt ammonia pressure leach process

Powder Properties and Applications

Currently, there are four types of nickel powders produced by thermally decomposing nickel tetracarbonyl—single spikey particles, filaments, high-surface-area powders, and high-density semismooth particles. These powders exhibit uniform size and structure and high surface areas, and the method of production affords a high degree of purity with respect to other metallic elements.

Spikey particles are a general-purpose type of nickel powder (Inco type 123). They are fine and regular in shape. with rough surface projections (Fig. 2). The powder typically contains 700 to 900 mL/m^3 (700 to 900 ppm) oxygen, 3 to 5 mL/m^3 (3 to 5 ppm) iron, 1 mL/m^3 (1 ppm) sulfur or less, and 600 to 700 mL/m^3 (600 to 700 ppm) of graphitic carbon. The Fisher subsieve size is 3 to 7 μm, and apparent density is 1.8 to 2.7 g/cm^3. Surface area is on the order of 0.4 m^2/g (BET).

Filimentary nickel powders (Inco types 255, 287, etc.) are characterized by their unique, spikey, chainlike structure of relatively fine particles, which makes them fluffy in nature, with low apparent density (0.5 to 1.0 g/cm^3) and large specific surface area (0.6 to 0.7 m^2/g). The structure and asymmetrical filament shape of this powder type are illustrated in Fig. 3.

Extra fine nickel powders (e.g., Inco type 210) also have filamental morphology, but the filaments are finer and specific areas range from 1.5 to 6 m^2/g depending on the grade.

Semi smooth high-density nickel powders are available in fine and coarse sizes. The fine is 10 to 20 μm in diameter (Fig. 4); the coarse is −16+40 mesh. Powder apparent density ranges from 3.5 to 4.2 g/cm^3.

The properties of these high-purity nickel powders have been tailored to meet the demanding requirements of the chemical, energy, and metal industries. Currently, these products provide the basic nickel source for a wide range of specialized products. Application areas for these powders include:

- Batteries and fuel-cell electrodes
- P/M parts
- Ferrites
- Carbide cutting tools
- Catalysts
- Welding products
- Pigments and coatings
- Chemicals
- Electronic alloys
- Getters
- Conductive resins and plastics
- Electromagnetic shielding

Production of Nickel Powder by Hydrometallurgical Processing

Gavin Freeman, Sherritt International Corporation

The production of nickel powders by hydrometallurgical processing incorporates leaching, solution purification, and metal recovery operations. The Sherritt (Corefco) refinery in Fort Saskatchewan, Alberta, Canada, provides an example of hydrometallurgical processing used in the production of nickel powder. This refinery commenced operations in 1954, at which time the feedstock for the plant was nickel concentrate from Sherritt's Lynn Lake mining operations in Manitoba. The Sherritt refinery was the first successful commercialization of a hydrometallurgical process for the treatment of a nickel sulfide concentrate. The refinery has undergone process and equipment modifications in recent years to increase its capacity and enable the acceptance and treatment of different feedstocks. At present, the primary feed to the refinery is a high-grade nickel cobalt sulfide from Moa, Cuba.

A flow diagram of the Sherritt process is shown in Fig. 5. The Sherritt process is adaptable to handle a wide variety of different feedstocks. It has been adopted by Western Mining Corporation in a refinery at Kwinana, Western Australia. Further information regarding the Sherritt process is contained in Ref 4 to 8.

Feed to Process. At present, the principal feed to the Sherritt refinery is a nickel and cobalt sulfide produced in Moa, Cuba. The Moa Bay lateritic deposits contain ore that is high in iron and also relatively rich in cobalt. The processing of this ore is accomplished through pressure leaching with sulfuric acid. The key step in this process is the separation of nickel and cobalt from iron. Iron, which typically constitutes 50% of the ore, is separated by precipitation at high temperature and pressure in an acid leaching step. It is possible to achieve the selective dissolution of nickel and cobalt with less than 1 g/L iron in the leach liquor. The leach liquor is neutralized using limestone mud, and the metal values in the leach liquor are then precipitated with hydrogen sulfide. The chemical analysis of the cobalt-nickel sulfide produced at Moa is given in Table 1. This material is packaged, as a wet filter cake, in bulk bags and transported to the refinery in Fort Saskatchewan. A schematic flowsheet for the Moa Bay process is shown in Fig. 6.

Leaching. At Fort Saskatchewan, the first stage in the refining process is the leaching of the metal values from the sulfide feeds into solution. The fine sulfide concentrate feed materials are blended with an ammonium sulfate leaching liquor in a repulp tank and are then fed into the leach autoclaves. In the leaching process, air, ammonia, and the sulfidic minerals react in a series of interrelated steps. The leaching process is carried out at elevated temperature and pressure in a continuous process. The typical conditions in the leach autoclaves are given in Table 2.

The net leach reaction is the oxidative dissolution of the metal sulfides to soluble ammine complexes with the simultaneous oxidation of sulfide sulfur (S^{2-}) to form a series of soluble

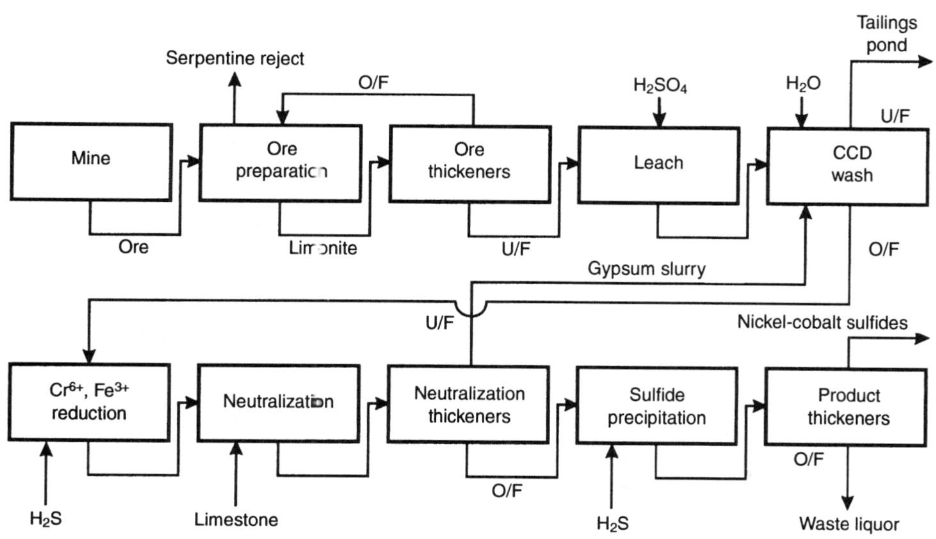

Fig. 6 Flowsheet for the Moa Bay process. O/F, thickener overflow; U/F, thickener underflow; CCD, counter current decantation

Table 1 Chemical composition of Moa sulfides

Component	Analysis, wt %
Nickel	51–56
Cobalt	5.0–6.0
Iron	0.5–0.8
Copper	0.1–0.2
Sulfur	32–37
Zinc	1.0–1.3

Table 2 Leaching conditions

Parameter	Value
Temperature, °C	90–95
Pressure, psig	110–120
Ammonia, g/L	80–110
Ammonium sulfate,g/L	150–200

thio salts including sulfate (SO_4^{2-}). The general equation for the ammoniacal leaching of sulfides may be represented as:

$$MS + nNH_3 + 2O_2 \rightarrow M(NH_3)_n^{2+} + SO_4^{2-}$$

where M = nickel, cobalt, iron, copper, zinc, and n = 2 to 6.

The leaching process takes advantage of the differing stability of the ammine complexes. The iron ammine complex is the least stable and is completely hydrolyzed to a hydrated iron oxide, $Fe_2O_3{\cdot}H_2O$ (hematite), which is reprecipitated. The leach slurry passes to a lamella thickener where the leach liquor is separated from the solid residue.

In fact, the actual reaction chemistry of the leaching process is much more complex due to the behavior of sulfur in alkaline solutions. The reaction sequence of the oxidation of sulfur to form sulfate, sulfamate, and a range of ammonium thiosalts may be represented as follows:

$$S^{2-} \rightarrow S_2O_3^{2-} \rightarrow S_nO_6^{2-} \rightarrow SO_4^{2-} + SO_3NH_2^-$$

where n = 2 to 6. At any given time, the leach liquor will contain sulfur in several forms. The amounts of these species present depend on the leach conditions and duration (Fig. 7.). The leach solution typically contains 60 to 70 g/L Ni, 6 to 9 g/L Co, 1 to 2 g/L Cu, 130 g/L NH_3, 200 g/L $(NH_4)_2SO_4$, and varying amounts of other sulfur oxyanions.

Before it is possible to reduce the nickel from the leach solution as the pure metal, several solution purification steps must be completed.

Cobalt Separation. The leach liquor is transferred to the nickel-cobalt separation plant. Here cobalt is separated from the nickel as pure crystalline cobaltic hexammine salt. The Sherritt cobalt refining process is further discussed in the article "Production of Cobalt-Base Powders" in this Volume. After the removal of cobalt, the nickel-rich, or "pregnant," solution proceeds through the copper-removal circuit.

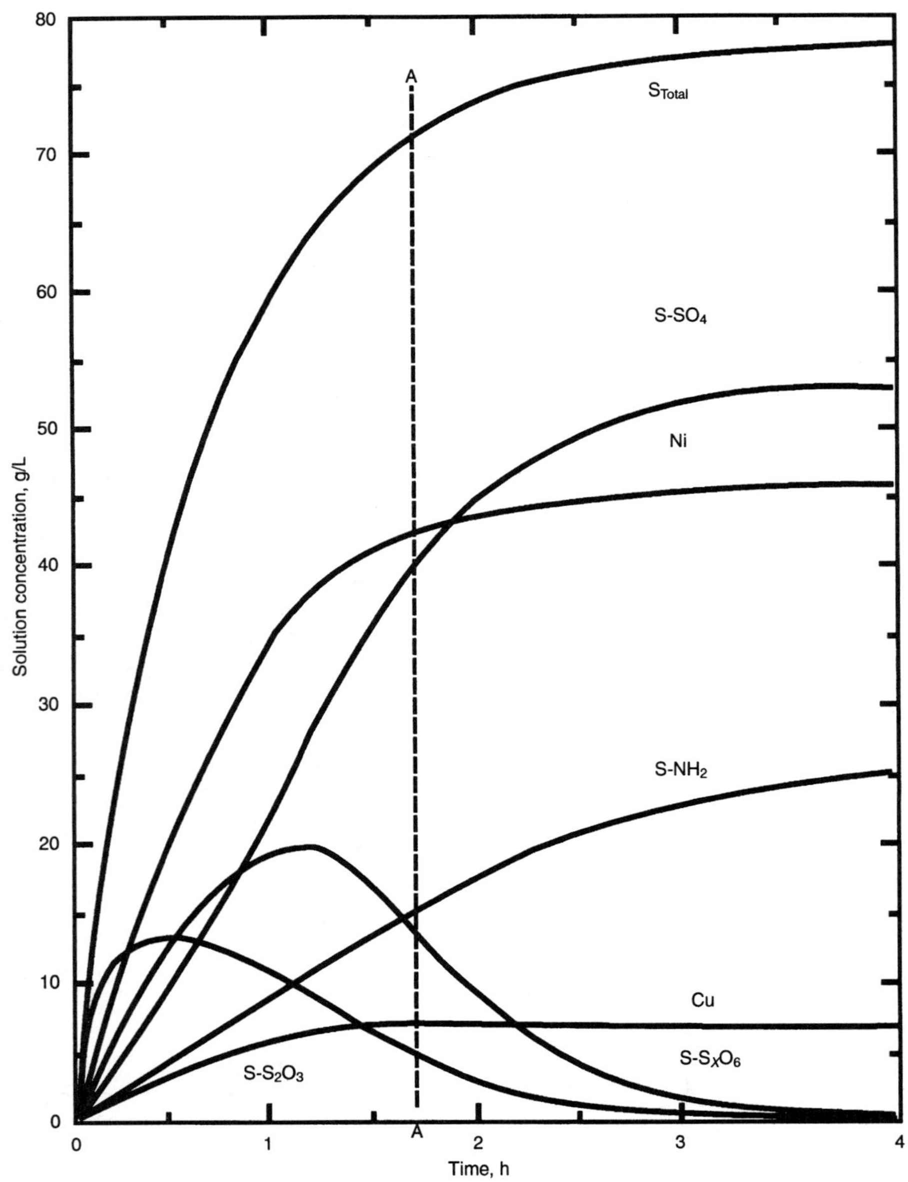

Fig. 7 A representation of the leach solution composition as determined in batch leaching tests

Copper Removal. Copper is removed from the leach liquors and pregnant solution from cobalt separation in a unit operation referred to as the "copper boil." In the copper boil, the combination of lowering the ammonia level by boiling to distill off the free ammonia and the addition of elemental sulfur and sulfur dioxide at elevated temperature results in the precipitation of copper sulfide. The copper sulfide is a mixture of cupric sulfide (CuS) and cuprous sulfide (Cu_2S) and is prepared by the reaction of soluble copper with thiosulfate ($S_2O_3^{2-}$):

$$Cu^{2+} + S_2O_3^{2-} + H_2O \rightarrow CuS + SO_4^{2-} + 2H^+$$
$$2Cu^+ + S_2O_3^{2-} + H_2O \rightarrow Cu_2S + SO_4^{2-} + 2H^+$$

The selectivity for copper is achieved by the fact that copper sulfide is less soluble than nickel sulfide. This copper sulfide by-product is filtered, washed, and then sold to a copper smelter.

Oxidation and Hydrolysis. The solution remaining after copper removal contains appreciable amounts of sulfamate ($SO_3NH_2^-$) and other forms of unsaturated sulfur. The removal of all unsaturated sulfur compounds is necessary before the nickel can be recovered. If these sulfur compounds are not fully converted to sulfate, the nickel powder product would contain unacceptably high levels of sulfur. These are removed by hydrolyzing the sulfamate to sulfate at 245 °C (473 °F) and oxidizing the thionates to sulfate with air at 4800 kPa (700 psi) in one combined step referred to as "oxydrolysis." The major reactions taking place in oxydrolysis are:

$$(NH_4)_2S_2O_3 + 2O_2 + H_2O + 2NH_3 \rightarrow 2(NH_4)_2SO_4$$
$$(NH_4)_2S_3O_6 + 2O_2 + 2H_2O + 4NH_3 \rightarrow 3(NH_4)_2SO_4$$
$$(NH_4)SO_3(NH_2) + H_2O \rightarrow (NH_4)_2SO_4$$

Ammonium sulfamate is a herbicide and must be hydrolyzed to sulfate to avoid contamination of the ammonium sulfate fertilizer. Ammonium sulfate is crystallized from the reduction end solution, following sulfide precipitation of the minor amounts of metal remaining in solution after hydrogen reduction.

Metal Recovery. In the final step of the Sherritt process, nickel is precipitated from solution as a metal powder through a process referred to as hydrogen reduction. The objective of the hydrogen reduction process is to precipitate the nickel ions present in the purified solution to nickel powder using hydrogen gas as a reductant. The reduction of metals from solution by hydrogen requires that the hydrogen potential of the system should exceed the electrode potential of the metal ions, in which case hydrogen will go into solution and the metal precipitates. The hydrogen potential is governed by the pH of the solution and by the partial pressure of hydrogen applied as shown in Fig. 8. The electrode potential of the metal ions is, however, little affected by concentration.

The reduction process is carried out in mechanically agitated horizontal autoclaves at elevated temperatures and pressures. This is a batch process consisting of two distinct stages, initial nucleation, and subsequent densifications. In the initial nucleation stage, fine nickel seed powder is formed in the autoclave. The densification refers to the process whereby nickel from solution is reduced onto the seed powder, which then grows in size. The reduction process, where 1 mole of hydrogen reduces 1 mole of nickelous ions to metallic nickel and yields 2 moles of ammonium ions as by-product, can be written:

$$Ni^{2+} + 2NH_3 + H_2 \rightarrow Ni^{o} + 2NH_4^{+}$$

The presence of free ammonia is necessary to neutralize the hydrogen ions produced during reduction. At low pH and lower hydrogen potentials the reduction of the metal ions effectively ceases. During the reduction process, the slurry in the autoclave is agitated by impellers; however, once reduction is complete the agitation is stopped and the nickel powder is allowed to settle. The spent solution (or reduction end solution) is then pumped out and the autoclave is filled with fresh nickel solution, the agitators are restarted and hydrogen pressure is applied. This batch reduction process is repeated 50 to 60 times to build up consecutive coatings of nickel on the powder particles. The overall process is stopped when the powder has reached the desired size. During the final discharge, the agitators are kept running and the entire contents of the autoclave are pumped to the flash tank.

In addition to nickel, the feed solution to reduction also contains significant quantities of cobalt and zinc. Zinc does not present a problem with respect to contamination of the product powder, because zinc will not be reduced to the metallic state under the conditions employed. Cobalt on the other hand has a similar standard reduction potential to that of nickel (–0.267 V for cobalt compared to –0.241 V for nickel). In order to effect the preferential reduction of nickel, the ammonia concentration is adjusted in the nickel reduction feed solution to give a NH_3-to-total metals molar ratio of 1.9 to 1.

In practice, the reduction feed solution contains 65 g/L Ni, 2.5 g/L Co, and 2 g/L Zn. The reduction of nickel ions to metal occurs rapidly only as a heterogeneous reaction, in which nickel is precipitated onto an existing surface. Figure 9 shows the change of particle size and apparent density with the number of reduction cycles or densifications. Figure 10 shows a scanning electron micrograph of a Sherritt produced nickel powder. The analysis of Sherritt's standard grade nickel powder is given in Table 3.

The addition of various organic additives to the densification reductions can also be used to provide the means of controlling the size, shape, and surface morphology of the powder produced.

The slurry of powder and reduction end solution is discharged from the autoclave and collected in flash tanks. The solution is allowed to overflow into a holding tank and is then pumped to sulfide precipitation. The nickel powder is then drawn off from the flash tank cone as a 90% solids slurry and is filtered and washed on a rotary pan filter. From the pan filter the wet powder (typically 5% moisture) is fed by a screw feeder into a gas-fired rotary dryer.

The dried-and-cooled nickel powder is either packaged as a powder product, or it can be directed to the briquetting operation. For briquetting, the nickel powder is first blended with an organic binder in a pug mill and is then compressed using a roll-briquette device. The briquettes are fed through a sintering furnace to effect carbon and sulfur removal and, through sintering at high temperature, impart structural integrity to the briquettes. The removal of carbon from the briquettes relies on the oxidation of carbon and its release as either CO or CO_2. In contrast to this, the mechanism for sulfur removal depends on the maintenance of reductive conditions through hydrogen injection, with the release of H_2S. The sintering furnace therefore consists of three distinct zones: the oxidizing "preheat zone" for carbon removal, the reducing "hot zone" for sulfur removal, and finally a "cooling zone." After exiting the furnace, the briquettes are packaged as the finished product (Fig. 11).

Sulfide Precipitation. The liquor from hydrogen reduction contains residual nickel, cobalt, and zinc. These metal ions are precipitated using hydrogen sulfide and recovered as sulfide precipitates. The general precipitation reaction is:

$$M^{2+}(aq) + H_2S \rightarrow MS(s) + 2H^{+}(aq)$$

where M = nickel, cobalt, zinc. Metals can be selectively removed from solution by controlling pH of the sulfide precipitation reaction. The pH is controlled by the addition of ammonia to neutralize the acid generated in the precipitation reaction:

$$NH_3 + H^{+} \rightarrow NH_4^{+}$$

The nickel and cobalt sulfides are then returned to the leach, the zinc sulfide (which is precipitated independently of nickel and cobalt sulfide) is sold as a by-product, and the barren liquor (metal free ammonium sulfate solution) proceeds to ammonium sulfate recovery.

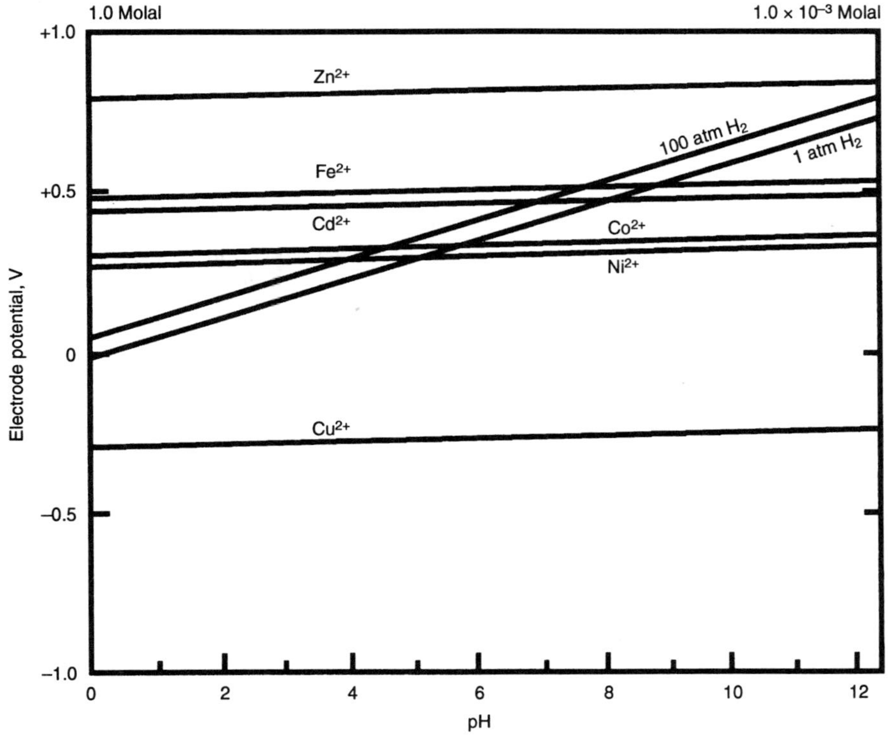

Fig. 8 Hydrogen and metal electrode potentials in solution

Table 3 Typical properties of Sherritt's standard grade nickel powder

Chemical composition, %	
Nickel	99.9
Cobalt	0.08
Copper	0.003
Iron	0.010
Sulfur	0.025
Carbon	0.006
Mesh screen size, %	
+100	0–10
–100+150	5–30
–150+200	20–45
–200+250	10–25
–250+325	10–35
–325	5–25
Apparent density, g/cm³	3.5–4.5

Ammonium Sulfate Recovery. The solution obtained after removal of the metal values as metal sulfides, contains only ammonium sulfate. This solution, referred to as "barren solution" is then evaporated and ammonium sulfate, which is formed as an overall refinery by-product, is recovered in a crystalline form. The ammonium sulfate recovery consists of crystallization followed by centrifuging, drying, and screening. The crystalline ammonium sulfate is then sold as a fertilizer.

Applications. Most hydrometallurgically produced nickel powder is compacted into briquettes or rondelles for consumption as an alloying additive in the steel industry. Smaller amounts of powder are roll compacted into nickel strip or dissolved to produce various nickel salts.

Table 4 Compositions and melting ranges of typical nickel-base hardfacing alloys

Nominal composition	Melting range °C	Melting range °F	Deposit hardness, HRC
1.5B-2.8Si-Ni	940–1260	1725–2300	19–24
1.7B-0.35C-7.5Cr-3.5Si-Ni	994–1152	1820–2105	35–42
2.4B-0.45C-11Cr-4Si-Ni	976–1063	1790–1945	49–52
3.5B-0.8C-15.5Cr-4.3Si-Ni	964–1003	1770–1840	59–62

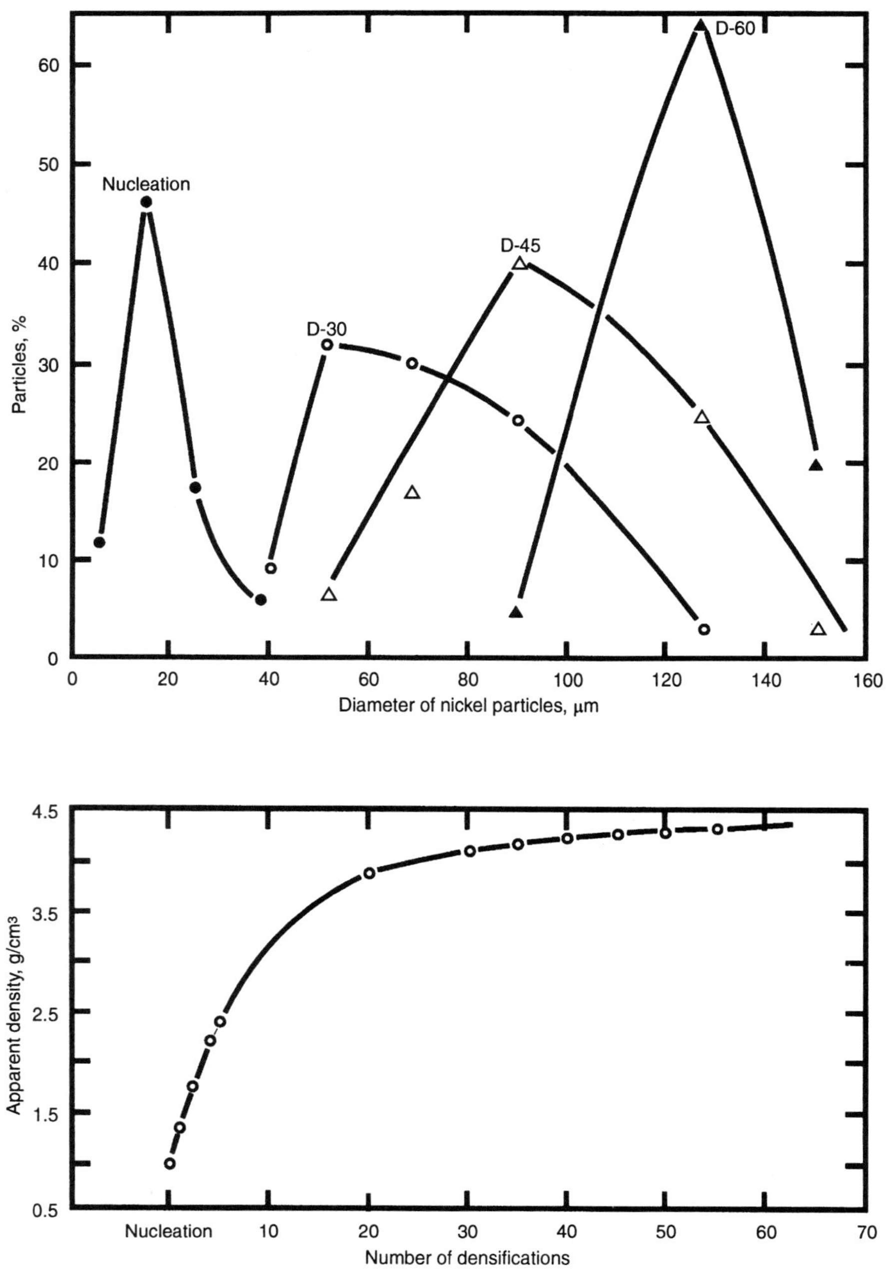

Fig. 9 Physical characteristics of nickel powder produced in a commercial operation

Atomization

K.M. Kulkarni,
Advanced Metalworking Practices, Inc.

Atomized nickel-base alloy powders are used primarily for hardfacing or aerospace components. Materials for hardfacing processes include various types of hardfacing processes, equipment, materials, and types of applications that are described in greater detail in the article "Welding and Hardfacing Powders" in this Volume. Atomized nickel-base powders are also described below with additional information in the article "Powder Metallurgy Superalloys."

Nickel-base hardfacing powders are produced by gas and water atomization. Unlike other alloy systems, nickel-base hardfacing powders are predominantly spherical in shape, even when produced by water atomization. Oxygen content is somewhat higher than gas-atomized powders, but is still below 1000 mL/m^3 (1000 ppm). Generally, melting practice and control during each manufacturing step affect product quality more than the atomization process.

Most nickel-base hardfacing powders are of the Ni-Cr-B-Si type and are self-fluxing during deposition because of the presence of boron and silicon. These elements also influence the melting temperature range of any specific alloy; melting range is an important factor for hardfacing powders. Table 4 gives typical compositions of several common grades.

The industrial use of Ni-B-Si and Ni-Cr-B-Si alloys for thermally sprayed and subsequently fused coatings developed beginning around 1955. During the period 1956 to 1960, most of these metal powders were produced by crushing processes; however, since about 1960 atomization from a melt became more widely utilized. Most of the atomization installations for these types of alloys utilized high-pressure water atomization as was also used for the production of powders.

The influence of cooling rate during atomization can have important effects on the metallurgical structure of the powders and on their behavior during thermal spraying. For example, there is a clear tendency toward the presence of unstable intermetallic nickel-boron compounds, such as Ni_3B_4 and Ni_4B_3, with increased atomization cooling rates (Ref 9). The quantity of these unstable phases has an influence on the self-fluxing reactions and on the wetting properties during fusing. This is due to the much higher velocity of the reaction of these unstable phases with the oxides present on the base metal surface. For these phases this velocity is estimated to be more than twice that for the more stable phase, NiB.

When chromium boride particles are present (i.e., from a lower atomization cooling rate), they have little influence on wetting during fusing. An increase in the content of stable phases (such as Cr_2B or CrB) increases hardener stability and ductility because less boron is used for the self-fluxing reaction. The crack resistance during bending also is distinctly improved for coatings produced with lower-cooling-rate powders.

The final boron content of the fused coatings produced with rapidly cooled powders may be up to 0.2% less than for the coatings produced with slower-cooled powders. The overall finer structure produced with more rapidly cooled powders also provides improved resistance to wear in the final coating. These same features are expected to contribute to improved corrosion resistance.

Superalloy Powders (Adapted from Ref 10). In aerospace applications, several P/M superalloys have replaced forged turbine disk alloys. The P/M alloys are characterized by a high homogeneous concentration of both solid-solution strengthening elements and the γ'-forming elements aluminum and titanium. These factors would limit forgeability of conventionally cast and wrought alloys. Compositions are listed in Table 5.

Various means of producing superalloy powders are summarized in Table 6. Inert gas atomization is the most common technique. An ingot is first cast, typically by vacuum induction melting, in order to minimize oxygen and nitrogen contents. In some cases remelting may be carried out by electron beam heating, arc melting under argon, or plasma heating. Atomization is carried out by pouring master melt through a refractory orifice; a high-pressure inert gas stream (typically argon) breaks up the alloy into liquid droplets, which are solidified at a rate of about 10^2 K/s.

The spherical powder is collected at the outlet of the atomization chamber. The maximum particle diameter resulting from this process depends on the surface tension, γ, viscosity, η, and density, ρ, of the melt, as well as the velocity, ν, of the atomizing gas. The principal factor is gas velocity. Oxygen contents are of the order of 100 ppm, depending on particle size. Finer particle sizes are obtained by screening.

Another important powder production method, the soluble gas process, is based on the rapid expansion of gas-saturated molten metal, resulting in a fine spray of molten droplets that form as the dissolved gas, usually hydrogen, is suddenly released (Fig. 12, Ref 12). The droplets solidify at a rate of about 10^3 K/s, and the cooled powder is collected under vacuum in another chamber, which is sealed and backfilled with an inert gas. This method is capable of atomizing up to 1000 kg (2200 lb) of superalloy in one heat and produces spherical powder that can be made very fine. This method has been successfully employed for LC Astroloy, MERL 76, and IN-100.

The third method of powder preparation is based on centrifugal atomization. The melt is accelerated and disintegrated by rotating under vacuum or in a protective atmosphere. One example of this method is the rotating electrode process (REP) used in the early production of IN-100 and René 95 powder. In this process, a bar of the desired composition, 15 to 75 mm (0.6 to 3 in.) in diameter, serves as a consumable electrode. The face of this positive electrode, which is rotated at high speed, is melted by a direct current electric arc between the consumable electrode and a stationary tungsten negative electrode. The process is carried out in helium. Centrifugal force causes spherical molten droplets to fly off the rotating electrode. These droplets freeze and are collected at the bottom of the tank, which is filled with helium or argon. A major advantage of this process is the elimination of ceramic inclusions and the lack of any increase in the gas content of the powder relative to the alloy electrode.

A variant on the REP process is the plasma rotating electrode process (PREP). Instead of a tungsten electrode, a plasma arc is used to melt the superalloy electrode surface. Cooling rates are higher, up to 10^5 K/s for IN-100 powder. On average, particle sizes are nearly twice as large in these processes as in gas atomization. More information on the REP process is described in the article "Rotating Electrode Process" in this Volume.

When solidification rates of powder exceed 10^5 K/s, the process is referred to as rapid solidification rate (RSR). For superalloys, the objective of the high rates is to obtain a microcrystalline alloy rather than an amorphous material. Apart from extremely fine grain size, such powders display nonequilibrium solubilities and very uniform compositions because of the very fine dendritic arm spacings resulting from rapid solidification. Both conventional superalloys and new alloys based on Ni-Al-Mo-X alloys, where X = tantalum or tungsten, have been prepared by RSR.

Rapid solidification processing can be done by centrifugal atomization with forced convective cooling. In this method, the alloy is vacuum in-

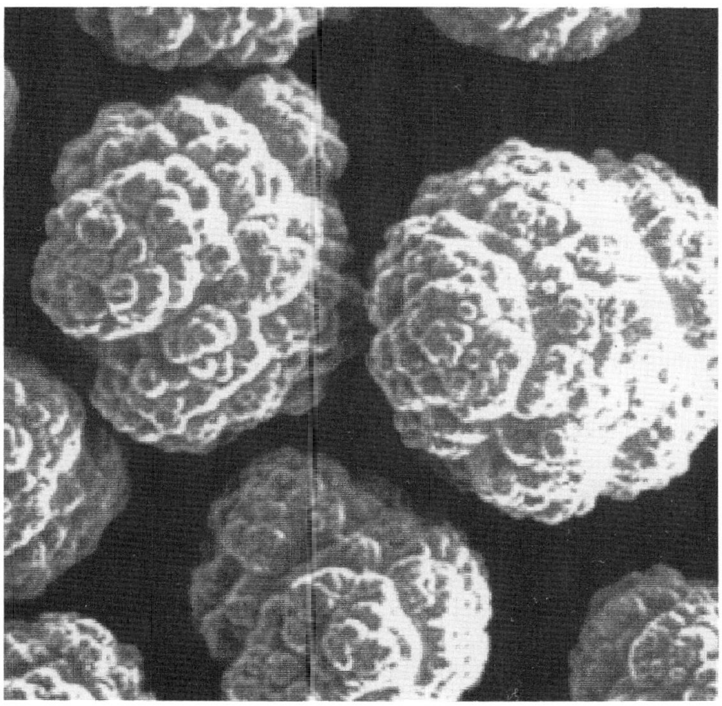

Fig. 10 Scanning electron micrograph (1000×) of nickel powder

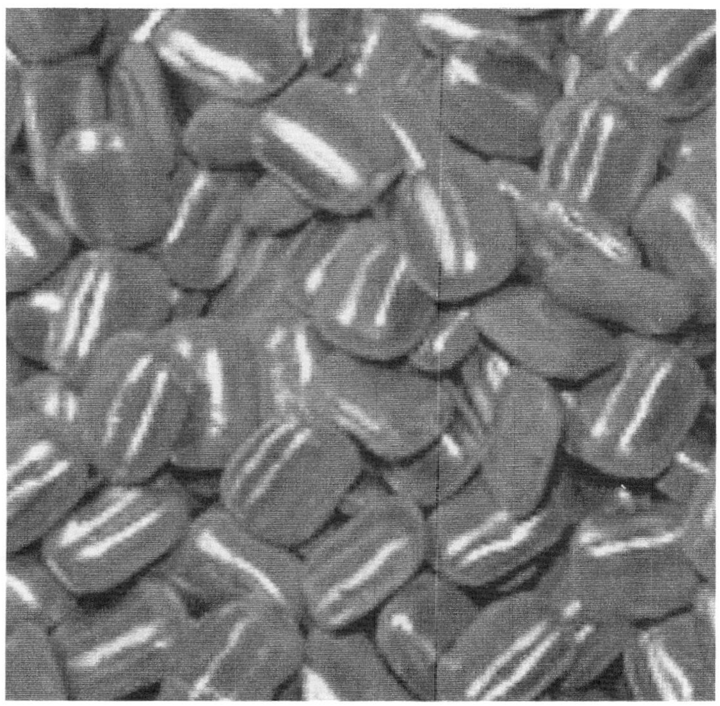

Fig. 11 Nickel briquettes (each briquette is ~38 by 25 by 18 mm, or 1.5 by 1.0 by 0.7 in.)

duction melted in the upper part of a chamber. The chamber is then backfilled with helium, and the alloy is poured in a preheated tundish. The liquid is poured through the tundish nozzle onto a rotor that is spun at 24,000 rev/mm. The melt is accelerated to rim speed and then ejected longitudinally as droplets. Further atomization and cooling of the droplets is accomplished by the injection of helium gas through annular nozzles. Spherical powder in the 10 to 100 μm (400 to 4000 μin.) diam range is produced; the cooling rate typically varies between 10^5 and 10^7 K/s, the higher rates being achieved with the smaller particles.

Rapidly solidified powders can also be prepared from melt spun ribbon that is pulverized after solidification. The ribbon is produced by pouring the melt through an orifice under pressure and impinging it on a rotating wheel of, for example, copper, that acts as a heat sink. Typical cooling rates are approximately 10^6 K/s, and ribbon thicknesses are less than 25 μm (1000 μin.). The ribbon must be mechanically pulverized, and this method is generally limited to small quantities of experimental alloys.

Mechanical Alloying*

Mechanical alloying (MA) was originally developed for the manufacture of nickel-base superalloys strengthened by both an oxide dispersion and γ′ precipitate. Now in its third decade of advancement since the first week of Benjamin (Ref 15), MA provides a means for producing a full range of complex dispersion-strengthened nickel-, cobalt-, iron-, titanium-, and aluminum-base powder systems.

*Adapted from Ref 13 and 14

Mechanical alloying is a dry, high-energy ball-milling process for producing composite metallic powders with a controlled, fine microstructure. It is carried out in a highly agitated ball charge by repeated cold welding and fracturing (Fig. 13) of a mixture of metal powders to which some nonmetal powders may be added. Its widest use has been in the production of dispersion-strengthened nickel- and iron-base superalloys for service at temperatures of 1000 °C (1830 °F) and above.

Unlike mechanical mixing processes, mechanical alloying produces a material whose internal homogeneity is independent of starting powder particle size. Thus, ultrafine dispersions (less than 1 μm interparticle spacing) can be obtained with relatively coarse initial powder (50 to 100 μm average diameter).

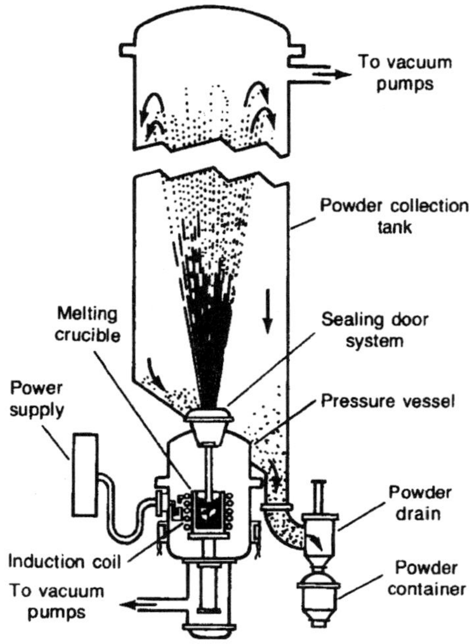

Fig. 12 Soluble gas atomization system for producing superalloy powder. Source: Ref 12

Equipment. The machinery used for mechanical alloying consists of one of several types of high-energy ball mills. These are selected on the basis of given processing times, ranging from hours to tens of hours. The types of ball mills employed include shaker mills, vibratory mills, stirred ball mills, centrifugal ball mills, and conventional ball mills with diameters greater than about 1 m. The restriction on conventional ball mills arises from the relatively low-energy density of operation of smaller mills, which leads to excessive processing times.

Unlike the procedure used in ball milling for comminution, the ratio of balls to powder in mechanical alloying is relatively high. These ratios range from 6:1 by weight to as high as 30:1, but most commonly are in the range of 10:1 to 20:1. The balls themselves range from 4 to 20 mm (0.16 to 0.8 in.) in diameter, but are usually 8 to 10 mm (0.32 to 0.4 in.) in diameter and are made of a through-hardened steel, such as 52100. The environment within the grinding machine is controlled wherever practical; water cooling and atmosphere control are employed. The milling atmosphere consists of either nitrogen or argon with measured trace amounts of oxygen. Liquids can also be used.

Production of Oxide-Dispersion-Strengthened (ODS) Superalloy Powders. The raw materials used for mechanically alloyed dispersion-

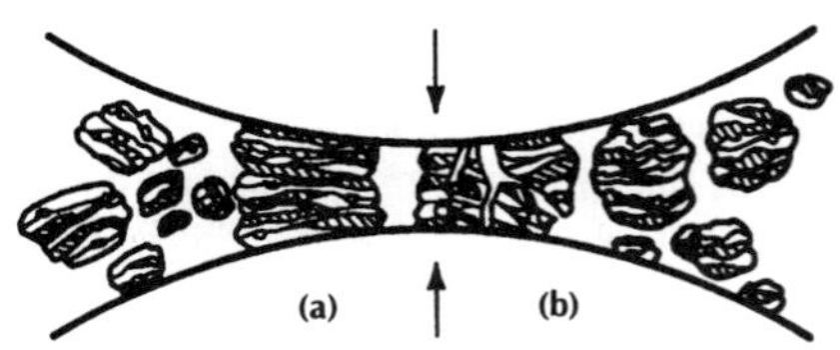

Fig. 13 Ball-powder-ball collision of powder mixture during mechanical alloying. (a) Cold welding. (b) Powder fracture. Courtesy of International Nickel Co., Inc.

Table 5 Chemical compositions of some nickel-base superalloys produced by powder metallurgy

Alloy	Composition, %													
	C	Cr	Co	Mo	W	Ta	Nb	Hf	Al	Ti	V	B	Zr	Ni
IN-100	0.07	12.4	18.5	3.2	...	...	...	...	5.0	4.3	0.8	0.02	0.06	bal
LC Astroloy	0.023	15.1	17.0	5.2	...	...	...	...	4.0	3.5	...	0.024	<0.01	bal
Waspaloy	0.04	19.3	13.6	4.2	...	...	...	...	1.3	3.6	...	0.005	0.048	bal
NASA II B-7	0.12	8.9	9.1	2.0	7.6	10.1	...	1.0	3.4	0.7	0.5	0.023	0.080	bal
René 80	0.20	14.5	10.0	3.8	3.8	...	...	...	3.1	5.1	...	0.014	0.05	bal
Unitemp AF2-1DA	0.35	12.2	10.0	3.0	6.2	1.7	...	...	4.6	3.0	...	0.014	0.12	bal
MAR-M 200	0.15	9.0	10.0	...	12.0	...	1.0	...	5.0	2.0	...	0.015	0.05	bal
IN-713 LC	0.05	12.0	0.08	4.7	...	...	(2.0)	...	6.2	0.8	...	0.005	0.1	bal
IN-738	0.17	16.0	8.5	1.7	2.6	1.7	0.9	...	3.4	3.4	...	0.01	0.1	bal
IN-792 (PA 101)	0.12	12.4	9.0	1.9	3.8	3.9	...	...	3.1	4.5	...	0.02	0.10	bal
AF-115	0.045	10.9	15.0	2.8	5.7	...	1.7	0.7	3.8	3.7	...	0.016	0.05	bal
MERL 76	0.025	12.2	18.2	3.2	...	...	1.3	0.3	5.0	4.3	...	0.02	0.06	bal
René 95	0.08	12.8	8.1	3.6	3.6	...	3.6	...	3.6	2.6	...	0.01	0.053	bal
Modified MAR-M 432	0.14	15.4	19.6	...	2.9	0.7	1.9	0.7	3.1	3.5	...	0.02	0.05	bal
New alloys														
RSR 103	...	...	...	15.0	...	...	...	...	8.4	...	...	...	...	bal
RSR 104	...	...	...	18.0	...	...	...	...	8.0	...	...	...	...	bal
RSR 143	...	...	...	14.0	...	6.0	...	...	6.0	...	...	...	...	bal
RSR 185	0.04	...	...	14.4	6.1	...	...	...	6.8	...	...	...	...	bal

Source: Ref 11

strengthened superalloys are widely available commercially pure powders that have particle sizes that vary from about 1 to 200 μm. These powders fall into the broad categories of pure metals, master alloys, and refractory compounds. The pure metals include nickel, chromium, iron, cobalt, tungsten, molybdenum, and niobium. The master alloys include nickel-base alloys with relatively large amounts of combinations of aluminum, titanium, zirconium, or hafnium.

These master alloys are relatively brittle when cast and easily comminuted to powder. In addition, because they consist of relatively exothermic intermetallic compounds, the thermodynamic activity of the reactive alloying elements, such as aluminum and titanium, is considerably reduced compared to that of the pure metals.

A typical powder mixture may consist of fine (4 to 7 μm) nickel powder, –150 μm chromium, and –150 μm master alloy. The master alloy may contain a wide range of elements selected for their roles as alloying constituents or for gettering of contaminants. About 2 vol% of very fine yttria, Y_2O_3 (25 nm, or 250 Å) is added to form the dispersoid. The yttria becomes entrapped along the weld interfaces between fragments in

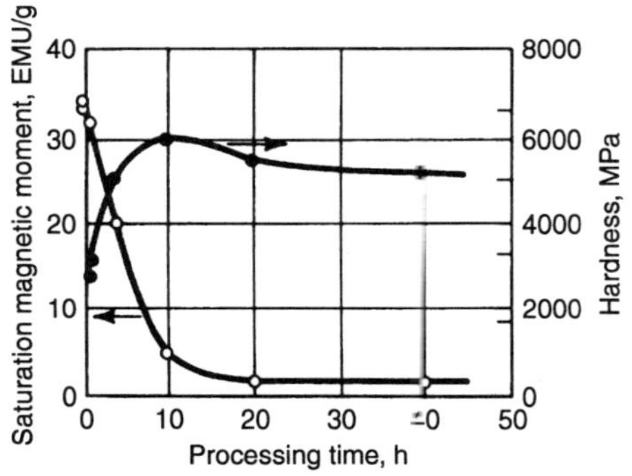

Fig. 14 Change of hardness and magnetic response during mechanical alloying. Courtesy of International Nickel Co., Inc.

the composite metal powders. After completion of the powder milling, a uniform interparticle spacing of about 0.5 μm (20 μin.) is achieved.

The oxygen contents of the commercially pure metal powders and the master alloys range from 0.05 to 0.2 wt%. The refractory compounds that can be added include carbides, nitrides, and oxides. For the production of dispersion-strengthened materials, such additions are limited to very stable oxides, such as yttria, alumina, or less frequently thoria. These oxides, which are prepared by calcination of oxalate precipitates, consist of crystallites of about 50 nm agglomerated into pseudomorphs of about 1 μm.

The only restriction on the mixture of powder particles for mechanical alloying (other than the particle size range mentioned above and the need to minimize excessive oxygen) is that at least 15 vol% of the mix should consist of a compressibly deformable metal powder. The function of this component, which can consist of any one or all of the pure metals, is to act as a host or binder for the other constituents during the process.

Other similar metals, such as copper, zinc, and magnesium, are suitably ductile but not normally added to superalloys. For dispersion-strengthened superalloys, the amount of refractory oxide added ranges from about 0.4 to 1.5 wt% (1 to 2.7 vol% for yttria).

Powder Characteristics. A uniform distribution of submicron refractory oxide particles must be developed in a highly alloyed matrix for the production of ODS alloys. A given sample of mechanically alloyed superalloy powder may contain particles ranging from 10 to 500 μm, with an average particle size between 50 and 200 μm. The internal structure of the powder is independent of particle size once the steady state is achieved.

Because of the severe plastic deformation that occurs during mechanical alloying, very high hardnesses are achieved in the powders. Hardness increases almost linearly during the initial stages of the process (Fig. 14), reaching a saturation value, after which time it is presumed that work softening balances further cold work.

It has also been noted by magnetic measurements (Fig. 14) that true alloying occurs, as evidenced by the loss of saturation magnetic moment, with increasing processing time in a high-nickel composition. This may be due to a combination of factors. First, the extreme amount of cold work leads to low activation energy and pipe diffusion. Second, the temperature is moderately elevated to approximately 150 °C (300 °F) during the process. Third, relatively short interdiffusion distances (less than 1 μm) exist toward the end of the process.

Commercial Alloys. The most common mechanically alloyed ODS alloys include MA 754, MA 758, MA 956, MA 6000, and MA 760.

Alloy MA 754 was the first mechanically alloyed ODS superalloy to be produced on a large scale. This material is basically a Ni-20Cr alloy strengthened by about 1 vol% Y_2O_3 (see Table7). It is comparable to TD NiCr (an earlier ODS material strengthened by thoria, ThO_2) but has a nonradioactive dispersoid, and because of its higher strength, has been extensively used for aircraft gas-turbine vanes and high-temperature test fixtures.

Alloy MA 758 is a higher-chromium version of MA 754 (see Table 7). This alloy was developed for applications in which the higher chromium content is needed for greater oxidation resistance. The mechanical properties of this alloy are similar to those of MA 754 when identical product forms and grain structures are compared. This alloy has found applications in the thermal processing industry and is also used in the glass-processing industry.

Alloy MA 956. The production of alloy MA 956 demonstrates the ability to add large amounts of metallic aluminum by mechanical alloying (see Table 7). This material is a ferritic Fe-Cr-Al alloy, dispersion-strengthened with yttrium aluminates formed by the addition of about 1 vol% of yttria. Because of its generally good hot and cold fabricability, MA 956 has been produced in the widest range of product forms of any MA ODS alloy. In sheet form, this alloy is produced by a sequence of hot and cold

Table 6 Powder production methods

Step	Inert gas atomization(a)	Soluble gas process	Rotating electrode process(b)	Plasma rotating electrode process	Centrifugal atomization with forced convective cooling (RSR)(c)
Melting 1	VIM; ceramic crucible	VIM; ceramic crucible	VIM, VAR, ESR	VIM, VAR, ESR	VIM; ceramic crucible
Melting 2	...	...	Argon arc	Plasma	...
Melt disintegration system/environment	Nozzle; argon stream	Expansion of dissolved hydrogen against vacuum and Ar + H_2 mixture	Rotating consumable electrode; argon or helium	Rotating consumable electrode; argon	Rotating disk; forced helium convective cooling

(a) VIM, vacuum induction melting. (b) VAR, vacuum arc remelting; ESR, electroslag remelting. (c) RSR, rapid solidification rate. Source: Ref 11

Table 7 Nominal composition of selected mechanically alloyed materials

Alloy designation	Composition, %											
	Cr	Al	Ti	W	Mo	Ta	Y_2O_3	C	B	Zr	Ni	Fe
MA 754	20	0.3	0.5	...	...	...	0.6	0.05	...	...	bal	...
MA 758	30	0.3	0.5	...	...	...	0.6	0.05	...	...	bal	...
MA 760	20	6.0	...	3.5	2.0	...	0.95	0.05	0.01	0.15	bal	...
MA 6000	15	4.5	2.5	4.0	2.0	2.0	1.1	0.05	0.01	0.15	bal	...
MA 956	20	4.5	0.5	...	...	...	0.5	0.05	...	...	...	bal

Table 8 Physical properties of selected mechanical alloying ODS materials

Alloy	Melting point		Modulus of elasticity		Mass density,	Coefficient of expansion, at 20 to 980 °C (70 to 1800 °F)	
	°C	°F	GPa	10^6 psi	g/cm³	μm/m · K	μin./in. · °F
MA 754	1400	2550	151	22	8.3	16.9	9.41
MA 956	1480	2700	269	39.0	7.2	14.8	8.22
MA 6000	1296–1375	2365–2507	203	29.4	8.11	16.7	9.3

working, which yields large pancake-shape grains following heat treatment. This grain structure ensures excellent isotropic properties in the plane of the sheet. MA 956 is used in the heat treatment industry for furnace fixturing, racks, baskets, and burner nozzles. It also is used in advanced aerospace sheet and bar components, where good oxidation and sulfidation resistance are required in addition to high-temperature strength properties.

Alloy MA 956 has a very high melting point (1480 °C, or 2700 °F), a relatively low density (7.2 g/cm³) compared to competitive materials, and a relatively low thermal expansion coefficient (see Table 8). This combination of properties makes the alloy well suited for sheet applications such as gas-turbine combustion chambers.

Alloy MA 6000 has a composition based on an alloy-development philosophy similar to that of the more sophisticated cast and wrought superalloys. This is because it contains a critical balance of elements to produce strength at intermediate and elevated temperatures, along with oxidation and hot-corrosion resistance. Alloy MA 6000 combines γ' hardening from its aluminum, titanium, and tantalum content for intermediate strength, with ODS from the yttria addition for strength and stability at very high temperatures. Oxidation resistance comes from its aluminum and chromium contents, while titanium, tantalum, chromium, and tungsten act in concert to provide sulfidation resistance. The tungsten and molybdenum also act as solid-solution strengtheners in this alloy. MA 6000 is an ideal alloy for gas-turbine vanes and blades where exceptional high-temperature strength is required.

Alloy MA 760 is an age-hardened nickel-base alloy with a composition designed to provide a balance of high-temperature strength, long-term structural stability, and oxidation resistance. Its primary use is expected to be for industrial gas turbines. The composition of this alloy is shown in Table 7. It is similar to MA 6000 in that its strength is supplemented by γ' age hardening. Its properties also benefit from zone annealing to give coarse elongated grains. The stress-rupture properties of alloy MA 760 exceed those of MA 754 but are exceeded by those of MA 6000 (see the article "Powder Metallurgy Superalloys" in this Volume).

REFERENCES

1. L. Mond, C. Langer, and F. Quinke, *Proc. Chem. Soc.*, Vol 86, 1890, p 112
2. W.M. Goldberger and D.F. Othmer, The Kinetics of Nickel Carbonyl Formation, *Ind. Eng. Chem., Process Des. Dev.*, July 1963, p 202–209
3. H.E. Carlton and W.M. Goldberger, Fundamental Considerations of Carbonyl Metallurgy, *J. Met.*, June 1965, p 611–615
4. D.G.E. Kerfoot, *CIM Bull.*, Vol 82 (No. 926), June 1997, p 136–141
5. M.T. Anthony and D.S. Flett, *Min. Ind. Int.*, Jan 1997, p 26–42
6. D.G.E. Kerfoot, *Ullmann's Encyclopedia of Industrial Chemistry*, Vol A 17, VCH, 1991, p 157–219
7. J.R. Boldt and P.E. Queneau, *The Winning of Nickel*, Longman Group, 1967
8. D.J.I. Evans, "The Production of Metals by Gaseous Reduction from Solution, I. Processes and Chemistry," Symposium on Advances in Extractive Metallurgy, Institution of Mining and Metallurgy, London, April 1967
9. P.A. Kammer, W. Simm, and H.T. Steine, The Influence of Atomization Process and Parameters on Metallic Powders and Coating Properties, *Thermal Spray Research and Applications*, ASM International, 1990, p 773–776
10. N.S. Stoloff, Wrought and P/M Superalloys, *Properties and Selection: Irons, Steels, and High-Performance Alloys*, Vol 1, ASM Handbook, ASM International, 1990, p 950–975
11. K.C. Anthony, Wear Resistant Cobalt-Base Alloys, *J. Met.*, Feb 1983, p 52–60
12. S. Reichman and D.S. Chang, *Superalloys II*, C.T. Sims, N.S. Stoloff, and W.C. Hagel, Ed., John Wiley & Sons, 1987, p 459
13. J. de Barbadillo and J. Fischer, Dispersion-Strengthened Nickel-Base and Iron-Base Alloys, *Properties and Selection: Nonferrous Alloys and Special Purpose Materials*, Vol 2, ASM Handbook, ASM International, 1990, p 722
14. P. Gilman and J.S. Benjamin, Nickel- and Iron-Based Dispersion Strengthened Alloys, *Powder Metallurgy*, Vol 7, ASM Handbook, American Society for Metals, 1985, p 722
15. J.S. Benjamin, Dispersion Strengthened Superalloys by Mechanical Alloying, *Metall. Trans.*, Vol 1, 1970, p 2943–2951

Production of Cobalt-Base Powders

Gavin Freeman, Sherritt International Corp.

COBALT POWDER is used primarily in the P/M industry as an alloying element in the production of permanent magnets, superalloys (high-temperature, creep-resistant alloys), wear-resistant hardfacing alloys, tool and die steels, and as a binder in cemented carbides. Approximately 60% of the world production of cobalt originates in Zaire and Zambia.

Cobalt is an important material to the P/M industry due to its ability to enhance the high-temperature properties of superalloys, the cutting ability of sintered carbide and high-speed tool steels, and the toughness of high-strength steels. This article discusses the three methods of cobalt powder production: hydrometallurgical processing, reduction of oxides, and atomization. Additional references are found at the end of this article.

Hydrometallurgical Processing

The hydrometallurgical processing of cobalt includes three main processing steps: chemical dissolution (leaching), solution purification, and metal recovery. The production of cobalt powder at the Sherritt (Corefco) refinery in Fort Saskatchewan, Alberta, Canada, currently amounts to ~2400 tonnes of cobalt per year, representing roughly 10% of the world's supply. In addition, this process is an integral part of the solution purification steps of the nickel refining process as the same feedstock to the plant contains both nickel and cobalt values. A flow diagram of this process is shown in Fig. 1. More information on the process is contained in Ref 1 to 3.

Leaching. The production of cobalt at the Sherritt International refinery begins with the leaching of a mixed nickel and cobalt sulfide obtained from Moa, Cuba. The leaching process is based on the ammoniacal pressure leaching of the sulfides to produce soluble nickel and cobalt hexammine sulfate complexes. The details of the leaching process are outlined in the article "Production of Nickel-Base Powders" in this Volume.

The hexammine leach solution contains a variety of metal ammine sulfate complexes. Leaching conditions are carefully controlled to maximize the formation of the cobaltic hexammine complex. This is crucial as the efficiency of the subsequent cobalt separation depends on the fact that cobalt exists primarily as the hexammine complex in the leach solution.

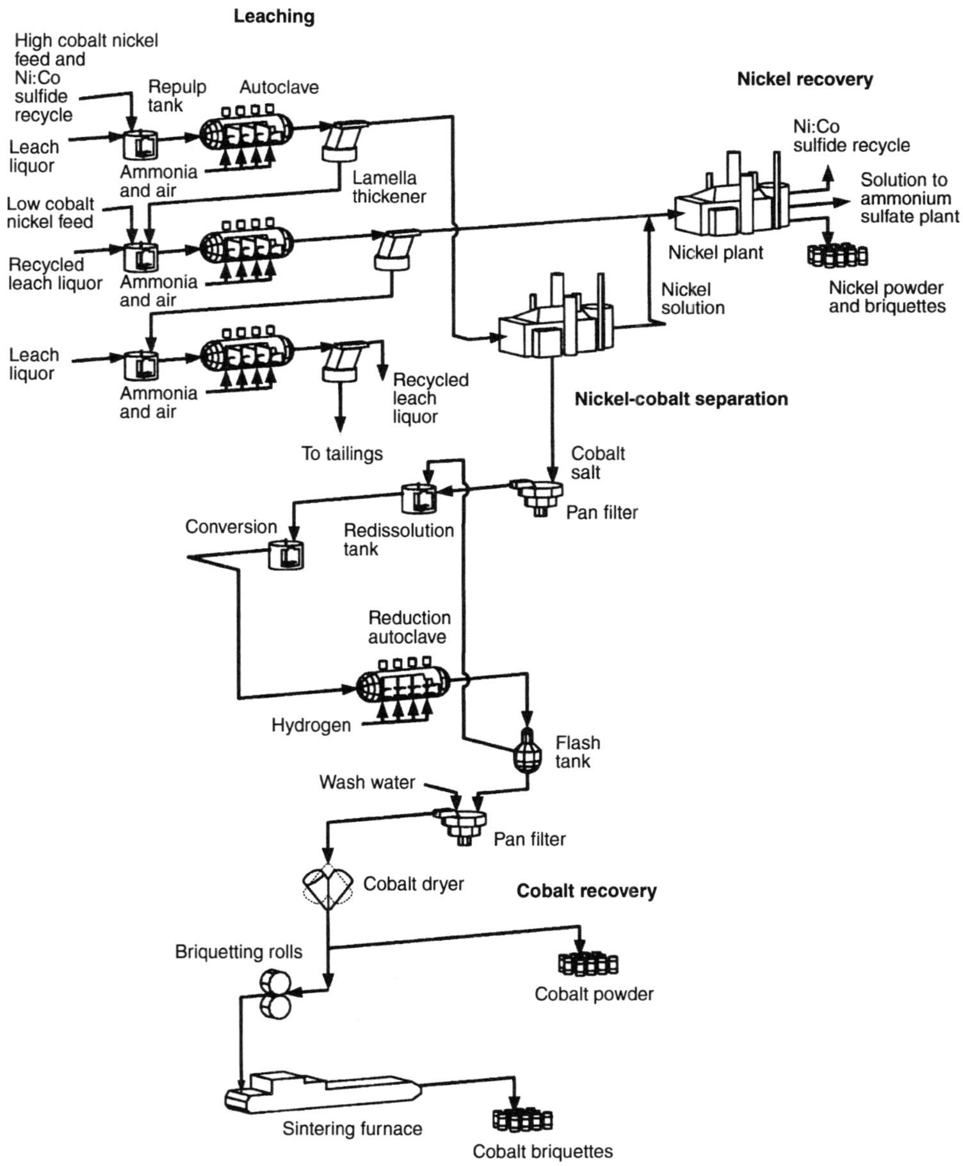

Fig. 1 Flowchart of the Sherritt cobalt refining process

The leach solution is then transferred to the cobalt separation plant, where cobalt is separated as a pure crystalline cobaltic hexammine salt. The nickel rich solution remaining after the selective removal of cobalt is directed to the nickel plant for further solution purification steps and ultimately nickel powder production.

Nickel-Cobalt Separation. The basis for the separation of cobalt from nickel is the separation and purification of cobalt (III) hexammine sulfate, $[Co(NH_3)_6]_2(SO_4)_3$. The initial step of this separation involves the precipitation of a cobalt-nickel hexammine salt. This precipitation is effected by the addition of ammonium sulfate and ammonia to the leach solution as shown in the following reaction:

$$[Co(NH_3)_6]_2(SO_4)_3\,(aq) + 2\,Ni(NH_3)_6(SO_4)\,(aq) + (NH_4)_2SO_4 \rightarrow 2(NH_4)[Co(NH_3)_6][Ni(NH_3)_6](SO_4)_3\,(s)$$

The cobalt-nickel salt is then dissolved in water, followed by precipitation of the relatively insoluble cobaltic hexammine sulfate, while the more soluble nickel hexammine sulfate remains in solution. This selective nickel removal from the cobalt-nickel hexammine salt results in the isolation of a purified cobalt salt.

$$2(NH_4)[Co(NH_3)_6][Ni(NH_3)_6]\,(SO_4)_3\,(s) + xs\,H_2O \rightarrow 2(NH_4)^+\,(aq) + 2[Ni(NH_3)_6]^{2+}\,(aq) + 2[Co(NH_3)_6]^{3+}\,(aq) + 6(SO_4)^{2-}\,(aq)$$

$$2[Co(NH_3)_6]^{3+}\,(aq) + 3(SO_4)^{2-}\,(aq) \rightarrow [Co(NH_3)_6]_2(SO_4)_3\,(s)$$

The cobalt hexammine salt is then recrystallized to further remove any nickel contamination. After washing and filtration, the pure cobalt salt is redissolved in an ammonium sulfate solution.

Cobalt Recovery. Reduction of the cobalt (III) present in the pure cobaltic hexammine solution to cobalt (II) is completed in a step referred to as "conversion." The conversion of cobaltic hexammine to a cobaltous ammine is a disproportionation reaction, and it is accomplished by the controlled addition of metallic cobalt as defined by the following stoichiometry.

$$[Co(NH_3)_6]_2(SO_4)_3 + Co\,(powder) \rightarrow 3[Co(NH_3)_4]SO_4$$

Sulfuric acid is added to neutralize the free ammonia and maintain the molar ratio of ammonia to cobalt at approximately 2.3 to 1.

$$2NH_3 + H_2SO_4 \rightarrow (NH_4)_2SO_4$$

After conversion and adjustment of the ammonia concentration, this solution is suitable to be fed to the reduction autoclaves. The objective of the reduction process is the chemical reduction of cobaltous ammine to cobalt powder. The reduction of cobalt is a batch operation that occurs in a reduction autoclave under hydrogen atmosphere at elevated temperature and pressure.

Cobalt powder is produced in a cycle comprising one nucleation reduction, in which fine cobalt powder of sufficient surface area to function as catalyst is prepared, and up to 60 densifications. Nucleation reduction refers to the hydrogen reduction process by which fine cobalt seed powder is produced with the aid of a sodium sulfide/sodium cyanide catalyst. At the completion of the nucleation reduction, the solution is decanted from the seed powder, which remains in the autoclave to act as a catalyst during subsequent densification operations. In the course of a cycle, the originally fine seed powder coarsens or becomes densified.

The purpose of the densification reduction is to deposit reduced cobalt onto a seed particle until desired powder density, chemical composition, and screen fraction have been achieved. Each densification is a hydrogen reduction operation in which the autoclave is filled with the purified cobaltous solution, and hydrogen is added under pressure. Metallic cobalt then precipitates from solution depositing on the seed powder, by the following reaction:

$$[Co(NH_3)_2]SO_4 + H_2 \rightarrow Co\,(metal) + (NH_4)_2SO_4$$

The rate of the reduction reaction and its completeness depend on several factors including the ammonium sulfate concentration, the ratio of ammonia to cobalt, temperature, the hydrogen partial pressure, and the amount of seed material present. At the end of each densification, the powder is settled, the spent solution is decanted, and a fresh charge of cobaltous solution is introduced.

The slurry is then discharged from the autoclave to a flash tank where the initial separation of cobalt powder from the end solution is accomplished. The cobalt powder is transferred from the flash tank to a vacuum pan filter. After the cobalt is washed to remove ammonium sulfate, and dried, the product is packaged as powder or compacted into briquettes and sintered. Typical analysis of the Sherritt cobalt powder is given in Table 1.

Table 1 Typical properties of the Sherritt cobalt powder

Chemical composition, %	
Co	99.9
Ni	0.02
Cu	0.005
Fe	0.005
S	0.03
C	0.05
Mesh screen size, %	
+100	0–15
−100 +150	5–25
−150 +200	5–15
−200 +250	5–15
−250 +325	20–45
−325	10–50
Physical property	
Apparent density, g/cm³	2.5–3.5

Thermal Decomposition of Cobalt Oxide

One of the major applications of cobalt powder is its use as a binder in the production of cemented carbides. Most of this powder is produced by the reduction of cobalt oxide with hydrogen at relatively low temperatures, generally below 800 °C (1470 °F) to obtain the necessary degree of fineness.

Powder grades used in cemented carbides should pass through 325 mesh. Small particle sizes are required, because the cobalt powder must be intimately mixed with tungsten carbide particles by ball milling. Typical properties of reduced cobalt powders are the following:

Chemical analysis, %	
Cobalt(a)	99.60
Nickel	0.08
Iron	0.08
Silicon	0.035
Calcium	0.020
Manganese	0.020
Carbon	0.015
Zinc	0.010
Sulfur	0.008
Copper	0.001
Lead	0.003
Hydrogen loss	0.20
Physical properties, %	
Apparent density, g/cm³	1.8
Tap density, g/cm³	3
Screen analysis(b), %	
+100 mesh	0.01
−100+200 mesh	0.04
−200+300 mesh	0.15
−300+400 mesh	0.20
−400 mesh	99.60

(a) Hydrogen loss not included. (b) Average particle size, 5 μm Source: Ref 4

Table 2 Composition and hardness of selected cobalt-base hardfacing alloys

AWS designation or trade name	Nominal composition	Nominal macrohardness DPH	Nominal macrohardness HRC	Approximate hardness of microconstituents: Matrix, DPH	Hard particles: Type	Hard particles: DPH
Alloy 21	Co-27Cr-5Mo-2.8Ni-0.2C	255	24–27	250	Eutectic	900
RCoCrA	Co-28Cr-4W-1.1C	424	39–42	370	Eutectic	900(a)
RCoCrB	Co-29Cr-8W-1.35C	471	40–48	420	Eutectic	900(a)
RCoCrC	Co-30Cr-12W-2.5C	577	52–54	510	M_7C_3	900(a)
					M_6C	1540
Alloy 20	Co-32Cr-17W-2.5C	653	53–55	540	M_7C_3	1700
					M_6C	900
Tribaloy T-800	Co-28Mo-17Cr-3Si	653	54–64	800(b)	Laves phase	1100

(a) Matrix and M_7C_3 eutectic. (b) Matrix and Laves phase eutectic

Atomization

Atomized powders of cobalt-base alloys find wide usage in high-temperature applications, although this section discusses only hardfacing powders. Cobalt-base alloys for hardfacing applications typically are the cobalt-chromium-tungsten-nickel-carbon types. These powders are gas atomized to minimize oxygen pickup.

Some manufacturers prefer vacuum melting and inert-gas atomization over air melting followed by inert-gas (typically nitrogen or argon) atomization. Oxygen content in these atomized powders is significantly <1000 ppm, although the nitrogen level may vary from 600 to 2000 ppm. Typical melting and atomizing practice requires strict control to produce a quality product.

Compositions and hardnesses of several typical alloys are given in Table 2. Detailed information on cobalt hardfacing alloys can be found in the article "Welding and Hardfacing Powders" in this Volume.

REFERENCES

1. Kerfoot, U.S. Patent 5,468,281, 1995
2. Kunda and R. Hitesman, in *Hydrometallurgy,* Vol 4, 1979, p 347–375
3. Mackiw and T.W. Benz, *Interscience,* 1961, p 503–534
4. A.R. Powder, Ed., *Handbook of Metal Powders,* Reinhold, 1966, p 109

SELECTED REFERENCES

- L. Cheng, Y. Zhang, D. Sun, X. Xiong, and K. He, Mechanical Alloying of Cobalt and Aluminum Powders, Chin. *J. Met. Sci. Technol.,* Vol 8 (No. 3), 1992, p 219–221
- A. Corrias, G. Ennas, G. Licheri, G. Marongiu, and G. Paschina, Mechanical Alloying of Cobalt and Boron Powders, *Mater. Sci. Eng. A,* Vol 145 (No. 1), 25 Sept 1991, p 123–125
- J. Eckert, L. Schultz, and K. Urban, Glass Formation and Extended Solubilities in Mechanically Alloyed Cobalt-Transition Metal Alloys, *J. Less-Common Met.,* Vol 166 (No. 2), 1 Nov 1990, p 293–302
- M.D. Faris and D.C. Irving, "Automation and Modelling of the Cobalt Hydrogen Reduction Process," Conf. on Modelling, Simulation, and Control of Hydrometallurgical Processes (Quebec City, Quebec), Canadian Institute of Mining, Metallurgy and Petroleum, 24 Aug–21 Sept 1993
- C. He, C. Lin, Y. Wang, and Y. Shi, Process for Preparing Rare Earth Containing Hard Alloy, U.S. Patent 5,248,328, 16 July 1991
- S.A. Makhlouf, E. Ivanov, K. Sumiyama, and K. Suzuki, Structural and Magnetic Properties of Nanocrystalline BCC Cobalt Particles Obtained by Leaching of Mechanically Alloyed Co-Al, *J. Alloy. Compd.,* Vol 189 (No. 1), 20 Nov 1992, p 117–121
- Y. Mitrani and E.Y. Gutmanas, Design of Dispersion Strengthened Cobalt Based Alloys for Elevated Temperatures, *PM '90–Int. Conf. on Powder Metallurgy*, Vol 3, Institute of Metals, 1990, p 190–194

Production of Precious Metal Powders: Silver, Gold, Palladium, and Platinum

Howard D. Glicksman, DuPont Company

Precious metals are used extensively in such advanced material applications as electronic components, fuel cell electrodes, and high-performance catalysts. Fine precious metal powders are used in the electronics industry for the manufacture of fired conductors, multilayer ceramic capacitors, and conductive adhesives using thick-film technology. Multilayer ceramic capacitors are small electronic chips built up of alternating thin layers of nonconducting ceramic material sandwiched between an equal number of thin layers of the conducting electrode layer. The precious metal powders most commonly used are silver, gold, palladium, platinum, and combinations such as silver-palladium and gold-platinum. Excellent conductivity, corrosion resistance, good solderability, and stability in air during firing of the metal ceramic films make precious metals the preferred electronic material for high-performance applications (Ref 1). The cost of using these more expensive materials is more than outweighed by their ease of firing, subsequent stability in air, and ease of attachment by soldering and welding (Ref 2).

Precious metal pastes are used in a variety of sensors including systems for measuring oxygen, humidity, dew point, wind speed, flow rate, pressure, and temperature. These sensors make use of the catalytic characteristics of the metals as well as their moisture resistance, heat resistance, and conductivity. Gold, platinum, rhodium, and ruthenium are the principal precious metals used (Ref 3).

The conductor pastes used in thick-film applications generally contain four components consisting of the metal powder, a glass powder included as a bonding agent for joining the metal to the substrate, an organic vehicle containing an organic resin that imparts the needed rheology to the paste for screen printing, and a solvent to control the viscosity and solids content. The particle sizes for the metal powders range from 5 μm down to less than 0.1 μm. Uniform size distributions and spherical morphology are prerequisites for a good conductor paste. Precious metal particles with two-dimensional flake morphology are also used and are preferred for the manufacture of conductive adhesives and for polymer thick-film applications where the precious metal powders are not sintered.

The majority of these fine precious metal powders are prepared by chemical precipitation from an appropriate aqueous salt solution (Ref 1). Homogeneous nucleation is the prerequisite for fine, smooth, and dispersed metal particles made by chemical precipitation. Any heterogeneous nucleation needs to be suppressed. In addition to controlling the nucleation mechanism, the prevention of aggregation or coagulation of the fine particles precipitating out of the solution is equally important. Van der Waals attraction forces are responsible for coagulation, while electrostatic repulsion forces arising from the electric double layers surrounding the particles act against coagulation. Without controlling the attraction forces, the particles will become flocculated, making it very difficult to control the size of the powders. Generally, changing the surface potential of the particles via adsorption of surfactants is the method used to prevent aggregation. By precipitating the metal powders directly from an organic phase, which can provide uniform and enhanced steric stabilization with no stabilization by electrostatic repulsion, finer and smoother particles are obtained (Ref 1, 2).

Chemical precipitation begins with the preparation of the metal-soluble salt solution by dissolving the metal in an inorganic acid, such as nitric acid or aqua-regia (hydrochloric acid and nitric acid). When aqua-regia is used, the nitrosyl compound must be removed by continuous boiling with the addition of more hydrochloric acid. In the manufacture of binary or ternary metal powders, such as palladium-silver or platinum-palladium-silver, the metal salts are dissolved in a common solvent (can be water) prior to precipitation.

Platinum-group metal salts in aqueous solution when treated with reducing agents such as formaldehyde, sodium formate, hydrazine, and so forth, precipitate a finely divided metal powder. Most often the reduction is carried out in alkaline solution, although acidic reductions have been used (Ref 4).

By applying wet, continuous comminution to precious metal powders, in the presence of a proper lubricant and solvent, platelike precious metal flake particles can be successfully produced. Various types of milling equipment can be used including ball mills, vibratory mills, and attritor mills. The size, shape, and other physical characteristics of the milled particles are primarily controlled by the morphology of the starting metal powder, the lubricant chosen, the solvent, the milling media, and the milling conditions such as temperature, revolutions per minute, the ratio of powder to solvent, and overall time.

Production of Silver Powders

Silver powders are used in a variety of electrical, electronic, and industrial applications. In powder metallurgy, silver powders are used in the manufacture of electrodes for primary batteries and storage cells, and as mixtures with other metals that are used in the preparation of electrical contacts and other P/M parts containing silver. These mixtures include silver-tungsten, silver-iron, silver-tungsten carbide, silver-molybdenum, silver-cadmium oxide, silver-graphite, and silver nickel (Ref 5). These electrical applications generally require a dense, coarse silver powder of high purity with good compressibility. Average particle sizes vary from approximately 10 to 100 μm.

In the electronics industry, silver powders are used in conductive inks, pastes, and adhesives. Conductive silver inks were probably the first thick-film inks to be developed and are used in the manufacture of capacitors, potentiometers, printed circuits, and other electronic parts. Silver pastes for glass substrates are heated up to 650 °C and are used in such applications as fluorescent character display tubes, plasma displays, heaters for kettles, hot trays, automobile antennas, and heated automobile windows (defoggers). Lower-temperature silver pastes containing cured polymers are used as conductive films for electromagnetic shields, undercoat for copper plating, membrane switches, flexible substrates, and amorphous solar cells. Silver pastes are also used as conductive adhesives for connecting leads to terminals and components on substrates or components to components. Examples include die-attach materials, lead wire outlets of quartz oscillators, silicon die bonding, and bonding of sliding carbon brushes to the lading springs (Ref 2).

The value of using silver is that it is the least expensive precious metal that is compatible with the normal thick-film process and can be made to have good bond strength with very high conductivity. Although its leach resistance to solder can be poor, slight modifications to the solder can help overcome this problem. Silver compositions are also easily compatible with resistor and dielectric systems. The major disadvantage of silver is its strong tendency to migrate over the surface of insulators and resistors when subjected to electrical fields under conditions of higher humidity. The rate of migration is considerably reduced in silver alloys, but migration is always a concern with silver-bearing conductors (Ref 6).

Generally, silver powders used in electronic processes are chemically precipitated, exhibit high purity, and are small in size. Average particle sizes range from 10 μm to below 1 μm. Silver powders that are mechanically milled into flat powders or silver flakes are also used to provide more surface coverage of the substrate. These flakes are well suited for conductive coatings and shielding materials and are mixed with silver powders in conductive ink systems.

Other industrial applications of silver powders include water purification, catalyst manufacture, dental amalgams, and preparation of silver alloys. In addition, silver powders are used as a protective coating against hostile industrial environments and as a source of very-high-purity silver (99.99+% pure).

Most of the methods currently used to manufacture metal powders can be applied to the production of silver powders. Chemical, physical (atomization and mechanical), thermal decomposition, and electrochemical processes are used industrially to make silver powders. These powders can be characterized by physical properties including surface area, tap density, particle size, particle size distribution, and shape.

Chemical Processes. Most silver salts are easily reduced from solution, with limitations imposed only by their solubilities. The most common salts are silver nitrate and silver oxide. Organic reducing agents such as alcohol, sugar, or aldehydes are used to manufacture silver powder from silver oxide. For example, dextrose can be used to produce a very fine silver powder:

$$2AgNO_3 + 2NaOH \rightarrow Ag_2O_{(S)} + 2NaNO_3 + H_2O$$

$$12Ag_2O + C_6H_{12}O_6 \rightarrow 24Ag_{(S)} + 6CO_2 + 6H_2O$$

Organically precipitated silver powders have small particle sizes with average sizes ranging from 5 to less than 0.1 μm. Unfortunately, these silver powders tend to agglomerate into aggregates of 10 μm or larger. To avoid agglomeration, the powder can be precipitated in the presence of a surfactant or deagglomerated after the precipitation using mechanical processes such as high shear mixing, ball milling, or jet milling. It was found that the addition of a surfactant such as gelatin can control the particle size of silver powder produced from the reduction of silver oxide with formaldehyde. For this reaction, it is important to control the amount of sodium hydroxide and gelatin, the formation of the silver oxide, and the temperature (Ref 7):

$$2AgNO_3 + 2NaOH \rightarrow Ag_2O_{(S)} + 2NaNO_3 + H_2O$$

$$Ag_2O_{(S)} + HCHO \rightarrow 2Ag_{(S)} + HCOOH$$

Inorganic reducing agents such as hydrazine, potassium sulfite, and sodium formate are used to reduce silver nitrate to silver powder. Two examples of the reduction process are:

$$2AgNO_3 + 2K_2SO_3 \rightarrow 2Ag_{(s)} + K_2SO_4 + 2KNO_3 + SO_{2(g)}$$

$$AgNO_3 + NaCO_2H \rightarrow Ag_{(s)} + NaNO_3 + CO_{2(g)} + \tfrac{1}{2}H_{2(g)}$$

These powders are coarse in size, resembling those made by electrochemical methods. Average particle sizes vary from less than 1 to 20 μm. These powders exhibit good flow and good sintering characteristics. Chemically precipitated silver powders can be made in a range of sizes and shapes. Examples are shown in Fig. 1 to 3.

Electrochemical Processes. Several electrochemical processes are used to manufacture silver powders. Galvanic reduction precipitates silver powder from a solution by a more anodic metal such as zinc, copper, or iron (Fig. 4). This process produces an irregularly shaped powder with a free bulk density ranging from 1.5 to 4.0 g/cm^3. Particle sizes of 100 μm or more are produced by galvanic reduction. Some galvanically reduced powders exhibit good flow characteristics.

In electrolytic reduction, an electrical current passes through an electrolyte at room temperature to produce a crystalline silver. Current densities vary from 215 to 2150 A/m^2 (20 to 200 A/ft^2), depending on the desired particle size. Silver anodes, usually as cast silver bars or granulated silver, are dissolved. Crystallized silver is then grown on a cathode in an experimental setup such as that shown in Fig. 5. This crystalline silver powder is dendritic and can be very large, with particle sizes ranging from 1 μm to as large as several millimeters, depending on processing conditions and current densities (Ref 8).

Atomization and Aerosol Processes. Atomization of a molten stream of silver with gas or liquid is used to produce silver powder. The atomized powder generally approaches a spheroidal shape, with a very smooth surface, minimal porosity, and a very high powder density. Average particle sizes generally are 10 to 50 μm, and physical classification is usually required to provide a narrow size distribution. The average particle size can be shifted by adjustment of the gas velocity, design of the atomization die, and by control of the metal feed (Ref 5).

Aerosol routes can be used to produce silver powders. Solid, spherical, dense silver powders can be produced by spray pyrolysis (aerosol decomposition). An aerosol of a silver salt solution such as silver nitrate is heated to temperatures as low as 600 °C in nitrogen to decompose the silver salt and produce a fully dense silver powder. Spherical silver powders of different sizes and density can be made by controlling reaction temperature, carrier gas, solution concentration, residence time, and droplet size. Particle size of the silver powders range from 5 to smaller than 0.5 μm (Ref 9).

Another spray pyrolysis technique for making silver powders involves the spraying of a silver nitrate solution into a flame formed by a coaxial burner. Control of the flame temperature and the spray conditions allow for the control of particle size and shape. By adjusting the flame temperature, particles can be made nonspherical (temperatures below the melting point), spherical (temperatures above the melting point), and ultrafine particles by having the flame temperature sufficiently hot enough to evaporate the silver (Ref 10).

Mechanical Milling Processes. Mechanical comminution can be used to alter most types of silver powders with processes such as ball milling, vibratory milling, or attrition milling. These methods are time consuming and care needs to be taken to keep from introducing significant levels of impurities from the equipment and media. Many types of milling media can be used including glass balls, zirconia balls, and metal balls. Sizes of media range from 10 to less than 1 mm depending on the size of the starting silver powder and the size of the resulting silver flake desired. Initially, larger silver particles and silver agglomerates are broken down during the milling process. Continued milling produces flat, two-dimensional silver flake as shown in Fig. 6. Usually an organic surfactant is introduced during the milling process to prevent the cold weld-

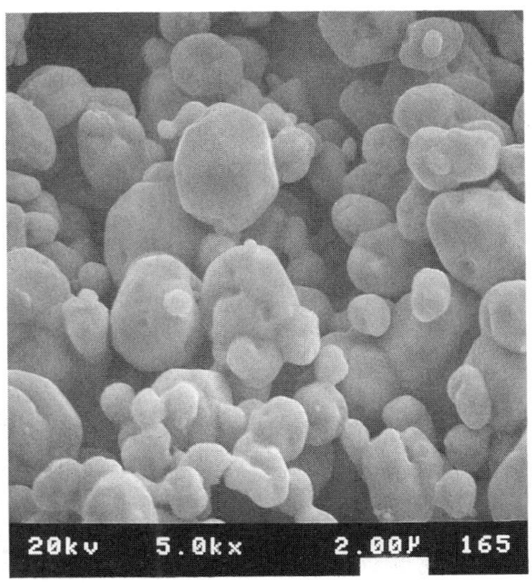

Fig. 1 Irregular-shaped silver powder

ing of the particles to each other during the grinding process and to control the size and shape of the silver flake. Typical surfactants include saturated and unsaturated fatty acids and long-chain aliphatic molecules. By altering the surfactant, solvent, milling media, and milling conditions, specific silver flakes can be designed for particular applications (Ref 11).

Thermal Reduction. Heat is used to reduce silver salts to produce silver powders. Silver oxide and silver carbonate can be reduced to silver powder by thermal reduction in an inert atmosphere at temperatures greater than 250 °C (480 °F) as follows:

$$2Ag_2O + \text{heat} \rightarrow 4Ag_{(s)} + O_{2\,(g)}$$

$$Ag_2CO_3 + \text{heat} \rightarrow 4Ag_{(s)} + 2CO_{2(g)} + O_{2(g)}$$

These powders tend to be spongy, agglomerated, and very porous and therefore require subsequent mechanical treatment to provide usable silver powder.

Production of Gold Powders

The reliability of gold conductors makes them highly desirable for a variety of applications in electronics and communication systems. In hybrid microcircuitry it is the bonding behavior of gold as well as its electrical conductivity that makes it desirable. The reliability of gold is largely due to its superior resistance to oxidation and tarnish. Using gold initially may cost more, but much more would be sacrificed in terms of lost operating time and repair expenses if the electrical properties of the nongold substitute were to deteriorate due to the formation of oxide or sulfide films (Ref 6, 12).

Pure gold conductors have high conductivity, excellent wire bondability, excellent migration resistance, and are generally compatible with other components of thick-film materials. This makes gold particularly suitable for resistor terminations, wire- and die-attach pads, and highly reliable electrical interconnections (Ref 13). Applications for these devices include medical equipment, computers, satellite telecommunications, automotive control devices, and defense-related electronic components.

Pure gold conductors are not readily solderable using conventional solders because of poor solder leach resistance. The gold conductors can be made to be solderable through the addition of platinum. This imparts solderability at the expense of conductivity. Where high reliability and solderability is required and cost is not an issue, platinum-gold conductors are the industry standard (Ref 13, 14).

Many processes for making gold powder use an aqueous solution of chloroauric acid ($HAuCl_4$) as the starting salt solution. This gold chloride solution is made by dissolving gold grain or shot in aqua-regia (a 3-to-1 mixture of hydrochloric acid and nitric acid) followed by heating to decompose and remove the nitric acid.

A variety of reducing agents can be used to reduce gold to gold powder including other metal powders (zinc, magnesium, lead, aluminum, and iron), sulfur compounds (sulfur dioxide, sodium sulfite, potassium sulfite, and ferrous sulfate), hydrazines (hydrazine hydrate, hydrazine hydrochloride, and hydrazine sulfate), aldehydes (formaldehyde), oxalates (oxalic acid and potassium oxalate), hydrogen peroxide, hydroquinone, hydroxylamine, and hypophosphorous acid (Ref 15, 16).

The morphology of the gold particles is determined by the conditions of the precipitation process. Using reducing agents without the addition of particle morphology modifiers produces gold particles with nodular or irregular-shaped particles. Heating the solution increases the rate of reduction. The presence of free hydrochloric acid can retard or even prevent the reduction. Solutions that are basic can increase the rate of reduction (Ref 17). Sodium hydroxide, sodium carbonate (or bicarbonate), and amines (including ammonia) can be used to make the solutions basic and regulate the pH during the reaction.

Reproducibility and morphology are the keys to the applications for gold powder and are the driving forces for choosing one process over another. Gold powder can be precipitated in a wide range of particle size distributions and shapes. Different reducing systems produce different types of gold powders. Modifiers can be added to control the nucleation stage to closely control particle size, and colloidal materials can be added to inhibit particle-to-particle aggregation (Ref 18).

The shape characteristics of gold powder are almost entirely spherical where the reductants are sulfur dioxide, sulfites, and related compounds. Potassium sulfite or sodium sulfite can be directly added to aqueous gold chloride solution to produce spherical gold particles with controlled size and density (Ref 19):

$$2HAuCl_4 + 3Na_2SO_3 + 3H_2O \rightarrow 2Au_{(S)} + 3H_2SO_4 + 6NaCl + 2HCl$$

This reaction is very dependent on pH, the ratio of gold chloride to sulfite, the mixing, and the temperature. If high temperatures are used, a mixture of spheres and flat plates can be made (Ref 16).

Gold powder containing a mixture of gold flakes and gold spheres can be made by reducing an aqueous gold chloride solution with a mixture of hydroquinone and oxalic acid in the presence of a protective colloid such as gum arabic, methylcellulose, sodium alginate, gelatin, and so forth (Ref 20). Fatty acids, fatty acid alcohols, and fatty acid amines can also be used as coatings for the gold particles to control the particle size and act as lubricating agents when using the gold powders (Ref 21).

An example of spherical gold powder is shown in Fig. 7, and a gold powder containing a mixture of spheres and flakes is shown in Fig. 8.

Gold powder can also be made mechanically. Sheets of gold are reduced in thickness by rolling and beating to produce a foil between 0.001 and 0.002 cm thick that is then beaten to produce a gold 0.2 to 0.5 μm thick. These leaves are

Fig. 2 Small-sized silver powder

Fig. 3 Spherical-shaped silver powder

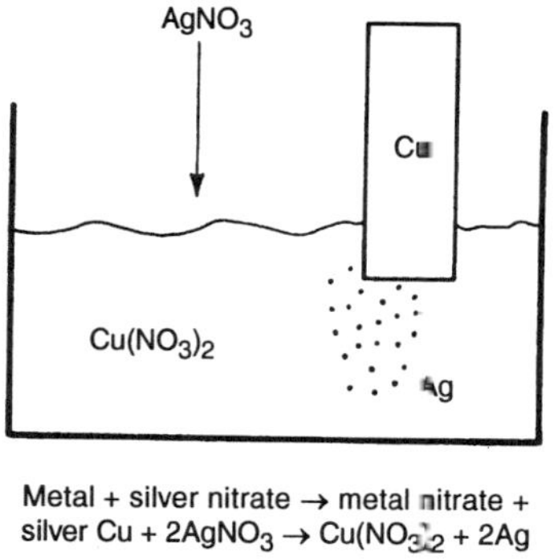

Fig. 4 Galvanic reduction

broken down further during a dispersion process to enable them to be used as fine gold powder. Larger gold powders and gold flake can be made through standard ball milling of gold powder (Ref 18).

Palladium Powders

Palladium is used in the electrical and electronics industries in electrical contacts, multilayer ceramic capacitors, thick-film paste conductors, resistors, thermocouples, and heating pads. The corrosion resistance, high melting point, contact resistance, and reasonable electrical conductivity of palladium account for its use as an electrical contact material where long life and reliability are essential. Palladium can be preferred for low-current, long-life relays operating at low contact forces (Ref 22, 23).

Palladium powders, mixtures of silver and palladium powders, and palladium-silver alloy powders are used extensively as the electrode material in multilayer ceramic capacitors. Palladium pastes are used in multilayer ceramic capacitors made at high temperatures, whereas silver-palladium is used in capacitors prepared at lower fired temperatures. The properties of the powders intended for the internal electrodes of multilayer ceramic capacitors are extremely important as compatibility is required between the metal powder and the organic medium of the thick-film paste and between the paste itself and the surrounding dielectric material (Ref 24).

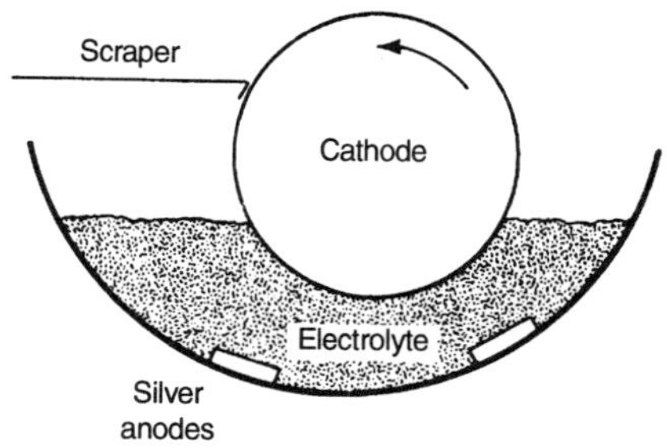

Fig. 5 Electrochemical reduction

Other thick-film conductor applications for palladium powders include chip resistors, resistor networks, and hybrid conductors for industrial, automotive, telecommunications, and even high-technology consumer applications (Ref 25). The presence of palladium serves to inhibit the solder leaching of silver from the conductor circuit line, and it reduces the migration of the silver into the dielectrics or resistors in hot or humid conditions, especially when high electric fields are present between closely adjacent conductor tracks held at different electric potentials (Ref 13).

As with other precious metal powders, soluble palladium salts such as chloropalladous acid, palladium nitrate, and palladium bromide are chosen as starting materials. Chloropalladous acid is the most frequently used. However, for the production of precious metal powders involving silver by coprecipitation, chloride salts should be avoided because they immediately precipitate the silver ions to form insoluble silver chloride. To prevent the precipitation of insoluble silver chloride, the nitrate salts of palladium and silver are used when making coprecipitated silver and palladium. Prior to reduction, the pH of the aqueous solution of the reactants may be adjusted by the addition of ammonium hydroxide, sodium, or potassium hydroxide, sodium or potassium carbonate, or similar compounds.

Although the basic reaction for making palladium powder remains the same:

$$Pd^{2+} + 2e^- \rightarrow Pd^{\circ}$$

subtle variations in the reaction parameters can have a large effect on the physical characteristics of the metal powder formed. Reducing agents that are effective for gold or platinum precipitations are applicable for the manufacture of palladium powder. Hydrazine, formaldehyde, hypophosphorous acid, hydroquinone, sodium borohydride, formic acid, and sodium formate are used. Important variables in the reduction process for making palladium powder include the type of palladium salt used as the source of palladium, the type of complexing agent (if any), the pH, the choice of the reducing agent, the concentration of the reactants, the reaction temperature, the mode and degree of agitation of the solution, the solution viscosity, and the presence of additives or surfactants. Subtle changes to the precipitation parameters can produce different sizes ranging from <0.1 to greater than 5 μm and different morphologies including nodular, dendritic, irregular, and spherical-shaped palladium particles (Ref 24). Examples of palladium powders are shown in Fig. 9 and 10.

Metal powders such as zinc and copper can be used to produce palladium powder from acidic solution. This process is known as "cementation" or "galvanic precipitation." The reaction is:

$$H_2PdCl_4 + Zn^{\circ} \rightarrow Pd^{\circ} + ZnCl_2 + 2HCl$$

Generally, products prepared by this technique require special treatment to reduce the amount of impurities present. In addition, these palladium

Fig. 6 Silver flake

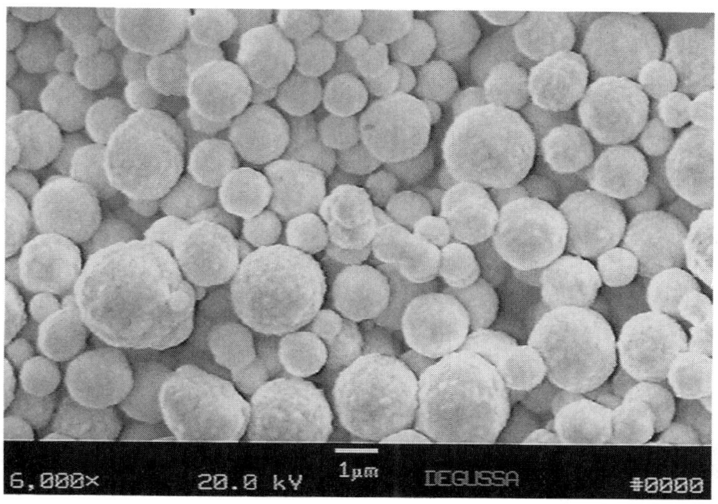

Fig. 7 Spherical gold powder

powders are large in size, irregular shaped, and tend to be agglomerated.

Spherical, dense palladium powder can be produced using spray pyrolysis. A precursor palladium salt solution (such as palladium nitrate) is atomized by an aerosol generator, the droplets are evaporated and reacted in a furnace, and the resulting spherical palladium powder is collected. The parameters controlling the particle properties include the choice of palladium salt, reactor temperature, solution concentration, aerosol droplet size, reaction atmosphere, and the residence time in the reactor. By controlling these parameters, spherical, dense palladium particles can be made with average sizes ranging from 0.1 to about 5 μm (Ref 26).

Platinum Powders

Platinum is the noblest of all of the metals. Despite its great expense, it is the best known of the precious metals and is widely used. Platinum has a very high density, does not corrode, and is very chemically resistant. With its low, but temperature-dependent resistance, platinum is used for temperature measurement, corrosion-resistant electrical contacts, electrodes, and sensors for gas detection such as oxygen and toxic gases (Ref 27). Oxygen sensors are used in automobiles to monitor the exhaust gases to ensure that the autocatalysts are working and that the fuel-to-air ratios are optimal. Platinum sensors can be used as flow-rate sensors because they have a quicker response than the conventional wire-type sensors (Ref 3). Platinum is also used for applications that need to endure severe conditions so that the added expense relative to palladium or silver is necessary to provide the reliability and durability (Ref 22).

Platinum powders are used in thick-film pastes as spherical or flake-shaped particles with size ranges from 0.1 to 20 μm. An example of platinum powder is shown in Fig. 11. These pastes are sprayed, brush-painted, or screen printed onto inert surfaces that are then fired to 500 to 1500 °C to form strongly bonded conductors that can be soldered (Ref 22). In addition, small amounts of platinum powders can be added to silver thick-film pastes to improve the bondability and leach resistance of the formed conductor film. Platinum powders are added to gold-powder thick-film compositions to impart excellent solder leach resistance. These platinum gold conductors are the standard for high-reliability, solderable conductor applications where cost is less of an issue (Ref 13).

An aqueous salt solution of chloroplatinic acid (H_2PtCl_6) is the most common starting material for the manufacture of platinum powder. It can be prepared by dissolving platinum sponge in aqua regia. After the nitric acid is driven off by continuous boiling with the addition of hydrochloric acid, the solution is then precipitated by a suitable reducing agent at the proper pH level. Reducing agents that are used include sodium formate, formic acid, sodium borohydride, sodium hydrosulfite, hypophosphorous acid, and hydrazines. The pH of the solution can be ad-

Fig. 8 Gold powder containing flakes and spheres

Fig. 9 Small-sized palladium powder

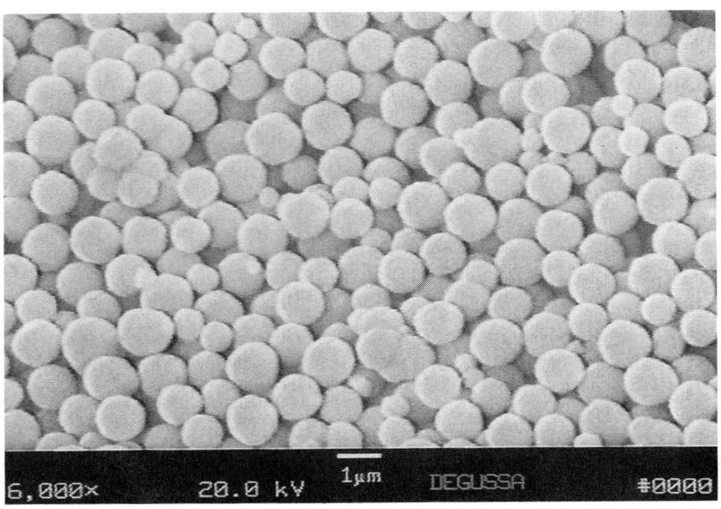

Fig. 10 Spherical palladium powder

Fig. 11 Platinum powder

justed by the addition of reagents such as sodium, potassium, or ammonium hydroxides, or their carbonates. Sodium borohydride can be used in the preparation of platinum alloys including ruthenium-platinum, platinum-palladium, platinum-rhodium, and platinum-iridium (Ref 4).

Very small sized platinum particles are very active and commercially important as catalysts. One method of making platinum powder with a size of about 0.030 μm is the reduction of chloroplatinic acid with sodium borohydride. The reaction is as follows:

$$H_2PtCl_6 + NaBH_4 + 3H_2O \rightarrow Pt_{(s)} + H_3BO_3 + 5HCl + NaCl + 2H_{2(g)}$$

An organic protective agent such as poly(vinylpyrrolidone) is generally added to prevent coagulation of these small platinum particles (Ref 28).

ACKNOWLEDGMENT

Scanning electron micrographs were provided by Degussa Corporation Metal Group, S. Plainfield, NJ.

REFERENCES

1. G. Demopoulos and G. Pouskouleli, Hydrochemical Preparation of Fine Precious Metal Powders, *J. Met.,* Vol 40 (No. 6), 1988, p 46–50
2. R.G. Loasby and P.J. Holmes, Development of Thick Film Technology, *Handbook of Thick Film Technology,* P.J. Holmes and R.G. Loasby, Ed., Electrochemical Publications, Ltd., 1976, p 5
3. L.S. Benner et al., Ed., *Precious Metals Science and Technology,* International Precious Metals Institute, 1991, p 319–333, 584–585
4. Platinum Suppl. Vol A1, *Gmelins Handbook of Inorganic Chemistry,* 1986, p 126
5. A. Butts and C.D. Coxe, Ed., *Silver—Economics, Metallurgy, and Use,* Krieger Publishing, 1975, p 441–445
6. J. Savage, Conductor Materials, *Handbook of Thick Film Technology,* P.J. Holmes and R.G. Loasby, Ed., Electrochemical Publications, Ltd., 1976, p 106
7. T. Kubota, *J. Appl. Phys. (Jpn.),* Vol 39 (No. 9), 1970, p 861–868
8. F.V. Lenel, *Powder Metallurgy Principles and Applications,* Metal Powder Industries Federation, 1980, p 39–40
9. T.C. Pluym, Q.H. Powdell, A.S. Gurav, T.L. Ward, T.T. Kodas, L.M. Wang, and H.D. Glicksman, *J. Aerosol Sci.,* Vol 24 (No. 3), 1993, p 383–392
10. K. Nagashima, Y. Morimitsu, and A. Kato, *Nippon Kagaku Kaishi,* Vol 12, 1987, p 2293–2300
11. E.M. Jost, K. McNeilly, and T.J. Santala, *ISHM '87 Proc.,* International Society for Hybrid Microelectronics, 1987, p 548–553
12. R.R. Davies, New Development in the Use of Gold in Electronics and Communications, *Precious Metals,* R.O. McGachie, and A.G. Bradley, Ed., 1981, Pergamon, p 193–199
13. W. Borland, Thick Film Hybrids, *Electronic Materials Handbook,* Vol 1, ASM International, 1989, p 332–353
14. M.L. Topfer, *Thick-Film Microelectronics,* 1971, p 46
15. B.P. Block, *Inorganic Synth.,* Vol IV, 1953, p 15
16. D.J. Arrowsmith and K.J. Lodge, *Trans. IMF,* Vol 65, 1987, p 120–126
17. T.K. Rose and W.A.C. Newman, *Metallurgy of Gold,* 7th ed., Met-Chem Research, 1986, p 70
18. N. Collier, Advances in Gold Powder Technology, *Gold Bull.,* Vol 10 (No. 3), 1977, p 62–66
19. O.A. Short, U.S. Patent 3,771,996, 1973
20. O.A. Short and R.V. Weaver, U.S. Patent 3,725,035, 1973
21. V.R. Driga, U.S. Patent 3,843,379, 1974
22. C.F. Vermaak, *The Platinum-Group Metals,* Mintek, 1995, p 165–180
23. E.M. Wise, *Palladium, Recovery, Properties, and Uses,* Academic Press, 1968
24. G. Ferrier, A. Berzins, and N. Davey, The Production of Palladium Powders for Electronic Applications, *Platinum Met. Rev.,* Vol 29 (No. 4), 1985, p 175–179
25. P.D. Gurney and R.J. Seymour, The Platinum Group Metals in Electronics, *Chemistry of the Platinum Group Metals,* F.R. Hartley, Ed., Elsevier, 1991, p 594–616
26. T.C. Pluym, S.W. Lyons, Q.H. Powell, A.S. Gurav, T.T. Kodas, L. Wang, and H.D. Glicksman, *Mater. Res. Bull.,* Vol 28, 1993, p 369–376
27. F.R. Hartley, The Occurrence, Extraction, Properties and Uses of the Platinum Group Metals, *Chemistry of the Platinum Group Metals,* F.R. Hartley, Ed., Elsevier, 1991, p 6
28. P. Rheenen, M. McKelvy, R. Marzke, and W. Gaunsinger, *Inorganic Synth.,* Vol 24, 1986, p 238–242

Production of Refractory Metal Powders

THE REFRACTORY METALS—tungsten, molybdenum, niobium, tantalum, and rhenium—are distinguished by several common characteristics, including high density, high melting point, and superior resistance to wear and acid corrosion. Tungsten, for example, has a density over twice that of iron and a melting point of 3410 °C (6170 °F), the highest of any element. These metals have body-centered cubic crystal structures (with the exception of rhenium, whose crystal form is hexagonal). All are subject, however, to severe oxidation above 500 °C (930 °F) and must be protected for service by coatings or nonoxidizing atmospheres.

The refractory metals are extracted from ore concentrates, processed into intermediate chemicals, and then reduced to metal, which can be in the form of powder. The pure or alloyed powders are pressed, sintered, and subsequently worked. This article discusses the raw materials, processing steps, powder properties, and finishing techniques (blending, handling, inspection and packaging) involved in the production of refractory metal and carbide powders. For information on applications of these materials, see the article "Refractory Metals" in this Volume.

Production of Tungsten and Tungsten Carbide Powders

B. Lux, Vienna University of Technology
B. Zeiler, Wolfram Bergbau- und Huetten-GmbH Nfg KG

Tungsten gained major industrial importance at the beginning of the 20th century due to its technical application as alloying element in high-speed steel and as filament wire in incandescent lamps (Ref 1–3). For both applications, pure tungsten was produced in powder form. From 1900 to 1912 only pure tungsten powders were used in high amounts for high-speed steels containing 16 to 24% tungsten (Ref 1). This application dominated the tungsten consumption at the time, but around 1915 ferrotungsten replaced the tungsten metal powder in this application.

Later (1910–1920), diamonds were needed in large numbers to produce drawing dies for tungsten lamp wire fabrication. Due to their high costs, there was a strong driving force to find cheaper alternatives. At this time, the very hard tungsten carbides (WC and W_2C), studied by Henri Moissan at the end of the 19th century during his attempts to make artificial diamond (Ref 3, 4), were considered as appropriate alternatives. However, all trials to compact tungsten carbide powders resulted in porous and brittle products unusable for the anticipated drawing dies application. In 1922, researchers of the OSRAM study group combined the brittle tungsten carbide with a ductile metal binder (iron, cobalt). After sintering the respective powder mixtures, a material of both high strength and toughness was obtained. Thus the "hardmetal" also named "cemented carbide" was created. OSRAM used the new material mainly for wire drawing. Since they apparently did not yet realize the great importance of their invention, patents were applied for in only Germany, Great Britain, and the United States, while in all other countries, priority deadlines were missed (Ref 3).

In 1925 KRUPP, a company experienced in steel technology and tool manufacturing, realized the enormous potential of the hardmetals and bought the patent rights. In 1927 they introduced hardmetals under the brand name WIDIA (an abbreviation for WI-e DIA-mant, which in German means "similar to diamond") (Ref 3). Since then, cemented carbides have changed the tungsten consumption pattern completely; today, hardmetals (and thus tungsten carbide powders) account for more than half of the world's tungsten consumption (Ref 5–8).

Uses and Annual Consumption. A breakdown of tungsten usage in the three most important economic regions— United States, Japan, and Europe— is shown in Fig. 1 (Ref 5–8). Over the last 20 years, annual global consumption of tungsten has been in the range of 40,000 to 50,000 metric tons per year (Ref 9). As noted, over half of tungsten consumption is for tungsten carbide powder for hardmetal production. However, important products are also based on tungsten metal powders, such as:

- Heavy metal alloys (W/Fe/Cu/Ni/Co-alloys)
- Electrical contact materials (W/Cu, W/Ag)
- Ductile tungsten (including lamp filaments)

Tungsten, as an alloying element for steels and superalloys, is predominantly used as ferrotungsten, or as melting base alloys, which are prepared from either ore concentrates or tungsten-

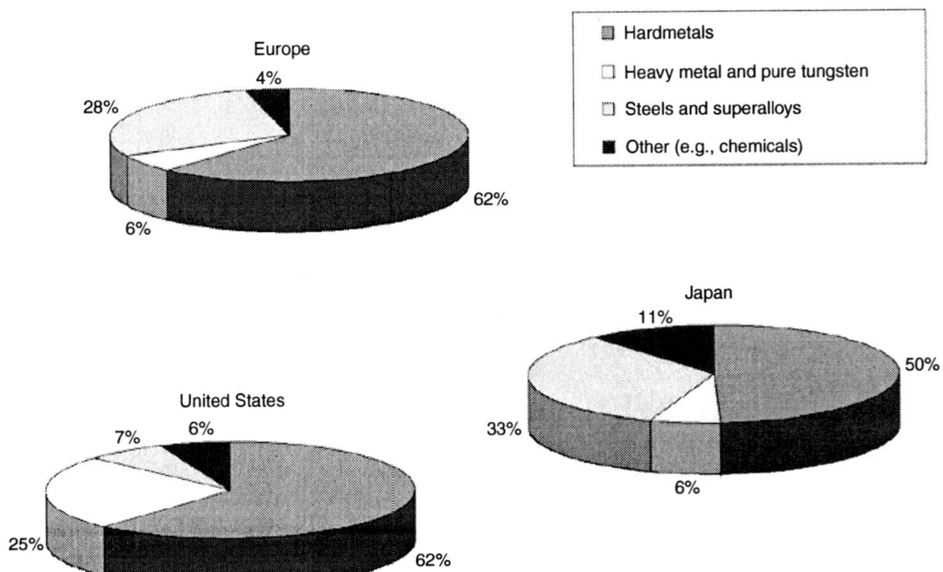

Fig. 1 Breakdown of tungsten consumption for different economic regions. Source: Ref 5–8

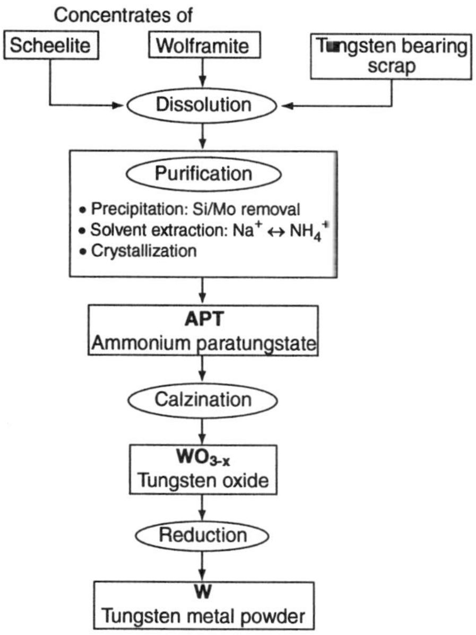

Fig. 2 Production of tungsten metal powder

Fig. 3 Multitube push type furnace. Courtesy of Wolfram Bergbau- und Huetten-GmbH Nfg KG, Austria

bearing scrap. Tungsten powders are hardly used for this application. A minor part of the tungsten consumption is used in the form of various chemical products, such as catalysts, pigments, solid lubricants, and heavy liquids (Ref 10).

Tungsten Metal Powder

Production

Industrially relevant tungsten-bearing minerals are Wolframite ([Mn, Fe]WO_4), Ferberite ($FeWO_4$), Hübnerite ($MnWO_4$) and Scheelite ($CaWO_4$). The tungsten concentration of workable ore starts at 0.1 to 0.3 wt% WO_3. Concentrations of more than 2 wt% are rare. The average WO_3 content is usually about 0.5 wt% (Ref 11–13). Depending on the mineral type, different techniques are used to concentrate the ore up to typical WO_3 contents about 50 to 70%, such as gravity methods and flotation. An important alternative tungsten source to the natural mineral resources is tungsten-bearing scrap, which usually has a high tungsten content.

Process. Figure 2 schematically shows the process from the concentrate or scrap via the formation of the high-purity ammonium paratungstate (APT) intermediate to tungsten metal powder (Ref 11, 14, 15).

Dissolution Step. The tungsten-bearing raw materials are treated with NaOH (Wolframite and oxidized scrap) or Na_2CO_3 (Scheelite) to form water-soluble sodium tungstate (Na_2WO_4).

Purification steps are as follows:

1. Silicon, phosphorus, fluorine, aluminum, and arsenic are removed by precipitation.
2. Sodium is exchanged by NH_4 and an ammonium tungstate solution ($(NH_4)_2WO_4$) is formed. This is commonly done by solvent extraction (SX), which is a liquid ion exchange (LIX) process, where tungsten is extracted by means of an aliphatic amine dissolved in isodecanol and kerosene. Reextraction into aqueous ammonia solution finalizes this step.
3. Ammonium Paratungstate (APT) is crystallized by evaporation of ammonia and water. APT is the intermediate product of highest purity for tungsten metal and tungsten carbide powder production. Today other intermediates, such as tungstic acid, have practically no industrial relevance and have almost disappeared completely.

Calzination Step (Ref 16). By heating the APT to 400 to 900 °C, mainly in rotary furnaces, ammonia and water are removed leading to tungsten oxide. If carried out in excess of air, the resulting oxide is yellow (WO_3- tungsten yellow oxide). Under exclusion of air (oxygen), a slightly reduced blue colored oxide is formed ($WO_{3-x,}$ tungsten blue oxide).

Reduction Step (Ref 2, 11). The tungsten oxide is reduced by hydrogen to tungsten metal powder. Two different furnace designs for the reduction process are industrially used today, the multitube push type (Fig. 3) and the rotary furnaces (Fig. 4).

Push Type Furnace. The tungsten oxide is filled in flat metal "boats," which are pushed through the furnace tubes. This results in a static powder layer throughout the whole reduction process. Layer height and push cycle can be varied. Through individual heating zones temperature profiles may be applied. Typical reduction temperatures range from 600 to 1100 °C. The flow of hydrogen is usually in the countercurrent direction. Removal of the water vapor formed during reduction occurs from the powder layer surface and is effected by the hydrogen flow rate. By variation of reduction parameters like temperature and powder layer height (oxide load of the boat), the particle size of the tungsten metal powder can be controlled in the range between 0.1 and 100 µm. The throughput for production of fine powders is lower than for manufacturing of coarser ones. The push type furnace is very flexible with respect to the production of different particle sizes and can quickly be adjusted to produce the desired powder properties.

Rotary Furnace. The tungsten oxide is directly introduced into the rotating furnace tube. As a consequence, there is no static but a dynamic powder layer. Depth of the layer is influenced by the feed rate, rotational speed, incline, and lifters inside the tube. The temperature range is comparable to that of push type furnaces, and the hydrogen flow direction is also in most cases countercurrent. Concerning changes between particle sizes of the tungsten metal powder to be produced, the rotary furnace is not as flexible as the push type furnace. A typical application is the continuous production of medium sized, 1 to 3 µm powders.

Reduction in General. Although the technical process of tungsten oxide reduction is carried out in one single process step, it proceeds through several intermediate oxide phases, as illustrated in Fig. 5 (Ref 17). The sequence in which oxides occur depends on the applied temperature and the local humidity within the static (push type furnace) or the dynamic (rotary fur-

Fig. 4 Rotary furnace. Courtesy: Wolfram Bergbau- und Huetten-GmbH Nfg KG, Austria

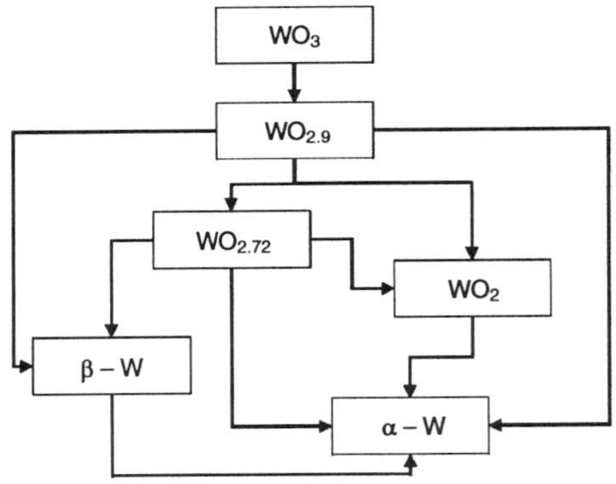

Fig. 5 Tungsten oxides occurring in the temperature range 500–1000 °C. Source: Ref 17

nace) powder bed. The intermediate oxides differ not only with respect to their color but also to their morphology (Ref 17, 18), which is shown in Fig. 6 (Ref 18). The reason for these morphological changes is the transport of tungsten via the gas phase chemical vapor transport (CVT) during the reduction sequence (Ref 19, 20). $WO_2(OH)_2$ has been identified as the most volatile compound in the system, W/O/H, and is responsible for the CVT (Ref 19–22). Figure 7 shows the CVT reaction from WO_2 to tungsten metal (Ref 23). If the partial pressure ratio, p_{H_2O}/p_{H_2}, is low enough, the reduction from WO_2 to tungsten (at a certain temperature) starts. From a thermodynamic point of view, $WO_2(OH)_2$ may form from both WO_2 and tungsten, but the equilibrium pressure is lower above tungsten, which causes the CVT of $WO_2(OH)_2$ from WO_2 to tungsten, where this vapor deposits as tungsten metal, thus leading to growth of the tungsten crystal. The CVT mechanism enables the control of the W-metal powder particle size through the interplay between temperature and humidity (partial pressure ratio, p_{H_2O}/p_{H_2}), which is shown schematically in Fig. 8 for the push type furnace (Ref 11, 19, 20, 24).

Properties

Tungsten has two outstanding properties, a very high density of 19.3 g/cm^3 and the highest melting point of all metals at 3422 °C (Ref 25) (Table 1). It can not be put in shape by normal casting technologies. Consolidation of tungsten powders is based almost exclusively on powder metallurgical techniques. Thus, clear and extensive powder specifications are extremely important for any industrial consolidation in order to obtain the desired final properties of a consolidated tungsten product. Tungsten powder properties and their proper specification can be divided in physical and chemical aspects.

Physical Properties. The particle size of the tungsten powder is very important for its use. Today all sizes, from submicron up to one hundred microns, are produced on an industrial scale (Fig. 9); today's major tungsten metal powder consumption is within the range of 2 to 5 μm.

Particle sizes between 1 to 10 μm are measured by Fisher subsieve sizer (FSSS) (ASTM B 330, Ref 26) and correspond well to particle sizes observed directly by scanning electron microscopy (SEM). However below and above this range only SEM gives accurate values. SEM is also the best way to judge the overall grain morphologies and homogeneity.

Particle size distributions can be measured by sedimentation (Sedigraph) (ASTM B 761, Ref 26) or by laser scattering techniques (ASTM B 822 (Ref 26)). Appropriate regulations of reduction parameters permit to obtain both, narrow or broad, size distributions and to fulfill almost any specific requirement on tungsten powders needed for various applications.

Different powder density measurements, such as apparent or Scott density (ASTM B 212 and B 329, Ref 26), tap density (ASTM B 527, Ref 26) as well as green density are of great importance for predicting and controlling the behavior of a powder during the subsequent consolidation steps (i.e., pressing and sintering).

Chemical Properties. Generally, tungsten metal powders, in comparison with most other metal powders used for powder metallurgical products, have a relatively very high purity, as shown by a

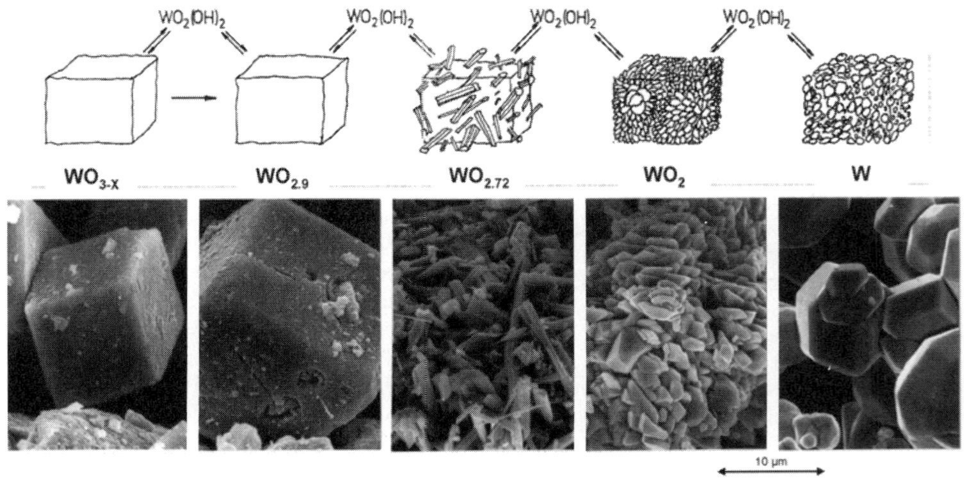

Fig. 6 Morphological changes during reduction at 1000 °C. Source: Ref 18

typical product analysis given in Fig. 10. Trace elements, such as molybdenum, aluminum, silicon, phosphorus, arsenic, sodium, potassium, calcium, magnesium, and sulfur are already removed during APT production. The uptake of elements like nickel, chromium, iron, and cobalt, typical alloying elements of reduction boats or rotary furnace tubes, has to be prevented during metal powder production.

Tungsten Applications and Special Products

Ultrahigh Purity (UHP) Powders. Sputter targets for semiconductor manufacturing need ultrahigh purity (5N and 6N) tungsten metal powders. In particular, the incorporation of alpha emitters such as uranium and thorium, is very critical and has to be kept below 1 ppb. The mobile alkali ions (e.g., sodium and potassium) have to be kept below 0.1 ppm. Purification of preselected APT by multiple extraction and crystallization steps guarantees the required purity, where all production steps must be carried out and can be achieved only under clean room conditions (Ref 11, 27, 28).

Pure Tungsten Materials. For wire and sheet production, the tungsten powders are consolidated to rods by pressing and sintering the powder. In order to get a suitably high densification, high temperatures between 2500 and 3100 °C (Ref 2, 11, 29) are applied. Thereby, most impurities still present at this stage are largely removed by evaporation. The sintered tungsten bars thus obtained are swaged or rolled at elevated temperatures in order to produce either rods, which are further drawn to produce wires, or sheets. Such pure tungsten materials are widely used for high-temperature applications, for example in high-temperature furnace construction (Ref 2, 11).

Nonsag Tungsten. The "creeping" of tungsten wires in incandescent lamps during very high operation temperatures must be avoided. The filament should not sag. Accidentally and involuntarily discovered decades ago, this avoidance can be achieved by incorporating traces of potassium into the tungsten wires, thus successfully preventing the sagging of the tungsten filaments in light bulbs (Ref 2, 30).

Today's technology is to add a mixture of aluminum, silicon, and potassium into tungsten oxide prior to the reduction to tungsten metal. During this reduction, a certain part of the Al-, Si-, K-doping elements is enclosed into the tungsten metal powder particles. In the subsequent sintering of the pressed tungsten powder, aluminum and silicon evaporate, while potassium is too big to diffuse through the tungsten lattice. This procedure leads to an incorporation of potassium droplets as inclusions. During the wire drawing, the tiny potassium droplets are elongated and finally break into a multitude of tiny potassium bubbles in rows allocated parallel along the wire axis, which leads to an elongated interlocked grain structure. During recrystallization at elevated temperatures, the potassium bubbles stabilize very effectively by an effect analog to dispersion hardening, the movement of dislocations and thus the migration of the grain boundaries perpendicular to the wire axis. This particular structure is the reason for the nonsag properties of the tungsten wires in light bulbs (Ref 2, 11, 30, 31).

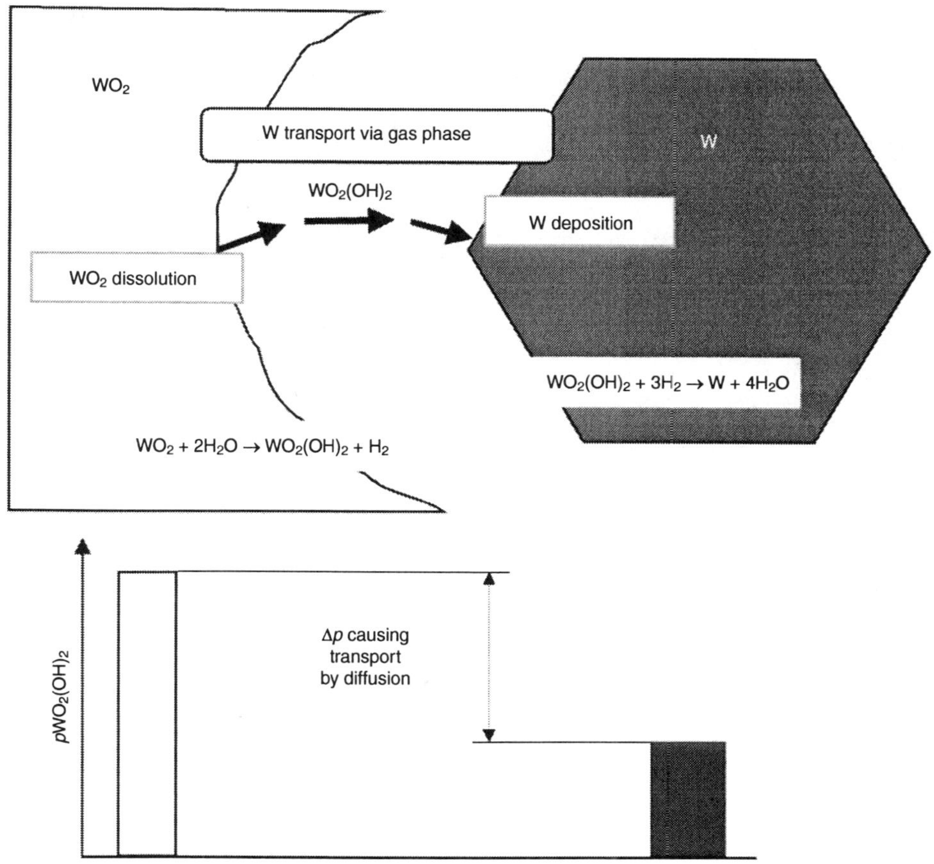

Fig. 7 Chemical vapor transport (CVT) of W in the WO_2 to W reduction step. Source: Ref 18

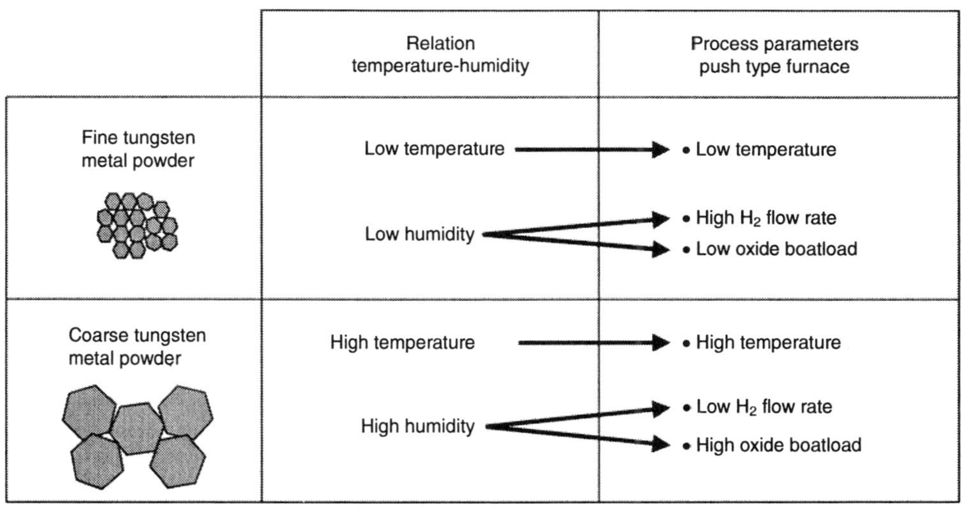

Fig. 8 Control of the W-metal powder particle size in the push type furnace. Source: Ref 11, 19, 20, 24

Tungsten with Oxide Dispersions. Tungsten metal powder and thorium-containing compounds (such as ThO_2 or Th-nitrate) are mixed, pressed, sintered, and further processed to wire or sheets. They have two major applications (Ref 2, 11):

- Welding electrodes, whereby the high electron emission capability of Th is used for easier electric arc formation and its stabilization (Ref 32, 33)

Fig. 9 Tungsten metal powder particle size range. Courtesy: Wolfram Bergbau- und Huetten-GmbH Nfg KG, Austria

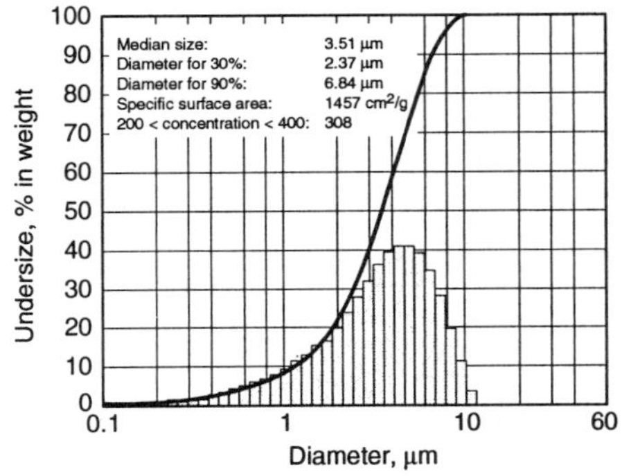

(a)

(b)

Physical analysis of tungsten, 2–5 μm	
BET surface, m²/g	0.18
Grain size FSSS, μm	
As supplied	3.27
Lab milled	2.93
Porosity	
As supplied	0.661
Lab milled	0.497
Scott density, g/in.³	52
Trace impurities, ppm	
Aluminum	2
Carbon	11
Calcium	<2
Cobalt	<5
Chromium	<10
Copper	<2
Iron	12
Potassium	<4
Magnesium	<1
Molybdenum	<20
Sodium	9
Nickel	<5
Oxygen	210
Phosphorus	<20
Sulfur	<5
Silicon	<10

Fig. 10 Analysis of a typical medium sized tungsten metal powder: Courtesy: Wolfram Bergbau- und Huetten-GmbH Nfg KG, Austria

- Improved physical properties (dispersion hardening) in tungsten sheets and wires for (e.g., vibration-resistant automotive lamps) (Ref 34)

Due to environmental health and safety regulations associated with the use of the radioactive thorium, alternatives have been developed. Lanthanum and cerium have proved to be potential candidates to substitute thorium in welding electrodes (Ref 32, 33). For certain lighting applications, however, suitable substitutes are still missing (Ref 32, 34).

Tungsten Composite Materials. Tungsten metal powder mixed with other powders, such as iron, nickel, copper, and cobalt (both individually or in combination) can be liquid phase sintered at about 1500 °C, forming so called "heavy metals." Applications are based on the high density (17–19 g/cm^3) and ductility of this composite and include counter weights, high energy penetrators, and radiation shields (Ref 11).

A porous skeleton of pressed tungsten metal powder can be infiltrated with copper alloys, or silver, leading to composite materials. W/Cu (and W/Ag) composites are used as electrical contacts (Ref 35, 36) due to their high electrical conductivity and their stability against damage caused by electrical arc formation. Recent developments (Ref 37, 38) describe the production of W/Cu-composite powders, which can be directly pressed to shape and sintered, thus preventing the infiltration and machining steps, which is of particular importance with respect to the application as heat sink in the semiconductor industry.

Recycling

Actually almost all kind of tungsten products, if available in large enough amounts and economically feasible, are recycled today (Ref 11, 39). Even unsorted low grade scraps containing enough tungsten can be used as a raw material for the chemical APT production process or can be used for alloying in the steel industry (Ref 40). Nevertheless, any options or possibility to use sorted tungsten scrap directly is of course preferably applied for economic reasons (Ref 40). Typical examples include:

- Sorted production scrap of pure tungsten can simply be crushed in high energy mills (planetary mills) and can be directly reused for applications with lower quality demands.
- Turning chips of heavy metal, obtained during the shaping operations, can be converted into reusable powder by oxidation and subsequent reduction. Too large solid pieces can be machined into smaller parts to enable further recycling. The oxidation step leads to a total disintegration of the solid chips into oxide powder containing WO_3 and tungstates of iron, nickel, or cobalt. Subsequent hydrogen reduction leads to a metal powder, whereby all components (W/Fe/Co/Ni) of the original heavy metal are still present (Ref 40).
- Composite scrap mixtures, such as W/Cu, are difficult to oxidize and are therefore not suitable for chemical recycling as described above. Also the recycling via the steel industry cannot be recommended due to the detrimental effect of copper in steel. A possible alternative is salt melt digestion with $NaNO_2$, $NaNO_3$, and Na_2CO_3 mixtures. However, this method suffers from environmental problems due to the NO_x formation (Ref 40). An alternative would be the electrochemical recovery of tungsten (Ref 41).
- W/Th recycling has specific problems due to the radioactive radiation. Recycling in the steel industry has become almost impossible during the last years. Chemical recycling in the conventional process leads to insufficient separation of thorium and is therefore also not applicable. The only way to recycle this material is still under development and based on the electrolytic dissolution of tungsten in ammonia (Ref 42). The difficulty is to run the process in a way to prevent dissolution of any of the radioactive fission products.

Fig. 11 Single-tube push type tungsten carbide powder production furnace. Courtesy: Wolfram Bergbau- und Huetten-GmbH Nfg KG, Austria

Tungsten Carbide Powder

Conventional production of tungsten carbide powders accounts world wide for the major part of the tungsten carbide powder. This flexible process can produce tungsten carbide powders with particle sizes between 0.15 and 50 μm (Ref 43). This range offers products for almost all industrial applications and is briefly described below along with alternative methods.

Production

Conventional. The main production steps involve the use of tungsten metal powder, which is mixed with high-purity carbon black, filled into graphite boats, and pushed through a high-temperature furnace under hydrogen at temperatures above 1300 °C. In contrast to the reduction step for tungsten, temperatures are much higher, and single tube push type furnaces (Fig. 11) are used. The particle size of the tungsten metal powder, used for carburization, determines directly the particle size of the tungsten carbide powder (Fig. 12).

After cooling the tungsten carbide powder is deagglomerated by means of ball mills, jet mills, or crushers. The deagglomeration step just separates particles from sticking to each other and is not used to adjust the particle size (Ref 43). The final process steps are homogenization (in double cone blenders) and sieving.

Fine Grained Powder Production. Recently, there has been an increasing demand for tungsten carbide powders finer than 1 μm. Together with the limitations of the conventional process concerning the lower efficiency of fine grained tungsten metal powder production (Ref 44) there has been a strong motivation to develop alternative production routes.

Tokyo Tungsten has developed the direct carburization of tungsten oxide (WO_3) as an alternative to the conventional process (Ref 45–47). This process is carried out today on an industrial scale. The main production steps are as follows:

- Production of pellets from a mixture of tungsten oxide with carbon black
- Subsequent reaction in rotary furnaces at temperatures up to 1600 °C (Ref 47) to tungsten carbide powder in two steps: first the direct reduction of the WO_3 to tungsten by carbon under nitrogen, followed by carburization to WC under hydrogen. Particle sizes of 0.1 to 0.7 μm can be produced (Ref 47).

More recently another direct carburization process has been developed by Dow Chemical (Ref 48–50):

- A mixture of WO_3 and carbon black falls through a vertical reactor at temperatures

Fig. 12 Correlation between tungsten metal and tungsten carbide particle size for ultrafine and coarse powders. Courtesy: Wolfram Bergbau- und Huetten-GmbH Nfg KG, Austria

around 2000 °C. Thereby the reaction time is typically a few seconds, leading to an intermediate mixture of W, W_2C, WC, and C.

- This product has to be subjected to a second, more conventional, carburization step.

The typical particle size range for these direct carburization processes is between 0.1 and 0.8 µm, which is identical with the lowest limits of feasibility of the conventional process (Ref 43).

To produce powders finer than 0.1 µm (100 nm), so-called nanopowders, two processes were developed. The spray conversion process has been promoted by Nanodyne Inc. (Ref 51, 52). Aqueous solutions of tungsten and cobalt salts are "spray dried," subsequently reduced, and carburized with gas mixtures like H_2/CH_4 or CO_2/CO in a fluidized bed reactor. The product is a relatively large (about 75 µm), hollow WC/Co composite powder particle, which consists of WC crystallites with a grain size of about 30 nm. This product must be milled in order to get a powder usable for being compacted (Ref 52).

The chemical vapor reaction (CVR) process, developed and patented by H.C. Starck (Ref 53) is based on the gas phase reaction of metal halides with different gas mixtures, to produce nanosized powders (5 to 50 nm). Production of tungsten carbide powder is possible in principle, although the process seems to be better suited for production of nitrides or carbides of other group IV and V refractory metals (e.g., TiN).

Coarse-Grained Powder Production. The conventional tungsten carbide powder production route is usually limited by the application of carburization temperatures around 2000 °C. With these temperatures, it is possible to carburize 50 to 100 µm tungsten metal particles, but the resulting product is polycrystalline (Ref 43).

The growth of tungsten carbide single crystals can be highly boosted by the use of a liquid phase, which is able to dissolve both tungsten and carbon. The Menstruum process (also called macro process) involves the formation of tungsten carbide within the melt of auxiliary metal, such as Fe (Ref 54, 55). Mixtures of tungsten ore concentrates, Fe_3O_4, Al, CaC_2 and/or C react exothermally (after ignition by using a starter) to form WC, Fe, CaO, and Al_2O_3. Temperatures of

Table 1 Properties of tungsten

Property	Value
Crystal structure	Body centered cubic
Density at 20 °C, g/cm^3	19.3
Hardness at 0 °C, HV	450
Modulus of elasticity at 0 °C, kN/mm^2	407
Melting point, °C	3422
Coefficient of thermal expansion at 500 °C, mm/m	2.3
Specific electrical resistivity at 20 °C, µΩ · m	0.055
Thermal conductivity at 0 °C, W/m · K	129.5
Vapor pressure at 1700 °C, Pa	1×10^{-10}

Source: Ref 11, 25

Table 2 Properties of WC

Property	Value
Crystal structure	Hexagonal close packed
Density, g/cm^3	15.7
Microhardness, dN/mm^2	2080
Modulus of elasticity, kN/mm^2	696
Melting point, °C	Decomposition 2600
Linear coefficient of thermal expansion (20–400 °C), $10^{-6}/K$	5.2
Electrical resistivity, μcm	22

Source: Ref 56

about 2500 °C are reached and complete reduction and carburization of about 22 tons of WC is achieved in 60 minutes. To get the tungsten carbide in powder form, the solidified auxiliary metal has to be dissolved, which is typically done with hot hydrochloric acid. Separation of the obtained tungsten carbide powder according to particle size is done by sieving operations. Menstruum tungsten carbide powders have a high extent of well faceted single crystal particles, contain approximately 0.2% Fe, and have typical particle sizes of 40 mesh (about 400 μm) and below.

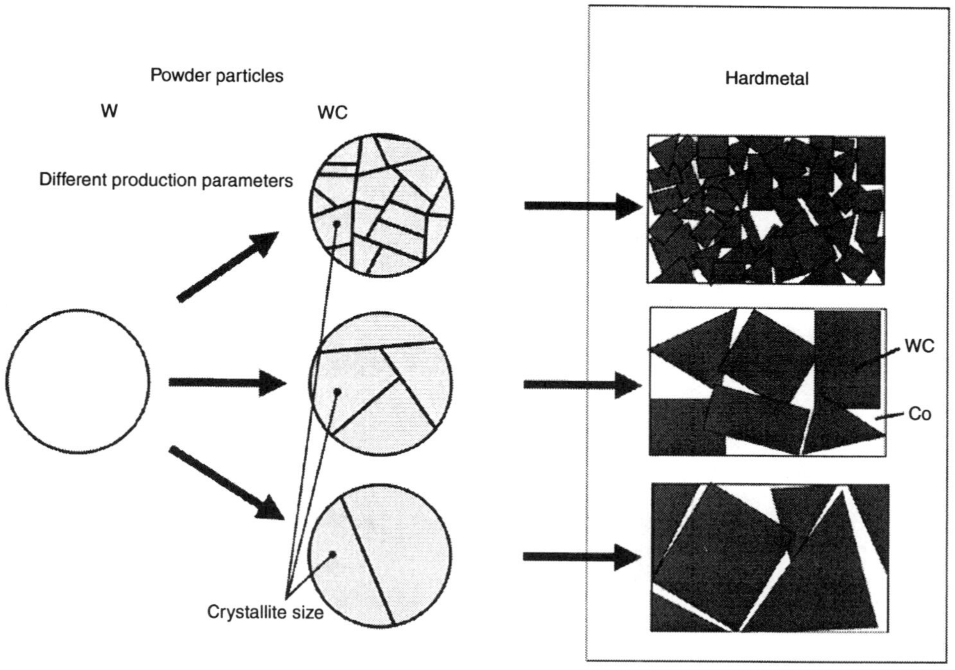

Fig. 13 Effect of crystallite size of tungsten carbide particles on their behavior during hardmetal manufacturing

Properties and Applications

One of the most important properties of pure tungsten carbide is its high hardness (for properties see Table 2), which is the basis for the application in hardmetals.

Hardmetals, also called "cemented carbides," achieve an optimized balance between the high hardness of the tungsten carbide and the high toughness of a ductile binder metal, such as cobalt, nickel, or iron, whereby optimal interfacial bonding is achieved (Ref 56, 57).

To produce hardmetals, WC and cobalt powders are mixed, pressed, and liquid phase sintered. The excellent wetting of WC by cobalt is one of the key properties. By variation of the tungsten carbide grain size (from below one micron to tens of microns) and the percentage of cobalt (between 3 and 30%), tailor-made properties of the hardmetals, (i.e., a great variety of different hardness to toughness combinations) can be obtained (Ref 56–58). A general rule thereby is the finer the carbide and the lower the cobalt content, the harder, but the less tough the composite is (Ref 59). Additions of other carbides such as TiC, TaC or carbo-nitrides (TiCN) are used industrially in particular for steel cutting grades (Ref 58). For more detailed information see the article "Cermets and Cemented Carbides" in this Volume. Tungsten carbide powder properties can be divided in physical and chemical aspects.

Physical Properties. There are a number of technologically measurable and characteristic values, giving information about the particle size and size distribution, quite similar to those described in the section "Properties" for tungsten metal powder. However, until now, none of these testing methods considered the fact that the tungsten carbide powder particles can be single or polycrystalline. Figure 13 shows that tungsten carbide particles with equal particle size may quite well differ concerning the size of the single crystal domains within the individual particles. Depending on the powder manufacturing process, the same particle can either be single or polycrystalline. This fact has indeed a strong impact on the behavior of the powder during the manufacturing of the hardmetal.

Polycrystalline particles may separate during sintering in their individual subgrains. Thus the higher the number of subgrains per WC particle, the finer the microstructure of a hardmetal can be. This is very important for all powders with particle sizes larger than one micron. Below one micron and especially in the range of ultrafine tungsten carbide powders, there is the problem that even the measurement of the particle size becomes more difficult. It could be shown by scanning electron microscopy that powders with an FSSS particle size around 0.6 μm in reality were already as fine as 0.3 μm (Ref 43). Also the determination of the particle size through measurement of the BET surface area can be misleading for characterization of ultrafine tungsten carbide powders because of its sensibility for surface roughness. A coarse particle with high surface roughness may give a high surface area measurement, leading to the wrong conclusion, that it is much finer (Ref 43). In these cases, all the properties that can be measured in the powder can be taken to check the lot to lot reliability of one production process, but not to estimate the properties of the sintered hardmetal. These have to be evaluated by a test sintering. Other physical properties, such as Scott, tap, and green density are important to control or predict the shrinkage during sintering, which is essential because most of the hardmetal parts are sintered to size (near net shape).

Chemical Properties. The most important chemical aspect of tungsten carbide powder is its carbon content. Stoichiometric pure WC has a total carbon content (C_t) of 6.135 wt%. From a thermodynamic point of view, WC has a very narrow stability range (Rudy, 1969 in Ref 2, p 386). Thus, already small carbon deficiency leads to W_2C subcarbide formation and excess of carbon to the formation of free carbon (C_f). In industrially produced WC powders, there is always a small portion of free carbon present (about 0.03% at a stoichiometric C_t level). Even if the total carbon is below the stoichiometric level, the coexistence of traces of W_2C and C_f can occur. The carbon balance is quite important for hardmetal manufacturing, because a lack of carbon leads to the formation of the brittle eta-phases (e.g., Co_3W_3C) and any excess in carbon leads to graphite precipitation; both cases result in a detrimental effect on the mechanical properties of the final product (Ref 56, 58).

Because of their grain growth inhibiting effect during hardmetal sintering, elements such as vanadium, chromium, tantalum, and niobium are added intentionally to very fine WC, mainly below 1 μm (Ref 60), in amounts around 0.1 to 1.5 wt%. Such additions can be done as carbide to the tungsten carbide powder, or as oxide or carbide prior to the carburization step. Trace elements like nickel, iron, and cobalt can be tolerated in certain amounts (100 ppm and more). However, there are other impurities that should be kept in the low ppm range due to their detrimental effect on hardmetal properties. Typical representatives are calcium, sulfur, silicon, and phosphorus (Ref 61–63). Figure 14 shows a typical example of an ultrafine tungsten carbide powder as commonly characterized.

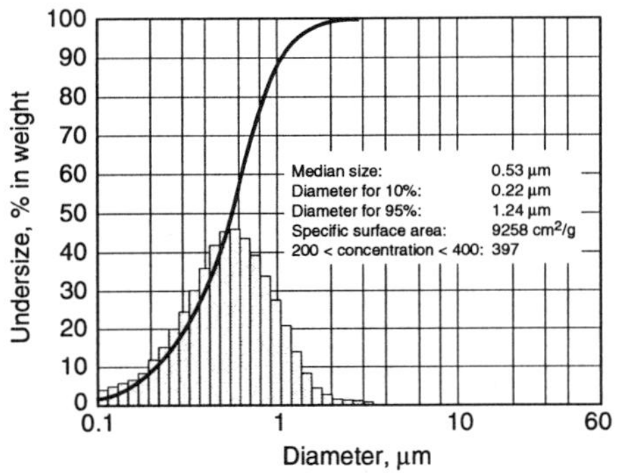

(a)

(b)

Physical analysis of WC05D	
BET surface, m^2/g	1.90
Grain size FSSS, µm	
As supplied	0.53
Lab milled	0.56
Porosity	
As supplied	0.698
Lab milled	0.654
Scott density, g/in.^{3}	24
Composition, %	
C	6.23
C	0.04
Cr_3C_2	0.64
VC	0.32
Trace impurities, ppm	
Aluminum	10
Calcium	3
Cobalt	<5
Copper	<2
Iron	46
Potassium	<4
Magnesium	<1
Molybdenum	<20
Sodium	3
Nickel	<5
Oxygen	1440
Phosphorus	<20
Sulfur	6
Silicon	<10

Fig. 14 Analysis of ultrafine tungsten carbide powder. Courtesy: Wolfram Bergbau- und Huetten-GmbH Nfg KG, Austria

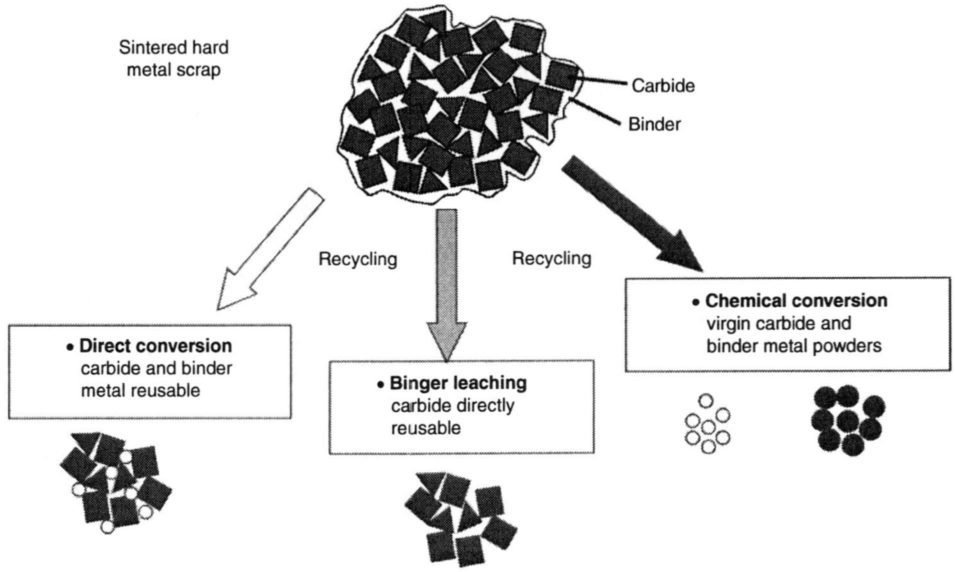

Fig. 15 Different hardmetal recycling processes. Source: Ref 39, 65

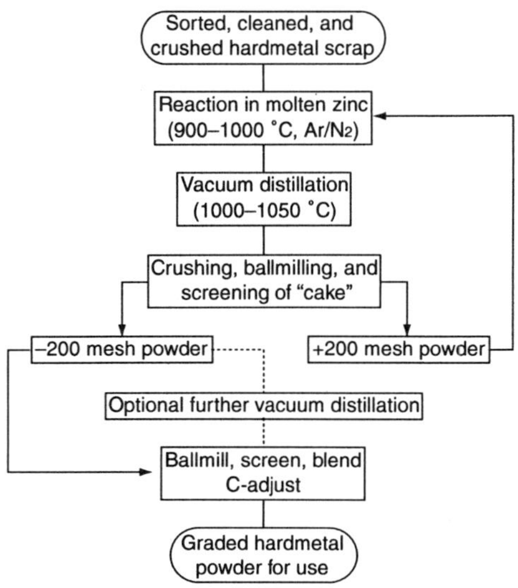

Fig. 16 The zinc process. Source: Ref 39

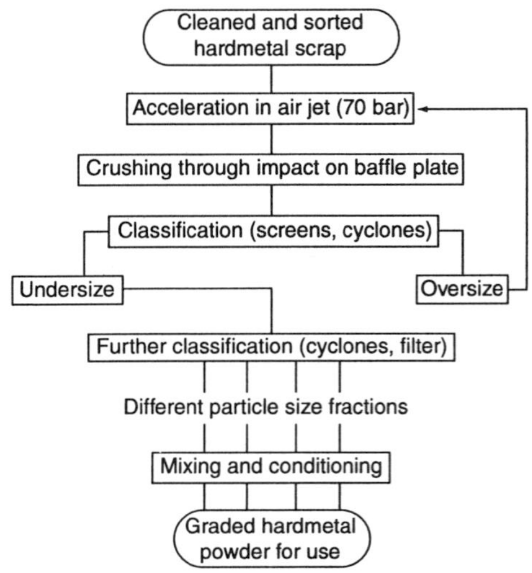

Fig. 17 The coldstream process. Source: Ref 70

Other Applications

Hardfacing. Carbide/metal mixtures very similar to hardmetal formulations are applied onto machine parts to improve their wear resistance. This can be done by flame or plasma spraying of powders or by welding electrodes. For this application, not only tungsten carbide powder as described above is used, but also so-called cast carbide, which is basically a crushed W_2C/WC mixture (Ref 56).

Electrical Contacts. A rather small quantity is used for the production of WC/Ag contact materials.

Diamond Tools. In diamond tool manufacturing, tungsten carbide powder (together with certain metal powders) are sometimes used to adjust the properties of soldering alloys in which the diamonds are embedded (Ref 56).

Catalysts. Some indications in the literature describe the use of very fine tungsten carbide with high surface area as catalyst in electrochemical reactions (Ref 64).

Recycling

Due to the fact that the majority of tungsten carbide powders are used for hardmetal production, this section focuses on recycling these products (Ref 65). Tungsten carbide containing hardmetal scrap can be divided into two groups:

- Soft scrap, which is powder scrap, such as hardmetal grinding sludge, filter dusts, floor sweepings, and broken green parts
- Hard scrap, which represents sintered parts, consisting of either production scrap or used parts

Another criteria for classification is the contamination of the scrap with impurities. While powder scrap frequently contains impurity levels that cannot be accepted for direct reuse, hard scrap, even if heavily contaminated, can be sorted, thus making a direct reuse easier.

Processes for hardmetal recycling (Fig. 15) can be divided into different categories (Ref 39, 65):

- Direct conversion into graded powder ready for pressing and resintering can be done with sorted hard scrap only. The most important process is the so-called "Zn process," shown schematically in Fig. 16 (Ref 39, 66–69). Also used on an industrial scale, but not as flexible from an economic point of view, is the cold-stream process, shown in Fig. 17 (Ref 70). For more brittle hardmetals with low binder contents, mechanical crushing is also an alternative (Ref 65).
- Leaving the carbides as is, but removing the binder chemically is another way of hardmetal recycling (Ref 65). The composition of the hardmetal scrap used for the leaching process determines the quality of the resulting carbide residue.
- All contaminated scraps—either soft or hard—can be treated in the chemical conversion process (Fig. 18) to gain back virgin powders of the hardmetal constituents (Ref 65).

Due to the inherent value of the hardmetal constituents and due to the increasingly more severe legislation concerning waste disposal and preservation of natural resources, recycling has today a considerable importance, which will increase in the future (Ref 39, 71).

Other Refractory Metals

Powder production of other refractory metals is briefly described in this section. More detailed information on fabrication and applications is given in the article "Refractory Metals" in this Volume.

Production of Molybdenum Powder

Molybdenum disulfide or molybdenite (MoS_2) is the major source of raw material. Molybdenum disulfide is concentrated by flotation and converted to an impure technical oxide (molybdenum trioxide, MoO_3) by roasting, which oxidizes the sulfur and removes it as gaseous sulfur dioxide (SO_2) (Ref 72).

Further purification commonly is done by taking advantage of the sublimation characteristics of molybdenum trioxide. Above 550 °C (1020 °F), it sublimes readily and can be distilled from its impurities and condensed again as "pure" molybdenum trioxide. Solvent extraction procedures are also available (Ref 73)

Additional purification can also be achieved by dissolving molybdenum trioxide in ammonia to form ammonium molybdate, $(NH_4)_2MoO_4$. The ammonium molybdate or the pure molybdenum trioxide is reduced in furnaces similar to those used for tungsten reduction. Typical purification and reduction reactions are:

Molybdenite (MoS_2) + $\frac{7}{2}O_2$ + heat = $MoO_3 + 2SO_2$

$MoO_3 + 2NH_4OH = (NH_4)_2MoO_4 + H_2O$

$(NH_4)_2MoO_4 + 3H_2 = Mo + 2NH_4OH + 2H_2O$

$(NH_4)_2MoO_4 + H_2 = MoO_3 + NH_4OH$

$MoO_3\ 3H_2 = 3H_2O + Mo$

Molybdenum recovery processes have been reviewed extensively in Ref 72–74.

Reduction is made more difficult by the volatility of molybdenum trioxide. Molybdenum dioxide is more easily controlled and can be formed from molybdenum trioxide below the sublimation temperature. Thus, the reduction sequence should always be $MoO_3 \rightarrow MoO_2 \rightarrow$ Mo (Ref 75).

Properties of Molybdenum Powder. The particle size of molybdenum is affected by the same variables that influence tungsten production, but to a lesser extent. Thus, there are fewer particle sizes available over a smaller range—typically, 1 to 6 μm. Typical compositions are:

Element	Content, ppm
Aluminum	5–25
Calcium	3–15
Chromium	5–25
Copper	5–10
Iron	10–100
Lead	5–10
Magnesium	1–10
Manganese	1–5
Nickel	5–50
Potassium	10–30
Silicon	5–250
Sodium	5–10
Tin	15–50
Tungsten	100–300
NVM (summation of Ca, Na, K, Si)	30–400
Carbon	10–50
LOR (a)	500–1000
Molybdenum: 99.9%, exclusive of gases	

(a) The combination of oxygen plus water is referred to as "loss on reduction" (LOR).

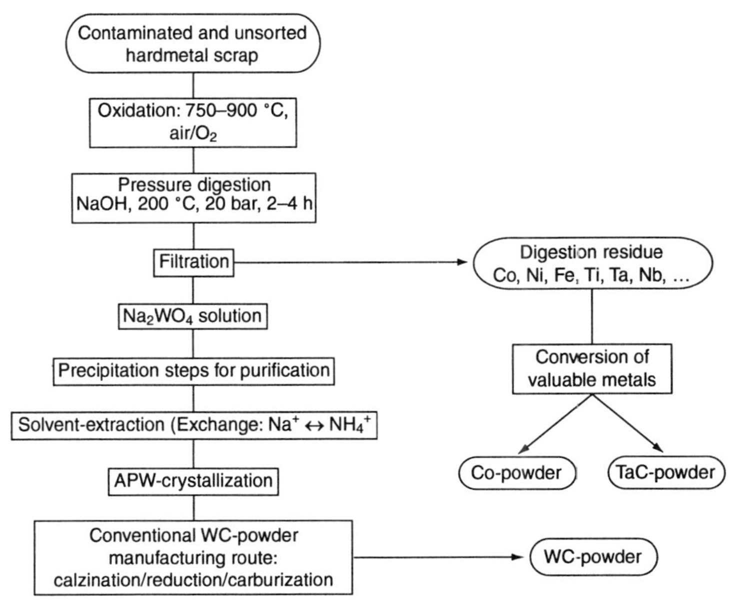

Fig. 18 The chemical conversion process. Source: Ref 65

Tantalum and Niobium Powders

Tantalum and niobium frequently are found in close association in their ores. The most important tantalum-bearing minerals are tantalite and columbite, which are variations of the same compound (Fe, Mn)(Ta, Nb)$_2$O$_6$. The tantalum mineral is called tantalite when the tantalum pentoxide (Ta_2O_5) content exceeds the niobium pentoxide (Nb_2O_5) content. Columbite is a tantalum ore in which the reverse relationship is true. Cassiterite (SnO_2) deposits found in Thailand, Malaysia, Australia, and other countries also contain significant amounts of tantalum and niobium. Byproduct slags from tin smelting operations are frequently used as tantalum feedstock.

Tantalum and niobium are separated from the source ore and from one another by digestion of the ore in aqueous hydrofluoric acid, followed by liquid-liquid solvent extraction with methyl isobutyl ketone (MIBK), as shown in Fig. 19. The liquid-liquid-solvent extraction process depends on promoting solute ion exchange between immiscible liquids in intimate contact, typically by varying the concentration of other solution components. In the first contacting, only the fluorides of tantalum and niobium are soluble in the MIBK; they are therefore separated from the contaminants iron, manganese, titanium and zirconium, which remains in aqueous solution. The solubility of tantalum in MIBK is high over an extensive range of acidity; niobium is soluble only at high acidity. In a series of subsequent steps, the MIBK is contacted with water of varying acidity to produce separate streams of tantalum fluoride and niobium fluoride.

Commercially important deposits of niobium also are found in the mineral pyrochlore [(Na, Ca)$_2$(Nb, Ta, Ti)$_2$O$_4$(OH, F) · H_2O], which is mined in Canada and Brazil. These ores are al-

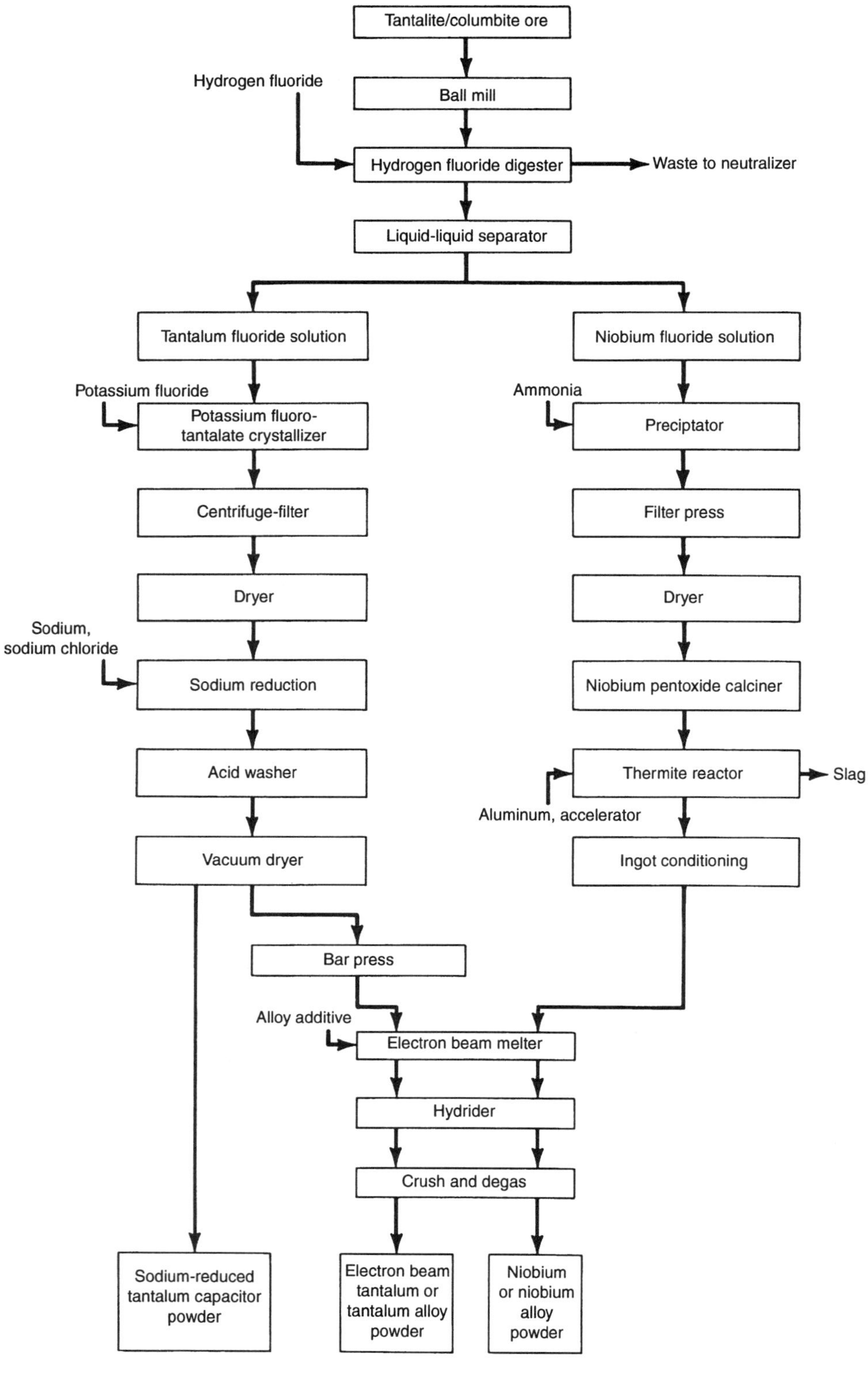

Fig. 19 Tantalum/niobium production flowchart. Source: Ref 76

Fig. 20 Particle shape of tantalum powder anode produced by sodium reduction of potassium tantalum fluoride. 2600×

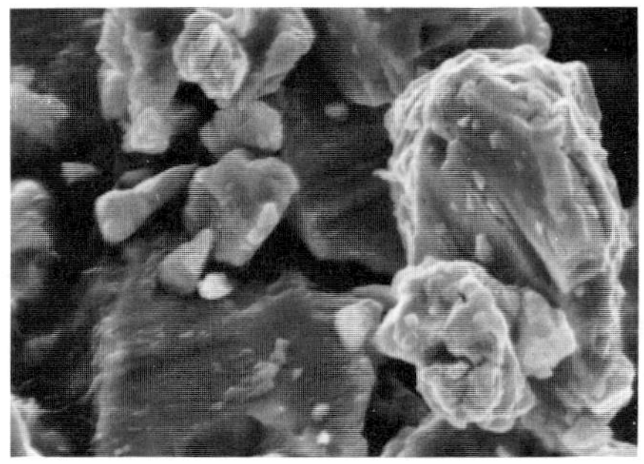

Fig. 21 Particle shape of tantalum powder made by electron beam melting, hydriding, crushing, and degassing. 2900×

most tantalum free and are processed to concentrates containing 55 to 60% niobium pentoxide by a series of operations including size reduction, flotation, and leaching. Extraction of niobium by high-temperature chlorination of the ore is also a suitable means of obtaining niobium (Ref 77).

Tantalum

Usually, the starting material for producing tantalum powder is the double salt, potassium tantalum fluoride (K_2TaF_7), which is precipitated by adding potassium fluoride to the aqueous tantalum solution. The metal is freed from the compound by chemically reacting with molten sodium in a stirred retort containing molten sodium chloride diluent under inert atmosphere. After cooling, the resulting salt cake is removed from the retort and crushed. The salt-encased tantalum powder is recovered by thoroughly washing with water, then acid.

Particles, as shown in Fig. 20, are usually spherical in shape, with a tendency to form grape-like clusters during reduction. Individual particles in such clusters range in size from 1 to 10 μm, depending on reduction variables such as temperature, agitation, and salt purity. Electronic capacitor applications normally require fine particle sizes with high surface area, whereas P/M mill products use the easier to handle coarse particle fractions.

In one modification of the process, potassium tantalum fluoride, sodium chloride, and sodium metal are mixed into a paste and are reacted without premelting (Ref 78). Platelet-shaped tantalum particles are produced by this technique. Because platelets have a higher surface-to-volume ratio than spheres, they are well suited for capacitor applications.

Purified tantalum powders also are produced by a process that involves electron beam melting as the means of purification. Sodium-reduced powders are pressed into bars that are subsequently electron beam melted. The resulting ingot is placed in an evacuated and purged furnace into which hydrogen is introduced. The tantalum is fully hydrided on slow cooling from 800 °C (1470 °F) under hydrogen. The brittle hydride is crushed, ground, and classified to yield powder with a mean particle size ranging

Table 3 Typical compositions of sodium-reduced and electron beam melted degassed-hydride tantalum powders

Element	Powder analysis, ppm	
	Sodium-reduced	Electron beam melted degassed-hydride
Aluminum	10	5
Calcium	10	5
Carbon	100	45
Chromium	25	5
Cobalt	10	5
Copper	10	5
Hydrogen	30	30
Iron	50	30
Lead	10	5
Magnesium	10	5
Manganese	10	10
Nickel	50	5
Niobium	50	20
Nitrogen	90	40
Oxygen	2400	1650
Silicon	25	10
Sodium	10	5
Tin	10	5
Titanium	10	5
Tungsten	25	25
Vanadium	10	5
Zirconium	10	5

Source: Fansteel, Inc.

from 3 to 6 μm. Tantalum powder with angular particles, as shown in Fig. 21, is obtained after dehydriding (degassing).

For capacitor manufacture, small tantalum pellets are pressed. It is desirable to press the powders to as low a density as possible, while retaining sufficient green strength for handling. Because capacitance is directly proportional to the tantalum surface area accessible to an electrolyte, a porous pellet structure is desirable. In general, sodium-reduced powders can be pressed to lower densities than electron beam melted, degassed-hydride powders. However, electron beam melting purifies the tantalum, thereby allowing capacitor devices to operate at higher voltages. The advantages of the two basic types of tantalum powders when used for capacitor manufacture are:

Sodium-reduced powder

- Higher capacitance per gram of powder
- Better green strength for pressed compacts

Electron beam melted, degassed-hydride powder

- Higher purity
- Better flow characteristics
- Higher voltage capability in capacitor devices

Table 3 shows typical compositions of sodium-reduced and electron beam melted powders.

The "hydride-crush-dehydride" process is also a convenient method for reclaiming tantalum metal scrap and converting it to powder. Platelet particle shapes also can be made by ball or vibratory milling of the pure, ductile particles obtained after degassing.

Table 4 Typical composition of niobium and C-103 niobium alloy powder made by the hydride-dehydride process

Element	Analysis, ppm	
	Niobium	C-103
Niobium	99.7 + %	87.2 + %
Oxygen	1820	1980
Tantalum	800	2800
Hafnium	<20	9.8%
Zirconium	<20	1800
Titanium	20	0.91%
Carbon	500	194
Iron	100	200
Aluminum	<20	<20
Nitrogen	197	62
Silicon	30	<20
Copper	<40	<40
Cobalt	<10	<10
Boron	<1	<10
Hydrogen	150	50
Nickel	<20	<20
Molybdenum	<20	100
Tungsten	<50	1100
Other elements(a)	<20	<20

(a) Other elements include cadmium, chromium, magnesium, manganese, lead, tin, vanadium, and zinc. Source: Fansteel, Inc.

Although there are few current applications for alloy tantalum powder, a Ta-7.5%W powder has found use as a corrosion-resistant spring material. Such powders usually are produced by the melting and hydriding method. The tungsten is added during electron beam melting. The homogeneous ingot is then hydrided, crushed, and dehydrided. Except for the tungsten addition, the chemistry and properties of this alloy powder are typically the same as for unalloyed degassed-hydride (Table 3).

Niobium

Currently, virtually all niobium metal is extracted and purified by aluminothermic reduction of the oxide:

$$3Nb_2O_5 + 10Al \rightarrow 6Nb + 5Al_2O_3$$

During the exothermic reaction, oxide impurities slag from the molten niobium. The metal is further purified by electron beam melting. Subsequently, the ingot is converted to powder by hydriding, crushing, and dehydriding. The particle structure is angular and completely analogous to degassed-hydride tantalum. The niobium particle size is coarser than tantalum because its use does not require the generation of high surface area

Alloy powders are made by adding alloying agents during melting. Low-volatility metals such as tungsten and tantalum can be added during electron beam melting. More volatile alloying agents, such as titanium, hafnium, or zirconium, are frequently added during arc remelting of the electron beam melted ingot prior to hydride-dehydride processing. Typical compositions of niobium and C-103 alloy (Nb-10Hf-1Ti) powder made by this method are given in Table 4.

Table 5 Physical properties of thermally agglomerated, sodium-reduced and electron beam melted, degassed-hydride tantalum powders

	Sodium-reduced powder	Electron beam melted, degassed-hydride powder
Sieve analysis, %		
−35 + 80	0–20	15–25
−80 + 200	10–25	15–25
−200 + 325	10–20	5–10
−325	45–70	45–60
Physical properties		
Scott bulk density, g/cm^3	1.6–2.0	3.5–4.0
Apparent Fisher subsieve particle size, μm	2.5–3.5	6.5–8.5
Hall flow index(a), s	130	50

(a) ASTM B 213

Physical Properties

Tantalum is unusual among powder materials because its most important use is in capacitors, and optimum capacitor performance requires very low pellet density after sintering. Therefore, many of the conventional metallurgical techniques for improving flowability cannot be used, and the green strength of low-density pellets is a serious concern. Production of a "popcorn ball" structure through vacuum thermal agglomeration of powders results in significant improvements in both characteristics (Ref 79). In fact, unagglomerated capacitor powders are almost obsolete because of inferior flow and green strength. The agglomeration step is carried out by heating the loose powder somewhat below the temperatures used for subsequent pellet sintering. The lightly sintered product is then recrushed and screened to a predetermined sieve analysis. Table 5 compares typical properties of agglomerated sodium-reduced and degassed-hydride tantalum powders.

ACKNOWLEDGMENTS

B. Lux and B. Zeiler would like to express their gratitude to Prof. W.D. Schubert of Vienna University of Technology for helpful and constructive discussions in preparing the section "Production of Tungsten and Tungsten Carbide Powders." Dr. M. Spross, Managing Director of Wolfram Bergbau- und Huetten-GmbH Nfg KG, is gratefully acknowledged for his kind and benevolent support. Mr. C. Keller's help in preparing the figures is also appreciated.

The section "Other Refractory Metals" was adapted from articles by John Lambert, Robert Droegkamp, and Peter Elott in *Powder Metallurgy,* Volume 7, *Metals Handbook,* 9th ed., 1984.

REFERENCES

1. P. Borchers and W. Müller, *Proc. of the 5th Int. Tungsten Symp.* (Budapest), MPR Publishing

Services Ltd., in cooperation with the International Tungsten Industry Association, 1990, p 40–55

2. S.W.H. Yih and C.T. Wang, *Tungsten Sources, Metallurgy, Properties, and Applications,* Plenum Press, 1979
3. H. Kolaska, *Powder Metall. Int.,* Vol 24 (No. 5), 1992, p 311–314
4. J. Gurland and J.D. Knox, *3rd Int. Conf. on Tungsten and Refractory Materials,* 1996, p 219–227
5. M. Maby, *Met. Powder Rep.,* April 1991, p 14–15
6. R. Bunting, *Proc. of the 5th Int. Tungsten Symp.* (Budapest), MPR Publishing Services Ltd., in cooperation with the International Tungsten Industry Association, 1990, p 87–94
7. O. Hedebrant, *Proc. of the 5th Int. Tungsten Symp.* (Budapest), MPR Publishing Services Ltd., in cooperation with the International Tungsten Industry Association, 1990, p 95–98
8. K. Miyake, *Proc. of the 5th Int. Tungsten Symp.* (Budapest), MPR Publishing Services Ltd., in cooperation with the International Tungsten Industry Association, 1990, p 99–109
9. Technical data, International Tungsten Industry Association
10. ITIA, International Tungsten Industry Association, 1997
11. E. Lassner, W.D. Schubert, E. Lüderitz, and H.U. Wolf, Tungsten, Tungsten Alloys, and Tungsten Compounds, *Ullmann's Encyclopedia of Industrial Chemistry,* Vol A27, VCH Verlagsgesellschaft GmbH, 1992
12. M. Spross, *Berg-und Huettenmaennische Montashefte (BHMMAM): Journal of Mining, Metallurgical, Materials, Geotechnical, and Plant Engineering,* Vol 141 (No. 8), 1996, p 359–362
13. M. Spross, BHMMAM, Vol 129, 1984, p 352
14 E. Lassner, *Österr. Chem. Z.,* Vol 80, 1979, p 111–114
15. B. Zeiler, BHMMAM, Vol 142 (No. 4), 1997, p 167–169
16. E. Lassner and W.D. Schubert, *Int. J. Refract. Met. Hard Mater.,* Vol 13, 1995, p 111–115
17. W.D. Schubert, *Int. J. Refract. Met. Hard Mater.,* Vol 9, 1991, p 178
18. R. Haubner, thesis, TU-Vienna, Austria, 1982
19. R. Haubner et al., *Int. J. Refract. Met. Hard Mater.,* Vol 2, 1983, p 108
20. R. Haubner et al., *Int. J. Refract. Met. Hard Mater.,* Vol 2, 1983, p 156
21. W. Sahle and S. Berglund, *J. of Less Common Met.,* Vol 79, 1981, p 271–288
22. T. Millner and J. Neugebauer, *Nature,* 1949, p 601–602
23. W.D. Schubert, B. Lux, and B. Zeiler, *Int. J. Refract. Met. Hard Mater.,* Vol 13, 1995, p 119–135
24. H. Hellmer, W.D. Schubert, E. Lassner, and B. Lux, *Proc. 11th Plansee Seminar,* Vol 3, H. Bildstein and H. Ortner, Ed., Verlaganstalt Tyrolia, Austria, 1985, p 43–86
25. A. Zezairliyan, *High Temp. Sci.,* Vol 4, 1972, p 248–252
26. ASTM Standards, American Society for Testing and Materials
27. R. Huenert, G. Winter, W. Kiliani, and D. Greifendorf, *Int. J. Refract. Met. Hard Mater.,* Vol 11, 1992, p 331
28. P. Wilhartitz, H.M. Ortner, and R. Krismer, Proc. *12th Plansee Seminar,* Vol 3, H. Bildstein and H. Ortner, Ed., Verlaganstalt Tyrolia, Austria, 1989, p 575
29. J.A. Mullendore, *The Metallurgy of Doped Non-Sag Tungsten,* E. Pink and L. Bartha, Ed., Elsevier, 1989, p 61–81
30. E. Pink and L. Bartha, Ed., *The Metallurgy of Doped Non-Sag Tungsten,* Elsevier, 1989
31. Special issue: The Chemistry of Non-Sag Tungsten, *Int. J. Refract. Met. Hard Mater.,* L. Bartha, E. Lassner, W.D. Schubert, and B. Lux, Ed., Vol 13, 1995
32. J. Resch and G. Leichtfried, *Proc. of the 7th Int. Tungsten Symp.* (Goslar), International Tungsten Industry Association, 1996, p 294–311
33. O. Prause, *Proc. of the 7th Int. Tungsten Symp.* (Goslar), International Tungsten Industry Association, 1996, p 289–293
34. F. Mertens, *Proc. of the 7th Int. Tungsten Symp.* (Goslar), International Tungsten Industry Association, 1996, p 277–288
35. H. Kippenberg, *Proc. of the 7th Int. Tungsten Symp.* (Goslar), International Tungsten Industry Association, 1996, p 312–316
36. L. Zehnder, *Proc. of the 7th Int. Tungsten Symp.* (Goslar), International Tungsten Industry Association, 1996, p 317–333
37. D.E. Jech, J.L. Sepulveda, and A.B. Traversone, Process for Making Improved Copper/Tungsten Composites, U.S. Patent 5,686,676, 1997
38. L. Desrosiers, S. Nigarura, and G. Daigneault, Advances in Powder Metallurgy and Particulate Materials, Vol 1, Metal Powder Industries Federation, 1997, p 7.65–7.80
39. B.F. Kieffer and E. Lassner, BHMMAM, Vol 139 (No. 9), 1994, p 340–345
40 B.F. Kieffer and E. Lassner, *Proc. of the 4th Int. Tungsten Symp.* (Vancouver), MPR Publishing Services Ltd., in cooperation with the International Tungsten Industry Association, 1987, p 59–67
41. J. Lebok, W.D. Schubert, and B. Lux, *Int. J. Refract. Met. Hard Mater.,* Sept 1988, p 156–160
42. P. Leichter, H. Trombitas, and A. Vesely, AT-Patent AT 403,060 (Austria), 1997
43. B. Zeiler, *Proc. 14th Plansee Seminar,* Vol 4, G. Kneringer, P. Rödhammer, and P. Wilhartitz, Ed., RWF Werbegesellschaft, Austria, 1997, p 265–276
44. W.D. Schubert, A. Bock, and B. Lux, *Proc. 13th Plansee Seminar,* Vol 4, H. Bildstein and R. Eck, Ed., RWF Werbegesellschaft, Austria, 1993, p 283–305
45. M. Miyake and A. Hara, *J. Jpn. Soc. Powder Metall.,* Vol 26, 1979, p 16–21
46. J. Yamamoto, A. Matsumoto, S. Honkawa, and N. Shigaki, *Proc. of 1993 Powder Metallurgy World Congress,* Y. Bando and K. Kosuge, Ed., The Japan Society of Powder Metallurgy, 1993, p 785–788
47. Y. Yamamoto, A. Matsumoto, and Y. Doi, *Proc. 14th Plansee Seminar,* Vol 2, G. Kneringer, P. Rödhammer, and P. Wilhartitz, Ed., RWF Werbegesellschaft, Austria, 1997, p 596–608
48. S.D. Dunmead, W.G. Moore, A.W. Weimer, G.A. Eisman, and J.P. Henley, Method for Making Submicrometer Carbides, Submicrometer Solid Solution Carbides, and the Material Resulting Therefrom, U.S. Patent 5,380,688, 1995
49. D.F. Carroll, *Proc. 14th Plansee Seminar,* Vol 2, G. Kneringer, P. Rödhammer, and P. Wilhartitz, Ed., RWF Werbegesellschaft, Austria, 1997, p 168–182
50. C. Conner, D. Carroll, G. Cochran, S. Dunmead, Z. Jovanovic, K. vanHeel, and A. Weimer, *Proc. of the 7th Int. Tungsten Symp.* (Goslar), International Tungsten Industry Association, 1996, p 171–180
51. L.E. McCandlish, B.H. Kear, and B.K. Kim, Carbothermic Reaction Process for Making Nanophase WC-Co Powders, U.S. Patent 5,230,729, 1993
52. P. Seegopaul, L.E. McCandlish, and F.M. Shinneman, *Int. J. Refract. Met. Hard Mater.,* Vol 15, 1997, p 133–138
53. T. König and D. Fister, H.C. Starck GmbH & Co. KG, Goslar, Germany, EU Patent EP 0 568 861 B1, 1993
54. E.N. Smith, *Met. Powder Rep.,* Vol 35, 1980, p 53–54
55. P.M. McKenna, U.S. Patent 3,379,503, 1968
56. W. Schedler, *Hartmetall für den Praktiker,* VDI-Verlag GmbH, 1988
57. A.T. Santhanam, *The Chemistry of Transition Metals Carbides and Nitrides,* S.T. Oyama, Ed., Blacky Academic and Professional, 1996, p 28–52
58. K.J.A. Brookes, *World Directory and Handbook of Hardmetals and Hard Materials,* 6th ed., International Carbide Data, Hertfordshire, U.K., 1996
59. W.D. Schubert, H. Neumeister, G. Kinger, and B. Lux, *Proc. 14th Plansee Seminar,* Vol 4, G. Kneringer, P. Rödhammer, and P. Wilhartitz, Ed., RWF Werbegesellschaft, Austria, 1997, p 249–264
60. A. Bock, W.D. Schubert, and B. Lux, *Powder Metall. Int.,* Vol 24, 1992, p 20–26
61. W.D. Schubert and E. Kübel, *Proc. 12th Plansee Seminar,* Vol 2, H. Bildstein and H. Ortner, Ed., Verlaganstalt Tyrolia, Austria, 1989, p 869–909
62. B. Uhrenius, H. Brandrup-Wognsen, U. Gustavsson, B. Lethinen, and H. Manninen, *Proc. 12th Plansee Seminar,* Vol 2, H. Bildstein and H. Ortner, Ed., Verlaganstalt Tyrolia, Austria, 1989, p 71–96
63. G. Friedbacher, M. Grasserbauer, H.M. Ortner, A. Virag, and P. Wilhartitz, *Proc. 12th Plansee Seminar,* Vol 2, H. Bildstein and H. Ortner, Ed., Verlaganstalt Tyrolia, Austria, 1989, p 321–350
64. S.T. Oyama, Ed., *The Chemistry of Transition Metals Carbides and Nitrides,* Blacky Aca-

demic and Professional, 1996
65. B. Zeiler, *Proc. Hagener Symp. Pulvermetallurgie,* R. Ruthardt, Ed., Werkstoff-Informationsgesellschaft mbH, 1997, p 283–303
66. B.F. Kieffer and E.F. Baroch, *Extractive Metallurgy of Refractory Metals,* H.Y. Sohn, O.N. Carlson, and J.T. Smith, Ed., AIME, 1981, p 273–294
67. B.F. Kieffer, *Proc. of the 2nd Int. Tungsten Symp.* (San Francisco), Mining Journal Books Ltd., in cooperation with the International Tungsten Industry Association, 1982, p 102–114
68. K. Yajima and T. Yamaguchi, *Powder Metall. Rep.,* Vol 14, 1982, p 90–94
69. B.F. Kieffer, *Int. J. Refract. Met. Hard Mater.,* Vol 5, 1986, p 65–68
70. J. Walraedt, Proc. *7th Plansee Seminar,* Metallwerk Plansee GmbH, 1971
71. G.R. Smith, American Metal Market—Tungsten, 31 Aug 1993
72. R.R. Dorfler and J.M. Laferty, Review of Molybdenum Recovery Processes, *J. Met.,* Vol 33 (No. 4), 1981, p 48–54
73. M. MacInnis, T. Kim, and J. Laferty, *1st Int. Conf. Chem. Uses of Molybdenum,* AMAX, Greenwich, CT, 1973, p 56–58
74. Molybdenum Ore Treatment—1980 to December 1982, NTIS Published Search, PB83-857250, Dec 1982
75. M.J. Kennedy and S.C. Bevan, *1st Int. Conf. Chem. Uses of Molybdenum,* AMAX, Greenwich, CT, 1973, p 11–13
76. R.E. Kirk and D.F. Othmer, Ed., *Encyclopedia of Chemical Technology,* Vol 22, 3rd ed., John Wiley & Sons, 1983, p 541–564
77. B.J. Lerner, U.S. Patent 3,294,482, 1966
78. P. Borchers et al., Ed., *Extractive Metallurgy of Refractive Metals,* Conf. Proc., Metallurgical Society of the American Institute of Mechanical Engineers, 22–26 Feb 1981
79. J.A. Pierret, U.S. Patent 3,418,106, 1968

Production of Beryllium Powders

James M. Marder, Brush Wellman

BERYLLIUM has a high elastic modulus (303.4 GPa, or 44 × 10^6 psi) and a low density (1.85 g/cm^3, or 0.067 lb/in.^{3}). It is used extensively in aerospace structures, weight- and inertia-critical structures, precision optics, and nuclear applications. Powder metallurgy is the primary fabrication technology used in the manufacture of beryllium products. This article discusses the techniques used to produce beryllium powder and the resulting material properties.

The production of beryllium powder begins with a vacuum-melted ingot, followed by chipping and mechanical or pneumatic pulverization. The mechanical properties of the final consolidated product depend on grain size and oxygen content, as well as on the method of powder production because it affects the crystallographic texture of the billet. Additional information on the properties, consolidation, and applications of beryllium can be found in the article "P/M Beryllium" in this Volume.

Powder metallurgy is used primarily to produce beryllium mill products, because the casting process does not produce parts with adequate structural integrity and mechanical properties. In making a beryllium structural casting, the degree of superheat that produces sufficiently low molten metal viscosity required for filling a mold cavity of any complexity also results in a very large as-cast grain size. Excessive grain size decreases strength and ductility. High superheat conditions can cause an excessive metal-mold reaction; insufficient superheat can result in a lack of mold fill, excessive porosity, or cold shuts. In addition, the reactivity and viscosity of molten beryllium exaggerate problems encountered when casting conventional materials.

The ingot/wrought billet manufacturing sequence also poses serious manufacturing difficulties. Columnar grain growth, which results from low-temperature ingot pouring, causes exaggerated planes of weakness at the impingement of the sets of columnar grains. Subsequent mechanical breakdown operations, therefore, result in cracking along the weak planes. Ultrasonic vibration (Ref 1) and the use of inoculants (Ref 2) are ineffective grain-refinement techniques. Zero-gravity experiments aboard Spacelab using beryllia and alumina as inoculants have also been ineffective (Ref 3). Powder metallurgy, therefore, is the only viable technique for manufacturing strong, homogeneous billets.

Physical Properties

Beryllium has a close-packed hexagonal crystal structure with a = 0.2286 nm and c = 0.3584 nm and a c/a ratio of 1 to 1.56. This value is low compared to other metals such as titanium (1 to 1.58). Partly as a result of this c/a ratio, slip occurs almost exclusively on basal planes, with a Burger's vector of $b = \frac{1}{3}\langle 11\bar{2}0\rangle$. Slip also can occur—with much greater difficulty, however—on first-order prism planes ($b = \frac{1}{3}\langle 11\bar{2}0\rangle$), and with even more difficulty on second-order pyramidal planes ($\{11\bar{2}2\}$ with $b = \frac{1}{3}\langle 11\bar{2}3\rangle$) (Ref 4).

Basal slip is the predominant deformation mode, however. Basal cleavage is the predominant fracture mode. These crystal structure characteristics affect the shape of powder particles and subsequently affect billet properties.

Processing Sequence

Manufacturing a consolidated, fully dense beryllium product involves ingot melting and casting, powder manufacture, and consolidation. The consolidated product usually is a right circular cylinder that is made by vacuum hot pressing. This vacuum hot pressed billet can be readily machined into parts for satellites, aircraft, optical systems, or guidance instruments. Beryllium is particularly well suited to these applications because of its superior strength, modulus, and density. The ductility afforded by P/M processing also facilitates forging, rolling, extruding, and other hot-working processes.

Raw Material. Extraction begins with the mining of bertrandite ore, $Be_4Si_2O_7(OH)_2$, and/or beryl ore, $3BeO\text{-}Al_2O_3\text{-}6SiO_2$, from domestic and international sources. This is con-

Fig. 1 Scanning electron micrograph of flakelike attritioned powder. 270×

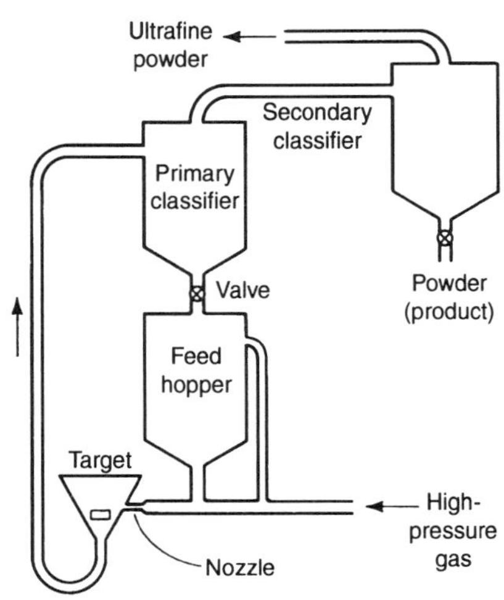

Fig. 2 Impact grinding system

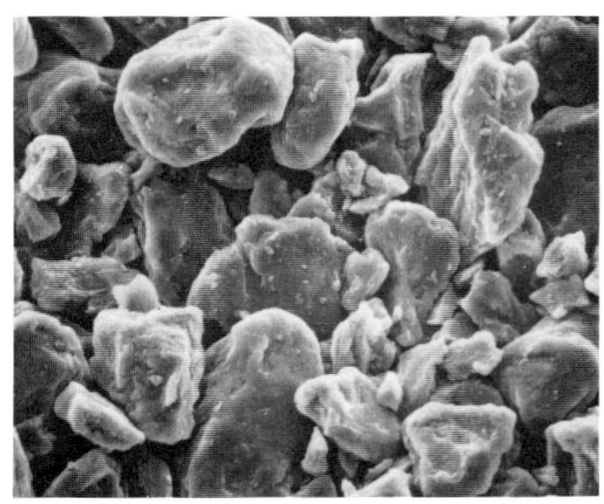

Fig. 3 Scanning electron micrograph of blocky impact ground powder. 270×

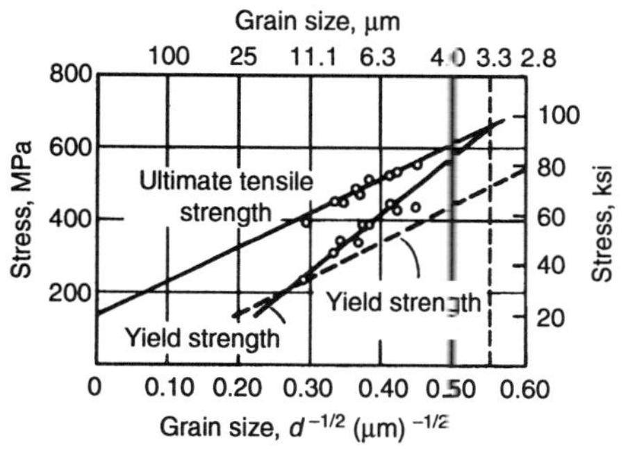

Fig. 4 Hall-Petch diagram of strength versus grain size of vacuum hot pressed beryllium with intermediate purity. Dashed line is yield strength values from Ref 6.

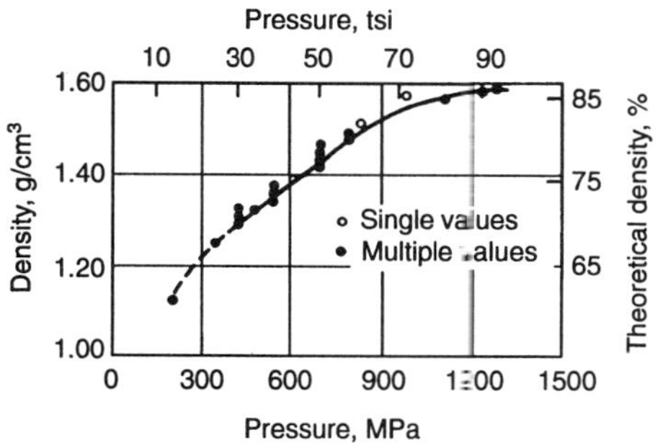

Fig. 5 Green density versus compacting pressure in uniaxially cold pressed beryllium powder. Source: Ref 7

verted to crude beryllium hydroxide product, purified, converted to anhydrous beryllium fluoride, and reduced with magnesium to produce "pebble" or primary beryllium. Residual reduction slag containing these beryllium pebbles is purified by vacuum melting. Vacuum melting uses a magnesium oxide crucible and a graphite mold to produce an ingot weighing 180 to 200 kg (400 to 450 lb). The ingot is "skinned" (machined) to remove the high-carbon material adjacent to the mold wall. Next, a multihead cutting tool is used to reduce the ingot to chips. Standard P/M techniques are used for powder production at this stage.

Powder Manufacture

Chips produced by the multihead cutter are relatively coarse and must be ground or comminuted to fine particle sizes before consolidation into a homogeneous, fine-grained billet. Before discussing grinding and comminution, however, it is important to recognize the effects that the close-packed hexagonal structure and chemical reactivity have upon the powder-making processes and the physical and mechanical properties of the final product.

During the late 1960s and early 1970s, mechanical attritioning and ball milling were used almost exclusively to reduce machining chips to powder. Mechanical attritioning is a rotary grinding procedure similar to the grinding of corn in a grist mill. Chips enter an area between a stationary and rotating beryllium plate, where they are ground. The particles then exit the mill at its periphery.

This process results in the occurrence of basal plane cleavage. Therefore, the powder particles produced by attritioning tend to be flakes, with large, flat surfaces corresponding to {0002} basal planes. A scanning electron photomicrograph of attritioned powder is shown in Fig. 1. Ball-milled powders have a similar shape. As a result of their flat particle shape and lack of a truly random crystallographic texture, consolidated billets exhibit anisotropy of both physical and mechanical properties. During vibratory die loading for consolidation, the flat faces of many particles align, forming extended volumes with a common crystallographic feature. The basal plane corresponds to the flat faces.

During consolidation—usually uniaxial vacuum hot pressing—the imposed stress causes these extended areas to rotate so that the *c*-axis poles align parallel to the stress axis. In billets made with attritioned powder, *c*-axis poles exist with two to three times the frequency in the longitudinal as compared to the transverse direction. The net result is substantial anisotropy of mechanical and physical properties. For example, ductility is substantially reduced in the longitudinal direction.

Impact grinding is used to minimize anisotropic effects. Figure 2 schematically illustrates the impact grinding process. Chips and powder are introduced into a high-pressure gas stream and are impacted against a beryllium target. The high speed of this pulverization process, coupled with relatively low temperature due to adiabatic gas cooling, activates the secondary and tertiary (nonbasal) cleavage systems.

Resultant particles are blocky, rather than flakelike (Fig. 3). Consequently, there is a substantially reduced correlation between crystallographic basal planes and flat powder particle faces, leading to reduced anisotropy and an improvement in longitudinal tensile elongation

Table 1 Comparison of billets using impact ground and attritioned powder

Powder lot(a)	Ductility and crystallographic orientation effects: Test direction	Tensile elongation, %	Elongation T/L ratio	{0002} pole density T/L ratio
Attritioned powder				
Lot A	Transverse	4.4	2.93 to 1	2.62 to 1
	Longitudinal	1.5	...	...
Lot B	Transverse	5.0	3.3 to 1	2.5 to 1
	Longitudinal	1.6	...	...
Impact ground powder				
Lot A	Transverse	5.0	1.7 to 1	1.48 to 1
	Longitudinal	2.9	...	...
Lot B	Transverse	5.4	1.5 to 1	1.45 to 1
	Longitudinal	3.6	...	...

(a) Powder lots A and B are produced from separate melts.

Table 2 Characteristics of powder used for commercial beryllium billets

	Grade: SP-65B	SP-200F	IP-70	IP-220
Composition, %				
Beryllium assay, min	99.0	98.5	99.0	98.0
Beryllium oxide, max	1.0	1.5	0.7	2.2
Aluminum, ppm max	600	1000	700	1000
Carbon	1000	1500	700	1500
Iron	800	1300	1000	1500
Magnesium	600	800	700	800
Silicon	600	600	700	800
Boron	2	...	...	...
Cadmium	2	...	...	...
Calcium	100	...	...	...
Chromium	100	...	...	...
Cobalt	10	...	...	...
Copper	150	...	...	...
Lead	20	...	...	...
Lithium	3	...	...	...
Manganese	120	...	...	...
Molybdenum	20	...	...	...
Nickel	300	...	...	...
Nitrogen	300	...	...	...
Silver	10	...	...	...
Other metallic impurities(a), ppm max	200	400	400	400
Particle size	98% –325 mesh	98% –325 mesh	98% –325 mesh	98% –325 mesh

(a) Each; determined by normal spectrographic methods

Table 3 Properties of beryllium hot pressed block

Property	Grade S-200F	S-65B	I-70A	I-220B
Chemical composition, %				
Beryllium assay (min)	98.5	98.5	99.0	98.0
Beryllium oxide (max)	1.5	1.0	0.7	2.2
Aluminum (max)	0.1000	0.06	0.07	0.10
Carbon (max)	0.15	0.12	0.07	0.15
Iron (max)	0.13	0.12	0.10	0.15
Magnesium (max)	0.06	0.08	0.07	0.08
Silicon (max)	0.06	0.06	0.07	0.08
Other metal impurities (max), %	0.04	0.04	0.04	0.4
Powder type	Impact ground	Impact ground	Impact ground	Impact ground
Minimum density, g/cm^3	1.84	1.84	1.84	1.84
Theoretical density, %	99	99	99.3	99
Average grain size (max), μm	25	15	20	25
Ultimate tensile strength (min), MPa (ksi)	324 (47)	289.6 (42)	241.3 (35)	379 (55)
Yield strength, 0.2% offset (min), MPa (ksi)	241 (35)	206.8 (30)	172.3 (25)	275 (40)
Elongation (min) in 25 mm (1 in.), %	2	3	2	2

Note: Pressing (billet) sizes can range from 18 to 183 cm (7 to 27 in.) in diameter and 15 to 168 cm (6 to 66 in.) in length, depending on grade and composition. (a) Special grades available with 9.000 permissible exposure limit (min) or 1.878 g/cm^3 (min)

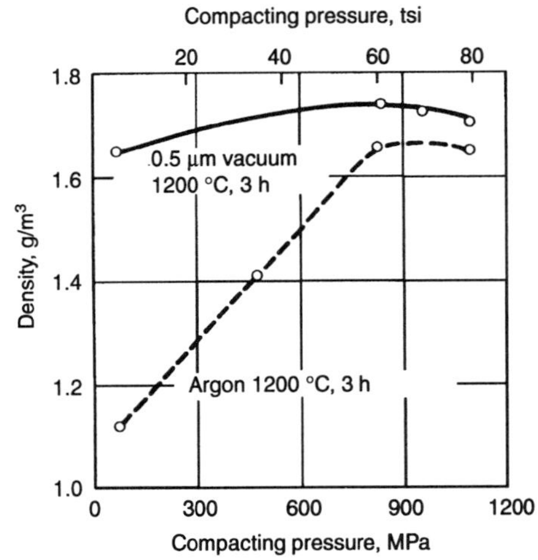

Fig. 6 Effects of compacting pressure on density after sintering in argon and in vacuum. Source: Ref 7

Table 4 Comparison of beryllium specification mechanical properties for VHP and HIP

Property	S-200F HIP	S-200F VHP	I-70 HIP	I-70 VHP	I-220 HIP	I-220 VHP
0.2% yield strength, MPa (ksi)	296 (43)	241 (35)	207 (30)	172 (25)	345 (50)	276 (40)
Ultimate tensile strength, MPa (ksi)	414 (60)	323 (47)	345 (50)	241 (35)	448 (65)	379 (55)
Elongation, %	3	2	2	2	2	2

(Ref 5). These effects are readily seen in Table 1, which illustrates the relationship between the method of powder comminution, {0002} pole density, and ductility.

The transverse-to-longitudinal (T/L) basal pole density ratio and ductility T/L ratio for attrition milled powder are about 3 to 1. For impact-ground powder, these values are approximately 1.6 to 1. The isotropy of the impact-ground powder billet is improved, as is the minimum elongation value in any direction. This demonstrated correlation between crystal texture and ductility is causing attritioning to be gradually replaced by impact grinding.

The relationship of powder particle size to oxide content and grain size also affects billet properties. Beryllium obeys a Hall-Petch relationship, as shown in Fig. 4. As the size of the powder particles decreases, the oxygen content of the powder increases due to the increase in specific surface area. The mean free path between oxide particles also is reduced. Because oxides are a primary grain-boundary pinning agent, fine particle size enhances higher oxide content, finer grain size, and higher strength. The strength imparted by fine grain size and the potential embrittling effect of excessively high oxide content must be balanced against each other. Characteristics of powder used to produce commercial beryllium billets are given in Table 2.

Gas Atomization. Inert gas atomization has been used to produce beryllium powder in semi-production quantities (Ref 8). The major advantage of this process over conventional impact ground powder is the increased isotropy of the consolidated product. This can be important in optical systems, in which isotropy is a primary factor in component performance, particularly in cryogenic systems wherein thermal contractions can cause distortion in anisotropic components. Economic benefits may also be realized if this technique is brought to production scale. As yet, the use of inert gas atomization has been confined to the advanced development stage, and no specification has been written for this material. The chemistry of a typical heat of inert gas atomized beryllium powder compares as follows with S-200F:

Impurity	Spherical powder, wt %	S-200 F, wt %
Beryllium oxide	0.32	1.2
Carbon	0.07	0.15
Iron	0.09	0.10
Aluminum	0.03	0.10
Silicon	0.02	0.06
Magnesium	0.01	0.06

Consolidation

Vacuum hot pressing is the traditional method by which beryllium components have been fabricated. Other methods, such as cold pressing and sintering and vacuum sintering, have also been used successfully to consolidate beryllium powder.

Vacuum hot pressing (VHP) typically is carried out at 1000 to 1200 °C (1830 to 3630 °F) and at a compacting pressure of approximately 8.3 MPa (1200 psi) or less. This process yields a product that is essentially 100% dense. Billets up to 183 cm (72 in.) in diameter have been manufactured by this technique. Properties of commercial vacuum hot pressed block are given in Table 3.

Hot isostatic pressing (HIP) is the consolidation method used for an increasing proportion of beryllium products and now approaches vacuum hot pressing in terms of the volume of material processed.

In the HIP process, the powder is loaded into a mild steel canister, degassed to remove air and water vapor, and the canister is sealed. Hot isostatic pressing takes place at a pressure of 103 MPa (15 ksi) with temperature in the range of 850 to 1000 °C (1560 to 1830 °F), depending upon the desired properties. A processed density in excess of 99.75% of theoretical is achieved in production. As a result of the greater latitude in consolidation temperature allowed by the use of HIP process, mechanical property improvements over conventionally vacuum hot pressed materials can be achieved. The specification mechanical properties of three grades of HIPed beryllium are compared to their vacuum hot pressed counterparts in Table 4.

A cold pressing/sintering/coining technique has been used successfully for specific applications, such as the production of aircraft brakes, heatsinks, rotors, and stators. The relationship of green density to pressure depends strongly on the specific powder used; a typical curve is shown in Fig. 5. After cold pressing to obtain a high green density, sintering is carried out in vacuum. An argon sintering atmosphere severely restricts final density, as shown in Fig. 6.

REFERENCES

1. I.G. Polotzki, T.Y. Benieva, and Z.L. Khodor, Effect of Supersonic Vibrations on the Crystallization Processes, *Akad. Nauk*, Vol 6, 1953
2. A.E. Bibb and S.M. Bishop, "Grain Refinement of Cast Beryllium," KAPL-1917, General Electric Co., 1950

3. S.H. Gelles and R.K. Malik, Process Development for Producing Fine-Grain Casting in Space," Report CR-120739, National Aeronautics and Space Administration, April 1975
4. F. Aldinger, Flow and Fracture of Single Crystals, *Beryllium Science and Technology*, Vol 1, D. Webster and G. London, Ed., Plenum Press, 1977
5. R.M. Paine and A.J. Stonehouse, "Investigation into Effects of Microalloying and Thermal Treatment on the Properties of Beryllium," Report BW-TR-549, Brush Wellman, 1974
6. G.J. London, et al., *Met. Eng. Q.,* Vol 16 (No. 4), 1976
7. H.H. Hausner and N.P. Pinto, *Trans. ASM,* Vol 43, 1951, p 1052
8. D. Hashiguchi and J.M. Marder, Structure and Properties of Beryllium Powder, *Modern Developments in Powder Metallurgy,* Vol 20, Metal Powder Industries Federation, 1988, p 109

Sampling and Classification of Powders

Terence Allen

WHENEVER POWDER is analyzed, whether for physical or chemical assay, the quality of the measurement depends upon how representative the sample is of the bulk from which it is drawn. The measurement may be made on a few milligrams taken from a bulk of many tons, so the chances of measuring a nonrepresentative sample are considerable. Therefore, sampling regimes and/or methods are needed to obtain representative samples for analysis.

In addition, classification methods are used to exclude certain powder sizes from a powder distribution, or to obtain particular powder distributions. Sieving methods, for example, are used to obtain particular powder distributions, or to obtain narrow size ranges of a powder.

This article briefly summarizes sampling, classification, and sieving methods for powders. The properties of metal powders are statistical in nature, so the material to be tested should be properly blended or mixed. An ideal mix is one in which all particles and constituents are distributed uniformly throughout the mixture. This condition is accomplished only through proper application of the variables involved in the mixing process, such as type of blender, volume of metal powder, rotational speed of the blender, and blending time (see the articles "Blending and Premixing of Metal Powders and Binders" and "Effect of Blending Techniques on Properties of Metal Powder Mixes" in this Volume).

Sampling of Powders

The object of sampling is to gain knowledge of the characteristics of the whole from measurements impracticable to apply to the whole: Bias, at any of the reduction stages, adversely affects the final analysis. Problems arise due to inhomogeneities in the parent distribution. If the bulk powder is homogeneous, or can be mixed prior to sampling in order to generate a homogeneous powder, sampling problems do not arise. Frequently it is not possible to mix the powder and a sampling campaign has to be prepared in order to obtain representative samples.

In order to establish homogeneity it is necessary to examine samples taken from the bulk, either at random or according to some predetermined pattern. If homogeneity is established, a single sample is representative of the bulk. The definition of homogeneity requires specification of the portions or sample sizes between which the variability is sufficiently small to be neglected.

The total sampling error is made up of errors due to the primary sampling, errors in subsequent sample dividing, and errors in the analysis itself. Sampling is said to be accurate when it is free from bias, that is, the error of sampling is a random variable about the true mean.

Two types of sampling errors are possible. *Errors due to segregation of the bulk* depend on the previous history of the powder and can be minimized by suitable mixing and building up a sample from a large number of increments.

Statistical errors cannot be prevented. Even for an ideal random mixture, the quantitative distribution in samples of a given magnitude is not constant but is subject to random fluctuations. This is the only sampling error that cannot be suppressed and occurs in ideal sampling. It can be estimated beforehand and reduced by increasing the sample size.

Samples are withdrawn from a population in order to estimate a certain characteristic of that population and to establish confidence limits for that characteristic. The characteristic may be particle size, composition, or quality; a measure of the spread of the distribution may also be required. The objective may be to set up limits between which the quality of a final product is acceptable, to decide whether the characteristics of a given lot meet preset criteria, or to estimate the variability within a lot or between lots.

If the material comes in containers, or can be viewed as discrete units, the objective may be to estimate the number of units outside of specification. The value of the estimate is largely dependent on the sampling technique adopted. It is of little value, and could impart false information, if it is biased or imprecise.

It is usually impossible to measure the size distribution of a bulk powder, so it is necessary to carry out measurements on a sample extracted from the bulk. This sample is itself frequently too large and has to be further subdivided. The stages in reducing from bulk to measurement samples are bulk or process stream (10^nkg), gross sample (>kg), laboratory sample (<kg), test sample (g), and measurement sample (mg).

The gross sample can be one of a series of spot samples, which have been extracted in order to determine the variability of the bulk or process stream with location or time, or it can be made up of subsamples to be representative of the bulk as a whole. Historically the bulk sample has been obtained manually, but automatic in-line sampling is becoming more common.

In some cases the gross sample is too large to send to the laboratory and has to be reduced in quantity. This reduction needs to be carried out in such a way that the laboratory sample is fully representative of the gross sample. When this reduction is unnecessary the gross sample is also the laboratory sample.

The laboratory sample may be required for a number of tests, so it is sometimes necessary to further subdivide it into test samples.

The test sample may be used in its entirety or further subdivided to form measurement and reference samples.

Powders may be classified as free-flowing or cohesive powders. Free-flowing powders tend to segregate during handling and storage, so spot samples are rarely representative. Cohesive powders tend to retain their characteristics during handling, so that if they are segregated during manufacture or packaging they will tend to remain segregated.

For these reasons it is recommended that, wherever possible, the *golden rules of sampling* should be followed:

- A powder should always be sampled when in motion.
- The whole of the stream of powder should be taken for many short increments of time, in preference to part of the stream being taken for the whole of the time.

Observance of these rules, coupled with an understanding of the manner in which segregation takes place, leads to the best sampling procedure. Any method that does not follow these rules should be regarded as a second-best method likely to lead to errors. Finally, the need for care and skill in abstracting samples cannot be overemphasized. There are a very large number of possible systems from which a gross sample has to be extracted, so it is impossible to lay down instructions that will address all situations. Essentially, the solution depends on whether the powder is stationary or moving and whether it is cohesive or free-flowing.

Sampling Stored Material

It is usual to assume that the powder was mixed before storage. If this assumption is not true, then the homogeneity of the powder will depend on its previous history. Thus, a nonflowing material that has been segregated during manufacture, or prior to storage, will remain segregated. For a free-flowing material, segregation can, and often does, occur during transfer from the mixer to the storage container.

Sampling Stored Nonflowing Material

Nonflowing material is a term for very fine cohesive powders, sticky material, moist material, or fibrous solids. These may be stored in small containers such as drums or bags, or in large containers such as trucks or railway wagons. Surface sampling is usually carried out with a scoop because of its simplicity. A presupposition is that the powder at the sampling point is representative of the bulk (i.e., that the powder was mixed before storage). Accuracy is increased by taking more than one sample, and the samples should be analyzed separately and combined in later analyses if the variation between samples is at an acceptable level.

Sampling Spears. Sampling accuracy is improved if samples from the body of the material are included, and this may be carried out with the aid of a sampling spear (thief). Three types are available (Fig 1). Type 1 consists of an inner and outer tube with a closed end and longitudinal window so that the sampling chamber runs the full length of the spear; this generates a core sample. In type 2 the sampling chamber is at the end of the spear; this generates a spot sample. In type 3 there are separate chambers along the length of the spear; this generates several spot samples simultaneously. With a type 2 sampling spear it is usual to extract several samples at different depths, typically top, middle, and bottom, and blend these to obtain a composite sample. This procedure presupposes homogeneity within the container and should only be used when this assumption has been verified (i.e., the three samples should be analyzed separately a sufficient number of times to establish an acceptable level of statistical confidence).

The spear is thrust into the powder with the inner chamber closed off, and when in position the outer tube is rotated to allow powder to fall into the inner chamber. When the chamber is full, the inner tube is turned to the closed position and the spear is withdrawn. Possible segregation throughout the bed may be investigated with type 3, an average value for the length of the spear with type 2, and a spot sample with type 3. Frequently the spears are vibrated to facilitate filling, and this can lead to an unrepresentative quantity of fines entering the sample volume. The sampling chambers also have a tendency to jam if coarse particles are present, because they can get lodged between the inner and outer chambers.

Coning and Quartering. In industry it is common practice to sample small heaps by coning and quartering. The powder is formed into a heap, which is first flattened at the top and then separated into four equal segments with a sharp-edged board or shovel. The segments are drawn apart and, frequently, two opposite quadrants are recombined and the operation is repeated until a small enough sample has been generated. This practice is based on the assumption that the heap is symmetrical, and since this is rarely so, the withdrawn sample is usually nonrepresentative.

Sampling from Trucks or Wagons. Consignment sampling is carried out on a single consignment (e.g., a truck or wagon load). In sampling from a truck or a wagon it is recommended that eight samples be extracted (Ref 1). No increment should be taken at less than 12 in. below the surface; this avoids the surface layer, in which segregation can have occurred due to vibration (Fig. 2). Care needs to be taken to prevent powder sliding down the slope created due to removal of surface material.

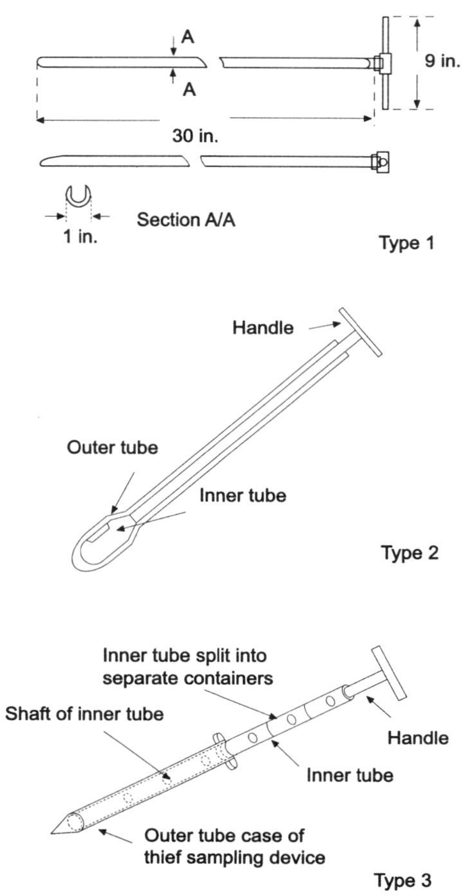

Fig. 1 Sampling spears

Sampling Stored Free-Flowing Material. It is practically impossible to representatively sample stationary free-flowing powder because of the severe segregation that has almost certainly occurred. There is only one sound piece of advice to give regarding sampling such material: Don't! If there is no alternative, several samples should be taken and analyzed separately so that an estimate can be made of the degree of segregation.

Sampling from Containers. Suppose an analysis is required from several tons of material that is available in bags or small containers. Several of these containers should be selected systematically or, preferably, using a table of random numbers. The recommended number of samples depends on the number of containers (Table 1). The whole of each bag should then be sampled using a full stream or Vezin-type splitter so that the golden rules of sampling are obeyed. This is the only way to obtain a representative sample from each bag. Where this is not possible a sampling thief may be used. It is preferable to obtain a sample as the containers are being filled or emptied.

Sampling Flowing Streams

Most powder systems are transported at some time during their manufacture as flowing streams: Hoppers are emptied by screw or belt conveyors, powders are transferred to bagging operations by screw or pneumatic conveyors, and many solids are transported through pipes. A general rule in all sampling is that whenever possible, the sample should be taken while the powder is in motion. This is usually easy with continuous

Table 1 Recommended number of containers to be sampled from a packaged lot

Number of containers in lot	Number of containers to be sampled
1–5	all
6–11	5
12–20	6
21–35	7
36–60	8
61–99	9
100–149	10
150–199	11
200–299	12
300–399	13

Note: For every additional 100 containers, one additional container should be sampled. Source: ISO 3954

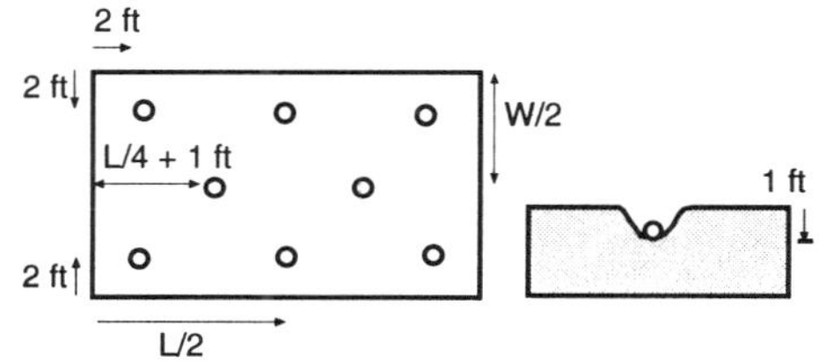

Fig. 2 Sampling points for a wagon or a truck

processes; with consignment sampling it may be possible during filling and emptying of storage containers.

Sampling from a Conveyor Belt. When a sample is to be collected from a conveyor belt, the best position for collecting the increments is where the material falls in a stream from the end of the belt. If access at such a point is not possible, the sample must be collected from the belt. The whole of the powder on a short length of the belt must be collected. The particles at the edge of the belt may not be the same as those at the center, and particles at the top of the belt may not be the same as those at the bottom. If the belt can be stopped, the sample may be collected by inserting into the stream a frame consisting of two parallel plates shaped to fit the belt; the whole of the material between the plates is then swept out. A scoop can be used to scoop out an increment, but this operation can be hazardous if the belt is moving.

When sampling from a continuous stream, the sampling may be continuous or intermittent. In continuous sampling, a portion of the flowing stream is split off and frequently further divided subsequently. In intermittent sampling, the whole stream is taken for many short increments of time at fixed time intervals. These increments are usually compounded and samples for analysis are taken from this gross sample. Continuous sampling is deprecated because if there is segregation on the belt, the extracted sample may not be representative.

It is common practice, in sampling from a blender, to extract three samples: the first after the blender has been emptying for a few minutes, the second when the blender is half empty, and the third when the blender is almost empty. Note that blenders sometimes have a "heel" of unmixed material that is first out of the blender. Powder from the whole cross section of the blender discharge steam should be collected for each sample. This practice should only be used after the mixing efficiency of the blender has been established for each product by taking multiple samples and analyzing these separately.

Point Samplers. Samples can be extracted from the product stream by projecting a sample tube, containing a nozzle or orifice, into the flow. The particles impact the tube and fill the open cavity. The sampling head is out of the stream when not sampling Snorkel-type samplers are available for vertical or inclined applications and can be preprogrammed for sampling frequency. It is not possible to sample nonhomogeneous streams representatively with this type of device. With the auger-type sampler, a slot inside the process stream is rotated to capture a cross section of the process stream, which is then delivered into a sample container. This type of device does not collect a representative sample unless the stream is homogeneous, and it has the added disadvantage that it obstructs flow.

Sampling from Falling Streams. In collecting from a falling stream of power, care should be taken to offset the effects of segregation. Each increment should be obtained by collecting the whole of the stream for a short time. Care must be taken in putting the sampler in and out of the stream. Figure 3 shows correct and incorrect ways of doing this. Unless the time during which the receiver is stationary in its receiving position is long compared with the time taken to insert and withdraw the sampler, the method shown in Fig. 3(a) will lead to an excess of coarse particles, because the surface region of the stream, usually rich in coarse particles, is sampled for a longer time than the rest of the stream. The method shown in Fig. 3(b) is not subject to this objection. If this method is not possible due to some obstruction, the ratio of stationary to moving time for the receiver should be made as large as possible. In many cases it is not possible to collect the whole of the stream as this would give too large an amount to be handled. The best procedure in this case is to pass a sample collector of the form shown in Fig. 3(c) through the stream.

The width of the receiver, b, should be chosen to give an acceptable weight of sample but must not be made so small that the biggest particles have any difficulty in entering the receiver. Particles that strike the edges of the receiver are likely to bounce out and not be collected, so that the effective width is $(b–d)$, where d is the particle diameter. The effective width is therefore greater for small particles than for large ones. To reduce this error to an acceptable level, the ratio of receiver width to the diameter of the largest particle should be made as large as possible with a minimum value of 3:1. The depth, a, must be great enough to ensure that the receiver is never full of powder. If the receiver fills before it finishes its traverse through the powder, a wedge-shaped heap will form that is size selective. As more powder falls on top of the heap, the fine particles will percolate through the surface and be retained, whereas the coarse particles will roll down the sloping surface and be lost. The length of the receiver, c, should be sufficient to ensure that the full depth of the stream is collected.

Stream Sampling Ladles. Powder may be manually withdrawn from a moving stream of

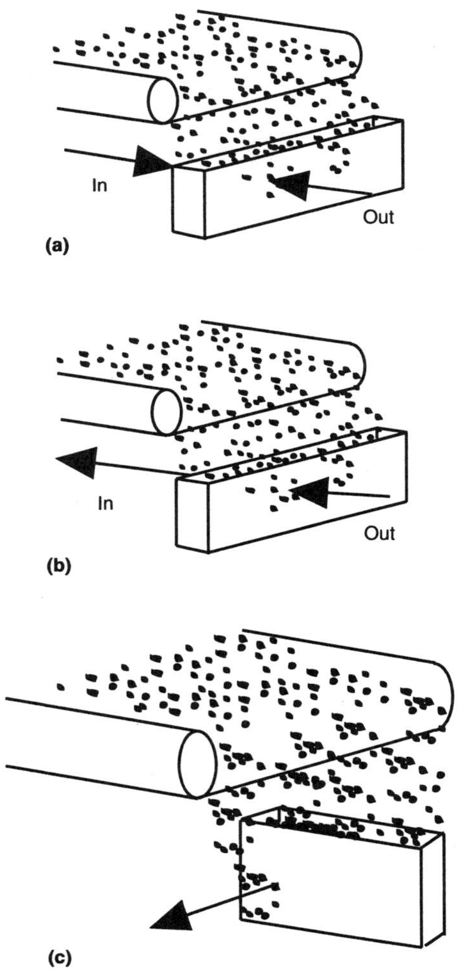

Fig. 3 Sampling from falling streams. (a) Bad sampling technique. (b) Good sampling technique. (c) Sampling procedure to be adopted for high mass flow rate

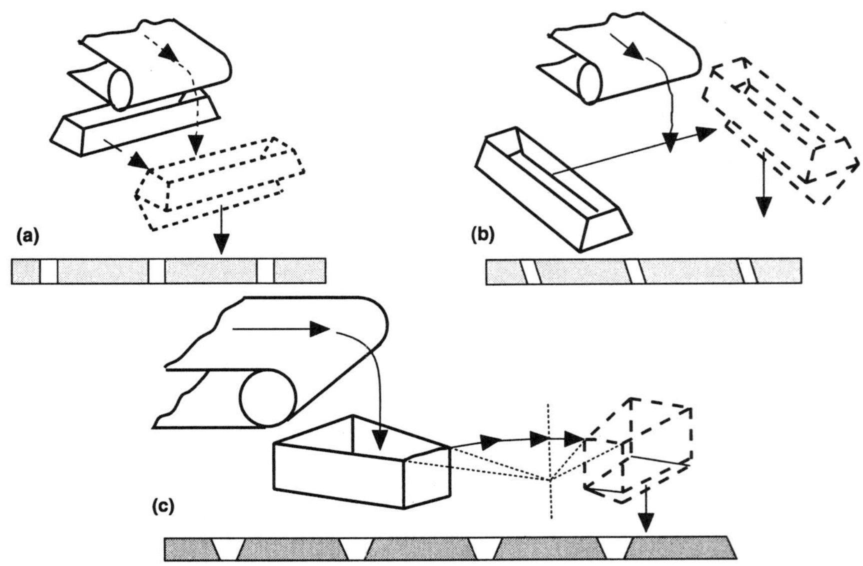

Fig. 4 Traversing cutters. (a) Straight path action, in line. (b) Cross line. (c) Oscillating or swinging arc path

powder using one of the several commercially available ladles. These are suitable for occasional use, but automatic on-line stream sampling samplers are preferred for frequent applications.

Traverse Cutters. With large tonnages, samples taken from conveyors can represent large quantities of material that need to be further reduced. With the action shown in Fig. 4(a) and (b), uniform increments are withdrawn to give a representative sample, but with the action shown in Fig 4(c), a biased sample results if the inner and outer arcs of the container are significantly different and the powder is segregated horizontally on the belt.

Often, a traversing cutter is used as a primary sampler, and the extracted sample is further cut into a convenient quantity by a secondary sampling device. The secondary sampler must also conform with the golden rules of sampling.

A traversing cutter is satisfactory for many applications, but it has limitations that restrict its use:

- Although a traversing cutter is comparatively readily designed into a new plant, it is frequently difficult and expensive to retrofit an existing plant because of the space requirements.
- The quantity of sample obtained is proportional to product flow rate, and this can be inconvenient when the plant flow rate is subject to wide variations. On the other hand, where the daily average of a plant is required, this is a necessary condition.
- It is difficult to enclose the sampler to the extent required to prevent the escape of dust and fumes when handling dusty powders.

Sampling Dusty Material. Figure 5 shows a sampler designed to sample a dusty material, sampling taking place only on the return stroke. This is suitable provided that the trough extends the whole length of the stream and does not overfill. The radial cutter or Vezin sampler shown in Fig. 6 is suitable for sample reduction. These samples vary in size from a 15 cm laboratory unit to a 152 cm commercial unit.

Diverter Valve Sampler. The diverter valve shown in Fig. 7 is suitable for online intermittent sampling when there is limited head room. It can also be operated manually.

Sample Reduction

The gross sample is frequently too large to be handled easily and may have to be reduced to a more convenient weight. Obviously, the method employed must conform with the two golden rules mentioned above. The amount of material to be handled is usually small enough that getting it in motion poses no great difficulty. There is a natural tendency to remove an aliquot with a scoop or spatula, and this must be avoided because it negates the effort involved in obtaining a representative sample from the bulk. Placing the material in a container and shaking it to obtain a good mix prior to extracting a scoop sample is not recommended.

To obtain the best results, the material should be made as homogeneous as possible by premixing. It is common practice then to empty the material into a hopper, and this should be done with care. A homogeneous segregating powder will segregate when fed to a hopper from a central inlet, because in essence it is being poured into a heap. In a core flow hopper the central region (which, with improper feeding, will be rich in fines) empties first, followed by the material nearer the walls, which has an excess of coarse particles. The walls of the hopper should have steep sides (at least 70°) to ensure mass flow, and the hopper should be filled in such a way that size segregation does not occur. This can best be done by moving the pour point about so that the surface of the powder is more or less horizontal. Several sample-dividing devices are discussed briefly below.

Scoop sampling consists of plunging a scoop into the powder and removing a sample. This method is particularly prone to error because the whole of the sample does not pass through the sampling device, and because the sample is taken from the surface, where it may not be representative of the mass. For powder in a container, it is usual to shake the sample prior to sampling in an attempt to achieve a good mix. However, the method of shaking can promote segregation.

Coning and quartering consists of pouring the powder into a heap and relying on its radial symmetry to give identical samples when the heap is flattened and divided by a cross-shaped cutter. This method is no more accurate than scoop or thief sampling, which are simpler to carry out, but gross errors are to be expected. Coning and quartering should never be used with free-flowing powders.

Table Sampling. In a sampling table the material is fed to the top of an inclined plane in which there is a series of holes. Prisms placed in the path of the stream break it into fractions. Some powder falls through the holes and is discarded, while the powder remaining on the plane passes on to the next row of holes and prisms, and more is removed, and so on. The powder reaching the bottom of the plane is the sample. The objection to this type of device is that it relies on the initial feed being uniformly distributed, and on a complete mixing after each separation, and in general these conditions are not achieved. As it relies on the removal of part of the stream sequentially, errors are compounded at each separation, hence its accuracy is low.

Chute Splitting. The chute splitter consists of a V-shaped trough, along the bottom of which is a series of chutes that alternately feed two trays placed on either side of the trough. The material is repeatedly halved until a sample of the desired size is obtained. When carried out with great care this method can give satisfactory sample division, but it is particularly prone to operator error, which is detectable by unequal splitting of the sample.

The above methods are all popular because the samplers contain no moving parts and are consequently inexpensive.

Rotatory Sample Divider. The rotary sample divider conforms to the golden rules of sam-

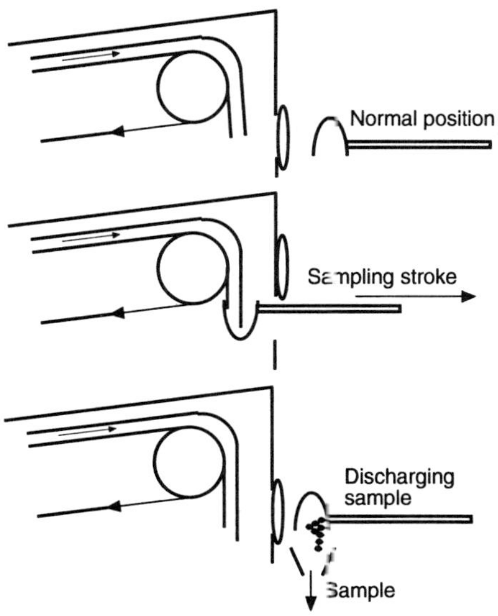

Fig. 5 Full-stream trough sampler

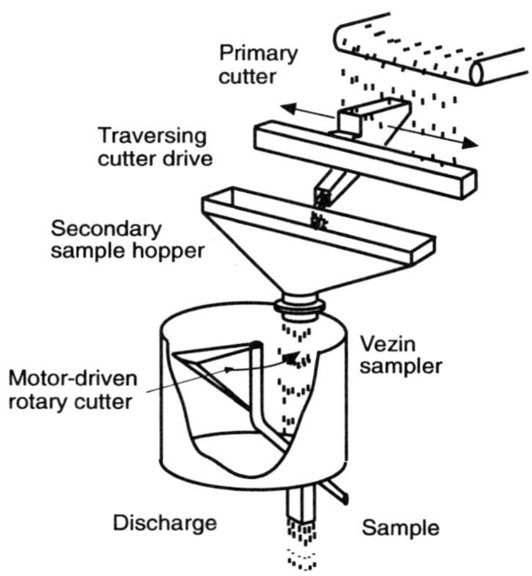

Fig. 6 Schematic of a primary and secondary system based on Denver Equipment Company's type C and Vezin samplers

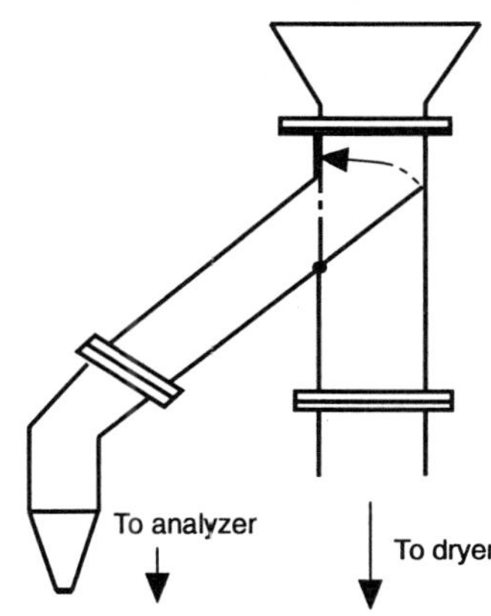

Fig. 7 Diverter valve sampler

pling. The preferred method of using this device is to fill a mass flow hopper in such a way that segregation does not occur. The table is then set in motion and the hopper outlet is opened so that the powder falls into the collecting boxes. The use of a vibratory feeder is recommended to provide a constant flow rate. Several versions of this instrument are available, some of which were designed for free-flowing powders, some for dusty powders, and some for cohesive powders. They handle quantities from 40 L down to a few grams.

Slurry Sampling

Slurry process streams vary in flow rate, solids concentration, and particle size distribution. Any sampling technique must be able to cope with these variations without affecting the representativeness of the extracted sample. For batch sampling, automatic devices are available where a sampling slot traverses intermittently across a free-falling slurry. Unfortunately, it is difficult to improvise with this technique for continuous sampling, because such samplers introduce pulsating flow conditions into the system.

Evaluation of Sampling

Experimental tests of sampling techniques are compared in Table 2 as an example. Binary mixtures of coarse and fine sand (60:40 ratio) were examined (Ref 2) using various laboratory sampling techniques. In every case, 16 samples were examined to give the standard deviations shown in column 2. It may be deduced that very little confidence can be placed in the first three techniques and that the rotary sample divider is so superior to all other methods that it should be used whenever possible.

Weight of Sample Required

Gross Sample. Analyses are carried out on a sample extracted from the bulk, which, irrespective of the precautions taken, never represents the bulk exactly. The limiting (minimum) weight of the gross sample may be calculated, using a simple formula to give an error within predesignated limits, provided that the weight of the gross sample is much smaller than that of the bulk. The limiting weight is given by:

$$M_s = \frac{1}{2}\left(\frac{\rho}{\sigma_i^2}\right)\left(\frac{1}{w_1} - 2\right)d_1^3 \times 10^3 \qquad \text{(Eq 1)}$$

where M_s is the limiting weight in grams, ρ is the powder density in $g \cdot cm^{-3}$, σ_i^2 is the variance of the tolerated sample error, w_1 is the fractional mass of the coarsest size class being sampled, and d_1^3 is the arithmetic mean of the cubes of the extreme diameter in the size class in cubic centimeters. This equation is applicable when the coarsest class covers a size range of not more than and $\sqrt{2}$:1 and w_1 is less than 50% of the total sample. Table 3 gives sample values.

Sample by Increments. For sampling a moving stream of powder, the gross sample is made up of increments. In this case the minimum incremental weight is given by:

$$M_i = \frac{\mu_0 w_0}{v_0} \qquad \text{(Eq 2)}$$

where M_i is the average mass of the increment, μ_0 is the average rate of flow, w_0 is the cutter width for a traversing cutter, and v_0 is the cutter velocity. If w_0 is too small, a biased sample deficient in coarse particles, results. For this reason w_0 should be at least 3*d*, where *d* is the diameter of the largest particle present in the bulk.

ISO 3081 suggests a minimum incremental mass based on the maximum particle size in millimeters. These values are given in Table 4. Secondary samplers then reduce this to analytical quantities.

Gy (Ref 3) proposed an equation relating the standard deviation, which he calls the fundamental error σ_F, to the sample size:

$$\sigma_F^2 = \left(\frac{1}{w} - \frac{1}{W}\right)Cd^3 \qquad \text{(Eq 3)}$$

where *W* is the mass of the bulk; $w = n\omega$ is the mass of *n* increments, each of weight ω, that make up the sample; C is the heterogeneity constant for the material being sampled; and *d* is the diameter of the coarsest element.

For the mining industry (Ref 3), Gy expressed the constant C in the form C = *clfg* where:

$$C = \frac{1-P}{P}\rho \qquad \text{(Eq 4)}$$

P is the investigated constant. ρ is the true density of the material. *l* is the relative degree of homogeneity where for a random mixture $l = 1$, and for a perfect mixture $l = 0$. *f* is a shape factor assumed to be equal to 0.5 for irregular particles and 1 for regular particles. *g* is a measure of the width of the size distribution where $g = 0.25$ for a wide distribution and $g = 0.75$ for a narrow distribution (i.e., $d_{max} < 2d_{min}$).

For the pharmaceutical industry, Deleuil (Ref 4) suggested C = 0.1*lc* with the coarsest size being replaced by the 95% size.

For $W \gg w$, Eq 4 can be written:

$$w\theta^2 = 0.1l\left(\frac{1-P}{P^3}\right)\rho t^3 d^3 \qquad \text{(Eq 5)}$$

where $\theta = t\frac{\sigma F}{P}$ and $t = 3$ (99.9% confidence level) for total quality.

- For $d_{95} = 100$ µm, $\rho = 1.5$, $P = 10^{-3}$ (1000 ppm), $\theta = 0.2$, $l = 0.03$ (random), and $w = 1000$ g.
- For $d_{95} = 100$ µm, $\rho = 1.5$, $P = 0.05$, $\theta = 0.05$, $l = 1$ (homogeneous), and $w = 4$ g.
- For $d_{95} = 20$ µm, $\rho = 1.5$, $P = 10^{-4}$ (100 ppm), $\theta = 0.05$, $l = 0.03$ (random), and $w = 8000$ g.

Powder Classification

Classification methods are used to obtain particular powder distributions or to exclude certain powder sizes from a distribution. The general process for separating dispersed materials is known as "air classification" or "fluid classification," where powder classification is based on the movement of the suspended particles at different points under the influence of a force. The fluid is usually water or air, and the field force may involve gravity or centrifugal or coriolis forces. The other forces of importance are the drag forces due to the relative flow between the particles and the flow medium, and the inertia forces due to accelerated particle movement. The classification process is defined in terms of sorting and sizing. The former includes processes such as froth flotation, where particles are separated on the basis of chemical differences and particle density. The latter, which is covered here, is based only on differences in particle size.

In an ideal system the cut size is well defined and there are no coarse particles in the fine fraction, and vice versa. In practice, however, there is always overlapping of sizes. The cut size may be predicted from theory, but this usually differs from the actual cut size due to the difficulty of accurately predicting the flow pattern in the system. It is therefore necessary to be able to predict the future performance of classifiers based on their past performance.

Table 2 Reliability of selected sampling methods using a 60:40 sand mixture

Sampling technique	Standard deviation, %
Coning and quartering	6.81
Scoop sampling	5.14
Table sampling	2.09
Chute slitting	1.01
Rotary sample dividing	0.146
Random variation	0.075

Table 3 Minimum sample mass required for sampling from a stream of powder

Upper sieve size, µm	Lower sieve size, µm	Mass % in class ($100w_1$)	Sample weight required, g
600	420	0.1	37,500
420	300	2.5	474
300	212	19.2	14.9
212	150	35.6	1.32

Table 4 Minimum incremental mass required for sampling from a stream of powder

Maximum particle size, mm	Minimum mass of increment, kg
250–150	40.0
150–100	20.0
100–50	12.0
50–20	4.0
20–10	0.8
10–0	0.3

Basic Variables

Consider a single stage of a classifier where W, W_c, and W_f are the weights of the feed, coarse stream, and fine stream, respectively; $F(x)$, $F_c(x)$, and $F_f(x)$ are the cumulative fraction undersize of feed, coarse stream, and fine stream, respectively; and x is particle size. Then:

$$W = W_c + W_f \qquad \text{(Eq 6)}$$

$$W\frac{dF(x)}{dx} = W_c\frac{dF_c(x)}{dx} + W_f\frac{dF_f(x)}{dx} \qquad \text{(Eq 7)}$$

The total fine efficiency may be defined as:

$$E_f = \frac{W_f}{W} \qquad \text{(Eq 8)}$$

and the total coarse efficiency as:

$$E_c = \frac{W_c}{W} \qquad \text{(Eq 9)}$$

so that $E_f + E_c = 1$. The total efficiency has no value in determining the effectiveness of a classification process, because it only defines how much of the feed ends up in one or other of the two outlet streams, not how much of the desired material ends up in the correct outlet stream. To discover this, it is necessary to determine the grade efficiency, which is independent of the feed, provided that the classifier is not overloaded:

$$\text{Grade efficiency at size } x = \frac{\text{Amount of desired material in product of size } x}{\text{Amount of desired material in feed of size } x} \qquad \text{(Eq 10a)}$$

$$\text{Coarse grade efficiency, } G_c(x), \text{ at size } x = \frac{\text{Amount of coarse product of size } x}{\text{Amount of feed of size } x} \qquad \text{(Eq 10b)}$$

$$G_c(x) = W_c\frac{dF_c(x)}{dx} \div W\frac{dF(x)}{dx} \qquad \text{(Eq 10c)}$$

$$G_c(x) = \frac{W_c}{W}\frac{dF_c(x)}{dF(x)} \qquad \text{(Eq 10d)}$$

$$G_c(x) = E_c\frac{dF_c(x)}{dx} \qquad \text{(Eq 10e)}$$

Similarly, the fine grade efficiency is defined as:

$$G_f(x) = \frac{W_f}{W}\frac{dF_f(x)}{dF(x)} \qquad \text{(Eq 11a)}$$

$$G_f(x) = E_f\frac{dF_f(x)}{dx} \qquad \text{(Eq 11b)}$$

Hence, from Eq 7, 10, and 11:

$$G_c(x) = 1 - G_f(x) \qquad \text{(Eq 12)}$$

These equations can be used to evaluate the grade efficiency of a classifier, provided that the total efficiency and the size distribution of two of the streams are known. Results are usually plotted as grade efficiency curves of $G_c(x)$ or $G_f(x)$ against x. The classifier separates on the basis of Stokes diameter, so it is preferable to carry out the size determinations on the same basis.

The grade efficiency curve is best determined by plotting $F_c(x)$ against $F(x)$ and differentiating (Fig. 8b), because this allows experimental errors to be smoothed out. The tangent at $F_c(x) = 100\%$ in Fig. 8(b) has a slope of $dF_c(x)/dF(x) = 100/60$; hence, $E_c = 60/100$. Because this tangent merges with the curve at $x = 58$ μm, all particles coarser than 58 μm are collected with the coarse fraction. Differentiating this curve at selected values of $F(x)$ and multiplying by 60 gives $G_c(x)$, the relevant diameters being determined from size distribution data. The 50% size on the grade efficiency curve is called the *equiprobable* size because particles of this size have an equal chance of being in either the coarse or the fine stream. Figure 8(a) shows how the feed is split between the coarse and fine fractions, i.e., $F_f(x) + F_c(x) = F(x)$.

The grade efficiency is often expressed as a single number. This number is known as the sharpness index, ψ, and is a measure of the slope of the grade efficiency curve:

$$75\psi 25 = \frac{x_{75}}{x_{25}} \qquad \text{(Eq 13)}$$

where x_{75} and x_{25} are the particle sizes at which the grade efficiency is 75% and 25%, respectively. For perfect classification ψ = 1, while values above 3 are considered poor. Alternatively 10ψ90 has been used.

These ratios are not always adequate to define the sharpness of cut. In many cases it is important to keep the amount of fines in the coarse or the amount of coarse in the fines as small as possible. For these cases a measure of the effec-

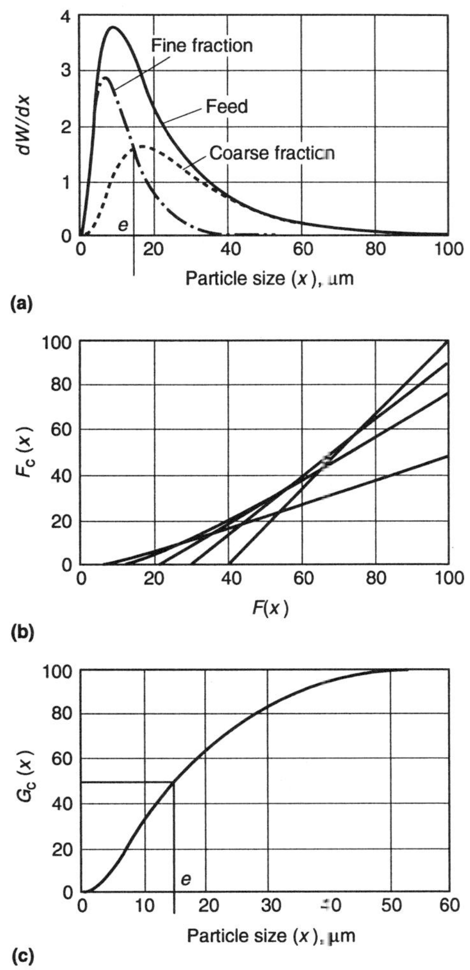

Fig. 8 Graphical determination of grade efficiency curve

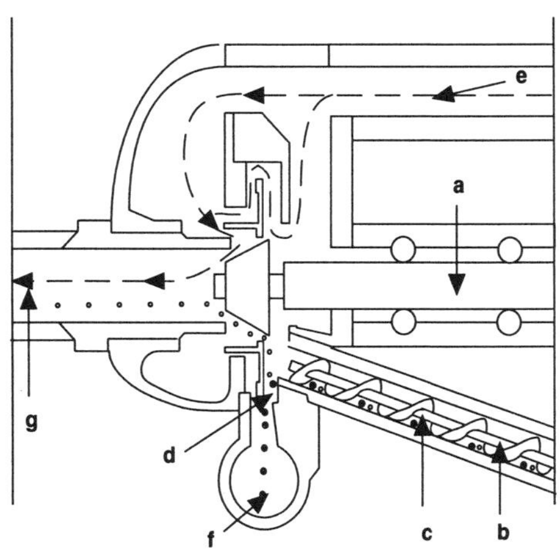

Fig. 9 The Alpine zig-zag centrifugal laboratory classifier

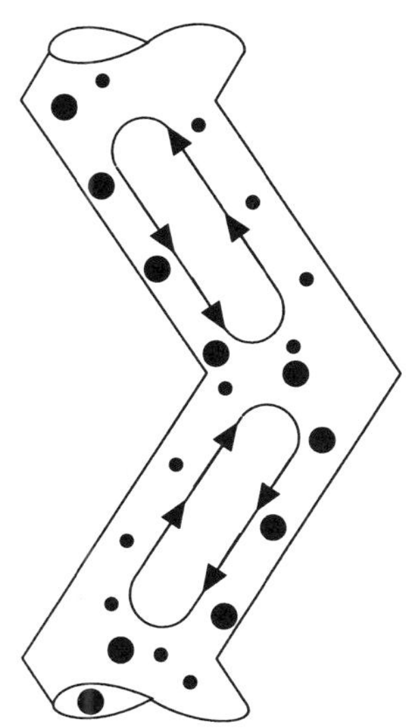

Fig. 10 Mode of action of a zig-zag classifier. The small balls are the fine material; the large balls are the coarse material. Bolts indicate the material guide motion

tiveness of a separation process is given by the following for the coarse yield:

$$\psi_c = \frac{\text{Weight of particles coarser than } e \text{ in the coarse fraction}}{\text{Weight of particles coarser than } e \text{ in the feed}} \quad \text{(Eq 14a)}$$

$$\psi_c = E_c \frac{\sum_{e}^{x_{max}} F_c(x)}{\sum_{e}^{x_{max}} F(x)} \quad \text{(Eq 14b)}$$

Similarly, for the fine yield:

$$\psi_f = E_f \frac{\sum_{x_{min}}^{e} F_c(x)}{\sum_{x_{min}}^{e} F(x)} \quad \text{(Eq 15)}$$

Systems for Powder Classification

Classifiers may be divided into two categories: counterflow equilibrium and crossflow separation.

Counterflow can occur in either a gravitational or centrifugal field. The field force and the drag force act in opposite directions and particles leave the separation zone in one of two directions according to their size. At the "cut" size, particles are acted on by two equal and opposite forces; hence, they stay in equilibrium in the separation zone. In gravitational systems these particles remain in a state of suspension, while in a centrifugal field the equilibrium particles revolve at a fixed radius, which is governed by the rate at which material is withdrawn from the system. They would therefore accumulate to a very high concentration in a continuously operated classifier, if they were not distributed to the coarse and fine fractions by a stochastic mixing process.

In a crossflow classifier, the feed material enters the flow medium at one point in the classification chamber, at an angle to the direction of fluid flow, and is fanned out under the action of field, inertia, and drag forces. Particles of different sizes describe different trajectories and so can be separated according to size.

Counterflow Equilibrium Classifiers in a Gravitational Field: Elutriators. Elutriation is a process of grading particles by means of an upward current of fluid, usually air or water. The grading is carried out in one or a series of containers, the bodies of which are cylindrical and the bases of which are inverted cones. The cut size is changed by altering the volume flow rate and the cross-sectional areas of the elutriation chambers. The flow medium is usually air, although water is used occasionally.

In air elutriators, air containing particles sweep up through the system at a preset flow rate. Particles with a settling velocity lower than the air flow rate are carried out with the air stream, whereas larger particles are retained in the elutriation chamber. The separation is very slow but can be speeded up by the use of zig-zag classifiers, which act as a succession of elutriators in series.

Zig-Zag Classifiers. Several versions of the zig-zag classifiers are available (Fig. 9). These may be categorized as gravitational or centrifugal counterflow classifiers. A feed rating worm (b) feeds the unclassified material (c) into a classifying chamber. Radially arranged blades on the outer face of the classifier rotor (d) speed the inflow of material up to the peripheral velocity of the rotor suspending it, extra air being admitted through (e). The dust-air mixture is then sucked in to the zig-zag-shaped rotor channels where classification takes place (Fig. 10). Fine material is sucked into the classifier center (g), where it leaves via a cyclone. The coarse material (f) is expelled by centrifugal force. At the periphery it is flushed by the incoming air before being discharged. Gravitational instruments operate in the 1 to 100 μm size range, centrifugal instruments from 0.1 to 6 μm.

Cross-flow gravitational classification is performed with the Warmain Cyclosizer (Ref 5), which is a hydraulic cyclone elutriator (Fig. 11). Using inverted cyclones as separators with water as the flow medium, samples of between 25 to 200 g are reduced to five fractions having cut sizes (for quartz) of 44, 33, 15, and 10 μm. The cyclones are arranged in series, and during a run the oversize for each cyclone is trapped and subjected to elutriating action for a fixed time period. At the end of a run the trapped materials are extracted by opening the valves at the apex of each cyclone in turn, and after decantation, the solids are recovered by filtration and evaporation.

Counterflow Centrifugal Classifiers. The Bahco Classifier is a centrifugal elutriator (Fig. 12). The sample is introduced into a spiral air current created by a hollow disc rotating at 3500 rpm. Air and dust are drawn through the cavity in a radially inward direction against centrifugal forces. Separation into different size fractions is made by altering the air velocity, which is effected by changing the air inlet gap by the use of spacers. Instrument calibration is necessary. For the sample, 5 to 10 g of powder are required, which can be graded in the size range 5 to 100 μm.

Crossflow Centrifugal Classifiers. The principle of a crossflow centrifugal laboratory classifier is illustrated in Fig. 13. A vaned rotor produces a centrifugal field, while at the same time air is drawn into the center of the rotor. All but about 5% of the air intake, induced by a positive displacement pump downstream of the classifier, enters the classification zone through a very narrow gap formed between the rotor and the stator. This leads to a very high turbulence in the preclassification zone. The material enters the classifier through a venturi-type nozzle with the remaining 5% of air. Between planes 1 and 2 the ratio of centrifugal force to drag force is kept very nearly constant by a diverging radial cross section. This zone is the classification zone. The smaller particles are carried out through the middle and the larger ones move toward the stator, where they undergo disaggregation until they reach the exit. The cut size of these machines ranges from 0.5 to 50 μm.

Crossflow Elbow Classifier. In the crossflow air classifier (Fig. 14) the main air is introduced at a_1 and secondary air at a_2. Both streams are bent around a solid wall (b), and the resulting flow follows the bend without leaving the wall

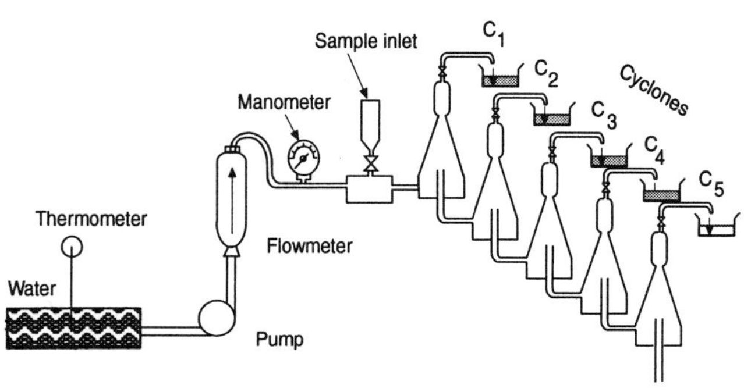

Fig. 11 Principle of the Warmain Cyclosizer

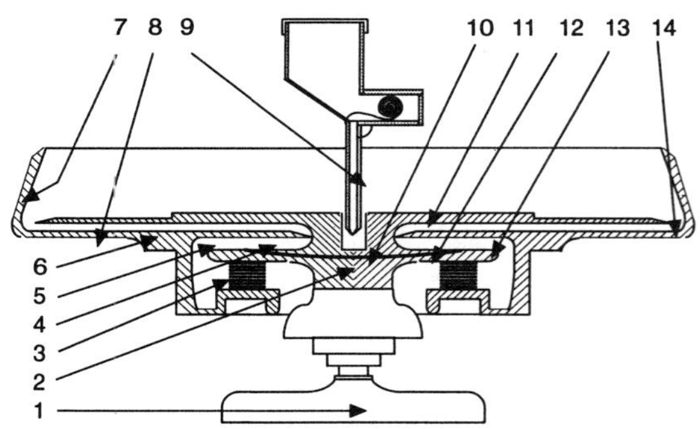

Fig. 12 Simplified schematic diagram of the Bahco microparticle classifier. 1, electric motor; 2, threaded spindle; 3, symmetrical disk; 4, sifting chamber; 5, container; 6, housing; 7, top edge; 8, radial vanes; 9, feed point; 10, feed hole; 11, rotor; 12, rotary duct; 13, feed slot; 14, fanwheel outlet

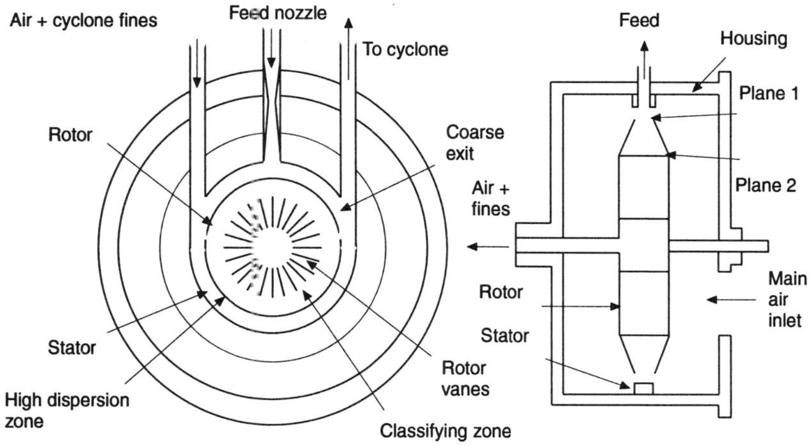

Fig. 13 A cross-flow centrifugal laboratory classifier

or forming vortices. The so-called Coanda effect helps to maintain the flow around the bend for approximately 90°, and this is enhanced by the application of suction.

Sieving Methods

Sieving methods are used to obtain particular powder distributions or to obtain narrow size ranges of a powder. Sieving is a particularly useful procedure because particles are sorted into categories solely on the basis of size, independently of other properties (density, surface, etc.). It can be used to classify dry or wet powders and generates narrowly classified fractions. With micromesh sieves, near-monodisperse powders can be generated in the range of 1 to 10 μm.

Sieving consists of placing a powder sample on a sieve containing openings of a fixed size and agitating the sieve in such a way that particles that can pass through the openings do so. To speed up the analysis, several sieves are stacked on top of each other, with the sieve containing the coarsest openings on top. This "nest" of sieves is vibrated until the residue on each sieve contains particles that could pass through the upper sieve but not through the lower sieve.

A variety of sieve apertures are currently available, ranging in size from about 20 μm to millimeters for woven wire sieves, down to 5 μm or less for electroformed sieves, and greater than 1 mm for punched plate sieves. Woven wire sieves generally have pseudosquare apertures (i.e., the weaving process generates trapezoidal apertures in three dimensions), but punched plate and electroformed sieves are available with round and rectangular apertures. A variety of other shapes are also readily available.

Sieve analysis presents three major difficulties. With woven wire sieves the weaving process produces three-dimensional apertures with considerable tolerances, particularly for fine-woven mesh (Ref 2). The mesh is easily damaged in use (Ref 3). The particles must be efficiently presented to the sieve apertures.

Fractionation by sieving is a function of two dimensions only, maximum breadth and maximum thickness, because unless the particles are excessively elongated, the length does not affect the passage of particles through the sieve apertures (Fig. 15). Particles having two dimensions smaller than the openings will pass through when the sieve is vibrated, whereas larger particles will be retained. The sieve size d_A is defined, on the basis of woven wire sieves, as the minimum square aperture through which a particle can pass.

The sieve surface consists of a go/no-go barrier: particles much smaller in size pass through rapidly, and larger particles pass through more slowly. The apertures have a range of sizes, so the final particle that can pass through will pass through the largest opening when its two smaller dimensions are in a preferred direction. Because this takes a very long time, sieving is usually deemed complete when not more than 0.2% of the original weight of sample passes through in a 2 min sieving operation (Ref 6).

Sieves are often referred to by their mesh size, which is the number of wires per linear inch. American Society for Testing Materials standards range from 635 mesh (20 μm) to 5 mesh (125 mm). The apertures for the 400 mesh are 37.5 μm; hence, the wire thickness is 26 μm and the percentage open area is 35.

Sieve analyses can be highly reproducible even when different sets of sieves are used. Although most of the problems associated with sieving have been known for many years and solutions have been proposed, reproducibility is rarely achieved in practice due to failure to recognize these problems (Ref 7).

Sieve Types

Woven-Wire and Punched Plate Sieves. Sieve cloth is woven from wire, and the cloth is soldered and clamped to the bottom of cylindrical containers (Ref 8). Although the apertures are described as square, they deviate from this shape due to the three-dimensional structure of the weave. Heavy-duty sieves are often made of perforated plate, giving rise to circular holes (Ref 9). Various other shapes, such as slots for sieving fibers, are also available. Fine sieves are usually woven with phosphor bronze wire, medium sieves with brass, and coarse sieves with mild steel. Special-purpose sieves are also available in stainless steel, and the flour industry uses nylon or silk. A variety of electroformed sieves are also available.

The cylindrical sieve cloth containers (sieves) are formed in such a way that they will stack, one on top of another, to give a snug fit (Fig. 16).

A variety of sieve aperture ranges are currently used, and these may be classified as

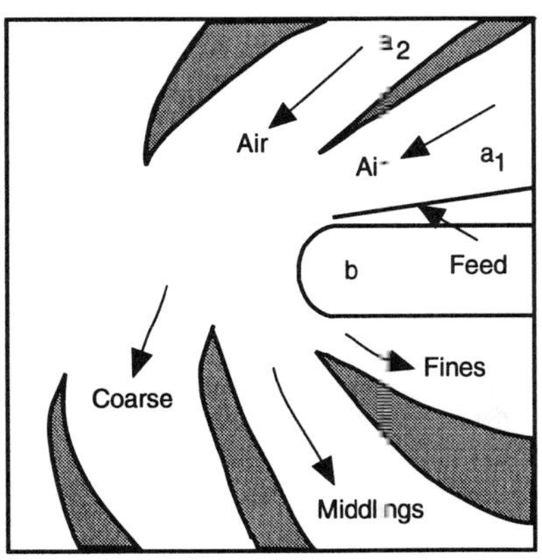

Fig. 14 Principle of the cross-flow elbow classifier

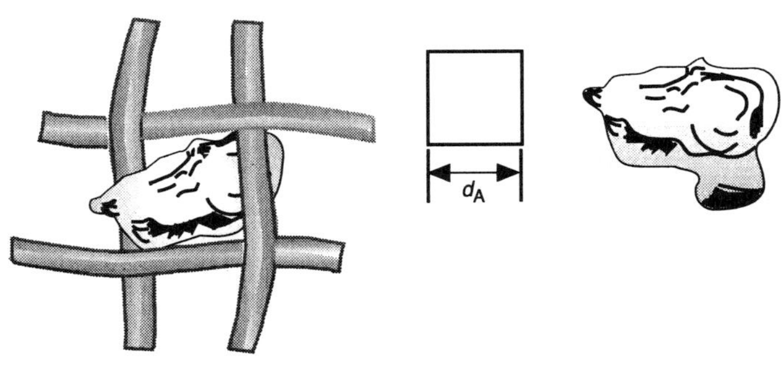

Fig. 15 Sieve size of an irregularly shaped particle

coarse (4 to 100 mm), medium (0.2 to 4 mm), and fine (less than 0.2 mm). The fine range extends down to 20 μm with woven wire mesh and to 5 μm or less with electroformed sieves.

Large-scale sieving machines, to take a charge of 50 to 100 kg, are needed for the coarse range. A wide range of commercial sieve shakers are available for the medium range, and these usually classify the powder into five or six fractions with a loading of 50 to 100 g. Special sieving techniques are used with the finer micromesh sieves.

Due to the method of manufacture, woven wire sieves have poor tolerances, particularly as the aperture size decreases. Tolerances are improved and the lower size limit is extended with the electroformed micromesh sieves.

Electroformed Micromesh Sieves. Micromesh sieves (Ref 10), generated by a photoetching process, are available in a wide range of sizes and aperture shapes.

Basically, the photoetching process is as follows. A fully degreased metal sheet is covered on both sides with a photosensitive coating and the desired pattern is applied photographically to both sides of the sheet. Subsequently the sheet is passed through an etching machine and the unexposed metal is etched away. Finally, the photosensitive coating is removed. A supporting grid is made by printing a coarse line pattern on both sides of a sheet of copper foil coated with photosensitive enamel. The foil is developed and the material between the lines is etched away. The mesh is drawn tautly over the grid and nickel plated onto it. The precision of the method gives a tolerance of 2 μm for apertures from 300 to 500 μm, reducing to 1 μm for apertures from 5 to 106 μm. For square mesh sieves the pattern is ruled onto a wax-coated glass plate with up to 8000 lines per inch, with each line 0.0001 in. wide, and the grooves are etched and filled. The lower limit for off-the-shelf sieves is about 5 μm but apertures down to 1 μm have been produced. Apertures are normally square or round. The percentage of open area decreases with decreasing aperture size, ranging from 2.4% for 5 μm aperture sieves to 31.5% for 40 μm aperture sieves. This leads to greatly extended sieving time when the smaller aperture sieves are used.

Some round aperture sieves have apertures in the shape of truncated cones with the small circle uppermost. This reduces blinding but also reduces the open area and therefore prolongs the sieving time. When sturdier thicker sieves are required, they are subjected to further electrodeposition on both sides to produce biconical apertures.

The tolerances with micromesh sieves are much better than those for woven-wire sieves, the apertures being guaranteed to ±2 μm of nominal size except for the smaller-aperture sieves. Each type of sieve has advantages and disadvantages; for example, sieves having a large percentage of open area are structurally weaker but measurement time is reduced.

Size and shape accuracy are improved by depositing successive layers of nickel and copper on a stainless steel plate, followed by etching through a photolithographic mask additional layers of copper and nickel. The holes are filled with dielectric, after which the additional nickel is removed down to the copper layer.

Dry sieving is often possible with the coarser micromesh sieves, and this may be speeded up, and the lower limit extended, with air jet and sonic sifting. Agglomeration of the particles can sometimes be reduced by drying or adding about 1% of a dispersant such as stearic acid or fumed silica. If this is unsuccessful, wet methods may have to be used.

Ultrasonics are frequently used as an aid to sieving or for cleaning blocked sieves, The delicate mesh can be ruptured under these conditions, and this readily occurs at low frequencies. The rate of cavitation is less in hydrocarbons than in alcohol and about six times greater in water than in alcohol. Saturating the alcohol with carbon dioxide reduces the rate. An upper limit of 40 kHz and a power level of 40 W is recommended. Still higher frequencies reduce the amount of erosion. ASTM E 161-70 recommends cleaning sieves, 15 to 20 s at a time, with a low-power 40 kHz ultrasonic bath containing an equivolume mixture of isopropyl or ethyl alcohol and water with the sieve in a vertical position.

Standard Sieves. The United States Standard Sieve Series described by ASTM (Ref 8) ranges from 20 μm to 5 in. According to ASTM, the nominal 75 μm sieve has a median aperture size in the range of 70 to 80 μm, and not more 5% of the apertures shall fall in the intermediate to maximum size range of 91 to 103 μm. This implies that there is a probability of having a 103 μm aperture in a nominal 75 μm sieve. It is clear that oversize apertures are more undesirable than undersize ones, because the latter are merely ineffective while the former permit the passage of oversize particles. The relative tolerance increases with decreasing nominal size, leading to poor reproducibility when analyses are carried out using different nests of uncalibrated sieves. Electroformed sieves with square or round apertures and tolerances of ±2 μm are also available.

Complete instructions and procedures on the use and calibration of testing sieves are contained in Ref 11. This publication also contains a list of all published ASTM standards on sieve analysis procedures for specific materials or industries. Standard frames are available with nominal diameters of 3, 6, 8, 10, or 12 in. Nonstandard frames are also available.

Most test sieves are certified and manufactured to ISO 9002 standards. Woven-wire sieves, having apertures ranging from 20 μm to 125 mm, are readily available in 100, 200, 300, and 450 mm diameters as well as 3, 8, 12, and 18 in. diameters. Microplate sieves are available in 100 and 200 mm diameters with round or square apertures ranging from 1 to 125 mm.

Micromesh electroformed sieves are available in 3, 8, and 12 in. diameter frames, as well as custom-made sizes. Sieves have apertures ranging from 5 to 500 μm, and other apertures are available. Etched sieves have round apertures ranging from 500 to 1200 μm.

Fig. 16 Stacked sieves on a shaker with rotary and tapping action

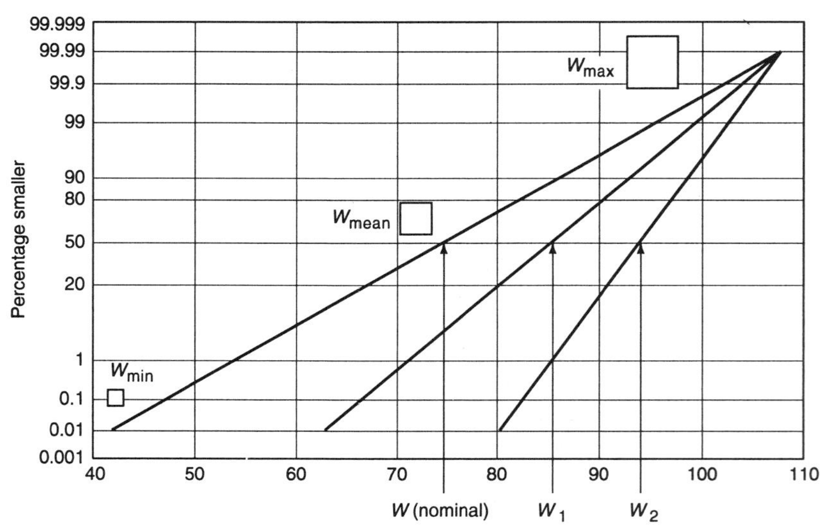

Fig. 17 Effective aperture width distribution with increasing sieving time

Process Variables

End Point of the Sieving Process. The nominal wire thickness for a 75 μm sieve is 52 μm. Hence, at the commencement of a sieving operation, the nominal open area comprises 35% of the total area [i.e., $(75/127)^2$], with apertures ranging in size from 42 to 108 μm (Fig. 17). As sieving progresses, the number of particles that can pass through the smaller aperture decreases, as well as the percentage of available open area. At the same time the effective sieve size increases, rising, in the example given above, to 84 μm, then to 94 μm, and eventually to the largest aperture in the sieve cloth. Thus, the mechanism of sieving can be divided into two regions with a transition region in between: an initial region that relates to the passage of particles much finer than the mesh openings, and a second region that relates to the passage of near-mesh particles (Fig. 18). Near-mesh particles are defined as particles that will pass through the sieve openings in only a limited number of ways, and the ultimate particle is the one that will pass only through the largest aperture in only one orientation.

The end point of sieving should be selected at the beginning of region 2. This can be done by plotting the time-weight curve on log-probability paper and selecting the end point. It is difficult to do this in practice, and an alternative procedure is to use a log-log plot and define the end point as the intersection of the extrapolation for the two regions (Fig. 18).

Using the conventional rate test, the sieving operation is terminated sometime during region 2. The true end point, when every particle capable of passing through a sieve has done so, is not reached unless the sieving time is unduly protracted.

Near-mesh particles are defined as particles that pass through the sieve openings in only a limited number of ways relative to the many possible orientations with respect to the sieve surface. The passage of such particles is a statistical process; that is, there is always an element of chance as to whether a particular particle will or will not pass through the sieve. In the limit, the sieving process is controlled by the largest aperture through which the ultimate particle will pass in only one particular orientation. In practice there is no end point to a sieving operation, so this is defined in an arbitrary manner.

The rate method is fundamentally more accurate than the time method, but it is more tedious to apply in practice, and for most routine purposes a specified sieving time is adequate.

Conventional dry sieving is not recommended for brittle material because attrition takes place and an end point is difficult to define. The rate of passage of particles may fail to decrease with sieving time due to particle attrition, particle deagglomeration, or a damaged sieve.

Calibration of Sieves. It is not widely realized that analyses of the same sample of material by different sieves of the same nominal aperture size are subject to discrepancies, which may be considerable. These discrepancies may be due to nonrepresentative samples, differences in the amount of time the material is sieved, operator errors, humidity, different sieving actions, and differences in the sieves themselves.

Ideally, sieves should have apertures of identical shape and size. However, due to the methods of fabrication, woven-wire sieves have a range of aperture sizes and weaving gives a three dimensional effect. Fairly wide tolerances are accepted and specified in standards, but even these are sometimes exceeded in practice. In order to obtain agreement between different sets of sieves it is therefore necessary to calibrate them and thenceforth to monitor them to detect signs of wear.

One way of standardizing a single set of sieves is to analyze the products of comminution. It is known that the products are usually log-normally distributed; hence, if the distribution is plotted on a log-probability paper a straight line should result. The experimental data are fitted to the best straight line by converting the nominal sieve aperture to an effective sieve aperture.

The traditional method of determining the median and spread of aperture sizes for a woven-wire sieve is to size a randomly selected set of apertures using a microscope. Due to the method of manufacture the measurements for the warp and weft will tend to differ. The limiting size may also be determined by using spherical particles. These are fed onto the sieve, the sieve is shaken, and the excess is removed. Many spheres will have jammed into the sieve cloth and may be removed for microscopic examination.

Sieves may be monitored using glass beads, available from the National Bureau of Standards, every few months to find out whether the sieve apertures are deteriorating or enlarging. The whole sample is placed on the sieves and shaken in the manner in which tests will be made on test samples. The percentage retained on each sieve is determined and the effective opening on each sieve is determined from calibration data supplied with the bead set. Standard polystyrene spheres are also available for calibration purposes from laboratory supplies (such as Gilson and Duke Standards).

Sieves may also be calibrated by a counting and weighing technique applied to the fraction of particles passing the sieve immediately prior to the end of an analysis. These will have a very narrow size range and the average particle size may be taken as the cut size of the sieve. A minimum number (n) of particles need to be weighed to obtain accurate volume diameters (d_v). Let this weight be m and the particle density be ρ, then:

$$d_v = \left(\frac{6m}{\rho n}\right)^{1/3} \quad \text{(Eq 16)}$$

Particles larger than 250 μm can be easily counted by hand and, if weighed in batches of 100, d_v is found to be reproducible to three significant figures. For particles between 100 and 250 μm, it is necessary to count in batches of 1000 using a magnifier. For sizes smaller than this the Coulter principle can be used to obtain the number concentration in a known suspension. Sieve analyses are then plotted against volume diameter in preference to the nominal sieve diameter. The method is tedious and time consuming, and the Community Bureau of Reference in Brussels has prepared quartz samples by this method for use as calibration material. The quartz is fed to a stack of sieves, and the analytical cut size is read off the cumulative distribution curve of the calibration material.

Sieving Errors. Hand sieving is the reference technique by which other sieving techniques should be judged. For instance, the French standard NFX 11-57 states: "If sieving machines are used, they must be built and used in such a way that the sieve analysis must, within the agreed tolerances, agree with the analysis obtained by hand sieving."

The apertures of a sieve may be regarded as a series of gages that reject or pass particles as they are presented at the aperture. The probability that a particle will present itself to an aperture depends on:

- *The particle size distribution of the powder.* The presence of a large fraction of near-mesh particles reduces the sieving efficiency. An excess of fines has the same effect.
- *The number of particles on the sieve (load).* The smaller the sieve loading, the more rapid the analysis. Too low a load, however, leads to errors in weighing and unacceptable percentage losses.
- *The physical properties of the particles.* These include adhesion and other surface phenomena.
- *The method of shaking the sieve.* Sieve motion should minimize the risk of aperture blockage and preferably include a jolting action to remove particles that have wedged in the sieve mesh.
- *Particle shape.* Elongated particles sieve more slowly than compact particles.
- *The geometry of the sieving surface* (e.g., the fractional open area). Whether or not the particle will pass the sieve when it is presented at the sieving surface will depend on its dimensions and the angle at which it is presented.

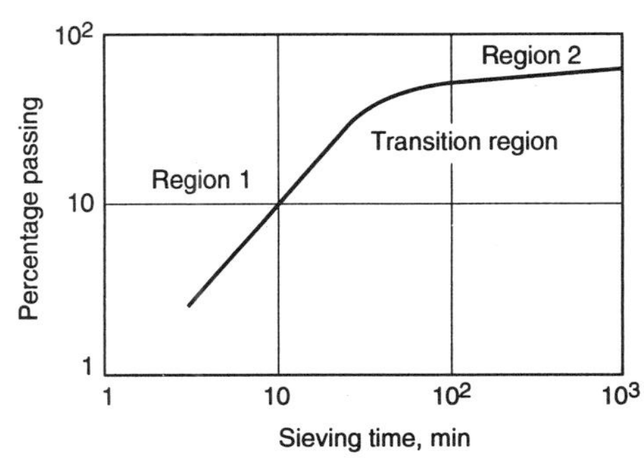

Fig. 18 The rate at which particles pass through a sieve

The size distribution given by the sieving operation also depends on:

- Duration of sieving
- Variation of sieve aperture
- Wear
- Errors of observation and experiment
- Errors of sampling
- Effect of different equipment and operation

It is evident that a reduction in sieve loading is more effective than prolonging the sieving. It is therefore recommended that the sample should be as small as is compatible with convenient handling. Although differences between analyses are inevitable, standardization of procedure more than doubles the reproducibility of a sieving operation. Sieve calibration increases this even further.

Methods of Sieving

Large-scale sieving machines, requiring a charge of 50 to 100 kg of powder, are used for the coarsest size range. A range of commercial sieve shakers are available for medium-sized-aperture sieves, and they usually classify the powder into five or six fractions, with a charge of 50 to 100 g.

Dry sieving is used for coarse separation, but other procedures are necessary as the powder becomes finer and more cohesive. Machine sieving is performed by stacking sieves in ascending order of aperture size and placing the powder on the top sieve. The most aggressive action is performed with Pascal Inclyno and Tyler Ro-tap sieves, which combine gyrating and jolting movements, although a simple vibratory action may be suitable in many cases. Automatic machines use an air jet to clear the sieves or ultrasonics to effect passage through the apertures. The sonic sifter combines two actions, a vertically oscillating column of air and a repetitive mechanical pulse. Wet sieving is frequently used with cohesive powders.

Amount of Sample Required. In determining the amount of sample to be used, it is necessary to consider the type of material, its sievability, and the range of sizes present. Two opposing criteria must be met: it is necessary to use sufficient material for accurate weighing and a small enough sample that the sieving operation is completed in a reasonable time. The natural tendency is to use too large a sample, although in practice, the smaller the sample, within limits, the more reproducible the data.

As a rough guide, the amounts recommended for 8 in. diameter sieves are given in Table 5. Alternatively, the sample weight may be based on the median particle size (Table 5), but this neglects to take into account that the narrower the distribution, the smaller the sample required. A more detailed table is presented in Ref 12.

Table 5 Amount of sample required for 8 in. diameter sieves

Basis	Sample required
Particle density, g · cm^{-3}	
<1.5	25
1.5–3.0	50
>3.0	100
Median diameter of particle, mm	
>2	500
2–1	200
1–0.5	100
0.5–0.25	75
0.25–0.075	50
<0.075	25

Hand sieving is time consuming but necessary for dependable dry sieving data. A representative sample is obtained and the whole of the sample is used in the analysis. The preferred method of sampling is with a spinning riffler or, failing that, a chute splitter. Coning and quartering induces segregation and should never be used with free-flowing powders.

The smallest-aperture sieve to be used should be rested on a catch pan, the tared sample placed on the mesh, and the whole sealed with a lid. The sieve should be slightly inclined to the horizontal and tapped with a cylindrical piece of wood about 20 cm long and 3 cm in diameter; this should be wrapped in duct tape to eliminate splintering. (The heel of the hand is recommended in some standards and is an acceptable alternative if the hand can take it!) The rate of tapping should be about 150 taps per minute and the sieve should be rotated ⅛ of a turn every 25 taps. After about 10 min the residue is transferred to the coarsest sieve, which is nested in a second catch pan for subsequent weighing. A white card should be placed on the bench (approximately 60 cm square) so that accidental spillage may be recovered. The process is repeated in 5 min cycles until less than 0.2% of the original charge passes through the sieve. The powder is not normally removed from the sieve unless excessive blinding is occurring; both the sieve and the residue are weighed and the residual weight is determined. At the end of the sieving operation the sieve is upturned on a white sheet of paper, and the fine particles adhering to it are removed with a soft brush and added to the sieve residue.

The process is repeated with sieves of increasing fineness and the residue weights are collated. Finally, the process is repeated with the finest sieve and the fines are added to the dust collected initially. Brushing is not recommended for sieves with apertures less than 200 μm due to the possibility of damage. Sieves should be washed and dried after use. Ultrasonics should be used to remove particles clogging the apertures, or these may be leached out if possible without damage to the sieve.

The results may be expressed in terms of the nominal size, although it is preferable to use calibrated sieves. A reference set of sieves should be used after every fiftieth analysis for comparison purposes in order to detect wear. In essence, the smaller the sieve loading, the more rapid the sieving operation. The low weights, however, lead to errors in weighing and intolerable percentages losses.

It is recommended that for a dry sieving operation, the fines should be removed prior to the sieve analysis. This is effected by presieving, usually by hand, on the finest sieve to be used in the subsequent analysis. If this is not done the fines will have to pass through the whole stack of sieves, thus increasing the time of the sieving operation and increasing the risk of high powder loss. Because small particles often adhere to large ones, presieving may be carried out with water (with wetting agent if necessary) or with some other liquid in which the powder is insoluble.

The time of sieving is closely related to sieve loading, a reduction in the latter resulting in a reduction of the former. It is usual for routine analysis to machine sieve for 20 min.

If more accurate data are required it is preferable to sieve for 10 min and weigh the residues. Repeated 2 min sieving should then be carried out until the amount passing any given sieve is less than 0.2% of the initial load. Care should be taken, with brittle material, that particle breakage does not occur. Granular particles pass through the sieve more rapidly than elongated particles, although spherical particles are likely to block the apertures.

Some materials tend to form granules when sieved; this effect can often be reduced by coating the particles in order to reduce cohesiveness. Powders may, for example, be shaken in a container with 1% fatty acid (stearic acid is often used) or fumed silica. Alternatively the powder may be sieved wet. The addition of 0.1% subsieve carbon black is useful (although rather messy) for eliminating electrostatic charge.

Brittle powders are best sieved using a gentle sieving action, as is found, for example, with the Air Jet Sieve (Micron Powder Systems, Summit, NJ) or the Sonic Sifter.

For reliable data, about 5% of the sample should be retained on the coarsest sieve and a similar amount should pass through the finest. Unless the size distribution is narrow, alternate sieves (a root two progression in aperture size) are recommended.

Machine sieving is carried out by stacking the sieves in ascending order of aperture size and placing the powder on the top sieve. A closed pan, a receiver, is placed at the bottom of the stack to collect the fines and a lid is placed at the top to prevent loss of powder. A stack usually consists of five or six sieves in a root two progression of aperture size. The stack of sieves is clamped on to a test sieve shaker that is vibrated for a fixed time, and the residual weight of powder on each sieve is determined. Results are usually expressed in the form of a cumulative percentage of the nominal sieve aperture.

The three essentials required in machine sieving are an effective sieving action in order that an end point is reached, as short a time as possible to reach the end point, and reproducible results.

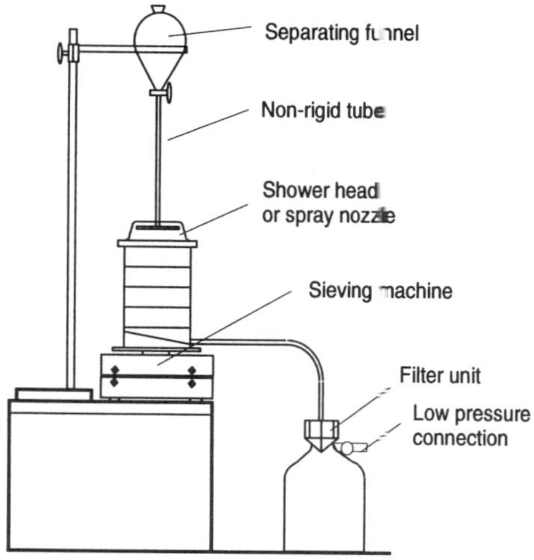

Fig. 19 The Retsch wet sieving machine

For routine control purposes it is usual to machine sieve for 20 min. In ASTM 452, a 20 min initial sieving period is recommended, followed by 10 min periods during which the amount passing should be less than 0.5% of the total feed. For coarse aggregates, sieving is deemed to be complete when the rate falls below 1% per min (ASTM C 136).

It is generally recommended that if losses during sieving exceed 0.5% of the total feed, the test should be discarded. Preliminary hand sieving on the finest sieve should be carried out for the removal of dust; this dust would otherwise pass through the whole nest of sieves and greatly prolong the sieving time. It would also percolate between sieves in the nest and increase powder loss.

The sieving action of some commercial machines is highly suspect, and frequently subsequent hand sieving will produce a sieving rate far greater than that produced on the machine. For the vibratory test sieve shaker, a rapid vertical movement is needed to help keep the apertures clear and prevent blinding. ASTM B 214 suggests 270 to 300 rotations per min for granular materials combined with 140 to 160 taps to reduce blinding of sieve apertures. A sieve enclosure to reduce noise levels may also be necessary.

Modifications to the methods may be necessary for materials that are not free flowing, are highly hygroscopic, are very fragile, have abnormal particle shapes, or have other properties that cause difficulty in sieving. For example, in ASTM C 92 the fines are first removed by washing through the finest sieve; the residue is then dried and analyzed in the dry state.

Manual Wet Sieving. Several methods of wet sieving using micromesh sieves have been described. An 80 kHz, 40 W ultrasonic bath can be used with the micromesh sieve resting on a support, which in turn rests on a beaker in the bath. Sieving intervals are 2 min with an initial load of 1 g. Sieving continues until no further powder can be seen passing through the sieve.

Powder (0.5 to 1 g) is dispersed in a small volume of the liquid, and the suspension is rinsed through the sieve with a suitable liquid into a glass beaker. Sieving continues by moving the sieve up and down in a glass beaker filled with the same liquid so that the direction of flow through the sieve is continuously reversed. If necessary, ultrasonics may be used. After a standardized time the operation is repeated using a second beaker containing fresh sieving liquid. Sieving is deemed to be complete when the amount passing through the sieve is visibly negligible.

For woven-wire and coarse micromesh sieves, the sieves may be mechanically rapped to facilitate sieving. The residual powder is then rinsed off the sieve and weighed, or the pretared sieve is dried and weighed. With fine micromesh sieves an ultrasonic probe may be necessary, and the sieving liquid must have a low surface tension (e.g., acetone), or it will not pass through the sieve apertures. The procedure is then repeated with sieves of increasing fineness.

Ultrasonics are frequently used as an aid to sieving or for cleaning blocked sieves. Under these conditions, rupture of the delicate mesh readily occurs at low frequencies, about 50 Hz, and sometimes at frequencies as high as 20 kHz. Cavitation is less in hydrocarbons than in alcohol and about six times greater in water than in alcohol. Saturating the alcohol with carbon dioxide reduces the rate.

ASTM E 161-70 recommends that cleaning be carried out, 15 to 20 s at a time, with a low-power 40 kHz ultrasonic bath containing an equivolume mixture of isopropyl or ethyl alcohol and water with the sieve in a vertical position.

Wet Sieving by Machine. In most automated wet sieving methods a stack of sieves is filled with a liquid and the sample is then fed into the top sieve. Sieving is accomplished by rinsing, vibration, reciprocating action, vacuum, ultrasonics, or a combination. Commercial equipment is available in which the sample is placed in the top sieve of a stack of sieves and sprayed with water while the stack is vibrated (Fig. 19).

A wet sieving device is commercial available for the size range 10 to 100 μm that includes a sieve vibrator of variable amplitude (Fig. 20).

The Hosokawa Mikropul Micron Washsieve is one version of a wet sieving process where water is sprayed onto the surface of a vibrating sieve. The machine consists of a sprinkler section, a sieving section, and an electromagnetic section. The sprinkler rotates through the force of water to give an even spray while the sieve is vibrated to prevent blockage.

Air-Jet Sieving. The principle of operation of the Air Jet Sieve (Fig. 21) is that air is drawn upward, through a sieve, from a rotating slit so that material on the sieve is fluidized. At the same time a negative pressure is applied to the bottom of the sieve, which removes fine particles to a collecting device (a filter paper). With this technique there is less tendency to blind the apertures, and the action is very gentle, making it suitable for brittle and fragile powders. Sieving is possible with some powders down to 10 μm in size, but with others balling occurs.

The reproducibility is much better than by hand or machine sieving. Size analyses are performed by removal of particles from the fine end of the size distribution by using single sieves consecutively.

The end point of sieving can be determined by microscopic examination of the cleanness of the

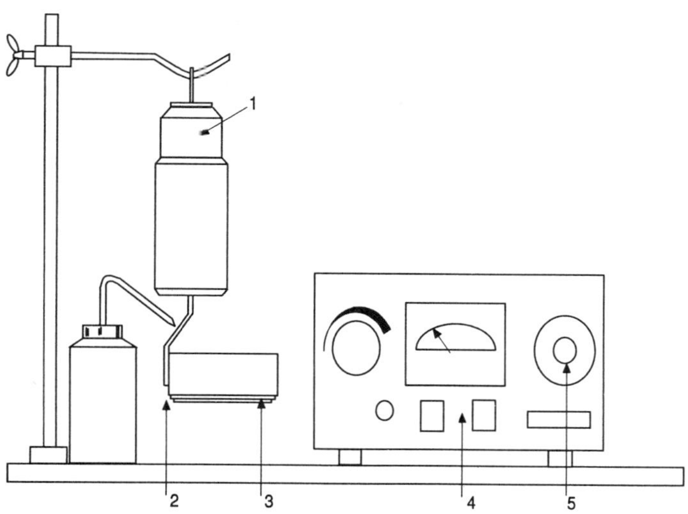

Fig. 20 The Alpine wet sieving device. 1, variable-amplitude vibrator; 2, support; 3, microsieve; 4, regulator with voltmeter for adjusting amplitude; and 5, timer

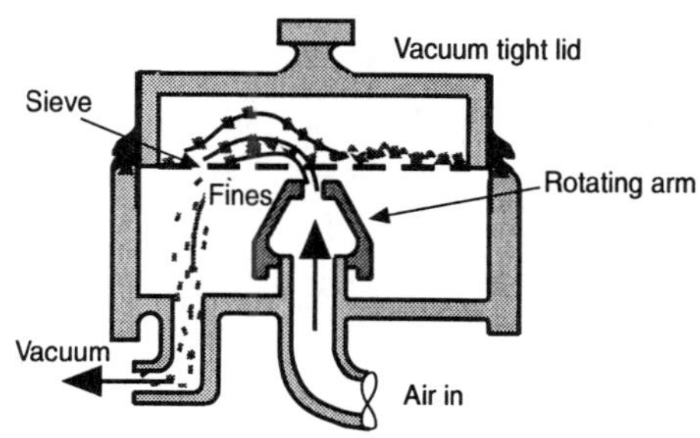

Fig. 21 Mode of action of an Air Jet Sieve

sample or by adopting the same criteria as are used in conventional dry sieving. Because only one sieve is in operation at any time, a full analysis may be unduly protracted, although with sieves coarser than 30 μm the run time is less than 3 min. Sieving is more protracted with finer mesh sieves, and a sieving time of 20 min is usual with a 1 g load on a 3 in. diameter 20 μm sieve.

The Sonic Sifter. The ATM Sonic Sifter is produced as a laboratory model and as an industrial model. It is said to be able to separate particles in the 2000 to 20 μm size range for most materials and 5660 to 5 μm in some cases. It combines two motions to provide particle separation, a vertical oscillating column of air and a repetitive mechanical pulse. The Sonic Sifter moves the air in the sieve stack (Fig. 22). The oscillating air sets the sample in periodic vertical motion, which reduces sieve blinding and breaks down aggregates and yet produces very little abrasion, thus reducing sieve wear and particle breakage.

The Gilson GA-6 Autosiever uses a unique double tapping action together with sonic sifting. The GA-6 features a memory for storing up to ten programs.

A downward air flow has been found to improve sieving rate for 10, 25, and 45 μm electroformed sieves using a sonic sifter operating at 107 dB and 78 Hz. The sieving rate also increased with decreasing feed rate, especially for sub-10 μm sieves.

The Seishin Robot Sifter is an automated version of the ATM in that it includes a robot arm that transports the empty sieves to a balance where they are weighed and the data are stored. At the end of this operation a nest of sieves are assembled on the balance ready to receive a sample of powder on the top sieve. The robot arm then transfers the nest of sieves to the sieving position where sieving is automatically performed. Upon completion of sieving the sieves are transferred to the balance and reweighed. The data are stored in a computer and can be printed out as required (Fig. 23).

Automated Systems. Several types of sieve machines are automated for particle size analysis.

Gradex Particle Size Analyzer. Gradex uses an octagonal two-compartment drum that can be arranged with up to 14 testing sieves. The operator selects the preset cycle for the sample to be analyzed and then loads the sample into the drum, via a feed chute, onto the sieve having the smallest opening. The operator then initiates the automatic cycle, which imparts a reciprocating motion to the drum, separating the smallest fraction from the sample. At the completion of the cycle for this separation, the reciprocating motion stops and the material that passed through the sieve is weighed. The sample then passes to the next coarsest sieve and the operation is repeated. Up to six samples can be loaded into an automatic dispenser on top of the analyzer.

The cycle continues for each sieve in the cycle and the material passing through each sieve is accumulated on an electronic scale. A computer reads and stores the weight of each fraction, then tares the balance before the cycle proceeds to the next sieve. The computer uses this data to calculate and print out the size distribution. When the computer controlled cycle is selected the analysis is stopped when the throughput rate decreases to a set value. Test times vary from 5 to 15 min, depending on the analysis required.

Labcon Automatic Sieve System. The Labcon P114 computer system consists of an Epsom Pine computer mounted in a custom-made portable case containing the necessary power supplies and interface components to allow control of a sieve shaker and electronic balance. Software is included to provide a variety of methods of data handling and presentation procedures.

Gilson Compu-Sieve Analysis System includes a computer with an interface for an electronic scale or a balance for direct data collection. Any electronic balance with an RS-232 interface can be connected to the system. A record of test data is produced by a color plotter. Fifteen sieves and pans can be handled.

Ultrasonic Machine Sieving. The Retsch ultrasonic sieving apparatus was developed for accurate sieving with micromesh sieves in the 5 to 100 μm range. High-intensity ultrasonic energy (50 W) produces small bubbles that continuously collapse and reform. The resulting cavitation keeps the particles in constant motion and simultaneously prevents blocking of the mesh. Sieve loading is 100 to 500 mg unless the sub-5 μm fraction is small, in which case loading can be increased to around 2 g.

The sieves are held in a stack and sealed from each other with radial packing rings. Gravitational action carries the suspension through the stack through progressively smaller meshes.

The Vibrosonic sieve employs ultrasonic vibration of the separator mesh for industrial separation at high rates down to 400 mesh.

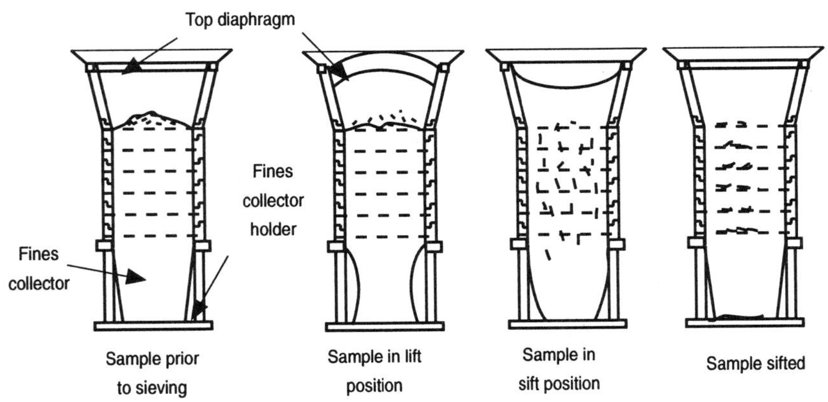

Fig. 22 Mode of action of the Sonic Sifter

Appendix: Dispersion of Powders in Liquids

Ralph D. Nelson, Jr., DuPont Co., Wilmington, DE

Preparation of a dispersion is not a trivial matter. Special equipment and chemicals have been developed to facilitate the three steps of preparing a stable dispersion:

- Wetting powder clumps into the liquid
- Breaking up the wetted clumps
- Preventing flocculation of the dispersed particles

The particle size distributions (PSDs) measured for successive dispersions made from aliquots of dry powder taken from a well-mixed powder sample will be comparable only if the procedures for these three steps are optimized and carefully followed and if the chemicals used (including the liquid) are consistent. This appendix discusses the factors that affect the measured PSD in a dispersion prepared from a dry powder and indicate how they may be controlled in an optimum dispersion procedure—one for which the measured PSD is relatively insensitive to small changes in procedure.

Wetting Powder Clumps into the Liquid

The primary consequence of inadequately wetting the clumps into the liquid is that air retained in the submerged clumps may remain in the system as small submerged bubbles whose sizes are included in and bias the reported PSD. A secondary consequence is that particles may remain attached to bubbles that rise to the top and thus be carried out of the analysis region, biasing the reported PSD.

The wet-in process for a powder starts when a clump is dropped onto the liquid surface, continues as the air is displaced from the interstices (the interparticle regions or "pores" of the clump), and is completed when the clump is fully submerged with no air in the interstices. The rate at which the liquid can flow into a network of interstices is proportional to the square of the length traveled and inversely proportional to the mean interstitial diameter. Thus clumps wet fastest if they are small and are made up of large and loosely packed particles. The best way to wet a batch of powder into liquid is to sprinkle the powder as small clumps across the entire liquid surface at a rate just below that which would cause buildup of a layer of dry solids. A wetting liquid hastens the wetting by pulling itself into the interstices thus increasing the submersion force on the clump. If the liquid is nonwetting, the addition of a wetting agent can dramatically speed the wet-in process.

When a clump of buoyant (less dense than the liquid) particles is placed in a "Neutral" liquid that has no tendency to wet or to reject the solid (contact angle near 90°) it will submerge to a depth at which the downward force of gravity equals the upward buoyancy force of the displaced liquid. A wetting liquid will pull the clump farther into the liquid; a nonwetting liquid will push it farther out.

A clump of nonbuoyant (more dense than the liquid) particles will submerge fully in a neutral or wetting liquid, but may remain on the surface if the liquid has sufficiently nonwetting force to overcome the (submerged) weight of the particle. Because water is nonwetting for perfluoroethylene polymers (or oiled needles), they will "float" on the surface of water even though they are nonbuoyant.

How can one determine if a liquid is wetting for a particular solid? When a drop of liquid is placed on a flat, horizontal solid surface it spreads out until it attains an equilibrium shape. The contact angle is the angle from the solid surface through the liquid to the gas surface. It is related to the balance of surface tension forces at the drop perimeter where the three phases meet (the "shoreline"). A liquid is called "wetting" if the contact angle is less than 90° and "nonwetting" if it is over 90°. A liquid is most likely to be wetting for a solid that has a similar chemical structure or a similar value of interfacial energy density ("like wets like").

The surface tension (or interfacial energy density) is a measure of the force that a material exerts (in a vacuum) to pull smooth a scratch or bump on its surface. It is lowest for hydrocarbons and fluorocarbons (as low as 20 mJ/m^2), moderate for metal oxides (up to 100 mJ/m^2), and highest for metals (up to 1000 mJ/m^2) (Ref 13). *Note:* Older texts report values of surface tension in dyne/cm (equivalent to erg/cm^2); 1 mJ/m^2 = 1 dyne/cm.

Breaking up Wetted Clumps

In a powder that is dry (nothing beyond a monolayer of surface moisture) or non-saturated (interparticle regions not filled with water) the individual particles are held together in clumps by forces acting at or near the points where the particles contact one another:

- Polarizability attraction
- Surface tension of the liquid located at these contacts
- Sinter bonds due to pressure on malleable materials
- Precipitation bonds from drying of solution after solid-liquid separation

The strength of bonds holding clumps together may increase with storage time due to the surface migration of material *into* the high energy contact region as a result of solubilization and mobilization due to adsorbed vapors or surface treatment chemicals or *out of* the high-pressure contact region to a low-pressure non-contact region. Some powders sold commercially have been coated with a solid or liquid to decrease the likelihood of such changes and to keep the clumps "soft" (easy to deagglomerate).

Breaking clumps to fully disperse the submerged and wetted clumps as individual particles (often called "deagglomeration") may be achieved by high shear stress or mechanical impact (Ref 14–16). This generates only a small amount of new surface area (from breaking the contact points) and thus uses only a moderate energy per unit mass of material. The surface area newly created by deagglomeration may have higher adsorptivity and reactivity than the previously existing surface of the clump. *Note:* Deagglomeration is distinct from "milling," which employs similar equipment to fragment the fundamental particles, thus generating significant new surface area and requiring a significantly higher energy per unit mass of material.

High shear stress can be achieved in high-viscosity systems (for example molten plastic or systems with very high solids loading) using:

- Slowly turning mixing paddles
- Close clearances between the shearing elements
- High rotational speed of the moving element

For intermediate-viscosity systems (for example highly loaded inks) the dispersion may be:

- Passed through the gap between two mill-rolls that are rotating with different surface velocities
- Circulated past alternately up- and down-turned teeth at the periphery of a rapidly spinning disk impeller

For low-viscosity systems (for example mineral ore in water) dispersions may be:

- Circulated through the small gaps between a slotted stator outside a close-fitting slotted rotor moving at a high RPM or
- Pumped through an annular gap around a cone rotating in a housing machined so that the clearance decreases from entry to exit
- Forced at high pressure through a very small clearance in a spring-loaded valve

At any viscosity the heat generated by the shear may be a factor limiting the extent of the operating region and thus the deagglomeration that can be achieved.

Mechanical impact can be achieved using grinding media (usually spheres or cylinders) set in motion by:

- Stirring using cross-bars or disks attached to a driveshaft
- Vibration of the container holding the liquid-powder-media mixture
- Cascading by gravity within a container rotating on a horizontal axle
- Cascading by centrifugal force within (several, for balance) rotating containers mounted on a centrifuge with their axles parallel to and offset from the axis of centrifugation

Conventional mills have a grind limit (minimum achievable particle size) about $\frac{1}{20}$ the diameter of the media, but mills with high energy input can achieve a grind limit of $\frac{1}{200}$ the size of

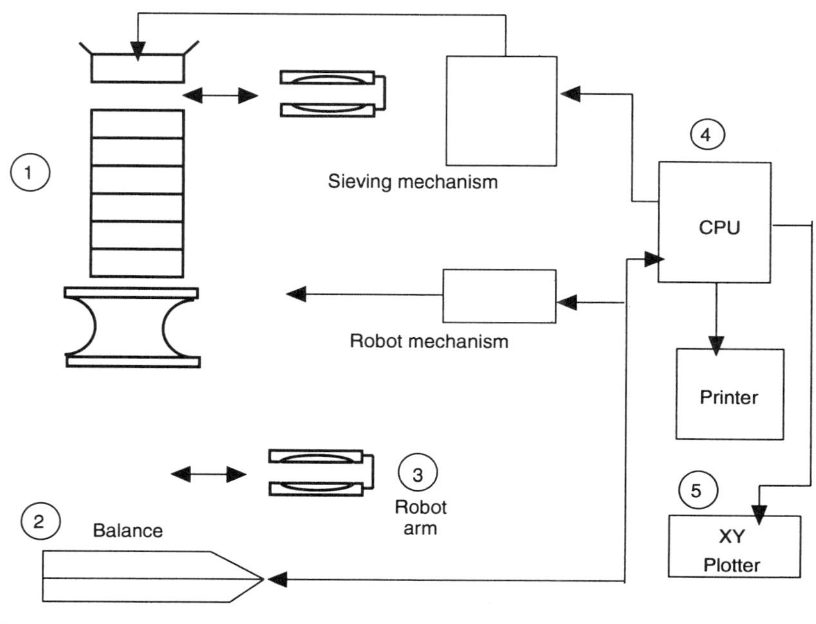

Fig. 23 The Seishin Robot Sifter

the media. It is hard to achieve high-energy impacts in a viscous medium, so impact methods are restricted to low-viscosity systems.

Autogenous Milling. If contamination by wear fragments from the media is a problem, autogenous milling may be preferable. In this process particles are ground through impact with one another caused by:

- Directing one high-velocity jet of dispersion against another
- Vigorously stirring a highly loaded dispersion

The number of collisions is maximized if the solids loading is above 40 volume percent so that there is little liquid separating the particles.

Ultrasonic baths or probes are commonly used to prepare laboratory dispersions in low-viscosity liquids. Piezo-electric crystals attached to the container wall produce pressure waves that alternately create a vacuum (that can pull the liquid apart to form evacuated cavities) and a pressure (causing the cavities to collapse) at 10^4–10^7 Hz. The cavities formed next to particle surfaces are hemispheric in shape. The jets of liquid that form as these hemispheres collapse impinge on the particle surface and focus stress that can shatter the bonds holding a clump together. Because the cavitation energy is strongly attenuated as it propagates through a dispersion, cavitation occurs in only a limited volume (the "active zone") near the vibrating wall. It is important to pass all the dispersion through the active zone and to have a sufficient total residence time there to achieve full deagglomeration without overheating the dispersion or breaking the fundamental particles. Since the presence of vapor in the cavity will decrease the intensity of collapse, the closer the liquid is to the boiling point, the less effective is ultrasonication.

Preventing Flocculation of the Dispersed Particles

Particles attract one another according to how much their interfacial energy densities differ from the intervening liquid. In an unstable dispersion, these attractive forces dominate and cause flocculation. Most size analysis techniques respond to a floc as though it were a single particle, so the particle sizes reported for a flocculated dispersion will be larger than those reported for a well-dispersed dispersion. A dispersion will be stable against flocculation if the particles repel one another or cannot approach each other closely.

Particles repel one another if they bear charges of the same sign *(electrostatic stabilization)* or are covered with molecules that prevent close approach *(steric stabilization)*. The repulsion between particles may be increased by changing solution conditions or using additives called "dispersing" agents to modify particles so that they will repel one another.

Electrostatic repulsion is the simplest and generally least expensive way to stabilize a dispersion against flocculation. Surface charge can be modified by:

- pH changes that alter ionization or adsorption of surface acid and base groups
- Chemical reactions that create or deactivate ionizable surface groups
- Adsorption of ions from solution
- Ionization of an adsorbed dispersing agent
- Changing the concentration of dispersing agent to shift the adsorption equilibrium
- Using a strongly adsorbing dispersing agent to displace an unwanted adsorbate that is adsorbed less strongly

Electrostatic stabilization is most often used in aqueous systems; ionization is difficult to achieve in organic liquids.

The zeta potential characterizes the effectiveness of the charge on a particle. If the zeta potential of particles in an aqueous system (with ionic strength below 0.1 mol/L) is larger than 30 mV (positive or negative) the dispersion is generally stable against flocculation. The ionic strength is half the sum of the concentration of each ion multiplied by the square of the number of charges on that ion. It is a measure of the ability of ions in solution to form a counter-ion atmosphere about an oppositely charged particle and thus offset the effectiveness of its zeta potential.

The zeta potential for acid or base surface groups changes as a function of pH. At low pH the surface may adsorb or react with hydrogen ions to become positively charged. At high pH the surface may adsorb or react with hydroxide ions to become negatively charged. The *isoelectric pH* is the pH at which the zeta potential is zero. Isoelectric pH values range from 2 for silica to 6.4 for aluminum trihydroxide (gibbsite) to 9 for zinc oxide. Surface chemistry depends on the residual stresses at the surface, so the isoelectric point for a particle may depend upon its prior heat treatment, impurities, grinding, and aging in contact with chemicals. Since the zeta potential may drop by as much as 50 mV per pH unit near the isoelectric pH, it is important to keep the pH of a dispersion at least 1 pH unit away from the isoelectric pH to ensure stability against flocculation.

Since multiply charged ions are more strongly attracted to a charged surface than singly charged ions are, the effect of a millimole of $MgSO_4$ in solution is four times that of a millimole of NaCl in diminishing the effectiveness of ionic stabilization. A change from distilled to tap water or from tap water to river water can have disastrous consequences on a process if the change introduces small quantities of multiply charged ions into a slurry that is stabilized by particle charge.

Steric repulsion arises as particles covered by a steric dispersing agent approach closely. Effective steric agents contain one or more soluble ("solvated") polymeric chain segments ("loops" or "tails") and one or more strongly adsorbed anchor ("head") groups. The typical molar mass of the soluble chain segments is 5,000 to 15,000 g/mol. When two such "hairy" particles come close together, the unfavorable energetics of trying to put both sets of solvated and moving loops and tails into the same region of solvent prevent the particles from coming close enough for the attractive force to become significant. The magnitude of steric repulsion depends on polymer-liquid interactions that influence polymer chain configurations. The hairy shell may expand or contract with temperature and liquid composition (pH, ionic strength, cosolvents). An expanded configuration generally provides the highest degree of stabilization.

Selecting a Dispersing Agent

The selection process suggested by Nelson (Ref 17) will soon be superseded by an ISO Standard 14887, "Dispersing Procedures for Powders in Liquids," currently in preparation with the author as convenor of the working group. A brief summary of the considerations involved in finding an effective dispersing agent is given below for the major chemical categories of particles.

Particles having a *natural surface charge* may be electrostatically stable to flocculation with no further treatment. If the natural charge is inadequate to prevent flocculation, these charged particles readily adsorb ionized dispersing agents that have a charge opposite that of the surface (to give a neutral particle) or multiply charged ions (which may reverse the charge on the particle).

Metal particles readily adsorb dispersing agents with highly polarizable groups and can achieve steric stabilization if coated with gelatin. Many metals react with air to form surface oxides, and many metal oxides react with water to form surface hydroxides, in which case they may be electrostatically stabilized in water if the pH is one or more units away from the isoelectric pH. Strong magnetic dipole attractions create special problems for dispersing transition metal oxide particles, which commonly have unpaired electrons that make them paramagnetic or ferromagnetic. The best approach to stabilization of magnetic particles is to provide a thick coating (for example from protein albumin) that prevents the close approach of two particles.

Highly polar materials readily adsorb dispersing agents with highly polarizable groups. A prospective dispersing agent for dispersion in aqueous systems would be sulfonated naphthalenes; for dispersion in organic systems it would be polar-substituted polyesters (acid, ester, or amine).

Nonoxide refractories consist of nonmetallic atoms held together by covalent bonds. Refractories have low values of density, polarizability, surface energy, reactivity, and adsorptivity. Because of their low polarizability, the particles are more easily dispersed in organic liquids than in water. They do not form strong flocs. A prospective dispersing agent would be a polyoxyisopropylene chain grafted to a poly-oxyethylene chain.

Polar polymers have strongly dipolar, hydrogen-bonding, or reactive groups along the chain, so their surface energies and polarizabilities are high. The particles are often incompatible with water and strongly attracted to one another, so flocculation can be a serious problem. A prospective dispersing agent would be a hydrophilic polar polymeric surfactant, which would adsorb at random points along its polymeric chain and be solvated along the rest of its chain.

Nonpolar polymers have no strongly polar groups on their repeating units. They do not require dispersing agents in nonpolar liquids. Since they have low polarizabilities, they do not form strong flocs in water, but they are not wet by water. Good dispersing agents for nonpolar polymers in water have a long nonpolar chain (that adsorbs on the polymer particles) grafted to a water-soluble chain. A prospective would be a hydrocarbon chain grafted to a polyoxyethylene chain.

REFERENCES

1. "Sieve Analysis of Granular Mineral Surfacing for Asphalt Roofing and Shingles," D 451–63, American Society for Testing and Materials, 1963
2. T. Allen, *Particle Size Measurement,* Chapman & Hall, 1997
3. P. Gy, *Sampling of Particulate Matter: Theory and Practice,* Elsevier, Amsterdam, 1982
4. M. Deleuil, Powder Technology and Pharmaceutical Processes, *Handbook of Powder Technology,* Vol 9, D. Chulia, M. Deleuil, and Y. Pourcelot, Ed., Elsevier, 1994
5. D.F. Kelsall and J.C.H. McAdam, *Trans. Inst. Chem Eng.,* Vol 41, 1963, p 84–94
6. "Methods for the Use of Fine-Mesh Sieves," S 1796, British Standard Specifications, 1952
7. K. Leschonski, in *Proc. Particle Size Analysis Conf.,* M.J. Groves, Ed., Heyden, London, 1977, p 205–217
8. "Standard Specification for Wire-Cloth Sieves for Testing Purposes," E 11-87, American Society for Testing and Materials
9. "Standard Specifications for Perforated Plate Sieves for Testing Purposes," E 323-80, American Society for Testing and Materials, reapproved 1990
10. "Standard Specification for Precision Electroformed Sieves (Square Aperture)," E 161-87, American Society for Testing and Materials, reapproved 1992
11. *Manual on Testing Sieving Methods,* STP 447B, American Society for Testing and Materials
12. *Test Sieving Manual,* Endecotts, London
13. R.D. Nelson, Dispersing Powders in Liquids, in the article "Powder Handling" in *"Kirk-Othmer Encyclopedia of Chemical Technology,"* Fourth ed., John Wiley, New York, 1996
14. R.F. Conley, *Practical Dispersion,* VCH Publishers, 1996
15. E. Jarvis, *Chem. Eng.,* (London), Vol 387, 1980, p 477–481
16. I.A. Manas-Zloczower, A. Nir, and Z. Tadmor, "Dispersive Mixing in Rubbers and Plastics," *Rubber Chemistry and Technology,* Vol 57, 1984, p 583–620
17. R.D. Nelson, *Dispersing Powders in Liquids,* Elsevier, 1988

Bulk and Surface Characterization of Powders

R. Iacocca, Materials Characterization P/M Lab, The Pennsylvania State University

CHEMICAL ANALYSIS of metal powders and sintered metal components can be a complicated process, depending upon whether the analysis is to be carried out on the powder or the sintered part, and whether quantitative or qualitative information is required. These complications arise not out of difficulty in obtaining the data, but in selecting the appropriate instrument and interpreting the data correctly.

Figure 1 is a chart that can be utilized in the selection of the proper analytical tool. The first question that should be satisfied is whether powder or a sinter-bonded part is to be examined. Starting first with powder, one should determine whether the chemical information is to be quantitative or qualitative. Beyond that, one should select a technique that provides surface or bulk chemistry.

Quantitative determination of the bulk alloy chemistry is the most sought after information for metal powder. For such measurements, atomic absorption spectroscopy (AAS) and inductively coupled plasma optical emission spectroscopy (ICP-OES) are widely employed. The former can provide the parts per billion (ppb) for some elements. AAS, however, detects one element at a time. ICP-OES has a lower resolution, typically 0.2 to 0.5 wt% for dissolved solids; however, the analysis provides quantitative data for twenty or more elements at a time, and can detect a total number of 70 elements or more (depending on the particular instrument). For both AAS and ICP-OES, the sample should be dissolved in solution.

X-ray fluorescence offers the same broad element detection of ICP-OES, however, the material can be analyzed as a bulk solid rather than dissolving the metal in an aggressive solution. For trace analyses, this method has a lower detection limit of approximately 100 ppm. This, of course, is also a function of the element to be detected.

For carbon, sulfur, nitrogen, and oxygen detection, bulk chemical analyses can be obtained using high temperature combustion and inert gas fusion, respectively. Provided great care is taken in sample preparation, resolution in the ppm range can be achieved. When considering the relative abundance of these four elements in the ambient environment, reproducibility in the ppm range is difficult to achieve.

Because of the small scale of metal powders, surface analysis should be approached with great care. Typically, microfocus electron beam techniques are the only solution for this. The scanning electron microscopy (SEM) equipped with an energy dispersive spectroscopy (EDS) system can provide a qualitative analysis of a metal particle. If the instrument can be calibrated with known standards, the analysis can be semiquantitative. Should quantification be required, the scanning electron microprobe, which is a scanning electron microscope equipped with wavelength dispersive detectors, can provide a quantitative analysis. Microprobe analysis is referred to as microchemical analysis because of the small sampling volume that is used to obtain the data. For metal powders, this small sampling volume is required to prevent the generation of spurious data.

For surface analysis of metal powder, x-ray photoelectron spectroscopy (XPS) and auger electron spectroscopy (AES) can be utilized. To interpret the data correctly, one should understand the electron beam-sample interaction. For example, when using AES, the chemical composition obtained can in no way represent the bulk chemical composition due to the tendency of certain elements to segregate.

X-ray diffraction does not supply elemental chemical analysis for metal powders. It is one of the few methods that can be used to quantitatively determine the phases that are present in a powder sample. If specific crystallographic information is needed for a particular phase that is present, transmission electron microscopy should be employed.

For a sinter-bonded part, all of the techniques described above are available, in addition to several others. Secondary ion mass spectrometry (SIMS) can be used to measure composition and trace impurity levels as a function of depth, having a detection limit in the ppb to ppm range for many elements. The sample should be flat for this method to work, thereby eliminating the analysis of powder.

By no means is Fig. 1 an exhaustive list for the characterization of particulate materials; however, the techniques listed offer a spectrum of analytical methods that are available. In the sections that follow, a more in-depth presentation is given for these techniques.

Microanalysis Methods

For many years, metallurgy has involved chemical analysis on a macroscopic scale with

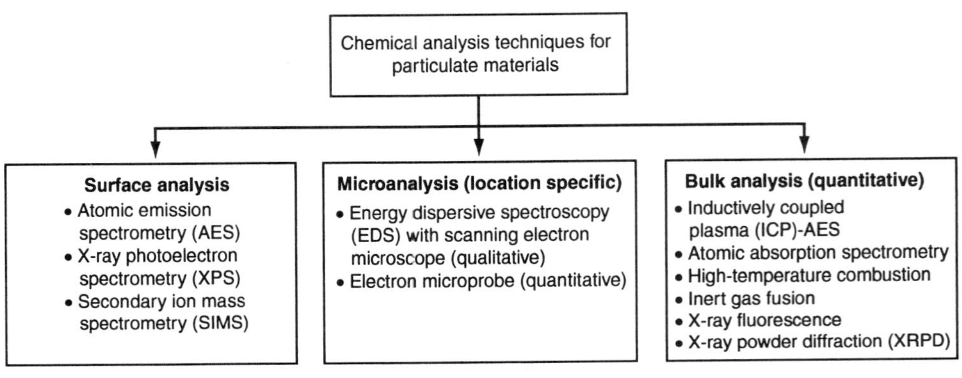

Fig. 1 Material characterization methods for particulate materials

Table 1 General capabilities and applications of SEM and electron probe microanalysis

Method	General capabilities	Applications
Scanning electron microscopy (SEM)	Imaging of surface features at 10 to 100,000×. Resolution of features down to 3 to 100 nm, depending on sample When equipped with a backscattered detector, microscope allows observation of grain boundaries on unetched samples, domain observation in ferromagnetic materials, evaluation of the crystallographic orientations of grains with diameters down to 2 to 10 μm, and imaging of a second phase on unetched surfaces when the second phase has a different average atomic number	Evaluation of crystallographic orientation of features on a metallographically prepared surface; for example, individual grains, precipitated phases, and dendrites Identification of the chemistry of features down to micron sizes on the surface of bulk samples; for example, inclusions, precipitate phases, and wear debris Evaluation of chemical composition gradients on the surface of bulk samples over distances approaching 1 μm
Electron probe x-ray microanalysis	Qualitative and quantitative elemental analysis of solids for elements with atomic number of 11 (sodium) or greater, detection limits of approximately 100 ppm, and at a lateral spatial resolution of approximately 1 μm Qualitative elemental analysis for light elements with atomic numbers from 5 (boron) to 10 (neon) Elemental compositional mapping of areas with dimensions as large as millimeters with spatial resolution to 1 μm	Compositional analysis of individual phases at the microstructural level in multiphase samples, for example, analysis of individual inclusions in steels and other alloys Analysis of compositional gradients at boundaries Determination of compositional homogeneity or heterogeneity at the micrometer scale in single-phase materials Compositional mapping of heterogeneous specimens to produce maps of elemental location and concentration

microscopic investigation of structure by means of metallography. Microanalysis is a method that combines structural (location specific) and compositional analysis in one operation. Microanalysis thus provides information concerning specimen composition on a microscopic scale; hence, microanalytical processes relate to both chemical analysis and microscopy.

Microanalysis typically involves the use of a scanning electron microscope to provide chemical analysis of micron-size areas of a surface. When used in this mode, the scanning electron microscope is termed an electron probe microanalyzer. This function is covered briefly in the section "Electron Probe X-Ray Microanalysis" in this article. A brief discussion of scanning electron microscopy (SEM) is also included. General capabilities and applications of the two methods are compared in Table 1. More detailed information on microscopic equipment and operation is contained in *ASM Handbook*, Volume 10, *Materials Characterization.*

SEM

As a tool for examining surfaces, the scanning electron microscope provides two outstanding improvements over the optical microscope: it extends the resolution limits so that picture magnifications can be increased from 1000 to 2000× up to 30,000 to 60,000×, and it improves the depth-of-field resolution more dramatically, by a factor of approximately 300. Commercial scanning transmission electron microscopes became available around 1973, and scanning Auger microscopes, with submicron resolution, appeared in the late 1970s. These two microscopes have much in common with the scanning electron microscope, and this section compares the three types of microscopes.

A primary use of the scanning electron microscope is to produce high-resolution and depth-of-field images of sample surfaces. A second use, perhaps just as important, is to provide chemical analyses of micron-size areas.

Three types of scanning electron beam instruments are currently available: the scanning electron, scanning transmission electron, and scanning Auger microscopes. These instruments all have in common the feature of obtaining information from the surface (or volume, for the scanning transmission electron microscope) of the sample by scanning an electron beam over a raster and analyzing the various signals generated. The scanning transmission electron microscope is closely related to the conventional transmission electron microscope. The scanning Auger microscope is designed to optimize information obtained from the Auger electron signal (see the section "Auger Electron Spectroscopy" in this article). The primary detector in a scanning electron microscope produces the high-resolution images from secondary electrons. Generally, scanning transmission electron and scanning Auger microscopes use a secondary electron detector; therefore, they can also operate in the SEM mode.

Samples for SEM analysis can be any solid or liquid having a low vapor pressure (greater than $\sim 10^{-3}$ torr, or 0.13 Pa). Sample size is limited by the scanning electron microscope available. Generally, samples as large as 15 to 20 cm can be placed in the microscope, but regions on such samples that can be examined without repositioning are limited to approximately 4 to 8 cm.

Sample preparation by standard metallographic polishing and etching techniques are adequate for electrically conducting materials. Nonconducting materials are generally coated with a thin layer of carbon, gold, or gold alloy. Samples should be electrically grounded to the holder, and fine samples, such as powders, can be dispersed on an electrically conducting film, such as a silver paint that has been thoroughly dried. Samples should be free from high vapor pressure liquids, such as water, organic cleaning solutions, and remnant oil-base films.

Limitations of SEM. Image quality on relatively flat samples, such as metallographically polished and etched samples, is generally inferior to the optical microscope below 300 to 400×. Feature resolution, although much better than the optical microscope, is inferior to the transmission electron microscope and the scanning transmission electron microscope.

Capabilities of related techniques are summarized as follows:

- *X-ray diffraction:* Provides bulk crystallographic information
- *Optical microscopy:* Faster, less expensive, and provides superior image quality on relatively flat samples at less than 300 to 400×
- *Scanning transmission electron microscopy, Auger electron microscopy:* Compared in Table 2.
- *Transmission electron microscopy:* Provides information from within the volume of material, such as dislocation images, small angle boundary distribution, and vacancy clusters. Superior resolution, but requires thin samples

Signals Generated by the Scanning Electron Beam. A picture of the scanned surface region can be taken using any signal generated by the electron beam. In the scanning Auger microscope, it is possible to use as many as five different signals to generate pictures. To understand the potential utility of these instruments, it is necessary to understand some elementary ideas on the nature of the signal generation when the electron beam interacts with sample surface.

When an electron beam strikes a solid surface, electrons and x-rays are emitted from the surface. The energy distribution of these signals is shown qualitatively in Fig. 2 (electromagnetic radiation of lower energies than x-rays is also emitted, which is termed cathodoluminescence). In addition to the secondary electron detector (which can also be used to detect backscattered electrons), most scanning electron microscopes are equipped with an x-ray detector, and specialized backscattered detectors are available at relatively low cost. The addition of an x-ray detector allows determination of the energy of the emitted characteristic x-ray shown in Fig. 2. Two types of x-ray detectors are used: wavelength-dispersive spectrometers and energy-dispersive spectrometers. These x-ray methods are discussed in the section "Electron Probe X-Ray Microanalysis" in this article.

In general, the electrons generated by the electron beam can be partitioned into three types: secondary, Auger, and backscattered (Fig. 3). The intensity scale shown in Fig. 3 has been increased to reveal the details not apparent in

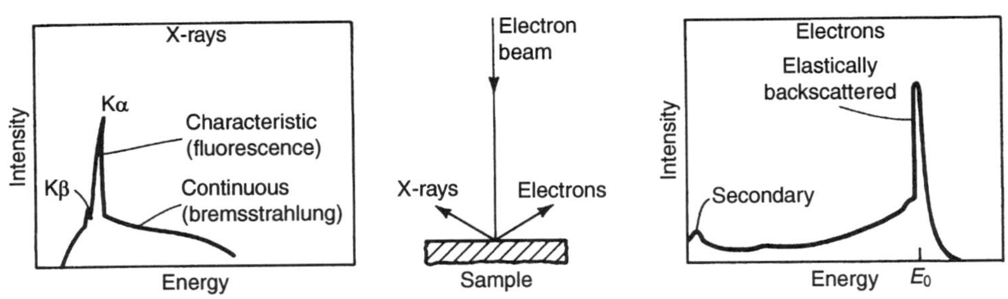

Fig. 2 Energy distribution of signals generated by the electron beam

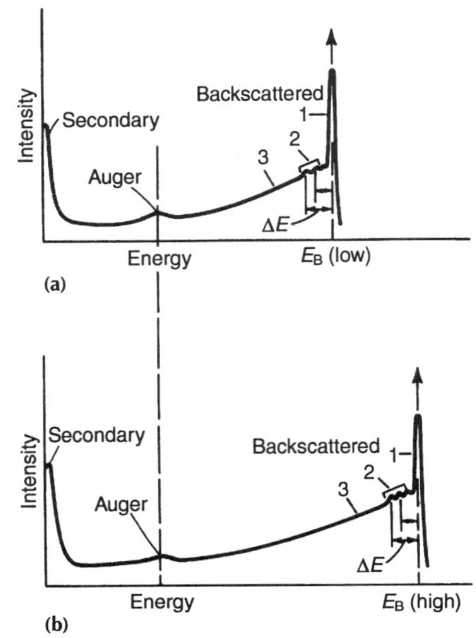

Fig. 3 The energy distribution of emitted electrons at (a) low beam energy (around 1 keV) and (b) a higher beam energy (around 5 keV)

Fig. 2. The backscattered electrons can be further partitioned into three types:

- *Type 1:* Elastically scattered
- *Type 2:* Plasmon and interband transition scattered
- *Type 3:* Inelastically scattered

Elastically scattered electrons emerge with essentially the same energy as the beam energy; inelastically scattered electrons generally undergo many scattering interactions and emerge with a spectrum of energies lower than the beam energy. The type 2 electrons are scattered by interactions that produce a plasmon oscillation of the electrons in the sample material or a transition of sample electrons between different energy bands.

Comparison of Fig. 3(a) and (b) illustrates that the energies of the secondary and Auger electrons are fixed, but the backscattered electrons shift their energy values as the primary beam energy is changed.

Summary. Tables 2 and 3 provide an overview of the three types of scanning electron beam instruments and summarize the source of the signals used. Table 2 compares the scanning transmission electron microscope and scanning Auger microscope with regard to top surface analysis. The scanning transmission electron microscope enables direct probing through the sample. With the scanning Auger microscope, surface films can be probed through by ion sputtering. In general, these instruments are multifunctional.

To interpret correctly the physical significance of the various signals used in the three scanning electron beam instruments, the volume below the surface from which the signal is originating must be known. In general, it is not possible to analyze quantitatively particles with diameters less than approximately 1 to 2 μm using the electron probe microanalyzer, although these particles appear to be huge on the scanning electron microscope screen. It is important to realize that the x-ray sample volume and shape vary with the electron beam voltage and the sample atomic number. Table 4 lists some estimates of the sample volume for the remaining signals. The inelastic backscattered electrons emerge from a volume roughly one-half the depth of the scattered electron range. The Auger electrons are collected from sample depths of 0.5 to 3 nm below the surface, depending on their energy. The Auger electron energies are relatively low, and only those electrons near the sample surface escape without suffering additional energy loss.

Electron Probe X-Ray Microanalysis

As previously noted, scanning electron microscopes are available that have been designed specifically to optimize chemical analysis with x-ray detectors. These instruments are often termed electron microprobes. Commercial electron microprobes were available before the commercial scanning electron microscope. Recent models of these machines, which provide structural resolution nearly equivalent to the less expensive conventional scanning electron microscope, are generally superior in their function as an electron probe microanalyzer, and they include an optical system that enables viewing of the sample in situ at approximately 400×. They are, however, generally more expensive than a scanning electron microscope equipped with equivalent x-ray analyzers.

Electron probe microanalysis (EPMA) is based on the measurement of characteristic x-rays emitted from a microscopic part of a solid specimen bombarded by a beam of accelerated electrons. The electrons are focused into a beam of less than 1 μm in diameter, so that their action can be limited to the chosen microvolume at the surface of the specimen.

The ionization leading to the emission of a characteristic x-ray photon requires that the exciting electron possess or exceed a minimum energy that is also characteristic of the element and line involved. The energy required sets a minimum to the usable beam energy, which typically is between 10 and 20 keV; to keep the reactive volume within one or a few micrometers. the energy of the electron beam should not exceed 20 keV.

The x-rays generated within the specimen suffer attenuation on their way to the specimen surface. When low-energy x-rays are used in EPMA for elements of low atomic number ($Z < 15$), the corresponding signal loss and the concomitant loss of accuracy can only be limited by maintaining the beam energy toward the lower end of the range indicated above.

X-ray spectra can be obtained for all elements except hydrogen. The emission from the first ten elements consists of bands in the low-energy region, where the losses of emission by absorption in the specimen are large. Chemical effects, related to binding energies between atoms, can modify the intensity, position, and shape of these lines, thus further complicating their use for quantitative procedures. Therefore, these elements, including such important ones as oxygen, nitrogen, and carbon, are often determined by difference or stoichiometry, and the quantitative analysis for elements of low atomic number is a very specialized field (Ref 1).

Microanalysis provides information concerning specimen composition on a microscopic scale; hence, microanalytical processes relate to both chemical analysis and microscopy. Electron probe microanalysis (EPMA) produces highly element-specific signals that can be used for quantitative measurements of high accuracy. The sensitivity of the method reaches into the parts per million region for most elements; the sensitivity of EPMA is moderate only, being in most cases much poorer than that of secondary ion mass spectrometry. The key feature is the possibility of local analysis of a small region on the specimen surface. The physical separation of microscopic specimen fragments from their surroundings, which would be impracticable in most instances, is unnecessary because in EPMA the element-specific x-rays are emitted from a small region surrounding the point of impact of the electron beam, and are therefore independent of the composition of the rest of the specimen.

Two types of x-ray spectrometers are used for EPMA measurements: the energy-dispersive spectrometer or the wavelength-dispersive spectrometer. In both the energy-dispersive spectrometry (EDS) and wavelength-dispersive spectrometry (WDS) detection processes, the photoelectric effect is employed for detection.

Table 2 Comparison summary of scanning electron beam instruments equipped with secondary electron and x-ray detectors

Instrument	Features optimized	Minimum area: Surface pictures	Minimum area: Microchemical analysis	Comments
Scanning electron microscope	Surface pictures: above ~500× on polished and etched samples; at all magnifications on high depth-of-field surfaces; accuracy and sensitivity of microchemical analysis	4–5 nm (conventional scanning electron microscope) 2–3 nm (in-lens)	1–3 μm (EDS and/or WDS)	Can be equipped with a WDS x-ray detector that maximizes sensitivity and light-element analysis. WDS: elements with Z > 4, EDS: elements with Z > 10
Scanning transmission electron microscope	Small area microanalysis of thin films; small area diffraction	2–3 nm (SEM mode, in-lens)	5–30 nm (EDS)	Samples must be thinned; generally also functions as a transmission electron microscope; allows chemical analysis of particles characterized by transmission electron microscope observation
Scanning Auger microscope	Chemical analysis of monolayers on surfaces made by in situ fracturing, and low-Z elements on surfaces cleaned by in situ ion etching	~100 nm (Auger) 10 nm (SEM mode)	~100 nm (Auger) 1–3 μm (EDS)	Requires UHV and careful surface preparation; can also detect electron loss signal

Table 3 Comparison summary of signals used in scanning electron beam instruments

Signal type	Type	Energy	Source	Use
X-ray	Characteristic (fluorescent)	Discrete values; different for each element: Cu Kα = ~8000 eV; Si Kα = ~1800 eV	Interband transitions: L → K = Kα, M → K = Kβ; Kα: lose K electron, L → K, photon ejects	Chemical analysis from micro areas in SEM, STEM, and SAM
	Continuous	Continuous	Decleration electron	None (background noise)
Electron	Auger	Discrete values; different for each element range: 100–1500 eV; Si LMM = ~100 eV; Cu LMM = ~900 eV	Interband transitions; LMM: lose L electron, M → L, M electron ejects	Monolayer surface analysis in SAM
	Backscattered (elastic)	Essentially same as beam energy	Beam electron scattered back after elastic collision	Atomic number contrast, channeling contrast, channeling patterns, and magnetic contrast in SEM
	Backscattered (inelastic)	Energies less than beam energy	Beam electron scattered back after inelastic collision	Atomic number contrast, channeling contrast, channeling patterns, and magnetic contrast in SEM
	Backscattered (plasmon and interband transition interactions)	1–1000 eV less than beam energy	Beam electron scattered back after collision producing plasmon oscillations or interband transition	Surface analysis in SAM; light-element analysis in STEM where scattering is in forward direction
	Secondary	~5 eV	Loosely bound electrons scattered from surface	Main signal for image formation in SEM

SEM, scanning electron microscopy; STEM, scanning transmission electron microscopy; SAM, scanning Auger microscopy

Table 4 Estimation of the volume of the various signals produced in iron by a 20 keV electron beam

Signal	Approximate volume dimensions: Diameter	Depth
X-ray	≈1 μm	≈1 μm
Backscattered electrons		
Inelastic (Type 3)	≈0.8 μm	≈0.5 μm
Energy loss (Type 2)	≈d_B	...
Elastic (Type 1)	≈d_B	...
Auger electrons	≈1.1d_B	≈0.5–2 nm
Secondary electrons	≈1.2d_B	≈10 nm

The x-ray photon is absorbed by an atom with the subsequent ejection of photoelectron, which creates change carriers in the detection device. These charge carriers are used in different ways in the EDS and WDS detectors. Additional information can be found in the article "Electron Probe X-Ray Microanalysis" in *Materials Characterization,* Volume 10, *ASM Handbook.*

Sensitivity is 100 ppm with x-ray measurement by wavelength-dispersive spectrometry; 1000 ppm (0.1 wt%) is typical with x-ray measurement by energy-dispersive spectrometry. Poorer sensitivity occurs for light elements in a heavy matrix.

Lateral and depth spatial resolution of EPMA are approximately 1 μm, limited by electron scattering in specimen and not the focused electron beam diameter. Quantitative analysis is limited to flat, polished specimens. Unusual geometries such as fracture surfaces, individual particles, and films on substrates can be analyzed, but with greater uncertainty.

Surface Analysis

Elemental analysis of surfaces is performed by several techniques including Auger electron spectroscopy, x-ray photoelectron spectroscopy (also known as electron spectroscopy for chemical analysis), and ion-scattering spectroscopy. Higher sensitivities are achievable by using secondary ion mass spectrometry (SIMS). Microanalysis is also done with some SIMS instrumentation. Surface analysis has various applications in solving problems related to lubrication, catalysis, adhesion, corrosion, wear, and fracture applications in which resultant properties are determined by the composition at the surface of the solid. However, little has been reported concerning the use of such techniques in powder metallurgy.

Some specific features and characteristics of the various surface analysis techniques are summarized in Table 5. This section briefly reviews the fundamentals and provides an aid to selecting techniques for surface analysis of P/M products. More detailed information on these methods is provided in *Materials Characterization,* Volume 10, *ASM Handbook.*

Auger Electron Spectroscopy

Auger electrons are produced whenever incident radiation—photons, electrons, ions, or neutral atoms—interacts with an atom with an energy exceeding that necessary to remove an inner-shell electron (K, L, M, ...) from the atom. This interaction, or scattering process, leaves the atom in an excited state with a core hole, that is, a missing inner-shell electron. These excited atoms are unstable, and de-excitation occurs immediately. The energy thus released is either emitted as a photon (x-ray fluorescence) or given to another outer shell electron, which is emitted if the energy is great enough (called the Auger transition).

In Auger electron spectroscopy (AES), the sample to be analyzed is bombarded with electrons. If an incident electron (Fig. 4a) strikes a K-shell (core-level) electron with sufficient energy to free it (Fig. 4b), the atom is left in a singly ionized state with a core-level electron vacancy. If the atom is near the surface, both the incident electron and the core-level electron are emitted from the sample as backscattered electrons, with energies below approximately 25 eV. The singly ionized atom undergoes electron rearrangement to achieve a more stable energy

Table 5 Surface analytical techniques

Testing parameters	Auger electron spectroscopy/ scanning Auger microscopy	X-ray photoelectron spectroscopy or electron spectroscopy for chemical analysis	Secondary ion mass spectroscopy	Ion-scattering spectroscopy
Principle	Emission of Auger electrons by core-hole recombination	Emission of photoelectrons by core-level excitation	Sputtering of surface atoms by ion beam	Elastic ion backscattering
Probe	~1 to 30 keV electrons	X-ray, photons	~1 to 3 keV ions	~0.5 to 3 keV ions
Signal	Derivative of electron emission energy	Electron current vs. electron binding energy	Ion current vs. mass	Ion current vs. energy
Elements detected	All except H and He	All except H	All	All except H and He
Sensitivity, general	Variable	Variable	Variable	High
Sensitivity, low atomic number	High	High	High	Low
Spatial resolution	High	Low	High	Low
Spectral shift due to chemical bonding	Yes	Yes	No	Generally no
Molecular information	Moderately effective	Yes	Usually no	Not effective
Quantitative analysis	Yes	Yes	Probably no	Yes
Beam-induced damage	Yes	No	Slight	Slight
Materials to be studied	Excellent for metals, damaging to polymers	Good for metals and polymers	Good for metals and polymers	Good for metals and polymers
Analysis, depth	<2 nm (20 Å)	~1 to 3 nm (~10 to 30 Å)	Monolayer	Monolayer
Depth-profiling analysis versus sputter	Simultaneous or sequential	Usually sequential	Usually sequential	Usually sequential

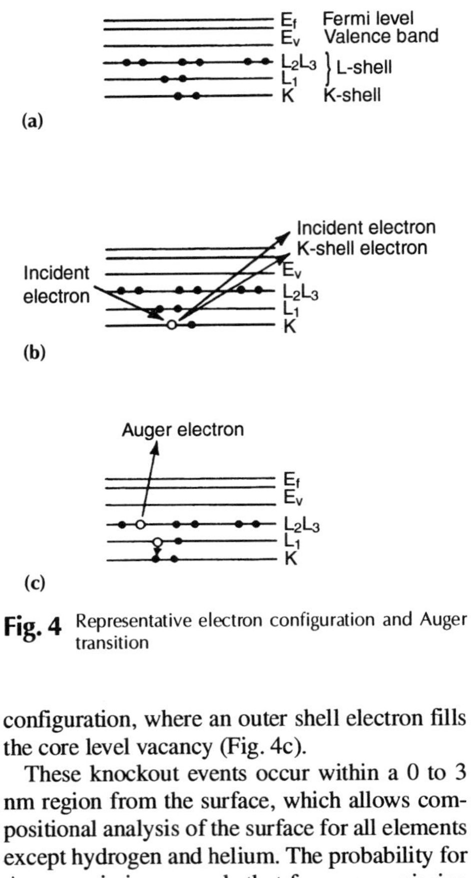

Fig. 4 Representative electron configuration and Auger transition

configuration, where an outer shell electron fills the core level vacancy (Fig. 4c).

These knockout events occur within a 0 to 3 nm region from the surface, which allows compositional analysis of the surface for all elements except hydrogen and helium. The probability for Auger emission exceeds that for x-ray emission as atomic number decreases. This is one of the reasons why Auger analysis has some advantages for light-element analysis. As with the characteristic x-ray emission, the energy of the Auger electrons is different for each element; therefore, analysis of Auger energies yields information on chemical identity. In addition, as with energy loss electrons, the Auger energy levels sometimes shift when an atom becomes oxidized, nitrided, and so on; therefore, information on the chemical state of the surface atoms can sometimes be obtained from Auger analysis.

The instrumentation typically used in AES includes an electron gun for primary electron excitation of the sample, an electron spectrometer for energy analysis of secondary electrons, a secondary electron detector for secondary electron imaging, a stage for sample manipulation, and an ion gun for sputter removal of atoms from the sample surface.

The electron spectrometer is usually the central component of an AES system. Various types of analyzers are in use and include sector analyzers, retarding field analyzers, hemispherical analyzers, and cylindrical mirror analyzers. For electron-excited AES, the cylindrical mirror analyzer generally offers superior signal-to-noise performance, which is associated with its high point transmission. Consequently, it remains the most prevalent spectrometer.

Detection sensitivity for most elements (except hydrogen and helium) is from 0.1 to 1.0 at%. The accuracy of quantitative analysis is limited to ±30% of the element present when calculated using published elemental sensitivity factors. Better quantification (±10%) is possible by using standards that closely resemble the sample.

Samples can be solid metal, ceramic, and organic materials with relatively low vapor pressures ($<10^{-8}$ torr at room temperature). Higher vapor pressure materials can be handled by sample cooling. Similarly, many liquid samples can be handled by sample cooling or by applying a thin film onto a conductive substrate.

Sample size can be individual powder particles as small as 1 μm in diameter. The maximum sample size depends on the specific instrument.

Interpretation of Auger spectra, particularly spectra of irregular surfaces such as powders or porous structures, is difficult due to the number of variables present, including surface contamination and roughness, dissociation effects of chemical species due to the incident electron beam itself (beam damage), and differences in sputtering rates of different chemical species. Thus, conversion of peak heights into atomic concentrations presumes an accuracy that may not exist. However, even as a qualitative or semiquantitative tool, Auger electron spectroscopy/scanning Auger microscopy has already proved invaluable in the solution of many problems.

XPS Surface Analysis

X-ray photoelectron spectroscopy (XPS) uses x-rays to remove an electron from a core level to the Fermi (1s) level. Knowing the photon energy and measuring the kinetic energy with the electron spectrometer permit direct determination of the binding energy of electrons, which is unique and can be used to identify specific elements. Because the mean free path of the electrons is very small, only electrons that originate in the top atomic layers are detected.

X-ray photoelectron spectroscopy is a surface-sensitive technique, and an inert environment such as an ultrahigh vacuum system—for example, 10^{-10} torr—is required for measurements. A typical x-ray photoelectron spectroscopy analytical chamber consists of an x-ray source, an electron energy analyzer, an electron multiplier, a pulse-counting detector, and a computer system. An ion gun is employed for sputtering and compositional depth profiling. For good resolution during profiling, the ion beam—for example, argon—must be defocused over a surface area of approximately 1 cm^2 (0.15 $in.^2$), because photoelectrons are collected from a surface area of several square millimeters during analysis.

Surface sensitivity of XPS is comparable to other surface analysis techniques. Data collection is slow compared with other surface analysis techniques, but analysis time can be decreased substantially when high resolution or chemical state identification is not needed. The requirement for ultrahigh vacuum conditions may necessitate an overnight vacuum pumpdown before analysis. Qualitative analysis can be performed in 5 to 10 min. Quantitative analysis requires 1 h to several hours, depending on information desired.

SIMS Analysis

In secondary ion mass spectroscopy (SIMS), an energetic beam of focused ions is directed at the sample surface in a high or ultrahigh vacuum (UHV) environment. The transfer of momentum from the impinging primary ions to the sample surface causes sputtering of surface atoms and molecules. Some of the sputtered species are ejected with positive or negative charges; these are termed secondary ions. The secondary ions are then mass analyzed using a double-focusing mass spectrometer or an energy-filtered quadrupole mass spectrometer. The principles of SIMS are represented schematically in Fig. 5.

This method can be used to acquire a variety of information about the surface, near-surface, or bulk composition of the sample, depending on the instrumental parameters. If the rate of sputtering is relatively low, a complete mass spectrum can be recorded to provide a surface analysis of the outermost 5 nm of the sample. This is often termed static SIMS. Although a useful mode of operation, it is not yet a routine analytical technique. Alternatively, the intensity of one or more of the peaks in the mass spectrum can be continuously recorded at it higher sputtering rate to provide an in-depth concentration profile of the near-surface region. At very high sputtering rates, trace element or impurity analysis in the bulk is possible. Finally, a secondary ion image of the surface can be generated to provide a spatially resolved analysis of the surface, near surface, or bulk of the solid. This article will focus on the principles and applications of high sputter rate dynamic SIMS for depth profiling and bulk impurity analysis.

In general, secondary ion mass spectroscopy can provide characterization of solid samples with high spatial and in-depth resolution; due to the inherent sensitivity of mass spectroscopy, this is usually attained with high detection sensitivity as well. The extent to which these capabilities can be realized, especially in a quantitative fashion, depends on the nature of the specimen, the instrument design, and the particular instrumental parameters and methods used in the analysis. The interpretation of SIMS spectra and depth profiles can be difficult; thus, SIMS is not yet applicable for chemical analysis of unknowns in the true sense of the word. Rather, SIMS has unique capabilities for answering specific questions about specimens whose stoichiometry and matrix structure are already characterized.

P/M Applications. Secondary ion mass spectroscopy has been applied to P/M materials on a limited basis. Surface analysis of a material in powder form presents a unique problem. The surface roughness of a powder specimen generates problems with uniform sputtering for depth profiling. Roughness influences the area of sample surface that is excited and the quantity of the secondary ions that can reach the detector.

One application of secondary ion mass spectroscopy in powder metallurgy is the measurement of adsorbed or chemisorbed species on a powder surface. In this case, the technique can distinguish whether a molecular species is adsorbed intact or has dissociated at the surface. For example, secondary ion mass spectroscopy was used to determine whether carbon monoxide on the surface of nickel powder was present as molecular or dissociated carbon monoxide (Ref 2). The presence of the former leads to $NiCO^+$ and Ni_2CO^+ species, while the latter leads to NiC^+ and NiO^+.

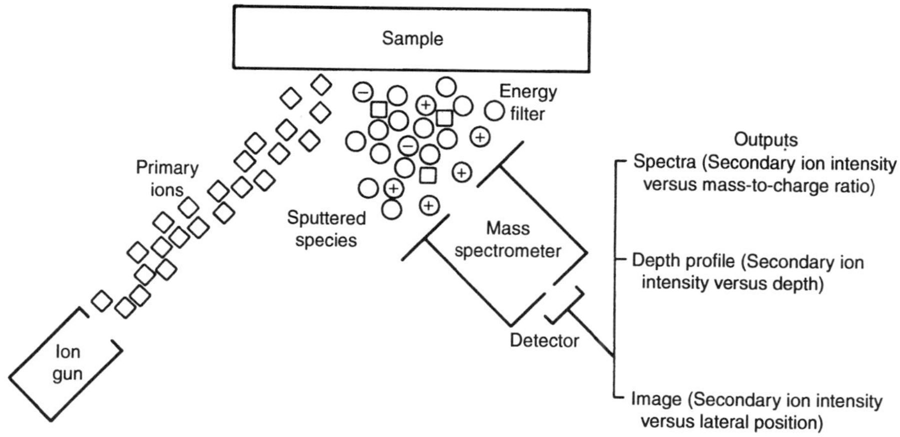

Fig. 5 Schematic representation of the principles of SIMS

Instrumentation. There is a wide variety of instrument designs for SIMS, and the quality and applicability of the analysis depends strongly on the details of the instrumentation. Secondary ion mass spectroscopy instruments can be categorized into three broad classifications in which the distinctions refer to versatility, mass resolution, primary beam characteristics, and, in particular, imaging capability.

The simplest SIMS instrument, sometimes called an add-on, macro-, or broad-beam instrument, is intended primarily for surface analysis and qualitative depth profiling and less so for quantitative elemental analysis, microanalysis, or imaging. This is seldom a dedicated unit, but rather a set of components often used in conjunction with an Auger or x-ray photoelectron spectrometer. Figure 5 shows a simple SIMS system. The instrument uses a standard electron-impact inert ion gun that is often the same sputter gun used for sample cleaning or Auger sputter-profiling. The mass spectrometer is usually a quadrupole type, but to select a portion of the wide energy distribution of sputtered secondary ions, it requires an energy prefilter. Other than the addition of the energy prefilter, the quadrupole mass spectrometer and the ion detector are very standard.

A dedicated SIMS instrument often incorporates a more intense, finely focused primary ion beam suitable for probe imaging. Because it is unnecessary to provide the flexibility for performing analyses by AES or XPS in this type of instrument, the sample orientation and the ion collection system can also be optimized.

These instruments, often called ion microprobes, are designed primarily for quantitative in-depth profiling and microanalysis and less so for surface analysis. A typical ion microprobe uses a duoplasmatron ion source to generate inert, as well as reactive, primary ion beams. Most commercially available units use an energy-filtered quadrupole mass spectrometer, but in principle, a magnetic analyzer could also be used.

A direct-imaging SIMS instrument, usually called an ion microscope, creates a secondary beam with the ejected secondary ions. The secondary beam is mass analyzed in a double-focusing electrostatic/magnetic sector; the transmitted secondary ions can then be counted or used to create a direct image of the sample surface on a microchannel plate. Direct-imaging instruments are designed primarily for microstructural analysis and imaging, quantitative in-depth profiling, and trace element analysis. They are not suitable for true surface analysis, and due to the precise ion collection optics, the ability to treat or manipulate the specimen in situ is essentially lost.

Bulk Analysis

Many analytical methods are available to determine the chemical composition of powders. Most of these procedures are geared toward the analytical chemist, and a brief description of methods used in powder analysis are summarized in Table 6 and described in this section.

If a pressed or sintered part is to be chemically analyzed, milling, drilling, or crushing can be used to obtain a representative sample. If the part contains oil, it should be removed by using a Soxhlet extractor with a solvent as detailed in ASTM B 328. Oil can also be removed by heating in a protective atmosphere at 705 to 815 °C (1300 to 1500 °F). However, this method cannot be used for materials such as sintered aluminum because they melt.

X-Ray Powder Diffraction

X-ray powder diffraction (XRPD) techniques are used to characterize samples in the form of loose powders or aggregates of finely divided material. These techniques cover various investigations, including qualitative and quantitative phase identification and analysis, determination of crystallinity, microidentification, lattice-parameter determinations, high-temperature studies, thin film characterization, and, in some cases, crystal structure analysis. The powder method, as it is referred to, is perhaps best known for its use as a phase characterization

Table 6 Comparison of bulk analysis method for powders

Method	Samples	General uses	Limitations
XRPD(a)	Powder samples (1 mg usually adequate) of crystalline solids	Identification of crystalline phases contained in unknown samples Quantitative determination of the weight fraction of crystalline phases in multiphase materials Characterization of solid-state phase transformations	Must be crystalline for phase identification which requires existence of standard patterns
Inert gas fusion	Solids, chips, or powders of usually 2 g or less, depending on material type and the expected amount of gases present	Quantitative determination of oxygen, nitrogen, and hydrogen in ferrous and nonferrous materials	Special precautions for metals with low boiling points Materials with stabile nitrides or oxides require addition of fluxes Method is destructive to the material
High-temperature combustion	Solids, chips, or powders of 1 g or less, depending on type of material	Determination of carbon and sulfur in metals and organics	Specimen must be homogenous Graphite-bearing specimens require special handling Method is destructive
ICP-AES(b)	Liquids, gases, and solids; liquids are most common; sample size of 5 to 50 mL for solutions; 10 to 500 mg for solids	Simultaneous multielement analysis Quantitative and qualitative analysis for over 70 elements with detection limits in the parts per billion (ng/mL) to parts per million (μg/mL) range Determination of major, minor, and trace elemental components	Detection limits parts per billion to parts per million Cannot analyze for noble gases Halogens and some nonmetals require vacuum spectrometer and optics Sensitivity poor for alkali elements, especially rubidium; cannot determine cesium
AAS(c)	Solids, solutions, and gaseous (mercury) size depends on technique used—from a milligram (solids by graphite furnace atomic absorption spectrometry) to 10 mL of solution for conventional flame work	Quantitative analysis of approximately 70 elements	Detection limits range from subparts per billion to parts per million Cannot analyze directly for noble gases, halogens, sulfur, carbon, or nitrogen Poorer sensitivity for refractory oxide or carbide-forming elements than plasma atomic emission spectrometry Basically a single-element technique

(a) XRPD, X-ray powder diffraction. (b) ICP-AES, inductively coupled plasma atomic emission spectroscopy. (c) AAS, atomic absorption spectroscopy. Source: *Materials Characterization*, Vol 10, *ASM Handbook*

tool partly because it can routinely differentiate between phases having the same chemical composition but different crystal structures (polymorphs). Although chemical analysis can indicate that the empirical formula for a given sample is $FeTiO_3$, it cannot determine whether the sample is a mixture of two phases (FeO and one of the three polymorphic forms of TiO_2) or whether the sample is the single-phase mineral $FeTiO_3$ or ilmenite. The ability of XRPD to perform such identifications more simply, conveniently, and routinely than any other analytical method explains its importance in many industrial applications as well as its wide availability and prevalence.

In general, an x-ray powder diffraction characterization of a substance consists of placing a powder sample in a collimated monochromatic beam of x-radiation. The diffraction pattern is recorded on film or using detector techniques, then analyzed to provide x-ray powder data that can be used to solve such problems as those listed in Table 6.

In XRPD analysis, samples usually exist as finely divided powder (usually less than 44 μm in size) or can be reduced to powder form. The particles in a sample comprise one or more independently diffracting regions that coherently diffract the x-ray beam. These small crystalline regions are termed crystallites. Consolidated samples, such as ceramic bodies or as-received metal samples. will likely have crystallites small enough to be useful for powder diffraction analysis, although they can appear to have considerably larger particle sizes. This occurs because a given grain or particle can consist of several crystallites (independently diffracting regions). Although larger grain sizes can sometimes be used to advantage in XRPD, the size limitation is important because most applications of powder diffraction rely on x-ray signals from a statistical sample of crystallites. The angular position, θ, of the diffracted x-ray beam depends on the spacings, *d*, between planes of atoms in a crystalline phase and on the x-ray wavelength λ:

$$n\lambda = 2d \sin \theta$$

The intensity of the diffracted beam depends on the arrangement of the atoms on these planes.

X-ray powder diffraction techniques usually require some sample preparation. This can involve crushing the sample to fit inside a glass capillary tube, rolling it into a very thin rod shape for the Debye-Scherrer camera technique, spreading it as a thin layer on a sample holder, or packing it into a sample holder of a certain size for other XRPD techniques. In some cases, samples compatible with metallographic examination can be accommodated in powder diffractometers, but some form of sample preparation will usually be necessary. Preparation will depend on the equipment available and the nature of the examination.

A diffraction pattern can be recorded using film, analog, or digital methods. Whether film, analog, or digital data collection is used, the final data can be displayed as a graph of intensity, as a function of interplanar distance *d*, or as a function of diffraction angle 2θ. Many modem automated powder diffractometers can provide further data reduction, including peak finding. a tabular listing of peak intensity versus interplanar spacing, search/match software, and other computer utilities.

A powder pattern from a single-phase material will have maxima at positions dictated by the size and shape of its unit cell and will increase in complexity as the symmetry of the material decreases. For example, many metal patterns and those from simple compounds that tend to be mostly of cubic symmetry and have small unit cell edges will produce powder patterns having fewer lines or maxima than would be expected from a compound of lower symmetry or one having a very large unit cell. A pattern of a mixture of phases in which all the individual patterns are superimposed will produce a complex experimental pattern, especially when the number of phases present in the mixture exceeds approximately three or when the phases constituting the mixture are all of very low symmetry or have very large unit cell dimensions.

Phase identification using XRPD is based on the unique pattern produced by every crystalline phase. Much as a fingerprint is unique for each person, the diffraction pattern can act as an empirical fingerprint for that phase, and qualitative identification of phases can be accomplished by pattern-recognition methods that include established manual techniques and the newer methods that use computers. most of which implement programs based on heuristic algorithms. All of these methods make use of the database maintained by the JCPDS International Centre for Diffraction Data (Ref 3).

Inert gas fusion is used to determine the quantitative content of gases in ferrous and nonferrous materials. These gases, such as hydrogen, nitrogen, and oxygen, are introduced to a material by physical and chemical adsorption. Surface metals deposited on base materials physically adsorb atmospheric gases to form metal-bearing compounds. The type of metal atom available and its compatibility with the types and concentrations of atmospheric gases available under the temperatures and pressures to which the material is exposed determines the types of metallic compounds formed. Although the bond at low temperatures is unstable, the metallic compounds formed become more stable as temperature increases and the physical adsorption bond becomes a chemical adsorption bond.

As base materials are smelted, the adsorbed gases are chemically absorbed into the melt. The less stable compounds are reduced in the melt of the material, allowing each to recombine and form a more complex stable compound. The metallic elements will absorb atmospheric gases based on the affinity of the metal for each gas at various temperatures during the melt.

Processed materials are again subject to gas adsorption as the material is initially cooled and worked. Stages of drawing, rolling, heat treating, or annealing provide favorable conditions for physical and chemical adsorption of atmospheric gases. Strict control of hydrogen, nitrogen, and oxygen levels minimizes their adverse effects on material strength.

Hydrogen causes internal cracks that generally appear during cooling in such processes as drawing, rolling, or forging of the material. Materials with large cross sections can break under high or continuous stress due to internal cracks. Hydrogen is adsorbed and diffused often during working of the material. Hydrogen that is adsorbed will generally diffuse during cooling or aging. Materials used under high temperatures and pressures and exposed to high-hydrogen environments can develop structural problems due to hydrogen embrittlement.

Nitrogen in some materials can provide strength. For example, nitrogen is added to austenitic manganese steels to increase yield strength. A material can be subjected to nitriding to increase hardenability. Nitrogen may decrease ductility.

The effects of oxygen on a material are similar to those of hydrogen in that inclusions and blowholes can appear in the material. Oxygen, when combined with the carbon and nitrogen in a material, will also cause increased hardness with age. Oxygen is often the hardest to control because it is available from various sources and reactive with many metals. Smelting and processing regulate the levels of gases introduced into materials by controlling temperature, pressure, and environment, which create adsorption and absorption of gases.

Gases introduced into the material are commonly quantitatively determined using inert gas fusion. Inert gas fusion reverses the physical and chemical bonding between the gases and the metals to dissociate the gases and sweep them from the fusion area with an inert carrier gas. Resistance or induction heating of a sample in a pure graphite crucible dissociates the gas/metal bonds. Because they are formed over a wide range of temperatures, bonds can be broken only by heating the specimen above the highest temperature at which the gas/metal bonding occurred. More information on these operations can be found in "Inert Gas Fusion" in *Materials Characterization,* Volume 10, *ASM Handbook.*

Inductively Coupled Plasma Atomic Emission Spectroscopy

Inductively coupled plasma atomic emission spectroscopy (ICP-AES) is an analytical technique for elemental determinations in the concentration range of major to trace, based on the principles of atomic spectroscopy. In theory, the technique applies to all elements except argon, and samples can be introduced as liquids, gases, or solids. In practice, favorable analytical results are obtained for approximately 70 elements, with detection limits usually attainable at the parts per billion level, and most samples are introduced in liquid form as aqueous solutions. The technique has found widespread application in the metallurgical, geological, environmental, agricultural, industrial, and biological fields and is an important technique in the modern analytical laboratory.

The ICP is an excitation source for atomic emission spectroscopy. It is an argon plasma operated at atmospheric pressure and sustained by inductive coupling to a radio frequency (RF) electromagnetic field. Argon gas flows axially through a quartz tube surrounded by three or four turns of an induction or work coil connected to an RF generator. The standard frequencies of operation are 27.12 MHz or, less commonly, 40.68 MHz, the frequencies allowed by the Federal Communications Commission for scientific and medical instrumentation. Power output is generally 1 to 5 kW.

The high-frequency current of up to 100 A flows in the water-cooled copper induction coils, generating oscillating magnetic fields whose lines of force are oriented axially inside the quartz tube and follow closed elliptical paths outside the tubes. If free electrons are present inside the tube, the induced magnetic fields cause the electrons in the gas to flow in oscillating closed annular paths inside the quartz tube space. This electron flow is termed the eddy current, and the electrons are accelerated by the time-varying magnetic field, causing collisions that result in further ionization of the argon gas and resistive heating. These electrical and magnetic fields responsible for the plasma are represented in Fig. 6.

The energy transmission in the plasma is similar to an electrical transformer in which the induction coils are the primary winding and the ionized gas is the secondary. Because the argon gas is initially neutral and nonconducting, the plasma must be initiated by seed electrons. usually generated by a brief tesla discharge. With RF power applied, the plasma ignites instantaneously, then is self-sustaining. The resulting plasma is a highly ionized gas with temperatures in the proximity of 10,000 K.

Analytical Characteristics. The ICP as an analytical technique provides the capability of performing simultaneous multielement analysis for as many as 60 elements within 1 to 2 min. It is applicable to most elements in the periodic table and has a large linear dynamic range (calibration curves that are linear over three to six orders of concentration magnitude). This enables determination of trace, minor, and major components in a single analysis.

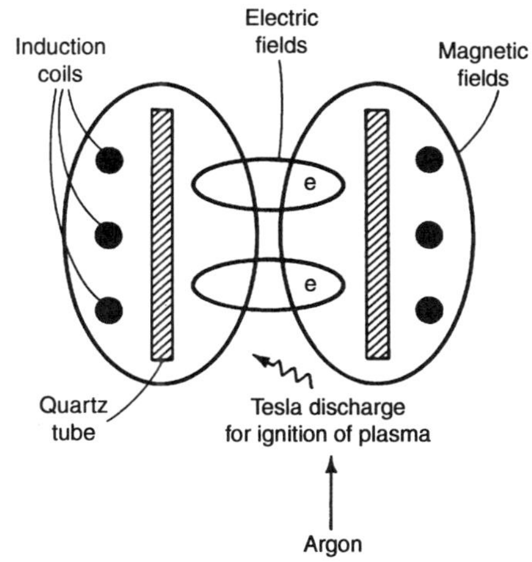

Fig. 6 Electric and magnetic fields of the inductively coupled plasma

Detection limits are in the parts per billion (ng/mL) range for most elements; precision and accuracy on the order of 1%; and relative freedom from chemical interferences. Detection limits for the ICP are determined by first establishing a calibration curve (plot of signal intensity at a given wavelength versus concentration for a series of standard solutions). The detection limit is calculated as the concentration that would correspond to an analytical signal equal to two (or three, according to choice of definition) times the standard deviation (noise) of repeated measurements of a blank at that wavelength. This concentration is the lowest value measurable with any certainty. These detection limit values should be considered extreme limits, because they are determined under ideal situations. Practical detection limits will be somewhat higher. Some sample preparation procedures, such as dissolution and dilution, necessarily degrade the achievable detection limits of the elements in the original sample material.

Precision and Accuracy. The precision of the ICP technique is usually determined by making several consecutive measurements, then calculating the standard deviation of the replicates as a percentage of the mean value. Major causes of signal fluctuations in the ICP are small variations in the RF power applied to the plasma and changes in the nebulization process. Precision can be increased to less than 1% by close regulation of the RF power (most new instrumentation accomplishes this to ±0.1%), by improved nebulization techniques, or by use of an internal standard. Nebulization is stabilized by using a mass flow controller to regulate the nebulizer gas flow and by use of improved nebulizer designs, such as the high-pressure crossflow nebulizer. The accuracy of the ICP technique is essentially limited by the precision and by systematic errors, such as interference effects, but is usually shown to be comparable to the precision.

Atomic Absorption Spectrometry

Atomic absorption spectrometry (AAS) is generally used for measuring relatively low concentrations of approximately 70 metallic or semimetallic elements in solution samples. The basic experimental equipment used is essentially the same as that of 30 years ago—enhanced by modern electronics, background-correction schemes, and alternate types of atomizers. The predominance of AAS in general-purpose trace-metal analysis has recently been somewhat eclipsed by modern atomic emission spectrochemical methods designed to permit solution analysis. However, its ruggedness and relatively low equipment costs keep AAS competitive. Atomic absorption spectrometry performed using the graphite-tube furnace atomizer usually remains the method of choice for ultra-trace-level analysis.

Figures 7 and 8 show the relationship between the flame atomizer versions of AAS and the related techniques of atomic fluorescence spectrometry (AFS) and atomic emission spectrometry (AES). Several features are common to all three techniques. The first is a sample-introduction/atomization system consisting of a sample sprayer (referred to as the nebulizer) and the flame. The flame desolvates, vaporizes, then atomizes (dissociates to free atoms) the fine sample droplets produced by the nebulizer. Next is the monochromator, which isolates a wavelength of light characteristic of a particular quantized transition between electronic energy levels of the outer electrons of the selected analyte element. The third component is the light intensity-to-electrical signal transducer, usually a photomultiplier tube (PMT). Finally, an electronic data-reduction system converts this electrical signal to an analytical response proportional to the concentration of analyte in the sample solution.

It should be noted that within typical atmospheric pressure atomizers, the intrinsic width of the absorption/emission lines of the elements are typically from 0.002 to 0.008 nm. A comparison of these linewidths with the approximately 600 nm-wide working range of the spectrometers normally used in atomic spectroscopy indicates that approximately 10^5 resolution elements are potentially available to determine the approximately 100 elements of the periodic table. The relatively high ratio of the number of possible resolution elements to the number of chemical elements (approximately 1000:1) explains why atomic spectroscopic methods tend to be more specific than most other analysis techniques. The ease of applying this intrinsic specificity in actual practice differs substantially among the three types of atomic spectroscopy.

In atomic emission spectrometry (AES) (Fig. 7a and 8a), the flame serves an additional function not required in AAS or AFS: excitation. To produce the desired signal, hot flame gases must thermally (collisionally) excite a significant fraction of the free atoms produced by dissociation in the atomizer from the relatively populous ground-state level to one or more electronically excited states. The excited atoms emit light at discrete wavelengths corresponding to these differences in energy levels when they spontaneously relax back to the lower states. That is, the instrument "sees" the excited-state population of analyte atoms, not the ground-state population.

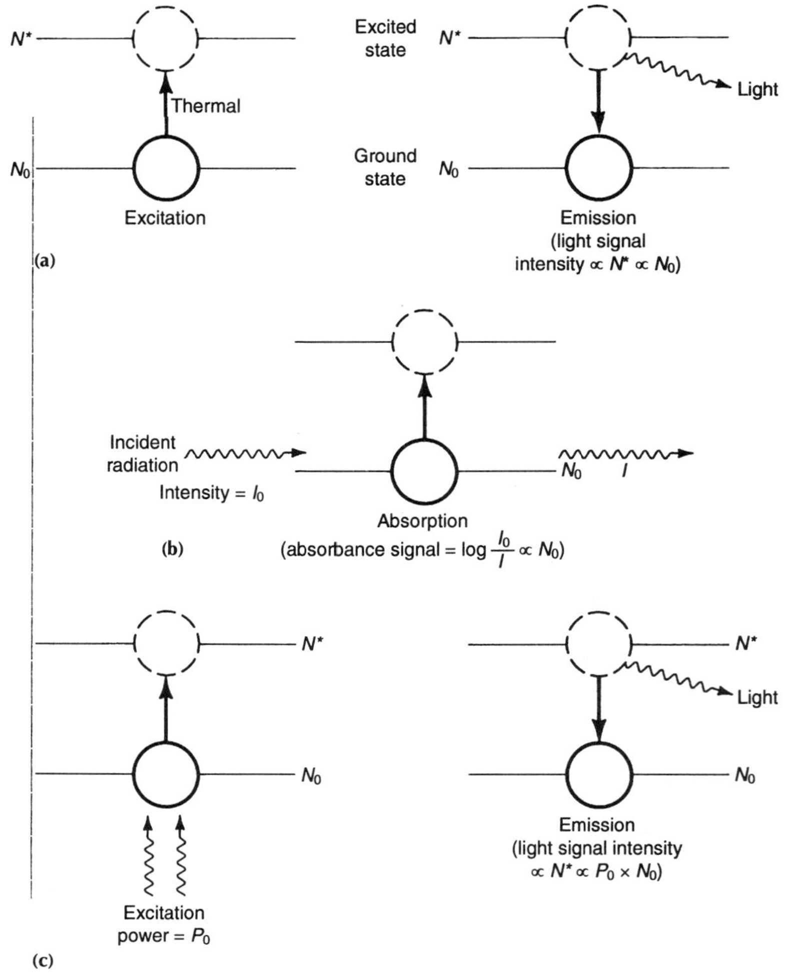

Fig. 7 Energy-level transitions of the atomic spectrometries. N^*, number of atoms in the excited state; N_0, number of atoms in the ground state; I_0, light intensity measured without the analyte present; I, light intensity measured with the analyte present. (a) Atomic emission spectrometry. (b) Atomic absorption spectrometry. (c) Atomic fluorescence spectrometry

The ratio of the population of atoms in a thermally excited state to that in a lower energy state follows the Boltzmann distribution; that is, the logarithm of the ratio is directly proportional to the absolute temperature and inversely proportional to the difference in energy between the states. Therefore, the absolute magnitude of emission signals is temperature dependent. At typical atomizer temperatures, only a small fraction of atoms are excited to levels capable of emitting visible or ultraviolet radiation; most remain at or very near to the ground-state energy level.

Because flames do not specifically excite only the element of interest, a monochromator able to resolve close lines while maintaining a reasonable light intensity through-put should be used to reduce the probability of spectral interferences. Additional broad band-like light emitted by similarly thermally excited molecular species, for example, OH or CH flame radicals or matrix-metal oxides, complicates isolation of the desired signal from the background.

Atomic emission spectrometry usually necessitates scanning the monochromator completely over the analytical spectral line to obtain the background signal values necessary for the calculation of a correct analytical response. Reasonably inexpensive high-resolution monochromators capable of automatic correction of background emission are available.

In atomic absorption spectrometry (AAS) (Fig. 7b and 8b), radiation from a lamp emitting a discrete wavelength of light having an energy corresponding to the difference in energies between the ground state and an excited state of the analyte element is passed through the atomizer. This light is generated by a low-pressure electrical discharge lamp containing a volatile form of the analyte element. Free analyte atoms within the atomizer absorb source-lamp light at wave-

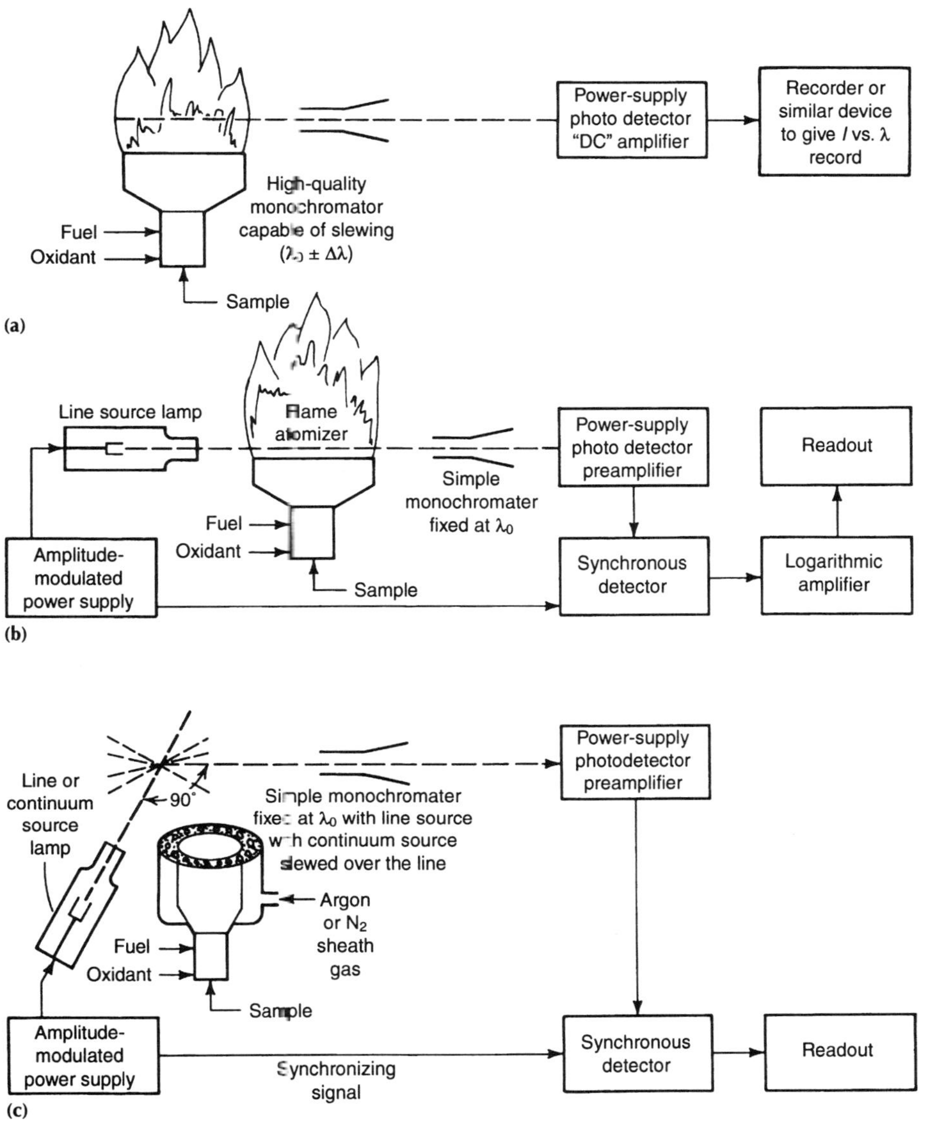

Fig. 8 Comparison of (a) flame atomic emission spectrometry, (b) flame atomic absorption spectrometry, and (c) flame atomic fluorescence spectrometry

lengths within their absorption profiles. The absorption lines have a bandwidth approximately twice as wide as the emission profiles of the same elements in the low-pressure source lamp. In contrast to AES, ground-state (not excited state) atomic populations are observed.

The source light not absorbed in the atomizer passes through the monochromator to the light detector, and the data reduction/display system of the spectrometer outputs an absorbance response directly proportional to the concentration of analyte in the sample solution. Absorbance is the logarithm (base 10) of the ratio of the light intensities measured without (I_0) and with (I) the analyte atoms present in the light path (absorbance = log I_0/I). In practice, the intensity of the source lamp is amplitude modulated at a specific frequency to permit subsequent electronic isolation of the ac light signal of the lamp from the dc light caused by the emission from species thermally excited by the atomizer. Direct current light is invariant relative to time.

Only the relatively highly populated ground-state population of the same element in the atomizer that is in the source lamp can contribute to the signal. Therefore, the analytical response of atomic absorption spectrometers is element-selective and not as sensitive to atomizer temperature variations as that of atomic emission spectrometers. In addition, the electronic lamp modulation/signal demodulation system renders the spectrometer blind to extraneous light sources. The monochromator serves only to isolate the desired analytical line from other light emitted by the one-element source lamp. Consequently, a less sophisticated monochromator suffices in AAS than is usually required for general-purpose AES. The major error signal encountered in AAS is the nonselective absorption or scattering of source-lamp radiation by undissociated molecular or particulate species within the atomizer. Several different types of background correction systems will be discussed later in this article.

Atomic fluorescence spectrometry (Fig. 7c and 8c), an emission technique, relies on an external light beam to excite analyte atoms radiatively. The absorption of light from the light source creates a higher population of excited-state atoms in the atomizer than that predicted by the Boltzmann equation at that temperature. Consequently, the absolute sizes of the atomic emission signals detected are larger than those seen in AES experiments performed with the same concentration of analyte atoms within the atomizer. A source-lamp modulation/signal demodulation scheme similar to that applied in atomic absorption spectrometers isolates the atomic fluorescence response from that emitted by thermally excited analyte, flame, or matrix species.

Atomic fluorescence spectrometry spectra tend to be much simpler than AES spectra. This is true even when bright continuum light sources, such as xenon-arc lamps, are used instead of line sources for excitation, because only those atomic emission lines originating from energy levels whose populations are enhanced by the initial atomic absorption step can contribute to an AFS response. When line-source excitation lamps are used, this initial excitation step is very selective, and the AFS spectrum becomes extremely simple. Monochromators are often not used; the only concession to spectral isolation is the use of a photomultiplier insensitive to room light.

Atomic fluorescence spectrometry has two major sources of error. The first is chemical scavenging or de-excitation (termed quenching) of the nonequilibrium excited-state analyte atom population (that in excess of the thermally excited population) before a useful light signal can be emitted. The magnitude of this error signal depends on the concentration of the quenchers in the gas phase, which depends on the chemical makeup of the sample matrix accompanying the analyte element. Consequently. quenching introduces a potential source for matrix effects in AFS not found in AAS.

The second source of error is scatter of the exciting radiation by particulate matter within the atomizer. Some refractory metals, such as zirconium and uranium, if present in high concentrations in the sample, are apt to be incompletely dissociated or even gassified in conventional atomizers. This scatter signal is sometimes compounded with molecular fluorescence emission signals from naturally present gaseous flame species, a condition especially troublesome when continuum-type excitation sources are used.

Advantages of Atomic Absorption Spectrometry. The AAS method has been one of the most widely used trace element analysis techniques, largely because of the degree of specificity provided by the use of an analyte-line light source. This reduces the probability of "false positives" caused by matrix concomitants and serves to enhance greatly the reliability of AAS

Key: Element — top value: Absorbance of 1 µg/mL solution with flame atomizer (a); bottom value: Absorbance of a 25 µL aliquot of a 1 µg/mL solution with furnace atomizer (b)

0.26 Li 37	0.26 Be 1100											0.0005 B –	 C –	 N –	 O –
1.5 Na 275	1.5 Mg 157											0.006 Al 275	0.003 Si 3.7	0.00003 P 	 S –
0.44 K 19	0.21 Ca 2.8	0.013 Sc –	0.003 Ti 3.7	0.005 V 7.3	0.08 Cr 220	0.18 Mn 110	0.071 Fe 366	0.067 Co 37	0.066 Ni 44	0.11 Cu 137	0.49 Zn 550	0.004 Ga 210	0.003 Ge 277	0.006 As 18	0.009 Se 11
0.11 Rb –	0.10 Sr 110	0.0009 Y –	0.0005 Zr –	0.0002 Nb –	0.013 Mo 18	 Tc –	0.006 Ru –	0.029 Rh 11	0.048 Pd 5.5	0.122 Ag 220	0.40 Cd 366	0.012 In 10	0.009 Sn 18	0.015 Sb 14	0.017 Te 16
0.04 Cs –	0.022 Ba 27	(see La row below)	0.0004 Hf –	0.0004 Ta –	0.0008 W –	0.0004 Re –	0.004 Os 0.41	0.006 Ir 0.65	0.004 Pt 2.4	0.034 Au 110	0.020 Hg 11	0.016 Tl 23	0.04 Pb 61	0.02 Bi 22	 Po –

0.0009 La 1.9	– Ce –	0.0002 Pr –	0.0005 Nd –	– Pm –	0.0006 Sn –	0.011 Eu 12	0.002 Gd –	0.0004 Tb –	0.007 Dy 2.1	0.005 Ho 1.2	0.007 Er 2.4	0.015 Tm –	0.057 Yb 85	0.0006 Lu –
– Ac –	– Th –	– Pa –	0.00004 U –											

Fig. 9 Typical analytical sensitivities obtained using flame or graphite furnace atomic absorption spectrometry. (a) Results obtained by Varian Techtron Ltd., Melbourne, Australia. (b) Results obtained by Allied Analytical Systems, Waltham, MA

determinations performed on "unknown" samples. A background-corrected atomic absorption instrument is also one of the most reliable, albeit slow, tools available for qualitative analysis. The need for simple monochromators has maintained the cost of AAS equipment well below that of AES instrumentation having similar capabilities.

Line-excited AFS has the specificity of AAS as well as other desirable characteristics, such as a greater dynamic range and potentially better detection limits. However. obtaining appreciable improvement over AAS detection capabilities with AFS usually requires more attention to optimizing the optics, atomizer, and electronics of the system. In addition, because correction for "scatter" signals is fundamentally more important and more difficult to accomplish than is background absorption correction in AAS, no flame- or furnace atomizer-based AFS unit is commercially available.

Atomic emission spectrometry has found limited acceptance in the instrument marketplace. Although most of the better atomic absorption instruments sold during the 1960s and 1970s could be used for flame-excited AES, the instrument requirements for the two techniques are so different that the results achieved using these spectrometers did not reflect the true potential of the method. Only the recent introduction of electrical plasma emission sources, such as inductively coupled plasma (ICP) or direct current plasma (DCP), designed for the routine analyses of solution samples has prompted commercial production of fairly inexpensive and compact spectrometers optimized for AES. However, these instruments remain considerably more expensive than basic atomic absorption spectrometers.

Atomic Absorption Spectrometry Sensitivities. The periodic table shown in Fig. 9 lists typical analytical sensitivities obtained using representative atomic absorption spectrometers with either a flame or the more sensitive graphite furnace atomizer. The entries in Fig. 9 represent the magnitude of the atomic absorbance signal expected when a 1 ppm solution of the element is continuously aspirated into a flame atomizer or introduced as a discrete 25 µL aliquot into a graphite furnace.

In practice, the performance of reliable analysis requires signal magnitudes ranging from 0.01 to 1.0 absorbance unit. This is a consequence of the signal-to-noise considerations involved in measuring small differences in two relatively large light signals. This rather limited dynamic analytical range often necessitates the concentration or dilution of sample solutions before analysis.

The reasons for the extreme differences in AAS sensitivities noted in Fig. 9 can be divided into three basic categories, the first two of which affect AAS, AFS, and AES nearly equally. First, because the number of atoms within the light path at a given time fundamentally determines the instantaneous signal, the mass-based sensitivities in Fig. 9 are biased in favor of the lighter elements.

Second, a substantial number of elements do not possess ground-state lines in a region of the spectrum that is accessible with normal spectrometers and to which the gases present within normal atomizers are transparent. Less sensitive alternative analytical lines may sometimes be used, for example, with mercury and phosphorus; for other elements (most of the fixed gases), no good lines are available. Further, many elements possess a multitude of atomic energy states near the absolute ground-state level. These low-lying levels are thermally populated to some degree at the working temperature of the atomizer, which tends to reduce the fraction of analyte atoms available at any one energy level to absorb a specific wavelength of light emitted by the source lamp. This reduces the sensitivity achievable by any atomic spectroscopic technique for many of the transition, lanthanide, and actinide elements.

Lastly, none of the atomizers commonly used in atomic absorption spectrometers provides conditions capable of substantially dissociating some of the more commonly encountered forms of some chemically reactive analyte elements. For example, boron forms stable nitrides, oxides, and carbides. No practical adjustment of the operating conditions of any of the conventional atomic absorption atomizers can provide an environment that is simultaneously sufficiently free of nitrogen, oxygen, and carbon to give a favorable degree of boron dissociation. However, the combination of the much higher temperatures and inert gas environments found in electrical plasma AES sources makes boron one of the most sensitive elements determined by modern AES instruments.

Atomizers. The sensitivity of AAS determinations is determined almost wholly by the characteristics of the light source and the atomizer, not by the optics or electronics of the spectrometer. Simple, inexpensive AAS instruments have the same sensitivity as more sophisticated models. Because the line-source lamps used in AAS have essentially the same line widths, most of the sensitivity differences noted between instruments can be attributed to differences in the atomizers.

ACKNOWLEDGMENT

Portions of this article were adapted from the following articles in *Materials Characterization,* Volume 10, *ASM Handbook,* 1986:

- L.M. Faires, Inductively Coupled Plasma Emission Spectroscopy, p 31–42
- D.D. Siemer, Atomic Absorption Spectrometry, p 43–59
- R.B. Fricioni and L. Essig, High-Temperature Combustion, p 221–225
- R.B. Fricioni and L. Essig, Inert Gas Fusion, p 226–232
- R.P. Goehner, X-Ray Powder Diffraction, p 333–343
- J.D. Verhoeven, Scanning Electron Microscopy, p 490–515
- K.F.J. Heinrich and D.E. Newbury, Electron Probe X-Ray Microanalysis, p 516–535
- A. Joshi, Auger Electron Spectroscopy, p 549–567
- J.B. Lumsden, X-Ray Photoelectron Spectroscopy, p 568–580
- C.G. Pantano, Secondary Ion Mass Spectroscopy, p 610–627

REFERENCES

1. G.F. Bastin and H.G.M. Heigliers, *Microbeam Analysis—1985,* San Francisco Press, 1985, p 1
2. H. Hopster and C. Brundle, *J. Vacuum Sci. Technol.,* Vol 16, 1979, p 548–551
3. Joint Committee on Powder Diffraction Methods, International Centre for Diffraction Data, Swathmore, PA

Particle Size and Size Distribution

Ronald G. Iacocca, The Pennsylvania State University

PARTICLE SIZE AND SIZE DISTRIBUTION have a significant effect on the behavior of metal powders during processing; thus to a considerable extent, they govern the properties of the final products made from powder. Consequently, characterization of such properties is essential.

In the P/M industry, the traditional and most widely used method of particle size measurement is sieving. Sieves or screens are used not only for particle size measurement, but also for separation of powders into different sieve fractions. This twofold use of sieves, in addition to the fact that most P/M powders are −80 mesh (smaller than about 180 μm in diameter, with only minor amounts smaller than 10 μm) has been well suited to industrial applications. For powders with large percentages of −400 mesh (38 μm) particles, sieve distribution data are often complemented with other particle size measuring instruments.

Several electronic methods of particle size analysis have been developed. These generally provide higher measurement speed, resolution, and small-size sensitivity than sieves, saving time and labor while yielding more precise data. However, there has also been some confusion; particle size data for the same powder can vary greatly depending on the instrument and method used.

There are several reasons for such discrepancies. One is the phenomenon of particle shape. A definition of particle size is simple for spherical particles in which particle diameter generally is used to characterize size. Most metal powders, however, are irregularly shaped. Determination of particle size and size distribution of such powders usually is based on the measurement of certain properties, or a combination of powder particle properties—length, volume, mass, settling rate, total or projected surface area, light scattering, and electric field perturbation, for example.

From such measurements, so-called equivalent spherical particle diameters can be calculated. Equivalent spherical diameters are the diameters of spheres that have the same, or equivalent, lengths, volumes, masses, or settling rates as the irregular particles themselves.

Various methods of particle size analysis provide identical or similar analysis results for spherically shaped particles only. Typically, the various principles used to measure particle size respond differently to changes in particle shape. Very different results can be achieved for the same particle size, if particle shape deviates from spherical.

Another factor is the method of measurement. Recent advances in instrumentation and computer technology, combined with the industrial need for more accurate and reproducible data

Table 1 Typical methods of particle size and size distribution measurement

Measuring principle	Method	Approximate useful size range, μm
Mechanical or ultrasonic agitation	Sieving	5 ~ 125 mm
Microscopy	Optical	0.5 ~ 100
	Electron	0.001 ~ 50
Electrical resistivity	Coulter counter	0.5 ~ 800
	Electrozone	0.1 ~ 2000
Sedimentation	Sedigraph	0.1 ~ 100
	Roller Air Analyzer	5 ~ 40
	Micromerograph	2 ~ 300
Light scattering	Microtrac	2 ~ 100
Light obscuration	HIAC	1 ~ 9000
Permeability	Fisher subsieve sizer	0.2 ~ 50
Surface area	Gas adsorption (BET)	0.01 ~ 20

BET, Brunauer-Emmett-Teller

Table 2 Methods of particle size analysis and nominal size ranges

Method	Nominal particle size μm	Nominal particle size mil	Measurement diameter
Sieving			
Dry	>10	>0.4	Geometrical
Wet	>2	>0.08	
Field scanning			
Optical microscopy	0.5–1000	0.02 to 40	Image
Electron microscopy	0.01–10	4×10^{-4} to 0.4	
Stream scanning			
Resistivity	0.05–500	2×10^{-3} to 20	Dynamic/Stokes
Optical	1–500	0.04 to 20	
Ultrasonic attenuation	100–1000	4 to 40	
Column hydrodynamic chromatography	0.1–1.0	4×10^{-3} to 0.04	
Sedimentation field-flow fractionation	0.01–1.0	4×10^{-4} to 0.04	
Laser Doppler velocimetry	0.01–3.0	4×10^{-4} to 0.12	
Gravity sedimentation			
Pipette	1–100	0.04 to 4	Stokes
Photoextinction	0.5–100	0.02 to 4	
X-ray absorption	0.1–130	4×10^{-3} to 52	
Centrifugal sedimentation			
Photoextinction	0.05–100	2×10^{-3} to 4	Dynamic/Stokes
Mass accumulation	0.05–25	2×10^{-3} to 1	
X-ray absorption	0.1–5	4×10^{-3} to 0.2	
Other			
Gas absorption	0.005–50	2×10^{-4} to 2	Equivalent spherical
Mercury intrusion	0.01–200	4×10^{-4} to 8	
Gas permeability	0.1–40	4×10^{-3} to 16	
Cascade impaction	0.05–30	2×10^{-3} to 12	
Brownian motion	0.01–3	4×10^{-4} to 0.12	Geometrical

Source: Ref 5

have dramatically increased the availability of commercial particle size analysis equipment over the past several years. Many different techniques, combined with dozens of commercial suppliers, have resulted in hundreds of different particle size measuring instruments and methods (Ref 1–4).

Common particle size measuring techniques include:

- Sieve analysis
- Quantitative microscopy
- Fraunhofer diffraction
- Neutron/x-ray scattering
- Time-of-flight
- Sedimentation
- Electrical zone sensing
- Mie diffraction
- Centrifugal sedimentation
- Packed bed permeability (Fisher Subsieve Analyzer)
- Photon correlation spectroscopy
- Elutriation
- Field-flow fractionation
- Size exclusion chromatography

This list is not exhaustive; however, it does represent the breadth of available techniques. Several of the methods listed, such as field-flow fraction, are specific to only a few industries and would not be encountered in companies producing components from technical metal and ceramic powders. Typical methods and effective size ranges are listed in Table 1.

With these different methods, it is important to understand that measurement methods are based on different analytical models (see, for example, the grouping of "measurement diameters" in Table 2 for a wide variety of methods. In reality no method measures the "true" particle size, because it is impossible to describe the size of a three-dimensional particle with a single number, therefore, all analytical methods measure an index of particle size such as geometric diameter, aerodynamic diameter, equivalent circle diameter, and so on. Because of this, even if the mathematical algorithm is accurate and the hardware is properly developed, one can obtain different "correct" particle size data on two different instruments.

With the large number of available size measuring methods and instruments, the challenge is to select the proper instrument for a required measurement. One method does not provide more accurate results than another method but one method may indeed be more appropriate or discriminatory in a given application.

Sample Preparation

A key step in the measurement of particle size distributions is the preparation of powder samples for particle size measurement. Powder dispersion is more important than the measurement methods, and instrument limitations or inaccuracies can be secondary to sample preparation.

Various sample methods include batch, continuous, or in situ measurements as outlined in Fig. 1. The procedures for the collection of representative samples and the preparation of dispersed powders or slurries are critical steps when measuring particle size distributions and estimating errors (as briefly discussed below). More detailed information of sampling and dispersion is discussed in the article "Sample and Classification of Powders" in this Volume.

Sampling Errors. Sampling is distinguished as either a probabilistic or nonprobabilistic process where:

- *Probabilistic sampling* is when all elements of the lot are submitted, along with a probability of being selected.
- *Nonprobabilistic* sampling is deterministic rather than being based on probability, for example, as in grab sampling.

Probabilistic sampling is the preferred method.

In terms of errors, both sampling and analysis are error-generating steps, with the consequence that the overall estimation error (OE) is the sum of the total sampling error (SE) and analytical (or measurement) error (AE). Sampling and analysis involve several preparation stages, alternated with selection stages, all of which can potentially generate errors. Therefore, total overall estimation error is represented by:

$$OE = \sum_n (PE_n + SE_n) = AE_n$$

where PE_n and SE_n are preparation and selection errors, respectively, of each stage, n (n = 1, 2, ...). Various errors described by Allen and Davies (Ref 6) are shown in Fig. 2. Relative particle sizes and errors depend on specific powders. For submicrometer-sized ceramic powders, dispersion-related errors appear to dominate.

Dispersion. Powder dispersion includes deagglomeration and formation of a stable suspension. Of the several methods used for deagglomeration (ultrasonic probe/bath, stirring, tumbling),

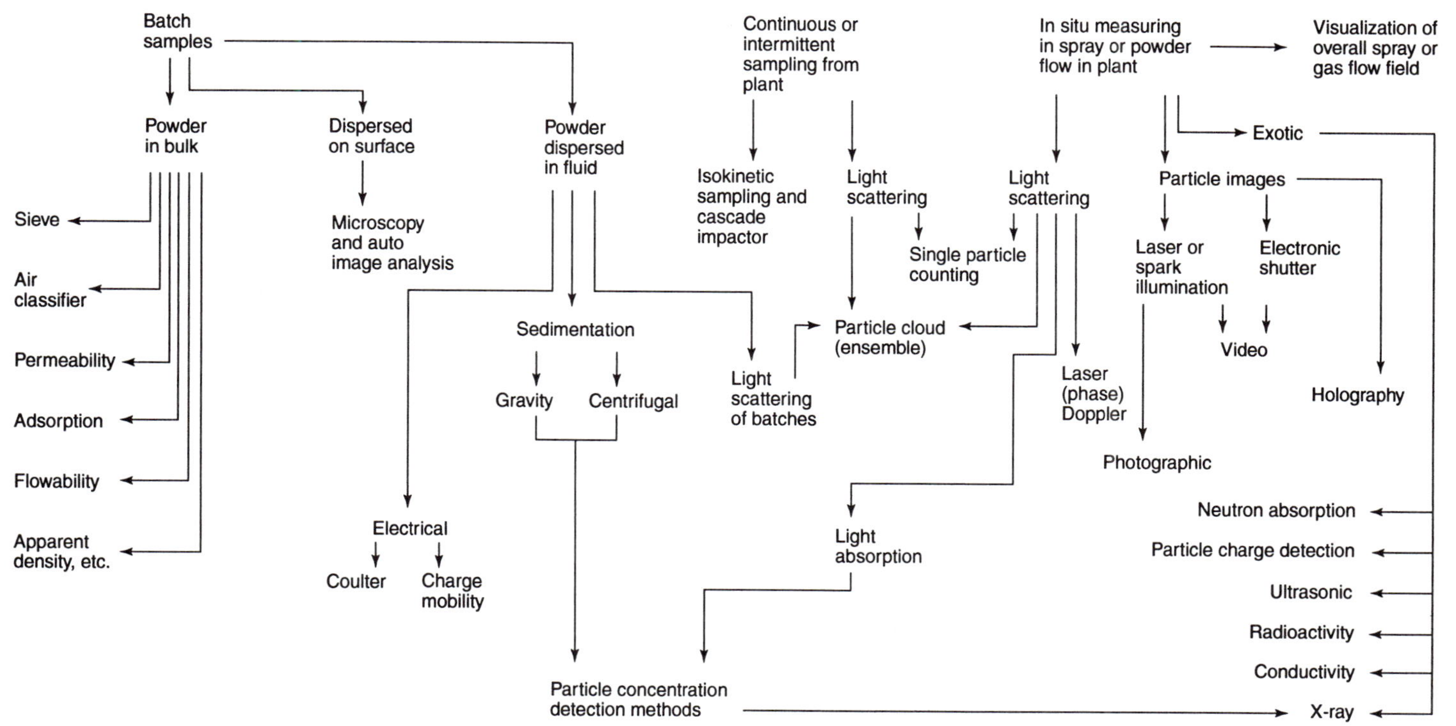

Fig. 1 General methods of powder sampling and particle size measurements

ultrasonication with a probe is the most effective in achieving the separation of particles held together by weak forces. An optimal level of power input to the suspension is required to achieve enhanced data reproducibility. The probe diameter, volume of suspension, power input, time of ultrasonication, and rate of power input are some of the parameters that affect deagglomeration.

The surface chemistry of the solvent-powder interface also controls the preparation of a stable suspension. A repulsive force of sufficient strength to overcome attractive forces between the particles in order to disperse them is required. The major attractive force that hinders dispersion is the van der Waals type. Electrical double-layer forces are the counterforces that can effectively provide repulsive forces. The optimal range of electrical double-layer forces in polar liquids is provided by high surface charge, moderate electrolyte concentration, and an adsorbed layer of polyelectrolyte of an appropriate type and concentration. A combination of electrical double-layer repulsion and steric stabilization has been found to be highly effective for dispersing powders in polar solvents. In nonpolar solvents, steric stabilization alone is often sufficiently effective (Ref 7) and represents the only choice available.

Particle Size Measurement

Although hundreds of methods have been developed or reported for particle size measurement (Ref 1–4), there are essentially two basic methods of defining particle size. The first method is to inspect the particles and make actual measurements of their dimensions. Microscopic techniques, for example, measure many dimensional parameters from particle images. The second method utilizes the relationship between particle behavior and its size. This often implies an assumption of an equivalent spherical size, developed using suitable relationships. For example, sedimentation is used to measure size according to Stokes's law and light scattering is used to measure the size according to the Fraunhofer-Mie theories.

With the large number of available size measuring techniques, the challenge for practitioners working in this area is to select the proper instrument for the job. Table 3 and the brief summaries below introduce the most widely used characterization methods and the associated advantages and disadvantages. More detailed descriptions are in the articles that follow.

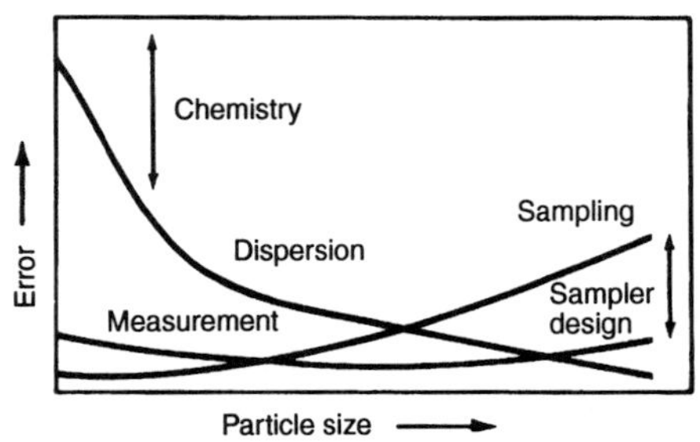

Fig. 2 Dispersion error related to particle size

Sieve Analysis. Over the years, sieving has almost become an industry standard because of its low cost and ease of operation. This method, however, has multiple disadvantages with respect to technical ceramic and metal powders. First, it is not useful in determining particle size smaller than −325 mesh (45 μm). Screens do exist with finer meshes; unfortunately, electrostatic cohesion and van der Waals forces cause the particles to adhere to the screen. Several techniques have been developed to overcome these attractive forces. By placing small rubber cubes on the sieves, a larger fraction of fine particles will pass through the mesh. Likewise, by placing the particles in an air stream (i.e., aerodynamic sieving), sieving will work more efficiently with smaller particles. Additionally, if a negative pressure is introduced underneath the sieve(s), or if the particles are sieved in a slurry, particles as small as 5 μm in diameter can be analyzed.

There are other inherent problems with sieving. Particle shape can greatly skew the distribution to the coarse or fine end. For example, if a powder has an irregular shape or ligmental shape, some of the particles will pass through the screen if they are perpendicular to the mesh, while others that do not have this orientation will not pass. Third, even if the articles have a spherical shape, the information obtained in screening is limited. For example, the designation −270 +325 for a powder indicates that the powder has passed through a 270 mesh screen, but has not passed through a 325 mesh screen. No information is available about the actual distribution of the particles within this band.

Subsieve Analysis. The Fisher Subsieve Analyzer is an instrument that has also gained great popularity in determining average particle size, especially in the carbide and refractory metal industries. With this device, compressed air is forced through a packed bed of powder. The pressure drop across this bed is measured. Knowing the flow rate of the air and the pressure drop, it is then possible to calculate the porosity of the bed, which in turn can be converted into a surface-dependent mean particle size. This number may or may not agree with an average particle size given by a different device. One of the major flaws of this instrument is that it is assumed that the particles are monosized spheres. Furthermore, no information is provided on the particle size distribution. The instrument is effective in providing a relative ranking of powders in terms of an "effective" particle size, typically in a production environment where the powder chemistry or size shouldn't change significantly from lot to lot. In no instance should such a number be mistaken as the true mean particle size.

Quantitative Image Analysis. Optical and electron microscopy are used to directly observe and measure individual particle size and shape.

Manual particle counting is a very tedious method to obtain a particle size distribution. In years past, particles were measured from micrographs (optical or SEM). A variety of measurements were then taken from these micrographs. Such manual measurements are extremely time consuming. Despite the time intensive nature of this method, it is still widely used because of its accuracy. However, automat-

Table 3 Advantages and disadvantages of common particle-size measurement methods

Method	Particle size range	Advantages	Disadvantages
Optical particle counting	nm to mm	Extremely accurate Reproducible Direct measurement Inexpensive Can be automated	Time consuming if done manually
Sieve analysis	44 μm to mm	Widely used Relatively inexpensive Easy to use	Lose accuracy over time Particle shape can greatly skew results Particles can adhere to screen If done properly, process can become expensive
Permeability (Fisher Subsieve Analyzer)	0.25 to 25 μm	Widely used in certain industries Easy to use Inexpensive	Does not give a true particle size
Laser diffraction	0.01 μm to mm	Fast Reproducible Easy to use Many instruments commercially available	
Traditional sedimentation	1 μm to 50 μm	Widely used in certain industries Does not need to be calibrated Relatively easy to use	
Centrifugal sedimentation	10 nm to 300 μm	Ability to measure submicron particles Analysis time is greatly reduced from nongravitational methods Does not need to be calibrated Easy to use	Expensive Problems with dispersion Must experiment to obtain optimal operating conditions
Photon correlation spectroscopy	5 nm to 5 μm	Easy to use Absolute method; does not require calibration Measures extremely small particles	Complicated algorithm used to calculate particle size distribution Agglomeration can be a real problem Hard to verify results

ic counting and image analysis techniques have advanced significantly with computer technology. Recent trends in image analysis are described in the article "Particle Image Analysis" in this Volume.

Even though quantification of the images is a direct method of observing particle size, it requires the selection of an appropriate index. Mean intercept length, area equivalent diameter, Feret's diameter, maximum and minimum caliper diameter, and so forth, are all valid size measurements. With the advance of computer electronics, few people rely on manual particle counting as a primary analysis tool. Usually, it is done only to verify the accuracy of another instrument.

Laser Diffraction (Coherent Light Scattering). One of the most versatile instruments to measure particle size operates on the principle of coherent light scattering. Particles, while suspended in a slurry, are pumped in front of a laser beam. From the angle and intensity of the diffracted beam, the particle size distribution is calculated. When first developed, these instruments had an operating range of approximately 1 to 250 μm. With more sensitive detectors and more advanced technology, these light scattering devices can effectively measure particles in the range of 0.01 to more than 5000 μm. This encompasses all of the powders one would currently encounter in powder injection molding. The limits of the unit are dependent on whether the software has been programmed to include Mie scattering theory, or whether it measures Fraunhofer diffraction alone. The former permits the measurement of particles from 0.5 to 0.01 μm. At larger particle sizes, the mathematics of Mie scattering theory converge to the basic Fraunhofer theory.

Several issues must be addressed in using coherent light scattering for particle size analysis. Care must be taken to ensure that the powder is adequately dispersed in the slurry. This is accomplished by adding a surfactant to the slurry. The appropriate surfactant is a function of the powder. Table 4 shows suspension liquid/surfactant pairs that provide adequate dispersion. Note that in many instances, if the appropriate suspension liquid is chosen, a surfactant is not necessary. If the as-received powder is severely agglomerated, ultrasonic sound waves must be used in conjunction with the surfactant to break up the aggregates. If excessive ultrasonic energy is applied, the particles will fracture, producing a size distribution that is not representative of the original sample.

Table 4 Surfactant/suspension pairs for liquid powder dispersion

Powder	Surfactant	Suspension liquid
Alumina	Sodium pyrophosphate	Water
Copper	Alkylphenol/ethylene oxide condensate (Lissapol)	Water
Aluminum	Sodium, oxalate, sodium tartarate, or sodium hexametaphosphate	Water
Iron	None necessary	Rape oil and acetone; 1:1 soy bean oil and acetone
Molybdenum	None necessary	Ethanol, glycerol, acetone
Steel	Alkylphenol/ethylene oxide condensate (Lissapol)	Water
Tungsten carbide	None necessary	Ethylene glycol, oil
Tungsten	None necessary	Water, ethanol, glycerol, acetone plus rape oil

The most recent advance in coherent light scattering has been to substitute a dry powder handling system for the slurry/pump system. With the dry system, powder is delivered in front of the light source via a stream of gas. The problem of agglomeration persists even in dry-dispersing systems.

Sedimentation Methods. In certain industries, devices that characterize particle size and size distribution based on particle settling in a liquid have been widely accepted. Such industries include mineral processing, technical ceramics, carbides, and refractory metals. The basic principle behind these instruments involves the projection of a beam of electromagnetic radiation, (usually white light, monochromatic light, or x-rays) through a column containing a homogeneous suspension. Either the radiation source scans the settling column measuring the amount of light or x-rays absorbed, or the source remains at a fixed height and the flux of particles passing in front of the beam is measured. This method of particle size analysis does have advantages over other techniques in that the attenuation of the beam can be measured accurately, the sample size is small, and there is no probe inserted into the measurement that could possibly influence the measurement. As particle size decreases, however, this method becomes unreliable because of the turbulence and Brownian motion of the particles.

Stokes's Law, the principle behind sedimentation instruments, assumes nonturbulent settling of the particles. As particle size increases, turbulence increases. Therefore, there is an upper limit on the particle size that can be accurately measured with these devices. The severity of the turbulence is also affected by the density difference between the particles and the settling medium (usually water or alcohol). With most metal powder used in powder injection molding, these restrictions impose a useful upper limit of approximately 50 μm. Some equipment manufacturers have extended the lower detection limit of sedimentation instruments by introducing a centrifuge into the settling chamber; however, this also decreases the maximum particle size that the instrument can analyze.

Aerodynamic time-of-flight is a recently developed method for particle size measurement. Particles are dispersed in a stream of flowing gas, which is directed between two sensors (typically laser diodes or a split laser beam) separated by a known distance. Knowing the true density of the material (as measured by helium pycnometry or some other technique) and the time needed to traverse the fixed distance between light beams, the particle size is calculated.

Again, as with all techniques, an index of particle size is measured. Particles with aspect ratios greater than one may preferentially align in the gas stream, which may bias the data slightly. The benefit of this method is that it is an individual particle counter. Each particle that is accelerated in the gas stream is measured and recorded. Other methods such as laser diffraction produce more of an "average" particle size distribution. For spherical powders, both laser diffraction and time-of-flight report very similar data.

One of the advantages of time-of-flight is that the optical properties of the powder, that is, the relative refractive index, is not required. The material density is needed; however, that is an easier measurement to obtain.

Electrical Zone Sensing. This is one of the more established methods for measuring particle size, as well as one of the more accurate techniques. Particles are suspended in an electrolytic fluid and are pumped through an orifice in a glass tube, across which a bias is applied. As the particle passes, the resistivity of the fluid changes, which is registered as a change in potential. The change in potential is directly proportional to the cross-sectional area of the particle. No information is needed about the particles. If the particles are extremely conductive, then the instrument must be tuned to a known standard.

The major disadvantage of this method is the width of the particle size distribution that can be detected. Each orifice can accommodate a 25× difference in particle size. If the larger particles exceed this limit, the hole will be blocked. If the particles are too small, the voltage change that is produced will not be detected. Within the operating range of a given sample cell, the measurements are extremely accurate. Like time-of-flight, electrical zone sensing measures the size of individual particles as they pass through the opening. In this method, the size limits may be extended from 400 to 0.6 μm using a multiple tube technique. An advantage of the method is its independence from particle density, which allows powder mixtures to be sized.

Photon Correlation Spectroscopy. Instruments are available that utilize laser light scattering and digital electronics to measure particle size. This spectroscopy method is used to extract information on the dynamics of a system of scatterers by the calculation of the correlation function of the scattered radiation. The correlation function of a scattered light signal with itself, that is, the autocorrelation function, is the Fourier transform of the power frequency spectrum of the input signal. This information is analyzed to quantify the dynamics of the scattering process that generated the signal. Some of these instruments have the capability to measure ζ potentials of the particles, by the application of laser Doppler velocimetry.

REFERENCES

1. R. Irani and C. Callis, *Particle Size Measurement, Interpretation, and Application,* John Wiley & Sons, 1963
2. C. Bernhardt, *Particle Size Analysis,* Chapman and Hall, 1994
3. H. Barth, Ed., *Modern Methods of Particle Size Analysis,* John Wiley & Sons, 1984
4. T. Allen, *Particle Measurement Size,* Chapman and Hall, 1997
5. S.G. Malghan and A.L. Dragoo, Characterization of Ceramic Powders, *Ceramic and Glasses,* Vol 4, *Engineered Materials Handbook,* 1991, p 66
6. T. Allen and R. Davies, Modern Aspects of Particle Size Analysis, *Ceramic Powder Science,* Vol 21, *Advances in Ceramics,* G.L. Messing et al., Ed., American Ceramic Society, 1987, p 721–746
7. G.D. Parfitt, Chapt. 1–3, *Dispersion of Powders in Liquids,* 3rd ed., *Applied Science,* 1981

SELECTED REFERENCES

- R. DeHoff and F. Rhines, *Quantitative Microscopy,* McGraw-Hill, 1968
- M. Fayed and L. Otten, Ed., *Handbook of Powder Science and Technology,* Chapman and Hall, 1997
- R. Iacocca and R. German, A Comparison of Powder Particle Size Measuring Instruments, *Int. J. Powder Metall.,* Vol 33 (No. 8), 1997, p 35–42
- K. Kurzdłowski and B. Ralph, *The Quantitative Description of the Microstructure of Materials,* CRC Press, 1995
- K. Linoya, K. Gotoh, and K. Higashitani, Ed., *Powder Technology Handbook,* Marcel Dekker, 1991

Sieve and Fisher Subsieve Analysis

Reviewed by Scott Nushart, ATM Corporation

TWO COMMON METHODS of particle size measurement in the P/M industry are sieve analysis and the Fisher subsieve sizer. Sieve analysis is the most widely used method of determining particle size distribution of metal powders down to ~5 μm. The test sieve is frequently the first quality control check on the receiving dock or the last verification of product quality before shipment.

The Fisher subsieve sizer also is used extensively throughout industry, especially for process control purposes where only relative values are required. This method can measure particle sizes between ~0.2 to 50 μm. These two common methods are briefly described in this article.

Sieve Analysis

Standard sieves have been in industrial use for over a hundred years. Particle size distribution is controlled and certified by the powder producer and is frequently checked by the end user. Typically, a series of sieves is selected that spans the full range of particle sizes present in a powder. Sieves are stacked in order, with the largest mesh size at the top and a pan at the bottom (Fig. 1). An appropriate sample weight of metal powder is spread on the top sieve and covered. The stack of sieves is agitated in a prescribed manner (shaking, rotating, or tapping) for a specified period of time. The powder fractions remaining on each sieve and contained in the bottom pan are weighed separately and reported as percentages retained or passed by each sieve.

Table 1 Standard U.S. sieve series

Sieve designation, mesh	Sieve openings	
	μm	in.
30	600	0.0232
40	425	0.0164
50	300	0.0116
60	250	0.0097
80	180	0.0069
100	150	0.0058
140	106	0.0041
200	75	0.0029
230	63	0.0024
325	45	0.0017

Source: ASTM E 11

Sieving Equipment

Sieves. Wire cloth sieves are woven in a square mesh pattern, usually from phosphor bronze or stainless steel wires. Distributions of opening sizes and wire diameters are closely controlled to determine the nominal sieve opening.

In 1970, the American Society for Testing and Materials (ASTM), the American National Standards Institute (ANSI), and the International Standards Organization (ISO) agreed on a standard sieve series (see Table 1). Wire diameters and manufacturing tolerances are specified in ASTM standard E 11 and ISO standards 565 and 3310/1. Sieves can be purchased with the following types of certification:

- Statement by the sieve manufacturer that the wire cloth has been made in compliance with ASTM E 11
- Certification from the manufacturer listing the optical measurements at several random spots on the sieve
- Certification from the National Bureau of Standards that the individual sieve has been optically examined and meets the qualification of ASTM E 11

The National Bureau of Standards supplies calibrated glass spheres for evaluating the effective openings of testing sieves. Sieve manufacturers supply "matched sieves," where extra precision is desired for comparing the results of several locations. Frequently, a laboratory performing many sieve analyses purchases a matched pair of sieves. One sieve is used as the routine working sieve, and the other sieve is kept as a "master" to periodically check the accuracy of the working sieve.

Sieve Agitators. Several methods of advancing the powder particles through the sieve openings have been devised that rely on one or more of these mechanisms:

- Hand shaking
- Gyratory motion
- Gyratory motion with tapping
- Mechanical shaking
- Mechanical vibration
- Electromagnetic vibration
- Oscillating air column
- Vacuum

Precision and accuracy must be assessed for each test method. Microscopic checks of the end sieve fractions and reproducibility must be evaluated. The length of sieving time must be established for each method as well. Usually, the sieving period ends when the quantity passing through the sieve retaining the largest fraction of the test portion in 1 min is <0.1% of the test portion.

Standard Test Method

Testing standards have been established for the sieve analysis of metal powders. These include ASTM standard B 214, published by the American Society for Testing and Materials; MPIF standard 05, formulated by the Metal Powder Industries Federation; and ISO standard 4497, developed by the International Standards Organization. These standards are very similar and specify the following requirements:

- Standard wire cloth sieves conforming to ASTM E 11, ISO 565, or ISO 3310/1
- Mechanical sieve shakers combining a rotary motion and tapping action
- Test specimen size of ~100 g (3.53 oz), a nominal catch weight, for metal powders with an apparent density >1.50 g/cm^3 and a 50 g (1.76 oz) test specimen for lower apparent density powders
- Mechanical sieve shaking for 15 min or another specified time
- Weighing the powder retained on each sieve and in the bottom pan to the nearest 0.1 g (0.0035 oz) and totaling all weights
- Calculating and reporting the percentages of each sieve fraction to the nearest 0.1%

Sieving Problems

Overloaded Sieves. In this condition, the sieve surface becomes crowded with oversized and near-mesh particles after material finer than one half the size of the mesh opening has passed through. On an overloaded sieve, the weight of the oversized material tends to wedge the near-mesh particles into the openings, thus blocking passage of additional particles.

Blinded sieves (blanked sieves) result when near-mesh size particles become entrapped in the openings. When brushing fails to remove the particles, an ultrasonic wash with a wetting agent must be used to dislodge them.

Damaged sieves may be torn or have stretched and distorted mesh openings. A microscope should be used to inspect sieves regularly. Epoxy has been used to repair small defects.

Irregularly shaped particles may pass through sieve openings in a specific orientation only. For these particles to pass successfully through the openings, vibrating or shaking should be sufficient to change the orientation of all particles. Some methods of sieve agitation are more efficient than others in achieving this goal, so they must be evaluated individually.

Agglomeration of particles prevents fine particles from passing through the mesh openings. Electrostatic charges may cause fine particles to be attracted to one another, frequently forming small balls. Fine particles also can attach themselves to larger particles. The use of sieving aids, such as powdered magnesium carbonate, tricalcium phosphate, hydrophobic fumed silica, activated carbon, or commercial antistatic agents, may alleviate agglomeration. Sieving aids are usually very fine powders that pass through the sieve and collect in the pan. Thus, the amount of sieving aid used can be subtracted from the –325 mesh fraction collected.

Wet Sieving

Some difficult-to-screen materials can be handled successfully by wet sieving techniques. Apparatuses are available to handle a single sieve or a stack of sieves. Ultrasonic wet sieving devices are also available. Wet sieve analyses are cumbersome to perform and require the sieve fractions to be dried before weighing. Difficult-to-screen materials are usually very fine powders, such as silicon or aluminum, that agglomerate poorly and may not separate completely using many dry sieving techniques.

Micromesh Sieving

Micromesh sieves are produced by photo-etching and electrodeposition techniques. Electrodeposition nickel produces precise square openings with planar surfaces. This thin mesh is supported by a coarser square-etched grid of nickel-plated cupronickel. These sieves are mounted in 75 or 203 mm (3 or 8 in.) stainless steel rings.

Sieve openings may be as small as 3 μm at frequent intervals. ASTM E 161 specification permits a maximum deviation of only ±2.0 μm for sieves ranging from 125 to 5 μm.

Sieving techniques include the same mechanical shakers and vibrators used for wire cloth sieving. However, for sieves finer than 325 mesh (45 μm), most dry sieving techniques are inadequate. For these smaller size sieves, a vacuum-type siever can be used (Fig. 2). This type of apparatus uses suction coupled with a rotating blowback nozzle. Fine particles are suctioned through the mesh, while the blowback nozzle redistributes the powder sample on the sieve, thus breaking up agglomerates and purging the mesh openings.

The sieving operation is rapid and allows sieving down to 10 μm. Due to the smaller sieve diameter (75 mm, or 3 in.) and the smaller amount of open area of the finer micromesh sieves, samples of only 1 or 2 g (0.035 or 0.070 oz) are used to prevent overloading. It is also possible to perform wet sieve analyses using micromesh sieves.

Sieve Analysis Variations*

Although sieve analysis is the most common method of measuring particle size by both producers and buyers, variation among results for different sieving tests of the same material is common. For example, Table 2 presents statistical results (in percentage of sample weight retained on a given sieve) for the 88 sieving tests (Ref 2). As the table shows, the values ranged widely, despite the controlled conditions of the test.

Table 3 also shows the number of variables for each variation source (4 batches, 88 samples, 2 sieve stacks, 2 sieving machines, and 4 operators) and the average variation (in percent) each source contributed to the total variation (Ref 2). Sampling (19.4%) and sampling reduction (33.5%) together contributed most of the total variation. The variation is largely due to the natural tendency of silica sand to segregate, which makes it difficult to take a representative sample and then split the sample into smaller representative samples. Table 3 also shows that using two sieve stacks contributed 29.9% of the total variation.

*Portions adapted with permission from *Powder and Bulk Engineering*, Vol 1 (No. 8), 1987, p 14–19

Most variations in sieve analysis, both in house and between plants, can be traced to variations in sample preparation, condition of the test equipment, and testing procedure. To minimize variations in sieve analysis, it is necessary to reduce or identify the sources of variation in sampling, sample reduction, sieve stacks, sieving machines, and operators (Ref 1).

Sample Preparation. The initial conclusions drawn from sieve analysis data, which deviates from expected values, is that either the entire shipment is bad, or the test equipment is out of calibration. While either or both conclusions may be valid (if only partially), the basic tenet of securing a representative sample is often critical.

Accurate sampling is extremely difficult. At the heart of accurate sampling is the topic of

Table 2 Particle size distribution determined by dry sieving tests of 88 silica sand samples

Mesh size(a)	Mean particle size distribution (percentage of sample weight retained)	Standard deviation, %	Variation from the mean (percentage of sample weight retained): Maximum	Minimum
30	0.08	0.051	0.20	0
40	7.03	1.281	9.80	5.10
50	36.08	1.514	40.30	33.20
70	26.38	0.417	27.20	25.50
100	19.37	0.521	20.60	18.10
140	9.28	0.396	10.00	7.90
200	1.59	0.111	1.80	1.20
270	0.12	0.044	0.30	0

(a) U.S. standard sieve series. Source: Ref 2

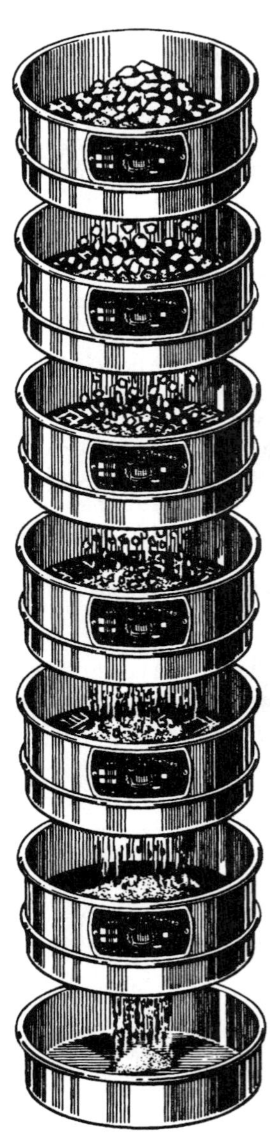

Fig. 1 Schematic of sieve series stacked in order of size

segregation. If the powder is made up of individual particles of only one size (that is, no real distribution), the effects of segregation are minimal. In the real world, however, most process powders exist as distributions of particle sizes. Segregation takes place as fine particles settle down through the spaces formed between larger particles. The disparate particle sizes in a mixture cause segregation.

To counteract segregation effects, representative samples should be obtained from a moving stream of powder. The entire stream of powder also should be taken for many short increments of time in preference to part of the stream being taken for the entire time (Ref 3). Sampling methods that do not follow these rules are second-best methods likely to cause errors. Alternative sampling methods—such as scoop sampling, coning and quartering, and riffling—are not precluded, just noted to be less than ideal. More information on the critical aspect of sampling is discussed in detail in the article "Sampling and Classification of Powders" by Terence Allen in this Volume.

Great care also must be taken to ensure uniform sampling. Not only is it important to sample correctly, but similar methods are needed to ensure similar results. For example, when the entire sample must be reduced or split, using one person and method to reduce each sample minimizes variation. Proper gross sample reduction should be performed as directed in the ASTM *Manual on Test Sieving Methods* and other industry-specific publications.

Condition of the Test Equipment. Understanding the process of a test sieve helps demonstrate the importance of keeping the equipment in top condition. Regardless of whether the sieve and contents are hand or mechanically agitated, the powder is constantly being reoriented in relation to the sieve openings. Those particles that are considerably smaller than the openings will clear first, followed by particles closer to the size of the openings. Eventually, the end point is approached—that point at which a negligible percentage of particles continues to pass through the openings.

Unfortunately, this description assumes that the test sieve and the powder being sieved are both perfect; the test sieve is assumed to have perfectly uniform square openings of the same size, while the powder is assumed to have regular, blocky, or spherical particles with a perfectly even distribution. These assumptions, of course, are inaccurate. The unique characteristics of individual powder samples and individual testing sieves combine to create another variable.

Standards and Specifications. Because test sieves are fabricated from a woven mesh material, variations in the weave are common. The chances of locating two sieves of identical distribution are extremely remote. This adversely affects the reproducibility of test results between sieves. To reduce testing variations, the dimensional integrity of the sieve openings must be closely controlled. The stringent standards imposed by ASTM, the International Standards Organization (ISO), and other regulatory bodies have created tolerance factors that allow for variations in the weave while striving to maintain a level of uniformity in test grade sieve cloth.

Each sieve should comply, as a minimum requirement, with the ASTM E 11 specification on "Wirecloth Sieves for Testing Purposes." This specification addresses the topics of nominal opening size, permissible variations in the opening size and wire diameter, general sieve frame construction parameters, and calibration procedures. In addition, appendices to the specification provide guidelines on how these dimensional values should be checked.

By compliance with standards, both powder suppliers and users have a common starting point, a basis of comparison. To further reduce performance variations, sieves characterized or physically measured as being near the nominal opening size or centerline of the tolerance should be specified.

Care and Maintenance. Test sieves, like any other piece of analytical laboratory equipment, require regular care to maintain their performance standards. The sieves should be kept clean and dry at all times and be stored in the cardboard carton provided or a suitable cabinet.

For the test sieve to perform properly, the wire cloth must be taut and free from variations in opening size. Regular cleaning, preferably in a low-power ultrasonic cleaner filled with a room temperature solution of mild detergent and water, will prevent the buildup of particles trapped in the mesh and prolong the useful life of the sieve. Cleaning should be done regularly to ensure good results.

Any brushing done on the woven wirecloth increases the risk of wirecloth variations, especially in mesh sizes finer than 80, where the fine wires are more likely to bend, distort, or even break. Practicality necessitates brushing in most instances, but utmost care should be exercised.

Agitation methods. The method for shaking or agitating the sieve must also be carefully monitored to reduce interlab variations. Hand sieving generally exhibits the greatest incidence of variations between operators. The amount of shaking and tapping force imparted to a sieve stack by an assortment of operators is difficult to quantify.

For this reason, many labs have introduced mechanical sieving devices (sieve shakers) into their regular programs to help minimize variations between operators. Unfortunately, machines from different manufacturers can provide clearly divergent data after a similar amount of agitation time. Variations can also be seen in test runs on similar machines from the same manufacturer. Standardization tests should be undertaken to reduce variations and to ensure reproducibility within each lab and correlation between labs.

Data formats for sieve calibration are also important. For example, if a sample is run through two different sieves, it is desirable to compare the sieves by both percentage of sample weight and cumulative percentage of sample weight retained (Ref 2), as shown in Table 4. The percentage of sample weight passing a given sieve more accurately shows whether each sieve meets the standards than does the cumulative percentage of sample weight retained, which is influenced by the sieve above it. Though a tight or loose sieve can pass either fewer or more

Table 3 Number of variables and average percentage of total variation for each variation source in sieving test series

Variation source	Number of variables	Average of total variation, %
Sampling	4 batches	19.4
Sample reduction	88 samples (22 from each batch)	33.5
Sieve stack	2	29.9
Sieving machine	2	11.4
Operator	4	5.8

Source: Ref 2

Table 4 Sieve calibration results for two sieve stacks (in percentage of sample weight passing and cumulative percentage of sample weight retained)

	Sample weight passing, %		Cumulative sample weight retained, %	
Mesh size(a)	Sieve stack 1	Sieve stack 2	Sieve stack 1	Sieve stack 2
20	99.5	99.6	0.5	0.4
30	76.3	83.5	23.7	16.5
40	58.7	58.8	41.3	41.2
50	32.2	32.6	67.8	67.4
70	13.8	14.1	86.2	85.9
100	6.3	6.4	93.7	93.6
140	2.3	2.6	97.7	97.4
200	0.0	0.1	100.0	99.9
Pan	0.0	0.0	100.0	100.0

(a) U.S. standard sieve series. Source: Ref 2

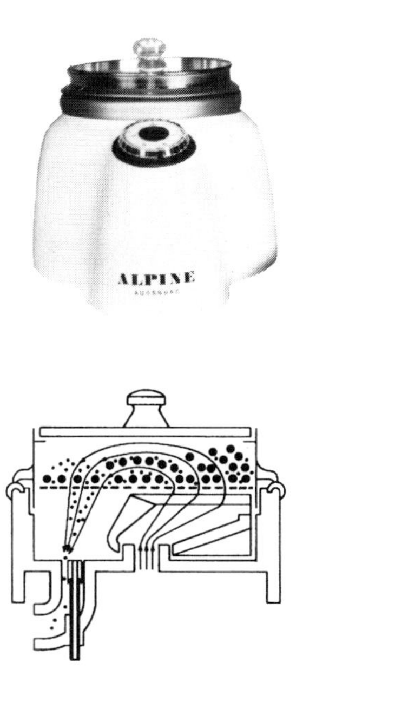

(a)

(b)

Fig. 2 Micromesh vacuum siever. (a) Apparatus. (b) Sectional view particles to the next sieve, very few near-size particles are likely to pass to the next sieve. Thus, presenting calibration results in both percentage of sample weight passing and cumulative percentage of sample weight retained will help eliminate the effect of the preceding sieve on the sieve in question (Ref 2).

Test Procedure. How a test procedure is performed can provide valuable insight into why variations occur. Specific documentation and adherence to a given procedure ensure good interlab correlation. The importance of proper documentation of test procedure is highlighted by the following example.

Example. When the analysis numbers did not match between a referee lab and a powder supplier, the referee lab asked about the specific test methods. The vendor said the sample was washed through a 400 mesh sieve using a fifty-fifty mixture of acetone and deionized water. The residue was then oven dried for 60 min at 300 °F, run through a demagnetizing coil, and put through the rest of the sieves. When the customer and the referee lab tried the procedure, they found complete correlation.

Summary. Good laboratory procedure is essential. Particle size analysis with a stack of testing sieves is an analytical procedure like any other used to determine the physical or chemical properties of a substance. The level of care taken during all phases of an analysis is directly proportional to the success ratio of the testing.

Communication between vendor and buyer can eliminate a great deal of misunderstanding and expense. If both parties have a full understanding of what is expected, questions regarding sampling, equipment, and procedures will be satisfactorily answered before any particles are analyzed.

Common sense is important. If the sampling techniques, particle size measuring techniques, and accepted testing procedures are reasonable to both parties, there will be no surprises. No phase should be expected to provide a higher level of test accuracy than any other phase. If the sampling phase is highly accurate, the equipment and procedural phases of the test should parallel that level of accuracy.

Fig. 3 Fisher subsieve sizer

Fisher Subsieve Sizer*

The Fisher subsieve sizer is another common industrial method that allows particle size measurements in the range of 0.2 to 50 μm. It is a simple method to measure particle sizes below the range of sieve analysis.

The Fisher subsieve sizer is a simple, inexpensive permeameter, which measures the flow of air through a packed powder bed. From this measurement, average particle size and porosity can be calculated from a chart that is an integral part of the instrument. The Fisher subsieve sizer is used extensively throughout industry, especially for process control purposes where only relative values are required.

Frequently, however, the results are often misinterpreted to designate an exact particle size. Consequently, analysis results refer to a Fisher "number" rather than a size. Fisher subsieve numbers should not be compared with size results from other measuring techniques, as frequently no correlation exists. Figure 3 illustrates a typical Fisher subsieve sizer, and Fig. 4 provides a schematic of the method.

Permeameters accurately and precisely measure permeability. They do not measure surface area or particle size. The flow through a porous compact depends on the structure of the pores. Flow, therefore, depends on the porosity, particle shape, particle size, particle size distribution, and method of compaction, all of which affect pore structure. Conversion of the measured permeability to a surface area is accomplished by the Kozeny-Carman equation:

$$K = E^3/(1.0 - E^2)BS_w^2\rho^2 \qquad \text{(Eq 1)}$$

and

$$E = (V_{bed} - V_{powder})/V_{bed}$$

*From article by Donald Parsons in *Metals Handbook*, 9th ed., Vol 7, 1985, p 230–232

where K is permeability of the powder bed; E is porosity of the powder bed; V is volume, cm^3; ρ is true, or absolute, density of the powder material, g/cm^3; S_w is surface area per unit weight of powder, cm^2/g; and B is constant.

From Eq 1, it can be seen that permeability is related to three macroscopic quantities: bed porosity, surface area, and particle density. Average particle diameter (d_m), also known as the Sauter mean diameter, the surface mean diameter, or the volume surface mean particle size, is calculated from the surface area by:

$$d_m = 6/\rho S_w$$

For powders of the same composition and of similar size distributions, Eq 1 is probably satisfactory. However, if particles can be packed differently and exhibit the same porosity, but with different pore structures, this relationship may not hold. For instance, two powders of the same chemical composition with different particle size distributions have been given the same Fisher subsieve numbers. Comparison of results from the Fisher subsieve sizer with those from other surface area and size distribution techniques may show no agreement and should not be undertaken.

When particle shape, size, and distribution are essentially the same and do not vary significantly from one production lot to another, Fisher subsieve analysis provides adequate comparisons for quality control purposes. The major limitations include constant sample packing and instrument calibration.

Operation. ASTM B 300 standard, "Average Particle Size of Powders of Refractory Metals and Compounds by the Fisher Subsieve Sizer," provides calibration and operating instructions for Fisher subsieve analysis. Standard ASTM C 721, "Average Particle Size of Alumina and Silica Powders by Air Permeability," also provides guidelines. These specifications are based on the Fisher subsieve sizer instruction manual (Ref 4).

The output of an air pump is regulated to a pressure head of 50 cm (19.7 in.) of water by the water level in the standpipe. At this constant pressure, the flow through the packed powder bed is measured by the water manometer. Calibration is accomplished by varying the leakage rate through the calibrating valves.

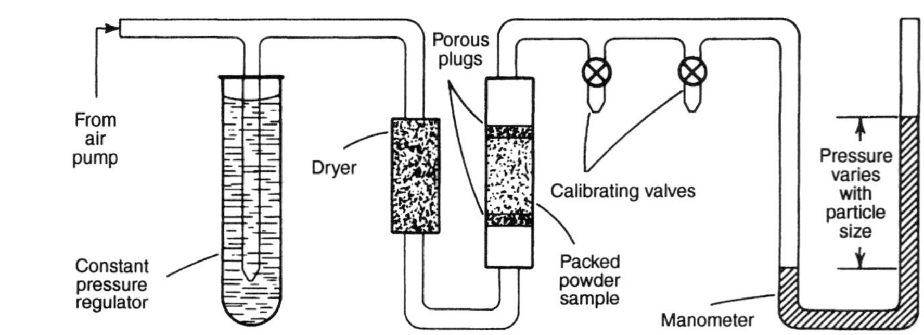

Fig. 4 Schematic of Fisher subsieve sizer

The powder sample is weighed to 0.1 g (0.003 oz) and placed in the sample tube. Using a rack and pinion, it is compacted with a 222 N (50 lb) force applied with a torque wrench. This technique minimizes sample-to-sample packing variation. Porosity is determined from the compacted sample height and the instrument calculator chart.

The sample tube is then placed in the air system, and after equilibrium is reached, the manometer water height is measured. The instrument calculator chart is used to obtain the average size. If the sample height is such (high or low) that the calculator chart cannot be used, the actual height (in centimeters) above the base line is measured for both sample height and manometer water height. These values are used in the following equation to provide calculated values. The calculator chart uses a modification of Eq 1:

$$\text{Porosity} = [LA - (M/\rho)]/LA$$

and

$$\text{Fisher number} = CL(M/\rho)\sqrt{2H/(P-2H)[AL-(M/\rho)]^3}$$

where L is sample height after compaction, cm; A is cross-sectional area of sample tube, cm^3 (1.267); M is weight of sample, g; ρ is true, or absolute, density of powder material, g/cm^3, H is height of water column above the base line, cm; P is overall air pressure, cm (50 cm, or 19.7 in., of water); and C is instrument constant (3.80).

If the sample weight is made equal to the true density, the equation is simplified:

$$\text{Porosity} = (LA - 1.0)/LA$$

and

$$\text{Fisher number} = CL\sqrt{2H/(P-2H)(AL-1.0)^3}$$

Standardization. The Fisher subsieve sizer must be standardized daily (or more frequently, if necessary) according to the procedures given in the instrument manual (Ref 4). Monitoring of the sample post height, which sets the chart base line as the instrument zero, is frequently overlooked. This can be verified easily by placing two filter paper sheets between the two sample tube porous plugs located on top of the sample post. The rack is brought down in contact with the upper porous plug, and the pointer is checked to see that it is on the base line. If it is not, post height is adjusted until the measurements coincide. This is an infrequent but critical adjustment.

A method of standardization uses a Fisher ruby orifice mounted in a sample tube, which is calibrated by the manufacturer against the primary standard, the National Bureau of Standards Cement No. 114. Operating the Fisher subsieve sizer with the calibrating tube in place and the chart positioned to the proper porosity produces a water column of a predetermined height. This indicates a size on the chart that is equal to that engraved on the side of the tube. If it does not, the internal calibration valves are adjusted until the correct value is obtained. Other secondary standards may be used and include any uniform powder that has an accepted Fisher subsieve number.

Results. Because analysis results are obtained by interpolation of the pointer position between lines on the calculator chart, they should be reported with varying intervals and places after the decimal point, as indicated in ASTM C 721, Section 9.1. Reading and chart line intervals for given particle ranges are shown in the following:

Particle range, µm	Chart line interval, µm	Reading interval, µm
0.2–4	0.1	±0.2(a)
4–8	0.2	±0.1
8–15	0.5	±0.1(b)
15–20	1.0	±0.2
20–25	5.0	±1.0

(a) For porosities less than 0.55, the reading interval for the 0.5–4 µm range is +0.5 µm. (b) For porosities greater than 0.6, the reading interval for the 11–15 µm range is ±0.2 µm.

In one laboratory, a five-year weekly evaluation of the ASTM B 459 tungsten powder gave an average of 1.23 µm and a standard deviation of 0.012 µm. Fisher subsieve numbers are generally agreed to be precise within ±5%. Because of the care taken in making the measurements, many laboratories report ±3% precision.

REFERENCES

1. S. Nushart, Sieve Analysis: Making Your Numbers Come Out Right, *Powder Bulk Eng.*, Vol 1 (No. 8), 1987, p 14–17
2. S.D. Cowley and S.J. Morris, Understanding and Minimizing Variation Sources in Dry Sieving Tests, *Powder Bulk Engr.*, Feb 1991, p 34–41
3. T. Allen, *Particle Size Measurement*, Chapman & Hall, 1981
4. Fisher Subsieve Sizer Instructions Manual, Fisher Scientific Co., Pittsburgh, PA

Sedimentation Methods

SEDIMENTATION is a mechanism for classifying metal powders according to their settling rate in a fluid. This measurement is based on Stokes's law of fluid dynamics, which states that, at low velocities, the frictional force on a spherical body moving through a fluid at constant velocity is proportional to the product of the velocity, the fluid viscosity, and the radius of the sphere.

The Stokes equation is a relationship between settling velocity, U_{st}, and particle diameter, D:

$$U_{st} = \frac{(\rho_s - \rho_f)\, gD^2}{18\eta}$$

where ρ_s is solid density, ρ_f is fluid density of viscosity η, and g is gravitational acceleration (Ref 1). The particle size distribution can be determined by examining a sedimenting suspension of the powder. Changes with time in the concentration of density of the suspension at known depths are determined, and the size distribution is determined from these data (Ref 2).

Sedimentation is routinely used to determine particle size and size distribution. Among the number of sedimentation methods available, only a few are commonly used for metal powders—the micromerograph and light and x-ray turbidimetry. Another analyzer using the sedimentation method is the roller air analyzer; however, it is no longer manufactured, and its standards are being withdrawn.

To obtain valid results, convection currents must be avoided in the suspending fluid, and the relative rate of motion between the fluid and the powder particles must be slow enough to ensure laminar flow. This generally restricts the particle sizes that can be determined to the subsieve range.

Particles should be large compared with inhomogeneities in the fluid, which for sedimentation in air restricts the available methods to particles >5 μm. For sedimentation in liquids, particle sizes down to 0.1 μm can be determined.

Particles in the suspension must be perfectly dispersed, and the suspension must be diluted enough to guarantee independent motion, which translates to a maximum concentration of ~1 vol% of particles in the suspending medium. Finally, wall effects should be minimized; consequently, the inside diameter of the sedimentation chamber should be sufficiently large to overcome this phenomenon.

Micromerographs

This equipment is a sedimentation balance occasionally used for determining the particle size distribution of subsieve metal powders. The powder is suspended in air by projecting the sample with a burst of nitrogen through a deagglomerating device consisting of a conical annulus into the settling chamber.

The chamber consists of a thermally insulated vertical aluminum tube with a 10 cm (3.94 in.) inside diameter that is 2.5 m (8.20 ft) high. The pan of an automatic balance that weighs the amount of powder settling on it is located at the bottom of the chamber. A recorder calculates the cumulative weight of powder settled as a function of time. Particle size distribution is calculated from this value on the basis of Stokes's law. Particle size distribution ranging from 2 to 100 μm can be determined with this instrument. One limitation of the method is the tendency of powder particles of various sizes to cling to the wall of the column.

Light and X-Ray Turbidimetry

Turbidimetry methods are widely used to determine the particle size distribution of refractory metal powders, such as tungsten and molybdenum, and of refractory metal compound powders, such as tungsten carbide. The turbidimeter is standardized in ASTM B 430. Because turbidimetry is used for refractory metals and compounds, several additional factors must be considered to obtain reproducible size and size distribution data. The particles of fine refractory metal powders are often agglomerated by sinter bonds. In ASTM B 430, determination of particle size distribution of the powder is described in the as-supplied condition and after deagglomeration by rod milling (laboratory milling). To obtain consistent, reproducible results by turbidimetry, not only the turbidimetric procedure, but also the deagglomeration procedure, must be standardized. Dispersion of the powder in the liquid before turbidimetric analysis must also be standardized.

Figure 1 shows a typical turbidimeter. The procedure for determining particle size distribution of refractory metal powders with this instrument is included in ASTM B 430. A sample of powder dispersed in a liquid is poured into a glass cell and allowed to settle. A collimated beam of light is passed through the cell at a level having a known vertical distance, h, from the liquid level. The intensity of the light beam is determined by the current generated in a photocell. The current is passed through a potentiometer, where the voltage drop across it is measured by a recording millivoltmeter.

The reading of the millivolt recorder for the intensity of the light beam shining through the clear solution is adjusted to 100%, while the concentration of the suspension is adjusted so that the reading for the intensity of the light beam through the suspension before any settling has occurred ranges from 20 to 40% of that through the clear solution. As the suspension settles, the projected area of the particles in the suspension decreases, and the intensity of the light beam increases.

At the beginning of the settling, all particle sizes are uniformly distributed through the volume of the sedimentation cell. As settling proceeds, large particles settle faster than small particles. After a given time (t_x), all particles larger in diameter than x have settled below the level of the light beam. The concentration of particles at the light beam level is now equal to the original

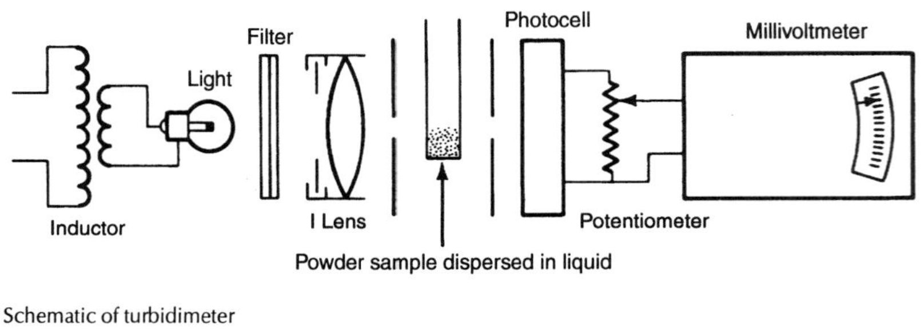

Fig. 1 Schematic of turbidimeter

concentration of particles, minus all particles with diameters equal to or larger than x. The projected surface of the particles at time, t_x, is therefore smaller than that of the particles in the original suspension, and the intensity of the transmitted light is greater.

Determination of Particle Size Distribution. To obtain information on particle size distribution from the plot of light intensity versus time, use is made of the relationship between the total weight of n particles of size x, which is proportional to nx^3, and the projected surface area of the n particles, which is proportional to nx^2. Therefore, the cumulative weight of particles up to a given particle size x_{lim}, which is:

$$\int_0^{x_{\text{lim}}} dW$$

is proportional to

$$\int_0^{x_{\text{lim}}} x dS$$

which is the integral of the product of particle size and projected surface area integrated from 0 to size x_{lim}.

Turbidimetric measurements using white light are relatively inexpensive and have proven quite reproducible and useful. They are used in research and in routine analyses for comparing different lots of refractory metal powders. These methods generally are used for comparative work on the same type of material.

When x-rays instead of white light are used to determine particle size distribution of a subsieve particle suspension, the attenuation of the x-ray beam intensity is proportional to the mass of the powder particles rather than their projected area.

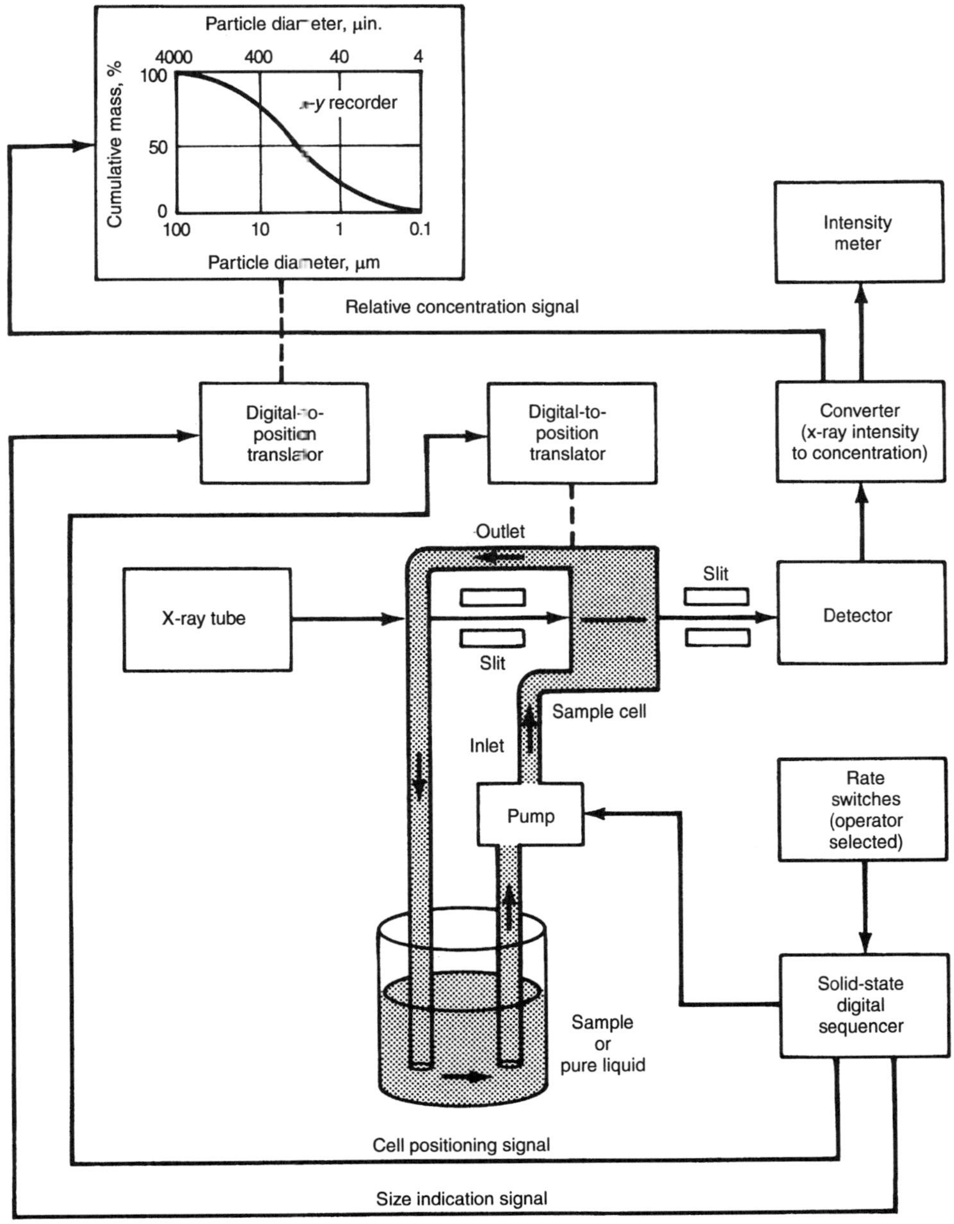

Fig. 2 Sedigraph schematic

Roller Air Analyzers

This apparatus determined the particle size distribution of metal powders by suspension in a stream of air. Using this apparatus, a powder is classified into particle size fractions that range from 5 to 40 µm. A stream of high-velocity air flowing through a nozzle of suitable size impinges the powder sample contained in a U-tube, so that the powder becomes dispersed within the stream. The velocity (v) of the air stream through the chamber balances the settling velocity of particles with diameter and density determined by a Stokes law calculation:

$$v = 29.9 \times 10^{-4} \rho x^2$$

where v is the velocity of the air stream, cm/s; ρ is the particle density, g/cm^3; and x is the particle diameter, µm.

Given this velocity, particles with a size smaller than x are carried through the settling chamber into the collecting system, which consists of an extraction thimble. Larger particles fall back into a U-tube. By using a series of vertical settling chambers with diameters in the ratio 1 to 2 to 4 to 8 and a constant volumetric rate of flow, the powder could be classified into particle size fractions with the maximum sizes in a ratio of 1 to 2 to 4 to 8 (for example, 5, 10, 20, and 40 µm). As previously noted, this sedimentation is no longer in use.

Other Sedimentation Methods

Besides gravity sedimentation, other methods of particle size measurement also employ the Stokes principle.

Andreasen's Pipette. In the pipette method of size analysis, the concentration changes that occur within a settling suspension are followed by using a pipette to draw off definite volumes. Stokes's law is used to calculate the size distribution.

In photosedimentation, a narrow horizontal beam of parallel light is projected through a suspension at a depth, h, onto a photocell. In a homogeneous suspension, if the particles are allowed to settle, the number of particles that leave the light beam will be balanced by the number entering it from above. However, after the largest particle, D_m, in the suspension has fallen from the surface to the measurement zone, the emergent light flux will begin to increase, because no more particles of this size will enter the measurement zone from above. Hence, the concentration of particles in the light beam at any time, t, will be the concentration of particles smaller than D_{st}, where D_{st} is given by the Stokes equation (Ref 3). The particle size distribution is calculated from the relationship between the attenuation of the light beam and the projected surface area of the particles.

X-Ray Sedimentation. Similar to the use of a light beam, x-rays can be used in combination with the gravity settling of particles. In this case, x-ray density is proportional to the weight of the powder in the beam:

$$I = Io \exp(-BC)$$

where B is a constant and C is the concentration of powder in the beam. The x-ray density, X_d, is defined as:

$$X_d = \log_{10}\left(\frac{I}{Io}\right)$$

The sedigraph (Ref 4) is a well-known instrument based on the gravitational settling and x-ray absorption analysis. This instrument automatically presents the cumulative percentage frequency, and the sedimentation cell is driven in such a way that the concentration is recorded directly as a function of the Stokes' diameter. The x-rays from a 22.6 keV source are collimated into a narrow beam and pass through a fixed thickness of suspension. The sedimentation cell (Fig. 2) is filled and emptied with the suspension by a built-in pump. The transmitted radiation is detected as pulses by a xenon-filled scintillation detector. The counting electronics give a voltage that is proportional to the powder concentration. Particle size distribution is calculated by the application of Stokes' law to these data.

Centrifugal Techniques. Because the centrifuging process speeds up the rate of settling of particles, it overcomes one of the serious disadvantages of the gravitational sedimentation techniques, which take an unduly long time for particles <5 μm (200 μin.). In addition, most sedimentation devices suffer from the effects of convection, diffusion, and Brownian motion. Centrifugal sedimentation can be carried out by either a two-layer or homogeneous suspension technique, either cumulatively or incrementally. The two-layer techniques give rise to streaming, a phenomenon most likely to affect cumulative analyses, in which the fraction sedimented against time is determined (Ref 2).

REFERENCES

1. L. Svarovsky, *Solid-Liquid Separation,* Butterworths, 1977, p 15–23
2. T. Allen, *Particle Size Measurements,* 3rd ed., Chapman and Hall, 1981
3. T. Allen and R. Davies, Modern Aspects of Particle Size Analysis, *Advances in Ceramics,* Vol 21, *Ceramic Powder Science,* G.L. Messing, et al., Ed., American Ceramic Society, 1987, p 721–746

Electric and Optical Sensing Zone Analysis

PARTICLE SIZE MEASUREMENTS by sensing zone analysis are based on dynamic/Stokes principles. These methods are not widely used for metal powders, but they are used for ceramic powders. The two main methods are electrozone (electrical sensing zone) size analysis and optical sensing zone size analysis.

Sensing zone methods involve the suspension of powders in a liquid medium, and there are important factors to be considered when interpreting particle size distribution results for metal powders. Most metal powders have a higher specific gravity than the particulates for which the obscuration-principle apparatus, as well as other particle size analyzers, was originally designed. As a result, a greater tendency exists for particles to settle in the sample reservoir before being sampled, thus causing the distribution to be skewed toward the fine side. This phenomenon is greater for coarser and more regularly shaped particles. Good reproducibility of a distribution does not guarantee the absence of particle settling, as settling of coarse particles also occurs and is reproducible.

Electrozone Analysis

Electrical sensing zone analysis is based on the Coulter principle. This method determines the number and size of particles suspended in an electrolyte by causing them to pass through a small orifice with an electrode immersed on either side. The changes in resistance as particles pass through the orifice generate voltage pulses with amplitudes proportional to particle volumes. In this method, the size limits can be extended from 400 to 0.6 μm (16 to 0.024 mils), using a multiple tube technique. An advantage of the method is its independence from particle density, which allows powder mixtures to be sized.

The electrozone method detects the volumes of individual particles in a liquid suspension from the electrical pulse amplitudes that represent individual particle volumes. The pulses are processed via microcomputer to yield particle size distribution data in whatever form may be required. Figure 1 shows the main components of a computerized electrozone system.

The powder to be evaluated is suspended in a conducting liquid. The suspension is circulated through a sampling chamber, which contains a sensing orifice with immersed electrodes on either side. Within the sensing orifice, particles displace the electrolyte to cause a resistance change (and consequently a voltage pulse) proportional to the "electrical envelope" volume. Because this voltage pulse is independent of particle shape, diameter is expressed as that of a sphere of equal volume, or equivalent spherical diameter.

The size distribution measured by a given testing setup depends on orifice diameter, orifice current, amplifier gain, and logarithmic distribution. The latter provides a distribution selection from 2 to 3% up to 50 to 75% of the orifice diameter, which is full range for a given orifice. Overall, the electrozone method is capable of measuring particle diameters ranging from 0.1 to 2000 μm, depending on particle density. For wide distributions, the computer program can blend data sets, which is facilitated by logarithmic scales.

Collection of size distribution data normally requires several minutes and consists of transfer of the digitized values for many thousands of particle pulses to their proper locations in computer memory. The initial data format is thus a frequency histogram, usually of 128 channels.

Data processing includes smoothing, scale normalizing or converting (both size and quantity axes), marking, editing, extrapolating and blending, subtracting, and creating ratios. Throughout data acquisition and processing, a videoscope display provides interactive, operative monitoring of data and associated items.

Sample preparation requires a particle-free, filtered electrolyte. Many electrolytes have the required characteristics to adequately test various materials. Powdered metals are best analyzed by aqueous solutions of common salts, such as sodium or potassium phosphates (basic) or chlorides, sulfates, or nitrates (neutral). Electrolyte conductivity (strength) may be approximately 2000 mmho/orifice diameter (1% sodium chloride conductivity is ~20 mmho/orifice diameter).

Anionic or nonionic dispersants are also used, sometimes with ultrasonic dispersants as needed. Solvating organic electrolytes are available for oily samples. To avoid excessive spurious pulses from coincidence in the sensing zone, sample concentration is limited to 1 to 50 ppm by volume, depending on distribution and shape of the size distribution.

Metal Density. Metal powders have relatively high density, which limits the maximum particle size for uniform electrozone sample suspension. Even if the suspension has efficient agitation and high viscosity (50% glycerol, for example), electrozone data generally will not be acceptable if

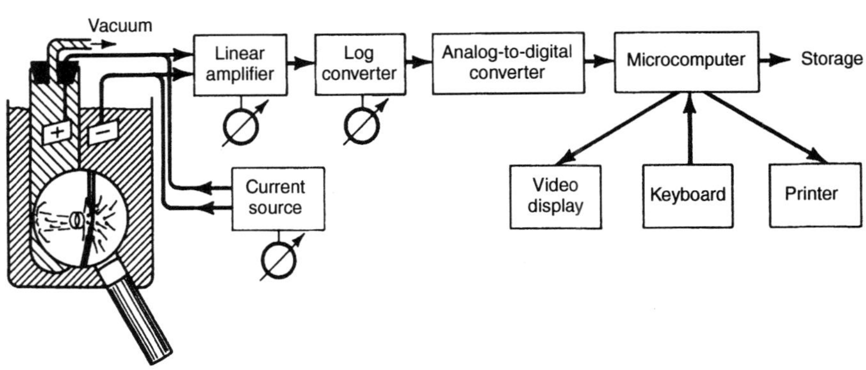

Fig. 1 General schematic of key instruments in an electrozone system

the particle size-density function exceeds 500, expressed by:

$$(d_m)|\rho - \rho_0|^{-1.5} > 500$$

or

$$(D_m)_{max} \leq 500\ |\rho - \rho_0|^{-1.5}$$

where d_m is diameter, μm, and ρ and ρ_0 are densities of particle and liquid, respectively. Typical values are:

Density, g/cm^3	2	5	10	17
d_m (max), μm	500	62	19	8

Below a value of 100 for this function, sieving becomes increasingly effective. Thus, some samples may require both sieve and electrozone analyses, with blending of the two data sets in the overlap region of 100 to 500 for this size-density function.

Metal Conductivity/Magnetism. The oxide surface layer present on most metal particles causes them to appear to be nonconductive, in effect creating a barrier impedance. Electrozone currents high enough to produce 1 to 3 V across the surface layer overcome this barrier impedance, thus causing conduction for only part of the pulse cycle (prior to the occurrence of sulfur polarization and ion depletion). Consequently, the particle pulse will erroneously appear to be smaller.

This effect can be detected by test observation of the total relative volume of particles per unit volume of suspension with increased current value. Samples may be treated with coating reagents or surfactants to increase surface layer impedance. Orifice current settings then must be kept below the level where this effect appears. Similarly, porous particles appear to be too small, due to the conductance of the pores aligned with the electric field, unless they are plugged with nonconductive material.

Magnetic powders may be dispersed for a sufficient time to permit electrozone measurement if they are degaussed for 20 min at temperatures as high as 750 °C (1380 °F) (but below the sintering point, however) and/or spatulated with viscous honey to inhibit reagglomeration. A microscope slide covered with the honey suspension is then dipped into the electrolyte in the stirred sample beaker. In this manner, data acquisition is initiated instantly, while the suspension is being formed from the dissolving honey.

Advantages. The electrozone method rapidly measures particle size distribution with excellent low-end sensitivity and with good resolution (volume response) and precision (minimal side effects).

By contrast, photozone (photicsensing zone) methods, which include light-beam scattering, blockage, and diffraction, have much lower resolution and precision. This is due to their area response and several significant side effects, including photic properties of particles, high coincidence levels, and beam/sensor instabilities. Sieving and sedimentation methods have similar limitations to response and side effects and are much slower in producing data.

Optical Sensing Zone

The light blockage principle (obscuration) was used by Carver in 1958 to develop a method for counting and sizing particulate contamination in hydraulic fluids (Ref 1, 2). To date, this principle has found application in other particle-fluid systems and has been used for the measurement of metal powder particles.

In an apparatus using the light-obscuration principles, particles are suspended in a fluid with a refractive index that differs from the refractive index of the particles. The suspension is then passed through a restricting orifice—the sensing zone—across which a collimated white light beam passes. Figure 2 illustrates the sensing zone. The beam of light falls on a photodetector, which measures the intensity of the light passing through the zone.

Large particles block the light beam, thus lowering the intensity with which the beam strikes the sensor. Small particle sizes (<3 μm) do not block the light rays; the rays bend around the particles. The sensor can be designed so that a slight bend that is equal to an index of refraction change of 0.05 causes the light to miss the detector, thereby detecting particles.

Instrumentation. Figure 3 is a schematic of the particle size analyzer based on the obscuration principle. The suspension is prepared in the reservoir cup, which is equipped with a stirrer. A vacuum pump pulls the suspension up through the sampling tube and through the sensing zone located in the back of the sampler stand. Once analyzed, the suspension is collected in a waste chamber.

The sensing zone is designed to prevent laminar flow, which causes each particle to tumble through the zone. The sensor output is a voltage signal, which is proportional to the amount of light blocked by a particle at a given instance, superimposed on a constant baseline voltage. The height of the pulse generated above the baseline as a particle traverses the sensing zone is proportional to the particle's largest cross section.

The relationship between the size of the particle and amplitude of the pulse produced can be determined theoretically. In practice, however, calibrated spheres are used to determine this relationship and to calibrate the sensor (Ref 3, 4). Sensors are available in the following particle size ranges:

- 1 to 45 μm
- 1.5 to 67.5 μm
- 2 to 90 μm
- 2.5 to 112.5 μm
- 5 to 225 μm
- 10 to 450 μm
- 16.7 to 750 μm
- 41.7 to 1875 μm
- 150 to 6750 μm

The first five ranges are the most useful for studying metal powders used in the P/M industry. The actual sensor dimension (height in Fig. 2) is one-third larger than the largest particle size for which the sensor is rated; consequently, sensor blockage is minimized (Ref 3, 4).

The microprocessor scales the sensor output pulses and sorts them into 23 size ranges, or channels, in a geometric progression of 1 to 1.18 by using 24 analog comparators. Within the limits of the physical size of the sensing zone, the thresholds can be adjusted to change the widths of the channels to provide more or less detail of certain areas of the particle size distribution, or to extend the range slightly. The threshold adjustment circuits are housed on a printed circuit board that is changed when the sensor is changed to monitor a new particle size range, for example.

The system memory accumulates the number of pulses generated between each pair of thresholds, and the central processing unit reduces the data, calculates the output data, and controls the printer and plotter output functions.

Accuracy and Interpretation. Instrument-related factors must be considered, even with

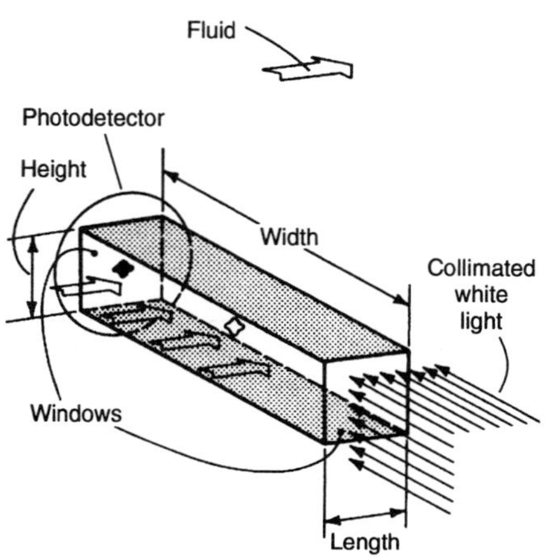

Fig. 2 Sensing zone for (1 to 45 μm) particle sensor based on light obscuration

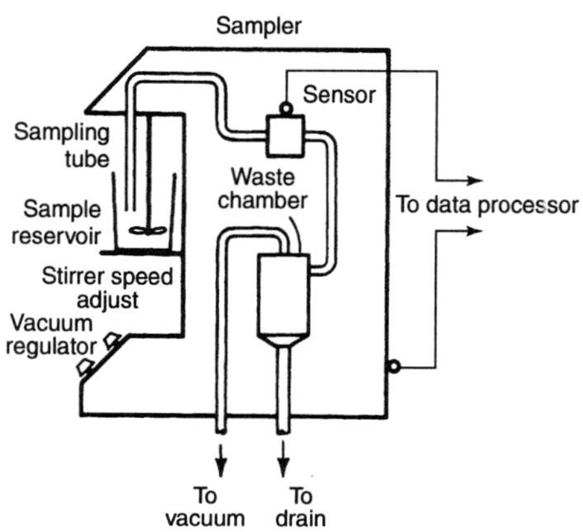

Fig. 3 Schematic of particle size analysis apparatus based on light obscuration

perfectly dispersed, spherical particles (Ref 5). The oversized particle indicator has another purpose in addition to detecting large particles that exceed the range of the sensor—it indicates an overconcentrated solution. If several particles overlap while passing through the sensor, the oversize indicator flashes, and the oversize counter is incremented. If this tendency occurs frequently (more than once or twice per analysis), the solution may need to be diluted, or the sensor may be blocked. If dilution does not eliminate the oversize indication, but does reduce its frequency by an amount proportional to the dilution, the large particles are most likely part of the actual distribution.

Even if dilution eliminates the oversize indication, small particles may overlap and skew the distribution toward the coarse side. To eliminate this occurrence, a maximum recommended number of particles per cubic centimeter of liquid is given for each sensor. High-concentration sensors that allow twice the concentration to be analyzed by restricting the volume of fluid, and therefore the occurrence of two particles in the sensing zone at the same time, are available.

To increase detail in some parts of the distribution and to extend the analysis range, channel thresholds can be adjusted. Adjustable thresholds, however, can slip out of adjustment. The appearance of a sharp peak and an adjacent dip in the same place on several different derivative distributions may indicate the need for threshold adjustment. When this condition occurs, counts are being placed in the channel above or below the proper one. For the cumulative percent under or over distributions, the effect is cancelled when both of these channels are passed. Threshold adjustment can be performed with a printed circuit extender board and a digital voltmeter. The overall detection level of the instrument should be checked periodically using calibrated spheres.

Settling of particles in the sample reservoir can be reduced by using a more viscous fluid such as glycerine. Pure glycerine, however, may be too viscous for the vacuum to draw and also may prevent the dispersion of particles that tend to agglomerate. Glycerine can be thinned with water, but antisettling benefits are diminished.

A more effective way to eliminate settling is to use reservoir shapes, such as round-bottom beakers, and stirrer configurations that maximize turbulence and prevent "dead" spots in the stirring current, without the generation of foam or bubbles that appear as particles in the sensor. Even if reservoir suspension is ideal, particle settling can still occur in the sampler tube (the tube connection between the reservoir and the sensor).

The higher the linear velocity of the fluid, the larger the size of the particles that can be entrained and carried along. The linear velocity of the fluid in the sampler tube is determined by the rate of flow through the restricted sensing zone and by the diameter of the sampler tube. The small sensor has a smaller sensing zone; consequently, the flow rate through the sampler tube is lower, causing the larger particles to settle before reaching the sensor. This tendency can be controlled by fitting the sensor with a smaller sampler tube. This increases the linear velocity of the fluid and keeps the larger particles entrained.

Small particles may have a tendency to loosely agglomerate due to moisture, or van der Waal's forces, for example. Stirring the water suspension of such a powder may be only partially successful in separating these particles. Addition of a surfactant or type IC dispersing aid to the powder before water is added helps the water wet the particles and allows them to separate with the stirring action. Ultrasonically vibrating the solution for several minutes also helps separate particles.

Effect of Lubricants. Many metal powders for P/M use are prelubricated by the powder manufacturer. Typical lubricant additions range from 0.5 to 1.0 wt%. While the presence of lubricants or other chemical additives or particle surfaces may inhibit wetting and cause agglomeration, lubricants have another significant effect. Lubricants or other chemical additives are much less dense than metal particles. Consequently, a small amount of lubricant by weight may contribute greatly to the particle count (population) distribution.

When the volume distribution is calculated from the population distribution, the small amount of lubricant (by weight) is weighted the same (per particle) as the metal powder. In this case, the volume distribution is not the true weight distribution. The same occurs during the analysis of a powder mix in which the constituents have different specific gravities. Single-component powders provide the best test results, unless only the population distribution is desired. Powders should be unlubricated, or the lubricant should be removed prior to analysis.

REFERENCES

1. G.C. West, "The Application of Light Obscuration Principle to the Particle Size Analysis of Powder Materials," Third Particle Size Analysis Conference, University of Salford, England, Sept 1977
2. H.S. Kupfer, Automatic Particle Size Distribution Analysis of Powdered Material by the Light Blockage Principle, *Powder Metall.*, Vol 10 (No. 2), 1979, p 96–97
3. T. Allen, *Particle Size Measurement,* 3rd ed., Chapman and Hall, 1981
4. "Operation and Maintenance Manual for HIAC Model PA-720," HIAC/Royco Instruments, Menlo Park, CA
5. H.N. Frock and P.E. Plantz, Correlation Among Particle Sizing Methods, *Proceedings of the Technical Program* (Rosemont, IL), International Powder and Bulk Solids Handling and Processing, May 1982, p 77–88

Light Scattering

Michael C. Pohl, Horiba Instruments, Inc.

LIGHT-SCATTERING measurements are capable of providing particle-size analysis on a broad range of particle sizes in a rapid, convenient fashion. Current technology is capable of measuring from several tens of nanometers to several millimeters in less than one minute. The instruments have proven themselves capable of solving a variety of problems associated with the production and handling of these powders.

Previous versions of these instruments had much more limited size range capabilities (Ref 1). The advent of modern microprocessors has led to the development of more advanced hardware and software designs. These advances have led to many new features that can be beneficial to the user of the equipment. Table 1 lists some of the more important benefits. The positive benefits have led to increasing use of this technique.

Light-scattering instruments are now capable of determining subtle differences in distributions that are not reflected in a change of the median size of the distribution (Ref 2). In many cases these subtle changes can be correlated to a change in the performance of the particles of interest. This is an important characteristic of the generated data. As correlations are understood and are related back to the processing conditions in the plant, knowledge is gained about process variations. This knowledge can be used in process control and is one reason for the continuing use of the technique.

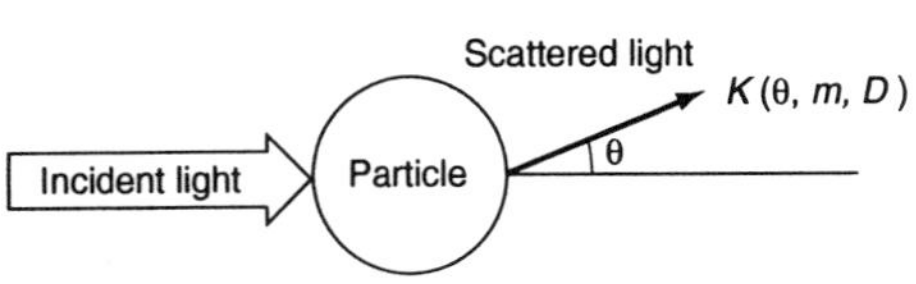

Light scattering intensity obtained from Mie theory

$$K(\theta, m, D) = \frac{\lambda^2}{8\pi^2 R^2} \left[i(\theta, m) + i(\theta, m, D) \right]$$

$\alpha = \pi D/\lambda$ Size parameter
D: Diameter
λ: Incident light wave length
R: Radial distance

$$m = \frac{\mu_1}{\mu_2} = \frac{\text{Refractive index of particle}}{\text{Refractive index of dispersing medium}}$$

Fig. 1 Laser-scattering principle. Calculation of scattering intensity based on the Mie theory

General Theory

The general theory of light scattering was developed in the early 1900s (Ref 3) while practical application occurred somewhat later (Ref 4). The fundamental concept is the Mie theory (Ref 5), a complete equation that explains the scattering phenomenon. In the past this was difficult to use, and simplifying assumptions were made. The most common assumption is that the particles are opaque, circular obstructions in the beam of light (Ref 6). This allows the Mie equation to be simplified to Fraunhofer diffraction theory. This greatly reduces the amount of input data and the complexity of the calculations to be performed. However, it does have a cost. The accuracy of particle-size measurements (especially for small-sized particles with size approaching that of the wavelength of the incident radiation) is lowered when the Mie equation is simplified in this way.

Modern computers make it possible to easily solve the more complex Mie equation. Figure 1 represents this equation. A number of the parameters are determined by the design of the system and are constants. These include λ and R. Other variables are refractive indices (μ_1 and μ_2), which also must be known to ensure accurate measurements.

Refractive Indices

Refractive indices are well-characterized values for most particle systems. In many cases it is merely a matter of referring to standard reference literature (Ref 7). In some cases there are actual tables that are being prepared by International Standards Organization committees (Ref 8). In the rare example of newly developed materials or mixed dispersant systems, these values can be measured experimentally. In a liquid medium, the values could easily be measured by a refractometer. These are available from a number of commercial vendors. In the case of particles, the measurement can be performed by a microscopic-immersion method.

Microscopic-Immersion Method. In this approach the particle is immersed in a liquid of known refractive index, and the medium refractive index is matched to the solid in a trial and error approach. This approach is best described in the following paragraph (Ref 9):

"The refractive index of any solid may be determined microscopically by successfully immersing fine particles of the solid in liquids of known refractive index until, in one such liquid, the particles become invisible. When this occurs, the refractive index of the invisible particles must be identical with that of the liquid medium in which they are immersed. Similarly, the index of the solid may be determined by varying the index of the mounting liquid until it matches that of the solid and then measuring the index of the liquid. The index may be varied by mixing liquids, by varying the temperature and/or the wavelength of light and by dissolving varying amounts of solid in a liquid."

Becke Test. The microscopic-immersion method can be a very slow laborious task if the selection of fluids is done in a totally random fashion. Fortunately there are approaches available to ac-

Table 1 General features of light-scattering particle-size analysis

Feature	Possible range of features	User benefit
Expanded range	0.02–2000 μm	More types of samples can be analyzed
Automatic analysis	Autosampler for 24 samples	Unattended operation of the instrument
Dry analysis	Samples can be run wet or dry	Samples can be run as they are normally used
PC control of the instrument	All standard programs, such as Windows, can be used	Easy storage, retrieval, and analysis of data
Flexible analysis of data	Narrow modes, multimodes, or broad distributions can be analyzed	Broader range of samples can be analyzed quickly
Mie scattering equations can be used	More complicated algorithms are used to analyze data	More correct particle size data are presented

celerate this process. The most popular approach is the Becke test, where the Becke line is a bright halo near the boundary of a transparent particle that moves with respect to that boundary as the microscope is focused up and down. The halo will always move to the higher refractive index medium as the position of focus is raised. The halo crosses the boundary to the lower refractive index medium when the microscope is focused. The particle must be illuminated with a narrow cone of axial light, which may be obtained by closing the iris diaphragm of the condenser until only a small aperture remains (Ref 9).

Absorptive Component of Refractive Index. The real component of the refractive analysis can be found or measured. Unfortunately in the case of the particles, there is typically also an imaginary or absorptive component. As Annex IV (Ref 8) shows, this term has also been determined in the case of many particles. In the event it is not known, there are several analytical techniques that may be used (Ref 4). Beyond these approaches, there are also empirical approaches that may be used (Ref 10). Thus by a variety of approaches, the liquid and solid refractive index can be determined. The relative refractive index is then determined by dividing the value for the particle by that for the liquid. This value is then input to the instrument for determination of the particle-size distribution.

Instrument Design

A typical laser light-scattering instrument consists of a light source, a particle-dispersion and delivery system, a detector to measure the scattering pattern, and a computer to control the system and determine the particle-size distribution. Older instruments were only capable of measuring the scattering pattern at low angles (<20°) and were limited to measuring particles >1 μm. More modern instruments collect scattering information at more and larger angles. This can be accomplished by the use of a converging beam, more or larger lenses, a second light-generating system, or the use of more detectors. This allows particles <0.1 μm to be measured accurately. These newer developments form the basis for this new description of light-scattering measurements (Ref 8).

Figure 2 shows the typical configuration for a laser-diffraction instrument. In the conventional setup, a laser is used to generate a monochromatic, coherent, parallel beam. A beam processing unit, usually a beam expander with an integrated filter, produces a beam to illuminate the dispersed particle collective. The direct, unscattered beam is then measured by the obscuration detector. The sample is passed through the laser beam in the measuring zone. The scattered light encounters the Fourier lens which focuses the

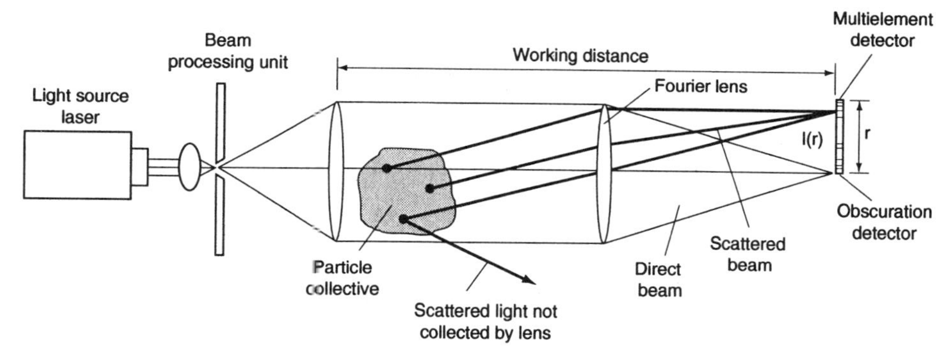

Fig. 2 Example of a laser diffraction instrument setup

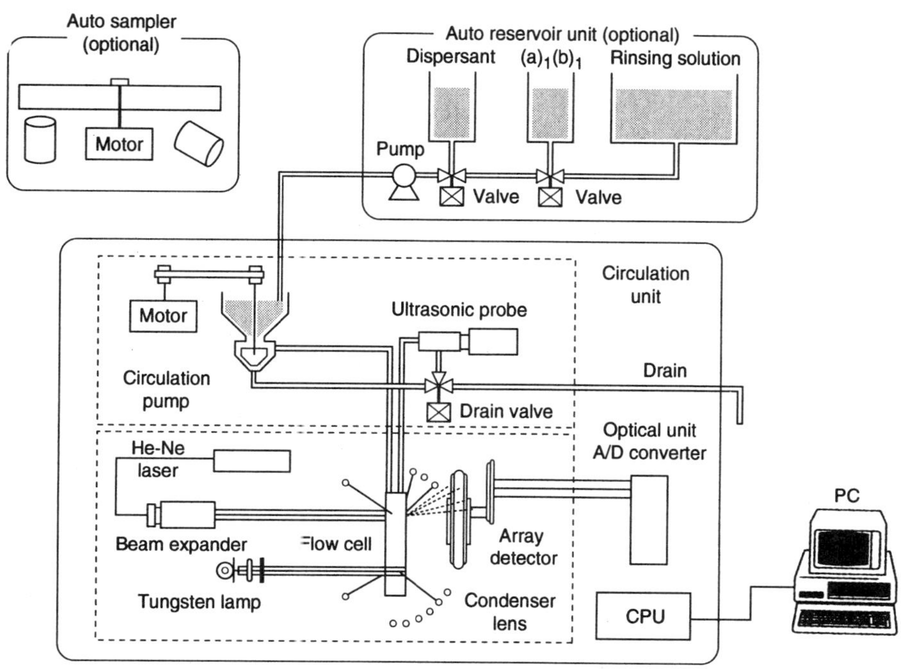

Fig. 3 Flow schematics

$$g(N) = \sum_{j=1}^{m} K(N, D)\, f_j\, (D)\Delta D$$

$g(N)$: Output from the Nth detector
$K(N, D)$: Response function of the Nth detected volume
$f_j(D)$: Particle size distribution based on volume
D: Typical particle size
ΔD: Interval between particle sizes

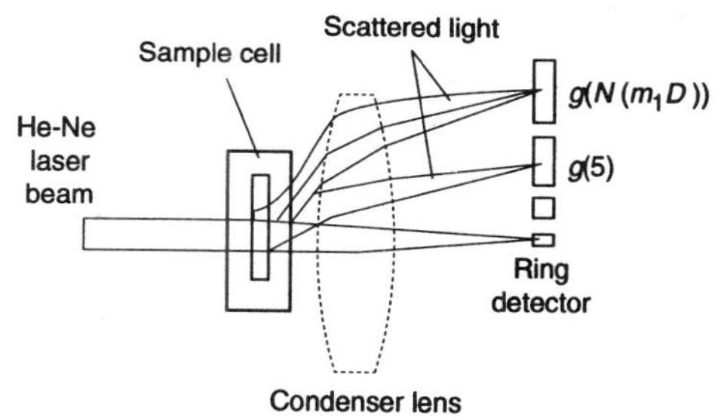

(a) Accumulation of scattered light at the detector
(b) Relationship between the scattering intensity, $g(N)$, of the scattered light accumulated at the detector and the particle size distribution, $f(D)$

Fig. 4 Relationship between position of the detector and scattering intensity, $g(N)$

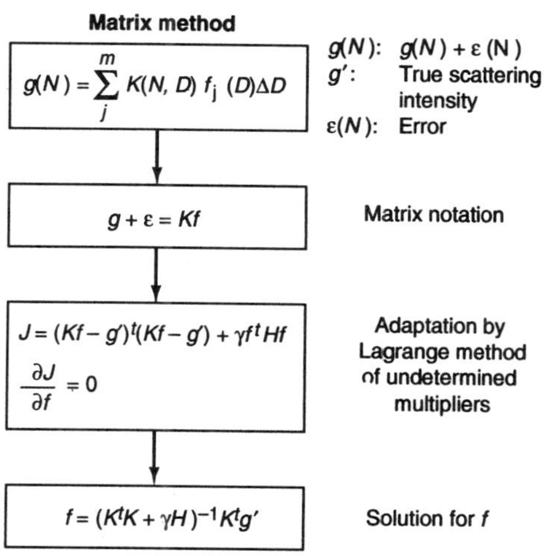

Fig. 5 Calculation of the particle-size distribution, $f(D)$, by the matrix method

scattered beam and eliminates the scatter from inappropriate particles. The focused scattered light is then collected on the multielement detector. This light represents the information that is processed to determine the particle-size distribution.

This type of system can be compared to the new designs available. Figure 3 shows a good example of such a system. In the heart of such a system is the more conventional laser system that is measuring the forward scattering from the larger particles. In this particular arrangement, the helium-neon laser is also being used to measure the scattered-light intensity at a variety of narrow angles. A second optical system featuring a tungsten lamp is also incorporated. It is filtered to a much shorter wavelength and is optimized to measure the light scattered at very wide angles. This is an ideal system for measuring the smallest particles.

Several other features are prominent in Fig. 3. Dispersion and proper delivery of particles are critical to the success of these newer instruments. Because they analyze a broad range of sizes (0.02 to 2000 μm), the pump must be capable of delivering particles uniformly to the laser beam. Also because many particles, especially small ones, tend to agglomerate, dispersion of the sample plays a key role. The powerful, ultrasonic probe is incorporated into this system to meet that challenge. Another key area is the addition of new components to better control the feed of the sample into the instrument. In this case an auto sampler and auto reservoir unit are shown as optional to this system. Improving instrument performance of these units is the focus of new development. As a result, the units are discussed in a later section.

Algorithm Calculations

The hardware of the system allows measurement of the scattered-light intensity at a series of different angles with several different wavelengths. This information is then used in a series of algorithms to calculate the particle-size distribution. Figure 4 shows the basic equation governing this relationship. This emphasizes the idea that several more pieces of information must be available to solve for the desired result ($f_j(D)$). There are two general approaches to solving this problem.

The older approach, called the matrix method, is illustrated in Fig. 5. When the Lagrange method is applied, the $f(D)$ can be determined. This method suffers from a number of limitations that have caused it to fall from favor. The limitations include problems with accuracy on monodispersed samples, the ability to properly recognize multimodal samples, and susceptibility to electronic noise in the system. As a result, a newer approach was developed. This method was developed from the solution of a somewhat unrelated analytical problem (Ref 10). This method, shown in Fig. 6, is called the iteration method.

Figure 7 shows a schematic representation of the steps involved in the iteration method. This method works by constantly comparing the measured scattering pattern to the proposed theoretical particle-size distribution. As more iterations are performed, this difference A, sometimes called χ^2 should become progressively smaller. As this happens, the calculated particle-size distribution becomes ever closer to the true value. While this process is repeated, the attempt

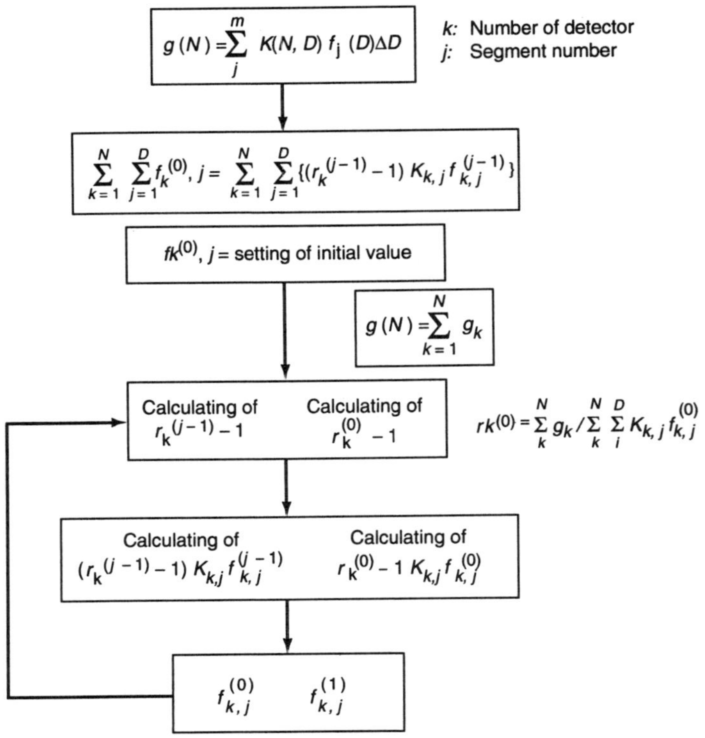

Fig. 6 Calculation of the particle size distribution, $f(D)$, by the nonlinear iteration method

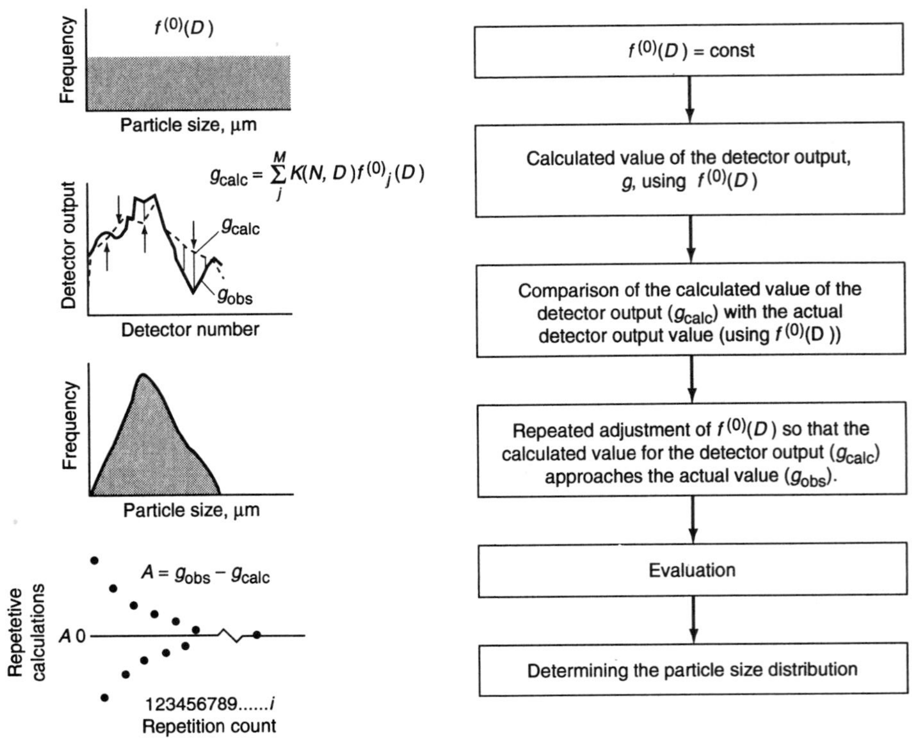

Fig. 7 Schematic representations of the steps involved in particle-size distribution, $f(D)$, by the nonlinear iteration method

is made to reduce it to zero. Practically this does not happen, so criteria for an acceptable A value is established, and the iterations are stopped at that point. Control over the number of iterations is determined depending upon the system being analyzed.

In this analysis, the missing major component is the relative response of each size particle at each detector. Figure 8 illustrates this problem. This relative response must be taken into account to properly solve for the particle-size distribution. A generalized approach has been to calculate a kernel function for a well-characterized system and then apply it to the equation to solve for $f(D)$. A very common starting point is to use a series of well-characterized polystyrene latex (PSL) samples. These samples are available in a wide variety of sizes and have been sized by a variety of techniques, including microscopy. From the known information, plus the fact that the relative refractive index is (1.59 – 0.00i)/1.33, it is possible to calculate the response factors.

New Applications

The basic physics of the light-scattering process have been understood for quite some time. Instrument design using this technique has enabled measurement of particles >5 orders of magnitude in size. As the size range has rapidly reached its theoretical limitation, other issues have become paramount. These issues include proper introduction of the sample into the laser beam, proper dispersion of the sample prior to analysis, the introduction of more exotic samples into the beam, and automation to simplify and extend the use of technology. These issues have come to the forefront recently and are expected to remain critically important in the future.

Improved Pumping Systems. One of the primary issues in a laser light-scattering experiment is presenting a representative sample to the laser for analysis. In most cases only a very small fraction of the sample can be measured at any given time. As a result this is a prime issue to consider, as larger and larger particles are being analyzed. These developments have caused the pumping system to be treated as a separate module that is added to the analysis system. In many cases, older peristaltic pumps that were used for particles in the several hundred micrometer range, have had to be replaced. A variety of approaches to this problem have been implemented.

One of the most popular approaches has been to incorporate a centrifugal pumping system. A typical example of such a pumping system is shown in Fig. 9. This type of pump provides adequate power to pump even 2000 μm particles around the system in a uniform, smooth fashion. Similar systems are incorporated into all newer instruments.

Another different approach is to totally remove the pumping module from the system. This option typically works only with smaller particles as the larger particles tend to become sediment below the laser beam. In spite of this limitation, this approach is very good for analyzing particles in obnoxious fluids. It minimizes the use of the solvent and simplifies the retrieval of the sample. The approach has proven useful in the analysis of a variety of particles, including emulsions and polymers.

Dispersion Accessories. The major obstacle to measuring an accurate, reproducible particle size is dispersion. In older instruments this need was typically not met by the analyzer. It was assumed that dispersion would be performed external to the instrument. The instrument was designed to handle stable, well-dispersed particles only. This, however, required a large laboratory space to house the instrument and the auxiliary equipment needed to disperse the sample. It also meant that the fundamental particle size could not be determined for less stable dispersions. These issues and the desire for quick, simple analysis procedures led to the incorporation of the dispersion accessories into the instrument.

A typical example of this is seen in Fig. 9. In this system, the powerful ultrasonic probe is placed directly into the stream of flowing liquid. This forces the particles to impinge directly on the tip of the horn and helps to maximize the energy of the imploding bubbles to effect the best possible dispersion of the sample. It is also a great aid in removing bubbles prior to the introduction of the sample. The rapid flow rate and its position immediately in front of the cell where

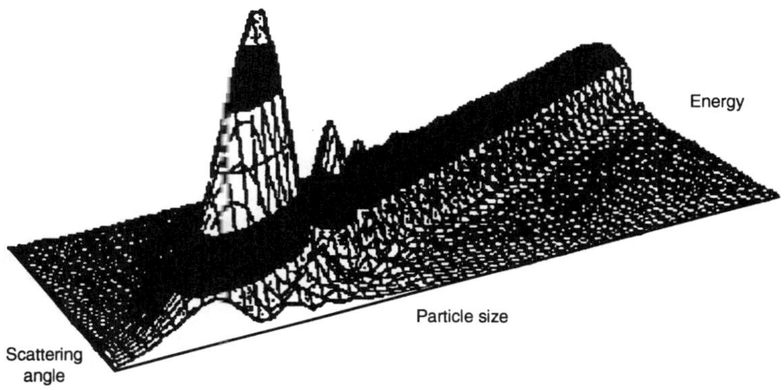

Fig. 8 Light energy scattering patterns for arbitrary detector geometry against particle size and scattering angle for equal volumes of particles (Mie theory, latex particles RI 1.60–0.0i, in water RI 1.33)

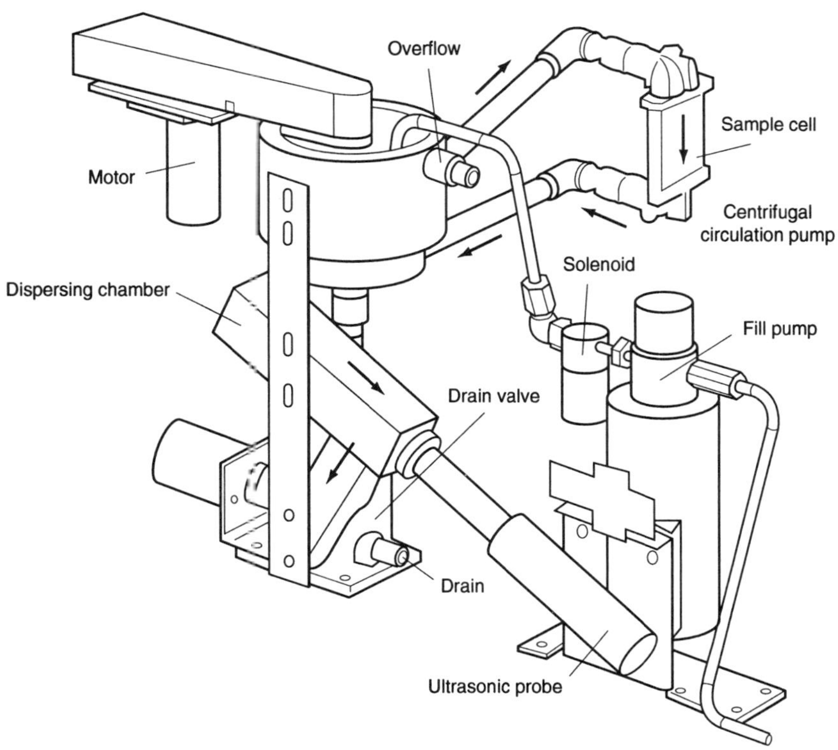

Fig. 9 LA-920 pumping system

the laser beam is focused make it possible to analyze even strongly agglomerating samples. Variable power outputs and duty cycles make it possible to handle even delicate particles. These accessories have greatly expanded the range of samples that the instrument can analyze.

Dry Analysis. Many particles are prepared dry and remain dry in their end use. These particles include, in some cases, pharmaceutical tablets, food products, dry pigments, and airborne particulates. In these cases, the introduction of a fluid medium serves to complicate matters significantly, because it introduces serious constraints on the analysis. A variety of dispersion liquids for the laser diffraction method are available. They should:

- Be transparent at the laser wavelength (e.g., for HeNe 633 nm)
- Be compatible with the materials used in the instrument (O-rings, tubing, etc.)
- Not dissolve or swell the particulate material
- Favor easy and stable dispersion of the particulate material
- Have a known refractive index that differs significantly from that of the particulate material
- Have low or medium viscosity to enable recirculation
- Not be hazardous to health and safety

Wet Dispersion. When working in the wet state, many aids are available to assist in the dispersion of the sample. The presence of large numbers of solvent molecules aids in dispersion, and often the separation effect of the liquid is augmented by the use of viscous media. For some examples, dispersants or surfactants are added to the liquid to improve dispersion. In many cases, shear energy or ultrasonic energy are applied to aid in the dispersion. In the dry phase, none of these approaches are applicable, so a totally different approach has to be taken. These technologies have greatly advanced in the last several years.

Dry Dispersion. Figure 10 (Ref 8) illustrates three approaches for dry dispersion. The two basic approaches of collisions and shear stress use very different technology to address dispersion. In the two collision methods, it is assumed that the energy of the collision is transferred to the breaking of the "bonds" holding the particles together. In cases where the particles are hard and not susceptible to fracture, this may work. In some cases the energy is used to break the individual particles instead. In the use of velocity gradients, the force is typically less violent and less prone to fracturing particles. This has come to be the most accepted approach for dispersing dry particles.

Figure 11 illustrates the principle of operation. The desired velocity gradient is produced by the introduction of compressed air to the vacuum stream at the jet nozzle. The shock of going from a vacuum condition to >80 psi in a short distance provides the desired force. Usually this is sufficient energy to disperse the samples.

Comparison of Wet and Dry Analysis. The increasing popularity of dry analysis has caused comparisons of wet and dry analysis. In countless cases, these two analyses have produced comparable results. In cases where differences have persisted, issues have been identified to explain the differences. In some cases, powders have flowed so poorly that they could not be fed properly. The addition of a small amount of silica enabled proper feeding of the material. In the case of strongly agglomerated particles, the particles actually behaved that way in the dry process. As a result, the dry, agglomerated results were actually more correct. Despite problems with dry dispersion, dry analysis is a firmly established technique.

Automated Features. Most of the automation that has been developed to date has been in the area of wet analyses. This can be seen in areas such as the simplification of the analysis through personal computer (PC) programs or the use of sophisticated programs to allow one button analysis. Automatic alignment of the optical bench of the instrument is also a feature. Although these are important developments, they typically make operation more convenient but not necessarily faster. The major improvement in this area has been in the form of automated feeding of the sample or the liquid used to dilute the sample. These systems can take on a variety of configurations depending on the type of sample that is being analyzed.

Liquid feeding can be achieved through a variety of mechanisms. In the simplest configuration, a container of fluid is placed at a high elevation and allowed to flow into the instrument. In most cases, a source of fluid at a positive pressure fluid is available to introduce the liquid at an appropriate rate and in a correct volume.

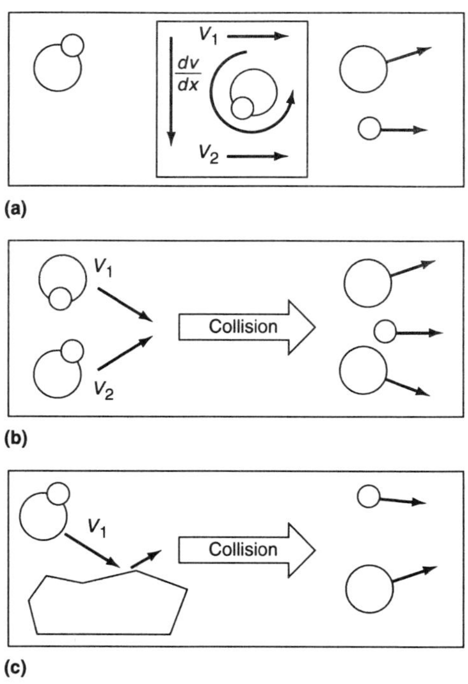

Fig. 10 Process involved for dry dispersion of powders. (a) Velocity gradients caused by shear stress. (b) Particle to particle collisions. (c) Particle to wall collisions

Reservoir systems also afford the opportunity to select a variety of liquid and surfactants to be fed into the instrument. This versatility increases the speed of sample preparation, sample analysis, and sample clean up.

Sample feeding is a major innovation for automation. A variety of configurations are available and perform quite differently. The best systems are those that are specifically designed for particle size analyzers rather than those that are "borrowed" from a gas chromatograph or similar analyzer. In specially designed systems, different samples are fed to the instrument, and a scheduler program allows for many types of experiments to be run. These include dispersion curves, kinetic studies, stability experiments, reproducibility tests, etc. All of these can be completed in an unattended fashion with increased time savings and sample throughput.

On-Line Analysis. Typically, the sample is in a jar or beaker for analysis, which means an operator mechanically withdrew the sample to be analyzed. This is an extra step requiring additional time and effort and is a possible source of additional error. To overcome these issues, laser light-scattering instruments also are used for on-line process analysis (Ref 12). While this marks the evolution of all analytical technologies, it is a very recent development in the particle sizing arena.

Historically the acceptance of this technology for plant use has been limited by many problems. They include the lack of ability to obtain a representative sample, lack of ability to keep the sample dispersed, lack of adequate cleaning for the instrument, and lack of ruggedness in the instrument design. Recently these problems have been solved for some of the well understood and better behaved processes. Many discussions of this analysis are now encountered in the literature (Ref 13). On-line analysis of dry powders is becoming more common. Examples of on-line

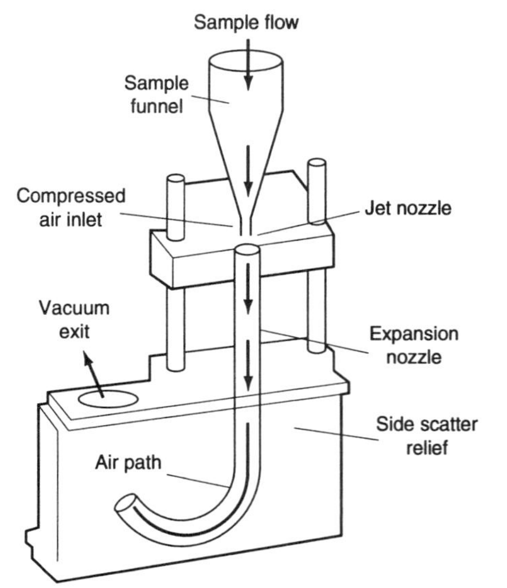

Fig. 11 Powder jet systems

measurement of wet systems also are becoming quite common (Ref 15).

Future Developments. In the past 10 to 12 years, an explosive growth has occurred in the popularity of laser light scattering as a technology for performing particle-size analysis. This popularity has led to the development of many new features and capabilities. The range of analysis has been greatly expanded as have the types of samples that can be analyzed. The analysis has been automated, and the reproducibility and repeatability have been improved. Even the algorithms used to determine the particle size have been improved. There are some indications that it can be used routinely in an on-line application. All of these trends are expected to continue well into the future. As new types of particles are developed and new particle systems are introduced, adaptations will develop to feed them to a laser light-scattering instrument to measure the particle size.

REFERENCES

1. H.N. Frock, in *Powder Metallurgy,* Vol 2, *ASM Handbook,* American Society for Metals, 1984, p 216
2. E. Orr, New Pigment Control Techniques: Part 2, *Am. Paint Coat. J.,* 14 Aug 1995, p 43
3. G. Mie, *Ann. Phys.,* Vol 25, 1908, p 377
4. M. Kerker, *The Scattering of Light and Other Electromagnetic Radiation,* Academic Press, 1969
5. T. Allen, *Particle Size Measurement,* 4th ed., Chapman & Hall, 1990, p 484
6. G. Row, Particle Sizing Technologies, *Int. Ceram.,* in Issue 1, March 1997, p 57
7. *CRC Handbook of Chemistry and Physics,* CRC Press
8. "Determination of Particle Size Distributions—Laser Diffraction Methods," ISO TC 24/SC4 Draft 2, to be published
9. W.C. McCrone, *The Particle Atlas,* Vol 1, *Instrumentation and Techniques,* Ann Arbor Science, 1973, p 72
10. D. Pugh, *Am. Ceram. Soc. Bull.,* Vol 76 (No. 7), July 1997, p 44
11. S. Twomey, On the Numerical Solution of Fredholm Integral Equations, *J. Assoc. Comp.,* 10 March 1963, p 97
12. V.A. Hackley, Measuring Particle Size of Slurries with Sound Waves, *Ceram. Ind.,* June 1997, p 52
13. A.P. Malcolmson and M.P. Bonin, Moving Laser Diffraction Particle Size Analysis from Lab to Line, *Ceram. Ind.,* April 1996, p 106
14. T.L. Harvill, J.H. Hoog, and D.J. Holve, In-Process Particle Size Distribution Measurements for Pharmaceutical Applications, *Powder Handl. Process.,* Vol 7 (No. 2), April 1995, p 139
15. M. Bumiller and T. Oja, Measuring Particle Size at High Concentration, *Ceram. Ind.,* July 1997, p 24

Time of Flight

Brian H. Kaye, Department of Physics and Astronomy, Laurentian University

THE GALAI INSTRUMENT is one of the first time-of-flight instruments used to characterize the size of powder grains (Galai Production Ltd., Ha Hameak, Israel). For a time this instrument was widely sold in the United States by the Brinkmann Company, so it is sometimes referred to as the Brinkmann size analyzer. Figure 1 shows the basic principles of the Galai instrument. Light from a laser is focused to a fine beam, and the beam is rotated by the rotating optical wedge. As the laser tracks a circular path in the cell containing the particles to be characterized, the size of the profile is measured by the time required for the laser beam to track across the profile of the powder grain. The instrument determines when a particle is intercepted at its maximum width, and it is calibrated using standard fine particles. The instrument also incorporates a video camera to view the particles being characterized at right angles to the rotating laser (Ref 1).

Figure 2 demonstrates the accuracy of the time-of-flight, or time-of-transit method, by the agreement in the measurement of the powder-size distribution measured by the Galai instrument, with the known size distribution of a standard powder prepared and distributed by the European Commission on Standard Materials, Brussels, Belgium. The sample studied in Fig. 2 is BCR-67.

Another system that uses time-of-transit measurements to obtain particle size information is manufactured by Lasentec (Ref 2). This system is often used in on-line monitoring of material. Figure 3 shows the basic system used in their equipment. In technical literature (Ref 2) Lasentec states:

> "The focused beam will cross the particle structure on a straight line between any two points on the edge of the structure. The distance between two points is a chord length. Thousands of chords are typically counted per second. The number of chords measured over a specific time period are sorted by chord length into a 38 channel distribution. The resulting chord length distribution tracks changes in particle geometry, a function of the shape and dimension of the particles, and particle structures as they naturally exist in process."

Two time-of-flight instruments that study the size of fine particles in the aerosol format are available from TSI Inc., St. Paul, MN, and Amherst Process instruments Inc., Amherst, MA. Figure 4(a) shows the basic system of the instrument manufactured by TSI, known as the aerodynamic particle sizer (APS) analyzer. After the powder to be characterized is made into an aerosol, the aerosol is fed into the equipment. As shown in the Fig. 4(a), some of the aerosol under study is filtered to create a clean air sheath, which confines the aerosol being created to the center of the measurement zone. The two laser beams are 125 μm apart and ~200 μm downstream from the nozzle orifice. The air feed to the interrogation zone is used to accelerate the particles, which respond to the accelerating

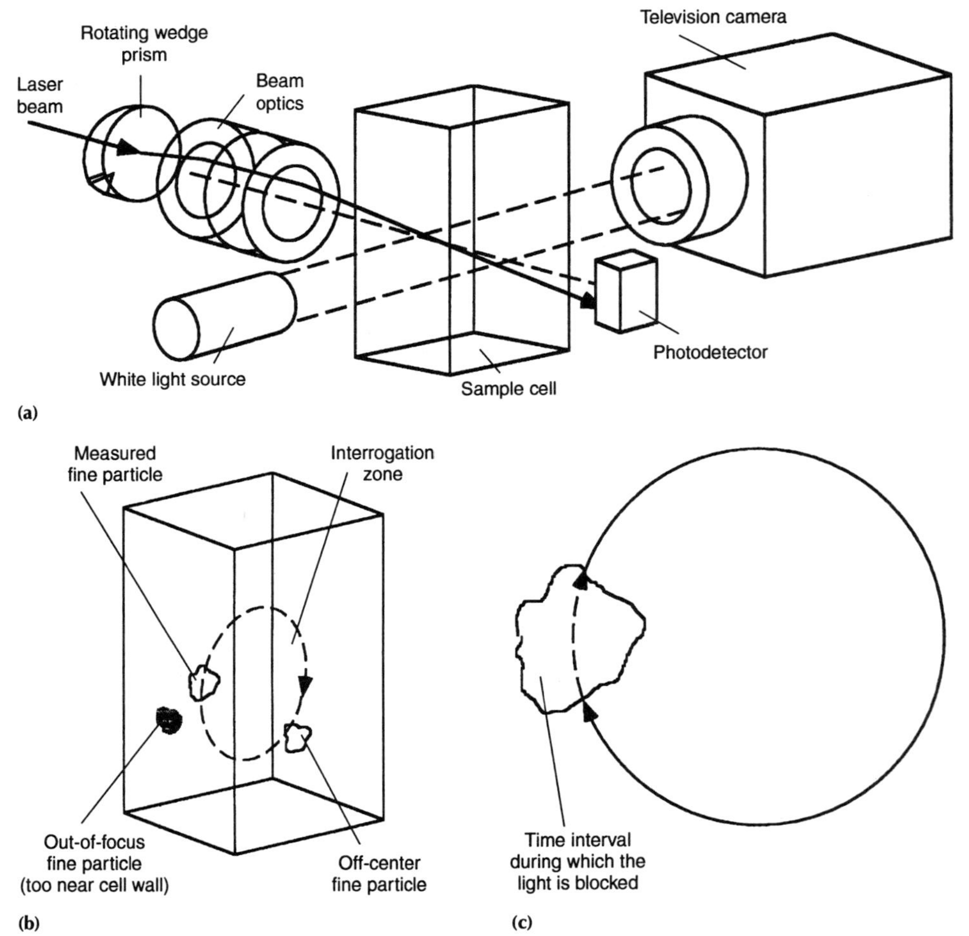

Fig. 1 Galai particle size analyzer. (a) Basic layout of the Galai instrument. (b) Laser beam tracing a circular path within the cell and the logic of the instrument rejecting any particles that are off center or out of focus. (c) Laser beam blocked from reaching the photodetector

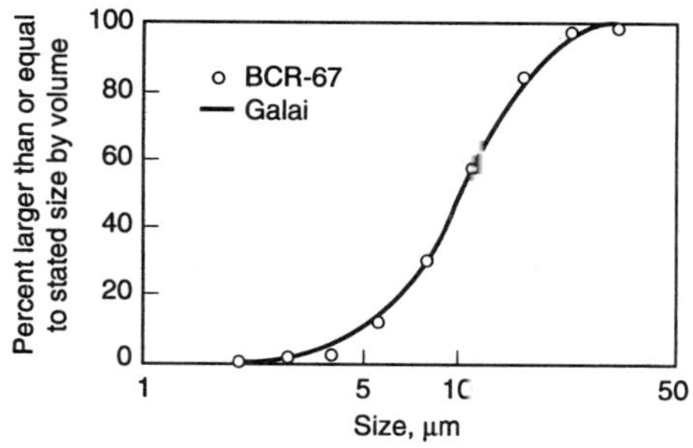

Fig. 2 Data for BCR-67, standard quartz test powder, obtained form the Galai particle size analyzed compared with known size distribution

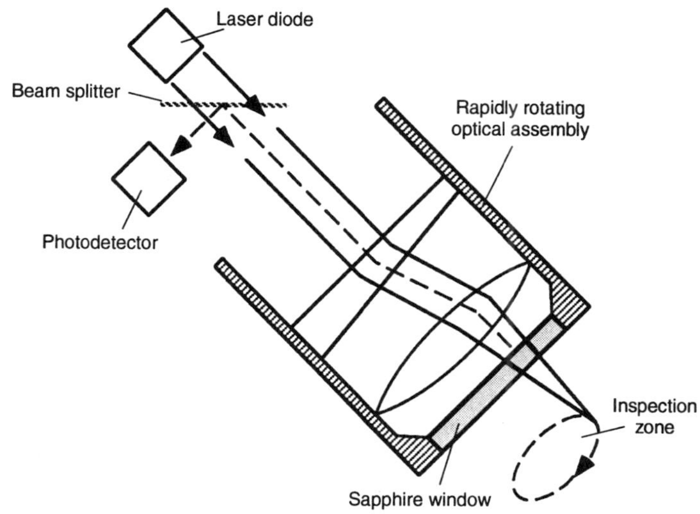

Fig. 3 Lasentec focused beam reflectance measurement system

forces at varying rates depending on their mass. The time-of-flight between the two laser beams is measured and used to calculate the particle velocity. The instrument is able to measure particles in the region of 0.5 to 30 μm. The instrument is calibrated using standard polystyrene spheres. Figure 4(b) shows typical results for the instrument.

The parameter of the powder grains measured by this instrument is known as the aerodynamic diameter of the particles. The aerodynamic diameter of a particle is defined as the size of a sphere of unit density having the same velocity as the particle under the same flow conditions. This aerodynamic diameter differs from diameters measured in other methods of particle size, and data need to be interpreted in a knowledgeable manner. Several publications deal with the relationship between particle sizes measured by time-of-flight and other well-known methods of size analysis, such as microscope image analysis and diffractometers (Ref 3).

The TSI Inc. system works at subsonic velocities and uses one photocell. In 1994, TSI Inc. announced an advanced version of their instrument, the Model 3320 aerodynamic particle size spectrometer, using double crested optics. TSI Inc. claims that the new model sizes particles from 0.5 to 20 μm aerodynamic size and can operate at 1000 particles/cm^3.

Figure 5 shows the system used in the Aerosizer, manufactured by Amherst Process Instruments Inc. Note that one of the problems in using instruments such as a time-of-flight aerosol spectrometer is ensuring that the powder under study is efficiently dispersed as an aerosol (Ref 4). The Amherst instrument has an auxiliary piece of equipment, the Aero-Disperser, for dispersing powders that enable the investigator to vary the shear rate applied to the powder being aerosolized (Fig. 6). The equipment is used to

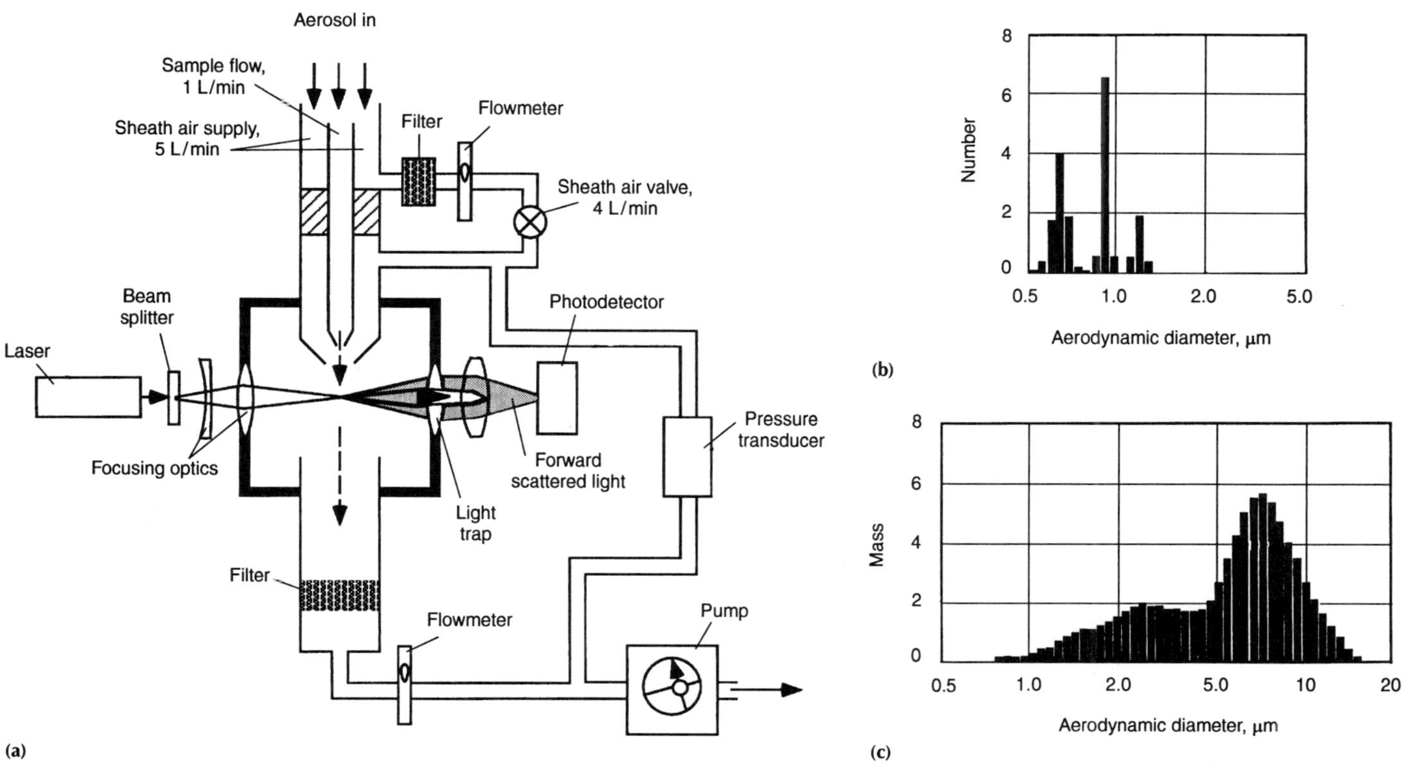

Fig. 4 The TSI aerodynamic particle sizer. (a) Basic layout. (b) Results from a mixture of three sized of polystyrene latex spheres. (c) Results for a therapeutic aerosol

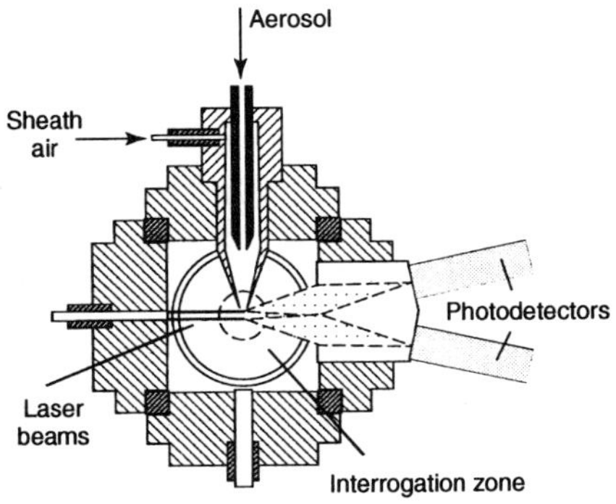

Fig. 5 Schematic diagram of the inspection zone of the Amherst Process Instruments Inc. Aerosizer time-of-flight particle sizer

carry out investigations of various shear rates applied to powder as it is aerosolized, until further increase in the shear rate does not result in a difference in distribution measurements. The aerosol being interrogated is surrounded by sheath air as the particles under study are accelerated into the interrogation zone. As shown in Fig. 5, the interrogation zone differs from that of the APS equipment in that two photocells are used to measure the optical energy scattered by the particles as they move through the two beams. A sophisticated electronic editor distinguishes which signal arriving at the photodetectors is associated with the initial signal from a given particle as it enters the first laser beam. The Aerosizer operates at sonic velocities and can count fine particles at a very high rate. The equipment measures the aerodynamic diameter and is calibrated using particles of known size. The output can be either a measure of the number of particles of a given aerodynamic size (Fig. 7a) or as a transformation into geometric size by volume (Fig. 7b). The geometric size by volume is a transformation of the aerodynamic diameter and takes into account the density of the particles being measured.

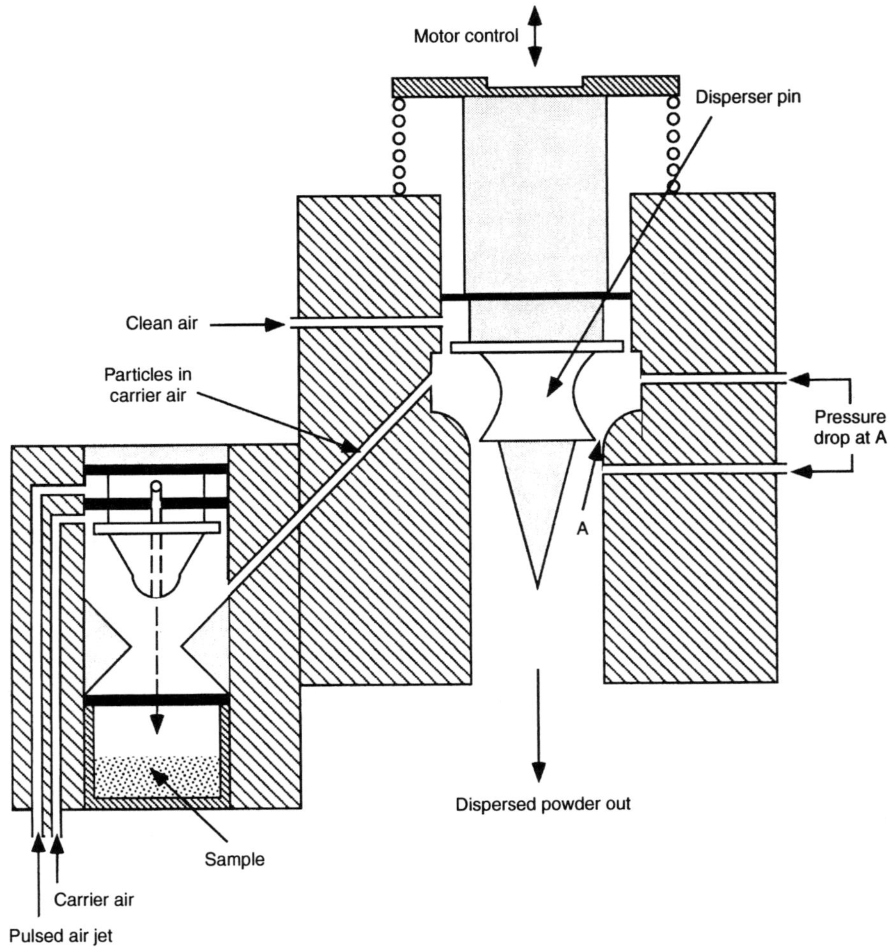

Fig. 6 Aero-Disperser used with the Aerosizer for aerosol preparation using controlled shear rates to disperse the sample and ensure that there is no agglomeration

REFERENCES

1. O.B. Weiner, W. Tscharnuter, and N. Karasikov, Improvements in Accuracy and Speed Using the Time-of-Transition Method and Dynamic Image Analysis for Particle Sizing; Some Real World Examples, *Proc. of the Symposium on Particle Size* (Orlando, FL), American Chemical Society, 1996
2. Technical literature, Lasentec, Redmond, WA
3. F.M. Etzler and M.S. Sanderson, Particle Size Analysis: A Comparative Study of Various Methods, *Part. Part. Syst. Charact.*, Vol 12, 1995, p 217–224
4. B.H. Kaye, Generating Aerosols, *KONA*, (No. 15), 1997, p 68–80

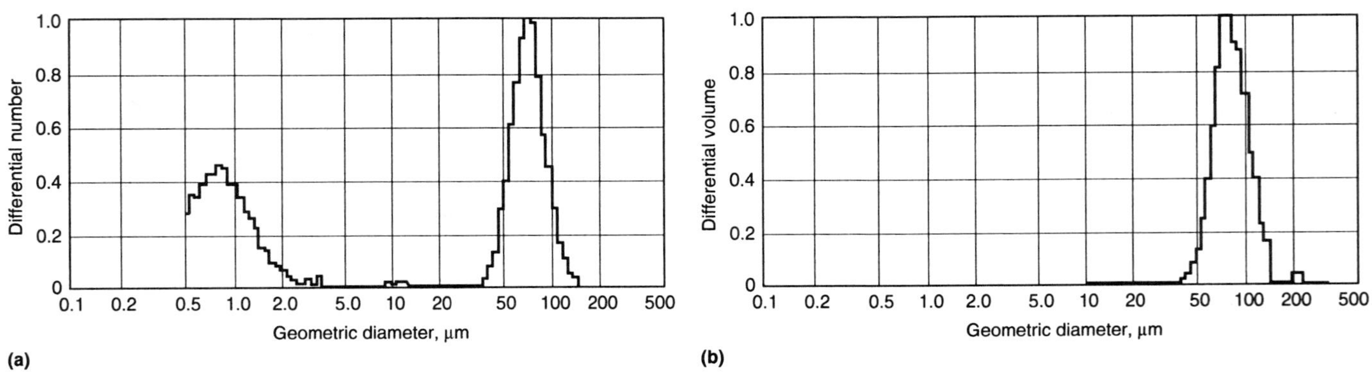

Fig. 7 Output of measurements for the size distribution of an irregular iron powder. (a) Number size distribution of the iron powder. (b) Volume size distribution of the iron powder

Particle Image Analysis

MICROSCOPY is the most definite method of particle size analysis, because individual particles are observed and measured. However, optical and electron microscopy also afford the opportunity for particle shape analysis. Particle shape, like particle size, is a primary powder characteristic that must be considered when determining the optimum utilization of a P/M material. Behavioral characteristics of a metal powder, such as flow rate, apparent density, compressibility, and sinterability, are all influenced by particle shape and size.

This article briefly reviews the common methods of optical and electron microscopy for image analysis of particle size distributions and particle shape characterization. Image analysis of particle size distribution is a very precise quantitative practice, while particle shape analysis is more qualitative in nature. Advancements in qualitative shape analysis are based on the use of the scanning electron microscope for determination of particle shape. In addition to its high magnification capability, this device produces an image with a three-dimensional appearance. Because of its advantages, scanning electron microscopy (SEM) has replaced optical microscopy to a large extent in the area of shape and topographical analyses of metal powders.

Quantitative shape analysis has posed numerous difficulties. However, with the advent of advanced computers and graphics systems, interest in quantitative analysis has increased. Through the use of optical and scanning electron microscopy, qualitative descriptors have been developed to label different particle shapes. As computational technology advanced, mathematical algorithms were written to describe particle shape. Fourier analysis and fractal geometry are two mathematical disciplines that are being applied to describe particle shape. Because of the computational intensity of these disciplines, most of the work has been confined to a two-dimensional space. The current challenge is to develop three-dimensional modeling programs that can accurately describe the shape of irregularly shaped particles.

Image Analysis of Powder Size

Microscopy is a very precise method of particle size analysis, but some of the techniques used are more art than science. However, if the basic principles of sampling, preparation, and counting are followed, a precise count can be made with a thorough understanding of the nature of the particles being studied. ASTM standard E 20 details the use of microscopy for particle sizing.

Size Measurements

Various techniques are used to measure the size of irregularly shaped particles when viewed through a microscope. This has resulted in different measurements, which are used to classify these two-dimensional particle images in terms of an equivalent spherical particle. Accepted measurements (Fig. 1) include:

- *Feret's diameter (F)*: The maximum length of the particle measured in a fixed direction
- *Martin's diameter (M)*: The length of a line that bisects the area of the particle image; all particles measured in the same direction
- *Project area diameter (d_a)*: The diameter of a circle with the same area as the two-dimensional image of the particle
- *Longest dimension:* The maximum Feret's diameter for each particle: no set direction
- *Perimeter diameter (d_p)*: The diameter of a circle having the same circumference as the perimeter of the particle
- *Maximum horizontal intercept*: The length of the longest line that can be drawn through the particle in a fixed direction

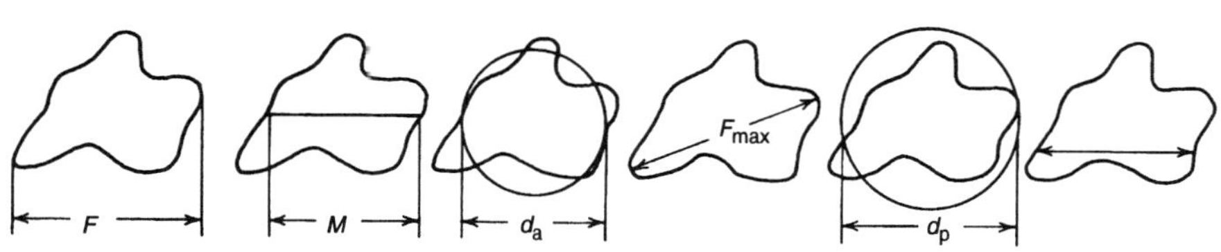

Fig. 1 Techniques for measuring sizes of irregularly shaped particles. See text for identification of variables.

Feret's diameter is the easiest to measure manually. The average Feret's diameter is related to the perimeter of the convex hull of the particle, as shown in Fig. 2, by the relationship:

$$F_{av} = \frac{P_{CH}}{\pi}$$

where P_{CH} is the perimeter of the convex hull. Martin's diameter is related to the specific surface S_v of the particle by:

$$M = \frac{4}{S_v}$$

The projected area diameter gives the best estimate of the true cross-sectional area of the particle.

Generally, the three diameters above are related to one another by the expression $M < d_a < F$. Their ratios remain fairly constant for a given material. The expression F/M has been used as a shape function, which is equal to 1 for spherical particles and increases in magnitude as particle shape becomes more acicular. Perimeter diameter and maximum horizontal intercept (or longest chord) are measured easily by some of the automatic particle sizing analyzers.

Sampling Techniques

The most important step in any particle size analysis, but especially by microscopic techniques, is sampling from the bulk. Because an extremely small quantity of material is used to determine the particulate size, an accurate analysis cannot be obtained if the bulk material is not properly sampled. Particles tend to segregate according to size. If handling has caused vibration of the sample, coarse material tends to collect near the surface. When free-flowing material is poured into a pile, the coarse portion collects near the outside of the pile, and the fines concentrate in the center. The following sampling techniques are recommended for obtaining a valid representation of particle sizes in a laboratory sample.

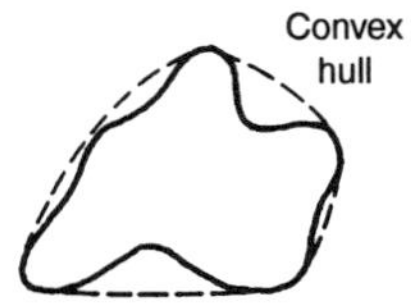

Fig. 2 Feret's diameter

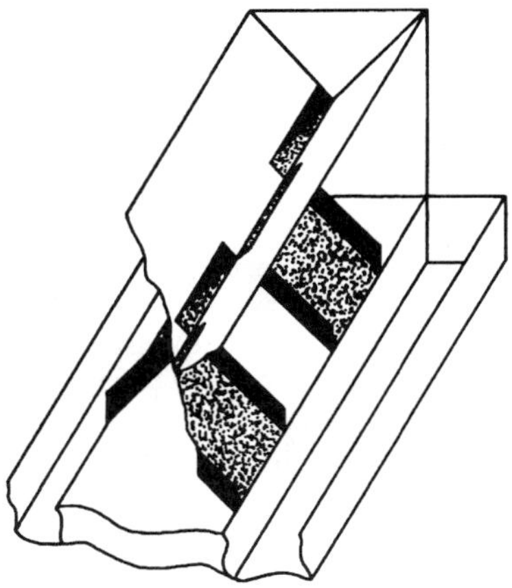

Fig. 3 Chute riffler

Chute rifflers (Fig. 3) consist of V-shaped hoppers that feed a series of chutes, which alternately feed into trays on opposite sides of the hopper. A sample that is properly poured into the hopper is halved, and the process is repeated until a suitable sample size is obtained. Sampling efficiency depends on an even feed of the sample into the hopper and on the width of the chutes. Numerous narrow chutes provide more accurate sample statistics than a few wide chutes, provided the powder can flow easily through them.

Spinning rifflers (Fig. 4) divide a sample by feeding it into a number of containers that rotate on a table under the feed. This method is efficient, provided the table rotates at least 100 times during the flow of the sample.

Oscillating Sampler. The feed hopper oscillates rapidly between two sampling cups, thereby halving the sample. Feed times must be long compared to oscillation times for optimum efficiency.

Cone and Cup Sampler. In this device, the sample is fed onto a cone that disperses the powder in all directions along the cone surface. A sample cup is rotated around the base of the cone, gathering portions of the sample at every position. The width of the cup and length of the cone periphery determine the sample reduction.

Sampling tables (Fig. 5) are used to reduce larger quantities of the sample. Sampling tables consist of an inclined plane with a series of holes that halve the sample in each of the four stages, leaving $^1/_{16}$ of the original volume as a final sample.

Cone and quartering involves pouring the sample into a cone-shaped pile on a glass plate, flattening the top, and dividing it into quarters

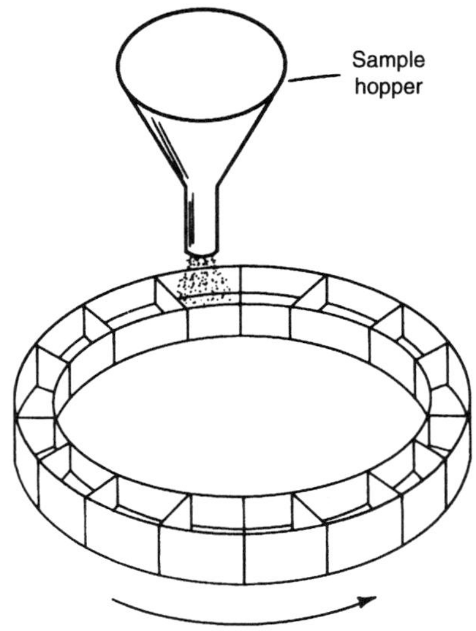

Fig. 4 Spinning riffler

with a thin metal blade. One of the quarters can then be mixed, repoured, and requartered until a suitably small sample is obtained. This method requires some technique to ensure symmetry of particle sizes in the pile and accurate quartering.

Scoop sampling is the least precise, but perhaps the most frequently used, sampling technique. The sample container is shaken, and a small portion is extracted with a scoop or spatula. However, segregation can be imparted to the sample by the shaking method. The scoop sampling technique that provides the least variance is shown in Fig. 6 and consists of shaking the container back and forth while rotating. Because particle sampling is the initial and most important step in particle size analysis, miniature sampling devices should be used whenever possible.

Optical Microscopy

The optical microscope is used to count and size particles ranging from 100 to 0.5 μm diameter. The resolution of an optical microscope is optimally about 0.25 μm, but diffraction effects at the edges of small particles result in significant errors in measurements within this size range. Resolution down to 0.1 μm can be improved by using ultraviolet light and quartz optics.

A severe limitation of the optical microscope is its small depth of field, which is about 10 μm at 100 and 0.5 μm at 1000×. This requires that the specimen powders be mounted carefully in one plane and eliminates the use of automatic counting devices on samples with wide particle size ranges. Another limitation is that due to the small number of particles usually counted, only the field of view can be characterized, not the entire lot from which the specimen was obtained.

Preparation of slides for optical microscopy is an art, and the quality of a dispersion depends more on the skill of the preparer than on the

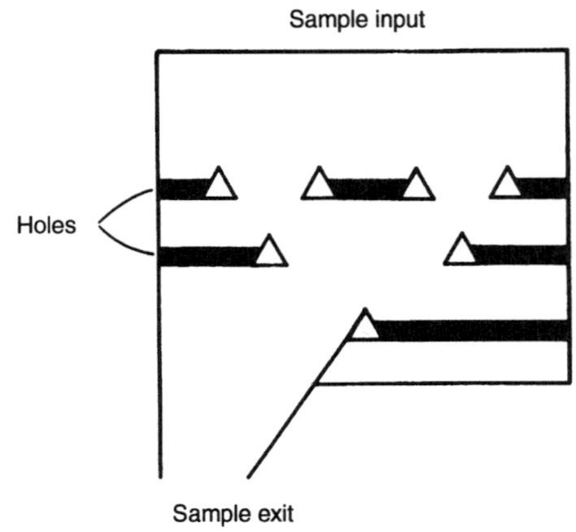

Fig. 5 Sampling table diagram

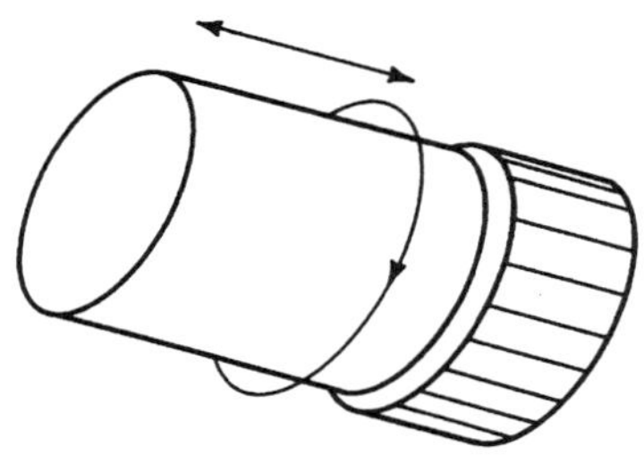

Fig. 6 Scoop sampling apparatus showing direction of shaking

particular technique used. Care must be taken to prevent segregation of particle sizes and to prevent the formation of agglomerates. Preparation techniques vary with the type of powder and depend on particle size, particle size range, and particle composition. Typical dispersion techniques are described below.

Dry Mounting. When a permanent slide is not required, and the powder exhibits good flow characteristics, slides can be prepared by "puffing" dry powder onto the slide or by applying it with a brush.

Dispersing Fluids. In applications that require shearing action to disperse the powder, a dispersing fluid such as cedar oil or glycerol can be used. The simplest technique involves placing a drop of dispersing fluid on a glass slide and placing a small amount of powder onto the drop. A cover slip is then placed carefully over the mounting fluid and pressed evenly with an eraser, while gently rotating the cover slip back and forth.

Another method involves placing some of the powder on a glass slide, adding a few drops of dispersing fluid, and working the mixture with a flexible spatula or soft brush. More fluid is added until the proper concentration is obtained. A drop of this mixture is then transferred to a new slide and covered with a cover slip.

Permanent Slides. A small amount of powder is placed in a beaker to which 2 or 3 mL (0.07 or 0.10 fluid oz) of a solution of 1 to 2% collodion in butyl acetate is added. The mixture is stirred vigorously, and a drop of the suspension is placed on the surface of distilled water. The drop spreads and forms a thin film as the solvent evaporates. The film is then picked up on a microscope slide and dried.

Other films can be produced by dissolving parlodion in amyl acetate, Canada balsam in xylene, and polystyrene in xylene. Films are formed by casting a 1% solution on water or a ½% solution directly on a glass slide. After the films dry, the powder is applied by spraying or by letting a drop of aqueous suspension dry on the surface. The particles sink into the medium, forming a permanent slide. Melted Canada balsam or glycerol jelly also can be used to form permanent samples.

Other Methods. Certain applications may require samples prepared by specialized techniques, such as heating magnetic particles above their Curie points for dispersion. These examples generally serve as guidelines for technique development.

Particle sizing usually is performed indirectly in the optical microscope, because the small depth of field frequently requires refocusing on individual particles. If the size range is narrow and the preparation is carefully made in a single plane, photographs or automatic particle counting devices can be used. Projection microscopes are also available that reduce operator fatigue for analyses that require numerous counts.

Transmission Electron Microscopy

The transmission electron microscope is used for counting particles that range from 0.001 to 5 μm in diameter. This instrument has a large depth of field; consequently, all particles in the field of view are in focus regardless of size. Particles are usually not counted directly from the viewing screen of the transmission electron microscope. Photographs are normally taken, and counts are made from prints or projected images using the negatives.

The electron beam is easily absorbed, and films greater than 100 to 200 nm (1000 to 2000 Å) are completely opaque. It is therefore necessary to produce very thin support films on which powders can be dispersed for counting. These films usually are made of carbon. One of the best techniques for producing strong, flat carbon support films involves cleaving high-quality mica and placing it immediately in a vacuum evaporation unit. After a hard vacuum has been obtained, carbon is evaporated onto the mica surface to a thickness of about 10 nm (100 Å).

Film thickness can be calculated by including a white porcelain slide containing a drop of vacuum oil in the vacuum chamber. When the porcelain that is not covered by the oil appears as a light chocolate shade, the carbon film is about 10 nm (100 Å) thick. The mica is then removed from the vacuum and slowly lowered at an angle into a beaker of distilled water. The carbon film floats off the mica onto the water surface, where it can be picked up on an electron microscope grid.

Another technique involves casting a thin film of parlodion onto the surface of distilled water by depositing two or three drops of 1% parlodion dissolved in amyl acetate onto a clean glass slide. The slide is allowed to drain and dry, and the film is floated from the slide onto distilled water. The plastic film is then picked up on transmission electron microscopy (TEM) grids, placed in a vacuum evaporator, and coated with carbon.

The parlodion can be carefully dissolved in amyl acetate by placing the grids on a fine stainless steel mesh bridge suspended in the center of a petri dish. Amyl acetate is added to the petri dish until the level of the liquid reaches the bottom of the mesh. The dish is then covered and left for a few hours, after which the grids are carefully removed.

If the parlodion film is thin enough, it can be left on the grids. However, this requires fresh, water-free solutions and care in handling, or holes will form in the thin plastic film as the solvent evaporates. Prepared carbon substrates on grids are available.

Powders are dispersed on the surface of support films by puffing from an aspirator or by allowing a drop of aqueous suspension to dry. Carbon films are hydrophobic; use of an aqueous suspension requires that they be rendered hydrophilic by either treatment with a thin solution of albumen or exposure to reactive oxygen in a low-temperature oxygen asher. Exposure to reactive oxygen must be done at a very low setting (4 W) for several seconds, or the carbon film will be destroyed.

Powders that require additional dispersion can be prepared by placing a small amount of powder on a clean glass slide and adding several drops of 1% parlodion in amyl acetate. This mixture is sandwiched between a second glass slide. As the amyl acetate evaporates, the mixture becomes more viscous. Immediately before the parlodion dries, the slides are sheared apart. This preparation can be viewed in the optical microscope to find the area of best dispersion. This area is then scribed into 3.2 mm (⅛ in.) squares and floated from the glass slide onto distilled water. The squares are picked up on electron microscope grids and coated with carbon to add stability to the electron beam.

For some samples that are difficult to disperse, ethylene glycol may be a suitable dispersing agent. A small amount of powder is added to a few drops of this viscous liquid on a glass slide and mulled until a good dispersion is obtained. A carbon substrate grid is then touched to the thin liquid film, which sticks to the grid. The grid is placed in a vacuum chamber and pumped for a few minutes, during which time the ethylene glycol sublimates, leaving the particles on the carbon substrate.

Particle thicknesses can be measured in the transmission electron microscope by "shadowing" particle dispersions in which the particles sit on a substrate and are not embedded in a plastic film. This is done by evaporating a small amount of metal placed at an angle to the substrate surface in a vacuum chamber. The metal coats the surface and particles in a line-of-sight fashion, leaving a "shadow" cast behind the particles. Commercially available precision-sized latex spheres can be included with the powder sample so that the shadow length to particle height ratio can be calculated.

Of the pure metals, platinum is best suited for shadowing because of its fine structure in an evaporated film. About 12 mm (0.5 in.) of 250 μm (10 mil) platinum wire, evaporated from a point source at an angle of 30° and a distance of 100 mm (4 in.), casts a good shadow.

Many other preparation techniques are used to observe powders in TEM, and most analysis methods are modified to suit the application and the operator's ability.

Scanning Electron Microscopy

The scanning electron microscope has a resolution of about 10 nm (100 Å) and is capable of very low magnification (about 10×) up to about 50,000×. It therefore can be used to count particles ranging in size from 1 mm to 0.1 μm. Particles smaller than 0.1 μm usually have too low a contrast with the background to be counted efficiently. The scanning electron microscope has about 300× the depth of field of an optical microscope.

The image in the scanning electron microscope usually is obtained by using the secondary electron output of the sample as it is scanned by a very narrow electron beam. The contrast of the image depends more on the topography of the sample than on differences in atomic number. Therefore, prepared powders must not be embedded in films, but dispersed on a smooth substrate. Any smooth surface can be used as a substrate. However, if energy dispersive x-ray analysis (EDXA) is to be performed for particle identification, a carbon or polystyrene surface is preferred.

An excellent substrate can be made by placing a polystyrene pellet on a glass slide and heating it on a hot plate until it softens. A second glass slide is then placed over this slide and pressed until the pellet forms a thin disk. The slides are removed from the hot plate and pressed together until the polystyrene sets. The disk thus formed is as smooth as the glass and contains no elements that may hinder EDXA. For sample preparations using aqueous suspensions, polystyrene surfaces can be rendered hydrophilic by a brief treatment in an oxygen asher at low power (5 to 10 W for 5 s).

While the substrates for SEM do not have to be as thin as those used for TEM, they must be conductive. Consequently, if glass or plastic surfaces are to be used, they must be coated with an evaporated metal (or carbon, for EDXA) film. This coating is usually applied after the particles have been dispersed on the surface.

Many of the dispersing techniques used for TEM can be applied to SEM. Particle dusting, drying from liquid suspensions, and mulling in liquids that can be sublimed in a vacuum are suitable dispersing methods, depending on the powder. If the technique of mulling in parlodion and amyl acetate is used, parlodion can be removed in an oxygen asher, thus leaving the particles on the substrate. Suitable substrates include glass or metal, because they are not affected by the ashing.

The prepared sample should always be placed in the scanning electron microscope with the surfaces normal to the electron beam so that the magnification, which changes with working distance, will be the same on all areas of the viewing screen. Particle counting can be done directly from the viewing screen, from photographs, or by using an automatic image analyzer.

Measurement Techniques

Direct Measurements. Particle diameters can be measured directly in an optical microscope with the use of a filar micrometer eyepiece. This eyepiece contains a scale and a movable cross hair that is operated by a calibrated knob on the side of the eyepiece. A particle is moved so that one side touches one of the fixed scale markings, and the cross hair is moved to touch the other side of the particle. The difference between the two readings is the Feret's diameter of the particle. The eyepiece is calibrated with an optical stage micrometer. This technique is time consuming, and because absolute measurements are not required for accurate size analysis, eyepiece graticules are usually used for direct measurements.

Figure 7 shows typical examples of eyepiece graticules. These patterns are etched on glass disks that are positioned in the back focal plane of the microscope ocular and are therefore in the same focal plane as the particle images. Using one of these gauges, the size range, into which the actual diameter of a particle falls, can be measured easily. Feret's diameter, projected area diameter, or perimeter diameter are measured conveniently with graticules. Where measurements are made from photographs, projected images, or cathode ray tube screens, simple scales or plastic overlay graticules can be used to measure any of the accepted diameters.

Image shearing eyepieces are used directly within the optical microscope. This device divides the optical beam into two parts by using mirrors and/or prisms. The distance between the two formed images is adjusted by shearing of the prisms, which is controlled by a micrometer dial on the side of the eyepiece. The maximum horizontal intercept is measured when the two images barely touch. With this device, a particle does not have to be moved to a particular position within the field of view to be measured.

The micrometer can be set to a given diameter, and the number of particles larger or smaller than this measurement can be counted, as determined by whether or not the images are touching. Red and green filters facilitate this by coloring the two images.

Semiautomatic Techniques. Many devices are available to shorten the time required to complete a particle size analysis. Most of these instruments record the number of particles in a size range, as judged by the analyst.

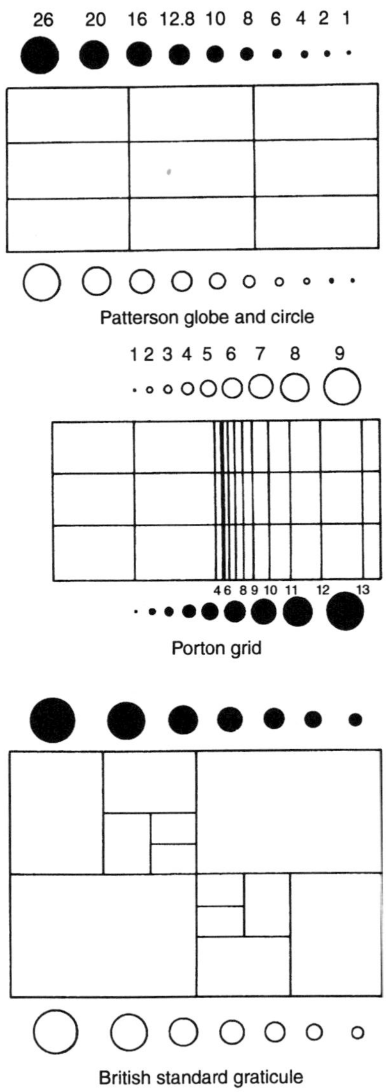

Fig. 7 Eyepiece graticules

Adjustable light spot analyzers use a circle of light projected onto a photograph. The diameter of the circle can be adjusted by a diaphragm; a foot switch causes one of a bank of registers to record the diameter according to the diaphragm setting. Each particle on the photograph is passed under the light, which is adjusted to measure the projected area diameter or the perimeter diameter.

A transparent electronic graticule is a plastic sheet on which various size circles have been drawn. Under each circle, an electrical contact is connected to one of a bank of registers. When a contact is touched by the analyst with a pencil contact, the register advances one number. This device can be used with photographs, back-projected images, and cathode ray tube screens. The analyst decides which circle best represents the diameter of a particle and touches the corresponding contact.

Recording calipers work in the same manner, with a bank of registers connected to the caliper spread adjuster. Any particles of the prescribed diameter can be recorded with this device.

Recording micrometer eyepieces use a similar recording mechanism, but work directly within the optical microscope. Two movable cross hairs that move toward or away from each other with the turn of a micrometer screw are used to bracket a particle; a button is subsequently pushed to record the diameter. This technique is particularly useful when the depth of field prevents photography of the particles, because each particle can be sharply focused before measuring.

Sensitive surfaces connected to computers facilitate counting from photographs. The computer can be programmed to record diameters when the opposite sides of particles are touched with a pointer, or specific areas of the particle are touched. The computer consequently compiles data to suit user requirements.

Automatic techniques are used for counting large numbers of particles directly within the microscope. The main limitations of these techniques are the inability to distinguish particles that are touching and the fastidious sample preparation that is required.

Spot scanning devices use the moving spot from a cathode ray tube, which is projected through the microscope onto the specimen. When the spot passes over a particle, the light beam is interrupted, and a photocell records the particle. Particles are sized by scanning with different spot sizes. This technique also has been used on photographs, so electron micrographs can be used.

The slit scanning method projects the microscope image onto a slit. The microscope slide is mechanically scanned, and the signal produced as the particle images pass the slit is recorded. The width of the slit can be varied to eliminate coincidence and overlap.

Quantitative image analyzers are computer-controlled devices that use television cameras for direct analysis within an optical microscope. Analysis of photographs can also be accomplished. A quantitative image analyzer can be connected directly to a scanning electron microscope to control the scanning system. Particles can be counted according to their maximum horizontal chords, vertical chords, perimeters, or areas.

Data Presentation. Table 1 shows the most precise form of presenting particle size data. However, graphical representations are more concise and visually show the mean and deviation from the mean. Some graphical presentations can yield specific values for descriptive constants such as the mean, median, and the standard deviation.

A histogram is a bar graph that illustrates the frequency of occurrence as a function of the size range. Figure 8 shows a typical histogram for a log-normal size distribution. The smooth curve drawn through the histogram is a valid size-

Table 1 Particle size data

Particle size range, µm	No. of particles	Particles, %	Cumulative percent less than size
1–2	0	0.0	0
2–3	3	0.6	0.6
3–4	8	1.6	2.2
4–5	15	3.0	5.2
5–6	79	15.9	21.1
6–7	163	32.9	54.0
7–8	121	24.4	78.4
8–9	64	12.9	91.3
9–10	28	5.7	97.0
10–20	13	2.6	99.6
20–30	2	0.4	100

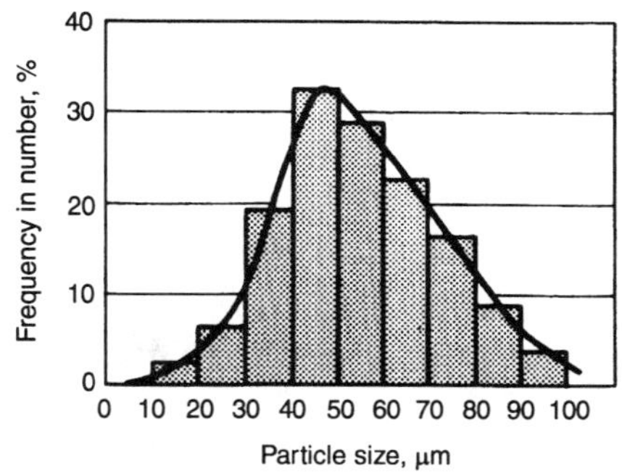

Fig. 8 Histogram and size frequency curve for log-normal size distribution

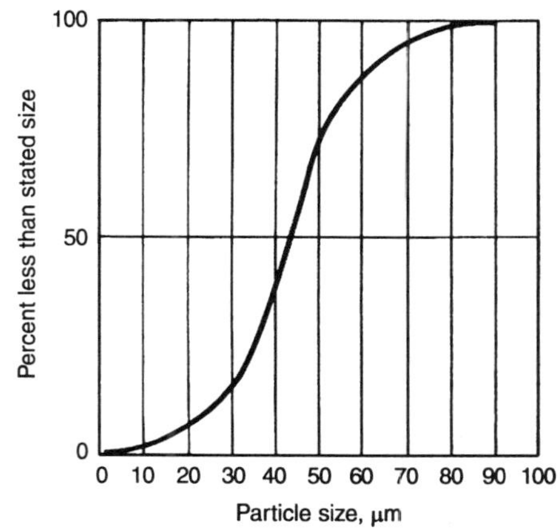

Fig. 9 Cumulative plot used in determining median particle size

frequency curve if sufficient particles are counted and the number of size intervals is at least ten.

In cumulative plots, such as Fig. 9, the median particle size can be noted easily. These plots are made by plotting the percentage of particles less than (or greater than) a given particle size against the particle size.

A special grid for plotting size-frequency data was developed so that the resulting curve is a straight line. This grid consists of a coordinate based on probability. Cumulative percentages are plotted on the probability scale against particle size on either an arithmetic or a log scale.

Symmetrical size distributions, characterized by a bell-shaped size-frequency curve symmetrical to a vertical line through the mode (apex) point, plot as a straight line on arithmetic-probability paper. Most natural particle distributions do not follow this normal probability, but follow a skewed or log-normal distribution. These distributions provide a symmetrical size-frequency plot if the particle-size scale is logarithmic and forms a straight line on log-probability paper. The median particle size is found at the 50% point on these probability plots and the standard deviation for arithmetic-probability plots can be calculated as:

$$\sigma = 84.13\%\ \text{size} - 50\%\ \text{size}$$
$$= 50\%\ \text{size} - 15.87\%\ \text{size}$$

and for log-probability plots can be calculated as:

$$\sigma = \frac{84.13\%\ \text{size}}{50\%\ \text{size}} = \frac{50\%\ \text{size}}{15.87\%\ \text{size}}$$

Particle Shape Characterization

Brian H. Kaye, Department of Physics and Astronomy, Laurentian University

Figure 10 shows a typical scanning electron micrograph of an array of iron particles for the type of powder grains encountered by the powder metallurgist in various areas of technology. Many of the techniques for characterizing the shape of powder grains make use of image analysis using pictures such as that shown in Fig. 10(a) (Ref 1). When an array of fine particles to be characterized is inspected, the first thing to be done is to decide what constitutes a separate powder grain. The economics of taking photographs using an expensive scanning electron microscope require working with a relatively dense field of view, which creates pseudoagglomerates from random juxtaposition. Thus, for the field of view shown in Fig. 10(a), the decision was made that the dashed lines represent separations between individual grains, which are close to each other.

Several commercial instruments are available for automated image analysis. Some automated image analysis systems have built-in logic procedures known as erosion logic. This procedure enables the device to strip ribbons of pixels away from an image to separate contiguous profiles to allow juxtapositioned particles to be evaluated as separate entities. Figure 10(b) shows the operation of erosion logic to separate profiles for subsequent image analysis. The software used in the work reported here was a program in which a television camera was used to obtain images through a light microscope.

The simplest method for characterizing the shape of profiles such as those shown in Fig. 10(a) is to evaluate what is known as geometric shape factors. In many cases, studies are carried out using fundamental measures on a set of profiles, such as those of Fig. 10(a), which are assumed to be in a position of maximum stability on the support system when they are photographed.

It should be noted that three-dimensional measurements on powder grain profiles can be used if the image is shadow cast with either evaporated metal or light illumination used in an appropriate manner. Thus, Fig. 11 shows a set of pieces of material as they might have been viewed by an optical microscope. The profiles are illuminated at a given angle to create shadows shown in the diagram (Ref 2).

The British Standards Institute has prepared a standard glossary of terms for use in the description of the appearance of powder grains (Ref 3):

- *Acicular:* Needle shaped
- *Angular:* Sharp edged or roughly polyhedral shaped. (Polyhedral derives from poly, meaning many, and hedra, meaning a base; therefore, polyhedral is understood to be a geometric shape having many faces, each of which can act as a base.)
- *Crystalline:* A geometric shape freely developed in a liquid
- *Dendritic:* A branched crystalline shape
- *Fibrous:* Regularly or irregularly threadlike
- *Flaky:* No formal definition in the British Standards. (It is assumed that a flaky fine particle is recognizable.)
- *Lamellar:* Platelike
- *Granular:* Approximately equidimensional but irregularly shaped
- *Irregular:* Lacking any symmetry
- *Modular:* Rounded, irregularly shaped
- *Spherical:* Globular shaped

The pioneer work on the development of shape factors was carried out by Heywood (Ref 4–6) and Hausner (Ref 7). Figure 12 shows the length, width, and thickness of a powder grain, as defined by Heywood. Using these measures, Heywood defined the elongation ratio as the length divided by breadth and the flakiness ratio as the breadth divided by thickness. Extensive reviews of procedures for deriving shape indices from geometric measurements have been prepared by Phal et al. (Ref 8) and Loveland (Ref 9).

Three-dimensional measurements of sugar grains have been used to develop three-dimensional graphs of a shape population variation within a powder (Ref 10).

The original shape factors described by Heywood are not currently in widespread use. They have, to some extent, been displaced by the aspect ratio and chunkiness, both of which are defined in Fig. 12. Figure 13(a) shows the simple silhouettes of the profiles shown in Fig. 11. Figure 13(b) shows the aspect ratio distribution for this population of profiles. When the population of aspect ratios are plotted in this way, it appears that there are two Gaussian populations of shapes in the powder. When plotted as a distribution function, the link of shape with size is lost in the calculation of the distribution of aspect ratios. In populations containing a wide range of

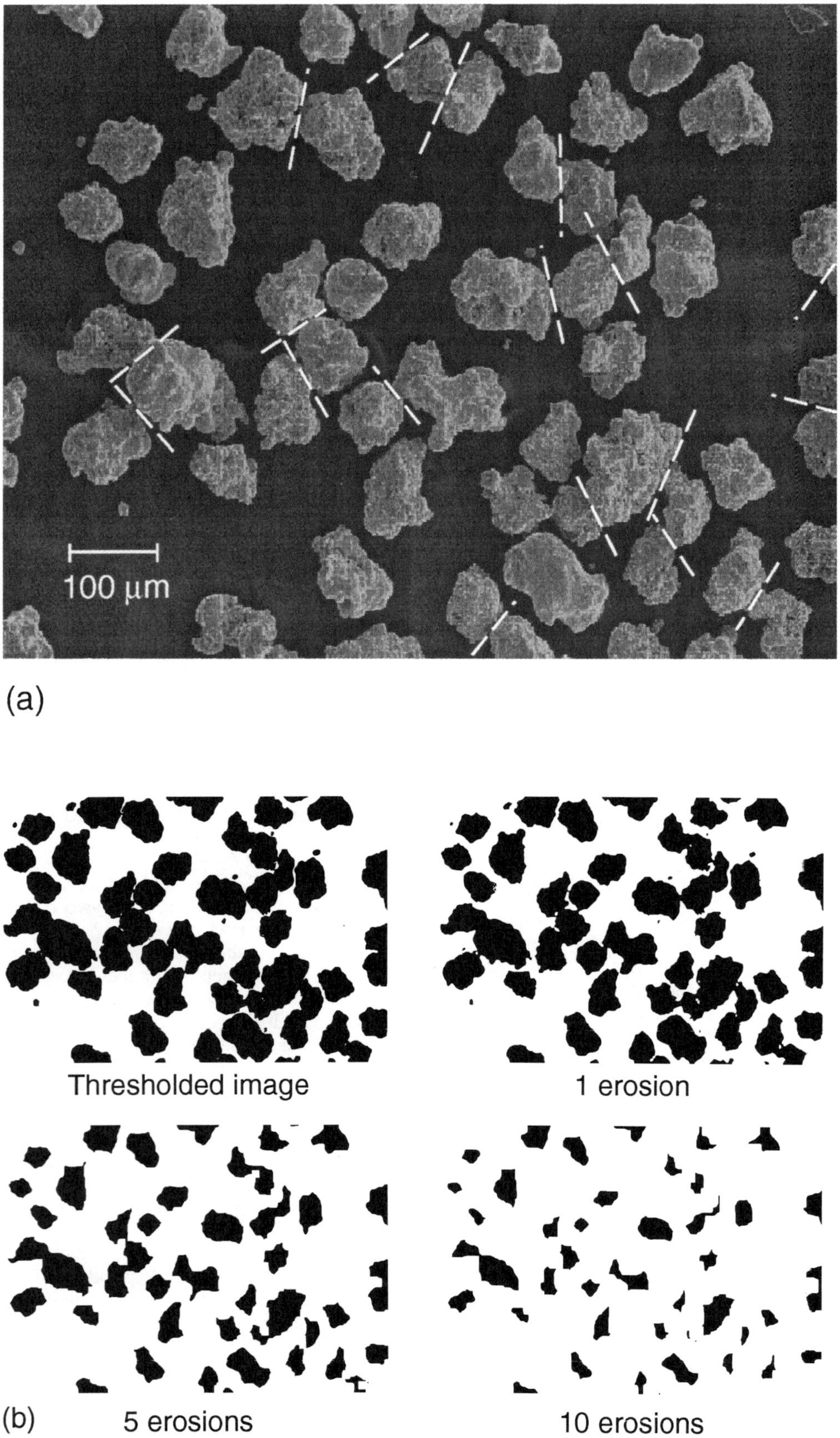

Fig. 10 Iron powder grains illustrate the type of shape characterization problems facing powder metallurgists. (a) Electron micrograph of an iron powder. Source: D. Alliet, Xerox Corporation. (b) Erosion of an image can be used to separate particles.

Fig. 11 Three-dimensional shape information determined by creating an image of a set of fine particles using shadow casting techniques.

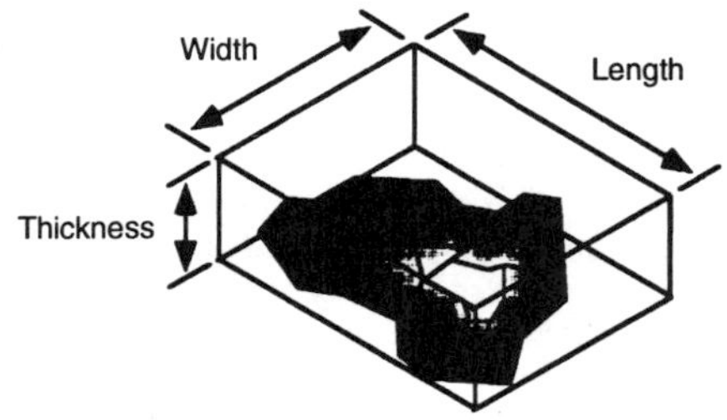

$$\text{Aspect ratio} = \frac{\text{Width}}{\text{Length}} = \text{Elongation}$$

$$\text{Chunkiness} = \frac{\text{Length}}{\text{Width}}$$

$$\text{Flakiness} = \frac{\text{Width}}{\text{Length}}$$

Fig. 12 Geometric shape indices defined on the basis of the magnitude of measurements made, either directly on the powder grain, or on images of the powder grains

shapes, the aspect ratio is not always useful, because in theory, it can vary from 1 to ∞, whereas the chunkiness can vary only between 0 and 1. In the graphical display known as the chunkiness-size domain, the shape of each profile can be plotted, as shown in Fig. 13(c). This type of plot retains the information on shape and size for each profile. Thus from the data display of Fig. 13, the technologist can see immediately that the shape increases with fineness of the powder, information that is not retrievable from Fig. 13(b).

Figure 14 shows the chunkiness size domain for the powders shown in Fig. 10(a), as determined by image analysis. This data display shows that powder is bimodal in size distribution. This probably arises from the fact that because the powder is magnetic, some fine powders were clinging to the coarser grains when the powder was made by a sieve fractionation process.

If measurements are carried out using the dimensional information obtained by shadow casting the images, a triangular grid can be used to plot the three-dimensional information on the profiles. Figure 15 illustrates the way measurements are made on the three-dimensional structure of a grain and then turned into a graphical display. To numerically quantify the scatter of

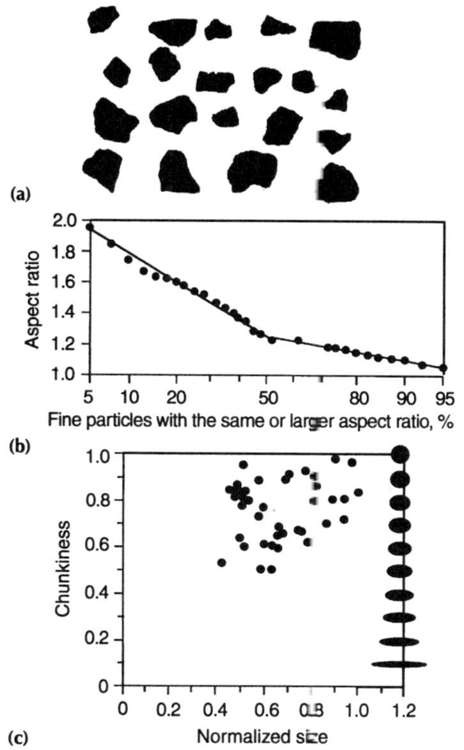

Fig. 13 Shape of fine particles can be characterized by aspect ratio or its reciprocal, chunkiness. (a) Silhouettes of the fine particles shown in Fig. 11. (b) Aspect ratio distribution of a larger set of fine particles like those of (a). (c) Chunkiness-size domain for the profiles described in (b)

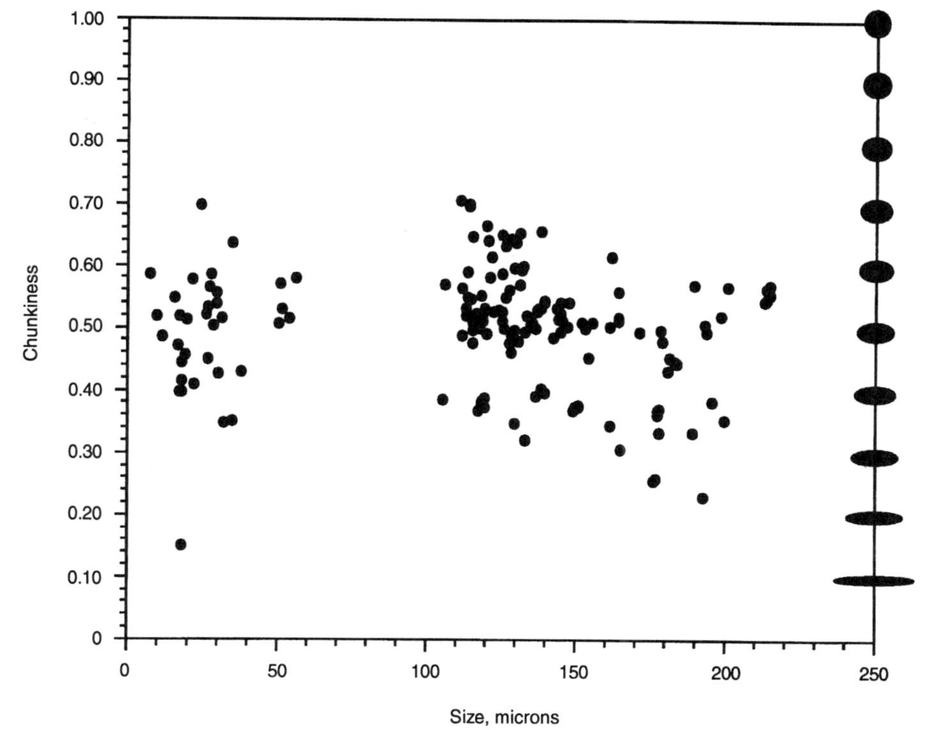

Fig. 14 Chunkiness-size domain summary of data generated by image analysis of the iron powder of Fig. 10 demonstrates a bimodal distribution indicating that the powder was probably sieved and that the smaller fine particles were most likely clinging to the larger fine particles.

data points on such a diagram, the points in space can be treated as unit masses. The centroid and the moment of inertia can then be calculated as a measure of the range of shapes present in the powder. Although several descriptions of this technique exist, no publications describe how to use such a population of shapes and relate it to the functional behavior of a powder (Ref 11, 12).

The gross shape information about a particle grain embodied in a geometric shape factor will govern the packing properties of a powder and is an important parameter to be considered because it contributes to the flow properties of a powder. When the surface reactivity of a powder and its sinterability is considered, the surface structure of the powder grains needs to be quantified.

Fourier Analysis. One method of characterizing the shape and texture of a powder grain is to generate and analyze what is known as the geometric signature waveform of a powder grain (Ref 13–15). A two-dimensional profile of the particle grain is needed to generate a geometric signature waveform. In one of the more frequently used techniques, the centroid of the profile viewed as a laminar shape is chosen as a reference point. The magnitude of the vector of the series of angles as shown in Fig. 16(a) is then recorded. The high resolution recording of the vector shows that the pattern generated by the rotated vector has the appearance of a complex wave, known as the signature waveform of the profile shown in Fig. 16(b). If the signature waveform is treated by Fourier analysis, the power spectrum of the waveform contains information on the shape and the texture of the profile. Flook (Ref 15) has analyzed the profile in Fig. 16 and shown that the first 5 harmonics of the Fourier analysis describe the gross shape of the profile, whereas the strength of the harmonics from 5 to 25 describe the texture of the profile as shown in Fig. 16(c) and 16(d). The use of signature waveforms and Fourier analysis were pioneered by Beddow et al. in the study of metal powder grains (Ref 16). Difficulties arise in choosing this straightforward geometric signature waveform if the profile of the particle has deep indentations, because in this situation the vector does not have a unique value crossing over the indenture in the profile. This can be overcome in modern data processing by creating the Fourier transform of the actual profile directly on computer logic, but this has not been widely used in powder metallurgy.

Fractal Analysis. Another technique for characterizing the texture for a metal profile is to measure the fractal dimension. Again current techniques available for measuring the fractal dimension are restricted to the use of two-dimensional images. The concept of a fractal dimension for measuring the ruggedness of a boundary was pioneered by Mandelbrot (Ref 17). Mandelbrot proposed that one way of describing the ruggedness of a boundary in space is to add a fractional number to the topological dimension of the system being studied to describe space-filling ability. In the description of the boundaries of powder grains, the topological dimension of a boundary in a two-dimensional space is one. Figure 17(a) shows basic concepts of systems put forward by Mandelbrot. There are several ways to characterize the fractal dimension of the profiles; one convenient technique is to explore the perimeter structure of various inspection resolutions and to construct a Richardson plot of the data generated.

The physical significance of these statements can be appreciated from the diagram summarized in Fig. 18. The first step in the evaluation of the fractal dimension of a profile is to digitize the profile as shown. A polygon from the perimeter of the profile is then constructed by counting off a number of steps around the profile. A polygon constructed in this way is an estimate of the perimeter at the resolution fixed by the number of steps taken around the perimeter. It can be shown that for an irregular perimeter that can be described by a fractal number, the plot of the estimates of the perimeter versus the size of the inspection stride (the term used to describe the step created by going a certain number of steps around the profile) on log-log graph paper generates a linear-data relationship. Furthermore, the absolute value of the slope is the fractional number to be used in the fractal dimension of the profile. This plot is known as the Richardson plot, named in honor of a pioneer scientist who studied the structure of irregular boundaries. Figure 18(b) and (c) show such a series of measurements and the data plot. A very rugged profile can have more than one data line on the graph. For example, in the case of powder formed by a fuming process such as a carbon black, a profile such as that of Fig. 19 can have two different data lines on the Richardson plot. The data line at high resolution represents the texture of the profile, whereas the data line of the coarser resolution is a measure of the overall gross structure of the profile. Figure 20 shows the shape, size, and fractal dimension of the population of powder grains from Fig. 10(a). Again the change of values of the three parameters can be quantified by treating the points representing the information as unit masses in space. This type of graph is used to look at the range of shapes in an abrasive powder.

Particle Shape Factors*

The most common approach to describe and differentiate particle shapes has been the use of qualitative concepts. Two fundamental concepts have been used: (a) the dimensionality of the particle and (b) the surface contour of the particle. By the use of these concepts, a model system of shape characterization is presented in Fig. 21. Photomicrographs of several types of loose powders described in the International Standards Organization standard ISO 3252 are shown in Fig. 22. Basic shapes are:

- *One-dimensional particles* (Fig. 22a and 22d). Two different types of one-dimensional particles can be considered on the basis of their surface contour. One particle is smooth and the other particle is roughened with an irregular type of surface.
- *Two-dimensional particles.* These are very flat in nature and the surface contour of such particles is usually irregular. The dendritic type (Fig. 22c) is characterized by a tree-like shape and is often associated with electrolytic powders. However, secondary mechanical treatments often destroy such a shape. Flake particles (Fig. 22e) are also considered two-dimensional particles.
- *Three-dimensional particles.* Most powders are three dimensional in nature. These powders can be equiaxed and nodular. The simplest type of particle in this category is the spherical type (Fig. 22i). By departing from this perfect shape and contour, irregular particles (Fig. 22g) and nodular types (Fig. 22h) are obtained.

Porous particles are different from irregular particles because of the presence of porosity, which can also be very irregular in both size and shape. Internal porosity may be isolated or interconnected. Large amounts of porosity make shape characterization difficult. It is best observed by examination of cross sections of mounted powder.

*Adapted from *ASM Handbook,* Volume 7, 1985, p 233–241

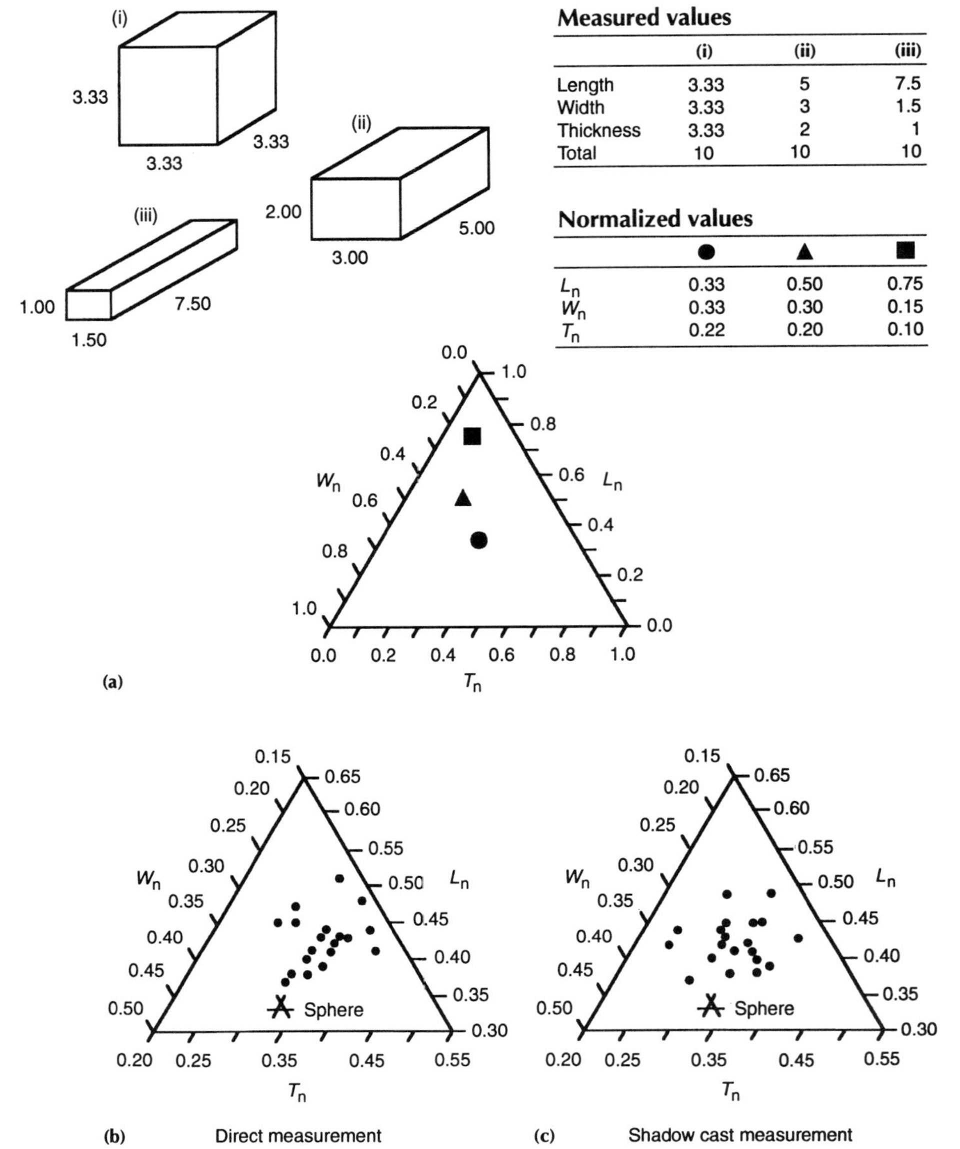

Measured values

	(i)	(ii)	(iii)
Length	3.33	5	7.5
Width	3.33	3	1.5
Thickness	3.33	2	1
Total	10	10	10

Normalized values

	●	▲	■
L_n	0.33	0.50	0.75
W_n	0.33	0.30	0.15
T_n	0.22	0.20	0.10

Fig. 15 Three-dimensional shape of a fine particle summarized using triaxial graph paper. (a) Demonstration of the use of triaxial graph paper to plot the dimensions of simple shapes. (b) Direct measurement of the three dimensions of the particles of Fig. 11. (c) Dimensions of the particles of Fig. 11 using shadow casting to determine the third dimension

Electron Microscopy

Qualitative SEM Shape Analysis. The most significant feature of the scanning electron microscope, in addition to its high magnification capability (useful magnifications beyond 10,000×), is its ability to produce images with a three-dimensional appearance. This ability derives from the fact that the depth of field for the scanning electron microscope is over 100× that of the optical microscope. This increased depth of field (ranging from 1 μm at 10,000× to 2 mm at 10×) also accounts for the extensive particulate analysis performed on the scanning electron microscope at magnifications that are within the capabilities of optical microscopy.

The imaging capabilities of the scanning electron microscope make it a useful tool for P/M applications concerned with all phases of powder production and processing. Scanning electron microscopy has been used to study all aspects of particle morphology, including size, shape, surface topography, surface structure (crystal, grain, and dendrite), coating or thin film characteristics (oxides), inclusion, void, and agglomeration characteristics, and satellite formation. The scanning electron microscope has also been used to study surface topography, effect of oxides or other coatings, porosity, inclusions, and other contaminants on P/M materials.

One of the more useful applications of the scanning electron microscope in powder metallurgy is qualitative particle characterization in terms of visual appearance. Despite the development of automated instruments for quantitative particle characterization, no substitute has been found for the interpretive capability of man. The use of the scanning electron microscope extends this capability into the microscopic domain.

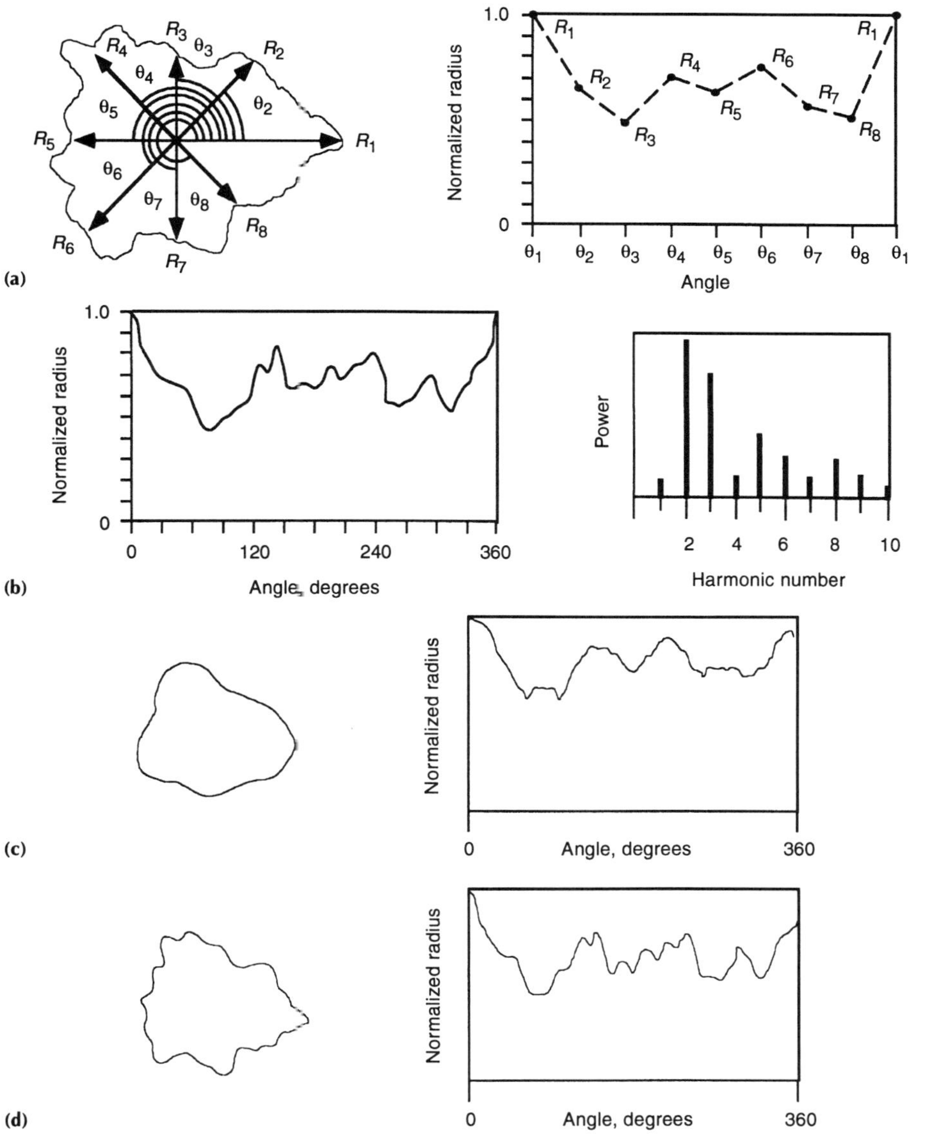

Fig. 16 Geometric signature waveform of a profile subjected to Fourier analysis to extract information on structure and texture. (a) Method used to obtain a simple geometric signature waveform. (b) More detailed signature waveform for the profile of (a) along with its harmonic spectrum. (c) First 5 harmonics used to reconstruct the profile show the gross morphology of the profile. (d) First 25 harmonics used to reconstruct the profile and add textural information to the structure

Inspection of powder particles to be used in P/M processing is essential, because individual and agglomerate characteristics can have significant effects on final material properties. Figure 23 shows representative alloy powder particles used in the direct rolling of aluminum P/M strip. The irregular shape promotes interlocking of particles, which provides sufficient green strength for strip processing.

Quantitative Analysis of Projected Images. In general, microscopists encounter two types of projected images. The first type of image results from a transmitted beam through the specimen, representing the features located within the three-dimensional space (such as by thin-foil transmission electron microscopy). In the second type, the projected image is generated by a reflected beam from the external surface of the specimen (such as by scanning electron microscopy).

Only the most rudimentary quantitative calculations can be made on images projected by the reflection techniques (Ref 18). In rough surfaces, the intensity levels depend on topography, and some features may be masked by others. Three-dimensional characterization is based on the photogrammetric analysis of stereopairs, for which automatic image-analyzing techniques are not yet available (Ref 19).

Quantitative statistical treatment of transmitted-beam images, however, has matured to a much greater extent. These analyses (Ref 20, 21) are too lengthy and complex to be treated here, but are described in the literature (see Ref 5).

One final topic will be included, because of its importance to the analysis of particulate systems. Figure 24 provides interrelated general equations of convex particles that express the important spatial parameters in terms of measurements made on the polish and projection planes. Application of the equations to specific particles is summarized in Table 2 for the sphere, the truncated octahedron (or tetrakaidecahedron), and convex particles in general. Tabulations of the type presented in Table 2 permit the microscopist to approximate microstructures with particles of known shape when other techniques are not feasible.

Conventional Shape Factors

Particle shape is a fundamental characteristic of powder particles and thus influences the properties of particulate systems. Various shape terms have been proposed to quantitatively represent particle shape. Early systems tended to measure one specific feature of a particle.

Table 3 (Ref 23–30) lists some of the most frequently used shape terms. The applicability and/or limitations of various shape factors, also referred to as shape parameters, are discussed in the following section of this article on stereological characterization of shape. For discussion purposes, the following method developed by Hausner (Ref 28) is used to demonstrate the applicability of shape factors in quantitative analysis of particle shape.

A rectangle of minimum area is drawn around the cross section of a particle (particle projection) as it is observed under the microscope (Fig. 25). The ratio of the rectangle side lengths permits calculation of particle elongation:

$$\text{Elongation factor, } x = \frac{a}{b}$$

The ratio of the area (A) of the projected particle to the area of the enveloping rectangle of minimum area ($a \times b$) indicates the bulkiness of the particle:

$$\text{Bulkiness factor, } y = \frac{A}{a \times b}$$

In this way, the cross-sectional area of the particle is correlated with some of its linear dimensions.

To characterize the surface configuration (or surface area), which is an essential factor of shape, the surface of the respective particle should be compared with the surface of a sphere of identical volume, or a cross section should be compared with that of the particle:

$$\text{Surface factor, } z = \frac{c^2}{12.6A}$$

where c is the perimeter of the projection profile of the particles. For a spherical particle, $z = 1$, and for particles of any other shape, $z > 1$.

Although the three ratios (x, y, and z) do not permit exact characterization of particle shape, they are nevertheless descriptive. It is possible, therefore, to correlate particle behavior with the three ratios—elongation factor, bulkiness factor, and surface factor. As indicated, however, there have been other shape factors proposed to characterize particle shape (see Table 3 and the next section of this article).

Stereological Characterization

Stereology refers to the study of the three-dimensional structure of materials from two-dimensional sections or projections. This discipline is primarily concerned with the geometrical evaluation of microstructural features, based primarily on geometrical probabilities. However, statistics, topology, and projection geometry also contribute to the overall quantitative characterization (Ref 20, 31). Quantitative characterization of the shape of microstructural features (e.g., pore space, precipitates, phase regions, and grains) and of particles on a substrate or embedded in a resin by means of data derived from planar images (sections or projections) is the prime consideration of stereology.

Stereology usually does not apply to the geometric nature of single particles, but rather to the geometric characteristics of a large number of particles or to the geometric characteristics of the "average particle" in a powder mass. Detailed information on the fundamentals, instrumentation, and applications of quantitative stereology can be found in Ref 18–22, 32, and 33.

Numerous shape parameters have been proposed, but only a few are practical. A useful stereological shape parameter (factor) must fulfill several requirements:

- *Shape sensitivity:* The value of a particular parameter must vary systematically with changes in shape—it should be sensitive to specific aspects such as elongation, bulkiness, and symmetry.
- *Independence of other geometric properties:* Size, size distribution, volume fraction, or other non-shape-related geometric characteristics should not influence the value for a given shape parameter.
- *Accessibility:* Quantities from which a stereological shape parameter is calculated must be available by simple measurements performed on planar images.

Other less significant requirements are that shape parameters must be dimensionless and independent of rotation or translation of the objects. Furthermore, it should be easy to visualize the significance of a shape parameter in terms of interpretable microstructural shape changes. The range for typical microstructural changes should be wide compared to statistical fluctuations and measuring errors.

From the parameters listed in standard textbooks (Ref 22, 32) and from parameters proposed in literature (Ref 34–37), none conforms to these requirements completely. In the following sections, some of the more useful stereological shape parameters, their limitations, and their practical alternatives (fingerprinting by means of simple two-dimensional parameters) are discussed.

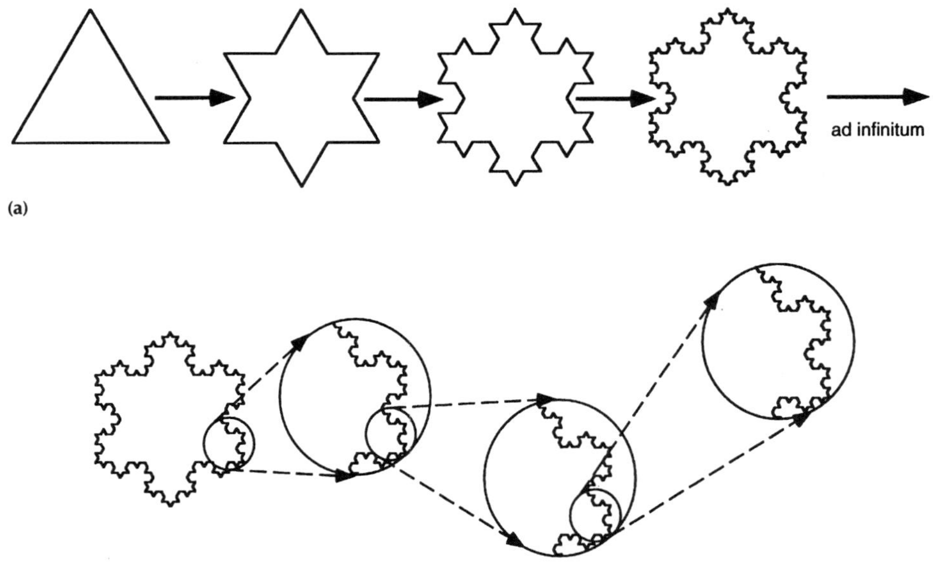

(a)

(b)

Fig. 17 Properties of the mathematical construction known as the Koch Triadic Island (a) Construction algorithm of the Koch Triadic Island. Each straight-line segment is divided by three, and the center segment is replaced with two segments of the same length at 60° to each other. (b) Triadic Island illustrating the property of self similarity. The magnification of an image of a boundary cannot be known because it looks the same at any magnification.

Image Data for Shape Characterization

With manual or semiautomatic data acquisition (in which counting and measurements are performed by the operator), only average quantities for the measured fields of view (field quantities), such as area fraction, total and mean intercept length, total and mean perimeter length, mean curvature, and number of features per unit area or unit length, can be obtained in a timely manner. Automatic instruments facilitate measurement of parameters describing the geometric properties of each individual planar feature, as seen in the fields of view at a high speed. From those parameters that can be used for shape description, the following have been shown to be of practical use: individual intersect areas, intercept length, tangent (Feret) diameters, as a function of direction and their extreme (maximum and minimum) for each closed planar feature; individual curvature; and moments of inertia.

Concise characterization of three-dimensional shape is possible only if the spatial coordinates of the particle surfaces or three-dimensional feature parameters (spatial Feret diameters, spatial moments of inertia, surface areas, and volumes of the particles) are available. Direct acquisition of three-dimensional data is possible by (a) serial sectioning, (b) stereometric measurements using SEM stereo-pair photographs and instrumented stereometers, and (c) additional specialized techniques, including x-ray topography, shadowgraphy, or densitometry (for semi-transparent materials). However, these techniques are time consuming and not yet well established.

Stereology of Shape

It is generally agreed that three-dimensional shape cannot be quantitatively assessed from planar images without severe limitations. For example, topological parameters, such as the number of separate parts (e.g., the number of pores in a porous particle) or the degree of connectivity

Table 2 Properties of a sphere, truncated octahedron, and convex particles

Property	Sphere, $D = 2r$	Truncated octahedron, edge length = a	General equations for convex particles
V	$4\pi r^3/3$	$11.314a^3$	$V = A' L_3 = AH'$
S	$4\pi r^2$	$26.785a^2$	$S = 4A' = 4V/L_3$
A'	πr^2	$6.696a^2$	$A' = S/4 = V/L_3$
H'	$2r$	$3.0a$	$H' = V/A = A'L_3/A$
A	$2\pi r^2/3$	$3.77a^2$	$A = V/H' = A'L_3/H'$
L_3	$4r/3$	$1.69a$	$L_3 = 4V/S = AN_A/N_1$
r, a, ρ(a)	$r = 2N_L/\pi N_A$	$a = 0.45N_L/N_A$	$\rho = H'/2 = N_A A/2N_L = A'/2L_2$
N_v	$\pi N_A^2/4N_L$	$0.744N_A^2/N_L$	$N_V = N_A/H' = N_L/A'$
	General equation: $V_V = N_V V + N_A A = N_L L_3$		

(a) ρ = half of mean tangent distance. Source: Ref 23

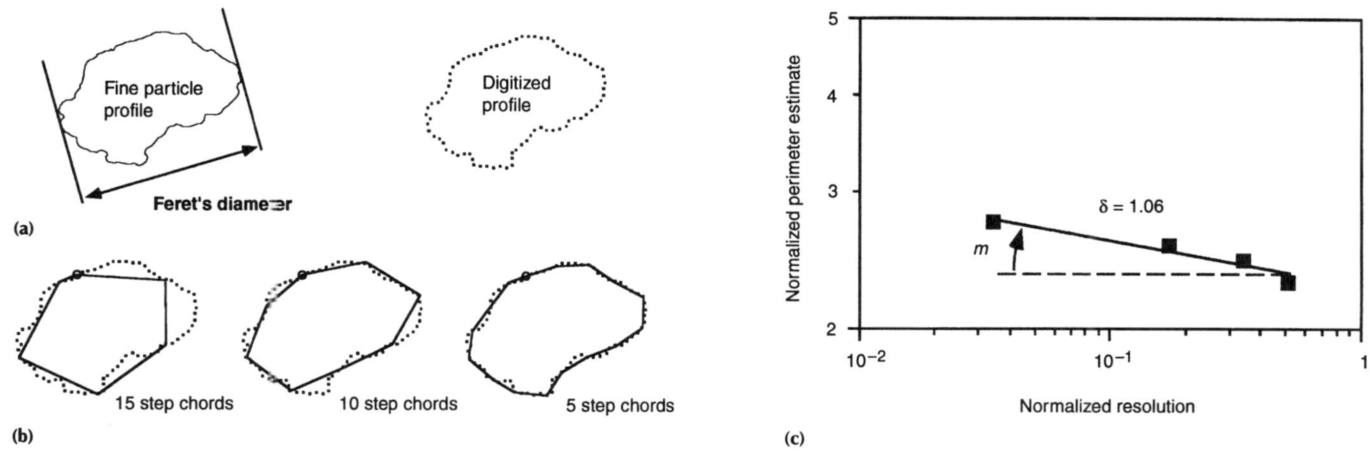

Fig. 18 Boundary fractal dimension of a profile—a dimensionless descriptor of the ruggedness of a fine particle boundary. (a) First step of the equipaced method of determining the boundary fractal. Profile is converted to a series of equally spaced (*x*,*y*) coordinates. (b) Estimates of the perimeter of the profile made at a series of decreasing step sizes. (c) Data normalized by dividing by the Feret's diameter and plotting perimeter estimate versus resolution of inspection on log-log graph paper. Fractal dimension (δ) = 1 + |*m*| where *m* is the slope of the best fit line through the data.

(the number of channels between the pores), cannot be obtained from a single cross section. From projections, especially from stereo-pair micrographs, a qualitative shape description can be easily derived. For stereological quantification, however, the data obtained from projections are generally less suitable than those obtained from the cross sections of particles embedded in a resin, a metal, or in a glass prepared by normal metallographic techniques.

Thorough analysis of stereological shape parameters obtained from a single cross section through a multitude of particles was proposed in the literature (Ref 38), intended to show that parameters that depend exclusively on the shape

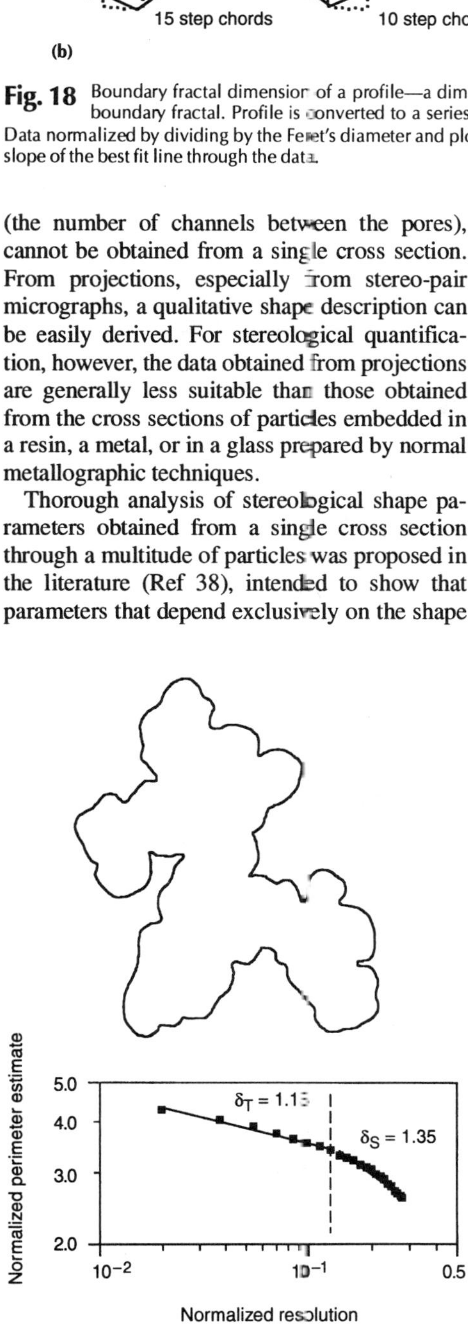

Fig. 19 Structural boundary fractal dimension (δ_S) relates to the gross morphology of the fine particle. Textural boundary fractal (δ_T) relates to the smaller features of the profile.

of the three-dimensional objects, and not on size or other non-shape-related features, can be traced to one simple combination of field data. This combination is the number of intersecting points between a measuring line and the perimeters of the particle cross section (P_L), which for convex particle features, is twice the number of particles per unit length of the measuring traverse—N_L ($P_L = 2N_L$)—the volume fraction (V_V) measured as area fraction (lineal fraction or point fraction) and the number of objects per unit area (N_A). Fischmeister (Ref 38) derives the universal shape parameter:

$$F_F = \frac{1}{6\pi} \cdot \frac{N_L^2}{V_V \cdot N_A} \quad \text{(Eq 1)}$$

The numerical factor aims to normalize the value to 1 for spheres. However, it can be easily demonstrated that this shape parameter is not independent of the size distribution, and deviations from a value of 1 for spheres of varying size are possible. A shape parameter that is actually independent of other geometric properties was proposed by De Hoff (Ref 39) in 1964:

$$F_D = \frac{1}{6\pi} \cdot \frac{N_L^2}{V_V \cdot N_A} \cdot \frac{M_1 \cdot M_3}{M_2^2} \quad \text{(Eq 2)}$$

where M_1, M_2, and M_3 are the first three moments of the spatial size distribution, which generally are not accessible from planar data without a priori information on shape (the shape of the particles must be simple and known in order to calculate three-dimensional size distributions from planar measurements). F_D reduces to F_F for uniform size, while logarithmic-normal size distributions with known standard deviation $ln\ \sigma$ yields:

$$F_D = \frac{1}{6\pi} \cdot \frac{N_L^2}{V_V \cdot N_A} \exp\,(\ln \sigma)^2 \quad \text{(Eq 3)}$$

The only other parameter that appears to meet the three basic requirements outlined above has

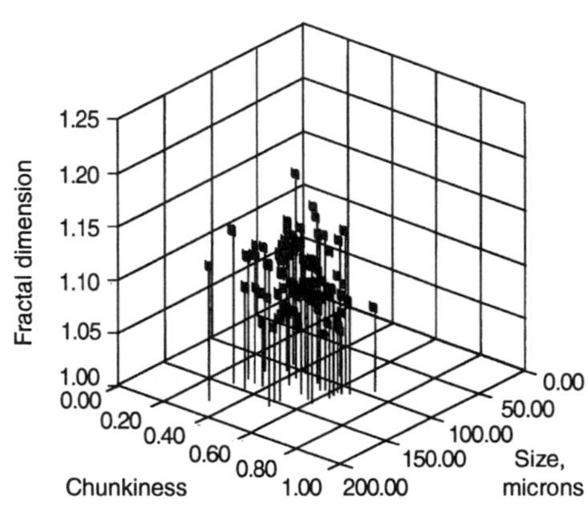

Fig. 20 Individual powder grain of Fig. 10(a) characterized by its fractal dimension and combined with size and chunkiness to form a single point in three-dimensional space. When a group of powder grains are treated in this way, they produce a cloud of points characteristic of the powder.

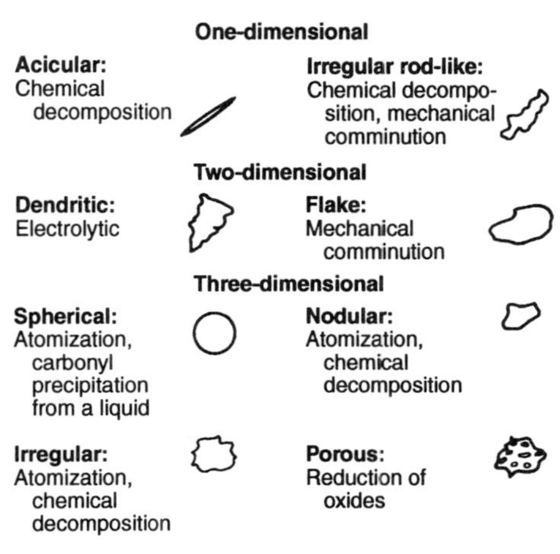

Fig. 21 Various shapes of powder particles and their methods of manufacture

been derived by Hilliard (Ref 40) from a geometric theorem relating the total surface and the total volume of three-dimensional objects to the first and the fourth moment of the intercept length distribution—$M_1(1)$ and $M_4(1)$ (see Ref 34). After normalizing to 1 for spheres again, this yields the shape parameter:

$$F_H = \frac{2}{3}\frac{N_L^2}{V_V}\sqrt{\frac{2}{3}M_4(1)} \quad \text{(Eq 4)}$$

N_L and V_V are defined above. The main limitation of F_H is that it fulfills Eq 3 in principle only; in practice, $M_4(1)$ is subject to large experimental errors due to the fact that the longest intercepts, which usually are present with a low numerical frequency, contribute most to $M_4(1)$.

Numerous other shape parameters have been proposed by Underwood (Ref 35, 36), Ministr (Ref 41), and others. However, most equations are in pronounced contradiction to one or more of the basic requirements for true shape factors. Due to the problems outlined in this section, two-dimensional shape characteristics provide practical alternatives and as such are used almost exclusively in practical work.

Quantitative Description of Planar Shape

Two-dimensional shape parameters provide a means to monitor shape changes occurring in microstructures of P/M products due to variations in processing parameters and a means to compare the particle shape of powders obtained with different materials and production techniques. For this method, Fischmeister (Ref 39) uses the term "fingerprinting," because if applied properly, it yields indirect but accurate information on three-dimensional shape. The basic requirement outlined previously, however, must also be observed for two-dimensional shape factors.

Hausner's shape parameters, described earlier in this article in the section on conventional shape parameters, are an example of "fingerprinting," as are other methods of planar shape description, such as Fourier coefficients and fractals of the sectioned or projected particle circumferences. In sectioned or projected particle circumferences, parameters are described that are sensitive to particular shape aspects and are easily determined with semiautomatic or fully automatic image analyzers. In the Hausner method, the fitting of the minimum area rectangle requires repeated fitting, which in some cases leads to ambiguous results.

By far the most frequently used planar shape parameter (implemented in most image analyzing devices) combines the area and the perimeter of a planar feature into a dimensionless number normalized to 1 for circles of the same size:

$$f_{PL} = 4\pi a/b^2 \quad \text{(Eq 5)}$$

where a and b are the area and the perimeter of the planar features (intersects or projections), re-

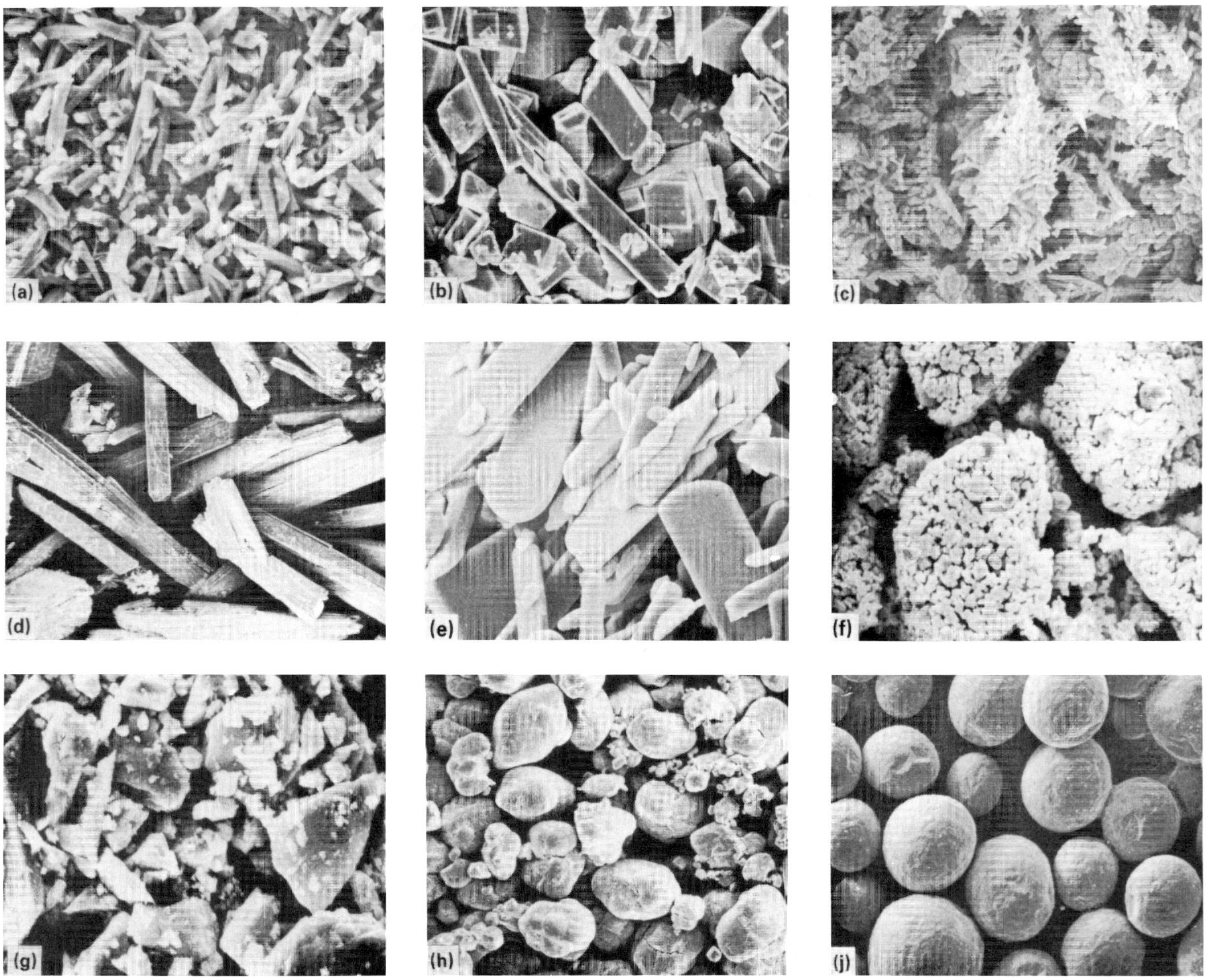

Fig. 22 Common particle shapes as depicted in ISO 3252. (a) Acicular powder particles. (b) Angular powder particles. (c) Dendritic powder particles. (d) Fibrous powder particles. (e) Flaky powder particles. (f) Granular powder particles. (g) Irregular powder particles. (h) Nodular powder particles. (j) Spheroidal powder particles

spectively. Averaging can be completed easily if a and b are measured individually for each feature:

$$F_L = \sum_{i=1}^{n} f_{PL}/n$$
$$= 4\pi \sum_{i=1}^{N} (a/b^2)/n \qquad \text{(Eq 6)}$$

where n is the total number of planar features. However, if a and b are averaged, or if total area and perimeter length are used (as provided by field analysis), the average shape parameters obtained depend on size distribution, with values deviating from unity for a system of circles of varying diameter.

If Eq 6 is used, deviations from unity are interpreted as deviations from circularity. Elongation (elliptical deformation), as well as concave deformations of the perimeter ("rugged" outlines), yields smaller values approaching zero for highly elongated features or highly rugged perimeters.

Thus, F_1 combines different aspects of shape. To differentiate between these aspects, simple combinations of moments of inertia, Feret diameters, or curvature have been derived by Schwarz (Ref 42). Exner and Hougardy (Ref 34) propose the following parameters for individual features, which also can be averaged for any number of features in the same way as f_{PL} (see Eq 6):

$$\text{Elongation,} \quad f_{EL} = \sqrt{i_1/i_2} \qquad \text{(Eq 7)}$$

$$\text{Compactness,} \quad f_{COM} = \frac{1}{2\pi} \cdot \frac{a^2}{\sqrt{i_1^2 + i_2^2}} \qquad \text{(Eq 8)}$$

$$\text{Ruggedness (waviness),} \quad f_{RU} = b_c/b \qquad \text{(Eq 9)}$$

where i_1 and i_2 are the two principal moments of inertia, b_c is the convex perimeter, and b is the actual perimeter.

These parameters are sensitive to the respective shape aspects, are easily measured by computer-aided image analysis, and are independent of other geometric properties; their interpretation through visualization is easily accomplished.

Consideration should be given to the method of averaging, however. Averaging on a volume basis, rather than on a number basis as in Eq 6, may be appropriate in some practical applications. If three-dimensional measurements are performed, similar shape parameters can be derived for the various aspects of spatial shape.

Multidimensional Shape Characterization

In order to analyze differences in shape precisely, more than one shape aspect must be monitored. Depending on the number of aspects considered using mutually independent shape parameters, two-, three-, or multidimensional representation of shape can be useful for comparison, and cluster analysis may be appropriate to quantify the significance of differences. Figure 26 shows a two-dimensional shape characterization of idealized particles in a ruggedness/elongation diagram.

These procedures require a computer with a large capacity, but which is modest compared to Fourier analysis, for example. The assessment of the individual parameters involves straightforward arithmetic calculation, and cluster analysis is performed on a few numbers for each feature measured.

Simple shape characterization, normally accomplished using Eq 6, prevails in practice, because the multidimensional procedures referred to previously are not yet implemented in com-

Table 3 Shape terms and their definitions

Term	Symbol	Definition
Volume specific surface (S_V)		
$S_v = \frac{S}{V}$		V: Particle volume
		S: Particle surface area
Heywood ratios (Ref 24)		
Elongation ratio (n)		
$n = L_h/B_h$	B_h	Breadth–the minimum distance between two parallel planes that are perpendicular to planes defining T_h
Flakiness ratio (m)		
$m = B_h/T_h$	T_h	Thickness—minimum distance between two parallel planes tangential to the particle maximum stability plane
	L_h	Length—distance between two parallel planes that are perpendicular to both T_h and B_h planes
Heywood shape factor (ϕ_h) (Ref 25)		
$\phi_h = f/k$		
where $f = \frac{S}{d_a^3}$	d_a	Diameter of a circle having the same projected area
	S:	Particle surface area
$k = \frac{V}{d_a^3}$	V:	Particle volume
	For spheres: $n = m = 1, \phi_h = 6$	
	For spheres: $n = m = 1, \phi_h = 6$	
Wadell		
Sphericity (ψ_s) (Ref 26)		
$\psi_s = (d_v/d_s)^2$	d_v:	Diameter of the sphere that has the same volume of the particle
$\approx 4.84\,(k/f)^{2/3}$	f, k:	Diameter of the sphere having the same surface area as the particle
	f, k:	Defined in Heywood ratios
Roundness (ψ_r) (Ref 27)		
$\psi_r = \frac{\sum_{i=1}^{N} r_i}{NR}$	r_i	ith radius of curvature along the particle profile which is smaller than the radius of the largest inscribed circle
	N:	Total number of radii of curvature smaller than the inscribed circle
	R:	Radius of the largest inscribed circle
Krumbein (Ref 28)		
Sphericity (ψ_k)	L_k:	Longest dimension of the particle
$\psi_k = \sqrt[3]{\left(\frac{T_k}{B_k}\right)\left(\frac{B_k}{L_k}\right)^2}$	B_k:	Breadth—measured perpendicular to L
	T_k:	Particle thickness
Hausner (Ref 29)		
Elongation ratio (x)	a:	Length of the enveloping rectangle that has the minimum area
$x = a/b$		
Bulkiness factor (y)	b:	Width of the rectangle
$y = A/ab$	A:	Projected area of the particle
Surface factor (z)	c:	Perimeter of the projected profile
$z = c^2/12.6A$		
Church shape factor (ϕ_c) (Ref 30)		
$\phi_c = E(d_M)/E(d_F)$	d_M	Martin's diameter—length of the chord which divides the profile into two equal areas with respect to a fixed direction
	$E(d_M)$:	Expectation of Martin's diameter
	d_F	Feret's diameter, the distance between a pair of parallel tangents of the particle profile with respect to a fixed direction
	$E(d_F)$:	Expectation of Feret's diameter
Centroid aspect ratio (CAR) (Ref 31)		
$\text{CAR} = \frac{d_m}{d_p}$	d_m	Longest chord passing through the centroid
	d_p	Length of the chord passing through the centroid and perpendicular to the one defining d_m

mercial image analyzers. At present, the best approach to the difficult problem of quantitatively characterizing the shape of a system of particles in a powder mass is to use one or several shape parameters carefully adjusted to the problem under investigation.

Morphological Analysis

Morphological analysis is primarily the characterization of particle shape. Particle size may also be obtained in the analysis (Ref 30). Typically, a shape analyzer is used to conduct the analysis. This instrument, which consists of a high-quality imaging (graphics) system that is augmented with computer equipment, must provide a clear image of the object for effective and accurate morphological analysis.

The image is digitized to obtain the (x, y) data set of the profile. This data set is then converted to the polar form. The polar data set (R, θ) constitutes a line, and the Fourier equation that represents this line can be constructed and the set of coefficients extracted from this equation. In turn, the coefficients are converted to invariant forms called "morphic features." Statistical properties of these features are also determined.

The rigorous definition of various morphological descriptors are given in Eq 10 to 14, and statistical properties are given in Eq 15 to 17. The size term is defined as the equivalent radius, R_0, of the particle profile, such that R_0^2 is the area of the circle of equivalent area to that of the particle profile.

$$R_0 = \sqrt{a_0^2 + \frac{1}{2}\sum^{\infty} (a_n^2 + b_n^2)} \qquad \text{(Eq 10)}$$

The mean radius a_0 and R_0 are related as follows:

$$L_0 = \frac{a_0}{R_0} \qquad \text{(Eq 11)}$$

$$L_1(n) = 0 \text{ for all } n \qquad \text{(Eq 12)}$$

The next set of terms, called the $L_2(n)$ terms, where n is the order of the Fourier coefficient, is useful in describing morphic features such as aspect ratio, triangularity, squareness, roughness, and other shape aspects.

$$L_2(n) = \frac{1}{2R_0^2}(a_n^2 + b_n^2)$$

$$(n = 1, 2, 3, \ldots N \text{ is the order of the coefficient}) \qquad \text{(Eq 13)}$$

$$L_3(m, n) = \frac{3}{4R_0^3}(a_m a_n a_{n+m} - b_m b_n a_{n+m} + a_m b_n b_{n+m} + b_m a_n b_{n+m}) \qquad \text{(Eq 14)}$$

The mean, standard deviation, and skewness of the radial distribution of the particle are defined below. They are standard statistical properties and each has a distinct physical meaning. For example, the standard deviation is an indicator of the not-roundness of the particle profile.

$$\mu_0 = L_0 R_0 \text{ (mean radius)} \qquad \text{(Eq 15)}$$

$$\mu_2 = R_0^2 \sum^{\infty} L_2(n)$$

$$\text{(second moment about the mean)} \qquad \text{(Eq 16)}$$

$$\mu_3 = R_0^3 \sum_{m=1}^{\infty} \sum_{n=1}^{\infty} L_3 \,(m, n)$$

$$\text{(third moment about the mean)} \qquad \text{(Eq 17)}$$

Morphological Descriptors and Statistical Terms. Morphological descriptors are unique in that they can be used to regenerate the form of the original particle profile. A set of descriptors therefore, contains all of the information in the original profile. Because morphological analysis has been developed to solve real problems, it is important that the morphological features can be identified with a corresponding physical meaning wherever possible.

The equivalent radius R_0 is defined so that πR_0^2 is the area of the particle. Therefore, R_0 can be interpreted as a size term; a_0 is the mean radius of the particle.

The series of terms $L_2(n)$ are defined in terms of the Fourier coefficients a_n and b_n, and they are normalized by dividing by R_0^2 to make them independent of size. Therefore, $L_2(n)$ terms are

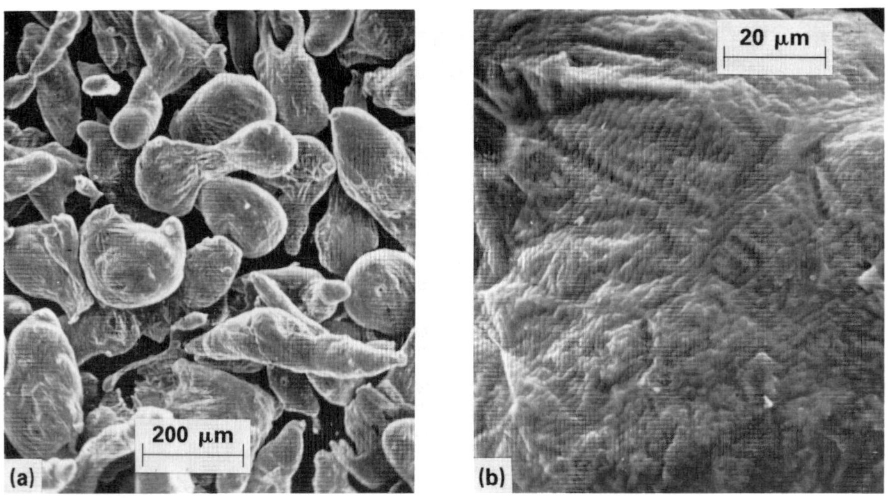

Fig. 23 Scanning electron micrographs of 7091 aluminum alloy particles used in direct powder rolling of strip. (a) 70×. (b) 700×

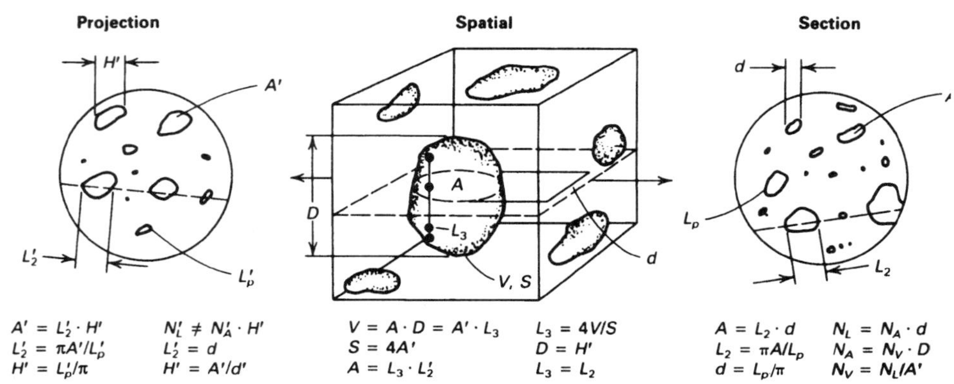

Fig. 24 Relationships among convex particles in space, their sections, and their projections. Projected quantities are indicated by primes. Source: *ASM Handbook*, Volume 9, p 134

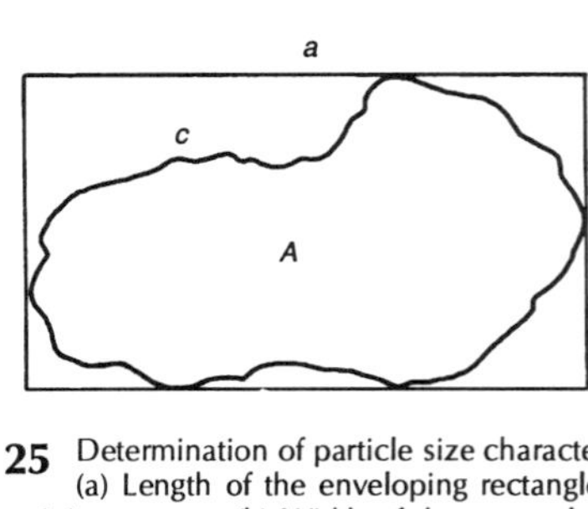

Fig. 25 Determination of particle size characterization. (a) Length of the enveloping rectangle, which has a minimum area. (b) Width of the rectangle. (c) Circumference of the projected particle. A, surface area of the projected particle

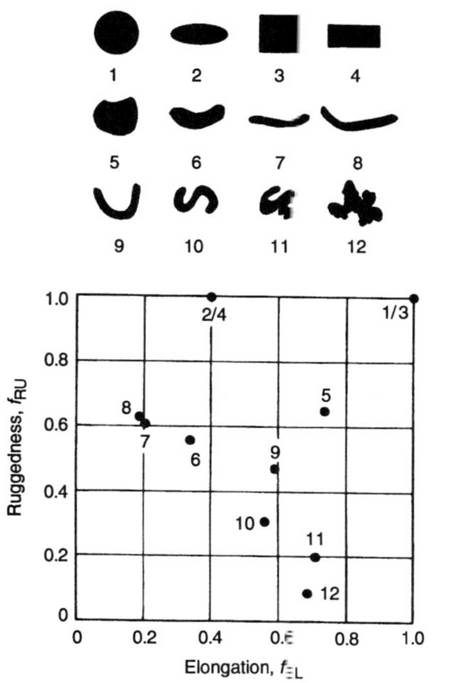

Fig. 26 Characterization of the shape of twelve typical particle projections (or sections) in a two-dimensional shape space

the equivalent of the Fourier coefficients A_n. In terms of their physical meaning, an indication of the relative strength of the $L_2(3)$ term can be obtained by dividing the value of $L_2(3)$ by the sum of $L_2(n)$. This value provides an indication of the angularity (triangularity) of the particle.

There is a relationship between the $L_3(m,n)$ terms and the radial skewness as shown in Eq 17. Therefore, if all of the $L_3(m, n)$ terms are summed and multiplied by R_0^3, the skewness of the radial distribution can be obtained. A pear-shaped drop or a teardrop shape, for example, would provide a high value of skewness. However, the physical meaning of the individual $L_3(m, n)$ terms is not understood at present.

The statistical properties of the morphological features defined in Eq 15 and 17 have been discussed above. The second moment of the radial distribution (the radiance), which is defined in Eq 16, is interpreted as indicating the not-roundness (out-of-roundness) of the particle profile.

The equivalent radius squared minus the mean radius squared is equal to the sum of all the $L_2(n)$ terms multiplied by R_0^2. Equation 10, therefore, relates intrinsic and extrinsic properties of the particle profile quantitatively. In addition, the radance (the sum of the $L_2(n)$ terms) can be equated to a difference between the square of the size minus the square of the mean radius.

This discussion indicates that there is a basis for the physical interpretation of the morphological features of a particle profile. The exception to this is the physical meaning of the $L_3(m, n)$ terms. This problem requires further study.

REFERENCES

1. B.H. Kaye, Particle Shape Characterization, Chap. 2, *Handbook of Powder Science and Technology,* 2nd ed., M.E. Fayed and L. Otten, Ed., Chapman & Hall, 1997, p 35–52
2. T. Allen, *Particle Size Analysis,* 4th ed., Chapman & Hall, 1996
3. *British Standards 2955 Glossary of Terms Relating to Powders,* British Standards Institute, 1965
4. H. Heywood, Size and Shape Distribution of Lunar Fines Sample 12057, 72, *Proc. of Second Lunar Science Conf.,* Vol 13, 1971, p 1989–2001
5. H. Heywood, Numerical Definitions of Particle Size and Shape, *Chem. Ind.,* Vol 15, 1937, p 149–154
6. H. Heywood, Particle Shape Coefficients, *J. Imp. Coll. Eng. Soc.,* Vol 8, 1954, p 25–33
7. H.H. Hausner, Characterization of the Powder Particle Shape, *Proc. of the Symp. on Particle Size Analysis* (Loughborough, England), The Society for Analytical Chemistry, 1967, p 20–77
8. M.H. Phal, G. Schädel, and H. Rumpf, Zusammenstellung von Teilchenform Beschreilbungsmethoden, *Aufbereit.-Tech.,* Vol 5, 1973, p 7–11
9. R.P. Loveland, *ASTM Committee E29.02 on Fineparticle Shape,* Report, ASTM, 1975
10. B.H. Kaye, Chap. 10, *Direct Characterization of Fineparticles,* John Wiley & Sons, 1981
11. B.H. Kaye, G.G. Clark, and Y. Liu, Characterizing the Structure of Abrasive Fineparticles., *Part. Part. Syst. Charact.,* Vol 9 (No. 1), 1992, p 1–8
12. R. Davies, A Simple Feature Based Representation of Particle Shape, *Powder Technol.,* Vol 12, 1975, p 111–124
13. H.P. Schwartz and K.C. Shane, Measurement of Particle Shape by Fourier Analysis, *Sedimentology,* Vol 13, 1969, p 213–231
14. R. Ehrlich and B. Weinberg, An Exact Method for Characterization of Grain Shape, *J. Sediment. Petrol.,* Vol 40 (No. 1), March 1970, p 205–212
15. A.G. Flook, A Comparison of Quantitative Methods of Shape Characterization, *Acta Stereol.,* Vol 3, 1984, p 159–164
16. J.K. Beddow, M.D. Nasta, and G.C. Philip, Characteristics of Particle Signatures, *Proc. of the Int. Conf. on Powders and Bulk Handling Systems,* May 1977 (Chicago, IL), International Powder Institute, 1977
17. B.B. Mandelbrot, *Fractals; Form, Chance and Dimension,* Freeman, 1977
18. J.E. Hilliard, *J. Microsc.,* Vol 95 (Part 1), Feb 1972, p 45–58
19. T.O. Johari, *Res. Develop.,* Vol 22 (No. 7), 1971, p 12
20. E.E. Underwood, The Stereology of Projected Images, *J. Micros.,* Vol 95 (Part 1), Feb 1972, p 25–44
21. E.E. Underwood, The Mathematical Foundations of Quantitative Stereology, *Stereology and Quantitative Metallography,* STP 504, ASTM, Philadelphia, 1972, p 3–38
22. E.E. Underwood, *Quantitative Stereology,* Addison-Wesley, 1970
23. H. Heywood, Symposium on Particle Size Analysis, *Trans. Inst. Chem. Eng.,* Vol 25, 1947, p 14–24
24. H. Heywood, *J. Imperial Coll. Eng. Soc.,* Vol 8, 1954
25. H. Wadel, *J. Geol.,* Vol 40, 1932, p 343–357
26. H. Wadel, *J. Geol.,* Vol 43, 1935, p 250–280
27. W.C. Krumbein, *J. Sedimen. Petrol.,* Vol 11 (No. 2), 1941, p 64–72
28. H.H. Hausner, *Planseeber,* Vol 14 (No. 2), 1966, p 75–84
29. T. Church, *Powder Technol.,* Vol 2, 1968–1969, p 27–31
30. J.K. Beddow, *Particulate Science and Technology,* Chemical Publishers, 1980
31. E.E. Underwood, Stereological Analysis of Particle Characteristics, *Testing and Characterization of Powders and Fine Particles,* J.K. Beddow and T.P. Meloy, Ed., 1980, p 77
32. E.R. Weibel, *Stereological Methods,* Vol 2, Theoretical Foundations, Academic Press, 1980
33. J. Serra, *Image Analysis and Mathematical Morphology,* Academic Press, 1982
34. H.E. Exner and H.P. Hougardy, Quantitative Image Analysis of Microstructures, *A Practical Guide to Techniques, Instrumentation and Assessment of Materials,* Deutsche Gesellschaft für Metallkunde, Oberursel, 1984
35. E.E. Underwood, Quantitative Shape Indices by Stereological Methods, *Quantitative Analysis of Microstructures in Medicine, Biology and Materials Development,* Riederer-Verlag, 1975, p 223–241
36. E.E. Underwood, "Three-Dimensional Shape Parameters from Planar Sections," Proc. Fourth Int. Cong. Stereology, NBS Special Publication 431, National Bureau of Standards, 1976, p 91–92
37. E.E. Underwood, Stereological Analysis of Particle Characteristics, *Testing and Characterization of Powders and Fine Particles,* Heyden and Sons, London, 1980, p 77–96
38. H.F. Fischmeister, Shape Factors in Quantitative Microscopy, *Z. Metallkd.,* Vol 65, 1974, p 558–562
39. R.T. DeHoff, The Determination of the Geometric Properties of Aggregates of Constant Size Particles from Counting Measurements Made on Random Plane Sections, *Trans. AIME,* Vol 230, 1964, p 764–769
40. J.E. Hilliard, private communication, 1975
41. Z. Ministr, The Determination of Parameters for Two and Three Dimensional Microstructures with Aid of Specific Perimeter or Specific Surface Area of the Microstructural Features, *Practical Metallogr.,* Vol 8, 1972, p 333–359, p 407–423 and Vol 12, 1975, p 244–258
42. H. Schwarz, Two Dimensional Feature-Shape Indices, *Proc. Fifth Int. Cong. Stereology,* Mikroskopie, Vol 37 (Suppl.), 1980, p 64–67

Surface Area, Density, and Porosity of Powders

A LARGE NUMBER of industries deal with powders at some point during their processing procedures. The characterization of surface area, porosity, density, and particle size is of particular importance. Especially in the case of metal powders, those characteristics are important factors for understanding and controlling material properties and processing behavior. Knowledge about the surface area of the powder is helpful in understanding the sintering behavior, because the reduction of surface area is the essential driving force for the sintering process. Porosity and density are closely related, and they may influence, for example, the hardness of the material.

Different techniques can be used to determine these powder characteristics. The methods used to determine surface area, density, and porosity are not specific to metal powders: many have been developed for testing other materials in powder form. The major characterization techniques are summarized below. Expanded discussions of several test methods can be found in subsequent sections of this article, including testing parameters, specifications, and instrumentation. When these techniques are used, however, the results of each test method can vary from the results obtained by a different technique. This is only partly due to sample variations. Yet, each test method determines those properties in a very specific way. Pore size and even surface area are ambiguous parameters. For example, the size of an ideal cylindrical pore would be defined as the radius or diameter of this cylinder. However, real pores are hardly ideal cylinders. The odd-shaped geometry of pores makes it difficult to define the size of such a pore, and each measuring technique has its own way of looking at those pores and determining an average pore size of the sample. The pore size also depends on whether the actual void size or the entrance size or neck of the pore is of interest. The latter could be the restricting factor for transport processes. It is also impossible to clearly define the surface area. Depending on the "yardstick" of the observer, a sample may have a larger or smaller surface area. An optical microscope may not show cracks and pores in a sample, whereas the same sample can show a substantial surface roughness under the electron microscope or when analyzed by gas adsorption techniques. Correspondingly, the value of the calculated surface area will be different.

- *Gas adsorption method:* Determines the surface area of a powder sample by measuring the amount of gas adsorbed on the sample surface at low temperatures. The data are then used to calculate the monolayer capacity from which the surface area is calculated using the "known" size of the adsorbed molecules. Nitrogen at liquid nitrogen temperature is the most common gas used; krypton at liquid nitrogen temperature is used for samples with an area of less than 1 m^2. Gas adsorption can also be used to determine pore volume and pore size distribution ranging from one Angstrom to about 100 or 200 nm.
- *Permeametry:* Measures the resistance to fluid flow through a compacted powder bed. This information is used to determine related properties of a powder, such as pore size distribution, specific surface area, or the average particle size for a packed powder sample.
- *Pycnometry:* Determines density by measuring the difference between the specific and bulk volumes of a sample. This method is based on the displacement principle, using the powder as the solid body and helium, water, or mercury as the displaced medium. Pycnometry can be used to determine total pore volume or density. However, it does not provide quantification of the pore size or the pore size distribution.
- *Mercury porosimetry:* Measures the volume of mercury intruded into the pores of a powder sample as a function of the pressure applied to the mercury. This method gives pore size and distribution over a wide range: 0.3 mm to 3 nm depending on the capability (pressure range) of the apparatus used. The method is based on the nonwetting behavior of mercury toward most materials. However, certain metals, for example, gold or copper, can react with mercury, and before those samples are to be analyzed specific precautions have to be taken.

Although surface area, density, and porosity of powder are interrelated, a given powder sample may require the use of several testing methods to provide a complete analysis of these characteristics, because each technique is more or less sensitive for a specific sample. A comprehensive treatment of each of the methods is given in this article.

Gas Adsorption

H. Giesche, School of Ceramic Engineering and Sciences, Alfred University

The Brunauer-Emmett-Teller (BET) (Ref 1) method of measuring specific surface area is based on the determination of the amount of gas that is adsorbed on the surface of the sample. The specific surface area (m^2/g) determined by this method includes the external as well as internal (pores) surface area. The surface area of closed pores cannot be determined because the adsorbing gas molecules have no physical path to that surface.

Theory of BET Method

The BET model is based on a kinetic model of the adsorption process that was described first by Langmuir in 1916 (Ref 2, 3). Langmuir regarded the solid surface as an array of adsorption sites, and a state of dynamic equilibrium was postulated in which the rate molecules arrive from the gas phase and condense on the bare sites is equal to the rate at which molecules evaporate from occupied sites. In 1938 Brunauer, Emmett, and Teller extended Langmuir's kinetic monolayer adsorption theory to a multilayer adsorption theory. The following relationship was derived to calculate the monolayer capacity, V_m:

$$\frac{V}{V_m} = \frac{C\,[1-(n+1)\,(P/P_0)^n + n(P/P_0)^{n+1}]}{[(P_0/P)-1]\,[1+(C-1)\,(P/P_0) - C\,(P/P_0)^{n+1}]}$$

In the actual experiment the adsorbed amount of gas, V, is measured as a function of gas pressure, P. P_0 is the saturation pressure of the liquefied gas at the corresponding temperature of the adsorption measurement and the ratio P/P_0 is often referred to as the relative pressure. The parameters C and n are part of the BET theory. The parameter C is:

$$C = \frac{a_1 \nu_1}{a_2 \nu_2} \exp\left\{\frac{q_1 - q_c}{RT}\right\} \approx \exp\left\{\frac{q_1 - q_c}{RT}\right\}$$

The parameters a, ν, q, R, and T are the activity coefficient, vibration frequency, heat of adsorption, gas constant, and temperature, respectively (Ref 4). The parameter C can be regarded simply as a measure of the additional or specific heat of adsorption in the first adsorbed layer (q_1) compared to the heat of condensation, q_c, which is assumed to be valid for the second and following adsorption layers. The parameter n corresponds to the maximum number of adsorbed layers on the surface. Only when pore sizes are very small is there a physical limitation of the number of layers and the factor, n, has to be considered. Usually n is assumed to be infinite and the "standard" BET equation results, which can then be rearranged to yield a linear relation:

$$\frac{P}{V(P - P_0)} = \frac{1}{V_m C} + \frac{(C-1)\,P}{V_m C P_0}$$

A plot of $P/[V(P - P_0)]$ versus P/P_0 generally gives a straight line in the relative pressure range from $P/P_0 = 0.06$ to 0.3. From the slope and intercept of this line, the amount of gas adsorbed in a monomolecular layer (V_m) is determined as well as the value of C. By using the specific size of the adsorbing molecule, the total surface area of the sample can then be calculated. It should also be noted that the value of C is quite sensitive to errors in the measurement and it can vary by 50% or more depending how accurately data points are taken and whether or not certain data points are omitted in the calculation. Frequently, C values for metal surfaces are quite large (>100 or 200); under those conditions, the intercept of the BET line can be taken as 0. The latter approach is used in the so-called single-point BET method, when the monolayer capacity, V_m, is directly calculated from one single data point using the following equation:

$$\frac{P}{V(P - P_0)} = \frac{(C-1)\,P}{V_m C P_0} \approx \frac{P}{V_m P_0}$$

or $V_m = V(1 - P/P_0)$.

Different gases can be used for the surface area determination. The different experimental results can be compared if the size of the adsorbed gas molecules is known. Different values for the so-called cross-sectional area (size) of the adsorbate molecule have been reported in the literature. The following tables lists values as used by Davis, DeWitt, and Emmett (Ref 5):

Adsorbate gas and temperature, K	Size according to the density of the liquid, nm^2	Revised valued recommended in Ref 5, nm^2
Nitrogen, 78	0.162	0.162
Krypton, 78	0.152	0.208
Krypton, 195	0.297	0.434
Butane, 273	0.321	0.469
Butane, 195	0.247	0.375
Freon-21 ($CHCl_2F$), 273	0.264	0.401

However, even with the use of such modified cross-sectional areas, anomalies between different samples could be diminished, but not completely eliminated, as shown in the data in Table 1 (Ref 5). The interpretation of surface area results also has to be carefully evaluated when the BET-C parameter is very large. A value of several hundred is a strong indication of either the presence of microporosity, adsorption on active sites at the surface, or even a chemisorption effect. As shown in the table below, very clean metal surfaces can produce extremely large C values and the adsorption of the gas molecules occurs in those cases in highly localized positions on the surface. As a result, the calculated surface area may not reflect the true area of the sample.

Values of the BET-C parameter for krypton adsorption at a temperature of 77 K (Ref 4) for various solids are:

Material	C-parameter
Organic materials	10–70
Glass	20–80
Silica	25–75
Ferric oxide	30–75
Nickel oxide	70–120
Silica gel	80
Ferrites	60–200
Micas	100–130
Tungsten powder	215, 290
Carbon black	230
Nickel film, contaminated	400, 1000
Nickel film, clean	1200, 2300

BET Apparatus

Figure 1 shows the adsorption apparatus originally developed by Emmett. Commercial instruments are now totally automated, and instead of the burette-type mechanism to measure the adsorbed gas volume very precise pressure transducers are used to calculate the adsorbed amount of gas from the known volume of a calibration chamber and the pressure drop that occurs when the valve between sample and calibration chamber is opened. Other instruments use a mass-flow controller to monitor the amount of gas introduced to the sample. In flow through cells, the adsorbing gas is passed over the sample using a carrier (helium, which is assumed to be nonadsorbing at the specific conditions). The composition of the gas stream before and after is analyzed to provide information about the total amount of adsorbed gas. A third technique uses a microbalance to measure the weight change as gases adsorb on the sample. Each technique has advantages and disadvantages. Further details

Table 1 Surface area determinations with different size values of the adsorbed gas

Adsorbate solid	Gas (Temperature, K)	Area per gram using liquid density values, m^2/g	Area per gram using revised values (see text), m^2/g	C-parameter
Glass spheres (7 μm)	N_2 (78)	0.434	0.434	150
	KR (78)	0.322	0.441	32
	C_4H_{10} (195)	0.333	0.489	7
	$CHCl_2F$ (195)	0.315	0.479	106
Tungsten powder	N_2 (78)	2.69	2.69	81
	Kr (78)	1.96	2.68	290
	C_4H_{10} (273)	1.67	2.43	26
	$CHCl_2F$ (273)	1.73	2.62	21
Zinc oxide	N_2 (78)	9.40	9.40	155
	Kr (78)	6.82	9.34	150
	C_4H_{10} (273)	6.93	10.1	52
	$CHCl_2F$ (273)	6.63	10.1	215
Silver foil (geometrical area, 1.56 m^2/g)	Kr (78)	1.56	2.14	19
	C_4H_{10} (195)	1.22	1.78	6
	$CHCl_2F$ (195)	1.13	1.72	11
Monel ribbon (geometrical area 1.56 m^2/g)	Kr (78)	0.456	0.622	13
	C_4H_{10} (195)	0.652	0.952	4
	$CHCl_2F$ (195)	0.577	0.878	7

Source: Ref 5

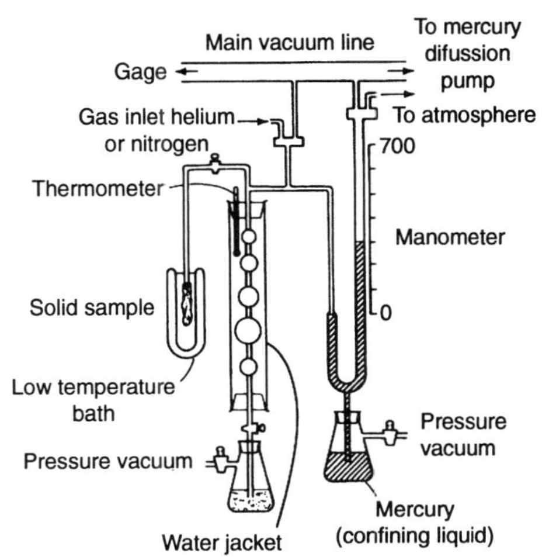

Fig. 1 Brunauer-Emmett-Teller apparatus for determining specific surface area

are described in all standard textbooks (Ref 6–9).

The isotherm is usually constructed point by point by the admission or withdrawal of known amounts of gas, with adequate time allowed for equilibration at each point. A small amount of heat is produced during the adsorption process, which has to be removed from the sample before final equilibrium is achieved. The amount of adsorbed gas is very sensitive to temperature changes. Prior to the determination of the isotherm all physisorbed material has to be removed. The exact conditions required to obtain a "clean" surface depend on the nature of the sample.

For the determination of the surface area and the mesopore size distribution by nitrogen adsorption, outgassing to a residual pressure of about 10^{-4} torr ($\approx 10^{-2}$ Pa) is considered acceptable. The outgassing conditions and the determination of the outgassed mass of the sample is one major source of error (or at least the reason for discrepancies between different experimenters). Inorganic oxides are usually outgassed at temperatures $\approx$150 °C, while microporous carbons and zeolites require higher temperatures $\approx$300 °C. Depending on the synthesis or other previous processing steps of the metals powders, the sample should be cleaned by washing with an appropriate solvent or treated at sufficiently high outgassing temperatures.

Usually nitrogen gas is adsorbed on the sample at liquid nitrogen temperature. The BET surface area determined from those measurements is used as a standard reference point. However, it might be appropriate or advantageous to use a different adsorbing gas in other applications. For example, nitrogen adsorption is limited to samples having a total surface area of more than 1 m^2. However, small surface areas can be measured much more precisely by using krypton as the adsorptive gas. A total sample surface area of down to 50 cm^2 can be measured with many commercial instruments.

The increased sensitivity of the krypton analysis is due to the fact that the saturation pressure or krypton is only about 2.5 torr ($\approx$300 Pa) at liquid nitrogen temperature. Due to this lower saturation pressure, the ratio between the actual amount of adsorbed gas on the sample surface against the amount of gas that remains in the surrounding gas phase is much higher for krypton compared to the same data point in a nitrogen isotherm and correspondingly the amount of adsorbed krypton can be determined with a much higher precision. The next section demonstrates using a simple calculation why krypton measurements are so much more sensitive for low surface area samples.

However, krypton at liquid nitrogen temperature has other problems. The saturation pressure under those conditions cannot be unambiguously defined. According to the phase diagram (Fig. 2), krypton could either exist as a solid or as an undercooled liquid at that temperature. Most publications refer to the supercooled liquid state. This uncertainty severely limits the krypton analysis. A similar condition is also true for argon or xenon at liquid nitrogen temperature. In the case of argon, the experiment could be performed at a different temperature. For example, liquid oxygen would create a temperature of about 90 K, which is above the triple-point temperature of argon (83.8 K). However, liquid oxygen is much more problematic to handle and not as commonly available as liquid nitrogen.

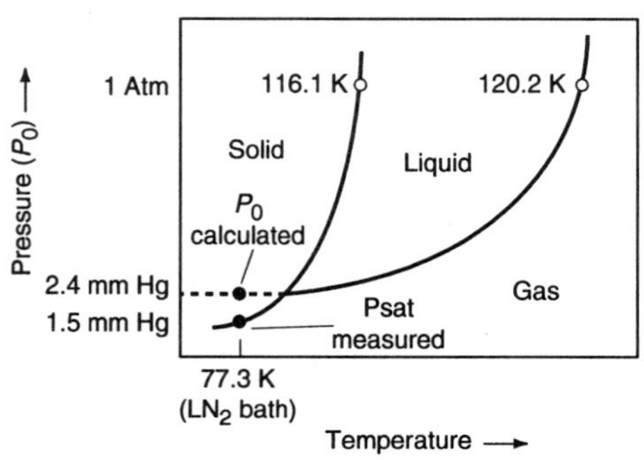

Fig. 2 Krypton phase diagram

Krypton for Low Surface Area Determination

The use of krypton or argon rather than nitrogen for adsorbate improves the accuracy of low specific surface area measurements.

Assume the sample has a total surface area of 1 m^2 and the volume of the sample tube is 20 cm^3. The saturation pressure for nitrogen is about $P_{0(N2)} = 760$ torr and for krypton it is $P_{0(Kr)} = 2.5$ torr at liquid nitrogen temperature (assuming a super-cooled liquid state for the krypton).

The single-point BET equation for those gases yields:

$$V_m = 4.35\ V_a\,(1 - P/P_0)\ \text{for nitrogen}$$

and

$$V_m = 5.64\ V_a\,(1 - P/P_0)\ \text{for krypton}$$

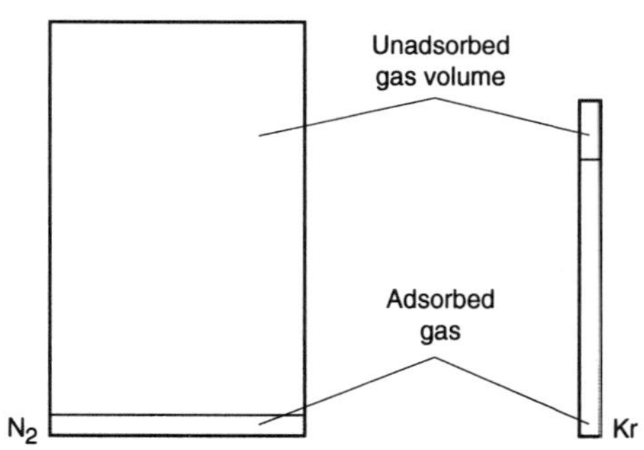

Fig. 3 Ratio of adsorbed versus remaining gas demonstrates more accurate measurements in the krypton analysis compared to N_2 analysis for low surface area samples

At a relative pressure of $P/P_0 = 0.2$ the adsorbed volume, V_a, is then calculated to:

$$1.0 = 4.35\,V_a\,(1 - 0.2)$$

or

$$V_a = 0.287\ \text{cm}^3\ \text{(STP nitrogen)}$$

$$1.0 = 5.64\ V_a\,(1 - 0.2)$$

or

$$V_a = 0.222\ \text{cm}^3\ \text{(STP krypton)}$$

The remaining volume of gas in the sample cell can be calculated as follows:

$$20\ \text{cm}^3\ \frac{(760\ 0.2)\ \text{torr}}{760\ \text{torr}}\ \frac{273\ K}{77.2\ K}$$

$$= 14.144\text{cm}^3\ (\text{STP for } N_2)$$

$$20\ \text{cm}^3\ \frac{(2.5\ 0.2)\ \text{torr}}{760\ \text{torr}}\ \frac{273\ K}{77.2\ K} = 0.046\ \text{cm}^3\ (\text{STP for Kr})$$

As shown in Fig. 3, it becomes obvious that the ratio of adsorbed versus remaining gas allows for a much more accurate measurement in the krypton analysis.

Accuracy

The high degree of accuracy attainable in some branches of chemistry and physics is out of the question where evaluation of specific surface from adsorption data is concerned. Even in the favorable case of nitrogen or argon, a divergence of at least ±10% from the actual area of the solid should be expected (owing to theoretical factors not yet supported by accurate quantitative assessments). In addition, there are several nonnegligible experimental uncertainties, which can be strikingly illustrated by results of a round-robin test. The project embraced thirteen laboratories, all well experienced in the field (Ref 10). The participants determined detailed isotherms of nitrogen (often 30 to 40 points), calculating both the specific surface, based on $a_m(N_2) = 16.2$ Å^2 and the value of the BET *C*-parameter.

Solid	Surface area and standard deviation, m^2/g	No. of labs
Graphitized carbon (Vulcan 3G-2700)	71.3 ± 2.7	6
Graphitized carbon (Sterling FT-2700)	11.00 ± 0.8	5
Plasma produced silica (TK 800)	165.8 ± 2.1	4
Mesoporous silica gel (Gasil (1))	286.2 ± 3.5	3

Pore Size Analysis

Gas adsorption/desorption experiments can also be used to study the pore size distribution in a sample. Pores of up to 100 or 200 nm can be determined. The measurement is based on the capillary condensation effect, which essentially means that a gas will condense into a liquid phase below its saturation point when it is confined in small pores. The basic relationship between relative pressure and pore size is given by the Kelvin equation:

$$\ln\left(\frac{P}{P_0}\right) = -\frac{2\gamma_{lv}V_l}{RTr}$$

where γ_{lv} is the interfacial tension of the liquid-gas interface, V_l is the molar volume of the liquid gas phase, and the other factors are gas constant, R, temperature, T, and the radius of curvature of the liquid-gas interface, r. Numerous theories and models have been developed for the pore size analysis. The most commonly accepted theory was developed by Barrett, Joyner, and Halenda (Ref 11). Frequently, the adsorption-desorption isotherm shows a hysteresis and the shape as well as the relative pressure range where the hysteresis occurs provides information about the pore shape and pore size. Several theories have been developed to explain and analyze the hysteresis effect in more detail including deriving additional information about network and connectivity effects of the pore system in a sample. Details are described in several textbooks (Ref 6–9).

Permeametry

Peter J. Heinzer, Imperial Clevite Technology Center

Permeametry is the measurement of resistance to fluid flow through a compact powder bed. Its main purpose in powder metallurgy is not so much to quantify resistance as it is to measure the related properties of a particle population—namely, specific surface area and average particle size. Fluids may be in liquid or gaseous form. Liquid permeametry was prevalent in the early stages of development and is the simplest method if the minimum diameter of any appreciable size fraction is 5 μm (Ref 12). Settling and segregation, aggregation, and the difficulty of removing bubbles make liquid permeametry unsatisfactory for smaller-sized particles. Gas permeametry is now the preferred method because it extends the size measurement capability down to 0.1 to 0.5 μm.

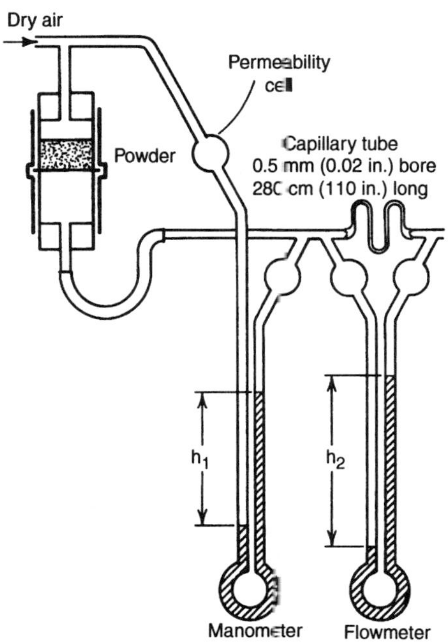

Fig. 4 Lea and Nurse permeability apparatus with manometer and flowmeter

Commercial permeametry is applicable in the following ranges (Ref 13, 14):

Specific surface area, cm^2/g ($in.^2/g$)	70–20,000 (10.8–3100)
Particle size, μm	0.5–50

In production practice, these ranges are subdivided into smaller segments to improve accuracy, because the behavior of fluid flow through powder beds with diverse characteristics changes considerably.

A typical sample occupies a volume of about 5 cm^3 (0.30 $in.^3$) and, depending on density, weighs from 5 to 20 g (0.18 to 0.70 oz). Reproducibility in the surface area range cited for commercial instruments is ±1%. Accuracy varies with the type of sample, as discussed later in this section.

Inherent problems in the use of permeametry for particle size measurement involve its use of semiempirical relationships with parameters that can only be evaluated indirectly. Additionally, permeametric methods provide only an average particle size—not the more useful size distribution information. Despite these disadvantages, permeametry is popular due to its simplicity of operation, reproducibility, speed of analysis (2 to 15 min compared to 1 h or more with other methods), and low cost (one-fifth to one-tenth of the equipment cost of competing methods).

History and Theory

The original study of fluid flow through compacted particulate matter is attributed to D'Arcy (Ref 15), who examined water flow rates from the public fountains of Dijon, France, through sand beds of varying thicknesses in 1856. He formulated the basic principle behind permeability, showing that the average flow rate is proportional to the pressure gradient and inversely proportional to the thickness of the bed.

In 1927, Kozeny (Ref 16) published the first derivation showing correlation among porosity, permeability, and particle surface area. Following the lead of Blake (Ref 17), Kozeny treated the flow of fluid through a particulate bed as being equivalent to the flow of fluid through a comparable volume of parallel pipe channels of circular cross section. This simplification resulted in determining an equivalent diameter of pipe channel to characterize flow rate through the powder bed.

In 1938, Carman (Ref 18) and Dallavalle (Ref 19) independently proposed the determination of specific surface area for powders using permeability methods. Carman published related experimental work in 1941 (Ref 20). He developed a liquid flow technique to determine surface area for coarse materials by taking into consideration (1) the dependence of permeability on the number of permeable pores of the particle bed, (2) the pore or void volume fraction contribution to total bed volume, (3) the friction of the gas or liquid flowing through the bed, and (4) the adsorption of immobile liquid layers that effectively reduce the capillary diameter, thus causing less permeability and therefore a greater apparent surface area.

The Kozeny-Carman equation, given below, has served as the most widely used basis for all permeability variations:

$$S^2 = \frac{L^2}{2\eta v L_e^2 \rho^2} \frac{\Delta P}{L} \frac{f_v^2}{(1-f_v)^3}$$

where S is surface area per unit weight of the powder, ΔP is the pressure drop across the powder bed, f_v is void fraction of packed sample, v is velocity of fluid flow, ρ is density of the powder material, η is viscosity of the fluid, L is length of the powder bed, and L_e is average path length through the powder bed.

Apparatus

Lea and Nurse (Ref 21) developed the apparatus shown in Fig. 4 to provide permeability measurements. The powder was compacted in the sample cell to a predetermined porosity. Air was permitted to flow through the bed, and the pressure drop (h_1) was measured on the first ma-

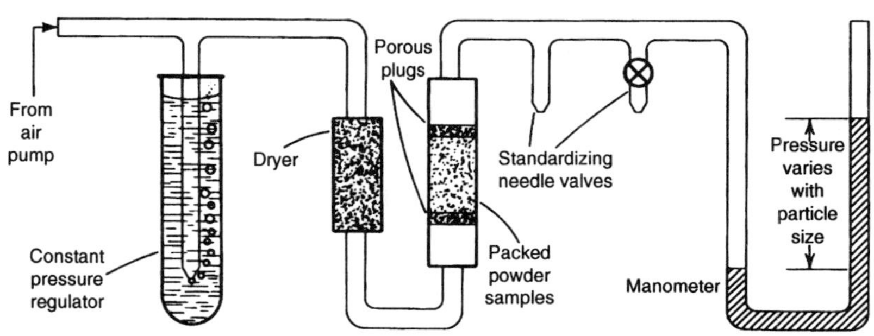

Fig. 5 Fisher subsieve sizer operation

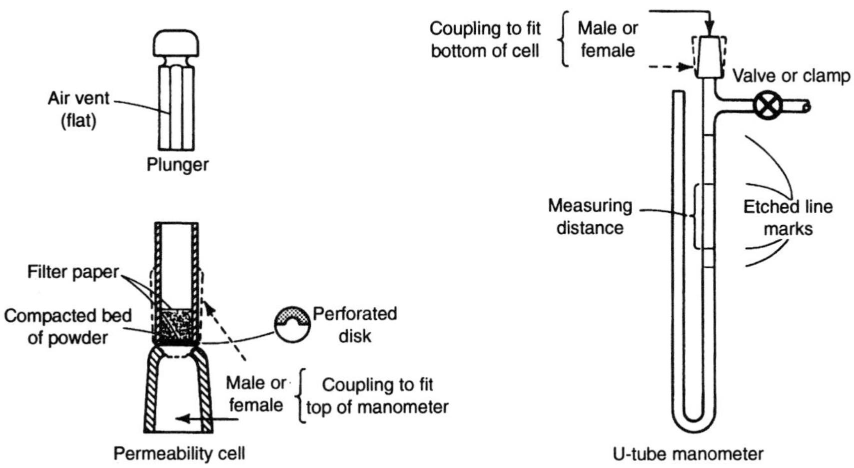

Fig. 6 Blaine air permeability apparatus. Source: Ref 23

nometer; the air then passed through a capillary flowmeter, across which another pressure drop was measured as h_2 on a second manometer.

The capillary permitted the system to operate under a constant pressure. The volume rate of flow through the flowmeter, the pressure drop across the bed as measured by the manometer, and the constants associated with the apparatus permitted determination of the specific surface area (surface area per unit volume).

Gooden and Smith (Ref 22) added a self-calculating chart to a modified Lea and Nurse apparatus to enable direct readout of the specific surface. The commercial version of their modification is known as the Fisher subsieve sizer (Fig. 5).

A simplified version of the air permeameter, known as the Blaine permeameter (Fig 6) (Ref 23), relied on a variable pressure technique (Ref 24). A vacuum was used to displace the oil in a U-tube connected in series with the powder cell. The resultant pressure caused air to flow through the powder bed, and the time required for the displaced oil to fall back to its equilibrium position was measured. This method resulted in a measured specific surface area, which decreased with porosity. Usui (Ref 25) showed that log t and the void fraction exhibited a linear relationship and that a plot of these parameters gave a value for surface area.

Limitations

For very fine powders, the basic Kozeny-Carman equation is not accurate. This is because the laminar flow assumption on which it is based is no longer valid. Compressed fine particles result in a powder bed with very small channel widths. If these widths are comparable to the mean free path length of the gas molecules, laminar flow conditions are not maintained.

Such a situation, involving molecular flow or diffusion conditions, is known as Knudsen flow (Ref 26) and can occur with very fine powders or with coarser particles at low pressures. Figure 7 shows a typical apparatus used to measure find particles under molecular flow conditions (Ref 27, 28). In some powders, both laminar and molecular flow may be significant. This is known as the transitional region.

Evaluating surface areas with steady-state flow conditions historically excluded noninterconnected blind pores. A method for including blind pores by utilizing transient-state flow measurements is the principle behind the apparatus shown in Fig. 8(a) (Ref 29, 30). A typical flow rate curve, showing extrapolation of the steady-state portion to determine the time lag is given in Fig. 8(b). The basic principle of the permeability analysis has not changed much during the last several years; even so, instruments now use more sophisticated pressure transducer and flowmeter or flow controller. Combined with a computerization of the instrument and an improved data analysis software, the instrument offer a wider variety of analysis tasks and computational methods.

Pycnometry

Peter J. Heinzer, Imperial Clevite Technology Center

Pycnometry is used to determine the true density of P/M materials. Based on the displacement principle, pycnometry is actually a method of determining the volume occupied by a solid of complex shape, such as a powder sample. For commercial pycnometers, typical sample sizes range from 5 to 135 cm^3 (0.30 to 8.24 $in.^3$). A properly prepared specimen can be analyzed in 15 to 20 min.

The pycnometric determination of density can be quite useful in P/M applications. In addition to its primary use in measuring the true density of a P/M part or product, it can be used to distinguish among different crystalline phases or grades of material, different alloys, compositions, or prior treatments.

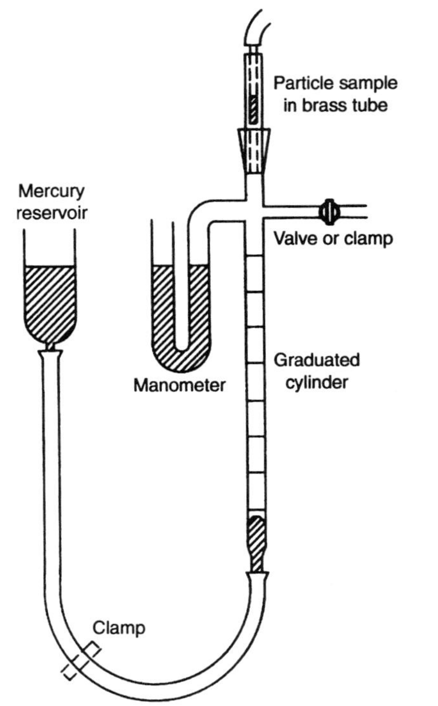

Fig. 7 Modified Pechukas and Gage apparatus for fine powders. Source: Ref 27

Information on the porosity of a material can be obtained from pycnometry if the sample has a uniform geometry, or if the bulk volume is known. Pore volume is the difference between the bulk volume (1/bulk density) and the specific volume (1/true density). Finally, pycnometry can be useful in determining properties that relate to density. Often, P/M materials have no solid counterpart to use for measuring true density, making percentage of theoretical density measurements questionable. Pycnometric measurements of the true density of the powder have provided a good point of reference.

Density is one of the most important properties of P/M materials. Critical processing parameters, such as applied force and pressure, and properties of the resulting P/M product, such as strength and hardness, usually depend on the density of the materials being processed. Standard industry practice compares the achieved density of a P/M product with the full, or theoretical, density.

Theory and Apparatus

Archimedes devised the first method for determining true density by using the displacement principle. Modern pycnometry represents a refinement of the displacement principle and uses either a liquid or gaseous substance as the displaced medium.

Absolute densities of solids can be measured by the displacement principle using either liquid or gas pycnometry. In liquid pycnometry, vol-

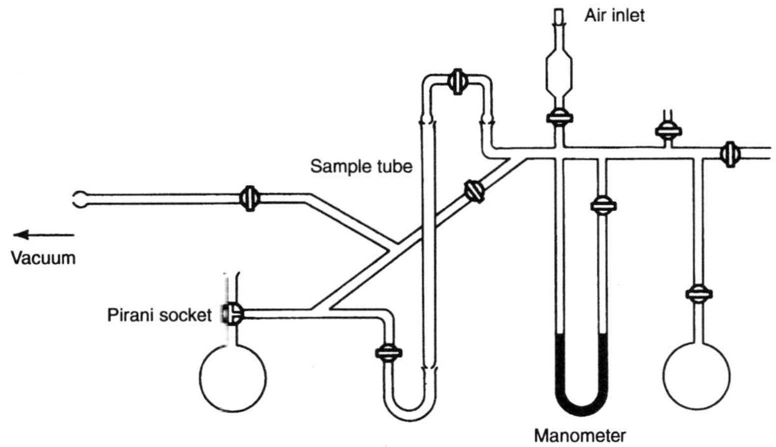

Fig. 8(a) Transient flow apparatus

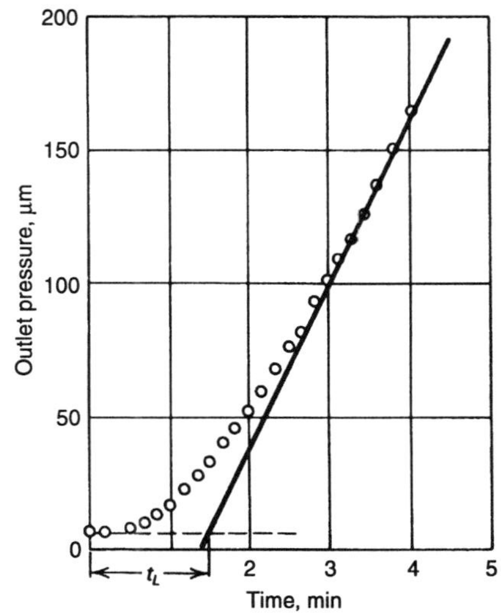

Fig. 8(b) Flow rate curve for the transient flow apparatus

ume displacement is measured directly, as liquids are incompressible. Inability of the liquid to penetrate pores and crevices, chemical reaction or adsorption onto the sample surface, wetting or interfacial tension problems, and evaporation contribute to errors in density measurement. Therefore, gas pycnometry is usually preferred for P/M applications.

In gas pycnometry, volume displacement is not measured directly, but determined from the pressure/volume relationship of a gas under controlled conditions. Gas pycnometry requires the use of high-purity, dry, inert, nonadsorbing gases such as argon, neon, dry nitrogen, dry air, or helium. Of these, helium is recommended because it:

- Does not adsorb on most materials
- Can penetrate pores as small as 0.1 nm (1 Å)
- Behaves like an ideal gas

In commercial pycnometers, the sample is first conditioned or outgassed to remove contaminants that fill or occlude pores and crevices, thus changing surface characteristics. This is accomplished by evacuating the system and heating to elevated temperatures, following by purging with an inert gas such as helium.

The helium-filled sample system (Fig. 9) is "zeroed" by allowing it to reach ambient pressure and temperature. At this point, the sample cell and reference volume are isolated from each other and from the balance of the system by valves.

The state of the system can then be defined by:

$$PV = nRT \quad \text{(Eq 1)}$$

for the sample cell and:

$$PV_R = n_R RT \quad \text{(Eq 2)}$$

for the calibrated reference cell. In these equations, P is the ambient pressure, Pa; V is the volume of the sealed empty sample cell, cm^3; V_R is volume of a carefully calibrated reference cell, cm^3; n is moles of gas in the sample cell volume at P; n_R is moles of gas in the reference cell volume at P; R is the gas constant; and T is ambient temperature, K.

A solid sample of volume (V_s) is then placed in the sample cell:

$$P(V - V_s) = n_1 RT \quad \text{(Eq 3)}$$

where n_1 is moles of gas occupying the remaining volume in sample cell at P. The system is then pressurized to P_2, about 100 kPa (15 psi) above ambient:

$$P_2(V - V_s) = n_2 RT \quad \text{(Eq 4)}$$

where n_2 is moles of gas occupying the remaining volume in the sample cell at P_2. The valve is then opened the connect the sample cell with the calibrated reference volume, and the pressure drops to a system equilibrium P_3:

$$P_3(V - V_s) + P_3 V_R = n_2 RT + n_R RT \quad \text{(Eq 5)}$$

Substituting PV_R from Eq 2 for $n_R RT$ in Eq 5 and substituting $P_2(V - V_s)$ from Eq 4 for $n_2 RT$ results in:

$$P_3(V - V_s) + P_3 V_R = P_2(V - V_s) + PV_R \quad \text{(Eq 6)}$$

Simplifying:

$$(P_3 - P_2)(V - V_s) = (P - P_3)V_R \quad \text{(Eq 7)}$$

$$V - V_s = \frac{(P - P_3)\,V_R}{(P_3 - P_2)} \quad \text{(Eq 8)}$$

$$V_s = V + V_R/[1 - (P_2 - P)/(P_3 - P)] \quad \text{(Eq 9)}$$

Because P is "zeroed" at ambient before pressurizing, the working equation becomes:

$$V_s = V + V_R/[1 - (P_2/P_3)] \quad \text{(Eq 10)}$$

Over the last few years, the gas pycnometer has been further improved by using a more accurate

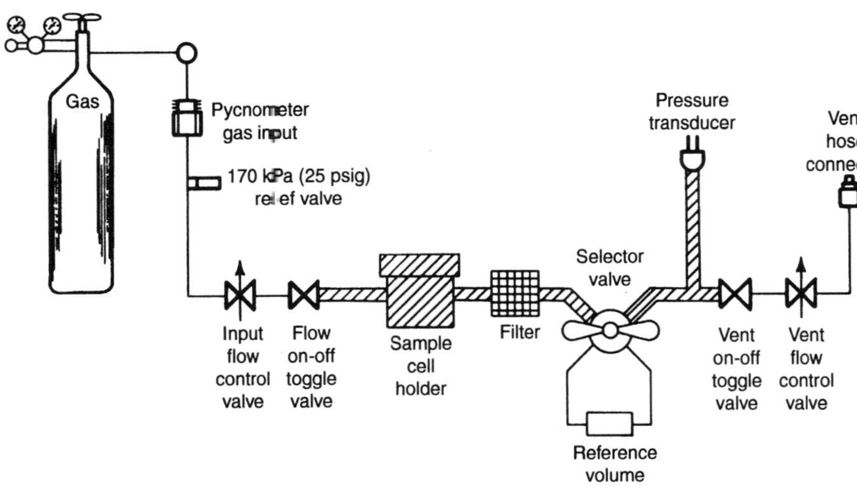

Fig. 9 Flow chart for typical pycnometer

pressure transducer, a better temperature control of the entire system, and by an automation (computerization) of the actual analysis process. Modern pycnometers can now reach an accuracy of 0.01%.

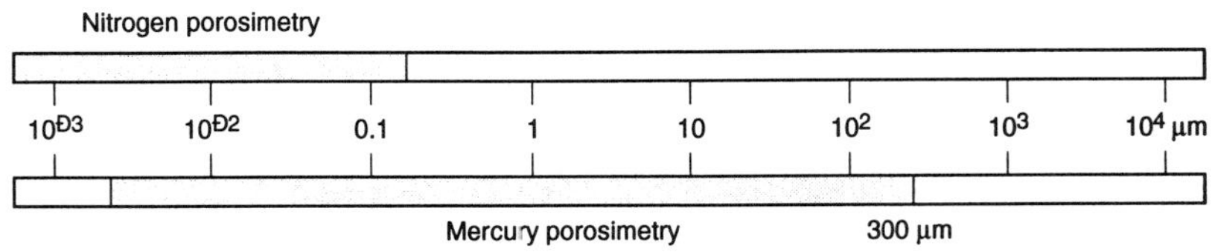

Fig. 10 Pore radii ranges covered by gas and mercury porosimetry

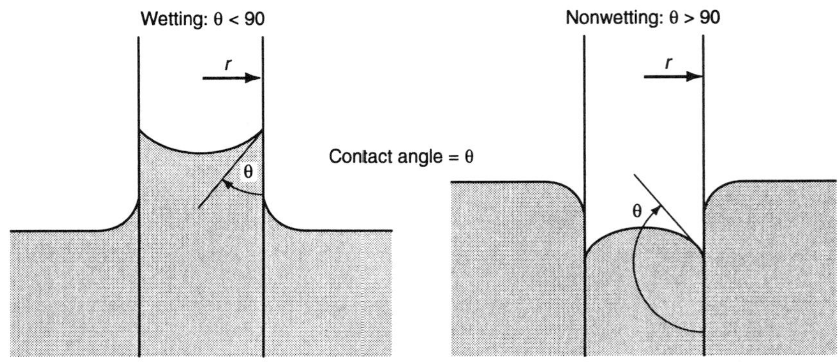

Fig. 11 Contact angle (θ) of a liquid in a capillary

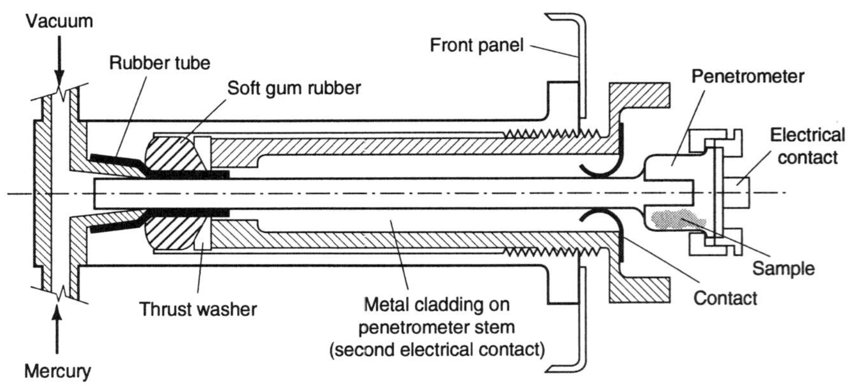

Fig. 12 Low-pressure mercury porosimeter. Source: Ref 34

Mercury Porosimetry

H. Giesche, School of Ceramic Engineering and Sciences, Alfred University

Many commercially important processes involve the transport of fluids through porous media and the displacement of one fluid, already in the media, by another. The role played by pore structure is of fundamental importance in understanding of these processes. The quality of powder compacts is also affected by the void size distribution between the constituent particles. For these reasons, mercury porosimetry has long been used as an experimental technique for the characterization of pore and void structure.

Gas and mercury porosimetry are complementary techniques with the latter covering a much wider size range from 0.3 mm to 3.5 nm (Fig. 10). Mercury porosimetry consists of the gradual intrusion of mercury into an evacuated porous medium at increasingly higher pressures followed by extrusion as the pressure is lowered. The simplest pore model is based on parallel circular capillaries that empty completely as the pressure is reduced to zero. This model fails to take into account the real nature of more porous media, which consist of a network of interconnecting noncircular pores. The network effects lead to hysteresis and mercury retention during the extrusion cycle.

General Description

Relationship between Pore Radii and Intrusion Pressure. Mercury porosimetry is based on the capillary rise phenomenon whereby an excess pressure is required to cause a nonwetting liquid to enter a narrow capillary. The pressure difference across the interface is given by the equation of Young (Ref 31) and Laplace (Ref 32), and its sign is such that the pressure is less in the liquid than in the gas (or vacuum) phase if the contact angle θ is greater than 90° or more if θ is less than 90°. If the capillary is circular in cross section, and not too large in radius, the meniscus will be approximately hemispherical. The curvature of the meniscus can be related to the radius of the capillary, and the Young-Laplace equation reduces to the Washburn equation (Ref 33):

$$\Delta P = \gamma_{lv}\left(\frac{1}{r_1}+\frac{1}{r_2}\right)=\frac{2\gamma_{lv}}{r_p}\cos\theta \qquad \text{(Eq 11)}$$

This is the Young-Laplace and Washburn equation where γ_{lv} is the surface tension of the liquid (e.g., for mercury, 0.485 N/m), r_1 and r_2 are mutually perpendicular radii of a surface segment. The angle θ is the angle of contact between the liquid and the capillary walls and is always measured within the liquid (Fig. 11). r_p is the capillary radius.

Equipment Fundamentals. The sample is placed into the penetrometer assembly; it is then evacuated to a set vacuum level for a specific time, before the sample cell is filled with mercury. Air is admitted to the low-pressure chamber, and the increasing pressure forces the mercury to penetrate the largest pores of the sample. The amount or volume of mercury penetrating into the sample is recorded at each pressure (or pore size) point; the first reading usually is taken at a pressure of 0.5 psi (0.003 MPa), although readings at a pressure of 0.1 psi (0.7×10^{-4} MPa) are possible. The pressure is then increased to 1 atm, or in some instruments the pressure is actually increased to a slight overpressure (up to 50 psi in some cases). After the low-pressure run is finished, the penetrometer is then inserted into a high-pressure port and surrounded with a special grade of high-pressure oil; it is special with respect to the dielectric constant and viscosity of the oil under high-pressure conditions. The pressure is increased up to a final pressure of 60 ksi (400 MPa). Commercial instruments work either in an incremental or continuous mode. In the former, the pressure is increased in steps and the system allowed to stabilize at each pressure point before the next step. In the continuous mode, the pressure is increased continuously at a predetermined rate. Schematics of low-pressure and high-pressure systems are shown in Fig. 12 and 13, respectively (Ref 34).

Measurement Techniques

Measuring Displacement Volumes (Pore Volume). Mercury volume displacements may be measured by direct visual observation of the mercury level in a glass penetrometer stem (Fig. 14) with graduated markings (Ref 35). However, most (if not all) instruments on the market will measure this volume automatically by one of the following techniques:

- *Precision capacitive bridges:* measure changes in the capacitance between the column of mercury in a dilatometer stem and a coaxial sheet surrounding the column
- *Mechanical transducer:* indicate the change in height of the mercury column by moving a contact wire and measuring the displacement of the mercury interface in the stem
- *Submerged wires:* measure changes in resistivity corresponding to the change in length of the mercury column

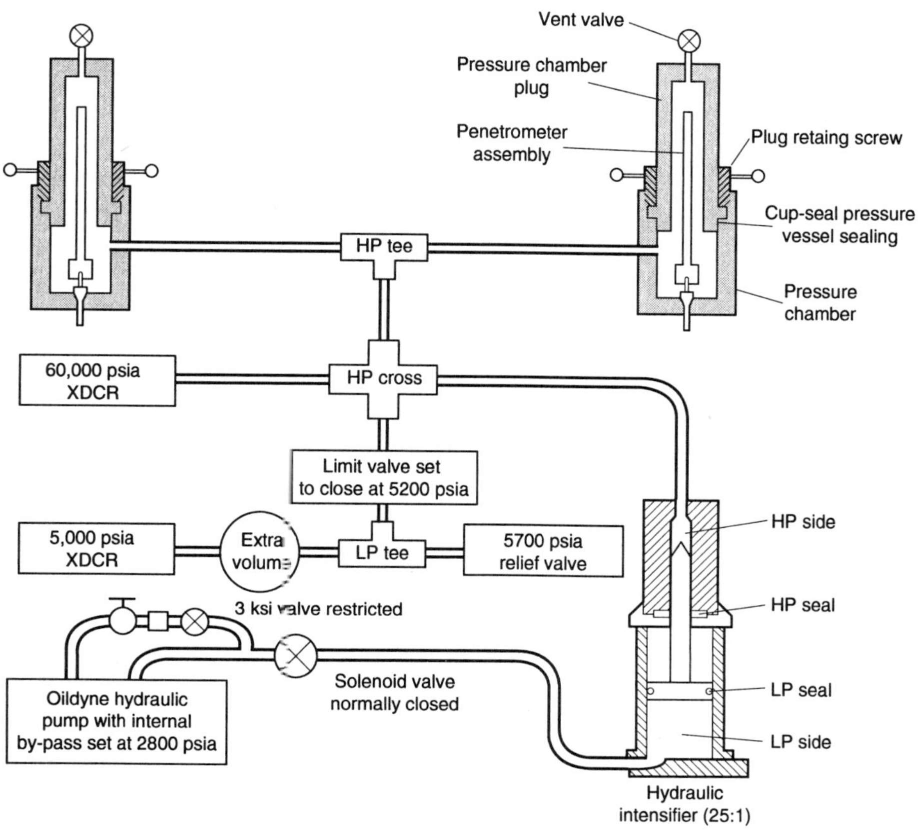

Fig. 13 Micromeritics high-pressure mercury porosimeter

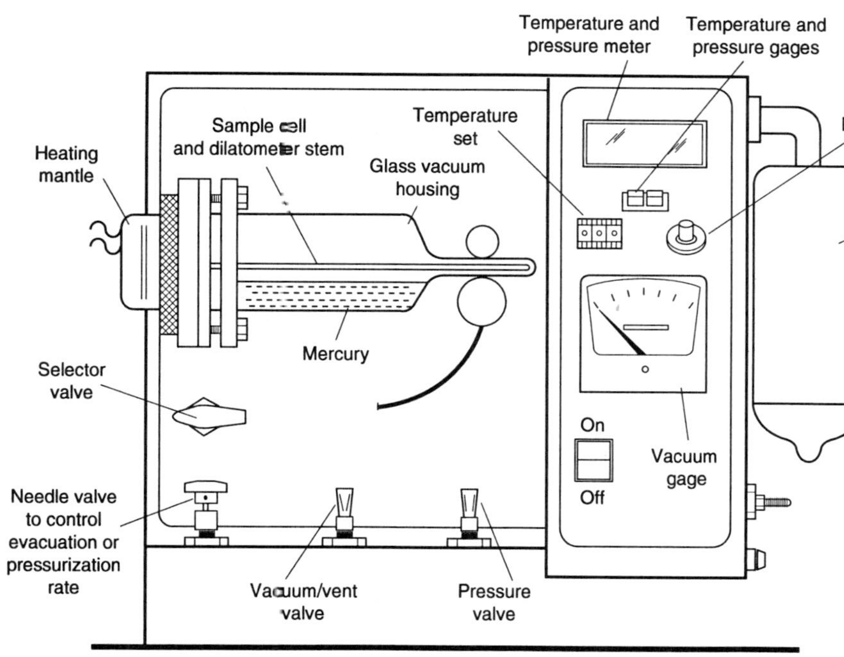

Fig. 14 Quantachrome filling mechanism and low-pressure porosimetry system. Source: Ref 35

From a practical point of view, the sample mass (pore volume) and the stem volume of the penetrometer should be adjusted to use the instrument to its highest resolution. In general, larger samples are preferred because they provide a better representation of the overall sample.

Pressure. The corresponding pressure at which mercury is filling the pore system is usually measured with electronic pressure transducer or with Heise-Bourdon manometer, used in older manual setups. A series of those pressure transducers ensures that accurate data are determined over the entire range from 0.1 psi (0.7×10^{-4} MPa) to 60 psi (0.4 MPa).

Looking at the Washburn equation (Eq 11) makes it obvious that two additional parameters play a critical role in the calculation of pore size from the applied pressure: contact angle, θ, and the surface tension, γ_{Hg}.

Contact Angle Determination. Various techniques are available to determine the contact angle:

- A drop of mercury can be placed on the flat surface of the sample, and the resulting contact angle is visually observed. Problems related to "micro" and "macro" contact angles have been reported (Fig. 15) (Ref 36). Brashforth and Adams (Ref 37) published tables that allow calculation of the contact angle as well as the surface tension of liquids from the shape of a drop of mercury on the substrate surface (Fig. 16). A simplified formula can be used when the maximum height of the drop is reached (Ref 35, 37, 38):

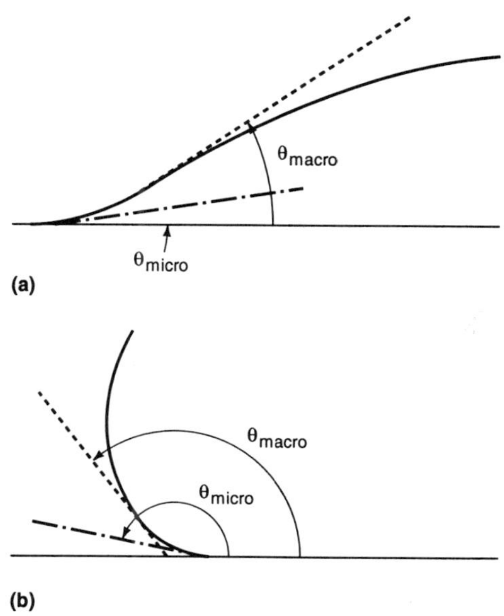

Fig. 15 Differences between microscopic and macroscopic measurement of the contact angle (θ) under conditions of (a) wetting and (b) nonwetting

$$\cos\theta = 1 - \frac{\rho g h_{max}^2}{2\gamma_{lg}}$$

with g, the acceleration of gravity, and ρ, the density of the liquid.

- A powder compact can be pressed in such a way that a well-defined hole is created in a disk. Mercury is now placed on top of this disk, and the contact angle can be calculated from the necessary pressure to force the mercury through this cylindrical pore.
- The Willhelmy plate method (Fig 17) can be used to determine the contact angle (Ref 39). Figures 17 and 18 illustrate the critical observation of an advancing and receding mercury interface. Surface roughness (Fig. 18) or the change in surface composition during the contact with mercury can explain the presence of this difference. (Note: there is no thermodynamic reason or explanation for any contact angle hysteresis.) No surface roughness effects are assumed below pore sizes of about 100 nm. This effect also emphasizes the importance of clean samples and clean mercury.

The contact angle between mercury and the sample being tested is frequently assumed to be 130 or 140°. This assumption is probably the largest source of error. Contact angles of different materials may differ significantly, as shown in Table 2 (Ref 40).

The Washburn equation is directly proportional to the cosine of the contact angle; the respective pore size errors for iron ($\theta = 115°$) and glass ($\theta = 153°$), using the values from Table 2 versus a constant value of 130° for θ, would be:

$$52\% = \frac{\cos 130 - \cos 115}{\cos 153}100$$

or

$$28\% = \frac{\cos 153 - \cos 130}{\cos 153}100$$

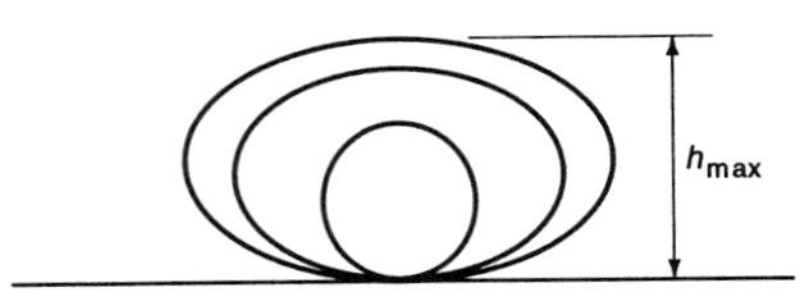

Fig. 16 Change of mercury-drop shape with size

However, published contact angles differ widely between different research groups, even when presumably the same material was studied.

Some materials might react with mercury, for example, zinc, silver, or lead samples. This severely changes the nonwetting behavior of mercury with that sample and may even lead to a contact angle of <90°. In those cases a protective film of stearic acid can be applied to the sample/pore surface and thus prevent the reaction. Though, for several reactive metals, such as copper, their natural oxide layer on the surface is a sufficient protection (Ref 41).

Surface Tension of Mercury. Values for the surface tension of mercury can vary with atmosphere, temperature, and purity of the mercury used. The purity of mercury has a significant effect on surface tension. Reported values vary by up to 0.1 N/m (100 dynes/cm). Mercury is unusually prone to contamination, and this probably accounts for the lack of reproducibility to be found in the values of surface tension in earlier publications. Later work, however, showed very consistent results (Table 3) (Ref 6). The effect of temperature is minimal, because the temperature coefficient of the surface tension of mercury is only 2.1×10^{-4} N/m · °C. Another error is caused by neglecting the change of surface tension for very small radii of surface curvature. The following correction has been suggested by Kloubek (Ref 42):

$$\gamma_{corr} = \gamma - 2.66 \times 10^{-4}\ \text{m} \cdot \Delta P$$

For $\Delta P = 200$ MPa, the correction term gives an error of 12% [$\gamma_{corr} = (0.485 - 0.053)$ N/m].

Compressibility of Mercury. Due to the slight compressibility of mercury, the measured pore volume of a porous material appears larger than its actual volume. Using a well-balanced combination of the compressibility of the glass sample cell, the mercury and the changes of dielectric properties especially of the high-pressure fluid minimizes this blank effect. In general, the larger the sample and pore volume of the sample in comparison to the amount of mercury in the penetrometer, the smaller the errors due to compressibility. Moreover a blank run correction can be used to correct for compressibility effects. However, the total effect is a combination of all components in the system: the mercury, the high-pressure oil, the glass penetrometer, and the sample.

Restrictions and Limitations (Ref 43)

Results obtainable by mercury porosimetry are limited a priori in three ways:

- The volume of pore space filled with mercury is limited by the maximum pressure. This aspect is related to the question of what is to be counted as pore space; this is not self-evident for all materials (e.g., cement stone). Mercury cannot penetrate very small pores.
- At the other end of the scale, the penetration of very large pores is limited by the height of the sample, which determines a minimum pressure. Hence, very large pores may go unnoticed. (This was observed, e.g., for porous nickel plaques.)
- The sample size is finite and usually quite small. This boundary condition determines that results are not necessarily representative of an infinite pore space, as large pore openings (at the surface) are more easily accessible in a smaller sample. Empirical support for this has been found by carrying out experiments with jacketed and unjacketed samples.

Table 2 Contact angle between mercury and select P/M materials

Powder	Angle, degrees
Aluminum	140
Copper	116
Glass	153
Iron	115
Zinc	133
Tungsten carbide	121
Tungsten	135

Source: Ref 40

Table 3 Surface tension of mercury in vacuum

Temperature, °C	Surface tension, mN/m	Method used
25	484 ± 1.5	Sessile drop
25	484 ± 1.8	Sessile drop
20	485 ± 1.0	Drop pressure
25	483.5 ± 1.0	Sessile drop
25	485.1	Sessile drop
16.5	487.3	Pendant drop
25	485.4 ± 1.2	Pendant drop
20	484.6 ± 1.3	Pendant drop
20	482.5 ± 3.0	Bubble pressure

Source: Ref 6

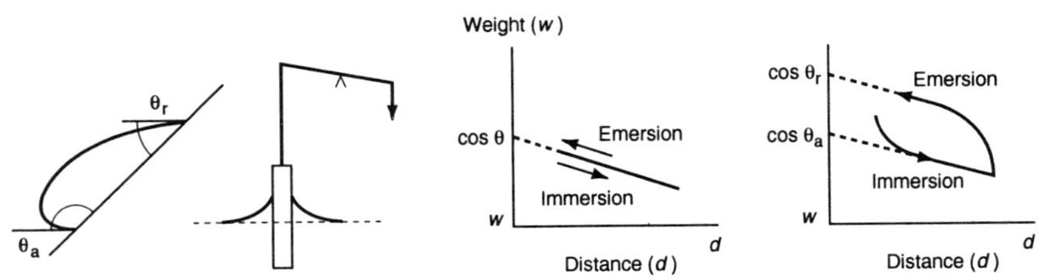

Fig. 17 Wilhelmy plate method showing the effect of contact angle hysteresis for emersion and immersion. Adapted from Ref 39

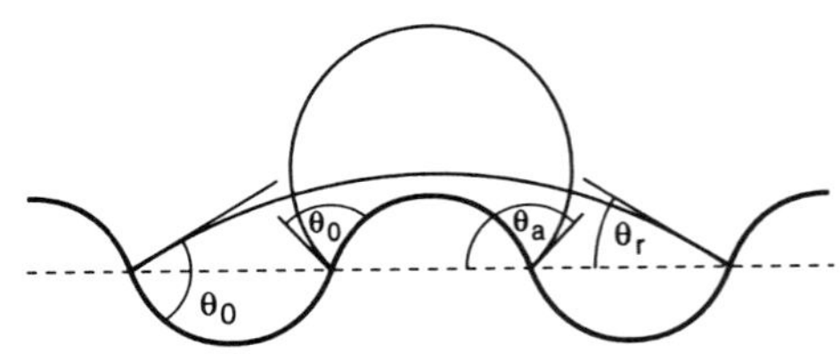

Fig. 18 Effect of surface roughness on contact angles

Geometrical properties of the sample can affect the reproducibility and can cause difficulty in giving an unambiguous interpretation of the result:

- In many cases, a distinction is to be made between the interparticle and intraparticle voids. In a packing of nonporous particles there is only an interparticle pore space. However, in many applications—for example, sorbents—the prime concern is in the intraparticle void space. In such cases, a judgment must be made as to which part of the measured pore volume belongs to the interparticle voids and which part belongs to the intraparticle porosity. If the sample exists as larger particles, the pore size distribution frequently shows a bimodal pore size distribution. The larger pore size fraction can then be attributed to the interparticle porosity. It may also help to use a narrow size fraction of these particles for the porosity analysis. A narrow particle size distribution causes a distinct peak in the pore size distribution at about 20 to 50% of the particle size, which can be clearly attributed to the interparticle voids.
- A special problem is caused by the roughness of the surface of the particles or lumps that are measured. Reich (Ref 44) already pointed out that characterization of samples should be based on those with the same surface roughness. He found a significant difference for pieces of brick broken in different ways. Corrections for the "part of the surface" that belongs to the particle pore space have been made, but it remains difficult to separate pores at the surface from pores in the inside if a cross-sectional area through the surface is different from a cross-sectional area through the interior of the porous medium. Sometimes these differences are very great, and the presence of a characteristic surface layer can actually be observed.
- Pretreatment of the sample may involve comminution. This may change the internal pore space in two ways. First, if closed pores are present, some of these will be broken open. Second, particles tend to break along large pores; hence the relative volume of larger pores decreases.
- Similar problems arise for nonparticulate materials such as paper. Sheetlike material should be packed in a controlled way to eliminate artificial pores between the sheets.

The points mentioned up to now are basically limitations of the technique. If they are overlooked, they can lead to errors of interpretation. The following is a list of factors that can upset the accuracy and the reliability of the results in a more direct way.

- Even if clean mercury is used at the beginning, the surface may soon be contaminated by components that were adsorbed on the surface of the sample. It is known that impurities can change the surface tension of mercury by as much as 30%.
- Before mercury enters the dilatometer, the sample should be outgassed. Evacuation of the sample can change the contact angle of the sample (compared with the contact angle on the sample in its original condition). It has been stated that strong outgassing is necessary if the smallest pores are to be measured. However, the error caused by compression of residual air seems to become significant only if the evacuation does not reach a pressure below 10 torr.
- A serious problem arises when the sample is mechanically destroyed by the applied pressure during the analysis. Fragile porous materials may be subject to a breakdown of pores during pressurization, or sol-gel materials are

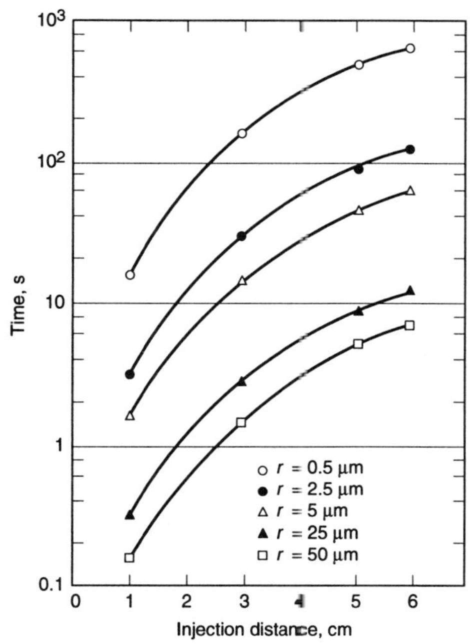

Fig. 19 Distance-to-time ratio for mercury advancing in tubes of five different radii (r). Source: Ref 49

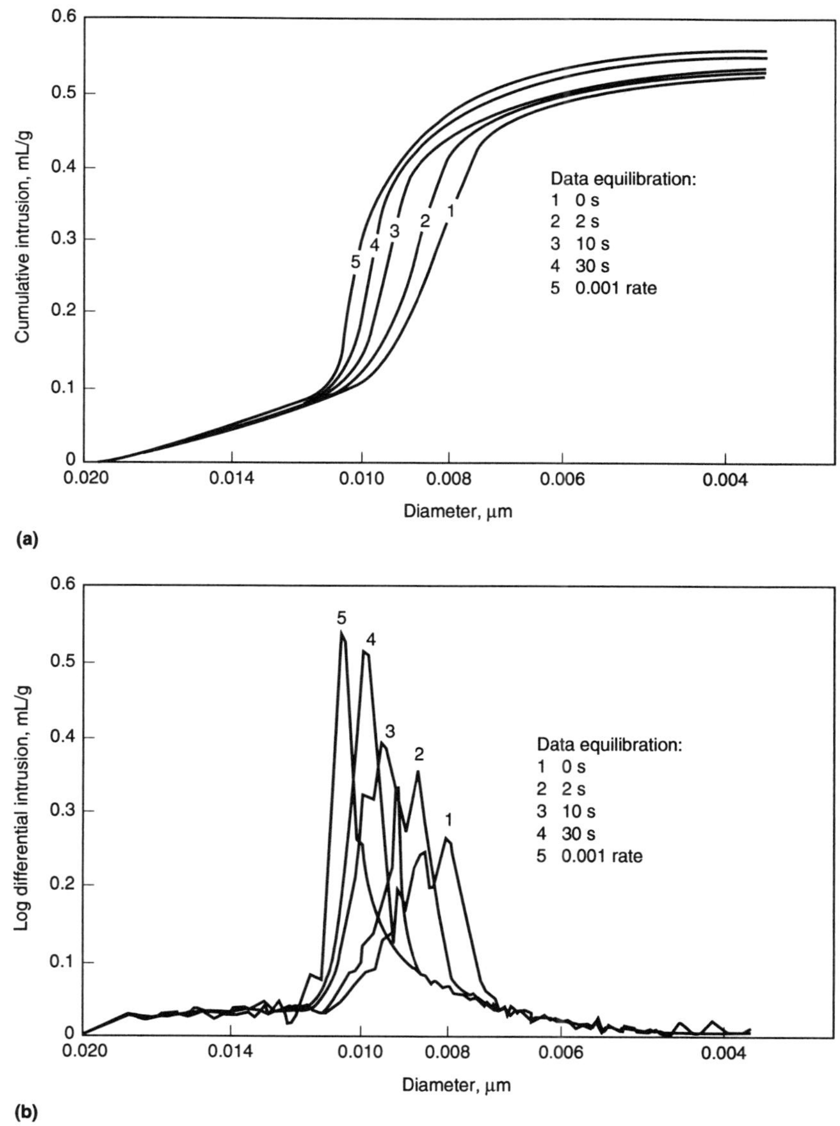

Fig. 20 Effect of equilibration routines on measurement of (a) cummulative intrusion and (b) log differential intrusion with alumina extrudate. Source: Courtesy of D. Smith, Micromeritics and Ref 53

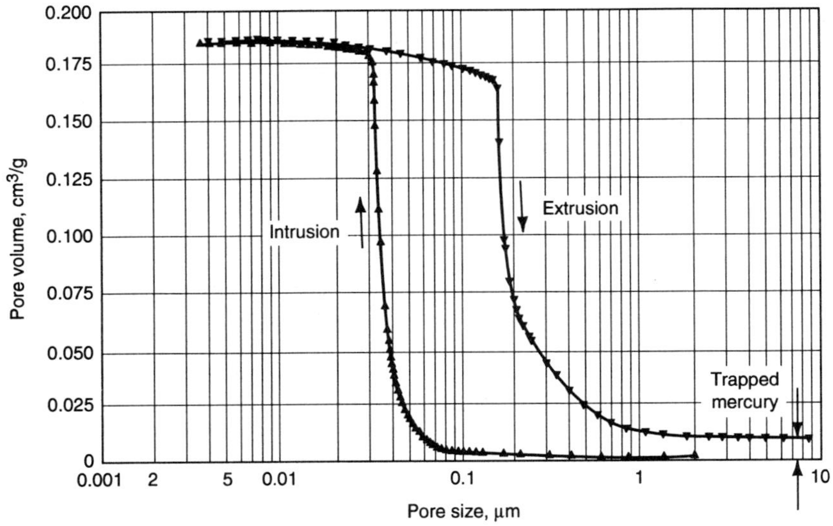

Fig. 21 Mercury porosimetry analysis of an ordered packed sphere structure. Source: Ref 48

Table 4 Comparison of surface area measurements with N_2 gas adsorption and Hg porosimetry

Sample	Surface area, m^2/g	
	Hg-porosimetry	N_2-adsorption
Tungsten powder	0.11	0.10
Iron powder	0.20	0.30
Zinc dust	0.34	0.32
Copper powder	0.34	0.49
Silver iodide	0.48	0.53
Aluminum dust	1.35	1.14
Fluorspar	2.48	2.12
Iron oxide	14.3	13.3
Anatase	15.1	10.3
Graphitized carbon black	15.7	12.3
Boron nitride	19.6	20.0
Hydroxyapatite	55.2	55.0
Carbon black (Spheron-6)	107.8	110.0

frequently compressed reversibly or irreversibly during the mercury porosimetry analysis at high pressure. Knowledge of the compressibility or fracture strength of the material to be analyzed is desirable to properly estimate whether mercury intrusion will occur before deformation or fracture of the porous material occurs. In other cases, a powder compact can be further compacted during the measurement; for example, a powder was compacted at a pressure of 10 MPa; however, mercury does not penetrate the pores before a pressure of 20 MPa is reached. In this case, the sample experiences a second compaction and the results are not related to a compaction pressure of 10 MPa but rather 20 MPa.

- Corrections have to be made for the compressibility of the dilatometer and the mercury, as well as compressibility effects of the sample. Due to the compression of the mercury and oil, the temperature in the dilatometer may rise considerably when a high pressure is applied. It has been estimated that the temperature rise could be as much as 15°. Although it has occasionally been mentioned that a cooling fan is used, this point seems to have gone largely unnoticed. A change in temperature then changes the volume of the mercury and the dilatometer and thus causes an artificial pore volume effect.
- The so-called kinetic hysteresis effect is related to the time that is required for the mercury to flow into pores. If a volume reading is taken before equilibrium has been reached, this will result in a shift of the pore size distribution toward smaller values.

Moreover, the rate of advance of mercury interfaces in horizontal, cylindrical capillaries was computed and also verified experimentally. For a given sample size and pore size, the time required for equilibrium to be achieved during injection can be calculated.

For a horizontal, cylindrical capillary, mercury enters the capillary at the threshold pressure, as given by the Laplace equation, but does not continue to advance. A finite rate of advance is dependent on an excess pressure (ΔP) above the threshold pressure and the distance-to-time relationship for the advancing mercury front is given by the following equation, which can be derived from the Poiseulle equation (see Ref 39, p 286, and Ref 49):

$$t = \frac{l^2 4\eta}{\Delta P r_p^2}$$

where t is time, l is distance, η is viscosity, r is radius, and ΔP is pressure applied in excess of the injection pressure.

The distance-to-time relationship for mercury in tubes of five different sizes is shown in Fig. 19. A total applied pressure of 110% of the injection pressure—that is ΔP in the previous equation is equal to 10% of injection pressure given by the Laplace equation—was used. The graph shows that for a tube of 0.5 µm radius, the time required to travel 3 cm is in excess of 100 s.

The measured data are not reliable if they do not refer to an equilibrium situation. In older publications, for example, kinetic hysteresis is stressed and the effect of mechanical vibrations ("tapping") is noted. In practice, those limitations only apply to very large samples or for extremely small pore sizes. An equilibration time of 1 to a maximum of 5 min should be sufficient in all samples.

In general, effects due to different intrusion rates are not very severe. However, an example where pore size as well as pore volume was greatly influenced is shown in Fig. 20. Five samples of an alumina extrudate were analyzed with different equilibration routines. One sample was scanned (equilibration by time for 0 s). Three others were analyzed at different equilibration intervals (2, 10, and 30 s), and a fourth was analyzed at an equilibration rate of 0.001 µL/g · s. The data from the five experiments are summarized in Fig. 20.

Surface Area Determination

Under the assumption of a specific pore geometry, one can calculate the surface area of a sample from mercury porosimetry measurements. Usually the cylinder pore model is used.

Rootare and Prenzlow (Ref 45) derived the following equation:

$$\text{Area} = -\int_0^V \frac{P\,dV}{\gamma_{lv}\cos\theta}$$

This equation is equivalent to the cylinder pore model.

The later derivation does not contain an assumption about the pore geometry; it does however, assume that the movement of the mercury meniscus is reversible. For an interconnected pore system, this is not the case and the equation will always calculate higher values for the surface area. Surprisingly many publications report a relatively good correlation between surface area values derived from mercury porosimetry measurements and corresponding N_2-gas adsorption BET values (Ref 6, 45) (Table 4).

The value of the specific surface area is often used to determine ("adjust") the contact angle of mercury on a specific material by modifying the contact angle until the surface area calculated from mercury porosimetry measurements correlates with the values determined from nitrogen adsorption data. For materials that do not indicate a severe effect of compressibility or pores smaller than about 10 nm, this method might be a valid alternative to determine the contact angle.

Hysteresis and Entrapped Mercury in the Sample. Interpretation of the extrusion data from mercury porosimetry measurements has been mostly neglected because the interpretation was very vague and questionable. Traditionally, three theories are used to explain the hysteresis between the intrusion and extrusion curve in mercury porosimetry measurements (Fig. 21)

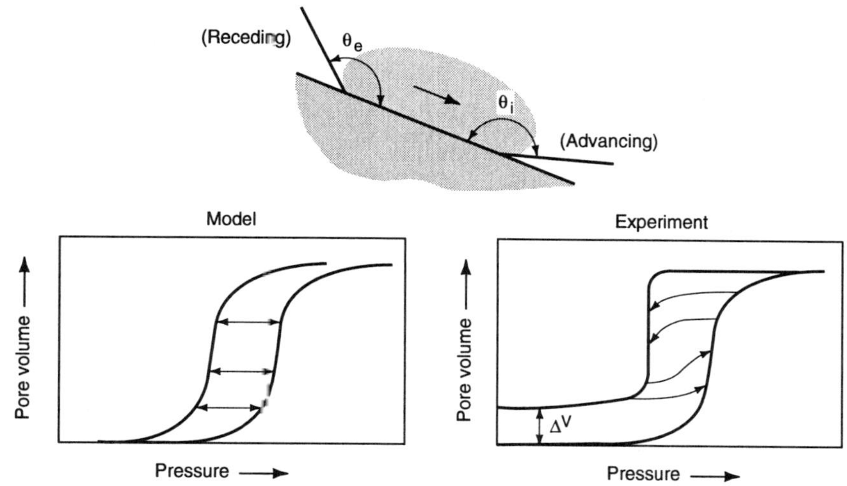

Fig. 22 Hysteresis model based on differences in advancing and receding angles

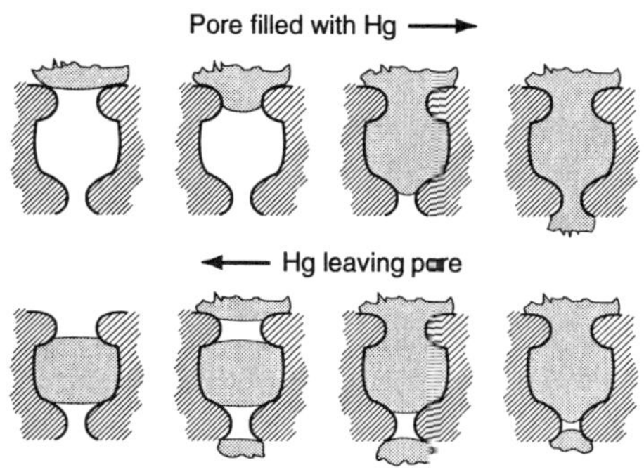

Fig. 23 The ink-bottle theory of hysteresis when mercury enters and leaves the pores

(Ref 48): contact angle hysteresis, ink-bottle theory, and connectivity model.

Contact Angle Hysteresis. Numerous authors have reported different values for advancing and receding contact angles (Fig. 22). The problem, however, is much more complicated because thermodynamic reasoning does not support such a difference. In addition, the observed intrusion and extrusion curves do not fit as perfectly as they should according to the contact angle hysteresis model. Moreover, the model does not provide any explanation for the trapped mercury, which may remain in the sample even after complete depressurization.

The ink-bottle theory accurately describes the situation in which mercury enters and leaves pores (Fig. 23). It explains why some of the mercury remains trapped in the sample. However, it explains the principal hysteresis only in part. For most samples, the ink-bottle theory would predict a much larger amount of mercury remaining in the sample than actually observed in the measurement. The contradiction has not been resolved.

The connectivity model uses a network of pores. In order for a pore to become filled with mercury, it is essential to be larger than the corresponding pore size at the applied pressure, but it is also necessary that a continuous path of mercury leads to the pore. Large internal voids that are surrounded by smaller pores are not filled unless the pressure is sufficient to fill the smaller pores. During the extrusion process, the reverse process can occur, and certain pores or islands of pores remain filled with mercury.

The connectivity model probably best describes the real situation, but conflicting results have been reported as well. Several new approaches have been studied recently, and further details are given in the corresponding publications (Ref 46–49). Analysis of the extrusion curve as well as a back-and-forward pressure scanning within the hysteresis is presently studied. This technique can provide a better understanding of the pore and network structure of the sample.

Conclusions

Mercury porosimetry is a very valuable technique, especially because it covers an extremely wide pore size range (Fig. 10). It certainly has its limitations, but carefully set experimental process conditions and a sensible data analysis provide valuable information about the sample. It is recommended that standard procedures be followed, for example, ASTM standards D 4284 (Ref 50) and C 493 (Ref 51). A good overview on mercury porosimetry with respect to different materials and other aspects of the technique is given in several articles in Ref 52.

It is recommended that close attention be paid to a correct and meaningful data presentation. The differential plot of the pore size distribution can easily lead to a misinterpretation of the results if the graph is calculated with the "wrong" axis:

- A Δpore volume/Δlog (pore size) should be used to represent the differential distribution equivalent to the cumulative distribution when using a logarithmic axis for pore size.
- Using Δpore volume/Δ pore size leads to an overestimation of small pores.
- Plotting the pore volume versus applied pressure on a linear pressure axis allows determination and recognition of compressibility effects of the sample or the mercury.

REFERENCES

1. S. Brunauer, P.H. Emmett, and E. Teller, *J. Am. Chem. Soc.*, Vol 60, 1938, p 309
2. I. Langmuir, *J. Am. Chem. Soc.*, Vol 38, 1916, p 2221
3. I. Langmuir, *J. Am. Chem. Soc.*, Vol 40, 1918, p 1361
4. K.S.W. Sing and D. Swallow, *Proc. Br. Ceram. Soc.*, Vol 39 (No. 5), 1965
5. R.T. Davis, T.W. DeWitt, and P.H. Emmett, *J. Phys. Chem.*, Vol 51, 1947, p 1232
6. S.J. Gregg and K.S.W. Sing, *Adsorption, Surface Area and Porosity,* Academic Press, 1982
7. S. Lowell and J.E. Shields, *Powder Surface Area and Porosity,* Chapman & Hall, 1991
8. T. Allen, *Particle Size Measurement,* Vol 2, *Surface Area and Pore Size Determination,* Chapman & Hall, 1997
9. A.W. Adamson, *Physical Chemistry of Surfaces,* John Wiley & Sons, 1982
10. D.H. Everett, G.D. Parfitt, K.S.W. Sing, and R. Wilson, *J. Appl. Chem. Biotechnol.*, Vol 24, 1974, p 199
11. E.P. Barrett, L.G. Joyner, and P.H. Halenda, *J. Am. Chem. Soc.*, Vol 73, 1951, p 373
12. T. Allen, Permeametry, *Particle Size Measurement,* Chapman and Hall, 1968
13. C.F. Callis and R.R. Irani, Miscellaneous Techniques, *Particle Size: Measurements, Interpretation, and Application,* John Wiley & Sons, 1963
14. "Permaran Specific Surface Area Meter," Outokumpu Oy, Instrument Division, Tapiola, Finland, 1973
15. H.P.G. D'Arcy, *Les Fontaines Publiques de la Ville de Dijon,* Victor Dalmont, 1856
16. J.J. Kozeny, *Akad. Wiss. Wein. Math. Naturwiss. K.I.,* Sitzungsbor, Abstr. IIA, Vol 136, 1927, p 271–306
17. F.C. Blake, *Trans. Am. Inst. Chem. Eng.,* Vol 14, 1922, p 415
18. P.C. Carman, *J. Soc. Chem. Ind. Lond.,* Vol 57, 1938, p 225–234; *Trans. Inst. Chem. Eng.,* Vol 15, 1932, p 150–166
19. J.M. Dallavalle, *Chem. Met. Eng.,* Vol 45, 1938, p 688
20. P.C. Carman, *Symposium on New Methods for Particle Size Determination in the Sub-Sieve Range,* ASTM, 1941, p 24
21. F.M. Lea and R.W. Nurse, *J. Soc. Chem. Ind.,* Vol 58, 1939, p 277–283; Symposium on Particle Size Analysis, *Trans. Inst. Chem. Eng.,* (suppl.), Vol 25, 1947, p 47–56
22. E.L. Gooden and C.M. Smith, *Ind. Eng. Chem. Anal. Ed.,* Vol 12, 1940, p 479–482
23. K. Niesel, External Surface of Powders From Permeability Measurements—A Review, *Silicates Industrials,* 1969, p 69–76

24. R.L. Blaine, *ASTM Bull.,* No. 123, 1943, p 51–55; also see *ASTM Bull.,* No. 108, 1941, p 17–20
25. K. Usui, *J. Soc. Mater. Sci. Jpn.,* Vol 13, 1964, p 828
26. M. Knudsen, *Ann. Physik,* Vol 28, 1909, p 75–130
27. A. Pechukas and F.W. Gage, *Ind. Chem. Eng. Anal. Ed.,* Vol 18, 1946, p 37
28. P.C. Carman and P.R. Malherbe, *J. Soc. Chem. Ind.,* Vol 69, 1950, p 134
29. R.M. Barrer and D.M. Grove, *Trans. Faraday Soc.,* Vol 47, 1951, p 826, 837
30. G. Kraus, R.W. Ross, and L.A. Girifalco, *J. Phys. Chem.,* Vol 57, 1953, p 330
31. T. Young, *Miscellaneous Works,* Vol 1, Murray 1855, p 418
32. P.S. Laplace, *Mecanique Celeste,* Suppl. Book 10, 1806
33. E.W. Washburn, *Proc. Nat. Acad. Sci. U.S.A.,* Vol 7, 1921, p 115
34. AutoPore II 9220 Operator's Manual, Micromeritics, 1993
35. Lowell and J.E. Shields, *Powder Surface Area and Porosity,* Chapman & Hall, 1991
36. R.J. Good and R.Sh. Mikhail, *Powder Technol.,* Vol 29, 1981, p 53
37. F. Brashforth and J.C. Adams, *An Attempt to Test the Theory of Capillary Action,* Cambridge University Press, 1883
38. L.I. Osipow, *Surface Chemistry,* Reinhold, 1964
39. P.C. Hiemenz and R. Rajagopalan, *Principles of Colloid and Surface Chemistry,* Marcel Dekker, 1997
40. H.M. Rootare, A Review of Mercury Porosimetry, *Advanced Techniques in Powder Metallurgy,* Vol 5, *Perspectives in Powder Metallurgy,* Plenum Press, 1970
41. M. Svata, *Powder Technol.,* Vol 29, 1981, p 145
42. J. Kloubek, *Powder Technolg.,* Vol 7, 1981, p 63, 162
43. J. van Brakel, S. Modry, and M. Svata, *Powder Technol.,* Vol 29, 1981, p 1
44. B. Reich, *Chem. Ing. Tech.,* Vol 39 (No. 22), 1967, p 1275
45. H.M. Rootare and C.F. Prenzlow, *J. Phys. Chem.,* Vol 71, 1967, p 2733
46. G.P. Matthews, C.J. Ridgway, and M.C. Spearing, *J. Colloid Interface Sci.,* Vol 171, 1995, p 8
47. G.P. Matthews, A.K. Moss, and C.J. Ridgway, *Powder Technol.,* Vol 83, 1995, p 61
48. H. Giesche, *Mater. Res. Soc. Symp. Proc.,* Vol 431, 1996, p 251
49. N.C. Wardlow and M. McKellar, *Powder Technol.,* Vol 29, 1981, p 127
50. "Pore Volume Distribution of Catalysts by Mercury Intrusion Porosimetry," D 4284, *Annual Book of ASTM Standards,* Vol 5.03, ASTM
51. "Bulk Density and Porosity of Granular Refractory Materials by Mercury Intrusion Porosimetry," C 493, *Annual Book of ASTM Standards,* Vol 15.01, ASTM
52. *Powder Technol.,* Vol 29, 1981
53. P. Webb and C. Orr, *Analytical Methods in Fine Particle Technology,* Norcross, 1997, p 165

Bulk Properties of Powders

John W. Carson and Brian H. Pittenger, Jenike & Johanson, Inc.

THE P/M INDUSTRY has grown considerably in the past decade. As a result of this growth, more critical components in the automotive, aircraft, tooling, and industrial equipment industries are being considered for manufacture using this technology. This is placing increasingly stringent quality requirements on the final P/M part. Variations in part density, mechanical properties including strength, wear, and fatigue life, as well as in aesthetic appearance and dimensional accuracy are no longer tolerated. As a result, metal powder producers and P/M part manufacturers must continually improve their capabilities to ensure the delivery of a consistent, uniform product. Research has demonstrated that these part qualities are significantly affected by changes (variations) in the particle size distribution, particle shape, and consequently the uniformity of powder blends (a combination of one or more particle sizes of a single powder) and mixes (a combination of one or more types of powders) (Ref 1, 2).

Powder Flow

This article reviews the general factors of powder flow, and the following properties are discussed, along with examples of their applications in equipment selection:

- Cohesive strength
- Frictional properties
- Bulk density
- Permeability and flow rate
- Sliding at impact points
- Segregation tendency
- Angle of repose

The flow of metal powders in bins, hoppers, feeders, chutes, and conveyors is not always reliable or uniform. This often results in the press having to operate at lower cycle times, wasted product due to composition or apparent density variations, and operational nightmares. The powder may form a stable arch or rathole; particle segregation may occur, resulting in unacceptable variations in the bulk density of the powder supplied to the feed shoe, or the powder may flood uncontrollably.

Bulk Properties. One of the main reasons that powder flow problems are so prevalent is lack of knowledge about the bulk properties of various powders. For many engineers, the name of a powder, such as atomized aluminum, is thought to connote some useful information about its handling characteristics. While this may be true in a general sense, it is not a reliable tool. Unfortunately, major differences in flowability often occur between different grades and types of powders with the same name.

For those who go beyond the generic name of a powder, one or more of the following four attributes are often relied on in trying to predict the behavior of metal powders and other bulk solids. However, these attributes rarely provide engineers with direct assistance during the design or specification of a bin, hopper, feeder, chute, or conveyor.

Angle of Repose. Determining the angle of repose is relatively easy: simply form a pile of material and measure its slope. Knowing what to do with the data is the difficult part.

For most materials, the angle of repose varies significantly, depending on how the pile was formed. Furthermore, the mechanics of pile formation bear little resemblance to the formation of an arch or rathole in a bin or hopper, uniformity of die fill, powder homogeneity, or to the other key parameters needed when designing a material handling system. In general, the angle of repose of a material is not an accurate measure of its flowability.

Flow Rate. The Hall and Carney flowmeters (described later in this article) are widely used in the P/M industry to characterize powder flowability. However, there are two major flaws with this approach:

- If a powder will not flow through the funnel, no information on its flowability can be determined.
- Even if a powder does flow well, the value obtained (s/50 g) cannot be extrapolated to predict limiting press speed, limiting flow rate through a feed hopper, or other rate-limiting phenomena.

The attempt to combine measurements of two material flow properties (minimum orifice size and flow rate) result in a method that does not measure either one very well.

Apparent Density or Tap Density. Neither of these parameters, nor their ratio (the Hausner Ratio), is a direct indicator of powder flowability. They do not, for example, assist in sizing hopper outlets or calculating appropriate hopper angles.

Free-Flowing versus Nonfree-Flowing. Whether or not a metal powder is considered free-flowing depends to a large extent on the size and shape of the die cavity into which it is expected to flow. For example, a powder that flows through a Hall flowmeter might be considered free-flowing; however, that same powder may have difficulty completely filling a die cavity for a thin-wall part. Thus, these terms are relative and not absolute indicators of powder flowability.

Flow Pattern Considerations. There are two flow patterns that can develop in a bin: funnel flow and mass flow. Both patterns are shown in Fig. 1.

In funnel flow, an active flow channel forms above the outlet with nonflowing material at the periphery. As the level of material in the hopper decreases, layers of the nonflowing material may or may not slide into the flowing channel, which can result in the formation of stable ratholes. In addition, funnel flow can cause product caking, provide a first-in last-out flow sequence, and increase the extent to which sifting segregation impacts the discharging material.

In mass flow, all of the material is in motion whenever any is withdrawn from the hopper. Material from the center as well as the periphery moves toward the outlet. Mass flow hoppers provide a first-in first-out flow sequence, eliminate stagnant material, reduce sifting segregation, provide a steady discharge with a consistent bulk density and a flow that is uniform and well controlled. Requirements for achieving mass flow include sizing the outlet large enough to prevent arching and ensuring the hopper walls are sufficiently smooth and steep enough to promote flow at the walls.

Useful Bulk Flow Parameters. Armed with information about the bulk properties of the powder, engineers can optimize the selection of storage and handling equipment. These same

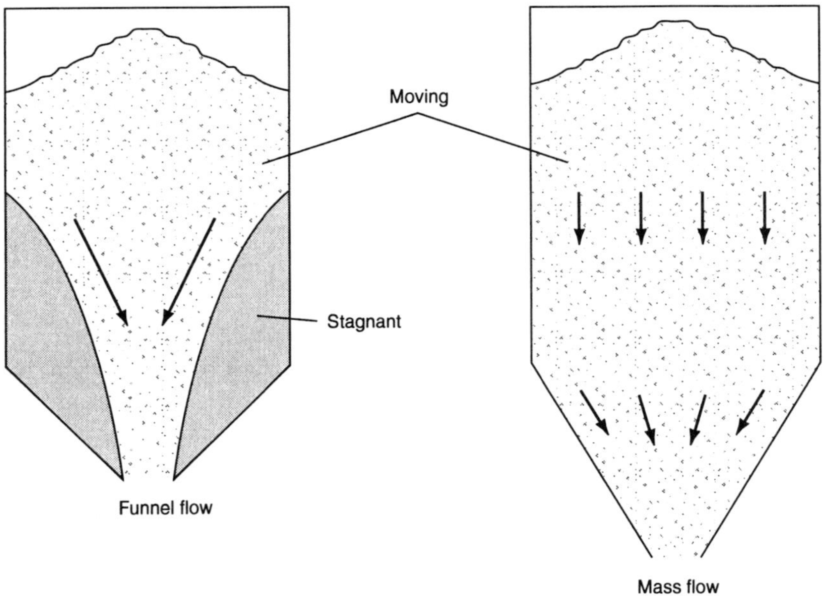

Fig. 1 Two flow patterns that can occur in a bin: funnel flow and mass flow

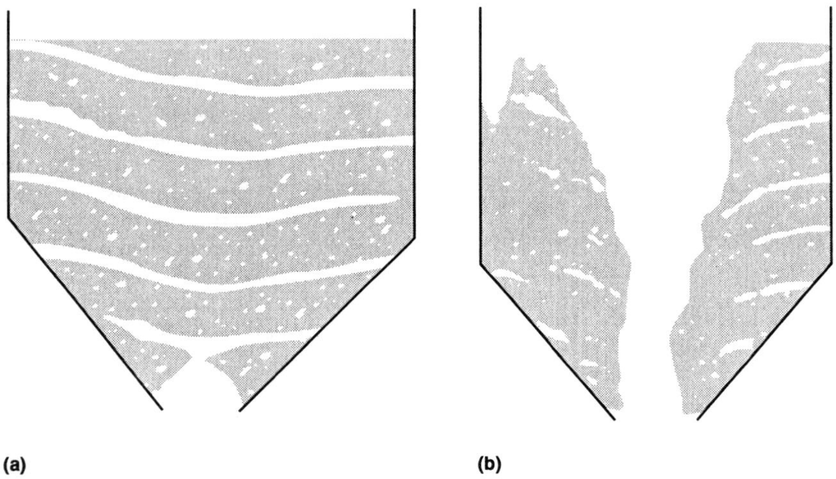

Fig. 2 Examples of no-flow situations where the darkened areas represent material within the bin. (a) Cohesive arch at the outlet of a bin. (b) Stable rathole formed within bin

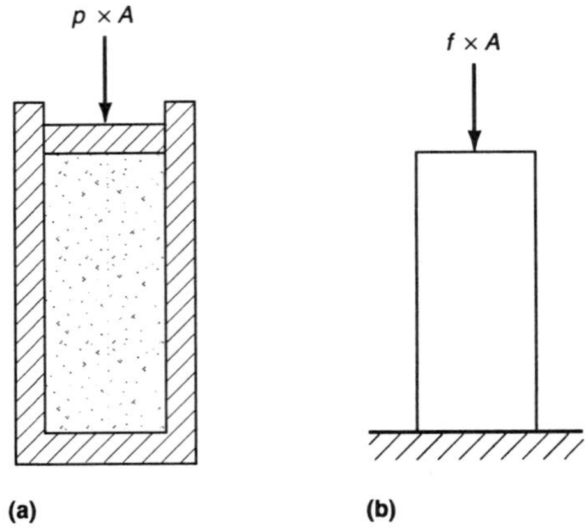

Fig. 3 An idealized flow function test. (a) Consolidation. (b) Failure

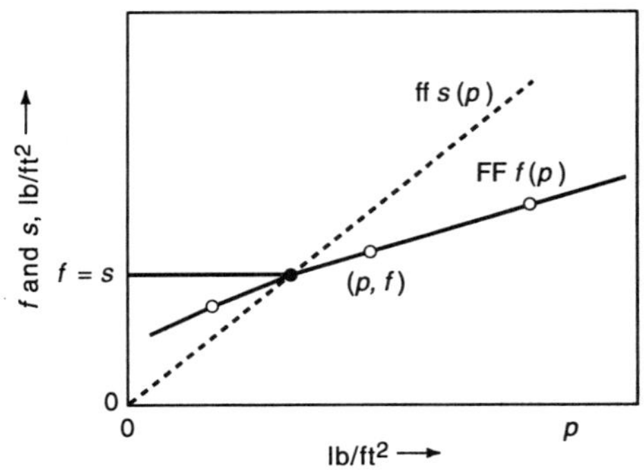

Fig. 4 Solids flow function (FF) and hopper flow factor (ff). See text for details.

properties can be used to retrofit existing processes to correct flow problems.

Discussed below are several bulk solids handling properties that are relevant to predicting flow behavior. The direct application of these parameters has been proven over the last 30 years in numerous installations handling the full spectrum of powders used in the P/M industry, including metal powders, fine chemical additives, polymers and waxes, and graphites/carbons (Ref 3, 4).

Cohesive Strength

Many metal powders and other bulk solids, when poured from a box, flow like a liquid. Under these conditions, such a material has no cohesive strength. However, when squeezed in the palm of one's hand, the material may gain enough strength to retain its shape once the hand is opened.

A similar range of conditions occurs inside bins, hoppers, and containers. Consolidation pressures range from zero at the surface to relatively large values at increasing depth within the container. If a powder gains cohesive strength because of the pressures applied to it, an arch or rathole can form (Fig. 2).

An arch (also called a bridge or dome) is a stable obstruction that forms over the point of narrowest cross section of the storage vessel (usually the discharge outlet). The arch supports the rest of the bin contents, preventing discharge.

A rathole is a stable pipe or vertical cavity that empties out over the outlet. Material is left stranded in stagnant zones that usually remain in place until an external force is applied to dislodge them.

Cohesive strength can be measured as a function of the applied consolidation pressure. This relation, which is of primary importance in the analysis of flow, can be described as follows: suppose that a quantity of a metal powder has been placed in a cylindrical mold of cross-sectional area A, with frictionless walls (Fig. 3), and consolidated under a force ($p \times A$) applied to the piston. Now suppose that, without disturbing it, the consolidated cylinder of powder is removed from the mold and placed on a table, and a compressive force is applied to it (Fig. 3). The force is increased from zero until the cylinder collapses at some value of the force ($f \times A$). The experiment is repeated for several values of p. For each value of p, a corresponding value of f is obtained. Points (p, f) are plotted in Fig. 4, and a smooth line is drawn through them. This relation is called the flow-function (FF) of the solid.

This compression test serves well as an illustration of the concept, but is not practical for a number of reasons; for instance, a mold cannot be made frictionless, and it is difficult to obtain uniform consolidation of a powder in a relatively tall cylinder. In addition, most metal powders have such low cohesive strength that the compacted cylinder of powder would fall apart when it was removed from the cylindrical mold.

A more accurate and controlled procedure is described in ASTM D 6128 (Ref 5). In a laboratory, a sample of the powder is placed in a Jenike

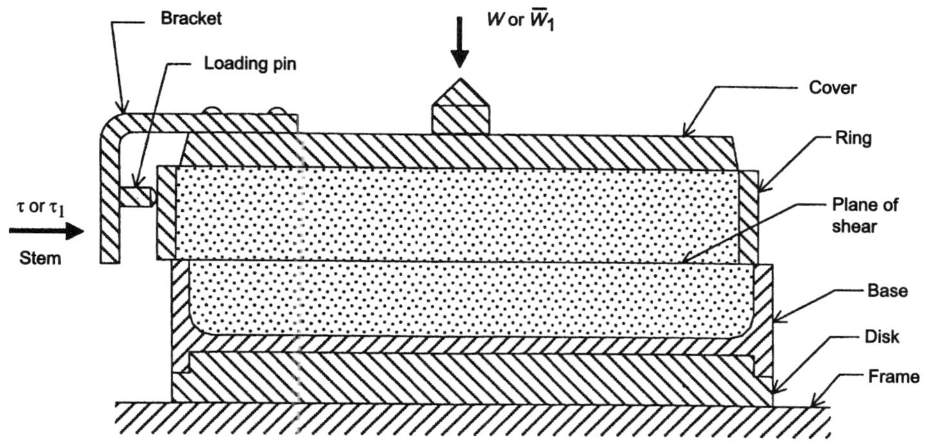

Fig. 5 Jenike shear cell in initial offset shearing position

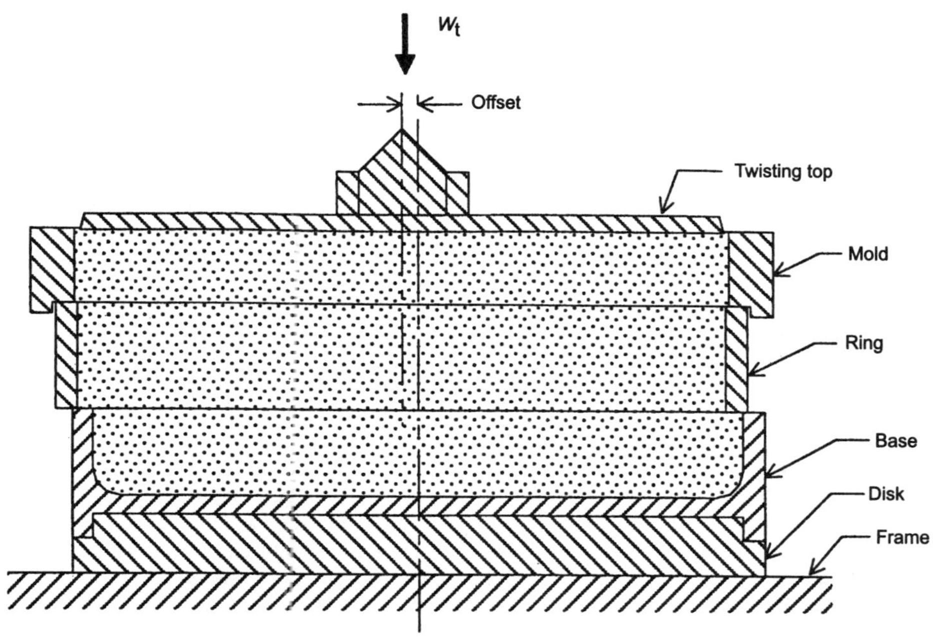

Fig. 6 Jenike shear cell with mold ring and consolidation lid set up for pre-consolidation

shear cell (Fig. 5), and both compressive and shear loads are applied to simulate flow conditions in a container.

The shear cell, Fig. 5, is composed of a base located on the frame of the machine, a ring resting on top of the base, and a cover. The bottom of the cover and the inside of the base are roughened to increase friction with the tested powder. The base and the ring are filled with the powder to be tested, and a vertical load is applied to the cover. A horizontal shearing force is applied by means of a stem, which acts on a bracket attached to the cover. This shearing force, which acts in the plane of contact between the ring and the base, is partly transferred from the bracket to the ring through a loading pin. This ensures a sufficiently uniform distribution of the shearing force across the cell. The standard shear cell is 95 mm (3.75 in.) inside diameter, but a 65 mm (2.5 in.) diameter is also often used.

A shear tester is equipped with a shear cell, a gravity vertical loading system, and an electronic shearing force applicator. The applicator in the Jenike shear tester has a shearing rate of 2.54 mm/min (0.10 in./min). The shearing force necessary to maintain the strain rate is continuously recorded on a strip-chart recorder. This arrangement produces a permanent record of the stress-strain relations for each test.

Following the simplified example (described above) of testing using a cylindrical mold, shear testing is a two-step process involving consolidation (also called preshear) and shear.

Consolidation is carried out in two stages. The purpose of the first stage, called preconsolidation, is to prepare a uniform specimen. With the cover off the test cell, a packing mold is placed on top of the ring, and both the mold and the ring are placed in an offset position on the base, as shown in Fig. 6. A sample of the tested powder is then placed in the cell. One layer after another is slightly packed with the fingers, up to the top of the mold. The excess material is scraped off level with the top of the mold, and a twisting top is placed over the powder. A vertical force is applied to the top by means of a weight hanger. This force causes a vertical pressure σ_t in the material. By means of a special wrench, a number of oscillating twists is applied to the cover. This preconsolidates the powder and ensures a uniform specimen.

Consolidation is completed in the second stage by causing the specimen to flow under given stresses until a steady state is reached, or closely approached. This is attained in the following way: The twisting load is taken off, the twisting top and the mold are removed, the excess material is scraped off level with the top of the ring, and the test cover is placed on the material. A smaller load is now placed on the cover, and the stem of the shearing device is advanced against the bracket (Fig. 5). This smaller load compacts the sample to a preshear normal stress of σ_p.

As shearing proceeds, a condition is reached when a layer of the powder across the whole specimen is caused to flow plastically: the recorded shearing stress reaches a steady value τ_p. Consolidation determines point (σ_p, τ_p), (Fig. 7).

Shear. When consolidation is completed, the stem of the shearing force device is retracted. The preshear normal stress σ_p is replaced by a smaller normal stress σ_s to locate a useful point (σ_s, τ_s) of the yield locus (Fig. 7). The sample is now sheared until a failure plane has developed. This fact is indicated on the recorder by the stress τ_s passing a maximum value. After shearing, the plane of failure of the specimen is checked. It should roughly coincide with the plane of shear of the cell. If the planes deviate, it means that the measured point (σ_s, τ_s) does not lie on the yield locus, and the test is repeated.

The determination of one yield locus requires the measurement of three to five points of the locus $(\sigma_s, \tau_s)_1$, $(\sigma_s, \tau_s)_2$, $(\sigma_s, \tau_s)_3$ and so forth (Fig. 7). For each point, the specimen is first consolidated and then sheared. The value of shear normal stress σ_s typically ranges between 25 and 80% of the preshear normal stress σ_p. It is necessary to obtain the values of these points for the same steady consolidating shear each time. This is accomplished by running a sufficient number of tests to permit interpolation to a suitably selected value of τ_p.

The yield locus is now drawn and extrapolated toward the higher value of σ_p, and a Mohr semicircle is drawn through point (σ_p, τ_p) tangentially to the yield locus. The point of tangency (*E*) locates the terminus of the yield locus. The point of intersection of the semicircle with the σ-axis determines the value of the major consolidation stress, σ_1. The Mohr semicircle for the unconfined yield strength f_c is now drawn. The value of f_c is determined by the point of intersection of the circle with the σ axis, as shown in the Fig. 7.

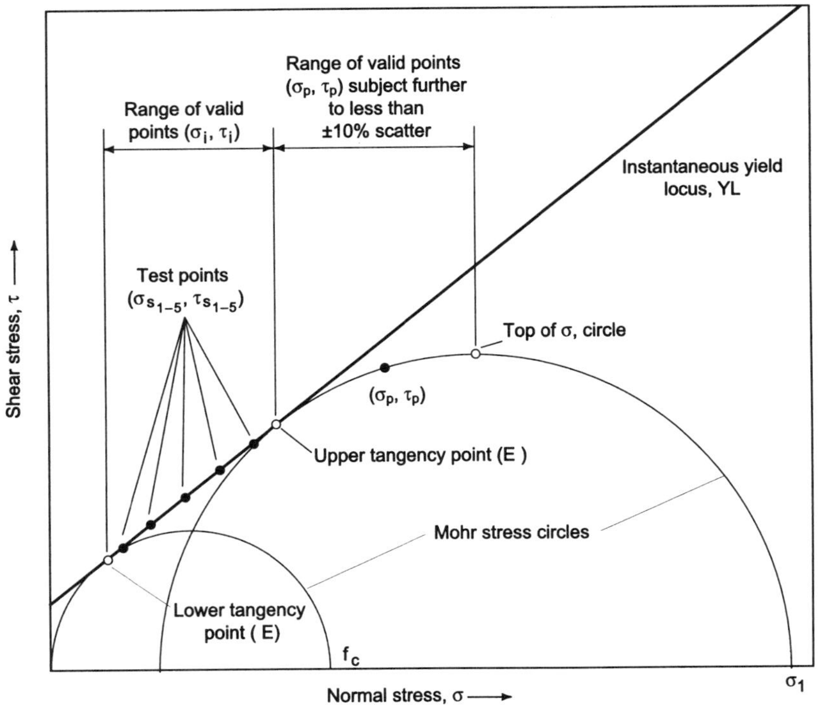

Fig. 7 Yield locus (YL) showing valid shear testing points

A plot of f_c versus σ_1 is the flow function of the powder under the conditions of particle size, moisture, and temperature tested.

To simulate time of storage at rest, the shear tester is used in conjunction with a six-cell consolidating bench (Fig. 8). The tester shown in Fig. 9 is used for temperature sensitive solids. Here, the consolidating bench is enclosed in a heated chamber that permits the control and recording of the temperature of the tested powder.

The flow function of a powder is used for a variety of engineering analyses, for example, to calculate minimum outlet dimensions required to prevent cohesive arches and ratholes from forming. Details can be found in Ref 3.

The cohesiveness of a bulk solid is a function of the following parameters:

- *Moisture:* Typically, cohesiveness rises as moisture content increases, although not in direct proportion. Hygroscopic materials can experience significant increases in moisture when exposed to humid air.
- *Particle size and shape:* There is no direct correlation between particle size, shape, and cohesiveness. Even so, in most cases, as a powder becomes finer, it also becomes more cohesive and difficult to handle. Angular or fibrous particles are often more cohesive than those that are rounded.
- *Temperature:* The temperature of a powder can affect its cohesiveness. For example, many thermoplastic blends (such as for PIM) become more difficult to handle as their temperatures rise. Some materials have more strength at constant temperatures, while others gain cohesive strength as their temperature changes during heating or cooling.
- *Time of storage at rest:* When a material resides in a bin or hopper for a period without moving, it can become more cohesive and difficult to handle. Such cohesion may be caused by settling and compaction, crystallization, chemical reactions, and adhesive bonding.
- *Chemical additives:* In some cases, adding a small amount of a chemical additive such as calcium, lithium, or zinc stearates can cause a cohesive powder to flow more easily.

Frictional Properties

Both internal and external friction values are important when characterizing the flow properties of a metal powder. Internal friction is caused by solid particles flowing against each other and is expressed by the angle of internal friction and the effective angle of internal friction. Both can be determined during the course of measuring cohesive strength with a Jenike shear cell (Fig. 5), as described in Ref 5.

External friction is expressed as the wall friction angle or coefficient of sliding friction. The lower the coefficient of sliding friction, the less steep the hopper walls need to be for powder to flow along them (mass flow). Also, the easier a feed shoe indexes to and from a die, the more uniform the flow of powder into the die cavity.

The coefficient of sliding friction can be measured by sliding a sample of powder across a stationary wall surface using a shear tester. The arrangement of the cell is shown in Fig. 10. In this case, a coupon of the wall material is placed on a filler so that the top surface of the coupon is the horizontal plane of the force measuring stem. The ring and packing mold are placed over the wall material coupon and filled with the powder.

After scraping off the excess material level with the top of the mold, a twisting top is placed over the powder. A vertical force is applied to the top by means of a weight hanger. This force causes a vertical pressure σ_t in the material. By means of a special wrench, a number of oscillating twists is now applied to the cover. This pre-

Fig. 8 Consolidation bench

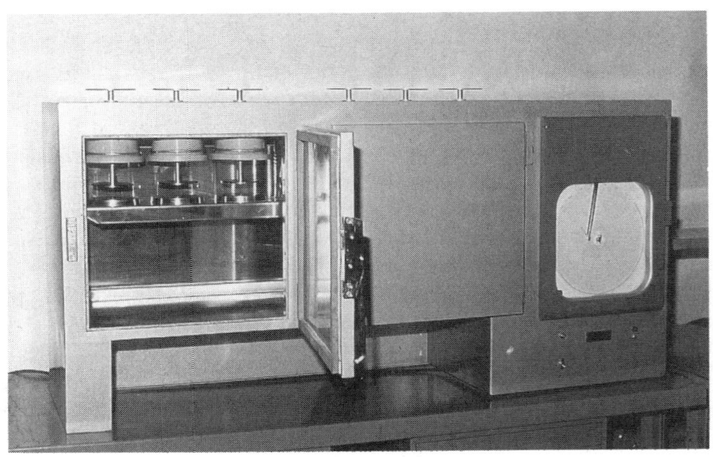

Fig. 9 Consolidation bench in heated chamber

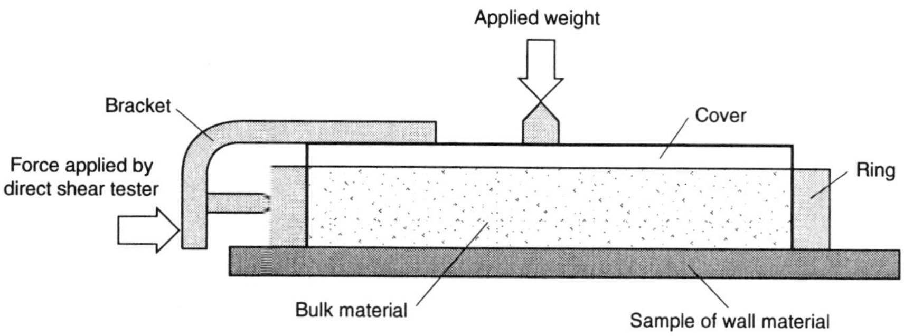

Fig. 10 Shear cell used in measuring wall friction properties. Test setup design allows shear stress (see horizontal arrow) to be measured as a result of applied weights (see vertical downward arrow).

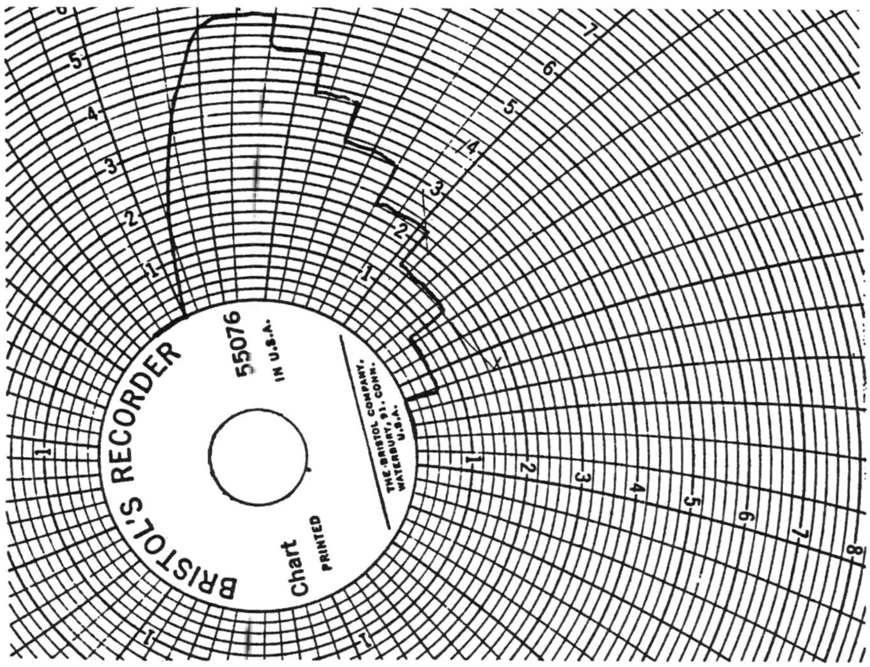

Fig. 11 Typical recorder chart in measurement of ϕ' (wall friction angle)

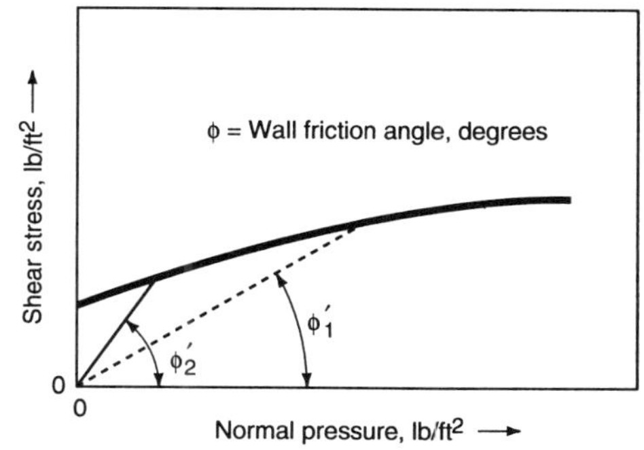

Fig. 12 Typical results of the test setup shown in Fig. 10 to help engineers determine wall friction angle (ϕ')

consolidates the powder and ensures a uniform specimen.

The twisting load is taken off, the twisting top and the mold are removed, the excess material is scraped off level with the top of the ring, and the test cover is placed on the material. A smaller load is now placed on the cover, and the stem of the shearing device is advanced against the bracket (Fig. 10). All the tests necessary to determine the coefficient of sliding friction are now run without replacing the powder.

Before the start of a test, the ring is twisted and manually lifted slightly off the wall material coupon to prevent it from dragging on the wall coupon. Several (say, six) one or two pound weights are placed directly on top of the cover of the shear cell to give the largest required normal stress σ_w. The stem is advanced. When the shear stress τ_w has leveled off, one weight is removed, after a while τ_w again levels off, another weight is removed and so on, until all the weights have been removed. The cover, ring, and the enclosed powder are then weighed. Their weight plus the superimposed weights determine the normal stresses σ_w.

A typical recorder chart is shown in Fig. 11. The points (σ_w, τ_w) are plotted in Fig. 12. A smooth line drawn through these points is the wall yield locus, WYL. Typically, the WYL is convex upward.

The coefficient of sliding friction is the ratio of the shear force required to cause sliding to the load applied perpendicular to the wall material coupon. The arc tangent of this value is the wall friction angle (Ref 3, 4).

The following variables can affect the internal and external friction values of a metal powder and are similar to those affecting cohesiveness:

- *Pressure:* Typically, as consolidating pressure increases, the effective angle of internal friction decreases. Similarly, the coefficient of sliding friction often decreases as pressure acting normal to the plate increases. However, the internal angle of friction is an intrinsic characteristic of the material that may increase, decrease, or remain the same as pressure acting on the material increases.
- *Moisture content:* As moisture increases, many bulk solids become more frictional.
- *Particle size and shape:* Typically, fine materials and those with a wide range of particle sizes are somewhat more frictional than coarse materials or those with a narrow particle size distribution; so the flow of the former is often more troublesome. Shape plays a role in that angular particles tend to interlock and also dig into a wall surface, thereby creating more friction.
- *Temperature:* For many materials, higher temperatures can cause particles to become more frictional.
- *Time of storage at rest:* If allowed to adhere to a wall surface, many powders experience an increase in friction between the particles and the wall surface. Such situations require steeper bin walls for unaided flow.
- *Wall surface:* The initial condition of a surface can play a major role in how materials slide along it. Smoother surfaces are typically less frictional, although this is not always true. Also, as a carbon steel container ages, corrosion can roughen the walls, making sliding more difficult.

Friction data are used to:

- *Design a mass flow hopper:* Values of both wall friction angle and effective angle of internal friction are required to design a mass flow hopper. Using these angles along with design charts given by Jenike (Ref 3), one can determine allowable hopper angles required to promote mass flow.
- *Anticipate sliding on chutes:* A chute is used to transfer material from one point to another in a bulk handling system. By definition, the cross section of a chute is only partially full at any given time, and the discharge rate of a powder is equal to the chute filling rate.

Beyond the impact point, the acceleration of a particle on a chute is directly related to the difference between the wall friction angle of the material and the chute angle. As long as the chute is steeper than the wall friction angle, particles will continue to accelerate. Otherwise they will slow down and may eventually block the chute.

Table 1 Effect of particle size on apparent density for several metal powders

Material	Average particle diameter(a), µm	Apparent density, g/cm³
Aluminum		
Atomized	5.8	0.62
	6.8	0.75
	15.5	0.98
	17.0	1.04
	18.0	1.09
	60% above 44 (+325 mesh)	1.22
	75% above 44 (+325 mesh)	1.25
Copper		
Electrolysis	90% min, −325 mesh	1.5–1.75
Hydrometallurgical	81.9%, −325 mesh	1.69
Oxide reduced	95% min, −325 mesh	2.10–2.50
Hydrometallurgical	49.1%, −325 mesh	2.42
Oxide reduced	50–65%, −325 mesh	2.65–2.85
Electrolysis	60–75%, +100 mesh	4.0–5.0
Atomized	70% min, −325 mesh	4.9–5.1
	50–60%, −325 mesh	4.9–5.5
Nickel		
Carbonyl	3.2	0.61
Precipitation	3.5	1.81
Carbonyl	3.8	1.87
	4.1	2.10
Precipitation	4.4	2.09
	8.0	2.60
	−40+325 mesh	3.60
Tungsten		
Oxide reduced	1.20	2.16
	2.47	2.52
	3.88	3.67
	6.85	4.40
	26.00	10.20
Stainless steel		
Atomized, spherical	−325 mesh	4.3
	−270+325 mesh	4.5
	−200+270 mesh	4.4
	−150+200 mesh	4.5
	−100+150 mesh	4.5
Iron		
Reduced	6	0.97
Carbonyl	7	3.40
Reduced	51	2.19
Electrolytic	53	2.05
	63	2.56
Reduced	68	3.03
Electrolytic	78	3.32

(a) From Fisher subsieve sizer for single values and screens for size fractions

Bulk Density

Bulk density measurements of powders include apparent density and tap density measurements as described below. A key factor is compressibility.

Apparent Density

Apparent density of a metal powder, or the weight of a unit volume of loose powder expressed in grams per cubic centimeter, is one of the fundamental properties of a powder. This characteristic defines the actual volume occupied by a mass of loose powder, which directly affects processing parameters such as the design of compaction tooling and the magnitude of the press motions required to compact and densify loose powder.

Table 2 Effect of mixture of spherical coarse and fine stainless steel particles on apparent density

Particle size (mesh)	Particles, %					
−100+150	100	80	60	40	20	...
−325	...	20	40	60	80	100
Apparent density, g/cm³	4.5	4.9	5.2	4.8	4.6	4.3

Table 3 Apparent densities and flow rates of electrolytic iron powders of three particle size distributions

Particle size (mesh)	Particles, % Powder A	Powder B	Powder C
+100	4	3	15
−100+150	11	26	10
−150+200	18	18	30
−200+250	16	6	25
−250+325	18	16	5
−325	33	31	15
Apparent density, g/cm³	2.6–2.8	3.2–3.4	3.8–3.9
Flow rate, s/50 g	29	24	20

In most compacting operations, dies are filled by volume measure, and presses operate either to a fixed position or a fixed pressure. If the press operates to a fixed position, pressure can be maintained at a constant level only if the apparent density of the powder does not change. If, however, the press operates to a fixed pressure, consistency in apparent density is necessary to ensure compacts of equal height. Small fluctuations in apparent density can be compensated for by adjustments of pressure or stroke of the presses, but large-scale compacting requires that the apparent density of the powder be controlled within close limits.

Factors Affecting Apparent Density. Apparent density of a metal powder depends on the density of the solid material, particle size, particle size distribution, particle shape, surface area and roughness of individual particles, and particle arrangement. Apparent density is strongly affected by particle size. It generally (1) decreases with decreasing particle size, (2) decreases as the particle shape becomes less spherical and more irregular, (3) decreases with increasing surface roughness, and (4) is frequently controlled by mixing various sizes of particles.

Particle Size. Decreasing particle size generally decreases apparent density. The smaller the particles, the greater the specific surface of the powder. This phenomenon increases the friction between particles and subsequently decreases the apparent density. Powder particles that exhibit very low friction because of their rounded shape, such as gas-atomized (spherical) stainless steel powder, do not demonstrate this characteristic. The effect of decreased particle size on density is particularly significant for particle sizes of less than 20 µm. Table 1 shows the effect of particle size on apparent density for several metal powders.

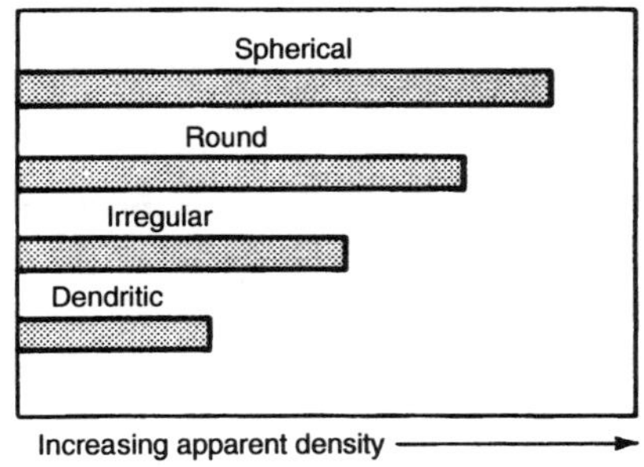

Fig. 13 Effect of particle shape on apparent density of a metal powder

Particle Shape. As particle shape becomes less spherical, apparent density decreases, due to both the increase in frictional surface area and less uniformity of powder particles during packing. Spherical powders, which are normally produced by atomizing, frequently have high apparent densities, about 50% of the density of the wrought metal. Spheres are most likely to pack without bridging or arching to create empty spaces; they tend to move easily past each other because of their smooth surfaces. At the other extreme in particle shape are flake powders, which often have apparent densities less than 10% of the wrought density. These powders are useful primarily as pigments, because their low apparent density aids in obtaining mixtures in paint.

Most powders used for compacting have irregular, somewhat equiaxed particle shapes with apparent densities that fall in the range between those of spherical and flake powders. Apparent densities of these particles range from 25 to 35% of the wrought density of the metal. Figure 13 illustrates the effect of particle shape on apparent density.

Surface Roughness. Decreasing surface area-to-volume ratios and decreasing surface roughness tends to reduce frictional forces between settling particles. This tendency thus increases apparent density by allowing the particles to move more effectively to fill the free spaces between previously settled particles.

Particle Size Distribution. An effective way to increase the apparent density of a powder is to fill the spaces between particles with smaller particles. Figure 14 shows the effects of adding differently shaped −325 mesh powder to a standard +325 mesh blend of stainless steel powder. Table 2 shows this effect for mixtures of fine and coarse spherically shaped stainless steel powders, where a mixture of about 60% coarse and 40% fine particles is optimal.

The addition of fine spherical powder effectively increases apparent density, while the opposite is true of flake powders. Distribution of a variety of particle sizes greatly affects apparent density. The relative amount of coarsest and finest particles and the percentage of particles between the two extremes determine apparent density. An example of this is shown in Table 3 for three particle size distributions.

Hall Flowmeter and Carney Funnel. The most common method for determining apparent

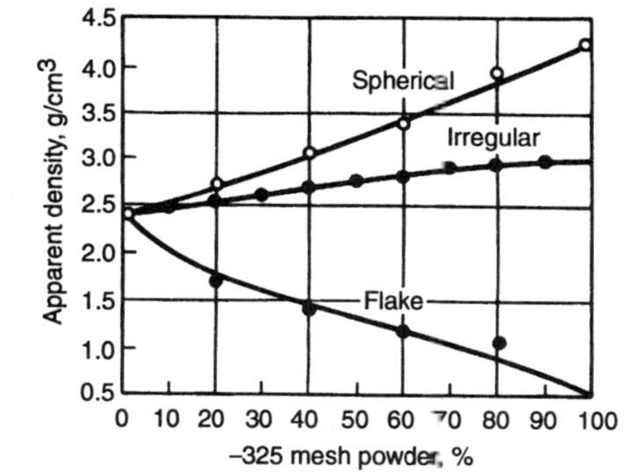

Fig. 14 Effect of three different shapes of −325 mesh powder addition to a +325 mesh distribution on apparent density of 316 stainless steel powder

density of metal powders uses the Hall flowmeter. Both ASTM B 212 and Metal Powder Industries Federation (MPIF) standard 04 describe this method.

Critical equipment dimensions are illustrated in Fig. 15(a), (c), and (d). Apparent density determinations are made by pouring powder into the funnel and allowing it to flow into the 25 cm^3 (1.5 $in.^3$) density cup. After the cup is fitted, the funnel is moved away and the excess powder is carefully leveled off using a spatula or straight edge. Care must be exercised to prevent physical densification of the powder in the cup when leveling. The apparent density in grams per cubic centimeter is then determined by weighing the powder in the cup in grams and dividing by 25 cm^3 (1.5 $in.^3$) (cup volume).

For powders that do not flow freely, a second method, described in ASTM B 417 and MPIF 28, has been devised. This is similar to the Hall flowmeter procedure, except that a different funnel, the Carney funnel, which has an orifice diameter twice that of the Hall funnel, is used (see Fig. 15a and b). This larger opening permits a greater variety of powders to flow. Powders that do not flow readily can be freed by poking a wire up and down in the hole. The wire must not enter the density cup at any time. This second method is fast and correlates well with the Hall flowmeter evaluation of free-flowing metal powders.

A wire is not used with the Hall flow-meter funnel because it may scratch the orifice and ruin the calibration of the funnel for flow tests. The Carney funnel is often used when measuring apparent density of lubricated powders because lubricant adhering to the smaller orifice of the Hall funnel temporarily affects the calibration of the Hall flowmeter.

Scott Volumeter. Another instrument frequently used for determining apparent density is the Scott volumeter, described in ASTM B 329, which was originally developed by Scott, Schaeffer, and White for the determination of the density of dry pigment for paint. As shown in Fig. 16, the device consists of:

- *A large brass funnel* with a metal screen and a smaller funnel with a straight stem for directing the powder into the baffle box
- *A baffle box* with glass sides and two wooden sides containing a series of glass baffle plates and a funnel at the bottom to collect the powder and direct it into the density cup
- *A square density cup* with a capacity of 16.4 ± 0.032 cm^3 (1 ± 0.002 $in.^3$) or a cylindrical cup with a capacity of 25.00 ± 0.05 cm^3 (1.5 ± 0.003 $in.^3$) with an inside diameter of 30.00 ± 2.00 mm (1.2 ± 0.08 in.)
- *A stand* to support the funnels and baffle box concentric with the density cup, so that the bottom of the baffle box funnel is 19 mm (0.75 in.) above the top of the density cup when the apparatus is assembled
- *A level, vibration free base* to support the funnels and baffle box
- *An analytical balance* having a capacity of at least 100 g (3.5 oz) and a sensitivity of 0.1 g (0.0035 oz)

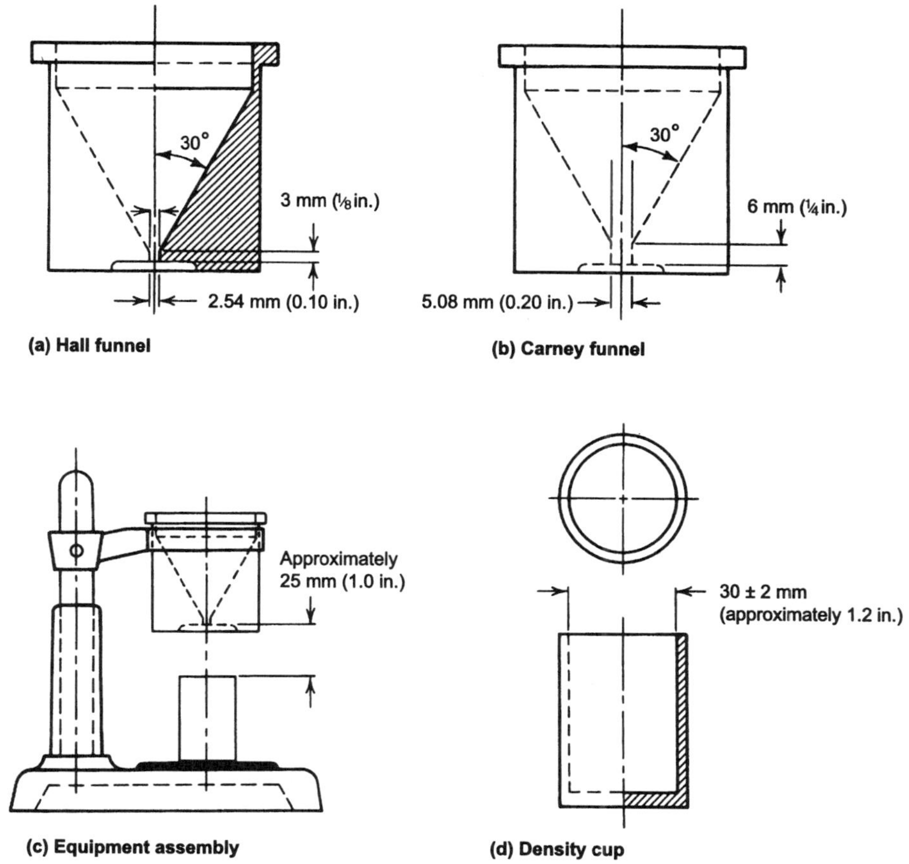

Fig. 15 Dimensions for apparent density equipment

Operating Procedure. The test specimen is carefully poured into the funnel. Ultrafine powders may require light brushing with a nylon brush to initiate powder flow through the screen in the funnel. Powder is allowed to run into the density cup until it completely fills and overflows the periphery of the cup. The funnel and baffle box should then be rotated approximately 90° in a horizontal plane to clear the cup.

Excess material should be removed from the cup by passing a spatula blade in flat contact with the top of the cup. The spatula is moved smoothly back and forth along the top of the cup until all excess powder has been removed. When insufficient powder is left for the first reverse pass to smooth the surface completely, powder on the spatula should be gently replaced on top of the cup. The spatula must be kept level at all times to prevent packing or pulling out of the powder.

After the leveling operation, the side of the density cup should be tapped lightly to settle the powder to avoid spilling. The powder is transferred to a balance and weighed to the nearest 0.1 g (0.0035 oz). The density of the powder in the density cup is given in grams per cubic centimeter if a metric cup is used, or grams per cubic inch if a nonmetric cup is used.

Arnold Meter. Another device developed to determine the apparent density of metal powders is the Arnold meter, described in ASTM B 703 and MPIF 48, which is designed to duplicate the action of a feed shoe filling a die cavity on a P/M press. A hardened, fully demagnetized steel block with a round hole having a volume of 20 cm^3 (1.2 $in.^3$) is placed on a sheet of glazed paper. A bronze bushing filled with powder is slid across the hole. The powder collected in the hole is then removed and weighed. Apparent density is calculated by dividing the weight by the volume of the hole.

Apparent densities obtained with this procedure, which takes less than 5 min to perform, are quite close to those values measured on powder that has filled a die cavity from a filling shoe in an automatic compacting press.

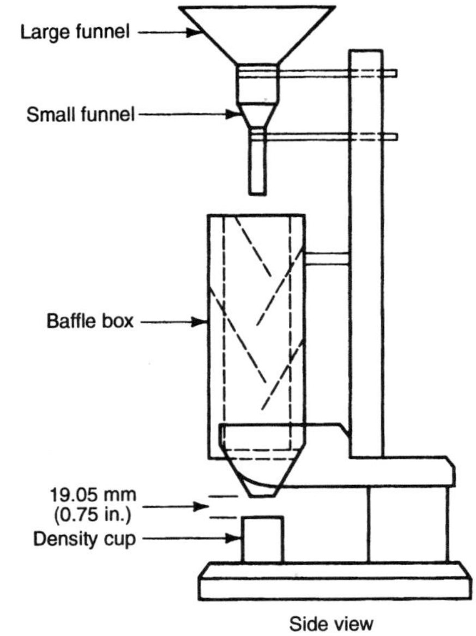

Fig. 16 Scott volumeter

Tap Density

Tap density is defined as the density of a powder when the volume receptacle is tapped or vibrated under specified conditions. Tapping or vibrating a loose powder induces movement and separation and lowers the friction between the powder particles. This short-term lowering in friction results in powder packing and in a higher calculated density of the powder mass. Tap density is always higher than the free-flow apparent density.

Tap density is a function of particle shape, particle porosity, and particle size distribution. It is commonly included as a control specification for metal powder, but is used in other industrial applications as a practical measure of the degree of powder packing that occurs in containers.

The amount of increase from apparent to tap density depends to a great extent on particle shape. Table 4 compares the density increases for three types of copper powders. Usually, the lower the apparent density, the higher the percentage increase in density on tapping.

Equipment and Test Procedures. Three pieces of equipment are needed to determine tap density:

- A balance with the capacity of weighing up to 100 g with an accuracy of 0.1 g
- A graduated glass cylinder with a capacity of 100 mL and an accuracy of 0.2 mL (or a smaller graduated cylinder may be used for high-density powders)
- A mechanical apparatus, such as the Tap-Pak volumeter, capable of tapping the graduated cylinder at a rate of 100 to 250 impacts per minute or, alternatively, a hard rubber slab approximately $100 \times 100 \times 5$ mm ($4 \times 4 \times \frac{1}{4}$ in.)

To determine tap density, a standard weight (usually 50 g) of powder is weighed to ±0.01 g.

Table 4 Effect of particle shape of copper powders

Particle size distribution is the same for apparent and tap density values

Particle shape	Apparent density, g/cm^3	Tap density, g/cm^3	Increase, %
Spherical	4.5	5.3	18
Irregular	2.3	3.14	35
Flake	0.4	0.7	75

Source: Ref 6

The powder is poured into a clean, dry graduated cylinder, taking care that a level surface of powder is obtained. For refractory metal powders that have high apparent densities (above 4 g/cm^3), it is preferable to use a reduced-volume graduated cylinder (25 mL) to improve the accuracy of the results.

The powder is settled in the cylinder by mechanical or hand tapping. If mechanical tapping (Fig. 17) is used, the filled cylinder is placed in the mechanical apparatus, which is operated until no further decrease in the volume of the powder is observed. If hand tapping is used, the base of the filled cylinder is tapped squarely on a hard rubber slab until no further decrease in volume is observed. Care must be exercised to avoid loosening the surface layers of the sample during this procedure.

The volume of the fully densified powder sample in the graduated cylinder is read and used in the following calculation of tap density:

$$\text{Tap density } \rho_t = \frac{m}{v}$$

where m is the mass of powder in grams and v is the volume of tapped powder in cubic centimeters. Results should be reported to the nearest 0.1 g/cm^3.

Representative tap densities of several metal powders and compounds are listed in Table 5, with their respective Fisher subsieve sizes. Mechanical tapping is performed in accordance with ASTM B 527, MPIF 46, or International Standards Organization (ISO) standard 3953. Manual tapping is performed in accordance with either the MPIF or ISO standard. Interlaboratory reproducibility of tap density values can be expected to fall within a standard deviation of about 3.5%.

Compressibility

In most cases, the bulk density of a material varies continuously as a function of the consolidating pressure acting on it. Therefore, it is not sufficient to describe a material simply in terms of its apparent density or tap density. Instead, this density-to-pressure relationship can be measured (Ref 7), and the results are often expressed as a straight line on a log-log plot (Fig. 18). In the bulk solids literature, this relationship is often called compressibility, although this term has another definition in the P/M industry.

The following variables can affect the bulk density of a material:

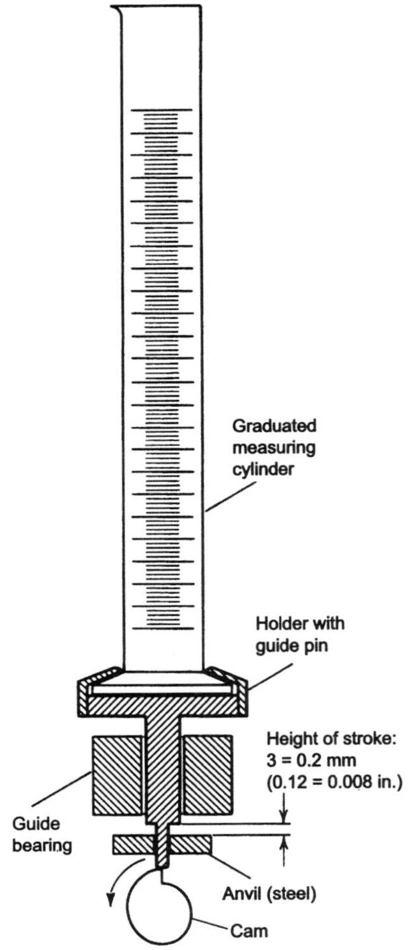

Fig. 17 Diagram of tapping apparatus

- *Moisture:* Higher moisture content usually makes a material more compressible.
- *Particle size and shape:* Often, the finer the bulk solid, the more dense and compressible it is. The shape of the particles can affect how they fit together, as well as their tendency to shear while being compacted.
- *Temperature:* Some materials become more compressible as their temperature increases.
- *Particle elasticity:* Elastic materials tend to deform significantly when they are compressed.

Some of the uses of compressibility data are:

- *Wall friction angle:* Bulk density values at various points in a hopper are used to calculate the pressures acting perpendicular to the hopper wall. After running a wall friction test, the wall friction angle is determined for a variety of pressures and used to calculate limiting angles in a mass flow hopper.
- *Feeder design:* To calculate the loads that act on a feeder or gate, one must know the bulk density of the material at the hopper outlet. Knowing this density also helps in sizing a volumetric feeder and choosing its speed.

Table 5 Typical tap densities of metal and metal carbide powders

Powder	Fisher subsieve size, μm	Tap density, g/cm³
Aluminum	5.05	1.30
Chromium	3.20	3.10
Chromium carbide (Cr_3C_2)	3.70	3.50
Cobalt	1.50	1.60
Hafnium carbide (HfC)	3.50	5.95
Iron	5.40	3.55
Manganese	3.40	3.05
Molybdenum	4.30	3.75
Molybdenum carbide (Mo_2C)	4.50	3.45
Nickel	3.00	1.90
Tantalum carbide (TaC)	2.65	8.00
Tin	2.45	3.15
Titanium carbide (TiC)	3.20	3.65
Tungsten		
Fine	1.15	4.45
Coarse	6.00	6.10
Tungsten carbide (WC)		
Fine	1.45	4.20
Coarse	6.50	6.50
Tungsten-titanium carbide (WC-TiC)	3.90	4.30
Vanadium carbide (VC)	4.50	2.70
Zirconium	3.70	2.50

Permeability and Flow Rate

Sizing a hopper outlet or feed tube to achieve the required discharge rate is more difficult, but no less important, than overcoming arching, particularly for fine powders. All bulk materials have a maximum rate at which they discharge through a hopper opening of a given size (e.g., a Hall flowmeter test). For example, for free-flowing bulk materials, a good approximation of this maximum discharge rate of a coarse material (e.g., 3 mm and larger particles), from a mass flow hopper can be predicted:

$$Q = \gamma A\{Bg/[2\,(1 + m)\tan\theta]\}^{1/2}$$

where Q is the maximum steady discharge rate, γ is the bulk density, A is the cross-sectional area of outlet, B is the outlet diameter or width, g is the acceleration due to gravity, m is 1 for circular opening and 0 for slotted opening, and θ is the flow channel angle (measured from vertical) in degrees. This equation can be modified to take particle size into account, but this modification is only important if the particle size is a significant fraction of the outlet size (Ref 8).

Obviously most metal powders cannot be considered "coarse" materials; therefore, the above equation rarely applies in P/M applications.

For fine powders, funnel-flow bins often exhibit high discharge rates, but controlling the flow rate is always a challenge because the flow channel is not likely to be stable. As a result, the actual size and shape of the stagnant region is neither well defined nor constant. The flow channel can change size radically or collapse, creating flow rates that range from no-flow conditions to complete flooding.

Fine powders are more easily handled in a mass flow bin, whose flow channel is stable and predictable. Because all of the material is constantly moving in a mass flow bin, the flow channel is set by the shape of the bin.

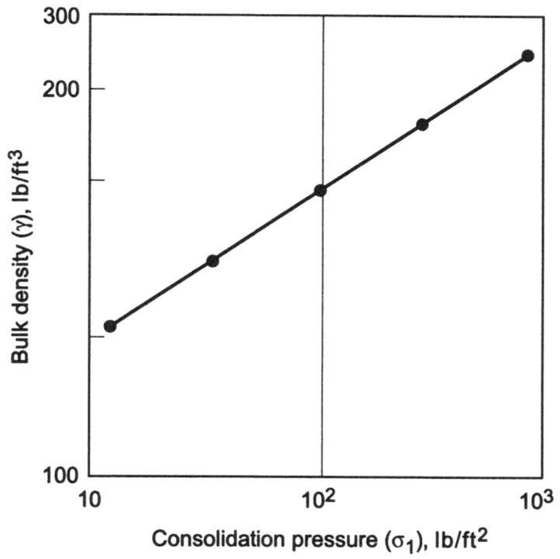

Fig. 18 Bulk density as a function of consolidation pressure (metal powder). Increasing consolidation pressure increases the bulk density.

However, it should be remembered that the maximum flow rate of a fine powder through the outlet of a mass flow bin is low compared with that of a coarse, granular solid. For fine materials, the expansion and contraction of voids during flow can create an upward air pressure gradient at the outlet of a mass flow bin. During discharge, this upward gradient acts against gravity, reducing the discharge rate. Such gradients do not usually form with coarser particle materials. Because coarse materials are more permeable than fine ones, air is allowed to flow freely into and out of the voids as they expand and contract.

This phenomenon can be analyzed by considering how gas flows through a bed of powder when a pressure differential occurs across the bed. When the gas velocity is low, flow through the bed is laminar, and Darcy's law can be used to relate gas velocities to gas pressure gradients within or across the bed. Darcy's law can be written in the following form:

$$u = -K\left[\frac{(dp/dx)}{\gamma}\right]$$

where K is the permeability factor of the bulk solid, u is the superficial relative gas velocity through the bed of solids, γ is the bulk density of the solid in the bed, and dp/dx is the gas pressure gradient acting at the point in the bed of solids where the velocity is being calculated. The permeability factor, K, has units of velocity and is inversely proportional to the viscosity of the gas. A permeability test is run by passing air (or other suitable gas) through a representative column of solids. The pressure across the bed is regulated, and the rate at which the gas flows is measured.

This approach allows the permeability of the bulk solid to be determined as a function of its bulk density. Figure 19 shows the test results for a sample of metal powder.

Because mass flow bins have stable flow patterns that mimic the shape of the bin, permeability values can be used to calculate critical, steady-state discharge rates from mass flow hoppers. Permeability values can also be used to calculate the time required for fine powders to settle in bins and silos and to design solids processing vessels to purge, heat, dry, or condition bulk solids.

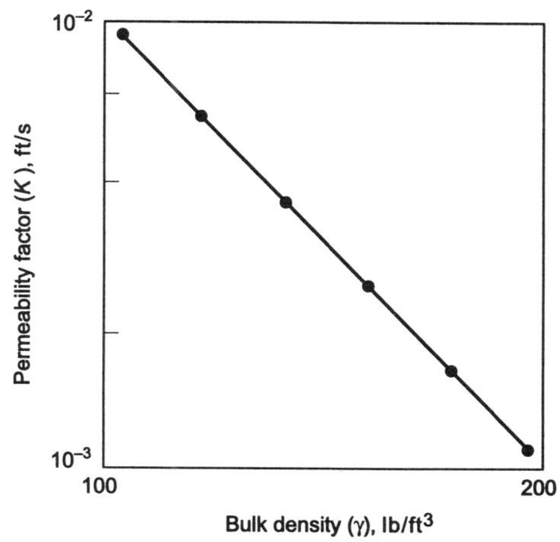

Fig. 19 Permeability as a function of bulk density (metal powder). Increases in bulk density reduce permeability of a material. Unless properly accounted for during bin selection, increased bulk density and reduced permeability can interrupt predictable flow.

Testing of Flow Rate

Flow of metal powders is determined by standard methods developed by ASTM and MPIF. Flow rate is the time required for a powder sample of a standard weight (50 g) to flow under atmospheric conditions through a funnel into the cavity of a container or mold. A determination of the flow rate of a powder is important in high-volume manufacturing, which depends on rapid, uniform, consistent filling of the die cavity. Poor flow characteristics cause slow and nonuniform press feeding and difficulty in ensuring even fills of the die cavity.

Before a powder is used in production, its flow characteristics must be known because some compacting tools require a free-flowing powder, while others can be used with a relatively poor-flowing powder. The term free-flowing refers to those physical properties of a powder—such as composition, particle fineness, and particle shape—that permit the powder to flow readily into the die cavity.

If a compacting tool is designed to handle a free-flowing powder, the use of a poor-flowing powder will necessitate modification. Compacting press manufacturers provide modified hopper designs and feeding shoe arrangements to accommodate finer, poor-flowing powders such as tungsten, molybdenum, or lighter aluminum powders.

The flow of powder from the feeding shoe into the die cavity can be increased by tapping or by changing the design of the filling device. These factors, however, are not taken into consideration in conventional flow rate test procedures, which consist of determining the time required for a given weight or volume of powder to flow through a standardized funnel-shaped cup with a small orifice at the bottom.

Table 6 Flow rate of metal powders through Hall and Carney funnels

Metal powder		Lubricant		Apparent density, g/cm³	Flow rate of 50 g (2 oz) powder, s		Flow rate of 25 cm³ (1.5 in.³) powder through Hall funnel, s	Weight of 25 cm³ (1.5 in.³) powder used in volumetric flow rate study		Calculated flow rate for 50 g (2 oz) powder based on volumetric flow rates, s
Material	Grade	Type	Additions, wt %		Hall funnel	Carney funnel		g	oz	
Iron	MP-35HD	Zinc stearate	None	2.81	25.77	4.62	37.74	70.27	2.46	26.86
		Zinc stearate	0.25	3.12	23.37	4.16	...	...	...	...
		Zinc stearate	0.50	3.05	25.93	4.30	...	...	...	...
		Zinc stearate	0.75	3.02	26.80	4.41	...	...	...	...
		Zinc stearate	1.00	3.00	27.57	4.59	40.55	74.52	2.61	27.21
Iron	MH-100	Zinc stearate	None	2.48	30.14	5.26	38.61	62.06	2.17	31.11
		Zinc stearate	0.25	2.97	23.23	4.14	...	...	...	...
		Zinc stearate	0.50	2.93	26.39	4.47	...	...	...	...
		Zinc stearate	0.75	2.86	28.97	4.80	...	...	...	...
		Zinc stearate	1.00	2.87	30.42	5.12	42.20	71.31	2.50	29.59
Iron	A-Met 1000	Zinc stearate	None	2.94	26.24	4.34	39.16	73.91	2.59	26.49
		Zinc stearate	0.25	3.27	23.89	4.04	...	...	...	...
		Zinc stearate	0.50	2.98	28.30	4.55	...	...	...	...
		Zinc stearate	0.75	3.18	25.46	4.41	...	...	...	...
		Zinc stearate	1.00	3.18	25.58	4.45	41.70	79.57	2.78	26.20
Stainless steel	304-L	Lithium stearate	None	2.61	30.62	4.92	39.65	65.45	2.29	30.29
		Lithium stearate	0.50	3.08	29.43	4.80	...	...	...	...
		Lithium stearate	0.75	3.01	33.20	5.42	...	...	...	...
		Lithium stearate	1.00	3.02	37.51	6.13	59.89	75.54	2.64	39.64
Premix bronze (90% Cu–10% Sn)	5099	Stearic acid	None	2.96	21.68	3.99	32.69	74.92	2.62	21.82
		Stearic acid	0.25	3.54	24.01	5.17	...	...	...	...
		Stearic acid	0.50	3.54	24.66	5.38	...	...	...	...
		Stearic acid	0.75	3.42	27.91	5.24	49.84	85.68	3.0	29.08
		Stearic acid	1.00	3.38	34.75	7.20	...	...	...	...
Brass	B-126	Lithium stearate	None	2.89	33.26	5.51	48.56	72.67	2.54	33.41
		Lithium stearate	0.25	3.06	33.52	5.77	...	...	...	...
		Lithium stearate	0.50	3.14	38.70	6.38	64.01	78.25	2.74	40.90
Aluminum	...	None	...	1.19	66.43	...	39.23	29.73	1.04	65.97

The following sections briefly review flow-rate test methods and variables that affect flow rate. Additional coverage is available in Ref 9 to 21.

Hall Flowmeter. The device most commonly used for measuring flow rate is the Hall flowmeter (Fig. 15a), taken from ASTM B 213 and MPIF 3 (equivalent standards include ISO 4490, Japanese standard JIS 7-2502-1966, and German standard 82-69). The test equipment consists of a funnel with a calibrated hole 2.5 mm (0.1 in.) in diameter. The funnel, which is made of aluminum alloy 6061-T6, is supplied with a smooth finish to minimize wall friction.

With the help of a stopwatch and weighing balance, the flow rate of metal powders can be easily determined. A dry 50 g weight sample is transferred to the funnel, the orifice of which is covered with the operator's fingertip. The stopwatch is started when the fingertip is removed and is stopped when the last quantity of the powder leaves the funnel. The flow rate (s/50 g) of the sample is reported as the elapsed time in seconds for 50 g of powder to flow through the orifice. A powder that does not flow through a 2.5 mm (0.1 in.) orifice Hall funnel, with or without an external impulse, is said to be a nonfree-flowing powder (as per ASTM B 213 and MPIF 3 method).

The Hall funnel is calibrated using a standardized powder (150-mesh Turkish emery grit), a sample of which is supplied with the equipment. The desired hole size is obtained by precisely honing a drilled hole until a satisfactory flow rate of emery powder is obtained.

A change in surface finish and the radius of the orifice (at the junction of the wall), buildup of material on the sidewalls of the orifice, or enlargement of the hole size due to continuous use can alter the standardization of the funnel. Verification of the calibration should be performed periodically by using the standardized emery powder. Calibration of the Hall funnel with Ballotini solid glass spheres, having particle size ranges of 0.090 to 0.102 mm (0.0036 to 0.004 in.) and 0.065 to 0.090 mm (0.0026 to 0.0036 in.) diameters, has yielded flow rates of 35.6 and 33.4 s/50 g, respectively.

The feed of powder to the die is handled on a volume basis. Thus, differences in apparent densities of powders can lead to considerable variations in the weight of material filling a given volume. A test for volumetric flow-rate determination is under investigation by ASTM. Table 6 lists the flow rates of metal powders for both weight and volume basis. The volumetric flow-rate tests were carried out using the Hall flowmeter. The data in Table 6 indicate that the readings for volumetric flow rate lie within a narrower range, as compared to the Hall flow rate.

Many factors can affect the accuracy of the results obtained with the Hall flowmeter, such as the moist finger of an operator (use of gloves can eliminate this effect), vibration of the surface on which the flowmeter is placed, humidity and temperature, condition of the sample, uniformity of the powder mix, and alternate use of the device for unlubricated and lubricated powders. Residual lubricant film left by a lubricated powder on the flow-meter wall or orifice can affect the subsequent flow test results of an unlubricated powder.

Other problems associated with its use include:

- *Segregation of the mixture:* Because only a small sample (50 g) is used, slight changes in particle size distribution can result in widely varying and misleading values.
- *Limited flowability information:* If a powder is too cohesive to flow through the funnel, no information on the flowability of the powder can be determined. It has been estimated that up to 40 to 50% of all powder mixes used in the P/M industry will not flow through the Hall flowmeter.
- *Meaningless results:* Even if a powder does flow well, the value obtained (s/50 g) is not meaningful for many design problems. It cannot, for example, be extrapolated to predict limiting press speed, limiting flow rate through the feed hopper, or other rate-limiting phenomena.

The attempt to combine measurements of two material flow properties (minimum orifice size and flow rate) results in a method that does not measure either one very well (Ref 22).

Carney Funnel. Certain characteristics of some metal powders, such as particle shape and size distribution and lower specific gravity, may affect the powders to such an extent that they will not flow through the Hall funnel. In this case, the Carney funnel (Fig. 15b), which has the same dimensions as that of the Hall, except for a larger orifice diameter of 5 mm (0.2 in.), can be used to obtain a relative measure of the flowability of nonfree-flowing metal powders. The use of the Carney funnel, which is further described in ASTM B 417 and MPIF 28, is not a standardized test method in that there is no standard calibration procedure. However, it is used in industry to compare flow rates through a 5 mm

(0.2 in.) orifice for a variety of materials. Because there is no correlating factor to relate the data obtained using the Carney funnel with that of the Hall, the user must establish an empirical relationship between the two methods.

Other Testing Methods. A number of other devices or methods for measuring the flow rates of metal powders have been developed. Efforts have been made to design test methods for powders that do not flow through either the Hall or Carney funnel and to reflect the shop floor conditions of these powders.

One method currently under consideration by the International Standards Organization determines the filling characteristics of metal powder into cavities of increasing sizes. A powder-filled shoe is slid back and forth once on the surface of a 40 mm (1.6 in.) thick plate (placed on a paper) into which bores with diameters of 5, 7, 10, 15, 20, 25, and 30 mm (0.2, 0.3, 0.4, 0.6, 0.8, 1.0, and 1.2 in.) have been drilled.

The plate is then lifted, and the powder that has fallen onto the paper from each of the seven die cavities is weighed. When the mass of the powder is divided by the volume of the respective cavity, the filling density for each cavity is obtained. The test provides the critical diameter, that is, the dividing line between cavity diameters which will be filled at a constant apparent density and those through which the powder may not flow at all or which may result in incomplete cavity fill.

In 1956, a second test method for determining the flow rate of metal powders that do not flow through the Hall funnel was developed by Chrysler Corporation as an internal standard. The test equipment consists of a powder shoe of 102 cm^3 (6.2 in.^{3}) in volume and a pivoted lever attached to the shoe at one end and to a cam follower at the other. A four-lobed cam is connected to a gearbox and is driven by a motor. A circular opening of 12.7 mm (0.5 in.) diam (with a 3.0 mm (0.12 in.) wide bar at center) drilled on a 2.4 mm (0.1 in.) thick metal plate acts as a die opening.

To conduct the test, the powder-filled shoe is moved back and forth over the opening four times. This is referred to as a one-cycle operation. The quantity of powder passing through the opening is collected on a balance pan and weighed. The test equipment simulates the action of a production press, and flow is measured in terms of the quantity of powder that passes through the opening in one filling cycle of a shoe.

In one production example, the normal range for a bronze powder mix was 80 to 95 g (2.8 to 3.3 oz) per cycle, whereas the data of all the lots ranged from 30 to 125 g (1 to 4 oz) per cycle of operation. The test method was used to check the flow of a production mix under conditions comparable to those encountered in the production presses. The test indicated the flowability of a powder mix accurately.

The die-filling operation of a production press can be simulated by a third test method in which actual bearing die cavities are used. For the production of bearings with varying wall thicknesses and complicated shapes, the tool design should accommodate the variations in filling properties of the powder mixes. The following test procedure was found to be especially useful for complex powder mixes that may or may not flow through the Hall flowmeter.

The equipment consists of a powder shoe, which slides over a metal plate with a recessed cavity in which a series of cups (with or without core rods) can be fitted. The sizes of the resulting bearing-shaped cavities vary between 3.3 to 8.2 cm^3 (0.2 to 0.5 in.^{3}) in volume with wall thicknesses ranging from 1 to 3 mm (0.04 to 0.12 in.) and corresponding core rod diameters from 19 to 15 mm (0.8 to 0.6 in.) with a constant height of 45 mm (1.8 in.). A reference cavity of the same height is coreless, with a volume of 16 cm^3 (1.0 in.^{3}).

With the selected size of the bearing die in position, the powder shoe is mechanically traversed through the die (or cup) opening and back to the original position. The cup is then removed and weighed. The quantity of powder is divided by the total volume of the cup, and the fill density of the powder is calculated. The process is repeated for the remaining die sizes by changing the core rod inserts. The filling density of the reference cavity is always the highest and is designated 100 for a given sample. This value is decreased with a reduction in wall thickness.

Tests with iron, nickel, copper, tin, and Turkish emery grit indicate that the results are consistent. The powders tested either flowed into the cup or did not enter at all. By evaluating filling densities in different die sizes, the test provides meaningful information in designing tools for a production shop. This test method was used to study blending variables and yielded close agreement with the performance of the production presses. Unfortunately, the test is not simple and requires the use of precise equipment.

Variables Affecting Flow Rate

Flow characteristics are dependent on several variables, including interparticle friction, particle shape and size, type of material, environmental factors, and weight of the bulk.

Characteristics of powder surfaces, such as surface oxide films and lubricant films, also affect flow characteristics. The presence of oxide films on powder particle surfaces alters the friction between particles and increases flow rate. Powders with lower surface oxide contents flow more slowly than powders with higher oxide levels. Minute additions of lubricants may increase the flow rates of metal powder, but further additions will reduce flow rates. For practical purposes, the higher the lubricant level, the slower the Hall flow rate (see Table 6).

In general, reduced flow rates are encountered with powders that exhibit one or more of the following characteristics: low specific gravity, low apparent density, high friction coefficient of fine particles, high specific surface area, a complex blend of different materials, and high moisture content.

Interparticle Friction. The resistance to flow depends primarily on the regions in which one particle hampers the free movement of other particles, either by direct contact or indirectly. This is mainly determined by the coefficient of interparticle friction. Particles may be prevented from moving separately by temporary adherence or interlocking. In this manner, clusters are formed that may occupy considerable volume. The phenomenon of cluster formation depends on the movement and type of powder, the flow varying markedly with the size and structure of the particles. If the particles were all truly spherical, they would generally roll readily into a die cavity. This can rarely be achieved in commercial powders; differences in size and shape are unavoidable.

Particle Size and Shape. Subsieve powders, those with particle sizes less than 44 μm, generally have poor flow rates. For this and other reasons, very fine powders are not used for compacts that are pressed on automatic presses. The particles of most powders used for compacting have irregular equiaxed shape, with flow rates between those of spherical (high flow rates) and flaky (low flow rates) powders.

Type of Material. Flow is influenced by the type of material, whether it be copper, aluminum, or iron. The major influence is the theoretical density. Other characteristics, such as adhesive and cohesive surface properties and magnetic or electrostatic interactions, are also factors.

Environmental Factors. Powders exposed to air containing high relative humidity absorb moisture on particle surfaces, resulting in reduced flow rate. On the other hand, as moisture content increases, many materials tend to agglomerate, which increases permeability and, consequently, increases discharge and settling rates. Seasonal changes in temperature do not affect the flow rate of metal powders considerably. Very low temperatures, however, can cause condensation of moisture; very high temperatures may partially melt the lubricants in the powder mix. Such extreme temperature conditions may cause interruptions in the flow of material through conveying systems. Also, because the permeability factor, K, is inversely proportional to the viscosity of the air or other gas in the void spaces, heating causes the gas to become more viscous, which makes the bulk solid less permeable.

Weight of the Bulk. A metal powder with a lower specific gravity, such as aluminum, generally exhibits slower flow rates compared to high specific gravity powders such as iron. At the same time, the higher the apparent density of a given material, the faster the flow. The ratio of apparent density to specific gravity can be used to correlate the bulk properties of various metal powders.

Sliding at Impact Points

Two key factors in chute design are the chute angle and the smoothness of the chute surface at the point of impact. Too shallow or too rough a surface in a chute impedes flow.

The required minimum chute slope can be determined by placing a ring-type device containing a sample of the powder on a representative sample of the chute wall surface and applying a predetermined vertical load to simulate impact. Once the weight has been removed, the plate is raised to determine the angle at which the particles begin to slide (Ref 23).

All of the factors affecting frictional properties can influence the optimal chute angle, except for time of storage at rest. Because it is uncommon for a chute to remain full without material moving through it, that parameter is generally not a design consideration.

Segregation Tendency

Providing a reliable, uniform mix or blend is critical in producing quality P/M parts. This requires not only proper mixing methods, but also proper handling techniques through compaction to ensure a uniform delivery of the powder to the die. As such, much greater emphasis is now being placed on appropriate storing, feeding, and transporting of powders.

The exact method of die fill depends on the specific compaction process being used (e.g., hydraulic, mechanical, rotary, isostatic, etc.) but generally a feed shoe is used. At the beginning of a compaction cycle, the feed shoe indexes over the die (lower punch) in the fill position and delivers a predetermined amount of powder (sometimes with the assistance of vibration). This is the final and most critical point where variations in the delivered mix directly impact part quality.

Much attention has been given to this specific step in trying to aid in powder filling such that a uniform fill is ensured (Ref 24). However if demixing has already occurred upstream in the process, then the compact will be nonuniform with resulting part-to-part variations. For example, lubricant variations will impact wall friction and subsequent compacting pressures. This, in turn, will cause some variation in compact strength, composition, and density.

Even more critical may be the variation in particle size as it impacts packing behavior and introduces significant density variations. Considering all the other critical variables that affect the compact density and strength (Ref 25) (e.g., compaction pressure, compaction velocities, temperature, wall surface variations and wear, etc.), ensuring that a proper final mix is delivered into the die becomes critical to reducing process complexity and ensuring a high-performance process.

Mechanisms of Powder Segregation. Five mechanisms have been identified as the primary cause of most segregation problems in particulate materials (Ref 26). Of these five, four are known to occur with P/M powders, and each of these mechanisms is described below. The conditions that tend to promote each mechanism are described in Ref 27. Methods to test the tendency for two of the mechanisms to segregate a particular powder blend or mix are also described below. More complete descriptions may be found in Ref 7 and 28.

Sifting. This common phenomenon occurs as smaller particles move through a matrix of larger ones. One of the most likely places in typical P/M processes where sifting segregation can occur is in the filling of a container. For example, during the discharge of the blender (or container) into a portable container (or surge hopper), a concentration of the fine particles will develop under the fill point while the larger particles will tend to roll or slide to the periphery of the pile.

The tendency of a material to segregate by sifting can be determined by running a sifting segregation test. First, a pile is formed under controlled conditions. The pile is then cored to gather a representative sample of particles from its center (under the fill point) and periphery (Fig. 20a). Each cluster is evaluated for particle size distribution, chemical content, and other relevant variables in order to determine the degree of segregation that has occurred. Figure 20(b) shows the results of a sifting segregation test, using a binary mixture.

Particle Velocity on a Surface. In general, the frictional drag on particles moving on a chute surface is higher for finer particles than for coarser particles. This results in a velocity difference between the coarse and fine particles, which translates to differences in particle trajectories off the end of a chute. The higher drag of the finer particles along with their position near the chute surface results in a concentration of the finer particles nearer to the end of the chute

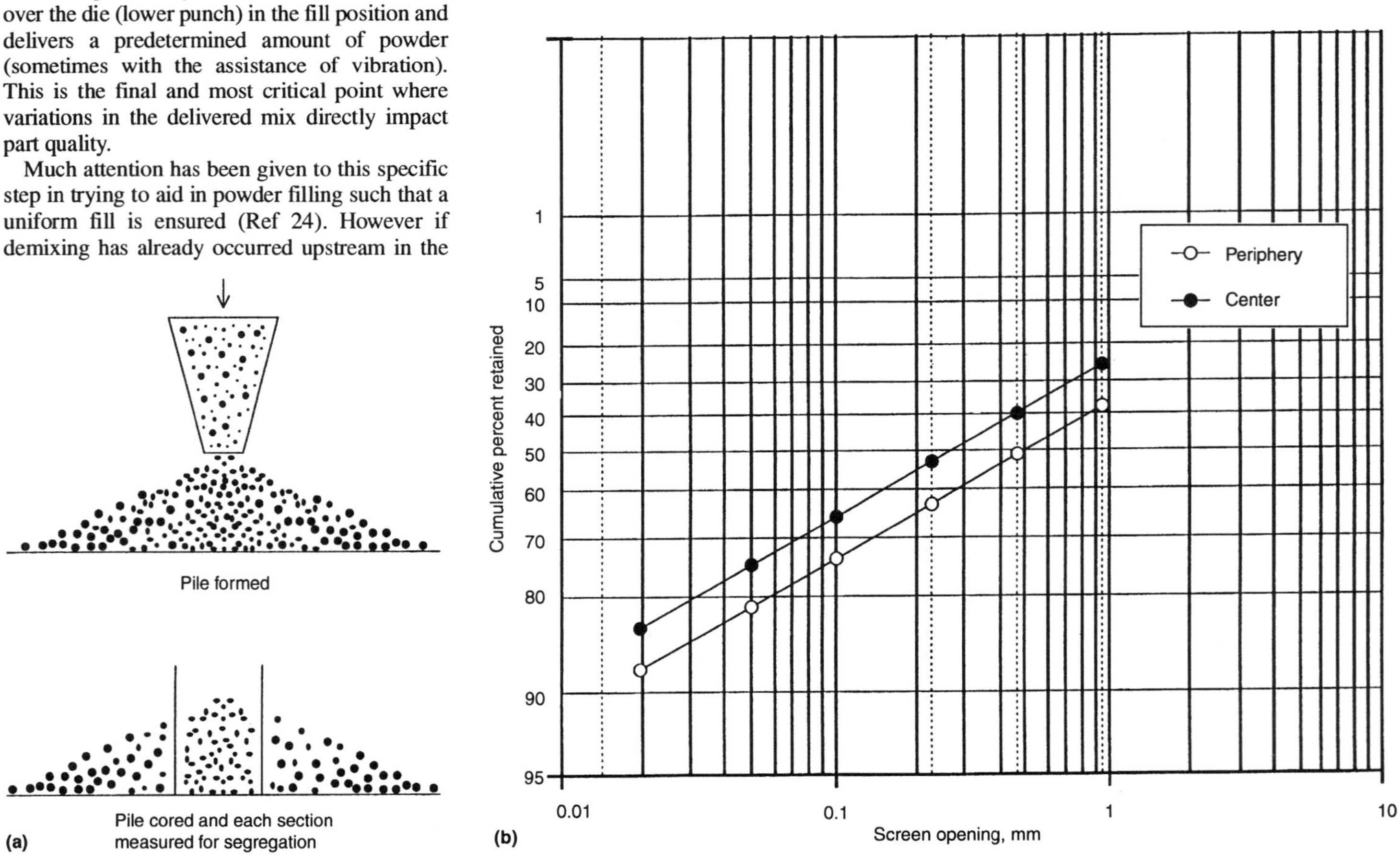

Fig. 20 Sifting segregation test. (a) Test setup. (b) Test results for binary mixtures

while the coarser particles come to rest at a much greater distance.

This can become particularly detrimental to the compacting process if portions of the powder stream go to multiple die cavity inlets or if the die cavity provides enough space for this trajectory segregation. In addition, if there are chutes within any of the powder transfer steps, there is the potential for this mechanism to occur.

Air Entrainment (Fluidization). In general, fine or light particles are less permeable than coarse or heavy ones. This allows the finer/lighter particles to retain air longer in their void spaces. Thus, when a mixture of coarse/heavy and fine/light particles is charged into a container or bin, down an empty vertical standpipe, or potentially into the die cavity, the finer particles remain fluidized longer while the coarser particles settle first (sometimes referred to as sedimentation). This results in a vertical segregation pattern within the powder bed.

To determine the likelihood of segregation by air entrainment, a tall cylinder containing a sample of the solid is aerated for a short period (Fig. 21). After the fluidizing air is turned off, sections of the cylinder are removed and the contents of each section are analyzed. If the material has segregated by air entrainment, fines will be located near the top of the cylinder and the coarser or heavier particles will be located at the bottom.

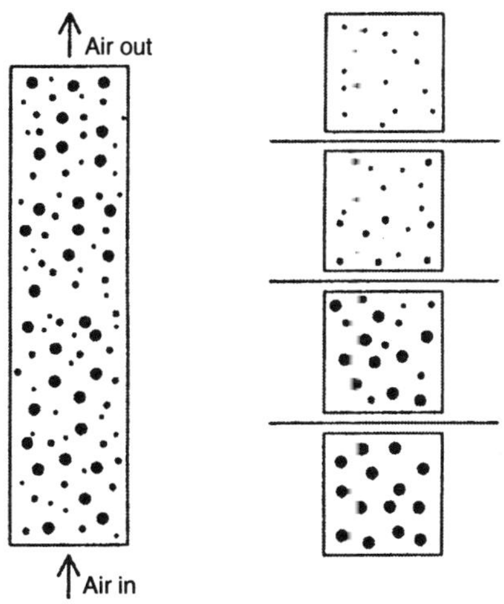

Fig. 21 Fluidization segregation test. (a) Column of material fluidized. (b) Column split and each section measured for segregation

Particle Entrainment. Fine/light particles tend to remain suspended in air longer than coarse/heavy particles. Thus, for example, air currents can carry airborne fines and light particles away from a fill point to certain areas of a bin, such as toward vents and dust collectors.

Variables that can affect the tendency of particles to segregate include:

- *Particle size distribution and shape:* Sifting segregation is most likely to occur when the material has a range of particle sizes and when there is interparticle motion during operation. Generally, particles greater than about 100 μm in diameter are most susceptible to sifting segregation. If most of the particles are smaller than 100 μm, segregation is more likely to occur by air or particle entrainment.
- *Cohesiveness:* The more cohesive a material, the less likely it is to segregate. Thus, for some materials, the potential for segregation can be reduced by adding binders. In general, this makes the material less free-flowing and causes fine particles to stick to coarse ones. Caution is advised, however, because the additive can rapidly change a situation from one of free flow to one of no flow.
- *Bin flow pattern:* The type of flow pattern that develops can significantly affect the segregation tendency of materials. Typically, funnel flow patterns exacerbate side-to-side segregation (such as that caused by sifting), whereas a mass flow pattern tends to minimize such problems.

Angle of Repose

The angle of repose of an aggregate, or the angle of the surface of an unconstrained pile of solids with the horizontal, is of practical interest rather than of theoretical concern. Of the various physical properties of bulk powders, the angle of repose is the easiest to obtain. It is related to interparticle friction and the flowability of cohesion-less material. This property is frequently used to characterize powdered materials.

The angle of repose of a powdered material does not always give satisfactory reproducibility and is often masked by other factors that are not inherent to the material (most frequently, the presence of a liquid). Thus, before a standard measuring method that provides reasonable reproducibility is developed, its usability as a measure of powdered material property must be established.

However, although the angle of repose is frequently used as a convenient characterization of powdered material, its ultimate reliability depends on the measuring method used. Therefore, the measuring method should be carefully selected, especially when the measured angle of repose is to be used to determine another property, so that it best reflects the property to be quantified.

Measurement Methods

Several methods have been used to measure the angle of repose of materials. Train (Ref 29) studied the angle of repose of a number of grades of glass spheres, lead shot, and silver sand using four different methods, as illustrated in Fig. 22. These methods are:

- *Fixed height cone:* The powder is carefully poured through a funnel at a fixed height until the apex of the heap formed by the powder reaches the tip of the funnel. The tangent of the angle of repose is the ratio of the height to the mean radius of the base of the powder heap.
- *Fixed base cone:* The powder is allowed to flow through a funnel, which is raised vertically until the heap covers a circular base of fixed size. The tangent of the angle of repose is calculated in the same manner as the fixed height cone method.
- *Tilting table:* A rectangular box filled with powder is tilted until the contents begin to slide. The angle that the upper surface of the box makes with the horizontal is equal to the angle of repose.
- *Rotating cylinder:* A sealed hollow cylinder half full of powder is rotated until the surface of powder exhibits its maximum angle with the horizontal. This maximum angle is the angle of repose.

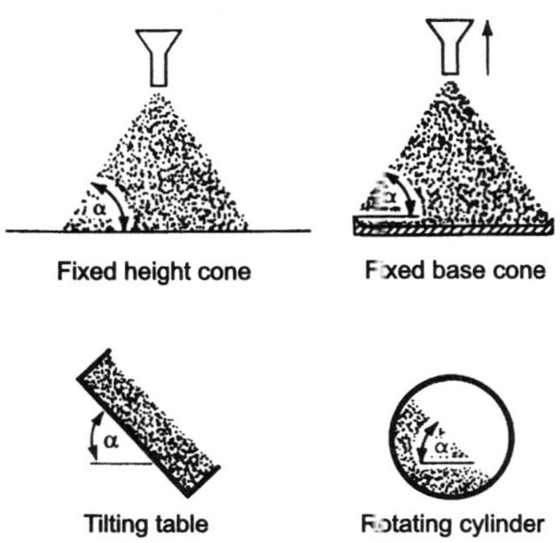

Fig. 22 Four methods used to measure the angle of repose. Source: Ref 30

Brown (Ref 31) reported three methods to measure the drained angle of repose (Fig. 23) and a rotating drum method to measure the dynamic angle of repose, as follows:

- *Ledge:* Material is first charged into a rectangular Perspex box that is 30 cm (12 in.) in

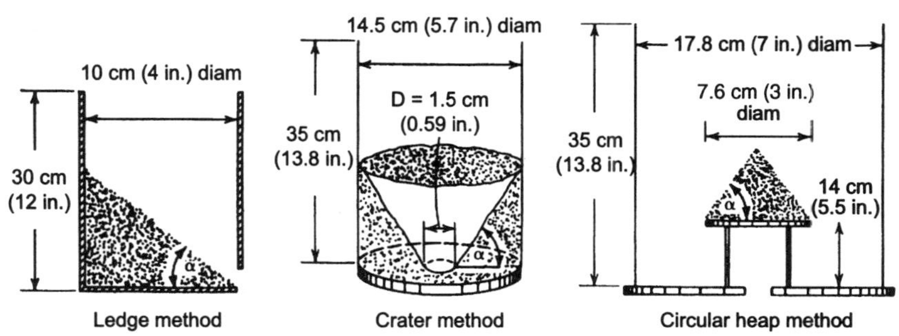

Fig. 23 Apparatus used to measure the drained angle of repose. Source: Ref 30

Table 7 Drained angles of repose (degrees to the horizontal)

Descriptive class	Material	Circular heap (±1½°), α	On a ledge (±1½°), α	Crater (±2°), α	Dynamic(a), α
Smooth, spherical	Beads	17½	25	27	25½
	Beads	20	23	21½	24
Rough, nearly spherical	Sand	32½(b)	34	35–35½	34½
	Tapioca	30	34	37½	32
	Rice	35	37	42	...
Angular	Sand	37	37	39	36½
	Sand	35½	37	38½	38½
	Sand	35½	36	37½	38½
	Durite	37	40	41	...
	Charcoal	...	38	...	...
	Charcoal	...	38½	...	...
	Charcoal	...	42½	...	...
	Charcoal	...	42½	...	...
	Coal	37½	37	41	34
	Coal	35½	37½	...	...
	Coal	36	38	...	...
	Coal	36½	38½	...	...
	Coal	38½	38½	...	...
Containing fine particles	Coal	...	52	...	...
	Coal	54	59–61	...	...
	Coal	...	47½	...	...
	Fine coal	...	67	...	...
	Limestone	...	64	...	...

(a) In a drum rotating at 6 rpm. (b) On a 5.1 cm (2 in.) diam platform in a 12.7 cm (5 in.) diam cylinder. Source: Ref 30

height with a 10 by 10 cm (4 by 4 in.) base. A slot at the base of one vertical wall can be closed by a board. The closure board is then removed to allow the material to flow slowly through the narrow slot. The angle with the horizontal of the surface of the material remaining when the flow stops is measured as the angle of repose.

- *Crater:* A circular Perspex tube with a 14.5 cm (5.7 in.) diam is place vertically on a flat, horizontal base plate having a 1.5 cm (0.59 in.) diam orifice in the center. The powder is discharged through the orifice. The height of the remaining material resting against the wall of the tube is measured at eight equidistant points around the circumference to determine the angle of repose.
- *Circular heap:* A circular platform 7.6 cm (3 in.) in diameter is supported horizontally over a circular hole in a flat base plate and surrounded by a cylindrical tube having 17.8 cm (7 in.) diam and sufficient height (35 cm, or 13.8 in.) to ensure that when it is filled with powder the platform and any heap that may form on it is completely immersed in the powder. The powder in the cylindrical tube is then allowed to flow slowly out of the circular hole in the base plate. The height of the resulting heap on the circular platform is then obtained. The tangent of the angle of repose is calculated as in the fixed height methods.
- *Dynamic angle of repose:* A drum 15 or 30 cm (6 or 12 in.) in diameter and 10.2 cm (4 in.) long with Perspex end faces and roughened internal surfaces is half filled with powder and slowly rotated counterclockwise, with its axis horizontal. Within a certain range of rotation speeds (usually 2.5 to 6 rpm), the surface of the powder in the drum becomes substantially steady. The angle of inclination of the surface to the horizontal is measured at various speeds to determine the angle of repose.

Table 7 lists the angles of repose measured using these methods. Henein et al. (Ref 32, 33) used a method similar to that used by Brown to determine the dynamic angle of repose and lower angle of repose (shear angle) of several materials. Rotating cylinders of 40 and 106 cm (16 and 42 in.) diam were lined with 24-3 grit type E silicon carbide abrasive paper. One of the two end plates of each cylinder were made of plexiglas to observe and photograph the bed. The maximum angle of bed inclination just before slump occurred was measured with a long-arm protractor, and this angle was designated the upper angle of repose, or the dynamic angle of repose. The angle relative to the horizontal of the shear plane that separated the slumping solids of the bed surface from the material moving with the cylinder wall was considered the lower angle of repose, or shear angle (see Fig. 24). Table 8 summarizes the upper and lower angles of repose of several materials.

Riley et al. (Ref 34) reported methods to measure the angles of repose of cohesive powders. They used both tilting box and fixed bed methods to measure the angle of repose of mixtures of glass ballotini and cohesive powders of different compositions and then extrapolated the results to obtain the angles of repose of pure cohesive material. Figure 25 shows a sample plot and gives the angle of repose of several cohesive powders.

Factors Affecting Angle of Repose

Internal factors are those inherent to powders, or characteristic of the nature of powders. These include particle size, particle shape, and cohesiveness. In general, larger particles have higher angles of repose. However, very small particles may exhibit cohesiveness due to the electrostatic effect, which increases the angle of repose (Ref 31–33). Because spherical particles have a greater tendency to roll, they typically have smaller angles of repose than irregularly shaped particles.

External factors are introduced by the environment and may include the method of measurement or the presence of other materials (either solid or liquid). The effect of measuring method is clearly evident as discussed above. In general, the more momentum introduced by the measuring method, the smaller the angle of repose. As a consequence, the drained angle of repose is higher than that obtained from the heap formation method.

The angle of repose of loosely packed dry powder may be increased by compacting, as well as by introducing liquid to the powder. Figure 26 demonstrates the significant effect these factors have on the angle of repose (Ref 30, 35).

Frydman (Ref 36) reported that the angle of repose of potash pellets may be increased by

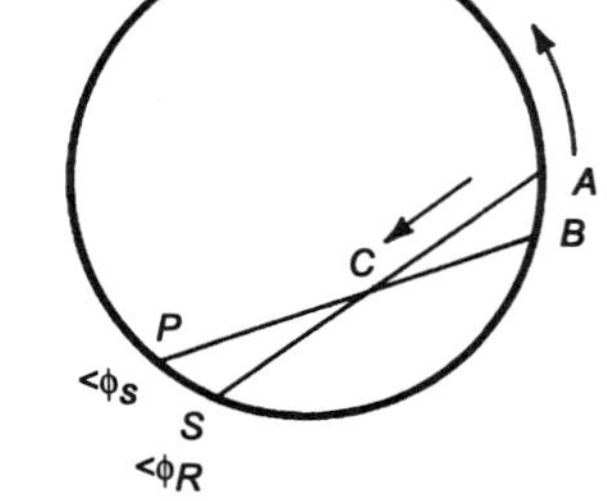

Fig. 24 Dynamic angle of repose and shear angle of material in a rotating cylinder. *B*, shear plane; ϕ_R, upper angle of repose (dynamic angle of repose). Angle ϕ_S, is lower angle of repose.

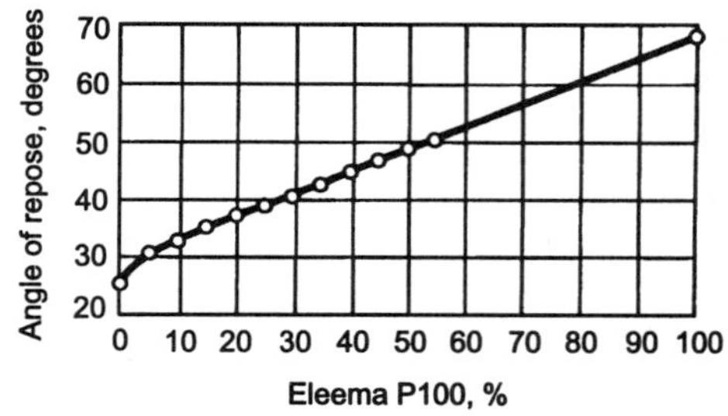

Powder	Tilting box	Fixed bed
Calcium carbonate	83.0	78.0
Eleema P100 (microfine cellulose)	64.5	68.0
Avieel PH101 (microcrystalline cellulose)	57.5	51.5
Methylcellulose 20 BPC	50.0	52.0
Methylcellulose 450 BP	63.0	55.0
Hydroxypropyl methylcellulose 4500	64.5	...
Hydroxypropyl methylcellulose 5000	59.5	58.5
Hydroxyethyl methylcellulose 3500	61.0	61.0
Magnesium stearate	69.0	66.0
Stearic acid	...	63.5

Source: Ref 34

Fig. 25 Typical plot of mixture composition versus angle of repose. Curve extrapolated to determine angle of pure material. Source: Ref 34

Table 8 Upper and lower angles of repose of several materials

Material	Average size, mm	Average size, in.	Particle shape	Particle density, kg/m³	Loose bulk density, kg/m³	Dense bulk density, kg/m³	Static angle of repose, degrees	Dynamic angle of repose, degrees	Cylinder diameter, m	Cylinder diameter, ft	Shear angle, degrees
Gravel	3.0	0.12	Angular	2870	1560	1690	40.7	37.5	0.40	1.3	34.7
								37.0	1.06	3.5	34.4
Iron oxide	11.6	0.46	Spherical	...	...	...	31.5	35.2	0.40	1.3	33.3
Limestone B	4.3	0.17	Irregular	2700	1450	1610	40.3	39.6	0.40	1.3	37.7
								36.5	1.06	3.5	34.5
Limestone C	1.5	0.06	Irregular	2690	1520	1600	37.8	36.0	0.40	1.3	33.6
									1.06	3.5	32.5
Limestone D	0.58	0.02	Irregular	2680	1490	1570	35.6	34.9	0.40	1.3	33.5
Limestone F	8.1	0.32	Angular	2690	...	...	42.8	41.5	1.06	3.5	38.5
Nickel oxide	4.9	0.19	Spherical	...	870	900	32.5	30.2	0.40	1.3	29.9
Sand B	0.50	0.02	Nodular	2660	1640	1740	33.4	33.6	0.40	1.3	32.2

Note: 1000 kg/m³ = 1 g/cm³. Source: Ref 33

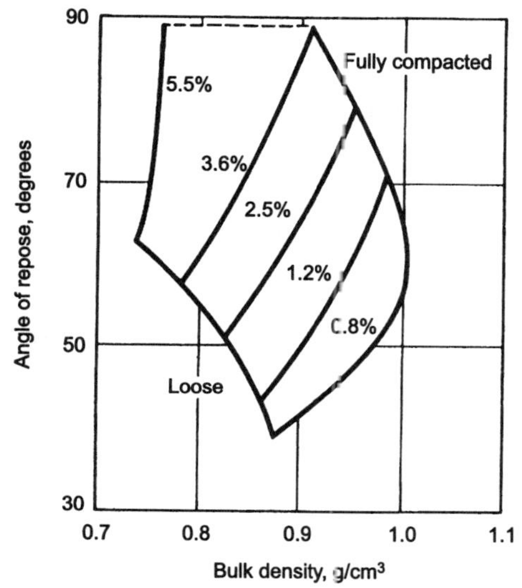

Fig. 26 Effect of moisture content on angle of repose of loose and compacted coal. Source: Ref 30

more than 100% by applying a light spray of moisture to the material surface. The effect of the presence of another solid is shown in Fig. 25.

REFERENCES

1. J.H. Bytnar, J.O.G. Parent, H. Henein, and J. Iyengar, Macro-Segregation Diagram for Dry Blending Particulate Metal-Matrix Composites, *Int. J. Powder Metall.,* Vol 31 (No. 1), 1995
2. *Properties and Selection: Irons and Steels,* Vol 1, *Metals Handbook,* American Society for Metals, 1978
3. A.W. Jenike, Storage and Flow of Solids, *Eng. Exp. Station Bull.,* No. 123, University of Utah, 1964
4. J. Marinelli and J.W. Carson, Solve Solids Flow Problems in Bins, Hoppers and Feeders, *Chem. Eng. Progr.,* May 1992, p 22–28
5. "Standard Shear Testing Method for Bulk Solids Using the Jenike Shear Cell," D 6128, ASTM, 1997.
6. A.R. Poster, Ed., *Handbook of Metal Powders,* Reinhold, 1966, p 17
7. J.W. Carson and J. Marinelli, Characterize Bulk Solids to Ensure Smooth Flow, *Chem. Eng.,* Vol 101 (No. 4), 1994
8. W.A. Beverloo, H.A. Leniger, and J. Van de Velde, The Flow of Granular Solids through Orifices, *Chem. Eng. Sci.,* Vol 15, 1961, p 260–269
9. J. Haertlein and J.F. Sachse, The Flow Rate of Metal Powders, *Handbook of Metal Powders,* Reinhold, 1966
10. A. Adler, Flow Properties of Metal Powders, *Int. J. Powder Metall.,* Vol 5 (No. 1), 1969, p 7–20
11. G. Matei, N. Claussen, and H.H. Hausner, Influence of Relative Humidity on Flow of Metal and Ceramic Powders, *Modern Developments in Powder Metallurgy,* Vol 8, Metal Powder Industries Federation, 1974, p 5–11
12. A.V. Zborovski, Correlation between the Apparent Density of Iron Powder and Its Flow Rate and Pressing Behavior, *Poroshk. Metall.,* Vol 84 (No. 12), Dec 1969, p 5–9
13. R.O. Grey and J. K. Beddow, On the Hausner Ratio and Its Relationship to Some Properties of Metal Powders, *Powder Technol.,* Vol 2, 1968–1969, p 323–326
14. M.J. Mahoney, "Method for Determination of Flow Rates of Metal Powders Blends," Laboratory Test 301–302, Amplex Div., Chrysler Corp., 1956
15. "Determination of Filling Characteristics," TC-119/SC-2N 234, International Standards Organization, 1981
16. P.E. Matthews, "Volumetric Flow Rate—Experimental," B 09:02, ASTM, 1980
17. J. Oakley, A Method of Assessing the Die-Filling Characteristics of Powders, *J. Inst. Met.,* Vol 87, 1958–1959, p 26–28
18. H.H. Hausner, Environmental Effects on the Behavior of Metal Powders, *Int. J. Powder Metall. Powder Technol.,* Vol 14 (No. 4), 1978
19. D. Yarnton, Effect of Lubricant on the Flow and Packing Density of Cu Powders, *Eng. Mater. Des.,* Sept 1970, p 1103–1107
20. "Flow Rate of Metal Powders," B 213, ASTM, 1977
21. "Determination of Flow Rate of Metal Powders Using the Hall Apparatus," No. 03, Metal Powder Industries Federation
22. J.W. Carson, Applying Bulk Solids Flow Principles to Metal Powders, *Int. J. Powder Metall.,* Vol 11 (No. 4), 1975
23. D. Stuart-Dick and T.A. Royal, Design Principles for Chutes to Handle Bulk Solids, *Bulk Solids Handl.,* Sept 1992, p 447–450
24. L.R. Lawrence and J.K. Beddow, Powder Segregation During Die Filling, *Powder Technol.,* Vol 2, 1968–1969
25. R.M. German, *Powder Metallurgy Science,* Metal Powder Industries Federation, 1984
26. J.W. Carson, T.A. Royal, and D.J. Goodwill, Understanding and Eliminating Particle Segregation Problems, Vol 6 (No. 1), *Bulk Solids Handl.,* 1986
27. B.H. Pittenger, H. Purutyan, and T.A. Royal, Eliminating Segregation Problems in Powdered Metal Processing, Reprinted from *Advances in Powder Metallurgy & Particulate Materials—1995,* Vol 1, Metal Powder Industries Federation, p 111–121
28. J.C. Williams, The Segregation of Particulate Materials, A Review, *Powder Technol.,* Vol 15, 1976
29. D. Train, Some Aspects of the Property of Angle of Repose of Powders, *J. Pharm. Pharmac.,* Vol 10 (Suppl), 127T–135T, Dec 1958
30. R.L. Brown and J.C. Richards, *Principles of Powder Mechanics,* Pergamon Press, 1970
31. R.L. Brown, Flow Properties, Powders in Industry, *Soc. Chem. Ind.,* Monograph No. 14, 1960, p 150–166
32. H. Henein, J.K. Brimacombe, and A.P. Watkinson, Experimental Study of Transverse Bed Motion in Rotary Kilns, *Metall. Trans. B,* Vol 14, 1983, p 191–205
33. H. Henein, J.K. Brimacombe, and A.P. Watkinson, The Modeling of Transverse Solids Motion in Rotary Kilns, *Metall. Trans. B,* Vol 14, 1983, p 207–220
34. G.S. Riley, S. Mann, and R.O. Jesse, Angles of Repose of Cohesive Powders, *J. Powder Bulk Solids Technol.,* Vol 2, 1978, p 15–18
35. E.F. Wolf and H.L. Hohenleiten, Experimental Study of the Flow of Coal in Chutes, *Trans. Am. Soc. Mech. Eng.,* Vol 67, 1945, p 585–599
36. S. Frydman, The Angle of Repose of Potash Pellets, *Powder Technol.,* Vol 10, 1974, p 9–12

Compressibility and Compactibility of Metal Powders

Steve Lampman, ASM International

COMPRESSIBILITY AND COMPACTIBILITY are terms that define the ability to form "green" (unsintered) compact from the die pressing of powders. In general terms, compressibility is quantified as the amount of compaction pressure that is required to achieve a given green density of a part. A more compressible powder can be compacted to a higher density and, consequently, permits the manufacture of parts possessing higher mechanical properties or the use of a smaller and less expensive compaction press. Thus, compressibility of a powder is a major factor in the design of pressing tools, the part density attainable, and the size of press needed.

Compactibility is a slightly different term related to the green strength of a powder compact, part complexity or fragility, and the ability to eject the compact from the die. For example, with all other factors equal, a simple part has more compactibility than a complex part, because the complex part would require higher ejection force (which increases the likelihood of breakage or cracking). Therefore, in order to increase the compactibility of complex or fragile parts, higher green strength may be needed in order to prevent breakage or cracking during ejection or subsequent handling.

The relation between compressibility and compactibility is complex, because some factors increase one at the expense of the other. In particular, particle shape and particle size are key factors that influence compressibility and compactibility (or green strength) in opposite ways. In terms of particle shape, for example, compressibility is reduced for a mix of irregularly shaped powder, because the particles do not rearrange themselves as easily as they would for spherical shapes. Thus particles with spherical shapes generally have higher compressibility than particles with irregular shapes. In contrast, green strength generally is increased by increasing the particle surface area. This can be achieved by increasing the particle surface roughness and/or reducing the average particle size, thus providing more sites for mechanical interlocking and a stronger compact. Coarse powders reduce the number of contact points between particles, which thus allows particle rearrangement for improved compressibility but also lowers green strength.

However, even though particle size and shape have competing effects on compressibility and green strength, there are factors that enhance both. For example, residual interstitial impurities or solid-solution alloying reduce both compressibility and green strength. This is shown in Fig. 1 for carbon contents of iron powder. Green strength and compressibility also improve with increased compaction pressure and die temperature as shown in Fig. 2 for a water-atomized iron powder (0.004 wt% C, 0.09 wt% O, 0.05 wt% Mn). Increasing temperature or compacting pressure promotes increased particle movement and deformation, which are the bases for good mechanical interlocking.

Table 1 shows the relationship among green strength, apparent density, compacting pressure, and green density for several types of iron powders. A comparison of compressibility and green strength is also shown in Fig. 3 for two types of iron powder. The powders are close in terms of compressibility (Fig. 3a), but the larger difference in green strength (Fig. 3b) is attributed to more interlocking for the atomized particles.

Compressibility

Compressibility of powders is generally expressed in terms of green density versus compaction pressure. Compressibility is also measured in terms of the *compression ratio,* which is the ratio of final pressed density to the apparent density of the powder.

Typical compressibility data for various kinds of metal powders are given in Fig. 4 as a percentage of theoretical density to demonstrate the relative compressibility of various powders. Because compressibility is so important in the effective and economical production of P/M parts, improvements in compressibility are a significant industry objective (Fig. 5).

Key Variables

Compressibility is determined by several factors including particle shape, density, composition, hardness, and to a lesser extent, particle size. Lubrication and compacting conditions also influence compressibility as described below.

Particle size and shape influence both apparent density and compressibility in a similar way. Small, spherical particles occupy more space than large, irregular ones, and higher compressibility and apparent density is expected for small, spherical particles. This is one reason for the approximate linear relation between compressibility and apparent density (Fig. 6). Highest compressibility is generally achieved with a wide spread in particle size, while powders of a uniform size exhibit relatively poor compressibility. This occurs because mixtures of particle sizes occupy more space and thus have better compressibility and higher apparent density.

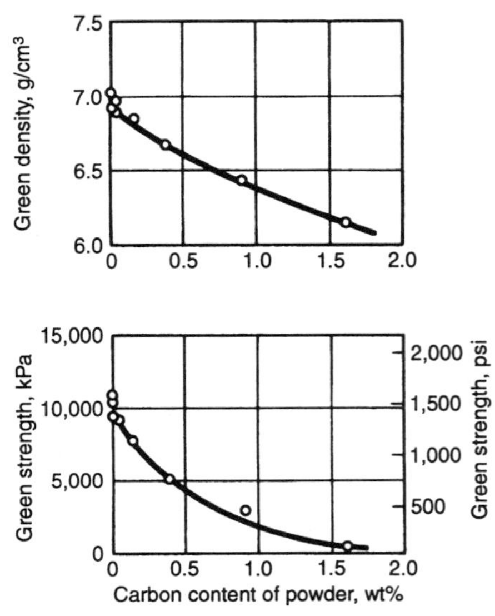

Fig. 1 Effect of residual carbon content on compressibility and green strength of water-atomized high-carbon iron powder. Pressed at 550 MPa (40 tsi) with 1% zinc stearate.

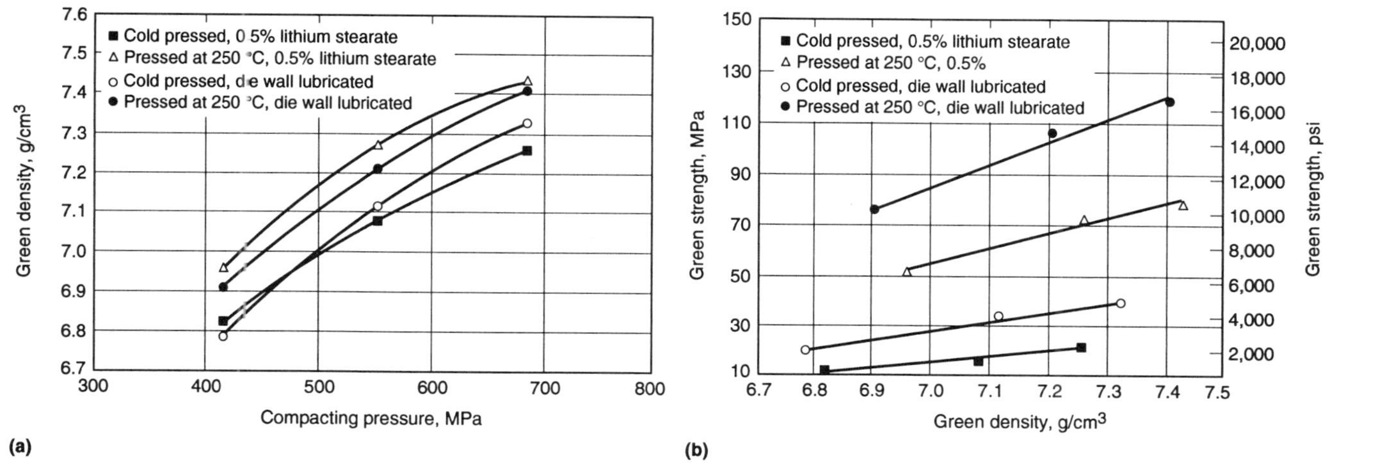

Fig. 2 Effect of compacting pressure and die temperature on (a) green density and (b) green strength of a high purity water atomized iron powder (0.004 wt% C, 0.09 wt% O, 0.05 wt% Mn). Source: Ref 1

Table 1 Green density and green strength for various types of iron powders

Powder	Apparent density, g/cm³	Compaction pressure MPa	Compaction pressure tsi	Green density, g/cm³	Green strength MPa	Green strength psi
Sponge(a)	2.4	410	30	6.2	14	2100
		550	40	6.6	22	3200
		690	50	6.8	28	4100
Atomized sponge(b)	2.5	410	30	6.55	13	1900
		550	40	6.8	19	2700
		690	50	7.0	...	...
Reduced(a)	2.5	410	30	6.5	16	2300
		550	40	6.7	21	3000
		690	50	6.9	24	3500
Sponge(a)	2.6	410	30	6.6	19	2700
		550	40	6.8	25	3600
		690	50	7.0	27	3900
Electrolytic(c)	2.6	410	30	6.3	32	4600
		550	40	6.7	43	6200
		690	50	6.95	54	7800

(a) Powders contained 1% zinc stearate blended in. (b) Powder contained 0.75% zinc stearate blended in. (c) Isostatically pressed. Source: Ref 2

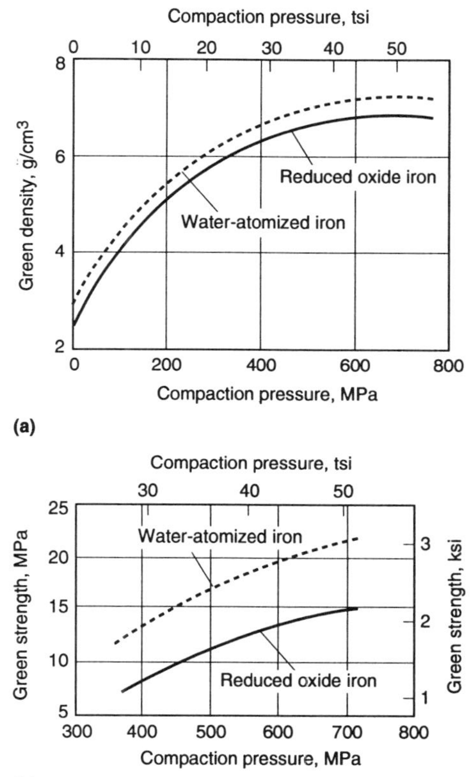

Fig. 3 Comparison of compressibility and green strength for two iron powders: (a) compressibility and (b) green strength. Source: Ref 3

Powder Hardness and Oxides. Because considerable deformation of particles occurs during pressing, particle hardness influences compressibility. In addition, some metals tend to work harden more than others. Nonmetallics such as unreduced oxides also reduce compressibility because of their hardness and low specific gravity.

Internal Porosity. Fine internal porosity in a powder tends to become noninterconnected during pressing, thus trapping air within particles. Although air is highly compressible, it takes up volume and contributes little to weight. Nonporous powders have the highest compressibility.

Particle Chemistry. Alloying additions or residual impurities generally degrade compressibility by substitutional or interstitial hardening. For iron powders, substitutional elements such as molybdenum, nickel, and manganese have a much smaller effect on compressibility (Fig. 7) than interstitial elements like carbon (Fig. 1). In particular, nickel and molybdenum produce only small reductions in the compressibility of prealloyed steel powders, while significantly higher compaction pressures are required for chromium, copper, and manganese. The reduction in compressibility from chromium and manganese is attributed not only to solid-solution strengthening, but also to the strong oxidation tendency of chromium and manganese during powder production. Residual oxygen is known to decrease iron powder compressibility as shown in Fig. 8.

Lubricants. The effect of lubricants on compressibility depends on the lubricant and the compaction pressures. Some lubricants perform better than others, but generally lubricants cause a reduction in green strength (Fig. 9). The reduction in compressibility from lubricants is more pronounced at higher compaction pressures (Fig. 10). At higher compaction pressures, lubricants reduce the green density because it occupies the available porosity.

Lubrication is used to assist ejection of the compacted specimen from the die. Failure to use a lubricant may result in tool seizure and excessive die wear, particularly at high compacting pressures. Common solid lubricants include zinc, stearate, amide wax, and stearic acid.

There are two lubrication techniques. One method is to blend a dry lubricant with the powdered metal. The other method, commonly referred to as die wall lubrication, is to lubricate the die walls and punches of the compacting tooling prior to introducing the powder metal into the die cavity. With the powder lubrication method, the level of lubricant addition may range from 0.5 to 1.5%. With die wall lubrication, the solid lubricant (for example, zinc stearate, 100 g) is mixed with a volatile organic liquid (for example, methylchloroform, 1 L) and is either painted or sprayed on the tooling. The organic liquid evaporates, leaving a thin film of

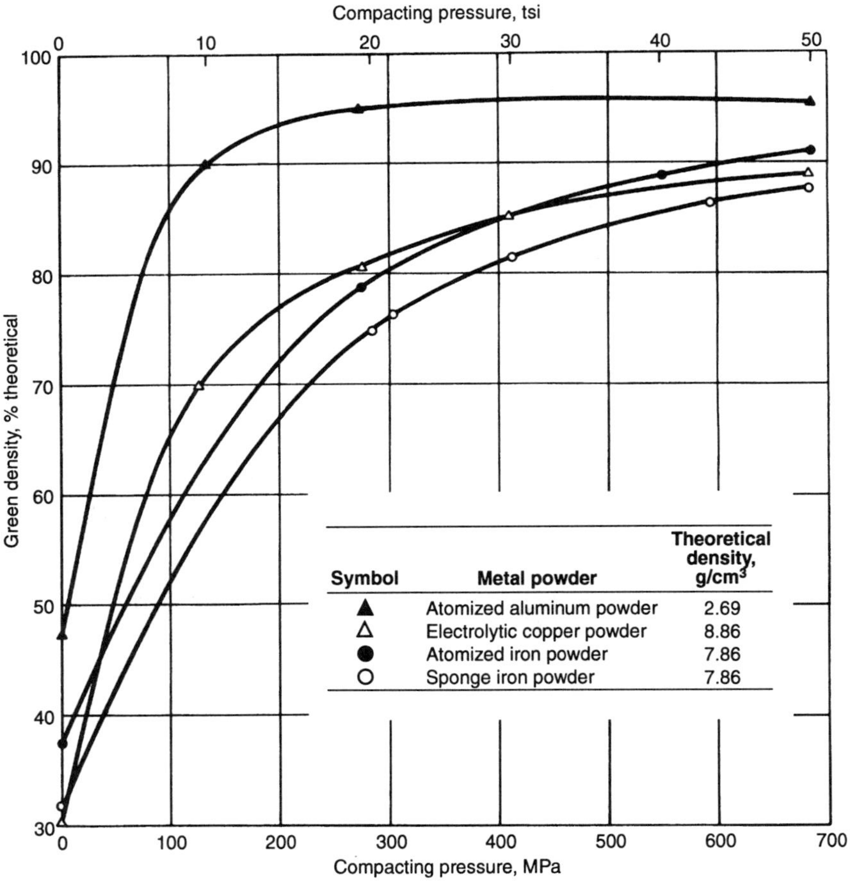

Fig. 4 Compressibility curves for various metal powders

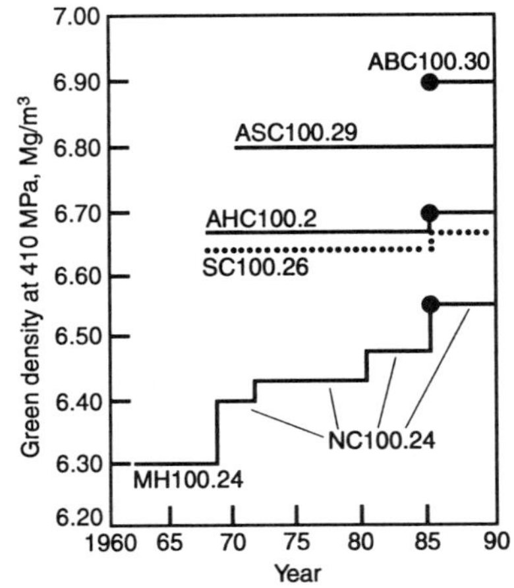

Fig. 5 Trend of improvements in iron powder compressibility. Source: Ref 4

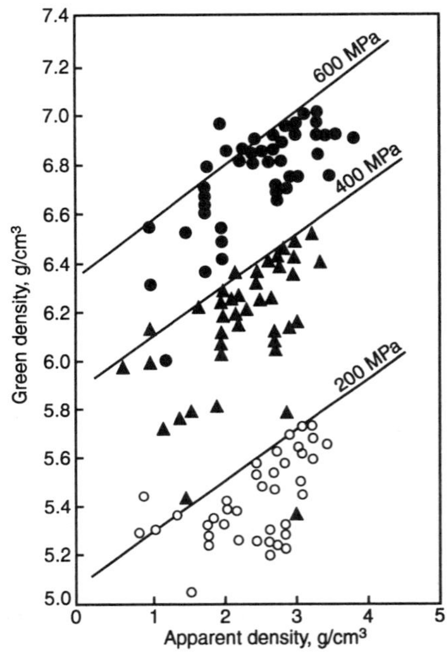

Fig. 6 Effect of apparent density on compressibility of iron powders produced by different methods. Compaction at 200, 400, and 600 MPa. Source: Ref 5

Table 2 Effect of processing on the green properties of ferrous compacts

Base material	Processing technique	Compaction pressure, MPa	Compaction pressure, tsi	Green density, g/cm³	Green strength, MPa	Green strength, psi	Peak die ejection force, MPa	Peak die ejection force, tsi
Ancorsteel 85HP(a)	Warm compaction	415	30	7.14	23.2	3370	29.6	2.15
		550	40	7.31	25.4	3685	33.5	2.43
		700	50	7.37	24.7	3580	32.0	2.32
	Cold compaction	415	30	7.00	9.9	1430	37.2	2.70
		550	40	7.19	12.2	1770	50.7	3.68
		700	50	7.29	13.4	1950	53.8	3.90
Distaloy 4800A(b)	Warm compaction	415	30	7.07	28.3	4100	27.4	1.99
		550	40	7.29	30.6	4445	31.7	2.30
		700	50	7.36	31.1	4515	32.3	2.37
	Cold compaction	415	30	6.93	12.2	1770	37.2	2.70
		550	40	7.15	15.0	2170	48.5	3.52
		700	50	7.26	16.9	2450	52.0	3.77

(a) Ancorsteel 85HP is a prealloyed steel powder containing 2.0% Ni, 0.85% Mo, 0.4% graphite, and 0.6% lubricant. (b) Distaloy 4800A is a diffusion-alloyed steel powder containing 4% Ni, 1.5% Cu, 0.50% Mo, 0.5% graphite, and 0.6% lubricant. Source: Hoeganaes Corporation

dry lubricant on the working surfaces of the die cavity and punches.

Density of the ejected test specimen varies with the method of lubrication (Fig. 9), as well as with the type and amount of lubricant used. These parameters should be specified when reporting density values that are to be used to determine the compactibility of a metal powder.

Warm compaction of the ferrous P/M parts is a recent innovation that uses heated tooling and powder during the compaction step. The powder and tooling are typically heated between 130 and 150 °C (260 and 300 °F). In order for the powder premix to perform at these temperatures, proprietary lubricant systems have been developed that can withstand the higher die temperatures in warm compaction. Polymer binders in the premix also limit segregation and provide enhanced flow characteristics of the powder premix.

Improved compaction of metal powders (such as aluminum, copper, and iron) at higher temperatures has been used for many years, but the problem with ferrous parts has been that available lubricants for proper ejection could not withstand warm die temperatures. This problem has been solved with improved lubricants (such as the Densmix polymer lubricant first developed and patented by Hoeganaes Corp.), that allow warm compaction. By utilizing warm compaction technology, the green density of the consolidated part can be increased from 0.10 to 0.25 g/cm^3 over traditionally processed (single pressed/single sintered) materials. The green strength is typically increased between 50 and 100%. Table 2 compares the green properties of warm compacted and cold compacted P/M parts. This increase in green strength provides advantages such as a reduction in green chipping and cracking due to part handling prior to sintering and makes possible the crack-free compaction of complex multiple-level parts. Additionally, the

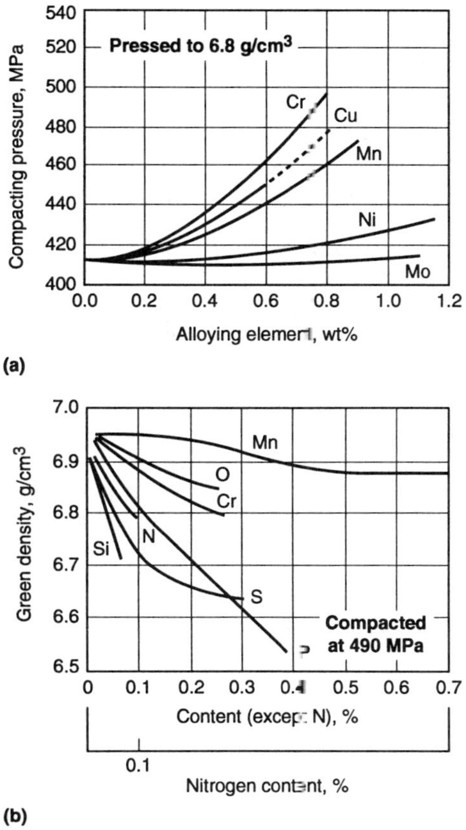

Fig. 7 Effect of alloying or compressibility and green strength of steel compacts. (a) Compressibility of water-atomized prealloyed powders (prealloyed powder samples mixed with 0.5% graphite – 0.75% zinc stearate and pressed to 6.8 g/cm^3. Source: Ref 6. (b) Green strength of steel compacted at 490 MPa. Source: Ref 7

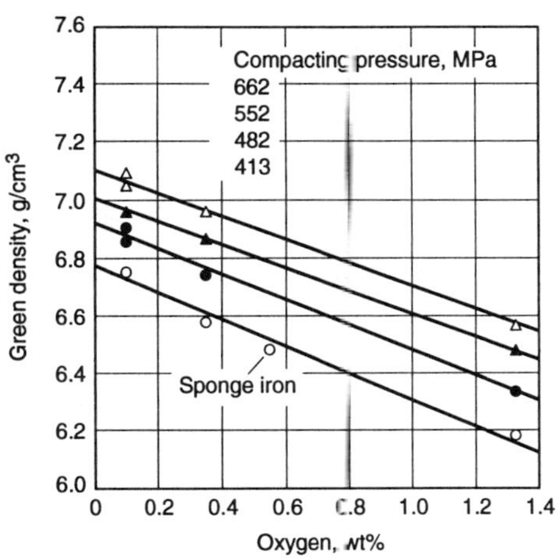

Fig. 8 Effect of oxygen content on compressibility of water atomized iron powder (<0.2 wt% Mn, 0.01 wt% Si) blended with 0.75 wt% Acrawax C and 0.4 wt% graphite. Data at 0.1 wt% O includes results from iron powder with 0.6 wt% Mn. Source: Ref 8

higher green strength provides an opportunity to machine the P/M part in the green state. This capability is important in the use of high-performance alloys that achieve high hardness in the sintered condition.

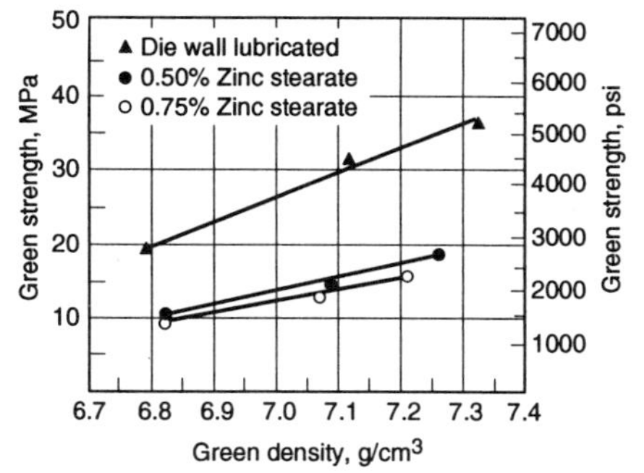

Fig. 9 Effect of lubrication method on green strength. Source: Ref 1

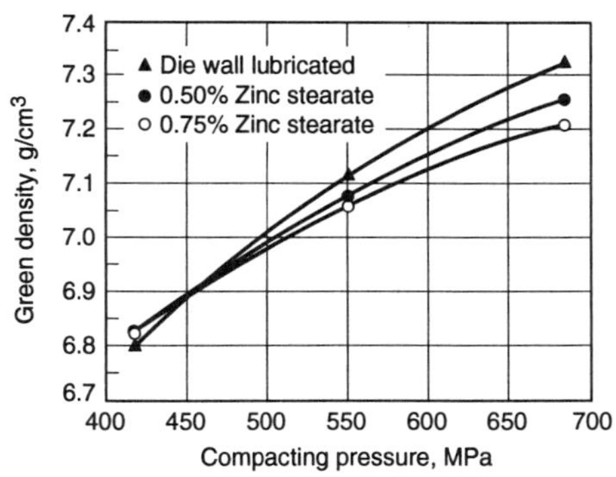

Fig. 10 Effect of lubrication on green density. Source: Ref 1

Annealing of metal powders has become common heat treating practice, because many powder-producing processes yield powders containing residual amounts of carbon, oxygen, and/or nitrogen that reduce green strength and compressibility. Thus, in metal powder production the term "annealing," or the stress relieving and recrystallizing of a cold worked material to increase ductility, includes chemical reactions such as oxygen reduction, decarburization, and denitriding.

Annealing also is used to remove thin surface oxide films from powders that tarnished during prolonged storage or exposure to humidity. Use of annealing for this purpose is more prevalent with relatively noble metal powders such as iron and nickel, which are easier to reduce but which frequently oxidize more readily and completely. Generally, the lower the reduction temperature used to reduce oxide films, the less stable the powder toward reoxidation. Lubrication and/or protective surface treatments improve the stability of powders.

Annealing of Alloy Steel Powders. High-speed tool steel powders require an annealing treatment after atomization. Martensitic and ferritic stainless steels sometimes are annealed in either a nitrogen or dry reducing atmosphere such as hydrogen or dissociated ammonia to improve their compacting properties. Annealing temperatures are kept as low as possible to minimize sintering. Austenitic stainless steel powders are generally used in the as-atomized condition. The effect of annealing on the compressibility (green strength) of 434-L ferritic stainless steel is illustrated in Fig. 11.

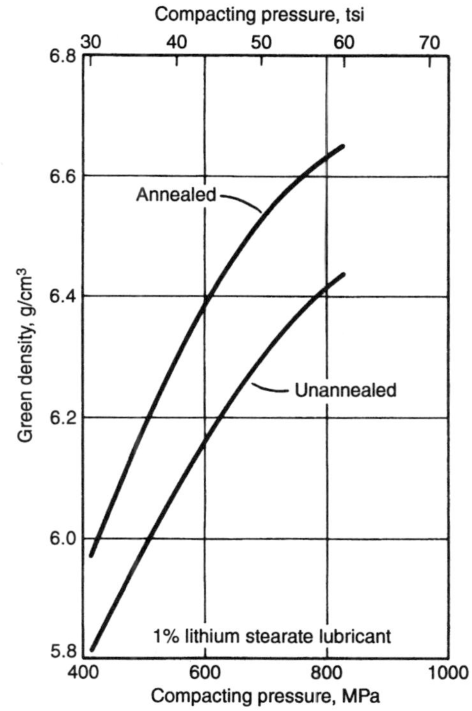

Fig. 11 Compactibility (green strength) of ferritic stainless steel 434-L powder in the annealed and unannealed condition

Water-atomized high-speed tool steel powders are annealed either in vacuum or hydrogen. Martensite and retained austenite are thereby transformed to a mixture of ferrite and spheroidal carbide. This renders the powder softer and more ductile and suitable for compaction.

Generally, the annealing treatment for tool steels combines deoxidation and temper annealing. The particles of as-atomized tool steels are as hard as their fully hardened wrought counterparts, with apparent hardnesses in the range of 63 to 65 HRC, which renders the powder virtually impossible to compact, and oxygen levels of ~2000 to 4000 mL/m^3 (2000 to 4000 ppm). The annealing treatment reduces the hardness to 23 to 25 HRC and oxygen to ~800 to 1000 mL/m^3 (800 to 1000 ppm). A small amount of decarburization is a by-product of annealing.

Compressibility Testing

Standard test methods for compressibility have been issued as:

- ASTM B331, "Compressibility of Metal Powder in Uniaxial Compaction," by the American Society for Testing and Materials
- MPIF 45, "Determination of Compressibility of Metal Powders," by the Metal Powder Industries Federation
- ISO 3927, "Metallic Powders Excluding Powders for Hardmetals—Determination of Compactibility (Compressibility) in Uniaxial

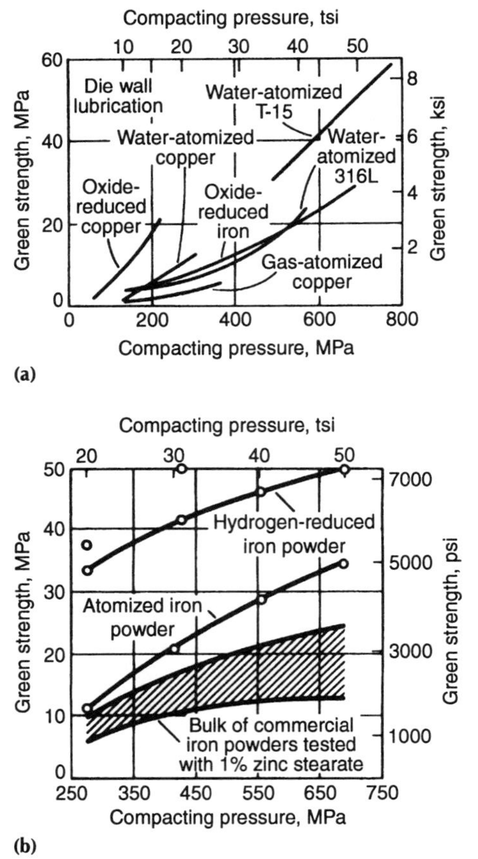

Fig. 12 Green strength versus compaction pressure for various types of powder. (a) Copper, iron and steel powders. (b) Iron powders

Compression," by the International Standards Organization

Typically, a cylindrical or rectangular test piece is made by pressing powder in a die, with pressure applied simultaneously from top and bottom. The pressure required to achieve a specified density is a measure of compressibility. Compressibility can also be specified as the density achievable at a given pressure. By plotting the density obtained by a series of increasing levels of pressure against these pressures, a compressibility curve is developed.

Two shapes of test pieces have been standardized. One is a cylinder having a diameter of 25 mm (1 in.) and a height of 12.7 to 25 mm (0. 5 to 1 in.). The other is a rectangular bar 12.7 mm (0.5 in.) wide by 31.8 mm (1.25 in.) long and 5 to 7 mm (0.2 to 0.3 in.) high. Sufficient powder to form the test bar is loaded into the appropriate die, and a pressing load is applied to the opposing punches. When the specified pressure is reached, the part is ejected from the die and its height and peripheral dimensions measured to determine volume. The weight of the part divided by its volume is a measure of the density achieved at the specified pressure. Alternatively, a given weight of powder is pressed to a predetermined thickness, thus to a specific density. The pressure needed to reach this density is the measure of compressibility.

Tooling used to make these specimens consists of a die, preferably of cemented carbide, or alternatively of tool steel, and two steel punches for producing either cylindrical or rectangular compacts. Exact tool dimensions required for making the cylindrical test piece and the rectangular bar are described in MPIF 45 and ISO 3927.

Green Strength

Green strength refers to the mechanical strength of a cold pressed powder compact. Adequate mechanical strength of a green part permits cold pressing of the powder, ejection of the green part from the die, and transfer of the green part to the sintering furnace without breakage. Strength of green compacts results mainly from mechanical interlocking of irregularities on the particle surfaces, which is promoted by plastic deformation during pressing. Figure 12 shows green strengths of various powders versus compaction pressure.

For most applications, green strength values of about 5.5 MPa (800 psi) or more permit pressing and safe handling of green parts. Lower green strength values may produce insufficient strength and horizontal lamination in the part. These laminations appear when the part is ejected from the die as a result of a small expansion of the part, due to the release of elastic stresses. This type of defect can be minimized or avoided by providing the die with a slight taper (draft angle) or a small radius at the upper rim. Parts with sharp contours or thin sections, or parts made from a blend of several powders in which some of the components detract from the overall strength of the part, require powders with higher green strength characteristics.

Testing for Green Strength

Although metal powder compacts are seldom used in the green condition, green compacts must be strong enough to resist abrasion and breakage while being transferred from a compacting press to a sintering furnace. This is particularly important for thin parts, thin sections of large parts, low-density parts, and part edges. Thus green strength is a key factor that determines the compactibility of powders.

The Rattler test, adapted from a method used to test paving bricks, was used in the early P/M industry to determine green strength. Five compacts of predetermined weight and size were placed in a bronze-screened cylinder and rotated for a certain number of revolutions for a specified time. The compacts were then removed from the cylinder, weighed, and the weight loss determined. The Rattler test provided information on the abrasion resistance of a compact and its ability to retain its shape (edge stability). The Rattler test (MPIF Standard 15-517) is still widely used in Japan.

Transverse Bend Test. Current test methods to determine green strength of an unsintered compact are specified in ASTM B 312 and MPIF 15. The standard green strength test is a transverse bend test of a 12.7 by 31.7 mm (0.50 by 1.25 in.) rectangular specimen 6.35 mm (0.25 in.) thick. A simple rectangular die with an upper and lower punch is needed to produce this transverse-rupture specimen. Prior to pressing the compact, the die wall is lubricated with a mixture of zinc stearate in a solvent. The die is then drained, dried, and filled with the test powder. Alternatively, the powder and lubricant may be mixed prior to compaction.

Comparisons between the two methods are not recommended, because green strength, green density, compressibility, and rupture characteristics vary depending on whether the die or the powder particles are lubricated. ASTM B 312 and MPIF 15 provide all necessary information on test procedures, die dimensions, transverse-rupture test fixture dimensions, transverse-rupture bar dimensions, and required formulations. The term "green strength" in these standards defines the stress calculated from the flexure formula required to break the specimen, and it relates to the resistance of a pressed compact to abrasion and breakage.

Variables Affecting Green Strength

Many theories have been proposed to explain the precise mechanics of green strength. Because the strength of green compacts results mainly from mechanical interlocking of particle surface irregularities, particle shape is the most important factor contributing to green strength. Powders with irregularly shaped particles produce compacts of higher green strength than powders with spherical particle shape. Spherical particles provide the lowest degree of mechanical strength because of low initial surface contact between adjacent particles and a low surface-to-volume ratio.

Generally, green strength is increased by increasing the powder surface area. This can be achieved by increasing the particle surface roughness and/or reducing the average particle size, thus providing more sites for mechanical interlocking. These characteristics also result in decreased apparent density. Figure 13 illustrates the dependence of green strength on apparent density for various iron powders.

Green strength also is increased when oxidation and contamination of particle surfaces are reduced. Optimum mechanical interlocking is impeded when particles are covered with heavy oxide films and adsorbed gases.

Green strength also is affected by other variables such as green density and compressibility. Increasing green density or compacting pressure promotes increased particle movement and deformation, which are the bases for mechanical interlocking. Particle porosity also increases green strength (Fig. 14a), although this also reduces green density (Fig. 14b).

Effects of Lubricants. Increasing the amount of additives also decreases green strength. The addition of some alloying elements, such as adding graphite to iron, or the addition of lubricants can reduce the green strength of a compact.

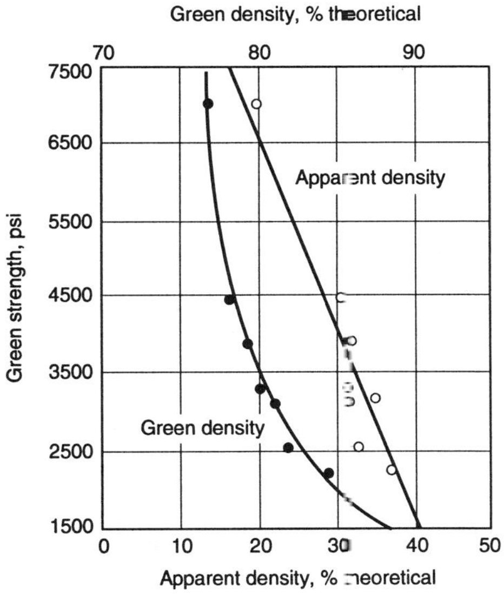

Fig. 13 Effect of apparent density and green strength on green strength of various iron powders. Source: Ref 9

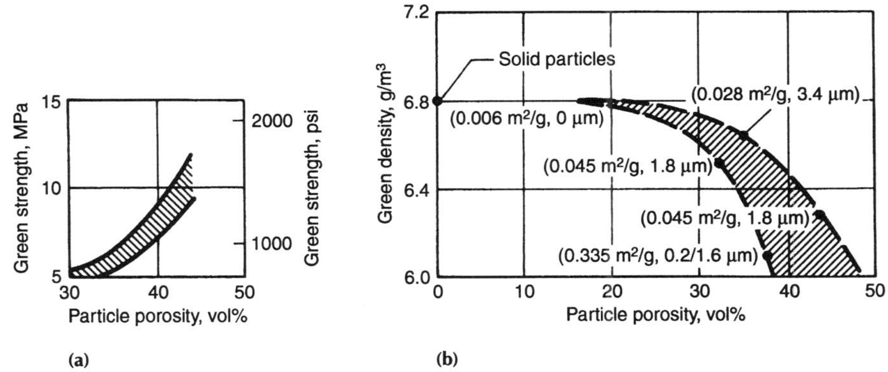

Fig. 14 Effect of internal powder porosity on (a) green strength and (b) green density. Solid and porous iron powders pressed at 414 MPa (30 tsi) using die wall lubrication. Figures in parentheses signify BET-specific surface areas and average intraparticle pore sizes of powders.

Table 3 Effect of lubricant on density and strength of iron and copper compacts pressed at 550 MPa (40 tsi)

Material	Metal stearate lubricant, %	Green density, g/cm³	Green strength MPa	psi
Reduced iron	None	6.47	32	4600
	1	6.57	23	3300
Electrolytic copper	None	7.97	67	9700
	1	8.11	35	5100

Source: Ref 10

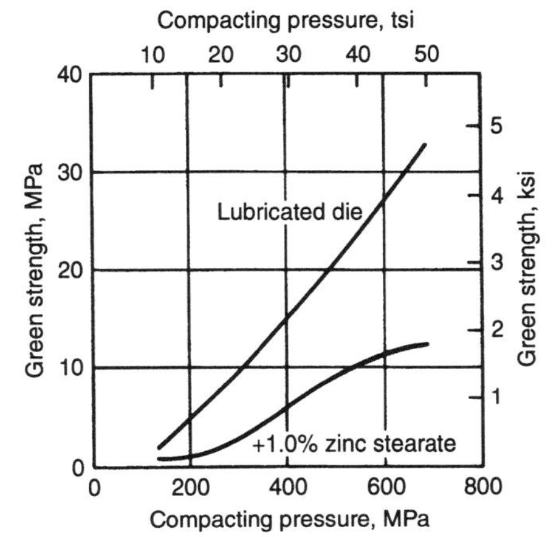

Fig. 15 Effect of admixed lubricant on green strength of water-atomized 4600 low-alloy steel powder

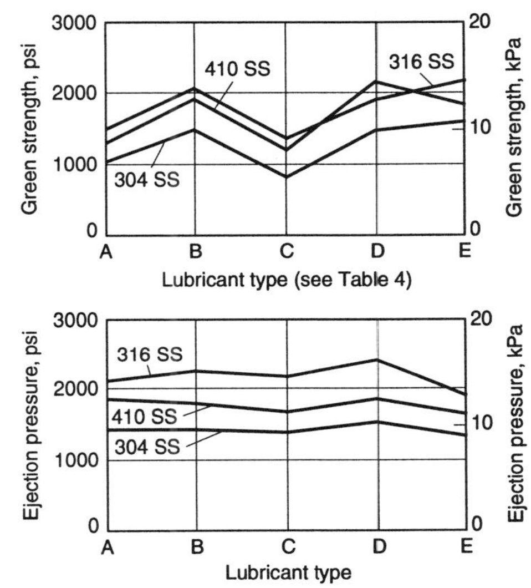

Fig. 16 Green strength and ejection pressures for various stainless steels and lubricants listed in Table 4. Source: Ref 9

Compacts from mixtures of base metal powders and a lubricant generally have considerably lower green strength than those from the powder alone, since the lubricant, which has been added to facilitate ejection from the die wall, does not act as a binder. Instead, the lubricant interferes with particle-to-particle binding. Therefore, green strength tests should be performed on the powder-lubricant mixture as well as on the powder alone.

Table 3 illustrates the effect of a lubricant on green strength and green density of iron and copper compacts. Using a different lubricant, such as an amide wax rather than a metal stearate, can also improve green strength values. Die wall lubrication also improves green strength over that of mixed lubrication (Fig. 15).

The effect of lubricants on green strength is complex and depends on various factors such as production speed, part size, and lubricant type. Figure 16 and Table 4 illustrate the differences for five different kinds of lubricants. These data are meant to show that different lubricants may help if there are difficulties in the production of green parts.

Green Strength Theories

The two mechanisms used to explain the origin of green strength are cold welding and mechanical interlocking of particles. Cold welding refers to the formation of metallic contacts between neighboring particles during pressing; mechanical interlocking refers to their entangling in the manner of a three-dimensional jigsaw puzzle.

Because both of these mechanisms respond similarly to many powder properties, it is difficult to provide conclusive experimental evidence for either theory. For instance, the important influence of a low apparent density of a powder and the accompanying increase of shearing and heat development during compaction is beneficial to the formation of metallic bonds, as well as to improved keying and interlocking of particles. Nevertheless, some of the features and relationships believed to be specific to each of the two mechanisms are summarized below.

Proponents of the cold welding theory argue that the well-known fact that clean metal surfaces form strong attractive junctions when joined (Ref 11) also applies to the pressing of parts from metal powders. In fact, experimental evidence of such interatomic or metallic bonding forces between powder particles has been presented (Ref 12).

Large and similar reductions in both green strength and electrical conductivity of copper compacts pressed from pure and oleic acid-coated copper powders (Fig. 17) were interpreted to indicate that green strength, like electrical conductivity, depends on the number and size of metallic microcontacts formed between particles. Thin oxide films on copper powder particles gave similar results (Ref 13).

High green strength powders pressed and sintered to the same density as low green strength powders of the same chemical composition produce significantly stronger parts. Because the strength of sintered parts is explainable in terms of metallic bonding and microstructural features, such as grain size and pore structure, and because proportionality exists between green and sintered strength, it is reasonable to assume that bonding forces responsible for strength are identical for both green and sintered parts.

Proponents of the interlocking theory of green strength argue that scanning electron micrographs of fractured surfaces of green parts do not show any evidence of metallic contacts and that parts pressed from spherical powders have

Table 4 Comparison of lubricants in Fig. 16

Lubricant type	Particle size	Melting point		Ash content,
		°C	°F	%
Ethylenebisstearmide atomized	0.1% +325 mesh	143	289	0.05
Ethylenebisstearmide ground	1% max 100 mesh	143	289	0.05
Ethylenebisstearmide with zinc stearate and stearic acid	0.1% +325 mesh	137	279	1.0
Ethylenebisstearmide	0.5% +325 mesh	140	284	0.05
Palmitic and stearic	5.0% max 100	66	151	0.05

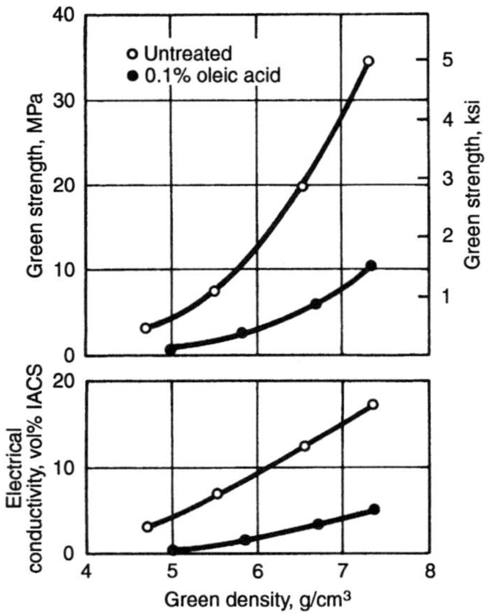

Fig. 17 Effect of thin film of oleic acid on green strength and electrical conductivity of copper parts

no measurable green strength, in spite of substantial deformation and nominal contact areas between particles. Furthermore, roll-compacted green strip has good mechanical strength, yet is quite flexible (Ref 12). Some evidence shows that interlocking may be prevalent at low densities and that cold welding may be dominant at high densities (Ref 13, 14).

The theories on friction welding and interlocking of particles explain the mechanisms responsible for the strength of a green part, but these theories do not permit prediction of green strength from basic powder characteristics. They do, however, provide principles for the development of powders with specific green strength characteristics. The large body of empirical green strength data that is available may be conveniently separated into intrinsic, surface-related, and geometric powder properties (Ref 14).

Intrinsic Powder Properties. Figure 12 shows green strength compacting pressure curves of several metal powders. These curves are typically linear or concave upward (with respect to the pressure axis) and illustrate that green strength increases rapidly with increasing compacting pressure. These curves may be converted into green strength/green density curves by means of the corresponding compressibility (green density/compacting pressure) curves of the powders.

At high compacting pressures or with high-speed compaction, green strength may be lower than expected due to air entrapment during compaction. Such parts may have lamination defects with an appearance similar to those produced by differential expansion during ejection from the die.

In general, green strength increases with the increasing intrinsic softness, or plasticity, of a powder. This relationship is, however, often masked by the influence of geometric powder characteristics. Annealing of a work-hardened powder raises its green strength compacting pressure curve; it may lower its green strength/green density curve, because the annealed powder requires a lower compacting pressure to achieve a given density.

If compaction exceeds the work-hardening capacity of a powder, particle fracture may occur and may cause green strength irregularities.

Surface-Related Factors and Powder Mixtures. The most important commercial examples in this category include the use of mixtures of powders (such as copper and tin or the addition of copper and nickel to iron), the addition of graphite and lubricants, and the presence of oxide films due to tarnishing. The green strength of mixtures of similar metal powders may be estimated from the green strength of its components by applying the rule of mixtures.

The addition of graphite and most lubricants reduces green strength to a much greater extent than indicated by the rule of mixtures, however. The effect of a 1% addition of zinc stearate on the green strength of a water-atomized low-alloy steel powder is shown in Fig. 15. The detrimental effect of the lubricant is more pronounced at high compacting pressures, because it causes the shape of the curve to change from concave to convex.

Because lubricants affect many other powder properties, most notably compressibility and powder flow, selection of amount and type of lubricant addition is made on the basis of the best compromise for a given application. Typically, lubricants that are less detrimental to green strength reduce both powder flow and compressibility.

Most metal powders possess thin oxide films—either directly from manufacturing, as in the case of water-atomized powders, or from subsequent tarnishing in air. The effect of these oxide films on green strength appears to depend on whether these films break down during compaction. For both iron and copper powders, tarnishing or surface oxidation produces significant losses in green strength. Figure 18 illustrates the extent of such losses for copper powder and also shows that oxide films on water-atomized stainless steel powders do not impair green strength. These oxide films were produced by low-temperature oxidation with air in a fluid bed.

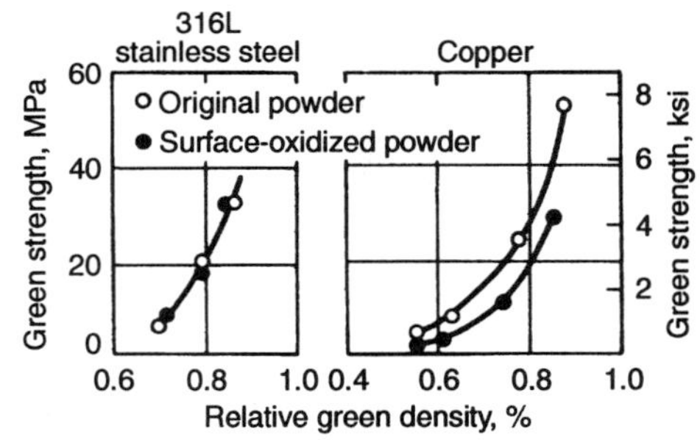

Fig. 18 Effect of surface oxide films on green strength of copper and type 316L stainless steel powders

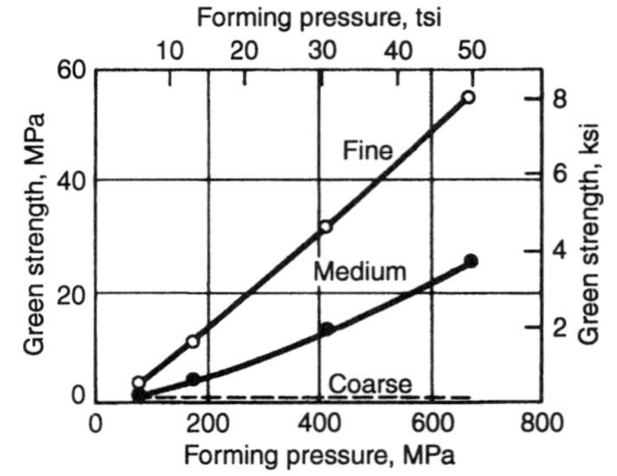

Fig. 19 Effect of particle size on green strength of isostatically pressed electrolytic iron powder. Fine: 100% –325 mesh, 90% 10 to 44 μm. Medium: 22% –325 mesh, 78% –65 + 325 mesh. Coarse: 100% –42 + 100 mesh. Source: Ref 16

Geometric powder properties include particle shape, particle size, and particle porosity. Spherically shaped solid particles (typical of many gas-atomized powders) have such low green strength that they cannot be used in conventional die compaction. Many water-atomized powders have an irregular particle shape with a solid bulk structure. Such powders typically provide intermediate green strength. Oxide-reduced powders are of irregular particle shape and possess internal particle porosity. Such powders give maximum control over green strength. Green strength increases with increasing particle porosity and with decreasing particle pore size.

Other geometric factors that are beneficial to green strength include decreasing particle size and increasing specific surface area of a powder. Figure 19 shows the effect of particle size for isostatically compacted iron powders. The true effect of particle size is sometimes masked by other factors. Figure 20, for example, shows a maximum for the green strength of a water-atomized steel powder of intermediate particle size. In this instance, particle shape becomes more regular with decreasing particle size; oxygen content also varies with particle size. This is typical of many water-atomized powders. In such cases, the apparent density of a powder is frequently a useful measure of particle shape and green strength. All geometric factors that are

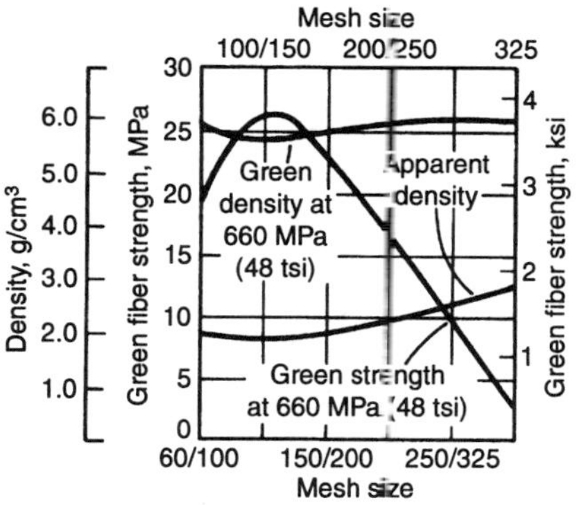

Fig. 20 Green strength, green density, and apparent density of water-atomized steel powder. Source: Ref 17

beneficial to green strength are detrimental to compressibility.

ACKNOWLEDGMENT

The author gratefully acknowledges Fritz Lenel, John Davidson, and Carl Evans. Significant portions of this article were adapted from their articles in *Metals Handbook,* 9th ed., Vol. 7, *Powder Metallurgy,* 1985, p 302–304, 286–287, and 288–289.

REFERENCES

1. F. Chagnon, C. Gélinas, and Y. Trudel, Development of High Density Materials for P/M Applications, *Advances in Powder Metallurgy and Particulate Materials,* Vol 3, Metal Powder Industries Federation, 1994, p 199–206
2. J.S. Hirschhorn, *Introduction to Powder Metallurgy,* American Powder Metallurgy Institute, 1969, p 140
3. R. German, *Powder Metallurgy of Iron and Steel,* John Wiley & Sons, 1998
4. Y. Morioka, Recent Trends in Powder Metallurgy Industry and Technology, *J. Jap. Soc. Powder Powder Metall.,* Vol 8 (No. 8), 1993, p 755–762
5. A. Šalak, *Ferrous Powder Metallurgy,* Cambridge Interscience Publishing, 1995
6. Y. Trudel, M. Gagné, and F. Racicot, Design Criteria of Low-Alloy Steel Powders, *Adv. Powder Metall.,* Vol 5, 1991, p 45–58
7. M. Nitta et al., *Proc. Jap. Powder Powder Metall.,* Vol 54, 1979
8. S. Kaufman, Oxygen in Iron and Steel Powders, *Advances in Powder Metallurgy and Particulate Materials,* Vol 1, Metal Powder Industries Federation, 1994, p 135–149
9. W.V. Knopp, Effect of Type of Lubricant on Green Strength, *Advances in Powder Metallurgy and Particulate Materials,* Vol 2, Metal Powder Industries Federation, 1993, p 27–33
10. H.H. Hausner and M.K. Mal, *Handbook of Powder Metallurgy,* 2nd ed., Chemical Publishing, 1982, p 160
11. F.P. Bowden and D. Tabor, *Friction and Lubrication of Solids,* Clarendon Press, Oxford, 1954
12. K.E. Easterling and A.R. Thölén, The Role of Surface Energy and Powder Geometry in Powder Compaction, *Powder Metall.,* Vol 16 (No. 31), 1973, p 112–118
13. E. Klar and W.M. Shafer, On the Nature of Green Strength, *Int. J. Powder Metall.,* Vol 5 (No. 2), 1969, p 5–10 and Vol 5 (No. 4), 1969, p 5–16
14. J.A. Lund, Origin of Green Strength in Iron Compacts, *Int. J. Powder Metall. Powder Technol.,* Vol 18 (No. 2), 1982, p 117–127
15. E. Klar and W.M. Shafer, Green Strength and Compressibility in Metal Powder Compaction, *Modern Developments in Powder Metallurgy,* Vol 9, Metal Powder Industries Federation, 1976, p 91–113
16. C.E. van Buren and H.H. Hirsch, Hydrostatic Pressing of Powders, *Powder Metallurgy,* W. Leszynski, Ed., Interscience, 1961, p 403–441
17. Technical Bulletin on Green and Sintered Properties of 316L and Other High Alloys, Hoeganaes Corp., Riverton, NJ, 1969

Shaping and Consolidation Technologies

Powder Shaping and Consolidation Technologies

B. Lynn Ferguson, Deformation Control Technology, Inc.
Randall M. German, The Pennsylvania State University

POWDER METALLURGY PARTS are formed by several methods that range from high-pressure die compaction to pressureless shaping methods, such as slip casting. Die compaction at high pressures where densification and shaping occur simultaneously is the dominant approach. However, shaping of P/M products is also accomplished by various low-pressure methods using binders to maintain part shape for further consolidation by sintering. Another alternative is full-density processing, where the application of high pressure at high temperatures leads to P/M components with mechanical properties approaching and sometimes exceeding those of wrought forms. Finally, P/M part technology also includes spray forming of metal powders into high-density parts.

In all of these methods, the basic purpose is to consolidate powders into useful forms. As simple as this sounds, however, shaping and consolidation of powders is a complicated subject. The selection of a shaping/consolidation method involves many decisions, which may include:

- The type of powder to be used (i.e., spherical, sponge, flake)
- The chemistry of the powder (i.e., prealloyed, elemental blend, partially alloyed)
- The use of rigid or flexible tools
- The use of a binder or lubricant
- The use of a directed powder spray to directly consolidate the powder as it is being made

This article briefly introduces and compares powder shaping and consolidation methods, which are described in more detail in subsequent articles of this Section. The emphasis here is to introduce methods for shaping of powder forms and to briefly describe common powder compaction methods, such as those listed in Table 1. The differences between compaction and shaping technologies are determined by many production and application factors, such as part size, desired final density, and shape complexity.

Powder Shaping Technologies

General categories of powder shaping methods are as follows (Ref 1):

- Binder-assisted extrusion: long structures, small powders, constant cross section, relatively simple shapes
- Injection molding: complex, small components, high-performance materials
- Slip casting: very large structures, constant wall thickness, low precision
- Tape casting: flat sheets, small powders, very simple shapes

In general these methods employ binders that hold the particles together in the desired shape; the particles are then consolidated to higher densities by sintering. The part is shaped at relatively low pressures (compared to die compaction) with green (unsintered) porosity ranging from ~40 to 60% (ignoring the binder that is sacrificial) (Ref 1). Most binders are polymers such as mineral oil or polyethylene. The following binders are used in powder shaping:

Extrusion

- 56% water, 25% methyl cellulose, 13% glycerine, 6% boric acid
- 72% water, 12% hydroxypropyl methyl cellulose, 8% glycerin, 4% ammonium polyacrylate, 4% ammonium stearate
- 65% polyethylene glycol, 30% polyvinyl butyryl, 5% stearic acid

Injection molding

- 69% paraffin wax, 20% polypropylene, 10% carnauba wax, 1% stearic acid
- 75% peanut oil, 25% polyethylene
- 50% carnauba wax, 50% polyethylene
- 55% paraffin wax, 35% polyethylene, 10% stearic acid

Slip casting

- 96% water, 4% sodium lignosulfonate, trace calcium nitrate
- 93% water, 4% agar, 3% glycerine
- 99% water, 1% ammonium alginate
- 97% water, 3% polyvinyl alcohol

Tape casting

- 77% water, 9% polyacrylate emulsion, 9% glycerin, 3% ammonium polyacrylate, 2% ammonium hydroxide
- 80% toluene, 13% polyethylene glycol, 7% polyvinyl butyral
- 47% mineral spirits, 24% isopropanol, 8% polyvinyl butyral, 8% dibutyl phthalate, 5%

Table 1 Compaction options

Process	Cold isostatic compaction	Cold forging	Die compaction	Explosive compaction	Roll compaction	Warm compaction
Pressure	Moderate, 400 MPa	Very high, >800 MPa	High, 700 MPa	Very high, >1 GPa	Low	High, 700 MPa
Temperature	Ambient	Ambient	Ambient	Very high	Ambient	Warm
Tooling	Soft	Hard	Hard	Soft	Hard	Hard, heated
Deformation rate	Low	High	High	Very high	Low	High
Continuous	No	No	No	No	Yes	No
Pressing direction	3	1	1	1	1	1
Shape complexity	Moderate to high	Moderate	High	Low	Low	Moderate
Polymer level	None	None	Low, 0.5%	None	Low, 0.1%	Low, 0.6%
Precision	Low	Moderate	High	Low	High	High
Use	Moderate	Low	Extensive	Very low	Moderate	Low

Source: Ref 1

polyethylene glycol, 5% stearic acid, 3% menhaden fish oil

In most cases, the shaping process requires four steps:

- Formation of a powder-binder feedstock
- Shaping of feedstock using customized tooling
- Extraction of the binder
- Sintering densification

Feedstock formulation involves small powders and selected binders that are mixed in a ratio dictated by the desired rheology. Shaping may be conducted by several techniques, but powder lubrication by the polymer is key to fabrication of the desired shape. Low-pressure molding concepts, such as slip casting, require an excess of binder. On the other hand, high-pressure shaping is possible with high-viscosity mixtures that contain more powder. If necessary, the polymer viscosity can be controlled by either heat or solvent.

After shaping, the binder is removed (debinding), and the remaining powder structure is sintered. These last two steps can be combined into a single thermal cycle. Since significant shrinkage is associated with sintering densification, the final dimensions rely on uniform shaping to hold tight final tolerances.

Slip Casting. In this method, a low-viscosity slurry (typically from 10 to 50 Pa · s) is prepared with a liquid carrier and powder. The slurry mixture is then poured into a porous mold usually fabricated from plaster of paris. The carrier is evaporated and absorbed into the porous mold, and a free standing shape is thus achieved. In the case of cements, a chemical bond can be created during the drying or curing process. For many powders, sintering is required to achieve particle bonding and to produce a part having useful mechanical properties. This process is more applicable to ceramic powders, and the most popular application is the production of ceramic bathroom fixtures. Slip casting is not commonly used for metal powders. For steels, the largest application is rapid fabrication of prototype shapes and pilot tool steels (Ref 1). Slip casting is very cost effective for large shapes due to inexpensive tooling. More information on this method can be found in the article "Slip Casting of Metals" in this Volume.

Tape casting uses a similar low-viscosity mixture of powder, water, and polymer, but the forming step is via deposit of the mixture on a moving Mylar film. As the water evaporates, the particles are glued together into a thin sheet. It is used to generate homogeneous long, thin, and flat structures, such as sheets. Many of the characteristics of the binder and powder are the same as encountered in slip casting. After forming, the solvent or water is evaporated, leaving residual binder to hold the particles in place.

Tape casting is widely used to form battery electrodes, brazing layers, microelectronic substrates, and small production levels of steel sheets. In tape casting, the powder-binder slurry is fed onto a moving paper or plastic sheet that passes through a controlled opening. A doctor blade levels the slurry to form a continuous thin sheet (Fig. 1). Subsequently, one constituent of the binder is evaporated, leaving bonds of binder behind to provide strength during separation of the tape from the substrate before final sintering. Binders for tape casting are similar to those discussed earlier. Acrylics, waxes, polyvinyl alcohol, and polyvinyl butyryl are common components.

Freeze Casting. A novel shaping technique relies on water and freezing to form a shape. The water contains polyvinyl alcohol or latex. After casting into hard tooling, the powder-binder mixture is frozen in the tool set. Subsequently, the ice is extracted by vacuum evaporation (freeze-drying or sublimation) from the compact, leaving behind a polymer component to hold the particles in place for subsequent handling. Sublimation of the water is slow and requires at least a few hours. Finally, the powder is sintered to a high density. The process is known by several names, including adiabatic forming, freeze-firing, and quick set. This freeze-sublime-sinter approach is mostly used for large structures; stainless steel is a preferred material. Current production targets components >l kg (2.2 lb) in mass with modest production volumes—typically a few per day. This is dictated by the time to freeze the component in the tooling, which limits the cycle time to ~1 h.

Injection Molding of Metals and Ceramics. Injection molding is now the most widely employed shaping technique for steel powders. The molding process competes with many other shaping technologies. It overcomes the property limitations inherent in plastics, the shape limitations of traditional powder compaction, the costs of machining, the productivity limits of isostatic pressing and slip casting, and the defect and tolerance limitations of casting.

Part production by injection molding is used with various metals, alloys, and ceramic materials, such as the following (Ref 2):

Elements

- Beryllium
- Copper
- Iron
- Gold
- Molybdenum
- Nickel
- Niobium
- Silver
- Titanium
- Tungsten

Alloys

- Bronze
- Cobalt-base
- Copper-base
- Gold-base
- Hastelloy
- Inconel
- Invar
- Iron-nickel
- Iron-silicon
- Kovar
- Nickel-base
- Niobium-base
- Stainless steel
- Steel
- Stellite
- Sterling silver
- Superalloys
- Tool steel
- Tungsten heavy alloy

Ceramics and compounds

- Alumina (Al_2O_3)
- Alumina-chromia (Al_2O_3-Cr_2O_3)
- Aluminum nitride (AlN)
- Ferrites ($MnFe_2O_4$)
- Hydroxyapatite ($Ca_{10}(PO_4)_6(OH)_2$)
- Mullite ($3Al_2O_3$-$2SiO_2$)
- Nickel aluminide (Ni_3Al)
- Silica (SiO_2)
- Silicon carbide (SiC)
- Silicon nitride (Si_3N_4)
- Spinel ($MgAl_2O_4$)
- Titania (TiO_2)
- Yttria (Y_2O_3)
- Zirconia (Zr_2O_3)

Cermets and composites

- Al_2O_3-SiC
- Al_2O_3-ZrO_2
- Mo-Cu
- NbC-Ni
- Ni_3Al-Al_2O_3
- Si_3N_4-SiC

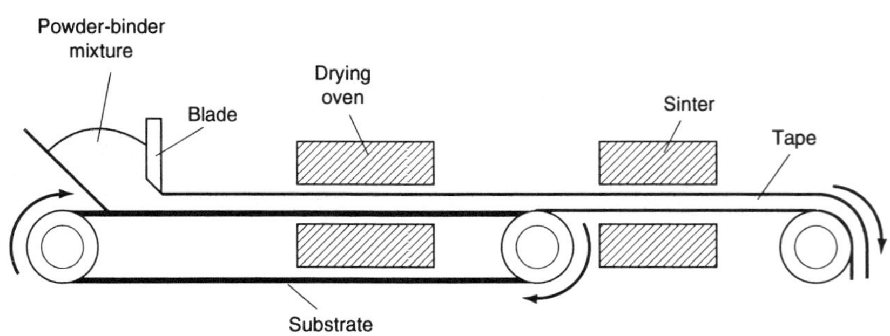

Fig. 1 In tape casting, a thin layer of powder-binder mixture is formed on a moving Mylar substrate, which carries the tape into a drying oven prior to sintering. The technique is used to form thin sheets. Source: Ref 1

- SiO_2-Si
- TiC-Ni-Mo
- W-Cu
- WC-Ni
- WC-TaC-Co
- ZrO_2-MgO
- ZrO_2-Y_2O_3

Noticeably absent are magnesium and aluminum; these reactive metals develop powders with surface oxides that are difficult to remove during sintering. Stainless steel is the most widely used material for injection molded parts, as shown by the relative ranking in Fig. 2. Stainless steel 316L is the most commonly used alloy. Ferritic and duplex stainless steels are also used.

Injection molding moved quickly from a laboratory development to a commercial process for two reasons. First, a precision shape could be achieved at a significant cost reduction compared to a conventional manufacturing process. Second, the use of fine powders promoted densification during sintering, and high properties could be achieved. The high potential of the process added a third reason that sustained growth. Availability of fine powders and pellet feedstock from suppliers increased as the process showed early successes. There is still a great deal of development work toward more efficient and economical production of fine powders (<20 μm) specifically to supply the injection molding industry.

In the as-molded state, the green part is typically ~65 vol% powder and ~35 vol% binder. Therefore, upon removal of the binder, considerable densification is required to achieve density levels of ~95 to 98% of full density. This densification must be accompanied strictly by size change because unwanted shape distortion would render the part useless. Therefore, uniform particle distribution in the binder is essential. Also, to minimize contamination, the binder must either be removed completely before sintering, or it must participate in alloying during sintering.

A binder/plasticizer has also been used to extrude long shapes rather than just mold individual shapes. A heated barrel plasticizes the binder, and a die gives shape to the powder being extruded. With this technique it is possible to form fluted shapes, hollow tubes, tubes with multiple cavities, and other complex shapes. More information is in the article "Powder Injection Molding" in this Volume.

Rapid prototyping technology includes quick methods for making templates, models of parts, prototype molds and dies, and fully functional components. Rapid prototyping methods rely on variants of powder metallurgy capabilities such as selective sintering, powder spraying, ink jet techniques for directed spraying, and deposition. Several methods are discussed in the article "Powder Metallurgy Methods for Rapid Prototyping" in this Volume.

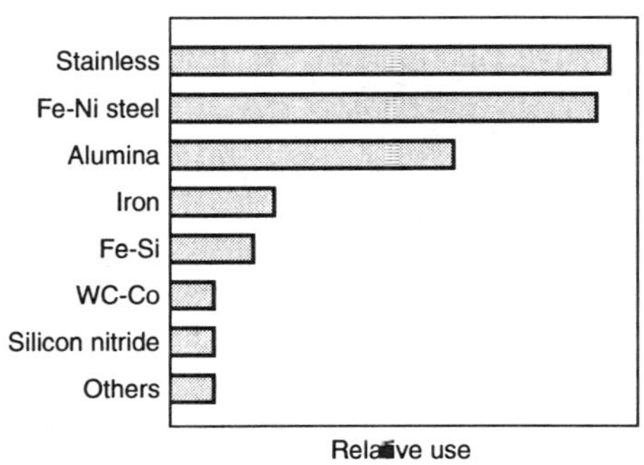

Fig. 2 Relative material utilization in powder injection molding (PIM) processing on a weight basis, showing the dominant position of steels and stainless steels. Source: Ref 2

Powder Compaction Methods

Powders are compacted under high pressure by various methods (Table 1) where the application of pressure packs the powders and reduces porosity. Unlike shaping methods, compaction techniques cause particle deformation. Many compaction methods are used, but the most prevalent method for P/M parts production is uniaxial compaction in a rigid die. This method is cost effective with relatively straightforward tooling.

However, compaction methods inherently produce parts with density variations that result in dimensional variations of the final sintered part. This occurs because the transmittal of stress through a mass of powder particles is not uniform, even in isostatic pressing. In the case of compaction in a rigid die, the uniformity of stress transmitted through the powder mass and the density distribution in the green compacts produced in rigid dies are much less uniform than in isostatic compacting. This can be understood by considering flow around corners. When a liquid is subjected to hydrostatic pressure inside a confined, rigid die, the stress transmitted by the liquid upon the interior surfaces of the die is uniform, regardless of whether the liquid must flow around corners. This is not so when a powder is pressed; it flows only in the direction of the applied pressure and not around corners. This phenomenon is illustrated in Fig. 3, which schematically shows the compaction of powder in a die with a sidearm. If pressure is applied only to the top punch, the powder is compacted only in the vertical section of the die, but remains loose in the horizontal section. Compaction in the sidearm is obtained when pressure is applied to both the top punch and the side punch.

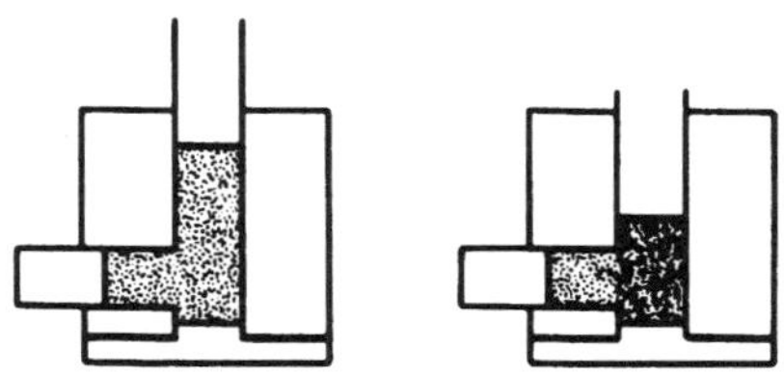

Fig. 3 Density distribution of compacts pressed in dies with a sidearm

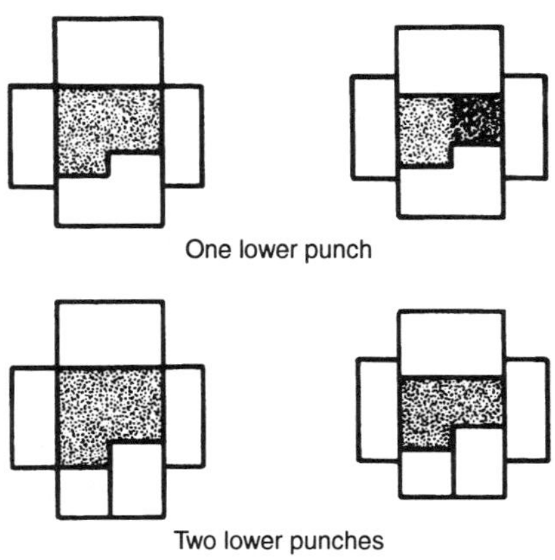

Fig. 4 Density distribution in a two-level compact

Uniaxial Die Compaction

In uniaxial compaction of metal powder, pressure is applied to the powder only with punches that move in the vertical direction. Compacting metal powder parts in rigid dies, where the punches move only in the vertical direction, imposes limitations on the shape of the parts that can be readily produced. Parts with re-entrant angles and with holes at an angle to the vertical direction generally are not produced by compacting.

Another consequence of powder flowing only in the direction of the applied pressure during compaction is that, when parts with different levels of thickness in the direction of pressing are compacted with only a single lower punch, they develop different green densities in the different levels. Individual punches for each level are necessary for a more uniform density. These punches must travel in such a way that the ratio of the height of the loose powder to that in the green compact (the compression ratio) is the same for all levels. This type of punch arrangement is illustrated in Fig. 4. Pressing from both top and bottom also reduces density variations (Fig. 5).

Even compacts with only one level in the direction of pressing show variation in density in the pressing direction due to the friction between the die wall and the powder. This friction is reduced as much as possible by lubricants. Usually, lubricants are added to the metal powder in powder form. Lubricants reduce friction between the powder and the side wall, as well as between individual powder particles. They also may be applied as a thin coating to the walls of rigid dies. In large-scale production, this die wall lubrication is not commonly used.

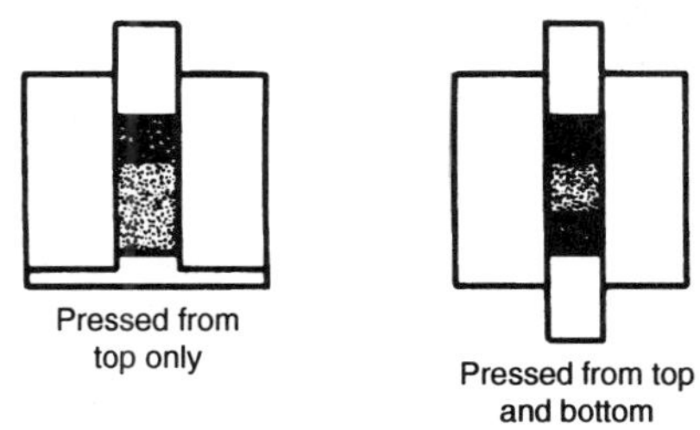

Fig. 5 Density distribution in compacts

Mathematical relationships and modeling of pressure-density relationships in powder compaction have important practical value, especially with greater emphasis on concurrent engineering. Modeling of compaction is discussed in the article "Mechanical Behavior of Metal Powders and Powder Compaction Modeling" in this Volume.

Cold pressing in rigid dies is the most commonly used compaction process, and the concept of die pressing is straightforward. Powder is poured into a die cavity, a movable punch seals the die cavity, and a load is then applied via the advancing punch. In the most simple case there is only one moving punch, and the die is stationary. However, a density gradient in the compact occurs as a consequence of die wall friction, with the highest density being next to the punch face. A floating die table reduces the density gradient by moving the die to offset the friction effect. The powder is densified from both top and bottom planes, and the middle plane has the lowest density. As more features are added to the compact, additional punches are required to produce an acceptable green compact.

Typical green densities for die compacted parts are 75 to 85% of full density. Green density is related almost exponentially to the applied load. At low densities, a small increase in load causes a major increase in density, while at high density levels, a large increase in applied load is required to get a small increase in density. The required compaction pressure to achieve a desired level of density is a function of the following:

- The powder shape (i.e., sponge, flake, water atomized, reduced)
- Particle size and size distribution
- Powder chemistry (i.e., prealloyed, blended master alloy)
- Lubrication practice

For steel powders, commonly used pressures are in the 400 MPa (30 tsi) range to achieve green densities from 80 to 85% of full density. The use of "low compressibility" powders, which truly do require less pressure to achieve the same density as prealloyed powders, allows either larger parts to be made on the same pressing equipment, or parts with higher green density to be made at the same compaction pressure. More information on compressibility is provided in the article "Compressibility and Compactibility" in this Volume.

Cold pressing in rigid dies has advantages of dimensional control due to the well defined cavity, high compaction pressures due to mechanical or hydraulic pressing equipment, process repeatability due to mechanization and improved powder consistency on a batch-to-batch basis, and high rates of production. Limitations of the process include size restrictions due to press capacity, height-to-diameter limitations due to die wall friction, ejection cracking problems with compacts pressed from powders with poor green strength, and the natural limitations of re-entrant angles and undercuts. Creative toolmakers circumvent or overcome these limitations daily.

Warm compaction is a rather recent development in production die pressing. In warm compaction, a plasticizer is added to the powder whereby the application of moderate temperatures causes the plasticizer to soften or melt and friction is substantially reduced. Friction here refers to friction between the powder and the rigid die wall and to friction between powder particles. With this process, significantly higher green densities can be achieved, even >92% of full density. This technology opens new applications to pressed and sintered parts because of the achievable density and the improved mechanical and physical properties. In addition, higher green strengths allow machining of green compacts.

Usually warm compaction involves the use of a polymer addition that helps bond particles together. The polymer-coated powder is more costly than typical die-compaction grades, unless a simple lubricant is admixed with the powder. Various common stearates or other lubricants work, including, Teflon (E.I., duPont de Nemours & Co., Inc., Wilmington, DE). Depending on which polymer is selected for coating the powder, ejection forces can be highly variable. Close temperature control is necessary, since product uniformity suffers if the polymer is too hot. In tests with various powders, the green density usually increases by 0.15 g/cm^3 over room-temperature compaction. After cooling to room temperature, the warm-compacted powder is stronger because chilling the polymer adds strength to the compact. However, there is no evidence of greater strength during ejection, which means that green cracks from ejection stresses are not reduced by warm compaction. Consequently, a hold-down pressure is required during ejection to avoid cracking.

Heating of the die and punches requires modifications to the compaction press, and a heater is required in the powder feed mechanism. Both microwave and hot-oil heaters are available for heating the powder. A typical temperature for the powder and tooling is ~150 °C (300 °F), and compaction pressures are usually in the range of 700 MPa (50 tsi) for steels. The major role of warm compaction is to lower the pressure required for attaining densities of >7.0 g/cm^3 of ferrous P/M compacts.

Hot Pressing. The production of large billets can be accomplished by compacting powder in heated dies. The use of elevated temperatures and long dwell times allows densities of >95% of full density to be achieved at compaction pressures that are one third to one half those needed for cold pressing to lower density levels. Full density is usually not achieved, and 3 to 5% porosity remains in the billet. This porosity somewhat limits the use of hot pressed billets because the properties are reduced from those of fully dense material. For this reason hot pressed billets are often used as stock for upset forging, closed die forging, and other deformation processes that can eliminate this residual porosity. Some metals, such as beryllium, are routinely processed by hot pressing, and acceptable performance levels for many applications are achieved.

Isostatic Pressing

Isostatic pressing allows more uniform density compared to uniaxial compaction in rigid dies. These methods rely on flexible molds for application of pressure in all directions, which reduces friction and allows compaction of compact shapes.

Cold Isostatic Pressing. CIP uses a flexible membrane to isolate the powder from a liquid medium that is pressurized to cause densification of the powder. Typical mold materials are latex, neoprene, urethane, polyvinyl chloride, and other elastomeric compounds. Because the mold moves with the powder as it densifies, friction effects are minimized. Also, because the pressure is applied uniformly around the mold, there is no theoretical size limit. Height-to-diameter and overall size are limited by the pressure vessel size. Often a rigid mandrel is part of the tooling; and, because powder must slide along this mandrel, it is coated with a friction reducing material.

In comparison to die pressing, cold isostatic pressing can achieve more uniform densities due to minimized friction effects. Pressure vessels are typically limited to pressures of 415 MPa (60 ksi) although units with twice this capacity have been produced. Isostatic pressing equipment can be automated (i.e., dry bag CIP units), but the production rates are lower than those of die pressing. Dimensional control is generally not as tight as with die pressing due to flexible tooling. As stated, however, rigid members can be incorporated into the flexible mold assembly to produce accurate surfaces where desired.

Hot isostatic pressing (HIP) is a versatile near-net shape process that has found niches in the production of aerospace structure and engine markets, high alloy and tool steel mill shapes and individual components, titanium hardware, and monolithic and composite alloy components for the energy industry. The process fundamentals and manufacturing steps are covered in detail in the article "Hot Isostatic Pressing of Metal Powders" in this Volume.

The aim of hot isostatic pressing is a near-net shape and full density. Powder is hermetically sealed in a container that is flexible at elevated temperatures; the "canned" powder is heated within a pressurized vessel and held for a specified time. Commercially used containers include low carbon steel sheet formed into a container, stainless steel sheet, and even glass. The pressurized medium is usually an inert gas such as argon, and pressures range between 100 and 300 MPa (15 and 45 ksi). The temperature for HIP is material dependent, of course, but typical production equipment can heat parts ~1000 to 1200 °C (2000 to 2200 °F). HIP units for ceramics and carbon-base materials may heat up to 1500 °C (2700 °F). Densification mechanisms active during HIP include bulk deformation (limited amount), sintering, and creep, with the latter accounting for a significant portion of densification. Densities >98% of full density are typical, and full density is routinely achievable with care during powder sealing and strict control of time, pressure, and temperature.

The powders used in hot isostatic pressing are usually spherical in shape and very clean. The particle surfaces are free of contaminants, such as oxide films. The sphericity facilitates can loading and handling, and the particle surface cleanliness facilitates particle bonding. Powder handling and avoidance of contamination is critical to the success of the process and considerable investment in facilities and equipment, followed up by attention to operating procedures and "good housekeeping," is required.

In comparison to hot pressing where only billet shapes are produced, hot isostatic pressing is capable of producing complex shapes. As in cold isostatic pressing, the achievable dimensional tolerances are at best near-net due to the flexible mold. Some net surfaces may be achieved if rigid members are incorporated into the mold. Figure 6 compares HIP capabilities with other compaction methods.

Bulk Deformation Processes

Powder Forging. In powder forging, a preform shape is cold pressed to between 75 and 85% of full density, sintered, heated to a forging temperature, and then forged in trapped dies using one blow to produce a fully dense net or near-net shape. The sintering step is optional, but it is normally included as a particle surface cleaning step (deoxidation) and to improve the workability of the porous preform. Powder forging fundamentals and general applications are presented in the article "Powder Forging and Hot Pressing" in this Volume. More detailed information is also contained in the article "Powder Forged Steel."

Normally, powder forging is performed hot (1000 to 1200 °C, or 1800 to 2200 °F, for steel powders), but it can also be conducted at warm or cold forging temperatures. Typical steel powder forging pressures are 550 to 950 MPa (40 to 70 tsi). Because the workability of a porous preform is poor, the design of the preform is critical to the process in terms of avoiding defects. Local surface tensile stresses and internal hydrostatic tension must be avoided. Nonetheless, with a correct preform shape and a well controlled process, powder forged parts have sound microstructures, good hardenability, and performance that meets or exceeds cast-wrought part performance.

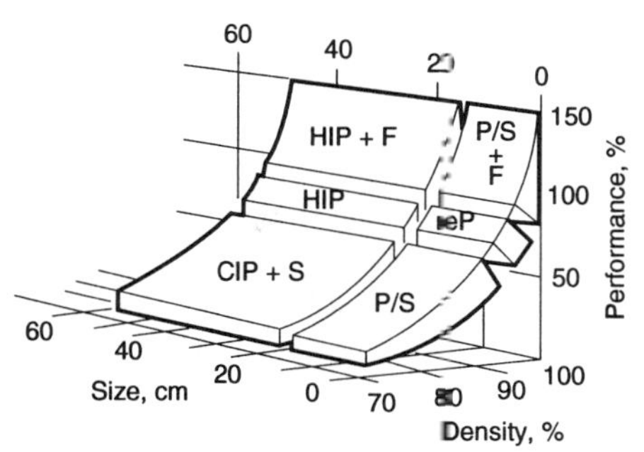

Fig. 6 Application areas of HIP based on part size, complexity, and level of densification three variables that dictate the P/M approach are size, density, and performance (as a percentage of wrought). This behavior corresponds to ferrous-base P/M systems, but is representative of many P/M materials. P/S, press and sinter; reP, press, sinter and repress; P/S + F, press and sinter and forge; CIP + S, cold isostatically press and sinter; HIP, hot isostatic press; HIP + F, hot isostatic press plus forge.

There are two classes of forging practice. In repressing, the preform shape is nearly identical to the forged shape. In true forging, considerable shear deformation is involved since the preform is different in shape from the forging. The dynamic properties of toughness and fatigue resistance are higher for forged parts than for repressed parts.

Powder Extrusion. Loose powder can be containerized and extruded to full density, either with or without heat. Extrusion ratios of at least 9:1 have been shown to produce full density, and many materials are extruded to full density using much higher extrusion ratios. Extrusion at such high reductions subjects the powder particles to high levels of shear deformation and compression as they pass through the die. The result is sound particle bonding. As in hot isostatic pressing, the powder is usually hermetically sealed in a container prior to extrusion. In some alloy systems, residual air provides oxygen for selection oxidation of alloy ingredients prior to extrusion, and an oxide dispersion strengthened material is produced. In some other cases, particulates or chopped fibers may be added to the matrix powder, and extrusion produces a composite material. In still other cases, combinations of powders or powder plus wrought pieces are extruded to produce multimetal parts, e.g., bimetallic tubing. These and other uses of extrusion are discussed in the article "Extrusion of Metal Powders" in this Volume.

Powder rolling or roll compaction of powders consolidates loose powder(s) into a porous strip as it passes through a roll gap. It is possible to produce a monolithic strip or a multilayer strip. Further processing may include sintering of the strip and additional rolling to densify the strip. Applications of this type of processing include such diverse products as clad metal for coin stock, automotive sleeve bearings, and electrode stock.

Sintering

For sintering to be effective, the powder particles must be in intimate contact. For this reason, sintering is typically performed on compacted or molded powder and not on loose powder. While most powder parts are sintered during their manufacture, the densification stage of sintering may not be utilized, and in these cases the primary use of sintering is to achieve metallurgical bonding of particles. Die pressed parts typically fall into this category because dimensional control is of primary importance.

Process routings that rely on sintering specifically to achieve high densities usually fall into three categories. A very fine particle size is used so that bulk diffusion paths are minimized at the last stage of sintering, and there is a high initial surface area to drive sintering in its early stages. This is one reason for the use of ultrafine powders in injection molding where sintering is the primary mechanism for development of high property levels. A liquid phase, either permanent or transient, is present during sintering so that particle rearrangement is promoted. A high temperature is used so that local melting may accompany sintering.

Liquid Phase Sintering. The use of blended powders allows the combination of a low melting powder with a higher melting powder so that a liquid is present during sintering. This liquid can aid powder rearrangement and densification of a green compact. Cermets are typically produced by this technique. Nearly pore-free microstructures may be achieved by this technique.

Transient Liquid Phase Sintering. A subclass of liquid phase sintering is transient liquid phase sintering. Here, a blend of powders contains a lower melting point phase, as above. Only in this case, the liquid alloys with the solid phase which remains solid, and the liquid is present only for a brief period. This process is not used as the primary densification method, but it is used to achieve another increment in density. For example, it may be used in high alloy steel or tool steel production to increase the density of a green compact from 80% to >95% of full density.

High-Temperature Sintering. If the temperature of the green compact is raised to levels very close to the solidus temperature of the equilibrium diagram, small pools of liquid can form due to microsegregation of alloying elements. This small amount of liquid accelerates densification during sintering. Although this represents a somewhat extreme use of high temperature sintering, this technique is useful for a wide range of alloys, and it is used especially for tool steels and other high alloy steel components. The article "High-Temperature Sintering of Ferrous Powder Metallurgy Components" in this Volume contains more information on high-temperature sintering.

Reactive Sintering. By taking advantage of the flexibility of powder metallurgy, a blend of powders may be compacted and then sintered so that during sintering a self propagating reaction occurs (see "Combustion Synthesis of Advanced Materials " in this Volume). This reaction, if controlled properly, can densify the compact rapidly while forming the new alloy. If not controlled, the component may revert to powder!

Compaction to Higher Density

Many methods have been investigated and developed for pressing metal powders to higher densities. Most of the commercial methods involve high-temperature compaction (such as hot isostatic pressing and powder forging, as described in detail in the section "Hot Consolidation and Higher-Density Applications" in this Volume). In general, these methods are used to develop fully dense or nearly fully dense P/M products. For example, conditions for full-density iron and steel by HIP are summarized in Table 2.

However, other methods besides high-temperature compaction are used to achieve higher densities in green or consolidated form. Some of these miscellaneous methods are described here. These methods are not necessarily full-density methods, and none have reached any significant commercial significance. They are described for general reference. Many experiments also have been reported that attempt to press compacts to higher densities or to produce a more uniform density and stress distribution in pressed compacts. Although none of the techniques developed has led to large-scale industrial use, they are discussed here.

Die Barrel Rotation. Rotating the die barrel while the powder is being pressed using a fine atomized aluminum powder was reported by Hammond and Schwartz (Ref 3). They lubricated the die barrel with a suspension of lithium stearate in didecyl alcohol. Annular compacts, 12.7 mm (½ in.) high, were pressed with outer diameters of 38 mm (1½ in.) and inner diameters of 25 mm (1 in.). The core rod was stationary, but the die barrel could be rotated during compression.

Compaction with a stationary and a rotating die barrel was compared. While in static compression, 20% of the applied stress was consumed in die wall friction, while only 2% of the stress was consumed when the die barrel was rotated. In addition, the pressure necessary to eject the compact from the die was reduced to approximately one half of the pressure for compacts pressed with a stationary die barrel. Similar experiments on the effect of die barrel rotation on the density of iron powder compacts were reported by Rutkowski et al. (Ref 4).

Triaxial compression by simultaneous isostatic and uniaxial compression is obtained by applying pressure to the circumference of a cylindrical specimen confined in a flexible envelope while an axial load is superimposed by a vertical piston. With this method, the level of pressure necessary to obtain a given density is less than with isostatic or uniaxial compression alone. For example, to compact an atomized iron powder (Ancor 1000) to a relative density of 85%, uniaxial compression at 540 MPa (78 ksi) or isostatic compression at 415 MPa (60 ksi) is necessary. The same density can be obtained by combining a confining (isostatic) pressure of 83 MPa (12 ksi) with a uniaxial pressure of 470 MPa (68 ksi). The principles involved in this method of compaction have been reviewed by Broese van Groenou (Ref 5).

High-Energy-Rate Compacting. The rate at which pressure is applied in compacting in a hydraulic press is slow. Compacting in certain mechanical presses is somewhat faster. The effects of the rate of pressure application in compacting were studied by Davies and Elwakil (Ref 6). They found that somewhat higher densities can be obtained for a given pressure and for a given energy input when compacts from iron powder are pressed in high-speed presses (petroforge presses). They also determined the effects of multiple blows during pressing.

In the fabrication of sheet metal products, techniques were developed that formed sheet metal at rates considerably higher than those obtained in fast-acting mechanical presses. This is the high-energy-rate forming technique, which generally uses explosives (see also Ref 1).

The success of high-energy-rate forming in fabricating sheet metal led to extensive experimental work on high-energy compacting of metal powders. The most common means to achieve high velocity is by the use of explosives. In one experiment, for example, compaction was done in a rigid die with pressure applied by a projectile propelled by an explosive charge that moves through a barrel (Ref 7). In a similar experiment compressed gas actuated the projectile (Ref 8). This experiment showed that the density of copper powder is not so much a function of projectile velocity, but depends primarily on the kinetic energy of the projectile, which can be varied by proper selection of gas pressure and projectile mass (Ref 8). For copper compacts with a volume of 0.86 cm^3 (0.052 in.^{3}), relative densities of ≥95% were obtained with energies of 150 J (110 ft · lb).

The most widely used method of explosive compacting is shown schematically in Fig. 7. The powder to be compacted is placed in a steel tube, which is closed at each end by steel plugs. The steel tube is surrounded by an explosive that is set off by a detonator located so that on detonation the tube collapses uniformly inward. Experiments by Lennon et al. (Ref 9) showed that density is a function of energy. They developed the equation:

$$D_c = D_T - \Delta D^{-(\beta E^{\gamma})}$$

where D_c is the compacted density, D_T is the full density of the material, ΔD is the difference between full and compacted density, β and γ are constants, and E is the net energy absorbed by the powder. The highest relative densities obtained for iron, nickel, copper, and aluminum powders and the corresponding net energies absorbed per unit volume of compact in their experiments were:

Powder	Density(a), %	Net energy, J/cm^3
Iron	98.1	261
Nickel	98.1	556
Copper	98.5	285
Aluminum	99.0	182

(a) Percent of theoretical density

This method of explosive compacting is not necessarily confined to cylindrical compacts. Cones and hollow cylinders have also been explosively compacted by this technique (Ref 10).

Vibratory Compacting of Powders. Vibration can be very effective in obtaining higher packed densities in powders. The relative densities of powders vibrated under carefully controlled conditions are much higher than those obtained by simply pouring the powder into the container. Therefore, much lower compaction pressures are required to reach a given density for a vibrated powder than for a poured powder. This is illustrated in Fig. 8 for a carbonyl iron powder. The density of 5.53 g/cm^3 (71% relative density), reached by compacting under a pressure of 245 MPa (35 ksi), is due to the plastic deformation of the iron powder particles, while the 5.37 g/cm^3 (69% relative density) obtained by vibrating at 167 oscillations per second is due mainly to vibratory packing. Plastic deformation during the simultaneous compacting at 2.4 MPa (0.36 ksi) is minimal. The method of consolidating powders by vibrating and simultaneous compacting is, therefore, primarily applicable to hard

Table 2 Hot isostatic pressing conditions for full-density iron and steel

Material	Pressure, MPa	Temperature, °C	Time, h
4 μm carbonyl iron	200	805	1
75 μm sponge iron	98	1000	1
70 μm low-alloy steel	150	800	1
190 μm maraging steel	210	1200	3
100 μm austenitic stainless steel	160	1150	3
120 μm martensitic stainless steel	150	1150	3
65 μm Fe-10Al-5Si	200	1000	1
80 μm tool steel	100	1100	1
165 μm iron superalloy	69	1200	3

Source: Ref 1

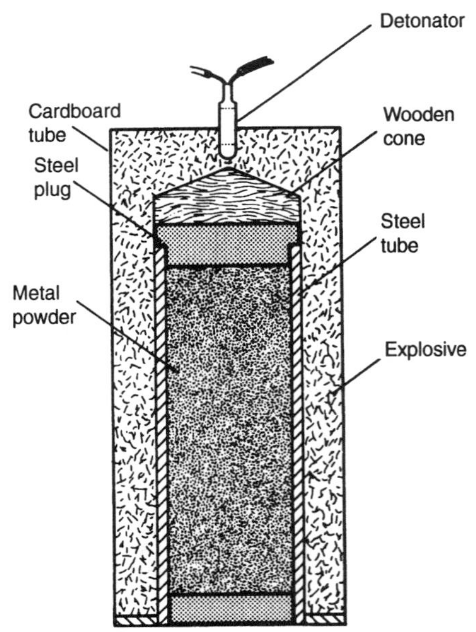

Fig. 7 Explosive compacting with powder contained in a steel tube

powders, such as refractory metal and cemented carbide powders, which can be densified relatively little by pressure application alone.

Melt-spray deposition of powders encompasses a wide variety of materials and product forms. It can be used to produce monolithic shapes by build up of the spray deposition, or it can be used to form coatings by deposition of a thin layer. One of the largest applications for melt-spray deposition is welding, including hardfacing and plasma spraying techniques for coatings.

Spray forming is also a method for producing preforms by a buildup of sprayed metal powder. These preforms and billets subsequently can be consolidated into various mill shapes. Of these processes, the Osprey process, developed in Wales by Osprey Metals Ltd., and the controlled spray deposition process are in commercial use. Several other methods are being developed; plasma spray buildup has high commercial potential. Laser techniques, such as laser glazing, also have commercial potential, especially when combined with rapid solidification technologies.

Osprey Process. Facilities for production of preforms made by the Osprey process consist of induction melting equipment and a preform production unit. In the Osprey process, an alloy is melted and subjected to gas atomization under inert conditions (usually nitrogen or argon is used). The atomized droplets are collected in a mold or group of molds, in which final solidification occurs. Molds are normally copper cooled by water. High-temperature ceramics offer other material options for molds. During solidification, welding of particles causes buildup of alloy in the mold. A preform having a density >96%, and normally >99%, of theoretical is generated by this buildup of alloy. The preform then can be consolidated to full density and formed into a mill or near-net shape part.

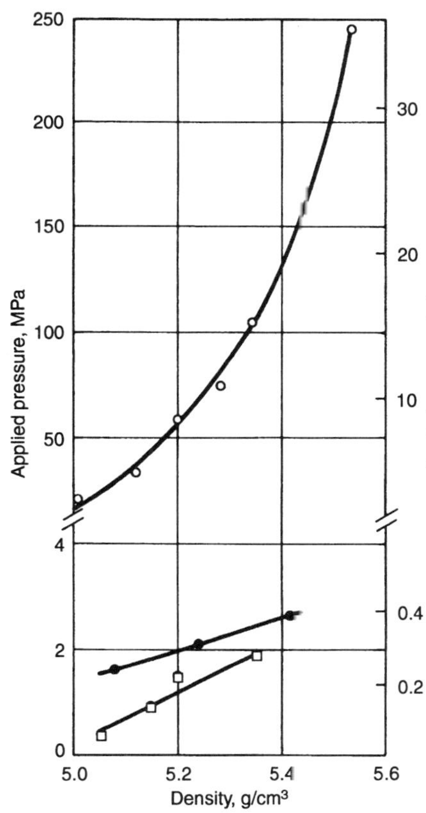

Fig. 8 Effect of powder vibration on densities of carbonyl iron compacts. ○ Obtained in static pressing. ● Vibratory compacting at a frequency of 233 oscillations per second. ♦ Vibratory compacting at a frequency of 167 oscillations per second

Alloys that have been processed by the Osprey process include stainless steels, high-speed steels, and nickel-base superalloys, although many materials appear to be compatible with the process. Alloy development has centered on high-alloy ferrous metals, Stellite alloys, superalloys, and composite materials. Because an inert atmosphere is maintained during spraying, oxygen levels similar to conventional ingot metallurgy products are attained, typically 20 to 40 ppm for superalloys. The high preform density ensures that no interconnected porosity is present in the preform, preventing internal oxidation during transfer of the material to subsequent consolidation and forming operations.

The Osprey process is used to produce a wide variety of preform shapes and sizes. Typical preform shapes are tubes, rings, cylinders, disks, or simple billets. Size is dictated by economics, with the melt facility, atomizer, and inert chamber sized for a particular product line. The largest preform size produced weighed 540 kg (1200 lb). Typical deposition rates range from 10 to 90 kg/min (20 to 200 lb/min). More information on the process is described in the article "Spray Forming" in this Volume.

The controlled spray deposition process is similar to the Osprey process in principle, but uses different machinery. Controlled spray deposition uses centrifugal atomization, while the Osprey process involves gas atomization. This process is used for production of mill shapes from high-alloy steel, which utilizes the enhanced workability of P/M workpieces. Highly alloyed metals may suffer from macrosegregation, which reduces the material workability. Elimination of segregation on a macroscale, coupled with a uniform distribution of fine carbides (2 to 3 μm range for M-2 high-speed steel), results in improved workability for P/M workpieces produced by spray deposition. These billets are processed subsequently into mill shapes and sheet products.

By atomizing liquid metal into droplets 0.5 to 1.5 mm (0.02 to 0.06 in.) in diameter, solidification rates three or more orders of magnitude higher than those of conventional ingot solidification are achieved. Impacting liquid droplets of metal onto a cooled substrate increases the solidification rate, resulting in solidification rates of 10,000 to 1,000,000 °C/s (18,000 to 1,800,000 °F/s). Controlled spray deposition relies on this type of splat solidification to build up a solidified deposit that becomes a workpiece for subsequent deformation processing. As the thickness and temperature of the built-up material increases, the solidification rate decreases, but it remains much higher than that of conventional ingot solidification. Heating prior to hot working can remove any microstructural variations that may exist throughout the thickness of the built-up deposit.

Along with the metallurgical benefits of controlled spray deposition, economic advantages of direct spraying of powder into preform shapes are attained by eliminating sieving, blending, and other powder preparation steps. Also, primary compaction of powder into a green shape is eliminated. Controlled spray deposition proponents claim that these reductions in equipment needs and processing steps allow more efficient utilization of energy, compared to conventional pressing and sintering P/M technology.

Selecting a Process

Selecting the proper process for consolidating powder to produce a part requires making many decisions. Using the performance requirements as the accept-reject criteria, the wide range of possible processes can be narrowed down, but usually the designer/manufacturer still finds that several methods are possible for producing a particular part. The final selection is then usually based on the availability of equipment, the experience with particular processes, and the marketplace requirements of cost, delivery, and quantity.

Some guidelines and corollaries to these guidelines are presented here as an example of process selection decision making. These are not hard-and-fast rules for several reasons. First, the craft or skill aspect of parts making will alter these guidelines for each and every company. Second, new methods, new powders, and new equipment constantly require that any such set of guidelines be adjusted. Third, the marketplace itself dictates changes and additions to such guidelines. The reader is encouraged to take these guidelines and corollaries as a starting point, then modify and add new statements with quantified details wherever possible so that the decision making process within each organization is captured.

Guideline 1: Control of porosity is the most important aspect of a consolidation process. The size, shape, distribution, and volume fraction of pores is the singlemost important property of a powder metallurgy part. The performance of the part is directly dependent on these aspects of porosity. Therefore, the partsmaker must control porosity in order to achieve a usable part, and the consolidation process is critical to controlling part porosity.

Corollary: Dynamic properties such as toughness and fatigue resistance improve dramatically as porosity is eliminated. Higher part stresses during application require higher densities. If a powder metallurgy part must compete with a wrought part, full density must be achieved in all critically stressed regions. This is especially true for applications that require high toughness or where cyclic loading is a life limiting condition.

Guideline 2a: Control of dimensions is the second most important aspect of the consolidation process. The successful implementation of parts production using powder metallurgy techniques is due to the minimization or elimination of machining operations. Therefore, the consolidation process must offer substantial control of the final part dimensions so that secondary operations are minimal.

Corollary: The use of rigid tooling provides superior control of dimensions in comparison to flexible tooling. Therefore, die pressing, injection molding, and powder forging are preferred over cold and hot isostatic pressing in terms of dimensional control. The shape of specific sections of isostatically pressed components can be better controlled by the selective use of rigid fixtures such as mandrels or pressing plates.

Corollary: The effect of friction on porosity distribution must be taken into account in order to maintain control of dimensions. Porosity gradients due to friction can cause nonuniform dimensional change during sintering. This is especially critical in die compaction where friction along vertical die walls and punch faces must be minimized through the use of admixed lubricants or die wall lubricants.

Corollary: Uniform powder loading is required to achieve a uniform dimensional change during sintering. Because the sintering rate is density dependent, a nonuniform distribution of powder will produce a nonuniform dimensional change during sintering. For injection molding, this means that the volume fraction of binder should be uniform throughout the green part. For multilevel die pressed compacts, this means that multiple pressing motions should be used so that the proper powder distribution is achieved before consolidation starts. In die pressing there is negligible transfer of powder between cavity sections while pressure is applied.

Corollary: Prior to sintering but after debinding, injection molded parts must be supported because particle bonding at this point is simple cohesion. After ejection from the mold, an injection molded shape is held together by the binder. Prior to sintering, the binder is removed, leaving a time period when the only force holding the powder mass together is particle cohesion plus residual binder. A supporting bed of an unreactive matter such as alumina spheres may be necessary.

Guideline 2b: Ignore Guideline 2a if powder metallurgy is the only method that can be used to produce a particular material. In some cases powder metallurgy offers a way to process a unique material. In these cases, the uniqueness is the material often outweighs the economics of dimensional control so that substantial machining or secondary processing is acceptable.

Guideline 3: Metallurgical bonding of particles must be achieved during the consolidation process. After porosity, the next most important characteristic for achieving a usable part is metallurgical bonding or welding of the particles. Sintering is the most common process for accomplishing this bonding. The alternative to sintering is shear deformation of particles while under a general state of compression. In processes such as forging and extrusion, the particle deformation and local pressures are sufficiently high that strong welds develop between particles.

Corollary: Spherical particles should not be cold pressed if they are to serve a structural function. Cold pressing of spherical powders does not result in significant mechanical bonding of particles and green strength is poor. With careful handling, cold pressing of spherical powder can be accomplished, followed by sintering to achieve particle bonding; P/M filters are an example.

Corollary: The particle surface quality directly effects the strength of the particle bond. Sintering is not effective in improving mechanical properties if particle surface contaminants such as carbides or oxides block diffusion of alloy species. Therefore, sintering requires clean surfaces or it must include a surface cleaning step such as deoxidation.

Corollary: The mode of densification affects the quality of the interparticle bonding. The strength of the bond between powder particles is improved by shear deformation in comparison to repressing or isostatic compaction. Shear plus pressure breaks up surface contaminants and promotes particle bonding. As surface cleanliness is improved toward being contaminant free, there is less difference between consolidation processes involving high or low amounts of shear.

REFERENCES

1. R. German, *Powder Metallurgy of Iron and Steel,* John Wiley & Sons, 1998
2. R. German and A. Bose, *Injection Molding of Metals and Ceramics,* Metal Powder Industries Federation, 1997
3. L.F. Hammond and E.G. Schwartz, The Effect of Die Rotation on the Compaction of Metal Powders, *Int. J. Powder Metall.,* Vol 6 (No. 1), 1970, p 25–36
4. W. Rutkowski, D. Bialo, and J. Duszczyk, Problems of Increase of Homogeneity and Density of Powder Metal Compacts Produced under Conditions of Pressing Process Assisted by Additional Movements of the Die, *Planseeberichte für Pulvermetallurgie,* Vol 28, 1980, p 194–203
5. A. Broese van Groenou, Pressing of Ceramic Powders, A Review of Recent Work, *Powder Metall. Int.,* Vol 10, 1978, p 206–211
6. R.I. Davies and S. Elwakil, "Comparison of Slow Speed, High Speed and Multiple Compaction in Ferrous Powders," Proc. of 17th Machine Tool Design and Research Conf. sponsored by University of Birmingham, Vol 3, 1976, p 483–488
7. J.W. Hagemeyer and J.A. Regalbuto, Dynamic Compaction of Metal Powders with a High Velocity Impact Device, *Int. J. Powder Metall.,* Vol 4 (No. 3), 1968, p 19–25
8. R.M. Rusnak, Energy Relationship in High Velocity Compacting of Copper Powder, *Int. J. Powder Metall.,* Vol 12 (No. 2), 1976, p 91–99
9. C.R.A. Lennon, A.K. Bhala, and J.D. Williams, Explosive Compacting of Metal Powders, *Powder Metall.,* Vol 21, 1978, p 29–34
10. S.W. Porembka, Explosive Compacting, *Ceramic Age,* Dec 1963

Powder Treatments and Lubrication

Erhard Klar, Consultant, and C.B. Thompson, OMG Americas

MOST POWDERS receive at least one treatment prior to compaction. These treatments are tailored to the use of the powder and may include (a) particle size distribution adjustment through screening and/or air classifying, (b) annealing for the purpose of improving compacting properties, (c) lubricant addition for compacting grade powders, (d) mixing of different powders for premixes, and (e) blending of powders and powder mixes to homogenize their various components. These treatments are usually performed by the powder producer. The quality of these treatments can greatly affect the uniformity and consistency of sintered part properties. With increasing emphasis on zero defect manufacture, the treatments have received more attention in recent years.

Classifying/Screening

Many powder production processes yield relatively broad particle size distribution, and because of the many manufacturing variables, the distributions and average particle sizes may exhibit marked lot-to-lot variations, which contribute to the variation of important powder and sintered part properties. Classifying and screening are used to render the particle size distributions of powders more uniform with well-defined upper particle size limits. Powder producers often manufacture series of powders that differ only in particle size distribution. Such powders exhibit "graded" differences in dimensional change during sintering. For filter manufacture, classifying and screening are used to generate narrowly sized powder fractions for controlled pore size in filters and flow restrictors.

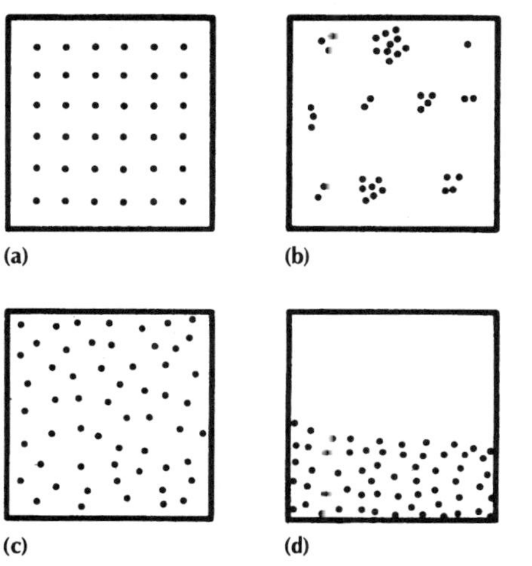

Fig. 1 Schematic representation of particle patterns in a powder mixture. (a) Ordered. (b) Agglomerated. (c) Statistical (random) distribution. (d) Demixed or segregated

Mixtures and Segregation

Most P/M powders are multicomponent systems and, therefore, are subject to segregation. Segregation is even possible in a one-component metal powder if, for instance, coarse and fine powder particles "demix" as a result of vibration. The opportunity for powder segregation exists in processes such as shipping and the filling of hoppers and compaction dies, where individual components exhibit different flow rates due to differences in particle size, shape, density, surface roughness, and other properties.

The most widely used P/M compacting grade powders are mixtures of iron with graphite, copper, nickel, and/or molybdenum. The preferred use of powder mixtures rather than prealloyed powders is related to several factors. The most important one is that elemental powder mixtures generally possess significantly higher compressibilities than their prealloyed counterparts. Secondly, powder mixtures are usually less expensive than prealloyed powders. Finally, powder mixtures more often can be formulated to provide transient liquid phases during sintering, which can reduce sintering times and improve mechanical properties.

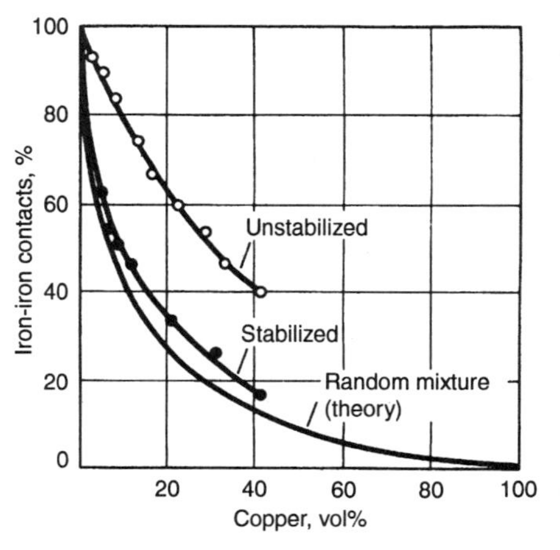

Fig. 2 Effect of stabilizer on iron-iron contact formation in binary iron-copper system. Lower curve represents theoretical random mixture.

In most powder mixtures, the base powder typically comprises 90% or more of the powder mixture. Since the other components are present in only small amounts, and their alloying effects are powerful, it is very important that they are uniformly distributed; homogeneity of the powder must be preserved during shipping and handling at the parts producer's site until compaction is completed. Without such precautions, the properties of the sintered parts are not optimal and the standard deviation, that is, the scatter of the properties, can be quite large.

Stabilization of Powder Mixtures

Because of the wide use of powder mixtures and their sensitivity to demixing, much work has been done towards the stabilization of powder mixtures.

Stabilizers. A powder mixture is optimally mixed if its components approach a random (statistical) distribution free of agglomeration (Fig. 1). The quality of mixing can be measured by the number of particle contacts (in green compacts) between identical or different powder components as illustrated in Fig. 2 for iron-copper mixtures, or by the chemical analysis of appropriate samples taken from the powder mixture.

Blenders and mixers that rely mainly on gravity (tumblers) are suitable for powders that mix readily. More intense mixing is accomplished with low-shear agitated-type blenders that use ribbons, slow-speed paddles, screw-type augers, or other means of motion. Figure 3 illustrates how spherical powders mix quite readily, but also are subject to demixing or overblending. Consequently, mixing should be stopped once a near-random distribution has been achieved. The variability coefficient in Fig. 3 represents the standard deviation of the measured degree of mixing divided by the average value of the measured property. The quality of the mixture improves with decreasing variability coefficient.

Demixing is often caused by the accumulation of electric charges, which frequently can be dissipated by the addition of a small amount of water. Surfactants and stabilizers are sometimes used to improve the flow of materials. They consist of wetting liquids and oils, which have no negative effects on the sintering process. Figure 4 shows the beneficial effect of oleic acid (dissolved in benzene) for 90%Fe-10%Al mixtures. In this example, the best mixture results from the addition of the optimized amount of a stabilizer after eight minutes of mixing, when the mixture has approached a random distribution (curved).

Diffusion-Bonded Powders. In a diffusion-bonded powder, a powder mixture is heated in a reducing atmosphere to yield a sintered powder cake, which after grinding, produces a powder that consists of agglomerates of the components of the powder. The process is controlled so that the degree of alloying among the different components is only small. This retains much of the desirable high-compressibility characteristics of the original elemental powder mixture and minimizes or eliminates the demixing tendency of such a powder. As a bonus, green strength is usually increased.

Binder-Treated Powder. In recent years powder producers have developed proprietary binders that are used in small amounts to bond fine graphite onto the coarser iron particles. Such powders reduce the dusting of a powder, and improve several important engineering properties of the sintered parts due to the more uniform concentration of carbon in the final parts.

Prealloyed Powders. Despite the widespread use of powder mixtures, the use of prealloyed powders in the structural parts industry has been increasing. In the low-alloy steel segment, nickel, molybdenum, and manganese are used as alloying elements to provide greater hardenability than is possible with admixed copper or nickel. Admixed nickel requires economically prohibitive sintering times (even at high sintering temperatures) for complete alloying. Higher alloyed powders, such as stainless steels, tool steels, and superalloys, are all prealloyed powders. The requirement of complete alloying in these alloy classes makes the elemental powder approach impractical. The problem of low compressibility, that is, low green density and/or low green strength, is solved by high-temperature and liquid phase sintering, and through the use of special consolidation methods, such as hot isostatic pressing and extrusion.

Annealing

Although the term annealing refers to a heat treatment that renders a metal soft by the removal of strain, in P/M, the term is used more broadly and includes the softening of a powder by the removal of interstitials, particularly carbon, nitrogen, and oxygen. Powders that have been work hardened through milling; powders that contain residual amounts of carbon, oxygen, or nitrogen; and powders that contain objectionable surface oxide films are most often subjected to this type of annealing. Examples include sponge and atomized iron powders, carbonyl and electrolytic iron powders, tool steel powders, and ferritic stainless steel powders. In most cases, annealing is performed in a reducing atmosphere, which not only protects the powder against oxidation, but also removes surface and interior oxides. Annealing improves the compressibility and green strength of a powder due to the softening of the metal. Frequently, annealing also produces slightly agglomerated powders with lower apparent densities and further increases in green strength.

Lubrication

Lubrication is essential to reduce friction between the pressed compact and the rigid tool components when compacting metal powders in steel or carbide tooling. Extremely high part ejection pressures and, ultimately, seizing and excessive tool wear result without proper lubrication. In addition, the surface quality of the part is detrimentally affected. Although much work has been done with systems where the tooling itself is lubricated, these systems generally have not proved effective high-volume P/M parts production. Parts produced by isostatic compaction do no require lubrication, due to the absence of high frictional forces during this type of processing.

Compacting grade metal powders are lubricated to facilitate part ejection from the die and to minimize die wear. Lubricants can strongly affect compacting and sintering properties of a powder. For this reason, powder producers try to optimize amount, composition, and structure of a lubricant to not only reduce friction during powder compaction, but also to protect the powder against oxidation and to obtain consistent apparent density and powder flowability, proper dimensional change during sintering, and maximum and consistent mechanical properties of the sintered parts.

Although a lubricant largely evaporates and/or decomposes in the preheating zone of a sintering furnace, its interactions and reactions with the metal powder can have critical consequences. Potential problems range from lubricant segregation, burnout, and blistering to carbonization and lubricant residue interference. Lubricant performance also depends on part density, metal powder composition, and particle size.

Lubricant Selection

In the majority of P/M applications, the lubricant is premixed directly with the metal powder.

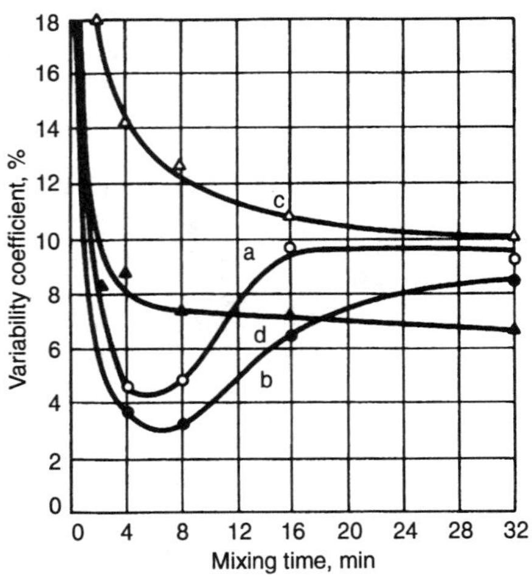

Fig. 3 Effect of particle size and shape of components of 90%Fe-10%Cu mixtures on degree of blending. Quality of blending improves as variability coefficient decreases. Particle size and shape for components: (a) Cu, 200 to 300 μm; Fe, <63 μm of spherical particle shape. (b) Cu 200 to 315 μm; Fe, 100 to 200 μm of spherical particle shape. (c) Cu, 200 to 315 μm; Fe, <63 μm of irregular particle shape. (d) Cu, 200 to 315 μm; Fe, 100 to 200 μm of irregular particle shape

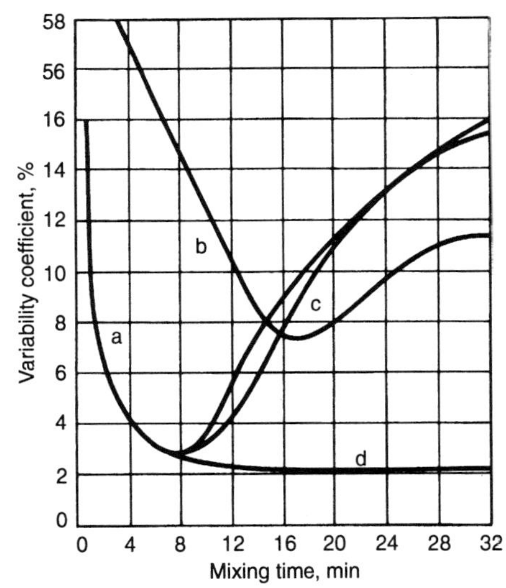

Fig. 4 Effect of stabilizer on degree of blending of 90%Fe-10%Al mixtures. Particle size for both components: 100 to 200 μm. Stabilizer: benzeneloleic acid. (a) Without stabilizer. (b) With 1.0% stabilizer. (c) 1.0% stabilizer added after 8 min of blending. (d) 3% stabilizer added after 8 min of blending

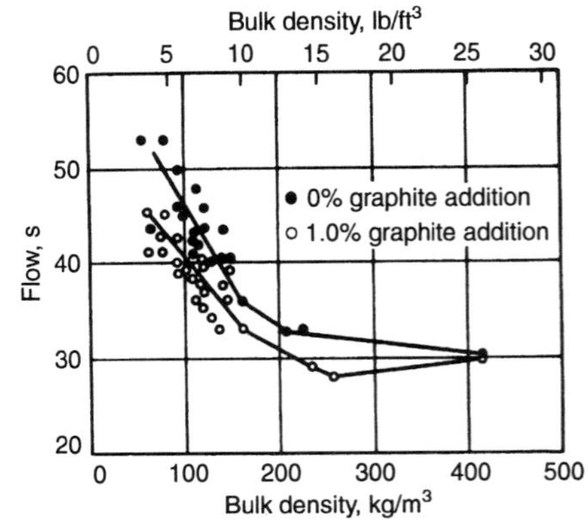

Fig. 5 Relationship of mix flow and bulk density of Ancor MH-100 iron powder with 1.0% zinc stearate lubricant. Source : Ref 1

Stearic acid, zinc stearate, lithium stearate, and synthetic waxes, such as Acrawax, are the most popular lubricants. Lubricant selection is based primarily on the ability of the lubricant to adhere to metal particle surfaces. The amount of lubricant added depends on many factors, including composition of the metal powder, type of tooling, compacted density, and complexity of the part. Amounts generally vary from 0.5 to 1.5 wt%.

Lithium stearate is widely used in stainless steel powders because of its beneficial effect on compressibility. Lithium stearate and zinc stearate mixtures are used in brass and nickel silver powders to produce stainfree parts. Synthetic waxes with their low moisture and ash contents are widely used for high-density parts where ash residues are objectionable. Multicomponent lubricants permit the powder producer to adjust or influence the packing, pressing, and sintering characteristics of powder. The use of die wall rather than admixed powder lubrication is still in the development stage.

Ferrous Materials

Although lubricants are a necessary addition to compacting metal powders, they can have some critical, and often deleterious, effects on the ultimate premix. Hoeganaes Corporation (Ref 1) has made a comprehensive study of lubricants and their influence on the properties of iron premixes. Lubricant properties, such as bulk density and particle size, are very important; for example, high-bulk density and large particle size provide good flow and low stripping pressure, as illustrated in Fig. 5 and 6. To minimize premix segregation, the particle size of the lubricant must be smaller than the size of the largest particle of iron powder; for example, a 150- to 200-mesh (104- to 75-μm) lubricant should be used with 100-mesh (150-μm) iron powder.

The melting point of the lubricant must be high enough to prevent melting or softening from heat developed during mixing. If this occurs, flow deteriorates, apparent density decreases, and the mix tends to agglomerate.

Higher lubricant content yields poorer flow, lower apparent density, and lower stripping pressure. Figure 7(a) shows the effect of lubricant content on apparent density of atomized stainless steel powder. The effect of increasing amounts of three lubricants on the flow of iron powder is shown in Fig. 7(b) and 8.

Stripping pressure is the initial amount of pressure required to start the ejection process of a green compact from the die after compaction. The sliding pressure is lower than the stripping pressure and can be defined as the pressure required to complete the compact ejection cycle. Both stripping and sliding pressures are dependent on the compaction pressure and type of lubricant, as shown in Fig. 9 and 10. Thus, different lubricants may be required for steel and carbide tooling materials.

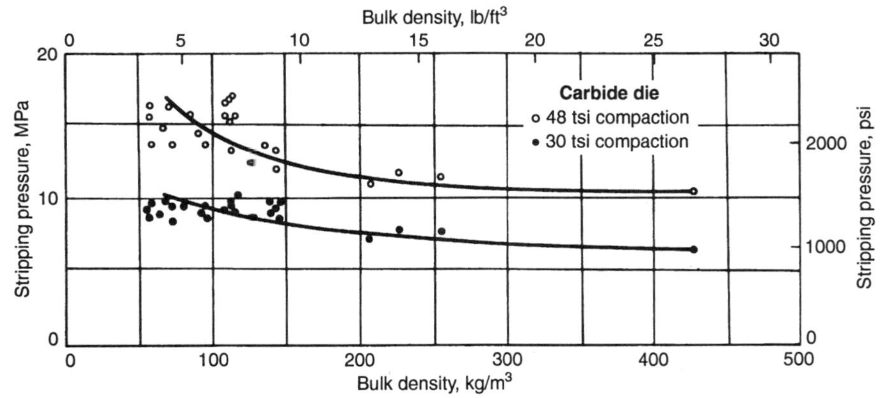

Fig. 6 Relationship of stripping pressure and bulk density of iron powders with 1.0% zinc stearate lubricant. Source: Ref 1

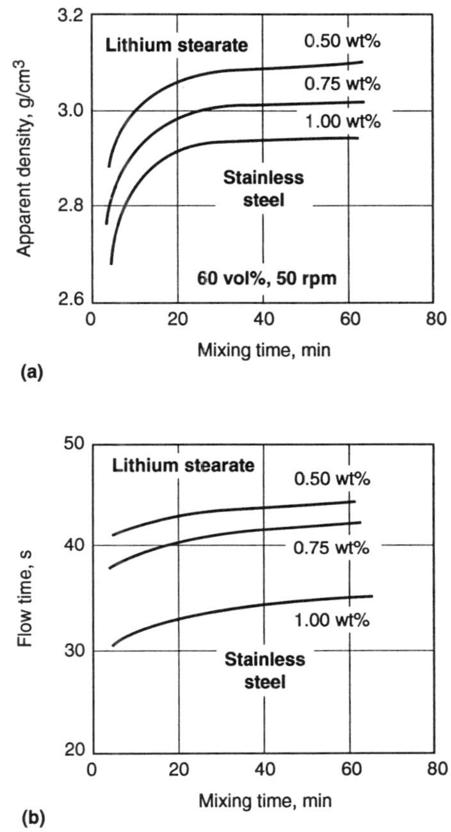

Fig. 7 Effect of lubricant additions and mixing time on (a) apparent density and (b) flow time of water-atomized stainless steel powder

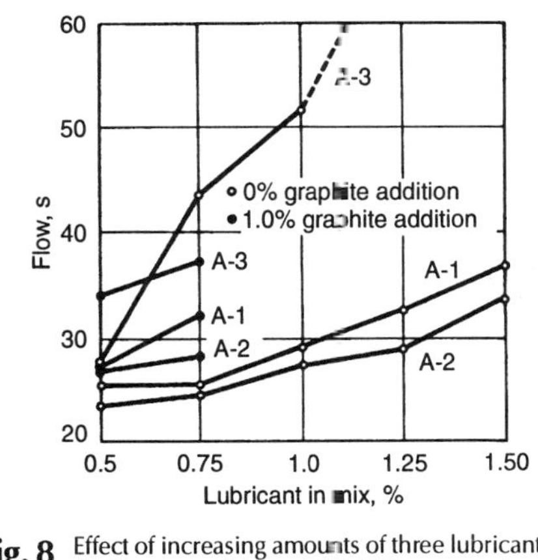

Fig. 8 Effect of increasing amounts of three lubricants on the flow rate of Ancor MH-100 iron powders. Lubricant A-1: 82% stearic acid, 13% palmitic acid, 1.0% oleic acid; lubricant A-2: 49% stearic acid, 50% palmitic acid, 0% oleic acid; lubricant A-3: 41% stearic acid, 51% palmitic acid, 6% oleic acid. Mixing time of 30 min. Source: Ref 1

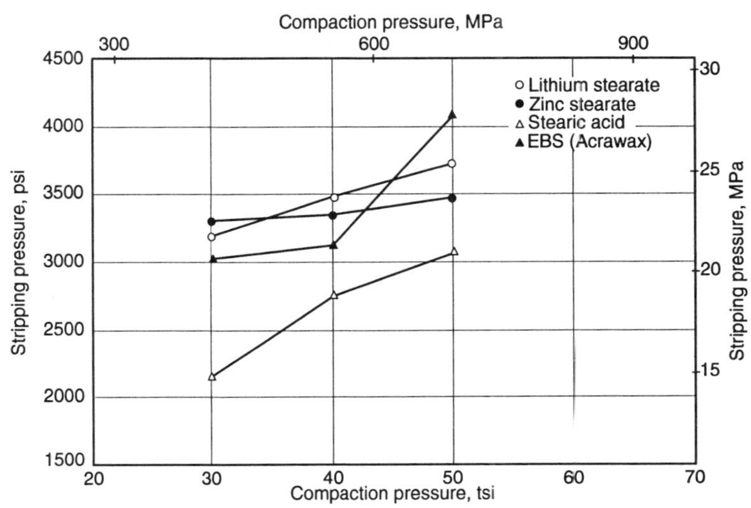

Fig. 9 Effect of compaction pressure and 0.75% lubricant on stripping pressure of Ancorsteel 45P (Ancorsteel 1000B + 2.90 wt% ferro-phosphorus)

Stripping strength can vary considerably with the type of lubricant used. When compared to parts pressed without lubricant, zinc stearate and stearic acid decrease sintered strength only slightly, as illustrated by Fig. 11 and 12. However, several lubricants, such as calcium and barium stearates, added to pure iron or to an iron and 1% graphite mixture cause significantly large decreases in sintered strength, as shown in Fig. 13. The appearance of P/M parts after sintering also can be affected by lubricants. Although the effect on iron parts is not as significant as with brass and nickel-silver powders, spotty surfaces and discoloration can be a major concern. Lubricants that vaporize completely during sintering generally leave a clean surface. Dark mottled surfaces occasionally result from zinc and calcium stearate lubricants, due to condensed zinc metal or calcium oxide deposits.

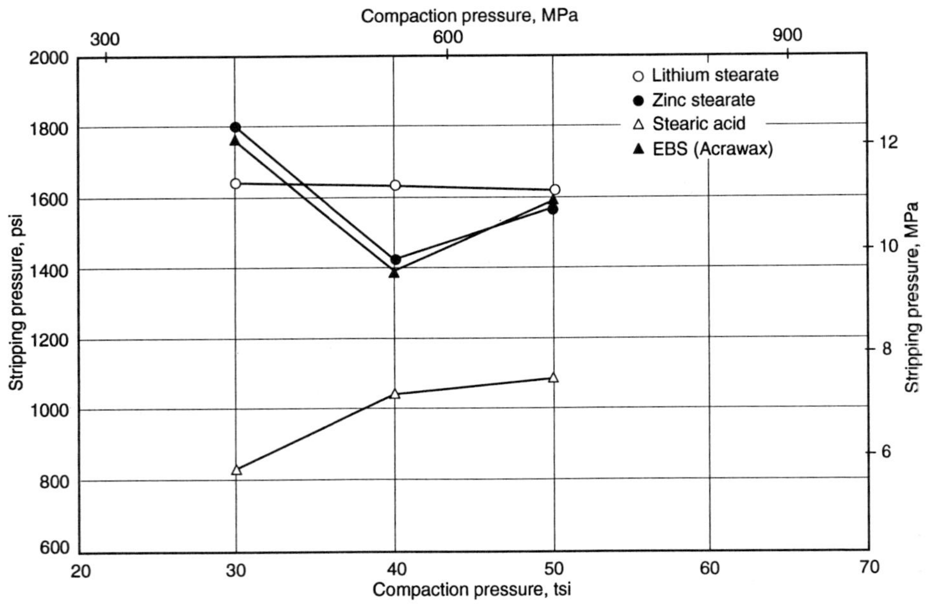

Fig. 10 Effect of compaction pressure and 0.75 wt% lubricant on the sliding pressure of Anchorsteel 45P (Anchorsteel 1000B + 2.90 wt% ferro-phosphorus)

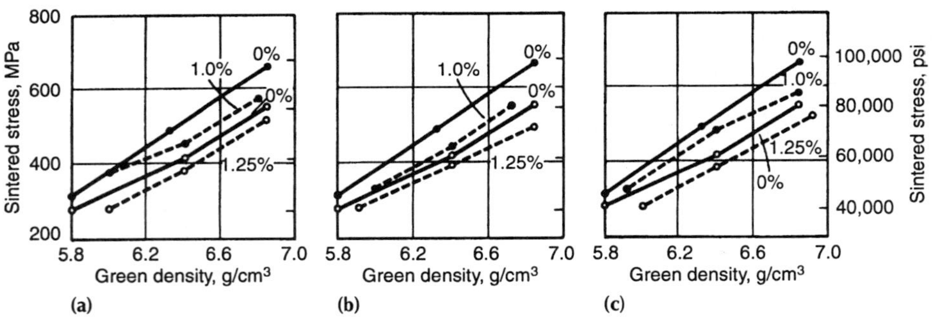

Fig. 11 Effect of zinc stearate additions on sintered strength of Ancor MH-100 compacted powders sintered 40 min at 1120 °C (2050 °F) in purified exothermic gas. Preheat of 650 °C (1200 °F). Bulk density: (a) 141 kg/m^3 (8.8 lb/ft^3); (b) 256 kg/m^3 (16.0 lb/ft^3) (c) 91 kg/m^3 (5.7 lb/ft^3). Particle size (a) 1.9 μm; (b) 4.5 μm; (c) 1.0 μm. Graphite addition of 0%, open circle data point; graphite addition of 1.0% closed circle data point. Without lubricant, solid line; with lubricant, dashed line. Source: Ref 1

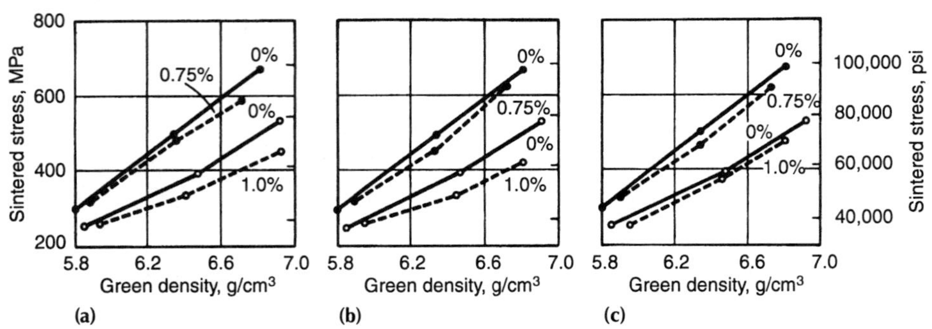

Fig. 12 Effect of zinc stearate additions on sintered strength of Ancor MH-100 compacted powders sintered 40 min at 1120 °C (2050 °F) in purified exothermic gas. Preheat of 650 °C (1200 °F). Bulk density: (a) 354 kg/m^3 (22.1 lb/ft^3); (b) 434 kg/m^3 (27.1 lb/ft^3); (c) 384 kg/m^3 (24.0 lb/ft^3). Softening temperature: (a) 64 °C (147 °F); (b) 55 °C (131 °F); (c) 54 °C (129 °F). Graphite addition of 0%, open circle data point; graphite addition of 1.0%, closed circle data point. Without lubricant, solid line; with lubricant, dashed line. Source: Ref 1

Lubricant removal, prior to the actual sinter operation, is vitally important for several reasons. Generally, this removal is accomplished during the sintering preheat operation and is referred to as "burn-off." Because of highly adverse effects of residual lubricant carbon deposits when sintering stainless steel, resulting in lowered corrosion resistance, Moyer (Ref 2) conducted an extensive study of the burn-off characteristics of common lubricants (lithium stearate, zinc stearate, Acrawax C, and Nopco Wax) in 316L powder compacts. Burn-off temperatures of 370, 425, and 480 °C (700, 800, and 900 °F) were used in both air and dissociated ammonia atmospheres. This study revealed that the waxes leave no residue at 425 °C (800 °F) or higher; the stearates, however, leave approximately 15% residue even when burned at temperatures up to 540 °C (1000 °F). Maximum lubricant burn-off is achieved at about 425 °C (800 °F). Burn-off is less controllable in dissociated ammonia, and the amount of lubricant removed decreases as the compacting pressure is increased. Carbon burn-off is incomplete when compacts are burned off in a dissociated ammonia atmosphere.

Nonferrous Materials

Many lubricants frequently used in iron-base powder systems have not been effective in all copper-base prealloyed systems. Without lubricant system development, primarily lithium stearate, it is unlikely that brass and nickel-silver P/M parts would hold the strong market position that they do today (Ref 3). The use of lithium stearate overcame many early sintering problems and provided consistently high properties in brass P/M parts.

Lubricants such as zinc stearate, lithium stearate, stearic acid, and Acrawax have been used successfully with conventional premixed bronze materials (90%Cu-10%Sn). A recent study (Ref 4) comparing the two stearates and Acrawax revealed slightly better compressibility and lower ejection pressures with zinc stearate. Furthermore, green strength increased with decreased amounts of lubricant, and Acrawax provided superior green strength zinc and lithium stearate.

Bi-lubricant systems, such as lithium-zinc stearate, provide the advantages of high physical properties and relatively clean surface appearance. Lithium stearate and lithium-zinc stearate lubricant additions in the range of 0.5 to 0.75% are used most often in brass and nickel-silver parts. Although waxes are excellent for maintaining green strength, their use results in lower

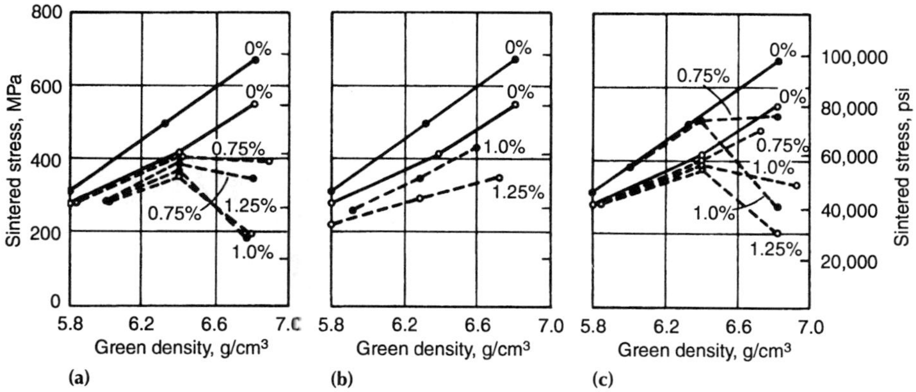

Fig. 13 Effect of calcium and barium stearate additions on sintered strength of Ancor MH-100 compacted powders sintered 40 min at 1120 °C (2050 °F) in purified exothermic gas. Preheat of 650 °C (1200 °F). (a) and (b) Calcium stearate. (c) Barium stearate. Bulk density: (a) 104 kg/m³ (6.5 lb/ft³); (b) 431 kg/m³ (26.9 lb/ft³) (c) 178 kg/m³ (11.1 lb/ft³). Particle size (a) 11 μm; (b) 18.8 μm; (c) 1.2 μm. Graphite addition of 0%, open circle data point; graphite addition of 1.0%

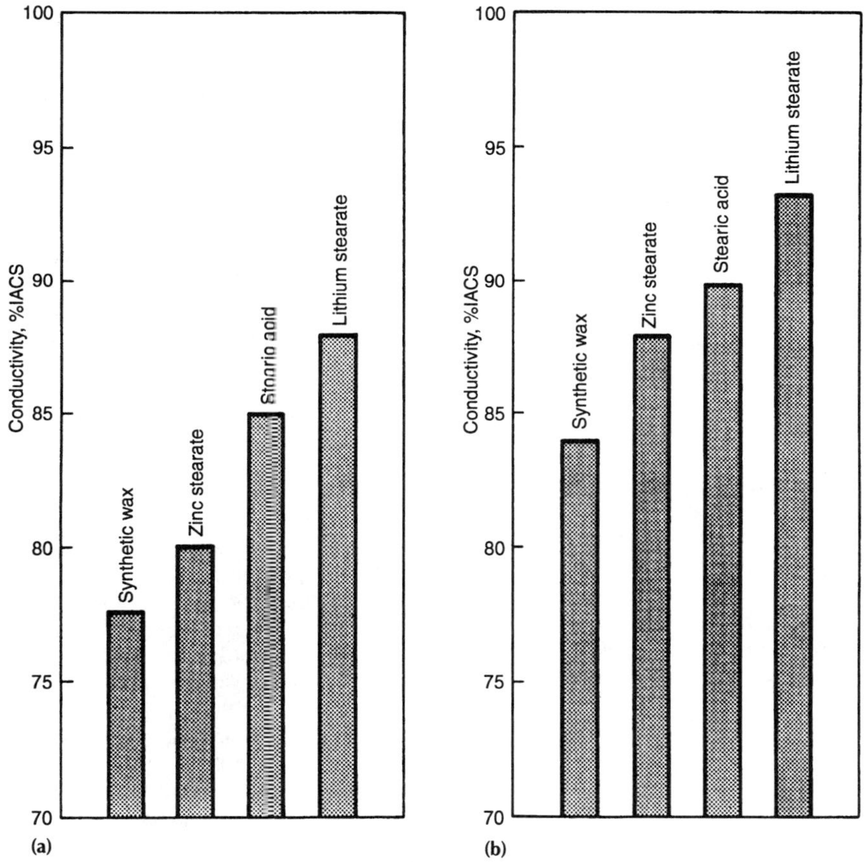

Fig. 14 Effect of lubricant on electrical conductivity of copper P/M parts. Compacting pressure: (a) 276 MPa (20 tsi); (b) 414 MPa (30 tsi)

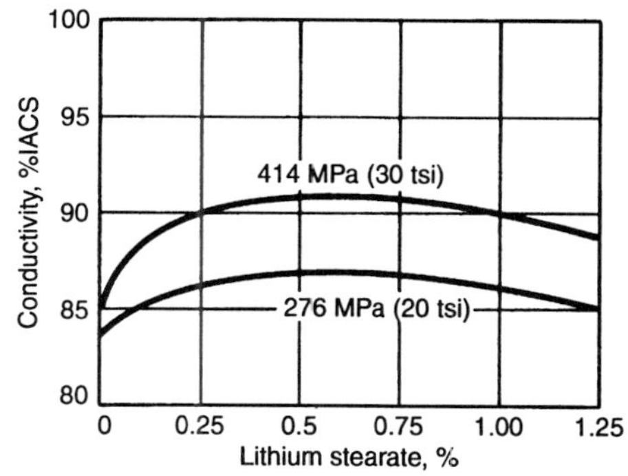

Fig. 15 Effect of lubricant quantity on the electrical conductivity of copper P/M parts

apparent densities, higher briquetting pressures, and lower sintered strengths in brass parts.

Frequently, a bi-lubricant system of zinc stearate and stearic acid at 0.75 wt% is employed for conventional bronze premix blends; these two lubricants complement each other. Stearic acid has excellent lubricating characteristics, but its spherical morphology tends to promote segregation, therefore, the irregular shape of the zinc stearate particles provides a supportive interlocking effect. Acrawax often is used alone to increase green strength, although it can lower the apparent density and retard the flow of bronze premixes.

Lubricants play a unique role in the production of copper P/M parts for high electrical conductivity applications (Ref 5). Lithium stearate allows the highest electrical conductivity among lubricated compacts, as illustrated in Fig. 14. Compacts pressed at 275 MPa (20 tsi), using a synthetic wax lubricant, have conductivities of about 74% IACS; similar compacts produced with lithium stearate exhibit conductivities of about 86%. The effect of the quantity of lubricant on electrical conductivity of copper P/M parts is shown in Fig. 15. The electrical conductivity reaches a maximum with approximately 0.6% lithium stearate in compacts pressed at 275 MPa (20 tsi) and with approximately 0.5% at 415 MPa (30 tsi).

Synthetic organic waxes, which have low moisture and ash content, are employed in the production of aluminum P/M premixes (Ref 6). The addition of 1.5% lubricant (nominally) permits the consistent production of high-density aluminum P/M parts with fine surface quality and eliminates tool problems with seizing and cold welding. High sintered properties are achieved because these lubricants leave virtually no ash to interfere with particle bonding during sintering.

ACKNOWLEDGMENT

Portions of this article are based on C.B. Thompson's article "Lubrications of Metal Powders" from *Powder Metallurgy,* Vol 7, *ASM Handbook,* 1984.

REFERENCES

1. E. Geijer and R.B. Jamison, "Lubricants for Powder Metallurgy Parts Manufacturing," Hoeganaes Report No. 142 1M, June 1965
2. K.H. Moyer, The Burn Off Characteristics of Common Lubricants in 316L Powder Compacts, *Int. J. Powder Metall.,* Vol 7 (No. 3), 1971, p 33–43
3. P.E. Matthews, Brass and Nickel Silver Powders, *Copper Base Powder Metallurgy,* Vol 7, Metal Powder Industries Federation, 1980
4. J. McGraw and M.J. Koczak, A Laboratory/Production Comparison of Powder Compacting and Ejection Response, *Int. J. Powder Metall.,* Vol 16 (No. 1), 1980
5. P.W. Taubenblat, Production and Properties of High Conductivity Copper P/M Parts, *Copper Base Powder Metallurgy,* Vol 7, Metal Powder Industries Federation, 1980
6. J.H. Dudas and W.A. Dean, The Production of Precision Aluminum P/M Parts, *Progress in Metallurgy,* Vol 25, Metal Powder Industries Federation, 1969

Mechanical Behavior of Metal Powders and Powder Compaction Modeling

J.R.L. Trasorras and R. Parameswaran, Federal-Mogul, Dayton, Ohio
A.C.F. Cocks, Leicester University, Leicester, England

CONSOLIDATION OF METAL POWDER into useful products generally follows a three-step process: (1) powder transfer into a container or die, (2) powder compaction through applied force, and (3) sintering. Particular processes combine two of these steps into one. For example, hot isostatic pressing combines steps two and three through the simultaneous application of pressure and temperature in an autoclave. This article describes step two, the mechanical response of the powder to the applied compaction forces. Density of the powder transferred into the container depends on the properties of the powder and on the method of transfer, which may include tapping and vibrating. Consideration of packing density is beyond the scope of this article; however, a comprehensive review of this topic can be found in Ref 1. Proper design of compaction processes requires a description of the mechanical or thermomechanical behavior of the powder. Issues related to applications of practical interest include:

- *Compaction in rigid dies:* Density distribution in the compact, effect of tooling motion histories on density, tooling stresses and tooling deflections, stresses in the compact during ejection and fracture likelihood
- *Cold isostatic pressing:* Geometric distortion of the pressed component, effect of container dimensions and material on distortion
- *Hot isostatic pressing:* Geometric distortion of the pressed component, effect of pressure and temperature cycles on distortion, effect of container design on distortion
- *Powder forging:* Density distribution in the formed part, forging loads, and tooling stresses

Unfortunately, the complex nature of powder mechanical behavior, coupled with complex geometry and boundary conditions, precludes analytical solutions to these problems. During the late 1980s, the maturing of numerical solutions to finite strain plasticity allowed initial attempts at modeling powder compaction processes. Powder compaction modeling has not achieved widespread industrial use yet. However, many useful and encouraging results have been published, and it seems likely that during the next decade, modeling will become central to the design of compaction processes.

Formally, powder compaction modeling requires the solution of a boundary value problem—a set of partial differential equations representing:

- *Balance laws:* conservation of mass, momentum balance, and conservation of energy
- *Constitutive laws:* stress-strain relation and friction laws

The finite element method (FEM) is the numerical technique most widely used to solve these equations. Modern FEM codes incorporate constitutive models for a wide variety of materials. Some of them are suited to modeling metal powders. Furthermore, these codes also provide facilities for the user to implement a particular constitutive law.

Constitutive Models for Metal Powder Compaction

This section examines the response of powder compacts under mechanical loading histories that are typical of those experienced during the manufacture of powder metal parts. Deformation of a powder compact occurs as the result of rearrangement and plastic deformation of the particles. Frictional resistance, as particles slide over each other, can also contribute to the macroscopic response. If the packing is initially loose, then the initial stages of compaction are dominated by particle rearrangement, with little or no plastic deformation occurring between the contacting particles. If, however, the initial apparent density of the powder is relatively high, further macroscopic plastic deformation under constrained compressive stress states is dominated by plastic deformation of the particles themselves.

Knowledge of the major physical processes that occur when a powder compact is deformed can be used to guide the development of constitutive models of the compaction process. At the microlevel, the constitutive response of fully dense materials governs plastic deformation between the contacting particles. The macroscopic response can then be determined by combining the contributions from individual contacts. There are general theoretical techniques that can be employed to achieve this and determine what general statements can be made about the structure of the macroscopic constitutive laws. Micromechanical models have been developed using these techniques. It may not be appropriate to precisely follow this route, but, instead, to develop empirical constitutive laws based on observations made at the macroscopic level. The results then provide a suitable structure and general guidelines for the construction of these models. As the material approaches full density, these models should reduce to classical incompressible models. Therefore, these models serve a second purpose; they provide a starting point from which more general constitutive relationships can be developed.

Throughout the article, the response of an element of material in a particle of the compact (referred to as a "microscopic element") is identified in terms of the local stress σ_{ij} and local strain ε_{ij}. When evaluating the response of a powder compact, a representative "macroscopic element" of material, which contains a large number of particles but is small compared to the size of a pressed component, is identified. The response of such an element is expressed in terms of a macroscopic average stress Σ_{ij} and strain E_{ij}.

Deformation of Powder Compacts: Experimental Observations

Before examining the structure of constitutive laws for powder compaction, it is instructive to identify the major physical processes that occur within the compact as it is deformed and to determine how these are likely to influence the macroscopic behavior. When the powder is initially poured into a die or mold, the particles are arranged randomly with particles only in point contact with each other.

Consider the situation where the compact is densified hydrostatically. As the pressure is increased, plastic flow occurs in the vicinity of the contacts. As a result, the contact zones spread, the centers of the particles move closer together, and the material densifies. During the initial stages of this process, the porosity remains connected, consisting of a network of interconnected channels threading through the material. In the micromechanical models, this is referred to as stage 1 compaction. As the material is densified further, the channels pinch off, leaving a distribution of isolated pores (stage 2). Throughout this process, because of the initial random structure, the porosity is randomly distributed throughout the material, particularly during stage 1 compaction. If, however, the sample is much larger than the mean particle size, no preferential orientations or distributions of pores develop, and the structure remains macroscopically isotropic. A single state variable can then be used to describe the structure and macroscopic response. A convenient state variable is the relative density D, which is defined as the ratio of the density of the compact to the density of the material, so that at full density $D = 1$.

In practical compaction processes, the densifying material can see complex stress states and histories. As a result, the structure that develops is no longer isotropic, and it may not be appropriate to describe this structure and the macroscopic response using a single state variable. For example, consider the case of frictionless closed-die compaction. As a cylinder is compacted along its axis, there is no straining in the transverse direction. The contact patches that develop between the deforming particles are therefore larger normal to the direction of loading than along the axis. Also, the size of these patches is different from those in a compact that has been densified hydrostatically to the same relative density. Thus, the relative density does not uniquely describe either the microstructure or the macroscopic response.

Any constitutive law that is developed needs to be validated experimentally. An examination of the types of experiments that have been performed and how the results from these tests are commonly presented follows. The analysis is based on the identification of a yield surface for the material (which is a surface in stress space within which the material responds elastically; plastic deformation can only occur for stress states on the yield surface). The size and shape of the yield surface is a function of the history of loading. The yield surface can be determined using a number of different methods. A number of samples can be prepared in the same way (for example, hydrostatically compacted to the same relative density). Each sample is then loaded along a different path in stress space until it yields. The yield surface is then formed by connecting the different yield points. Alternatively, a single specimen can be used. After compacting to the desired state, a series of probing experiments can be performed by unloading and reloading along a range of different stress paths until yield occurs. The yield surface can be mapped out in exactly the same way as before. When using this method, it is important to ensure that only a small amount of plastic straining occurs during each probe so that there is no significant change of microstructure (i.e., state) over the series of probing tests.

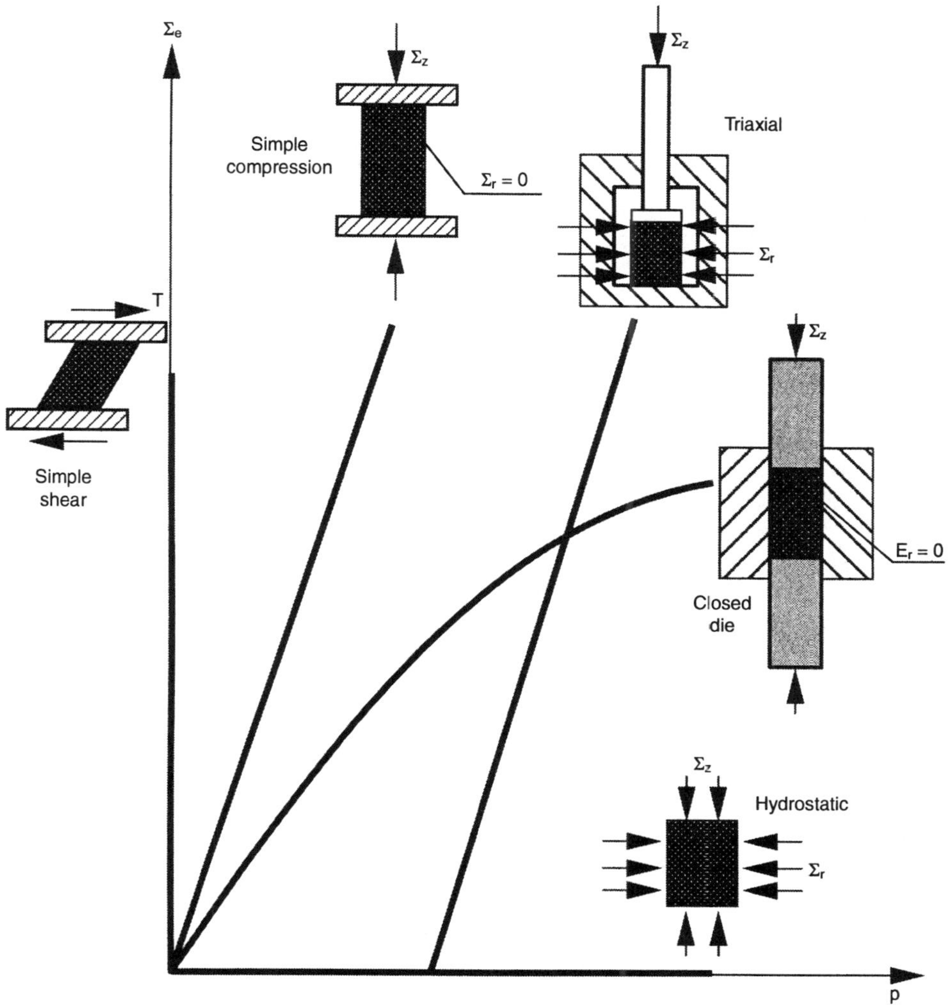

Fig. 1 Schematic representation of the load paths corresponding to the different test procedures. Source: Ref 31

When examining a range of different stress states, it proves convenient to present the results in terms of global measures of stress. Two convenient quantities are the von Mises effective stress, Σ_e, and the mean (or hydrostatic) stress, Σ_m. If Σ_1, Σ_2, and Σ_3 are the principal stresses,

$$\Sigma_e = \sqrt{\frac{1}{2}\{(\Sigma_1 - \Sigma_2)^2 + (\Sigma_2 - \Sigma_3)^2 + (\Sigma_3 - \Sigma_2)^2\}} \quad \text{(Eq 1)}$$

and

$$\Sigma_m = \frac{1}{3}(\Sigma_1 + \Sigma_2 + \Sigma_3) \quad \text{(Eq 2)}$$

More general definitions of these stresses are provided in the following section.

Alternatively, instead of using Σ_m, the material response can be presented in terms of the hydrostatic component of stress (or pressure) $P = -\Sigma_m$. These definitions are used interchangeably throughout the article.

Many of the experimental methods used today to evaluate the behavior of metal powders have been inspired by techniques used to test soils. Most published data has been generated using such prodedures. Before examining the material data in detail, it proves instructive to first examine the general features of the most common testing methods.

It is convenient to represent the yield function, as well as the stress paths during testing, in the $P - \Sigma_e$ plane. In this representation, the closer a point is to the ordinate (Σ_e axis), the more shear experienced in the material; the closer it is to the abscissa (P axis), the more hydrostatic pressure experienced. The typical powder compaction process involves stress

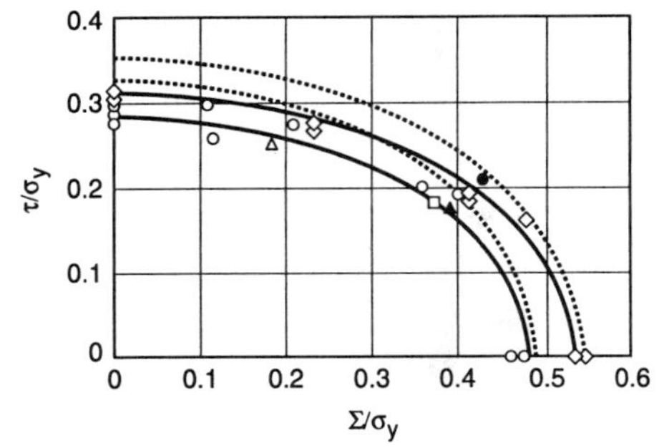

Fig. 2 Yield surfaces determined by Kim et al. (Ref 5) from tension/torsion tests on sintered iron powders

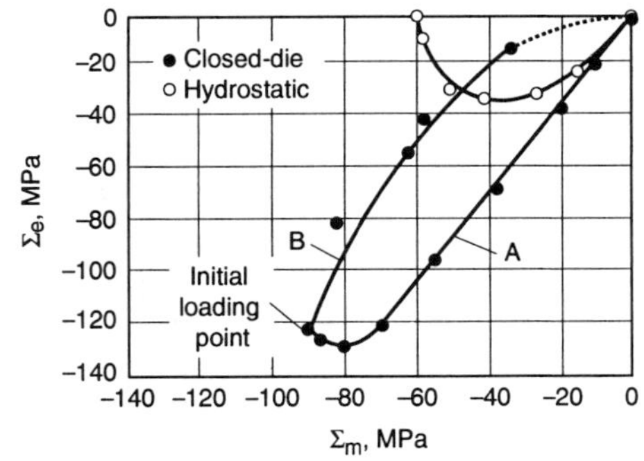

Fig. 3 Yield surface for spherical copper powders determined by Akisanya et al. (Ref 10) for isostatically pressed and close-die compacted samples

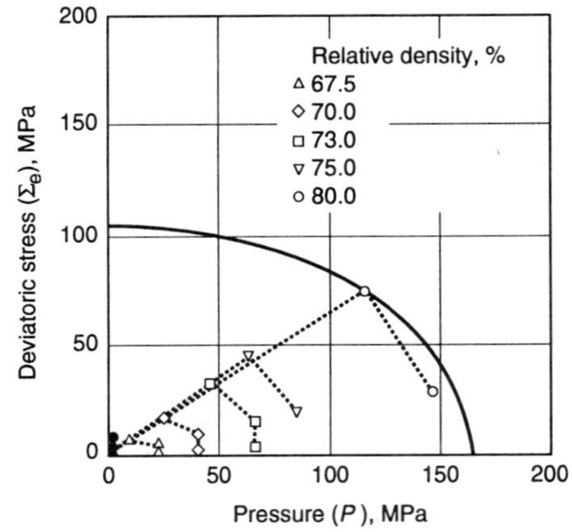

Fig. 4 Isodensity surfaces for spherical copper powders tested by Brown and Abou-Chedid (Ref 11)

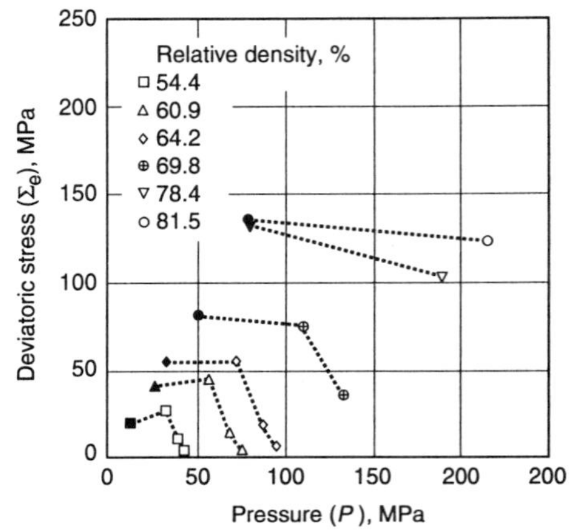

Fig. 5 Isodensity surfaces for irregular iron powders tested by Brown and Abou-Chedid (Ref 11)

states that have a high-pressure component, a natural consequence of the confinement of the powder in a die cavity.

Figure 1 shows a schematic representation of the load paths corresponding to the different test procedures:

- *Simple shear test:* This is marked by a complete absence of any hydrostatic load. This test has little value in the context of loose powders and can only be applied to sintered powders or porous materials.
- *Simple compression test:* This is once again a shear-dominated test, with friction at the interface of the powder and dies assumed absent. As with the previous case, this test can realistically only be used for sintered powders or porous materials.
- *Triaxial test:* The primary advantage of this test is that a variety of stress paths can be examined through a combination of compressive stresses in axial Σ_z and radial Σ_r directions. A detailed description of this test is provided in the section "Experimental Determination of Powder Metal Constitutive Properties and Functions" in this article. The effective and hydrostatic components of stress for this loading situation are given by $\Sigma_e = |\Sigma_z - \Sigma_r|$ and $P = \frac{1}{3}(2\Sigma_y + \Sigma_z)$.
- *Closed die compaction:* The state of stress in the powder in this test is possibly the best description of the actual stress state in a typical compaction. This test does not lend itself to parameter extraction procedures because friction is present at the die walls. It is possible, however, to simulate frictionless closed die compaction in a triaxial cell.
- *Hydrostatic compaction:* The state of stress during this test is one of pressure alone with shear components completely absent. This test is easy to set up and provides the compressibility of the material.

All the cases illustrated in Fig. 1 involve the testing of cylindrical specimens. The inherent symmetry in the shape and loading can be exploited to simplify the analytical treatment required for the extraction of material parameters.

It should be noted that results from a particular testing procedure might be used to extract material parameters for different constitutive models. Simplifying assumptions are an essential ingredient in such determinations. For instance, friction is inevitably present in most testing procedures. However, it often becomes necessary to ignore the effects of friction in the extraction procedure, because not doing so may seriously impair the ability to obtain reasonable values for material parameters.

Kim et al. (Ref 5) cold isostatically pressed tubes of iron powder (Hoeganaes ASC 100-29) to relative densities in the range D = 0.8 to 0.85. These tubes were then sintered at 1150 °C for 1 h in a hydrogen atmosphere. The tubes were then tested in combined tension, Σ, and torsion, T. Typical yield surfaces obtained from these experiments are shown in Fig. 2, where the results are normalized with respect to the yield strength of the fully dense material, σ_y. The sintering process has bonded the particles, imparting significant tensile strength to the compact. A number of studies for determining the mechanical properties of powder compacts employ the same type of procedure. For example, the experimental studies of Kuhn and Downey (Ref 6) and Shima and Oyane (Ref 7), which form the basis for a widely used constitutive law for powder compaction, were performed on sintered compacts. Brown and Weber (Ref 8) demonstrated that the mechanical response of compacted and then sintered powders is very different from that of identical powders that have only been compacted. Watson and Wert (Ref 9) compacted aluminum powders hydrostatically and by using closed-die compaction. They performed uniaxial tensile and compressive tests on these samples to construct the yield surface.

Akisanya et al. (Ref 10) used the probing method to determine yield surfaces for samples of spherical powders, which were either hydrostatically compacted or compacted by simulating closed-die compaction. These tests were performed using a triaxial cell in which hydrostatic pressure and uniaxial compression could be applied independently. Yield surfaces obtained for a relative density of 0.8 are shown in Fig. 3 using the axes of $\Sigma_s = \Sigma_z - \Sigma_r$ and Σ_m. Note that in these experiments $\Sigma_r > \Sigma_z$ and thus Σ_s is negative with $\Sigma_s = -\Sigma_e$. It is important to note that the shape of the yield surfaces obtained by compacting the samples using these two different methods are significantly different from each other. Thus, the states of the materials are different even though the densities are the same.

Another convenient way of presenting experimental data is in the form of isodensity plots. These plots are constructed by loading samples proportionally (either by maintaining the stresses or the strains proportional to each other) and then determining the stress states that correspond to a given density. If the material response can be described using the relative density as the only state variable, then these surfaces would be equivalent to the yield surfaces of the material. It is evident from the results of Fig. 3 that more than one state variable is required to describe the response. However, in many practical situations, an element of material experiences near proportional loading and isodensity or similar plots provide valuable information about the constitutive behavior. Brown and Abou-Chedid (Ref 11) constructed a series of isodensity plots in Σ_e – Σ_m space for spherical copper and sponge iron (Hoeganaes MH-100) powders (Fig. 4, 5). The solid symbols in these figures represent the yield strength in uniaxial compression following biaxial compression to the desired density. The major observations here are the different forms of the surface for the different powders, and the fact that the spherical powder has limited strength in uniaxial compression. These conclusions are consistent with the experiments of Akisanya et al. (Ref 10) presented in Fig. 5. The irregular powders, however, exhibit significant strength in uniaxial compression. Also, the irregular powders tested by Watson and Wert (Ref 9) had a significant yield strength when loaded in tension, while samples of spherical powder compacts can readily be broken in tension, often by hand, due to the lack of interlocking of the particles. These plots will be discussed and evaluated in the following sections. There are however, three important points to note at this stage:

- It is possible to identify a yield surface for a powder compact.
- The state of the material is a function of the prior history of loading. This influences the size and shape of the yield surface.
- The yield behavior depends on the morphology of the powder particles. Irregular particles interlock, giving tensile strength, while compacts of smooth spherical particles can be readily pulled apart.

Elastoplastic Constitutive Behavior

The constitutive behavior is discussed within the framework of small deformation theory of plasticity, which allows the decomposition of the total strain into elastic and plastic components and development of relatively simple expressions for the constitutive response. In practice, large plastic straining of a body can occur during the compaction process. The general forms of expression are still applicable, provided appropriate measures of stress, strain, stress-rate, and deformation-rate for these problems are defined. At this stage, it is not necessary to precisely determine what these definitions are. All necessary information can be determined by making the simple assumption of small strains. Start by considering the response of a microscopic continuum element of material (i.e., the response of the individual particles that form the compact).

Assume that the total strain, ε_{ij}, a body experiences can be decomposed into elastic and plastic components, such that:

$$\varepsilon_{ij} = \varepsilon^{e}_{ij} + \varepsilon^{p}_{ij} \qquad \text{(Eq 3)}$$

Where appropriate, the index notation for tensorial quantities is used to represent the constitutive response. The indices i and j can have a value in the range of 1 to 3. A repeating suffix implies summation over this range (thus, $\varepsilon_{ii} = \varepsilon_{11} + \varepsilon_{22} + \varepsilon_{33}$).

Elastic Behavior. The elastic strains depend on the current stress. For a linear elastic material,

$$\varepsilon^{e}_{ij} = C_{ijkl}\sigma_{kl} \qquad \text{(Eq 4)}$$

where C_{ijkl} is the elastic compliance matrix. If the material response is isotropic, it is more convenient to decompose the strain into deviatoric, ε^{ed}_{ij}, and volumetric, ε^{e}_{v}, components, such that:

$$\varepsilon^{e}_{ij} = \varepsilon^{ed}_{ij} + \frac{1}{3}\varepsilon^{e}_{v}\delta_{ij} = \frac{s_{ij}}{2G} + \frac{\sigma_m}{3K}\delta_{ij} \qquad \text{(Eq 5)}$$

where G is the shear modulus, K is the bulk modulus, $s_{ij} = \sigma_{ij} - \frac{1}{3}\sigma_{kk}\delta_{ij} = \sigma_{ij} - \sigma_m\delta_{ij}$ are the deviatoric components of stress, $\sigma_m = \frac{1}{3}\sigma_{kk}$ is the mean stress, and δ_{ij} is the Kronecker delta function ($\delta_{ij} = 0$ for $i \neq j$; $\delta_{ij} = 1$ for $i = j$). For a fully dense material, G and K are material constants. If the material is porous, both these moduli can be a function of the relative density of the material. This point is revisited later in the article in an examination of the constitutive response of powder compacts.

Classical Plasticity for Incompressible Materials. Central to the development of constitutive laws for plastically deforming materials is the concept of a yield surface. Here a more precise definition of ways in which the changing state of the material can be modeled is presented. The discussion is limited to classical plasticity concepts for fully dense materials, where plastic flow occurs at constant volume. Porous bodies are presented in subsequent sections.

A yield surface is shown schematically in Fig. 6 in principal plane stress space, in which there are two measures of stress σ_1 and σ_2. Mathematically, this surface can be defined by the equation:

$$f(\sigma_{ij}) = 0 \qquad \text{(Eq 6)}$$

where $f(\sigma_{ij})$ is a function of stress. For example, for an isotropic incompressible material, a suitable form of this expression is:

$$f(\sigma_{ij}) = \sigma_e - \sigma_y \qquad \text{(Eq 7)}$$

where σ_y is the yield strength of the material in uniaxial tension and σ_e is the von Mises effective stress, given by:

$$\sigma_e = \sqrt{\frac{3}{2} s_{ij}s_{ij}} \qquad \text{(Eq 8)}$$

In principal plane-stress space,

$$\sigma_e = \sqrt{\sigma_1^2 + \sigma_2^2 - \sigma_1\sigma_2} \qquad \text{(Eq 9)}$$

Equation 9 then represents the von Mises ellipse shown in Fig. 6. If $f < 0$, the material response is elastic, and if $f = 0$ (i.e., if the stress state is on the yield surface), the material is able to deform plastically. The increments of plastic strain at yield are related through the associated flow rule:

$$d\varepsilon^{p}_{ij} = \lambda\frac{\partial f}{\partial \sigma_{ij}} \qquad \text{(Eq 10)}$$

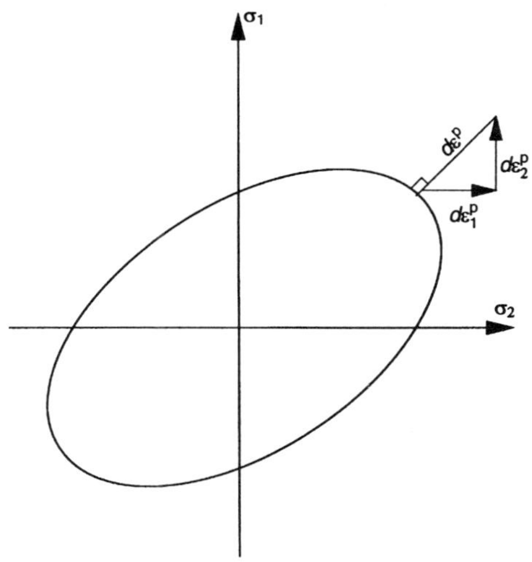

Fig. 6 Schematic of yield surface in plane-stress space, showing strain increment vector normal to the yield surface

where λ is a plastic multiplier. This equation simply states that the strain increment vector is normal to the yield surface in stress space. Additional information is needed to determine the exact value of λ.

For the yield condition of Eq 7,

$$\frac{\partial f}{\partial \sigma_{ij}} = \frac{\partial \sigma_e}{\partial \sigma_{ij}} = \frac{\partial \sigma_e}{\partial s_{ij}} = \frac{3}{2}\frac{s_{ij}}{\sigma_e} \qquad \text{(Eq 11)}$$

Now,

$$d\varepsilon^{p}_{ij} = \lambda\frac{3}{2}\frac{s_{ij}}{\sigma_e} \qquad \text{(Eq 12)}$$

Also,

$$d\varepsilon^{p}_{ij}\, d\varepsilon^{p}_{ij} = \lambda^2\frac{9}{4}\frac{s_{ij}s_{ij}}{\sigma_e^2} = \frac{3}{2}\lambda^2 \qquad \text{(Eq 13)}$$

Thus,

$$\lambda = \sqrt{\frac{2}{3} d\varepsilon^{p}_{ij}\, d\varepsilon^{p}_{ij}} = d\varepsilon^{p}_{e} \qquad \text{(Eq 14)}$$

where $d\varepsilon^{p}_{e}$ is an effective strain increment. The volumetric component of the strain increment at yield is:

$$d\varepsilon^{p}_{v} = d\varepsilon^{p}_{ii} = \frac{3}{2}\frac{s_{ii}}{\sigma_e} = 0 \qquad \text{(Eq 15)}$$

which is consistent with the initial assumption of incompressibility.

This manipulation of equations is standard, but there are some important points to note:

- The effective stress, which drives the deformation, is only a function of the deviatoric stresses (i.e., it is independent of the magnitude of the mean stress).

- The strain increments at yield are proportional to the deviatoric components of stress.
- The deviatoric component of stress, and therefore the effective stress, drives changes of shape of an element.
- Volume changes are driven by the mean stress.

In the above description the magnitude of the strain increment at yield has not been determined; only the relative magnitudes of the strain components have been determined. To determine the magnitude, the evolution of the state of the material with time must be understood. In Eq 7, the state is represented in terms of the yield strength of the material. Use of a single state variable of this type implies isotropic behavior. Below, two particular forms of isotropic models are examined.

Perfect Plasticity. The simplest assumption is that the yield strength remains constant, in other words, the yield surface retains the same size and shape in stress space. The plastic strain increment $d\varepsilon_{ij}^{p}$ can then have any value. In a structural problem, the exact value of the strain increment is determined either from the full elastoplastic response of the body or from any resulting geometry changes.

Isotropic Hardening. As the material is deformed plastically, the yield stress can increase. Consider the situation where yield strength is a function of the accumulated plastic strain ε_{e}^{p}:

$$\varepsilon_{e}^{p} = \int_{0}^{\varepsilon_{e}^{p}} d\varepsilon_{e}^{p} \quad \text{(Eq 16)}$$

that is,

$$\sigma_{y} = \sigma_{y}(\varepsilon_{e}^{p}) \quad \text{(Eq 17)}$$

and

$$d\sigma_{y} = \frac{d\sigma_{y}}{d\varepsilon_{e}^{p}} d\varepsilon_{e}^{p} = h(\sigma_{y}) d\varepsilon_{e}^{p} \quad \text{(Eq 18)}$$

where $h(\sigma_{y})$ is the slope of a plot of yield strength against plastic strain in a standard uniaxial material test.

Plastic flow can only occur if the stress state is on the yield surface, in other words, plastic flow only occurs as the stress is increased from σ_{ij} to $\sigma_{ij} + d\sigma_{ij}$ if $f = 0$ throughout the increment and $(d\sigma_{ij})(\partial f/\partial \varepsilon_{ij}) > 0$. Then, differentiating (Eq 7) gives:

$$df = \frac{\partial f}{\partial \sigma_{ij}} d\sigma_{ij} + \frac{\partial f}{\partial \sigma_{y}} d\sigma_{y} = \frac{\partial f}{\partial \sigma_{ij}} d\sigma_{ij} - d\sigma_{y} = 0 \quad \text{(Eq 19)}$$

Combining Eq 18 and 19 gives

$$d\varepsilon_{e}^{p} = \frac{1}{h(\sigma_{y})} \frac{\partial f}{\partial \sigma_{ij}} d\sigma_{ij} \quad \text{(Eq 20)}$$

The individual strain increments can be obtained by combining Eq 20 with Eq 12 and 14.

In the above model, it is assumed that the material remains isotropic as it deforms plastically, so that the yield surface maintains the same shape and simply changes in size as plastic strain is accumulated. In practice, the material response can become anisotropic. One means of representing anisotropy is through the use of a kinematic hardening model, wherein the center of the yield surface in stress space is allowed to translate according to some rule. See Ref 12 for further information. As a different approach towards anisotropy, Hill (Ref 13) presents a generalized theory of plasticity that allows for a change in the shape of the yield surface, while the basic requirements of convexity and incompressibility are maintained.

There are many other types of constitutive laws that have been developed to model the response of plastically deforming materials. As in the models described here, these alternatives mainly consist of identifying a yield surface and developing appropriate laws for the expansion, translation, and change in shape of the surface as the material deforms plastically. These laws can contain many state variables, together with complex evolution laws. The choice of constitutive law depends on the type of loading histories being evaluated. If the stress history experienced by the material point is simple, then relatively simple constitutive laws adequately describe the material response. For example, for monotonic proportional loading, the kinematic and isotropic hardening models predict the same response provided they are fit to the same base data. A detailed knowledge of the shape and orientation of the entire yield surface and how these evolve with time is only required if complex loading histories are being considered. It should also be remembered that the more complex the constitutive law, the wider the range of tests that need to be devised to determine all the material parameters and the evolution equations for each of the state variables. When selecting the structure of a constitutive law, it is therefore important to first determine the type of stress history that each material point is likely to experience and to select the equations capable of capturing the major feature of the response.

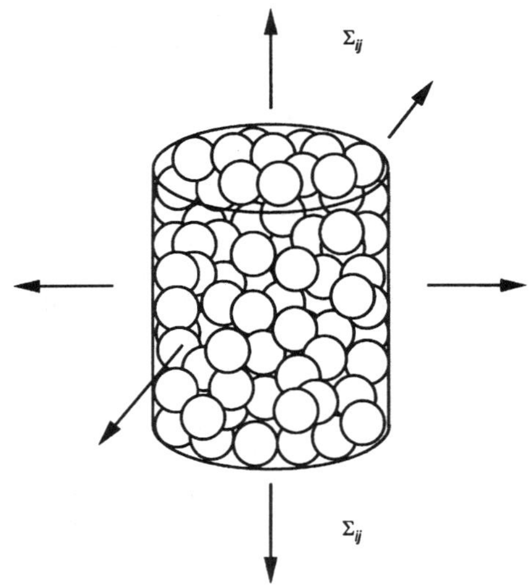

Fig. 7 Macroscopic element of powder compact subjected to stresses Σ_{ij}

In the following section, assume that the individual particles of a compact deform according to the constitutive models described in this section. This understanding provides a structure for the development of the constitutive laws for the compact.

The Structure of Constitutive Laws for Powder Material

Consider an element of a powder compact (Fig. 7), subjected to a macroscopic stress state Σ_{ij}, and experiencing strains E_{ij}. The strain can be decomposed into macroscopic elastic and plastic components, as before:

$$E_{ij} = E_{ij}^{e} + E_{ij}^{p} \quad \text{(Eq 21)}$$

Assume that the elastic response is isotropic. The elastic constitutive law can then be expressed in the form of Eq 5:

$$E_{ij}^{e} = E_{ij}^{ed} + \frac{1}{3} E_{v}^{e}\delta_{ij} = \frac{S_{ij}}{2G} + \frac{\Sigma_{m}}{3K}\delta_{ij} \quad \text{(Eq 22)}$$

Table 1 Experimental techniques for the determination of elastic properties of powder compacts

Experimental technique	Elastic property	Comments	Reference
Triaxial test unloading	*E*(*D*)	Can determine *E* at low density (*D* is ~0.65 for atomized steel powders). Experimental apparatus is involved.	Ref 14
Resonant frequency	*E*(*D*)	Very accurate and repeatable. Simple experimental setup. Simple sample preparation. Commercial laboratories available that use this technique. Minimum density *D* is ~0.75 for atomized steel powders.	Ref 15
Ultrasound	*E*(*D*)	Higher scatter than resonant frequency. Able to determine anisotropic properties in a transverse rupture bar.	Ref 15
Hydrostatic test unloading	*K*(*D*)	Can determine *K* at low density (*D* is ~0.65 for atomized steel powders). Experimentally difficult at very high density.	Ref 14

where G and K are functions of the state of the material. Several techniques for determining the elastic moduli G and K are summarized in Table 1. It is sometimes more practical to measure Young's modulus E instead of shear modulus G. Shear modulus can be determined using $G = 3KE/9(K - E)$.

The variables E and K can be measured during the unloading of a triaxial compaction test and a hydrostatic compaction test, respectively. Another technique used to determine E is resonant frequency. The powder is compacted into a beam, typically a transverse rupture bar. The beam is excited through an impact load, and the natural vibration frequency is measured. This frequency can be used to compute E. Ultrasound has also been applied to the determination of E in compacted specimens.

Watson and Wert (Ref 9) measured the variation of G and K as a function of D for samples which had been compacted in closed die. They found that:

$$G = G_m \exp(8.15(D - 1)) \quad \text{(Eq 23)}$$

$$\chi = \chi_m \exp(5(D - 1)) \quad \text{(Eq 24)}$$

with

$$K = \chi + \frac{2}{3}G \quad \text{(Eq 25)}$$

and G_m and χ_m representing the shear modulus and Lamé constant for the fully dense material, respectively.

If a microscopic element deforms plastically according to any of the models described, and there are no other dissipative mechanisms operating, then it is possible to identify a yield surface in stress space:

$$F = F(\Sigma_{ij}, S_\alpha) = 0 \quad \text{(Eq 26)}$$

which is a function of stress and the state of the material, described in terms of a number of state variables, S_α. Plastic flow can only occur if the stress state is on the yield surface. It can further be shown that associated flow at the micro-level guarantees associated flow at the macroscopic level, thus:

$$dE_{ij}^p = \Lambda \frac{\partial F}{\partial \Sigma_{ij}} \quad \text{(Eq 27)}$$

where Λ is the macroscopic plastic multiplier.

If frictional sliding contributed to the internal dissipative processes, then it is still possible to identify a yield surface, but associated plastic flow cannot be guaranteed. Fleck (Ref 16) has examined the two extremes of frictional behavior: free sliding; and sticking, whereby sliding can only occur by shear yielding in the vicinity of a contact. In each case, there is no frictional dissipation and associated flow can be guaranteed. The yield surface for intermediate frictional conditions should lie between the yield surfaces for these two situations. Fleck (Ref 16) found that there is only a small difference between the surfaces for a given assumed state, indicating that only a small amount of energy is dissipated by shearing between the particles. Thus, even if frictional sliding occurs it will only have a small influence on the macroscopic response, and it is appropriate to assume that an associated flow rule is valid.

During compaction the stress point remains on the yield surface, therefore, following Eq 19:

$$dF = \frac{\partial F}{\partial \Sigma_{ij}} d\Sigma_{ij} + \frac{\partial F}{\partial S_\alpha} dS_\alpha = 0 \quad \text{(Eq 28)}$$

In order to complete the model, evolution laws for the state variables in terms of the stress, strain, and the state of the material are needed. In the models, the state could be described in terms of suitable measures of strain, or alternatively, plastic strains can be considered the state variables. Then Eq 28 becomes:

$$dF = \frac{\partial F}{\partial \Sigma_{ij}} d\Sigma_{ij} + \frac{\partial F}{\partial E_{ij}^p} dE_{ij}^p = 0 \quad \text{(Eq 29)}$$

Combining Eq 29 with Eq 27 gives

$$\Lambda = -\frac{(\partial F/\partial \Sigma_{ij})d\Sigma_{ij}}{(\partial F/\partial E_{kl}^p)(\partial F/\partial \Sigma_{kl})} \quad \text{(Eq 30)}$$

with the individual strain components given by (Eq 27).

To simplify constitutive models, isotropic behavior is often assumed. This consideration dictates the use of mechanical and material descriptions independent of a particular choice of coordinate system. This freedom allows one to use rotation invariant or coordinate system independent quantities in the constitutive model. For the mechanical stress or strain, which are mathematically represented as tensors of the second order, these quantities are the first and second tensor invariants. The yield function can then be expressed as a function of $\Sigma_e = \sqrt{3/2\, S_{ij} S_{ij}}$, the macroscopic von Mises effective stress in the powder aggregate, a form of the second invariant of the macroscopic stress and $\Sigma_m = 1/3\Sigma_{kk}$, the mean stress in the powder aggregate, a form of the first invariant of the macroscopic stress. Equation 26 then assumes the form: $F = F(\Sigma_e, \Sigma_m, S_\alpha) = 0$.

If appropriate forms for the yield function in terms of the stress and accumulated plastic strain can be found, the above equations can be employed to determine the full constitutive response. In the following sections, different possible forms of the yield function are examined.

Yield Functions from Micromechanical Models. The section "Deformation of Powder Compacts: Experimental Observations" states the nature of the porosity changes as a material densifies. Initially, the porosity is open and distinct necks exist between the contacting particles (stage 1). At high relative densities ($D > 0.9$), the porosity is closed, which is referred to as stage 2. Different forms of micromechanical models have been developed for these two stages. In the stage 1 models, it is assumed that there is no interaction between the deformation zones that form in the vicinity of the different contacts. Analyses of the contact of two isolated bodies can then be used to obtain the appropriate contact law for a given microplasticity model. The macroscopic response is then determined by combining the contributions from each contact. Fleck et al. (Ref 16) employed a perfectly plastic microplasticity model and assumed that the material was isotropic and could be described using a single state variable, which they took to be the relative density, D. This assumption is equivalent to assuming that the material had previously been compacted isostatically to a given density. From their analysis, Fleck et al. (Ref 16) propose an approximate yield function:

$$F = \left(\frac{\sqrt{5}\,\Sigma_m}{3P_y}\right)^2 + \left(\frac{5\Sigma_e}{18P_y} + \frac{2}{3}\right)^2 - 1 = 0 \quad \text{(Eq 31)}$$

where P_y is the yield strength in hydrostatic compression;

$$P_y = 3D^2 \frac{(D - D_0)}{(1 - D_0)} \sigma_y \quad \text{(Eq 32)}$$

where D_0 is the initial dense random packing density, which is generally taken to be 0.64.

The full response can be obtained from Eq 26 to Eq 30 by noting that the densification rate is given by (neglecting elastic volume changes):

$$\dot{D} = -D\dot{E}_{kk}^p \quad \text{(Eq 33)}$$

In practice, the matrix material can strain harden as the contact zones deform plastically. Fleck et al. (Ref 16) demonstrate how the effects of strain hardening can be taken into account. A more compact form of model can be obtained by rewriting Eq 31 in the form of Eq 7, that is,

$$F = \bar{\Sigma} - \sigma_y = 0 \quad \text{(Eq 34)}$$

where $\bar{\Sigma}$ is an effective stress for the porous material, which is a function of the macroscopic stress and internal geometry described by D. Rearranging Eq 31 then gives:

$$\bar{\Sigma} = \left[\frac{1}{3}\Sigma_e + \sqrt{\frac{1}{4}\Sigma_e^2 + \Sigma_m^2}\right] \frac{(1 - D_0)}{3D^2 (D - D_0)} \quad \text{(Eq 35)}$$

For loading conditions on the yield surface, Eq 29 becomes:

$$dF = \frac{\partial \bar{\Sigma}}{\partial \Sigma_{ij}} d\Sigma_{ij} + \frac{\partial \bar{\Sigma}}{\partial D} dD - d\sigma_y = 0 \quad \text{(Eq 36)}$$

Fleck et al. (Ref 16) take σ_y to be the average yield strength in the plastically deforming material, and they define an average internal effective strain increment, $d\varepsilon_e^p$, by equating the internal and external work. Using the current terminology,

$$\Sigma_{ij} dE^{p}_{ij} = \bar{\Sigma}\Lambda = \sigma_y \Lambda = G(D)\sigma_y \, d\varepsilon^{p}_{e} \qquad \text{(Eq 37)}$$

where $G(D)$ is the effective volume fraction of plastically deforming material, given by Fleck (Ref 17) as:

$$G(D) = \frac{45}{\sqrt{3}} D^2 \left(\frac{D - D_0}{1 - D_0} \right)^{3/2} \qquad \text{(Eq 38)}$$

Combining Eq 36, 37, 33, and 18 gives:

$$\Lambda = d\bar{E} = \frac{(\partial\bar{\Sigma}/\partial\Sigma_{ij})d\Sigma_{ij}}{D(\partial\bar{\Sigma}/\partial D)(\partial\bar{\Sigma}/\partial\Sigma_{kk}) + [h(\sigma_y)/G(D)]} \qquad \text{(Eq 39)}$$

where $d\bar{E}$ is a macroscopic effective plastic strain increment not to be confused with the von Mises effective strain increment, dE_e. The advantage of this form is that it separates out the effects of geometric and material hardening and it leads to a more compact expression than the relationships proposed by Fleck et al. (Ref 16).

In the development of the above model, two major assumptions were made: the strengths of the contact zones were assumed to be the same in tension and compression, and the material was assumed to be isotropic. It was demonstrated in the section "Deformation of Powder Compacts: Experimental Observations" that the strengths of the contacts are likely to be different in compression and tension, particularly for smooth particles. Also, when a compact is loaded along stress paths other than hydrostatic, the microstructure (the distribution of necks) becomes anisotropic.

Fleck (Ref 17) has proposed a model in which the tensile strength of a contact patch is less than the compressive strength by a factor η. If P_y (Eq 33) is the magnitude of the pressure required to plastically deform a compact, then the magnitude of the hydrostatic tensile stress required to initiate plastic flow is ηP_y. Fitting a quadratic to Fleck's results and adopting the form of expression presented in Eq 34, the following equation is obtained:

$$\bar{\Sigma} = \frac{1}{\eta}\left[\frac{1}{6}\Sigma_e (1+\eta) + \tfrac{1}{2}\Sigma_m(1-\eta) + \sqrt{\tfrac{1}{36}\Sigma_e^2(1+7\eta+\eta^2) + \tfrac{1}{4}\Sigma_m^2(1+\eta)^2 + \tfrac{1}{6}\Sigma_e\Sigma_m(1-\eta^2)} \right] \frac{(1-D_0)}{3D^2(D-D_0)} \qquad \text{(Eq 40)}$$

This expression reduces to Eq 35 when $\eta = 1$. When $\eta = 0$, it becomes:

$$\bar{\Sigma} = -\frac{4\Sigma_m^2 + \frac{5}{9}\Sigma_e^2(1-D_0)}{4\Sigma_m + \frac{4}{3}\Sigma_e \, 3D^2(D-D_0)} \qquad \text{(Eq 41)}$$

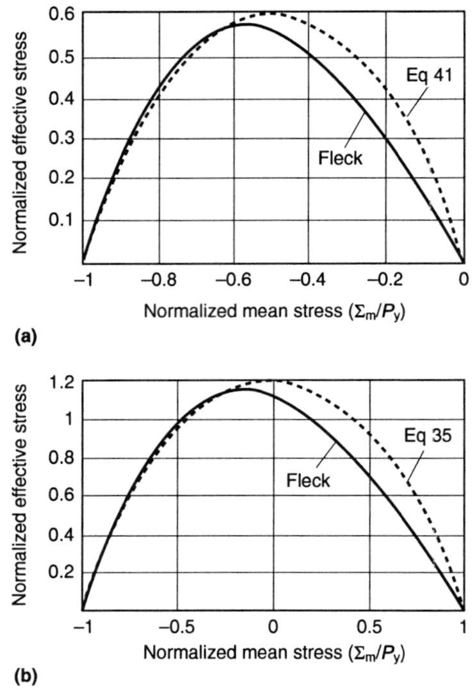

Fig. 8 Yield surfaces predicted by Fleck's (Ref 17) isotropic model for (a) $\eta = 0$ and (b) $\eta = 1$, compared with the prediction of Eq 41 and 35

The yield surfaces are plotted in Fig. 8 for $\eta = 0$ and 1, where they are compared with the surfaces obtained by Fleck (Ref 17). It should be noted that these surfaces have vertices where they meet the Σ_m axis. The direction of the strain increment vector is nonunique for pure hydrostatic stress states, which can lead to computational difficulties when implementing this model. In practice, this nonuniqueness has been bypassed by inserting a circular arc at the vertex to ensure a smooth continuous yield surface (Ref 19).

Fleck (Ref 17) has also examined the influence of material anisotropy by relating the size and number of contacts at a given orientation to the plastic strain E^p_{ij}. The surface for frictionless closed-die compaction, normalized by the hydrostatic stress, P, required to compact the material to the same density is shown in Fig. 9 for $\eta = 0$, where it is compared with the comparable surface for hydrostatic compaction. There are a number of important points to note:

- The surface for closed-die compaction is extended in the direction of loading and contracted in the transverse direction compared to the surface for hydrostatic compaction. Thus, the yield behavior cannot be described in terms of the relative density alone.
- The vertex on the yield surface is at the loading point. Similar results are obtained for crystal plasticity theory applied to fully dense materials.
- The yield surfaces expand in a self-similar manner. Thus, the different components of stress increase in proportion to each other when the straining is proportional.

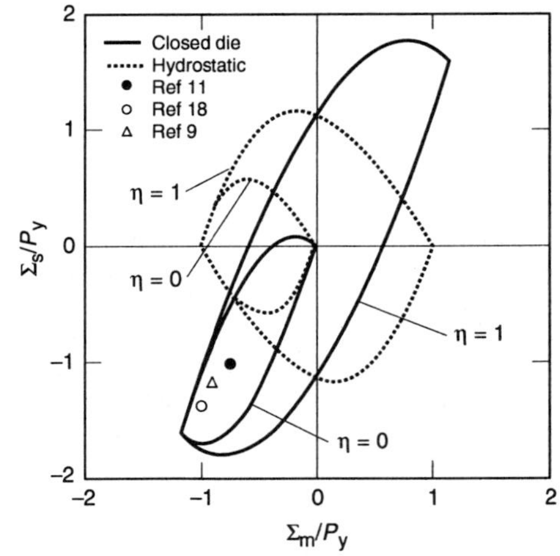

Fig. 9 Yield surfaces for isostatic and closed-die compaction predicted by the anisotropic model of Fleck (Ref 17)

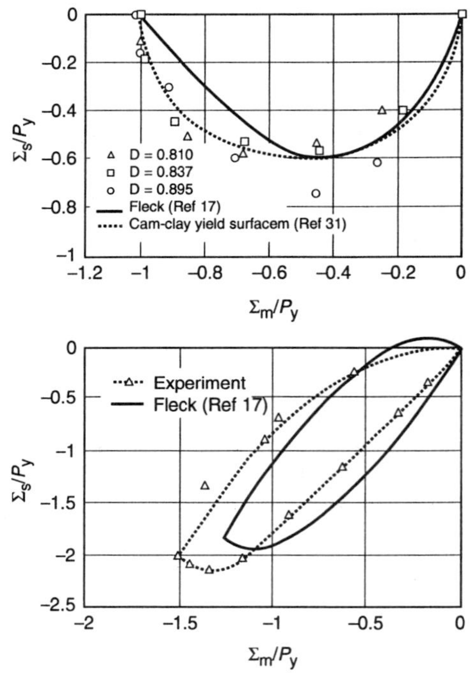

Fig. 10 Comparison of Fleck's (Ref 17) model with the experimental results of Akisanya et al. (Ref 10)

- It is not possible to obtain a closed form solution for arbitrary loading paths. Thus, the full micromechanical model is required in order to compute the constitutive response.

The predictions of this model are compared with the experimental results of Akisanya et al. (Ref 10), shown in Fig. 10 for hydrostatic and closed die compaction. It is evident that the general form of these surfaces is well represented by this model.

So far only the early stages of compaction, when distinct necks exist between the particles, have been discussed. In the latter stages of compaction ($D > 0.9$), the pores are isolated. Constitutive models employed in this regime have been taken from studies originally concerned with ductile failure. In these models, the pores are assumed to remain spherical. The most widely used model is developed by Gurson (Ref 20), which predicts a yield surface given by:

$$F = \left(\frac{\Sigma_e}{D\sigma_y}\right)^2 + \frac{2(1-D)}{D^2}\cosh\left(\frac{3\Sigma_m}{2\sigma_y}\right) - \frac{2(1-D)}{D^2} - 1 = 0 \quad \text{(Eq 42)}$$

This expression suffers from the same problem as Eq 31, in that the effects of material and geometric hardening are coupled. Cocks (Ref 21) proposed that a more appropriate form for the constitutive model, which does not differ significantly from that proposed by Gurson (Ref 20), is given by Eq 34, with

$$\bar{\Sigma} = \sqrt{\frac{\Sigma_e^2}{G_1(D)} + \frac{\Sigma_m^2}{G_2(D)}} \quad \text{(Eq 43)}$$

where

$$G_1(D) = \frac{3D^2}{5-2D} \quad \text{(Eq 44)}$$

$$G_2(D) = \left(\ln\left(\frac{1}{(1-D)^2}\right)\right)^2 \quad \text{(Eq 45)}$$

Equations 36 to 39 for the constitutive response again apply, where, now, $G(D) = D$.

In practice, the pores do not achieve a spherical shape during the latter stages of compaction. Instead, they maintain a cusp shaped profile. Liu et al. (Ref 22) and Akisanya et al. (Ref 23) have demonstrated that bodies with cusp-shaped pores have greater compressibility than bodies with spherical pores at the same density. Qian et al. (Ref 24) have suggested how the influence of the cusp-shaped pores can be incorporated into the constitutive models for creeping materials. They still assume that the body is isotropic macroscopically. Extending their analysis to a perfectly plastic material would simply result in different definitions of $G_1(D)$ and $G_2(D)$ in Eq 43.

Under general loading situations, the expectation might be that the pores would become squashed in the main direction of compaction. Ponte Castenada (Ref 25) has examined the effect of oriented ellipsoidal-shaped pores on the constitutive response, but there have not been any studies of the influence of distorted cusp-shaped pores on the material behavior. In all the stage 2 models, it has always been assumed that the contact patches between the distorted particles are able to support tensile tractions and the contacting particles are unable to slide over each other. There has been no attempt to analyze the effect of the strength of these contacts on the constitutive response.

The yield surfaces for stage 1 and stage 2 compaction can have different shapes. In order to ensure a smooth transition, a gradual change from one shape to another is assumed to occur over a given range of relative density. Assuming a linear variation from the yield expression for stage 1 to that for stage 2 over the range D_1 to D_2 gives:

$$\bar{\Sigma} = \left(\frac{D_2 - D}{D_2 - D_1}\right)\bar{\Sigma}_1 + \left(\frac{D - D_1}{D_2 - D_1}\right)\bar{\Sigma}_2 \quad \text{(Eq 46)}$$

where $\bar{\Sigma}_1$ is the effective stress for stage 1 compaction and $\bar{\Sigma}_2$ is the effective stress during stage 2. In practice, the transition is generally assumed to occur over the range $D_1 = 0.77$ to $D_2 = 0.9$.

Empirical Yield Functions. In the empirical models of powder compaction, the different form of porosity in stage 1 and stage 2 is not considered. Simple continuous expressions for the yield function are developed that satisfy the condition that the material response is incompressible when $D = 1$. The most widely used expressions are those due to Kuhn and Downey (Ref 6) and Shima and Oyane (Ref 7). These models have the same general structure as those described in the section "Yield Functions from Micromechanical Models." The yield function can again be represented in the form of Eq 34. For simplicity, a quadratic function of the von Mises effective stress, Σ_e, and mean stress, Σ_m, is assumed for $\bar{\Sigma}$. It is further assumed that the material response can be described in terms of a single state variable, the relative density, D. $\bar{\Sigma}$ can then be expressed in the form of Eq 43, where $G_1(D)$ and $G_2(D)$ are determined experimentally by assuming that the material yield strength σ_y remains constant. Shima and Oyane (Ref 7) determined these quantities by compacting samples in a closed die to a given density, sintering the samples and then performing uniaxial tensile and compressive tests to determine the yield properties. By measuring the uniaxial yield strength and Poisson's ratio, $G_1(D)$ and $G_2(D)$ could be determined at the chosen density; and by repeating these tests at different densities, the complete functions $G_1(D)$ and $G_2(D)$ could be determined. Their results for copper and iron powders are well approximated by:

$$G_1(D) = D^5 \quad \text{(Eq 47)}$$

$$G_2(D) = \frac{D^5}{6.25(1-D)} \quad \text{(Eq 48)}$$

The above form has been employed by a number of authors in experimental and computational studies of powder compaction. Kim and Suh (Ref 26) recognized that a material could harden as a result of shear deformation as well as a result of densification. They proposed a modified form of expression for the yield function:

$$F = \bar{\Sigma} - \kappa = 0 \quad \text{(Eq 49)}$$

with

$$\bar{\Sigma} = \sqrt{\Sigma_e^2 + 1.2(1-D)\Sigma_m^2} \quad \text{(Eq 50)}$$

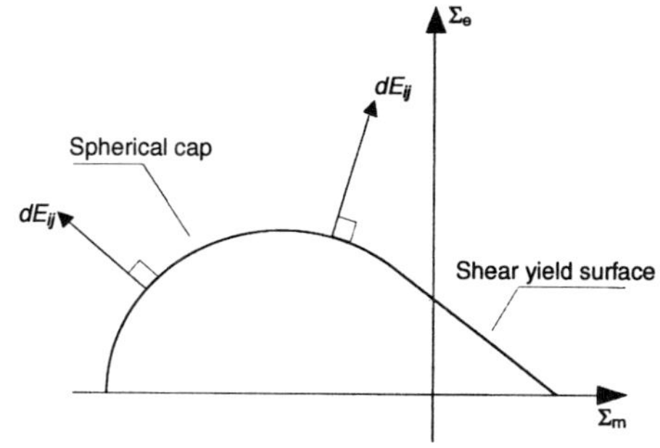

Fig. 11 Schematic of Watson and Wert's (Ref 9) two surface Drucker-Prager model

Here κ is the effective yield strength of the compact. The evolution law for κ is assumed to be of the form:

$$d\kappa = \beta_1\sigma_e d\varepsilon_e + \beta_2\sigma_m d\varepsilon_{kk} \quad \text{(Eq 51)}$$

which takes into account geometric and material hardening. Now, from Eq 28 and 51, find:

$$\Lambda = \frac{(\partial\bar{\Sigma}/\partial\Sigma_e)\, d\Sigma_e + (\partial\bar{\Sigma}/\partial\Sigma_m)\, d\Sigma_m}{D\,(\partial\bar{\Sigma}/\partial D)\,(\partial\bar{\Sigma}/\partial\Sigma_m) + \beta_1(\partial\bar{\Sigma}/\partial\Sigma_e)\,\Sigma_e + \beta_2(\partial\bar{\Sigma}/\partial\Sigma_m)\,\Sigma_m} \quad \text{(Eq 52)}$$

and the individual plastic strain increments are given by Eq 27. The yield function of Eq 49 was found to give a slightly better fit of the tension/torsion experiments of Kim et al. (Ref 5) on sintered iron compacts discussed earlier than the model of Shima and Oyane (Ref 7).

These models were validated by performing experiments on sintered compacts. The effect of the sintering process is to anneal the particles and to bond them together. As a result, the yield strength is expected to be the same in tension and compression, as observed by Shima and Oyane (Ref 7), and it is not unreasonable for a quadratic function, such as those presented in Eq 43 and 50, to adequately describe the shape of the yield surface. A consequence of this choice of yield function is that it will always predict a negative volumetric strain-rate when the mean stress is negative (i.e., compaction will always occur under net compressive stress states). In experiments on compacted, but not sintered, powders, Brown and Abou-Chedid (Ref 27) and Watson and Wert (Ref 9) observed that the volumetric strains were positive in uniaxial compression.

Based on their experimental observations, Watson and Wert (Ref 9) proposed the adoption of Drucker and Prager's (Ref 28) two part yield surface, which has been employed to model the response of soils. This surface is shown schematically in Fig. 11 and consists of a shear surface and a spherical cap. Watson and Wert (Ref 9) assumed associated flow for both surfaces (i.e., the strain-increment vector is normal to the surface). Thus, dilation occurs for stress states on the shear surface and compaction occurs if the loading point lies on the spherical cap.

Following Gurson and McCabe (Ref 29), Brown and Abou-Chedid (Ref 27) proposed a modified form of the Cam-Clay model (Ref 30, 31) to model their experiments. In this model the yield surface is represented by Eq 34, with

$$\bar{\Sigma} = \sqrt{A(D)\Sigma_e^2 + B(D)(\Sigma_m - C)^2} \quad \text{(Eq 53)}$$

Brown and Abou-Chedid (Ref 27) did not propose forms for the functions $A(D)$ and $B(D)$. More detailed experiments are required to determine the exact form of these expressions.

The macroscopic effective stress of Eq 53 is similar to the quadratic forms of Eq 43 and 50, but now the elliptic surface is centered on the point (0, C) in $\Sigma_e - \Sigma_m$ space. The quantity C represents the degree of cohesion between the particles. If they are perfectly bonded, $C = 0$, and the surface reduces to the symmetric expressions of Eq 43 and 50, which is appropriate for sintered materials. If the particles are smooth and the contacts are unable to support a normal tensile stress, then the value of C can be chosen so that the yield surface passes through the origin, as in the micromechanical model of Eq 41 for $\eta = 0$. Then,

$$C = -\frac{\bar{\Sigma}}{(B(D))^{1/2}} \quad \text{(Eq 54)}$$

and Eq 53 becomes

$$\bar{\Sigma} = -\frac{A(D)\Sigma_e^2 + B(D)\Sigma_m^2}{2(B(D))^{1/2}\,\Sigma_m} \quad \text{(Eq 55)}$$

Equation 55 is equivalent to the most common form of Cam-Clay model used to evaluate the response of soils (Ref 31).

These last two models were derived from constitutive laws that have been developed for soils. A wide range of models have been proposed in the literature for granular materials, in which the yield surfaces are allowed to adopt different shapes, as well as being able to translate in stress space, in a similar manner to the kinematic models (Ref 31).

A Constitutive Model for Metallic Powders with Ductile Particles

Several constitutive models have been applied to the numerical simulation of powder compaction in dies. The most widely used are variations of two types of models: models with an empirical quadratic yield function (Ref 32–34, 40, 41) and Cap models (Ref 35–36, 38–39).

The main practical advantages of a model of the first type as developed in Ref 34, 36, 40, 41 include:

- A reasonable representation of the behavior of metal powders with ductile particles under monotonic loading
- A continuous yield function that facilitates its numerical implementation
- Experimental calibration with a relatively small number of tests

Furthermore, this model has been implemented in the compaction modeling finite element code PCS (Ref 42) that is in use by several parts manufacturers in the United States. The overall strategy presented here can be employed for any of the models described in the section "Constitutive Models for Metal Powder Compaction." The only difference lies in the range of experiments that need to be performed in order to determine any unknown functions in the models.

Model Formulation

The model considered in detail here uses an empirical yield function and fits the general framework presented earlier with assumptions:

- As the powder aggregate is compacted, the particles deform plastically according to the behavior described by classical plasticity with isotropic hardening
- The powder compact displays macroscopic elastic-plastic behavior that is isotropic and independent of strain rate
- The total strain can be decomposed into elastic and plastic components
- The contributions of particle sliding to the overall deformation are negligible
- The state of the powder aggregate is represented by two state variables, the relative density, D, and the yield strength of the powder particle σ_y

Rigorously, for elastic-plastic models, a slightly different definition of the relative density D is necessary when D is used as a state variable. Earlier D was defined as the ratio of the density of powder to the density of the fully dense material. However, the plastic state of the material should not be affected by the changes in D that are due to elastic deformation. A more appropriate definition of D is the ratio of the density of unloaded powder to the density of the fully dense material, the relaxed relative density of the powder aggregate.

Following the general approach presented earlier, consider an element of a powder compact, which is subjected to a macroscopic stress state Σ_{ij} and experiences strains E_{ij}. The strain can be decomposed into macroscopic elastic and plastic components as per Eq 21:

$$E_{ij} = E_{ij}^e + E_{ij}^p \quad \text{(Eq 56)}$$

Here assume that the elastic response is isotropic. The elastic constitutive law can then be expressed by Eq 22:

$$E_{ij}^e = E_{ij}^{ed} + \frac{1}{3}E_v^e\delta_{ij} = \frac{S_{ij}}{2G} + \frac{\Sigma_m}{3K}\delta_{ij} \quad \text{(Eq 57)}$$

where G and K are functions of D.

An empirical yield function F of the general form of Eq 26, which is a function of the stress state and two state variables, is given by:

$$F = F(\Sigma_e, P, \sigma_y, D) = 0 \quad \text{(Eq 58)}$$

where $P = -\frac{1}{3}\Sigma_{kk} = -\Sigma_m$, is the pressure, a form of the first invariant of the macroscopic stress.

The specific form of the yield function is as defined in (Eq 34) and (Eq 43):

$$F = \frac{\Sigma_e^2}{G_1(D)} + \frac{P^2}{G_2(D)} - \sigma_y^2 = 0 \quad \text{(Eq 59)}$$

As the powder aggregate densifies and reaches high relative density, its response will approach the incompressible plastic behavior of fully dense metals. Therefore, the function $G_1(D)$ is expected to increase monotonically with an increase in relative density and the function $G_2(D)$ is expected to decrease monotonically with an increase in relative density.

The yield functions developed by Trasorras et al. (Ref 34) involved functions $b(D)$ and $c(D)$ that are expressible in terms of the functions $G_1(D)$ and $G_2(D)$ in the following manner: $G_1(D) = c(D)$ and $G_2(D) = 3c(D)/2b(D)$.

In the following development, the yield function is cast in the form used in Ref 34.

$$F = \Sigma_e^2 + \frac{3}{2}b(D)P^2 - c(D)\sigma_y^2 = 0 \quad \text{(Eq 60)}$$

The functions $b(D)$ and $c(D)$ are to be determined through experiments.

The associated flow at the macroscopic level (from Eq 27) yields:

$$dE_{ij}^p = d\Lambda\frac{\partial F}{\partial \Sigma_{ij}} = d\Lambda\left[\frac{\partial F}{\partial \Sigma_e}\frac{\partial \Sigma_e}{\partial \Sigma_{ij}} + \frac{\partial F}{\partial P}\frac{\partial P}{\partial \Sigma_{ij}}\right]$$
$$= 3d\Lambda\left[S_{ij} - \frac{1}{3}Pb(D)\delta_{ij}\right] \quad \text{(Eq 61)}$$

From Eq 61, the volumetric and deviatoric components of the plastic strain are given by:

$$dE_v^p = -d\Lambda\frac{\partial F}{\partial P} = -3d\Lambda Pb \quad \text{(Eq 62)}$$

$$dE_e^p = d\Lambda\frac{\partial F}{\partial \Sigma_e} = 2\Sigma_e d\Lambda \quad \text{(Eq 63)}$$

Eliminating $d\Lambda$ allows

$$dE_v^p\frac{\partial F}{\partial \Sigma_e} + dE_e^p\frac{\partial F}{\partial P} = 0 \quad \text{(Eq 64)}$$

The conservation of mass is expressed in the form of the continuity equation (from Eq 33):

$$\dot{D} = -D\dot{E}_{kk}^p \quad \text{(Eq 65)}$$

On integration, the continuity equation yields the following result for the evolution of relaxed relative density,

$$D = D_0 \exp(-E_{kk}^{p}) \quad \text{(Eq 66)}$$

The isotropic hardening law for the powder particle (from Eq 18) is:

$$d\sigma_y = h d\varepsilon_e^{p} \quad \text{(Eq 67)}$$

The particle hardening h may be constant (isotropic linear hardening), or may be a function of equivalent plastic strain in the particle.

If the energy dissipated by particle sliding is negligible, as assumed earlier, the external and internal plastic work of deformation can be equated (from Eq 37 with $G(D) = D$):

$$\Sigma_{ij} dE_{ij}^{p} = D\sigma_y d\varepsilon_e^{p} \quad \text{(Eq 68)}$$

Application to Powder Blends

In the previous sections, the constitutive models have assumed that the powder consists of an aggregate of metal particles of a single kind. In powder compaction in dies, the powder blend will contain lubricant, and in the case of ferrous alloys, often graphite and other alloying elements as well. The theories presented here are expected to be valid for powder blends, provided that the additions to the base metal powder occupy a small volume of the total aggregate (typically a few volume percent). That being the case, the models can be used with the constraints that:

- The relative density D be defined as the ratio of the total density of the powder to the pore-free density of the powder to enforce a plastically incompressible behavior when the material reaches the pore-free density, as opposed to the density of the fully dense metal particle.
- The elastic properties, which are strongly dependent on the presence of nonmetallic constituents, be determined experimentally for the specific blend.
- The constitutive functions $b(D)$ and $c(D)$ be experimentally determined for the specific blend.

Experimental Determination of Powder Material Constitutive Properties and Functions

In this section, the experimental procedures that can be applied to determine the constitutive parameters that appear in the model of the preceding section are presented. Techniques used to determine the elastic parameters and the form of the yield function have been described in the section "Deformation of Powder Compacts: Experimental Observations."

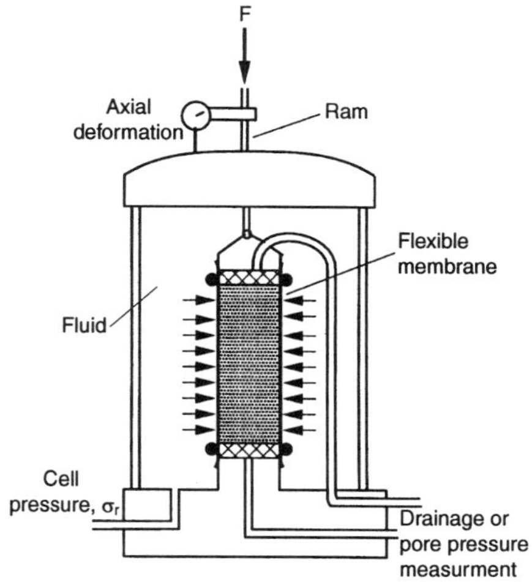

Fig. 12 Triaxial compaction cell. Source: Ref 31

Triaxial Tests

A brief description of the triaxial test and apparatus is presented in this section. For a more detailed treatment of the test, stemming from a soil mechanics perspective, see Ref 31. The triaxial test is important as it provides a convenient means of examining a variety of situations involving different ratios of the deviatoric and hydrostatic stress measures. A schematic of the triaxial test cell is shown in Fig. 12. The powder is placed in a container with a flexible wall, typically made of an elastomer, and immersed in a fluid under a pressure, p. Take care not to confuse this notation for the magnitude of fluid pressure with the one used elsewhere for the hydrostatic component of stress (P). A ram at the top of the apparatus provides a means of increasing the axial load on the powder while maintaining fluid pressure. Knowledge of the cell pressure as well as the applied ram load allows for the calculation of the resulting axial stress in the test sample. In addition, appropriate instrumentation on the apparatus ensures measurement of axial and radial dimensional changes. These measurements can then be translated into axial and radial strains. With the added assumption of a uniform stress state in the specimen and isotropic behavior of the material, it is reasonable to conclude that the radial and tangential strains, as well as radial and tangential stresses, are equal. Macroscopically, the stress state of the powder can be stated as:

$$\Sigma_z = -\left[\frac{F}{A} + p\left(1 - \frac{a}{A}\right)\right] \quad \text{(Eq 69)}$$

$$\Sigma_r = -p \quad \text{(Eq 70)}$$

The quantities A and a are the areas of cross section of the specimen and ram, respectively. The strains are calculated as:

$$E_r = \ln\left(\frac{r}{r_0}\right) \quad \text{(Eq 71)}$$

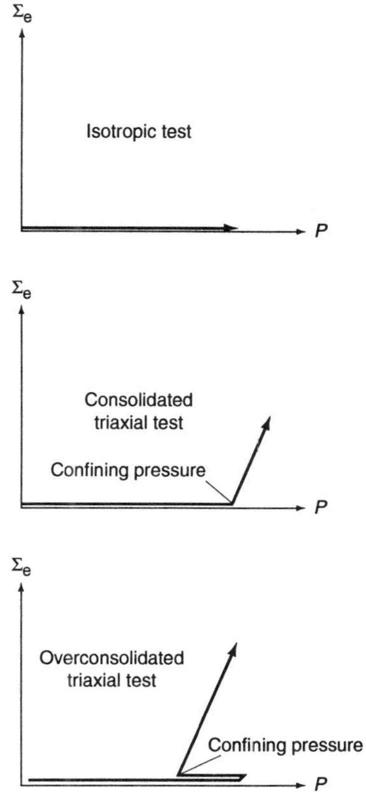

Fig. 13 Stress paths in isostatic compaction test, consolidated triaxial compaction test, and overconsolidated compaction test. Source: Ref 43

$$E_z = \ln\left(\frac{l}{l_0}\right) \quad \text{(Eq 72)}$$

where r_0 and l_0 are, respectively, the initial radius and length of the compact, and r and l, the current values of these quantities. The deviatoric and hydrostatic components of stress reduce to:

$$\Sigma_e = \sqrt{\frac{3}{2} S_{ij} S_{ij}} = \Sigma_r - \Sigma_z \quad \text{(Eq 73)}$$

$$P = -\frac{1}{3}(2\Sigma_r + \Sigma_z) \quad \text{(Eq 74)}$$

In soil mechanics, triaxial tests are often performed in one of three ways, as depicted in Fig. 13. The first, the isostatic compaction test, is used to characterize soils under hydrostatic loading and to determine the elastic bulk modulus. In this test, the ram load is maintained insofar as to ensure a hydrostatic state of stress in the test specimen. The second test, the consolidated triaxial test, involves initial pressurization of the test cell followed by an increase in ram load, the cell pressure being maintained constant. The path of stress on the P-Σ_e plane has a slope of 3 as the axial load increases at constant containing pressure. This path can be verified by taking the derivatives of the expressions for P and Σ_e while keeping p constant, and then calculating the slope using the ratio of the derivatives. Finally, the overconsolidated tests are employed to help determine the shape of the yield surface on the P-Σ_e plane. The test consists of subjecting the specimen to a history of deformation under pure pressure, following this by unloading to an intermediate pressure, and subsequently increasing the ram load while maintaining cell pressure to determine a different point on the yield surface.

As presented in the section "Deformation of Powder Compacts: Experimental Observations," the size and shape of the yield surface are functions of the history of loading. Thus, if the strain history only involved hydrostatic loading, subsequent probing of the yield surface in the manner of an overconsolidated triaxial test will only identify a yield surface defined by this initial loading. A different picture will emerge if the yield surface is identified following a stress path, such as the one used in a consolidation test, followed by unloading and subsequent probing of the yield surface (Fig. 3). This is of particular importance when considering tests to be used for material parameter extraction; the best tests shall be those involving stress paths as close to the real situation as possible. In the case of compaction in dies, the compact is constrained radially. The triaxial test setup should then include control of cell pressure to ensure zero radial strain in the sample to simulate actual compaction conditions.

Calibration of Material Parameters for an Iron Powder Blend

The procedures of the previous section are applied below to the calibration of the two-state variable model introduced earlier. The data is for a powder blend comprising 99.5% by weight of Distalloy AE, 0.5% by weight of graphite, and 1% wax Hoechst micropulver. This last component is admixed as internal lubricant. Distalloy AE is a diffusion alloyed iron powder with composition 4 wt% Ni, 1.5 wt% Cu, and 0.5 wt% Mo. Particle sizes for this powder range from 20 to 180 μm. The apparent density of the powder is 3.04 g/cm^3; the pore free density of the material is 7.33 g/cm^3. The data used are from experiments reported by Pavier and Doremus (Ref 14).

Consider the yield function of the model introduced in the section "A Constitutive Model for Metallic Powders with Ductile Particles":

$$F = \Sigma_e^2 + \frac{3}{2} b(D)P^2 - c(D)\sigma_y^2 = 0$$

To fully determine the yield function F the functions $b(D)$ and $c(D)$ must be identified. Figure 14 illustrates the yield surface for a lubricated atomized iron powder (Hoeganaes Ancorsteel 1000). The surface is plotted for different values of relative density and particle yield stress pairs. For this powder, the functions $b(D)$ and $c(D)$ were determined by Trasorras et al. (Ref 40) to be:

$$b(D) = \frac{0.13D^{-5.6}}{(D+0.62)^{1.35}} \quad \text{(Eq 75)}$$

$$c(D) = D^6 \quad \text{(Eq 76)}$$

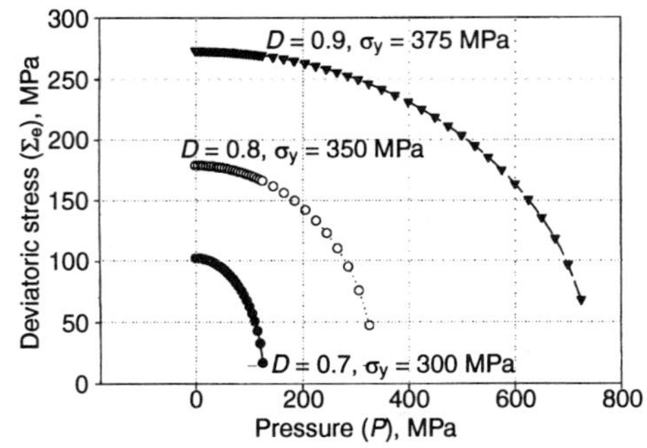

Fig. 14 Yield surface for lubricated atomized iron powder (Hoeganaes Ancorsteel 1000). Surface plotted for different values of relative density and particle hardening

The constitutive model has two-state variables and requires two different kinds of tests for its calibration.

Triaxial Consolidation Test for *b*(*D*). The state of stress in the test specimen is assumed to be uniform throughout (homogeneous). This assumption leads to the result that the radial Σ_r and tangential Σ_z stresses are equal and given by the cell pressure. If the specimen is cylindrical in shape, the components of stress in the powder aggregate can then be appropriately defined using a cylindrical coordinate system as follows:

$$[\Sigma] = \begin{bmatrix} -\Sigma_r & 0 & 0 \\ 0 & -\Sigma_r & 0 \\ 0 & 0 & -\Sigma_z \end{bmatrix} \quad \text{(Eq 77)}$$

The negative signs are assigned to indicate the compressive nature of the stresses in the aggregate. The deviatoric and pressure components are then:

$$P = -\frac{1}{3}(-\Sigma_r - \Sigma_r - \Sigma_z) = \frac{1}{3}(2\Sigma_r + \Sigma_z) \quad \text{(Eq 78)}$$

$$S = \Sigma + PI = \begin{bmatrix} -\tfrac{1}{3}\Sigma_r + \tfrac{1}{3}\Sigma_z & 0 & 0 \\ 0 & -\tfrac{1}{3}\Sigma_r + \tfrac{1}{3}\Sigma_z & 0 \\ 0 & 0 & \tfrac{2}{3}\Sigma_r - \tfrac{2}{3}\Sigma_z \end{bmatrix} \quad \text{(Eq 79)}$$

Calculating $\Sigma_e = \tfrac{3}{2}\sqrt{S_{ij}S_{ij}}$, find

$$\Sigma_e^2 = \tfrac{3}{2}\left(\tfrac{2}{9}(\Sigma_z - \Sigma_r)^2 + \tfrac{4}{9}(\Sigma_r - \Sigma_z)^2\right) \quad \text{(Eq 80)}$$

$$\Sigma_e = \sqrt{(\Sigma_z - \Sigma_r)^2} \quad \text{(Eq 81)}$$

The increment of plastic strain given by Eq 61 is:

$$\Delta E_r^p = 3\Delta\Lambda\left[\frac{1}{3}(\Sigma_z - \Sigma_r) - \frac{1}{9}(2\Sigma_r + \Sigma_z)b(D)\right]$$

$$\Delta E_\theta^p = 3\Delta\Lambda\left[\frac{1}{3}(\Sigma_z - \Sigma_r) - \frac{1}{9}(2\Sigma_r + \Sigma_z)b(D)\right]$$

$$\Delta E_z^p = 3\Delta\Lambda\left[\frac{2}{3}(\Sigma_r - \Sigma_z) - \frac{1}{9}(2\Sigma_r + \Sigma_z)b(D)\right] \quad \text{(Eq 82)}$$

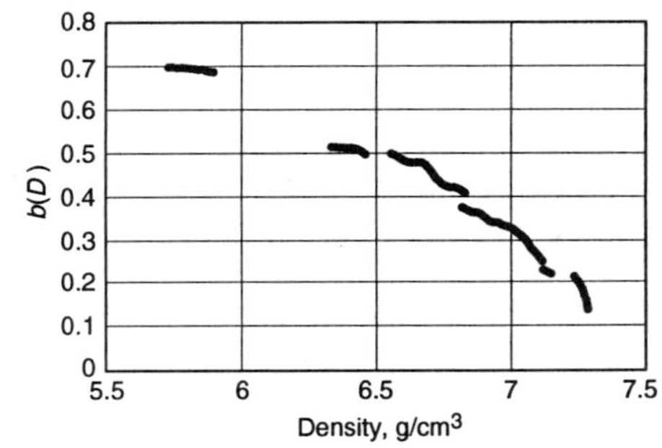

Fig. 15 Function *b*(*D*) for an iron powder blend comprising 99.5% by weight Distalloy AE, 0.5% by weight graphite, and 1% wax Hoechst micropulver

The deviatoric and volumetric components of plastic strain increment become:

$$\Delta E_v^p = 3\Delta\Lambda[-\tfrac{1}{3}(2\Sigma_r + \Sigma_z)b(D)] \quad \text{(Eq 83)}$$

$$\Delta E_z^p - \Delta E_r^p = -3\Delta\Lambda(\Sigma_z - \Sigma_r) \quad \text{(Eq 84)}$$

Eliminating $\Delta\Lambda$ obtain,

$$\frac{\Delta E_v^p}{\Delta E_z^p - \Delta E_r^p} = \frac{b(D)(2\Sigma_r + \Sigma_z)}{3(\Sigma_z - \Sigma_r)} \quad \text{(Eq 85)}$$

Solving for $b(D)$,

$$b(D) = \frac{3(\Sigma_z - \Sigma_r)\Delta E_v^p}{(2\Sigma_r + \Sigma_z)(\Delta E_z - \Delta E_r)} \quad \text{(Eq 86)}$$

Thus, the function $b(D)$ may be determined using data from the triaxial tests for values of axial stress Σ_z, axial plastic strain increment ΔE_z^p, and radial strain increment ΔE_z^p corresponding to a range of values of cell pressure Σ_r.

The results of applying Eq 86 to the experimental data of Pavier and Doremus (Ref 14) for the Distalloy AE blend are shown in Fig. 15. Because noise is inherently present in such information, a cursory smoothing using a moving average scheme was performed prior to computing the increments of strain. In addition, the elastic components of strain were assumed to be negligible. For low relative densities, this assumption is reasonable. However, as plastic incompressibility is approached, the validity of the assumption must be checked. A curve fit for $b(D)$ results in the following expression:

$$b(D) = \frac{1}{2.8287}\ln\left(\frac{7.6 - 7.33D}{0.2516}\right) \quad \text{(Eq 87)}$$

Isostatic Compaction Test for *c*(*P*). The state of stress in the isostatic compaction test is one of pure pressure, with no shear or deviatoric stress present. The isotropic test provides a convenient means of evaluating the evolution of relative density as a function of increasing cell pressure. Data are obtained by increasing cell pressure to a point before unloading the cell to determine the resulting powder relative density.

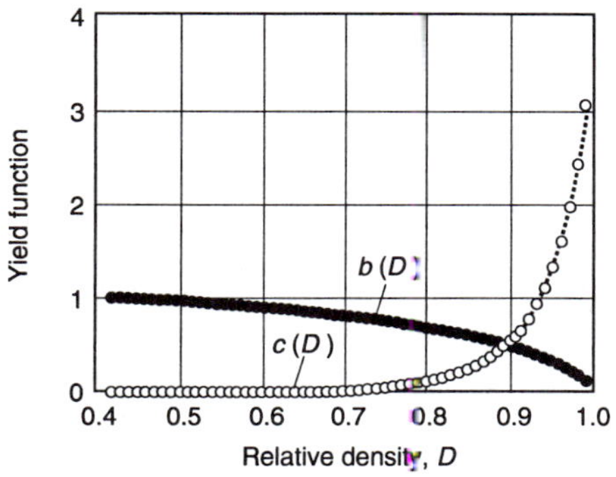

Fig. 16 Functions $b(D)$ and $c(D)$ for a iron powder blend comprising 99.5% by weight Distalloy AE, 0.5% by weight graphite, and 1% wax Hoechst micropulver

Following Pavier and Doremus (Ref 14), the relation between the relative density for the powder under hydrostatic loading and the applied pressure or hydrostatic stress was determined to be:

$$P_y = \frac{37.596}{1.0364 - D} - 46.2976 \qquad \text{(Eq 88)}$$

where P_y is expressed in MPa and represents the pressure at yield given a value of relative density, D in the isostatic test.

It is reasonable to assume that the material is continuously yielding with load. Further, since shear stress is completely absent, Eq 60 reduces to:

$$F = \frac{3}{2}P_y(D)^2 b(D) - \sigma_y^2\, c(D) = 0 \qquad \text{(Eq 89)}$$

If particle hardening is considered, the calculation of $c(D)$ involves the simultaneous integration of partial differential equations for the evolution of particle yield stress and the functions $b(D)$ and $c(D)$. However, for simplicity, assume the powder particles are perfectly plastic and obtain $c(D)$ from (Eq 89):

$$c(D) = \frac{3P_y(D)^2\, b(D)}{2\sigma_y^2} \qquad \text{(Eq 90)}$$

Using the forms for pressure and $b(D)$ from Eq 88 and 87, and assuming a mean value of 375 MPa for particle yield stress, the function $c(D)$ is calculated and plotted in Fig. 16.

Verification of Calibrated Model. This section evaluates the accuracy of the calibration performed for functions $b(D)$ and $c(D)$ for the Distalloy AE blend considered using triaxial test data. Analysis data were generated through a finite element simulation of the triaxial tests conducted. Because hardening of the powder particles is a likely event, it is considered here. Specifically, the material of the powder particle was modeled as having a yield stress of 250 MPa with a hardening of 333 MPa. These values were obtained from data reported by Trasorras et al. (Ref 36). Figure 17 shows a comparison of the predicted powder density as it evolves during the triaxial tests against data measured by Pavier et al. (Ref 14). There is very good agreement between simulation and experiment. Figure 18 shows a plot of axial strain versus radial strain for a triaxial cell pressure of 250 MPa. The strains are measured relative to the onset of shear in the specimen, marked by an increase in axial stress over the cell pressure. Figure 19 shows a plot of axial stress versus axial strain for the same triaxial cell pressure of 250 MPa. Again, the simulations compare well with the experiments.

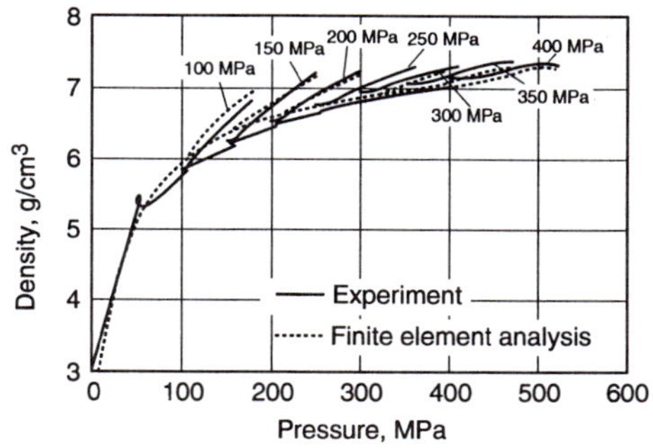

Fig. 17 Density evolution in a triaxial test: comparison of FEA results with experiment

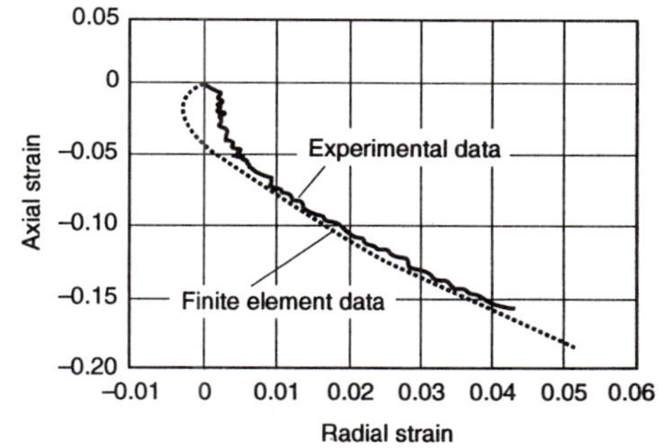

Fig. 18 Axial versus radial strain in a triaxial test with 250 MPa cell pressure: comparison of FEA results with experiment

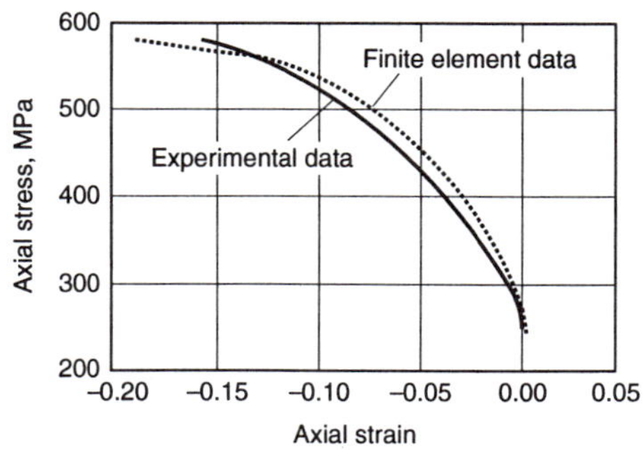

Fig. 19 Axial stress versus strain in a triaxial test with 250 MPa cell pressure: comparison of FEA results with experiment

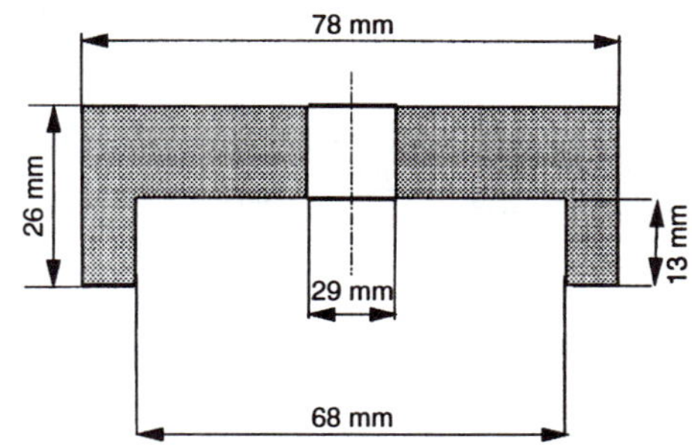

Fig. 20 Nominal dimensions of axisymmetric part used in compaction experiments

Numerical Modeling of Powder Compaction in Dies

In this section, practical illustrations of finite element analysis of powder compaction are presented. The first one is the compaction of a two-level component with emphasis on the effect of compaction kinematics on punch loads. The second example, the compaction of a long bushing, illustrates density predictions.

Compaction of a Two-Level Part

The model presented in the section "A Constitutive Model for Metallic Powders with Ductile Particles" is used to examine the compaction of an axisymmetric two-level component. This part was studied experimentally by Kergadallan et al. (Ref 4). The part was pressed with five different tooling motions; two of those cases, labeled as part 30 and part 34, are presented here. The compaction experiments and the finite element model are described, and numerical predictions are compared to experimental results.

Compaction Experiments. Following are the conditions of the compaction tests used to study an axisymmetric two-level component.

Part Geometry. The part is axisymmetric with a thin outer rim, a hub, and a bore. This geometry is representative of many common powder metal parts (e.g., engine camshaft timing pulleys, one-way mechanical diode clutch plates). Geometry and dimensions are shown in Fig. 20. The outer diameter is 78 mm and the overall height is 26 mm.

The powder blend is based on the diffusion alloyed iron powder, Hoeganaes Distalloy AE, as described in section "Calibration of Material Parameters for an Iron Powder Blend."

Press and Tooling. Figure 21 shows a schematic of the press and tooling. The part was pressed in a hydraulic COSMO press. The tooling consists of four moving components: an outer die and an inner core that move together, a top punch, and a lower inner punch. The tooling also includes a stationary lower outer punch. The press was instrumented with strain gauges to measure loads on tooling members (see Fig. 21) and potentiometer displacement transducers to measure tooling displacements.

Initial Conditions. The fill positions of the punches are given by R_1 and R_2 as shown in Fig. 22 and Table 2. The initial density in the die cavity was estimated by assuming uniform density within the rim section and the hub section of

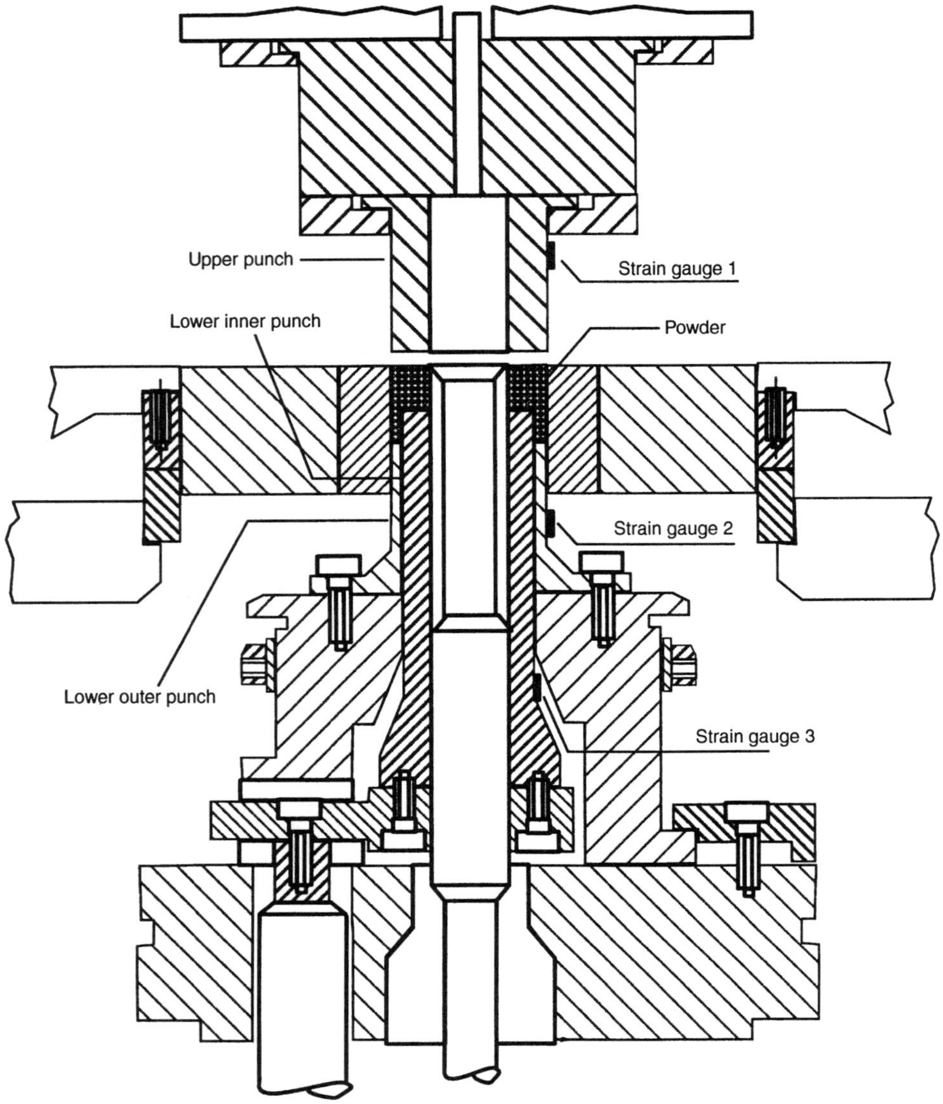

Fig. 21 Schematic of hydraulic COSMO press, tooling and strain gauges for load measurement. Source: Ref 4

the part, as shown schematically in Fig. 23. The initial density values (Table 2) were estimated by weighing the different sections of the part and assuming no mass transfer between the top and bottom sections.

Compaction Kinematics. The displacement histories for each of the moving tooling members are shown in Fig. 22. Tabulated information on punch displacements for part 30 and part 34 can be found in Ref 4. As explained later in this section, compaction of part 30 results in low density in the rim section. Also, cracks appear on the inner and outer surface of the rim. The aim of the compaction kinematics used with part 34 was to reduce the density imbalance between the hub and rim sections and to eliminate the cracks.

The Finite Element Model. The powder material model described in the section "A Constitutive Model for Metallic Powders with Ductile Particles" was implemented in the finite element code Abaqus/Standard (Ref 2) through a user-defined material subroutine. Details of the numerical implementation are beyond the scope of this article. Aravas (Ref 3) has described the numerical integration of constitutive models of this class.

Finite Element Model. An axisymmetric model of the compact was set up using four noded axisymmetric elements (Ref 2). Figure 24 shows a three-dimensional representation of the finite element mesh obtained by revolving the part about the axis of symmetry.

Tooling. For simplicity, the tooling components were modeled as rigid surfaces. In a more detailed model, the tooling can also be modeled assuming elastic material behavior. Tooling stresses can be computed this way. However, modeling the interfaces between the powder and the elastic tooling is numerically very intensive.

Material Model. Constant values were used for the elastic properties. A fixed value of Young's modulus (141 GPa), corresponding to the final average density, was calculated using the expression for elastic modulus determined by Pavier et al. (Ref 14). It was mentioned earlier that the elastic properties are strong func-

Table 2 Initial conditions for two-level compaction experiments

Conditions	Part 30	Part 34
Rim fill (R_1), mm	54.71	55.82
Hub fill (R_2), mm	27.23	25.18
ρ_{Hub}, g/cm^3	3.51	3.63
ρ_{Rim}, g/cm^3	3.21	2.97
Mass, g	498.09	482.42

Table 3 Measured and computed peak compaction loads

	Part 30 loads, MN		Part 34 loads, MN	
Punch	Measured	Predicted	Measured	Predicted
LIP	2.10	2.12	1.61	1.75
LOP	0.59	0.36	0.61	0.5
UP	3.28	2.85	2.55	2.48

Table 4 Measured sectional densities

	Density, g/cm^3	
Section of Fig. 26	Part 30	Part 34
1	7.01	6.9
2	7.07	6.93
3	7.09	7.03
4	6.90	6.91
5	6.89	6.97

tions of density. However, in an analysis such as this, severe numerical instabilities can result as a consequence of the soft elastic properties at low density. Further, the effect of elastic behavior is dominant only when the relative density approaches its pore-free value.

The friction coefficient between iron powder compacts and steel/carbide tooling varies with the normal stress (Ref 44, 45). For lubricated iron powders, the coefficient of friction varies in the range 0.1 to 0.2 for normal stresses in the range 100 to 700 MPa. For the present model, a Coulomb friction model was used and a coefficient of friction of 0.15 was assumed.

Compaction Kinematics. A piecewise linear approximation to the actual tooling motions was used as depicted in Fig. 22.

Part Fill Density. The initial fill density was set as represented schematically in Fig. 23.

Experimental and Modeling Results. Figure 25 shows the compaction forces measured for part 30. The forces exerted on the bottom and upper punches increase from the initial instant corresponding to the moment of contact between the powder and the upper punch. At instant t_{L1} (Fig. 22), the speed of the lower inner punch (LIP) is modified such that it descends at practically constant force over the rest of its travel through time t_{L2}. While the LIP is in motion, the upper punch (UP) and the lower outer punch (LOP) withstand forces that increase progressively up to their peak values. The rest of the cycle, for time $>t_{L2}$, continues while the LIP remains at a fixed position. The forces on the LOP and the UP increase rapidly to their maximum values. Compression finishes at time t_F, corresponding to the maximum displacement of the UP. The maximum loads recorded are shown

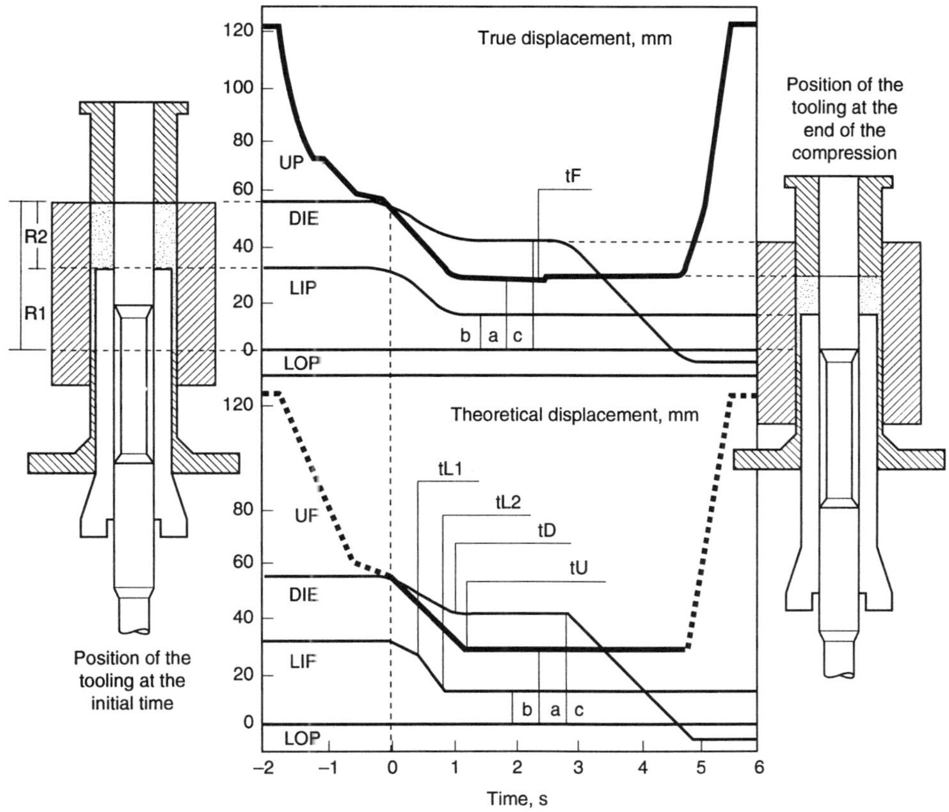

Fig. 22 Actual tooling motions for axisymmetric component with piecewise linear approximations. Source: Ref 4

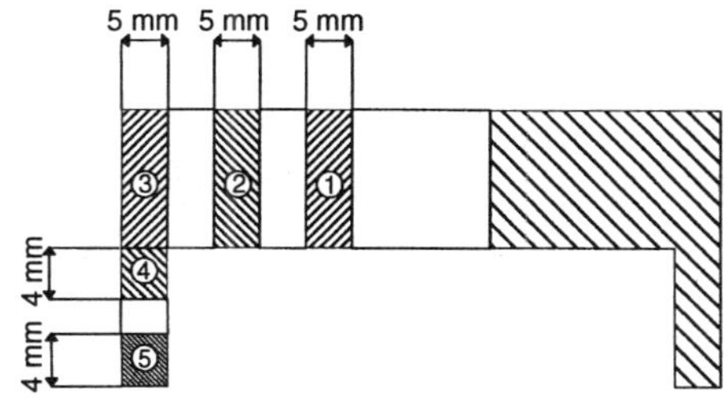

Fig. 26 Comparison of compaction tooling loads for part 34: predicted versus experimental

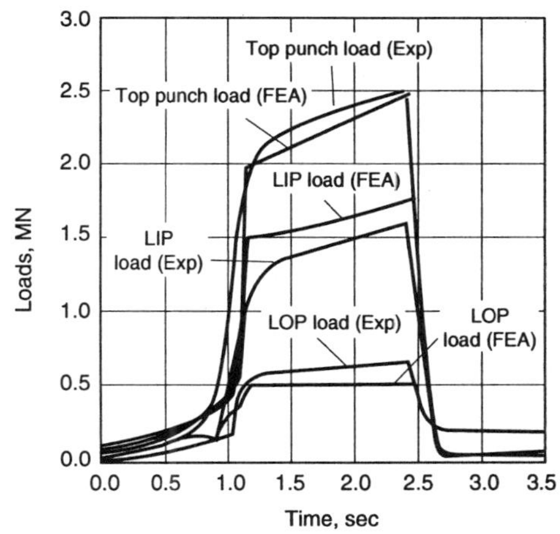

Fig. 27 Schematic of sections for density measurements for test parts

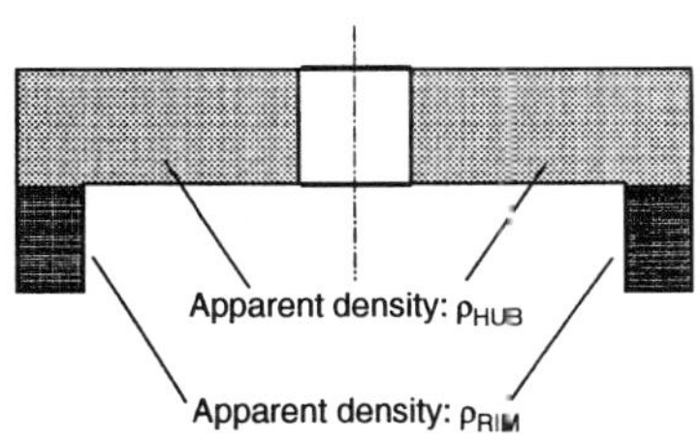

Fig. 23 Estimated initial fill density distribution for parts used in compaction experiments. Source: Ref 4

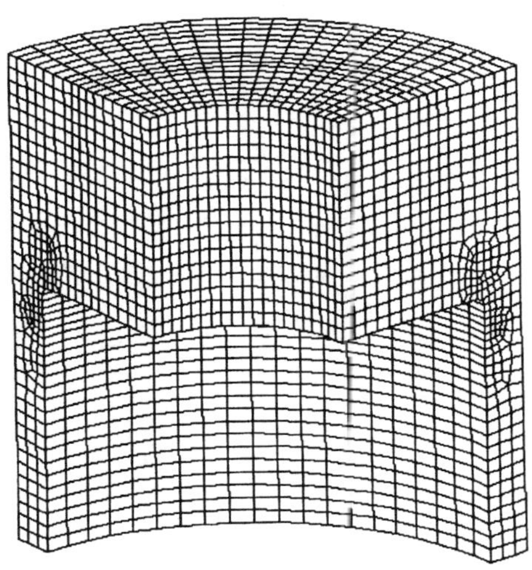

Fig. 24 Three-dimensional representation of the axisymmetric finite element discretization of the powder compact.

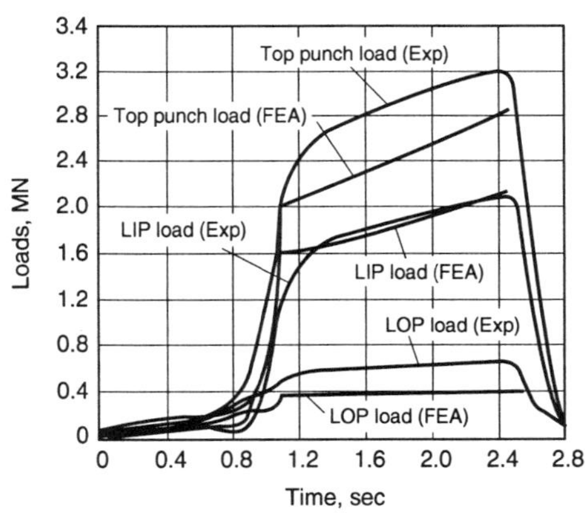

Fig. 25 Comparison of compaction tooling loads for part 30: predicted versus experimental

in Table 3. Densities were measured in five sections of the part as shown schematically in Fig. 26. Measurements are shown in Table 4. For part 30, the hub and rim densities are approximately 7.06 g/cm^3 and 6.90 g/cm^3.

After ejection, the parts were visually inspected for cracks. Part 30 presented a distinct crack around the rim ID, 2 mm below the hub. Very fine cracks were also present on the outer surface of the rim, close to the bottom end. Although it is not possible to state exactly when the cracks were formed, it is very likely that they appeared during load removal/ejection. For part 30, the maximum loads on the inner and outer lower punches are 2.1 MN and 0.59 MN, respectively. The high load on the lower inner punch results in high punch deflection. During ejection, the elastic recovery of the lower inner punch results in tension on the rim causing cracking.

To eliminate the cracks that appeared in part 30, part 34 was pressed with higher fill in the rim section and reduced fill in the hub section (Table 2). The tool motions applied were modified to accommodate the different fill positions, otherwise they were very similar to the motions used with part 30. The forces measured are shown in Fig. 27. The peak load values are given in Table 3. The load patterns for parts 30 and 34 are similar; however, for part 34, the lower inner punch load is reduced to 1.61 MN, while the outer lower punch force increases slightly to 0.61 MN. The hub and rim densities are approximately 6.97 g/cm^3 and 6.94 g/cm^3 (Table 3). The reduced load in the lower inner punch results in lower deflection and eliminates tension during load removal/part ejection. Part 34 was defect free.

Figure 28 illustrates the compaction process as predicted by the numerical model. The deformed mesh is plotted for part 34. The model shows that there is very limited transfer of powder between the top and bottom (rim) sections of the part. This behavior is expected because the tooling motions were designed to minimize powder transfer to avoid the formation of a defect at the corner between the two sections. Figure 29 plots the evolution of density in the compact during pressing of part 34.

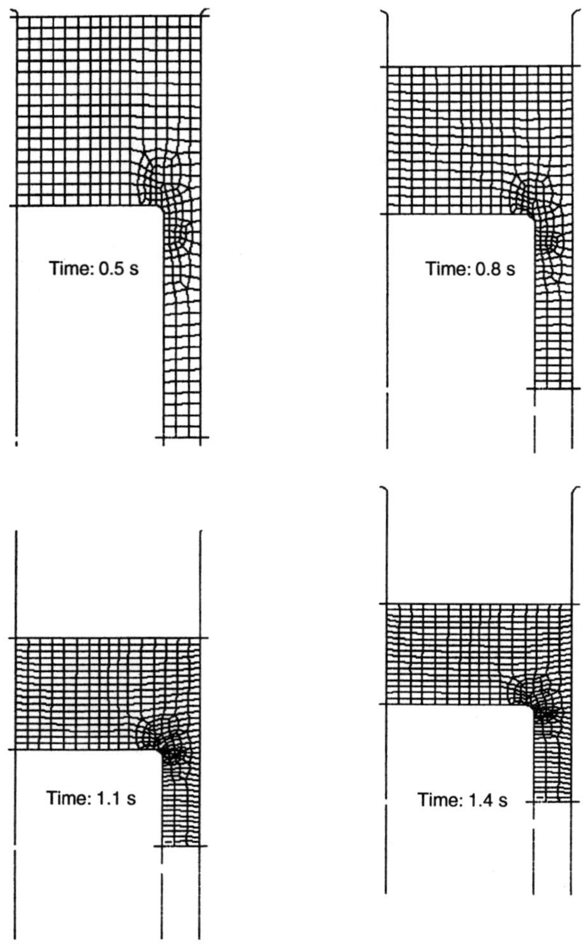

Fig. 28 Deformed mesh during stages of compaction of part 34

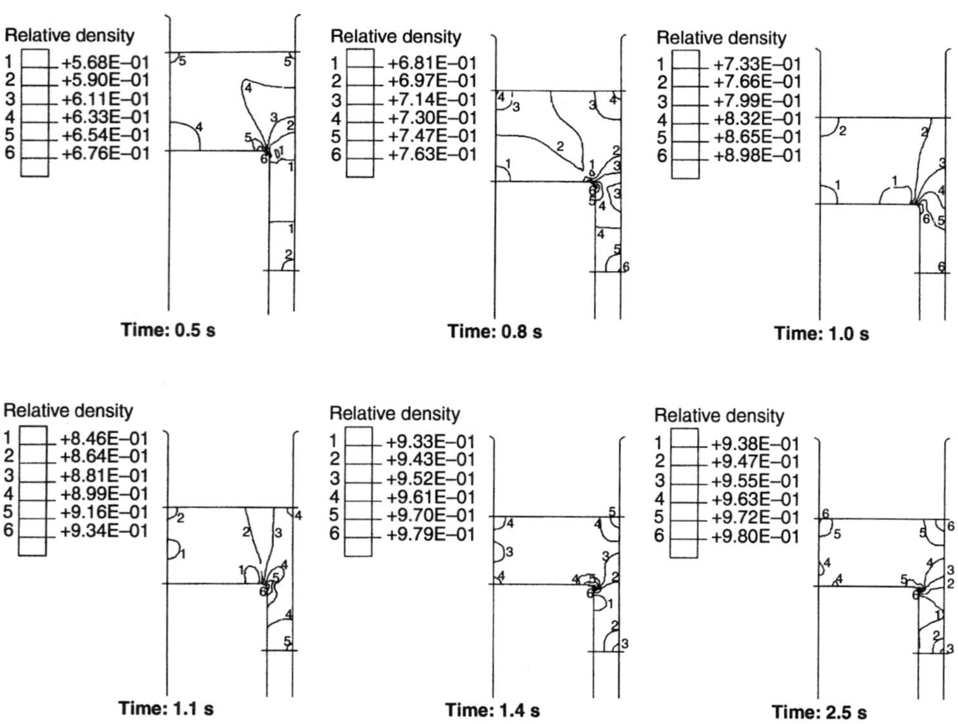

Fig. 29 Computed density fields at different stages of compaction of part 34

The loads computed for parts 30 and 34 are plotted in Fig. 25 and 27 and compared to the measured loads. There is good overall agreement between the experiments (Ref 4) and the calculations. The computed maximum load values for each punch are shown in Table 3. Again the agreement with the experiments is very good, showing that the model is capable of predicting the effects of punch kinematics and fill position on punch loads.

One observation on the nature of the load computations is pertinent. As the compact approaches its pore-free density, it becomes incompressible. Consequently, towards the end of the compaction stroke, minor variations in tool motions result in sharp changes in load. Thus, it is very important to ascertain the exact displacements of tooling members from their reference positions if accurate load predictions are sought.

Compaction of a Long Bushing

This example illustrates how the numerical model can be applied to predict density distributions. This part was studied experimentally by Trasorras et al. (Ref 40). A steel bushing was pressed in a 150 ton mechanical press. The green bushing dimensions are outside diameter (OD) = 19.05 mm, inner diameter (ID) = 12.7 mm, and height = 25.4 mm. The powder used was a blend of atomized steel powder (Ancorsteel 1000) with 0.75 wt% zinc stearate as lubricant. The punch motions comprise the following sequence. After powder filling, the top punch moves down to compact the powder, then rises and exits the die cavity. At the end of top punch motion, the die is stripped to eject the compact. Finally, the core rod is stripped. The lower punch and the die remain stationary throughout the compaction. As compaction starts, a density gradient develops in the bushing due to the friction between the compact and the tooling members. With continued top punch motion, the bushing densifies with the top always being at a higher density than the bottom. During ejection, there is some additional densification of the bottom end of the compact. Finally, the compact expands as it exits the die cavity thereby reducing the overall density. The axial density distribution in the bushing was determined by successively sectioning and weighing the compact.

Compaction of the bushing was modelled in PCS (Ref 42), a powder compaction modeling system based on the finite element code NIKE2D (Ref 46). Figure 30 shows the finite element discretization of the tooling and powder, with the punches shown at their fill position. The material model described earlier was used with the constitutive functions $b(D)$, $c(D)$, and elastic properties calibrated for the atomized iron powder. The initial apparent density of the powder was 3.2 g/cm^3. A complete model of the tooling was used (Fig. 30) and elastic isotropic behavior was assumed. The friction between the compact and the tooling members was assumed to follow Coulomb's model with a friction coefficient of 0.2. Figure 31 compares the axial density distribution predicted by the numerical model with

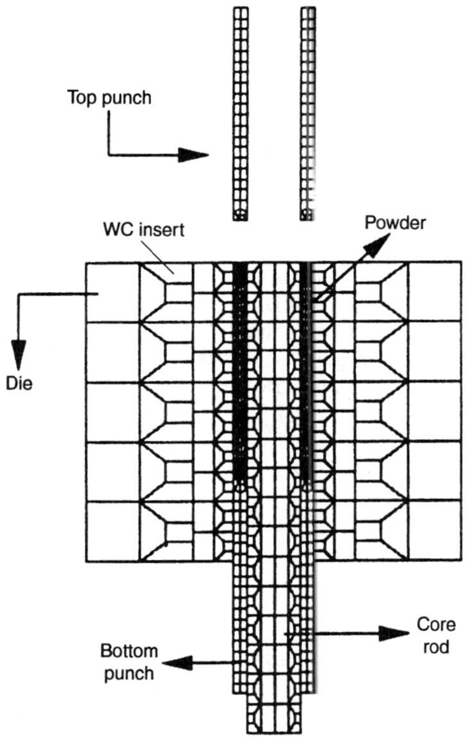

Fig. 30 Finite element discretization of powder and tooling used in the compaction of cylindrical bushing

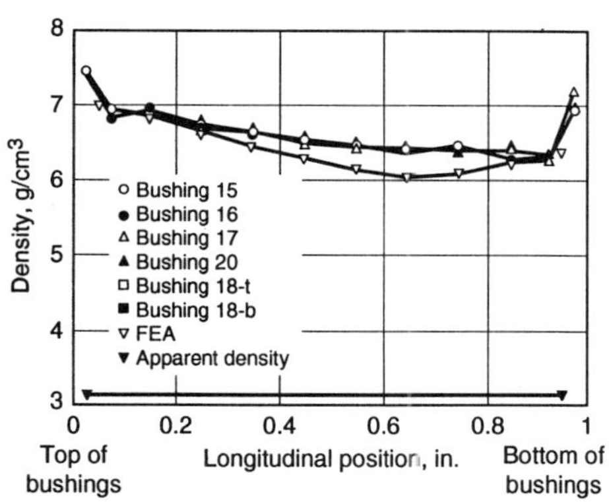

Fig. 31 Density distribution along axis of cylindrical bushing: FEA predictions versus experimental results

the experimental results. The model properly represents the densification that takes place during both compaction and ejection and the predicted final density distribution is in good agreement with the experiments. The experiments show a sharp increase of density at the powder layer in contact with the top and bottom punches. The numerical model, with the discretization level used, was not able to predict this effect.

This article has examined the general structure of constitutive laws for the compaction of powder compacts and demonstrated how these material models can be used to model the response of real world components to a series of complex die operations. It identified the general structure of the constitutive law and described a number of models that have been proposed in the literature. This field is still evolving, and it is evident that there will be significant developments in this area over the next few years as a wider range of experimental studies are conducted, providing greater insights into the compaction process. At the current time, there is no universally accepted model. Therefore, a pragmatic approach and a relatively simple form of empirical model were adopted requiring, for the determination of the unknown functions, a limited range of experiments. This selection allowed an examination of the compaction of axisymmetric components in detail and a comparison of general features of the component response with practical measurements. Similar procedures could have been adopted for any of the methods described, although in general, more sophisticated experiments are required in order to determine any unknown function or coefficients, particularly if the shape of the yield function is not known, or assumed, a priori.

ACKNOWLEDGMENTS

The authors are grateful to a number of people for help in gathering the material for publication here; in particular, Janet Wilson and Kevin R. Holmes from Federal-Mogul Technology, Cawston, England, for their effort in preparing models for finite element analysis, E. Pavier and P. Doremus of Laboratoire Sols, Solides, Structures, Grenoble, France, for allowing extensive use of data toward the calibration of models for powder behavior, Gerard Puente of Federal-Mogul Sintered Products, Pont-de-Claix, France assisted in the analysis of the compaction experiments. Finally, Howard Sanderow of M.E.T., Dayton, Ohio, Dr. Sriram Krishnaswami of MARC Analysis Research Corp., Palo Alto, California, and Dr. Howard A. Kuhn of Concurrent Technologies Corp., Johnstown, Pennsylvania reviewed this article and contributed valuable comments.

REFERENCES

1. R. German, *Particle Packing Characteristics,* Metal Powder Industries Federation, 1989
2. *ABAQUS/Standard User's Manual,* Version 5.7, Vol 1–3, Hibbitt, Karlsson, & Sorensen, Inc., Providence, RI, 1997
3. N. Aravas, On the Numerical Integration of a Class of Pressure-Dependent Plasticity Models, *Int. J. Numer. Meth. Eng.,* Vol 24, 1987, p 1395–1416
4. Y. Kergadallan, G. Puente, P. Doremus, and E. Pavier, Compression of an Axisymmetric Part, *Proc. of the Int. Workshop on Modelling of Metal Powder Forming Processes* (Grenoble, France), 1997, p 277–285
5. K.T. Kim, J. Suh, and Y.S. Kwon, Plastic Yield of Cold Isostatically Pressed and Sintered Porous Iron under Tension and Torsion, *Powder Metall.,* Vol 33, 1990, p 321–326
6. H.A. Kuhn and C.L. Downey, Material Behavior in Powder Preform Forging, *J. Eng. Mater. Technol.,* 1990, p 41–46
7. S. Shima and M. Oyane, Plasticity Theory for Porous Metals, *Int. J. Mech. Sci.,* Vol 18, 1976, p 285–291
8. S.B. Brown and G.A. Weber, A Constitutive Model for the Compaction of Metal Powders, *Modern Developments in Powder Metallurgy,* Vol 18-21, 1988, MPIF, p 465–476
9. T.J. Watson and J.A. Wert, On the Development of Constitutive Relations for Metallic Powders, *Metall. Trans. A,* Vol 24, 1993, p 2071–2081
10. A.R. Akisanya, A.C.F. Cocks, and N.A. Fleck, The Yield Behaviour of Metal Powders (1996), *Int. J. Mech. Sci.,* Vol 39 (No. 12), 1997, p 1315–1324
11. S. Brown and G. Abou-Chedid, Yield Behaviour of Metal Powder Assemblages, *J. Mech. Phys. Solids,* Vol 42 (No. 3), 1994, p 383–399
12. W. Prager, *Proc. Inst. Mech. Eng.,* Vol 169, 1955, p 41
13. R. Hill, *The Mathematical Theory of Plasticity,* Oxford University Press, 1950
14. E. Pavier and P. Doremus, Mechanical Behavior of a Lubricated Powder, *Advances in Powder Metallurgy & Particulate Materials-1996,* Vol 2 (Part 6), Metal Powder Industries Federation, 1996, p 27–40
15. C.J. Yu, R.J. Henry, T. Prucher, S. Parthasarathi, and J. Jo, *Advances in Powder Metallurgy & Particulate Materials,* Vol 6, Metal Powder Industries Federation, 1992, p 319–332
16. N.A. Fleck, L.T. Kuhn, and R.M. McMeeking, Yielding of Metal Powder Bonded by Isolated Contacts, *J. Mech. Phys. Solids,* Vol 40, 1992, p 1139–1162
17. N.A. Fleck, On the Cold Compaction of Powders, *J. Mech Phys. Solids,* Vol 43 (No. 9), 1995, p 1409–1431
18. J. Gollion, D. Bouvard, P. Stutz, H. Grazzini, C. Levaillant, P. Baudin, and J.P. Cescutti, On the Rheology of Metal Powder during Cold Compaction, *Proc. Int. Conf. on Powders and Grains,* Biarez and Gourves, Ed., Clermont-Ferrand, France, 4–8 September 1989, p 433–438
19. R.M. Govindarajan and N. Aravas, Deformation Processing of Metal Powders, Part 1: Cold Isostatic Pressing, *Int. J. Mech. Sci.,* Vol 36, 1994, p 343–357
20. A.L. Gurson, Continuum Theory of Ductile Rupture by Void Nucleation and Growth, Part 1: Yield Criteria and Flow Rules for Porous Ductile Media, *J. Eng. Mater. Technol.,* Vol 99, 1977, p 2–15
21. A.C.F. Cocks, The Inelastic Deformation of Porous Materials, *J. Mech. Phys. Solids,* Vol 37 (No. 6), 1989, p 693–715
22. Y-M. Liu, H.N.G. Wadley, and J. Duva, Densification of Porous Materials by Power-Law Creep, *Acta Metall. Mater.,* Vol 42, 1994, p 2247–2260
23. A.R. Akisanya, A.C.F. Cocks, and N.A. Fleck, Hydrostatic Compaction of Cylindrical Parti-

cles, *J. Mech. Phys. Solids,* Vol 42 (No. 7), 1994, p 1067–1085

24. Z. Qian, J.M. Duva, and H.N.G. Wadley, Pore Shape Effects during Consolidation Processing, *Acta Metall. Mater.* Vol 44, 1996, p 4815
25. P. Ponté Castañeda and M. Zaidman, Constitutive Models for Porous Materials with Evolving Microstructure, *J. Mech. Phys. Solids,* Vol 42, 1994, p 1459–1497
26. K.T. Kim and J. Suh, Elastic-Plastic Strain Hardening Response of Porous Metals, *Int. J. Eng. Sci.,* Vol 27, 1989, p 767–778
27. S. Brown and G. Abou-Chedid, Appropriate Yield Functions for Powder Compacts (1992), *Scr. Metall. Mater.,* Vol 28, 1993, p 11–16
28. D.C. Drucker and W. Prager, *Q. Appl. Math.,* Vol 10, 1952, p 157–165
29. A.L. Gurson and T.J. McCabe, Experimental Determination of Yield Functions for Compaction of Blended Powders, *Proc. MPIF/APMI World Cong., on Powder Metallurgy and Particulate Materials* (San Francisco), Metal Powder Industries Federation, 1992
30. A. Schofield and C.P. Wroth, *Critical State Soil Mechanics,* McGraw-Hill, 1968
31. D.M. Wood, *Soil Behavior and Critical State Soil Mechanics,* Cambridge University Press, 1990
32. S. Shima, "A Study of Forming of Metal Powders and Porous Metals," Ph.D. thesis, Kyoto University, 1975
33. Y. Morimoto, T. Hayashi, and T. Takei, Mechanical Behavior of Powders in a Mold with Variable Cross Sections, *Int. J. Powder Metall. Powder Technol.,* Vol 18 (No. 1), 1982, p 129–145
34. J.R.L. Trasorras, S. Armstrong, and T.J. McCabe, Modeling the Compaction of Steel Powder Parts, *Advances in Powder Metallurgy & Particulate Materials-1994,* Vol 7, American Powder Metallurgy Institute, 1994, p 33–50
35. J. Crawford and P. Lindskog, Constitutive Equations and Their Role in the Modeling of the Cold Pressing Process, *Scand. J. Metall.,* Vol 12, 1983, p 271–281
36. J.R.L. Trasorras, T.M. Krauss, and B.L. Ferguson, Modeling of Powder Compaction Using the Finite Element Method, *Advances in Powder Metallurgy,* Vol 1, T. Gasbarre and W.F. Jandeska, Ed., American Powder Metallurgy Institute, 1989, p 85–104
37. B.L. Ferguson, et al., Deflections in Compaction Tooling, *Advanced in PM & Particulate Materials,* Vol 2, Metal Powder Industries Federation, 1992, p 251–265
38. H. Chtourou, A. Gakwaya, and M. Guillot, Assessment of the Predictive Capabilities of the Cap Material Model for Simulating Powder Compaction Problems, *Advances in Powder Metallurgy & Particulate Materials-1996,* Vol 2 (Part 7), Metal Powder Industries Federation, 1996, p 245–255
39. D.T. Gethin, R.W. Lewis, and A.K. Ariffin, Modeling Compaction and Ejection Processes in the Generation of Green Powder Compacts, *Net Shape Processing of Powder Materials,* 1995 ASME Int. Mechanical Engineering Congress and Exposition, AMD-Vol 216, S. Krishnaswami, R.M. McMeeking, and J.R.L. Trasorras, Ed., The American Society of Mechanical Engineers, 1995, p 27–45
40. J.R.L. Trasorras, S. Krishnaswami, L.V. Godby, and S. Armstrong, Finite Element Modeling for the Design of Steel Powder Compaction, *Advances in Powder Metallurgy & Particulate Materials-1995,* Vol 1 (Part 3), Metal Powder Industries Federation, 1995, p 31–44
41. S. Krishnaswami and J.R.L. Trasorras, Modeling the Compaction of Metallic Powders with Ductile Particles, *Simulation of Materials Processing: Theory, Methods and Application,* Shen and Dawson, Ed., Balkema, Rotterdam, 1995, p 863–858
42. Powder Compaction Simulation Software (PCS Elite) User's Manual, Concurrent Technologies Corp., Johnstown, PA
43. H-A. Haggblad, P. Doremus, and D. Bouvard, An International Research Program on the Mechanics of Metal Powder Forming, *Advances in Powder Metallurgy & Particulate Materials-1996,* Vol 2 (Part 7), Metal Powder Industries Federation, 1996, p 179–192
44. B.Wikman, H.A. Häggblad , and M. Oldenburg, Modelling of Powder-Wall Friction for Simulation of Iron Powder Pressing, *Proc. of the Int. Workshop on Modelling of Metal Powder Forming Processes* (Grenoble, France), July 1997, p 149–158
45. E. Pavier and P. Dorémus, Friction Behavior of an Iron Powder Investigated with Two Different Apparatus, *Proc. of the Int. Workshop on Modelling of Metal Powder Forming Processes* (Grenoble, France), July 1997, p 335–344
46. J. Hallquist, "NIKE2D-A Vectorized, Implicit, Finite Deformation, Finite-Element Code for Analyzing the Static and Dynamic Response of 2-D Solids," Technical report UCRL-19677, Lawrence Livermore National Laboratory, Livermore, California, 1993

Powder Metallurgy Presses and Tooling

Revised by John Porter, Cincinnati Incorporated

POWDER METAL COMPACTING PRESSES, equipped with appropriate tooling, frequently are used for producing P/M components. Although commonly called P/M presses, use is not limited to the pressing of metal powders. Almost any alloy or mixture of materials produced in powder form can be compacted into suitable end products. The majority of components fabricated by P/M presses, in number of pieces and pounds of product produced, consists of compacted metals. Ferrous-base metals constitute the largest usage. Powder metallurgy compacting presses usually are mechanically or hydraulically driven, but they can incorporate a combination of mechanically, hydraulically, and pneumatically driven systems.

Table 1 summarizes some of the developments for P/M presses in the last 40 years. Other recent improvements in compaction technology include:

- Split-die techniques to make multilevel parts having different peripheral contours at different levels
- Punch rotation capability to facilitate production of helical gears and other helical shapes
- Higher compaction pressures by using stronger tool materials, advanced pressure control methods, and die wall lubricants
- Better process control with computerized tool motion monitoring
- Warm compaction and improved "segregation-free" powders with enhanced flow characteristics

Compacting Press Requirements

Although P/M presses resemble stamping and forming presses, several significant differences exist. Press frames generally have straight sides. Gap-type or "C" frame presses are not suitable because the frame deflects in an arc under load, resulting in a slight out-of-alignment condition between the bed and side of the press. This arrangement produces a compacted part that is slightly out of parallel, top to bottom. Because P/M tooling clearances are generally ~0.025 mm/25 mm (0.001 in./1 in.) total, bending deflection can cause broken tooling or excessive tool wear.

Powder metallurgy presses apply sufficient pressure from one or both pressing directions (top and bottom) to achieve uniform density throughout the compact. Design should include provision for ejecting the part from the tooling. Pressing and ejection occur during each cycle of the press and must be accurately synchronized.

Presses need sufficient connected horsepower to compact and eject the part. In most press-working applications, the working stroke is a small portion of the total stroke of the press. In P/M presses, the working stroke during the compaction portion of the cycle is usually greater than the length of the part being produced, and the ejection portion of the cycle has a working stroke equal to or greater than the length of the part by a factor of approximately two or three. In some cases, the power required during the ejection cycle is greater than that required during compaction.

Presses should provide for adjustable die filling (the amount of loose powder in the tooling cavity). Automatic powder feeding systems that are synchronized with the compaction and ejection portion of the press cycle are desirable. Finally, P/M presses must meet federal, state, and local design and construction safety laws. Metal Powder Industries Federation (MPIF) standard 47 details safety standards for P/M presses.

Mechanical presses are available in top-drive and bottom-drive arrangements. In top-drive presses, the motor, flywheel, and gearing system are located in the crown or upper structure of the press. Presses with pressing capacities of ~1780 kN (200 tons) are floor mounted, requiring little or no pit. Top-drive presses with pressing capacities >1780 kN (200 tons) usually require a pit to maintain a convenient working height for the operator.

In bottom-drive presses, the drive mechanism, motor, and flywheel are located in the bed of the press. These presses usually are "pulled down"; that is, the top ram of the press is pulled downward by draw bars or tie rods. Bottom-drive presses with pressing capacities of >445 kN (50 tons) usually require pits. Top-drive and bottom-drive presses are comparable in terms of part-making capability, reliability, and equipment cost.

Press Tonnage and Stroke Capacity. Required press capacity to produce compacts in rigid dies at a given pressure depends on the size of the part to be pressed and the desired green density of the part, which in turn is determined by requirements for mechanical and physical properties of the sintered part. Compacting pressures can be as low as 70 to 140 MPa (5 to 10 tsi) for tungsten powder compacts or as high as 550 to 830 MPa (40 to 60 tsi) for high-density steel parts.

When a part is pressed from the top and bottom simultaneously, the press should apply the required load to the upper and lower ram of the press. To eject the pressed compact, an ejection capacity must be available that is sometimes divided into the load for the breakaway stroke (which is the first 1 to 12 mm ($\frac{1}{32}$ to $\frac{1}{2}$ in.) of the ejection stroke and the load for a sustained stroke). The load for a sustained stroke is generally one-fourth to one-half of the breakaway load.

Table 1 History of development in P/M presses

Years	Compacting press
1955–1959	Cam press, HP
1960–1964	Toggle press, MP
1965–1969	Large size HP (500 +), large size MP (500 +)
1970–1974	Multistepped MP, double die compacting press
1975–1979	Large size MP (750 +), tool holder quick change
1980–1989	NC press, multistepped HP (800 +), large size rotary press
1990–1994	Large size MP, automatic P/M manufacturing line
1995–present	Hybrid (mechanical/hydraulic) presses (800 tons)

HP, hydraulic press; MP, mechanical press; NC, numeric controlled

The stroke capacity of a press, or the maximum ram travel, determines the length of a part that can be pressed and ejected. In presses used for automatic compacting, the stroke capacity is related to the length available for die fill and ejection stroke.

Load Requirements. The total load required for a part is determined by the product of the pressure needed to compact the part to the required density and the projected area of the part. Compaction curves relate pressure, P, to the required density, q, and are usually obtained from compacting tests on cylindrical shapes with the height, L, equal to the diameter, D. For thicker parts the load must be increased, by as much as 25% for a length to diameter ratio of 4 to 1, to give the required density.

Required compacting pressures can be estimated with a correction factor, k, such that (Ref 1):

$$P = P_1 (1 + k)$$

where P is the compaction pressure for a larger part and P_1 is the compaction pressure for a "standard" part (i.e., $L = D$). The correction factor is:

$$k = (0.25/3)(L/D - 1) \quad \text{for } L/D > 1$$
$$k = 0 \quad \text{for } L/D < 1$$

For parts that are not cylindrical, an equivalent L/D ratio can be used:

$$L_e/D_e = (V \cdot p)/(2 \cdot A^2)$$

where V is the part volume and A is the projected area. The press load required is then obtained by multiplying the required compaction pressure by the projected area of the part.

Mechanical Presses

In most mechanical P/M compacting presses, electric motor-driven flywheels supply the main source of energy used for compacting and ejecting the part. The flywheel normally is mounted on a high-speed shaft and rotates continuously. A clutch and a brake mounted on the flywheel shaft initiate and stop the press stroke. To initiate a press stroke, the brake is disengaged and the clutch is engaged, causing the energy stored in the rotating flywheel to transmit torque through the press gearing to the final drive or press ram.

Clutch and brake systems should be of the partial revolution type that can be engaged and disengaged at any point in the pressing cycle. The clutch usually is pneumatically engaged with a spring release, and the brake is pneumatically released with a spring set, thereby providing full stopping ability in the event of loss of air pressure. An adjustable speed device normally is supplied with electric drive motor, providing production rate adjustment as indicated by pressing and ejection conditions.

On presses that have main motor capacities up to ~19 kW (25 hp), the adjustable speed drive is usually of the variable-pitch pulley or traction-drive type. Above 19 kW (25 hp), direct-current or eddy-current control devices are preferred. The motor and drive must be totally enclosed to prevent contamination by metal powder dust.

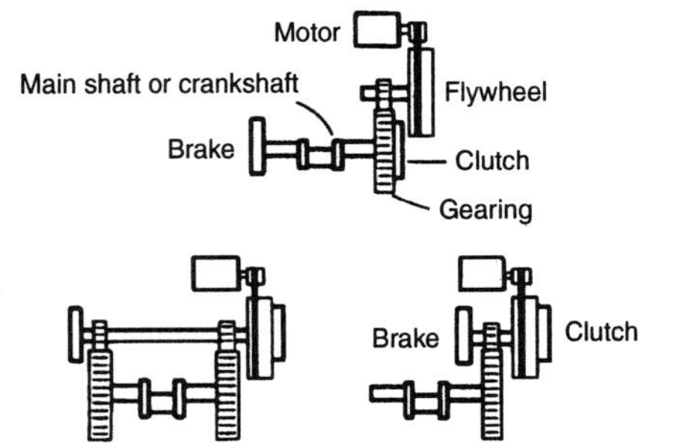

Fig. 1 Single-reduction gearing systems for P/M compacting press

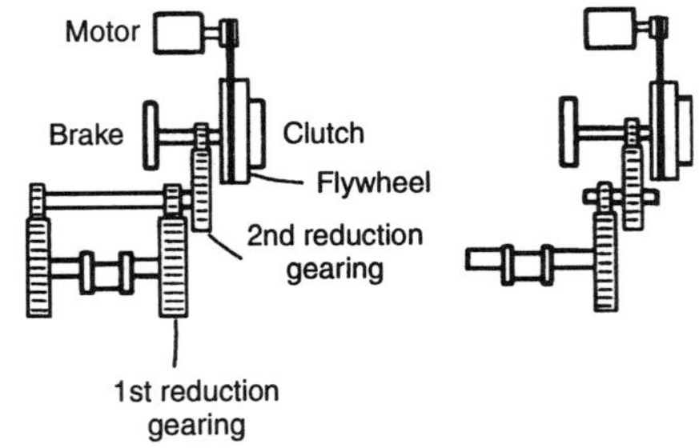

Fig. 2 Double-reduction gearing systems for P/M compacting press

Gearing systems usually are either single-reduction (Fig. 1) or double-reduction (Fig. 2) arrangements. Single-reduction gearing frequently is used in lower tonnage presses (≤445 kN, or 50 tons) that have stroking rates of ~50 strokes/min. Higher tonnage presses use double-reduction gearing and commonly have maximum stroking rates of 30 strokes/min.

The low-speed shaft of the press, normally called the main shaft, is linked to the press ram, causing motion of the tooling for the compacting and ejection cycles. Ram driving mechanisms can be either cam- or eccentric-driven arrangements.

Cam-driven presses generally are limited to pressing capacities ≤890 kN (100 tons). The main shaft of the press has two cams—one cam operates the upper ram, and the other cam operates the lower ram for compacting the part. The cam that operates the lower ram also controls the powder feed into the die and ejects the part from the die after compacting. Cams normally operate linkages that convert the main shaft rotary motion into the linear motion of the tooling.

Figure 3 shows a schematic of a cam-driven press. The cams in this type of press can be adjusted or arranged with removable sections, thus allowing cam motion to be varied to produce special motions to compact the part. Pressure can be applied either simultaneously or sequentially to the top and bottom of the compact. Anvil and rotary presses are types of cam-driven machines. These presses are described in more detail later in this article.

Eccentric-Driven Presses. Presses that have a final drive mechanism consisting of an eccentric or crank on the main shaft are the most widely used type of mechanical press. A connecting rod is used to convert the rotary motion of the main shaft into the reciprocating motion of the press ram. Generally, an adjustment mechanism is built into the connecting rod or press ram assembly, thus permitting the height position of the press ram to be changed with respect to the main shaft or press frame, thereby controlling the final pressing position of the ram. This adjustment mechanism can be used to control the length of the compacted part. Standard eccentric-driven presses have pressing capacities ranging from 6.7 to 7830 kN (0.75 to 880 tons).

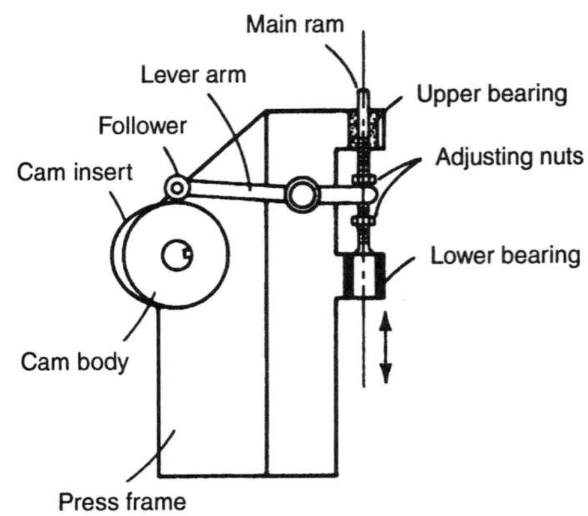

Fig. 3 Schematic of cam-driven compacting press

Hydraulic Presses

Hydraulically driven compacting presses are available with pressing capacities ranging from 445 to 11,100 kN (50 to 1250 tons) as standard production machines, although special machines with capacities ≤44,500 kN (5000 tons) have been used in production. Hydraulic presses normally can produce longer parts in the direction of pressing than mechanical presses, and longer stroke hydraulic machines are less expensive compared to an equivalent stroke produced in a mechanical press. The maximum depth of powder fill in mechanical presses is ~180 mm (7 in.), while 380 mm (15 in.) of powder fill is common in hydraulic presses.

The maximum production rate for hydraulic presses producing a single part per stroke is ~650 pieces per hour. The slower speed of a hydraulic press when pressing long parts is preferable, because the longer time during pressing permits trapped air within the powder to escape through the tooling clearances.

Most hydraulic presses are considered top-drive machines because the main operating cylinder is centrally located in the top of the press. This main cylinder provides the force for compacting the part. Hydraulic presses have three distinct downward speeds:

- *Rapid advance:* Produces minimal pressing force, used for rapid closing of the die cavity
- *Medium speed:* Pressing capacities ~50% of full-rated capacity, used during initial compaction when lower pressing force is required
- *Slow speed:* Maximum capacity available for final compaction

Two types of hydraulic pumping systems are commonly found in P/M presses: the high-low system and the filling circuit system. The high-low system has a double-acting main cylinder. A regenerative circuit is used for rapid approach. Initially, the piston of the cylinder is activated by a high-volume, low-pressure pump; the fluid from the bottom of the cylinder is directed into the top of the cylinder in addition to the low-pressure pump volume. At medium speed, the regenerative circuit is deactivated, while the piston remains activated by the low-pressure pump. In full-tonnage press, the low-pressure pump is deactivated, and a high-pressure pump activates the piston.

The filling circuit hydraulic pumping system has a single-acting main cylinder, and ram motion is controlled by small double-acting cylinders. The ram control cylinders are smaller than the main cylinder, so only a low flow rate of fluid is needed to cause rapid movement of the ram. During approach and return cycles, however, the fluid flow rate into and out of the main cylinder is high. The main cylinder is fitted with a large two-way valve that allows fluid to flow at low pressures (usually gravity feed). During pressing, the two-way valve is closed, and high pressure from the pump is applied to the main cylinder piston.

Ejection of the part usually is accomplished by a cylinder that is centrally located in the bed of the press. The cylinder either upwardly ejects the part or pulls the die downward from the part, depending on the type of tooling used.

When pressing parts to a given thickness, positive mechanical stops are used on hydraulic presses to control downward ram movement. When pressing parts to a desired density, downward ram movement is controlled by adjustment of the pressure to the cylinder. When the part is pressed to the desired unit pressure, the press ram stops and returns to the retracted position. Some types of P/M materials, such as P/M friction materials, are always pressed to density rather than size, because uniform density provides uniform friction and wear properties.

The drive-motor horsepower on a hydraulic press is considerably larger than on an equivalent mechanical press. A mechanical press has a flywheel from which energy is taken during the pressing and ejection of the part. Energy is restored to the flywheel during the die feeding portion of the cycle. The motor on a hydraulic press must supply energy directly during the pressing and ejection portion of the cycle.

Comparison of Mechanical and Hydraulic Presses

In terms of partmaking capability, no distinct advantage is gained by using either a mechanical press or a hydraulic press. Any part can be produced to the same quality on either type of machine. However, the following parameters influence press drive selection.

Production Rate. A mechanical press produces parts at a rate one and one-half to five times that of a hydraulic press as a result of inherent design of the energy transfer systems and stroke length.

Operating cost of a hydraulic press is higher, because the total connected horsepower of a hydraulic press is one and one-half to two times that of an equivalent mechanical machine. Theoretically, the required energy to compact and eject a part is the same for a hydraulic or a mechanical press, except that the overall efficiency of a mechanical press is slightly higher than that of a hydraulic press. Also, the kilowatt usage of the larger motor on a hydraulic press is greater than that of a mechanical press during the idle portion of the machine cycle.

Machine overload protection is an inherent feature of a hydraulic press. If the hydraulic system is operating properly, the machine cannot create a force greater than the rated capacity. Consequently, overload of the machine frame is not possible, even if a double hit or operator error occurs in adjusting the machine. Misadjustment or double hits can cause a mechanical press to overload, can damage the machine, or may cause tooling overload and failure if the tooling cannot withstand full machine capacity. Some new mechanical presses are equipped with hydraulic overload protection systems.

Equipment cost of a hydraulic press generally is one-half to three-quarters that of an equivalent mechanical press. Facility, foundation, installation, and floor space costs generally are comparable.

Die Sets. The mounting into which the tooling is installed is known as the die set. Generally, the die set must be well guided because of the close tooling clearances used. Guide bearings must be protected with boots or wipers to prevent powder particles from entering guiding surfaces. Tooling support team members should have high stiffness to minimize deflection.

The die set must be free of residual magnetism. The maximum acceptable level is ~2 G. To ensure press operator safety, die sets should be adequately guarded. In a complex tooling arrangement, as many as seven independent tooling members and supports are moving relative to one another during the pressing and ejection cycles.

Die sets can be classified as removable or nonremovable. Both types are used in mechanical and hydraulic presses. Nonremovable die sets are used throughout the entire tonnage requirements of available presses. Manually removable die sets are used primarily in presses with pressing capacities up to ~2670 kN (300 tons). Above this press size, the die set assembly is moved by a powered system, and removable die set presses with capacities of ≤17,800 kN (2000 tons) are available.

The major advantage offered by nonremovable die sets is flexibility in setup and operation. Presses equipped with nonremovable die sets usually have all adjustments required for setup and operation built into the press and die set, including:

- *Part length adjustment:* Any dimensions of the part in the direction of pressing can be quickly changed during production.
- *Part weight:* Material weight in any level of the part can be changed easily during production.
- *Tooling length adjustment:* Adjustments are provided to accommodate shortening of punch length due to sharpening or refacing.

Another advantage of nonremovable die sets is the greater space available for tooling, compared to the removable type. This space provides more freedom in tooling design. However, presses incorporating nonremovable die sets must be shut down during tooling changes or maintenance. Tooling change and setup time generally is from 1 to 4 hours—but sometimes substantially longer, depending on the complexity of tooling.

Nonremovable die sets are well suited for developing new P/M parts, because press and tooling adjustments can be made quickly to achieve the desired weight, density, and part dimension. Adjustment features of nonremovable die sets make them desirable on long production runs, where changes in powder quality among lots require frequent tooling adjustment to maintain part quality.

Users of removable die sets normally have two or more die sets per press. Tooling can be set up in a spare die outside the press. Removable die sets normally can be changed in less than 30 min, so loss of production time is minimal. On small presses where the die set is also small, the die set is restricted to a given set of tools and is considered semidurable tooling.

One disadvantage of many removable die sets is that pressing is controlled by pairs of pressing blocks made of hardened tool steel, such as D-2. The height of the pressing block controls the height of the part. If the part length dimension is changed due to design, or if the tooling length is changed due to repair, the pressing blocks must be changed accordingly. Removable die sets are ideally suited for shorter production runs. On newer presses with removable die sets, complete powder adjustment is available, even when the die set is outside the machine.

Part Classification

The Metal Powder Industries Federation has classified P/M parts according to complexity. Class I parts are the least complex, and class IV parts are the most complex. To better understand the types of commercially available P/M com-

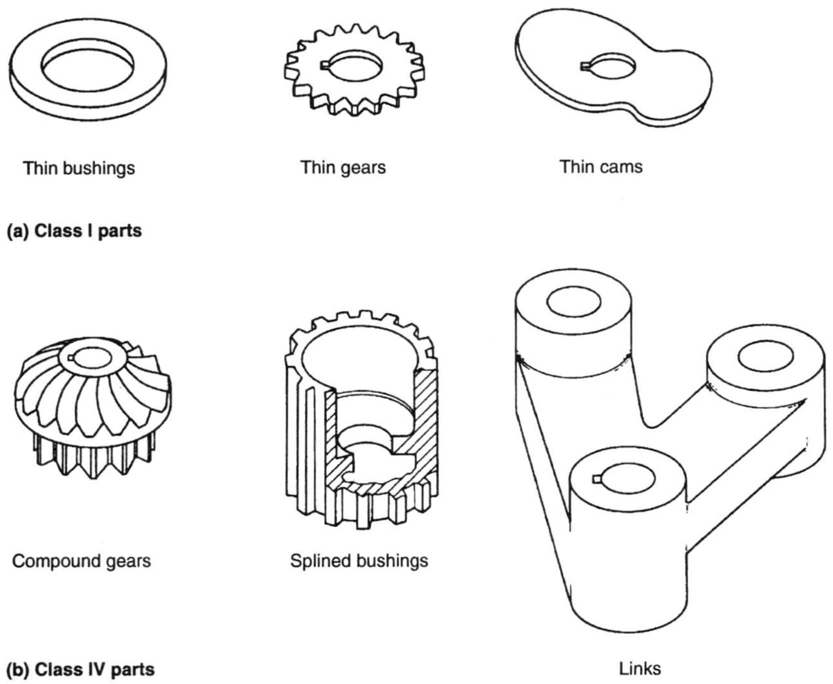

Fig. 4 Basic geometries of (a) MPIF class I (simple) and (b) MPIF class IV (complex) parts

pacting presses, and their advantages and limitations, an understanding of P/M part classification and tooling systems used to produce parts is necessary. Part thickness and number of distinct levels perpendicular to the direction of powder pressing determine classification—not the contour of the part.

Class I parts are single-level parts that are pressed from one direction, top or bottom, and that have a slight density variation within the part in the direction of pressing (Fig. 4a). The highest part density is at the surface in contact with the moving punch, and the lowest density is at the opposite surface. Parts with a finished thickness of ~7.5 mm (0.3 in.) can be produced by this method without significant density variation.

Class II parts are single-level parts of any thickness pressed from both top and bottom. The lowest density region of these parts is near the center, with higher density at the top and bottom surfaces.

Class III parts have two levels, are of any thickness, and are pressed from both top and bottom. Individual punches are required for each of the levels to control powder fill and density.

Class IV parts are multilevel parts of any thickness, pressed from both top and bottom (Fig. 4b). Individual punches are required for each level to control powder fill and density.

Shape of Rigid Tooling

Rigid tool compaction differs from roll compaction, isostatic compaction, hot isostatic pressing, and injection molding in that a quantity of powder (fill) is confined in a rigid die cavity at ambient temperature. The die cavity is entered by one or more punches, which apply compaction pressure to the fill powder. As a result of the compaction pressure, the fill powder densifies, develops green strength, and assumes the exact shape of the die cavity and punch faces. Following the pressure cycle, the shaped powder fill, now a piece part, is ejected (stripped) from the die cavity.

The physical size of parts made in rigid tool compaction systems is a function of press tonnage capacity, fill depth, and also the length of a green powder fill that can be effectively compacted in terms of a maximum density variance. Parts vary in size from those weighing ~1 g (0.035 oz) that are made in presses with capacities as small as 35 kN (4 tons) to those weighing 10 kg (22 lb) that are made in presses with capacities of 8900 kN (100 tons).

Rigid tools must also be constructed oversize, with exact linear dimensions, to compensate for the final volume change. Although theoretical computations are useful, most successful rigid tool sets are based on shrinkage allowances developed from existing tooling and the dimensional histograms developed for particular powders. However, shrinkage allowances can be complex depending on subsequent sintering and binder additives. For example, some metallic powders, such as the carbide and tool steel types, and some gas and centrifugally atomized specialty powders, such as spray-dried tungsten carbide, do not develop significant green strength, because their individual particles are predominantly spherical or they lack plasticity. To compact such powders in rigid tool systems, wax or wax-stearate binders are added, which can occupy up to 20 vol% of the green compacted shape. The development of full metallic properties during sintering also requires a volume shrinkage.

Powder Fill

The important consideration in P/M part production is the fill ratio required to produce parts to a density that is compatible with end use requirements. The fill ratios must remain constant for a given part to maintain dimensional reproducibility. Parts can be of single-level or multilevel design.

Single-level parts, designated as class I by the Metal Powder Industries Federation (MPIF), present the least difficulty to the tool designer, regardless of the size or part configuration. The main consideration is designing a die that is long enough guidance for the lower punch (usually 25 mm, or 1 in.) and providing adequate fill depth for compacting the powder to the required density. This challenge, coupled with the primary mechanical consideration of locating the center of mass in the press center, provides the best potential for producing a uniform quality part. Figure 4(a) shows basic geometries of MPIF class I parts.

Multilevel parts, with industry classifications II through IV, present two additional complications to the tool designer: powder fill and part ejection. Because metal powders tend to compact in vertical columns and generate little hydraulic flow, the tool designer must create fill levels in the tools that compensate for the thickness variations present in the final part configuration. Uniform density, neutral axis of compaction, and part ejection should be considered to determine the need to vary fill levels and the manner in which these variations are achieved. Excessive density variations contribute to green cracks and sintered distortion.

A common method of varying fill levels is by using multiple lower punches, which are timed to react to one another either through the use of springs or air, or by mounting on separate press platens. Other methods are less effective, because punches are not adjustable and are fixed on one of the tool members, such as the die or core rod.

Fixed levels are commonly referred to as die chokes, core rod steps, or splash pockets (Fig. 5). Fixed fills are sensitive to the apparent density of the material being compacted. In operations that control compacting pressure, such as in hydraulic pressing, fixed fills cause dimensional variations in part thickness. Because mechanical presses are set to operate to a fixed position relative to the die, the variation created by the apparent density of the powder causes overdensification or underdensification, resulting in a corresponding oversize or undersize peripheral area on the part. Green expansion occurs as a part is stripped from the die. Ideally, the part returns to die size through shrinkage during sintering.

When a part has more than one level in the compacting direction, the step height should be

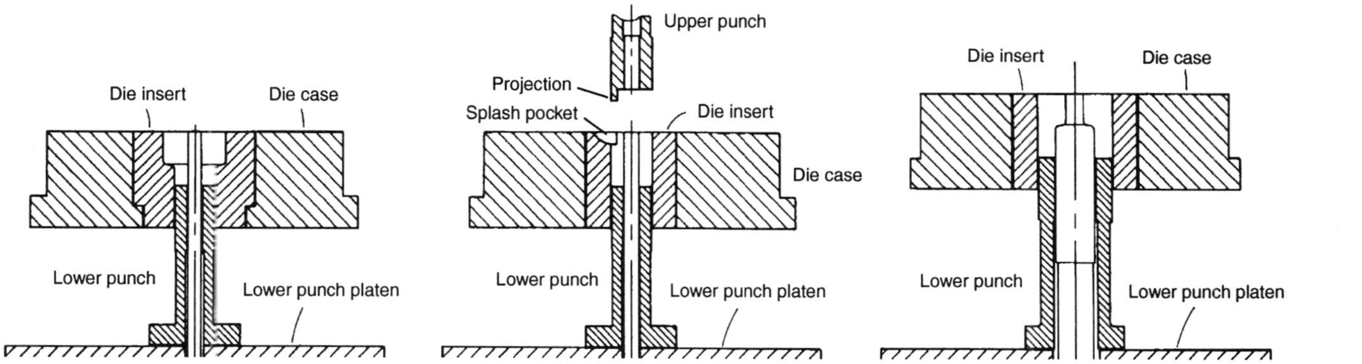

Fig. 5 Methods of achieving fixed fill levels. (a) Fixed fill on an upper level using a step die. (b) Fixed fill using a splash pocket to permit a projection feature on an upper punch. (c) Stepped core rod forming an internal shoulder

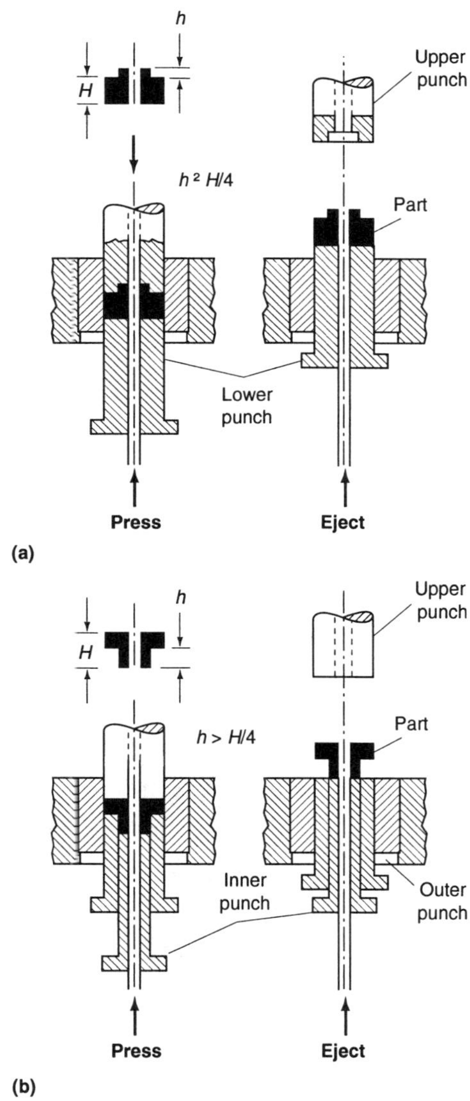

Fig. 6 Two-level compaction. (a) Single lower punch when $h \leq H/4$. (b) Double lower punches when $h > H/4$. Source:

limited to one-quarter of the overall height for a single punch (Fig. 6a). If a larger step is required, multiple punches should be considered (Fig. 6b).

Fill Height. The fill height is the depth of the loose powder required to give the required part thickness after compaction. The value is determined by the compressibility of the loose powder at the required density. The fill height, h_f, is obtained by multiplying the finished part height by the compression ratio of the powder:

$$h_f = tk_r$$

In this equation, t is the part thickness, k_r is the compression ratio, and $k_r = q/q_a$, where q is the part required compaction density and q_a is the apparent density of the loose powder. If the fill height is greater than the maximum fill height that can be accommodated in the press selected on the basis of the compacting load required, a larger capacity machine should be selected, which has the required fill height capacity.

Tooling Systems

High-production P/M compacting presses are available as standard production machines in a wide range of pressing capacities and production rate capabilities. Presses are designed to produce parts of a specific classification, as discussed previously.

Single-action tooling systems generally are limited to production of class I parts. During the compacting cycle, the die, core rod, and one of the punches (usually the lower punch) remain stationary. Compacting is performed by the moving punch, which is driven by the action of the press. One or more core rods may form any through holes in the part.

During ejection, the upper punch moves away from the formed part, and the part is ejected from the die by the lower punch. The core rod (Fig. 7) is stationary, and the part is ejected from the die and core rod simultaneously. On some presses, the core rod is arranged so that it is free to move upward (float) with the part as it is ejected. The compacted part experiences slight elastic expansion on ejection from the die, which causes the part to free itself from the core rod. The core rod is then free to move downward to the fill position. This floating core rod arrangement reduces ejection forces and core rod wear.

Double-action tooling systems primarily are used to produce class I and II parts. Force is applied to the top and bottom of the part simultaneously, because the punches have the same travel rate. The die and core rod are stationary. Densification takes place from the top and bottom, with the lowest density region near the center of the part. Although the core rod is fixed in this system, it can be arranged in a floating position. Figure 8 shows the compacting sequence of a double-action tooling system.

Floating die tooling systems are similar to double-action arrangements. As shown in Fig. 9, the die is mounted on a yielding mechanism (springs). However, pneumatic or hydraulic cylinders usually are used, because they offer an easily adjustable resisting force. As the upper punch enters the die and starts to compact the powder, friction between the powder and die wall causes the die to move down. This has the same effect as an upward-moving lower punch. After pressing, the die moves upward to the fill position, and the upward-moving lower punch ejects the part. The core rod can be fixed or floating.

Withdrawal tooling systems use the floating die principle, except that the punch forming the bottommost level of the part remains stationary and that the die motion is press activated rather than friction activated. The die and other lower tooling members, including auxiliary lower punches and core rods, move downward from the time pressing begins until ejection is complete.

Figure 10 shows the compacting sequence in a multiple-motion withdrawal tooling system. During compaction, all elements of the tooling system except the stationary punch move downward. The die is mounted on the top press member of the platen and is supported by pneumatic or hydraulic cylinders. Auxiliary punches are mounted on additional platens, which are similarly supported and have positive pressing stops. The stops control the finished length of each of the levels within the compacted part. Before ejection, these stops are released or disengaged so that the platens can be moved further downward. During ejection, the upper punch moves upward, away from the compact, while the die and lower punches move sequentially downward until all tool members are level with the top of the stationary punch. The compact is fully supported by the tooling members during ejection, resting on the stationary punch as the die and lower punches are lowered to release it.

The core rod can be provided with pressing position stops to allow a part to be produced with blind or counterbored holes. The core rod is held stationary until the part is free of all other tooling members before moving downward to the ejection position.

At this point in the machine cycle, the feeder moves across the die, pushing the compacted part from the die area and covering the die cavity. The die and auxiliary lower punches move upward to their respective fill positions. The core rod then moves upward, displacing the excess powder into the partially empty feed shoe. The feeder retracts, wipes the top fill level, and readies the press for the next cycle.

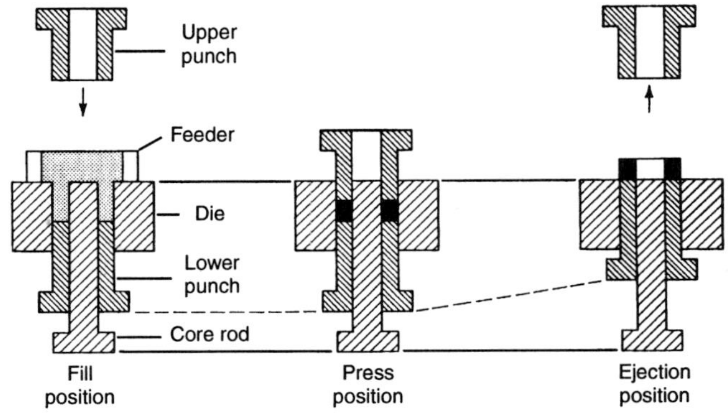

Fig. 7 Compacting sequence utilizing single-action tooling. Dashed line indicates motion of lower punch.

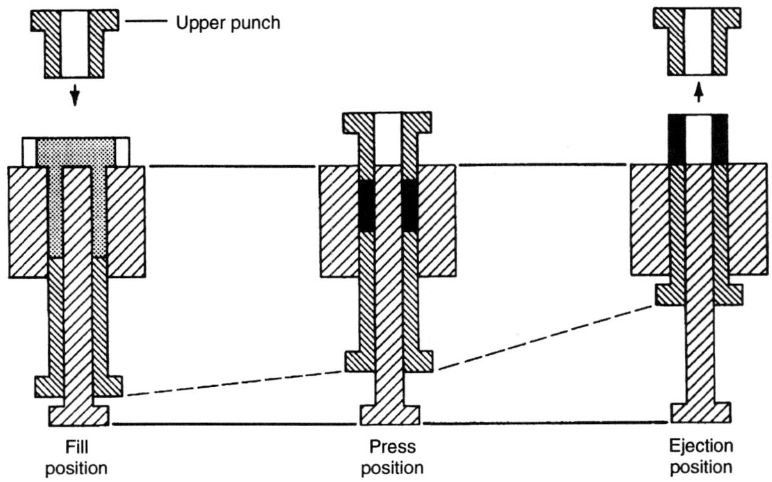

Fig. 8 Compacting sequence utilizing double-action tooling. Dashed line indicates motion of component parts.

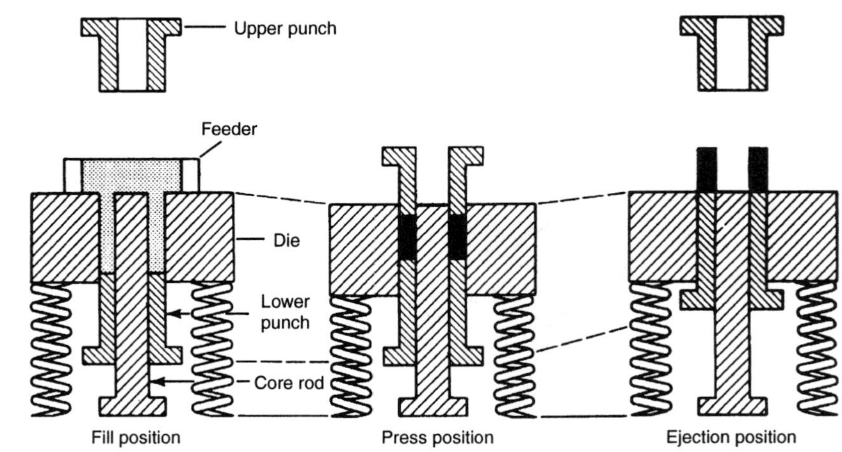

Fig. 9 Compacting sequence utilizing floating die tooling. Dashed lines indicate motion of component parts.

Types of Presses

Anvil presses generally are limited to compaction of class I parts in a single direction. Anvil presses do not have an upper punch; a moveable, solid, flat block seals the top of the die. Compacting is done by the lower punch, which, after the anvil is released and moved, moves farther to eject the compact from the die.

Anvil presses are available with pressing capacities ranging from 6.7 to 310 kN (0.75 to 35 tons), with maximum depth of fill ranging from 1 to 75 mm (0.040 to 3 in.). Multiple-cavity pressing frequently is used in anvil presses, with

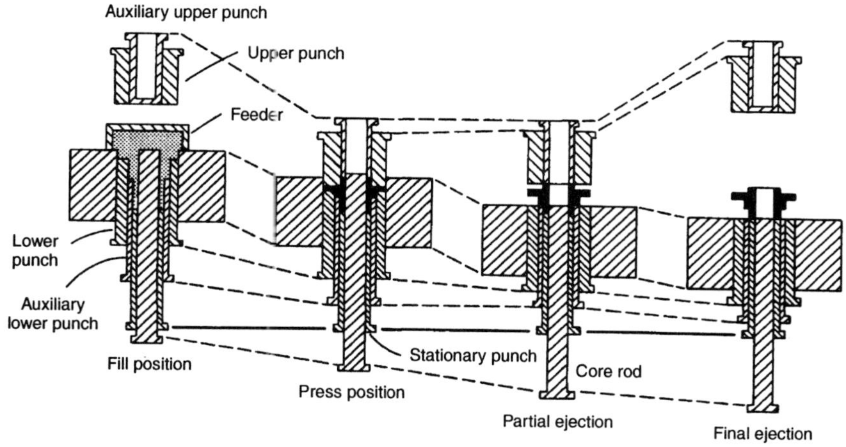

Fig. 10 Compacting sequence utilizing floating die withdrawal double-action tooling. Dashed lines indicate motion of component parts.

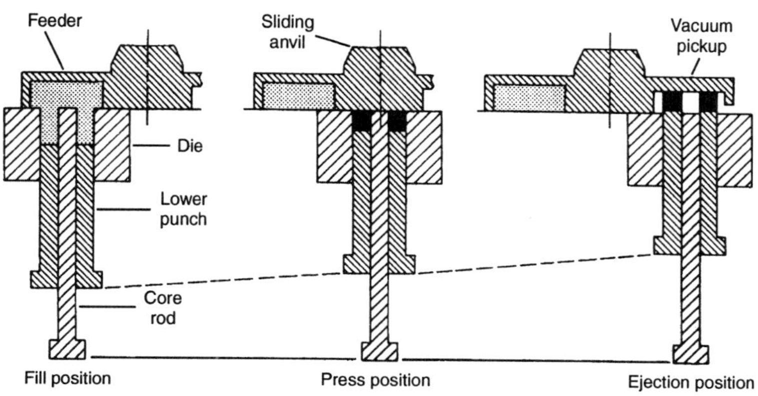

Fig. 11 Compacting sequence utilizing sliding anvil single-action tooling. Dashed line indicates motion of component parts.

possible production rates of >100,000 pieces per hour. Some anvil presses can be converted to double action, using an upper punch entry system. Anvil presses usually are mechanically driven. Figure 11 shows a schematic of an anvil press operation.

Rotary presses generally are limited to compaction of single-level class II parts, although some class III parts, such as flanged bushings, are produced. Rotary machines are available with pressing capacities ranging from 36 to 590 kN (4 to 66 tons), with a depth of fill up to 75 mm (3 in.). Production rates of >60,000 pieces per hour are possible, depending on machine size and the number of tooling stations. Rotary presses are mechanically driven.

Single-Punch Opposing Ram Presses. Like rotary presses, these machines are limited to production of class II and some class III P/M parts. These presses are available in top- and bottom-drive models, with pressing capacities ranging from 36 to 980 kN (4 to 100 tons) and with a maximum depth of fill up to 100 mm (4 in.).

Production rates of up to 3000 parts per hour are possible using mechanical presses with single-cavity tooling, although production rates of 900 to 1800 pieces per hour are more common. Hydraulic presses produce ~900 pieces per hour. Ejection of the part is accomplished by the lower punch moving upward. Mechanical and hydraulic presses are available.

Single-punch withdrawal presses have essentially the same partmaking capabilities as the single-punch opposing ram system in terms of pressing capacity, depth of fill, and production rate. The major difference is that floating dies are used to achieve top and bottom pressing. The die is moved downward to eject the part.

Multiple-motion die set presses can be designed to produce the most complex P/M parts. These presses use floating die and withdrawal tooling methods. Machines are available with either bottom- or top-drive arrangements. Pressing capacities range from 27 to 7830 kN (3 to 880 tons), with a maximum depth fill of 180 mm (7 in.). Production rates vary from more than 6000 pieces per hour on smaller machines to 1800 pieces per hour for 1960 kN (220 ton) presses.

In addition to producing complex parts, the removable die set (tool holder) minimizes press downtime for part changeover if the die set for the next part to be produced is set up outside the press and is ready for installation. Pressing position for each level being produced by a separate tooling member is controlled by fixed-height tooling blocks (stop blocks), which usually are ground to the proper height to produce a given dimension on the part. A small adjustment in the block mounting member allows for minor changes to part dimension. Full range adjustments are available on more recent presses.

Multiple-motion adjustable stop presses have the same partmaking capability as multiple-motion die set presses and use the same tooling methods. Pressing capacities range from 980 to 7340 kN (110 to 825 tons), with a maximum depth of fill of 150 mm (6 in.). These presses do not incorporate removable die sets; however, press stop positions are adjustable, and a change in any dimension of the part in the direction of pressing is easily accomplished.

Advanced Tool Motions

A common limitation of some rigid tooling systems is that part features not perpendicular to the direction of pressing cannot be compacted and stripped. Frequently, it is cost effective to form features such as cross holes and threads by machining. Other nonperpendicular features, notably helix shapes and hidden flanges, can be formed using complex tool motions. Another type of advanced tooling system permits production of complex shapes with magnetic orientation of the microstructure.

Helical shapes, typically helical spur gears, are produced in rigid compaction tool sets with punch rotation capability. In a simple system, a helical form lower punch is engaged in a die with a matching gear form. In such a system, the lower punch remains engaged in the die at all times, as is common practice for all rigid tool systems, so that indexing rotation of the punch to the die is avoided. The die acts as a guide. Rotation is carried out on a thrust bearing, which rests on the punch platen that supports the lower punch. An upper punch is not required, because the top of the die cavity is closed by an upper anvil, which does not enter the die cavity. Central core rods, with or without additional features such as splines and key forms, are commonly operated in this helical tool system.

Helical gears made in this manner are limited to helix angles of ~25° and a thickness of 32 mm (1¼ in.) due to fill limitations along the helix tooth form. More complex helical gear tooling systems have been developed for routine production using helical upper punches, driven by follower cams for indexed die entry, with inner and outer lower helical punches for stepped helical gears.

Split Die Systems. Another rigid tooling system that avoids some through-cavity limitations is known as the split die, or "double die," system. It enables the compaction of parts with completely asymmetric upper and lower sections in the pressing direction. Figure 12 shows typical tool motions in split die compaction. This system requires two die-holding platens to carry the upper and lower die. Each platen is controlled and moved independently.

Wet magnetic compaction (Fig. 13) has enjoyed wide usage in the production of magnetically oriented ferrite shapes. In this production process, a feed shoe is not required. Instead, the die cavity is injected with an aqueous slip (slurry) that has a high concentration of ferrite powder, with the addition of green binders as required. Typically, the die filling pressure is 35 MPa (5000 psi). By using an aqueous slip, many of the gravity die fill problems, such as attainment of uniform powder density and filling the areas that are difficult for the powder to reach, are avoided.

Following die fill injection, an orienting magnetic field is applied to the slip, resulting in magnetic polarization of the individual ferrite particles, which remain mobile at this point. The optimal orientation of the ferrite particles directly determines the quality of the finished permanent magnet. After magnetic orientation, the main pressing load is applied, densifying the ferrite mass and causing the suspending aqueous carrier to be expelled through drainage ports. The compact is imparted with the precision shape and dimensions of both the upper and lower dies, plus any core rods that may be inserted. The cycle is completed by separation of the press platens and ejection of the compacted ferrite shape.

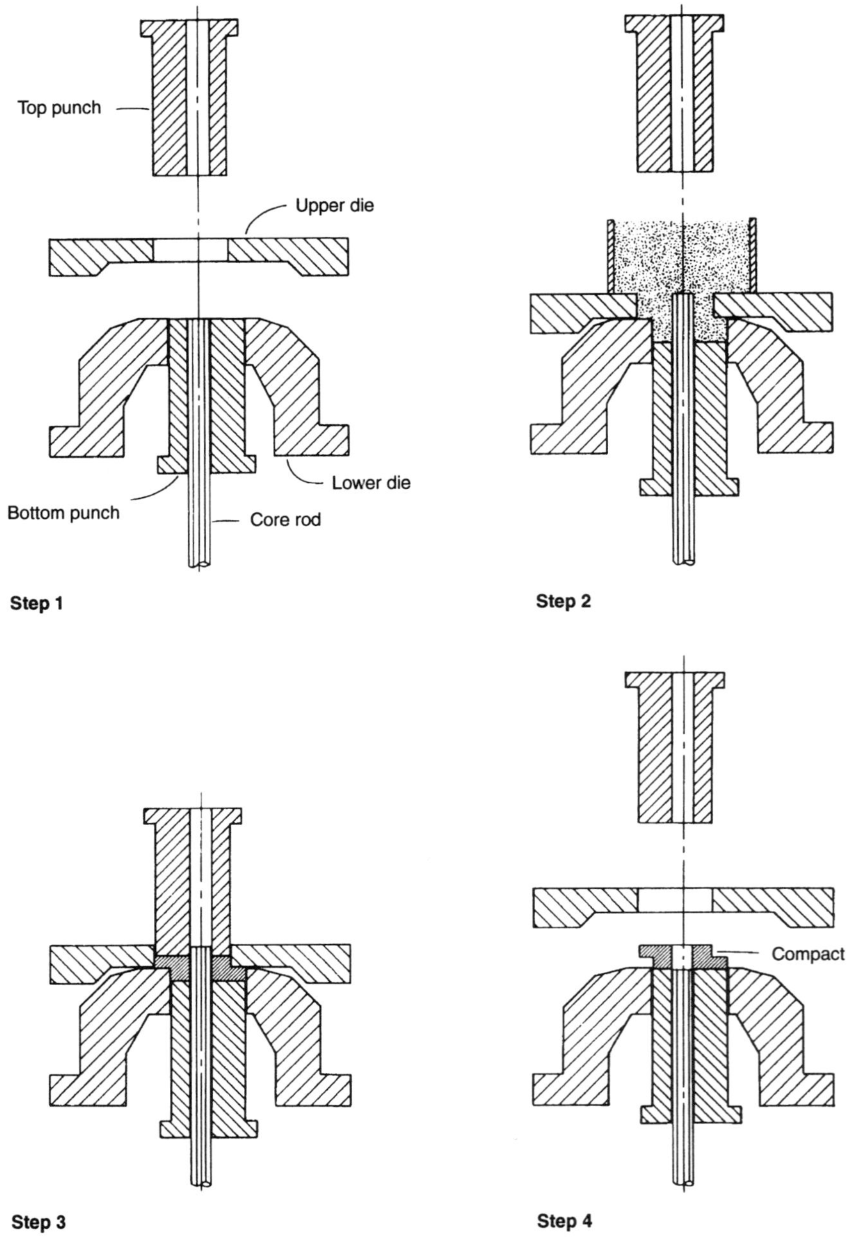

Fig. 12 Split die compaction sequence

Tooling Design

Traditionally, P/M tooling was designed on the basis of production experience. In simple parts, such as single-level class I and II parts, these determinations proved successful. As state-of-the-art materials and presses advanced to the production of complex, multilevel parts, the "cut-and-try" method of tool design became obsolete. The high cost of complex tooling and adapters, plus downtime to redesign and rebuild tooling, requires the partmaking system, including the press, to be carefully analyzed in terms of load, stress, and deflection.

Tooling layout is required to design a suitable set of tools and to determine the physical dimensions (length and thickness) of tooling members. A preliminary layout helps to determine fill, pressing, and ejection positions and to eliminate interference at these positions.

The die space drawing supplied with every compacting press, which usually starts with the ejection position, is the basis of the tooling assembly layout. Generally, tooling members are never closer than in the ejection position, which constitutes the minimum space available to contain all components and their adapters.

Die Design. Dies are commonly constructed by using inserts that are held in the die case by shrink fitting. The amount of interference between the insert and the die case depends on the inside and outside diameter of each member and on the compacting pressure used. The powder can be considered a fluid in a closed container that transmits the compacting pressure in all directions; therefore, the die must be designed as though it were a pressure vessel with internal pressure.

In actual practice, radial pressure on the die walls due to compacting rarely exceeds 50% of the compacting pressure. The interference fit of the die case and die insert should be such that the stress on the insert always remains in compression for round dies. However, for shaped dies such as gears, cams, and levers, the use of finite element analysis is the best method for accurately determining stress and deflection.

In P/M tooling, the die normally controls the outer peripheral shape and size of the piece part. Typically, it is constructed from materials such as tungsten carbide or high alloy tool steels, such as T15, D2, CPM-10V, or CPM-15V with high hardness and good wear resistance. Dies are usually constructed in one or more sections and compressed into a retaining ring made of a low-alloy steel, such as AISI 4340 or 6150.

Considerations in die design and material selection include initial tool cost, shear strength of the die material, and die shape. A large die may require tungsten carbide, which costs ten times as much as tool steel materials. Tungsten carbide may be the best material for a set of gear tools with a relatively steep helical angle. Sectional die construction may be required for specific shapes such as sharp corners or projections into the die cavity.

Die Wall Thickness. An exact calculation of the stress on die walls is almost impossible from a practical point of view because stress distributions in the compact are extremely complicated

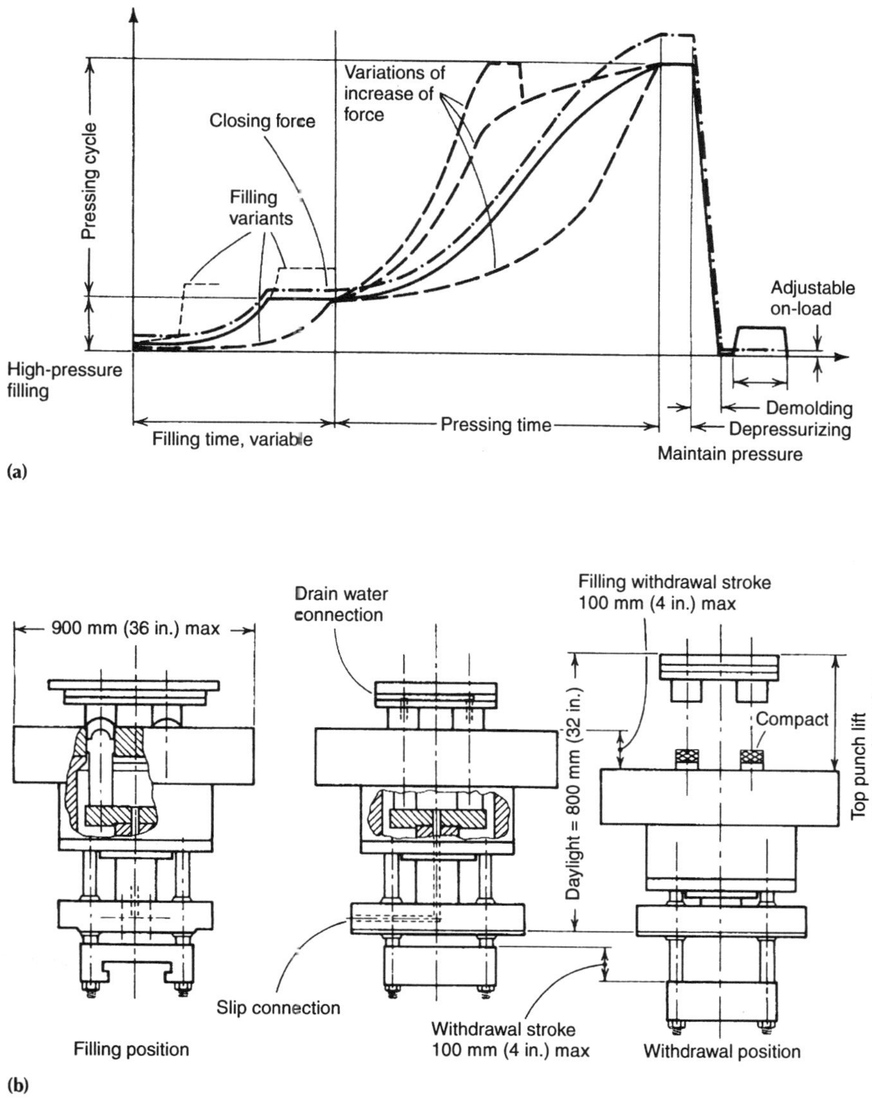

Fig. 13 Wet magnetic compaction. (a) Force-time diagram for magnet presses. (b) Schematic of press tool for chamber-filling method designed for withdrawal operation

and include variables such as part shape, particle size distribution, and other factors that affect transmission of compressive stress in the lateral direction (Ref 2). The vertical axial load can exert a horizontal force after a certain degree of consolidation has been attained. For example, when a simple shape is compacted at 400 MPa, as much as 120 MPa pressure can be exerted radially against the die walls.

If for purposes of simplification, the internal pressure is considered strictly hydrostatic in nature and the confined material is an incompressible liquid, then the die wall thickness for a cylindrical die could be determined by using Lame's formula:

$$D = d\sqrt{(S + p)/(S - p)}$$

where S is the maximum allowable fiber stress for the material of the die, D is the outer diameter of the die, d is the compact diameter, and p is the radial stress acting on the die wall. This is a simplification because during metal powder compaction the pressure is not hydrostatic and the material is not incompressible. Initially, the powder is compressed with a consequent reduction in the vertical height of the space filled by the powder. The compressed material begins to resemble a solid after a certain degree of compaction has been reached.

Poisson's ratio is 0.3 for fully dense and isotropic steel. While this wrought form value cannot apply to powder metal, it is assumed to be applicable in the fully compacted condition. Thus, the Poisson's ratio is introduced into the previous equation, and the following modified Lame's formula is used for estimating the die wall thickness for metal powder compaction.

$$D = d\sqrt{(S + \mu p)/(S - \mu p)}$$

where μ = Poisson's ratio = 0.3. This formula, however, does not take into consideration that the internal pressure acting over the length of the compact is balanced by the strength of the die having a larger length. The formula does address the friction at the tooling/powder interfaces resulting in nonuniform pressure distribution in the compact.

Generally speaking, the formula produces more conservative results than are necessary. The interference fit between the shrink ring and the die insert should be such that the stress on the insert always remains compressive for round dies. For shaped dies such as those used for production of gears and cams, the use of finite element analysis is the best method for accurately determining the stress and deflection.

Core Rods. Basically, the core rod is an extension of the die that controls the inner peripheral shape and size of the piece part. Tungsten carbide and M2 or M4 high-speed steels are the most common materials used for core rods. Primary factors in materials selection include wear resistance and hardness, which enable the core rod to resist the high compressive force exerted during compaction and the abrasive action sustained during part ejection. Core rods >25 mm (1 in.) in diameter or area are held to a base by mechanical means, such as a screw, while smaller core rods are held by means of silver solder or braze.

Punches can perform the function of a die or a core rod and carry the full load of the compressive force required to compact the P/M part. Wear resistance and toughness are the most important factors in materials selection. The most commonly used materials are A2, D2, S7, and H13 tool steels. Dimensional control, especially in areas such as concentricity and hole-to-hole location, depends on the amount of clearance that can be maintained between the punches, die, and core rods. Clearance should be calculated for each specific range and size of part. It is important to note that thermal size changes occur during operation, primarily because of the friction created by stripping the compacted part and the speed of the pressing cycle.

Punch Component Stress. Compacting powder causes compressive stress in the punch. This stress must be below the yield strength of the punch material. Calculation of buckling stability should be made for long, thin-walled punches.

Figure 14 shows the effect of axial compressive force on a tubular punch. A tubular punch is subjected to internal pressure during compacting of multilevel parts. In this case, the resulting circumferential tensile stress in the punch wall should be calculated. If the stress and accompanying deflection is excessive, tooling clearances should be designed so that when the outer punch wall expands, it is supported by the die wall before the stress reaches the yield limit (Fig. 15).

During ejection, the punch is subjected to compressive stresses by resisting the stripping action of the die and to tensile stresses from the stripping action of punch. These stresses normally are lower than compacting stresses. Components of the punch subjected to stress include the punch clamp ring and bolts, which should resist the ejection of the punch without perma-

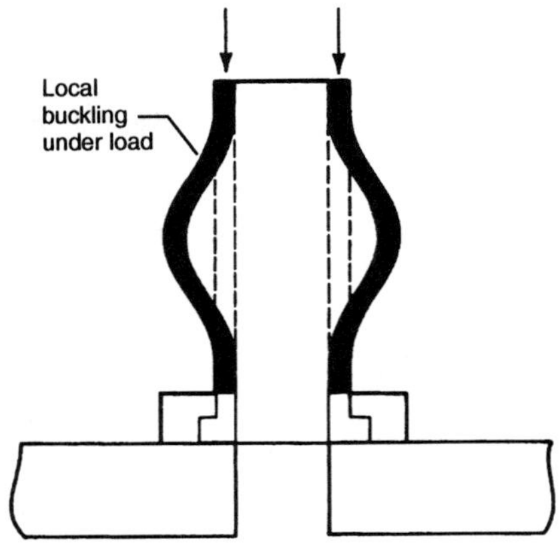

Fig. 14 Effect of compressive stress on tubular punch

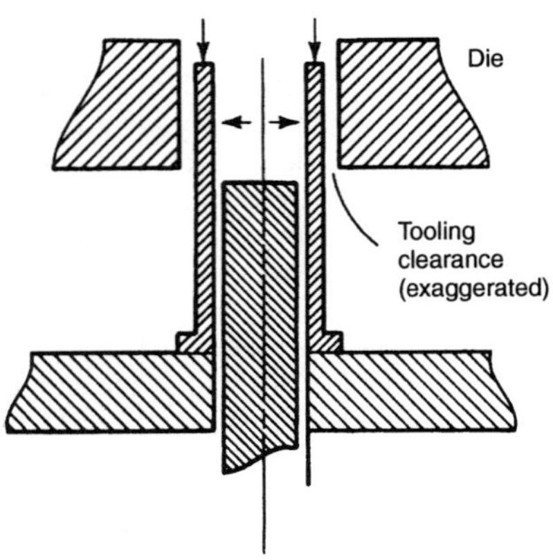

Fig. 15 Tensile stresses in a tubular punch during compacting. Large arrows indicate action of powder on walls of punch.

nent deformation. Punch adapters are subjected to bending loads that create a tensile stress around the center hole during compacting. This stress should not exceed the fatigue limit of the adapter material.

Tubular adapters must have sufficient cross-sectional area to withstand the pressing load without permanent deformation. A stepped core rod, or a core rod forming a blind hole, must not buckle during compacting. The base of the core rod must resist, without permanent deformation, whatever ejection loads are imposed on the core rod.

The core rod clamp ring and retaining bolts should be sized to withstand the ejection force on the core rod without permanent deformation. The core rod adapter generally is strong enough to resist both pressing and ejection loads, due to the size of the adapter when space is provided for clamp ring fasteners.

Deflection Analysis. Pressing of P/M parts at pressures >690 MPa (50 tsi) presents unique considerations for size and tolerance in multilevel parts. A variety of tool members should be utilized to establish proper fill ratios, and deflection and springback can occur. Deflection occurs because of the column loading effect on the

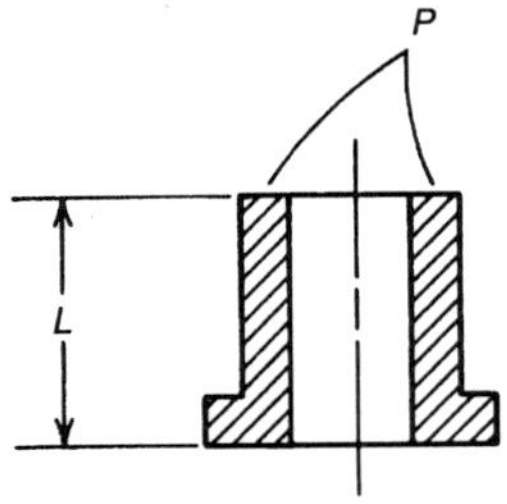

$$Y = \frac{PL}{AE}$$

Fig. 16 Punch compression. P is total punch load, L is length, Y is deflection, A is area of punch, and E is Young's modulus.

compacting tools during the briquetting cycle. For column load consideration, the bottom section of the lower punch is considered fixed, while the top section or working end of the lower punch can be considered free to rotate. The amount of deflection on the tool member will be determined by the column slenderness ratio of the punch and the adapter. When the column load is released after the press goes through the bottom dead center compaction point, the deflected punches will return to their original lengths, if their elastic material property limits have not been exceeded. This return movement is generally called springback and can be deleterious to the green part, depending on the fragileness of the green part section geometry involved.

Deflection can be minimized by strengthening the various tool members through changes in physical size or shape and/or by changes in material selection. The most common method of minimizing deflection effects is to equalize deflection using tool members and adapters that are designed to match the deflection characteristics of the most critical member. The ability of the tool designer to find the proper balance is paramount for production of crack-free parts.

When designing tools for production of parts other than single-level class I or II parts, deflection analysis of the tooling, tooling adapters, and press is desirable. These members are essentially stiff springs, each with a different spring rate or modulus. When the compacting load is applied, the parts deflect. When the load is released, they return to their original length. If the press contains two or more separate lower punches, the total deflection of each punch and the supporting members must be the same. Otherwise, the compacted part will move with the punch that has the greatest total deflection, leaving a portion of the part unsupported. This condition is likely to cause cracking during part ejection.

A punch under load normally is in pure compression and therefore will follow Hooke's law. If the punch has varying cross-sectional areas, each length having the same cross-sectional area is calculated individually. The total punch compression is the sum of these calculations. For a long, thin-walled punch, local buckling of the punch wall under load should be investigated. Compression of punches and their supporting members may be calculated using the equation given in Fig. 16.

Adapter Bending. The adapter, on which the punch is mounted, usually is a flat plate with the punch and the load positioned at the center, around a hole through which either another punch or a core rod passes. This plate, if supported at the outer edge, is subjected to the pressing load around the center hole. Two forms of deflection—bending and shearing—occur in this area. Adapter deflection is linearly proportional to force. Calculated adapter stress should be compared with the allowable adapter material stress to evaluate design suitability.

Press Deflection. Like the tooling and support, the press is subject to deflection. This tendency is considerably less than that of punch compression or adapter bending, but it must be considered in total tool design. Data regarding press deflection should be obtained from the press manufacturer. Deflection is linearly proportional to the amount of force exerted:

$$Y = C \times W$$

where C is the equipment constant, W is pressing force, and Y is deflection.

Tool Materials

Dies. In the most common type of die construction, wear-resistant inserts or liners are held in place by clamping or shrink fitting. The amount of interference between the insert and the die case depends on the inside and outside diameter of each member and on the compacting pressure used. The powder can be considered a fluid in a closed container that transmits the compacting pressure in all directions; therefore, the die must be designed as though it were a pressure vessel with internal pressure.

In actual practice, radial pressure on the die walls due to compacting rarely exceeds 50% of compacting pressure. The interference fit of the die case and die insert should be such that the stress on the insert always remains in compression for round dies. However, for shaped dies, such as gears, cams, and levers, finite element analysis is the best method for accurately determining stress and deflection.

Die inserts for compaction of carbide, ceramic, or ferrite powder most frequently are the medium- or coarse-grain 94WC-6Co grades of cemented carbide. Cemented tungsten carbide containing 12 to 16% Co can be used to make inserts for compacting metal powders in medium-to-long production runs.

The elastic moduli of carbides are considerably higher than those of steels, a fact that should be considered when designing composite steel and carbide die assemblies. Because carbide will deflect only 33 to 40% as much as steel, the steel portion generally should be designed with enough stiffness to support three times the expected loading in order to match the deflection of the carbide. The shrink-fit allowance should be ~1.0 mm/m (0.0010 in./in.).

Shrink rings and similar supporting parts of the tooling can be made from medium-carbon alloy steel, such as AISI/SAE 4340 or 6150, quenched and tempered to ~42 to 46 HRC. It is especially important that supporting parts for carbide tools provide sufficient support; otherwise, the carbide tools are likely to break in service.

Cemented carbides are relatively expensive, and shaping of parts to the required form must be done either by electrical discharge machining or by specialized methods of grinding.

Wear-resistant tool steel inserts are sometimes used instead of carbide inserts. Tool steel inserts are tougher and easier to fabricate than carbide inserts. Powder metallurgy tool steels such as CPM 10V, Vanadis 4, and Vanadis 10 are frequently chosen for medium-to-long production runs because they have wear resistance approaching that of carbides. Other wear-resistant tool steels, usually D2 or D3 or a high-speed steel such as M2 or M4, have been used for short-run applications. Tool steel inserts generally are heat treated to a working hardness of 62 to 64 HRC. For increased wear resistance, a nitrided case may be specified for dies made of CPM 10V, Vanadis V, Vanadis 10, or D2. For certain part designs, a solid die rather than an insert die is a more practical choice; an air-hardening 5% Cr tool steel such as A2 is generally used for such applications.

Punches. The stresses imposed on punches during service are such that toughness is a much more important material requirement than wear resistance, although wear resistance cannot be ignored. Type A2, and sometimes the shock-resisting type S7, are preferred for punches. Wear-resisting grades such as D2 and CPM 10V often lack the required toughness, particularly for solid punches. In type A2, which is deep hardening upon air quenching, an as-quenched hardness of 60 HRC can be developed in the center of a section 125 mm (5 in.) square, even though a solid punch this large would seldom be used. In contrast, for type S7 the maximum section size in which such hardness can be obtained is 65 mm ($2\frac{1}{2}$ in.) square. If sections >125 mm (5 in.) square are required, type A2 should be oil quenched from the austenitizing temperature to ~540 °C (1000 °F), then air quenched to 65 °C (150 °F) before tempering. S7 can be carburized or nitrided for added wear resistance.

For applications in which A2 or S7 punch faces become severely abraded, a more wear-resistant grade, such as CPM 10V, D2, D3, or M2, should be considered. Cemented carbide punches and core rods employ a higher cobalt grade (~11% Co) with a hardness of ~90 HRA. They can be made of solid carbide or a composite that uses tungsten carbide in the wear areas. More recently, fine-grain carbides with 10 wt% Co have also been employed in these applications.

Core Rods. Both toughness and wear resistance are important criteria in the selection of core-rod materials, but generally the primary consideration is wear resistance. Tungsten carbide and high-speed steels (M-grades) are the most common materials for core rods. For particularly abrasive conditions, CPM 10V has been used successfully, as have D2, M2, and A2 tool steels that have been nitrided or coated with tungsten carbide. Crucible P/M (CPM) tool steel CPM 15V has recently become popular for these applications.

Tooling support adapters normally are made from medium-carbon alloy steel, such as AISI/SAE 4140 or 6150, heat treated to a hardness sufficient to resist brinelling of the punches into the adapter surface without failing due to fatigue. Adapters should be heat treated to a minimum hardness of 28 to 32 HRC to reduce damage to critical mounting surfaces during handling.

Punch clamp rings normally are not highly stressed members, but they should be made from a heat-treatable steel to prevent damage during handling. Heat treating of the clamp ring is optional.

Operational Factors. Die working surfaces and core rods should be polished or lapped to a mirror-like surface finish, and final polishing should be done in a direction parallel to the axis of the tool. The faces and lands of the punches should also be given a fine finish. An exceptionally smooth surface finish reduces friction, thereby reducing some of the load on the tooling. It also makes it easier to eject the compacts, and it eliminates minute scratches and other stress raisers that could lead to premature fatigue failure.

Hard chromium plating is sometimes recommended to improve the life of tool steel punches and core rods, particularly when abrasive powders are involved. Some users claim that nitrided or chromium-plated die parts have up to ten times the wear resistance of untreated tool steel die parts; others claim that chromium plating is not very effective. Both nitrided and chromium-plated die parts are subject to chipping or flaking, especially at sharp edges. When this is a problem, a diffused surface layer such as that produced by chromizing may prove to be an effective alternative.

Tooling Clearances and Design

As in many other manufacturing operations, process variables (e.g., the type of materials being processed, the density of the part being produced, the amount and type of powder or die lubrication, and production rate) dictate operating conditions. Density and production rate greatly affect tool clearances during a continuous production run in which tooling temperature increases as compacted density and/or production rate increases. Temperature variations and corresponding dimensional changes within the various tooling members must be considered.

Standard tooling clearance is 0.016 mm/25 mm (0.0006 in./1 in.) on the diameter total. Minimum clearance should be used initially, because materials can always be removed from the punch or die to provide additional clearance as needed. In addition, the clearances must be smaller than the size of the powder particles to prevent their entrapment. Smaller clearances will also help reduce possible variations in the dimensions of the parts. The density of the compact produced and the production rate have a great influence on the determination of clearances. Compacting loads will be higher for increased densities. These, as well as higher production rates, increase the tooling temperature. As the temperature increases, dimensional changes occur in the tooling members. For this reason, the clearances must be sufficiently large to prevent seizure of the tools.

A representative value for the clearances between die walls and punches is 0.005 to 0.008 mm for precision parts and an upper limit of 0.013 mm for other parts. Minimum possible clearance should be used initially, because it can be increased as needed by removing material from the punch or the die.

The expansion ("pop out" or "springback") of the compact upon ejection makes it essential that the top edge of the die cavity be properly rounded or flared to allow the compact to make a smooth transition during ejection. Provision of a shallow chamfer is a more practical solution in the case of gears.

Die cavities and punch faces should be lapped and polished to a very high degree of surface finish, preferably <0.25 μm.

Shapes and Features. A shape or feature can be die compacted provided that it can be ejected from the tooling and the tools that form the feature have sufficient strength to withstand the repeated compaction loads. Due to the vertical closure of the tooling and the lack of tool motions perpendicular to the pressing direction, part removal from the tools controls many features. Examples of features that cannot be accommodated in die compaction, and therefore require secondary machining operations, include undercuts, reverse taper (larger on bottom than on top), annular grooves, and threads. The following guidelines provide assistance with many possible features in die compaction. Further details are provided in Ref 3.

Wall Thickness. Minimum wall thickness is governed by overall part size and shape. For parts of any appreciable length, walls should be not <1.5 mm (0.060 in.) thick. A maximum length-to-wall thickness ratio of 8 to 1 should be followed to ensure reasonable density uniformity and adequate tool life. Separate tool members (punches) should be used to provide density uniformity and proper ejection.

Steps. Simple steps or level changes not exceeding 15% of the overall part height (H) can be formed by face contours in the punches. A draft of 5° or more is needed to release this contour from the punch face during ejection. Features such as countersinks and counterbores can be similarly formed. This tooling method, as compared to multiple punches, will result in slight density variations from level to level. However, this approach offers the simplest tooling, lower-cost tooling, and closer axial tolerances than multiple punches.

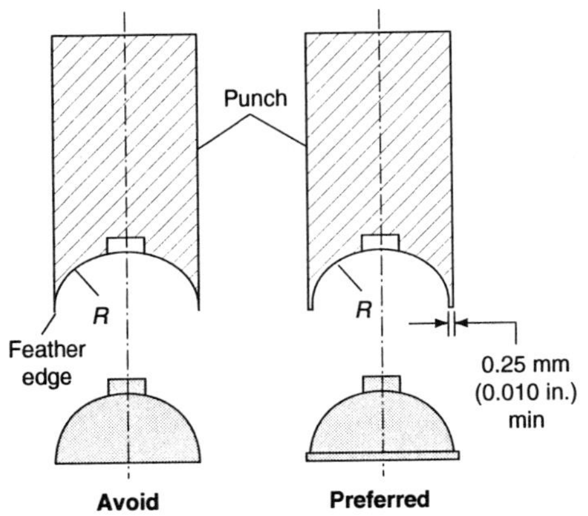

Fig. 17 Proper design of spherical shapes in P/M parts. Source: Ref 3

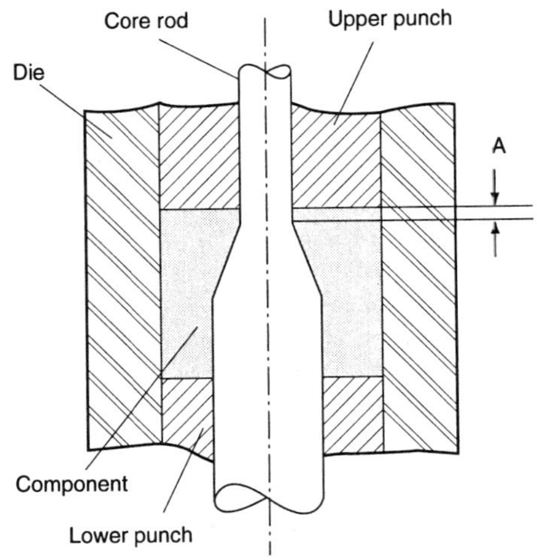

Fig. 18 Tapered hole design for P/M parts. Source: Ref 3

Spherical Shapes. Complete spheres cannot normally be made because the punches would have to feather to zero width (Fig. 17). Spherical parts require a flat area around a major diameter to allow the punch to terminate in a flat section (Fig. 17). Parts that must fit into ball sockets are repressed after sintering to remove the flats.

Taper and Draft. Draft is not generally required or desired on straight-through parts. While tapered side walls can be produced where required, the tools may demand a short straight surface (A in Fig. 18) at the end of the taper to prevent the punch from running into the taper in the die wall or on the core rod.

Holes. Through holes in the pressing direction are produced with core rods extending through the punches. Round holes require the least expensive tooling, but many other shapes, such as splines, keys, keyways, D-shapes, squares, and so forth can readily be produced. Blind holes, blind steps in holes, and tapered holes, are also readily pressed. For very large parts, lightening holes are added to reduce weight and the area of compacted surface.

Flanges. A small flange, step, or overhang can be produced by a shelf or step in the die. Separate lower punches are used if the amount of overhang becomes too great to permit ejection without breaking the flange.

Alphanumeric Characters. Numbers, lettering, logos, and other characters can be pressed into surfaces oriented perpendicular to the pressing direction. Recessed lettering is preferred because raised letters are fragile, easily damaged in the green compact, and prevent stacking of parts for sintering.

Chamfers, Radii, and Bevels. Chamfers are preferred rather than radii on part edges to prevent burring. It is common practice to add a 0.25 mm (0.010 in.) flat at a 45° chamfer; lower chamfer angles may not require the tooling flat.

Hubs and Bosses. Hubs or bosses that provide for drive or alignment rigidity in gears, sprockets, and cams can be readily produced. However, the design should ensure the maximum permissible material between the outside diameter of the hub and the root diameter gear or sprocket features.

Other Features. Additional information on these and other features (slots, grooves, knurls, studs, fillets, countersinks, etc.) can be found in Ref 3. Because shape complexity is a recognized limitation of die compaction, multipiece assembly is a useful alternative, especially where extensive machining would be required. Pulleys, spools, and sprockets have been produced using sinter bonding, brazing, and welding techniques.

REFERENCES

1. W.A. Knight, Design for Manufacture Analysis: Early Estimates of Tool Costs for Sintered Parts, *Annals of the CIRP,* Vol 40 (No. 1), 1991, p 131–134
2. S.D.K. Saheb and K. Gopinath, Tooling for Powder Metallurgy Gears, *Powder Metall. Sci. Technol.,* Vol 2 (No. 3), 1991, p 25–42
3. *Powder Metallurgy Design Manual,* 2nd ed., Metal Powder Industries Federation, 1995

Powder Injection Molding

Randall M. German, The Pennsylvania State University

INJECTION MOLDING is widely recognized as a manufacturing approach that can form complicated shapes from plastics. Since the 1920s, there has been a progressive evolution of injection molding from strict use on plastics to use with metal and ceramic powders. This new technology, known as powder injection molding (PIM), combines the productivity of injection molding with the ability to fabricate metals and ceramics. Thus, complicated shapes emerge from materials capable of operating at high temperature, or from materials that have desirable electrical, thermal, optical, or magnetic properties not available with polymers.

Much interest exists in PIM because of five key features: low production costs, shape complexity, tight tolerances, applicability to several materials, and high final properties. Many successful applications rely on particular combinations of these attributes. Examples include orthodontic brackets for straightening teeth, porous filters for treating hot waste water, magnets for controlling computer disk drives, small gears for electric hand tools and toothbrushes, cleats on sporting shoes, surgical tools such as scalpels, electrical connectors, handgun components, eyeglass and wristwatch components, golf clubs, and microwave filters for high-frequency microelectronics.

The technology has undergone widespread commercialization since the 1980s and today is practiced in many variants, reflecting different combinations of powders, binders, molding techniques, debinding routes, and sintering furnaces. An outline of the core production sequence is given in Fig. 1. In this simple form, four steps are required: formation of the feedstock, molding into the tooling, binder removal, and sintering.

As schematically illustrated in Fig. 2, the first step in the production sequence is to combine a small quantity of a polymer with the powder to form a feedstock. A thermoplastic polymer is selected to provide flow upon heating in the molding machine. Feedstock for the process is formulated by mixing small powders and thermoplastic binders to form pellets that easily flow in a molding machine. Today there are several suppliers of these precompounded feedstock pellets in a wide variety of compositions. These feedstock pellets are molded into tooling that defines the shape. Once the polymer is heated the mixture can flow by viscous flow. After molding, the binder is removed (debinding) and the remaining powder structure is sintered. These last two steps can be combined into a single thermal cycle because significant shrinkage is associated with sintering. The sintered product may be further densified, heat treated, coined, plated, or machined to complete the fabrication process. Most important, the sintered component has the precision of plastic injection molding in materials that deliver properties unattainable from polymers.

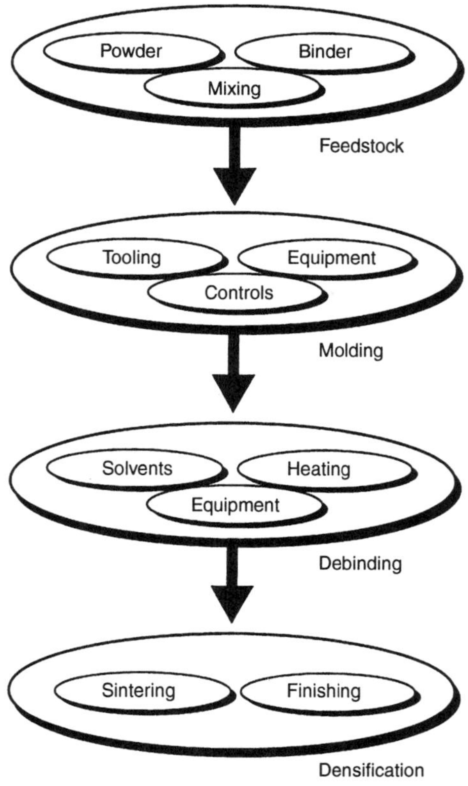

Fig. 1 Flow chart for the major activities making up the powder injection molding process. Processing tends to divide the operations into four major segments.

Materials and Equipment

Feedstock. PIM begins with the mixing of selected powders and binders. The particles are small to aid sintering, usually between 0.1 and 20 μm with near-spherical shapes. For example, a 5 μm carbonyl iron powder is widely used in the PIM process, as is a −16 μm gas-atomized stainless steel powder. Most common engineering alloys are used, including various steels, tool

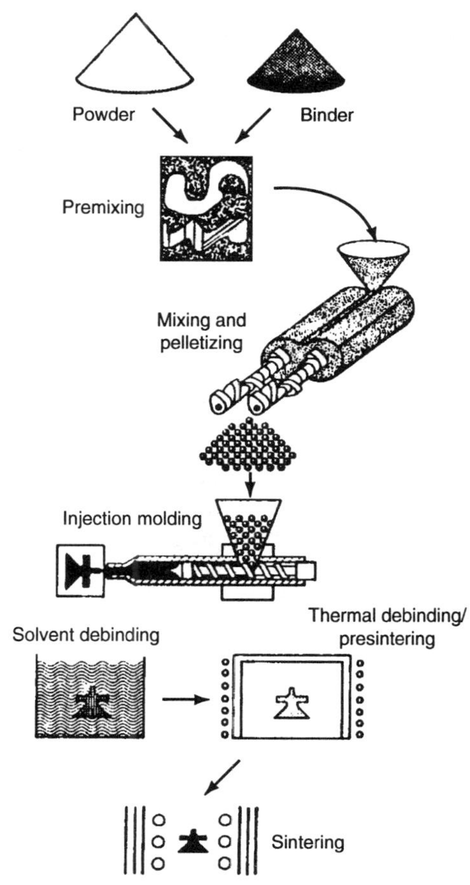

Fig. 2 A flow sketch of the powder injection molding process, showing the fabrication of a component from the original binder and powder, which are combined to form feedstock, then molded, debound, and sinter densified to produce the final component

steels, and stainless steels. Likewise, ceramics, refractory metals, and cemented carbides are processed in a similar manner. The binder is based on a common thermoplastic such as wax or polyethylene or wax-polypropylene, but food-grade polymers, polyacetal, cellulose, gels, silanes, water, and various inorganic substances are also in use. Usually the binder system consists of two or three components. An example binder, which is molten at 150 °C, consists of 65% paraffin wax, 30% polypropylene, and 5% stearic acid. A typical binder content is near 40 vol% of the mixture; for steel that corresponds to about 6 wt% binder. A few other binder systems are:

- 90% polyacetal, 10% polyethylene
- 69% paraffin wax, 20% polypropylene, 10% carnauba wax, 1% stearic acid
- 75% peanut oil, 25% polyethylene
- 50% carnauba wax, 50% polyethylene
- 55% paraffin wax, 35% polyethylene, 10% stearic acid

Feedstock is a term for the mixture of powder and binder. Many types of powders can be used, but great differences exist in mixing and molding, especially if the particle shape is nonspherical. The formulation of a successful feedstock balances several considerations. Sufficient binder is needed to fill all voids between particles and to lubricate particle sliding during molding. A viscosity similar to that of toothpaste is generally most desirable. Mixing is best achieved using a continuous twin screw compounder. Actually the viscosity depends on several factors. At too high a powder-to-binder ratio there is a high viscosity and insufficient binder to fill all void space between the particles. Consequently, it is hard to mold such a feedstock. Alternatively, too much binder is undesirable because component shape is lost during debinding. Inhomogeneities in the feedstock lead to defects in molding; thus, a high shear is required in mixing to force the binder among all particles. Consequently, special mixing practices are required to compound feedstock for most applications. The final step in feedstock preparation is to form pellets that are easily transported to the molding machine. Figure 3 shows both worms and pellets formed for molding. An important evolution in the technology has been the preparation of feedstock by major chemical companies, removing some of the licensing and technological barriers, allowing rapid growth in the field. Table 1 details the composition of a few common injection molding feedstocks, showing the binder, powder, and formulation details.

Pelletized feedstock is injection molded into the desired shape by heating it in the molding machine and hot ramming it under pressure into the tool cavity. By virtue of the binder, the feedstock becomes low enough in viscosity that it can flow into the die cavity under pressure. Cooling channels in the die extract heat and solidify the polymer to preserve the molded shape. The shaping equipment is the same as that used for plastic injection molding. It consists of a die filled through a sprue, runner, and gate from a heated barrel. Most popular is molding in a reciprocating screw machine. Here the screw in the barrel stirs the feedstock while it is melting and acts as a plunger to generate the pressure needed to fill the die. In the actual molding stroke, the molten feedstock is rammed forward to fill the cold die in a split second. Molding pressures depend on several parameters, but might be 60 MPa or more. Pressure is maintained on the feedstock during cooling until the gate freezes to reduce the formation of sink marks and shrinkage voids. After cooling in the die, the component is ejected and the cycle is repeated.

Tooling. The tool materials used in PIM are similar to those encountered in many metal working, plastic injection molding, and powder metallurgy operations. Table 2 identifies some of the common tooling materials. The tool material choice depends on the anticipated number of molding cycles and the required wear resistance. On the one hand, machining difficulty and material costs need to be considered. For molding tools, P-20 is the most common material, because of the combination of strength and cost. Yet wear concerns with PIM make the selection of higher-hardness tool steels most common. Rapid prototype tool materials, including epoxy, have been used in pilot production. Soft alloys of aluminum, zinc, or bismuth are used during tool development because of easy machining. Cemented carbides are useful where wear is a primary concern, but tool fabrication is expensive and tool damage is a problem because of the low toughness. Material cost varies by a factor of ten between these materials. Tool steels are best because of the combined strength, toughness, hardness, and machinability.

Tool fabrication occurs in a machining center via progressive removal of material from an initially oversized block of material. Most machining is computerized, but there is still the necessity to hand-finish critical components or dimensions in the tool set. A final surface roughness of 0.2 μm (8 μin.) is typical, but smoother finishes are used in selected applications. The desired tool hardness is typically more than 30 HRC, which is satisfied by many heat-treated stainless steels or tool steels. Under normal conditions, an injection molding tool set can mold up to one million parts. With soft tool materials like aluminum, the life is less, at 1,000 to 10,000 cycles.

The tool set has the cavity and further consists of the pathway for filling the cavity with ejectors for extracting the component from the cavity. In most instances, the tool set consists of a single

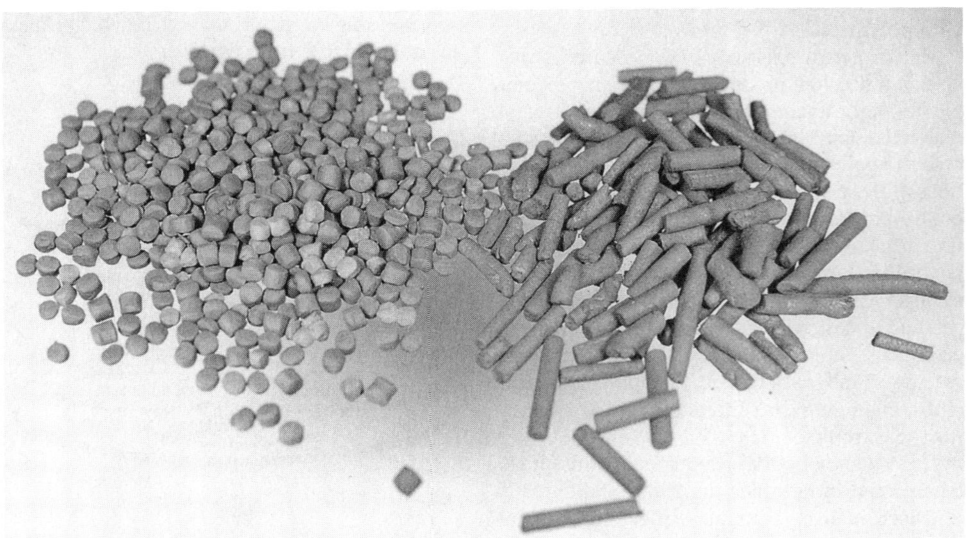

Fig. 3 Feedstock pellets and worms for molding

Table 1 Examples of powder injection molding feedstock

Powder	Binder, wt %	Solids loading, vol %	Density, g/cm³	Molding temperature, °C	Viscosity, Pa · s	Strength, MPa
4 μm Fe	60PW-40PE	58	4.90	120	35	5
4 μm Fe	55PW-45PP-5SA	61	5.12	150	19	22
4 μm Fe-2Ni	90PA-10PE	58	4.52	180	190	20
2.5 μm Mo	60PW-35PP-5SA	58	5.97	113	200	7
10 μm stainless	55PW-45PP-5SA	67	5.60	130	100	15
15 μm stainless	90PA-10PE	62	5.33	190	80	20
12 μm tool steel	90PA-10PE	62	5.33	190	180	20
8 μm W	65PW-30PP-5SA	56	11.22	142	...	5
1 μm W-10Cu	60PW-35PP-5SA	64	11.41	135	55	6

PA, polyacetal; PE, polyethylene; PP, polypropylene; PW, paraffin wax; SA, stearic acid cavity. This cavity captures the component shape, and it is oversized to allow for component shrinkage during sintering. Around the cavity are several tool parts needed for opening and closing the cavity, ejecting the component, aligning the die sections, moving inserts, cooling the component, and locating the sprue, runner, and gate. Figure 4 is a sketch of a molding tool set with ejector pins, ejector plate, and keyed slides to ensure proper closure of internal die components. Many operations use a three-plate mold for automated removal of the gate on mold opening.

A primary concern in designing injection molding tooling is component shrinkage. On a volume basis, the typical feedstock contains 60% solid and 40% binder. To attain the desired final component properties, the linear shrinkage during sintering may be 15%. The shrinkage in dimensions is known as the shrinkage factor Y, calculated from the solids loading, Φ, and the sintered fractional density, ρ/ρ_T:

$$Y = 1 - \left(\frac{\Phi}{\rho/\rho_T}\right)^{1/3} \qquad \text{(Eq 1)}$$

where ρ is the final density and ρ_T is the theoretical density for the material. This assumes isotropic shrinkage in sintering. For example, if a 13.8 mm dimension shrinks to 12 mm, then the shrinkage factor is 0.13 or 13%, calculated as the change in a dimension divided by the original dimension. Because the target is the final component size, each dimension of the tool cavity is oversized to accommodate shrinkage. If the desired final dimension is L_f at a fractional density ρ/ρ_T from a feedstock with a fractional solids content of Φ, then the initial dimension of the tooling is given in terms of the tool cavity expansion factor, Z:

$$Z = \frac{1}{1 - Y} \qquad \text{(Eq 2)}$$

As an example, if the shrinkage factor Y is 0.15 (15% shrinkage), then the tool expansion factor Z is 1.1764. Thus, to obtain a 12 mm final dimension requires a tool dimension of 14.11 mm (12 × 1.1764). Note that the shrinkage from 14.11 mm to 12 mm is 2.11 mm, so the measured shrinkage factor is 2.1 ÷ 14.11 mm, or 0.15 (15%).

At the end of the runner is the gate leading into the die cavity. It is a small opening designed to freeze before the cavity, runner, or sprue freezes. A solidified gate allows removal of the pressure at the machine while the mass in the cavity cools. Gates are usually near 3 mm in diameter. Actual gate size is determined by two factors: the filling shear strain rate and the section thickness. For the gate to freeze before the compact requires a thickness between 10 and 50% of the compact thickness.

In the typical tool set, the feedstock flow path is from the molding machine nozzle into the sprue, along the runner, through the gate, and into the mold cavity, as evident in the test ge-

Table 2 Construction materials for injection molding tools

Material	Composition	Hardness, HRC	Suggested applications
420 stainless	Fe-14Cr-1Si-1Mn-0.3C	50	Corrosion-resistant cavities, cores, inserts
440C stainless	Fe-18Cr-1Si-1Mn-1C	57	Wear-resistant, small inserts, cavities, cores
H13 tool	Fe-5Cr-1.5Mo-1Si-1V-0.4Mn-0.4C	50	Larger or intricate cavities, high toughness, low wear
M2 tool	Fe-6W-5Mo-4Cr-2V-0.3Mn-0.8C	61	Core and ejector pins
P20 steel	Fe-1.7Cr-0.8Mn-0.5Mo-0.4V-0.35C	30	General purpose, hot runner, large cavities
Cemented carbide	WC-10Co	80	High wear, compressively loaded small inserts

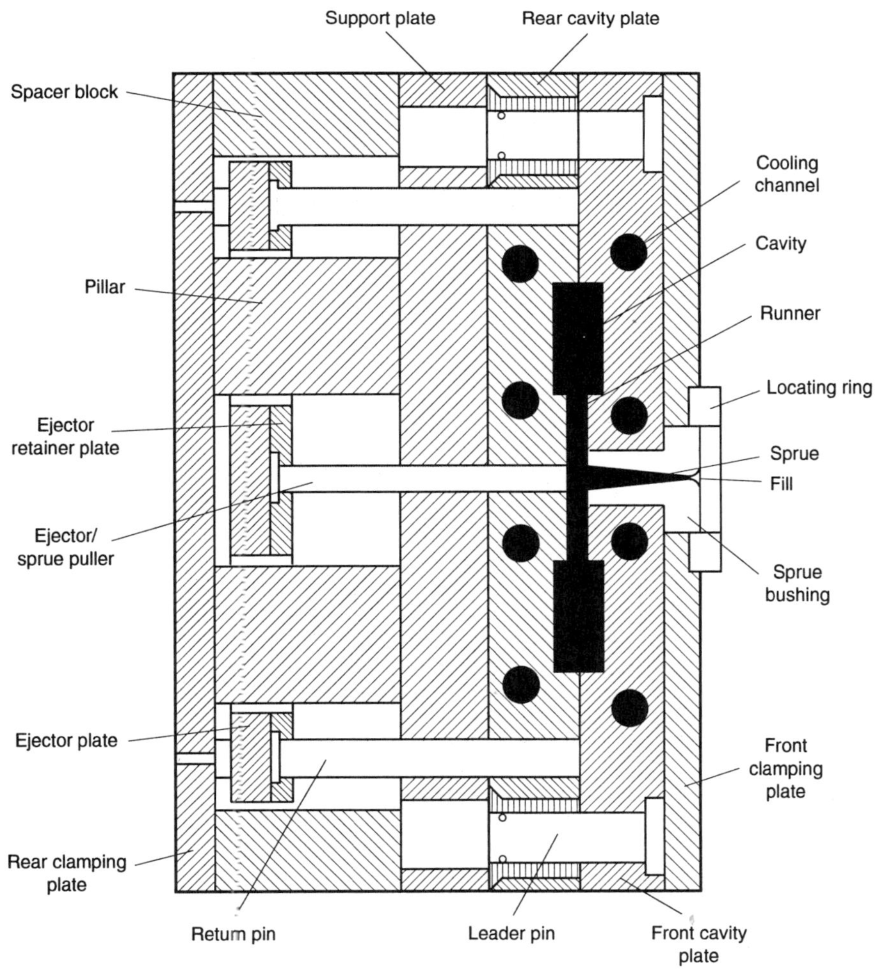

Fig. 4 A sketch of the tool set for powder injection molding, showing major components

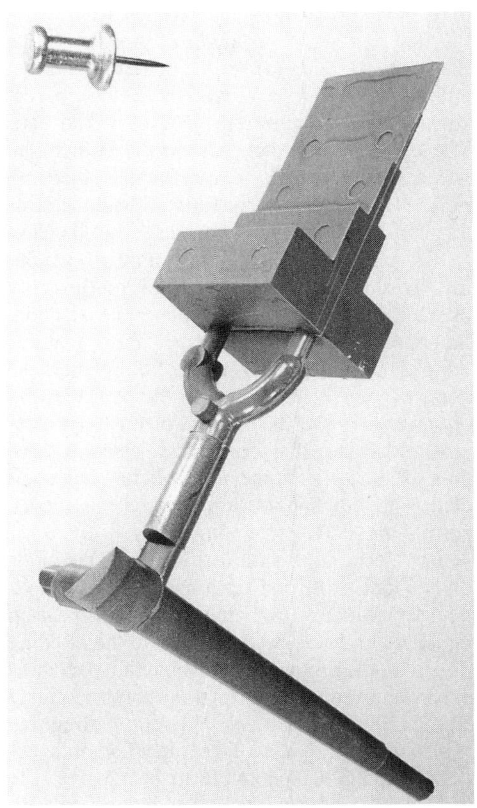

Fig. 5 A five-step test geometry with dual gates and attached runner and sprue

ometry shown in Fig. 5. This five-step geometry has two gates fed by the runner system from the sprue. This flow path is surrounded by various clamping plates, alignment and locating pins, and ejector components. Alignment of the components and their proper sequencing and smooth motions are important to successful PIM. Most of the tool components are available as premachined packages, so only the cavity needs to be custom machined. Also, within this geometry are the cooling or heating networks designed to control mold temperature. One key advantage of injection molding is the ability to fabricate complex shapes that cannot be produced by alternative techniques. Accordingly, complex tool designs are a necessary aspect of injection molding.

The number of cavities in the tool set depends on the number of components to be fabricated, the shot capacity of the molding machine, tool fabrication costs, and the available clamping force. Tool sets with up to 40 cavities have been used for high-volume production. Most injection-molded steel components are generated in tool sets with 1 to 16 cavities. A single cavity tool set is satisfactory for low production quantities, below about 200,000 parts. As higher productivity is required, more cavities may be justified, because production increases are gained without the purchase of new molding machines. As the number of cavities increases, the cost of manufacturing the tool set increases, but the net production cost per component decreases. Because tool cost is distributed over the production quantity, there is a minimum total cost that depends on the total production quantity increases.

After molding, the component is cooled in the die cavity. Cooling causes the binder to contract and this results in a progressively lower pressure, eventually allowing ejection of the part. The ejection force depends on the component shape and feedstock. To accomplish ejection, pins in the die body move forward with the ejector plate and push the component from the cavity. If inserts, internal cores, or threads are put into the shape, these must be retracted (possibly by motorized motions) to allow free ejection. Sometimes core pins, inserts, or other features are placed in the cavity before filling and these items become encapsulated in the part, ejecting out on each cycle. Ejector pins blemish the component, because they concentrate ejection force on a soft material. Some of the blemishes associated with molding include ejector pin marks, parting lines, and gate impressions.

For ejection, pins move from flush positions on the tool walls and push against the component to extract it from the cavity. To build in undercuts or holes perpendicular to the molding direction, the tooling must contain side-actuated cores or inserts. Using rotating cores to create threads or rifling patterns without a parting line is also possible. Unlike die compaction, in injection molding it is possible to design into the tooling perpendicular holes, undercuts, and indents using side cores that are mechanically actuated on mold opening and closing. A tool set can contain several such cores, which may be actuated using hydraulic pistons, electric motors, or mechanical motions.

Molding Machines. The three most common molding machines are reciprocating screw, hydraulic plunger, and pneumatic. Table 3 summarizes the attributes of a few such molding machines. Pneumatic machines simply apply gas pressure to move heated feedstock into the mold. They are inexpensive and effective for small components where internal flaws are not objectionable. However, voids form because the feedstock is under low pressure that fails to compensate for shrinkage on cooling. In a hydraulic molding machine, a plunger rams heated feedstock into the mold. Excess pressure is generated to compensate for the volume contraction normally encountered by feedstock cooling. This pressurization is important to forming defect-free compacts, but the control systems usually are not suitable for forming complicated shapes.

High-volume injection molding uses a horizontal reciprocating screw located inside a heated barrel. The screw is designed to compress and transport the feedstock to the die, and it becomes a plunger during mold filling. Figure 6 is a typical layout of a horizontal machine and Fig. 7 is a picture of a contemporary molding machine with vision system, robot, process controller, and data acquisition computer. The tooling is clamped in the center of the machine. The granulated or pelletized powder-binder feedstock is placed in the hopper for metering into the injection barrel. This is the beginning of the molding operation.

The heart of molding lies with the motions of the reciprocating screw. This needs to be wear resistant to withstand abrasion by the particles. It has a helical pitch, the design of which is adjusted for the viscosity of the feedstock, but generally it consists of gradual section changes along its length. Screw rotation is controlled via a hydraulic motor and heat is supplied by external heaters on the barrel.

An important role of the screw rotation is to de-air the feedstock and prepare the next charge for injection. This is termed *metering,* where the screw rotates to pressurize feedstock to the nozzle. During metering the screw acts as a mixer to ensure uniform powder-binder distribution and uniform heating. The screw has a check ring behind the tip that acts as a valve that allows feedstock flow into the front of the barrel. This ring seals on the screw during mold filling and forces molten feedstock to flow into the die cavity through the nozzle at the end of the barrel. Effectively, the screw becomes a plunger during mold filling. Control of the screw rotation, displacement, and pressure is important to the fabrication of precise components by injection molding.

During the molding cycle, the screw initially rotates, compressing the feedstock. Then, during the injection step, the screw moves forward,

Table 3 Sample attributes of powder injection molding machines

Attribute	Low pressure	Moderate tonnage	Intermediate tonnage	High tonnage
Type	Pneumatic	Reciprocating screw	Reciprocating screw	Reciprocating screw
Clamping force, ton	0.7	25	75	200
Clamp type	Pneumatic	Hydraulic	Toggle	Hydraulic
Platen size, mm	480	260	530	900
Drive motor, kW	2	5.5	15	30
Screw diameter, mm	None	18	28	50
Injection volume, cm^3	3000	35	70	350
Fill rate, cm^3/s	0.5	46	66	100
Injection stroke, cm	...	9	11	14
Plastication capacity, kg/h	...	12	20	40
Screw speed, mm/s	...	110	300	110
Molding pressure, MPa	0.5	190	230	200
Shortest cycle time, s	4.8	2	1.2	3.6
Maximum temperature, °C	150	200	200	200
Control	Open-loop	Adaptive	Closed-loop	Adaptive

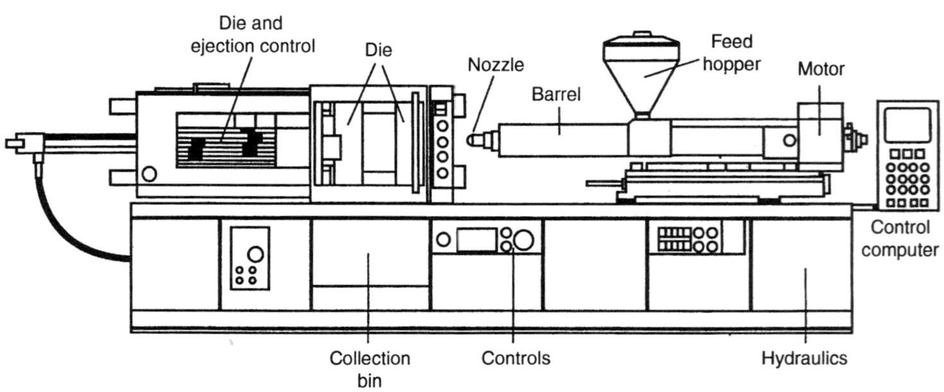

Fig. 6 Overview of a horizontal injection molding machine and key components

closing the check ring, and the shot is injected into the mold. A closed-loop control system with a quick response servohydraulic valve is required for screw position and pressure control. During the fill stroke, the volume of feedstock injected into the mold depends on the cross-sectional area of the screw and on the stroke length. A typical screw diameter is 22 mm, but it might range from 15 to 40 mm (0.6 to 1.6 in.), depending on the machine capacity.

Feedstock flow in molding depends on the applied pressure and viscosity. For a cylindrical runner, the volumetric feedstock flow rate, Q, varies with the runner diameter, D, to the fourth power according to Poiseuille's equation:

$$Q = \frac{P \pi D^4}{128 \eta L} \qquad \text{(Eq 3)}$$

where P is the applied pressure on the feedstock, L is the runner length, and η is the feedstock viscosity. The rate of mold filling is very sensitive to the injection pressure and runner diameter. Usually a high feedstock flow rate is needed in order to fill the die before the feedstock cools.

The barrel holds the rotating screw and is surrounded by heaters that control the mixture temperature. There are often multiple heater zones to ensure temperature control during mold filling. Because cold feedstock is abrasive, the first heater zone is geared to rapidly heat the feedstock, and subsequent zones might be at lower temperatures. The materials used in constructing the screw and barrel are critical to long service without contamination. Hard materials and close tolerances are required to reduce abrasive wear. Tool steels containing vanadium carbide and boride-clad tool steels prove most durable. Similarly, other machine components in the flow path can exhibit considerable wear, resulting in a loss of machine control.

All of the molding steps are controlled by a computer that might even correct errors during molding. Besides the molding machine, coordination is required with the peripheral operations needed in automation schemes. For example, robots are used to stage the compacts for debinding. Other automation features include conveyor systems, parts and tooling storage with automated retrieval, and continuous feedstock preparation and component debinding steps.

Process Description

Molding Cycles. A typical sequence for the injection pressure and screw position are shown in Fig. 8. Prior to filling, the screw rotates in the barrel and the external heaters bring the feedstock to temperature. The dwell time for the feedstock in the barrel must be sufficient to ensure thorough and uniform heating. Then, during molding, the screw plunges forward in a split second. This is traced by the screw displacement curve in Fig. 8. Three pressure curves are included to show that pressure is high at the source (hydraulic pressure) and lags at the cavity. The rapid rise in hydraulic pressure induces feedstock flow into the mold. Once the cavity is filled, the gate freezes and there is little effect of hydraulic pressure on the cavity pressure. The nozzle pressure is intermediate between these two.

Moldability is a measure of the ease by which a feedstock can be shaped to a given specification. A high moldability is desired from a feedstock. In practice, several combinations of temperature, fill rate, and pressure exist that produce defect-free components. This pressure is generated by the forward motion of the screw in the barrel as a plunger. A low pressure gives incomplete packing with cavities or sink marks on the compact surface. Excessive packing pressure causes the compact to stick in the die, resulting in severe ejection problems.

To compensate for the flow resistance and pressure gradients in the molding system, the screw position and hydraulic system pressure vary during molding in a coordinated manner. Friction along the flow path decreases pressure. After mold filling, pressure is held during the packing stage. The objective is to compensate for the thermal contraction of the feedstock during cooling. Finally, when the binder has sufficiently cooled to hold the component shape and withstand the ejection forces, the die is opened and the component is ejected. The ejector motion can be from the hydraulic system or a separate mechanical motion.

The time for molding relates to the cavity size, filling time, and cooling time. It can be as short as 5 s or as long as 1 min. Initial feedstock flow requires that the molding temperature be higher than the softening point of the binder. This is usually between 50 and 200 °C (122 to 392 °F). A low temperature results in short shots (incomplete mold filling), while a high temperature degrades the binder or causes flashing or powder-binder separation, and requires prolonged cooling. Molding pressure affects the mold filling rate and is usually limited by the machine design. Once the component has cooled, the final step is component ejection. Short shots occur if either the molding temperature or pressure is too low. At the higher pressures and temperatures the components stick to the die walls or open the die along the parting line, giving flash. Voids are captured in a part if the feedstock shoots across the cavity without pushing out all of the air. This is termed *jetting* and is very undesirable. Jetting occurs with too high a fill rate, so most molding machines are programmed for a fill rate that prevents jetting.

Table 4 gives the range of molding conditions and one example of a specific component and its molding conditions. That component is a steel trigger guard for a rifle formed from carbonyl iron powder and a binder based on paraffin wax,

Fig. 7 Picture of a research injection molding machine with attached vision system, robot, data acquisition computer, and control computer

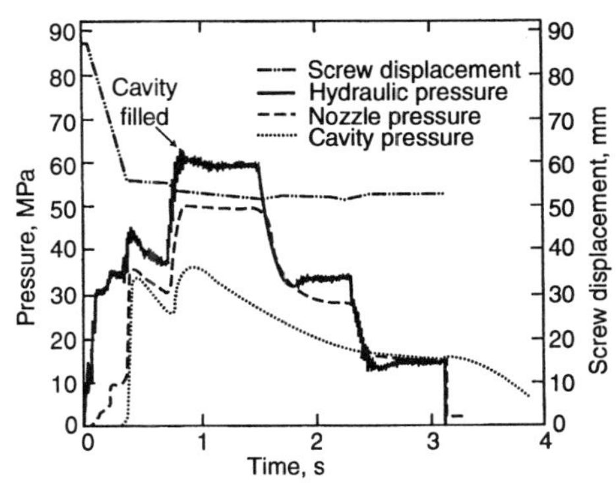

Fig. 8 Plots of the screw position and pressures involved in molding

polypropylene, and stearic acid. Once the component is molded, the polymer binder phase is extracted and the compact is sintered to near-full density. These steps are covered in the section "Sintering Technologies and Techniques" in this Volume.

Densification. Usually the binder is removed from the component prior to sinter densification. A wide array of options exist for binder extraction. Thermal debinding is the easiest to envision. The component is slowly heated to decompose the binder. As illustrated by Table 5, many variants exist. There are some differences in component precision and debinding rates. For this tabulation, the measured rates are compared on an equivalent basis of 5 μm steel powder and 10 mm section thickness. The most popular approach is to immerse the component in a solvent that dissolves some binder, leaving some polymer behind to hold the particles in place for subsequent handling. The remaining binder is thermally extracted as part of the sintering cycle. Newer binders are water soluble, so the debinding solvent is water. Major growth is occurring in the use of catalytic phase erosion for debinding. Most of the binder is attacked by a catalytic vapor, and the residual binder is removed during heating to the sintering temperature. This can now be performed as a continuous process at the beginning of sintering. Debinding is highly variable as to binder system, technique, and section thickness. Solvent and catalytic debinding ensure the best dimensional control, because the binder is kept rigid during extraction. For these techniques penetration rates of 1 to 2 mm/h are common.

The next step is sintering, which can be incorporated directly into a thermal debinding cycle. Sintering bonds the particles together, leading to densification. Often sintering serves the dual role of densification and chemical homogenization. In the latter process, mixed powders are molded and sintering causes them to form homogeneous alloys by long-range atomic motion. Usually sintering shrinkage is uniform and isotropic, so the molded component is oversized to deliver the desired final dimensions. For metals, sintering is performed in a protective atmosphere or vacuum at a peak temperature that causes rapid elimination of the pores that were previously filled with binder. Steels and stainless steels are sintered at 1120 to 1350 °C range for 30 to 120 min, with shrinkages of 12 to 18%. Table 6 details some common sintering cycles, giving the initial green density, heating rate, maximum temperature, hold time, sintering atmosphere, support material, and final density. If the powder structure is formed homogeneously, then sintering is uniform and final dimensions can be held to close tolerances.

After sintering, the component has excellent strength, with properties near or even superior to those available from other processing routes. Final densification can be assisted by both hot and cold deformation, including hot isostatic pressing. Other post-sintering steps include coining, drilling, reaming, machining, plating, passivation, and heat treatment. Options in heat treatment include tempering, precipitation hardening, nitriding, and carburization.

A typical PIM component is a trigger guard for a sporting shotgun, the curved piece that surrounds the trigger below the barrel. It is fabricated from a low-alloy iron-nickel steel, usually with a final weight of 40 g. A mixture of 5 μm Fe and 8 μm Ni powders are used. These are combined with wax and polyethylene to form feedstock that can be molded at 58 vol% solids. During molding, the nozzle temperature in the molding machine is 175 °C with a die temperature of 40 °C. The maximum pressure applied during mold filling is 20 MPa and a pressure of 8 MPa is held on the feedstock during cooling. The mold filling time is rapid (about 0.5 s), but the mold cooling time is 18 s, with a total cycle time of 37 s between parts.

Advantages and Limitations

Process Attributes. As emphasized already, PIM can produce a range of components from powders. A main attraction is the economical production of complex parts from high-performance materials. Because of the high final density, the PIM products are often superior to those produced by other powder fabrication routes. Most materials are available, including all common ceramics and alloys: steel, stainless steel, tool steel, silicon nitride, cemented carbide, silicon carbide, copper, tungsten heavy alloys, nickel-base alloys, alumina, cobalt-base alloys, and composites that include tungsten-copper and molybdenum-copper.

Besides the primary advantages of shape complexity, low cost, and high performance, several other attributes are worthy of notice. Producing both internal and external threads in the molded component is possible, thereby avoiding post-sintering machining. Also, waffle patterns and insignias can be molded directly into the component. Furthermore, the surface finish is typically good.

PIM vs. Die Compaction. Some confusion exists about the difference between PIM and traditional die compaction. The latter process is

Table 4 Typical powder injection molding parameters

Parameter	Typical range	Trigger
Barrel temperature, °C	100–200	160
Nozzle temperature, °C	80–200	180
Mold temperature, °C	20–100	40
Screw rotation speed, rpm	35–70	35
Peak injection pressure, MPa	0.1–130	20
Packing pressure, MPa	0–10	8
Fill time, s	0.2–3	0.6
Packing time, s	2–60	3
Cooling time, s	18–45	20
Cycle time, s	8–360	37

Table 5 Comparison of debinding techniques and times

Binder system	Debinding technique	Conditions	Time
Wax-polypropylene	Oxidation	Slow heat 150 °C, hold, heat to 600 °C in air	60 h
Wax-polyethylene	Wicking	Slow heat to 250 °C, hold, heat to 750 °C in hydrogen	4 h
Wax-polymer	Supercritical	Heat in freon vapor at 10 °C/min to 600 °C under 10 MPa pressure	6 h
Wax-polyethylene	Vacuum extraction	Slow heat while passing low-pressure gas over compacts, heat to sintering temperature	36 h
Water-gel	Vacuum sublime or freeze dry	Hold in vacuum to extract water vapor from ice	8 h
Oil-polymer	Solvent immersion	Hold in ethylene dichloride at 50 °C	6 h
Water-gel	Air drying	Hold at 60 °C	10 h
Polyacetal-polyethylene	Catalytic debind	Heat in nitric acid vapor at 135 °C	4 h

Note: Section thickness, 10 mm; particle size, 5 μm; solids loading, 60 vol%

Table 6 Sample sintering cycles for powder injection molding materials

Material	*D*, μm	ρ_G, %	*dT/dt*, °C/min	*T*, °C	*t*, min	Atmosphere	Support	ρ_S, %
Ag	90	67	1400	900	60	Hydrogen	Stainless	70
Co-50Fe	5	50	50	1250	120	Hydrogen	Alumina	98
Cu	9	70	10	900	60	Hydrogen	Alumina	94
Fe	5	60	10	1200	60	Vacuum	Alumina	100
Fe-49Co-3V	6	58	5	1350	240	Vacuum	Alumina	96
Fe-2Ni	5	64	15	1250	60	Hydrogen	Alumina	99
Fe-2Ni-0.8C	4	58	4	1200	60	Hydrogen	Alumina	97
Fe-50Ni	5	60	10	1250	60	Hydrogen-nitrogen	Alumina	96
Fe-29Ni-17Co	7	60	10	1250	240	Hydrogen-argon	Alumina	97
Fe-3Si	8	60	10	1350	180	Hydrogen	Alumina	97
Mo-15Cu	9	30	10	1400	60	Hydrogen	Alumina	86
Ni_3Al	14	52	10	1340	60	Hydrogen	Alumina	99
316L stainless	15	62	10	1325	120	Vacuum	Alumina	97
W-10Cu	8	50	10	1350	60	Hydrogen	Alumina	96
W-5Ni-2Fe	2	55	10	1500	30	Hydrogen	Alumina	100

D, particle size; ρ_G, green density; *dT/dt*, heating rate; *T*, sintering temperature; *t*, time; ρ_S, sintered density widely employed in forming powders into squat (low-height) shapes that can be ejected from compression tooling. Die compaction usually employs high forming pressures but gives less shape complexity. Most important, uniaxial die compaction results in density gradients in the compact, unlike PIM, which is hydrostatic. Density gradients result in distortion during sintering. Consequently, components fabricated from pressed powder are either sintered at lower temperatures, where sintering shrinkage is avoided, or machined after sintering. Otherwise, dimensional scatter becomes unacceptably large. Because of forming pressure gradients, sintering temperature differences, and differing performance levels, PIM and traditional powder compaction are usually not applied to the same structures. The PIM approach is suited for complex shapes sintered to near-full density, while die compaction is suited to simple shapes, sintered at lower temperatures to lower performance levels. The low porosity in PIM materials gives a high strength, toughness, ductility, conductivity, magnetic response, and so on.

Advantages. Besides the traditional materials, PIM can also produce specialty materials such as nickel superalloys, intermetallics, precious metals, refractory metals, and ceramic-fiber reinforced ceramic composites. Co-molding is another possibility, where two materials are combined to make a laminated structure. Such components can be joined in the green condition. This option has merit for creating corrosion barriers, wear surfaces, electrical interconnections, or high-toughness structures.

For the producer, injection molding is a desirable option because of manufacturing ease, including process control, flexibility, and automation. Inherently, injection molding is associated with large production volumes. Various components are produced at rates approaching 100,000 per day. On the other hand, small production runs are possible, with as few as 5000 parts per year being economical. This flexibility fits well with the current desires for quick response in manufacturing.

As with all technologies, the essence centers on economics. Powder injection molding is cost advantageous for the more complex shapes. The largest advantage comes from the elimination of the secondary operations, such as grinding, machining, drilling, or boring, that typically are required for precision components. Also, since the feedstock material (runners, sprues, and damaged moldings) can be recycled, material use is nearly 100%. This is especially important for costly raw materials such as refractory metals, specialty ceramics, and precious metals.

Process Limitations. Generally, PIM is viable for all shapes that can be formed by plastic injection molding. Still, for shapes with simple or axial-symmetric geometries, it is not competitive with traditional screw machining or die compaction and sintering.

Table 7 General criteria for powder injection molding

Attribute	Minimum	Maximum	Typical
Thickness, mm	0.2	25	10
Thickness variation	None	100×	2×
Longest dimension, mm	2	1,000	100
Tolerances, % (standard deviation)	0.03	2	0.3
Number of dimensional specifications	20	1,000	100
Mass, g	0.02	20,000	40
Material	Simple element	Composites	Alloys
Properties	Unimportant	Highest attainable	Handbook
Cost per part, $	0.20	400	2
Production quantity per year	200	20,000,000	150,000

In some cases another limitation is the component size. Large components require more powder (a large expense in some compositions), and large molding and sintering devices, which are more expensive and difficult to control. Small components with simple geometries can be more economically produced by standard machining, die compaction, or casting techniques.

Debinding is a key problem with PIM because the time for binder removal depends on the section thickness. Consequently, various manufacturers have set upper limits on section thicknesses, ranging from 10 to 50 mm. On the other hand, PIM has been used to form section thicknesses less than 0.5 mm. In practice, dimensional tolerances are typically within 0.3% of a target, although holding tighter tolerances on critical dimensions is possible. For better dimensional control, machining or coining is required after sintering. Density gradients can result from uneven filling, thickness variations, or direction changes during molding, and they often cause component warpage in sintering. Thus, manufacturers minimize changes in section thickness and if possible hold the variation in thickness within a factor of two. However, 10- to 100-fold section thickness changes are possible. Maximum sizes depend on several factors, including tool costs, powder costs, and equipment capacities. Typically, the largest dimension is below 100 mm, with the total component volume being below 100 cm^3. However, much larger components are in production.

The small particles used in PIM are more expensive than larger powders or wrought materials. This becomes a barrier to large component fabrication, because powder cost becomes a larger fraction of the production cost. On the other hand, as consumption increases, the powder cost continues to decline.

Substantial problems facing the technology are the lack of knowledge on the part of end users, missing property data, and the shortage of personnel trained in the basic process. These problems are acute with the ceramic materials because of the high sensitivity of ceramics to microstructural flaws induced during molding. Further, a lack of design guidelines has inhibited PIM from being adapted as a replacement for other production routes. These problems are being addressed via seminars and educational literature. However, the many process variants create initial confusion.

Design Guide

Certain component characteristics are best suited to PIM. Early identification of a match with these characteristics ensures technical and economic success. The considerations involve the materials, properties, component size and shape, tolerances, production costs, production quantity, and delivery schedule. Powder injection molding is best applied to the fabrication of complex, small components, especially those that would otherwise require multiple assembly or forming operations in alternative production routes. Further advantages occur when the material is expensive or difficult to process by alternative techniques. However, the shape complexity that can be produced is limited. Various shapes require secondary machining operations or complicated tooling that increase the manufacturing expense.

Tolerances are another concern. Precise tolerances prove difficult to achieve; production variations on a dimension are ±0.1% for special cases, and ±0.3% is more typical. Besides dimensional variability, successful component design must allow for noncritical placement of the parting line, ejector pin marks, and gate. It is also desirable to include a slight draft or taper of 0.5 to 2° to assist in ejection from the mold. Because precision tooling is expensive, the production quantity must justify the initial tooling cost. These factors need to be evaluated early to ensure feasibility. Greatest success is attained by early discussions of the manufacturing approach so that features can be incorporated that will aid PIM processing.

To emphasize the contemporary fabrication range, Table 7 summarizes the minimum, maximum, and typical features. Some explanation is in order. Components are characterized by the geometric aspects, including the largest dimension and the wall thickness. For PIM, the wall thickness is usually small, and it might be less than 10% of the largest dimension. Further, the variation in thickness over the part is usually small. The blades of scissors are an example where the length is much greater than the thickness, and the thickness is not highly variable. It is thickness that determines debinding time and is the main limitation of PIM.

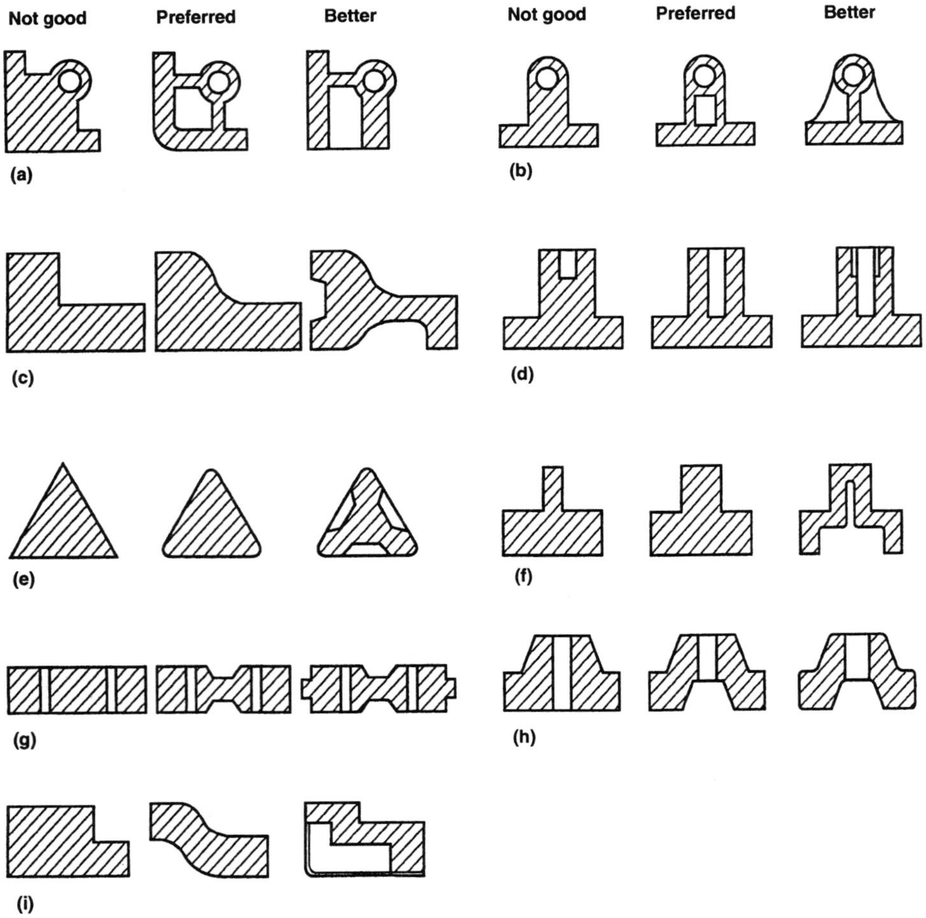

Fig. 9 Recommended designs for powder injection molding components and suggestions for improved processing

Table 8 Nominal powder injection molding component design guidelines

Restrictions
No inside closed cavities
Corner radius greater than 0.075 mm
Smallest hole diameter 0.1 mm
Weight range 0.02 g to 20 kg
No undercuts on internal bores
2° draft on long parts
Minimum thickness 0.2 mm
Desirable features
Gradual section thickness changes
Weight less than 100 g
Assemblies in one piece
Small aspect ratio geometries
Largest dimension below 100 mm
Wall thickness less than 10 mm
Flat surfaces for support
Allowed design features
Holes at angles to one another
Stiffening ribs
Protrusions and studs
D-shaped and keyed holes
Hexagonal, square, blind, and flat bottom holes
Knurled and waffle surfaces
External or internal threads
Part number or identification in die

Another gage of PIM candidates is through the information content, which is akin to the number of dimensions on the engineering drawing. A simple shape, like a flat washer for a bolt, has only a few features, while a microcomputer circuit has many millions of features; both would be poor applications for PIM. Common PIM successes involve several dimensions in the 20 to 40 g range—a wristwatch case is one example that matches well with the technology. The typical mass is low, and similar attributes are evident in components for electronic packaging, computer disk drives, surgical tools, and firearms. On the other hand, metal, ceramic, and carbide bodies have been fabricated that are several kilograms in mass and up to 1 m in maximum length. These are somewhat lower-precision structures used in electric power turbines, steel mills, and furnace construction. The general production limit for precise components is 250 g, although several special processes have evolved to produce precise components as large as 1 kg.

Only in a few instances are mechanical or physical properties unimportant, such as in jewelry and wristwatch cases. More typically the goal is to attain near-handbook properties for the selected material. Likewise, only rarely are properties pushed to the limits for any material. In these cases secondary hot isostatic pressing or other treatments might be employed to ensure maximum properties. From a cost consideration, typical prices are moderate, but that assumes a high production volume to justify the cost of tool construction, and dimensional tolerances that are compatible with the inherent process variability. Thus, a general processing window exists that proves most feasible. The typical parts are small enough to be held in a hand, but can be as small as the tip of a pencil. This implies that very large objects are difficult or expensive to produce by PIM, as contrasted with other forming technologies.

Geometric Considerations. In a simple sense, if a shape can be formed from plastic, then PIM is viable. Tables 7 and 8 provide general quantitative and qualitative component design guidelines.

The particle size is a dictate for the minimum section thickness. The powder size needs to be less than one-tenth of the smallest thickness. Particle size limits the sharpness of all corners, because the minimum particle radius is the edge radius, but more typically several particles are needed to fill out the dimension. A conceptual limit on an edge radius is 0.05 mm and a corner radius is 0.1 mm.

Table 8 also lists options in the design or production of PIM compacts. These options include square holes, flat-bottomed holes, knurled surfaces, twists, curls, helical geometries, and external or internal threads. Although these features are not needed to make a PIM compact viable, they are possible and add to the design flexibility. Certain attributes, such as component size and mass, will increase as experience and manufacturing optimization occur. Already components up to 10 kg and 1,000 mm maximum length have been fabricated by PIM. These require special efforts and are not routine products. With respect to changes in the section thickness, a ratio of 2 is typical; however, compacts have been successfully fabricated with thickness changes of 100:1.

Component design has a great influence on the success of PIM. Identification of injection molding as a production process in the design stage allows changes that increase the ease and lower the manufacturing cost. Possible design shifts that give essentially the same functional features are sketched in Fig. 9. Two alternatives are shown in each case. Both alternatives achieve

Table 9 Typical standard deviation of powder injection molding production components

Feature	Best possible	Typical
Angle	0.1°	2°
Density	0.2%	1%
Weight	0.1%	0.4%
Dimension	0.05%	0.3%
Absolute dimension	0.04 mm	0.1 mm
Hole diameter	0.04%	0.1%
Hole location	0.1%	0.3%
Flatness	0.1%	0.2%
Parallelism	0.2%	0.3%
Roundness	0.3%	0.3%
Perpendicularity	0.1% or 0.1°	0.2% or 0.3°
Average roughness	0.4 μm	10 μm

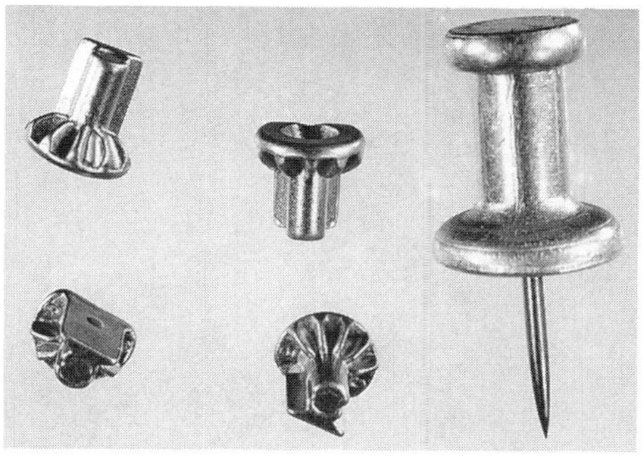

Fig. 10 Stainless steel gears for an electric toothbrush

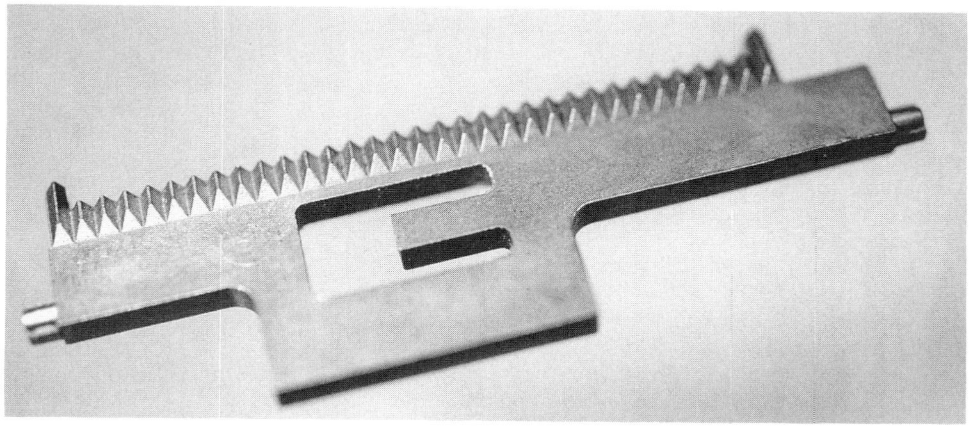

Fig. 11 Paper tape cutter for a postage meter, formed out of tool steel

the same design objectives, but one is better aligned with the specifics of the PIM approach.

Table 9 summarizes the minimum and typical tolerances possible in PIM processing without secondary operations. Generally, the ability to hold tolerances increases with production experience. With little effort, dimensional tolerances can be held within 0.5% (one standard deviation). A controlled process with solvent, wicking, or catalytic debinding allows dimensions to be held with a deviation of 0.3% or less, and critical dimensions can be held within 0.1%. With newer binder systems and integrated process controls these values can be improved to 0.05%. In the production of jet engine ceramic casting cores, the mass ranges from 10 to 1200 g and part lengths range from 2 to 50 cm. These cores are produced with final dimensional scatter of ±0.05% on length (one standard deviation) in production quantities over 50,000 per week. Larger dimensions exhibit more scatter. Fortunately, the component density can be held within ±1% and weight variation to ±0.1%.

Materials and Applications

The availabilities of small powders that match the needed characteristics for molding determine the production materials used in PIM. Most of the classic engineering materials are available except aluminum, glass, lead, and tin. Typically ferrous alloys and alumina-base ceramics are used most often. For the stainless steels, the 316L composition is used frequently, because of its combined strength and corrosion resistance. Another favorite is 17-4 PH stainless steel, which is precipitation hardenable. Other compositions with more chromium are available, and for easier sintering, two-phase compositions that have high levels of molybdenum (up to 6%), chromium (up to 22%), or silicon (up to 3%) are preferred. These are sometimes called duplex stainless steels, because of the two-phase final microstructure. Stainless steels, iron-nickel compositions, and alumina are the materials most often used. Figures 10 to 14 show a few examples of PIM components.

Properties. The properties attainable with PIM are generally equivalent to those possible via other production techniques. Although many PIM components are selected for thermal, optical, or wear applications, the bulk of the property evaluation has focused on mechanical properties. A few sample mechanical properties of PIM materials are given in Table 10. However, there is considerable variability. For example, there have been several reports on 316L stainless steel formed by PIM. Sintered densities ranged from 93 to 100% of theoretical, with a reported mean yield strength of 220 MPa, but a range from 170 to 345 MPa. The lowest strength occurred with the lowest density. Likewise, a range from 18 to 81% elongation has been reported, with a mean of 45% and a standard deviation of 17%. The scatter reflects variations between vendors, especially in controlling impurities such as carbon, oxygen, and nitrogen during debinding and sintering. These same impurities degrade corrosion resistance.

In the Fe-2Ni steels, when the carbon level is almost zero the sintered yield strength is about 190 MPa and the fracture elongation is 30%. When the carbon level is increased to 0.5%, the heat-treated material has a yield strength of 1230 MPa with 1% fracture elongation. These dramatic shifts in strength and ductility are due to the retained carbon. For ductile systems, the strength will typically have a standard deviation

Fig. 12 An injection-molded iron-nickel housing used in personal computers

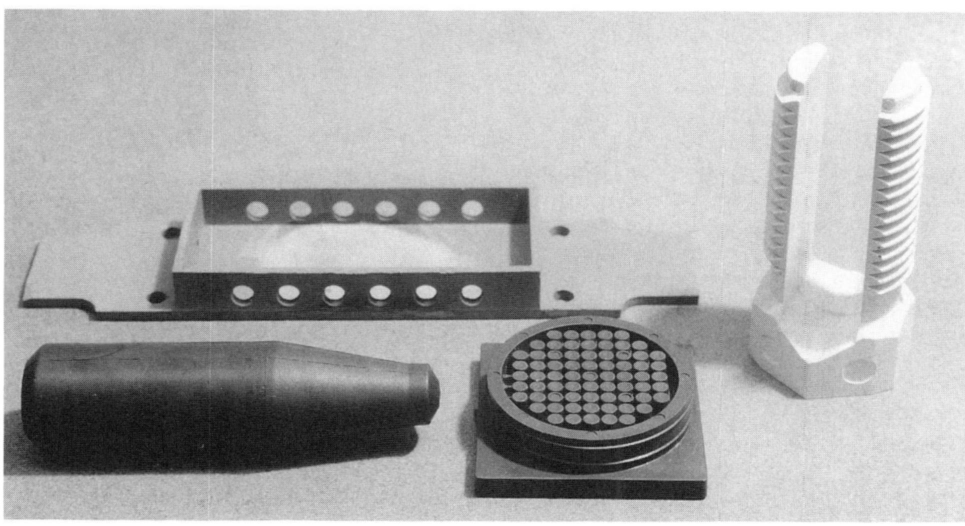

Fig. 13 Collection of parts and materials showing some of the possibilities with powder injection molding: a steel microelectronic package, tungsten bullet, tungsten-copper computer heat sink, and alumina split bolt

of 20 MPa and the elongation will have a standard deviation of approximately 1%.

The dynamic properties, such as fatigue and impact toughness, depend on the pore structure. In the case where the final pores are small and spherical, there is competitive resistance to crack propagation. Unfortunately, dynamic properties are not commonly tested. Early reports gave fatigue endurance strengths of 219 MPa for Fe-7Ni, 237 MPa for Fe-7Ni-0.5C, 575 MPa for 4640 steel, and 517 MPa for 17-4 PH stainless steel. Improved fatigue strength comes with case hardening. Hot tensile tests on PIM 316L stainless steel have shown a steady decline in strength as temperature increases, with yield strengths of 258, 177, 121, 71, and 62 MPa at temperatures of 180, 300, 500, 700, and 900 °C. The ductility remained fairly high, over 25% up to 900 °C.

Impact toughness tends to be low when measured with pre-notched samples, largely because of residual porosity effects on cracking. Fracture toughness has been evaluated for only a few PIM materials, but these are similar to more conventional processing routes.

Corrosion resistance of the PIM stainless steels is a concern. When properly processed without contamination, the PIM products are corrosion resistant, and they are often superior to wrought materials for thin sections. Corrosion is highly dependent on impurities, density, and final thermal cycles. However, as a simple generalization, often the sintered corrosion properties match or exceed the typical properties observed with alternative processing routes.

The soft magnetic characteristics are of interest for the ferrous systems such as iron, Fe-2Ni, Fe-3Si, Fe-0.45P, Fe-0.6P, and Fe-50Ni. Of these, the Fe-50Ni alloy has the most attractive combination of high magnetization and low coercive force. Experience shows considerable variability in the coercive force. Part of this variation is directly due to differences in sintered density, but impurity control is also a major factor. Like corrosion and strength, the magnetic characteristics are sensitive to contaminants in the compacts.

Economics. Cost is the critical parameter in determining the feasibility of PIM production. It depends on many factors, not the least being the initial tooling cost. Other factors include the number of cavities in the tool set, production quantity, powder cost, details of the fabrication steps, surface roughness, packaging requirements, labor rates, and required tolerances. The best applications for PIM involve high production quantities of complex parts formed from materials that are difficult to fabricate. A desirable production quantity is more than 50,000 per year, although at quantities over approximately 100,000 per year the tooling cost per compact is not significantly less, because of needed tooling refinishing. For production of 250,000 per year or more, the use of multiple cavity tooling is typical, and up to 16 cavities are typical. New products are planned to reach production rates of 300,000 to 500,000 per day. Current industry growth rates are near 22% per year, after reaching sustained rates of 32% per year through the 1990s.

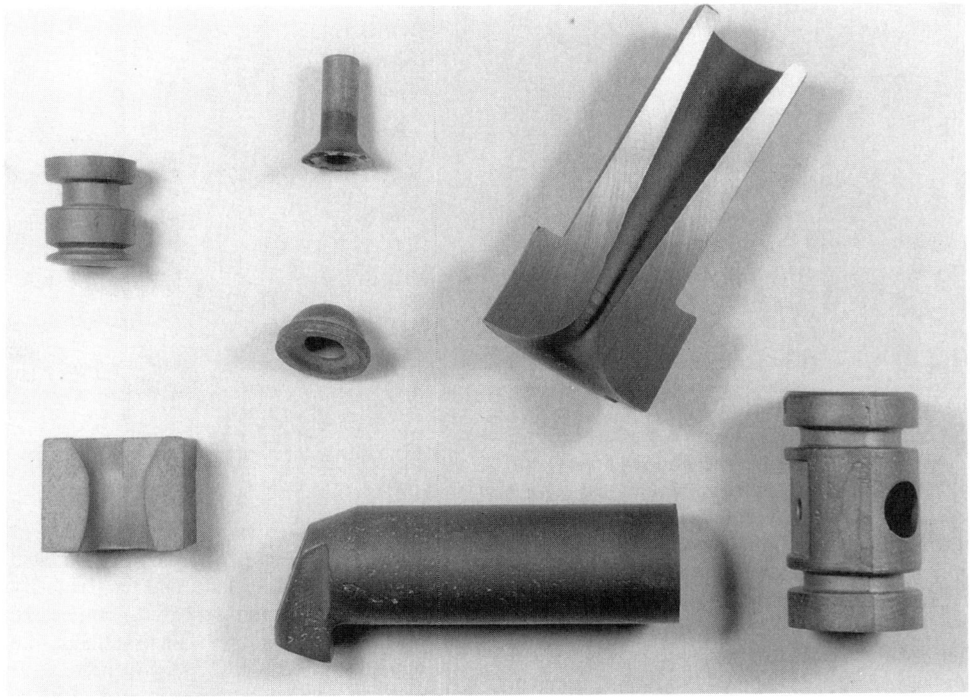

Fig. 14 Some wear components formed by powder injection molding out of cemented carbides

Table 10 Selected mechanical properties of powder injection molding metals and alloys

Material	Composition, wt%	Density, %	Yield strength, MPa	Ultimate tensile strength, MPa	Elongation, %	Hardness
4140 (HT)	Fe-1Cr-0.4C	93	1240	1380	2	40 HRC
4640 steel (HT)	Fe-2Ni-1Mo-0.4C	97	1400	2000	3	30 HRC
Iron-copper steel	Fe-2Cu-0.8C	95	...	700	10	92 HRB
Iron-chromium steel	Fe-1Cr-0.5C	94	...	600	10	90 HRB
Iron-nickel	Fe-50Ni	96	170	420	20	50 HRB
Iron-nickel steel (HT)	Fe-2Ni-0.5C	94	1230	1230	1	45 HRC
Iron-silicon	Fe-3Si	99	345	520	25	85 HRB
Kovar or F15	Fe-29Ni-17Co	98	350	520	42	60 HRB
Stainless 17-4 PH (HT)	Fe-16Cr-4Ni-4Cu	96	965	1140	12	35 HRC
Stainless 316L	Fe-17Cr-12Ni-2Mo-2Mn	96	220	510	45	75 HRB
Stainless 420 (HT)	Fe-13Cr-1Mn-1Si	92	690	1440	6	47 HRC
Stellite	Co-28Cr-4W-3Ni-1C	99	...	1020	3	40 HRC
Super Invar	Fe-32Ni-5Co	96	285	440	40	65 HRB
Ti-6-4	Ti-6Al-4V	98	800	880	12	35 HRC
Tool steel	Fe-6W-5Mo-4Cr-2V-1C	99	...	2000	0	66 HRC
Tungsten heavy alloy	W-4Ni-1Fe	99	650	1000	20	50 HRA

HT, heat treated

SELECTED REFERENCES

- H.H. Angermann and O. Van Der Biest, Binder Removal in Powder Injection Molding, *Rev. Part. Mater.*, Vol 3, 1995, p 35–69
- P.H. Booker, J. Gaspervich, and R.M. German, Ed., *Powder Injection Molding Symposium—1992*, Metal Powder Industries Federation, Princeton, NJ, 1992
- R.M. German, *Powder Injection Molding*, Metal Powder Industries Federation, Princeton, NJ, 1990
- R.M. German and A. Bose, *Injection Molding of Metals and Ceramics*, Metal Powder Industries Federation, Princeton, NJ, 1997
- R.G. Iacocca, A Critical Assessment of the Characterization Tests Needed to Support Powder Injection Molding Component Fabrication, *Rev. Part. Mater.*, Vol 2, 1994, p 269–313
- K.S. Hwang, Fundamentals of Debinding Processes in Powder Injection Molding, *Rev. Part. Mater.*, Vol 4, 1996, p 71–103

Binder-Assisted Extrusion

David E. Alman and Jeffrey A. Hawk, U.S. Department of Energy, Albany Research Center

BINDER-ASSISTED EXTRUSION is a plastic forming process in which a highly viscous feedstock, that is, a powder mixed with a binder and other rheology modifiers (also termed the paste), is forced through a die to form a shaped product. The binder-assisted extrusion process is used commercially to produce ceramic parts, such as furnace tubes, bricks, insulators, pipes, tiles, tubular capacitors, catalyst supports, magnets, heat exchangers, wires, and springs (Ref 1–8). However, binder-assisted extrusion has also been used to fabricate metals and high-temperature superconductors (Ref 9, 10), as well as being a technique for aligning reinforcement in the processing of advanced composites (Ref 11–13). Depending on the powders being shaped, the extrusion process can be performed at room temperature with the utilization of a solvent-based binder, or at elevated temperatures with a plastic binder system.

The basic principles of binder-assisted extrusion are similar to powder injection molding (PIM) in that both processes consist of four very similar steps: (1) feedstock or paste preparation; (2) product shape forming (powder extrusion in one case, powder injection in the other); (3) binder removal; and (4) the consolidation step, that is, sintering. During the feedstock/paste preparation stage, the powders are mixed with the constituents of the binder to obtain the proper flow characteristics. Feedstock preparation may consist of several steps (e.g., dry mixing, wet mixing, and high-shear mixing) to remove agglomerations, to produce a uniform mixture, and to coat the powder particles with the binder. During the forming step, the feedstock/paste is extruded through the appropriately shaped die. At this point in the process, the part must possess enough structural integrity to be manipulated while maintaining its shape. In the finishing step the part is cut to size, dried, and thermally processed. Table 1 summarizes the steps generally followed in making a part using the binder-assisted extrusion process. Knowledge and control of each step in the process are crucial to the production of a sound part.

Binder-assisted extrusion differs from PIM in that the extrusion usually consists of a constant cross section. Although parts with complicated cross sections can be produced, a selection of which is shown in Fig. 1.

Binder-assisted extrusion and its derivatives have many advantages and some important disadvantages when compared to other powder processing techniques, for example, powder compaction in dies, isostatic pressing, slip and tape casting, and so forth. According to Benbow and Bridgwater (Ref 6), the primary advantages of using the binder-assisted extrusion processes include:

- The formation of complicated cross sections
- The small amount of binder/liquid added to the solids
- The ability to shape very hard powders
- A uniform density distribution
- The construction of long, thin sections

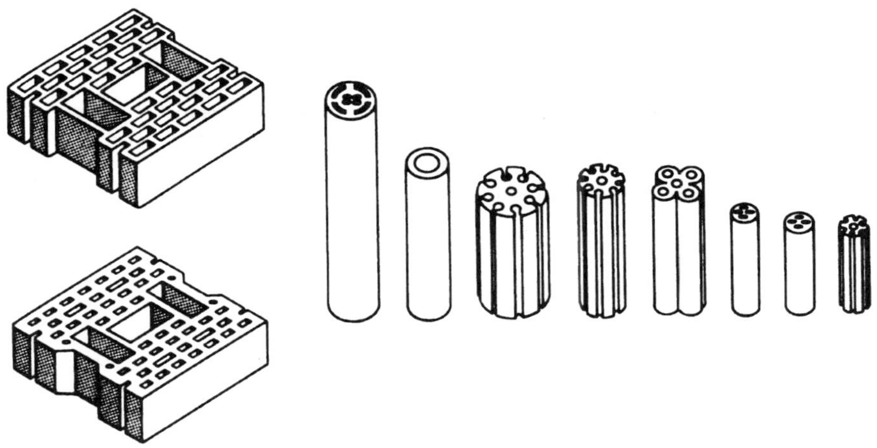

Fig. 1 Schematic of typical cross sections produced by binder-assisted extrusion. Source: Ref 2

Table 1 Stages of binder-assisted extrusion

Processing stage	Purpose	Potential difficulties
Dry mixing	Uniformly mix all solid components	Regions of poorly mixed solids
Wet mixing (for solvent-based binders systems)	Uniformly distribute particles in liquid	Air entrapment; liquid rheology may prevent mixing; adherence to surfaces
High-shear mixing	Break down agglomerates; ensure binder surrounds individual powder particle	Binder—powder separation; incomplete deagglomeration
Degassing	Remove entrapped gas	Hard to achieve in some equipment
Extrusion	Shape feedstock	Binder—powder separation; excessive pressure drop; lamination flow defects between confluent streams; surface defects
Extrudate handling and cutting	Remove product from extruder and cut to desired length	Damage to weak product; surface damage; closure of internal holes during cutting
Binder removal	Remove binder; aid shape retention	Dry cracks from solvent-based binder system; incomplete binder burnout resulting in residual contaminates in product
Sintering	Generate strength in product; remove porosity	Temporary weakness in extrudate

Adapted from Ref 6

- Competitive capital and operating costs relative to other processes

The disadvantages include:

- The final structure is dependent on the powder properties.
- The strength is not imparted during the extrusion process itself.
- Structural defects may be introduced.
- Structure or strength are not generally adjustable by changing the extrusion conditions.
- Binder and rheology modifiers can adversely affect product properties, and the rate at which the binder and rheology modifiers can be removed from the extrudate is slow.

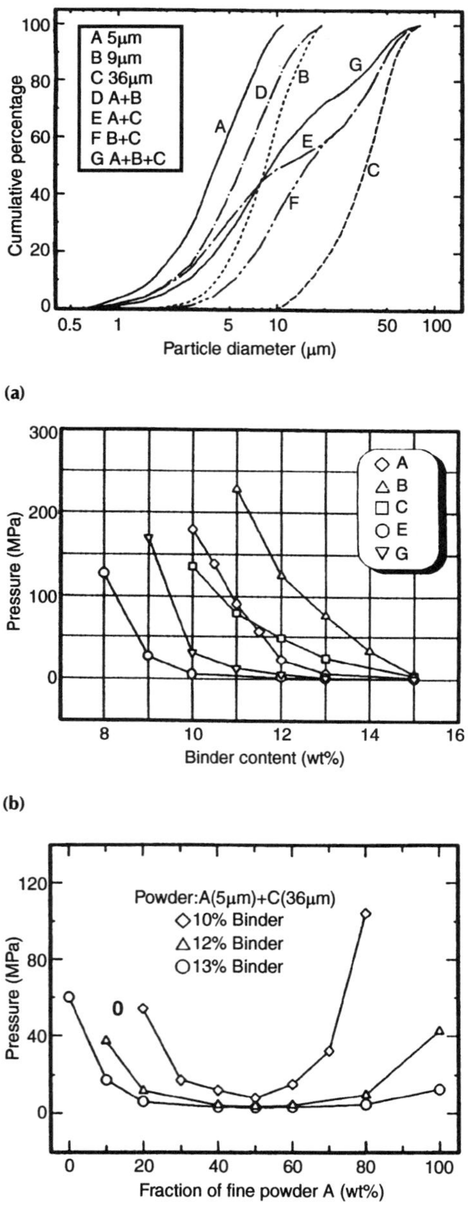

Fig. 2 Effect of binder (water soluble hydroxypropyl methylcellulose) content and 304L stainless steel powder size distribution on extrusion pressure. (a) Stainless steel powder particle size distributions. (b) Extrusion pressure versus binder content for stainless steel powders of different particle size distributions (reduction ratio = 10; extrusion velocity = 10 mm/min). (c) Extrusion pressure versus utilization of powders with a bimodal particle size distribution (mixtures of powders A and C). Source: Ref 16

Feedstock/Paste

The feedstock, or paste, is the moldable material that is subsequently used to make the part. It is important that during preparation the feedstock is able to retain its shape long enough to allow handling and further processing. Correct feedstock formulation and preparation is thus crucial for a defect-free, successful extrusion. The required characteristics of the feedstock include the following (Ref 2):

- The feedstock must be sufficiently plastic to flow under pressure into the desired cross section, while retaining enough rigidity to resist deformation due to its own weight, that is, slumping.
- The feedstock must have enough fluidity so that it does not stick to the die or other tooling, and that upon exit from the die it possesses a smooth surface.
- Phase separation within the feedstock (i.e., separation of the binder and powder) must not occur during extrusion.

Fundamentals of Flow

Fluids can be broadly classified according to the relation between applied shear stress and the rate of deformation of the fluid. Fluids in which the shear stress is directly proportional to the rate of deformation are Newtonian, that is:

$$\tau = \eta \left(\frac{d\gamma}{dt} \right) \quad \text{(Eq 1)}$$

For fluids where this proportionality does not hold, that is, where they possess nonlinear behavior, they are broadly classified as non-Newtonian. Some well-known categories of non-Newtonian fluid flow are dilatant, viscoelastic, and Bingham.

When discussing the flow characteristics of the feedstock, only viscous, plastic, and elastic effects are considered (Ref 6). In the extrusion process, to facilitate flow of the feedstock in the extruder and die assemblies, it should be viscoelastic, that is, the feedstock should possess an initially high strain rate that decreases over time as elastic stresses are developed. This type of behavior is more commonly referred to as "strain-thinning" behavior, or pseudoplasticity. Paints possess a strain-thinning characteristic; that is, they have a low viscosity at high shear rates (smooth paint flow during brushing) and a high viscosity at low shear rates (particles remain in suspension during storage) (Ref 2). Thus, the characteristic of major import in the extrusion process is the decrease in viscosity of the feedstock with increasing shear strain rate. Consequently, the flow stress (τ) for a pseudoplastic material, having a yield stress (τ_0), is given by:

$$\tau = \tau_0 + \eta\,(\dot{\gamma})^n \quad \text{(Eq 2)}$$

where $\dot{\gamma}$ is the shear rate equal to $d\gamma/dt$, η is the shear rate factor, or the viscosity, and n is the exponent of shear rate. In the case of pseudoplastic materials, the value of the exponent is less than 1.

During compression, when the feedstock flows through the barrel, shear strain rates are low and high viscosity prevents phase separation and promotes consolidation. By contrast, during shaping, when the feedstock deforms as it flows through the die, shear strain rates are high and a low-viscosity feedstock minimizes the required extrusion pressure. A feedstock possessing pseudoplastic behavior, therefore, possesses both a high viscosity at low shear strain rates and a low viscosity at high shear strain rates, which is why it is the desirable feedstock flow behavior for binder-assisted extrusion.

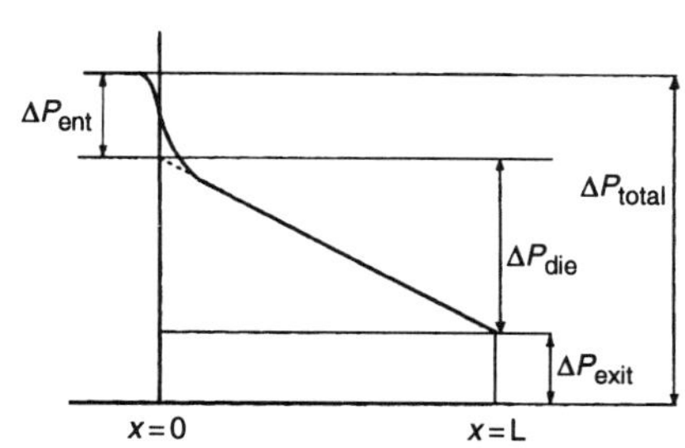

Fig. 3 Pressure profile in the extrusion die. Source: Ref 5

Feedstock Formulation

The characteristics of both the binder and powders need to be considered in feedstock formulation. Binder content is one of the most important parameters in the formulation process. The binder is used to provide the proper rheological behavior for the shaping of the powders. The binder physically separates the individual powder particles and acts as lubrication between the powder/binder mixture and the die walls during the extrusion process. There is a narrow range of binder volume fractions that produce a feedstock with rheological characteristics suitable for producing good extrusions (Ref 1, 2, 5, 6, 14–16). If insufficient binder is present in the feedstock, it will not flow easily and high pressures will be needed to extrude the feedstock (Ref 6, 14–16). This situation is shown in Fig. 2(b), which shows the pressure needed to extrude stainless steel powders with different weight fractions of binder (Ref 16). Notice that as the binder weight fraction decreases, the extrusion pressure increases for all powder configurations. Conversely, if too much binder is added to the powder, the feedstock will be too soft and pliable, and the extruded part will most likely slump. To prevent slumping, the strength of the extruded part must exceed $\rho g h$, where h is the height of the extruded part, ρ is the density of the feedstock, and g is the acceleration due to gravity. Slumping can be prevented by adding coagulants to the binder formulation, or by utilizing binders that will gel as a result of heat input during subsequent processing steps (Ref 1–6). Generally, the minimum amount of binder required to prevent slumping of the part is what should be used. Typically, a feedstock for hot extrusion will contain between 30 and 40 vol% plastic binder. A solvent-based feedstock for room-temperature extrusion can contain upwards of 50 vol% of a liquid phase.

The powder particle characteristics, such as size, shape, and size distribution, can influence the extrusion process in several important ways. For instance, the largest powder particle that can be processed must be less than the smallest dimension found in the extruded part. This criteria may be significant during production of parts that contain thin sections, like honeycombs where the wall thicknesses can be as thin as 100 μm. In practice, a good rule of thumb is that the absolute maximum powder particle dimension should be less than one-tenth that of the smallest die dimension (Ref 6). Powder particle size distributions (e.g., whether it is bimodal or widely distributed) that allow efficient packing will result in feedstock formulations with minimal binder content (Ref 6, 14–16). Proper extrusion also depends on having some excess binder, over and above that which is necessary to fill the interparticle voids, to lift and separate the powder particle surfaces, and to facilitate flow. For a given weight fraction of binder, extrusion pressures will be lower for powders packed efficiently, as shown in Fig. 2. The reduced pressure arises because a larger percentage of the binder is being used for filling the space between the powder particles, rather than for facilitating flow as in the case where the powders are not efficiently packed. Also, an extruded part, with the minimum amount of binder needed to hold its shape, will be easier to handle, and it will ultimately lead to a part with a higher density (i.e., minimum porosity).

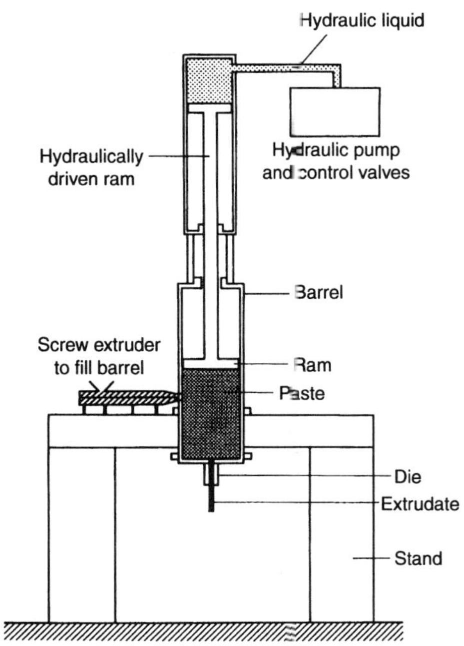

Fig. 4 Ram extruder. Source: Ref 6

Binders

The selection of an appropriate binder system is dependent on the physical and mechanical properties of the powders being extruded. It is important to select a fugitive binder that can be completely removed, as residual contaminants can degrade the mechanical properties of the finished product. To facilitate proper rheological behavior, binder formulations may contain several components, including solvents, surfactants, deflocculants, coagulants, lubricants, plasticizers, and preservatives.

Solvent-Based Binders for Room-Temperature Extrusions

Solvent-based binder systems are used for room-temperature extrusion. Typically, a water soluble clay and/or an organic polymer binder is dissolved in water (Ref 1, 3, 5,17). The water uniformly disperses the binder throughout the powder and provides the rheological characteristics necessary for proper extrusion. The organic binder is used to increase the viscosity of the feedstock and to change the flow characteristics of the powder/water feedstock from Newtonian to the desired pseudoplastic flow required for good extrusions. Nonaqueous binder systems are also used. Table 2 lists some of the more common aqueous and nonaqueous binders used in the extrusion process (Ref 5,17). Organic binders are available in a wide range of viscosities and are typically sold as a function of the molecular weight or the strength of their molecular bonds. Typically, the viscosity of the feedstock for extrusion should be about 10 Pa · s (Ref 17).

Table 2 Typical binder systems for room temperature extrusions

Aqueous binders	
Binder type	**Viscosity grade(a)**
Gum arabic	VL
Lignosulfonates	VL
Dextrins	VL to L
Polyvinyl alcohol	VL to M
Acrylates	VL to M
Polyethylene-imine	L to H
Methylcellulose	L to H
Sodium carboxymethylcellulose	M to VH
Polyethylene oxide	M to VH
Natural gums	H to VH
Alginates	M to VH
Nonaqueous binders	
Binder	**Solubilities**
Acrylics (PMMA, PMA, PBMA, PEMA)(b)	Ketones, alcohols, glycol ethers, hydrocarbons
Polystyrene	Ketone, hydrocarbons
Cellulosics (ethyl, methyl, acetate, butyrate)	Wide
Polyethylene oxide	Polar organics

(a) Viscosity grade is defined as concentration of binder in aqueous solution required to give a solution viscosity of 2 Pa · s. Very low (VL) > 10 wt%; Low (L) 3–10 wt%; Medium (M) 1–3 wt%; High (H) 0.3–1 wt%; Very high (VH) < 0.3 wt%. (b) PMMA, polymethyl methacrylate; PMA, polyethyl methaceylate; PBMA, polybutyl methaceylate; PEMA, polymethacrylic acid; Adapted from Ref 5

Plastic Binders for Hot Extrusion

Plastic binders, utilized for hot extrusion, are blends of organic and metall-organic polymers, with a small percent of modifying agents (Ref 7). Typically, the feedstock contains 65 vol% powder, 35 vol% binder, and has a viscosity between 200 and 500 Pa · s, within the temperature range of 80 to 220 °C, and at a shear rate of between 100 and 1000 s^{-1}. Many of the exact binder/feedstock formulations utilized for hot extrusion are proprietary. However, many binder/feedstock combinations are based on waxes (Ref 7, 9, 13, 16, 18–21), and those suited for PIM should work just as well for binder-assisted extrusion. A list of several binder/feedstock systems utilized in PIM can be found in Ref 22.

Feedstock Preparation—Mixing

The purpose of mixing the powder and the binder is to produce a uniform feedstock suitable for extrusion where all of the powder particles have been coated with the binder. Areas of inhomogeneity (e.g., pockets of uncoated powder and/or pure binder) can result in strength-limiting defects in the final product. Obtaining uniformity of the feedstock can be problematic because of the stiffness of the final extrudable mixture. Mixing of the powder and binder consists of several steps. For a solvent-based binder system for extrusions made at room temperature, the solid components (i.e., the powders, clays, organic polymers) are first mixed in the dry state. The liquid solvent is then added to the thoroughly mixed powders. To maximize homogeneity of the feedstock, wet mixing occurs first by slurry mixing and then through high-shear mixing. During this stage, it is common practice to add extra solvent to lower the viscosity of the feedstock in order to facilitate homogeneity during mixing. The excess solvent is then removed by the partial drying of the feedstock mixture prior to extrusion. In an alternative approach, the liquid solvent and powders are mixed to form a low-viscosity slurry, to which the solid component of the binder is incrementally added until the mixture possesses the desired viscosity.

High-shear mixing can be performed in a pug-type mill, in which two rows of shafts rotate in opposite directions. The purpose of this type of mixing is to break down the agglomerates inherent in the starting powders and any formed during subsequent mixing. It is important to avoid separation of the liquid and solid phases during high-shear mixing.

In the hot-extrusion process, the specific components that make up the plastic binder are compounded. The powder and binder are then dry mixed. The material is then run through a high-shear, screw-type pug mill, where the powder/binder is mixed, then melted and remixed under high-shear conditions. It is finally pumped through a die to form a strand. The strands are pelletized and fed into the extruder. Similar procedures employed during feedstock preparation for PIM can be applied in the feedstock preparation for hot extrusion.

Extrusion

Extrusion is the pressing of the feedstock into and through a die of size and cross-sectional shape appropriate to that of the product. The specific steps (Ref 1) used in binder-assisted extrusion are: (1) feedstock/paste delivery into extruder, (2) consolidation and flow of feedstock/paste through the barrel, (3) deformation and shaping of feedstock/paste as it flows through the die, (4) flow of feedstock/paste through the die land, and (5) part ejection. For the feedstock to flow through the die, the pressure generated by the extruder must be greater than the resistive forces associated with the deformation of the feedstock at the die entrance due to the reduction in diameter and the drag forces along the die wall. The pressure drop along the axis of the extruder reflects the amount of work required to overcome the resistive forces. The pressure is highest in the barrel and lowest at the exit of the die landing, as shown in Fig. 3. There is a pressure drop at the exit of the die, which tends to be small in magnitude compared to the drop associated with the deformation and motion of the feedstock. However, if pressure drop upon exiting is too high, there can be significant "springback," or elastic rebound, of the extruded part. Springback can result in axial cracking and warpage of the part after it has left the die. Slower extrusion rates, longer extrusion dies, and the incorporation of short relief sections at the end of the die (i.e., a negative taper) all contribute to minimize the effect of springback (Ref 5).

Table 3 Comparison of single and twin-screw extruders

Factor	Single screw	Twin screw
Energy loss	Viscous flow	Heat transfer
Feedstock transport	Wall friction dependent	Positive displacement
Throughput	Pressure dependent	Independent
Shear forces	Larger	Smaller
Power/kg of material	Higher	Lower
Temperature gradients	Higher	Lower
Air evacuation	Simple	Difficult
Capital cost	Lower	Higher

After Ref 1

Equipment

Two common extruders are used in the binder-assisted extrusion process: the screw, or auger-type, extruder and the piston, or ram, extruder. In piston extrusion (Fig. 4), the feedstock, commonly a degassed billet, is inserted in the barrel, and the piston compresses it, forcing the feedstock down the barrel and through a die of the desired shape. As the ram contacts and compresses the feedstock, the pressure increases (Fig. 5). As the feedstock is pushed along barrel by the piston, the pressure falls as a result of drag forces along the barrel wall. As the feedstock enters the die, the pressure once again rises. Constant pressure in the barrel can only be maintained if inverted extrusion is employed (Ref 5). (In inverted extrusion the die moves instead of the feedstock.)

Two advantages of the piston extrusion process are (Ref 1–6): (1) very high pressures can be generated independent of the feedstock properties and (2) there is precise control over output flow. As a consequence, piston extruders are commonly used for the production of intricately shaped components. However, a drawback to the use of piston extruders is the intermittent nature of the process and the low capacity inherent with batch-type processes. However, with the aid of a screw extruder to load the piston barrel with the feedstock, as depicted in Fig. 4, piston extrusion can be made to perform semicontinuously.

Screw extruders are used to produce parts in large batches in a continuous process. Potential capacities of screw extruders can be as high as 100 tonnes/h (Ref 1–6). The screw extruder typically consists of several sections (shown schematically in Fig. 6). The feedstock is fed into a pug mill, which supplies high-shear mixing. The feedstock is then forced through a perforated plate where it is shredded. This shredded material falls into the main extruder chamber. Degassing can occur at this stage, making use of a vacuum system attached between the two chambers. In the extruder chamber, the feedstock is consolidated and forced through the forming die. Screw extruders may use single- or twin-screw configurations. When using a twin-screw system, the two screws are able to rotate in either the same or the opposite direction. Twin-screw extruders force the feedstock through the barrel using positive displacement, whereas in the single screw extruder, the movement of the feedstock is entirely dependent on the frictional forces that develop among the screw, the feedstock, and the barrel (Ref 1–6). For conveying, feedstock properties are less significant with a twin-screw extruder than for a single-screw extruder. A comparison of the operating parameters for the single- and twin-screw extruders is given in Table 3.

In a single-screw extruder pressure builds up as the segmented feedstock is conveyed and compressed, as shown in Fig. 7 (Ref 23). The segmented feedstock becomes continuous in the metered zone, and the pressure continues to rise. Finally, the feedstock is conveyed into the die, where it is deformed. It is at this point that the highest pressures are generated. The pressure in the die is dependent on the screw geometry and the feedstock rheology (Ref 1). Larger and/or tapered screws may be used to generate higher pressures. The number of flights of the screw controls the number of feed columns that are

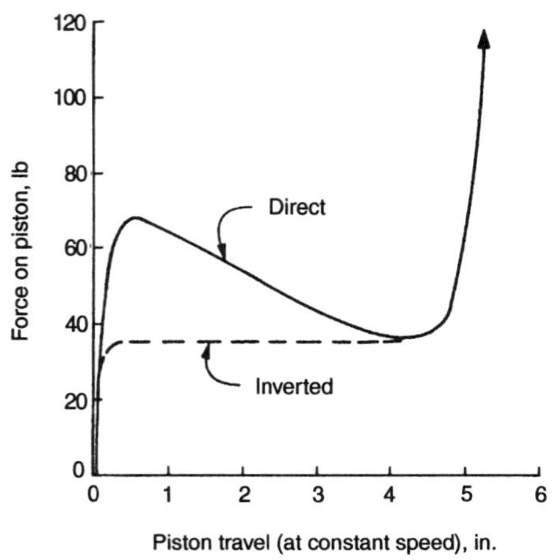

Fig. 5 Pressure profile in the barrel of a ram extruder. Source: Ref 5

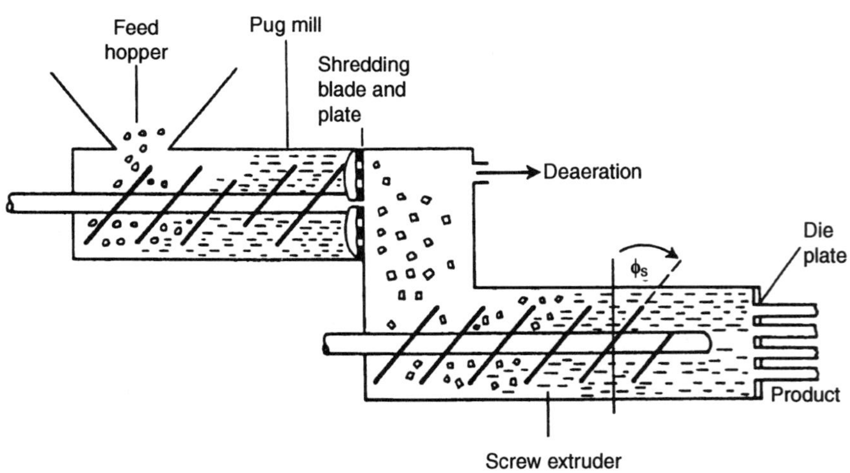

Fig. 6 Screw extruder. Source: Ref 6

displaced. Larger helical angles increase the potential delivery rate, but reduce the compressive thrust on the feedstock. The required ratio of screw diameter to part diameter increases as the yield strength of the feedstock increases or when the die land area increases. The length of the feed section is commonly five times that of the screw diameter, and the length of the metering section is ten times that of the screw diameter. Typical screw geometries include (Ref 1, 2, 3, 7) a helical angle between 5 and 20°, radial clearance between the screw and barrel of 0.1 %. of screw diameter, and a channel depth ratio of about 3 to 1. Generally, these dimensions are selected based on standard screw geometries developed for commercial extrusions of other products, such as thermoplastics and food articles. However, recent experiment results indicate that for optimal binder-assisted powder extrusion, larger helical angles should used (Ref 24).

Flow through the Extrusion Die

The flow through the die is composed of two parts: (1) flow from the barrel into the die land (i.e., flow through the die entry) and (2) flow through the die land (Ref 1, 5, 6,14–16, 24–28). The feedstock deforms—that is, it extends in the axial direction and decreases in cross-sectional area—as it flows through the entrance to the die. For a square entry die, the work required to deform and extend the feedstock in the axial direction results in a pressure drop (ΔP_1) at the entrance to the die. This pressure drop is given by the following equation (Ref 6, 25):

$$\Delta P_1 = \sigma_b \ln\left(\frac{A_o}{A}\right) \quad \text{(Eq 3)}$$

where σ_b is the internal yield stress of the feedstock, and A_o and A are the barrel and die cross-sectional areas, respectively. For a circular square entry die (Fig. 8), the pressure drop at the entrance is represented by the following equation (Ref 6, 25):

$$\Delta P_1 = 2\,\sigma_b \ln\left(\frac{D_o}{D}\right) \quad \text{(Eq 4)}$$

where D_o and D are the initial and final diameters, respectively.

The flow through the die land is resisted by drag forces along the die wall, that is, friction force between the feedstock and the die wall. The drag force acts over the perimeter of the die land. The pressure drop (ΔP_2) required to cause flow through the die land can be derived from a simple force balance (Ref 6):

$$\Delta P_2 A = \tau_f ML \quad \text{(Eq 5)}$$

where τ_f is the extruded part-wall shear stress (assumed to be independent of the position on the perimeter), and A, M, and L are the cross-sectional area, perimeter, and length of the die land, respectively. For flow through a die with a circular cross section, a geometry used for wire and bar production, the pressure drop is given as:

$$\Delta P_2 = 4\tau_f\left(\frac{L}{D}\right) \quad \text{(Eq 5a)}$$

where D is the diameter of the die land. For flow through a die land with a square cross section, a geometry used to make brick and tile, the following equation represents the pressure drop:

$$\Delta P_2 = \tau_f\left(\frac{b+c}{bc}\right)L \quad \text{(Eq 5b)}$$

where b and c are the depth and width of the rectangular cross section. Finally, for flow in the space between two annular cylinders, a geometry used to make tubular products, the equation for the pressure drop is given by:

$$\Delta P_2 = 4\tau_f\left(\frac{L}{D_1 - D_2}\right) \quad \text{(Eq 5c)}$$

where $D_1 - D_2$ are the outer and inner diameters of the tube, respectively.

The total pressure drop (ΔP) required for feedstock flow through a circular, square entry die is (Ref 6, 25):

$$\Delta P = \Delta P_1 + \Delta P_2 = 2\tau_b \ln\frac{D_o}{D} + 4\tau_f\left(\frac{L}{D}\right) \quad \text{(Eq 6)}$$

where σ_b and τ_f are related to the flow characteristics of the feedstock. Experimentally (Ref 25), it has been found that $\sigma_b = (\sigma_o + \alpha v^n)$ and $\tau_f = (\tau_o + \beta v^m)$, which is very similar in form to Eq 1. Here, σ_o is the yield stress, extrapolated to zero velocity; α characterizes the effect of velocity on the yield stress; τ_f is the initial wall shear stress; β is a factor accounting for the velocity

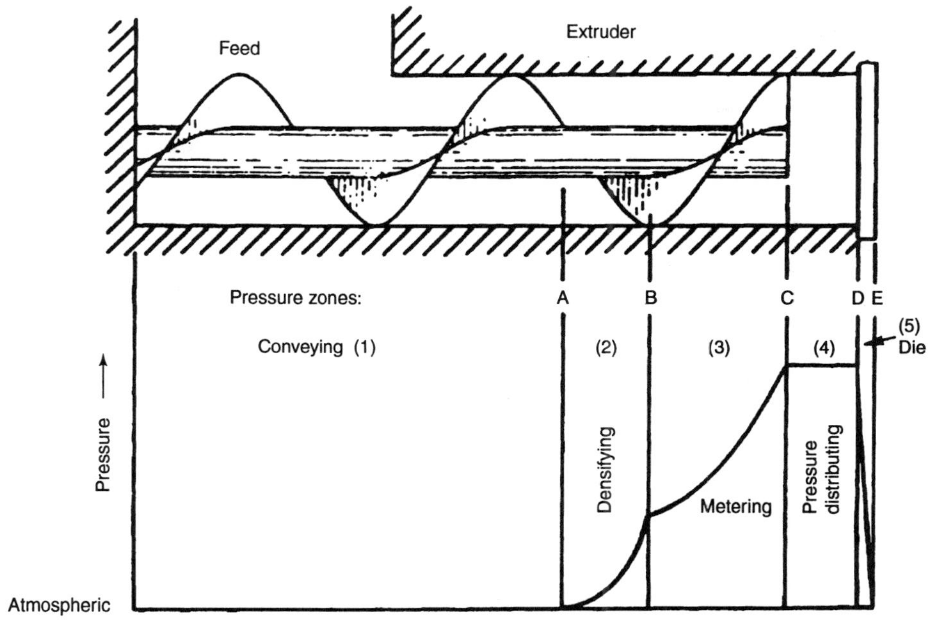

Fig. 7 Pressure profile in the barrel of a screw extruder. Source: Ref 23

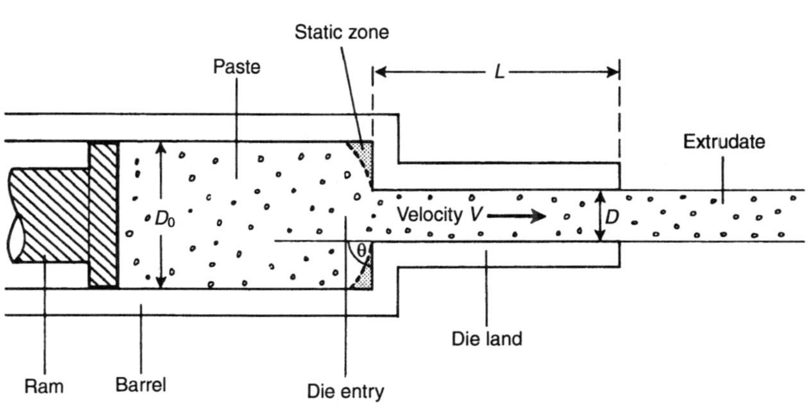

Fig. 8 Circular square entry die. Source: Ref 6

dependence of wall shear stress; n and m are shear-thinning exponents; and v is the feedstock velocity. The values of feedstock parameters (σ_o, τ_o, α, β, m, and n) are determined from ram extrusion experiments in which extrusion pressures are measured through two dies of differing die-land lengths and at a range of extrusion velocities. For a circular square entry die the extrusion pressure can be calculated as:

$$\Delta P = 2\,(\sigma_o + \alpha v^n)\ln\frac{D_o}{D} + 4\,(\tau_o + \beta v^m)\ln\frac{L}{D} \qquad \text{(Eq 7)}$$

The pressure calculated using Eq 7 is compared to the measured pressure for an alumina extrusion in Table 4. Differences in calculated and measured pressures may be attributed to the die-entry velocity being a function of the shear rate factor and shear-thinning exponent (Ref 27).

Conical or tapered dies are frequently used in extrusion processes. For flow through a conical die entry, the feedstock is subjected to both compressive and shear stresses, with the generation of significant drag forces along the die wall. The pressure drop (ΔP_1) for flow through a die with a conical entry can be written as (Ref 6):

$$\Delta P_1 = (\sigma_o + \alpha_1 v^n + \tau_o \cot\theta)\ln\left(\frac{D}{D_o}\right) + \left(\frac{\beta_1 v^m \cot\theta}{m}\right) \qquad \text{(Eq 8)}$$

where θ is the die entrance angle (note that for a square entry die $\theta = 90°$ and the cot ($\theta = 0$; thus Eq 8 reverts to Eq 4). The flow velocity in the tapered die is a function of both radial and axial location. The material nearest the central axis experiences more acceleration, and experiments utilizing striated feeds of different colors (Ref 30), and monitoring extrusion with nuclear magnetic resonance (Ref 29), indicate that the differential flow velocity between the center and wall of the die increases with increasing die entrance angle (as shown in Fig. 9).

Analysis of the preceding equations reveal that the extrusion pressure will be high when the extrusion velocity is high and when the ratio of the die land length to die diameter *(L/D)* is large, which is consistent with practice (Fig. 10). The extrusion pressure will also be high when the reduction ratio (D/D_o) is large and when the die entry angle (θ) is small (Fig. 11). Finally, the extrusion pressure will be high when the friction in the die land is high (Fig. 12). From Fig. 12, the addition of lubricants decreases the frictional force along the die land, and thus reduces the work required for feedstock flow.

A more generalized form of Eq 7 and 8 is given by Eq 9 (Ref 6, 26):

$$\Delta P = f(v)\ln\frac{A_o}{A} + g(v)\frac{M}{A} \qquad \text{(Eq 9)}$$

where $f(v)$ and $g(v)$ are used to characterize the feedstock flow behavior. In this case, $f(v)$ is a function of the change in cross-sectional area of the feedstock due to extension into the die entry, and $g(v)$ is a function of the shear flow, or drag of the feedstock along the die land. In Eq 9, A_o and A are the cross-sectional areas of the barrel and die, respectively, while M is the perimeter distance of the die land.

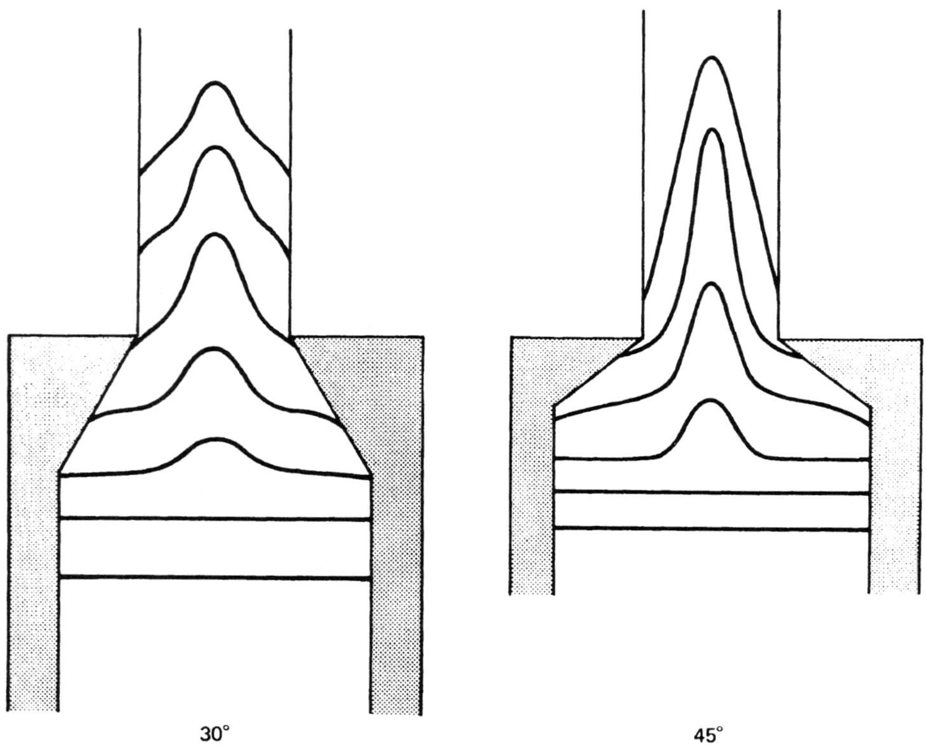

Fig. 9 Influence of die taper on the extrudate velocity profile. Source: Ref 30

Practical Concerns for Extrusion Die Design

A large die entrance angle may result in a zone of static material (analogous to the dead-metal zone in metal extrusion) at the periphery of the die. Static material may be a source of defects

Table 4 Comparison of calculated and measured extrusion pressures for extrusion of alumina powder

Extrusion velocity, mm/s	Extrusion pressure, kPA	
	Measured	Calculated using Eq 7
Square entry—circular die land (D_o = 25.4 mm; D = 9.5 mm; L = 50.8 mm)		
1.3	5.05	5.0
2.7	6.26	5.6
5.3	7.74	6.5
13.3	11.00	9.6
Square entry—square die land (A_o = 506 mm^2; A = 9.0 mm^2; b = 3.0 mm; c = 3.0 mm; L = 3 mm)		
4.7	1.89	1.23
18.7	3.58	2.27
46.7	5.21	4.53

Note: Feedstock: 83 wt% Al_2O_3, 3.5 wt% clay, 6 wt% water, 7.5 wt% glucose. σ_o = 24 kPa, α = 18 kPa · s/mm; τ_o = 19 kPa; β = 18 kPa · s/mm. Adapted from Ref 26

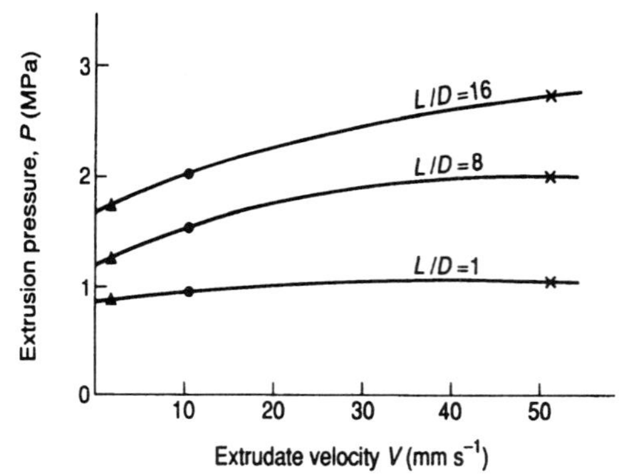

Fig. 10 Influence of extrudate velocity and die land aspect ratio (*L/D*) on extrusion pressure (extrusion of Al_2O_3 powder; binder is clay, starch, and water; D_o = 25.4 mm, D = 3.2 mm. Source: Ref 6

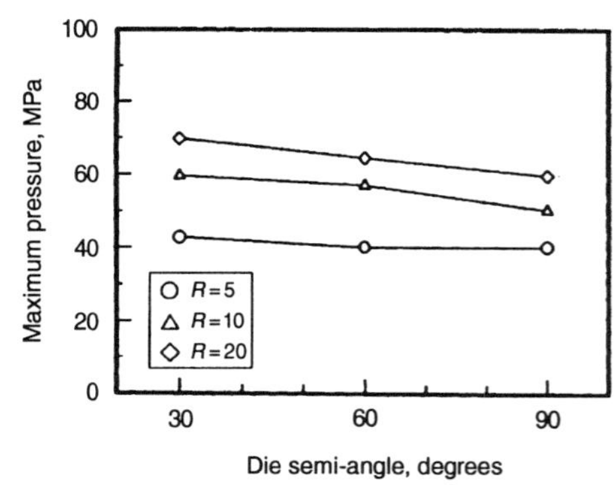

Fig. 11 Influence of die semiangle and reduction ratio on extrusion pressure (extrusion of ZrO_2 powder; binder is hydroxypropyl methylcellulose and water. Source: Ref 28

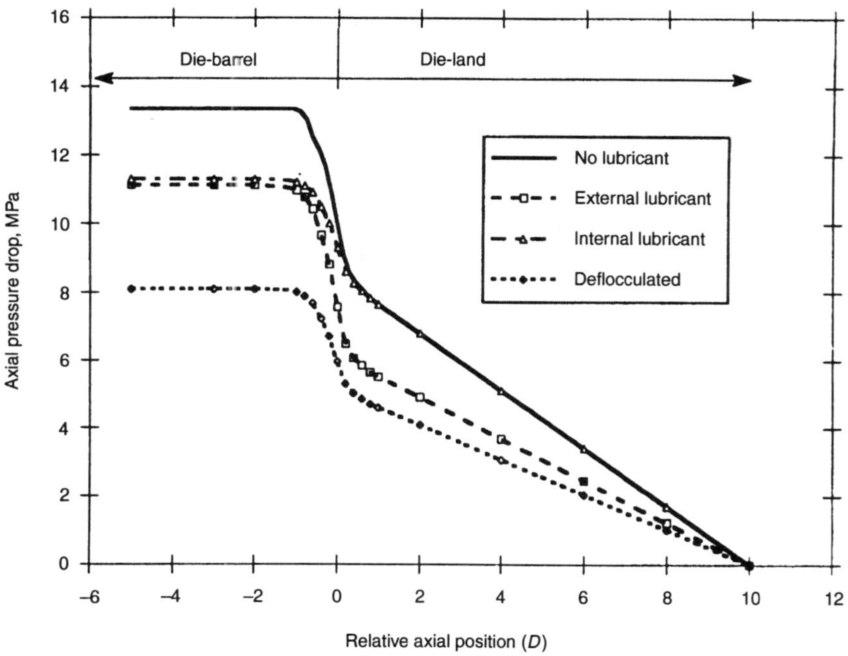

Fig. 12 Influence of lubrication on extrusion pressure. Source: Ref 1

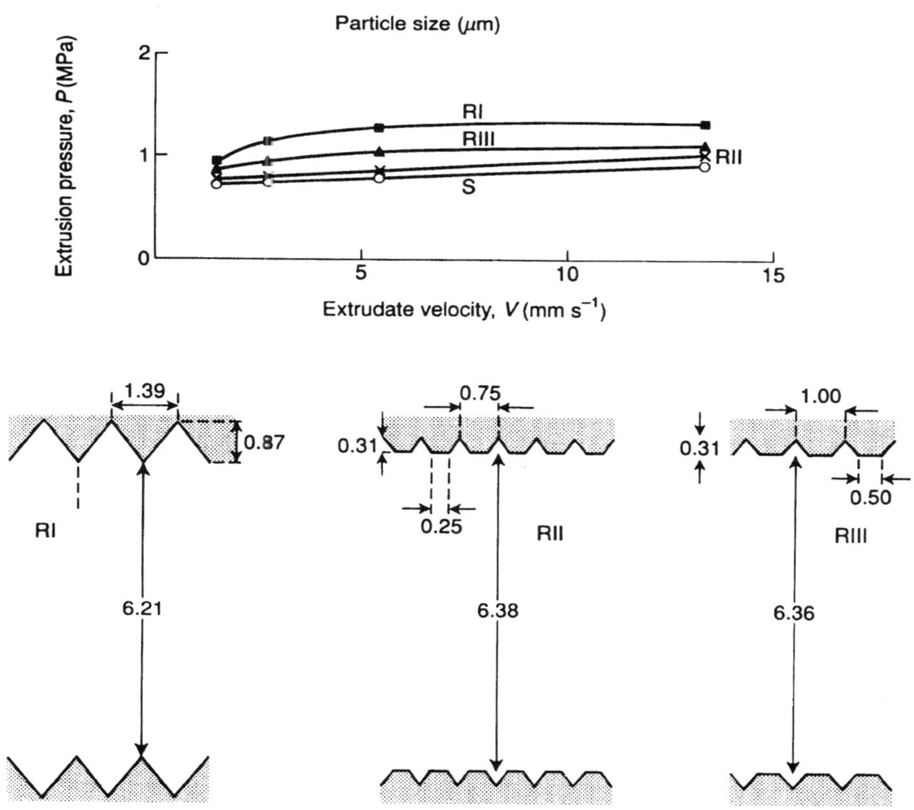

Fig. 13 Influence of die land surface roughness on extrusion pressure. Top, extrusion pressure versus extrudate velocity for a variety of die land surface roughness. Bottom, roughness profiles (all dimensions in millimeters; s, smooth die had an internal diameter of 6.38 mm). (Extrusion of Al_2O_3 powder; binder is clay, starch, and water; D_o = 25.4 mm, D = 6.35 mm, L = 25.4 mm. Source: Ref 6

and should be avoided (Ref 1–6). For industrial extrusions, die entrance angles are usually less than 25° and the *(L/D)* ratios range from 2-to-1 to 4-to-1 (Ref 1–6). The tooling must have a smooth finish and should be inspected for wear, as extrusion pressure will increase with increasing surface roughness of the die (Fig 13). Finally, uniform cross sections are an important aspect of tool and part design. Nonuniform cross sections can cause flow-rate differentials that will affect the flow characteristics of the part during extrusion, as well as causing variations in the final product (e.g., flaws, density gradients, etc.). Therefore, parts with sharp edges should be avoided, and for tube production the spider should be designed with a "bullet point" at the die entrance (Ref 1–6).

Extrusion Flow Defect

There are three principal types of flow defects that can occur during extrusion (Ref 1, 5, 6, 30): lamination, surface fracture, and phase separation. Lamination-type flow defects occur when two adjacent flow streams do not properly join together. This will be the source of a structural flaw in the final product. These defects occur during the forming of tubes or multihole extrusions and are caused by flow of the feedstock around the supports of the spiders. Typically, the spider is supported in the center of the die by three prongs at 120° to each other. The feedstock must squeeze around these prongs and reunite on the other side to form, for example, a continuous hollow cylinder prior to exiting the die. Lamination defects can also form in the barrel of a screw extruder. At the point where the feedstock leaves the screw, the trailing face of one twist comes into contact with the leading face of the preceding twist. It is here where lamination defects can occur, mitigated by the reduction in the cross section of the die after streams are brought together.

Surface fractures (or "feathering") are imperfections or tears that form at the surface of the extruded part (shown in Fig. 14). They arise from drag forces that build up parallel to the surface of the extruded part. Surface fractures tend to occur when the die land has sharp corners (i.e., for dies with square or triangular cross sections) and at high extrusion velocities. The advent of surface fractures can be mitigated by (Ref 6):

- Employing a die that consists of a large die land length to diameter *(L/D)* ratio, resulting in a high pressure drop in the die land, suppressing the effects that occur at the junction of the die entry and land
- Utilizing a tapered die
- Using lubricants in the feedstock or by designing the die so that lubricants can be injected into the die land during extrusion (this may also be accomplished by warming the die-wall surface)
- Formulating the feedstock in such a manner that the strength of the feedstock is significantly greater than the initial wall shear stress

Phase separation or seepage is another common problem that occurs during extrusion. This situation occurs when the phases move at significantly different velocities as a result of the pressure gradient. This condition arises when the permeability through the packed powder particles is high and the binder viscosity is low. As a result, the feedstock becomes dry and the pressure needed for the extrusion of the feedstock rises (Fig. 15). Evidence of phase separation in the feedstock include:

- Accumulation of liquid or binder at the die exit, or wet and dry patches on the extruded part surface
- High extrusion pressure at low extrusion rates rather than at high extrusion rates (i.e., the slower the extrusion speed, the more time available for phase separation)
- Extrusion pressure increases rather than decreases as the feedstock flows through the die.

Phase migration is affected by both the powder properties and binder rheology. The migration of the binder is a function of the interparticle void space. Therefore, employing fine powders (to a first approximation the interparticle void space is equivalent to the powder diameter, assuming that the powders do not agglomerate), or powder mixtures that pack efficiently, will reduce the interparticle void space and decrease the propensity for phase separation during extrusion (Ref 6, 14, 15). Selection of a binder system with the appropriate rheological characteristics will also help mitigate phase separation. Thus, a feedstock with a high initial yield stress (τ_0, in Eq 2) will lessen phase separation.

Postextrusion Processing

Cutting of the extruded part to the desired length can be accomplished by fixing the cutting device at the die exit. For continuous and large throughput extrusions, the ratio of the cutting speed to the extruded part speed will be a significant factor. As the ratio of speeds increase, the angle between the cutter and extruded part will approach 90° (Ref 6). It is important that the cutting operation not cause any surface tears or close any small internal holes found in complex parts, such as the holes in catalysts supports.

Binder Removal

The binder needs to be removed from the part prior to final consolidation. For solvent-based binders, removal occurs in two steps: (1) drying to remove the solvent and (2) thermal treatment to remove the organic component of the binder. Drying occurs by solvent evaporation at the surface. Solvent dispersed in the interior of the part must be able to diffuse to the surface to be removed. Solvent evaporation at the surface and subsequent diffusion through the part are accelerated by temperature. Shrinkage will occur during the drying process; hence, care must be taken to ensure that the solvent has enough time to diffuse through and then evaporate, as differential drying rates will result in differential shrinkage rates. Differential shrinkage rates can generate tensile stresses in the part, which may be relieved by the generation of a crack in the extruded part.

Most solid-phase organic binders used in binder-assisted extrusion can be burned off in an appropriate environment (e.g., inert, oxidizing, or reducing atmosphere, depending on the chemistry of the binder) at a temperature between 300 and 1000 °C. The burnoff must be slow to allow the carbonaceous gases generated by the decomposition of the organic species to escape through the interconnected pores. Too rapid of a burnoff rate can crack or fracture the part. If the organic binders used for solvent-based extrusion are not burnt off properly, they will decompose, leaving residual carbon and other contaminants trapped within the part, which will subsequently degrade the performance over time. The exact thermal treatment (i.e., its temperature/time profile and environment) will depend on the actual binder used in the feedstock. Plastic binders for hot extrusion are removed via thermal treatments. The techniques used in PIM can also be utilized for plastic binder removal in the hot-extrusion process.

Consolidation

Consolidation of the powders, of course, occurs through sintering. The time/temperature profile and sintering environment depend on the

Table 5 Examples of extruded ceramics

Product	Ceramic
Furnace tubes	Alumina, mullite, silicon carbide, zirconia
Insulators	Alumina, beryllia, steatite
Tubular capacitors	Barium titanate
Catalyst supports	Cordierite, alumina, silica, aluminosilicates
Resistance heaters	Barium titanate
Electronic substrates	Alumina, cordierite
Magnets	Ferrites
Heat-exchanger tubes	Silicon carbide, mullite

After Ref 5

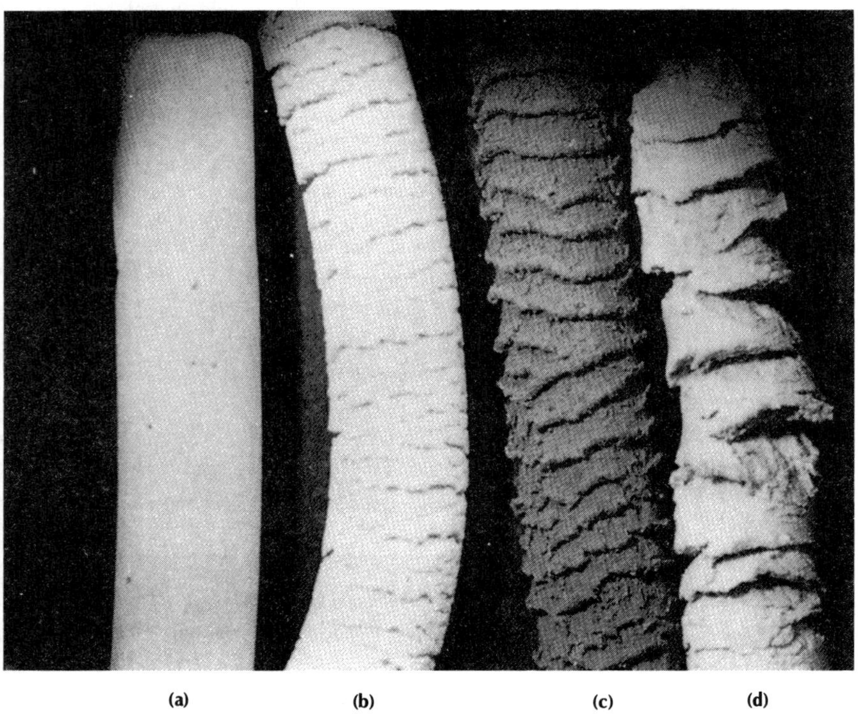

Fig. 14 Examples of surface flaws in extruded alumina (feedstock in wt%): 75% Al_2O_3 powder, 3% bentonite clay, 17% glucose solution; 5% water; D_0 = 25.4 mm, D = 3.2 mm, L = 9.5 mm). (a) No flaws (v = 0.23 mm/s). Surface flaws: (b) v = 1.2 mm/s, (c) v = 2.3 mm/s, and (d) v = 6.0 mm/s. Source: Ref 6

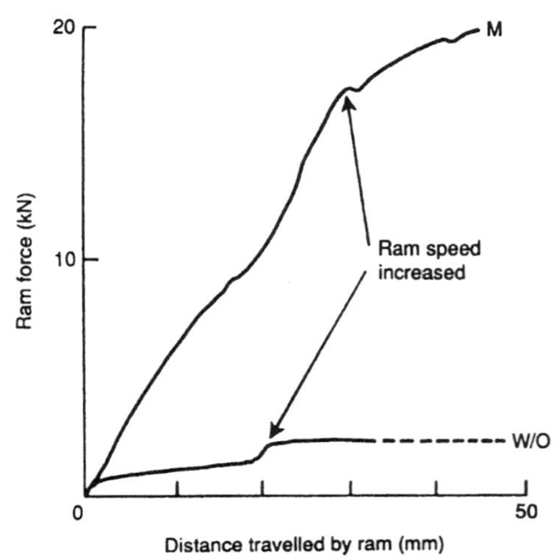

Fig. 15 Effect of binder powder phase separation on extrusion pressure. M, phase migration or separation; W/O, no phase migration. Source: Ref 14

materials being consolidated. For some materials, sintering may be an extension of the binder-removal operations, provided the environment for organic-binder removal does not interfere with the sintering process for the material in question.

Examples of Binder-Assisted Extrusion

Ceramics

Binder-assisted extrusion is most commonly used to form ceramic parts (Ref 1–5). Room-temperature extrusion, utilizing water-based solvent binder systems, has been used to produce a variety of ceramic parts, a selection of which are listed in Table 5. Hot extrusion is commonly used to produce advanced structural ceramics (e.g., Si_3N_4), or where the powders tend to be water sensitive or are extremely small (<1 μm) in size (Ref 7).

A typical solvent-based binder formulation for extruding alumina (in vol%) is 46% Al_2O_3 powder, 4% ball clay, 2% methyl cellulose, and 48% water (Ref 1). However, clays can cause contamination problems for nonoxide ceramics, such as silicon carbide (Ref 1). For these materials, a high-molecular-weight organic polymer, such as cellulose ether, can be substituted for the clay. The polymer is dissolved in a liquid phase and yields a solution with a high viscosity at low stresses and a low viscosity at high stresses. A typical feedstock formulation for room-temperature extrusion of silicon carbide (in vol%) is 50% SiC powder, 6% hydroxyethyl cellulose, 42% water, and 2% polyethylene glycol (Ref 1).

High-Temperature Superconducting Wires

Ceramic superconductor (e.g., $YBa_2Cu_3O_{7-x}$) wires have been fabricated via binder-assisted extrusion (Ref 8, 9, 18, 19, 31, 32). Both solvent-based and polymer binders have been used, and extrusions have been made at room and elevated temperatures. For solvent-based binder systems, care must be taken to select a solvent that does not contaminate the ceramic superconducting powders. A binder system that has been used in the extrusion of $YBa_2Cu_3O_{7-x}$ powders at room temperatures is (Ref 19): ethyl cellulose as the organic binder, glycerol as the solvent, and stearic acid as a lubricant. Other solvents that are compatible with $YBa_2Cu_3O_{7-x}$ powders include methylethylketone, methanol, and xylene (Ref 32). Polymers, such as polyethylene, polystyrene, and waxes, have been used as binders for hot extrusion of the $YBa_2Cu_3O_{7-x}$ powders (Ref 9, 18, 19). Wires with diameters of between 0.2 and 20 mm, with lengths well over 15 m, have been produced by binder-assisted extrusion. Because the extruded wire is flexible prior to binder removal, coils can be made by wrapping the wire around a mandrel.

With $YBa_2Cu_3O_{7-x}$ powders that possess a platelike morphology, the apparent viscosity of the feedstock can significantly influence the critical current density for superconductivity (Ref 19). This is shown in Fig. 16, where the platelike particles align unidirectionally during

Table 6 Comparison of powder forming techniques

Forming method	Shape of product	Advantages of method	Potential difficulties and disadvantages
Die pressing	Two-dimensional; flat objects	Economics; readily automated; products can be directly sintered	Limited sizes and shapes; potential density gradients
Doctor blade	Thin sheets	High productivity; secondary operations can be used to produce laminates and multilayered substrates	Binder removal may be difficult
Slip casting	Thin-walled products; irregular shapes	Utilizes simple equipment; can create relatively complex shapes	Distortions occur during castings; drying of slurry and slurry control can be difficult
Binder-assisted extrusion	Long objects with regular cross sections	Continuous production possible; no limitation on length of product	Objects with large cross sections require large equipment; binder removal can be difficult
Injection molding	Complex shapes	Mass produce, complex parts; readily automated; dimensional control; good surface finish	Economics (mold costs are high); binder removal can be difficult

Adapted from Ref 39

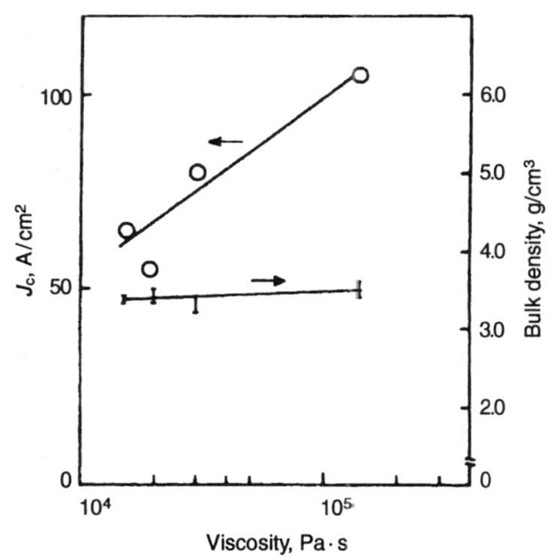

Fig. 16 Influence of feedstock viscosity on the critical current density (J_c) for and bulk density for binder-assisted extruded superconducting wires. Source: Ref 19

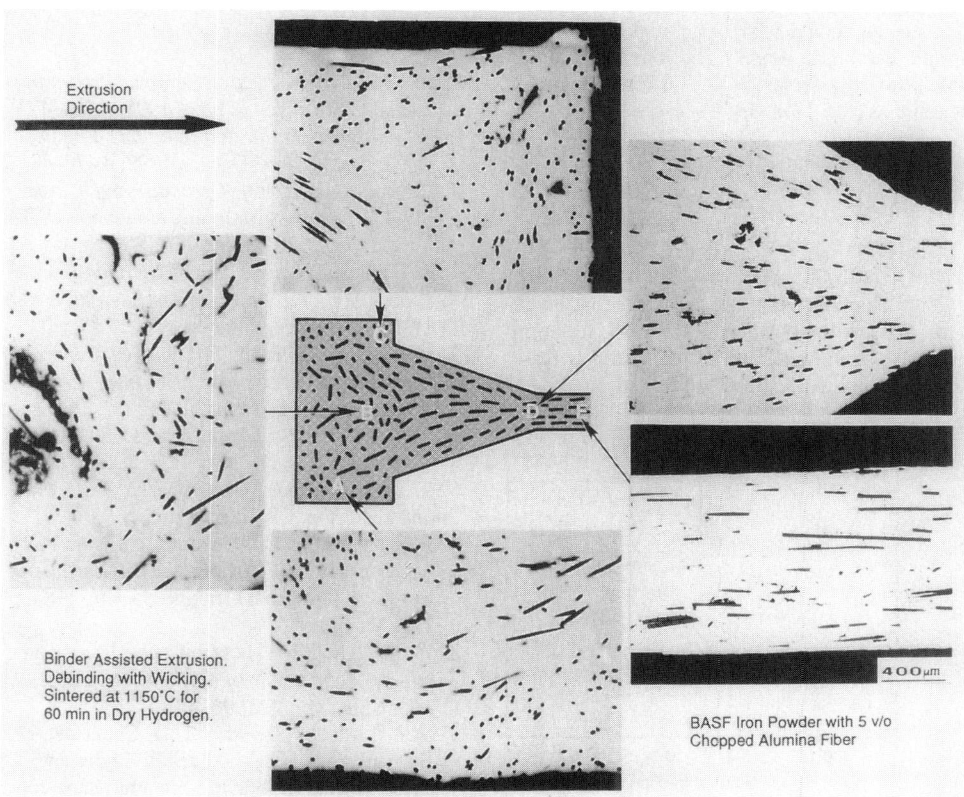

Fig. 17 Fiber alignment during binder-assisted extrusion of short-fiber-reinforced composites. Courtesy of A. Bose

the extrusion process as a result of the high apparent viscosity of the mixture.

Metals

Metals, for example, stainless steels (Ref 16, 28), high-speed tool steels (Ref 33, 34), and tungsten alloys (Ref 10), have been produced using binder-assisted extrusion. Again, both room-temperature and hot extrusions have been made. Acrawax C has been used as a plastic binder for the hot extrusion of a high-speed tool steel. A water-soluble hydroxypropyl methylcellulose has been employed as the binder system for the room-temperature extrusion of stainless steel powders. A solvent-based binder has been employed in the extrusion of the tungsten powders. The hot-extrusion technique has also been employed to fabricate 1.5 mm diam wires of a two phase Ni_3Al and NiAl (Ni-36.8 at.% Al) intermetallic alloy (Ref 21).

Composites

As mentioned previously, when particles possess a platelike morphology, they will tend to align under certain extrusion conditions. The same is true for chopped fibers mixed with the powder (Ref 11–13, 20, 35–38). Fibers tend to align parallel to the flow direction when extruded from a converging (i.e., constricting) nozzle, as shown in Fig. 17. When the flow occurs in a diverging nozzle, the fibers tend to align perpendicular to the flow direction. However, for alignment like this to occur, the maximum powder particle size cannot be greater than the edge-to-edge separation between the fibers. The relationship between fiber diameter, fiber volume fraction, and powders size can be derived from a simple geometric model. The maximum allowable powder particle size (λ) for fiber alignment is given as (Ref 13):

$$\lambda = \left(\frac{\pi D^2}{2\sqrt{3}\, v_f}\right) - D \qquad \text{(Eq 10)}$$

where D is the fiber diameter and v_f is the fiber volume fraction. Analysis of this equation reveals that smaller powders are required for alignment as the volume fraction of fibers increase. This behavior is illustrated in Fig. 18.

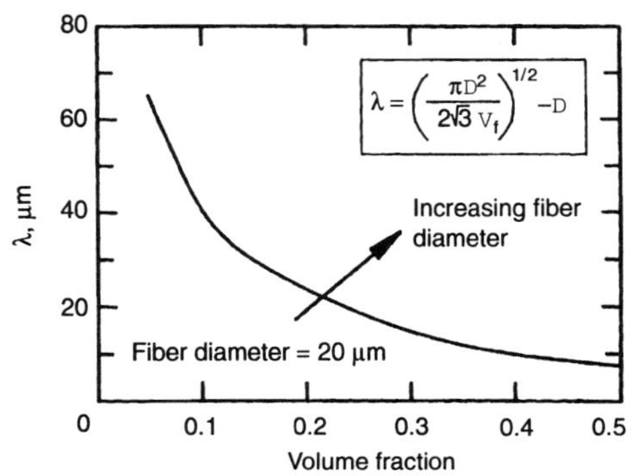

Fig. 18 Maximum powder particle size (λ) for fiber alignment in a powder matrix. Source: Ref 13

Conclusions

Binder-assisted extrusion is a commercially used powder forming technique. In this technique, powders are mixed with a binder to produce a feedstock with rheological characteristics that facilitate extrusion. The process consists of four basic operations, each step is critical for producing a defect free product. The steps include (1) feedstock formulation and preparation, (2) extrusion for shaping, (3) binder removal, and (4) sintering. Feedstock preparation involves mixing the powders with the binder and is a critical step in producing homogeneous mixtures. Both the characteristics of the powders and the binder influence the rheology of the feedstock. The feedstock is then extruded through a die to form a part with a particular cross section. Care must be taken to avoid flow defects during extrusion. Postextrusion processing includes binder removal and consolidation. At this stage, it is important to completely remove the binder so residual contaminants are not left behind that can degrade the physical and mechanical properties of the final product. Finally, binder-removal cycles should be designed to prevent cracking.

Binder-assisted extrusion is used to produce a variety of industrial parts. These parts may be ceramic or metallic in nature. This technique is amenable to producing long objects such as tubes, bricks, and honeycombed structures, as long as the parts have regularly shaped cross sections. Table 6 compares and contrasts the various powder forming approaches.

REFERENCES

1. J.S. Reed, *Principles of Ceramic Processing*, 2nd ed., John Wiley & Sons, 1995, p 450–473
2. D.W. Richerson, *Modern Ceramic Engineering*, 2nd ed., Marcel Decker, 1992, p 478–488
3. I. Ruppel, Extrusion, *Ceramics and Glasses*, Vol 4, *Engineered Materials Handbook*, ASM International, 1991, p 166–172
4. T.A. Ring, *Fundamentals of Ceramic Powder Processing and Synthesis*, Academic Press, 1996, p 643–653
5. M.A. Janney, Plastic Forming of Ceramics: Extrusion and Injection Molding, *Ceramic Processing*, R.A. Trepstra, P.P.A.C. Pex and A.H. deVries, Ed., Chapman and Hall, 1995, P 174–194
6. J.J. Benbow and J. Bridgewater, *Paste Flow and Extrusion*, Clarendon Press, 1993
7. B.C. Mutsuddy, Hot Ceramic Extrusion Process, *Forming of Ceramics*, Vol 9, *Advances in Ceramics*, American Ceramic Society, 1984, p 212–219
8. J.K. Wright, R.M. Thomson, and J.R.G. Evans, On the Fabrication of Ceramic Windings, *J. Mater. Sci.*, Vol 25, 1990, p 149–156
9. M.A. Lusk, J.A. Lund, A.C.D. Chaklader, M. Burbank, A.A. Fife, S. Lee, B. Taylor, and J. Vrba, The Fabrication of a Ceramic Superconducting Wire, *Super. Sci. Technol.*, Vol 1, 1988, p 137–140
10. R.B. Neal, B.S. Lankford, D.E. White, A.L. DeMint, and W.G. Northcutt, Jr., High-Density Tungsten Product by a Powder Metallurgy Extrusion Process, *Advances in Powder Metallurgy: 1989*, Vol 2, T.G. Gasbarre and W.F. Jandeska, Jr., Ed., MPIF/AMPI, 1989, p 371–383
11. S. Blackburn and H. Bohm, Silicon-Carbide Fiber Reinforced Alumina Extrusion, *J. Mater. Res.*, Vol 10, 1995, p 2481–2487
12. Y. Goto and A. Tsuge, Mechanical Properties of Unidirectional Oriented SiC-Whisker-Reinforced Si_3N_4 Fabricated by Extrusion and Hot-Pressing, *J. Am. Ceram. Soc.*, Vol 76, 1993, p 1420–1424
13. D.E. Alman, N.S. Stoloff, A. Bose, and R.M. German, Structure and Properties of Aligned Short Fiber-Reinforced Intermetallic Matrix Composites, *J. Mater. Sci.*, Vol 30, 1995, p 5251–5258
14. J.J. Benbow, S.W. Jazayeri, and J. Bridgewater, Ceramic Extrusion Mechanisms: The Effect of Paste Formulation and Liquid Phase Rheology on Die-Flow Resistance, *Proc. First Inter. Conf. on Ceramic Powder Processing Science*, G.L. Messing, E.R. Fuller, Jr., and H. Hausner, Ed., The American Ceramic Society, 1988, p 624–634
15. J.J. Benbow, E.W. Oxley, and J. Bridgewater, The Extrusion Mechanisms of Pastes—Influence of Paste Formulations on Extrusion Parameters, *Chem. Eng. Sci.*, Vol 42, 1987, p 2151–2162
16. Z. Chen, J.-X. Xie, T. Murakami, and K. Ikeda, Effects of Particle Size Distribution on Extrusion Characteristics of Stainless Steel Powders Containing a Binder, *J. Jpn. Soc. Technol. Plast.*, Vol 37, 1996, p 653–658
17. G.Y. Onoda, Jr., The Rheology of Organic Binder Solutions, *Ceramics before Firing*, G.Y. Onoda, Jr. and L.L. Hench, Ed., John Wiley & Sons, 1978, p 235–251
18. S.R. Su, M. O'Connor, M. Levison, and P.G. Rossoni, Microstructural Effects on Superconducting Properties of Sintered $YBa_2Cu_3O_{7-x}$ Wires, *Physica C*, Vol 178, 1991, p 81–88
19. S. Suzuki, S. Asada, and M. Takahashi, Effect of Extrusion on the Critical Current Density of Superconducting Wires, *J. Am. Ceram. Soc.*, Vol 74, 1991, p 714–717
20. D.E. Alman and N.S. Stoloff, Powder Fabrication of Monolithic and Composite NiAl, *Int. J. Powder Metall.*, Vol 27, 1991, p 29–41
21. D.E. Alman, Reaction Synthesis of Ni-36.8 at.% Al, *J. Mater. Sci. Lett.*, Vol 13, 1994, p 483–486.
22. R.M. German, *Powder Injection Molding*, MPIF/AMPI, 1990, p 99–124
23. J.R. Parks and M.J. Hill, Design of Extrusion Auger and the Characteristic Equation of Ceramic Extrusion Machines, *J. Am. Ceram. Soc.*, Vol 42, 1959, p 2–6
24. A.S. Burbridge and J. Bridgewater, The Single Screw Extrusion of Pastes, *Chem. Eng. Sci.*, Vol 50, 1995, p 2531–2543
25. J.J. Benbow, The Dependence of Output Rate on Die Shape During Catalysts Extrusion, *Chem. Eng. Sci.*, Vol 26, 1971, p 1467–1473

26. J.J. Benbow, T.A. Lawson, E.W. Oxley, and J. Bridgewater, Prediction of Paste Extrusion, *Ceram. Bull.*, Vol 68, 1989, p 1821–1824
27. J. Zheng, W.B. Carlson, and J.S. Reed, Flow Mechanics on Extrusion through a Square Entry Die, *J. Am. Ceram. Soc.*, Vol 75, 1992, p 3011–3016
28. Z. Chen, J.-X. Xie, T. Murakami, and K. Ikeda, Extrusion Characteristics of Zirconia and Stainless Steel Powder-Binder Compounds, *J. Jpn. Soc. Technol. Plast.*, Vol 37, 1996, p 546–551
29. J. Gotz and H. Buggisch, NMR Imaging of Pastes in a Ram Extruder, *J. Non-Newtownian Fluid Mech.*, Vol 49, 1993, p 251–275
30. G.C. Robinson, Extrusion Defects, *Ceramics before Firing*, G.Y. Onoda, Jr. and L.L. Hench, Ed., John Wiley & Sons, 1978, p 391–407
31. D. Ponnusamy and K. Ravi-Chandar, Extrusion of Superconducting Wires of $Yba_2Cu_3O_{7-x}$, *J. Mater. Res.*, Vol 8, 1993, p 268–273
32. R.B. Poeppel, M.T. Lanagan, U. Balachandran, S.E. Dorris, J.P. Singh, and K.C. Goretta, High-Temperature Superconductors, *Ceramics and Glasses*, Vol 4, *Engineered Materials Handbook*, ASM International, 1991, p 1156–1160
33. H. Nakamura and Y. Mochida, Production and Characteristics of Sintered Hollow High Speed Tool Steel Blanks by Binder Assisted Powder Assisted Extrusion Process, *J. Jpn. Soc. Powder and Powder Metall.*, Vol 38, 1991, p 689–693
34. T. Takada and K. Minagawa, Extrusion Forming of SKH51 High Speed Steel Fine Powder-Wax Mixture, *J. Jpn. Soc. Powder and Powder Metall.*, Vol 35, 1998, p 636–640
35. R.M. German and A. Bose, Fabrication of Intermetallic Matrix Composites, *Mater. Sci. Eng.*, Vol A107, 1989, p 107–116
36. A. Bose, *Advances in Particulate Materials*, Butterworth-Heinemann, 1995, p 391–399
37. R. Lenk and J. Adler, SiC Platelet Orientation in a Liquid Phase Sintered SiC Composite Formed by Thermoplastic Forming Techniques, *J. Europ. Ceram. Soc.*, Vol 17, 1997, p 197–202
38. S. Blackburn and H. Bohm, The Influence of Powder Packing on the Rheology of Fibre-Loaded Pastes, *J. Mater. Sci.*, Vol 29, 1994, p 4157–4166
39. S. Somiya, *Advanced Technical Ceramics*, Academic Press, 1989, p 76–77

Warm Compaction

Francis G. Hanejko, Hoeganaes Corporation

THE FERROUS P/M INDUSTRY continues to grow because of developments in raw materials and part production processes enabling the manufacture of components with greater complexity and higher levels of performance. Advances in raw materials include higher compressibility iron powders, molybdenum prealloyed steels, diffusion-alloyed powders, and the use of binder-treated iron powders. These new powders and premix technologies offer P/M users greater flexibility in mechanical properties at traditional part densities, typically less than 7.1 g/cm^3. However, end users of P/M parts are demanding still higher levels of mechanical properties available solely through higher part densities. Traditional methods used to achieve higher densities include the use of copper infiltration, double-pressing/double-sintering (DP/DS), and powder forging. Because these techniques involve the use of secondary processing, significant cost penalties are encountered, often negating the potential cost savings realized by powder metallurgy. Warm-compaction process is a technique to achieve DP/DS densities and mechanical properties utilizing a single compaction process. The process incorporates the use of heated powder and heated tooling in standard compacting presses to achieve higher green and sintered densities.

Temperatures above room temperature and below the hot-forging range are used extensively in wrought steels to lower forging loads and minimize distortion. Work at MeriSinter AG in the mid-1980s found certain advantages to be present when ferrous, bulk lubricated powders were heated in the vicinity of 100 °C. It was of particular interest that compressibility improved with heating as compared to the some powders in an unheated condition (Ref 1). The practical application of warm compaction of powders was realized in 1994 with the introduction of Hoeganaes Inc. Ancordense and Densemix powders (Ref 2). Figure 1 schematically illustrates the warm-compaction process. The powder and die temperatures used vary from 75 to 150 °C, with every 100 °C rise in compaction temperature resulting in a 0.08 g/cm^3 increase in green density (Ref 3). Experimental work at Höganäs AB showed a 30% decrease in the compressive yield strength of iron powder when the powder was heated to 150 °C (Fig. 2) (Ref 4).

The maximum green density achieved via P/M techniques depends on the amount and type of premix additives used with iron powder. To calculate the maximum green and sintered densities attainable, it is useful to review the concept of pore-free density (PFD). Pore-free density is defined as the density of a green compact in which all the interparticle porosity is eliminated (Ref 2). This PFD can be calculated from the specific density and percentage of each additive in the premix. The calculation for PFD is given as:

$$\text{PFD} = 1/[\Sigma\ (\%\ \text{element in premix/specific gravity of additive})] \quad \text{(Eq 1)}$$

in which the percentage of each element is the weight percentage used and is expressed as a decimal.

Once the PFD is calculated, a practical upper limit of green density is 98% of the calculated value. The specific density of several common ferrous powders and typical premix additives are listed in Table 1. Additions of materials with specific densities higher than the base iron increase the PFD, while additions of materials with lower specific densities (lubricants and graphite) will lower the PFD.

Because of the reduced compressive yield strength at 150 °C, attaining 98% of the PFD is achieved with lower compaction pressures. In addition, compaction at 150 °C reduces the amount of lubricant between the particles while simultaneously increasing the amount of added lubricant reaching the die-part interface (Ref 5). This redistribution of the lubricant not only increases the green density, but reduces the ejection forces by 25 to 33% (Ref 2, 5). This enhanced lubricity implies that lower amounts of lubricant are necessary (typically 0.60% lubricant for warm compaction compared to 0.75% in conventional compaction), which again contributes to the attainment of higher green and sintered densities.

Advantages of the warm-compaction process include higher green densities, enhanced green strength, improved mechanical and soft magnetic properties, and greater uniformity of den-

Table 1 Specific density as measured by pycnometry

Material	Specific density, g/cm^3
Ancorsteel 1000B	7.841
Ancorsteel 4600V	7.844
Distaloy 4800A	7.896
Atomized copper	8.047
Inco nickel powder 123	8.846
Graphite	2.295
Lubricants	0.90–1.15

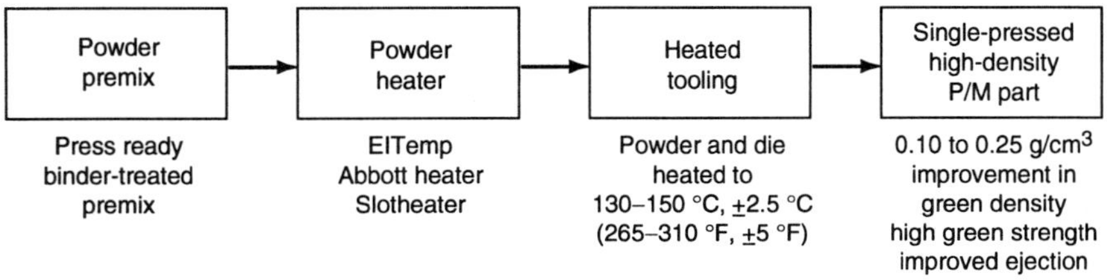

Fig. 1 Warm-compaction process

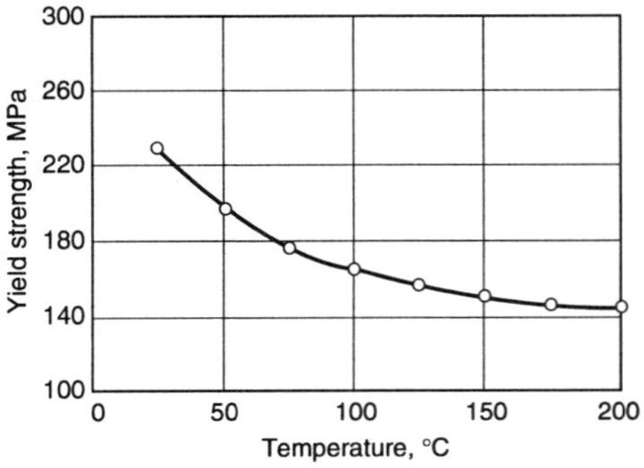

Fig. 2 Effect of temperature on the yield strength of pure iron powder. Source: Ref 4

sity throughout the as-sintered part. The balance of this article details the process and the improved properties resulting from warm compaction.

Effects on Green and Sintered Properties

Warm compaction results in a 0.10 to 0.25 g/cm^3 increase in the green and sintered densities of P/M parts (Ref 2). Figure 3 shows the improved green and sintered densities achieved with a diffusion-alloyed powder premixed with 0.6% graphite. At lower compacting pressures, the beneficial effect of warm compaction is greater than the improvement observed at higher compaction pressures. Figure 4 summarizes the transverse rupture strength results of the diffusion-bonded material compacted by both conventional and warm-compaction techniques under compaction conditions of 410 to 690 MPa.

Table 2 summarizes the as-sintered mechanical properties of various warm-compacted premix compositions (Ref 6). This processing is applicable to all iron and low-alloy powder compositions. The magnitude of the increase in sintered density depends on the material system and subsequent part processing. Premixes containing copper additions exhibit growth during the sintering process; this growth negates the beneficial effects of the warm-compaction processing. Consequently, copper-containing premixes are not considered ideal candidates for warm compaction (Ref 7).

Rotating bending fatigue testing was performed on a variety of warm-compacted materials in both the as-sintered and heat-treated conditions; Table 3 summarizes the available data (Ref 8, 9). As expected, increasing the density increased the fatigue endurance limit; however, higher-temperature sintering did not consistently improve the fatigue endurance limit. Reviewing Table 3, it is observed that no generalized correlation exists between the fatigue endurance limit and the tensile strength of P/M materials. It is recommended that designers use available data when specifying the fatigue endurance limit of P/M components.

Double pressing of conventionally compacted parts results in improved part densities and mechanical properties. In experimental work performed by Donaldson and others, warm-compacted P/M parts were presintered at 870 °C (1600 °F) and subsequently re-pressed at up to 690 MPa (50 tsi) at room temperature (Ref 10–12). Following re-pressing, the part was then sintered at either 1120 °C or 1260 °C, resulting in sintered densities ranging from 7.5 to 7.6 g/cm^3. Figures 5 and 6 present data on transverse rupture strength and impact energy from this study. These higher densities of the double-pressed/double-sintered (DP/DS) warm compacted materials produced an approximately 15% improvement in the transverse rupture strengths (Fig. 5), but more importantly resulted in a 50 to 80% improvement in the impact energy (Fig. 6) when compared to the 7.4 g/cm^3 density level (Ref 10). This study demonstrated the potential for significantly improved mechanical properties of P/M materials via DP/DS of a warm-compacted component. The resultant

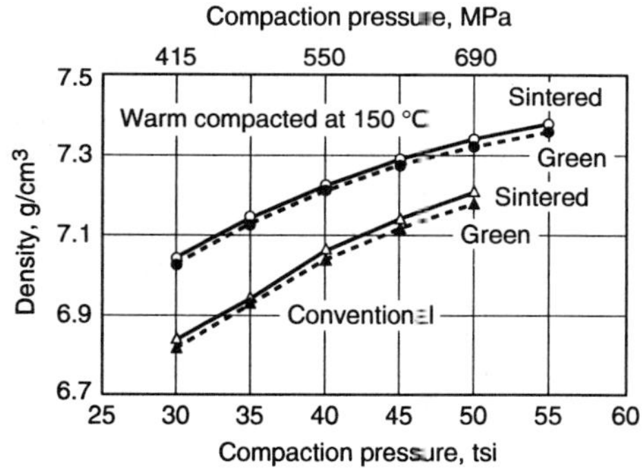

Fig. 3 Compressibility of diffusion-bonded 4% Ni, 1.5% Cu, and 0.5% Mo with 0.6% graphite and 0.6% lubricant

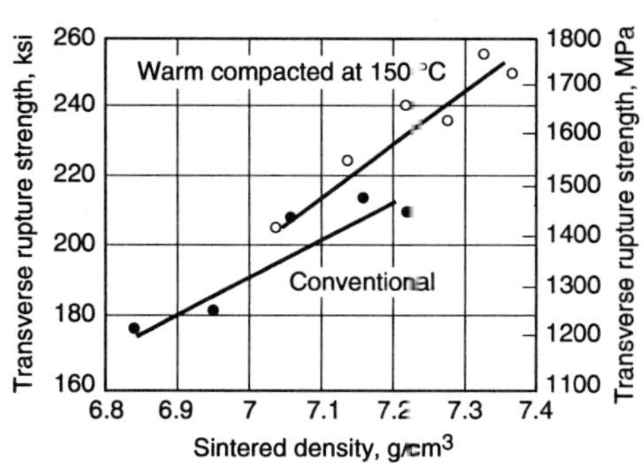

Fig. 4 Sintered transverse rupture strength of diffusion-bonded 4% Ni, 1.5% Cu, and 0.5% Mo with 0.6% premixed graphite

Table 2 As-sintered tensile properties of warm-compacted P/M materials, sintered at 1120 °C (2050 °F)

Composition(a)	Sintered density, g/cm^3	0.2% offset yield strength, MPa	0.2% offset yield strength, ksi	Tensile strength, MPa	Tensile strength, ksi	Elongation, %	Apparent hardness, HRB
FL-4405 with 0.6% graphite	7.37	273	40	471	69	3.5	77
FLN 2-4405 with 2% Ni and 0.6% graphite	7.44	444	65	628	92	2.8	87
FL-4205 with 0.6% graphite	7.24	417	61	506	74	1.7	81
Ancorsteel 150Mo (b) with 2% Ni and 0.6% graphite	7.40	533	78	718	105	1.3	93
FD-0405 with 0.6% graphite	7.25	425	62	117	800	2.6	97
Iron plus 0.45% phosphorus	7.39	267	39	422	62	25.2	67
FN 0250 with 0.6% graphite	7.37	267	39	452	66	3.52	79

(a) MPIF designations, based on MPIF Standard 35, 1997 edition. (b) Hoeganaes Corporation prealloyed powder with nominal 1.5% Mo

Table 3 Fatigue data of warm-compacted ferrous materials

Material	Sintering temperature, °C	Sintering temperature, °F	Heat treat	Density, g/cm^3	50% fatigue endurance limit, MPa	50% fatigue endurance limit, ksi	99% fatigue endurance limit, MPa	99% fatigue endurance limit, ksi	Tensile strength, MPa	Tensile strength, ksi
Fe-0.45 wt% P	1120	2050	No	7.23	207	30	185	26.8	365	53
				7.40	225	32.7	197	28.6	403	58
Fe-0.45 wt% P	1260	2300	No	7.33	216	31.4	210	29.7	403	59
				7.50	260	37.7	234	34	476	69
FC-0208	1120	2050	No	7.07	234	33.9	175	25.4	596	86
				7.17	243	35.3	193	28.0	621	90
FD-4805	1120	2050	No	7.19	230	33.3	181	26.3	710	103
				7.32	242	35.3	192	27.8	798	115
FD-4805	1260	2300	No	7.26	217	31.5	172	24.9	814	118
				7.37	227	32.9	185	26.9	925	134
FD-4805	1120	2050	Yes	7.20	399	57.9	317	46.0	249	181
				7.32	409	59.3	332	48.1	1327	193
A150HP, 2% Ni and 0.6 Gr	1120	2050	No	7.18	233	33.8	189	27.4	641	93
				7.34	262	38.0	241	35.0	693	101
A150HP, 2% Ni and 0.6 Gr	1260	2300	No	7.21	207	30.0	165	24.0	652	95
				7.37	256	37.1	201	29.2	710	103
FLN2-4405	1120	2050	No	7.31	253	36.7	222	32.2	632	91
FLN2-4405	1260	2300	No	7.35	247	35.8	219	31.7	672	98
FLN2-4405	1290	2350	No	7.20	239	34.6	227	32.9	621	90
A41AB	1290	2350	No	7.16	242	35.1	210	30.4	856	124
				7.27	270	39.2	234	34.0	917	133
A41AB	1290	2350	Yes	7.16	403	58.5	353	51.2	1211	176
				7.28	449	65.1	410	59.4	1349	196
FN0250	1120	2050	Yes	7.23	316	45.8	276	40.1	1193	176
FL-4405	1120	2050	Yes	7.17	330	47.9	283	41.1	1131	164
				7.30	336	48.8	279	40.4	1151	167
FD-0205	1120	2050	Yes	7.19	368	53.4	315	45.7	1192	173
				7.29	374	54.2	316	45.9	1303	189

mechanical properties of such parts are equivalent to the properties of ductile cast irons and machined carbon steel forgings.

Green Strength Enhancement. Warm-compaction processing provides improved green strength of the as-compacted component. This increase in the green strength results from the synergy of greater powder particle deformation with enhanced particle welding during compaction, plus the presence of the unique binder and lubricant utilized in the ANCORDENSE material (Ref 2, 5). The improved green strengths are realized at densities significantly below the pore-free density (Fig. 7). These data imply the potential of warm compaction being used in lower-density applications where the enhanced green strength reduces part breakage or part chipping of fragile features.

A consequence of the enhanced green strength via warm compaction is the ability to green machine the as-compacted part. This concept has been used in a commercial application of a P/M safety locker part (Ref 13). After compaction, the component was milled in the green condition, thus reducing the overall part cost. A machinability study utilizing a drilling test was conducted on a molybdenum prealloyed material premixed with 2% Ni, 0.5% graphite, and 0.6% lubricant (Ref 14). This study concluded that satisfactory surface finishes are achieved with machining conditions using high speeds and high feed rates. Additionally, modifications of the standard drill bit geometry from a standard 90° chisel bit to a 135 split-point drill bit enhanced the as-machined surface finish. Prior to establishing green-machining parameters, it is recommended that testing be conducted to examine the effects of tool-bit geometry, machining feed rate, and machining speed. Green machining of P/M parts in combination with sinter hardening offer the part designer greater flexibility in part design and material selection.

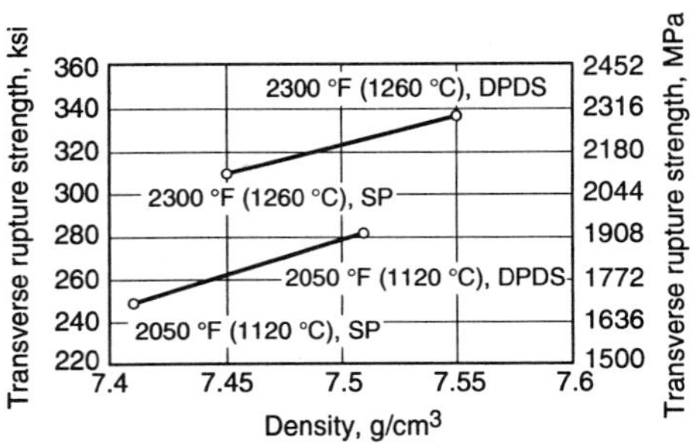

Fig. 5 Transverse rupture strength of diffusion-bonded 4% Ni, 1.5% Cu, and 0.5% Mo premixed with 0.3% graphite subsequently carburized and tempered

Table 4 Magnetic and mechanical properties of a warm-compacted iron-phosphorus steel versus AISI 1008

Property	Fe-0.45 wt % P	AISI 1008
Density, g/cm³	7.35	N/A
Sintering temperature, °C (°F)	1120 (2050)	N/A
0.2% yield strength, MPa (ksi)	285 (42)	285 (42)
Tensile strength, MPa (ksi)	405 (59)	383 (56)
Elongation, %	12	37
Maximum permeability	2700	1900
Coercive force, Oe	1.9	3.0
Saturation at 15 Oe, G	15,000	14,400

N/A, not applicable

Magnetic Applications

The use of warm compaction of P/M magnetic alloys effects higher sintered densities with corresponding higher saturation induction levels and higher permeabilities with no change in the coercive force (Ref 15). Sintered densities in excess of 7.4 g/cm^3 are possible using warm compaction of iron-phosphorus alloys. At this density level, the magnetic and mechanical performances of this family of materials are equal to the properties of a low-carbon steel forging. Table 4 summarizes the mechanical and magnetic property data for a 0.45% P/M phosphorus steel processed to 7.4 g/cm^3 and an AISI 1008 steel forging. From these data, the P/M material is a suitable replacement for the wrought steel.

Warm-compaction processing enables the introduction of a new class of P/M materials for use in alternating current (ac) magnetic applications (Ref 15–17). These materials utilize a high-strength polymer and warm-compaction processing to produce components that do not require sintering. The polymer acts to both electrically insulate the powder particles and provide strength without the need for sintering. As-compacted green densities in excess of 7.2 g/cm^3 are possible. Manufacturing flexibility can produce a variety of material options with unique magnetic performance. Applications for these materials include automotive ignition coils and stators for high-speed electric motors.

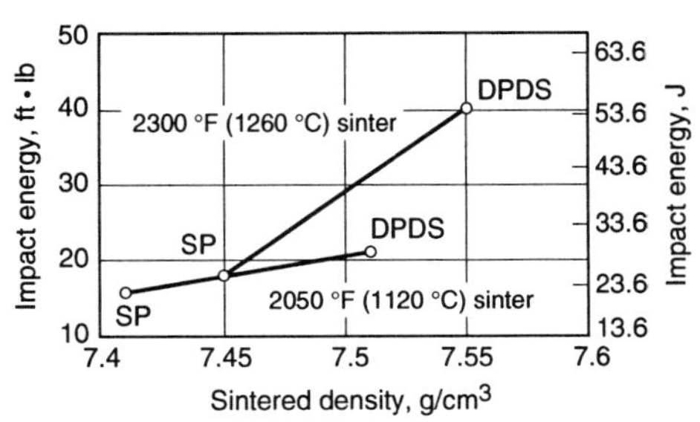

Fig. 6 Impact energy for diffusion-bonded 4% Ni, 1.5% Cu, and 0.5% Mo premixed with 0.3% graphite subsequently carburized and tempered

Table 5 summarizes these ac magnetic materials and their magnetic performance. These materials are ideally suited for applications with operating frequencies above 400 Hz. Optimizing the amount and type of insulation produces components that can operate at frequencies up to 50,000 Hz. The unique three-dimensional structure of these materials can be used to carry magnetic flux in any direction. The strength of these materials in the as-compacted condition is approximately 103 MPa (15 ksi) transverse rupture strength. Employing a 315 °C (600 °F) thermal treatment to the as-compacted part raises the transverse rupture strength to approximately 240 MPa (35 ksi).

Commercial Powder Heating and Delivery Systems

Successful utilization of the warm-compaction process necessitates that the powder, powder shuttle, and compaction tooling be heated to the proper temperature. Recommended temperature control of the heated powder and tooling is ±2.5 °C. It is imperative that the temperature of the powder not exceed 170 °C (320 °F); above this temperature the lubricant and binder degrade, resulting in diminished powder flow. Heating of the tooling is accomplished using cartridge heaters embedded in the stress ring of the die. Typically, eight to twelve 500 W cartridge heaters are required to heat the tooling to 150 °C (300 °F) in approximately 30 min. Heating of the powder shuttle is necessary to maintain the powder temperature during the transfer of the powder into the die cavity. Top-punch heating is recommended to eliminate the possibility of a tool binding between the top punch and core rod(s). Heating the core rod and lower punches is not necessary; where practical, incorporating a car-

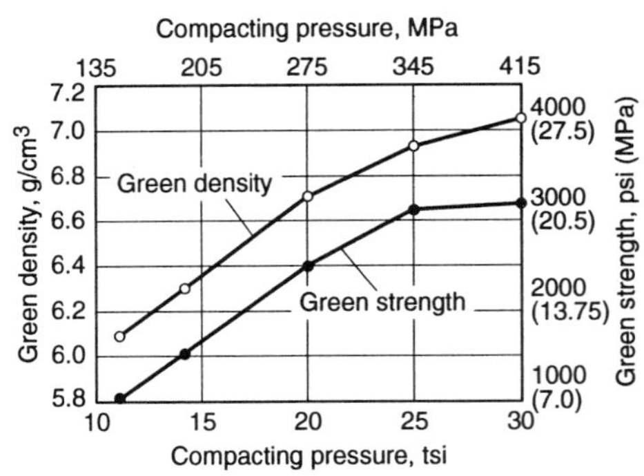

Fig. 7 Green density and green strength of warm-compacted FN0205 at low compacting pressures

Table 5 Magnetic performance of insulated iron powders

Material	Initial permeability	Maximum permeability	Coercive force, Oe	Induction at 40 Oe, G
Iron powder with 0.6% plastic	120	425	4.7	11,200
Iron powder with 0.75% plastic	100	400	4.7	10,900
Iron powder with oxide coating and 0.75% plastic	80	210	4.7	7,700

tridge heater in the core rod will provide greater temperature uniformity.

Currently, three commercial powder heating and delivery systems are available. Each system is capable of delivering heated powder at the proper temperature. Additionally, each has the capability of heating and controlling temperatures in the die, the punches, and the powder shuttle system. The three systems are:

- Cincinnati Inc. El Temp System
- Abbott Furnace Company Thermal Powder Processor
- Slotheater

The Cincinnati Incorporated El Temp system utilizes an auger to both heat and transport the powder from the powder feed hopper to the heated shuttle (Ref 18). The auger operates within a resistively heated shell; additionally, the auger is hollowed, allowing preheated air to provide for additional heating capability. The amount of powder heated is determined by the part weight and press speed. Production systems are available that heat up to a maximum of 9 kg/min (20 lb/min). A unique feature of the El Temp system is its direct interface with the Cincinnati computer operating system of the press, allowing for control of all press and heating functions from a single touch screen.

The Abbott Furnace Company Thermal Powder Processor, TPP 300 (patent pending) uses a low-pressure fluidizing air 35 kPa (5 psi) to heat the powder within a sealed reactor. Heating of the powder is accomplished in a stream of air that passes across resistively heated elements. As powder is withdrawn from the bed into the delivery system, additional powder is drawn into the reactor. This system uses a stand-alone programmable logic controller (PLC) controller to heat the powder, die, and powder shuttle. Units are available that can deliver powder up to 3.5 kg/min (8 lb/min) and 3.5 to 9 kg/min (8 to 20 lb/min). The TPP 300 is portable and can be adapted to any press. These units have no moving parts, thus minimizing maintenance.

The Slotheater uses the principle of direct contact of the powder with the heated surfaces of an oil-filled slotted heat exchanger (Ref 4). The powder flows via gravity from the press feeder hopper into the slot heater where it is heated and then flows into the powder delivery system. The temperature of the heated oil is controlled to a temperature approximately 4 °C (7 °F) hotter than the desired temperature of the powder. To achieve uniform temperature of the powder, the residence time of the powder in the heater must be at least 5 min. Commercial units are available that can deliver 3.5 kg/min (8 lb/min) of hot powder. However, the design is scalable to achieve up to 9 kg/min (20 lb/min).

Considerable attention has been given to the actual mechanism of heating the powder; however, attention must also be given to the powder shuttle system. Although no commercial systems exist, it is a relatively easy task to design and build a hot powder feed shoe. Heating of the feed shoe is accomplished by embedding cartridge heaters and a thermocouple in the aluminum feed shoe. Temperature control of the feed shoe is necessary to prevent any heat loss during the residence time of the powder in the shoe. Both a closed shoe and an open shoe have been successfully used. Unlike conventional powder shoes, the amount of powder in the feed shoe is critical. Excessive amounts of powder in the feed result in a long residence time within the feed shoe, possibly resulting in a temperature drop causing excessive part-to-part weight variations.

Tooling Design for Warm Compaction

The tooling design for warm compaction is essentially the same as for regular compaction with typical radial tooling clearance of 0.01 to 0.02 mm (0.0004 to 0.0008 in.). The choice of carbide inserts or tool steel inserts is not critical. Carbide inserts have proven to be successful; however, the designer is cautioned that additional interference fits are required to compensate for the differential thermal expansion of the carbide insert compared to the steel stress ring.

One word of caution in the design of tooling is the stress involved during the compaction to near-pore-free densities. As the density increases, the tooling loads increase rapidly. This increase in tooling pressure necessitates that thicker stress rings be used and the allowances made for the greater tool deflections. Shown in Fig. 8 is the green expansion as a function of the green density of powder compacts compacted using both conventional room-temperature compaction in addition to warm-compaction conditions. Note that the green expansion at equivalent density is lower for the warm-compacted material. The rationale for the lower green expansion for the warm compacted material is explained by the fact that lower compacting pressure was required to achieve this same density; thus the tooling load was decreased. However, as the green density increases to near-pore-free density, the green expansion increases dramatically. With this increased density, the tooling loads increase, resulting in greater expansion of the part.

This increased green expansion can cause microlaminations in the compacted part. These microlaminations are serious problems because they reduce the structural integrity of the sintered component. In multilevel parts, these microlaminations usually occur at the transition from one level to another. Incorporating top-punch hold-down during the ejection cycle often prevents these cracks from occurring. However, even top-punch hold-down is not sufficient to prevent microcracking if the density of the part exceeds 98% of the theoretical density.

Part Processing Considerations

Studies conducted by Hoeganaes, Presmet, QMP, and others demonstrated that the part-to-part variability of the warm-compaction process is equivalent to conventional compaction (Ref 10, 19, 20). Equal press speeds were achieved with the warm-compaction process compared to conventional compaction. The limiting feature in part production is the rated capacity of the powder heating system and the part weight. Although conventional compacting presses are used, attention must be given to prevent the heat generated in the tooling from reaching critical bearing components. Cincinnati Inc. recommends that stainless steel adapter plates be used to minimize the flow of heat to the critical components (Ref 18). Additionally, incorporating an air gap between the die body and die pot within the press minimizes the transfer of heat.

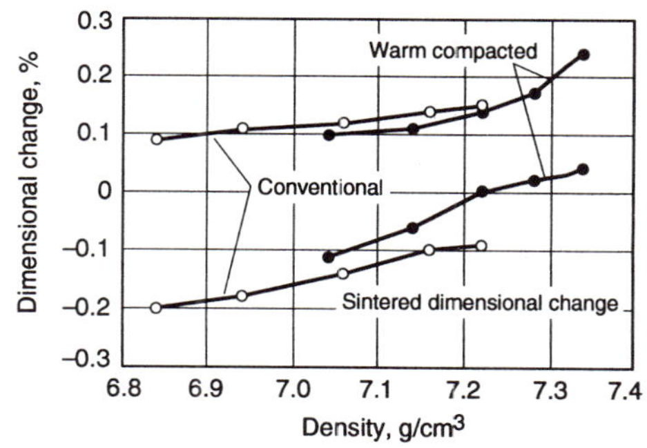

Fig. 8 Green expansion and sintered dimensional change of warm-compacted material relative to conventional compaction techniques

Effects of Prolonged Time at Temperature and Regrinding of Green Parts. Laboratory testing performed by Hoeganaes demonstrated that binder-treated powder can be reheated to warm-compaction temperatures a maximum of 4 cycles with minimal loss of powder flow and compressibility (Ref 4). Additionally, the powder can be held at temperature up to 4 h with no degradation of the apparent density, flow, green density, and green strength.

Although powder metallurgy is considered a low-scrap manufacturing process, nonusable parts are generated during the setup stage. To address this issue of potential green scrap, laboratory work was initiated at Hoeganaes Corporation to study the effects of adding reground warm-compacted powder into new premixes. This work demonstrated that additions up to 5% regrind can be successfully compacted without any degradation in the strength or flow characteristics of the premix. Although it is not recommended that regrind additions be made to critical components, this work demonstrated that additions can be made without any loss in powder or sintered properties.

Potential Applications of Warm Compaction

Because warm compaction is a single-press and single-sinter process, the process is ideal for complex multilevel P/M parts that require high mechanical properties that cannot be obtained at

Table 6 Density and process comparison between warm and cold compaction

Base powder	Graphite	Lubricant	Compaction	Sintering	Sintered density, g/cm3
Distaloy AE(a)	0.5 %	0.7% Kenolube	600 MPa cold compaction	1120 °C, 30 min, Endogas	7.07
		0.6% Densmix	600 MPa warm compaction	1120 °C, 30 min, Endogas	7.31
		0.6% Kenolube	650 MPa cold compaction	1120 °C, 30 min, 90% N_2/10% H_2	7.1
		0.6% Densmix	500 MPa warm compaction	1120 °C, 30 min, N_2/10% H_2	7.1
Distaloy DC(b)	0.5 %	0.6% Kenolube	650 MPa cold compaction	1120 °C, 30 min, 90% N_2/10% H_2	7.1
		0.6% Densmix	500 MPa warm compaction	1120 °C, 30 min, 90% N_2/10% H_2	7.1
Distaloy AE	0.8 %	0.6% Kenolube	600 + 500 MPa cold compaction (DPDS)(c)	750 + 1120 °C, 20 + 30 min, 90% N_2/10% H_2	7.3
		0.6% Densmix	700 MPa warm compaction	1120 °C, 30 min, 90% N_2/10% H_2	7.3

(a) Distaloy AE is a diffusion bonded powder utilizing pure iron with 4.0% Ni, 1.5% Cu, and 0.5% Mo. (b) Distaloy DC is a diffusion bonded powder utilizing a prealloy 1.50% Mo powder with 2.0% Ni. (c) DPDS, double-press double sinter. Source: Ref 21

Table 7 Comparison of warm-compacted materials to selected wrought and cast alloys

Material	Yield strength		Tensile strength		Elongation,	
	MPa	ksi	MPa	ksi	%	Hardness
AISI 1020	345	50	440	64	35	77 HRB
AISI 1050	427	62	745	108	20	96 HRB
AISI 8620	358	52	635	92	26	90 HRB
AISI 8620 Heat treat	1390	202	1482	215	10	45 HRC
Ductile iron 120-90-02	860	125	965	140	2	36 HRC
Powder forged F-0005	765	111	827	120	10	27 HRC
Powder forged FL-4605	1172	170	1455	211	9.5	47 HRC
FLN-4205 at 7.39 g/cm^3	1220	177	1503	218	1.9	42 HRC
FD-0405 at 7.33 g/cm^3	938	136	1248	181	1.7	41 HRC

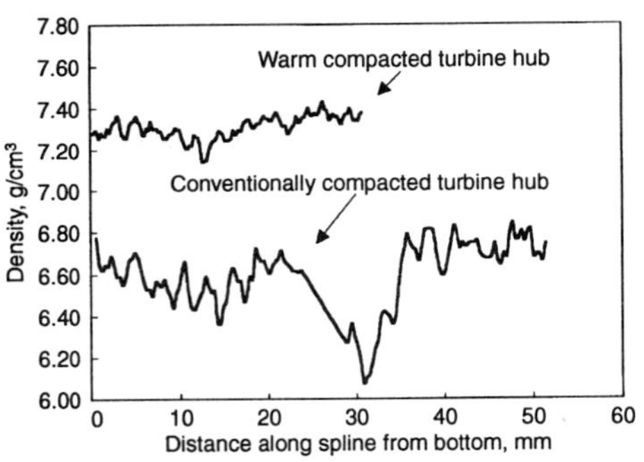

Fig. 9 Variation in sintered density and dimensional change of turbine hub processed by conventional P/M and warm compaction

conventional compaction densities. Higher density (or equivalent density at lower compaction pressures) can be achieved with warm compaction as compared with cold compaction (Table 6).

Recent articles have demonstrated the usefulness of the warm-compaction process in the fabrication of an automotive turbine hub for high-performance engines (part weight 1100 g), the manufacture of helical gears with gear densities in excess of 7.3 g/cm^3, lock components (part weight 27 g), and gearing with complex gear forms or spiral gears requiring high gear densities (Ref 22–24). The current production parts made by warm compaction are parts with a complex shape that are not adaptable for double pressing and double sintering. Warm compaction offers a simplified manufacturing process with resulting mechanical properties that met or surpassed the part specification.

Mechanical properties of warm-compacted steel powders were compared to selected wrought and forged alloys (see Table 7). Note that the yield and tensile strengths of the warm-compacted alloys were equivalent to those of wrought alloys. Thus it would seem that components made from these alloys are suitable candidates for the warm-compaction process. It must be noted that the elongation of the P/M materials is significantly lower than the wrought alloys chosen (except for the heat-treated ductile iron). Thus, proper application of the warm-compaction process must consider the reduced elongation and impact energy of the P/M part.

One significant advantage of the warm-compaction process is the increased density uniformity achieved in the compacted part (Ref 22, 23). Quantitative metallographic techniques demonstrated this feature in both a helical gear and an automotive turbine hub. Figure 9 demonstrates the greater uniformity of sintered density achieved with a turbine hub compared to a conventionally compacted part. This enhanced density uniformity results in increased load-carrying capacity with reduced dimensional variations because of the uniform density.

Future applications of the warm-compaction process will exploit the ability to achieve higher green densities at lower compaction pressures, thus minimizing the tooling stresses. Additionally, with the increased interest in the sinter-hardening process, warm compaction offers the potential to green machine these components prior to sintering and subsequent hardening.

REFERENCES

1. G.F. Bocchini, *The Warm Compaction Process: Basics, Advantages, and Limitations,* Society of Automotive Engineers, 1998
2. H.G. Rutz and F.G. Hanejko, High Density Processing of High Performance Ferrous Materials, *Advances in Powder Metallurgy and Particulate Materials—1994,* Vol 5, Metal Powder Industries Federation, 1994, p 117–133
3. F. Chagnon and Y. Trudel, Effect of Compaction Temperature on the Sintered Properties of High Density P/M Materials, *Advances in Powder Metallurgy and Particulate Materials—1995,* Vol 2, Part 5, Metal Powders Industries Federation, 1995, p 106–126
4. U. Engstrom, B. Johansson, H. Rutz, F. Hanejko, and S. Luk, High Density Materials for Future Applications, *Advances in Powder Metallurgy and Particulate Materials—1995,* Vol 3, Part 11, Metal Powders Industries Federation, 1995, p 106–126
5. M. Gagne, "Behavior of Powder Mix Constituents During Cold and Warm Compaction," presented at the 1997 International Conference on Powder Metallurgy & Particulate Materials, 29 June to 2 July 1997, Chicago, IL
6. H.G. Rutz and T.M. Cimino, Advanced Properties of High Density Ferrous Powder Metallurgy Materials, *Advances in Powder Metallurgy and Particulate Materials—1995,* Vol 3, Part 10, Metal Powders Industries Federation, 1995, p 97–117
7. S. Luk, H. Rutz, and M. Lutz, Properties of High Density Ferrous P/M Materials—A Study of Various Processes, *Advances in Powder Metallurgy and Particulate Materials—1994,* Vol 5, Metal Powder Industries Federation, 1994, p 135–155
8. H.G. Rutz, T. Murphy, and T. Cimino, The Effect of Microstructure on Fatigue Properties of High Density Ferrous P/M Materials, *Advances in Powder Metallurgy and Particulate Materials—1996,* Vol 13, Metal Powder Industries Federation, 1996, p 375–390
9. R. O'Brien, "Fatigue Properties of P/M Materials," Technical Data, Hoeganaes Corporation, 1988
10. I. Donaldson and F. Hanejko, An Investigation into the Effects of Processing Methods on the Mechanical Characteristics of High Performance Ferrous P/M Materials, *Advances in Powder Metallurgy and Particulate Materials—1995,* Vol 5, Metal Powder Industries Federation, 1995, p 51–69
11. S.R. Sun and K. Couchman, Repressing of Warm Compacted Materials, *Advances in Powder Metallurgy and Particulate Materials—1996,* Vol 5, Metal Powder Industries Federation, 1996, p 109–125
12. I. Donaldson, An Evaluation of High Performance Materials Processed Using Warm Compaction Technology, *Advances in Powder Metallurgy and Particulate Materials—1996,* Vol 5, Metal Powder Industries Federation, 1996, p 235–246

13. U. Engstrom, B. Johansson, and O. Jacobson, "Properties and Tolerances of Warm Compacted PM Materials," presented at Euro '95 (Birmingham), Oct 1995
14. T. Cimino and S.H. Luk, Machinability Evaluation of Selected High Green Strength P/M Materials, *Advances in Powder Metallurgy and Particulate Materials—1995*, Vol 2, Part 8, Metal Powders Industries Federation, 1995, p 129–149
15. C.G. Oliver and H.G. Rutz, Powder Metallurgy in Electromagnetic Applications, *Advances in Powder Metallurgy and Particulate Materials—1995,* Vol 3, Part 11, Metal Powders Industries Federation, 1995, p 87–106
16. S. Pelletier, L.P. Lefebvre, and C. Gelinas, Resin Impregnation of Soft Magnetic Materials for Low Frequency Applications, presented at the 1997 International Conference on Powder Metallurgy and Particulate Materials (Chicago, IL), 29 June to 2 July 1997
17. D.E. Gay, Higher Performance Microencapsulated Powders for Various P/M Applications, *Advances in Powder Metallurgy and Particulate Materials—1995,* Vol 3, Part 11, Metal Powders Industries Federation, 1995, p 103–117
18. R. Unkel, Additional Applications of Cincinnati EL-Temp™, *Advances in Powder Metallurgy and Particulate Materials—1995,* Vol 2, Metal Powders Industries Federation, 1995, p 3–11
19. S. St.-Laurent and F. Chagnon, "Designing Robust Powders Mixes for Warm Compaction," presented at the 1997 International Conference on Powder Metallurgy and Particulate Materials (Chicago, IL), 29 June to 2 July 1997
20. F. Hanejko, H.G. Rutz, U. Engstrom, and B. Johansson, Properties of Diffusion Bonded Alloys Processed to High Densities, *Advances in Powder Metallurgy and Particulate Materials—1995,* Vol 3, Part 10, Metal Powders Industries Federation, 1995, p 77–97
21. O. Mars and U. Engström, Warm Compaction and Its Influence on the Dynamic Properties of Sintered Steels, *Powder Metallurgy in Automotive Applications,* P. Ramarkrishnan, Ed., Science Publishers Inc., 1998, p 147–155
22. J. Chemlar, B. Nelson, H. Rutz, M. Lutz, and J. Porter, An Evaluation of the ANCORDENSE Single Compaction Process and HPP Processing Technique on Fine Pitched Spur and Helical Gears, *Advances in Powder Metallurgy and Particulate Materials—1994*, Vol 5, Metal Powder Industries Federation, 1994, p 73–89
23. T. Miller and F. Hanejko, "Development of a Warm Compacted Automatic Transmission Torque Converter Hub," Paper 970428, Society of Automotive Engineers
24. Advertising Literature on Warm Compaction, Porite Taiwan Company, LTD

Cold Isostatic Pressing

Peter E. Price, Consultant

COLD ISOSTATIC PRESSING (CIP) is a materials processing technique in which high pressure is applied to metal powder in a sealed elastomer container shaped for the application. The powder is converted from a loose aggregate into a partially dense compact that has sufficient green strength to permit careful handling and transfer to the following process operation. Water treated with corrosion inhibitor is the usual pressurization medium. Compacting pressures range from 207 to 414 MPa (30 to 60 ksi), although pressures as high as 758 MPa (110 ksi) have been used. Compaction is performed at ambient temperature. The density of the loose powder poured into the elastomer mold is increased from 55 to 65% of theoretical, a nominal range for uncompacted powder, to 75 to 85% of the 100% theoretical density value of the metal being processed.

There are two variations of the CIP process depending on the way the elastomer mold is utilized. In the "dry-bag" process, the elastomer mold is fixed to the pressure vessel and the equipment is designed for high production rate of small parts more typically in ceramics than in metals. In the "wet-bag" process, the mold is filled outside the vessel, transferred in, pressure compacted, and removed. In this process, multiple molds can be processed simultaneously according to the size of the pressure vessel. The wet-bag method is illustrated schematically in Fig. 1. Dry-bag methodology has much greater application in ceramics processing than in powder metallurgy, however, and is not further discussed here.

Molded parts are designated as being in the "green" or unsintered state following common terminology used for die-compacted metal powder parts. The usual next step is sintering to increase density to a maximum value close to theoretical or to a designed lower value for intentionally porous structures. Cold isostatic pressing also can be used as a preliminary densification step in containerized hot isostatic pressing (HIP) where the objective is to increase the "fill factor" of the HIP capsule and permit increased dimensional control in that process. Post-sintering operations can include forging, extrusion, rolling, coining, and HIP. Certain powder metals such as titanium, which can be sintered to a closed-pore condition, are most suited to the widest selection of post-sintering process operations.

Cold isostatic pressing of metal powders was invented by H.D. Madden of the Westinghouse Lamp Company in 1913 (U.S. Patent 1,081,618) to process tungsten and molybdenum billets for subsequent conversion to wire. This application continues in current use in the refractory metals industry. A semiautomated process for production of ceramic parts using the dry-bag technique was patented by B. Jeffrey of the Champion Spark Plug Company in 1942 (U.S. Patent 2,290,910).

During World War II, the use of CIP was broadened to include applications in the pressing of explosives, ceramics, beryllium, and other special defense-related materials. In the postwar period, larger and higher pressure vessels were constructed to facilitate powder-based processes in metals and ceramics. In 1970, a vessel operating with a working chamber of 61 cm (24 in.) in diameter, 2.45 m (96 in.) long at a pressure of 551 MPa (80 ksi) was placed in operation in industry for processing beryllium powder. This vessel is still in commercial operation after transferring to new ownership and location. Currently, there are a number of primary metals and metal part production processes that utilize CIP as an essential process step. Applications include processing tungsten, molybdenum, beryllium, tantalum, tungsten carbide/cobalt, P/M high-speed steels, and P/M composites. These materials cannot be melt processed if desired manufacturing response and end-point functional properties are to be maintained. The historical foundation of the current process has been described in the literature (see the "Selected References" at the end of this article).

Process Characteristics

Unique aspects of CIP, as compared to die compaction, for P/M parts include:

- Hydrostatic application of pressure over all surfaces of the mold produces uniform powder density for simple and complicated shapes. Die-wall friction is not a factor in the densification process because of the elastic behavior of the mold.
- Organic binder or lubricant additions to the metal powders are not required to achieve useful green strength. In fact, these materials are detrimental because of the adverse influence on metal chemistry and related mechanical properties for many of the reactive metals commonly processed by CIP if all these materials were present.
- Shapes with high ratios (greater than 10) of length to characteristic diameter, can be densified and handled.

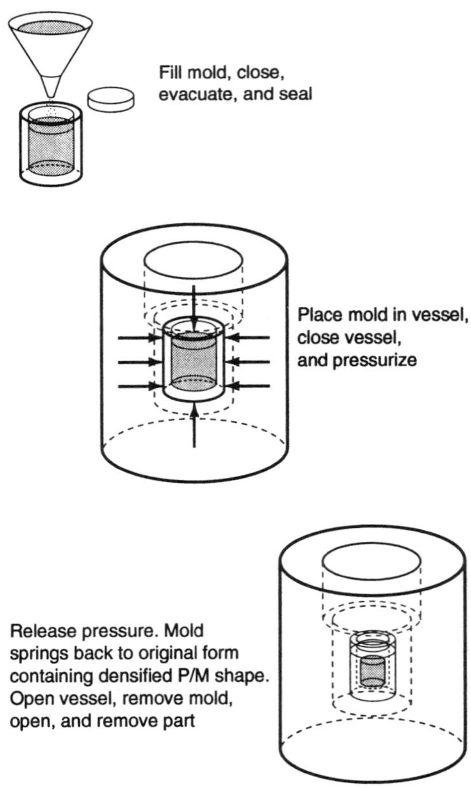

Fig. 1 Schematic of wet bag CIP process

- Parts with reentrant and three-dimensional curved geometries can be made.
- Elastomer tooling can be combined with metal tooling inserts to exactly control certain part dimensions such as cylinder bores or outside diameters.
- Thin-walled sections and parts can be pressed.
- Parts so large that comparable uniaxial die pressing capacity does not exist in industry can be made because of the large area by pressure product available in the largest CIP vessels. For the 24 by 96 in. by 80 ksi vessel mentioned above, this maximum product is 818 MN (92,000 tons).
- Elastomer tooling costs are low.

Recognized limitations of the process include:

- Dimensional control is generally less precise than with metal die compaction because of the absence of exactly dimensioned reference surfaces and variability of the local poured powder density within the mold.
- Surface finish of cold isostatically pressed parts is rough compared to die-compacted parts except for areas in contact with hard tool inserts in hybrid molds.
- Production rates are low compared to metal die compaction.
- Elastomer molds have a relatively short life because of abrasive wear.
- Leakage of a mold in the CIP vessel due to mold tearing or poor closure and sealing practice results in loss of material due to contamination by the working fluid. With reactive fine powders, such as aluminum, in large molds water leakage into the mold can produce a hazardous exothermic reaction.
- Not all metals and alloys of interest are readily available in the suitable particle size distribution and shape for application in the process.

Process Equipment

Equipment for the process includes a pressure vessel with design pressures ranging up to 758 MPa (110 ksi), fluid reservoir, low-pressure transfer pump, high-pressure intensifier, fluid filtration equipment, automatic and manual high-pressure valves, pressure gages, safety-relief valves, and a control system. The process flow diagram for a modern CIP system is shown in Fig. 2. Fluid filtration for removal of destructive abrasive metal powder particles is a significant modern improvement aimed at reducing maintenance costs.

Pressure Vessels. The pressure vessel is a unique and major component of the CIP system for powder metals. It is designed to meet all applicable engineering and regulatory codes for the size and pressure of the unit. Due to cyclic loading of the vessel, low-cycle fatigue life is a major engineering consideration, and a trade-off between desired operating life and initial cost is a factor in procurement. Design stress levels can be lowered to give increased operating life by increasing section thickness with consequent increase in cost. High-pressure vessels are manufactured as double- or even triple-shell designs wherein bore stresses are reduced by residual compressive stresses induced by a heat shrink assembly technique. Very large and very-high-pressure CIP vessels can be limited in design by the available manufacturing processes, transportation limitations due to size and weight, and engineering materials properties.

Closures and seals are critical design areas of pressure vessels. Closures may be full threaded, interrupted threaded, full resilient threads, or non-threaded support by an external yoke frame. Thread design, machining tolerances, and machine finish are important factors for long cycle life. For rapid opening, a quarter turn interrupted thread allows quick automated handling of the top closure of the vessel. A small CIP system with a quick opening top closure is shown in Fig. 3. The total length of thread in the interrupted design is approximately double that of the continuous thread design to accommodate equivalent load-bearing capacity. The bottom closure of a threaded vessel is usually full threaded because it is removed only for maintenance and inspection. A recent trend has been use of a closed-bottom end design even for two shell vessels. This innovation eliminates the need for a bottom-end bore seal, which in P/M processing can be subject to abrasive wear due to unavoidable contamination and settling out at the seal/bore interface crevice of metal powder particles from surfaces of elastomer tooling bags. An alternative to a closed-bottom design is a bottom-pressure seal plate supported by the bottom-threaded end closure. This design allows removal of the plate from the top for cleaning and seal change. All CIP vessels have fill and drain ports in the closures and a top air vent port to allow total displacement of air during fluid fill. Fill and drain ports are at the bottom. The air vent port is required to avoid air trapping that would otherwise result in slow pressurization/pumping cycles due to the air compressibility.

As an alternative to threaded closures, a yoke frame may be used to restrain end loads on the vessel closure plugs. Opening and closing a yoke frame unit requires moving the frame aside horizontally, extracting the end closure, and moving it aside for access to the working void space. User selection of yoke or threaded de-

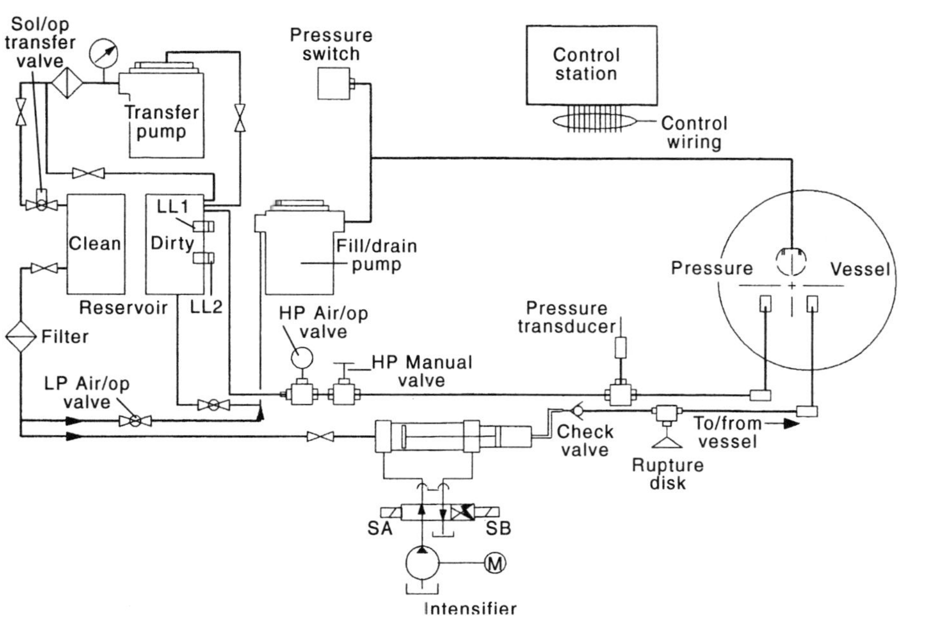

Fig. 2 Process flow diagram of modern CIP system. Source: EPS Inc. LP, low pressure; HP, high pressure; LL, liquid level; SA, side "A"; SB, side "B"; SOL/OP, solenoid to open; AIR/OP, air to open

Fig. 3 Small CIP system with quick-opening top closure. Courtesy of EPS Inc.

signs depends on factors including system size, design pressure, space availability, cost, inspectability, and preference.

Pressure Generators. After a low-pressure fluid-fill operation, pressure is increased in the system by air-driven or hydraulic-driven pumps and pressure intensifiers. Air-driven intensifiers are lower cost and slower than hydraulic units. The air units are used for small working void vessels and low production rates. Plant air is typically used for these pumps. Hydraulic intensifiers are installed for routine production, large vessels, and very high pressures. For wet-bag processing in most P/M applications, the pressurization medium is usually water with corrosion-inhibiting additives. An efficient filtering system is required to scavenge particulate contamination that would otherwise accelerate wear of the seals and pistons of these pumps.

Depressurization Design. The low compressibility of the usual working fluid, water, results in a rapid pressure drop in the vessel during let-down after only a small volumetric outflow. For large or convoluted-shape P/M parts, rapid release of the stored elastic energy and equalization of compressed gases (air) in the pressurized compact may cause fracture of the molded shape. First-stage depressurization is therefore preferably regulated with a manual or automatic control valve. With an automatic valve and computer controls, the depressurization rate can be controlled as a function of pressure allowing both a safe initial depressurization rate for molded parts and an efficient, fast, volumetric outflow rate from the vessel at lower pressures.

Control and Safety Features. The control system of a modern CIP unit functions to actuate the vessel closure, check off the locked condition of the closure, sequence valves, actuate a low-pressure pump for vessel filling, set control valves, accept and operate a pressure/time cycle protocol, actuate a high-pressure intensifier pump, provide information about system status (pumps, valves, pressure, etc.), and produce an analog/digital record of the process cycle. With a system with an automatic high-pressure control valve, the pressure let-down can be controlled according to cycle protocol. In addition to design and manufacture according to industry and regulatory standards of the vessel, valves, fittings, lines, and other items in the high-pressure flow path, safety-release overpressure rupture disks, analog and digital pressure indicators at the control station and at the equipment (analog type), closure lock indicator, instrument air supply indicator, and parallel/series manual valves for pressure let-down in the case of electrical or air supply outage are standard components of a modern CIP system.

Tooling. The simplest form of cold isostatic tooling is an elastomeric material mold or "bag" shown schematically in Fig. 1. The cavity dimensions of the mold take into account the volume of poured and loosely settled powder required to produce a compacted part. The mold cavity may be considered a three-dimensional equiaxial expansion of the cold isostatically pressed form. Additional complexity can be introduced into the tooling by use of inside-diameter, outside-diameter, or special-shape hard-tooling inserts in the bag to provide hard reference surfaces, shapes, and dimensions for the metal powder during compaction. With hybrid tooling, that is, elastomeric bags with hard tooling inserts, defined dimensions and shapes are produced from the "reference" surfaces, while the dimensions and surface smoothness of the "free" surfaces in contact with the elastomer are less exact. To prevent sag and bulging of large molds due to the weight of contained metal powder, external conformal support fixtures are used. Compression of powder can be either "outside-in" (collapsing-bag technique) or "inside-out" (dilating-bag technique). With the dilating-bag technique, there are stops or expansion limiting hard tooling components such as the inside diameter of a steel tube to define the limits of the mold cavity.

Tooling Material. Elastomer tooling is subject to abrasive wear and tearing. Tearing of a mold in the CIP vessel is a serious operational difficulty due to possible spillage of metal powder into the pressure vessel and subsequent introduction into the seal areas, valves, fittings, pumps, and filters. In such an event, immediate steps to drain, inspect, strip, and clean the system are required to avoid costly wear of critical system components. A control system for monitoring mold usage and condition as well as appropriate selection of mold material can prevent

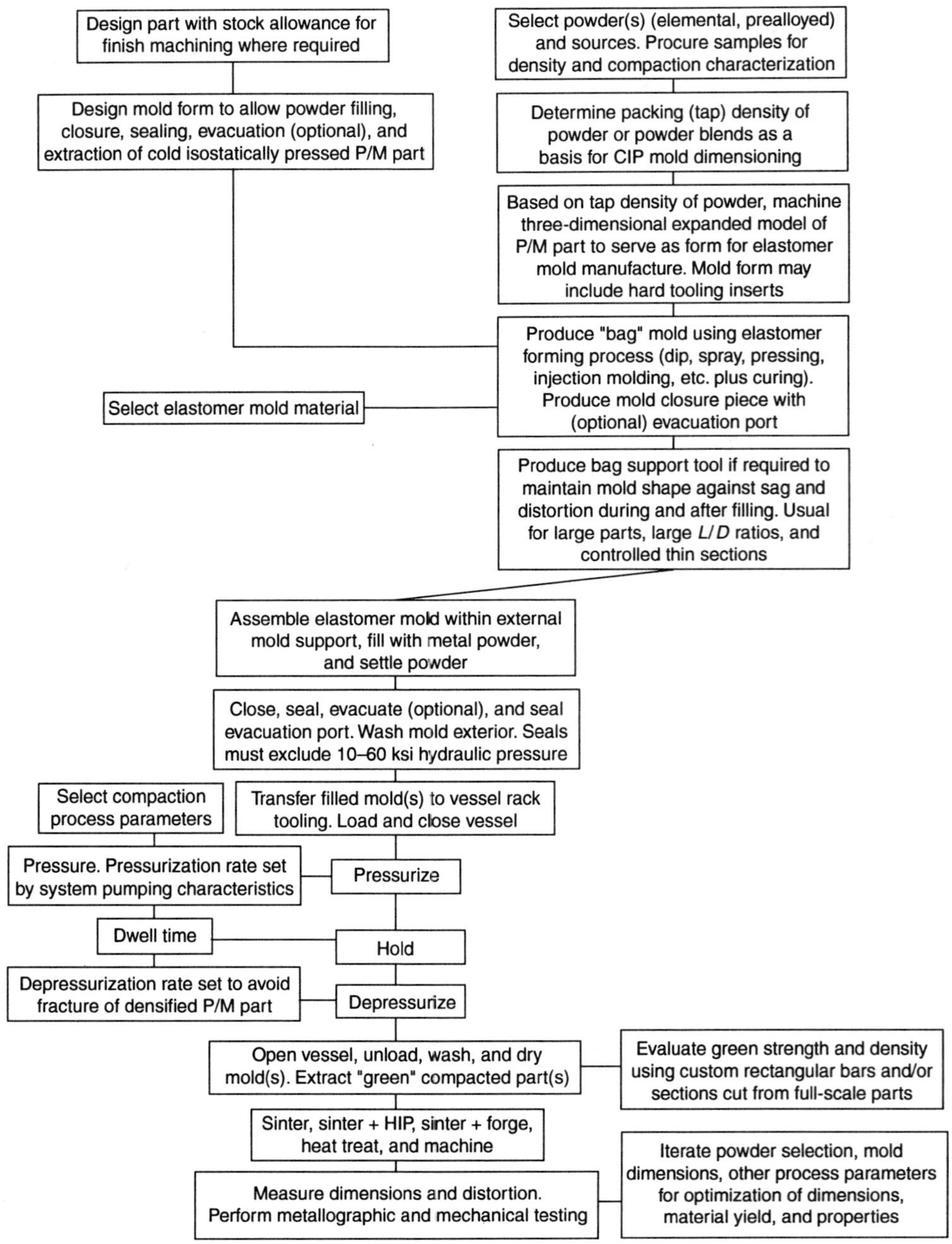

Fig. 4 Process flowchart for production of cold isostatically pressed parts

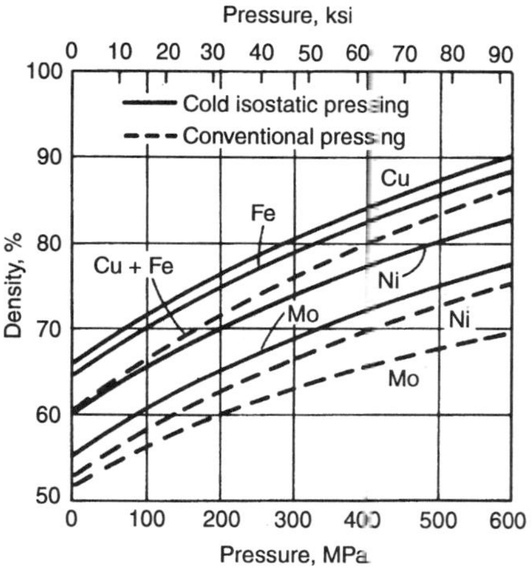

Fig. 5 Density as a function of pressure for cold isostatically pressed and die-compacted parts

such failures. Molds can be manufactured from a variety of elastomers including natural rubber, butyl rubber, polyvinyl chloride, polyurethane, neoprene, nitrile, and silicone rubbers. Abrasion resistance, hardness, and resilience are properties that are adjustable in the elastomer formulation. Polyurethane rubber meets these requirements well and can be produced in a variety of hardness or durometer specifications. Custom molds can be manufactured by commercial sources. Alternatively, a rubber formulation, dipping, and curing operation can be set up in-house when many and varied molds are required using commercially available materials and equipment.

Wet-Bag Isostatic Pressing

In wet-bag pressing, the mold is filled, evacuated (if this step is used), and sealed outside the pressure vessel. Because these operations occur in a powder-handling area, contamination of external mold surfaces with metal powder can easily occur. Introduction of dirty molds into the vessel results in working fluid contamination with long-term effects similar to mold tearing and gross metal powder spillage. Thorough mold washing as an integral part of the process flow chart is necessary to keep the process fluid system clean. Filled molds are stacked on a rack that serves as handling tooling for loading molds into the vessel. In normal practice, the vessel is not fully emptied of working fluid (most commonly water with corrosion preventative additive) between pressurization cycles. This means that even with an external filtering system for that portion of the working fluid that is cycled in and out of the vessel, a major fraction of the working fluid remains in the vessel and may accumulate traces of metal powder introduced in each process cycle. Depending on usage rate, periodic complete draining and cleaning of the vessel is necessary to maintain the highest level of working fluid cleanliness.

The wet-bag process is flexible in that it can be adapted for large shapes limited only by vessel dimensions or multiple small shapes that can be stacked in a loading rack for a single process run. With a given metal powder that establishes the required process pressure, a mix of mold sizes and shapes can also be processed as a single batch, thereby utilizing the flexibility of the process. The requirement for mold support tooling must be determined for each configuration. Currently, when CIP is used for P/M applications the wet-bag process is usually selected.

Part Size and Shape

The upper limitation on the size of a P/M part is the diameter and internal height of the available CIP pressure vessel minus the dimensional space required for the rack tooling required to lift the mold (plus mold support tooling) into and out of the vessel. Alternatively, multiple small shapes can be processed using stacked layers of parts on rack-type tooling.

Billets are simple shapes including cylinders, rectangular solid slabs, and sheet bars. After CIP, these shapes undergo further processing including typically sintering, hot deformation, and cold deformation with intermediate annealing for conversion to final form.

Forging preforms have a shape related to the final forging shape depending on whether it is a closed-die or open-die forging. The preform must have sufficient total mass to fill the closed-die cavity or finish the open-die part with sufficient excess stock to clean up in machining. Sintering following CIP is necessary to ensure metal flow without premature fracture. The advantage of a forging preform is elimination of one or more intermediate forging steps, particularly in closed-die applications. Forging acts to increase the density of the preform, refine the sintered grain structure, and produce deformation texture allowing production of a part with mechanical properties dependent primarily on chemistry and heat treatment. Both forging preforms and near-net-shape parts produced by CIP facilitate production of prototype parts with reduced initial tooling costs.

Near-net shapes are produced using elastomeric molds with and without hard tooling inserts. The mold cavity is dimensioned so that after CIP and sintering final dimensions require only a minimal amount of finish machining. The utilization of expensive metal powders is efficient. Blends of powders with disparate properties such as melting point can be processed by CIP to produce unique functional properties. Also, blends of powders with significantly different mechanical properties, that is, soft and ductile versus hard and brittle can be cold isostatically pressed to net shape to produce parts that could not otherwise be produced by deformation processing. Sintering can produce parts with densities in the range 90 to 97% of theoretical, which is sufficient for designated applications. Intentionally low-density net-shape

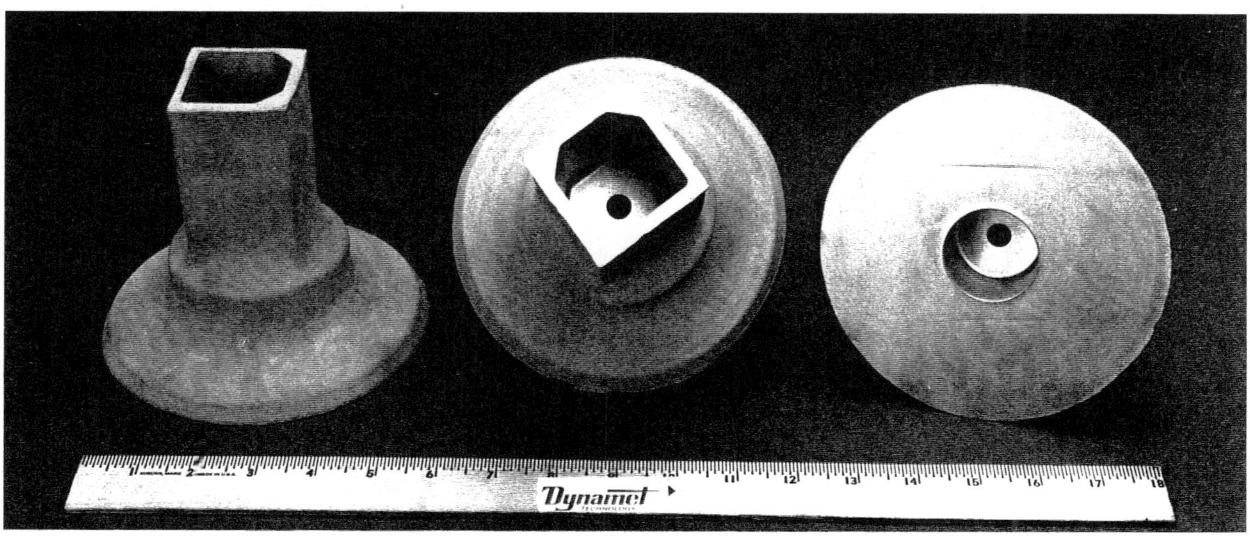

Fig. 6 Complex cold isostatically pressed P/M part made using internal round and square hard-tooling inserts. Courtesy Dynamet Technology

Fig. 7 Closed one end precision bore cylinder. Material-efficient use of blended elemental titanium and alloying powders. Courtesy Dynamet Technology

parts for applications such as filters can be produced by selection of CIP pressure and sintering schedule. Alternately, for select metal powders such as titanium and its alloys CIP to net shape followed by sintering to a closed-pore condition allows a final step of hot isostatic pressing to be employed to produce substantially fully dense parts. Shrinkage allowances for CIP compaction, sintering compaction, and hot isostatic compaction must be factored into the CIP mold design to realize optimal material utilization and minimal final machining. Final dimensions of sintered or sintered and hot isostatically pressed near-net-shape parts depend critically on mold dimensions, powder properties, and the thermal process cycle.

Powder Properties

Metal powder characteristics directly determine the parameters of the CIP process. In mold filling, particle size distribution and particle shape are controlling factors in determining tap density of the filled mold. For billet molds, where final shape is determined by deformation processing operations, exact fill is not as important as in near-net-shape parts. Nevertheless, reproducible particle size and shape factors would be required to ensure repeatable post-CIP processing. In the case of net-shape parts, the mold must be filled to a uniform and repeatable tap density at all locations in the mold to yield predictable green part dimensions. This requires exact control and reproducibility of the above-indicated factors. German (Ref 1) has presented the existing knowledge base covering particle packing characteristics. In the case of blends of powders with disparate characteristics, that is, tungsten plus copper, for example, particle characteristics including size, shape, and density control the extent of separation during handling and mold filling (Ref 1).

Powder blending represents the most adaptable method for creating alloy compositions using CIP. Elemental powders can be blended with master alloy powders to create ranges of compositions. Separation of the blend must be avoided by selecting appropriate size and shape factors for the constituents (Ref 1).

In CIP compaction, a series of events takes place including particle rearrangement, localized deformation, homogeneous deformation, and elastic bulk compression. These phases of compaction depend on yield strength, strain hardening rate, and elastic properties, all of which are generally inferred from bulk properties of the related metal or alloy. Hardness is a shorthand indicator of compaction response because it is a property derived from the plastic flow properties of the bulk material. The softer the bulk metal is, the lower will be the required compaction pressure. With blends of powders with dissimilar properties, that is, hard and soft, compaction response is governed by the constituent with the greatest connectivity, that is, that phase of the blend for which a continuous path between like particles can be traced throughout the structure without intersection in the alternate phase.

Although produced regularly in tonnage quantities in a variety of compositions, powders atomized from the melt have a generally spherical

Fig. 8 Cold isostatically pressed parts with designed regions of controlled density and porosity. Courtesy Dynamet Technology

particle shape and have been found difficult to use in CIP processing. This behavior is attributed to absence of shear and interlocking mechanisms during compaction processes that occur in chemically produced metal powders with irregular shapes. Atomized powders can be altered in particle shape by mechanical processing such as milling or attrition.

Process Parameters

Figure 4 outlines a simple flowchart for the CIP process. Critical elements of the overall process for producing uniform and reproducible parts include: part design, mold design, reproducibility of metal powder density in filled mold, mold support, and CIP pressure. Operations that can lead to process failure if not properly executed include mold closure, mold sealing, evacuation, sealing of mold evacuation port, depressurization rate, and handling.

Dwell pressure for producing "green" strength that allows handling after CIP and maximum density if full density (as opposed to intentionally porous) parts are desired is determined by metal and powder type. Typically, pressures in excess of 200 MPa (30 ksi) are required for the process. Applicable ranges of dwell pressures for a variety of powder metals are:

	Pressure range	
Powder	MPa	ksi
Aluminum	55–140	8–20
Copper	140–275	20–40
Iron	310–415	45–60
High-speed tool steel	240–345	35–50
Stainless steel	310–415	45–60
Titanium	310–415	45–60
Tungsten	240–415	35–60
Tungsten carbide	170–205	25–30

For isostatic compaction, lower pressures are required than for die compaction due to the absence of die-wall friction effects. Nominal CIP pressure for a metal powder is also substantially independent of part size, again because of the isostatic nature of the process. Figure 5 shows density as a function of pressure for select cases of isostatically and die-compacted powders.

Dwell time is generally only a few minutes for the process. At room temperature (the usual process temperature), there is negligible time-dependent particle deformation for the materials listed above. Thus, density follows pressure very closely in time, and no further increase occurs at constant pressure.

Depressurization rate is a unique process parameter for CIP. Under full process pressure, the densified metal part stores elastic strain energy in the solid and the $P \cdot V$ (pressure · volume) energy due to compression of air and gases entering into the mold in the void space of the low-density metal powder. Rapid release of process pressure can result in cracked parts. Entrapped gas can be eliminated by evacuation and sealing of the filled mold, but this process option, which adds complexity to the process and mold design, is frequently bypassed. Because of the relatively low compressibility of water (most frequently used pressurization medium), a very small volume release at high pressure causes a large pressure drop in the CIP system. Therefore, controlled depressurization is achieved by use of a high-pressure flow-control valve in the overall system. A depressurization schedule that avoids part cracking should be determined experimentally.

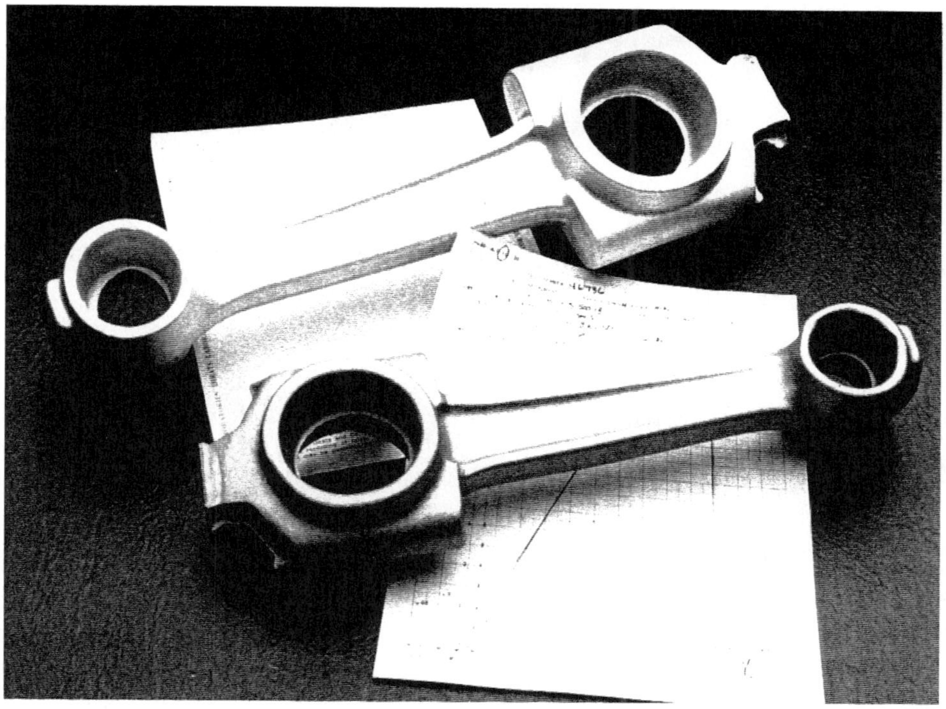

Fig. 9 Forging preforms cold isostatically pressed with titanium alloy and aluminum alloy powders. Courtesy Dynamet Technology

Evaluation of green strength and density of a cold isostatically pressed part may be carried out for a given metal powder and process cycle in the development or qualification phase of a powder and/or part by molding rectangular bars and carrying out three-point bend tests according to the protocol of ASTM B 312 (1997). These same bars may be measured and weighed before the bend test to determine approximate green density. Because the mold is elastomeric and not a hard die as specified in the ASTM procedure, oversize parts can be produced and machined green to standard dimensions for the development and control tests. Uniformity of density and green strength for larger parts can be determined using cut-ups taken along various directions of interest.

Thermal processing of cold isostatically pressed parts usually starts with sintering. Sintering cycles may be designed to maximize density or produce a specified percentage of porosity. Sintering atmospheres are based on metal chemistry requirements and include vacuum, reducing, and chemically active atmospheres such as carburizing and nitriding. Full density of cold isostatically pressed and sintered P/M parts can be achieved by hot isostatic pressing following sintering if a closed-pore condition can be established. This is common for titanium alloy parts and metal compositions that can be liquid-phase sintered such as tungsten carbide/cobalt, iron-bonded TiC, and tungsten/copper. Low-pressure liquid-phase sintering can be used as a complete alternative to HIP for selected materials. More complicated thermal processing sequences can be invoked such as sinter plus HIP plus forge to produce a fully dense forging with desired grain flow structure. This sequence is material efficient and eliminates the requirement for multiple forging steps, dies, and reheats.

Applications

Parts that are candidates for application of CIP originate within the sets (not all inclusive). They are expensive materials such as titanium, tungsten, molybdenum, and tantalum; materials difficult to machine such as tungsten carbide/cobalt, titanium carbide/iron, titanium, and tool steels; complex shapes; hybrid structures such as layered structures of multiple materials or combined porous/dense structures; and intersections of the above. Using titanium and its principal alloy Ti-6Al-4V as materials for illustration, application examples of CIP are shown in the following. Figure 6 shows the shape complexity feature of CIP where a nearly axisymmetric part 125 mm (5 in.) in diameter with round and rectangular inside-diameter features was produced using hard-tooling core mandrels and a net-shape external elastomer mold. The hard-tooled surfaces have evolved as smooth compared to

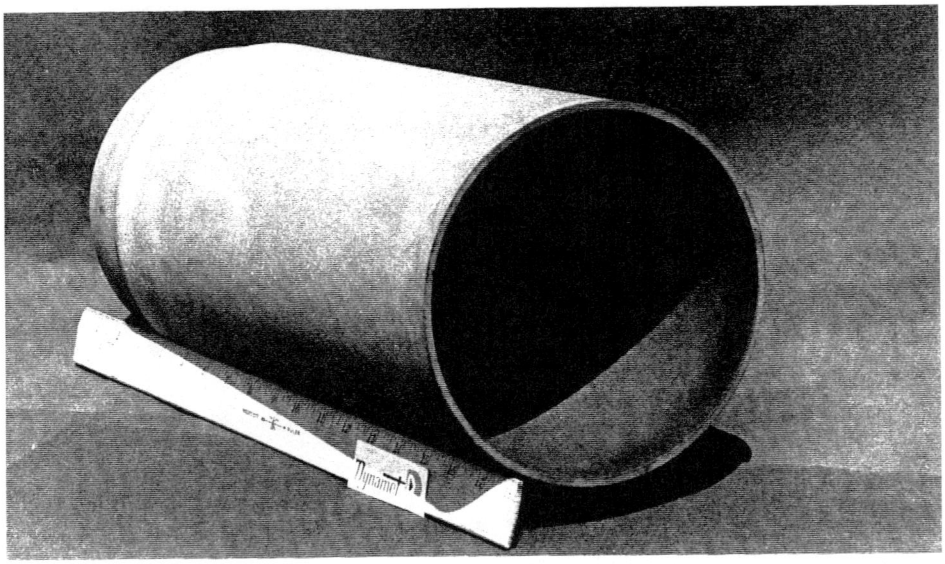

Fig. 10 200 mm *D* by 500 mm *L* thin-wall tube cold isostatically pressed in Ti-6Al-4V powder blend. Courtesy Dynamet Technology

the elastomer mold interface surfaces following CIP plus HIP. The surface produced by the elastomer mold reflects the particle size of the blended powder. In Fig. 7, CIP has been used with blended elemental powders of Ti-6Al-6V-2Sn and Ti-6Al-4V to produce with high material efficiency a hollow cylindrical cup with isotropic (texture-free) mechanical properties using a hard-tool core mandrel. The part has a high surface-area-to-volume ratio. The alternative to the use of CIP is machining from stock with a resulting low part-weight-to-stock-weight yield ratio. In Fig. 8, a series of parts with variable geometry and integral regions of controlled density have been produced using CIP. Powder size distribution can be selected as well as multiple-step CIP to facilitate production of the design. Porosity can range from zero to slightly greater than one minus the tap density of the metal powder, that is, approximately 1 – 0.55/0.65, where 0.55/0.65 is a normal range of powder density, when low-pressure CIP plus low-temperature sintering increases the original powder density slightly. Filters, filter supports, vibration damping, and heat transfer control devices are typical applications for porous parts. Figure 9 illustrates forging preforms in an aluminum alloy (lighter-color part) and a titanium alloy produced by CIP plus sintering. A final part can be produced with a single die set and forging stroke. Preform dimensions can be modified so that the forging operation produces desired effects including: amount and location of plastic strain, final densification, microstructural homogenization, and property enhancement, particularly fracture toughness. In the case of the connecting rods, overall length is maintained in forging. For tubular or solid parts with length-to-diameter (*L/D*) ratios of 2 or greater CIP can produce parts such as shown in Fig. 10. This part has an *L/D* of 2.25 with wall thickness of ~6 mm (~0.25 in.) with a closed bottom. Similar parts such as tubes, pipes, hydraulic pressure fittings, and vessel liners with thin walls, high *L/D* ratios, tapered wall thickness, and stepped bores are all practical with CIP technique.

REFERENCE

1. R.M. German, *Particle Packing Characteristics,* Metal Powder Industries Federation, 1989

SELECTED REFERENCES

- S. Abkowitz, Isostatic Pressing of Complex Shapes from Titanium and Titanium Alloys, *Production to Near Net Shape Source Book,* American Society for Metals, 1983
- C.B. Boyer, J.H. Peterson, and F.D. Orcutt, Equipment for Hydrostatic Pressing of Metal Powders, *Proc. National Conference on Fluid Power,* Vol XXV, 27th Annual Meeting (Chicago), Fluid Power Society and National Fluid Power Association, 19–21 Oct 1971
- C.B. Boyer and J.C. Wahl, "Equipment for Cold Isostatic Pressing," No. 74-PVP-21, Conf. Proc., 24–28 June 1974, American Society of Mechanical Engineers
- E.L.J. Papen, Isostatic Pressing, *High Pressure Technology,* I.L. Spain and J. Paauwe, Ed., Marcel Dekker, 1977
- P. Popper, *Isostatic Pressing,* Heyden & Sons, 1976
- J.T. Winship, Upsurge in Isostatics, Special Report 765, *Am. Mach.,* April 1984

Roll Compacting of Metal Powders*

Reviewed by Walter V. Knopp, P/M Engineering and Consulting Co.

ROLL COMPACTING, or powder rolling, refers to the continuous compaction of metal powders by a rolling mill. In this process, metal powders are fed from a hopper to a set of compacting rolls that produce a continuous green (unsintered) strip or sheet. These materials undergo further processing by sintering (heating) and re-rolling to produce an end product with the desired material properties. Material can be produced fully dense or with a desired porosity and pore size.

Ideal powder characteristics for roll compaction are comparable to those desired in powders for die compaction; flowability is important to ensure consistency in the final product; softness and irregularity of powder particle shape promote green strength; and particle size and size distribution influence green density. Metal powders suitable for roll compacting include elemental powders, blends of elemental powders or elemental and alloy powders that are homogenized during sintering, blends that contain additives that are nonreactive and that cannot be produced by standard melting and casting processes, and prealloyed atomized powders.

Atomization has the capability to produce powder particles that are individually uniform in composition and unique in internal structure. Strip produced by roll compacting of powders subsequently produces materials that are clean and completely uniform throughout the coil, with a fine grain size.

Production Procedures

Roll Position. The type and position of the rolls vary with the type of system chosen to produce the desired finished strip. Rolls can be positioned vertically (as in conventional rolling mills), horizontally, or at an inclined angle, as shown in Fig. 1.

Roll position selection is determined by several factors: the characteristics of the material, the amount of strip to be produced, the desired end product, and special structural requirements. Horizontal roll position (Fig. 1a) using gravity feed offers the most flexibility in controlling powder feed and is a convenient way of introducing several powders for multiple-layer strip. A conventional vertical mill can also be used to roll compact powder. However, vertical roll position (Fig. 1c) requires a very tall sintering furnace, or the green strip must be turned horizontally to enter a conventional furnace. This action could disrupt green bonds and inhibit sintering. The inclined angle position (Fig. 1d) offers a compromise between vertical and horizontal roll positions. Using this approach, a conventional mill can usually be used and powder can be gravity fed.

Powder Feeding. The initial step in the roll compacting process is the feeding of powder to the rolls. The green strip should exhibit uniform density from edge to edge; otherwise, problems may be encountered during processing, and nonuniform conditions cannot be rectified. If strip has a higher density on one side, it may exit the re-rolling mill at an angle, exhibiting camber or a long edge. However, if the center of the strip has a lower density, it may crack on re-rolling due to densification and elongation of the edges. If the edges of the strip have a lower density, the higher density center area may densify and elongate during re-rolling, causing edge cracks.

Two methods of feeding powder to the rolls are illustrated in Fig. 2. The powder is fed to the hopper by a belt or a vibrating feeder from the

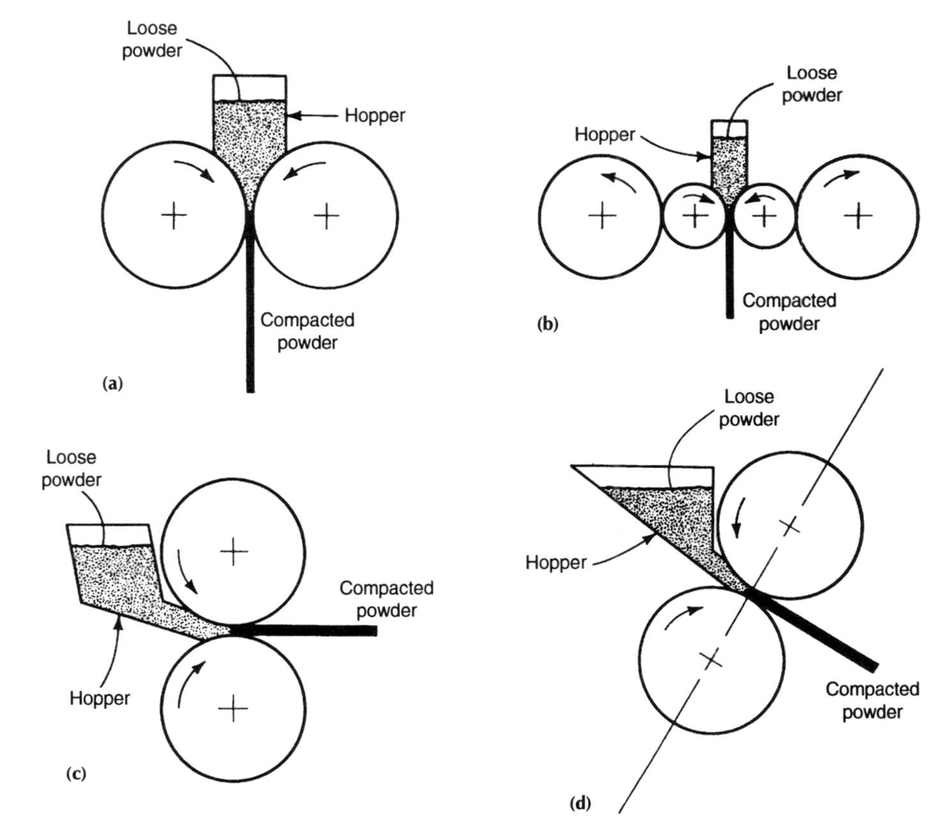

Fig. 1 Typical roll positions. (a) and (b) Horizontal positioning. (c) Vertical positioning. (d) Inclined angle positioning

*Reviewed and adapted from previous edition article by Walter V. Knopp, P/M Engineering & Consulting Co., Inc.; William R. Duncan, Superintendent, Fabricated Metal Products, Sherritt Gordon Mines Ltd.; Alan J. Moses, Plant Manager, Pfizer, Inc.; Milton W. Toaz, Program Manager, Engine Parts Division, Imperial Clevite Inc.

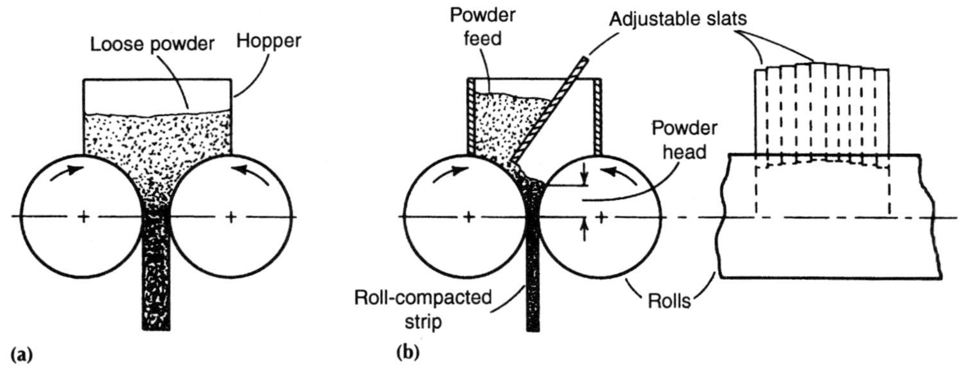

Fig. 2 Methods of feeding powder to the rolls. (a) Schematic of roll compaction with saturated feed, rolling mil rolls arranged horizontally. (b) Roll compaction with unsaturated feed controlled by adjustable gates

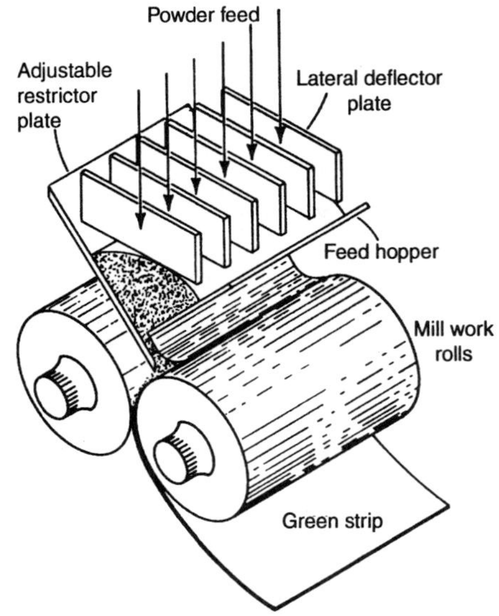

Fig. 3 Typical powder compaction mill feeder arrangement

storage bin. During saturated feeding (Fig. 2a), the powder falls directly from the hopper into the roll gap. The head of powder in the hopper should be held constant, because it applies pressure on the powder in the roll gap. A variation in pressure at the gap affects the feed rate of the powder, which results in varying density along the strip length. In the unsaturated, or "starved," feeding system (Fig. 2b), the amount of powder entering the roll gap is controlled by adjustable gates. Variations in compaction density can be made by adjusting the gates. Adjustable gates also can be used for saturated feeding.

Cover plates, formed to fit the curvature of the roll diameter, are available in sections so that powder feed can be varied at different areas of the roll surface as required. Figure 3 illustrates a typical powder feeder. This design incorporates a lateral deflector plate that distributes the powder evenly across the powder hopper. At present, strips of any roll-compacted powder can be produced by both saturated and unsaturated feeding in widths up to 53 cm (21 in.) and thicknesses of 2.5 to 3.2 mm (0.100 to 0.125 in.). Lateral deflector plates in the powder feeder are not necessary to achieve these results in all cases.

Edge Control. The green strip rolled from powder should have uniform thickness and density across the width of the strip, and its edges should be well formed and as dense as the center of the strip. Therefore, a method of containing the powder being fed to the compacting rolls is essential to process control. Without edge control, powder escapes from the roll gap. Consequently, a strip with low-density edges is produced, which necessitates edge trimming after the strip is re-rolled.

Figures 4 and 5 illustrate typical methods of retaining powder in the roll gap. The method shown in Fig. 4(a) uses floating flanges that are attached to one roll and that overlap the other roll. Pressure is applied to the flanges as they approach the roll gap, thus preventing powder loss from the gap. A continuous belt that covers the gap at the edge of the rolls also is effective in preventing powder loss (Fig. 4b and 5). Other systems are available that produce strip with a tight edge that requires minimal edge trimming.

Roll Diameter. In powder rolling, strip thickness is mainly determined by the roll diameter used. Roll diameter varies from 12.7 mm (0.5 in.) to 920 mm (36 in.). Figure 6 illustrates the effect of roll diameter. The nip, or gripping, angle for conventional metal rolling is the same for powder rolling (approximately 7 to 8°). Note that on the larger diameter roll, a larger arc is included by the nip angle on the roll surface. Therefore, a larger amount of powder is pulled into the roll gap than in the small-diameter rolls. For a given powder, a thicker strip can be produced with the larger diameter rolls.

Figure 7 shows the strip thickness that can be obtained with different roll diameters with two types of nickel powder. The atomized nickel is soft, while the hydrometallurgical nickel powder is hard. The ratios of roll diameter to optimum green strip thickness vary widely for different powders. They depend on the flow and apparent density of the powder, the coefficient of friction between the powder and roll, roll temperature, and whether the powders produced are hard or soft.

If the coefficient of friction between the powder and the rolls is increased by roughening the roll surface, a larger amount of powder will be pulled into the roll gap. For the same roll setting, higher compacted densities will be obtained. An increase in roll temperature, achieved through preheating or increased pressure, will also increase the coefficient of friction with the same results.

The roll gap is normally set to provide a green density approximately 80 to 90% of the theoretical density at a thickness that will provide full density when processed to final gage. Hot and

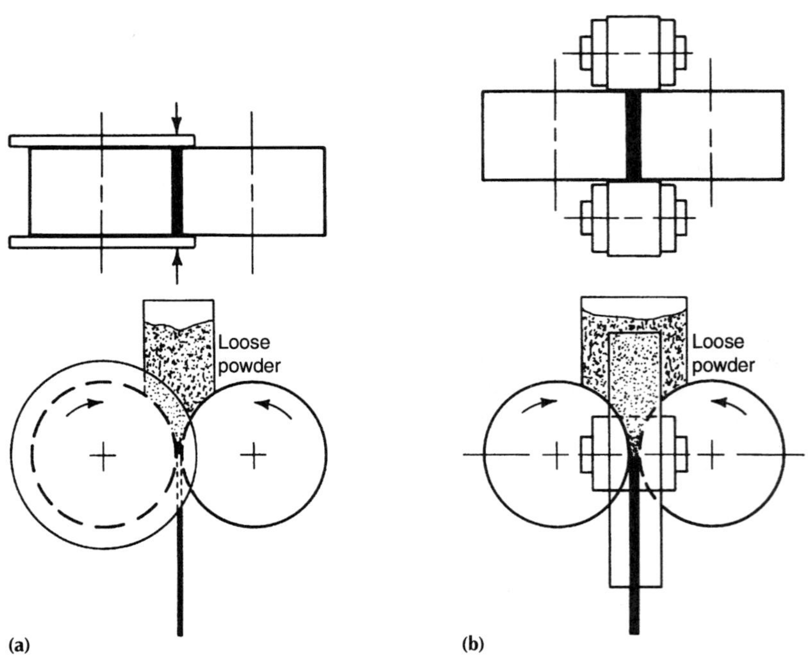

Fig. 4 Methods of controlling powder feed to compacting rolls. (a) Flange edge control. (b) Belt edge control

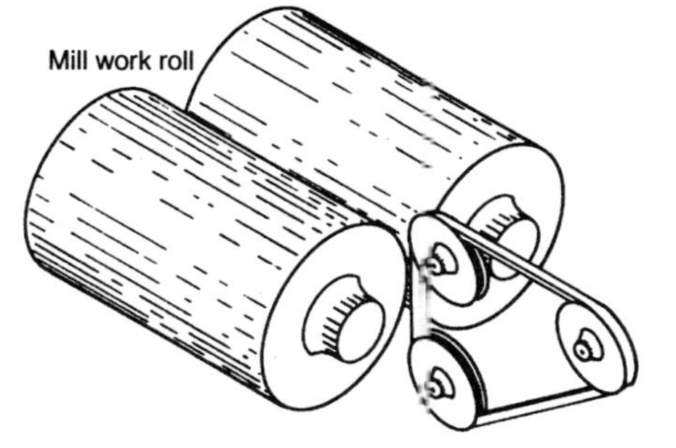

Fig. 5 Edge restriction device. V-belts and pulleys are used to contain powder at the edge of roll face and in the roll gap of a powder compaction mill

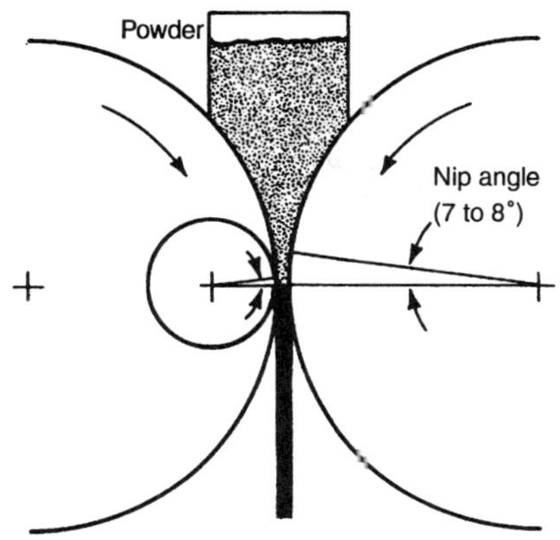

Fig. 6 Effect of roll diameter

cold workability of the material will determine the compacted gage.

Finishing. Uniform green strip produced by the compacting mill must be sintered before further densification. Green density usually is between 75 and 90% of theoretical density. A higher density renders the green strip hard and brittle, while a lower density produces a weak strip that is difficult to handle. Finishing of strip can be accomplished by several methods, depending on the type of material being produced, quantity required, sintering temperature and time, atmosphere requirements, and hot or cold working characteristics.

Figure 8 shows two methods of handling strip when densification is achieved by cold rolling. In Fig. 8(a), strip is coiled green and then sintered in a bell-type furnace. In Fig. 8(b), strip is passed through a sintering furnace, cooled, and

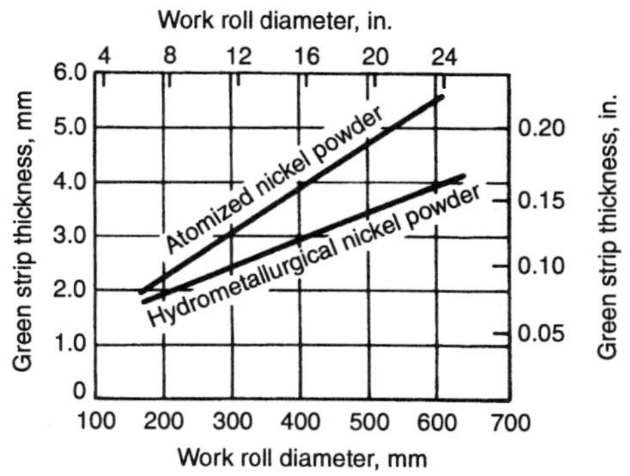

Fig. 7 Effect of work roll diameter on green strip thickness for two types of nickel powder

then coiled. If necessary, additional sintering and re-rolling steps are completed in the same manner.

In Fig. 9, strip from the compacting mill is passed through the sintering furnace. While still hot, strip is hot rolled to full density. Strip is then cooled before leaving the furnace and subsequently coiled.

Commercial Production

Powder rolling was described in a number of early patents (Ref 1), but the first in-depth analysis of the process was undertaken by Naeser and Zirm (Ref 2), who documented their work on rolling RZ iron powder into strip. Powder rolling of copper powder is described in Ref 3 and 4.

From 1958 to 1968, powder rolling processes received considerable attention in Great Britain, the United States, Canada, the Soviet Union, and Japan. Principal interest centered around combining low-cost processes for producing metal powders, such as copper, nickel, cobalt, iron, and aluminum, with the powder rolling process to develop methods of fabricating thin sheet or strip more economically.

Nickel Powder Strip. The Sherritt Gordon process of rolling nickel powders into strip and sheet was one result of these efforts (Ref 5). This strip was used to make coinage blanks for the Canadian Mint. Typical mechanical properties of roll compacted nickel strip at various stages of manufacturing are given in Table 1.

After nickel powder is compacted into a green strip, the strip is sintered between 1000 and 1200

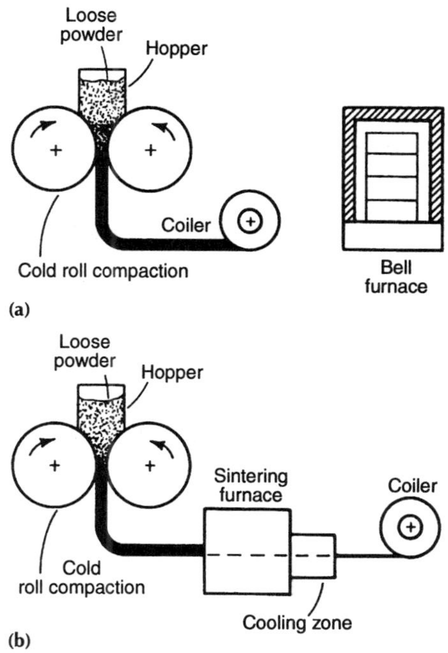

Fig. 8 Powder rolling processes. (a) Cold rolled green strip is coiled and then sintered in a bell-type furnace. (b) Cold rolled strip is reeled into individual coils after first sintering treatment

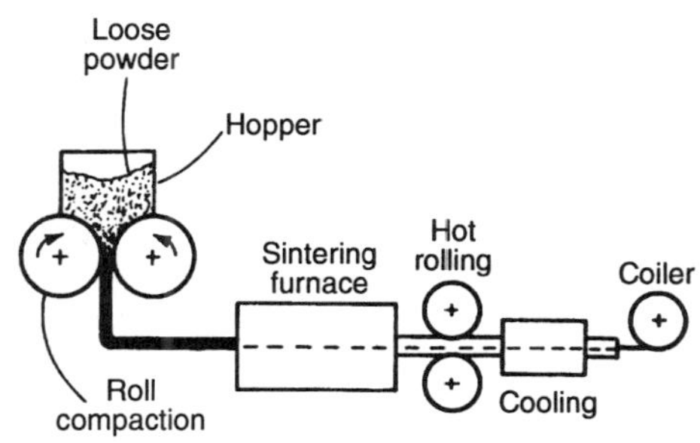

Fig. 9 Powder rolling process. Shown are horizontal arrangement of rolls, sintering, hot rolling, cooling, and coiling of strip

°C (1830 and 2190 °F) in a muffled furnace. Hydrogen, dissociated ammonia, or even less reducing gases provide a suitable atmosphere. Strip is conveyed through the furnace on a mesh belt or rollers.

Densification of the roll-compacted nickel strip can be achieved by hot or cold rolling. Hot rolling of nickel strip is performed above 800 °C (1470 °F), with a reduction in thickness of approximately 50%. Because of its porous nature, the sintered strip should be protected from oxidation when heating for hot rolling. This can be accomplished by using an inert or exothermic atmosphere.

Cold rolling and annealing cycles also may be used to densify the sintered strip. To be successful, sintering must be performed at 1100 °C (2010 °F) or higher. Limited reduction is achieved on the first cold mill pass. The strip must then be

Table 1 Typical mechanical properties of hydrometallurgical powder rolled nickel strip

Strip	Strip thickness		Strip density, %	Ultimate tensile strength		Yield strength		Elongation, %
	mm	in.		MPa	ksi	MPa	ksi	
Green strip	4.0	0.158	79	4	0.6	4	0.6	0
Sintered strip	4.1	0.161	79	138	20	136	19	0
Hot rolled strip	2.1	0.084	100	358	52	165	24	38
Cold rolled strip	1.3	0.052	100	579	84	572	83	5
Annealed strip	1.3	0.052	100	362	53	83	12	48

Note: Compacting roll diameter, 560 mm (22 in.). Roll speed, 2.2 rpm. Roll gap (green strip), 3.5 mm (0.140 in.)

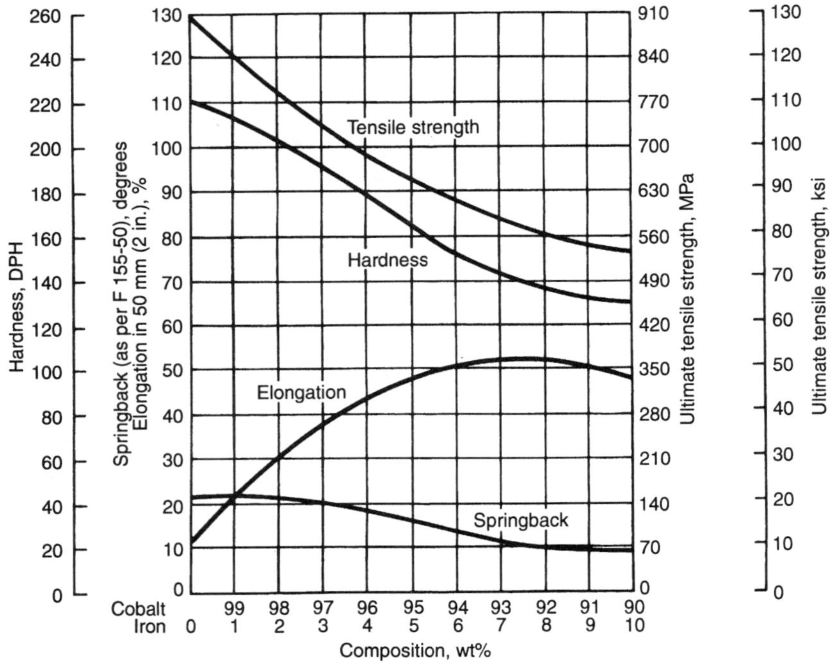

Fig. 10 Properties of cobalt-iron alloys. Manufactured by powder rolling of blended cobalt and iron powders

Table 2 Typical mechanical properties of hydrometallurgical powder rolled cobalt strip

Strip	Strip thickness, mm	Strip thickness, in.	Strip density, %	Ultimate tensile strength, MPa	Ultimate tensile strength, ksi	Yield strength, MPa	Yield strength, ksi	Elongation, %
Green strip	2.1	0.084	86	22	3	22	3	0
Sintered strip	2.1	0.084	86	201	29	195	28	5
Hot rolled strip	1.2	0.048	100	758	110	413	60	15
Cold rolled strip	0.9	0.036	100	1103	160	1100	159	1
Annealed strip	0.9	0.036	100	793	115	345	50	20

Note: Compacting roll diameter, 254 mm (10 in.). Roll speed, 6.0 rpm. Roll gap (green strip), 1.5 mm (0.06 in.)

fed directly to a furnace without coiling, where it is annealed and cold rolled to full density. Reductions in excess of 35% are necessary to achieve full density.

Several advantages are gained from producing high-purity nickel strip by roll compacting. Lower electrical resistivity is possible (73 to 79 $\times 10^{-6}\ \Omega \cdot m$, or 44 to 48 $\Omega \cdot$ cir mil/ft). Stability can be maintained at ±2% throughout the coil and from heat to heat. Wrought nickel offers an erratic ±6% tolerance. Work hardening rates for roll-compacted nickel strip are 25% less than for wrought nickel strip.

The lower softening (annealing) temperature coupled with high purity makes this nickel useful in clad metal combinations. In these applications, low and closely controlled annealing temperatures are required to minimize interdiffusion and to prevent incipient melting reactions.

Finished nickel strip produced from powder is virtually indistinguishable from strip produced from an ingot. Differences in physical properties are the result of compositional variances, rather than the method of fabrication. Typical physical properties for strip made from nickel powder are:

Density	8.90 g/cm^3
Coefficient of thermal expansion at 20–100 °C	14 μm/m · °C
Coefficient of thermal expansion at 20–500 °C	15 μm/m · °C
Thermal conductivity	86.23 W/m · K (0.206 cal/cm · s · °C)
Cold working capacity	Good
Hot forming capacity	Good
Hot work temperature	800–900 °C (1470–1650 °F)
Annealing temperature	700–900 °C (1300–1650 °F)
Magnetic properties	
Curie temperature	353 °C (667 °F)
Initial permeability	130
Maximum permeability	1240
Saturation induction	6.05 T
Remanance	3.25 T
Coercivity	23.87 A/m
Magnetostriction (soft) 1590 A/m	0.000032 mm/mm

Cobalt Powder Strip. Other metal powders can be roll compacted using methods similar to that of nickel powder. Powder characteristics and material properties alter processing parameters and conditions, however. For example, cobalt has a close-packed hexagonal crystal structure, which changes the surface morphology of the cobalt powder. Roll-compacted cobalt powder produces a green strip that is stronger and denser than the comparable green nickel strip. Typical mechanical properties of powder rolled cobalt strip at each stage of the production process are given in Table 2.

After sintering at 1100 to 1150 °C (2010 to 2100 °F), cobalt may be densified by hot or cold rolling. Cold rolling of pure cobalt is inhibited by the close-packed hexagonal crystal structure, which characteristically produces rapid work hardening and restricts cold reduction to approximately 25% between anneals.

The use of high-purity cobalt powders for the feed to a powder rolling mill results in a relatively ductile strip. Use of controlled conditions during annealing results in increased amounts of cubic phase (usually associated with cobalt above 445 °C, or 835 °F), which is retained at room temperature. The retained phase is stable and shows little tendency to transform to the hexagonal structure at room temperature.

The rapid work hardening characteristics of cobalt make it an ideal candidate for powder rolling. The initial strip can be rolled to a thickness very close to the required final gage. Because initial strip thickness is thin (1.5 to 2.0 mm, or 0.06 to 0.08 in.), cold rolling is minimized. Typical physical properties of powder rolled cobalt strip are:

Density	8.85 g/cm^3
Coefficient of thermal expansion at 20–100 °C	13 μm/m · °C
Coefficient of thermal expansion at 20–500 °C	14 μm/m · °C
Thermal conductivity at 70 °C (160 °F)	115.9 W/m · K (0.277 cal/cm · s · °C)
Cold working capacity	Poor
Hot forming capacity	Good
Hot work temperature	600–800 °C (1110–1470 °F)
Annealing temperature	800–1000 °C (1470–1830 °F)
Magnetic properties (Unworked high-purity powder rolled cobalt strip)	
Curie temperature	1121 °C (2049 °F)
Initial permeability	11.6
Maximum permeability	29.2
Saturation induction	1.9 T
Saturation field strength	310,000 A/m
Remanance	0.3 T
Coercivity	3,600 A/m
Coefficient of friction at 70 °C (160 °F) cobalt/cobalt	0.3

Production of heterogeneous alloys is possible with roll compacting processes. In cobalt-iron materials, for example, the presence of the iron promotes a suppression of the cubic to hexagonal phase transformation in cobalt. The ductility of the cobalt-iron mixture increases with increased iron content. The effects of adding iron powder to cobalt powder on the properties of cobalt strip are shown in Fig. 10.

Specialty Applications

Roll compacting facilitates the production of small coils in which minor compositional changes can be made during blending. Because roll compacting produces materials with uniform and re-

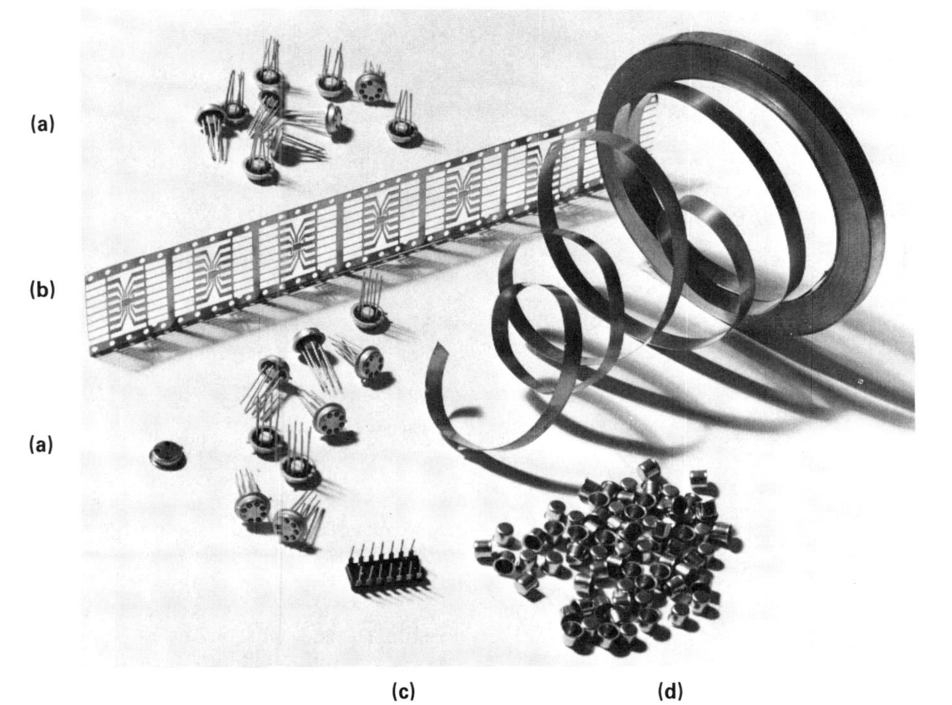

Fig. 11 Typical roll-compacted nickel strip applications. (a) Pure nickel semiconductor caps. (b) Nickel-iron lead frame. (c) Nickel-iron Dual Inline Package integrated circuit. (d) Nickel-iron resistor end caps

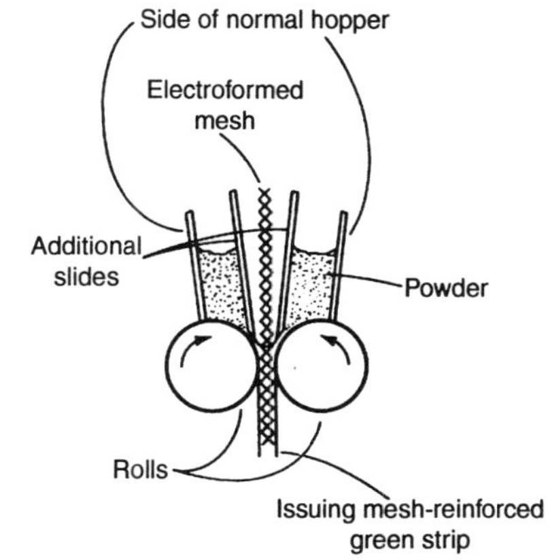

Fig. 12 Rolling of strip from nickel powder. Designed to incorporate mesh into strip

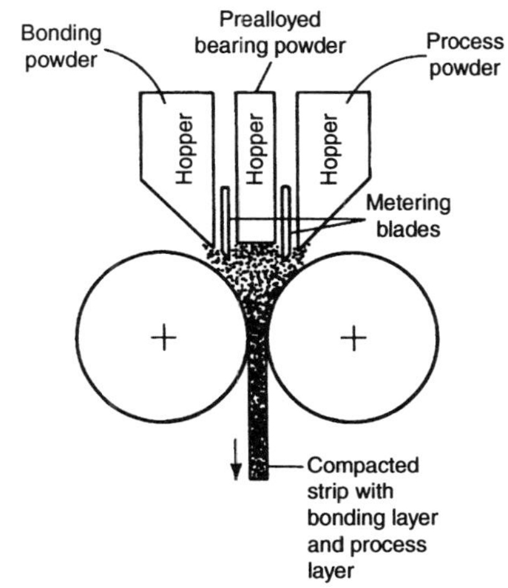

Fig. 13 Rolling of strip for bearings. Using bonding powder, prealloyed bearing powder, and process powder

liable properties, many specialty materials are produced by these methods (Ref 6).

Specialty P/M strip produced by roll compacting includes various compositions of nickel-iron strip produced for controlled expansion properties and special copper-nickel-tin alloys. The latter must meet stringent property requirements, such as strength, formability stress relaxation, electrical conductivity, solderability, and corrosion resistance. Casting of this material produces elemental segregation during solidification, which can be avoided by using P/M fabrication techniques.

Typical applications of roll-compacted strip materials include:

- High-purity nickel strip parts for heart pacemaker batteries
- Nickel-iron strip for resistor caps (Fig. 11)
- Nickel-iron lead frames (Fig. 11)
- Ni-Fe-Co cans for glass-sealed electronic parts
- Al-SiC composite strips (Ref 7)
- Ti-Al foils for composite manufacture (Ref 8 and 9)
- Ni-Ti strip for shape memory (Ref 10)

Figure 12 illustrates a method for producing porous strip of pure nickel for alkaline battery and fuel cell electrodes. In this procedure, an electroformed nickel screen or mesh is fed into the center of the powder hopper and incorporated into the rolled strip to produce highly porous electrodes with adequate green strength. The same system can be used to feed wires into a green strip for extra strength.

Composite Bearings

Roll compacting also can be used for producing composite, or "sandwich," materials. An example of such a roll-compacted composite material is bimetallic strip used in producing main and connecting rod bearings (Ref 11). A change in automotive emission standards led to the use of these new materials to replace traditional copper-lead sleeve-bearing components. The rolled strip consists of a layer of Al-8.5Pb-4.0Si-1.5Sn-1.0Cu prealloyed powder "sandwiched" to a pure aluminum layer.

The production setup for roll compacting such a composite structure is shown in Fig. 13. Three powder hoppers are required, as well as powder flow blade that controls the flow of the powders into the roll gap. The coil of the composite strip is then sintered, and eventually the pure aluminum layer is roll bonded to a steel backing material—an operation that is common in the bearing industry.

Powder Melting. An induction melting furnace is charged with elemental aluminum, copper, and silicon (Ref 11). At an intermediate elevated temperature, lead and tin are added to the melt. Furnace temperature is raised to the single-phase temperature (925 °C, or 1070 °F), and 38 °C (100 °F) of superheat is added to provide a safety margin. The induction current of the furnace creates a stirring action that ensures complete dissolution of the lead and tin to form a true single-phase solution.

From the induction furnace the molten alloy is poured through a launder into a gas-fired tundish furnace. An extension on the bottom of the tundish crucible holds a ceramic nozzle that meters the molten alloy at a controlled rate.

Atomization. The thin metal stream falls vertically into the atomizing chamber, where it is disintegrated into discrete particles and rapidly solidified. Powder particles fall into a 6 m (20 ft) high by 1 m (4 ft) diam collector. They then pass through a cyclone separator to remove the fines and a powder screen to separate oversized particles.

Small, finely dispersed lead-tin alloy particles are contained in a hypoeutectic aluminum-silicon-copper matrix. Because of solubility considerations, the lead-tin bearing constituent remains intact throughout subsequent processing. Controlled distribution of the lead-tin ensures the development of desired mechanical properties.

Consolidation. A consolidation process is selected to produce a strip of prealloyed aluminum-lead alloy that can be ultimately roll bonded to a low-carbon steel backing strip.

Previous experience with wrought aluminum-based bearing materials has proven them unsuitable for direct bonding to a steel liner. Ultimate fatigue resistance of the bearing material depends on both the intrinsic strength of the alloy

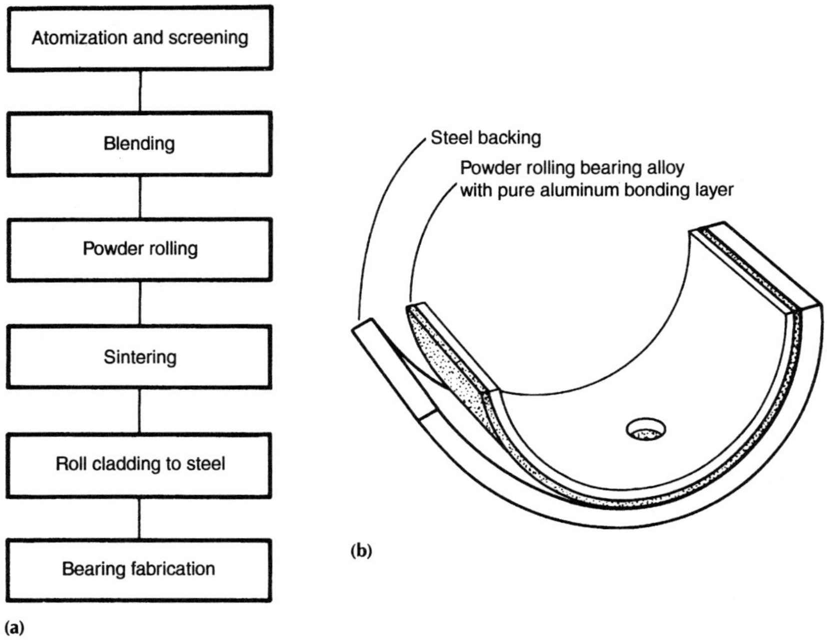

Fig. 14 Fabrication of powder-rolled sleeve bearing. (a) Flowchart. (b) Sleeve-bearing component

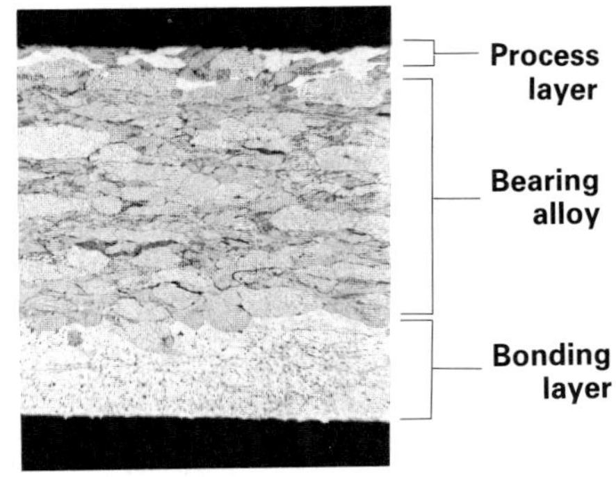

Fig. 15 Cross section of unsintered powder rolled aluminum-lead strip. Original powder particle boundaries are visible. 25×

and the integrity and strength of the bond between alloy layer and steel backing.

When aluminum alloys with soft phases such as lead or tin are bonded directly to steel, the bond interface necessarily contains microscopic discontinuities that occur where lead or tin precipitates are in direct contact with the steel. These imperfections create brittleness in the interface and act as sites for accelerated fatigue crack propagation and bond separation under service conditions. Consequently, wrought aluminum alloys containing appreciable amounts of soft bearing phase are almost always used with a pure aluminum or nickel bonding layer next to the steel. Powder rolling provides a convenient means of incorporating a pure aluminum bonding layer during the consolidation of the aluminum-lead prealloyed powder.

A third layer on the opposite side of the prealloyed layer from the bonding layer is necessary to ensure that the strip exits cleanly from the rolls of the powder rolling mill. This process layer, which is made from a blend of aluminum and lead-tin particles, is machined off in subsequent bearing finishing operations.

The three distinct compositions are fed into the compacting mill, as shown in Fig. 13. The work rolls compact these powders into a three-layer strip, which is close to 100% of theoretical density as it leaves the roll nip. A flowchart of the fabrication of powder rolled sleeve bearings is shown in Fig. 14. The production mill used to fabricate these bearings resembles a conventional four-high rolling mill (Fig. 1b). A cross-sectional view of the unsintered structure of the aluminum-lead alloy strip is shown in Fig. 15.

Sintering. The green strip is then sintered, which improves the morphology of one of the phases (silicon) and provides the necessary diffusion across particle boundaries to achieve excellent strength and ductility. The finely distributed lead phase is unaffected. Further processing consists of cladding the aluminum strip to a steel backing, followed by blanking, forming, and machining of the bearing halfshells.

Bearing materials must possess a unique combination of properties, including fatigue strength, seizure resistance, wear resistance, and corrosion resistance. The performance characteristics of several bearing alloys are given in Table 3.

Recent Developments

Roll compacting of metal powders into strip is a relatively new process. Research and development currently is being conducted on many different materials to produce a superior material as well as to achieve properties not possible with conventional ingot metallurgy (Ref 12).

Roll compacting is especially useful for the continuous and semicontinuous preparation of special-type strip and sheet materials including the following:

- Al-base sheet and strip (Ref 6, 13)
- Ni-base strip applications
- Ti-base sheet and strip (Ref 6 and 10)
- Composites (Ref 7)
- Stainless steel and copper strip (Ref 6)

Iron strip made from high-purity iron powder has produced a low-carbon sheet with properties equivalent to those of conventional low-carbon sheet steel for production of containers for consumer goods. The following are typical properties of thin-gage roll-compacted low-carbon iron powder:

Gage	0.30 mm (0.012 in.)
Yield strength	290 MPa (42 ksi)
Tensile strength	330 MPa (48 ksi)
Elongation	34%
ASTM grain size	11–12
Limiting draw ratio	2.10
Strain ratio	1.3–1.6
Earing	3%
Work hardening exponent	0.220

Source: Ref 14

Table 3 Typical properties of common bimetal bearing alloys

Alloy composition	Fatigue (Underwood life)	Surface action at 150 °C (300 °F): Coefficient of friction	Wear scar width, mm	Wear scar width, in.
Al-8Pb-5Si-1.5Sn-2Cu (Clevite 66)	200 h at 49 MPa (7 ksi)	0.21	4.52	0.178
Cu-10Pb-10Sn (SAE 792)	200 h at 69 MPa (10 ksi)	0.22	5.15	0.203
Cu-23Pb-3Sn-2Zn (SAE 794)	200 h at 49 MPa (7 ksi)	0.17	5.08	0.200

Note: The Underwood test simulates the cyclic loading of a rod bearing resulting from the powder stroke in an internal combustion engine. Pressure is based on unit load. Wear scar and coefficient of fraction are obtained from the standard LFW-1 test machine, Faville LeVally Corp.

REFERENCES

1. W.D. Jones, *Fundamental Principles of Powder Metallurgy*, Edward Arnold, 1960, p 924
2. G. Naeser and F. Zirm, *Stahl Eisen*, Vol 70, 1950, p 995–1003
3. H. Franssen, *Z. Metallkd.*, Vol 45, 1954, p 238
4. P.E. Evans and G.C. Smith, "The Continuous Compacting of Metal Powders," Special Report 58, *Symposium on Powder Metallurgy*, The Iron and Steel Institute, London, 1956, p 131–136
5. M.H.D. Blore et al., Pure Nickel Strip by Powder Rolling, *Met. Eng. Quart.*, Vol 6 (No. 2), 1966, p 54–60

6. M. Roman, Roll Compaction of Metal Powders in Manufacturing Special Materials, *First Int. Conf. on Materials and Manufacturing Technologies (Matehn '94),* Technical University of Cluj-Napoca Press, Romania, 1994, p 215–220
7. C. Scorey, Aluminum-Silicon Carbide Composite and Process for Making the Same, U.S. Patent No. 5,384,087, 1995
8. I. Morgenthal, B. Kieback, G. Hubner, and D. Nerger, Preparation of Ti-Al Foils by Roll Compaction of Elemental Powders, *Powder Metallurgy World Congress (PM '94),* Vol 2, Editions de Physique, 1994, p 1247–1250
9. S.C. Jha and J.A. Forster, Producing Titanium Aluminide Foil from Plasma-Sprayed Preforms, *JOM,* Vol 45 (No. 7), July 1993, p 57–59
10. K. Kusaka and T. Shimizu, Production of Ni-Ti Alloy Strip by Powder Rolling and Its Shape Memory Behaviour, *Denki Seiko (Electr. Furn. Steel),* Vol 58 (No. 4), Nov 1987, p 226–234
11. M.L. Mackay, Innovation in P/M: An Engine Bearing Material, *Met. Prog.*, Vol 111 (No. 6), 1977, p 32–35
12. D.H. Ro, M.W. Toaz, and V.J. Moxson, The Direct Powder Rolling Process for Producing Thin Metal Strip, *J. Met.,* Vol 35 (No. 1), Jan 1983, p 34–39
13. D.H. Ro and M.W. Toaz, Direct Powder Rolling of High Strength Aluminum Alloy Strip, *Processing of Metal and Ceramic Powders,* R.M. German and K.W. Lay, Ed., TMS/AIME, 1981
14. A.J. Klein et al., A Modern Mini-Mill Process for Powder Sheet Rolling, *Prog. Powder Metall., Vol 31, 1981, p 329–345*

Spray Forming

Alan Lawley and Roger D. Doherty, Department of Materials Engineering, Drexel University

SPRAY FORMING is a near-net-shape solidification process that takes place by capturing a spray of gas atomized metal or alloy droplets on a moving substrate. The metal flow rates are high, typically 0.2 to 2.0 kg/s, and multiple sprays produce even larger rates. On the basis of experimental studies and process modeling (Ref 1), it is known that at the moment of impact the heat content of the spray is equivalent to a solid fraction in the range 0.5 to 0.8. This is material, in the form of small droplets, that has been rapidly solidified by high velocity gas while in flight to the substrate in a few milliseconds. If the process is performed correctly, a large fraction of the sprayed material, typically 0.8 or more, is captured on the substrate where it forms a deposit with the geometry of a billet, tube, sheet/plate, or other shape with dimensions ranging from a few to hundreds of millimeters. The remaining solidification then occurs within the deposit at a much slower rate over times that correlate with the size of the deposit, ~10 to 100 s or more (Ref 1, 2). The resulting as-sprayed deposits exhibit low porosity and a homogeneous, fine-grained structure. Given appropriate control of the spray and substrate movement, deposits with well-controlled shapes can be produced (Ref 3). Thus, the technology is capable of producing shaped castings with microstructures comparable to or superior to those characteristic of wrought ingot alloys.

Grant (Ref 4) provides an overall review of all aspects of the process and Lavernia and Wu (Ref 5) give an extensive overview of both atomization and deposition, the two major processes integral to spray forming. Their monograph provides a complete coverage of publications in the area up until 1995. The reports of three international conferences on spray forming (Ref 6–8) give an account of the range of application of the process and show how the subject has evolved in recent years with increasing sophistication of process modeling.

In this article, a description is given of the spray forming process and the attendant metallurgical characteristics of the product. This description is followed by a review of the commercialization and application of spray forming illustrated in terms of several product forms. The article concludes with a detailed analysis of modeling and evolution of microstructure in spray forming.

Process Description

In spray forming in the Osprey mode, a source of molten metal is converted into a spray of droplets by gas atomization. The droplets are collected on a substrate where they aggregate to form a high density preform (Ref 5, 9), as shown in Fig. 1. The preform can be used in the as-sprayed condition or after conventional working operations. The production of billets, strip, and tubing by spray forming is illustrated schematically in Fig. 2.

For commercial viability in spray forming in the Osprey mode, close tolerances in shape and dimensions, consistency in microstructure, and a high product yield are essential. This viability requires an understanding of and control over the effects of several independent process parameters: melt superheat, metal flow rate, gas pressure, spray motion, spray height (distance between the gas nozzles and the substrate), and substrate motion (substrate rotation speed, withdrawal rate, and tilt angle). Several parameters can be selected prior to, but not during, the process, such as the diameter of the metal delivery nozzle, the atomizer design, and the substrate material.

Primary process parameters that can be changed during the operation (i.e. on-line) are listed in Table 1. Advantages and disadvantages of the Osprey process are cited as follows:

Advantages

- Densities >98% of theoretical
- Fine equiaxed grains
- No macroscopic segregation
- Absence of prior particle boundaries
- Enhanced mechanical properties
- Material/alloying flexibility
- High rate of deposition, ≤2 kg/s

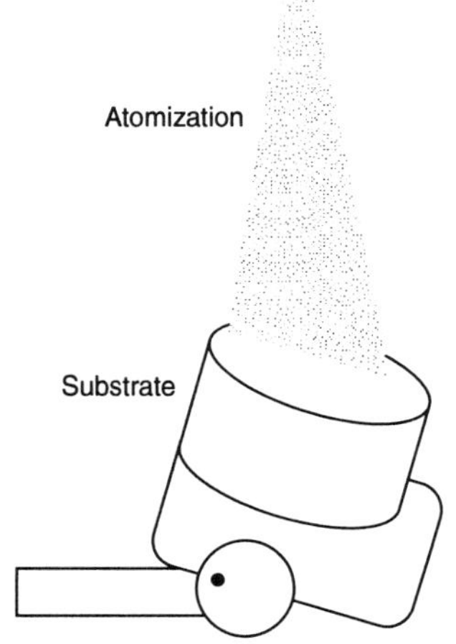

Fig. 1 A schematic of spray forming in the Osprey mode

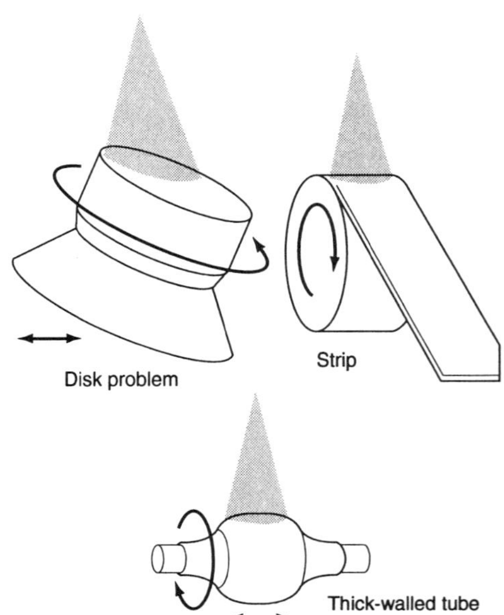

Fig. 2 A schematic of the production of billet, strip, and tubing by spray forming

Table 1 Representative on-line parameters in Osprey spray forming

Melt superheat, °C	10–200
Melt flow rate, m^3/s	10^{-5}–10^{-3}
Atomizing gas	N_2, Ar, He, or mixed gases
Gas pressure, MPa	0.5–1.0
Scanning frequency, Hz	0–10
Spray height, mm	≥350

Table 2 Tensile properties of stainless steel

	Yield strength, N/mm^2	Tensile strength, N/mm^2	Elongation, %
Spray formed (as-deposited)	260	618	57
Ingot metallurgy (as-extruded)	290	632	58
Spray formed (cold Pilger rolled)	980	1058	15
Ingot metallurgy (cold Pilger rolled)	931	1007	17

Sanicro 28: Fe-26Cr-30Ni-3.5Mo. Source: Ref 9

Limitations

- Overall yield
- Shape control
- Compositional control in microalloying

Metallurgical Characteristics

Metallurgically, the mechanical properties of spray cast alloys normally meet or exceed those of the counterpart ingot alloy (Ref 9). This fact is illustrated in Table 2 for stainless steel tubing in the spray cast condition and after cold working.

The advantages of spray forming in its application to aluminum alloys are vested in low oxygen and hydrogen levels and a refined uniform microstructure (Ref 9). In the absence of oxide films or particle boundaries, minimal hot working is required to optimize mechanical properties. In the standard 2000 and 7000 series aluminum alloys, spray forming enhances damage tolerance. This is attributed to refinement of the grain structure (~50 μm) and of the primary phase distribution (~1 μm), particularly in alloys that are sensitive to the presence of iron-containing intermetallics or to the distribution of transition element phases. This refinement results in attractive increases in ductility, toughness, and fatigue resistance. Fracture toughness data for spray formed 7075, as a function of yield strength, are given in Fig. 3; corresponding data for the direct chill (DC) cast alloy are included for comparison.

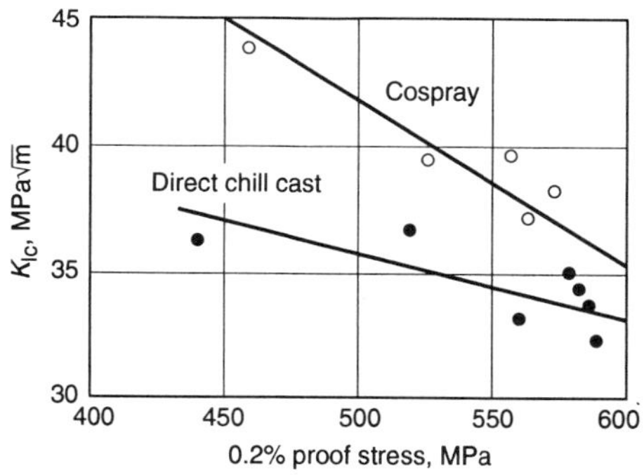

Fig. 3 Fracture toughness of 7075 aluminum alloy as a function of strength and mode of fabrication. Source: Ref 9

Alloy modification via spray forming has led to further strengthening of the 7000 series aluminum alloys (Ref 9). For example, strengths from 700 to 800 MPa have been achieved by increasing the levels of zinc and magnesium. In the T6 condition, fracture toughness is 27 to 32 MPa$\sqrt{m}$, with attendant improvements in fatigue resistance compared to 7075. Significantly, a 30% increase in strength has been achieved with this new alloy composition compared to 7050 at the same toughness level. Thus the new alloy has the potential for replacing 7075 and 7050 extrusions and 7050 forgings.

Spray Forming Applications

Metallurgically, the benefits of spray forming in the Osprey mode can be realized commercially only if the resulting products are cost effective in the marketplace. To this end, research and development programs have focused on technical improvement and cost reduction in spray forming (Ref 10, 11). To compete with conventional fabrication technologies (ingot casting, forging, rolling, etc.) spray plant throughput has been maximized, and process efficiency has been improved (Ref 11). Throughput has been improved via a larger melt size and preform size, higher production rates, and increased productivity. Improved process efficiency is reflected in lower gas consumption rates, higher yields of deposition, and improved dimensional control in spray formed products.

A review of the commercialization of spray forming in the Osprey mode is given here for several product forms.

Rolls

Sumitomo Heavy Industries Foundry and Forging Company (SHIFF) has been manufacturing spray formed rings in high-chromium cast irons and high-carbon, high-speed steels since 1991. After downstream processing, the rings are fitted to a roll mandrel for use in round bar, flat bar, wire rod, and section mills, and in finishing or intermediate stands. The rings exhibit a fine, rapidly solidified carbide microstructure leading to improved thermal fatigue resistance and a longer life compared to conventionally cast products. Examples of spray formed rolls are given in Fig. 4; roll sizes up to 800 mm diam by 0.5 m length have been fabricated.

More recently, interest has centered on the direct manufacture of large clad rolls for use in hot and cold strip mills. Participants in this joint project, funded partially by the U.K. Department of Trade and Industry, were Forged Rolls (U.K.),

Fig. 4 Spray formed rolls. Source: Ref 11

Fig. 5 Plant for spray forming clad rolls. Source: Ref 11

British Rollmakers, Sheffield University, and Osprey Metals. Arbors up to 400 mm diam by 1 m length have been clad with cold and hot mill roll alloys and with high-speed steel. Figure 5 illustrates the facility with a hot spray clad roll in place. Key to the process is a sound metallurgical bond at the arbor-spray deposit interface.

Tubing

In Sweden, Sandvik produces stainless steel tubes up to 400 mm diam by 8 m length with a wall thickness ≤50 mm by spray forming. The spray process configuration for tube fabrication using a single atomizer is shown in Fig. 6. By using multiatomizers, it is possible to minimize porosity on the internal bore of the tube. Spray cast tubes of In625 are shown in Fig. 7.

A potential market exists for bimetallic tubing with a corrosion resistant layer sprayed onto the outside of a collector mandrel. Projected applications include boilers, incinerators, and waste heat recovery plants. To this end, Sandvik has developed a corrosion-resistant spray forming alloy Fe-Cr-Ni-Mo (Sanicro 65) for service in waste incinerators.

Rings

In the United States, Howmet operates a plant capable of spray forming superalloy ring blanks up to 800 mm diam by 0.5 m length. These are used for ring/casing components in high thrust gas turbine engines. A study of Udimet 720 has shown that the ring preforms can be processed subsequently by ring rolling or hot isostatic pressing.

Billets

Round billets can be fabricated by spray forming in vertical (Fig. 8) or horizontal (Fig. 9) plants. The two techniques differ only in relation to the collector movement, which is either vertical or horizontal. The former is used currently on aluminum and copper alloy billet plants. Osprey Metals utilizes the horizontal configuration for processing high alloy steels.

With a melt size of 1.2 tonne, tool steel and precipitation hardened stainless steel billets have been sprayed at diameters up to 400 mm by 1.2 m length (Fig. 10). Significant cost reductions have been achieved by the use of twin atomizers, higher metal dispensing rates, a reduction in gas consumption, and an increase in yield of the deposit.

Aluminum alloy billets are normally spray formed in the vertical mode (Fig. 8); Sumitomo Light Metals (Japan) uses this configuration on a commercial scale. The spray formed billets are subsequently extruded or forged into the final product configuration. Current billet size is 250 to 400 mm diam by 1 to 1.4 m length. Smaller diameter billets can be extruded without machining because of the tight tolerances achieved. Primary applications for spray formed aluminum-silicon alloys are in the automotive field (e.g., cylinder liners, pistons, connecting rods).

Other spray formed aluminum alloys in billet form are at an advanced stage of development. These include aluminum-zinc alloys, ultralow density aluminum-lithium alloys, and Al-Si-X ternary alloys, which exhibit a low coefficient of thermal expansion (Ref 11).

Copper alloy billets up to 300 mm diam by 2.2 m length are being spray formed on pilot plants in operation at Wieland (Germany) and Swissmetal (Switzerland). Production levels are not high because the emphasis has been on alloy development. Swissmetal produces Cu-Ni-Sn alloys that combine high strength with acceptable levels of electrical conductivity, copper-bronzes with a high tin content that exhibit enhanced levels of hot and cold workability, Cu-Cr-Zr alloys for high-temperature stability, and copper-carbon alloys in which the graphite particles enhance wear properties.

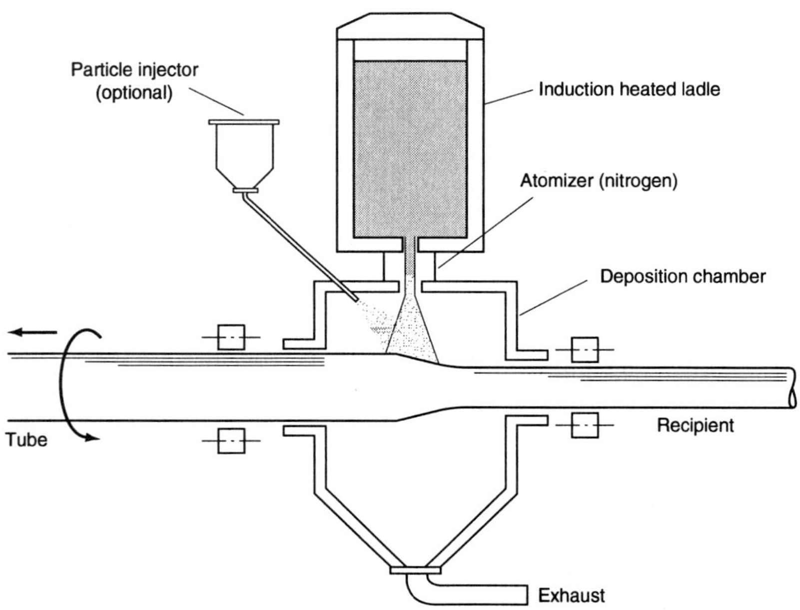

Fig. 6 Tube fabrication by spray forming. Source: Osprey Metals, Ltd.

Modeling and Microstructure

The process of spray forming involves a series of steps in which a molten alloy at a selected superheat is disintegrated by high velocity gas atomization into a spray of rapidly cooled droplets possessing a range of sizes and thermal conditions. These droplets then impact on a substrate to give a high yield of a partially solid deposit of controlled shape. This deposit is cooled by the gas stream, and solidification is completed at much slower rates (comparable to those in ingot casting) than the initial cooling rates in the spray. The steps in the process are complicated, and many of them are not yet fully understood, which makes them unpredictable at least from first principles. The process has, however, been modeled, and some understanding of the microstructure characteristics of spray formed alloys has been obtained (Ref 2).

The successful models, which include those of Mathur et al. (Ref 1, 12), Lavernia et al. (Ref 13, 14), Trapaga et al. (Ref 15), Tsao and Grant (Ref 16), Grant et al. (Ref 17, 18) and others (Ref 19–21), combine detailed numerical models (for those processes in which the physics is well understood) with empirical input parameters for the other processes or steps that cannot, as yet, be modeled. The model developed at Drexel University is illustrated in Fig. 11 (Ref 3, 12). It shows the preset process parameters (e.g., metal superheat, gas pressure, metal flow rate, and gas to metal ratio [GMR]) and the measured (empirical) parameters (e.g., the gas velocity field, the particle size distribution [PSD], the radial mass flux distribution [RMF], and the droplet sticking efficiency [SE]). At the present time, none of the last four parameters can be predicted reliably from first principles. The atomization stage is similar to gas atomization in powder production and has been reviewed in detail by Lawley (Ref 22, 23), Yule and Dunkley (Ref 24), and Lavernia and Wu (Ref 5). Modeling methods are available for predicting gas flow fields using numerical techniques (Ref 19); physics-based methods are available for estimating the mean particle size in gas atomization (Ref 25); and empirical methods are available for predicting the PSD, such as those of Lubanska (Ref 26). However, these methods, when tested in spray forming studies, do not appear to be sufficiently reliable to be useful in modeling the detailed thermal processes intrinsic to spray forming. In addition, the current models of gas atomization do not address the angular dependence of particle density in the spray.

To overcome these difficulties, it has been found necessary in experimental studies of spray forming to measure all the required empirical parameters in gas atomization. One such parameter is the gas velocity field in the spray chamber, usually determined by Pitot tube measurements in the absence of a stream of molten metal (Fig. 12). Results of empirical determinations of the measured PSD and RMF of the alloys investigated at Drexel University and at the Naval Surface Warfare Center (NSWC) were re-

Fig. 7 Spray formed tubes of In625 alloy. Source: Osprey Metals, Ltd.

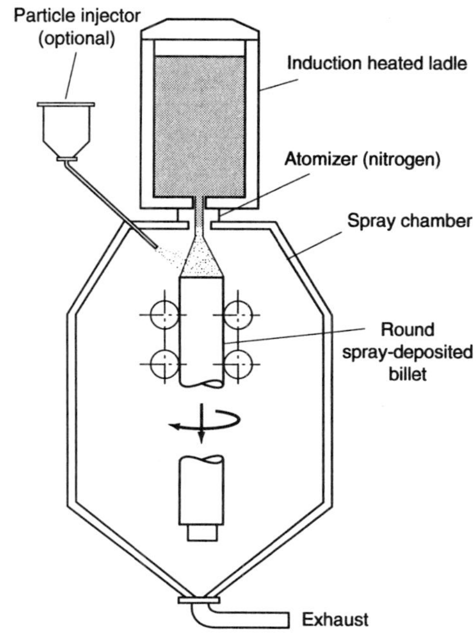

Fig. 8 Vertical mode billet fabrication by spray forming. Source: Osprey Metals, Ltd.

ported and analyzed by Cai et al. (Ref 3, 27). The method used was to spray the alloys (nickel-base alloy In625, high-strength low-alloy steel, and Cu-6Ti) under a range of gas pressures, metal flow rates, and superheats in a modified Osprey facility at NSWC. The height of the chamber was extended to ensure that solidification of the largest droplets was complete before capture in a series of concentric tubular collectors (Fig. 13). Concurrently, this experiment yielded measurements of the RMF (Ref 3). Analysis of these observations showed a general tendency for the mass mean particle size to fall with increased radial distance from the axis of the spray, as shown in Fig. 14 (Ref 27). Empirical results such as these can then be fitted to descriptive equations in order to model the spray conditions of the alloys in the same facility at NSWC and in a similar facility at Drexel University. A comparable experimental investigation of PSD and RMF for tin, steel, and copper was conducted by Uhlenwinkel and Bauckhage (Ref 28).

With current heat and mass transfer techniques, it is possible to predict the velocity and cooling rate of an individual droplet of a given size in flight to the substrate (Fig. 11) (Ref 5–8, 12–18). By integration of the resultant heat content of all the droplets in the PSD, it is then possible to calculate the enthalpy of the spray at the moment of impact onto the growing deposit. If it is assumed that thermal and chemical equilibrium are achieved rapidly after deposition, an equivalent value of the effective fraction of liquid in the spray, $f_l(s)$, can be calculated. This calculated value of $f_l(s)$ is a more meaningful physical parameter than the integrated enthalpy in the spray from which $f_l(s)$ is derived. With the use of empirical values of the PSD and gas flow fields, the predicted values of $f_l(s)$ should be reliable. The fraction of liquid on the surface of the deposit, $f_l(d)$, under the spray for a billet shaped preform is predicted to be very close to the calculated fraction of liquid in the spray. However, for other geometries such as tubes, the fraction of liquid on the deposit can be less than that in the spray by virtue of cooling of the deposit when it rotates out of the spray and into the gas stream.

To calculate the rate of growth of the deposit, various inputs are needed. The first of these is the rate of arrival of metal from the spray at each point on the substrate surface. Cai and Doherty (Ref 3, 29) have developed a full three-dimensional shape model that can predict the rate of growth at all points on the deposit, assuming a known sticking efficiency. Examples of the predicted shapes are reproduced in Fig. 15. The

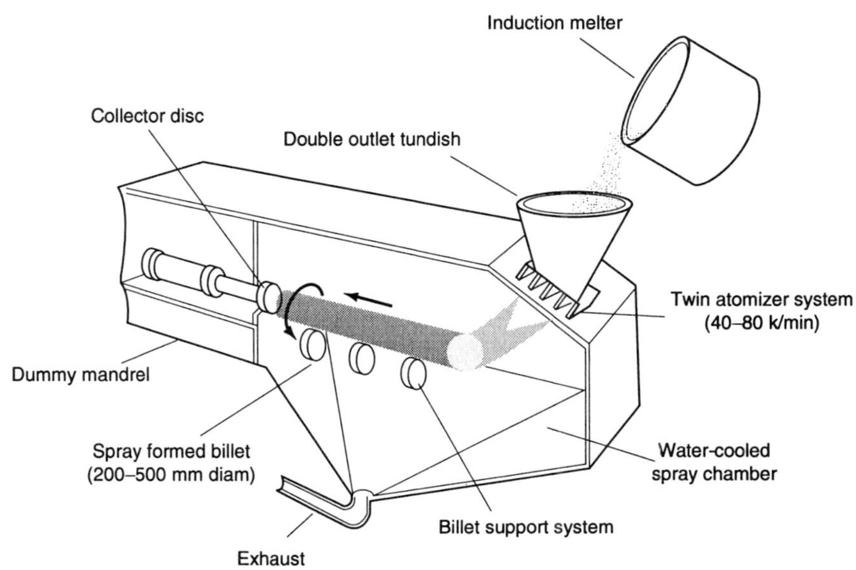

Fig. 9 Horizontal mode billet fabrication by spray forming. Source: Osprey Metals, Ltd.

Fig. 10 Horizontal mode spray formed tool steel billet. Source: Ref 11

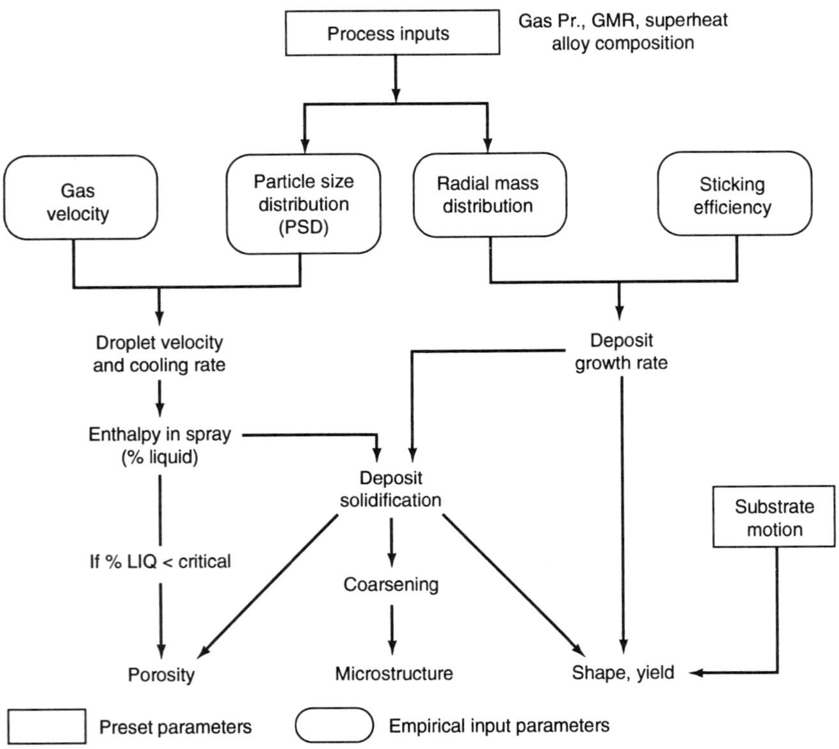

Fig. 11 Flow chart of the Drexel University model of spray forming illustrating empirical inputs and predicted outputs. Source: Ref 3

shape model can be developed further to delineate the substrate motion required to give specific shapes—including those of complex geometry such as a curved tube where, by altering the substrate motion, a uniform wall thickness was designed (Ref 3, 29). These predictions do, however, require a knowledge of the sticking efficiency of the droplets onto the deposit.

Deposit Cooling and Solidification

One of the early predictions of the spray models was the result that successful spray deposition requires a partially solid-partially liquid spray in which $1 \geq f_l(s) \geq 0$ (Ref 1). The droplets arriving at the surface of the deposit at velocities ~50 to 100 m/s are predicted to be fully solid if small, fully liquid if large, and for the majority of the mass of the spray of intermediate size, partially solid and partially liquid on impact. Given the high thermal conductivity of metallic alloys, rapid thermal equilibration of all the deposited droplets is expected with a partially solid and partially liquid structure on the surface of the deposit. Initial models of the thermal behavior of the deposit were one dimensional and predicted that after deposition of the partially molten alloy, most of the subsequent cooling was caused by the high velocity gas, with only limited cooling taking place through the substrate—at least for material deposited some distance away from the substrate. The model predicted complete solidification times (10–200 s), that were confirmed by embedded thermocouple measurements (Ref 5). Later models, such as those of Grant et al. (Ref 17, 18) and Cai et al. (Ref 3), confirm the qualitative form of these results, as shown in Fig. 16 and 17. Models of thermal deposition were two dimensional, incorporating gas cooling from the side and top surfaces of the deposited billets (Ref 3, 29). For billets, the two-dimensional thermal models appear to capture the major features of the cooling and solidification process (Ref 29). After a short period of deposition, the temperature at the top surface of the deposit achieves a value corresponding to a partially molten surface. The depth of this incompletely solidified region increases, and at the center of the billet time required for complete solidification is tens of seconds. There is, however, a critical need for a three-dimensional thermal model for more complex shapes, such as a rotating tube (Ref 27, 29). In tube deposition, which is one of the major commercial applications of the technology (Ref 6–11), the thermal conditions are complex because the tube is subjected to additional cooling as it rotates into and out of the spray in the presence of the cooling gas stream. As a result, it is expected that for the same enthalpy in the spray, $f_l(s)$, there will be a lower fraction of liquid on the surface of the deposit during spray forming of tubes than of billets.

An initial two-dimensional model for the spray forming of tubes (Ref 27) captured one significant aspect of this geometry. In addition to the usual buildup of surface temperature with time after overcoming the initial chilling effect of the substrate, there was a second but opposite effect in the tube. As the length of the deposited tube grew, the cooling effect of the gas stream on the as-deposited tube resulted in a decrease in the surface temperature under the spray with time as shown in Fig. 18 (Ref 27). Also shown in Fig. 18 is the chilling effect of the tubular substrate, which causes rapid cooling of the deposit in contact with the mandrel even after extensive deposition. The limitations of the two-dimensional model for this geometry are clear from Fig. 18. The thermal model of a billet shows that the fraction of liquid on the surface of the deposit, $f_l(d)$, approaches that in the spray, $f_l(s)$ (Fig. 17).

It should also be noted that with a rapidly rotating substrate, particularly in tubes, embedded thermocouple measurements are difficult to make (Ref 5). However, for plates sprayed under linear nozzles, embedded thermocouple measurements can be made, and these show close agreement with the thermal model (Ref 30) shown in Fig. 19 and 20. In particular two features of the process are shown in Fig. 20. First, the cooling rate in the solid-liquid phase field in the presence of a preheated steel substrate (to offset the chilling effect of a cold steel substrate) is low. The second feature is the thickness in the deposit (10% at 400 °C and 40% at 30 °C) during which the chilling effect of the substrate causes the deposit to form without the temperature rising above the solidus temperature. Under these conditions, the only liquid present is the transient liquid in the spray as it arrives at the surface of deposition. This transient liquid will disappear rapidly during thermal equilibration with the deposit, and the partially liquid region at the deposit will form only after the deposit is relatively thick (Fig. 18). This result is of major significance in the formation of a deposit of low porosity. Although this study pertains to continuous deposition, the experimental and modeling methods are applicable if the spray is oscillated in the growth direction, thereby allowing for additional cooling at the deposit surface when the spray is temporarily elsewhere. Similar conditions are inevitable in the manufacture of other shapes, most significantly tubes.

Consolidation at the Surface of the Deposit

The two critical unknowns in understanding the spray forming process are the sticking efficiency of droplets onto the growing deposit and the level of porosity in the final deposit (Fig. 11). As discussed, the droplet velocity and thermal models of the conditions in the spray can predict the velocities and the enthalpies of the individual droplets at the moment of impact. In most of the present models, the enthalpy of the spray is averaged over the full radial angle of the spray, ignoring the slightly different gas velocities and droplet size distributions in different radial directions. A correction for the different values of $f_l(s)$ at different angles from the spray axis could be handled in the numerical models, but to-date, the resulting improved predictions have not been readily usable given the difficulties of experimental observation. Thus, although reasonable insight into the thermal condition of the spray is available, at present there is no physically based model that attempts to describe the complexity of the impact of a very high density of high-velocity, partially liquid droplets onto a similar partially liquid surface.

How much of the material arriving at the surface of deposition will stick? Also, how will the sticking efficiency vary with droplet size, impact angle, and the thermal condition (fraction of liquid) of the individual droplets and of the deposit surface? Other important and incompletely answered questions include what determines the resultant microstructure of the deposit, in particular the level of porosity and the grain size? These parameters are likely to be determined, at least in part, by the processes occurring during impact of the high velocity spray onto the deposit surface. Although not yet predictable, each of these phenomena are capable of investigation experimentally.

Sticking Efficiency

Because of the fine partially molten dendritic structure in the majority of the particles, it is believed that dendrite fragmentation will occur, breaking up the solid into small pieces, such as one to a few dendrite arm spacings (Ref 2–5). Typical solidification and flight times in spray deposition are only a few milliseconds and from the standard model of dendrite arm coarsening this leads to a dendrite arm spacing (cell size) of ~1 μm (Ref 31–32). This result is characteristic of the fine scale in the dendritic microstructure observed in fully solidified powder droplets, such as the overspray particles that failed to hit or stick to the deposit (Ref 2). Fragmentation of the dendrites in the partially molten droplets as they impact the growing deposit at velocities of ~100 m/s appears to provide the basis for the fine grain structure characteristic of spray formed materials (Ref 2–4).

This grain fragmentation process produced by high velocity impact is probably similar to that found in stircasting (rheocasting) (Ref 33). In stircasting, solidification of the dendritic struc-

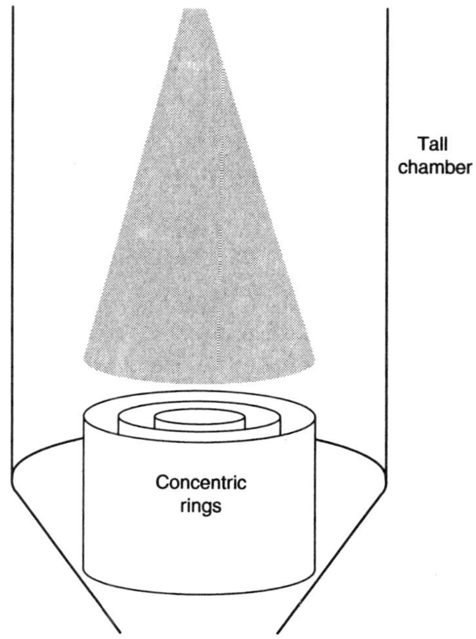

Fig. 13 Schematic of system for collecting fully solidified particles from a spray cone for measurement of radial mass flux and particle size distribution. The distance of fall (standoff distance) has been increased from that used in spray forming to ensure a fully solidified spray at the collector. Source: Ref 3, 27

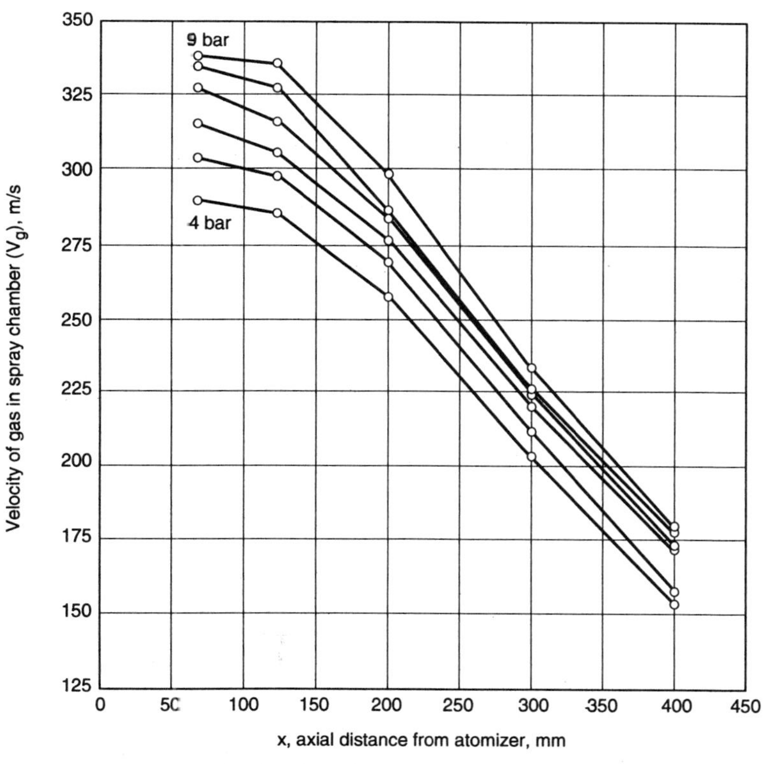

Fig. 12 Axial gas flow velocities as a function of distance measured in the Osprey facility at Drexel University for nitrogen gas at different preset atomization gas pressures

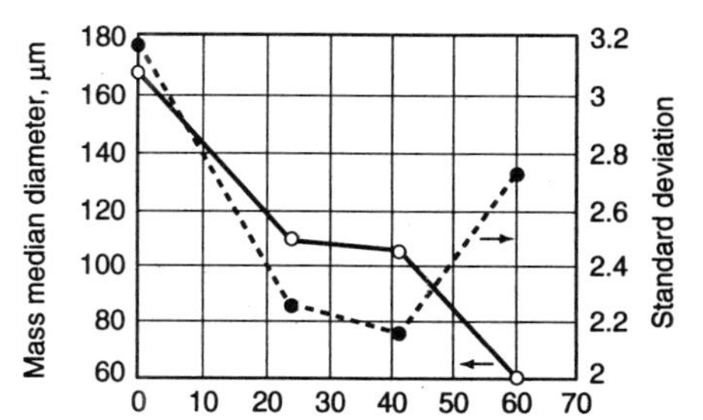

Fig. 14 Measured mass median particle diameter and standard deviation of the log normal distribution as a function of radial distance for Cu-6Ti sprayed at a GMR of 0.68, metal flow rate of 0.32 kg/s and superheat of 120 °C. Source: Ref 27

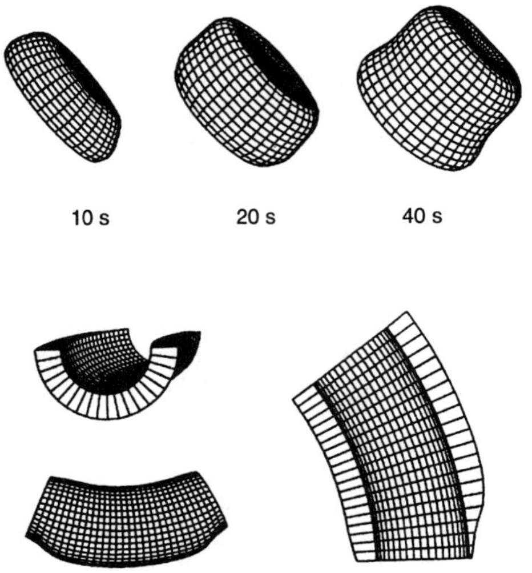

Fig. 15 Predicted shapes for spray deposited billets as a function of deposition time and for a curved tube sprayed onto a 100 mm diam mandrel with appropriate substrate motion using the three-dimensional shape model. Ref 27

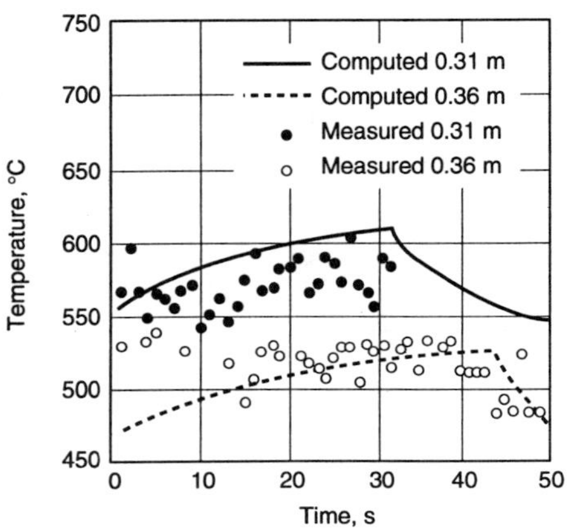

Fig. 16 Numerically modeled and experimentally observed deposit temperatures (using infrared imaging) as a function of deposition time for Al-4Cu at two axial spray heights. Source: Ref 17

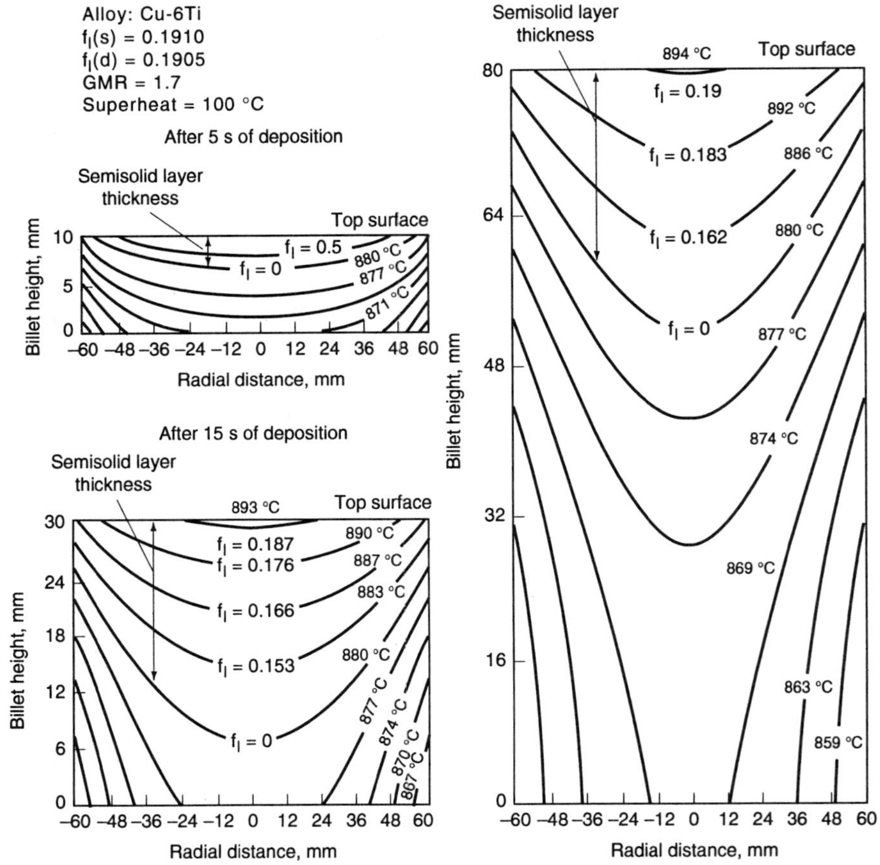

Fig. 17 Predicted temperature contours for a Cu-6Ti cylindrical billet as a function of deposition time. The enthalpy of the spray corresponds to a fraction of liquid (f_l) in the spray of 0.19. Note the depth of the partially solid region. Source: Ref 27

tures under high shear rates causes break up of the dendrites to give small rosettelike particles or with faster stirring and longer times, spherical particles. The model for dendrite fragmentation found to be valid for stircasting is plastic deformation of the dendrites via dendrite arm bending, based on the low yield strength of low solute metal at temperatures close to the liquidus of the alloy. Under this condition, when dendrite arms are bent through large angles (>20°) at high temperatures, rapid dislocation recovery occurs producing high angle grain boundaries. These high angle grain boundaries are then wetted by liquid causing separation between the grains because, as shown by Gundoz and Hunt (Ref 34), the energy of a high angle grain boundary, γ_{gb}, with a misorientation of 20° or more, is equal to that of two solid liquid interfaces, $2\gamma_{sl}$. Clear evidence for this phenomenon was shown in stircast aluminum alloys (Ref 33), and a similar process has been proposed for spray casting (Ref 35). Thus the fragmentation of metallic dendrites appears to occur not by fracture but by deformation, recrystallization, and melting.

Irrespective of the microstructural origin of the fine spherical grains found in most spray formed deposits, the partially solid material at the surface of the deposit is expected to exhibit thixotropic mechanical properties (Ref 33). These properties include a large increase in viscosity in semisolid metallic systems when the stirring rate falls. In consequence, the behavior of the material can change rapidly from fluidlike to that of a solid as bonds build up at grain boundaries between the particles (Ref 33). For spray formed materials, this suggests that, at the deposit surface with its high frequency of high velocity impacts (≈100 m/s), the material will be rather fluid. This fluidlike character will, however, decay rapidly at a given position as the deposit grows and the material just below the surface is no longer agitated by the spray. Videotapes of the surface of aluminum deposits at high liquid fractions (Ref 36) do show rapid surface movement similar to that seen in puddles during a heavy rainfall. This liquidlike behavior at the surface should terminate rapidly beneath the surface of the deposit where mixing by droplet impact will not penetrate deeply into the viscous semisolid material. The strength of the deposit is confirmed by the fact that, even with high rotation rates in spray formed billets and tubes, the shape of the deposit usually remains stable. If this were not the case, the process would be unworkable.

With reference to Fig. 11, the three parameters have been examined (Ref 37) in which the thermal model of the process was used to design and analyze a set of experiments which attempted to understand the process parameters that control the resulting output parameters. In performing these experiments, it was assumed that the enthalpy of the spray at the moment of impact on the deposit (usually expressed as the equivalent fraction of liquid in the spray, $f_l(s)$) was likely to be the critical control parameter. A series of experiments were run on copper-titanium alloy billets, and previous data on In625 tubes (Ref 38–40) were reanalyzed using the model. For the billets, the process parameters were selected using the model to design experiments with a wide range of $f_l(s)$ in the spray. For the tubes, the experiments were analyzed to determine the operating values of $f_l(s)$ in the spray at the moment of impact. In addition, using Fig. 11, the final size and shape of each spray formed billet and tube were predicted assuming 100% sticking efficiency (SE) (Ref 3, 27). The sizes predicted were then compared with the measured deposit

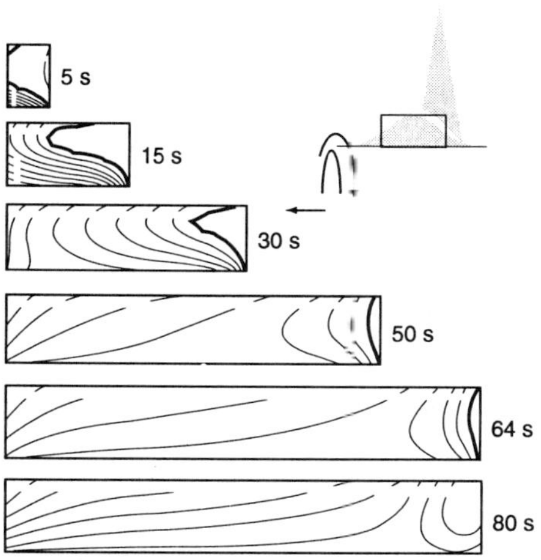

Fig. 18 Isotherms modeled for In625 tube under a fraction of liquid in the spray of 0.46. Limitations of the two-dimensional model are clear, and acceleration of cooling in the later stages of deposition (due to the cooling effect of the gas stream on the as-deposited tube) are seen from the rapid decline of the partially solid region, even before termination of the spray which occurred after 64 s. Source: Ref 27

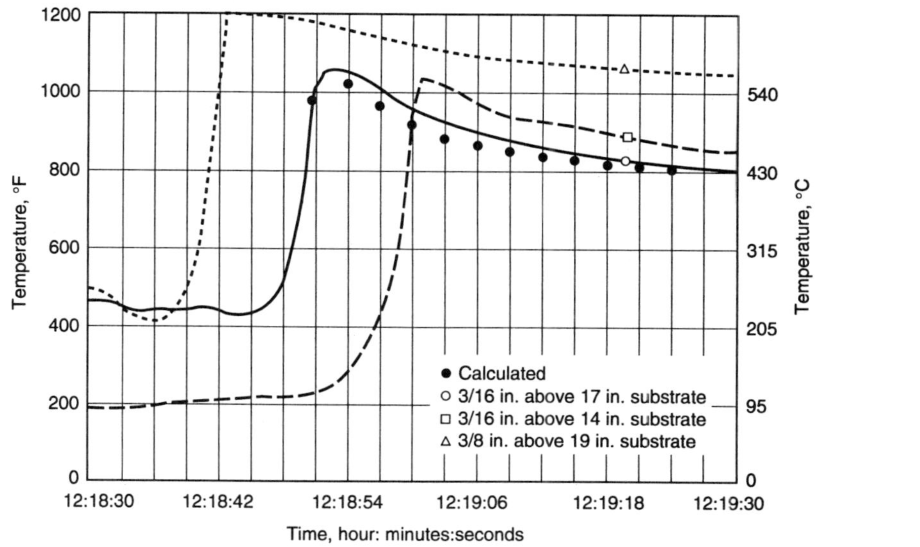

Fig. 19 Calculated and measured temperature versus time plots in spray formed plates using a linear nozzle. Source: Ref 30

sizes to give a first estimate of the sticking efficiency. These estimated values of SE were then used in revised runs of the three-dimensional shape model, and the comparison was repeated until, for each spray formed deposit, the predicted and measured shapes of the deposit agreed with the fitted value of SE.

Figures 21 and 22 show the resulting measured values of SE in the two studies as a function of the modeled values of f_l(s). The two plots show optimum values of f_l(s) to achieve maximum sticking efficiency but at very different values in the two experiments. For the copper-titanium billets, the highest SE was ~0.9 (by weight) at an optimum f_l(s) of ~0.3. In contrast, for the In625 tubes, the optimum SE occurred at a much higher value of f_l(s): namely ~0.5. The sticking efficiency for the In625 tubes was also enhanced by reducing the centrifugal force, resulting from a slower substrate rotational speed (Fig. 22). Some material is spun off from the rotating deposit because the sticking efficiency is decreased at higher rates of rotation. The decrease in SE at fractions of liquid higher than the optimum value is also explained by this effect.

Porosity

The other key characteristic of the spray formed microstructure is a fine scale porosity that, at best, is ~1% by volume and exists primarily in the form of small voids (~1 μm) between grains (Ref 2). In cold regions of the deposit, such as against a cold substrate and at the edges of the deposit, larger amounts of interconnected porosity are usually found (Ref 2, 30, 41). In addition, in hot deposits (high fractions of liquid), spherical pores can be found that are much larger than the grain size of the deposit (Ref 2, 5, 41). These three types of porosity appear to originate from different causes.

Interconnected porosity appears to arise when the chilling effect of a cold substrate causes the fraction of solid at the surface of the deposit to approach 100% (Ref 1). Under these circumstances, true spray deposition does not occur because filling of the pores between solid particles by excess liquid cannot take place. Recent thermal modeling studies (Ref 30) demonstrate clearly this idea (Fig. 20). The absence of a calculated solidification cooling rate before 0.4 of the normalized thickness for the unheated steel substrate and 0.1 of the normalized thickness for the heated substrate, reflect the minimum thicknesses modeled before any retained liquid is predicted to occur on the deposit surface. These distances match closely the experimental thickness of the deposit containing a high density of interconnected pores (Ref 30, 41). The results of this study (Ref 30) confirm the importance of achieving a rapid buildup of a partially molten (incompletely solidified) surface on the deposit, if a spray deposit of low porosity is to be produced. The results also confirm a well-established empirical feature of successful spray forming, namely the need to have a deposit surface that does not rapidly chill the spray. Empirical studies have shown repeatedly that it is not desirable to have too fine a droplet size in the spray because this leads to a rapid buildup of solid in the spray. It is also important to use a thermally insulating substrate, or alternatively, if the substrate is metallic, then it has to be preheated. This requirement is critical in the spray forming of thick coatings on a metallic substrate, such as spraying tool steel onto metallic iron rolls or spraying stainless steel onto mild steel (Ref 11).

It is generally considered, but not completely confirmed, that the fine scale pores at grain boundaries are ubiquitous to spray formed deposits (Ref 2) and arise from trapped gas from the atomization step. Indirect evidence for this comes from the observation that in the deposition of high-temperature alloys such as copper- and nickel-base alloys deposits are either essentially pore-free or become so after light thermomechanical processing. This change only happens if a chemical reaction occurs between the nitrogen (atomization) gas and a reactive nitride forming element in the alloy such as chromium or zirconium in copper (Ref 42) or nickel (Ref 43). Spraying nickel-base alloys with an inert gas such as argon always produces fine scale porosity that is not readily or permanently removed by subsequent processing (Ref 43). Heavy deformation can close up the pores, but if the gas remains unreacted, then the porosity may return on subsequent annealing.

The coarser spherical porosity that is often reported in deposits produced under hot spray conditions with a high fraction of liquid in the spray occurs as spherical pores that are much larger than the grain size (Ref 2, 41). It appears likely that these pores may form by the coalescence of smaller pores in the presence of a high liquid content. This allows more fluid motion, and the growing pores can push the solid particles apart. Irrespective of the mechanism(s) in-

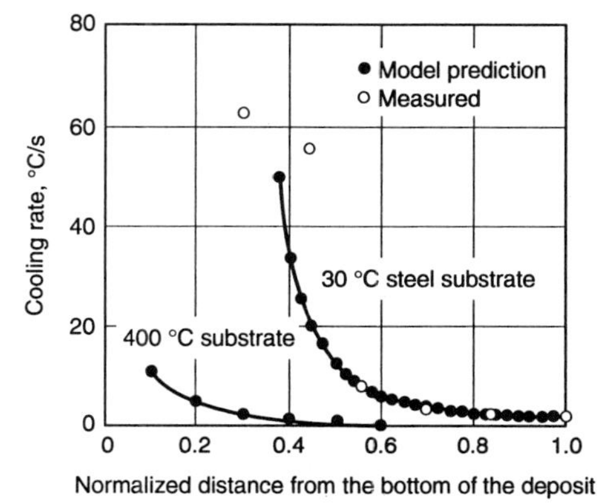

Fig. 20 Calculated cooling rates in spray formed plates showing the chill layer. Source: Ref 30

volved, a simple thermodynamic analysis (Ref 41) suggests that growth of the pores will produce a much larger volume fraction of gas. For finely divided pores with a radius of curvature, r, of a few micrometers, the interfacial energy, γ, and induced pressure, ΔP, can reduce significantly the volume of trapped gas. For example with copper ($\gamma = 1.5$ J/m^2) and $r \approx 3$ μm:

$$\Delta P = 2\gamma/r \approx 1 \text{ MPa} \approx 10 \text{ bar} \quad \text{(Eq 1)}$$

When a number, m, of these small voids of initial radius, r_1, coalesce, the radius increases to r_2 and the pressure falls, giving a larger volume of gas. Use of the ideal gas law ($PV = nRT$) shows (Ref 37) that:

$$r_2 = r_1 \cdot m^{1/2} \quad \text{(Eq 2)}$$

and that the volume, V_1, of the m small pores increases to a volume V_2, where:

$$V_2 = V_1 \cdot m^{1/2} \quad \text{(Eq 3)}$$

Thus, extended coalescence giving the large spherical bubbles ($m \gg 1$) will result in a significant increase in the porosity observed. If, as is often observed, the coarse gas-filled voids are much larger ($r_2 \approx 60$ μm) than the fine grain boundary voids ($r_1 \approx 3$ μm) (Ref 2), then relaxation of the interfacial induced pressure (Eq 1) will for the same volume of trapped gas increase the porosity 20 times (Eq 2). This analysis gives a moderate overestimate of the apparent increase in porosity following coalescence because the model ignores the effect of atmospheric pressure. The large spherical gas voids described in Ref 2 were produced in a copper-titanium billet sprayed under conditions of $f_l(s) = 0.6$, which should allow for more pore coalescence.

Equation 1 also shows how heavy thermomechanical processing can close up the gas pores even with insoluble gases. If deformation fragments the pores, it will reduce the pore size leading to increased pressure and a decrease in gas volume ($m \ll 1$, in Eq 3). However, unless the gas in the very small pores reacts chemically, it is always capable of increasing its volume fraction by coalescence during high-temperature solid state processing.

Recently a series of quantitative studies (Ref 37) confirmed fine scale porosity in copper-titanium and the nickel-base superalloy In625. The average porosity in the deposits was measured in copper-titanium billets and In625 tubes. Additional experiments were also performed on copper-titanium tubes and In625 billets. The billets were sprayed onto oxide substrates with a low thermal conductivity. Figure 23 shows the porosities measured in the copper-titanium billets and in the In625 tubes as a function of the modeled fraction of liquid in the spray (Ref 38–40).

A steady fall in the level of porosity with increased values of the fraction of liquid in the spray $f_l(s)$ can be seen. The two alloys show very similar results, but the behavior is very different in the two geometries. The billets show porosity levels only as high as 8 vol% in very cold sprays ($f_l(s) \approx 0.1$), falling to 1 vol% as $f_l(s)$ reached 0.55. In the spray formed tubes, there is a decrease in porosity from a very high value of 25 vol% at a $f_l(s)$ of 0.25 to porosity levels 2 to 3 vol% at high values of $f_l(s)$, such as 0.5 to 0.6. Lavernia and Wu (Ref 5) reviewed the published data on porosity and, as with the results reported here, found high levels of porosity in sprays deposited under cold conditions; the level of porosity decreased as the temperature of the spray increased. Results have also been reported showing the opposite effect. An increase in porosity at high fractions of liquid (Ref 2, 4, 29, 30), particularly in aluminum alloys (Ref 29, 30) and in alloys sprayed with an inert gas in which there is no interaction between the gas and the alloy constituents has been shown. A minimum in porosity is indicated and present results suggest that this may occur at ~0.3 fraction of liquid in the spray in billets (Ref 4). Clearly the optimum value of $f_l(s)$ is higher for tubes than for billets.

The different results between the levels of porosity in billets and tubes show clearly that to obtain equivalent behavior in terms of sticking efficiency and porosity, the spray must be much hotter for tubes than for billets. The difference between these two geometries arises because in billets the surface of deposition remains essentially under the spray at all times. As shown in Fig. 17, the thermal model predicts that the fraction of liquid on the billet surface under the spray rapidly approaches the fraction of liquid in the spray, $f_l(s)$. The geometry of tube spraying requires that the deposit rotates into and out of the spray with each revolution. Thus, at most, deposition occurs ~50% of the time, and the rate of arrival of the spray falls to zero at the "3 and 9 o'clock" orientations. At these positions, high velocity gas is sweeping past the deposit surface, and the mean fraction of liquid on the surface of the deposit, $f_l(d)$, should be significantly less than that in the spray, $f_l(s)$. The difference suggests that the control parameter is likely to be the fraction of liquid on the surface of the deposit under the spray, $f_l(d)$, rather than the fraction of liquid in the spray. Since deposition involves both the spray and the deposit, it is possible that some combined parameter involving both $f_l(d)$ and $f_l(s)$ is critical. At present, no reliable three-dimensional thermal models of tube deposition are available. Given such a reliable and validated thermal model, $f_l(d)$ can be modeled along with the modeled values of $f_l(s)$, and the experimental results on sticking efficiencies and porosity can be correlated with these two parameters. The current results suggest that the thermal conditions in the spray and the deposit are the critical parameters in controlling sticking efficiency and porosity. A wider range of studies is needed, particularly for other geometries such as sheet and plate and with other atomization geometries such as the linear nozzle (Ref 27). The current results do, however, illustrate the value of combining experimental studies with modeling.

Grain Structure in Spray Formed Deposits

The defining feature of the microstructure of spray deposited materials is the fine equiaxed grain structure with minimum solute segregation (Ref 2–4). It is generally agreed that the fine grain size arises from dendrite fragmentation on impact of the spray followed by grain coarsening during solidification. The times for solidification increase with the fraction of liquid in the spray at a given geometry (Fig. 24). However the coarsening is slower than that predicted using standard coarsening theory, which results in a small grain size (Fig. 25). A detailed investigation of coarsening under conditions of spray forming was carried out by Annavarapu and Doherty (Ref 32) using aluminum AA2014 (Fig. 26) and copper-titanium (Fig. 27) reheated into the solid-

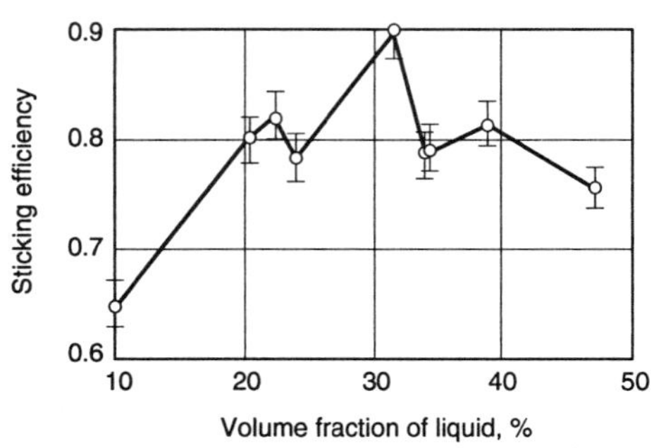

Fig. 21 Sticking efficiency obtained by matching experimental results with the shape model for spray formed copper-titanium billets compared to the modeled fraction of liquid in the spray. Source: Ref 37

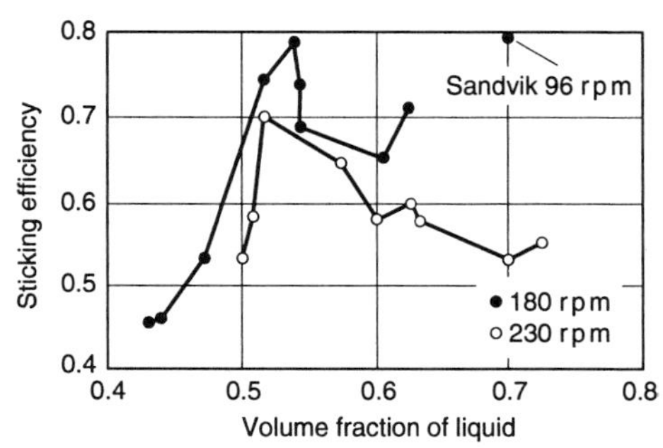

Fig. 22 Sticking efficiency obtained by matching experimental results with the shape model for spray formed tubes of In625 compared to the modeled fraction of liquid in the spray for two different rotational speeds. Source: Ref 37, 38–40

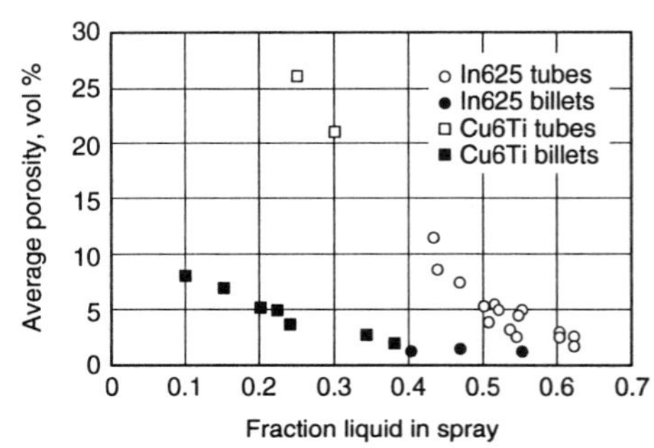

Fig. 23 Average porosity in billets and tubes of copper-titanium and In625 as a function of fraction of liquid in the spray. Source: Ref 37, 38–40

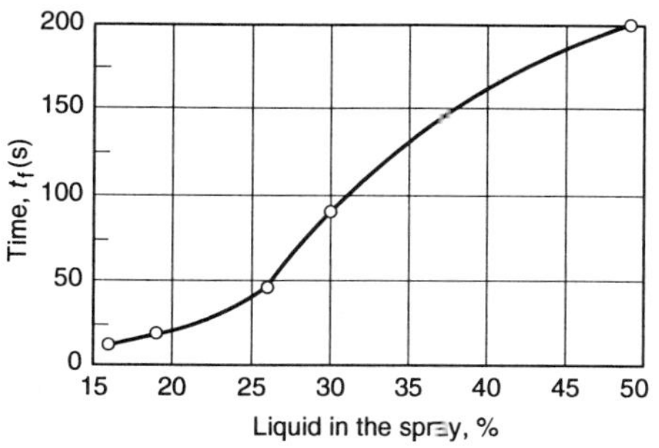

Fig. 24 Thermally modeled total local solidification times, t_f(s), in copper-titanium billets at an axial location 25 mm from the base of the deposit for different fractions of liquid in the spray. Source: Ref 3

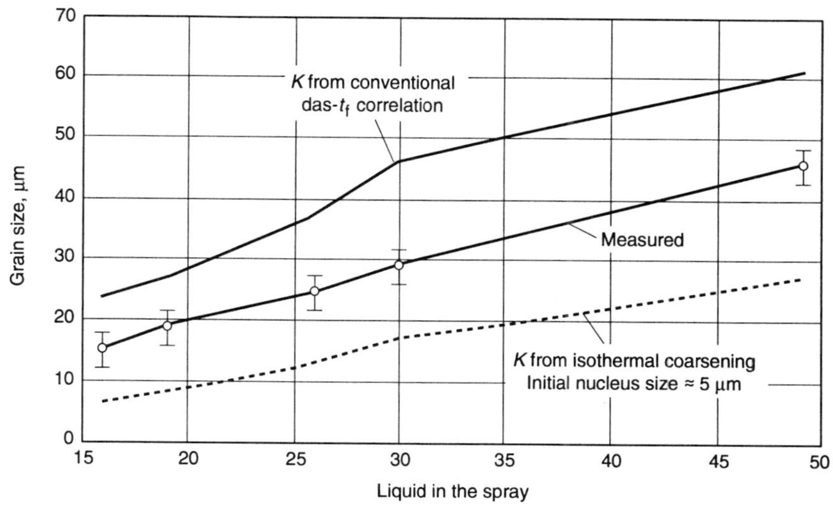

Fig. 25 Measured and thermally modeled grain sizes in copper-titanium billets as a function of the fraction of liquid in the spray. das-t_f, dendrite arm spacing-solidification time correlation. Source: Ref 3

liquid phase field. It was found that the measured coarsening rates were much slower than those expected for dendritic structures, and the coarsening rate fell as the fraction of solid in the mixture increased. This is in qualitative but not quantitative agreement with the experimental results for copper-titanium (Fig. 14). Other studies on grain coarsening in spray formed structures (Ref 44, 45) yield similar results. Slower coarsening in solid-liquid mixtures with increasing solid fraction is, however, in direct conflict with experimental studies (Fig. 28) and modeling of coarsening in high purity two phase solid-liquid systems (Ref 32). In these studies, coarsening is accelerated as the volume fraction of solid increases due primarily to shorter diffusion distances. At the highest volume fraction of solid where the liquid is present as small droplets trapped at the boundaries, coarsening is controlled apparently by the drag of these droplets (Ref 32). This conflict appears to arise from the presence of solid second phase impurity particles in the spray formed material which inhibit grain boundary movement (Ref 32, 45). Currently this is established qualitatively, and a detailed quantitative investigation of this phenomenon is required to allow the thermal modeling of deposit solidification to yield reliable predictions of the resulting grain size in different parts of the deposit. Figure 29 shows the result of an experimental study of the grain size variation in a spray formed deposit (Ref 37).

The physical processes that give rise to the fine grain structures in spray forming also give rise to another desirable characteristic, low microsegregation and macrosegregation. Insights into these phenomena were reviewed by Biloni and Boettinger (Ref 46). Rapid solidification to a high fraction of solid (giving fine scale solid particles that are then allowed to cool more slowly with significant coarsening) is expected to result in decreased microsegregation. In addition, the fact that the deposit forms with a high fraction of solid reduces significantly the solidification shrinkage within the deposit, which combined with the restricted permeability of the fine structure, should limit severely the fluid flow processes that cause macrosegregation. These ideas appear to fit qualitatively with the experimental observations of homogeneous spray formed deposits (Ref 4). However few attempts have been made to test the ideas of the expected reduced microsegregation and macrosegregation quantitatively. This is an area that needs additional study.

Modeling: Current Status and Future Needs

Current models of the spray process, which rely on a combination of empirical inputs and

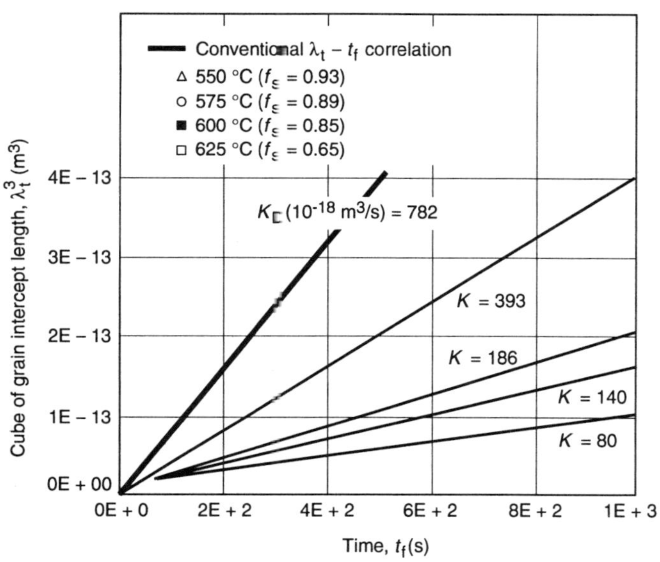

Fig. 26 Experimentally measured coarsening rates for 2014 (Al-Cu-Mg-Fe-Si) both as conventional dendrite arm coarsening (K_D = 782 μm^3/s) and for reheated spray cast Al 2014 at f_s = 0.65 (625 °C, K = 393 μm^3/s), f_s = 0.85 (600 °C, K = 186 μm^3/s), f_s = 0.89 (575 °C, K = 140 μm^3/s), and f_s = 0.93 (550 °C, K = 80 μm^3/s). Source: Ref 32

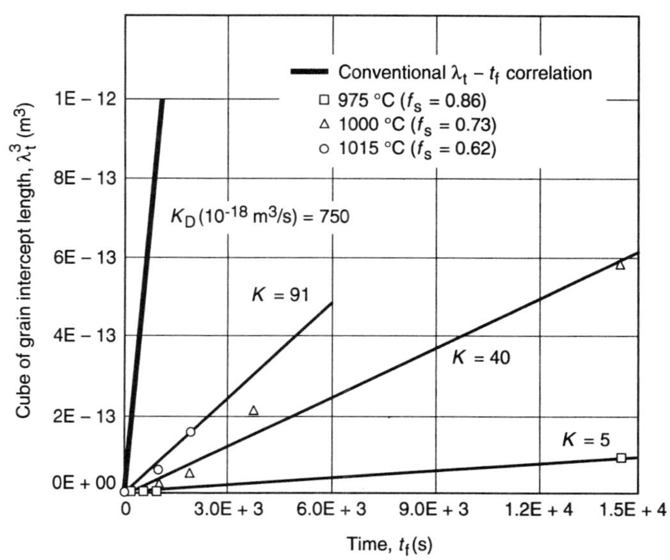

Fig. 27 Experimentally measured coarsening rates for copper-titanium both as conventional dendrite arm coarsening (K_D = 750 μm^3/s) and for reheated spray cast copper-titanium billet at f_s = 0.62 (1015 °C, K = 91 μm^3/s), f_s = 0.73 (1000 °C, K = 40 μm^3/s), and f_s = 0.86 (975 °C, K = 5 μm^3/s). Source: Ref 32

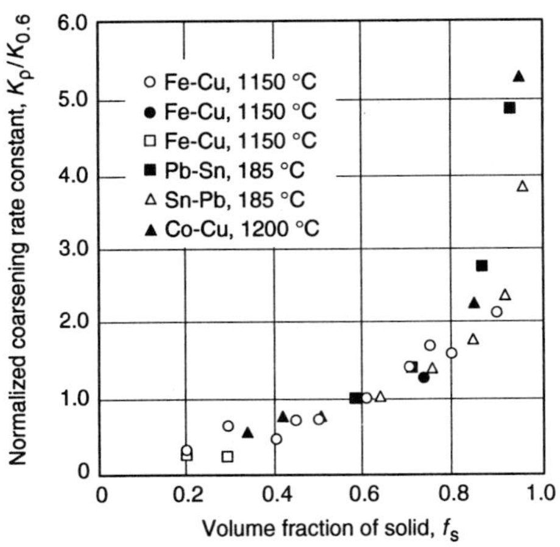

Fig. 28 Measured normalized coarsening rates; K_p normalized by the value of K_p at a fraction solid, f_s of 0.6, as a function of f_s. This is the situation in high purity binary alloys containing one solid phase. Source: Ref 2, 32

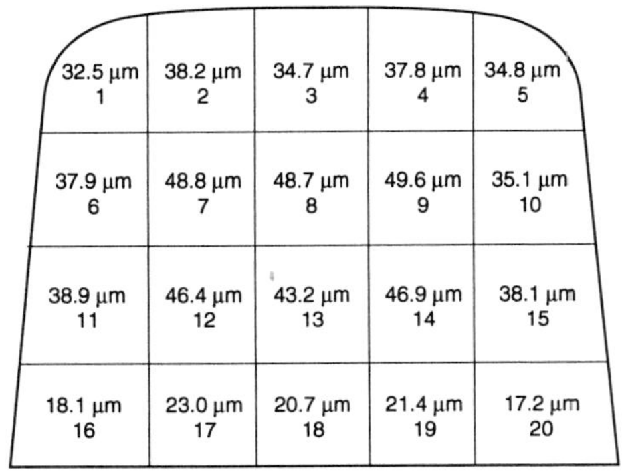

Fig. 29 Measured grain sizes as a function of location in a spray deposited billet of In690. Source: Ref 37

physically modeled processes, appear to describe the shape (Ref 3, 27) of all types of spray formed deposits and the thermal conditions in billets (Ref 15, 27) and plates (Ref 27). The modeled thermal conditions in the spray have been found to control the two major unknowns in the process: sticking efficiency, which determines the yield, and trapped gas porosity, which is the major structural defect (Ref 34). These successes were achieved by use of carefully and expensively measured intermediate process parameters—notably the gas velocity field, PSD, and RMF (Ref 3, 27). If each modeling attempt requires this level of prior investigation, then the models will probably have limited application. Renewed experimental and theoretical studies that can lead to reliable predictions of these parameters in different spray forming chambers and with alloys of interest in spray forming are needed. It may be possible and more desirable to make progress directly in such studies by direct investigative techniques in actual gas atomization sprays. Techniques using lasers (Ref 47, 48), high-speed photography, and thermal imaging (Ref 15) appear to offer potential for direct measurements of droplet sizes, velocities, and deposit temperatures from which a process model should be able to back-calculate the needed process parameters and their variation with process variables, such as atomization, gas pressure, and metal flow rate.

The three-dimensional shape models need to be developed further to give thermal models that, once tested and validated, will facilitate improved process design to avoid undesirable features, such as near substrate porosity, and to promote finer and perhaps more uniform grain sizes. The features that limit grain coarsening in the high solid fractions of spray formed deposits also need further study. This is desirable both as a scientific puzzle (Ref 29) and as a potential means of further refining spray formed grain sizes. Given the growing commercial successes of the process (Ref 1), further detailed scientific analysis of all aspects of spray forming is warranted.

REFERENCES

1. P.C. Mathur, D. Apelian, and A. Lawley, Analysis of Spray Deposition, *Acta Metall.*, Vol 37, 1989, p 429
2. S. Annavarapu and R.D. Doherty, Evolution of Microstructure on Spray Forming, *Int. J. Powder Metall.*, Vol 29, 1993, p 331
3. C. Cai, S. Annavarapu, and R.D. Doherty, Modeling-Based Microstructural Control in Spray Forming, *Proc. of the Second Int. Conf. on Spray Forming,* ICSF II (Swansea, U.K.), J.V. Wood, Ed., Woodhead Publishing, Cambridge, U.K., 1993, p 67
4. P.S. Grant, Spray Forming, *Prog. Mater. Sci.*, Vol 39, 1995, p 497
5. E.J. Lavernia and Y. Wu, *Spray Atomization and Deposition,* John Wiley & Sons, 1996
6. J.V. Wood, Ed., *Proc. of the First Int. Conf. on Spray Forming,* ICSF I (Swansea, U.K.), Osprey Metals, Neath, U.K., 1990
7. J.V. Wood, Ed., *Proc. of the Second Int. Conf. on Spray Forming,* ICSF II (Swansea, U.K.), Woodhead Publishing, Cambridge, U.K., 1993
8. J.V. Wood., Ed., *Proc. of the Third Int. Conf. on Spray Forming,* ICSF III (Cardiff, U.K.), Osprey Metals, Neath, U.K., 1997
9. A.G. Leatham and A. Lawley, The Osprey Process: Principles and Applications, *Int. J. Powder Metall.*, Vol 29, 1993, p 325
10. A.G. Leatham, A.G.W. Ogilvy, and L. Elias, The Osprey Process: Current Status and Future Possibilities, *Powder Metallurgy: Aerospace Defense and Demanding Applications,* F.H. Froes, Ed., Metal Powder Industries Federation, 1993, p 165
11. A.G. Leatham, Commercial Spray Forming: Exploiting the Metallurgical Benefits, *Mater. World,* Vol 4, 1996, p 317
12. P.C. Mathur, S. Annavarapu, D. Apelian, and A. Lawley, Spray Casting: An Integral Model for Process Understanding and Control, *Mater. Sci. Eng. A*, Vol 142, 1991, p 261
13. E.J. Lavernia, E.M. Gutierrez, G.M. Trapaga, J. Szekely, and N.J. Grant, A Mathematical Model of the Liquid Dynamic Compaction Process: I Heat Transfer in Gas Atomization, and II Formation of the Deposit, *Int. J. Rapid Solid.*, Vol 4, 1988, p 89, 125
14. E.M. Gutierrez, E.J. Lavernia, G.M. Trapaga, J. Szekely, and N.J. Grant, A Mathematical Model of the Spray Deposition Process, *Metall. Trans. A*, Vol 20, 1989, p 71
15. G.M. Trapaga, E.F. Matthys, J.J. Valencia, and J. Szekely, Fluid Flow, Heat Transfer and Solidification of Molten Metal Droplets Impinging on Substrates: Comparison of Numerical and Experimental Results, *Metall. Trans. B,* Vol 23, 1992, p 701
16. C.Y.A. Tsao and N.J. Grant, Modeling of the Liquid Dynamic Compaction Process, *Int. J. Powder Metall.*, Vol 30, 1994, p 323
17. P.S. Grant, B. Cantor, and L. Katgerman, Modeling of Droplet Dynamics and Thermal Histories during Spray Forming, *Acta Metall. Mater.*, Vol 41, 1993, p 3097
18. P.S. Grant, P.P. Maher, and B. Cantor, Heat Flow in Spray Formed Al-4Cu, *Mater. Sci. Eng. A*, Vol 179, 1994, p 72
19. D.T. Gethin, J.O. Medwell, and N. Muhamad, A Comparison of Lumped Parameter Numerical Techniques for Liquid Metal Sprays, *Proc. of the First Int. Conf. on Spray Forming*, ICSF I (Swansea, U.K.), J.V. Wood, Ed., Osprey Metals, Neath, U.K., 1990
20. U. Fritschung, H. Zhang, and K. Bauckhage, Modeling of Thermal Histories and Solidification in the Spray Cone and Deposit of Atomized and Compacted Metals, *Proc. of the Second Int. Conf. on Spray Forming,* ICSF II (Swansea, U.K.), Woodhead Publishing, Cambridge, U.K., 1993, p 35
21. J. Forrest, S. Lile, and J.S. Coombs, Numerical Modeling of the Osprey Process, *Proc. of the Second Int. Conf. on Spray Forming,* ICSF II (Swansea, U.K.), Woodhead Publishing, Cambridge, U.K., 1993, p 117
22. A. Lawley, An Overview of Powder Atomization Processes and Fundamentals, *Int. J. Powder Metall.*, Vol 13, 1977, p 169
23. A. Lawley, Atomization: The Production of Metal Powders, *Monographs in P/M,* Metal Powder Industries Federation, 1992
24. A.J. Yule and J.J. Dunkley, *Atomization of Melts for Powder Production and Spray Deposition,* Clarendon Press, Oxford, U.K., 1994
25. D. Bradley, On the Atomization of a Liquid by High Velocity Gases, *J. Appl. Phys.*, Vol 6, 1973, p 1724, 2267
26. H. Lubanska, Correlation of Spray Ring Data for Gas Atomization of Liquid Metals, *JOM,* Vol 20, 1970, p 45
27. C. Cai, "A Modeling Study for Design and Control of Spray Forming," Ph.D. dissertation, Department of Materials Engineering, Drexel University, 1995
28. V. Uhlenwinkel and K. Bauckhage, Mass Flux Profile and Local Particle Size in the Spray Cone During Spray Forming of Steel, Copper and Tin, *Proc. of the Second Int. Conf. on Spray Forming,* ICSF II (Swansea, U.K.), Woodhead Publishing, Cambridge, U.K., 1993, p 25

29. R.D. Doherty, "Modeling for Microstructural Control," Final Report N00014-89-J-P0001, Office of Naval Research, Arlington, VA, 1996
30. S.J. Pien, H. Ding, and M.K. Chyu, Modeling of the Multi-Phase Transport Phenomena and Solidification in a Spray Forming Process with a Linear Nozzle, *Sprayforming,* Vol 2, K. Bauckhage and V. Uhlenwinkel, Ed., Kolloquium des SFB, University of Bremen, 1997, p 17
31. M.C. Flemings, *Solidification Processing,* McGraw-Hill, 1974
32. S. Annavarapu and R.D. Doherty, Inhibited Coarsening of Solid-Liquid Microstructures in Spray Casting at High Volume Fraction Solid, *Acta Metall. Mater.,* Vol 43, 1995, p 3207
33. R.D. Doherty, E.A. Feest, and H.-I. Lee, Microstructure in Stir-Cast Alloys, *Mater. Sci. Eng.,* Vol 65, 1984, p 181
34. M. Gundoz and J.D. Hunt, The Measurement of the Solid-Liquid Interfacial Energies in the Al-Cu, Al-Si, and Pb-Sn Systems, *Acta Metall.,* Vol 33, 1985, p 1641
35. S. Ashok, Comments on Evolution of Microstructure on Spray Forming, *Int. J. Powder Metall.,* Vol 30, 1994, p 141
36. R. Kozarek, Alcoa Technical Center (Alcoa, PA), private communication
37. R.D. Doherty, C. Cai, and L.K. Warner-Kohler, Modeling and Microstructural Development in Spray Forming, *Int. J. Powder Metall.,* Vol 33, 1997, p 50
38. R.D. Payne, M.A. Matterson, and A.L. Moran, Application of Neural Networks in Spray Forming, *Int. J. Powder Metall.,* Vol 29, 1993, p 345
39. M.A. Matterson, R.D. Payne, C.J. Madden, and A.L. Moran, An Approach to Modeling the Spray Forming Process with Artificial Neural Networks, *Proc. of the Second Int. Conf. on Spray Forming,* ICSF II (Swansea, U.K.), J.V. Wood, Ed., Woodhead Publishing, Cambridge, U.K., 1993, p 105
40. A.L. Moran, C.J. Madden, and R.E. Rebis, Industrialization and Certification of Large Diameter Spray Formed Superalloy Components for Military Applications, *Proc. of the Second Int. Conf. on Spray Forming,* ICSF II (Swansea, U.K.), J.V. Wood, Ed., Woodhead Publishing, Cambridge, U.K., 1993, p 165
41. M.G. Chu, Microstructure of Aluminum Alloy Deposits Produced by Spray Forming with a Linear Nozzle, *Sprayforming,* Vol 2, K. Bauckhage and V. Uhlenwinkel, Ed., Kolloquium des SFB, University of Bremen, 1997, p 115
42. W.G. Watson, Thermal and Microstructural Characterization of Spray Cast Copper Alloy Strip, *Proc. of the First Int. Conf. on Spray Forming,* ICSF I (Swansea, U.K.), Osprey Metals, Neath, U.K., 1990, p 19
43. M.G. Benz, T.F. Sawyer, W.T. Carter, R.J. Zabala, and P.L. Dupree, Nitrogen in Spray Formed Superalloys, *Proc. of the Second Int. Conf. on Spray Forming,* ICSF II (Swansea, U.K.), J.V. Wood, Ed., Woodhead Publishing, Cambridge, U.K., 1993, p 171
44. P.S. Grant, R.P. Underhill, W.T. Kim, K.P. Mingard, and B. Cantor, Grain Growth in Spray Formed Al Alloys, *Proc. of the Second Int. Conf. on Spray Forming,* ICSF II, J.V. Wood, Ed. (Swansea, U.K.), Woodhead Publishing, Cambridge, U.K., 1993, p 45
45. R.P. Underhill, P.S. Grant, and D.B. Cantor, Grain Growth in Spray-Formed Ni Superalloys, *J. Mater. Synth. Process.,* Vol 3, 1995, p 171
46. H. Biloni and W.J. Boettinger, Solidification, *Physical Metallurgy,* 4th ed., R.W. Cahn and P. Haasen, Ed., Elsevier Science B.V., 1996, p 670
47. K. Bauckhage, H.H. Floegel, U. Fritsching, and R. Hiller, The Phase-Doppler-Difference-Method: A New Laser-Doppler Technique for Simultaneous Size and Velocity Measurements, *Particle Charac.,* Vol 5, 1988, p 66
48. B. Bewley and B. Cantor, Modeling of Spray Deposition: Measurements of Particle Size, Gas Velocity, Particle Velocity, and Spray Temperature in Gas Atomized Sprays, *Metall. Trans. B,* Vol 21, 1990, p 899

Thermal Spray Forming of Materials

Richard Knight and Ronald W. Smith, Center for the Plasma Processing of Materials (CPPM), Drexel University, Philadelphia, PA

THERMAL SPRAY is a "family" of particulate/droplet consolidation processes capable of forming metals, ceramics, intermetallics, composites, and polymers into coatings or freestanding structures. During the process, powders, wires, or rods are injected into combustion or arc-heated gas jets, where they are heated, melted or softened, accelerated, and directed toward the surface, or substrate, being coated. On impact at the substrate, the particles or droplets rapidly solidify, cool, contract, and incrementally build up to form a deposit. The thin "splats" undergo very high cooling rates, typically in excess of 10^6 K/s for metals (Ref 1).

Increased attention is being paid to thermal spray processes today because of their ability to consolidate virtually any material that has a stable molten phase, producing coatings or deposits exhibiting relatively homogeneous, fine-grained microstructures (Ref 1, 2). Thermal spray is very versatile and has also been proposed for several new applications, including rapid part prototyping (Ref 3), production of fiber-reinforced intermetallic-matrix composites (IMCs) (Ref 4), and for producing so-called "functionally gradient materials" (FGMs) (Ref 5). Thermal spray is, nevertheless, a mature process technology, having been established now for more than a century as an economic, efficient method for coating and recoating surfaces, repairing worn components and spray forming. The latter has been practiced for at least 50 years as a means of producing ceramic crucibles for the containment of molten or reactive metals, nozzles for hot gases (Ref 6), production of forging dies (Ref 7), and the structural forming of difficult-to-process particulate materials such as high-temperature alloys, refractory metals, intermetallics, and ceramics (Ref 8). Thermal spray processing has become an important powder consolidation technique, and new developments are resulting in more novel material synthesis routes and processing of unique material combinations. This article reviews thermal spray processing advances made in the last 25 years, focusing on the recent innovations that are driving its application in metallic-, ceramic-, and intermetallic-matrix composites (Ref 9), materials synthesis, and FGMs. Thermal spray processes and their characteristics are described, together with descriptions of the properties and characteristics of thermally sprayed materials. The first part describes the different classes of thermal spray processes and the materials systems that can be processed, either as coating or as spray-formed deposits. The second part focuses on thermal spray forming techniques and applications.

Thermal Spray Processing—Introduction

The term "thermal spray" describes the family of four or five processes that use the heat energy generated by chemical (combustion) or electrical (plasma or arc) methods to melt, or at least soften, and accelerate fine dispersions of droplets to velocities in the range 50 to >1000 m/s. The high particle temperatures and velocities achieved result in significant droplet deformation on impact at a surface, producing thin layers or lamellae, often called "splats," which conform and adhere to the substrate surface. Solidified droplets build up rapidly, particle by particle, as a continuous stream of droplets impact to form continuous rapidly solidified layers. Individual splats are generally thin (~1 to 20 μm), and each droplet cools at very high rates (>10^6 K/s for metals, Ref 1) to form uniform, very fine-grained, polycrystalline coatings or deposits. Figure 1 shows a schematic of a generic thermal spray powder consolidation process, illustrating the key features and a typical deposit microstructure.

Sprayed deposits usually contain some level, typically between 0 and ~10%, of porosity, some unmelted or partially melted particles, fully melted and deformed "splats," and varying amounts of other phases—as a result of oxidation of some hot particles traveling in the spray jet due to entrainment of, and reaction with, the surrounding air.

Thermal spray jets heat large numbers (millions) of particles simultaneously, and because of the nonuniform temperature distributions of the spray jets used, and differing particle sizes, many particles do not experience the same degree of heating. Some may be completely unmelted and can cause porosity or trapped "unmelts" in coatings. The level of these coating defects varies depending on the particular thermal spray process used, the operating conditions

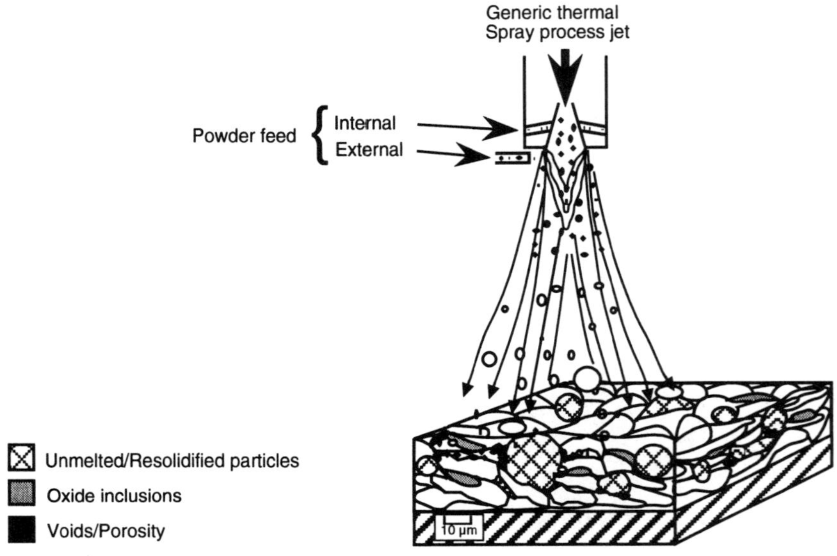

Fig. 1 Schematic representation of a typical thermal spray powder consolidation process

selected, and the material being sprayed, as described later. Figure 2 shows a photomicrograph of a thermally sprayed (high-velocity oxygen fuel, or HVOF) 80/20 NiCr alloy coating showing the characteristic lamellar splat structure. The splats and coating layers were built up as each droplet impacted at the substrate. The microstructure in Fig. 2 includes partially melted particles and dark, oxide "inclusions" that are characteristic of many metallic coatings sprayed in air. As described later, the thermal spray process selected for spraying a material can have a significant effect on the microstructure and composition of the resulting coating and hence on its properties. Figure 2 also shows the characteristically wide range of variation in oxide content, structure, and porosity that can be obtained for the same material as a function of thermal spray technique.

Thermal spray consolidation of particulate materials, as described above, is essentially a thermomechanical forming process, involving a combination of solid (unmelted or partially resolidified) and liquid phases. Some material may also be vaporized, but this is generally <0.1% by weight. The thermal energy of the process jet heats injected particles close to their melting points and also heats the coating/substrate surface during deposition. Hot working of the particles may also occur owing to the high kinetic energy imparted to the particles by the high-velocity gas jet used to accelerate them, in many cases to velocities exceeding 300 m/s. Consolidation thus occurs through a combination of liquid flow, solid redistribution, solid-phase deformation, recrystallization, and diffusion. In many cases, the continuous bombardment of newly deposited surfaces by subsequent impacting particles locally densifies the underlying deposits through a "peening" process of plastic deformation. All of these consolidation steps are driven by the thermokinetic energy characteristic of the various thermal spray processes. Thermal spray processes are, in practice, microcasting/microthermomechanical forming techniques in which the temperature and velocity of the particles determine the final microstructure and overall degree of consolidation. Unlike other spray casting techniques, thermal spray solidification events are very localized, on the scale of each droplet, and hence there is no long-range solidification structure. Such localized solidification leads to very fine grains, typically 0.1 to 10 μm, and no longer-range segregation.

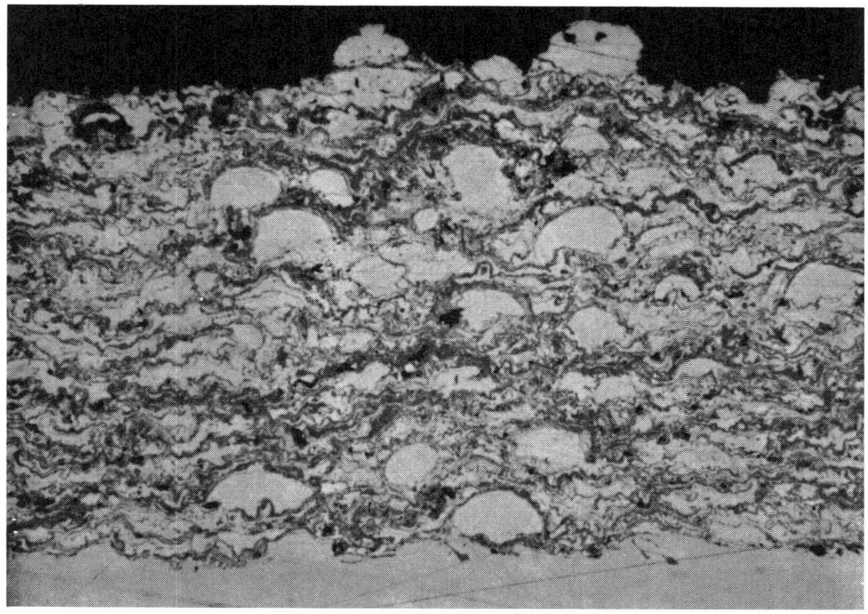

Fig. 2 Photomicrograph showing the microstructure of a thermally sprayed 80/20 NiCr alloy

Thermal Spray Processes and Techniques

All thermal spray methods use chemical or electrical energy to heat feed materials injected into hot-gas jets to create a stream of molten droplets that are accelerated and directed toward the substrates being coated. Figure 3 shows how the various thermal spray processes can be broken down according to the energy source used (combustion versus electrical), feed material form (powdered or solid), and surrounding environment (air, low pressure [vacuum], inert gas, etc.). The thermal spray "family" of processes are typically grouped into three major categories (Ref 10):

- Combustion spray (flame—powder/wire/rod, HVOF)
- Wire arc or arc spray (air or inert gas)
- Plasma (direct current air, vacuum or inert atmosphere plasma spray; RF induction)

Figure 4 shows typical schematics of the major thermal spray processes.

Combustion spray burns mixtures of fuels such as acetylene and oxidizing agents (air or oxygen) that are used to subsequently heat and accelerate particles of material injected into the hot expanding gas jets. Combustion spraying has, during the last 15 years, been subdivided into two distinct categories: (a) Flame spray, which uses low pressure (70 to 140 kPa, or 10 to 20 psi) fuel/oxygen mixtures, burning unrestricted and external to the spray gun itself in ambient air, and (b) HVOF spraying, which utilizes internal combustion at higher gas flow rates and pressures (up to ~1.5 MPa, or several hundred psi) generating supersonic jet exit velocities. The powder particles are also injected internally in HVOF, such that the heat and momentum transfer to the particles, and hence particle velocity, are substantially higher (>1000 m/s) than in conventional flame spray.

Wire arc spray utilizes two electrically conducting wires as feedstock, rather than powders. A continuous direct current (dc) arc melts the tips of the advancing wires and a high flow (typically >1500 slm) gas jet atomizes the molten material and accelerates the resulting small droplets toward the substrate being coated. A typical wire arc spray gun consists of a nozzle that directs a cool, high-velocity air, nitrogen or, in special cases argon, gas jet at the wire tips and a feedback-controlled wire-feed mechanism that

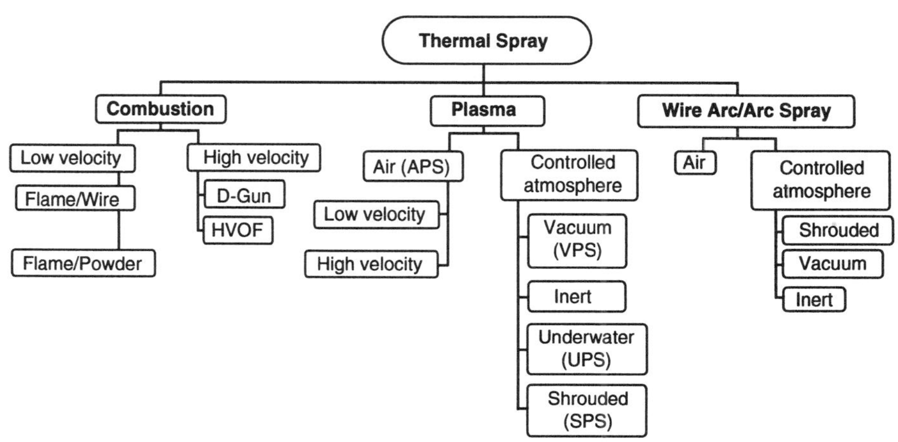

Fig. 3 General types of thermal spray processes

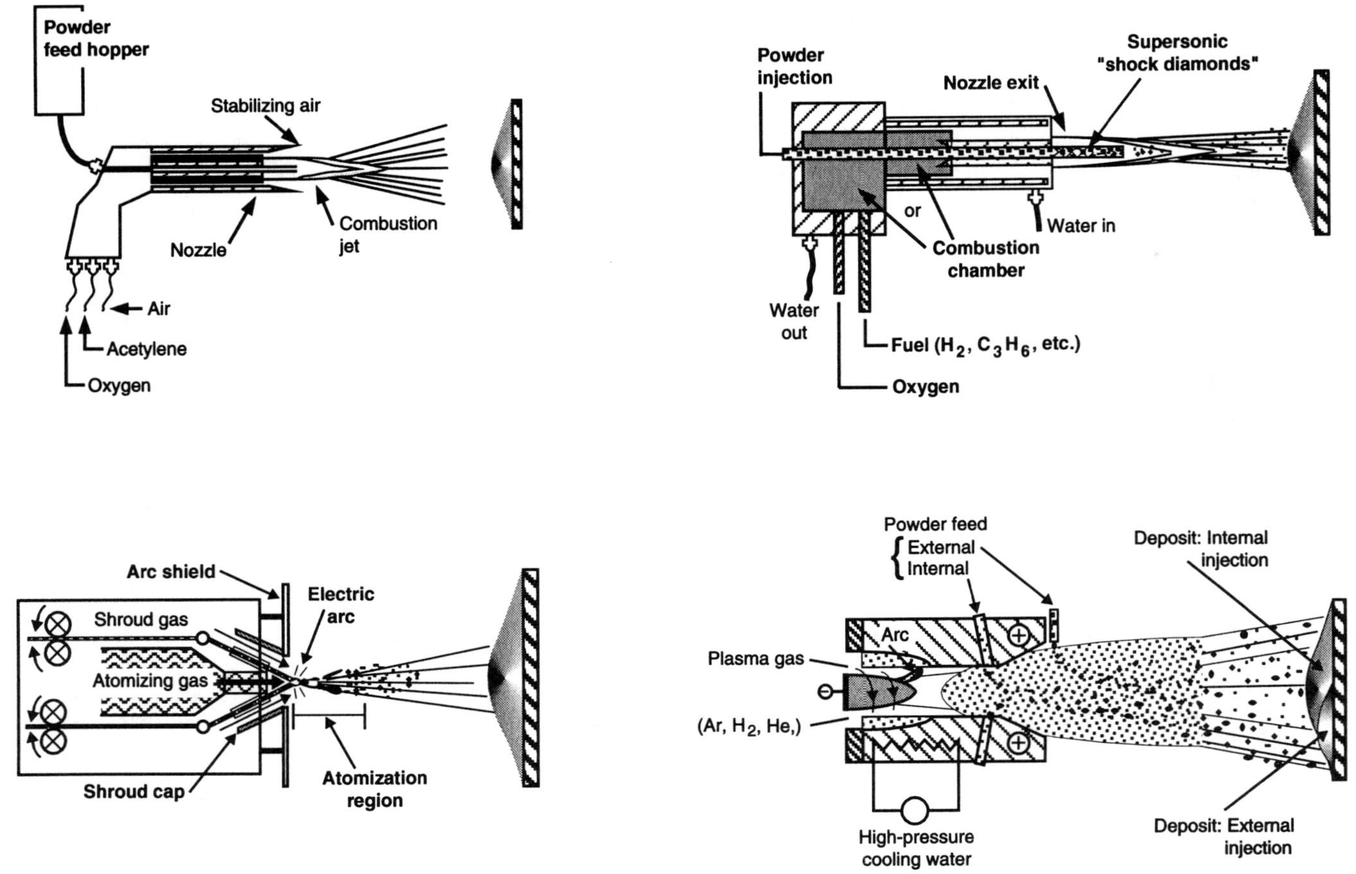

Fig. 4 Major thermal spray processes

continuously positions and advances the wires. Melt rate and interelectrode voltage determine the arc gap and size distribution of the atomized droplets leaving the tips of the wires and hence affect the final coating microstructure. Because cool gas jets are used to atomize the molten tips of feed wires in wire arc spray, the components being coated are not heated as much as in other thermal spray processes, an advantage when thermally sensitive substrate materials are being coated. Even thin sheets of paper can be coated by the wire arc process. Wire arc spraying is, however, limited to electrically conductive materials (metals) as feedstock, but has the lowest processing temperatures (at the substrate) of all the processes. Particle velocities vary, but are generally higher than in conventional flame spray and lower than HVOF or plasma spray. Plasma spraying (see the section that follows) has the highest particle-heating potential, but its particle velocities, although higher than conventional flame and wire arc spray, are generally lower than those of HVOF spraying. In other cases, accelerating gas cap extensions have been added to conventional wire arc spray systems to increase particle velocity and produce finer droplets, both leading to increased deposit densities, and reduced differences between the microstructures and properties of wire arc sprayed deposits and those produced by other thermal spray processes.

Plasma spray can be broken down into three major categories: (a) air plasma spray (APS), (b) "vacuum" plasma spray (VPS) or low-pressure plasma spray (LPPS), and (c) controlled atmosphere plasma spray (CAPS), which also includes inert gas shrouded plasma jets, depending on the environment used. These distinctions affect the level of interaction between the process jet and materials being sprayed with the surrounding atmosphere and thus control the microstructure and properties of the sprayed materials. All current commercially significant plasma spray processes use a nontransferred (i.e., no arc is established to the workpiece or substrate) dc electric arc to heat gases to peak temperatures in excess of 25×10^3 K, producing jets of ionized "plasma" with temperature ranges of 3×10^3 to 15×10^3 K. At these temperatures, the plasma gases (Ar, H_2, He, or N_2) are dissociated and ionized into an equilibrium mixture of positive ions and electrons as energy is pumped into them by the confined arc discharge. The plasma state gives the process both its name and its ability to melt any material exhibiting a stable melting point. A typical plasma spray device consists of a cylindrical, thoriated tungsten cathode, which emits electrons when heated by an electric arc, located concentrically and coaxially inside a cylindrical water-cooled copper nozzle. Electrons emitted from the cathode flow to the nozzle anode wall under the influence of the applied electric field, typically 30 to 80 V. Gases injected into this interelectrode region are heated, dissociated, and ionized by the electric arc and subsequently expanded to atmosphere through the nozzle as a subsonic or supersonic plasma jet, depending on the nozzle design and pressure ratio. The gas velocity may, in some cases, exceed 1.5×10^3 m/s, depending on the power input, nozzle design, gas composition and flow rate. Powders are injected into the plasma jet as fluidized streams of "carrier" gas and powder, either externally, by means of a tube that directs the stream of powder particles radially into the hot gas jet immediately after it exits the nozzle, or internally through powder feed ports through the nozzle wall itself. Internal injection produces longer particle heating or "dwell" times and improves melting efficiency, enabling higher melting temperature materials to be sprayed.

Plasma spray devices generally use argon or mixtures of argon plus helium, hydrogen (H_2), or sometimes nitrogen (N_2) to generate the hot plasma jets used for spraying. Coatings sprayed at low pressures (VPS), in inert atmospheres in an enclosed pressure vessel, or using an inert gas shroud, have very low oxide inclusions and porosity levels. This is achieved by eliminating interaction between the molten particles and oxygen being entrained into the jets from the surrounding air. The sprayed coatings generally exhibit low oxide content, such that in many cases VPS materials have the same, or better, properties than cast materials. This characteristic has enabled VPS processes to be used to spray form net shapes of "difficult-to-process" materials, including ceramics, refractory metals, intermetallics, and composites.

Inert gas shrouded plasma spray has the potential of achieving vacuum plasma quality coatings under air plasma spray conditions, thereby lowering the total processing cost. Several effective inert gas shroud designs have been developed and reported, including porous metal nozzle inserts (Ref 11) and discrete, multijet, parallel flow designs located downstream of the nozzle exit on a conventional plasma spray gun (Ref 12).

As stated above, the structures and properties of sprayed materials can vary widely, depending on the process used, classes, and the processing conditions. The different processes each result in different thermal and velocity histories for the injected particles. Particle melting and deformation are thus different in each thermal spray class, leading to the differences observed in the performance of spray deposited materials. Key features of the different processes are as follows:

- Conventional flame spray has the lowest heating potential and lowest mean particle velocity of all thermal spray processes, but is nevertheless used in many applications.
- High-velocity oxygen fuel processes typically have the highest particle velocity, but its gas jet temperatures limit the maximum attainable particle temperatures.
- Wire arc spray has lower processing temperatures and produces intermediate particle velocities, but—owing to its low gas temperatures—can be used to coat thermally sensitive substrates.
- Plasma spray, with the highest processing temperatures, is most flexible with respect to materials selection, and because of its ability to operate under inert gas conditions, is commonly chosen for consolidating oxidation-sensitive and reactive materials.

Thermal spray has been defined as a particulate/droplet consolidation mechanism capable of processing virtually any material and most materials combinations. Despite the widespread use of thermal spray to produce coatings, and, to a lesser extent for spray forming, thermal spray has not been widely used for the synthesis and production of advanced materials. Developments in thermal spray over the last 25 years, including controlled atmosphere plasma spraying, HVOF, reactive plasma spraying, improved wire arc spray methods, new material and powder production methods, and particularly improved process controls have, however, resulted in many new opportunities for the technology in the area of powder consolidation.

The range of materials that can be sprayed, their possible structures, and the application of thermal spray processing are limited only by the imagination of the users and the economic viability of the application. Thermal spray can process many materials, in many forms, to produce deposits on many substrate materials, thus making thermal spray an ideal technique for processing many advanced materials and structures. Many challenges, however, still remain:

- Cost/economics
- Compositional and structural control
- Deposition rate
- Control of deposit properties

Thermal spray is now, owing to recent advances in materials, powder production techniques, process understanding, and control, at the leading edge of novel powder consolidation methods. Some recent advances are reviewed below that present the current state of the technology for implementation as a viable powder consolidation process.

Materials for Thermal Spray

Thermal spray is capable of processing materials with jet temperatures ranging from 500 to 15×10^3 K, which enables virtually any material, or combination of materials, to be processed. Because only small volume powder particles are heated in a given time and because, on solidification, only small volumes, typically 5 to 100 μm in diameter, are being cooled, segregation is not a limiting factor. On the other hand, residual stress can have detrimental effects on the properties and performance of sprayed coatings, its significance varying from process to process and material to material. Also, because materials are deposited in relatively thin (12 to 50 μm, or 0.0005 to 0.002 in.) layers, many unique combinations of materials can be produced.

Three characteristic types of deposit can be sprayed: (a) Monolithic materials, such as metals, alloys, intermetallics, ceramics, and polymers; (b) composite particles such as cermets (WC/Co, Cr_3C_2/NiCr, NiCrAlY/Al_2O_3, etc.), reinforced metals, and reinforced polymers; and (c) layered or graded materials. Some examples of these, and their particular advantages and associated technical challenges, are described below.

Monolithic Materials

Metals. Tungsten, molybdenum, rhenium, niobium, superalloys (nickel, iron, and cobalt base), zinc, aluminum, bronze, cast iron, mild and stainless steels, NiCr and NiCrAl alloys, cobalt-base Stellites, cobalt/nickel-base Tribaloys, and NiCrBSi "selffluxing" Colmonoy have all been successfully thermal spray consolidated either as coatings or structural deposits. Recently Tribolite (FeCrNiBSi) and AmaCor (amorphous) alloys have also been developed for spraying and exhibit excellent wear and corrosion resistance (Ref 13). Monolithic alloys have advantages due to their similarity to many base metals requiring repair, their high strength, and their corrosion, wear and/or oxidation resistance. Applications include automotive/diesel engine cylinder coatings; piston rings or valve stems; turbine engine blades, vanes, and combustors; protection of bridges and other corrosion-prone infrastructure; petrochemical pumps and valves; and mining and agricultural equipment. Except in the case of controlled-atmosphere spraying (VPS, inert chamber, and shrouded jets) thermally spraying these metals and alloys produces microreinforced composites of monolithic alloys due to their varying levels of oxide inclusions. Figure 2 shows a range of microstructures for thermally sprayed monolithic metals. These coatings exhibit characteristic lamellar microstructures with the long axis of impacted splats oriented parallel to the substrate surface, together with a distribution of similarly oriented oxides. Oxide content varies from relatively thick layers to finely distributed, particulate, intersplat phases, depending on whether the coatings are wire arc, plasma, or HVOF sprayed. The progressive increases in particle velocity of these processes leads to differing levels of oxide, and differing degrees of oxide breakup on impact at the surface. Oxides may increase coating hardness and may also provide lubricity. Conversely, excessive and continuous oxide networks can lead to cohesive failure of a coating and contribute to excessive wear debris. It is thus important, when selecting materials, coating processes, and processing parameters, that oxide content and structure be controlled to acceptable levels.

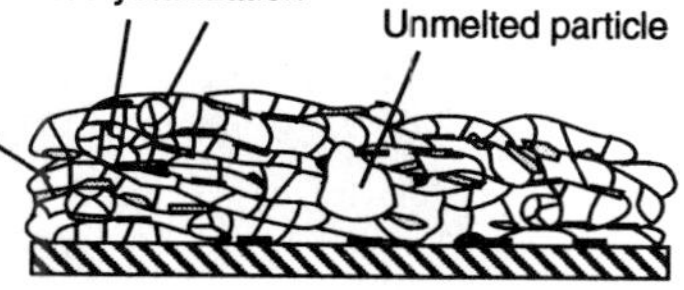

Fig. 5 Schematic microstructure of an "as-sprayed" thermally sprayed metal deposit

Thermal spray coatings may, depending on the spray process, particle velocity and size/size distribution, and spray distance, also contain varying levels of porosity and unmelted particles. High levels of porosity may lead to early deposit failure owing to poor intersplat cohesion. Conversely, low levels of porosity (<5%) may be beneficial in tribological applications through retention of lubricating oil films. Lamellar oxide layers can also lead to lower wear and friction due to the lubricity of the oxide. The porosity of thermal spray coatings is typically <5% by volume. The retention of some unmelted and/or resolidified particles can lead to lower deposit cohesive strength, especially in the case of "as-sprayed" materials with no postdeposition heat treatment or fusion.

Other key features of thermally sprayed deposits are their generally very fine grain structures and microcolumnar orientation. Thermally sprayed metals, for example, have reported grain sizes of <1 μm prior to postdeposition heat treatment. Grain structure across an individual splat thus normally ranges from 10 to 50 μm, with typical grain diameters of 0.25 to 0.5 μm, due to the high cooling rates achieved (~10^6 K/s) (Ref

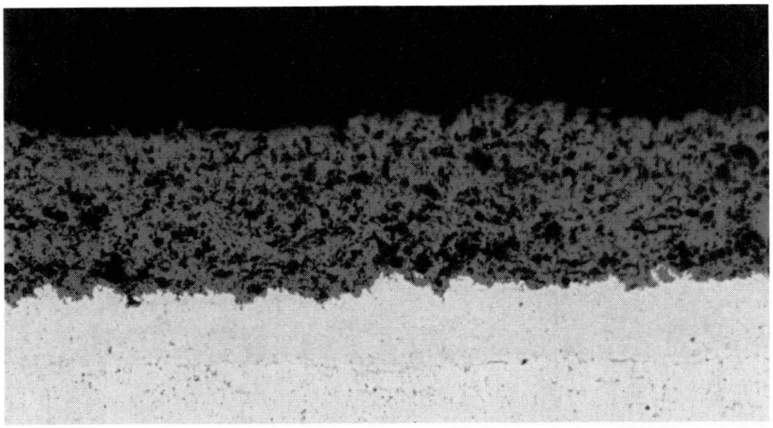

Fig. 6 Typical microstructure of a thermally (plasma) sprayed ceramic (yttria-stabilized zirconia)

1). Such rapid cooling rates, known to form fine-grained martensitic microstructures in steels, contribute to the high strengths exhibited by thermally sprayed materials. The "as-sprayed" microstructure of a typical metallic coating is shown schematically in Fig. 5.

The tensile strengths of "as-sprayed" deposits can range from 10 to 60% of those of the fully cast or wrought material, depending on the spray process used. Spray conditions leading to higher oxide levels and lower deposit densities result in the lowest strengths. Controlled-atmosphere spraying leads to ~60% strength, but requires postdeposition heat treatment to achieve near 100% values. Low "as-sprayed" strengths are related to limited intersplat diffusion and limited grain recrystallization during the rapid solidification characteristic of thermal spray processes. The microstructures of thermally sprayed metals are typically very uniform and exhibit excellent tensile properties. After heat treatment at >0.8 T_m (melt temperature), much of the characteristic lamellar structure recrystallizes, but depending on the alloy, grain growth may be limited to <100 μm. The fine-grained structures, found in highly alloyed, grain-growth-stabilized superalloys, have been found to improve thermal fatigue properties, but to increase creep rates. After postdeposition heat treatment (vacuum or hot isostatic pressing), the high-temperature creep properties of these alloys have been found to be lower than cast alloys due to the fine grain sizes and the oxygen interstitials. In this reported work, final grain size in heat-treated microstructures was found to be substantially retained, due to very stable oxide and/or carbide networks formed during solidification after spraying, many times originating from the alloy feedstock powders. Final deposit ductilities were also lowered by the retention of relatively high oxygen contents originating from the surfaces of the original powders (Ref 14).

On the beneficial side, the addition of nitrogen to steel matrices by promoting, rather than minimizing, atmospheric interactions during air plasma spraying has been shown to increase the strength of thermally sprayed steels and is a viable strengthening mechanism for selected monolithic alloys (Ref 15). Grain nucleation is suppressed in alloys containing high levels of silicon and boron, yielding so-called "amorphous" coatings (Ref 16). These "micrograined" or amorphous microstructures contribute to the high strengths and toughnesses of thermally sprayed metals at low to intermediate temperatures. Fatigue failure is also harder to propagate in such structures, except through coating defects such as oxide inclusions. The role of grain size effects on the wear performance of thermally sprayed coatings is understood, but has not been independently measured; hence it is hard, for example, to determine its contribution to sliding-wear performance.

Ceramics. Oxides such as Al_2O_3, ZrO_2 (stabilized with MgO, CeO, Y_2O_3, etc.),TiO_2, Cr_2O_3, and MgO; carbides such as Cr_3C_2, TiC, Mo_2C and SiC (generally in a supporting metal matrix) and diamond; nitrides such as TiN and Si_3N_4, and spinels or perovskites such as mullite and superconducting oxides, have all been thermally spray deposited. Sprayed deposits of these materials are used to provide wear resistance (Al_2O_3, Cr_2O_3, TiO_2, Cr_3C_2, TiC, Mo_2C, TiN, and diamond), thermal protection (Al_2O_3, ZrO_2, MgO), electrical insulation (Al_2O_3, TiO_2, MgO), and corrosion resistance. With the exception of SiC (which sublimes), diamond, and SiC or Si_3N_4 (which must be in a metallic binder), ceramics are particularly suited to thermal spraying, with plasma spraying being most suitable due to its high jet temperatures.

The processing and materials flexibility and high temperatures gives plasma spraying a leading role in the spraying of thermal barrier coatings (TBCs), although the use of HVOF is also being investigated. Thermal barrier coatings are, after wear and corrosion coatings, likely to be one of the largest growth markets for thermal spray, with increased use in the automotive, metalworking, and chemical industries. Such broad usage, however, will require a more thorough understanding of the behavior of these materials during high-temperature service. Thermal spray processing of ceramic materials that melt, rather than decompose, is essentially the same as for metals; however, the higher mean melting temperatures and low thermal conductivity of ceramics limits the selections of thermal spray processes that can successfully melt these materials. Combustion spray methods generally have insufficient process jet enthalpies and/or particle dwell times to efficiently spray ceramic powders. Most HVOF systems are also limited in their efficient spraying of ceramics, where the powder sizes required are either too small or deposit efficiencies are too low for the processes to be economically viable. Wire arc spray is excluded because it needs conductive materials, leaving only plasma spraying as an economic method for spraying ceramic powders. There are, however, some specialized flame spray techniques such as the "Rokide" process (Norton Company), which uses a combustion jet to melt the tips off pressed-and-sintered ceramic rods and a secondary gas jet that atomizes the molten material from the melting rod tip. Air plasma spraying is, however, most widely used to deposit ceramics, finding applications in TBCs, wear coatings on printing rolls (Cr_2O_3), and for electrical insulators (Al_2O_3).

The microstructures of sprayed ceramics are similar to those of metals, with two important exceptions: grain orientation and microcracking. Rapid solidification of small droplets (generally <50 μm) results in very fine grains as in metals; however, owing to the low thermal conductivity of ceramics, multiple grains may exist though a splat thickness, whereas in metals a single grain may cross an entire splat and even grow into an adjacent one. Intersplat and intrasplat microcracking is widespread in ceramic coatings, resulting from the accumulation of highly localized, residual cooling stresses. Figure 6 illustrates the typical microstructure of a thermally sprayed yttria-stabilized zirconia ceramic coating, showing fine grains, characteristic lamellar structure, and a network of porosity and microcracks.

Splats normally exhibit through-thickness cracking owing to the very low ductility of most ceramics, but these cracks do not usually link up through the whole deposit thickness, at least not until an external stress is applied. Microcracking of splats is a major contributor to the effectiveness of TBCs, even under high-temperature gradients and moderate strains, conditions under which conventionally formed bulk ceramics would fail.

Intermetallics. Usually produced from powders due to their intrinsically low ductility, intermetallics are generally consolidated either by pressing and sintering or hot isostatic pressing. Over the last 15 years researchers have reported on the thermal spray consolidation and forming of intermetallic powders. The high heating and cooling rates of the thermal spray process reduce the segregation and residual stresses that ordinarily limit the formability of these brittle materials. Thermal spray processes are also able to deposit materials onto mandrels, building up thin layers of material and thus forming near-net shapes and providing the opportunity for "engineered microstructures" and functionally graded structures. Researchers have also reported on the plasma spraying of TiAl, Ti_3Al, Ni_3Al, NiAl, and $MoSi_2$ with excellent deposit characteristics and properties. Improved ductilities have been obtained, with tensile strengths equal to, or better than, those of materials consolidated by other powder processing techniques.

Thermal spray processing of intermetallics, unlike ceramics, is generally not an application for plasma spraying. Plasma spray in controlled atmospheres is, however, the method of choice for the production of bulk deposits. High-velocity oxygen fuel and other combustion spray techniques are normally only used to spray compatible intermetallics, such as semiconducting, insulating, or corrosion- and wear-resistant coatings. Most intermetallics are very reactive at high temperatures and very sensitive to oxidation, hence the preference for using inert atmosphere plasma spray. Plasma spray forming is also well suited for the net-shape forming of brittle intermetallics. Investigators have found that plasma spray forming can actually *increase* the ductility of intermetallics such as $MoSi_2$ (Ref 17) and NiAl/Ni_3Al (Ref 18). These improvements were linked to a decrease in grain boundary and/or splat interface contaminants that limit localized plastic flow under an applied strain and lead to early crack linking, thus lowering the overall plastic flow, measured as ductility. It has been shown, for example, that reactive plasma species reacting at the surfaces of $MoSi_2$ powder particles in flight actually reduce the residual SiO_2 content below that of commercial $MoSi_2$ powders and hence reduce the SiO_2 "pest" reaction that degrades the properties of pressed-and-sintered or hot isostatically pressed $MoSi_2$.

The as-sprayed microstructures of thermally sprayed intermetallics are very similar to those of metals. Figure 7 shows the microstructure of thermal spray consolidated $MoSi_2$, showing the characteristic lamellar structure and fine grains. Intermetallics are generally somewhat more porous than sprayed metals owing to their limited ductility and low plasticity, which translates into a narrower "processing window" of plasma spray parameters for the production of high-density intermetallics, requiring tighter process control than for spraying metals. Higher-velocity plasma spray processes are preferred because the increased particle velocities result in more complete deformation of individual splats and denser deposits overall.

Polymers. Polymeric materials can also be successfully thermally sprayed, provided they are available in particulate form. Thermal spraying of polymers has been commercially practiced for 20 years, and a growing number of thermoplastic polymers and copolymers have now been sprayed, including urethanes, ethylene vinyl alcohols (EVAs), nylon 11, polytetrafluoroethylene (PTFE), polyetheretherketone (PEEK), polymethylmethacrylate (PMMA), polyimide, and copolymers such as polyimide/polyamide, Surlyn, and Nucryl (Du Pont), and polyvinylidene fluoride (PVDF). Conventional flame spray and HVOF are the most widely used thermal spray consolidation methods used to date (Ref 19), although use of plasma spray has also been reported. Figure 8 shows a typical dense, well-bonded, HVOF-sprayed nylon coating. Consolidation of polymers to full densities and high cohesive strengths, unlike metals, relies to some extent on continued heat input to the polymer layers after deposition onto the substrate. This may be due to the more complex, long-range-ordered nature of the molecular bonding in polymers, compared to the simpler, short-range-order bonding in metals. Thermal spray, a high-temperature, rapid-heating/rapid-solidification process, has been found to decompose polymeric materials, lowering the molecular weight and producing "charring." In many cases, however, this does not impair the functionality of the sprayed coating, although it does change certain polymer properties compared to polymers consolidated by conventional means. Despite thermal spraying of polymers having been practiced for more than 20 years, little research into the material structures and properties has been carried out until recently. Commercial interest in this application is growing, for applications such as aqueous corrosion protection and as a solventless processing alternative to environmentally hazardous volatile organic compound (VOC)-based techniques.

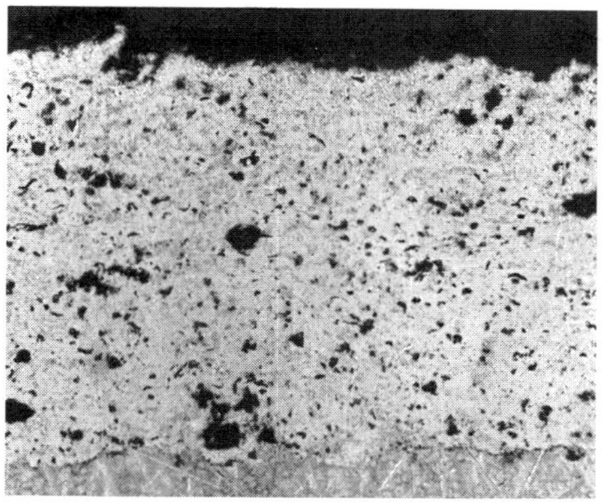

Fig. 7 Microstructure of a plasma spray consolidated intermetallic material ($MoSi_2$)

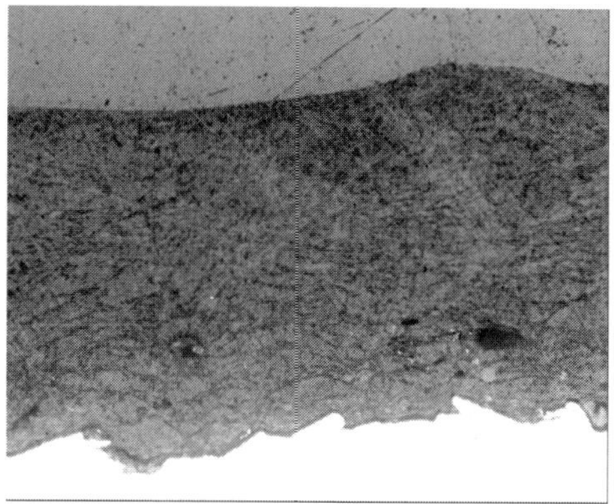

Fig. 8 Microstructure of an HVOF-sprayed nylon coating

Composite Materials

Thermal spraying, either as a coating or as a bulk structural consolidation process, has clearly demonstrated advantages for the production of composites. Difficult-to-process composites can be readily produced by thermal spray forming, with vacuum plasma spray being the process of choice for the most reactive matrix materials. Particulate-, fiber-, and whisker-reinforced composites have all been produced and used in various applications. Particulate-reinforced wear-resistant coatings such as WC/Co, Cr_3C_2/NiCr, and TiC/NiCr are the most common applications and comprise one of the largest single thermal spray application areas. Figure 9 shows schematically the diverse forms of composites that can be thermally spray formed.

Whiskers of particles can be incorporated using so-called "engineered" powders, mechanical blending, and by coinjecting different materials into a single spray jet. Mechanical blends and coinjection, although useful, have been found to result in segregation of the reinforcing phase and, in many cases, degradation of the second-phase whiskers or particles. Thermal spray composite materials can have reinforcing-phase contents ranging from 10 to 90% by volume, where the metal matrix acts as a binder, supporting the reinforcing phase. The ability to consolidate such fine-grained, high reinforcing phase content materials is a major advantage of thermal spray over other methods used to produce composites.

Thermal spraying of composite materials with discontinuous reinforcements, such as particulates or short fibers, is usually accomplished by spraying powders or powder blends. Investigators have developed techniques for the production of continuous fiber-reinforced materials that overcome the "line-of-sight" limitations of thermal spray processes. This includes "monotape" fabrication techniques, where continuous fibers are prewrapped around a mandrel and a thin layer of a metal, ceramic, or intermetallic matrix material is sprayed (Ref 20). Plasma, HVOF, and wire arc spray have been used, although in the cases of intermetallics and high-temperature alloys, controlled atmosphere plasma spray (VPS)

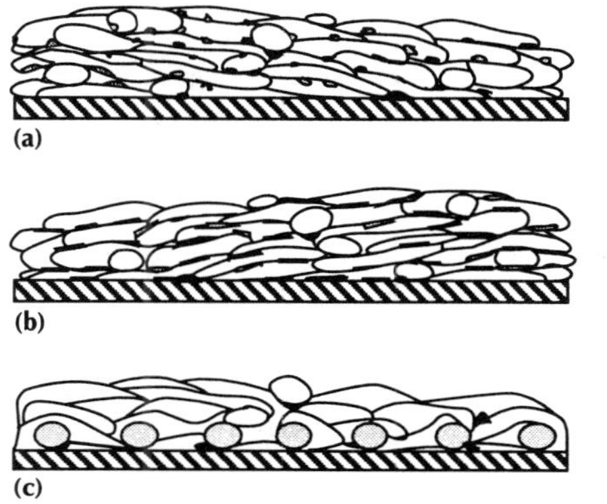

Fig. 9 Schematic microstructures of possible thermally spray formed composites. (a) Deposit with a particulate-reinforced second phase. (b) Deposit with a whisker-reinforced second phase. (c) Deposit with a continuous-fiber-reinforced second phase

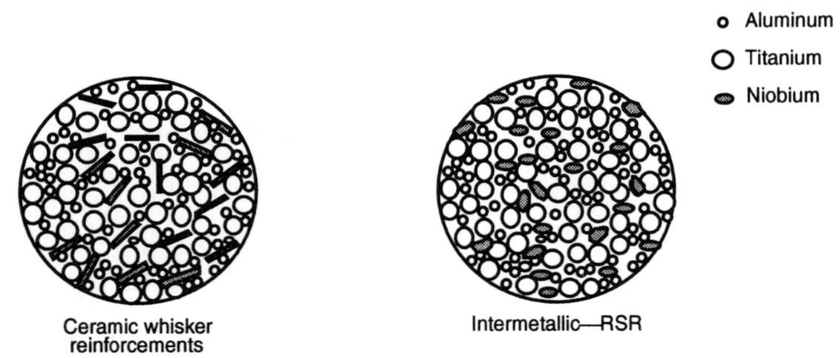

Fig. 10 Schematic representations of typical "composite" thermal spray powders

has generally been used. The fibers are thus encapsulated within thin monolayer tapes and are subsequently removed for consolidation to full density by hot pressing with preferred fiber orientations, producing continuously reinforced bulk composites.

Powders for Sprayed Composites

Discontinuously reinforced composites produced using thermal spray techniques use either composite powders or direct reactive synthesis approaches, as described below. Powders can be produced mechanically, chemically, thermomechanically, or by using high-temperature synthesis. "Engineered powders" defines powders in which different phases are incorporated to produce a "microcomposition" of the final desired structure. Typically these powders contain the desired sizes, size distributions, and morphologies of the equilibrium phases. These powders also permit the introduction of higher concentrations of phases than those normally achievable through conventional melt or reaction processing. Figure 10 shows two types of powder that could be produced in this way: short ceramic fibers within a metal matrix and an intermetallic-matrix material reinforced with more ductile phases. The latter has been found to be a viable approach for increasing the fracture toughness (K_{Ic}) of intermetallics. Varying reinforcing phase combinations, compositions, morphologies, and distributions can be produced. The rapid heating and cooling experienced by these powders during thermal spray forming limits dissolution and degradation of the phases, which remain relatively unchanged after consolidation, although some microstructural refinement and solutioning can take place. Generally, more significant changes occur during conventional powder consolidation processes because of the longer processing times.

Mechanical Blends. Blended powders are produced by mixing the required proportions of a binder phase (a metal, an intermetallic, or a ceramic) together with a reinforcing phase. The matrix and reinforcing phases may segregate over time during shipping, handling, and feeding into a thermal spray process and often yield poor "as-sprayed" deposit uniformity. Many mechanical blends can, however, be "agglomerated" and fixed using an organic binder, which reduces their sensitivity to handling. Spherical agglomerates are known to flow and feed better, and spray drying has been used to produce such materials. Agglomerated powders have enabled thermally sprayed deposits with improved uniformity to be produced, although the high viscosity, shear, and thermal forces acting on injected agglomerated particles tend to break the agglomerates, leading to segregation and direct exposure of the second phase to the thermal spray jet. Physically and thermally stable agglomerates are thus desirable.

Agglomeration, Sintering, and Melt Densification. The mechanical stability of agglomerated composite particles can be improved by agglomeration and sintering, which is a solid or rapid "matrix" melting and cooling process. Metal-matrix/ceramic hard-phase-reinforced powders (e.g., WC/Co, Cr_3C_2/NiCr, NiCrAlY/Al_2O_3, etc.) can be sintered in the solid state by heating in a protective atmosphere furnace, sometimes using a fluidized bed, followed by a gentle milling to break any weak interagglomerate bonds. Agglomerated and sintered powders retain a level of porosity, typically 5 to 10% by volume, depending on the reinforcing phase/matrix combination, the sintering time, and temperature. Retained porosity, although not generally a source of phase segregation in thermally sprayed deposits, can lead to higher as-sprayed porosity and increases the spray jet melting enthalpies required to melt the powders because the thermal conductivity of the individual particles is lowered by the porosity. Melting of metallic binding phases and retention of spherical particle morphologies can be achieved by processing the powders through a thermal plasma jet. Known as "plasma densification," this process produces spherical powders exhibiting near 100% particle density and uniform distributions of reinforcing phases. Plasma-densified powders result in the most uniform thermal spray-consolidated deposits, but at significantly increased cost owing to the additional handling, processing steps, and lower process yields.

Self-propagating high-temperature synthesis (SHS) is a composite powder production method now finding increasing application in thermal spray forming. The SHS process exploits the high heats of formation released from exothermic reactants which, when mixed and ignited, produce a self-sustaining combustion-type reaction that propagates until all reactants are consumed. Self-propagating high-temperature synthesis processing uses mixtures of reactant powders placed in a reactor vessel, generally in an argon atmosphere. The reactant mixture is then ignited using an incandescent filament or laser beam. Self-propagating high-temperature synthesis has been developed to the point where it is now capable of producing composite materials with well-controlled distributions of carbides or other reacted phases in metallic or intermetallic matrices (Ref 21, 22). Self-propagating high-temperature synthesis produces low-density, porous "ingots" that must then be further processed by crushing, milling, and screening to yield powders in size distributions compatible with thermal spraying. Researchers have successfully thermally sprayed SHS-produced $MoSi_2$, SiC-reinforced $MoSi_2$, FeCr/TiC, and NiCr/TiC materials. The sprayed deposits reportedly yielded equivalent, and in many cases superior, proper-

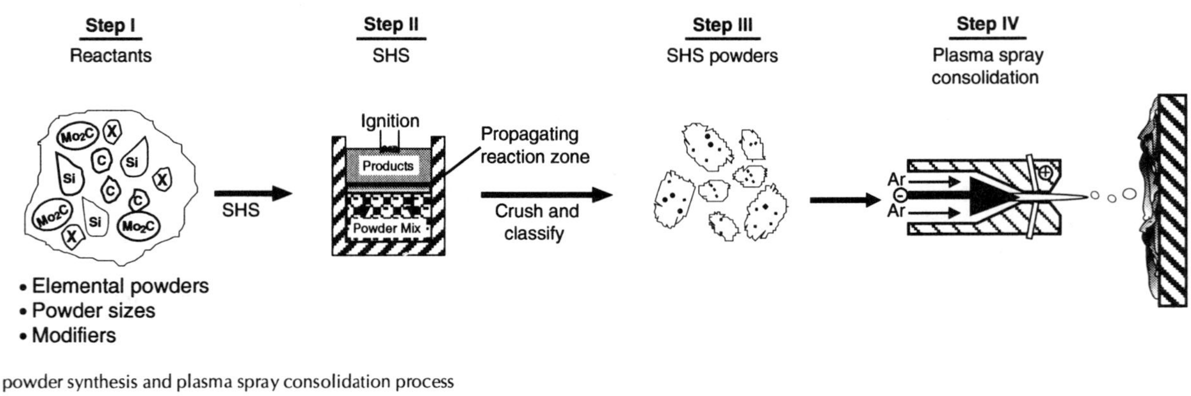

Fig. 11 SHS powder synthesis and plasma spray consolidation process

ties to deposits produced from powders produced by other synthesis routes Some examples of the improved properties obtained include increased wear resistance in TiC/metal-matrix composites and increased fracture toughness in intermetallic composites such as $MoSi_2/SiC$ (Ref 23). Figure 11 illustrates the SHS powder production route, including powder synthesis and consolidation via plasma spray forming.

Thermal Spray Forming

While the thermal spray technology described above originated primarily as a coating or surfacing process, with materials being deposited as thin (<250 μm), nonload-bearing coatings for dimensional restoration, wear, corrosion, and thermal protection, it has also evolved, during the last 50 years, into a process capable of spray forming free-standing materials onto mandrels at thicknesses of >100 mm. Early applications included zirconia λ-oxygen sensors and bulk ceramic materials for crucibles. Historically "difficult-to-form" materials such as the refractory metals tungsten, hafnium, molybdenum, and tantalum; superalloys; and intermetallics have now all been deposited with varying degrees of success, with the ultimate mechanical properties of the sprayed deposit being the main limitation. Process technology advances such as HVOF, which can produce sprayed deposits with lower tensile stresses and degrees of thermal degradation than other processes and with increased deposit density, controlled-atmosphere plasma spray, and improved powder formulations and morphologies have all contributed to the successful development of thermal spray forming of monolithic and composite materials.

Thermally sprayed coating deposits are usually <1000 μm thick, more typically ~400 μm. Thermal spray forming, however, can yield deposit thicknesses in excess of 25 mm. Freestanding shapes can be produced by spraying onto sacrificial mandrels, which are mechanically or chemically removed after spraying.

The incremental nature of thermal spray processes, using particulate-based feedstocks, and with deposited layers typically ~15 to 25 μm thick, enables graded and laminated structural materials to be formed. Also, because of the high processing temperatures and high localized forming energies (high particle kinetic energies at impact), traditionally difficult-to-form materials can be processed into near-net-shape components. Virtually all common and refractory metals, many intermetallics, ceramics, and combinations of these have been spray formed as "composite" materials. The properties (apart from the ductility of metals) of these deposited materials, after postdeposition heat treatment, have been reported to be close to, if not exceeding, those of cast or wrought materials. The ductility of metals is limited by contamination sources occurring either within the feedstock powders used, or in the spray process itself, due to oxidation of the material during heating under atmospheric conditions. Nickel-base superalloys and other heat-resistant alloys have all been thermally spray formed during the last 20 years, either as preforms or as a component repair technique. Plasma spray in low pressure inert/controlled atmosphere (VPS) chambers has been found to be most suitable for forming these alloys; however, recent investigations using HVOF have shown that this too may be a viable alternative to plasma spraying in controlled atmospheres, at least for some component repair applications. Despite the technical capability of thermal spray, however, economics and process/coating reliability concerns related to material structure, uniformity, and material properties still limit the use of thermal spray forming.

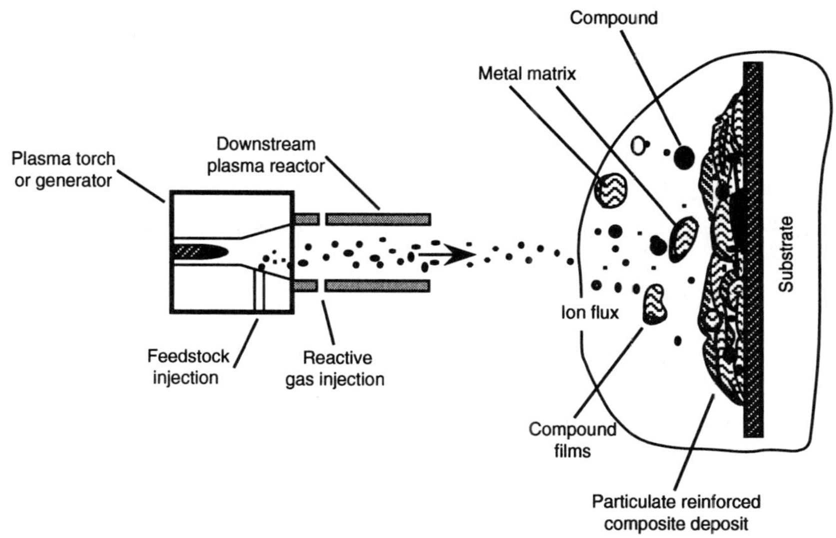

Fig. 12 Reactive plasma spray process for the synthesis of composite materials

The current major advantages of thermal spray forming techniques are in the forming of refractory metals, both in graded or layered structures and in composites. Some of the more innovative spray forming application developments are described below, together with some insights into emerging technologies that may ultimately lead to expanded implementation of thermal spray forming as a materials processing technique.

Reactive Plasma Spray Forming

Research in recent years has demonstrated that the synthesis of advanced materials by chemical routes is more promising than many "conventional" processing methods—casting, forming, and possibly powder metallurgy (Ref 24). Plasma spray forming, a rapid particulate consolidation, high-temperature process, has an inherent suitability for chemical synthesis and is thus a strong candidate for synthesizing multiphase, advanced materials. As described in the literature (Ref 25), plasma spray coating and forming have been practiced for many years and are now practical and economical methods for processing difficult-to-form intermetallic, ceramic, and composite materials (Ref 1, 2, 26). The development of plasma spraying in controlled atmosphere chambers (VPS or LPPS) in the early 1970s (Ref 27–29) was a key factor behind its advancement from a coating to a forming technology. Controlled-atmosphere plasma spraying, adapted for *reactive* plasma spraying, has recently been developed to combine controlled dissociation and reactions in thermal plasma jets, for the in situ forming of new materials or to produce new phases in sprayed deposits. Reactive plasma spray forming is consequently emerging as a viable method for producing a wide range of advanced materials. The process, a logical evolution from conventional plasma spraying, allows "reactive" precursors to be injected into the particulate and/or hot gas streams. These reactive precursors may be liquids, gases, or mixtures of solid reactants which, on contact with the high-temperature plasma jet, decompose or dissociate to form highly reactive and ionic species that can then react with other heated materials within the plasma jet to form new compounds. Figure 12 illustrates the reactive plasma spray concept. Chemical reactions rely on plasma-induced dissociation of the injected precursors, for example gaseous methane (CH_4), which decomposes into elemental or ionic species such as CH_3^-, C^{4-} or even atomic carbon. These species can react with other elements to produce carbides such as TiC and WC, or under certain conditions, diamond or diamondlike carbon (DLC) films. Figure 12 shows the plasma heating of gases, using either a nontransferred electric arc or an inductively coupled plasma (ICP) radio frequency (RF) discharge; injection of reactive precursors into the plasma jet; and injection of powders in a carrier gas stream. Injection locations may vary depending on the materials or reactions desired and can be directly in the plasma generator (torch or gun) or into a reactor located immediately downstream. The primary requirements are that the precursors dissociate into reactive "species" and that the reaction times and temperatures are sufficiently long for the desired prod-

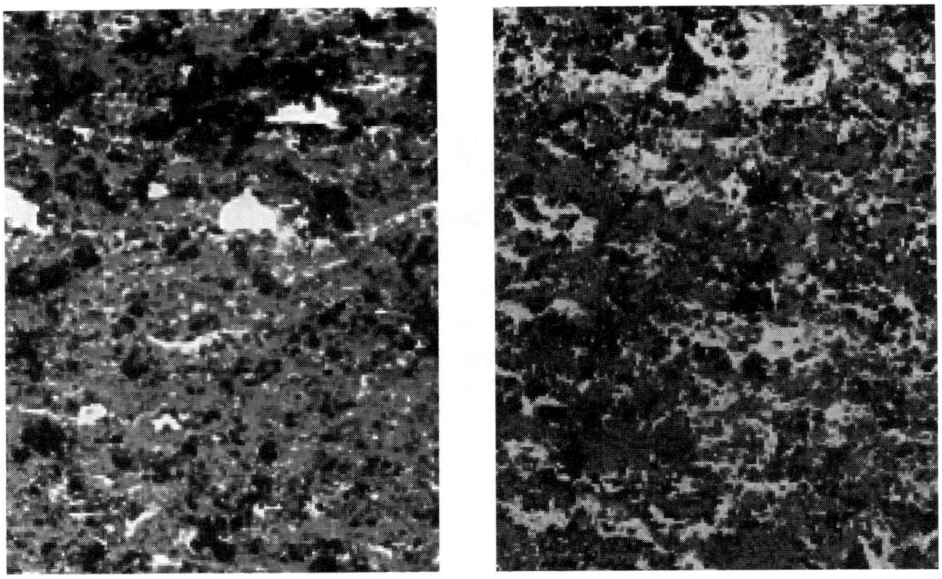

(a) (b)

Fig. 13 Microstructures of reactively plasma sprayed (a) Al_2O_3/SiAlON and (b) $MoSi_2$/SiC

ucts or phases to form. Reactive plasma spray forming has several key components, as shown in Fig. 12:

- A thermal plasma generator
- A method of injecting the reactants or precursors
- A reaction region where compounds can form and nucleate, either on the surfaces of particles or on a substrate
- A gas jet to transport the heated reactants and products to the surface(s) on which the products are formed

Reactive plasma spray applications include the synthesis of composite materials, shaped brittle intermetallic alloys, reinforced or toughened ceramics, and tribological coatings with in situ formed hard or lubricating phases. Reactive plasma spray forming enables a wide range of materials to be produced, for example, aluminum with AlN, Al_2O_3, or SiC; NiCrTi alloys with TiC or TiN; intermetallics such as TiAl, Ti_3Al, $MoSi_2$, and other ceramics with oxides, nitrides, borides, and/or carbides. All of these have been produced in situ in reactive thermal plasma jets.

Figure 13 shows typical microstructures of some reactively plasma sprayed materials, demonstrating the versatility of the process for the in situ forming of metal/carbide, metal/nitride, and ceramic composites.

Research indicates, however, that significant technical challenges still remain because, despite the promises and advantages indicated previously, the yields and quality of the final products depend strongly on the starting precursors, their injection methods, and the atmospheres in which they are sprayed (Ref 24, 30). Monolithic SiC, SiN, and AlN powders have also been synthesized in plasma jets (Ref 31, 32). The uniformity of these particular materials has considerable variability, however, and has limited the practical use of many plasma spray processes for forming monolithic compounds, although commercial plasma production of ZrO_2 and TiO_2 powders in plasma jets has been reported (Ref 9).

Recent investigations have demonstrated that reactive plasma spray forming utilizing incomplete or nonequilibrium reactions in plasma jets to form mixtures of phases can produce functional composites with the reactant products embedded in metallic, ceramic, and even intermetallic matrices. These materials have been shown to be very hard and have potential application as wear-resistant coatings (Ref 33). These reactively plasma sprayed coatings were, in practice, metal-matrix composites with in situ reacted hard phases formed when the reactive gas mixtures reacted with molten metal particles. Fine TiC, WC, TaC, and other refractory carbides, oxides, silicides, or nitrides have been spray formed in situ as particulate phases in near-net-shape metallic, intermetallic, and ceramic matrices, thus eliminating many difficult post-forming operations. Lightweight structural materials such as Ti_3Al, TiAl, NiAl, and $MoSi_2$ have also been produced by reactive spray forming. In these cases, aluminum metal powders were injected into $TiCl_4$ plasmas to produce TiAl. Molybdenum particles have also been injected into plasma jets seeded with disilane (Si_2H_6) to form $MoSi_2$.

The utility and potential of reactive plasma spray synthesis is further illustrated by two high potential examples presented below. Research and development in these areas is receiving increased attention as the potential of thermal plasma processing of these high value materials is realized.

Spray Forming of Superconducting Oxides (High T_c Ceramics)

Researchers have reported on the deposition and synthesis of high critical temperature (T_c) ceramic superconductors using dc and RF induction coupled plasma jets (Ref 34, 35). Dense, thick deposits have been produced using prealloyed 1-2-3 $YBaCu_xO_y$ powders, although the high T_c phase was only formed after a postdeposition heat treatment in oxygen. Researchers have not yet been able to directly plasma spray form the superconducting phase by dc plasma spraying prealloyed powders because the rapid cooling of the "1-2-3" materials from the liquid state suppresses direct formation of the high T_c phase. The microstructures of the deposits were also found to contain many small grains and pores as well as fine cracks that limited the maximum critical currents to very low values. Researchers at the University of Minnesota have developed a method of overcoming this limitation by using a vapor phase, plasma-assisted, reactive plasma spray process (Ref 34). This system uses an RF, induction-coupled, plasma heater in combination with liquid feeding of oxalate precursors of the copper, yttrium, and barium components. The liquids were ultrasonically atomized to generate an aerosol, subsequently carried by an oxygen carrier gas into the RF plasma jet. The longer heating, dissociation, and reaction times available in RF plasma jets have been shown to be able to produce the high-temperature 1-2-3 T_c phase directly. Thin, epitaxially deposited, films of the "1-2-3" high-T_c superconducting oxides were formed without the need for any postdeposition heat treatments or annealing in oxygen. The films were single crystals that exhibited none of the critical current limitations characteristic of the directly plasma sprayed materials.

Diamond Synthesis and Deposition

Several researchers have synthesized diamonds using either dc or RF thermal plasma jets (Ref 35–37) and have shown that generation of CH_3^- radicals and ionic hydrogen, which typically occurs under many plasma jet conditions, leads to the formation of diamond films. In these cases, precursor gases such as methane and hydrogen were injected into the plasma jets, where the high temperatures (>10,000 K) produced the required ionic concentrations such that on collision with a heated surface at ~1000 °C amorphous carbon and diamond phases were formed. It is generally accepted that the ionic hydrogen assists in preferentially etching away the amorphous carbon phase, leaving diamond as the surface film. High growth rates (>1 μm/min) can be achieved; however, the exact morphology and bonding of the deposited diamond films varies and can be a limiting factor. A wide range of thermal plasma processing conditions with a wide range of environments have now been used to produce diamonds. The critical elements of diamond growth clearly occur in the thin, plasma-enhanced, reaction zone created immediately above the substrate surface when the thermal plasma jet impinges on the substrate surface. The precise formation mechanisms are not yet completely understood, but research is now showing that it is the thermal plasma jet conditions that are primarily responsible for producing diamonds at high rates.

Functionally Gradient Materials

Functionally gradient materials (FGMs) are a rapidly growing application area with significant promise for the future production of (a) improved materials and devices for use in applications subject to large thermal gradients, (b) lower-cost clad materials for combinations of corrosion and strength or wear resistance, and (c) perhaps improved electronic material structures for batteries, fuel cells, and thermoelectric energy conversion devices. The most immediate application for FGMs is in thermally protective claddings, where large thermal stresses could be minimized and component lifetimes improved by "tailoring" the coefficients of thermal expansion, thermal conductivity, and oxidation resistance. These FGMs could, and indeed are, finding use in turbine components, rocket nozzles, chemical reactor tubes, incinerator burner nozzles, or other critical furnace components. Figure 14 illustrates an example of a thermally sprayed FGM proposed for the protection of copper using a layered FGM ceramic structure. Successful fabrication of this would have application as improved burner nozzles, molds, and furnace walls (Ref 38).

Thermally protective FGMs may also be used in applications where reduced weight is needed, for example in aircraft or other transportation systems, or for the thermal protection of lightweight polymeric materials using insulating materials. The economic production of graded metallic, oxide, and intermetallic structures may also enable advanced batteries, solid oxide fuel cells (SOFCs) (Ref 39), and thermoelectric devices consisting of alternating layers of semiconducting materials ($FeSi_2$) to be produced.

Thermally sprayed graded materials are now also being considered for the production of oxide/metal/air-type electrode/electrolyte systems. This has been proven by research and development on SOFCs and electrode developments (Ref 40) using thermal spray methods. Spray-formed battery cells promise to be both lightweight and significantly more environmentally acceptable than current lead/acid and nickel/cadmium batteries, thus opening up large potential markets in consumer, computer, and automotive electronics. These products would all use FGMs, which promises to increase performance and lower the cost. Thermal spray forming of such gradient structures has been proposed because of the unique ability this approach has of being able to deposit thin, individual layers of a wide range of materials—metals, intermetallics, and ceramics—thereby enabling layered or continuously graded structures to be produced. The forming of composite gun barrels by spraying tubular laminates and superalloy composites

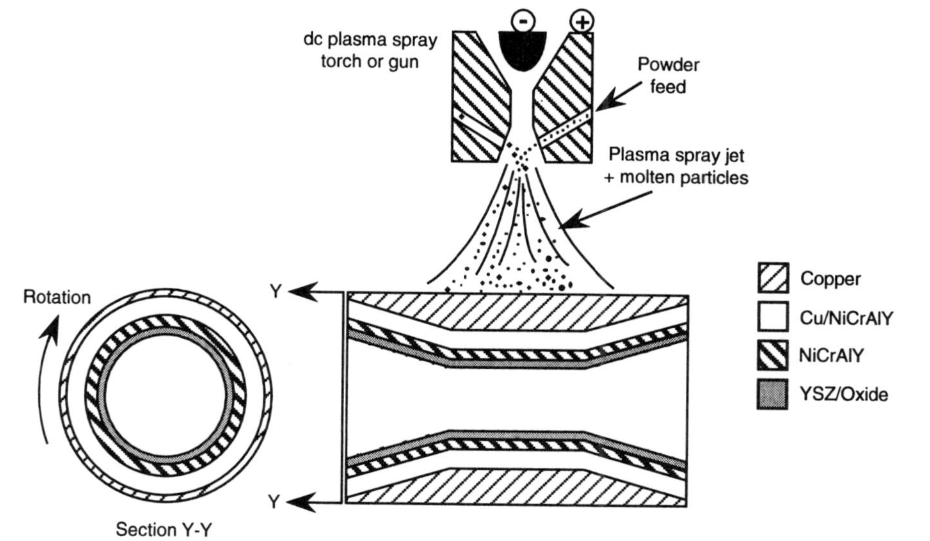

Fig. 14 Schematic of a thermally sprayed FGM for burner nozzle applications

(Ref 41), intermetallics, and refractory metals is one example of completed work. Thick thermally sprayed coatings with graded structures are being developed for several commercial applications, including:

- The ceramic outer air seals in aircraft gas turbines and other "clearance-control" coatings in rotating machinery
- Thick, multilayer, TBCs for heavy-duty diesel engine pistons
- High-performance dielectric coatings for electronic devices
- Wear-resistant coatings for diesel engine piston rings
- Oxidation-resistant coatings for high-temperature conditions

Future growth of FGM applications will ensure that thermal, particularly plasma, spray forming will increase and develop, provided that material properties can be controlled and processing costs can be optimized.

Thermal spray forming can produce continuous gradations of metals, ceramics, and intermetallics, either in alternating stepwise layers or as microlaminates. Materials could be sprayed to replicate the mechanical behavior of a material, for example making tungsten shear locally in high-impact applications. Normally under high impact, tungsten and its alloys behave in a ductile fashion, exhibiting significant plastic deformation, or "mushrooming." When localized shear behavior is required, however, very thin, alternating layers of intermetallic materials can be deposited concurrently while spray forming the bulk tungsten alloy. Thermal spray forming thus enables microlaminate structures to be formed that can be used to locally modify the overall mechanical behavior of the system. The

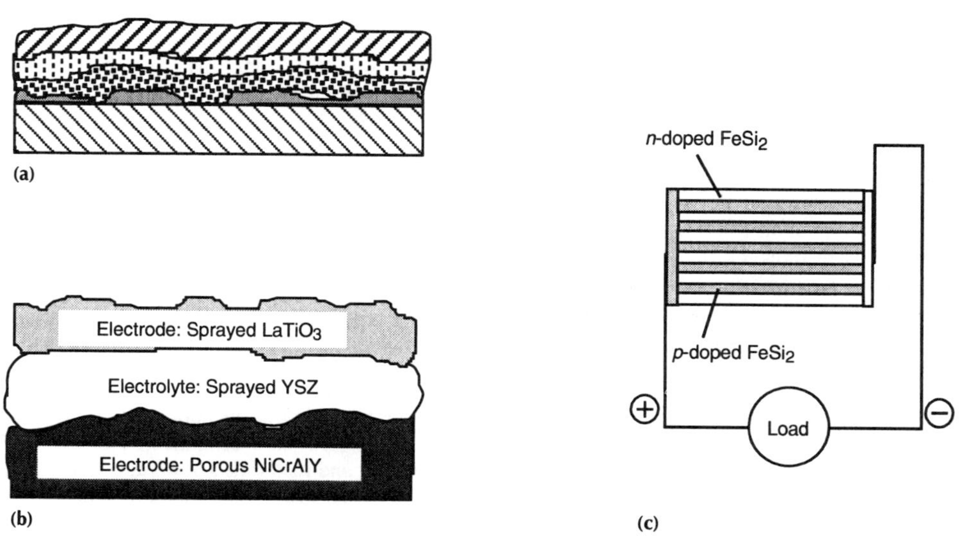

Fig. 15 Other examples of thermally sprayed FGMs. (a) Multilayer coating for strain control. (b) Sprayed electrodes for SOFCs. (c) Sprayed alternating layers for thermoelectric devices

same approach could also be used to modify the electrical or thermal properties of a material.

Thermally sprayed gradient structures have also been proposed as strain controlled coatings and/or structures and for electrical devices, as shown schematically in Fig. 15.

The powder production route for materials permits the production of oxides and/or brittle intermetallics that have unique electrical and thermal properties. Figure 15 shows that oxide-based fuel cells can be spray formed by depositing alternating layers of oxide and metal electrodes, and oxide electrolytes. The figure also shows the spraying of alternating layers of thermoelectric materials ($FeSi_2$) and the concurrent deposition of metallic electrodes. This processing combines the grading capability of thermal spray together with its ability to spray materials with widely varying melting points, thus truly realizing the advantages of thermal spray forming.

Thermal Spray Forming of Tooling

Two growing, and related, areas for thermal spray forming of materials are the direct production of solid free forms and tooling manufacture and repair.

Solid free-form fabrication has been demonstrated using a CAD model in conjunction with thermal spray deposition to fabricate free standing solid parts. Successive planar layers have been deposited using the wire arc spray process to incrementally build up three-dimensional shapes without the need for preformed mandrels (Ref 3). The wire arc spray process has also been used to replicate metallic, wooden or polymeric tool and die patterns produced using CAD/rapid part prototyping techniques such as stereolithography (SLA). Materials sprayed for this application include zinc and aluminum alloys. Zinc is particularly attractive because of its low shrinkage on solidification, but it is limited by compatibility of the final application with relatively soft tooling. More demanding applications require the spraying of tool steel, which can damage thermally sensitive polymeric SLA models. Wire arc sprayed Kirksite-type alloys as the tool surface have also been proposed (Ref 42) as a means of reducing tooling costs for many plastics molding processes, including structural foam, injection molding, compression molding, and vacuum forming. This approach is attractive for short production runs in which the cost of traditionally machined tooling can be very high. Coating thicknesses of 1.5 to 2 mm (60 to 80 mil) are typical, with reported cost savings ranging from a half to a twentieth of that of conventionally produced tooling.

A recent development in this area (Ref 43) was the unique combination of spray forming of tool steel using nitrogen as the atomizing gas with in situ shot peening to incrementally eliminate distortion due to stresses in the sprayed deposit. Tools and dies have been produced from 0.8% C steel, 13% Cr steel, M-2 tool steel, and 18/8 stainless steel using this technique.

Thermal spray forming of materials is also being used for the repair of tooling subsequently used in other metal forming processes. Air plasma sprayed (APS) coatings consisting of 70 vol% yttria-stabilized zirconia + 30 vol% CoNiCrAlY have been successfully applied to the surfaces of cast centrifugal casting molds (Ref 44). Service life was reportedly extended from 20 to 200 runs, with reduced product rejection rate.

Conclusions

Thermal spray is a unique materials processing tool capable of processing the widest range of materials into useful forms: coatings of widely varying thickness, monolithic spray-formed materials, and spray-formed composite or graded structures. Thermal spray processing commonly uses powders as feedstock, and as described here both the form and characteristics of the starting powders play key roles in the development of the final thermally sprayed microstructure. The process heat source, combustion versus plasma, also has a significant influence on the microstructure of the coating or spray-formed deposit. The ultimate properties of sprayed materials, for example in coatings, are quantified in terms of the density/porosity, hardness, wear and/or corrosion resistance, thermal conductivity, and electrical resistivity. Key physical properties of spray-formed structures can also include coefficient of thermal expansion, tensile and creep strength, and fracture toughness. The structures of thermally sprayed materials therefore depend on the materials processing route used, including powder processing, heating rates and dwell times, particle velocity, and on the environmental exposure during processing.

Advances in thermal spray will continue to stimulate its growth as an important powder consolidation technique, especially for difficult-to-form materials. Important advances to date include:

- Emergence of the SHS route as an economical means of producing a wide range of improved composite feedstock materials for thermal spray forming
- Ever-growing knowledge and application bases for reactive plasma spray processing for synthesizing a wide range of advanced materials
- Growth of thermal spray as a method of producing a wide range of FGMs for thermal, mechanical, chemical, and electrical applications

Thermal spraying, developed just over a century ago, is now on the verge of many new and important forming applications. Thermal spray forming is now benefiting from the increased processing and materials knowledge that will allow increased use of its structures and characteristics. Further understanding of the potential for thermal spray consolidation of powders and the potential of this route for forming, however, necessitates that the influences of the processes on material microstructures be studied and understood. The last 25 years of research have been dedicated to this, and now new applications for the forming of materials for molds, dies, tooling, functionally graded structures, refractory metals, and composites are being established. The future challenges facing thermal spray are cost effectiveness, together with improved structure/property control, using highly reliable, repeatable processes. The potential is there, and many recent advances in materials and process control have ensured that industry is using thermal spray as both a coating and a forming process.

ACKNOWLEDGMENTS

The authors would like to acknowledge the support of Drexel University, The Ben Franklin Partnership, and RWTH Aachen.

REFERENCES

1. M. Jackson, J. Rairden, J. Smith, and R.W. Smith, *J. Metals,* Vol 1, 1981, p 146
2. S. Sampath and H. Herman, *Thermal Spray Technology,* ASM International, 1988, p 1–8
3. L. Weiss, F. Prinz, and D. Adams, Solid Free-form Fabrication by Thermal Spray Shape Deposition, *Proc. 13th International Thermal Spray Conf.* (Orlando, FL), ASM International, 1992, p 847–851
4. N. Stoloff and D.E. Alman, Innovative Processing Techniques for Intermetallic Matrix Composites, *Proc. Symposium on Intermetallic Matrix Composites* (Pittsburgh, PA), MRS, 1990, p 31–43
5. Y. Kimura, T. Yoshioka, and M. Kaazawa, On the Effects of Sealing Treatment and Microstructural Grading upon Corrosion Characteristics of Plasma-Sprayed Ceramic Coating, *Proc. Seventh National Thermal Spray Conf.* (Boston), ASM International, 1994, p 527–532
6. T. Nguyentat, K.T. Dommer, and K.T. Bowen, Metallurgical Evaluation of Plasma Sprayed Structural Materials for Rocket Engines, *Proc. 13th International Thermal Spray Conf.* (Orlando, FL), ASM International, 1992, p 321–325
7. K.M. McHugh, Materials Processing with De Laval Spray Forming Nozzles: Net-Shape Applications, *Proc. Seventh National Thermal Spray Conf.* (Boston), ASM International, 1994, p 477–483
8. R.W. Smith and R. Novak, Advances and Applications in U.S. Thermal Spray Technology, *Powder Metallurgy International,* Vol 23-3 and 23-4, Springer Verlag, 1991
9. M. Thorpe, *Chem. Eng.,* Nov 1991, p 54–57
10. R.W. Smith, Equipment and Theory, *Lesson from Thermal Spray Technology, Course 51,* Materials Engineering Institute, ASM International, 1992
11. H.C. Chen, Z. Duan, J.V.R. Heberlein, and E. Pfender, Influence of Shroud Gas Flow and Swirl Magnitude on Arc Jet Stability and Coat-

ing Quality in Plasma Spray, *Proc. Ninth National Thermal Spray Conf.* (Cincinnati, OH), ASM International, Oct 1996 p 553–561

12. M. Mohanty, R.W. Smith, R. Knight, W.L.T. Chen, and J.V.R. Heberlein, Shrouded Air Plasma Processing of Lightweight Coatings, *Proc. Ninth National Thermal Spray Conf.* (Cincinnati, OH), ASM International, Oct 1996, p 967
13. R.W. Smith, P. Kangutkar, R. Drossman, and R. Krepski, A New Iron Base Thermal Spray Coating for Wear Resistance, *Proc. 13th International Thermal Spray Conf.* (Orlando, FL), ASM International, 1992, p 653–659
14. K. Tanaka et al., *J. Mater. Sci.*, Vol 22, 1987, p 2192–2198
15. K. Ishizaki et al., *J. Mater. Sci.*, Vol 24, 1989, p 3553–3559
16. S. Matsumoto, T. Kobyashi, and M. Hino, *Proc. Eighth International Symposium on Plasma Chemistry*, Vol 4, IUPAC, 1987, p 2458
17. R.G. Castro et al., Ductile Phase Toughening of Molybdenum Disilicide by Low Pressure Plasma Spraying, *Mater. Sci. Eng.*, Vol A155, 1992, p 101–107
18. S. Sampath, H. Herman, and S. Rangaswamy, Ni-Al Re-evaluated, *Proc. First National Thermal Spraying Conf.* (Orlando, FL), ASM International, 1987, p 47–53
19. L.S. Schadler, K.O. Laul, R.W. Smith, and E. Petrovicova, Microstructure and Mechanical Properties of Thermally Sprayed Silica/Nylon Nanocomposites, *J. Thermal Spray Technol.*, Vol 6 (No. 4), 1997, p 475–485
20. L.J. Westfall, Composite Monolayer Fabrication by an Arc-Spray Process, *Proc. First National Thermal Spray Conf.* (Orlando, FL), ASM International, 1987, p 417–426
21. Z.A. Munir and V. Anselmi-Tamburini, Self-Propagating Exotherm Reactions: The Synthesis of High-Temperature Materials by Combustion, *Mater. Sci. Rep.*, Vol 3 (No. 7/8), 1989, p 277
22. S. Sampath et al., Synthesis of Intermetallic Composite Powders via Self-Propagating Synthesis, *Proc. 1993 Powder Metallurgy World Congress* (Kyoto, Japan), Japan Society Powder and Powder Metallurgy, 1993, p 401–403
23. R.G. Castro, H. Kung, and P.W. Stanek, Reactive Plasma Spraying of $MoSi_2$ using an Ar-10% CH_4 Powder Carrier Gas, *Mater. Sci. Eng.*, Vol A185, 1994, p 65–70
24. R.W. Smith, Reactive Plasma Spray Forming for Advanced Materials Synthesis, *Powder Metall. Int.*, Vol 25 (No. 1), 1993, p 9–16
25. R.W. Smith and R. Knight, Thermal Spraying I: Powder Consolidation—From Coating to Forming, *JOM*, Vol 47 (No. 8), 1995, p 32–39
26. P.G. Tsantrizos, The Reactive Spray Forming Production of Titanium Aluminides in the Tail Flame of a DC Plasma Torch, *Proc. 13th International Thermal Spray Conference* (Orlando, FL), ASM International, 1992, p 195–199
27. E. Muehlberger, *Proc. Seventh International Thermal Spray Conf.* (London), The Welding Institute, 1974, p 245
28. R.W. Smith, W.F. Schilling, and L.G. Peterson, *Plasma Processing and Synthesis of Materials,* J. Szekely and D. Apelian, Ed., Materials Research Society, 1984, p 30–217
29. R.W. Smith, M.R. Jackson, J.R. Rairden, and J.S. Smith, *J. Met.*, Vol 33 (No. 11), 1981, p 23
30. R.W. Smith and Z.Z. Mutasim, Reactive Plasma Spraying of Wear-Resistant Coatings, *J. Thermal Spray Technology*, Vol 1 (No. 1), 1992, p 57–63
31. T.N. Meyer, A.J. Becker, J. Edd, F.N. Smith, and J. Liu, *Proc. Eighth International Symposium on Plasma Chemistry*, IUPAC, 1987, p 2006–2011
32. S. Matsumoto, T. Kobayashi, and M. Hino, *Proc. Eighth International Symposium on Plasma Chemistry*, IUPAC, Vol 4, 1987, p 2458
33. R.W. Smith, E. Lugscheider, P. Jokiel, U. Mueller, J. Merz, and M. Wilbert, Synthesis of Composite Materials by Reactive Plasma Spray Processing, *Proc. Fifth National Thermal Spray Conf.* (Anaheim, CA), ASM International, 1993, p 439–444
34. H. Zhu, Y.C. Lau, and E. Pfender, *Mater. Sci. Eng.*, Vol A139, 1991, p 352
35. C. Tsai, J. Nelson, W. Gerberich, J. Heberlein, and E. Pfender, *Proc. Third International Conf. on New Diamond Science and Technology (Diamond '92)* (Heidelberg, FRG), COMST, Sept 1992
36. S.L. Girschick et al., Diamond Deposition by Atmospheric-Pressure Induction Plasma: Effects of Impinging Jet Fluid Mechanics on Film Formation, *Diamond and Related Materials*, Vol 2, 1993, p 1090–1095
37. U. Müller and E. Lugscheider, *Proc. Third International Conf. on New Diamond Science and Technology (Diamond '92)* (Heidelberg, FRG), COMST, Sept 1992
38. A.H. Bartlett, R.G. Castro, D.P. Butt, H. Kung, J.J. Petrovic, and Z. Zurecki, Plasma Sprayed $MoSi_2/Al_2O_3$ Laminate Composite Tubes as Lances in Pyrometallurgical Operations, *Ind. Heat.*, Vol LXIII (No. 1), 1996, p 33–36
39. E. Fendler, R. Henne, and M. Lang, The Production of Porous Layers for the Solid Oxide Fuel Cell by Vacuum Plasma Spraying, *Proc. Eighth National Thermal Spray Conf.* (Houston, TX), ASM International, 1995, p 533–537
40. R. Henne, A. Kayser, V. Borck, and G. Schiller, Production of Mo-Doped Raney Nickel Electrodes Applying Vacuum Plasma Spraying, *Proc. 13th International Thermal Spray Conf.* (Orlando, FL), ASM International 1992, p 817–824
41. L.J. Westfall, Thermal Spray: Advances in Coating Technology, *Proc. Second National Thermal Spray Conf.* (Cincinnati, OH) ASM International, 1988, p 417
42. L.J. Grant, Metal Spray Tooling: A Low Cost Alternative, *UTECH Asia '93* (Singapore), Dec 1993
43. A New Low Cost Replication Technique for Making Tool Steel Dies, *European Commission Brite-Euram Feasibility Study,* Project No. FA-89-57, Sprayforming Developments Ltd.
44. T. Oka, J. Takeuchi, Y. Harada, and H. Nakahira, Extending the Service Life of Metal Moulds for Centrifugal Casting by Atmospheric Plasma Spraying (APS), *Proc. First Plasma-Technik-Symposium* (Lucerne, Switzerland), Vol 1, May 1988, p 173–170

Slip Casting of Metals

R.A. Haber and C.A. Paredes, Ceramic Casting Technology Program, Rutgers University

SLIP CASTING is a very common process for consolidating ceramic powders. It can be described as the filtration of ceramic particles suspended in a liquid, at room temperature, into a porous mold. The porous mold removes the solvent by means of either capillary forces or an external pressure, leaving the compacted piece in the mold. Figure 1 shows the critical process steps involved in slip casting of ceramics and the different parameters that must be controlled to obtain the best properties of the final product (Ref 1).

Traditionally, pressure compacting of metal powders has been used to produce metal parts. This technique presents several disadvantages. Hausner (Ref 2) lists some of these disadvantages:

- Restriction on size of parts being made due to capacity limitations of presses and relatively high die cost
- Shape restrictions of complicated pieces due to nonuniform pressure distribution in a mass of metal powders, which is caused by friction conditions prevailing during application of pressure
- Economic restrictions, because this process needs to be a relatively large-scale production due to the relatively high cost of the dies

The earliest reference to slip casting of metal powders is a German patent granted to Reichmann in 1936 (Ref 3). This method consists basically of pouring a suspension of metal powder and water into a plaster mold. The mold has the negative shape of the desired form, and it absorbs all the water from the suspension. The slip should have a low enough viscosity to enable an easy pour into the mold. The green body should be firm enough to retain its shape after stripping the piece from the mold. Figure 2 shows schematic of slip casting of metals (Ref 5). The two most common variations of slip casting are solid casting and drain casting (Fig. 3). Drain casting is a method where the slip is poured into the mold and after a determined time the slip is drained from the mold, leaving a deposited cast layer inside the mold. After some time is allowed for the piece to become rigid, the part is demolded, dried, and fired. Solid casting is similar to drain casting, but the slip is allowed to sit in the mold during dewatering, leaving the solid piece inside.

Colloidal Stability

Because slip casting of metals is a wet process, some understanding of the interaction between the material and water must be established. Most materials when immersed in a polar solvent become charged. The different mechanisms by which the surface acquires such charge depend greatly on the type of material used. These mechanisms include ion dissolution and ion adsorption (Ref 4, 7). Ionic materials acquire their surface charge through preferential dissolution. A very well known example is AgI, where the solubility product gives the amount of soluble species Ag^+ and I^- When an excess of I^- ions is present, the colloid is negatively charged, and vice versa. These ions are called potential-determining ions (PDI). For the case of metal

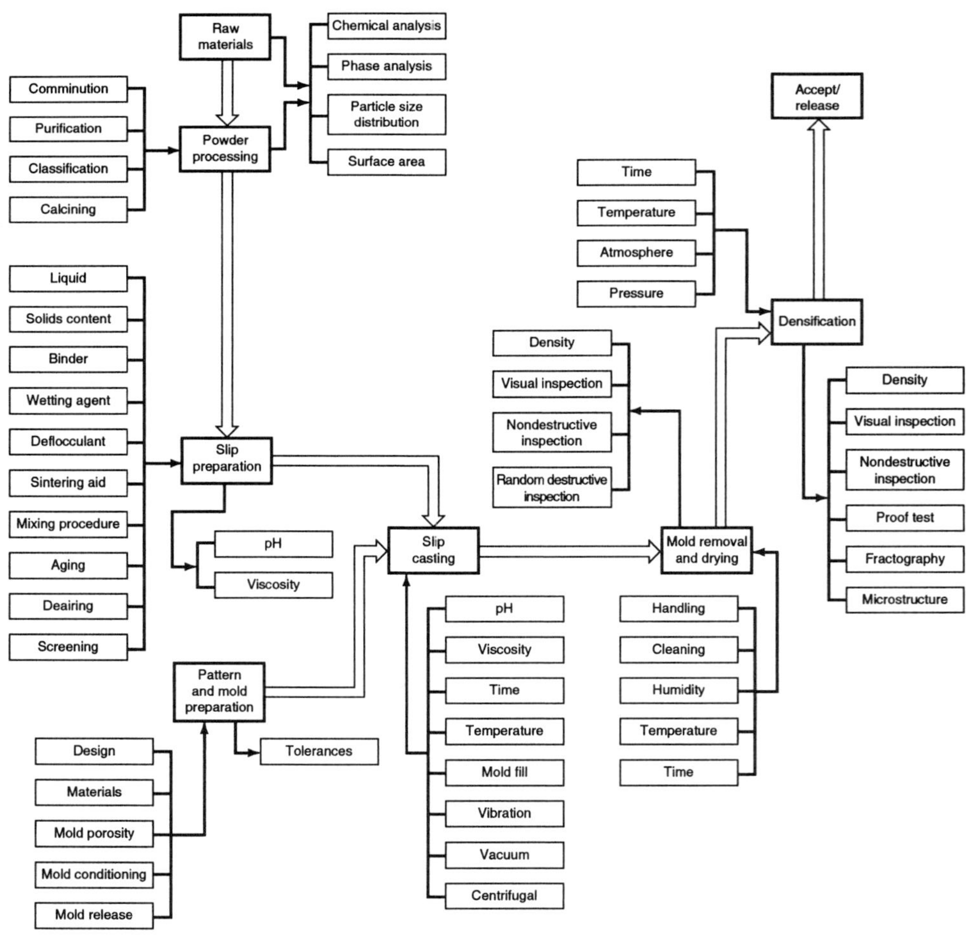

Fig. 1 Process steps in slip casting and parameters that must be controlled. Source: Ref 1

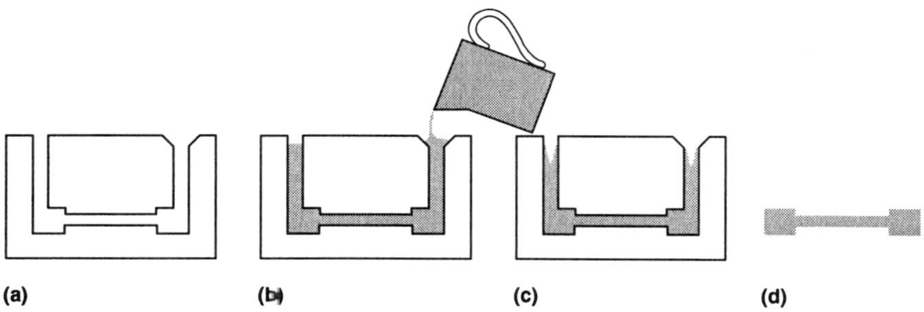

Fig. 2 Slip casting of metal powders. (a) Assembled mold (Source: Ref 4). (b) Filling the mold. (c) Absorbing water from the slip. (d) Finished piece, removed from the mold and trimmed

oxides, hydronium and hydroxyl ions become PDIs. This process is illustrated by the next reaction:

$$MOH + H^+ \rightarrow MOH_2^+$$

$$MOH + OH^- \rightarrow MO^- + H_2O$$

A material can also gain surface charge by unequal adsorption of ions in the surrounding solution. If a larger quantity of positive ions is adsorbed on a particle surface, then a positive surface charge will result, and vice versa.

Electrical Double Layer. It is necessary to understand the behavior of the potential and charge distribution around the vicinity of colloidal particles. For this reason, it is very important to understand the source of the charge distribution. The electrical double layer is thought to be composed of two parts, the inner portion of the double layer and the diffuse double layer.

Helmholtz was the first to relate the electrical double layer to a metal surface (Ref 7). He assumed that the charge on the particle surface was balanced by an equal and opposite charge located near the particle surface in the surrounding solution. This theory was then modified by French and British scientists and the model was called the Gouy-Chapman model. This model is the simplest quantitative treatment for the diffuse part of the double layer. The assumptions are (Ref 4):

- The particle surface is assumed to be planar with a uniform surface charge.
- Ions of the diffuse portion of the double layer are distributed according to the Boltzmann distribution.
- The solvent is assumed to influence the double layer only by its dielectric constant.
- A symmetrical one-to-one electrolyte is assumed.

A schematic of this model for a particle medium interface is shown in Fig. 4. This model gives the mathematical function that describes the potential distribution at a distance from the particle surface:

$$\Psi = \Psi_o \exp[-\kappa x] \quad \text{(Eq 1)}$$

where Ψ is the potential at a distance x from the particle surface, Ψ_o is the surface potential, and κ is the Debye parameter, or the inverse double layer thickness. For low surface potential, the potential can be related to the surface charge through:

$$\sigma_o = \varepsilon\kappa\Psi_o \quad \text{(Eq 2)}$$

where σ_o is the surface charge and ε is the medium dielectric constant. The surface potential depends on both the surface charge density and the Debye parameter, which is in turn affected by the ionic strength of the medium. As a result, when the double layer is compressed, κ increases and both the surface charge and the surface potential are affected. Surface charge must increase and surface potential must decrease, respectively, or both can occur.

In the Gouy-Chapman treatment, the ions on the diffuse layer are considered point charges. Stern modified the double-layer theory to account for specific adsorption of ions as well as for the fact that ions have a finite size. This theory proposed a division of the double layer into two parts by a plane called the Stern plane. Ions with centers located within an ionic radius of the particle surface are in the inner part of the double layer. The ions located beyond the Stern plane are in the diffuse portion of the double layer, where the Gouy-Chapman treatment is considered to be applicable. The potential changes from Ψ_o to Ψ_d. Figure 5 shows schematic of Stern's representation of the electrical double layer.

In the presence of specific adsorption, either co-ion or counter-ion adsorption can occur. The Stern potential is either reduced or reversed with respect to the surface potential, as shown in Fig. 6. The Stern potential can be determined by electrokinetic measurements. Electrokinetic behavior depends on the potential at the surface of shear between the charged surface and the electrolyte solution (Ref 4). This potential is called the zeta potential [ζ]. The exact location of the shear surface is unknown, but it is accepted that it is located very close to the Stern plane. Therefore, it is usual to assume identity between Ψ_d and ζ. The zeta potential is a property that can be readily measured, while the Stern potential cannot. For this reason, zeta potential values are used in determining material surface properties.

DVLO Theory. The stability of colloidal-sized particles depends essentially on the balance of attractive and repulsive forces surrounding the suspended particle. A quantitative theory enabling determination of stability was devel-

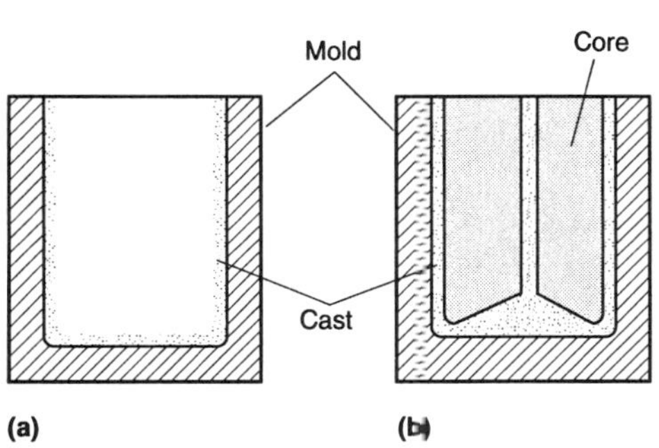

Fig. 3 Drain casting (a) and solid casting (b). Source: Ref 6

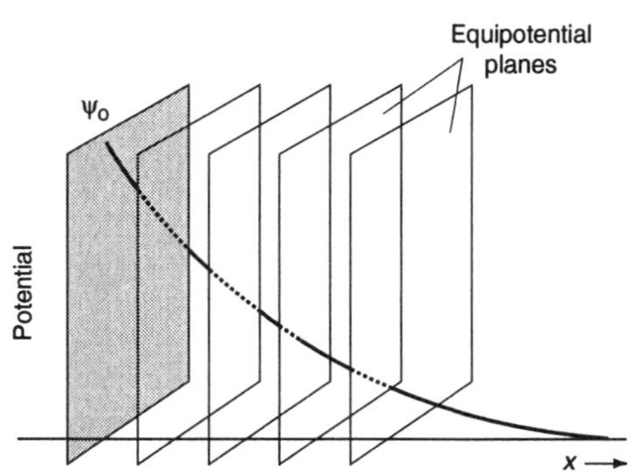

Fig. 4 Electrostatic potential distribution in parallel planes near a metal surface. Source: Ref 7

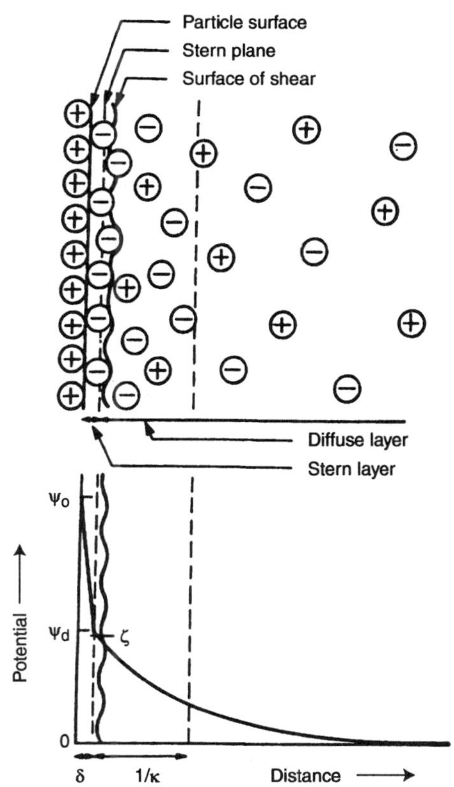

Fig. 5 Representation of Stern's electrical double layer. Source: Ref 4

oped independently by Deryagin and Landau and Verwey and Overbeek and is called the DVLO theory. According to this theory, the total interaction between particles is the sum of two contributions, V_r, the repulsive part due to the overlapping of double layers, and V_A, the attractive portion resulting from Van der Waals attraction (Ref 4, 7, 8):

$$V_T = V_A + V_R \quad \text{(Eq 3)}$$

Figure 7 shows total interaction energy curves. The repulsive forces are an exponential function that is effective on the range of the double layer. It is positive for all values of distances between the two particles. The attractive forces, on the other hand, decrease as the inverse power of the distance between the particles. At very small distances, the attractive potential becomes very negative, forming the potential energy well described as the primary minimum. When particles lie at this distance, they typically form very strong aggregates that cannot be broken up under normal shear conditions. As the distance between these particles increases, the repulsive forces start taking over and a maximum energy barrier results. The magnitude of this barrier depends on the concentration of electrolyte present. A secondary minimum develops at large separation distances between particles. The attractive forces dominate at this distance, leading to a potential well. In order for flocculation of particles to occur, the energy barrier between particles should be either very small or very big. One way in which an energy barrier may be reduced is the presence of a significant amount of electrolyte that will cause a compression of the electrostatic double layer between particles and allow for Van der Waals forces to dominate between particles. A more detailed review of this theory can be found in Ref 4 and 7.

Process Considerations: Slip Control

There are variables that should be monitored and controlled at all times to prevent agglomeration of the particles as well as settling. Some of these variables are:

- Water-to-powder ratio (solids loading)
- Type and amount of deflocculant
- Viscosity of the slip
- pH of the slip
- Surface chemistry of the metal particles
- Temperature of the slip
- Particle size distribution

All these factors have a direct effect on the stability of the suspension. The ratio of densities of the metal particles to the fluid affects the nature of the suspension. As this ratio increases, the nonuniform settling rate of the metal particles starts taking place. The various sizes and shapes of the particles also affects the stability of the slip. These settling forces can be balanced by using powders of small and fairly uniform particle size and by increasing the viscosity of the suspension. The use of finer particles has a further advantage because their high surface-area-to-volume ratio causes them to sinter faster.

Two very important processing variables in preventing flocculation mentioned above are viscosity and pH of the suspension. Tamura (Ref 9) reported that slip casting $MoSi_2$-base heating elements was best achieved using a water-to-metal ratio of 0.180, pH 4.05, and a viscosity range of 1000 to 1800 cp. In 1939, Hauser and Hirshon (Ref 10) suggested that the charge on the particles might originate as a result of either dissociation of different ions for the particle surface or by preferential adsorption of ions from the surrounding medium of the particles. In either case, the charges in the metal surfaces are balanced by counter-ions, which form a double layer. Jones (Ref 11) proposed the mechanism by which the metal particles repel each other. He proposed a formation of either an oxide film or an adsorbed film in the metal surface. The development of either film can be controlled by pH adjustment. Figure 8 shows how viscosity decreases when pH increases up to a minimum value, after which the viscosity starts increasing again. St. Pierre (Ref 12) showed this curve using molybdenum powder suspended in 5% aqueous polyvinyl alcohol solution.

Deflocculants are chemical additives used to control the flow properties of casting slips. In ceramics, common deflocculants are sodium silicate, sodium carbonate, and lately polyacrylates. For dispersing nonclay slips, strong acids or strong bases are often used. There is a strong correlation between the water-to-metal ratio and the amount of the deflocculant in the slip (Ref 5). An increase of solids loading requires an increase in the amount of deflocculant necessary to maintain the pH and viscosity. Lidman and Rubino (Ref 13, 14) presented two papers on slip casting of stainless steel powder. The slip

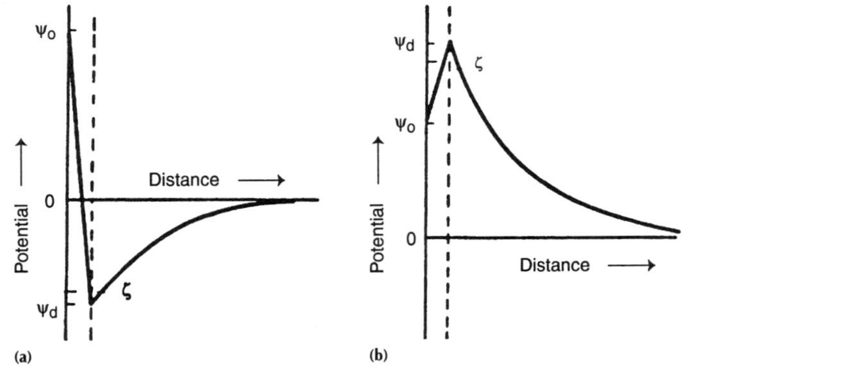

Fig. 6 (a) Reversal of charge due to adsorption of surface-active counter- ions. (b) Adsorption of surface-active co- ions. Source: Ref 4

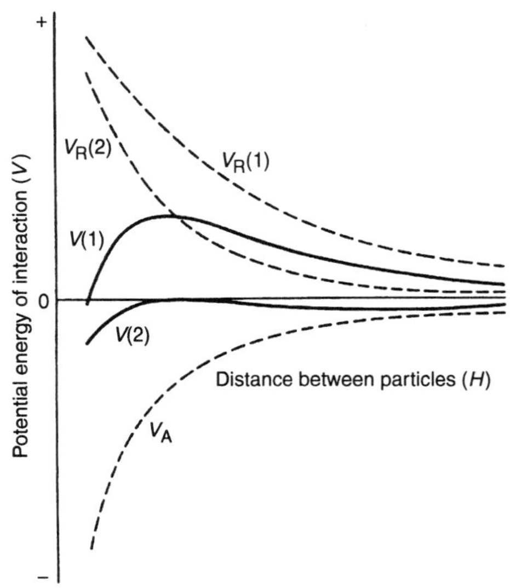

Fig. 7 Total interaction energy curves $V(1)$ and $V(2)$, obtained by the summation of an attractive curve, V_A, with different repulsion curves, $V_R(1)$ and $V_R(2)$. Source: Ref 4

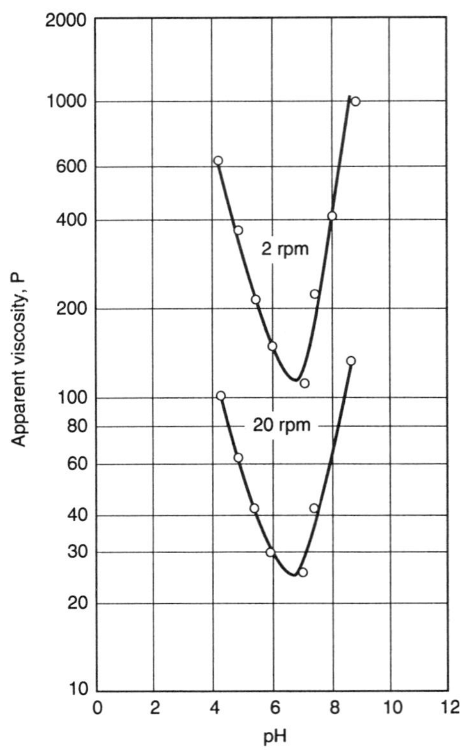

Fig. 8 pH-viscosity relations for a fine molybdenum powder suspended in 5% polyvinyl alcohol solution. Source: Ref 12

was prepared using ammonium alginate for bonding the powder particles, polyvinyl alcohol for elasticity, and water for fluidity. Rempes et al. (Ref 15) conducted slip casting experiments using different powder materials, metals, ceramics, and cermets. They found that for slip casting chromium and stainless steel 302 a combination of a sodium salt of alginic acid (Keltex of Kelco Co., used as a suspension agent and binder) and an acrylic acid polymer (Carbopol 934, Goodrich Chemical Co.) deflocculants was required. For the case of titanium and molybdenum, only a solution of Keltex was necessary to achieve slip stability.

Mold Control

The main components of the slip casting process are the slip and the mold. Both components play an important part in the production of pieces. Slip casting is a filtration process in which the mold serves as the filter and also is the driving force for filtration. Plaster mold properties are extremely important to the overall casting system. The ideal mold material is one that is inexpensive and can reproduce intricate detail, remain dimensionally stable, and provide good dewatering capacity indefinitely. Gypsum has proven to be the best material for these needs. Norton (Ref 16) lists why gypsum plaster is the best mold material:

- Fine detail may be reproduced.
- The mold is stable chemically and physically.
- The absorption can be varied over wide limits for any desired use. The porosity allows release of the clay.
- A smooth, durable surface can be readily formed.
- Uniform physical and chemical properties can be maintained.
- The pores are not easily sealed by colloids.
- The cost is moderate.

Plaster Production. Gypsum is found in many parts of the world, typically in deposits of marine origin. It is the most valuable and widely used form of calcium sulfate ($CaSO_4{\cdot}2H_2O$). Partially dehydrating gypsum results in the two forms of hemihydrate gypsum, α and β. Both forms are widely used in the production of molds for slip casting because of their ability to rehydrate and form rigid porous molds. Molds made from the α hemihydrate-derived molds are known for high strength, while the β hemihydrates are known for their excellent absorptive power. The primary difference between the two materials is the hemihydrate crystal morphology, as controlled by the calcination process. Table 1 summarizes the known forms of calcium sulfate.

Table 1 Forms of calcium sulfate

Material	Formula	Preparation
Dihydrate	$CaSO_4{\cdot}2H_2O$	Naturally occurring gypsum rock
α-hemihydrate	$CaSO_4{\cdot}\frac{1}{2}H_2O$	By dehydrating in saturated steam
β-hemihydrate	$CaSO_4{\cdot}\frac{1}{2}H_2O$	By dehydrating in dry atmosphere
αβ-soluble anhydrite	$CaSO_4$	By dehydrating α-hemihydrate
β-soluble anhydrite	$CaSO_4$	By dehydrating β-hemihydrate
Insoluble anhydrite	$CaSO_4$	By high-temperature processing

Source: Ref 16

Plaster Setting. The set rate of plaster is important because it can affect the compressive strength of the mold and in turn its durability. The basic driving force for the setting of plaster is the room-temperature solubility difference between hemihydrates and gypsum. The hemihydrate is far more soluble than gypsum. As it comes in contact with water, it dissolves to reach its solubility equilibrium. Simultaneously, gypsum precipitates because its solubility limit is exceeded by a considerable amount (Ref 17). There are two mechanisms regarding the initiation of gypsum crystallization being argued: the growth from colloidal gypsum present due to supersaturation or the growth from gypsum crystal nuclei present as an impurity (Ref 18 and 19). Though only 18.6 g of water per 100 g of hemihydrate is required for hydration from hemihydrate to gypsum, excess water is added to afford fluidity of the suspension. The excess water creates the porosity present after drying. Concentrated suspensions set to form hard masses due to the interlocking of the growing crystals.

Setting expansions are caused by crystal growth, which actually exerts pressure on container walls. The growth of needlelike crystals forms a network where crystals lie against one another, transferring force and displacement throughout the setting mass.

The set time can be affected by different factors such as plaster-to-water ratio and tempera-

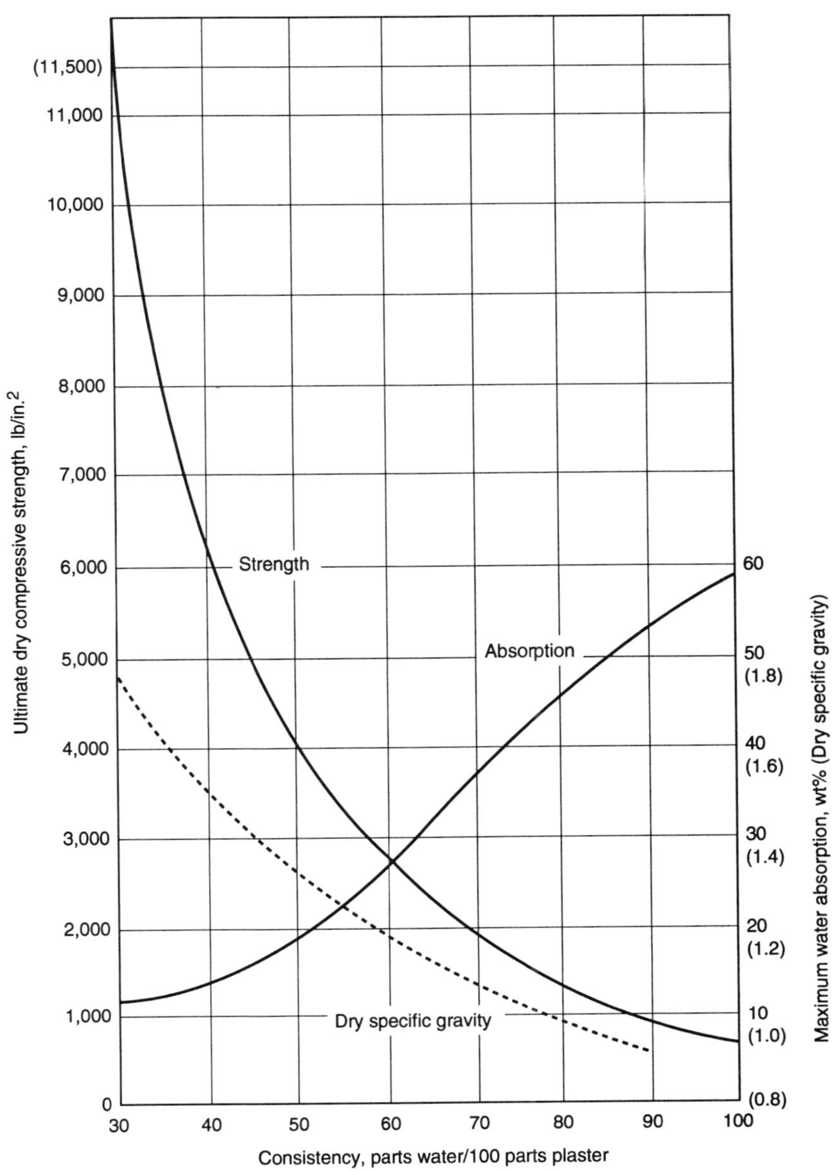

Fig. 9 Variation of strength, density, and total absorption as a function of consistency. Source: Ref 20

ture of the plaster/water mixture or through use of set accelerators and retardants. It is clear that control of all these parameters is important for maintaining consistent plaster properties on set time and the resulting microstructural dependence.

Plaster Mold Production. Once the ideal mold properties are obtained, they are only reproducible through accurate control of mold production procedures. Often these procedures are overlooked.

The term consistency is commonly used for the water-to-plaster ratio, or number of grams to 100 g of plaster. It is probably the single most powerful means of adjusting plaster properties. Nearly all the properties that this variable controls can be fine-tuned by proper preparation procedures. Figure 9 shows a plot of strength, density, and total absorption as a function of consistency (Ref 20).

Mold durability in casting is influenced by the overall strength of the mold (Ref 21). For this reason, the mold used is one having the maximum strength allowable concurrent with good permeability and capillarity.

Foaming of the plaster slurry decreases the density of the mold while maintaining reasonable strength. Adjustment in particle size of the plaster or by mixing the α and β hemihydrates can result in a stronger mold material.

Mixing of the plaster slurry can be considered the most crucial step in mold production. The reaction of hydration is dependent on solubility, the number of seed crystals, and energy, due to the exothermic nature of the reaction. Vigorous mixing can accelerate solubility equilibrium. Breaking of the dissolving crystals during mixing can increase the hemihydrate surface area, helping also the attainment of the solubility equilibrium. Breaking of gypsum crystals increases the number of nucleation sites and, therefore, the hydration rate. Excessive mix energy can increase the temperature and initiate the related effects.

There are many other factors that should be considered very carefully to obtain the best properties for the plaster mold. Mixing time has an inverse effect on the compressive strength and water absorption rates—the longer the mixing time, the higher the rates. Aging of calcined plaster during storage can also alter the water demand, set time, and appropriate mixing requirements for a given plaster batch (Ref 22). While in storage, the character of the hemihydrate may change to gypsum, affecting the total hydration and the number of nucleation sites. A humid environment may also serve to fuse hemihydrate crystals, lowering crystal disintegration, surface area, and solubility, thereby reducing water demand and set rate.

The drying of plaster molds is very important in the efficient production of pieces by the slip casting process. The ideal mold moisture for casting is between 8 and 17%. A mold that is too dry will cast quickly at first, forming a dense layer that slows further casting. A mold that is too wet will lead to a slow cast and poor dewatering shrinkage, which complicates demolding (Ref 21). Moisture distribution is also important, so the mold must be allowed to equilibrate between casts to prevent moisture concentration at the casting surface, giving a very slow second cast.

Mechanics of Filtration

Cake filtration involves the separation of fine solid particles from liquid in which they are dispersed by collecting the solid particles on a surface of a medium that is permeable only to the liquid. In slip casting, the filter medium is either a gypsum or synthetic mold. As the casting process continues, additional particles are collected on the surface of the initial layer of solids, thereby building up a cake. At any instant of time during the casting of a cake, the liquid will flow through the previously deposited solid layers.

The driving force for slip casting in plaster is the capillarity of the mold. The opposing force in the system is the resistance of both the mold and the cake to the flow of water. Permeability is the inverse of the resistance. Plaster molds absorb water according to the parabolic rate law where the rate of absorption decreases linearly with the square root of time. Assuming a constant driving force, the decreasing rate is caused by the increasing volume of the mold that is being penetrated, thus contributing to the resistance. This balance of forces in the mold leads to an increase in absorptive rate with increasing porosity. The decrease in mold resistance outweighs the decrease in capillary suction (Ref 23).

Addition of a cake to the system yields a more complex relationship where the absorptive rate is inevitably decreased dramatically. There is a new balance between the one driving force and two resisting forces.

Mold and Cast Cake Permeability. Though the mold and cake are both permeable, the permeability values are low enough to serve as the resistance to capillary suction of the mold. Deeg first modeled slip casting as a diffusional process, and later it was more correctly described by Deen as a filtration process (Ref 24).

Flow through a porous material was first quantified and described by D'Arcy, who noted that the rate of liquid flow through a sand bed was proportional to the pressure gradient across the bed. Kozeny then related the flow rated per unit area to the viscosity, porosity, and internal surface area of the porous medium. Carman combined these analyses and related the properties to the permeability (Ref 25, 26). Adcock and McDowall (Ref 27) bridge the gap between ceramics and this area of fluid mechanics. Ceramists now use the Adcock and McDowall equation for permeability:

$$K = \frac{\eta \, dv/dt \, L}{PA} \quad \text{(Eq 4)}$$

where dv/dt is the volumetric flow of the filtrate having a viscosity η, as forced through a bed of thickness L and area A, by the pressure gradient P.

The permeability equation is commonly used to calculate the permeability of the cast cake, and more recently the plaster mold. Dal and Berden incorporated the permeability of the plaster into D'Arcy's equation:

$$\frac{dv}{dt} = \frac{PA}{\eta[(L_c/K_c) + (L_m/K_m)]} \quad \text{(Eq 5)}$$

where the subscript c stands for the cake and m for the mold. Now it is apparent that the volumetric flow is dependent on both permeabilities, as a pair of resistance in series (Ref 24).

The casting rate is dependent on both the volumetric flow rate and the total porosity of both the cake and mold. A mold of low porosity will have the highest penetration distance of the waterfront, representing the effective thickness, L. Consequently, the greater effective filter thickness and the lower permeability as dictated by the lower porosity will reduce the rate of fluid flow. The casting rate and cake permeability are similarly dependent on the void fraction of the cake.

Pressure Casting

Pressure casting is rapidly being implemented in the production of different components. The need for shorter casting times, reduced costs, and improved physical properties of the cast has contributed to its development. However the level of science and fundamental knowledge is still quite limited. Most of the factories run a pressure casting machine at the suggested pressures and cycle times specified by the supplier of the equipment. Little variation is usually done in either the cycle time, ramping rate, and pressure levels.

Basically, pressure casting refers to the application of pressure to the slip to increase the casting rate. Originally, plaster molds were used for this purpose, but because of its lack of strength and limitation of the applied pressures, porous plastic molds are used now. Development of these molds allowed the pressure to be increased tenfold. Blanchard (Ref 28) reported results for different bodies for a 6 mm thick compact. He observed a decrease in time from 45 to 15 min by increasing the pressure from 0.025 to 0.4 MPa for a porcelain formulation.

There are two types of pressure casting systems: medium pressure (0.3 to 0.4 MPa) and high pressure (up to 4 MPa). Pressure casting is currently being used in different applications such as in sanitaryware in the case of medium-pressure systems and dinnerware in the case of high-pressure systems.

There are several advantages of pressure casting over slip casting. Complex shapes become easier to cast with applied pressure because the quality of the greenware is improved and the fluctuations reduced. The filtration rate of plaster molds used for slip casting is slow, thus requiring a large number of molds and available space. The casting cycle using pressure casting is shorter, reducing labor costs and improving productivity.

REFERENCES

1. D.W. Richerson, *Modern Ceramic Engineering: Properties, Processing and Use in Design,* 1st ed., Marcel Dekker, 1992, p 444–460
2. H.H. Hausner, Slip Casting of Metal Powders, *Proc.: Annual Meeting—Metal Powder Association,* Vol 14, 1958, p 79–90
3. R. Reichmann, "Method of Fabricating Sintered Bodies, Especially Hollow Shapes of High Refractory Metals," German Patent 627,980, 26 March 1936
4. D.J. Shaw, *Colloid and Surface Chemistry,* 4th ed., Butterworth-Heinemann, 1992
5. F.V. Lenel, *Powder Metallurgy: Principles and Applications,* 1st ed., Metal Powder Industries Federation, 1980, p 309–312
6. G.W. Phelps, private communication
7. R.J. Hunter, *Introduction to Modern Colloid Science,* 1st ed., Oxford University Press, 1993
8. R. Hogg, Flocculation Phenomena in Fine Particle Dispersions, *Advances in Ceramics,* Vol 21, Ceramic Powder Science.
9. K. Tamura, Fabrication of $MoSi_2$ Base Heating Element by Slip Casting, *Funtai Oyobi Funmatsuyakin,* Vol 10, 1963, p 44–50; *See Chem. Abs.,* Vol B59, 1963, 15759g
10. E.A. Hauser and S. Hirshon, Behavior of Colloidal Suspension with Electrolytes, *J. Phys. Chem.,* Vol 43 (No. 8), 1939, p 1015–1036
11. W.D. Jones, *Fundamental Principles of Powder Metallurgy,* 1st ed., Edward Arnold Publishers, 1960, p 372–385
12. P.D. St. Pierre, *Research,* Vol 12 (No. 12), 1959, p 460–466
13. W.G. Lidman and R.V. Rubino, Slip Casting of Stainless Steel Powder—Part 1, *Precis. Met. Molding,* Vol 14, (No. 8), 1956, p 40–41, 83
14. W G Lidman and R.V. Rubino, Slip Casting of Stainless Steel Powder—Part 2, *Precis Met. Molding,* Vol 14, (No. 9), 1956, p 64–66, 98
15. P.E. Rempes, B.C. Weber, and M.A. Schwartz, Slip Casting of Metals, Ceramics and Cermets, *Ceram. Bull.,* Vol 37 (No. 7), 1958
16. F.H. Norton, *Fine Ceramics: Technology and Applications,* 2nd ed., McGraw-Hill, 1978
17. C.H. Desch, The Setting of Plaster, *Trans. Brit. Ceram. Soc.,* Vol 15, 1981
18. W.C. Hansen, The Setting and Hardening of Gypsum Plasters, *Mater. Res. Stand.,* Vol 3 (No. 5), 1963, p 359
19. Power, Transient Solubilities in the Calcium Sulfate-Water System, *J. Chem. Eng. Data,* Vol 9 (No. 3), 1964, p 437
20. R. Hamilton, How They Test Gypsum Cements and Plasters for the Ceramic Industry, *Ceram. Ind.,* Vol 71 (No. 1), 1958
21. B.W. Nies and C.M. Lambe, Movement of Water in Plaster Molds, *Ceram. Bull.,* Vol 35, 1956, p 319
22. J. Camm, Technical Control of Mould Making, *InterCeram,* Jan 1939, p 35
23. E.G. Walker, Plaster and the Sanitary Ware Industry, *Trans. Brit. Ceram. Soc.,* Vol 316, 1961
24. P.H. Dal and W.J. Berden, The Capillary Action of Plaster Molds, *Science Ceram.,* Vol 4, 1968, p 113
25. P. Carman, The Determination of the Specific Surface of Powders I, *J. Soc. Chem. Ind.,* Vol 58 (No. 1), July 1938
26. P. Carman, Some Physical Aspects of Water Flow in Porous Media, *Trans. Inst. Chem. Eng.,* Vol 15, 1937, p 150–166
27. D.S. Adcock and I.C. McDowall, The Mechanism of Filter Pressing and Slip Casting, *J. Am. Ceram. Soc.,* Vol 40 (No. 10), 1957, p 355–362
28. E G Blanchard, Pressure Casting Improves Productivity, *Am. Ceram. Soc. Bull.,* Vol B67 (No. 19), 1988

SELECTED REFERENCES

- H.H. Hausner and D.P. Ferriss, Metal Powder Slip Castings, *Mater. Methods,* Vol 43 (No. 5), 1956, p 132–134
- J.T. Jones and M.F. Berard, *Ceramics: Industrial Processing and Testing,* 1st ed., Iowa State Press, 1972
- G.W. Phelps, S.G. Maguire, W.J. Kelly, and R.K. Wood, *Rheology and Rheometry of Clay-Water Systems,* Cyprus Industrial Minerals Co., Sandersville, GA
- E. Posnjak, The System, $CaSO_4$-H_2O, *J. Am Chem. Soc.,* Vol 35a, 1938, p 247
- J.S. Reed, *Introduction to the Principles of Ceramic Processing,* 4th ed., John Wiley & Sons, 1988
- United States Gypsum, "Testing Gypsum Cements and Plasters," Bulletin No. IG-501
- E.G. Walker, The Role of the Mould in Casting, *Brit. Ceram. Res. Assoc. J.,* Vol 64, 1965, p 233

Powder Metallurgy Methods for Rapid Prototyping

DIRECT FABRICATION is a manufacturing innovation whereby engineers integrate design and manufacturing operations for structural materials via computer software to fabricate components in an automated fashion. Several techniques are under development, and their use for rapid prototyping (RP) can eliminate the need for costly redesign after a part has gone into full-scale production. Traditional prototyping can itself be costly and time consuming, often delaying product introduction. Rapid prototyping technologies were developed in response to increasing competitive pressures to bring quality products to market quickly. A key element in any RP process is the use of sophisticated computer codes used to generate computer numerical control (CNC) motion and process control data from "slices" taken from solid models.

The following section in this article summarizes the principal RP methods. Later sections provide more detailed information about the following P/M RP methods: soft tooling, low-pressure molding, selective laser sintering, and laser-based direct fabrication.

Rapid Prototyping Methods

Leander F. Pease III, Powder-Tech Associates, Inc.

If every designer had immediate access to the required raw material, computer-controlled machining centers, and the facilities for heat treating and finishing parts, the prototype process would already be irreducibly short. It is the real-world lack of immediate access to raw material and machining that has inspired the search for desktop manufacturing (DTM). In its most elegant form, the engineer's CAD system drives a clean, quiet machine in the corner of the office to produce the required prototype part the same day, and, for some classes of plastic materials, this kind of system is in use today. A laser system traces out the part in a bath of photocurable plastic. In the past, it was not common practice to machine a plastic prototype part from a monolithic block. Rather, the usual procedure was to build tooling and then injection mold the part, a process that could take up to 12 weeks.

The rapid development of laser-processed plastic prototypes has inspired the search for a similar system to work with metals. The output of such a system might be fully dense parts to simulate the product of ingot metallurgy, or it might be less than fully dense materials that simulate P/M components. Whatever the system, it must offer advantages over the current machining methods. Such advantages can be in saving development time or reducing machining costs. It is important to understand the current methods used in P/M part prototyping so as to be able to assess the real value of the developments in DTM.

Powder Metallurgy Prototypes

Powder metallurgy prototypes are now produced either by building hard tooling and making limited runs of parts or by machining the prototypes from monolithic blocks of the required sintered materials (Ref 1). The hard tooling route requires 4 to 12 weeks, a few thousand dollars worth of tooling, and then the use of valuable production press time to make the parts. If the design must then be changed, the whole process must be repeated. Alternatively, blocks of the required P/M materials are readily available from P/M parts producers. Such prototype materials are produced in all the alloy designations found in MPIF Standard 35, ASTM B 783 or ISO 5755, using powders in inventory. They are available in two days to two weeks. The P/M blocks have the same density and mechanical and physical properties as the materials listed in the standards. Just as in the case of ingot-based metals, the P/M materials are then fashioned into prototype parts on machining centers or by wire electrical discharge machining. The parts have the required tolerances and surface finish. Heat treating is performed subsequent to the machining. The time required for the machining operation depends on the facilities at hand: a captive model shop may be available, or an outside cus-

Fig. 1 Actuator prototype after wire electric discharge machining from a sintered block. Courtesy of Airtech, Inc.

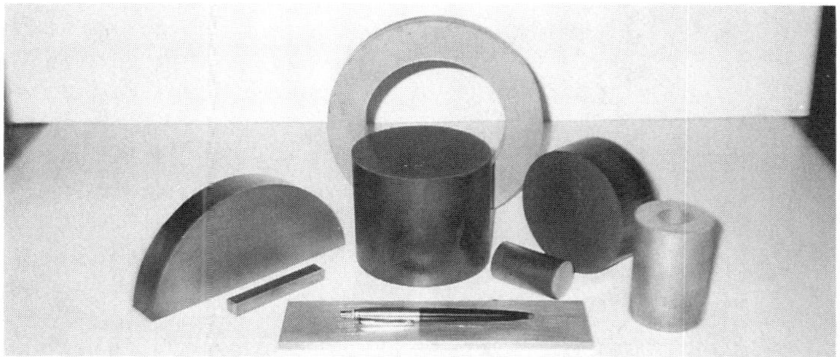

Fig. 2 Examples of monolithic blocks of sintered P/M materials used to machine prototype parts. Courtesy of Powder-Tech Associates

tom machine shop used. The times range from a few days to a few weeks.

Realistically, the new rapid-prototyping systems must compete with the existing P/M methods. This means that in some cases, the machined P/M prototypes will be available in as little as 1 week, and they will meet the drawing requirements for material properties, dimensions, and surface finish. Figure 1 shows an actuator insert electrical discharge machined and further machined from a simple block of P/M material such as those shown in Fig. 2. In one version of the insert, the 32 teeth were wire electrical discharge machined at a cost of $192. In a second version, the teeth were ground, requiring 1.5 h.

Soft Tooling. The fabrication of conventional tool steel molds for metal injection usually requires 10 to 12 weeks. The use of soft tooling allows limited numbers of metal-injection molded (MIM) parts in only 1 to 2 weeks. Soft tools are constructed of cross-linked polymers such as epoxy resin for rigid tooling and silicone for resilient tools. The rigid materials withstand metal/binder injection at a few thousand pounds per square inch while the silicone tools are used for low-pressure molding.

To create a rigid tool, a model of the proposed part is first machined. The part is covered in a parting agent and cast in epoxy resin. The resin may be reinforced with fillers. The model must have a parting line film or sheet attached, so as to create the two halves of the mold. Prior to curing, the epoxy is de-aired by vacuum. After curing, the model is removed from the two mold halves. The molds can be used in the MIM process to make 10 to 1000 pieces

Soft tooling also refers to tooling machined from aluminum, which can run up to 50,000 pieces at normal injection molding pressures. Such tools can be made in 2 to 4 weeks.

Ink-Jet Technology. In this process, an ink-jet printer (Ref 2) deposits 80 μm droplets of liquid acrylic copolymer binder onto a bed of metal powder. At each pass of the print head, a bonded path 175 μm (0.007 in.) wide and 75 μm (0.007 in.) deep is created. The printer follows computer-generated instructions to construct a part layer by layer. After one layer is bonded, the powder bed is lowered and a new 175 μm (0.007 in.) layer of powder is rolled into place. This layer is then bonded to the layer below, gradually building a three-dimensional object.

In making injection molding tooling, internal water passages are created that conform well to the shape of the cavity. Such carefully placed water passages provide optimal cooling during MIM and result in faster cycle times and less residual stress in the molded part. There are no ingot-based metallurgical or machining techniques that can create arbitrary cooling passages inside an object.

When the object is complete, it is removed from the bed and the loose powder blown off and collected. The powder inside the cooling passages is removed by water immersion of the green part. The part is then heated at 100 °C (212 °F) to cure the acrylic and strengthen the structure. The acrylic is removed in a thermal debinding step, and the part is then easily handled.

In an example with 60 μm (0.002 in.) stainless steel powder, the debound part is placed on a stainless steel substrate, which in turn rests on zirconia pellets in an alumina crucible. A piece of 90/10 Cu/Sn bronze is placed on the substrate adjacent to the porous part. In a furnace containing argon + 5% H_2, the temperature is raised to 980 °C (1795 °F). The bronze melts, wicks along the stainless substrate, and enters the side of the porous part. This continues until the bronze fills all the pores by capillary action. The zirconia pellets keep the stainless steel substrate from being brazed to the alumina crucible. The stainless substrate is machined off the bottom of the fully dense mold. The cores and cavities are then machined and polished as necessary.

Other materials produced in the same fashion include iron, stainless steel, tungsten, tungsten-nickel, tungsten carbide, and tungsten carbide cobalt. It is possible to locally alloy a part by depositing fine metal powders along with the acrylic binders. The fine powders then diffuse into the base metal during the sintering cycle. In terms of capital cost, thermal printers are much less expensive than lasers. One can imagine a whole room full of printers constituting a small factory for low-volume production runs.

Selective laser sintering (SLS) was developed by DTM of Austin, TX (Ref 3) and uses laser energy to transform metal powders into useful tooling and parts. A CAD model of the part is numerically transformed into thin slices. Loose, polymer-coated 55 μm (0.002 in.) metal powder is laser fused into the required shape, as it traces out the area of the computer-generated slice. The workstation lowers this fused layer by 75 to 250 μm (0.003 to 0.10 in.), and a new layer of powder is rolled into place, atop the fused layer. Laser fusing of this second layer bonds the particles to each other and to the layer below. In this way, a part is built up, layer by layer. Molds 15 by 25 cm (6 by 10 in.) by a few inches thick have been produced by this method.

After the fusing step, the part has a green strength of about 2.7 MPa (400 psi) and requires care in handling. The unfused powder falls away from the fused part and is recycled. The fused green part is impregnated with acrylic resin containing a cross-linking polymer and then dried at 50 °C (112 °F). The dried part is placed on an alumina plate in a graphite crucible, and the porous preform is infiltrated with molten copper, in an atmosphere-controlled furnace. The furnace then returns to room temperature, and the assembly is removed.

The infiltrated part contains about 60% Fe, 39% Cu, and 0.8% C. During infiltration, the part shrinks 2.5% linearly, and this is compensated at the initial part design. The infiltrated parts have a tolerance of ±0.25 mm (±0.010 in.) over distances of 13 cm (5 in.).

In all, the process takes about 5 days and results in a mold that is ready for plastic injection molding or metal injection molding. Such a mold can produce 50,000 plastic parts. The main use for SLS via the DTM process is making tooling for the injection molding of plastics. The infiltrated material so produced does not conform to MPIF Standard 35 and so could not be used in functional prototypes.

In principle, other alloy steels could be SLS to form 60% dense preforms that could then be sintered and shrunk to the correct density to serve as P/M prototype parts. This would involve linear shrinkages of 7 to 10% with difficulties in maintaining tolerances. In general, parts sintered to such a degree would not have the same mechanical properties as the standardized P/M grades.

Direct Laser Sintering (Laser Free-Form Fabrication). A method has been devised to directly laser sinter metal powders and intermetallic compounds, without the use of a polymeric binder film (Ref 4). They use a three-axis machine to aim both a fiber-optically guided Nd-YAG laser beam and a stream of metal powder at a substrate as shown in Fig. 3. The laser is under computer control to develop the part as a series of layers. The powder falls into the focal zone of the laser where it sinters to itself and to the underlying layer. This is done in an atmosphere of argon with <5 ppm oxygen. The original machine creates $2\frac{1}{2}$ dimensional parts in an envelope of 9.5 by 9.5 by 38 mm ($\frac{3}{8}$ by $\frac{3}{8}$ by 1.5 in.) high. These are thin-walled parts 1.5 to 3.2 mm (0.060 to 0.125 in.) wall. The minimum wall thickness is 0.8 mm (0.030 in.). Powder is deposited at a rate of 2 g/min with a 2 kW laser. The powder can be formed into a fully dense

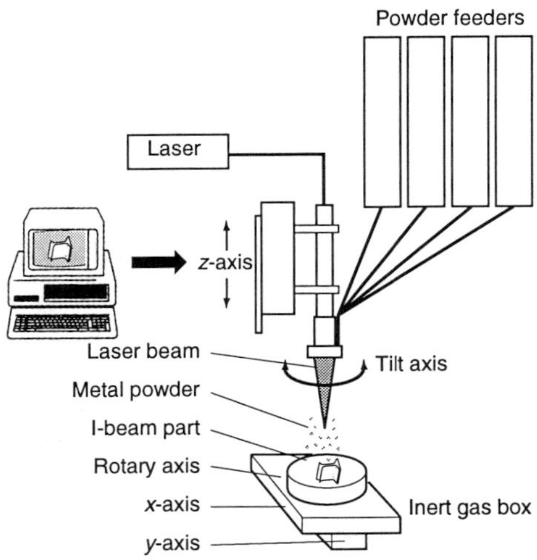

Fig. 3 Laser-sintering apparatus. Courtesy of G. Lewis, Los Alamos National Laboratory

object or into one that has controlled porosity. Parts with residual porosity showed particle boundaries, and their mechanical properties are not yet known. A new five-axis machine makes parts within a working envelope of 60 by 60 by 60 cm (24 by 24 by 24 in.). It deposits powder at the rate of 10 to 20 g/min. The powders cool at 100 K/s (180 °F/s) in rod forms and 10,000 K/s (18,000 °F/s) in flat plate forms.

The process is a near-net-shape process with tolerances of ±0.25 μm (±0.010 in.). The first layer of the part sinters to a base plate, which must later be machined off. Thick-walled parts are harder to make because of the longer time needed to make a massive part. Also, it is hard to form sharp corners on thick-walled parts. To a degree, this problem is addressed by the five-axis machine, which can be angled, and is able to make special finishing passes and overhangs. The usual powder is −100 down to −325 mesh, but 10 μm (0.0004 in.) powders give better surface finish. Powders processed to date include P20 tool steel, 316 and 410 stainless steels, tungsten, $MoSi_2$, and NiAl. No work has been done on low-alloy steels. The carbon must be in solution in the steel, or the laser beam will vaporize it. This suggests that only special prealloyed steel powders with carbon in solution will be used. The excess powder can be recycled and reused. A laser can be used to form powder directly into a part, as shown in Fig. 4. This work (Ref 5) has resulted in usable tool steel molds.

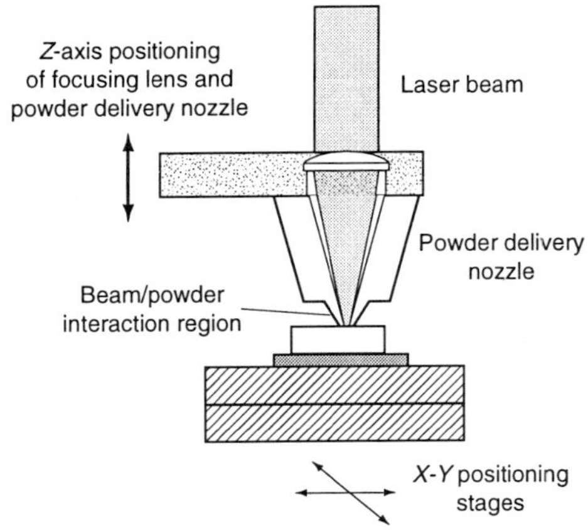

Fig. 4 Laser-sintering apparatus. Courtesy of D. Keicher, Sandia National Laboratory

Free-form laser-sintering work is being researched in order to develop a way to laser sinter titanium alloy powders into the 150 kg (330 lb) bulkheads for F16 aircraft. The starting powder is Hunter-Kroll unpurified sponge titanium fines at −40/+100 mesh and vanadium-aluminum master alloy. The powders are in a bed fluidized by argon with <15 ppm oxygen. The original chamber is 0.3 by 0.3 by 0.3 m (1 by 1 by 1 ft) and is for concept demonstration. Laser-sintered titanium samples that weighed about 454 g (1 lb) were produced to determine mechanical properties.

The laser is 7 kW with a spot size of 12 mm (½ in.). This results in the melting of the titanium and the alloying of the vanadium and aluminum master alloy. The melting tends to create a flat surface. At the same time, with one or two passes, any residual chlorine or sodium in the Hunter-Kroll fines is evaporated. Kroll fines, which are made by reducing $TiCl_4$ with Mg, are undesirable as a feedstock because residual Mg salts have been observed in the pores of laser deposited material. Hydrogen levels are low, but can be further reduced by vacuum annealing after laser forming. The first layer is attached to a striking plate, and the object is built up layer by layer. The fluidized bed allows one layer to be submerged in the bed, in preparation for the next layer to be sintered. The final object is usually machined 3 mm (⅛ in.) on each surface. The striking plate tends to warp about 3 mm (⅛ in.) in 100 mm (4 in.). Because the object is almost, but not completely, pore-free the part is subsequently hot isostatically pressed without a can or forged to full density.

The second-generation machine has a working area of 3 by 3 m (10 by 10 ft), and a bed depth of 1 m (3 ft). A 14 kW laser rides on a gantry. This allows the manufacture of an F16 bulkhead that is 3 by 1 by 0.1 m (9 by 3 ft by 3 in.) thick. A finished bulkhead weighs about 150 kg (330 lb). The laser covers 45 cm^2 (7 $in.^2$) to a depth of 1 mm (0.040 in.) in 1 min. This is 4.6 cm^3/min (0.28 $in.^3$/min), and for titanium at 4.5 g/cm^3, 20.6 g/min (0.05 lb/min).

Table 1 summarizes the capabilities of several free-form fabrication processes in terms of tolerance, size, and speed. Current applications of laser-sintered metal powders are injection molding tools. Expected applications include laser-sintered titanium alloy bulkheads, tooling, prototype parts, and one-of-a-kind or short runs. A shipboard manufacturing center that will use computer-stored information on part geometry to laser sinter and manufacture replacement parts while at sea is also expected as well as other, similar objectives. Laser sintering could fabricate some interesting composite materials and might be a production manufacturing method rather than just the prototype method. One example might be intermetallic valves for automotive engines.

It seems likely that laser free-form fabrication using metal powders will provide a rapid progression from design to prototype parts. The process readily makes fully dense objects. It must be determined if the tolerances so produced are adequate for fit and function tests, or whether additional machining will be needed. In some cases the machining might be minimal, but in other cases it might be more cost effective to start with a simple block of material and machine it all over. The laser-formed materials will have advantages when working with expensive or difficult-to-machine raw materials.

For less than fully dense P/M prototypes, of the type covered by MPIF Standard 35, the utility of the laser free-form fabrication process is

Table 1 Comparison of P/M-based rapid-prototype/DTM processes

Tolerances(a)		Surface finish		Size		Deposition rate			
μm	in.	μm	in.	cm	in.	mm^3/h	$in.^3$/h	Materials	Ref
±50 X-Y and ±375 Z	±0.002 X-Y and ±0.015 Z	4 RMS	0.00016 RMS	7.5 × 10 × 15	3 × 4 × 6	16	1	Full density	5
±250	±0.010	50–100	0.002–0.004	5	2	8; Aim, 98	0.5, Aim is 6	Full density	4
<±375	<±0.015	75–125	0.003–0.005	22.5 × 12.5 die	9 × 5 die	1-245	0.06–15	Full density	6
±375 X-Y	±0.015 X-Y	...	...	15 × 15 × 30	6 × 6 × 12	50	3	Infiltrated	3
100	0.004	...	...	15 × 30 × 2.5	6 × 12 × 1	245	15	Infiltrated	2
±1500	±0.060	±750	±0.030	Aim 210 × 90 × 7.5	Aim 84 × 36 × 3	295	18	Full density	7

RMS, root mean square. (a) For dimensions up to about 15 cm (6 in.)

yet to be determined. In addition to competing with the tolerances and surface finish of parts machined from simple blanks, there is the difficulty of matching the physical and mechanical properties of P/M parts. Many P/M materials are essentially composites, rather than homogeneous alloys, for example, nickel steels and diffusion alloys. The most common P/M material is FC-0208, iron + 2% Cu + 0.8% combined C. Laser sintering does not work well with loose graphite, so special prealloyed Fe-0.8% steels would have to be produced. The copper could not be prealloyed because normal P/M sintering practice does not diffuse copper to the center of the iron particles.

Low-pressure molding (LPM) is done at <0.7 MPa (<100 psi), usually in soft tools. To obtain a low viscosity that will pour into tooling, the mixtures are prepared under conditions of low shear rate, which lowers viscosity by tenfold. The mixtures solidify in the molds, and after removal, parts are thermally debound and then sintered. There are two classes of binder materials: dilute polymer solutions and polymerizable systems.

Dilute polymers consists of 0.5 to 5.0% aqueous solutions of polysaccharides, vinyl polymers, waxes, alginates, and so forth. The mixtures are poured into porous molds and dried at room temperature (slip casting). As dried, the parts are already 99% debound, and the remaining binder is removed thermally at 500 °C (932 °F) in protective atmosphere. In one example, stainless steel is sintered to 96% of the pore-free density. In another dilute polymer method, Quickset (Ref 8), the mixture of powder and dilute polymer is frozen. The water is then removed by sublimation. The binder comes out during vacuum sintering.

An example of a polymerizable system is furfuryl alcohol and powder. It is poured into the mold, and heat and catalysts are applied. In 1 to 5 min, it cross links and hardens. The water is expelled as gas, leaving a porous skeleton that later allows for the removal of the rest of the binder, at 400 to 500 °C (752 to 932 °F). In a variant called gelcasting, 10 to 20% acrylamide in water is mixed with metal powder. The mixture is poured into soft tools and allowed to gel for 10 to 60 min. After removal of the part, it is dried for a few hours. It is debound at 350 to 500 °C (662 to 932 °F). After high-temperature sintering and substantial shrinkage, 17-4 PH steel reached 7.5 g/cm^3.

Soft Tooling

S.V. Atre and J.A. Thomas,
The Pennsylvania State University

Soft tooling is a general term that is used to describe molds fabricated from polymers. The principle utilizes the low viscosity of polymer precursors to accurately form the reverse pattern of a given desired shape. The solidified reverse pattern can then be used to replicate the original shape many times.

The development of soft tooling gained impetus as a consequence of the material and labor shortages during World War II. However, its widespread acceptance has only occurred more recently as the economic benefits of reduced lead times from "art to part" have become apparent. This increased awareness has coerced dramatic changes in the field, to the extent that physical reverse patterns can now be directly created from virtual desired shapes stored in a computer. To date, the development of soft tooling can be attributed to traditional manufacturing techniques such as injection molding and thermoforming for plastics, sheet metal forming for metals, and hand lay-up for composites (Ref 9). The current accelerated growth of binder-assisted P/M forming techniques has made it crucial for the parts manufacturer to take advantage of the quickly fabricated, low-cost, lightweight tooling obtained from polymers. This section outlines the materials, methods, attributes, and applications of soft tooling.

Soft-tooling materials are usually cross-linked polymers (Ref 10). They can be categorized on the basis of the consistency of their final structure under two classes: rigid and resilient materials.

Rigid materials include thermosetting resins such as epoxies, polyesters, phenolics, and polyurethanes. Epoxy resin is the most common resin used for constructing rigid molds. The mold is shown in Fig. 5. It is used for fabricating M2 tool steel parts by MIM.

Resilient materials include thermosetting resins such as silicones, polysulfides, polyurethanes, and polyureas, as well as aqueous network structures such as biopolymer gels based on agar or gelatin. Silicone is the most commonly used class of materials for constructing resilient molds. The mold is shown in Fig. 6. It is used for fabricating 316L stainless steel watch cases by LPM.

The mechanical properties and tooling performance of the polymers are dependent on their physical and chemical structure as well as fabrication conditions. As a result, considerable property variations occur even within a given class of polymers. Resin manufacturers usually provide detailed information on the properties of the polymers to aid the tooling engineers in the selection of the correct material for their application. The hardness of the common resins used in soft tooling (Ref 9, 11, 12) and metals (Ref 12,

Fig. 5 A rigid mold fabricated from filled epoxy for the production of M2 tool steel parts. Courtesy of J.A. Thomas and A. Griffo, The Pennsylvania State University

Table 2 Materials used in fabricating molds, arranged in order of decreasing surface hardness

Material	Surface hardness(a)
G30 cemented carbide	71 HRC
Steel, nitriding	>68 HRC
Steel, carburizing	60–65 HRC
D3 tool steel	62 HRC
Nickel-cobalt alloy	45–52 HRC
Copper-beryllium 275	46 HRC
Aluminum bronze	188 BHN
Steel, low alloy and carbon	180 BHN
Kirksite	80–105 BN1
Brass	50 BN1
Sprayed metal	<50 BN1
Epoxy, metal filled	85 Shore D
Epoxy, unfilled	85–90 Shore D
Polyurethane, hard	70–85 Shore D
Polyurethane, soft	40–65 Shore A
Silicone rubber	15–65 Shore A

(a) HRC, Rockwell hardness, C scale; BHN, Brinell number 3000 kg load; BN1, Brinell number 500 kg load. Source: Ref 9, 11–13

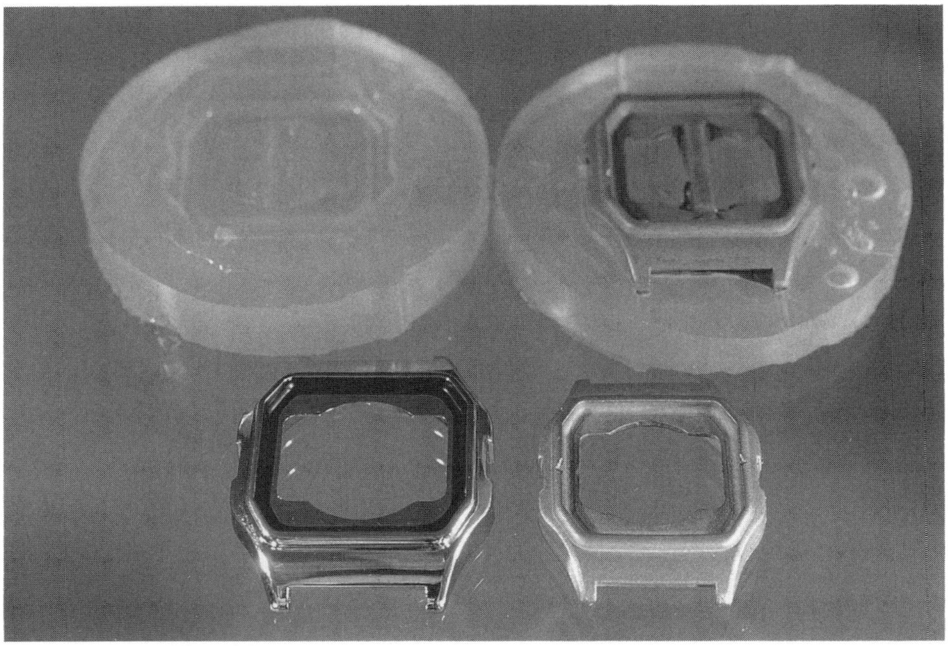

Fig. 6 A flexible mold fabricated from RTV silicone. Clockwise from bottom left: master model, top half of mold, bottom half of mold and green part, final part after debinding and sintering. Courtesy of C.D. Turner, The Pennsylvania State University

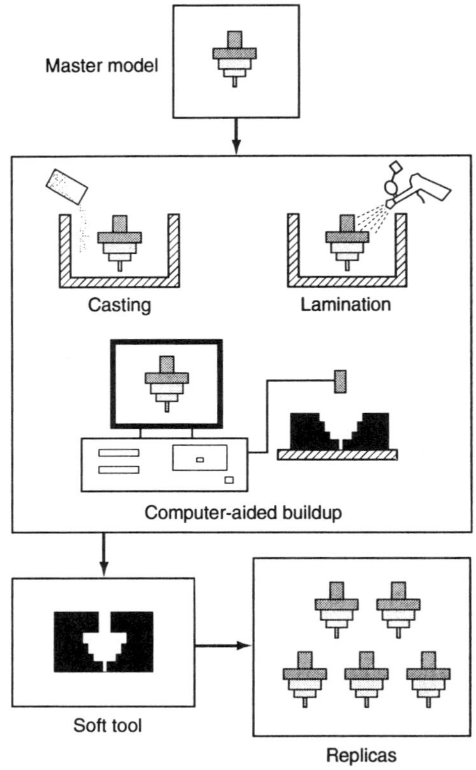

Fig. 7 An outline of the soft tooling process

Table 3 A comparison of attributes of metals and polymers for tooling

Comparison variable	Result (for polymers)
Capital investment	Favorable
Labor cost	Favorable
Lead time	Favorable
Large tool cost	Favorable
Tool weight	Favorable
Repair and design changes	Favorable
Short runs and prototypes	Favorable
Ease of duplication	Favorable
Corrosion resistance	Favorable
Surface finish	Favorable
Skilled supervision	Approximately equal
Material cost	Approximately equal
Thermal conductivity	Unfavorable
High-temperature strength	Unfavorable
Long production runs	Unfavorable

Source: Ref 9

13) used in traditional "hard" tooling are given in Table 2. The much lower hardness observed for polymers is the rationale behind the use of the term "soft" tooling.

Methods. Figure 7 outlines the key steps involved in the manufacture of parts from a soft tool. A master model is fabricated by machining or other rapid-prototyping methods.

Recent developments directly use a CAD file stored in the computer to be used as the master model (Ref 14). Using the master model, a soft tool is fabricated from procedures such as casting, lamination, and computer-aided buildup.*

*Lamination refers to the fabrication of alternate layers of glass cloth or chopped fiber and a resin. Computer-aided buildup is used here to describe all techniques that include a CAD file to directly create a three-dimensional object. Some examples are stereolithography (SL), SLS, fusion deposition modeling (FDM), three-dimensional printing (3DP), and ballistic particle manufacturing (BPM).

The master is separated from the cured tool and is used to make replicas by a forming technique of choice.

Several rapid-prototyping techniques for fabrication of plastic objects have been covered elsewhere in this Volume. Casting is briefly described in this section. Most thermosetting materials used in casting are also excellent adhesives. Therefore, a release or parting agent must be applied to the master model before application of the tooling resin. Particulate fillers such as aluminum, silicon carbide, and sand are often added to the resin in order to increase thermal conductivity, improve surface hardness, and reduce costs, respectively. The high viscosity of filled suspensions makes it important to ensure the removal of air bubbles entrapped during the mixing or pouring stages (Ref 15). In order to fabricate large tools, the soft tool can be cast onto a core of materials such as stainless steel, aluminum, kirksite, or resin-fiber glass laminates in a procedure known as backing (Ref 9). More recently, an additional stage known as fronting has been developed in order to create a thin metal film on the surface of the soft tool in order to improve thermal properties and hardness (Ref 16).

Attributes and Applications. Complex shapes can be cast or laminated to final dimensions using soft-tooling techniques. Lower costs can be achieved compared to conventional tooling because fabrication times are short and expensive machining is often unnecessary. The time for fabrication is on the order of days as opposed to weeks or months. There are two main limitations of soft tooling. The poorer thermal properties result in an increase in molding cycle times and impose restrictions on the maximum processing temperature. The inferior mechanical properties, in general, reduce the lifetime of the tooling to the order of 10 to 1000 parts. Table 3 summarizes the comparisons between the attributes of soft tooling and conventional hard tooling.

From the above discussion, it is evident that polymers are most useful for the quick and inexpensive fabrication of tooling for low-volume prototyping runs. Further, the tooling is useful when a high level of surface finish is required. Soft tooling is applicable to binder-assisted P/M forming techniques such as powder injection molding and LPM.

Low-Pressure Molding

S.V. Atre, The Pennsylvania State University

There are several shape-forming techniques that utilize powder-binder suspensions (Ref 17). In recent years, a number of methods have emerged in which molding occurs under low shear rate conditions. These methods are collectively referred to as low-pressure molding (LPM). Figure 8 shows the generic steps involved in LPM. A well-mixed feedstock consisting of the powder and binder constituents is introduced into a mold at pressures below 0.7 MPa (100 psi). Generally, the viscosity of the feedstock is such that the material can flow at room temperature (1 to 10^3 Pa · s). The feedstock is

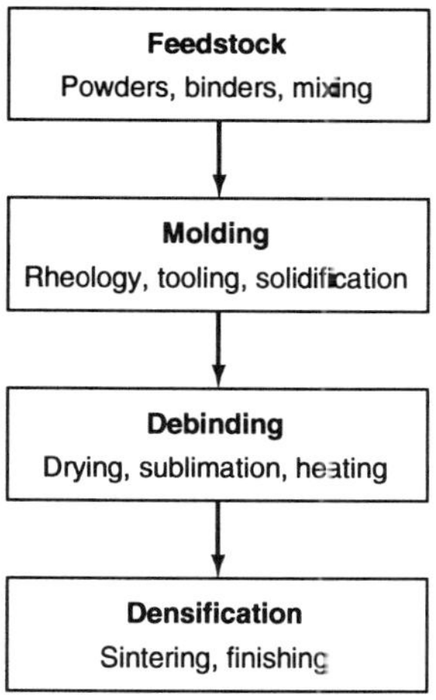

Fig. 8 Flow chart for the LPM process, showing the four major stages and highlighting the important parameters within each

solidified in the mold to the desired shape. The green part is removed from the mold and is subjected to thermal debinding and sintering cycles to high final densities (90 to 100% of the theoretical value). The main variations in the process occur in the binder composition and in the solidification stage.

A critical requirement for any molding operation is that the feedstock viscosity allow complete and uniform mold filling (Ref 18). The viscosity of suspensions depend on many variables such as temperature, shear rate, particle characteristics, and solids loading (Ref 19, 20). For example, thermoplastic polymers and waxes soften and flow easily above their melting point. Certain polymers and suspensions display a shear thinning behavior, wherein the viscosity decreases with increasing shear rate. Particle characteristics such as shape, size distribution (mean and width), packing characteristics, and surface chemistry have all been recognized to play a role on the suspension viscosity. A semiempirical equation (Ref 18) relating the suspension viscosity, η, to some of these variables is:

$$\eta = \eta_b A(1 - \phi/\phi_m)^{-n}$$

where η_b is the viscosity of the binder, ϕ is the solids loading, ϕ_m is the maximum particle packing density, and n is a constant usually close to 2. A is a function that accounts for shear rate, temperature, and particle size effects. In principle, some or all of the above factors can be controlled to allow molding at low shear rate conditions. In LPM practice, however, the suspension viscosity is usually reduced by lowering η_b. Typical examples of LPM techniques are discussed next.

Methods

Frequently, LPM methods for shape forming with metal powders are extensions of processes developed for ceramics. The majority of LPM processes can be classified based on the binder system under two categories: dilute polymer solutions and polymerizable systems. Variations under each of these categories are discussed.

Processes Using Dilute Polymer Solutions. A straightforward approach to reducing binder viscosity is to use polymer solutions. The viscosity of a polymer solution is lowered with decreasing concentration and molecular weight of the polymer (Ref 10). Both aqueous and nonaqueous systems have been practiced, although the former is preferred for environmental and safety reasons, provided the metal system is not reactive toward water. Two examples are presented using this method.

Slip casting is a classic example of a forming method used primarily for ceramics (Ref 17). Although the method had been explored several decades ago for metal powders (Ref 21), it was rarely practiced until more recently (Ref 22, 23). Slurries consisting of aqueous polymer solutions as binders and metal powders have been used for slip casting. Polysaccharides, vinyl polymers, waxes, and various organometallic compounds have been used in solutions (concentration varying between 0.5 and 5 wt%) for slip casting (Ref 24). A homogeneous feedstock is obtained by the low shear mixing of the ingredients at room temperature. The viscosity of the feedstock can be reduced by more than an order of magnitude under conditions of low shear (e.g., 450 Pa · s at a shear rate of 40 s^{-1} for a 316L stainless steel feedstock, solids loading of 60 vol%). The feedstock is then poured into molds at ambient conditions of temperature and pressure and dried before ejection of parts. The use of porous tooling helps achieve green densities in the range of 70 to 74% of the theoretical density (Ref 23). The parts obtained are nearly 99% debound and have a green strength of 9.5 MPa (1380 psi). The remaining binder is rapidly removed in a thermal debinding step (500 °C). Because the amount of carbon in the starting feedstock is low, carbon control, especially for the fabrication of thick sections, is relatively easy. Carbon analysis performed on 17-4 PH stainless steel parts formed by slip casting resulted in 0.045 wt% residue (Ref 23). Final sintering is conducted under standard conditions. Starting with identical powders and solids loading, comparable properties can be obtained for 316L stainless steel parts fabricated by slip casting and powder injection molding (Ref 25), as shown in Table 4. The latter process is an example of a high-pressure molding process. Depending on the mold geometry, pressures in the range of 3.5 to 70 MPa (500 to 10,000 psi) are typical for powder injection molding (Ref 18). In a separate study, a sintered density of 7.74 g/cm^3 and hardness of 40 HRC have been reported for slip cast 17-4 PH stainless steel (Ref 23). This powder is unsuitable for use in the fabrication of parts by die compaction due to its hard and spherical nature. Powder injection molded parts for the same material are 7.5 g/cm^3 and 35 HRC, respectively (Ref 26).

The Quickset process is another variation of binder systems using aqueous polymer solutions (Ref 8). Here, the parts are frozen in the mold,

Table 4 Sintered properties of 316L stainless steel parts obtained by slip casting and powder injection molding

Property	Slip casting(a)	Powder injection molding(b)
Shrinkage after sintering, %	14.8	13.3
Sintered density, % theoretical	95.6	91.0
Ultimate density strength, MPa	492	448
Elongation to fracture, %	38	34
Hardness, HRB	59	57

Feedstock: water-atomized 316L stainless steel (d_{50} = 14.3 μm) at a solids loading of 62 vol% was used in both processes. The binder system used in slip casting was an aqueous alginate solution. The binder system used in powder injection molding consisted of paraffin wax, polypropylene, and stearic acid. Sintering conditions for both processes: 1350 °C, 1 h, H_2. (a) Ref 22. (b) Ref 25

and the water molecules are removed by sublimation. Additives are introduced in the feedstock to minimize volume changes during freezing. The residual binder is removed in a thermal debinding step. For Sialon, flexural strength of 975 MPa and Weibull modulus of 19.4 have been reported using the process. Comparative runs with isopressed parts resulted in values of 785 MPa and 8.7, respectively, for the flexural strength and Weibull modulus.

Polymerizable Systems. The viscosity and melting point of a polymer increase with its molecular weight (Ref 10). Therefore, a logical alternative to using polymer solutions is to use low-molecular-weight polymer precursors (referred to as monomers, oligomers, or prepolymers) in order to obtain low binder viscosities at room temperature (typically <1 Pa · s). A catalyst is used to initiate the polymerization reaction, often in combination with heat. The polymers are of the thermosetting type because a cross-linking agent is usually added to the mixture to solidify the part in the mold. This approach owes its origins to fabrication techniques used in polymer-matrix composites. Two examples of this approach are presented.

The debinding reaction injection molding (D-RIM) process uses a low viscosity (0.2 to 1.0 Pa · s), thermosetting liquid polymer to fluidize the powder (Ref 27). Solids loading of 60 to 65% for metal powders can be achieved. The furfuryl alcohol-based polymer undergoes solidification as a result of a cross-linking reaction initiated by the addition of heat or a catalyst. A green strength of 6.5 MPa (950 psi) is typical. Water molecules are evolved in the gaseous state as a by-product of the reaction and leave behind an open pore network. Nearly 50% of the initial binder is removed in this manner at the end of the molding stage. The remaining binder is thermally decomposed in air, hydrogen, or vacuum at 400 to 500 °C. Mold residence times depend on part dimension and typically vary from 1 to 5 min. An important aspect of processes using thermosetting resins is that thermal decomposition occurs without any liquid phase being formed. The porous structure formed in the present process as a result of the condensation reaction provides a convenient pathway for the es-

Fig. 9 An example of rapid tooling produced by LPM. Courtesy of J.A. Thomas, The Pennsylvania State University

cape of the gaseous decomposition products. This minimizes the formation of internal stresses that could lead to defects such as cracks and blisters during the thermal debinding stage. Tensile strength for carbonyl iron D-RIM parts was reported to be 215 MPa. A value of 220 MPa reported for powder injection molded processed iron (Ref 26).

Gelcasting is another variation based on the use of cross-linkable monomers (Ref 28). A monomer solution provides a low-viscosity vehicle for suspending the powder. For example, the viscosity of a 62 vol% slurry of alumina has been reported to be 1.8 Pa · s at a shear rate of 10 s^{-1}. The monomers used are a mixture of acrylamides (10 to 20%), and the solvent is usually water (80 to 90%). This is introduced into a mold made of aluminum or polyethylene where it is cross linked to form a solid polymer-solvent gel. The cross-linking reaction can be performed by the addition of heat (50 to 60 °C) and a suitable catalyst. The gelled part can be removed from the mold without distortion and dried to remove the solvent. The mold residence time depends on part geometry and varies from 10 min to an hour. Drying is dependent on the part and occurs outside the mold cavity in a matter of hours. The residual binder was removed at 350 to 500 °C. The process was originally developed for processing ceramics (e.g., alumina and silicon nitride) and has been extended to metal powder systems (e.g., 17-4 PH stainless steel and H13 tool steel). A sintered density of 7.58 g/cm^3 and hardness of 37 HRC have been reported for gelcast 17-4 PH stainless steel. Values for powder injection molded parts for the same material are 7.5 g/cm^3 and 35 HRC, respectively.

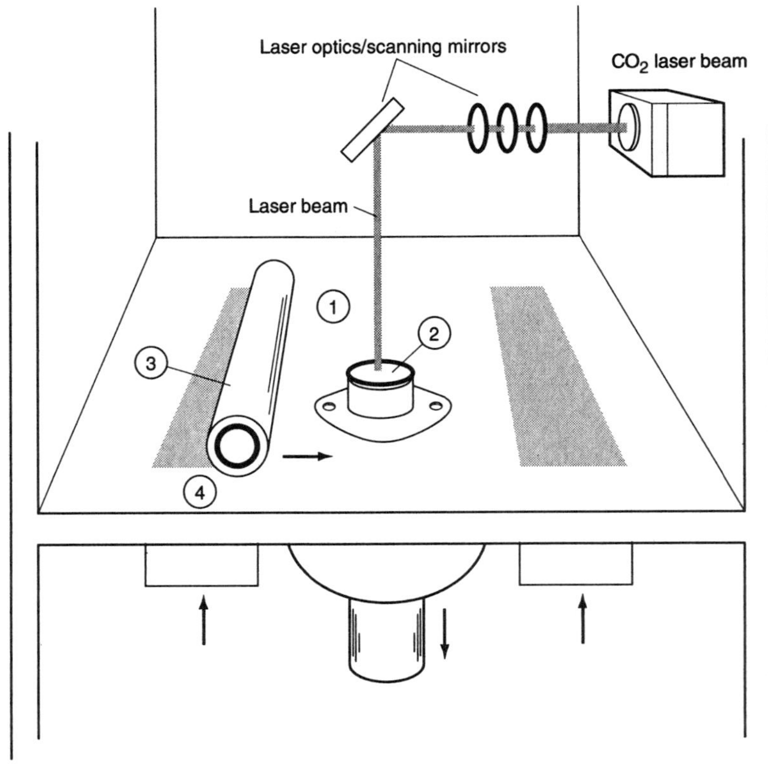

Fig. 10 Process for producing a selective laser-sintered tool. Numbered steps in the figure are described in the text. Source: Ref 3

Advantages

The lower shear and room-temperature conditions required for the mixing and molding of LPM feedstocks afford considerable savings in equipment and energy costs. In addition, the extent of wear of tooling and equipment is reduced, as is the accompanying contamination problems in the feedstock. Further, the low shear conditions during molding allow the use of inexpensive mold materials such as aluminum, plaster, epoxy resin, and silicone rubber. An added advantage of using soft tooling such as epoxy resin or silicone rubber is that the surface finish of the product is greatly enhanced. The low molding pressures imply that the forces required for clamping the mold are also reduced. Because a major portion (50 to 99%) of the binder removal is accomplished either during the molding cycle or by drying, the problems associated with the storage, handling, and disposal of solvents (in the case of solvent debinding methods) or corrosive acids (as in the case of catalytic debinding methods) present in powder injection molding are completely avoided. The residual binder is rapidly removed in a thermal debinding step. With the ease of debinding, carbon control, especially for the fabrication of thick sections is easier. Density variations of ±0.1% at the 3σ level has been reported for the LPM of silicon nitride in an automated press (Ref 29). Data on dimensional tolerance of parts produced by LPM are limited. One study reports a 3σ tolerance of 0.2 to 0.3% for sintered parts (Ref 8). More work related to dimensional tolerance and surface finish needs to be performed for other LPM methods.

Applications

Low-pressure molding methods can currently use a number of metal powders such as stainless steels, tool steels, composites, and superalloys. In cases where the metal surface is reactive toward the binder components (e.g., iron), a passivating coating using silanes has been applied prior to the mixing of the feedstock (Ref 27). Depending on the extent of automation, LPM techniques are ideally suited for situations involving low- to intermediate-volume runs (1 to 10,000). The rapid production of tooling is a very important application. Figure 9 shows an example of P/M tooling produced by LPM technology. Applications for LPM have been demonstrated in traditional P/M areas such as the aerospace, automotive, microelectronics, biomedical, and consumer industries.

ACKNOWLEDGMENT

The author wishes to thank Dr. Anthony Griffo for providing helpful suggestions during the preparation of the manuscript.

Selective Laser Sintering

Leander F. Pease III, Powder-Tech Associates, Inc.

Selective laser sintering (SLS), developed by DTM of Austin, Texas (Ref 3), uses laser energy to transform metal powders into useful tooling and parts. Figure 10 is a diagram of the first part in the process, wherein a loose, polymer coated metal powder is laser fused into the required shape. A CAD model of the part is numerically transformed into thin slices, 75 to 250 μm (0.003

to 0.010 in.). In the machine shown in Fig. 10, a very thin layer of heat-fusible powder is deposited on top of the build cylinder (step 1). Then, the powder layer is fused by the laser, as it traces out the computer-generated slice (step 2). The layer so formed is held together by the strength of the fused plastic coating on the 55 μm (0.002 in.) particles. The work stations lower this fused layer by 75 to 250 μm (0.003 to 0.010 in.), and a new layer of powder is rolled into place (step 3) atop the fused layer. Laser fusing of this second layer bonds the particles to each other and to the layer below. The support platform moves the part downward a layer at a time (step 4), and the process repeats itself until the part is fully formed, that is, built up layer by layer. The chamber size is 38 by 33 by 41 cm (15 by 13 by 16 in.); parts and molds 15 by 25 cm (6 by 10 in.) thick have been produced.

After the fusing step, the part has a green strength of about 2.7 MPa (400 psi) and requires care in handling. The unfused powder falls away from the fused part and is recycled. The fused green part is impregnated with acrylic resin containing a cross-linking polymer and then dried at 50 °C (122 °F). The dried part is then placed on an alumina plate in a graphite crucible. Blocks of copper are also placed in the crucible to serve as an infiltrant for the porous laser-formed part. The assembly is heated in 70%N_2-30%H_2 at 0.5 °C/min (1 °F/min) to 1120 °C (2048 °F). At 300 °C (572 °F) the polymer burns out, but the cross-linked polymer remains to strengthen the part and prevent slumping. At 700 °C (1292 °F) the iron particles begin to sinter and bond. At 1083 °C (1981 °F) the copper melts and runs laterally into the porous steel preform. It infiltrates from the sides and then runs up into the part by capillary action. There is enough copper present to fill all the pores, but the copper does not adhere to the alumina plate on which the infiltrated part now rests. The furnace then returns to room temperature, and the assembly is removed.

The infiltrated part contains about 60% Fe, 39% Cu, and 0.8% C. Its properties are:

Ultimate tensile strength, MPa (ksi)	475 (69)
Yield strength, MPa (ksi)	255 (37)
Elongation in 25 mm (1 in.), %	15
Hardness, HRB	75
Elastic modulus, GPa (10^6 psi)	207 (30)
Coefficient of thermal expansion, mm/mm · °C (10^6 in./in. · °F)	14.4 (7.99)

The thermal conductivity is important for injection molding tools and is 185 W/m · K (107 Btu/ft · h · °F) or about six times that of P20 steel. During infiltration, the part shrinks 2.5% linearly, and this is compensated at the initial part design by making it over size. The infiltrated parts have a tolerance of ±0.25 mm (±0.010 in.) over distances of 13 cm (5 in.). Walls can be as thin as 1 mm (0.040 in.). All parting lines and shutoffs for molding tools must be machined and therefore have 0.5 mm (0.02 in.) added at laser forming for that purpose. The as-infiltrated surface roughness is 0.25 μm (10 μin.) R_a and may be polished to 2.5 to 25 nm (0.1 to 1 μin.). Some polishing may also be done in the green state.

In all, the process takes about 5 days and results in a mold that is ready for plastic injection molding or MIM. Such a mold can produce 50,000 plastic parts in its useful lifetime. The main use for SLS via the DTM process is making tooling for the injection molding of plastics.

The infiltrated material produced by SLS does not match properties of pressed-and-sintered materials (such as in MPIF Standard 35), and so could not be used in functional prototype parts. In principle, other alloy steels could be selective laser sintered to form 60% dense preforms that would then be sintered and shrunk to the correct density to serve as P/M prototype parts. This would involve linear shrinkages of 7 to 10% with difficulties in maintaining tolerances. In general, parts sintered to such a degree would not have the same properties as the standardized P/M grades. (The prolonged sintering would result in erroneously high elongation and toughness.)

Laser-Based Direct Fabrication

Eric J. Whitney, Applied Research Laboratory, The Pennsylvania State University
John E. Smugeresky, Sandia National Laboratories
David M. Keicher, Optomec Design Company

Many rapid prototyping methods are under development, and some use metals or ceramics separately, while others are exploring using metals and ceramics together. In one process, metal powder can be deposited at a rate of about 0.1 to 10 g/min in precise amounts and with positional accuracy such as a net-shape part results. In another process, metal powder is deposited at a higher rate, about 10 to 100 g/min, resulting in a near-net-shape part. Both of the processes are completely computer controlled such that there are no molds used in the fabrication of a shape. Generally, the processing atmosphere is controlled to limit interstitial contamination from oxygen and nitrogen. The processes completely melt and fuse the metal powder, often achieving full theoretical density. These approaches eliminate or reduce the need for nonfunctional prototyping, such as the traditional trial-and-error prototyping approach using wood, clay, or plastic and the new rapid-prototyping (RP) processes that integrate computer-aided engineering, CAD, and CAM techniques with polymer deposition processes. These RP processes are also known as stereolithography or selective laser sintering (SLS).

During the period that RP was being developed, several laser laboratories experimented with using CNC-controlled laser-cladding systems to build structural parts (Ref 30–32). These laboratories concentrated on obtaining structurally sound material and addressed making only rudimentary shapes. The importance of this early work, however, was the demonstration that structurally useful material could be deposited using laser technology. More recently, several laboratories have taken advantage of the RP software concepts to push laser-based direct-fabrication technology to a much higher level of shape definition (Ref 33–38). While there are a number of terms used to describe this technological evolution, for example, rapid manufacturing (RM), flexible fabrication (FF), and solid free-form fabrication (SFF), they are laser-based direct-fabrication techniques for structural rather than model or surrogate materials.

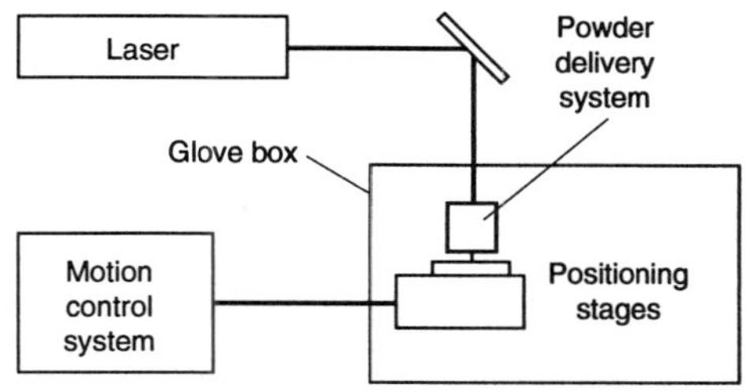

Fig. 11 Basic approach to laser-depositing metallic materials. Source: Ref 5

The material deposited by these processes must have structurally useful properties, such as high tensile strength, adequate ductility, good fracture toughness, and good fatigue life. At a minimum, the properties must be equivalent to conventionally cast material of the same composition. The potential of the process, however, is the ability to make materials that have unique characteristics, in shapes that are not constrained to have a single composition throughout the part.

For example, turbine engine parts often operate with severe temperature gradients. The limiting mechanical property in the high-temperature region may be creep while at the lower-temperature region the limiting property may be tensile or fatigue strength. In designing such a chemically homogeneous component manufactured using current processes, a compromise between creep resistance and high fatigue must be made. However, with the laser-based direct-fabrication techniques, parts are formed by depositing a small volume of material and building the part a layer at a time. This allows for composition changes on a scale equivalent to the width and height of the deposit such that gradients in the composition can be intentionally built into the part. Appropriate selection of the composition gradients provides the opportunity to adjust the properties as a function of position, creating so-called functionally graded materials. Thus, a design engineer may be able to increase the life of a component by metallurgically optimizing each region of the part for the anticipated thermomechanical environment. Both thermal profiling and composition profiling, in principle, can be engineered into a part to optimize its design.

Lasers are an ideally suited heat source for laser-based direct-fabrication processes. The la-

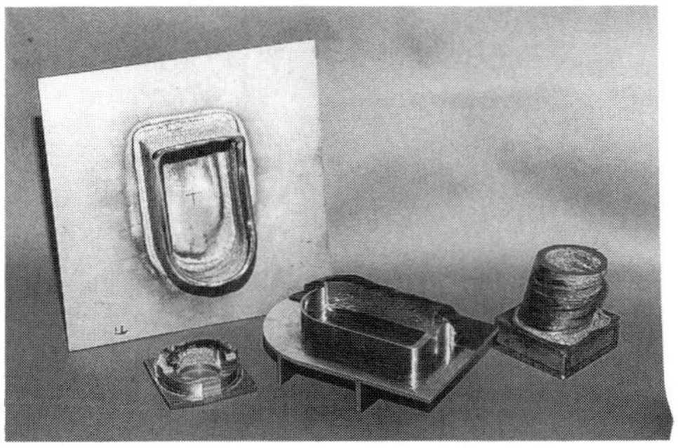

Fig. 12 Thick-walled Ti-6Al-4V laser-based direct fabrication. Source: Ref 41

Fig. 13 Fine-grained ductile 316 stainless steel produced under nonoptimized conditions. Left: optical macrostructure, showing layer geometry. Source: Ref 5. Right: transmission electron micrograph showing ultrafine grain size. Source: Ref 42

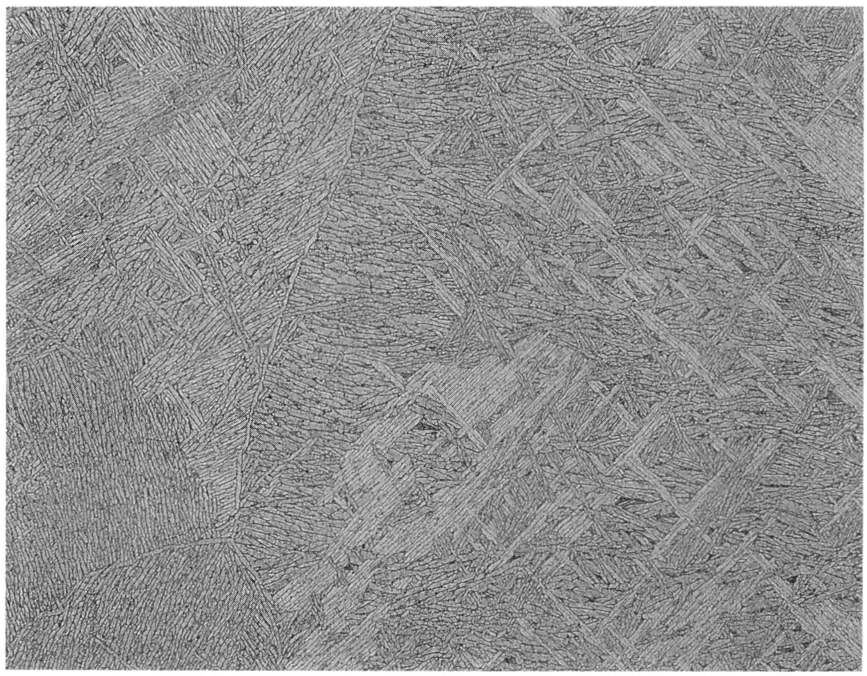

Fig. 14 Microstructure of laser-based direct-fabricated Ti-6Al-4V

ser provides an intense heat source that is easily configured into highly automated systems. Laser-based direct fabrication has much in common with laser cladding and surfacing techniques. Laser power level (energy density), powder deposition rate, and travel speed are parameters that can be scaled to suit the requirements for thermal profiling, size and composition of the deposit to control microstructure, mechanical properties, dimensional tolerances, and surface finish. Figure 11 (Ref 5) shows a schematic of a simple laser-based direct-fabrication setup.

Many types of lasers can be used, depending on the size and sophistication of the desired part and the type of material being processed (Ref 39). Selection of laser type depends on the light absorption, thermal transport, and thermodynamic properties of the material being deposited. For small intricate parts, a laser power of 1 kW or less can be used (Ref 40). For larger parts, such as those shown in Fig. 12, with thick-wall sections multikilowatt lasers must be used to achieve economical deposition rates. Adjustments to processing parameters are also required to meet dimensional and surface finish specifications.

Laser-Based Direct Fabricated Microstructures. A wide variety of microstructures, some of which are unique, can be achieved with laser-based direct fabrication, depending primarily on the power level of the laser utilized. At power levels less than about 5 kW, a small molten pool of material is created on a substrate, the added feedstock increases the size of the pool momentarily, and solidification occurs when the laser beam is traversed. Solidification rates are generally high enough, due to the heat-sink effect of the substrate, that metastable microstructures characteristic of rapid-solidification technology are readily generated. This corresponds to ultrafine, extended solid-solution microstructures capable of producing enhanced mechanical properties. In welding terms, cellular dendritic microstructures are produced where the secondary dendrite arm spacing can be used to estimate the solidification rate, with rates of approximately 10^4 s^{-1} being realized. Figure 13 (Ref 42) shows 316 stainless steel produced by two different laser-based direct-fabrication processes.

In making thick-section parts, high-power lasers in the 10 to 45 kW range are generally used. Here the solidification rates are slower, possibly approaching those of thin-walled castings. Figure 14 shows the microstructure for a laser-based direct-fabrication-processed Ti-6Al-4V using a 14 kW CO_2 laser and a thin-walled conventional casting.

The chemical and mechanical properties of laser-based direct-fabricated materials are excellent. The following alloys were deposited and tested: Ti-6Al-4V, Ti-5Al-2.5Sn, 316 stainless steel, alloy 625, alloy 690, alloy 718, H13 tool steel, and nickel-aluminum-bronze. Some representative data are presented here.

Ti-6Al-4V. It is well known that titanium alloys are subject to embrittlement by oxygen and nitrogen, impurities commonly introduced during conventional processing. Table 5 shows a comparison of the composition of laser-based direct-fabricated deposited Ti-6Al-4V and the starting powder composition. The data show that, under properly controlled conditions, the increase in oxygen and nitrogen can be held very low. Tensile and hardness properties of laser-based direct-fabricated Ti-6Al-4V material are also within ASTM B 367 Grade C5.

The data shown in Tables 5 and 6 are from laser-based direct-fabricated Ti-6Al-4V material made from prealloyed hydride-dehydride powder and using a 14 kW CO_2 laser. After forming, the material was heat treated in vacuum at 788 °C (1450 °F) for 2 h. Fatigue testing has shown that laser-based direct-fabricated Ti-6Al-4V is

Table 5 Chemical composition of laser-based direct-fabricated Ti-6Al-4V and starting powder

Sample(a)	Composition, wt % O	N	H	Fe	Al	V
Starting powder	0.23	0.032	0.012	0.199	6.01	3.95
Laser-based direct-fabricated material(b)	0.23	0.048	0.0072	NC	5.83	3.75

NC, not checked. (a) Powder sample average of two tests. (b) Vacuum mill annealed

Table 6 Tensile properties of laser-based direct-fabricated Ti-6Al-4V

Test	ASTM B 367 C5	Laser-based direct-fabricated Ti-6Al-4V
Hardness, HRC	39 max	36
Ultimate tensile strength, MPa (ksi)	896 (130)	1007 (146)
Yield tensile strength, MPa (ksi)	827 (120)	876 (127)
Elongation, (in 25.4 mm, or 1 in.)	6 min	8.5

equivalent to cast and hot isostatically pressed material.

Alloy 718 is a widely used γ''-strengthened superalloy that is available in cast and wrought forms. The alloy generally exhibits good weldability. Table 7 shows the 538 °C (1000 °F) tensile properties of laser-based direct-fabricated alloy 718 after postform aging.

Alloy 625 and 316 stainless steel are austenitic materials that are based on nickel and iron, respectively. The alloys are available in wrought and cast forms. Alloy 625 is a high-temperature material that is corrosion and oxidation resistant. It is used in the chemical, petroleum, paper, nuclear, and aerospace industries and is generally regarded as the most widely used nickel-base alloy in the world. At lower temperature, 316 stainless steel is used in many of the same industries as alloy 625. Both alloys represent a significant portion of the premium metals market.

Excellent properties for 316 stainless steel, where yield strengths double that of wrought annealed bar with no sacrifice in ductility, are routinely achieved by laser-based direct-fabrication processing (Table 8) (Ref 42). The properties of similarly processed alloy 625 exceed those of annealed bar. Properties are dependent on processing parameters, and work on 316 stainless steel demonstrated that mechanisms as simple as grain refinement over conventionally processed materials lead to Hall-Petch strengthening. Figure 14 shows a grain size of approximately 10

Table 7 Tensile properties of laser-based direct-fabricated Alloy 718

Test	Typical cast value	Laser-based direct-fabricated alloy 718
Ultimate tensile strength, MPa (ksi)	860 (125)	931 (135)
Yield tensile strength, MPa (ksi)	725 (105)	793 (115)
Elongation, %	9	15

μm compared to 50 μm for wrought 316 stainless steel. Hardness values have been reported (Ref 42) from 180 to 232 on the Knoop scale, equivalent to Rockwell B values of 85 to 96. Rockwell B values in the 80 to 85 range for stainless steels correspond to tensile yield strengths in annealed bar of about 240 MPa. Tensile yield strengths greater than 480 MPa have been reported (Ref 43) for similarly processed material, for samples with Knoop hardness in the 230 to 240 range (96 HRB). In that study, the grain size within the fabricated structures was reported to be on the order of 5 to 10 μm, whereas the grain size for the annealed 316 stainless steel is typically around 100 μm. This difference in grain size is believed to be the primary cause of the improved strengths for the laser-based direct-fabricated structures and is consistent with the Hall-Petch relationship between yield strength and grain size. Consequently, full-size parts can be made with a wide range of strengths.

H13 Tool Steel. The fabrication and maintenance of tools is a huge cost element for many manufacturing operations. Laser-based direct-fabrication of tools is of immense economical importance. Utilizing the unique metallurgical qualities of laser-based direct fabrication to increase the useful life of tools may also reduce costs. H13 tool steels processed by laser-based direct fabrication can have both a coarse and fine structure (Ref 44), depending on processing conditions. Figure 15 shows the tempering response of laser-based direct-fabricated H13 as compared to a conventionally prepared alloy of the same composition.

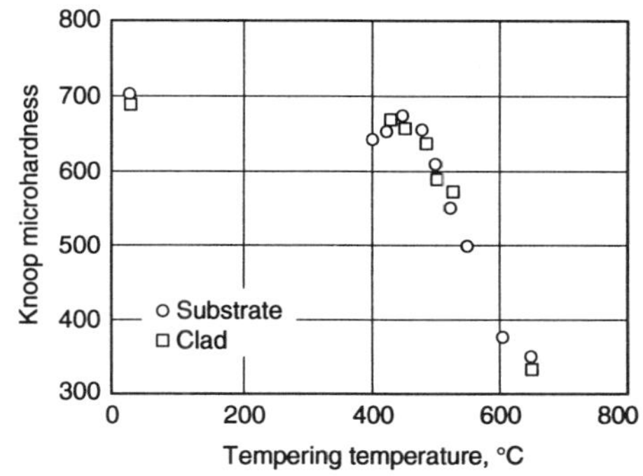

Fig. 15 Tempering response of H13 tool steel

Table 8 Mechanical tensile test data for laser-based direct-fabricated 316 stainless steel and alloy

Plane orientation with respect to tensile direction	Ultimate strength MPa	ksi	Yield strength MPa	ksi	Elongation (in 25 mm, or 1 in.)
316 stainless steel perpendicular	790	115	450	65	66
316 stainless steel parallel	805	117	590	86	33
316 stainless steel anneal bar	585	85	240	35	50
Inconel 625 parallel	930	135	635	92	38
Inconel 625 perpendicular	930	135	515	75	37
Inconel 625 annealed bar	835	121	400	58	30

REFERENCES

1. *Powder Metallurgy Design Manual,* 2nd ed., Metal Powder Industries Federation, 1995
2. E. Sachs, S. Allen, et al., "Three Dimensional Printing, a Desktop Manufacturing Process," presented at ANTEC '95 (Boston), 7–11 May 1995
3. Product literature, DTM Corp., Austin, TX, 1997
4. G.K. Lewis, "Directed Light Fabrication," ICALEO 1994 Conf. (Orlando, FL), 1994
5. D.M. Keicher et al., The Laser Forming of Metallic Components Using Particulate Materials, *JOM,* May 1997, p 51–54
6. J. Mazumder, private communication
7. F.G. Arcella, E.J. Whitney, et al., Materials Characterization of Laser Formed Titanium, *Powder Metallurgy and Particulate Materials Conf., PM2TEC96,* Metal Powder Industries Federation, 1996
8. B.E. Novich, C.A. Sundback, and R.W. Adams, Forming Science and Technology for Ceramics, *Ceram. Trans.,* Vol 26, M.J. Cima, Ed., American Ceramic Society, 1991, p 157
9. W.P. Benjamin, *Plastic Tooling: Techniques and Applications,* McGraw Hill, 1972
10. F.W. Billmeyer, *Textbook of Polymer Science,* 3rd ed., John Wiley & Sons, 1994
11. Technical literature, Ciba Geigy
12. J.H. DuBois and W.I. Pribble, *Plastics Mold Engineering Handbook,* 4th ed., Van Nostrand Reinhold, 1987
13. K. Stoeckhert, *Mold-Making Handbook for the Plastics Engineer,* Hanser, 1983
14. P.F. Jacobs, *Rapid Prototyping and Manufacturing: Fundamentals of Stereolithography,* Society of Manufacturing Engineers, 1992
15. E.A. Sheard, *Elastomerics,* Vol 112 (No. 10), 1990, p 49–57
16. M. Wilson and M. Yeung, *Rapid Prototyping,* Vol 2 (No. 1), 1996, p 20
17. R.M. German, *Powder Metallurgy Science,* 2nd ed., Metal Powder Industries Federation, 1994
18. R.M. German, *Powder Injection Molding,* Metal Powder Industries Federation, 1990

19. A.B. Metzner, *J. Rheology,* Vol 29, 1985, p 739
20. H.A. Barnes and S.A. Holbrook, *Processing of Solid-Liquid Suspensions,* P.A. Shamalou, Ed., Butterworth-Heinemann, 1993, p 222
21. P.E. Rempes, B.C. Weber, and M.A. Schwartz, *Ceram. Bull.,* Vol 37, 1958, p 334
22. S.V. Atre, L. Cai, and R.M. German, *Advances in Powder Metallurgy and Particulate Materials,* Vol 19, Metal Powder Industries Federation, 1996, p 243
23. A. Griffo, K. Sivaraman, T. Potter, J.A. Thomas, and R.M German, *Advances in Powder Metallurgy and Particulate Materials,* Metal Powder Industries Federation, 1997
24. R. Moreno, *Am. Ceram. Soc. Bull.,* Vol 71, 1992, p 1521, 1647
25. L. Cai and R.M. German, *Advances in Powder Metallurgy and Particulate Materials,* Vol 4, Metal Powder Industries Federation, 1994, p 157
26. R.M. German, *Injection Molding of Metals and Ceramics,* Metal Powder Industries Federation, 1997
27. G.M. Brasel, S.J. Brasel, and J.E. Duke, *Powder Injection Molding—1992,* P.H. Booker, J. Gaspervich, and R.M. German, Ed., Metal Powder Industries Federation, 1992, p 67
28. M.A. Janney, W. Ren, and S. Viswanathan, *Advances in Powder Metallurgy and Particulate Materials,* Vol 17, Metal Powder Industries Federation, 1996, p 227
29. J.A. Mangels, *Am. Ceram. Sci. Bull.,* Vol 73 (No. 5), 1994, p 37
30. F. Arcella and G. Lessmann, U.S. Patent 4,818,465, 1989
31. C.O. Brown, E.M. Breinan, and B.H. Kear, U.S. Patent 4,323,756, 1982
32. V. Pratt, W. Scheidt, and E. Whitney, U.S. Patent 5,038,014, 1991
33. D.M. Keicher, J.L. Jellison, L.P. Schanwald, et al., *SAMPE 1995,* 1995, p 1009–1018
34. D.M. Keicher, J.E. Smugeresky, J.A. Romero, et al., *SME Conf.,* Society of Manufacturing Engineers, 1996
35. G.K. Lewis, R. Nemec, J. Milewski, et al., *ICALEO 1994 Conf.,* Vol 17f, 1994, p 17
36. G.K. Lewis, J.O. Milewski, R.B. Nemec, et al., Los Alamos National Laboratory, Los Alamos, NM, 1995
37. J. Mazumder, J. Koch, K. Nagarthnam, et al., *Powder Metallurgy and Particulate Materials Conf.,* Vol 15, *Intermetallics and Lasers,* Metal Powder Industries Federation, 1996, p 12
38. M. McLean, F. Shannon, and W. Steen, *Powder Metallurgy and Particulate Materials—1997, PM2TEC97,* Metal Powder Industries Federation, 1997
39. S.S. Charschan, Laser Institute of America, Orlando, FL, 1993
40. G.K. Lewis, D.J. Thoma, R.B. Menec, et al., *Powder Metallurgy and Particulate Materials Conf.,* Vol 15, *Intermetallics and Lasers,* Metal Powder Industries Federation, 1996, p 15-65, 15-76
41. J.T. Schriempf, E.J. Whitney, P.A. Blomquist, et al., *Powder Metallurgy and Particulate Materials—1997, PM2TEC97,* Vol 3, R.A. McKotch and R. Webb, Ed., Metal Powder Industries Federation, 1997, p 21-51, 21-59
42. J.E. Smugeresky, D.M. Keicher, J.A. Romero, et al., *Powder Metallurgy and Particulate Materials—1997, PM2TEC97,* Vol 3, R.A. McKotch and R. Webb, Ed., Metal Powder Industries Federation, 1997, p 21-33, 21-42
43. D.M. Keicher, J.A. Romero, C.L. Atwood, et al., *Powder Metallurgy and Particulate Materials—1996, PM2TEC96,* American Powder Metallurgy Institute, 1996
44. J. Mazumder, J. Choi, K. Nagarathnam, et al., *JOM,* 1997, p 55–60

Consolidation Principles and Process Molding

Revised by Randall M. German, Pennsylvania State University

CONSOLIDATION of metal powders generally involves sintering, whereby an assembly of loose or compacted particles metallurgically bond into a coherent body at elevated temperatures. During the process, sintering powder particles form strong bonds and alloy any admixed elements. Temperatures below the melting point of the major constituent are usually used, although liquid phase sintering is the most common technique in production.

Although sintering has been the subject of extensive experimental and theoretical work, there is no universally applicable theory of sintering. Much is not understood about the process, and no single applicable explanation or understanding of the process has emerged. Sintering generally consists of six distinct, but not necessarily sequential, stages:

- Initial particle bonding
- Neck growth
- Pore rounding
- Pore channel closure
- Densification and pore shrinkage
- Pore coarsening

Material transport mechanisms involved include surface diffusion, volume diffusion, evaporation and condensation, and grain boundary diffusion. Several factors affect sintering. However, the most significant are time and temperature, with temperature being the most important single variable. Particle size, compact porosity, and prealloying also affect the sintering process.

This article discusses the important changes in density, microstructure, and mechanical properties that occur when metal powder compacts in the as-pressed condition (green compacts) are sintered. These changes are described phenomenologically and from the point of view of the driving forces and the material transport mechanisms that are important in the modeling and characterization of changes in density, microstructure, and mechanical properties from sintering green compacts. The mechanisms of homogenization during sintering and of activated and liquid phase sintering also are examined. In addition, process modeling is discussed at the end of this article.

Changes in the dimensions and density of compacts are important considerations in sintering. Changes in density during sintering are primarily due to changes in phases, in alloying, or in volume (dimensions), but changes in mass, which may be caused by the loss of lubricant added during compacting, volatilization of a component such as zinc from brass, and the reduction of oxide skins on the powder particles during sintering should also be considered.

In studies of the fundamentals of sintering, dimensional changes are determined as the difference in dimensions between green and sintered compacts, parallel and perpendicular to the direction of pressing, and are expressed in percent of the green dimensions. From a technological standpoint, the dimensional change during sintering of greatest importance is the difference between the dimensions of the die and the sintered compact in the radial direction, that is, the direction perpendicular to that of pressing.

Another approach to studying shrinkage and densification during sintering is to determine the dimensional changes with a dilatometer. A compact is heated at a constant rate to a given sintering temperature and then cooled at this rate from the maximum temperature to room temperature. For example, a dilatometer curve is shown in Fig. 1. This curve represents the dimensional changes in the axial direction of the same type of copper compacts as Fig. 2. Compacts were pressed at a pressure of 138 MPa (20,000 psi) and were heated and cooled at a uniform rate of 3.9 °C/min (7 °F/min). This type of experiment is particularly well suited for studies of the rate of densification as a function of temperature.

Consolidation of Compacts

Consolidation of metal powders generally entails compacting, followed by sintering. However, metal powders sometimes are sintered without compacting, which is called "loose powder sintering." This process is suitable for certain specialized P/M applications such as bronze filters and porous electrodes, as well as for model experiments to study the mechanism of sintering. Additionally, all powder injection molded structures begin sintering from this same condition once the binder is removed.

The simplest type of sintering is for compacts of a single metal powder, or compacts of homogeneous solid-solution alloy powders (alpha brass powder or an austenitic stainless steel powder). However, powder metallurgy alloys also can be produced by blending two or more constituent powders in proportions required to

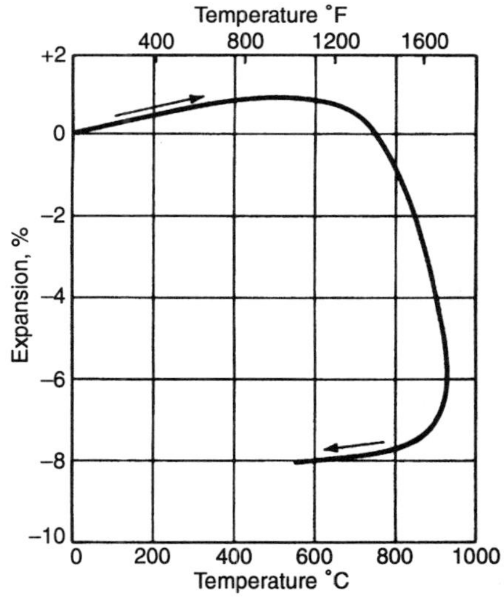

Fig. 1 Dilatometer curve of dimensional changes in a copper P/M compact from sintering. The first portion of the curve up to 315 °C (600 °F) is identical to that of a specimen of solid copper and represents the thermal expansion of the metal with increasing temperature. When sintering begins, shrinkage counteracts thermal expansion, and the curve deviates from that for solid copper. As the temperature increases, the rate of shrinkage increases and the compact contracts. Immediately after reaching the maximum sintering temperature of 925 °C (1700 °F), the compact is cooled at the same rate at which it was heated. During cooling, contraction initially is faster than normal thermal contraction. Eventually, the dilatometer curve slope follows normal contraction. Compacts from −74+43 μm copper powder pressed at 138 MPa (20,000 psi) and heated at a rate of 3.9 °C/min (7 °F/min) to 925 °C (1700 °F) and then cooled at the same rate.

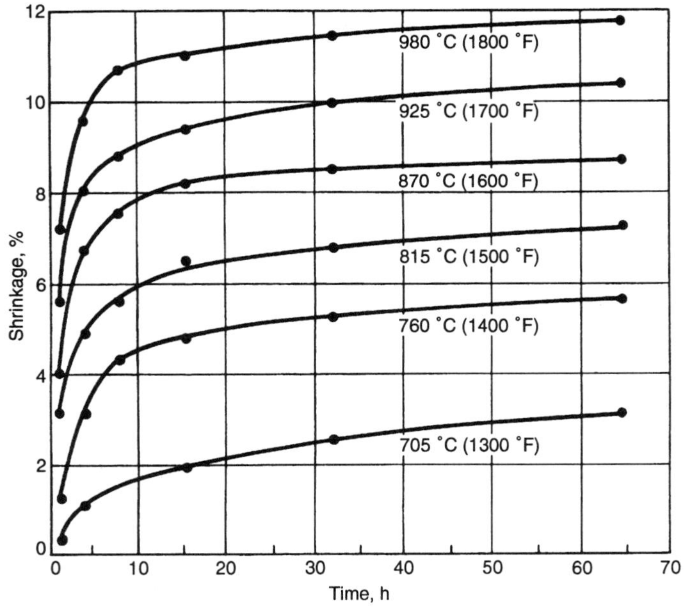

Fig. 2 Shrinkage of copper powder compacts from –74+43 μm sieve fraction. Compacted at 138 MPa (20,000 psi) and sintered at various temperatures as a function of sintering time

achieve the desired overall alloy composition. Constituent powders of the mixture may be composed of pure elements (elemental powders) or alloys containing two or more elements (master alloy powders). If a homogeneous alloy is to be produced from a mixture of powders, sintering must be controlled to achieve the required degree of homogenization.

Another alloying method is the use of prealloyed powders in which each powder particle has the same composition as the final part produced by compacting and sintering. Such prealloyed powders typically are produced by atomization of a molten alloy by impingement of a high-pressure gas or liquid stream. The fluid disperses the liquid alloy into small droplets to solidify into powder particles.

Homogenization of elemental powders occurs through interdiffusion of the chemical elements among the constituent powders of different composition. Often, it is a solid-state diffusion process. Complete homogenization during sintering is achieved during the sintering of mixtures of iron and graphite powder, where graphitic carbon goes into solid solution in gamma iron (austenite) at the sintering temperature.

Homogenization circumvents some of the disadvantages of prealloyed powder processing. Hardness and strength of prealloyed powder particles result in powder compacts with low densities and low green strengths when pressed at room temperature. Mixture of powder with the major component of the alloy present as an elemental powder (usually soft and ductile) typically exhibit high densities and high green strengths in the as-compacted condition. Furthermore, adjustments to alloy composition can be made readily by varying the proportions of constituent powders in the mixture, thus eliminating the need for additional powders of different composition. Mixtures composed of elemental powders normally can be obtained with high purity. Prealloyed powders, conversely, may contain elements other than the main components of the alloy because of pickup of impurities during melting prior to atomization.

The disadvantages of homogenization processing center around the need to remove compositional heterogeneities by thermal processing (often in conjunction with mechanical working) subsequent to compaction. Homogenization often involves more extensive elevated-temperature exposure than normal sintering; this added processing expense must be minimized by limiting processing conditions to levels that provide the required homogeneity. Also, in many cases, one element diffuses much faster as compared with the other, resulting in swelling and dimensional growth during sintering.

Activated Sintering. Sometimes, compacts pressed from a mixture of a high-melting-point metal with a small addition of a lower melting point metal are sintered, such as compacts of tungsten powder to which small amounts of nickel powder have been added. During sintering at temperatures below the melting points of both metals, the compacts densify at much lower temperatures than those of the high-melting-point metal. This is one type of activated sintering (Ref 2), which is discussed later in this article and elsewhere in this Volume.

Liquid-Phase Sintering. Compacts of mixtures of metal powders also may be sintered at temperatures that form a certain amount of liquid phase. This results in a much faster process, which is very popular in industry, because of the decreased sintering times. The process is widely used in sintering iron, copper, tungsten, cobalt, and cemented carbide compositions. The quantity of the liquid phase must be small enough to be held by capillary force within the skeleton of the remaining solid phase so that the compacts retain their shape. In this process, the liquid phase may be present through the duration of the sintering cycle. This is the case in the sintering of compacts from a mixture of tungsten, nickel, and iron powders (the so-called "heavy alloys"), and in the sintering of cemented carbides (combinations of tungsten and other refractory metal carbides with cobalt). Because the solution of small particles of the solid phase in the liquid and its reprecipitation on the large particles during sintering plays an important role, this method often is referred to as the solution and reprecipitation process of liquid-phase sintering.

In transient liquid-phase sintering, however, a liquid is formed when the compact is heated to the sintering temperature. The liquid is transformed into a solid by interdiffusion while the compact is at the sintering temperature, such as in the sintering of compacts from mixtures of copper and tin powders, in which an alpha bronze solid solution is formed.

Pressure-Assisted Sintering. The simplest form of pressure-assisted sintering is that which occurs by uniaxial hot pressing. A refractory mold (die) is inserted into a furnace, and both the powder and the die are heated while the punches are driven by an external pressure source. Hot isostatic pressing is another widely used process. This process is also called gas pressure bonding, because a high-pressure gas assists densification. In hot isostatic pressing, the powder is enclosed in a thin-walled container that is flexible but pressure-tight at sintering temperatures. A pressurized gas is uniformly applied to the container in a high-pressure chamber. The chamber contains a furnace to heat both the material and the gas. The hot consolidation of metal powders, where sintering and densification are combined by applying pressure to the powder at elevated temperature, is not discussed in detail in this article. Hot consolidation processes are discussed elsewhere in this Volume.

Shrinkage and Densification

During sintering, most compacts pressed from a single-phase metal powder shrink rather than expand. The higher the sintering temperature, the longer the sintering time, the smaller the particle size of the powder from which the compacts are pressed, and the lower the green density of the compacts, the greater the shrinkage. Compacts from powders of body-centered cubic metals (molybdenum and tungsten) of a given particle size pressed to a given density and sintered a given length of time at a given homologous temperature (ratio of sintering temperature to melting temperature of the metals in degrees kelvin) show higher shrinkage than compacts from face-centered cubic metals (copper and nickel) sintered under corresponding conditions.

Compacts made of iron powder generally are sintered in the temperature range where iron has the gamma (face-centered cubic) structure. One reason is because carbon, an important strength-

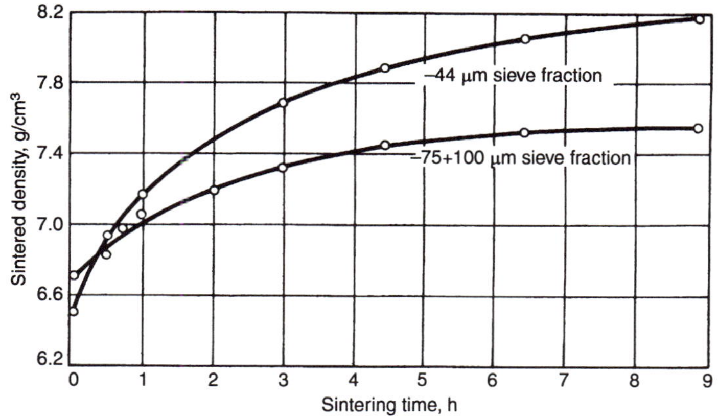

Fig. 3 Density of compacts from electrolytic copper powder. Two particle size fractions: –75+100 μm and –44 μm sintered at 865 °C (1590 °F) as a function of sintering time

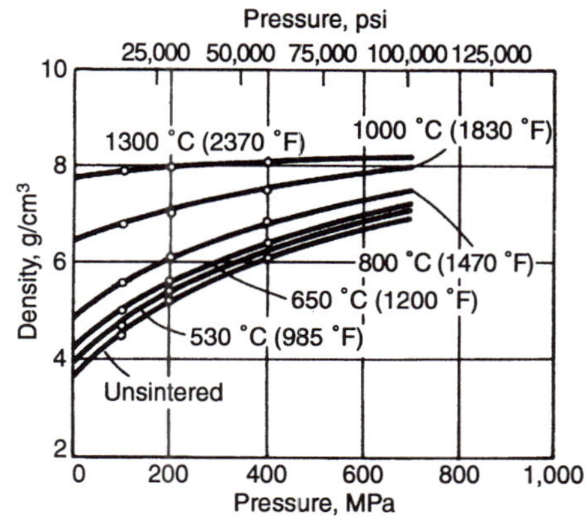

Fig. 4 Density of carbonyl nickel powder compacts. Sintered at a series of temperatures for 2 h as a function of compacting pressure

ening element, is soluble in the face-centered cubic form of iron, but is insoluble in the body-centered cubic phase of iron. Compacts made of iron powder with conventional commercial particle size distribution, pressed to green densities in the range from 6.6 to 7.1 g/cm^3 and sintered at temperatures ranging from 1100 to 1200 °C (2010 to 2190 °F) for ½ to 1 h, experience minimal shrinkage, often less than 1% with some alloying additions the compacts will expand. Despite this small amount of shrinkage, iron powder compacts are sintered adequately with regard to their mechanical and physical properties. This is important in the fabrication of structural parts from iron and steel because of the need for control of dimensions. Large amounts of shrinkage during sintering may be undesirable, thus rendering control of final dimensions of the sintered part difficult.

Shrinkage of copper powder compacts from a –74+43 μm sieve fraction, pressed at 138 MPa (20,000 psi), as a function of sintering time and temperature is shown in Fig. 2. These compacts were heated rapidly to the isothermal temperature and, after being held at this temperature, cooled rapidly to room temperature. These data represent the "isothermal" shrinkage of the compacts, even though shrinkage during heating and cooling is included. As shown in Fig. 2, shrinkage of the copper powder compacts is primarily a function of sintering temperature. The higher the temperature, the greater the shrinkage. Shrinkage also increases with increasing sintering time. The rate of shrinkage is initially quite high, but then decreases with increasing sintering time. The higher the sintering temperature, the more rapid the decrease in shrinkage rate. Accordingly, high sintered densities can be obtained more readily by increasing sintering temperature than by increasing sintering time.

In other P/M applications, large amounts of shrinkage are desirable to obtain sintered densities near those of the solid metal. For instance, compacts from very small tungsten powder (body-centered cubic) sintered at temperatures near the tungsten melting point increase in density from 11 g/cm^3 in the green condition to 18 g/cm^3 in the sintered condition; that is, they shrink 15% linearly. Because the sintered compacts are subsequently swaged and drawn into wire, close control of sintered dimensions is not required.

Effect of Powder Size on Shrinkage. Another factor affecting shrinkage is the particle size of the powder. As shown in Fig. 3, an increase in density during sintering of copper powder compacts pressed at 276 MPa (40,000 psi) and sintered at 865 °C (1590 °F) is a function of sintering time. The rate of densification of compacts from –44 μm powder is considerably faster than that of compacts from a –75+100 μm sieve fraction.

Effect of Green Density on Sintered Density. A final factor affecting the densification of compacts from a single-phase metal powder is the green density of the compact or the pressure at which the compact is pressed, which determines green density. As shown in Fig. 4, density of compacts from carbonyl nickel powder is plotted as a function of the compacting pressure for green compacts and for compacts sintered at five different temperatures after pressing. The higher the compacting pressure, the smaller the sintering shrinkage, or the change from green to sintered density. However, the sintered density is much improved by the higher green density.

Expansion during sintering is common in compacts made of soft metal powders, such as copper, that are pressed at high pressures. This expansion is due to gases entrapped in closed-off pores of the compact. Gases may be entrapped during compacting or may be formed by chemical reactions, such as the formation of water vapor by reaction of oxide skins on the surface of the particles with the hydrogen of the sintering atmosphere which diffuses into the closed pores. Such expansion or growth generally is not observed in compacts that are made of harder metal powders.

Microstructure

The microstructure of green metal powder compacts reveals the outlines of the original powder particles and the pores. Depending on the particle size distribution of the powder from which the compact is pressed, powder particles

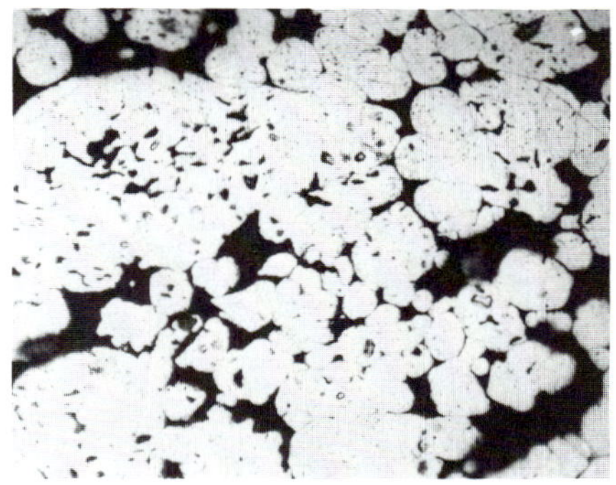

Fig. 5 Microstructure of green compact. Pressed from copper powder at 180 MPa (26,000 psi). Unetched. 450×

and pores that are visible in the microstructure may have narrow or wide size distributions. If the powder particles are polycrystalline, grain boundaries may be observed within the particles. Particles may be flattened and distorted, depending on the nature of the powder and the compacting pressure. Typically, uniaxial compaction creates flattened pores that subsequently contribute to anisotropic dimensional change in sintering.

The microstructure of a green compact pressed from copper powder at a pressure of 180 MPa (26,000 psi) in the unetched condition is shown in Fig. 5. This microstructure indicates that the copper powder had a fairly wide particle size distribution and that some of the particles had interior porosity. Most of the boundaries between particles can be readily discerned. When this compact was sintered 15 min at 705 °C (1300 °F), a temperature too low for adequate sintering, the microstructure shown in Fig. 6 was obtained. The principal difference between the microstructures of the green compact and the one sintered at 705 °C (1300 °F) is in the shape of the pores within the powder particles, which have become rounded during sintering. The sintering temperature is too low to produce adequate bonding between powder particles; however, significant strengthening occurs by this temperature.

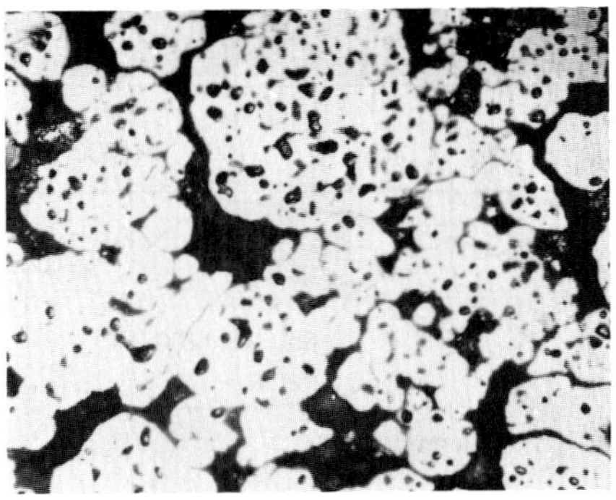

Fig. 6 Microstructure of copper powder compact. Pressed at 180 MPa (26,000 psi) and sintered 15 min at 705 °C (1300 °F). Unetched. 450×

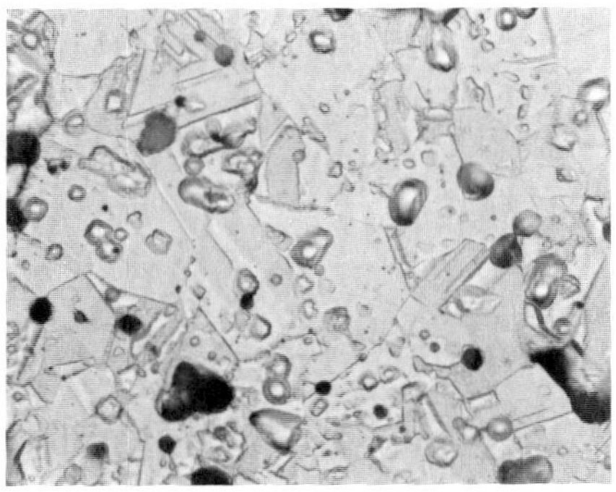

Fig. 7 Microstructure of copper powder compact. Pressed at 180 MPa (26,000 psi) and sintered 30 min at 980 °C (1800 °F). Potassium dichromate etch. 500×

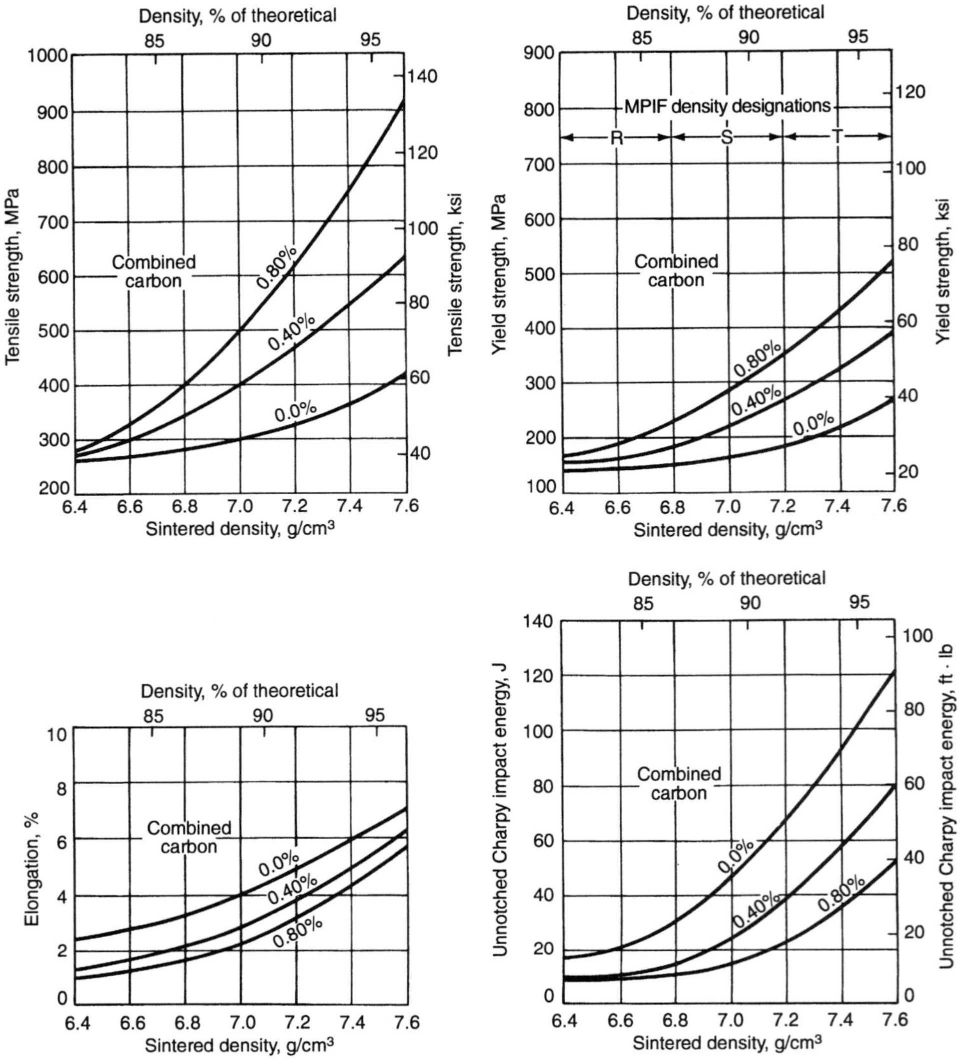

Fig. 8 Effects of density on mechanical properties of as-sintered 4% nickel steel (FN-04XX)

Figure 7 shows the microstructure of the copper powder compact sintered 30 min at 980 °C (1800 °F). The sample was etched to reveal the grain boundaries in the microstructure. Boundaries between particles can no longer be discerned; instead, a network of grain boundaries similar to those in wrought and annealed copper is illustrated. Pores of different sizes filled with the medium used to mount the specimen for microstructural examination are seen. These pores are no longer cusp-shaped and irregular, but have become rounded. In compacts made of small powders, complete spheroidization of the pores often is observed.

The transition in microstructure from that of a green compact to that of a well-sintered compact is a function of sintering temperature and time. During annealing of cold worked wrought material, a recrystallization temperature is reached, where nuclei of strain-free grains are formed that grow into the recrystallized structure. Subsequently, grain growth of the recrystallized structure occurs. This temperature for recrystallization and grain growth is a function of the time during which the cold worked material is annealed and the amount of prior cold work. The temperature range in which the typical microstructure of a well-sintered powder compact is developed is much higher than the recrystallization temperature of the wrought material. Compacting pressure frequently has a relatively minor effect on the temperature range at which the sintered structure is developed.

Development of the typical well-sintered structure of a metal powder compact requires that grain growth occur across prior particle boundaries. This grain growth is restricted until material transport that occurs during sintering has progressed to a point where a substantial increase in the contact area between particles has taken place. This increase in contact area is impeded by the network of pores. It is not responsive to the effects of strain hardening during compacting, which are relieved at temperatures below those at which the development of a sintered microstructure is observed. The principal effect of higher compacting pressures on grain growth during sintering is to facilitate extensive contact between particles.

Mechanical Properties

Mechanical properties of P/M structural materials depend on the composition, density, and heat treatability of the material, as well as processing and design considerations. Final density, which is expressed in g/cm^3 as well as percent of theoretical density—the ratio of the density of a P/M material to that of its wrought counterpart—has the greatest effect on properties of P/M materials. Powder metallurgy parts with theoretical densities less than 75% are considered to be low density, those above 90% are high density, and those in between are classified as medium density. Generally, structural parts have densities ranging from 80% to above 98%. Oil-impregnated bearings have densities of approximately 75%; densities of filter parts can be as low as 35%.

As shown in Fig. 8, mechanical properties generally improve with increases in density. Figure 8 illustrates the effects of density on the mechanical properties of as-sintered nickel steels. The data of Fig. 8 apply only when electrolytically produced iron-base powder is used and the processing includes presintering and repressing of all materials above 6.6 g/cm^3. Other iron powders and processing methods produce different results. One option is based on specified minimum tensile properties, where the fabrication suppliers may select suitable powders, density, and processing to meet minimum

Table 1 Typical density designations and ranges of ferrous P/M materials

Density of pure iron is 7.87 g/cm³.

Designation MPIF density suffix	ASTM type(a)	SAE type	Density, g/cm³
N	I	1(b)	Less than 6.0
P	II	2	6.0 to 6.4
R	III	3	6.4 to 6.8
S	IV	4	6.8 to 7.2
T	V(c)	5(c)	7.2 to 7.6
U	...	...	7.6 to 8.0

(a) ASTM B 426 only; different density ranges used in ASTM B 310 and B 484. (b) Density range of 5.6 to 6.0 g/cm³ is specified. (c) Minimum density of 7.2 g/cm³ is specified.

strength levels. The density designations in Fig. 8(b) are defined in Table 1.

Transverse-Rupture Strength. Changes in mechanical properties of compacts that occur during sintering may be illustrated by comparing the properties of green compacts with those of sintered compacts. The strength of green compacts usually is determined by the transverse-rupture strength of rectangular compacts according to the American Society for Testing and Materials standard ASTM B 312 and the Metal Powder Industries Federation standard MPIF 15. The level of strength determined by this test is quite low, generally below 15 MPa (2000 psi). Strength must be adequate enough to allow handling of green compacts during transfer from the compacting press to the sintering furnace.

The transverse-rupture strength of sintered metal powder specimens may be determined by a test very similar to the green strength test. The test is only applicable to brittle P/M materials and is often mistakenly applied to ductile alloys, leading to high, but invalid values. It is described in ASTM B 528 and MPIF 35. Strength values obtained are much higher than those for green specimens. The values vary, depending on the type of material (lower for weak metals and alloys and higher for strong metals and alloys) and specimen density (ratio of sintered compact density to that of the solid material). The highest values will be achieved for the densest sintered compacts. The sintered strength values are usually at least 20 times the transverse-rupture strength of green compacts. For many sintered materials, transverse-rupture strength is 1.5 to 2.2 times the ultimate tensile strength.

Although transverse-rupture strength is not a design consideration, it is often used to measure the strength of specimens from a lot of metal powder and to compare its strength with a standard strength value. The transverse-rupture strength test of sintered metal powder specimens is applicable only to a relatively brittle material whose permanent deflection, as a result of testing, is less than 0.25 mm (0.010 in.). However, low-density, high-carbon P/M steels generally are relatively brittle. The transverse-rupture strength test is therefore applicable to such P/M steels.

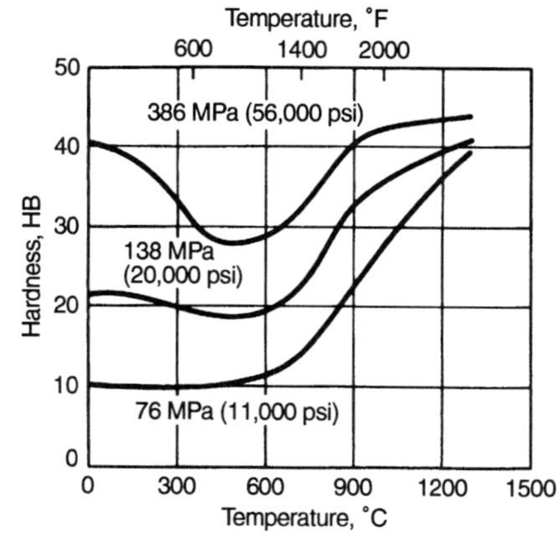

Fig. 9 Brinell hardness of compacts from carbonyl iron powder. Pressed at 76, 138, and 386 MPa (11,000, 20,000, and 56,000 psi) and sintered for 2 h as a function of sintering temperature

Strength and Toughness. Materials with "near-full" density have strength properties comparable to those of wrought materials of the same composition. Strength decreases with decreasing density, but the rate of decrease is not nearly as rapid as the decrease in ductility (elongation and reduction in area) and particularly impact energy. These properties are generally much lower than properties of corresponding wrought materials. Only sintered materials with densities that closely approximate wrought materials (on the order of 99% or more of theoretical density) have ductility and impact energy properties that approach properties of corresponding wrought materials. However, the mechanical properties of many cast products are equaled or exceeded.

Hardness. The large difference in mechanical properties between green and sintered specimens does not apply to conventional indentation hardness testing, particularly Brinell hardness. Green compacts pressed at high pressures often exhibit high Brinell hardness values, which decrease when the compacts are sintered because the strain hardening introduced during compacting is relieved.

At still higher sintering temperatures, compacts shrink and hardness increases due to their higher density. Conversely, the hardness of compacts pressed at low pressures increases with increasing sintering temperature, because the effect of increase in compact density predominates. This is illustrated in Fig. 9, in which the hardness of compacts made of carbonyl iron powder pressed at three different pressures is plotted as a function of sintering temperature for a sintering time of 2 h.

Sintering of Single-Metal Compacts

As discussed above, important changes in the dimensions of compacts occur when green compacts are heated to the sintering temperature. In many sintering operations, green compacts shrink and may even reach the density of corresponding wrought materials. Even when compacts shrink only minimally or not at all, changes in interior geometry occur. In pressed compacts, the approach of the surfaces of the individual powder particles toward each other to distances on the order of metal lattice dimensions is quite limited.

One of the important occurrences in the early stages of sintering is the increase in the contact strength between particles. This may occur without significant changes in the overall density of the compact. Another change within the green compact that does not necessarily involve large density change is the rounding and spheroidization of the pore structure, which is best observed by metallography. Changes in mechanical properties during sintering are, of course, a direct consequence of these geometrical changes.

During compacting, powder particles are subjected to strain hardening that increases with higher pressure. At low temperatures, sintering causes annealing, during which strain hardening is relieved, primarily through recrystallization. A small portion of the geometrical changes that occur in compacts during sintering results from recrystallization. However, the fact that a well-developed sintered microstructure is formed only at temperatures above the recrystallization temperature range indicates that the major driving force for sintering is the reduction in surface energy, not cold work.

The principal driving forces are capillary forces due to the surface and interfacial tensions, also called specific surface and specific interfacial free energies, of the free surfaces and the interfaces between grains (Ref 1). These forces cause stresses that are related to the curvature of the surfaces and interfaces. Stresses due to surface and interfacial tension forces tend to decrease the surface and the interfacial areas and are the principal driving forces in sintering.

In single-phase systems, the existing interfaces are the free surfaces between the solid and gas phase and the interfaces between grains with different orientations. The simplest case would be a single-phase, monocrystalline system, for which the relationship between surface tension (specific surface free energy) and stress is given by the Laplace equation:

$$\sigma = \gamma\left(\frac{1}{r_1} + \frac{1}{r_2}\right) \quad \text{(Eq 1)}$$

where σ is the stress, γ is the surface tension, and r_1 and r_2 are the radii of curvature of the surfaces. Under convex surfaces for which the curvature is negative, the surface stress is tensile ($\sigma < 0$). For concave surfaces with positive curvature, the surface stress is compressive. Experimental methods for readily determining the surface tension of metals and alloys at sintering temperatures are available (Ref 2). The stress defined by the Laplace equation causes a gradient in chemical potential between surfaces with different radii of curvature:

$$\Delta\mu = \sigma\Omega \quad \text{(Eq 2)}$$

where Ω is the atomic volume.

Single-phase metal powders shrink rather than expand during sintering, except for soft metal powders compacted at high pressures. Gases entrapped in these compacts may expand when heated during sintering. Stress caused by gas pressure, which increases with increasing temperature, counteracts surface tension stresses that cause shrinkage.

In this discussion of the driving forces that cause sintering—or, more specifically, material transport during sintering—compacts are assumed to be sintered without the application of external pressure. If external pressure is applied, the capillary forces are of secondary importance compared to external forces, particularly in metallic systems.

Several possible mechanisms have been identified that cause material transport under the influence of a chemical potential gradient. In single-phase systems, the chemical potential gradient is due to the stress gradient between surfaces and interfaces.

One of the early mechanisms postulated is evaporation and condensation. Atoms evaporate from surfaces that have a higher vapor pressure and are transported through the gas phase to highly curved convex surfaces that have a lower vapor pressure, where they condense. Because most metals do not have high vapor pressures at temperatures near their melting points, evaporation and condensation contribute nominally to material transport in the most common metallic systems.

Diffusional flow is the most important mechanism of material transport (Ref 3). It is based on the concept that a certain concentration of vacancies exists in the crystal lattice of a metal. This concentration is a function of temperature and may reach values on the order of one vacancy for every 10,000 occupied lattice positions at temperatures near the melting point of the metal. Vacancy concentration also is a function of the chemical potential or stress to which the surface of the metal is subjected. Consequently, a gradient of vacancies exists between a highly curved convex surface, which has a higher vacancy concentration, and an adjacent flat surface, which has a lower vacancy concentration.

Studies of diffusional flow have concentrated on the first or initial stage of sintering, where the small necks between particles become larger. This causes rounding of the pores and some decrease in total pore volume. However, the original particles are distinguishable. The studies were based on theoretical analysis and on experiments on models, such as those in which a sphere is sintered to a flat surface (Ref 4) or in which two spheres are sintered together (Ref 5).

Figure 10(a) schematically illustrates a cross section of a two-sphere model. The radius, ρ, is formed at the end of the neck between the two particles in Fig. 10(a). Figure 10(b) shows the geometry near this radius in more detail. Assuming that the two particles are single crystals, with different orientations, a grain boundary is formed at the neck. The difference in curvature at the neck with ρ and the adjacent flat surface causes a difference in stress and chemical potential between the two points, which in turn produces a gradient in the concentration of vacancies between the highly curved neck surface, which has a high vacancy concentration, and the adjacent flat surface, which has a lower concentration.

The difference in vacancy concentration under surfaces with different radii of curvature can be expressed mathematically (Ref 3). This difference causes a flux of vacancies away from the highly curved surface to the flat surface, which is equivalent to a diffusional flow of atoms in the opposite direction.

There are two types of diffusional flow possible under the influence of vacancy concentration differences between flat and highly curved surfaces. They are volume diffusion, indicated schematically by path *a* in Fig. 10(b), and surface diffusion, indicated by path *b*. As shown in Fig. 10(b), these two types of diffusional flow cause rounding and spheroidization of irregular pores in a compact. However, they do not explain shrinkage, which would be represented by a decrease in the distance between the centers of the two spherical particles.

To explain shrinkage, flow of atoms from the grain boundary to the neck must be considered. This is indicated by path *c*, which represents grain-boundary diffusion, and path *d*, which represents volume diffusion from the interior of the grain boundary to the neck. When atoms flow from the grain boundary to the neck, vacancies flow in the opposite direction and are thus eliminated at the grain boundaries. This causes the centers of the two particles to approach each other, and shrinkage occurs. The rates of the diffusional mechanisms (volume diffusion, surface diffusion, and grain-boundary diffusion) vary significantly, and the geometries that govern transport also vary and are quite complex even for simple model systems, such as sintering a sphere to a sphere. Nevertheless, the calculated rates of pore rounding and shrinkage in model systems, based on known values of specific surface and interface free energy, and the various types of diffusivities closely approximate those observed experimentally (Ref 6).

Rates of material transport during the initial stage of sintering can be calculated for model systems such as the two-sphere model discussed above. The presence of neighboring particles has constraining effects on neck growth and shrinkage (Ref 7).

Mathematical models have been constructed not only for the initial stage of sintering, but also for the intermediate and final stages. During the intermediate stage of sintering, the original particles are less distinguishable, because the pore channels between particles are pinched off and become closed. Grain boundaries between the original particles migrate due to grain growth. The pores form a more or less connected continuous network throughout the compact. They become isolated in the third or final stages of sintering. Models for the intermediate and final stages of sintering are complex (Ref 8).

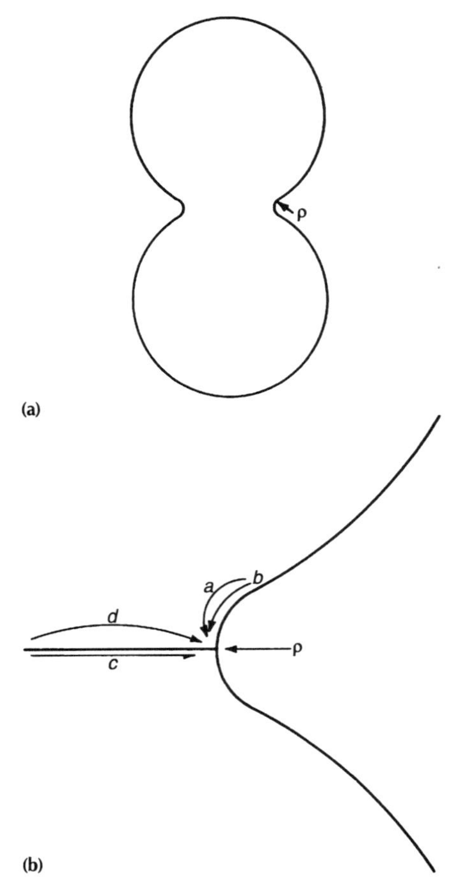

Fig. 10 (a) Schematic cross section through two spheres sintering together. Radius, ρ, is at the end of the neck. (b) Types of material transport when two single crystal spheres with a grain boundary at the interface sinter together. Path *a*, volume diffusion from the flat surface to the neck. Path *b*, surface diffusion from the flat surface to the neck. Path *c*, grain-boundary diffusion from the grain boundary to the neck. Path *d*, volume diffusion from the grain boundary to the neck

Homogenization During Sintering

Alloys may be produced by P/M techniques either by sintering compacts formed from homogeneous prealloyed powder particles or by first mixing different powders for shaping and sintering. Typically, powder mixtures exist as solute-rich particles (either elemental or master alloy) surrounded by a matrix of solvent particles (normally elemental particles of the base element of the alloy). The density to which the powder is pressed determines the degree of residual interparticle porosity.

Dispersion of the solute-rich particles facilitates rapid homogenization and can be optimized through use of proper mixing procedures and determination of the best particle size ratio (the finer the matrix particle size, the greater the dispersion of solute-rich particles). Overall alloy composition and composition of the solute-rich particles also can affect the dispersion, but sometimes are specified and cannot be considered independent variables. Density differences between two different types of particles often cause mixing difficulties. Such problems are

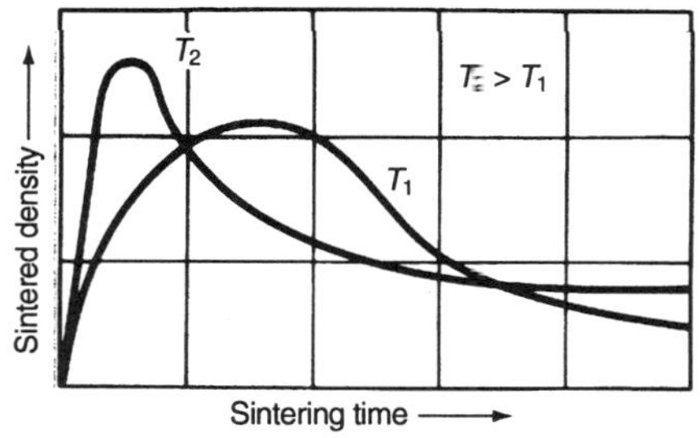

Fig. 11 Variation of sintered density during homogenization in nickel-copper compacts

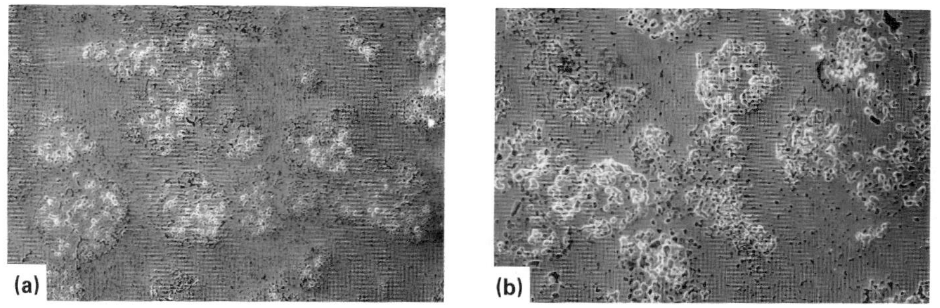

Fig. 12 Formation and coarsening of Kirkendall porosity in compacts made from a blend of nickel (–325+400 mesh) and copper (–100+140 mesh) powders. Mean composition: 0.19 atom fraction copper. 100×. (a) T = 850 °C (1560 °F); t = 1.0 h. (b) T = 850 °C (1560 °F); t = 24.0 h

minimized by varying the size ratio of the particles used.

Sintering causes solid-state diffusion (lattice diffusion) of the solute atoms into the solvent matrix and diffusion of solvent atoms into the solute-rich particles. Homogenization, as a function of time at temperature, may be described in terms of concentration/distance profiles. Over time, the local composition approaches the average compact composition as the concentration gradients are eliminated via interdiffusion. Generally, the compact behaves as a collection of many small diffusion couples that are allowed to undergo interdiffusion to the point where atomic fluxes become negligible and homogenization is achieved.

Variation in sintered density with sintering temperature and time during homogenization is frequently quite different compared to the variation that exists if compacts are composed of particles of identical composition. This is due to two phenomena, which have opposing effects on the densification of the compact. The matrix densifies due to pore shrinkage. Additionally, a net flow of vacancies from one particle to the other occurs if there is a difference in the fluxes of the atomic species to and away from a specific type of particle. These vacancies can accumulate in the form of pores, which is called Kirkendall porosity (Ref 9). The formation of Kirkendall porosity results in expansion of the compact.

The variation of sintered density with time for compacts made from a blend of nickel and copper powders is shown schematically in Fig. 11. Density of the compact first increases and then decreases until a minimum is reached, at which point the density appears to stabilize. In the initial stages of sintering (short sintering times), the densification mechanism dominates, and density increases sharply. However, a point is reached where density begins to decrease, indicating that the formation of Kirkendall porosity (expansion) has begun to dominate.

With increased sintering time, density reaches a minimum, after which it remains almost stable (Ref 9). The time at which the density reaches a minimum indicates the point at which the two opposing mechanisms have balanced each other and that an advanced level of homogenization has been reached. The formation of Kirkendall porosity and its eventual stabilization are shown in Fig. 12(a) and (b), respectively. At higher sintering temperatures, the observed changes in density occur earlier.

The formation of austenite during sintering of compacts from a mixture of iron and graphite powders is an example of complete homogenization. This is due to the high diffusivity of carbon in gamma iron at the sintering temperature. In this case, compacts generally exhibit only slight shrinkage during sintering. At the end of the sintering treatment at temperatures near 1100 °C (2010 °F), compacts consist of homogeneous austenite. When cooled to room temperature at a moderate rate, the austenite transforms. The microstructure of a sintered steel of eutectoid composition (0.8% C content), which consists of the typical lamellae of ferrite and cementite (pearlite). Except for residual pores, this microstructure is similar to that of an annealed wrought eutectoid steel.

The following additional processing parameters can affect the rate and extent of densification or expansion during homogenization:

- Difference in the diffusivities of the particles in the compact
- Size of the particles in the compact
- Mean composition of the compact
- Compacting pressure
- Impurity content of the starting powders
- Entrapped gases within particles

Thus, the form of the density-time relationship for a specific homogenization treatment may vary, and it is frequently more complicated for multiphase systems.

Powder, Alloy System, and Homogenization Variables

Characteristics of the powder particles that compose the mixture exert a significant effect on homogenization kinetics (Ref 10). The sizes of the particles in the mixture establish the distance over which interdiffusion must occur to achieve homogeneity. In alloy mixtures where the elemental solvent particles form a continuous matrix around the dispersed solute-rich particles, particle size of the dispersed particles becomes the important parameter in establishing the interdiffusion distance (assuming that the proportion of the powders in the mixture is already established by the alloy composition desired). Very small powders, such as those produced by reduction of submicron mixtures of oxides or alternatively by mechanical alloying, result in diffusion distances that are small enough to preclude the consideration of homogenization.

The composition of the dispersed particles also can influence homogenization kinetics; for example, solvent alloyed into the solute-rich particles reduces the extent of inhomogeneity in the mixture and can shorten the interdiffusion distance. If the dispersed copper particles in a nickel matrix was 50% Cu, the amount of nickel needed in the mixture to achieve the same overall composition would be less, thus shortening interdiffusion distance.

Alloy systems with more than two components that contain more than a single phase are more difficult to analyze in terms of homogenization kinetics. Interdiffusion in single-phase ternary systems must be described by four diffusion coefficients to consider the diffusion of both solutes. Each is influenced by its own concentration gradient and that of the other solute.

Multiphase binary systems must be described by the solute solubilities and interdiffusion coefficients of each of the phases present (Ref 10–12). In such complexities, the parameters required to analyze the homogenization process are not available, and reasonable simplifications (assumption of binary and/or single-phase behavior) are made. Similarly, the presence of a transient liquid phase at the outset of the homogenization process also is difficult to analyze, and reasonable approximations are necessary.

The duration of sintering required to achieve a desired degree of homogeneity is critically dependent on temperature, because interdiffusion coefficients ($\tilde{D}$) (units of length squared time) are exponentially dependent on temperature:

$$\tilde{D} = D_0 e^{-Q/RT} \qquad \text{(Eq 3)}$$

where D_0 is the preexponential factor, Q is the activation energy, R is the universal gas constant, and T is the absolute temperature.

For example, the interdiffusion coefficient for most solid-solution elements in nickel doubles for every 50 K increase in temperature near the liquidus. The total duration (t) of elevated-temperature treatment to achieve essentially complete homogenization is inversely proportional to $\tilde{D}$:

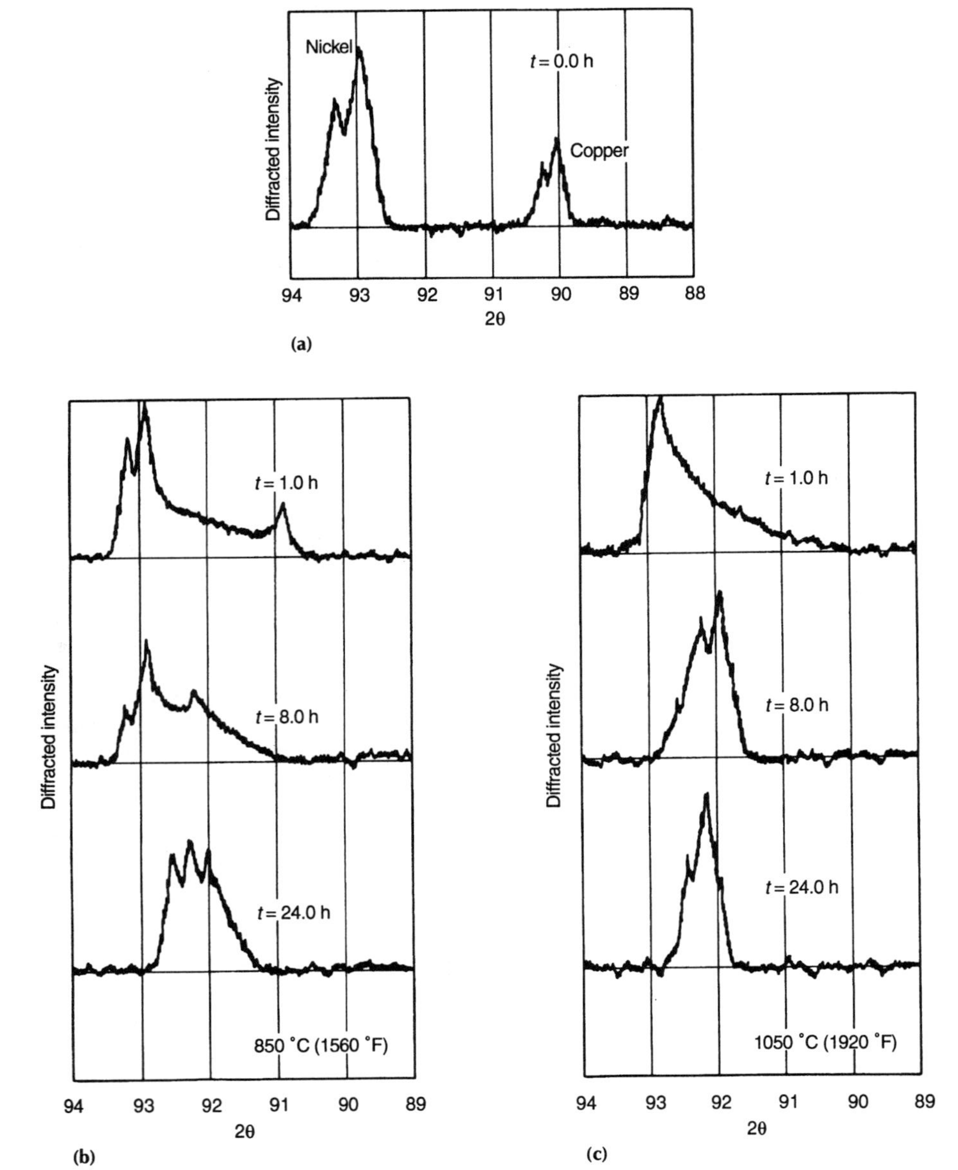

Fig. 13 X-ray diffraction peaks (311) for compacts of blended nickel and copper powders. Mean composition: 0.28 atom fraction copper. Nickel: –400 mesh. Copper: –200+270 mesh. (a) Initial condition. (b) and (c) Various partially homogenized conditions

$$t=k\,l2\tilde{D} \quad \text{(Eq 4)}$$

where l is the average diameter of the dispersed particles and k is a constant, depending on alloy system parameters, powder particle compositions, and overall alloy composition. For example, k is approximately 0.2 for a 4-to-1 mixture of elemental powders that exhibits complete solid solubility (Ref 10). Thus, it can be seen that both the homogenization temperature and the size of the dispersed particles are the most significant variables in controlling homogenization kinetics.

Mixing and mechanical working during powder fabrication of an alloy also can influence homogenization kinetics (Ref 10, 13). Ideal mixing is assumed in Eq 4 to predict homogenization time. Inadequate blending technique or incompatible powder particles (significantly different sizes and/or densities) result in poor mixing. The greatly enlarged interdiffusion distances dictate longer homogenization times. The serious problems of poor mixing stem from homogenization time being proportional to the square of the interdiffusion distance.

Effective interdiffusion distance may be reduced and homogenization kinetics increased by mechanical working (Ref 10). Such working has the same effect as a reduction in particle size of the dispersed particles. However, the effect of a given mechanical working reduction is less than that indicated by the same reduction in l in Eq 4, because the spherical diffusion flux geometry in a compacted mixture of powders is changed to the less efficient unidirectional flux (by rolling) or cylindrical flux (by extrusion). Any enhancement of homogenization kinetics will result only if mechanical working reduces the thicknesses of the solute-rich regions, little, if any, increase in homogenization kinetics is exhibited during subsequent thermal processing.

Determination of the Degree of Homogeneity

Direct experimental techniques for the determination of the progress of homogenization or the evaluation of the state of homogeneity include microscopy (qualitative and quantitative), compositional analysis of the microstructure via electron microprobe analysis or scanning electron microscope instrumentation, magnetic, electrical, or x-ray compositional line broadening techniques (Ref 10–12, 14).

X-ray compositional line broadening is a technique for the rapid determination of the degree of homogeneity. It displays the spectrum of compositions existing in a given inhomogeneous phase from the shape of a diffraction peak broadened by a range of lattice parameters in the phase. This technique may be used for the analysis of single-phase or multiphase binary alloys. The experimental procedure involves obtaining x-ray diffractometer patterns of the specimen whose degree of homogeneity is to be determined.

The progress of homogenization is seen clearly, although qualitatively, in Fig. 13. The Ni (311) peak occurs at a diffraction angle $2\theta \approx 93°$, and the Cu (311) peak occurs at $2\theta \approx 90°$. Figure 13(b) indicates that the initial stages of homogenization (low temperatures and short times) proceed by forming the complete range of compositions from nickel to copper. This is the manifestation of the initial interdiffusion across the nickel-copper interparticle boundaries.

As shown in Fig. 13(c), later stages of homogenization (high temperatures and long times) result in the formation of a single peak at a diffraction angle between those of pure nickel and pure copper. This corresponds to the approach to a single homogeneous phase having the mean composition of the system. The diffraction angle of the single peak is determined by this mean composition. Quantitative analysis of peak shapes can be used to obtain the spectrum of phase compositions within a specimen.

Microscopy is also useful to determine the degree of homogeneity, particularly for multiphase systems that are difficult to analyze using x-ray compositional line broadening techniques. Figure 14 shows the development of microstructures and the progress of homogenization in compacts made from a blend of nickel and tungsten powders. In a partially homogenized compact (Fig. 14), three distinct regions are apparent:

- Dispersed light gray particles (the undissolved tungsten-rich phase that originated as tungsten particles)
- Dark gray unetched regions surrounding the tungsten-rich particles (the nickel-rich phase that has tungsten concentrations ranging from approximately 0.05 atom fraction of tungsten

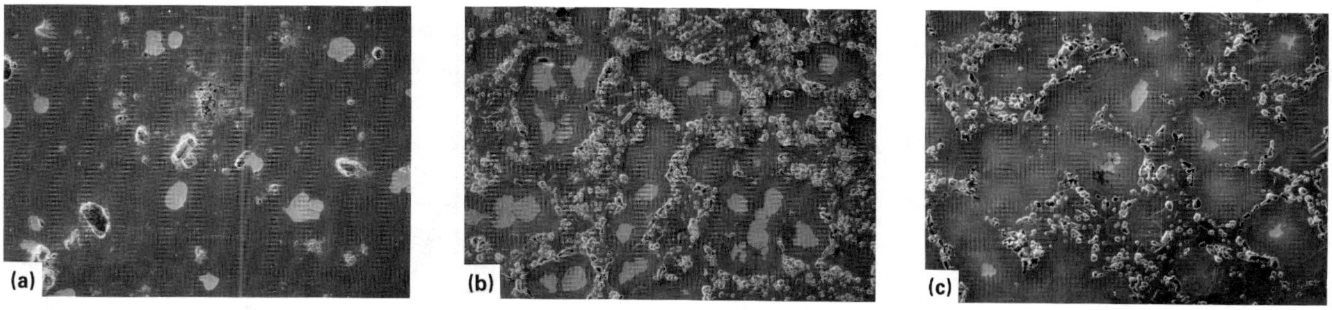

Fig. 14 Scanning electron micrographs showing the progress of homogenization in compacts made from a blend of nickel (–400 mesh) and tungsten (–325+400 mesh) powders. Mean composition: 0.05 atom fraction tungsten. 75×. (a) $t = 0.0$ h. (b) $T = 1150$ °C (2100 °F); $t = 10.0$ h. (c) $T = 1150$ °C (2100 °F); $t = 48.0$ h

at the interface between the etched and unetched regions up to 0.17 atom fraction of tungsten at the interface between the unetched region and the tungsten-rich particles)
- Etched matrix (the nickel-rich phase that has tungsten concentrations less than 0.05 atom fraction)

Quantitative microscopy can be used to determine the extent of homogenization in multiphase systems. This technique is compatible with x-ray compositional line broadening (XCLB) in providing data on the presence of second phases and may be performed on the same test specimens used for XCLB measurements.

Activated and Liquid-Phase Sintering

In activated sintering, the rate of sintering is enhanced compared to that observed in compacts of a single metal powder or a homogeneous alloy powder. The process to which the term activated sintering is applied are the solid-state sintering processes. In contrast, liquid-phase sintering refers to processes in which a liquid phase is formed. Of the two types of liquid-phase sintering discussed in this article, the solution and reprecipitation process frequently is used to provide enhanced liquid-phase sintering. Transient liquid-phase sintering, however, may be used for applications where rapid densification is undesirable, or where alloy formation is desired.

Enhanced sintering, whether achieved by activated or liquid-phase sintering, is commonly used with refractory metals. This is due to the difficulties associated with the extremely high sintering temperatures of these metals. The problems in understanding conventional sintering are compounded in enhanced sintering by the presence of a second phase or supplemental treatment. Most enhanced sintering studies have been conducted with little or no theory to predict beneficial treatments. The eventual success of this approach cannot be denied. A theory to explain such behavior has evolved (Ref 15).

Sintering enhancement generally results from an increased driving force, through physical or chemical treatments. Many such processes are well known. While most attention has focused on tungsten, several other materials, including molybdenum, rhenium, iron, tantalum, uranium, tin, copper, aluminum, titanium, and several ceramic materials, have been investigated.

Enhancement of the sintering process generally is attributed to one or more changes in the fundamental material properties resulting from a special treatment. The strongest effects are those associated with changes in the interfacial properties (higher surface energy or lower grain-boundary energy). Alternatively, a less common means is to induce the operation of a normally dormant mass transport mechanism.

To the ceramist, many of these phenomena are commonplace. Impurities and stoichiometry departures can provide enhanced sintering of many ceramic materials. Likewise, the sintering atmosphere can have a profound influence on sintering rate, as well as the sintering mechanism. Any change in a material that induces an enhanced defect concentration or higher atomic mobility or that promotes the operation of new mass transport processes is considered enhanced sintering.

Typically, enhanced mass flow during the sintering cycle is beneficial. However, in the fabrication of filters, porous bearings, and flow restrictors, enhanced sintering can be detrimental. For most P/M materials fabricated for structural, magnetic, radiation, thermal, or electrical applications, improved service properties are associated with greater mass flow during sintering. Hence, any technique that delivers a greater degree of sintering is beneficial to these applications.

Usually, superior sintered properties are ensured by a higher sintered density. Pressing to high densities (above 90% of theoretical) before sintering is difficult. Consequently, specific techniques have evolved to enhance densification of powder compacts during sintering. Although the mathematics describing densification are somewhat formidable, they provide a concise description of the effects of the various process parameters. In a qualitative sense, there are some useful concepts worthy of review. For example, a smaller grain size (or smaller particle size) aids sintering densification and final properties. Higher sintering temperatures have a significant effect (because of an exponential term), thus increasing the rate of densification.

In a similar manner, a lower process activation energy has the same effect as an increase in sintering temperature. By contrast, a large pore size inhibits densification. Sintering time has a nominal effect on densification; generally, prolonged sintering is not advantageous, because grain size is increased. The grain size increase reduces the amount of grain-boundary area, thus reducing the beneficial effects of the enhanced diffusion rates at grain boundaries. Furthermore, time at temperature is expensive and usually is avoided for economic reasons.

Phase Changes. As previously discussed, compacts made from powders of body-centered cubic metals exhibit more rapid sintering than compacts made from face-centered cubic metals. This is directly related to the higher diffusivities in metals with a body-centered cubic lattice structure compared to metals with a face-centered cubic lattice structure.

In sintering compacts of iron-base compositions, it may be desirable to achieve maximum shrinkage rather than dimensional control. The body-centered cubic phase of iron (ferrite) is unstable at the usual sintering temperatures of 1000 to 1300 °C (1830 to 2370 °F). However, this phase may be stabilized in iron alloys with silicon or molybdenum. This is achieved by sintering compacts of mixtures of iron powder and ferrosilicon or molybdenum powders. Densification increases with the amount of ferrite stabilized at the sintering temperature. Increased densification is most likely due to the higher diffusivity and the fact that the phase boundary is a good vacancy sink. Additionally, the mixed phase microstructure resists grain growth during sintering.

Activated Sintering

Activated sintering refers to any of several techniques that lower the activation energy for sintering. Several techniques have been developed to achieve this goal, including chemical additions to the powder and the use of special atmospheres. In this respect, the sintering treatments described previously in this article that achieve stabilization of the body-centered cubic crystal structure in iron are considered a form of activated sintering.

Many of the detailed investigations of activated sintering have been conducted on tungsten. Tungsten without special treatment exhibits a high level of porosity and is quite weak. Treatment with 0.4 wt% Pd promotes activated sinter-

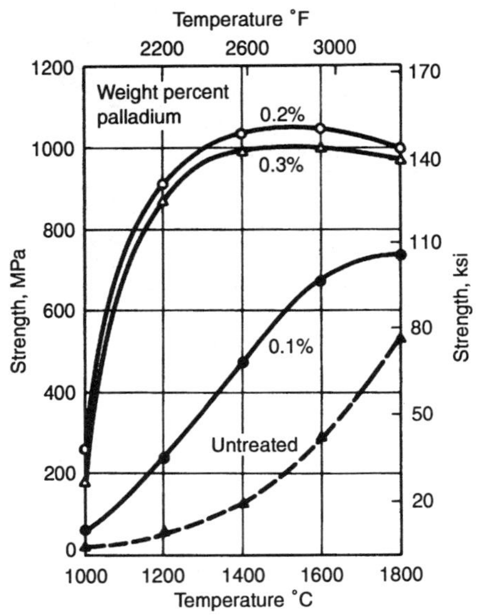

Fig. 15 Compressive strength of compacts from 0.8 μm tungsten powder. Treated with varying amounts of palladium as a function of sintering temperature

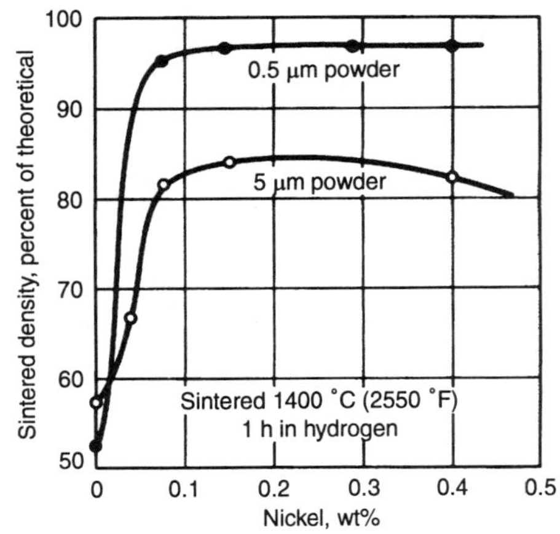

Fig. 16 Density of compacts of two tungsten powders treated with nickel with 0.5- and 5-μm particle size. Sintered 1 h at 1400 °C (2550 °F) as a function of the amount of nickel

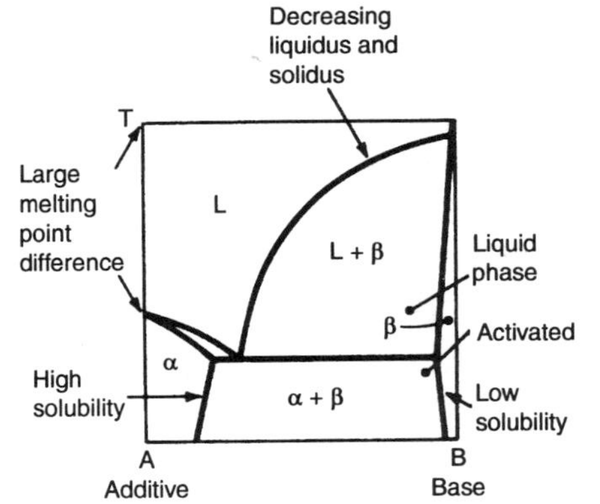

Fig. 17 Idealized phase diagram showing favorable conditions for enhanced sintering by activated and liquid-phase sintering

ing. Diminished porosity produced with enhanced sintering increases strength significantly.

Fine tungsten powder coated with a uniform layer of certain transition metals undergoes rapid densification at unusually low temperatures. The amount of additive required to promote low-temperature sintering is equivalent to one atomic layer on the powder surface. Additive contents above this amount produce minimal further enhancement in sintering and tend to lessen the degree of activation. Densification occurs in two stages. The second, slower stage begins with the onset of grain growth. Thus, the importance of grain boundaries is well demonstrated in activated sintering studies.

The effect of palladium as an activator in the sintering of tungsten on the compressive strength of tungsten compacts is shown in Fig. 15. The strength of the compacts is plotted as a function of palladium content and sintering temperature. In addition to the amount of activator and sintering temperatures, particle size of the powder is an important parameter. In Fig. 16, the density of sintered compacts of two types of tungsten powder with 0.5 and 5 μm particle size doped with nickel are plotted as a function of nickel content.

The type of additive that proves successful as an activator must meet several criteria. First, it must form a phase that has a lower melting temperature than the base metal being sintered. Second, the activator must have a high solubility for the base metal, while the base metal should have a low solubility for the activator. The function of the activator is to remain segregated to the interparticle interfaces during sintering. Such a segregated layer provides a high diffusivity path for rapid sintering. A lower melting temperature ensures a lower activation energy for diffusion, while the solubility ensures that the activator is not dissolved into the base metal during sintering. Typically, an activator that decreases the liquidus and solidus of the base metal remains segregated to the interface between particles. Figure 17 shows an ideal phase diagram for activated as well as liquid-phase sintering systems. Note that at temperatures slightly above the activated sintering range, a liquid forms. The formation of a liquid phase is another means of enhancing sintering.

The kinetics of activated sintering are dependent on the rate of diffusion of the base metal through the thin activator layer. It is necessary to form sufficiently thick layers to provide significant diffusion fluxes by sintering activation. Concentrations above this level frequently do not prove beneficial. The measured activation energies for shrinkage in activated systems closely approximate those for self-diffusion in the activator. Because the process has a low activation energy for diffusion, temperature is the most sensitive process control. The mechanism resembles grain-boundary diffusion-controlled sintering, thus shrinkage initially depends on the cube root of time. During the later stages of sintering, the rate of grain growth appears accelerated because of the low porosity and the high grain-boundary motion. Consequently, rapid grain growth during long-term sintering actually degrades the sintering rate and sintered properties.

Activated sintering also refers to sintering processes in which activation is produced through control of the sintering atmosphere. For example, the addition of a halide to the sintering atmosphere aids transport during sintering by the formation of high-vapor-pressure molecules. Consequently, major changes in pore shape are possible. In such cases, the sintered product has greater strength and significantly higher ductility.

Table 2 compares the strengths and the ductilities for iron compacts sintered in hydrogen. The addition of 1% hydrogen chloride to a hydrogen atmosphere results in improved properties. Metallographic examination indicates that the main effect of the hydrogen chloride is in promoting more rapid vapor phase transport by iron chloride molecules. Improvement in mechanical properties has been demonstrated in other systems. It is thus possible to achieve unique benefits from the sintering atmosphere in addition to control of oxide or carbon content.

Chemical additions are the most successful means of effecting activated sintering. Other processes, such as radiation treatments, have been successful in promoting rapid sintering. However, there has been little interest in this process. Sintering activation treatments can be classified as means of altering either the kinetics or the driving force of sintering. In radiation bombardment, sintering kinetics are altered by the creation of a vacancy excess. This eliminates the vacancy formation energy from the activation energy for diffusion. Alternatively, treatments such as cyclic heating of a material such as iron through the polymorphic phase transformation represent a change in the driving force.

Table 2 Effect of hydrogen chloride on iron sintered in hydrogen

Temperature		Time,	Atmosphere, % hydrogen	Density,	Strength		Elongation,
°C	°F	min	chloride	g/cm³	MPa	ksi	%
950	1740	30	0	6.20	131	19	6
		30	1	6.30	159	23	10
		120	0	6.30	138	20	6
		120	1	6.30	159	23	10
1375	2505	30	0	7.00	193	28	11
		30	1	7.20	234	34	20
		120	0	7.50	234	34	17
		120	1	7.80	283	41	25

In the latter case, cyclic heating generates an internal stress, which effectively raises the driving force. Such a treatment is analogous in some respects to external stresses with hot pressing.

Liquid-Phase Sintering

In systems involving mixed powders, liquid formation is possible because of the differing melting ranges of the components. Common systems involving liquid-phase formation during sintering include copper-cobalt, tungsten-copper, tungsten-nickel-iron, tungsten-silver, copper-tin, iron-copper, tungsten carbide/cobalt, and copper-phosphorus. Because of the faster sintering and superior properties, liquid phase sintering is applied to 70% of all P/M alloys and accounts for 90% of the value of all sintered products.

Heating of a mixed-phase system begins with a mixture of elemental powders. When the liquid forms, it flows and wets the solid particles and begins acting on the solid phase. Melt penetration between the solid particles can cause swelling of the compact, especially if the green density is high, and the particles are coarse.

However, along with melt penetration, there is possible rearrangement of the particles. The combination of wetting, liquid flow, and particle rearrangement contributes to a rapid change in the volume of the compact. With continued heating in the presence of a liquid phase, the solid phase begins to dissolve. Depending on the solubility limit, amount of liquid, phase diagram, and composition of the alloy, one of two actions occurs. If a solid has a high solubility in the liquid, the liquid composition may recross a solidus boundary of the phase diagram and solidify. This is termed transient liquid-phase sintering.

A commercial application of transient liquid-phase sintering is the production of self-lubricating bronze bearings by sintering compacts from a mixture of copper and tin powders. In these bearings, porosity must be controlled. The goal of this process is not to achieve rapid densification, but to control carefully the changes in density during sintering. This is achieved by proper selection of the initial particle size distribution of the powders, compacting pressure, sintering temperature, and rate of heating the compacts.

An alternative type of liquid-phase sintering is the solution and reprecipitation method, in which a limited solubility of the liquid phase in the solid is evident. In this type of sintering, the amount of liquid grows until saturation for the base phase is achieved. With continued time at temperature, the liquid phase becomes a carrier for the solid-phase atoms. Solution and reprecipitation does not change the amount of liquid and solid; however, it may lead to coarsening of the solid phase. Accordingly, the solid phase achieves a higher packing density through particle shape accommodation, with a flattening of faces.

An example of phase accommodation in liquid phase sintering is given in Fig. 18, which depicts a tungsten-nickel-iron alloy containing 95% tungsten. This material has achieved a high tungsten packing density (and full densification) by shape accommodation of the tungsten grains. The grains are several times larger than the tungsten particles in the original powder mixture.

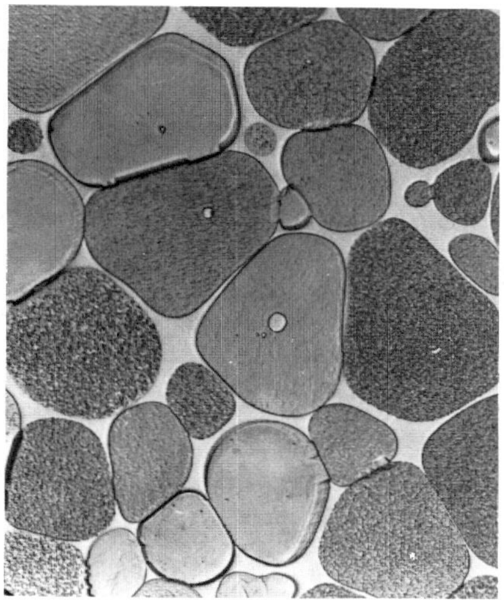

Fig. 18 Compact of 95 wt% W, 3.5 wt% Ni, 1.5 wt% Fe. Sintered 1 h at 1470 °C (2680 °F). Murakami's etch. 350×

Another example of liquid-phase sintering with solution and reprecipitation is shown in Fig. 19, which represents a cemented tungsten carbide with 10% cobalt. In this example, the solid-phase tungsten carbide grains are triangular and rectangular, rather than rounded. Their size is only slightly larger than that of the tungsten carbide powder particles in the powder mixture, indicating a slow rate of coarsening.

The degree of dimensional change in the solution and reprecipitation type of liquid-phase sintering can be quite large. A schematic plot of density versus sintering time is given in Fig. 20. Initially, the liquid forms and contributes to rapid densification through rearrangement. Subsequently, the rate of sintering slows as the solution and reprecipitation stage begins. During sintering, the solid phase forms an interconnected network, which further reduces the densification rate. The use of small particles and large volume contents of liquid phase aids the achievement of high sintered densities in short times. With prolonged sintering, coarsening of the microstructure results, and shape accommodation is evident. Generally, coarsening should be avoided, because it degrades the mechanical properties, although it is often favored in magnetic components.

Liquid-phase sintering kinetics generally are more sensitive to temperature than to sintering time. For solution and reprecipitation shrinkage, the transport path is through the liquid phase. In the final stage, solid-state sintering is dominant, and the rate of densification is quite low. Hence, the major focus is on the solution and reprecipitation stage, where larger amounts of liquid and higher solubilities aid densification. Increasing the liquid content up to approximately 35 vol% aids initial densification. Beyond this amount, little benefit is gained, as full density has been achieved already. Such large amounts of liquid phase are not used, because most alloys lack sufficient rigidity during sintering to maintain dimensional or shape control.

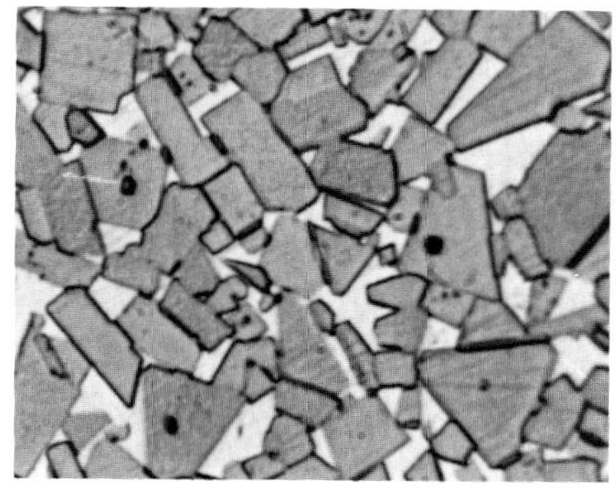

Fig. 19 Microstructure of tungsten carbide/cobalt alloy with 10 wt% Co. Murakami's etch. 1500×

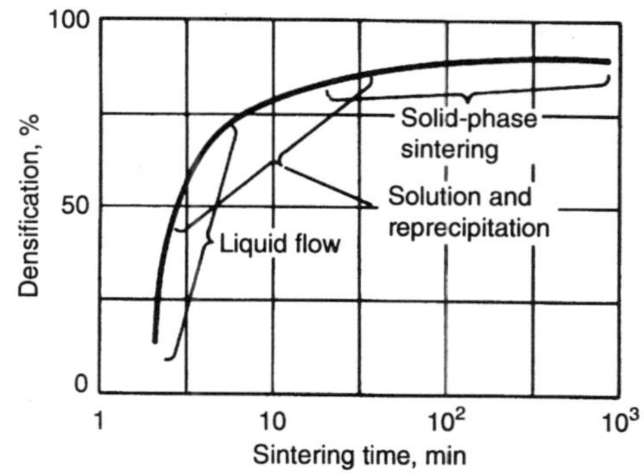

Fig. 20 Density versus sintering time for liquid-phase sintering

Green density and initial particle size can have a profound effect on densification. Coarse particle size and high green density offset the favorable effects of the liquid. In many systems, swelling rather than densification results. The melt penetrates along the interparticle junctions and causes particle separation. Expansion occurs if there is not rapid dissolution of the solid phase.

Throughout liquid-phase sintering, material is transferred through the liquid phase. When the interfacial tension between the solid and the liquid phase is isotropic (liquid solidifies to same crystal as solid), a microstructure of smooth, rounded grains of the solid phase is produced in the matrix, as shown in Fig. 18 for a tungsten-nickel-iron alloy. When the interfacial tension between the solid and the liquid phase is highly anisotropic (the liquid precipitates at least one crystal type different from the solid), as in the tungsten carbide/cobalt system, a microstructure of rectangular or triangular particles is produced in the matrix, as shown in Fig. 19. The initial green compacts exhibit considerable porosity and irregular particle shape. After liquid-phase sintering, the microstructure shows essentially full density.

In many applications of the solution and reprecipitation type of liquid-phase sintering, the solid phase coarsens considerably with increasing sintering time, as shown in Fig. 18. In other cases, such as the tungsten carbide/cobalt system, change in the size of the solid-phase particles is minimal once characteristic shape has been established. In general, higher sintering temperatures cause increased microstructural coarsening.

The occurrence of swelling in liquid-phase sintering has been the subject of extensive study. In the iron-copper system, the onset of swelling is best explained by melt penetration along the grain boundaries within the iron particles. As a consequence, iron particles disintegrate and separate. If the amount of liquid is large, green density is high, and the particles are coarse, then swelling is more likely. Swelling can be controlled by selecting a finer particle size, lower compaction pressure, and slower heating rate. In the iron-copper system, the addition of carbon also offsets swelling. Carbon inhibits melt (copper) penetration of the iron grain boundaries.

The solution and reprecipitation types of liquid-phase sintering and activated sintering are similar in many respects. Both involve the use of a second phase at the sintering temperature to provide rapid mass transport by an effective short circuit path. As shown in Fig. 17, the phase diagram characteristics applicable to both enhanced sintering techniques are similar. Furthermore, the use of fine initial particle sizes and short sintering times produces the best sintered particles. Grain coarsening via rapid diffusional processes is experienced with both forms of enhanced sintering. Closing of large pores is slow by either technique.

A major difference between liquid-phase and activated sintering is the amount of second phase sintering is the amount of second phase present at the sintering temperature. Liquid-phase sintering systems typically have several times more second phase than activated sintering systems. In many respects, this is beneficial, because the second phase is the continuous phase in the sintered product and is better able to sustain strain and deformation, leading to improved mechanical properties.

Process Modeling

The modeling of powder processing is quickly advancing, especially as numerical simulation tools become widely available. Die compaction, hot isostatic compaction, mixing, sintering, injection molding, hot forging, and extrusion are now computer simulated on a routine basis in production operations. Unfortunately, most of these models are proprietary, and details are closely guarded. Two areas where there is much information on powder processing models is in sintering. The activities started in the 1960s, but computation speed was slow. Over the next several years, the sophistication of the models progressed, and the computer speeds advanced, making it possible to computer simulate real sintering cycles in seconds. Thus, experimentation is more rapid via simulation, as compared to experimentation. Many powder processing operations have embraced these new technologies as tools to improve production control and process specification. As a few examples, the basic concepts in modeling die compaction and solid-state sintering are considered. Further information can be found by referring to the Selected References.

Die Compaction

The major difficulty with die compaction traces to die wall friction. This friction inhibits ejection and, more importantly, causes density gradients in the green compact. Punch motion against the powder is similar to plowing snow. Close to the punch, the packing is dense, but far removed from the punch, the powder is unaffected. This pressure decay with distance is because the powder spreads load to the die wall in the form of friction.

Green density increases with compaction pressure. Consequently, density declines with distance from the punch. When the compact length is approximately six times the width, then the powder is poorly compressed. This is among the reasons why most die compacted parts are squat, to minimize density gradients. Also, density gradients are reduced when both the upper and the lower punch move toward the compact center. Double action pressing is a term used to describe the application of stress from both the top and bottom punches. In one variant, the die floats with respect to the lower punch, giving double action even though the lower punch remains stationary. Positive die motion ensures uniformity in green density. Inside the compact, the point where the lowest density occurs is the neutral pressure plane, and in most instances, the desire is for the neutral plane to be in the center of the compact.

As the compaction pressure and density increase, the die wall friction also increases. The die wall friction is proportional to the radial pressure when the powder is compressed uniaxially. By the end of the compaction stroke, wall pressures exceed 50% of the applied pressure. Thus, the higher the applied pressure is, the higher the radial pressure and concomitant die wall friction.

If pressure decays with depth in a compact due to wall friction and green density depends on local pressure, then die compaction results in green density gradients. Besides friction with the die wall, an additional factor is friction on the punches and core rods. Consequently, an objective of computer simulations is to calculate the green density contours in a multiple level compact pressed under a variety of press or tool options. If there are large density gradients, then the compact will exhibit nonuniform shrinkage or swelling in sintering, resulting in component distortion and loss of precision. Low density regions tend to change dimensions more in sintering as compared to high density regions.

Minimized density gradients are important to final dimension control. Accordingly, much study has been focused on reducing density gradients. Unfortunately, most cures possible by changing the powder or alloy are ineffective, but a few concepts are important. However, computer simulation techniques show some important options. These include keeping the compact thin in the pressing direction (thickness), lubricating the die wall, and double action pressing. One alternative being studied is to move the top punch in a circular motion, which causes powder flow in a sideways motion to help eliminate low density regions. The need for minimum density gradients increases as the component becomes more complex. Since die wall friction is the culprit, the more die wall there is per unit volume, the greater the problem. This is one of the major barriers to the production of precise gears to P/M, since gear teeth have a large surface area leading to more die wall friction and larger density gradients.

Because of density gradients, problems occur with local weak spots, cracking on ejection, and subsequent component warpage in sintering. Thus, much effort is being applied to calculation of density gradients in pressing. Usually the friction between the powder, die, and punches is the main cause of density contours, but there is possibly a large effect from the die filling operation, especially if powder separation or nonuniform flow occurs. Most of the calculations are performed using a concept of finite element analysis. In this approach, the component is numerically subdivided into hundreds of tiny computation cells. The placement of these calculation cells is visible on the component as a fine mesh. Each cell compresses and deforms based on the forces from the neighboring cells using equations that relate pressure to density. Computer calculation of what happens in each cell then determines the flow of mass and stress through the compact, resulting in the plotted density contours. Soon, most tooling design and compaction setups will be guided by such computer simulations, with the goal being reduction of density gradients and weak spots in the compact. The aim is to make the tooling correct with minimal trial and error. Most tool design is assisted by computer-aided engineering packages. These packages allow accurate drawings and analysis of form, fit, and function. Also, finite element analysis is used to assess the stresses on the press and in the tool sets. Accordingly, the leading press manufacturers provide codes for analysis of press frame deflection. Of course, such efforts require knowledge on the part size and shape, and powder characteristics.

Most presses operate in an open-loop control mode. The position of the various tool members is adjusted at the beginning of a production run, but no monitor is provided for following operation conditions. Parts are selectively checked for weight and dimensions, but the preservation of the desired tool motions, pressures, fills, and other characteristics of compaction is not ensured.

A few advanced presses include pressure sensors, displacement transducers, strain gages, temperature sensors, position sensors, and other monitors needed to watch for consistent production. Rarely does this advance to in situ control where instantaneous corrections are made; although a few hybrid computer controlled presses are now installed. If sensors are used, they usually measure the die temperature and strain, upper and lower punch loads (via in-line transducers), and relative displacement of the tool components. Hydraulic and computer numerical controlled presses prove easier to instrument and control. For example, in-line load cells can be used to adjust pressure on the hydraulic cylinder via a servo-hydraulic valve. This results in one form of a hydraulic computer numerically controlled press, best suited to complex, multiple level part. The result is more consistent densities. Likewise, a few presses have multiple step feed shoes. The first step weighs the powder charge to ensure the proper amount is used in filling.

Computerization of die compaction has two objectives. One objective is to record the operating data, including set-up parameters and production data. This allows initialization at the same conditions each time the tooling is installed. In one variant, the tool setup is fully automated using computerized electric motors. The second variant is to monitor trends in parts. Part-to-part variations and differences between production lots are both concerns, which require monitoring systems and logic systems. Robots might be coupled to the press for component extraction, placement on sintering trays, or inspection. These too are coupled into the logic and control system. Pick and place robots are used on large volume production items, but much of the P/M industry relies on human intervention should there be a problem.

Monitoring is largely through load measurements in the pressing direction. Logic systems then decide what changes are appropriate when drifts in product quality are detected. Most important, these systems can stop production in cases where tool damage might occur. For example, if the powder fill changes, then excessive powder in the cavity might deform a punch. A load monitor during compaction would detect such a change and halt production. Generally only three monitors are used to detect product quality drift at the press: changes in weight, changes in compaction load, or changes in dimensions. The latter parameter might be manually input to the control computer via an operator.

Solid-State Sintering

The growth of the sinter bond from an initial loose powder contact is characterized as the initial stage of sintering. In this stage the neck size is sufficiently small that neighboring necks grow independent of one another. The initial stage ends when the necks begin to impinge. Various laws link the neck size ratio to other useful sintering parameters, including shrinkage, surface area, and density.

Computer analysis of initial stage sintering has been a popular theoretical topic. Early codes assumed sintering occurs between two equal spheres with conservation of volume by a single mass transport mechanism. Thus, any change in the neck size must have a corresponding functional relation to the volume of material deposited at the neck. There is a gradient in the curvature over the sintering geometry. The curvature gradient drives the mass flow, by any of the mechanisms presented, to smooth the surface and possibly densify the structure. At any point on the neck, there is a surface contour defined by the function $S(v)$, where v is a parameter describing the spacial relation between the x-y coordinate system and surface position. Consider surface transport controlled sintering where at any point on the surface curve the following exists:

$$\frac{dS}{dv} = \left[\left(\frac{dx}{dv}\right)^2 + \left(\frac{dy}{dv}\right)^2\right]^{1/2} \quad \text{(Eq 5)}$$

The instantaneous change in the neck profile depends on the normal motion at the surface, which can be described in the following manner:

$$\left(\frac{dS}{dv}\right)^{-1}\left(\frac{dx}{dv}\mathbf{Y} - \frac{dy}{dv}\mathbf{X}\right) \quad \text{(Eq 6)}$$

where **X** and **Y** are the unit axis vectors. At point v on the neck profile, the principle radii of curvature are expressed according to the following:

$$R_1 = y\frac{dS/dv}{dx/dv} \quad \text{(Eq 7)}$$

and

$$R_2 = \frac{dS/dy}{\left(\frac{d^2y}{dv^2}\right)\left(\frac{dx}{dv}\right) - \left(\frac{d^2x}{dv^2}\right)\left(\frac{dy}{dv}\right)} \quad \text{Eq 8}$$

This allows determination of the local curvature κ:

$$\kappa = \frac{1}{R_1} + \frac{1}{R_2} \quad \text{(Eq 9)}$$

It is the curvature gradient from point to point that drives the flux of atoms during initial stage sintering. Comparing the curvature at a position v and neighboring positions $v - 1$ and $v + 1$ gives a basis for calculation of the flux. According to Fick's first law, the flux depends on the curvature gradient at each point and the mobility of atoms, meaning the neck volume change depends on the arrival rate for mass at the sinter bond:

$$\frac{dV}{dt} = JA\Omega \quad \text{(Eq 10)}$$

where J is the atomic flux, A is the bond area over which the new mass is distributed, and Ω is the volume of a single atom or molecule. The change in curvature with position gives a chemical gradient that directs mass flow into the sinter bond between the particles. The deposited or removed atoms change the neck size and shape. In turn the new shape and curvature gradient influence the flux. High temperatures promote faster mass transport and thereby contribute to faster neck growth. Many concurrent deposition processes react simultaneously to the same driving force. Consequently, accurate calculations of sintering rates require numerical techniques.

Sintering models with simplified forms are available that provide an estimate of the relative rate of neck growth from any transport mechanism. These result in the following initial stage sintering model for the neck size ratio X/D as a function of sintering time t under isothermal conditions:

$$\left(\frac{X}{D}\right)^n = \frac{Bt}{D^m} \quad \text{(Eq 11)}$$

where X is the neck diameter, D is the particle diameter, t is the isothermal sintering time, and B is a term that collects material and geometric constants that are specified in Table 3. The values of n, m, and B change with time and degree of sintering, but for many analyses, they have been assumed constant with the mechanism of mass transport. Many experiments have been analyzed using the exponent n to identify the sintering mechanism. However, in cases where the grain boundary energy is relatively high (dihedral angle is small), neck growth will be paced by grain growth ($X = G \sin \phi/2$) and will not follow Eq 11. In these cases, there will be a groove at the neck.

There are many different solutions to the initial stage sintering problem, so the forms given

Table 3 Initial stage sintering equation for spheres

$(X/D)^n = Bt/D^m$

Mechanism	n	m	B
Viscous flow	2	1	$3\gamma/(2\eta)$
Plastic flow	2	1	$9\pi\gamma\mathbf{b}D_V/(kT)$
Evaporation-condensation	3	2	$(3P\gamma/\rho^2)(\pi/2)^{1/2}(M/(kT)^{3/2}$
Volume diffusion	5	3	$80D_V\gamma\Omega/(kT)$
Grain boundary diffusion	6	4	$20\delta D_b\gamma\Omega/(kT)$
Surface diffusion	7	4	$56D_s\gamma\Omega^{4/3}/(kT)$

Symbol	Definition
b	Burgers vector
D_b	Grain boundary diffusivity
D_s	Surface diffusivity
D_v	Volume diffusivity
k	Boltzmann's constant
M	Molecular weight
P	Vapor pressure
T	Absolute temperature
γ	Surface energy
δ	Grain boundary width
η	Viscosity
ρ	Theoretical density
Ω	Atomic volume

in Table 3 are only representative of the efforts. For example, depending on the assumptions, various models for surface diffusion controlled initial stage sintering give n values of 3, 5, 6, 7, and 7.5. The simplified model represented by Eq 11 gives a rough approximation up to a neck size ratio of 0.3. The diffusion coefficient is embedded in the parameter B and follows an Arrhenius temperature dependence. The required frequency factor and activation energy for several common materials are collected in Table 4 for surface, volume (lattice), and grain boundary diffusion. Equations such as those introduced here for initial stage solid-state sintering are typically in error by 10 to 20% over actual numerical solutions. Hence, little precision can be assigned to calculations based on these integral models.

Although not very exact, Eq 11 does illustrate some key processing factors in sintering. A high sensitivity to the inverse particle size means smaller particles give more rapid sintering. Surface diffusion and grain boundary diffusion have the highest sensitivity to particle size changes; thus they are enhanced relative to the other processes by a smaller particle size. In all cases, temperature appears in an exponential term, which means small temperature changes can have a large effect. Finally, time has a relatively small effect in comparison to temperature and particle size.

Bulk transport processes decrease the interparticle spacing as neck growth processes, which results in compact shrinkage and the formation of additional, new necks. It is easier to measure the compact dimensional change instead of the neck size. Shrinkage is approximately related to the neck size by a simple model as follows:

$$\frac{\Delta L}{L_0} = \left(\frac{X}{2D}\right)^2 \quad \text{(Eq 12)}$$

where the shrinkage $\Delta L/L_0$ is the compact length change divided by the initial length. Shrinkage is actually a negative value, but the sign is usually ignored. Actual measurements of both shrinkage and neck size reveal that the shrinkage is slightly larger than predicted by Eq 12. On this basis, shrinkage during initial stage sintering follows a kinetic law similar to Eq 11:

$$\left(\frac{\Delta L}{L_0}\right)^{n/2} = \frac{Bt}{2^n D^m} \quad \text{(Eq 13)}$$

The parameter B is exponentially dependent on temperature:

$$B = B_0 \exp\left(-\frac{Q}{kT}\right) \quad \text{(Eq 14)}$$

where k is Boltzmann's constant, T is the absolute temperature, and B_0 consists of material parameters (such as surface energy, atomic size, atomic vibration frequency, and system geometry as defined in Table 3).

One of the benefits of such computer simulations is recognition that strength and other mechanical, thermal, or physical properties are dependent on the neck size. For example, strength data are directly related to the neck size ratio squared. A major complication arises due to multiple mechanisms contributing simultaneously to particle bonding. Accurate calculation of the sintering rate is necessary to build a basis for sintering predictions. Unfortunately, most of the models are only approximations to an underlying complexity that cannot be treated by simple geometric applications. Consequently, computer simulation is enlisted to address the sintering problem. This is an important direction to provide realistic modeling capabilities. Much of the early work focused on accurate predictions of the sintering geometry and mass flow kinetics by a single transport mechanism, for example grain boundary diffusion.

Only recently, have computer simulations embraced realistic conditions. A typical shortfall of the early models was a failure to incorporate a grain boundary dihedral angle in the sinter neck. Other problems are associated with the assumption of monosized spheres in ideal packing geometries with instantaneous heating. A serious deficiency in early simulations was a loss of mass or volume. For example, simulating the coalescence of spheres by surface diffusion resulted in nearly a 5% gain in volume. These problems limit the usefulness of the early simulations. In contrast, actual sintering experiments involve a wide range of particle sizes, packing inhomogeneities, and multiple sintering mechanisms during slow heating to the final temperature. A real difficulty is in using too few initial particles in the simulation. It is not uncommon in practice to see a final grain size that is ten times larger than the initial particle size. Thus, the initial simulation must start with over 1000 particles for every final grain in the sintered structure. This is a major computational burden. Furthermore, there are many situations where 10^9 particles are consumed to form one sintered grain, which is far beyond current simulation capabilities.

The variety of computer simulations for sintering processes is large. Approaches include numerical solution to the different equations, finite element analysis, finite difference solutions, and Monte Carlo simulations. A review of the field is provided in the Selected References. Although there are over 200 sintering simulations in circulation, many focus on single mechanisms; even so they have provided important checks on the models and a basis for multiple mechanism simulations.

Multiple mechanism assessments of sintering are useful in determining dominant processes. For example, volume diffusion versus surface diffusion controlled sintering can be assessed by taking the ratio of neck growth rates. These assessments show that early sintering favors surface diffusion and grain boundary diffusion, while the latter stages tend to shift to volume diffusion control.

Most of the modern simulations recognize that sintering involves the delivery of mass to the interparticle bond by several simultaneous transport mechanisms. A typical assumption is that the instantaneous sintering mass flux is simply the sum of the contributions from each mechanism. At any instant, the rate of neck growth is the sum of the individual fluxes:

$$\left.\frac{dX}{dt}\right|_{\text{total}} = \Sigma \left.\frac{dX}{dt}\right|_i \quad \text{(Eq 15)}$$

This assumes that each of the individual mechanisms, represented by the subscript i, provides an instantaneous contribution to the total neck growth. Accordingly, computer simulations are employed to continuously calculate the contributions from each transport mechanism at a given time. Once the rates of mass flow are determined, the sum is used to enlarge the neck by determining its new volume and shape for the next instant of time. With reposition of the transported mass, the geometry is recalculated. Time is advanced by a small step size, and the corresponding microstructure gradients are used to recalculate the fluxes and total mass flow. This process goes on with millions of small steps to eventually determine the geometry after a reasonable sintering time. Only in the past few years, has the speed of such calculations for just two particles exceeded the speed of actual sintering. (In the 1970s it would have taken 10 times longer to simulate the process than the actual experimental time; computer times of 600 min would be required for simulation of 60 min of sintering for two spheres.)

In calculating the rate of sintering, the evolution has been from simple geometric approximations to sophisticated multiple mechanism computer simulations. The usefulness of such simulations is evident in several of the proprietary codes, where the agreement with the experiment is excellent. Surface diffusion dominates sintering, while evaporation condensation is a minor contributor.

A consequence of these computer-based sintering calculations is the realization that the process is very involved and not easily treated using simple models. The logical extension has been the synthesis of sintering diagrams that provide guidelines on the interplay of the processing parameters with rigorous geometric models.

A sintering diagram proves useful in condensing and representing sintering behavior. These are based on computer simulations, wherein multiple mechanisms contribute to mass flow, and accurate geometric representations are used for the microstructure. Contributions are included from each of the mass transport mechanisms, and the geometric approximations are minimized to accurately reflect microstructure evolution during sintering. The more recent simulations include a change in grain size with time and recognize densification results in new particle contacts. A sintering diagram can have any of several forms, but typically shows the

Table 4 Physical properties and diffusion factors for various materials

Material	Density, g/cm^3	Crystal type(a)	Atomic diameter, nm	Melting point, °C	Boiling point, °C	Heat capacity, J/(kg · °C)	Heat of fusion, kJ/mol	Heat of vaporization, kJ/mol	Solid-vapor surface energy, J/m^2	VD frequency(b), m^2/s	VD activation energy(b), kJ/mol	GBD frequency(c), m^3/s	GBD activation energy(c), kJ/mol	SD frequency(d), m^2/s	SD activation energy(d), kJ/mol
Alumina (Al_2O_3)	3.98	H	...	2054	2980	775	...	...	1.12	0.19	636	3×10^{-3}	477	3×10^{-3}	536
Aluminum	2.70	fcc	0.286	660	2467	917	10.7	294	1.14	2×10^{-4}	142	3×10^{-14}	60	5×10^{-2}	142
Antimony	6.69	R	0.364	630	1635	209	20.9	68	0.45	1×10^{-4}	165	...	93	...	...
Beryllia (BeO)	2.86	H	...	2570	3787	1046	...	...	1.42	...	530	...	...	...	...
Beryllium	1.85	HCP	0.228	1289	2970	1880	9.8	308	1	6×10^{-5}	161	...	...	...	...
Brass (Cu-30Zn)	8.50	fcc	...	965	...	377	...	...	...	1.6×10^{-6}	126	...	...	...	...
Bronze (Cu-10Sn)	8.88	fcc	...	957	...	377	...	...	1.7	1.6×10^{-10}	207	6×10^{-10}	105	2.5	205
Chromia (Cr_2O_3)	5.21	H	...	2440	3000	670	...	...	...	16×10^{-3}	423	1×10^{-15}	240	...	...
Chromium	7.23	bcc	0.250	1875	2665	461	15.0	342	2.2	2×10^{-5}	308	1×10^{-13}	198	83	215
Cobalt	8.90	HCP	0.250	1495	2870	423	15.0	382	2.1	8×10^{-5}	292	2×10^{-14}	117	4×10^{-5}	130
Cobalt oxide (CoO)	6.46	NaCl	0.426	1935	...	703	...	...	...	5×10^{-3}	398	1×10^{-13}	200	...	...
Copper	8.96	fcc	0.256	1083	2578	386	13.0	307	1.75	6×10^{-5}	213	2×10^{-14}	107	2.6	205
Erbium	9.07	HCP	0.356	1529	2863	168	17.2	293	...	4.5×10^{-4}	302	...	...	...	...
Europium	5.24	bcc	0.398	822	1597	182	10.5	176	...	1×10^{-4}	144	...	...	...	...
Gadolinium	7.90	HCP	0.358	1313	3266	300	15.5	312	...	1×10^{-6}	138	...	...	...	...
Germanium	5.32	Diamond	0.244	937	2830	310	34.7	224	0.7	7.8×10^{-4}	287	1×10^{-17}	172	...	...
Glass-borosilicate (SiO_2-13B_2O_3)	2.40	...	...	1140	...	335	...	...	...	1×10^{-7}	84	...	...	...	...
Gold	19.30	fcc	0.288	1063	2807	343	13.0	343	1.37	6×10^{-6}	172	3×10^{-16}	110	1×10^{6}	234
Hafnium	13.31	HCP	0.334	2227	5197	147	25.5	661	...	1.21×10^{-7}	163	...	...	...	...
Hematite (Fe_2O_3)	5.24	H	...	1570	...	628	...	...	...	2×10^{-4}	326	4×10^{-13}	210	...	...
Ice	0.92	H	...	0	100	4217	...	...	0.1	9×10^{-4}	59	8.3×10^{-13}	38	3.6×10^{2}	59
Indium	7.31	fct	0.332	156	2080	243	3.3	226	0.6	3×10^{-4}	79	...	...	...	...
Invar (Fe-36Ni)	8.05	fcc	...	1425	...	515	...	...	...	2.5×10^{-3}	310	...	...	...	...
Iron	7.87	bcc	0.248	1536	2750	456	15.0	340	1.95	2×10^{-4}	251	1×10^{-17}	128	1.1×10^{1}	239
Iron oxide (FeO)	5.70	NaCl	...	1370	3414	803	...	...	...	1×10^{-2}	326	1×10^{-13}	195	...	...
Kovar (Fe-29Ni-17Co)	8.36	...	...	1450	...	460	...	...	...	1.5×10^{-3}	315	...	...	...	...
Lanthanum	6.15	HCP	0.374	921	3457	200	10.0	400	0.7	1×10^{-6}	103	...	...	...	...
Lead	11.10	fcc	0.350	328	1740	130	5.0	178	0.61	1×10^{-4}	109	2×10^{-13}	68	2.6×10^{-4}	109
Lithium	0.53	fcc	0.304	180	1347	3517	4.6	135	0.4	2.3×10^{-5}	55	...	...	...	...
Lithium fluoride (LiF)	2.64	NaCl	...	846	1717	1631	...	...	...	7.4×10^{-3}	214	...	...	...	...
Magnesia (MgO)	3.58	NaCl	...	2825	3260	920	...	...	0.5	1.4×10^{-6}	460	1.4×10^{-15}	230	1.5×10^{3}	460
Magnesium	1.74	HCP	0.320	649	1090	1038	9.0	129	0.57	1×10^{-4}	135	5×10^{-12}	92	...	...
Molybdenum	10.22	bcc	0.272	2610	4612	251	28.0	590	2.2	5×10^{-5}	418	6×10^{-14}	263	10	241
Mullite ($3Al_2O_3$-$2SiO_2$)	3.17	O	...	1828	...	765	...	...	0.66	3.6×10^{6}	810	0.0036	434	1×10^{7}	405
Nichrome (Ni-20Cr)	8.50	fcc	...	1400	...	430	...	...	...	1.6×10^{-4}	285	2.8×10^{-15}	115	...	...
Nickel	8.90	fcc	0.250	1453	2732	452	18.0	375	1.86	1.4×10^{-4}	276	4×10^{-16}	108	2×10^{-2}	164
Nickel aluminide (NiAl)	5.86	Ordered	...	1647	...	843	...	...	...	4.8×10^{-5}	335	1.6×10^{-15}	290	2.2	325
Nickel trialuminide (Ni_3Al)	7.25	Ordered	...	1380	...	123	...	...	2	4×10^{-4}	306	2×10^{-13}	152	1×10^{4}	306
Niobium	8.57	bcc	0.284	2468	4927	268	27.0	697	2.3	1×10^{-4}	402	5×10^{-14}	263	1	235
Palladium	12.02	fcc	0.274	1552	3140	247	17.2	393	2.1	2×10^{-5}	274	...	...	...	...
Platinum	21.45	fcc	0.278	1769	3827	134	19.7	511	1.3	3×10^{-5}	294	...	...	4×10^{-7}	108
Potassium	0.86	bcc	0.462	63	774	754	2.4	78	0.12	3.2×10^{-5}	41	...	...	...	...
Rhenium	21.02	HCP	0.274	3180	5625	138	33.1	707	...	...	...	...	...	9×10^{-5}	221
Rhodium	12.41	fcc	0.268	1966	3727	243	21.6	495	...	...	...	...	...	4×10^{-6}	173
Silicon	2.33	Diamond	0.234	1410	2355	729	39.6	383	0.9	0.9	496	1×10^{-15}	300	...	...
Silicon carbide (SiC)	3.15	H	...	2700	2972	628	...	...	3	0.05	696	2.2×10^{-11}	557	7×10^{4}	696
Silicon nitride (Si_3N_4)	3.18	Cubic	...	2564	...	400	...	...	0.5	0.05	636	2×10^{-11}	509	7.6×10^{-2}	636
Silver	10.50	fcc	0.288	961	2212	234	11.3	255	1.14	4×10^{-5}	185	6×10^{-15}	90	5×10^{3}	266
Sodium	0.92	HCP	0.380	98	883	1240	2.6	89	0.19	2.4×10^{-5}	44	...	...	...	...
Sodium chloride (NaCl)	2.16	NaCl	...	801	1465	870	14.0	323	0.28	0.025	217	6.2×10^{-10}	155	5.1×10^{4}	217
Spinel ($MgAl_2O_4$)	3.58	Cubic	...	2135	...	812	...	...	...	8.9×10^{-5}	439	1×10^{-14}	264	...	...
Stainless steel, 304L	8.00	fcc	...	1400	...	500	...	...	2.2	4×10^{-5}	280	2×10^{-13}	167	0.5	220
Stainless steel, 316L	8.05	fcc	...	1375	...	500	...	...	2	4×10^{-5}	280	2×10^{-13}	167	0.4	250
Steel 1018	7.86	bcc	0.252	1525	...	464	15.0	340	2.1	2×10^{-5}	270	8×10^{-14}	159	300	220
Superalloy (Ni-base)	8.52	Mixed	...	1400	...	413	...	...	1.7	2×10^{-4}	285	2×10^{-6}	115	4.8	285
Superconductor ($YBa_2Cu_3O_7$)	...	...	...	1200	...	...	...	...	...	8×10^{-4}	278	5×10^{-18}	139	...	...
Tantalum	16.65	bcc	0.228	2996	5427	142	31.0	758	2.8	1.2×10^{-4}	425	5.5×10^{-12}	300	1.1×10^{-3}	326
Thoria (ThO_2)	10.00	CaF_2	...	3250	4400	230	...	...	...	3.5×10^{-5}	625	1×10^{-14}	375	...	...
Thorium	11.72	fcc	0.360	1750	3850	100	19.2	544	...	10^{-5}	349	...	...	...	...
Tin	7.30	Tetra.	0.316	232	2270	226	7.2	290	0.68	9×10^{-4}	105	3×10^{-15}	40	0.07	64
Titania (TiO_2)	4.17	H	...	1830	...	799	...	...	...	4.6×10^{-6}	240	...	...	...	...
Titanium (hcp)	4.50	HCP	0.294	1668	3286	528	20.9	429	1.7	6×10^{-12}	123	4×10^{-16}	97	4×10^{-3}	150
Titanium (bcc)	4.00	bcc	0.294	1668	3286	418	20.9	429	1.7	1×10^{-4}	251	5×10^{-17}	153	6×10^{-3}	153
Titanium aluminide (TiAl)	3.83	Ordered	...	1457	...	650	...	...	1.76	8×10^{-7}	200	8×10^{-13}	180	2.4	150
Titanium carbide (TiC)	4.93	NaCl	0.300	3065	...	564	83.6	501	1.19	4.4	737	10^{-13}	543	2.2×10^{-4}	288
Ti-6Al-4V	4.46	...	...	1650	...	544	...	...	...	1.3×10^{-6}	180	...	...	...	...
Tool steel (Fe-4Cr-6W-5Mo-2V-0.9C)	8.00	...	...	1410	...	450	...	...	2.2	4×10^{-5}	280	2×10^{-13}	167	0.5	220
Tungsten	19.30	bcc	0.274	3410	5657	138	35.0	824	2.8	8×10^{-6}	520	3×10^{-13}	385	10^{-3}	293
Urania (UO_2)	10.96	CaF_2	...	2827	...	243	...	...	0.74	10^{-5}	452	2×10^{-15}	293	0.1	452
Uranium	19.10	O	0.276	1132	3745	117	15.5	423	...	2×10^{-7}	123	3×10^{-11}	184	...	...
Vanadium	6.11	bcc	0.264	1887	3377	498	18.0	460	2.5	4×10^{-5}	317	5×10^{-14}	209	...	...
Zinc	7.13	HCP	0.266	420	906	394	6.7	115	0.8	2×10^{-5}	94	10^{-14}	60	9×10^{-6}	27
Zirconia-yttria (ZrO_2-$3Y_2O_3$)	6.08	Mono.	...	2600	4548	456	...	...	0.3	2.7×10^{-5}	423	2×10^{-12}	309	0.04	430
Zirconium	6.50	bcc	0.316	1852	4377	264	23.0	567	1.9	1.3×10^{-4}	272	6×10^{-16}	184	0.06	180
Zirconium carbide (ZrC)	6.73	NaCl	...	3532	5100	368	...	...	...	0.1	720	5×10^{-12}	468	...	...

(a) Crystal types: bcc, body-centered cubic; H, hexagonal; fcc, face-centered cubic; R, rhombohedral; O, orthorhombic; fct, face-centered tetragonal, Tetra., tetragonal; Mono., monoclinic. (b) VD, volume diffusion. (c) GBD, grain boundary diffusion. (d) SD, surface diffusion. Source: Ref 15 density or neck size versus temperature for isothermal sintering at various times. The underlying simulations rely on material data to combine mechanisms and determine the effects of the main process variables (time, temperature, green density, particle size, and grain size). Data needed for these calculations are collected in various books on sintering theory. The diagrams are sensitive to input material characteristics. The output might be density, neck size, dimensional change, or final compact shape (includes distortion from green density gradients) as a function of compaction conditions, sintering conditions, composition, or powder characteristics. The graphical output is most impressive, since the plots show the interactions between the key processing variables.

The effect of simultaneous transport mechanisms is to increase the overall rate of neck growth, but possibly with a decreased rate of shrinkage. One merit of computer simulated diagrams is that such combined effects are included in the calculations. When the sintering diagram is based on density, the surface transport contributions are less evident.

Today, the complexity of sintering is recognized, and realistic sintering calculations rely on sophisticated computer calculations. Early models failed to appreciate the multiple mechanisms and complex geometric changes associated with solid-state sintering. In modern industrial sintering situations, the systems are complex, and many proprietary codes exist that include nonlinear heating, wide particle size distributions, liquid formations, atmosphere interactions, and the behavior of mixed powders. Sintering models have only recently embraced some of the practical realities.

REFERENCES

1. H.E. Exner, Principles of Single Phase Sintering, *Rev. Powder Metall. Phys. Ceram.*, Vol 1 (No. 1–4), 1979
2. H. Udin, A.J. Shaler, and J. Wulff, The Surface Tension of Solid Copper, *Trans. AIME*, Vol 185, 1949, p 186–190
3. R.M. German, *Powder Metallurgy Science*, 2nd ed., Metal Powder Industries Federation, 1994
4. G.C. Kuczynski, Self-Diffusion in Sintering of Metallic Particles, *Trans. AIME*, Vol 185, 1949, p 169–178
5. G.H. Gessinger, F.V. Lenel, and G.S. Ansell, Continuous Observation of the Sintering of Silver Particles in the Electron Microscopy, *Trans. ASM*, Vol 61, 1968, p 598–604
6. D.L. Johnson, New Method of Obtaining Volume, Grain Boundary, and Surface Diffusion Coefficients from Sintering Data, *J. Appl. Phys.*, Vol 40, 1969, p 192–200
7. P.C. Eloff and F.V. Lenel, The Effect of Mechanical Constraints Upon the Early Stages of Sintering, *Modern Developments in Powder Metallurgy*, Vol 4, H.H. Hausner, Ed., Plenum Press, New York, 1971, p 291–302
8. G.C. Kuczynski, Statistical Approach to the Theory of Sintering, *Sintering and Catalysis*, G.C. Kuczynski, Ed., New York, 1975, p 325–337; G.C. Kuczynski, Statistical Theory of Sintering, *Z. Metallkd.*, Vol 67, 1976, p 606–610
9. B. Fisher and P.S. Rudman, Kirkendall Effect Expansion During Sintering in Cu-Ni Powder Compacts, *Acta Metall.*, Vol 10, 1962, p 37–43
10. R.W. Heckel and M. Balasubramaniam, The Effects of Heat Treatment and Deformation on the Homogenization of Compacts of Blended Powders, *Metall. Trans.*, Vol 2, 1971, p 379–391
11. R.A. Tanzilli, R.D. Lanam, and R.W. Heckel, Techniques for the Study of Homogenization in Compacts of Blended Powders, *Advanced Experimental Techniques in Powder Metallurgy (Perspectives in Powder Metallurgy)*, Vol 5, J.S. Hirschhorn and K.H. Roll, Ed., Plenum Press, 1970, p 139–188
12. R.W. Heckel, Diffusional Homogenization of Compacted Blends of Powders, *Powder Metallurgy Processing*, H.A. Kuhn and A. Lawley, Ed., Academic Press, New York, 1978, p 51–97
13. M.S. Masteller, R.W. Heckel, and R.F. Sekerka, A Mathematical Model Study of the Influence of Degree of Mixing and Particle Size Variation on the Homogenization Kinetics of Compacted Blends of Powders, *Metall. Trans. A*, Vol 6, 1975, p 869–876
14. P.S. Rudman, An X-Ray Diffraction Method for the Determination of Composition Distribution in Inhomogeneous Binary Solid Solutions, *Acta Crystallogr.*, Vol 13, 1960, p 905–909
15. R.M. German, *Sintering Theory and Practice*, John Wiley & Sons, 1996

SELECTED REFERENCES

- M.F. Ashby, A First Report on Sintering Diagrams, *Acta Metall.*, Vol 22, 1974, p 275–289
- K. Breitkreutz and K. Haedecke, Calculated Simulation of Non-Isothermal Sintering Kinetics, *Powder Metall. Int.*, Vol 22 (No. 6), 1990, p 14–15
- R.M. German, Overview of Key Directions and Problems, *Computational and Numerical Techniques in Powder Metallurgy*, D.S. Madan, I.E. Anderson, W.E. Frazier, P. Kumar, and M.G. McKimpson, Ed., Minerals, Metals and Materials Society, 1993, p 1–15
- R.M. German, *Sintering Theory and Practice*, John Wiley & Sons, 1996
- S.H. Hillman and R.M. German, Constant Heating Rate Analysis of Simultaneous Sintering Mechanisms in Alumina, *J. Mater. Sci.*, Vol 27, 1992, p 2641–2648
- K.S. Hwang, R.M. German, and F.V. Lenel, Analysis of Initial Stage Sintering Through Computer Simulation, *Powder Metall. Int.*, Vol 23 (No. 2), 1991, p 86–91
- A. Jogota and P.R. Dawson, Simulation of the Viscous Sintering of Two Particles, *J. Am. Ceram. Soc.*, Vol 73, 1990, p 173–177
- H.J. Leu, T. Hare, and R.O. Scattergood, A Computer Simulation Model for Particle Sintering, *Acta Metall.*, Vol 36, 1988, p 1977–1987
- L.R. Madhavrao and R. Rajagopalan, Monte Carlo Simulations for Sintering of Particle Aggregates, *J. Mater. Res.*, Vol 4, 1989, p 1251–1256
- H. Riedel and J. Svoboda, A Theoretical Study of Grain Growth in Porous Solids During Sintering, *Acta Metall. Mater.*, Vol 41, 1993, p 1929–1936
- C.M. Sierra and D. Lee, Modeling of Shrinkage During Sintering of Injection Molded Powder Metal Compacts, *Powder Metall. Int.*, Vol 20 (No. 5), 1988, p 28–33
- F.B. Swinkels and M.F. Ashby, A Second Report on Sintering Diagrams, *Acta Metall.*, Vol 29, 1981, p 259–281

Sintering Furnaces and Atmospheres

DURING SINTERING, compacted metal powders are bonded or sintered by heating in a furnace to a temperature that is usually below the melting point of the major constituent. Sintering occurs in a series of overlapping but balanced phases, all of which depend on temperature, time, and atmospheric composition, flow, circulation, and direction.

Furnaces used for sintering typically are similar to protective atmosphere furnaces used in brazing and heat treating of steel. However, the significant differences between heat treating and sintering require careful selection of furnace atmospheres for sintering. Primarily, P/M compacts are porous; therefore, a much greater surface area is exposed to the furnace atmosphere than with solid parts. Sintering temperatures are considerably higher than heat treating temperatures (1120 °C, or 2050 °F, for iron and steel compared to 900 °C, or 1650 °F, for carburizing and neutral hardening of steel).

This combination of greater exposed surface area and higher temperatures enhances chemical reactivity between the surface and atmosphere during sintering. The trend toward higher sintering temperatures also places additional requirements on furnace capability and tight atmosphere control. These factors are briefly discussed for both conventional and high-temperature sintering.

Continuous Furnaces

Tim Raffeinner,
Abbott Furnace Company

Continuous production sintering furnaces for powder metal compacts come in several designs including the mesh-belt conveyor furnace, the ceramic-belt conveyor furnace, the roller-hearth furnace, the pusher furnace, the walking-beam furnace, and the continuous vacuum furnace (Fig. 1). A typical continuous sintering furnace has four distinct areas: the preheat, delube, or burn-off area; the high heat or sintering area; the slow cool or transition area; and the final cooling area (Fig. 2).

Furnace Characteristics

Preheat Area. The function of the preheat area is to remove the compacting lubricants from the green powder metal part. Hence the preheat area is also called the delubing or burn-off zone.

Lubricants commonly used in compacting are zinc stearate, lithium stearate, Acrawax, and par-

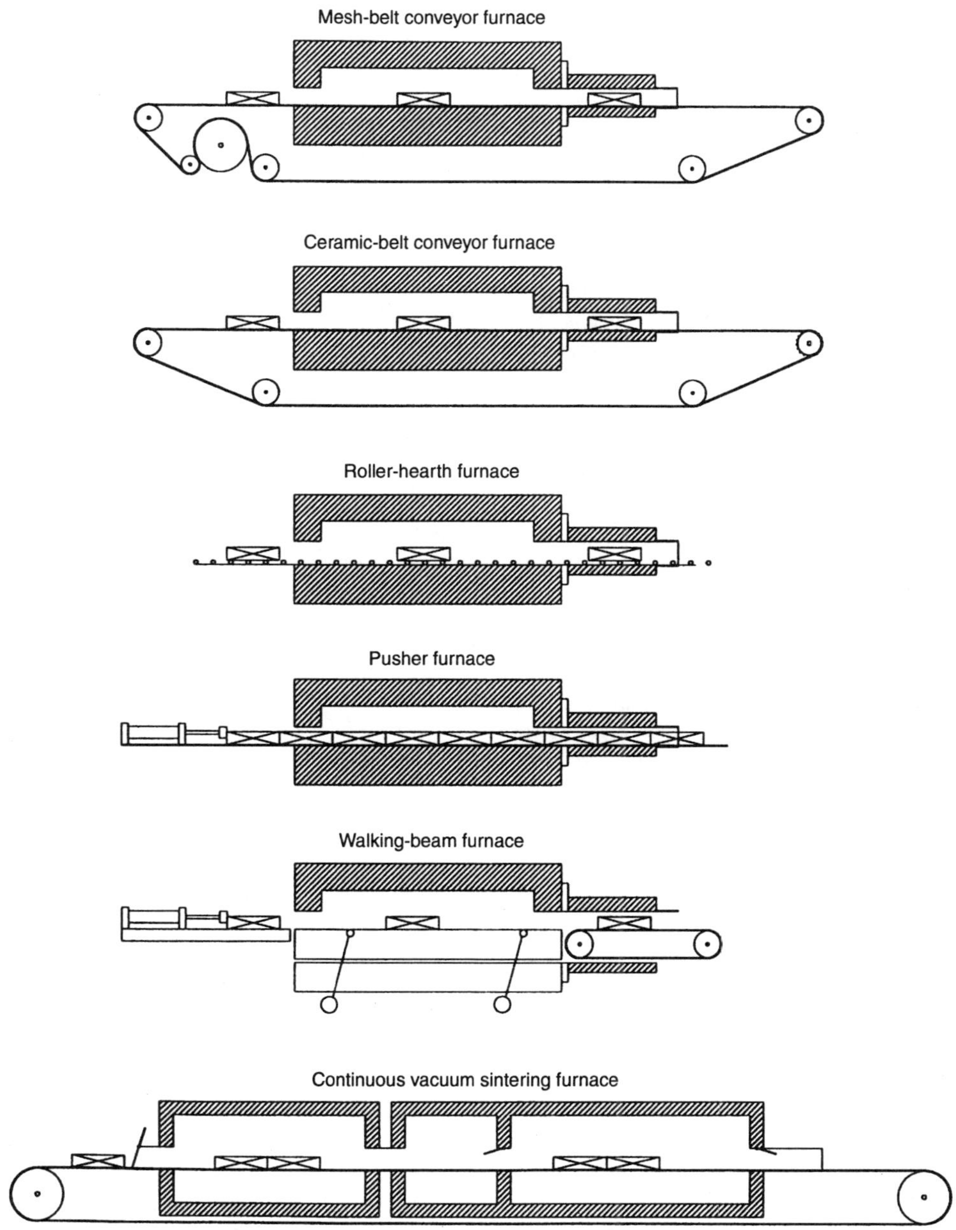

Fig. 1 Schematic of various continuous furnaces

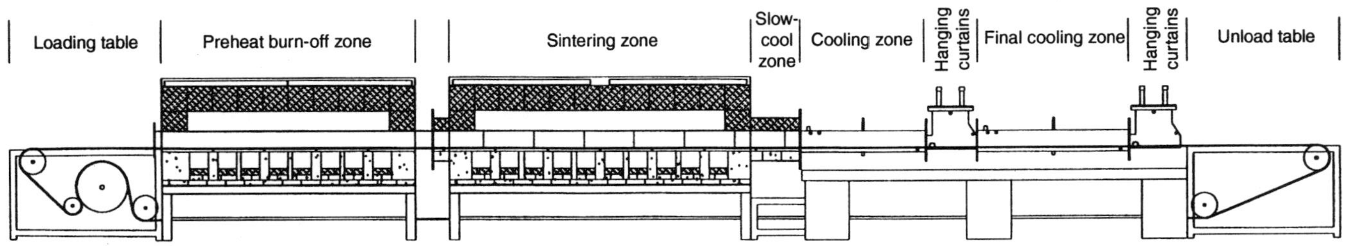

Fig. 2 Schematic of a muffled continuous mesh-belt sintering furnace

affin (Table 1). Because these lubricants have relatively low melting points up to ~220 °C (430 °F), it is assumed that the maximum operating temperature of the preheat would only need to be slightly above this. However, because the difficulty in removing lubricants increases as density and cross-sectional area increases, actual maximum operating temperature of the preheat area typically is ~1010 °C (1850 °F).

It is important for the green compact to have a controlled heating rate to avoid part blistering because the lubricant liquefies and progresses through the resulting porosity toward the surface of the part where it vaporizes. Length of the preheat is typically 50 to 100% of the high-heat section.

Most preheat areas contain a muffle that forms a narrow passageway for the vaporized lubricant to flow out of the front of the furnace. The muffle also prevents lubricant from condensing on other internal furnace components and reducing their life.

Preheat areas can be heated by either electric or gas. Wire elements are commonly used in place of silicon carbide elements due to the lower operating temperatures.

Nitrogen moisturizing systems, also known as bubblers, help to increase the dew point in the preheat area to aid in the delubing process. Nitrogen is fed into the bottom of a sealed, heated container filled with water, and then it is collected at the top of the container and injected into the preheat after it has picked up moisture.

Enhanced lubricant removal systems are specially designed for parts with large cross-sectional areas or heavy furnace loads. The length of the enhanced lubricant removal area varies depending on whether it is added as a separate unit previous to the preheat section or a combination thereof.

The enhanced delubing process works on the principle that the combustion products of a gas burner combined with the regular furnace atmosphere provide the proper combination for more efficient lubricant removal. On some units, once the gas burner is set, it fires at the same rate even as furnace loading changes. Wire-wound elements under the furnace hearth provide any extra heat needed as loading increases. Burner velocity does not increase, which means that internal furnace pressures remain unchanged. Consistent pressure yields a steady frontal flow, which results in consistent delubing results (Fig. 3).

High-Heat Area. The function of the high-heat area is to sinter the individual powder particles together, while at the same time increasing density due to shrinkage of the part.

The design of the high-heat area should be such that the P/M parts reach sintering temperature in the quickest time possible and then remain there until particles are adequately bonded and alloying elements fully diffused into the matrix. The maximum operating temperature of the high-heat area for conventional ferrous powders is usually 1177 °C (2150 °F) and 1300 °C (2400 °F) for high nickel/chrome alloys.

Length of the high heat is based on the average part cross section and belt loading. The high heat can be 100 to 200% of the length of the preheat area.

The high heat can be an open-chambered design or can contain a muffle. Open-chambered furnaces expose the brickwork and heating elements to reaction with the furnace atmosphere and do not offer as much carbon control as muffled furnaces. Muffles can be made of either an alloy or ceramic. Ceramic muffles can be silicon carbide or alumina, with silicon carbide offering better thermal conductivity, which in turn yields better thermal shock resistance. Ceramic muffles come in sections that are fitted together with a male/female joint structure (Fig. 4). Each ceramic muffle section can be a single piece design or can be a two-piece, base and lid, design. Because of the slight amount of porosity of ceramic muffles, each furnace section containing the ceramic muffle sections is sealed and purged with nitrogen gas to ensure that none of the P/M parts are subjected to outside air.

High-heat sections can be heated with gas or electricity. When heating with gas, a refractory alloy muffled design is preferable to keep burner off gases from entering the furnace atmosphere. Electric elements can be composed of silicon carbide or molybdenum, depending on the maximum operating temperature of the high-heat section.

Slow-Cooling Area. The function of the slow-cooling zone can be to aid in microstructure development or to act as a shock zone or transition zone, to lessen the thermal stress between the heated and cooled sections of the

Table 1 Commonly used compacting lubricants

Lubricant	Chemical formula	Melting point °C	Melting point °F
Zinc stearate	$Zn(C_{18}H_{35}O_2)_2$	130	266
Lithium stearate	$LiC_{18}H_{35}O_2$	221	430
Paraffin	$C_{22}H_{46}$ to $C_{27}H_{56}$	40–60	104–140
Acrawax	...	140–143	284–290

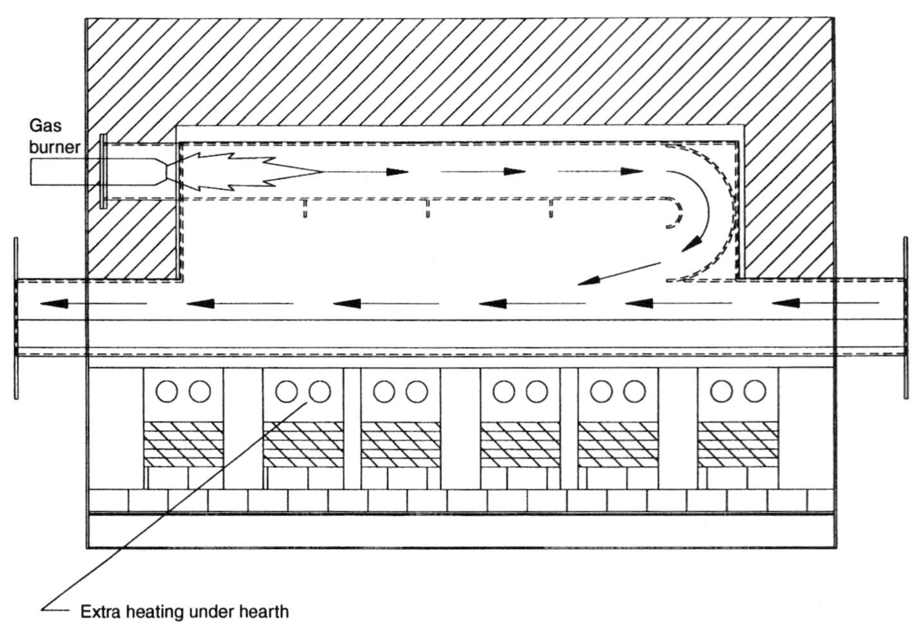

Fig. 3 Schematic of an enhanced delubing system

furnace. This section can be an open-chambered design, or it can contain an alloy or ceramic muffle.

Final Cooling Area. The function of the final cooling section is to cool the P/M parts below their oxidation point under atmosphere before they exit the furnace into the outside air.

Typically, the final cooling area consists of cool water, which surrounds a muffle-like design to keep the parts from being exposed directly to the water. This type of unit relies mostly on radiant cooling, with limited conduction through the product transfer mechanism. As the water collects heat, it is pumped to an area where it is cooled, then recirculated to the final cooling area to accept more heat. Length of the final cooling area is typically 150 to 250% that of the sintering section.

A water-jacketed cooler can be any one of several designs. It can be a simple inner and outer steel jacket with water flowing in between. The cooler can have an open top that eliminates pressure buildup. It can contain inserts welded into place between the inner and outer jacket to help direct the flow of water. The cooler can contain multiple injection locations that oppose each other to create a turbulent effect between the two jackets around the entire cooler. Uniform water temperature in the multiple injection port cooler eliminates thermal stress caused by uneven expansion associated with the other types of water-jacketed coolers (Fig. 5).

Another type of final cooling area now gaining rapid acceptance utilizes convection cooling to remove heat from the P/M product. This type of cooling system consists of an area that removes existing heated atmosphere from the furnace, a heat exchanger to cool that atmosphere, and another area where the cooled atmosphere is reintroduced into the furnace. These units contain features to increase atmosphere velocity and turnover, which in turn effect the British Thermal Unit (BTU) exchange rate.

Some systems offer a variable speed blower, the speed of which can be adjusted depending upon the cooling needs of each individual product line. It can also be completely water jacketed, allowing the convection portion to be shut off and the unit to be used as a conventional cooling chamber.

Uses of the convection cooling chamber can be to cool heavy loads or parts with large cross-sectional areas, to shorten the overall length of the furnace in areas with restricted floorspace, or to sinterharden certain materials, which eliminates subsequent heat treatment operations.

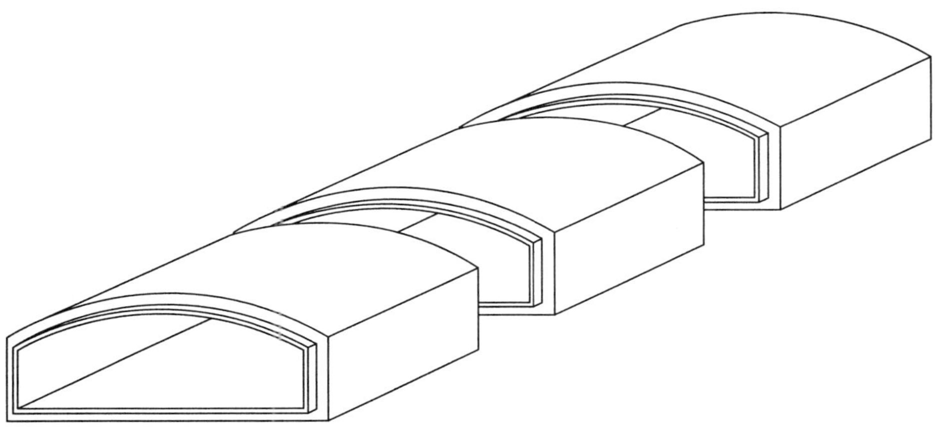

Fig. 4 Schematic of ceramic muffle sections

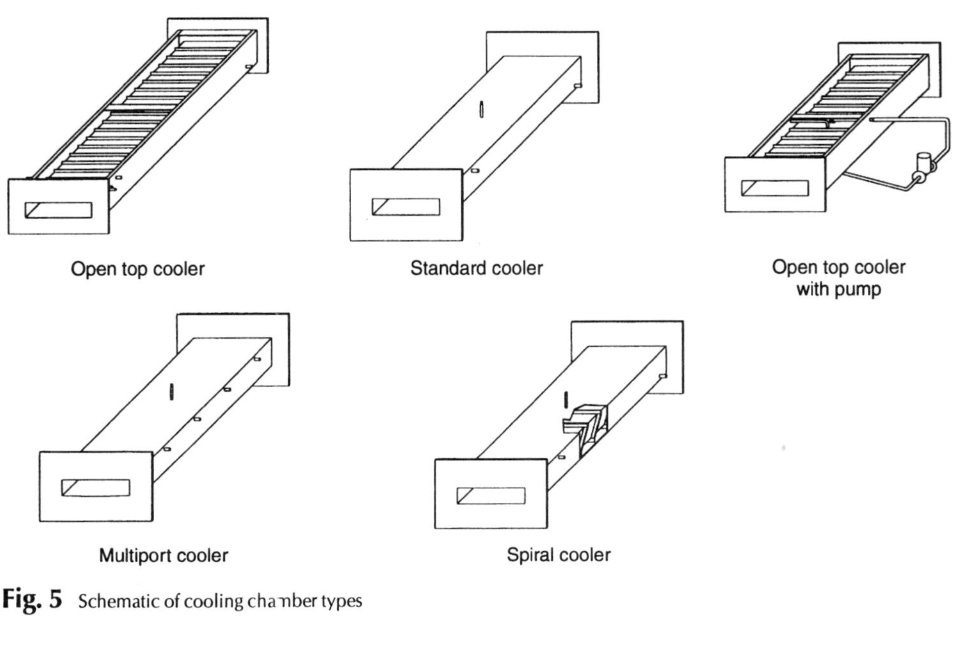

Fig. 5 Schematic of cooling chamber types

Continuous Furnace Types

Mesh-belt Conveyor Furnaces. Mesh-belt conveyor furnaces used for sintering P/M compacts are the most common type of production sintering furnace. These furnaces provide a continuous, reproducible time-temperature-atmosphere profile.

The mesh-belt furnace consists of a load table, a preheat area, a high-heat area, a slow-cool area, a final cooling area, and an unload or discharge table. The belt is driven by a drum system located on the charge table, which frictionally drives the belt by pinching it between the drums.

The preheat and high-heat areas of the furnace can contain a muffle or can be of an open-chambered design. Muffled furnaces provide a narrower channel for better atmosphere flow, as well as protection for other furnace internals against reactions with the furnace atmosphere and lubricant byproducts.

The belt is made of mesh woven from alloy wire. Temperature limitations are typically ~1177 °C (2150 °F).

Parts can be placed directly on the belt at the charge end of the furnace. Finer mesh screens or trays are used when parts will fall through the mesh size of the belt, or when certain part surface conditions must be met.

Mesh-belt conveyor furnaces have typical loading capabilities of 73 kg/m^2 (15 lb/ft^2). With heavier mesh belts and drive mechanisms, loading can be increased up to 122 kg/m^2 (25 lb/ft^2).

Ceramic-belt conveyor furnaces are used for sintering P/M compacts at a higher temperature than mesh-belt conveyor furnaces. Like their lower-temperature counterparts, they provide a continuous, reproducible time-temperature-atmosphere profile.

Ceramic-belt furnaces consist of a load table, a preheat area, a sintering area, a slow-cooling area, a final cooling area, and an unload table. The belt is driven by a sprocket located on the discharge table.

The high-heat area of the furnace contains a ceramic muffle instead of the conventional alloy design. Muffle material can be silicon carbide or alumina to eliminate the distortion at higher temperatures.

The belt can be made of silicon carbide or alumina. Silicon carbide offers better thermal conductivity and specific heat characteristics, which yield better thermal shock resistance. The belt consists of pressed links and spacers, held together by a crossrod. Parts are placed on ceramic trays to prevent contact and subsequent reaction with the belt.

The ceramic-belt conveyor furnace has production capabilities up to 136 kg/h (300 lb/h), with belt loading of 49 kg/m^2 (10 lb/ft^2). With a special roller belt design, which reduces friction inside the furnace, belt loading can be increased above the 49 kg/m^2 (10 lb/ft^2) rate.

Humpback furnaces are basically a mesh-belt sintering furnace adapted to using high-purity, low-density atmospheres such as dissociated ammonia or hydrogen. They are especially useful to conserve atmosphere when the need for a large charge opening exists.

Humpback furnaces consist of a load table, an upward inclined preheat area, a horizontal sintering area, a horizontal slow-cool area, a downward inclined, final cooling area, and an unload table (Fig. 6). The humpback furnace can be an open-chambered or muffled design. Muffled designs can contain a refractory alloy or ceramic muffle.

The belt is made of alloy wire woven into a mesh design. Parts can be placed directly on the belt but can also be placed on trays to assist them on the inclined areas of the furnace.

The humpback furnace typically has belt loading capabilities of 73 kg/m^2 (15 lb/ft^2). With a special helper belt to assist the main belt up the incline, loading can be increased to ~122 kg/m^2 (25 lb/ft^2).

Pusher Furnaces. In a pusher furnace, parts are loaded on trays or ceramic plates that are pushed through the stationary hearth furnace. Generally, this type of furnace has a lower output than a belt furnace. Furnaces can have a manual, hydraulic, or mechanical pusher system. The pusher mechanism can operate intermittently or continuously. When it is continuous, a reproducible thermal profile of the work can be achieved. Pusher furnaces operate at temperatures up to ~1650 °C (3000 °F).

Roller-Hearth Furnaces. In a roller-hearth furnace, parts are carried on trays through the furnace on driven rolls. Roller-hearth furnaces can be made to greater lengths and widths than most conveyor-belt furnaces, so they have large output potential. Generally, these furnaces are capable of heavier loading (pounds per square foot) on the hearth than mesh-belt furnaces. The roller-hearth design allows for fast entry and exit of the work, which minimizes open door time and reduces atmosphere consumption. This furnace also allows a continuous, reproducible thermal profile. The maximum achievable temperature is ~1150 °C (2100 °F) because of the limitations imposed by available alloys for rolls.

Atmosphere quality in a roller-hearth furnace may be poor. The large cross-sectional area of this furnace gives the lowest forward velocity of atmosphere, which can allow lubricant burn-off residue to diffuse backward into the sintering area.

Walking-beam furnaces are particularly well suited for applications in which sintering temperatures are above the limitations of the mesh-belt conveyor and roller-hearth furnaces. These furnaces operate up to 1650 °C (3000 °F).

The work to be conveyed through the furnace is placed on ceramic carrier plates. In a walking-beam furnace, the middle section of the hearth is a movable beam system that extends the entire length of the furnace. This beam can be raised above the hearth level, lowered below the hearth level, and moved longitudinally. To convey the work through the furnace, the beam is raised, lifting the ceramic carrier plates off the stationary hearth section. The beam then moves forward a fixed distance, taking the carriers with it, and drops below the stationary hearth level, where it retracts to the initial position. With this rectilinear movement, the beam intermittently moves the carriers containing the work through the furnace and cooling zones.

Design of the walking-beam furnace permits all-ceramic construction of the furnace interior, the walking beam, and the carrier plates, thereby facilitating operation at high temperatures. In addition, walking-beam furnaces are capable of carrying extremely heavy loads at temperatures ≤1650 °C (3000 °F), which is unattainable in mesh-belt conveyor, roller-hearth, and pusher furnaces.

Fig. 6 Humpback sintering furnace. Source: Abbott Furnace Co.

Batch-Type Furnaces

Bell Furnace. The base of the bell furnace is permanently installed on the floor. Work to be processed is loaded onto the base and then covered with a heat-resistant alloy retort to contain the protective atmosphere. The furnace bell then is lifted and placed over the retort and base.

The elevator furnace has the furnace bell above the mill floor on a fixed structure. The base and work load are covered with a retort and rolled on tracks under the furnace. An elevator system then raises the base, work, and retort into the furnace.

Bell and elevator furnaces are commonly used for sintering friction materials. They can be equipped with work-pressing devices to apply heavy pressure to loads such as clutch and brake facings. They also are useful for sintering heavy loads or large pieces and for applications that require long programmed heating and cooling cycles. These furnaces can also be used when extremely pure atmosphere conditions are required.

Vacuum Furnaces

Vacuum sintering can be conducted in either batch-type or continuous furnaces. Vacuum sintering does not differ from conventional atmosphere sintering, except for one modification: a vacuum furnace operates in the absence of an internal atmosphere.

Batch-type vacuum sintering furnaces were first developed for sintering compacts from refractory and reactive metals such as tantalum, niobium, titanium, zirconium, and their alloys (which react with any but the noble gases, such as argon or helium), for sintering cemented carbides containing titanium carbide and tantalum carbide and for sintering alnico magnets.

Generally, a batch-type vacuum furnace consists of an outer vacuum-tight cylindrical casing that contains a furnace with radiation shields or other types of insulation, work support, and heating elements. This casing is fitted with roughing and diffusion pumps to achieve the desired vacuum levels.

Vacuum furnaces are designed for use at a variety of temperatures. Units capable of reaching 2760 °C (5000 °F) are available. Because all heat transfer in a vacuum is accomplished by radiation, cooling cycles can be quite long. Therefore, in many cases, the furnace is designed for back filling with nitrogen or argon, which when combined with natural convection shortens cooling time.

Continuous vacuum sintering of iron and steel compacts and silicon-iron magnetic material is accomplished in automated furnaces. The furnace typically is comprised of an external loading table, followed by an atmosphere delubrication chamber, a transfer station, and an "atmosphere to vacuum" vestibule section, followed by a heating chamber, vacuum cooling chamber, a combination fan cooling and "vacuum-to-atmosphere" vestibule, and an unloading table. The advantages of a continuous vacuum sintering furnace are a more flexible temperature range and a choice of slow cooling or rapid cooling in a pressure quench. Most importantly, there is absolute carbon control and no need to continuously monitor an atmosphere for dew point or carbon concentration.

Sintering Atmospheres

H.S. Nayar, BOC Gas Inc.

Atmospheres used for sintering must perform several functions (Ref 1–2). During sintering, atmospheres must:

- Prevent air from entering the furnace
- Delube or dewax the parts
- Reduce surface oxides on the powder particles
- Control carbon on the surface and in the core of steel parts
- Remove carbon in special applications
- Provide controlled oxidation during cooling in special applications
- Convey or remove heat efficiently and uniformly

Because furnace atmospheres affect the sintering process and the material being treated, sintering is never performed in air or in an oxygen-rich atmosphere. A basic function of a sintering atmosphere is to protect metal parts from the effects of contact with air. Atmospheres must be maintained at a sufficient pressure and flow rate to prevent infiltration of air through furnace openings. In addition to protecting the part, atmospheres provide sufficient conduction convection for uniform heat transfer to ensure even heating or cooling within the various furnace zones (Fig. 7).

Delube and Preheat Atmospheres

Delubing involves the burning and removal of hydrocarbon lubricants used to compact the powdered metal. Removal occurs early in the sintering process in the low-temperature delube zone of the furnace. Small amounts of oxidizing gases in the atmosphere cause effective lubricant burning without oxidizing parts or decarburizing steel parts. The most effective oxidants are controlled amounts of water and carbon dioxide.

Without such oxidizing gases, lubricants tend to decompose thermally into hydrogen and carbon >540 °C (1000 °F). Hydrogen is given off as a gas, but carbon remains as a sooty deposit. If a suitable oxidant is present, however, a chemical reaction (oxidation) rather than thermal decomposition occurs. The carbon in the lubricant combines with the oxidant and is removed as a gaseous mixture of carbon monoxide and carbon dioxide. For example, assuming C_xH_y is a wax-like lubricant and water is the oxidizing agent, the following reaction applies:

$$C_xH_y + H_2O \xrightarrow{\text{approx } 550\ °C\ (1020\ °F)} CO + CO_2 + H_2$$

Delubing in a sintering atmosphere is performed most efficiently when a wet (high dew point) atmosphere is used. Atmosphere circulation should also be provided; atmosphere flow toward the furnace entrance is desirable. Ideally, the maximum temperature of the P/M compacts during delubing should range from 425 to 650 °C (800 to 1200 °F).

Sooting is primarily caused by improper delubing. The majority of sintering problems are caused by incomplete or improper delubrication. Sooting occurs when hydrocarbon vapors from the lubricants decompose thermally by the following reaction:

$$CH_4\ (g) \rightarrow 2H_2\ (g) + C\ (s)$$

Sooting can also be explained by the following reaction:

$$2CO\ (g) \rightarrow C\ (s) + CO_2\ (g)$$

Sooting problems can be classified into four types to help take corrective action as described in the following paragraphs (Ref 3).

Adherent soot appears as a black stain and is not easily rubbed off. It can occur on the top, bottom, and the sides of the P/M part. The lubricant oozes to the surface and does not fully vaporize or oxidize before the part enters the hot zone. The causes can be any one or more of the following:

- Dew point too low in the preheat
- The preheat length too short
- Part density too high
- Incorrect preheat temperature
- Belt loading too high
- Incorrect temperature profile

The short term solution is to increase the preheat dew point using a humidified nitrogen system. Reducing belt loading and belt speed can also help. Ideally, the preheat should be lengthened or a rapid burn off unit installed to accelerate the delubing process.

Granular soot looks like black snow and appears mostly on the top surface of the part. The flaky black soot can be easily blown off. It is caused when lubricant vapors thermally decompose into black soot and rains down onto the part. Granular soot is caused when the following conditions occur:

- The dew point is too low in the preheat
- Lubricant vapors migrate into the hot zone
- The forward velocity of atmosphere is too low
- The atmosphere profile shifts due to plant drafts
- Improper design of exhaust stacks

Solutions for granular sooting include increasing total atmosphere flows, redirecting the flow, and redistribution of flows into different inlets. It is also helpful to control up and down drafts in the exhaust stacks.

Soot deposits on belts are indicative of granular sooting. Correcting sooting problems in general will help get rid of soot on the belts. Even small amounts of sooting tend to accumulate along the belt edges over extended periods of time. Soot on belts can lead to carburization and embrittlement of the belts.

Shiny soot appears as a black uniform and shiny coating on all exposed surfaces of the part. It is caused by the catalytic cracking of natural gas on the P/M part surface in the hot zone. The obvious solution is to decrease the amount of natural gas addition. Changing the location of the natural gas addition can also help solve this problem.

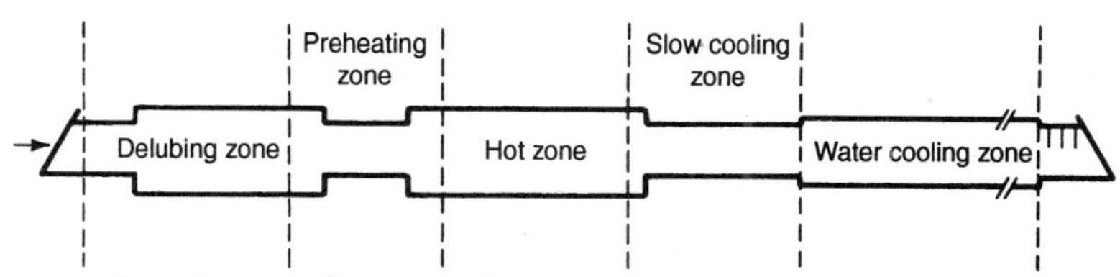

	Delubing zone	Preheating zone	Hot zone	Slow cooling zone	Water cooling zone
Atmosphere functions →	• Convey heat quickly and uniformly • Burn and sweep out lubricants to front exit	• Reduce surface oxides • Carbon diffusion	• Copper melting, coating, or infiltrating • Bonding • Carbon control	• Carbon control • Cooling rate control	• Cooling • Prevent oxidation or controlled light oxidation
Atmosphere composition →	• Lightly oxidizing	• Highly reducing • Neutral to carbon	• Reducing • Neutral to carbon preferred	• Reducing • Neutral to carbon preferred	• Slightly reducing or neutral or slightly oxidizing
Temperature range (steel) →	• 425 to 650 °C (800 to 1200 °F)	• 650 to 1040 °C (1200 to 1900 °F)	• 1040 to 1120 °C (1900 to 2050 °F)	• 1120 to 815 °C (2050 to 1500 °F)	• 815 °C to ambient (1500 °F to ambient)

Fig. 7 Typical furnace schematic for sintering steel

Blistering/rippling can be seen on the P/M part and sometimes the part crumbles into powder at sharp edges. Blistering is an extreme case of sooting, which occurs within the P/M part. It is commonly seen in nickel-containing grades and P/M parts with very high densities. Another cause is too rapid a heating rate in the preheat zone. Solutions include decreasing belt speed and belt loading rates. Increasing wet N_2 flows and adding a pre-preheater can be helpful.

Preheating Zone and Oxide Removal. As the sintering process progresses, the furnace performs another function in addition to delubrication. At ~650 °C (1200 °F), generally near the end of the delube zone and through the hot zone (~1120 °C, or 2050 °F for iron and steels), the atmosphere begins to strip surface oxides on the metal particles that comprise the part.

In the preheat zone, the atmosphere must reduce surface oxides on the metal particles to ensure a clean, metallic surface on the particles comprising the P/M compact. This permits admixed graphite to diffuse into the iron particles, thus creating a pearlitic microstructure during cooling. Graphite diffusion into the particles is essentially completed when parts reach ~1040 °C (1900 °F).

Clean particle surfaces further improve bonding of the particles at >1100 °C (2010 °F). Thus, the greater the oxide reducing effect of the atmosphere, the stronger the sintered bond. Furthermore, porosity between the particles becomes rounded, thus improving the structural integrity and toughness of the part.

Oxidation

Under most operating conditions, metal oxides *(MO)* are reduced to metal *(M)* primarily by the hydrogen component of the atmosphere as shown by the following reaction:

$$MO + H_2 \rightarrow M + H_2O$$

The effectiveness of an oxide-reducing atmosphere can be determined by the ratio of hydrogen to water, as shown for iron in Fig. 8(a) and (b). In the P/M industry, water content is generally determined by measuring the dew point of the atmosphere. Figure 9 shows water content as a function of sintering atmosphere dew point. As the ratio of hydrogen to water increases, the reducing reaction is more effective. The ratio of hydrogen to water, not the absolute levels of these constituents, determines the reducing effectiveness of the atmosphere. Furthermore, for a given dew point, the hydrogen content of the atmosphere becomes more reducing to oxides as temperature increases.

Oxidation of P/M parts can occur in many forms and to varying degrees. Oxidation of iron can be caused by air, moisture, or carbon dioxide.

Oxidation by air appears as a light gray to dull black color. It is generally a function of the oxygen partial pressure, temperature, and the metal. Oxygen contents >25 ppm in the cooling section will oxidize a steel part (Ref 3). Oxidation by moisture (H_2O) and carbon dioxide (CO_2) for a given temperature and material, is a function of the ratios H_2/H_2O and CO/CO_2.

The color of the oxidation is an indication of the degree of the problem. Given the color of oxidation and the temperature profile of the furnace, it is possible to approximate where the oxidation occurred. Table 2 gives colors for steel.

Surface oxidation occurs in the cooling zones and, in general, has only a surface effect with the other properties unaffected. Oxidation problems in sintering furnaces can be differentiated based on the source of oxidation.

Air Infiltration from the Exit End. The oxidation caused by air ingress from the exit end of the furnace will appear randomly and erratically. The oxidation color will change with part weight. It can occur preferentially along one edge of the belt and can show only on the leading edge of the part, or after a gap in the belt loading. Possible causes include:

- Insufficient atmosphere flow
- Atmosphere flow imbalance
- Pressure differential at the ends of the furnace
- Up and down drafts in exhaust stacks
- Improper design of exit door curtains

The solutions include increasing atmosphere flow rates, controlling the up and down drafts in the exhaust stacks, fixing or replacing exit door curtains, relocating fans at front and exit ends to avoid crosswinds, and keeping the exit door as low as possible with the front door kept 13 mm (0.5 in.) higher than the exit end door.

Air Leaks in Water-Cooled Sections. The oxidation in this case will be continuous and will get progressively worse. The oxidation will be on all surfaces, and belt marks may be seen on the bottom surface. Shielded surfaces may appear bright and clean. The causes include the following:

- Loose flanges due to creep of bolts
- Air leaks in the atmosphere piping
- Cracks in the joints between water-cooled sections

Solutions are to find and fix the leaks. An atmosphere profile will help pinpoint the leaks. The flanges should be checked for leaks. It is desirable to have the flanges welded. The incom-

Table 2 Oxidation colors for steel

Temperature(a) °C	°F	Color
205	400	Faint straw
225	440	Straw
245	475	Deep straw
270	520	Bronze
280	540	Peacock
310	590	Full blue
340	640	Light blue
340–390	640–730	Gray to black

(a) Higher temperature oxidation results in scaling or etching.

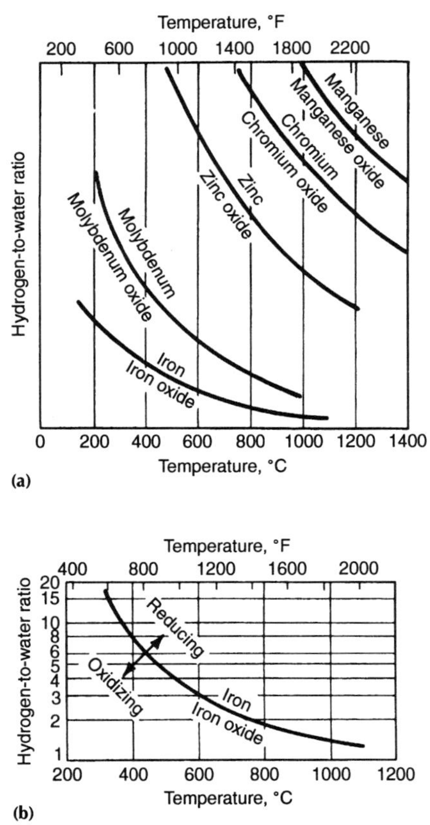

Fig. 8 (a) Metal-to-metal oxide equilibria for common elements used in powder metallurgy. Copper, lead, cobalt, nickel, and tin oxide are easier to reduce than iron oxide. (b) Iron-to-iron oxide equilibria

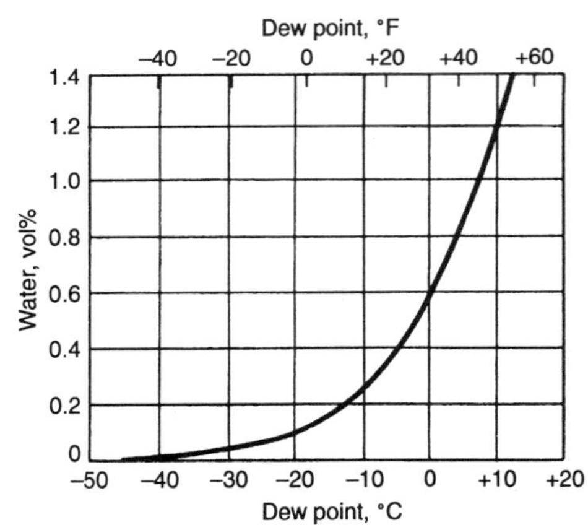

Dew point °C	°F	Water, vol%
−80	−112	0.00005
−60	−76	0.00106
−50	−58	0.00388
−40	−40	0.01270
−30	−22	0.03760
−20	−4	0.1020
−10	14	0.2570
0	32	0.6020
10	50	1.2120
20	68	2.3080

Fig. 9 Water content as a function of dew point of the atmosphere

Table 3 Typical physical properties of various gaseous feedstocks at 20 °C (68 °F) and 1 atm

Property	Natural gas (CH_4)	Ammonia vapor(a)	Air	Propane gas	Methanol vapor(a)	Nitrogen gas	Hydrogen gas
Specific gravity(b)	0.59	0.60	1.00	1.55	1.11	0.97	0.07
Gas density, kg/m^3 (lb/ft^3)	0.72 (0.045)	0.74 (0.046)	1.20 (0.075)	1.91 (0.119)	1.33 (0.083)	1.17 (0.073)	0.08 (0.005)
Specific volume, m^3/kg (ft^3/lb)	1.39 (22.2)	1.35 (21.7)	0.83 (13.3)	0.52 (8.43)	0.75 (12.05)	0.85 (13.8)	12.50 (192.0)
Molecular weight	16.08	17.03	28.95	44.00	32.04	28.01	2.02
Specific heat vapor, J/kg · °C × 10^{-3} (Btu/lb · °F)	2.48 (0.593)	2.20 (0.525)	1.01 (0.241)	1.69 (0.404)	1.38 (0.33)	1.04 (0.249)	14.3 (3.42)
Caloric value, J/m^3 × 10^7 (Btu/ft^3)	3.837 (1,030)	1.814 (487)	...	9.501 (2,550)	3.029 (813)	...	1.211 (325)
Caloric value, J/kg × 10^7 (Btu/lb)	5.343 (22,970)	2.549 (10,958)	...	5.001 (21,500)	2.279 (9,800)	...	14.210 (61,095)
Air/fuel ratio at 100% combustion	10.3	3.6	...	24.0	7.14	...	2.4

(a) 1 kg of liquid ammonia yields 1.4 m^3 of ammonia vapor (1 lb produces 22.5 ft^3). Ammonia vapor doubles in volume on dissociation; 1.4 m^3 of ammonia vapor becomes 2.8 m^3 of nitrogen and hydrogen mixture (1-to-3 ratio). 1 kg of liquid methanol yields 0.68 m^3 of methanol vapor (1 lb produces 11 ft^3). Methanol vapor triples in volume on dissociation. 0.68 m^3 of methanol vapor becomes 2.0 m^3 of carbon monoxide and hydrogen mixture (in 1-to-2 ratio). (b) Air equals 1.0.

ing atmosphere should be checked to ensure that the O_2 content is <10 ppm.

Water Leaks in Cooling Section. This results in oxidation of a darker color, which is more adherent. The oxidation is continuous and gets worse with time. The causes are the following:

- Holes or cracks in the water-cooled section
- Incoming atmosphere too wet

Solutions to this problem involve differentiating between a water and an air leak. This can be done by performing the copper/steel test and by taking air and dew point readings.

A copper/steel test is performed to determine if the oxidation occurring in the cooling section is caused by an air leak or by a water leak. The test is performed as follows:

- Reduce the furnace temperature to 1040 °C (1900 °F)
- Place a piece of bright copper strip and a steel strip side by side on the belt and send them through the furnace
- Copper will be oxidized by oxygen but not by water
- Steel will be oxidized by both oxygen and moisture

Heavily oxidized parts occur when O_2 or moisture seeps into the hot zone from the front or the exit end or from leaks in the muffle. The part appears heavily oxidized (black) and has poor sintered strength. Adverse conditions in the preheat and water-cooled sections can compound the problem. Solutions include the following:

- Find and fix source of oxidant
- May require higher hydrogen percentage
- Incoming atmosphere may be too wet
- Incorrect pressure differential between the ends of the furnace

Frosted parts exiting the furnace will appear to have a dull matte finish on surfaces that normally have a shiny finish, such as the sides of the part. The part also tends to be decarburized and have lower hardness. This phenomenon occurs when oxidation occurs in the preheat zone followed by reduction in the hot zone. Causes include:

- Dew point too high in the preheat zone
- Air entering the preheat zone from the front end or hole in the muffle
- Down draft at front exhaust
- Shifts in atmosphere profile

Frosty parts can be avoided by reducing the front end oxidation, fixing imbalances in the exhaust system, and redirecting the atmosphere flows.

Belt Oxidation. Belts for sintering furnaces (314 stainless steel), contain chromium and will oxidize at lower oxidation potentials than regular steel parts. If the belt appears dark and parts look bright, then marginal air infiltration from the exit end is occurring. Increasing atmosphere flow at the exit end will dilute the O_2 contaminant. If the belt is clean and adherent scale is seen, then preheat oxidation followed by hot zone reduction is occurring. Solutions are the same as those for frosted parts.

Carbon Control

Tight control of carbon levels in powder metal parts is one of the distinguishing attributes of the high-quality P/M producer. The carbon content affects many of the properties of the finished part such as dimensional tolerances and mechanical properties. The pick up of carbon also decreases the corrosion resistance of stainless steel.

Carbon control is achieved primarily through a fine tuning of the sintering atmosphere, as it relates to temperature, admixed carbon, carbon and oxygen content of the powder, and delubrication. The control of this atmosphere is greatly improved through the use of synthetic (N_2H_2) versus generated atmospheres.

The reactions relating to carbon control occur primarily in the hot zone and the slow-cool zone of the furnace (Fig. 7). If the atmosphere is very active—an endothermic atmosphere with 60% combustibles (hydrogen and carbon monoxide) at ~5 °C (40 °F) dew point, for example—a steel surface with ~0.8% combined carbon is decarburized to ~0.2% combined carbon (Fig. 10) in the hot zone (typically 1120 °C, or 2050 °F) to a depth of 0.5 to 2.5 mm (0.02 to 0.1 in.).

The degree and depth of decarburization depend on the dew point of the active endothermic atmosphere, the processing time and temperature in the hot zone, and the porosity of the steel part. Higher values for these parameters lead to a greater degree of decarburization.

During slow cooling from ~1120 to 815 °C (2050 to 1500 °F) under an active endothermic atmosphere, carbon is more or less restored to

Table 4 Typical physical properties of various liquid feedstocks at 20 °C (68 °F) and 1 atm

Property	Propane	Methanol	Ammonia	Nitrogen	Hydrogen
Liquid density:					
kg/m^3 × 10^2	5.09	7.94	6.09	8.09	0.71
lb/gal	4.25	6.63	5.08(a)	6.75	0.5906
lb/ft^3	31.8	49.5	38.0	50.5	0.071
Specific gravity(b)	0.509	0.794	0.609	0.809	4.42
Molecular weight	44.0	32.04	17.03	28.01	2.02
Freezing point, °C (°F)	−123.3 (−190)	−97.8 (−144)	−77.8 (−108)	−209.9 (−345)	−252.1 (−434.5)
Boiling point, °C (°F)	−46.1 (−51)	64.5 (−148)	−33.3 (−28)	−195.6 (−320)	−252.7 (−423)
Vapor pressure:					
At 20 °C (68 °F), kPa (mm Hg)	855.2 (6,413)	13.3 (100)	889.5 (6,670)	Very high	Very high
At 50 °C (120 °F), kPa (mm Hg)	1,692.4 (12,691)	53.5 (400)	2,033.9 (15,252)	...	...
Specific heat (liquid), J/kg · °C × 10^{-3} (Btu/lb · °F)	2.46 (0.588)	2.55 (0.61)	4.73 (1.13)	2.00 (0.475)	7.32–9.76 (1.75–2.33)
Heat of vaporization at boiling point, J/m^3 × 10^8 (Btu/gal)	2.19 (786)	8.75 (3,136)	8.36 (2,997)	1.62 (580)	0.318 (114)
Heat of dissociation, J/m^3 × 10^8 (Btu/gal)	12.05 (4,318)	31.71 (11,360)	16.59 (5,944)	...	...
Heat of combustion, J/m^3 × 10^8 (Btu/gal)	255.03 (91,375)	179.49 (64,310)	155.37 (55,670)	...	100.71 (36,083)

(a) At 889 kPa (129 psi). (b) Water equals 1.

the surface of the part. Decarburization in the hot zone followed by restoration of carbon in the slow-cooling zone also occurs in other active atmospheres, such as nitrogen-diluted endothermic or nitrogen-diluted dissociated methanol, with a combustible content of 20% or more hydrogen and carbon monoxide. The slow-cooling section of the furnace is designed to aid carbon restoration.

If the atmosphere is neutral to carbon (a low dew point synthetic nitrogen-base atmosphere with <10% combustibles such as hydrogen, for example), decarburization in the hot zone and subsequent carbon restoration in the slow-cooling zone do not occur at all or occur to a much less extent than they do under active endothermic atmospheres. Carbon potential for neutral atmospheres is unrelated to temperature. Consequently, neutral atmospheres, unlike endothermic atmospheres, are suitable for high-temperature sintering of steel P/M parts.

Furthermore, parts with different carbon levels can be sintered in the same tray or in random tray sequence under neutral atmosphere without the need for furnace conditioning. Carbon content in the steel part depends on the amount of admixed graphite in the powder and the dew point of the neutral atmosphere. With a lower dew point, less carbon is lost.

In some specialty sintering applications, such as those used for magnetic and stainless steels and high-ductility iron, carbon must be reduced to the lowest possible level during sintering. In such applications, atmospheres with carbon-containing gases such as carbon monoxide should be avoided. A relatively high dew point is required to decarburize the part without adversely affecting the sintered quality of the part.

In some applications, an atmosphere that is slightly oxidizing to sintered parts during cooling at <540 °C (1000 °F) is desired to provide controlled oxidation for improved corrosion resistance and increased service life.

Types of Atmospheres

Conventional atmospheres used in sintering are endothermic gas, exothermic gas, dissociated ammonia, hydrogen, and vacuum. According to the Metal Powder Industries Federation, the most common atmosphere, endothermic gas, was used by ~83% of the industry in 1976 for sintering carbon steels (Ref 4). Endothermic gas is generated directly from partial combustion of natural gas. Exothermic gas, also a product of incomplete combustion of natural gas, was used by 5% of the industry for sintering nonferrous materials, such as copper-base metal components (Ref 4).

Dissociated ammonia and hydrogen are indirectly derived from natural gas. These atmospheres are used for sintering stainless steels, magnetic alloys, and other specialty materials.

Among the more recently adopted innovations, synthetic nitrogen-base atmospheres were used by ~30% of the industry in 1983. Recent data show that nitrogen-base atmospheres, which are more versatile and relatively independent of natural gas, are becoming more economical and thus more widely used.

Tables 3 and 4 list typical physical properties of various gases and liquids used to produce sintering atmospheres. Table 5 summarizes the composition of conventional and synthetic nitrogen-base atmospheres. Various chemical requirements—burning of lubricant vapors, reduction of surface oxides, and carburization and decarburization, for example—are controlled by the ratios of various atmospheric constituents, as shown in Table 5.

Endothermic Atmospheres (Ref 5–7)

Endothermic, or endo gas is generated by reacting relatively rich mixtures of hydrocarbon gas (usually natural gas, predominantly methane) and air over a clean catalyst such as nickel oxide in an externally heated chamber (Fig. 11). The volumetric reaction between natural gas and air is the following:

$$2CH_4 + 4.8\,\text{air}\,(O_2 + 3.8N_2) \rightarrow 2CO + 4H_2 + 3.8N_2$$

The theoretical ratio of air to natural gas is 2.4 to 1; however, generator ratios are generally between 2.5 and 3 to 1. This reaction is endothermic. Consequently, the generator must be heated by an external source. Operating temperature in the catalyst must range from 980 to 1000 °C (1800 to 1830 °F); typically, the generator operates at ~1060 to 1100 °C (1960 to 2010 °F).

Air-to-gas ratios and the temperatures in the generator must be controlled to prevent sooting of the catalyst. Insufficient air or too low temperature can cause carbon monoxide to decompose to carbon dioxide and amorphous carbon (soot). Soot deposits destroy the efficiency of the catalyst and cause high levels of unreacted methane, carbon dioxide, and water in the output atmosphere. When a methane-rich atmosphere is introduced to the furnace, the methane deposits soot on the parts.

A sooted catalyst also causes loss of control over the ability of the generator to maintain a definite carbon potential in the atmosphere. When higher concentrations of carbon monoxide are required, endo gas can be generated from propane instead of natural gas. Commercial propane containing >5% unsaturated hydrocarbons does not react properly and causes sooting.

After passing over the catalyst, the reaction is "frozen" by rapid chilling of the gases to below 315 °C (600 °F) in a water-jacketed zone surrounding the top of the retort. This prevents the reaction from reversing and forming carbon (soot) and carbon dioxide from carbon monoxide:

$$2CO \rightarrow C + CO_2$$

The reaction in the direction indicated previously predominates at 700 to 480 °C (1300 to 900 °F). The soot-free reacted gas can be further cooled for metering and distribution.

Changes in the air to natural gas ratio also change the composition of the output gas and hence the dew point and carbon potential. Sulfur (as H_2S) reacts with the catalyst to deactivate it. More than 0.65 g (10 grains) of sulfur per 2.83 m^3 (100 ft^3) of gas renders the catalyst less effective; consequently, carbon monoxide and hydrogen contents of the endo gas are lowered, and water, methane, and carbon dioxide contents are increased.

A higher concentration of hydrogen is found in endo gas generated by the hydrocarbon steam process. In this generator, natural gas and water are reacted to produce carbon monoxide and hydrogen:

Table 5 Compositions of principal furnace atmosphere constituents

Atmosphere	AGA class	Air-to-natural gas ratio	Dew point °C	Dew point °F	Nominal composition, vol %: Nitrogen	Hydrogen	Water	Carbon monoxide	Carbon dioxide	Methane	Key ratios(a): Hydrogen to water	Carbon monoxide to carbon dioxide	Hydrogen to carbon monoxide
Lean exothermic	101	9.0	20(a)	68(b)	84.7	1.2	2.5	1.4	10.2	...	0.5	0.1	0.9
Rich exothermic	102	6.0	20(a)	68(b)	69.8	12.2	2.5	10.2	4.9	0.4	4.9	2.1	1.2
Endothermic	302	2.5	5	40	38.2	40.4	0.8	19.8	0.3	0.5	51	66	2.0
Dissociated methanol	...	...	15.5	60	...	65.6	1.7	32.4	0.4	...	39	81	2.0
Dissociated ammonia	601	...	−50	−60	25.0	75.0	0.004	...	...	...	19,000	...	...
Hydrogen	...	...	−60	−80	...	100	0.001	...	...	...	100,000	...	...
Nitrogen	...	...	−60	−80	100	...	0.001	...	...	...	...	...	...

Note: For nitrogen-base atmospheres, composition, dew point, and key ratios can be synthesized over a broad range, depending on which of the above generated atmospheres is blended with the high-purity nitrogen gas. (a) Control carburization or decarburization, oxidation or reduction, or combination of these. (b) Dew point is ~6 °C (10 °F) above temperature of cooling water; dew point may be reduced to 5 °C (40 °F) by refrigeration or to −45 °C (−50 °F) by absorbent tower dehydration. Hydrogen-to-water changes accordingly.

$$CH_4 + H_2O \rightarrow CO + 3H_2$$
(–12 °C, or 10 °F, dew point)

The result is a higher proportion of hydrogen and an absence of nitrogen.

Regardless of the process or feedstock used to produce the endo gas, it is monitored by measuring dew point and carbon dioxide content upon exiting the generator. Oxygen probes are also used to monitor the composition of the endo gas.

Variations in composition and peak shaving of the natural gas feedstock make control difficult. Consequently, input ratios of air and methane or propane are adjusted to maintain the dew point and carbon dioxide at a constant level. Frequent monitoring and adjustment are required to maintain a consistent endo gas composition. Automatic ratio adjusters are available to ensure consistent composition of the endo gas (Ref 8). However, these controls require careful calibration and maintenance.

Endo gas is typically 40% nitrogen, 40% hydrogen, and 20% carbon monoxide, with small amounts of water, carbon dioxide, and methane (see Table 5). Typically 60% of the gaseous constituents are combustibles (carbon monoxide, hydrogen, and methane), which are oxide-reducing agents. Carbon monoxide and methane are carburizing; carbon dioxide and water decarburize, oxidize, and delube. The carbon dioxide and water content of endo gas generally burns off the lubricant vapors by chemical reaction in the delubing section of the furnace. Other atmospheres containing higher oxidant levels also are capable of achieving optimum results.

Typically, hydrogen-to-water ratio determines the reduction-oxidation function of an atmosphere (Fig. 8). Endo gas is reducing to iron at >260 °C (500 °F) because the typical 40% hydrogen and 0.8% water (dew point: 5 °C, or 40 °F) contents provide a hydrogen-to-water ratio of 50 to 1. Although this ratio achieves acceptable oxide reduction, other atmospheres—notably those based on hydrogen or dissociated ammonia or synthetic nitrogen-base atmospheres—can provide higher ratios.

The ability of typical endo gas to transfer carbon depends on the temperature and water or carbon dioxide content (Ref 9). Figure 10 shows carbon potential at various temperatures for three different dew points. Typically, a high-carbon part loses carbon at and near the surface as it approaches the peak hot zone temperature (1120 °C, or 2050 °F) in the heat-up cycle. At this peak temperature, the carbon potential corresponds to ~0.2% C, depending on the dew point of endo gas in the furnace. As the part slowly cools, the carbon potential increases, as shown in Fig. 10. The carbon thus tends to be restored.

The extent of decarburization and recarburization determines the finished carbon content at or near the surface of the P/M part. Because the sintering temperature profile is determined by the sintering process, carbon potential must be controlled by the dew point or the carbon dioxide content in the endo gas. Most sintering under endothermic atmosphere is done at dew points ranging from –5 to 15 °C (23 to 59 °F).

High-carbon steel parts are sintered at low dew points, and low-carbon parts are processed at high dew points. Endo generators tend to operate soot free at higher dew points; consequently, natural gas is sometimes added to the high dew point endo gas atmosphere to secure a lower dew point in the furnace for sintering high-carbon parts.

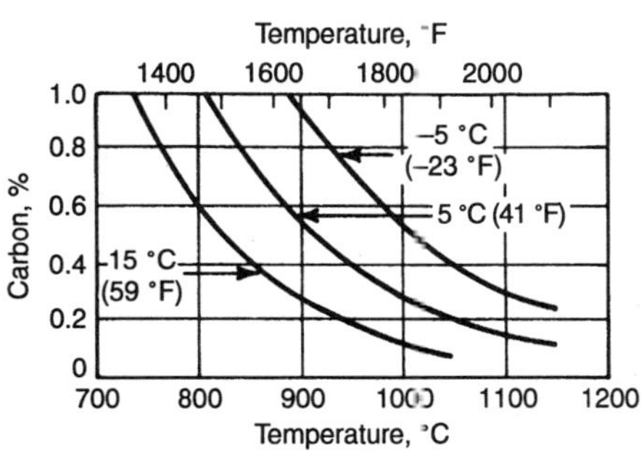

Fig. 10 Carbon equilibrium in endo gas at different dew points. Endo gas composition: 40% hydrogen, 20% carbon monoxide, and 40% nitrogen

Exothermic Atmospheres (Ref 5–7)

Exothermic, or exo, atmospheres typically contain 67 to 87% nitrogen. The remainder is carbon monoxide, hydrogen, carbon dioxide, and water. Exothermic atmospheres are produced by the partial combustion of hydrocarbon fuels (generally natural gas) with air. A 6-to-1 ratio of air to natural gas produces rich exo gas, American Gas Association (AGA) class 102, which is high in carbon monoxide and hydrogen.

In contrast, lean exo gas (AGA class 101), produced from a 9-to-1 ratio of air to gas, contains low percentages of carbon monoxide and hydrogen, but correspondingly higher percentages of carbon dioxide and nitrogen (Table 5). Partial combustion in a gas generator equipped to control the ratio of air to fuel gas is capable of producing any desired composition between rich and lean exothermic gases.

The exothermic gas generator is a refractory-lined, water-jacketed combustion chamber that operates at 1100 to 1425 °C (2010 to 2600 °F). The internal temperature of the chamber depends on (a) the ratio of air to fuel gas, (b) the volume of gas being burned, and (c) the effectiveness of the heat transfer from the inside of the chamber to the water-jacketed exterior. Figure 12 shows a typical exo gas generator.

The gas mixture enters the two-zoned converter, which is the main portion of the generator. In the first zone, where combustion occurs, most of the fuel and air react to produce hydrogen, carbon monoxide, and nitrogen. In the second zone, which is heated by the earlier combustion, the unburned hydrocarbon and combustion products are cracked over the catalyst into hydrogen and hydrogen byproducts. The turn-down volume (expressed as a percentage of the rated capacity) is limited by the minimum exo gas output that will maintain the chamber at 1030 °C (1890 °F) or above. Lean ratios permit lower turn-down volumes than rich ratios.

The hot gases are quenched in the hot gas cooler to separate water as condensate from the gas stream for removal through a condensate trap. The amount of moisture in the cooled gas varies according to the temperature of the incoming cooling water; the temperature of the exit gas should be within ~5 °C (10 °F) of the cooling water.

More moisture can be removed by refrigerating to a dew point of ~4 °C (40 °F) or by adsorbent tower dehydration to a dew point of –45 °C (–50 °F) when relatively high oxide reducing capability (hydrogen-to-water ratio) is desired.

Most exo gas (lean or rich ratio) has a typical dew point of 21 °C (70 °F), corresponding to 2.5% water. It contains 1.2 or 12.2% hydrogen, depending on whether it is a lean or rich exo gas. The hydrogen-to-water ratios for lean and rich

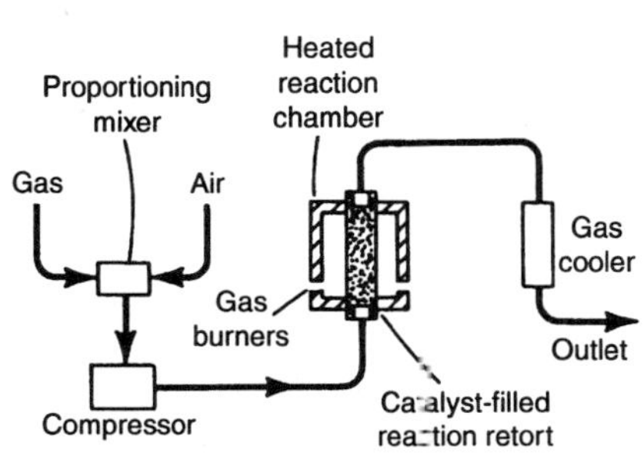

Fig. 11 Schematic of endothermic gas generator

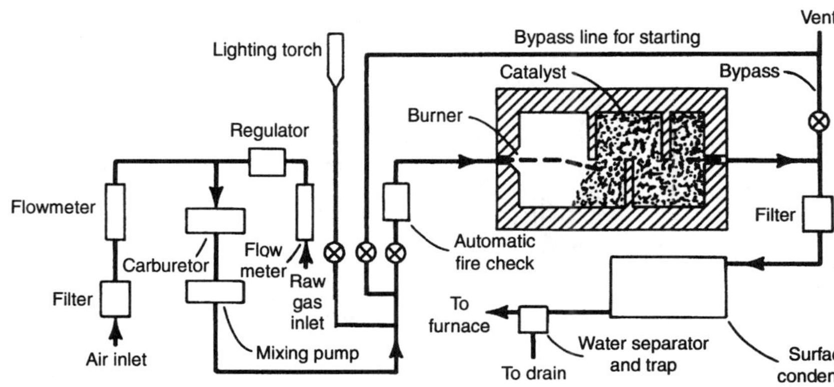

Fig. 12 Schematic of exothermic gas generator

exo gases are 0.5 and 4.9, respectively (Table 5). These atmospheres are significantly less oxide reducing than endo gas or synthetic nitrogen-base atmospheres.

Exo gas is highly decarburizing to steel above ~700 °C (1300 °F). Because of the high water and carbon dioxide content, it is a very effective delubing agent.

Exo gas generally is used to sinter bronze and some other copper-base parts. Copper and most copper-base alloys do not require high oxide-reducing power (Fig. 8). Exo gas requires much less natural gas than endo gas; consequently, it is one of the most economical atmospheres.

Dissociated Ammonia (Ref 5–7)

Heating metallurgical-grade anhydrous ammonia in the presence of a catalyst causes the ammonia to dissociate into a mixture of hydrogen and nitrogen molecular gases in a 3-to-1 ratio (75% hydrogen and 25% nitrogen):

$$2NH_3 \xrightarrow[\text{catalyst}]{\text{heat}} 3H_2 + N_2$$

Dissociated ammonia is pure, consistent, and dry; the dew point is typically less than –51 °C (–60 °F). Residual undissociated ammonia is generally well below 250 ppm.

Although ammonia molecules begin dissociating at 315 °C (600 °F), 980 °C (1800 °F) provides a practical operating compromise between dissociated rate and equipment life. Higher temperatures must be used to ensure that virtually no undissociated ammonia molecules remain. Generally 1 kg of liquid ammonia yields ~1.4 m^3 of undissociated ammonia vapor or ~2.8 m^3 of dissociated ammonia (DA) (hydrogen and nitrogen in a 3-to-1 ratio). In English units, 1 lb of liquid ammonia yields 22.5 standard ft^3 of undissociated ammonia vapor or 45 standard ft^3 of DA.

A typical ammonia dissociator produces 2 to 140 m^3/h (70 to 4950 ft^3/h) of dissociated ammonia at an outlet pressure of 35 to 100 kPa (5 to 15 psi). Liquid ammonia is fed from a storage tank into a vaporizer, which transforms it into ammonia vapor. A heater in the storage tank maintains gas operating pressure if the temperature of the liquid ammonia falls below the recommended level.

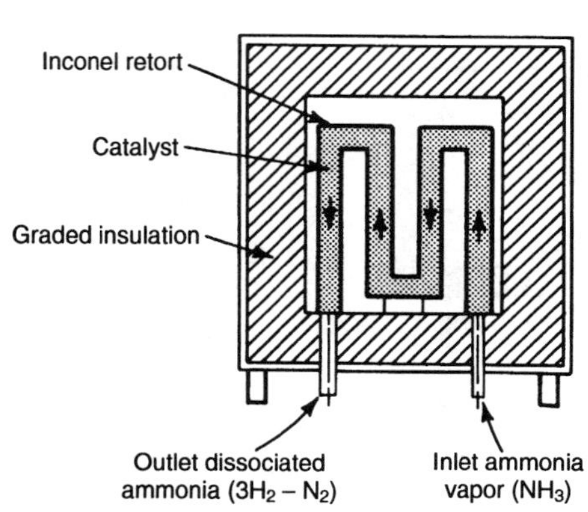

Fig. 13 Schematic of ammonia dissociator

The vapor then passes through a pressure regulator, is preheated, and enters an externally heated retort (typically made from Inconel) that contains a nickel- and iron-bearing catalyst. The ammonia molecules dissociate into molecular hydrogen and nitrogen at 925 to 1040 °C (1700 to 1900 °F). The resulting mixture leaves the retort and passes through a heat exchanger; heat is subsequently transferred to the incoming ammonia vapor. If necessary, the dissociated ammonia can be pressure regulated and piped to the furnace.

Output can be easily controlled between 0 and 100% of capacity by regulating flow at any point along the processing line. This control feature is unavailable on exo and endo gas generators, which typically operate between 70 and 100% of their rated capacity. Some recent designs of endo generators allow greater output reduction.

The most important component of the dissociator is the catalyst chamber or retort, which is generally fabricated from Inconel. This chamber is heated either electrically or by natural gas. If electrically heated, a dissociator requires 0.5 to 1.0 kW (depending on the size of dissociator) of electricity per cubic meter (14 to 28 W/ft^3) of dissociated ammonia produced. Figure 13 shows a schematic of a typical ammonia dissociator. Ammonia dissociators are relatively maintenance free compared to endo and exo gas generators.

Because dissociated ammonia is very dry (low dew point), it is highly reducing to surface oxides, as the high hydrogen-to-water ratio given in Table 3 indicates. Consequently, it is frequently used where the high reducing capability of stable oxides is desirable, such as in sintering of stainless steel. Thermodynamically, the high hydrogen content in dissociated ammonia is decarburizing to carbon steel; kinetically, however, the decarburizing reaction is very slow when the dew point in the furnace remains low (below –40 °C, or –40 °F).

The decarburizing reaction can be further slowed by diluting dissociated ammonia with a lower dew point inert gas such as nitrogen when sintering carbon steel that cannot have any decarburization at the surface. Because dissociated ammonia is very dry, it does not provide optimum burning of the lubricant during delubing in the preheating zone. However, because dissociated ammonia is 75% hydrogen, it has a thermal conductivity higher than nitrogen-rich atmospheres.

Hydrogen (Ref 10)

Hydrogen, the lightest of all the gases listed in Table 3, is produced for the most part by steam reforming natural gas and light hydrocarbon liquids over a catalyst. Metallurgical hydrogen is refined from excess hydrogen-gas feedstocks from petrochemical steam-reforming plants. The gas supplier removes hydrocarbons and moisture from the raw gas stream and liquefies the partially refined gas at –253 °C (–423 °F). Other impurities such as nitrogen and argon are removed during the liquefaction process.

Hydrogen is delivered either as compressed gas or liquid to volume and purity requirements. Commercial grades of compressed hydrogen gas, with typical pressure of 17 MPa (2500 psig), are delivered by trailer in cylinders with capacities between 1000 and 4000 m^3 (35,000 and 140,000 standard ft^3). Smaller quantities of compressed hydrogen usually are available in cylinders containing 7 m^3 (250 ft^3).

Most liquid hydrogen is delivered by trailer to liquid receivers or storage vessels with capacities between 5000 and 75,000 L (1500 and 20,000 gal). Liquid hydrogen usually is vaporized as required. One liter of liquid hydrogen yields 0.85 m^3 (113.4 ft^3) of hydrogen gas (or one gallon yields 133.4 ft^3 of hydrogen gas). Smaller quantities of liquid hydrogen are available in 1150 L (40 gal) cylinders.

Hydrogen specifications vary, depending on the grade. Typical hydrogen is 99.999% pure. Commercial-grade hydrogen meets the following specifications:

Hydrogen	99.995% min
Water	8 ppm max at –68 °C (–90 °F)
Oxygen	1 ppm max

Because hydrogen is highly flammable, care must be taken to ground major stationary and mobile equipment to remove static electricity and the potential for spark ignition of the hydrogen. Electrical equipment must meet applicable national electric codes and fire prevention standards.

As indicated in Table 5, hydrogen is the most effective reducing atmosphere available because it has the largest hydrogen-to-water ratio. It is essentially nondecarburizing, like dissociated ammonia, to carbon steels as long as the dew point remains below approximately –40 °C (–40 °F). It is similarly inefficient at burning lubricant vapors because of a low dew point. Of all atmospheres commonly used in the P/M industry, hydrogen is the most heat conductive.

Vacuum (Ref 11–12)

Vacuum is essentially a lack of atmosphere. It is used mainly for sintering stainless steels, tool steels, carbides, magnetic alloys, and metals such as titanium, zirconium, uranium, tantalum, and other refractory metals and compounds that react with hydrogen-, nitrogen-, and carbon-monoxide-bearing atmospheres.

Vacuums are being used increasingly for high-temperature sintering of conventional ferrous P/M parts. Under vacuum, care must be taken not to lower the pressure in the furnace below the vapor pressure of the constituents of the alloy to be sintered so that depletion does not occur.

Vacuum cannot be used in the conveyor furnaces conventionally used by the sintering industry. Currently, almost all vacuum sintering furnaces are batch types. Some continuous (compartmentalized) conveyor vacuum furnaces are being introduced to the P/M industry.

Green compacts of carbides, refractory metals, and other materials are sometimes heated under a protective or reducing atmosphere in a separate furnace to drive off volatile lubricants before entering the vacuum furnace for sintering. This minimizes the contamination of heater elements and heater element supports that are frequently associated with vacuum furnaces.

Nitrogen-Base Atmospheres (Ref 13–22)

The main constituent of the recently developed nitrogen-base systems is molecular nitrogen. Molecular nitrogen is obtained from air, which consists of ~78% nitrogen, 21% oxygen, 0.93% argon, 0.03% carbon dioxide, and small amounts of rare gases such as neon and helium.

Nitrogen is produced commercially by cryogenic and noncryogenic methods. Air is first filtered to remove particulates and is then compressed and passed through a heat exchanger for cooling to remove water and carbon dioxide. The air is then rapidly depressurized to an ultracold liquid that is distilled to separate out the high-purity molecular nitrogen gas. The gas is either piped directly to the sintering plant or liquefied for long-distance shipping. One cubic meter of liquid nitrogen equals 697 m^3 of nitrogen gas.

Nitrogen is also produced in noncryogenic generators located at the sintering plant. In noncryogenic production, a stream of compressed air is passed through a bed of special adsorbent, such as a carbon molecular sieve maintained at ambient temperature. Oxygen, water vapor, and carbon dioxide are preferentially retained on the adsorbent, but nitrogen (with a small amount of oxygen) flows through it. As the adsorbent bed becomes nearly saturated with oxygen, water, and carbon dioxide, the system automatically switches to a second bed without interrupting nitrogen delivery.

The system regenerates the first bed by reducing the bed pressure and releasing the adsorbed bases. The regenerated bed is then ready to repeat the cycle. All adsorption and desorption activities occur at ambient temperatures rather than at the low temperatures used in cryogenic plants. To reduce the oxygen content of the nitrogen, the product stream with added hydrogen is passed through a catalytic deoxidizer and a dryer to produce high-purity nitrogen. Powder metallurgy fabricators with a nitrogen consumption greater than ~100 m^3/h (3500 ft^3/h) could use an on-site noncryogenic plant for economic reasons. These plants have been introduced into the market only recently.

Whether produced cryogenically or noncryogenically, nitrogen is readily available and economical. Additionally, it provides a consistent source of high technical quality, high-purity product—maximum oxygen content is 10 ppm (typically 2 ppm) and dew point is –65 °C (–85 °F). Because nitrogen is molecular rather than atomic, it is essentially inert to steel and other compositions that are usually sintered in P/M plants.

Inert nitrogen has a density close to that of air (Table 3); consequently, it effectively prevents air from entering the furnace, thus protecting porous P/M parts from harmful contact with air during sintering. Assuming an adequately airtight furnace, an atmosphere of 100% nitrogen is suitable for sintering aluminum and some less critical ferrous and nonferrous metal parts. Nitrogen alone, however, will not reduce particle surface oxides or effectively control surface carbon of steel parts.

With small, controlled additions of active gases, nitrogen atmospheres can perform all functions required of sintering atmospheres. Furthermore, the amount and type of active ingredients can be varied to change the level of reactivity of the atmosphere. The most important active ingredient required for sintering of commonly produced P/M parts is hydrogen.

Systems that use low dew point hydrogen sources and that maintain other oxidants at low levels require only small amounts of reducing gas (hydrogen) to create an effective and efficient nitrogen-base atmosphere. Such atmospheres are potentially highly reducing to metal oxides. Typically, nitrogen-base systems consist of essentially inert nitrogen and small amounts of one or more active gaseous ingredients.

Conventional atmosphere generators are not required in such systems; instead, feedstock gases (nitrogen and active gases) are piped through a flow panel into the furnace. The gases entering the panel regulation system exit as the atmosphere–the metering panel becomes, in effect, an atmosphere generator.

These systems exhibit great flexibility by enabling the ratio and types of active ingredients to be varied easily to suite the various sintering requirements within the plant. Nitrogen systems can also provide different atmospheres in different sections of a furnace, as shown in Fig. 14. The volume output reduction for nitrogen is 100%, considerably more than for endo and exo atmospheres. Nitrogen-base systems contain three basic types of active ingredients, as discussed below.

Oxide-Reducing Agents

Hydrogen is the most desirable ingredient for reduction of particle surface oxides, which thus improves bonding and provides maximum integrity to the sintered part. It is supplied by compressed hydrogen tube trailers or liquid hydrogen storage tanks, dissociated ammonia, endo gas, or dissociated alcohols such as methanol. Each supply source has advantages and disadvantages. Technically, hydrogen sources can be divided into the following two systems.

System 1. Hydrogen is supplied from endo gas or dissociated methanol. The total amount of combustibles (primarily hydrogen and carbon monoxide) in system 1 (nitrogen-endo or nitrogen/dissociated methanol) atmospheres varies between 15 and 30%, typically 24%.

Hydrogen is accompanied by carbon monoxide in a 2-to-1 ratio. The dew point is relatively high. Consequently, the hydrogen-to-water ratio is relatively low (40 to 50, as shown in Table 3). Although this ratio achieves acceptable oxide reduction in steel parts, other atmospheres—nota-

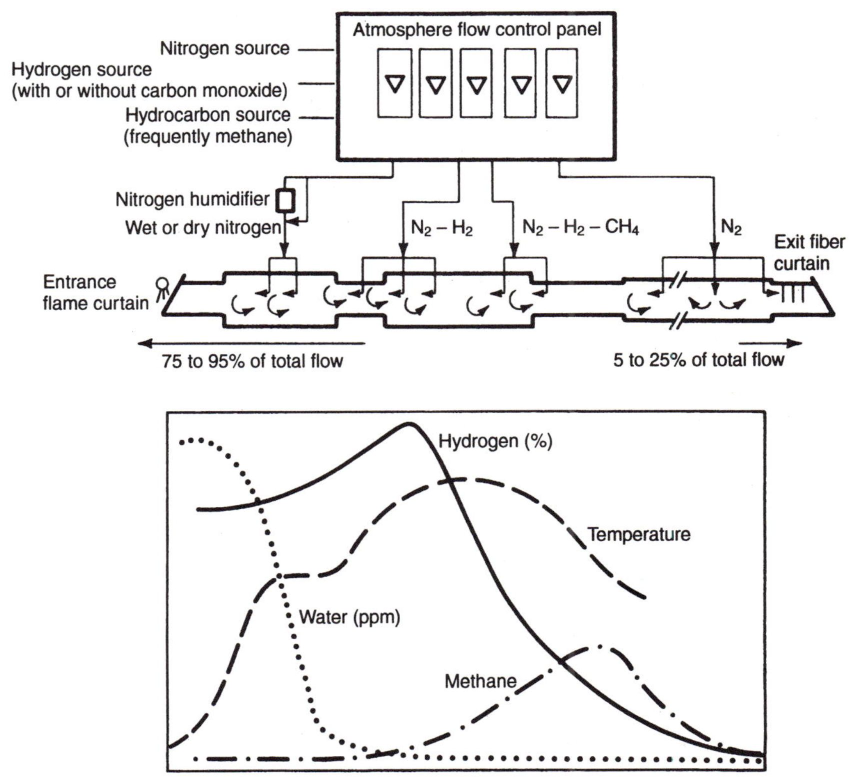

Fig. 14 Schematic of nitrogen-base zoned atmosphere system. Atmosphere and temperature profiles corresponding to the nitrogen-base atmosphere system are shown.

bly system 2, which uses dissociated ammonia in a nitrogen base—provide better ratios and consequently greater oxide reduction.

Methanol, as a source of hydrogen, is mixed with nitrogen before introduction into the furnace. Methanol molecules dissociate in the hot zone of the furnace as:

$$CH_3OH \xrightarrow{\text{heat}} 2H_2 + CO$$
$$+ \text{small amounts of } H_2O, CO_2, \text{and } CH_4$$

For this reaction to occur, furnace temperatures must be >800 °C (1470 °F), preferably >900 °C (1650 °F). The reaction is endothermic and requires ~5 kW of heat energy per gallon of methanol dissociated.

One liter of methanol produces ~1.7 m^3 of hydrogen and carbon monoxide in a 2-to-1 ratio (1 gal produces 225 ft^3). Typical water, carbon dioxide, and methane contents in dissociated methanol (without nitrogen additions) are 1.7, 0.4, and 0.2%, respectively. The carbon potential of dissociated methanol (without nitrogen additions) is 0.2 to 0.3% C at 925 °C (1700 °F). Composition control is generally improved when hydrogen is derived from methanol rather than from endo gas.

Because hydrogen produced from endo gas or dissociated methanol is always accompanied by carbon monoxide, use is limited to sintering P/M compositions that are not adversely affected by the presence of the gas. Endo gas or methanol is not recommended as a hydrogen source in P/M plants that use only one basic nitrogen system and have capabilities to sinter not only carbon steels but also bronze (at relatively low temperatures), magnetic steel, stainless steel, high-ductility iron parts, and high-carbon high-density parts that blister, or "ripple," during sintering.

System 2. Hydrogen is supplied from tube trailers, liquid storage tanks, or from dissociated ammonia. This atmosphere does not contain carbon monoxide and is pure and dry (very low dew point). The total amount of combustibles (primarily hydrogen) in system 2 (nitrogen-hydrogen or nitrogen/dissociated ammonia) atmospheres varies between 3 and 10%, typically 5%.

Hydrogen-to-water ratios are much higher than in endo gas or dissociated methanol atmospheres (Table 5). Tube trailers and storage tanks are the most flexible hydrogen supply sources, but are approximately two to three times as expensive as dissociated ammonia per unit of furnace atmosphere.

A system 2 using dissociated ammonia provides the optimum combination of oxide-reducing power, long-range economics, technical flexibility, operator safety, furnace life, and ease of operation. Because these atmospheres do not contain carbon monoxide, a broader range of P/M compositions can be sintered compared to system 1.

Carbon-Control Agents. Systems using hydrogen or dissociated ammonia (system 2) as the only active ingredient in nitrogen-base atmospheres do not experience surface decarburization of P/M steel parts if the dew point in the furnace is kept below approximately −30 °C (−20 °F). However, if the dew point in nitrogen-hydrogen (dissociated ammonia) atmosphere is slightly higher, small amounts of noticeable surface decarburization occur, and wear resistance of the finished part is lowered. In such situations, small amounts of hydrocarbon, preferably from natural gas, can be used with an oxide-reducing agent (hydrogen or dissociated ammonia) in system 2 to prevent decarburization or to add small amounts of carbon to the surface of the P/M part.

Generally, the amount of carbon monoxide present in system 1 is adequate to prevent decarburization of P/M steel parts. Small amounts of hydrocarbon derived from natural gas can be added to nitrogen-endo gas or nitrogen/dissociated methanol (system 1) atmospheres if additional carbon is desired in the part surface.

Oxidants. Oxidizing agents enhance burning of lubricants in P/M parts. Water and/or carbon dioxide can be added or generated within the furnace in controlled amounts for use in the delube section of the furnace to accomplish this purpose.

Oxygen from air is a relatively strong oxidant compared to water and carbon dioxide. Generally, adding oxygen or air to the atmosphere entering the delube section is not recommended for burning of lubricants.

Small amounts of oxidizing agents such as water can be added to the atmosphere entering the hot zone to achieve limited surface decarburization of parts requiring subsequent pressing and sizing. Small amounts of oxygen (air) or water can be added to nitrogen-base atmospheres with some circulation in the cooling section of the furnace to provide controlled but limited surface oxidation of the sintered parts. This procedure improves the appearance of the parts by imparting a straw, blue, or gray finish, depending on the amount of air or water added and the temperature of the part in contact with the oxidant-containing atmosphere. This also improves corrosion resistance and service life of sintered parts.

Furnace Zones Concept (Ref 23–28)

Typical sintering furnaces can be thought of as having three or more interconnected zones (depending on the material being sintered), each with a separate function. The sintering process consists of several sequential phases, each requiring a unique combination of temperature, time, and atmosphere composition; flow; direction; and circulation. Each phase of the sintering process occurs in a specific zone of the furnace. Separating these zones and phases conceptually enhances design flexibility. A close match between the temperature and atmosphere of each zone and the function of each phase results in an optimum overall sintering process. In a single system, the base nitrogen can be modified with other gases or active ingredients to produce an appropriate and optimum atmosphere composition for each sintering phase before introduction into the proper furnace zone.

To understand the optimum combination of temperature, time, and atmosphere for each sintering phase, the processing sequence of a P/M part through the sintering furnace must be visualized. If the first sintering phase is completed effectively and efficiently in the first furnace zone, then the second phase in the second furnace zone, and each successive phase, is likely to be successful. The net result is improved productivity (production rate, quality, consistency, safety, versatility, and cost). Figure 7 shows a schematic of the processing sequence for sintering steel.

For example, if a P/M part composed of iron, 1% graphite, 2% copper, and 1% lubricant was processed in this furnace, at least five phases would be required to complete the sintering process. These would occur in five different zones of the furnace. The requirements for each sintering phase and furnace zone are outlined in Fig. 7 and are summarized in Table 6.

Prime objectives in sintering this green part include the absence of sooting (internal or external). Admixed graphite should be effectively and uniformly diffused into iron particles as combined carbon. This results in an optimum pearlitic microstructure during cooling with minimum carbon loss. Admixed copper, after melting, should be uniformly distributed over all particle surfaces in the P/M part.

High structural integrity (maximum possible number of bonds or necks between particles, high average bond or neck area, and relatively rounded pores) is desirable. Surface carbon should be approximately equal to the carbon content in the core or interior. Emerging parts should be clean and bright or should have a controlled thin surface oxide to improve corrosion resistance. To meet these technical objectives economically, the total flow of atmosphere should be held to a minimum, with the least amount of added active gases.

For the purposes of the example, the atmosphere introduced into the furnace is nitrogen-base system 2 (carbon monoxide free), with 5% hydrogen and 0.5% methane. If the furnace is conventional, all of the atmosphere is introduced at one inlet—generally in the slow-cooling section between the hot zone and the water-cooled section. As a consequence, atmosphere composition (water, hydrogen, and methane content) is generally uniform throughout the different zones or sections as the atmosphere flows from the inlet toward both ends. This means that each active ingredient is at the same level whether or not the atmosphere is in a zone where a particular ingredient is needed. For the furnace to function, it is necessary to compromise between the atmosphere needed in one part of the furnace and that needed in another.

In the atmosphere system used in a zoned sintering furnace, however, different atmospheres (composition, flow, direction, and circulation) are introduced to different zones of the furnace (Fig. 14) to obtain optimum results. Wet nitrogen is introduced near the loading end of the furnace, hydrogen in the center, and dry nitrogen near the exit. Custom-designed injectors provide proper flow, circulation, and direction to these atmospheres. Injectors are also designed to act as gas barriers and reduce overall atmosphere flow. As a result of this zone, the constituents peak where they are needed (Fig. 14). In this example, water (dew point) is at the highest level near the entrance in the preheat zone, where delubing takes place. Hydrogen is introduced at a peak level at the mouth of the hot zone and in the hot zone, where it efficiently reduces surface oxides. Methane is added in sufficient amounts to reach maximum levels at the end of the hot zone and at the beginning of the slow-cooling zone, where it is needed for carbon control. Only nitrogen is added in the cooling zone, constituting nearly 100% of the atmosphere near the exit end of the furnace; furthermore, there is no need for a flame at the exit end. Therefore, exiting parts are cooler, and operators are safer and more comfortable.

In contrast to the conventional system, no compromise is needed to balance atmosphere makeup and furnace function; the zoned sintering atmosphere facilitates optimum sintering regardless of makeup or other part specifications. In zoned systems, atmosphere is introduced as needed with respect to composition, quantity, and location. Atmosphere zoning, combined with suitable temperature and time, allows each phase of the sintering process to be completed effectively, efficiently, and economically at the lowest possible atmosphere flow rates.

Nitrogen-base zoned atmosphere systems are compatible with existing furnaces, and the constituents can be varied to permit sintering of a wide variety of materials. A well-designed nitrogen-base system can meet the needs of a plant that sinters a wide range of ferrous and nonferrous parts. Technically, zoned atmospheres can achieve the same results as endo, dissociated ammonia, and exo atmospheres, while being more energy efficient, easier to operate, safer, and more flexible.

Nitrogen-base zoned atmosphere systems are capable of meeting the varying requirements imposed by furnace components. Atmospheres must be mild to furnace components. Conveyor belt alloys, for example, oxidize easily at sintering temperatures. However, under very dry and reducing atmospheres, the protective oxide layers are removed, and belt links tend to weld together. The belt "crackles" and gradually loses strength. Furthermore, belts, heating elements, and muffles become brittle and have short service lives under carburizing atmospheres. In addition, insulating brick oxides are reduced if the atmosphere is too reducing, which shortens service life.

Safety Precautions (Ref 29–35)

Sintering atmospheres (endo, exo, dissociated ammonia, dissociated alcohols, and nitrogen-base) normally consist of a mixture of gases, which can be flammable, toxic, asphyxiant, or a combination of these. Explosion, fire, and poisoning are potential hazards. National Fire Protection Association standard 86C covers various safety considerations pertaining to continuous conveyor belt furnaces used in the P/M industry.

Four intrinsic dangers are associated with producing and using the common atmosphere gases. A mixture of atmosphere gas and air can accumulate in a confined area and explode. Relatively small quantities of atmosphere gas can unexpectedly burn or flash out of control. Personnel can be poisoned by carbon monoxide, ammonia, or methanol. "Simple" asphyxiation is possible when asphyxiants are present in high concentrations.

Table 7 describes the characteristics of the common sintering atmosphere constituent gases. Most of the major constituents are flammable; four are toxic and four are simple asphyxiants. Carbon monoxide, ammonia, and methanol are both flammable and toxic. The percentage by volume of the flammable ingredients ranges from 75% hydrogen in dissociated ammonia atmosphere to only a few percent in purified exothermic and nitrogen-base atmospheres.

Dangers are present even when heat treating with atmospheres that contain relatively small percentages of hazardous ingredients because these gases may accumulate or concentrate. Under typical operating conditions, however, the active ingredients in nitrogen systems are diluted below the level that is flammable, even if all the nitrogen is replaced by air. Explosion, burning, poisoning, and asphyxiation hazards are thus reduced, but not eliminated.

The explosive ranges of typical atmosphere constituents are the following:

Atmosphere constituents	Concentration in air, %
Hydrogen	4.0–74
Carbon monoxide	12.5–74
Methane	5.3–14
Ammonia	15.0–28
Methanol	6.7–36

Any mixture between the high and low limit of flammability will burn when ignited and, under certain conditions, detonate or explode. The destructive power of the pressure wave from an ignited flammable mixture depends on the amount of gas and the heat of combustion of the fuel gas, the combustion mode (deflagration or

Table 6 Furnace zones, sintering phases, and atmosphere requirements to sinter an iron, 2% copper, 1% graphite, and 1% lubricant part

Furnace zone	Temperature, °C	Temperature, °F	Time, min	Sintering phase	Sintering subphases	Atmosphere requirement	Atmosphere composition(a), %: Hydrogen(b)	Methane	Water	Flow	Circulation
1 delube	425–650	800–1200	10–30	Delubing	Lubricant vaporizing, vapor burning, flushing vapors to entrance	Fast, uniform heat transfer, slightly oxidizing	2–7	0–0.1	0.5–1.5	High toward entrance	Highly desirable
2 preheat	650–1070	1200–1960	5–15	Elimination of particle surface oxides	Oxide reducing, graphite diffusing, copper melting, and coating particle surfaces	Highly reducing to surface oxides, neutral to carbon fast, uniform heat transfer	5–15	0.1–0.3	0.01–0.03	High toward entrance	Desirable
3 high temperature	1070–1150	1960–2100	10–30	Particle bonding	Copper diffusing, neck (bond) growing, pore rounding	Neutral to carbon reducing	3–8	0.2–0.5	0.01–0.02	Medium toward entrance	Desirable
4 slow cool	1150–815	2100–1500	5–15	Carbon restoring	Carbon transferring, homogenizing	Uniform slow cooling, slightly carburizing	2–7	0.3–1.0	0.01–0.02	Medium toward entrance	Highly desirable
5 cool down	815–50	1500–120	30–90	Cooling down, optionally oxidizing	Pearlite forming, part cooling down, preventing oxidation or controlling light oxidation	Slightly reducing or neutral (optionally oxidizing) to iron, uniform fast cooling	0–2	0–0.1	0.01–0.02	Low to medium partially toward exit	Highly desirable

(a) Balance is essentially nitrogen with or without small amounts of carbon monoxide and carbon dioxide. (b) Hydrogen can be derived from hydrogen storage vessel, dissociated ammonia, endothermic gas, or dissociated methanol. (c) If slightly oxidizing atmosphere is required, water is relatively higher or a small amount of air is introducd along with nitrogen in water-cooled section at a selected location.

Table 7 Potential hazards and functions of heat treating atmosphere constituent gases

Gas	Potential hazard			Atmosphere function
	Flammable	Toxic	Simple asphyxiant	
Nitrogen	...	...	Yes	Inert
Hydrogen	Yes	...	Yes	Strongly reducing
Carbon monoxide	Yes	Yes	...	Carburizing and mildly reducing
Carbon dioxide	...	Yes	Yes	Oxidizing and decarburizing
Natural gas	Yes	...	Yes	Strongly carburizing and deoxidizing
Ammonia	Yes	Yes	...	Strongly nitriding
Methanol	Yes	Yes	...	Carbon monoxide and hydrogen generating

detonation), and the configuration of the confinement space. The energy released is either absorbed by the surroundings or destroys them.

The explosive potential of the large volumes of flammable gases used in sintering furnaces poses an important safety consideration. To ensure the safety of sintering and heat treating, the equipment and systems that handle gases must be designed, operated, and maintained to prevent accumulation of explosive mixtures. Dangerous accumulations that cannot be readily detected by personnel can occur. Properly designed safety systems, maintained and operated by well-trained and competent personnel, substantially reduce explosion hazards.

Generally, sufficient volumes of atmosphere gas flowing through the furnace leave the charge and discharge doors, mix with air, and burn uniformly and completely. However, several conditions can exist that prevent the atmosphere gas from burning off properly. Sometimes, a combustible mixture of air and gas can form in the furnace throats or vestibules. It can burn rapidly and forcefully exhaust, or flash, flames and hot gases through the furnace door. Unprotected personnel in the vicinity can be burned. Eyes are particularly sensitive to damage.

Although flashes of this type are more common than explosions, they are potentially less destructive. Nonetheless, personnel who approach these areas without safety glasses, protective face shields, gloves, and flameproof clothing risk serious flash burns. If safe methods are established and used for clearing jam ups, inspecting the furnace interior, and removing products from the furnace, this hazard can be substantially reduced.

Table 8 Physiological effects of ammonia

Concentration, ppm	Physiological effects
20	First perceptible odor
40	Slight eye irritation in a few individuals
100	Noticeable irritation of eyes and nasal passages after a few minutes of exposure
400	Severe irritation of the throat, nasal passages, and upper respiratory tract
700	Severe eye irritation; no permanent effect if the exposure is limited to less than ½ h
1700	Serious coughing, bronchial spasms; less than ½ h of exposure may be fatal
5000	Serious edema, strangulation, asphyxia; almost immediately fatal

Protective guards and shields used for routine operation are no substitute for the required protection of personnel working in the vicinity of the furnace doors. However, the possibility of flash back under nonflammable, nitrogen-rich atmospheres is remote.

Ammonia, carbon monoxide, and methanol, which are highly toxic, are used routinely in the sintering and heat treating industries. Liquid ammonia produces dissociated ammonia or ammonia vapor for nitriding, and carbon monoxide is a constituent of exothermic, endothermic, and dissociated methanol gases. Methanol is used to produce dissociated methanol, which consists of hydrogen and carbon monoxide. Concentrations of <½% ammonia or carbon monoxide in the air are considered fatal in <½ h of exposure. Tables 8 and 9 give the physiological effects of various concentrations of these gases.

Threshold limit values, published by the American Conference of Governmental and Industrial Hygienists, list carbon monoxide, carbon dioxide, ammonia, and methanol as commonly used toxic chemicals. Concentrations as small as 35 ppm of ammonia, 400 ppm of carbon monoxide, and 250 ppm of methanol are harmful; therefore, only short exposures are allowed.

Protection from poisoning by raw ammonia or methanol vapor or from contact with liquid ammonia or methanol is provided partly by the design of the storage and delivery system. Suppliers are excellent sources of detailed safety information on ammonia and methanol systems and ammonia dissociators.

Carbon monoxide is not quite as toxic as ammonia; however, because it is odorless, it poses a greater safety hazard. Lethal concentrations can collect in isolated areas and remain undiscovered until personnel are overcome by fumes. According to Ref 30, "... in concentrations far short of the asphyxiation level, carbon monoxide is still dangerous, especially if exposure is prolonged. Carbon monoxide (poisoning) can be a contributing factor to death that ultimately results from other causes, such as fatal accidents, and is a health as well as safety problem." Carbon monoxide interferes with the ability to breathe and, as a result, the body's ability to utilize oxygen. It is over 200 times more attractive than oxygen to the blood hemoglobin that delivers oxygen throughout the body. A relatively small amount of carbon monoxide, therefore, depletes the body of a large amount of oxygen.

Table 9 Physiological effects of carbon monoxide

Concentration, ppm	Physiological effects
100	Allowable for an exposure of several hours
400	Can be inhaled for 1 h without appreciable effect
600	Causes a barely appreciable effect after 1 h of exposure
1000	Causes unpleasant symptoms, but not dangerous after 1 h
1500	Dangerous for exposure of 1 h
4000	Fatal for exposure of less than 1 h

To ensure personnel safety, piping between the generator and the furnace must be leaktight, and all gas that enters the furnace must be either burned off or properly vented. The highly toxic nature of carbon monoxide requires use under carefully controlled conditions. The safest and most convenient disposition of carbon monoxide is to ensure complete mixing of the furnace atmosphere effluent with air to combustible proportions and subsequent ignition of the mixture. Products of combustion are far less toxic, but should still be properly vented.

In some sintering applications, carbon monoxide cannot be burned directly because it has been diluted in inert gas to a level that is nonflammable when mixed with air. Special precautions must be taken to vent and dilute it to below toxic limits. To ensure the continued effectiveness of the venting, carbon monoxide levels near the furnace setup must be monitored continuously. Consequently, apparatus using or emitting gas containing carbon monoxide that will not burn when mixed with air should not be operated unless the user is willing to accept the added liability. Additionally, special gas monitoring and disbursing equipment should be installed.

Personnel should also be familiar with the early symptoms of carbon monoxide poisoning and trained in appropriate first aid. Early symptoms include slight dizziness, weakness, or headache. In the later stages of poisoning, the victim's lips and skin turn a characteristic cherry red.

Finally, the danger of asphyxiation is not intrinsic to the gases used in the P/M industry, but it is a concern. Asphyxiation can be caused by a harmless gas (for example, a gas that has no significant physiological effect) if it is present in high concentrations in the air. Normal oxygen content in air is ~21 vol%. Minimum oxygen content should be 18 vol% under normal atmospheric pressure.

ACKNOWLEDGMENT

The editors gratefully acknowledge Thomas Philips for portions of Ref 3 adapted for this article.

REFERENCES

1. R.M. German, *Metallurgy of Iron and Steel*, John Wiley & Sons, 1998
2. R. Heckel, Effects of Gas Composition on Gas-Solid Equilibrium During Sintering of Ferrous Powders, *Prog. Powder Metall.*, Vol 35, 1979
3. T. Philips, Trouble-Shooting Guide for Sintering Furnace Atmospheres, *Int. J. Powder Metall.*, Vol 26 (No. 3), 1990, p 245–6250
4. H.S. Nayar, Status Report on Energy Consumption and Distribution in Powder Metal Industry, *Prog. Powder Metall.*, Vol 34, 1978
5. *Furnaces Atmospheres and Carbon Control*, American Society for Metals, 1964
6. *Metals Handbook*, Vol 4, 9th ed., American Society for Metals, 1981
7. H.S. Nayar and J. Drew, Furnace Atmospheres: Feedstock, Make-Up, Properties, *Heat Treat.*, July–Sept 1980
8. R.W. Stevenson, C. Alabrese, and P.E. Fredette, Generator Control System Gives Endo an Edge, *Heat Treat.*, March 1982
9. S. Kaufman, The Three-Zone Sintering Furnace, *Prog. Powder Metall.*, Vol 33, 1977
10. *Matheson Unabridged Gas Data Book, Hydrogen*, Matheson Gas Products, 1974
11. D.W. Doak, State of the Art in Vacuum Sintering Furnace Technology, *Carbide Tool J.*, March 1980
12. W.J. Doekler, Sintering Powder Metallurgy Compacts in Vacuum Furnaces, *Prog. Powder Metall.*, Vol 35, 1979
13. U.S. Patent 4,106,921, 1978
14. H.S. Nayar, Nitrogen-Based Sintering Atmospheres, *Modern Developments in Powder Metallurgy*, Vol 9, Metal Powder Industries Federation, 1976
15. H.S. Nayar, Sintering and Infiltration of Steel P/M Parts in Nitrogen-Based Atmospheres, *Prog. Powder Metall.*, Vol 33, 1977
16. H.S. Nayar, Endo Vs. Nitrogen, Some Technical Considerations, *Heat Treat.*, March-April 1980
17. J.S. Becker, Comparison of Endothermic and Nitrogen-Based Sintering Atmospheres, *Prog. Powder Metall.*, Vol 35, 1979
18. J. Muzik, R. Smith, and J. Marsden, Comparative Results of Nitrogen-Based and Conventional Sintering, *Modern Developments in Powder Metallurgy*, Vol 12, Metal Powder Industries Federation, 1980
19. A.T. Sibley, J.G. Marsden, and D.M. Buck, High Temperature Sintering with a Nitrogen-Methanol Atmosphere, *Prog. Powder Metall.*, Vol 38, 1982
20. M. Kostelitz, Heat Treatment Processes in Nitrogen-Based Atmospheres, *Powder Metall. Int.*, Vol 13 (No. 3), 1981, p 140–144
21. G.W. Gaines and R.H. Kohler, Powder Metallurgy Sintering with a Nitrogen Methanol Atmosphere, SAE Paper No. 810245, Society of Automotive Engineers, 1981
22. S.M. Adams and H.S. Nayar, Advantages of Nitrogen as a Brazing Atmosphere, *Heat Treat.*, April 1981
23. H.S. Nayar and D. Schaeffer, How Furnace Zoning Can Optimize Atmosphere Efficiency, *Heat Treat.*, March 1981
24. U.S. Patent 4,294,395, 1981
25. H.S. Nayar, "The Concept of Furnace Zoning: Its Use in Developing Highly Effective Sintering Atmospheres," Int. Powder Metallurgy Conf. (Florence, Italy), Associazione Italiana di Metallurgica, June 1982
26. H.S. Nayar, Atmosphere Zoning to Optimize Productivity During Sintering of Steel—Part 1, *Prog. Powder Metall.*, Vol 39, 1983
27. R. Frey, Effects of Furnace Atmosphere Compositions on Sintering Ferrous P/M Parts, *Prog. Powder Metall.*, Vol 38, 1982
28. M. Nowotarski, Reduction of Atmosphere Requirements in Continuous Sintering Furnaces, *Prog. Powder Metall.*, Vol 38, 1982
29. W. McKinley and H.S. Nayar, Safety Considerations in Sintering Atmospheres, *Prog. Powder Metall.*, Vol 35, 1979
30. J.T. Holtzberg, Requirements for Monitoring Carbon Monoxide, *Independent. Heat.*, March 1980
31. "Industrial Furnaces Using a Special Processing Atmosphere," ANSI/NFPA 86C, National Fire Protection Association, Quincy, MA, Dec 1983
32. "Threshold Limit Values of Substances in Workroom Air," American Conf. of Governmental and Industrial Hygienists (Cincinnati), 1979
33. Safety and Health Standard 29CFR 1910, Occupational Safety and Health Administration, June 1981
34. F.T. Bodurtha, *Industrial Explosion, Prevention and Protection*, McGraw-Hill, 1980
35. *Handbook of Industrial Loss Prevention*, Factory Mutual Engineering Co., McGraw-Hill, 1967

SELECTED REFERENCES

- P.E. Johnson, H.S. Nayar, and W.O. Judge, Microstructure and Quality Control of P/M Parts Heat Treated under Various Atmospheres, *Progress in Powder Metallurgy*, 1984, Vol 39, p 281–306
- G. Lei, R.M. German, and H.S. Nayar, Corrosion Control in Sintered Austenitic Stainless Steels, *Progress in Powder Metallurgy*, 1984, Vol 39, p 391–410
- H.S. Nayar and T. Philips, Oxide Coating Process to Improve Corrosion Resistance of Sintered Steel Parts, Industrial Heating, April 1992
- H.S. Nayar, Productivity and Quality Improvements Via Evolution in Equipment Design and Process Changes in Conventional Sintering, *Industrial Heating*, May 1994
- T. Philips and H.S. Nayar, A Troubleshooting Guide for Sintering Furnace Atmospheres, *Industrial Heating*, Vol 64 (No. 8), Aug 1994, p 33–37

Production Sintering Practices

SINTERING is a complex high-temperature process that consolidates and strengthens loose or compacted particles into a more dense coherent body. During sintering powder particles form coherent bonds and densify by pore shrinkage. There is no universally applicable theory of sintering, and much is not understood about the process; but sintering generally consists of six distinct, but not necessarily sequential, stages:

- Initial particle bonding
- Neck growth
- Pore channel closure
- Pore rounding
- Densification or pore shrinkage
- Pore coarsening

A comprehensive and systematic theory of sintering has not been established due to the multitude of variables that influence sintering. Nonetheless the major factors that influence sintering response can be classified in terms of transport mechanisms, reaction types, and sintering parameters. These major factors include the following variables:

Mechanism of material transport
- Viscous flow
- Surface diffusion
- Lattice or volume diffusion
- Grain boundary diffusion
- Evaporation and condensation

Types of sinter reactions
- Single-component sintering of metals, oxides, or hydrides
- Multiple component sintering of solid solutions, intermetallic, or mechanically bonded systems
- Activated sintering with additives

Sinter Conditions
- Rate of heating and cooling
- Sintering time
- Sintering atmosphere
- Sintering temperature

Sintering time and temperature are the most significant factors from a practical perspective, with temperature being the most important variable. Particle size, compact porosity, and powder type (mixed, prealloyed, and diffusion-alloyed) also influence sintering practices.

Table 1 lists typical sintering temperatures for various P/M materials and ceramics. Sintering is generally performed at temperatures around $2/3$ to $4/5$ of the absolute melting point or solidus of the material for a single-component system. Multicomponent powder mixtures are generally sintered near the melting point of the constituent with the lowest melt temperature.

Sintering times are typically 20 to 60 min under a protective atmosphere. Widely used furnace atmospheres include endothermic gas, exothermic gas, dissociated ammonia (DA), hydrogen, hydrogen-nitrogen mixtures, and vacuum. The main function of the atmosphere is to protect a part from oxidation or nitridation, as might occur when heating in air. Frequently, however, critical aspects of a sintering atmosphere also include reducing and carburizing power and capability for efficient removal of the lubricant. Some sintering furnaces contain so-called rapid burn-off zones for rapid and efficient removal of the lubricant. This usually entails an atmosphere with a higher oxidation potential, that is, a higher concentration of gases such as steam and carbon dioxide.

Inadequate sinter generally is indicated by low strength, low hardness, and improper dimensions. Causes of inadequate sintering are often related to the atmosphere, but several factors may be involved, such as:

- Sintering temperature too low
- Insufficient reducing agent
- Dew point too high in the hot zone
- High O_2 content in hot zone
- Incorrect green density
- Incorrect belt speed or time at temperature

Corrective action includes:

- Measure and control dew point and O_2 content in the hot zone
- Measure and increase H_2 content
- Check and correct belt speed and/or powder compositions

Further information on atmosphere control for sintered steel is discussed in the preceding article "Production Sintering Atmospheres and Furnaces" in this Volume.

Sintering of Ferrous Materials

Pressed and sintered iron-base materials represent the largest segment of the P/M parts industry. Ferrous P/M parts are made from iron powders, alloy steel powders, or mixtures of (a) iron and graphite powders, (b) iron (or steel) and copper powders, or (c) iron, copper, and graphite powders (with or without powders of other metals). The overall sintering process consists of several stages and temperature regimes (Table 2). Excluding lubricant burnoff, the following reactions occur during sintering of ferrous powders:

- Metallurgical bonds develop from the mechanical interlocks between powders in the compact.
- Metal oxides are reduced by reaction with the carbon from the blended graphite powders.

Table 1 Sintering temperatures for powder metal alloys and special ceramics

Sintered material	Sintering temperatures, C°
Aluminum alloys	590–620
Bronze	740–780
Brass,890–910	
Iron, carbon steels, low-alloyed steels (Cu, Ni)	1120–1150
Low-alloyed steels (Cu, Ni, Mo; Distaloy)	1120–1200
High-alloyed ferritic and austenitic steels (Cr, Cr-Ni)	1200–1280
Hard magnets (Alnico)	1200–1350
Hard metals (cemented carbides)(a)	1350–1450
Molybdenum and molybdenum-alloys	1600–1700
Tungsten(b)	200–2300
Heavy metal (W alloy)	~1400
Ferrites (soft and hard)	1100–1300
Silicon nitride (with different additives)(c)	1750–2000
Silicon carbide (with different additives)(d)(e)	1750–2100
Alumina(e)	1400–1800
Zirconia (with different additives)(e)	1400–1750

(a) TiC-based hard metals (cermets) up to 1600 °C. (b) ~3000 °C, when direct sintering is used. (c) Highest temperature under pressured N_2 atmosphere or in powder bed. (d) Low temperatures for liquid phase sintering. (e) Low temperatures for highly active powders. Source: Ref 1

Table 2 General temperature regimes during steel sintering

Process stage(a)	Temperature range, °C
Lubricant removal, in the liquid state	150–200
Lubricant removal, by gaseous decomposition	300–600
Refining (reduction of iron oxide)	750–1000(b)
Carbon diffusion	900–1050
Diffusion of alloy additions	1050–1300
Carbon pick-up by the atmosphere (in some cases)	1000–1300
Carbon restoration	1050–800
Microstructure formation	950–600
Final cooling, under reducing or "inert" conditions	Cool from 600 to 50

(a) Not all stages apply and some stages may overlap. (b) Reduction of metal oxides other than iron oxide may require higher temperatures depending on the element and furnace atmosphere. Source: Ref 2

Table 3 Sintering temperatures for various iron-base alloys

Material	Sintering temperature(a), °C
Fe, Fe-Cu, Fe-Cu-Ni	1120–1280
Fe-C, Fe-Cu-C, Fe-Cu-Ni-C	1120–1150
Fe-Cu-Ni-Mo (Distaloy)-C	1120–1200
Fe-Mn, Fe-Mn-C	1120–1280
Fe-Cr, Fe-Cr-C	1150–1280
Fe-Mn-Cu	1120
Fe-Cr-Cu	1200–1280
Fe-Mo, Fe-Mo-C	1120–1280
Fe-Cr-carbide	>1280
Fe-V carbide	>1280
Fe-Ti-carbide	1200–1280
Fe-W-carbide	1200–1280
Fe-Mn-Cr-Mo-C, Fe-Mn-V-Mo-C	>1280
Fe-Cr-Ni	>1280

(a) Temperatures represent the range of typical or possible sintering temperatures. Source: Ref 3

- Carbon alloying of the iron powder is obtained by diffusion of carbon from graphite powder.
- Densification of P/M compacts is achieved by shrinkage and the reduction of porosity.

Conventional sintering generally is conducted at 1120 °C (2050 °F) in mesh belt conveyor furnaces with conveyor mesh belts made from nickel-chromium alloy wire. Mesh belt furnaces can be used at temperatures up to 1150 °C (2100 °F) only, due to the temperature limitation of the belt material. For high-temperature sintering applications, walking beam furnaces have been successfully used for continuous operation at temperatures up to 1315 °C (2400 °F). Furnace temperatures rarely exceed 1350 °C (2460 °F) in ferrous P/M operations (Ref 3).

Sintering temperatures depend on the material and desired properties. Alloy steels typically require higher sintering temperatures (Table 3) and times to promote homogenization. Typical sintering practices for several ferrous materials, with an emphasis on the types of atmospheres used and the resulting as-sintered microstructures and properties, are described in this section. In practice, sintering parameters are usually determined empirically from a combination of technical requirements and economic factors (and furnace type).

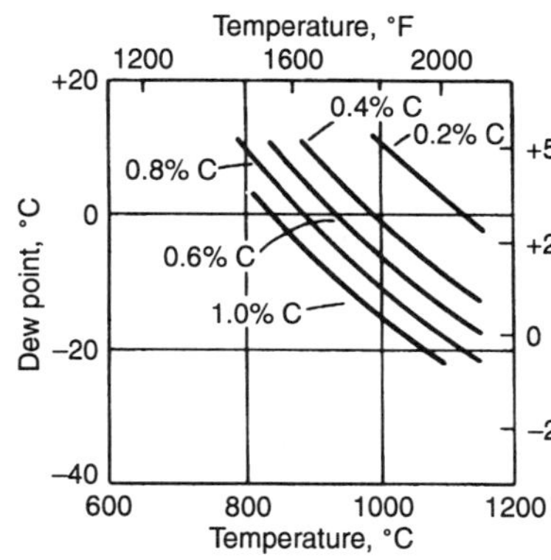

Fig. 1 Carbon potential of endothermic gas as a function of dew point and temperature

Sintering Atmospheres

Sintering atmospheres primarily control chemical reactions between the materials being processed and the furnace surroundings. Additionally, atmospheres may be the source of one or more chemical elements that alloy with the material being sintered. An example of this is the addition of carbon to iron in a carburizing furnace atmosphere. Sintering atmospheres may also prevent the loss of alloying elements existing in the material being processed. Atmospheres are routinely used to flush the decomposition products of the admixed lubricants from the furnace to prevent deposits on furnace interiors, where they rapidly degrade the performance of the equipment.

In common operating practice, the most important functions of select atmospheres are to aid in the reduction of oxides on the surfaces of the metal particles in the compact and to control carburization and decarburization of iron and iron-base compacts caused by oxygen, water vapor, and carbon dioxide when present in improper proportions with respect to the hydrogen and carbon monoxide contents of the sintering atmosphere. Iron oxides are reduced by hydrogen, carbon monoxide, and carbon. Carburization is caused by carbon monoxide and by hydrocarbons such as methane.

The most frequently used atmospheres in commercial sintering of P/M iron and steel materials are endothermic, exothermic, dissociated ammonia, pure hydrogen, and nitrogen-base atmospheres. Of these, endothermic gas is the most widely used, followed by dissociated ammonia.

Endothermic and Exothermic Atmospheres. Although other hydrocarbon gases may be used, methane and propane are the most commonly used bases for the production of endothermic gas and exothermic gas atmospheres. These furnace atmospheres consist of mixtures of nitrogen, water vapor, carbon monoxide, hydrogen, carbon dioxide, and methane. The degree of combustion that occurs is controlled by the amount of air admitted to the process. This air-to-gas ratio determines the properties of the resultant gas. With methane, endothermic gas is produced with air-to-gas ratios of ~2.4, while exothermic gas ratios are usually above the 9.0+ value required for complete combustion (10-to-1 is the air-to-gas ratio for complete combustion).

For an endothermic-type gas produced from methane, the reaction can be approximated by:

$$2CH_4 + O_2 + 3.8N_2 \rightarrow 2CO + 4H_2 + 3.8N_2$$

This gas is strongly reducing to iron oxide. Control of carbon level, however, is difficult because the carbon potential of endothermic gas varies with temperature and dew point, as shown in Fig. 1. The carbon potential of the endothermic gas with −1 °C (30 °F) dew point varies from about 0.2% at 1120 °C (2050 °F) to 1.0% at about 850 °C (1560 °F). Dew point is a measure of the dryness of the atmosphere gas. The lower the dew point, the lower the water vapor content of the atmosphere. Generally, low dew point gases increase the reduction potential of the atmosphere. Figure 2 shows the relationship between dew point and water vapor content.

When generating exothermic-type atmospheres, the reaction is self supporting and generates heat, which in the presence of a nickel catalyst promotes cracking of any of the hydrocarbons remaining unburned into hydrogen and carbon monoxide. The two reactions that occur are:

$$CH_4 + 2O_2 \rightarrow 2H_2O + CO_2 + \text{heat}$$

$$2CH_4 + O_2 + \text{heat} \rightarrow 4H_2 + 2CO$$

Exothermic atmospheres are not strongly reducing to iron and are decarburizing at normal sintering temperatures. Their use in sintering iron-base materials is limited to applications that do not require a residual carbon content. Removal of water and carbon dioxide can improve the properties of this protective atmosphere.

Nitrogen-Base Atmospheres. Protective atmospheres based on nitrogen obtained from the fractional distillation of liquid air are enjoying increased usage, not only in sintering, but in heat treating as well. This is primarily due to the increasing cost of hydrocarbon fuels.

Basically, nitrogen can be taken from a bulk storage facility or directly from the distillation tower and can be used as a replacement for all, or nearly all, of the carbon monoxide and/or hydrogen-containing gases previously used. Typical usage consists of utilizing nitrogen to extend either endothermic gas or dissociated ammonia through dilution. Hydrogen content can be reduced to a few percent of the total, versus a normal 40% presence in conventional endothermic gas.

Because there is less hydrogen available to reduce the oxide present on the surface of powder particles, greater demand is placed on the graphite present to take part in the reduction process. As a result, greater amounts of admixed graphite are required if the sintered carbon content is to remain the same.

Dissociated Ammonia Atmospheres. Dissociated ammonia, consisting of 75 vol% hydrogen and 25 vol% nitrogen, is produced from gasified ammonia by the reaction:

$$2NH_3 \Rightarrow N_2 + 3H_2$$

Dissociated ammonia directly from the dissociator is very dry, generally yielding dew points of −40 °C (−40 °F) or lower.

It is common practice to generate dissociated ammonia with higher nitrogen contents by reacting ammonia with air either catalytically or through partial burning. The water formed in such reactions is removed from the gas by absorption.

Hydrogen Atmospheres. Typical methods used to produce metallurgical-grade hydrogen include the electrolysis of aqueous solutions and the reaction of hydrocarbon gases with steam or oxygen to form mixtures of hydrogen and carbon monoxide. Carbon monoxide is subsequently converted to carbon dioxide and scrubbed from the mixture. Hydrogen is used primarily in the sintering of iron-base magnetic materials. Alnico permanent magnets, for example, are frequently sintered in an atmosphere of very dry hydrogen at 1300 °C (2370 °F).

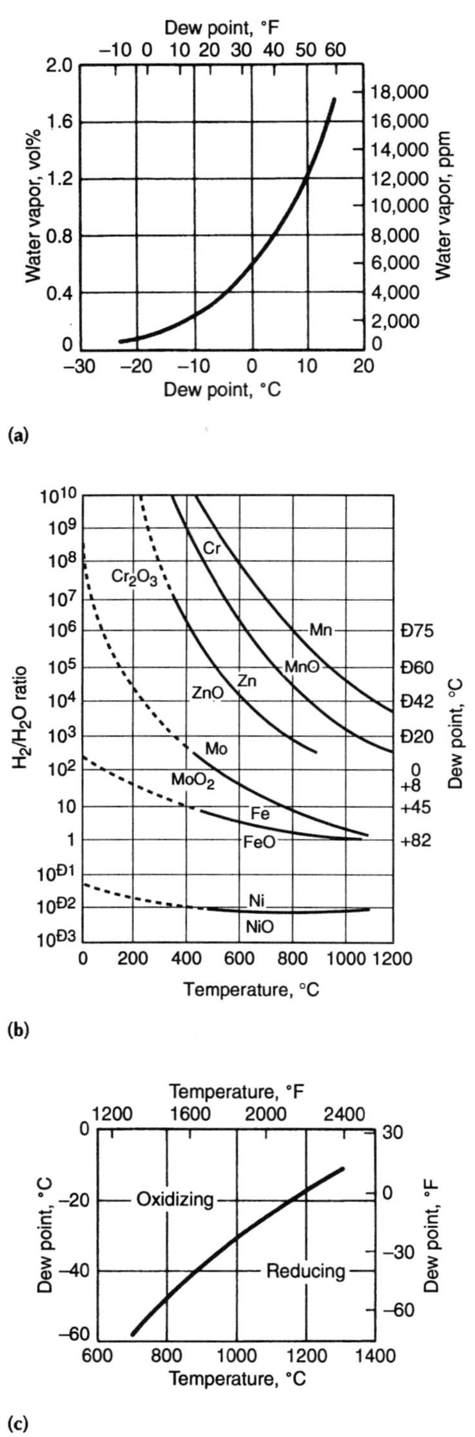

Fig. 2 Relationship between (a) dew point and water vapor content. (b) Equilibrium values of H_2/H_2O ratio and dew point for selected oxides. (c) Oxidation-reduction of chromium oxide for pure chromium in hydrogen.

Iron and Iron-Graphite Powder

Sintering of plain iron powder sequentially involves the establishment and growth of bonds between the particles of powder at their areas of contact, grain growth and migration of the grain boundaries formed at the bonds, spheroidization of the pores between the particles, and the elimination of small pores (and, possibly, the growth of large pores). The formation of bonds is opposed by residual material from the lubricant, by impurities and surface oxides, and by poor contact.

As the sintering temperature increases, porosity decreases and shrinkage increases. Bonds form between the particles during sintering, and the number of particle bonds increases as the temperature increases. In addition, some grain growth and spheroidization of pores occur as sintering temperatures increase.

Mixtures of iron and graphite powders are sintered by (a) establishment and growth of iron-to-iron bonds, (b) diffusion of carbon into and combination of carbon with iron, and (c) spheroidization of pores. The oxide content of iron is an important factor in determining the reactivity of iron with graphite.

During sintering, the major portion of the oxide must be reduced by the graphite and the sintering atmosphere before the iron and graphite can combine. To ensure high reactivity with the iron for rapid reduction of oxide and fast carburization, fine graphite powder—free of silicon carbide and having low ash content—usually is used.

Although iron powder of low oxide content is not required, determination of the oxide content (hydrogen loss value) is essential, because the graphite addition must be large enough to allow for reaction with the oxide and still provide a sufficient quantity to produce the desired iron-carbon alloy.

Effect of Combined Carbon. Aside from bonding, the main factor affecting the properties of a sintered steel part is the amount of combined carbon formed in the steel. Overall expansion during sintering is directly proportional to the amount of combined carbon. As with wrought steel, the strength of sintered steel increases rapidly with increasing combined carbon content. For wrought steel, strength increases rapidly up to 1.0% combined carbon, then gradually reaches a maximum at the highest carbon content of 1.6%, as shown in Fig. 3.

Figure 4 shows a similar plot for sintered steel. Combined carbon was varied by adding increasing amounts of graphite. Test bars were pressed to a density of 6.3 g/cm^3, then sintered 30 min at 1120 °C (2050 °F) in dissociated ammonia. Figure 4 shows transverse rupture strength as a function of combined carbon. As with solid steel, there is a rapid initial increase in strength with increasing combined carbon content, but a maximum is reached at about the eutectoid composition of 0.85% C. Between 0.90 and 1.00%, strength drops markedly.

As the amount of pearlite that contains the combined carbon increases, strength and hardness increase correspondingly until a combined carbon content of 0.8% is reached, at which point the structure consists entirely of fine pearlite. At 1.0% combined carbon, free cementite

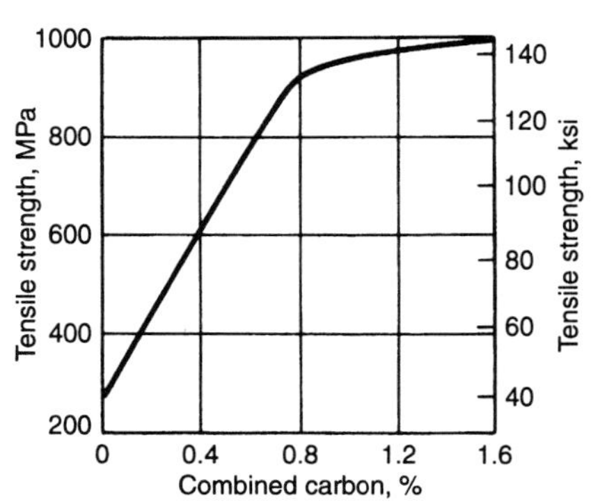

Fig. 3 Effect of combined carbon content on the tensile strength of wrought (rolled) steel. Source: Ref 4

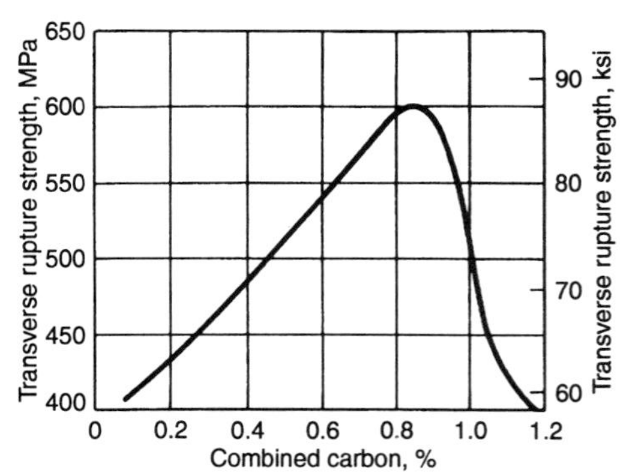

Fig. 4 Effect of combined carbon content on the transverse rupture strength of sintered steel. Test bars were pressed to a density of 6.3 g/cm^3, then sintered for 30 min at 1120 °C (2050 °F) in dissociated ammonia. Source: Ref 4

forms at the grain boundaries, which reduces the transverse rupture strength. At 1.2% combined carbon, the network of free cementite becomes continuous, further reducing the transverse rupture strength.

Thus, most sintered steel has a combined carbon content of 0.8 to 0.9%, and a graphite addition greater than 0.9%. Carbon loss is controlled by maintaining the carbon potential of the sintering atmosphere at 0.7 to 0.9%. The furnace atmosphere must be controlled at a high enough carbon potential to prevent excessive loss of carbon to the atmosphere, because decarburization results in correspondingly poor properties.

As the sintering temperature or time is increased, spheroidization of the pores causes the strength of a compact to continue to increase after carburization of the iron to an all-pearlite structure is complete. The resulting densification of the compact causes a reduction in overall growth of the part. The effect of sintering temperature on strength is shown in Fig. 5 for typical steel compact.

Effect of Atmosphere Composition. The tendency of parts to oxidize or reduce and to carburize or decarburize, as well as the rates at which these reactions occur during sintering, depends on the sintering temperature and the proportion of various gases contained in the sintering atmosphere. The sintering atmosphere has a strong effect on the amount of combined carbon formed during sintering. If the carbon potential is not controlled, excessive carbon may be lost to the atmosphere, or excessive carburization may occur.

The carbon potential is determined by the ratio of water vapor to hydrogen, carbon dioxide to carbon monoxide, and methane to hydrogen. The amount of graphite in the parts also contributes to the carbon potential in the furnace atmosphere. Figure 6 illustrates equilibrium ratios at various temperatures for oxidation-reduction and decarburization-carburization reactions. Typically, the carbon dioxide to carbon monoxide ratio can be fairly high at all temperatures without causing oxidation. However, it is necessary to keep the carbon dioxide content low to prevent decarburization.

Carbon dioxide content is kept low by maintaining a low enough dew point (water vapor content) to balance the otherwise continuously reversible "water-gas" reaction:

$$CO_2 + H_2 \rightarrow H_2O + CO$$

A dew point of −4 to −1 °C (25 to 30 °F) at the generator is usually low enough to maintain a 0.70 to 0.90% carbon potential during continuous sintering of iron-graphite mixes. If lower combined carbon is desired, then a smaller graphite addition can be made or a higher dew point can be used to raise the carbon dioxide content.

The ratio of methane to water vapor also determines the tendency of parts to gain or lose carbon. At normal sintering temperatures, even small amounts of methane in the furnace atmosphere increase carburizing tendency. Increased amounts of methane further increase this tendency.

Atmospheres having better than equilibrium ratios are required unless long sintering times or high sintering temperatures are permitted. Also, nitrogen content should be kept as low as possible. For example, purified exothermic gas (high-nitrogen content) can be used, but frequently larger graphite additions are required to obtain the desired combined carbon. Endothermic gas (low-nitrogen content), however, promotes rapid carbon pickup, and less graphite is required.

Effect of Temperature. Although sintering temperature has a profound effect on the amount of combined carbon formed for a given sintering time, graphite combines readily under normal sintering temperatures and times. Figure 5 illustrates the effect of sintering temperature on the transverse rupture strength of bars made of iron plus 1.25% graphite, pressed to a density of 6.1 g/cm^3 and sintered for 30 min at temperature. No strengthening occurs as the sintering temperature is raised from 900 to 1010 °C (1650 to 1850 °F). Above this temperature, however, a substantial increase is noted.

Figure 7 illustrates the changes in microstructure. At low sintering temperatures, no carbon is combined, as evidenced by the complete absence of pearlite. At 1010 °C (1850 °F), some carbide has begun to form (Fig. 7a), while at 1040 °C (1900 °F), the maximum combined carbon content is achieved (Fig. 7b). Additional strengthen-

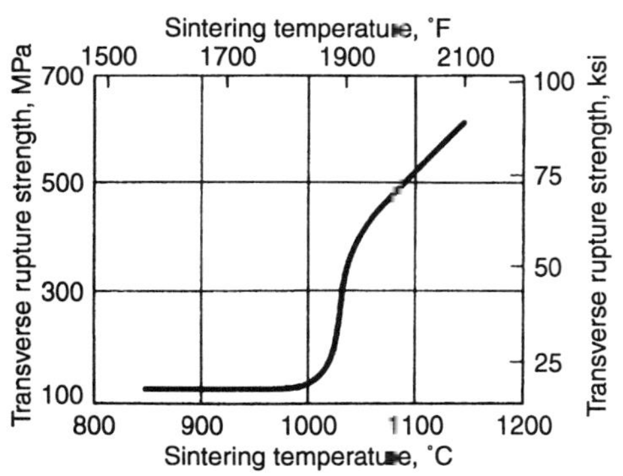

Fig. 5 Effect of sintering temperature on transverse rupture strength of iron plus 1.25% graphite test bars. Compacted to a density of 6.1 g/cm^3 and sintered for 30 min at temperature. Source: Ref 4

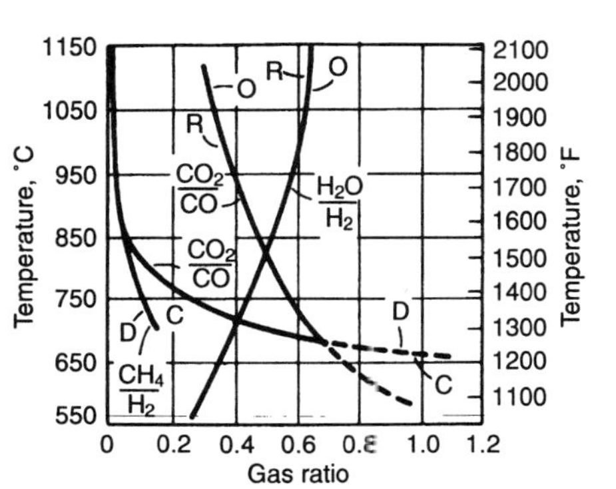

Fig. 6 Equilibrium ratios at various temperatures for oxidation-reduction and decarburization-carburization reactions. Equilibrium reactions: R, reducing; C, carburizing; O, oxidizing; D, decarburizing. Source: Ref 4

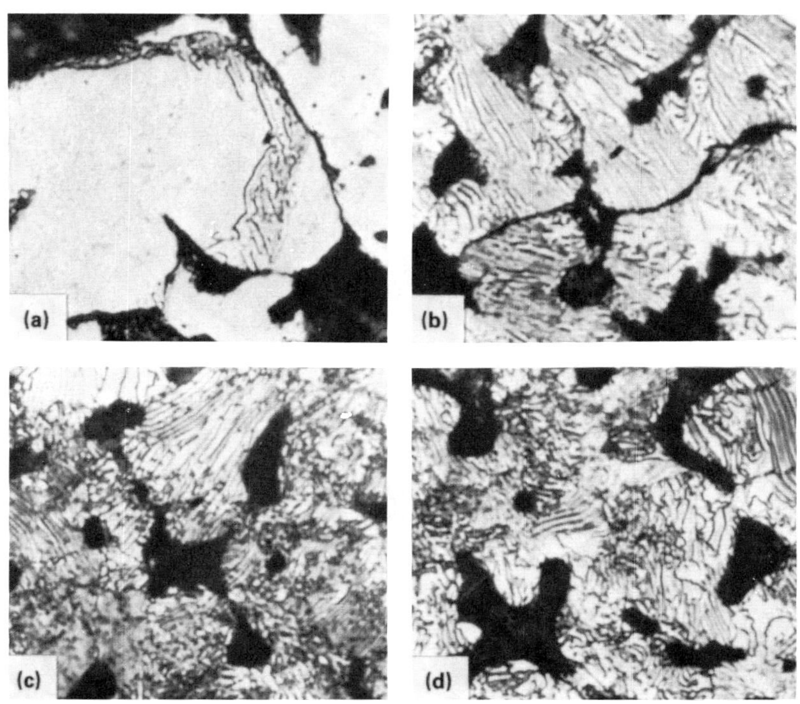

Fig. 7 Effect of sintering temperature on the microstructure and transverse rupture strength of iron plus 1.25% graphite test bars. Etchant: 4% picral plus 0.5% nitric acid. 800×. (a) Sintering temperature: 1010 °C (1850 °F). Combined carbon: 0.1%. Strength: 138 MPa (20 ksi). (b) Sintering temperature: 1040 °C (1900 °F). Combined carbon: 0.75%. Strength: 352 MPa (51 ksi). (c) Sintering temperature: 1120 °C (2050 °F). Combined carbon: 0.75%. Strength: 552 MPa (80 ksi). (d) Sintering temperature: 1175 °C (2150 °F). Combined carbon: 0.75%. Strength: 655 MPa (95 ksi)

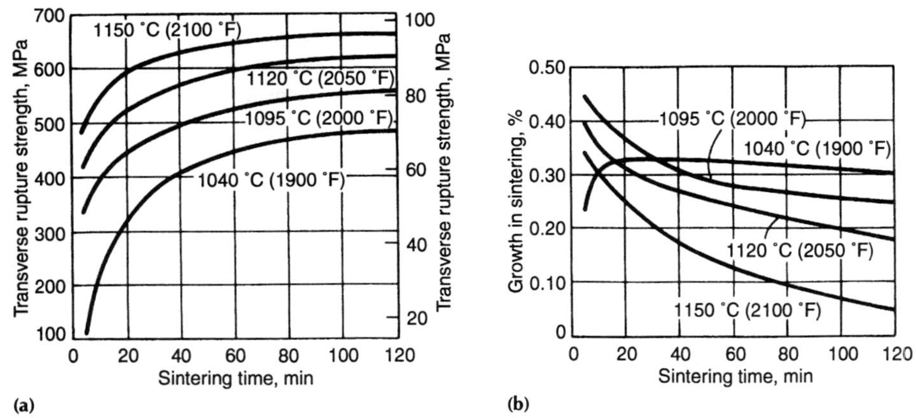

Fig. 8 Effect of sintering time on properties of iron-graphite test bars pressed to a density of 6.1 g/cm^3. (a) Transverse rupture strength. (b) Dimensional change. Source: Ref 4

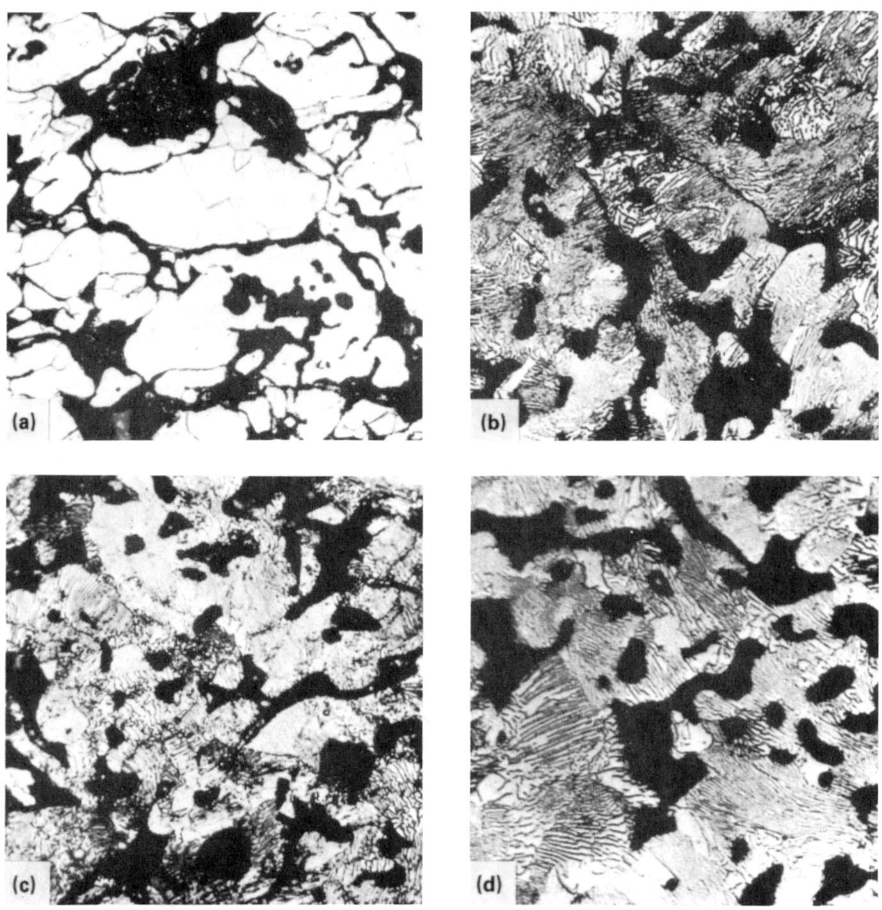

Fig. 9 Effect of sintering time on the microstructure and transverse rupture strength of iron-graphite test bars pressed to a density of 6.1 g/cm^3. (a) As-pressed condition. Combined carbon: 0%. Strength: ~124 MPa (18 ksi). (b) Sintering time: 5 min at 1120 °C (2050 °F). Combined carbon: 0.7%. Strength: ~421 MPa (61 ksi). (c) Sintering time: 30 min at 1120 °C (2050 °F). Combined carbon: 0.7%. Strength: ~552 MPa (80 ksi). (d) Sintering time: 120 min at 1120 °C (2050 °F). Combined carbon: 0.7%. Strength: ~627 MPa (91 ksi)

ing at higher temperatures is caused by increased sintering, as evidenced by the elimination of grain boundaries and spheroidization of pores—shown in Fig. 7(c) at 1120 °C (2050 °F) and Fig. 7(d) at 1175 °C (2150 °F).

Effect of Sintering Time. Time of sintering also affects the amount of combined carbon formed. At the sintering times normally used, however, the maximum amount of combined carbon is usually formed. A noticeable effect on strength and dimensional change during sintering was apparent for a series of test bars sintered at several temperatures for times varying from 5 to 120 min. Figure 8(a) is a plot of sintering time versus transverse rupture strength. A rapid increase in strength occurs up to 30 min, followed by smaller increases at longer times.

Figure 8(b) is a plot of sintering time versus dimensional change for several temperatures. These are typical size change curves showing a high growth at relatively short times, followed by lesser growth at longer times. Figure 9 illustrates how microstructure is influenced by sintering time at 1120 °C (2050 °F). Figure 9(a), a microstructure of an unsintered bar, shows the individual particles, grains within particles, and graphite located primarily at the pores formed between particles. Figure 9(b) shows almost complete formation of pearlite in 5 min; numerous grain boundaries are visible, and the porosity is quite angular. Figure 9(c) shows some disappearance of grain boundaries and slight spheroidization of pores after 30 min, while Fig. 9(d) shows an almost complete absence of grain boundaries and substantial spheroidization of pores after 120 min. Sintering for 10 min at 1200 °C (2200 °F) typically results in strength equal to that attained after 120 min at 1120 °C (2050 °F).

Effect of Cooling Rate. The influence of cooling rate on strength in wrought steel is well known. Iron-graphite samples similar to those used in the study of effect of time and temperature were sintered for 30 min at 1120 °C (2050 °F), followed by cooling at different rates. Table 4 shows the three rates used and the resulting mechanical properties. Typically, faster cooling rates cause increased strength and hardness.

Iron-Copper and Iron-Copper Graphite

The sintering of iron-copper mixtures involves (a) solid bonding of iron to iron, (b) solid bonding of copper to iron, (c) melting of copper, (d) solution and diffusion of copper in solid iron, and (e) solution and precipitation of iron in liquid copper.

At typical sintering temperatures of 1095 to 1120 °C (2000 to 2050 °F), 7.5 to 9.0% Cu is soluble in iron. However, with this copper content and typical sintering conditions, some of the molten copper remains undissolved, or free, and can dissolve ~3% Fe.

The solubility of copper in iron decreases with decreasing temperature and is less than 0.1% at room temperature. Therefore, copper dissolved in iron at the sintering temperature must precipitate as the compact is cooled, thus hardening it. Faster cooling rates lower the temperature at which the precipitate forms, which makes the precipitate finer and increases the hardening effect.

The solution of copper in iron causes growth of the compact; the solution of iron in the free copper causes shrinkage. These processes go on simultaneously, with solution of copper predominating only in the early stages of sintering. For example, the addition of 7.5 to 10% Cu causes significant amounts of growth, but a 20%

Table 4 Effect of of cooling rate on iron-graphite samples

Pressed to a density of 6.1 g/cm3 and sintered 30 min at 1120 °C (2050 °F).

Sample No.	Cooling rate		Transverse rupture strength		Hardness, HRB	Pearlite spacing
	°C/min	°/min	MPa	ksi		
1	1.9	3.5	462	67	37	Very coarse
2	64	115	565	82	50	Medium
3	125	225	600	87	57	Very fine

Source: Ref 5

Table 5 Alloying properties of selected elements in iron

Alloying element	Hardening Grossman factor(a)	Diffusion coefficient (D_x/D_{Fe})(b)	Enthalpy of oxide formation (ΔG), KJ/g-atom of oxygen
Copper	1.7	1	–150
Nickel	1.4	0.5	–250
Cobalt	2	0.5	–270
Tungsten	...	4.5	–270
Molybdenum	3.7	5	–310
Iron	...	...	–350
Chromium	3.1	5	–540
Manganese	4.5	2.5	–500
Vanadium	...	5	–620
Boron	...	...	–650
Silicon	1.7	10	–680

(a) Grossman factor for 1 w% alloy. (b) Diffusion coefficient of alloying element in γ-iron (D_x), self-diffusion coefficient of γ-iron (D_{Fe})

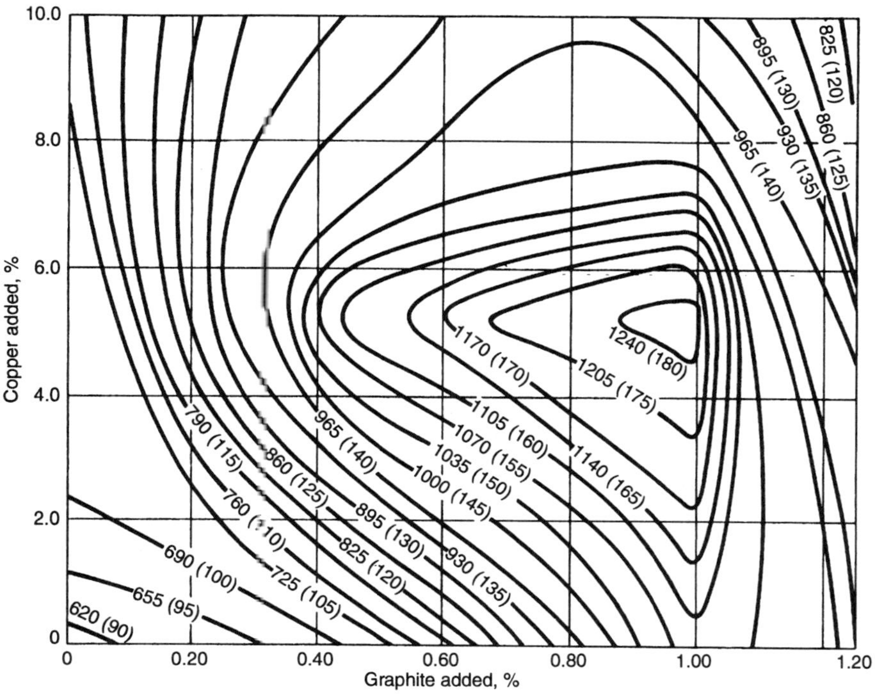

Fig. 10 Transverse rupture strength of iron, copper, and graphite powder compacts. Sintered to a density of 6.8 g/cm^3 in endothermic gas. Lines represent compositions having the same transverse rupture strength, given in MPa with ksi equivalent values in parentheses; combined carbon in alloys is ~80% of amount of graphite included in mixture.

addition increases the amount of free copper to the extent that growth is no greater than for a 7.5% copper addition.

The sintering of iron-copper-graphite mixtures involves the mechanisms associated with both the iron-graphite and iron-copper mixtures mentioned previously. Diffusion of carbon usually is complete before the melting point of copper is reached. Carbon has little effect on the solubility of copper in solid iron, but it decreases the rate of solution. Thus, there is usually more free copper remaining when carbon is present, which acts to reduce the amount of growth.

Figure 10 illustrates the combined effects of graphite and copper powder additions on iron compacts sintered in an endothermic atmosphere to a density of 6.8 g/cm^3. Transverse rupture strength is plotted as a function of the amounts of graphite and copper powder added to the iron powder. The amount of combined carbon in these compacts is ~80% of the graphite added. The highest value of transverse rupture strength of 1240 MPa (180 ksi) is obtained with additions of 0.9% graphite and 5% copper.

A steel of this composition is quite brittle and is used in applications that do not require superior toughness. The high strength of iron-copper-carbon compacts in the as-sintered condition is essentially due to the combined effect of copper causing precipitation strengthening and increasing the hardenability of the steel.

Sintering of Alloy Steels

Sintered alloy steels are made from various powder types that may be elemental, prealloyed, diffusion alloyed, or a hybrid mix of these types. The most common alloying method is mixing and compaction of elemental powders, which then alloy during sintering by diffusion. Mixing of elemental powders is advantageous because elemental powders have high compressibility due to the absence of solid solution strengthening. However, in order to achieve adequate homogenization, uniform composition and good diffusion rates must be maintained during sintering. Ferrous master alloys are sometimes used for base powders in admixed systems. Powders manufactured from ferrous master alloys can also be advantageous when the alloy elements are strong oxidizers (like chromium or manganese).

In contrast, prealloyed steel powders are alloyed during atomization and do not require homogenization during sintering. This allows a higher level of chemical and structural homogenity in sintered parts. Prealloyed powders are more expensive and have lower compressibility than elemental mixes, but they are used in high-performance parts that require structural integrity in demanding applications. Diffusion alloyed powders are an intermediate powder form where alloy elements are partially bonded onto base iron particles. Because the elements are not fully alloyed, the powders still have good compressibility, while the bonding improves the level of chemical homogeneity. Diffusion bonded powders typically involve iron alloying with nickel, molybdenum, and/or copper and can be sintered like elemental mixes. Prealloyed powders include a wider variety of possible compositions.

Sintering of Ferrous Alloys with Copper, Nickel, Molybdenum, and/or Phosphorus. A major constraint in conventional P/M alloy development is that typical production furnaces and atmospheres require elemental mixes and diffusion-alloyed powders based on elements with a low affinity for oxygen (copper, nickel, molybdenum, and phosphorus). Nickel, molybdenum, phosphorus, and copper have low affinity with oxygen (Table 5) which thus allows reduction of their oxides in typical mesh-belt furnaces (which cannot exceed 1150 °C and have typical atmosphere dew points around –10 to –40 °C). Under these typical production conditions, elements with high-oxygen affinity (chromium, manganese, and silicon) may not be adequately reduced during sintering. This limitation on alloying can be overcome to some extent by vacuum sintering, high-temperature sintering, or by using prealloyed powders.

Table 6 Properties of chromium alloyed steels prepared from elemental powders

Chromium powder size, µm	Sintering parameters(a)	Density, g/cm^3	Dimensional change, %, linear	Ultimate tensile strength MPa	Yield strength (0.2 offset), MPa	Elongation, %	Impact toughness, J	HV30
Fe-2%Cr-0.7%C								
<25 µm	1 h/H_2	7.23	0.83	937	740	2.9	38	274
	2 h/vac	7.36	0.26	863	546	10.9	58	230
45–63 µm	1 h/H_2	7.26	0.65	643	544	3.5	10	273
	2 h/vac	7.34	0.3	755	490	9.2	38	210
Fe-1.0%Cr-0.5%Mo-1.0%C								
<25 µm	1 h/H_2	7.16	0.92	927	776	3.4	38	278
	2 h/vac	7.35	0.18	923	635	8.0	29	266
Bondaloy	1 h/H_2	7.29	0.38	948	800	4.2	29	285
	2 h/vac	7.45	–0.33	961	648	5.1	29	275
Fe-1.0%Cr-0.5%Mo-1.0%C								
<25 µm	1 h/H_2	7.17	0.89	855	718	1.0	1.9	278
	2 h/vac	7.41	0	897	773	2.7	10	335

(a) Compacted at 1200 MPa, sintered at 1 h at 1270 °C/H_2 or 2 h at 1270 °C/vacuum as listed. vac, vacuum

Table 7 Chemical composition of master alloys for alloying powder steels with manganese for liquid-phase sintering

Chemical composition (wt %)							T_l	T_s	ΔT
Mn	Ni	Cr	Mo	Fe	Cu	Si	(°C)	(°C)	(°C)
40	30	15	5	10	...	...	1171	999	172
44	25	...	11	19	...	...	1204	1166	38
55	18	3	8	14	...	2	1129	943	186
56	24	3	6	11	...	...	1157	1032	125
47	20	13	6	14	...	2.5	1210	1166	45
75	25	...	...	...	...	...	1054	982	72
74	12.5	...	...	...	12.5	1	1060	927	133
36	30	18	6	10	...	...	1207	1096	111
41	25	18	6	10	...	...	1216	1077	139
38	23	18	6	15	...	...	1229	1093	136
64	16	...	10	10	...	...	1188	1132	56
56	14	...	15	15	...	...	1260	1093	167
56	14	15	5	10	...	...	1227	1102	125
59	11	15	5	10	...	...	1249	1116	133
53	17	15	5	10	...	...	1216	1093	123
56	14	22	8	...	...	...	1335	1049	286
50	20	15	5	10	...	...	1204	1143	61
46	24	15	5	10	...	...	1204	1088	116
72	14	...	...	...	14	2	1043	966	77

T_l, liquidus temperature; T_s, solidus temperature. Source: Ref 3

Nickel-Steel. Sintering of mixed iron-nickel powders is subject to inhomogeneous alloying due to the low diffusion rate of nickel into iron (Table 5). Fine (carbonyl) nickel powder should be used when mixing with iron powder, and the sintering temperature should be at least 1100 °C (2010 °F) or preferable up to 1200 °C (2190 °F).

Cu-Ni-Mo Steels. Copper-nickel steels are sintered at temperatures up to 1150 °C, and the temperature must be increased to at least 1200 °C when molybdenum is added (Ref 3).

Phosphorus Steel. Unlike wrought forms (where phosphorus causes grain boundary embrittlement), phosphorus is a common alloying element in P/M Steels. phosphorus P/M steels provide a unique combination of strength and ductility, which is attributed to the formation of a eutectic liquid phase in the initial stages of sintering (starting at 1050 °C) and the stabilization of ferrite even at normal sintering temperatures of 1120 °C (2050 °F).

Sintering of Ferrous Alloys with Chromium, Manganese, and Silicon. Chromium, manganese, and silicon have a higher affinity with oxygen (Table 5), which thus requires cleaner atmospheres and high enough temperatures for adequate oxide reduction (Fig. 2b) during sintering. A nitrogen atmosphere with a small amount of hydrogen and hydrocarbons can ensure oxide reduction (Ref 3). Another general approach is to tie up the reactive alloying elements in compound form by adding alloy powders from either ferroalloys (ferrochromium, ferromanganese, and ferrosilicon) or as carbides. Addition of carbides (or carbon ferroalloys) has a low-alloying cost, but carbides increase tool wear and require sintering temperatures above 1200 °C (2190 °F) in typical production atmospheres (Ref 3). This practice is not prevalent.

Chromium Steel Sintering (Ref 3). Chromium steel parts are sintered from powder mixes, prealloyed powders, or from special master alloys based on carbides, sigma-phase compositions, and other complex alloying with carbides. To obtain a homogenous structure from sintering of powder mixes, the sintering temperature should be above 1250 °C (2280 °F) with the formation of a liquid phase in the presence of carbon. Effective carbon control is a decisive factor, but high strength chromium alloy steel can be achieved (Table 6).

The problem of sintering chromium steels at high temperatures can also be solved by adding chromium to the iron powder in the form of low-melting master alloy. Higher hardness was achieved in a Fe-1.5Cr-0.25C sintered steel based on a low-melting master alloy (33%Cr5%C-8%MN+Ni+Mo+P) as compared to chromium carbide or ferrochromium additions. With the low-melting master alloy, the material can be effectively sintered at 1120 °C (Ref 3).

Manganese Steels. Manganese is one of the least expensive alloying elements for iron, but is not widely used in P/M steels due to high reactivity with oxygen. It can be added to iron in elemental form or as ferromanganese alloys. It is also added with other elements as low-melting master alloys (Table 7).

Manganese steels can be sintered efficiently at 1120 °C in typical production atmospheres due to self gettering of the atmosphere. Manganese sublimation occurs at sintering temperatures and manganese vapor reacts with oxygen in the furnace atmosphere, which in effect cleans oxygen from the furnace atmosphere (Ref 3). Table 8 presents mechanical properties of manganese steels prepared from atomized iron powder and electrolytic manganese sintered at different temperatures. The properties change only slightly if ferromanganese is used instead of electrolytic manganese. Dimensional stability was obtained at 2 to 3% Mn.

Manganese Silicon Steels. Iron silicon materials are not used in the fabrication of P/M parts due the excessive shrinkage of parts during sintering. However, the shrinkage is reduced by the addition of manganese in elemental form or by master alloys. To increase activity, additions for liquid phase sintering are also considered (Ref 3).

High-Temperature Sintering

Interest in high-temperature sintering of ferrous components continues to grow in the P/M industry (see the article "High-Temperature Sintering of Ferrous Powder Metallurgy Components" in this Volume). Improvements in both production rates and properties are possible as sintering temperatures increase above 1120 °C (2050 °F). Rounding of the porosity results in increased strengths, especially impact strength. Higher diffusion rates also increase the strength and hardenability of admixed powder compositions. The greater oxide reduction that occurs at higher temperatures proves valuable in powder forging and other applications.

To demonstrate the feasibility of high-temperature sintering of ferrous materials on a pro-

Table 8 Effect of sintering temperature and time on mechanical properties of Fe-Mn steels

Alloy	Sintering: Temperature, °C	Sintering: Time, min	Density, g/cm³	Yield strength, %	Ultimate tensile strength, %	Elongation, %
Fe-2Mn	1200	50	7.03	170	300	11.0
	1200	100	7.05	220	340	10.0
	1250	50	7.06	200	320	10.0
	1280	100	7.06	200	330	15.0
Fe-4Mn	1200	50	6.9	340	490	3.5
	1200	100	6.91	320	530	4.0
	1280	50	6.92	330	530	4.0
	1280	100	6.92	330	550	4.0
Fe-6Mn	1200	50	6.79	370	600	3.2
	1200	100	6.81	420	650	4.0
	1280	50	6.8	450	610	2.5
	1280	100	6.87	440	620	4.0

Compacted at 600 MPa. Source: Ref 3

Table 9 Effect of sintering temperature on transverse rupture strength of test bars pressed to 6.8 g/cm³

Alloy	Transverse rupture strength at 1120 °C (2050 °F), MPa	ksi	Transverse rupture strength at 1290 °C (2350 °F), MPa	ksi
Fe-0.9C	693	100.5	728	105.6
Fe-4Ni-0.7C	910	132.0	924	134.0
Fe-2Cu-0.9C	975	141.5	964	139.8

Transverse rupture strength of bars sintered for 32 min at 1120 °C (2050 °F). Average transverse rupture strength for all atmosphere tests of bars sintered for 10 min at 1290 °C (2350 °F). Source: Ref 5

Table 10 Effect of sintering temperature on dimensional change of test bars pressed to 6.8 g/cm³

	Change at 1120 °C (2050 °F), %	Change at 1290 °C (2350 °F), %
Fe-0.9C	+0.25	+0.04
Fe-4Ni-0.7C	0.0	−0.2
Fe-2Cu-0.9C	+0.4	+0.4

Dimensional change of bars sintered for 32 min at 1120 °C (2050 °F) and average for all atmosphere tests of bars sintered for 10 min at 1290 °C (2350 °F). Source: Ref 5

Table 11 Effect of sintering temperature on the apparent hardness of test bars pressed to a density of 6.8 g/cm³

Alloy	Apparent hardness, 1120 °C (2050 °F)	HRB, 1290 °C (2350 °F)
Fe-0.9C	57.9	63.2
Fe-4Ni-0.7C	73.1	75.9
Fe-2Cu-0.9C	75.6	81.3

Apparent hardness of bars sintered for 32 min at 1120 °C (2050 °F) and 10 min at 1290 °C (2350 °F) in comparable atmospheres. Source: Ref 5

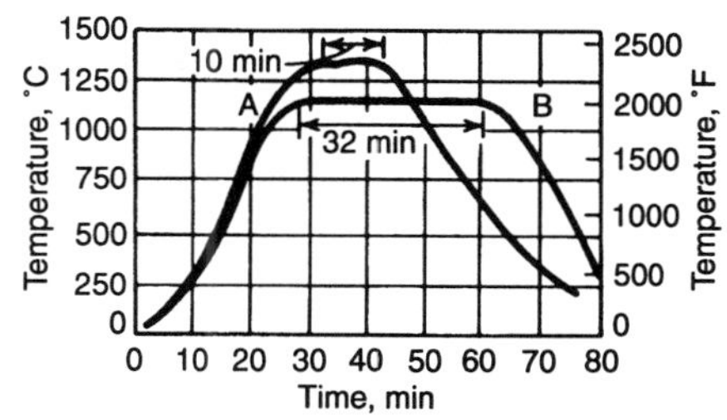

Fig. 11 Temperature profile of sintering cycles. Profile A shows the high-temperature cycle used in study. Profile B shows the conventional sintering cycle used as a standard. Source: Ref 5

duction basis, a study was conducted on three iron-base compositions sintered in a nitrogen-methanol atmosphere (Ref 5). Alloys tested had the following nominal compositions: Fe-0.9C, Fe-4Ni-0.7C, and Fe-2Cu-0.9C.

Transverse rupture bars were pressed to a density of 6.8 g/cm³. Methanol was blended with nitrogen to form the following compositions: 1% carbon monoxide and 2% hydrogen, remainder nitrogen; 3% carbon monoxide and 6% hydrogen, remainder nitrogen; and 5% carbon monoxide, 10% hydrogen, remainder nitrogen. Methane additions of 0, 0.25, and 0.5% were made to each of the atmospheres. Atmosphere composition in the hot zone of the furnace was monitored by infrared analysis, gas chromatography, and dew point analysis.

Lubricant was burned off in a mesh belt furnace at 760 °C (1400 °F) for 35 min. An atmosphere of 90% nitrogen and 10% hydrogen with a dew point of −12 °C (10 °F) was used for lubricant burn off. Sintering was performed in a pusher furnace with a ceramic muffle and wound molybdenum heating elements.

A temperature cycle was selected (Fig. 11) that held the parts above 1290 °C (2350 °F) for 7 min and above 1280 °C (2340 °F) for 10 min. The maximum temperature reached was 1301 °C (2374 °F). For comparison with conventional sintering, a set of test bars was sintered at 1123 °C (2053 °F) for 32 min (Fig. 11). A single nitrogen methanol atmosphere forming 5% carbon monoxide, 10% water vapor, 0.25% methane, and the remainder nitrogen was chosen for conventional temperature sintering.

The sintered bars were tested for transverse rupture strength, apparent hardness, and dimensional change. Surface and core carbon contents were determined metallographically and by combustion and thermal conductivity analysis.

Effect of Sintering Temperature. The 10 min sinter at 1290 °C (2350 °F) exhibited equivalent or slightly better properties than the longer 30 min sinter at 1120 °C (2050 °F) in a comparable atmosphere. Transverse rupture strengths were very similar, with only a 1 to 1.5% variation in the Fe-Ni-C and Fe-Cu-C alloys and 5% for the Fe-C steel (Table 9).

Metallographic analysis revealed a slightly more spheroidized pore structure and increased sintering in the 1290 °C (2350 °F) specimen. Dimensional change (Table 10) was more negative by approximately 0.2% for the Fe-C and Fe-Ni-C samples sintered at 1290 °C (2350 °F). Consequently, increased particle diffusion may have occurred at the higher temperatures even though a shorter sintering time was used. No difference in dimensional change was evident for the Fe-Cu-C samples.

Apparent hardness increased slightly for all three alloys at the higher sintering temperature (Table 11). Generally, the 10 min sinter at 1290 °C (2350 °F) was found to be an adequate substitute for a 30 min sinter at 1120 °C (2050 °F). This type of cycle can be used if an increased production rate that yields equivalent mechanical properties is desired.

Effect of Atmosphere Composition. Variations in methanol (forming carbon monoxide and hydrogen) and methane additions to the nitrogen carrier gas did not result in any significant change in transverse rupture strength for any of the alloys. This is attributed to the consistent carbon content (0.6 to 0.7%) at the core of the samples for all atmospheres tested.

Dimensional change of the alloys was unaffected by the atmosphere composition. The Fe-Cu-C and Fe-Ni-C alloys were consistent for all atmospheres tested. Some variations were evident in the Fe-C compacts, although no trend was evident.

Atmosphere composition was found to significantly affect surface carbon content and apparent hardness. Methanol enrichment without methane additions resulted in the most decarburization. Increases in methanol alone did not significantly increase the surface carbon. The surface of an Fe-0.9C sample was decarburized to a combined carbon content of 0.45 to 0.55% when sintered for 10 min at 1290 °C (2350 °F) in 15% dissociated methanol. Methane additions were required to increase the surface carbon.

Although some decarburization was still evident, acceptable results were obtained with an atmosphere containing 15% dissociated methanol (5% carbon monoxide and 10% hydrogen) with 0.5% methane additions. The dissociation of methanol and the resulting carbon dioxide and water vapor levels were found to be affected by localized temperature, residence time, and availability of local catalytic surfaces. When introduced properly, the methanol dissociation was demonstrated to be very efficient at 1290 °C (2350 °F).

Concentrations of carbon dioxide and water vapor were low, and the formation of methane was too low to be detected by gas chromatography. The atmosphere formed with nitrogen-methanol compared favorably to an equivalent atmosphere formed by blending nitrogen and endothermic gas.

These results indicate that efficient dissociation of methanol was achieved and that methane additions should be used to further reduce carbon dioxide and water vapor levels. As a result, nitrogen-methanol-methane atmospheres can be considered as viable alternatives to nitrogen-hydrogen-carbon monoxide and nitrogen-hydrogen-methane atmospheres for high-temperature sintering of ferrous alloys.

Sintering of Stainless Steel

E. Klar and P.K. Samal, OMG AMERICAS

Sintering is the most critical step in processing stainless steel parts. It is also more complex than the sintering of most other materials. Correctly sintered stainless steel parts will approach the corrosion resistance of wrought stainless steel of the same composition. Incorrectly sintered parts will have inferior corrosion resistance, frequently inferior by one to two orders of magnitude.

The majority of the studies dealing with the corrosion resistance of sintered stainless steels are based on immersion and/or salt spray testing in 5% aqueous NaCl, using the length of time until the appearance of rust as a quantitative measure of corrosion resistance (Ref 6). Some are based on weight loss measurements in H_2SO_4, HNO_3, and HCl of various concentrations. Metallography (Ref 7) and potentiostatic anodic polarization tests (Ref 8) have been used to elucidate the various mechanisms of corrosion, as well as for troubleshooting.

Microstructure-corrosion resistance correlations of sintered stainless steels have shown that for maximum corrosion resistance, it is important to minimize or avoid the occurrence of precipitates of chromium carbides, chromium nitrides, and of silicon and chromium oxides in the structure (Ref 6). Surface depletion of chromium during vacuum sintering can also be a cause for low-corrosion resistance (Ref 9). Furthermore, certain sintered densities can cause severe crevice corrosion in a neutral salt environment (Ref 10).

Requirements for optimum sintering of stainless steels are described in this section for the most commonly used sintering atmospheres. The typical problems present during processing, particularly during sintering, are discussed together with measures necessary to avoid the previous problems. The precautions necessary to avoid the previous problems differ for the various sintering atmospheres. Moreover, to guarantee maximum corrosion resistance, quality control must begin powder production.

Powder Quality

Contamination-Free Stainless Steel Powder. Widely used commercial grades of stainless steel powders are summarized in Table 12. Those grades containing tin have superior corrosion resistance over their tin-free equivalents on account of an enrichment of tin on the surfaces of the sintered parts.

When powder producers manufacture stainless steel powders in a facility or with equipment where cross contamination with iron or low-alloy steel powder is possible, the sintered parts are likely to have low-corrosion resistance because the electrochemically less noble contaminants will form so-called galvanic couples with the stainless steel particles in which the less-noble particles oxidize (corrode) (Ref 11). Austenitic stainless steel powders contaminated with ferritic stainless steel powders also form galvanic couples, that is, the corrosion resistance of a sintered austenitic stainless steel is significantly reduced if it is contaminated with ferritic stainless steel particles. The only exception to the above statement is when the contaminant particles dissolve during sintering, which is possible for fine particles (< 38 μm) and high sintering temperatures (>1260 °C, or 2300 °F).

Contamination can also occur at the part producer's facility if stainless steel and carbon steel processing are done under one roof.

It is important to assure that a stainless steel powder is contamination free prior to pressing and sintering because after sintering, the identification of contamination as the cause for inadequate corrosion resistance can be rather difficult and time consuming. The authors have found that as little as 100 ppm of iron contamination in 316L can reduce corrosion resistance in aqueous NaCl by 50%. It is for this reason that stainless steel powders as well as green parts should be subjected to the so-called copper sulfate or Ferroxyl test prior to compacting and, additionally, prior to sintering if contamination in the part producer's facility is possible.

Copper Sulfate Ferroxyl Tests. For the detection of contamination of a stainless steel powder with a less-noble metal powder, two tests are available.

The copper sulfate test (ASTM 380) is based on the plating out of elemental copper from a concentrated copper sulfate solution onto particles that are less noble than copper, in accordance with:

$$CU^{++} + Fe \rightarrow Cu + Fe^{2+}$$

The plated-out copper can be seen within minutes with the naked eye or preferably with a low-power microscope. A serious shortcoming of the copper sulfate test is that lubricants prevent the copper sulfate solution from properly wetting the powder or green compact.

The Ferroxyl test is based on the development usually within minutes, of a blue precipitate, Turnbull's blue, when elemental iron or ferrous ions react with hexacyanoferrate (III) in accordance with:

Table 12 Compositions of stainless steel powders

	Composition(a), %									
	Cr	Ni	Si	Mo	Cu	Sn	Mn	C	S	P
Austenitic grades										
303	17–18	12–13	0.6–0.8	...	...	...	0.3(b)	0.03(b)	0.1–0.3	0.03(b)
303LSC	17–18	12–13	0.6–0.8	...	2(c)	1(c)	0.3(b)	0.03(b)	0.1–0.3	0.03(b)
304L	18–19	10–12	0.7–0.9	...	...	...	0.3(b)	0.03(b)	0.03(b)	0.03(b)
Ultra 304L	19(c)	11(c)	0.8(c)	...	0.8(c)	2(c)	0.2(c)	0.02(c)	0.01(c)	0.01(c)
316L	16.5–17.5	13–14	0.7–0.9	2–2.5	...	...	0.3(b)	0.03(b)	0.03(b)	0.03(b)
316LSC	16.5–17.5	13–14	0.7–0.9	2–2.5	2(c)	1(c)	0.3(b)	0.03(b)	0.03(b)	0.03(b)
317L	19(c)	14(c)	0.8(c)	3(c)	...	...	0.3(b)	0.03(b)	0.03(b)	0.03(b)
SS100	20(c)	17(c)	0.8(c)	5(c)	...	...	0.2(c)	0.02(c)	0.03(b)	0.02(c)
Martensitic grade(d)										
410L	12–13	...	0.7–0.9	...	...	...	0.3(b)	0.03(b)	0.03(b)	0.03(b)
Ferritic grades										
430L	16–17	...	0.7–0.9	...	...	...	0.3(b)	0.03(b)	0.03(b)	0.03(b)
434L	16–18	...	0.7–0.9	0.5–1.5	...	...	0.3(b)	0.03(b)	0.03(b)	0.03(b)

(a) Oxygen content, 1000–2500 ppm; nitrogen content, 200–500 for all listed grades. (b) Maximum. (c) Typical. (d) With C <0.03%, this grade is ferritic. Graphite addition to the low carbon grade powder renders it martensitic upon sintering.

$$K^+(aq) + Fe^{2+}(aq) + [Fe(CN)_6]^{3-}(aq) \rightarrow KFe^{III}Fe^{II}(CN)_6(s) \text{ (Turnbull's blue)}$$

Ref 12 summarizes optimum solution strengths for performing this test. This test is also mentioned in ASTM Standard for Cleaning and Descaling Stainless Steel Parts (Ref 13).

Delubrication

The delubing of stainless steel compacts influences oxygen and carbon balance, which can affect corrosion resistance of the sintered part by precipitation of chromium carbides. Carbide precipitation (sensitization) depletes chromium from grain boundaries.

Sensitization. For a number of austenitic stainless steels, the maximum carbon content allowable to avoid sensitization is 0.03%. Higher carbon contents are tolerable only if the cooling rate after sintering is high enough to prevent sensitization and if the parts are not subjected to elevated temperatures again such as in welding. Critical cooling rates can be estimated from time-temperature-sensitization curves for wrought austenitic stainless steels. Carbide precipitates can be readily shown with optical microscopy. A properly sintered low-carbon austenitic structure (Fig. 12a) is free of carbide precipitates and has clean and thin grain boundaries with evidence of twinning. Higher carbon structures show necklace type (Fig. 12b) and continuous precipitates (Fig. 12c) of chromium-rich carbides in the grain boundaries.

For ferritic stainless steels, the maximum amounts of C + N allowable to avoid sensitization are much lower, namely ~100 to 150 ppm (Ref 14). Because such low contents of interstitials are difficult to obtain in practice, niobium and/or titanium are often used as stabilizers in wrought ferritics. For P/M versions of ferritic stainless steels, niobium is the stabilizer of choice as niobium becomes less oxidized during water atomization than titanium. More information on sensitization conditions and corrosion is discussed in the article "Corrosion-Resistant Powder Metallurgy Alloys" in this Volume.

Atmospheres. In the early years of stainless steel parts production, much of the sintering was done in dissociated ammonia at 2050 to 2100 °F (1120 to 1150 °C). Under these conditions, lubricant removal in the preheat zone of the furnace occurs under partial decomposition such that residual carbon contents typically exceed 0.03%. As this residual carbon is not sufficiently removed during low temperature (<1205 °C, or 2200 °F) sintering, the sintered parts have low-corrosion resistance due to the formation of chromium-rich carbides upon cooling. Delubing was, therefore, often done separately in air, which resulted in lower-carbon contents because of the more complete combustion of the lubricant. While air delubing reduced the carbon content to more acceptable levels, particularly for austenitic stainless steels, it also was at the expense of increased oxidation. Oxidation begins before complete lubricant removal (Fig. 13). It

(a) (b) (c)

Fig. 12 Microstructures of type 316L stainless steel sintered in hydrogen at 115 °C (2100 °F) (Glyceregia). (a) C is 0.015%, clean and thin grain boundaries. (b) C is 0.07%, necklace type chromium-rich carbide precipitates in grain boundaries. (c) C is 0.11%, continuous chromium-rich carbide precipitates in grain boundaries.

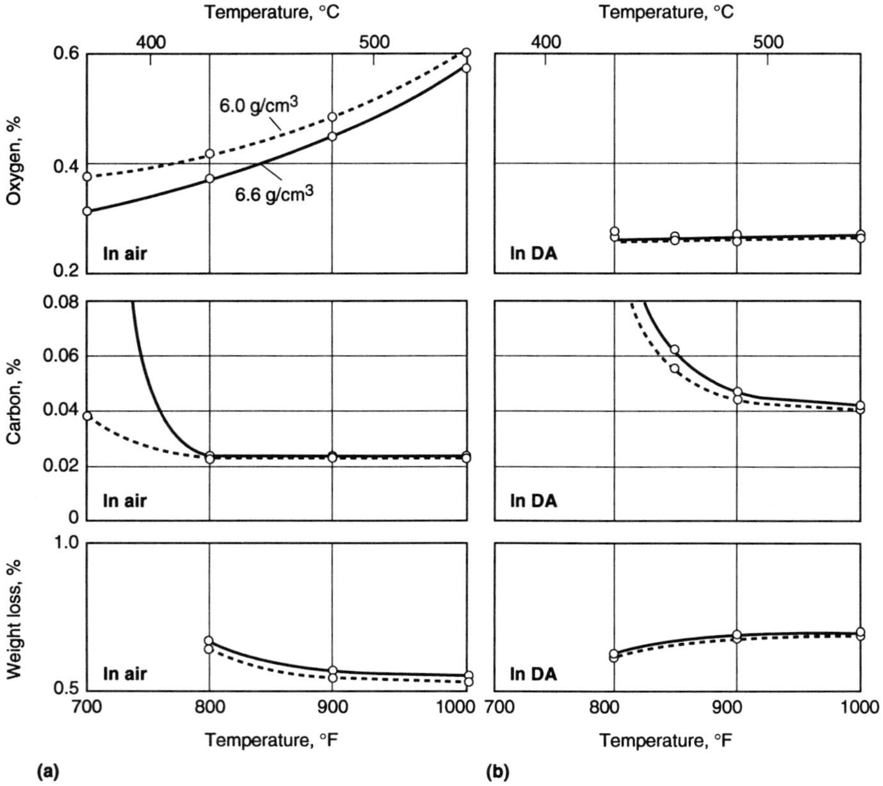

Fig. 13 Effect of delubrication temperature on oxygen and carbon contents, and weight loss of 316LSC transverse rupture strength compacts of two densities—6.0 g/cm^3, dashed lines, and 6.6 g/cm^3, solid lines—lubricated with Acrawax and delubed for 30 min in (a) air and (b) dissociated ammonia

appears impossible to obtain maximum carbon removal without additional oxidation. The oxides formed during delubing are not always removed during sintering and can impair the corrosion resistance of the sintered parts.

With higher sintering temperature (>1205 °C, or 2200 °F), the reaction between residual oxygen and carbon is more complete, and delubing is, therefore, preferably completed in a reducing atmosphere. As Fig. 14(a) shows for several stainless steels, lubricated with 1 % Acrawax and pressed to green densities of 6.5 to 6.7 g/cm^3, delubing in dissociated ammonia prevents any significant oxidation up 510 to 538 °C (900 to 1000 °F). Although carbon removal under these conditions is not yet at its maximum, and still >0.03% (Fig. 14b), a higher sintering temperature effectively lowers the carbon content to below 0.03%. It is partly because of these relationships that stainless steel parts sintered at high temperatures often exhibit better corrosion resistance than those sintered at lower temperatures.

For high-temperature vacuum sintering, the reaction between residual carbon and oxygen is even greater so that carbon can even be added, in the form of graphite, to some stainless steels to achieve lower-oxygen contents at acceptable carbon contents (Ref 9, 15,16). Lower-oxygen contents are beneficial to both corrosion resistance (Ref 7, 9) and dynamic mechanical properties.

Sintering

Most commercial sintering of stainless steel parts is completed in belt, pusher, walking beam, and vacuum furnaces. A wide range of processing parameters is common. Typical sintering atmospheres include hydrogen, dissociated ammonia, H_2-N_2 mixtures, and all of "low" dew point, as well as vacuum. Sintering temperatures range from 1120 to 1344 °C (2050 to 2450 °F), and sintering times range from 20 to 60 minutes. Insufficient sintering, either too short a time or at too low a temperature, will result in sintered parts showing insufficient bonding, original particle boundaries, and sharp, angular pores as shown in Fig. 15 for a 316L stainless steel part sintered for 30 min at 1093 °C (2000 °F) in H_2. Such sintering produces parts with a high concentration of interstitials (carbon and oxygen), low-corrosion resistance, and inferior mechanical properties.

The significance of the various sintering parameters is discussed in the following paragraphs.

Sintering in Hydrogen. In spite of higher cost, the use of hydrogen for the commercial sintering of stainless steel parts has been increasing at the expense of dissociated ammonia. The main reasons for this change are that with present sintering furnaces, it is very difficult to obtain good corrosion resistance in dissociated am-

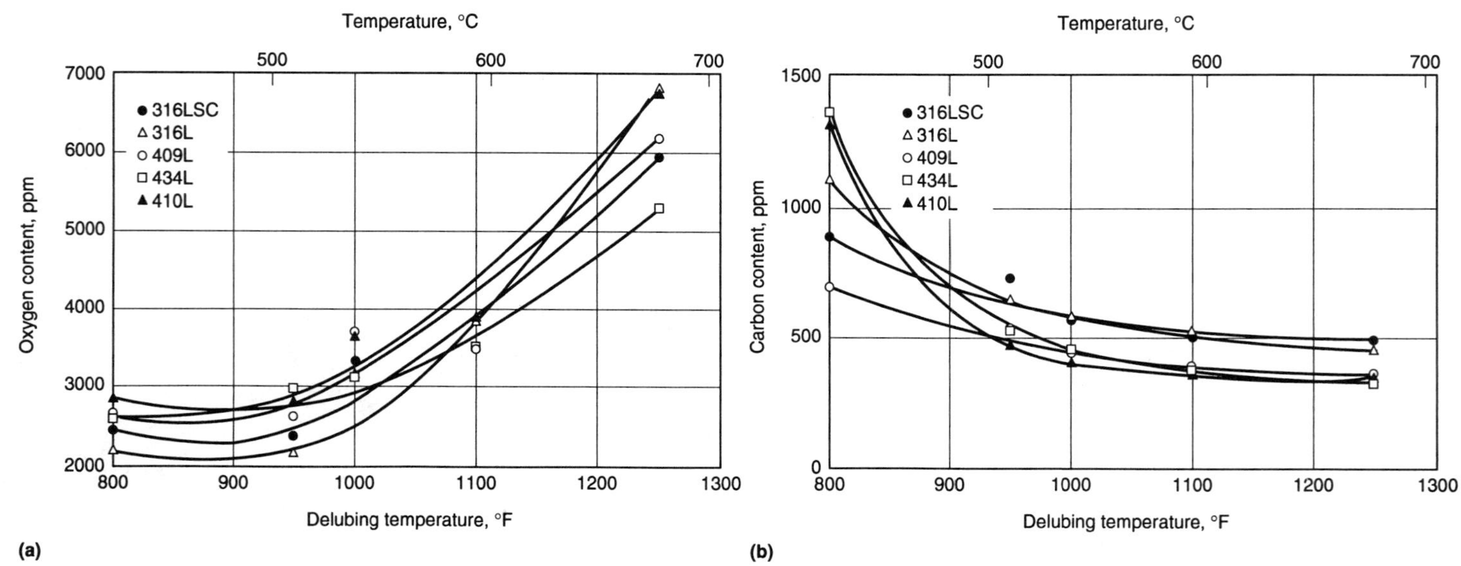

Fig. 14 Effect of delubing temperature on (a) oxygen content and (b) carbon content of stainless steel transverse rupture strength bars (6.5 to 6.7 g/cm^3) lubricated with 1 % Acrawax and delubed for 30 min in dissociated NH_3

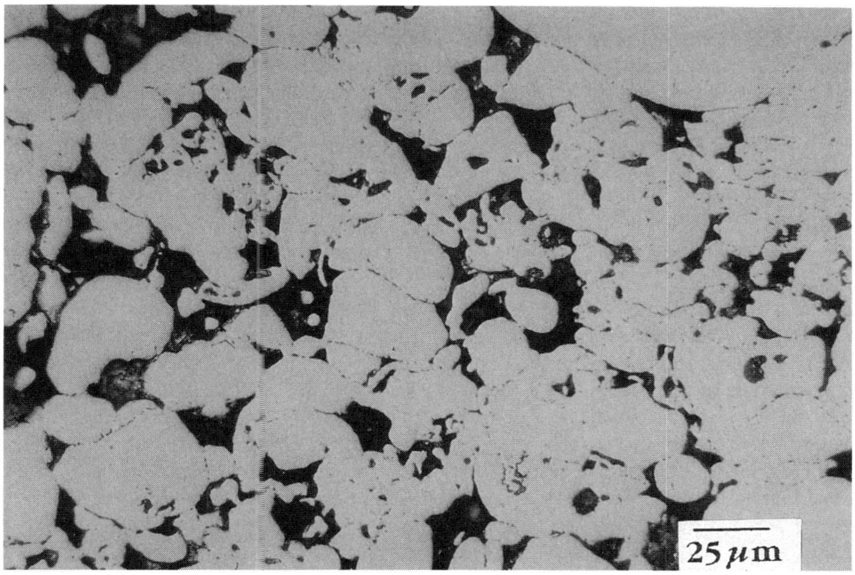

Fig. 15 Insufficiently sintered 316L showing original particle boundaries (1) and sharp, angular pores (2). As-polished cross section

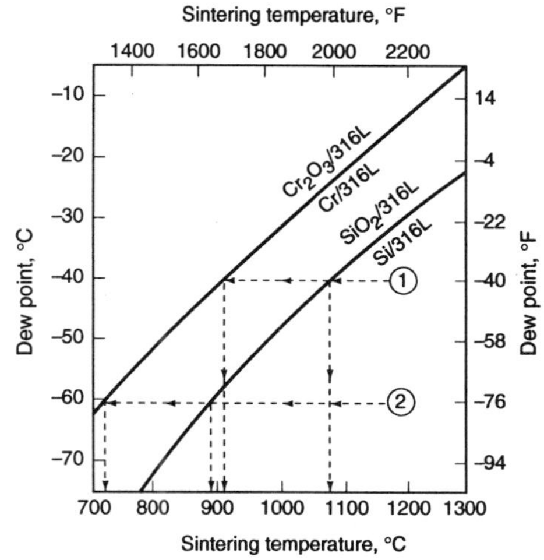

Fig. 16 Redox curves for chromium and silicon in 316L in H_2 at atmospheric pressure. Source: Ref 17

monia, and the increasing use of ferritic stainless steels in magnetic applications also requires the absence of nitrogen in the sintered part. Nevertheless, sintering in hydrogen also requires careful control to obtain maximum corrosion resistance.

Figure 16 shows the redox curves for chromium and silicon in 316L in H_2 of various dew points at atmospheric pressure (Ref 17). Chromium and silicon are the two constituents in 316L with the highest oxygen affinities. Sintering with dew-point sintering temperature combinations to the left of the two curves of Fig. 16 will result in selective oxidation of chromium and silicon, respectively, and produce parts of inferior corrosion resistance. Thus, for the lowest sintering temperatures employed in commercial sintering of stainless steels (1120 °C, or 2050 °F), the hydrogen should have a dew point not higher than approximately –40 °C (–40 °F) to ensure reduction of chromium oxides in 316L stainless steel (Fig. 16). A similar relation is shown in Fig. 2(c) for pure chromium.

What happens after sintering, that is, during cooling, is even more important. Figure 16 shows two scenarios. For both scenarios, the sintering temperature is 1200 °C (2192 °F). In the first scenario, the dew point of the sintering atmosphere is –40 °C (–40 °F); in the second scenario, it is –60 °C (–76 °F). In the first scenario, the stainless steel part crosses the SiO_2/Si redox curve (Fig. 16) upon cooling at ~1070 °C (1960 °F). At this temperature, the rate of oxidation of silicon is quite rapid and, therefore, rapid cooling is necessary to prevent or minimize, the formation of silicon oxides on the surface of the stainless steel part. Figure 17 shows a scanning electron microscopy (SEM) of such oxide precipitates for 316L. These precipitates do not cause the part to discolor, and they are visible only under a microscope. When tested in aqueous $FeCl_3$, in accordance with ASTM G 46 (20 °C), such parts exhibit inferior corrosion resistance due to pitting. Higher-alloyed stainless steel powder such as SS-100 (20Cr-17Ni-0.8Si-5Mo) appear to be more immune to this type of corrosion (Ref 10).

With the second scenario in Fig. 16, the lower dew point of –60 °C (–76 °F) causes the parts to cross the SiO_2/Si redox curve in the cooling zone of the furnace at the much lower temperature of ~890 °C (1632 °F). At that temperature, the rate of silicon oxidation either is very slow, or any oxides formed at that temperature have little effect on the corrosion resistance of the part. Figure 18 shows tentative critical cooling rate-dew point combinations as a function of dew point for three hydrogen sintered austenitic stainless steels. Upper critical cooling temperatures, that is, the lowest high temperatures where rapid cooling is to commence, are shown in Fig. 19.

Processing to the right of the cooling rate-dew point curves produces maximum corrosion resistance. Processing to the left results in rapid deterioration of corrosion resistance as shown schematically in Fig. 19 for 316L. It is clear from these relationships that for maximum corrosion resistance, sintering in hydrogen requires very low dew points and/or rapid cooling after sintering. Mechanical properties of hydrogen sintered stainless steels are given in the article "Powder Metallurgy Stainless Steels" in this Volume.

Sintering in Vacuum. In the early years of commercial sintering of stainless steel parts, vacuum furnaces were said to be good alternatives to other types of sintering because of their low consumption of gas. After years of experience, parts producers learned that in addition to high initial capital cost, vacuum furnaces also were costly to maintain. Nevertheless, it is clear from the previous section on sintering in hydro-

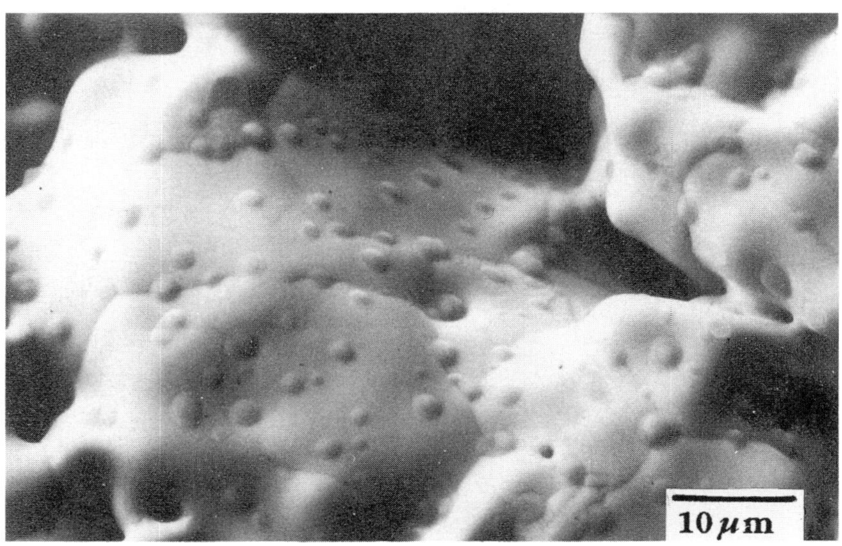

Fig. 17 Spheroidal silicon oxide particles formed on 316L part on cooling

gen that with the typical furnaces (belt, pusher, and walking beam) used presently in the industry, the number one property of stainless steel, superior corrosion resistance, is for most presently used P/M stainless steels not attainable to a sufficient degree. Therefore, future P/M opportunities that require excellent corrosion resistance cannot be realized until furnace manufacturers construct furnaces that are capable of lower dew points and parts producers equip their furnaces with (already available) rapid cooling devices. This is where vacuum furnaces are used. With a state of the art vacuum furnace, it is much easier to maintain a low dew point and to obtain rapid cooling than it is with a typical atmosphere furnace. Nevertheless, certain precautions are necessary.

For maximum corrosion resistance of vacuum sintered stainless steel, surface depletion of chromium due to high-vapor pressure and the presence of original surface oxides must be minimized. Sintering under a partial pressure of nitrogen or argon of ~1500 µm of mercury effectively reduces chromium losses. Reference 9 shows that after high temperature sintering (>1205 °C, or 2200 °F), where chromium losses are more severe, holding the parts prior to cooling for a short time at a lower temperature, or by increasing the partial pressure of argon to 1 at the lower temperature, significantly improves corrosion resistance. Both measures allow the parts to replenish (from the interior) surface chromium that was lost at the high-sintering temperature.

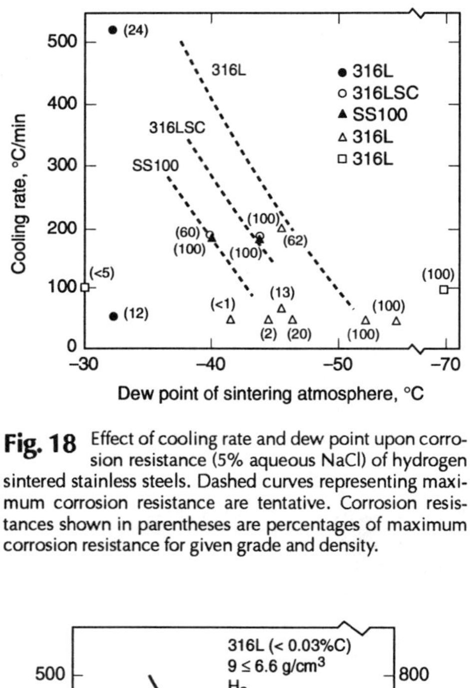

Fig. 18 Effect of cooling rate and dew point upon corrosion resistance (5% aqueous NaCl) of hydrogen sintered stainless steels. Dashed curves representing maximum corrosion resistance are tentative. Corrosion resistances shown in parentheses are percentages of maximum corrosion resistance for given grade and density.

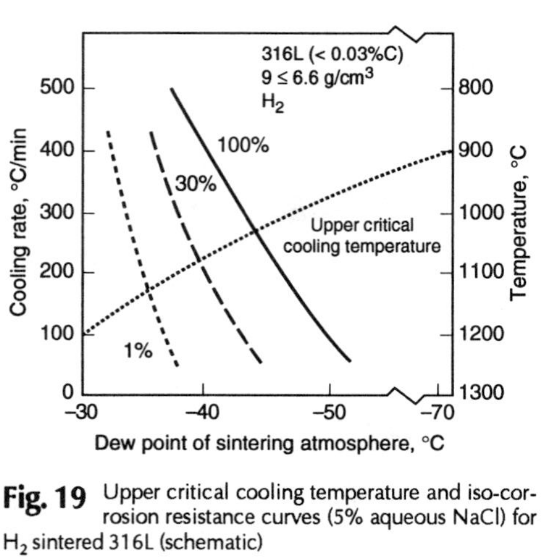

Fig. 19 Upper critical cooling temperature and iso-corrosion resistance curves (5% aqueous NaCl) for H_2 sintered 316L (schematic)

In spite of the high-oxygen contents of water atomized stainless steel powders, vacuum sintered stainless steel parts usually are bright. This is because some of the surface oxides, during sintering, diffuse into the interior.

In the absence of an external-reducing gas atmosphere, vacuum-sintered stainless steel parts have relatively low carbon and oxygen contents due to the reaction between carbon and oxygen (or oxides) particularly at high-sintering temperatures, to form carbon monoxide (Fig. 20) (Ref 18 and 19). There is evidence, however, that the typical amounts of carbon present in a stainless steel part after delubrication are insufficient for removing most of the original oxide particles present on the outer surfaces of a part. These unreduced original oxide particles give rise to pitting corrosion. Admixing small amounts of graphite to the stainless steel powder, overall oxide reduction, particularly at the higher-sintering temperatures, is greatly enhanced and is also sufficient to reduce surface oxides. Alternatively, a small partial pressure of H_2 should accomplish the same.

When making graphite additions to a stainless steel powder, it should be kept in mind that the carbon content of the sintered part will increase. Thus, the optimum graphite addition is the maximum addition that produces no chromium carbide precipitates in the cooling zone of the sintering furnace. It depends, among other factors, on the composition of the stainless steel, the oxygen content of powder, the sintering temperature, and the cooling rate employed.

Vacuum-sintered stainless steels should be rapidly cooled in a non-oxidizing gas to prevent the formation of deleterious surface oxides. Cooling in nitrogen will result in the formation and precipitation of chromium nitrides on the surfaces of the parts. The attendant chromium depletion will cause the parts to have very low-corrosion resistance. Mechanical properties of vacuum sintered stainless steels are shown in the article "Powder Metallurgy Stainless Steels" in this Volume.

Sintering in H_2-N_2 Gas Mixtures. Sintering at 1120 °C (2050 °F) in dissociated ammonia was the most widely used method for sintering stainless steels in the 1960s and 1970s. Dissociated NH_3 was not only less expensive than H_2, but it also increased the strength of the sintered parts, although at some reduction in ductility, to levels comparable to wrought stainless steels of the same composition, at densities of ~85 to 90% of theoretical. The strengthening is the result of nitrogen absorption during sintering. The amount of nitrogen absorbed follows known phase equilibria in accordance with Sievert's law, that is, nitrogen absorption is proportional to the square root of the partial pressure of nitrogen in the sintering atmosphere. (See the article "Corrosion-Resistant Powder Metallurgy Alloys.") The relationships for 304L are shown in Fig. 21 (Ref 20), showing both the amount of nitrogen absorbed as a function of sintering temperature and sintering atmosphere (dissociated ammonia and nitrogen) and the strength increase due to nitrogen absorption.

The problems with nitrogen absorption and chromium nitride (Cr_2N) precipitation during cooling (after sintering), and sensitization and loss of corrosion resistance, when sintering is done in dissociated ammonia, are described in Ref 21. They were not appreciated for many years, in part because corrosion resistance demands were modest and/or corrosion resistance was not assessed. Later, with increasing demands for improved corrosion resistance, and as more quantitative information on the effect of sintering in dissociated ammonia became available, recommendations were made to limit nitro-

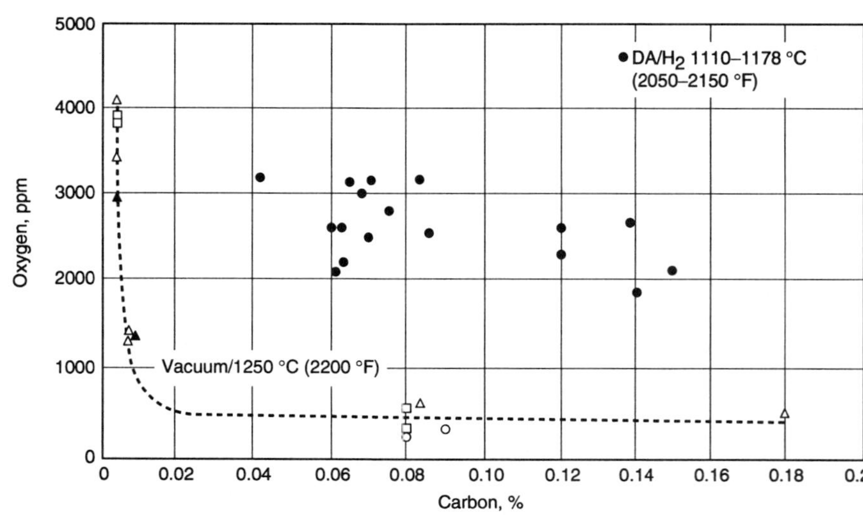

Fig. 20 Oxygen versus carbon contents of vacuum and atmosphere sintered P/M austenitic stainless steels of varying compositions

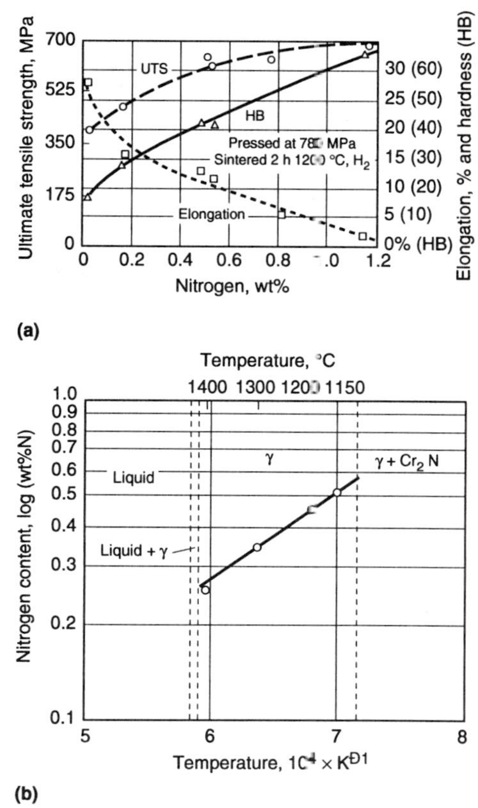

Fig. 21 (a) Effect of nitrogen content on ultimate tensile strength and elongation of 304L stainless steel. (b) Effect of sintering temperature on amount of absorbed nitrogen for 304L. Source: Ref 20

gen absorption to some 3000 ppm in austenitic stainless steels. While this limitation seemed to satisfy some corrosion resistance requirements in an acidic environment, it was unsatisfactory for parts tested in aqueous NaCl. Even with small amounts of Cr_2N precipitates, rust spots would form in a short time. Figure 22 shows examples of Cr_2N precipitates in sintered austenitic stainless steels.

In an early study, Sands et al. (Ref 22) pointed out that 316L sintered in dissociated NH_3 required a cooling rate of 200 °C/min for preventing nitrogen absorption and precipitation of Cr_2N. More recently, Frisk et al. (Ref 23) determined in a laboratory study that sintering of 316L in dissociated NH_3 at 1250 °C (2280 °F) required cooling rates of >450 °C/min (Fig. 23). The higher critical cooling rate of Frisk et al. can probably be ascribed to their much lower dew point (–100 °C versus –40 to –60 °C for Sands), which allows for more rapid nitrogen absorption during cooling as illustrated in Fig. 24 (Ref 24) for the bright annealing of stainless steel strip. The deleterious reactions of increasing nitrogen absorption with decreasing dew point, and of increasing oxidation with increasing dew point, leave a relatively narrow dew point window for sintering in dissociated ammonia.

These cooling rate requirements are even higher than those for sintering in hydrogen to prevent oxidation during cooling, and the cooling-rate dew point relationship appears to be reversed. It is, therefore, not surprising that stainless steel parts producers are increasingly shifting towards hydrogen sintering at the expense of sintering in dissociated NH_3.

Sintering in an atmosphere of $10\%N_2$-$90\%H_2$ may be a more practicable compromise that greatly reduces the high-cooling rate requirements of dissociated NH_3 to more manageable levels, while still benefiting substantially from the solid solution strengthening obtainable with the lower nitrogen concentration. Good corrosion resistance for such conditions were reported by Larsen (Ref 25) and Mathiesen (Ref 26).

The positive effect of nitrogen on corrosion resistance, as documented for wrought stainless steels, is expected to apply equally to sintered stainless steels. However, this beneficial effect has not yet been well documented for. sintered stainless steels, probably because of the overshadowing negative effect of excessive nitrogen absorption on the surface of parts from the sintering atmosphere during cooling. Mechanical and other properties for 316L and 434L sintered in dissociated NH_3 are given in the article "Powder Metallurgy Stainless Steels" in this Volume.

Sintered Density

For many years, inferior corrosion resistance of sintered stainless steel parts has been and continues to be mistakenly attributed to the presence of pores in accordance with the mechanism of crevice corrosion. That this hypothesis is untenable follows from widely available evidence (Ref 7) that sintered stainless steel parts of identical composition and similar pore volumes, pore sizes, and pore shapes, but sintered under varying conditions, may have corrosion resistances (in 5% aqueous NaCl) that can vary by two orders of magnitude. Furthermore, a comparison of wrought and sintered (85% of theoretical density) type 316L for susceptibility to crevice corrosion in 10% $FeCl_3$, in accordance with ASTM G 48, showed that the wrought part was actually more severely attacked than the porous part (Ref 7, 27). More recent studies have shown that most cases of inferior corrosion resistance of sintered stainless steels are due to incorrect sintering, as previously described.

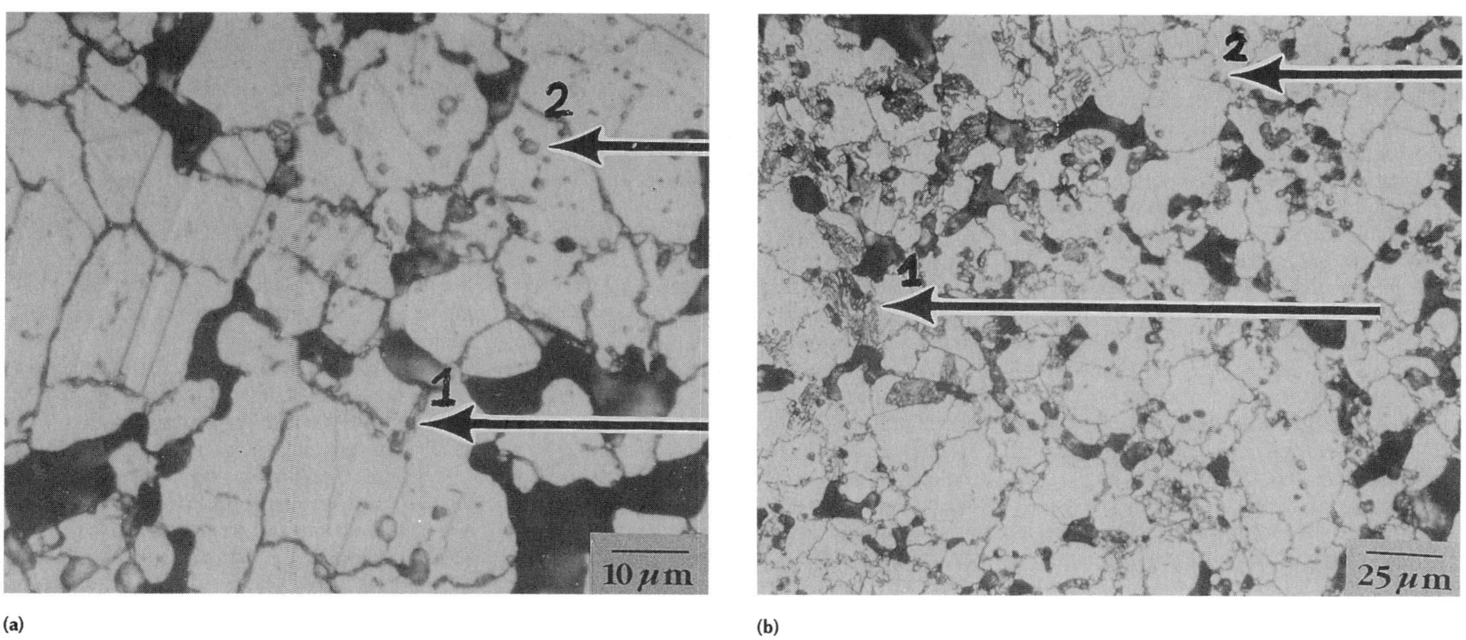

Fig. 22 Chromium nitride precipitates in 316L (a) sintered at 2100 °F (1150 °C) in dissociated NH_3; 4500 ppm N_2; Cr_2N precipitates along grain boundaries (1), and within grains (2). (b) sintered at 1120 °C (2050 °F) in dissociated NH_3 and slowly cooled; 6500 ppm N_2; Cr_2N in lamellar form near surface (1) and as grain boundary precipitates in the interior (2).

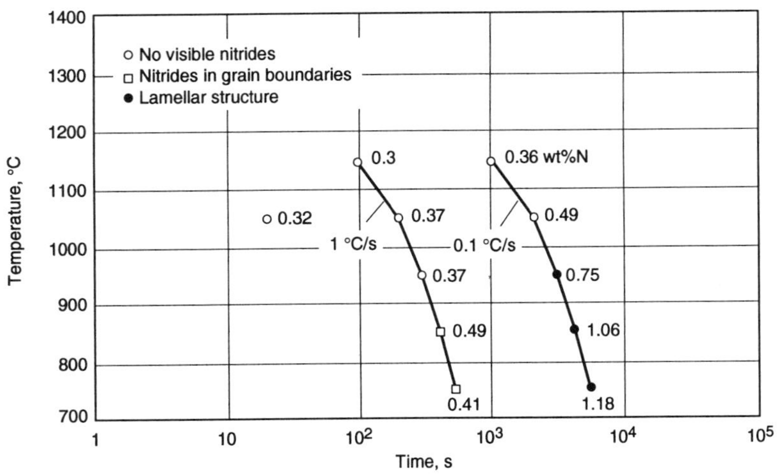

Fig. 23 Effect of cooling rate on presence of chromium nitrides in microstructure of 316L parts sintered at 1250 °C in dissociated NH_3. Source: Ref 23

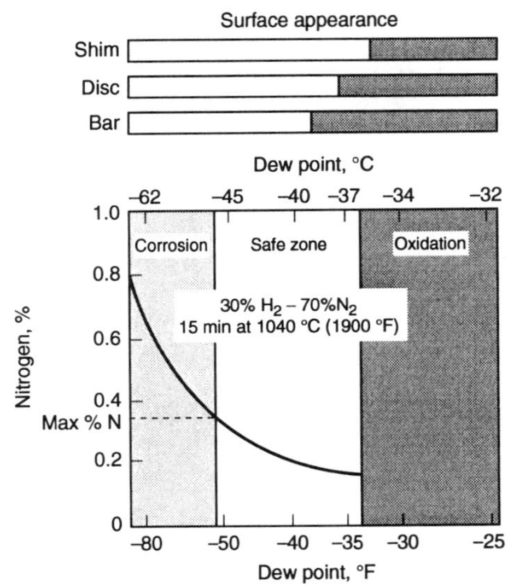

Fig. 24 Effect of dew point on nitrogen absorption and oxidation of 316L shim, disk, and bar stock annealed for 15 min at 1038 °C (1900 °F) in 30% H_2-70%N_2. It took 2.3; 2.8; and 4.7 min respectively to cool the three materials from 1038 °C (1900 °F) to 538 °C (1000 °F). Source: Ref 24

Also, for many years, there existed controversy as to the effect of sintered density upon corrosion resistance. Corrosion testing in an acidic environment, such as dilute H_2SO_4, HCl, and HNO_3, always showed that corrosion resistance improved with increasing density. Testing in a neutral salt solution, however, showed the corrosion resistance to decrease with increasing density (Ref 17, 28–31), when corrosion testing was done in long-term immersion or salt spray tests, whereas higher density was found to be beneficial in short-term potentiodynamic polarization tests (Ref 32). Maahn and Mathiesen (Ref 8) attributed the failure to observe this important relationship between corrosion resistance and density in short-term, potentiostatic anodic polarization tests to the lack of time for the time-consuming build up of localized attack within the pores, in analogy to the mechanism of crevice corrosion. By using slow stepwise polarization, the expected relationship, that is, a decrease of the stepwise initiation potential with increasing density (equivalent to deteriorating corrosion resistance), was observed (Ref 33).

Recent studies, with austenitic stainless steels (Ref 10, 17, 31) showed that the corrosion resistance in 5% aqueous NaCl (by immersion) can be reduced by up to two orders of magnitude due to the presence of porosity. The negative effect appears at sintered densities of ~6.7 g/cm^3, reaches a minimum corrosion resistance between ~6.9 and 7.2 g/cm^3, depending on pore morphology, and thereafter disappears at densities of ~7.4 g/cm^3 (Fig. 25). To the left of the minimum, corrosion resistance decreases with decreasing pore size due to increasing oxygen depletion and failure to maintain the passive layer. To the right of the minimum, corrosion resistance improves as pores become closed off and inaccessible to the surface. This type of corrosion can be reduced by impregnating the pores with a resin, by metallurgical modification of the pore surfaces, or by the use of higher-alloyed stainless steels, particularly those containing higher concentrations of molybdenum.

Another approach to avoiding the problem of long-term corrosion in a neutral salt solution due to the presence of certain size pores is to make use of the various forms of liquid-phase sintering (transient, persistent, and supersolidus) and to achieve sintered densities >7.4 g/cm^3. For austenitic stainless steels, silicon additions of several percent (Ref 34, 35) or smaller amounts of boron (Ref 17), have been used. For ferritic stainless steels phosphorus additions (Ref 35) have been used. The large shrinkage occurring during liquid-phase sintering is often accompanied by increasing loss of dimensional stability. Sizing is usually employed to establish dimensional accuracy. Also, depending on the composition of the liquid-phase sintering additive, secondary phases may be present after sintering, which can have a negative effect on corrosion resistance (Ref 17).

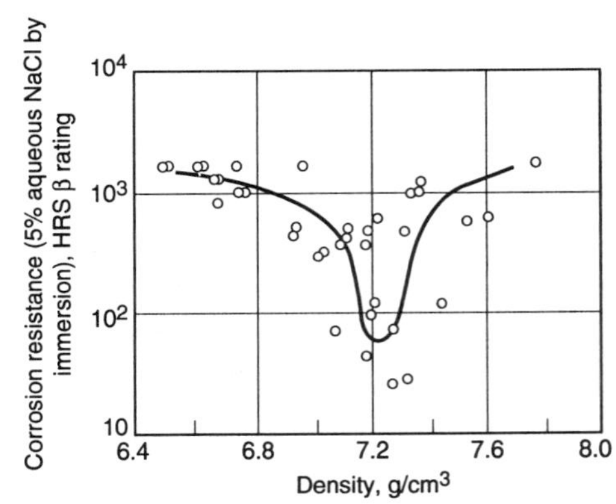

Fig. 25 Effect of density on corrosion resistance (B rating) of pressed, sintered, repressed, and annealed 317L parts

Sintering of High-Speed Steels and Tool Steels

Powder metallurgy processing offers several advantages to costly and highly alloyed tool steel materials. These advantages include uniform and finer microstructure, improved grindability, improved cutting performance, and capabilities of high-speed steels and tool steel alloys that cannot be made by conventional ingot metallurgy.

The use of conventional press-and-sinter technology offers the additional advantage of net shape or near-net shape capability. Cutting tools, bearings, and wear parts are being produced commercially by fully dense sintering. Wear parts also are produced by conventional P/M techniques to densities of 80 to 90%.

Current commercial fully dense sintering uses high green strength and compressible water-atomized tool steel powders that are compacted in rigid dies using uniaxial pressing or in flexible rubber molds using cold isostatic pressing to make green tools and parts. These parts are then sintered in a microprocessor-controlled vacuum furnace near the solidus temperature of the alloy to virtually full density (at least 98% and frequently 99+% of theoretical).

The flexibility of this process allows production of pressed and sintered tool and die parts in various forms. Generally, parts are pressed to 70 to 85% of theoretical density before sintering to full density. Pressing to lower green densities tends to require longer times at temperature to obtain full density and results in coarser microstructures. Pressing at higher pressures, which is required for increased green densities, results in increased tool wear and breakage. Parts with high green density also may require extra care in sintering. The surface can sinter rapidly to high density and entrap gases from the center of the part.

Sintering Mechanisms

Important contributions to sintering arise from diffusion and viscous flow. Diffusion rates increase with increasing temperature because of the increased number of vacancies that promote diffusion of substantial alloying elements. Sintering temperature must be held very close to the solidus temperature to attain full density in a reasonable time. A sintering temperature 5.5 to 11 °C (10 to 20 °F) above the solidus reportedly forms a small amount of liquid, which allows high-diffusion rates for enhanced sintering. These results are based on relatively rapid densification at high temperatures. However, metallographically, there is no resemblance between a successfully sintered high-speed steel and a typical liquid-phase sintered material, such as tungsten carbide or tungsten heavy metal. Additional research is required to fully understand high-temperature sintering.

Factors Affecting Sintering. Sintering is a series of complex processes, of which densification is only one phase. As green parts are heated to the sintering temperature, gases can be adsorbed (nitrogen, hydrogen, or oxygen) or evolved (nitrogen, hydrogen, or carbon monoxide). Admixed carbon dissolves and homogenizes, while carbides dissolve and grow. Grain growth also occurs as pores shrink and virtually disappear.

Sintering Time and Temperature. Increasing the sintering temperature decreases the amount of time required to achieve full density. Higher temperatures also reduce the time between reaching full density and oversintering. These relationships are shown schematically in Fig. 26. A sintering curve for M2 high-speed steel is shown in Fig. 27. Table 13 provides typical properties of M2 tool steel for various sintering temperatures.

Effect of Carbon Content. Carbon content has a significant effect on sintering temperature. Increasing carbon content reduces the sintering temperature by lowering the solidus temperature.

Carbon can be controlled closely by blending graphite or lampblack into the powder before pressing. Carbon dissolves rapidly, and its high diffusivity enables rapid homogenization above 980 °C (1800 °F). However, blending of more than 0.15 to 0.2% C can cause several detrimental effects, such as nonuniform distribution of carbon, variable response to sintering, variable part-to-part density, and part distortion.

More than 0.15% C can be admixed with 0.05 to 0.30% blending oil, such as mineral oil. However, blending oils may create powder flow difficulties and can lead to excessive outgassing during vacuum sintering.

Figure 28 shows the effect of carbon content on microstructure for M2 high-speed tool steel. The amount of mixed carbon critically affects the microstructure. For a given sintering cycle, insufficient carbon produces incomplete sintering, and excessive carbon produces oversintering. Table 13 gives the effect of carbon additions for five heats of M2 tool steel and three sintering temperatures.

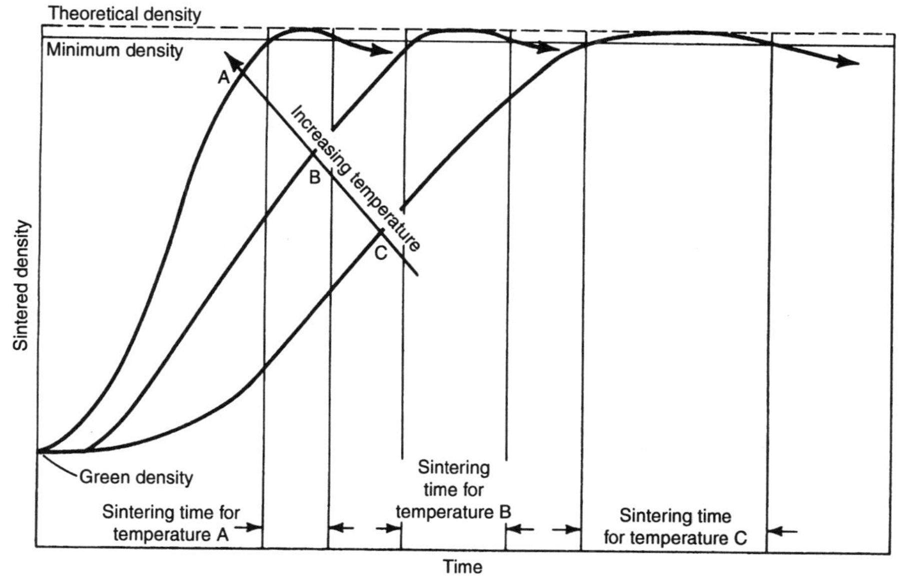

Fig. 26 Relationship among sintered density, sintering time, and sintering temperature

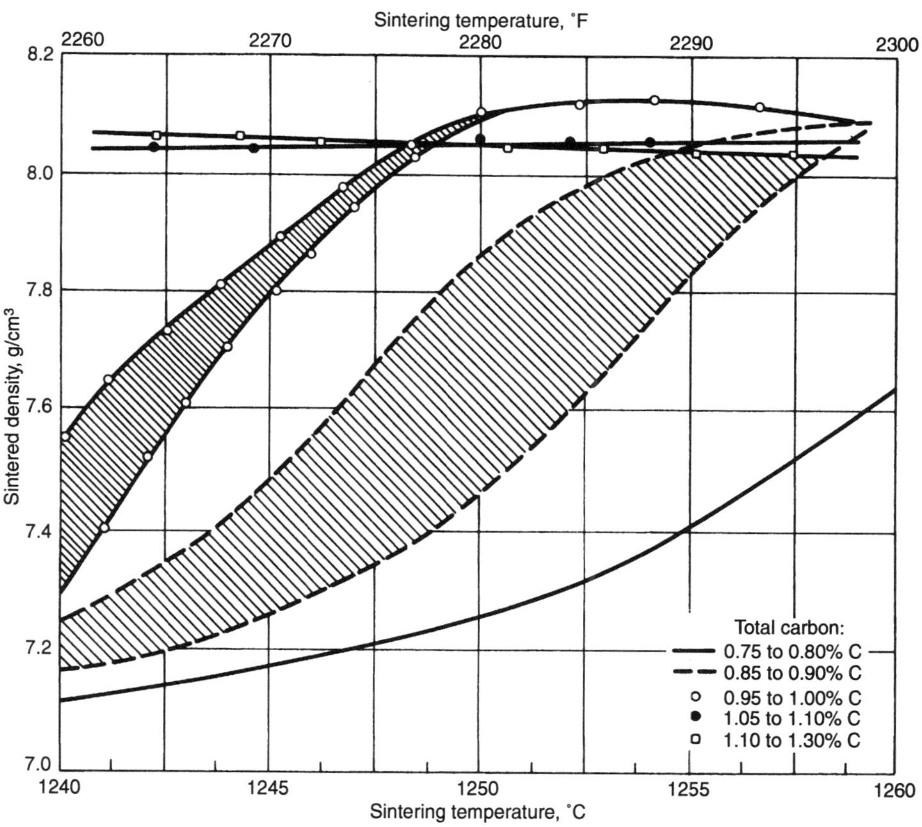

Fig. 27 Sintering curves for compacted M2 high-speed steel. Sintering time for all materials is 1 h.

Effect of Oxygen Content. Oxygen affects the sintering of tool steels by reducing carbon content during sintering. Oxygen from the annealed powder reacts with carbon during sintering to form carbon monoxide. Approximately 0.01% C is lost in the atmosphere for every 0.01% of oxygen present in the annealed powder. This reaction reduces the oxygen to less than 200 ppm during vacuum sintering.

Effect of Silicon Content. Silicon aids sintering by suppressing the melting point of tool steels. A carbon equivalent (CE) for estimating sintering temperature can be calculated as:

Table 13 Sintering data for M2 tool steel

Testing was conducted on five lots. All compacts were pressed at 827 MPa (60 tsi) and sintered for 60 min.

Total carbon(a), %	Added graphite, %	Sintered properties: At 1240 °C (2260 °F) Density, g/cm³	At 1240 °C (2260 °F) Characteristics	At 1250 °C (2280 °F) Density, g/cm³	At 1250 °C (2280 °F) Characteristics	At 1260 °C (2300 °F) Density, g/cm³	At 1260 °C (2300 °F) Characteristics
Lot 1							
0.77	0.0	7.19	MLP, FG	7.29	FG, SP	7.47	MSP, VFG
0.87	0.1	7.25	VFG, LP	8.03	FP, LP	8.04	FP, SP
0.97	0.2	7.48	VFG, LP	8.11	FP	8.09	FP
1.07	0.3	8.01	FG	8.09	FP, SP	8.07	E, LP
Lot 2							
0.79	0.0	7.00	VFG, MP	7.25	FG, MP	7.77	MP
0.89	0.1	7.18	VFG, MP	7.76	MP	8.08	FP
0.99	0.2	7.58	MP, VFG	8.08	FP	8.06	FP, LG
1.09	0.3	8.07	VFP	8.00	LG, SP	8.05	E, LP, FP
Lot 3							
0.86	0.0	7.17	VFG, MP	7.47	MP	8.09	FP
0.96	0.1	7.29	VFG, MP	8.08	MLP	8.08	FP, LP
1.06	0.2	7.84	VFG, MP	8.09	SP	8.06	E, LP, LG
1.16	0.3	8.07	FG, VFP	8.07	SP, E	8.07	E, LG
Lot 4							
0.87	0.0	7.22	VFG, MP	7.87	MP	8.09	FP, LP
0.97	0.1	7.47	FG, MP	8.08	FP, SP	8.07	SP
1.07	0.2	8.07	FP	8.06	FP, SP	8.06	E, MSP
1.17	0.3	8.08	FP	8.06	E, SP	8.05	E, MSP
Lot 5							
1.00	0.0	7.82	VFG	8.09	FP, SP	8.06	E, FP, LG
1.10	0.1	8.08	FG	8.07	FP, SP	8.05	E, EP, LG
1.20	0.2	8.07	VFP	8.06	LG, E	8.05	E, FP, LG
1.30	0.3	8.06	LG	8.06	LG, E	8.04	E, FP, LP

M, many; V, very; FG, fine grain; LG, large grain; SP, small pores; LP, large pores; FP, few pores; MP, many pores; E, eutectic. (a) wt% C in powder plus wt% admixed graphite; does not account for carbon loss during sintering due to deoxidization

Table 14 Typical sintering temperatures and compositions for several high-speed steels

Alloy	Sintering temperature(a) °C	Sintering temperature(a) °F	Composition, % C	Cr	Mo	W	V	Co	Relative sinterability
M2	1245	2270	0.85	4.2	5.0	6.3	1.9	...	3
M2 (high carbon)	1240	2260	1.00	4.2	5.0	6.3	1.9	...	3
M3 type 2	1255	2290	1.20	4.1	5.0	6.0	3.0	...	2
M4	1260	2300	1.32	4.5	4.5	5.5	4.0	...	3
M35	1225	2240	1.15	4.2	5.1	6.4	2.0	5.0	4
M42	1220	2230	1.10	3.8	9.5	1.5	1.2	8.0	5
T15	1265	2310	1.60	4.4	...	12.5	4.8	5.0	1

(a) Approximate sintering temperature for powders annealed at atmospheric pressure and pressed at 830 MPa (60 tsi). (b) 1 represents the easiest, 5 is the most difficult.

$$\text{CE} = \%\text{C (powder)} + \%\text{C (graphite)} + \frac{\%\text{Si}}{30}$$

Admixed silicon does not homogenize rapidly because of low diffusivity; consequently, it usually causes incipient melting.

Effect of Boron Content. Boron forms a low-melting-point eutectic with steels. Only small amounts can be added without changing carbide morphology. Boron can be added to powder as an alcoholic solution of boron oxide, followed by drying at ~95 °C (200 °F). This results in an even coating of the particle with boron oxide. The treated powder thus sinters uniformly.

Effect of Carbide Formers. The amount of carbon required for sintering and heat treating increases with the concentration of tungsten, molybdenum, and vanadium. Tungsten, molybdenum, and vanadium combine with carbon to form carbides that deplete the matrix of carbon. Vanadium is especially potent, because the vanadium-rich MC carbide is very stable and resists dissolution at high-sintering temperatures.

The stability of the vanadium carbide effectively retards grain growth. Generally, increasing vanadium content (when properly balanced with carbon) facilitates sintering. High-speed steels containing 3% or more vanadium (M3 type 2 and T15) can be sintered to full density with microstructures finer than wrought.

In high-speed steels, chromium content does not affect sintering of alloys due to the diminished stability of chromium carbide compared to molybdenum, tungsten, and/or vanadium carbides. Chromium carbide is formed only in annealed material. Chromium content affects the sintering of alloys that contain only minor amounts of vanadium, molybdenum, or tungsten, such as the high-chromium die steels.

Vacuum Sintering to Full Density

The prime object of vacuum sintering to full density is to uniformly expose all parts to sufficient temperature and time. This is typically done to produce the minimum required density and to prevent overheating and/or excessive carbide growth.

Sintering Cycles. Figure 29 shows typical sintering cycles. These sintering cycles consist of a lower temperature hold, followed by one or more high-temperature holds to densify compacts.

Deoxidation. Heating rates to the deoxidation temperatures are not critical. Parts should be heated at the maximum practical heating rate to minimize furnace time, provided the temperature of the load does not greatly surpass the deoxidation soak temperature. Batch-type vacuum furnaces can be backfilled with inert gas or hydrogen to >46 kPa (350 torr) to improve the heating rate and temperature uniformity of the load.

The primary function of the deoxidation step is to react oxygen with carbon to form carbon monoxide, which is removed by the vacuum pumps. Oxygen, nitrogen, and other gases that may be present in the powder compact must be removed before sintering closes the interconnected porosity to the surface. Once the porosity is no longer interconnected, gas evolution results in blistering.

The deoxidation soak or hold also provides an opportunity for load temperature to become uniform. Temperature in any part of the load should not vary more than ± 9 °C (± 16 °F) before heating to the sintering soak. The load may be soaked at 1040 °C (1900 °F) for several hours to improve temperature uniformity without seriously affecting the microstructure.

Heating to sintering temperature is slow and closely controlled to ensure temperature uniformity. Typical ramp rates are 0.5 to 5.5 °C/min (1 to 10 °F/min).

Sintering Soak. Sintering treatments may consist of an isothermal soak or a very slow ramp (typically 3 to 33 °C/h, or 5 to 60 °F/h). Temperature uniformity is essential for successful sintering. Temperature range should not vary by more than 8 °C (15 °F) at any location within the load. Table 14 gives typical sintering temperatures for two-soak sintering cycles for several alloys.

Cooling. Generally, cooling from the sintering temperature is performed as rapidly as possible by backfilling the furnace with inert gas and using forced convection cooling (also known as fan cooling or gas quenching) to minimize furnace time. Typically, the hardnesses of gas-quenched fully dense tool and high-speed steels are 50 to 60 HRC. These hardnesses are substantially lower than the austenitized and quenched hardnesses of 55 to 65 HRC because of excessive dissolution of carbides, which results in excessive retained austenite.

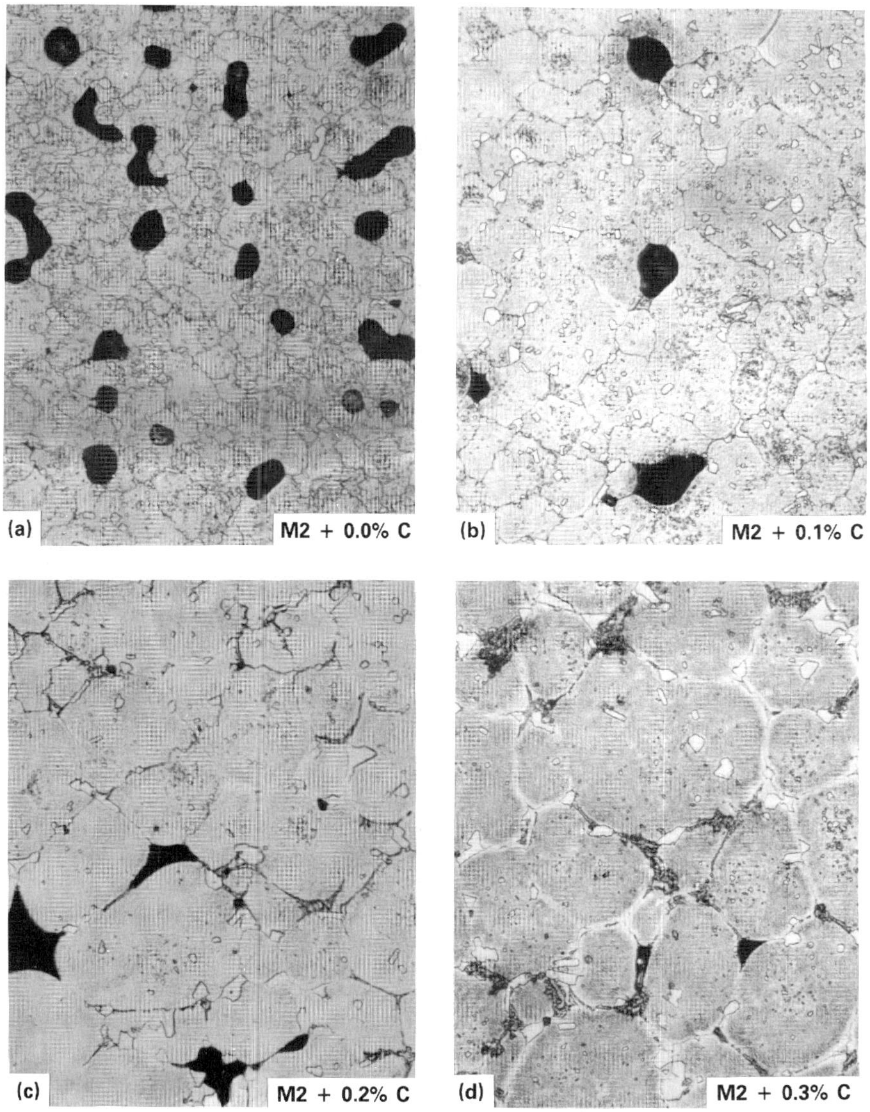

Fig. 28 Effect of increasing carbon content on the sintered microstructure of M2. All samples were sintered in the same run. Carbon was increased by blending in graphite. Note the microstructure changes from being undersintered (round pores) for 0.0 and 0.1% C to being oversintered (sharp angular pores, eutectic) for 0.2 and 0.3% C. 250×

Nitriding. Tool steels and high-speed steels can be alloyed with nitrogen by maintaining a nitrogen partial pressure after deoxidation, but before sintering. Significant alloying does occur with partial pressure at >133 Pa (1 torr). Increasing the partial pressure increases the nitrogen content. Depending on the alloy, nitrogen contents from 4000 ppm (M2) to over 8000 ppm (T15) can be produced by backfilling to atmospheric pressure.

Nitrided cases can be produced by introducing a suitable nitrogen partial pressure after sintering has closed off the interconnected porosity. Use of nitrogen as the quenching gas during forced convection cooling does not result in any significant nitriding.

Heat Treatment

Heat treatment of sintered high-speed steels is similar to the heat treatment of wrought counterparts. In both cases, heat treatment consists of hardening an annealed structure, followed by tempering to achieve desired properties. The finer microstructure of P/M high-speed steels may require slightly lower temperatures than wrought components to optimize performance.

Annealing. Sintered parts should be annealed before austenitizing. This treatment provides grain refinement and transforms large amounts of retained austenite. Sintering temperatures and times are higher and longer than optimum austenitizing heat treatment times. Consequently, parts quenched from the sintering temperature contain large amounts of retained austenite, which lower material properties. A suitable annealing cycle for sintered parts is heating to 900 °C (1650 °F) for 4 h and cooling at a rate of 50 °C/h (90 °F/h) to 500 °C (930 °F), followed by rapid cooling to ambient temperature.

Austenitizing. The austenitizing temperature of high-speed steels is influenced by the exact composition of the alloy (particularly carbon), as

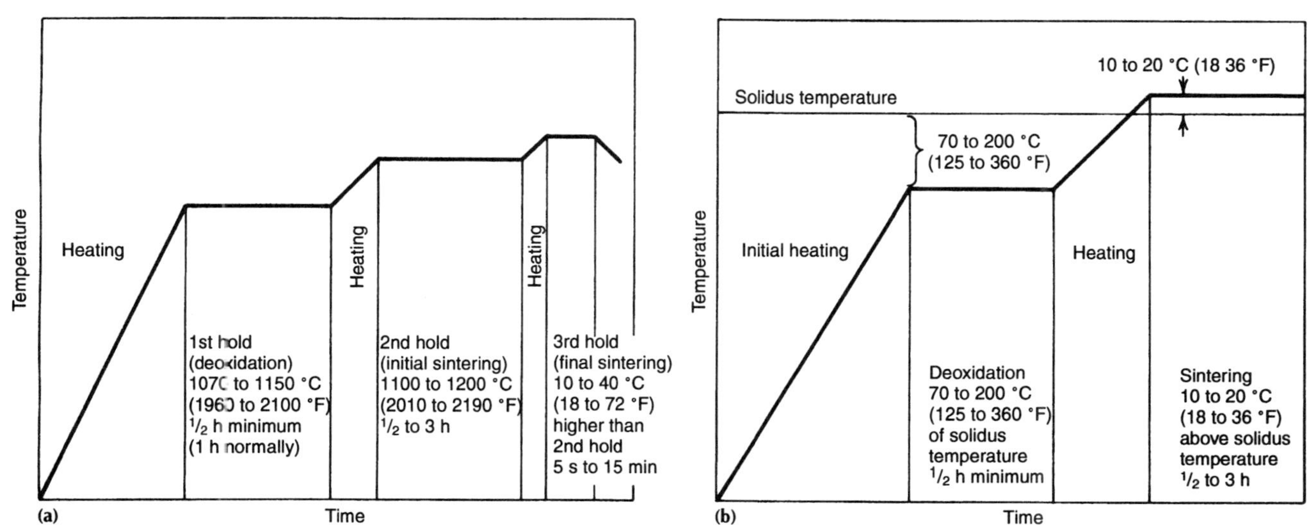

Fig. 29 Sintering cycles for fully dense sintering of high-speed steels. (a) British Patent 1 562 788. (b) U.S. Patent 4 063 940

Table 15 Recommended conditions for salt bath hardening of sintered high-speed steels

Alloy	Requires hardness		Hardening temperature(a)		Tempering temperatures(b)		Intercept grain size
	HRC	DPH	°C	°F	°C	°F	
M2	63–64	790–815	1170	2138	570	1058	...
	64–65	815–840	1180	2156	560	1040	...
	65–66	840–870	1200	2192	550	1022	...
M3 type 2	64–65	815–840	1170	2138	570	1058	15
	65–66	840–870	1180	2156	570	1058	15
	66–67	870–905	1200	2192	560	1040	15
	67–68	905–940	1220	2228	550	1022	14
M4	63–64	790–815	1180	2156	570	1058	10.5
	64–65	815–840	1200	2192	560	1040	8
	65–66	840–870	1200	2192	540	1004	8
M15	63–64	790–815	1180	2156	560	1040	17
	64–65	815–840	1200	2192	580	1076	17
	65–66	840–870	1200	2192	570	1058	17
	66–67	870–905	1210	2210	560	1040	15
M35	63–64	790–815	1180	2156	580	1076	12
	64–65	811–840	1180	2156	560	1040	12
	65–66	840–870	1200	2192	560	1040	12
T15	65–66	840–870	1170	2138	570	1058	17
	66–67	870–905	1180	2156	565	1049	17
	67–68	905–940	1200	2192	550	1022	16
	68–69	940–980	1220	2228	520	968	13.5
T42	66–67	870–905	1160	2120	570	1058	15
	67–68	905–940	1180	2156	570	1058	15
	68–69	940–980	1210	2210	560	1040	14

Based on 20 g (0.7 oz) specimen size. (a) All samples were hardened for 150 s soak time. (b) All samples were triple tempered (× 3 × 1 h)

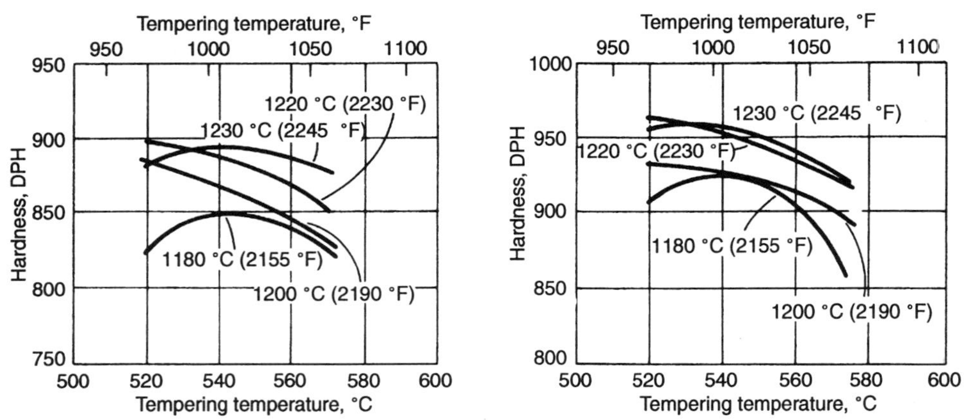

Fig. 30 Tempering curves for high-speed steels at varying hardening temperatures. (a) Mean hardening response of sintered M2. (b) Mean hardening response of sintered T15

well as the carbide size. Fine carbides dissolve more rapidly than coarse carbides; consequently, parts with fine carbides should be heat treated at a lower austenitizing temperature. Table 15 gives hardness values, hardening and tempering temperatures, and the average Snyder-Graaf intercept grain size for several high-speed steels.

The Snyder-Graaf method for determining intercept grain size is based on an actual count of the grains. In the as-quenched condition, the grain boundaries of high-alloy tool steels are clearly revealed by deep etching in nital. The test method is conducted on a metallograph, with the structure shown on a ground-glass screen at a magnification of 1000×. A 127 mm (5 in.) line drawn on the ground glass represents a length of 127 μm (0.005 in.) on the sample. The number of grains crossed or touched by this 127 mm (5 in.) line is counted; the average of ten readings at random points on the sample gives the intercept grain size.

Tempering temperatures of P/M parts are similar to those used for wrought materials. Depending on property requirements, temperatures range from 540 to 595 °C (1000 to 1100 °F). Figure 30 shows tempering curves for M2 and T15. Heat treated properties of sintered M2, M35, and T15 are given in Table 16.

Sintered Microstructures

Fully dense sintered tool steel and high-speed steel metallurgy and microstructures are similar in most respects to wrought counterparts. Lower significant differences in carbide size and uniformity of carbide distribution may have significant consequences. Sintered T15 is capable of finer carbides and grain size than the wrought alloy. Sintered M2 exhibits carbide and grain sizes that generally are not as fine as sintered T15 and are frequently coarser than wrought M2. T15 contains a large amount of vanadium-rich carbides (MC) that are very stable and do not coarsen readily. These carbides also inhibit grain growth during sintering.

Table 16 Typical mechanical properties of commercially sintered M2, M35, and T15 high-speed steels

Property	Grade M2	M35	T15
Density, g/cm^3	8.05–8.2	8.05–8.2	8.15–8.3
Ultimate tensile strength(a), MPa (ksi)	750–800 (108–116)	770–820 (112–119)	770–830 (112–120)
Elongation(a), %	12–14	6–9	3–6
Ultimate tensile strength(b), MPa (ksi)	750–2000 (108–290)	770–2000 (112–290)	770–2000 (112–290)
Hardness, HRC	62–65	63–66	64–67

(a) Fully annealed. (b) Depending on heat treatment

Sintered D2 microstructure exhibits somewhat finer carbides than wrought D2. Wrought D2, like wrought T15, frequently has a segregated or "banded" structure in which large carbides group together. Improved properties of P/M tool steels are attributed to their finer and more uniform microstructure.

As-sintered microstructures consist of untempered martensite within prior austenite grain boundaries and large amounts of retained austenite. Etching in nital reveals the presence of prior austenitic grain boundaries during sintering. Etching also can reveal gray or white areas in the matrix. Gray areas are reported to be caused by the very fine carbide dispersions that result from marginal or insufficient vacuum quenching. White areas indicate locations in which precipitation has not occurred.

Figure 31 shows undersintered, correctly sintered, and oversintered microstructures. Undersintered microstructures (Fig. 31a) have fine or very fine carbides and grains in addition to porosity. Porosity is rounded (without sharp ends) and can be irregular in shape when sintered density is low. Correctly sintered microstructures have little or no porosity, no eutectic structure or evidence of melting, uniformly dispersed carbides, and uniform grain size, as shown in Fig. 31(b).

The oversintered microstructure has large carbides and grains. Porosity with sharp ends, usually at grain triple points, is evident. A eutectic structure from localized melting and a continuous carbide network around grains may also be present in oversintered microstructures (Fig. 31c). Oversintered parts have slightly lower sintered densities (~0.2 to 0.5 g/cm^3) than correctly sintered parts because of the formation of porosity at the grain triple points.

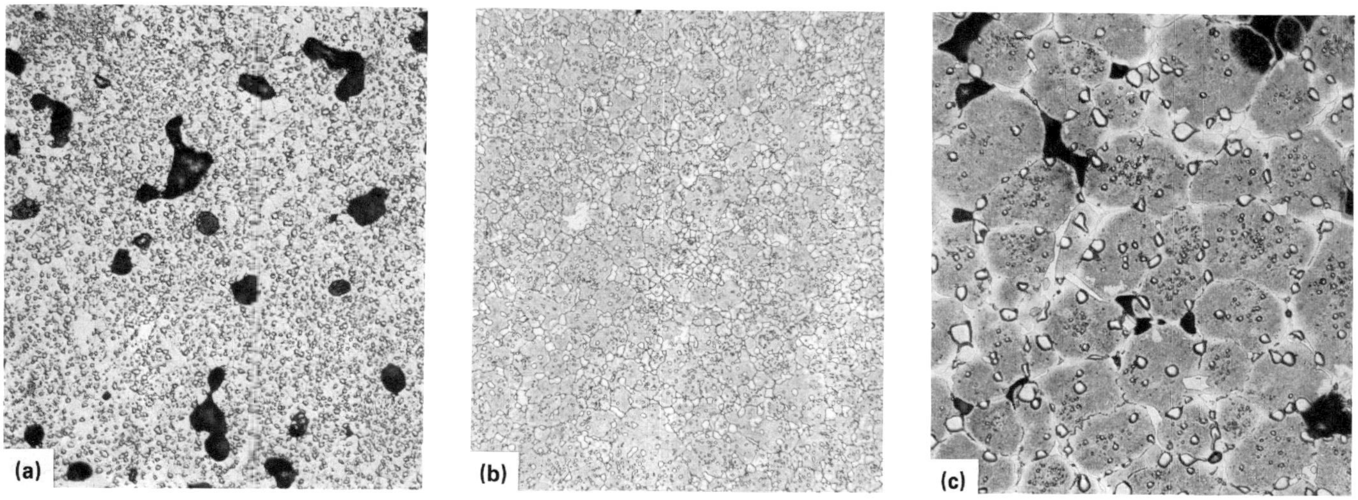

Fig. 31 Photomicrographs of as-sintered T15 high-speed steel. (a) Undersintered structure. 190×. (b) Correctly sintered structure. 190×. (c) Oversintered structure. 190×

Contamination. Metallic contamination can be readily detected in sintered tool steel microstructures. Contamination appears as carbide-free areas with porosity and/or carbide networks. Such defects occur when lower melting alloy particles or alloy particles that react with carbon from the tool steel powder to form low-melting alloys are present in higher melting tool steel alloys. Iron, stainless steel, and low-alloy steel powders form such defects in M2, T15, and other high-speed steels. These defects also occur when particles of T15 are present in M2 high-speed steel.

Tool steel powder producers have significantly reduced typical levels of contamination by using powder cleaning, compaction, and sintering equipment only for tool steel powders and by keeping powder-producing environments meticulously clean. Tool steel powder users must also clean powder hoppers and blenders to maintain high-quality P/M part production.

Atmospheric Pressure Sintering

Although vacuum is the preferred mode of sintering, wear parts made of tool steels and high-speed steels can be sintered to conventional P/M densities or to full density at atmospheric pressure and in atmospheres with dew points below −40 °C (−40 °F). Atmosphere-sintered parts have higher oxygen and nitrogen levels than vacuum-sintered parts.

Nitrogen-based, dissociated ammonia, and hydrogen atmospheres are viable alternatives to vacuum. Atmosphere composition has little effect, however, on sintered density. Parts sintered in pure nitrogen or nitrogen-base atmospheres are nitrided. Increasing the nitrogen content decreases the sintered transverse rupture strength. Consequently, as-sintered transverse rupture strength is lowest for parts sintered in nitrogen and highest for parts sintered in hydrogen. Sintered high-speed steels and tool steels should be double tempered to maximize transverse rupture strength.

Sintering of Copper-Base Alloys

Alain Marcotte, United States Bronze Powders, Inc.

Copper-base P/M materials rank second only to iron-base parts in terms of commercial applicability. As with other P/M materials, the final properties and related performance of copper-base parts depend on successful sintering techniques. This section reviews sintering practices for Cu-Sn, Cu-Zn-Pb, and Cu-Ni-Zn alloys.

Sintering of Bronze

Sintered bronze can be produced either from mixtures of copper powder and tin powder or from a prealloyed tin bronze powder. The nominal composition of 90Cu-10Sn can be complemented with other constituents such as dry organic lubricants, graphite, lead, and iron, depending on the specified grade.

Premix/Diffusion Alloyed Bronzes

Premix or partially diffused bronzes are used extensively in the manufacture of porous, self-lubricating bushings and bearings and for more complex structures requiring superior bearing and mechanical strength. Self-lubricating bushings and bearings are produced at nominal densities (oil impregnated) of 5.8 to 7.2 g/cm^3, with oil contents ranging from 24 to 11 vol%, respectively. Corresponding radial crush values (K strength constant) are approximately 69 MPa (10 ksi) of the lower density, increasing to 228 MPa (33 ksi) at the highest nominal density.

The basic manufacturing procedure consists of compacting the powder shapes to the appropriate green density and sintering to achieve a homogeneous metallurgical alpha bronze structure, followed by oil impregnation. A sizing operation completes the process to ensure dimensional precision and general surface integrity.

Dimensional Change. Several methods are used in the industry to produce bearings and structural parts from sintered 90Cu-10Sn bronze. For practical purposes, either pure copper and pure tin or prealloyed bronze and tin powder can be mixed. There are advantages and disadvantages to both methods. Mixed powders possess relatively good pressing properties, but there is always a risk of segregation. Premix powder containing prealloyed bronze (e.g., 94Cu/6Sn + 4Sn) have less liquid phase during sintering, which accounts for a comparatively lower sintered strength at similar green density. However, "partially prealloyed" bronze minimizes the risk of segregation and still maintains acceptable sintered strength.

Partially diffused bronze is possible when premix bronze is presintered at a temperature range from 400 to 750 °C (750 to 1380 °F), such that a metallurgical bond between tin and copper powder is created. The sintered cake is then crushed, ground, and screened. This treatment minimizes segregation and gives similar sintered properties to premix bronzes.

To include all dimensional patterns with related absolute dimensional change values for each commercially available premix system is prohibitive. A variety of premixed powders are available with specific sintered dimensional patterns to satisfy customer design and tooling needs. Despite the dimensional magnitude of the particular premix being sintered, compositions of this type exhibit a common sintered dimensional pattern. Absolute sintered dimensional characteristics typically are unique to a specific source of copper and tin powders. For example, sintered dimensional consistency can be obtained by blending two or more copper-base

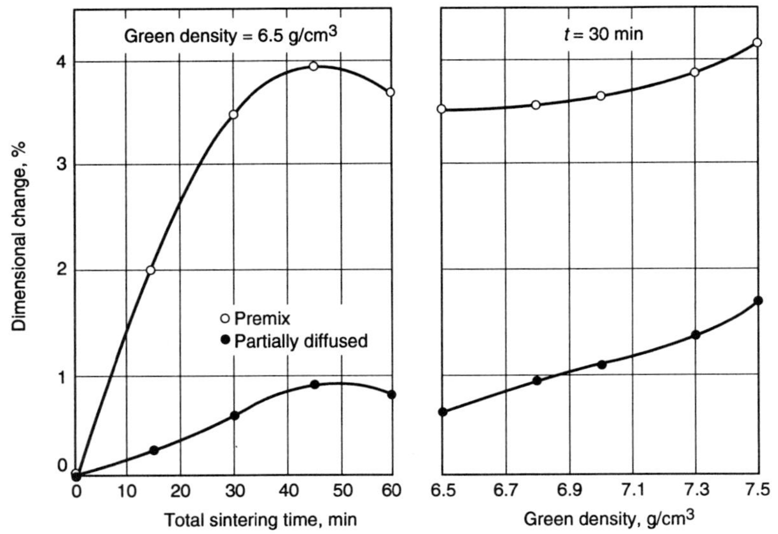

Fig. 32 Effect of properties on dimensional change of bronze. Sintered at 820 °C (1500 °F), under dissociated ammonia atmosphere.

powders that exhibit different growth characteristics and/or by use of tin powders that also exhibit different growth characteristics.

Generally, copper-tin blends composed of relatively coarse powder sinter to higher growth values than a blend composed of finer powders. After powder blends have been tested and adjusted to provide an approximation of target dimensions, final adjustments are made during production sintering to obtain dimensional precision. For discussion purposes, Fig. 32 shows the relationship between dimensional change, sintering time, and green density of a premix bronze and a partially diffused bronze, using the same source of powders and a similar particle size distribution (see Ref 36).

Factors affecting the ultimate, or peak, dimensional values include physical characteristics of the constituents and compacted density. Control of sintered dimensions in premix systems is achieved by manipulating sintering time and/or temperature.

Sintering Time and Temperature. Typical sintering furnace temperatures for bronze range from 815 to 860 °C (1500 to 1580 °F); total sintering time within the hot zone ranges, from 15 to 30 min, depending on the furnace temperature selected, required dimensional change, and most importantly, the presence of an optimum alpha grain structure. Figure 33 shows the typical microstructure of a premix bronze part.

Sintering atmospheres should be protective and reducing to facilitate sintering. Reduction of the copper oxides that may surround each copper powder particle and reduction of tin oxide formation allow for increased diffusion rates. Consequently, faster sintering rates and more homogeneous structures can be obtained.

Prealloyed Bronzes

Sintered bronze alloys are rather uncommon in powdered metal usage. This is primarily due to their relatively high cost compared to low-alloy steels. However, sintered bronze properties can be advantageous for non-magnetic applications that require very good corrosion resistance, good mechanical strength, and excellent ductility.

Prealloyed 80Cu-9Sn-2Zn bronze powders with a select lubricant are intended for the fabrication of high-density P/M structural components. Unlike many elemental copper-tin premixes, the sintering of prealloyed bronze results in the attainment of high sintered densities (85 to 90% of theoretical), that provide correspondingly high strengths and hardnesses.

Sintering properties. Prealloyed bronze powders are relatively easy to work due to the excellent ductility obtained on sintering. For example, an 89Cu-9Sn-2Zn bronze pressed at 414 MPa (30 tsi) to 85% of theoretical density, exhibits >30% elongation thus allowing for a substantial degree of cold working. Compared with sintered brass, bronze powders reach higher yield strength and hardness levels when compacted under similar conditions. For example, pressing at 550 MPa (40 tsi) results in hardness of 45 HRB for an 89Cu-9Sn-2Zn-2Fe bronze and of 90 HR for a 70Cu-30Zn brass, respectively (after 30 min sintering at 840 °C (1550 °F) under dissociated ammonia).

The addition of 2% of a select grade iron in he composition improves the sintered structure, which has a more uniform grain size and results in comparatively higher yield strength and hardness. Figure 34 shows the microstructure of a prealloyed bronze. Sintering temperature and atmospheres of prealloyed bronzes are similar to

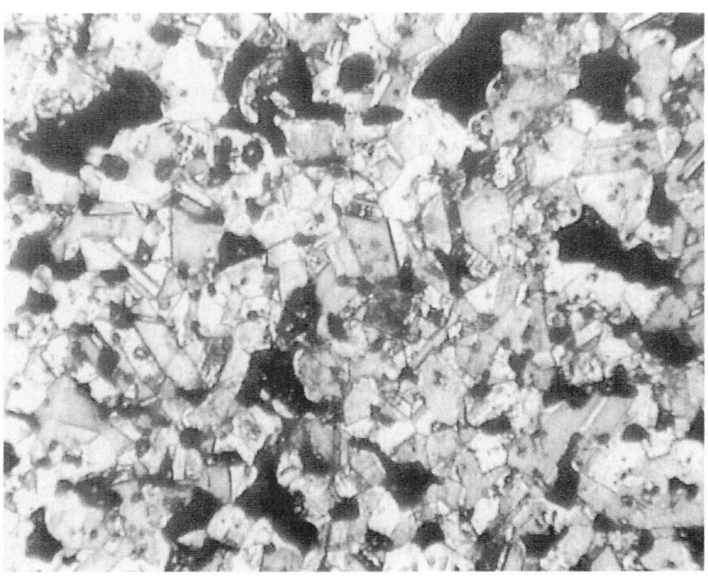

Fig. 33 Typical microstructure of sintered premix bronze. 90-10 bronze, etched in $K_2Cr_2O_7$ solution. 175×

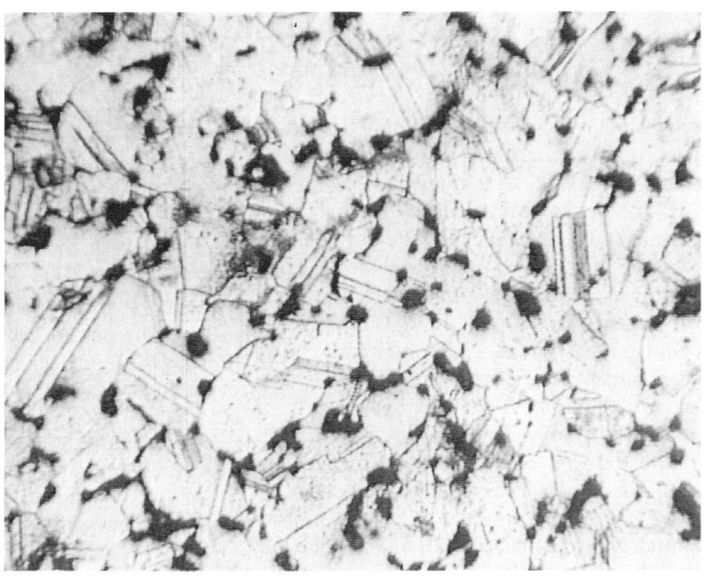

Fig. 34 Typical microstructure of sintered prealloyed 89Cu-95-2Zn-2Fe bronze, etched in $K_2Cr_2O_7$ solution. 350×

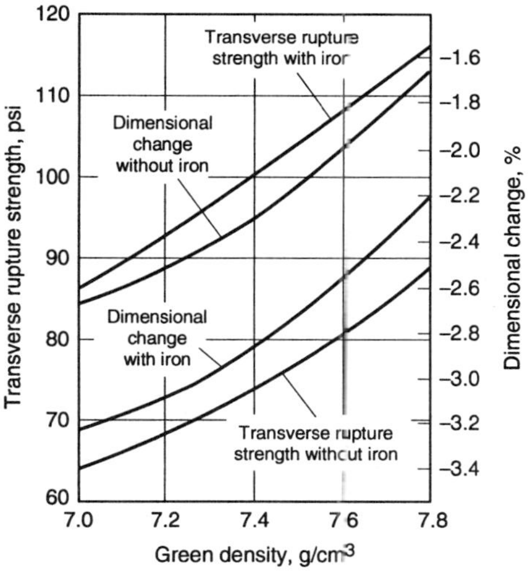

Fig. 35 Properties of prealloyed bronzes. Sintered 30 min at 840°C (1550 °F), under DA

those of bronze premix. Figure 35 shows the dimensional change and transverse rupture strength of prealloyed bronzes as a function of green density.

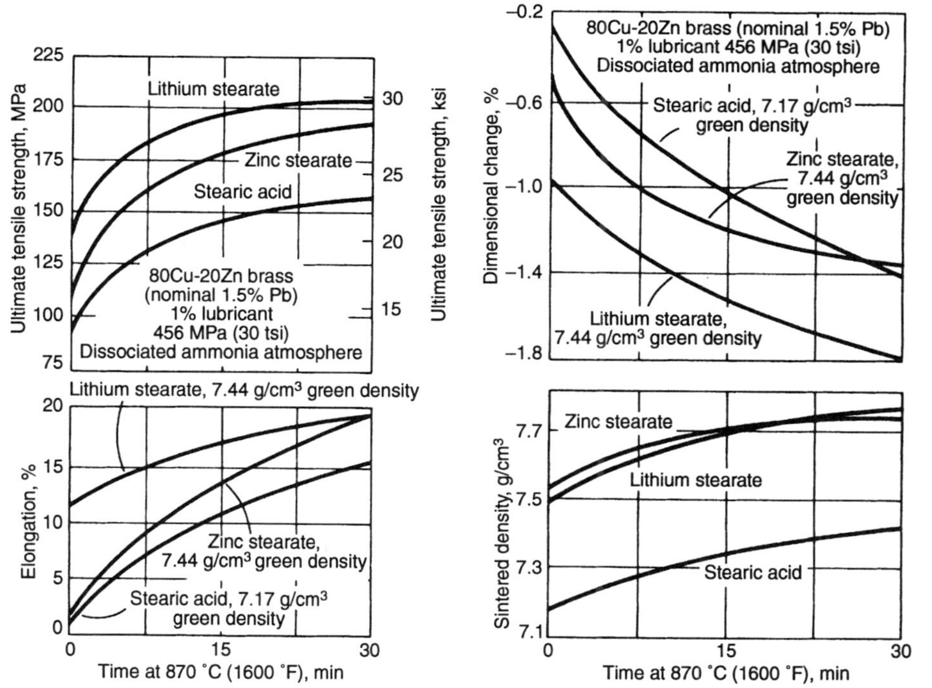

Fig. 36 Effects of lubricants and sintering time at temperature on tensile properties, sintered density, and dimensional change of brass compacts

Sintering of Brass and Nickel Silvers

Powders of brasses and nickel silvers are prealloyed, single-phase (alpha) powders that, on sintering, yield moderate mechanical strength, excellent ductility, and good corrosion resistance. Parts can be subsequently burnished to improve surface finish. The various alloy compositions produced also provide suitable color, or shade, selection for applications that require a high degree of surface finish and appearance.

During alloy preparation, lead can be added to improve machinability of the sintered forms. Typical machining operations include drilling, tapping, turning, threading, and grinding. Excellent sintered ductility also facilitates secondary operations, such as sizing, cold densification, swaging, and staking. By using multiple pressing and sintering operations, the yield strength and hardness of the P/M structure may approach those of its wrought alloy counterpart (Ref 37).

Standard prealloyed brass and nickel silver powder compositions are controlled to conform to existing materials standards. This conformance precludes the additions of foreign metallic constituents, such as tin and iron, that affect sintered mechanical and dimensional characteristics. In spite of the various compositions comprising these standard alloy powders, most exhibit similar characteristics. After blending with lubricant, powders densify approximately 10% more than their as-atomized apparent density. For example, an as-atomized powder with an apparent density of 3.0 g/cm^3 usually blends to a density of 3.3 g/cm^3 minimum with the addition of dry lubricant. Compressibilities are excellent, as lubricated powders compact to 85% of wrought counterpart densities at 414 MPa (60 ksi). Compression ratios of lubricated powders range from 2.0-to-1 to 2.2-to-1.

Effect of Lubricant. Blending of powders with dry organic lubricants is normally accomplished in a double cone-type blending unit. To minimize the inclusion of large lubricant agglomerates and other undesirable particles, the material is passed through a 40 mesh sieve, or screening is recommended prior to blending. The primary lubricant employed with brasses and nickel silvers is lithium stearate.

Lithium stearate provides an apparent scavenging or cleansing effect that enhances the sinterability of these powders. It may also result in spotty or speckled superficial stains on sintered surfaces. These phenomena affect the appearance of the components, but are not detrimental to mechanical properties. To minimize staining, lithium stearate additions of less than 0.5 wt% are recommended, and zinc stearate can be added to provide the additional required lubricity.

Different types of lubricant have a marked effect on the physical and mechanical properties of nonferrous prealloyed powder (Ref 37). To illustrate the effects on mechanical properties, Fig. 36 shows data for three frequently used lubricants in the P/M industry—lithium stearate, zinc stearate, and stearic acid. The beneficial effect of 1 wt% lithium stearate and the deleterious effect of 1 wt% stearic acid on mechanical properties are shown.

Compacting of lubricated powders is performed with standard types of compacting presses employing steel or carbide dies and punches. Excellent compressibility and good green strength permit compacting to 75% of theoretical density at pressures as low as 207 MPa (30 ksi).

Although powders are normally free of gangue and other inclusive, abrasive material that may cause tool wear, their relative softness does cause tooling difficulties. If powders are not adequately lubricated, fines within tool clearances gall the die wall and adjacent punch areas, thereby requiring tool removal and cleaning. The amount of lubricant added to the powder should be proportional to the total surface area of the die assembly requiring lubrication during forming and ejection. Low-profile, minimum die wall contact parts may only require 0.5 wt% added lubricant, whereas a high die wall contact part having core rods for holes or internal cavities may require 1.0 wt% lubricant.

Generally, there are no restrictions on the compacted configuration of these powders. In tool design, particularly with regard to die fill, consideration must be given to the relatively higher apparent densities of lubricated brass and nickel silver powder. Typical apparent densities for lubricated powders range from 3.3 to 3.6 g/cm^3.

Sintering of brasses and nickel silvers typically is not difficult; however, basic sintering practices do differ from those employed with other common alloy systems, such as elemental copper-tin blends and iron powder blends. These differences include sintering temperature, time at sintering temperature, and atmosphere protection.

Sintering temperatures for standard brasses range from 760 to 925 °C (1400 to 1700 °F). Temperature selection depends on the brass alloy being sintered and the mechanical properties

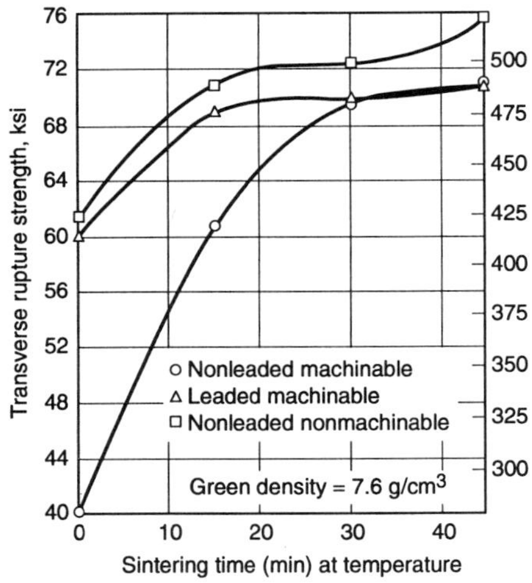

Fig. 37 Transverse strength of 80Cu-20Zn brasses. Sintered in hot zone at 870 °C (1600 °F) in DA

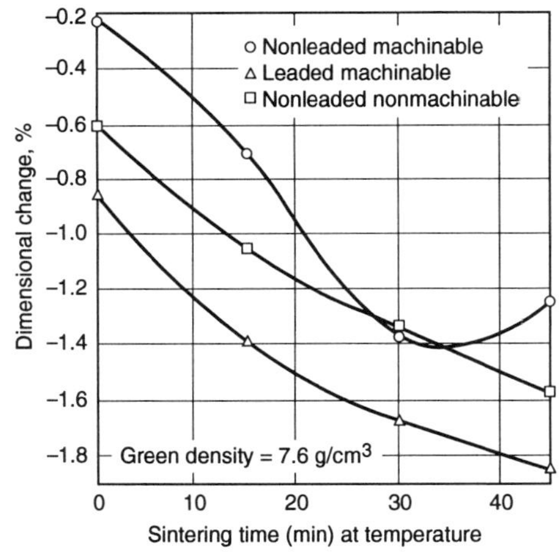

Fig. 38 Dimensional change of 80Cu-20Zn brasses. Sintered in hot zone at 870 °C (1600 °F) in DA

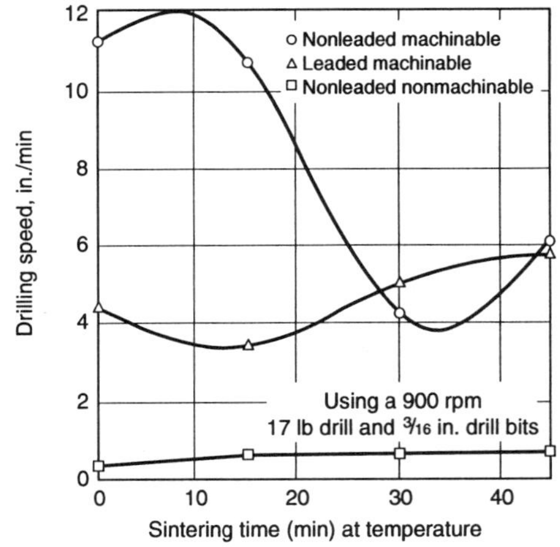

Fig. 39 Machinability of 80Cu-20Zn brasses. Sintered in hot zone at 870 °C (1600 °F) in DA

desired after sintering. Lower brasses with higher zinc contents and lower melting points are sintered at the lower temperature. Generally, a starting temperature of 100 °C (180 °F) below the solidus temperature (as determined from any copper-zinc binary alloy constitutional diagram) is suitable.

Nickel silver can be sintered at 870 to 980 °C (1600 to 1800 °F). Currently, only one base alloy is used for the manufacture of P/M structural parts; it has nominal composition of 64Cu-18Ni-18Zn. The leaded alloy composition contains 1.5% Pb. Sintering characteristics are similar to those of the brasses; therefore, responses to sintering parameters that affect dimensional and mechanical properties of brass are equally applicable to nickel silver.

Sintered Properties. Dimensional and mechanical properties of brasses and nickel silvers are primarily affected by compact density and the amount of time at temperature, as well as the sintering temperature itself. As mentioned previously, other elements that affect dimensional and mechanical properties usually are not added to powders. However, sintered properties, especially dimensional change, can be effectively controlled by manipulation of sintering time at the appropriate temperature. Each alloy exhibits unique dimensional characteristics—a 90Cu-10Zn brass compacted at 414 MPa (30 tsi) and sintered for 30 min at 870 °C (1600 °F) may shrink 0.5%, while a 70Cu-30Zn brass similarly treated may shrink 2.5%.

Figures 37 and 38 show typical property relationships that can be controlled through manipulation of time at temperature. The leaded 80Cu-20Zn brass shown in Fig. 37 and 38 is commonly used for structural parts fabrication. The density of 7.6 g/cm^3 is "average" for compacting lubricated prealloyed powders containing 0.375% lithium stearate and 0.375% zinc stearate at 414 MPa (30 tsi). As shown, close dimensional control can be obtained with a minimum reduction in mechanical properties after 15 min at temperature. Ductility is increased for subsequent forming operations, such as sizing, cold repressing for densification, or coining, by increasing sintering time.

Non-leaded machinable brasses have been recently introduced on the market to reduce or eliminate lead. It has been shown (Ref 38) that by replacing lead with select alloy additions, brass parts still maintain similar mechanical and physical properties, while having improved machinability. Figures 37 and 38 illustrate the properties of non-leaded brass, leaded brass, and a non-leaded brass containing select alloy additions. Figure 39 also shows the machinability for those different types of 80Cu-20Zn brasses.

Atmosphere protection is required for sintering brasses and nickel silvers to prevent oxidation and to ensure effective sintering. Use of lithium stearate as the base lubricant allows the use of most common sintering atmospheres over a wide range of dew points. Although dry hydrogen or dissociated ammonia provides the best sintering atmosphere, comparable properties can be obtained with nitrogen-base or partially combusted hydrocarbon gas atmospheres.

When sintering, compacts should be protected from direct impingement of furnace flame curtains and atmosphere gases by partially or fully covering loaded trays to minimize zinc loss. Because it has a high vapor pressure at standard sintering temperature (boiling point of pure zinc is 906 °C, or 1663 °F), zinc may be lost to the atmosphere as it diffuses through to the particle surfaces. Loss of excessive surface zinc results in a change in surface composition. In the case of brasses, pink copper or zinc-depleted areas are apparent. Although superficial zinc losses do not adversely affect sintered properties, surface finish is diminished; finished parts may be rejected because of color differences.

Furnace Design. A variety of protective sintering tray arrangements can be used, including graphite trays with full covers and graphite-base plates with steel covers, for example. Sintering of brasses on open belts is not common practice for several reasons; for example, overheating can melt brass parts that can fuse to the mesh belt. Also, excessive heat and direct impingement of flame curtain gas through the mesh can impair effective sintering due to oxidation and premature partial lubricant removal. Direct exposure to the sintering atmosphere can also cause objectionable surface dezincing.

Conversely, extremely gastight protective sintering setups should be avoided, because lubricant decomposition products may not be completely volatilized. Excessive residual lubricant within the compact inhibits sintering; consequently, mechanical properties are adversely affected. Typically, compacts with excessive internal lubricant residuals cannot be salvaged by additional passes through a full sintering cycle. To avoid or minimize the deleterious effect of residual lubricants, covered tray setups should facilitate free venting of lubricant gases to the flowing furnace atmosphere.

Sintering of Aluminum and Aluminum Alloys*

Conventionally pressed and sintered aluminum powder metal parts have been commercially available for many years. Sintered aluminum P/M parts are competitive with many aluminum castings, extrusions, and screw machine products that require expensive and time-consuming finishing operations. In addition, sintered aluminum P/M parts compete with other metal powder parts in applications where some of the attractive physical and mechanical properties of aluminum can be used.

Commercially available aluminum powder alloy compositions consist of blends of atomized aluminum powders mixed with powders of vari-

Adapted from *Properties and Selection: Non-Ferrous Alloys and Special-Purpose Materials,* Vol 2, *ASM Handbook,* 1990 ous alloying elements such as zinc, copper, magnesium, and silicon as follows:

Grade	Composition, %				
	Cu	Mg	Si	Al	Lubricant
601AB	0.25	1.0	0.6	bal	1.5
201AB	4.4	0.5	0.8	bal	1.5
602AB	...	0.6	0.4	bal	1.5
202AB	...	0.6	0.4	bal	1.5
MD-22	2.0	1.0	0.3	bal	1.5
MD-24	4.4	0.5	0.9	bal	1.5
MD-69	0.25	1.0	0.6	bal	1.5
MD-76	1.6	2.5	...	bal	1.5

Aluminum P/M parts can be sintered in a controlled, inert atmosphere or in vacuum. Sintering temperatures are based on alloy composition and generally range from 595 to 625 °C (1100 to 1160 °F). Sintering time varies from 10 to 30 min. Nitrogen, dissociated ammonia, hydrogen, argon, and vacuum have been used for sintering aluminum; however, nitrogen is preferred because it results in high as-sintered mechanical properties (Table 17). It is also economical in bulk quantities. If a protective atmosphere is used, a dew point of –40 °C (–40 °F) or below is recommended. This is equivalent to a moisture content of 120 mL/m^3 (120 ppm) maximum.

Aluminum preforms can be sintered in batch furnaces or continuous radiant tube mesh or cast belt furnaces. Optimum dimensional control is best attained by maintaining furnace temperature at ± 2.8 °C (± 5 °F). Figure 40 illustrates typical heating cycles for aluminum parts sintered in various furnaces.

Mechanical properties are directly affected by thermal treatment. All compositions respond to solution heat treating, quenching, and aging in the same manner as conventional heat-treatable alloys. Table 18 summarizes some typical properties of aluminum P/M alloys.

Sintering Atmospheres

Sintering in Nitrogen. Nitrogen at high purity and low dew point is particularly well suited for sintering aluminum P/M parts because of ready availability and moderate cost. Special handling is not required, nor are a generator and adsorbent dryer required to convert nitrogen to a dry, gaseous form. The highest sintered strength in both 601AB and 201AB parts is achieved in nitrogen.

Dissociated ammonia is used in many P/M applications for sintering brass and bronze parts. It can be used for aluminum as well (Ref 39).

Dissociated ammonia contains high concentrations of flammable hydrogen; consequently, care must be taken in handling, particularly for aluminum processing during which sintering temperatures are not high enough to ensure self-ignition upon contact with air. One precautionary method involves purging the furnace with an inert gas such as nitrogen prior to introducing dissociated ammonia.

Conditions and production rates for aluminum P/M parts sintered in dissociated ammonia were similar to those sintered in nitrogen; however, tensile properties of 601AB and 201AB alloys sintered in dissociated ammonia (Table 19) tend to be lower than for nitrogen sintered parts. The lower properties of parts sintered in dissociated ammonia appear related to the presence of hydrogen and/or undissociated ammonia in the sintering atmosphere. The tendency of hydrogen to cause gassing of aluminum has been well documented (Ref 40). Also, ammonia reacts with aluminum to liberate hydrogen. When 100% ammonia vapor was used as a heat treating atmosphere for aluminum, the tensile strength of 2024 alloy sheet was reduced 29%, and the elongation was reduced 82% (Ref 40).

Dimensional Change During Sintering. Dimensions of sintered aluminum P/M parts are affected by compact density, sintering atmosphere, temperature, and dew point. The effects of green density and atmosphere on sintered dimensions of 601AB and 201AB alloys are illustrated in Fig. 41(a) and (b). Dimensions increased with increased green density in all atmospheres. Shrinkage or a lack of growth was exhibited by 85% green density parts, whereas high-density compacts exhibited growth when sintered in dissociated ammonia and vacuum. Nitrogen-sintered parts experienced shrinkage over the full range of densities—except for 95% density 601AB, in which no change was noted.

Table 17 Typical properties of nitrogen-sintered aluminum P/M alloys

Alloy	Compacting pressure		Green density		Green strength		Sintered density		Temper	Tensile strength(a)		Yield strength(a)		Elongation,	Hardness
	MPa	tsi	%	g/cm^3	MPa	psi	%	g/cm^3		MPa	ksi	MPa	ksi	%	
601AB	96	7	85	2.29	3.1	450	91.1	2.45	T1	110	16	48	7	6	55–60 HRH
									T4	141	20.5	96	14	5	80–85 HRH
									T6	183	26.5	176	25.5	1	70–75 HRE
	165	12	90	2.42	6.55	950	93.7	2.52	T1	139	20.1	88	12.7	5	60–65 HRH
									T4	172	24.9	114	16.6	5	80–85 HRH
									T6	232	33.6	224	32.5	2	75–80 HRE
	345	25	95	2.55	10.4	1500	96.0	2.58	T1	145	21	94	13.7	6	65–70 HRH
									T4	176	25.6	117	17	6	85–90 HRH
									T6	238	34.5	230	33.4	2	80–85 HRE
602AB	165	12	90	2.42	6.55	950	93.0	2.55	T1	121	17.5	59	8.5	9	55–60 HRH
									T4	121	17.5	62	9	7	65–70 HRH
									T6	179	26	169	24.5	2	55–60 HRE
	345	25	95	2.55	10.4	1500	96.0	2.58	T1	131	19	62	9	9	55–60 HRH
									T4	134	19.5	65	9.5	10	70–75 HRH
									T6	186	27	172	25	3	65–70 HRE
201AB	110	8	85	2.36	4.2	600	91.0	2.53	T1	169	24.5	145	24	2	60–65 HRE
									T4	210	30.5	179	26	3	70–75 HRE
									T6	248	36	248	36	0	80–85 HRE
	180	13	90	2.50	8.3	1200	92.9	2.58	T1	201	29.2	170	24.6	3	70–75 HRE
									T4	245	35.6	205	29.8	3.5	75–80 HRE
									T6	323	46.8	322	46.7	0.5	85–90 HRE
	413	30	95	2.64	13.8	2000	97.0	2.70	T1	209	30.3	181	26.2	3	70–75 HRE
									T4	262	38	214	31	5	80–85 HRE
									T6	332	48.1	327	47.5	2	90–95 HRE
202AB Compacts	180	13	90	2.49	5.4	780	92.4	2.56	T1	160	23.2	75	10.9	10	55–60 HRH
									T4	194	28.2	119	17.2	8	70–75 HRH
									T6	227	33	147	21.3	7.3	45–50 HRE
Cold-formed parts (19% strain)	180	13	90	2.49	5.4	780	92.4	2.56	T2	238	33.9	216	31.4	2.3	80 HRE
									T4	236	34.3	148	21.5	8	70 HRE
									T6	274	39.8	173	25.1	8.7	85 HRE
									T8	280	40.6	250	36.2	3	87 HRE

(a) Tensile properties determined using powder metal flat tension bar (MPIF standard 10-63), sintered 15 min at 620 °C (1150 °F) in nitrogen

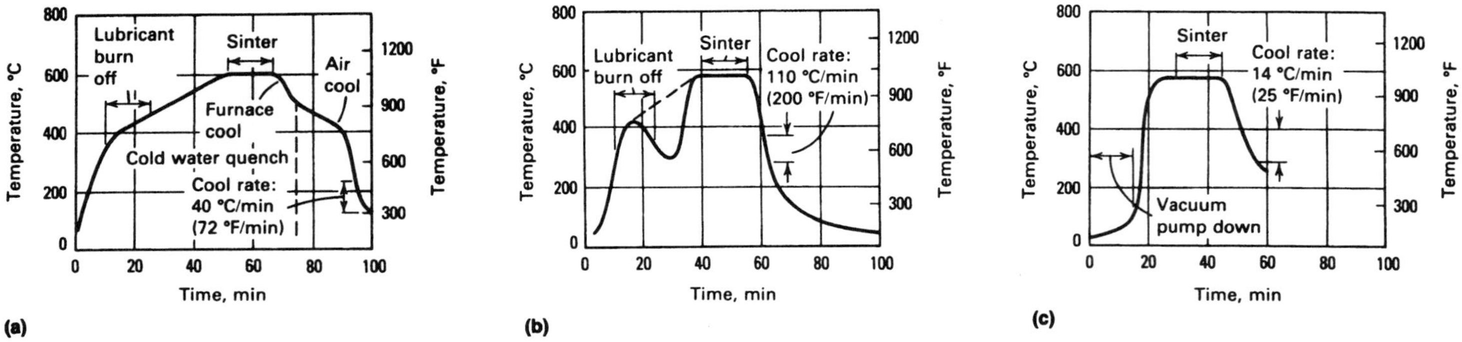

Fig. 40 Typical heating cycles for aluminum P/M parts sintered in (a) batch furnace, (b) continuous furnace, and (c) vacuum furnace

Table 18 Typical heat-treated properties of nitrogen-sintered aluminum P/M alloys

Heat-treated variables and properties	Grades: MD-22	MD-24	MD-69	MD-76
Solution treatment				
Temperature, °C (°F)	520 (970)	500 (930)	520 (970)	475 (890)
Time, min	30	60	30	60
Atmosphere	Air	Air	Air	Air
Quench medium	H_2O	H_2O	H_2O	H_2O
Aging				
Temperature, °C (°F)	150 (300)	150 (300)	150 (300)	125 (257)
Time, h	18	18	18	18
Atmosphere	Air	Air	Air	Air
Heat-treated (T6) properties(a)				
Transverse-rupture strength, MPa (ksi)	550 (80)	495 (72)	435 (63)	435 (63)
Yield strength, MPa (ksi)	200 (29)	195 (28)	195 (28)	275 (40)
Tensile strength, MPa (ksi)	260 (38)	240 (35)	205 (30)	310 (45)
Elongation, %	3	3	2	2
Rockwell hardness, HRE	74	72	71	80
Electrical conductivity, %IACS	36	32	39	25

(a) T6, solution heat treated, quenched, and artificially age hardened

Table 19 Properties of P/M aluminum alloys sintered in dissociated ammonia

Alloy	Green density %	Green density g/cm^3	Temper	Tensile strength MPa	Tensile strength psi	Yield strength MPa	Yield strength psi	Elongation in 25 mm (1 in.), %
601AB(a)	90	2.42	T1	93	13,500	76	11,000	2.5
			T4	108	15,700	88	12,700	3.5
			T6	159	23,100	...	...	1.0
	95	2.55	T1	121	17,600	87	12,600	3.5
			T4	146	21,200	99	14,300	5.0
			T6	207	30,100	205	29,700	1.5
201AB(b)	90	2.50	T1	161	23,300	141	20,500	2.0
			T4	198	28,800	163	23,700	2.5
			T6	247	35,800	...	...	0.5
	95	2.64	T1	174	25,200	152	22,000	2.0
			T4	221	32,000	180	26,100	3.0
			T6	288	41,800	287	41,600	1.0

(a) Sinter 10 to 30 min at 620 °C (1150 °F) at a dew point of –40 to –50 °C (–40 to –60 °F). (b) Sinter 10 to 30 min at 595 °C (1100 °F) at a dew point of –40 to –50 °C (–40 to –60 °F)

These dimensional changes were consistent as long as the sintering temperature was constant and the dew point in the furnace was at least –40 °C (–40 °F). Higher than normal temperatures caused excessive shrinkage and distortion or even melting in extreme cases. Lower than normal temperatures produced parts with increased dimensions and reduced properties.

Thus, for example, lowering the sintering temperature of 201AB alloy from 595 to 570 °C (1100 to 1060 °F) caused 95% density specimens sintered in dissociated ammonia to change from 0.25% shrinkage to 1.0% growth. At 580 °C (1080 °F), 0.25% growth was observed. Mechanical properties were less affected, although a reduction in tensile strength of 2 to 10% was observed, depending on thermal treatment. High dew points in the furnace resulted in excessive part expansion and a significant reduction in properties.

Sintering of Cemented Carbides

Revised by Mack Greenfield and George Wolfe, Kennametal, Inc.

Cemented carbides consist of various mixtures of carbides and iron-group metals, such as tungsten carbide/cobalt, tungsten carbide/titanium carbide/tantalum carbide/niobium carbide/cobalt, or titanium carbide/molybdenum carbide/nickel. These materials typically are sintered to essentially 100% of theoretical density using a liquid-phase sintering process. The successful application of cemented carbides to high-stress operations such as metal cutting, oil well drilling, or metal forming dies depends on achieving low levels of residual porosity during sintering. Additionally, sintering must be controlled carefully to obtain desired microstructure and chemical composition.

Cemented carbides are used in many applications in the as-sintered condition. Frequently, the as-sintered surface acts as the critical wear/stress-bearing surface. In most metal-cutting applications, the tool is considered worn out when the wear scar depth exceeds 0.2 to 0.4 mm (0.008 to 0.016 in.). Consequently, control of surface properties is essential.

Sintering Methods

Two basic methods are used to sinter cemented carbides. Hydrogen sintering uses a hydrogen-base atmosphere at atmospheric pressure to dynamically control part composition. Vacuum sintering uses a vacuum or reduced-pressure environment to control composition through slowing of the reaction kinetics. The vacuum method enjoys wide commercial usage. The use of over-pressure sintering (i.e., sinter HIP) and hot isostatic pressing (HIP) are significantly influencing the production of cemented carbides for specialty applications.

Sintering of cemented carbides consists of removal of pressing lubricant, oxide reduction, densification, and microstructure development. Each of these phases is discussed below for hy-

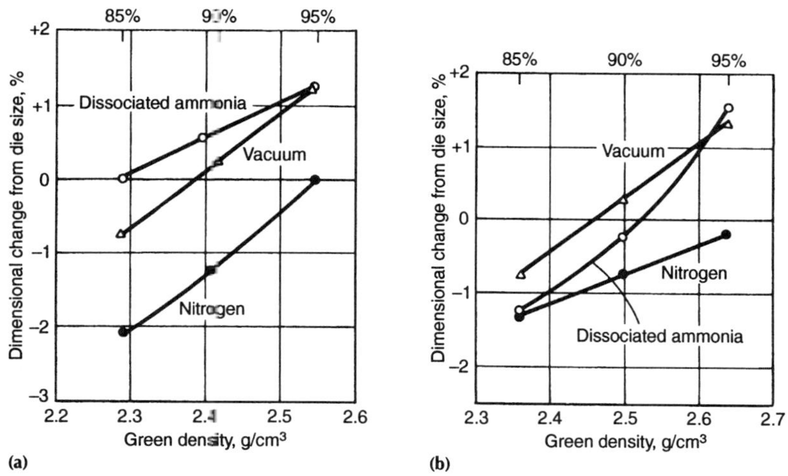

Fig. 41 Effect of green density and atmosphere on sintered dimensions. (a) 601AB. (b) 201AB

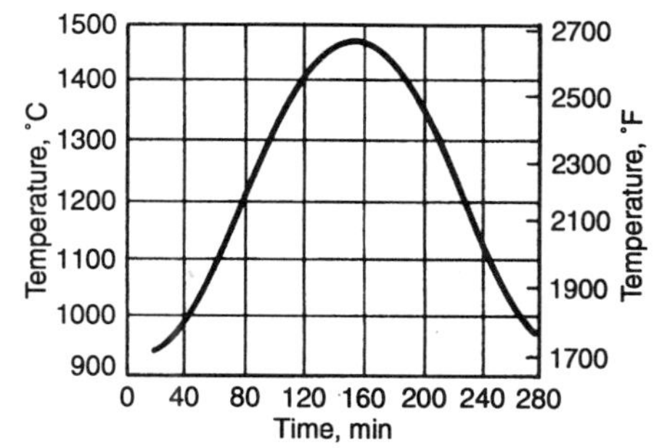

Fig. 42 Typical time-temperature cycle for hydrogen stoking furnace

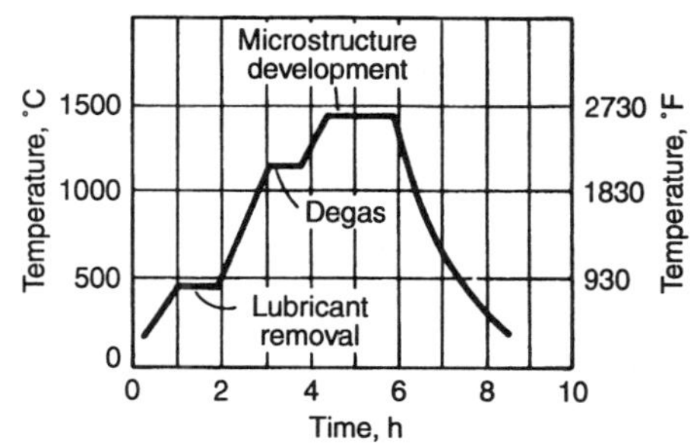

Fig. 43 Cemented carbide vacuum sinter time-temperature cycle

drogen and vacuum sintering modes, sinter HIP, and the effects of HIP.

Hydrogen Sintering. During hydrogen sintering, an atmosphere of hydrogen provides a reducing environment which, when subsequently modified by reactions with the furnace walls and carrier devices, provides the correct carburizing and oxidizing potentials to maintain thermodynamic equilibrium with the cemented carbide. In conventional hydrogen sintering, carbon content of the carbide powder is adjusted to the nominal theoretical value, and the hydrogen atmosphere subsequently maintains that value throughout the sintering cycle. For example, a 94WC-6Co composition may enter the furnace with a carbon content of ~5.70 to 5.80 wt% and exit the furnace with 5.76 ± 0.04%.

The atmosphere control capability of the hydrogen process is adequate for straight tungsten carbide/cobalt grades, but the oxidizing potential usually is too high to provide high-quality compositions containing titanium carbide, tantalum carbide, and/or niobium carbide additions suitable for steel machining. Such compositions usually are sintered in vacuum to achieve lower oxygen contents. Hydrogen sintering frequently is accomplished by mechanically stoking the parts through the furnace; a presintering furnace is required to remove pressing lubricant to prevent contamination from evolved lubricant vapors during the high-temperature stage of sintering. Presintering can also be used to partially sinter parts to green strengths that permit green forming operations, such as lathe turning or drilling. Peak presintering temperatures range from 500 to 800 °C (930 to 1475 °F), depending on the level of green strength desired.

Figure 42 shows a typical time-temperature curve for a mechanically stoked hydrogen sintering run. The combined hot zone profile and stoking rate resulted in a heating rate of ~6 °C/min (11 °F/min) up to 1350 °C (2460 °F), holding above 1350 °C (2460 °F) for 75 min, and a peak temperature of ~1460 °C (2660 °F).

Vacuum Sintering. Vacuum sintering is the predominant industrial process for sintering cemented carbides. Compared to hydrogen atmosphere sintering, vacuum sintering has several major advantages. This process allows superior control of product composition. At pressures of 1.3 to 133 Pa (10^{-2} torr to 1 torr), the rate of carbon and oxygen exchange between the atmosphere and the cemented carbides is very low. the main factor controlling composition is the oxygen content of the carbide powder, not the rate of reaction with the atmosphere.

Because of the low oxidizing potential of the vacuum sintering atmosphere, the vacuum process is superior to the hydrogen process for the sintering of titanium carbide, tantalum carbide, and niobium carbide compositions that are sensitive to oxidizing potential. While capable of attaining low oxidizing potentials, most commercial hydrogen sintering furnaces have relatively high oxidizing potentials due to air leakage and reactions with ceramic furnace parts and fixturing.

Secondly, vacuum sintering offers greater flexibility and control of the thermal cycle, particularly the heating cycle. This is due to the fact that vacuum sintering is a batch process, while most hydrogen sintering is a continuous process. Slow heating rates are essential, usually followed by an intermediate temperature hold to obtain optimum product quality when titanium carbide, tantalum carbide, and niobium carbide additives are used in the carbide composition.

This allows sufficient time for the carbon-oxygen reactions to occur; a sufficient open-pore network exists to allow escape of the carbon monoxide gas that forms. If the temperature is raised too rapidly, gas becomes entrapped in the structure, thus forming pores. The heating cycle can be adjusted easily for vacuum furnaces; for mechanically stoked hydrogen furnaces, however, it is fixed by the hot zone design and the peak temperature time required to obtain the desired microstructure.

Conventional vacuum sintering offers low operating cost. Previous vacuum equipment used relatively small induction-heated furnaces that were characterized by high power consumption, long cooling times, and the need for separate lubricant removal prior to sintering. Commercial vacuum sintering furnaces currently cost less to operate than hydrogen furnaces due to:

- Large batch size
- Resistance heating
- Forced gas cooling
- Built-in lubricant removal

Figure 43 shows a typical vacuum process time-temperature cycle. The most significant features are three temperature holds—the first at ~500 °C (930 °F) for lubricant removal, the second at ~1200 °C (2190 °F) for degassing, and a third at 1400 to 1500 °C (2550 to 2730 °F) for microstructure development.

Sinter Hot Isostatic Pressing. Overpressure sintering, more commonly referred to as pressure sinter or sinter HIP consists, in general, of a sintering furnace contained in a vessel of intermediate pressure capability. Argon pressures in the range of 1. 5 to 10 MPa are employed at sintering temperature to reduce, or eliminate remnant porosity by the application of isostatic pressure to closed porosity parts. A single cycle includes delubing, oxide reduction, and sintering, followed by low pressure HIP with chamber pressure being increased to a higher level only after the carbide has sintered to the closed porosity range. The closure of clean porosity and macrovoids has been reported for straight WC-Co grades, that is, 3 to 6 wt% Co in the temperature range of 1420 to 1460 °C (2590 to 2660 °F) utilizing sinter HIP pressures of ~0.2 MPa, while temperatures in the 1430 to 1480 °C (2610 to 2700 °F) range yielded similar results for grades containing cubic carbides (Ref 41). The cost of single cycle sinter HIP units is substantially

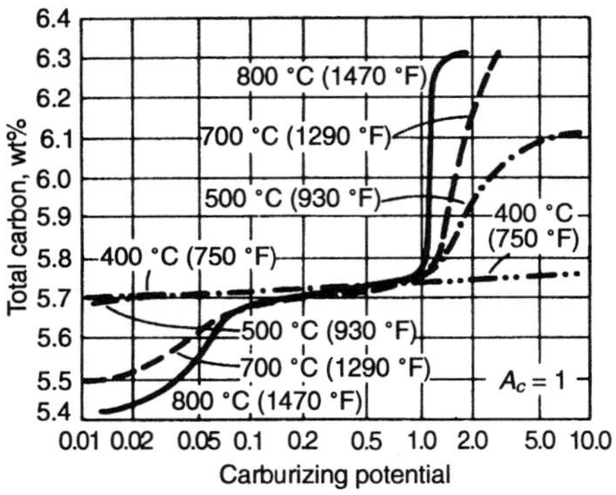

Fig. 44 Effect of carburizing potential of hydrogen-methane mixtures on carbon content of WC-6Co. 1 h hold; 1 atm pressure. A_c is the thermodynamic carbon activity. Total carbon content is the sum of the compounded carbon in the carbide phase plus any graphite present. It does not include carbon present in the lubricant.

Table 20 Atmosphere compositions required to maintain neutral carburizing potentials for tungsten carbide/cobalt as a function of temperature and total oxygen content

Temperature °C	Temperature °F	Carbon activity	Hydrogen, %	Carbon monoxide, %	Water, ppm	Carbon dioxide, ppm	Methane, %
1450	2640	0.6–0.25	99.944–99.976	None	None	None	0.056–0.023
1450	2640	0.6–0.25	99.45–99.47	0.5	3.4–8	4.7×10^{-3}–1.1×10^{-2}	0.055–0.023
1450	2640	0.6–0.25	89.96–89.98	10	60.0–144.0	1.7–4.0	0.045–0.019
1200	2190	0.6–0.25	99.84–99.93	None	None	None	0.24–0.10
1200	2190	0.6–0.25	99.27–99.40	0.5	12.2–29.2	3×10^{-2}–7.2×10^{-2}	0.24–0.10
1200	2190	0.6–0.25	89.79–89.87	10	220.0–570.0	12.2–29.2	0.19–0.08
1000	1830	0.5–0.2	99.56–99.82	None	None	None	0.44–0.18
1000	1830	0.5–0.2	99.05–99.29	0.5	114.0–290.0	0.34–0.85	0.44–0.18
1000	1830	0.5–0.2	89.43–89.34	10	2080.0–5200.0	135.0–338.0	0.36–0.14

At pressure of 1 atm

more expensive than conventional vacuum sinter units of similar capacity.

Hot isostatic pressing is completed as a post-sintering operation in a specially designed high-pressure vessel usually employing argon at pressures in the range of 100 MPa and a temperature in the same general range as that used in sintering. It is normally employed as a separate operation on cemented carbides after normal sintering has already removed most of the remnant porosity. Hot isostatic pressing can also be applied, in certain cases, to clad green preforms, that is, presintered product. Hot isostatic pressing units represent major capital investments, and their use as a post sintering operation adds to materials handling costs, utilities usage, and production lead times (Ref 42). The application of either sinter HIP or HIP to the manufacture of cemented carbide materials employing the appropriate time/temperature/pressure criteria results in increases in strength relative to both hydrogen or vacuum sintered product. This strength increase is greater for (a) fine-grain sized materials, (b) low cobalt grades, that is, 3 to 8 wt% Co, and (c) grades containing low additions of "cubic carbides" (Ref 43).

Lubricant Removal

Pressing lubricants used for cemented carbides range from low levels (0.5 to 2.5 wt%) of simple paraffins for pressed parts up to high levels of various synthetic polymers for extruded, injection molded, and slurry cast parts. The heating rate used for lubricant removal in a hydrogen atmosphere depends on the type and amount of lubricant used and the size of the pressed part. Heating rates as high as 15 to 20 °C/min (25 to 35 °F/min) to 500 °C (930 °F) can be used for pressed cutting inserts containing 1 to 2% paraffin. Much slower rates (0.5 to 4 °C/min, or 0.9 to 7 °F/min) are required for larger parts or for parts with higher lubricant contents.

Commonly used lubricants, such as paraffin, evaporate readily in vacuum at 100 to 250 °C (210 to 480 °F). Consequently, heating rate should be controlled in this temperature range to prevent excessive pressure within the compact due to evolving lubricant vapors. A condenser is used in the vacuum line to remove the portion of the lubricant vapor that condenses at ~10 to 90 °C (50 to 195 °F). The interior walls of the furnace are sometimes used to condense paraffin vapor to liquid, which is subsequently drained to the bottom of the furnace to enter the main condenser. Other furnaces employ carrier gases such as H_2, argon, or N_2 to completely remove paraffin vapors to external condensers. Carbowax is another common lubricant used in lieu of paraffin and is removed under atmospheric H_2.

Carbon Control

Carbon content of cemented carbides must be controlled to within very narrow limits, typically ± 0.04 wt%, to prevent formation of brittle lower carbides, such as the eta phase in tungsten carbide alloy systems on the low carbon side and carbon precipitates on the high carbon side.

Because carbon content can be changed substantially during sintering by reactions with oxygen-containing phases in the powder and by carbon exchange reactions with the furnace atmosphere, control of the furnace atmosphere carbon potential and the oxygen content of the powder is essential to produce high-quality P/M sintered parts.

During hydrogen presintering, reactions between the hydrogen atmosphere and cemented carbides are relatively sluggish at the lower end of the peak temperature range (400 to 600 °C, or 750 to 1110 °F), but reaction rates are sufficiently high at temperatures >600 °C (1110 °F) to cause significant changes in the carbon content of the part if the atmosphere is not properly adjusted.

This phenomenon is illustrated in Fig. 44, which shows the effect of varying the carburizing potential of the hydrogen atmosphere by methane additions on the carbon content of a 94WC-6Co composition held 1 h at the peak presinter temperature. The carburizing potential can be described as:

$$\frac{1}{k} \cdot \frac{CH_4}{H_2^2}$$

where k is the equilibrium constant for the reaction $2H_2 + C \leftrightarrow CH_4$, and CH_4/H_2^2 is a partial pressure ratio.

At temperatures up to 500 °C (930 °F), reactions causing carbon loss are too slow to noticeably change compositions. At 700 and 800 °C (1290 and 1470 °F), however, large carbon losses occur when the carburizing potential (relative to pure carbon) is less than ~0.05.

Thus, even pure hydrogen can cause large carbon losses above ~600 °C (1110 °F). When the carburizing potential exceeds 1.0 (saturation value), large carbon increases occur at 500 °C (930 °F) and above, thus allowing buildup of the carbon-containing volatilized lubricant in the peak temperature portion of the furnace.

Above ~900 °C (1650 °F), hydrogen-base atmospheres can change the carbon content of cemented carbides substantially by:

$$3WC + 3Co + 2H_2O \pm W_3CO_3C + 2CO$$
$$+ 2H_2WC + Co + CH_4 \pm WC + CO + C + 2H_2$$

To prevent the formation of undesirable carbon reaction phases, the carburizing potential (carbon activity) of the atmosphere must be carefully controlled. Table 20 shows experimentally determined activities required to maintain the desired carbon contents in tungsten carbide/cobalt cemented carbides at 1000 to 1450 °C (1830 to 2640 °F), with examples of typical gas compositions required to provide such carbon activities (Ref 44). The three major constituents of cemented carbide sintering atmospheres are hydrogen, carbon monoxide, and methane.

Typical commercial practice for hydrogen sintering carbon control is based on partial reaction of the incoming gas (hydrogen plus impurity levels of water vapor and oxygen), which has zero carbon activity. Graphite fixturing is used to carry the carbide parts through the furnace to increase the atmosphere carbon activity. Fixturing may thus contain openings that are empirically sized to react the incoming gas to the proper degree with the graphite to obtain the desired carbon activity.

Alternately, the graphite carrier may completely enclose the parts, thus causing carbon activity to increase to near unity. Modifiers such as aluminum oxide sand are included within the carrier to react with the atmosphere to thus lower the carbon activity to the desired level. A com-

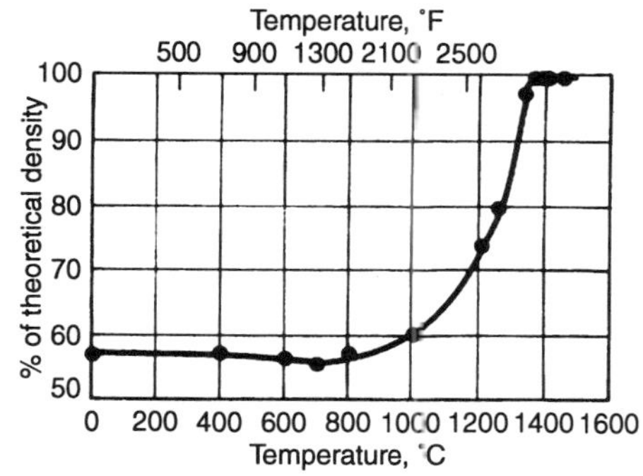

Fig. 45 Density-temperature relationship for WC-10Co. 30 min hold at each point

mon practice uses a packing mixture of aluminum oxide sand and a small amount of carbon that surrounds the cemented carbide parts inside graphite carriers. By adjusting the carbon addition, carbon activity can be adjusted to the desired level.

Because the rates of carbon and oxygen exchange between the atmosphere and cemented carbides are low during vacuum sintering, the final carbon content of the cemented carbide when vacuum sintered is determined primarily by the initial carbon content of the powder and the amount of carbon lost during heating due to reactions between carbon and chemisorbed oxygen in the powder. Much of the oxygen present as oxides of tungsten and cobalt reacts with carbon in the powder during the vacuum heating cycle and evolves as carbon monoxide and carbon dioxide gases, whereas in hydrogen sintering, those oxides are reduced by hydrogen and evolve as water vapor. Studies have shown that the cobalt binder phase is reduced of chemisorbed oxide between 550 to 650 °C (1020 to 1200 °F); tungsten oxides are reduced around 900 °C (1650 °F), while reduction of the oxides of the cubic carbides (Ta_2O_5 and TiO_2) occurs at 1000 °C (1830 °F) (Ref 45). When titanium carbides and tantalum carbides are present, much of the oxygen contained in these compounds is also removed by carbon reduction. To compensate for this carbon loss, vacuum-processed powder usually contains about 0.1 to 0.3 wt% added carbon than is used for hydrogen-processed powder.

Densification

During hydrogen sintering, densification of pressed cemented carbide parts usually is accomplished by mechanically stoking through a muffle-type electrically heated furnace held at a peak temperature of ~1400 to 1500 °C (2550 to 2730 °F). A densification curve for WC-10Co composition (Fig. 45) shows that significant densification begins at ~1000 °C (1830 °F) and that greater than 90% theoretical density can be obtained at 1300 °C (2370 °F). Liquefaction of the cobalt phase occurs between 1280 and 1350 °C (2335 and 2460 °F) as a result of the formation of a tungsten carbide/cobalt eutectic or eutectic-like phase (approximately 35 wt% tungsten carbide that forms during heating, as discussed in Ref 46). Essentially, the part becomes fully dense soon after the liquid phase forms.

During densification, the pressed part undergoes a volumetric shrinkage of ~40 to 45% (16 to 18% linear shrinkage). Complete densification is readily accomplished for typical cemented carbide compositions that contain ~3 to 25 wt% Co (5 to 40 vol%). Complete densification is more difficult to achieve for cobalt contents of <3%. For cobalt contents greater than ~25%, part shape is difficult to maintain due to the presence of high liquid-phase fraction, which allows the part to sag.

Densification proceeds during vacuum sintering at about the same rate as it does for hydrogen sintering, reaching nearly 100% of theoretical density soon after the cobalt or nickel phase liquefies at 1280 to 1350 °C (2385 to 2460 °F); the exact temperature of liquefaction depends on the carbon content. When titanium carbide, tantalum carbide, and/or niobium carbide are present in the cemented carbide, the carbon-oxygen degassing reactions must be completed before densification closes the pores. If this does not occur, the evolving gases become entrapped in the interior of the compact, thus causing high residual porosity. Holding at ~1100 to 1250 °C (2010 to 2280 °F) for ~30 to 90 min ensures proper degassing of titanium carbide, tantalum carbide, and niobium carbide compositions.

Microstructure

At the beginning of peak temperature holding, the microstructure is nearly fully dense, and the carbide particles are nearly unchanged in size and shape from the original milled powder state. They are relatively small, irregularly shaped, and poorly dispersed, tending to agglomerate. The cobalt phase is also poorly dispersed, and many pools or lakes may be present (Fig. 46a). A noticeable amount of residual porosity also exists (Fig. 46a).

The main purpose of the final stage of the sintering operation is to develop the microstructure by holding at a temperature above the cobalt melting point for a time sufficient to develop a more uniform carbide structure with good cobalt phase dispersion and minimal residual porosity (Fig. 46b). This is usually accomplished by holding for 30 to 90 min above 1350 °C (2460 °F), reaching a peak temperature at ~1400 and 1600 °C (2550 and 2910 °F). During this period, the cobalt phase, driven by capillary forces, disperses more evenly. This process also improves carbide particle distribution.

Carbide distribution is further improved during holding by the dissolution of small particles into the liquid phase, with subsequent precipitation onto the larger particles during cooling. This results in a gradual increase in average particle size (Fig. 47a). Increasing the sintering temperature has a similar influence on grain growth (Fig. 47b).

One undesirable consequence of grain growth by the solution and precipitation process is the tendency of large tungsten carbide grains to grow at a disproportionately high rate. This discontinuous grain growth (Fig. 46c) occurs more readily at lower cobalt contents (3 to 6 wt%) with finer average particle size mixes. It is most pronounced when tungsten carbide is the only carbide phase present. Small additions (0.1 to 0.5%) of group VB carbides (vanadium carbide, niobium carbide, and tantalum carbide) undergo

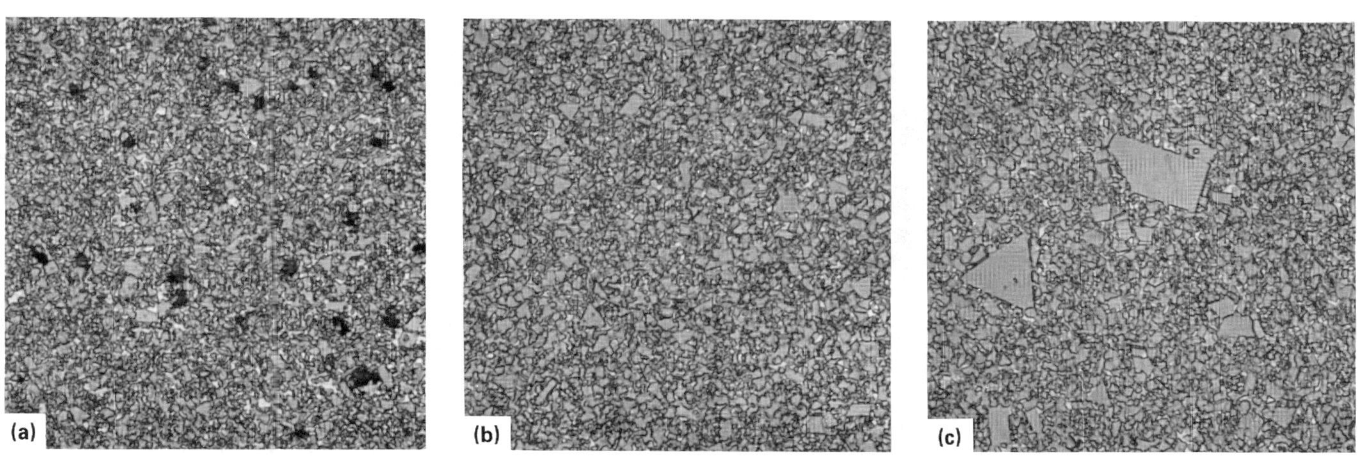

Fig. 46 Microstructure of WC-6Co. (a) Undersintered. (b) Normal structure. (c) Discontinuous grain growth

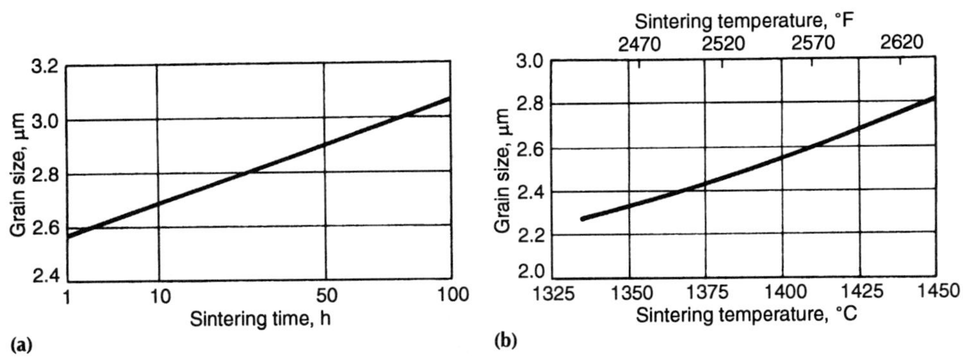

Fig. 47 Effect of sintering on grain growth of WC-25Co compacts. (a) Effect of sintering time at 1400 °C (2550 °F). (b) Effect of sintering temperature (1 h hold)

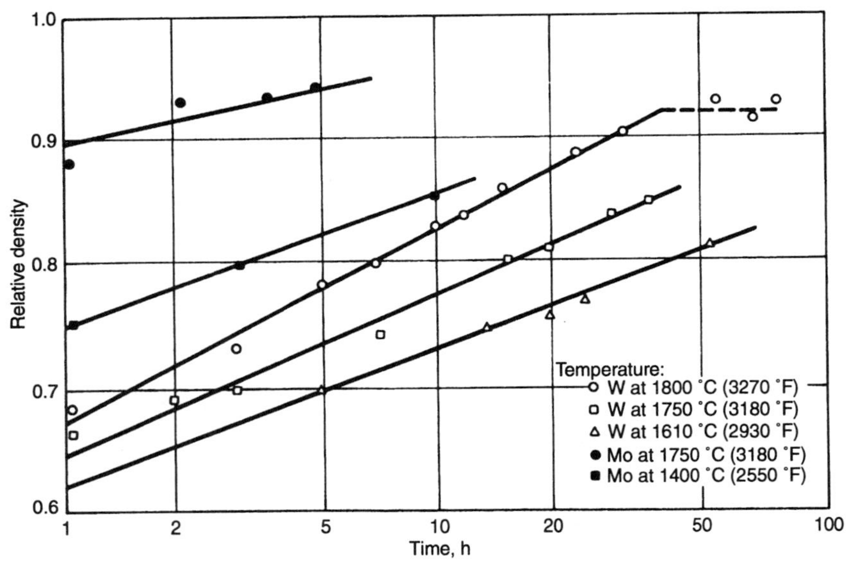

Fig. 48 Relative density versus the logarithm of sintering time for 4 µm tungsten and molybdenum powders

significant grain growth in fine-grained tungsten carbide/cobalt compositions. Titanium carbide is also a strong grain growth inhibitor (Ref 46), as well as chromium carbide.

During vacuum sintering, cobalt losses by evaporation should be controlled. If uncontrolled, as much as 10 to 20% of the cobalt content of the part may be lost, thus resulting in a loss of mechanical strength and the formation of a rough, coarse-grained surface structure caused by the precipitation of tungsten carbide from the evaporating cobalt (Ref 47 and 48). Cobalt evaporation can be minimized by completely enclosing the pressed parts in graphite fixturing, the walls of which are maintained at the sintering temperature.

The enclosure causes a buildup of cobalt vapor pressure around the cemented carbide parts, which reduces the evaporation rate. Cobalt losses can also be controlled by operating at pressures of at least 66 Pa (0.5 torr), rather than in the range of 1.3 to 13 Pa (0.01 to 0.1 torr) that mechanical vacuum pumps and blowers are capable of maintaining. Higher pressures can be maintained by placing a throttling valve between the furnace and the pump or by injecting an inert gas into the furnace.

Sinter HIP results in the closing of pores and macrovoids with microstructure typical of the surrounding material. Similar results can be obtained through the use of HIP provided that judicious selection of time and temperature is employed in conjunction with HIP pressure for the specific binder level of the material undergoing post-sintering treatment. Early attempts at HIP resulted in the extrusion of binder material into the pores and macrovoids once the melting point had been reached and the formation of cobalt "lakes" or "pools" in the microstructure (Ref 49).

Cooling

After holding at sintering temperature, loads containing small parts can be cooled to room temperature in ~2 to 4 h by circulating inert gas (argon or helium) through the load at or near atmospheric pressure. Cooling of large parts is prolonged to minimize thermal stresses. When all significant material and process parameters, such as powder preparation, chemical composition, heating rate, holding times and temperatures, and cooling procedures, are properly adjusted, the as-sintered surfaces of all pieces in the load achieve the same composition and structure as the interiors of the pieces. This eliminates scrap or rework associated with compositional imbalances and reactions with the furnace atmosphere.

Sintering of Tungsten and Molybdenum

Richard Cheney, GTE Products

Tungsten and molybdenum sintering practices are similar; consequently, they will be described together in this section. The achievement of high density is a primary goal when sintering tungsten and molybdenum, because high density is frequently required for performance. It is also required if metalworking operations are necessary, in which case at least 90% of theoretical density is essential. High-density values indicate that compact shrinkage has been significant, because pressed densities are seldom >75% of theoretical. In addition to ensuring high density, purification occurs during sintering, and the resultant microstructure is established.

Sintering Variables

Several factors influence the sintering of tungsten and molybdenum, including time, temperature, particle size, atmosphere, purity of powders and compacts, compacted density, thermal gradients, heating rate, cooling rate, particle size distribution, compact weight, gravity, and friction, and additives such as dispersed oxides and sintering activators. These variables and their interrelationships are discussed in the following sections.

Time and temperature are the fundamental determinants of the sintering kinetics for tungsten and molybdenum—in addition to particle size, which is discussed below. Mass transport by diffusion is very slow for refractory metals at most conventional sintering temperatures that are economically or technically feasible. Therefore, the effects of time, temperature, and particle size are more significant for these metals than for metals where temperatures up to and exceeding their melting points are readily attainable.

Figure 48 shows sintering rates, in terms of relative density, achieved for tungsten and molybdenum at temperatures from 1400 to 1800 °C (2550 to 3270 °F) for particle sizes of 4 µm. The temperature of 1800 °C (3270 °F) is a practical limit for furnaces that use molybdenum or tungsten heating elements and aluminum oxide refractory materials. Note the long sintering time (50 h at 1800 °C, or 3270 °F) required to attain

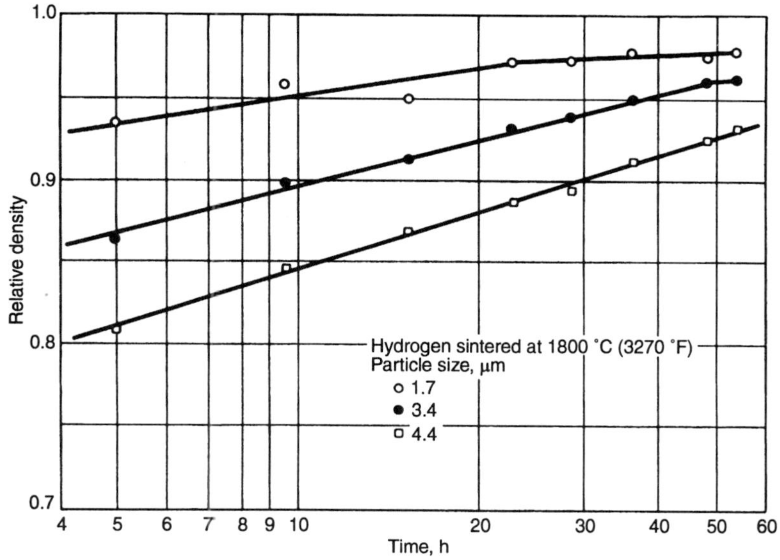

Fig. 49 Relative density versus the logarithm of sintering time for several tungsten powder sizes

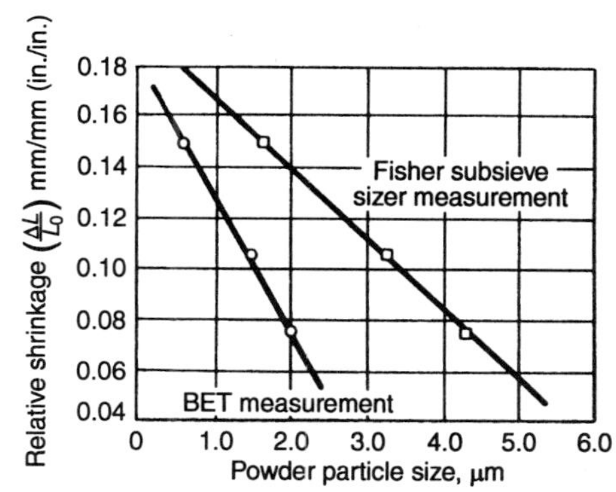

Fig. 50 Relative shrinkage versus tungsten powder particle size. Relative shrinkage ($\Delta L/L_0$) is the change of length over the original length.

92% density for tungsten. At the same temperature, molybdenum sinters to a similar density in 1 to 2 h.

Powder particle size is extremely important to the sintering of tungsten, but less so for molybdenum. Figure 49 shows the effect of particle size on the density of tungsten sintered in hydrogen at 1800 °C (3270 °F). Figure 50 compares the effect of several particle sizes on the sinterability of tungsten, as expressed by the relative shrinkages of compacts pressed to the same starting density and sintered at 1800 °C (3270 °F).

Induction and tungsten mesh heating can be used to attain temperatures up to 2400 °C (4350 °F). At these temperatures, sintering can proceed to high density in 1 to 2 h if other factors are not rate limiting. These other factors include heating and cooling rates and billet outgassing.

Self-resistance heating permits attainment of temperatures >3000 °C (5430 °F), and complete sintering cycles require only 20 to 30 min, depending on compact size. Compact size is limited by the available electrical power; the largest practical compact has a maximum cross section of 645 mm^2 (1 in.^{2}) and ranges from 610 to 915 mm (24 to 36 in.) in length.

Atmosphere. Tungsten and molybdenum are always sintered in a hydrogen atmosphere. Vacuum sintering generally is unsuccessful. The only exception to this rule is initial exposure of compacts to hydrogen up to ~1200 °C (2190 °F) before furnace chamber evacuation. Tungsten and molybdenum are highly reactive with carbon and oxygen at almost all sintering temperatures, and the presence of either is usually catastrophic to sintering and billet properties.

Dissociated ammonia generally can be used as a source of hydrogen with no detrimental effects. Nitride formation appears nonexistent for either pure molybdenum or tungsten. However, in some applications, entrapped or dissolved nitrogen may cause a reduction in properties in worked rods and wire.

Powder and Compact Purity. Impurities greatly affect the sintering of tungsten and molybdenum. Because tungsten and molybdenum are sintered at such high temperatures, most impurities are molten, and many exhibit high vapor pressures. Other impurities, however, may react with the sintering atmosphere to form new compounds, while some may alloy with the base metal—all leading to undesirable influences on properties.

Tungsten and molybdenum powders are fine, typically with particle diameters of 1 to 10 μm and have 0.1 to 0.2% of adsorbed gases (oxygen, nitrogen, and water vapor) on particle surfaces. Entrapment of gases prevents achievement of high density. Additionally, densification does not proceed satisfactorily until surfaces are purged of these gases. Heating and holding periods of sintering schedules should be designed to allow sufficient time for entrapped gases to be released, particularly for large billets. An adequate hold at ~1200 °C (2190 °F) ensures uniformly low gas concentrations and thus uniform density and microstructure. Large compacts and loads require significantly longer hold times.

Control of Carbon Content. Typically, carbon should be avoided. Carburization begins at 800 °C (1470 °F) for molybdenum and at 1100 °C (2010 °F) for tungsten. Small amounts of carburization greatly inhibit densification and may greatly affect the ductility and/or workability of the part. Common carbon sources include vacuum pump oils and heating elements.

Metallic elements are used as sintering activators, or they may form alloys or oxides. As liquids, metallic elements such as nickel, cobalt, and iron activate localized grain growth and may leave large voids where the contaminant was originally located. As alloys or oxides, they usually retard local grain growth and frequently inhibit sintering. Generally, contaminants make uniform densification impossible and further working difficult.

Compacted Density. Generally, higher compacted densities are desirable, as long as entrapped gases do not result. Figure 51 shows the effect of cold isostatic compaction pressure on the compacted density of three tungsten powders of different sizes. Figure 52 shows the effect of higher compacted density on the sintered density for a 4.4 μm tungsten powder after various sintering times in hydrogen at 1800 °C (3270 °F). Higher compacted density significantly shortens the sintering times needed to produce a given density. From a practical viewpoint, pressures greater than 517 MPa (75 ksi) are economically difficult to obtain, and compacted billet densities therefore usually range from 65 to 75% of theoretical.

Thermal gradients and heating rates are particularly important when sintering large furnace loads, which is frequently the case with tungsten and molybdenum. Because sintering rates are so sensitive to temperature, thermal gradients greatly affect the density attained in a given time. This in turn affects property and dimensional control. From empirical rate equations, it can be calculated that sintering at 1775 °C (3225 °F) rather than at 1825 °C (3315 °F) will require 110 h (4.6 days) rather than 40 h (1.66 days) to reach 92% density, a difference of 275%.

Temperature gradients are particularly detrimental when sintering large compacts such as 130 mm (5 in.) diam by 250 mm (10 in.) long tungsten billets for forging. At 1800 °C (3270 °F), radiation is the predominant mode of heating, and billet surfaces reach sintering temperatures rapidly compared to billet interiors. Sintering thus begins sooner at the billet perimeter; if uncontrolled, a thin high-density shell forms, which effectively restricts further densification of the interior of the billet.

Furnace heating rate is the primary method of controlling exterior shell formation. Proper choice of powder particle size is also important—coarser sizes decrease the effect of temperature gradients at the expense of lowering the initial rate of sintering.

The rate of outgassing of impurities is also related to temperature gradients in large com-

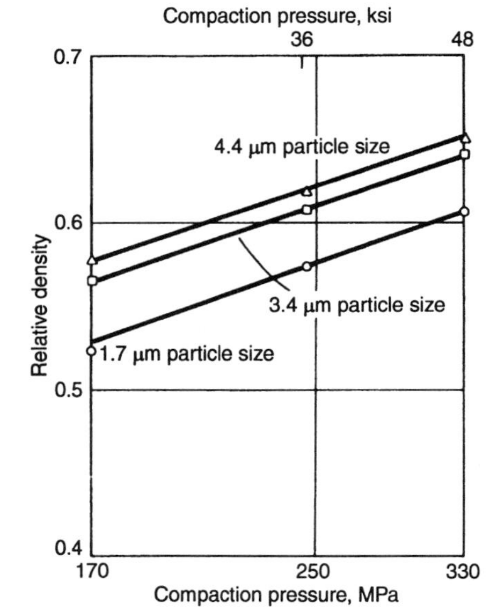

Fig. 51 Relative compact density versus the logarithm of compaction pressure for tungsten powders of various sizes

pacts. Surface outgassing proceeds more rapidly than outgassing from interior volumes. The inhibiting effect of these gases allows surface volumes to sinter more rapidly, thus creating a high-density shell. Proper control of furnace heating prevents shell formation.

Cooling rate is particularly important with very large parts (305 mm, or 12 in. diam or more). Excessively rapid cooling causes cracking due to thermal stresses.

Particle size distribution is of secondary importance to the sintering variables mentioned previously. It is also difficult to quantify these effects to provide general guidelines. Consequently, they will not be discussed in this section.

Compact weight, gravity, and friction are important considerations when large or heavy parts are being sintered. Tungsten cylinders weighing 23 kg (50 lb) or more shrink less at their bases than at their tops when sintered on end. This tendency is mainly due to restricting friction forces. A 12.7 mm (0.5 in.) thick pedestal of compacted tungsten placed under the cylinder alleviates this problem.

Gravity is a significant factor in only a few sintering applications. Rounds may become slightly elliptical when sintered on their sides. Rods suspended from their top ends shrink more across top sections and less along lengths because of their suspended weight. These shrinkage disparities diminish as the bottom of the rod is approached.

Furnace Loading. The use of radiative heating necessitates loading of furnaces to minimize nonuniform heating. Heating rates are important. Parts partially shielded by other parts heat slower and with less uniformity. Parts partially buried in a refractory sand also experience nonuniform heating.

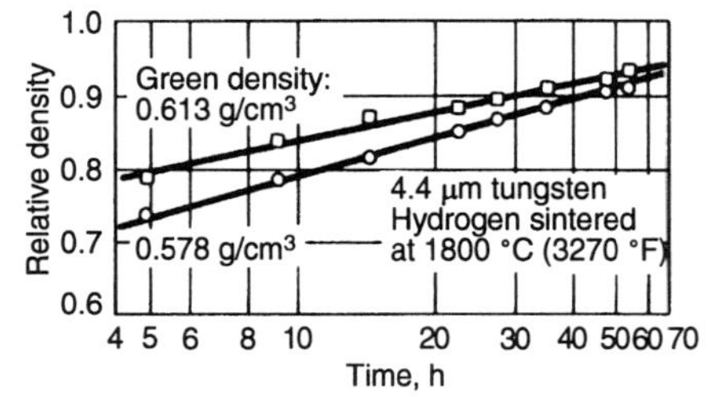

Fig. 52 Relative density versus the logarithm of sintering time for tungsten powders compacted at different pressures

Oxide-Dispersed Additives. Typically, these additives, or dopants, may be thorium dioxide or the so-called KAS dopant, which is a combination of potassium oxide, aluminum oxide, and silicon dioxide. Their presence inhibits sintering, which can be useful in promoting outgassing of other impurities, in minimizing the effects of thermal gradients, and in controlling microstructure. The silicon dioxide and KAS compounds usually vaporize when sintering occurs at ≥2000 °C (3630 °F).

Sintering Activators. The sintering of tungsten and molybdenum can be activated to a considerable degree, but at the expense of significant losses of ductility and workability (Ref 50). Nickel additions to tungsten ranging from 0.2 to 0.5% result in densities of >90% of theoretical in only several hours at temperatures as low as 1100 °C (2010 °F), compared to 50 h for unactivated tungsten at 1800 °C (3270 °F). Molybdenum is also activated, but to a lesser extent. The activating elements for both tungsten and molybdenum include palladium, nickel, cobalt, iron, rhenium, ruthenium, iridium, and osmium.

Effects on Microstructure. Sintered grain size is primarily a function of sintered density and starting powder size, as shown in Fig. 53 for tungsten. Grain growth typically proceeds at a more rapid rate above about 95% density, as shown by the lower line. This is due to the decreased effectiveness of grain boundary pinning by the pores. Once the grain boundaries break away from the pores, sintering rates decrease significantly.

The workability of compacts of tungsten and molybdenum is highly dependent on sintered density and, to a lesser extent, on grain size. Densities should be a minimum of 90% of theoretical, and values ranging from 92 to 95% are more desirable. Typical grain sizes are 10 to 30 μm.

Sintering Equipment

Self-resistance heating has been used extensively for tungsten because of the extremely high temperatures required for practical sintering. Typically, rods with cross sections less than 645 mm^2 (1 in.^{2}) are heated to 3000 °C (5430 °F) or higher. Densification to greater than 90% of theoretical can occur in less than ½ h.

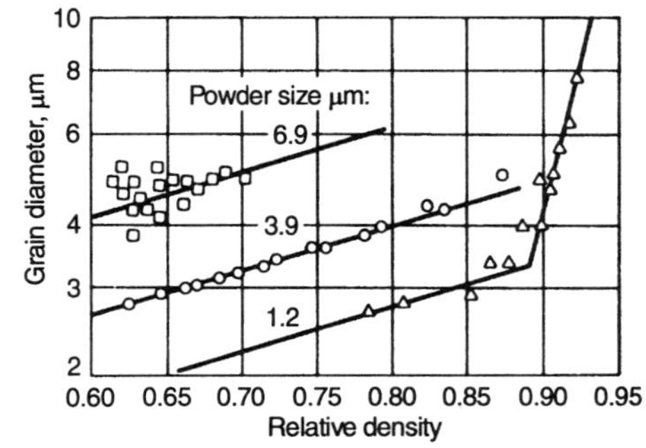

Fig. 53 Logarithm of grain diameter versus relative density for tungsten powders of various sizes. Data cover a range of sintering temperatures (1600 to 1800 °C, or 2910 to 3270 °F)

Sintering is done under a hydrogen atmosphere in a water-cooled, copper-jacketed enclosure. Tungsten contacts are required, and one contact must be able to move as the bar shrinks. Bar cross sections must be uniform or sintering temperature will vary along the bar length. Temperatures and densities are always low within several inches of the clamped ends because of the heatsink effect of the contacts. Sintering of molybdenum by this method has been replaced by resistance and induction heated furnaces.

Resistance element heating furnaces are frequently used for hydrogen sintering of molybdenum and tungsten at 1800 °C (3270 °F). Furnaces having uniformly hot zones up to 1.2 m (4 ft) long and 0.14 m^2 (1.5 ft^2) are available. Loads of up to several thousand pounds can be sintered.

A typical furnace consists of water-cooled entrance and exit zones and a central zone of alumina refractory brick that supports molybdenum heating element rods. Furnace loads are supported on molybdenum slabs that are conveyed through the entrance and exit zones by powered rolls. The load is moved through the hot zone by a walking beam that oscillates above and below the hearth and that carries the load forward during the portion of the cycle that is above the hearth. Varying the frequency of oscillation varies the rate of forward movement of the parts in the furnace. Smaller furnaces with lighter loads can be operated by mechanically stoking a series of boats through the furnace.

Induction radiant heating is used when temperatures above 1800 °C (3270 °F) are required and also when very large (greater than 405 mm, or 16 in. diam) parts are involved. Molybdenum alloy billets weighing up to 4500 kg (10,000 lb) have been sintered with this type of heating mode.

A typical furnace consists of a large-diameter water-cooled chamber that contains a cylindrical susceptor ring of tungsten or molybdenum. The susceptor is inductively heated, and the compact is heated by radiation from the susceptor. Depending on the construction materials and loads, sintering temperatures ≤2400 °C (4350 °F) can be obtained.

Sintering of Tungsten Heavy Alloys

Tungsten heavy alloys are W-Ni-Fe or W-Ni-Cu alloys containing 90 to 98% tungsten. Typical nickel-to-iron ratios range from 1-to-1 to 4-to-1, while nickel-to-copper ratios generally range from 3-to-2 to 4-to-1. Minor amounts of other metals can also be added. Because of the liquid-phase sintering, these alloys achieve essentially full theoretical density. They exhibit good machinability and can exhibit substantial ductility. For example, a 90W-7Ni-3Fe alloy may exhibit elongation as high as 40%. Copper-containing alloys are less ductile and are more difficult to sinter to full density, especially in larger cross sections. One advantage of copper-containing alloys is that they are nonmagnetic. Primary uses for tungsten heavy alloys are as counterweights, radiation shielding, and kinetic energy penetrators.

Tungsten heavy alloys are made from blends of elemental powders. Homogeneity is of prime concern, so fine tungsten powders in the 2 to 8 μm range are used. Standard production procedures for tungsten powder yield fine particle sizes. Carbonyl powders are used in iron and nickel blends, and fine electrolytic powder is used in copper blends. As a precaution against impurities and agglomerates, powders are screened to –200 mesh prior to blending. Any standard powder blending procedure can be used.

Likewise, almost any P/M procedure for compacting powders can be used. If die pressing is used, a binder must be added to the powder. Like pure tungsten powder, high pressed densities cannot be achieved in tungsten heavy alloys. Consequently, allowance must be made for considerable shrinkage during sintering. A die-pressed piece typically is ~55% dense, while an isostatically pressed piece is ~65% dense.

Sintering Furnaces and Atmospheres. Sintering can be carried out in any furnace capable of sustaining the temperature and atmosphere required. Generally, the pieces are supported on aluminum oxide sand or aluminum oxide fixtures. Good control of temperature and heating rates is essential to produce consistent, high-quality tungsten heavy alloys.

Tungsten heavy alloys usually are sintered in pure hydrogen, but hydrogen and nitrogen mixtures, dissociated ammonia, or vacuum also can be used. Argon or mixtures of argon and hydrogen are not used, because the presence of argon during sintering results in a lower density product. With some tungsten powders, blistering of the heavy alloy parts occurs during sintering. The use of wet hydrogen through the sintering schedule usually corrects this problem.

Regardless of the atmosphere used during sintering, a cleanup step in hydrogen at ~1000 °C (1830 °F) is advised. For small pieces, only a few minutes at temperature are required; large pieces may require an hour or more.

During vacuum sintering, hydrogen cleanup is essential. For die-pressed pieces, hydrogen cleanup must be used to remove any residual carbon remaining from lubrication removal. If substantial carbon is still present when the liquid phase forms, serious density and mechanical property problems result.

Solid-State Sintering. At 1200 to 1430 °C (2190 to 2605 °F), it is possible, and sometimes desirable, to solid-state sinter W-Ni-Fe alloys to near-theoretical density. Solid-state sintering is not possible with W-Ni-Cu alloys because of the low melting point of copper. Once near-theoretical density is achieved, only minimal holding at liquid-phase temperatures is required to complete sintering.

Liquid Phase Sintering. During liquid-phase sintering, alloys densify as tungsten particles grow by dissolution and subsequent precipitation and coalescence of the tungsten grains. Tungsten particles grow from 3 to 8 μm to rounded grains 50 to 150 μm in diameter. As the tungsten content of the alloy increases, grain size tends to increase.

In W-Ni-Cu alloys, the low melting point of the copper and the wide liquidus temperature range of the matrix make it difficult to achieve full density in large parts. Normally, these alloys are sintered at 1380 to 1450 °C (2515 to 2640 °F). Very low heating rates that begin near the melting point of copper are required to properly sinter these alloys. Sintering times vary from less than an hour for small parts to several hours for large billets. One advantage of sintering these alloys is that they do not have the strong tendency to slump (distort) as do W-Ni-Fe alloys.

Liquid-phase sintering of W-Ni-Fe alloys is carried out at 1450 to 1600 °C (2640 to 2910 °F). Generally, sintering temperature increases with tungsten content. Heating rates are not as critical as with W-Ni-Cu alloys. With large parts, however, heat should proceed slowly to the temperature where the liquid phase first forms (1450 to 1560 °C, or 2640 to 2840 °F). Large temperature gradients within the part resulting from too rapid heating can cause compositional variations and porosity.

Sintering times are relatively short; 30 min at temperature is sufficient in most cases. Some large parts and alloys with tungsten contents >97% may require sintering times as long as 2 h. One limitation experienced in sintering of W-Ni-Fe alloys is slumping of the parts. As tungsten content decreases, slumping increases. For example, a 97% tungsten alloy experiences minimal slumping, while a 90% tungsten alloy may exhibit severe slumping. Slumping can be controlled by using the lowest sintering temperature possible and by supporting the parts during sintering.

Cooling. On cooling from liquid-phase temperatures, some problems may be experienced. If a long bar is directionally cooled, the composition of both the solute and the resulting solid are likely to change during solidification, thus causing end-to-end variations in composition. Gases released during solidification also can collect near the tail end of the bar, which is the last section to solidify.

If small pieces are cooled rapidly from the liquid phase, a solid shell forms around the part, thus causing the liquid matrix to be drawn from the center of the part in an effort to compensate for solidification shrinkage. Porosity is subsequently produced at the centers of thicker sections.

Heat Treatment. For maximum tensile elongation, W-Ni-Fe alloys must be heat treated after sintering. Typical heat treatment involves heating in nitrogen, vacuum, or inert atmosphere at 900 to 1300 °C (1650 to 2370 °F). Time varies with temperature and cross section. One hour at 1200 °C (2190 °F) is sufficient for a 12.7 mm (½ in.) diam bar, but a 50 mm (2 in.) diam bar requires ~24 h.

Mechanical properties of tungsten alloys vary greatly, depending on processing. For example, a MIL-T-21014B specification calls for 2% minimum elongation for a 95W-Ni-Fe alloy. However, if this alloy is properly processed, it exhibits a typical elongation of 28 to 32%.

As-sintered hardness and yield strength are affected only slightly by tungsten content. Typical hardness is 28 to 31 HRC, and yield strength ranges from 550 to 620 MPa (80 to 90 ksi). Elongation ranges from a maximum of 40% for a 90% tungsten alloy to 18% maximum for a 97% tungsten alloy.

Sintering of Titanium

Sintering practice (as opposed to hot pressing techniques) generally is applied only to fine titanium powders, which are the by-products of the production of titanium sponge. Such powders, referred to as "sponge fines," are produced by the reduction of titanium tetrachloride with sodium or magnesium. Particles are soft and irregularly shaped, thus well suited to cold compaction. Although sodium-reduced sponge fines are preferred for titanium P/M sintering in the United States, similar powders made by molten salt electrolytic reduction are reported in Russian literature.

Another class of titanium powder, prealloyed powder, is made from alloy ingot by ultraclean processes such as the plasma-rotating electrode process. Such powders are spherical and are not compactible at room temperature by ordinary means. These powders can, however, be sintered in loose form to produce a porous structure or coating. This section concentrates primarily on the sintering practices and resultant microstructures and property relationships for parts manufactured from sponge fine types of powders.

Titanium Powder and Compacts

Titanium is very reactive, forming very stable oxides and nitrides. Unlike many metals, titanium is capable of dissolving its own surface oxides at sintering temperatures. Consequently, these oxides do not have to be reduced through the use of a reactive atmosphere. The formation of additional oxides should be avoided; consequently, the use of high vacuum or high-purity inert gas atmosphere is common practice.

Titanium sponge fines used for P/M parts typically range from −80 to −100 mesh. Sodium-reduced sponge fines contain ~0.23% residual sodium chloride entrained in the particles. Although it is impossible to remove this residual salt by leaching, some salt usually is liberated at high temperatures and condenses in cooler parts of the sintering system.

Sponge fines can be compacted at room temperature in rigid dies or isostatically pressed in elastomeric tooling. Due to the reactive nature of titanium, admixed lubricants are not used, but die-wall lubrication is required with rigid dies. The irregular particle shape of sponge fines provides compacts with excellent green strength, and green densities can range from 60 to 85%, the latter at die pressures of 552 MPa (40 tsi) or isostatic pressures of 414 MPa (60 ksi).

When additional sintered strength is required, alloying powders can be elemental or master alloys; in the case of Ti-6Al-4V, for example, a master alloy powder of 60Al-40V is frequently used. Sponge fine powders exhibit strengthening from sintering at temperatures as low as 1000 °C (1830 °F), but higher temperature and sufficient time must be used when alloying powders are added to allow for complete interdiffusion and to obtain the strengthening benefit of the alloy additions. Temperatures in the range of 1200 to 1260 °C (2200 to 2300 °F), with holding times of 1 to 4 h, are sufficient to homogenize admixed alloy powders.

Table 21 Comparative properties of P/M and wrought titanium

Property	Commercially pure P/M	Commercially pure Wrought grade II	Ti-6Al-4V Cold pressed, sintered blend	Ti-6Al-4V Cold pressed, sintered blend	Ti-6Al-4V Wrought
Density, %	95.5	100	94	99	100
Modulus of elasticity, GPa (10^6 psi)	103 (15)	102.7 (14.9)	103 (15)	110 (16)	110 (16)
Reduction in area, %	14	35	3	23	43
Ultimate tensile strength, MPa (ksi)	414 (60)	345 (50)	754.3 (109.4)	960.5 (139.3)	948 (137)
Yield strength, MPa (ksi)	324 (47)	345 (50)	643.3 (93.3)	884.6 (138.3)	869 (126)
Elongation, %	15	5	3	12	19

Vacuum Sintering

The use of cold-walled vacuum furnaces generally is preferred for sintering of titanium powders. The furnace should be equipped with a diffusion pump that has a sufficient capacity to accommodate outgassing during heat up and maintenance of high vacuum (133 MPa, or 1×10^{-3} mm Hg) at peak temperatures. Heating elements of molybdenum or graphite cloth with a temperature limit of at least 1245 °C (2275 °F) can be used.

The use of cold traps is necessary to condense vaporized sodium chloride and other impurities that may impair diffusion pump performance. Because significant outgassing occurs during heating, a control system that cuts power to the furnace when a preset pressure level (400 MPa, or 3×10^{-3} mm Hg) is reached should be used. Heating may be restored when higher vacuum is achieved again.

This type of system automatically compensates for variations in furnace load, lubricants, or compact density, for example. The use of an inert gas recirculating system for cooling is optional, except for alloys requiring rapid cooling through the β-α temperature range. While a cooling system can improve furnace through put, the added difficulties and costs of additional valving (sources of leaks) and maintenance of a water-cooled heat exchanger should be considered carefully before making the installation.

Furnace Loading

Due to reactivity, titanium adheres to most support materials after sintering. The preferred support or racking material is molybdenum, which has a very low solubility in titanium and is able to maintain strength at high temperatures. Ti-6Al-4V plate can also be used by placing sheets of felted ceramic between the work and support. At lower temperatures (1000 to 1080 °C, or 1830 to 1975 °F), high-density graphite, coated with a wash of yttrium oxide, can be used. It must be dried separately and outgassed prior to sintering. To help ensure against contamination of the parts being sintered, some loose titanium powder can be included with each load to act as a "getter" for gases. Molybdenum foil placed on top of the load can be used to contain the powder.

Leak Testing. Prior to heating, the furnace should be leak tested for each load. This can be accomplished by pumping the furnace down at room temperature to diffusion pump levels, then isolating the furnace chamber from the pumping system. The leakage, or drop in vacuum, is read on the tank vacuum gauge. The leakage rate constant should not exceed a value of 40 calculated by multiplying the observed leakage rate (μm Hg/h) by the furnace chamber volume (ft^3). Thus, for a furnace 3 ft in diameter by 3 ft long, with a volume of 21.2 ft^3, the maximum allowable leakage rate would be 1.89 μm Hg/h. This procedure helps to ensure against contamination of the work load, as a high-capacity pumping system can maintain a good vacuum even with a leak. However, continuous introduction of a stream of air over the work rapidly saturates the getter and begins to contaminate the work.

Sintering Cycles. Care must be taken during heating to avoid excessive outgassing of the work, furnace walls, and elements, unless the system pumping capacity is sufficiently large compared to the furnace volume. Heat transfer in vacuum furnaces is primarily by radiation; therefore, a "shadowing" effect occurs with workpieces close to the heating elements—they heat before pieces in the center of the load are heated. Because a heating cycle can be relatively long (~8 h), peak temperature should be maintained for 2 to 3 h to ensure equalization of the load temperature before cooling.

The use of load-monitoring thermocouples at several locations in the load is recommended, particularly with large loads. Sintering temperatures of 1000 to 1300 °C (1830 to 2370 °F) are commonly used. Lower temperatures are usually used for porous and commercially pure products, and higher temperatures are used to produce alloy products.

Alternative Practices

While vacuum sintering is the most widely used sintering practice for titanium, an alternative method uses purified argon for sintering powder-rolled titanium strip product. Argon is purified by passing it over titanium chips heated to 800 to 900 °C (1470 to 1650 °F).

The use of temperature cycling through the α-β transformation—880 °C (1620 °F) for unalloyed titanium—can promote densification; the effect is the most significant when heating from α to β, which lasts for ~2 min.

Because large thermal lags exist in most vacuum furnaces, this technique is not practical for large production loads. In addition to normal sintering, a supplementary containerless HIP cycle can be used to improve density and mechanical properties (Ref 35, 39). Because the as-sintered density of P/M Ti-6Al-4V is 94 to 96% of theoretical, the porosity is not interconnecting, and densification can be achieved through the application of gas pressure and heat.

Microstructure and Properties

Because diffusion rates are higher in the beta-phase, body-cubic centered (bcc) region, sintering is generally completed at >880 °C (1620 °F). The microstructure following slow cooling is a transformed beta microstructure, which is considered an inferior structure for conventional wrought material. In the case of commercially pure titanium parts, this limitation is not significant, because static strength and corrosion resistance are of prime importance. Only fatigue and fracture toughness-related properties are affected by this microstructure.

Static properties of pressed-and-sintered P/M titanium are comparable to those of wrought material, as shown in Table 21. Fatigue resistance is not quite as good as for wrought alloys, primarily due to the presence of salt-related residual porosity and to a lesser extent the transformed beta microstructure. More current information of P/M titanium alloys is in the article "Titanium Powder Metallurgy Alloys" in this Volume.

Sintering of Nickel and Nickel Alloys

Sintered nickel-base materials enjoy wide commercial usage due to their unique properties, such as corrosion resistance, wear resistance, mechanical strength at low and elevated temperatures, thermal expansion, electrical conductivity, and magnetic permeability. Commercial products represent a wide range of shapes, sizes, and microstructures. Sintered products with ≤90% porosity are technologically as important as fully dense products. Sintering practices aimed at achieving such diverse properties represent a variety of manufacturing technologies.

Depending on the end product, sintering operations can be performed on a loose powder bed, a thin layer applied to a substrate, a lightly compacted coating, a roll-compacted strip, or die-compacted preforms. Sintering temperatures range from those at which the material is completely solid to those at which ≤50% of the material is in the liquid state.

Conventional sintering practices generally are useful, although more recent techniques such as HIP are required in specialized applications. Sintering atmospheres are generally reducing; in some applications, however, nonreactive environments such as vacuum or an inert gas are required. Physical and chemical characteristics of the starting powders, such as moisture content, degree of oxidation, segregation, morphology, and purity, are equally as important as the green compact microstructure prior to sintering.

Sintered porous nickel products of commercial interest include electrodes of alkaline nickel-cadmium, nickel-zinc, and nickel-iron rechargeable batteries, electrodes of alkaline fuel cells and metal-air (zinc-air or iron-air) cells, electrolyzers, and filter elements. Table 22 lists the functions and structural requirements these components must meet for efficient operation.

Highly porous electrodes for alkaline rechargeable batteries are made from nickel powders that have low bulk densities (<1 g/cm^3). These powders are produced by carbonyl processes. Slurries made from these powders are coated onto a support strip and then sintered. Alternatively, loosely packed beds of powder can be sintered. For all other porous structures, a carbonyl nickel powder with higher bulk density is suitable. Finer nickel powders can also be used by lightly compacting in a die or by roll compacting to produce the desired high green density.

Sintering temperatures between 850 and 1050 °C (1560 and 1920 °F) are commonly used to produce porous nickel products. The reducing conditions required for sintering are obtained easily with nitrogen-hydrogen gas mixtures or a burnt natural gas with low combustibles.

During sintering, loosely packed beds or compacted preforms of carbonyl nickel powders do not densify at <600 °C (1110 °F). However, considerable strengthening of compacted materials occurs at <600 °C (1110 °F) due to surface diffusion. As the temperature increases above 600 °C (1110 °F), volume diffusion increases. At ~950 °C (1740 °F), filaments in the porous structures become smooth and broadened, while in consolidated structures the pores become rounded. To attain adequate strength without excessive loss of porosity, sintering temperatures between 950 and 1000 °C (1740 and 1830 °F) are optimum.

Table 23 gives typical porosity, strength, and electrical resistivity values for loose sintered and pressed-and-sintered materials. Table 24 shows reduction in surface area as a result of sintering. Figure 54 shows pore size distributions of sintered porous nickel structures.

Sintered Dense Nickel Products. Sintered high-density nickel products generally are used as preforms for further working operations. Additional densification occurs as the preform is hot and/or cold worked into a useful commercial shape. Preforms can be round or square billets or strips. Conventional compacting and sintering of individual small components is also used commercially.

A typical example of a high-density billet for further hot working involves extremely high-purity (99.97% Ni), inclusion-free nickel for electronic applications. Commercial melting and casting practices previously used could not guarantee the high purity and freedom from inclusions required for this application. Powder metallurgy processing, however, provides extremely high-purity nickel powders produced by techniques other than atomization. As a result, various mill shapes such as plates, rod, bar, and tube were manufactured from sintered billet stock.

The optimum starting powder for this application is a high-purity nickel powder with a carbon content below 0.1%, oxygen content below 0.15%, and sulfur content below 7 ppm. A high degree of compressibility is desirable. Powder is typically compacted by cold isostatic pressing to a density of approximately 6 g/cm^3. Sintering is carried out in a sulfur-free, reducing atmosphere such as hydrogen, dissociated ammonia, or a mixture of 90% nitrogen and 10% hydrogen. The prime objectives of the sintering operation

Table 22 Functional, service, and structural requirements for sintered porous nickel products

Product	Function	Service conditions of service requirements	Structural requirements
Electrodes of alkaline nickel-cadmium, nickel-zinc, and nickel-iron rechargeable batteries	Hold active mass (nickel hydroxide or cadmium hydroxide) of the battery	Pores in the electrodes are impregnated with the active mass before battery assembly. Electrodes are in contact with a highly caustic solution of potassium hydroxide. Contraction and expansion of the active mass during discharge and charge cycles leads to ~25 vol% change.	75–85% porosity for holding the active mass, mechanical strength to withstand handling during assembly and volume changes during battery operation
Electrodes of alkaline fuel cells and metal-air cells	Provide a barrier between the electrolyte and gases	Must provide the largest electrolyte/gas interface. Must avoid complete wetting of pores or bubbling of gases through pores	Small pore sizes; very narrow pore size distribution; total pore volume of about 48%
Filter elements	Filter highly caustic solutions	Depends on specific applications	Depends on specific applications. Both pore volume and pore size distribution are important.

Table 23 Typical physical properties of sintered porous nickel structures

Material	Condition	Porosity, %	Electrical resistivity, μΩ · cm	Bend strength, MPa	Bend strength, psi
Nickel 287	Loose sintered	83	150–200	5.8	840
Nickel 255	Slurry sintered	82	100–150	12.7	1,840
Roll compacted strip	Sintered (49% porosity)	38	30–40	137	19,865
Nickel 123	Loose sintered	60	60–70	24.5	3,550

All materials were sintered at 950 °C (1740 °F) for 5 min.

Table 24 Surface area of sintered porous nickel structures

Powder type	Surface area, m^2/g: Loose powder	As sintered at 850 °C (1560 °F) for 5 min	As sintered at 950 °C (1740 °F) for 5 min
Nickel 255	0.6	0.18	0.16
Nickel 123	0.3	0.11	0.10

Surface area determined by BET method using krypton

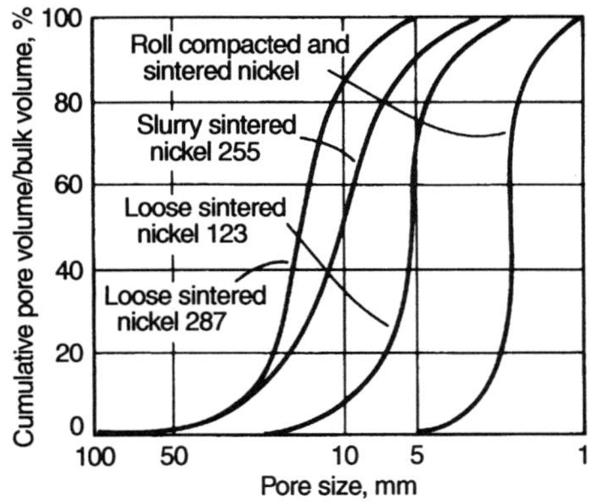

Fig. 54 Pore size distributions in sintered porous nickel structures

are densification, development of strength (for subsequent hot working operations), and reduction of carbon and oxygen levels to below 0.01% and 0.005%, respectively.

During heat up of the green billet, gases adsorbed on powder surfaces are liberated. Carbon and oxygen in the powder then begin to react and produce gaseous products such as carbon dioxide and carbon monoxide. Sufficient time must be allowed at low and intermediate temperatures to permit the escape of these gases and the resulting decarburization before densification begins. For billets to reach a density above 96% of theoretical, sintering temperature is raised to 1100 to 1200 °C (2010 to 2190 °F) following decarburization. Depending on sintering temperature, sintering time may vary from 4 to 12 h.

Coinage strip represents another application in which conventional P/M operations are combined with conventional metalworking operations to yield a unique product. While coin blanks can be compacted from powders using high-speed presses, considerations such as economy and productivity may favor punching of the coin blanks from powder-rolled strip. Green strip is produced by roll compacting nickel powder that is sintered in a continuous sintering furnace to a reasonably high density. The density must be high enough (generally >80% of theoretical) to facilitate handling and coining operations. Following sintering, the strip is hot and/or cold worked to full density and to the desired thickness of the final product. See the article "Roll Compacting of Metal Powders" in this Volume for additional information.

Nickel powders with consistent high flow rates and compressibility are required so that high volume production units with automatic powder-feeding mechanisms can be used. Water-atomized nickel powders usually are used for this purpose, although powders produced by other techniques are also used commercially. Water-atomized nickel powder typically contains ~0.1 to 0.2% oxygen. A major portion of this oxygen is present on or near the surfaces of the powder particles.

Conventional reducing atmospheres (hydrogen, dissociated ammonia, or burnt natural gas) can be used for sintering. The purpose of sintering is to increase the strength of the roll-compacted strip. Sintering time-temperature cycles are designed to obtain the desired strength and density, while matching the speed of the roll compacting operation.

The same considerations that apply to high-purity, electronic-grade nickel (high purity and freedom from inclusions) are responsible for commercial interest in the production of magnetic and low-expansion P/M alloys. Soft magnetic alloys such as Ni-50Fe, Ni-17Fe-4Mo, and Ni-17Fe-5Cu-4Mo are commercially produced from high-purity elemental metal powders. Carbonyl nickel and iron powders and high-purity copper and/or molybdenum powders are mechanically blended, isostatically compacted, and sintered in hydrogen at 1200 to 1400 °C (2190 to 2550 °F). During sintering, the density increases from ~70% in the as-pressed condition to 90 to 95% in the sintered condition. Bulk liquid hydrogen gas with an incoming dew point of approximately–73 °C (–99 °F) is used for sintering. Sintered billet is subsequently converted to a wrought product by hot and cold working.

Selection of Sintering Atmosphere. Composition and dew point of a sintering atmosphere determine whether reducing or oxidizing conditions prevail during sintering of a given material. The Richardson free energy chart should be consulted for proper conditions. For example, in sintering of nickel at 1000 °C (1830 °F), according to the Richardson free energy chart, the hydrogen-to-water and carbon monoxide-to-carbon dioxide ratios must be >0.01 to prevent oxidation of nickel. A reducing gas containing approximately 5% hydrogen, 5% carbon monoxide, 8% carbon dioxide, and the remainder nitrogen easily meets these requirements.

However, if nickel is alloyed with reactive elements such as silicon, manganese, chromium, vanadium, titanium, aluminum, zirconium, magnesium, and calcium, much higher ratios must be maintained to prevent oxidation of the reactive elements. Thus, for instance, the hydrogen-to-water and carbon monoxide-to-carbon dioxide ratios must be greater than 8×10^3 to prevent oxidation of chromium. Very pure hydrogen with a dew point of –45 °C (–49 °F) or lower is required for sintering nickel-chromium alloys. Also, time and temperature during heat up to sintering temperature must be carefully controlled so that the material being sintered is surrounded by an atmosphere with the required ratios. Additionally, alloys containing nitride-forming elements (chromium, titanium, and zirconium, for example) must be sintered in nitrogen-free atmospheres. In some cases, vacuum sintering may be necessary.

Nonconventional Sintering Practices. Many modifications to conventional pressing and sintering to produce dense bodies from metal powders are commercially important, as discussed below.

Hardfacing and brazing rods of nickel-base alloys classified by the American Welding Society as AWS BNi-1 (13 to 15% Cr, 2.75 to 3.50% B, 4.0 to 5.0% Si, 4.0 to 5.0% Fe, 0.6 to 0.9% C, remainder Ni), AWS BNi-3 (2.75 to 3.50% B, 4.0 to 5.0% Si, 0.5% Fe, remainder Ni), AWS BNi-4 (1.5 to 2.2% B, 3.0 to 4.0% Si, 1.5% Fe, remainder Ni), and low-carbon variations of AWS BNi-1 are customarily cast products.

These alloys can be produced more economically from water-atomized powders of these compositions. The process consists of packing graphite molds with powders to a packed bed density of 50 to 60% and heating them in a reducing atmosphere to a temperature between the solidus and liquidus temperatures—977 to 1030 °C (1790 to 1886 °F) for BNi-1, 982 to 1038 °C (1800 to 1900 °F) for BNi-3, and 982 to 1066 °C (1800 to 1950 °F) for BNi-4. Typically, 40 to 60% mushy liquid is produced at the sintering temperature. Sintering cycles are adjusted so that cooling begins as soon as the mushy liquid is produced. The sintered product is typically 85 to 95% dense. Sintered rods up to 1350 mm (53 in.) long are routinely produced. In some applications, further mechanical working, such as extrusion and wire drawing, is required.

Production of high-temperature corrosion- and creep-resistant P/M nickel-base alloys is also an established practice. The presence of large amounts of highly reactive elements in these alloys makes it difficult to produce fully dense shapes by the conventional press-and-sinter method. Powders of these alloys are produced by inert gas (argon) atomizing a vacuum-induction-melted alloy, so as to exclude any oxygen and nitrogen. In addition, they are sized and encapsulated in an inert atmosphere such as argon.

Consolidation by HIP to produce fully dense, complex near-net shapes components, such as gas turbine engine disks, has become commercially viable. Other consolidation practices for nickel-base superalloy powders include hot isostatic pressing plus forging, hot extrusion, and extrusion plus rolling. Techniques of producing thin, hollow shapes of reactive metals by plasma spraying onto a disposable mandrel are being applied to production of large integral gas turbine engine components. Alloy powder, usually –400 mesh, is sprayed onto acid-leachable steel preforms by low-pressure plasma spraying.

REFERENCES

1. F. Thümmlee and R. Oberacker, *Introduction to Powder Metallurgy,* Institute of Materials, 1993
2. P. Ramakrishman, *Powder Metallurgy in Automotive Applications,* Science Publishers, Inc., 1998
3. A. Šalak, *Ferrous Powder Metallurgy,* Cambridge International Science Publishing, 1995
4. P. Gummeson and A. Stosuy, "Iron-Carbon Behavior During Sintering," Technical Bulletin D 164, Hoeganaes Corp., 1972
5. D.M. Buck, J.G. Marsden, and A.T. Sibley, High Temperature Sintering With a Nitrogen-Methanol Atmosphere, *Progress in Powder*

Metallurgy 1982, Vol 38, Metal Powder Industries Federation, 1983, p 53–64

6. E. Klar and P.K. Samal, Powder Metals, *ASTM Manual 20: Corrosion Tests and Standards*, ASTM, 1995, p 551–557
7. E. Klar, Corrosion of Powder Metallurgy Materials, *Corrosion Metals Handbook*, 9th ed., ASM International, 1987, p 823–845
8. E. Maahn and T. Mathiesen, "Corrosion Properties of Sintered Stainless Steel," presented at UK Corrosion 91 (Manchester), 1991
9. E. Klar and P.K. Samal, Optimization of Vacuum Sintering Parameters for Improved Corrosion Resistance of P/M Stainless Steels, *Advances in Powder Metallurgy and Particulate Materials*, Vol 7, Metal Powder Industries Federation, 1994, p 239–251
10. E. Klar and P.K. Samal, Effect of Density and Sintering Variables on the Corrosion Resistance of Austenitic Stainless Steels, *Advances in Powder Metallurgy & Particulate Materials*, Vol 3, Metal Powder Industries Federation, 1995, p 11–3 to 11–17
11. M.A. Pao and E. Klar, Corrosion Phenomena in Regular and Tin Modified P/M Stainless Steel, *Progress in Powder Metallurgy*, Vol 39, H.I. Nayar, S.M. Kaufman, and K.E. Meiners, Ed., Metal Powder Industries Federation, p 431–444
12. E. Klar and P.K. Samal, On Some Practical Aspects Related to the Corrosion Resistance of Sintered Stainless Steels, *Powder Metallurgy World Congress* (Paris), Vol III, Editions de Physique, 6–9 June 1994, p 2109–2112
13. Annual Book of ASTM Standards Section 1, Vol 01. 03., A 380–78
14. J.J. Demo, Structure and Constitution of Wrought Ferritic Stainless Steels, *Handbook of Stainless Steels*, D. Peckner and I.M. Bernstein, Ed., McGraw Hill, 1977
15. T. Tunberg, L. Nyborg, and C.X. Liu, Enhanced Vacuum Sintering of Water-Atomized Stainless Steel Powder by Carbon Addition, *Advances in Powder Metallurgy and Particulate Materials*, Vol 3, 1992, p 383–396
16. R.M. Larsen and K.A. Thorsen, Removal of Oxygen and Carbon During Sintering of Austenitic Stainless Steels, *Proc. of 1993 Powder Metallurgy World Congress* Part 1 (Kyoto, Japan), p 377–380
17. E. Maahn, S.K. Jensen, R.M. Larsen, and T. Mathiesen, Factors Affecting the Corrosion Resistance of Sintered Stainless Steel, *Advances in Powder Metallurgy & Particulate Materials*, Vol 7, 1994, p 253–271
18. C. Lall, Fundamentals of High Temperature Sintering: Application to Stainless Steels and Soft Magnetic Alloys, *Int. J. of Powder Metall.*, Vol 27 (No. 4), 1991, p 315–329
19. E. Klar, M. Svilar, C. Lall, and H. Tews, Corrosion Resistance of Austenitic Stainless Steels Sintered in Commercial Furnaces, *Advances in Powder Metallurgy & Particulate Materials*, Vol 5, Metal Powder Industries Federation, 1992, p 411–426
20. N. Dautzenberg, Paper No. 6.18, Eigenschaften von Sinterstählen aus Wasserverdüsten Unlegierten und Fertiglegierten Pulvern, *2nd European Symposium on Powder Metallurgy*, 1968, EPMA, Vol II
21. M.A. Pao and E. Klar, On the Corrosion Resistance of P/M Austenitic Stainless Steels, *Proceedings of the International Powder Metallurgy Conference* (Florence, Italy), Associazone Italiano di Metallurgia, 1982
22. R.L. Sands, G.F. Bidmead, and D.A. Oliver, The Corrosion Resistance of Sintered Stainless Steels, *Modern Developments in Powder Metallurgy*, Vol 2, H.H. Hausner, Ed., Plenum Press, 1966, p 73–85
23. K. Frisk, A. Johanson, and C. Lindberg, Nitrogen Pick up During Sintering of Stainless Steel, *Advances in Powder Metallurgy & Particulate Materials*, 1992, Vol 3, Metal Powder Industries Federation, p 167–179
24. R.H. Shay, T.L. Ellison, and K.R. Berger, Control of Nitrogen Absorption and Surface Oxidation of Austenitic Stainless Steels in H-N Atmospheres, *Progress in Powder Metallurgy*, Vol 39, H.S. Nayar, S.M. Kaufman, and E.E. Meiners, Ed., Metal Powder Industries Federation, 1983, p 411–430
25. M. Larsen, "Debindering and Sintering in Powder Metallurgy Processes," Ph.D. thesis, Technical University of Denmark, 1994 (in Danish)
26. T. Mathiesen, "Corrosion Properties of Sintered Stainless Steels," Ph.D. thesis, Technical University of Denmark, 1993 (in Danish)
27. D. Ro and E. Klar, Corrosion Behavior of P/M Austenitic Stainless Steels, *Modern Developments in Powder Metallurgy*, Vol 13, H.H. Hausner and P.W. Taubenblat, Ed., Metal Powder Industries Federation, 1980, p 247–287
28. M. Svilar and H.D. Ambs, P/M Martensitic Stainless Steels: Processing and Properties, *Advances in Powder Metallurgy*, Vol 2, 1990, p 259–272
29. S.K. Chatterjee, M.E. Warwick, and D.J. Maykuth, The Effect of Tin, Copper, Nickel, and Molybdenum on the Mechanical Properties and Corrosion Resistance of Sintered Stainless Steel (AISI 304L), *Modern Developments in Powder Metallurgy*, Vol 16, E.N. Aqua and C.I. Whitman, Ed., Metal Powder Industries Federation, 1984, p 277–293
30. F.M.F. Jones, The Effect of Processing Variables on the Properties of Type 316L Powder Compacts, *Progress In Powder Metallurgy*, Vol 30, Metal Powder Industries Federation, 1970, p 25–50
31. R.M. Larsen, T. Mathiesen, and K.A Thorsen, The Effect of Porosity and Oxygen Content on the Corrosion Resistance of Sintered 316L, *Powder Metallurgy World Congress* (Paris), Vol III, Les Editions de Physique, 1994, p 2093–2096
32. G. Lei, R.M. German, and H.S. Nayar, Corrosion Control in Sintered Austenitic Stainless Steels, *Progress in Powder Metallurgy*, Vol 39, H.S. Nayar, S.M. Kaufman, and K.E. Meiners, Ed., Metal Powder Industries Federation, 1984, p 391–410
33. T. Mathiesen and E. Maahn, Corrosion Behavior of Sintered Stainless Steels in Chloride Containing Environments, *12th Scandinavian Corrosion Congress* (Helsinki), 1992, p 1–9
34. W.F. Wang and Y.L. Su, Powder Metallurgy, Vol 29, 1986
35. N.S. Mikkelsen and M. Jensen, "Sinterwerkstoff", European Patent Publication 0564778 A1, 1993
36. E. Peissker, *Modern Developments in P/M*, Vol 7, Metal Powder Industries Federation, 1974, p 597–613
37. P.E. Matthews, Brass and Nickel Silver Powders, *Copper Base Powder Metallurgy*, Vol 7, Metal Powder Industries Federation, 1980
38. T.W. Pelletiers, *Advances in P/M*, Vol 3, Metal Powder Industries Federation, 1991, p 147–158
39. Paul Matthews, *Int. J. Powder Metall.*, Vol 4, 1968, p 39–46
40. K.R. Van Horn, *Aluminum*, Vol 3, American Society for Metals, 1967, p 26–28, 313–314
41. R.C. Leuth, Mouldless Hot Pressing of Cemented Carbides, *Refractory and Hardmetals Journal*, June 1985, p 87–91
42. M.S. Greenfield, B. North, and W.R. Pfouts, Selecting Pressure Sinter Hot Isostatic Pressing or Vacuum Sinter for Cemented Carbide Applications, *Advances in Powder Metallurgy and Particulate Materials*, Vol 6, 1991, p 465–479
43. B. North, M.S. Greenfield, W.C. McCoy, and J.S. VanKirk, Effect of Pressure Sinter and Post-Treatment on the Strength of Cemented Carbides, *Advances in Powder Metallurgy and Particulate Materials*, Vol 8, 1992, p 111–123
44. T.E. Hale, private communication
45. R. Hack, F. Kiefer, D. Schuler, and N. Muller, "Monitoring the Gas Composition in a Sintering Furnace," Application Report, Balzers AG, Furstentum, Liechtenstein
46. D. Jaffrey, J.W. Lee, and J.D. Browne, Co-WC Pseudobinary Eutectic Reaction, *Powder Metall.*, (No. 3), 1980, p 140–144
47. H. Tullhoff, On the Grain Growth of WC in Cemented Carbides, *Modern Developments in Powder Metallurgy*, Vol 14, H.H. Hausner, et al., Ed., Metal Powder Industries Federation, 1980, p 269–277
48. O. Rudiger and H. Rottger, On the Problem of Evaporation of the Auxiliary Metal During Vacuum Sintering of Hard Metals, *International Symposium on Powder Metallurgy* (Paris), June 1964
49 K. Tamai, R. Fukumoto, T. Watanabe, and Y. Kondo, Sinter HIP Processing of Cemented Carbides, *Proceedings of Powder Metallurgy '90*, p 251–256
50. P.E. Zovas, et al., Activated and Liquid Phase Sintering—Progress and Problems, *J. Metals*, 1983, p 28–33

Consolidation of Ultrafine and Nanocrystalline Powder

David L. Bourell, The University of Texas at Austin
Joanna R. Groza, University of California at Davis

Ultrafine or nanocrystalline powder has either particle size or grain size less than about 100 nm (0.1 μm). The synthesis techniques are described in the article "Ultrafine and Nanophase Powders" in this Volume. Ultrafine powder forms are of two types. The first is particulate, generally a single crystal, whose size is less than about 100 nm (Fig. 1). Synthesis routes include evaporation/condensation, sol-gel, and other precipitation techniques. The other powder form has conventional powder size but is composed of a fine, nanocrystalline grain structure. This structure is obtained through a deformation processing technique, such as high-energy, attrition ball milling (Ref 2), or severe plastic deformation consolidation.

Consolidation of these materials into bulk form with retention of the fine structure is the subject of this article. Particle curvature is the dominant feature affecting consolidation. It produces a large driving force for densification and enables surface/boundary diffusion mechanisms to operate at substantial rates. Densification takes place at temperatures consistently below those of larger grained powders by up to hundreds of degrees, and the activation energies for sintering are lower. Low sintering temperature may result in retention of fine grain size, elimination of sintering aids, avoidance of undesirable phase transformations, and absence of deleterious interfacial-reaction decomposition.

Understanding and controlling grain growth and particle coarsening are critical to successful consolidation of nanocrystalline powder. Pressureless sintering has been used to produce bulk materials with nanocrystalline structure, but control of particle agglomeration is critical to its success. Other manufacturing processes involve applied pressure of the order of the curvature stresses or greater: sinter forging, HIP, hot/warm pressing, extrusion, and high-pressure consolidation. Molecular dynamic simulations enhance our understanding of the fundamentals of sintering, including particle rotation and neck formation as sintering mechanisms.

Consolidation Processes

Nanocrystalline powder synthesis, consolidation, and properties are the subjects of several reviews (Ref 3–8). Densification of both types of nanocrystalline powder (as previously described) generally follows the same rules of consolidation and sintering applicable in conventional powder processing if the effects of fine powder size and powder agglomeration are incorporated. Fine powder possesses extremely large surface area, given for spherical particles as:

$$S = 6/\rho_{th}D \quad \text{(Eq 1)}$$

where S is the specific surface area (per unit mass), ρ_{th} is the material theoretical density, and D is the particle diameter. For example, a powder mass of 25 μm iron powder has a surface area of 0.03 m^2/g, compared to a 10 nm ceramic powder that might have a specific surface area as high as 260 m^2/g. Associated with this high surface area is large particle curvature, which, along with optional applied pressure, acts as a driving force for densification. Ashby (Ref 9) developed an equation for the effective surface driving pressure, P_s, due to particle curvature during initial consolidation, based on a work argument for spherical particles:

$$P_s = \frac{6\gamma}{D}\Delta^2\left(\frac{2\Delta - \Delta_o}{1 - \Delta_o}\right) \quad \text{(Eq 2)}$$

where γ is the material surface free energy, Δ is the relative density (ρ/ρ_{th}) at some point during the consolidation process, and Δ_o is the relative density (ρ_o/ρ_{th}) at the onset of consolidation. This relation yields surface pressures of approximately 1 to 10 MPa for nanocrystalline powder.

Particulate nanocrystalline powder is usually loosely agglomerated. Magnetic nanocrystalline powder and β-SiC align in a one-dimensional fashion typified in Fig. 2 (Ref 10, 11). Because the particle mass to surface area ratio is small for nanocrystalline powder, interparticle friction effects dominate, producing low tap densities of approximately 10 to 30% of theoretical density (Ref 5, 12, 13). These effects have a negative impact on consolidation. For pressureless sintering, agglomerates densify quickly, leaving large

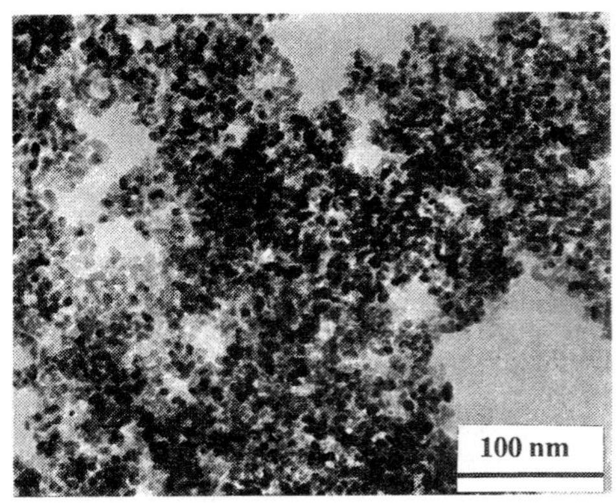

Fig. 1 Nanocrystalline particulate zirconia. Source: Ref 1

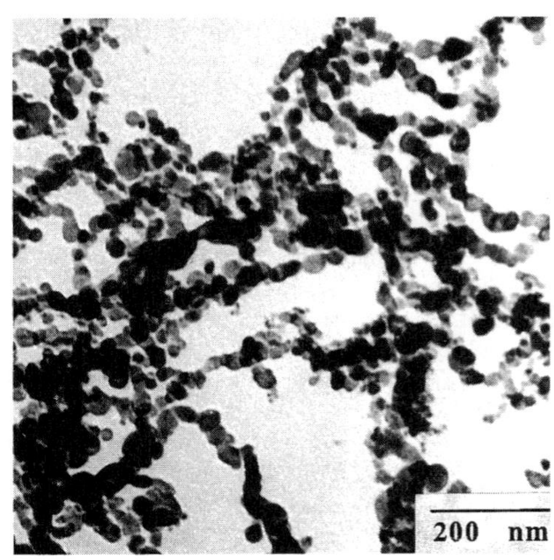

Fig. 2 Chainlike or necklace particulate orientation of nanocrystalline iron. Source: Ref 10

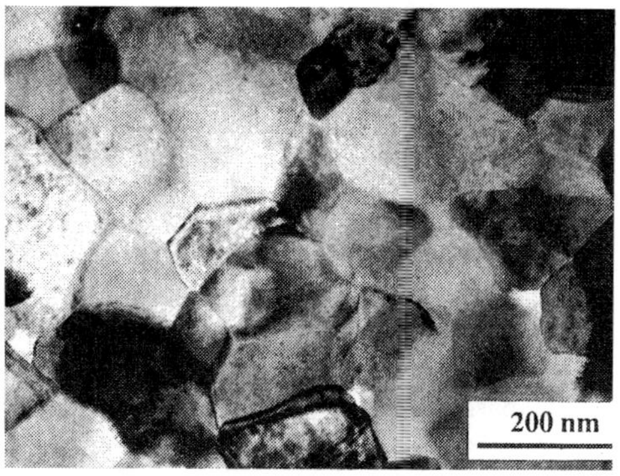

Fig. 3 Grain growth of 10 nm particulate yttria-stabilized zirconia (Fig. 1a). Powder was HIP at 1500 K for 1 h. Source: Ref 1

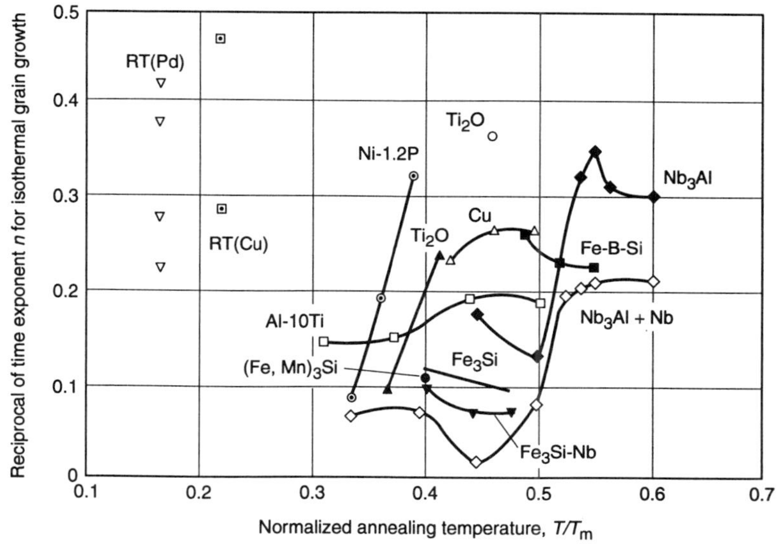

Fig. 4 Reciprocal of the grain coarsening exponent (1/*n*) as a function of temperature for nanocrystalline materials. RT, room temperature. Source: Ref 14

interagglomerate porosity that densifies at much reduced rates. For pressure-assisted methods, pressures exceeding 1 GPa are required to overcome the large interparticle friction effects.

Grain Growth. The greatest obstacle to consolidation of nanocrystalline powder is retention of the fine structure in the bulk. Grain growth and particle coarsening are thermally activated processes. Therefore, the use of thermal excursions to assist in consolidation are always accompanied to some degree by particle coarsening. Figure 3 shows the effects of sintering on grain size in nanocrystalline zirconia (Ref 1). Pressure-assisted consolidation methods are tailored to minimize coarsening during consolidation.

Grain growth/particle coarsening in nanocrystalline materials can be described:

$$D^n - D_0^n = (k_0 t)\exp\left(-\frac{Q}{RT}\right) \quad \text{(Eq 3)}$$

where D is the grain/particle size after time t at temperature T, D_0 is the initial grain/particle size, k_0 is a constant, Q is the activation energy for coarsening, R is the universal gas constant, and n is a constant ranging from 2 to 10 (Ref 6, 7, 14). Figure 4 shows the reciprocal of n as a function of homologous temperature for many nanocrystalline materials. The activation energy for coarsening in nanocrystalline materials is associated with grain boundary diffusion, approximately half the value for lattice self-diffusion (Ref 14). When external pressure assists densification, grain boundary sliding is also suggested as an assisting mechanism (Ref 15, 16).

Experimentally, a strong correlation exists between the closure of open porosity and the onset of exaggerated grain growth (Ref 17–25). As shown in Fig. 5 for nickel and TiO_2, accelerated grain growth occurs at densities above about 90% when the pores become isolated or closed. Often, the final grain size upon full densification is close to 1 μm. In contrast, grain growth is much less exaggerated at high densities for nonagglomerated Al_2O_3 powders (Ref 26). Abnormal grain growth, localized excessive growth of a few grains, has been observed in alumina-zirconia nanocomposites (Ref 27), silicon (Ref 28), and iron (Ref 10). It is illustrated for iron in Fig. 6.

Fine pore size throughout the sintering process is also critical in controlling the final grain size. For these purposes, a small and uniform pore population is desired in the green compact. Most often, this pore distribution is associated with a high green density in nonagglomerated or weakly agglomerated powders. For example, it has been shown that deagglomeration of a commercial ZrO_2 may reduce sintering from 4 h at 1775 k to 1 h at 1375 K (Ref 29). Efforts to produce nonagglomerated powders have intensified. A good summary of the progress for ceramics can be found in Ref 5. The effects of fine pore size and narrow pore size distribution on obtaining high densities and fine structure are well correlated (Ref 18, 26, 30–34).

Considerable coarsening can occur in nanocrystalline materials during heating to and cooling down from the pressureless sintering temperature. A relationship has been proposed to describe the extent of coarsening based on a modification of Eq 3 (Ref 35, 36):

$$D^n - D_0^n = \left[\frac{k_0 R}{v_0 Q}\right]\left[T_f^2 \exp\left(\frac{-Q}{RT_f}\right) - T_i^2 \exp\left(\frac{-Q}{RT_i}\right)\right] \quad \text{(Eq 4)}$$

where v_0 is the heating or cooling rate, T_f is the final sintering temperature, and T_i is the initial temperature. This relationship is verified in heating/quenching experiments for nanocrystalline zirconia (Ref 36). In general, the degree of coarsening occurring during heating in time t to the sintering temperature is approximately equivalent to one third of the grain size obtained for an isothermal anneal (Eq 3) at the sintering temperature for time t (Ref 35).

Pressureless Sintering. Pressureless sintering of nanocrystalline powder occurs generally by surface-assisted and boundary-diffusion-assisted mechanisms. One feature of pressureless sintering is rapid densification of agglomerated regions followed by removal of interagglomerate porosity. Densification rates for the latter stage are associated with sintering of polycrystalline powder whose particle size is associated with the

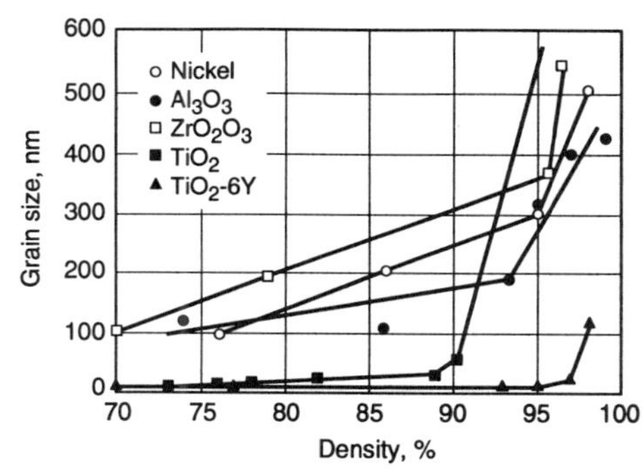

Fig. 5 Grain size as a function of density for pressureless sintering of selected nanocrystalline materials

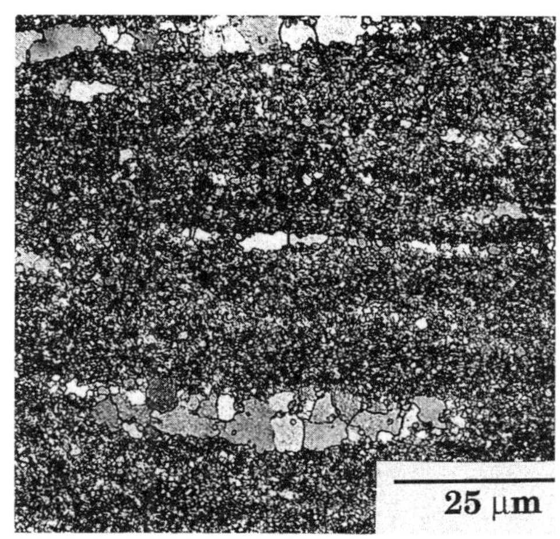

Fig. 6 Abnormal grain growth in 15 nm iron close-die sintered at 810 K (10 MPa, 1 h) followed by pressureless sintering at 1080 K for 1 h. Source: Ref 10

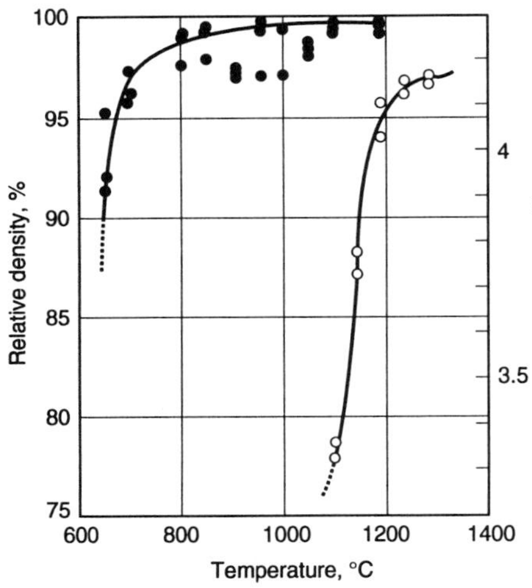

Fig. 7 Pressureless sintering curves for TiO_2 (2 h), showing accelerated densification of 40 nm particles (closed circles) compared to commercial, fine-particulate material (0.2 μm, open circles). Source: Ref 40

densified agglomerate. The overall effect is rapid initial densification that slows to conventional rates as the agglomerates become fully dense. This effect is mitigated by application of external pressure.

Nanocrystalline powders densify at sintering temperatures significantly lower than conventional powders, as exemplified in Fig. 7 for titania. Generally, densification occurs at 0.2 to 0.4 T_m, compared with 0.5 to 0.8 T_m for conventional powder. To rationalize this decrease in the sintering temperature, different scaling laws have been applied. For particles under 100 nm, Alymov et al. (Ref 37) developed an empirical relationship for the dependence of the sintering onset temperature, T_s, on the mean particle size D:

$$\ln T_s = 1/D \quad \text{(Eq 5)}$$

They calculated a ratio of sintering onset to melting temperature that is in good agreement with

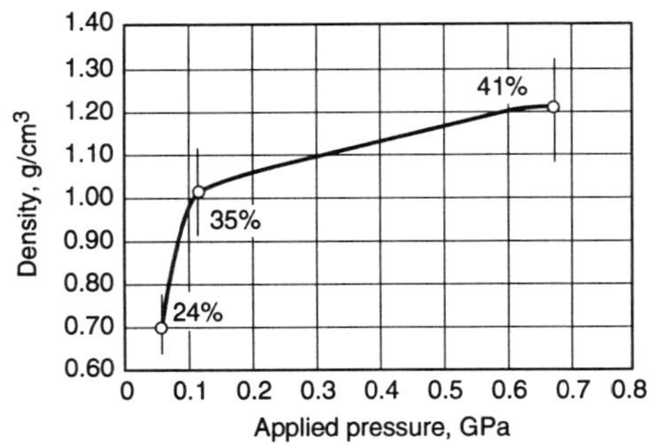

Fig. 8 Effect of compaction pressure on the density of a 17 nm silicon nitride preform. Density after precompaction in WC-Co piston-cylinder die. Source: Ref 40

experimental data for sintering of nanocrystalline metals. In 1950, Herring developed the classical scaling rule for the effect of the particle size on sintering time (Ref 38). Assuming operation of a single sintering mechanism, the time t to achieve the same sintering condition is related to the powder particle size D by:

$$t_1/t_2 = (D_1/D_2)^n \quad \text{(Eq 6)}$$

where n is the exponent that has a value of 3 if the main sintering mechanism is volume diffusion and a value of 4 if the main sintering mechanism is grain boundary diffusion. Considering an Arrhenius expression for temperature, the sintering temperature dependence on the particle size becomes:

$$n\ln(D_1/D_2) = Q/R[(1/T_1) - (1/T_2)] \quad \text{(Eq 7)}$$

where Q is the activation energy for the predominant sintering mechanism, R is the universal gas constant, D_1 and D_2 are the different powder particle sizes, and T_1 and T_2 are the respective sintering temperatures. Reasonable agreement of experimental and calculated data has been found for Al_2O_3 and TiO_2 (Ref 26, 39).

Pressure-Assisted Consolidation. Classical temperature-pressure interactions offer distinct advantages to nanocrystalline powder consolidation due to the ability to restrict grain growth by lowering consolidation temperature. Illustrations of the pressure effect on nanopowder densification are shown in Fig. 8 to 10. Applied pressure adds an additional component to the curvature related driving pressure for densification described earlier. At elevated temperature, it activates new plasticity-driven densification mechanisms, such as local yielding, creep, and stress-assisted diffusion. Applied pressure can induce particle rearrangement and large pore collapse, increasing the number of particle contacts and thereby accelerating densification.

The role of room temperature compaction pressure, P, on relative density, Δ, has been studied extensively (Ref 43). The generally accepted form is:

$$\ln[1/(1-\Delta)] = -\ln(\varepsilon) = kP + B \quad \text{(Eq 8)}$$

where k and B are constants, and ε is the fractional porosity (Ref 44). Changes in densification mechanism, such as particle rearrangement, plastic deformation, and agglomerate fracture, effect changes in the constants as demonstrated in Fig. 9.

Pore collapse scales with the shear stress level. The shear component of stress is minimal in hot isostatic pressing (HIP) and is present increasingly in quasi-isostatic pressing, hot pressing, sinter forging, and extrusion (Ref 45). Additionally, shear stress is beneficial for mechanical disruption of surface oxide layers, which improves interparticle bonding.

Most pressure-assisted consolidation methods have been applied to nanocrystalline powder: HIP, forging and hot pressing, extrusion consolidation, cold sintering (high-pressure consolidation), dynamic consolidation, and field activated sintering. Details of these techniques are discussed in other articles.

The classical pressure-assisted sintering diagrams developed by Ashby and coworkers (Ref 46–48) have been used to describe nanocrystalline powder consolidation (Ref 49–51). Densification maps illustrate both the change in

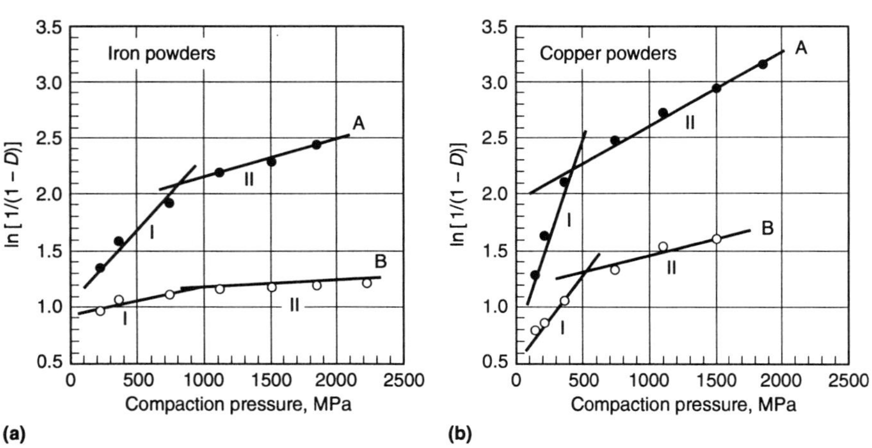

Fig. 9 Effect of compaction pressure on density of (a) iron and (b) copper powder. In each figure, "A" refers to micrometer-sized powder (2 μm Fe, 50 μm Cu) and "B" to nanocrystalline powder (50 nm Fe, 58 nm Cu). Two regions of consolidation are evident from the plots. Low-pressure Region I is associated with particle sliding and elastic deformation between adjacent particles. High-pressure Region II is associated with plasticity at interparticle necks. Source: Ref 41

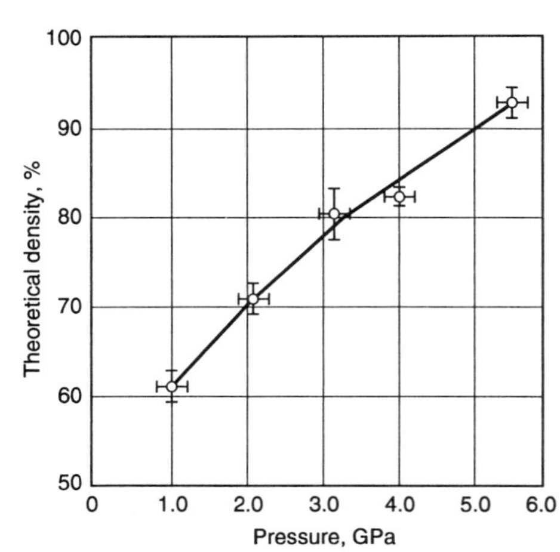

Fig. 10 Cold compaction of 13 nm γ-alumina. Source: Ref 42

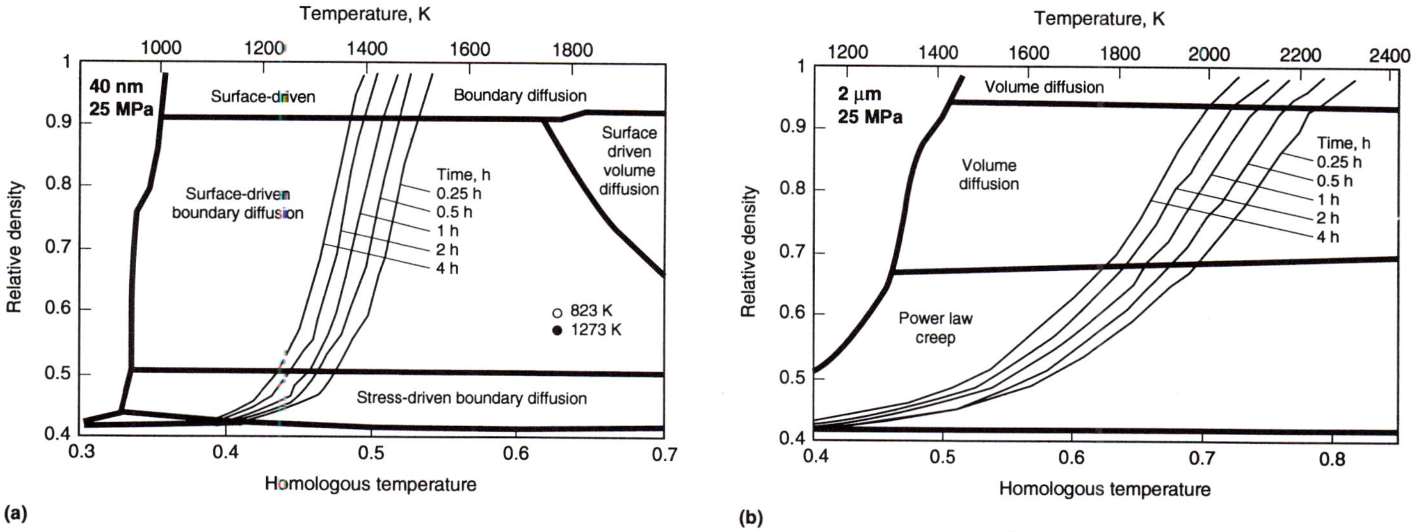

Fig. 11 Densification maps computed for zirconia HIPed with 25 MPa pressure. (a) 40 nm particulate. (b) 2 μm particulate. Symbols represent experimental data for 1 h HIP of 40 nm zirconia calcined at 823 K and 1273 K. Source: Ref 50

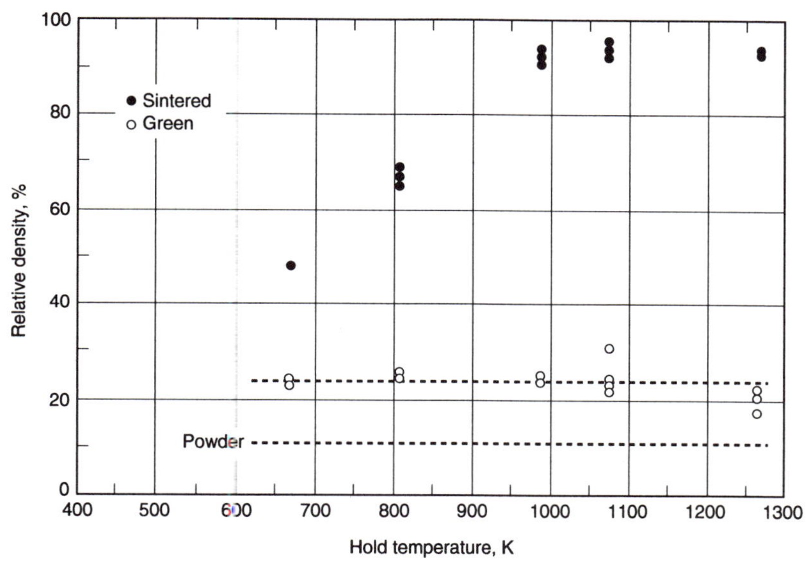

Fig. 12 Effect of sintering temperature on density for 15 nm iron powder hot pressed at 10 MPa for 1 h. Dashed lines show the powder density before (11%) and after (23%) compaction. Source: Ref 10

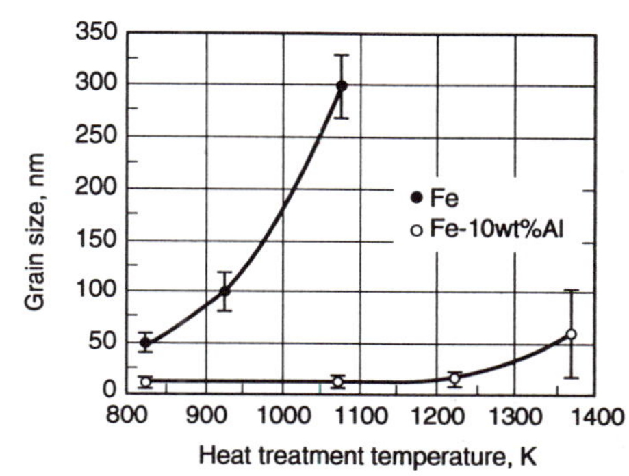

Fig. 13 Effect of fine-particle AlN as a grain-growth retardant for nanocrystalline iron cryomilled in nitrogen. Source: Ref 56

mechanism and accelerated densification relative to conventional powders. This is shown for HIP of nanocrystalline zirconia in Fig. 11(a) (Ref 50). The primary densification mechanism is boundary diffusion driven primarily by curvature effects, even though 25 MPa external pressure was applied. Densification was complete after 1 h at 1500 K. Figure 11(b) shows the computed densification behavior under identical conditions for 2 μm zirconia. Densification is controlled by creep and volume diffusion driven by the applied pressure, and full density is not obtained below about 2200 K.

Metals

Nanocrystalline metallic powder can be densified using conventional sintering. The largest departure from nonnanocrystalline powder sintering behavior is seen in the initial sintering stages. Here, lower activation energies and sintering onset temperatures are observed. For instance, an activation energy of 134 kJ/mol was reported for 40 nm tungsten (Ref 52), which is lower than the value for volume diffusion (580 kJ/mol) and surface diffusion (~300 kJ/mol) (Ref 53). Two apparent activation energy values were calculated in nanocrystalline iron: a low value for low-temperature sintering (125 kJ/mol) and a high value (248 kJ/mol) corresponding to high-temperature densification (Ref 54). The latter value is close to that of volume diffusion in iron and is similar to the value for coarse-grained iron. The low-temperature value is less than the grain boundary diffusion value.

The sintering onset temperature is particle-size dependent (Eq 5). In pure tungsten, the onset is 1100 K (0.3 T_m) for 31 nm particles and 900 K (0.24 T_m) for 9 nm particles (Ref 55). The onset temperature for densification of 40 nm iron powder is as low as 370 K (0.21 T_m), as compared to ~900 K (0.5 T_m) for 2 μm iron powder (Ref 10, 56), as shown in Fig. 12. Alloying iron with aluminum improves fine structure after sintering (Fig. 13) due to formation of AlN dispersoids (Ref 56).

Sintering of iron-nickel nanocrystalline powder is initiated when oxide layers are reduced (Ref 57). This occurs above 500 K in hydrogen. When vacuum is used, no sintering takes place up to 725 K. To avoid further contamination, as-produced nanocrystalline powder can be either consolidated in situ or handled in a controlled environment prior to sintering. The oxygen role in nanocrystalline, evaporation-condensation, iron sintering is important (Ref 10). Although sintering is carried out in a reducing hydrogen atmosphere, oxygen is present in the final product as fine, grain-refining oxides. This is due to adsorbed oxygen entrapment at pore closure.

Iron, nickel, copper, and aluminum have been consolidated using high-pressure densification.

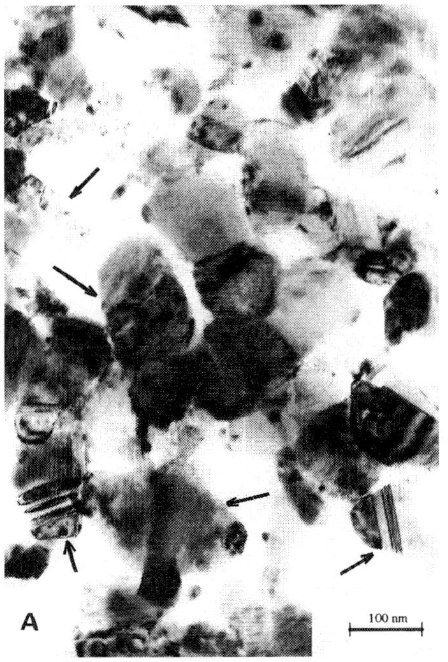

Fig. 14 Bright-field transmission electron microscopy image of 58 nm copper powder cold compacted at 750 MPa. Arrows indicate deformed particles presenting flat surfaces. Source: Ref 41

The data conform to the compaction law given as Eq 8 and illustrated for iron and copper in Fig. 9. The transition point at which the slope changes (Fig. 9) is associated with agglomerate failure. When agglomerates are strong, they do not break and consequently, no transition point is seen. Rate controlled sintering of pure nickel resulted in 99% density with 70 to 80 nm grain size (Ref 58). Densities up to 97% at room temperature and 99% at 575 K in nickel (65 nm), iron (30 to 50 nm), and copper (58 nm) have been obtained using 3 GPa in a high pressure cell, as seen in Fig. 14 (Ref 41, 59). Historically, high pressure consolidation has been the preferred in situ consolidation method for classical nanocrystalline powders synthesized using inert gas condensation (Ref 60–64). A dense product with 22 nm final grain size has been achieved in Al-Fe (Mo, Si, B) alloys by high pressure consolidation (3 GPa at 1073 K) of alternating amorphous layers of constituent metals (Ref 65). Fully dense iron, nickel and Cu-50Ag specimens with less than 20 nm grain size were produced by severe plastic deformation consolidation in which torsional true strains of the order of 7 are imposed (Ref 66–68).

Sinter Forging. In numerous cases, sinter forging is shown to be very effective in achieving full or near full densities and grain sizes less than 100 nm in metals, as shown in Fig. 15 (Ref 49, 69–71). The composite Fe-10Cu has been densified by sinter forging (Ref 49). The benefits of sinter forging in comparison to HIP are evident, as sinter forging produced material with 45 nm grain size at 800 K and 525 MPa compared to HIP at 975 K and 170 MPa with a final grain size of 130 nm (Ref 49, 72). Sinter forging of

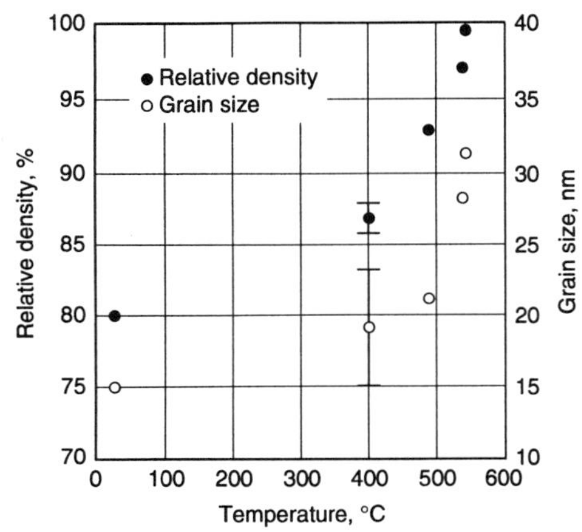

Fig. 15 Effect of sinter-forging temperature on the density and grain size of mechanically milled Fe-29Al-2Cr. The initial particle size was 30 nm and forging was effected with 1.25 GPa for 5 h. Source: Ref 71

ceramics is equally effective (see the section "Zirconia" in this article).

Shock wave consolidation has been applied to consolidation of metallic nanoparticles. For example, fully dense specimens were obtained from ball-milled Fe-C and Fe-N solid-solution, powders by shock wave consolidation (Ref 73, 74). The ball-milled iron powder (20 nm grain size) was consolidated by dynamic compaction (12–19 GPa) followed by 1 h anneal at 500 °C (930 °F).

Hot extrusion usually involves high stresses that can be applied at relatively lower temperatures than in other pressure-assisted techniques, such as HIP. A compromise is usually sought between the strength, which requires high temperatures, and final grain size, for which low temperatures are desired. Hot extrusion has been primarily used to consolidate metallic nanopowders (Ref 75–78). Grain sizes less than 100 nm have been achieved at 1120 K using a 0.5 GPa stress in nickel and iron (Ref 75), Al-Ni-Zr with mischmetal additions under an extrusion ratio of 10 to 1 (Ref 70), and in Mg-10Y-5Cu at 323 to 773 K with extrusion ratios of 5 tol to 10 tol (Ref 70).

Ceramics and Intermetallics

Alumina. Room-temperature phases other than α-Al_2O_3 are usually observed in the nanocrystalline Al_2O_3 system. These include the γ polymorph (Ref 79, 80), as well as bayerite, pseudoboehmite, and θ-Al_2O_3 (Ref 80). Nanocrystalline alumina is pressureless sintered by a surface-diffusion mechanism (Ref 81–83). Colloidally processed alumina sintered to 99.5% density at 1425 K (Ref 80). Early densification activation energy for pressureless densification of 13.5 nm Al_2O_3 was reported to be 234 kJ/mol (Ref 80), compared to 500 kJ/mol for volume diffusion of bulk alumina. Toward the end of the densification cycle, activation energies correspond to volume diffusion (Ref 26).

Densification maps for HIP of nanocrystalline Al_2O_3 powders confirmed boundary diffusion as the primary densification mechanism (Ref 51). However, comparison with experimental HIP results indicated that the calculated densification rates were consistently higher than experimental rates. For instance, 20 nm Al_2O_3 HIPed for 1 h at 1625 K under 300 MPa pressure resulted in a dense sample of only 84% compared to a full-density prediction for the same time at 1170 K and 20 MPa. Among the possible reasons is retarded densification due to large interagglomerate pores.

Both alumina (Ref 84, 85) and zirconia-toughened alumina (Ref 86) have been sinter forged successfully. Hot pressing has been performed (Ref 80, 87, 88), but under certain conditions it posed difficulties due to the formation of a high-density surface layer that reduced the pressure internally (Ref 80). In cold isostatic pressing at very high pressures up to 5.6 GPa, densities in excess of 90% have been achieved (Fig. 10) (Ref 42). Dynamic magnetic compaction (DMC) is an electric pulsed power method that induces short high pressure pulses up to 5 GPa (Ref 89, 90). Alumina powders were consolidated using DMC to 83% density while retaining a nanocrystalline structure (Ref 89). This method has also been applied to sintering of large parts of nanocrystalline alumina (Ref 90).

Microwave sintering is useful in densification of nanocrystalline alumina because grain boundaries are the primary sites for electric dipoles and resultant heating (Ref 91–95). In these materials, densities of about 95% have been achieved with retention of fine structure.

Titania. Nanocrystalline TiO_2 has been densified by pressureless sintering with restrained grain growth through improvements in nanocrystalline powder synthesis, particularly the control of agglomeration (Ref 5, 96, and 97). A sintering onset below 950 K was reported for titania with 40 nm initial particle size (Ref 39). The sintering onset was later measured to be 823 K (0.4 T_m) for 12 to 14 nm TiO_2 (Ref 17). Elimination of small pores by rapid surface diffusion occurred between 425 K and 775 K. Densification was complete at 1300 K as compared to more than 1670 K for 1.3 μm TiO_2 (Fig. 7). For gel-derived TiO_2 powders, near full density is achieved at 1075 K (Ref 98).

A large green density may not always reflect a uniform pore size and distribution. When large agglomerates are present, the green density can be larger than in nonagglomerated powders. For instance, a relatively high density of 75% in as-compacted nanocrystalline TiO_2 at 425 K (2 GPa pressure, 2 h) is explained due to easier packing of large agglomerates (50 nm) rather than individual smaller particles (14 nm) (Ref 17).

The removal of large interagglomerate porosity based on vacancy diffusion requires high temperature and long sintering times. Mayo et al. (Ref 5) compiled evidence for three TiO_2 powders showing that sintering temperature scales with agglomerate size, rather than with size of the individual powder particle (Fig. 16).

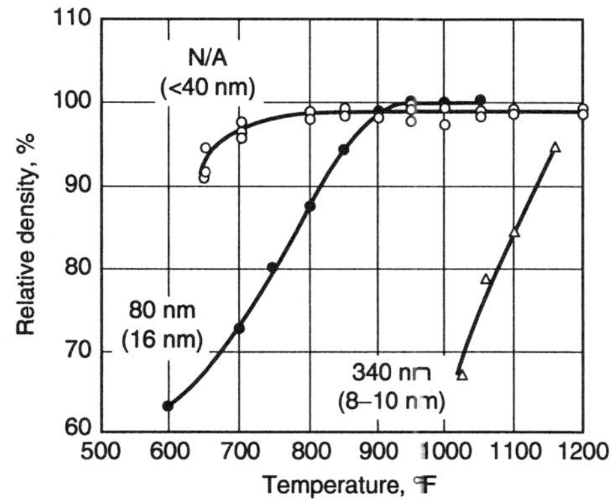

Fig. 16 Effect of agglomerate size on pressureless sintering of nanocrystalline TiO_2. Bold type indicates agglomerate size and the number in parentheses is the particle size. "N/A" indicates that the powder was not agglomerated. Source: Ref 5

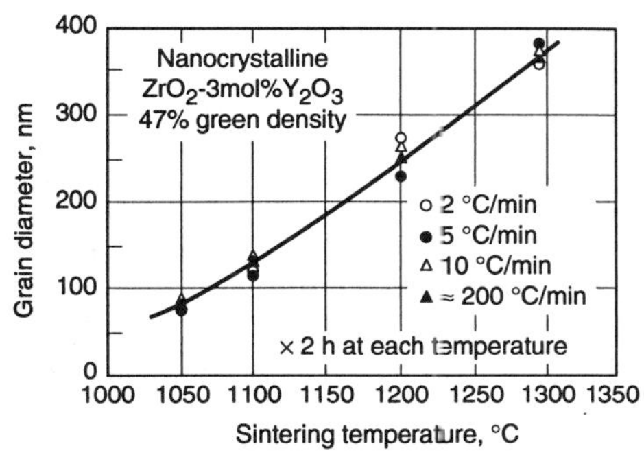

Fig. 17 Grain growth in pressureless sintered 15 nm zirconia, showing the independence of heating rate. Source: Ref 117

Table 1 Pressureless sintering of metastable nanocrystalline ceramics

	Sintering parameters				Final properties		
Material	Initial grain size, nm	Temperature, K	Time, h	Atmosphere	Density, %	Grain size, nm	Reference
ZrO_2	6–9	1400	1.3	Air	Full	80	31
ZrO_2	6–9	1250	0.67	Vacuum	Near full	60	31
ZrO_2 – 3mol%Y_2O_3	N/A	1270	0.67	Vacuum	Full	<100	31
ZrO_2 – 3mol%Y_2O_3(a)	<10	1373	1	NR	99.9	80	5
ZrO_2 – 5.8mol%Y_2O_3	8	1273	100	Air	>95	80	15
ZrO_2 – 5.8mol%Y_2O_3	8	1323	7	Air	96	60	15
CeO_2	10–15	1523	2	Air	Near full	~100	164
CeO_2, – 6at.%Ca	10–15	1623	10 °C/min	Air	Full	30	164
TiO_2	14	1173	15–20	Oxygen	~90	50	17
TiO_2(b)	~6	873	N/A	N/A	99	<60	98
TiO_2 – 6%Y	~15	1173	15–20	Oxygen	~98	30	25
TiAl	10–20(c)	723–773	2	Vacuum	>95	15–20	128–129

N/A, not applicable; NR, not reported. (a) Centrifugally consolidated. (b) Sol-gel prepared. (c) Partial amorphous structure

The agglomerate size scales inversely with particle size. Nonagglomerated powders sinter at the lowest temperature, despite their largest nanoparticle size.

In TiO_2, the quantitative agreement between observed and calculated grain growth (Eq 7) is better when grain boundary diffusion is considered the sintering mechanism (Ref 39). The fitting value for the activation energy used (139 kJ/mol) is about half of the activation energy for lattice diffusion (251 kJ/mol). Activation energies as low as 96 kJ/mol have been reported for 11.5 nm TiO_2 (Ref 82).

A grain growth exponent (Eq 3) for nanocrystalline TiO_2 of about 3 has been reported (Ref 5, 15, 99). Solute additions, such as yttrium (Fig. 5), calcium, or tin, inhibit grain growth in near-full density nanocrystalline TiO_2 (Ref 100). Alternatively, sintering temperatures up to 600 K lower than conventional minimize or even eliminate the need for sintering additives or grain growth inhibitors (Ref 101).

The curvature part of the total sintering pressure during hot pressing (Eq 2) increases when the particle size is reduced and can reach high values as the particles become increasingly small. The applied pressure must be higher than this curvature driven pressure to be effective. Consistent with this threshold effect, no external pressure effect is observed in TiO_2 unless a certain pressure level is achieved (Ref 20, 21, 102, and 103). Hot pressing gives distinct advantages in comparison to pressureless sintering in achieving full densities and minimal grain growth (Ref 17, 104, 105). Generally, the pressures used in hot pressing span a large range from low (<100 MPa) (Ref 80, 106), to moderate (100 to 500 MPa) (Ref 31, 105, 107) to high levels (>0.5 GPa) (Ref 17, 18, 70, and 108). Final grain size is inversely proportional to the applied pressure. Nanocrystalline TiO_2 is sintered to theoretical density at 725 to 825 K (~0.35 T_m) using 1 GPa with minimal grain growth (Ref 18). Application of high pressure (>1 GPa) resulted in grain size less than 60 nm in TiO_2 sintered to 95% density at 675 K (Ref 109).

Sinter forging has been extensively applied to consolidation of nanocrystalline titania (Ref 20, 103, 110, and 111). As an illustration, TiO_2 has been sinter forged with a final grain size of 87 nm, compared to pressureless sintering which gave a 400 nm grain size for the same final density (91%) (Ref 20, 21). Grain sizes less than 50 nm have been retained in TiO_2 densified at 0.43 T_m (Ref 103).

Processing time in microwave sintering is reduced as compared to conventional heating from external sources due to the direct energy coupling with electric dipoles within the heating body (Ref 91, 92). These shorter times produce energy and final property benefits. Temperature gradients are reduced and an overall short sintering time minimizes grain growth. Because grain boundaries are the primary sites for electric dipoles, microwave sintering appears particularly attractive for the densification of nanocrystalline powders. The method has been applied to nanocrystalline TiO_2 (Ref 93, 112, 113). Densities no greater than 95% have been achieved for grain size less than 100 nm.

Zirconia. The pressureless sintering onset of 8 to 9 nm zirconia is 870 to 920 K (0.3 T_m) as compared to 1370 K (0.5 T_m) for 70 nm powder (Ref 18, 31). Commercial powders sinter above 1945 K (Ref 18). Rhodes sintered centrifugally cast ZrO_2-6.5 mol Y_2O_3 to 99.5% density at 1375 K with a final grain size of 200 nm (Ref 29). Similar attempts to sinter ZrO_2 achieved greater than 98% density at 1493 K when Y_2O_3 was added and 93% at 1208 K when the additive was Bi_2O_3 (Ref 116), as shown in Table 1. In both cases, the final grain size was 600 to 700 nm. Pressureless sintering of yttria-stabilized zirconia resulted in grain growth independent of heating rate (Fig. 17) (Ref 117). Surface diffusion is the dominant mechanism initially (Ref 15, 114, 115). Thuenissen et al. (Ref 15) found an atypically low activation energy (100 kJ/mol) in the early sintering of 15 nm yttria-stabilized tetragonal zirconia. In contrast, when aggregate size is 50 nm, the activation energy has the usual value of 275 kJ/mol. Intensified efforts on sintering nanocrystalline ZrO_2 have resulted in fully dense ceramics with restrained grain growth. The sinterability of ceramic nanopowders has been greatly enhanced by improvements in nanopowder synthesis, particularly the control of nanopowder agglomeration (Ref 5, 96, and 97). Examples of dense zirconia ceramics that have retained nanosized grains by pressureless sintering are provided in Table 1. As seen in this table, full densification without grain coarsening has been achieved predominantly in the ZrO_2 system. The explanation for this behavior has been found to be twofold (Ref 5, 15, 18, 29, and 118). First, ZrO_2 has the slowest coarsening tendency among ceramics. Second, the pore size does not seem to affect grain coarsening (Ref 5).

In HIP under identical conditions, a change in the sintering mechanism from creep and volume diffusion in coarse-grained zirconia to grain boundary diffusion in nanocrystalline zirconia occurs (Ref 51), as seen in Fig. 11. Full density processing is also possible by hot pressing (Ref 31, 51).

Wet processing provides significantly improved packing uniformity of nanocrystalline zirconia (Ref 5). For instance, a green density close to the ideal value for monosize spheres (74%) is achieved in 12.5 nm Y_2O_3-doped ZrO_2 by centrifugation from a slip suspension (Ref 29).

Pressure-assisted consolidation of nanocrystalline zirconia synthesized by precipitation routes is hampered by formation of strong agglomerates. Upon cold pressing, they do not

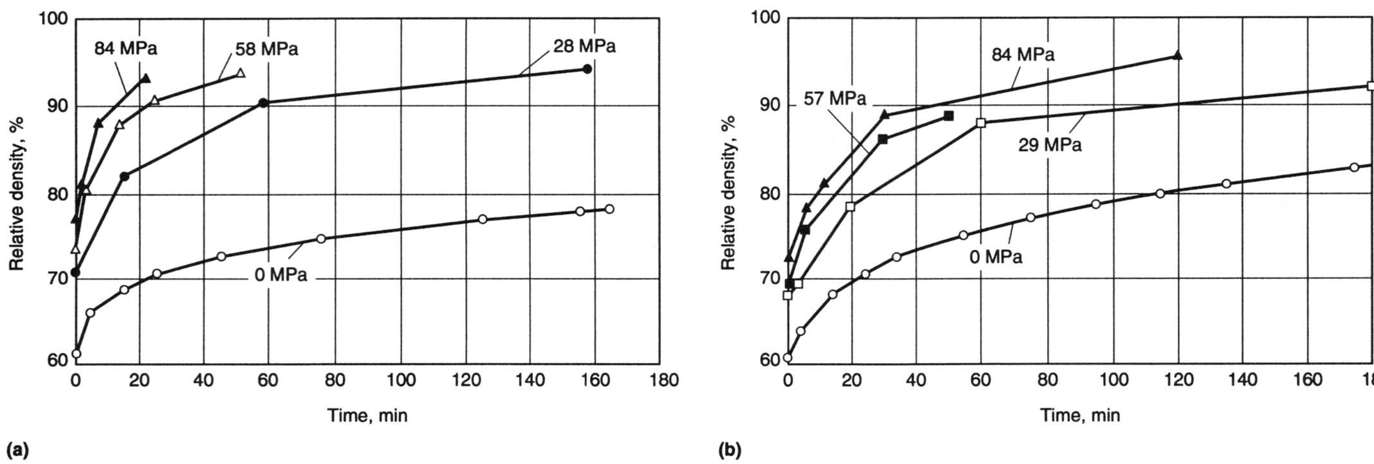

Fig. 18 Increase in relative density with time at 1370 K during pressureless sintering and sinter forging of (a) $95ZrO_2$-$5Y_2O_3$, and (b) $92ZrO_2$-$4Y_2O_3$-$4CeO_2$. Source: Ref 124

break, and consequently, no transition point is observed as seen for metals (Fig. 9). This is the case for agglomerates in calcined water-washed, yttria-stabilized zirconia that survive a cold isostatic pressing at 400 MPa (Ref 119). Particle rearrangement is hindered and lower green densities are likely to be achieved as compared to conventional micron size powders. Bourell et al. (Ref 51) achieved 45% density for 70 nm yttria-partially stabilized zirconia and only 29% for 40 nm particle size. Conversely, when particle sliding is facilitated, high green densities are obtained. This can be accomplished by using lubricants or coatings. Deagglomeration of a commercial ZrO_2 reduced sintering from 4 h at 1775 K to 1 h at 1375 K (Ref 29). Efforts to produce nonagglomerated powders have intensified; Mayo (Ref 5) produced a good summary of the progress. The benefits of a small pore size and narrow size distribution in reaching high densities and reduced grain coarsening have been shown by several researchers for nonagglomerated ZrO_2 (Ref 18, 31, 32).

Sinter forging has been applied to nanocrystalline zirconia, producing materials with grain size less than 50 nm at 0.4 T_m (~1200 K) (Ref 31, 86, 120–124) (see Fig. 18). Precautions should be considered to avoid ceramic specimen fracture at high strain rates and low temperatures.

Other Ceramics and Intermetallics. The benefits of fine grain size in ceramic sintering were reported almost 30 years ago when the sintering onset for very fine grained MgO was found to be as low as 625 K (Ref 125). Examples of dense ceramics and intermetallics with retained nanocrystalline grains up to 100 nm by conventional sintering are provided in Table 1.

Low sintering temperatures during conventional sintering are common. The sintering onset for 4 nm Y_2O_3 is 600 K (<0.25 T_m) (Ref 18). Si_3N_4 has been fully sintered 700 K lower than coarse-grained material after cold pressing (Ref 40, 126), as seen in Fig. 8. TiC powder of 140 to170 nm particle size sintered to 91% density at 1900 K as compared to 5 μm powder with the same density at 3070 K (Ref 127). Densification of nanocrystalline TiN starts at 1170 K and is completed below 1823 K, whereas microcrystalline powder is only 63% dense at 1823 K (Ref 34). Macroscopically, stabilization of particle surfaces (i.e., lower surface energy values when oxides are formed) results in a decrease of the driving force for sintering. This has been observed in nanocrystalline TiN at higher oxygen contents (Ref 34). TiAl was pressureless sintered to greater than 95% density by conventional sintering at low temperatures (0.4 T_m) with grain size less than 15 to 20 nm (Ref 128, 129).

Green densities up to 65% can be reached by conventional cold compaction of 70 to 80 nm TiN. At very high pressures (7 GPa) the density obtained was approximately 80% (Ref 130). Subsequent warm compaction at 675 K resulted in densities exceeding 90%. High pressure compaction (1 GPa) at 525 K also resulted in high densities for the intermetallics TiAl (Ref 110), Fe_3Al, and Ni_3Al (Ref 70).

HIP has been applied to nanoceramic densification in only a few cases where densities of 91 to 92% have been reached, such as in SiC at 2175 K (Ref 131) and Si_3N_4 at 1925 K (Ref 132). The final grain size exceeded the nanometer range. Generally, the final grain size in dense materials obtained by HIP of nanocrystalline powders falls in the micron size range (Ref 131–133). However, by careful control of HIP parameters, particularly temperature, it is possible to produce dense products with grain size less than about 100 nm (Fig. 19). This has been demonstrated for ball-milled TiAl, and Ti_3Al (Ref 134, 135), and FeAl (Ref 136). Another limitation in using HIP for ceramic/intermetallic densification is the lack of adequate encapsulation materials with softening temperatures close to ceramic sintering temperatures (Ref 5).

Theoretical or near-theoretical densities and grain sizes less than 100 nm have been achieved by hot pressing of mechanically alloyed Al-10Ti (Ref 137), Fe_3Al, Ni_3Al (Ref 70, 138) and TiAl (Ref 140).

In cold isostatic pressing at very high pressures (up to 5.6 GPa), densities in excess of 80% have been achieved for SiO_2 (Ref 42). A factor in high pressure approaches in ceramic powders is the high level of residual stress that can result in fracture upon subsequent handling (Ref 5).

Sinter forging has been effectively applied (Fig. 18) in full or near full density processing of ceramics (Ref 25, 31, 32, 87, and 104). Generally, the stress levels required for densification by sinter forging are lower than in hot pressing or HIP. This method received substantial attention both theoretically and experimentally as applied to ceramic densification (Ref 84, 120, 140–142). The most attractive benefit of sinter forging is the capability to densify green compacts with large interagglomerate pores.

In the Ti-Si system, shock consolidation yielded 30 to 40 nm grain size of crystalline $TiSi_2$ and Ti_5Si_3 phases (Ref 143). Only limited grain coarsening took place upon subsequent annealing at 1070 K for 1 h. Full densification was also reported in mechanically alloyed TiAl specimens with a final grain size of 15 nm (Ref 144). These results are compared with HIP, in which full consolidation was achieved at 1075 K (207 MPa for 2 h) but with a grain size of about 100 nm.

Composites and Cermets

Handwerker et al. (Ref 83) summarized the sintering behavior of ceramic composites. Hot pressing gives distinct advantages in comparison to pressureless sintering in achieving full densities and minimal grain growth as shown, for example, in nanograined Fe-(Fe, Mo)$_6$C (Ref 145, 146) and ZrO_2-Al_2O_3 (Ref 87). A grain size of only 16 nm has been retained in cryomilled Fe-10Al which was hot pressed at 823 K and heat treated for 1 h at 1223 K (Ref 107). This unusual stability is attributed to nanometer dispersoids of γ-Al_2O_3 and AlN particles. Similar densification results with final nanocrystalline structures have been reported by hot pressing of tungsten-

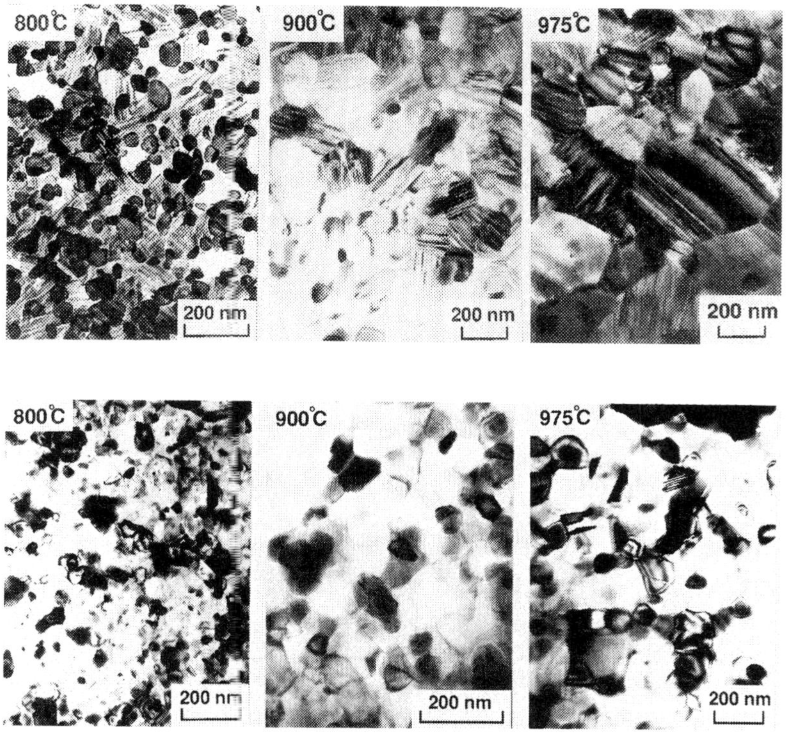

(a)

(b)

Fig. 19 Microstructures of ball-milled and HIPed (a) Ti-24Al-11Nb, and (b) Ti-55Al. Initial particle size was 10–15 nm. HIP was performed at the temperature shown for 2 h with 207 MPa pressure. Source: Ref 134

titanium and metal-nitride composites (Ref 147, 148). Hot pressing of nanoceramic composites, such as ZrO_2/Al_2O_3 (Ref 149), Si_3N/SiC (Ref 150), and Al_2O_3/Ni (Ref 151), achieved full densities and grain sizes below 100 nm. Mechanically alloyed Al-10.7Ti powder was densified to 98% under 2 GPa at 573 K with virtually no grain growth and retention of the initial aluminum supersaturation (Ref 109).

Nanograined WC-Co powders have been fully consolidated with minimal grain growth by conventional sintering, but the final grain size is approximately 100 to 200 nm (Ref 152–154). This final grain size is mainly attributed to the liquid phase mechanism that typically promotes grain growth. One approach in restricting grain growth in WC-Co is to use inhibitors (Ref 152).

Computational Studies

Molecular dynamics simulations have been performed for two to nine nanocrystalline metallic particles: copper composed of 4,688 atoms sintered at 700 K (Ref 16, 155–157), nickel composed of 767 atoms sintered between 50 K and 945 K (Ref 158), and 3 nm palladium clusters (Ref 159). In all cases, two-particle interactions were typified by particle rotation to a low-energy orientation accompanied by neck growth. The time required for these processes to occur was 10 to 500 picoseconds. Rapid neck formation has been attributed to dislocation activity driven by contact Hertzian stresses that exceed the ideal shear strength (Ref 156). Dislocations are generated in the neck and move along the usual slip systems in the fcc crystals. Particle rotation during sintering has been observed experimentally (Ref 114, 160, and 161). Further, particle rotation at least partially explains twin formation in nanocrystalline metallic particle pairs (Ref 115, 162, 163).

REFERENCES

1. G.B. Prabhu and D.L. Bourell, Processing of Nanocrystalline Zirconia and Zirconia Alumina Composite Powder by Hot Pressing and HIP, *Synthesis and Processing of Nanocrystalline Powder,* D.L. Bourell, Ed., TMS, 1996, p 181–191
2. C.C. Koch and Y.S. Cho, Nanocrystals by High-Energy Ball Milling, *Nanostructured Mater.,* Vol 1, 1992, p 207–212
3. A.C.F. Cocks, The Structure of Constitutive Laws for the Sintering of Fine Grained Materials, *Acta Metall. Mater.,* Vol 42 (No. 7), 1994, p 2191–2210
4. H. Gleiter, Nanocrystalline Materials, *Prog. Mater. Sci.,* Vol 33 (No. 4), 1989, p 223–315
5. M. Mayo, Processing of Nanocrystalline Ceramics from Ultrafine Particles, *Int. Mater. Rev.,* Vol 41, 1996, p 85–115
6. C. Suryanarayana, Nanocrystalline Materials, *Int. Mater. Rev.,* Vol 40, 1995, p 41–64
7. J. Weismuller, Nanocrystalline Materials—An Overview, *Synthesis and Processing of Nanocrystalline Powder,* D.L. Bourell, Ed., TMS, 1996, p 3–19
8. J. Groza, Nanocrystalline Powder Consolidation Methods, *Nanostructured Mater.,* Carl Koch, Ed., Noyes Publisher, 1998
9. M.F. Ashby, "Sintering and Isostatic Pressing Diagrams," Department of Engineering, University of Cambridge, Cambridge, U.K.
10. D.L. Bourell and W.A. Kaysser, Nanocrystalline Iron Sintering Behavior and Microstructural Development, *Metall. Mat. Trans. A,* Vol 25, 1994, p 677–685
11. R.J. Conder, C.B. Ponton, and P.M. Marquis, Processing of Alumina/Silicon Carbide Nanocomposites, *Nanoceramics,* R. Freer, Ed., The Institute of Materials, 1993, p 105–116
12. R.A. Andrievski, Compaction and Sintering of Ultrafine Powders, *Int. Powder Metall.,* Vol 30, 1994, p 59–66
13. W. Chen, A. Pechenik, S.J. Dapkunas, G.J. Piermarini, and S.G. Malghan, Novel Equipment for the Study of Compaction of Fine Powders, *J. Am. Ceram. Soc.,* Vol 77, 1994, p 1005–1010
14. T.R. Malow and C.C. Kock, Grain Growth of Nanocrystalline Materials—A Review, *Synthesis and Processing of Nanocrystalline Powder,* D.L. Bourell, Ed., TMS, 1996, p 33–44
15. G.S.A.M. Thuenissen, A.J.A. Winnubst, and A.J. Burggraaf, Sintering Kinetics and Microstructure Development of Nanoscale Y-TZP Ceramics, *J. Eur. Ceram. Soc.,* Vol 11, 1993, p 319–324
16. R.S. Averback, H. Zhu, R. Tao, and H.J. Höfler, Sintering of Nanocrystalline Materials: Experiments and Computer Simulations, *Synthesis and Processing of Nanocrystalline Powder,* D.L. Bourell, Ed., TMS, 1996, p 203–216
17. H. Hahn, J. Logas, and R.S. Averback, Sintering Characteristics of Nanocrystalline TiO_2, *J. Mater. Res.,* Vol 5, 1990, p 609–614
18. H. Hahn, Microstructure and Properties of Nanostructured Oxides, *Nanostructured Mater.,* Vol 2, 1993, p 251–265
19. R.A. Andrievski, Compaction and Sintering of Ultrafine Powders, *Int. Powder Metall.,* Vol 30, 1994, p 59–66
20. M. Mayo, D.C. Hague, and D.-J. Chen, Processing Nanocrystalline Ceramics for Applications in Superplasticity, *Mater. Sci. Eng. A,* Vol 166, 1993, p 145–159
21. M.J. Mayo, Superplasticity in Nanostructured Ceramics, *Mechanical Properties and Deformation Behavior of Materials Having Ultra-Fine Microstructures,* M. Nastasi, D.M. Parkin, and H. Gleiter, Ed., Kluwer Academic Press, Dordrecht, Netherlands, 1993, p 361–380
22. V.A. Hackley, M.A. Anderson, and S.A. Spooner, Small-Angle X-ray Scattering Study of Microstructure Evolution During Sintering of Sol-gel-Derived Porous Nanophase Titania, *J. Mater. Res.,* Vol 7, 1992, p 2555–2571
23. T.L. Wen, V. Hebert, S. Vilminot, and J.C. Bernier, Preparation of Nanosized Yttria-Stabilized Zirconia Powders and Their Characterization, *J. Mater. Sci.,* Vol 26, 1991, p 3787–3791

24. J.-M. Wu and C.-H. Wu, Sintering Behavior of Highly Agglomerated Ultrafine Zirconia Powders, *J. Mater. Sci.,* Vol 23, 1988, p 3290–3299
25. H. Hahn, R.S. Averback, H.J. Höfler, and J. Logas, Sintering and Deformation of Nanocrystalline Ceramics, *Clusters and Cluster-Assembled Materials,* R.S. Averbach, J. Bernholc, and D.L. Nelson, Ed., Materials Research Society, 1991, p 569–580
26. G.L. Messing and M. Kumagai, Low-Temperature Sintering of α-Alumina Seeded Bohemite Gels, *Am. Ceram. Bull.,* Vol 73, 1994, p 88–91
27. G.B. Prabhu and D.L. Bourell, Abnormal Grain Growth in Alumina-Zirconia Nanocomposites, *Nanostructured Mater.,* Vol 5 (No. 6), 1995, p 727–732
28. K. Hayashi and H. Kihara, Sintering Behavior of Ultrafine Metal Powders, *Fundamentals of Diffusion Bonding,* I. Yshida, Ed., Elsevier, 1987, p 501–509
29. W.H. Rhodes, Agglomerate and Particle Size Effects on Sintering Yttria-Stabilized Zirconia, *J. Am. Ceram. Soc.,* Vol 64, 1981, p 19–22
30. S.J. Wu and L.C. De Jonghe, Sintering of Nanophase γ-Al_2O_3 Powder, *J. Am. Ceram. Soc.,* Vol 79, 1996, p 2207–2211
31. G. Skandan, Processing of Nanostructured Zirconia Ceramics, *Nanostructured Mater.,* Vol 5, 1995, p 111–126
32. G. Skandan, H. Hahn, B.H. Kear, M. Roddy, and W.R. Cannon, Processing of Nanostructured Zirconia Ceramics, *Molecularly Designed Ultrafine/Nanostructure Materials,* K.E. Gonsalves, G.-M. Chow, T.D. Xiao, and R.C. Cammarata, Ed., Materials Research Society, 1994, p 207–212
33. J.-P. Ahn, M.-Y. Huh, and J.-K. Park, Effect of Green Density on Subsequent Densification and Grain Growth of Nanophase SnO_2 Powder During Isothermal Sintering, *Nanostructured Mater.,* Vol 8, 1997, p 637–643
34. T. Rabe and R. Wasche, Sintering Behavior of Nanocrystalline Titanium Nitride Powders, *Nanostructured Mater.,* Vol 6, 1995, p 357–360
35. D.L. Bourell and W. Kaysser, Effect of Nonisothermal Heating or Cooling on Grain Growth, *Acta Metall. Mater.,* Vol 41 (No. 10), 1993, p 2933–2937
36. G.B. Prabhu and D.L. Bourell, Synthesis, Sintering Behavior and Grain Coarsening Characteristics of Nanocrystalline Yttria Stabilized Tetragonal Zirconia, *Novel Techniques in Synthesis and Processing of Advanced Materials,* J. Singh and S. Copley, Ed., TMS, 1995, p 91–102
37. M.I. Alymov, E.I. Maltina, and Y.N. Stepanov, Model of Initial Stage of Ultrafine Metal Powder Sintering, *Nanostructured Mater.,* Vol 4, 1992, p 737–742
38. C. Herring, Effect of Change of Scale on Sintering Phenomena, *J. Appl. Phys.,* Vol 21, 1950, p 301–303
39. M.F. Yan and W.W. Rhodes, Low Temperature Sintering of TiO_2, *Mater. Sci. Eng.,* Vol 61, 1983, p 59–66
40. A. Pechenik, G.J. Piermarini, and S.C. Danforth, Fabrication of Transparent Silicon Nitrides from Nanosize Particles, *J. Am. Ceram. Soc.,* Vol 75, 1992, p 3283–3288
41. O. Dominguez, Y. Champion, and J. Bigot, A Microstructural Study of the Consolidation of Fe and Cu Nanocrystalline Powders, *Synthesis and Processing of Nanocrystalline Powder,* D.L. Bourell, Ed., TMS, 1996, p 193–201
42. M.R. Gallas, A.R. Rosa, T.H. Costa, and J.A.H. da Jornada, High Pressure Compaction of Nanosize Ceramic Powder, *J. Mater. Res.,* Vol 12, 1997, p 764–768
43. A.R. Cooper and L.E. Eaton, Compaction Behavior of Several Ceramic Powders, *J. Am. Cer. Soc.,* Vol 45, 1962, p 97–101
44. I. Shapiro and M. Kolthoff, The Compressibility of Silver Bromide Powders, *J. Phys. Colloid Chem.,* Vol 51, 1947, p 483–493
45. J.R. Groza, Non-Conventional Pressure-Assisted Powder Consolidation Processes, *J. Mater. Eng. Perform.,* Vol 2, 1993, p 283–290
46. E. Arzt, M.F. Asby, and K.E. Easterling, Practical Applications of Hot-Isostatic Pressing Diagrams: Four Case Studies, *Metall. Trans. A.,* Vol 14, 1983, p 211–221
47. F.B. Swinkels, D.S. Wilkinson, E. Arzt, and M.F. Ashby, Mechanisms of Hot-Isostatic Pressing, *Acta Metall.,* Vol 31, 1983, p 1829–1840
48. M.F. Ashby, *Operating Manual for HIP 487,* Cambridge University, 1990
49. G.R. Shaik and W.W. Milligan, Consolidation of Nanostructured Metal Powders by Rapid Forging: Processing, Modeling, and Subsequent Mechanical Behavior, *Metall. Mater. Trans. A,* Vol 28, 1997, p 895–904
50. D.L. Bourell, Parimal, and W.A. Kaysser, Sol-Gel Synthesis of Nanophase Yttria-Stabilized Tetragonal Zirconia and Densification Behavior Below 1600 K, *J. Am. Ceram. Soc.,* Vol 76, 1993, p 705–711
51. M.G. McKimpson, Densification Maps for Nanosized Powders, *Mater. Manuf. Process.,* Vol 11, 1996, p 935–949
52. L.I. Trusov, V.N. Lapovok, and V.I. Novikov, Problems of Sintering Ultrafine Powders, *Science of Sintering,* D.P. Uskokovic, H. Plamour III, and R.M. Spriggs, Ed., Plenum Press, 1989, p 185–192
53. R.M. German, *Sintering Theory and Practice,* John Willey & Sons, 1996
54. T.R. Malow and C.C. Koch, Grain Growth in Nanocrystalline Iron Prepared by Mechanical Attrition, *Acta Mater.,* Vol 45, 1997, p 2177–2186
55. O. Dominguez and J. Bigot, Materials Transport Mechanisms and Activation Energy in Nanometric Fe Powders Based on Sintering Experiments, *Nanostructured Mater.,* Vol 6, 1995, p 877–880
56. R.J. Perez, B. Huang, A.A. Sharif, and E.J. Lavernia, Thermal Stability of Cryomilled Fe-10wt%Al, *Synthesis and Processing of Nanocrystalline Powder,* D.L. Bourell, Ed., TMS, 1996, p 273–280
57. M. Eldrup et al, Reduction of Surface Oxide on Ultrafine FeNi Particles in *Mechanical Properties and Deformation Behavior of Materials Having Ultra-fine Microstructures,* M. Nastasi, D.M. Parkin, H. Gleiter, Ed., Kluwer Academic, 1993, p 571–577
58. A.V. Ragulya and V.V. Skorokhod, Rate-Controlled Sintering of Ultrafine Nickel Powder, *Nanostructured Mater.,* Vol 5, 1995, p 835–843
59. E.Y. Gutmanas, L.I. Trusov, and I. Gotman, Consolidation, Microstructure and Mechanical Properties of Nanocrystalline Metal Powders, *Nanostructured Mater.,* Vol 4, 1994, p 893–901
60. R.W. Siegel, Synthesis and Processing of Nanostructured Materials, *Mechanical Properties and Deformation Behavior of Materials Having Ultra-Fine Microstructures,* M. Nastasi, D. Markin, and H. Gleiter, Ed., Kluwer Academic Publ., Netherlands, Dordrecht, 1993, p 509–538
61. W. Nieman, J.R. Weertman, and R.W. Siegel, Mechanical Behavior of Nanocrystalline Cu and Pd, *J. Mater. Res.,* Vol 6, 1991, p 1012–1027
62. G.W. Nieman, J.R. Weertman, and R.W. Siegel, Mechanical Behavior of Nanocrystalline Metals, *Nanostructured Mater.,* Vol 1, 1992, p 185–190
63. N.P. Kobelev, Y.M. Soifer, R.A. Andrievski, and B. Gunther, Microhardness and Elastic Properties of Nanocrystalline Silver, *Nanostructured Mater.,* Vol 2, 1993, p 537–544
64. A. Kumpmann, B. Gunther, and H.-D. Kunze, Thermal Stability of Ultrafine-Grained Metals and Alloys, *Mater. Sci. Eng. A,* Vol 168, 1993, p 165–169
65. B. Yao, B.Z. Ding, A.M. Wang, D.J. Li, and Z.Q. Hu, Preparation of a Single Phase Bulk Al-Fe (Mo, Si, B) Nanostructured Alloy Under High Pressure, *Mater. Lett.,* Vol 22, 1995, p 81–86
66. R.Z. Valiev, R.S. Misra, J.R. Groza, and A.K. Mukherjee, Processing of Nanostructured Nickel by Severe Plastic Deformation of Ball Milled Powder, *Scr. Metall. Mater.,* Vol 34, 1996, p 1443–1448
67. H. Shen, Z. Li, B. Gunther, A.V. Korznikov, and R.Z. Valiev, Influence of Powder Consolidation Methods on the Structural and Thermal Properties of a Nanophase Cu-50wt-Percent Ag Alloy, *Nanostructured Mater.,* Vol 6, 1995, p 385–388
68. R.Z. Valiev, Processing of Nanocrystalline Materials by Severe Plastic Deformation Consolidation, *Synthesis and Processing of Nanocrystalline Powder,* D.L. Bourell, Ed., TMS, 1996, p 153–161
69. L. He and E. Ma, Processing and Microhardness of Bulk Cu-Fe Nanocomposites, *Nanostructures Mater.,* Vol 7, 1996, p 327–339
70. L. He and E. Ma, Nanophase Metallic Alloys Consolidated from Powders Prepared by Mechanical Alloying, *Mater. Sci. Eng. A,* Vol 204, 1995, p 240–245

71. L. He, J.Y. Huang, and E. Ma, Retention of Nanostructure in Sinter-Forged Metallic Alloys, *Synthesis and Processing of Nanocrystalline Powder,* D.L. Bourell, Ed., TMS, 1996, p 111–118
72. J.E. Carsley, W.W. Milligan, S.A. Hackney, and E.C. Aifantis, Glasslike Behavior in a Nanostructured Fe/Cu Alloy, *Metall. Mater. Trans. A,* Vol 26, 1995, p 2479–2481
73. C.P. Dogan, J.C. Rawers, R.D. Govier, and G. Korth, Mechanical Processing, Compaction, and Thermal Processing of Alpha-Fe Powder, *Nanostructured Mater.,* Vol 4, 1994, p 631–644
74. G.E. Korth, T.M. Lillo, and J.C. Rawers, Thermal Stability of Dynamically-Consolidated Ball-Milled Nanocrystalline Fe-2C Powder, *Synthesis and Processing of Nanocrystalline Powder,* D.L. Bourell, Ed., TMS, 1996, p 263–272
75. M.I. Alymov and O.N. Leontieva, Synthesis of Nanoscale Ni and Fe powders and Properties of Their Compacts, *Nanostructured Mater.,* Vol 6, 1995, p 393–395
76. H. Nagahama, K. Ohtera, and K. Higashi, Mechanical Properties of Rapidly Solidified Aluminum Alloys Extruded from Amorphous or Nanocrystalline Powders, *Philos. Mag. Lett.,* Vol 67, 1993, p 225–230
77. M.A. Morris and D.G. Morris, Microstructural Refinement and Associated Strength of Copper Alloys Obtained by Mechanical Alloying, *Mater. Sci. Eng.,* Vol 111, 1989, p 115–127
78. A. Kato, T. Suganuma, H. Horikiri, Y. Kawamura, A. Inoue, and T. Masumoto, Consolidation and Mechanical Properties of Atomized Mg-based Amorphous Powders, *Mater. Sci. Eng. A,* Vol 179–180, 1994, p 112–117
79. J.M. McHale, A. Navrotsky, and A.J. Perrotta, Effects of Increased Surface Area and Chemisorbed H_2O on the Relative Stability of Nanocrystalline γ-Al_2O_3 and α-Al_2O_3, *J. Phys. Chem. B,* Vol 101, 1997, p 603–613
80. J.-F. Kuo and D.L. Bourell Sol-Gel Synthesized and Hot Pressed Alumina and Alumina-8 vol% Zirconia Composite, *Proc. of the Symp on Synthesis and Processing of Nanocrystalline Powder,* D.L. Bourell, Ed., TMS Annual Meeting (Anaheim, CA), TMS, 4–8 Feb 1996, 237–248
81. J.E. Bonevics and L.D. Marks, The Sintering Behavior of Ultrafine Alumina Particles, *J. Mater. Res.,* Vol 7, 1992, p 1489–1500
82. P. Vergnon, M. Astier, and S.J. Teichner, Initial Stage for Sintering of Ultrafine Particles (TiO_2 and Al_2O_3), *Fine Particles,* W.E. Kuhn, Ed., The Electrochemical Society, 1974, p 299–307
83. C.A. Handwerker, J.E. Blendell, and R.L. Coble, Sintering of Ceramics, *Science of Sintering,* D.P. Uskokovics, H. Palmour III, and R.M. Spriggs, Ed., Plenum Press, 1989, p 3–37
84. O.-H. Kwon, C.S. Nordahl, and G.L. Messing, Submicrometer Transparent Alumina by Sinter-Forging Seeded γ-Al_2O_3 Powders, *J. Am. Ceram. Soc.,* Vol 78, 1995, p 491–494
85. C.S. Nordahl and G.L. Messing, Transformation and Densification of θ-Alumina during Sinter Forging, *J. Am. Ceram. Soc.,* Vol 79, 1996, p 3145–3154
86. A.J.A. Winnubst, M.M.R. Boutz, Y.J. He, A.J. Burggraaf, and J.H. Verweji, Plasticity of Nanocrystalline Zirconia Ceramics and Composites, *Ceram. Int.,* Vol 23, 1997, p 215–221
87. G.B. Prabhu and D.L. Bourell, Synthesis and Sintering Characteristics of Zirconia and Zirconia-Alumina Nanocomposites, *Nanostructured Mater.,* Vol 6, 1995, p 361–364
88. R.S. Misra, C.E. Lesher, and A.K. Mukherjee, High-Pressure Sintering of Nanocrystalline γAl_2O_3, *J. Am. Ceram. Soc.,* Vol 79, 1996, p 2989–2992
89. V. Ivanov, Y.A. Kotov, O.H. Samatov, R. Bohme, H.U. Karow, and G. Schumacher, Synthesis and Dynamic Compaction of Ceramic Nanopowders by Techniques Based on Electric Pulsed Power, *Nanostructured Mater.,* Vol 6, 1995, p 287–290
90. B. Chelluri, J.P. Barber, E.J. Gonzales, and N.N. Thadani, Nano Grain Size Component Fabrication Using Dynamic Magnetic Compaction (DMC) Process, presented at 1997 Int. Conf. on Powder Metallurgy and Particulate Materials (Chicago), MPIF, 1997
91. D.E. Clark and W.H. Sutton, Microwave Processing of Materials, *Annu. Rev. Mater.,* Vol 26, 1996, p 299–331
92. J.D. Katz, R.D. Blake, and C.P. Scherer, Microwave Sintering of Titanium Diboride, *Ceram. Eng. Sci. Proceed.,* Vol 10, 1989, p 857–867
93. A.W. Fliflet, R.W. Bruce, A.K. Kinkead, R.P. Fischer, D. Lewis, III, R. Rayne, B. Bender, L.K. Kwihara, G.-M. Chow, and P.E. Schoen, Application of Microwave Heating to Ceramic Processing: Design and Initial Operation of a 2.45-GHz Single-Mode Furnace, *IEEE Trans. Plasma Sci.,* Vol 24, 1996, p 1041–1049
94. J. Freim, J. McKittrick, J. Katz, and K. Sickafus, Microwave Sintering of Nanocrystalline γ-Al_2O_3, *Nanostructured Mater.,* Vol 4, 1994, p 371–385
95. G. Link, V. Ivanov, S. Paranin, V. Khrustov, R. Bohme, G. Muller, G. Schumacher, M. Thumm, and A. Weisenburger, A Comparison of mm-Wave Sintering and Fast Conventional Sintering of Nanocrystalline Al_2O_3, *Microwave Processing of Materials V,* M.F. Iskander, J.O. Kiggans, Jr., and J.-C. Bolomey, Ed., Vol 430, Materials Research Society, 1996, p 157–162
96. M.M.R. Boutz, R.J.M. Olde Scholtenhuis, A.J.A. Winnubst, and A.J. Burggraaf, A Hydrothermal Route for Production of Dense, Nanostructured Y-TZP, *Mater. Res. Bull.,* Vol 29, 1994, p 31–40
97. M.G.S. Murray, C.B. Ponton, and P.M. Marquis, Processing of Ultrafine Grained Hydroxyapatite Bioceramics, *Nanoceramics,* R. Freer, Ed., Institute of Materials, London, 1993, p 25–30
98. K.-N.P. Kumar, K. Keizer, A.J. Burggraaf, T. Okubo, H. Nagamoto, and S. Morooka, Densification of Nanostructured Titania Assisted by a Phase Transformation, *Nature,* Vol 358, 1992, p 48–51
99. J.A. Eastman, Microstructural Development in Nanophase TiO_2 During Sintering, *J. Appl. Phys.,* Vol 75, 1994, p 770–779
100. C.D. Terwilliger and Y.-M. Chiang, The Effect of Calcium Segregation on Grain Growth in Nanocrystalline TiO_2, *Nanostructured Mater.,* Vol 4, 1994, p 651–661
101. R.W. Siegel, S. Ramasamy, H. Hahn, Z. Li, T. Lu, and R. Gronsky, Synthesis, Characterization, and Properties of Nanophase TiO_2, *J. Mater. Res.,* Vol 3, 1988, 1367–1372
102. H.J. Höfler and R.S. Averback, Sinter-Forging of Nanocrystalline TiO_2, *Nanophase and Nanocomposite Materials,* S. Komarneni, J.C. Parker, and G.J. Thomas, Ed., Vol 286, Materials Research Society, 1993, p 9–14
103. M. Uchic, H.J. Höfler, W. Flick, R. Tao, P. Kurath, and R.S. Averbach, Sinter-Forging of Nanophase TiO_2, *Scr. Metall. Mater.,* Vol 26, 1992, p 791–796
104. H. Hahn and R.S. Averback, High Temperature Mechanical Properties of Nanostructured Ceramics, *Nanostructured Mater.,* Vol 1, 1992, p 95–100
105. C.D. Terwilliger and Y.-M. Chiang, Characterization of Chemically- and Physically-Derived Nanophase Titanium Oxide, *Nanostructured Mater.,* Vol 2, 1993, p 37–45
106. J. Rawers, G. Slavens, D. Govier, C. Dogan, and R. Doan, Microstructure and Tensile Properties of Compacted, Mechanically Alloyed Nanocrystalline Fe-Al, *Metall, Mater. Trans. A,* Vol 27, 1996, p 3126–3134
107. R.J. Perez, B. Huang, and E.J. Lavernia, Thermal Stability of Nanocrystalline Fe-10 wt%Al Produced by Cryogenic Mechanical Alloying, *Nanostructured Mater.,* Vol 7, 1996, p 565–572
108. H. Araki, S. Saji, T. Okabe, Y. Minamino, T. Yamane, and Y. Miyamoto, Consolidation of MA Al-10.7 at% Ti powder at Low Temperature and High Pressure of 2 GPa, *Mater. Trans., JIM,* Vol 36, 1995, p 465–468
109. S.-C. Liao, K.D. Pae, and W.E. Mayo, High Pressure and Low Temperature Sintering of Bulk Nanocrystalline TiO_2, *Mater. Sci. Eng. A,* Vol 204, 1995, p 152–159
110. R.S. Averback, H.J. Höfler, and R. Tao, Processing of Nano-Grained Materials, *Mater. Sci. Eng. A,* Vol 166, 1993, p 169–177
111. D.C. Hague and M.J. Mayo, The Effect of Crystallization and a Phase Transformation on the Grain Growth of Nanocrystalline Titania, *Nanostructured Mater.,* Vol 3, 1993, p 61–67
112. D. Lewis, R.J. Rayne, B.A. Bender, L.K. Kurihara, G.-M. Chow, A. Fliflet, A. Kincaid, and R. Bruce, Conventional and High Frequency Microwave Processing of Nanophase

Ceramic Materials, *Nanostructured Mater.,* Vol 9, 1997, p 97–100

113. Y. Bykov, S. Gusev, A. Eremeev, V. Holopstev, N. Malygin, S. Pirarunas, A. Sorokin, and A. Shurov, Sintering of Nanophase Oxide Ceramics by Using Millimeter Wave Radiation, *Nanostructured Mater.,* Vol 6, 1995, p 855–858
114. J. Rankin and B.W. Sheldon, In Situ TEM Sintering of Nano-Sized ZrO_2 Particles, *Mater. Sci. Eng. A,* Vol 24, 1995, p 48–53
115. M. Kusunoki, K. Yonemitsu, Y. Sasaki, and Y. Kubo, In Situ Observation of Zirconia Particles at 1200 °C by High-Resolution Electron Microscopy, *J. Am. Ceram. Soc.,* Vol 76, 1993, p 763–765
116. A.J.A. Winnubst, K. Keizer, and A.J. Burggraaf, Mechanical Properties and Fracture Behavior of ZrO_2-Y_2O_3 Ceramics, *J. Mater. Sci.,* Vol 18, 1983, p 1958–1966
117. M.J. Mayo and D.-J. Chen, Densification vs. Grain Growth in Nanocrystalline ZrO_2-3mol%Y_2O_3: How to Win, *Synthesis and Processing of Nanocrystalline Powder,* D.L. Bourell, Ed., TMS, 1996, p 227–236
118. D.-J. Chen and M. Mayo, Densification and Grain Growth of Ultrafine 3 mol % Y_2O_3-ZrO_2 Ceramics, *Nanostructured Mater.,* Vol 2, 1993, p 469–478
119. C.D. Sagel-Ransijn, A.J.A. Winnubst, B. Kerkwijk, A.J. Burggraaf, and H. Verweij, Production of Defect-Poor Nanostructured Ceramics of Yttria-Zirconia, *J. Eur. Ceram. Soc.,* Vol 17, 1997, p 831–841
120. D.C. Hague and M.J. Mayo, Sinter-Forging of Nanocrystalline Zirconia: 1, Experimental, *J. Am. Ceram. Soc.,* Vol 80, 1997, p 149–156
121. D.M. Owen and A.H. Chokshi, An Evaluation of the Densification Characteristics of Nanocrystalline Materials, *Nanostructured Mater.,* Vol 2, 1993, p 181–187
122. M.M.R. Boutz, R.J.M. Olde Scholtenhuis, A.J.A. Winnubst, and A.J. Burggraaf, Preparation, Microstructural Control and Superplasticity of Nanostructured Yttria/Ceria Stabilised Tetragonal Zirconia Ceramics, *Nanoceramics,* R. Freer, Ed., The Institute of Materials, London, 1993, p 75–86
123. G. Skandan, H. Hahn, B.H. Kear, M. Roddy, and R.W. Cannon, The Effect of Applied Stress on Densification of Nanostructured Zirconia during Sinter-Forging, *Mater. Lett.,* Vol 20, 1994, p 305–309
124. M.M.R. Boutz, L. Winnubst, A.J. Burggraaf, M. Nauer, and C. Carry, Low-Temperature Sinter Forging of Nanostructured Y-TZP and Ce-TZP, *J. Am. Ceram. Soc.,* Vol 78, 1995, p 121–128
125. P.E.D. Morgan, Superplasticity in Ceramics, *Ultrafine-Grained Ceramics,* J.J. Burke, L.R. Norman, and V. Weiss, Ed., Syracuse University Press, 1968, p 251–271
126. A. Pechenik, G.J. Piermarini, and S.C. Danforth, Low Temperature Densification of Silicon Nitride Nanoglasses, *Nanostructured Mater.,* Vol 2, 1993, p 479–486
127. S.A. Panfilov, O.V. Padalko, and N.G. Mitrenko, Behavior of Ultrafine Titanium Carbide Powders during Annealing and Sintering, *Poroshk. Metall.,* Vol 11, 1985, p 45–49
128. H. Chang, H.J. Höfler, C.J. Alstetter, R.S. Averbach, Synthesis, Processing and Properties of Nanophase TiAl, *Scr. Metall. Mater.,* Vol 25, 1991, p 1161–1166
129. H. Chang, H.J. Höfler, C.J. Alstetter, R.S. Averbach, Synthesis, Processing and Properties of Nanophase Aluminide, *Mater. Sci. Eng. A,* Vol 153, 1992, p 676–678
130. R.A. Andrievski, V.I. Torbov, and E.N. Kurkin, Nanocrystalline Titanium Nitride, *Proc. of the 13th Int. Plansee Seminar,* H. Bildstein and R. Eck, Ed., Vol 3, Metallwerk Plansee, 1993, p 649–655
131. A. Kaiser, R. Vassen, D. Stover, and H.P. Buchkremer, Heat Treatment of Ultrafine SiC Powders to Reduce Oxygen Sensitivity and Grain Growth, *Nanostructured Mater.,* Vol 4, 1994, p 795–802
132. S.C. Danforth, W. Symons, and K.J. Nilsen, Processing of Silicon Nitride Powders, *Advanced Ceramic Processing and Technology 1,* J.G.P. Binner, Ed., Noyes, Park Ridge, NJ, 1990
133. I. Borner and J. Eckert, Grain Size Effects and Consolidation in Ball-Milled Nanocrystalline NiAl, *Mat. Sci. Forum,* Vol 235–238, 1997, p 79–84
134. C. Suryanarayana, G.E. Korth, F.H. Froes, and J. Hebeisen, Synthesis of Nanostructured Titanium Aluminides by Mechanical Alloying and Hot Isostatic Pressing, *Synthesis and Processing of Nanocrystalline Powder,* D.L. Bourell, Ed., TMS, 1996, p 133–141
135. C. Suryanarayana, G.E. Korth, G.-H. Chen, A. Frefer, and F.H. Froes, Thermal Stability of Nanostructured Titanium Aluminides, *Nanostructured Mater.,* Vol 2, 1993, p 527–535
136. F.H. Froes, C. Suryanarayana, D.K. Mukhopadhyay, D.K. Brand, G. Korth, D. Zick, P. Tylus, and J. Hebeisen, Compaction of Nanograined Gamma TiAl by Hot isostatic Pressing, *Advances In Powder Metallurgy & Particulate Materials,* M. Phillips and J. Porter, Ed., Vol 3, Part 9–13, Metal Powder Industries Federation, 1995, p 12–63
137. L. Guoxian, M. Qingchang, Li Zhichao, and W. Erde, Consolidation of Nanocrystalline Al-Ti Alloy Powders Synthesized by Mechanical Alloying, *Nanostructured Mater.,* Vol 5, 1995, p 673–678
138. L. He and E. Ma, Full-Density Nanocrystalline Fe-29Al-2Cr Intermetallic Consolidated from Mechanically Milled Powders, *J. Mater. Res.,* Vol 11, 1996, p 72–80
139. K. Aoki, Y. Itoh, J. Park, Y. Kawamura, A. Inoue, and T. Masumoto, Preparation of Nanocrystalline Bulk TiAl by a Combination of HDDR and Hot Pressing, *Intermetallics,* Vol 4, 1996, p 5103–5111
140. R.S. Averback, H.J. Höfler, H. Hahn, and J.C. Logas, Sintering and Grain Growth in Nanocrystalline Ceramics, *Nanostructured Mater.,* Vol 1, 1992, p 173–178
141. D.C. Hague and M.J. Mayo, Modeling Densification During Sinter-Forging of Yttria-Partially-Stabilized Zirconia, *Mater. Sci. Eng. A,* Vol 204, 1995, p 83–89
142. B.J. Kellet and F.F. Lange, Experiments on Pore Closure during Hot Isostatic Pressing and Forging, *J. Am. Ceram. Soc.,* Vol 71, 1988, p 7–12
143. S.C. Glade and N.N. Thadhani, Shock Consolidation of Mechanically Alloyed Amorphous Ti-Si Powders, *Metall. Mater. Trans. A,* Vol 26, 1995, p 2565–2569
144. C. Suryanarayana, G.E. Korth, and F.H. Froes, Compaction and Characterization of Mechanically Alloyed Nanocrystalline Titanium Aluminides, *Metall. Mater. Trans. A,* Vol 28, 1997, p 293–302
145. R.K. Sadangi, B.H. Kear, and L.E. McCandlish, Hot Pressing of Nanograined Fe-(Fe, $Mo)_6C$ Composite Powders, *Nanostructured Mater.,* Vol 2, 1993, p 563–569
146. R.K. Sadangi, B.H. Kear, L.E. McCandlish, Pressureless Sintering of Nanograined Fe-(Fe, $Mo)_6C$ Composite Powders, *Powder Metall.,* Vol 37, 1994, p 277–282
147. R.L. Axelbaum, J.I. Huertas, C.R. Lottes, S. Hariprasad, and S.M. Sastry, Nano-Phase W and W-Ti Composite via Gas-Phase Combustion Synthesis, *Mater. Manuf. Proc.,* Vol 11, 1996, p 1043–1053
148. J. Secondi, O. Drbohlav, and R. Yavari, Metal-Nitride Composites Obtained by Alloy Milling under Nitrogen Gas, *Mater. Sci. Forum,* Vol 179, 1995, p 287–294
149. S. Bose, Application of Nanocrystalline Materials as Matrices of Composites: Processing and Performance Advantages, *Mater. Sci. Eng. A,* Vol 196, 1995, p 105–109
150. Z. Lences, A. Bellosi, and F. Monteverde, Factors Influencing the Crystallization and the Densification of Ultrafine Si/N/C Powders, *Mater. Chem. Phys.,* Vol 41, 1995, p 46–54
151. T. Sekino, T. Nakajima, and K. Niihara, Mechanical and Magnetic Properties of Nickel Dispersed Alumina-Based Nanocomposite, *Mater. Lett.,* Vol 29, 1996, p 165–169
152. L. Wu, B.K. Kim, B.H. Kear, and L.E. McCandlish, Grain Growth Inhibition in Sintering of Nanostructured WC-Co Alloys, *Proc. 13th Int. Plansee Seminar,* H. Bildstein and R. Eck, Ed., Vol 3, Metallwerk Plansee, 1993, p 667–679
153. Z. Fang and J.W. Eason, Study of Nanostructured WC-Co Composites, *Int. J. Refractory & Hard Met.,* Vol 13, 1995, p 297–303
154. L.E. Candlish, B.H. Kear, and B.K. Kim, Processing and Properties of Nanostructured WC-Co, *Nanostructured Mater.,* Vol 1, 1992, p 119–124
155. H. Zhu and R.S. Averback, Sintering Process of Two Nanoparticles: A Study by Molecular Dynamics Simulations, *Philos. Mag. Lett.,* Vol 73, 1996, p 27–33

156. H. Zhu and R.S. Averback, Sintering of Nano-Particle Powders: Simulations and Experiments, *Mater. Manuf. Process.,* Vol 11, 1996, p 905–923
157. H. Zhu and R.S. Averback, Molecular Dynamics Simulations of Densification Processes in Nanocrystalline Materials, *Mater. Sci. Eng. A,* Vol 204, 1995, p 96–100
158. H.L. Heinisch, Computer Simulations of Nanoparticle Interactions, *Synthesis and Processing of Nanocrystalline Powder,* D.L. Bourell, Ed., TMS, 1996, p 299–310
159. C.-L. Liu, J.B. Adams, and R.W. Siegel, Molecular Dynamics Simulation of Consolidation Processes During Fabrication of Nanophase Palladium, *Nanostructured Mater.,* Vol 4 (No. 3), 1994
160. N. Guillou, L.C. Nistor, H. Fuess, and H. Hahn, Microstructural Studies of Nanocrystalline CeO_2 Produced by Gas Condensation, *Nanostructured Mater.,* Vol 8, 1997, p 545–557
161. M. Yeadon, J.C. Yang, M. Ghaly, D.L. Olynick, R.S. Averbach, and J.M. Gibson, Sintering of Sputtered Copper Nanoparticles on (001) Copper Substrates, *Nanophase and Nanocomposite Materials II,* S. Komarneni, J.C. Parker, and H.J. Wollenberger, Ed., Materials Research Society, 1997, p 179–184
162. F. Bodker, S. Morup, and S. Linderoth, Surface Effects in Metallic Iron Nanoparticles, *Phys. Rev. Lett.,* Vol 72, 1994, p 282–285
163. Y. Champion and J. Bigot, Preparation and Characterization of Nanocrystalline Copper Powders, *Scr. Mater.,* Vol 35, 1996, p 517–522
164. M.N. Rahaman, Sintering and Grain Growth of Ultrafine CeO_2 Powders, *Sintering Technology,* Marcel Dekker, 1996, p 93–100

Reactive Sintering

David E. Alman, Albany Research Center, U.S. Department of Energy

COMBUSTION SYNTHESIS (also termed variously in the literature as self-propagating, high-temperature synthesis (SHS) and reactive synthesis) has been used to produce advanced ceramic and intermetallic compounds, such as $MoSi_2$ and WC, in particulate form (Ref 1–3) since the late 1940s and early 1950s. Quite simply, combustion synthesis consists of forming compounds or composites from elemental or subcompound components:

$$xA + yB \rightarrow A_xB_y \quad \text{(Eq 1a)}$$

or

$$AB_y + xC \rightarrow AC_x + yB \quad \text{(Eq 1b)}$$

The driving force for the reaction is the negative heat of mixing of the forming compound. This results in a release of energy (as heat), which sustains and propagates the reaction through the body of the reactants. A complete description of the thermodynamics, wave propagation, and phenomenological aspects of combustion synthesis can be found in the article "Combustion Synthesis of Advanced Materials" in this Volume. Commonly, the product of a combustion synthesis reaction is porous and therefore easily milled into powder.

Evolution of Porosity During Combustion Synthesis. Porosity is induced during combustion synthesis from several factors (Ref 4–11):

- *Kirkendall and Frenkel effects.* The diffusivity of the elements is unbalanced (i.e., element A diffuses more rapidly into element B than B diffuses into A; as a consequence, after the reaction propagates through the reactants a pore forms at prior particle A powder sites).
- *Gas evolution.* The heat generated by the initiation of the combustion synthesis reaction may be great enough to volatilize residual impurities attached to the surfaces of the starting powders. The evolution of gaseous species may be violent enough to prevent densification of the product or can be entrapped within the product resulting in porosity.
- *Molar volume changes.* The product phase has a different molar volume than the reactants. Typically, the product phase is denser than the reactants. For aluminides, the density increase is typically less than 12%, while for ceramics, the increase is quite substantial, ranging from 15 to 30% (see Table 1, Ref 8). The resulting volumetric change is intrinsic, as it is a direct result of the inherent atomic structure and bonding of the resultant product versus those of the starting constituents. Unless the change in the dimensions of the product (e.g., shrinkage due to sintering) is equal to the intrinsic volume change resulting from the reaction, porosity will be generated in the product.
- *Initial porosity.* The starting green compact of pressed powders contains 20 to 50 vol% porosity. This must to be eliminated during the sintering operation for product densification to occur.

The aim of reactive sintering is to form dense intermetallic and ceramic compounds, alloys, or composites directly from elemental powders. As discussed below, during combustion synthesis of certain compounds, the formation of porosity can be eliminated by careful control over the reaction and/or through the application of external pressure during reactive sintering by reactive hot pressing (RHP) or reactive hot isostatic pressing (RHIP).

Reactive Sintering. Coble (Ref 12) defines seven classes of reactions that can occur during sintering:

$A_1 \rightarrow A_2$
Crystallographic phase change (1)
$A + B \rightarrow \alpha$
Solid-solution formation (2)
$A + B \rightarrow \alpha + \beta$
Solid-solution formation (3)
$A + B \rightarrow \alpha + \text{liquid}$
Liquid phase sintering (4)
$A + B \rightarrow AB$
Compound formation (5)
$A_1 + \text{liquid} \rightarrow A_2 + \text{liquid}$
$\beta \rightarrow \alpha Si_3N_4$, with liquid (6)
$A_{solid} + B_{liquid} \rightarrow AB_{solid}$
Disappearing liquid phase, reactive infiltration (7)

Reactions 2 and 3 are simple eutectic systems at temperatures below the eutectic temperature. The final product is either single phase (α) or two phase ($\alpha + \beta$). The fourth reaction is analogous to the third reaction, except at temperatures above the eutectic temperature. The sixth reaction is for phase transformation via solution in a liquid. An example of this is $B \rightarrow \alpha Si_3N_4$ transformation in magnesium silicate liquid. This article is concerned with sintering when reactions of type 5 occur and both starting constituents are in solid form. Reactive liquid infiltration of powder preforms is not within the scope of this article.

In many instances, the combustion synthesis reaction is accompanied by the formation of a transient liquid phase (Ref 13, 14). As with any liquid phase sintering operation, densification occurs due to capillary effects associated with the formation of the liquid. Figure 1 illustrates the reactive sintering process for forming the compound AB from a mixture of A and B powders (Ref 13). The initial compact is composed of mixed elemental powders, which diffusionally interact during heating. When the lowest liquidus temperature is reached (e.g., a eutectic or the melting point of one of the constituents), partial melting of the compact occurs. The liquid flows into pores between the solid particles resulting in densification. Also, the liquid is a rapid

Table 1 Intrinsic volume and density changes during formation of ceramic and intermetallic products from elemental powders

Product	Volume change (ΔV), %	ρ_p/ρ_r(a)
$MoSi_2$	−40.6	1.39
SiC	−28.4	1.39
$TiSi_2$	−27.5	...
TiC	−24.4	1.32
WC	−23.8	1.31
TiB_2	−23.3	1.29
NiAl	−12.6	...
TiAl	−5.3	...
Ni_3Al	−5.2	...
ReHf	−2.2	...
$3TiC + 2Al_2O_3$	−22 to −28	1.28 to 1.39
$3TiB_2 + 5Al_2O_3$	−27 to −28	1.37 to 1.39
$9Fe + 4Al_2O_3$	−22	1.28
$3Fe + Al_2O_3$	−19	1.23

(a) ρ_p/ρ_r is theoretical density of product(s)/theoretical density of reactants (i.e., density of nonporous compact of reactants). Source: After Ref 8 diffusion path between the elements, which facilitates compound formation. The final product is single phase and dense AB.

Formation of the transient liquid phase during processing does not guarantee a dense product. Shrinkage porosity may develop during the solidification of the liquid phase (Ref 8). Control over the formation of the liquid phase, hence the reaction, through judicial selection of the characteristics of the starting powders and the process parameters is paramount for producing dense articles with useful microstructures by reactive sintering. Further difficulties can arise during reactive sintering when the adiabatic flame temperature of the reaction exceeds the melting point of the forming compound. Nevertheless, monolithic aluminides, ceramics, and composites have been fabricated by reactive sintering and its variants, as discussed in detail below.

Densification during Reactive Sintering

Figure 2 shows a microstructure of Ni_3Al produced by reactive sintering nickel and aluminum powders at 750 °C for 15 min in vacuum. The elemental powder mixture was die pressed at room temperature to a green density of about 70% and heated at a rate of 20 K/min to the sintering temperature. As is evident, near-fully dense (>97% theoretical) Ni_3Al can be produced utilizing processing cycles consisting of short sintering times and low homologous sintering temperatures ($T_H = 0.60$ and is defined as the sintering temperature, 1023 K, divided by the melting temperature of compound, 1660 K). Densification coincides with the initiation of the reaction, as shown by the dilatometer and differential thermal analysis scans illustrated in Fig. 3. The high final density is a result of:

- The relatively small volume change (–5.2%, see Table 1) that occurs when the nickel and aluminum powders transform to Ni_3Al
- The presence of a transient liquid phase during the sintering cycle

The processing parameters that influence the reaction between the constituent powders and alter the amount, distribution, and duration of the liquid phase impact the density of the final product. These processing parameters include: powder particle sizes, composition, green density, heating rate, and sintering atmosphere. Proper manipulation of these parameters can enhance the density of the reactively sintered product.

Manipulation of the Liquid Phase with Process Parameters. The size ratio of the reactant powder particles plays an important role in the distribution of the liquid phase and therefore impacts the final density of several reactive sintered compounds (Ref 15–22), as illustrated in Fig. 4 for Ni_3Al. By selecting appropriate nickel and aluminum powder particle sizes, the initial green microstructure will consist of two interpenetrating and interconnected nickel and aluminum powder networks. In this configuration, the contact surface between the nickel and aluminum will be maximized, which mitigates pore formation when the transient liquid reacts and flows into the solid phase. Interconnectivity of the particulate phases is necessary for long-range densification by means of capillary action. If the aluminum particles exist in an isolated configuration, pools or pockets of liquid form upon melting and porosity will form at the prior aluminum sites when the liquid flows from the pockets into the solid phase during the reaction.

Aluminum constitutes 34 vol% of the solid powder mixture in stoichiometric Ni_3Al, and at this volume fraction interconnectivity requires a

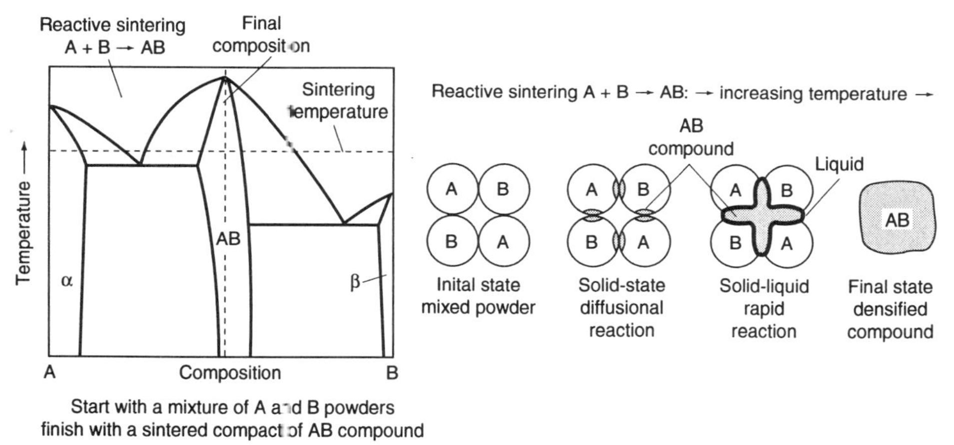

Fig. 1 A binary phase diagram and the typical sequence of events expected in reactive sintering of compound AB from a mixture of A and B powders. Source: Ref 13

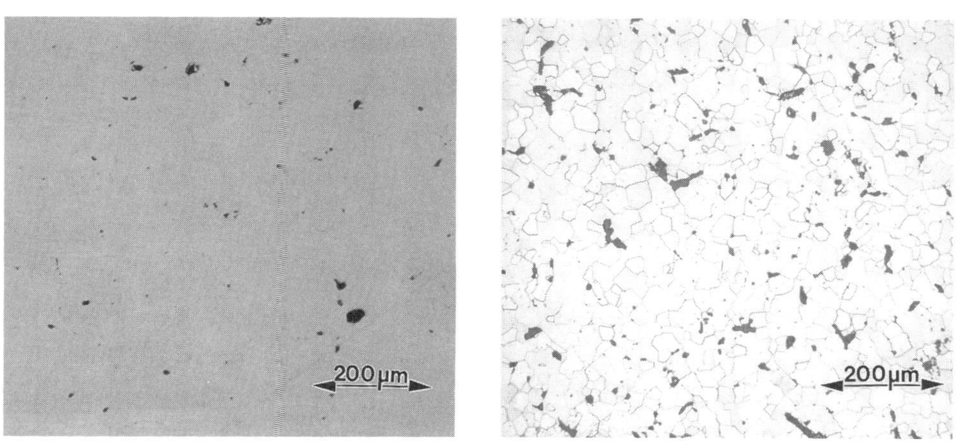

(a) (b)

Fig. 2 Microstructure of reactive sintered Ni_3Al (750 °C, 15 min). (a) As-polished. (b) Etched with Kallings solution

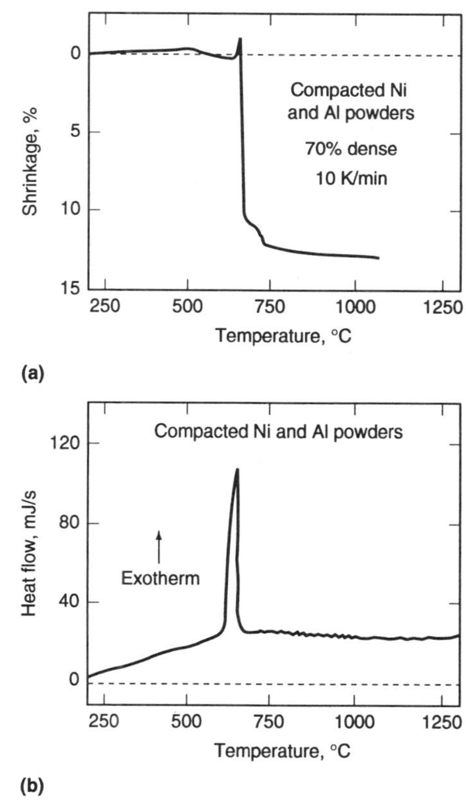

Fig. 3 Dilatometry (a) and differential thermal analysis (b) of compacted nickel and aluminum powders mixed at the Ni_3Al stoichiometry. Densification (shrinkage on the dilatometry scan) corresponds with the reaction (as indicated by the exothermic reaction on the DTA scans). Source: Ref 14

nickel-to-aluminum powder size ratio of 2.4 to 1. German and coworkers (Ref 15, 17–20), reactively sintered Ni_3Al to 97% of its theoretical density, using 45 μm agglomerated Ni and 15 μm Al powders (i.e., a nickel-to-aluminum particle size ratio of 3 to 1). Nishimura and Liu (Ref 21), using similar powders (agglomerated 45 μm Ni and 15 μm Al), obtained 98% dense Ni_3Al by reactive sintering at 620 °C for 15 min in a vacuum. Experiments using other nickel-to-aluminum particle size ratios produced Ni_3Al microstructures with substantially more porosity (Ref 15–20), as shown in Fig. 5(a). Powder size ratio also influenced the density of reactive sintered $NbAl_3$, as shown in Fig. 5(b) (Ref 22). In this case, aluminum constitutes 74 vol% of $NbAl_3$, and interconnectivity requires a niobium-to-aluminum particle size ratio of 1 to 3. Dense $NbAl_3$ (95%) was reactively sintered using a niobium-to-aluminum particle size ratio of 1 to 3 (10 μm Nb and 30 μm Al powders). The pores in this specimen were less than 10 μm in diameter, indicating that capillary action caused densification of the structure. Specimens produced using other niobium and aluminum particle size ratios (3 to 1, 1 to 1.5, and 1 to 9.5) swelled upon sintering. The microstructure of these specimens contained large pores. The large pores formed when the disconnected liquid aluminum pools were consumed by the reaction (the liquid aluminum diffuses into the niobium powders, leaving pores). Unfortunately, similar particle size adjustments did not mitigate swelling during reactive sintering of TiAl, Al_3Ta, and Fe_3Al (Ref 18, 23, 24).

The influence of heating rate and sintering atmosphere on the density of reactive sintered Ni_3Al is shown in Fig. 6 (Ref 15, 17–20). The role of sintering atmosphere is explained by heat conduction and entrapped gas effects. Heat is carried away from the compact during reaction by the higher thermal conductivity of gas versus vacuum. This lowers the maximum temperature obtained during the reaction, which impedes densification. Furthermore, because of the speed of the reaction there is no time for absorbed vapors and atmosphere captured in the pores to escape. Therefore, the gas becomes entrapped in pores and prevents densification. Heating rate effects on transient liquid phase sintering have been explained based on solid-state interdiffusion prior to liquid formation. With slow heating rates, there is more solid-state diffusion and intermediate compound formation prior to the reaction initiation. This inhibits liquid formation and results in low sintered densities.

Control over Excessive Liquid Formation by Adding a Diluent. Shape retention is lost if too much liquid forms during reactive sintering. This can occur when the adiabatic flame temperature of the reaction exceeds the melting point of the forming compound. This situation occurs during combustion synthesis of NiAl (T_m = 1640 °C), as illustrated in Fig. 7(a) (Ref 25). Control over the reaction can be gained by adding a high melting phase to act as a ballast and dilute the reaction. This phase can be a ceramic reinforcement phase, such as TiB_2 or Al_2O_3, or prealloyed NiAl, as is shown in Fig. 7(b). The addition of the high-melting and inert diluent phase absorbs some of the heat generated by the reaction and lowers the adiabatic flame temperature (Ref 26).

The diluted combustion synthesis reaction is as follows:

$$xA + xB + (1-x)D \rightarrow xAB + (1-x)D \qquad \text{(Eq 2)}$$

The amount of diluent $(1-x)$ required to prevent slumping, due to the adiabatic flame temperature (T_{ab}) of the reaction exceeding the melting point of the compound (T_m), can be determined from a simple energy balance:

$$x\left[\Delta H_f^T\right]_{AB} = x\left[\int_T^{T_m} C_{p_{AB}}\,dT\right] + (1-x)\left[\int_T^{T_m} C_{p_D}\,dT\right] \qquad \text{(Eq 3)}$$

where ΔH_f^T is the heat of formation for the compound at a reference temperature, T, such as the reaction initiation temperature or 298 K. $C_{p_{AD}}$ and C_{p_D} are the heat capacity of the compound, AB, and the diluent, D, phase, respectively. The values for ΔH_f^T, $C_{p_{AD}}$, and C_{p_D} are obtained from thermodynamic handbooks. From Eq 3, the mole fraction of the product phase AB, (x), can be determined using numerical integration techniques. The reaction utilizing prealloyed NiAl powder as the diluent for the combustion synthesis of nickel and aluminum powders is as follows:

$$x\text{Ni} + x\text{Al} + (1-x)\text{NiAl}_{\text{prealloyed}} \rightarrow x\text{NiAl}_{\text{reaction}} + (1-x)\text{NiAl}_{\text{prealloyed}} \rightarrow \text{NiAl} \qquad \text{(Eq 4a)}$$

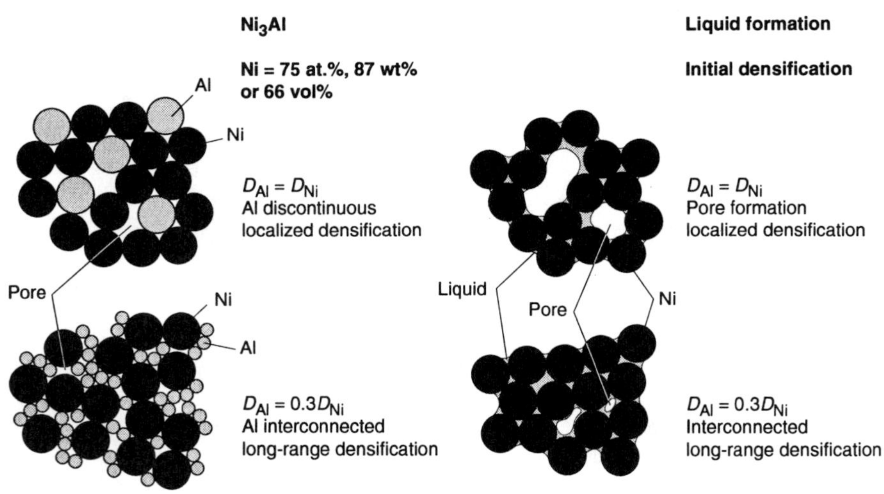

Fig. 4 A schematic diagram depicting the effect of particle size on reactive sintering of Ni_3Al. Source: Ref 15

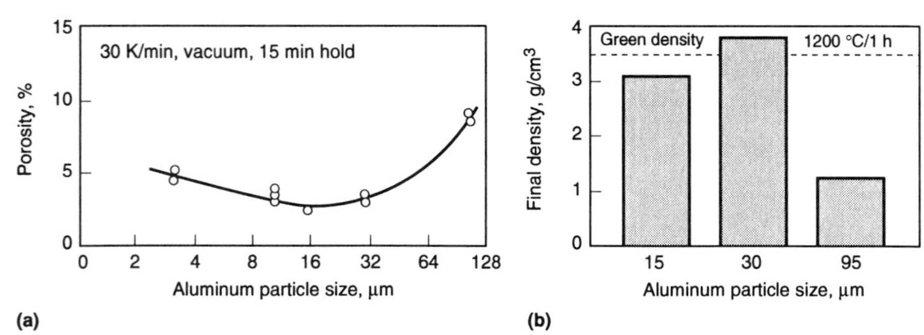

Fig. 5 Influence of aluminum particle size on the density of: (a) reactive sintered Ni_3Al, using 45 μm Ni powders. Source: Ref 17 and (b) $NbAl_3$, using 10 μm Nb powders. Source: Ref 22

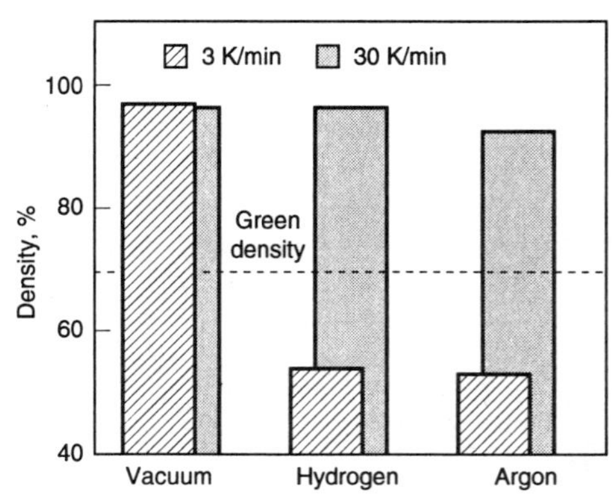

Fig. 6 Density of reactive sintered Ni_3Al for two heating rates and three sintering atmospheres. Source: Ref 17

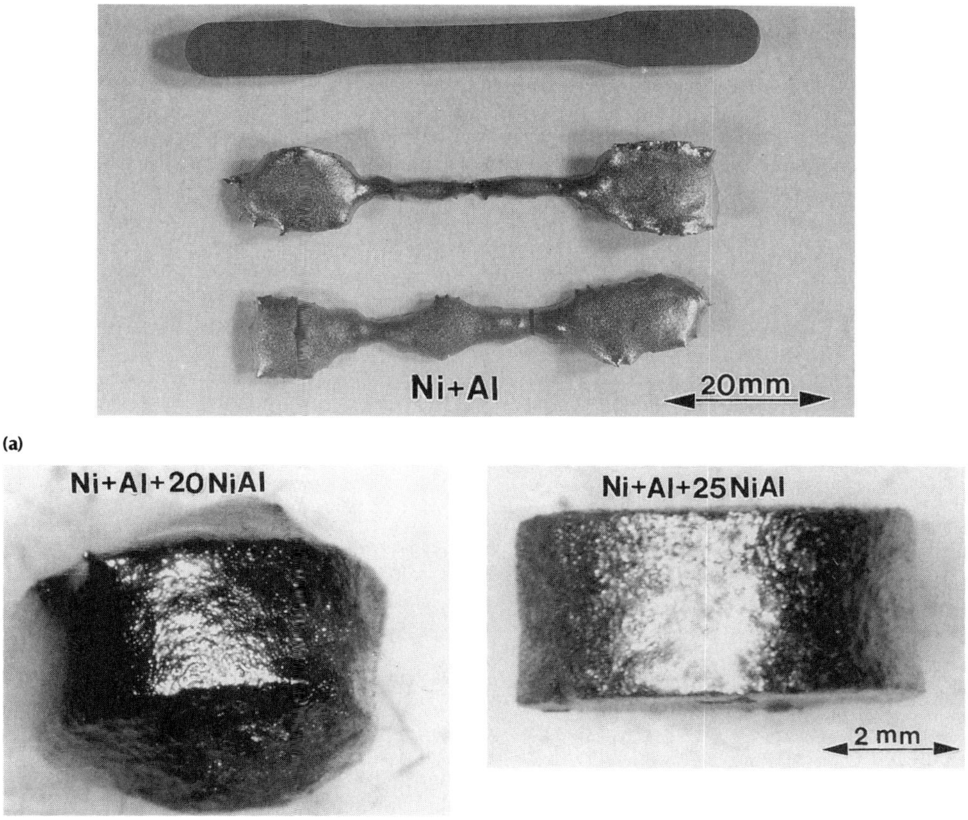

Fig. 7 Macrophotographs of pressed and reactive sintered NiAl (700 °C, 15 min, vacuum). (a) Loss of shape retention in undiluted reaction (i.e., Ni + Al → NiAl). (b) Shape retention using NiAl as diluent (i.e., xNi + xAl + (1 − x)$NiAl_{prealloyed}$ → $x NiAl_{reaction}$ + (1 − x)$NiAl_{prealloyed}$ → NiAl)

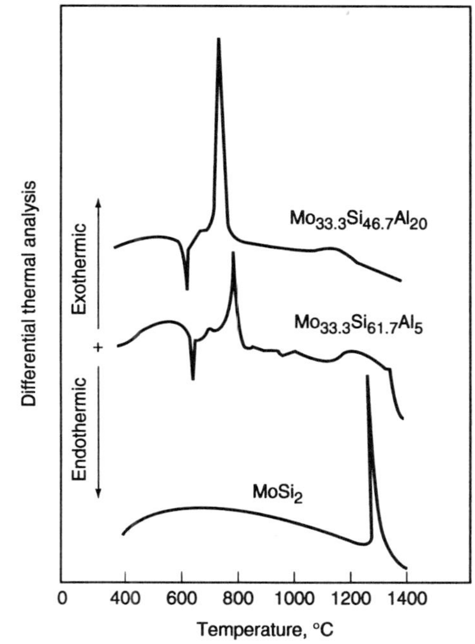

Fig. 8 Differential thermal analysis scans for reactive synthesis of $Mo(Si,Al)_2$ from elemental molybdenum, silicon, and aluminum powders. Source: Ref 27. The addition of aluminum augments the liquid phase formation during reactive sintering.

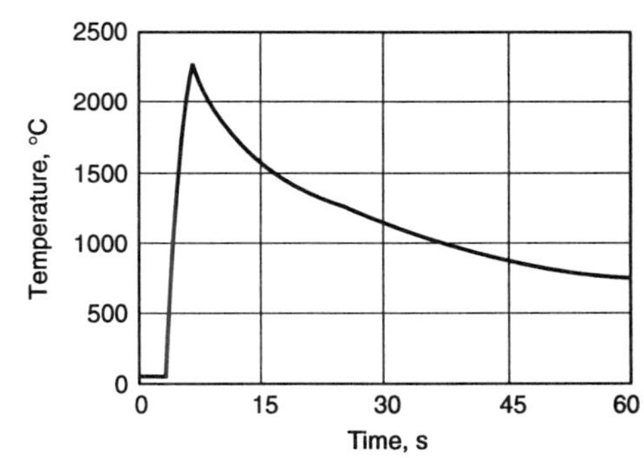

Fig. 9 Time-temperature profiles for the reactive synthesis of Ti + C + 25 wt% Ni → TiC-Ni. Source: Ref 29

The subsequent energy balance is:

$$x\left[\Delta H_{\text{f}}^{298}\right]_{\text{NiAl}} = \left[\int_{298}^{1913} C_{p_{\text{NiAl}}}\, dT\right] \qquad \text{(Eq 4b)}$$

Using Eq 4(b), x was determined to be 78 mol%, which corresponds to 22% NiAl prealloyed powder required to prevent slumping during combustion synthesis of NiAl. This is consistent with experimental results (Fig. 7, Ref 25). In practice, it is important not to overdilute the reaction as this will inhibit liquid formation and impede densification.

Augmenting Liquid Phase Formation by Alloying. The addition of alloying elements that melt during combustion synthesis prolongs the duration, distribution, and amount of the liquid phase during processing. These elements may or may not participate in the actual reaction. An example of the former is the reactive synthesis of aluminum-modified $MoSi_2$, from the following reaction:

$$\text{Mo} + 2(\text{Si} + \text{Al}) \rightarrow \text{Mo(Si,Al)}_2 \qquad \text{(Eq 5)}$$

The aluminum melts (or forms an aluminum-silicon eutectic) prior to the initiation of the combustion synthesis reaction (Ref 27), as illustrated in the differential thermal analysis (DTA) scans of Fig. 8. The presence of this "extra" liquid acts as a rapid diffusion path, and as a consequence the initiation of reaction associated with the formation of the ternary compounds occurs at a lower temperature than the reaction to form binary $MoSi_2$. The "extra" liquid phase can also be used to enhance densification during RHP. Similar results have been observed for silicon-modified aluminide compounds, such as Ti(Al,Si) (Ref 28).

Another example of the use of alloying to augment the liquid phase is the combustion synthesis of cermets. The reaction to form the TiC-Ni cermet is:

$$\text{Ti} + \text{C} + x\text{Ni} \rightarrow \text{TiC} + x\text{Ni} \qquad \text{(Eq 6)}$$

and a temperature-time profile for this reaction is shown in Fig. 9 (Ref 29). The reaction temperature exceeds the melting point of nickel (1453 °C) for about 20 s. Liquid nickel is present during this time period, and in fact the slight shoulder observed on the profile upon cooling at 1450 °C corresponds to the solidification of the nickel phase. Similar behavior is also noted for the simultaneous reaction of TiC and NiAl and TiC and Ni_3Al from titanium, carbon, nickel, and aluminum powder mixtures (Ref 30).

Alloying additions can also influence the diffusion rates between the reacting elements and minimize Kirkendall porosity. An example of this is the reactive sintering of nickel-modified iron aluminides (Ref 31). Without nickel additions, iron aluminide swells upon reactive sintering. The addition of nickel (substituting for iron) results in a high-density reactively sintered product (Fig. 10). The nickel inhibits the rate of aluminum diffusion in the iron, thus eliminating the pores that develop at the prior aluminum particle sites (Ref 31). However, this only occurs in Fe-Al-Ni powder billets rolled to an 80% reduction prior to reactive sintering at 650 °C for 15 min. This effect does not occur in similar material that was die pressed. This is due to differences in the green structure that develop during rolling compared to die pressing, as discussed below.

Influence of Green Compaction. As mentioned previously, Fe-Al-Ni powder billets rolled

Table 2 Conditions and tensile properties of select compounds, alloys, and composites fabricated by RS, RHP, and RHIP

Material	Process	Microstructure	Room-temperature tensile properties: Yield strength (σ_{ys}), MPa	Ultimate tensile strength (σ_{UTS}), MPa	Elongation to fracture (ε_f), %	Ref
Nickel aluminides						
Ni_3Al	Ni + 3Al: RS: 700 °C, vacuum	Ni_3Al 97% dense	270	270	1	18
Ni_3Al + B	Ni + 3Al + B doped: RS: 700 °C, vacuum	Ni_3Al 97% dense	353	682	12	18
	Ni + 3Al + B doped: RHIP: 800 °C, 0.5 h, 104 MPa	Ni_3Al, Ni_5Al_3 100% dense	265	727	10	18
	Ni + 3Al + B doped: RHIP: 1100 °C, 1 h, 170 MPa	Ni_3Al, Ni_5Al_3 100% dense	494	677	2	18
	Ni + 3Al + B doped: RHIP: 1100 °C, 1 h, 170 MPa + heat treatment	Ni_3Al 100% dense	591	827	5	18
Ni_3Al	Ni + 3Al; RS: 620 °C, 15 min (heat rate 60 K/min)	Ni_3Al: 98% dense	...	...	...	21
	Ni + 3Al; CS + HP: 550 °C, 25 MPa initiated reaction with a Ti + B at 550 °C	Ni_3Al: 98% dense	...	...	...	37
Ni_3Al + B	Ni + 3Al + B doped; RHP: 1250 °C	Ni_3Al 98.6% dense	401	401	0.3	34
Ni_3Al + B + 3.56 vol% Al_2O_3 particles	Ni + 3Al + B doped + Al_2O_3; RHP: 1250 °C	Ni_3Al, Al_2O_3 99% dense	521	625	4.7	34
Ni_3Al + B + 6.88 vol% Al_2O_3 particles	Ni + 3Al + B doped + Al_2O_3; RHP: 1250 °C	Ni_3Al, Al_2O_3 98.7% dense	486	486	0.9	34
Ni_3Al + 20 vol% Y_2O_3 particles	Ni + 3Al + Y_2O_3: RHIP: 800 °C, 1 h, 170 MPa	Ni_3Al, Y_2O_3 100% dense	...	464	...	18
NiAl	Ni + Al + 25 wt% NiAl: RS: 700 °C, 15 min vacuum	NiAl 100 dense	...	...	...	25
	Ni + Al + 20 wt% NiAl: RS + HIP: 1200 °C, 172 MPa, 1 h	NiAl 100% dense	135	154 (at 800 °C)	14	25
NiAl + 20 vol% TiB_2 particles	Ni + Al + 20 wt% NiAl + TiB_2: RS + HIP: 1200 °C, 172 MPa, 1 h	NiAl, TiB_2 100% dense	344	344 (at 800 °C)	0	25
NiAl + 10 vol% Al_2O_3 aligned sort fibers	Ni + Al + Al_2O_3 RS + HIP: 1200 °C, 172 MPa, 1 h	NiAl, Al_2O_3 100% dense	142	163 (at 800 °C)	3	38
Iron aluminides						
Fe_3Al	Fe + 3Al; RHIP: 1000 °C, 140 MPa	Fe_3Al 100% dense	759	759	0	39
Fe_3Al + 5% Cr	Fe + 3Al + 5Cr: RHIP + 24 h/1100 °C/ + HIP 1100 °C/207 MPa/ + 2 h/750 °C/oil quench	Fe_3Al 100% dense	857	1095	7	39
Fe_3Al	Fe + 3Al; RHP 1250 °C	Fe_3Al 98.2% dense	404	521	0.3	34
Titanium aluminides						
TiAl	Ti + Al: extruded (*R* = 350) + RS: 1350 °C, 6 h	TiAl 98% dense	...	...	...	33
Ti-50Al	Ti + Al; extruded (*R* = 14) + RHIP: 1100 °C, 2 h, 150 MPa	TiAl; Ti_3Al 100% dense	450	750 (at 700 °C)	...	40
Ti-49Al	Ti + Al; extruded (*R* = 14) + RHIP: 1000 °C, 3 h, 200 MPa + HT: 1250 °C, 3 h	TiAl; Ti_3Al 100% dense	380	710	...	40
	Ti + Al; extruded (*R* = 14) + RHIP: 1000 °C, 3 h, 200 MPa + HT: 1250 °C, 3 h	TiAl; Ti_3Al 100% dense	300	750 (at 700 °C)	...	40
Ti-48Al-2Cr	Ti + Al + Cr; extrude + RHIP: 1100 °C, 3 h, 125 MPa	TiAl; Ti_3Al 100% dense	700	1500	0.2	41
	Ti + Al + Cr; extrude + RHIP: 1100 °C, 15 min, 125 MPa	TiAl; Ti_3Al 100% dense	650	1550 (at 500 °C)	0.25	41
TiAl-Ti_5Si_3	Ti + Al + Si: RHP: 1300 °C, 15 min, 6.5 MPa	TiAl; Ti_3Al; Ti_5Si_3	...	...	...	42
TiAl-TiB_2	Ti + Al + B: RHP: 1300 °C, 15 min, 6.5 MPa	TiAl; Ti_3Al; TiB_2	...	...	...	42
TiAl-TiC	Ti + Al + C: RHP: 1300 °C, 15 min, 6.5 MPa	TiAl; Ti_3Al; TiC	...	...	...	42
$(Al,Cr)_3Ti$	Al + Ti + Cr: RHIP: 1250 °C, 2 h, 173 MPa	$Al_{66}Cr_9Ti_{25}$ 100% dense	470	1360 (in compression)	14	43
Other intermetallics						
$MoSi_2$	Mo + 2Si: RHP: 1600 °C, 45 min, 36 MPa	$MoSi_2$ 89% dense	...	...	...	44
$Mo(Si,Al)_2$	33Mo + 46.7Si + 20Al: RHP: 850 °C, 15 min, 30 MPa	$Mo(Si,Al)_2$, Mo, Mo_3Si, 90% dense	...	...	...	27
$NbAl_3$	Nb + 3Al: RHIP: 1200 °C, 4 h, 173 MPa	$NbAl_3$, $NbAl_2$ >98% dense	...	...	...	22
$NbAl_3$	Nb + 3Al: RS: 1200 °C, 1 h	$NbAl_3$ 95% dense	...	...	...	22
Ceramics and cermets						
TiC	Ti + C: combustion synthesized	TiC 50% dense	...	...	...	11
	Ti + C: RHP: 1600 ° C, 27.6 MPa	TiC 95% dense	...	...	...	11
SiC	Si + C: high-pressure combustion synthesis 3 GPa	βSiC 90% dense	...	...	...	45
TiB_2	Ti + 2B: high-pressure combustion synthesis 3 GPa	TiB_2 95% dense	...	...	...	46
TiB_2/SiC-Ni	TiH_2 + Si + B_4C + (2 wt%)Ni: RHP: 2000 °C, 1 h, 30 MPa	TiB_2, SiC, Ni 99% dense	...	K_c(SENB) = 6.6 MPa$\sqrt{m}$ σ_F (three-point bend) = 496 MPa	...	47
TiB_2/SiC	TiH_2 + Si + B_4C: RHP: 2000 °C, 1 h, 30 MPa	TiB_2, SiC, Ni 99% dense	...	K_c(SENB) = 8.7 MPa$\sqrt{m}$ σ_F (three-point bend) = 330 MPa	...	47
TiB_2-Al	TiH_2 + AlB_2: RHP: 1500 °C, 34 MPa	TiB_2-Al	...	...	...	48
TiC-NiAl	Ti + C + Ni + Al: RHP: 21 MPa	TiC, NiAl 94–100% dense	...	...	...	30

CS, combustion synthesis; K_c, critical stress intensity; σF, fracture stress; SENB, single-edge notched bend test

to an 80% reduction prior to reactive sintering at 650 °C for 15 min were dense; however, similar composition of elemental powders, die pressed at more than 550 MPa, swelled upon sintering (Ref 31). Similar behavior is observed during reactive sintering of TiAl. Die-pressed titanium and aluminum powders swell upon reactive sintering. However, Dahms and coworkers (Ref 5, 6, 32, 33) were able to produce 97% dense TiAl by reactive sintering extruded billets of titanium and aluminum powders, as illustrated in Fig. 11. The extrusion ratio played an important role in the densification of TiAl, with higher extrusion ratios resulting in denser (less porous) pressureless sintered billets.

It is evident that the high pressures that can be generated during rolling and extrusion influence the green structure in a beneficial manner. The deformation of the elemental powders that occurs during extrusion and rolling results in increased surface contact area, decreased initial pore size, and improved aluminum distributions in the green microstructure. This results in minimal pore formation during reactive sintering.

Pressure-Assisted Reactive Sintering

Even though many of the examples discussed above can be reactively sintered to densities greater than 97% theoretical, this small amount of porosity can be detrimental to the mechanical properties—specifically tensile and fracture behavior—of these compounds. Commonly, a postreactive sintering routine, such as hot isostatic pressing (HIP), is used to fully densify these materials.

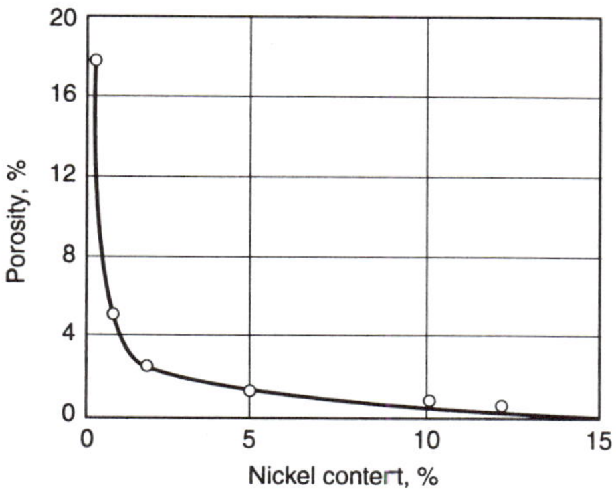

Fig. 10 Effect of nickel content on the porosity of reactive sintered (650 °C, 15 min) iron-aluminides (elemental powder mixtures corresponding (in wt%) to 86.4(Fe + Ni) + 13.6Al. Source: Ref 31. Compacts of elemental powder mixtures were rolled to 80% reduction prior to sintering.

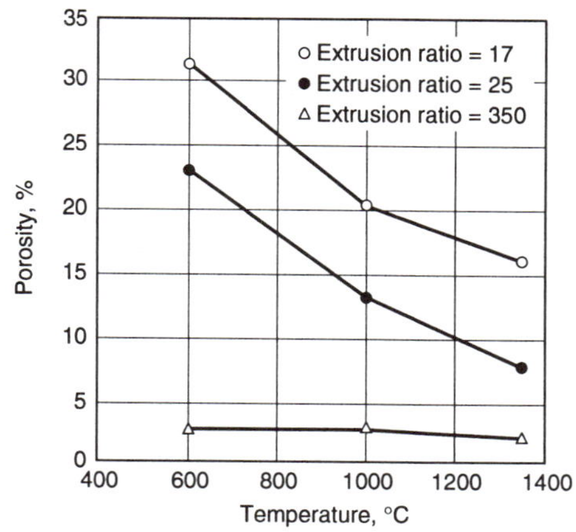

Fig. 11 Porosity of reactive sintered TiAl specimens as a function of sintering temperature and extrusion ratio. Source: Ref 33. Compacts of elemental titanium and aluminum were extruded prior to sintering.

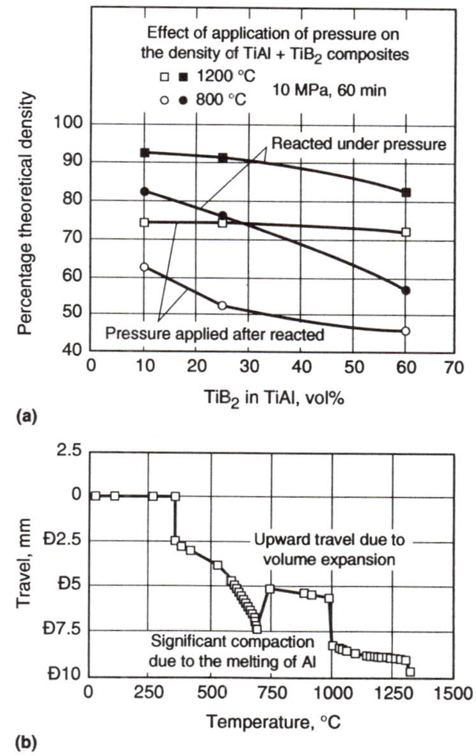

Fig. 12 Influence of the application of pressure on reactive sintering. (a) Densities of TiAl-boride in situ composites fabricated by reactive hot pressing (RHP) and reactive sintering followed by hot pressing (RS + HP). Source: Ref 28. (b) Hot press ram travel during RHP of Ni_3Al. Source: Ref 34

Alternatively, the powders can be sintered under pressure by RHP or RHIP. Greater densities can be achieved in the final product if the reaction is initiated under pressure (i.e., RHP), as opposed to applying pressure subsequent to the reaction (i.e., RS + HP), under identical hot pressing conditions. This is illustrated in Fig. 12(a) (Ref 28) for a variety of RHP in situ TiAl-boride composites. The composites that were reacted under pressure were considerably more dense than similar composites in which the pressure was applied subsequent to the reaction (e.g., after the hold temperature was reached). During RHP, compaction of the powders (as indicated by ram travel, Fig. 12b) occurs as the hot-press temperature nears the reaction initiation temperature and the transient liquid phase forms (Ref 34). The applied pressure forces the transient liquid phase into the voids between the solid powder particles, and as a consequence the compact densifies. Further, the heat liberated by the reaction raises the temperature of powder compact. At this temperature, the product may be soft and deform under the applied pressure, resulting in densification (Ref 35).

While the simultaneous application of pressure to the reacting powders promotes densification, it can result in inhomogeneous microstructures. This is particularly true for RHP aluminides. Alman and coworkers (Ref 25, 27, 28) noted that microstructures of RHIP-NiAl and RHP-TiAl contained "hot" spots of aluminum-rich intermetallic phases. However, the microstructure of similar aluminides, produced by RS + HIP or RS + HP, were homogeneous. The microstructural inhomogeneities that developed during RHP were attributed to the same mechanism for the enhancement in density. The transient liquid phase is forced into the voids between the solid particles (presumably nickel or titanium). The liquid is presumably aluminum rich and, upon solidification, results in the microstructural inhomogeneity. Liu and coworkers (Ref 21, 36) found that the application of pressure has a negative impact on the density and microstructure of reactive sintered Ni_3Al. In their studies, single-phase and dense (98%) Ni_3Al was produced by RS + HP. However, porous (70% dense) and multiphase (Al_3Ni, Al_3Ni_2, and nickel) microstructures resulted when the nickel and aluminum powders were RHP under stresses of 5 and 50 MPa. The low densities and inhomogeneity were attributed to the heat transfer associated with the contact of the ram and the powder compact, quenching the reaction. This was further supported by a size effect of the powder compact: the thinner the powder compact hot pressed, the more inhomogeneous was the resultant microstructure.

Conclusions

Reactive sintering (RHP and RHIP) have been utilized to produce a variety of ceramics, intermetallics, and composite materials. Table 2 summarizes the processing conditions and tensile properties of select materials produced by reactive powder processing. This is not intended to be an all-inclusive list, but representative of materials produced by these methods.

The microstructure that develops during reactive sintering is influenced by a variety of process parameters, including the green structure, the sintering conditions, and product composition. An interconnected configuration of the reactant powders minimizes pore formation during reactive sintering. The distribution of the reactants in the green structure can be adjusted by altering the powder particle characteristics (e.g., particle size ratio) or the green processing method (e.g., extrusion, rolling versus compaction). The parameters of the sintering cycle that influence reactive sintering, include heating rate, sintering atmosphere, and application of pressure. Heating rates and sintering atmospheres that limit interdiffusion and intermediate compound formation prior to the reaction between the powders are preferred. Application of pressure during combustion synthesis results in dense products, but the microstructure of these products may be inhomogeneous, particularly if large amounts of transient liquid form during processing. Both the densification and inhomogeneity can be attributed to the applied pressure forcing the transient liquid into pores between solid powders. Finally, composition of the product can affect the reaction, by the addition of alloying elements and diluent phases (inert, high melting, reinforcement phases). These can affect the amount, duration, and distribution of the transient liquid phase and can also affect the diffusion rates between the reacting elements. With the judicial selection of all these parameters, dense and useful components can be produced by reactive sintering.

REFERENCES

1. R.A. Long, "Fabrication and Properties of Hot-Pressed Molybdenum Disilicide," RM-E50F22, National Advisory Committee for Aeronautics, Washington, DC, 1950
2. J.B. Huffadine, *Special Ceramics,* P. Popper, Ed., Academic Press, 1960, p 220
3. E.N. Smith, Macro Process for Direct Production of Tungsten Mono-Carbide, U.S. Patent No. 3,379,503, 1968
4. A.P. Savitskii, *Liquid Phase Sintering of Systems with Interacting Components,* Russian Academy of Sciences, Tomsk, 1993

5. G.X. Wang and M. Dahms, *Powder Metall. Int.*, Vol 24 (No. 4), 1992, p 212
6. M. Dahms and F. Schmelzer, *Z. Metallk.*, Vol 84 (No. 5), 1993, p 351
7. K. Shibue, *Sumitomo Light Met. Tech. Rep.*, Vol 32, 1992, p 32
8. R.W. Rice and W.J. McDounough, *J. Am. Ceram. Soc.*, Vol 68, 1985, p C122
9. A. Bohm, T. Jungling, and B. Kieback, *Advances in Powder Metallurgy and Particulate Materials—1996,* Part 10, T.M. Cadle and K.S. Narasimhan, Ed., MPIF/AMPI, 1996, p 51
10. D.C. Dunand, *Mater. Manuf. Process.*, Vol 10 (No. 3), 1995, p 373
11. J.B. Holt and Z.A. Munir, *J. Mater. Sci.*, Vol 21, 1986, p 251
12. R.L. Coble, *Sintering Theory and Practice,* D. Kolar, S. Pejovnik, and M.M. Ristic, Ed., Elsevier, 1982, p 145
13. R.M. German, *Liquid Phase Sintering,* Plenum Press, 1985, p 157
14. R.M. German, *Sintering Theory and Practice,* John Wiley and Sons, 1996, p 395
15. A. Bose, B. Moore, R.M. German, and N.S. Stoloff, *J. Met.*, Vol 40 (No. 9), 1988, p 14
16. K.A. Philpot, Z.A. Munir, and J.B. Holt, *J. Mater. Sci.*, Vol 22, 1987, p 159
17. D. Sims, A. Bose, and R.M. German, *Prog. Powder Met.*, Vol 43, 1987, p 575
18. R.M. German, A. Bose, D. Alman, J. Murray, P. Korinko, R. Oddone, and N.S. Stoloff, *P/M 90,* Vol 1, The Institute of Metals, 1990, p 310
19. A. Bose, B.H. Rabin, and R.M. German, *Powder Metall. Int.*, Vol 20 (No. 3), 1998, p 25
20. B.H. Rabin, A. Bose, and R.M. German, *Combustion and Plasma Synthesis of High-Temperature Materials,* Z.A Munir and J.B. Holt, Ed., VCH, 1990, p 114
21. C. Nishimura and C.T. Liu, *Acta Metall.*, Vol 41, 1993, p 113
22. J.C. Murray and R.M. German, *Metall. Trans. A,* Vol 23, 1992, p 2357
23. R.R. Oddone and R.M. German, *Advances in Powder Metallurgy,* Vol 3, T.G. Gasbarre and W.F. Jandeska, Ed., MPIF/AMPI, 1989, p 475
24. B.H. Rabin and R.N. Wright, *Metall. Trans. A,* Vol 22, 1991, p 277
25. D.E. Alman and N.S. Stoloff, *Int. J. Powder Metall.*, Vol 27 (No. 1), 1991, p 29
26. W.R. Wrzesinski and J.C. Rawers, *J. Mater. Sci. Lett.*, Vol 9, 1990, p 432
27. D.E. Alman and R.D. Govier, *Scr. Metall.*, Vol 34, 1996, p 1287
28. D.E. Alman, J.A. Hawk, and M. Ziomek-Moroz, *In situ Reactions for the Synthesis of Composites, Ceramics and Intermetallics,* E.V. Barrera, S.G. Fishman, F.D.S. Marquis, N.N. Thadhani, W.E. Frazier, and Z.A. Munir, Ed., TMS, 1996, p 141
29. S.D. Dunmead, D.W. Readdy, C.E. Semler, and J.B. Holt, *J. Am. Ceram. Soc.*, Vol 72, 1989, p 2318
30. S.D. Dunmead, Z.A. Munir, J.B. Holt, and D.D. Kingman, *J. Mater. Sci.*, Vol 26, 1991, p 241
31. X.Q. Wang, G. Fair, and J.V. Wood, *Powder Metall.*, Vol 36 (No. 3), 1993, p 187
32. G.-X. Wang and M. Dahms, *JOM,* Vol 5, 1993, p 52
33. G.-X. Wang and M. Dahms, *Metall. Trans. A,* Vol 24, 1993, p 1517
34. S.C. Deevi and V.K. Sikka, *High Temperature Ordered Intermetallic Alloys VI,* J.A. Horton, I. Baker, S. Hanada, R.D. Noebe, and D.S. Schwartz, Ed., MRS Symp. Proc., Vol 364, *MRS,* 1995, p 917
35. J.B. Holt, *MRS Bull.*, 1 Oct/15 Nov 1987, p 60
36. S. Miura and C.T. Liu, *Intermetallics,* Vol 2, 1994, p 297
37. W.C. Williams and G.C. Strangle, *J. Mater. Res.,* Vol 10, 1995, p 1736
38. D.E. Alman, N.S. Stoloff, A. Bose, and R.M. German, *J. Mater. Sci.,* Vol 30, 1995, p 5251
39. B.H. Rabin and R.N. Wright, *Metall. Trans. A,* Vol 23, 1992, p 35
40. M. Dahms, J. Seeger, W. Smarsly, and B. Wildhagen, *ISIJ Int.*, Vol 31, 1991, p 1093
41. G.-X. Wang. A. Bartels, and M. Dahms, *Mater. Trans. JIM,* Vol 34, 1993, p 228
42. J.C. Rawers and W.R. Wrzesinski, *J. Mater. Sci.,* Vol 27, 1992, p 2877
43. M. Otuski and N.S. Stoloff, *Scr. Metall. Mater.,* Vol 26, 1992, p 325
44. S.C. Deevi and N.N. Thadhani, *Mater. Sci. Eng.,* A192/193, 1995, p 604
45. O. Yamada, Y. Miyamoto, and M. Koizumi, *Am. Ceram. Soc. Bull.,* Vol 64 (No. 2), 1985, p 319
46. Y. Miyamoto, M. Koizumi, and O. Yamada, *J. Am. Ceram. Soc.,* Vol 67, 1984, p C-224
47. G.J. Zhang, Z.Z. Jin, and X.M. Yue, *J. Mater. Sci.,* Vol 32, 1997, p 2093
48. T.P. DeAngelis and D.S. Weiss, in *Combustion and Plasma Synthesis of High-Temperature Materials,* Z.A. Munir and J.B. Holt, Ed., VCH, 1990, p 144

Combustion Synthesis of Advanced Materials*

A. Varma and A.S. Mukasyan, University of Notre Dame

COMBUSTION SYNTHESIS (CS) is an attractive technique for synthesizing a wide variety of advanced materials, including powders and near-net-shape products of ceramics, intermetallics, composites, and functionally graded materials. This method was discovered in the late 1960s in the former Soviet Union (Ref 1), and its development has led to a new scientific endeavor that incorporates aspects of both combustion and materials science. At about the same time, some work concerning the combustion aspects of this method was done in the United States (Ref 2–4), although the full potential of CS in the production of advanced materials was not utilized. The scientific and technological activity in the field increased during the 1980s. The significant results of combustion synthesis have been described in a number of review articles (Ref 5–12).

There are two modes by which combustion synthesis can occur: *self-propagating high-temperature synthesis* (SHS) and *volume combustion synthesis* (VCS). A schematic diagram of these modes is shown in Fig. 1. In both cases, reactants may be pressed into a pellet, typically cylindrical in shape. The samples are then heated by an external source (e.g., tungsten coil, laser) either locally (SHS) or uniformly (VCS) to initiate an exothermic reaction.

The characteristic feature of the SHS mode is, after initiation locally, the self-sustained propagation of a reaction wave through the heterogeneous mixture of reactants. The temperature of the wave front can reach quite high values (2000 to 4000 K). In principle, if the physicochemical parameters of the medium are known, along with the instantaneous spatial distributions of temperature and concentration, then the combustion velocity and reaction rate throughout the mixture can be calculated. Thus, the SHS mode of reaction can be considered a well-organized wave-like propagation of the exothermic chemical reaction through a heterogeneous medium, followed by the synthesis of desired condensed products.

During VCS, the entire sample is heated uniformly in a controlled manner until the reaction occurs essentially simultaneously throughout the volume. This mode of synthesis is more appropriate for weakly exothermic reactions that require preheating prior to ignition, and it is sometimes referred to as the *thermal explosion* mode. However, the term "explosion" used in this context refers to the rapid rise in temperature after the reaction has been initiated, not to the destructive process usually associated with detonation or shock waves. For this reason, volume combustion synthesis is perhaps a more appropriate name for this mode of synthesis.

From the viewpoint of chemical nature, three main types of CS processes can be distinguished. The first, *gasless combustion synthesis from elements,* is described by:

$$\sum_{i=1}^{n} X_i^{(s)} = \sum_{j=1}^{m} P_j^{(s,l)} + Q \qquad \text{(Eq 1)}$$

where $X_i^{(s)}$ are elemental reactant powders (metals or nonmetals), $P_j^{(s,l)}$ are products, Q is the heat of reaction, and the superscripts (s) and (l) indicate solid and liquid states, respectively. Perhaps the most popular example of this type of reaction is carbidization of titanium:

$$\text{Ti} + \text{C} = \text{TiC} + 230\ \text{kJ/mol}$$

The second type of CS, called *gas-solid combustion synthesis,* involves at least one gaseous reagent in the main combustion reaction:

$$\sum_{i=1}^{n-p} X_i^{(s)} + \sum_{i=1}^{p} Y_i^{(g)} = \sum_{j=1}^{m} P_j^{(s,l)} + Q \qquad \text{(Eq 2)}$$

where $Y_i^{(g)}$ are the gaseous reactants (e.g., N_2, O_2, H_2, CO) that, in some cases, penetrate the sample by infiltration through its pores. This type of CS is also called *infiltration* (or *filtration*) *combustion synthesis.* Nitridation of titanium and silicon are common examples:

$$\text{Ti}^{(s)} + 0.5\text{N}_2^{(g)} = \text{TiN}^{(s)} + 335\ \text{kJ/mol}$$

$$3\text{Si}^{(s)} + 2\text{N}_2^{(g)} = \text{Si}_3\text{N}_4^{(s)} + 750\ \text{kJ/mol}$$

The third main type of CS is *reduction combustion synthesis,* described by:

$$\sum_{i=1}^{n-q-r} (MO_x)_i^{(s)} + \sum_{i=1}^{r} Z_i^{(s)} + \sum_{i=1}^{q} X_i^{(s)} = \sum_{j=1}^{m-k} P_j^{(s,l)} + \sum_{j=1}^{k} (ZO_y)_j^{(s,l)} + Q \qquad \text{(Eq 3)}$$

where $(MO_x)_i^{(s)}$ is an oxide that reacts with a reducing metal $Z_i^{(s)}$ (e.g., Al, Mg, Zr, Ti), resulting in the appearance of another, more stable oxide $(ZO_y)_j^{(s,l)}$, and reduced metal $M_i^{(s,l)}$. This reaction may be followed by the interaction of $M_i^{(s,l)}$ with other elemental reactants $X_i^{(s)}$ to produce desired products $P_j^{(s,l)}$. Thus, in general, the reduction combustion synthesis may be considered a two-step process, where the first step is a *thermite* reaction:

$$\sum_{i=1}^{n-q-r} (MO_x)_i^{(s)} + \sum_{i=1}^{r} Z_i^{(s)} = \sum_{j=1}^{k} (ZO_y)_j^{(s,l)} + \sum_{j=1}^{l} M_j^{(s,l)} + Q_1 \qquad \text{(Eq 4a)}$$

and the second step is the synthesis from elements similar to Eq 1:

$$\sum_{j=1}^{l} M_i^{(s,l)} + \sum_{i=1}^{q} X_i^{(s)} = \sum_{j=1}^{m-l} P_j^{(s,l)} + Q_2 \qquad \text{(Eq 4b)}$$

with the total heat release $Q = Q_1 + Q_2$. An example of this type of CS is:

$$\text{B}_2\text{O}_3^{(s)} + 2\text{Al}^{(s)} + \text{Ti}^{(s)} = \text{Al}_2\text{O}_3^{(l)} + \text{TiB}_2^{(s,l)} + 700\ \text{kJ/mol}$$

*This article is based in part on Ref 12.

where TiB_2 is the desired product and Al_2O_3 can be removed (e.g., by centrifugal separation) and used separately, or a ceramic composite material ($Al_2O_3 + TiB_2$) can be produced. In some cases, the reducing reactant (Z_i) is the same as that used for the synthesis (X_i), for example:

$$2B_2O_3^{(s)} + 5Ti^{(s)} = 3TiO_2^{(s)} + 2TiB_2^{(s)} + 140\,\text{kJ/mol}$$

The number of products synthesized by CS increased rapidly during the 1970s and 1980s, and currently there are more than 400 different compounds. Specifically, these materials include carbides (TiC, ZrC, SiC, B_4C, etc.), borides (TiB_2, ZrB_2, MoB_2, etc.), silicides (Ti_5Si_3, $TiSi_2$, $MoSi_2$, etc.), nitrides (TiN, ZrN, Si_3N_4, BN, AlN), and intermetallics (NiAl, Ni_3Al, TiNi, TiAl, CoAl, etc.).

Throughout this work, a greater emphasis is placed on the SHS mode of synthesis, because more information is available for it than for the VCS mode. Also, the production of powders by gas-phase combustion synthesis processes (Ref 13, 14) is not considered in this article.

Table 1 Adiabatic and measured combustion temperatures for various reaction systems

System	Adiabatic combustion temperature, K	Measured combustion temperature, K	Lowest melting point on the phase diagram, K
Carbides			
Ta + C	2700	2650	3295 (Ta)
Ti + C	3290	3070	1921 (eut)
Si + C	1690	2000(a)	1690 (Si)
Borides			
Ta + B	2728	2700	2365 (B)
Ti + 2B	3193	3190	1810 (eut)
Ti + B	2460	2500	1810 (eut)
Silicides			
Mo + 2Si	1925	1920	1673 (eut)
Ti + 2Si	1773	1770	1600 (eut)
5Ti + 3Si	2403	2350	1600 (eut)
Intermetallics			
Ni + Al	1912	1900	921 (eut)
3Ni + Al	1586	1600	921 (eut)
Ti + Al	1517	N/A	933 (Al)
Ti + Ni	1418	N/A	1215 (eut)
Ti + Fe	1042	N/A	1358 (eut)
Nitrides(b)			
$2Ta + N_2$	3165	2500	3000 (Ta)
$2Nb + N_2$	3322	2800	2673 (NbN)
$2Ti + N_2$	3446	2700	1943 (Ti)
$2Al + N_2$	3639	2300	933 (Al)
$3Si + 2N_2$	2430	2250	1690 (Si)
$2B + N_2$	3437	2600	2350 (B)
Thermite type			
B_2O_3 + Mg	2530	2420	1415 (eut)
B_2O_3 + Mg + C	2400	2270	N/A
$B_2O_3 + Mg + N_2$	2830	2700	N/A
SiO_2 + Mg	2250	2200	1816 (eut)
SiO_2 + Mg + C	2400	2330	N/A
La_2O_3 + Mg + B_2O_3	N/A	2400	N/A

eut, eutectic. N/A, not applicable. (a) With preheating. (b) T_c^{ad} calculated for 1 atm

Theoretical Considerations and Phenomenology

Thermodynamics

For any reaction system, the chemical and phase composition of the final product depends on the *green mixture composition,* gas pressure, reactive volume, and initial temperature. CS reactions can be represented in the following general form:

$$\sum_{i=1}^{N_X} \nu_i X_i^{(s)} + \sum_{i=1}^{N_Y} \nu_i Y_i^{(g)} + \sum_{i=1}^{N_{MO_X}} \nu_i (MO_x)_i^{(s)} + \sum_{i=1}^{N_Z} \nu_i Z_i^{(s)} = \sum_{j=1}^{N_P} \nu_j P_j^{(s)} + \sum_{j=1}^{N_M} \nu_j M_j^{(s)} \quad \text{(Eq 5)}$$

where $X_i^{(s)}$ are the metal or nonmetal solid reactants, $Y_i^{(g)}$ are the gas-phase reactants, $(MO_x)_i^{(s)}$ are the oxide reactants, $Z_i^{(s)}$ are the reducing metal reactants, $P_i^{(s)}$ are the solid products (e.g., carbides, borides, nitrides), and $M_i^{(s)}$ are the reduced metal products. In limiting cases, Eq 5 leads to the three main classes of CS reactions:

- Gasless combustion (Eq 1): $N_Y = 0$, $N_{MO_X} = 0$, and $N_M = 0$
- Gas-solid combustion (Eq 2): $N_{MO_X} = 0$, and $N_Z = 0$, and $N_M = 0$
- Reduction-type combustion (Eq 3): $N_{MO_X} \neq 0$, and $N_Z \neq 0$

Thermodynamic calculations can identify the adiabatic combustion temperature, as well as the equilibrium phases and compounds present at that temperature. The composition of the equilibrium final products is determined by minimizing the thermodynamic potential. For a system with $N^{(g)}$ gas and $N^{(s)}$ solid number of components, at constant pressure, this may be expressed as:

$$F(\{n_k\},\{n_l\}) = \sum_{k=1}^{N^{(g)}} n_k \left(\ln \frac{p_k}{p} + G_k \right) + \sum_{l=1}^{N^{(s)}} n_l G_l \quad \text{(Eq 6)}$$

where p_k is the partial pressure of the k-th gas-phase component, n_i is the number of moles of component i, and G_l is the molar Gibbs free energy of component l (Ref 15). The adiabatic combustion temperature, T_c, is determined by total energy balance:

$$\sum_{i=1}^{N_0} H_i(T_0) = \sum_{k=1}^{N(g)} n_k H_k(T_c^{ad}) + \sum_{l=1}^{N(s)} n_l H_l(T_c^{ad}) \quad \text{(Eq 7)}$$

where the enthalpy of each component is:

$$H_i(T) = \Delta H_{f,i}^0 + \int_{T_0}^{T} C_{p,i}\, dT + \sum \Delta H_{s,i} \quad \text{(Eq 8)}$$

and $\Delta H_{f,i}^0$ is the heat of formation at 1 atm and reference temperature T_0, $C_{p,i}$ is the heat capacity, and $\Delta H_{s,i}$ is the heat of s-th phase transition for component i.

Thermodynamic calculations of adiabatic combustion temperatures have been made for a variety of systems and compared with experimentally measured values (Ref 16, 17). Under conditions that lead to full conversion, good

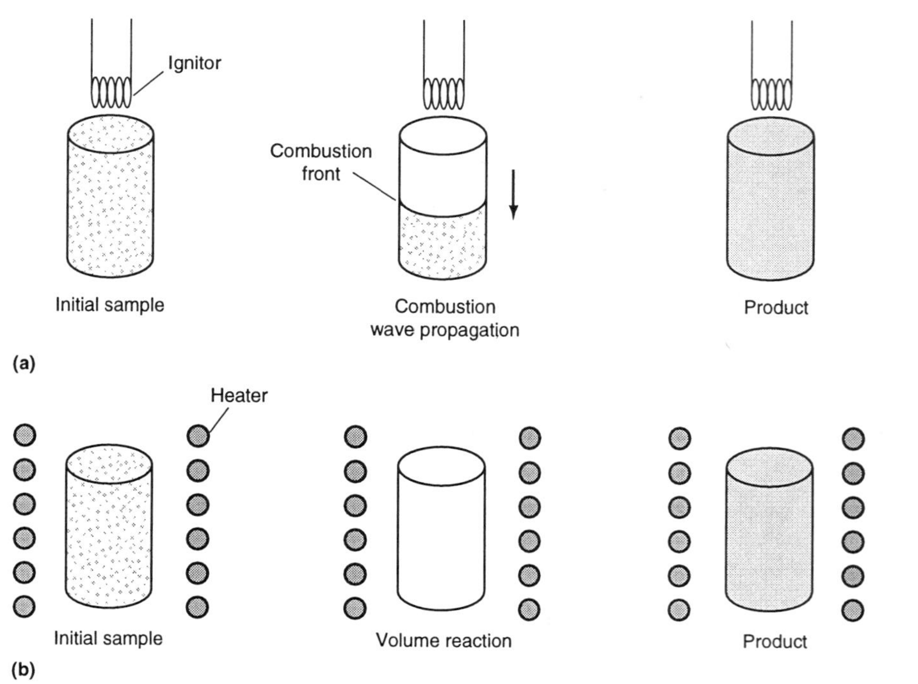

Fig. 1 The modes of combustion synthesis. (a) Self-propagating high-temperature synthesis. (b) Volume combustion synthesis

agreement between theoretical and experimental values has generally been obtained (Table 1).

Combustion Wave Propagation Theory

The application of combustion theory to combustion synthesis processes is based on a comparison of the mass and thermal diffusivities in the solid reactant mixture. In these types of systems, the Lewis number is very small:

$$\mathrm{Le} = \frac{\mathcal{D}}{\alpha} << 1 \qquad \text{(Eq 9)}$$

where $\mathcal{D}$ and α are the mass and thermal diffusivities, respectively, indicating that heat conduction occurs much faster than mass diffusion. As a result, mass transfer by diffusion at the *macroscopic* scale may be neglected, and an average concentration of the reactants in any local region of the heterogeneous mixture may be used. Thus, the physical and thermal properties (e.g., density, thermal conductivity, heat capacity) are the average of reactant and product values. In this case, the combustion synthesis process is controlled only by heat evolution from the exothermic reaction and heat transfer from the reaction zone to the unreacted mixture

Propagation of combustion wave through the reactant mixture with a velocity U can be described by the set of energy continuity and chemical kinetic equations (Ref 18):

$$\rho C_p \frac{dT}{dt} + \rho C_p U \frac{dT}{dx} = \frac{d}{dx}\left(\lambda \frac{dT}{dx}\right) + Q\rho\phi(\eta,T)$$

$$\frac{d\eta}{dt} + U\frac{d\eta}{dx} = \phi(\eta,T) \qquad \text{(Eq 10)}$$

where ρ, C_p, and λ are respectively the density, heat capacity, and thermal conductivity of the reaction mixture. The term $\phi(\eta,T)$ represents the reaction rate, and x is the spatial coordinate moving with the reaction front. Assuming that the temperature (T) and conversion (η) profiles do not change with time (t), i.e. *constant-pattern propagation,* then the time derivatives can be set equal to zero. Then Eq 10 takes the form:

$$\rho C_p U \frac{dT}{dx} = \frac{d}{dx}\left(\lambda \frac{dT}{dx}\right) + Q\rho\phi(\eta,T) \qquad \text{(Eq 11a)}$$

$$U\frac{d\eta}{dx} = \phi(\eta,T) \qquad \text{(Eq 11b)}$$

along with the boundary conditions:

$$T = T_0,\ \frac{dT}{dx} = 0,\ \eta = 0 \text{ at } x \rightarrow -\infty \qquad \text{(Eq 12a)}$$

$$T = T_c,\ \frac{dT}{dx} = 0,\ \eta = 1 \text{ at } x \rightarrow +\infty \qquad \text{(Eq 12b)}$$

The solution of Eq 11 and 12 was first obtained in the context of gas flames (Ref 19) and was later adapted for solid mixtures (Ref 20, 21) to yield the following formula for reaction wave velocity:

$$U^2 = \frac{\lambda}{\rho Q}\,\frac{RT_c^2}{E}\,\frac{k_0 \exp\left(\frac{-E}{RT_c}\right)}{\int_0^1 \frac{1-\eta}{\varphi(\eta)}\,d\eta} \qquad \text{(Eq 13)}$$

where E is activation energy and R is a gas constant. This expression can be solved for various types of kinetics. For example, for zero-order reaction, then:

$$U^2 = \frac{2\lambda}{\rho Q}\,\frac{RT_c^2}{E}\,k_0 \exp\left(\frac{-E}{RT_c}\right) \qquad \text{(Eq 14)}$$

This formula is widely used to determine effective kinetic constants from experimental data, because:

$$\ln\left(\frac{U}{T_c}\right) = -\frac{E}{2RT_c} + \text{constant} \qquad \text{(Eq 15)}$$

Thus by varying T_c (by dilution or by changing initial temperature, T_0) and plotting $\ln(U/T_c)$ vs. $1/T_c$, the activation energy can be obtained readily. In this manner, the effective values of E have been measured for various SHS systems (Ref 5).

While providing a simple method for analyzing the redistribution of energy in the combustion wave, the model discussed above does not account for the local structural features of the reaction medium. *Microstructural models* account for details such as reactant particle size (d) and distribution, product layer thickness, and so on, and correlate them with the characteristics of combustion to give expressions of the form:

$$U = f(d)\,F(T) \qquad \text{(Eq 16)}$$

The first microstructural models were developed independently and essentially simultaneously (Ref 18, 21–23). For these models, the elementary reaction cell, which accounts for the details of the microstructure, consists of alternating lamellae of the two reactants (A and B), which diffuse through a product layer (C), to react (see Fig. 2a). Assuming that the particles are flat allows one to neglect the change in reaction surface area during synthesis. The characteristic particle size, d, is equivalent to the layer thickness, and the relative thicknesses of the initial reactant layers are determined by stoichiometry.

A simple analytical solution for the case of diffusion-controlled kinetics has been developed, and yields $f(d) \propto 1/d$ (Ref 21). A more accurate expression, based on the same physical geometry, was developed using the following kinetic function (Ref 22, 23):

$$\phi(\eta, T) = k_0 \exp\left(\frac{-E}{RT}\right) e^{-m\eta}\,\eta^{-n} \qquad \text{(Eq 17)}$$

which describes linear (m = 0, n = 0), parabolic (m = 0, n = 1), cubic (m = 0, n = 2), and exponential ($m > 0$, n = 0) kinetic dependencies. For example, the solution of Eq 11 and 12 for power-law kinetics (i.e., m = 0, $n \geq 0$) can be written as:

$$U = \frac{(n+1)(n+2)}{d^{(n+1)/2}}\,F(T) \qquad \text{(Eq 18)}$$

For the next two types of theoretical models, the elementary reaction cell consists of a spherical particle of one reactant surrounded by a melt of the other reactant. In the first case, the product layers (C) grow on the surface of the more refractory particles (B) due to diffusion of atoms from the melt phase (A) through the product layer (see Fig. 2b). At a given temperature, the

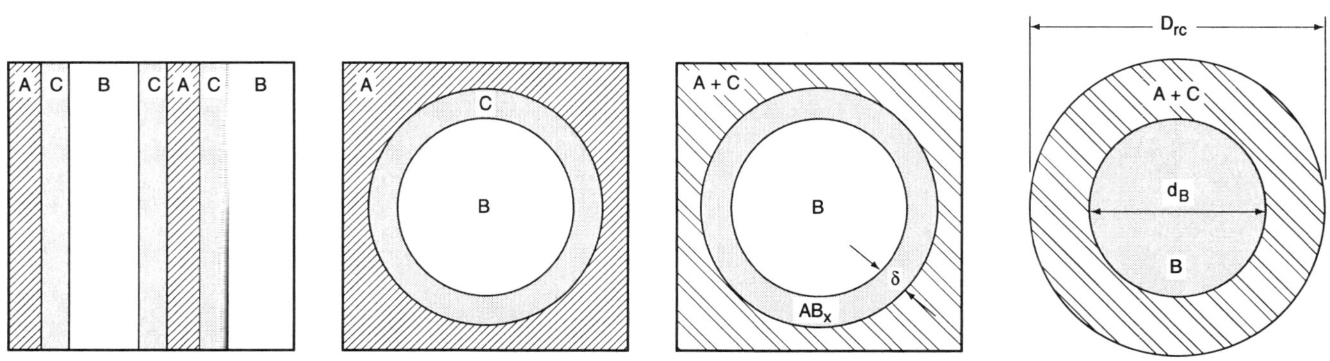

(a) A(s) + B(s) → C(s) (b) A(l) + B(s) → C(s) (c) A(l) + B(s) → C(1) (d) A(l) + B(1) → C(1)

Fig. 2 Geometry of the reaction cells considered in the theoretical models concentrations at the interphase boundaries are determined from the phase diagram of the system (Ref 24, 25). The second model (Fig. 2c) assumes that upon melting of reactant A, a layer of initial product forms on the solid reactant surface. The reaction proceeds by diffusion of reactant B through this layer, whose thickness is assumed to remain constant during the reaction (Ref 26, 27). The final product (C) crystallizes in the volume of the melt after saturation. Based on this model, an analytical expression for the combustion wave velocity has been reported (Ref 28):

$$U=\left[\frac{1}{d}\left(\frac{2}{d}+\frac{1}{\delta}\right)\right]^{1/2} F(T) \qquad \text{(Eq 19)}$$

where δ is the thickness of the initial product layer.

For the case where both reactants melt in the preheating zone and the liquid product forms in the reaction zone, a simple combustion model using the reaction cell geometry presented in Fig. 2(d) has been developed (Ref 29). After both reactants melt, their interdiffusion and the formation of a liquid product occur simultaneously. Numerical and analytical solutions were obtained for both kinetic- and diffusion-controlled reactions. In the kinetic-limiting case, for a stoichiometric mixture of reactants (A and B), the propagation velocity does not depend on the initial reactant particle sizes. For diffusion-controlled reactions, the velocity may be written as:

$$U=\frac{\bar{b}}{d}\cdot F(T) \qquad \text{(Eq 20)}$$

where $\bar{b}$ is a constant that depends on the mixture composition.

The microstructural models described above represent theoretical milestones in gasless combustion. Using similar approaches, other models have also been developed. For example, in Ref 30 the solid-liquid model (Fig. 2c) was used to determine the combustion velocity as a function of stoichiometry, degree of dilution, and initial particle size. Calculations for a variety of systems compared favorably with experimental data. In addition, an analytical solution was developed for diffusion-controlled reactions, which accounted for changes in λ, ρ, and C_p within the combustion wave, and led to the conclusion that $U \propto 1/d$ (Ref 31).

Phenomenological Aspects

As stated above, the characteristic feature of combustion synthesis, as compared to conventional powder metallurgy, is that the process variables, such as combustion wave velocity, U, and temperature, T_c, are strongly related. For example, a small change in T_c may result in a large change in U (see, e.g., Eq 13) and hence the characteristic time of synthesis. The process parameters (e.g., green mixture composition, dilution, initial density, gas pressure, reactant particle characteristics) influence the combustion velocity and the temperature-time profile, and in turn can be used to control the synthesis process.

Based on the analysis of literature data and incorporation of additional details, some general relationships for gasless combustion synthesis of materials from elements have been outlined as shown schematically in Fig. 3. Both characteristic features of the process, U and T_c, have maximum values when the composition of the green mixture corresponds to the most exothermic reaction for a given system (Fig. 3a). In general, U and T_c decrease with increasing initial reactant particle size and with addition of an inert (nonreactive) diluent to the green mixture (Fig. 3b, c), while increasing significantly with increasing initial sample temperature (Fig. 3d). Different trends have been observed when the initial sample density is varied. With increasing ρ_0, the combustion front velocity either increases monotonically or goes through a maximum, while the combustion temperature generally remains constant (Fig. 3e). A decrease in the sample size (e.g., sample diameter, D) does not influence U and T_c when the size is larger than a critical value D^*, because heat losses are negligible as compared to heat release from the chemical reaction. Below the critical sample size, both the combustion velocity and temperature decrease due to significant heat losses (Fig. 3f). Many exceptions to the dependencies discussed above have been observed, however, even for the simplest case of gasless combustion synthesis from elements. The combustion wave behavior becomes more complicated in gas-solid and reduction-type reactions. All of these effects are discussed in greater detail in Ref 8, 10, and 12.

Within the region of optimal experimental parameters, the combustion wave velocity remains constant and the temperature profile $T(t)$ has the same form at each point of the reaction medium. This regime is called *steady propagation* of the combustion synthesis wave, or *steady SHS process*. As the reaction conditions move away from the optimum, where the heat evolution decreases and/or heat losses increase, different types of *unsteady propagation* regimes have been observed. These include the appearance of an *oscillating combustion synthesis* regime, where macroscopic oscillations of the combustion velocity and temperature occur. The reaction may also propagate in the form of a hot spot that, for example, may move along a spiral pattern in cylindrical samples, and is called the *spin combustion* regime of CS. The combustion regime has great importance in the production of materials, because it influences the product microstructure and properties.

Methods for Large-Scale Synthesis

In general, methods for the large-scale production of advanced materials by combustion synthesis consist of three main steps: (a) preparation of the green mixture, (b) high-temperature synthesis, and (c) post-synthesis treatment. A schematic diagram of these steps is presented in Fig. 4. The first step is similar to the procedures commonly used in powder metallurgy, where the reactant powders are dried (e.g., under vacuum at 80 to 100 °C), weighed into the appropriate amounts, and mixed (e.g., by ball mixing). For some applications, cold pressing of the green mixture is necessary, especially for the production of low-porosity or poreless materials. Typically, no plasticizer is used, and the density of the cold-pressed compacts varies from 40 to 80% of the theoretical density. The final procedure in sample preparation determines the type of product to be synthesized: a powder product results from loose powder reactants, while sintered products are yielded from cold-pressed compacts. Pressing the green mixture into special molds or machining pressed initial compacts yields complex-shaped articles.

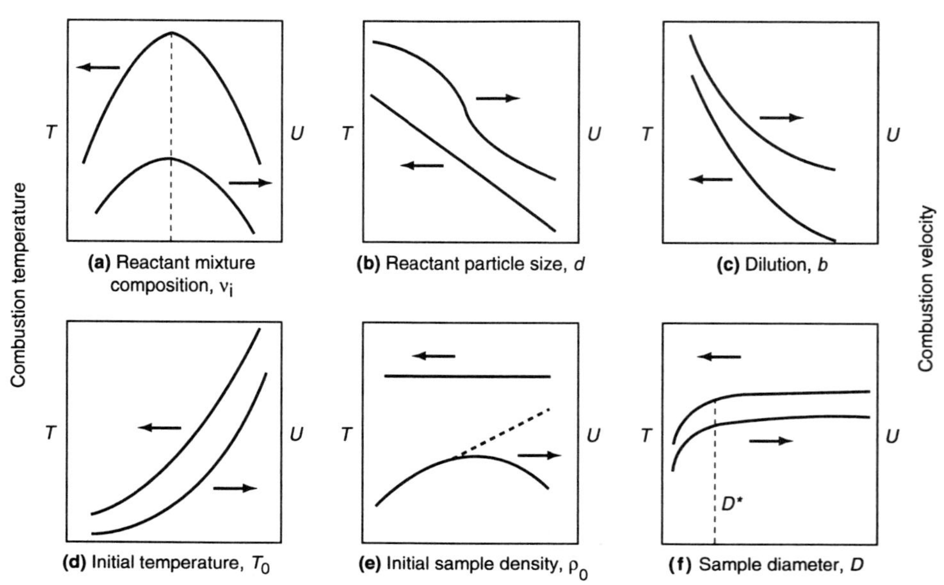

Fig. 3 Dependencies of combustion velocity, U, and maximum combustion temperature, T_c, on various combustion synthesis parameters

The main production technologies of combustion synthesis by the SHS mode are presented in the second block of Fig. 4. They may be classified into several major types: powder production and sintering, densification, casting, and coating.

The third main step of combustion synthesis technologies is post-synthesis treatment. Not all products require this step. Powder milling and sieving are used to yield powders with a desired particle size distribution. Annealing at elevated temperatures (800 to 1200 °C) removes residual thermal stress in brittle products. The synthesized materials and articles may also be machined into specified shapes and surface finish.

The VCS mode is used primarily for the synthesis of weakly exothermic systems. Various types of heaters, mostly commercially available furnaces but also spiral coil and foil heaters, are used to preheat the sample to the ignition point. To date, VCS materials have been produced only in laboratories.

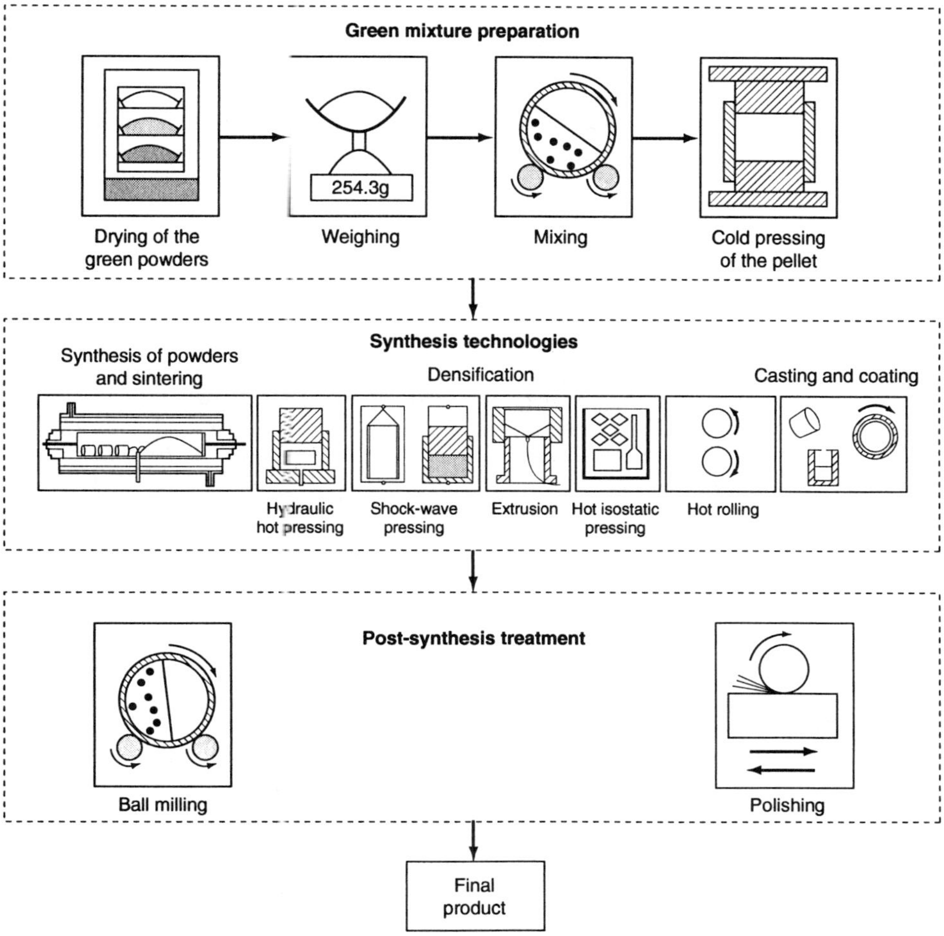

Fig. 4 Generalized schematic diagram of combustion synthesis technologies

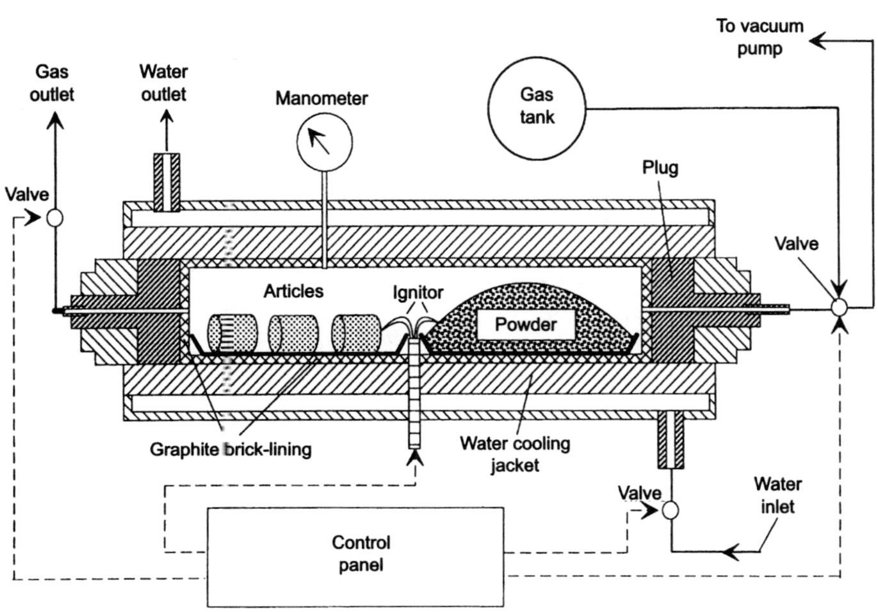

Fig. 5 Schematic diagram of self-propagating high-temperature synthesis reactor

Powder Production and Sintering

The design of a commercial reactor for large-scale production of materials is shown in Fig. 5. Typically it is a thick-walled stainless steel cylinder, with volume up to 30 L, that can be water cooled. The inner surface of the reactor is lined with an inert material to protect the vessel from extreme reaction temperatures. Graphite is generally used for lining during carbide, boride, or silicide synthesis, while boron nitride and silicon nitride provide protection during nitride synthesis. The green mixture or pressed compacts are loaded inside the vessel, which is then sealed and evacuated by a vacuum pump. After this, the reactor is filled with inert or reactive gas (Ar, He, N_2, O_2, CO, CO_2). Alternatively, a constant flow of gas can be supplied at a rate such that it permeates the porous reactant mixture. Two different types of reactors are used, depending on the product synthesized. The first type can maintain pressures up to 150 atm and is widely used for production of powders in gasless and gas-solid systems. Carbides, borides, silicides, intermetallics, chalcogenides, phosphides, and nitrides are usually produced in this type of reactor. The second type, a high-pressure reactor (up to 2000 atm), is used for the production of nitride-based articles and materials, because higher initial sample densities require elevated reactant gas pressures for full conversion. For example, well-sintered pure BN ceramic with a porosity about 20 to 35% was synthesized in 100 to 5000 atm N_2 (Ref 32).

SHS with Densification

The application of an external force during or after combustion is generally required to produce a fully dense (i.e., poreless) material. A variety of techniques have been investigated for applying the external force, such as static, dynamic, and shock-wave loading (see Fig. 6).

The oldest method uses a relatively slow (static) loading provided by a hydraulic press along with a specially designed die (Ref 33–37). Owing to the high temperature of the combustion products being densified, several approaches have been used to isolate the reacting sample from the die. One possible solution is the use of steel dies lined with graphite or BN ceramics (Ref 38). Another possibility is to use a pressure-transmitting medium such as SiO_2 (sand) or Al_2O_3, as shown in Fig. 6(a) (Ref 39). Essentially poreless (porosity less than 1%) and large-scale (up to 1 m in diameter) ceramic and metal-ceramic composite materials (Fig. 7) can be produced by SHS using the static pressing method (Ref 40).

Another method of SHS densification involves high-speed loading by an explosion shock. Two possible schemes of *high-speed shock-wave pressing* are presented in Fig. 6(b). The first method provides axial densification; the other, radial densification. The dynamic axial densification method has been used to synthesize dense materials including TiC, TiB_2, and HfC (Ref 41–44). The radial shock-wave densification technique has been applied to produce several materials, including dense high-temperature ceramic superconductor, $YBa_2Cu_3O_{7-x}$ (Ref 45–46).

In the techniques described above, combustion synthesis was first initiated, and then dynamic loading was applied to the hot product immediately after the combustion front propagated through the sample. Alternatively, shock compression applied to the reaction mixture may result in a rapid increase in temperature, initiating the chemical reaction with supersonic propagation rates. This approach, called *shock-induced synthesis,* has recently been developed (Ref 47–49). A variety of dense silicides (e.g., $MoSi_2$, $NbSi_2$, Ti_5Si_3, $TiSi_2$) and aluminides (NiAl, $NiAl_3$) have been synthesized using this method. Although the boundary between combustion and shock-induced synthesis is not well defined, the difference between them is addressed in a review article that describes different shock-induced methods to synthesize materials (Ref 50).

Another method used for the production of fully dense materials by combustion synthesis is SHS with *extrusion,* shown schematically in Fig. 6(c) (Ref 51). In this case, a powder compact of the reaction mixture is placed in the mold, and the process is initiated locally by a heated tungsten wire. After the combustion wave has propagated through the sample, appropriate pressure is applied to the plunger, extruding the products through the hole of the conic die. A high plasticity at the elevated reaction temperature allows the formation of long rods of refractory materials. The form and size of the die hole determine the extruded product configuration. The best developed application of SHS extrusion to date is the production of TiC-based cermet electrodes used for electric spark alloying (Ref 52).

A promising method for densification of combustion-synthesized products is a combination of SHS with *hot isostatic pressing* (HIP). This idea was first applied to the synthesis of TiB_2 ceramics under a pressure of 3 GPa, provided by cubic anvil press, resulting in 95% dense material (Ref 34, 53). The relatively low exothermic reaction of SiC from elemental powders was also carried out under these conditions (Ref 54), and 96% conversion to β-SiC was achieved as compared to 36% conversion when the reaction was initiated locally.

Another approach is to use high gas pressure for densification of the product simultaneously with combustion synthesis. The method of SHS + HIP with pressing by gas has been developed by using a so-called "chemical furnace." A schematic drawing of this *gas pressure combustion sintering* method is shown in Fig. 6(d) (Ref 34, 35). The green mixture is placed in evacuated glass containers, which are surrounded by a highly combustible mixture (e.g., Ti + C) that is enveloped by high-pressure gas (e.g., argon at 100 MPa). After ignition, the combustible mixture acts as a chemical furnace that heats the samples in the containers to their ignition point. The heat evolved from the chemical furnace also heats the glass to its softening temperature, where it becomes plastic and easily deformable. The material synthesized inside the containers is then pressed isostatically by the surrounding high gas pressure to zero porosity. A variety of ceramics, cermets, and functionally graded materials have been produced using this method, such as TiC, TiB_2, TiB_2-Ni FGM (Ref 56, 57),

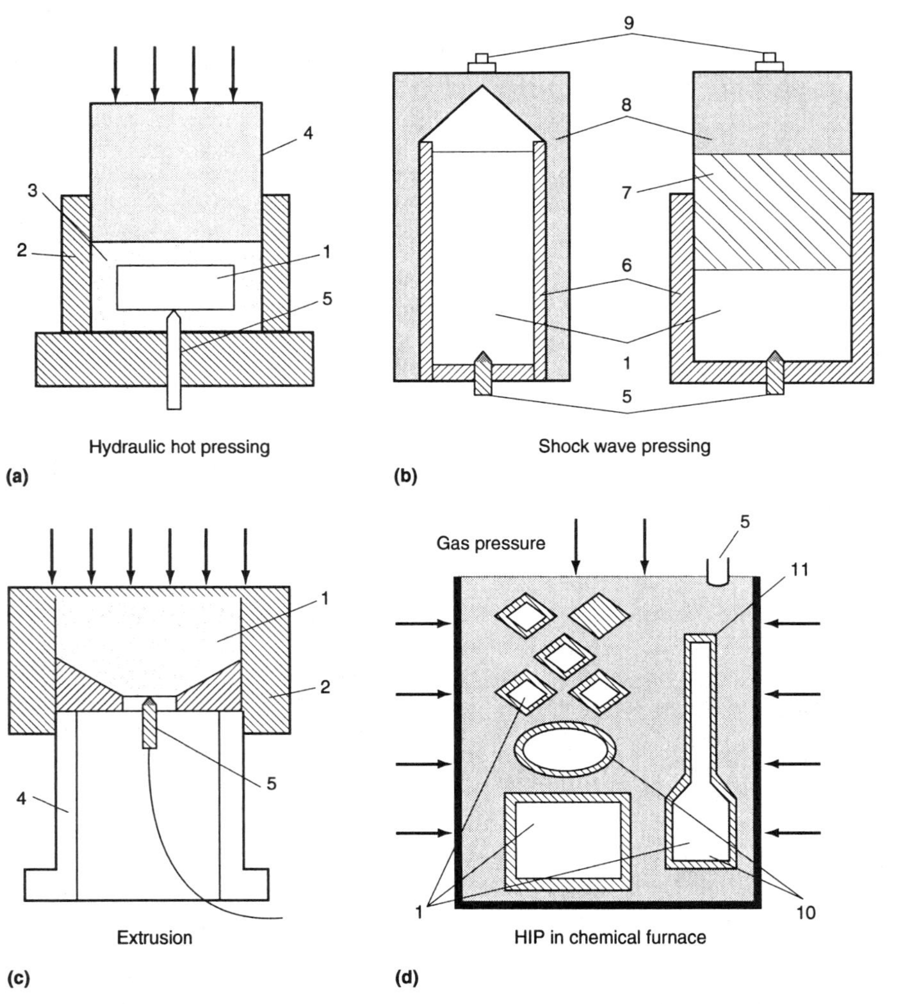

Fig. 6 Schemes of self-propagating high-temperature synthesis densification. 1, sample; 2, press die; 3, pressure-transmitting medium; 4, pressing body; 5, ignitor; 6, metal container; 7, massive piston; 8, explosive; 9, electric fuse; 10, glass containers; 11, "chemical furnace" mixture

Fig. 7 Commercially produced hard-alloy articles. Reprinted with permission. Source: Ref 6

$MoSi_2$-SiC/TiAl FGM (Ref 58), and Cr_3C_2-Ni (Ref 59).

Along with the various methods of combustion product densification, the *hot rolling* technique has been investigated (Ref 60, 61). It has been shown that simultaneous synthesis and hot rolling of intermetallic and ceramic materials (e.g., TiAl, $TiC_{0.47}$) under vacuum yields articles with porosity in the range of 5 to 50% (Ref 61).

SHS with Casting

Combustion synthesis of highly exothermic reactions (typically reduction-type) often results in completely molten products, which may be processed using common metallurgical methods. Casting of CS products under inert gas pressure or centrifugal casting has been used to synthesize cermet ingots, corrosion- and wear-resistant coatings, and ceramic-lined pipes.

Casting under gas pressure is similar to conventional SHS production. The reduction-type initial mixture (e.g., CrO_3 + Al + C, WO_3 + Al + C) is placed in a casting die and the reaction is initiated under an inert gas pressure (0.1 to 5 MPa) to prevent product sputtering by gas evolution from the thermite reaction (Ref 62, 63). Also, the molten product may consist of two immiscible phases, either molten oxides (e.g., Al_2O_3, MgO) with dispersed non-oxides (e.g., metals carbides or borides) or molten metals with dispersed ceramic droplets. Owing to the difference in densities, the two phases may separate. By controlling the characteristic times of cooling (t_{cool}) and phase separation (t_{ps}), different distributions of phases may result. Thus, it is possible to obtain cermets with uniform ($t_{cool} << t_{ps}$), gradient ($t_{cool} \sim t_{ps}$), and layered ($t_{cool} >> t_{ps}$) products.

Centrifugal casting allows for greater control of the distribution of phases by controlling the time of separation. The two types of centrifugal casting equipment are shown in Fig. 8 (Ref 64). *For radial centrifuges* (Fig. 8a), the sample is placed at a fixed radial position from the axis of rotation, and the applied centrifugal force is parallel to the direction of propagation. The influence of centrifugal acceleration on the degree of phase separation, η_{ps}, is shown in Fig. 9. This parameter characterizes the phase distribution in the final product:

$$\eta_{ps} = m_s / m_t$$

where m_s is the mass of the non-oxide phase (e.g., metal, carbide, boride) fully separated from the oxide matrix, and m_t is the total mass of this phase produced by the reaction. Thus, η_{ps} = 0 for the cermet product with uniformly distributed oxide and non-oxide phases, while η_{ps} = 1 for multilayer materials or cases in which the metal-type ingot is totally separated from the oxide slag layer. Finally, the ingots have a gradient distribution of phases when $0 < \eta_{ps} < 1$.

The second type of centrifugal casting apparatus, called an *axial centrifuge* (Fig. 8b), is used for production of ceramic, cermet, or ceramic-lined pipes. The axial centrifuge casting method was developed further for production of long pipes with multilayer ceramic inner coatings (Ref 65, 66). In this process, a thermite mixture (e.g., Fe_2O_3/2Al) is placed inside a rotating pipe and ignited locally. A reduction-type combustion reaction propagates through the mixture, and the centrifugal force results in separate layers of metal and ceramic oxide, with the latter forming the innermost layer (Fig. 10). The process is carried out in air under normal pressure, and pipes with diameters up to 30 cm and 5.5 m long have been obtained.

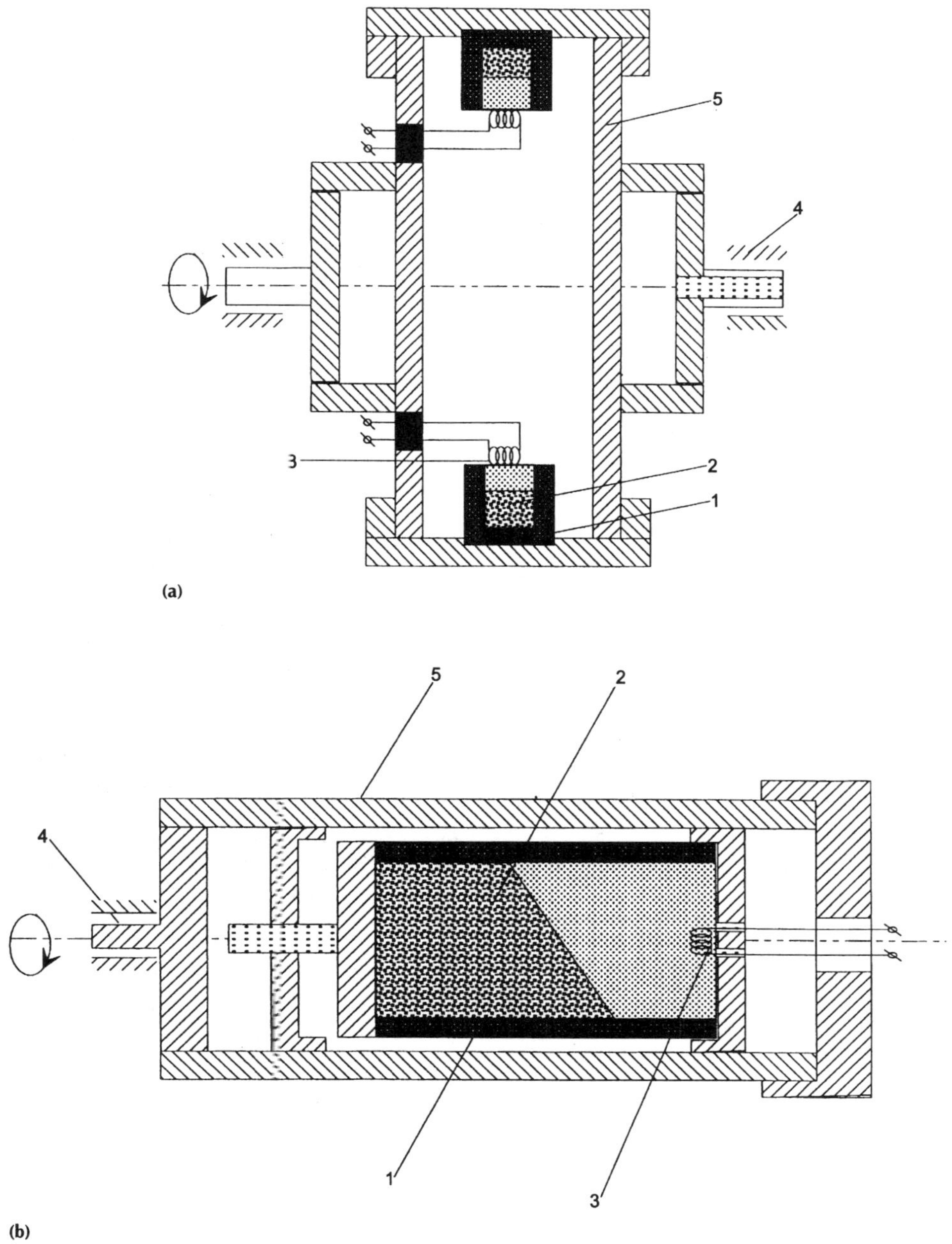

Fig. 8 Scheme of self-propagating high-temperature synthesis plus centrifugal casting. (a) Radial centrifuge. (b) Axial centrifuge. 1, sample container; 2, reactant mixture; 3, ignitor; 4, axle; 5, reactor

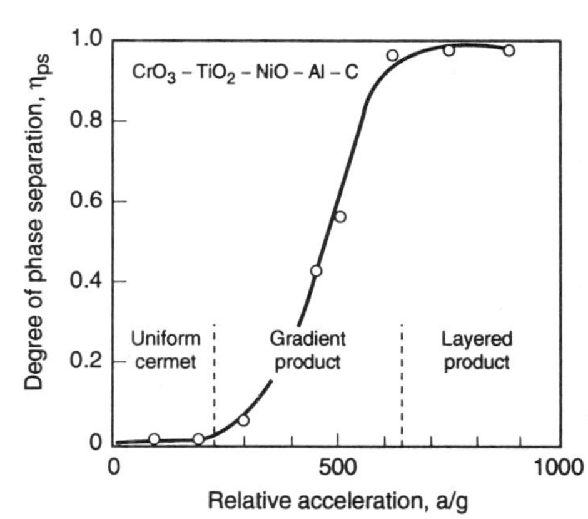

Fig. 9 The degree of phase separation as a function of centrifugal acceleration, *a*, where *g* is acceleration due to gravity. Source: Ref 64

Classes and Properties of Synthesized Materials

The synthesis of several hundred materials, by both the SHS and VCS modes, has been reported in the literature. The types of compounds produced include carbides, borides, intermetallics, silicides, aluminides, composites, nitrides, hydrides, and oxides. This section describes the synthesized compounds, as well as the materials (i.e., powders, poreless materials, and functionally graded materials) and articles produced.

Gasless Combustion Synthesis from Elements

Group IV Metal Carbides. Group IVa transition metal carbides (TiC, ZrC, HfC) were among the first compounds synthesized by the SHS method (Ref 67). With high melting points, high hardness, and good chemical stability, these materials can be used as abrasives (economical substitutes for diamond powders and pastes) or as components of cermet and ceramic materials (Ref 68).

These compounds have relatively high heats of formation (e.g., –185 kJ/mol for TiC and –198 kJ/mol for ZrC). The heat evolved upon initiation of the reaction allows the materials to be synthesized in the SHS mode, with a high degree of conversion. For example, the amount of unreacted carbon has been reported to be as low as 0.09 wt% for TiC and 0.01 wt% for ZrC and HfC (Ref 67). Emission spectrochemical analysis of the powders (Ref 16) has confirmed the high purity of the combustion-synthesized TiC, where the largest impurity constituents were Al (up to 2000 ppm), Ar (300 ppm), Si (100 ppm), Fe (100 ppm), B (80 ppm), and Ca (30 ppm). The impurities in the product were typically 2 to 7 times lower than in the initial mixture. Owing to the high temperature of combustion, the material self-purifies by purging any volatile impurities from the reaction mixture. Some properties of SHS powders are presented in Table 2.

The high heat of formation of TiC and specific features of the Ti-C phase diagram make it possible to produce, by CS, non-stoichiometric carbides, TiC_x (x = 0.38 to 1.0) (Ref 69), self-binding materials that consist of non-stoichiometric TiC and titanium located along the carbide grain boundaries, and cemented carbides (Ref 70). A wide variety of metal additives have been also used to produce dense cermets, where combustion-synthesized TiC grains are joined by metal (e.g., up to 60 wt% Ni or Fe) or intermetallic binders. Static pressing of hot TiC_x immediately after synthesis yields dense ceramic or cermet (self-binded carbide) materials, with product densities as high as 99% of theoretical (Ref 16, 39, 70). Dense self-binded TiC_x-Ti cermets with residual porosity ~1% have also been produced by pseudo-HIP (Ref 71, 72). By applying SHS with isostatic pressing to a Ti + 0.42C green mixture, dense (4.54 g/cm³) cermets with high hardness (86 to 87 HRA) and bending strength (652 MPa) have been synthesized.

The production of TiC-based materials with different binders has been reported: TiC + Ni (Ref 73, 74), TiC + (Ni-Al) (Ref 75, 76), TiC + (Ni-Mo) (Ref 70, 77), TiC + (Ni-Mo-Cu) and (TiC-Cr_3C_2) + Ni (Ref 70), (TiC-Cr_3C_2) + steel and (TiC-TiN) + (Ni-Mo) (Ref 32), TiC + (Ti-Al) (Ref 78), and TiC-Al (without pressing) (Ref 79). Some properties of combustion-synthesized and static-pressed cemented carbides are listed in Table 3, and characteristic microstructures of TiC-Ni cermets are shown in Fig. 11.

Group Va Carbides. To a lesser extent, combustion synthesis from elements of transition (Group Va) metal carbides has also been investigated, especially for TaC (Ref 80, 81) and NbC (Ref 82). Some properties of reaction-synthesized NbC powders are presented in Table 2.

Nonmetal Carbides. Among non-metallic carbides, SiC synthesized from elemental silicon and carbon powders has attracted the most attention. Silicon carbide has a relatively low heat of formation (69 kJ/mol), and additional heat input is required for combustion to occur. For example, locally igniting pressed Si + C samples, placed on a ribbon heater, yielded ~64% conversion to β-SiC, while increasing the area of ignition with a specially designed carbon-sleeve heater improved the conversion to ~99% (Ref 83). Dense SiC ceramics have been produced by combining the carbon-sleeve heater with hot pressing (Ref 54). Another method of heating and ignition is to pass an electric current through the sample, yielding fine stoichiometric SiC powders (Ref 53). Fine-grained powders (0.2 to 0.5 μm) with high specific surface area (6.2 m²/g) have been synthesized in the VCS mode by inductively preheating the Si + C mixture 1300 °C (Ref 84, 85). Solid SiC and SiC-C blocks have also been produced by infiltration of arc-melted silicon into porous carbon, during which the SHS reaction occurred (Ref 86). A characteristic microstructure of SiC powder synthesized by the SHS method is shown in Fig. 12.

An alternative method of activating the reaction between silicon and carbon without preheating has been reported (Ref 87), which involves using potassium nitrate (KNO_3) as an activating agent. With this method, fine (<μm) SiC powders containing 28 to 29 wt% bound carbon and 0.5 wt% free carbon have been produced.

Borides. Most of the transition metal borides can be synthesized in the SHS mode (Ref 88, 89). Early works reported synthesis of relatively pure (with residual unreacted boron less than 0.3 wt%) ZrB_2, TiB_2, HfB_2, and MoB (Ref 4, 67). Combustion-synthesized borides also include TiB, TaB_2, LaB_6, and NbB_2 (Ref 7). Similar to TiC for carbides, the combustion synthesis of transition metals borides from elements has focused primarily on TiB_2.

The high heat of formation for these borides (e.g., –279 kJ/mol for TiB_2, –160 kJ/mol for TiB) makes it possible to add chemically inert metals and alloys (or weakly exothermic mixtures) to the metal-boron system. The subsequent plasticity of the products at the combustion temperature allows for various methods of densification to be applied, yielding poreless boride ceramics or boride-based cermets. For example, dense TiB-Ti self-binding borides (Ref 90) and TiB_2 ceramics with 95% theoretical den-

Table 2 Some characteristics of self-propagating high-temperature synthesis powders

Powder	Chemical composition, wt %	Main impurities, wt %	Average particle size, μm or specific surface area, m²/g
TiC	C_{fixed} = 19.2–19.5	C_{free} < 0.8	3–5 μm
TiC (titanium carbide abrasive)	Ti = 77.0–79.5 C_{fixed} = 18.0–19.8	C_{free} < 0.2–2.0 Mg = 0.06 Fe = 0.1	Polydispersed 1–200 μm
NbC	C_{total} = 11.0	C_{free} < 0.1 O < 0.2	Polydispersed 1–100 μm
SiC (β - phase)	C_{total} = 27.0–27.8	C_{free} < 0.3–0.8 O < 1.0 Fe < 0.2	8–10 m²/g
BC_4	B_{total} = 74 C_{total} = 20.0	B_2O_3 < 0.5 Mg < 0.5	3.9 m²/g
TiB_2 (polydispersed)	Ti = 69.6 B_{total} = 29.2	O < 0.5	1–200 μm
TiB_2 (finely dispersed)	Ti = 66.7 B_{total} = 30.5	B_2O_3 < 0.2 Mg < 0.3	1–10 μm
$MoSi_2$	Mo = 61.9–62.5 Si_{total} = 35.5–36.5	Si_{free} < 0.3 O < 0.3 Fe < 0.1	1–50 μm
$TiSi_2$	Ti = 44.3 Si_{total} = 52.0	O < 0.1 Fe < 0.1	1–100 μm

Source: Ref 32, 163

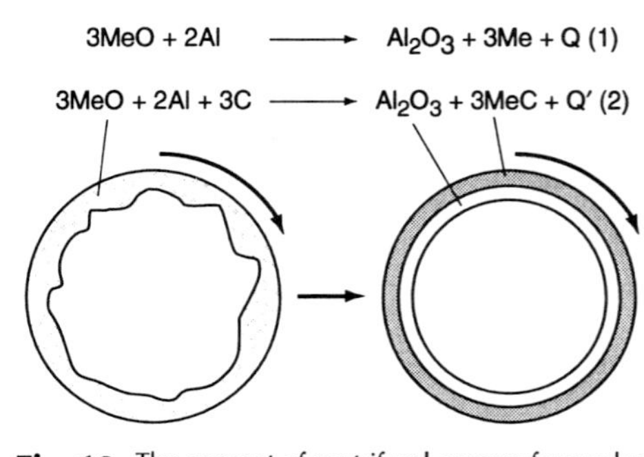

Fig. 10 The concept of centrifugal process for production of ceramic-lined steel pipes. Source: Ref 66

sities (Ref 36) have been formed by hydraulic pseudo-isostatic pressing and uniaxial pressing in a graphite-lined die, respectively. The hydrostatic pressing method has also been applied to synthesize TiB_2-TiNi composites with 93% relative density and molar content of TiNi ranging from 0 to 95% (Ref 91). Dense TiB_2 ceramic materials have been formed by high-pressure combustion sintering under 3 GPa in a cubic anvil press (Ref 34), as well as by gas HIP at 100 MPa using a chemical furnace (Ref 92). Synthesis of composite products, such as TiB_2-Al (Ref 93) and TiB_2-Fe (Ref 94), has also been reported.

TiC-TiB_2 Composite Ceramics. In some cases, several refractory compounds can result from two or more parallel reactions occurring simultaneously in the combustion wave. A typical example is the Ti-C-B system, where both the Ti + C and Ti + 2B reactions affect the combustion synthesis and structure formation processes (Ref 95). By adjusting the contents of carbon and boron powders in the reactant mixture, either carbide- or boride-based ceramics may be obtained.

The Ti-C-B system has been used for a series of ceramic poreless composites produced by the SHS + HIP method (Ref 70). Near-eutectic 40TiC-60TiB_2 (wt%) compositions were primarily investigated, while various metal additives were tested (e.g., Co, Cu) as the binder phase, with contents varying from 0 to 12 wt%. The materials demonstrated good mechanical properties (Table 3) and may be used for cutting tools. Materials with higher TiC content (80 wt%) possess higher bending strength.

Silicides. Compounds of metals with silicon were synthesized from elements by both the SHS (Ref 96–99) and VCS modes (Ref 100). Powders and sintered samples of various chemical compositions were obtained, including Ti_5Si_3, TiSi, $TiSi_2$, Mo_3Si, Mo_5Si_3, $MoSi_2$, Zr_2Si, Zr_5Si_3, ZrSi, and $ZrSi_2$. Among them, titanium and molybdenum silicides have received the most attention due to their high melting points, excellent oxidation resistance, relatively low density and high electrical and thermal conductivities (Ref 101, 102). The properties of some silicide powders are shown in Table 2.

Combustion-synthesized $MoSi_2$ powders have been used to produce high-temperature heating

Table 3 Some properties of self-propagating high-temperature synthesis cermet alloys

Cemented carbide alloy	Basic composition	Density, g/cm^3	Average grain size, μm	Hardness, HRA	Ultimate flex strength, kg/mm^2	Application
STIM-1B/3	TiC/TiB_2 + Cu	4.94	5–7	93.5	70–80	Cutting edges, targets
STIM-2	TiC + Ni	5.50	5–7	90.0	100–120	Wear-resistant coatings
STIM-2A	TiC + (Ni-Mo)	6.40	1–2	87.0	160–180	Press tools
STIM 3B/3	TiC-Cr_3C_2 + Ni	5.37	3–4	92.5	80–100	Cutting edges
STIM-3B	TiC-Cr_3C_2 + steel	5.40	2–4	92.5	100–120	Corrosion-resistant articles
STIM-4	TiB + Ti	4.20	1–2	86.0	100–120	Thermal-shock-resistant articles
STIM-5	TiC-TiN + Ni-Mo	5.0	1–2	91.0	120–140	Cutting edges

Source: Ref 163

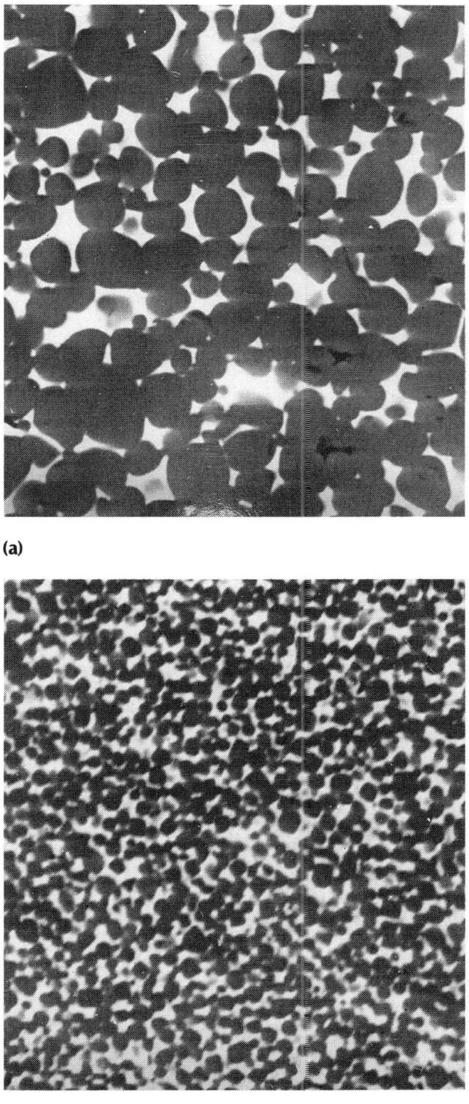

(a)

(b)

Fig. 11 Characteristic microstructure of self-propagating high-temperature synthesis Ti-C-Ni cermets (1800×). (a) 20 wt% Ni. (b) 40 wt% Ni

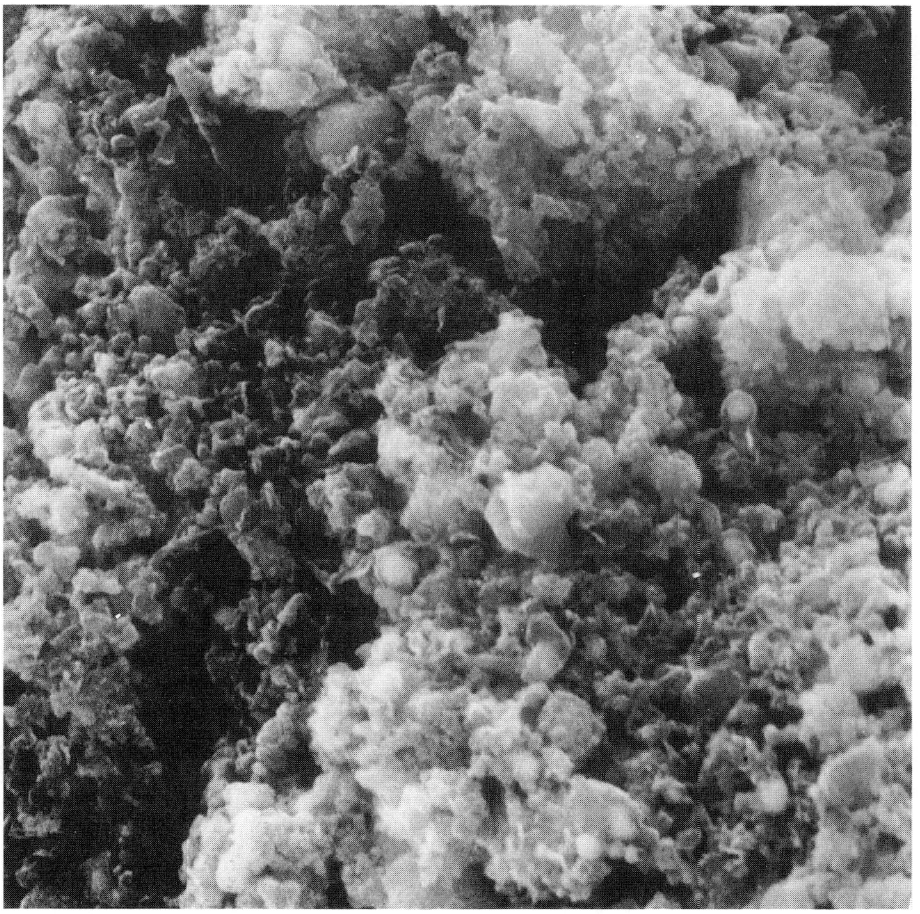

Fig. 12 Characteristic microstructure of self-propagating high-temperature synthesis β-SiC powder conglomerate (16,000×)

elements (Ref 6). The application of the SHS + extrusion method for one-step production of $MoSi_2$ heaters has been reported (Ref 51). Combustion synthesis of molybdenum alumosilicide (Mo-Al-Si) has also been demonstrated (Ref 103).

Aluminides. Combustion synthesis of nickel aluminides in the VCS mode was first demonstrated during sintering of a Ni-Al mixture (Ref 104). Shortly thereafter, the SHS mode of combustion synthesis was reported for a variety of aluminides, including those of Zr, Ti, Cr, Co, Mo, Cu, Fe, Nb, etc. (Ref 105–110). However, most attention has been paid to nickel aluminides (Ref 111–113) and titanium aluminides (Ref 114–116).

As discussed above, aluminides have been used as binders for carbide- and boride-based cermets, for example by adding nickel and aluminum powders to exothermic mixtures of titanium with carbon or boron. On the other hand, some intermetallic compounds (e.g., NiAl, Ni_3Al, TiAl) possess high enough heats of formation that composites with intermetallic matrices can be produced either in the VCS or SHS regimes. The ceramic components are either added in the green mixture or synthesized in situ during the reaction.

For example, Ni_3Al-matrix composites reinforced by Al_2O_3, SiC, or B_4C whiskers have been formed in the SHS mode (Ref 117, 118). A typical microstructure of Ni_3Al/Al_2O_3 composite is shown in Fig. 13. Ni_3Al-TiB_2 composites have been synthesized in situ under microgravity conditions ($a = 10^{-4}$ m/s^2). The kinetics of growth of TiB_2 particles in the Ni_3Al matrix, for both normal and microgravity environments, are shown in Fig. 14, indicating that finer microstructure results in reduced gravity synthesis (Ref 119). NiAl-matrix composites reinforced with TiB_2 particles (10 and 30 wt%) have been synthesized using SHS + HIP hot pressing (Ref 120). The TiAl matrix was formed in the SHS mode, while reinforcing SiC particles were added in the initial Ti + Al mixture (Ref 121). Also, aluminum and $TiAl_3$ matrix composites, reinforced by TiC, TiB_2, and TiC + TiB_2 particles synthesized in situ, were fabricated recently in the SHS regime (Ref 122).

Owing to their high plasticity at the combustion temperature, poreless aluminides can be produced using various methods of densification. Dense titanium materials have been obtained in the VCS regime with pseudo-HIP (Ref 123), while hydraulic hot pressing combined with SHS has been used to form NiAl-matrix composites with porosities of 1% or less (Ref 119). Hot rolling under vacuum has been applied to form combustion-synthesized TiAl (Ref 61). Other materials have been synthesized by the extrusion of nickel-rich NiAl (Ref 109). Shock-induced synthesis with dynamic densification has been reported for titanium, niobium, and nickel aluminides (Ref 47, 124, 125). Densification of iron aluminides and Fe_3Al-matrix composites reinforced with Al_2O_3 particles has been performed by pressing in graphite-lined dies at 10 to 70 MPa (Ref 113), as well as by HIP and pseudo-HIP (Ref 126). It has been shown that a

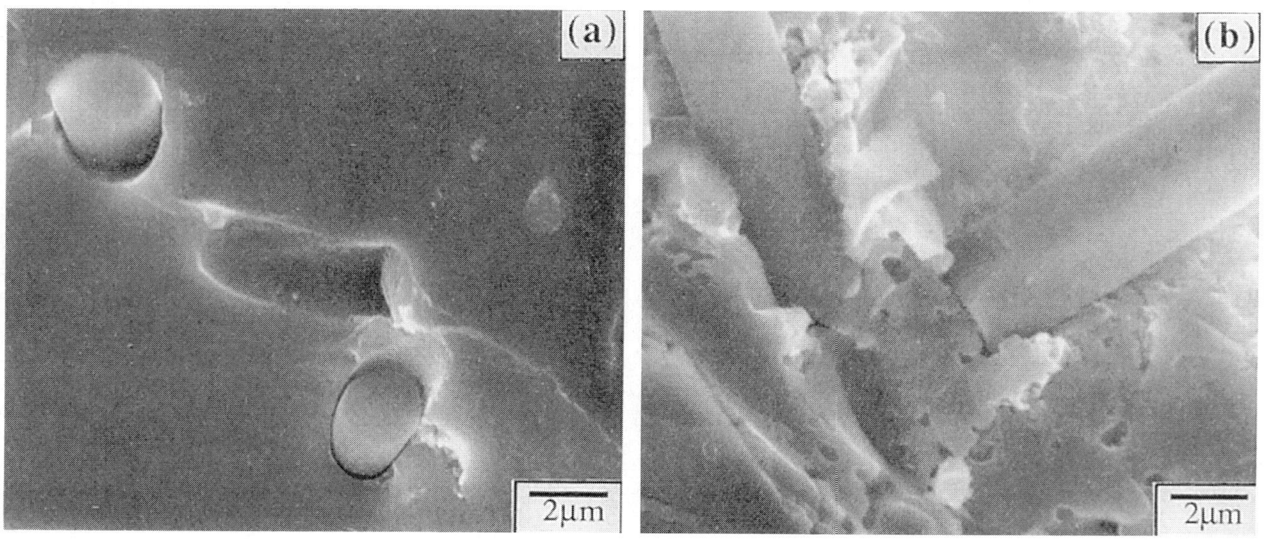

Fig. 13 SEM view of fracture surface from nickel aluminide/Al_2O_3 composite. Source: Ref 118

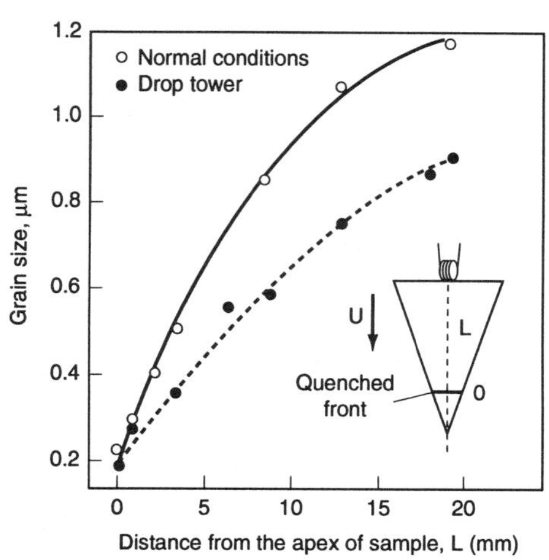

Fig. 14 Evolution of TiB_2 grain size in (3Ni + Al)/(Ti + 2B) quenched sample. Source: Ref 119

Fig. 15 Hot-rolled coil (70 kg) of NiTi wire (9.5 mm diameter) after forging. Reprinted with permission. Source: Ref 132

99.7% dense Fe_3Al-Nb alloy billet can be prepared by SHS followed by superplastic consolidation (Ref 127).

Compounds of Titanium. While intermetallic Ti-(Ni, Co, Fe) compounds cannot be synthesized from elemental powders without preheating, a relatively low increase above room temperature is generally sufficient to initiate a self-sustained reaction. Almost all compounds of the Ti-Ni, Ti-Co, and Ti-Fe systems have been produced by this method (Ref 128–130). Intermetallic TiNi compounds are especially attractive for their shape memory characteristics. Combustion-synthesized Ti-Ni powders and materials have been used in the industrial production of sheets, tubes, and wires with shape memory in Russia (Ref 131) and in Japan (Ref 132). Cold-deformed binary alloys have retained up to 90% of their original shape, while 99.5% of the original shape may be restored for ternary alloys (e.g., Ti-Ni-Fe and Ti-Ni-Al). A 70 kg coil of hot-rolled TiNi wire produced by the CS + HIP technique is shown in Fig. 15 (Ref 132).

Diamond Composites. At normal pressures, diamond is a metastable phase of carbon. At temperatures higher than 1300 K (either in vacuum or inert gas), it converts to graphite. The rate of conversion is extremely rapid at temperatures higher than 2000 K (Ref 133). Thus, graphitization during sintering of diamond composites makes it difficult (if not impossible) to produce refractory-matrix composites with diamond particles. Due to the short time of interaction, SHS is an attractive method for producing diamond-containing composite materials with refractory and wear-resistant matrices. SHS combined with pressing (pseudo-HIP) has been used to obtain TiB-diamond composites from elemental Ti + B with up to 20 wt% synthetic diamond (Ref 134). By decreasing the combustion temperature from 2200 to 1900 K, the extent of graphitization was reported to decrease from 6 to 0.5% (Ref 135). Further investigation of the combustion synthesis of TiB- and NiAl-matrix diamond-containing materials has shown that under certain conditions, diamond particles retain their shape, surface quality, and mechanical properties after synthesis (Ref 136, 137). The characteristic microstructure of the diamond-containing portion of an Ni-Al/diamond composite synthesized by SHS is shown in Fig. 16.

Functionally Graded Materials. The concept of functionally graded materials (FGMs) to tailor nonuniform distribution of components and phases in materials, and hence combine mechanical, thermal, electrical, chemical, and other properties that cannot be realized in uniform materials. For example, the material structure may have a smooth transition from a metal phase with good mechanical strength on one side, to a ceramic phase with high thermal resistance on the other side (see Fig. 17). With a gradual variation in composition, FGMs do not have the intermaterial boundaries found in multilayer materials, and hence they exhibit better resistance to thermal stress.

Combustion synthesis has been applied successfully in the production of TiB_2-Cu FGM, with the amount of TiB_2 varying from 0 (metal side) to 100% (ceramic side) for cylindrical samples, 30 mm in diameter and 5 mm thick (Ref 138). A predetermined distribution of elements in the green mixture was created using an automatic powder stacking apparatus. During synthesis, simultaneous densification by hydrostatic pressing was also applied. Similar methods were used to produce TiB_2-Cu FGM samples from cylindrical green compacts, 17 mm in diameter and 10 mm in height (Ref 140).

The method of gas-pressure combustion sintering was applied to produce a TiC-Ni FGM (Ref 56). The Vickers microhardness of this material gradually changed from 2 GPa at the nickel side to 23 GPa at the TiC side, where the TiC grain size decreased with increasing local nickel content. Figure 18 shows the microstructure of a Cr_3C_2/Ni FGM, consolidated by the SHS + HIP method from a green compact consisting of powder layers with different Cr-C-Ni compositions.

Another approach is to prepare a green mixture with relatively few (typically two or three) layers, but so that desired gradients in the product appear during the SHS process due to infiltration and migration of melts, diffusion, and other transport phenomena (Ref 141). This approach has been used to produce poreless functionally gradient cermets, including TiC-Ni and (Ti,Cr)C-Ni, by combining SHS with pseudo-HIP. An example of a multilayer material synthe-

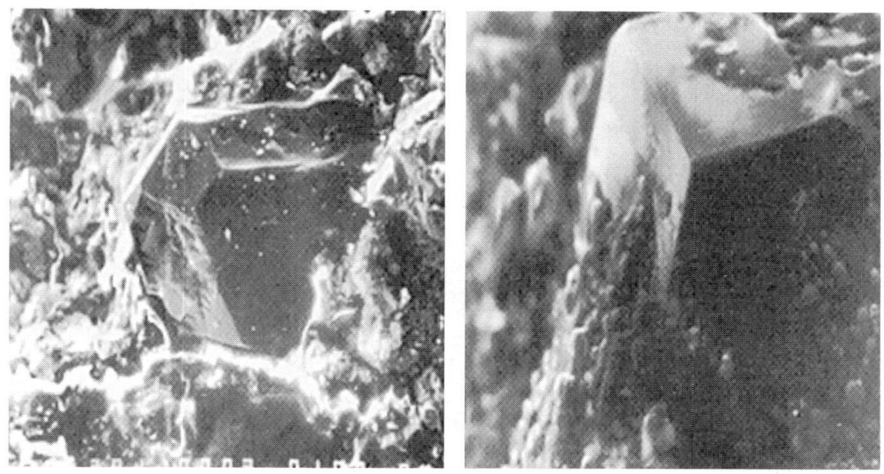

Fig. 16 Characteristic microstructure of the diamond-containing region in a Ni-Al diamond composite produced by self-propagating high-temperature synthesis. Reprinted with permission. Source: Ref 137

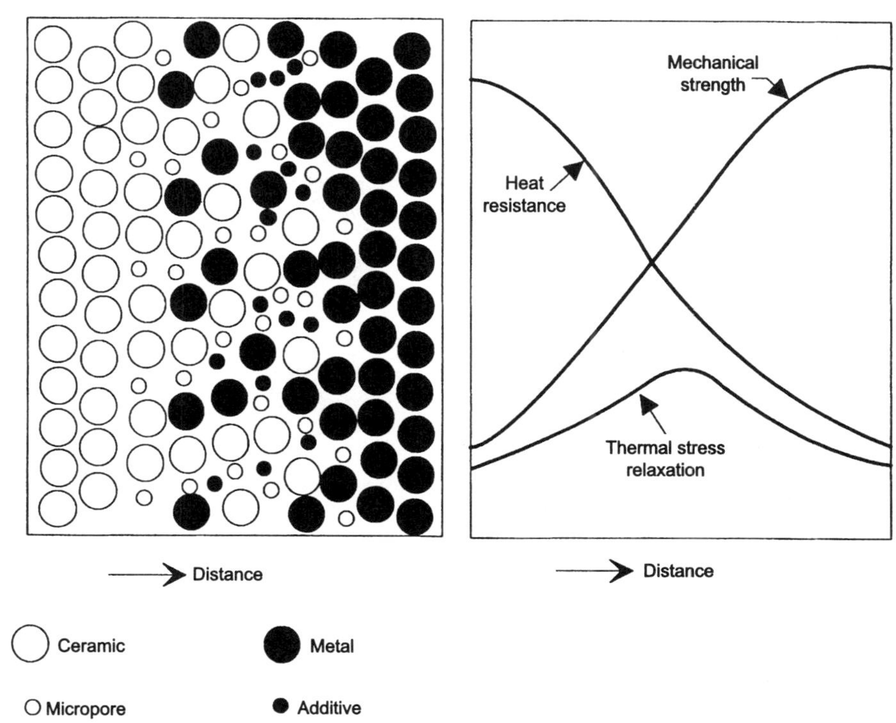

Fig. 17 The concept of functionally graded materials. Source: Ref 138

sized by CS + hot pressing is the TiB_2-Al_2O_3-ZTA (zirconia toughened aluminum) composite. The microstructure of this layered material is shown in Fig. 19 (Ref 142). The layers can have different thicknesses and are well bonded to one another. The production of diamond-containing FGMs has also been reported (Ref 143), where properties of materials are determined by the gradient distribution of diamond in the combustion-synthesized matrix.

Other Classes of Products. Chalcogenides, including sulfides and selenides of Mg, Ti, Zr, Mo, and W represent another group of compounds that can be synthesized from elements during the combustion process. Experimental synthesis of luminescent ZnS samples was also carried out (Ref 144). Combustion reactions between chalcogenides and alkali metal compounds have been applied to produce GaAs and GaP from solid-state precursors (Ref 145). In addition, compounds of metals (Ni, Cu, Zr, etc.) with phosphorus were obtained (Ref 146, 147). Also, phosphides of Al, Ga, and In were formed by direct combustion synthesis from elemental powders (Ref 148). An interesting example of the combustion synthesis of aluminum iodide Al_2I_6 from a mixture of aluminum powder with crystalline iodine has been reported (Ref 149). It should be noted, however, that synthesis of chalcogenides, phosphide, and iodides represents a process intermediate between gasless and solid-gas processes due to intense vaporization of the nonmetal reactant during combustion.

Among intermetallic compounds not mentioned previously, Mg_2Ni is a promising material for storing hydrogen. It has been shown that the combustion synthesis of this material is feasible, in spite of a relatively low heat of reaction. For example, high-purity Mg_2Ni has been obtained in the SHS mode (Ref 150).

Combustion Synthesis in Solid-Gas Systems

Nitrides and nitride-based ceramics produced by combustion synthesis in solid-gas systems are described below.

Metallic Nitrides. A wide variety of metallic nitrides, including TiN, ZrN, NbN, HfN, TaN, and VN, have been produced by combustion synthesis. Among these, titanium and zirconium nitrides have been investigated most thoroughly. These nitride powders can be used to form refractory corrosion-resistant coatings and conducting materials in electronic applications. Depending on the nitrogen pressure, titanium and zirconium nitrides with a wide range of nitrogen compositions have been synthesized, ranging from solid solutions with a minimum amount of nitrogen ($MeN_{0.1}$) to single-phase supersaturated solutions ($MeN_{0.34}$ to $MeN_{0.45}$), as well as stoichiometric MeN. Full conversion to the stoichiometric compound can be achieved by dilution of the initial metal powder with the corresponding nitride product (Ref 151, 152).

Other metallic nitrides have been synthesized by combustion synthesis. SHS-vanadium nitride (Ref 153) is the basis of ferrovanadium (VN + Fe), a commonly used alloying agent for the manufacture of steels with low-temperature applications (Ref 154). A unique cubic structure of TaN, which has the highest microhardness among all transition metal nitrides (3200 kg/mm^2) and can be used as a hard alloy component, has been formed during combustion of tantalum powder in liquid nitrogen (Ref 155). Niobium nitride powders, with a transition temperature to superconductivity at 14.3 K, have been synthesized by SHS. In addition to powders, thin superconducting plates and wires of NbN have been obtained using Nb + NbN + N_2 mixture as a chemical furnace (Ref 156). Both SHS and VCS modes have been used for the synthesis of magnesium nitride, Mg_3N_2 (Ref 157). The characteristics and properties of some combustion-synthesized transition metal nitrides are presented in Table 4.

Nonmetallic nitrides including Si_3N_4, BN, and AlN, have important applications as high-temperature structural ceramics. As discussed above, two methods may be used to produce nitrides by CS: elemental reactions (Si + N_2, B + N_2, Al + N_2) and reduction reactions (e.g., B_2O_3 + Mg + N_2). Tables 5 to 7 compare the chemical composition and specific surface area of powders produced by CS and by conventional method. Powders synthesized by SHS (see Fig. 20) have nitrogen contents that are comparable to, if not higher than, those of powders produced by conventional methods. The metal impurities (mainly iron) are the consequence of either inappropriate mixing/grinding equipment or impure raw materials, both of which may be remedied. The mechanism of combustion SHS nitrides synthesized from elements has also been investigated (Ref 158).

The industrial ceramics produced by conventional sintering of β-phase Si_3N_4 powders synthesized by SHS (Ref 159) can be used as high-temperature articles with attractive dielectric properties (tan $\delta = 4.4\ 10^{-3}$ at $f = 10^6$ Hz, and dielectric strength of 9.2 kV/mm). Also, silicon nitride powders with a relatively high α-phase content (≥80%) have been used for the production of advanced structural ceramics with good mechanical properties (σ_b = 900 MPa, K_{Ic} = 6.7 to 7.2 mN/M$^{3/2}$, and HV = 13.3 to 14.2 GPa). Because their mechanical properties are maintained at high temperatures (up to 1700 K), sili-

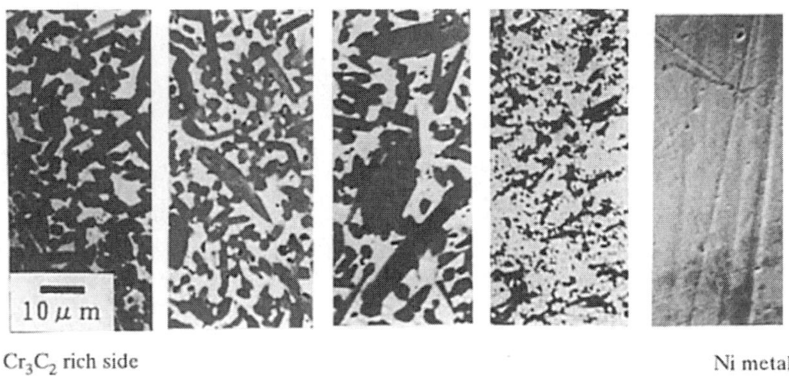

Fig. 18 SEM image of Cr_3C_2/Ni functionally graded material microstructure. Reprinted with permission. Source: Ref 139

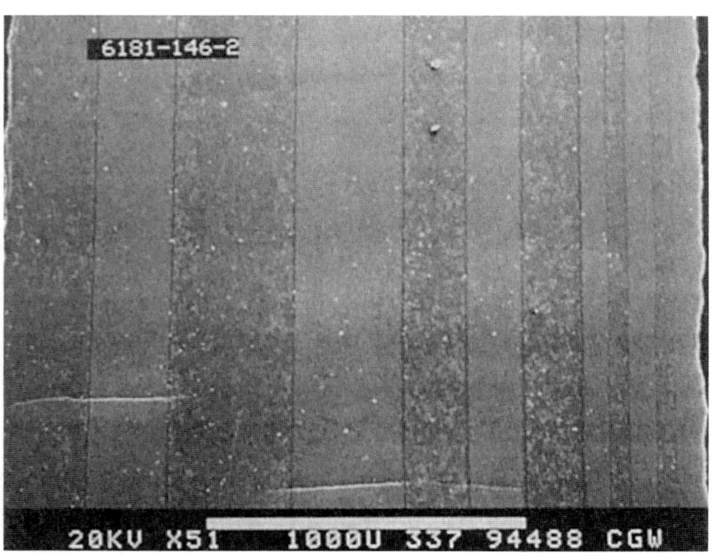

Fig. 19 SEM photomicrograph of multilayer TiB_2/Al_2O_3 with zirconia-toughened alumina. Reprinted with permission. Source: Ref 142

Table 4 Some characteristics and properties of metallic self-propagating high-temperature synthesis nitride powders

Products	Chemical composition, wt%	Impurities, wt%	Particle size and specific surface area	Other properties
Titanium nitride (cubic, $a = 4.24$)	Ti = 77.0–78.5, N = 2.03–21.5	O = 0.3–0.5, Fe < 0.2	1–200 μm, easily grindable	$H = 1800$ kg/mm^2
Hafnium nitride (cubic, $a = 4.510$)	Hf = 92.0–94.0, N = 5.9–6.0	O = 0.3–0.5	1–50 μm	$T_{sc} = 6.6$ K
Tantalum nitride (cubic, $a = 4.323$)	Ta = 91.7–92.2, N = 7.3–7.6	O = 0.1–0.3, C = 0.05–0.1	1–10 μm, 0.6 m^2/g	$H = 3200$ kg/mm^2
Zirconium nitride (cubic, $a = 4.576$)	Zr = 87.3, N = 12.6	O = 0.3–0.5, Fe < 0.2	1–200 μm	$H = 1570$ kg/mm^2, $T_{sc} = 9.6$ K
Niobium nitride (cubic, $a = 4.385$)	Nb = 86.9, N = 6.1	O = 0.1–0.3, C = 0.05–0.1	1–50 μm	$H = 1670$ kg/mm^2, $T_{sc} = 14.3$ K

H, Microhardness; T_{sc}, transition temperature to superconducting state. Source: Ref 67, 163

Table 5 Some characteristics and properties of Si_3N_4 powders

Parameter	SHS β-phase	SHS α + β-phase, (α ≥ 80%)
Basic compound, Si_3N_4	98–99	98–99
N, wt%	38.5–39.0	38.5–39.0
O, wt%	0.5–1.0	0.5–1.0
Fe, wt%	0.1–0.3	<0.01 to 0.3
C, wt%	<0.1	≤0.1
Silicon free, wt%	<0.1	<0.1
Specific surface, m^2/g	1–3	3–11

Source: Ref 32

con nitride materials are attractive for use in high-temperature structural applications in chemically aggressive environments. For example, Table 8 shows some properties of a Si_3N_4-based SHS ceramic (Si_3N_4 + SiC + TiN, Ref 160) that is suitable for ceramic engines.

The dielectric materials produced by hot pressing SHS AlN powders have about the same thermal conductivities, but higher bending strength, compared to those produced in conventional furnaces (Ref 161). For this reason, they are suitable for fabrication of panels in the microelectronics industry. A poreless composite material, TiAl-TiB_2-AlN, has been obtained by reacting a Ti + (0.7–0.95)Al + (0.05–0.50)B mixture under 30 to 100 atm N_2 pressure (Ref 162). The use of high-pressure nitrogen gas was found to be effective for simultaneous synthesis and consolidation of nitride ceramics with dispersed intermetallic compounds (e.g., TiAl). Dense, crack-free products with uniform grains (approximately 10 mm in size) were obtained. Also, aluminum nitride powders synthesized by SHS have been used as a heat-conducting component in silicon-organic adhesive sealants. This glue has a thermal conductivity about twice that of alternative adhesives (Ref 32).

Combustion synthesis of boron nitride powder, BN, was reported in one of the earliest works on SHS. More recently, finely dispersed hexagonal BN powder has been obtained from reduction-type reactions (Ref 163). The main applications for BN materials take advantage of their dielectric properties, which are maintained at high temperatures (even above 2000 K) in nonoxidizing atmospheres. Some properties of BN materials are listed in Table 9 (Ref 160). The high temperatures present in the CS process make it possible not only to produce powders, but also to sinter materials and form net-shape articles (see Fig. 21) in the combustion wave.

Hydrides. The heats of formation for transition metal hydrides are much lower than those for nitrides or oxides. However, the high diffusivity of hydrogen within the reaction medium makes it possible to organize self-sustained combustion (Ref 165). More than thirty hydrides and deuterides of scandium, titanium, vanadium, and the major lanthanide metals have been obtained by gas-solid combustion synthesis (Ref 166). Hydrides of the transition metal alloys and intermetallic compounds (e.g., $ZrNiH_3$, Ti_2COH_3) have also been obtained (Ref 167). The production of complex hydrocarbides and hydronitrides is of special interest (Ref 168). For example, multicomponent single-phase hydronitrides of the following compositions may be synthesized: $Zr_{0.9}$-$Nb_{0.1}$-C_x-N_{1-x}-$H_{0.19}$ to $Zr_{0.4}$-$Nb_{0.6}$-C_x-N_{1-x}-$H_{0.26}$.

Oxides. Combustion synthesis has been used to produce complex oxide materials, such as ferrites and ferroelectric piezoelectric, and superconducting materials. In these systems, the heat required for SHS is supplied by the heat released from the reaction between oxygen and metal. Oxygen can be supplied to the reaction by either an external source (oxygen gas) or an internal source (e.g., reduction of oxides in the starting reaction mixture) or by a combination of the two.

High-Temperature Superconductors. One of the promising oxide materials produced by SHS is the $YBa_2Cu_3O_{7-x}$ (Y_{123}) ceramic superconductor (Ref 6). This material has a number of applications, including targets for magnetron spraying, superconducting shields, wires and cables. The compound is produced by SHS using the following reaction:

$$3Cu + 2BaO_2 + \tfrac{1}{2}Y_2O_3 + (1.5 - x)/2O_2 \rightarrow YBa_2Cu_3O_{7-x}$$

The effects of processing conditions, including green density, oxygen pressure, particles sizes of copper, BaO_2, and Y_2O_3 and the use of Cu_2O and CuO as starting materials have been investigated (Ref 169-172).

Table 6 Some characteristics and properties of BN powders

Parameter	SHS product: Ultrapure	SHS product: Technically pure	Furnace product: ORPAC Grade 99	Furnace product: Denka (Japan)
Basic compound, BN, wt%	>99.5	97.3	98–99	>98
N, wt%	>55.7	54.9	54–55	54.5
O, wt%	<0.5	1.5	1.5	1.5
C, wt%	<0.01	0.3	...	...
Metal impurities (Mg, Fe), wt%	<0.2	0.3	...	...
Specific surface, m^2/g	11.0	8–14	10	...

Source: Ref 32

Table 7 Some characteristics and properties of AlN powders

Parameter	SHS product: Ultrapure	SHS product: Technically pure	Furnace product: ART USA A-100	Furnace product: Starck (Grade B) (Germany)
Basic compound, AlN, wt%	99.7	98.8	99.0	98.1
N, wt%	33.9	32.7	33.0	33.3
O, wt%	0.3	0.6	1.0	2.3
Fe, wt%	0.07	0.12	0.005	100 ppm
Specific surface, m^2/g	2.0–20	1.5	2.5–4.0	1.0–8.0

Source: Ref 32

Table 8 Some properties of self-propagating high-temperature synthesis silicon nitride ceramics

Property	Value
Density,	3.34
Porosity, %	<1 to 15
Elastic modulus, GPa	180–250
Rockwell hardness, HRA	85–93
Vickers hardness, GPa	6–14.5
Bend strength, MPa, T = 1750 K	270–650
Critical stress intensity factory, MN/M$^{3/2}$	2.5–5
thermal conductivity (873–1371 K), W/m · K	15–20

Source: Ref 160

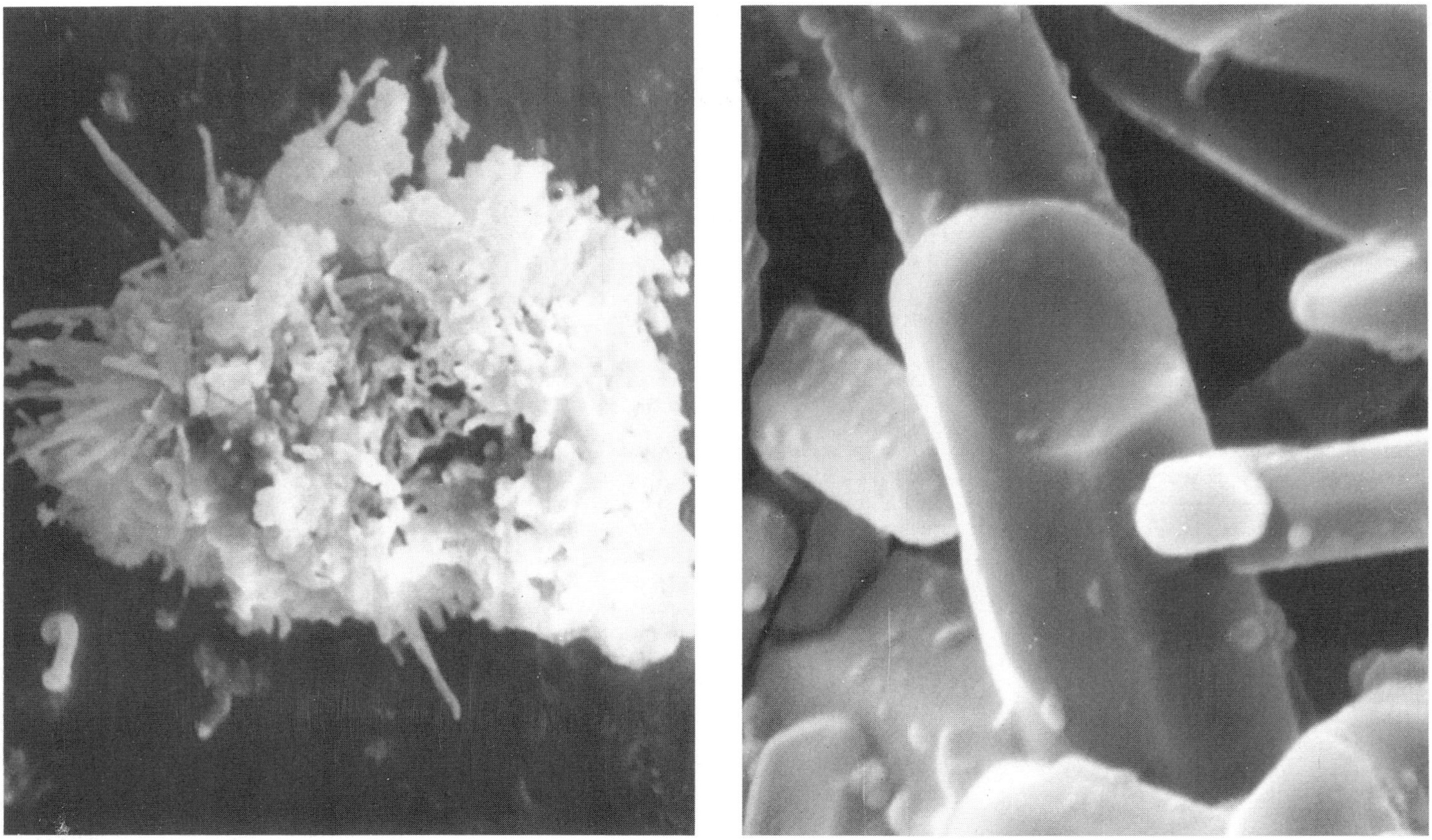

(a) (b)

Fig. 20 Characteristic microstructure of self-propagating high-temperature synthesis Si_3N_4 powders. (a) α-Si_3N_4. 8,625×. (b) β-Si_3N_4. 15,000×

Table 10 compares Y_{123} powders produced by SHS (SHS-J1 and SHS-J2) with those produced by furnace synthesis. The oxygen content, transition temperature, orthorhombic phase Y_{123} content, basic impurities, and mean particle sizes of SHS powders compare favorably with those produced by furnace synthesis. However, the time of Y_{123} production for a conventional furnace is an order of magnitude longer than that for SHS. Items of relatively high density (~90%) with superconducting transition temperature, T_{sc} of 92 K have been fabricated by SuperConco Company (U.S.) by sintering SHS-J2 powders (Ref 173). Thick films (1 mm) made from this powder exhibit a critical current density up to 10^6 A/cm^2. Materials of $Bi_2Sr_2CaCu_2O_y$ and $Tl_2Ba_2Ca_2Cu_3O_y$ compositions with T_{sc} = 125 and 131 K, respectively, have also been synthesized by SHS (Ref 169).

Ferroelectric Materials. The properties of electronic ceramics produced by SHS, including chemical and phase compositions as well as electromagnetic properties, are strongly affected by the maximum combustion temperature (Ref 173). This feature suggests that the properties of the final product can be tailored for a specific application by changing the combustion temperature, which can be achieved by adjusting the processing conditions.

Ferroelectric laminated bismuth compounds, which have applications as high-temperature piezoelectrics and semiconductors, can be synthesized by SHS. Table 11 presents the properties of several ferroelectric materials along with their potential applications. In addition, $LiNbO_3$ and $LiTaO_3$ can be obtained by SHS with densities two times higher than those produced by conventional methods, by varying the Li_2O/Me_2O_5 ratio in the product (*Me* is Nb or Ta). The formation of rare-earth (lithium and terbium) molybdate materials can also be achieved by SHS. All of the above materials have electronic applications. During conventional synthesis, difficulties arise when forming single crystals of lithium and terbium molybdates due to the high volatility of molybdenum oxide. This difficulty has been overcome by using SHS-produced stoichiometric powders, which are characterized by minimal losses during heating, melting, and crystallization.

Products of Thermite-Type SHS

It is impossible to describe, even briefly, all of the products obtained by thermite-type self-propagating reactions. Some examples involving magnesium metal are presented in Table 1. The use of magnesium leads to MgO product, which can be easily leached, leaving behind the desired product composition in powder form. In this section, some selected products reported in the literature are considered, specifically those whose phase distribution and properties are controlled by gravitational or centrifugal forces. Owing to high heats of formation, the products are completely molten at the combustion temperature. As a result, gravity-induced processes of convection and buoyancy may affect phase separation and crystallization, as well as combustion wave propagation.

Fig. 21 Ceramic materials and net-shape articles produced by combustion synthesis

Table 9 Some properties of self-propagating high-temperature synthesis BN ceramics

Properties	BN	BN + SiO_2
Chemical composition, wt%	N = 55; O < 0.5; B_{free} < 0.5; B_2O_3 < 0.3; C < 0.5	BN = 74 SiO_2 = 26
Density, g/cm^3	1.5	1.85
Dielectric strength, kV/mm	25	19
Dielectric permeability	3.1	6.0
Tangent of dielectrical loss at 1 MHz	0.0034	0.036
Resistivity, Ω	2×10^8	1.3×10^8
Thermal conductivity, W/M · K		
T = 400 K	20	...
T = 900 K	8	8
Bend strength, MPa	25	50

Source: Ref 160

Centrifugal SHS casting has been used to produce oxide-carbide ceramics by the following reaction: $MO_x + Al + C \rightarrow Mc_y + Al_2O_3$ (Ref 174). In general, titanium, chromium, molybdenum, and tungsten carbides are synthesized with an alumina melt in the combustion wave (e.g., Cr_3C_2-Al_2O_3, WC-Al_2O_3, (Ti,Cr)C-Al_2O_3, MoC-Al_2O_3. Also, addition of metal (iron group) oxides (NiO, CoO, Fe_2O_3) results in formation of cermet composites containing a metal binder (e.g., (Ti,Cr)C-Ni), in addition to the carbide phase. The lighter oxide phase can be fully separated from the cermet by centrifugal acceleration. A centrifugal overload of 150 g is sufficient to separate MoC from Al_2O_3, while 300 and 1000 g are required to separate WC-Co cermet and WC, respectively, from the alumina melt (Ref 64). Lower centrifugal overloads result in incomplete separation and the formation of graded materials. This phenomenon has been used to obtain refractory and chemically stable protective coatings on various substrates (see, e.g., Fig. 22) that possess relatively complex compositions and structures (Ref 62, 175).

An interesting application of thermite-type SHS reaction with centrifugal overload is the fabrication of multilayered composite pipes (Ref 176–178). The simple and well-known composition Fe_2O_3 + Al was used to produce a two-layer lining inside the steel pipe; see Fig. 10 (Ref 65). The iron outer layer provides good adhesion of the coating to the inner surface of the steel pipe, while the inner layer (Al_2O_3) provides protection from corrosion and abrasion. Separation of the layers was achieved by rotating the pipe (centrifugal acceleration between 50 and 200 g). Some properties of the ceramic layer are presented in Table 12. The effects on the structure of the pipe lining layer of adding SiO_2 and Al_2O_3 to the Fe_2O_3 + 2Al mixture have been investigated (Ref 179). More complex structures have been formed using multicomponent mixtures (Ref 64). For example, the Ti-C-Ni-Al-O reaction system yielded multiphase gradient layers, where the composition gradually changed from TiC-Ni cermet to the Al_2O_3 + TiO_2 ceramic. The local density changed from 6.8 g/cm^3 at the cer-

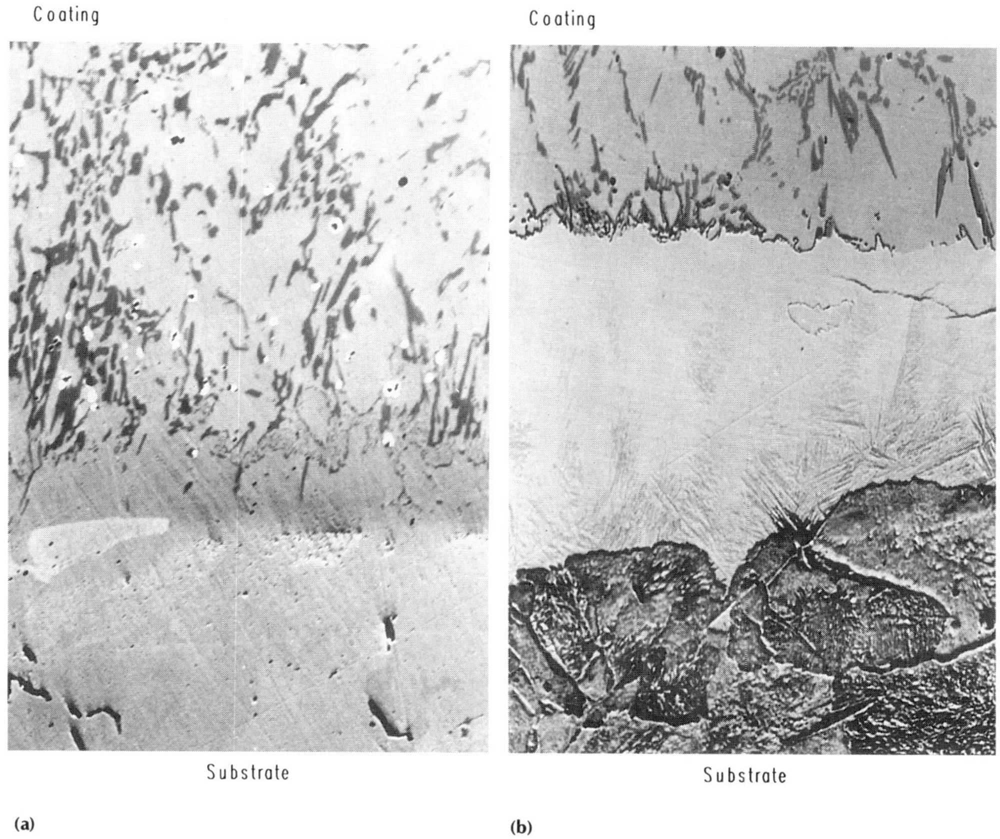

Fig. 22 Characteristic microstructures of protective coatings on steel produced by combustion synthesis (400×). (a) Titanium-chromium carbide coating. (b) Chromium-nickel carbide coating

Table 10 Some characteristics and properties of superconductive powders ($YBa_2Cu_3O_{7-x}$) produced by conventional methods and self-propagating high-temperature synthesis

Powder	Oxygen content (7–x)	TSC, K	Orthorhombic (Y_{123}) phase, %	Basic impurities	Average particle size, µm
SC5-S(a)	6.92	93.5	98	CuO	40.0
SC5-P(a)	6.87	93.5	~99	CuO	8.2
SC5-6.5(a)	6.88	90.5	97	$BaCuO_2$	6.8
SC5-7(a)	6.85	92.5	~99	CuO	5.7
CPS A-1203(a)	6.85	92.0	~99	$BaCuO_2$	3.0
SSC 03–0065(a)	6.89	92.0	~99	...	6.5
SHS-J1(b)	6.90	92.0	97	CuO	9.0
SHS-J2(b)	6.92	93.5	~99	...	8.0

T_{sc}, transition temperature to superconducting state. (a) Conventional methods. (b) Combustion synthesis. Source: Ref 169

Table 11 Some characteristics and properties of ferroelectric materials produced by self-propagating high-temperature synthesis

Composition	Resistivity, ρ, (Ω · cm)	Dielectric constant, ε	Dielectric dispersion, tan δ	Applications
$Bi_4V_2O_{11}$	4.7×10^9	63	0.089	Gas sensor (ethanol, acetone)
$BiFeO_3$	4.2×10^9	46	0.169	Gas sensor (ethanol, acetone, gasoline)
$Bi_2Fe_4O_9$	6.0×10^9	10	0.004	Gas sensor (ethanol, acetone)
$Bi_4Fe_2O_9$	2.8×10^9	40	0.109	Temperature-sensitive resistance, gas sensor (ethanol, acetone, gasoline)
$BaBi_2Ta_2O_9$	4.0×10^{11}	49	0.0022	Gas sensor (ethanol, acetone, natural gas, H_2O)
Bi_3TiNbO_9	1.9×10^{10}	70	0.0047	Moisture sensor
$BaBi_2Ni_2O_9$	3.5×10^{11}	92	0.0081	Moisture sensor
$Na_{0.5}Bi_{4.5}Ti_4O_{15}$	1.45×10^9	82	0.0059	Moisture sensor

Source: Ref 173

Table 12 Properties of the alumina-based ceramic layer produced by the centrifugal-thermite process

Property	Value
Porosity	5.6%
Bulk density	3.86 g/cm^3
Linear expansion (at 293–1273 K)	8.57×10^{-6} K^{-1}
Thermal conductivity (at 293 K)	13.9 W/m · K
Resistivity (at 293 K)	2×10^7 Ω
Vickers hardness	1200 MPa
Compressive strength	> 850 MPa
Bending strength	150 MPa
Bulk modulus	2.2×10^5 MPa

Source: Ref 65

met side to 3.9 g/cm^3 at the ceramic side. The production of steel pipes lined with a porous alumina-zirconia ceramic layer by reaction of Fe_2O_3 + Al + Zr + additives has also been described (Ref 180).

ACKNOWLEDGMENT

The authors gratefully acknowledge financial support from the National Science Foundation (CTS 95–28941).

REFERENCES

1. A.G. Merzhanov, V.M. Shkiro, and I.P. Borovinskaya, Soviet Union Certif. [Patent] 255221, 1971
2. F. Booth, *Trans. Farad. Soc.*, Vol 49, 1953, p 272
3. J.D. Walton and N.E. Polous, *Am. Ceram. Soc.*, Vol 42, 1959, p 40
4. A.P. Hardt and R.W. Holsinger, *Combust. Flame*, Vol 21, 1973, p 91
5. Z.A. Munir and U. Anselmi-Tamburini, *Mater. Sci. Rep.*, Vol 3, 1989, p 277
6. A.G. Merzhanov, *Combustion and Plasma Synthesis of High-Temperature Materials*, Z.A. Munir and J.B. Holt, Ed., VCH Publishers, 1990, p 1
7. J.B. Holt and S.D. Dunmead, *Annu. Rev. Mater. Sci.*, Vol 21, 1991, p 305
8. R.W. Rice, *J. Mater. Sci.*, Vol 26, 1991, p 6533
9. A. Varma and J.-P. Lebrat, *Chem. Eng. Sci.*, Vol 47, 1992, p 2179
10. A.G. Merzhanov, *Int. J. SHS*, Vol 2, 1993, p 113
11. J.J. Moore and H.J. Feng, *Prog. Mater. Sci.*, Vol 39, 1995, p 243
12. A. Varma, A.S. Rogachev, A.S. Mukasyan, and S. Hwang, *Adv. Chem. Eng.*, Vol 24, 1998, p 79
13. H.F. Calcote, W. Felder, D.G. Keil, and D.B. Olson, in *Twenty-Third Symposium (International) on Combustion*, The Combustion Institute, 1990, p 1739
14. K. Davis, K. Brezinsky, and I. Glassman, *Combust. Sci. Tech.*, Vol 77, 1991, p 171
15. J.M. Prausnitz, R.N. Lichtenthaler, and E.G. de Azevedo, *Molecular Thermodynamics of Fluid-Phase Equilibria*, Prentice-Hall, 1986
16. J.B. Holt and Z.A. Munir, *J. Mater. Sci.*, Vol 21, 1986, p 251
17. I. Glassman, K.A. Davis, and K. Brezinsky, in *Twentv-Fourth Symposium (International) on Combustion*, The Combustion Institute, 1992, p 1877
18. A.P. Aldushin and B.I. Khaikin, *Combust. Explos. Shock Waves*, Vol 10, 1974, p 273
19. Y.B. Zeldovich and D.A. Frank-Kamenetskii, *Zh. Fiz. Khim.*, Vol 12, 1983, p 100
20. B.V. Novozilov, *Phys. Dokl.*, Vol 141, 1961, p 836
21. A.P. Hardt and P.V. Phung, *Combust. Flame*, Vol 21, 1973, p 77
22. A.P. Aldushin, A.G. Merzhanov, and B.I. Khaikin, *Dokl. Phys. Chem.*, Vol 204, 1972, p 475
23. A.P. Aldushin, T.M. Martem'yanova, A.G. Merzhanov, B.I. Khaikin, and K.G. Shkadinskii, *Combust. Explos. Shock Waves*, Vol 8, 1972, p 159
24. E.A. Nekrasov, V.K. Smolyakov, and Y.M. Maksimov, *Combust. Explos Shock Waves*, Vol 17, 1981, p 513
25. E.A. Nekrasov, V.N. Tkachenko, and A.E. Zakirov, *Combust. Sci. Tech.*, Vol 91, 1993, p 207
26. V.V. Aleksandrov, M.A. Korchagin, and V.V. Boldyrev, *Dokl. Phys. Chem.*, Vol 292, 1987, p 114
27. V.V. Aleksandrov and M.A. Korchagin, *Combust. Explos. Shock Waves*, Vol 23, 1998, p 557
28. G. Cao and A. Varma, *Combust. Sci. Tech.*, Vol 102, 1994, p 181
29. E.V. Okolovich, A.G. Merzhanov, B.I. Khaikin, and K.G. Shkadinskii, *Combust. Explos. Shock Waves*, Vol 13, 1977, p 264
30. A. Makino and C.K. Law, *J. Am. Ceram. Soc.*, Vol 77, 1994, p 778
31. M.G. Lakshmikantha and J.A. Sekhar, *J. Mater. Sci.*, Vol 28, 1993, p 6403
32. A.G. Merzhanov, in *Particulate Materials and Processes: Advances in Powder Metallurgy and Particulate Materials*, Metal Powder and Industries Federation, 1992, p 341
33. I.P. Borovinskaya, V.M. Vishnyakova, V.M. Maslov, and A.G. Merzhanov, in *Combustion Processes in Chemical Engineering and Metallurgy*, A.G. Merzhanov, Ed., USSR Academy of Sciences, 1975, p 141 (in Russian)
34. Y. Miyamoto, M. Koizumi, and O. Yamada, *J. Am. Ceram. Soc.*, Vol 67, 1984, p C224
35. S. Adachi, T. Wada, T. Mihara, Y. Miyamoto, M. Koizumi, and O. Yamada, *J. Am. Ceram. Soc.*, Vol 72, 1989, p 805
36. P.D. Zavitsanos, J.J. Gebhardt, and A. Gatti, in *Combustion and Plasma Synthesis of High-Temperature Materials*, Z.A. Munir and J.B. Holt, Ed., VCH Publishers, 1990, p 170
37. S.D. Dunmead, B.J. Holt, and D.D. Kingman, in *Combustion and Plasma Synthesis of High-Temperature Materials*, Z.A. Munir and J.B. Holt, Ed., VCH Publishers, 1990, p 186
38. T. Nishida and K. Urabe, *Int. J. SHS*, Vol 1, 1992, p 566
39. A.G. Merzhanov, I.P. Borovinskaya, V.I. Yukhvid, and V.I. Ratnikov, *Scientific Principles of Material Science*, Nauka, Moscow, 1981, p 193
40. V.L. Kvanin, V.L. Gorovoi, N.T. Balikhina, I.P. Borovinskaya, and A.G. Merzhanov, *Int. J. SHS*, Vol 2, 1993, p 56
41. L.J. Kecskes, T. Kottke, A.J. Niiler, *J. Am. Ceram. Soc.*, Vol 73, 1990, p 1274
42. H.A. Grebe, A. Advani, N.N. Thadani, and T. Kottke, *Metall. Trans.*, Vol 23A, 1992, p 2365
43. B.H. Rabin, G.E. Korth, and R.L. Williamson, *Int. J. SHS*, Vol 1, 1992, p 336
44. G.A. Adadurov, I.P. Borovinskaya, Y.A. Gordopolov, and A.G. Merzhanov, *J. Eng. Phys. Thermophys.*, Vol 63, 1992, p 1075
45. Y.A. Gordopolov, N.A. Zotov, A.S. Mukasyan, and R.M. Shikhverdiev, *Int. J. SHS*, Vol 5, 1996, p 191
46. V.M. Fedorov, Y.A. Gordopolov, and A.G. Merzhanov, *J. Eng. Phys. Thermophys.*, Vol 63, 1992, p 1166
47. S.J. Work, L.-H. Yu, N.N. Thadhani, M.A. Meyers, R.A. Graham, and W.F. Hammetter, in *Combustion and Plasma Synthesis of High-Temperature Materials*, Z.A. Munir and J.B. Holt, Ed., VCH Publishers, 1990, p 133
48. K.S. Vecchio, L.-H. Yu, and M.A. Meyers, *Acta Metall. Mater.*, Vol 42, 1994, p 701
49. M.A. Meyers, L.-H. Yu, and K.S. Vecchio, *Acta Metall. Mater.*, Vol 42, 1994, p 715
50. N.N. Thadani, *Prog. Mater. Sci.*, Vol 37, 1993, p 117
51. V.V. Podlesov, A.V. Radugin, A.M. Stolin, and A.G. Merzhanov, *J. Eng. Phys. Thermophys.*, Vol 63, 1992, p 1065
52. V.V. Podlesov, A.M. Stolin, and A.G. Merzhanov, *J. Eng. Phys. Thermophys.*, Vol 63, 1992, p 1156
53. O. Yamada, Y. Miyamoto, and M. Koizumi, *J. Am. Ceram. Soc.*, Vol 70, 1987, p C206
54. O. Yamada, Y. Miyamoto, and M. Koizumi, *J. Am. Ceram. Soc. Bull.*, Vol 64, 1985, p 319
55. M. Koizumi and Y. Miyamoto, in *Combustion and Plasma Synthesis of High-Temperature Materials*, Z.A. Munir and J.B. Holt, Ed., VCH Publishers, 1990, p 54
56. Y. Miyamoto, Y. Nakanishi, I. Tanaka, T. Okamoto, and O. Yamada, in *Proc. First US-Japanese Workshop on Combustion Synthesis*, National Research Institute for Metals, 1990, p 173
57. Y. Miyamoto, T. Takakura, K. Tanihata, I. Tanaka, O. Yamada, M. Saito, and H. Takahashi, in *Proc. First Int. Symp. on Functionally Gradient Materials*, Functionally Gradient Materials Forum, 1990, p 169
58. Y. Matsuzaki, J. Fujioka, S. Minakata, and Y. Miyamoto, in *Proc. First Int. Symp. on Functionally Gradient Materials*, Functionally Gradient Materials Forum, 1990, p 263
59. K. Tanihata, Y. Miyamoto, K. Matsushita, X. Ma, A. Kawasaki, R. Watanabe, and K. Hirano, *Ceramics Trans.*, Vol 34, 1993, p 361

60. R.W. Rice, W.J. McDonough, G.Y. Richardson, J.M. Kunetz, and T. Schroeter, *Ceram. Eng. Sci. Proc.,* Vol 7, 1986, p 751
61. E.Y. Osipov, Y.A. Levashov, V.N. Chernyshev, A.G. Merzhanov, and I.P. Borovinskaya, *Int. J. SHS,* Vol 1, 1992, p 314
62. V.I. Yukhvid, *Pure Appl. Chem.,* Vol 64, 1992, p 977
63. A.G. Merzhanov, V.I. Yukhvid, and I.P. Borovinskaya, *Dokl. Chem.,* Vol 255, 1989, p 503
64. A.G. Merzhanov and V.I. Yukhvid, in *Proc. First US-Japanese Workshop on Combustion Synthesis,* National Research Institute for Metals, Japan, Vol 1, 1990
65. O. Odawara, *Int. J. SHS,* Vol 1, 1992, p 160
66. O. Odawara and J. Ikeuchi, *J. Am. Ceram. Soc.,* Vol 69, 1986, p C80
67. A.G. Merzhanov and I.P. Borovinskaya, *Dokl. Chem.,* Vol 204, 1972, p 429
68. E.K. Storms, *The Refractory Carbides,* Academic Press, 1967
69. V.M. Shkiro and I.P. Borovinskaya, in *Combustion Processes in Chemical Engineering and Metallurgy,* A.G. Merzhanov, Ed., USSR Academy of Sciences, 1975, p 253 (in Russian)
70. I.P. Borovinskaya, V.I. Ratnikov, and G.A. Vishnyakova, *J. Eng. Phys. Thermophys.,* Vol 63, 1992, p 1059
71. I.P. Borovinskaya, *Pure Appl. Chem.,* Vol 64, 1992, p 919
72. R.V. Raman, S.V. Rele, S. Poland, J. LaSalvia, M.A. Meyers, and A.R. Niiler, *J. Metals,* Vol 47, 1995, p 23
73. A.S. Rogachev, V.M. Shkiro, I.D. Chausskaya, and M.V. Shvetsov, *Combust. Explos. Shock Waves,* Vol 24, 1988, p 720
74. Z.Y. Fu, W.M. Wang, H. Wang, R.Z. Yuan, and Z.A. Munir, *Int. J. SHS,* Vol 2, 1993, p 307
75. S.D. Dunmead, B.J. Holt, and D.D. Kingman, in *Combustion and Plasma Synthesis of High-Temperature Materials,* Z.A. Munir and J.B. Holt, Ed., VCH Publishers, 1990, p 229
76. B. Mei, W. Wang, R. Yuan, and Z. Fu, *Int. J. SHS,* Vol 3, 1994, p 79
77. J.C. LaSalvia and M.A. Meyers, *Int. J. SHS,* Vol 4, 1995, p 43
78. H.E. Maupin and J.C. Rawers, *J. Mater. Sci. Let.,* Vol 12, 1993, p 540
79. Y. Choi and S.-W. Rhee, *J. Mater. Sci.,* Vol 28, 1993, p 6669
80. V.M. Shkiro, G.A. Nersisyan, and I.P. Borovinskaya, *Combust. Explos. Shock Waves,* Vol 14, 1978, p 455
81. V.M. Shkiro, G.A. Nersisyan, I.P. Borovinskaya, A.G. Merzhanov, and V.S. Shekhtman, *Sov. Powder Metall.,* Vol 18, 1979, p 227
82. V.M. Martynenko and I.P. Borovinskaya, in *Combustion Processes in Chemical Engineering and Metallurgy,* A.G. Merzhanov, Ed., USSR Academy of Sciences, 1975, p 127 (in Russian)
83. O. Yamada, Y. Miyamoto, and M., Koiizumi, *J. Mater. Res.,* Vol 1, 1986, p 275
84. R. Pampuch, L. Stobierski, and J. Lis, *J. Am. Ceram. Soc.,* Vol 72, 1989, p 1434
85. R. Pampuch, L. Stobierski, J. Lis, and M. Raczka, *Mater. Res. Bull.,* Vol 22, 1987, p 1225
86. S. Ikeda, K. Urabe, M. Koizumi, and H. Izawa, in *Combustion and Plasma Synthesis of High-Temperature Materials,* Z.A. Munir and J.B. Holt, Ed., VCH Publishers, 1990, p 151
87. S.L. Kharatyan and H.H. Nersisyan, *Int. J. SHS,* Vol 3, 1994, p 17
88. J.B. Holt, D.D. Kingman, and G.M. Bianchini, *Mater. Sci. Eng.,* Vol 71, 1985, p 321
89. R.W. Rice, G.Y. Richardson, J.M. Kunetz, T. Schroeter, and W. J. McDonough, *Adv. Ceram. Mater.,* Vol 2, 1987, p 222
90. V.A. Shcerbakov, A.N. Gryadunov, and A.S. Shteinberg, *J. Eng. Phys. Thermophys.,* Vol 63, 1992, p 1111
91. N. Yanagisawa, O. Asano, N. Sata, and N. Sanada, in *Proc. First US-Japanese Workshop on Combustion Synthesis,* National Research Institute for Metals, Japan, 1990, p 157
92. K. Urabe, Y. Miyamoto, M. Koizumi, and H. Ikawa, in *Combustion and Plasma Synthesis of High-Temperature Materials,* Z.A. Munir and J.B. Holt, Ed., VCH Publishers, 1990, p 281
93. Y. Taneoka, O. Odawara, and Y. Kaieda, *J. Amer. Ceram. Soc.,* Vol 72, 1989, p 1047
94. Z.Y. Fu, H. Wang, W.M. Wang, R.Z. Yuan, and Z.A. Munir, *Int. J. SHS,* Vol 2, 1993, p 175
95. V.A. Shcerbakov and A.N. Pityulin, *Combust. Explos. Shock Waves,* Vol 19, 1983, p 631
96. A.R. Sarkisyan, S.K. Dolukhanyan, and I.P. Borovinskaya, *Sov. Powder Metall.,* Vol 17, 1978, p 424
97. T.S. Azatyan, V.M. Mal'tsev, A.G. Merzhanov, and V.A. Seleznev, *Combust. Explos. Shock Waves,* Vol 15, 1979 p 35
98. S.C. Deevi, *J. Mater. Sci.,* Vol 26, 1991, p 2662
99. S.C. Deevi, *Mater. Sci. Eng.,* Vol A149, 1992, p 241
100. J. Trambukis and A.Z. Munir, *J. Amer. Ceram. Soc.,* Vol 73, 1990, p 1240
101. R. Rozenkranz, G. Frommeyer, and W. Smarsly, *Mater. Sci. Eng.,* Vol 152A, 1992, p 288
102. D.M. Shah, D. Berczik, D.L. Anton, and R. Hecht, *Mater. Sci. Eng.,* Vol 155A, 1992, p 45
103. H.G. Hakobian and S.K. Dolukhanyan, *Int. J. SHS,* Vol 3, 1994, p 299
104. Y.S. Naiborodenko, V.I. Itin, and K.V. Savitskii, *Sov. Phys. J.,* Vol 11, 1968, p 19
105. Y.S. Naiborodenko, G.V. Lavrenchuk, and V.M. Filatov, *Sov. Powder Metall.,* Vol 21, 1982, p 909
106. V.I. Itin, A.D. Bratchikov, and L.N. Postnikova, *Sov. Powder Metall.,* Vol 19, 1980, p 315
107. B.H. Rabin and R.N. Wright, *Metall. Trans.,* Vol 22A, 1991, p 277
108. V.M. Maslov, I.P. Borovinskaya, and M.K. Ziatdinov, *Combust. Explos. Shock Waves,* Vol 15, 1979, p 41
109. C.R. Kachelmyer and A. Varma, *Mater. Res. Soc. Symp.,* Vol 350, 1994, p 33
110. L.L. Wang, Z.A. Munir, and J.B. Holt, *Metall. Mater. Trans.,* Vol 21B, 1990, p 567
111. K.A. Philpot, Z.A. Munir, and J.B. Holt, *J. Mater. Sci.,* Vol 22, 1987, p 159
112. J.-P. Lebrat and A. Varma, *Combust. Sci. Tech.,* Vol 88, 1992, p 211
113. D.E. Alman, *J. Mater. Sci. Let.,* Vol 13, 1994, p 483
114. H. Kuroki and K. Yamaguchi, in *Proc. First US-Japanese Workshop on Combustion Synthesis,* National Research Institute for Metals, Japan, 1990, p 23
115. L. Ho-Yi, Y. Hong-Yu, M. Shu-Xia, and Y. Sheng, *Int. J. SHS,* Vol 1, 1992, p 447
116. Y.D. Hahn and I.H. Song, *Int. J. SHS,* Vol 4, 1995, p 293
117. J.-P. Lebrat, A. Varma, and P.J. McGinn, *J. Mater. Res.,* Vol 9, 1994, p 1184
118. J.-P. Lebrat, A. Varma, and A.E. Miller, *Metall. Trans.,* Vol 23A, 1992, p 69
119. A.S. Mukasyan, A.E. Pelekh, A. Varma, A.S. Rogachev, and A. Jenkins, *AIAA Journal,* Vol 35, 1997
120. W. Wang, B. Mei, Z. Fu, and R. Yuan, *Int. J. SHS,* Vol 2, 1993, p 183
121. J.C. Rawers, W.R. Wrzesinski, E.K. Roub, and R.R. Brown, *Mater. Sci. Tech.,* Vol 6, 1990, p 187
122. I. Gotman, M.J. Koczak, and E. Shtessel, *Mater. Sci. Eng.,* Vol 187A, 1994, p 189
123. P.H. Shingu, K.N. Ishihara, F. Ghonome, T. Hayakawa, M. Abe, and K. Taguchi, in *Proc. First US-Japanese Workshop on Combustion Synthesis,* National Research Institute for Metals, Japan, 1990, p 65
124. A. Ferreira, M.A. Meyers, and N.N. Thadani, *Metall. Trans.,* Vol 23A, 1992, p 3251
125. A.J. Strutt, K.S. Vecchio, L.-H. Yu, and M.A. Meyers, *AIP Conf. Proc.,* Vol 309, 1994, p 1259
126. B.H. Rabin, R.N. Wright, J.R. Knibloe, R.V. Raman, and S.V. Rale, *Mater. Sci. Eng.,* Vol 153A, 1992, p 706
127. A. Dutta, *Int. J. SHS,* Vol 4, 1995, p 309
128. V.I. Itin, V.N. Khachin, A.D. Bratchikov, V.E. Gyunter, E.F. Dudarev, T.V. Monasevich, D.B. Chernov, G.D. Timonin, and A.P. Paperskii, *Russ. Phys. J.,* Vol 12, 1977, p 1631
129. V.I. Itin, V.N. Khachin, V.E. Gyunter, A.D. Bratchikov, and D.B. Chernoy, *Porosh. Metall.,* Vol 3, 1983, p 156
130. H.C. Yi and J.J. Moore, *JOM,* Vol 42, 1990, p 31
131. V.I. Itin and Y.S. Naiborodenko, *High Temperature Synthesis of Intermetallic Compounds,* Tomsk University Publishing, Russia, 1989 (in Russian)
132. Y. Kaieda, M. Otaguchi, and N. Oguro, in *Combustion and Plasma Synthesis of High-Temperature Materials,* Z.A. Munir and J.B. Holt, Ed., VCH Publishers, 1990, p 106
133. V.R. Howers, *Proc. Roy. Soc.,* Vol 80, 1962, p 648
134. K.L. Padyukov, A.G. Kost, E.A. Levashov, I.P. Borovinskaya, and Y.V. Bogatov, *Int. J. SHS,* Vol 1, 1992, p 443
135. K.L. Padyukov and E.A. Levashov, *Diam. Relat. Mater.,* Vol 2, 1993, p 207

136. K.L. Padyukov, E.A. Levashov, I.P. Borovinskaya, and A.G. Kost, *J. Eng. Phys. Thermophys.*, Vol 63, 1992, p 1107
137. E.A. Levashov, I.P. Borovinskaya, A.S. Rogachev, M. Koizumi, M. Ohyanagi, and S. Hosomi, *Int. J. SHS,* Vol 2, 1993, p 189
138. N. Sata, N. Sanada, T. Hirano, and M. Niino, in *Combustion and Plasma Synthesis of High-Temperature Materials,* Z. A. Munir and I.B. Holt, Ed., VCH Publishers, 1990, p 195
139. Y. Miyamoto, K. Tanihata, Y. Matsuzaki, and X. Ma, *Int. J. SHS,* Vol 1, 1992, p 147
140. Z. Wang, C. Ge, and L. Chen, *Int. J. SHS,* Vol 3, 1994, p 85
141. A.N. Pityulin, Y.V. Bogatov, and A.S. Rogachev, *Int. J. SHS,* Vol 1, 1992, p 111
142. T.P. DeAngelis and D.S. Weiss, in *Combustion and Plasma Synthesis of High-Temperature Materials,* Z.A. Munir and J.B. Holt, Ed., VCH Publishers, 1990, p 144
143. E.A. Levashov, B.V. Vijushkov, E.V. Shtanskaya, I.P. Borovinskaya, M. Ohyanagi, S. Hosomi, and M. Koizumi, *Int. J. SHS,* Vol 3, 1994, p 287
144. I.E. Molodetskaya, V.P. Pisarskii, and O.O. Ulanova, *Combust. Explos. Shock Waves,* Vol 28, 1992, p 385
145. J.B. Wiley and R.B. Kaner, *Science,* Vol 255, 1992, p 1093
146. S.V. Muchnik, *Izv. Akad. Nauk SSR, Nearg. Mater.,* Vol 20, 1984, p 158
147. S.V. Muchnik, Y.F. Lomnitskaya, V.B. Chernogorenko, and K.A. Lynchak, *Sov. Powder Metall.,* Vol 32, 1993, p 865
148. N. Kanamaru and O. Odawara, *Int. J. SHS,* Vol 3, 1994, p 305
149. E.A. Shtessel, M.V. Kurylev, and A.G. Merzhanov, *Dokl. Phys. Chem.,* Vol 288, 1986, p 529
150. T. Akiyama, H. Isogai, and J. Yagi, *Int. J. SHS,* Vol 4, 1995, p 69
151. I.P. Borovinskaya and V.E. Loryan, *Dokl. Phys. Chem.,* Vol 231, 1976, p 1230
152. I.P. Borovinskaya and V.E. Loryan, *Sov. Powder Metall.,* Vol 11, 1978, p 851
153. Y.M. Maksimov, M.K. Ziatdinov, A.G. Raskolenko, and O.K. Lepakova, *Combust. Explos. Shock Waves,* Vol 15, 1979, p 420
154. Y.M. Maksimov, M.K. Ziatdinov, A.G. Merzhanov, L.G. Raskolenko, and O.K. Lepakova, *Combust. Explos. Shock Waves,* Vol 20, 1984, p 487
155. I.P. Borovinskaya, A.G. Merzhanov, A.A. Butakov, A.G. Rabin'kin, and V.S. Shehtman, *Izv. Bull. 9,* Certif. 264365, Patent 1311660, 1970
156. M. Ohyanagi, M. Koizumi, K. Tanihata, Y. Miyamoto, O. Yamada, I. Matsubara, and H. Yamashitaa, *J Mater. Sci. Let.,* Vol 12, 1993, p 500
157. J.-T. Li, Y.-L. Xia, and C.-C. Ge, *Int. J. SHS,* Vol 3, 1994, p 225
158. A.S. Mukasyan and I.P. Borovinskaya, *Int. J. SHS,* Vol 1, 1992, p 55
159. V.Y. Petrovskii, E.I. Gorvits, I.P. Borovinskaya, and V.M. Martinenko, in *Problems of Technological Combustion,* Russian Academy of Sciences, 1981 p 5
160. A.S. Mukasyan, M.Y. Blinov, I.P. Borovinskaya, and A.G. Merzhanov, in *Proc. Int. Conf. on Engineering Ceramics,* Institute of Inorganic Chemistry, Slovak Academy of Sciences, 1989, p 161
161. G.A. Gogotsi, V.P. Savada, and F.Y. Kharitonov, *Probl. Prochn.,* Vol 12, 1984, p 7 (in Russian)
162. O.Y. Yamada, *Soc. Mater. Sci. Japan,* Vol 43, 1994, p 1059
163. I.P. Borovinskaya, E.A. Levashov, and A.S. Rogachev, *Physical-Chemical and Technological Base of Self-Propagating High-Temperature Synthesis: A Series of Lectures,* Moscow Institute of Steel and Alloy Press, 1991
164. A.G. Merzhanov, "Combustion Processes That Synthesize Materials," paper presented at AMPT 93 International Conference on Advances in Materials and Processing Technology, Dublin, Ireland, 1993
165. S.K. Dolukhanyan, M.D. Nersesyan, A.B. Nalbandyan, I.P. Borovinskaya, and A.G. Merzhanov, *Dokl. Akad. Nauk,* SSSR, Vol 231, 1976, p 675
166. S.K. Dolukhanyan, H.G. Hakobyan, and A.G. Aleksanyan, *Int. J. SHS,* Vol 1, 1992, p 530
167. S.K. Dolukhanyan, A.G. Akopyan, and A.G. Merzhanov, *Combust. Explos. Shock Waves,* Vol 17, 1981, p 525
168. N.N. Agadzhanyan and S.K. Dolukhanyan, *Combust. Explos. Shock Waves,* Vol 26, 1990, p 739
169. A.G. Merzhanov, in *Ceramic Transactions: Superconductivity and Ceramic Superconductors,* K.M. Nair and E.A. Giess, Ed., American Ceramic Society, 1990, p 519
170. J.-P. Lebrat and A. Varma, *Physica C,* Vol 184, 1991, p 220
171. J.-P. Lebrat and A. Varma, *Combust. Sci. Tech* , Vol 88, 1992, p 177
172. S.-C. Lin, J.T. Richardson, and D. Luss, *Physica C,* Vol 233, 1994, p 281
173. P.B. Avakyan, M.D. Nersesyan, and A.G. Merzhanov, *Am. Ceram. Soc. Bull.,* Vol 75, 1996, p 50
174. V.I. Yukhvid, A.R. Kachin, and G.V. Zakharov, *Int. J. SHS,* Vol 3, 1994, p 321
175. Y.M. Grigor'ev and A.G. Merzhanov, *Int. J. SHS,* Vol 1, 1992, p 600
176. O. Odawara, U.S. Patent 4363832, 1982
177. O. Odawara, *J. Am. Ceram. Soc.,* Vol 73, 1990, p 629
178. A.R. Kachin and V.I. Yukhvid, *Int. J. SHS,* Vol 1, 1992, p 168
179. R. Orru', B. Simoncini, P.F. Virdis, and G. Cao, *Int. J. SHS,* Vol 4, 1995, p 137
180. J. Li, M. Zhou, and J. Wang, in *Proc. Third Int. Symp. on Self-Propagating High-Temperature Synthesis (Book of Abstracts),* Wuhan University of Technology, 1995, p 48

Infiltration

Claus G. Goetzel, Stanford University, and Joanna Groza, University of California, Davis

DURING INFILTRATION, a liquid, metal mass or front moves through or penetrates the pore system of a solid-phase powder compact. This process is similar to liquid-phase sintering, in which one of the constituent powders of a compact melts and disperses throughout the compact in situ. Voids are filled by shrinkage and particle rearrangement. During infiltration, liquid externally contacts the porous solid, and capillary forces draw it inward. Reduction of total surface-free energy of the system is a prerequisite during infiltration and determines whether a particular solid-liquid phase system is suitable for infiltration.

A variation of capillary-action infiltration at ambient pressure consists of penetration of pores by liquid assisted by external force. Impregnation treatment of this type is widely used to fill pores of sintered products with hydrocarbons, organics, and nonmetallic fillers. For a solid-liquid phase system with suitable surface free energy, infiltration can be used to attain:

- Full density without high compacting or repressing pressures or subsequent forging, rolling or extrusion, or hot pressing
- Precise and complex shapes, including large sizes, using conventional P/M operations, tooling, and equipment
- Layered P/M products of different composition, or of one P/M section and a cast or forged element
- Surface characteristics that permit standard joining and coating methods
- Improved machinability due to uninterrupted chip forming, facilitated by minimization of pores or due to favorable chip-breaking caused by a duplex microstructure
- Reasonable control of phase distribution that may result in a uniform or a purposely graded microstructure
- Good mechanical properties due to a minimum of angular stress-raising pores
- Increased strength through subsequent effective heat treatment

Mechanism of Infiltration

The principles of infiltration mechanisms are discussed in Ref 1 to 6. Infiltration rate is related to Poisseuille's law defining the rate of flow or rise of a liquid in a glass capillary (Ref 2, 3). The parabolic rate law was confirmed for the early penetration stages of infiltration for a number of systems. However, bent and twisted channels in powder compacts made comparison with the straight capillaries of the model almost impossible. Furthermore, in many systems involving metals, penetration is followed by a second infiltration stage when the pore filling liquid reacts with the solid.

Successful infiltration of a porous, skeletal solid with a liquid phase requires that the total surface-free energy of the system after infiltration be lower than the total surface-free energy before infiltration. The total includes the surface-free energies of the solid and the liquid phases, as well as the interfacial energy between the solid and liquid. This relationship is expressed by:

$$\gamma_{s/l} = \gamma_s - \gamma_l \cos \theta$$

Where $\gamma_{s/l}$ is the specific surface-free energy of the solid-liquid interface, γ_s and γ_l are the respective surface-free energies of the solid and liquid, and θ is the liquid-solid contact or wetting angle. A low contact angle is essential for infiltration. The rate and depth or height of infiltrant penetration of the compact pore system can be enhanced by lowering the contact angle with the use of detergents (Ref 7). Solid, liquid, or vapor-forming surfactants mixed with the powder or passed through the compact prior to infiltration are suitable additives.

With a sufficiently low contact angle, surface tension forces in many systems cause the liquid metal to spread on the outside faces of the skeleton body. Because of diminished resistance, complete envelopes may form before capillary forces draw the liquid into the interior.

A number of conditions other than a low liquid-solid contact angle must be fulfilled to ensure successful infiltration and useful end products. These conditions have been discussed by many investigators (Ref 8–10) and are reviewed below.

The Skeleton. The compacted powder matrix, or skeleton, should consist of a network of solid particles or grains throughout the infiltration process. It should provide a system of interconnected pores and channels of a size range that permits unimpeded capillary force action. Closed-off pores must be avoided; they cannot be penetrated by the liquid infiltrant under normal circumstances. An attempt should be made to remove oxide or nitride films prior to infiltration.

Infiltrant must have a melting temperature below that of the skeleton. High fluidity in the liquid state is desirable because it aids the driving force of the surface tension. Thermal expansion characteristics of the infiltrant affect ultimate strength of the infiltrated body. If the infiltrant skin surrounding the infiltrated skeleton expands during rapid solidification, the strength of the finished part may be decreased; however, the opposite reaction occurs if the skin contracts on cooling from the liquid state. This phenomenon was found for iron infiltrated with copper and bismuth, respectively (Ref 11).

System Compatibility. The ideal liquid infiltrant should have a contact angle with a solid that is near zero, and reaction between solid and liquid should be kept to a minimum. Some refractory metal-conductor metal systems fulfill this condition. If the reaction products formed during infiltration (intermetallic compounds, eutectics, or solid solutions) have a specific volume equal to or greater than the combined initial specific volumes of the skeleton and infiltrant, penetration of the liquid is halted before completion, or it is blocked completely. A similar effect is obtained if the reaction products are dissolved in the liquid phase and the infiltrant becomes less fluid. In either case, external infiltrant residue and unfilled pore space remain.

Solubility. When solid solution formation between components is minimal at low temperatures and when equilibrium conditions at infiltration temperature produce minimal dissolution of the skeleton in the liquid infiltrant, the effects of solubility generally are beneficial. Maximum operating temperature may be lower and kept at only slightly above the liquidus of the infiltrant alloy. Strong bonds at the interface between the two phases are formed, and nearly full density of the product is obtainable.

If the infiltrant dissolves more than traces of the skeleton material, the infiltration rate may be lowered, requiring longer times for complete penetration. Also, at the point of contact with the liquid infiltrant mass, the exterior surface of the skeleton body can erode. Presaturation of the infiltrant with the skeleton metal or alloying with a solubility-inhibiting element eliminates this problem. Prealloying of copper with iron for infiltration into iron skeletons is a typical example.

The stability of the compact becomes impaired as the solubility of the solid in the liquid increases. Consequently, wide pore channels and short infiltration times should be utilized. Diffusion barriers deposited on the pore walls of the skeleton can cause sufficient delay in dissolving the solid surfaces to permit penetration of the liquid phase through the entire pore system.

In applying copper films on the inner surfaces of iron skeletons by precipitation from a copper sulfate solution, carbon diffusion through the resulting copper-base solid solution can be slowed sufficiently to prevent solidification of a cast iron infiltrant before the entire compact is penetrated. Unless these precautions are taken, excessive solubility of the solid in the liquid rapidly leads to the collapse of the skeleton and to a meltdown of the partly or fully penetrated compact.

Microstructure. Structural integrity of the infiltrated body is closely associated with the microstructure. If the dihedral angle at the intersection of the boundary between two solid grains or particles and a liquid phase approaches zero, the liquid metal tends to penetrate along the grain boundary. An example of this is the W-Ni-Cu heavy alloy system. During the penetration stage of infiltration of the nickel-copper alloy, closed-off pores can become filled. This mechanism can lead to a loosening of the contacts between the solid tungsten grains and a loss of coherence of the skeleton. Consequently, volume expansion can occur during infiltration, while some of the original pores remain unfilled. New pores also can be generated by particle rearrangement (Ref 12), or as a result of liquid-solid phase reactions.

Control of dimensional changes of the compact during infiltration is difficult in systems where the dihedral angle is zero and may even require the confinement of the skeleton in a mold. Also, the ultimate properties of the infiltrated body can be affected by the cohesive strength level of the skeleton that remains after boundary penetration by the liquid phase.

Vacuum. Capillary infiltration can be enhanced by subjecting the system to a vacuum. Volatile impurities are more readily removed from the infiltrant melt and the free surface of the solid under vacuum. For alloy infiltrants, the solidus temperature may be raised, and premature freezing in the skeleton channels can occur if the composition is changed as a result of a high vapor pressure of one of the elements. In that event, vacuum operations should be avoided, and pressure should be applied in a sealed vessel during infiltrant melting and melt penetration of the skeleton.

Pressure Gradient. If a vacuum is applied to a confined skeleton on the side opposite the face contacting the infiltrant, penetration of the liquid is enhanced. By increasing the pressure gradient further with the aid of an external force acting on the liquid, compacts with a less controlled pore structure or wider pore spectrum can be penetrated. Infiltration becomes progressively less dependent on capillary action and good wetting as the pressure gradient becomes larger.

Systems in which the liquid forms high contact angles with the solid can be infiltrated if the pressure differential is sufficient, but the liquid cannot be retained by the skeleton and may be exuded partially or completely. In that case, the skeleton functions as a filter.

Infiltration Techniques

Throughout the development of metal infiltration technology, numerous methods have been employed. All acceptable methods for production work utilize, at least in part, procedures, tooling, and equipment common to the P/M industry. However, depending on the system involved and the quality specifications of the product, specific tools, devices, fixtures, supports, or molds generally are required. For example, the shape complexity of a component or its potential for reaction with either the skeleton or liquid metal can determine whether a carbon base or ceramic container or mold is used during infiltration.

The type of furnace used must permit control of time and temperature for both stages of infiltration—penetration of the liquid through the pore system and reaction of the liquid with the skeleton after filling of the pores. The particle size distribution of the powder must be selected or custom blended, and the compacting pressure must be adjusted to produce a large number of interlocking pores, as well as free paths between them, required to receive and pass the liquid metal.

Capillary-Dip Infiltration. The skeleton body is partly immersed into a molten metal bath contained in a crucible, where it acts as a wick (Fig. 1a). Liquid is drawn in by capillary forces and rises, expelling the gas volume contained in the pores. The bath may be metered to fill exactly the available pore volume of the solid.

Full-Dip Infiltration. The skeleton is completely submerged in the melt (Fig. 1b). Penetration occurs from all sides toward the core; therefore, the gas volume can be displaced only by diffusion through the liquid. To avoid gas entrapment, immersion of the skeleton must proceed slowly or in stages. The melt may be poured onto the porous body from one side with a tilting bath container. Application of a vacuum facilitates degassing. Excess molten metal is unavoidable, but can be used as a cast metal extension of the infiltrated compact.

Contact Infiltration. The infiltrant is initially placed in juxtaposition with the free-standing skeleton. After melting, a liquid film is formed that penetrates the pores. If the infiltrant is placed on top of the skeleton (Fig. 1c), surface tension can spread the film over all faces of the skeleton before penetration into the core occurs. Gas displacement from the pores is aided by placing the infiltrant beneath the skeleton or by applying a vacuum. Exact allotment of the metal required to fill the pore volume is possible, thus eliminating the need for liquid metal containers. This method is an adaptation of conventional sintering and elevates infiltration to a production process.

Gravity-Feed Infiltration. In this method, capillary forces are augmented by external pressure exerted by the high head of an infiltrant melt reservoir above the skeleton, which in turn is contained in an investment-casting inert mold. Auxiliary force is a function of the height of the liquid metal head; if the mass of the head is large enough, more than one skeleton can be fed, and cluster arrangements similar to those of the investment-casting process are possible.

Separation of the infiltrated product from excess infiltrant of the head or sprues is aided by suitable gating, made of coarse powder com-

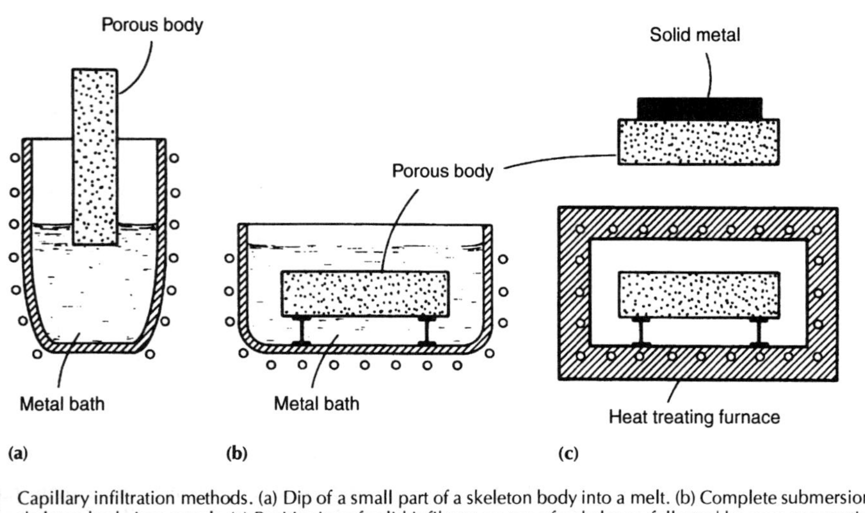

Fig. 1 Capillary infiltration methods. (a) Dip of a small part of a skeleton body into a melt. (b) Complete submersion of a skeleton body into a melt. (c) Positioning of solid infiltrant on top of a skeleton, followed by pore penetration on melting. Source: Ref 8

pacts or fiber felts to facilitate passage of the molten metal. If gate and skeleton materials are the same composition, the liquid can dissolve some of the gate and become partially saturated with the alloying element, thus preventing surface erosion of the infiltrated product. This process is suitable for producing precision-shaped and graded cermet turbine blades. Figure 2 shows the infiltration mold assembly used in this technique.

External Pressure Impregnation. When capillary forces are ineffective because of poor wetting, unsuitable pore size and distribution, or high viscosity of the liquid, the solid body can be impregnated by the molten metal only with the help of a substantial external force. This force can be supplied by a pressurized gas or liquid, a dead load, or a piston inside a cylinder, and must exert pressure on the melt. The infiltrant mass may be limited to the exact amount required for filling the pores, providing a continuous film surrounding the end product, thereby eliminating the need for cropping excess material.

Molten infiltrant can make contact with the skeleton at any desired face, although the top or bottom usually is the most practical contact surface. The liquid metal must be contained in a suitable pressure vessel before and during impregnation. This method is applicable to systems in which the molten infiltrant tends to solidify before penetration is completed, due to diffusion-induced changes in composition.

An example of such diffusion-solidification can be found in iron/cast iron systems. High-speed steel has been produced by impregnating molten high-carbon iron into porous low-carbon steel compacts made from shotted granules or wire clippings with 0.5 MPa (75 psi) pressure (Ref 14). Figure 3 shows the closed mold arrangement used for this process.

Centrifugal pressure impregnation is an adaptation of external pressure-aided impregnation of liquid metal into pores of a solid. This process uses the centrifugal investment casting method, but requires special features such as more refractory, chemically inert, and tight-fitting ceramic molds and atmospheric control. The centrifugal force on the molten infiltrant can cause segregation due to exudation of a heavy liquid from the skeleton. This phenomenon may require special mold adapters or inserts.

Vacuum Infiltration. Two methods can be employed to create a vacuum in metal infiltration. One method creates a pressure gradient by applying suction to the liquid phase through the interconnected pore system of the skeleton, allowing atmospheric pressure on the molten infiltrant to be the driving force. A closed-mold system that contains the molten infiltrant and the skeleton is required. The vacuum then is applied to the end of the skeleton, away from contact with the melt.

The second method simply provides for placement of the entire infiltration arrangement in a vacuum furnace. This method is useful for systems with strong outgassing components, but can cause difficulties in controlling the infiltration cycle of monitoring skeleton and infiltrant temperatures.

Infiltration Systems

Even if essential infiltration conditions are strictly met, the possible binary systems in which a high-melting single metal or compound can be infiltrated by a lower melting single metal are many. Table 1 lists systems that are partially or fully immiscible in the liquid state (Ref 8), but also includes workable systems that form terminal or even complete solid solutions (e.g., iron-copper and nickel-copper). Systems for which infiltration has gained acceptance for past production tasks or where infiltration is presently in use by industry are shown by closed circles, open circles indicate systems capable of infiltration in the laboratory or systems possessing infiltration potential on the basis of established criteria.

The material combinations listed in Table 1 are incomplete, because the possibilities for infiltration combinations are expanded if binary or more complex alloys are used for the skeleton and/or infiltrant. It is possible to produce nonequilibrium alloys of a final composition that cannot be made by melting and casting; if the respective phase diagram indicates some solubility relationships in the liquid or solid state, diffusion heat treatments just below or above the melting temperature of the infiltrant can result in truly equilibriated alloys, usually distinguished in their microstructure by rounded grains embedded in a matrix (Ref 8).

In principle, homogeneous alloys can be produced by infiltration if solubility of the skeleton in the infiltrant (or vice versa) is unlimited in the liquid or solid state. The nickel-copper system is an example; Monel compositions were successfully produced with 63 to 67% Ni by complete penetration of nickel skeletons with copper and binary alloys of copper with 10% Al or 10% Si.

Another example is the capillary infiltration of iron or low-carbon steel powder compacts of predetermined porosity with molten cast iron under conditions of close time, temperature, and carbon-diffusion control. When a homogeneous alloy is produced, process variables such as pore

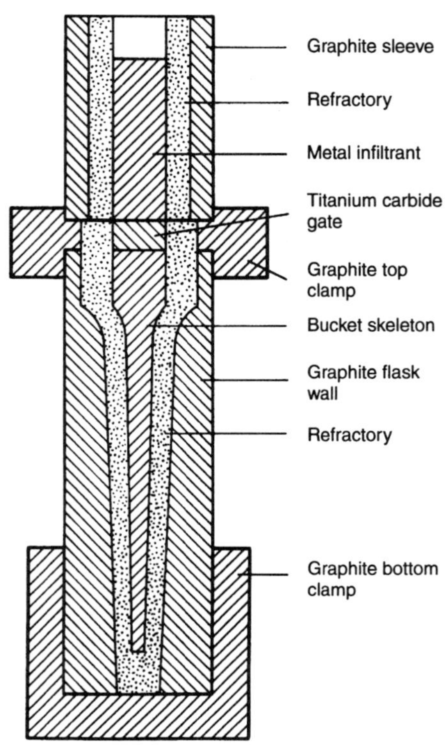

Fig. 2 Cermet turbine blade infiltration mold assembly. Source: Ref 13

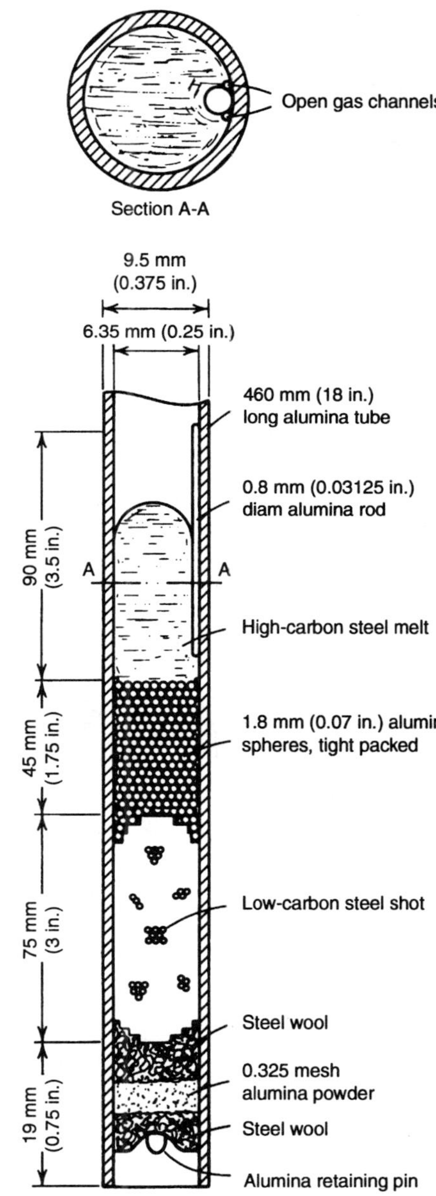

Fig. 3 Mold arrangement for pressure impregnation of low-carbon steel compacts with a high-carbon steel melt (experimental process). Compact dimensions: 6.3 mm (0.25 in.) diam, 80 mm (3 in.) in length. Source: Ref 14

and capillary channel, size of the powder compact, time for melting the infiltrant and its complete penetration of the skeleton, and type of diffusion inhibitor must be carefully controlled to ensure homogeneity.

After the penetration stage has been successfully completed, sintering with the liquid phase in place can proceed at the infiltration temperature, and full homogenization of the alloy can be accomplished by subsequent heat treatment in the solid state. The principal advantage of this procedure lies in the potential for obtaining alloys of nearly full density without mechanical working. Unfortunately, process control problems have made this approach impractical.

Refractory metal-based composite structures of tungsten-copper, tungsten-silver, molybdenum-copper, and molybdenum-silver systems are among the oldest P/M products produced by infiltration. The solubility of refractory metals in molten silver is minimal, and solubility in molten copper is practically nil. Along with excellent wetting characteristics, these properties constitute ideal conditions for infiltration.

The ratio of the two phases that can be combined by this method can vary considerably (from about 35 to 40 vol% up to 85 to 90 vol% of the refractory metal). The upper limit is set by the requirement of an interconnected pore structure in the compacted and sintered refractory metal, while the lower limit can be controlled by selection and treatment of the powder, which is piled loosely in a suitable container (e.g., graphite).

Generally, refractory metal powder is reduced oxide type with particle sizes ranging from about 1 to 50 μm. However, coarser powders ranging from 50 to about 400 μm with a more sharp-edged crystalline particle shape, such as that obtained by mechanical comminution of sintered or swaged scrap metal, have produced sintered skeleton structures with improved erosion and wear resistance (Ref 15). The desired porosity in the loose pile is obtained by tapping, vibration, jolting, or ramming, whereas porosity of compacted powders is controlled by the static pressure of the press tools.

The refractory metal skeleton usually is sintered prior to infiltration to improve strength and cohesion. Sintering temperature must be selected in conformity with powder particle size and shape, compacting pressure, and specific final composition; it is usually about 100 °C (180 °F) above the melting point of the infiltrant metal. Higher temperatures tend to produce severe grain growth and shrinkage and can even destroy the interconnected pore system needed for successful infiltration, especially in very fine starting powders. Figure 4 schematically presents the growth of tungsten grains with rising sintering temperature and the change in the distribution of a subsequently infiltrated copper phase, especially above 1600 °C (2900 °F).

Infiltrated bodies that have a continuous skeleton phase above approximately 65 vol% can be shaped only by machining, while those having a smaller proportion of refractory metal dispersed as loose grains in the ductile metal matrix are plastically deformable at elevated temperatures (Ref 8). Soldering, brazing, or plating of the infiltrated product is aided by generally smooth surface films or high contents of infiltrant metal.

The preceding binary systems illustrate combinations of two metals of widely differing melting temperatures that can be advantageously produced to near-net shape and full density by infiltration. Nickel also can be infiltrated into tungsten, but equilibrium at liquid-phase temperature causes severe attack of the refractory metal and requires careful process control to prevent incomplete penetration due to diffusion solidification. Coarse tungsten powder helps to produce compacts with larger capillary channels for better penetration. High heating rates, especially after about 90% of the absolute melting temperature of nickel is reached, improve infiltration conditions.

If nickel is alloyed with copper, the solubility of tungsten in the liquid phase decreases as the copper percentage is increased. Infiltration is more practical (Ref 8) for heavy alloy systems because tungsten powder and process control requirements are less stringent. If nickel is alloyed with iron, however, the solid-liquid phase interaction is similar to the binary tungsten-nickel system, and infiltration again is more difficult.

The same principle applies to the more complex refractory metal systems with nickel-chromium and cobalt-chromium alloys. Nevertheless, skeleton bodies of tungsten and molybdenum, as well as bodies of binary 85W-15Cr and 75W-25Cr alloys, can be successfully infiltrated with superalloys of the Nichrome-V, Hastelloy-C, Stellite, and Vitallium compositions into shapes simulating mechanical and engine test specimens (Ref 17).

Table 1 Binary metal infiltration systems

	Infiltrant(a)																		
Skeleton	Aluminum	Antimony	Bismuth	Cadmium	Calcium	Cobalt	Copper	Gold	Iron	Lead	Magnesium	Manganese	Mercury	Nickel	Silver	Sodium	Thallium	Tin	Zinc
Aluminum			○	○						●						○	○		
Beryllium	○										○								
Chromium	○		○	○		○	●			○			○		○		○	○	
Cobalt			○				○			○			○		○		○	○	
Copper		○	●							●			○				○	●	○
Iridium								○							○				
Iron	○	○	○	○	○		●			●	○		○		○	○	○	○	○
Lead																○		○	
Magnesium																○			○
Manganese			○							○					○		○		
Molybdenum	○		○	○	○		●	○		○		○	○	○	●			○	○
Nickel			○				○			○	○		●		●		○	○	
Niobium				○						○			○		○				○
Platinum								○							○				
Rhodium								○							○				
Silicon		○	○	○						○							○	○	
Silver			●										○						
Tantalum				○				○		○			○		○				○
Titanium	○						○											○	
Titanium carbide						●	○		●					●					
Tungsten	○	○	○	○	○	○	●	○		●		○	○	●	●				○
Tungsten carbide						●	○		○					●	●				
Vanadium							○								○				
Zinc			○							○						○	○	○	
Zirconium							○			○					○				

(a) Open circle (○) indicates experimental and potential combinations; closed circle (●) indicates industrially significant combinations. Source: Ref 8

Several other refractory metal-based composite structures can be readily produced by infiltration. These include the high-density tungsten-lead system to produce materials suitable for shielding against radiation and the chromium-copper system to produce compositions for welding electrodes (Ref 8). To retain the low liquid-solid contact angles in these systems, a strong reducing atmosphere is necessary to prevent oxide films on the molten lead or on the solid chromium. A free metallic surface requires sintering of the skeleton above 1250 °C (2280 °F) to reduce any oxide film on the solid chromium.

Carbide-Based Systems. Liquid-phase sintering of tungsten carbide/cobalt or titanium carbide/nickel systems capitalizes on the eutectics of the two phases. A limited solubility of the carbide in the matrix metal facilitates bonding; carbon and metal diffusion through the liquid are less important in densification than reactions at the carbide-metal phase boundaries. It is unknown whether the carbide particles in these systems form a rigid skeleton, but the interfacial tension between crystals of the carbide and the liquid metal appears to be anisotropic (Ref 18).

During cooling from the sintering temperature, some or most of the carbon and metal dissolved in the liquid precipitates on grains that remained solid during the process. This mechanism can be altered somewhat if the rigid skeleton is formed first and the liquid phase subsequently infiltrates into the pore system. By first saturating the matrix metal with carbon and skeleton metal, the liquid phase dissolves less skeleton materials, and shape distortion is diminished. Carbide coalescence and grain growth also are decreased.

The earliest attempts to produce cemented carbides were made by infiltrating carbide skeletons with unalloyed binder metals (Ref 19). Later, binder metal prealloyed with elements of the skeleton to inhibit contact face erosion was used in a broad investigation of infiltrating single and double carbides with many cobalt and nickel alloys (Ref 20). Table 2 lists some of the alloys used. The large number of feasible combinations includes several noteworthy successes, especially for titanium carbide-based stems with Nichrome and Vitallium infiltrants. A feasibility study (Ref 17) produced similar results for the same infiltration systems and laid the foundation for an extensive development program to utilize these materials for heat-resistant applications.

The structure of infiltrated carbides reflects infiltration mechanics on a macroscale. In zones penetrated by the infiltrant, fully dense regions and slight expansion of the skeleton due to carbide grain separation are observed. Substantial porosity and some shrinkage occur in areas inadequately penetrated by the liquid alloy, such as the side opposite the contact face in unidirectional infiltration, or in the center for infiltration from opposite sides. Subsequent heat treatment is ineffective in eliminating porosity. Erosion at the contact faces is greatly diminished by infiltrating skeletons composed of tungsten-base multicarbides, so that saturating the infiltrant with skeleton elements by prealloys frequently is not required.

The microstructure reflects the crystallographic characteristics of the carbide, and fully infiltrated regions do not differ in grain size and morphology from material whose liquid phase was sintered in situ. Rectangular and triangular grains are retained in infiltrated tungsten carbide, whereas for infiltrated titanium carbide, the cubic lattice is reflected by distinctly rounded grains. Grains of solid-solution carbides of tungsten and titanium or titanium and molybdenum are slightly rounded at the corners. Graphite precipitates accompany porosity in poorly infiltrated regions for tungsten carbide and titanium carbide skeletons, especially if starting powders contain more than a trace of free carbon.

Ferrous-Base Systems. The thermodynamic affinity between solid iron and liquid copper offers the potential for virtually pore-free P/M products by infiltration. Moreover, the excellent

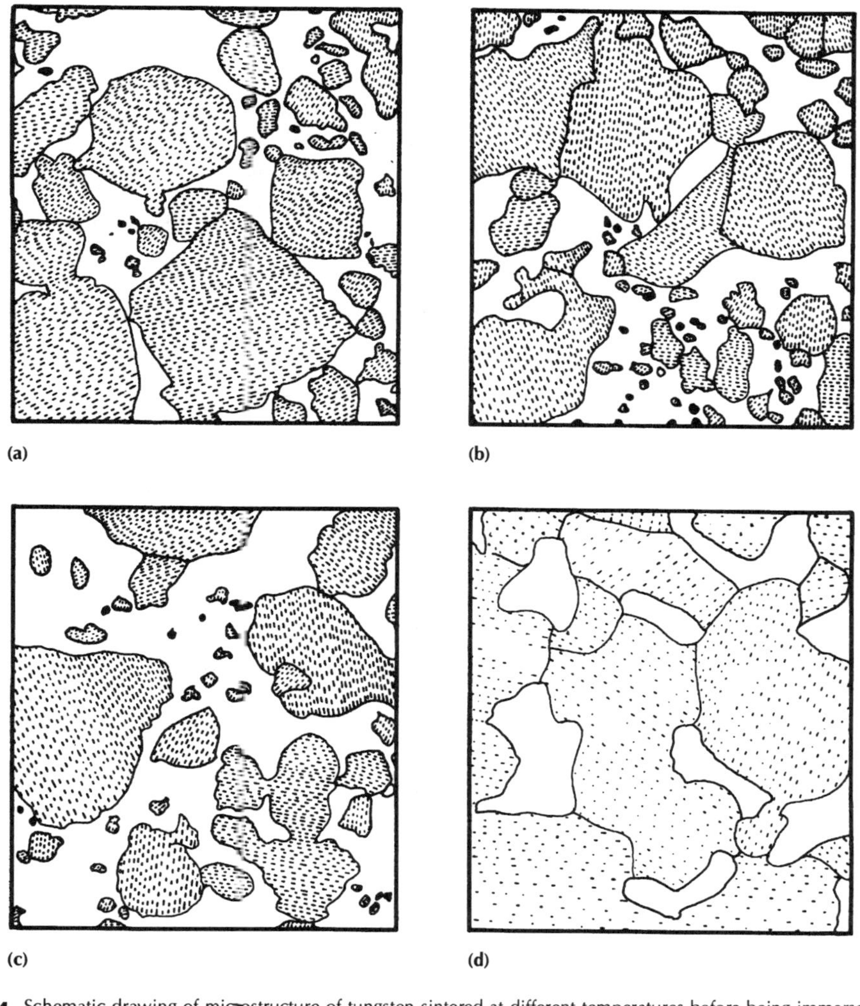

Fig. 4 Schematic drawing of microstructure of tungsten sintered at different temperatures before being immersed into molten copper. (a) 1100 °C (2010 °F). (b) 1350 °C (2460 °F). (c) 1600 °C (2910 °F). (d) 2800 °C (5070 °F). 500×. Source: Ref 16

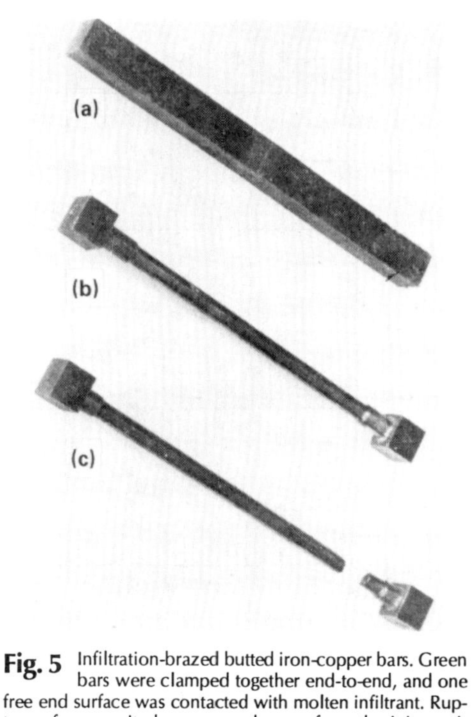

Fig. 5 Infiltration-brazed butted iron-copper bars. Green bars were clamped together end-to-end, and one free end surface was contacted with molten infiltrant. Rupture of composite bar occurred away from the joint, evidence of the high strength of the brazed bond. (a) Butt joined bars after infiltration. (b) Machined tensile bar. (c) Tested tensile bar

wetting characteristics that exist in the brazing of steel can also be utilized for joining disparate bodies during infiltration. A powder compact and a casting or forging, or halves of complex or offset configurations, can be joined. Figure 5 (Ref 10) illustrates this self-brazing capacity without strength degradation. Finally, the generally smooth cupric film surrounding the infiltrated body serves as a base for surface coating or plating.

The ability of the infiltration process to combine major proportions of normally unalloyable industrial premier metals (iron and copper) was recognized as early as World War I (Ref 21, 22), but it was only in the late 1940s, through refinements in technique, that sound products could be made (Ref 10, 23–25). These products, in turn, have culminated in the present advanced state of the art. While copper content must be higher than for most commercially sintered iron-copper alloys because of the need to maintain an interconnected pore system for complete infiltration, good mechanical properties can be realized.

This is apparent from the tensile strength-elongation data given in Fig. 6 for commercial iron powder with and without graphite additions (Ref 24). Strength is enhanced because the infiltrated structure, with a minimal amount of isolated pores, is virtually free of internal notches. The iron-copper system permits a precipitation-strengthening mechanism. If carbon is diffused into the iron to produce a hypoeutectoid structure of the skeleton, hardening by martensite transformation is possible. Where other metals are alloyed with the copper, solid solution strengthening of the matrix can be achieved.

In the binary iron-copper system, about 3.8 to 4% Fe is dissolved by the liquid copper under equilibrium conditions at an infiltration temperature of 25 to 50 °C (45 to 90 °F) above the peritectic temperature of 1090 °C (2000 °F) at the copper side of the phase diagram, while the γ-iron dissolves about 8 to 8.5% of the copper. At 900 °C (1650 °F), the solubility of α-iron in copper is about 1.5%; it diminishes to less than 0.04% at room temperature, at which point the solubility of copper in iron is equally low.

These thermodynamic relations form the basis for precipitation hardening. However, macrodispersion of the two phases in the infiltrated alloys causes concentration of precipitates in thin zones at the phase boundaries, as shown in Fig. 7 (Ref 26). Consequently, conventional hardness tests are not precise enough to show a noticeable increase in macrohardness after a precipitation treatment, such as quenching from 900 °C (1650 °F) followed by prolonged tempering at 600 °C (1110 °F). The precipitation mechanism produces increases in strength, elongation, and impact resistance (Ref 8) and also can be traced through changes in the electrical conductivity (Ref 25).

If carbon is diffused into the iron skeleton with copper infiltrated afterward, two processes compete with one another during cooling. During quenching, normal martensitic transformation occurs inside the skeleton structure. At the same time, however, precipitation of the dissolved iron in the copper and copper in iron, respectively, is suppressed in the phase boundary zones. During reheating, martensite decomposition causes a decrease in hardness of the steel skeleton structure, with simultaneous increases in hardness and strength of the boundary zones.

Table 3 shows the effect of such a heat treatment on the mechanical properties of a copper-infiltrated 0.3% carbon steel. Water quenching produces an appreciable increase in hardness and brittleness. With increasing reheating tem-

Table 2 Carbide infiltration test matrix and evaluation

Specimen No.	Skeleton(a) Composition	Skeleton(a) Density, g/cm^3	Skeleton(a) Pore volume, %	Infiltrant composition, %	Infiltration(b) Type	Infiltration(b) Temperature, °C	Infiltration(b) Temperature, °F	Infiltration(b) Time, min
1a	WC (6.1% C)	11.05–13.35	29.3–14.6	100Co	Contact, opposing sides(d)	1500	2730	15
1b				95Co-5WC		1460	2660	5
1c				75Co-25WC		1390	2530	5
1d				60Co-40WC		1350	2460	5
2a	80WC-20TiC(e)	6.65–8.34	37.1–21.3	100Co	Contact, opposing sides(d)	1500	2730	15
2b				95Co-5WC		1460	2660	15
3a	TiC (18.8% C)	3.01–3.63	33.2–19.3	80Co-20Cr	Contact, opposing sides(d)	1500	2730	15
3b				66Co-28Cr-6Mo		1450	2640	15
3c				72.7Co-17.3Cr-10TiC		1400	2550	15
3d				80Ni-20Cr		1450	2640	15
4a	97TiC-3Mo_2C(e)	3.38–4.03	25.7–11.4	80Co-20Cr	Contact, one side(d)	1500	2730	5
4b				66Co-28Cr-6Mo		1450	2640	5
4c				72.7Co-17.3Cr-10TiC		1400	2550	5
4d				80Ni-20Cr	Capillary dip in molten infiltrant	1550	2820	3
5a	95TiC-5Mo_2C(e)	3.46–4.05	24.8–11.6	80Co-20Cr	Contact, one side(d)	1500	2730	5
5b				66Co-28Cr-6Mo		1450	2640	5
5c				72.7Co-17.3Cr-10TiC		1400	2550	5
5d				72.7Ni-17.3Cr-10TiC	Capillary dip in molten infiltrant	1550	2820	3
6	90TiC-10Mo_2C(e)	3.54–4.14	26.0–14.6	80Ni-20Cr	Contact, opposing sides(d)	1400	2550	15
7	70TiC-30Mo_2C(e)	4.09–4.75	22.9–9.9	80Ni-20Cr	Contact, opposing sides(d)	1400	2550	15
8	50TiC-50Mo_2C(e)	4.69–5.58	21.3–8.3	80Ni-20Cr	Contact, opposing sides(d)	1400	2550	15

(a) All skeletons were presintered at 950 °C (1740 °F) and high sintered at 1500 °C (2730 °F) for 2 h in a carbon tube resistor furnace under hydrogen, except No. 1 to 3, which were high sintered in vacua in a carbon susceptor induction furnace. (b) In vacuum induction furnace. (c) A qualitative assessment of resistance against fragmentation by hammer blows. (d) Infiltrant mass was 40 to 45% of mass of infiltrated product. (e) Solid solution. Source: Ref 19 peratures, however, the material becomes softer and tougher, without loss in strength (Ref 8).

When a copper alloy is used as an infiltrant, other benefits can accrue. When precipitates form in the matrix, such as copper alloys with beryllium, chromium, or silicon, the resulting strength increase during heat treating of the infiltrated body reinforces iron-copper precipitation zones at the boundaries. The strengthening effect of precipitates is independent of heat treatment and augments strength increases obtained with unalloyed copper infiltrant in solid-solution alloys. However, some solid-solution alloy infiltrants minimize erosion at the point of initial contact with the iron skeleton, because these alloys melt slowly over a range of temperatures. Brass containing 20% Zn exhibits this phenomenon.

A copper alloy with manganese in amounts up to 5%, especially if it also contains sufficient iron to inhibit a severe attack of the skeleton contact faces, has a different beneficial effect. The manganese oxidizes preferentially in the commercial atmospheres where infiltration takes place. A nonadhering porous crust results that can be removed during finishing more easily than the tenacious residue formed by a binary copper-iron infiltrant (Ref 27, 28).

Many ferrous metal-base infiltration systems have been explored experimentally. Kieffer and Benesovsky (Ref 8) have investigated the iron-gold, iron-bismuth, iron-cadmium, iron-lead, iron-antimony, and iron-tin systems for bearings, and the iron-cobalt-silicon, iron-copper-silicon, and iron-manganese-silicon systems for magnetic or structural parts. Alloys of the iron-zinc system have also been produced by infiltration, but treatment in a pressure vessel is required to overcome the high vapor pressure of the zinc (Ref 29).

Austenitic stainless steel skeletons infiltrated with silver possess excellent corrosion resistance, thus making them suitable for food processing applications. Ferritic stainless steel and high-manganese steel compacts with varying carbon contents also display improved corrosion resistance when infiltrated with cupric alloys. These alloys also exhibit extraordinary hardness and wear resistance, coupled with a considerable toughness. Table 4 lists mechanical and technological properties of several ferrous-base infiltrated materials.

Nonferrous-Based Systems. The major nonferrous metals with higher melting points are thermodynamically compatible with many low-melting metals in the liquid state. Consequently, skeleton bodies of cobalt and nickel can be readily infiltrated with gold, as well as with many of the low-melting heavy metals, such as bismuth, lead, or antimony (Ref 8). Copper can also be infiltrated into cobalt and nickel skeletons; however, because of the formation of solid solution in all proportions, infiltration into nickel powder compacts requires a narrow particle size range, wide capillaries within a pore volume not exceeding about 35%, short infiltration time, and a vacuum to assist the capillary forces. Mercury wets nickel well without forming an amalgam and is easily impregnated into skeleton bodies, provided the pore structure prevents exudation of the heavy liquid metal (Ref 30).

Copper is another skeleton metal with pores that can be readily filled with liquid low-melting-point metals, such as lead (Ref 31, 32) or bismuth (Ref 33). Vacuum impregnation is suitable for incorporating lead-base alloys, such as those containing 15% Sb and 5 to 10% Sn, into spongy structures of nickel-copper or nickel-iron supported by steel backing (Ref 34).

Combinations of aluminum or aluminum alloys with low-melting metals such as bismuth, lead, thallium, or thallium-lead by means of infiltration in hydrogen or a vacuum have been proposed (Ref 33, 35). However, strict control of powder characteristics, especially particle shape and surface condition, to maximize wetting appears to put infiltration at a disadvantage over in situ liquid-phase sintering. To overcome this problem, zinc or cadmium can be added to the aluminum of the skeleton, followed by cleaning

Table 2 (continued)

Analysis, %	Hardness, HRA	Infiltrated product: Surface condition	Infiltrated product: Microstructure	Infiltrated product: Bench test(c)	Specimen No.
	86–87	Contact face erosion, slight residue	Porous in core, graphite precipitates, fairly uniform grain size	Very tough	1a
					1b
	85–86	No erosion, heavy residue		Tough	1c
					1d
	86.5–87	Contact face erosion	Uniform phase distribution, some porosity	Tough	2a
					2b
	87.5–88	Slight erosion		Very tough	3a
	89.5–90	Slight porosity		Fairly tough	3b
	88	No erosion, smooth		Very tough	3c
24.6Ni, 6.1Cr, bal TiC	83.5–85	Slight erosion	Dense, uniform	Very tough	3d
	88+	Heavy contact face erosion and residue	Higher matrix concentration near contact face, porosity increasing toward far end	Very tough	4a
	89.5			Tough where dense, brittle where porous	4b
	88+	Less contact face residue than in 4a			4c
22.5Ni, 5.7Cr, 2.1Mo_2C, bal TiC	84.5–85	Alloy skin becoming heavier toward bottom, forming excess on bottom end	Uniform phase distribution, generally dense	Tough	4d
	88+	Contact face erosion, some residue	Similar to 4a–c	Tough	5a
	90				5b
	88.5	Slight contact face residue			5c
	85	Similar to 4d	Similar to 4d	Tough	5d
22.9Ni, 5.5Cr, 7.1Mo_2C, bal TiC	85–86	Slight contact face erosion, small residue, slightly porous	Porous in core, less uniform phase distribution than in 4d	Less tough than 4 and 5	6
22.6Ni, 5.6Cr, 21.4Mo_2C, bal TiC	86–87	Similar to 6, but more porous	More porous in core, less uniform phase distribution than in 6	More brittle than 6	7
22.3Ni, 5.7Cr, 35.8Mo_2C bal TiC	86–87	Similar to 7, but more porous	Very porous, nonuniform phase distribution	More brittle than 7	8

(a) All skeletons were presintered at 950 °C (1740 °F) and high sintered at 1500 °C (2730 °F) for 2 h in a carbon tube resistor furnace under hydrogen, except No. 1 to 3, which were high sintered in vacua in a carbon susceptor induction furnace. (b) In vacuum induction furnace. (c) A qualitative assessment of resistance against fragmentation by hammer blows. (d) Infiltrant mass was 40 to 45% of mass of infiltrated product. (e) Solid solution. Source: Ref 19 and activating the free surface of the pores by evaporation of the lower boiling metal before proceeding with the impregnation of a metal such as lead (Ref 36).

Infiltration Products

Electrical Contacts and Electrodes. Composite materials for heavy-duty service contacts consist of (a) a refractory metal component to supply the necessary hardness, wear resistance, burning and material transfer through arcing, and high-temperature strength, and (b) a matrix of an electrically and thermally conducting metal. Tungsten and molybdenum are the principal refractory metals, but tungsten carbide, tungsten-niobium double carbide, and titanium carbide have also been used for this purpose (Ref 8). Conductor metals are copper and silver.

Because the two metal types do not alloy with one another, the rule of mixture can be applied to determine the density of a specific composition. Figure 8 shows the effect of increasing tungsten content on density, hardness, and electrical conductivity for copper-tungsten contact material (Ref 37). The straight-line relationship between volume ratio and conductivity is typical for this system. Figure 9 shows the change in thermal expansion coefficient with tungsten content; principal mechanical properties versus tungsten content are plotted in Fig. 10 (Ref 38). The boldface portions of the curves represent composition ranges of material that can be readily produced by infiltration with a low-melting metal content of approximately 10 to 40 vol% (5 to 25 wt%).

Table 5 lists mechanical and electrical property data for tungsten-silver and molybdenum-silver contact materials (Ref 39). Also included are data for nickel-silver contact materials that can be produced by infiltration.

In infiltrated materials, the soft conductor metal matrix is the minor constituent, and extrusion for profile generation is not practical except for nickel-silver composites. All materials are readily machinable into a variety of shapes, especially caps, platelets, and rings. Usually, the heavy-duty contact metal, as a facing, is joined by clamping or brazing onto a support structure. The facing also may be penetrated by an infiltrant with a major portion, that on cooling, solidifies like a casting; an example is copper that

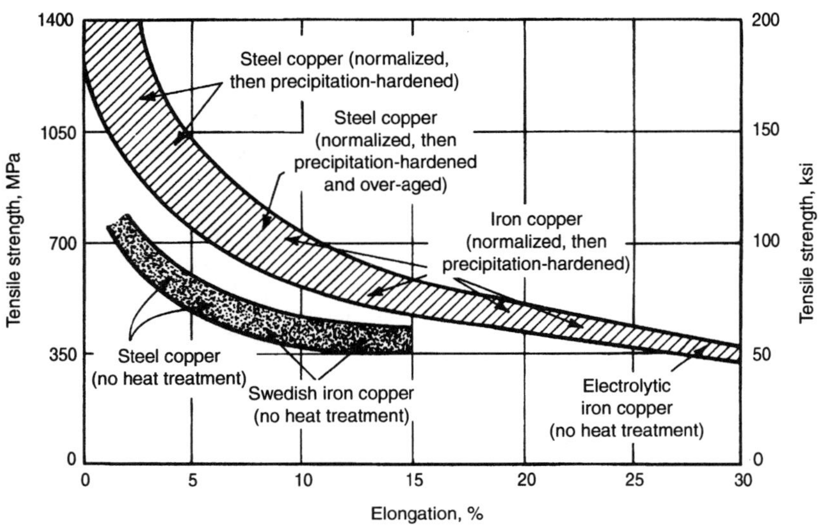

Fig. 6 Mechanical property ranges of copper-infiltrated iron and hypoeutectoid steel compacts before and after heat treatment

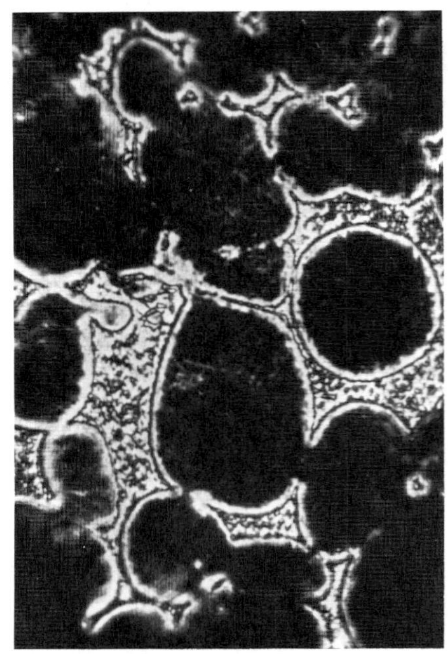

Fig. 7 Microstructure of 25Cu-Fe compact sintered at 1100 °C (2010 °F) for 30 min. Large, rounded, dark areas are α-iron, separated by a diffusion layer from the light copper phase containing the fine, dark, iron-rich precipitate. 1000×. Source: Ref 26

Table 3 Effect of heat treatment on the mechanical properties of 0.3% carbon steel infiltrated with 11 vol% Cu

Treatment(a)	Hardness, HV	Ultimate tensile strength, MPa	Ultimate tensile strength, ksi	Elongation, %	Impact resistance (b), J	Impact resistance (b), ft · lb
Copper-infiltrated and furnace cooled	262	704	102.1	7.3	22.16	16.35
Reheated to 900 °C (1650 °F) and water quenched	437	790	114.6	5.2	11.76	8.68
Reheated for 2 h at 400 °C (750 °F)	360	...	...	...	14.71	10.85
Reheated for 2 h at 500 °C (930 °F)	302	771	111.8	7.6	23.14	17.07
Reheated for 2 h at 600 °C (1110 °F)	255	750	108.8	13.5	62.75	46.30

(a) Hametag iron powder pressed to 88% of theoretical density, sintered at 1220 °C (2230 °F) for 1 h in hydrogen, and infiltrated with electrolytic copper at 1100 °C (2010 °F) for 30 min. (b) Unnotched test bar of 1 cm^2 cross section. Source: Ref 8

Table 4 Properties of some ferrous metal-based infiltrated alloys

Skeleton composition, %	Infiltrant composition, %	Infiltrant, vol%	Density, g/cm³, Calculated	Density, g/cm³, Determined	Hardness, HV	Ultimate tensile strength, MPa	Ultimate tensile strength, ksi	Elongation, %	Impact resistance(a), J	Impact resistance(a), ft · lb	Comment
100Fe	100Pb	10	8.15	7.95	93.5	251	36.4	14	...	...	Free machining, extrudable
99Fe-Cu	100Ag	11	8.09	8.00	178	378.5	54.9	11	...	...	...
100Fe	80Cu-20Ni	17	8.04	7.76	213	419.5	60.9	8	21.57	15.91	...
100Fe	65Cu-35Mn	13	7.92	7.63	256	446	64.7	10	31.37	23.14	...
93.2Fe-6Mn-0.8C	100Cu	13	7.90	7.87	740	...	...	...	...	...	Naturally hard, wear resistant
87.2Fe-12Mn-0.8C	100Cu	9	7.89	7.69	310	562	81.5	6	...	...	Wear resistant, work hardening
93.5Fe-3Cr-3Mn-0.5C	100Cu	14	7.96	7.93	502	957	138.8	4	...	...	...

(a) Unnotched test bar of 1 cm^2 cross section. Source: Ref 8 is strengthened by small amounts of beryllium, chromium, or nickel and is cast into the form of a backing, arm, or other support structure.

Primary applications for infiltrated tungsten-copper composites are resistance welding electrodes and make-and-break contact facings in oil or air circuit breakers or transformer taps. Tungsten-silver contacts are used in switch gear and low-voltage regulators (Ref 37). Low contact resistance and nonsticking properties of nickel-silver contacts make them suitable for high-voltage disconnect switches (Ref 39). Porous, cup-shaped nickel compacts have been impregnated with mercury as part of a short circuiting safety switch contact in proximity fuses for radio-controlled detonation (Ref 40, 41).

Rocket Nozzles. The manufacture of silver-infiltrated tungsten contacts formed the basis for the development and production of large billets, from which rocket nozzle throat liners have been fabricated. This material has proved successful in actual service in rockets of underwater-launched ballistic missiles (Ref 42) because:

- Extreme stability of the structure (virtually zero change in critical dimensions due to the high-temperature, high-pressure gaseous environment in the nozzle throat)
- High resistance to thermal shock during heating and cooling at the extremely high rates experienced by the throat surface in contact with rocket propulsion gas. This in turn requires high tensile strength and resistance to high hoop stresses at the outer perimeter of the throat liner, as well a good thermal conductivity to provide adequate heat transfer to the backup structure.
- Fabricability, especially machinability, on an economical production scale
- Optimum reliability and reproducibility of the nozzle throats in spite of their relatively large size, compared to similarly produced P/M parts in electrical switch gear

An additional advantage of using silver as the matrix metal is its relatively high vapor pressure. Silver evaporation aids infiltration of large capillaries by forming deposits on the pore walls before the liquid metal enters individual pores. Also, the exposed surface region of the throat liner is kept at a temperature considerably below that of the propulsion gas due to evaporative transpiration cooling (Ref 43).

Figure 11 shows a typical silver-infiltrated tungsten ring machined from a large billet that weighed nearly 90 kg (200 lb). It contained about 80 vol% W; the conductor metal was uniformly dispersed throughout the entire cross sec-

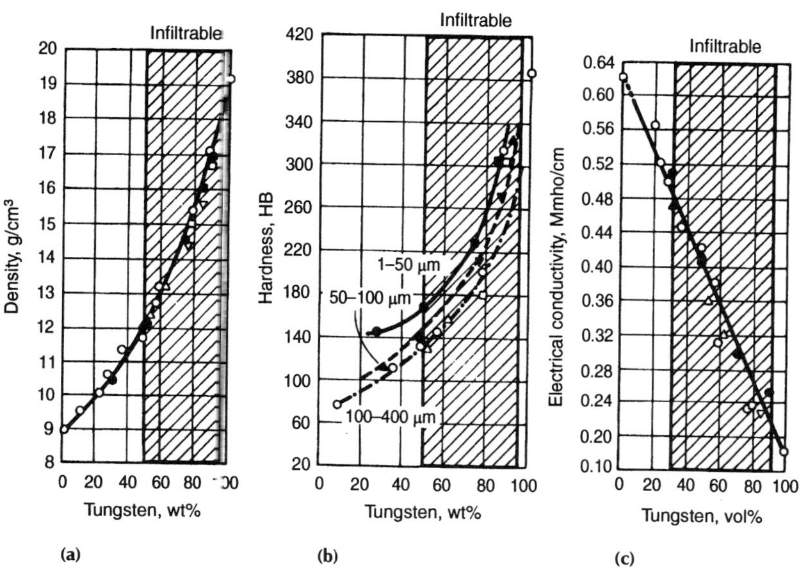

Production method	Particle size of tungsten powder, μm	Symbol
Mixing of the component elemental powders, briquetting, sintering, subsequent working	Coarse, 100–400	ϒ
	Medium, 50–100	▲
	Very fine, 1–50	●
Impregnation of the loosely filled tungsten powder in a mold, extrusion into shape	Coarse, 100–400	Δ
Dip impregnation of a pressed (and presintered) tungsten compact, machining to size	Coarse, 100–400	o
	Medium, 50–100	▼
	Very fine, 1–50	■

Fig. 8 Effect of composition on physical properties of tungsten-copper contact material. (a) Density. (b) Hardness. (c) Electrical conductivity. Heavy framed areas are composition ranges for liquid copper infiltration or sintering of compacts from powder mixtures. Source: Ref 37

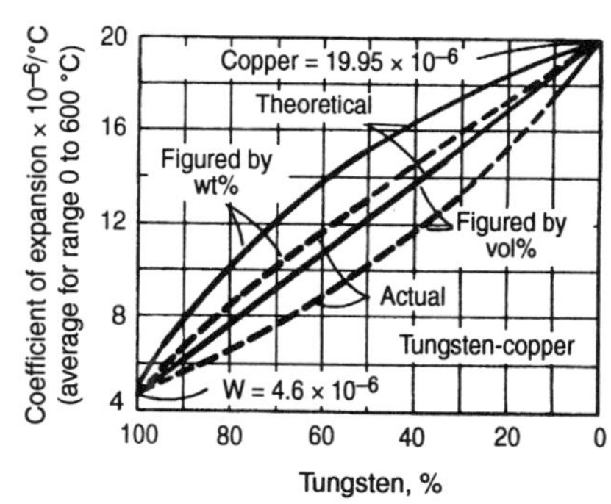

Fig. 9 Effect of composition on thermal expansion of tungsten-copper contact material. Boldface portion of curves designates liquid copper-infiltrated tungsten or sintered compacts from powder mixtures. Light portion of curves designates powder mixtures; compositions are not infiltrable. Source: Ref 38

Table 5 Properties of contact materials containing silver

Composition, wt%				Density, g/cm^3	Hardness, HB	Transverse rupture strength		Electrical conductivity		Estimated contact resistance(a)
Ag	W	Mo	Ni			MPa	ksi	Mmho/cm	%IACS	
10	90	...	...	17.5	230–250	1240–1400	180–200	0.20–0.23	35–40	1000
15	85	...	...	17.0	210–230	1100–1240	160–180	0.24–0.25	42–44	700
20	80	...	...	16.3	200–220	965–1100	140–160	0.26–0.28	45–49	500
25	75	...	...	15.8	160–180	860–965	125–140	0.28–0.30	49–53	350
30	70	...	...	15.2	140–160	790–825	115–120	0.30–0.33	53–57	250
20	...	80	...	10.25	190–210	1100–1240	160–180	0.28–0.30	49–53	500
25	...	75	...	10.27	180–190	965–1100	140–160	0.30–0.32	53–56	400
30	...	70	...	10.28	160–170	860–965	125–140	0.32–0.34	56–60	325
35	...	65	...	10.29	140–150	525–860	110–125	0.34–0.36	60–63	275
40	...	60	...	10.30	120–130	690–525	100–110	0.36–0.38	63–67	250
40	...	...	60	9.30	(b)	450	65	0.245	43	...
60	...	...	40	9.60	(c)	350	50	0.33	57	...

(a) As compared to bulk silver = 100. (b) 30T68 Hardness, Rockwell Superficial. (c) 30T46 Hardness, Rockwell Superficial. Source: Ref 42 tion, filling all interconnected pores (Fig. 12). Processing parameters, production details, and starting material characteristics are described in Ref 44 and 45.

Toensing and Zalsman (Ref 44) have also reported on the effect of these parameters on mechanical properties at ambient and elevated temperatures. Figure 13 shows the change in tensile strength with temperature for a tungsten skeleton with 20% pore volume before and after silver infiltration. Data represent average values, but strengths as high as 620 to 700 MPa (90 to 100 ksi) have been reported. As the melting temperature of the silver is approached, the strength of the composite converges with that of the tungsten skeleton. Table 6 summarizes mechanical and physical properties of 20 vol% silver-infiltrated material for different test temperatures within the operating temperature range experienced by the throat liner during rocket propulsion (Ref 42).

Jet Engine Components. The quest for materials to withstand the high temperatures and stresses imposed by combustion gases in the gas turbines of jet engines led to the development of cermets in the late 1940s and early 1950s. Cermets offered increased operating temperatures and engine efficiency when used in rotating blades. Their inherent brittleness prevented the use of cermets for this purpose, however, and their use for stationary gas-conducting nozzle vanes also failed to win acceptance. For more information, see the article "Cermets" in this Volume.

To overcome brittleness, the infiltration process was applied to the production of gas turbine components. Sintered and preformed titanium carbide skeletons were infiltrated with a nickel- or cobalt-base superalloy in a vacuum. Graded products that had the following structural characteristics were produced:

- Ductile and tough roots and airfoil tips containing nearly 100% superalloy that withstood combined tensile and bend stresses at moderately high temperatures
- Ductile and tough superalloy-rich leading and trailing edges that resisted impact from small solid particles at temperatures ranging from ambient to operational
- Strong, creep-resistant airfoil portions containing 60 to 80 vol% titanium carbide that withstood the centrifugal force-induced tensile stresses at the highest temperature zones near the radial centroid of the foil, about midway between the root and tip
- Oxidation-protective, ductile, superalloy-rich airfoil encasements integrally joined with the metallic matrix

Figure 14(a) shows a graded turbine bucket for a J-47 jet engine after infiltration. Figure 14(b) shows the graded turbine bucket after machining of the root configuration. The change in microstructure across the airfoil is shown in Fig. 15 (Ref 43). Table 7 gives stress-rupture for different zones of the graded bucket (Ref 43, 49). Other infiltrated titanium carbide turbine components are shown in Fig. 16.

While these infiltrated carbide turbine components, like their sintered cermet counterparts, did not reach commercial production, blades for several stages of the compressor for a J-33 jet en-

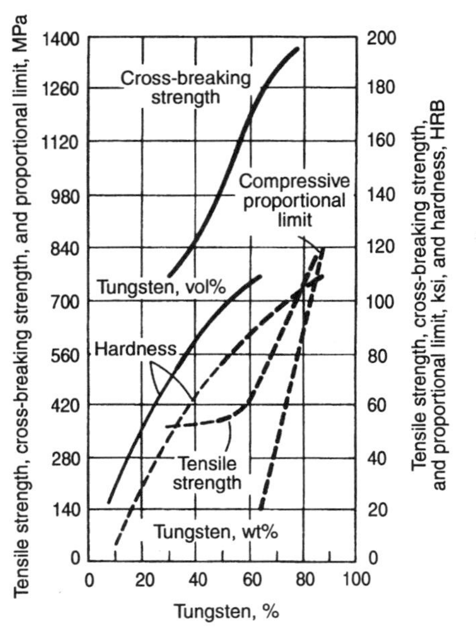

Fig. 10 Effect of composition on mechanical properties of tungsten-copper contact material. Boldface portion of curves designates liquid copper-infiltrated tungsten or sintered compacts from powder mixtures. Light portion of curves designates powder mixtures; compositions are not infiltrable. HRB. Source: Ref 39

Fig. 11 Silver-infiltrated tungsten billet for rocket nozzle throat liner. Source: Ref 42

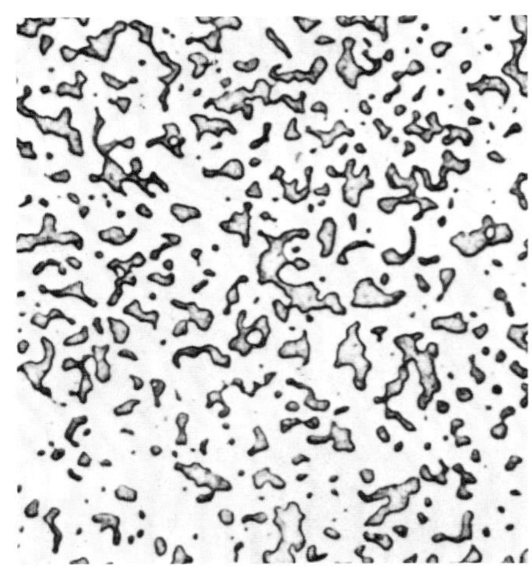

Fig. 12 Microstructure of silver-infiltrated 80W-20Ag billet. Angular microconstituents are silver. 500×. Source: Ref 42

Table 6 Physical and mechanical properties of silver-infiltrated tungsten

Determined at strain rate of 50 μm/mm/min (0.05 in./in./min) for 20 vol% Ag/80 vol% W+

Test temperature		Thermal expansion,	Thermal conductivity		Modulus of elasticity		0.2% offset yield strength		Ultimate tensile strength		Elongation in 31.75 mm	Reduction in area,	Ultimate shear strength	
°C	°F	μm/m	W/m · K	Btu·in./ft² · h · °F	GPa	ksi	MPa	ksi	MPa	ksi	(1.25 in.), %	%	MPa	ksi
25	75	...	...	...	262	38,100	...	...	510	74.0	0	0	294	42.7
260	500	1.1	206	1430	250	36,400	...	...	463	67.2	0	0	247	35.8
540	1000	2.7	138	960	193	28,000	308	44.7	331	48.0	9.3	11.2	140	20.3
815	1500	3.8	108	750	206	30,000	...	...	315	45.7	2.9	3.1	155	22.5
1095	2000	4.7	92	640	197	28,500	232	33.6	243	35.2	4.0	8.0	123	17.9
1370	2500	6.0	80	560	97.2	14,100	105	15.2	135	19.6	7.0	17.5	80.6	11.7
1650	3000	7.6	72	500	74.5	10,800	71.7	10.4	80.6	11.7	7.3	18.5	46.9	6.8
1925	3500	9.1	62	430	57.2	8,300	53.8	7.8	57.9	8.4	4.7	14.7	28.3	4.1
2215	4000	10.3	59	410	38.6	5,600	40.0	5.8	42.7	6.2	5.9	19.5	15.9	2.3

Source: Ref 46–48

gine were mass produced from copper alloy infiltrated steel compacts (Ref 50). These blades were heat treated to a yield strength of 620 MPa (90 ksi), with an elongation of 5%, and withstood three to four times as much vibration at an operating temperature of 370 °C (700 °F) as blades for these stages made from martensitic type 403 stainless steel precision forgings.

Tools. In parallel development with turbine blades, titanium carbide skeletons were infiltrated with a variety of liquid steel alloys to manufacture tools and wear-resistant parts. The infiltrant varied from simple low-carbon steel to alloy and high-speed steels (Ref 51). Angular or rounded titanium carbide grains resulted, depending on carbide content and matrix composition. Infiltrated materials can be heat treated to produce specific properties. For instance, hardness ranged from 90.1 to 90.6 HRA after water quenching and from 86.0 to 90.3 HRA after tempering. At temperatures up to 750 °C (1380 °F), hot hardness for the titanium carbide that was infiltrated with type T6 tungsten high-speed steel was equivalent to commercial grades of cemented carbides. This material had better oxidation resistance than cemented carbides up to 870 °C (1600 °F) and room-temperature transverse-rupture strengths up to 1500 MPa (220 ksi). Cutting speeds were twice those of molybdenum high-speed steel, one and one-half times those of Stellite, but only one fourth those of commercial steel-cutting grades of cemented carbides.

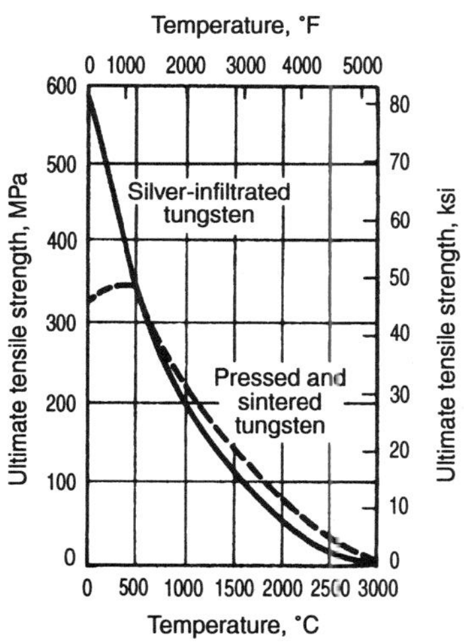

Fig. 13 Effect of temperature on tensile strength of porous uninfiltrated and silver-infiltrated (20 vol%) tungsten. Source: Ref 44

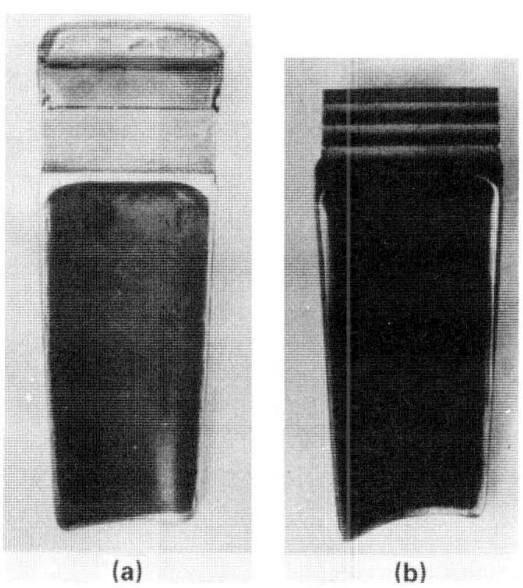

Fig. 14 Graded cermet turbine blade. (a) After superalloy infiltration of titanium carbide skeleton. (b) After machining of root

Because the solubility of titanium carbide in liquid steel is high, the rate of infiltration is slowed so that penetration of the liquid tends to be confined to the regions near the original contact face. The remaining porosity cannot be filled because of diffusion solidification, regardless of the time allowed for penetration of the liquid. As a result, severe size limitations are required for the infiltrated product to be sound and uniform in structure. This disadvantage has caused the substitution of in situ liquid-phase sintering of titanium carbide and steel powder mixtures in the production of such tool materials. For more information, see the article "Cermets" in this Volume.

Mechanical Parts. Infiltration is widely used in the production of ferrous structural parts requiring densities in excess of 7.4 g/cm^3 and mechanical properties superior to those obtained by compacting, sintering, and coining. Depending on the application, porous skeletons of iron or steel can be fully or partially infiltrated with copper alloy.

There are several advantages to infiltration of iron-base structural parts with copper alloys (Ref 52):

- *Increased mechanical properties.* Higher tensile strengths and hardnesses, greater impact energies, and fatigue strengths are obtained through infiltration. Figure 17 shows the effect of infiltration on strength.
- *Uniform density.* Parts that contain non-uniform and/or heavy sections can be infiltrated to obtain more uniform density; infiltration tends to even out density variations.
- *Higher density.* Infiltration is a useful method to increase sintered part weight without increasing the size of the part. Given the press size limitations and restrictions in pressing technique and powder compressibility, it is often easier to obtain high density through infiltration. Certainly, when considering normal P/M operations, it would be difficult to obtain densities in excess of 7.2 g/cm^3 without resorting to additional pressing and sintering operations. Infiltration makes densities in

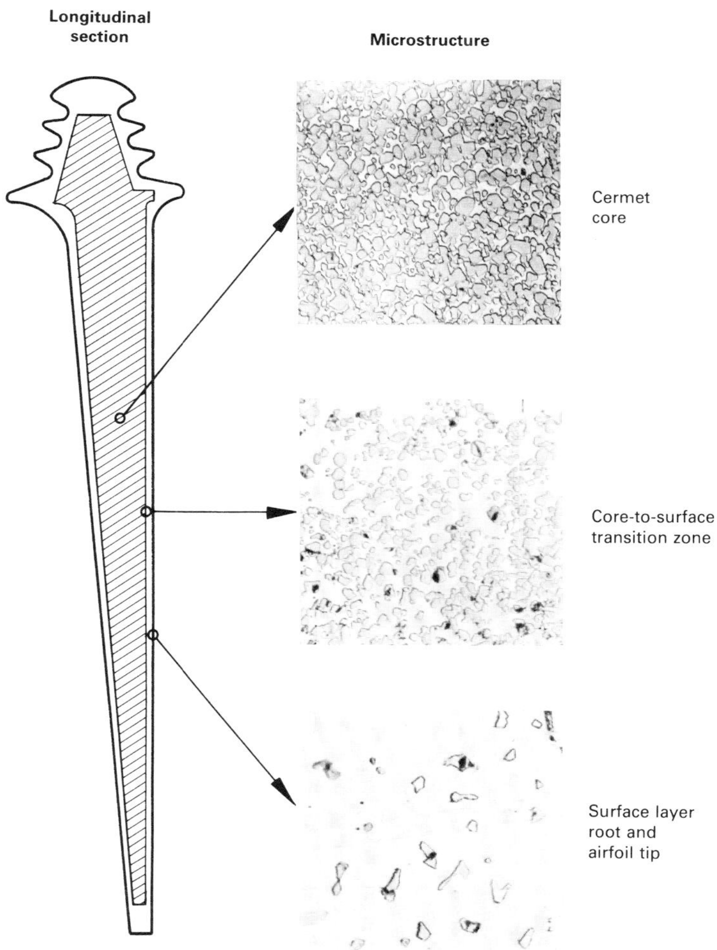

Fig. 15 Microstructure of graded turbine blade across airfoil. Angular and rounded microconstituents are titanium carbide. 150×. Source: Ref 43

excess of 7.2 g/cm^3 possible in a single pressing and sintering operation.

- *Removal of porosity for secondary operations.* Infiltration can be used in place of impregnation as a method to seal surface porosity so that secondary operations such as pickling and plating can be performed without damaging the interior of the part and creating subsequent "bleeding" problems. It is also a method of sealing a part used for applications in which no porosity is desired.
- *Selective property variation.* It is possible by infiltrating only selected areas of a part to obtain, within limits, a controlled variation of properties in the part—for example, variations in density, strength, and hardness. This is known as localized infiltration. Infiltration to considerably less than the full density in the part (e.g., 7.1 g/cm^3) is known as starve infiltration.
- *Assembly of multiple parts.* Different sections of the final part, pressed separately, can be assembled by sintering the individual pieces together and bonding the pieces into one part through common infiltration.

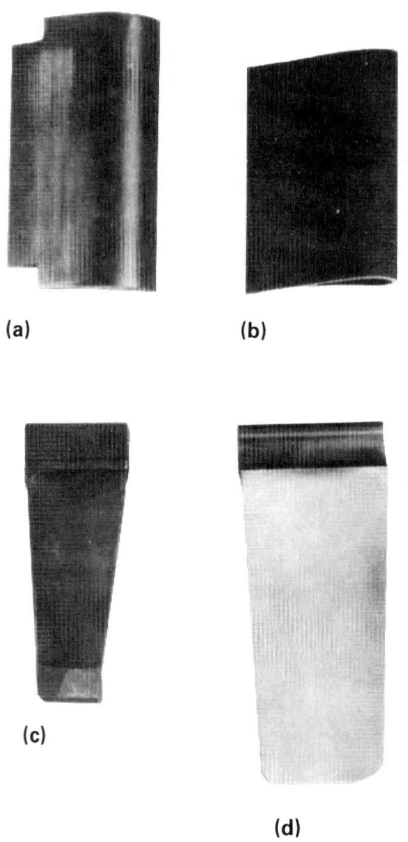

Fig. 16 Infiltrated titanium carbide cermet jet engine turbine components. (a) J-35 nozzle vane. (b) Experimental hollow vane. (c) J-57 turbine bucket. (d) J-47 turbine bucket

Table 8 gives composition and typical properties of P/M infiltrated steels.

The usual method of infiltrating iron and steel skeletons is to place a compact pressed from the powder of the infiltrant material next to the skeleton. The compact of infiltrant powder can be positioned on top or underneath the skeleton compact, or two infiltrant compacts can be used—one on top, the other underneath the skeleton compact. The exact amount of infiltrant needed can be compacted in the same die in which the powder for the porous skeleton is pressed.

After the skeleton has been sintered, the green compact or compacts of infiltrant powder are positioned next to the skeleton and the assembly heated to the infiltration temperature. Sintering of the skeleton and infiltration can be combined into one operation, in which the green compact or compacts of the infiltrant are positioned next to the green compact of the skeleton. By controlling the rate of heating, the skeleton compact will be adequately sintered by the time the melting point of the infiltrant is reached. This operation is called "sintrating."

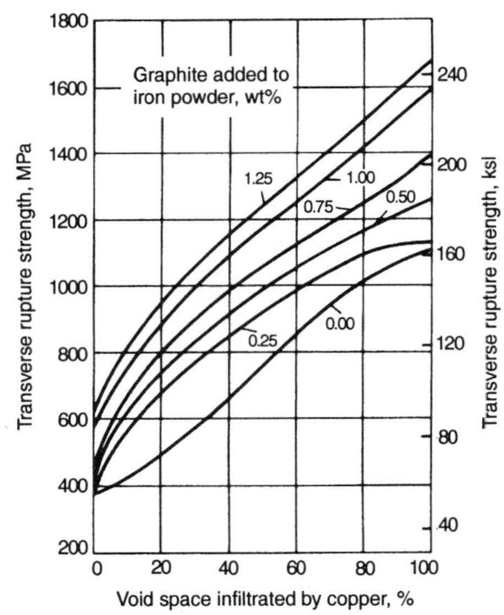

Fig. 17 Effect of infiltration on transverse-rupture strength of iron-carbon alloys sintered to a density of 6.4 g/cm^3. Combined carbon in alloys was about 80% of graphite added to iron powder; amount of copper infiltrant was adjusted to fill various fractions of void space. Source: Ref 53

If only part of the porous skeleton is to be infiltrated—for example, the teeth of an infiltrated gear—the skeleton can be positioned in a graphite container in which space is provided adjacent to the gear teeth to be preferentially infiltrated. This space is then filled with the appropriate amount of infiltrant in powder form. Because iron has some solubility in copper, the liquid infiltrant will attack the surface of the skeleton when it first comes in contact with it, and severe erosion can take place at this point. One method to minimize erosion is to use a copper alloy (e.g., 80%Cu-20%Zn brass) as infiltrant. Because the brass does not melt at one temperature, but over a range of temperatures, less erosion takes place.

Another method is to use an alloy of copper with iron as an infiltrant. However, it is difficult to specify the exact amount of iron; too little iron can cause erosion, and too much iron can cause an undesirable adherent deposit on the infiltrated part. A third method is to use an alloy of copper, iron, and a third alloying constituent, such as manganese, which oxidizes in the atmosphere in which infiltration takes place. In this case, a crust containing an oxide of the oxidizing alloying ingredient of the infiltrant is formed, which does not adhere to the skeleton but which can be more or less readily removed.

Figure 18 shows a small foot holder that required strength, machinability for tapping a thread, and plateability. These characteristics were achieved by infiltrating a sintered iron-rich skeleton (7% Cu) with a copper-rich alloy containing 5% Mn and 5% Fe. The compacts were infiltrated under furnace conditions similar to those used for sintering, using infiltrating slugs previously pressed. The infiltrated parts were electroplated with chromium to a satin finish. Processing details for the two-step procedure are given in the table that accompanies Fig. 18.

The control of dimensions during infiltration can cause problems. It depends on the composition of the steel; carbon-free iron shows the greatest growth during infiltration. It also depends on the exact temperature of infiltration and the time during which the infiltrated compact is above the liquidus temperature of the infiltrant.

Bearings. The best example for the industrial use of infiltration in the bearing field is strip-backed steel, precision-type main, and connecting rod bearings for automobiles. In this application, the anti-friction babbitt alloy is used to infiltrate the pores of a skeletal cupronickel strip that is bonded to the steel backing and produces an extra layer on the surface facing the crankpin (Ref 7, 34).

Infiltration techniques can also be applied to high-temperature bearing materials, such as bearing retainers in gas turbines for supersonic aircraft. To combat the high wear of these cages,

Table 7 Stress-rupture properties of infiltrated graded cermet bucket

Bucket section	Material	Average service temperature		Average service stress		Stress-rupture life (100 h)			
						Temperature		Stress	
		°C	°F	MPa	ksi	°C	°F	MPa	ksi
Root	Superalloy-rich region	650–760	1200–1400	140–170	20–25	705	1300	455	66
Airfoil tip	Superalloy-rich region	870–930	1600–1700	55	8	870	1600	138	20
Airfoil body	Superalloy-infiltrated TiC	1000	1800	70–100	10–15	980	1800	83	12

Source: Ref 44

an ideal structure consists of a hard, load-bearing phase intertwined with a liquid metallic lubricant. Structures that closely approach the ideal concept and operate up to 370 °C (700 °F) have been developed (Ref 53).

Materials exhibiting improved anti-friction properties against steel under simulated service conditions involving very high rotational speeds are efficiently fabricated by P/M techniques that include an infiltration step. These materials are characterized by a duplex structure (a soft, metallic phase of about 15 vol%) that is uniformly dispersed throughout the hard matrix. The matrix can be either Monel or a 48Ni-48Cr-4Si alloy, with 10 wt% molybdenum disilicide added in some cases to reduce wear. The soft phase consists of silver that is infiltrated after sintering the hard alloy powder compact.

Skeleton		Infiltrating slug	
Reduced iron	92.0%	Copper	89.0%
Copper	7.0	Manganese	5.0
Zinc stearate	1.0	Iron	5.0
		Lubricant	1.0

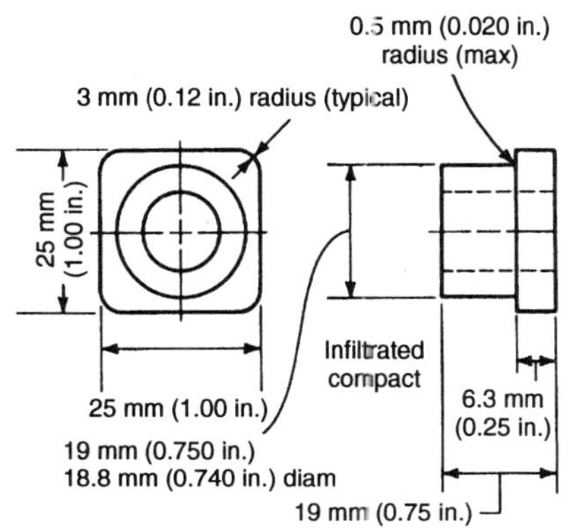

Press and tool details	
Type of press	Mechanical(a)
Press capacity	45.4 metric tons (50 tons)
Die material	Tungsten carbide (6% Co)
Punch material	D2 tool steel
Core-rod material	Tungsten carbide (6% Co)
Processing details	
Compacting pressure:	
Skeleton	415 MPa (60 ksi)
Infiltrating slugs	276 MPa (40 ksi)
Preheating treatment	15 min at 1100 °C (2040 °F)
Sintering treatment	15 min at 1100 °C (2040 °F)
Infiltrating treatment	15 min at 1100 °C (2040 °F)
Atmosphere for sintering and infiltrating	Endothermic

(a) Two movements above and three below

Fig. 18 Small compact infiltrated with copper alloy to provide machinability, strength, and plateability

Infiltration Update

Recently, significant advances have been made into the understanding and control of infiltration processes by developing models for liquid metal flow, heat transfer, wetting and spreading phenomena, and pressure effects. A short summary of such analytical and experimental treatments follows in the section "Mechanism of Infiltration." On the practical side, new ideas concerning infiltration techniques have been developed, the range of infiltration systems has extended, and, most notably, infiltration applications and final products have diversified to cover traditional and new areas. A literature survey on the latter topics is given in "Selected References" at the end of this article.

Mechanism of Infiltration

Analytical treatments to describe fluid flow and heat transfer during infiltration of porous preforms by a liquid metal have been addressed by different groups at MIT (Ref 54–61) and Colorado School of Mines (Ref 62) in the United States, also in England (Ref 63), France (Ref 64), Japan (Ref 65), China (Ref 66), and Jordan (Ref 67). General expressions have been derived to describe heat, mass, fluid flow, and kinetics during the infiltration process. A review of theoretical factors and mechanisms dealing with infiltration of a porous body by a liquid metal is given in Ref 68. Factors that control the matrix grain size, interfacial reactions, and morphology of the fiber-matrix interface and its stability have been predicted (e.g., initial fiber temperature, preform compressibility, reaction heat, external cooling, or reaction kinetics) (Ref 54, 57, 59, 70). Infiltration kinetics, size of the remelting region, and temperature distributions can be calculated (Ref 54, 58, 61, 62, 65, 70).

Models for pressure effects in conventional or new infiltration processes (squeeze casting or chemical vapor infiltration) have been developed (Ref 54, 57, 60, 61, 67, 70–80). Based on the analysis of physical phenomena that govern the infiltration process, good practical guidelines to optimize the processing techniques and the materials produced can be found in Ref 75.

The mechanism of wetting and spreading of the liquid metal has been studied (Ref 54, 78, 81, 82). These studies reveal the formation of a precursor film ahead of the bulk liquid front by adsorption, condensation, and surface migration on the solid wall (Ref 82). A model for chemical vapor infiltration is presented in Ref 83 and 84.

Sensors and models apparatuses (e.g., noninvasive capacitance technique), model experiments, and instrumented casting facilities have been developed to confirm theoretical predictions (Ref 55, 56, 58, 60, 64, 78, 81, 85–87). Modeling of magnetic field assisted and ultrasonic infiltration also has been attempted (Ref 88–90). A validation of the Lorenz force model has been experimentally obtained (Ref 88).

REFERENCES

1. P. Schwarzkopf, The Mechanism of Infiltration, *Symposium on Powder Metallurgy, 1954,* Special Report No. 58, The Iron and Steel Institute, London, 1956, p 55–58
2. K.A. Semlak, C.W. Spencer, and F.N. Rhines, Rate of Capillary Rise of Liquid Metal in a High Melting Metal Powder Compact, *Trans. AIME,* Vol 209, 1957, p 63–64
3. K.A. Semlak and F.N. Rhines, The Rate of Infiltration in Metals, *Trans. AIME,* Vol 212, 1958, p 325–331
4. W.D. Jones, *Fundamental Principles of Powder Metallurgy,* Edward Arnold, 1960, p 505–512
5. C.G. Goetzel and A.J. Shaler, Mechanism of Infiltration of Porous Powder Metallurgy Parts, *J. Met.,* Vol 16 (No. 11), 1964, p 901–905
6. A.J. Shaler, Theoretical Aspects of the Infiltration of Powder Metallurgy Products, *Int. J. Powder Metall.,* Vol 1 (No. 1), 1965, p 3–14
7. T. Kimura, J.C. Kosco, and A.J. Shaler, Detergency During Infiltration in Powder Metal-

Table 8 Composition and properties of P/M infiltrated steels

MPIF designation	Condition(a)	MPIF composition limits(b), %			Tensile strength		Yield strength		Elongation in 255 mm (1 in.), %	Impact energy(c)		Apparent hardness	Elastic modulus	
		Carbon	Copper	Iron	MPa	ksi	MPa	ksi		J	ft · lb		GPa	10^6 psi
FX-1005	AS	0.3–0.6	8.0–14.9	80.5–91.7	570	83	440	64	4.0	19	14	75 HRB	135	20
	HT	0.3–0.6	8.0–14.9	80.5–91.7	830	120	740	107	1.0	9.5	7.0	35 HRC	135	20
FX-1008	AS	0.6–1.0	8.0–14.9	80.1–91.4	620	90	515	75	2.5	16	12	80 HRB	135	20
	HT	0.6–1.0	8.0–14.9	80.1–91.4	895	130	725	105	60.5	9.5	7.0	40 HRC	135	20
FX-2000(d)	AS	0.3 max	15.0–25.0	70.7–85.0	450	65	...	...	1.0	20	15	60 HRB	...	...
FX-2005(e)	AS	0.3–0.6	15.0–25.0	70.4–84.7	515	75	345	50	1.5	12.9	9.5	75 HRB	125	18
	HG	0.3–0.6	15.0–25.0	70.4–84.7	790	115	655	95	<0.5	8.1	6.0	30 HRC	125	18
FX-2008(f)	AS	0.6–1.0	15.0–25.0	70.0–84.4	585	85	515	75	1.0	14	10	80 HRB	125	18
	HG	0.6–1.0	15.0–25.0	70.0–84.4	860	125	740	107	<0.5	6.8	5.0	42 HRC	125	18

Note: All materials have a density range of 7.2 to 7.6 g/cm^3. (a) AS, as sintered; HT, heat treated (typically austenitized at 870 °C (1600 °F)), oil quenched and tempered 1 h at 200 °C (390 °F). (b) MPIF Standards require that the total amount of all other elements be <2.0%. (c) Unnotched Charpy. (d) ASTM B 303, Class A; SAE 870. (e) ASTM B 303, Class B. (f) ASTM B 303, Class C; SAE 872.

lurgy, *Proc. of 15th Annual Meeting,* Metal Powder Industries Federation, 1959, p 56–66
8. R. Kieffer and F. Benesovsky, The Production and Properties of Novel Sintered Alloys (Infiltrated Alloys), *Berg- und Hüttenmännische Monatshefte,* Vol 94 (No. 8/9), 1949, p 284–294
9. C.G. Goetzel, Infiltration Metallurgy, *Research,* Vol 4 (No. 12), 1951, p 555–561
10. F.V. Lend, *Powder Metallurgy,* Metal Powder Industries Federation, 1980, p 313–319
11. G. Matsumura, Stress Infiltration in Two-Phase Alloys, *Planseeberi. Pulvermetall.,* Vol 8 (No. 3), 1960, p 110–118
12. W.A. Kaysser, S. Takajo, and G. Petzow, Skeleton Dissolution and Skeleton Formation During Liquid Phase Sintering of Fe-Cu, *Modern Development in Powder Metallurgy,* Vol 12, Metal Powder Industries Federation, 1981, p 473–482
13. H.W. Lavendel and C.G. Goetzel, Recent Advances in Infiltrated Titanium Carbides, *High Temperature Materials,* R.F. Hehemann and G.M. Ault, Ed., John Wiley & Sons, 1959, p 140–154
14. G. Langford, High Speed Steel Made by Liquid Infiltration, *Mater. Sci. Eng.,* Vol 28, 1977, p 275–284
15. C.G. Goetzel, *Treatise on Powder Metallurgy,* Vol 2, Interscience, 1950, p 196
16. K. Schröter, Border Regions of Metallography, *Z. Metallkd.,* Vol 23 (No. 7), 1931, p 197–201
17. J.M. Krol and C.G. Goetzel, "Refractory Metal Reinforced Super Alloys," USAF Technical Report 5892, ATI No. 57154, May 1949
18. F.V. Lend, *Powder Metallurgy,* Metal Powder Industries Federation, 1980, p 383–400
19. R. Kieffer and F. Kölbl, Production of Hard Metals by Infiltration, *Berg. Hüttenmämann. Monatsh.,* Vol 95 (No. 3), 1950, p 49–58
20. H. Baumhauer, Hard Tools and Process for Making Them, U.S. Patent 1,512,191, 1924, German Patent 443,911, 1927
21. L. Reimann, Production of Bodies from Metallic Compounds, German Patent 300,669, 1917; L. Reimann and H. Leiser, Metallic Alloy, British Patent 148,533, 1921
22. C.L. Gebauer, Process of Producing Metal Bodies, U.S. Patent 1,342,801, 1920; C.L. Gebauer, Production of a Composite Metallic Article, U.S. Patent 1,395,269, 1921
23. F.P. Peters, Cemented Steels—A New High-Strength Powder Metallurgy Product, *Materials and Methods,* Vol 23 (No. 4), 1946, p 987–991
24. E.S. Kopecki, Cemented Steels, *Iron Age,* Vol 157 (No. 18), 1946, p 50–54
25. C.G. Goetzel, Cemented Steels—Infiltration Studies with Pure Iron and Copper Powders, *Powder Metall. Bull.,* Vol 1 (No. 3), 1946, p 37–43
26. L. Northcott and C.J. Leadbeatèr, Sintered Iron-Copper Compacts, *Symposium on Powder Metallurgy,* Special Report No. 38, The Iron and Steel Institute, 1947, p 142–150
27. P. Schwarzkopf, Infiltration of Powder Metal Compacts with Liquid Metal, *Met. Prog.,* Vol 57 (No. 1), 1950, p 64–68
28. G. Stern, The Effect of Infiltration on Physical Properties of Sinterings, *Prec. Met. Mold.,* Vol 11 (No. 6), 1953, p 92–102
29. J. Schramm and A. Mohrnheim, Precipitation Hardening of Iron-Zinc and Cobalt-Zinc Alloys, *Z. Metallkd.,* Vol 39, 1948, p 71–78
30. F.R. Hensel, Treatment of Bearing, U.S. Patent 2,364,713, 1941
31. E. Fetz, Bearings from Metal Powder—A New Art, *Metals and Alloys,* Vol 8 (No. 9), 1937, p 257–260
32. E. Fetz, Manufacture of Composite Bearings, U.S. Patent 2,234,371, 1941
33. F.R. Hensel, Impregnation of Metallic Composition with Bismuth, British Patent 590,412, 1947
34. A.L. Boegehold, Copper-Nickel-Lead Bearings, *Powder Metall.,* J. Wulff, Ed., American Society for Metals, 1942, p 520–529
35. F.R. Hensel and E.I. Larson, Sintered Porous Aluminum-Base Bearings, U.S. Patent 2,418,841, 1947
36. F.R. Hensel, Method of Making Porous Bearing Surfaces, U.S. Patent 2,447,980, 1948
37. R. Kieffer and W. Hotop, *Powder Metallurgy and Sintered Materials,* Springer-Verlag, 1943, p 324–326, 329
38. F.R. Hensel, E.I. Larsen, and E.F. Swazy, Physical Properties of Metal Compositions with a Refractory Metal Base, *Powder Metall.,* J. Wulff, Ed., American Society for Metals, 1942, p 483–492
39. C.G. Goetzel, *Treatise on Powder Metallurgy,* Vol 2, Interscience, 1950, p 207–209, 216–217
40. P. Schwarzkopf, *Powder Metallurgy,* Macmillan, 1947, p 167–168
41. C.G. Goetzel, *Treatise on Powder Metallurgy,* Vol 2, Interscience, 1950, p 539–540, 634
42. C.G. Goetzel and J.B. Rittenhouse, The Influence of Processing Conditions on the Properties of Silver-Infiltrated Tungsten, *Symposium sur la Métallurgie des Poudres, Edictions Métaux* (Paris), 1964, p 279–288
43. C.G. Goetzel and H.W. Lavendel, Infiltrated Powder Components for Power Plant and Propulsion Systems, *Metals for the Space Age, Plansee Proceedings 1964,* F. Benesovsky, Ed., Springer-Verlag, 1965, p 149–162
44. C.H. Toensing and S. Zalsman, "Silver-Infiltrated Tungsten Characterization Study," Firth Sterling, Inc., 31 Dec 1962
45. J.J. Warga and R.E. Matt, Infiltration of Tungsten Using Silver, *Materials Science and Technology for Advanced Applications,* Vol 2, American Society for Metals, 1964, p 552–568
46. J.R. Kattus, Report, 5525–1428 II, Southern Research Institute, 19 Sept 1962
47. J.R. Kattus, Report, 6253–1526 VI, Southern Research Institute, 9 Dec 1963
48. J.R. Kattus, Report, 6606–1498 XI, Southern Research Institute, 31 Jan 1964
49. H.W. Lavendel and C.G. Goetzel, "A Study of Graded Cermet Components for High Temperature Turbine Applications," Report, WADC-TR 57–135, May 1957
50. G. Stern and J.A. Gerzina, Making Jet Engine Compressor Blades by Powder Metallurgy, *Iron Age,* Vol 165 (No. 8), 1950, p 74–77
51. C.G. Goetzel and L.P. Skolnick, Some Properties of a Recently Developed Hard Metal Produced by Infiltration, *Sintered High-Temperature and Corrosion-Resistant Materials Plansee Proc.,* F. Benesovsky, Ed., Pergamon Press, 1956, p 92–98
52. C. Durdaller, "Copper Infiltration of Iron-Based P/M Parts," Hoeganaes Corp., Riverton, NJ, 1969
53. J.T. Burwell, Wear Behavior of High Temperature Bearing Materials, *Prec. Met. Mold.,* Vol 14 (No. 10), 1956, p 40, 41, 87, 88, 90, 91
54. A. Mortensen, L.J. Masur, J.A. Cornie, and M.C. Flemings, Infiltration of Fibrous Preforms by a Pure Metal, Part I: Theory, *Metall. Trans. A,* Vol 20, 1989, p 2535–2547
55. J. Masur, A. Mortensen, J.A. Cornie, and M.C. Flemings, Infiltration of Fibrous Preforms by a Pure Metal, Part II: Experiment, *Metall. Trans. A,* Vol 20, 1989, p 2549–2557
56. A. Mortensen, Infiltration of Fibrous Preforms by a Pure Metal, Part III: Capillary Phenomena, *Metall. Trans. A,* Vol 21, 1990, p 2257–2263
57. A. Mortensen and V.J. Michaud, Infiltration of Fiber Preforms by a Binary Alloy, Part I: Theory, *Metall. Trans. A,* Vol 21, 1990, p 2059–2072
58. V.J. Michaud and A. Mortensen, Infiltration of Fiber Preforms by a Binary Alloy, Part II: Further Theory and Experiments, *Metall. Trans. A,* Vol 23, 1992, p 2263–2280
59. R.B. Calhoun and A. Mortensen, Infiltration of Fibrous Preforms by a Pure Metal, Part IV: Morphological Stability of the Remelting Front, *Metall. Trans. A,* Vol 23, 1992, p 2291–2299
60. V.J. Michaud, L.M. Compton, and A. Mortensen, Capillarity in Isothermal Infiltration of Alumina Fiber Preforms with Aluminum, *Metall. Trans. A,* Vol 25, 1994, p 2145–2152
61. A. Mortensen and J.A. Cornie, On the Infiltration of Metal Matrix Composites, *Metall. Trans. A,* Vol 18, 1987, p 1160–1163
62. G.P. Martins, D.L. Olson, and G.R. Edwards, Modeling of Infiltration Kinetics for Liquid Metal Processing of Composites, *Metall. Trans. B,* Vol 19, 1988, p 95–101
63. R.M.K. Young, Liquid Metal Infiltration Model of Unidirectional Fiber Preform in Inert Atmospheres, *Mater. Sci. Eng. A,* Vol 135, 1991, p 19–22
64. E. Lacoste, M. Aboulfatah, M. Danis, and F. Girot, Numerical Simulation of the Infiltration of Fibrous Preforms by a Pure Metal, *Metall. Trans. A,* Vol 24, 1993, p 2667–2678
65. M. Yokota, Basis and Application of Infiltration Techniques for Powder Metallurgy, *J. Jpn. Soc. Powder Powder Metall.,* Vol 38, 1991, p 464–471
66. L. Hu, S. Luo, and W. Huo, Determination of Threshold Pressure for Infiltration of Liquid

Aluminum into Short Alumina Fiber Preform, *Trans. Nonferrous Met. Soc. China,* Vol 6, 1996, p 133–137

67. A.E.M. Assarand and M.A. Alnimr, Fabrication of Metal Matrix Composites by Infiltration Process, Part I: Modeling of Thermodynamic and Thermal Behavior, *J. Compos. Mater.,* Vol 28, 1994, p 1480–1490
68. R. Lumpkins, Theoretical Review of the Copper Infiltration of P/M Components, *Powder Metall. Int.,* Vol 17, 1985, p 120–123
69. S.G. Warrier and R.Y. Lin, Interaction between SiC Fibers and a Titanium Alloy during Infrared Liquid Infiltration, *Metall. Trans. A,* Vol 26, 1995, p 1885–1894
70. X. Tong and J.A. Kahn, Infiltration and Solidification/Remelting of a Pure Metal in a Two-Dimensional Porous Preform, *J. Heat Transfer (Trans. AIME),* Vol 118, 1996, p 173–180
71. T.W. Cline and J.F. Mason, Squeeze Infiltration Process of Metal-Matrix Composites, *Metall. Trans. A,* Vol 18, 1987, p 1519–1530
72. S. Nourbakhsh, F.-L. Liang, and H. Margolin, Calculation of Minimum Pressure for Liquid Metal Infiltration of a Fiber Array, *Metall. Trans. A,* Vol 20, 1985, p 1861–1866
73. N.H. Tai and T.W. Chou, Modeling of an Improved Chemical Vapor Infiltration Process for Ceramic Composites Fabrication, *J. Am. Ceram. Soc.,* Vol 73, 1990, p 1489–1498
74. Z.H. Xia, Y.H. Zhou, Z.Y. Mao, and B.L. Shang, Fabrication of Fiber-Reinforced Metal Matrix Composites by Variable Pressure Infiltration, *Metall. Trans. B,* Vol 23, 1992, p 295–302
75. A. Mortensen, V.J. Michaud, and M.C. Flemings, Pressure-Infiltration Processing of Reinforced Aluminum, *JOM,* Vol 45, 1993, p 36–43
76. H. You, M.G. Bader, Z. Zhang, S. Fox, and H.M. Flower, Heat Flow Analysis of the Squeeze Casting of Metal-Matrix Composites, *Compos. Manuf.,* Vol 5, 1994 p 105–112
77. S. Long, Z. Zhang, and H.M. Flower, Infiltration and Wetting of Liquid Infiltration of Unidirectional Fiber Arrays by Squeeze Casting, *Acta Mater.,* Vol 42, 1994, p 1389–1397
78. T.R. Jonas, J.A. Cornie, and K.C. Russell, Infiltration and Wetting of Alumina Particulate by Aluminum and Aluminum-Magnesium Alloys, *Metall. Trans. A,* Vol 26, 1995, p 1491–1497
79. S. Long, Z. Zhang, and H.M Flower, Characterization of Liquid Metal Infiltration of a Chopped Fiber Preform Aided by External Pressure, Parts I–III, *Acta Mater.,* Vol 43, 1995, p 3489–3509; Vol 44, 1996, p 4233–4240
80. T. Yamauchi and Y. Nishida, Consideration on Suitable Infiltration Conditions for Molten Metal into Fibrous Preforms, *J. Jpn. Inst. Light Met.,* Vol 45, 1995, p 409–414
81. J.-H. Ahn, N. Terao, and A. Berghezan, Experimental Factors on Wetting and Infiltration Fronts in Metals, *Scr. Mater.* Vol 22, 1988, p 793–796
82. J.-H. Ahn and A. Berghezan, Scanning Electron Microscopy of Liquid Metal Infiltration of Capillaries, *Mater. Sci. Technol.,* Vol 7, 1991, p 643–648
83. R.P. Courier and S.M. Valone, Time-Dependent Solution to the Tai-Chou Chemical Vapor Infiltration Model, *J. Am. Ceram. Soc.,* Vol 73, 1990, p 1758–1759
84. T.L. Starr and A.W. Smith, 3-D Modeling of Forced-Flow Thermal-Gradient CVI for Ceramic Composite Fabrication in Chemical Vapor Deposition on Refractory Metals and Ceramics, T.M. Besmann and B.M. Gallois, Ed., *Mater. Res. Soc.,* 1990, p 55–60
85. T.R. Fletcher, J.A. Cornie, and K.C. Russell, Capacitance Technique for Studying Pressure Infiltration, *Mater. Sci. Eng. A,* Vol 144, 1991, p 159–163
86. R. Asthana and P.K. Rohatgi, Melt Infiltration of Silicon Carbide Compacts, *Z. Metallkd.,* Vol 83, 1992, p 887–892
87. Z. Zhang, S. Long, and H.M. Flower, Light Alloy Composite Production by Liquid Metal Infiltration, *Composites,* Vol 25, 1994, p 380–392
88. R.M. Andrews and A. Mortensen, Lorentz-Force-Driven Infiltration by Aluminum, *Mater. Sci. Eng. A,* Vol 144, 1991, p 165–168
89. H. Nakanishi, Y. Tsunekawa, N. Mohri, and I. Niimi, Ultrasonic Infiltration in Alumina Particle/Molten Aluminum System, *J. Jpn. Inst. Light Met.,* Vol 43, 1993, p 14–19
90. Y. Tsunekawa, H. Nakanishi, M. Okumiya, and N. Mohri, Application of Ultrasonic Vibration to Molten Aluminum Infiltration, *Key Eng. Mater.,* Vol 104–107, 1995, p 215–224

Selected References

Infiltration Techniques

- Lorentz-Force-Driven Infiltration by Aluminum, R.M. Andrews and A. Mortensen, *Mater. Sci. Eng. A,* Innovative Inorganic Composites Symp., 8–11 Oct 1990 (Detroit, MI), Vol A144, 1991, p 165–168. A new process for infiltrating fibrous preforms with liquid metal that eliminates the need for pressurization of the metal or preform chemical pretreatment is presented. The molten matrix material and preform are held in a ceramic crucible so that the melt forms an annulus around the preform. The material and preform are then subjected to an intense high frequency magnetic pulse. The eddy currents induced in the melt interact with the magnetic pulse to propel the metal into the preform at a high speed, infiltrating the preform. The infiltration length is governed by the nature and number of discharges. The predicted electromagnetic field is then used together with Ergun's equation to calculate the infiltration distance.
- The Preparation of Oxidation Resistant Silicon Carbide-Silicon Carbide Composites by Chemical Vapor Infiltration (CVI), D. Bashford, *New Materials and Their Applications 1990,* Proc. of the 2nd Int. Symp., 10–12 April 1990 (Warwick, U.K.), D. Holland, Ed., IOP Publishing, Bristol, U.K., 1990, p 239–245. Continuous fiber-reinforced ceramics need to be prepared so that they possess benign fracture characteristics to give a good balance between strength and fracture toughness. CVI is particularly suitable for preparing SiC-SiC composite shapes, provided specific attention is given to preform construction, fiber to matrix interfaces, infiltration techniques, and overall composite microstructure.
- Processing of Carbon Fiber-Reinforced Aluminum Composite Using K_2ZrF_6 Treated Carbon Fibers: A Degradation Study, S.N. Patankar, V. Gopinathan, and P. Ramakrishnan, *J. Mater. Sci. Lett.,* Vol 9 (No. 8), 1990, p 912–913. The infiltration of molten aluminum in an array of carbon fibers by application of K_2ZrF_6 onto the fiber surface is discussed. The presence of K_2ZrF_6 causes instantaneous infiltration of carbon fiber torus; however, fiber degradation occurs if the molten aluminum temperature and time of contact during infiltration are not regularly checked.
- Microstructural Evolution during the Infiltration Treatment of Titanium Carbide-Iron Composite, K.-W. Chae, D.-I. Chun, D.-Y. Kim, and Y.-J. Baik, *J. Am. Ceram. Soc.,* Vol 73 (No. 7), 1990, p 1979–1982. The microstructural evolution of grain-matrix interfaces in TiC-Fe cermet has been observed and explained in terms of equilibration reaction and interfacial energy minimization. The coherency strain energy is assumed to be the driving force for enhanced dissolution of certain specific planes, which results in a very irregular grain shape. The equilibrium shape appears after heat treatment for a long period of time.
- Synthesis of SiC Platelet Reinforced 2014 Aluminum Alloys by a Pressure Infiltration Technique, R. Asthana and P.K. Rohatgi, *J. Mater. Sci. Lett.,* Vol 10 (No. 4), 1991, p 230–234. A technique that employs countergravity pressure is used to infiltrate prepacked columns of SiC platelets. This liquid metallurgy technique is applicable to synthesize aluminum alloy composites containing relatively high (50% or more) volume fractions of platelet-shaped silicon carbide reinforcement.
- Fabrication of Cast Particle Reinforced Metals via Pressure Infiltration, E.M. Klier, A. Mortensen, J.A. Cornies, and M.C. Flemings, *J. Mater. Sci.,* Vol 26 (No. 9), 1991, p 2519–2526. A new casting process for fabrication of particle-reinforced metals whereby a composite of particulate reinforcing phase in metal is first produced by pressure infiltration is presented. This composite is then diluted in additional molten metal to obtain the desired reinforcement volume fraction and metal composition. This process produces a pore-free as-cast particulate metal matrix composite. It was demonstrated for Mg-SiC particles (3, 10, 30 μm) and compared to compocasting (which produces unacceptable porosity).
- Vapor-Phase Fabrication and Properties of Continuous-Filament Ceramic Composites, T.M. Besmann, B.W. Sheldon, R.A. Lowden,

and D.P. Stinton, *Science,* Vol 253 (No. 5024), 1991, p 1104–1109. The continuous-filament ceramic composite is becoming recognized as necessary for new high-temperature structural applications. Yet because the filaments are susceptible to damage from traditional methods for the preparation of ceramics, vapor-phase infiltration has become the fabrication method of choice. The chemical vapor infiltration methods for producing these composites are now being studied in earnest, with the complexity of filament weaves and deposition chemistry being merged with standard heat and mass-transport relationships. Two of the most influential effects on the mechanical properties of these materials are adhesion and frictional force between the fibers and the matrix, which can be controlled by a tailored interface coating. A variety of materials are available for producing these composites, including carbide, nitride, boride, and oxide filaments and matrices. SiC-based materials are by far the most advanced and are already being used in aerospace applications.

- Chemical Vapor Infiltration (CVI) of Silicon Carbide Fiber Preforms, R. Lundberg, L. Pejryd, and G. Loof, *J. Phys. (France) IV,* Vol 1 (No. C2), 1991, p 491–495. An alternative fabrication technique for ceramic matrix/ceramic fiber composites involving filling a fiber fabric preform with an Si/Si_3N_4 slurry and subsequent reaction bonding is presented. For the first processing step, chemical vapor deposition (CVD) is evaluated as a means of both binding together and protecting the fibers. SiC (Nicalon) fiber preforms consisting of through-stitched two-dimensional fabrics are infiltrated. The preform thickness is more than 4 mm, and a 3 µm TiC/TiN layer is subsequently deposited on all the fibers, even in the center of the preform after less than 8 h. The purpose of the CVI layer is not only to bind the fibers together but also to protect the fiber/matrix interfacial carbon film from reaction with the subsequently infiltrated matrix. The remaining porosity of the preform is filled with a Si/Si_3N_4 powder slurry, which is nitrided to form a reaction bonded Si_3N_4 (RBSN) matrix. The advantage of this combined CVI/RBSN route is a substantial cut in processing time compared to a process where most of the porosity is filled using CVI.
- The Fabrication of Metal Matrix Composites by a Pressureless Infiltration Technique, M.K. Aghajanian, M.A. Rocazella, J.T. Burke, and S.D. Keck, *J. Mater. Sci.,* Vol 26 (No. 2), 1991, p 447–454. A novel technique for fabricating metal matrix composites by the spontaneous (pressureless) infiltration of filter preforms with molten aluminum alloys is described. Numerous reinforcing materials, including Al_2O_3 and SiC of various configurations, such as particles, agglomerates, and fibers, have been incorporated as fillers. The effects of processing variables, such as alloy chemistry, process temperature, and filler material, on the infiltration kinetics and resultant microstructures are discussed. Comparisons with existing infiltration technology and preliminary composite properties are presented.
- High Pressure Infiltration Casting: Manufacturing Net Shape Composites with a Unique Interface, R.B. Bhagat, *Mater. Sci. Eng. A,* Vol 144, 1991, p 243–251. An overview of the major casting-based manufacturing techniques of metal matrix composites is presented (compocasting or rheocasting, squeeze casting, infiltration, investment casting, and pressure casting or squeeze infiltration casting). Attention is focused on high pressure infiltration casting, which uses a rapid application of a relatively high pressure to force-infiltrate molten metal into fiber preforms. Reaction between fiber and matrix metal is negligible, and the cast composites are free from voids, gas porosity, and shrinkage cavities.
- Pressure Infiltration Casting of Metal Matrix Composites, A.J. Cook and P.S. Werner, *Mater. Sci. Eng. A,* Vol 144, 1991, p 189–206. In pressure infiltration casting, a pressurized inert gas forces the liquid metal into a preform of reinforcement material. The method utilizes an enclosed die chamber with controlled pressurization that makes possible the use of low-strength molds with high infiltration pressures. This is an inexpensive method to produce composite materials, prototypes, and near-net shapes. The parts can be infiltrated and directionally solidified.
- Combustion Synthesis of Ceramic-Metal Composite Materials: the TiC-Al_2O_3-Al System, H.J. Feng, J.J. Moore, and D.G. Wirth, *Metall. Trans. A,* Vol 23 (No. 9), 1992, p 2373–2379. One of the main disadvantages of combustion synthesis of ceramic and composite materials is the relatively high levels of porosity (≥50%) present in the product. The article discusses a novel application of combustion synthesis for producing ceramic-metal composites with reduced levels of porosity by allowing an excess amount of liquid metal, generated by the exothermic reaction, to infiltrate the pores. This application of combustion synthesis of ceramic-metal composite materials is discussed with respect to a model reaction system that utilizes an inexpensive oxide, i.e., TiO_2, reacted with carbon and an excess stoichiometric amount of aluminum. The aluminum is in the liquid state at the ignition temperature and is intentionally allowed to infiltrate the porous ceramic matrix (i.e., TiC-Al_2O_3) produced from the combustion synthesis reaction. This in situ process for producing ceramic-metal composites by the simultaneous liquid-metal infiltration of the pores in a ceramic matrix using the combustion synthesis approach provides considerable advantages over a conventional process.
- Observations on Infiltration of Silicon Carbide Compacts with an Aluminum Alloy, R. Asthana and P.K. Rohatgi, *J. Mater. Sci. Lett.,* 1 Oct 1992, Vol 11 (No. 19), 1992, p 1278–1281. The infiltration process consists of countergravity infiltration of suitably tamped and preheated compacts of SiC platelets under an external pressure in a special pressure chamber for a set period, followed by solidification of the infiltrant metal in the interstices of the bed at atmospheric pressure. Some experimental observations are presented on the infiltration behavior and matrix microstructures that form when porous compacts of platelet-shaped single crystals of α-SiC (hexagonal) are infiltrated under gas pressure with a liquid 2014 aluminum alloy (nominal composition Al-4.5Cu-0.8Si-0.8Mn-0.5Mg).
- Consolidation of Continuous Fiber Intermetallic-Matrix Composites, S.L. Semiatin, R.L. Goetz, and W.R. Kerr, *Intermetallic-Matrix Composites II Symp.,* 27–30 April 1992 (San Fransisco, CA), D.B. Miracle, D.L. Anton, and J.A. Graves, Ed., Materials Research Society, 1992, p 351–364. Processing routes for fabrication of continuous fiber, intermetallic-matrix composites, including conventional and hot isostatic pressing of layups of matrix material (e.g., foil or powder cloth) and fiber mats, consolidation of monotapes made by techniques such as arc, plasma, vapor, or electron beam deposition or tape casting, and liquid-metal infiltration-base methods are reviewed. The advantages and disadvantages of the various methods are discussed.
- Preparation and Properties of Reinforced Mg-Li Superlight Alloys by High Pressure Infiltration in an Autoclave, A. Schweighofer, E. Hornbogen, S. Kudela, and K. Schemme, *Magnesium Alloys and Their Applications,* 8–10 April 1992 (Garmisch Partenkirchen, Germany), B.L. Mordike and F. Hehmann, Ed., DGM, 1992, p 423–430. A high pressure infiltration method is used for the production of Mg-Li matrix composites. Most favorable materials are magnesium-lithium alloys with up to 8 wt% Li combined with coated carbon fibers. Other ceramic fibers such as SiC and Al_2O_3 are also suitable but raise the bulk density much more than carbon fibers.
- Near-Net Shape Long Fiber-Reinforced Intermetallic-Matrix Composites Produced by Reactive Infiltration Process, H. Chen, M. Kaya, and R.W. Smith, Ed., *Mater. Lett.,* Vol 13 (No. 4–5), 1992, p 180–183. A reactive melt infiltration technique is utilized for the processing of intermetallic (NiAl and Ni_3Al) matrix composites. The intermetallic compound matrix can be formed by in situ reaction during the flow of molten metal or alloy into a nickel-powder preform containing reinforcements. With this method, the macrosegregation of nickel can be controlled to produce materials with varying microstructural zones, which can enhance high-temperature mechanical properties.
- Discussion of a Liquid-Metal Pressure Infiltration Process to Produce Metal Matrix Composites (MMCs), J.T. Blucher, *J. Mater. Process. Technol.,* Vol 30 (No. 3), 1992, p 3821–3900. The two major processing techniques to manufacture MMCs are solid state diffusion bonding and squeeze casting. The

former has size limitations; it is suitable for flat sheets or plates, and the manufacturing rate is low and expensive. The latter, although suitable for large production of complex shapes, requires heavy and expensive tooling. The described pressure infiltration process was developed at the Massachusetts Institute of Technology. In this process, fiber, whisker, or particulate preforms are placed in inexpensive disposable containers, into which gas pressure forces the molten matrix material. The gas pressure quasi-isostatic, thus reducing the strength requirements for containers and making the process inexpensive. As the processing parameters, such as preform and melt temperatures, infiltration pressure, and cooling rate, can be accurately controlled, the process is particularly suitable for research and limited production.

- An Advanced Melt Infiltration Process for Net Shape Production of Metal Matrix Composites, O. Ottinger and R.F. Singer, *Z. Metallkd.*, Vol 84 (No. 12), 1993, p 827–831. An alternate process to squeeze casting to produce fiber-reinforced light metals for the automotive industry is proposed. This new process is expected to have advantages in certain cases, such as low volume production, intricate shapes, and low heat capacity metals. First results for the magnesium alloy MSR reinforced with 20 vol% Saffil short fiber are presented, indicating that mechanical properties are equivalent to composites obtained by squeeze casting.
- Graded Compositions and Microstructures by Infiltration Processing, B.R. Marple and D.J. Green, *J. Mater. Sci.*, Vol 28, 1993, p 4637–4643. Mullite/alumina particulate composites are fabricated by infiltrating porous alumina preforms with an SiO_2-containing sol, followed by a heating step for mullite formation and densification. Electron microprobe analysis is performed to obtain concentration profiles across sections of the sintered composites. These profiles indicate concentration gradients with mullite content decreasing with increasing distance from the surface of the bodies. An increase of the alumina grain size is also noticed in the same direction. These two effects (microstructural and compositional) have been related, and it has been concluded that while the mullite presence limits grain growth in alumina, the mullite content has to be at least ~5 wt% in order for grain growth to occur in a controlled fashion.
- Processing of Molybdenum Disilicide Using a New Reactive Vapor Infiltration Technique, N. Patibandla and W.B. Hillig, *J. Am. Ceram. Soc.*, Vol 76, 1993, p 1630–1634. In this new process, a loosely compacted molybdenum powder is exposed to a gaseous silicon precursor. Initially, a surface $MoSi_2$ layer forms that subsequently progresses inward. At 1200 °C, the silicide layer grows at a rate of about 20 µm/h. No excess silicon or molybdenum is present, unlike in chemically vapor deposited $MoSi_2$. Irrespective of the volumetric increases involved in siliciding molybdenum to $MoSi_2$, none of the samples silicided between 1100 and 1300 °C showed swelling or surface cracks.
- Reactive Melt Infiltration of Silicon-Niobium Alloys in Microporous Carbon, M. Singh and D.R. Behrendt, *J. Mater. Sci.*, Vol 9 (No. 7), 1994, p 1701–1708. Reactive melt infiltration of silicon-niobium alloys in microporous carbon preforms prepared by the pyrolysis of a polymer precursor is studied by modeling, differential thermal analysis (DTA), and melt infiltration experiments. Mercury porosimetry results indicate a very narrow pore size distribution with all open pores. The morphology of the residual phases ($NbSi_2$ and Si) in the infiltrated material can be tailored by the control of pore size and distribution of the carbon preform and alloy composition. The average room temperature four-point flexural strength of a reaction formed SiC (made by infiltration of medium pore size preforms with Si-5at%Nb) is 290 ± 40 MPa (42 ± 6 ksi), and the fracture toughness is 3.7 ± 0.3 MPa$\sqrt{m}$. The flexural strength decreases at high temperature due to relaxation of the residual thermal stresses and the presence of free silicon in the material.
- Silicon-Aluminum Network Composites Fabricated by Liquid Metal Infiltration, Y.Y. Chen and D.D.L. Chung, *J. Mater. Sci.*, Vol 29 (No. 23), 1994, p 6069–6075. A new method for fabricating interpenetrating silicon-aluminum network metal matrix composites is presented. The method involves infiltration of an aluminum-silicon alloy (Al-12Si-1Mg or Al-30Si-1Mg) liquid into a SiC (50 vol%) preform. The silicon particles are partially dissolved by the liquid alloy and contribute to the formation of a silicon network after solidification. The composites are metallurgically sound with no porosity. The thermal expansion coefficient is low (7.7×10^{-6} at 50–100 °C) with compressive strength up to 580 MPa, tensile strength up to 160 MPa, and Vickers hardness up to 390.
- Application of Ultrasonic Vibration to Molten Aluminum Infiltration, Y. Tsunekawa, H. Nakanishi, M. Okumiya, and N. Mohri, *Key Eng. Mater.*, Vol 104–107, 1995, p 215–224. A new method of infiltration at a very low applied pressure has been developed. Ultrasonic vibration has been applied to improve the wettability of Al_2O_3 particle or fibers by molten aluminum. This ultrasonic vibration allows infiltration to take place at a very low pressure. At the same time, noninfiltrated defects are minimized. The method has been successfully applied for the fabrication of in situ metal matrix composites starting with titanium powders and Al_2O_3.
- Development of the Technique of Extrusion Directly Following Infiltration for the Manufacturing of Metal-Matrix Composites, L.X. Hu, S.J. Luo, W.C. Huo, and Z.R. Wang, *J. Mater. Process. Technol.*, Vol 49, 1995, p 287–294. A new technique that involves extrusion directly after infiltration is aimed at providing a good bond between the matrix and the reinforcement and eliminating the voids. The new method combines melt infiltration, squeeze casting, and semi-solid extrusion and is particularly suitable for the manufacturing of bars, pipes, plates of fiber, or particulate MMCs.
- Reactive Melt Infiltration of Silicon-Molybdenum Alloys into Microporous Carbon Preforms, M. Singh and D.R. Behrendt, *Mater. Sci. Eng. A*, Vol 194 (No. 2), 1995, p 193–200. Reactive melt infiltration of Si-1.7at%Mo and Si-3.2at%Mo alloys into microporous carbon preform is studied by modeling, differential thermal analysis (DTA), and melt infiltration experiments. These results indicate that pore volume fraction of the carbon preform is an important parameter for the final composition of the reaction formed SiC and secondary phases. The liquid silicon-carbon reaction exotherm temperatures are influenced by pore and carbon particle size and the compositions of infiltrants. Room temperature flexural strength and fracture toughness of Si-3.2at%Mo alloy infiltrated into medium ore size preforms are presented. Various undesirable results (e.g., choking-off, specimen cracking, silicon veins, and lake formation) are presented in relationship to inadequate preform properties.
- Fabrication of Intermetallic Compound Matrix Composite by Spontaneous Infiltration and Subsequent In Situ Reaction Process, T. Choh, T. Mohri, and M. Kobashi, *J. Mater. Process. Technol.*, Vol 63 (No. 1–3), 1997, p 379–383. Titanium aluminides intermetallic compound matrix composites were in situ synthesized by spontaneous infiltration of liquid aluminum into a powder mixture of titanium and Al_2O_3.
- Fabrication of Al_2O_3 Composites by Reactive Melt Infiltration, W.P. Tai, *Am. Ceram. Soc.*, Vol 76 (No. 4), April 1997, p 86–89. Near-net shape Al_2O_3 composites containing a small amount of silicon metal are fabricated by infiltrating aluminum into Al_2O_3 particle preforms with the aid of borosilicate glass. The process is conducted under oxygen atmosphere at ambient pressure (1 atm) using a conventional furnace. The growth rate and mechanical properties of the composites are investigated by changing the processing factors.

Infiltration Systems

- Wetting and Infiltration of Porous Tungsten and Molybdenum by Liquid Gallium, A.V. Ivanov, T.S. Ivanova, L.I. Kostenetskaya, Y.I. Smirnov, and M.M. Churakov, *Sov. Powder Metall. Met. Ceram.*, Vol 23 (No. 56–58), 1984. Electrical contacts with a solid skeleton of tungsten or molybdenum have been produced by liquid gallium infiltration. Specimens are compacted under a pressure of 390 MPa and sintered for 2 h in hydrogen at 1400 and 1200 °C, respectively, with resulting porosities of 38 and 44%. Wetting experiments are carried out in helium, hydrogen, or vacuum. An intermetallic compound and sintered moldybdenum layers that formed at 600 °C prevented gallium infiltration.

- Kinetic Laws for Infiltration of Porous Titanium by Lead and Indium Melts, A.A. Kurilko, G.A. Kurshev, V.A. Rudyuk, and Y.V. Naidich, *Sov. Powder Metall. Met. Ceram.*, Vol 23, 1984, p 686–689. This process is similar to active soldering of nonmetallic oxide materials. A molten solder (e.g., tin, lead, or indium) infiltrates and spreads over its surface. The work is aimed at studying the kinetics of infiltration of titanium powders by two metallic melts: indium and lead.

Metal Matrix Composites (MMC)

- Casting Particulate and Fibrous Metal Matrix Composites by Vacuum Infiltration of a Liquid Metal Under an Inert Gas Pressure, J. Yang and D.D.L. Chung, *J. Mater. Sci.*, Vol 24 (No. 10), 1989, p 3605–3612. MMCs are fabricated by a combination of three techniques: vacuum infiltration, infiltration under an inert gas pressure, and squeeze casting. The particulate or fiber preform is placed in a mold with the matrix alloy above the preform. A gas under pressure (7 to 17.3 MPa or 1000 to 2500 psi) forces the melt to infiltrate the preform. The temperature and pressure are below those used in squeeze casting, thus minimizing the interfacial reactions. The method has been used to fabricate aluminum-matrix composites with short ceramic or continuous ceramic fibers (e.g., SiC whiskers) and particle reinforcements (e.g., SiC, Al_2O_3, graphite flakes).
- Infiltration of Aluminum into Silicon Carbide Compacts, P.B. Maxwell, G.P. Martins, D.L. Olson, and G.R. Edwards, *Metall. Trans. B*, Vol 21, 1990, p 475–485. Although liquid-metal processing of MMCs offers economic advantages, problems related to the nonwetting nature of the ceramic discontinuous reinforcement create obstacles to its ready implementation. Infiltration can occur only if a threshold pressure is applied to overcome the unfavorable interfacial forces in the system. Experiments were carried out to infiltrate SiC compacts with pure aluminum, Al-1wt%Mg and Al-1wt%Si. An incubation time was found to be necessary before infiltration can proceed, even though the threshold pressure is exceeded. It is suggested that the mechanism responsible for the incubation phenomenon might be related to a surface modification produced by either reaction of liquid aluminum with an oxide film on the surface of the particles or coverage of the surface by a capillarity-induced aluminum condensate.
- Ceramic Metal Composite Produced by Melt Infiltration, C. Toy and W.D. Scott, *J. Am. Ceram. Soc.*, Vol 73 (No. 1), 1990, p 97–101. A new ceramic-metal composite with continuous interconnected phases is fabricated from sintered AlN infiltrated with aluminum metal. A dense, lightweight, hard material with high strength is obtained. Advance of the infiltration front is linear with time. The activation energy for the process is high (330 kJ/mol) suggesting a chemical reaction rather than viscous flow as the rate controlling mechanism. The infiltration rate is inversely proportional to the pore size. The thermal and mechanical properties of the composite are promising.
- Interface Studies in Cast Fiber-Reinforced Metals, V.D. Scott and Y. Ming, *Advanced Structural Inorganic Composites, Proc. of the Satellite Symp. 2 on Advanced Structural Inorganic Composites of the 7th Int. Meeting on Modern Ceramics Technologies,* 27–30 June 1990 (Montecatini Terme, Italy), P. Vincenzini, Ed., Elsevier, 1991, p 95–108. The microstructures of a number of composite materials manufactured by liquid-metal infiltration of a fiber preform are described, with particular reference to the fiber/matrix interface. Examples are given whereby the interface is related to the type of bond developed (i.e., physical or chemical). The complexities introduced by the presence of matrix second phases, including impurities, are mentioned. The results, as well as demonstrating the importance of carrying out full and detailed microstructural studies of the fiber/matrix interface, emphasize the need to explore and to assess every aspect of composite structure.
- Directed Metal Oxidation and Pressureless Metal Infiltration: New Technologies for the Fabrication of Reinforced Ceramics and Metals, C.R. Kennedy, *Advanced Structural Inorganic Composites, Proc. of the Satellite Symp. 2 on Advanced Structural Inorganic Composites of the 7th Int. Meeting on Modern Ceramics Technologies,* 27–30 June 1990 (Montecatini Terme, Italy), P. Vincenzini, Ed., Elsevier, 1991, p 691–700. Key features and advantages of new technologies for fabricating reinforced ceramics and metals are reviewed. Microstructures and properties of composites tailored for a variety of applications are presented.
- Magnesium-Base Hybrid Composites Prepared by Liquid Infiltration, J. Schroeder and K.J. Kainer, *Mater. Sci. Eng. A*, Vol 135 (No. 1–2), 1991, p 33–36. Liquid infiltration of ceramic fiber preforms is an economical alternative to produce magnesium matrix composites with discontinuous fibers. The high cost stems from the preform and reinforcements (short fibers or whiskers).
- Two-Dimensional SiC/SiC Composites Processed According to the Isobaric-Isothermal Chemical Vapor Infiltration Gas Phase Route, R. Naslain, *J. Alloy. Compd.*, Vol 188, 1992, p 42–48. Chemical vapor infiltration (CVI) is a process according to which a ceramic material is deposited from a gaseous precursor in the open pores of a heated substrate. It has been used to elaborate ceramic matrix composites, starting from porous fiber preforms. SiC/SiC composites consist of an SiC CVI-matrix reinforced with Si-C-O expolycarbosilane yarn fibers. These composites exhibit nonbrittle mechanical behavior in tension loading when a thin layer of compliant interphase material (turbostratic pyrocarbon or hexagonal boron nitride) with a low shear failure stress is deposited onto the fibers prior to infiltration of the SiC-matrix. When exposed to oxidizing atmospheres, SiC/C/SiC composites are self-healing if their carbon-interphase is thin enough and the temperature is high. Their oxidation resistance can be improved by replacing the vulnerable carbon-interphase by a BN-interphase (low temperature exposures) and/ or adding a glass-former coating on the external surface of the composites.
- Effect of Pore Size on the Infiltration Kinetics of Aluminum in Titanium Carbide Preforms, D. Muscat, R.L. Harris, and R.A. Drew, *Acta Metall. Mater.*, Vol 42, 1994, p 4155–4163. Infiltration by capillarity is a simple and attractive way of fabrication metal/ceramic composites. Four different TiC preforms of approximately 58% density with different pore sizes ranging from 1 to 30 μm are produced. They are suspended in a thermogravimetric analyzer (TGA) and infiltrated with molten aluminum under flowing argon. Infiltration profiles are obtained by continuously monitoring the weight change of the preform at temperatures from 860 to 1300 °C. Kinetic analysis of the infiltration profiles yields activation energies ranging from 105 to 450 kJ/mol, depending on pore size. It is suggested that the driving mechanism changes from a diffusion controlled process for smaller pore size to one that is driven by a surface reaction at the interface of larger pores.
- Reactive Infiltration Processing of SiC-Metal and SiC-Intermetallic Composite, L. Hozer and Y.M. Chiang, *J. Mater. Res.*, Vol 11 (No. 9), 1996, p 2346–2357. Reactive infiltration is used to infiltrate carbon with silicon-aluminum and silicon-copper melts to prepare composites consisting of interpenetrating networks of β-SiC and a seconday phase assemblage containing ductile metal (silicon-aluminum) or a metal silicide (silicon-copper). The mechanisms of phase formation are characterized. A rapid initial reaction upon infiltration forms a largely stationary SiC network, within which the secondary phase evolves due to solute rejection and liquid phase diffusion processes. Both homogeneous composites and controlled composition gradients are synthesized.
- Fabrication and Analysis of an In Situ TiB_2Al Composite by Reactive Spontaneous Infiltration, M. Kobashi and T. Choh, *Scr. Mater.*, Vol 34 (No. 8), April 1996, p 1257–1265. An in situ reactive spontaneous infiltration is used to fabricate an Al-$MgTiB_2$, AlN composite using TiN, TiC_xN_{x-1}, boron powders and aluminum-magnesium ingot as starting materials. An experimental setup is also employed in order to study the infiltration process by using a precision multirecorder, which continuously moni-

tors the temperature change in the powder mixture of TiN or TiC_xN_{x-1} and B as the metal infiltrated.

Particulate MMC

- The Structure and Properties of Infiltrated Titanium Carbide Cemented Carbide, R.F. Mamleev and V.N. Antsiferov, *Sov. Powder Metall. Met. Ceram.*, Vol 29 (No. 2), 1990, p 153–156. The influence of the titanium carbide skeleton porosity and its temperature of infiltration with a multialloyed (9.5% Co, 9.5 Cr, 2.0 Mo, 10.0 W, 2.6 Ti, 5.4 Al, 0.3% Si) nickel alloy (SNS) on the structure and mechanical properties of the composite created at temperatures up to 1150 °C was studied. TiC powder to TU 6-09-492-75 with an average grain size of 3 to 4 μm and individual aggregates up to 20 μm is used to produce the skeletons. The powder is pressed with additions of 5% plasticizer under a pressure of 10 to 400 MPa and sintered in vacuum with a residual pressure of less than 10^{-2} Pa at 1400 to 1800 °C for 0.5 h. The infiltration is done by application of the metal to the skeleton in vacuum at 1450 to 1600 °C for 0.5 h. The yield strength in compression is established and the bend strength is determined by the three-point bend method.
- Pressureless Sintering of SiC-Whisker-Reinforced Al_2O_3 Composites II: Effects of Sintering Additives and Green Body Infiltration, H.-W. Lee and M.D. Sacks, *J. Am. Ceram. Soc.*, Vol 73 (No. 7), 1990, p 1894–1900. Pressureless sintering of SiC-whisker-reinforced Al/Al_2O_3 composites is investigated. The effects of Y_2O_3/MgO sintering additives and green body infiltration on the densification behavior and microstructure development are reported. Both sintering additives and green body infiltration result in enhanced densification. However, the infiltration approach is more effective for samples with high SiC whisker concentrations. Samples with 27 vol% whiskers could be pressureless sintered to approximately 93% relative density and approximately 3% open porosity. Fracture toughness values and microstructural features (e.g., grain size) for the infiltrated samples remained approximately the same as observed in the uninfiltrated samples.
- Squeeze Infiltration—A Potential Process to Develop Aluminum Ceramics Particulate Composites, R.S. Solanki, A.K. Sing, K. Basu, and C.B. Raju, *Advanced Structural Inorganic Composites, Proc. of the Satellite Symp. 2 on Advanced Structural Inorganic Composites of the 7th Int. Meeting on Modern Ceramics Technologies,* 27–30 June 1990 (Montecatini Terme, Italy), P. Vincenzini, Elsevier, 1991, p 747–756. All recent effort on metal matrix composites (MMCs) has considered the light metal, especially aluminum matrix composites (AMCs). The present investigation deals with the development of aluminum ceramic particulate composite using a low cost squeeze infiltration process. The defect-free AMCs were produced by a squeeze infiltration of suitably treated ceramic particulate porous preform by application of pressure up to 150 MPa. These AMCs were characterized by optical and scanning electron microscopes. Improvement in hardness value over AMCs produced by the Vortex method was observed.
- Properties of TaC-Based Metal-Matrix Composites Produced by Melt Infiltration, K. Shanker, L.T. Mavropoulos, R.A.L. Drew, and P.G. Tsantrizos, *Composites,* Vol 23 (No. 1), 1992, p 47–53. TaC/metal-matrix composites were developed for use as electrodes in plasma systems; in addition, they were observed to possess excellent mechanical properties. Composites of TaC/Al, formed by infiltrating TaC preforms with pure aluminum under optimal infiltration conditions of 1500 °C for 0.5 h, had yield strengths of approximately 100 to 125 MPa, ultimate tensile strengths of approximately 440 MPa, and elongations of 7% while containing 48 to 63% TaC. The composite consists of discrete TaC particles in a continuous aluminum matrix. Increasing the metal infiltration temperature results in a rapid decrease in the tensile properties of these composites. Composites of TaC/Cu were not as easy to fabricate as those of TaC/Al. Infiltration without distortion or cracking was feasible over a narrow TaC range (43 to 50%) and required the infiltration to be carried out in vacuum. Under these conditions, the composites had a yield strength of 90 to 135 MPa, ultimate tensile strength of approximately 760 MPa, and elongation of approximately 15.3%. All the composites had excellent machinability and could be machined using conventional methods and equipment.
- Al/TiC Composites Produced by Melt Infiltration, D. Muscat, K. Shanker, and R.A.L. Drew, *Mater. Sci. Technol.,* Vol 8 (No. 11), 1992, p 971–976. An Al/TiC composite has been developed using a melt infiltration technique. Green preforms of TiC were partially sintered to densities ranging from 50 to 85%. They were then infiltrated with molten aluminum in argon at atmospheric pressure at temperatures ranging from 950 to 1350 °C. The variations in mechanical properties of this composite material with ceramic content and fabrication conditions are presented and the resulting microstructures are discussed. The composite exhibited excellent tensile strength values of approximately 475 MN/m^2 and up to 5% elongation. Fracture behavior mechanisms are also described in relation to the stress-strain curves. High infiltration temperatures were found to be detrimental to the mechanical properties because the TiC particle network, which formed during sintering, was broken down by the metallic phase.
- Melt Infiltration of Silicon Carbide Compacts, R. Asthana and P.K. Rohatgi, *Z. Metallkd.,* Vol 83 (No. 12), 1992, p 887–892. Counter-gravity, pressure-assisted infiltration with a 2014 aluminum alloy of suitably tamped porous compacts of platelet-shaped single crystals of α (hexagonal) SiC was used to measure particulate wettability and infiltration kinetics under dynamic conditions relevant to pressure casting of composites. A threshold pressure (p_{th}) for ingression of the infiltrant was identified based on the experimental penetration length versus pressure profiles for a range of experimental variables that include infiltration pressure, infiltration time, SiC size, and SiC surface chemistry (plain, copper coated, and nickel coated). The results showed that p_{th} decreased whereas the penetration length increased with increasing SiC size and infiltration time. Copper coated SiC led to lower p_{th} (and lower estimated wetting angles) and larger (60 to 80%) penetration lengths compared to uncoated SiC under identical conditions. These observations have been discussed in the light of theoretical models of infiltration and the kinetics of wetting.
- Infiltration Processing of Metal-Matrix Composites: A Review, R. Asthana, P.K. Rohatgi, and S.N. Tewari, *Process. Adv. Mater.,* Vol 2 (No. 1), 1992, p 1–17. An overview of the current understanding of some fundamental materials phenomena relevant to infiltration processing of metal-matrix composites (MMCs) is presented. The fundamental aspects of wettability, chemical reaction-induced interfacial phenomena, capillarity, infiltration kinetics and fluid flow, and structure evolution in infiltrated MMCs are discussed in light of results of recent experimental and theoretical studies on some commercially important MMCs, such as Al_2O_3-Al and SiC/Al. The lack of a comprehensive theoretical framework suitable for rationalizing observations on processing and microstructural aspects of infiltrated MMCs is emphasized and potential areas for further research effort have been identified.
- Threshold Pressure for Infiltration in Mica Ceramic Particle/Aluminum Composite, H. Nakanishi, Y. Tsunekawa, M. Okumiya, M. Higashi, and M. Kanatani, *J. Jpn. Inst. Light Met.,* Vol 42 (No. 2), 1992, p 92–97. Threshold pressure method was used to evaluate wettability of mica-ceramic particles by molten aluminum. Three ceramic particle sizes were used (L, M, S). Pressure infiltration of the particle preforms were performed at 998 K in vacuum and air. The threshold pressures were found to be 80.9 (L), 77.8 (M), and 159 (S) in vacuum and 96.9 (L), 96.1, (M) and 162 (S) kPa in air. The lower threshold pressure in vacuum is attributed to the better wettability. Noninfiltrated defects with a meniscus configuration were observed in the vicinity of two-particle contact. Flat defects were noticed in L preforms and aggregate defects in S preforms. Modeling of defects with meniscus configuration has been performed, and a contact angle between mica-ceramic and molten aluminum of 2.39 rad was calculated. The relationship between the threshold pressure and contact angle is dependent on defect configuration.

- Diamond/Aluminum Metal Matrix Composites Formed by Pressureless Metal Infiltration Process, W.B. Johnson and B. Sonuparlak, *J. Mater. Res.*, Vol 8 (No. 5), 1993, p 1169–1173. The diamond particles were coated with SiC prior to pressureless infiltration to prevent the formation of Al_4C_3 compound. The measured thermal conductivity of the initial diamond-aluminum composite is as high as 259 W/m · K. The effects of coating thickness on the physical properties of the composite, including Young's modulus, 4-point bend strength, coefficient of thermal expansion, and thermal conductivity, are presented.
- Sintering by Infiltration of Loose Mixtures of Powders: a Method for Metal Matrix Composite Elaboration, V. Constantinescu, R. Orban, and H. Colan, *J. Phys.*, Vol 3 (No.7), 1993, p 1781–1786. A comparison between infiltration of loose powder mixtures and the classical melt infiltration is carried out. Factors that prevent excessive hydrostatic flow of the melt and, consequently, dispersion of the reinforcing particles and that promote wettability in both infiltration and liquid phase sintering were identified. Results for fusion tungsten carbide and diamond reinforced metal matrix composite are presented.
- Effect of Ceramic Particle Size, Melt Superheat, Impurities, and Alloy Conditions on Threshold Pressure for Infiltration of SiC Powder Compacts by Aluminum-Based Melts, S.Y. Chong, H.V. Atkinson, and H. Jones, *Mater. Sci. Eng. A*, Vol 173 (No. 1–2), 1993, p 233–237. An instrument heated pressure vessel was used to determine the effect of some experimental variables on the pressurized infiltration of SiC powder preforms by a series of aluminum-base melts. Threshold pressure for infiltration of 2014 aluminum alloy decreased with increasing SiC particle size and increasing melt superheating. Impurities in pure aluminum and copper addition (4.2 wt%) to 99.999% pure aluminum increased the threshold pressure, while the addition of an extra 1.0 wt% Mg to 2014 alloy lowered it. Superheating the melt reduced the incidence of entrapped porosity in compacts that have a tendency to increase in the infiltration direction. Partial SiC dissolution was noted during infiltration.
- Zirconia-Alumina Particulate Composites by Infiltration Processing, S. Riou, F. Queyroux, and P. Boch, *Ceram. Int.*, Vol 21 (No. 5), 1995, p 339–343. ZrO_2/Al_2O_3 particulate composites were prepared by infiltration of presintered alumina skeletons with a precursor of zirconia and subsequent sintering at 1600 °C. The composite specimens exhibit a gradient in composition; the zirconia content being higher in the superficial zones than in the core. The influence of the ZrO_2 content on the volume fraction of t-ZrO_2, the t to m transformation temperature, and the microstructural development were studied.
- Particulate Composites in the Al_2O_3-SiO_2-TiO_2 System by Infiltration Processing, S.J. Li, F. Queyroux, and P. Boch, *J. Eur. Ceram. Soc.*, Vol 13 (No. 1), 1994, p 3–9. Particulate composites in the Al_2O_3-SiO_2-TiO_2 system were prepared by infiltration of presintered alumina skeletons with precursors of silica and titania and subsequent reaction sintering at temperatures ≤1600 °C. The microstructure of composites is finer than that of the corresponding "pure" alumina. It is constituted of alumina grains surrounded by a mullitic, continuous phase.
- Fabrication of Particulate Aluminum-Matrix Composites by Liquid Metal Infiltration, S.W. Lai and D.D.L. Chung, *J. Mater. Sci*, Vol 29 (No. 12), 1994, p 3128–3150. The technology was developed to infiltrate high volume fraction (up to 75%) particulate preforms (AlN, SiC, Al_2O_3) with molten aluminum. A 0.1 wt% acid phosphate binder (with P/Al molar ratio of 23) was used for the preform, in contrast to higher binder amounts necessary for a whisker composite. After infiltration by a slurry that consists of the reinforcing particles, binder, and carrier (preferably acetone), a baking at 200 °C was applied for drying. Baking in air at 500 °C instead of 200 °C caused oxidation of AlN and subsequent decrease of composite thermal conductivity. The binder reinforcement reactivity was larger for AlN than SiC, but it did not affect overall composite properties due to the small amount of binder used. The best thermal conductivity and ductility were obtained in Al/AlN, and the poorest were obtained in Al/Al_2O_3 (due to Al_2O_3 particle clustering).
- Pressureless Infiltration of Aluminum Metal Matrix Composites, Y. Kajikawa, T. Nukami, and M.C. Flemings, *Met. Mater. Trans.*, Vol 26 (No. 8), 1995, p 2155–2159. The ceramic preforms are SiC with varying amounts of particulate aluminum, titanium, and nickel. The infiltrants used are aluminum and Al-12.5%Si. It is shown that a pressure differential within the preform is required for infiltration; measurements are made of pressure changes in the preform during infiltration. Results indicate that atmospheric pressure is essential for infiltration, but that capillarity can play a role as well.
- Synthesis of Composite Materials by Infiltration of IVa, Va Metals and BN Powder Mixtures with Molten Aluminum, M. Kobashi, N. Tonokura, and T. Choh, *J. Jpn. Inst. Light Metals*, Vol 46 (No. 12), 1996, p 638–643. Ceramic phases were in situ synthesized by the reaction: $2Al + 2BN + Me = MeB_2 + 2AlN$, where Me is a metal (Ta, Nb, Zr, or Hf). Each of these metals was mixed with BN powder. The spontaneous infiltration occurred with a 3600 s hold at 1473 K. Niether pores nor cavities were observed. TaB_2 and NbB_2 were completely formed together with AlN after holding for 3600 s in systems with tantalum and niobium. The in situ formed particles were uniformly distributed in the aluminum matrix. In systems with zirconium and hafnium, ZrB_2 and HfB_2, along with BN, were found. Differential thermal analysis indicated that the heat of Al-Ta and Al-Nb reactions can ignite a reaction involving BN. However, the Al-Zr and Al-Hf reactions that occur at 1000 K do not ignite a reaction with BN. Therefore, the processing time was longer for zirconium and hafnium systems. The Ta-BN powder mixture was infiltrated with molten aluminum at temperature above 1173 K, and both TaB_2 and AlN were found as a result of the in situ reaction.
- Fabrication of Metal Matrix Composites (MMCs) by Spontaneous Infiltration and Subsequent In Situ Reaction Processes, T. Choh, M. Kobashi, M. Nakata, and H. Kaneda, *Mater. Sci. Forum*, Vol 217–222, 1996, p 353–358. Fabrication of MMCs was developed by applying the in situ reaction followed by spontaneous infiltration of liquid aluminum into a mixture of ceramic and metal powders. Next, the spontaneous infiltration of liquid magnesium into SiC powder phase was completed using SiO_2 as an infiltration agent in the SiC powder phase.
- Precipitation during Infiltration of A201 Aluminum Alloy into Al-Fe-V-Si Preform, C.-C. Yang, Y.-C. Chen, and E. Chang, *Scr. Mater.*, Vol 34 (No. 7), 1996, p 1059–1065. A new near-net casting process is proposed in which liquid aluminum is infiltrated around the Al-Fe-V-Si particles to form a composite material. A preliminary study of the Al-Fe-V-Si particle reinforced A201 aluminum alloy demonstrates that the compression strength at 300 °C can be twice as high as A201 aluminum alloy.
- Metal Matrix Composites Fabricated by Pressure Assisted Infiltration of Loose Ceramic Powder, M.A. Taha and N.A. El-Mahallawy, *J. Mater. Process. Technol.*, Vol 73 (No. 1–3), 1998, p 139–146. Pressure was applied by two methods: centrifugal and squeeze casting. The structural features of the composite, including particle distribution, metal/ceramic interface, and part soundness, are shown. A comparison of the two techniques using a high volume fraction of Al_2O_3 (50–65%) is made.

Metal Matrix Composites (MMC) (Fiber)

- Characterization of Ceramic Matrix Composites Fabricated by Chemical Vapor Infiltration, D.P. Stinton, D.M. Hembree, Jr., K.L. More, and B.W. Sheldon, *Chemical Vapor Deposition on Refractory Metals and Ceramics Symp.*, 29 Nov–1 Dec 1989 (Boston, MA), T.M. Besmann and B.M. Gallois, Ed., Materials Research Society, 1990, p 273–280. A process for the preparation of fiber-reinforced SiC composites by chemical vapor deposition has been developed at Oak Ridge National Laboratory. Composites are prepared by infiltrating fibrous preforms with reactant gases that decompose at elevated temperatures to deposit silicon carbide between and around the fibers. Because the infiltration process utilizes both temperature and pressure gradients, SiC is deposited under conditions that vary considerably from the hot face to the cool face of the composite. Matrix characterization of composite samples by transmission electron microscopy and Raman spectroscopy is described.

- Carbon Coated Alumina Fiber/Glass Matrix Composites, R.L. Lehman and C.A. Doughan, *Compos. Sci. Technol.*, Vol 37 (No. 1–3), 1990, p 149–164. Carbon coated alumina fiber-reinforced borosilicate glass composites were fabricated by slurry infiltration and hot pressing. For fiber contents of 20 vol%, composite strengths of 118 and 263 MPa were observed for uncoated and carbon coated composites, respectively. Fiber pull-out was observed in fracture surfaces only for carbon coated composites, and pull-out lengths were approximately 30–60 μm. The fiber/matrix interface was studied to characterize the interfacial frictional stress. Direct measurements by indentation were compared with calculated values based on pull-out lengths and fiber strengths. Direct indentation measurements gave interfacial frictional stress values of 100 MPa, which were reduced by as much as 40% after a correction for Poisson's ratio effects. Values calculated from pull-out lengths suggested a value of 63–84 MPa, in good agreement with direct measurements. Elastic modulus and thermal expansion measurements confirmed the debonded character of the carbon coated fiber/matrix interface.
- Fiber-Reinforced Silicon Nitride Composites, R. Lundberg, R. Pompe, R. Carlsson, and P. Goursat, *Compos. Sci. Technol.*, Vol 37 (No. 1–3), 1990, p 165–176. Three possible processing routes to obtain silicon nitride reinforced with continuous fibers are identified and demonstrated. Slip-infiltrated, HIPed carbon fiber-reinforced material, slip-infiltrated, SiC fiber-reinforced nitrided Si_3N_4, and polysilazane solution infiltrated pyrolysed composites with SiC fibers have been successfully fabricated. Possible fiber/matrix reactions are discussed on the basis of scanning electron microscopy observations and the bend fracture behavior of the composites. All materials exhibited nonbrittle fracture and are thus potentially interesting composites for further development.
- Temperature and Concentration Dependence of SiC Deposition on Nicalon Fibers, T.M. Besmann, B.W. Sheldon, and M.D. Kaston, *Surf. Coat. Technol.*, Vol 3–4 (No. 1–3), 1990, p 167–175. Chemical vapor infiltration of Nicalon fiber bundles (approximately 500 filaments) was studied using methyltrichlorosilane to form the SiC matrix. The operating conditions were chosen to simulate those used in composite fabrication. It was determined that the deposition reaction has an activation energy of 66 ± 20 kJ/mol and is first order with respect to reactant concentration.
- Development of Advanced SiC/Al Composite Material, Manufacturing of Preform Wires by Liquid Metal Infiltration, Y. Imai, Y. Tanaka, H. Ichikawa, and T. Ishikawa, *J. Iron Steel Inst. Jpn.*, Vol 75 (No. 9), 1989, p 1555–1562. The continuous SiC fiber produced by polycarbisilane has outstanding properties as MMC reinforcement. SiC/Al (A1050) preform wires were continuously produced by the liquid infiltration method. After heating the SiC fiber yarn at 1373 K for desizing, it was dipped into molten aluminum at 953 K and then passed through a ceramic nozzle hole 0.25 mm in diameter. The tensile strength of the preform wire is 1560 MPa.
- Silicon Carbide/Silicon and Silicon Carbide/Silicon Carbide Composites Produced by Chemical Vapor Infiltration, M. Kmetz and S. Suib, *J. Am. Ceram. Soc.*, Vol 73 (No. 10), 1990, p 3091–3903. Composites of SiC/Si and SiC/SiC were prepared from single yarns of SiC. The use of carbon coatings on SiC yarn prevented the degradation normally observed when chemically vapor deposited silicon is applied to SiC yarn. The strength, however, was not retained when the composite was heated at elevated temperatures in air. In contrast, the strength of a SiC/C/SiC composite was not reduced after being heated at elevated temperatures, even when the fiber ends were exposed.
- Influence of Aluminum Alloy Matrices on the Strength of SiC_PC_S Fiber-Reinforced Preform Wires by Liquid Metal Infiltration, Y. Imai, Y. Tanaka, H. Ichikawa, and T. Ishikawa, *J. Iron Steel Inst. Jpn.*, Vol 76 (No. 1), 1990, p 65–72. The matrices used were aluminum-base binary alloys with 5wt% Si, Cu, and Mg, and commercial A2024, A5052, A6061, and A7075 alloys. No wire with high strength was obtained in any matrix. Silicon precipitated on the surface of SiC fiber in Al-5Si. $CuAl_2$ was observed in Al-$CuAl_2$ eutectic and magnesium-diffused into the fiber in Al-5Mg. Similar compounds were found when using industrial alloys. An Al-3Cu-2Ni alloy was designed and the fiber-melt contact was shortened. The tensile strength of the wire obtained was 1 GPa.
- Melt Infiltration and Reaction at the Fiber/Matrix Interface during the Brazing of a Fiber-Reinforced Ceramic to Metal, S. Karunanithy, *J. Am. Ceram. Soc.*, Vol 73 (No. 1), 1990, p 178–181. The presence of silica in the fiber/matrix interface of SiC fiber-reinforced alumina facilitates the brazing of these composites with metals. In addition, molten braze infiltrates the interfacial pores to form reaction layers with the matrix and the fiber. The chemical compatibility, wetting, and thermodynamics of the titanium-base braze leads to a potentially useful method of joining a fiber-reinforced ceramic to metal.
- Chemical Stability of Zirconia-Stabilized Alumina Fibers during Pressure Infiltration by Aluminum, J.A. Isaacs, F. Taricco, V.J. Michaud, and A. Mortensen, *Metall. Trans. A*, Vol 22 (No. 12), 1991, p 2855–2862. MMCs composed of high purity aluminum and continuous zirconia-stabilized polycrystalline Al_2O_3 fibers are fabricated by liquid metal infiltration using three different casting procedures. The microstructure of the composites showed discrete faceted $ZrAl_3$ particles at the interface that grow into the matrix in samples processed above matrix melting point for more than 13 minutes. This compound formation is in agreement with the thermodynamic stability calculations. There is also a reaction between solid aluminum and the fibers at 913 K, yielding a reaction product of the same morphology as with molten aluminum. When infiltration occurs in a preform heated below the metal melting point and solidification time is less than 1 min, no reaction products are seen in SEM. This leads to the conclusion that aluminum MMC can be cast with no apparent interfacial reaction product using these fibers provided that adequate processing parameters are chosen.
- Influence of Fiber Array on the Threshold Pressure on Infiltration in Alumina Fiber/Aluminum Composite, H. Nakanishi, Y. Tsunekawa, M. Okumiya, M. Higashi, and I. Niimi, *J. Jpn. Inst. Light Met.*, Vol 41 (No. 5), 1991, p 325–330. It was shown experimentally that the fiber arrangement greatly influences the threshold pressure of infiltration in the alumina fiber/molten aluminum system. The threshold pressure for the unidirectionally fiber array is 425 kPa, a value less than theoretical because of the slight disarray in the fiber alignment. By contrast, the experimental value for quasi-planar random array is 245 kPa. The relation between the threshold pressure and wettability is derived for infiltration with noninfiltrated defects at the contact of nonwetted two fibers. The threshold pressure of alumina fiber/molten aluminum system with defects is less than that without defects, and it decreases with decrease in contact angle of two fibers because of increasing defects.
- Influence of Processing Parameters on the Threshold Pressure of Infiltration in Alumina Fiber/Aluminum Composite System, H. Nakanishi, Y. Tsunekawa, M. Okumiya, and I. Niimi, *J. Jpn. Inst. Light Met.*, Vol 41 (No. 9), 1991, p 576–581. Although fiber arrangement in the preform does not affect the interfacial energies or contact angles, it affects the processing parameters like pressurizing rate, pressure holding at the maximum pressure applied, wettability, and therefore, threshold pressure. In this study, a δ-Al_2O_3/molten aluminum composite was selected to investigate such dependence. Pressurizing rate was changed from 0.20 to 16 kPa/s and pressure holding at maximum pressure was 300 s. No effect of these parameters on the threshold pressure was observed in Al_2O_3 preforms with quasi-planar random arrangement. Experimentally, it was shown that the threshold pressure obtained by extrapolation of infiltration distance and applied pressure agrees with that obtained from the Blake-Kozeny equation. The influence of molten aluminum temperature on the threshold pressure that was obtained by the pressure

drop was studied at the pressurizing rate of 2.4 kPa/s. The threshold pressure decreases from 259 to 162 kPa when molten aluminum temperature increased from 943 to 1163 K.

- Characterization of Metal-Matrix Composites Fabricated by Vacuum Infiltration of a Liquid Metal under an Inert Gas Pressure, J.-M. Chiou and D.D.L. Chung, *J. Mater. Sci.*, Vol 26 (No. 10), 1991, p 2583–2589. Silicon carbide whisker-reinforced aluminium was fabricated by vacuum infiltration of liquid aluminum into a porous whisker preform under an argon gas pressure, using an infiltration temperature of 665 °C. The volume fraction of whiskers ranged from 11 to 37%. No whisker pull-out was observed on the fracture surface for an infiltration temperature of 665 °C, but some whisker pull-out was observed for an infiltration temperature of 720 °C. Both the tensile strength and ductility decreased with increasing infiltration temperature above 665 °C. Tensile test results from room temperature to 300 °C are reported. They showed that the quality of these composites was comparable to that of composites made by powder metallurgy or squeeze casting. The coefficient of thermal expansion at 100 to 150 °C was decreased by 45% by the addition of 37 vol% whiskers.
- Synthesis of SiC Platelet Reinforced 2014 Al Alloys by a Pressure Infiltration Technique, R. Asthana and P.K. Rohatgi, *J. Mater. Sci. Lett.*, Vol 10 (No. 4), 1991, p 230–234. A counter-gravity pressure assisted liquid metal infiltration technique was used to infiltrate prepacked columns of SiC platelets by aluminum. The aluminum alloy composites contain relatively high (50% or more) volume fractions of SiC reinforcement.
- Aluminum-Matrix Silicon Carbide Whisker Composites Fabricated by Pressureless Infiltration, Y. Chen and D.D.L. Chung, *J. Mater. Sci.*, Vol 31 (No. 2), 1991, p 407–412. SiC whisker-reinforced aluminum composites were fabricated by pressureless infiltration of liquid Al-Mg or Al-Si-Mg alloy at 830 to 950 °C into a preform of nickel coated SiC fibers in nitrogen. Nickel coating was obtained by electroless plating. The composite exhibits a slightly lower tensile strength and modulus and slightly higher coefficient of thermal expansion than the same composite fabricated by pressure infiltration. The hardness decreased with increasing distance from the preform-melt interface much more significantly in pressureless composites as compared to pressure infiltrated counterparts. This hardness decrease, which was attributed to porosity increase, was larger in pressureless composites without prior evacuation than in those made by prior evacuation. The Al-SiC reactivity was larger in pressureless composites as compared to those made with pressure application because of a longer infiltration time.
- Effect of Fiber Distribution on Infiltration Processing and Fracture Behavior of Carbon Fiber-Reinforced Aluminum Composites, Z. Xia, Z. Mao, and Y. Zhou, *Z. Metallkd.*, Vol 82 (No. 10), 1991, p 766–768. Composites with bundles of carbon fibers were fabricated by a pressure-regulating infiltration technique. The molten metal infiltrates the preform in finger-like mode, which can make infiltration possible at a low pressure of 0.7 MPa in spite of poor wettability of fibers by the molten metal. The fiber arrangement in bundle resulted in high tensile strength and an increase in fracture energy of the composite.
- Infiltration of Fiber Preforms by an Alloy, Part III: Die Casting Experiments, P. Jarry, V.J. Michaud, A. Mortensen, A. Dubus, and R. Tirard-Collet, *Metall. Trans. A*, Vol 23 (No. 8), 1992, p 2281–2289. Preforms of 20 vol% Saffil alumina fibers are infiltrated with Al-4.4wt%Cu-0.3wt%Mg using a horizontal die casting machine. Fiber preform temperature is varied from 973 to 673 K. Increases in fiber volume fraction are observed in composites downstream of the infiltration path. It is proposed that these result from locking of the compressed fibers by solid metal present during infiltration. With this assumption, a good agreement was found between theory on solute concentration, fiber volume fraction distribution, matrix microstructure, and experiments. For preform temperature of 673 K, freckles, which are interpreted to result from the combined effects of pressure and significant enrichment in solute at the infiltration front, were found in the composite.
- Age-Hardening Behavior of 6061 Aluminum Alloy Reinforced with SiC Whiskers, H. Toda, T. Kobayashi, and M. Niinomi, *J. Jpn. Inst. Met.*, Vol 56 (No. 11), 1992, p 1303–1311. Cast reinforced composites containing 22% volume fraction of SiC whisker were fabricated by a high pressure infiltration technique. The age-hardening behavior of the 6061 aluminum alloy with and without SiC whisker was studied using hardness measurement, calorimetric technique, and transmission electron microscopy. The overall age-hardening sequence is not altered by the addition of SiC whisker. Moreover, the maximum hardness of 6061 with and without SiC whisker is associated with the same transition stage from the GP (II) zone to the intermediate β' phase. However, precipitation kinetics of the matrix alloy is affected by presence of the reinforcement. Times to achieve the peak hardness are shortened in the composites over the whole temperature ranges investigated. However, this tendency becomes marked with low aging temperature, suggesting the contribution from high diffusivity paths, such as dislocations and reinforcement-matrix interfaces. This acceleration is attributed to the drastic decrease of the activation energy required for the formation of the GP (II) zone and intermediate β' phase.
- High-Temperature Mechanical and Material Design for SiC Composites, Part A, N.M. Ghoniem, *J. Nucl. Mater.*, Vol 191–194, 1992, p 515–519. SiC fiber-reinforced composites (FRCs) are strong potential candidate structural and high heat flux materials for fusion reactors. A concise discussion of the main material and design issues related to the use of SiC FRCs as structural materials in future fusion systems is given in this paper. The status of material processing of SiC/SiC composites is first reviewed. The advantages and shortcomings of the leading processing technology, known as chemical vapor infiltration, are particularly highlighted. A brief outline of the design-relevant physical, mechanical, and radiation data base is then presented. SiC/SiC FRCs possess the advantage of increased apparent toughness under mechanical loading conditions. This increased toughness, however, is associated with the nucleation and propagation of small crack patterns in the structure. Design approaches and failure criteria under these conditions are discussed.
- Effect of Reinforcement Architecture on Mechanical Properties of a Short Fiber/Magnesium RZ5 MMC Manufactured by Preform Infiltration, D.J. Towle and C.M. Friend, *Mater. Sci. Eng. A*, Vol 188 (No. 1–2), 1994, p 153–158. Short fiber-reinforced metal-matrix composites (MMCs) produced by the liquid metal preform infiltration route exhibit a planar (two-dimensional) random fiber architecture. The mechanical properties measured in tension and compression of a Saffil fiber- reinforced RZ5 (ZE41) magnesium alloy, both parallel (P) and normal (N) to the fiber arrays, are presented. Generally, the mechanical properties are enhanced in the P orientation compared to the N orientation, and the properties in the N orientation are still better than those of unreinforced alloy. There is a significant effect of deformation mode on the mechanical properties both on the magnitude of the anisotropic effects and on the absolute strengths in tension and compression.
- Pressurized Infiltration of Aluminum Melt into Aluminum Borate Whisker Preform, T. Yamauchi and Y. Nishida, *J. Jpn. Inst. Met.*, Vol 59 (No. 5), 1995, p 564–570. The aluminum borate whiskers and alumina short fibers were used as reinforcements for fiber aluminum composites made by squeeze casting. The pressure was mechanically applied by a punch, and the pressure and punch speed were measured during the infiltration of molten aluminum. Hardness was measured along the infiltration direction. Based on these measurements, the distribution of volume fraction was calculated. The experimental results agree with theoretical analysis using D'Arcy's law. Deformation of fibrous preforms occurred due to compressive stress during infiltration with molten aluminum.
- Infiltration of Al_2O_3 Short Fiber Preforms during Squeeze Casting of A357 Al Alloys, H.-G. Kang, P.R.G. Anderson, and B. Cantor, *Proc. of the 1995 124th Annual Symp. Synthesis/Processing of Lightweight Metallic Materials* (Las Vegas, NV), TME annual meeting, TMS, 1995, p 153–174. Alloy A357 Al-

7wt%Si-0.6wt%Ma was infiltrated into 20 vol% Saffil preforms with discontinuous fibers with a diameter of 3 to 3.5 μm and 500 μm long by squeeze casting at 0 to 100 MPa pressure, melt temperatures of 700 to 900 °C, and die and preform temperature of 300 °C. It was found that presolidification of the molten alloy occurs during the time delay between die filling and pressurization. This way, a shell of coarse columnar grains of α-aluminum dendrites forms. The enriched remaining solute that infiltrates the preform produced a macrosegregation within the composite. Needle-like primary $FeSiAl_5$ and blocky $(Fe,Mn)_3Si_3Al_{1/5}$ intermetallics form due to solute enrichment at the composite/monolithic interface. Stresses build up within the preform leading to cracks and fragmentation. Shrinkage pores, cracks, and uninfiltrated regions are all force fed with eutectic liquid as the pressure increases at the end of the ram travel. High applied pressure leads to a fine grained structure.

- Microstructure and Properties of Al_2O_3/Al-1.5Mg Composite Manufactured by Extrusion Directly Following Liquid Infiltration, L. Hu, S. Luo, W. Huo, and Z. Wang, *Trans. Nonferrous Met. Soc. China,* Vol 5 (No. 4), 1995, p 146–150. The microstructure of the extruded composite after infiltration indicated a high density matrix with fine recrystallized grains without defects. The fibers are uniformly and quasi unidirectionally distributed. Mechanical properties (Young modulus, tensile and yield strength) and the thermal expansion resistance of the composite increase with fiber volume fraction, while ductility is still maintained at a high level. The corrosion resistance of Al_2O_3/Al-1.5Mg composite is similar to that of the Al-1.5%Mg alloy.
- Pressure Infiltration of Molten Aluminum Preform of Granulated Whisker, T. Yamauchi, Y. Nishida, and H. Nakae, *J. Jpn. Inst. Met.,* Vol 61 (No. 2), 1997, p 158–165. A granulated aluminum borate whisker preform was prepared to enhance permeability in squeeze casting. The permeability of granulated whisker was six times larger than that of uniform preforms with randomly distributed whiskers. When the pressure is applied, the molten metal first flows between the whisker granules to the bottom of the preform and then infiltrates into the granules of whiskers as the pressure increases. This way, the preform deformation was suppressed.
- Properties of Continuous Fiber-Reinforced Al- and Mg-Matrix Composites Produced by Gas Pressure Infiltration, H.P. Degisher, P. Schulz, and W. Lacom, *Key Eng. Mater.,* Vol 127–131, 1997, p 99–110. General steps in the production of continuous fiber-reinforced aluminum or magnesium or their alloys are described: preform manufacturing, liquid metal infiltration by gas pressure, solidification, and finishing. Elastic properties and tensile strengths of different fiber (carbon or ceramic multifilament tows)-light metal matrix systems are compared in longitudinal, and if available, transverse directions.

Other Infiltration Products

- Influence of Liquid Metal Infiltration on the Superconducting Characteristics of Niobium Nitride, L.T. Summers, J.R. Miller, M.J. Strum, R.J. Weimer, and D.E. Kizer, *Advances in Cryogenic Engineering,* Plenum Press, Vol 34, 1988, p 835–842. A fully stabilized multifilamentary BN superconductor was prepared using a combination of physical vapor deposition of NbN on graphite followed by liquid metal infiltration using copper or aluminum. The resulting conductor assumed a finely divided multifilamentary form embedded in a matrix of conducting copper or aluminum. The geometry provides high stability to flux jumps and high quench protection. The effects of liquid metal infiltration and process variables on the electrical properties were determined. Critical current dependence on field strength and stabilizer residual resistivity ratio are discussed.
- Anomalous Growth of the Complex Carbide Phase in Carbon-Deficient WC-Co Hard Metal during Infiltration of Eutectic Liquid, J.K. Park, K.Y. Eun, and D.N. Yoon, *Mater. Sci. Eng. A,* Vol 105–106, 1988, p 233–236. Large angular grains of η phase (Co_3W_3C) around WC grains formed during infiltration of the carbon-deficient 97WC-3Co with 25WC-75Co infiltrants. In contrast, in the uninfiltrated region, the fine η and CoW_3C phases were uniformly distributed. Some elongated η grains grew at the infiltration boundary. The anomalous growth of η phase is attributed to the composition change in the specimen from WC to η to potassium region to the WC-K liquid region during infiltration.
- Correlation Among Process Routes, Microstructures, and Properties of Chemically Vapor Deposited Silicon Carbide, R.F. Davis, *Chemical Vapor Deposition on Refractory Metals and Ceramics Symp.,* 29 Nov–1 Dec 1989 (Boston, MA), T.M. Besmann and B.M. Gallois, Materials Research Society, 1990, p 145–158. *Silicon carbide* is a generic term for a host of different materials produced by several process routes that yield a variety of microstructures and associated property characteristics. The route of chemical vapor deposition (CVD) is used primarily to deposit SiC for wear- and corrosion-resistant coatings and for diffusion barriers to and from the underlying substrate. Presently this technique is also being used to deposit monocrystalline semiconductor thin films of SiC and to infiltrate various high temperature woven fabrics. The paper describes the results of thermodynamic calculations to define SiC CVD diagrams using various precursor gas mixtures, discuss various CVD techniques, and detail the results of deformation, infiltration, and thin film deposition studies that have been recently conducted on vapor deposited SiC.
- Fabrication Aspects of Glass Matrix Composites for Gas Turbine Applications, D.A. Clarke, *New Materials and Their Applications 1990,* Proc. of the 2nd Int. Symp., 10–12 April 1990 (Warwick, U.K.), D. Holland, Ed., IOP Publishing, Bristol, U.K., 1990, p 173–183. A brief background to the development of glass matrix composites (GMCs) including the basic fabrication methods is presented. Some of the key problems involved in manufacturing production quantities of GMC gas turbine components are described and the need for highly repeatable, and hence largely automated, processing methods are highlighted.
- Processing and Properties of Silicon Carbide-Reinforced Aluminum Metal Matrix Composites for Electronic Applications, K.K. Aghajanian, *Proc. 1991 Int. Symp. Microelectronics,* International Society of Hybrid Microelectronics, Reston, VA, 1991, p 368–372. Alternative packaging materials with high thermal conductivity, high specific strength and stiffness, and low density are in demand for the microelectronic industry. SiC particulate reinforced aluminum MMCs satisfy these requirements. Using the Primex pressureless metal infiltration process, SiC loadings can be varied over a wide range, thus providing a family of composites with tailorable properties. The processing and properties, as well as the electronic applications of silicon carbide reinforced aluminum composites, are discussed.
- Microstructure and Properties of Metal Infiltrated Reaction-Bonded Silicon Nitride (RBSN) Composites, N.A. Travitzky and N. Claussen, *J. Eur. Ceram. Soc.,* Vol 9 (No. 1), 1992, p 61–65. RBSN-metal composites were fabricated using gas-pressure infiltration. Various RBSN types have been infiltrated with molten aluminum, an Al-Si-Mg alloy, a Ti-Al intermetallic, and silicon, resulting in considerable increase in mechanical properties when compared to uninfiltrated RBSN. For example, strength was raised to 510 from 227 MPa when infiltrated with a Ti-39wt%Al alloy and the toughness to >5 from 2.7 MPa$\sqrt{m}$ when pure aluminum was infiltrated. Silicon infiltration proved to be most effective in enhancing the wear resistance.
- Low Shrinkage Refractories by an Infiltration Technique, N. Lequeux, P. Larose, P. Boch, and N. Burkarth, *J. Eur. Ceram. Soc.*, Vol 14 (No. 1), 1994, p 23–27. Infiltration of presintered silica-zircon preforms with silica of alumina precursors and subsequent heating lead to the development of crystallized segregation (e.g., silica, alumina, or mullite), which decreases the sintering shrinkage. The decrease in shrinkage is very sensitive to the nature of segregations.
- Mechanical Properties of Al_2O_3/Si Composites Fabricated by Pressureless Infiltration Technique, N.A. Travitzky, E.Y. Gutmanas, and N. Claussen, *Mater. Lett.*, Vol 33 (No. 1–2), 1997, p 47–50. A microstress-induced strengthening concept was applied to Al_2O_3/

Si composites fabricated by pressureless infiltration of porous Al_2O_3 preforms with molten silicon. The mechanical properties (strength, hardness, fracture toughness) were superior to a material with the same amount of Al_2O_3 with aluminosilicate phase instead of silicon. For example, at 30% Si, the composite exhibits a bending strength of 320 MPa and K_{Ic} of 4.8 MPa$\sqrt{m}$ as compared to 230 MPa and 3.5 MPa$\sqrt{m}$, respectively, for a technical alumina AD-85. This strengthening is attributed to the compressive residual microstresses in an inherently weak silicon phase generated as a result of solidification-related expansion of silicon.

- Mullite/SiAlON/Alumina Composites by Infiltration Processing, M.P. Albano and A.N. Scian, *J. Am. Ceram. Soc.,* Vol 80 (No. 1), Jan 1997, p 117–124. The formation of mullite/SiAlON/alumina composites was studied by infiltrating a SiAlON/alumina-base composite with two different solutions, followed by thermal treatment. The base composite was prepared from a mixture of Al_2O_3 grains, fume SiO_2, and alumina powders. The mixture was pressed into test bars and nitrided in a nitrogen gas (N_2) atmosphere at 1480 °C. The infiltrants were prehydrolyzed ethyl polysilicate solution and ethyl polysilicate-aluminum nitrate solution. The composites were infiltrated under vacuum, cured at 100 °C, and precalcined in air at 700 °C. This infiltration process was repeated several times to produce bars that had been subjected to multiple infiltrations, then the bars were calcined in a N_2 atmosphere at 1480 °C to obtain mullite/SiAlON/alumina composites. The infiltration process increased the percentage of nitrogenous crystalline and mullite phases in the matrix; therefore, a decrease of the composite microporosity was observed. The infiltration increased the mechanical strength of the composites. Of the two composites, the one produced using prehydrolyzed ethyl polysilicate as the infiltrant had a higher mechanical strength, before and after being subjected to a severe thermal shock.
- Infiltration of Porous Aluminum Bodies with Solution Precursors, P. Honeyman-Colvin and F. Lange, *Am. Ceram. Soc.*, Vol 79 (No. 7), July 1996, p 1810–1814. Alumina powder compacts, partially densified with a low-temperature heat treatment and then cut into bars, were infiltrated with liquid precursors that decomposed to either mullite (Al_2O_3, SiO_2), fully stabilized zirconia (cubic $Zr(8Y)O_2$), or partially stabilized zirconia (tetragonal $Zr(4Y)O_2$). The specimens were repeatedly infiltrated and pyrolyzed to achieve a higher concentration of the precursor near the surface. The infiltrated bodies were then densified at 1500 °C for 2 h. Residual stresses developed during cooling from the densification temperature because of the higher concentration of the second phase near the surface and their differential thermal expansion relative to the matrix material. At least ten bars of each two-phase material were fractured in four-point bending to determine the effect of the second phase on strength. The alumina bars without a second phase had a larger grain size (≅7 μm) and a mean strength of 253 GPa. The intruded phases significantly reduced the Al_2O_3 grain size to ≅1 μm. Despite their higher concentration near the surface and apparent surface tensile stress, both of the $Zr(Y)O_2$ phases increased the mean strength to 413 MPa (cubic $Zr(8Y)O_2$) and 583 MPa (tetragonal $Zr(4Y)O_2$, an apparent toughening agent). The mullite second phase produced a high mean strength of 588 MPa, apparently due to its concentration gradient creating a compressive surface stress.

Liquid-Phase Sintering

Rajiv Tandon, Phillips Powder Metal Molding
John Johnson, Howmet Corporation

SINTERING is a process in which particles bond together when heated to a sufficiently high temperature. The driving force is the net reduction in the surface energy. Liquid-phase sintering (LPS) involves the formation of a liquid phase to promote higher densification rates and lower the sintering temperatures. Because of cost and productivity advantages, it is estimated that over 70% of sintering products are processed using liquid-phase sintering. Common systems for liquid-phase sintering include Cu-Co, W-Cu, W-Ni-Fe, W-Ag, Cu-Sn, Fe-Cu, WC-Co, TiC-Ni, and Fe-Cu-P. Key factors in determining whether a liquid phase forms during sintering of a specific material system include surface energies, solubilities, and diffusivities.

There are two main forms of LPS. When a liquid phase is obtained by inducing melting in a mixture of powders and is persistent throughout the high-temperature portion of the sintering cycle, the process is termed as persistent LPS. Classic LPS systems such as W-Ni-Fe and WC-Co are excellent examples of LPS with a persistent liquid. A persistent liquid can also be obtained by partially melting a prealloyed powder above its solidus temperature (termed as supersolidus LPS, or SLPS) and is widely used for processing tool steels, stainless steels, and superalloys. In some systems (for example W-Cu and Mo-Cu) with a low inter-solubility even in the presence of a persistent liquid, an activator can be used to enhance sintering. This is termed as activated liquid-phase sintering (ALPS). Alternatively, transient liquid-phase sintering (TLPS) involves liquid that disappears due to dissolution into the solid or formation of a new phase/compound. Although each material system has its own set of fabrication concerns, many general LPS concepts can be used to guide their processing. The following sections outline these concepts and provide specific examples of how they apply to industrially relevant materials.

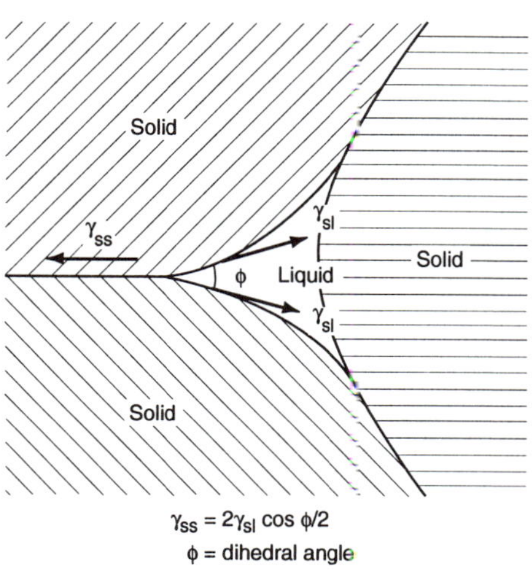

Fig. 1 The dihedral angle and surface energy equilibrium between two intersecting grains with a partially penetrating liquid phase

Thermodynamic and Kinetic Factors

Surface Energy. When a liquid forms during sintering, the microstructure consists of solid, liquid, and vapor phases. Successful LPS requires a reduction in surface energy as the liquid spreads across the surface of the solid particles. Thus, the solid-liquid surface energy must be less than the solid-vapor surface energy. In this case, the liquid "wets" the solid and provides a bonding force between the particles to aid densification.

Wetting is aided by solubility of the solid in the liquid, formation of intermediate compounds, and interdiffusion. Interfacial energies can be highly dependent on surface purity, so the wetting behavior can be drastically altered by contaminants or processing steps that clean the powder surface. Clean surfaces are especially important for metal-metal systems. Reactive metals tend to wet oxidized surfaces better than noble metals due to chemical reactions at the interface. Metal powders contain an oxide surface layer, so a reducing atmosphere or high vacuum is generally required to break down these surface oxides to obtain good wetting with less reactive liquid metals. In certain systems, the penetration of the liquid by capillary action is sufficient to mechanically break down oxide layers.

Surface energies also establish important microstructural features such as contiguity (the surface area of solid-solid contacts as a fraction of the total interfacial area), connectivity (the number of observed contacts per grain in a two-dimensional cross section), grain size, and dihedral angle. The dihedral angle characterizes the energy ratio between the grain boundaries and solid-liquid surfaces and is defined in Fig. 1. This ratio changes with sintering time and has a strong influence on microstructural development. Dissolution reactions during liquid formation decrease the solid-liquid interfacial energy below its equilibrium value, promoting particle rearrangement. With further sintering time, the interfacial energy increases back to its equilibrium value. The contiguity also increases with time as neck growth proceeds until an equilibrium neck size is reached. The contiguity is uniquely related to the solid volume fraction, dihedral angle, and grain size ratio (Ref 1). Connectivity is also strongly related to contiguity. A large dihedral angle gives a larger connectivity and contiguity for the same solid volume fraction. These microstructural parameters influence the electrical conductivity, strength, ductility, elastic behavior, dimensional stability, and thermal characteristics of LPS materials.

Solubility of the solid in the liquid is generally required for successful liquid-phase sintering, especially for systems with more than 65 vol% solid (Ref 2). This solubility permits solution reprecipitation and enables a more efficient packing of the grains, leading to higher sintered densities. On the other hand, solubility of the liquid in the solid is generally undesirable, because it can result in swelling as the liquid diffuses into the solid grains (leaving behind large pores that are typically difficult to eliminate during subsequent sintering). However, in certain systems, such as Cu-Sn and Cu-Zn, transient liquid phase can be controlled to yield unique properties. These two cases are schematically shown in Fig. 2.

The solubility parameter, which is defined as the ratio of the solid solubility in the liquid to the liquid solubility in the solid, provides a reasonable estimate of whether swelling or shrinkage occurs during liquid-phase sintering. Several examples are given in Table 1, which shows a potential for shrinkage at higher ratio values. The actual amount of dimensional change depends on processing conditions.

Phase diagrams are useful in identifying solubility parameters and other characteristics conducive to LPS (Ref 3–5). An ideal phase diagram for LPS is shown in Fig. 3. A deep eutectic is favorable due to the large reduction in sintering temperature with the formation of the liquid phase. The formation of intermediate compounds is generally unfavorable. High-temperature phases can lower diffusion rates, while brittle intermetallic phases that form during cooling can degrade mechanical properties. Phase diagrams also indicate the tendency of alloying elements and impurities to segregate to grain boundaries. The greater the separation of the solidus and the liquidus, the greater the solute segregation to interfaces. A downward sloping liquidus and solidus also indicate a tendency for solute segregation and lower surface energies (Ref 6). A sintering temperature just above the eutectic temperature is optimal for a composition in the L + β region. A classical LPS microstructure for such a case, as shown in Fig. 4, consists of relatively large, rounded grains suspended in a liquid matrix, with the degree of grain contact governed by the dihedral angle.

For systems that lack solubility of the solid in the liquid phase, a rigid skeletal structure is expected with densification governed by solid-state diffusion. High sintered densities are possible with the use of extremely fine starting powders or by using segregating activators that enhance solid-state sintering in the presence of the liquid phase. These techniques are covered in more detail in the section "Activated Liquid Phase Sintering" in this article.

Diffusivity. One of the benefits of LPS is the rapid densification and homogenization that occurs due to the high diffusion rates of base metal atoms in a liquid. Additives that have low melting temperatures are preferred due to their low activation energies and high diffusivities at lower temperatures. For certain systems, during heating, significant densification can occur due to enhanced solid-state sintering through the additive phase. In such cases, the diffusivity increases exponentially with temperature until the additive melts. Once the liquid phase forms, there is a rapid jump in both the solubility and diffusivity, due to the much weaker bonding associated with a liquid phase. An exponential increase in diffusivity accompanies further increases in temperature. Thus, higher temperatures are favorable for densification, but they also increase the vapor pressure of the liquid. Preferential evaporation of the liquid can change the overall composition of the sintered component. In addition, redeposition of vapors on furnace walls can result in damage to furnaces. Problems with preferential evaporation can be assessed through relative vapor pressures. A vapor pressure of 10^{-3} Pa provides a practical limit for vacuum sintering, while a vapor pressure of 1 Pa can lead to a measurable weight loss during sintering at atmospheric pressure.

In general, melting temperatures are good indicators of diffusion rates. Materials with lower melting temperatures have weaker atomic bonding and higher molar volumes, which allow faster diffusion at a given temperature. Liquid-phase sintering materials often involve mixed powders that have significantly different melting temperatures and thus unequal diffusivities. Unequal diffusion rates between the base alloy and the additive result in the Kirkendall porosity. For sufficient sintering times, the material will homogenize, and equilibrium factors, such as solubility, will dominate. Thus, unbalanced diffusivity ratios contribute to initial pore formation, but sustained shrinkage or swelling depends on the solubility ratio. Homogenization occurs more rapidly for small particles with a high initial homogeneity. Higher sintering temperatures and longer sintering times also aid microstructural uniformity of the sintered component.

Table 1 Solubility effects on densification in liquid-phase sintering

Base	Additive	Solubility ratio, at.%	Behavior
Al	Zn	0.004	Swell
Cu	Al	0.1	Swell
Cu	Sn	0.001	Swell
Cu	Ti	4	Shrink
Fe	Al	0.02	Swell
Fe	B	7	Shrink
Fe	Cu	0.07	Swell
Fe	Sn	0.01	Swell
Fe	Ti	3	Shrink
Mo	Ni	20	Shrink
Ti	Al	0.0003	Swell
W	Fe	5	Shrink

Table 2 Characteristics affecting initial-stage behavior in liquid-phase sintering

Factor	Swelling	Shrinkage
Solid solubility in liquid	Low	High
Liquid solubility in solid	High	Low
Diffusivities	Unequal	Equal
Additive particle size	Large	Small
Base particle size	Large	Small
Green density	High	Low
Contact angle	High	Low
Dihedral angle	High	Low
Temperature	Low	High
Time	Short	Long

Sequence of Stages

Densification during liquid-phase sintering commonly occurs in three stages after the liquid forms: rearrangement, solution-reprecipitation, and final-stage sintering. However, in many systems there is a strong solid-state diffusion contribution to the overall densification. A brief description of these stages follows.

Solid-State Sintering Prior to Liquid Formation. The same solubility, segregation, and diffusivity criteria that are favorable for LPS can lead to enhanced solid-state sintering prior to liquid formation. Powders prepared for LPS generally consist of admixed powders designed to lower the activation energy for diffusion, so solid-state sintering during heating can be

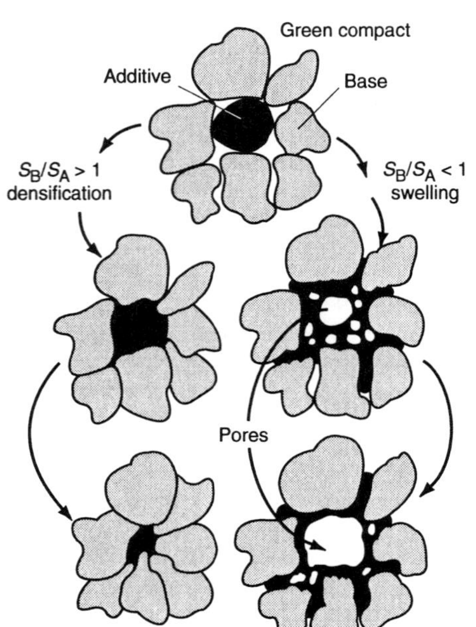

Fig. 2 A schematic diagram contrasting the effects of solubility on densification or swelling during liquid-phase sintering

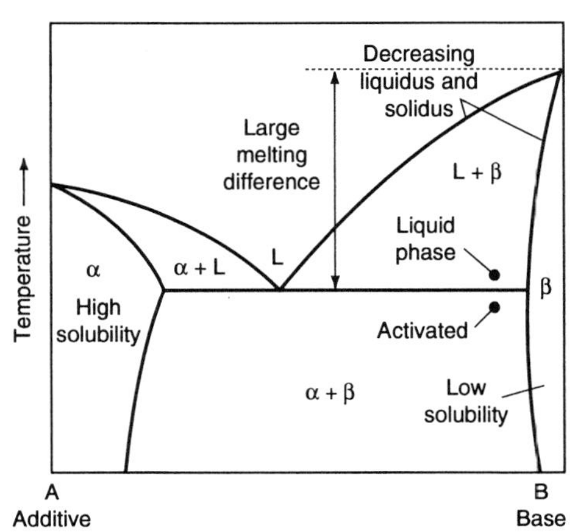

Fig. 3 Phase diagram of an ideal system for liquid-phase sintering

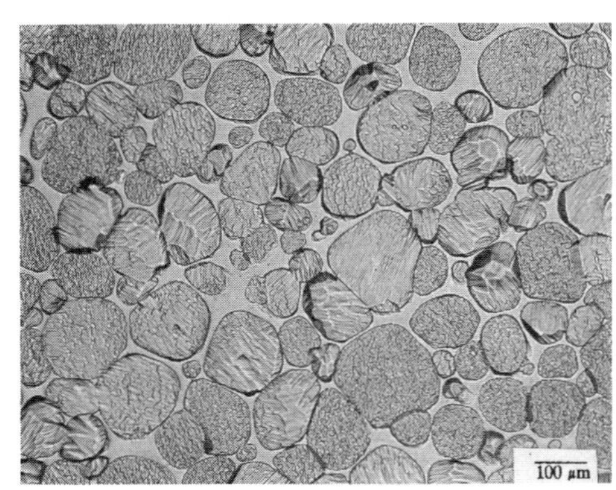

Fig. 4 Typical microstructure of a liquid-phase sintering system with the phase diagram characteristics shown in Fig. 3

greatly enhanced. This can result in significant densification prior to liquid formation. In fact, for systems such as the W-Ni-Fe heavy alloys, more than 95% of the total densification occurs prior to liquid formation (Ref 7). Thus, solid-state sintering is an important part of the overall LPS process. The relative contribution of solid-state sintering to total densification increases with increasing solubility of the base in the additive, increasing homogeneity of the additive compound, decreasing heating rates, and decreasing particle size. For the liquid to further assist densification, it must dissolve any solid-state sinter bonds that form during heating, and thus the solid must be soluble in the liquid. Indeed, even with the achievement of near-theoretical densities in the solid state, complete dissolution of solid-state sinter bonds has been demonstrated for the W-5Ni system (Ref 8).

Rearrangement. Once the liquid phase forms, typically the first step during classical LPS is rearrangement. At the onset of liquid formation, capillary forces pull the wetting liquid into particle necks and pores. These capillary forces also exert an attractive bonding force on the particles, resulting in rapid shrinkage. This is termed as primary arrangement. The liquid further enhances the packing by attacking and disintegrating clusters of particles by a process known as secondary rearrangement. Repacking and further densification occur by redistribution of the small particles between the large particles. Pore elimination occurs by viscous flow during this stage.

The dimensional behavior immediately after liquid formation is affected by both the processing conditions and material characteristics. The material and process variables that influence shrinkage and swelling are summarized in Table 2. A smaller base particle size is beneficial to rearrangement due to an increase in capillary forces (Ref 9). For systems with a small solubility ratio between the base alloy and additive, the amount of swelling can be reduced by the use of a smaller additive particle or by increasing additive homogeneity (Ref 10, 11). Particle shape is also important. Smooth, round particles tend to pack more homogeneously and are more easily rearranged, resulting in greater densification (Ref 12). Solubility also has a smoothing effect on the particle surfaces, decreasing interparticle friction and improving packing. High green densities and slow heating rates can restrict rearrangement due to a greater degree of bonding between the solid particles. A low contact angle and high liquid-vapor surface energy increase the capillary forces between the particles (Ref 13).

Solution-Reprecipitation and Grain Shape Accommodation. In most LPS systems, insufficient liquid is present to fill all the pores following rearrangement. Additional densification is accomplished by mass transport or solution reprecipitation in which the grains change shape and size distribution, resulting in a higher packing density. In this case atoms at point contacts between solid particles and at other convex surfaces have higher solubility in the liquid than atoms at concave surfaces. The atoms on the convex surfaces dissolve and diffuse through the liquid to neighboring concave surfaces where they reprecipitate. As the grains change shape, they are able to pack better and release liquid to fill any remaining pores (Ref 14, 15). Densification occurs as the centers of the grains grow closer to each other. The reduction in energy associated with the elimination of pores offsets the increase in the solid-liquid surface area. Equilibrium grain shapes are determined by the packing characteristics, surface energies, and liquid volume fraction (Ref 16).

Compared to the rearrangement stage, the shrinkage rate during the solution-reprecipitation stage is lower. However, for systems with only small amounts of liquid phase, solution reprecipitation can be the most significant densification mechanism. The kinetics of solution reprecipitation may be governed by interfacial dissolution or reprecipitation (reaction controlled) or by the rate of mass transfer through the liquid (diffusion controlled). In cases where the base material is a chemical compound, such as tungsten carbide, the kinetics are generally reaction controlled, while alloys are generally diffusion controlled. For diffusion-controlled cases, the densification rate is enhanced by a smaller particle size, higher solubility of the base in the liquid, and higher diffusivity of the base in the liquid. Concurrent with densification, there can be a significant degree of microstructural coarsening by the same diffusional processes.

Grain growth occurs concurrently with shape accommodation as small grains dissolve and reprecipitate on larger grains through Ostwald ripening or grain coalescence, or even solid-state grain growth of skeleton. The driving force is a decrease in interfacial energy through a reduction in both the amount and curvature of the solid-liquid interface. The rate of coarsening is generally described by:

$$G^n - G_o^n = K(\phi, T)\,(t - t_o) \quad \text{(Eq 1)}$$

where G_o is the initial grain size, G is the average grain size after time t, $K(\phi, T)$ is the growth rate constant that is dependent on the solid volume fraction (ϕ) and temperature T, and n is the growth exponent (the value of n varies between 2 and 5, depending on what mechanisms dominate mass transport). A cubic relationship is often observed in systems with rounded grains, such as the heavy alloys. In the case of reaction control, the exponent n is 2 and growth occurs at preferred crystal orientations, leading to a prismatic shape as in the cemented carbides. For a constant liquid volume fraction, an increase in temperature increases the value of K (Ref 17). In addition, the grain growth rate constant K increases with increasing solid fraction ϕ at a given temperature.

Generally it is desirable to avoid the microstructural coarsening stage of LPS, because it can degrade the mechanical properties. Pore coarsening is also a concern, especially if the sintering atmosphere or decomposition products become entrapped in closed pores (Ref 18).

Final Stage Sintering. The final stage of LPS is controlled by the slow densification of the solid skeleton structure. As noted, microstructural coarsening continues, and the residual pores enlarge if they contain trapped gas (Ref 18). If the entrapped gas is insoluble in the material, the gas pressure inside the pores restricts densification. If the entrapped gas has slight solubility in the matrix, large pores grow at the expense of smaller pores via an Ostwald ripening mechanism. This leads to compact swelling, which becomes more pronounced with large volume fractions of liquid. Pore coarsening due to gas entrapment is greatly reduced if the gas is highly soluble in the material; pore coarsening can be avoided with vacuum sintering.

Fabrication Concerns

Liquid-phase sintering is primarily limited to compositions with high solid volume fractions due to the poor dimensional stability associated with excess liquid. Although high liquid contents aid densification, they give distortion due to a lack of contact between the solid particles. However, the solid/liquid ratio needed to avoid distortion depends on the alloy system. Classic LPS systems, such as W-Ni-Fe, exhibit solubility of the solid in the liquid phase and generally display distortion at solid volume fractions of about 0.75 to 0.80 (Ref 19, 20). However, combinations of metals with little chemical similarities, such as tungsten and copper or molybdenum and copper, resist distortion for solid volume fractions as low as 0.2 (Ref 21). These systems generally have a much smaller grain size and are characterized by a high solid-liquid dihedral angle. An estimate of the critical solid volume fraction for structural rigidity as a function of the dihedral angle is plotted in Fig. 5.

The microstructure of LPS materials is also affected by gravity. For systems with density differences between the solid and liquid phases, gravity may cause solid-liquid segregation leading to compositional, microstructural, and property gradients. For example, the high-density tungsten grains in W-Ni-Fe alloys tend to settle to the bottom of the sample resulting in localized

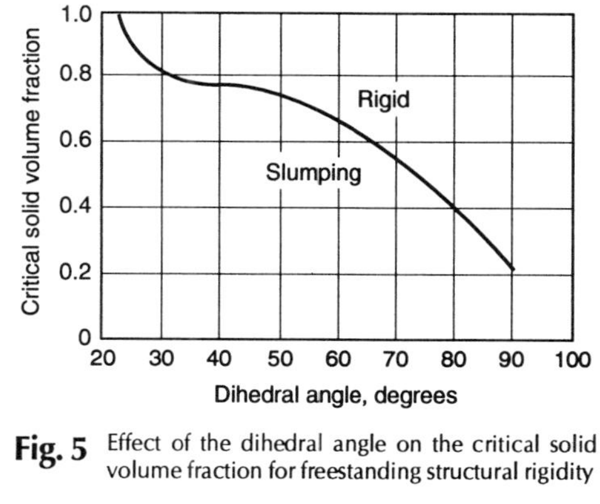

Fig. 5 Effect of the dihedral angle on the critical solid volume fraction for freestanding structural rigidity

Table 3 Typical properties of an Fe-10Cu-0.3C alloy after liquid-phase sintering

Property	6.4 g/cm^3, sintered	6.4 g/cm^3, heat treated	7.1 g/cm^3, sintered	7.1 g/cm^3, heat treated
Hardness	50 HRB	25 HRC	80 HRB	40 HRC
Yield strength, MPa	280	...	395	655
Tensile strength, MPa	310	380	550	690
Elongation, %	0.5	0.5	1.5	0.5
Fatigue strength, MPa	115	145	210	260
Impact energy, J	4	...	11	...
Elastic modulus, GPa	90	90	130	130

Source: Ref 27

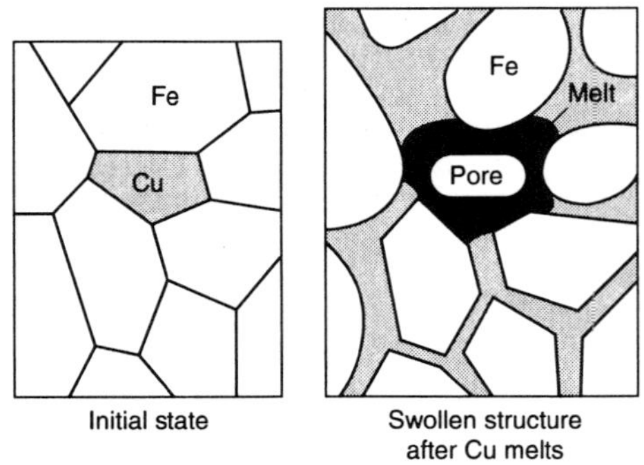

Fig. 6 Swelling in the Fe-Cu system associated with liquid copper penetration of the iron grain-boundaries, leading to a separation of the solid grains and formation of a pore at the prior copper site

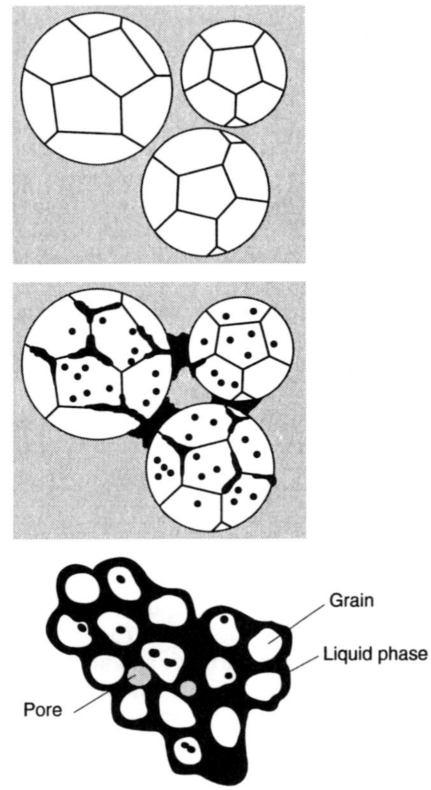

Fig. 7 Sequence of stages leading to densification via supersolidus sintering

Table 4 Typical properties of industrial cemented carbides after liquid-phase sintering

Type	Binder, wt%	Density, g/cm^3	Elastic modulus, GPa	Transverse strength, GPa	Hardness, HRA
TiC-Fe	55	6.6	300	1.9	87
WC-Co	14	14.1	520	2.8	87
WC-Co	11	14.4	550	2.7	90
WC-Co	6	15.0	630	1.9	91
WC-Co	3	15.2	680	1.5	93
WC-TaC-TiC-Co	5	12.0	500	1.5	92

Source: Ref 12

higher density (Ref 19, 20, 22–26). Increases in grain size, contiguity, and connectivity are associated with the higher solid volume fraction in the settled region. X-ray diffraction on this region indicates that there is no preferred orientation. However, a preferred orientation arises at the top of the sintered compact, indicating that grain rotation to a low-energy configuration occurs in this region (Ref 19).

Applications

Iron-Copper Alloys. Liquid-phase sintering is extensively used for Fe-Cu alloys in structural components for automobiles, household appliances, farming equipment, office machines, and electrical motors. An especially large market is for drivetrain components for the automotive industry. These parts require the high production rates, excellent mechanical properties, and dimensional control to within ±0.025 mm available through uniaxial die compaction and LPS (Ref 27). The Fe-Cu system is characterized by a high solubility of copper in iron and a low solubility of iron in copper. These characteristics result in swelling with the initial copper melt formation, as shown in Fig. 6. Swelling occurs due to the penetration of copper into the iron grain boundaries. Pores form at the prior-copper particle sites. The amount of swelling is dependent on the green density, amount of copper addition, particle sizes, internal powder porosity, copper distribution, amount of carbon, and atmosphere. Normally, swelling would be undesirable, but under controlled conditions, it enables the sintered dimensions to match the die dimensions to greatly simplify tooling design and achieve better dimensional tolerances. Zero dimensional change is accomplished by using the swelling to offset the die-to-green shrinkage and any other shrinkage that may occur during heating. Carbon and atmosphere control are critical because of their role on the surface energies and consequently the degree of penetration of the bonds between the iron particles. The solid solution-strengthening of copper in iron provides adequate mechanical properties, even with the presence of 10 to 15% porosity. Properties improve with increasing density or by heat treating, as shown in Table 3. Further increases in strength and ductility are possible by infiltrating the remaining pore space with copper.

Cemented Carbides. Another large commercial application of LPS is in the fabrication of cemented carbides for machining inserts, punches, dies, cutting tools, milling inserts, and saw blades. They are also used in rock drilling equipment, electrical contacts, armor-piercing projectiles, and in applications that require resistance to wear, erosion, cavitation, abrasion, and/or penetration. Cemented carbides consist of transition metal compounds of carbon in matrix phases, which are typically composed of cobalt, iron, or nickel alloys. Typical properties of industrial cemented carbides are given in Table 4.

Practically all commercial cemented carbides, including the carbides of titanium, tantalum, vanadium, and niobium, are liquid phase sintered. WC-Co is a classic example. Densification occurs during heating prior to liquid formation, due to a high stored strain energy and defect structure introduced by milling applied to the fabrication of these powders.

The anisotropic surface energy of the hexagonal close-packed tungsten carbide results in grain polygonization on heating. Additions of TiC or TaC are effective in controlling the grain shape and size during sintering. The peak sintering temperature is about 1400 °C under a vacuum of approximately 100 Pa. Carbon control is key in attaining the desired attributes, including hardness, strength, wear resistance, and fracture strength.

Supersolidus Liquid-Phase Sintering

Supersolidus LPS involves heating a prealloyed powder between the solidus and liquidus temperatures to form a liquid phase (Ref 28, 29). The fundamental difference between classic LPS and SLPS lies in the sequence of events leading to densification. A schematic of the stages in SLPS is shown in Fig. 7. In the first stage, liquid forms as a prealloyed powder is heated above the solidus temperature. The commonly observed liquid formation sites are the grain boundaries within a particle, the interparticle neck region, and the grain interior. These sites depend on several factors, such as powder microstructure, alloy chemistry, particle size, and the heating rate. As the liquid volume increases, at a critical temperature above the solidus a threshold amount of liquid exists along the grain

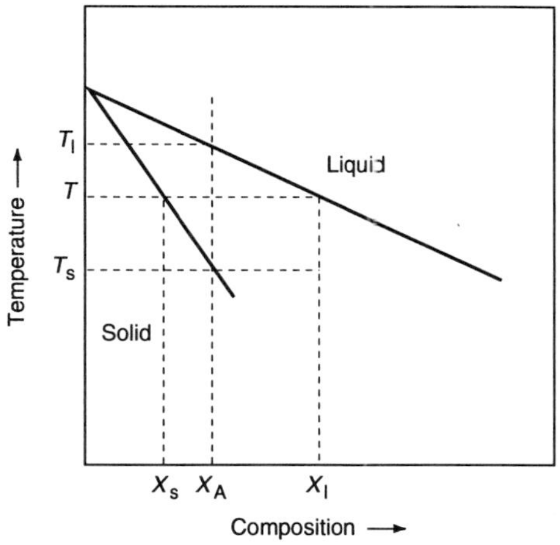

Fig. 8 Section of a hypothetical binary phase diagram showing the characteristics of a two-phase (solid + liquid) region that can influence supersolidus sintering. *T*, sintering temperature; T_s, solidus temperature; T_l, liquidus temperature; X_s, composition of solid phase; X_l, composition of liquid phase; X_A, alloy composition

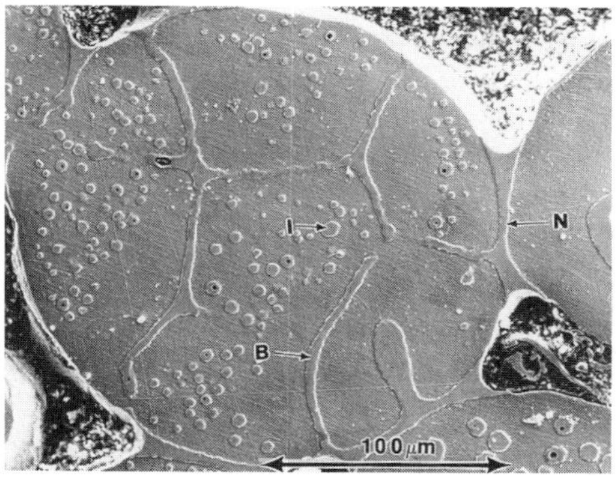

Fig. 9 Scanning electron micrograph showing the liquid formation and particle bonding in a prealloyed bronze water-quenched from 843 °C. The regions I, N, and B, represent intragranular neck, and grain-boundary liquid, respectively. Courtesy of Paul Murley

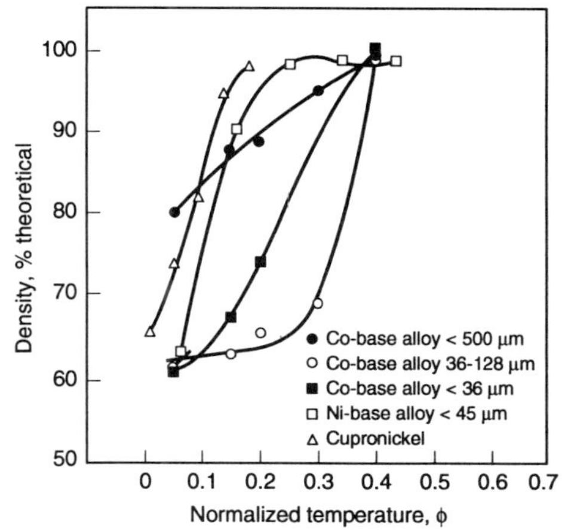

Fig. 10 Sintered density vs. normalized temperature for selected prealloyed powders, showing the dependence of particle size and alloy composition on the sintering behavior. Source: Ref 29, 39, 40

boundaries (Ref 28–31). Above this threshold, the grains possess enough mobility to rearrange, leading to particle fragmentation and capillary-induced rearrangement. Thus, densification during SLPS is analogous to viscous flow sintering, because the semi-solid particles turn mushy and flow once the liquid spreads along the grain boundaries. Subsequently, continued densification occurs by solution reprecipitation, grain shape accommodation, and pore removal, as in classic LPS.

Processing Variables. Both material and processing parameters affect SLPS. The former include the phase diagram characteristics and the chemical constituents of the alloy, especially with respect to the presence of additives or impurities. The latter include powder size, heating rate, sintering temperature, sintering time, and sintering atmosphere.

A phase diagram is an important tool for predicting the sensitivity of the liquid volume variation with respect to the sintering temperature. Figure 8 shows a section of a schematic binary phase diagram where T_s and T_l are the solidus and liquidus temperatures, respectively. A greater separation between the solidus and liquidus (i.e., $T_l - T_s$) is desirable, because the liquid volume fraction dependence on temperature is proportionate to the inverse of this separation. Similarly, a steep solidus and liquidus slope implies less sensitivity in the liquid volume with respect to an increase in temperature, thus simplifying the process control. In reality, atomized powders experience non-equilibrium solidification. Therefore, an equilibrium phase diagram can only be used as a qualitative predictor of sintering behavior.

During SLPS, the commonly observed sites for liquid melting are the neck region between particles, the grain boundaries within a polycrystalline particle, and intragranular isolated pools, as shown in Fig. 9. Each of these liquids contributes to densification in a unique way, depending on the region where it is present. For example, the liquid at the neck region provides the capillary force for rearrangement, as in classic LPS. The grain boundary liquid is responsible for particle fragmentation. However, the liquid in intragranular pools is essentially trapped and hence does not contribute directly to densification (Ref 32, 33). The optimal quantity of liquid for densification varies from 5 to 30%, depending on the system and sintering conditions. The liquid volume fraction is also dependent on the powder microstructure, heating rate, and powder chemistry. For example, large particles experience a slow cooling rate, resulting in greater segregation of the constituents, especially if the partition coefficient is small with respect to the base alloy. Therefore, the solidus temperature of large particles is expected to be slightly lower than that of fine particles. In reality, the difference depends on both the extent of segregation and the nature of additives in the large and small powders. Heating rate is also a factor, because a slow heating rate allows for homogenization of the segregated species, thus raising the solidus. Conversely, a rapid heating rate can lead to incipient liquid formation at a lower temperature, due to localized melting of the segregated additives.

The identification of sintering temperature for a given alloy composition is of fundamental concern in SLPS. Rapid densification occurs above a critical temperature due to viscous flow of the solid grains in a liquid matrix (Ref 28, 34, 35). The viscous flow is facilitated by the lowering of solid-liquid viscosity with an increase in temperature. Too high a sintering temperature results in degraded mechanical properties due to excessive grain coarsening (Ref 36–38), besides leading to compact slumping. Typically, the transition from an undersintered state to a fully sintered state occurs over a narrow temperature interval. In many cases, the sintering window is narrow (10 to 20 °C in the vicinity of the solidus temperature). Thus, for any alloy system the sintering temperature has to be optimized for a given starting particle size and heating rate. This is shown in Fig. 10, which is the plot of density versus normalized temperature for several different systems sintered via SLPS (Ref 29, 39, 40). The temperature is represented using a parameter ϕ, defined as:

$$\phi = \frac{T - T_s}{T_l - T_s} \quad \text{(Eq 2)}$$

where *T* is the sintering temperature.

A slow heating rate can induce significant densification by solid-state sintering prior to the formation of liquid phase. This implies that the liquid has to penetrate the solid-solid bonds to cause rearrangement, thus leading to restrained densification kinetics via particle rearrangement (Ref 41). A slow heating rate allows time for segregated species to homogenize. However, in most alloys that rely on the presence of trace amounts of segregated species (e.g., boron, phosphorous, sulfur, or carbon) to induce grain-boundary melting for enhanced densification, chemical homogenization can reduce liquid formation, thus restraining shrinkage.

The optimum sintering time is determined by the sintering temperature. For example, if sufficient liquid exists, complete densification can be achieved by rearrangement processes alone. In this case, the sintering time can be as low as 10 to 15 min, as shown in Fig. 11 (Ref 42). Alterna-

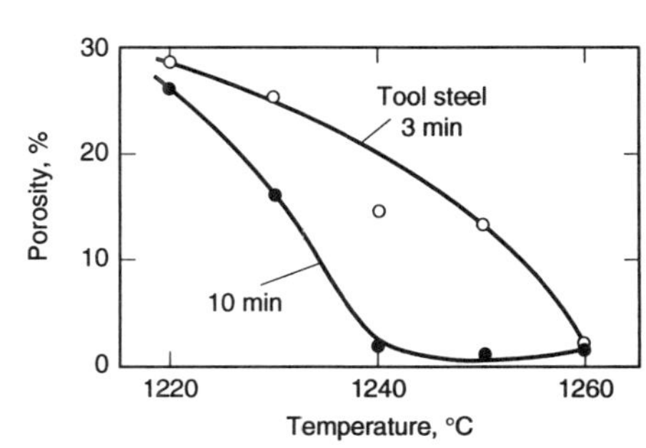

Fig. 11 Sintered porosity for a tool steel powder processed by supersolidus liquid-phase sintering for two hold times at the indicated sintering temperatures. Source: Ref 42

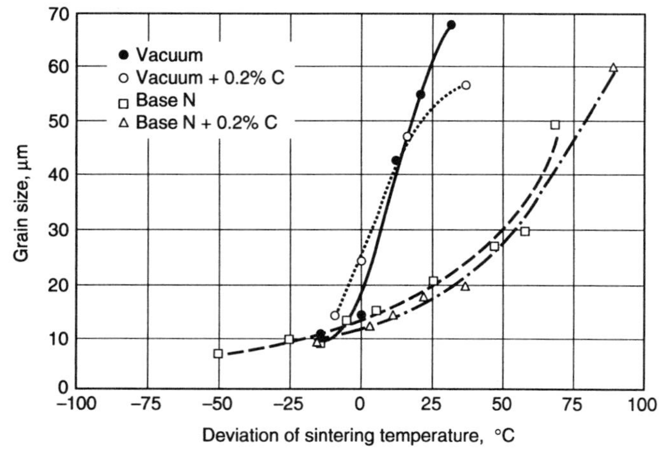

Fig. 12 Variation of grain size with the deviation from the optimum sintering temperature for different sintering atmospheres in T15 steel. Source: Ref 36

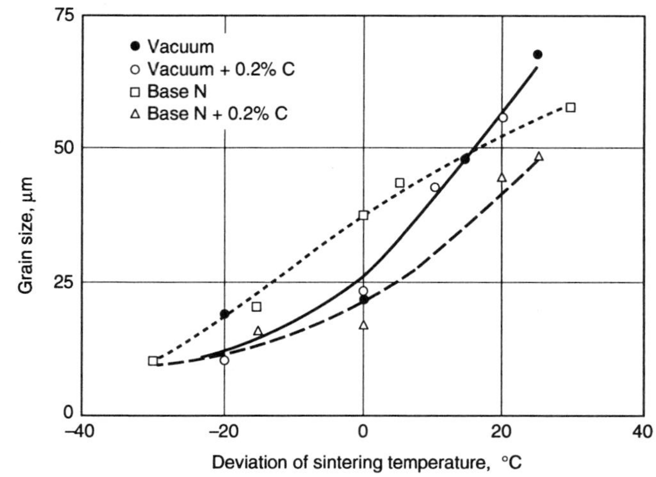

Fig. 13 Variation of grain size with the deviation from the optimum sintering temperature for different sintering atmospheres in M2 steel. Source: Ref 36

tively, a lower sintering temperature implies less liquid; hence, time-dependent mechanisms such as solution reprecipitation, grain shape accommodation, and pore removal are required to attain full density. These mechanisms are concomitant with grain coarsening.

The choice of sintering atmosphere does not affect the sintering mechanism. However, the degree of densification and the resulting mechanical properties can be significantly affected by the sintering atmosphere (Ref 36, 37). Vacuum sintering is widely employed for high-performance tool steels and nickel- and cobalt-base superalloys (Ref 40, 43–46). This is because vacuum avoids trapping gas in the pores at the final stage of densification, giving higher density and hardness. However, studies on T15 tool steel (Ref 36) have shown that sintering in a gas atmosphere of N_2, H_2, and CH_4 reduces the optimal sintering temperature and decreases the grain size as compared to vacuum. This is due to the tendency of nitrogen to substitute for carbon in vanadium nitride, which releases carbon into the matrix and lowers the solidus temperature. The vanadium carbonitride pins the grain boundaries and retards grain growth. Opposite results have been obtained for M2 (Ref 36), where the nitrogen dissolution in the matrix increases the solidus temperature. Thus the choice of sintering atmosphere depends mainly on the alloy system, in addition to economic considerations.

Grain coarsening during SLPS is analogous to that in classic LPS. However, in SLPS, the effects of temperature and liquid volume fraction are not separable, because the two are directly related. Simultaneously, the solubility of the solid and liquid continuously changes with temperature. Qualitative studies on T15 and M2 (Ref 36), BT42, and T42 (Ref 47) have shown that grain coarsening is rapid above the solidus temperature and is dependent on the sintering atmosphere, as shown in Fig. 12 and 13. Some alloys, such as boron-doped austenitic steels, exhibit a 20-fold increase in the grain growth rate constant between the subsolidus and supersolidus (Ref 48), as shown in Fig. 14. Therefore, factors such as sintering atmosphere, temperature, and time are important variables in SLPS processing.

Applications and Fabrication Concerns. Over the past 30 years, SLPS has been widely used for sintering both ferrous and nonferrous alloys. The former include stainless steels (including the 300 and 400 series), tool steels (e.g., T1, T2, T6, T15, T42, M2, M42), and precipitation-hardened steels such as 17-4 PH. The latter ranges from simple systems such as monels, brasses, and bronzes to complex nickel- and cobalt-base superalloys. Several existing applications based on SLPS include superalloy jet engine defect repair, self-lubricating bearings, corrosion-resistant components from stainless steels, high-conductivity solder pastes, wear surfaces from cobalt-chromium alloys, diamond bonds from molybdenum or cobalt alloys, gas turbine blades from martensitic stainless steels, and cutting tools from high-speed steels. Of these listed applications, the processing of tool steels and stainless steels are most important.

Current commercial full-density sintering of tool steel involves consolidation of water-atomized powders using either uniaxial die pressing or cold isostatic pressing to about 70 to 85% of theoretical density. Sintering is commonly performed in a vacuum furnace in the vicinity of the solidus temperature with tight temperature control. Typical sintering temperatures for tool steel powders such as M2, M3, M4, M35, M42, and T15 range from 1220 to 1260 °C with a recommended temperature uniformity of ±8 °C.

High-density processing of stainless steels such as 304, 316L, and 17-4PH to densities greater than 96% of theoretical is performed in a vacuum, batch-atmosphere, or continuous furnace at temperatures ranging from 1300 to 1350 °C. The use of a reducing atmosphere such as hydrogen, low dewpoint (less than –40 °C), and

Table 5 Examples of supersolidus liquid-phase sintering cycles

Material	Powder size, μm	Sintering temperature, °C	Time, h	Atmosphere	Density, % theoretical
316L	<20	1300 to 1350	1 to 2	Vacuum or hydrogen	96 to 99
316L + boron	<20	1230 to 1245	0.5 to 1	Hydrogen	98 to 100
17-4PH	<20	1300 to 1325	1 to 2	Vacuum or hydrogen	97 to 99
Bronze	100 to 200	850 to 880	0.5 to 1	Hydrogen	92 to 96
Ni superalloy	50 to 250	1250 to 1375	0.5 to 2	Vacuum	96 to 100
M2	<100	1265 to 1280	1 to 2	Vacuum or N_2-H_2-CH_4	99 to 100
T15	<100	1225 to 1270	1 to 2	Vacuum or N_2-H_2-CH_4	99 to 100
T42 modified	...	1150	1	N_2-H_2-CH_4	99 to 100
Tool steels + Cu_3P	<45	1160	0.5 to 1	Vacuum	99 to 100

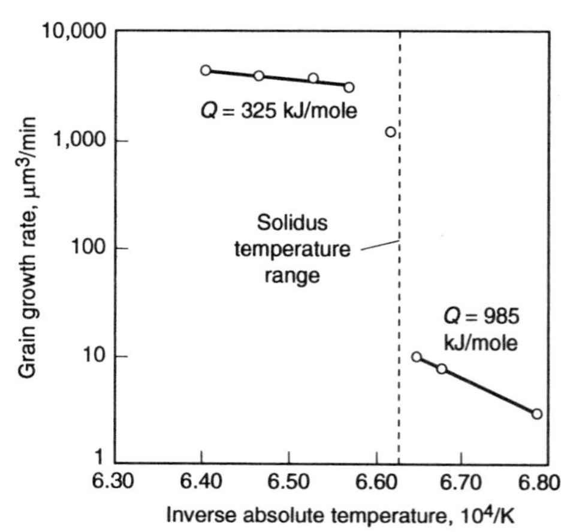

Fig. 14 Variation in the grain growth rate constant as a function of sintering temperature for a boron-doped austenitic stainless steel

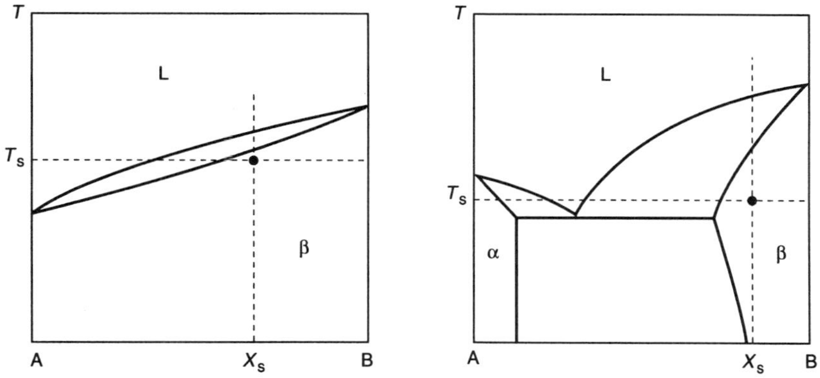

Fig. 15 Two binary-phase diagrams showing possible conditions where transient liquid-phase sintering occurs. The mixed powders form a temporary liquid, either by melting the low-temperature constituent or by eutectic formation. The final composition is in a single-phase region. *T*, sintering temperature; T_s, solidus temperature; X_s, composition of solid phase

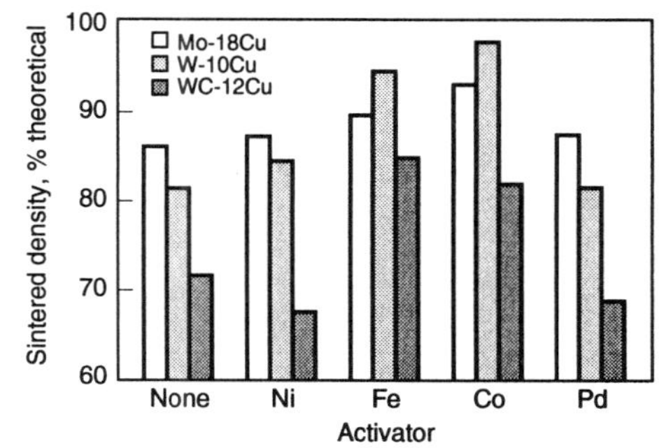

Fig. 16 Effects of transition elements on the sintered densities of W-Cu, Mo-Cu, and WC-Cu

fast cooling rate are some of the parameters that dictate the final properties.

From a practical consideration, temperature control plays the most important role during SLPS. Sintering below a threshold temperature results in incomplete densification due to insufficient liquid. The upper critical temperature is dictated by the shape distortion that results from the presence of excess liquid. Simultaneously, the temperature deviations that can be tolerated during the sintering cycle are alloy specific. In most commercial alloys, the sintering temperature has to be controlled to within ±5 °C. Some examples of commercial sintering cycles are listed in Table 5. In addition, the interactions between process parameters, such as the interaction between heating rate and sintering time, and the material parameters, such as powder chemistry, particle size, and grain size, are complex and are just beginning to be fully understood (Ref 49–51).

Transient Liquid-Phase Sintering

Transient LPS is observed in mixed powder systems where the liquid phase has a high solubility in the base material. Some examples of transient LPS include the Cu-Sn, Cu-Al, Ni-Cu, Fe-Mo-C, Fe-P, and Fe-Ti alloys. Figure 15 gives two sample phase diagrams of systems that could be processed using a transient liquid from mixed A and B powders. When a liquid forms, it initially results in melt penetration and shrinkage due to rearrangement. Unlike classic LPS, which involves the presence of a persistent liquid, the liquid in TLPS has a high solubility in the solid. As a consequence, the low-melting additive is replaced by a pore after liquid homogenization occurs. The advantage of a transient liquid is that the short liquid duration eliminates coarsening and results in uniform dimensional change. However, because the liquid content depends on several processing parameters, TLPS is highly sensitive to processing conditions such as the particle size, amount of additive, heating rate, and maximum temperature. Additionally, the solubility ratio necessary for the liquid to be transient can induce swelling during heating to the sintering temperature for certain systems. For applications such as porous bronze bearings this is beneficial, while for structural materials the porosity and swelling are detrimental. Some commercial applications for TLPS include dental amalgam based on silver and mercury, self-lubricating porous bronze bearings, and soft magnetic materials, such as Fe-Si and Fe-P.

Transient LPS of Fe-P and Fe-Si Magnetic Alloys. Ferro-phosphorus, Fe_3P, which is used as the source for P in the Fe-P system, exhibits a transient sintering response above its melting temperature near 980 °C (1800 °F). The Fe-0.45%P is the most common composition in this alloy system. Likewise, the ferrosilicon master-alloys (usually containing 9 or 17 wt% Si) exhibit a transient liquid phase above 1200 °C (2200 °F). A common alloy composition, Fe-3%Si, is used in applications involving high resistivity and low core loss, such as in relays and printer actuator mechanisms. Both the Fe-P and Fe-Si systems benefit from using high sintering temperatures that result in enhanced densities due to transient liquid phase sintering and a corresponding increase in the magnetic properties.

Activated Liquid-Phase Sintering

Activated LPS can be used to improve the sintering behavior of systems with little or no solubility of the solid in the liquid. Densification is enhanced by the addition of elements that segregate to grain boundaries and lower the activation energy for solid-state diffusion. Such elements can be identified by phase diagram characteristics. Limited additive solubility in the liquid helps to ensure segregation to grain boundaries, while the formation of low-melting temperature intermetallic phases with the base metal provides the high-diffusivity phase needed for enhanced solid-state sintering in the presence of the liquid. For liquid volume fractions of 35 vol% or more, high sintered densities can be achieved in systems that lack mutual solubility, because grain shape accommodation is not required. Thus, activators are only needed for compositions with low volume fractions of liquid. In this case, volume diffusion through a high-diffusivity interboundary phase provides the grain shape accommodation necessary to achieve high sintered densities.

Systems in which ALPS is particularly applicable are those in which high electrical or thermal conductivity is desired, such as W-Cu, Mo-Cu, and WC-Cu (Ref 52, 53). In the case of W-Cu, the base metal is practically insoluble in the liquid, but other transition elements, such as cobalt, iron, nickel, and palladium have substantial solubility for tungsten. Traditionally nickel, which has complete solubility with copper, is added to W-Cu to increase densification by solution reprecipitation through the Cu-Ni liquid phase, but the alloying of copper and nickel degrades the electrical and thermal conductivity. High conductivities can be maintained if the amount of activator (i.e., nickel) is small; however, this results in only a small increase in solubility in the liquid phase, which is not highly beneficial to improved density. On the other hand, cobalt and iron also have substantial solubility for tungsten but have only limited solubility in copper. As shown in Fig. 16, they are much more effective activators of W-Cu, Mo-Cu, and WC-Cu, which all have similar phase relationships with these four activators.

Increasing the amount of additive in a system that densifies by ALPS above a certain amount hinders densification and lowers properties, as shown in Fig. 17. The optimal amount of additive for ALPS corresponds closely with the amount of activator needed to coat the particle surfaces plus the amount of activator that remains in solution in the liquid phase. For systems that densify via solution reprecipitation, such as W-Cu-Ni, increasing the solubility of the liquid phase by increasing the nickel concentration improves the densification behavior.

Densification in an ALPS system is further improved by a high initial homogeneity and a fine particle size. In fact, with particle sizes below 1 μm, solid-state diffusion in the presence of the liquid phase can provide near-theoretical densities without the need for an activator (Ref

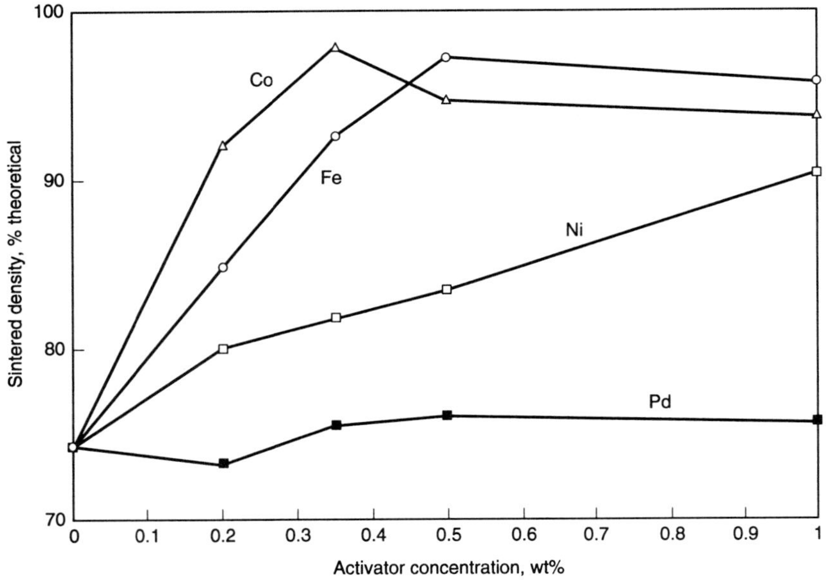

Fig. 17 Effects of activator concentration on W-10Cu for samples sintered at 1300 °C for 1 h in H_2

54). Low compaction pressures are beneficial because much of the densification occurs via rearrangement. When there are low volume fractions of liquid a greater portion of densification occurs through the surface layer, due to more contacts, and thus, there is more opportunity for solid-state sintering. Slow heating rates are detrimental to the densification behavior due to the formation of solid-state sinter bonds, which must be dissolved prior to rearrangement. Sintering temperatures and times can be optimized according to the desired properties and specific powder characteristics.

REFERENCES

1. R.M. German, The Contiguity of Liquid Phase Sintered Microstructures, *Metall. Trans.*, Vol 16A, 1985, p 1247–1252
2. W.D. Kingery, Densification during Sintering in the Presence of a Liquid Phase, *J. Appl. Phys.*, Vol 30, 1959, p 301–306
3. R.M. German, The Identification of Enhanced Sintering Systems through Phase Diagrams, *Modern Developments in Powder Metallurgy,* Vol 15, E.N. Aqua and C.I. Whitman, Ed., Metal Powder Industries Federation, 1985, p 253–273
4. R.M. German and B.H. Rabin, Enhanced Sintering through Second Phase Additions, *Powder Met.*, Vol 28, 1985, p 7–12
5. C.M. Kipphut and R.M. German, Alloy Phase Stability in Liquid Phase Sintering, *Sci. Sintering,* Vol 20, 1988, p 31–41
6. J.J. Burton and E.S. Machlin, Prediction of Segregation to Alloy Surfaces from Bulk Phase Diagrams, *Phys. Rev. Lett.*, Vol 37, 1976, p 1433–1436
7. S. Farooq, A. Bose, and R.M. German, Theory of Liquid Phase Sintering: Model Experiments on the W-Ni-Fe Heavy Alloy System, *Progress in Powder Metallurgy,* Vol 43, C.L. Freeby and H. Hjort, Ed., Metal Powder Industries Federation, 1987, p 65–77
8. S. Farooq and R.M. German, An Update on the Theory of Liquid Phase Sintering, *Sintering '87,* Vol 1, S. Somiya et al., Ed., Elsevier Applied Science, 1988, p 459–464
9. Y.V. Naidich, I.A. Lavrinenko, and V.A. Evdokimov, Densification during Liquid Phase Sintering in Diamond-Metal Systems, *Soviet Powder Met. Metal Ceram.*, Vol 11, 1972, p 715–718
10. W.J. Huppmann, The Elementary Mechanisms of Liquid Phase Sintering, Part I: Rearrangement, *Z. Metallkde.*, Vol 70, 1979, p 707–713
11. J.S. Lee, W.A. Kaysser, and G. Petzow, Microstructural Changes in W-Cu and W-Cu-Ni Compacts during Heating up for Liquid Phase Sintering, *Modern Developments in Powder Metallurgy,* Vol 15, E.N. Aqua and C.I. Whitman, Ed., Metal Powder Industries Federation, 1984, p 489–506
12. R.M. German, *Liquid Phase Sintering,* Plenum Press, 1985, p 82
13. W.J. Huppmann, Sintering in the Presence of a Liquid Phase, *Sintering and Catalysis,* G.C. Kuczynski, Ed., Plenum Press, 1975, p 359–378
14. W.J. Huppmann, The Elementary Mechanisms of Liquid Phase Sintering, Part II: Solution-Reprecipitation, *Z. Metallkd.*, Vol 70, 1979, p 792–797
15. W.A. Kaysser and G. Petzow, Present State of Liquid Phase Sintering, *Powder Metall.*, Vol 28, 1985, p 145–150
16. H.H. Park and D.N. Yoon, Effect of Dihedral Angle on the Morphology of Grains in a Matrix Phase, *Metall. Trans.*, Vol 16A, 1985, p 923–928
17. D. Uffelmann, W. Bender, L. Ratke, and B. Feuerbacker, Ostwald Ripening in Lorentz-Force Stabilized Cu-Pb Dispersions at Low Volume Fractions, Part I: Experimental Observations, *Acta Metall. Mater.,* Vol 43, 1994, p 173–180
18. R.M. German and K.S. Churn, Sintering Atmosphere Effects on the Ductility of W-Ni-Fe Heavy Metals, *Metall. Trans.,* Vol 15A, 1984, p 747–754
19. C.M. Kipphut and R.M. German, Gravity and Configurational Energy Induced Microstructural Changes in Liquid Phase Sintering, *Metall. Trans.,* Vol 19A, 1988, p 1905–1913
20. C.M. Kipphut and R.M. German, The Gravitational Effects on Liquid Phase Sintering, *Advances in Powder Metallurgy,* Vol 2, Metal Powder Industries Federation, 1989, p 415–429
21. A. Upadhyaya, J.L. Johnson, and R.M. German, Dimensional Stability in Liquid Phase Sintered W-Ni-Cu Alloys, *Tungsten and Refractory Metals—1995,* Metal Powder Industries Federation, 1995, p 3–18
22. A.N. Niemi and T.H. Courtney, Settling in Solid-Liquid Systems with Specific Application to Liquid Phase Sintering, *Acta Metall.,* Vol 31, 1983, p 1393–1401
23. S.C. Yang and R.M. German, Gravitational Limit of Particle Volume Fraction in Liquid Phase Sintering, *Metall. Trans.,* Vol 22A, 1991, p 786–791
24. S. Kohara and M. Hoshino, *Advances in Powder Metallurgy and Particulate Materials,* Vol 3, Metal Powder Industries Federation, 1994, p 295–302
25. R.M. German, Microstructure of the Gravitationally Settled Region in a Liquid Phase Sintered Dilute Tungsten Heavy Alloy, *Metall. Mater. Trans.,* Vol 26A, 1995, p 279–288
26. Y. Liu, D.F. Heaney, and R.M. German, Gravity Induced Solid Grain Packing during Liquid Phase Sintering, *Acta Metall. Mater.,* Vol 43, 1995, p 1587–1592
27. R.M. German, *Powder Metallurgy Science,* 2nd ed., Metal Powder Industries Federation, 1994, p 402
28. R.M. German, Supersolidus Liquid Phase Sintering, Part I: Process Review, *Int. J. Powder Metall.,* Vol 26, 1990, p 23–33
29. R. Tandon, "Densification Mechanisms and Microstructural Evolution Leading to High Density Processing of Prealloyed Powders in Supersolidus Liquid Phase Sintering," Ph.D. Thesis, The Pennsylvania State University, 1995
30. R.M. German, Supersolidus Liquid Phase Sintering, Part II: Densification Theory, *Int. J. Powder Metall,* Vol 26, 1990, p 35–43
31. R.M. German, A Quantitative Theory for Supersolidus Liquid Phase Sintering, *Powder Metall.,* Vol 34, 1991, p 101–107
32. M. Jeandin, S. Rupp, J. Massol, and Y. Bienvenu, Fundamental Study of the Later Stages of Liquid Phase Sintering of a Ni-base P/M Superalloy—Metallographic Observations on Quenched Supersolidus-Sintered Materials, *Sintering '85,* G.C. Kuczynski, D.P. Uskok-

ovic, H. Plamour, and M.M. Ristic, Ed., 1987, Plenum Press, p 179–188

33. R. Tandon, Y. Liu, and R.M. German, Role of Initial Powder Characteristics in Supersolidus Liquid Phase Sintering, *Advances in Powder Metallurgy and Particulate Materials,* Vol 3, Metal Powder Industries Federation, 1994, p 251–266
34. R. Tandon and R.M. German, Supersolidus-Transient Liquid Phase Sintering Using Superalloy Powders, *Int. J. Powder Metall.,* Vol 30, 1994, p 435–443
35. R. Tandon and R.M. German, Particle Fragmentation during Supersolidus Sintering, *Int. J. Powder Metall.,* Vol 33, 1997, p 54–60
36. S. Jauregi, F. Fernandez, R.H. Palma, V. Martinez, and J.J. Urcola, Influence of Atmosphere on Sintering of T15 and M2 Steel Powders, *Metall. Trans.,* Vol 23A, 1992, p 389–400
37. P. Maulik and W.J.C. Price, Effect of Carbon Additions on Sintering Characteristics and Microstructure of BT42 High Speed Steel, *Powder Metall.,* Vol 30, 1987, p 240–248
38. R. Tandon, Y. Liu, and R.M. German, High Density Processing of Ferrous Alloys via Supersolidus Liquid Phase Sintering, *Advances in Powder Metallurgy and Particulate Materials,* Vol 2, 1995, Metal Powder Industries Federation, p 5.37–5.49
39. S.R. Bala and J. Lund, Studies of the Supersolidus Sintering of Cupronickel Powder, *Z. Metallkd.,* Bd. 70, 1979, p 185–190
40. C. Guyard, C.H. Allibert, J. Driole, and G. Raisson, Liquid Phase Sintering of Prealloyed Powders of Co-Base Alloy, *Sci. Sintering,* Vol 13, 1981, p 149–163
41. C. Toennes and R.M. German, Density and Microstructure Control in a Martensitic Stainless Steel through Enhanced Sintering, *Powder Metall. Int.,* Vol 24, 1992, p 151–157
42. S. Takajo and M. Nitta, Observation of Liquid Phase Sintering of a High Speed Steel Powder, *Sintering '85,* G.C. Kuczynski, D.P. Uskokovic, H. Plamour, and M.M. Ristic, Ed., 1987, Plenum Press, p 189–196
43. E.J. Westerman, Sintering of Ni-Base Superalloys, *Trans. Metall. Soc. AIME,* Vol 224, 1962, p 159–164
44. K. Hajmrle and R. Angers, Sintering of Inconel 718, *Int. J. Powder Metall. Powder Tech.,* Vol 16, 1980, p 255–267
45. M. Jeandin, J.L. Koutny, and Y. Bienvenu, Liquid Phase Sintering of Nickel Base Superalloys, *Powder Metall.,* Vol 26, 1983, p 17–22
46. R. Kieffer, G. Jangg, and P. Ettmayer, Sintered Superalloys, *Powder Metall. Int.,* Vol 7, 1975, p 126–130
47. R.H. Palma, V. Martinez, and J.J. Urcola, Sintering Behavior of T42 Water Atomized High Speed Steel Powder under Vacuum and Industrial Atmospheres with Free Carbon Addition, *Powder Metall.,* Vol 32, 1989, p 291–299
48. R. Tandon and R.M. German, Processing Parameter Effects on the Mechanical Properties of a Boron-Doped Austenitic Stainless Steel, *Int. J. Powder Metall.,* 1998
49. Y. Liu, R. Tandon, and R.M. German, Modeling of Supersolidus Liquid Phase Sintering, Part I: Capillary Force, *Metall. Mater. Trans.,* Vol 26A, 1995, p 2415–2422
50. Y. Liu, R. Tandon, and R.M. German, Modeling of Supersolidus Liquid Phase Sintering, Part II: Densification, *Metall. Mater. Trans.,* Vol 26A, 1995, p 2423–2430
51. R.M. German, Liquid Phase Sintering of Prealloyed Powders, *Metall. Mater. Trans. A,* 1997, Vol 28A, 1997, p 1553–1567
52. J.L. Johnson and R.M. German, Phase Equilibria Effects on the Enhanced Liquid Phase Sintering of W-Cu, *Metall. Trans.,* Vol 24A, 1993, p 2369–2377
53. J.L. Johnson and R.M. German, Chemically Activated Liquid Phase Sintering of Tungsten-Copper, *Int. J. Powder Metall.,* Vol 30, 1994, p 91–102
54. J.L. Johnson and R.M. German, Solid-State Contributions to Densification during Liquid Phase Sintering, *Metall. Mater. Trans.,* Vol 27B, 1996, p 901–909

Cold Sintering—High Pressure Consolidation

Elazar Y. Gutmanas, Technion-Israel Institute of Technology, Haifa, Israel

COLD SINTERING or high pressure consolidation at ambient temperature, or at temperatures not exceeding 450 °C, of rapidly solidified and mechanically alloyed powders, as well as of very fine elemental powder blends, is an alternative method to achieve full density without compromising very fine and/or metastable microstructures (Ref 1–4).

Development of fine microstructures is a primary goal in the design of high performance materials (Ref 5, 6). Mechanical properties of materials with complex microstructures are controlled by the superposition of several strengthening mechanisms. Refinement of grain size, second phase precipitates or dispersoids, and the development of a homogeneous microstructure result in a combination of high strength and ductility.

Fine-scale microstructures can be obtained using advanced powder metallurgy processing techniques. Rapid solidification of powders yields supersaturated solid solutions, fine-scale precipitates, fine grain sizes, and homogeneity of composition and microstructure, as well as results in the retention of metastable phases including amorphous metal glasses (Ref 6–9). Alternatively, mechanical alloying of metal powders and ceramic particles results in extremely fine microstructures with a uniform distribution of dispersoids in the metal matrix (Ref 10–13). Similar features are observed in metal-oxide composite powders obtained by chemical methods (Ref 14–16) or by blending very fine micron/submicron elemental metal and ceramic powders (Ref 17, 18).

To produce high performance materials from rapidly solidified or other fine-scale powders, consolidation to full density is mandatory. Hot processing methods, such as extrusion, hot isostatic pressing (HIP), or powder forging, are normally employed for the consolidation of rapidly solidified or mechanically alloyed powders (Ref 7, 13, 16). The temperatures of hot processing are often higher than the subsequent service temperature. Thus, the refined microstructures obtained by rapid solidification or by mechanical alloying can coarsen, and the metastable constituents can dissolve in the matrix during hot consolidation. In the blends of very fine elemental powders, grain growth can take place.

Experimental Techniques of Cold Sintering

Most elemental powders and rapidly solidified alloys can be consolidated at ambient temperature to densities close to the theoretical (>95% TD) at the pressure of 3 GPa. With careful design, tool steel dies and punches can be used at this pressure for hundreds of cycles. Dies and punches made from CPM T15 high-speed steel have been used by the authors of the present article over the last 10 years. The T15 high-speed steel (12W-5V-5Cr-5Co-1.6C (wt%)), fabricated from the inert-gas-atomized powder by HIP, has a fine homogeneous structure and, as a result, a considerably higher strength than that of the ingot tool steel of the same composition (Ref 19). The high concentration of fine homogeneously dispersed carbides in the steel results in very good wear resistance. Some of the punches used for high pressure consolidation were prepared from the water-atomized T15 powder by cold sintering. Schematic of a pressure cell for cold sintering with punches, die walls, and punch supports made from high-speed tool steel is shown in Fig. 1. The inner and outer supporting rings are made of the high strength 4340 steel. Provided the alignment of the tooling is maintained, it is possible to use dies and punches made from CPM T15 steel up to ~3.5 GPa. The outer diameter of the die is usually only three times larger than the diameter of the part being consolidated (die opening). Typically, the compact is loaded to 2 GPa in a few seconds, further loading proceeding more slowly over the 10 to 15 s interval to achieve the pressure of 3 GPa. Pressures as high as 4.5 GPa are possible utilizing cemented carbide dies. These dies are more brittle, and high pressure consolidation of composite materials containing hard or super-hard particles often results in a failure of a cemented carbide die after several cycles.

Cylindrical dies with diameters ranging from 3 to 50 mm and rectangular dies with cross sections ranging from 2 by 18 to 40 by 42 mm have been used, the maximum size being limited mainly by the press capacity (Ref 3). Examples of dies and punches are shown in Fig. 2. A laboratory press with 5 MN capacity provides applied pressure (P) equal to 2.5 to 3 GPa for the largest dies. A large number of smaller size samples can be produced by placing each of them in a separate rigid cell (see Fig. 2b). Upscaling of the pressure die to an 88 by 88 mm opening has been done, and densities close to the theoretical have been obtained for aluminum, iron, and T15 high-speed steel employing a press with >25 MN capacity (Ref 4).

For some prealloyed, rapidly solidified, or mechanically alloyed powders, consolidation at am-

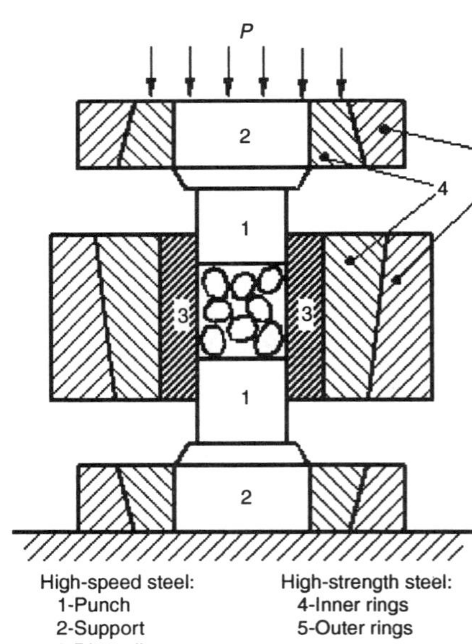

Fig. 1 Schematic of a pressure die for cold sintering

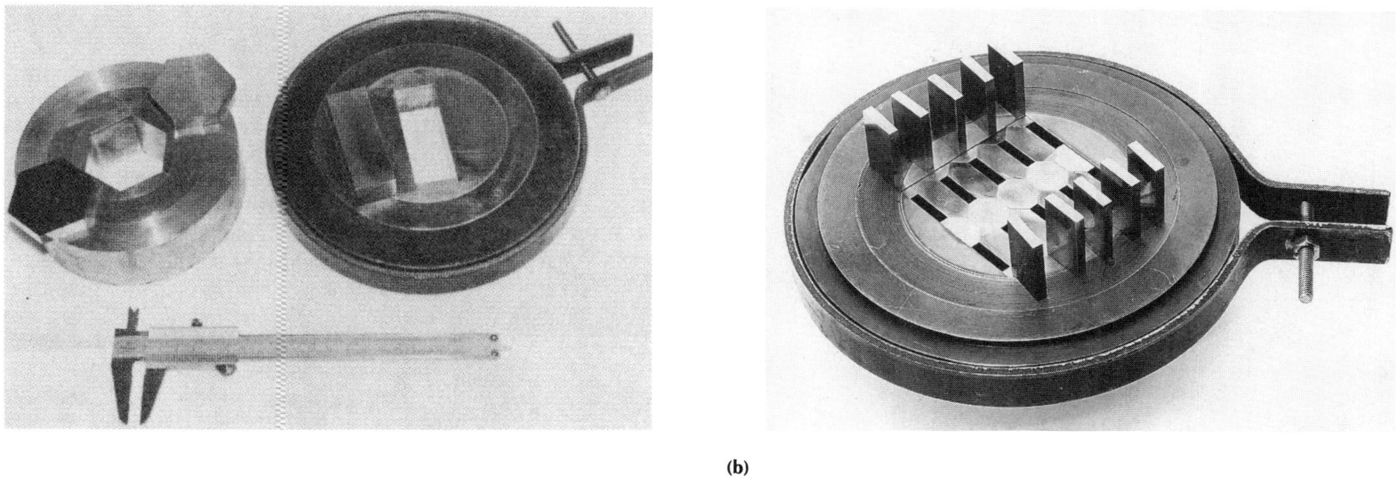

Fig. 2 Examples of dies and punches for cold sintering

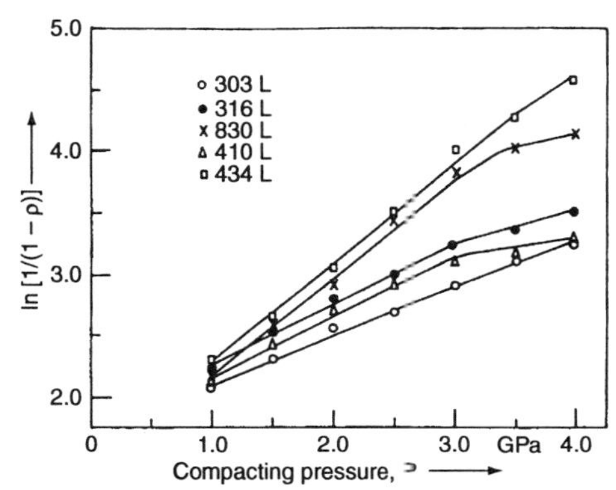

Fig. 3 Compaction curves ($\rho = f(P)$) of a stainless steel powder: $\ln[1(1 - \rho)]$ versus P

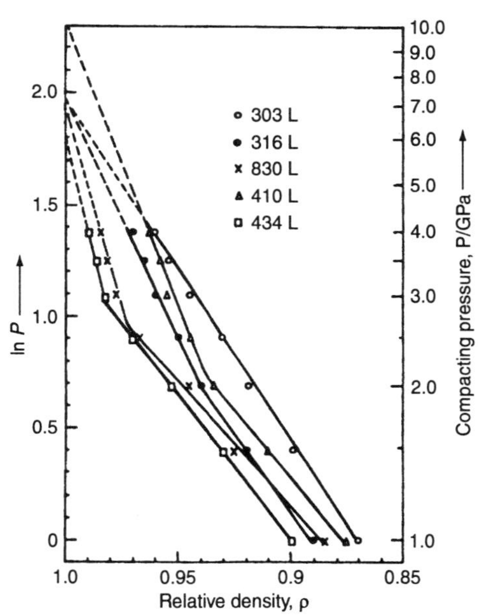

Fig. 4 Experimental data of Fig. 3 as ρ versus ln*P*

bient temperature and at $P = 3$ GPa results in densities 96 to 97% TD. Even lower densities have been obtained for some amorphous and microcrystalline rapidly solidified powders. For such powders, densities close to the theoretical can only be obtained by high pressure consolidation at temperatures above the ambient (warm consolidation). Thus, densities >99% TD were achieved at $T = 450$ °C and $P = 3$ GPa for amorphous powders (Allied Signal) and microcrystalline powders (Marco) with compositions corresponding to iron- and nickel-base wear-resistant alloys and high-speed steels (Ref 3). For some highly alloyed, rapidly solidified alloys, such as high-speed steels, stainless steels, and superalloys, consolidation at $T = 400$ °C and $P = 3$ GPa resulted in densities close to the theoretical and good bonding integrity of the powder particles. In many cases, annealing followed by repressing at $P = 3$ GPa provides consolidation to full density. At the pressure of 3 GPa, high-speed T15 steel pressure dies can be used at temperatures ≤450 to 500 °C.

Densities >96% TD were obtained by warm consolidation of amorphous ribbons of Fe-B-Si and Fe-B-Si-C alloys at 0.78 GPa and 2 GPa, respectively, at a temperature of 350 to 380 °C, which is below the glass transition temperature (Ref 20).

Densification of Powders under Pressure

Densification of powders under pressure was treated in the past by a number of authors as related to the compaction process, which is an important step in conventional powder metallurgy. Expressions used to describe density-pressure relationship were empirical in nature. For example, Heckel (Ref 21) used the following expression for compaction of powders in a rigid die:

$$\ln[1/(1 - \rho)] = kP + A$$

where ρ is relative density, P is applied pressure, and k and A are constants. The expression is in good agreement with the experimental results obtained for a number of powders at conventional pressures (<0.9 GPa) (Ref 6), as well as for the consolidation of powders under high pressures (Ref 22). The deviations from straight lines observed in $\ln[1/(1 - \rho)]$ versus P plots for stainless steel powders (see Fig. 3) were attributed to a martensite transformation at higher pressures. The empirical relationship developed by Balshin (Ref 23):

$$\ln P = K_1\rho + B$$

where K_1 and B are constants, is also in good agreement with experimental results. Figure 4 shows the experimental results of Fig. 3 in terms of Balshin's expression.

In the frames of the theory of plasticity, Torre (Ref 24) obtained an expression for the final density ρ_f as a function of the applied external pressure, P, and the yield stress of the material, σ_y, for spherical particles with spherical voids:

$$\rho_f = 1 - \exp(\tfrac{3}{2}P/\sigma_y)$$

According to this expression, $\rho_f = 87.7\%$ at $P = \sigma_y$, $\rho_f = 95.3\%$ at $P = 2\sigma_y$, $\rho_f = 98.9\%$ at $P = 3\sigma_y$, and $\rho_f = 99.75\%$ at $P = 4\sigma_y$. Experimental results obtained during high pressure consolidation of powders are in good agreement with this expression. For example, the data for pure iron and pure nickel powders in Fig. 5 fit the expression, if for iron $\sigma_y = 0.55$ and 0.7 GPa for iron and nickel, respectively. Experimental values for yield stress in compression of samples prepared from these powders at $P = 2$ GPa are $\sigma_y = 0.45$ and 0.5 GPa for iron and nickel, respectively.

Attempts were made to introduce yield stress and strain hardening of the material as densification parameters of powders (Ref 21). Such an approach usually fails because during powder densification, and especially at its final stage, the flow stress and strain hardening in the region of still open pores can be quite high and not typical

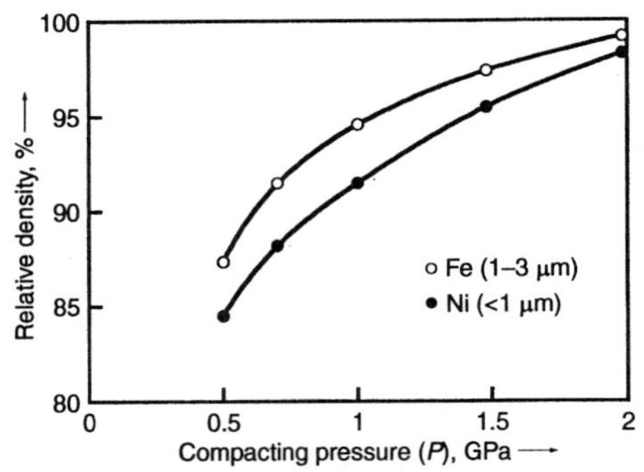

Fig. 5 Compaction curves for fine iron and nickel powders: ϒ - fine (1–3 μm) iron powders; λ - fine (<1 μm) nickel powders

for plastic deformation of a bulk sample in compression.

The final stage of densification of powders under pressure is worth additional discussion and requires a critical analysis of up-to-date results. In a number of materials densified by plastic flow, cusp-shaped pores <0.1 μm in size have been observed. The radius of material on the pore surface is much smaller than the particle radius. The material surrounding the pore has a hemispherical shape resembling the edge of a wire produced by extrusion or deep drawing (see Fig. 6). The pore radius decreases with deformation. Such cusp-shaped pores are less stable under applied pressure than the spherical pores considered by Torre (Ref 24) formed by diffusional flow. What in principle opposes the closure of cusp-shaped pores is the increasing surface tension force on the concave surfaces, resulting in LaPlace compressive stresses on these surfaces, $\sigma = \gamma_s / r_{curv}$ (γ_s is surface energy and r_{curv} is the radius of curvature). These stresses oppose tensile stresses caused by stress/pressure gradient. Enhanced surface diffusion from the concave surfaces to the convex parts of the cusp-shaped pores can be important for the final stage of densification even at a relatively low temperature.

What can be a more important problem for the final stage of densification by plastic flow is that generation of mobile dislocations is required in order to provide necessary plastic strain. Indeed, if we consider a powder particle under external pressure during the final stage of densification, the gradient of stress/pressure will result in shear stresses only in the regions adjacent to the surface of the pore. The inner part of the particle is under hydrostatic pressure. Regions adjacent to the pore become very small as the pore becomes smaller. For a pore ~0.1 μm in size, the particle region that can supply dislocations for plastic flow should be about the same size (~0.1 μm). Even for a microcrystalline material, this will most probably be a single grain. Most fine micron/submicron size metal particles are single crystals. For nanoscale particles <0.1 μm in size, the pore size will be <0.01 μm. The probability of the presence of an operative dislocation source in such a small region is very low. Thus, for the

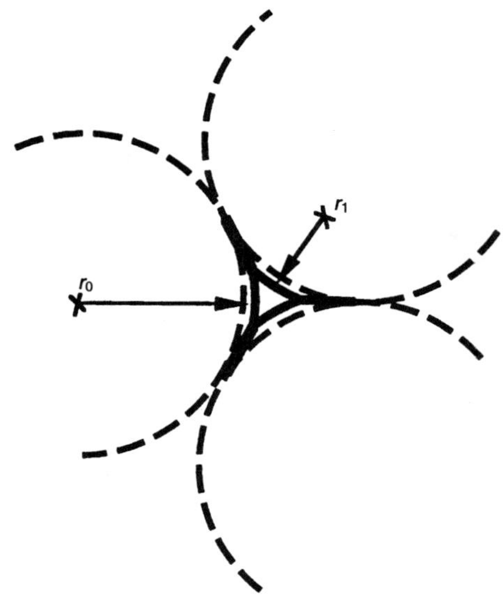

Fig. 6 Schematic of a cusp-shaped pore (triple point) in compacted spherical powders; r_1 is the radius of curvature of the pore after densification by plastic deformation.

pore closure by plastic flow at the final stage of densification, shear stresses exceeding the theoretical shear strength are required that are able to nucleate dislocations in a perfect lattice. This requirement can indicate a real change of the densification mechanism at the final stage.

Stresses necessary for dislocation nucleation in a perfect lattice are usually estimated to be in the range of $G/10$ to $G/30$, where G is shear modulus. If the shear stresses, σ_{sh}, acting on the material adjacent to the pore surface are assumed equal to the applied external pressure, the pressure $P = 3$ GPa yields $\sigma_{sh} = G/10$ for aluminum ($G = 28.6$ GPa) and $\sigma_{sh} = G/30$ for iron ($G = 86.9$ GPa). Higher shear stresses are expected around cusp-shaped pores. In the solution of the Hertzian problem for two spherical bodies of equal radius in contact under pressure (Ref 25), the maximum pressure is given by:

$$P_{max} = 0.388 \cdot 3\sqrt{2E^2P}$$

and the maximum shear stress for $\nu = 0.3$ is:

$$\sigma_{sh}^{max} = 0.31\ P_{max}$$

where P is applied pressure, E is Young's modulus, and ν is Poisson's ratio.

Maximum shear stresses for a number of metals calculated using a simplified approach that is based on the solution of the Hertzian problem and considers regions adjacent to pores as spheres being in contact under pressure are summarized in Table 1. The results indicate that increasing shear stress can result in nucleation of dislocations even in materials with high shear modulus, e.g. tungsten.

It should be noted that stresses required to move dislocations in high performance materials with high concentrations of very fine precipitates or in supersaturated solid solutions can exceed the stresses needed for dislocation nucleation in a perfect lattice of the pure metal, especially if superposition of different strengthening mechanisms takes place. According to the estimations made for iron, nickel, and cobalt base alloys (Ref 14), grain boundary strengthening for a nanocrystalline material with ~10 nm grain size is ~3 GPa, whereas dispersion strengthening due to the presence of fine particles with ~10 nm interparticle spacing is ~5 GPa. Because the high stresses required for the consolidation of prealloyed powders with very fine microstructures are rarely attainable in practice, another approach has been suggested, namely consolidation of very fine elemental powder blends. The flow stress of such blends will be affected by strain hardening only and will be controlled, at the final stage of consolidation, by stresses required to nucleate dislocation in a perfect lattice.

The Mechanism of Cold Sintering

Plastic deformation of powder particles leading to intimate contact between oxide- and contamination-free surfaces results in the formation of chemical bonds and adhesion. This has been proved by transmission electron microscopy for pure aluminum powder (Ref 26). Broken oxide layers exist as islands at powder particle interfaces. Similar observations were made for cold sintered copper, nickel, iron, and stainless steel powders. In a consolidated material, the interface between powder particles is transformed into a grain boundary. New grain boundaries formed at the former powder particle interfaces are unusual and worth special attention. Surfaces that are brought together by plastic deformation during high pressure consolidation have random orientation. Thus, the formed grain boundaries are mostly noncoherent. Usually, plastic deformation of polycrystals introduces defects in the lattice and thereby results in an increasing degree of disorder, becoming a driving force for recrystallization. In the case of consolidation of powders under high pressure

Table 1 Calculated shear stresses σ_{sh}^{max} in regions adjacent to a cusp-shaped pore

Material	Young's modulus (E), GPa	Shear modulus (G), GPa	σ_{sh}^{max}, GPa, at applied P, GPa: 1	2	3	σ_{sh}^{max}, in G, at applied P, GPa: 1	2	3
Aluminum	77.4	28.6	2.7	3.4	4.0	G/11	G/8	G/7
Iron	223	87	5.6	7.0	8.0	G/15	G/12	G/11
Copper	127	47	3.8	4.8	5.5	G/12	G/10	G/8
Tungsten	394	122	8.1	10.2	11.7	G/15	G/12	G/10

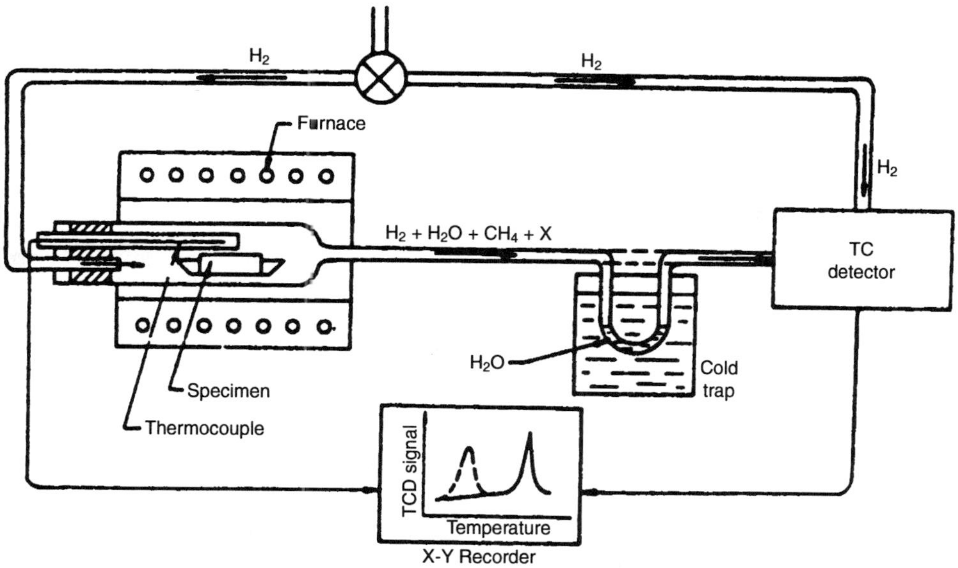

Fig. 7 Schematic of the apparatus used for investigating the reactions of hydrogen with powders and powder compacts and for construction of reduction diagrams

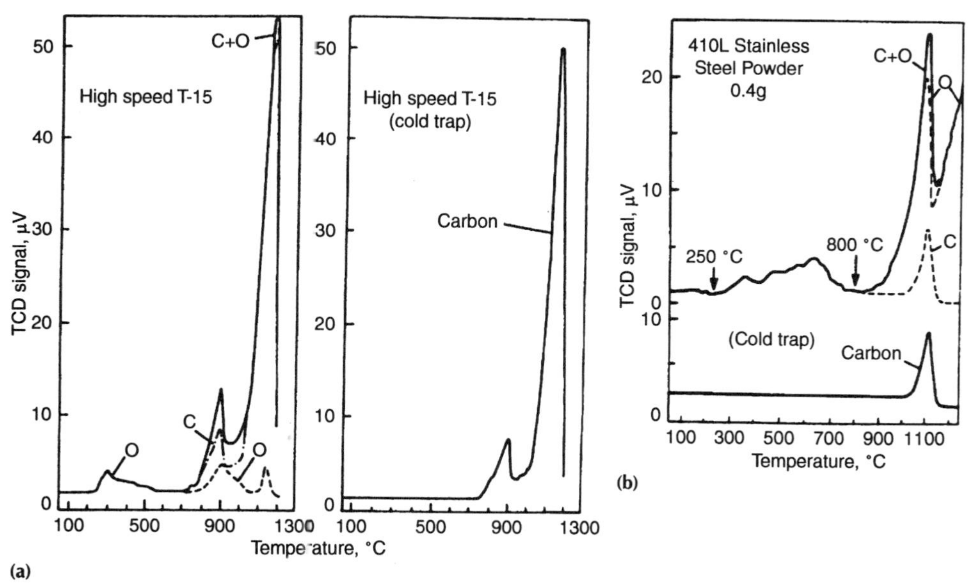

Fig. 8 Reduction diagrams for 80% dense compacts with an interconnected system of pores (water-atomized powders) (a) high speed steel T15, and (b) 410L stainless steel

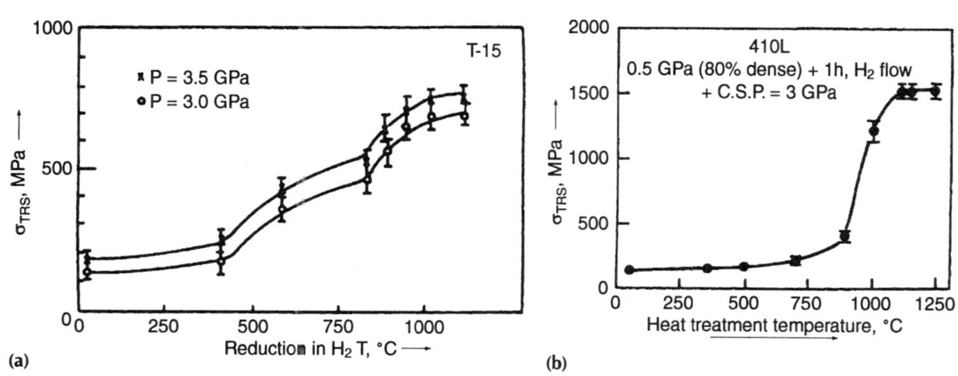

Fig. 9 Room temperature transverse rupture strength, σ_{TRS}, as a function of the reduction heat treatment temperature (1 h anneal in an H_2-flow)

and relatively low temperature, plastic deformation and surface diffusion at the interfaces can result in the partial ordering of the structure. Enhanced diffusion should also take place along interfaces in cold sintered samples. Such diffusion promotes adhesion and bonding integrity of consolidated powder particles.

Processing by Cold Sintering

Reduction of Surface Oxides

Consolidation to full density requires densification with the formation of strong adhesive bonds between the particles. When oxide- and contamination-free surfaces are brought into intimate contact, strong chemical bonds arise resulting in good integrity among the powder particles. Plastic deformation, through the breaking of oxide layers, can result in an intimate contact between the freshly formed surfaces. At higher temperatures, diffusional flow can result in dissolution or spheroidization of oxide layers on metal particle surfaces. Processing routes involving a pronounced shear deformation provide better conditions for breaking of surface oxide.

For a number of pure metals and alloys, oxide layers can be removed by heat treatment in hydrogen or in vacuum at relatively low temperatures, which guarantees the retention of the fine microstructure. Adhesion is considerably improved after such treatment. This has been demonstrated by the high bending strength (σ_{TRS} > 600 MPa) and high ductility (>20%) of cold sintered iron, nickel, and copper powders treated in hydrogen prior to consolidation (Ref 13). The formation of bonded chains of very fine oxide-free particles produced by vapor condensation was observed for Fe, Co, Ni, Ag, Cu, Mg, Zn, and Al by Tholen (Ref 27). For a number of alloys (e.g. aluminum and magnesium alloys), oxide layers cannot be removed by heat treatment. For stainless steels, chromium-containing high-strength or high-speed steels, superalloys and titanium alloys, reduction of surface oxides can require unduly high processing temperatures (Ref 28–30).

Good adhesion is also important to prevent partial fracture of adhesive bonds as a result of the elastic springback of consolidated powder particles. Elastic springback can be responsible for the lower effective density of consolidated powders after unloading, as compared to the density measured under load. A higher density was obtained in copper and nickel powders high pressure consolidated at 2 GPa, when the powders were treated in hydrogen at T = 200 °C and kept in vacuum prior to consolidation, as compared to the powders left in open air for ~1 h after the H_2 treatment (~99.5 versus ~98.5%). The effect of springback is even more pronounced when consolidation of nanoscale metal powders is considered.

Reduction diagrams can be used to establish temperature limits within which surface oxides on metal, carbide, or nitride powders are reducible (Ref 30, 31). An apparatus used for experi-

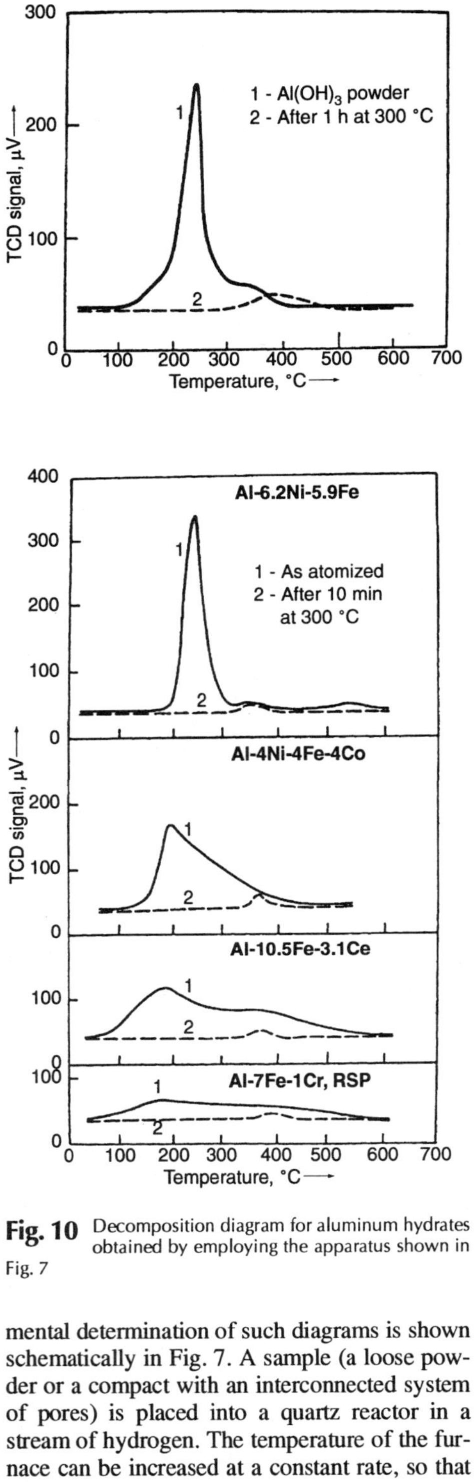

Fig. 10 Decomposition diagram for aluminum hydrates obtained by employing the apparatus shown in Fig. 7

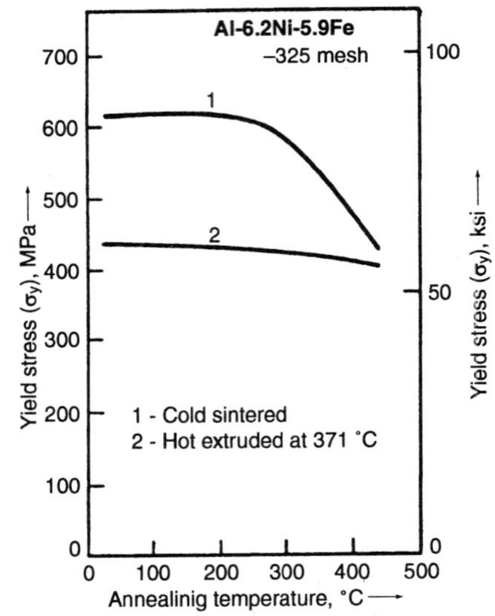

Fig. 11 Room temperature yield stress in compression as a function of annealing temperature (1 h exposure) for the cold sintered ($P = 3$ GPa) and hot extruded Al-6.2Ni-5.9Fe alloy (air-atomized powder)

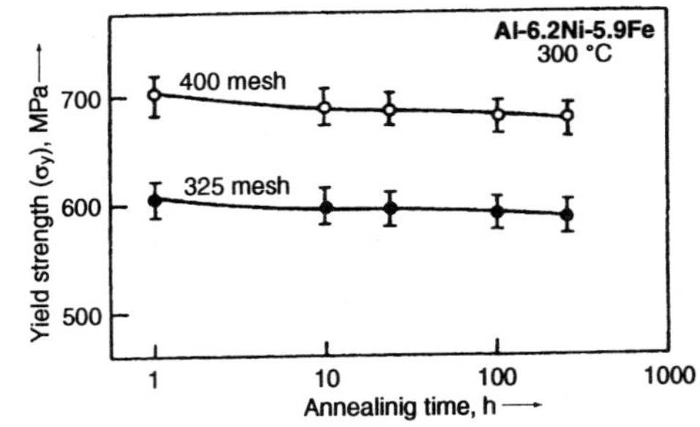

Fig. 12 Room temperature yield stress in compression as a function of annealing time at 300 °C for cold sintered Al-6.2Ni-5.9Fe alloy

mental determination of such diagrams is shown schematically in Fig. 7. A sample (a loose powder or a compact with an interconnected system of pores) is placed into a quartz reactor in a stream of hydrogen. The temperature of the furnace can be increased at a constant rate, so that the reduction behavior can be studied with a steady increase of temperature. In the hydrogen atmosphere, reduction of oxides, sulfides, carbides, etc. can take place, as well as reaction of hydrogen with dissolved carbon, nitrogen, or other elements with the formation of water vapor, hydrogen sulfide, methane, and other gases. The amount of the gaseous reaction product is proportional to the amount of the compound reacted with hydrogen. For example, the amount of water vapor is proportional to the amount of oxides reduced. Reaction kinetics can be studied by measuring the amount of gas released as function of time at a constant temperature. This is done by connecting the reactor to a thermal conductivity (TC) detector; the TC signal being proportional to the amount of evolved gases. The starting temperature for a reaction depends on thermodynamic parameters of the reagents and products. Thus, for each reaction, a peak of the TC signal versus temperature can be found within a specific range. Peaks corresponding to different reactions can be separated by using a number of cold traps at different temperatures. For example, if a dry ice cold trap is placed between the reactor and the TC detector, water vapor is removed and the peak corresponding to oxide reduction does not appear in the diagram. At the same time, peaks corresponding to reactions of hydrogen with carbon (CH_4), nitrogen (NH_3), etc. will remain. Examples of reduction diagrams for high-speed T15 steel and 410L stainless steel (12 wt% Cr) are shown in Fig. 8. In both cases, the first peak corresponds to the removal of iron oxide and the second to the reduction of mixed oxides (e.g. iron-chromium spinel, $FeCr_2O_4$). Figure 9 shows the room temperature transverse rupture strength, σ_{TRS}, of cold sintered T15 and 410L steel samples as a function of the preconsolidation reduction treatment temperature (80% dense compacts in hydrogen flow, 1 h). In both cases, an increase of σ_{TRS} was observed in agreement with the reduction peaks in Fig. 8. T15 samples cold sintered after reduction treatment at $T > 1000$ °C measured $\sigma_{TRS} > 600$ MPa, and a very high value of $\sigma_{TRS} > 1500$ MPa was obtained for 410 steel cold sintered after reduction treatment at $T > 1100$ °C. The high strength of these cold sintered steels is the result of good adhesion between the surface of oxide-free alloy particles.

The apparatus for reduction studies shown in Fig. 7 was also used for the investigation of decomposition of hydrates in aluminum alloys (Ref 4). Removal of hydrates formed on the surface of aluminum alloy powders in the process of atomization is important for improving the bonding integrity of consolidated materials (Ref 32, 33). Degassing leading to dehydration of atomized aluminum alloys is usually done at temperatures >400 °C, which is higher than the planned service temperature. At such temperatures the coarsening of fine metastable microstructures can take place (Ref 4). Decomposition diagrams of $Al(OH)_3$ and of hydrates obtained employing the apparatus with the TC detector for a number of aluminum alloy powders are shown in Fig. 10. It can be seen that, for a number of the aluminum alloys, degassing can be performed at <300 °C.

Determination of a temperature range necessary for the reduction of surface oxides is important for retention of fine microstructures of rapidly solidified/mechanically alloyed or of blends of very fine powders.

Processing of Aluminum Alloys for Elevated Temperatures

The main goal of the development of aluminum alloys for temperatures >200 °C was to replace high-strength aerospace materials for elevated temperatures based on iron or titanium. The disadvantages of the latter include a relatively high specific weight and high cost. A wide variety of alloy systems employing rapid solidification, mechanical alloying, or a combination of both techniques, followed by hot consolidation, were investigated. Alloying elements with low solubility and low diffusion rates in aluminum, such as Ni, Fe, Co, V, Mo, W, Ti, Zr, and Ce, were used in the development of alloys for elevated temperatures.

Normally, the processing temperatures for aluminum alloys are >400 °C, which is higher than planned service temperatures. Processing includes the removal of hydrates from the surface of rapidly solidified powders, hot forging, and hot extrusion. In many alloys, the partial coarsening of microstructure takes place during hot processing, resulting in deterioration of mechanical properties. As it has already been mentioned, for a number of aluminum alloys, especially those which do not contain magnesium, hydrites can be removed at $T < 300$ °C (Ref 31). Because cold sintering guarantees the retention

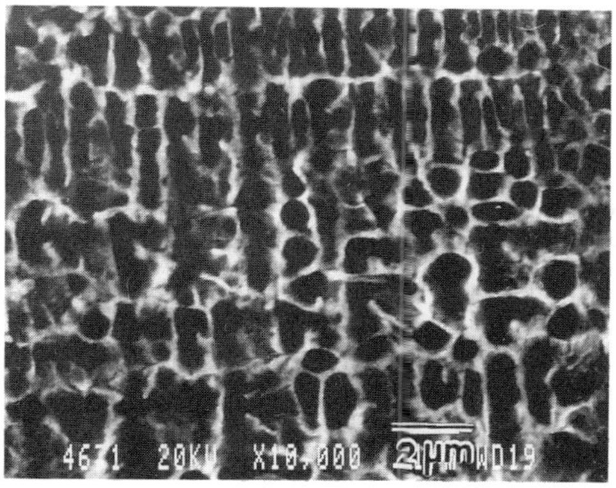

Fig. 13 A representative cell structure of carbides in a water-atomized high-speed steel powder cold sintered at 400 °C (SEM)

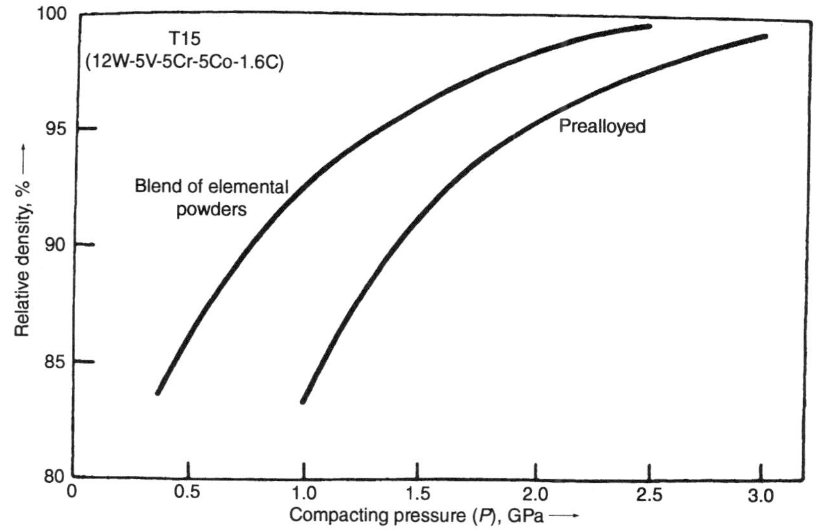

Fig. 14 Compaction curves for high-speed T15 steel

of metastable microstructures of rapidly solidified powders of aluminum alloys (Ref 26, 34–36), the method can be used for investigating the relationship between microstructure and properties, starting from the microstructure of an as-atomized powder.

The thermal stability of cold sintered samples can be monitored indirectly by measuring the room temperature yield stress, σ_y, after annealing at various temperatures. The dependence of σ_y on annealing temperature (1 h anneal) for the cold sintered (curve 1) and hot extruded at 371 °C (curve 2) Al-6.2Ni-5.9Fe (wt%) alloy is shown in Fig. 11. It can be seen that up to ~300 °C, there is no significant decrease in σ_y of the cold sintered samples, and that σ_y of the cold sintered samples is ~30% higher than that of the hot extruded samples. Pronounced coarsening of intermetallic precipitates is observed at temperatures >350 °C. The effect of annealing time at 300 °C on the room temperature yield stress, σ_y, of the cold sintered Al-6.2Ni-5.9Fe alloy is shown in Fig. 12. There is almost no drop in σ_y after long (up to 300 h) exposures, which is consistent with the thermally stable microstructure observed in TEM (Ref 18). Other rapidly solidified aluminum alloy powders for elevated temperatures (Al-Ni-Co, Al-Fe-Ce), as well as mechanically alloyed rapidly solidified Al-Fe-Ce alloys, were successfully consolidated to full density employing cold sintering.

Cold Sintering of Rapidly Solidified Iron- and Nickel-Base Alloys

Rapidly solidified high-speed steel powders (T15, M2, M35) were successfully consolidated to full density via cold sintering with retention of very fine microstructure (Ref 4, 30, 37, 38). Two processing routes can be used for full density consolidation of high speed steels: (a) high pressure consolidation at $P = 3$ GPa and temperature from 300 to 350 °C; and (b) cold sintering at ambient temperature at $P = 3$ GPa, followed by annealing at 900 to 1000 °C and repressing/cold sintering at room temperature. Both routes result in finer carbide dispersions compared to similar high speed steels prepared by HIP (Ref 4).

Rapidly solidified high-speed tool steel powders are characterized by a fine cellular structure of carbides (Ref 4), as shown in Fig. 13. Such structure of carbides results in high microhardness and should result in a high elastic modulus, which can be advantageous for some applications. However, exposure to the high processing temperatures (>1200 °C) during consolidation or/and austenizing treatment results in the breaking up of the cell structure and in the partial coarsening of carbides. High pressure consolidation at 400 °C provides full density without destruction of the fine cellular structure. Rapidly solidified nickel-base superalloys can be successfully consolidated to full density using the same cold sintering routes (Ref 4, 29).

Processing of High Performance Materials by Cold Sintering of Fine Elemental Powder Blends

Materials with fine microstructure can be produced by cold sintering of very fine elemental powders, followed by a homogenization treatment. The main advantage of this approach, compared to consolidation of prealloyed (rapidly solidified or mechanically alloyed) powders, is the higher compressibility of elemental powders. To utilize high compressibility, all the preconsolidation heat treatments (removal of surface oxides, annealing, etc.) should be performed at relatively low temperatures in order to prevent interparticle diffusion which occurs in alloying and strengthening. Compaction curves of a prealloyed water-atomized high-speed T15 steel powder and of the fine elemental powder blend of the same overall composition are shown in Fig. 14. Significantly higher densities are achieved at the same pressure for the blended powders. The same applies to 410 stainless steel, nickel- and cobalt-base superalloys, nickel, cobalt, and titanium aluminides, and nickel, cobalt, and iron titanides (Ref 39–42). When a blend of two or more metal powders is consolidated, a metal with a lower flow stress and higher ductility flows around the harder particles, the shape

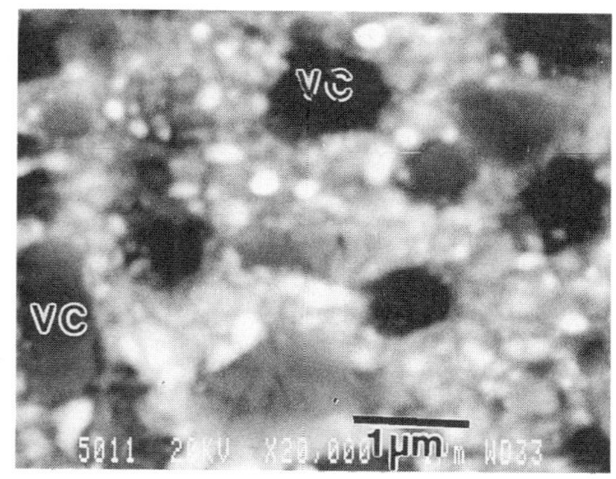

Fig. 15 A representative microstructure (SEM) of cold sintered and heat treated Fe-25VC-5Cr_3C_2 alloy prepared from the fine elemental powder blend

of the latter remaining practically unchanged (Ref 4, 43).

Prior to cold sintering, ~80% dense compacts with an interconnected system of pores are prepared and annealed in a reducing atmosphere (H_2) at temperatures $T < 400$ °C, which is high enough for the removal of surface oxides from elemental nickel, cobalt, iron, and copper powders. Titanium may be added as very fine TiH_2, which decomposes in vacuum at temperatures >400 °C. In this case, reduction treatment can be replaced by vacuum degassing where hydrogen evolved during the decomposition of TiH_2 will promote the reduction of metal oxides. After cold sintering, the dense samples are annealed to homogenize the material by diffusion. This can be followed by heat treatment to optimize mechanical properties (aging, quenching, tempering, etc.). Homogenization can lead to diffusion alloying, precipitation hardening, and solid state reactions, resulting in the synthesis of intermetallics or amorphization. In some cases, the reactions of new phase (intermetallic) formation are accompanied by a substantial volume decrease

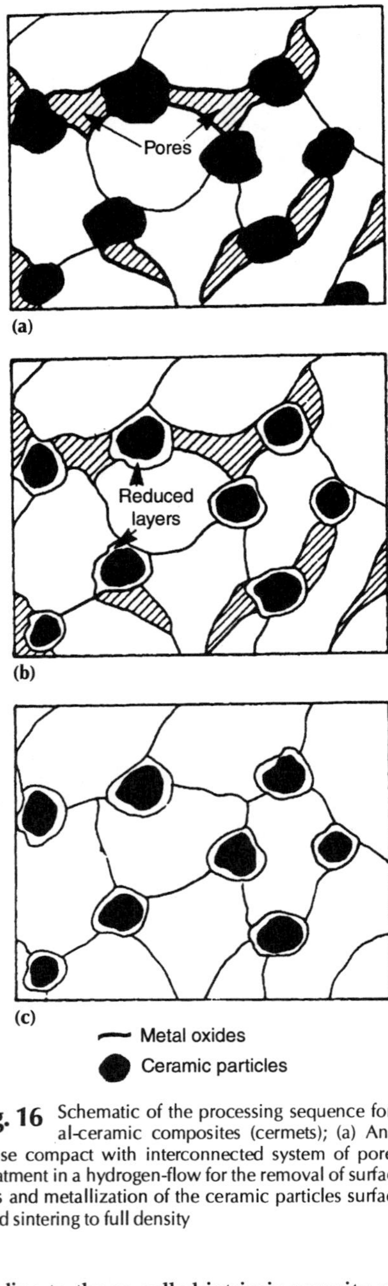

Fig. 16 Schematic of the processing sequence for metal-ceramic composites (cermets); (a) An 80% dense compact with interconnected system of pores. (b) Treatment in a hydrogen-flow for the removal of surface oxides and metallization of the ceramic particles surface. (c) Cold sintering to full density

Fig. 17 SEM micrographs of the fracture surface of a 4640 steel-20 wt% VC composite cold sintered after reduction treatment at 950 °C (1 h). (a) Cleaved VC particle. (b) The region marked by an arrow at a larger magnification

leading to the so-called intrinsic porosity of the final product. In such cases, external pressure should be applied during postconsolidation heat treatment in order to obtain a fully dense product.

Fine elemental metal powders are often agglomerated. To obtain homogeneous distribution of different metal powders and ceramic particles in the blend, high energy attrition milling has been used for the blending of powders (Ref 3, 44, 45). By choosing a low grinding media (steel balls) to powder ratio (1:1 to 2:1), and by adding neutral organic liquid, welding and mechanical alloying (by diffusion) during milling can be avoided. The much higher 5:1 to 10:1 media-to-powder ratios are typically used for mechanical alloying. Homogeneous distribution of powder particles on a microscopic scale (areas <10 by 10 μm) is normally achieved after 1 to 2 h blending.

Processing of Composites by Blending of Very Fine Powders

Preparation of iron-base materials by blending of elemental powders offers wide possibilities in the design of thermally stable microstructures by adding carbides or nitrides that are resistant to coarsening. This applies to the carbides and nitrides of elements which have high affinity for carbon and nitrogen and are strong formers and stabilizers of carbides and nitrides (e.g., V, Nb, and Ta). This approach has been utilized for the design of wear-resistant iron-base materials containing unusually large volume fractions of thermally stable carbides, nitrides, and carbonitrides (Ref 3, 38). For example, the abrasive wear resistance of cold sintered Fe-25VC-5Cr_3C_2 and Fe-50VC compositions containing large volume fractions of a hard VC carbide is two times higher than that of the wear-resistant T15 high-speed tool steel. Representative microstructures of the cold sintered and heat treated Fe-25VC-5Cr_3C_2 material with the homogeneous distribution of carbides are shown in Fig. 15. The initial chromium carbide, Cr_3C_2, has been dissolved at a relatively high austenizing temperature (1200 °C) and has reprecipitated in the form of fine mixed carbides. Vanadium carbide particles ~1 μm size have been partially dissolved, but they are stable to coarsening at austenizing temperatures <1300 °C. For Fe-50VC, carbides form a skeleton after austenizing treatment at $T > 1200$ °C. Still finer microstructures can be obtained if finer starting powders of elemental iron and milled carbides, nitrides, etc. are used. It should be mentioned that compositions containing higher volume fractions (up to 80 vol%) of micron-size carbides in micron-size iron- and nickel-base matrices have been successfully consolidated to full density employing cold sintering.

Cold Sintering of Metal Matrix Composites with Partial (Surface) Reduction of Ceramic Particles

Bonding integrity between a metal matrix and ceramic particles can be improved by partial (surface) reduction or metallization of the ceramic particles prior to consolidation of the composite to full density. Such an approach has been applied to the processing of iron-base matrix composites (cermets) reinforced with Cr_2O_3 and VC particles (Ref 31, 46). A schematic of the process is shown in Fig. 16. Metallization of ceramic particle surfaces result in an increased composite strength. The fracture path moved from the particle-matrix interface (in the untreated cermet) to the metal matrix. Some of the coarser ceramic particles were cleaved, which indicates a very good adhesion to the metal matrix (see Fig. 17). An improvement in strength was also obtained for Co-WC composites prepared according to the scheme in Fig. 16. Surface metallization of ceramic particles by such a reduction treatment can only be utilized for oxides, carbides, nitrides, etc., that are reducible at relatively low temperatures when no coarsening of the matrix microstructure takes place. In principle, ceramic particles (or fibers) can be metallized by reduction treatment before being added to the metal matrix. In this case, higher reduction temperatures can be employed and matrices with metastable structures can be used; these useful structures will not be destroyed during cold sintering.

For oxides or other ceramics that are impossible or difficult to reduce, coating by a reducible layer can be used as an intermediate step. Such coatings are produced by annealing an Al_2O_3 plate placed in contact with a nickel, cobalt, or chromium powder in an oxidizing atmosphere followed by partial reduction/metallization of the formed surface layer ($NiAl_2O_4$, $CoAl_2O_4$, and $(AlCr)_2O_3$ for nickel, cobalt, and chromium, respectively) (Ref 47).

Coating of non-oxide ceramic particles and fibers employing Powder Immersion Reaction Assisted Coating (PIRAC) (Ref 48–50) results in the formation of conducting metal-covalent carbide, nitride, etc. surface layers that can lead to a better chemical bonding between ceramic reinforcements and a metal matrix in cold sintered products. Ceramic particles or fibers with conductive PIRAC surface layers can further be coated by electrodeposition before introducing them into a metal matrix. This approach has

Fig. 18 Examples of parts produced by cold sintering

been successfully used for the fabrication of metal bonded diamond composites for stone cutting (Ref 51).

Cold Sintering of Nanocrystalline Powders and Nanocomposites

Dense nanocrystalline materials can be synthesized by assembling atom clusters that are obtained by physical or chemical methods or by mechanical attrition. Because it is difficult to produce bulk materials by physical or chemical vapor deposition or by electrochemical deposition methods, consolidation of nanosize powders is the only practical way to fabricate bulk nanocrystalline materials (Ref 52–54). Until recently consolidation of nanoscale powders has resulted in densities ~85% TD and not >95% TD. Porous nanocrystalline materials find a variety of useful applications as catalysts or extra fine molecular filters/sieves. However, the potential of nanoscale materials in the field of high performance structural and functional applications, as well as the properties of dense nanoscale materials, remain practically unexplored due to inability to achieve full density.

Cold sintering performed immediately after reduction treatment has been successfully used for consolidation of nanocrystalline aluminum, nickel, and iron powders as well as Ni-TiC nanocomposites, and densities >99% have been obtained (Ref 55). It should be emphasized that the effect of surface oxides is especially pronounced for nanoscale powders characterized by an unusually high grain boundary density. These films are formed during the manufacturing of powders, because contamination of nanosize particles cannot be completely prevented at this stage. It is believed that the relatively low densities reported in most publications dealing with consolidation of nanocrystalline metal powders (e.g., 95% TD for untreated nickel powder consolidated at 5 GPa and 900 °C (Ref 55) compared to >99% TD obtained by cold sintering in vacuum (10^{-3} Torr) at 3 GPa and 300 °C of a similar powder heat treated in a hydrogen flow) are the result of poor adhesion caused by the presence of surface oxides.

High strength and microhardness values were obtained in cold sintered nanocrystalline materials (Ref 55). Thus, σ_y = 600 and 1100 MPa were obtained for pure nanocrystalline nickel (65 nm) and iron (30 nm), respectively. This is considerably higher than the strength of these metals cold sintered from 1 μm size powders: σ_y = 200 and 300 MPa for nickel and iron, respectively.

Fabrication of Net Shape Parts

Because cold sintering uses rigid dies and achieves full density, parts with very close tolerances can be produced: <0.01 mm for a 10 mm opening (Ref 18). Thus the method can be used for the fabrication of parts with final shape and dimensions that do not need any additional machining. Elastic expansion after unloading (pressure release) should be taken into account. Wear resistant parts, tools, parts made of difficult to machine superalloys, stainless steels and titanium alloy, as well as parts made of metal glasses are of primary consideration. Examples of parts produced by cold sintering are shown in Fig. 18. Rectangular (3.2 by 10 by 40 mm) segments for stone-cutting tools were fabricated from metal-bonded (cobalt alloy or tool steel) diamond composites employing cold sintering followed by heat treatment at temperatures lower than the typical hot processing temperatures. Up to 28 segments could be produced in one pressure die, the processing cycle duration being 3 to 4 min. The performance of the cold sintered segments in field tests was better than that of similar segments manufactured by hot processing.

REFERENCES

1. E.Y. Gutmanas, A. Rabinkin, and M. Roitberg, On Cold Sintering of Metal Bonded Diamond Composites, *Scripta Metall.*, Vol 13, 1979, p 11–15
2. E.Y. Gutmanas, Cold Sintering under Pressure—Mechanisms and Application, *Powder Metall. Int.*, Vol 15, 1983, p 129–132
3. E.Y. Gutmanas, Consolidation of Powders under High Pressure—Cold Sintering, *New Materials by Mechanical Alloying Techniques,* DGM Informationgesellschaft Verlag, Oberursel, 1989, p 129–142
4. E.Y. Gutmanas, Materials with Fine Microstructures by Advanced Powder Metallurgy, *Prog. Mater. Sci.*, Vol 34, 1990, p 261–366
5. R.F. Decker, Alloy Design Using Second Phases, *Metall. Trans.*, Vol 4, 1973, p 2495–2517
6. M. Cohen, Progress and Prospects of Metallurgical Research, *Advancing Materials Research*, National Academy Press, 1986, p 51–110
7. N.J. Grant, *Advances in Powder Metallurgy,* American Society for Metals, 1982, p 1–21.
8. A. Lawley, *Atomization,* Metal Powder Industries Federation, 1992
9. W.J. Boettinger, *Mater. Sci. Eng.*, Vol 98, 1988, p 123–130
10. J.S. Benjamin, *Metall. Trans. A,* Vol 1, 1970, p 2943–2951.
11. J.S. Benjamin and M.J. Bomford, *Metall. Trans. A,* Vol 8, 1977, p 1301–1305
12. P.S. Gilman and W.D. Nix, *Metall. Trans. A,* Vol 12, 1981, p 813–824
13. P.S. Gilman and J.S. Benjamin, *Ann. Rev. Mater. Sci.*, Vol 13, 1983, p 279–300
14. F.J. Anders, G.B. Alexander, and W.S. Wartel, *Met. Prog.*, Vol 82, 1962, p 88–91, 118–121
15. R.W. Fraser, B. Meddings, D.J.I. Evans, and V.N. Machiw, *Modern Developments in Powder Metallurgy*, Vol 2, H.H. Hausner, Ed., Plenum Press, 1966, p 87–111
16. F.V. Lenel, *Powder Metallurgy Principles and Applications*, Metal Powder Industries Federation, 1980
17. E.Y. Gutmanas and A. Lawley, Design and P/M Processing of New Hierarchical Material Structures, *Modern Developments in Powder Metallurgy*, Vol 20, Metal Powder Industries Federation, 1988, p 467–479
18. E.Y. Gutmanas and A. Lawley, Processing of Specialty Alloys, Composites and Intermetallics from Fine Elemental Powders, *Advances in*

Powder Metallurgy, Vol 2, Metal Powder Industries Federation, 1990, p 1–14

19. A. Kasak and E.J. Dulis, *Powder Metall.*, Vol 21, 1978, p 114–123
20. H.H. Lieberman, *Mater. Sci. Eng.*, Vol 46, 1980, p 241–250
21. R.W. Heckel, *Trans. AIME,* Vol 221, 1961, p 1001–1008
22. E.Y. Gutmanas, High Pressure Compaction and Cold Sintering of Stainless Steel Powders, *Powder Metall. Int.*, Vol 12, 1980, p 178–182
23. Y.M. Balshin, Theory of Compaction, *Vestnik Metaloprom,* Vol 18, 1938, p 137–147
24. C. Torre, *Montan Hochschulle Leoben,* Vol 93, 1948, p 62–67
25. S. Timoshenko and Y.N. Goodier, *Theory of Elasticity*, McGraw-Hill, 1951
26. D. Shechtman and E.Y. Gutmanas, Transmission Electron Microscopy of Metallic Powders, *Prakt. Metallogr.*, Vol 18, 1981, p 587–592
27. A.R. Tholen, *Acta Metall.*, Vol 27, 1979, p 1765–1783
28. D.B. Goldman and E.Y. Gutmanas, Effect of Heat Treatments on Cold Sintering of 410 Stainless Steel, *Powder Metall. Int.*, Vol 17, 1985, p 269–272
29. D.B. Goldman, E.Y. Gutmanas, and D. Zak, Reduction of Oxides and Cold Sintering of Water Atomized Powders of Ni, Ni-20Cr, and Nimonic 80A, *J. Mater. Sci.*, Vol 4, 1985, p 1208–1211
30. D.B. Goldman, J.B., Clark, S. Hart, and E.Y. Gutmanas, Effect of Heat Treatment in Reducing Atmosphere on Cold Sintering of High Speed Steel, *Modern Development in Powder Metallurgy*, Vol 17, Metal Powder Industries Federation, 1985, p 427–440
31. E.Y. Gutmanas, D.B. Goldman, S. Hart, and D. Zak, Cold Sintered Steel-Vanadium Carbide Composites, *Powder Metall. Int.*, Vol 18, 1986, p 401–404
32. Y.-W. Kim, W.M. Griffith, and F.H. Froes, *J. Met.*, Vol 37, 1985, p 27–34
33. G. Stanek, D. Shechtman, J. Ramon, and W. Bunk, *Powder Metall. Int.*, Vol 22, 1990, p 7–12
34. E.Y. Gutmanas, M. Premkumar, and A. Lawley, Microstructure and Mechanical Properties of Cold Sintered P/M Aluminum Alloys, *Progress in Powder Metallurgy*, Vol 39, Metal Powder Industries Federation, 1983, p 669–682
35. O. Botstein, E.Y. Gutmanas, and A. Lawley, Stability and Mechanical Behaviour of Cold Sintered P/M Aluminum Alloys, *Modern Developments in Powder Metallurgy*, Vol 15, Metal Powder Industries Federation, 1985, p 761–773
36. O. Botstein, E.Y. Gutmanas, and A. Lawley, High Pressure Consolidation of a Dispersion Strengthened Aluminum Alloy, *Progress in Powder Metallurgy*, Vol 14, Metal Powder Industries Federation, 1986, p 123–137
37. E.Y. Gutmanas and D. Zak, Mechanical Behavior of Cold Sintered High Speed Steel-Carbides Composites, *Modern Developments in Powder Metallurgy*, Vol 20, Metal Powder Industries Federation, 1988, p 421–429
38. I. Gotman and E.Y. Gutmanas, Diffusion and Microstructural Changes at 1200 °C in Cold Sintered T15–4640 Joints, *J. Mater. Sci. Lett.*, Vol 6, 1987, p 1303–1306
39. E.Y. Gutmanas, I. Gotman, L. Farber, and E. Paransky, Solid State Synthesis of Intermetallics via Cold Sintering of Fine Elemental Powders, *Proc. 1993 Powder Metallurgy World Congress* (Kyoto, Japan), Japan Soc. of Powder and PM, 1993, p 1148–1151
40. M. Boshmann, G. Petzov, L. Farber, and E.Y. Gutmanas, Synthesis of NiAl from Ultrafine Powders via Solid State Reaction and Reactive Sintering, *PM '94, Powder Metallurgy World Congress,* Vol II, Éditions de Physique, Paris, 1994, p 1485–1488
41. E. Paransky, E.Y. Gutmanas, I. Gotman, and M.J. Koczak, Pressure Assisted Reactive Synthesis of Titanium Aluminides from Dense 50 Al-50 Ti Elemental Powder Blends, *Metall. Mater. Trans.*, Vol 27A, 1996, p 2130–2139
42. L. Farber, E.Y. Gutmanas, I. Gotman, and M.J. Koczak, Dense CoAl Based Alloys with Improved Ductility, *Metall. Mater. Trans.*, Vol 27A, 1996, p 2140–2150
43. L. Farber, L. Klinger, and I. Gotman, *Mater. Sci. Eng.*, Accepted, 1998
44. E.Y. Gutmanas, Design of Iron Based Wear Resistant Materials by Advanced Powder Metallurgy, *P/M into 1990's*, Vol 2, The Institute of Materials, London, 1990, p 359–363
45. Y. Mitrani and E.Y. Gutmanas, Design of Dispersion Strengthened Cobalt Based Alloys for Elevated Temperatures, *P/M into 1990's*, Vol 3, The Institute of Materials, London, 1990, p 190–194
46. E.Y. Gutmanas, D.B. Goldman, J.B. Clark, and S. Hart, Cold Sintered Stainless Steel-Chromium Oxide Composites, *Progress in Powder Metallurgy*, Vol 41, Metal Powder Industries Federation, 1986, p 631–640
47. I. Gotman and E.Y. Gutmanas, Joining of Al_2O_3 to P/M Iron and Nickel Based Alloys by Cold Sintering, *Modern Developments in Powder Metallurgy*, Vol 18, Metal Powder Industries Federation, 1998, p 75–88
48. E.Y. Gutmanas, I. Gotman, and W.A. Kaysser, Coating of Non-Oxide Ceramics by Interaction with Metal Powders, *Mater. Sci. Eng.*, Vol 157A, 1992, p 233–241
49. I. Gotman and E.Y. Gutmanas, Microstructure and Thermal Stability of Coated SiC and Si_3N_4, *Acta Metall. Mater.*, Vol 40, 1992, p 121–131
50. I. Gotman, E.Y. Gutmanas, and P. Mogilevsky, Interaction between SiC and Ti Powder, *J. Mater. Res.*, Vol 8, 1993, p 2725–2733
51. J. Karwan-Baczewska, I. Gotman, and E.Y. Gutmanas, Processing of Metal-Bonded Coated Diamond Composites, *Non-Ferrous Metals '97, 3rd Int. Conf.*, Univ. Mining and Metallurgy, Kracow, 1997, p 350–353
52. H. Gleiter, Nanocrystalline Materials, *Prog. Mater. Sci.,* Vol 33, 1989, p 223–315
53. R.W. Siegel, *Mater. Sci. Eng.*, Vol 168A, 1993, p 189–197
54. V.G. Griaznov and L.I. Trusov, Micromechanics of Nanocrystals, *Prog. Mater. Sci.,* Vol 37, 1993, p 289–401
55. E.Y. Gutmanas, L.K. Trusov, and I. Gotman, Consolidation, Microstructure and Mechanical Properties of Nanocrystalline Metal Powders, *Nanostructured Mater.*, Vol 8, 1994, p 893–901

Field-Activated Sintering

Joanna R. Groza, University of California

IN POWDER CONSOLIDATION, the imposition of an electric current is known to enhance sintering kinetics and therefore high densities are achieved at lower temperatures or in shorter times as compared to conventional sintering methods. This way, full or near-full densification may be realized with minimal undesirable microstructural changes. A shorter processing time usually results in productivity gains. In addition, there is less sensitivity to initial powder preparation, such as no need for sintering aids and subsequent debinding steps. Most of the time, air sintering is appropriate; that is the need for controlled environment is eliminated. Finally, interparticle bonding is improved. As a consequence, sintered parts of higher quality may be expected.

Numerous processes have been developed that apply an external current to assist powder consolidation. The initial idea to use resistance sintering of metal powders started with Taylor in 1933 in hot pressing of cemented carbides (Ref 1). Cremer patented a field-assisted sintering method in 1944 (Ref 2). He used a 60 Hz current for 1 to 2 cycles at a current density of 62 kA/cm^2 under 70 to 140 MPa pressure to sinter copper, brass, bronze, and aluminum. In late 1950s, resistance sintering under pressure was applied to metal powders by Lenel using equipment similar to spot welding machines (Ref 3). Presently, resistive hot pressing is commonly used and consists of a low-voltage (5 to 40 V), high-amperage (up to 25 kA) current passing through the powders with a simultaneous pressure application. The electrically conductive powders are contained in an internally insulated die. The passage of the electric current provides the resistive heating of the powders by Joule effect; that is, the electric power is used as the heating source for the consolidating compact. Therefore, the processing times are extremely short, on the order of seconds. Pressure levels are low to moderate (15 to 340 MPa).

In another case of field-assisted sintering, large electric currents induce magnetic forces that contribute to powder compaction due to a pinch effect. By concentrating the pulse in the compaction zone, the magnetic field forces may reach high values, up to an equivalent to 5 GPa pressure (Ref 4). Examples of such consolidation methods are dynamic magnetic compaction (DMC) (Ref 5) and indirect high-energy, high-rate processing (HEHR) (Ref 6, 7).

In *field-activated sintering,* an initial activation of powder particles is achieved by the application of electrical discharges on either conductive or nonconductive materials. (It is also sometimes called electroconductive sintering for conductive materials.) These discharges are produced by pulsed electric power. The electrical discharge per se does not densify powders and, therefore, additional energy is required to increase the final density. This extra energy may be mechanical, as an applied pressure, and/or thermal energy. The electrical discharge only activates the sintering mechanism (i.e., neck formation) for densification by these other means.

In one of these methods, known as electrical discharge compaction (EDC), the electrical energy is suddenly released by discharging a capacitor bank through the powder preform contained in an insulating container (Ref 8–13). Usually, there is only one discharge. Repeating up to three discharges increases activation and thus contributes to final density. No further density improvement is observed beyond 4 to 5 discharges. The discharge period is on the order of hundreds of μs at a high voltage (up to 30 kV) and current density (~10 kA/cm^2). After discharge, a high amperage current is passed through the metal powder column for Joule heating. In this stage, an external pressure may also be applied.

Neck formation between powder particles takes place in the discharge stage. The experimental measurement of the powder column electrical resistance has shown an initial rapid drop in the discharge step. The reduction of initial high resistivity of powder column was first attributed to the neck formation between metal powder particles (Ref 8, 11). Later, Okazaki and coworkers demonstrated that the resistance drop is caused by oxide removal rather than only neck buildup. To verify this, they documented oxide film disruption on nickel powders by electron microscopy studies (Ref 8, 10). Some quantitative oxygen measurements in initially heavily oxidized powder and final EDC consolidated part indicated that most of the nickel oxide was dissociated by the sudden heat generated by discharge. This explosive disruption of surface oxides on metal powders due to electric discharge contributes to neck formation. The actual pore elimination and powder consolidation to high densities is achieved by Joule heating and pressure application in the second stage of the process. The method has been applied to compact a variety of metal powders in England (Ref 11–13), Japan (Ref 14), and the United States (Ref 8–10). In many cases, electrical discharge compaction was not the final fabrication step. It was aimed to produce at a high-density compact capable of withstanding subsequent processing such as forging or rotary swaging (Ref 8, 11, 12). The initial EDC compacts may be obtained as bars with the same or stepped various cross sections, L-shaped, or hexagonal (Ref 12). More details about EDC may be found in Ref 8.

A sustained activation of the powder particle surface may be achieved by the application of multiple power pulses that produce repeated electrical discharges. This activated consolidation has been pioneered in the spark-sintering technique, which was developed in the 1960s (Ref 15–17). This technique comprises two steps. The first step consists of the application of an alternating current of 500 to 1000 Hz frequency at a low pressure for 15 to 30 s. The spark discharges generated in this step activate powder particle surfaces. The next stage proceeds with resistance heating of the powders under the same or increased pressure. An increase of pressure in the second step provides the best results. All tools—die and plungers—are made of high-density graphite. Extensive experimental spark-sintering work was performed on various materials such as aluminum, beryllium, copper, nickel, iron, titanium, and their alloys; superalloys; refractory metals; carbides; borides; oxides; and beryllides in the late 1960s and early 1970s (Ref 18–21). Bars up to 250 mm (~10 in.) in diameter and complex near-net-shape parts have been processed. Multilevel components that involve sinter bonding of various parts were also fabricated by spark sintering. In the latter case, the main attraction was the solid metallurgical bonding achieved while minimizing side effects such as grain coarsening or other microstructural alterations involved in joining materials. The process was advanced to an industrial scale using a semiautomatic system in the commercial production of beryllium and titanium parts (Ref 18, 19). Mechanical properties were comparable to those of wrought metals, particularly when postconsolidation forging was applied. The main difficulty reported was the temperature measurement and control. Other problems were related to the mechanical strength of the graphite die, powder reaction with graphite, tooling cost for complex parts, and in some cases, low productivity. Frustration due to the lack of understanding of the physics of electrical activation often surfaced.

A similar method termed impulse resistance sintering was used in the mid 1970s (Ref 22). The main efforts were directed toward achieving balanced heating and densification by the control of heating rate and distribution and of heat losses. A large difference in temperature between the specimen center and the die wall due to the cooling of the edges by the die was documented. In some cases (e.g., for conductive ceramics such as TiB_2), the die temperature was only 50 to 60% of that of the densifying specimen. A mushroom-shaped punch was used to minimize heat losses at the water-cooled contact with the copper electrodes (small cross section) and thus to provide the highest temperatures at the powder-punch interface (large cross section).

In the last decade, the benefits of a pulse discharge have been recognized more widely, and numerous field-activated sintering techniques have been developed. Such techniques are instrumented pulse electrodischarge consolidation, spark/plasma sintering, or plasma-activated sintering developed in Japan (Ref 23–28), resistance/spark sintering under pressure in Korea (Ref 29), pulsed electrical discharge with pressure application in Russia (Ref 30, 31), HEHR processing in the United States (Ref 6, 7) and plasma sintering in Brazil (Ref 32). All these methods are similar concepts to pulse discharge and subsequent resistancesintering. The pulsed electrodischarge stage may be built in as an added option for powder activation in other nonfield sintering techniques such as the piston-cylinder high-pressure method (Ref 33). This article addresses only pulsed-electrical-discharge activated-pressure sintering that is generally called field-activated sintering technique (FAST). Because an in situ plasma generation may be possible among discharged particles, plasma-activated sintering (PAS) is another common name given to such field-activated densification. A distinction has to be made in the latter case to differentiate PAS from plasma sintering techniques where the specimen is "bathed" in an external plasma environment such as in microwave sintering.

Process Description

All electrical field-activated techniques use pulsed electrical discharge combined with rapid heating and pressure application to achieve fast powder sintering. A schematic of the FAST process is shown in Fig. 1. The equipment consists of a mechanical device capable of uniaxial pressure application and the electrical components to apply the pulsed and steady current. The loose powders are directly loaded into a punch-and-die unit. Graphite die and graphite punches are commonly used. This limits the pressure levels to low values, generally <100 MPa. The graphite confinement provides a reducing component to the sintering environment. Spark discharges may have the ability to clean the powder particle surfaces. For sintering more sensitive powders, the machines are equipped with chambers for vacuum or controlled environment.

The consolidation process consists of two stages: (1) an initial activation through the application of a pulsed voltage and (2) the subsequent heating and densification using direct current (dc). These stages may be applied sequentially or simultaneously. Throughout the process, a uniaxial pressure is applied. The pressure may be constant or changed in the second stage. Sometimes, the initial pressure level is light (~10 to 15 MPa) and then increased during the heating stage.

The FAST processing parameters are shown in Table 1 and Fig. 2. A typical pulse discharge is achieved by the application of a low voltage (~30 V) and a 600 to 1000 A current. In PAS machines, an alternating current (ac) rectifier is used to achieve a controlled dc pulsing. The result is a square-wave current with adjustable on and off pulses. The duration of each pulse may be varied between 1 and 300 ms. The total pulsing time is typically 30 s, but may go up to 300 s. In some FAST methods, an ac current is used for powder particle activation stage (Ref 29). In the second stage, when regular sintering takes place, the current is essentially dc at a level that depends on the powder type. The conductive powders are mainly heated due to the Joule effect. For nonconductive powders, heating occurs through heat transfer from the die and plungers. In this case, the die and punches are heated through their own resistance.

The sintering time is dependent on the type of powder. To determine sintering time, the PAS machines are equipped with a linear gage that

Table 1 Typical FAST processing parameters

Parameter	Value
Applied pressure, MPa	10–65
Pulse stage:	
Voltage, V	30
Current, A	750
Pulse on-stage duration, ms	30–80
Pulse off-stage duration, ms	30–80
Total pulsing time, s	30–60
Resistance heating:	
Current, A	1000–4000
Duration (hold time), min	3–5
Atmosphere	Air or controlled
Total time, min	<10

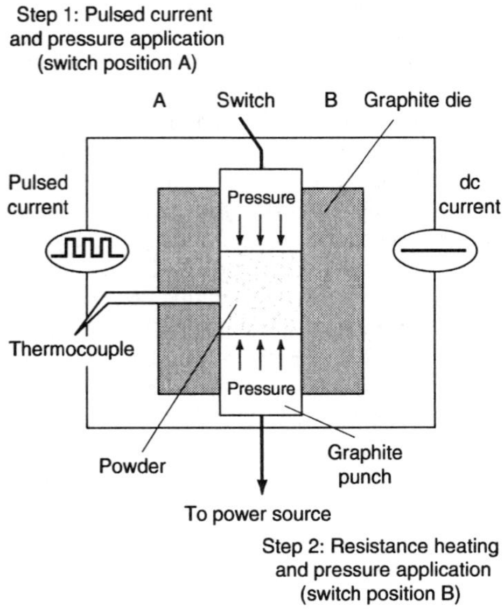

Fig. 1 FAST process

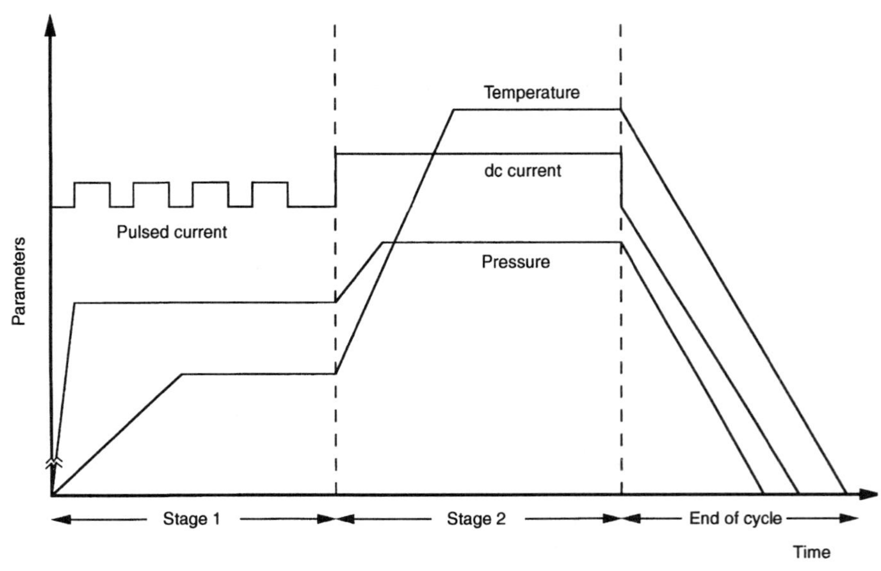

Fig. 2 Processing parameters in FAST sintering

monitors the powder shrinkage and shrinkage rate during densification. The shrinkage decreases continuously while the shrinkage rate is at a maximum and then decreases to essentially zero. The densification is considered complete when the shrinkage rate reaches the zero value. When sintering is completed, the temperature is allowed to ramp down and the compact is removed. Stripping the compact out of the punch-and-die unit is easier for a simple configuration such as a circular or square disk. Generally, the part is cooled in the die because ejection of hot compact is difficult. Cooling rates are usually not controlled, although fans can be used or hot specimens can be ejected from the die and quenched afterwards. The entire operation, from the loading of the die to the ejection of the finished part, may be accomplished in less than 10 min.

Temperature measurements are difficult in all field-assisted sintering techniques. The process is fast, and the steady-state conditions are extremely short. These conditions make the use of conventional pyrometers and thermocouples troublesome. Insertion of a thermocouple in the compact is mechanically difficult. A pyrometer sighting of the compact is possible only for transparent dies. Therefore, die rather than part temperature is generally measured using a pyrometer or a thermocouple. The latter is inserted in the die cavity as close to the heating specimen as possible. Large temperature differences between die and compact have been noticed, particularly in the heating stage. For the sintering step, that is, steady heating, these differences have been measured to be 70 to 100 °C for metal powders (Ref 34). Because current passes through the powders, they reach a higher temperature than the external graphite die. For nonconductive powders that are heated from the dies, powders are at a lower temperature than the die.

Process Mechanisms

Because loose powders are directly loaded into the sintering machines, the use of binders to achieve the strength of the green part becomes unnecessary. Consequently, the debinding step commonly used to burn out the volatiles is rendered superfluous. This is not only helpful in eliminating processing steps and increasing productivity, but part purity also may be enhanced.

The accelerated densification by electric field application may be partly explained by the ability to remove the oxides and impurities from the particle surfaces. In general, the remaining oxides on powder particles are known to cause consolidation difficulties and low fracture toughness of the sintered part. If direct boundary contacts between grains can be achieved, this permits full advantage to be taken of the intrinsic strength of the material. Oxide removal may be attributed to phenomena ranging from electric breakdown of the oxide layer to a possible plasma generation among powder particles.

As in any sintering process, field-activated densification starts with a highly porous body in which pores form an interconnected network. The initial pressure application proceeds with neck formation. Because the pressure is at a low level, most of these necks are only interparticle point contacts. At this stage, the pulsed current is applied, and the goal is to achieve a uniform current path rather than local channels that may concentrate all passing current. Usually, the initial 1 to 2 pulse applications ensure a uniform current path, particularly for high resistance powders. Next, the current is forced to choose the path of the least resistance, that is, through the contact points to complete the current path through powders. The powder particles have an inherent oxide layer on their surface. This way, small capacitors are formed across any two particle contact points. Electrical discharges are generated across these capacitor gaps. The interfering surface oxide films are "broken" when a

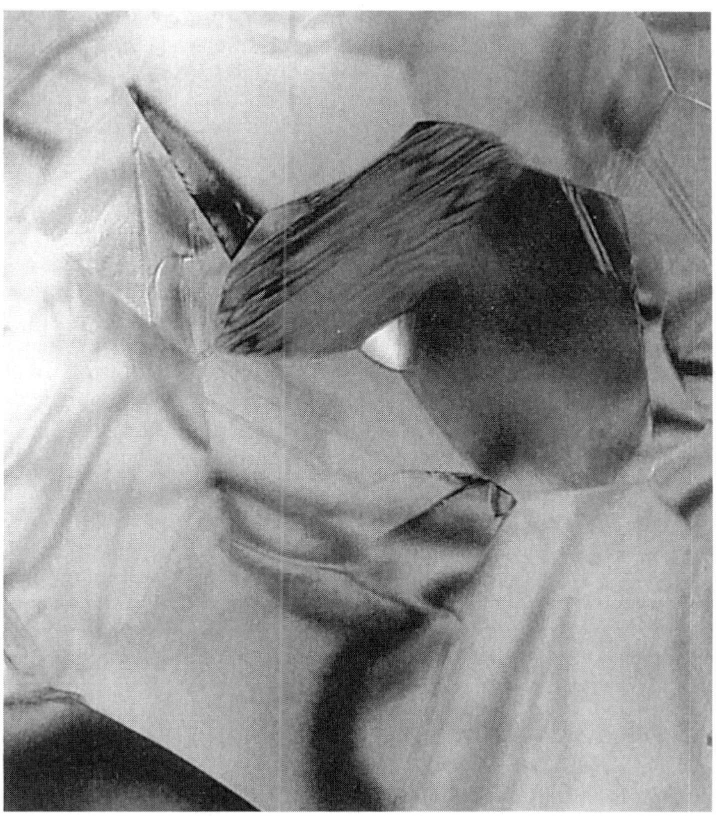

Fig. 3 Transmission electron micrograph of aluminum nitride sintered by PAS. Courtesy of T.A. Guiton, Dow Chemical Company. 69,000×

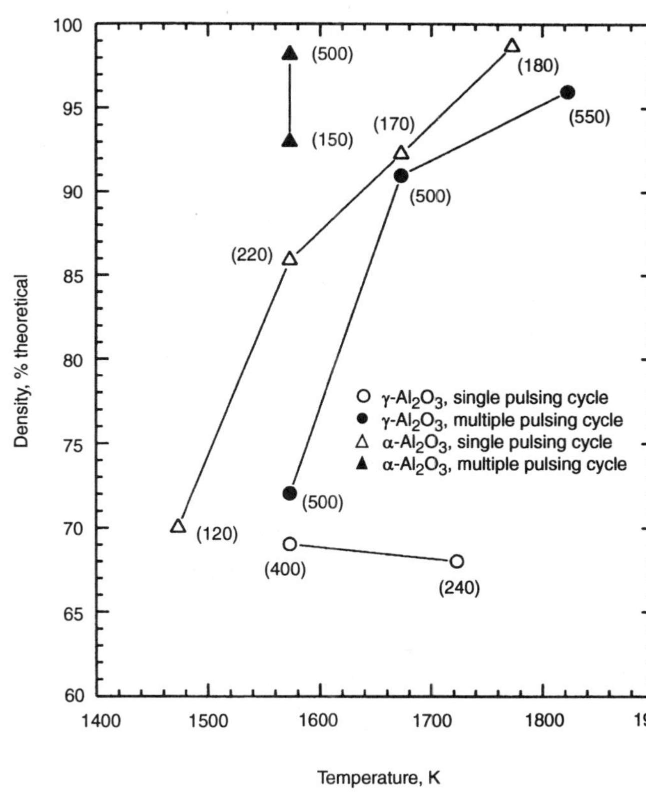

Fig. 4 Effect of multiple pulsing on the densification of α-Al_2O_3 and γ-Al_2O_3 powders. The single pulse cycle was applied at 293 K. The pulse cycle current was 750 A, pulse duration 60 ms for a total of 60 s. The multiple pulse cycle involved 4 cycles, one each at 293, 473, 773, and 1273 K. The pulsing parameters were the same as single pulse cycle and kept constant for all the cycles. The numbers in bracket denote holding time at sintering temperature for each run. The initial pressure was 40 MPa and increased to 66 MPa in the sintering stage. Courtesy of R.S. Misra (U.C. Davis)

certain voltage level is achieved. This level, or electrical breakdown voltage, is dependent on the dielectric strength of the oxide layer. The mechanisms of electrical failure that take place at the breakdown voltage involve arcing across adjacent points on particles and electrical breakdown of dielectric film on the powder particle surface.

Alternatively, the electrical discharges across the sample may generate a plasma, that is, an ionized gas among powder particles. It appears that plasma generation occurs only in some conditions that are not yet well defined. If plasma is generated, surface oxide removal and elimination of surface adsorbates take place. The resultant "clean" surfaces provide a better intergranular cohesion, thus enhancing the neck formation. Consequently, enhanced sinterability is observed upon further densification. The dc current applied in the second stage actually promotes diffusion bonding of these clean interparticle contacts.

Presently, the type of surface effects caused by electric field activation has not been unambiguously established. However, experimental evidence suggests that particle surfaces are "cleaned," and direct grain-to-grain contact is observed in the field-sintered specimens. For instance, a transmission electron microscopy (TEM) micrograph of aluminum nitride sintered under an electrical field application shows clean grain boundaries with no secondary or amorphous boundary layers (Fig. 3). A high-resolution electron microscopy image of a typical grain boundary in this aluminum nitride confirmed this direct grain-to-grain bonding (Ref 35). Furthermore, measurements of the oxygen content in electron energy loss spectra did not show any oxygen segregation at the grain boundary.

This result is in contrast to the conventional resistance-sintered specimens that do not involve an activation step. In this case, the higher resistance of the surface oxide films gives rise to higher temperatures than in the interior of the particles themselves and therefore causes melting at particle boundaries. This liquid-sintering mechanism explains the higher densities and shorter times for densification. When molten films develop, secondary phases are observed at grain boundaries of the sintered part (Ref 36).

Another indication of some special surface effects have been noticed in PAS densification of superconductors. Resistivity/temperature curves of Y-Ba-Cu-O (YBCO) superconductors consolidated by PAS showed new transition onsets from 240 to 278 K (Ref 37). Such transitions have not been observed in non-PAS consolidated superconductors. They can be explained by new surface phenomena (i.e., surface accumulation of electronic charge) that may be induced by a plasma generation.

After the particles have been connected, the second stage starts when a direct current is applied. Diffusional processes and plastic flow are the main contributors to the densification in this stage. The highest heating occurs at the thin necks formed by discharge. Diffusion is activated by the passage of the current. The highest temperatures achieved in the necks provide the highest diffusion rates and thus enhance matter transport toward the neck area. This is the area in which most of the matter transport is required for sintering. Therefore, field application intensifies the sintering rate. In addition, plastic flow of metal particles is also enhanced by the combination of applied pressure and high temperatures. The plastic flow may occur through the entire particle or, possibly, only localized plastic flow at the necks. In both cases, plastic flow significantly contributes to pore closure and therefore enhances overall densification rate. As a result, high density may be achieved and in shorter times or at lower temperature than in the absence of the applied current. The final result is a dense compact with good intergranular bonding.

Similar to metals, in ceramics electric charge is built up in the individual powders, which can be considered capacitors. A correlation between the dielectric properties of ceramics and the PAS sinterability has been established (Ref 38, 39). Ceramics with high dielectric constant and low dielectric strength sinter well under electrical field. When the dielectric constant of the ceramic powder is high, the powder surface can hold higher electric charges. This is the case for alumina and aluminum nitride. Conversely, no FAST sintering has been possible for low dielectric constant and high dielectric strength ceramics such as boron nitride and silicon nitride unless oxide additives are used. The favorable effect of a high dielectric constant on sinterability may be also noticed in sintering conditions that increase the dielectric strength of the material (Ref 39). For instance, the dielectric constant of alumina increases by heating. When pulses are applied at higher temperatures, alumina sinterability is enhanced. This has been achieved by multiple pulsing of powders (Fig. 4). The dielectric constant increases when alumina is heated up at increasingly higher temperatures before next pulse is applied.

Field sintering of alumina indicated a dependence on the pulse duration. Sintering is enhanced when longer pulse duration (30 to 160 ms) is used. The longer discharge times allow the ions in the ceramics more time for polarization and therefore more electric energy to be built up before electrical breakdown.

Process Fundamentals

The main parameters that control the field-assisted sintering process are pulsing duration and frequency, resistance-sintering current and time, pressure level, and sintering environment.

The effect of electrical discharge on the total sintering time was studied in PAS using pure aluminum powders and is shown in Fig. 5 (Ref 40). The sintering time reported is the duration required to attain final zero-shrinkage rate. The sintering time during resistance heating becomes shorter when a previous pulsed electrical discharge was applied. The effect of electrical discharge application on sintering time is more distinctly observed for currents between 750 and 1200 A. Sintering time is minimum for individ-

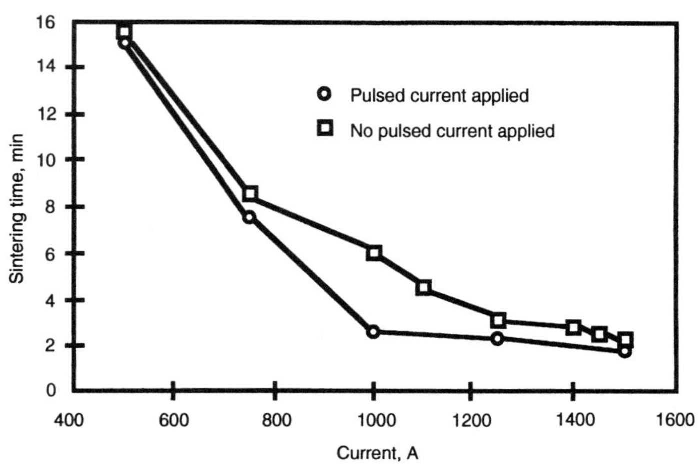

Fig. 5 Effect of pulsed current on the total sintering time for pure aluminum densified by FAST. Source: Ref 40

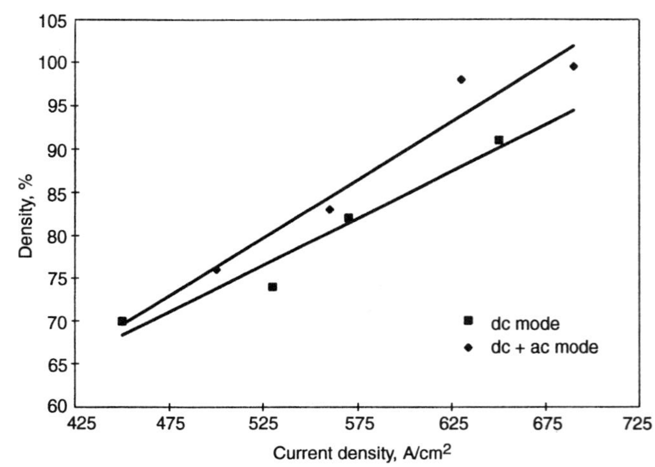

Fig. 6 Effect of alternating current plus direct current (ac + dc) superimposition on the density variation with current density for TiAl powders. Source: Ref 29

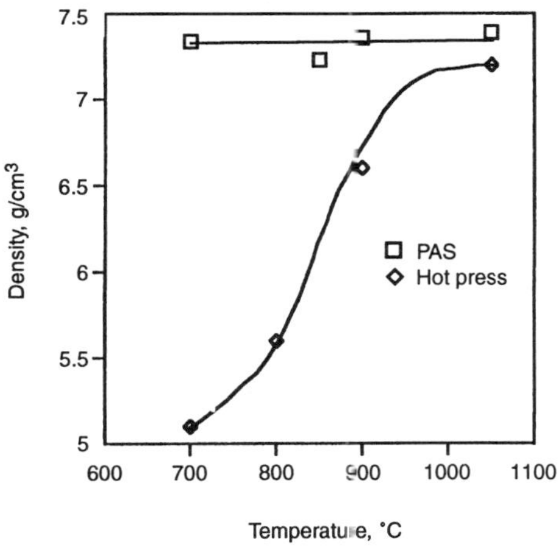

Fig. 7 Comparison of densities of Fe-2%C powders sintered by conventional hot pressing and FAST. Source: Ref 41

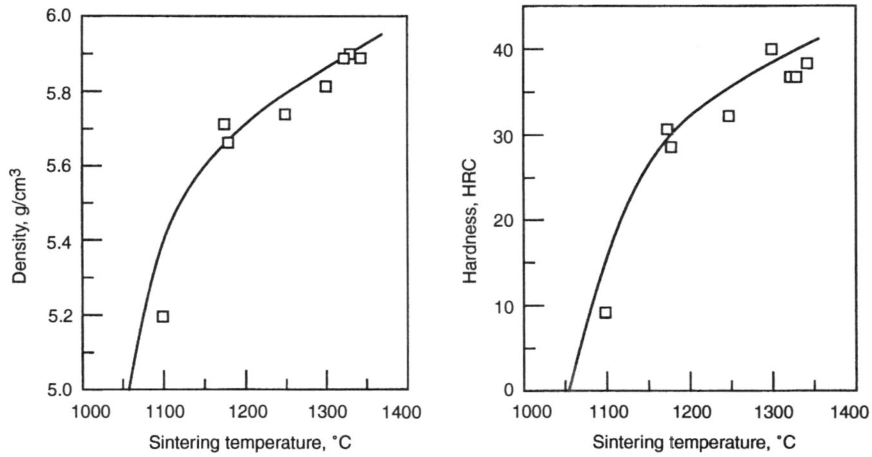

Fig. 8 (a) Density and (b) hardness variation with sintering temperature during FAST consolidation of atomized NiAl powders. Source: Ref 42

ual pulse duration of 30 to 80 ms. For pulses longer than 80 ms, sintering time increases, indicating that the efficiency of pulsed current diminished. Similarly, the density increases when alternating current is superimposed on the direct current for sintering in resistance/spark sintering under pressure (Ref 29) (Fig. 6).

Generally, the field effects in comparison with other sintering techniques are more noticeable when sintering proceeds at lower temperatures. A comparison of conventional hot pressing and PAS sintering of mechanically alloyed Fe-2%C powders is shown in Fig. 7 (Ref 41). The most significant difference in densities is observed at temperatures below 1273 K.

Beyond the electrical discharge stage, the sintering parameters in the resistance heating stage have an effect similar to those in conventional pressure sintering. As an example, the temperature effect is illustrated for the PAS sintering of atomized NiAl powders (Fig. 8) (Ref 42). Overall pressure effects are the same as in conventional hot pressing. The larger the pressure, the higher the density. Studies on PAS sintering of pure nickel powders indicated that the densification rate remains in overall agreement with the pressure-assisted models but becomes faster. A shift in densification behavior has been observed from only power-law creep to grain-boundary contribution to densification. This shift is probably due to field-enhanced diffusivity. Increased densification has been observed when pressure is increased in the resistance-sintering stage (Ref 41).

Grain growth during FAST sintering occurs and seems to be governed by temperature level and time at high temperature, similar to conventional sintering. However, the short-time high-temperature exposure characteristic of FAST processes usually minimizes grain growth.

Applications

Presently, industrial application of FAST sintering methods is in the production of both hard and soft magnetic materials and cutting tools in Japan (Ref 43, 44). For instance, manganese-zinc ferrites for high-frequency power supplies are PAS sintered using 60 s pulse current cycle followed by a heating current of 2000 A under 49 MPa pressure (Ref 43). Densities of >99% and grain size of 1 μm are obtained. This fine grain size minimizes the magnetic high-frequency core losses from 1800 kW/m^3 for a conventionally sintered 9 μm sample to only 720 kW/m^3. Commercial Nd-Fe-Co-B magnets are densified by PAS.

Laboratory-scale experiments have succeeded in field-activation sintering of a large variety of materials from metals to intermetallic compounds to ceramics and composites. Examples of metal powders that have been consolidated include pure metals (nickel, tungsten), intermetallics (TiAl, NiAl, Nb_3Al, $FeSi_2$), and composites (Ref 23, 27, 28, 41, 42, 45). The field-sintered ceramics were oxide (Al2O3, superconductors) and nonoxide-type (AlN, Si_3N_4) (Ref 25, 37–39, 46). The process parameters, density, grain size, and hardness values achieved in PAS-consolidated materials are shown in Tables 2 and 3.

To illustrate some advantages of field application, a comparison of conventional and FAST sintering of aluminum nitride powders should be considered. Aluminum nitride was field sintered to near-full density in 5 min at 2000 K without any additives (Ref 46). The resulting densities

Table 2 Field-activated sintering parameters and properties of metal powders

Material	Temperature, K	Holding time, min	Pressure, MPa	Density, g/cm^3(%)	Grain size, nm	Hardness
Nickel	1270	2	30	8.57 (96.2)	NA	228 HV
Tungsten	2163	6	61	17.4 (90)	~1500	NA
Fe-2%C(a)	973	3	47	7.22	52	41 HRC
Fe-2%Al(a)	973	3	47	7.34	38	43 HRC
Nb_3Al(a,b)	1423	3	30	7.57 (~100)	~200	1073 HV
NiAl	1616	1.5	30	5.88 (99.5)	NA	38 HRC
TiAl(c)	1602	<3	39	(100)	>100	NA
Fe-85vol%Fe_3C	723	3	63	(98)	45	1050 HV
WC-10%Co	1673	1	45	14.45	NA	1756 HV
Fe-Ni-5vol%TiC	1473	10	33	7.84	NA	249 HV
$FeSi_2$(d)	1123	5	59	(83.4)	NA	566 HV

All sintering tests were performed in air with pulse cycle time of 30 s, and on (off) durations of 30 ms. NA, not applicable. (a) Mechanically alloyed powders. (b) Nb_3Al compound synthesis occurred simultaneously with powder densification. (c) Pulse duration 100 ms. (d) Pulsing cycle time = 99 s

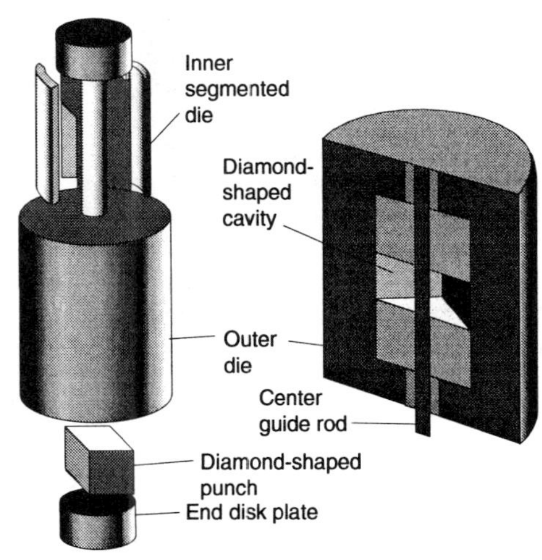

Fig. 9 Die-and-punch configurations for net-shape sintering of cutting tool insert. Courtesy of K. Yamazaki (U.C. Davis)

Table 3 Field-activated sintering parameters and properties of ceramic powders

Material	Temperature, K	Holding time, min	Pressure, MPa	Density, %	Grain size, μm
AlN	2005	5	50	99.3	0.8
α-Al_2O_3(a)	1573	<10	40	98.2	0.7
β-Al_2O_3(a)	1673	6	66	>99.5	1
β-Si_3N_4-5%Y_2O_3-2%MgO(b)	1823–1873	7–8	49	98.1–98.5	0.3
WC	2073	2	45	100	NA
TiN	1273	3	66	93.7	NA
BCCSO	1073	8.3	14	100	NA
YBCO	1173	15	15	100	NA

All specimens were sintered in air. (a) Multiple pulsing applied. (b) Sintered in a vacuum

ranged between 3.18 and 3.24 g/cm^3 (97.5 and 99.3%, respectively). For comparison, undoped aluminum nitride can be sintered to 95% at 2200 K for 30 h. When dopants are used, the densities achieved are 97 to 98% by conventional sintering at 2070 to 2220 K for 3 to 4 h.

Simultaneous synthesis and densification of $MoSi_2$ compound from elemental powders has been achieved by FAST (Ref 47). A density of 99.2% was obtained by using a 30 s pulse cycle at 60 MPa pressure followed by heating to 1700 °C in vacuum. This density is higher than that reported for conventional hot pressing of $MoSi_2$ powders (95 to 97%). The hardness values of field-densified $MoSi_2$ compare favorably with conventional specimens (>9.4 GPa).

Functionally graded materials combining refractory ceramics on a metal substrate, such as ZrO2 on NiCrAlY and TiAl, have been consolidated to high densities using FAST methods (Ref 48, 49). A temperature gradient was achieved by either using punches of different materials (i.e., one graphite and one tungsten) or using a stepped die. The temperature gradient was >100 K in the former case and >700 K for the latter.

Bonding capabilities of FAST process should be similar to those in spark sintering. However, only one attempt was reported on diffusion bonding of cubic boron nitride on metal substrates (Ref 50). The good adhesion obtained suggests that field sintering can be used for ceramic-to-metal joining.

Net-shape parts may be produced by FAST methods. One such example is a diamond-shaped cutting tool insert that was fabricated by PAS using a special punch-and-die set (Ref 51). The diamond-shaped cavity of the mold was formed by divided pieces of graphite to prevent mold destruction during sintering (Fig. 9).

The rapid rate of densification, no need for preliminary powder preparation steps such as cold compaction, additive use and debinding, and air sintering are some densification characteristics that make the FAST process economically competitive. This is especially true for materials that are difficult to cold press; oxygen-sensitive, metastable materials; or for critical bonding and purity requirements. Final-net-shape and diffusion bonding of different materials with minimal microstructural changes are high-potential capabilities of FAST consolidation that are worth future explorations.

ACKNOWLEDGMENT

Sincere appreciation goes to Dr. C.G. Goetzel, who generously shared information on spark-sintering technique.

REFERENCES

1. G.F. Taylor, Apparatus for Making Hard Metal Compositions, U.S. Patent 1,896,854, 7 Feb 1933
2. G.D. Cremer, Sintering Together Powders Metals Such as Bronze, Brass or Aluminum, U.S. Patent 2,355,954, Aug 1944
3. F.V. Lenel, Resistance Sintering under Pressure, *JOM,* Vol 7, 1955, p 158–167
4. V. Ivanov, Y.A. Kotov, O.H. Samatov, R. Bohme, H.U. Karow and G. Schumacher, Synthesis and Dynamic Compaction of Ceramic Nano Powders by Techniques Based on Electric Pulsed Power, *Nanostructured Mater.,* Vol 6, 1995, p 287–290
5. B. Chelluri, Dynamic Magnetic Consolidation (DMC) Process for Powder Consolidation of Advanced Material, *Mater. Manu. Process.,* Vol 9 (No. 6), 1994, p 1127–1142
6. H.L. Marcus, D.L. Bourell, Z. Eliezer, C. Persan, and W. Weldon, High-Energy, High-Rate Materials Processing, *JOM,* Vol 39, 1987, p 6–10
7. S.K. Raghunathan, C. Persad, D.L. Bourell, and H.L. Marcus, High-Energy, High-Rate Consolidation of Tungsten and Tungsten-Based Composite Powders, *Mater. Sci. Eng.,* Vol A131 (No. 2), 1992, p 43–53
8. K. Okazaki, Electro-Discharge Consolidation of Particulate Materials, *Rev. Partic. Mater.,* Vol 2, 1994, p 215–269
9. D.K. Kim and K. Okazaki, Characteristic Features of Electro-Discharge Compaction, *1991 P/M in Aerospace and Defense Technologies* (Tampa, FL, 4–6 March 1991), Metal Powder Industries Federation/American Powder Metallurgy Institute, 1991, p 365–372
10. D.K. Kim, H.-R. Pak, and K. Okazaki, Electrodischarge Compaction of Nickel Powders, *Mater. Sci. Eng.,* Vol A104, 1988, p 191–200
11. T.J. Davies and S.T.S. Al-Hassani, *Advances in Materials Technology in America,* Vol 2, I. LeMay, Ed., American Society of Mechanical Engineers, 1980, p 147
12. M. Shakery, S.T.S. Al-Hassani, and T.J. Davies, Electrical Discharge of Powder Compaction, *Powder Metall. Int.,* Vol 11 (No. 30), p 120–124
13. D.J. Williams and W. Johnson, Neck Formation and Growth in High-Voltage Discharge Forming of Metal Powders, *Powder Metall.,* Vol 25, 1982, p 85–89
14. S. Saito and A. Sawaoka, *Powder Metall. Int.,* Vol 5, 1973, p 70–75
15. K. Inoue, U.S. Patent 3,250, 892, May 1966
16. "Metallic Powder Activation," U.S. Patent 3,958,566, Aug 1971
17. G. DeGroat, One-Shot Powder Metal Parts, *Am. Mach.,* Vol 109 (No. 21), 1965, p 107–109
18. R.W. Boesel, M.I. Jacobson, and I.S. Yoshida, Spark Sintering Tames Exotic P/M Materials, *Mater. Eng.,* 1969, p 32–35
19. C.G. Goetzel, Electrical Discharge-Type Activated Pressure Sintering and Bonding of a Superalloy Composite Structure, *High Temp.—High Press.,* Vol 3, 1971, p 425–438
20. G. Goetzel and V.S. de Marchi, Electrically Activated Pressure-Sintering (Spark Sintering) of Titanium-Aluminum-Vanadium Alloy Powders, *Modern Developments in Powder Metallurgy,* H.H. Hausner, Ed., Plenum Press, 1971, p 127–150
21. "Spark Sintering of Parts and Preforms," Lockheed Missiles & Space Co., Sunnyvale, CA, 1969
22. L.A. Shepard and W.J. Croft, Impulse Resistance Sintering, *Powder Metall. Int.,* Vol 7 (No. 1), 1975, p 33–38
23. H. Kimura and S. Kobayashi, Synthesis of High Hardness Intermetallic Compound by Pulse Electro-Discharge Consolidation of Amorphous TiAl Powder, *J. Jpn. Inst. Met.,* Vol 58, 1994, p 201–207
24. H. Kimura, Synthesis of Nano-Structured High Temperature Titanium Aluminide by Instrumented Pulse Electro-Discharge Consolidation of Mechanically Alloyed Amorphous Powder, *J. Phy. IV* (France), Vol 3, 1993, p 423–428
25. T. Nishimura, M. Mitomo, H. Hirotsuru, and M. Kawahara, Fabrication of Silicon Nitride Nano-Ceramics by Spark Plasma Sintering, *J. Mater. Sci. Lett.,* Vol 14 (No. 15), 1 Aug 1995, p 1046–1047
26. K. Matsugi, T. Hatayama, and O. Yanagisawa, Effect of Direct Current Pulse Discharge on Specific Resistivity of Copper and Iron Powder Compacts, *J. Jpn. Inst. Met.,* Vol 59 (No. 7), 1995, p 740–745
27. S. Shiga, K. Masuyama, M. Umemoto, and K. Yamzaki, Plasma Activated Sintering (PAS) of β-Fe(Mn)Si(Al)$_2$ Prepared by Mechanical Alloying (MA), *Sintering Technology,* R.M. German, G.L. Messing, and R.G. Cornwall, Ed., Marcel Dekker, 1996, p 431–438
28. G. Jones, J.R. Groza, K. Yamazaki, and K. Shoda, Plasma Activated Sintering (PAS) of Tungsten Powders, *Mater. Manuf. Process,* Vol 9 (No. 6), 1995, p 1105–1114
29. G.-S. Choi, J.-Y. Kim, and D.-H. Lee, Resistance/Spark Sintering under Pressure of Intermetallic TiAl Powders, *J. Korean Inst. Met. Mater.,* Vol 30 (No. 7), 1992, p 840–847

30. A.I. Raichenko, M.Z. Kolchinskii, and D.A. Levina, Electric-Discharge Sintering of Oxidized Metal Powders, *Poroshk. Metall.,* Vol 10 (No. 10), 1976, p 19–26
31. V.N. Bazanov, S.A. Balankin, J.G. Grigoriev, V.V. Gunichev, S.V. Novikov, and V.A. Yartsev, Application of Electrical Discharge for Pulse Heating and Compaction of Powder Materials, *PM into the 1990's, International Conference on Powder Metallurgy,* The Institute of Metals, 1990, p 270–273
32. J.L.R. Muzart, V.J. Batista, C.V. Franco, and A.N. Klein, "Plasma Sintering of AISI 316L Stainless Steel: The Influence of the Cycle Processing on the Sample Density," presented at 1997 Int. Conf. Powder Met. and Particulate Materials, MPIF, 1997
33. R. Misra, C. Lesher, and J.R. Groza, personal communication, 1997
34. J.M. Doh, private communication, 1997
35. S.H. Risbud, J.R. Groza, and M.J. Kim, Clean Grain Boundaries in AlN Ceramics Densified without Additives by a Plasma Activated Sintering Process, *Philos. Mag.,* Vol 69, 1994, p 525–533
36. S.J. Hong and P.W. Kao, Reinforced Aluminum Composite Made by Resistance Sintering of Mechanically Alloyed Powders, *Mater. Sci. Eng.,* Vol A119, 1989, p 153–159
37. S.H. Risbud and C.-H. Shan, Resistivity Drops at >240 K and Diamagnetic AC Susceptibility up to 300 K in Rapidly Consolidated YBCO, *Mater. Lett.,* Vol 20, 1994, p 149–153
38. J.A. Schneider, R.S. Mishra, and A.K. Mukherjee, Plasma Activated Sintering of Ceramic Materials, *Proc. Second International Symposium on Advanced Synthesis and Processing, Ceramics Transactions,* R. Spriggs, Z. Munir, and K. Logan, Ed., 1996, p 143–151
39. R.S. Mishra, S.H. Risbud, and A.K. Mukherjee, Influence of Initial Crystal Structure and Electrical Pulsing on Densification of Nanocrystalline Alumina Powder, *J. Mater. Res.,* Vol 13, 1998, p 86–89
40. M. Yokogawa, K. Yamazaki, S.H. Risbud, J.R. Groza, H. Aoyama, and K. Shoda, Plasma Activated Sintering for Rapid Net Shape Manufacturing—Principle, Control, and Results, *Advancement of Intelligent Production,* E. Usui, Ed., Elsevier Science B.V./The Japan Society for Precision Engineering, 1994, p 582–587
41. S.H. Yoo, Field Effects in Plasma Activated Sintering, M.S. thesis, University of California at Davis, 1996
42. J.R. Groza, Consolidation of Atomized NiAl Powders by Plasma Activated Sintering Process, *Scr. Metall. Mater.,* Vol 30, 1994, p 47–52
43. S. Nagata, Y. Takahashi, M. Yorizumi, and K. Aso, "Fine Grained Mn-Zn Ferrite Produced by Plasma Activated Sintering Method (PAS)," presented at the Sixth Int. Conf. on Ferrites, Tokyo, Japan, 29 Sept–2 Oct 1992
44. K. Yamazaki, private communication, 1996
45. M. Tracy and J.R. Groza, Consolidation of Nanocrystalline Nb-Al Powders by Plasma Activated Sintering, *Nanostructured Mater.,* Vol 2, 1993, p 441–449
46. J.R. Groza, S.H. Risbud, and K. Yamazaki, Plasma Activated Sintering of Additive-Free AlN Powders to Near-Theoretical Density in 5 Minutes, *J. Mater. Res.,* Vol 7 (No. 10), 1992, p 2643–2645
47. I.J. Shon, Z.A. Munir, K. Yamazaki, and K. Shoda, Simultaneous Synthesis and Densification of $MoSi_2$ by Field-Activated Combustion, *J. Am. Ceram. Soc.,* Vol 79 (No. 7), 1996, p 1875–1880
48. H. Kimura and K. Toda, Design and Development of Functionally Graded Material by Pulse Discharge Resistance Consolidation with Temperature Gradient Control, *Powder Metall.,* Vol 39 (No. 1), 1996, p 59–62
49. S.K. Hur, S.H. Yoo, J.R. Groza, J.M. Doh, K. Yamazaki, and K. Shoda, Graded Coatings by Gradient Temperature Densification, *J. Mater. Res.,* May 1998
50. S. Yoo, J.R. Groza, T.S. Sudarshan, and K. Yamazaki, Diffusion Bonding of BN on Metal Substrates by Plasma Activated Sintering (PAS) Process, *Scr. Mater.,* Vol 34 (No. 9), 1996, p 1383–1386
51. K. Yamazaki, S.R. Risbud, J.R. Groza, and K. Shoda, "Potential Technologies for Processing High Performance and Innovative Materials Using Plasma Activated Sintering," presented at Int. Conf. on Manuf. Milestones towards the 21st Century, Japan, Society of Mechanical Engineering, 1997

Principles and Process Modeling of Higher-Density Consolidation*

W.B. Eisen, Crucible Research, J. C. Hebeisen, Bodycote IMT, Inc.

POWDER METAL PARTS are sometimes classified by density, where part production technologies may have low-density and high-density variants. Proponents of a particular technology for "high-density" parts may refer to the technology as a "high-density" or "full-density" process without explicit definition of the term for potential customers or comparison with other processes or materials. For the purposes of this article, "higher-density consolidation" means effectively 100% dense material that will compete with forgings in terms of mechanical properties. The most popular way to achieve 100% density in powder materials is with hot isostatic pressing, although there are a number of pseudoisostatic pressing techniques that can achieve high densities and at times at a cost savings. These are generally not important production processes at this time and are not included in this article.

Hot isostatic pressing (HIP) is a materials processing technique in which high isostatic gas pressure is applied to a powder part or compact at elevated temperatures to produce particle-to-particle bonding. This process usually results in the manufacture of a fully dense body, although partially dense bodies also can be intentionally produced. During processing, the compact is subjected to equal pressure from every side.

Elevated temperature, in reference to HIP of metal powders, ranges from approximately 480 °C (895 °F) for aluminum alloy powder to approximately 1700 °C (3090 °F) for tungsten powder. High-density argon gas is the most common medium used in the process, and pressures range from approximately 20 to 300 MPa (3 to 45 ksi), with 100 MPa (15 ksi) as the most typical pressure.

The HIP process was invented at Battelle Memorial Institute in 1955 by Saller et al. (Ref 1). Early designs utilized the "hot-wall" configuration; the furnace surrounded the pressure vessel. Material limitations precluded scale-up, and the development of a cold-wall vessel design, now used throughout industry, took place.

Hot isostatic pressing was initially used for diffusion bonding of clad nuclear fuel elements. Consolidation of beryllium metal powder "to-shape" was first carried out in 1964. High-volume hot isostatic compaction of high-speed tool steel was achieved in the United States and Sweden by 1972. The U.S. Air Force Materials Laboratories expanded HIP technology to include forging of preforms and near-net shapes of nickel-base superalloy and titanium alloy powders from 1970 to 1980. A comprehensive review of early HIP applications is given in Ref 2. During the 1970s it was also discovered that HIP could be used to permanently heal internal porosity in complex cast shapes without distortion of major casting features. This has remained a major use for HIP technology but is not discussed here.

Current applications of HIP technology in P/M processing include near-net shapes in steel and stainless steel alloys for power turbine components and for manifolds and valves for oil-field applications; nickel-base superalloys for aircraft engine turbine disks and shafts (such shapes are "squared-off" cross sections suitable for sonic inspection); nickel-base P/M forging and rolling preforms and nickel-base P/M integral pump and turbine impeller wheels; titanium alloy P/M billets, forging preforms, and shapes; tool steel billets (for mill processing), large die blocks, and composite structures; net shapes in P/M beryllium, niobium alloys, and other refractory metals; and dispersion and fiber-strengthened P/M aluminum alloys. Small parts processed by a combination of cold compaction of metal powder, sintering, and HIP include blended elemental titanium-base alloys, tool steel shapes, rare-earth magnets, and tools, dies, rolls, wear parts, and seals manufactured in tungsten carbide/cobalt and other carbide compositions. The iron- and nickel-base parts made for valve and manifold applications represent the leading edge of the technology in terms of size and shape complexity. Nickel-base P/M aircraft engine applications represent the highest technology level of the method in terms of mechanical properties, and tool steels represent the highest production tonnage. Hot isostatic pressing of P/M tungsten carbide/cobalt parts is still employed worldwide; however, most manufacturers have gone to a combined sinter/HIP process to reduce process time and costs.

Process Equipment

Typical HIP units consist of a pressure vessel, gas storage and handling system, furnace, tooling, power supply, controls, and instrumentation. Figure 1 is a schematic of a complete HIP system. An example of a large-sized HIP unit is shown in Fig. 2.

Pressure Vessel. Proper design of the pressure vessel is essential. The explosive energy stored inside an operating unit can be equivalent to several hundred pounds of high explosives in destructive capability, depending on unit size. Pressure vessels must be designed to meet appropriate American Society of Mechanical Engineers (ASME) codes and ASTM standards for materials.

Consideration must be given to the maximum allowable stress that the vessel can sustain for a designed service life. Typically, the general primary membrane stress intensity in the vessel wall cannot exceed one-third of the ultimate tensile strength. Due to the cyclic loading of the HIP unit, fatigue life of the vessel is an important design criterion. The possibility of plastic failure of the vessel caused by the applied test and operating pressures must be considered in the vessel design.

Provisions must be made for suitable ports into the pressure vessel to accommodate instrumentation, electrical power, and pressurization equipment. These penetrations are usually located at the bottom of the vessel in present designs.

*The first sections of this article through "World Wide HIP Capacity" have been adapted and updated from P.E. Price and S.P. Kohler, "Hot Isostatic Pressing of Metal Powders," *Powder Metallurgy,* Vol 7, *ASM Handbook,* American Society for Metals, 1984, p 419–443. The section "Modeling of Hot Isostatic Pressing" was adapted with permission from *Reviews in Particulate Materials,* Vol 4, Metal Powder Industries Federation, 1996.

The first HIP units were designed with the furnace surrounding the vessel. These hot-wall vessels were constrained to operate within a limited range of pressures and temperatures, up to about 207 MPa (30 ksi) and 790 °C (1460 °F). Currently, all HIP units are cold-wall designs (furnace inside the vessel), with pressure vessel components operating up to about 80 °C (150 °F). A heat shield is placed between the furnace and the vessel to limit heat losses and control vessel body temperature. Additionally, the wall of the pressure vessel is cooled by water or other suitable heat transfer fluids.

The cylindrical portion of the vessel can be cooled using several methods. Cooling fluid can be contained by a jacket and come into direct contact with the vessel wall, plate coils with heat transfer mastic can be clamped around the vessel, or an interior-cooled bore liner can be used. In all cases, careful control of the cooling fluid composition must be maintained to prevent corrosion or fouling of the cooling channels.

Hot isostatic pressing vessels are constructed by several methods: multiwall forged, monolithic forged, wire wound, or multiple ring liner. An example of a monolithic-forged pressure vessel is shown in Fig. 3. Pressure is sealed within the vessel by Bridgman seals, metal-to-metal seals, single or double O-rings, or a combination of these devices.

Closure of the pressure vessel is accomplished by either a threaded closure, pin closure, or nonthreaded closure supported by a yoke frame. In pressure vessels using threaded or pin-type closures, additional consideration must be given to the end loads carried by the vessel and closures. Threaded closures must be designed carefully to distribute the load evenly over all of the threads so that stress is not concentrated on the first two threads. Use of resilient threads or a specially designed thread form and undercut controls stress concentrations. For rapid opening of the vessel, interrupted threads are available that open the vessel in less than one rotation of the threaded closure.

In pin-type closures, a transverse pin is inserted through the pressure vessel and the end cap to close the vessel. This is accomplished quickly and easily; however, high unit stresses occurring at the pinholes in this type of design limit its application. Designs that reduce stress concentrations by using two pins have been proposed. Currently, this design is used primarily with smaller cold isostatic presses.

As an alternative to threaded-end closures, a yoke frame can be used (Fig. 4) to restrain the end loads. Opening and closing a pressure vessel with a yoke frame requires two actions. First the yoke frame (or the cylindrical body) is moved horizontally; then the end closure is extracted axially and moved to the side to allow access to the vessel. Large production units of both types (threaded closure and yoke frame) are currently in operation. User selection of design (yoke frame or threaded-end closure) depends on such factors as system size, design pressure, system application, space availability, cost, inspectability, and preference.

Gas System. The pressure medium used in HIP is gas, usually argon, although nitrogen and other mixtures can be employed. The gas supply system must be able to quickly deliver the gas at the desired high pressure. Additionally, argon must be delivered at a very high purity level to protect the workload and process equipment from detrimental chemical reactions with contaminants. Gas can be delivered from storage tanks and/or a liquid argon storage vessel.

Supplying argon from liquid storage (Fig. 5) ensures that the gas is of very high purity. Argon is delivered from liquid storage to a cryogenic pump by another smaller transfer pump or by gravity. Pressurized liquid argon moves from the cryogenic pump to a vaporizer, which delivers pressurized gas to the vessel.

Argon can be introduced into the system at a high rate. Unfortunately, there is no small-scale economic method of reliquefying argon and returning it to liquid storage after the HIP cycle. The used and possibly contaminated gas can

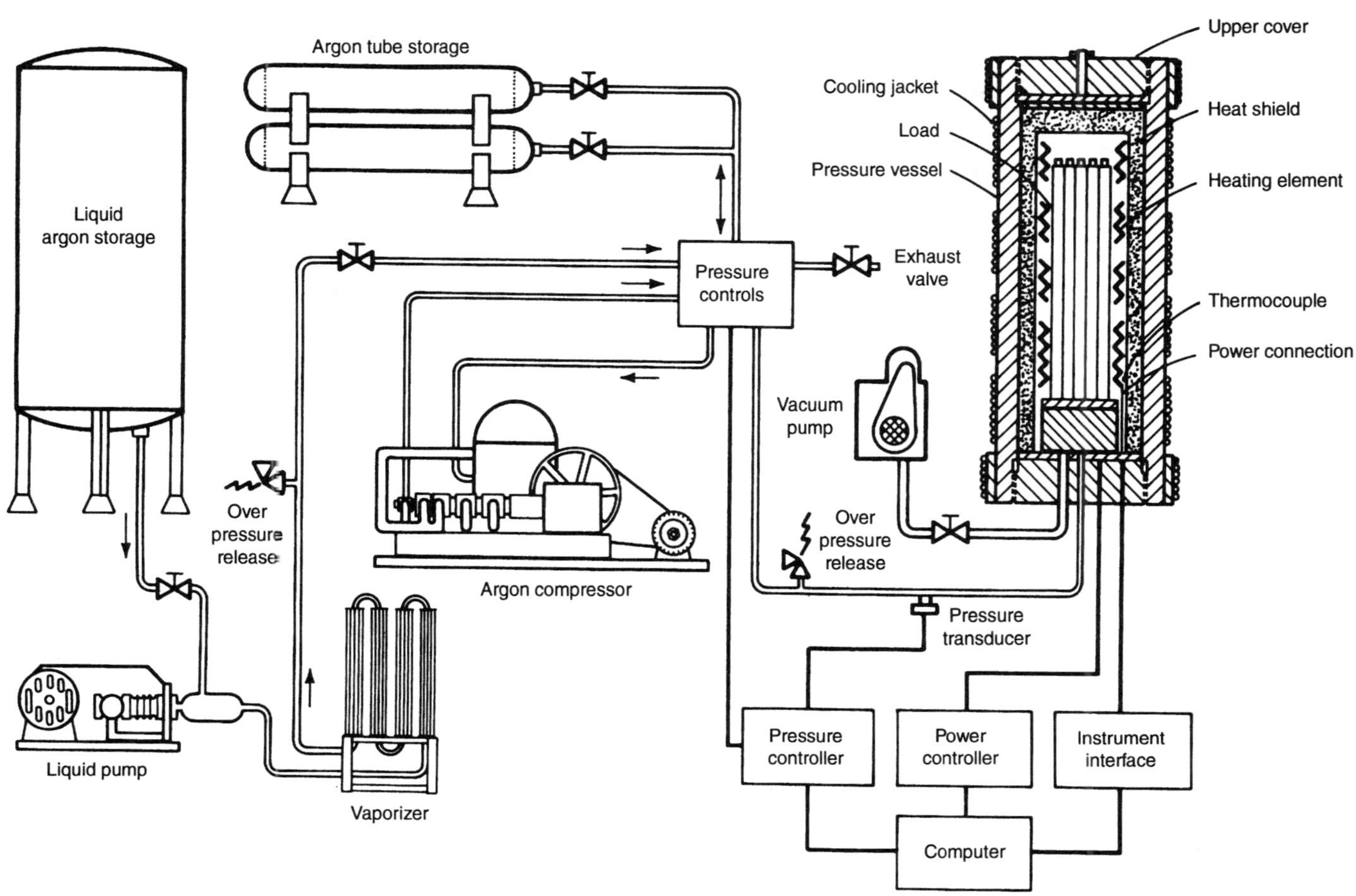

Fig. 1 Typical schematic of an HIP system

either be stored in a tube bank (Fig. 5) or vented to the atmosphere. Argon in gas tube banks must be pumped by a compressor for storage. Multiple-stage, oil-lubricated piston pumps have been used for this purpose. The use of multiple stages provides high-pressure gas at a high-volume rate. However, hydrocarbon contamination of the argon is possible with this type of compressor.

Diaphragm compressors are also used in HIP gas systems. They provide a very clean operation. Contamination of the gas can occur only after gross diaphragm failure. The use of multiple diaphragms and failure sensors and routine replacement of the diaphragm prevent this occurrence. Diaphragm compressors have a low volumetric rate capacity and low gain. High inlet pressure maximizes the mass pumping rate and outlet pressure.

Another type of compressor that has been introduced in the HIP industry is of relatively recent design. A double-acting hydraulic cylinder with a dry-lubricated, highpressure gas cylinder on each end has been used to compress argon to very high pressures (70 to 400 MPa, or 10 to 60 ksi) at a fairly high rate without contamination problems. Figure 6 shows such a typical high-pressure gas compressor.

The purity of the gas entering or leaving the HIP unit must be monitored to ensure that the workload and internal process equipment are not contaminated during a cycle. Possible contaminants include nitrogen, oxygen, water vapor, carbon monoxide, carbon dioxide, methane, and hydrocarbons. Analysis for these gases can be performed with an on-line gas chromatograph.

Different HIP applications require gases of varying purity levels. For example, when hot isostatically pressing powder in a can, protection of process equipment is the only factor that must be considered in determining acceptable gas purity.

Furnace. During HIP, a controllable hot-work zone is created and maintained by the furnace. The size of this zone is governed by vessel size and by furnace and insulation system design. Furnaces utilize primarily convection and radiation mechanisms to transfer heat. Convection dominates at lower temperatures, whereas radiation may be the dominant form of heat transfer at elevated temperatures. Presence of high-density gas as a heat transfer medium allows operation at a higher power density than is possible at atmospheric pressure or in a vacuum for comparable furnace elements.

Pressure vessel body temperatures are kept within design conditions by use of a heat shield between the furnace element and the vessel wall. This heat shield must meet several requirements. In addition to minimizing heat transfer to the vessel wall to reduce power consumption, it must shield the vessel wall as well. During cooldown, the shield should allow heat loss to reduce cooldown time. Density gradients in the argon (caused by thermal gradients between the vessel wall, furnace, and workload) can produce convection loops within the unit. These loops must

Fig. 2 Large HIP unit. 1.5 m (60 in.) diameter by 3.0 m (120 in.) work zone. Operating pressure: 100 MPa (15 ksi)

be controlled by the heat shield to maintain uniform operating temperature within the hot zone. The heat shield also must withstand the thermal gradient from the work zone to the vessel wall. These requirements must be satisfied with as little mass as possible to minimize heat-up and cooldown times and with the smallest volume to provide maximum space within the vessel for the workload. Multiple concentric metal shells, some of which are made from molybdenum or molybdenum alloys, are frequently used for the heat-shield elements.

Typically, the furnace consists of resistance heaters arranged in multiple, independently controlled zones below and beside the workload for optimal temperature control. A hot zone operating at 1750 °C (3200 °F) has been achieved with a single-level element beneath the workload. Natural convention in the high-density gas medium within the pressure vessel is the functional heat transfer mechanism in this design. Forced convection within the pressure vessel has been incorporated in this base heater design, producing an increased workload length as well as improved heating and cooling rates. A heat shield, radiation shields, and baffles are also used in this design.

Installation and loading of the furnace, mantle, and workload into the pressure vessel can be accomplished by several methods. The furnace and mantle can have the capability to automatically connect the power supply and load instrumentation within the vessel. The workload can be introduced into the vessel before, after, or along with the furnace. Some HIP units permit loading and unloading of the workload while it is still hot, thus requiring a shorter cycle time.

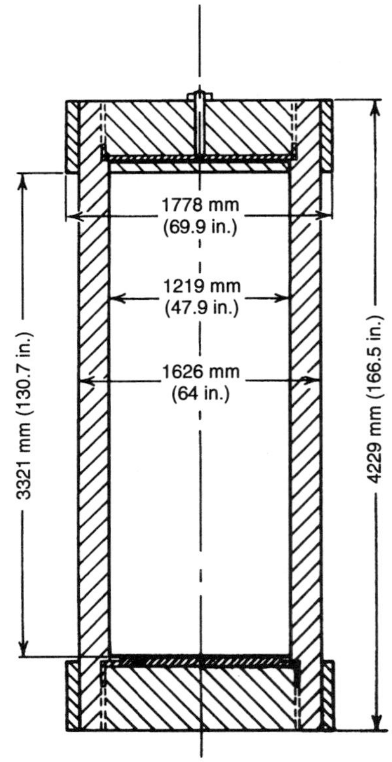

Fig. 3 Monolithic-forged pressure vessel design. Design conditions: 100 MPa (15 ksi) at 200 °C (400 °F). Material: ASTM A723, grade 3, class 3

The type of material used for furnace elements varies, depending on the range of operating temperatures. For applications up to 1230 °C (2250 °F), Kanthal A-1 (Fe-22/23Cr-4.5/5.7-Al-0.5/2.0-Co) (Kanthal Corp., Bethel, CT), or Hoskins 875 (Fe-22.5Cr-5.5Al-0.5Si, Hoskins Manufacturing Co., Detroit, MI) is used. Kanthal can be exposed to air at elevated temperatures; consequently, the workload can be hot loaded into a furnace of this type. However, this material is prone to oxidation and may embrittle. Broken elements can be repaired by conventional welding. Because Kanthal is a relatively poor thermal conductor, low-power densities must be used or overheating can occur, causing sagging and melt out. To compensate for this tendency, Kanthal frequently is manufactured and used in ribbon form.

Molybdenum furnace elements can operate in the temperature range of 500 to 1600 °C (930 to 2900 °F). They cannot be exposed to air at elevated temperatures because rapid oxidation occurs. Molybdenum usually is used as stranded wire or a band. Figure 7 shows the stranded molybdenum heater on a 1.5 m (60 in.) HIP system. During extended use, molybdenum recrystallizes, thus causing an embrittlement condition at ambient temperature. However, stranded molybdenum that becomes damaged can be repaired by lashing, a technique of splicing across the break using molybdenum wire wrapping. In the frequent presence of carbon monoxide in a hot isostatic pressure system, molybdenum case carburizes. The electrical resistivity of the wire becomes greater, increasing the heating capability of the wire.

Graphite is also used as both structural and resistance heating elements for HIP furnaces operating in the range of 400 to 2200 °C (750 to 3990 °F). Graphite is advantageous because of its excellent strength and dimensional stability at very high temperatures, its low cost compared to refractory metals, and its ready availability in various grades and stock sizes. Two forms of graphite are used: bulk and fiber reinforced. Bulk graphite is machined into the required furnace element shapes. Fiberreinforced graphite is used because it can be fabricated into thinner (and higher electrical resistance) sections while retaining good resistance to mechanical damage and breakage.

The principal shortcoming of graphite furnaces is that the system must be kept dry. This is frequently done by a combination of heating, purging, and/or evacuation. When moisture is present, carbon monoxide can be generated by reaction of water vapor and graphite. Carburization of workload pieces can occur unless proper care is taken in the evacuation cycle. Also, contact between graphite and most metals must be avoided to prevent melting reactions. Currently, graphite furnaces up to 75 cm (30 in.) in diameter are used in routine production at temperatures up to 1750 °C (3180 °F).

Controls and Instrumentation. The control system of the HIP unit must provide for the safe and reliable operation of the entire system, as well as accurate measurement and control of pressure and temperature parameters throughout the entire processing cycle. Pressure inside the vessel is measured by a Bourdon tube gage or a strain-gage pressure transducer. Temperature is measured with thermocouples that are distributed throughout the hot zone. With a multizoned furnace, each furnace element or zone must be monitored and controlled to adequately control the entire hot zone. Measurements are taken at several points on large workpieces, as well as throughout the working volume, to guarantee that process conditions are met. Permanent pressure and temperature records should be maintained.

The power supply to the furnace is usually a silicon-controlled rectifier system. Phase-angle firing is used to control power input for molybdenum element furnaces. Zone current, zone voltage, and total power are monitored to control furnace performance. Power consumption of a well-designed 25 cm (10 in.) diam HIP unit is about 15 kW; a 38 cm (15 in.) diam unit consumes about 40 kW, and a 102 cm (40 in.) diam unit uses about 200 kW. Ground-fault protection can be provided to prevent extensive furnace damage in the event of electrical insulation failure.

Other system parameters that should be monitored include gas purity, gas pressure, vessel temperatures, and water system conditions. If a system failure is detected, the control system should provide a warning, take corrective action, and/or initiate system shutdown.

Fig. 4 Hot isostatic pressing unit with a wire-wound pressure vessel and a yoke frame closure support. Courtesy of ABB-Autoclave Systems

Control equipment consists of automatic controllers, analog data presentation devices, and remotely operated valves and switches. Minicomputers are capable of automatically monitoring and controlling all system parameters. Minicomputers also can provide complete and reliable detection of process errors or system malfunction and take corrective action.

Processing Sequence. Figure 8 presents a basic process flowchart for production of various-sized encapsulated and unencapsulated P/M parts. This chart applies to both production and development routes, wherein interactive modification of process variables is required to obtain design objectives. Included is a provision for a subroutine where parts can be low-temperature hot isostatically pressed to a closed porosity condition, decanned, and hot isostatically pressed again, usually at higher temperatures. This option can be employed when the powder/container interaction (melting, alloying, contamination, etc.) is unacceptable at the preferred higher HIP temperature. This technique has been used, for example, for niobium alloys that are initially hot isostatically pressed at 1205 °C (2200 °F) in low-carbon steel containers, decanned, and hot isostatically pressed at 1595 °C (2900 °F) to circumvent an iron-niobium eutectic reaction at 1360 °C (2480 °F).

Part Size. The upper size limitation for densification of encapsulated parts is governed primarily by the processing unit uniform temperature working zone diameter and length. Tool steel billets as large as 86 cm (34 in.) in diameter and 300 cm (120 in.) long have been fully densified by HIP. Nickel-base superalloy P/M turbine disks greater than 1 m (3.3 ft) in diameter have been successfully densified. For sheet metal encapsulation of P/M parts weighing more than approximately 20 kg (44 lb), attachment of handling lugs is recommended. For large-diameter parts (greater than 0.5 m, or 1.6 ft in diameter) and weights greater than 100 kg (220 lb), sheet metal bending stresses due to enclosed powder

Fig. 5 Liquid argon storage and tube bank storage for argon gas. Courtesy of Bodycote IMT, Inc.

weight must be considered. Careful consideration must be given to the support of large parts in hot isostatic press tooling to prevent bending during heating prior to complete densification. For small net-shape parts (up to 1000 g, or 35 oz), particularly with thin sections, tooling that permits separate setting of each part is often required.

Production of small (less than 10 kg, or 20 lb) net-shape parts by HIP using encapsulation techniques is not generally economical because of the cost of container fabrication and leak checking. This applies particularly to P/M steel alloys. Exceptions include experimental parts and manufacture of specialty parts in P/M refractory metals, composites, and precious metals, where metal cost is a controlling factor. Small net-shape parts (less than 0.5 kg, or 1 lb) are best manufactured by containerless HIP, particularly for tool materials, provided satisfactory process procedures can be developed.

Worldwide HIP Capacity. It is estimated that there are nearly 1000 operational HIP units worldwide (Ref 6). Approximately 50 to 75% of these systems are research and development or laboratory units. About 35 to 40% of all operational HIP units are in the United States, with perhaps 75% of these being research and development units; Japan is currently home to around 300 operational HIP units, two-thirds of which are research and development/laboratory units; Western Europe houses about 165 units, probably 60+% of which are research and development/laboratory units; and there are between 30 and 95 units in Asian countries other than Japan. Eastern Europe may have as many as 85 HIP units, the majority of which (~80%) are production units.

The principal manufacturers of hot isostatic presses are ABB-Autoclave Systems, Inc. of Columbus, OH, and ABB of Vasteras, Sweden; Engineered Pressure Systems, Inc. (formerly National Forge); and Kobe. Together, these three companies have clearly dominated world HIP equipment production (and commercial HIP equipment technology development) in the 1990s. From 1990 to the present, ABB, EPSI, and Kobe have together sold no less than 162 HIP units worldwide.

Several other companies build HIP units, among these probably the most prominent being ABRA, American Isostatic Presses, Conaway, and Vitek. ABRA is a Swiss firm incorporated in 1988. Both American Isostatic Presses and Vitek are relatively new and small firms, and both have sold several units. Conaway is a second- or third-generation descendant of Conaway Pressure Systems, Inc., the company mainly responsible for one of the major technical innovations in HIP during the late 1970s and early 1980s—namely, the introduction of the self-contained, laboratory size "MiniHIPPER." Conaway today seems to be focused on the development of rapid, very-high-pressure HIP processing technology, an example of which is the Quick HIP system at Michigan Technological University.

General Process Parameters

Hot isostatic pressing of P/M material generally produces fully dense material with successful process parameters usually biased toward the shortest process time, using pressures within the equipment rating large enough to accommodate the parts being processed. Strict grain-size control also may be required for P/M parts subjected to subsequent isothermal forging or other hot-working operations.

Process Cycle. Several types of hot isostatic pressing cycles can be used. These vary primarily in the magnitude and sequence of pressure and temperature application, as shown in Fig. 9. Additional variations are possible with staged heating and cooling operations.

A type I cycle is carried out by cold pressurization to approximately one-fourth to one-third of final process pressure (T equals approximately 1100 to 1200 °C, or 2000 to 2200 °F), isolation of the pressure vessel, and heating to process temperature. Pressure rises due to increased gas temperature under constant volume constraint. This process constitutes a typical cold-loading cycle and is particularly useful for geometry control of large net shapes in sheet metal encapsulation.

Type II cycles involve heating workpieces to process temperature while maintaining a low vessel pressure (up to 7 MPa, 1000 psi). Pressurization is often accomplished by equalization from storage cylinders and pumping. This type of processing is equivalent to a hot-loading cycle and is used for glass-encapsulated powder and other fragile containers to prevent breakage.

Type III cycles involve full pressurization, followed by heating, while process gas is bled back to storage to maintain pressure within the equipment rating. This cycle tends to enhance recrystallization of powder particles through plastic deformation at the lowest possible temperature.

A type IV process cycle, with the accelerated cooling capabilities of HIP equipment, is the most cost-effective processing mode in terms of reduced process time for a cold-loading cycle. Pumping and heating occur concurrently to a level where sufficient pressure exists within the system to attain the required final pressure when the process temperature is achieved.

Selection of process parameters, such as temperature, time, and heating and cooling rates, to achieve material property and dimensional requirements usually is verified by processing through a matrix and testing. Guidelines to process temperatures for HIP of P/M parts can be obtained from published literature. Rapid diffusion and power-law creep, in which deformation rate is proportional to applied stress raised to a power (n), occur in most cases near these temperatures. General temperature guidelines indicate that metal powders can be densified at 100 MPa (15 ksi) pressure in 2 to 4 h, as indicated:

- *Nickel-base alloys:* 1100 to 1200 °C (2000 to 2200 °F)
- *Iron-base tool steels:* 1050 to 1200 °C (1950 to 2200 °F)
- *Titanium-base alloys:* 850 to 950 °C (1550 to 1750 °F)
- *Aluminum powders:* 480 to 530 °C (900 to 1000 °F)
- *Copper-base alloys:* 750 to 850 °C (1400 to 1550 °F)

- *Molybdenum and niobium fine powder:* 1200 to 1650 °C (2200 to 3000 °F)

Generally, the maximum pressure of the equipment (usually 100 MPa, or 15 ksi) should be used, unless intentionally porous parts are desired. This approach minimizes process time, but scale-up requirements must be considered to avoid requiring pressures not available in larger hot isostatic pressure vessels.

Process times for most powders listed above are 2 to 4 h at 100 MPa (15 ksi). This processing time is based on temperature after soak-through. Because of capacity and transfer lags, particularly for large parts weighing more than 25 kg (55 lb), workpiece temperature must be measured by closely coupled thermocouples (Ref 3). Figure 10 shows a recommended thermocouple installation within a HIP unit to eliminate overriding heat transfer effects from the hot high-pressure gas. Final-stage densification depends primarily on diffusion because pores shrink to radii where the applied-pressure driving force is insignificant compared with surface tension effects (Ref 4).

Evaluation of the time parameter effect on densification must be done on the basis of a structure-sensitive mechanical property, such as notched stress rupture. High-magnification microscopy of grain boundaries in polished sections, if grain-boundary structure is critical for the particular application, also discloses residual pores. For an alternate approach to process parameter selection, see Ref 5 for mechanism maps of HIP.

Special constraints on process parameter selection may relate to controlled grain-size requirements, avoidance of powder/container interactions, and solid-state phase transformations. Micron-sized (0.1 to 1.0 mm) powders (usually precompacted and partially sintered) can be densified at approximately 100 to 200 °C (180 to 360 °F) below coarser (100 mm mean particle size) powders at equivalent pressure and time. This effect is derived from greater contribution of grain-boundary sliding during creep deformation. It is useful in processing of refractory metal powders.

Modeling of HIP

Modeling of the HIP process has tried to define mathematically consolidation, and its objective has been to either predict the properties or the shape of the finished product. This work has been going on for more than 20 years, and the starting points for most models have been either microscopic, in which constitutive equations are based on the interaction of particles under high stress, or macroscopic, in which yield criteria were developed for the powder, treating it as a porous material continuum. A third technique, confined to shape prediction, results in an empirically derived model based on an analysis of actual production experience over an extended period of time. This section reviews all three of these methods and shows the current status and capabilities of all models.

Table 1 lists the nomenclature used in this section on HIP modeling. However, readers should be aware that often different symbols are used in literature for some properties or parameters. Another case is the material property parameters in constitutive equations developed by different authors. In some cases, the terms are identical, but have different mathematical definitions. In

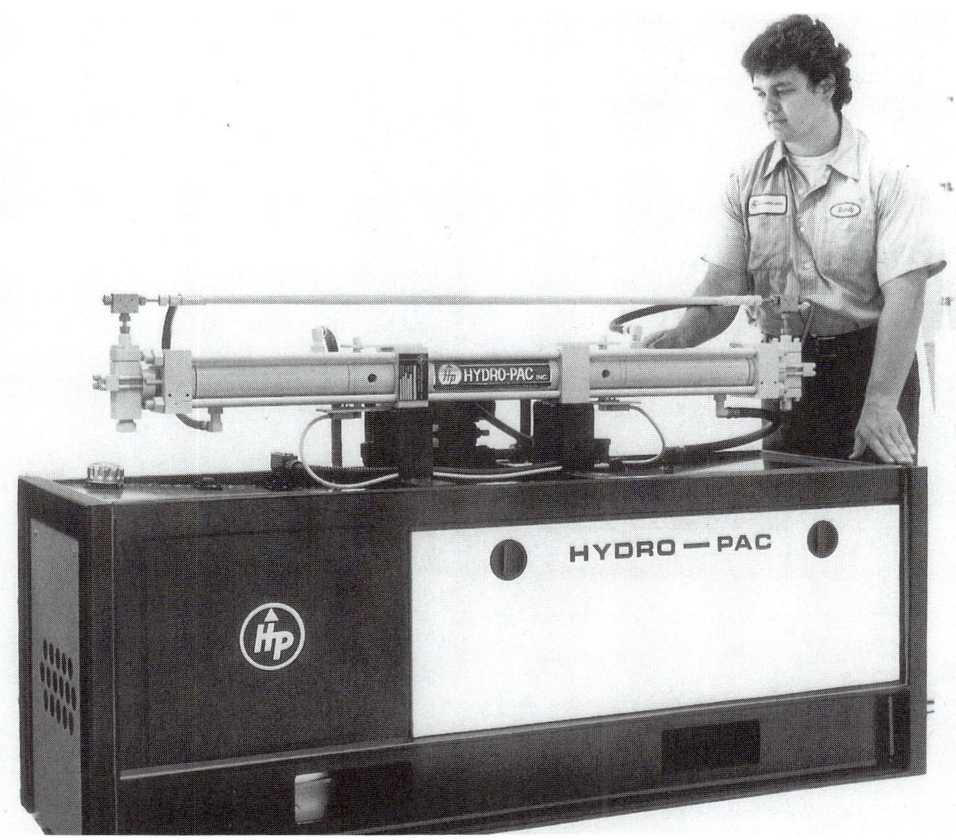

Fig. 6 High-pressure gas compressor. This compressor is rated for operation at 20 MPa (30 ksi) and is powered by a 20 hp electric motor. Courtesy of Hydro-Pac Inc.

Fig. 7 Large molybdenum HIP furnace. This 1.5 m (60 in.) diam heater consists of stranded molybdenum wire on a molybdenum shell. Courtesy of Bodycote IMT, Inc.

some cases, they are unique with their own mathematical description.

Empirical Models

The earliest models or estimates of size change were based on the starting packing density (fraction of theoretical) of the powder and on the estimate that shrinkage was the same in all directions. Total volumetric shrinkage was calculated according to the following analysis:

$$\frac{V_i - V_f}{V_i} = 1 - D_0 \qquad \text{(Eq 1)}$$

Assume that the volume can be approximated by the cube of the linear dimension:

$$l_i^3 = V_i \qquad \text{(Eq 2)}$$

$$l_f^3 = V_f \qquad \text{(Eq 3)}$$

Table 1 Nomenclature used in equations

Symbol	Definition
α	function of relative density, used in yield function with J_1
β	function of relative density, used in yield function with J_2'
ε	strain
$\dot{\varepsilon}$	strain rate
$\dot{\varepsilon}_0$	material property in power-law creep equation
$d\varepsilon_{eq}$	equivalent matrix strain increment
θ	temperature
$\delta\lambda$	nonnegative constant
σ	stress
$\sigma_1, \sigma_2, \sigma_3$	principal stresses
σ_{av}	average stress
σ_m	hydrostatic stress
σ_{eq}	equivalent stress
σ_0	material property parameter in creep equation
σ_y	yield stress of material
τ_s	shear stress
ϕ	function of relative density, used with hydrostatic stress in stress-strain equations
b	material constant used in various yield functions. May have different specific meanings
f, f_1, f_2, f	function used in various yield functions. May have different specific meanings
k	thermal conductivity of fully dense product
k_D	thermal conductivity as an exponential function of relative density
l_i	characteristic initial dimension of compact
l_f	characteristic final dimension of compact
m	material constant in various yield functions
n	power-law creep exponent
A	power-law creep coefficient
D	relative density of the powder at any time
D_0	initial relative density of the powder
$\dot{D}$	densification rate
F	yield function
J_1	$\sigma_1 + \sigma_2 + \sigma_3$, first invariant of the stress tensor
J_2'	$[(\sigma_1 - \sigma_2)^2 + (\sigma_2 - \sigma_3)^2 + (\sigma_3 - \sigma_1)^2]$, second invariant of the deviatoric stress sensor
P	pressure
P_{eff}	effective (contact) pressure on each powder particle
P_{lim}	external pressure that will cause yielding
Q	activation energy for creep
R	radius
R_0	universal gas constant
T	temperature
V_i	initial volume of powder container
V_f	final volume of powder container after HIP

The shrinkage is the same in all directions (isotropic behavior):

$$\frac{l_i^3 - l_f^3}{l_i^3} = 1 - D_0 \qquad \text{(Eq 4)}$$

$$\frac{l_f}{l_i} = D_0^{1/3} \qquad \text{(Eq 5)}$$

The final dimension is the initial dimension times the cube root of the initial relative density. In other words, isotropic shrinkage or the shrinkage of each linear dimension is equal to $(1 - D_0^{1/3})$.

From these equations, the initial dimensions, that is, the starting shape of the powder needed to end up with specific final dimensions, can be approximated if one knows the starting density and the desired finished dimensions. The assumption of isotropic shrinkage has limited validity. If a spherical container is hot isostatically pressed with no thermal gradient present, shrinkage is uniform, but all other starting shapes do not densify uniformly in all directions. This is, however, a satisfactory starting point when the initial packing density is greater than 0.65 and if an additional machining allowance is used to produce a near-net shape. A large allowance produces a satisfactory part all the time. This may be an uneconomical solution, however, because the additional weight of powder for the allowance may make the process too expensive. Shrinkage corrections can be made based on experience for nonspherical geometries, and one company, Crucible (Ref 7), has developed a model based on a continuing analysis of production data on near-net-shape parts. This is a significant refinement over the isotropic calculation and works very well when a complex part can be divided into simple cylinders, rectangular prisms, hollow cylinders, or spheres. The important variables in this model are starting density, aspect ratio, cylindrical area/end area, and the absolute value of the outside diameter. Because this concept involves the construction of the starting container based on an analysis of existing data, it may be thought of as an "expert system." Crucible's design procedure (using their shrink bus software) is shown in Fig. 11.

This practice has been in use for more than 10 years. Statistical analyses of production data show good repeatability on the same part, but there are some serious limitations to this model as shown by the question mark to the right of the hollow cylinder diamond in Fig. 11. It works

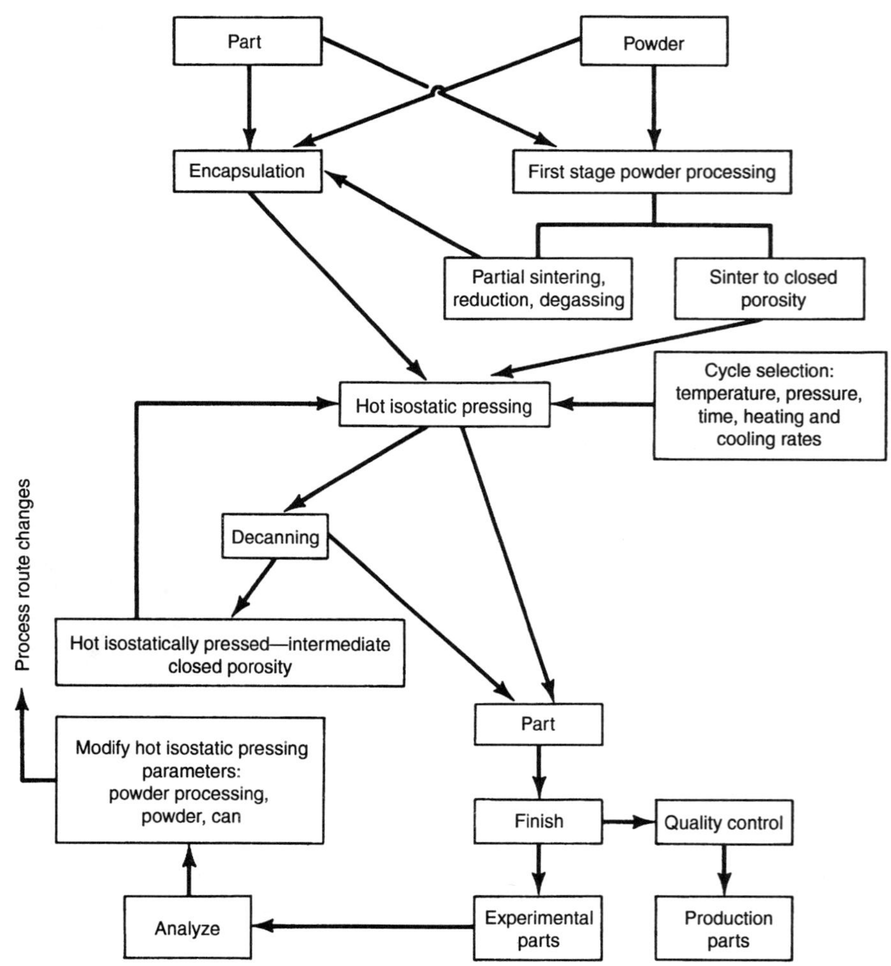

Fig. 8 Flowchart for P/M manufacture of metal parts by HIP

well with thin-walled containers and with relatively simple shapes, but initial attempts to predict how curved surfaces would shrink were incorrect. There are two other known empirical models. The Swedish Institute of Production Engineering described their work (Ref 8) using factors similar to Crucible's, also based on the analysis of production data. The third work is that of Powdermet Sweden which, like Crucible, is in the near-net-shape business. It was reported that they evaluated a finite element model, but are designing containers based on their production experience (Ref 9). A more fundamental understanding of the mechanics of densification and the effect of HIP cycles on shape change beyond the empirical models is needed to get more accurate shape control after HIP on complex parts.

Microscopic Models and Mechanisms

Mechanistic models of the microscopic mechanisms in the HIP process was pioneered by Ashby et al. (Ref 5, 10–13) in the early 1980s and is based on the interaction among powder particles during the consolidation process. Ashby and his collaborators did most of the early work on this subject and opened up many avenues for further study by later investigators. They divided the entire consolidation process into three stages, based on powder packing and how the porosity is located relative to the powder particles. Stage 0 describes the density reached by the packing of the loose powder. Stage 1 describes the early stages of densification (relative density less than 0.9) when the porosity is still connected. Stage 2 describes the final densification (relative density greater than 0.9) when the residual porosity is in the form of small holes. A schematic of these two stages is shown in Fig. 12.

Although this work assumes that the powder consists of spherical particles of a single size, other authors have attempted to determine densification rates with spherical particles (Ref 14–17) of nonuniform size distribution. Ashby recognized that the mechanism of densification depends on the temperature and pressure at any given time in the HIP cycle and derived equations in each stage for plastic yielding, power-law creep, Nabarro-Herring creep, and diffusion. For the complete derivation of all the equations used for each of these mechanisms, see Ref 16. This section presents the final results of this derivation.

Plastic Yielding Densification. Yielding occurs during the initial stage provided the effective pressure, P_{eff}, on the particle is equal to or greater than three times the yield stress of the material. The external pressure that just causes yielding is:

$$P_{lim} = 3D^2 \frac{D - D_0}{1 - D_0} \sigma_y \qquad \text{(Eq 6)}$$

Because yielding is a time-independent mechanism, the densification rate will be infinite if the effective pressure is greater than the limiting pressure for yielding. Likewise, the rate will be zero at effective pressures below the limiting pressure for yielding.

Power-Law Creep Densification. Densification also occurs by creep deformation. The constitutive equation for a solid exhibiting power-law creep is:

$$\dot{\varepsilon} = \dot{\varepsilon}_0 \left(\frac{\sigma}{\sigma_0}\right)^n \qquad \text{(Eq 7)}$$

σ_0, and n are material properties; $\dot{\varepsilon}$ and σ are the equivalent strain rate and stress, respectively.

In Stage 2, the densification rate becomes:

$$\dot{D} = \frac{3}{2}\left(\frac{\dot{\varepsilon}_0}{\sigma_0^n}\right) \frac{D(1 - D)}{([1 - (1 - D)^{1/n}]^n)} \left(\frac{3}{2n} P\right)^n \qquad \text{(Eq 8)}$$

Although densification in the latter part of Stage 2 does occur as Nabarro-Herring creep and diffusion, the relative changes in both density and geometry due to these mechanisms are usually small compared to plastic yielding and power-law creep. Equations have been developed for all these mechanisms and can be found in Ref 11 along with the derivations.

HIP Maps. The major use for these equations has been in the construction of HIP maps. These HIP diagrams identify the dominant mechanism that is responsible for densification in a particular temperature/pressure region and can be used as a tool to qualitatively predict densification rates and consolidation times as functions of pressure and temperature. The maps are constructed by evaluating the contribution of each densification mechanism and summing them to give the total densification rate. Like every mechanistic model, the map requires material property data. Data for many materials as well as how to construct and use these HIP maps are available from Ashby's software program, HIP 6.0 (Ref 18). The user manual describes in detail how to construct maps and analyze data from them. The output of this program is in one of three forms: temperature versus relative density, pressure versus relative density, or pressure versus temperature.

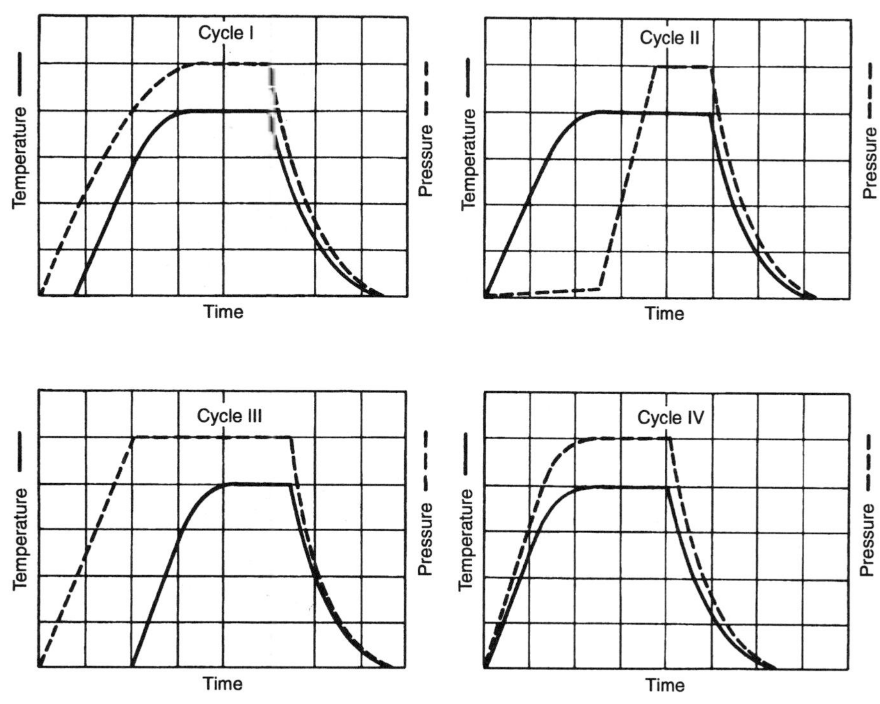

Fig. 9 Typical process cycles. Source: Bodycote IMT, Inc.

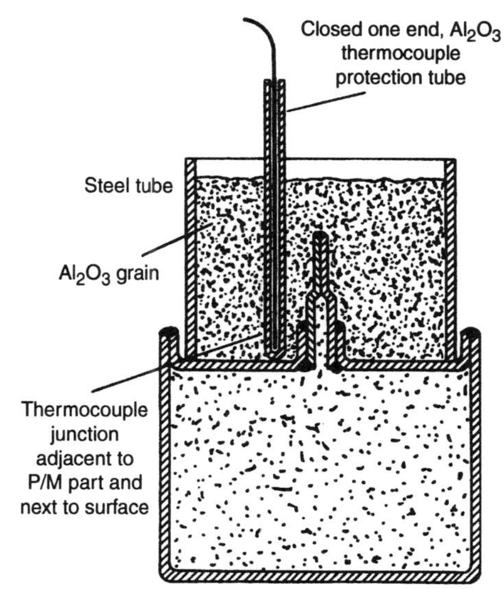

Fig. 10 Workpiece thermocouple installation for HIP cycles. This design is intended to minimize transfer and lag errors in timing P/M cycles. Source: Bodycote IMT, Inc.

A typical HIP map is shown in Fig. 13. It is a density versus temperature map of a tool steel starting with a relative density of 0.69 at a constant pressure of 100 MPa. The three times are 5, 30, and 180 min. The power-law creep exponent used is 7.5, which is the value from Ref 18. Because creep is the dominant mechanism in this consolidation, the densification rate is very sensitive to the power-law creep exponent. If 7.5 is not the correct value for this material an error will occur in the densification rate. Actual experimental data (Ref 19) indicate that full consolidation in large diameter parts takes place at about the rate that is predicted if the exponent is 3.5. This is the first indication that the correct material property parameters may be a major cause for inaccuracies in any HIP model.

Two points should be made: (1) although these maps do not provide a quantitative picture of what happens during a cycle, they are very useful in providing qualitative guidelines for densification changes at various times at either a given temperature or pressure, and (2) the dominant mechanism of densification is readily identified. One large advantage of HIP maps is that data can be input and the program run in about 1 min, so a great deal of information can be learned in a short amount of time by the creation of several maps.

Summary. Although the best use of the microscopic model is in qualitative evaluations, it has been used to determine shrinkage and size change during HIP on small cylindrical containers (up to 100 mm long by up to 200 mm in diameter) (Ref 20). This work showed that temperature gradients in the powder during HIP could cause densification fronts that are a major reason for unusual or anisotropic shape change. This same effect was also shown by Aren and Navara (Ref 21) who determined the effect of different geometries on densification. The contributions, significance, and use of the microscopically developed models can be summarized:

- Information about microscopic details of the HIP process enabled mathematical equations to be developed that established the mechanisms of densification in the different temperature/pressure regions of the process.
- Much of this work has provided education for those unfamiliar with the HIP process and has served as a means to qualitatively examine the process on an industrial scale.
- Computer programs for personal computers exist that can help to evaluate the effect that

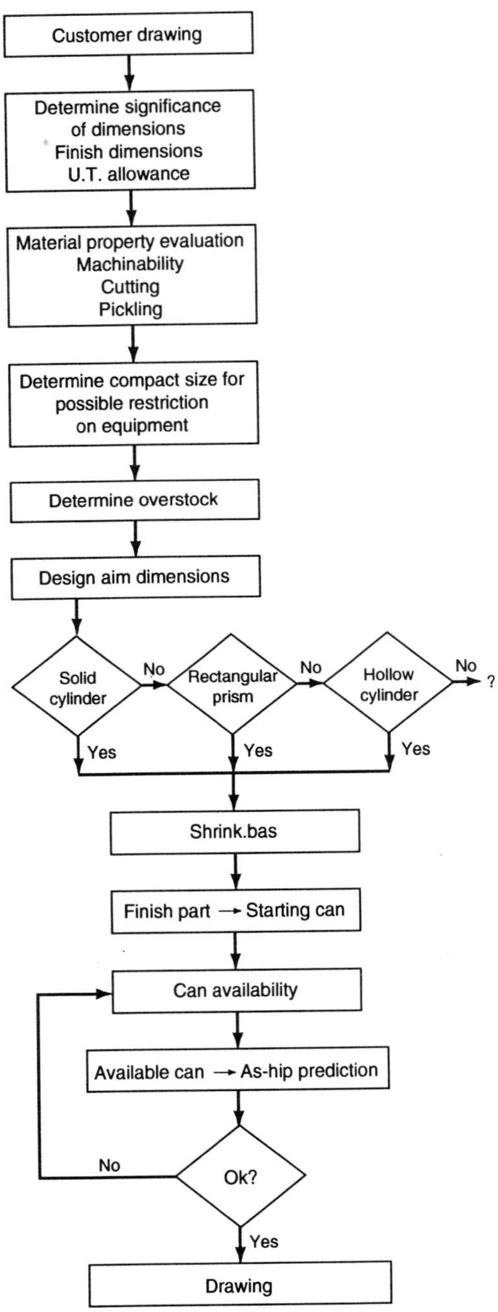

Fig. 11 Crucible design procedure. From Ref 7

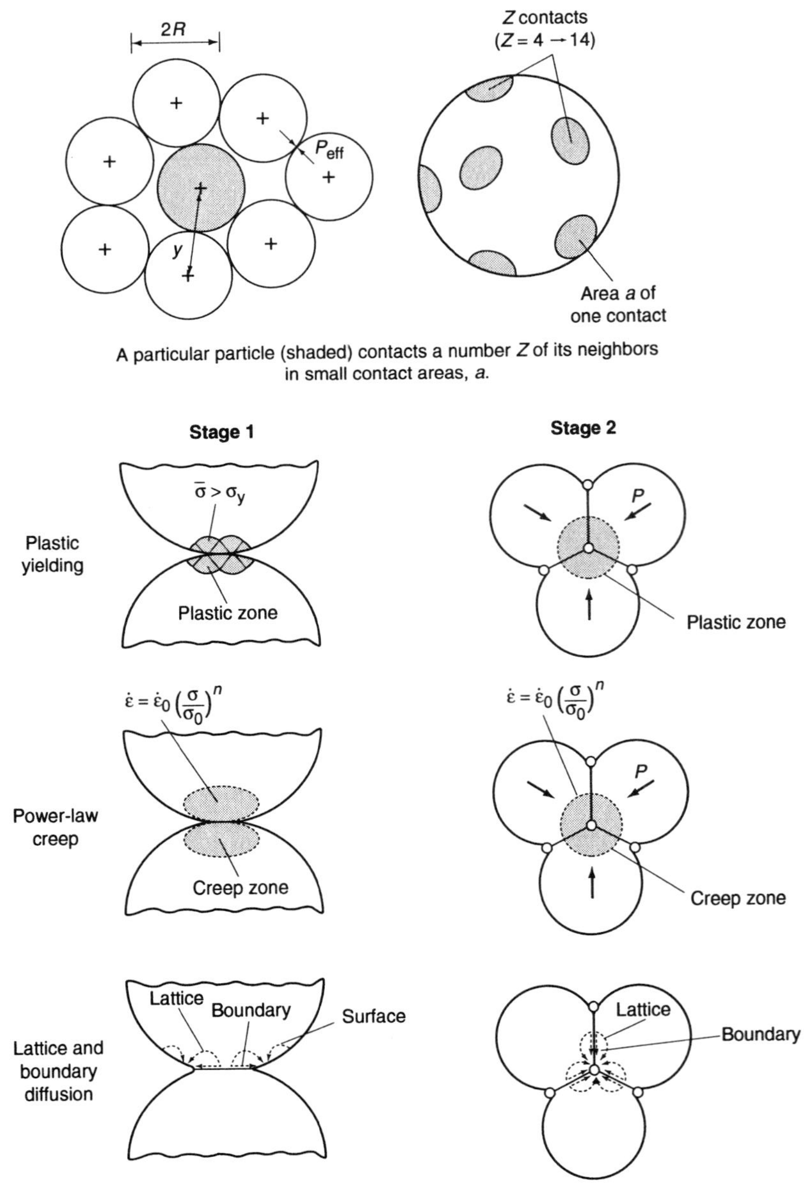

Fig. 12 Ashby stage 1 and stage 2 densification. From Ref 13

different HIP cycles may have on densification kinetics. These HIP maps have provided a basis for much of the knowledge of how metal powder is consolidated.

- The need to combine heat transfer and densification equations to quantitatively describe the HIP process was recognized.

Macroscopic Models of HIP

Developments Prior to about 1990. The earliest attempts to develop a macroscopic model were the work of Kuhn and Downey (Ref 22) and Green (Ref 23), who proposed a plasticity theory for porous metals. They independently developed yield criteria and stress-strain relationships for these materials. Both of these theories suggest that the yield criterion is a function of the first invariant of stress, J_1 and the second invariant of the deviatoric stress, J_2'. The relationship is expressed:

$$F = (\alpha J_1^2 + \beta J_2')^{1/2} \quad \text{(Eq 9)}$$

where α and β are functions of the relative density. They also defined the stress-strain relationship:

$$d\varepsilon_i = d\lambda(\sigma_i - \phi\sigma_m) \quad \text{(Eq 10)}$$

where φ is a function of relative density, $d\lambda$ is a nonnegative constant, and σ_m is the hydrostatic stress.

Shima and Oyane (Ref 24) built on this work and rewrote Eq 9 as follows:

$$F = \left[\frac{(\sigma_1 - \sigma_2)^2 + (\sigma_2 - \sigma_3)^2 + (\sigma_3 + \sigma_1)^2}{2} + \left(\frac{\sigma_m}{f}\right)^2\right]^{1/2} \quad \text{(Eq 11)}$$

where f represents the degree of influence of the hydrostatic stress component, σ_m, on the onset of yielding of porous bodies and is a function of relative density. Their expression for the yield criterion for porous materials as well as stress-strain relationships is:

$$\sigma_{eq} = \frac{1}{D^m}\left[\frac{(\sigma_1 - \sigma_2)^2 + (\sigma_2 - \sigma_3)^2 + (\sigma_3 + \sigma_1)^2}{2} - \left(\frac{\sigma_m}{f}\right)^2\right]^{1/2} \quad \text{(Eq 12)}$$

$$d\varepsilon_i = d\lambda\left[\sigma_i - \left(1 - \frac{2}{9}f^2\right)\sigma_m\right] \quad \text{(Eq 13)}$$

f and m in Eq 12 were determined by simple compression and tension tests and are given by:

$$f = \frac{1}{2.49}(1 - D)^{0.514} \quad \text{(Eq 14)}$$

$$m = 2.5 \quad \text{(Eq 15)}$$

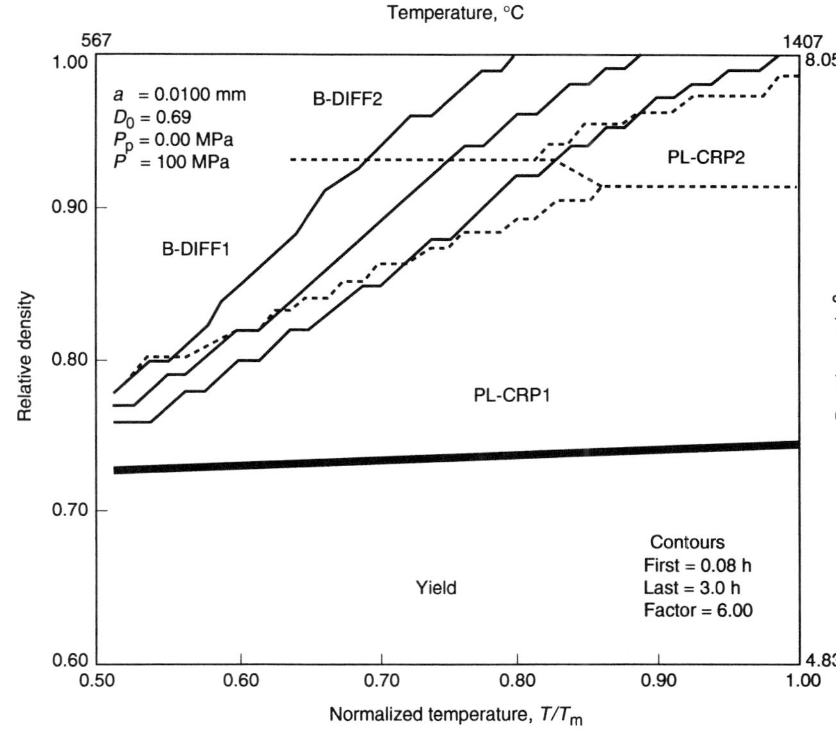

Fig. 13 Tool steel HIP diagram, creep exponent = 7.5. From Ref 18

$$d\lambda = \frac{3}{2}\frac{1}{D^{2m+1}}\frac{d\varepsilon_{eq}}{\sigma_{eq}} \quad \text{(Eq 16)}$$

These basic equations were applied to the analysis of closed-die compression, and the calculated pressure-relative density relationships for sintered copper agreed well with experimental results. Although this work did not deal with the HIP process itself, it has served as the basis for virtually all subsequent continuum models. The functions, f and m, in Eq 12 appear in similar forms in work done since this pioneering effort. These functions must be determined experimentally for each particular material.

The work of Shima and Oyane served as the basis for studies done by Soh, Nohara, and Nakagawa (Ref 25) and Nohara, Nakagawa, Soh, and Shinke (Ref 26). These investigators carried out uniaxial compression tests on MERL 76, a nickel-base superalloy and ended up with a comparison between the predicted shape change for an axisymmetric turbine disk and the actual shape change after a HIP cycle, as shown in Fig. 14 and 15. This particular HIP cycle is somewhat unusual because temperature is kept constant for a long time and then raised in a single step. This practice tends to eliminate temperature gradients during the HIP cycle. In most commercial cases, however, temperature and

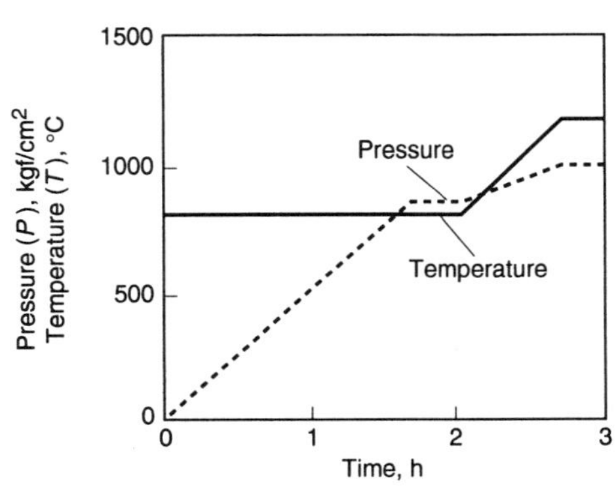

Fig. 14 HIP cycle for MERL 76 disk. From Ref 25

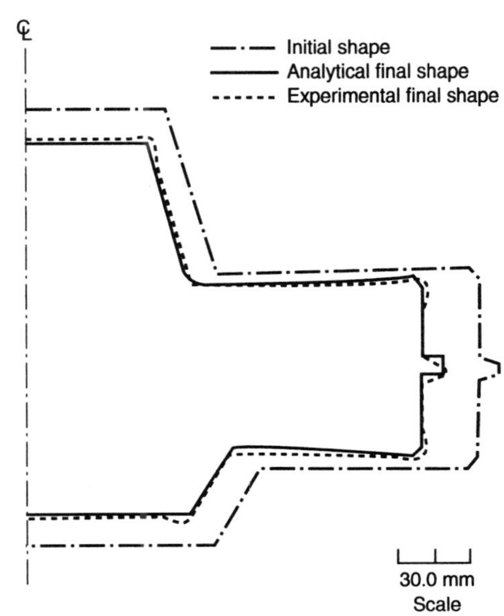

Fig. 15 Final shape of MERL 76 disk. From Ref 25

pressure increase together and thermal gradients do exist. How to calculate them and deal with them is a problem that still challenges all researchers who are working on the computer modeling. Work done in Ref 25 and 26 assumed a linear relationship between thermal diffusivity and relative density. This work was sponsored by Kobe Steel and has led to the finite element program, HIPNAS, which has been used in Japan.

Another important contribution to the understanding of the consolidation process was the work of Abouaf et al. (Ref 27) (early 1980s). This work also recognized that deformation or consolidation involved both plastic flow and time-dependent creep and from this developed a viscoplastic model. The form of Abouaf's constitutive equations are very similar to those of Shima and Oyane, but have an additional creep deformation term. The density-dependent functions were determined with simple compression tests at different temperatures on partially dense materials, while the creep parameters were determined on fully dense material tested at different temperatures. This is a technique that is currently used on most models to determine material properties.

Thermal characteristics were determined experimentally, and their results show an exponential relationship different from the Japanese investigators. Theoretical work of Argento and Bouvard (Ref 28) expressed diffusivity as an exponential function of relative density, but work done in 1997 (Ref 29) on the nickel-base alloy 625 using the laser flash technique confirmed the assumption of the Japanese investigators that the relationship is linear. There is yet no full agreement on how thermal diffusivity correlates with relative density. Because density does change during consolidation, this relationship has an important bearing on any calculation, and the lack of specific agreement on this relation adds to the difficulty of developing an accurate mathematical model.

A comprehensive review by Doraivelu et al. (Ref 30) defined a new yield function for porous bodies. The basis for their work was the large discrepancy between the experimental and theoretical dependence of compressive yield stress on relative density. Again, the yield function is tied in to the first invariant of the stress tensor and the second invariant of the deviatoric stress component. Their yield function is expressed only as a function of relative density:

$$F = \left[(2 + D^2) J_2' + \frac{(1 - D^2)}{3} J_1^2 \right]^{1/2} \quad \text{(Eq 17)}$$

This is in a form similar to Eq 9 and 11 and gives yield curves very close to those. This work also summarizes all previously developed expressions for yield curves proposed by several different researchers as shown in Fig. 16. Qualitatively all are similar, but quantitatively there are obvious differences. No HIP simulations came from this work, but three-dimensional yield surfaces were calculated by software developed from this mathematics by Brigham Young University.

Current Developments in HIP Modeling. Consolidation mathematics began to be integrated into computer code to create simulation models starting approximately 1990. The status in 1997 can be summarized:

- Constitutive equations based on plasticity and creep deformation have been developed for porous materials. All are of similar form and have functions in them dependent on relative density and material property parameters.
- Because yield strength of both the fully dense and the partially dense material are both functions of temperature as are the creep properties, it is necessary to develop the equations for heat transfer during the densification process so that the mathematical descriptions for deformation (consolidation) and heat transfer can be coupled into a single finite element code.
- Commercial models have been confined to relatively simple shapes, and insufficient data have been presented to show the quantitative relationship between the predicted and actual

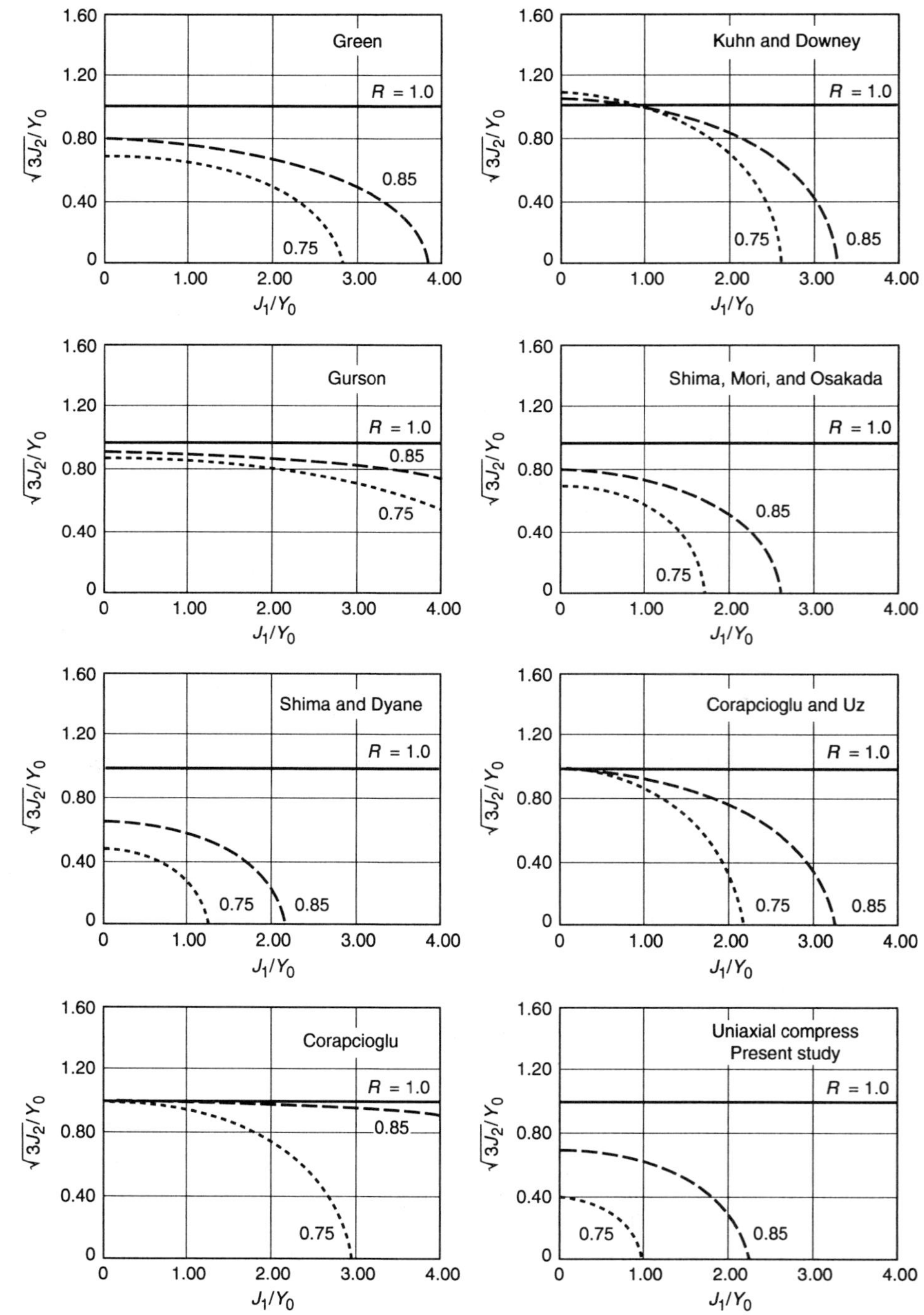

Fig. 16 Yield functions proposed by various researchers. From Ref 30

dimensions on any hardware to know whether any of the models are quantitatively accurate.

The utility of a model should be defined in terms of its capability to predict actual size and shape change during HIP, so that under perfect conditions using the model will eliminate the need for production trials. Work that is either a continuation of that already described or from new sources has recognized this. While there have been efforts from other investigators, specific attention will be paid to the results from four groups, because these are closest to realizing the goal. These four groups are LNT in Russia, CEA/CEREM in France, and MATSYS and CTC in the United States. LNT's model is plastic, while the other three employ a viscoplastic model. Each of these models are briefly described.

From a practical engineering and design approach, however, each of the HIP models suffers from lack of validation for any but the simplest of parts. The true test of the accuracy of any model can only be done on a relatively large and complex production part with curved surfaces and different cross sections. If any of these models is to gain the confidence of the production facility that wants to use it, additional validation experiments must be run. Moreover, the expected precision of size change predictions must be quantitatively defined if parts producers are to rely absolutely on the results predicted by the simulation alone. To make complex parts, either net or near-net shape, the designer must produce a prototype as a first iteration and then use the analysis of the results of that trial to make production hardware. Current HIP models for design systems can only work when some trial part is an integral part of the manufacturing process because the simulations are just not accurate enough to do the job alone.

LNT Model of HIP. Computer simulation of HIP started at the All Russia Institute of Light Alloys (VILS), which is now at the Laboratory of New Technology (LNT) in Moscow. Their work (Ref 31–34) was initially tied into the use of superalloy powder for as-HIP near-net-shape parts. The objective is to shorten the time it takes to manufacture a complex part from powder, and its model uses plastic yield criteria only. However, LNT research recognizes the existence of errors in both the mathematics of the constitutive equations as well as in the determination of material property parameters that are used in the model. Because of these errors, researchers believe the results of a simulation alone cannot be used directly in the manufacture of hardware. Their yield criterion is:

$$\frac{\theta_2}{f_1^2}+\frac{\sigma_{av}^2}{f_2^2}+\tau_s^2 \qquad \text{(Eq 18)}$$

where f_1 and f_2 are functions of density, σ_{av} is the average stress, τ_s is the shear stress, and θ is the temperature.

The LNT methodology is to determine yield curves and apply them to the manufacture of actual parts (Fig. 17). Temperature and pressure are raised together in a HIP cycle, and samples, S_1, S_2, ..., S_n, are obtained by interrupting the cycle. Relative densities of each sample are checked, and the compressive yield strength of each is determined. This gives the yield strength of the material at different densities at a specific temperature and gives yield strength as a function of relative density because subsequent HIP cycles have the same temperature, pressure, and time relationship as the test cycle. Their method assumes that the relative density in the simulation will change at the same rate as the samples used to measure yield strength. Temperature gradients do not exist. The variables f_1 and f_2 and the yield strength values are calculated and are implemented into a finite element code to simulate shrinkage. The next step is to design a capsule (container) and make a demonstrator part and compare the final actual geometry with that of the simulation. Because errors are expected, the model is then "fine-tuned" until the simulation accurately fits the dimensions of the demonstrator part. Finally, the model is used to determine the starting geometry for a production part. This is a practical engineering approach to a complicated problem, and LNT researchers use modeling in combination with the manufacture of a prototype part. Figure 18, which shows their flowchart for container design, divides the entire problem into four subproblems: modeling (simulation), parametric identification, capsule design, and stochastic analysis.

MATSYS Model of HIP. A second commercial finite element model of HIP was started under an Intelligent Processing of Materials initiative funded by ARPA. The HIP modeling was carried out by BDM (now MATSYS) and this viscoplastic model has been described in detail elsewhere (Ref 35–37). This work recognizes the different

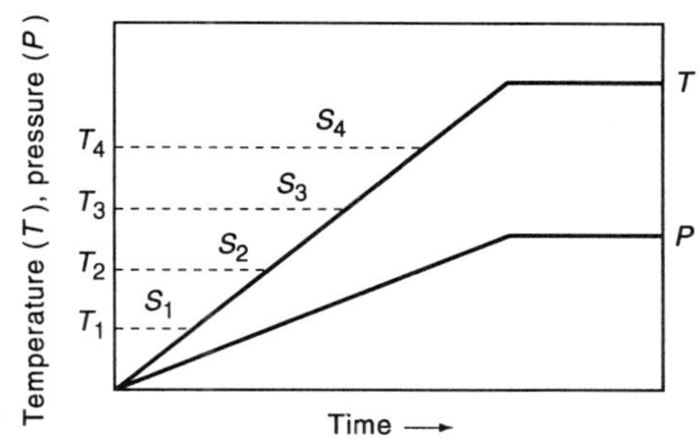

Fig. 17 LNT HIP cycle for calibration. From Ref 18

stages of consolidation as interpreted by Ashby and shown in Fig. 12, but defines them in terms of the relative density in each stage rather than in the geometry of the pores between powder particles:

- *Stage 1:* relative density is less than or equal to 0.75.
- *Stage 2:* relative density is greater than 0.90.
- *Transition region:* relative density is between 0.75 and 0.90.

Using these three ranges, they define yield curves in stage 1 based on the work of Fleck, Kuhn, and McMeeking (Ref 37) and in stage 2 based on the work of Gurson (Ref 38). In the transition region where relative density is between 0.75 and 0.90, they introduced a new concept, a yield function that is a weighted average of the two functions and dependent on the relative density at any given time:

$$F=\frac{D_2-D}{D_2-D_1}F_1+\frac{D-D_1}{D_2-D_1}F_2=0 \qquad \text{(Eq 19)}$$

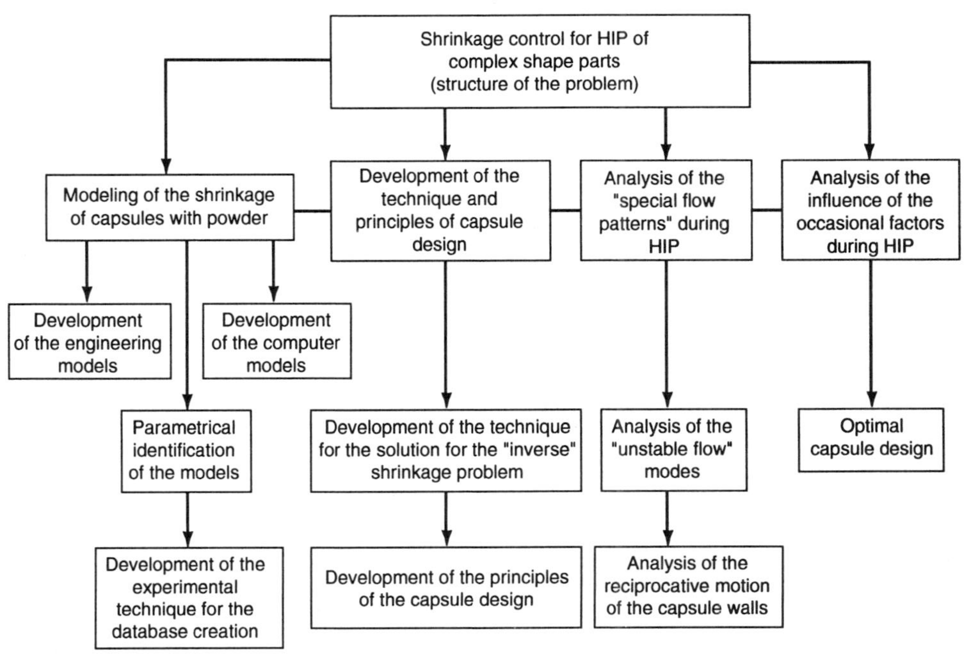

Fig. 18 LNT flowchart for container design. From Ref 32

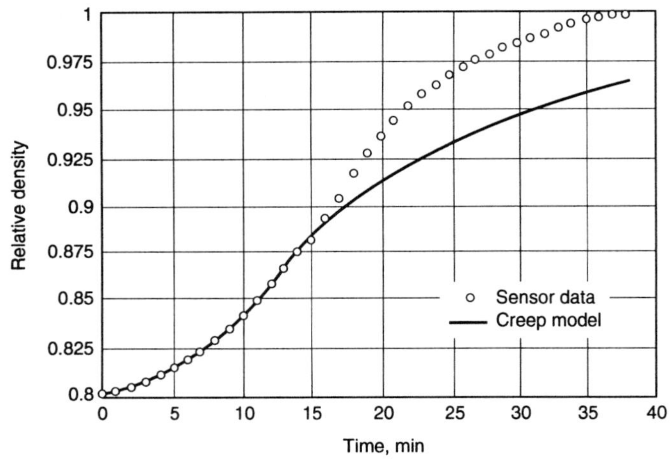

Fig. 19 BDM calculated versus senior shrinkage for TiAl. Estimation of γ-TiAl properties. Power-law creep is the active densification mechanism during stage 1.

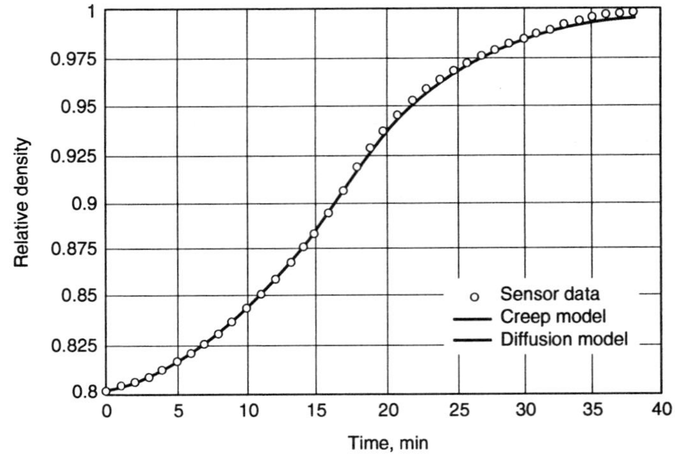

Fig. 20 BDM calculated versus sensor shrinkage for TiAl. Estimation of γ-TiAl properties. Power-law creep is active during stage 1. Diffusion is active during stage 2.

where F_1 is the yield criterion in stage 1, and F_2 is the yield criterion in stage 2.

Constitutive equations were also developed for deformation due to power-law creep in stage 1 and stage 2 with the transition region handled as in Eq 19. The models were implemented by determining the densification due to these two mechanisms, summing them, and calculating the size change that resulted. A third mechanism, diffusional creep has been added to their calculation for densification change. The determination of which mechanisms are active during densification is tied into the experimental method for calculation of material property parameters.

This method is quite different from that used by other model developers and is based on actual densification rates as experimentally determined using eddy-current sensors first described by Kahn, Mester, and Wadley (Ref 39). By measuring in real time the impedance of a small cylinder filled with powder during a HIP cycle and measuring density after the run, these two variables can be correlated so that the signal from the eddy-current sensor during a subsequent HIP cycle gives a dynamic measurement of actual density. Material property parameters in the densification equations are determined by initially estimating them in the densification change equations and comparing the calculated density changes with the actual density changes of the small samples. Nonlinear optimization techniques are then used to match the output of the model with the actual density changes. The end result of this match gives accurate determinations of the material parameters. Figure 19 is a plot of the actual sensor data superimposed over the creep densification mechanism densification rate for γ-titanium aluminide. There is an excellent fit between the two up to a relative density of about 0.90, but above that there is an obvious mismatch between the two curves, indicating that power-law creep is not the only densification mechanism. Figure 20 corrects this mismatch by using diffusional creep above a relative density of 0.90 to get a much better fit.

CTC Model of HIP. A third model, developed by Concurrent Technologies Corporation (CTC), is the offshoot of work started by Trasorras et al. (Ref 40) at Carnegie Mellon University. The specific goal of this U.S. Navy sponsored program at CTC is to develop a program that can accurately predict the final dimensions of a hot isostatically pressed component. Trasorras defined two different deformation processes: deformation of the powder aggregate idealized as a continuum and deformation of the individual particles. His model accounts for the plastic deformation of the powder aggregate as a whole, the stress response of the powder aggregate, heat transfer properties of the powder aggregate, and evolution of the matrix material state. This work uses both plastic yield and power-law creep as densification mechanisms and is thus another viscoplastic model. Most of the experimental work on this model has been done with commercially pure titanium (Ref 41–44). Trials using the CTC computer model shows it can be used in place of actual production trials (Ref 44). The trial was with a titanium part. The first trial of the model matched the actual production part in qualitative terms, but it did not meet the desired shape. The starting container shape was changed to determine what initial shape would produce the finish customer dimension. On the third computer trial, the simulated container produced the desired finished part shape. This system has been upgraded to coordinate solid modeling with mesh generation and finite element analysis. The methodology is similar to that of LNT with the only difference being the use of an actual part by CTC as opposed to a demonstrator part by LNT to "fine-tune" the model.

CEA/CEREM Model of HIP. Work by CEA/CEREM in Grenoble, France, is based on the Abouaf viscoplastic model. In this work (Ref 45, 46) an equivalent stress is defined as:

$$\sigma_{eq} = \left[fJ_1^2 + \frac{3}{2} bJ_2'^2 \right]^{1/2} \quad \text{(Eq 20)}$$

This form is similar to Eq 12, where f and b are functions of the relative density, D, and are material dependent. When the relative density equals 1, $f(1) = 0$ and $b(1) = 1$, this satisfies the von Mises criterion for incompressibility of a fully dense material. The equivalent stress and strain rate are also related through the creep law:

$$\dot{D} = A \exp\left(-\frac{Q}{R_0\theta}\right)\sigma_{eq}^n \quad \text{(Eq 21)}$$

The values of A and n are calculated using fully dense samples, while f and b are determined with porous samples with relative densities ranging from 0.77 to 1.0. Experimental work with T316LN powder was done to determine these functions (Ref 47, 48).

These investigators also have proposed a function in which thermal conductivity is an exponential function of relative density:

$$k_D = k\left(\frac{D - D_0}{1 - D_0}\right)^{1.46(1 - D_0)} \quad \text{(Eq 22)}$$

where k is thermal conductivity. The results of this work have been incorporated into a multimaterials finite element modeling code, PRECAD/M, that allows the simulation of HIP of axisymmetric and three-dimensional parts. The finite element software was taken from a commercial package and modified by CEA/CEREM specifically to handle the problem of dimensional change during HIP. Data were presented to show the comparison between initial and deformed geometry in a part made with Ti-6Al-4V powder with a solid core (Ref 46).

Validation tests of the different models by different collaborators have also been run for actual HIP cycles on production type parts, where beginning and final dimensions are reported so that the precision of the model can be evaluated (Ref 49–51). This work was done with 316L stainless steel powder from a single source so that the variable of powder type and size is kept constant.

Simulation results were reported by three investigators in a 1997 workshop with the following conclusions:

- Jeon and Kim (Ref 49) using the models of Abouaf and McMeeking concluded that it is not feasible to produce near-net shape parts under HIP when the part has an asymmetric shape.
- Eisen, Bouaziz, and Dellis (Ref 50) compared the prediction of the Crucible empirical model and the CEA/CEREM two-dimensional model using two different HIP cycles (one with a very rapid heat up and pressurization, the other with a slower ramp up for temperature and pressure) with the actual results. They found that the predictions of both models were qualitatively but not quantitatively accurate and that either model could be used for near-net-shape prediction. Errors (ratio of simulated final value minus measured final value to initial value minus measured final value) on the three major dimensions ranged from 4 to 32% for the Crucible model and from 7 to 42% for the CEA/CEREM model. They also showed that the isotropic predictions using Eq 5 gave an accuracy very close to actual results. On this basis, they hypothesized that in small parts and even in large parts that are heated slowly enough to minimize temperature gradients, isotropic shrinkage can be used as a first approximation.
- Dellis, Bouaziz, Baccino, and Moret (Ref 51) used a three-dimensional simulation on this same part and compared it to the actual shrinkage. Their error on the three main dimensions ranged between 1 and 10%, which is a significant improvement over the two-dimensional and empirical model. This three-dimensional model required a fine-mesh grid and took 72 h of central processing unit time compared with 4 h for the two-dimensional simulation.

It is also possible, however, that current models may not be accurate because the mathematics do not describe what is actually happening during the consolidation process. Work of DeLo and Piehler (Ref 52) using thin-walled containers holding Ti-6Al-4V powder that was hot isostatically pressed consolidated concluded that the vast majority of densification and shape change occurs early in the HIP cycle before steady state-temperature and pressure are achieved. They attributed particle motion as the reason for an underestimation of densification. It has also been suggested that because temperature and pressure are increased together in most HIP cycles, power-law creep may be the predominant mechanism during consolidation and that the plastic yielding contribution could be ignored. A model using power-law creep as the sole densification mechanism might work as well as any viscoplastic model.

REFERENCES

1. H.A. Saller et al., U.S. Patent 4,709,848, 1987
2. A.H. Clauer, K.E. Meiners, and C.B. Boyer, "Hot Isostatic Pressing," Report MCIC-82-46, Metals and Ceramics Information Center, Battelle Memorial Institute, Columbus, OH, 1982
3. R. Schumann, Jr., *Metallurgical Engineering Principles,* Vol 1, Addison-Wesley, 1952, p 277–282
4. R.L. Coble, Diffusion Model for Hot Pressing with Surface Energy and Pressure Effects as Driving Forces, *J. Appl. Phys.,* Vol 41 (No. 12), 1970, p 4798
5. E. Arzt, M.F. Ashby, and K.E. Easterling, Practical Applications of Hot Isostatic Pressing Diagrams: Four Case Studies, *Metall. Trans. A,* Vol 13, Feb 1983, p 211–221
6. "Hot Isostatic Pressing (HIP): Market Forecasts, Technology Assessments, and New Business and Product Opportunities to the Year 2000," Gorham Advanced Materials, Gorham, ME, 1997
7. W.B. Eisen, Advances in Near Net Shape, *High Performance Materials in Engine Technology,* Vol 9, *Advances in Science and Technology,* Techna, 1995, p 485–496
8. S. Fallman, L.L. Jaktlund, R. Tegman, and T. Garvare, Numerical Modelling of Shrinkage during HIP, *Proc. Int. Conf. Hot Isostatic Pressing,* Centek Publishers, 1987, p 89–96
9. J. Hebeisen, IMT, Andover, MA, 1995, private communication
10. D.S. Wilkinson and M.F. Ashby, Pressure Sintering by Power Law Creep, *Acta Metall.,* Vol 23, 1975, p 1277–1285
11. A.S. Helle, K.E. Easterling, and M.F. Ashby, Hot-Isostatic Pressing Diagrams: New Developments, *Acta Metall.,* Vol 33, 1985, p 2163–2174
12. W.-B. Li, M.F. Ashby, and K.E. Easterling, On Densification and Shape Change during Hot Isostatic Pressing, *Acta Metall.,* Vol 35, 1987, p 2831–2840
13. M.F. Ashby, The Modelling of Hot Isostatic Pressing, *Proc. Int. Conf. Hot Isostatic Pressing,* Centek Publishers, 1987, p 29–40
14. S.V. Nair and J.K. Tien, Densification Mechanism Maps for HIP of Unequal Sized Particles, *Metall. Trans. A,* Vol 18, Jan 1987, p 97–108
15. J.K. Tien and J.C. Borofka, Bimodal Modeling of HIP of Superalloys, *Proc. Int. Conf. Hot Isostatic Pressing,* Centek Publishers, 1987, p 41–50
16. R.D. Kissinger, S.V. Nair, and J.K. Tien, Influence of Powder Particle Size, Distribution, and Pressure on the Kinetics of HIP Consolidation of P/M Superalloy Rene 95, *Proc. Fifth Int. Symposium of Superalloys,* The Metallurgical Society of AIME, 1984, p 285–294
17. S.V. Nair, B.C. Hendrix, and J.K. Tien, Obtaining the Radial Distribution of Random Dense Packed Hard Spheres, *Acta Metall.,* Vol 34, 1986, p 1599–1605
18. M.F. Ashby, *HIP User Manual 6.0,* Cambridge University Engineering Dept., 1990
19. C. Dorsch, Crucible Research Center, Pittsburgh, PA, 1995, private communication
20. A. Nissen, L.L. Jaktlund, R. Tegman, and T. Garvare, Rapid Computerized Modelling of the Final Shape of HIPed Axisymmetric Containers, *Proc. Second Int. Conf. Hot Isostatic Pressing,* ASM International, 1989, p 55–62
21. B. Aren and E. Navara, Modelling the Shape Change of Parts Produced by Hot Isostatic Pressing of Powders, *Proc. Int. Conf. Hot Isostatic Pressing,* Centek Publishers, 1987, p 107–115
22. H.A. Kuhn and C.L. Downey, Deformation Characteristics and Plasticity Theory of Sintered Powder Materials, *Int. J. Powder Metall.,* Vol 7, 1971, p 15–25
23. R.J. Green, A Plasticity Theory for Porous Solids, *Int. J. Mech. Sci.,* Vol 14, 1972, p 215–224
24. S. Shima and M. Oyane, Plasticity Theory for Porous Metals, *Int. J. Mech. Sci.,* Vol 18, 1976, p 285–291
25. T. Soh, A. Nohara, and T. Nakagawa, HIP Process Simulation by the Finite Element Method, *Proc. Int. Conf. Hot Isostatic Pressing,* Centek Publishers, 1987, p 81–88
26. A. Nohara, T. Nakagawa, T. Soh, and T. Shinke, Numerical Simulation of the Densification Behavior of Metal Powder during Hot Isostatic Pressing, *Int. J. Numer. Methods Eng.,* Vol 25, 1988, p 213–225
27. M. Abouaf, J.L. Chenot, P. Bauduin, and G. Raisson, Prediction of the Deformation during the Production of Near-Net Shape Superalloy Parts by Hot Isostatic Pressing, *Second Int. Conf. Hot Isostatic Pressing,* MPR Publishing Services, 1982, p 9-1 to 9-24
28. C. Argento and D. Bouvard, Modeling the Effective Thermal Conductivity of Random Packing of Spheres thru Densification, *Int. J. Heat Mass Transfer,* Vol 39 (No. 7), 1996, p 1343–1350
29. R.E. Taylor and H. Groot, "Thermophysical Properties of Partially Dense P/M N625," Report to Crucible Materials Corporation, Feb 1997
30. S.M. Doraivelu, H.L. Gegel, J.S. Gunasekera, J.C. Malas, J.T. Morgan, and J.F. Thomas, Jr., A New Yield Function for Compressible P/M Materials, *Int. J. Mech. Sci.,* Vol 26, 1984, p 527–535
31. V.A. Goloveshkin, V.N. Samarov, D.G. Seliverstov, and A.M. Kasberovich, New Regularities of the Shape Changing of Hollow Parts During HIP, *Conf. Proc. Hot Isostatic Pressing Theory and Applications,* Elsevier Applied Science, 1991, p 281–294
32. E.P. Kratt, V.N. Samarov, and R.A. Haykin, Technology of HIPing Complex Shape Parts with Dual Chemical Composition and Properties from Metal Powders: Trends of Its Development in the USSR, *Conf. Proc. Hot Isostatic Pressing Theory and Applications,* Elsevier Applied Science, 1991, p 259–267
33. V. Samarov, Industrial Application of HIP for Near Net Shape Critical Parts and Components, *Hot Isostatic Pressing '93,* Elsevier Applied Science, 1993, p 171–184
34. D. Seliverstov, V. Samarov, V. Goloveshkin, S. Alexandrov, and P. Ekstrom, Capsule Design for HIP of Complex Shape Parts, *Hot Isostatic*

Pressing '93, Elsevier Applied Science, 1993, p 555–560
35. N.M. Wereley, T.F. Zarah, and F.H. Charron, Intelligent Control of Consolidation and Solidification Processes, *J. Mater. Eng. Perform.,* Vol 2, 1993, p 671–682
36. T.F. Zarah and F.H. Charron, Process Modeling and Design for Near-Net Shape Forming of P/M Materials, *Concurrent Engineering Approach to Materials Processing,* TMS, 1992, p 69–79
37. N.A. Fleck, L.T. Kuhn, and R.M. McMeeking, Yielding of Metal Powder Bonded by Isolated Contacts, *J. Mech. Phys. Solids,* Vol 40, 1991, p 1139–1162
38. A.L. Gurson, Continuum Theory of Ductile Rupture by Void Nucleation and Growth: Part I—Yield Criteria and Flow Rules for Porous Ductile Media, *Trans. ASME,* Vol 99, Ser. H, No. 1, 1977, p 2–15
39. A.H. Kahn, M.L. Mester, and H.N.G. Wadley, Eddy Current Techniques for Sample Dimension Measurements during Hot Isostatic Pressing, *Proc. Second Int. Conf. Hot Isostatic Pressing,* ASM International, 1989, p 341–348
40. J.R.L. Trasorras, and M.E. Canga, and W. Eisen, Modeling the Hot Isostatic Pressing of Titanium Parts to Near-Net Shape, *Advances in Powder Metallurgy & Particulate Materials—1994,* Vol 7, 1994, p 51–69
41. D. Bouvard and M. Lafer, Rheological Characterization of Metal Powder at High Temperature, *Advances in Powder Metallurgy,* Vol 1, Metal Powder Industries Federation, 1989, p 491–503
42. J.F. Zarzour, J.R.L. Trasorras, J. Xu, and J.J. Conway, Experimental Calibration of a Constitutive Model for Hot Isostatic Pressing (HIP) of Metallic Powders, *Advances in Powder Metallurgy & Particulate Materials—1995,* Vol 2, Metal Powder Industries Federation, p 5-89 to 5-112
43. J.R.L. Trasorras, R. Parameswaran, J. Xu, R. Anbajagane, and V.B.S. Rachakonda, Model Based Design of Hot Isostatic Pressing to Near Net Shapes, *Advances in Powder Metallurgy & Particulate Materials—1995,* Vol 2, Metal Powder Industries Federation, p 5-139 to 5-152
44. J.R.L Trasorras, "Development of a Hot Isostatic Pressing (HIP) Modelling System for Large Complex Parts," Technical Progress Report, U.S. Navy, 1994
45. C. Dellis, D. Bouvard, P. Stutz, M. Pierronnet, and G. Raisson, Modelling of the HIP Densification of Powders: Influence of Operating Parameters, and Rheology of the Porous Material, Fourth Int. Conf. Isostatic Pressing, MPR Publishing Services, 1990, p 39-1 to 39-12
46. C. Dellis, P. Le Gallo, D. Abondance, and F. Moret, 3D Modelling of Net Shape Forming by Hot Isostatic Pressing, *1995 European Conference on Advanced P/M Materials,* 1995, p 189–194
47. C. Dellis, F. Bernier, O. Bouaziz, R. Baccino, and F. Moret, Note Technique D.E.M. No. 01, CEA/CERAM/DEM Service of Material Engineering Grenoble, 1996
48. C. Dellis, D. Abondance, O. Bouaziz, Rheologie des Poudres Metalliques a Chaud, *Colloque sur les Traitements de Poudres et Leurs Consequences,* SF2M, Paris, 1996, to be published
49. Y.C. Jeon and K.T. Kim, Near Net Shape Forming of 316L Stainless Steel Powder under Hot Isostatic Pressing, *Proc. Int. Workshop of Modelling of Metal Powder Forming Processes,* 1997, p 315–324
50. W.B. Eisen, O. Bouaziz, and C. Dellis, 3D HIP Modeling: Results on Non-Axisymmetric Parts, *Proc. Int. Workshop of Modelling of Metal Powder Forming Processes,* 1997, p 325–334
51. C. Dellis, O. Bouaziz, R. Baccino, and F. Moret, HIPing Simulation Evaluation through Comparison of Measures on 2D and 3D Manufactured Parts with PRECAD Numerical Results, *Proc. Int. Workshop of Modelling of Metal Powder Forming Processes,* 1997, p 305–314
52. H.R. Piehler and D.P. Delo, Physical Modeling of the Early Stages of Metal Powder Compaction and Comparison to Existing Analytical Models for Hot Isostatic Pressing, *Proc. Int. Workshop of Modelling of Metal Powder Forming Processes,* 1997, p 203–210

Hot Isostatic Pressing of Metal Powders

J.J. Conway and F.J. Rizzo, Crucible Compaction Metals

HOT ISOSTATIC PRESSING (HIP) is a process involving the use of high-pressure gas isostatically applied to a part or workpiece at an elevated temperature performed in a specially constructed pressure vessel. When consolidating metal powders in pressure-tight, sealed compacts, the HIP process plastically deforms the powder, which closes up porosity and achieves 100% theoretical density in the part. Thousands of fully dense HIP P/M compacts (equivalent to >10,000 tons) are commercially produced each year. This includes net shapes, near-net shapes, and a variety of mill forms for subsequent thermomechanical processing. Monolithic and bimetallic or clad parts are part of the near-net shapes that are supplied in many different conditions such as: HIP, HIP plus heat treated, HIP plus thermomechanically processed, and so forth. The majority of individual parts are made as near-net shapes. Other parts are made as net shapes, or with some net surfaces and other surfaces that need finishing to match mating parts or to meet some other critical parameters. The elevated temperature in HIP ranges from approximately 480 °C (896 °F) for aluminum alloy powders to 1700 °C (3092 °F) for tungsten powders. Most of the commercial HIP activity is with steel and nickel alloys, which are commonly hot isostatically pressed between 1100 °C (2012 °F) and 1205 °C (2201 °F). High-density argon gas is the most common pressure medium used in the process, although other gases such as helium or nitrogen can also be used. Pressures ranging from 20 to 300 MPa (3 to 45 ksi) are possible with 100 MPa (15 ksi) being the most common.

The HIP process was invented at Battelle Memorial Institute in 1955 as a method to diffusion bond dissimilar materials where Zircaloy was clad to uranium oxide nuclear fuel elements (Ref 1, 2). As the technology evolved, applications other than diffusion bonding were discovered, namely consolidating encapsulated powder, porosity healing inside castings, and densification of presintered components. In the 1960s, HIP technology (sometimes referred to as gas pressure forging) was used to consolidate beryllium metal powder, refractory metal powder, ceramic powder, and cemented carbides. Not until the 1970s, however, did the HIP process expand to an industrial scale. For example, high-volume hot isostatic compaction of high-speed tool steel was achieved in the United States (Ref 3) and Sweden by 1972. By the late 1970s, the United States Air Force Materials Laboratories funded work to develop HIP technology to manufacture titanium and superalloy components for aircraft engines. This application currently represents the highest technology level of the process. Powder metallurgy tool steels represent the highest production tonnage.

Some of the current applications of HIP P/M parts include:

- *Tool steel billets* (for mill processing) for hot and cold working tools and dies
- *Nickel and titanium alloys* for high-temperature components on aircraft and marine gas turbine systems
- *Nickel* alloys in the oil and gas and petrochemical industries for corrosion-resistant components on wellheads and piping systems
- *Nickel alloys* in the nuclear power industry for corrosion-resistant components in reactors
- *Titanium and cobalt alloys* in the biomedical industry for prosthetic implants
- *Refractory metal alloys* for high-temperature service
- *Cemented carbides* for superior wear-resistant parts
- *Composite aluminum materials* for lightweight structural components

Although HIP technology is commonly used for sealing porosity in castings and P/M sintered compacts, this article focuses on the consolidation of metal powder and the diffusion bonding of dissimilar materials.

Process Selection

During the design phase of a particular component, the end user must select a material and process that will best suit the requirements of the application given the constraints of economics, availability, and time. There are several different processing routes available to designers, among them are ingot metallurgy (I/M) and powder metallurgy (P/M).

Ingot metallurgy processing involves casting molten metal into molds, solidifying at relatively slow cooling rates (0.1 °C/s), and subsequently converting to the final product form by thermomechanical processing. The as-cast ingot exhibits chemical compositional variations, porosity, and a nonuniform microstructure containing equiaxed and columnar grains. Postcasting processing (e.g., heat treatment, thermomechanical processing, etc.) helps to homogenize the materials and somewhat ameliorate these anomalies. However, in some cases, conventional ingot metallurgy restricts the material from attaining its fullest potential in terms of properties.

The P/M processing route enables materials to achieve higher alloyed compositions without encountering segregation. Once fully densified by HIP, the P/M material has a fine grain size with an equiaxed microstructure. This uniform microstructure with a consistent composition leads to isotropic mechanical and physical properties. Not only do the properties have less variation than those properties found in I/M products, but the mechanical properties are often enhanced due to the fine grain size and the fine dispersion of precipitates available for strengthening. Better properties extend product life cycles or allow lesser grade materials to be used in more stringent applications. In addition, machinability is enhanced due to the uniformity and fine microstructure of P/M materials.

One example where the P/M route was selected over ingot metallurgy is René 95, which initially started as a cast-and-wrought (C/W) nickel-base superalloy in 1972. This highly alloyed material used for rotating disks in jet engines is prone to microsegregation, which during forging (even under careful controls) ultimately causes severe cracking and an undesirable microstructure (Ref 4). Later, the process was modified to use HIP P/M (Ref 5) and more than 100,000 of these parts are in aerospace service today.

Another example of the beneficial use of P/M versus I/M is in highly alloyed tool steels (Ref 6, 7). Through the years, producers of these materials have increased the content of carbide-forming elements such as molybdenum, tungsten, and vanadium to achieve increased wear resistance. Increasing alloying additions has led to in-

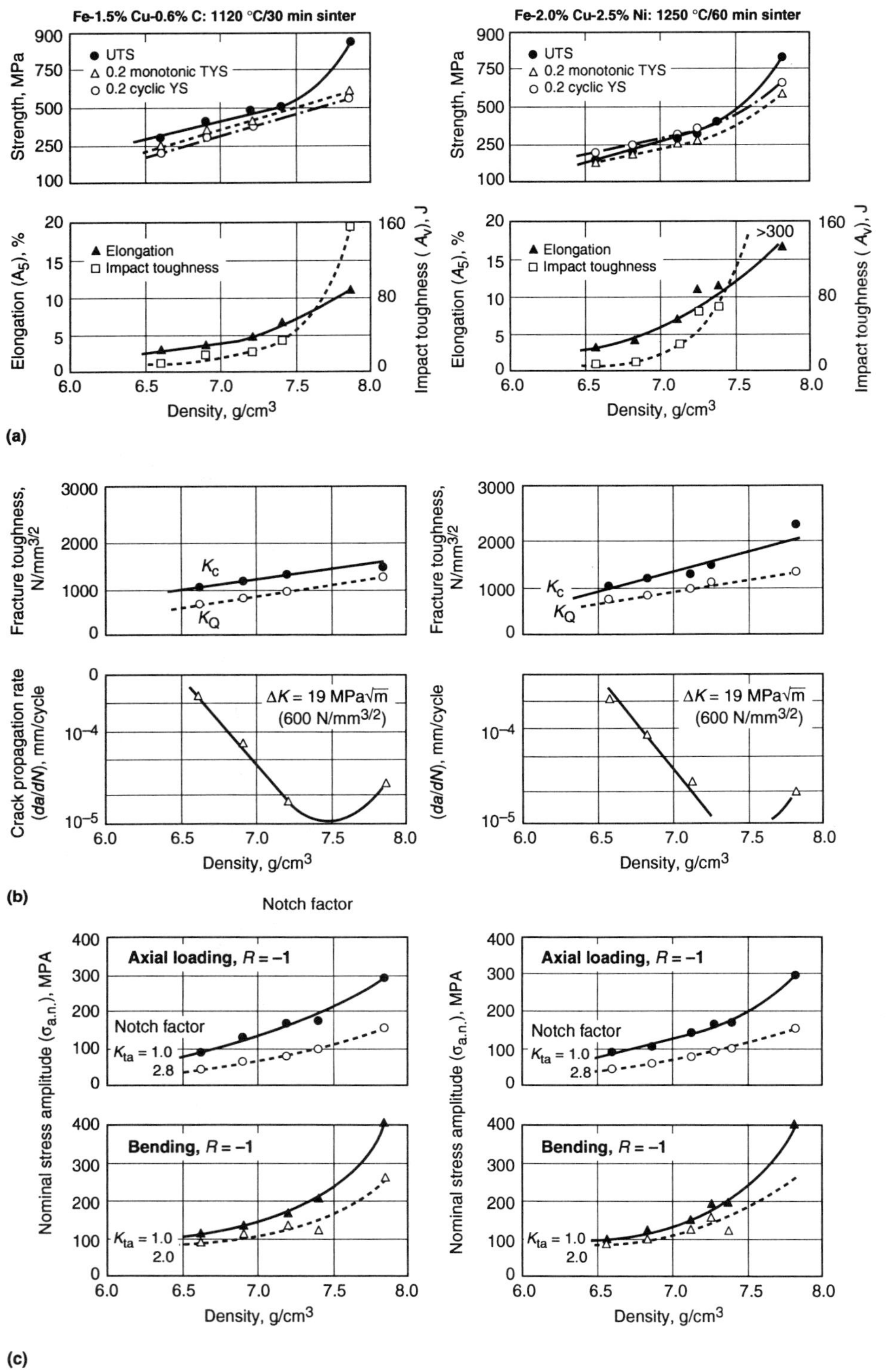

Fig. 1 Effect of density on mechanical properties of two P/M steels (Fe-1.5%Cu-0.6%C and Fe-2.0%Cu-2.5%Ni). (a) Strength, elongation, and impact toughness. (b) Fracture mechanic properties. (c) Nominal stress amplitude for 50% fatigue failure at 2×10^6 cycles. UTS, ultimate tensile strength; TYS, tensile yield strength; YS, yield strength. Source: Ref 8

creased segregation, which has resulted in lower product yields and lack of toughness. Due to the fine nature of the carbides in the P/M product and their high volume fraction, the wear resistance and toughness are greatly improved over I/M-produced tool steels.

HIP versus Other Powder Metallurgy Methods. Once powder metallurgy is established as the preferred processing route, other factors influence proper selection of a compaction process, namely:

- Properties required to meet component service requirements
- Size, shape, thickness, and complexity of the part

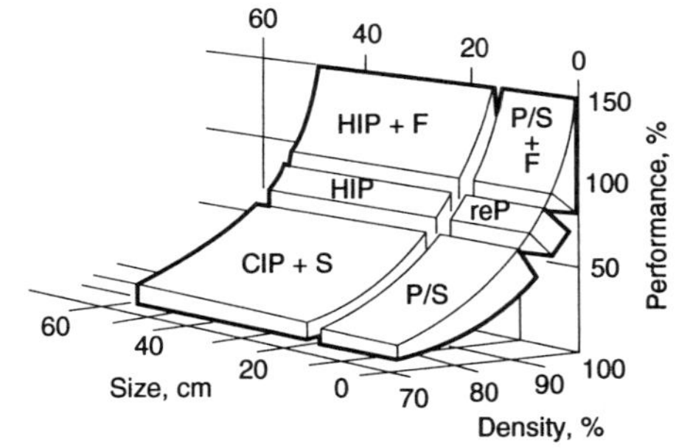

Fig. 2 Three variables that dictate the P/M approach are size, density, and performance (as a percentage of wrought). This behavior corresponds to ferrous-base P/M systems, but is representative of many P/M materials. P/S, press and sinter; reP, press, sinter and repress; P/S + F, press and sinter and forge; CIP + S, cold isostatically press and sinter; HIP, hot isostatic press; HIP + F, hot isostatic press plus forge. Source: Ref 9

- Tolerances and surface finish requirements
- Cost, availability, and lead time

Because mechanical performance improves with increasing density (Fig. 1) (Ref 8), applications requiring maximum properties generally employ HIP for larger-sized parts instead of pressing and sintering (Fig. 2). By utilizing the combination of high temperature and pressure, HIP can achieve a particular density at lower pressure when compared to cold isostatic pressing (CIP) or at lower temperature versus sintering. In addition to the processes described in Fig. 2, there are processes other than HIP that can be employed to achieve near-full or full density, each with their inherent advantages and disadvantages. Cold isostatic pressing combined with some subsequent thermal process (e.g., sintering, HIP, extrusion, etc.) can provide full density for most materials. The CIP approach can be used when it is metallurgically advantageous to begin densification prior to exposure to elevated temperature. Pressure-assisted sintering (PAS) involves the simultaneous application of pressure and heat to a powder mass typically created by hot pressing and pressurized sintering. Sinter plus HIP (containerless HIP) involves sintering a partially dense compact produced by die compaction or CIP and applying heat to sinter in an autoclave with subsequent pressure for the HIP phase. Sinter plus HIP has been used for tungsten carbide, silicon carbide, and silicon nitride for high-performance applications. Consolidation by atmospheric pressure (CAP) involves applying heat to a glass-enclosed, vacuum-tight compact to consolidate powder that has been chemically treated with a sintering activation agent. Rapid omnidirectional compaction (ROC), or fluid-die process, involves using a conventional forging press and a pot die to compact powder in a cavity of usually a thick-walled shaped "fluid" container. The Ceracon process (CERAmic CONsolidation) involves taking a heated preform and consolidating the material by pressure against a granular ceramic medium using a conventional forging press. Consolidation of metal powders via extrusion can be per-

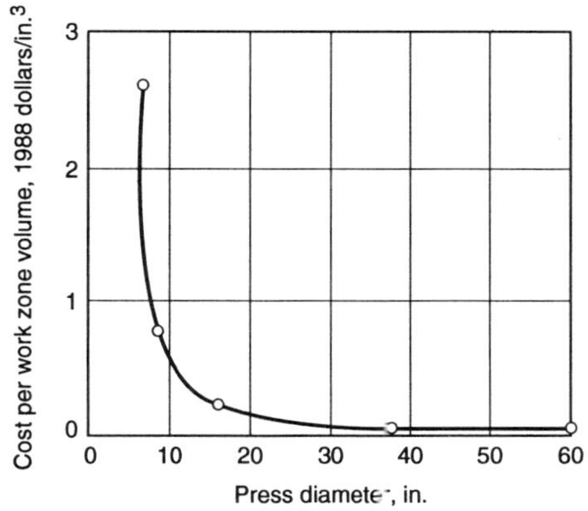

Fig. 3 HIP processing cost as a function of press size. The cost per unit work-zone volume to run a HIP cycle depends on the size of the HIP unit. Increasing the diameter of the work zone reduces the cost, but above 0.8 to 1 m diameter, further gains are small. Still larger HIP units are built to handle specific products. Source: Ref 10

formed by using either a blocker die or extruding loose powder to full density (see the article "Extrusion Consolidation" in this Volume).

Although these processes show promise, each has limitations. Some do not achieve full density (e.g. CIP and PAS). Others require subsequent processing (e.g., CIP and CAP), while others have not been scaled to production levels (e.g., Ceracon, ROC, and CAP). The HIP process is currently the most commercially viable method to produce fully dense P/M material. In excess of 10,000 tons of HIP P/M material is produced annually, far exceeding all of the other techniques.

When end users select the HIP P/M processing route to manufacture their parts, it is typically due to the following characteristics, which are desired for the final application:

- Full density of highly alloyed materials that usually cannot be produced by I/M processing
- Near-net-shapemaking capability
- Fine grain size
- Isotropic physical and mechanical properties
- Improved mechanical properties
- Extended life cycle
- Cost-effective components
- A range of possible part sizes from 1 to >10,000 lb
- Combinations of the above

Although cost of HIP P/M parts are often higher than the competing I/M or press-and-sinter P/M produced parts, HIP P/M can be an affordable method to increase part performance, reduce equipment weight (e.g., stronger material therefore thinner cross sections, etc.), maximize material usage by producing near-net shapes, or realize cost savings by increasing life cycles, eliminating downtime, and reducing maintenance and replacement costs. In general, the main cost elements for the HIP P/M process are the prealloyed, gas-atomized powder, the container (and container tooling if needed), HIP vessel efficiency (e.g., maximizing the number of parts during a single HIP run of the autoclave, Fig. 3), and postconsolidation processing. The combined considerations for customer requirements and the process cost elements determine if HIP P/M processing can be used for a given application. To properly assess each application, the end user should make a comparison with other production processes such as casting and forging. This should involve a total engineering evaluation that includes acquisition cost, finish part manufacturing cost, and component lifecycle cost.

HIP Technology

Whether it is consolidating powder in an encapsulated compact, bonding two surfaces together, or healing a casting or sintered compact, HIP is a process to remove porosity. The following describes the mechanisms and stages of pore closure for metal powders and HIP diffusion bonding of two surfaces.

Consolidation of Encapsulated Powder. As will be described in more detail later, powder particles are randomly packed inside a container that encapsulates the powder during the HIP process. Powder used for HIP is typically spherical with a broad particle size distribution. This is better for achieving high packing density when compared to irregularly shaped particles and/or monosized particles. With monosized particles or powder with narrow particle size distributions, there are no small particles to fill the interstices, thus decreasing packing density. However, even with the ideal size distribution, there will always be some percentage of void space due to the spherical morphology (Fig. 4) of the powder and possibly the presence of hollow particles (Fig. 5) that may have formed during the powder-making process.

To describe the consolidation of metal powders and thus the removal of pores, it is generally believed that the HIP process is similar to sintering where three physical mechanisms (Ref 11) are present, namely:

- Powder consolidation
- Neck growth
- Final densification

Powder in compacts prior to HIP is packed at a relatively low density (e.g., 60 to 80% of theoretical density). During the powder consolidation phase, powder is considered a cohesionless granular material where particles can slide freely without particle deformation. Particle rearrangement occurs, and some macroscopic deformation is observed. Once a particular density is attained and particle rearrangement discontinues, necks grow at the contacts between particles. At this stage, particles are beginning to bond, but the porosity is still interconnected. Various mechanisms describe this phenomenon as follows:

- Plastic deformation of the particles by dislocation
- Power-law creep or dislocation creep

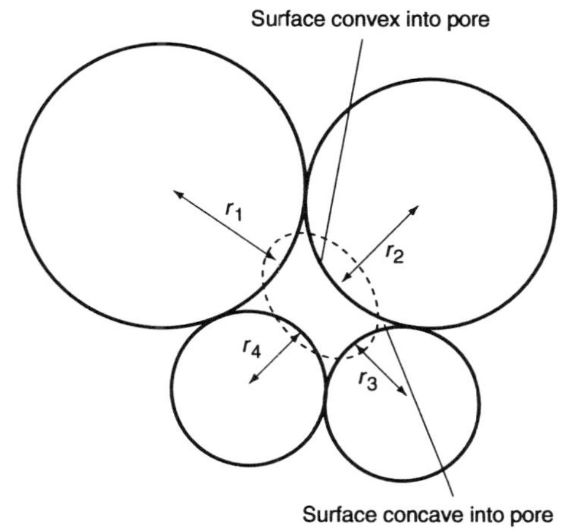

Fig. 4 Pore with varying radii of curvature around the surface. The broken curve shows the pore surface after some spheroidization has taken place by redistribution of material from convex surfaces to concave. Source: Ref 1

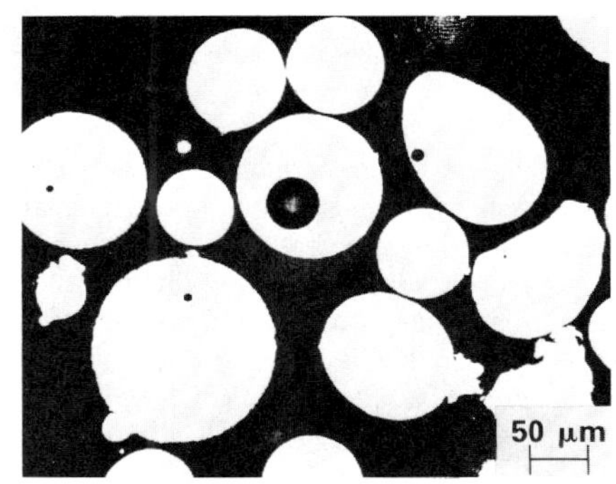

Fig. 5 Pores in –80 mesh Astroloy powder particles produced by argon gas atomization. Courtesy of Industrial Materials Technology, Inc.

- Nabarro-Herring creep or volume diffusional creep
- Coble creep or grain boundary diffusion creep

During final densification, the material may be considered a solid containing isolated pores connected by grain boundaries. The mechanisms controlling the final densification stage is the same as those controlling neck growth (Ref 11). A more detailed description of the mathematics is given in the article "Principles and Process Modeling of Higher-Density Consolidation" in this Volume and elsewhere (Ref 11–14).

Interface/Diffusion Bonding. Hot isostatic pressing technology was initially developed as a method to diffusion bond two dissimilar materials. Unlike fusion methods (e.g., welding, brazing, etc.), diffusion bonding depends on atomic transport across two mating surfaces to remove bond line pores. Fusion processes involve melting, which can cause segregation or solidification cracking. Unlike consolidation of a powder compact, components undergoing HIP diffusion bonding deform very little macroscopically. However, microscopically there is localized plastic flow across the bondline with the size of

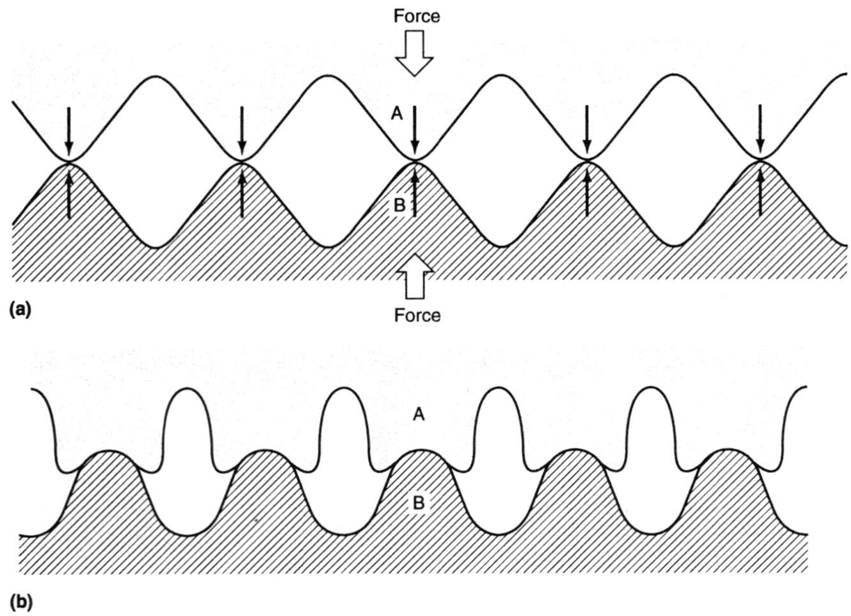

Fig. 6 (a) Magnified view of a region where two materials come into contact showing surface roughness. (b) Magnified view of the same region after microplastic deformation. (Material B is harder than material A.) Source: Ref 1

the initial pore being a function of the surface roughness (Fig. 6). Depending on the mismatch in physical and mechanical properties of the two materials, pores at the interface will spheroidize due to surface energy considerations. Afterward, previously described mechanisms take over until final densification is reached (Ref 1).

Process Description

The interaction of the HIP process and equipment must be understood to help explain the technology. There are two basic types of autoclave used for HIP. The first is known as a hot-loading cold-wall vessel where the parts to be densified are heated to the HIP temperature external to the high-pressure chamber. The second type is known as a cold-loading cold-wall vessel where the parts and furnace enter the high-pressure chamber at room temperature and are sealed in place. The heating and pressurizing of the parts and the vessel are then done according to a preprogrammed set of parameters.

A typical HIP cycle for a hot-loading cold-wall HIP unit is as follows:

1. The powder-filled compacts would be placed on fixturing designed to hold the load at elevated temperatures. Some compacts can be self-fixtured by attaching a lifting device to the container.
2. The compacts and fixtures are placed into a furnace and heated to the desired temperature at a controlled rate or progress through a series of preheat and high-temperature furnaces to a hold point. Because the thermal conductivity of loose powder is much less than that of fully dense material, long hold times are necessary to ensure the entire cross section of the largest compact in the load is thoroughly heated and ready for HIP.
3. Once heated, the compacts and fixture are lifted from the furnace and placed into the pressurizing vessel. The load may be protected with insulating material during the transfer to prevent cooling prior to placement in the HIP unit.
4. The lid is placed on the vessel and sealed.
5. A pressure valve is opened, and gas at high pressure is released from a reservoir into the densifying chamber until there is equalization in both tanks. The vessel is held at these conditions for a set period of time, which is usually short, and thus lends itself to high production rates.
6. The gas is pumped from the densification chamber back into the storage reservoir, and the connecting valve is closed.
7. The lid is removed from the HIP chamber, and the load is removed.
8. The load is cooled to room temperature either in air or in another chamber (e.g., a furnace) at a predetermined rate to protect the parts from undesirable reactions or microstructural phase transformations.

A typical HIP cycle for a cold-loading cold-wall autoclave is as follows:

1. The powder-filled compacts are placed on fixturing designed to withstand the load weight and elevated temperature to which it will be subjected. This is usually done at a station that can also accommodate the furnace. To maximize efficiency, the load is built in a duplicate base heater while the vessel is actively running a cycle.
2. The furnace is placed over/around the compacts and the base heater. The entire load is lifted and placed into the HIP vessel as a module. The base heater engages outlets in the bottom of the vessel, and the process is ready to start. The vessel is closed by putting the lid in place and sealing it properly via threads or a yoke.
3. The vessel is then evacuated and purged with inert pressurizing gas sufficient to remove all the traceable remnants of air and/or moisture from the vessel.
4. High-purity gas, usually argon, is discharged into the vessel from the gas storage tanks until the pressure is equalized.
5. The compressor is turned on, and vessel pressure is increased via pumping until the hold pressure is reached. For most cycles, heating takes place during high-pressure purging, contributing to the pressurization. In fact, gas expansion is responsible for the largest contribution to the high pressure.
6. The load is heated to the desired holding temperature previously determined and based on material properties and so forth at some predetermined rate. The duration of the hold is based on the nature and size of the load.
7. After the cycle completes the established hold conditions, the load can be cooled in three ways. First, the furnace can be turned off, and the entire system will cool naturally to a temperature low enough to protect the furnace elements and to enable the load to be handled safely. This rate will vary with vessel size and with the amount and configuration of the load. Second, the furnace can be programmed to cool slowly to prevent a material problem such as unwanted excess residual thermal stresses. Third, the furnace can be rapidly cooled by circulating new cooler gas to replace the high-temperature gas in the vessel (Ref 15). This process can be used to improve cycle turnaround time or affect a quench of the material in the HIP unit. Regardless of cooling cycle type, the gas can either be vented to the air or recycled for future use.
8. The HIP charge is removed from the vessel when the temperature is appropriately low. Another charge built during the cycling of the furnace charge replaces it in the vessel, and this process repeats itself.

The cycle times for a given vessel depend on the load and the size of the autoclave. However, a typical floor-to-floor time for medium- to large-size vessels is a matter of minutes in hot-loading units and 24 h or less in cold-loading units. The cycle time in a hot-loading unit is generally a matter of minutes. These run usually 30 min or less. However, to achieve maximum benefit from this type of system, the preheating furnace capacity must be extensive. The charge will generally require 12 to 24 h of heating before it is ready to be introduced into the HIP unit. Nearly all of the autoclaves used extensively in the HIP P/M industry are the cold-loading cold-wall style because they can achieve a more reproducible cycle, which is necessary for near-net shape processing. The hot-loading cold-wall vessels are used where high volume is needed and the product is a compact or preform for reprocessing.

Selection of Specific Cycle Parameters. Generally, HIP of P/M material produces fully dense material in the shortest process time, using pressures within the rating of equipment large enough to accommodate the parts being processed. Control of the fully densified microstructure may also be required for HIP P/M parts, especially if they will be subsequently hot worked. For most materials, the temperature of HIP is the primary controlling factor and is usually at least 70% of the melting point of the material. With most materials, the HIP temperature is specified by the end user because of its effect on the subsequent microstructure and properties of the product.

There are potentially an infinite number of HIP cycles available, and the rate of heating and pressurizing the vessel is of primary importance to the shape of the parts being produced. Work by Evans and Malley (Ref 16) showed that by varying how heat and pressure are applied, the shape of a simple cylinder can vary dramatically. All cycles can be classified into three basic categories: (1) heat up first and then pressurize, (2) pressurize first and then heat, or (3) heat and pressurize simultaneously. How the load of parts reaches the hold parameters determines the shape of the parts. This means that reproducibility of the HIP cycle is most important to the reproducibility of any given part and is a key variable in obtaining a desired shape.

Consolidation of Encapsulated Compacts

When metal powder is consolidated via HIP, encapsulation is required to act as a barrier between the high-pressure gas and the powder to provide the differential force required to create the plastic deformation of the powder. Powder is typically placed in a container or envelope made of metal or glass to a specified shape larger than final desired size. There are three main components to producing a HIP-encapsulated P/M part, namely: powder manufacturing and processing, container manufacturing, and compact manufacturing.

Powder Processing

Powder Manufacture. Although crushed/irregular, mechanically alloyed, and hydride/dehydride powders can be used for HIP processing, prealloyed, spherical powder produced via gas atomization is most commonly used throughout the HIP industry. The primary advantage of spherical powders is their high pre-HIP packing density (e.g., 60 to 80% or greater of theoretical density) and relatively low variation in packing, which provides repeatable packing density in production. As the initial packing density increases, the probability of predicting final fully dense shape is higher, thus the more reproducible the HIP P/M process becomes.

For HIP P/M parts, the most commonly used powder-making process is inert gas atomization, which entails creation of powder by impinging high-pressure inert gas on a molten metal stream to disintegrate the metal into tiny liquid droplets. Due to surface tension considerations, the liquid droplets form spheres that rapidly solidify into powder particles at a rate of 10^4 to 10^8 °C/s (Ref 17). This rapid solidification provides a powder particle with superior chemical homogeneity not possible with conventional ingot metallurgy. In general, several alloy systems can be gas atomized successfully, namely those based on iron, nickel, cobalt, niobium, titanium, aluminum, copper, silver, and so forth with the molten metal containment (i.e., refractory or skull melt) being the limiting factor.

In addition to gas atomization, spherical powder can be produced by other methods that utilize centrifugal force as the energy mechanism to break up liquid metal. The rotating electrode process (REP) and the plasma rotating electrode process (PREP) are two such methods. During REP, the face of a rotating bar is melted by a direct-current electric arc maintained between the electrically negative tungsten tip and the positive alloy bar (Ref 18). The liquid metal is centrifugally ejected from the outer edge of the rotating bar and solidifies in flight as spherical particles, which are collected in a sealed chamber. While REP is normally used to produce iron-base powders, applications requiring superior cleanliness (e.g., titanium, nickel alloys, etc.) can be produced by PREP. During PREP, the plasma torch uses a dc arc to melt the rotating bar with a water-cooled tungsten cathode that is protected by helium gas to minimize erosion (Ref 18). Another mechanical atomizing technique is the rotating disk process where a molten metal stream is impinged on a rapidly spinning disk. As with the first two processes, the liquid metal is spun off the disk and centrifugally atomized into spherical droplets, which solidify in flight. Particle cooling can be enhanced by blasting the emerging particles with a stream of helium (Ref 19). Centrifugally produced powder typically has a narrower particle size distribution and lower packing density than gas atomized powder.

Another process that is commonly used to make titanium powder is the hydride/dehydride process. Titanium sponge is charged with hydrogen and thus embrittled. The particles are mechanically crushed and dehydrided to yield fine powder particles. This powder is not as pure as that made by the PREP or inert gas atomization, nor is it spherical. However, it can be used to make HIP components. Powder of other refractory metals such as molybdenum and niobium can also be made in this way. It is also possible to make alloys of these elements using this process.

Powders can also be made by crushing and grinding ingot material. This process is usually used for highly alloyed materials and is not ideal for HIP P/M because of the inherently low packing density associated with this manufacturing method. These powders are often used with the HIP cladding process because they can be highly alloyed to achieve a specific property such as wear and/or corrosion resistance. These powders have a tendency to be extremely nonhomogeneous.

Powder Particle Size Classification. Classification may involve separating powder by particle size, typically via vibratory screening. Air classification through specially designed cyclones has also been employed. Bulk screening processes involve feeding raw powder onto the vibratory screener. Two exit ports from the screener are connected to separate collection bins to segregate the oversize and undersize fractions. The prime powder (usually the undersize) is used to manufacture HIP P/M parts, and the other fraction (usually the oversize) is reprocessed through melting.

Powder blending is performed to homogenize the particle size distribution. Hot vacuum blending is used to remove adsorbed gases from the surface of powder particles. Uniform particle size distribution is essential to HIP P/M near-net shape parts to minimize distortion during the consolidation step of the process. If a near-net shape is not being produced, then blending of the powder is not an essential step in the process and is often omitted. For some applications, blending can be used to achieve the desired chemistry of the end product. This can be as drastic as blending of elemental powders to make the entire composition or blending prealloyed powders that are close in chemistry as a refinement to meet a requirement for a specification. Blending for chemistry reasons is not considered a high-quality process and would not be utilized for technically demanding applications. After blending, powder is ready for loading into containers for HIP.

Container Manufacture

Container Design. There are a number of methods used to make both simple and complex shapes by the HIP P/M technology. The HIP component designer usually works with the end user to establish dimensional parameters that will meet the requirements of the application and still be compatible with the capability of the HIP process. The designer of the HIP container considers all the factors and then makes his decision on container type and material to give the best chance of successfully meeting the needs of the user. Figure 7 (Ref 20) displays a typical algorithm that designers of cylindrical HIP cans will employ. There may also be some consideration for metallurgical factors as well. As an example, section size of a part may be critical for cooling from heat treatment. These factors must be included in the part/container design process.

The basic requirements of the container are that it must:

- Be strong enough to maintain shape and dimensional control prior to and during processing

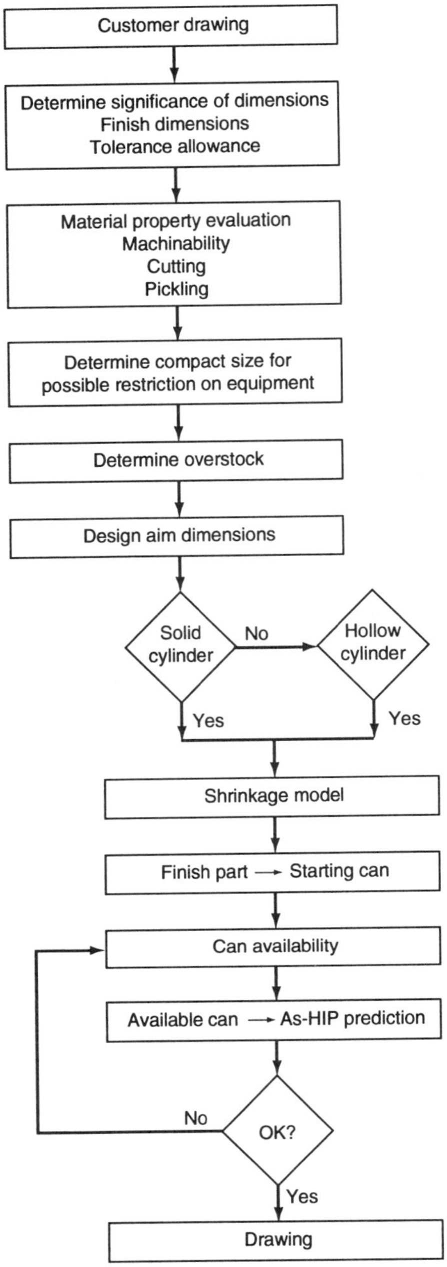

Fig. 7 Container design methodology for cylindrical shapes. Source: Ref 20

- Maintain leak tightness against the low and high pressures (typically up to 100 MPa, or 15 ksi) during evacuation, sealing, and densification
- Remain inert and plastic under applied temperature and pressure conditions
- Not contaminate the powder
- Be compatible with the material to be compacted and exhibit minimal integration with the powder by diffusion processes
- Be readily removable by machining, grinding, chemical dissolution, and so forth.

Currently, metal encapsulation is the most widely used containerization method for both experimental and production scale HIP P/M parts. In most cases, metal containers in the form of sheet, tubing, pipe, and so forth are used. Container components machined from heavy mill products (e.g., plate or bar) is not advised because pinhole piping originating from steel ingot solidification occurs frequently and persists through mill operations. This can result in a defective container that is likely to leak during the HIP cycle. Nickel and copper can be used, but thermodynamic reduction of sheet metal grain boundaries and weld metal oxides can occur in high-pressure argon during processing. Sheet materials commonly used for metal powder encapsulation include 1010, 1018, and 1020 low-carbon steel, extra-low-carbon-killed steel, type 304 austenitic stainless steel, commercially pure titanium, and Ti-6Al-4V alloy. Thicknesses less than 0.75 mm (0.030 in.) are impractical because of limited tensile ductility, whereas appreciable plasticity without through fracture may be required in certain container configurations that stretch or bend during the shrinkage that occurs during HIP. Special tooling (e.g., shear spinning tools) can be necessary to form the desired container shape. Large solid shapes, such as shown in Fig. 8, usually are produced with sheet metal containment. For shapes with internal cavities, sheet metal encapsulation with internal inserts that can be removed by machining or chemical etching is used (Fig. 9). Figure 10 illustrates the magnitude of the shrinkage a cylindrical compact experiences during HIP.

Fig. 8 Large as-HIP shapes. Source: Ref 21

The fill, evacuation, and seal tube attachment to the powder container is designed to prevent can buckling and tearing in the transition area. Thick-walled fill tubes are preferred, typically 19 mm outside diameter by 3 mm (¾ in. outside diameter by 11 gage) for compacts weighing 5 to 200 kg (11 to 440 lb) and 9.5 mm outside diameter by 2.3 mm (⅜ in. outside diameter by 13 gage) for smaller compacts. A thick-walled tube prevents buckling under the applied HIP pressure and facilitates metal flow during the forge welding crimping step.

A successful metal container weld design is the "matched-lip" style (Fig. 11), which accommodates severe deformation during consolidation and minimizes potential leaks. Applied process pressure creates a self-sealing action and eliminates shear loads on the weld. The small volume of accessible weld metal also permits repair by grinding and rewelding. When the matched-lip design is impractical, suitable fillet and butt welds can be employed so long as the welds are capable of passing radiographic inspection per ASME Section VIII (Ref 20).

Low-carbon steel parts made from plate or bar (free-machining grades are not recommended)

Fig. 9 HIP P/M shapes with internal cavities. Source: Ref 21

(a) (b)

Fig. 10 Cylindrical billet can. (a) After HIP. (b) Before HIP. Source: Ref 22

may be used for can interior shapes or cavity-forming components. As compact weight increases, container sheet gage should be increased to provide structural support during pre-HIP operations. Usually 14 gage (1.9 mm, or 0.07 in.) sheet is adequate for parts weighing approximately 200 kg (440 lb) and 12 gage (2.7 mm, or 0.1 in.) sheet is suitable for parts weighing approximately 1000 kg (2200 lb).

The presence of "continuous path" oxides in the metal container walls or welds can lead to subsequent leakage during densification and result in a nonusable product. Powder/container metal combinations must be evaluated or tested for detrimental interactions at processing temperatures. Binary phase diagrams can be consulted for potential low-temperature melting reactions. Commercially pure titanium encapsulation of refractory metals (molybdenum, niobium, tantalum, and tungsten) permits processing of these alloy systems between 1400 and 1650 °C (2250 and 3000 °F) without problems. However, any iron-containing capsules can only be used with titanium powders if the HIP temperature is kept below approximately 1000 °C (1830 °F). Solid-state reactions between can and powder include interdiffusion alloying and carbide formation.

In some boron- and carbon-containing materials (e.g., cemented carbide) thermodynamic activity of these elements may approach unity, leading to powder/container system melting points approximated by binary phase diagram eutectics. For example, the iron boron eutectic melts at 1149 °C (2100 °F). The latter case precludes steel canning of tungsten carbide/cobalt powder and HIP above approximately 1100 °C (2010 °F), because tungsten carbide/cobalt contains approximately 6 wt% C.

Glass that softens at the HIP temperature can be used as a container (e.g., Vycor glass for nickel alloys) (Ref 1). Glass containers can be fabricated by glass blowing or casting a slurry of ground glass to the required shape, drying, and then firing the shape. Typically, glass is used for high-temperature HIP cycles (e.g., refractory alloys, ceramics, etc.) where metal containers do not have the required properties. Currently, glass molds are not used extensively in production of HIP P/M components. Glass mold technology is described in detail elsewhere (Ref 23).

Another choice for container material is porous ceramic. The ceramic mold process (Ref 24) can produce solid net shapes with a complex geometry. The ceramic container is produced via the same principles as the investment-casting lost-wax process. A wax pattern is produced either by machining (usually experimental quantities) from wax billets or by wax injection molding in precision-machined tooling. A series of ceramic coatings are applied to the wax patterns and air hardened. The mold is then dewaxed and fired to leave the shape of the desired finished part expanded appropriately to accommodate the HIP shrinkage. Because the ceramic container is porous, it cannot be used directly as a container for HIP. The mold is filled with powder and placed into a secondary container and surrounded with a pressure transmitting medium (e.g., sand) and sealed except for an evacuation stem for outgassing. Once the entire assembly is outgassed, the secondary container stem is crimped and the compact is densified by HIP (Fig. 12). The ceramic mold process is particularly advantageous for production of complex precision shapes with compound curved surfaces and reentrant configurations.

Design Methodology. At first glance, the design of the HIP container would seem to be one of isotropic expansion of the finish shape given a particular packing density (Ref 25). However, in practice, it is rarely observed that a photographic reduction of the original shape is obtained (Fig. 10). Factors that can influence the dimensional distortion during HIP are:

- Stiffness of the container
- Thickness and variations in thickness and geometry
- Local variations in powder density
- Types and positions of welds

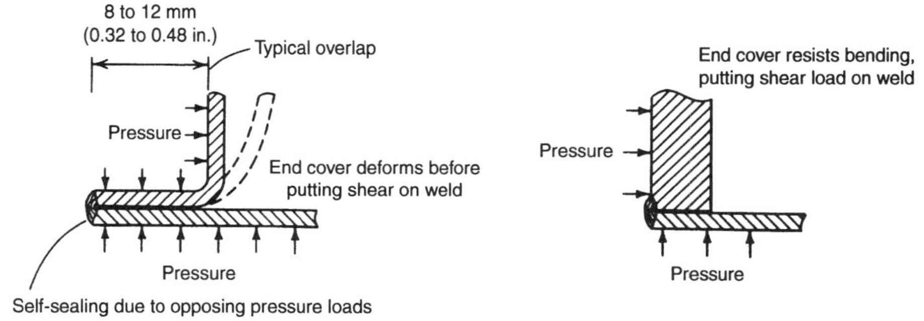

(a) (b)

Fig. 11 Typical can corner designs. (a) Recommended design. (b) Poor design. Source: Ref 22

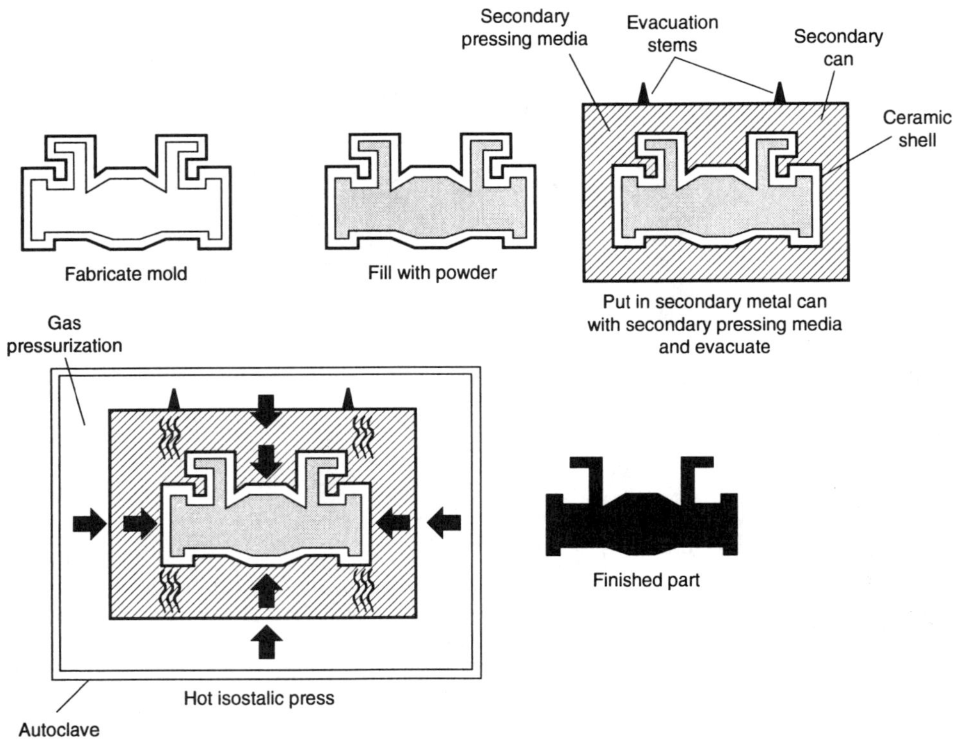

Fig. 12 Schematic diagram of ceramic mold process for producing fully dense P/M shapes. Source: Ref 24

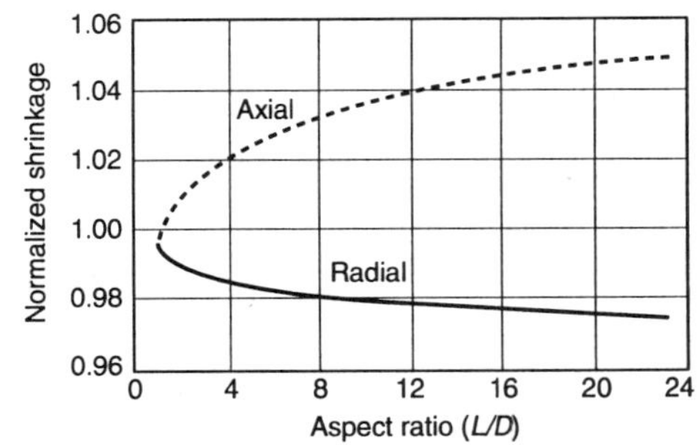

Fig. 13 Normalized shrinkage on solid cylinders (normalized shrinkage = actual shrinkage/isotropic shrinkage). Source: Ref 20

- Heating and pressurization rates during HIP consolidation
- Formation of densification fronts due to non-isothermal conditions of the powder (Ref 11, 26)

This dimensional anisotropy can be difficult to predict given the complex interrelationships between the properties of the powder and container materials as a function of temperature, density, and part geometry.

The designer of HIP cans starts with an end user drawing of the required part and a bill of material listing specifications to be met. The next step is to design a net or near-net shape to protect the dimensions of the part with considerations of machining allowance, powder/container interaction, bowing (particularly in thin sections), end effects, integral sample material allowance, and other secondary processing as described in Fig. 7 (e.g., nondestructive testing, machinability, ability to straighten, heat treatment, etc.). Once this is established, an aim configuration is determined with material allowances as required.

The container design may include several components fabricated from various techniques (e.g., shear spinning, stamping, hydroforming, etc.) and eventually put together to yield a capsule for loading. Because powder occupies only about 70% of the available volume inside the container, a considerable amount of shrinkage occurs. The science/art of designing a container must account for the packing density and the symmetry or lack thereof to achieve an acceptable part. During the initial production of a particular component, preproduction trials and/or iterations of the full-size shape may be necessary to determine the shrinkages empirically. However, this iterative approach is frequently costly and time consuming.

Through the years, HIP P/M part manufacturers have employed engineering intuition and previous experience to develop the starting can design. At this time, other approaches are being developed and used, namely, empirical and continuum mechanics/finite element modeling. Some of these are briefly described here.

Empirical Models. A large percentage of the HIP P/M compacts produced are either simple or hollow cylinders. An empirical model for these shapes was developed several years ago (Ref 26) by analyzing dimensional data from before and after HIP for a variety of cylindrical shapes. To eliminate the effects of the HIP cycle, alloy systems, and container thickness, the analysis focuses on cylinders made from nickel-base alloys consolidated in similar HIP cycles with a certain can thickness range. Best fit curves were generated for axial and radial shrinkage as a function of aspect ratio (length to diameter) and surface area ratio (area of cylindrical component to area of the lateral ends) as shown in Fig. 13. Based on these data, a computer program was generated to provide either the starting container dimensions to make a finished near-net shape or predict post-HIP dimensions given a specified starting container (Ref 26).

Engineering Models. One promising approach to perform a direct process simulation via the use of computer models is to combine a constitutive model of continuum mechanics equations solved by a finite element computational method. Several approaches ranging from simple plastic to compressible, viscoplastic constitutive models have been investigated as described in the article "Principles and Process Modeling of Higher-Density Consolidation" in this Volume. Even though some of these mathematical models are utilized in production, none has matured into a reliable modeling system for arbitrary geometries and HIP cycles.

Container Fabrication

Tooling and Container Component Fabrication. Once the design has been established, the metal container is fabricated. This is not a trivial step because the container must be producible in an economical fashion or the finished part cannot be manufactured. The most economical and easily formed container material is low-carbon steel; however, other materials (e.g., stainless steel, nickel alloys, titanium, etc.) can also be used. The process is constrained by existing metalforming techniques (e.g., metal spinning, hydroforming, stamping, hand forming, casting, machining, etc.) with each having its inherent advantages and limitations.

Tooling for the HIP P/M process refers to that which may be used to fabricate the container components. Usually, the quantity of parts to be made dictates the precision and cost committed to the tooling for HIP P/M containers. Large numbers of parts (1000 or more) would employ stamped containers. Anything less than this would be determined on a case-by-case basis. Cost of tooling must be amortized over the quantity of parts produced, so more expensive and more precise tooling can be cost effective only for sizable production runs. Because HIP P/M is most often used as a near-net shape process with small quantities of parts, container tooling is not made to be as precise as other processes where net shape is important.

Cleaning. Contamination of encapsulated powders will result, unless dirt, oxides, metalworking lubricants, and rust preventatives used to fabricate container components are removed. See *Surface Engineering,* Volume 5 of the *ASM Handbook* for cleaning procedures applicable to various metals. Proper cleaning, storage, and handling procedures immediately prior to any welding operation are necessary to prevent dirt entrapment or contamination on can surfaces.

Powder metallurgy alloys that are particularly sensitive to contamination (titanium, nickel-base alloys, and refractory metals) require controlled humidity and stringent cleanliness for final can preparation, assembly, welding, and filling. Electropolishing of stainless steel container components and nonchlorinated solvent cleaning (usually acetone, methylethylketone or methanol) of titanium container components represent typical cleaning processes for specialized applications. Carbon steel sheet metal containers should be supported carefully to prevent distortion during welding. Similar precautions are recommended for the outer sheet metal container used in a ceramic mold process.

Welding. A matched-weld-lip container configuration is designed to promote directional plane front solidification with good liquid metal feeding in the solidifying weld metal. Certain oxides (iron, nickel, and copper) can be reduced at high temperature in a high-pressure argon environment. This process may produce leaky containers during HIP if oxides extend through weld metal or container wall materials. Use of stainless steel filler metal for carbon steel container repairs is recommended because chromium oxides essentially are stable under processing conditions up to 1200 °C (2200 °F) in argon. Gas tungsten arc welding of nickel containers with stainless steel filler metal is also advised.

Containers for loose powder are assembled, welded, leak tested, and filled in sequence. Containers with interior spacers (mandrels), powder/solid composites (e.g., clad components), or precompacted and sintered P/M compacts can be filled with at least one cover removed. This procedure results in an extensive assembly weld area that cannot be leak tested in the vacuum mode because of the slow response time of helium through the interior of the filled container. Consequently, careful removal of loose powder from the weld area is necessary and use of precision and reproducible (preferably machine) weld techniques is required to prevent leakage. Leak tightness of HIP containers is a major process consideration.

Electron beam, gas tungsten arc welding, and stick welding are used for final container assembly. Argon dry box and electron beam welding are used for titanium alloy containers because nonoxidizing conditions are required. Gas tungsten arc welding with and without filler is used for carbon steel and stainless steel containers. Carbon steel containers may require a final reduction anneal after weld assembly, and some clad parts may need to be preheated prior to welding because of substrate material considerations or section size differences. Because weld metal is essentially a solidified casting, shrinkage and gas porosity are the fundamental causes of leakage at welds.

Leak Testing. Containerized HIP of metal powders can only be achieved successfully with leak-free containers. Location of leaks in a fully assembled container by use of valid leak-testing procedures and subsequent repair are fundamental requirements of HIP P/M technology. Leak detection is based on characteristics of helium and argon flow through small capillaries when compared at 1 atm (0.1 MPa) and 1000 atm (100 MPa) total pressure. Flow characteristics of a cylindrical capillary have been described by Guthrie and Wakerling (Ref 27):

$$Q = 1/L\,[C_1P^2 + C_2P + C_3 \ln(1 + C_4P)]_{P_2}^{P_1} \qquad \text{(Eq 1)}$$

where Q is the flow rate, cgs units; P_1 and P_2 are the exterior and interior pressure, cgs units; C_1, C_2, C_3, and C_4 are constants; and L is the capillary length, cm. For $P_2 = 0$ (evacuated container interior) and P_1 large:

$$Q = C_1P_1^2/L = \pi D^4P_1^2/256\ \eta\ L \qquad \text{(Eq 2)}$$

where η is the gas viscosity and D is capillary diameter, both in cgs units. This applies strictly in the viscous flow region, when Reynolds number (Re) is <1200:

$$\text{Re} = DV\rho/\eta < 1200 \qquad \text{(Eq 3)}$$

where V is the gas velocity, cgs units; and ρ is the gas density, cgs units.

Equation 2 indicates the relationship of container design, manufacturing, and leak testing. Leakage flow is proportional to the exterior pressure squared. Whereas leak testing is conducted with pressure differences across the container wall of one to several atmospheres, HIP typically uses 1000 atm (100 MPa). Thus, a leak occurring just below the detectability limit of a selected method permits leakage flow rates 10^6 times greater during HIP. Consequently, sensitivity of the leak-detection method is of utmost importance.

Capillary length (L) can be identified with container wall thickness, and with all other variables being constant, a capillary leaks ten times faster through a 0.25 mm (0.01 in.) wall than a 2.5 mm (0.1 in.) wall. The self-sealing, matched-weld-lip design is advantageous because capillary path length through the weld increases rapidly as weld flanges deform and solid-state bonding occurs. The fourth power dependence of leakage flow rate on capillary diameter indicates the necessity for procedures to eliminate weld porosity.

Leakage flow rates for helium at 1 and 10 atm and for argon at 1000 atm (100 MPa) for a set of capillary leak sizes and $D4/L$ parameters that could occur in practice are given in Table 1. This illustrates the major problem in leak testing of HIP containers: the leakage flow rate of argon through a capillary hole at 1000 atm (100 MPa) process pressure is approximately 10^5 times greater than the flow rate during a 1 atm (0.1 MPa) leak-testing procedure such as use of the helium mass spectrometer in the vacuum mode (i.e., evacuated container and/or atmospheric helium surrounding container exterior). This flow-rate difference defines a requirement for maximum sensitivity of the leak-testing method that is satisfied only by use of the helium mass spectrometer method in the vacuum mode.

Table 1 Leakage rate as determined by capillary (hole) diameter, gas type, and pressure

Capillary length $L = 0.1$ cm

Capillary diameter (D), μm	Leakage rate, cm^3/s, for: Helium at 0.1 MPa (1 atm)	Helium at 1 MPa (10 atm)	Argon at 100 MPa (1000 atm)
0.001	3.8×10^{-16}	3.8×10^{-14}	3.3×10^{-11}
0.01	3.8×10^{-12}	3.8×10^{-10}	3.3×10^{-7}
0.1	3.8×10^{-8}	3.8×10^{-6}	3.3×10^{-3}
1.0	3.8×10^{-4}	3.8×10^{-2}	3.3×10^{-1}

Note: Leakage rate is inversely proportional to the capillary length; that is, if the capillary length is twice as long, the leakage rate will be half as much. For example: if $D = 0.1$ μm, helium pressure = 1 MPa, and $L = 0.3$ cm, then leakage rate = $(3.8 \times 10^{-6}/3)\ cm^3/s = 1.3 \times 10^{-4} cm^3/s$. Source: Ref 22

Figure 14 illustrates a typical setup for a HIP container that is connected to a commercial helium mass spectrometer for leak testing in the vacuum mode. The effect of argon in powder compacts is estimated here to emphasize the importance of using only leak-free containers for powder encapsulation. Total leakage (in argon at standard temperature and pressure, assuming constant leak conditions and no appreciable pressure rise inside the container) for 1000 and 10,000 s flow times is given in Table 2. Distribution of this "leaked" argon within compacts of various sizes permits estimates of argon contamination in parts per million (ppm) by weight, as shown in Fig. 15.

Contained argon, although compressed during early powder densification stages in the HIP cycle, can limit end-point densification by "pressure balance" within small remaining pores. Regrowth of pores in subsequent heat treating operations, with related adverse effects on properties, can occur at levels as low as 0.1 mL/m^3 (0.1 ppm) for tool steels and 1 to 5 mL/m^3 (1 to 5 ppm) for superalloys. Leaks representing argon contamination at the 10 to 100 mL/m^3 (10 to 100 ppm) level generally prevent full densification, and larger leaks usually result in partial or no HIP densification.

Compact Manufacture

Loading. Filling of powder into the hermetically sealed, preshaped metal container can be performed in air, under inert gas, or under vacuum conditions, with the latter to aid in the removal of adsorbed gases. In some cases, powder is still loaded in open air as it was 25 years ago; however, most processing today is containerized to protect the product and prevent inhalation of the metal powder by operators (Fig. 16). Advanced filling systems have been developed to ensure clean, dry handling of powder for critical aerospace applications. Magnetic particle separation, screening, outgassing, and settling have been incorporated into these systems. In a production operation, there is a need for more so-

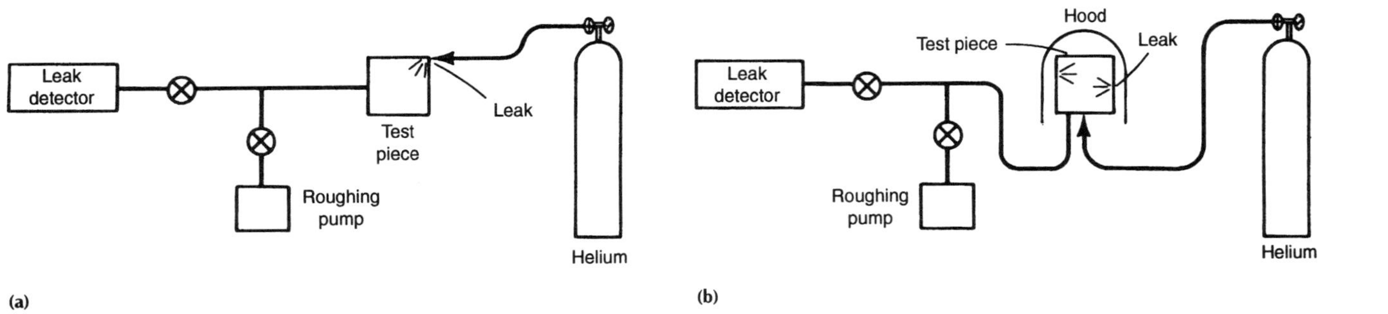

Fig. 14 Leak testing setup. Acceptance criterion: Q (flow rate) <10^{-9} standard cm^3/s. (a) Test piece evacuated and hooded with helium atmosphere to determine overall leakage rate. (b) Test piece evacuated; helium jet probe used to locate leak. Source: Ref 22

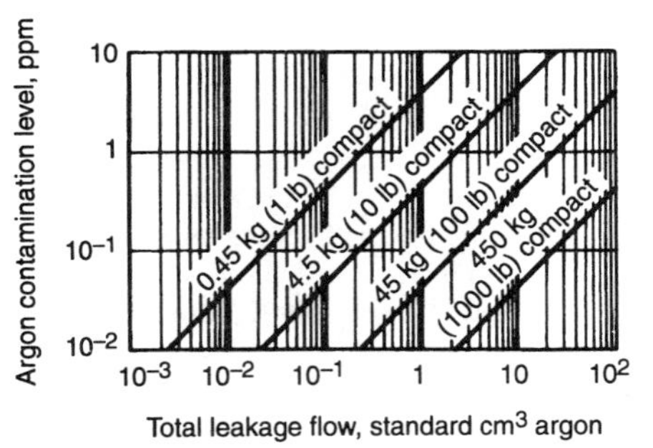

Fig. 15 Argon contamination level versus total leakage flow for various compact sizes. Source: Ref 22

Table 2 Total argon leakage flow at standard temperature and pressure

Capillary (hole) length L = 0.1 cm

Capillary diameter (D), μm	Leakage flow, cm^3 at a leakage time of: 1000 s	10,000 s
0.001	3.8×10^{-8}	3.8×10^{-7}
0.01	3.8×10^{-4}	3.8×10^{-3}
0.1	3.8×10^{-0}	3.8×10^{-1}

Note: Leakage flow is inversely proportional to capillary length; that is, if the capillary length is twice as long, the leakage flow will be half as much. For example, if D = 0.1 μm, argon pressure = 100 MPa (1000 atm), leakage time = 1000 s, and L = 0.3 cm, then leakage flow = $3.8 \times 10^0/3 cm^3 = 1.3 \times 10^0 cm^3$. Source: Ref 22

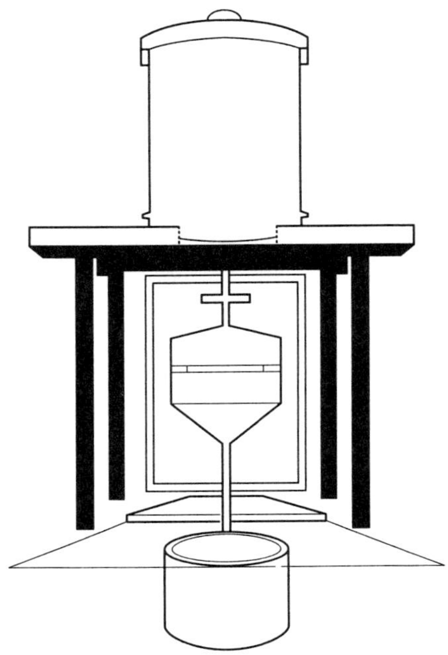

Fig. 16 Illustration of Modeen loading station

Fig. 17 Commercial degassing and capsule filling station

phisticated load stations that are automated to achieve maximum productivity. These are usually enclosed systems capable of operating with inert gas or vacuum conditions inside the container and the system. Some loading facilities are also capable of hot dynamic outgassing during the filling operation. If effective, this type of load station will prevent the need for subsequent outgassing once the compact is filled. Figure 17 illustrates a commercial degassing and capsule filling station.

The loading process must be performed in a manner to meet the desired packing density of the powder to obtain the proper post-HIP shape. Vibration to settle powder (i.e., packing) is usually employed to get the uniform distribution of powder throughout the inside volume of the container. This is important for good shape definition and for reproducibility among parts of the same configuration. It is also possible to load powder into the container first, and then by use of a large-amplitude low-cycle vibration pack the powder in the compact. This process is sometimes called "thumping." Organic materials such as rubber tubing are not recommended in the powder flow path as they are an obvious contamination source.

Particular attention should be paid to completion of loading. Figure 18 illustrates recommended and poor container filling practice. An incompletely filled container results in loss of shape control and may result in collapse and tearing of the container under external process pressure. Compacted powder in the fill tube provides integral contiguous material for testing and evaluation of the part.

Outgassing. The functional requirement of encapsulated powder vacuum/hot outgassing is to remove the atmosphere and water vapor (free and absorbed) from the packed powder bed to prevent formation of particle surface oxide and nitride films, which reduce workability and/or mechanical properties of the subsequent consolidated product. Behavior of powder in a heating/vacuum cycle (at temperatures up to approximately 400 °C, or 750 °F), for the purpose of defining process specifications, can be determined by thermogravimetry, combined with limited range mass spectrometry techniques. Vacuum outgassing does not remove gas entrapped in hollow powder particles originating from inert gas atmosphere atomization operations. Evacuation time for a packed powder bed can be estimated using viscous and molecular flow concepts. Elevated temperature is used to raise gas pressure within a bed and to promote desorption of water vapor. Packed metal powder beds are poor thermal conductors; therefore, an excessively high heating rate and temperature gradient in the compact during outgassing can result in redistribution of gas by chemisorption and reaction in the outer zone before all the gas is pumped out of the bed. This can occur because the center of the bed evolves gas at "low" temperature, which diffuses and reacts with the outer "high"-temperature portion of the bed before it leaves the compact.

The required practical end point for degassing a powder-filled hot isostatic pressing container can be estimated from the residual bed pressure (assuming air composition), which contributes oxygen and nitrogen levels ten times less than

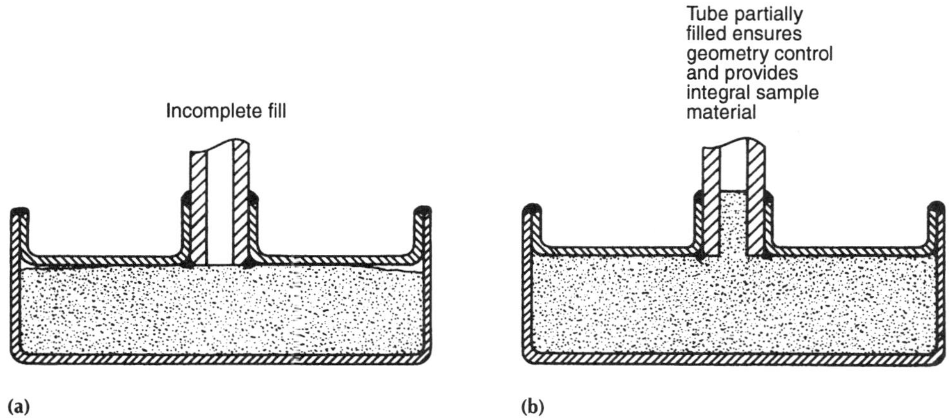

Fig. 18 Container filling practices. (a) Poor practice. (b) Recommended practice. Source: Ref 22

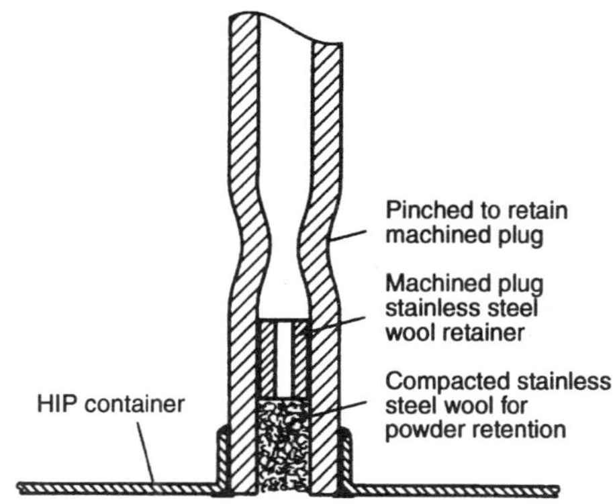

Fig. 19 Insertion of plugs to prevent loss of powder during evacuation. Source: Ref 22

the base level of the powder. Based on powder packing density, temperature, bed pressure, and ideal gas laws, parts per million by weight is given by:

$$\text{ppm} = 1.32\,Pf[(1 - \rho'/\rho)/\rho' RT]M \quad \text{(Eq 4)}$$

where P is bed pressure, μm Hg; ρ is full density of metal, g/cm^3; ρ′ is apparent (tap) density of powder, g/cm^3; f is fractional composition of gas, oxygen = 0.21 (air); M is molecular weight, g/mole; R is gas constant, 82.06 cm^3-atm/K; T is absolute temperature, K.

For $T = 600$ K, oxygen in air ($f = 0.2$) $\rho = 8.0$ g/cm^3, and $\rho' = 5.2$ g/cm^3 (65% packing density):

$$\text{ppm}_0 = 1.8 \times 10^{-4P} \quad \text{(Eq 5)}$$

and for $\text{ppm}_0 = 1$:

$$P \cong 5.5 \times 10^3\ \mu\text{m Hg} = 5.5\ \text{mm Hg} \quad \text{(Eq 6)}$$

The normal oxygen level of commercial superalloy powders ranges from 10 to 50 mL/m^3 (10 to 50 ppm) by weight. Thus, relatively high finishing evacuation pressures are acceptable in some cases. Other alloys, particularly refractory metals, may be more sensitive to residual gas. Carbon steel containers can be through oxidized in an air bake-out furnace with prolonged exposure. Container wall thickness, therefore, should be increased with increasing size and weight. For steel containers, grit-blast descaling is recommended after powder degassing. Stainless steel containers offer greater oxidation resistance during powder degassing and do not require descaling.

Loss of part dimensional control in large compacts (greater than about 25 kg, or 55 lb) also can occur during degassing, because the sheet metal container heats faster than the contained packed powder. New empty space is created inside the can at the bottom, into which powder flows from top areas. An oversize diameter at the bottom and uneven top geometry result from this type of powder movement during can heating without applied pressure. This particular problem also can occur in hot-loading HIP operations.

At the completion of the outgassing cycle, the fill/outgas stem is torch heated to approximately 982 °C (1800 °F) and sealed by use of a crimping tool. This leaves the powder-filled container sealed under vacuum and ready for consolidation by HIP. Loss of powder during evacuation and degassing (after can filling) can be prevented by inserting a stainless steel wool plug or a metal plug and partial crimping (Fig. 19).

Consolidation by HIP. The HIP of powder compacts is performed to consolidate the powder metal by plastic deformation to 100% of theoretical density. Details of the various HIP systems and cycles are discussed in previous sections of this article. In a cold-loading system (most commonly used for near-net-shape work), powder-filled compacts are fixtured and loaded into the autoclave (HIP) vessel. Pressure and temperature are increased at preprogrammed rates until the desired HIP hold parameters are reached. At the completion of the hold, the vessel is depressurized and the furnace is turned off. The fully dense parts are removed from the fixturing and sent to the next step in the process.

The upper size limitation for densification of encapsulated parts is governed primarily by the processing unit uniform temperature working zone diameter and length. Tool steel billets approximately 60 cm (24 in.) in diameter by 300 cm (120 in.) long and larger have been fully densified by HIP. Nickel-base superalloy P/M turbine disks greater than 1 m (3.3 ft) in diameter have been successfully densified. For sheet metal encapsulation of P/M parts weighing more than approximately 20 kg (44 lb), attachment of handling lugs is recommended. For large-diameter parts (greater than 0.5 m, or 1.6 ft, in diameter) and weights greater than 100 kg (220 lb), sheet metal bending stresses due to enclosed powder weight must be considered. Careful consideration must be given to the support of large parts in HIP tooling to prevent distortion during heating prior to complete densification. For small net-shape parts (1 to 1000 g, or 0.03 to 35 oz), particularly with thin sections, tooling that permits separate setting of each part is required.

Production of small (less than 10 kg, or 22 lb) net-shape parts by the HIP P/M process is not generally economical because it is labor intensive and the container fabrication costs are high. This applies particularly to the manufacturing of individual tools from P/M tool steels. Exceptions include experimental parts and manufacture of specialty parts in P/M refractory metals, composites, and precious metals, where metal cost is a controlling factor. Small net-shape parts (less than 0.5 kg or 1 lb) are best manufactured by containerless HIP, particularly for tool materials, provided satisfactory process procedures can be developed.

Postconsolidation Processing. After HIP consolidation, the compact fill stems are removed, and the components are dimensioned. The material in the fill stem is checked for density and microstructure. Failure to meet set criteria for any of these characteristics is cause for rejection of the compact. Many parts are further processed through heat treatment, container removal via chemical dissolution or machining, NDT, and mechanical testing prior to certification and shipment to the end user.

Because the HIP P/M process generally provides a near-net shape, the powder part manufacturer usually supplies a semifinished product. This could include material in any one or a combination of the following conditions:

- As-HIP
- HIP plus heat treated
- Rough machined
- NDT qualified
- A preform for thermomechanical processing
- HIP plus thermomechanically processed

The end users determine in what condition they want to receive their parts and specify all requirements such as dimensions, surface, thermomechanical history, properties, and so forth.

Applications

The ability of HIP to produce near-net shapes has been a primary impetus behind the develop-

Fig. 20 Large-sized cylindrical high-speed steel billet. Courtesy of Crucible Materials Research Center

Table 3 Heat treatments, grain size, and tensile properties of René 95 forms

Heat treatment/property	Extruded and forged(a)	Hot isostatic pressing(b)	Cast and wrought(c)
Heat treatment	1120 °C (2050 °F)/1 h AC + 760 °C (1400 °F)/8 h AC	1120 °C (2050 °F)/1 h AC + 760 °C (1400 °F)/8 AC	1220 °C (2230 °F)/1 h AC + 1120 °C (2050 °F)/ 1 h AC + 760 °C (1400 °F)/8 h AC
Grain size, μm (mils)	5 (0.2) (ASTM No. 11)	8 (0.3) (ASTM No. 8)	150 (6) (ASTM No. 3–6)
40 °C (100 °F) tensile properties			
0.2% yield strength, MPa (ksi)	1140 (165.4)	1120 (162.4)	940 (136.4)
Ultimate tensile strength, MPa (ksi)	1560 (226.3)	1560 (226.3)	1210 (175.5)
Elongation, %	8.6	16.6	8.6
Reduction in area, %	19.6	19.1	14.3
650 °C (1200 °F) tensile properties			
0.2% yield strength, MPa (ksi)	1140 (165.4)	1100 (159.5)	930 (134.7)
Ultimate tensile strength, MPa (ksi)	1500 (217.6)	1500 (217.6)	1250 (181.3)
Elongation, %	12.4	13.8	9.0
Reduction in area, %	16.2	13.4	13.0

(a) AC, air cooled. Processing: −150 mesh powder, extruded at 1070 °C (1900 °F) to a reduction of 7 to 1 in area, isothermally forged at 1100 °C (2012 °F) to 80% height reduction. (b) Processing: −150 mesh powder, HIP processed at 1120 °C (2050 °F) at 100 MPa (15 ksi) for 3 h. (c) Processing: cross-rolled plate, heat treated at 1218 °C (2225 °F) for 1 h. Source: Ref 28, 29

ment of HIP P/M parts. Conventional manufacturing methods for materials with high alloy content have low process yields and typically utilize only 10 to 30% of the material purchased in the final product; the remainder becomes scrap during machining. Hot isostatic pressing to near-net shape improves material utilization significantly during part manufacturing and finish machining. A hot isostatically pressed near-net shape part normally loses only 10 to 20% during final machining. The inability to provide nondestructive inspection of complex near-net-shape parts for certification has somewhat inhibited application of this technology, particularly for turbine engine applications.

High-Speed Tool Steels. The development of gas-atomized prealloyed steel powders in the 1960s (Ref 3) led to HIP P/M tool steels. This represented the first production application of HIP for a relatively low-cost material. Hot isostatic pressing improves the microstructure of tool steels by preserving the fine grain size and carbide distribution present in the atomized powder through the consolidation process. Increased homogeneity of the fine carbides throughout the material is an added benefit. Superior tool properties result from the improved microstructure. Shape stability during subsequent heat treatment is superior in HIP material. Grindability, wear resistance, and uniformity of hardness also are improved. Additionally, cutting performance of high-speed tool steels is improved by this processing treatment, due to the increased toughness related to fine austenite grain size. New high-alloy-content steels with enhanced material properties can be produced. High-speed tool steels are generally consolidated in billet form. A HIP high-speed steel compact is shown in Fig. 20.

Nickel-Base Superalloys. Starting with development in the early 1970s, nickel-base superalloys have evolved into one of the best applications for the P/M HIP technology. More than 5000 tons (4545 metric tons) of superalloy components are currently operating in commercial and military aircraft turbine engines. Hot isostatic pressing of forging preforms represents a significant portion of the current production, but there are approximately 100,000 as-HIP parts in service as well. The use of HIP P/M consolidation for superalloys is economically attractive because of its near-net-shape capabilities. High-alloy-content superalloys can be produced with attractive properties. Superalloys strengthened by a large volume fraction of second-phase γ' undergo severe segregation during ingot solidification. Such ingots would be virtually unworkable by conventional hot-working techniques for large-size parts. The division of the melt into small powder particles during atomization eliminates macrosegregation, and microsegregation is reduced because of high cooling rates during particle solidification. Hot isostatic pressing of these powders produces a homogeneous microstructure that improves mechanical properties and hot workability.

Superalloy powders are typically made by inert gas atomization or REP. Care must be taken in processing to avoid the presence of stable nonmetallic compounds on the surface of the powder particles because they can be detrimental to the properties of consolidated products. The article "Powder Metallurgy Superalloys" in this Volume discusses the properties of many nickel-base superalloys made via the HIP P/M process. A comparison of HIP properties with other forms is given in Table 3.

Heat treatment after HIP can have significant effects on material properties as shown in Table 4. Material response to post-HIP treatment depends on the processing conditions. Near-net-shape parts also may be subject to distortion during post-HIP heat treatment. If complex shapes are required, the ceramic mold process is suitable, particularly for static parts. If a carbon or stainless steel container is used for powder consolidation, a 0.5 mm (0.02 in.) diffusion zone may surround the part. This does not cause a problem in the final part because HIP envelopes usually exceed this dimension. Hot isostatic pressing conditions are alloy dependent. Processing temperatures may be keyed to the γ' solvus temperatures for purposes of grain size control in nickel-base superalloys.

Oxide-dispersion-strengthened superalloys also can be consolidated by HIP. Prior to processing, alloy powders, additives, and oxide dispersoids are put in a high-attrition ball mill and mechanically alloyed. This ensures fine grain size and uniform oxide distribution throughout the powder. Hot isostatic pressing produces fully dense material with these microstructural features maintained.

Titanium-Base Alloys. Powder production for titanium and titanium alloys requires special setups because of the reactivity of titanium. The hydride/dehydride process is the most common way to make titanium powders, but the particles resulting from this process are not spherical and thus do not work well for near-net-shape processing. The early method used to make spherical titanium powder was the REP. This was later supplanted by PREP to reduce contamination. Either of these processes depends on the ability to manufacture bar product of the alloy being made into powder. In the late 1980s, an inert-gas-atomizing technique was developed for titanium and its alloys (Ref 30). By the use of inert atmosphere or vacuum induction skull melting, the titanium alloy is brought to the molten state. The liquid is then poured through a metallic nozzle into a high-pressure gas stream. The metal breaks up and resolidifies as spherical titanium particles. The powder is collected in a cyclone system designed to cool the powder to prevent sintering.

There are any number of applications for titanium and titanium alloy powders. In the late 1970s and through the 1980s, the Air Force Materials Laboratory supported many programs to develop near-net shapes for military uses (Ref 31). For many reasons, this work never resulted in an ongoing production process, even though

Table 4 Mechanical properties of hot isostatically pressed plus conventionally forged Nimonic alloy AP1

Processing temperature		Size of sample disk			Tensile properties(a)								Stress rupture(b)			
					Yield point, 0.2% offset		Ultimate tensile strength		Elongation,	Reduction in area,	Notched tensile strength		Plain life,	Elongation,	Notch life,	Low-cycle fatigue(c),
°C	°F	mm	in.	Solution treatment	MPa	ksi	MPa	ksi	%	%	MPa	ksi	h	%	h	cycles
1150	2100	150	6	4 h at 1110 °C (2030 °F), air cool	971	141	1307	190	30.4	31.6	1869	271	42	30.1	195	>276,000
1150	2100	150	6	4 h at 1080 °C (1980 °F), oil quench	1120	162	1513	219	23.2	24.2	1992	289	64	15.3	159	>307,000
1150	2100	150	6	4 h at 1110 °C (2030 °F), quenched and aged(d)	1037	150	1381	200	30.4	46.7	1776	258	88	20.4	163	>214,000
1220	2230	150	6	4 h at 1110 °C (2030 °F), air cool	999	145	1328	193	28.6	32.7	1868	270	45	20.5	188	>155,000
1220	2230	150	6	4 h at 1080 °C (1980 °F), oil quench	1085	157	1463	212	23.2	23.4	1941	281	66	17.2	247	>228,000
1220	2230	150	6	4 h at 1110 °C (2030 °F), quenched and aged(d)	1052	153	1383	201	25.0	25.8	1844	267	74	16.9	315	>242,000
1150	2100	475	19	4 h at 1110 °C (2030 °F), air cool	952	138	1320	191	29.5	31.4	1521	221	85	22.9	>500	>35,000
1150	2100	475	19	4 h at 1080 °C (1980 °F), oil quench	993	144	1356	197	26.1	28.0	1785	259	113	20.3	>450	>100,000

(a) At 650 °C (1200 °F). (b) 760 MPa (110 ksi) at 705 °C (1300 °F). (c) 1080 MPa (157 ksi) at 600 °C (1110 °F). (d) 50% water-soluble polymeric compound, 50% water. All material aged 24 h; 650 °C (1200 °F); air cooled, 8 h; 760 °C (1400 °F); air cooled

there is still some experimental work being performed currently. All of the meaningful earlier work was conducted with PREP powder. When the gas-atomized powder became available, it was used for all subsequent activities. At that time, the emphasis changed to applications needing titanium aluminide powders. Because these can be easily made by the skull-melting/gas-atomization process, the bulk of the experimental work is currently being performed in this area. The powders are now being used to manufacture metal-matrix composites and intermetallic-matrix composites. The advantages of these products are their light weight, high strength, oxidation resistance, and creep resistance at high temperatures.

Cemented Carbides. Tungsten-carbide/cobalt tools are the premier example of containerless HIP to achieve full density by removing residual porosity. Superior transverse rupture strength results from HIP. The wear performance of cutting tools at high speeds is not significantly improved, however, because this behavior is governed by the hardness of the material rather than by its fracture properties. Low cobalt content (3%) alloys can be produced by HIP to give enough toughness for use in drawing dies.

Fully dense cemented carbide can be finished to give a perfectly smooth surface, which is required for high-quality rolls, dies, mandrels, and extrusion tools. Generally tungsten-carbide/cobalt tool materials are manufactured by CIP and sintering of blended powders, followed by HIP. Typical conditions for HIP are 1290 °C (2350 °F) at 100 MPa (15 ksi) for 1 h. Cemented carbide parts produced using HIP are shown in Fig. 21.

Refractory Metals. Consolidation of refractory metals by HIP is a two-step process. Processing these materials to net and near-net shape promotes conservation of these critical resources. Niobium alloy C-103 (Nb-10Hf-1Ti-5Zr) has been successfully hot isostatically pressed using a duplex cycle. Hydride/dehydride and PREP powders are consolidated in a plain carbon steel container filled with powder at 1260 °C (2300 °F) at 100 MPa (15 ksi) for 3 h. The container is then removed in a nitric acid solution and further chemically milled in a nitric-hydrofluoric acid solution to remove the alloy/container interaction layer. The material is finished in a HIP step at 1590 °C (2900 °F) at 100 MPa (15 ksi) for 3 h to a final density in excess of 99% of theoretical. Room-temperature and high-temperature (1650 °C, or 3000 °F) tensile strength and ductility properties compare favorably to wrought alloy properties. The ductile/brittle transition temperature is higher (–18 °C versus 160 °C, or 0 °F versus 320 °F, for standard products) in the HIP material due to increased oxygen content. Gas content of the hydride/dehydride material results in poorer weldability than the PREP powder. Hydrogen embrittlement also occurs in the hydride/dehydride alloy C-103. Vacuum baking at 870 °C (1600 °F) for 2 h eliminates embrittlement, and the alloy will fail in a ductile manner in tensile and Charpy tests.

Near-net shape forward bowls manufactured by consolidation of C-103 in the duplex HIP cycle are shown in Fig. 22. The diameter of the bowls was within 0.13 mm (0.005 in.) of final dimensions. The P/M net shape weighed 0.8 kg (1.8 lb). This, compared with rough forging weighing 1.7 kg (3.8 lb) and a final part weighing 0.6 kg (1.4 lb), illustrates the material savings achieved by HIP to near-net shape.

Included is a provision for parts to be low-temperature HIP to a closed porosity condition, decanned, and re-HIP usually at higher temperatures. This option can be employed when the powder/container integration (melting, alloying, contamination, etc.) is unacceptable at the preferred higher HIP temperature. This technique has been used, for example, for niobium alloys that are initially hot isostatically pressed at 1205 °C (2200 °F) in low-carbon-steel containers, decanned, and re-HIP at 1595 °C (2900 °F) to circumvent an iron-niobium eutectic reaction at 1360 °C (2480 °F).

Fig. 21 Tungsten-carbide/cobalt parts produced by HIP. Source: Ref 22

Stainless Steels. One of the most prominent applications of the HIP P/M technology is in the area of stainless steels. Both duplex and austenitic steels have been used extensively as P/M near-net shapes in the oil and gas and petrochemical industries. For example, valve bodies, fittings, and large complex manifolds for piping systems have been successfully produced in a cost-effective manner via HIP processing. Figure 23 (Ref 32) shows some of the typical fittings that have been made from 254 SMO material. Figure 24 (Ref 32) is a valve body that weighs more than 2 tons and was made from an austenitic stainless steel. Large manifolds with integral outlets hot isostatically pressed from a superduplex stainless steel have also been put in service in an offshore oil rig in the North Sea (Ref 32). In addition to the other benefits of a HIP P/M approach, the manifold can be fabricated in far less time and avoid costly welding processes. An analysis of the cost factors showed a greater than 20% savings over a similar manifold produced from fabricated cast and wrought components (Ref 32).

Fig. 22 Niobium forward bowls hot isostatically pressed to shape. Source: Ref 22

Fig. 23 Tees for underwater applications in the offshore industry hot isostatically pressed in 254 SMO grade. Weight: 155 kg/pc

Fig. 24 Hot isostatically pressed valve body in austenitic stainless steel. Weight: 2 t

Table 5 Diffusion bonding in comparison with other joining processes

	Fusion welding	Diffusion bonding	Brazing
Contacting method	Autogenous fusion, autogenous fusion and pressure, pressure and autogenous fusion	Pressure (no fusion)	Contact fusion, contact fusion and pressure, pressure and contact fusion
Bonding	Cohesive	Adhesive, diffusion	Cohesive, adhesive
Heating	Local	Local, total	Local, total
Temperature	Melting point of parent metal	0.5–0.7 of melting point of parent metal	Somewhat above melting point of braze
Surface preparation	Less exacting	Careful	Less exacting
Fit-up	Lenient	Precise	With capillary gap
Materials	Metals, alloys	Metals, alloys, nonmetals	Metals, alloys, nonmetals
Joint formation	Gradual	Simultaneous	Simultaneous, gradual
Edge preparation	Yes	No	Yes
Joining of dissimilar materials	Limited	Unlimited	Unlimited
Stepwise conduct of process	Limited	Unlimited	Unlimited
Susceptibility to solidification cracking	Strong	None	Weak
Porosity	Shrinkage, blowholes	None	Blowholes, shrinkage, diffusion
Overlapping with heat treatment	No	Unlimited	Limited
Warpage	Heavy	None	Light
Principle types of joint	Butt, lap	Flat (butt, lap, tapered plug in socket, between cylinders, spherical, curvilinear)	Butt, lap
Joining in hard-to-reach places	Limited	Unlimited	Limited
Product precision	Low	Fairly high	High
Disassembly of joint	No	No	Yes
Vibration survival	Low	Very high	High
Corrosion resistance	Satisfactory	Fairly high	Low
Strength	Close to that of parent metal	That of parent metal	That of braze
Air pollution and radiation emission	Yes	No	Yes

Source: Ref 1

Interface/Diffusion Bonding

Not only can HIP be used to consolidate loose powder, it can also be used to create a component of multiple bonded materials. Diffusion bonding by HIP can be performed on solid-to-solid, powder-to-solid, and in some cases, powder-to-powder surfaces. As with powder/metal container combinations, material compatibility must be evaluated to ensure no low-temperature melting reactions occur at the HIP temperature. If this does occur, interlayers can be used to alleviate this problem.

HIP Diffusion Bonding versus Other Joining Processes. As stated previously, HIP technology was initially developed as a method to diffusion bond two materials together. Table 5 shows various attributes of joining two materials when comparing diffusion bonding with fusion methods (i.e., welding and brazing). The major advantages of diffusion bonding are no melting of the parent metal and therefore no segregation or cracking problems, very little dimensional distortion, and stronger bonds due to the elimination of a low-melting-point filler.

When HIP diffusion bonding is compared with conventional diffusion bonding, there are several advantages, namely (Ref 1):

- Conventional diffusion bonding uses uniaxially applied pressure, which limits the geometry of the joint. For HIP, complex, shaped surfaces can be bonded together.
- Applied pressure must be low to prevent macroscopic plastic deformation with conventional diffusion bonding. For HIP, the plastic deformation is on the microscopic scale and therefore can be performed at a higher temperature.
- Powder and porous bodies can be simultaneously densified to a substrate with HIP diffusion bonding.

Encapsulation Methods. As with consolidating P/M compacts, components for HIP diffusion bonding must be encapsulated to ensure a differential pressure exists to create the driving force for bonding. One method is to simply weld the contact area between the two parts. Another is to seal only the contact area with a container component. Yet another is to encapsulate part or all of the substrate. Figure 25 shows the steps used to HIP diffusion bond a powder material with a solid substrate material (Ref 21).

HIP Parameters. The choice of HIP parameters is usually based on metallurgical and economic consideration. Diffusion bonding is typically enhanced by increasing temperature and pressure. The temperature will generally be 50 to 70% of the melting point of the lowest-temperature material in the system. The pressure shall be sufficient to close up all pores along the bond line as well as internal pores and pores created by interdiffusional pores (i.e., Kirkendall effect). The time at temperature should be kept to a minimum to decrease cost and potentially avoid any deleterious effects from formation of brittle intermetallics, excessive grain growth, and secondary recrystallization.

Use of Interlayers. An interlayer is sometimes used between surfaces to prevent the formation of deleterious brittle compounds and/or alleviate stresses due to thermal expansion mismatch. As described previously, interlayers must be compatible with each material that it contacts. The thickness must be sufficient enough to accommodate cooling stress and not so thick that the bond strength is decreased by the presence of a thick ductile interlayer. A 100 μm thick Ni-Cu-Ni interlayer was successfully used as a carbon diffusion barrier between BG42 tool steel and

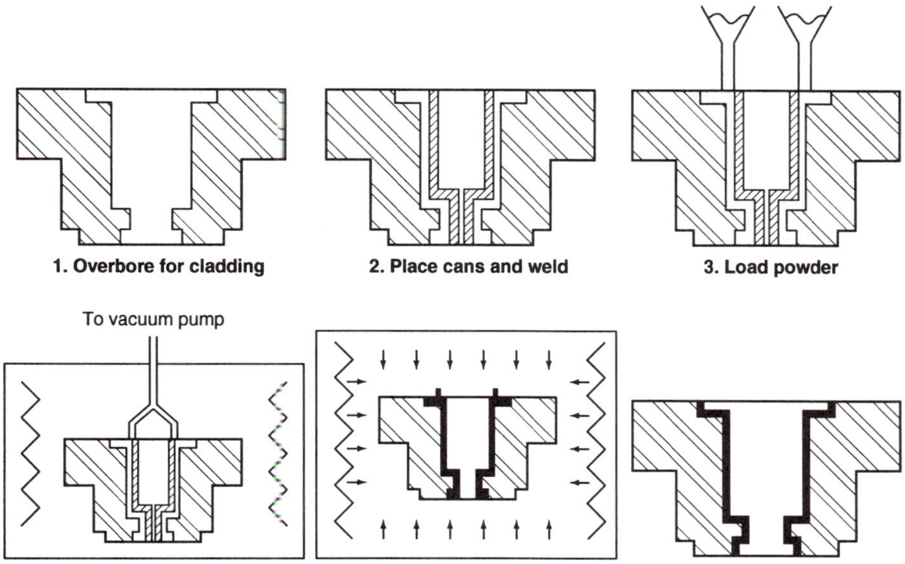

Fig. 25 HIP diffusion bonding of powder to solid. Source: Ref 21

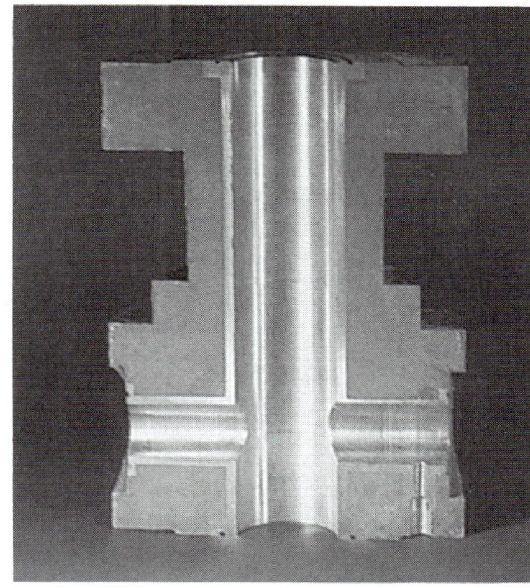

Fig. 26 Low-alloy steel HIP clad with alloy 625 for corrosion resistance

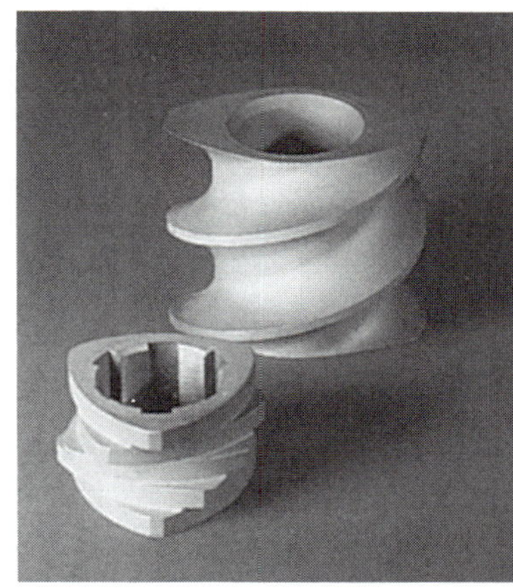

Fig. 27 Bimetallic wear-resistant screw segments for the plastic extrusion industry

17-4 stainless steel and a cobalt-base alloy and 17-4 stainless steel, thus maintaining a martensitic structure up to the interface (Ref 33). Another interlayer application was the use of refractory metal and ceramic interlayers during the fabrication of as-HIP foil of highly alloyed material (e.g., titanium, nickel, and niobium alloys) (Ref 34).

Applications. There have been several applications of bimetallic components that have utilized the HIP diffusion bonding process. Examples include:

- Corrosion-resistant alloy 625 clad to the interior of F22 steel (Fig. 26)
- Wear/corrosion resistant alloy (MPL-1) clad to 4140 steel (Ref 21)
- Alloy CPM 9V clad on the exterior of 4140 cylinders (Ref 20)
- Twin extrusion barrel internally clad with CPM 10V against 4140 steel (Ref 21)
- CPM 10V clad to low-carbon steel for segmented screws used inside the plastic extrusion barrel (Fig. 27)

Future Developments

Applications using HIP technology have evolved from diffusion bonding of dissimilar materials to consolidating encapsulated powder and sealing microporosity in castings. Hot isostatic pressing technology is continuing to grow with diversification into new areas. These areas include equipment improvements, mechanistic modeling of material undergoing HIP, and new applications of HIP.

Refinements of Batch Processing. One equipment refinement that is generating interest is "quick cool" or "HIP quenching." After the HIP cycle hold, furnace cooling on a cold-walled vessel can take several hours with cooling rates of about 100 °C to 200 °C/h depending on the vessel and size of the load. By utilizing a flow device (Ref 15) and the introduction of cold gas into the hot gas, the convective cooling is dramatically increased. One portion of the gas is forced to the outside of the thermal barrier for cooling while the other portion is circulating inside. To achieve the desired cooling rate, the proportion of the hot and cold gas can be computer controlled. The major driver of this technological improvement is to increase productivity, which ultimately increases capacity and decreases costs. In addition, there may be metallurgical enhancements of some materials, thus potentially eliminating some downstream processing steps.

HIP Modeling and Microstructure Prediction. As described in the article "Principles and Process Modeling of Higher Density Consolidation" in this Volume, there has been much work devoted in the 1990s (Ref 13) to predicting dimensional changes during HIP via continuum mechanics/finite element modeling. To predict shrinkage changes, an understanding is needed of the anisotropy of consolidation brought about by the complex interrelationships between the properties of the P/M and container materials as a function of temperature, density, and part geometry. With the development of the constitutive equations for the particles and powder aggregates to predict shrinkage, the underlying mathematics now exist to also predict microstructure of the HIP product (Ref 11, 12, 35). With computational power continually increasing at an affordable rate and material property characterization available from hot triaxial compaction tests (Ref 36), the ability to predict grain size (Ref 37) and other microstructural features (Ref 12, 38) may soon be possible.

HIP Modeling and Closing Porosity in Spray Formed Billets. To compete with ring-rolled products, there has been some interest in producing large nickel-base superalloy rings via spray forming followed by HIP (Ref 39, 40). For this process, metal is nitrogen-gas-atomized onto a low-carbon steel substrate to form a partially dense preform (typically, >90%). The resulting microstructure is determined by amount of liquid in the spray before impact and amount of liquid on the top surface of the deposit. As the amount of liquid is increased, an increase in deposit yield is observed (i.e., atomizing into a swamp); however, these slower solidification rates typically lead to a coarser grain size. If a finer grain size is required, the amount of liquid is decreased, but this typically increases the amount of unusable overspray that cannot be recycled due to increased nitrogen content concerns. Hot isostatic pressing of the preform increases the density to nearly 100% density with some interconnected surface porosity present.

ACKNOWLEDGMENTS

The authors wish to thank Peter Price and Steven Kohler for their efforts on the previous version of this article. Much of their work has been incorporated in this current article.

REFERENCES

1. H.V. Atkinson and B.A. Rickinson, *Hot Isostatic Pressing,* 10 P Publishing, 1991
2. E.S. Hodge, Elevated Temperature Compaction of Metals and Ceramics by Gas Pressures, *Powder Metall.,* Vol 7 (No. 14), 1964, p 168–201
3. C.S. Boyer, History: Development of a HIP Apparatus to Fulfill a Commercial Need, *Hot Isostatic Pressing Conf.,* ASM International, 20–22 May 1996
4. J.E. Coyne, W.H. Everett, and S.C. Jain, Superalloy Powder Engine Components: Controls Employed to Assure High Quality Hardware, *Powder Metallurgy Superalloys, Aerospace Materials for the 1980's,* Vol 1, Metal Powder Report Publishing, Shrewsbury, England, 18–20 Nov 1980, p 24-1 to 24-27
5. J.H. Moll and F.J. Rizzo, Production Applications of Rapidly Solidified Tool Steels, Superalloys, Titanium Alloys, and Corrosion-Resistant Alloys, *Rapid Solidification Processing Principles and Technologies III,* R. Mehrabian, Ed., 6–8 Dec 1982, p 686–691
6. W.B. Eisen, P/M Tool and High Speed Steel: A Comprehensive Review, *Proc. of the 5th Int. Conf. on Advanced Particulate Materials and Processes,* Metal Powder Industries Federation, 1997, p 55
7. W. Stasko, K.E. Pinnow, and R.D. Dixon, Particle Metallurgy Cold Work Tool-Steels Containing 3–18% Vanadium, *Proc. 5th Int. Conf. Advanced Particulate Materials and Processes,* Metal Powder Industries Federation, 1997, p 401
8. C.M. Sonsino, Fatigue Design for Powder Metallurgy, *PM-90, World Conf. Powder Metallurgy,* Vol 1, Institute of Materials, 1990, p 42–88
9. R.M. German, *Powder Metallurgy Science,* 2nd ed., Metal Powder Industries Federation, 1994, p 302–340
10. P. Hellman, Review of HIP Development, *Hot Isostatic Pressing—Theories and Application,* Centek Publishers, 1988, p 3–18
11. R.E. Smelser, J.F. Zarzour, J. Xu, and J.R.L. Trasorras, On the Modeling of Near-Net Shape Hot Isostatic Pressing—AMD, *Mechanics in Materials Processing and Manufacturing,* Vol 194, ASME, 1994, p 213–237
12. R.D. Kissinger, The Densification of Nickel Base Superalloy Powders by Hot Isostatic Pressing, *Diss. Abstr. Int.,* Vol 49 (No. 9), Mar 1989, p 1–39
13. W.B. Eisen, Modeling of Hot Isostatic Pressing, *Rev. Partic. Mater.,* Vol 4, 1996
14. E. Artz, M.F. Ashby, and K.E. Easterling, Practical Applications of Hot Isostatic Pressing Diagrams: Four Case Studies, *Metall. Trans. A,* Vol 13, Feb 1983, p 211–221
15. C. Bergman, J. Westerlund, and F.X. Zimmerman, HIP Quench Technology, *Hot Isostatic Pressing: Proc. Int. Conf. Hot Isostatic Pressing,* ASM International, 20–22 May 1996, p 87–90
16. D.J. Evans and D.R. Malley, "Manufacturing Process for Production of Near Net Shapes by Hot Isostatic Pressing of Superalloy Powder," Final Report on AFWAL-TR-83-4022, Air Force Wright Aeronautical Laboratories, June 1983
17. F.S. Biancaniello, J.J. Conway, P.I. Espina, G.E. Mattingly, and S.D. Ridder, Particle Size Measurement of Inert Gas Atomized Powder, *Mater. Sci. Eng. A,* Vol 124, 1990, p 9
18. P. Loewenstein, Superclean Superalloy Powders, *Met. Powder Rep.,* Vol 36 (No. 2), Feb 1981, p 59–64
19. U.S. Patent No. 4,078,873, 1978
20. J.J. Conway, F.J. Rizzo, and C.K. Nickel, Advances in the Manufacturing of Powder Metallurgy (P/M) Parts by Hot Isostatic Pressing, *Hot Isostatic Pressing, Proc. Int. Conf. Hot Isostatic Pressing,* ASM International, 20–22 May 1996, p 27–32
21. J.J. Conway and J.H. Moll, Current Status of Powder Metallurgy Near Net Shapes by Hot Isostatic Pressing, *Int. Third Conf. Near Net Shape Manufacturing* (Pittsburgh), ASM International, 27–29 Sept 1993, p 125–131
22. Product literature and data, Industrial Materials Technology, Inc.
23. U.S. Patent No. 3,622,313, Nov 1971
24. C.F. Yolton and J.H. Moll, Powder Metallurgy (P/M) Near-Net Shape Titanium Components from Prealloyed Powder, *Titanium 1986—Products and Applications,* Vol II, Ohio Titanium Development Association, 1987, p 783–800
25. G.S. Garibov, V.N. Samarov, and V.I. Geigin, Powder Metallurgy Industry, Economics, and Organization of Production, *Sov. Powder Metall.,* Vol 18 (No. 2), July 1979, p 136–140
26. J.J. Conway, "Final Shape Prediction of Hot Isostatic Pressed Powder Metallurgy (P/M) Compacts," MSE 298 Masters Project, University of Pittsburgh, 21 Aug 1990
27. A. Guthrie and R.K. Wakerling, *Vacuum Equipment and Techniques,* 1949, p 191
28. S. Reichman and D.S. Chang, *Superalloys II,* C.T. Sims, N.S. Stoloff, and W.C. Hagel, Ed., John Wiley & Sons, 1987, p 459
29. R.V. Miner and S. Gayda, *Int. J. Fatigue,* Vol 6 (No. 3), 1984, p 189
30. U.S. Patent No. 4,544,404
31. V. Peterson, V. Chandhok, and C. Kelto, Hot Isostatic Pressing of Large Titanium Shapes, *Powder Metallurgy of Titanium Alloys,* F. Froes and J. Smugeresky, Ed., AIME, 1980, p 251
32. C.G. Hjorth and H. Eriksson, New Areas for HIPing Components for the Offshore and Demanding Industries, *Hot Isostatic Pressing, Proc. Int. Conf. Hot Isostatic Pressing,* ASM International, 20–22 May 1996, p 33–38
33. M.A. Ashworth, M.H. Jacobs, G.R. Armstrong, R. Freeman, B.A. Rickinson, and S. King, HIP Diffusion Bonding for Gear Materials, *Hot Isostatic Pressing, Proc. Int. Conf. Hot Isostatic Pressing,* ASM International, 20–22 May 1996, p 275–285
34. A.M. Ritter, M.R. Jackson, D.N. Wemple, P.L. Dupree, and J.R. Dobbs, Processing of Metal Foil by Direct HIP of Powder, *Aeromat '96* (Dayton, OH), 5 June 1996
35. J.F. Zarzour, J.R.L. Trasorras, J. Xu, and J.J. Conway, Experimental Calibration of a Constitutive Model for Hot Isostatic Pressing (HIP) of Metallic Powders, *Advances in Powder Metallurgy and Particulate Materials—1995,* Vol 2, M. Phillips and J. Porter, Ed., Metal Powder Industries Federation, p 5–89
36. H.R. Piehler and D.M. Watkins, Hot Triaxial Compaction: Initial Results for Aluminum Compacts, *Advances in Powder Metallurgy,* Vol 1, E.R. Andreotti and P.J. McGeehan, Ed., Metal Powder Industries Federation, 1990, p 393–398
37. B.A. Hann, I. Nettleship, and S. Schmidt, An Investigation of Microstructural Evolution of PM Alloy N625 During Interrupted Hot Isostatic Pressing (HIP) Cycles, *Superalloys 718, 625, 706 and Various Derivatives,* E. Loria, Ed., TMS, 1997, p 781–789
38. M.C. Somani, N.C. Birla, Y.V.R.K. Prasad, and V. Singh, Deformation Behaviour and Process Modelling of Hot Isostatically Pressed P/M Alloy Nimonic AP-1 and Its Correlation with Microstructure, *Advances in Materials and Processes,* IBH Publishing, 1993, p 104–143
39. N. Paton, T. Cabral, K. Bowen, and T. Tom, Spraycast-X 718 IN718 Processing Benefits, *Superalloys 718, 625, 706 and Various Derivatives,* E. Loria, Ed., TMS, 1997, p 1–16
40. T.F. Zahrah, R. Dalal, and R. Kissinger, Intelligent HIP Processing of a Spraycast-X Superalloy for Aerospace Applications, *Hot Isostatic Pressing, Proc. Int. Conf. Hot Isostatic Pressing,* ASM International, 20–22 May 1996, p 163–166

Extrusion of Metal Powders*

B.L. Ferguson, Deformation Control Technology, Inc. and P.R. Roberts, American Superconductor Corporation

EXTRUSION is a relatively recent addition to metalworking as noted in a historical survey of extrusion and the development of the process (Ref 1). Notably, the inventive genius of Alexander Dick and the increasing availability of steels that could withstand higher working temperatures opened the way for the hot extrusion of copper alloys and laid the foundation for modern extrusion. Pearson and Parkins (Ref 1) and Lange and Stenger (Ref 2) have written comprehensively on the history, development, application, and mechanics of extrusion; these are recommended texts that provide an excellent background for understanding the process.

There are two main types of extrusion mechanisms, (a) direct and (b) indirect or inverted, as shown in Fig. 1. In direct extrusion, the ram pushes a workpiece forward through a die, causing a reduction in cross-sectional area of the workpiece. Conversely, in indirect extrusion, the workpiece remains stationary relative to the container and there is no friction between the workpiece and container. Both methods may be used to extrude metal powders, although direct extrusion is more widely practiced.

Powder extrusion provides a method to obtain a useful shape or form that may not be readily achieved by other means. It has been used to make seamless tubes, wires, and complex sections that would be difficult or impossible to fashion by any other process. Pioneering work in the extrusion of metal powders was conducted in the late 1950s to produce controlled ductility in beryllium, dispersions of nuclear fuels and control rod materials, and dispersion-strengthened aluminum (Ref 3).

The extrusion of metal powders occupies a special niche in extrusion technology for many reasons, which include:

- The ability to form shapes by extrusion from materials that are difficult or impossible to process by casting or working
- Improved properties and performance because of microstructural refinement and minimization of segregation that results from powder processing
- The dispersion of one species in another from the extrusion of powder mixtures
- The ability to form wrought structures from powder without the need for sintering or other thermal treatments
- Reduced extrusion pressures and wider temperature and ram velocity ranges for powder extrusion than for extrusion of cast billets

This article concentrates on direct extrusion processing where metal powders undergo plastic deformation, usually at an elevated temperature, to produce a densified and elongated form having structural integrity. Three main approaches to extrusion of powders are shown in Fig. 2. In the first instance where the principal material is poured loose into the extrusion container, the particle size is normally large. Lenel (Ref 4) mentions that this process is used to extrude magnesium alloy pellets (particle size 70 to 450 μm). A hot container supplies heat to the pellet charge, and extrusion is performed with no atmosphere protection.

The second method, shown in Fig. 2, relies on the use of a compacted billet as the extrusion workpiece. Precompaction is useful because a workpiece shape that can be handled is produced, supplying a form that is much easier to use and control in a manufacturing environment than loose powder, and compaction increases the density of the workpiece in comparison with that of loose powder. The higher density reduces the ram stroke and decreases the container length needed to produce the required extruded length. In this method, a powder that compacts readily is used; this type of powder has particles that are rough and jagged with multiple asperities and ragged protrusions, or a flake form. In loose form, such particles have low packing efficiency and a correspondingly low density, for example, 35 to 50% of theoretical. However, these particles can be pressed into a partially consolidated billet shape with densities of 70 to 85% of theoretical. This form is referred to as a "green" compact, and it has sufficient "green strength" to endure handling. Where greater resistance to crumbling during handling is needed, the porous compact may be sintered before extrusion. However, sintering is not a required processing step, and many materials are extruded without this additional processing. Alternatively, hot pressing may be used instead of cold pressing to produce the extrusion workpiece.

A more elaborate approach to powder billet preparation is that shown in the third variation in Fig. 2 and in more detail in Fig. 3, where powder is first partially densified directly in a can. This can may then be evacuated and sealed, as shown in Fig. 4, or it may be left open to the atmosphere. Canning is employed for the following reasons (Ref 5):

- Isolation of the principal material from the atmosphere and extrusion lubricants (clean extrusion technique)
- Isolation of toxic materials such as beryllium and uranium for safe handling
- Encapsulation of spherical and other difficult-to-compact powders to produce a billet form
- Improved lubricity and metal flow at the die interface by proper selection of the can material
- Isolation of the principal material from the extrusion die and region of highest shear, which is an important consideration for materials with limited ductility

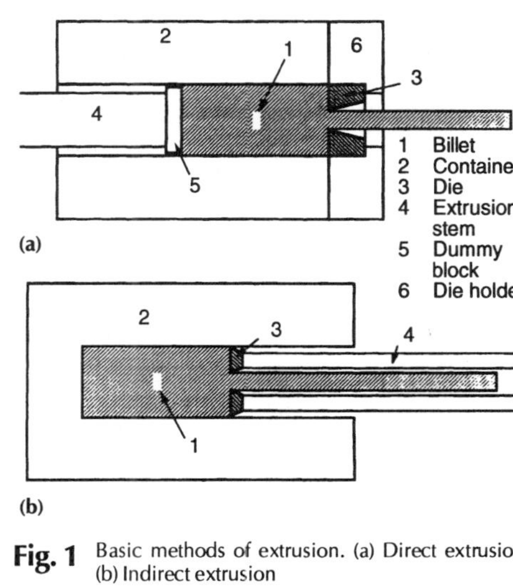

Fig. 1 Basic methods of extrusion. (a) Direct extrusion. (b) Indirect extrusion

*Adapted from article by P.R. Roberts and B.L. Ferguson, "Extrusion of Metal Powders," *International Materials Reviews,* Vol 36 (No. 2), 1991, p 62–79 with review by Peter W. Lee, The Timken Company and Donald Byrd, Wyman Gordon Forgings

- The ability to position powder and solid components within the can to produce unique and complex shapes (this is a variation of the filled billet extrusion technique that is discussed below)

When purity must be maintained, canning of the powder, including evacuation and sealing as shown in Fig. 4, is an essential step. Billet preparation for critical applications, such as gas turbine engine components, requires that filling and evacuation be performed with great care and refined practice. Procedures may include cleanroom container preparation and assembly, total isolation of powder from ambient air, evacuation at slightly elevated temperatures to drive off adsorbed gases from particle surfaces, and leak checking of sealed containers. The extruded product must be free from both prior particle boundary decorations and nonmetallic inclusions that degrade mechanical properties, especially fracture toughness and fatigue resistance.

With this introduction to the basic powder extrusion processes, it is pertinent to review briefly the mechanics of extrusion and to examine specific extrusion practices for the production of wrought material from powder stock.

Mechanics of Powder Extrusion

Typical pressure curves for direct and indirect extrusion of a conventional billet are shown in Fig. 5. Initially, pressure increases linearly with ram displacement as the billet upsets to fill the container and reaches a maximum value as the workpiece begins to flow through the die; this is known as the breakthrough pressure. Steady state is achieved as the ram advances. For direct extrusion, the pressure falls as the ram stroke continues, reflecting the decreasing frictional resistance as the contact area between the billet and container decreases. For indirect extrusion, the extrusion pressure is fairly constant because there is no relative movement and thus no friction between the billet and container. At the end of the stroke, a sharp rise in pressure may occur because of the increasing resistance to radial inflow of the residual billet material. This latter effect may be overcome by placing a follower of some disposable material between the billet and ram to ensure that the billet clears the die.

Extrusion pressure is a function of material flow stress, temperature, friction, strain rates, and deformation. Sachs and Eisbein (Ref 6) proposed an extrusion constant K such that:

$$P = K \ln R \quad \text{(Eq 1)}$$

where P is the required extrusion pressure, R the extrusion ratio A_i/A_f, which is the ratio of initial cross-sectional area to final cross-sectional area, and K the extrusion constant for the material.

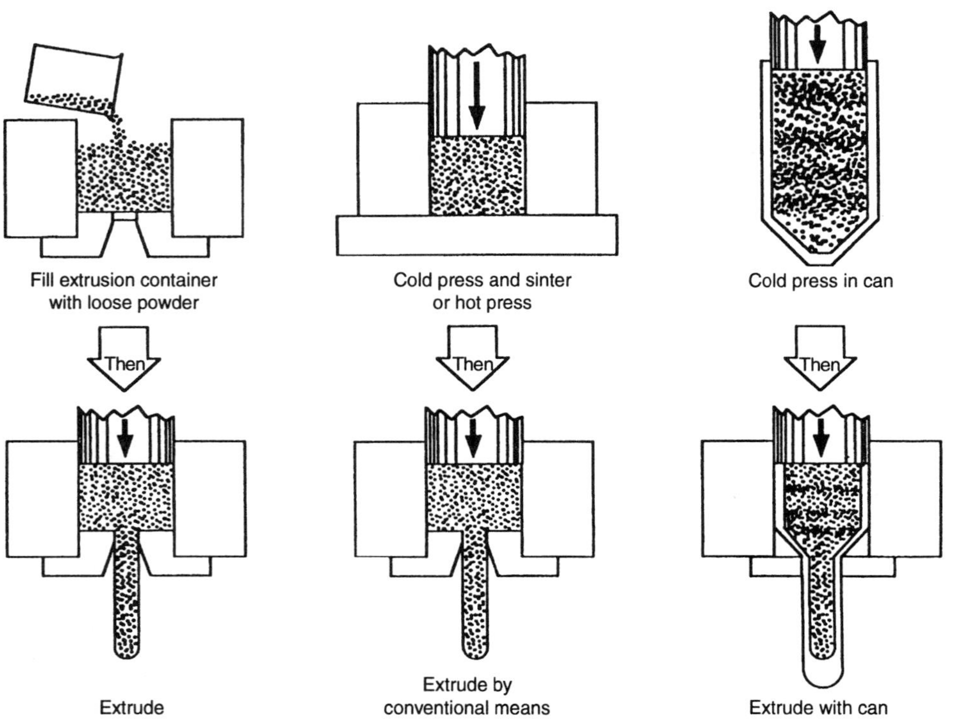

Fig. 2 Hot extrusion methods for metal powders

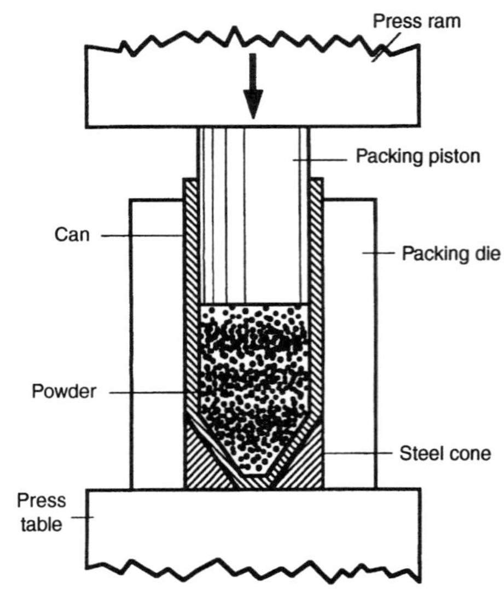

Fig. 3 Packing of powder in metal can

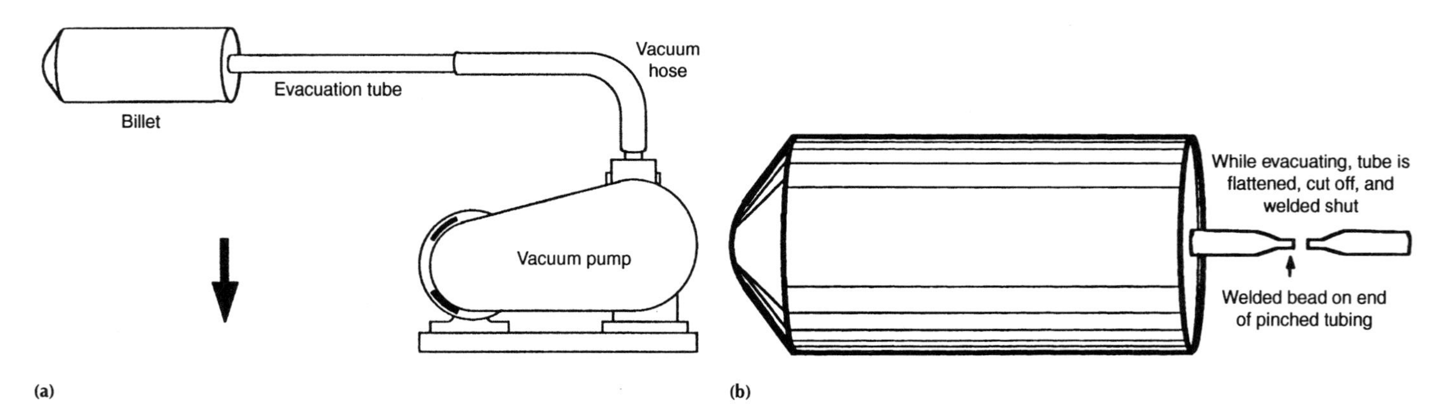

Fig. 4 Evacuation and sealing of powder extrusion billet. (a) Billet with evacuation tube leading to vacuum pump. (b) Billet with sealed tube

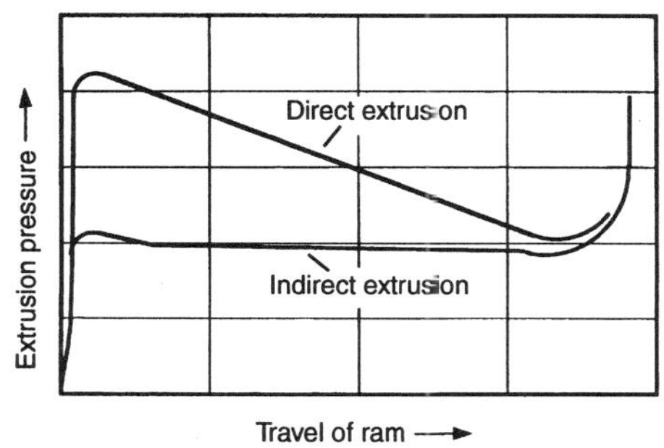

Fig. 5 Extrusion pressure as function of ram travel

The extrusion constant K combines flow stress, friction, and redundant work into one parameter. While K is dependent to some extent on particular extrusion practices, reported extrusion constant values may be broadly applied as an aid to the design of useful extrusions. Extrusion constant values for a variety of engineering materials are shown as a function of temperature in Fig. 6. These curves were compiled from many runs on various extrusion presses (Ref 7).

It is clear from Fig. 6 that K decreases with temperature, and it follows that hot extrusion requires lower extrusion pressure than cold extrusion. The effect of temperature on K is similar to the relationship between flow stress and temperature. For cast billets, K increases as the ram speed is increased, which reflects the relationship between flow stress and strain rate. Dieter (Ref 8) mentions that a tenfold increase in ram speed results in a 50% increase in the required extrusion pressure, as a general rule for cast/wrought billet stock.

In evaluating the dependency of extrusion pressure on temperature and strain rate, adiabatic heating, heat generated by friction, and conductive heat transfer must be taken into account. Proper extrusion speed and temperature are usually determined by experience, following general guidelines shown schematically in Fig. 7. This figure shows the relationship between pressure and temperature on deformation level and between strain rate and temperature on allowable deformation level. Together these relationships define an allowable range of processing conditions. With improved mathematical modeling capabilities, analytical methods may be used to verify selected extrusion conditions by computer simulation.

The discussion up to this point has only dealt with extrusion of cast/wrought billets. Extrusion of powder is mechanically different from extrusion of cast material. First, the billet stock for powder extrusion is usually compacted powder, with a density of ~75 to 85% of theoretical, or loose powder, with a density of 55 to 65% of theoretical. The presence of porosity and unbonded particulate changes the deformation mechanics significantly from that of cast/wrought material (Ref 3–5, 9–16).

Pressure versus ram displacement curves are shown in Fig. 8 for the extrusion of cold compacted Al-Mg-Si (as supplied) powder billets at two ram speeds (Ref 13). For comparison, a pressure versus displacement curve for a similar cast material is also shown. The initial pressure buildup is considerably different for powder and cast billets. Pressure increases linearly for a solid billet as it upsets to fill the container. For a powder billet, the pressure rise during the upsetting stage is both nonlinear and more gradual than for a cast billet. During this pressure buildup, the powder is being compacted to nearly full density. While densification is substantially completed before the onset of extrusion, powder particles remain poorly bonded and mechanical strength is low. The material does not attain measurable strength and ductility until it has traversed through the extrusion die and been subjected to a sufficient level of shear deformation. Shear deformation results in the formation of metallurgically sound bonds between particles and produces a "wrought" product.

In practice, back pressure or resistance must be provided initially to aid densification of the powder in the billet nose before the extrusion. Therefore, a solid nose or a pad is placed in front of the powder billet. The pad may also provide a measure of lubrication. Extrusion of green compacts without such assistance to densification has resulted in gross surface cracking (Ref 9). For loose powder, the pad is incorporated in the container design.

Experimental results for extrusion of compacted billets have shown that the extrusion pressure is less dependent on ram speed for powder than for cast material (Ref 9). Therefore,

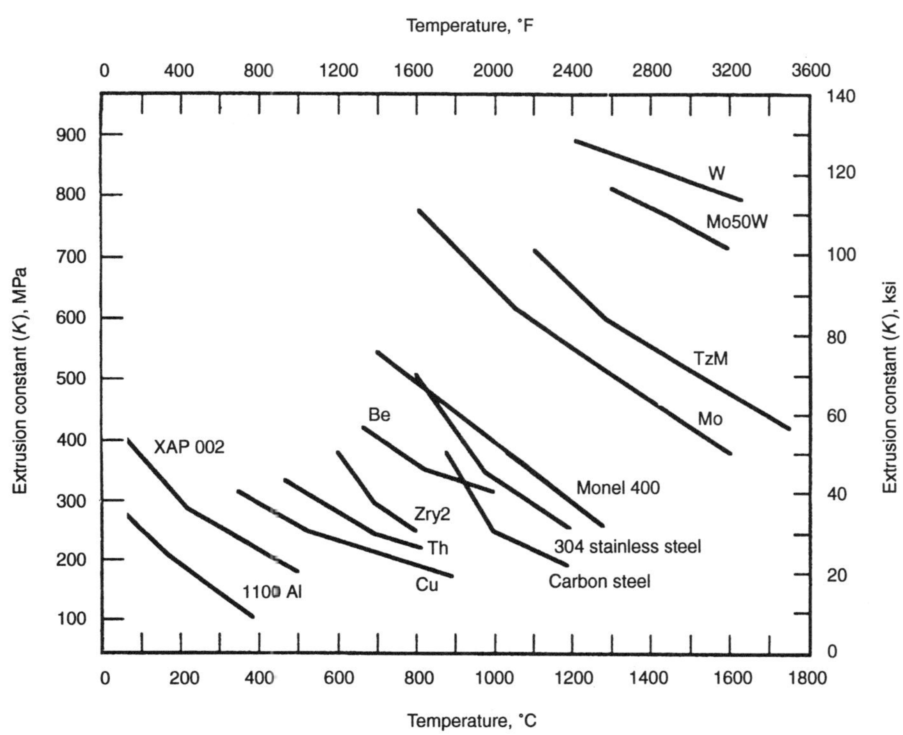

Fig. 6 Extrusion constant versus temperature for some engineering alloys

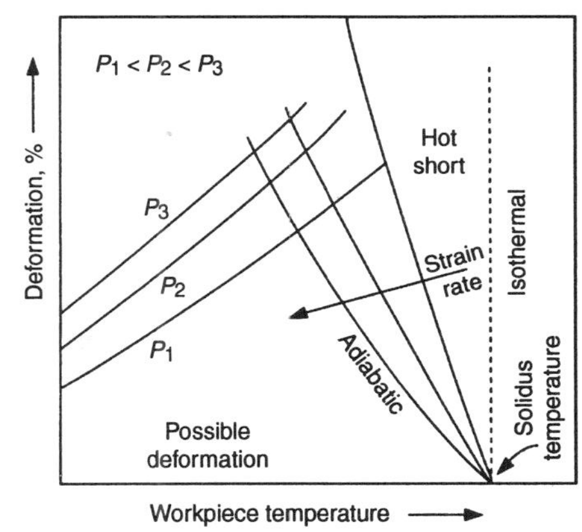

Fig. 7 General effects of pressure, strain rate, and temperature on allowable deformation. Source: Ref 8

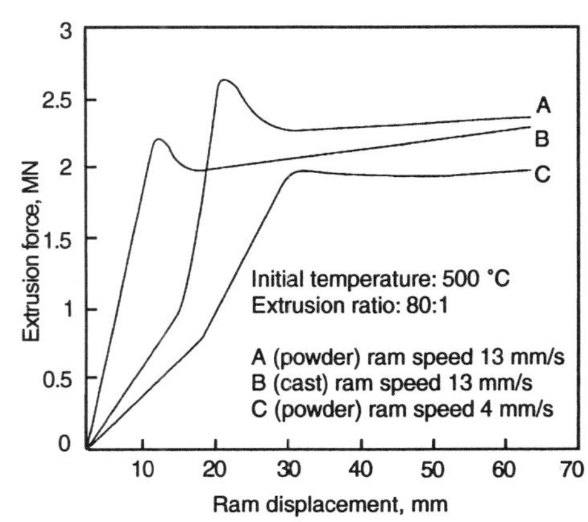

Fig. 8 Load versus displacement curves for extrusion of Al-0.33Mg-0.12Si (final) alloy powder (containing alumina) and cast billet. Source: Ref 12

Table 1 Extrusion pressure and reduction ratio relationships based on $P = a' + b' \ln R$

Powder, wt%	Extrusion relationship	a'/b'
Al-3.6Mn-0.95Al_2O_3	$P = 555 + 252.5 \ln R$	2.2
Al-2.4Fe-1.09Al_2O_3	$P = 475.5 + 252.5 \ln R$	1.9
Al-1.9Al_2O_3	$P = 277.5 + 252 \ln R$	1.1
Al-0.26Al_2O_3	$P = 187.5 + 252.5 \ln R$	0.74
99.5Al (cast)	$P = 20 + 287.5 \ln R$	0.07

Source: Ref 14

strain rate is of less importance in powder extrusion than in extrusion of cast materials. The major reason for this is thought to be the overwhelming influence of porosity and the formation of particle bonds during powder deformation as opposed to the importance of flow stress on extrusion pressure for cast/wrought material. However, the breakthrough load for powder extrusion remains a function of strain rate and increases as ram speed increases (Ref 13).

Theoretically, the pressure P needed to extrude a billet through a die has three components:

$$P = A + B \ln R + C \quad \text{(Eq 2)}$$

where A is related to redundant work, $B \ln R$ is the homogeneous work of deformation, C is the work needed to overcome friction, and R is the extrusion ratio.

For cast billets, Eq 2 is generally normalized by the material shear yield strength Y, as follows:

$$P/Y = a + b \ln R \quad \text{(Eq 3)}$$

For powder billets, Eq 2 does not hold because the yield strength of the material changes dramatically as it flows through the extrusion die. For extrusion of powder, Eq 4 is preferred:

$$P = a' + b' \ln R \quad \text{(Eq 4)}$$

In Eq 4 the term a' relates to redundant work and b' relates to homogeneous work. Table 1 contains experimentally determined constants a' and b' for extrusion of several aluminum powder alloys and cast aluminum at various reduction ratios. The ratio a'/b' indicates the relative importance of redundant work, and it is clear that redundant work is much more significant in powder extrusion than in extrusion of cast stock. For powder extrusion, the redundant work term includes the process of forming cold welds between particles, breaking welds, and rewelding, while particle deformation is occurring. Sheppard proposed that the high proportion of redundant work in extrusion of powder contributes to the formation of sound particle bonds and eliminates the need for subsequent sintering (Ref 9–15). A minimum level of extrusion reduction must be exceeded in order to produce sound bonds, with 9 to 1 being a typical minimum reduction ratio for extrusion of spherical powders.

Because the amount of redundant work that accompanies powder extrusion is a function of the number of welds formed, the extrusion pressure component related to redundant work is a function of particle size and size distribution. Fine powder, which has a higher surface area per unit volume than coarse powder and thus forms more welds, requires a higher extrusion pressure than coarse powder (Ref 9, 10).

Sheppard and coworkers have extended the upper-bound analysis to study powder extrusion (Ref 9–15). A typical deformation zone geometry, as defined by the upper-bound method, and the accompanying hodograph are shown in Fig. 9. This analytical technique is reviewed thoroughly by Sheppard et al. (Ref 9, 11, 12) and Avitzur (Ref 17).

For canned powders, an important consideration is the nature of the flow of material through the die. Modeling studies of material flow during extrusion have been conducted using such diverse materials as colored wax (Ref 18), mixtures of chalk, beeswax, and petroleum jelly with added coloring (Ref 19), and Plasticene (Ref 20). Green (Ref 20) has shown that a good working analogy to hot metal is afforded by such simple substances. Sachs and Eisbein (Ref 6) used small cylindrical tin billets that were sectioned longitudinally, with a grid scribed on the flat surface. The cylinder halves were held together and forced through conical dies of different included angles. The flow patterns were revealed in the distortion of the grid in a range of partial extrusions, as shown in Fig. 10. When the die is flat (180° included angle), a zone of dead metal forms in the front corner of the press container and flow at the periphery of the billet occurs by shear over a conically defined surface (Ref 6, 17). As the included cone angle decreases from the extreme case of the flat die, flow becomes more streamlined and the ogival characteristic of the transverse grid lines becomes less pronounced. Conical dies are used for the extrusion of canned powder to avoid the formation of a stagnant flow zone and the associated turbulence and shear experienced with flat dies. Conical dies work well with included angles of 90 to 120°, and good streamlined flow is obtained when there is effective lubrication and closely fitting extrusion tooling working in conjunction with the die. Canned extrusions have been limited to the fabrication of simple shapes such as circular sections, tubes, flats, and so forth, because of the necessity for a conical approach to the die land. Advances in analytical modeling of extrusion and in die-making techniques permit the extrusion of more complex cross sections.

Comparison of Extrusion with Cast and Powder Billets. Comparison of analysis results for cast billets with results for powder billets has led to some fundamental conclusions:

- Extrusion of powder is a continuous process of welding particles, breaking welds, and rewelding as particles are rearranged and deformed. Densification is nearly completed in the extrusion container, before actual extrusion. However, the achievement of adequate structural

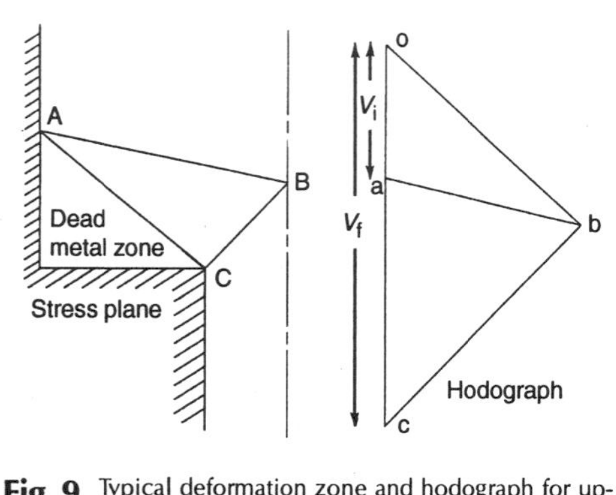

Fig. 9 Typical deformation zone and hodograph for upper bound modeling of extrusion of powder billet. *oa*, initial velocity V_i; *oc*, final velocity V_f. Source: Ref 11

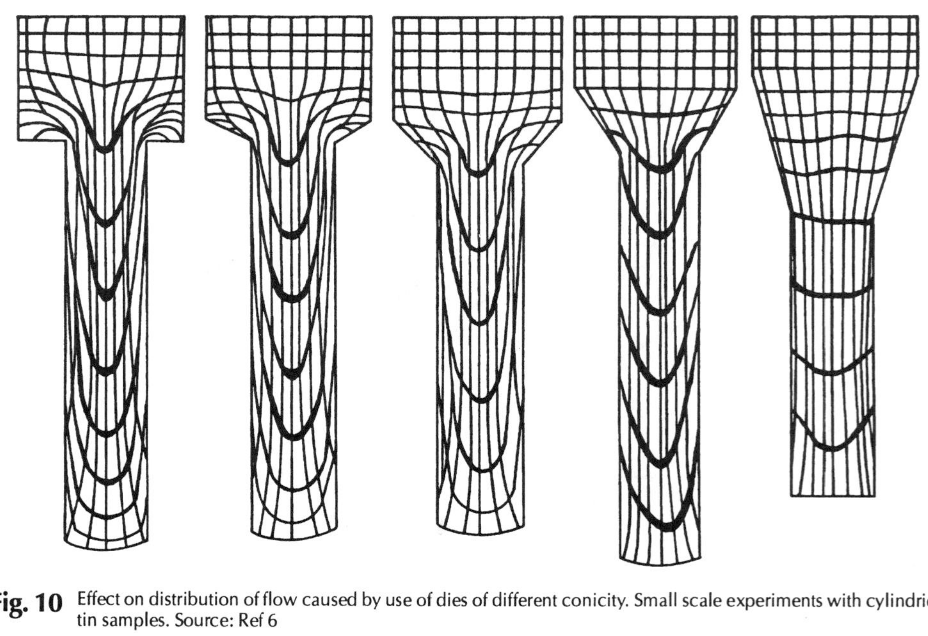

Fig. 10 Effect on distribution of flow caused by use of dies of different conicity. Small scale experiments with cylindrical tin samples. Source: Ref 6

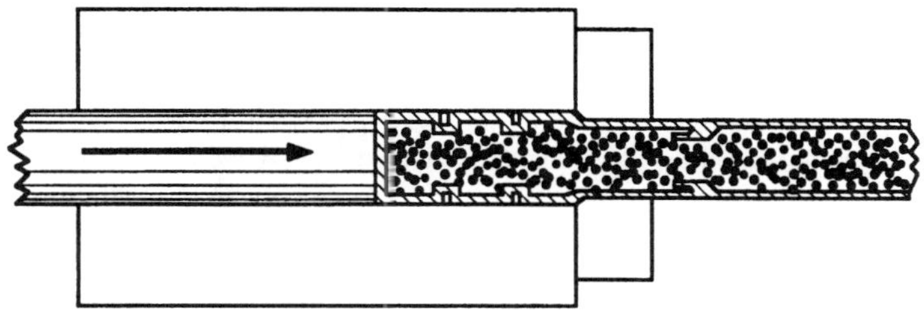

Fig. 11 Folding of metal can with loosely packed powder

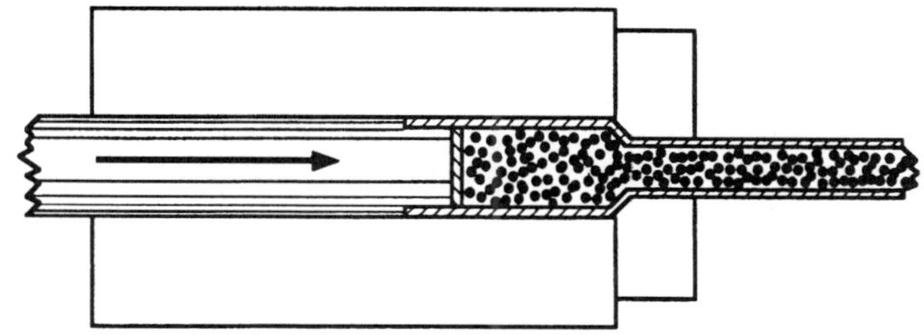

Fig. 12 Penetrator technique to avoid folding of can

integrity is dependent on shear deformation as the compacted powder is forced through the extrusion die.

- The presence of porosity and unbonded particles in the extrusion billet produce a deformation zone much different from that of the extrusion of a cast billet. The extrusion pressure for powder is much lower than for extrusion of a cast billet of similar composition. This results from the development of final properties in the powder extrusion only after the material has traversed through the deformation zone. Conversely, a cast billet exhibits higher flow stress before extrusion.
- The amount of redundant work is dependent on the surface area/volume ratio of the powder. As the powder particle size decreases, this ratio increases, and a greater number of welds are required for bonding. Consequently, the amount of redundant work increases.

Powder Extrusion Practice

Extrusion of canned powder can be considered a special case of filled-billet extrusion. Filled billet refers to an extrusion billet or workpiece that contains more than a single material. For canned powder, there are at least two components, the powder and the can. More typical filled-billet workpieces, discussed below, may have two or more different materials arranged in a geometric array within a container. In the extrusion of a filled billet, it is important that the materials of the various components are metallurgically compatible. A careful review of their mechanical and physical properties will avoid serious pitfalls, such as:

- The formation of eutectic phases
- Gross thermomechanical mismatch at extrusion temperature
- Excessive interdiffusion of can and workpiece material
- Extreme difficulty in removing the can material from the extruded product

These considerations are discussed in more detail by Roberts (Ref 5).

In the treatment of powder-filled billets, a further concern is the relative deformation of the canister and the powder contents during the upset of the billet. The can may buckle in the press container if the resistance of the contents of the container to upsetting is low in comparison with the buckling resistance of the can. The folds that result from buckling will carry through into the extruded rod as a defect. This is shown schematically in Fig. 11 and is discussed by Loewenstein et al. (Ref 3). The density of the powder and the wall thickness of the can must be balanced to avoid buckling. One solution to this problem is to use a ram that penetrates the canister, densifying the powder without axially loading the canister wall. This is shown in Fig. 12. Alternatively, the powder billet may be precompacted as in Fig. 3 or, more suitable for spherical powders, it may be jolted or vibrated during filling to increase the density of powder packing.

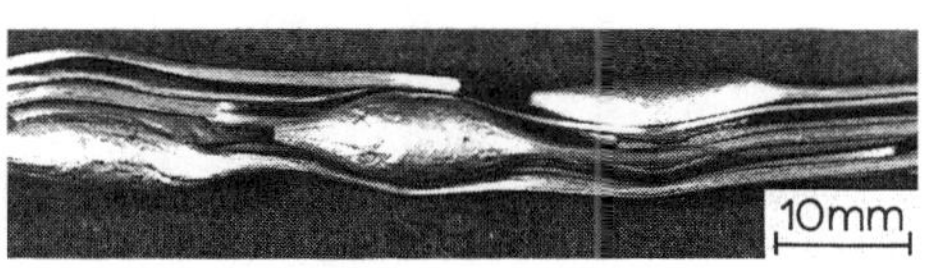

Fig. 13 Attempt to coreduce cast cobalt alloy rods in low-carbon steel matrix. See Fig. 14.

Because densification is largely completed during the upsetting stage, extrusion of the canned powder can be thought of as coreduction of two materials having a core-sheath relationship. The relative resistance to deformation of the two materials must be balanced in order to avoid severe instabilities in plastic flow and possible fracture of either the sheath (can material) or core (densified powder workpiece). Avitzur has presented an analysis used to predict conditions that cause fracture in either the cladding or the core based on the following parameters (Ref 17, 21):

- Overall reduction ratio
- Extrusion die angle
- Volume fraction of each material
- Relative flow stress of each material friction

Roberts and Roberts (Ref 22) cite examples and conditions for successful coreduction of metals by hot extrusion and cold wire drawing.

The extrusion constant K defined in Eq 1 can be used to design billets for either powder extrusion or extrusions of various arrays of solids within a can. The force required to coextrude a billet containing more than one material can be calculated by a rule of mixtures approach:

$$F = (K_1A_1 + K_2A_2 + \dots K_iA_i) \ln R \qquad \text{(Eq 5)}$$

where the various K and A terms are the respective extrusion constants and the cross-sectional areas of the components that sum to A, the cross-sectional area of the press container. A general guideline is to balance the required force components by selection of container material (K term) and relative area fractions of workpiece and container (A terms) for the required reduction ratio.

While precise matching of flow stress values is not practical, it serves as a simplified guide. Gross mismatch of flow stress values usually results in unstable metal flow manifested by large variations in component sections along the extruded rod. In extreme cases, discontinuities may occur. Figure 13 shows the results of uneven reduction obtained from the coextrusion of seven cast rods of a cobalt alloy (Coast Metals alloys No. 64, Co-30Cr-20W-1V-5Ni-1C) set in a drilled billet of low-carbon steel. The billet geometry is shown in Fig. 14. The reduction ratio and temperature used for this trial were 12 to 1 and 1065 °C. The cobalt alloy rods were exposed by etching away the steel matrix in hot dilute nitric acid. It is interesting to note that while this attempt to coextrude cast rods in a mild steel matrix was unsuccessful, much better

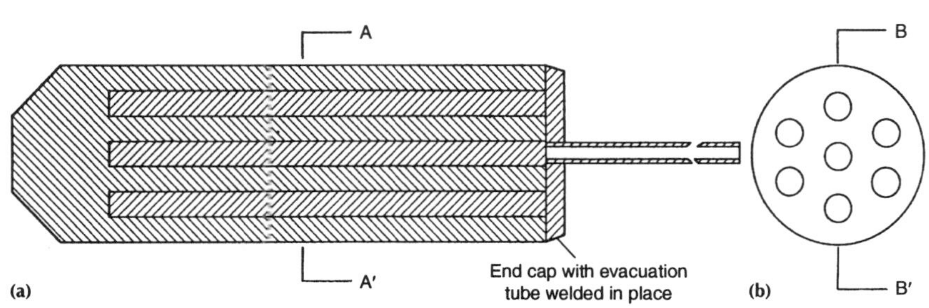

Fig. 14 Low-carbon steel drilled billet with seven cobalt alloy rods introduced (see Fig. 13). (a) Longitudinal section B-B′. (b) Cross section A-A′

results were obtained when the same cobalt alloy was introduced as –80 mesh (<175 μm) powder.

In certain instances, trials have been conducted successfully that violate the simple rule of matching flow stresses. An example involving a thin cladding of 10 carat gold alloy over a thin-walled tube of Inconel 600 with a central core of nickel rod and all contained within a steel can, extruded successfully with proportionate and uniform component sections in the extruded product (Ref 5). Had the components been presented in the same amounts but with a different geometry of distribution, such as in the form of parallel rods of these materials, much less uniform coreduction of section could be expected. The exact influence of component distribution is not well established, and a considerable amount of art is required for proper billet design.

Materials Processed by Powder Extrusion

The early development of powder extrusion processes concentrated on the search for materials with improved high-temperature performance and the development of methods for encapsulating and handling toxic materials. The use of an outer skin or container and a particulate workpiece also stimulated metalworkers to examine powder extrusion as a means for forming structural components from difficult-to-work material. These basic uses of powder extrusion still drive the industry. Examples are given below to illustrate the advantages of coupling powder metallurgy with a large deformation process such as extrusion for the production of unique materials.

The most critical applications of extruded powders today involve extrusion of canned material. The primary applications include tooling and aerospace markets, where the materials include tool steels, superalloys, titanium, copper, and aluminum. Common to the products produced from these materials are the requirements for structural integrity and high performance. These attributes are achieved only by proper powder processing and controlled extrusion practices.

It is well established that shear deformation in combination with pressure fractures oxides and other films on particle surfaces and produces sound bonds between particles. For this reason, extrusion of aluminum powders, which have substantial oxide contents, is common practice. On a more subtle scale, extrusion of "clean" powders, for example, with oxygen levels below 100 ppm, relies on the same mechanism to produce particle bonds of high integrity. Here, the environment within the sealed container is critical to the success of the process.

Thus far, little mention has been made of the mechanical properties or microstructure of the extruded products because the process may be used to form a wide variety of materials for many applications. A discussion of the versatility of the extrusion process as it applies to powder follows.

Dispersion-Strengthened Materials. In the ongoing search for materials with improved properties at elevated temperature, sintered aggregates of metals and nonmetals were investigated. This development started at the laboratories of Aluminium Industrie AG in Switzerland in 1946 as an accidental discovery when experiments to make extruded Al-C wires from powders showed much higher strengths than expected. The oxide skin that surrounds each aluminum powder particle inhibits sintering, and the result is a weak, brittle sintered product. However, this skin is easily broken to expose clean metal surfaces that bond metallurgically during extrusion deformation to produce sound products. Furthermore, extrusion tends to fragment and redistribute the oxide.

Subsequently, Irmann (Ref 23) reported details of a commercially manufactured material that came to be known as sintered aluminum product (SAP). Irmann's development of SAP encouraged further research in the 1950s and 1960s. A good account of the processing details and enhanced mechanical properties developed in the original SAP materials is found in Ref 24. Metallic systems with nonmetallic dispersion phases studied intensively included copper, cobalt, iron, nickel, silver, gold, platinum, and lead, using dispersions of oxides, carbides, and nitrides. Zwilsky and Grant (Ref 25) mixed electrolytic copper with fine alumina; pressed, sintered, and extruded rods were swaged; and it was found that recrystallization was affected significantly. A 10% addition of a 0.3 μm alumina to –325 mesh (<45 μm) copper powder so processed was found to suppress recrystallization completely and retain useful hardness up to 1000 °C. Cremens and Grant (Ref 26) mixed very fine alumina with fine carbonyl nickel. Pressed, sintered, and extruded rod stock showed a sevenfold increase over pure nickel for stress rupture at 100 h. In all of these experiments, extrusion was the primary densification process. Ansell (Ref 27) suggested a mechanism whereby the supply of dislocation sources was diminished by extrusion to explain the delay in the onset of creep and the often erratic creep behavior in these early oxide-dispersion-strengthened materials. He further suggested that extrusion or rolling may be essential manufacturing steps for these products.

Oxides of aluminum, magnesium, thorium, and the rare earths, such as yttrium, are generally preferred as dispersion phases because they are hard and stable at high temperatures. Nickel strengthened with a dispersion of thoria is an early example of these products. These early materials have been superseded by more well developed and precisely controlled dispersion-strengthened materials in recent years.

The main methods of producing current dispersion-strengthened metals are mechanical alloying and chemical reaction, for example, internal oxidation (Ref 28–35). Both of these techniques are based on powder metallurgy as a means of producing a uniform distribution of fine dispersoids, and both involve extrusion as the primary consolidation method. Examples of these materials are presented below.

Commercial grades of oxide-reinforced copper are available (C15715 with 0.3 wt% Al_2O_3 and C15760 with 1.1 wt% Al_2O_3) in wire and rod form produced from extruded bar (Ref 32). Primary applications include spot welding tips that require high electrical conductivity and resistance to deformation at elevated temperature. The oxide in this P/M product is produced by internal oxidation, and it is redistributed during the large area reduction of the extrusion process. At high temperatures, the material extrudes easily, and the extruded bar may be drawn to large reductions in section without intermediate annealing steps. Consequently, the product is fine grained with high strength and yet maintains electrical conductivity close to that of pure copper. The key to achieving a high elevated-temperature strength is the fine size (3 to 12 nm) and uniform spacing (50 to 100 nm) of the alumina dispersoid. The combination of powder metallurgy and extrusion provides these metallurgical features.

In the original aluminum SAP alloys, the oxide content was typically 12 to 15 wt%; high fractions of dispersoid were required because of the relatively coarse and uneven distribution. Commercial dispersion-strengthened aluminum alloys currently available contain a significantly lower dispersoid content, that is, <5 wt%, but these alloys are much more effective than SAP alloys because of the fine size and the uniform distribution of the dispersoid (Ref 29–31). IN-9021 (Al-4.0Cu-1.5Mg-1.2C-0.75O) and IN-9051 (Al-4.0Mg-0.7C-0.60O) mechanically alloyed aluminum alloys contain oxide and carbide dispersions of less than 5 wt%. Typical processing involves hot consolidation to densify the powder, followed by hot extrusion. Extruded forms may be rolled, swaged, or drawn.

Nickel alloy Inconel MA754 (Ni-20Cr-1Fe-0.3Al-0.5Ti-0.6Y_2O_3) is used for gas turbine vanes and is required to have a strong texture to provide good thermal fatigue and creep resistance (Ref 28, 33). Mechanically alloyed powder, attrited in a high-energy ball mill, is thermomechanically processed via extrusion, hot rolling, and zone annealing to provide a strong ⟨100⟩ grain-axis alignment parallel to the working direction in order to impart the desired properties. The directional working of extrusion and subsequent rolling or plane strain forging produces the microstructure needed to form a large, elongated grain structure during zone annealing to provide directional recrystallization.

Dispersion strengthening of silver by particulate cadmium oxide was developed to produce electrical contacts having improved resistance to arc erosion, low contact resistance, and antiwelding characteristics (Ref 34). While the CdO content may be high in comparison with the dispersoid content in copper (e.g., up to 25% CdO in Ag versus 1% Al_2O_3 in copper), the thermal and electrical conductivities of these materials are excellent. It is common practice to make rod stock in this alloy by pressing, sintering, and extruding coprecipitated or preoxidized powders or powder blends.

Platinum and its alloys, used in special furnace windings, heater tapes, and structural shapes that are required to operate close to their melting points, are fabricated from powder blends with thoria, yttria, or zirconia dispersions (Ref 35).

A dispersion-strengthened titanium powder produced by the plasma rotating electrode process (PREP) was consolidated by extrusion (Ref 36). The chemistry of this alloy was a modification of Ti-6246 with an erbium addition (Ti-6Al-2Sn-4Zr-2Mo-0.1Si-2Er). The rare earth was internally oxidized to Er_2O_3 in a fine dispersion of particles where the maximum size observed was 50 nm. The principal intent of this exercise was to obtain fine dispersoids in a rapidly solidified powder material. It is interesting to note that powder consolidation by extrusion did not compromise the useful metallurgical characteristics developed by rapid solidification processing of this material.

These are some of the many examples of materials that have had their properties significantly extended by dispersions, and extrusion is most often used to consolidate the precursor powder blends into bar form that may then undergo further working. Extrusion serves to densify the product and impart a uniform longitudinal and transverse distribution of the dispersed phase together with the development of preferred orientation in the metal matrix that assists further deformation by wire drawing or other methods for the reduction of cross section.

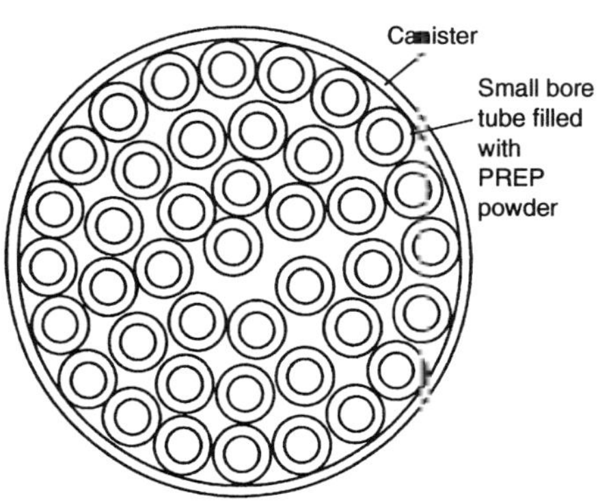

Fig. 15 Cross section of extrusion billet to fabricate superalloy weld wires

Superalloys. Extrusion is commonly used to consolidate superalloy powders into billet stock for subsequent forging operations (Ref 37, 38). Typical processing includes containerization of gas-atomized powder, hot compaction by forging to around 94% of theoretical density, and extrusion to full density at a reduction ratio of 5–7 to 1. After can removal and inspection, the extruded bar is sectioned into billets for forging. Extrusion is relied on to produce strong metallurgical bonds between powder particles by virtue of shear deformation and high pressure across particle interfaces. In addition, extrusion produces a fine grain size that facilitates forging. This is especially important in isothermal forging processes.

Superalloy powders are susceptible to carbide formation on particle surfaces because of the presence of strong carbide-forming elements. In high-carbon superalloys, such as René 41, carbides are relied on to provide grain size control and to enhance mechanical properties (Ref 38). In such cases, the extrusion temperatures must be controlled to avoid gross carbide formation before extrusion. High extrusion preheat temperatures result in excessive carbide formation along prior particle boundaries in the extruded form that reduces transverse strength and ductility to unacceptable levels. Furthermore, grain growth is limited to the powder particle size during high-temperature solution annealing, which results in low elevated-temperature properties. The use of a low extrusion temperature reduces the amount of particle-boundary carbide formation significantly. Subsequent solution annealing temperatures can then be controlled to produce either a fine-grained microstructure with more uniform properties or a large equiaxed microstructure with good elevated-temperature properties.

Unintentionally introduced organics are often the source of undesirable prior particle boundary decorations. Low-deformation consolidation processes, such as hot isostatic pressing, do not deform particles sufficiently to rupture surface films and present virgin surfaces for bonding. Extrusion, by virtue of its large deformation, effectively removes the deleterious effect on performance of such contaminants when potentially harmful particle surface films are fragmented and isolated within the product.

In a novel application of superalloy powder filled billet extrusion, a method has been developed that relies on a powder-solid configuration for fabricating thin-weld filler wire that is used for critical repair of turbine engine blades and vanes. Weld wire has been made in the cobalt and nickel alloys listed in Table 2; adjustment of component dimensions and process parameters were derived to provide wire at various diameters. While the technology is proprietary, the basic process is well known. Arrays of tubes filled with clean spherical powder made by PREP are packed within a larger tube that is capped, evacuated, and sealed. Because the billet contains both solid and powders, billet design must compensate for the differential in their packing densities and the effect this will have while the billet undergoes upset and the powder finally comes to full density. PREP powder is preferred because it produces slag-free welds due to virtual absence of nonmetallic contaminants. In addition, PREP spherical powders pack reproducibly to a density of 65% because particles are free of satellites. The number of tubes processed and the high aspect ratio preclude ramming or step compaction for powder fill. The billet is extruded to a diameter calculated to contain the desired wire diameter within the resultant rod. Weld wires are ultimately released from the rod by immersion in a reagent bath that attacks the container matrix but is inert to the weld wire. The cross section of a weld wire extrusion billet is shown schematically in Fig. 15. After extrusion, the rod section contains an array of evenly spaced wires consolidated to full density. The wire sections are not perfectly circular, but the total variation in cross-sectional area from wire to wire and along the length of any wire does not exceed 5%. An example of an extrusion section and some short samples of released wires are shown in Fig. 16.

Beryllium. Extrusion of beryllium serves as an excellent example of processing a toxic, difficult-to-work metal into useful structural shapes. Beaver and Wikle (Ref 39) have reported comprehensive mechanical properties for beryllium. This metal has a hexagonal close-packed crystal structure and shows the development of a considerable degree of anisotropy when hot-pressed compacts are hot worked by extrusion. Most notably, the ductility is enhanced in the working direction. Gelles et al. (Ref 18) described the texture developed by extrusion, which imparts the increased elongation properties in the extruded direction. Basal planes tend to lie parallel

Table 2 Weld wires fabricated by filled billet technique

Alloy	Normal composition, wt%													Wire diam, mm
	Ni	Co	Cr	Mo	W	Nb	Ti	Ta	Al	Zr	B	C	Others	
B-1900	bal	10	8	6	0.1 max	0.1 max	1	4.3	6	0.08	0.015	0.1	...	1.58
713C	bal	...	12.5	4.2	...	2	0.8	...	6	0.1	0.012	0.12	...	1.27, 0.89, 0.66
René 95	bal	8	14	3.5	3.5	3.5	2.5	...	3.5	0.05	0.01	0.16	...	0.89, 0.71
René 125	bal	10	9	2	7	...	2.5	3.8	4.8	0.05	0.015	0.11	1.5Hf	0.89, 0.71
IN-738LC	bal	8.5	16	1.8	2.5	1	3.5	1.7	3.5	0.05	0.01	0.1	...	1.58, 1.27, 0.89
PWA 1093A	15	bal	25	...	9	0.2	0.2	3	4	0.03	0.005	0.25	1Hf, 0.05Y	1.58
Coast Metals alloy No. 64	5	bal	28	...	20	...	...	...	...	...	0.01	0.8	1V	1.58, 1.27, 0.89, 0.81

Note: This technique, developed by Nuclear Metals Inc., involves coextrusion of spherical (PREP) powders with solid billet matrix components.

to the extrusion axis with the ⟨1010⟩ crystallographic direction aligned. Fracture occurs by cleavage on the (1120) planes.

The much-improved properties obtained by extrusion have resulted in the application of beryllium tubing in the construction of communication satellites. The combination of low density (1.85 g/cm^3) with a high modulus of elasticity and an elongation in excess of 10% in the longitudinal (extrusion) direction make this an ideal structural material for this application.

Loewenstein et al. (Ref 19) have concluded that while it is possible to extrude beryllium uncanned using a glass lubricant, after the method of Ugine-Se′journet (Ref 40, 41), cracking will occur in any section more complex than a simple round rod, unless considerable precautions are taken, especially in die design. In view of the toxicity of the metal and its proclivity to oxidize, any advantage that might be realized in dimensional tolerance control via bare extrusion would seem to be outweighed by the hazards inherent to this practice.

Normal canning for beryllium powder billet construction is low-carbon steel, and this is readily removed by nitric acid when the beryllium core is to be released. The use of pressed powder rather than cast material as the precursor beryllium charge produces much finer grain size in the extruded product (Ref 4). The grain refinement attained contributes to the improved ductility in the extruded product. A detailed review of beryllium extrusion practices is given by Loewenstein (Ref 42).

Ferrous Alloys. Although powder extrusion is intrinsically an expensive process, benefits in improved homogeneity and a more direct route from melt to a wrought, consolidated structure are realized for a number of materials that, until recently, have been wedded to conventional practice involving casting and extensive hot working. An example is the manufacture of fully dense stainless steel tubes for critical service, such as in the production of urea carbamate. Two stainless steel compositions, referred to as 724L and 724LN, are manufactured by Nyby Uddeholm AB, Torshalla, Sweden, and the compositions are given in Table 3 (Ref 43). Inert gas atomized powder is canned in mild steel, with a packing density of 70% of theoretical being achieved. The powder billets are cold isostatically pressed to densify the powder further to 85 to 90% of theoretical density. These are then extruded at 1200 °C using a glass lubricant to reduce friction. Tubing is fully densified for extrusion ratios at or above 4 to 1. The use of powder extrusion to produce tubing and pipe seems to be a logical extension of the technology developed to produce rolled sheet directly from stainless steel and other powders.

Fig. 16 Extrusion cross section and some released weld wires of cobalt alloy (Coast Metals No. 64-Co-30Cr-20W-5Ni-1V-1C). These wires were made from −80 mesh (<175 μm) powder, cf performance of cast material in Fig. 13.

Some high-speed steels and tool steels have been fabricated into rod form via extrusion of powders in order to overcome segregation and alloy limitations encountered during ingot metallurgy processing. These powders possess fine uniform grain sizes and provide improved homogeneity when consolidated. Spherical powders from gas atomizing are canned, and the container is evacuated, outgassed, sealed, and then extruded, as in Fig. 4. Water-atomized powders may be cold pressed and sintered and then extruded, as in Fig. 2. Typically, these materials are extruded at 1100 to 1200 °C, and extrusion pressures of 620 to 690 MPa are required (Ref 44). The improved properties and microstructures justify the expense of hot isostatic pressing or extrusion, and in many cases savings are achieved when the extensive and energy-intensive operations of rolling and forging from large ingot sections are bypassed (Ref 45).

Composite Materials. Powder metallurgy coupled with extrusion offers a unique method of fabrication of both macro- and microcomposite materials. Erich (Ref 46) reviewed P/M applications for the production of metal-matrix composites. Advantages of a P/M approach in comparison with fusion metallurgy methods include low processing temperatures, unique materials, use of particulates or whiskers, and a wider capability for in situ formation of strengthening phases. Canned extrusions have been developed to fabricate powder mixtures and multicomponent systems that include arrays of both solid sections and powders. The reinforcing agent may be formed in situ by metallurgical reaction, or it may be introduced as a separate component during powder blending. Extrusion serves to align the reinforcing agent as well as consolidate the product.

Intimate mixtures of metal powders sealed within an evacuated canister have been extruded to make fibers having high aspect ratios. Rebundling of the primary extrusion into a second can, such as the method of Klein (Ref 47), is capable of producing filaments that are extremely fine. Depending on the application, these may be released by differential chemical attack or the composite structure formed may be the desired product.

Composite rods extruded from copper-niobium and copper-tin bronze-niobium powder mixtures are described by Foner (Ref 48, 49). These rods were drawn to wires using large section reductions, and the niobium particles were coreduced proportionally to small diameters and long lengths. Appropriate heat treatment of the bronze matrix wires resulted in the formation of the intermetallic compound Nb_3Sn having a β tungsten (*A*15) structure and excellent hard superconducting properties.

Although the superconducting filaments are long, they are discontinuous and therefore might not be expected to impart superconducting properties to the wire. However, extensive reduction in cross section will provide small filament diameters with concomitantly finer interfilament spacing; a careful selection of particle sizes and section reduction schedules can provide interfilament spacing that is sufficiently small to permit electron pair tunneling as described in the theory of Bardeen et al. (Ref 50). The highly desirable fine diameters of the filaments together with this tunneling effect combine to produce excellent type II superconducting properties in the wire when it has received suitable processing. In addition to the fabrication of superconductors, this type of structure has other interesting applications. Foner (Ref 49) cites the possibility of making magnetic wire materials with this method, and indeed Levi (Ref 51) describes the formation of such wire structure involving filaments of iron in a copper matrix made by successive wire drawing and bundling steps. Although Levi uses neither powders or extrusion, the principle for successful coreduction of mechanically compatible materials is demonstrated, and structures offering a similar degree of reinforcement are feasible by the method of Klein (Ref 47).

Powder mixtures and powders mechanically mixed with chopped fibers can be extruded to produce fiber-reinforced structures. If all the components are metallic and possess similar flow strengths at the working temperature, they will elongate similarly as described by Roberts et al. (Ref 5, 22) and Klein (Ref 47). Ultimately, long fibers of one species in a matrix of another may provide a potent strengthening mechanism. Where the fibers and matrix have dissimilar elastic moduli, with that of the fiber being greater than that of the matrix, a complex stress distribution is developed during axial loading where the shear stress developed at the fiber/matrix interface and the axial tensile stress in the fiber may be represented graphically as shown in Fig. 17.

Table 3 Stainless steel compositions (wt%) extruded into tube by Nyby Uddeholm AB

Steel	C	Si	Mn	P	S	Cr	Ni	Mo	N_2	Fe
724L	0.015	0.45	0.80	0.02	0.01	17.5	14.5	2.35	0.09	bal
724LN	0.015	0.25	1.70	0.02	0.015	25.0	22.0	2.10	0.12	bal

Source: Ref 43

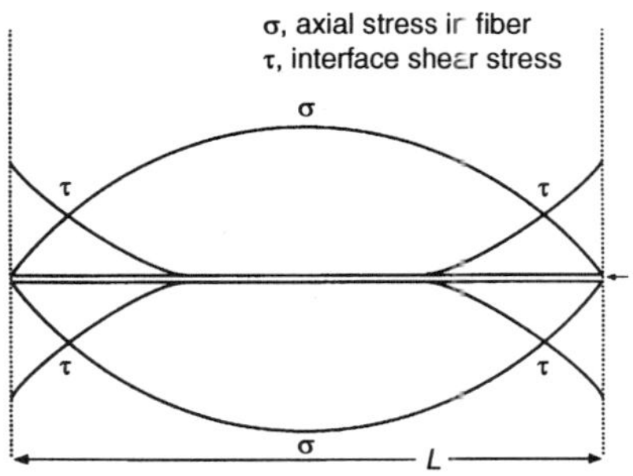

Fig. 17 Stress distribution along embedded fiber caused by uniaxial loading

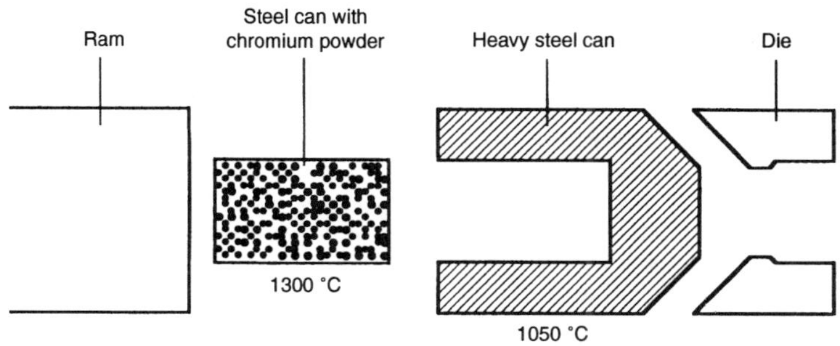

Fig. 18 Multitemperature extrusion of chromium

Essentially, this figure shows that the fiber carries a tensile load produced by the shear stress developed along the fiber matrix interface. High allowable interface shear stresses permit the fiber to carry high axial tensile stresses. Shear stresses above some critical value at the fiber ends cause the matrix to deform plastically. To obtain the best use of the high-strength fiber, the plastic zone in the matrix should not extend from the fiber ends to its midpoint before the fiber is overstressed. A critical fiber length L_c is given by:

$$L_c = \sigma_f d/2\tau_o \quad \text{(Eq 6)}$$

where σ_f is the fracture stress of a fiber of diameter d in a matrix having a shear yield stress τ_o. When the fiber length is greater than L_c, composite failure occurs by fiber fracture and the system has developed its full strengthening potential. The critical aspect ratio L_c/d for fibers determines the minimum coreduction of section ratio; for all extrusion processing this will be the minimum value for R, the extrusion reduction ratio.

Erich (Ref 46) cites an example of in situ composite fabrication by extrusion where tungsten powder was blended into a nickel powder matrix, compacted, and sintered. This blend can be sintered with a transient liquid phase technique. The sintered billet is then extruded and may be further worked by rolling or drawing to produce a composite with a microstructure of highly elongated tungsten grains in a nickel matrix. The resulting product is reported to equal the strength of a directionally solidified material of similar chemistry.

Innovative Applications

Extrusion processes have been developed that serve special purposes or produce special products. In this regard, they may be considered to be innovative applications of the generic extrusion process.

A most ingenious development reported by Loewenstein et al. (Ref 3) involved the fabrication of materials that are usually considered to be difficult to work and are therefore not made in wrought forms. They cite the manufacture of chromium tubes and rods by a technique referred to as multitemperature extrusion in which the total extrusion billet is assembled rapidly at the press from separate components that have been heated to different temperatures. The parts exhibit a similar stiffness when the assembly upsets and is extruded so that good proportional coreduction and dimensional control are obtained. The success of the procedure depends in large part on the speed of the press and the skill of the press crew. The billet must be assembled and pushed rapidly before the effects of temperature decline and normal thermal conduction are significant. The method is shown schematically in Fig. 18. It is sufficiently versatile that it has been used to consolidate ceramics such as UO_2 heated into the temperature range 1750 to 2000 °C required for plastic flow and coextruded within a relatively cold steel can heated to 700 °C.

In an early development of electric lamp filament manufacture, tungsten powders mixed with organic binders or amalgams of mercury, bismuth, and cadmium were forced under pressure through diamond dies to make long, fragile "strings." These strings were subsequently heated to drive off the volatile components and to sinter the material in preparation for hot swaging, which was the major means of densification and section reduction (Ref 52).

Recently, the new technology of metal injection molding (MIM) has also emerged for making precision metal parts via injection molding and sintering (Ref 53–57). Mixtures of metal powders and polymers are forced under pressure into accurately sized molds so designed and proportioned as to provide a component that will shrink to the desired size and density after a sintering cycle. These developments, which initiated in the military and aerospace markets, now have moved into the wider civilian sphere of computer peripheral devices, business machines, automotive, and biomedical industries. Superalloys, stainless steels, and cemented carbides have been processed into precision metal parts using conventional screw and piston plastic injection molding equipment.

Metal powders and organic polymer mixtures have been coprocessed to provide high-performance plastic parts that are lighter than metal for the automotive industry. In other instances, the physical properties of the plastic, such as electrical conductivity, thermal conductivity, electromagnetic shielding, and static discharge capabilities, may be modified. Various workers have reported on the influence of particle shape, size, and distribution on the performance of these composites (Ref 58–61). It is common practice to extrude or injection mold these materials because these methods provide continuous production and superior economics compared with compression molding.

Hydrostatic extrusion is a variant of the process that involves extrusion by the action of a liquid medium surrounding the billet being pressurized rather than direct axial loading by the ram. The liquid is pressurized to an extent that exceeds the flow stress of the billet material, and this is forced through the die. The principle was first expounded by Robertson (Ref 62) in 1894, but it was never applied until the work of Bridgman (Ref 63) and Pugh (Ref 64). The technique is shown schematically in its simplest form in Fig. 19, where the extrusion exits into the atmosphere. Later developments have the extrusion passing from a high-pressure to a low-pressure chamber, which permits the processing of brittle materials that might otherwise break up at the

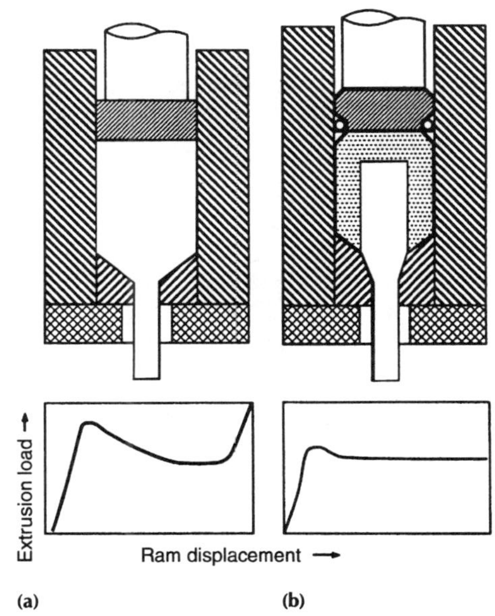

Fig. 19 Comparison of load profiles of simple hydrostatic extrusion and conventional extrusion. (a) Conventional. (b) Hydrostatic with conical die

die exit. The elevated pressure at the die exit in the two-chamber approach improves the ductility of the extrudant; however, in order for extrusion to proceed, the pressure in the first chamber must be increased over that which would be applied in the simple, single-chamber operation.

Originally this was a cold process using oils as working fluids. More recent developments have used "quasi-fluids" operating at higher temperatures. These are actually solids that are much softer than the billet at the deformation temperature. Examples of these media are NaCl, CaF_2, and assorted glass formulations operating just below their melting points (Ref 65). It is reported that extrusion reduction ratios of 30 to 1 and 40 to 1 at pressures of 1240 and 2068 MPa, respectively, have been obtained with a tool steel and a high-speed steel at 1000 °C. At room temperature, the best reduction ratio attained was 3 to 1 (Ref 65, 66).

No specific references have been found to the application of this technique to metal powders, but with proper canning there is no obvious reason why canned powders could not be processed by this method.

Continuous Extrusion Processes

The extrusion processes discussed up to this point are discontinuous; each involves a high-temperature, high-pressure operation performed on a single billet that is held in a closed-container system. Methods have been developed to provide continuous extrusion where the working principle employed involves driving a continuous supply of stock up to the die orifice by frictional gripping forces (Ref 67–69). The die is positioned at the far end of an open chamber admitting the continuous material feed. An ingenious method for making wires developed by Western Electric Company drives the feedstock with viscous drag forces exerted by a pumped viscous fluid (Ref 70, 71).

The "Conform" process (Fig. 20) achieves 160 to 1 reduction ratio in aluminum where 9.5 mm diam feedstock is driven by frictional contact with a grooved wheel up to a die arrangement to produce 0.75 mm diam wire (Ref 68). The machine capability to generate the wire at an exit speed of 3540 m/min was considered to be too fast for conventional spooling equipment. Pardoe (Ref 69) mentions that two of three developed Conform machines can accept either powder feedstock or rod stock. In this manner, either loose powder or compacted slugs may be fed into the machine.

Materials processed have included commercially pure aluminum and copper, alloys of aluminum, and oxide-dispersion-strengthened aluminum and copper. Feedstock forms included granules, wire clippings, and atomized powder. Finer powder sizes produced material with higher strength than coarse feedstock. Bowers (Ref 72) reports that the use of a powder feedstock reduces the pressure developed markedly as compared with solid metal feedstock. In addition, the use of powder provides the opportunity to feed blends of different metal powders for production of composite wires.

The "Linex" process (Fig. 21) as described by Black et al. (Ref 67) involves a similar approach where rectangular section stock is held and forced forward toward the die by opposed continuous loops of gripper blocks arranged like caterpillar treads. Reduction ratios of 30 to 1 are reported. Avitzur (Ref 72) presents an analytical review of these concepts, consolidating them into a group that he refers to as "Extrolling"—an amalgamation of rolling and open-chamber extrusion.

Conclusions

While elevated-temperature direct extrusion is now a mature metalworking technique, its utility has not diminished with time. Loose powders or precompacted slugs may be processed this way as well as cast billets, and the canned extrusion and filled billet approaches have extended the application of extrusion to powder metals for specific purposes. Examples of the benefits obtained from canned metal powder extrusions include the formation of composites and reinforced materials, near-net sections, sometimes complex in shape, that may be released by chemical removal of a coextruded matrix, sound rods and sections in materials of limited ductility, and the safe containment of materials that may be toxic or reactive.

The mechanisms for powder extrusion have been studied extensively and reported by Sheppard and coworkers using upper-bound analyses and hodographic depictions of the process as material traverses the extrusion deformation zone. They demonstrated that powder is compacted to near-full density before the onset of extrusion, although at this stage the particles remain poorly bonded. As the densified material passes through the extrusion deformation zone, redundant work is expended in welding, breaking welds, and reforming them between particles while they are deforming; the emergent material has suffered considerable shear deformation, and the formation of metallurgically sound interparticle bonds produces full density and wrought properties. The redundant work factor in the process is a function of particle size so that fine powder with a large surface-area-per-unit volume requires a higher extrusion pressure than coarse powder.

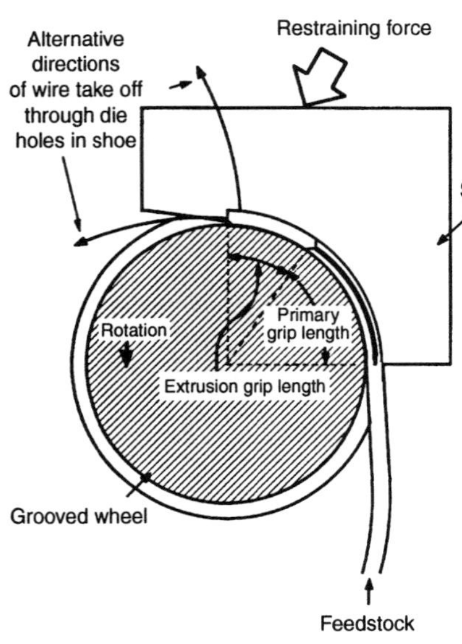

Fig. 20 Conform continuous extruder

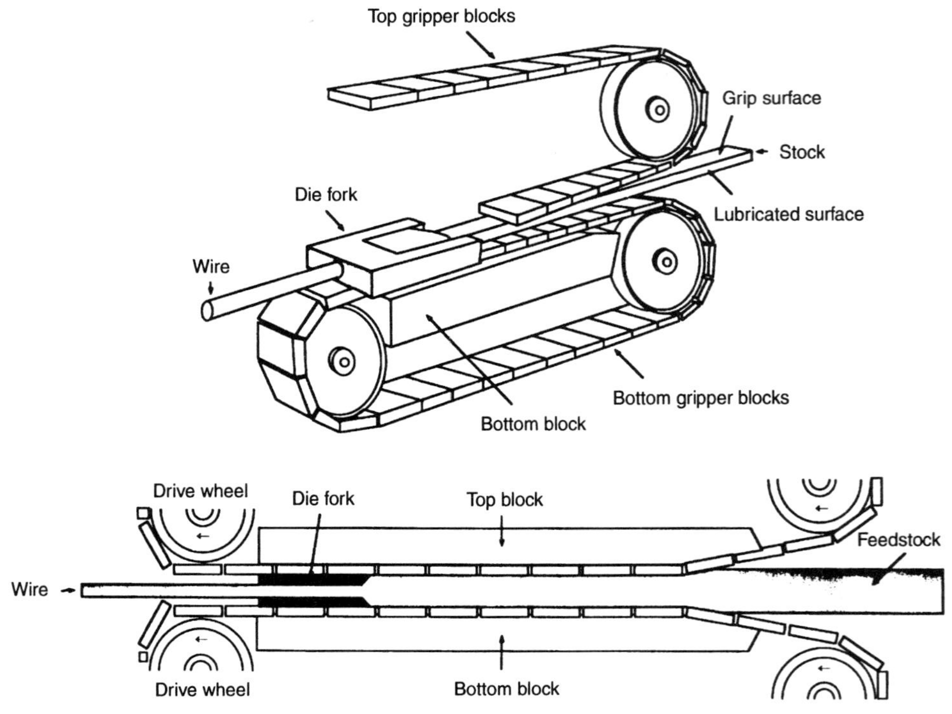

Fig. 21 Linex process

However, extrusion of powders generally requires lower pressures than cast billets of the same material because the final properties characteristic of the wrought material are developed only after it has passed through the deformation zone.

When powder mixtures or powder-solid combinations are to be extruded, successful coreduction of section depends on the metallurgical and thermomechanical compatibility of the various components. When flow stresses are well matched and the operating temperature does not cause the formation of liquid phases, uniform and proportionate coextrusion will be achieved. Within the broad confines of these simple criteria, it is possible to construct billet assemblies involving multiple components to make products that might otherwise be obtained with great difficulty, if at all.

The extrusion of metal powders is a useful industrial technique with a broad variety of applications that have been extended in part by inventive process and billet design. The obtainable material properties and product characteristics together with the relatively short process cycle involving melt-powder-extruded compact ensure that it will continue to be used well into the future.

REFERENCES

1. C.E. Pearson and R.N. Parkins, *The Extrusion of Metals,* 2nd ed., London, Chapman and Hall, 1960
2. K. Lange and H. Stenger, *Extrusion: Process, Machinery, Tooling,* American Society for Metals, 1981
3. P. Loewenstein, L.R. Aronin, and A.L. Geary, *Powder Metallurgy,* W. Leszyski, Ed., Interscience, 1961, p 563–583
4. F.V. Lenel, *Powder Metallurgy—Principles and Applications,* Metal Powder Industries Federation, 1980
5. P. Roberts, Tech. Paper MF 76-391, SME, 1976
6. G. Sachs and W. Eisbein, *Mitt. Deut. Mater. Pruf. Anst.,* Vol 16, 1931, p 67–96
7. unpublished data, Nuclear Metals, Inc.
8. G.E. Dieter, *Mechanical Metallurgy*, 3rd ed., McGraw-Hill, 1986
9. T. Sheppard and P.J.M. Chare, *Powder Metall.,* Vol 15 (No. 29), 1972, p 17–41
10. P.J.M. Chare and T. Sheppard, *Powder Metall.,* Vol 16 (No. 32), 1973, p 437–458
11. P.J.M. Chare and T. Sheppard, *Int. J. Powder Metall. Powder Technol.,* Vol 10 (No. 3), 1974, p 203–215
12. T. Sheppard and H. McShane, *Powder Metall.,* Vol 19 (No. 3), 1976, p 121–125
13. T. Sheppard and H. McShane, *Powder Metall.,* Vol 19 (No. 3), 1976, p 126–133
14. T. Sheppard, H.B. McShane, M.A. Zaidi, and G.H. Tan, *J. Mech. Work. Technol.,* Vol 8, 1983, p 43–70
15. A. Greasely and T. Sheppard, *Proc. 2nd Int. Conf. Consolidation of Particulate Materials* (Brighton), Aug 1975
16. A. Kumar, P.C. Jain, and M.L. Menta, *Powder Metall. Int.,* Vol 19 (No. 3), 1987, p 15–18
17. B. Avitzur, *Metal Forming: Processes and Analysis,* McGraw-Hill, 1968, p 250–292
18. S.H. Gelles, V. Nerses, and J.M. Siergiej, *J. Met.,* Vol 15 (No. 11), 1963, p 843–848
19. P. Loewenstein, E.F. Jordan, and J.M. Siergiej, "Extrusion of Unclad Beryllium," Internal Report, Nuclear Metals, Inc., 1961
20. A.P. Green, *Philos. Mag.,* Vol 42, 1951, p 365
21. B. Avitzur, *Wire J.,* Vol 3, Aug 1970, p 42–49
22. J.A. Roberts and P.R. Roberts, U.S. Patent 3,698,863, 1972
23. R. Irmann, *Tech. Rund.,* Vol 41 (No. 36), 1949, p 19
24. *Symposium on Powder Metallurgy,* The Iron and Steel Institute, London, 1954, 1956
25. K.M. Zwilsky and N.J. Grant, *J. Met.,* Vol 9 (No. 10), 1957, p 1197–1201
26. W.S. Cremens and N.J. Grant, *Proc. ASTM,* Vol 58, 1958, p 714–730
27. G.S. Ansell, *Trans. AIME,* Vol 215, 1959, p 249–250
28. J.S. Benjamin and T.E. Volin, *Metall. Trans.,* Vol 5 (No. 8), 1974, p 1929–1934
29. J.S. Benjamin and M.J. Bomford, *Metall. Trans.,* Vol 8A, 1977, p 1301–1305
30. G. Jangg, F. Kutner, and G. Korb, *Powder Metall. Int.,* Vol 9 (No. 1), 1977, p 24–26
31. K.H. Kramer, *Powder Metall. Int.,* Vol 9 (No. 3), 1977, p 105–112
32. A.V. Nadkarni and J.E. Synk, *Powder Metallurgy,* Vol 7, *ASM Handbook,* American Society for Metals, 1984, p 700–716
33. P.S. Gilman and J.S. Benjamin, *Powder Metallurgy,* Vol 7, *ASM Handbook,* American Society for Metals, 1984, p 722–727
34. J.E. Synk, *Powder Metallurgy,* Vol 7, *ASM Handbook,* American Society for Metals, 1984, p 716–720
35. Anon, *Powder Metallurgy,* Vol 7, *ASM Handbook,* American Society for Metals, 1984, p 720–722
36. F.H. Froes and R.G. Rowe, *Rapidly Solidified Alloys and Their Mechanical and Magnetic Properties,* B.C. Giessen et al., Ed., Symp. Proc., Vol 58, Materials Reserarch Society, 1986, p 309–334
37. B.L. Ferguson, *Powder Metallurgy,* Vol 7, *ASM Handbook,* American Society for Metals, 1984, p 646–656
38. G. Friedman, *Int. J. Powder Metall. Powder Technol.,* Vol 16 (No. 1), 1980, p 29–35
39. W.W. Beaver and K.G. Wikle, *Trans. AIME,* Vol 200, 1954, p 550–573
40. British patents, 607,285; 661,555; and 663,357
41. J. Séjournet, *Iron Coal Trades Rev.,* Vol 165, 1952, p 963
42. B. Loewenstein, *Beryllium Science and Technology,* D. Webster et al., Ed., Vol 2, Plenum, 1979, p 67–82
43. C. Aslund, *Prog. Powder Metall.,* Vol 39, 1983, p 543–558
44. N. Kawai and H. Takigawa, *Met. Powder Rep.,* Vol 37 (No. 5), 1982, p 237–240
45. R.S. Carbonara, State of the Art Review of Rapid Solidification Technology (RST), Report MCIC 81-45, Metals and Ceramics Information Center, 1982, p 36
46. D.L. Erich, *Int. J. Powder Metall.,* Vol 23 (No. 1), 1987, p 45–54
47. J.L. Klein, U.S. Patent 3,413,707, 1968
48. S. Foner, *Prog. Powder Metall.,* Vol 38, 1982, p 107–114
49. S. Foner, *Met. Powder Rep.,* Vol 39, 1984, p 583
50. J. Bardeen, L.N. Cooper, and J.R. Schrieffer, *Phys. Rev.,* Vol 108, 1957, p 1175
51. F. Levi, U.S. Patent 3,029,496, 1962
52. W.D. Jones, Fundamental Properties of Powder Metallurgy, Edward Arnold, 1960, p 242–602
53. R. Billiet, *Prog. Powder Metall.,* Vol 38, 1982, p 45–52
54. R. Billiet, *Prog. Powder Metall.,* Vol 41, 1985, p 723–742
55. A.R. Erickson and R.E. Wiech, Jr., *Powder Metallurgy,* Vol 7, *ASM Handbook,* American Society for Metals, 1984, p 495–500
56. L.F. Pease III, Gorham International Inc. Conf. (Monterey, CA), 1986
57. P. Crook, R.D. Rivers, and H.J. Klein, *Ind. Heat.,* May 1987, p 12–15
58. A. Milliaris and D. Turner, *J. Appl. Phys.,* Vol 42, 1971, p 614
59. L.E. Nielsen, *Mechanical Properties of Polymers and Composites,* Vol 1 (No. 2), Marcel Dekker, 1974
60. S.K. Bhattacharya and A.C.D. Chaklader, *Polymer Plast. Technol. Eng.,* Vol 19 (No. 1) 1982, p 21
61. F. Bueche, *J. Appl. Phys.,* Vol 43, 1972, p 4837
62. J. Robertson, British Patent, 19,356, 1894
63. P.W. Bridgman, *Studies in Large Plastic Flow and Fracture,* McGraw-Hill, 1952, p 362
64. H.Ll.D. Pugh and A.H. Low, *J. Inst. Met.,* Vol 93, 1964–1965, p 201–217
65. *Extrusion: Process, Machinery, Tooling,* American Society for Metals, 1981, p 100–101
66. D. Green, *J. Inst. Met.,* Vol 99, 1971, p 76–80
67. J.T. Black, W.G. Voorhes, and D. Breneiser, Tech. paper MF 76-141, SME, 1976
68. E. Hunter, Tech. paper MF 76-407, SME, 1976
69. J.A. Pardoe, *Powder Metall.,* Vol 22 (No. 1) 1979, p 22–28
70. F.J. Fuchs and G.L. Schmehl, *Wire J.,* Vol 6, Nov 1973, p 53–57
71. P. Andrus and G.L. Schmehl, Tech. paper MF 76-409, SME, 1976
72. J. Bowers, *J. Met.,* Vol 38 (No. 11), 1986, p 63
73. B. Avitzur, Tech. paper MF 75-140, SME, 1975

Forging and Hot Pressing

Howard A. Kuhn, Concurrent Technologies Corporation (CTC)

POWDER FORGING AND HOT PRESSING processes are similar in terms of their external physical characteristics. Each process involves the simultaneous application of pressure and heat to consolidate powders to full, or nearly full, density. Both processes occur within a die that gives shape to the final densified component.

At a more detailed level, forging and hot pressing are significantly different. Forging involves the deformation of a powder preform in which considerable lateral flow occurs (Fig. 1a). Forging is carried out on high-speed presses that are automated for mass production. Hot pressing, on the other hand, involves compression of loose powder or a powder compact fitted tightly in the die cavity, similar to repressing, resulting in very little lateral deformation, as shown in Fig. 1(b). Hot pressing generally is carried out at very low deformation rates in hydraulic presses.

At an even finer scale, viewing the localized deformation of a representative unit of material containing a pore (or void space) between particles, the lateral flow of forging leads to considerable shear, which breaks up any oxide surfaces on the particles and enhances bonding (Fig. 2a). In hot pressing, pores and voids simply collapse under axial deformation; see Fig. 2(b) (Ref 1). Hot pressing is similar in this respect to cold die compaction.

Because of the differences in speed and deformation mode between forging and hot pressing, the metallurgical mechanisms of densification are different. Forging relies primarily on plastic deformation to close the voids during the few milliseconds that the material is under pressure. Hot pressing involves plastic deformation, creep, and diffusion mechanisms to cause localized flow of material into the pores.

Forging involves only axial compressive stress while the material is undergoing lateral flow, or upsetting. When the forged material reaches the die wall, however, lateral pressure acts on the material much like repressing (Fig. 1a). In hot pressing, the material is subjected to lateral pressure p' in addition to axial pressure p, similar to repressing (Fig. 1b and 2b). The numerical difference between p and p' diminishes as full density is reached. Closure of pores by plastic deformation becomes more difficult as the lateral and axial pressures approach equality; i.e., a hydrostatic stress state. Therefore, it is important in forging that most of the densification occurs by shear before the material reaches the die walls.

While both powder forging and hot pressing can consolidate most powder materials to full density, forging is applied primarily to high production parts in ferrous alloys, such as automobile connecting rods, transmission parts, and hand tool components. Hot pressing is used primarily for specialty materials such as tungsten carbide, beryllium, and ceramics. Hot pressing is also used as a consolidation process to produce billets for further processing by conventional forging or extrusion of materials such as high strength P/M aluminum alloys, tool steels, and superalloys.

Successful application of forging and hot pressing involves careful consideration of powder preparation (compaction, heating, and atmosphere protection) and forming process parameters (preform design, heating, tooling, and equipment). Computer modeling of these processes has been introduced recently to accelerate the development of the proper process parameters and tooling for new applications and the solution of problems in existing processes. The important process features are described here for both powder forging and hot pressing, along with specific applications and materials.

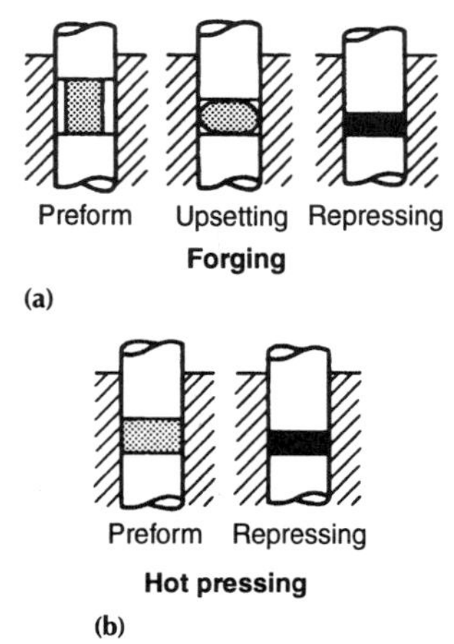

Fig. 1 Deformation mode for (a) forging and (b) hot pressing

Powder Forging

Powder Preparation

The most extensive application of powder forging is automotive drive train parts using high strength steels. Water-atomized low alloy steel 4620 is commonly used as well as iron-copper-carbon blended elemental powders (Ref 2). Powder characteristics, such as size and shape, are unimportant in the forging deformation process itself, but they may have important effects on the compaction and sintering processes prior to forging. For example, whether the powders are spherical or irregularly shaped, fine or coarse in size, they will lose their identity through the extensive deformation and accompanying densification during forging, and will have little effect on the deformation process. For the preform compaction process prior to forging, however, irregularly shaped powders are preferred because of their ability to interlock during compaction and produce compacts with sufficient green strength for handling. Coarse powders might also be preferred over fine powders because of ease of flow during compaction die filling, and because they present less surface area for contamination during sintering or heating for forging.

The chemical composition of the powders involved in powder forging is important, primarily

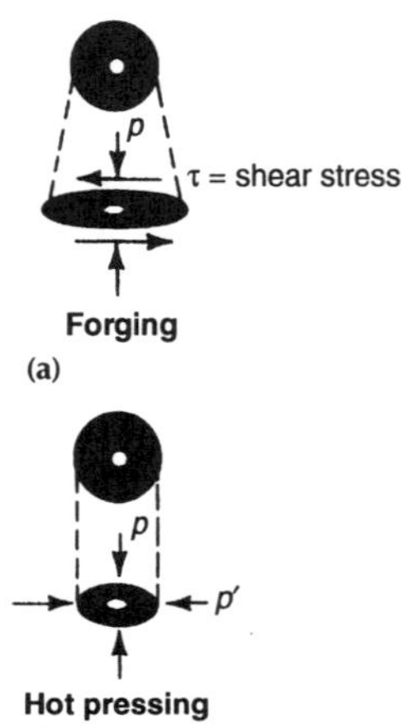

Fig. 2 Stress conditions for (a) forging and (b) hot pressing

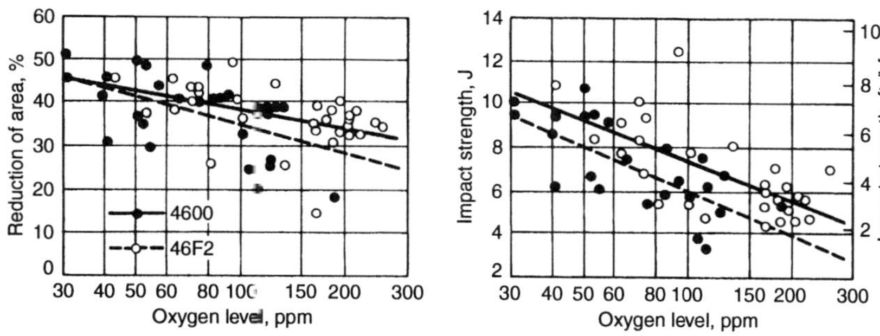

Fig. 3 Variation of reduction of area and impact strength as a function of oxygen content. Source: Ref 3

in terms of oxide content, because the volume fraction of contaminants strongly influences the final properties of the powder forged component. If the oxide appears on the surface of the particles, it can be readily reduced by proper sintering practice or broken up and dispersed by shear deformation during forging.

Preforms for forging are fabricated primarily by conventional die compaction, although dry bag isostatic compaction has also been used for some applications. The overall density of the preform has no significant effect on the density of the forged parts, but variations in density from region to region of the preform are often designed to control metal flow and avoid defects during the forging process. Another consideration is the influence of preform density on the sintering process. Preforms of low density contain extensive interconnected porosity, allowing the sintering furnace atmosphere to reduce larger amounts of surface oxides on the powder particles. On the other hand, lower density preforms are more prone to internal oxidation and carburization during exposure to air while the preform is transferred from the furnace to the forging press. For this reason, protective coatings are generally applied to the preform before heating.

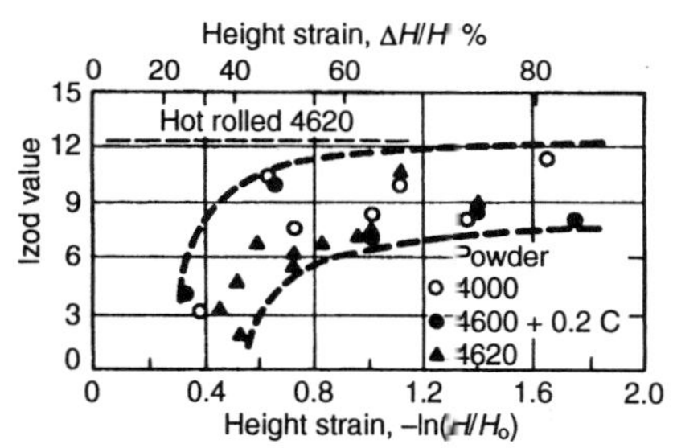

Fig. 4 Increase in impact resistance with increasing forging deformation of sintered nickel steel (4600 series) powder

Control of the preform weight is critical because hot forging is carried out in a trap die without flash, as shown in Fig. 1(a). Excessive preform weight may lead to tool breakage or stalling of the press. Conversely, underweight preforms will not achieve full densification by forging and may not fill extreme corners of the die.

Sintering is the final critical step in preparation of preforms for forging. Sintering not only bonds the powder particles by diffusion, it also provides a protective or reducing atmosphere to prevent contamination and reduce the levels of oxide contamination that may be present in the particles. The oxygen level of the sintered preform, which depends on the sintering temperature, sintering atmosphere, dew point, and furnace type, determines the oxygen level of a forged part. The oxygen level of sintered preforms decreases with higher sintering temperature and lower dew point. Mechanical properties in general, and dynamic properties (impact and fatigue strength) in particular, are strongly influenced by the oxygen level of the forged part. Low oxygen levels lead to higher dynamic properties because the metallurgical bonds developed between particles during forging are stronger if the particle boundaries are free of oxides; see Fig. 3 (Ref 3).

Forging Process

A major key to successful powder forging is proper preform design because it has a significant effect on the metal flow and distribution of stresses in the material during forging. These factors, in turn, affect the densification and probability of fracture. In general, sufficient metal flow must take place to achieve full density and good bonding across collapsed pores. However, increasing the amount of metal flow also increases the possibility of fracture. Thus, preform shape design involves a tradeoff between the lower limit of deformation to achieve the required properties, and the upper limit that would cause fracture.

As shown in Fig. 1(a) and 2(a), the deformation during forging involves considerable lateral flow and shearing of the powder particles. This shearing action causes any oxide films on the powder particles to be broken up, exposing clean metal, and enabling a strong metallurgical bond across collapsed pore interfaces. As a result, the dynamic properties of the material are enhanced. Figure 4, for example, shows that the impact energy of as-forged alloy steel powder preforms increases steadily with increasing deformation up to ~60% reduction. Similarly, Fig. 5 shows that the fatigue limit for powder forged 4620 steel increases with increased deformation (Ref 4).

Large deformation of sintered powder material can easily lead to fracture because tensile stresses usually develop, and pores in the material provide many sites for concentration of these stresses, leading to fracture. To overcome this limitation a fracture criterion has been developed for predicting fracture during powder forging. Based on the occurrence of fracture during

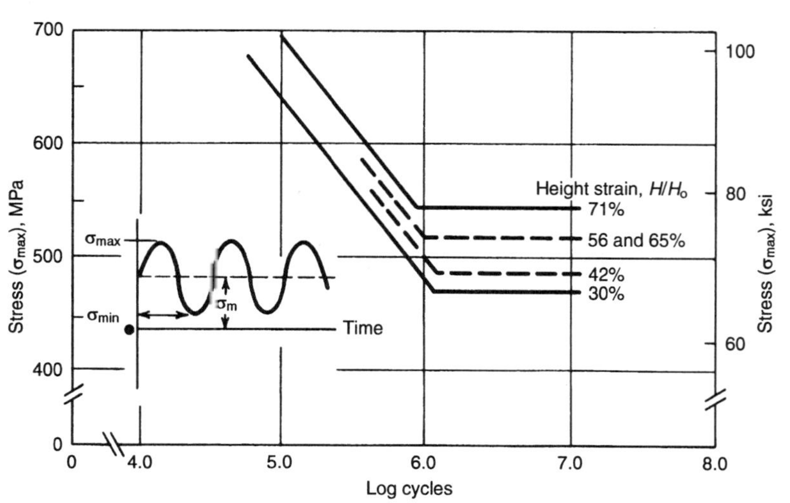

Fig. 5 Axial fatigue of P/M forged 4620 steel for various levels of forging deformation. Fatigue limit increases as deformation level (height strain) increases. Source: Ref 4

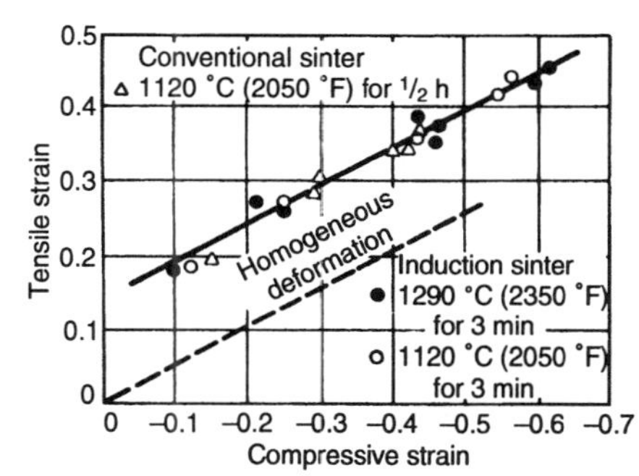

Fig. 6 Tensile strain vs. compressive strain at fracture of 4620 steel powder cylinders and sintered under various conditions. Dashed line represents homogeneous deformation (zero tensile stress).

upset compression testing, a locus of surface strains at fracture can be generated, as shown in Fig. 6 (Ref 5). For a given material, strain combinations in the material above this line will lead to fracture, while strain combinations below the line are safe. Such fracture lines can be generated for any material.

To illustrate the important role of preform design on defect-free powder forging, consider the part shown by shading within the tooling configuration in Fig. 7. This part is axisymmetric, consisting of an outer rim (upper section) formed by the upper punch, and a hub (lower section) formed by the lower punch. A flange is formed between both punches, and an outer die and core rod complete the tool set. A preform containing the rim and hub features is not considered because metal flow would be limited, leading to incomplete densification. Rather, the preform for this part would simply be a cylinder with a hole at the center. Forging the part from a cylindrical preform will involve various combinations of back extrusion of metal into the rim, forward extrusion into the hub, and lateral (radial) flow in the flange. However, as seen in Fig. 8, for a given volume of material, the inside and outside diameters of a cylindrical preform can have a variety of combinations. Using the fracture criterion in Fig. 6 and knowledge of the stresses in the various modes of deformation, the proper preform can be determined.

In Fig. 8(a), the preform leads to successful forging because radial expansion of the outer surface is prevented, forcing inward flow to form the hub. This inward flow is in compression and does not lead to fractures. Meanwhile, the rim forms by backward extrusion without radial expansion and is also safe from fracture. The preform in Fig. 8(b) allows considerable radial expansion and flow of material around the upper corner radius into the rim, which easily leads to fracture on the inside of the rim. In Fig. 8(c), the preform is constrained from lateral flow on both the inner and outer diameters; the resulting forward extrusion into the hub section causes large tension on the surface of the hub, leading to fracture. The preform in Fig. 8(d), like Fig. 8(b), leads to fracture on the inside surface of the rim as material flows around the upper corner of the punch into the rim.

It is clear that, even though a cylindrical preform is the obvious choice for this part, the specific dimensions of the preform must also be determined. Subtle changes in the preform dimensions mark the difference between successful and unsuccessful forging. This observation is true in preform design for nearly every part.

For more complex shapes, preform design is a major challenge, since a wide variety of deformation modes and combinations may occur. Preform design for connecting rods, for example, requires specification of the preform dimensions for the circular sections at the pin end and the crank end, as well as the beam section connecting these ends. Lateral flow and extrusion occur locally in all three sections, and material flow may occur across the intersections as well. Preform design for this complex case requires careful determination of the weight of material in each section and detailed analysis of the localized metal flow in each section. Unlike most preforms, it is necessary to specify different densities in each of the three sections of the preform to ensure full densification and avoid defects in the finished connecting rods.

The tooling for powder forging is based on the trap die concept, an example of which is shown in Fig. 7. The material is completely trapped between the punches and die, with no flashland. While the trap die concept produces a near-net shape that does not require removal of excess metal in a flashland, it does lead to high stresses on the tooling. Combined with the high temperatures of forging, these stresses and metal flow lead to high wear rates on the punches and die. The most commonly used material for the dies is H13, a hot-working die steel. With proper preform design and cooling of the tools between press strokes, 50,000 to 100,000 parts can be produced before die refurbishment is required.

The temperature of the preform as it enters the forging process influences the mechanical properties of the forged parts as well as the life of the forging tooling. Higher forging temperatures enhance densification of the forged part because plastic flow occurs more easily, as shown in Fig. 9 (Ref 6). On the other hand, higher forging temperatures lead to greater die wear. Typically, steel powder preforms are forged at 980 °C (1800 °F) as the optimum temperature to minimize die wear and ensure part densification.

The speed of forging also affects the quality of parts produced by powder forging. Since the preform is porous, rapid cooling of the hot preform surface by contact with the colder die surface will lead to residual porosity at these surfaces. To minimize this effect, the contact time between the preform and die must be as short as possible. For this purpose, screw presses with a very stiff frame are widely used for powder forging. The screw press motion ensures rapid withdrawal of the forging punch from the die, and the stiff frame minimizes contact time between the part and die at the peak load. The total contact time for forging by a screw press is ~30 ms, while the contact time for a typical mechanical crank press is 80 ms. This difference has a very large effect on local cooling of the preform and, consequently, residual surface porosity in the forged part.

Applications

Forging of P/M preforms to near-net shape or net shape parts has several advantages: material utilization is increased by eliminating or substantially reducing machining operations, energy efficiency is greater than conventional forging because operations such as hot rolling of bar stock are eliminated, and powder forging uses lower temperatures than conventional forging for the same alloys. As an example, a machine gun accelerator is conventionally produced by hot forging billets of 4340 steel to a rough shape in six operations, followed by 27 machining operations. Powder forging of the same part consists of compacting a preform from atomized 4600 steel powder, sintering the preform, and hot forging, followed by just seven machining steps. This leads to a net cost savings of >50% compared to the conventional approach.

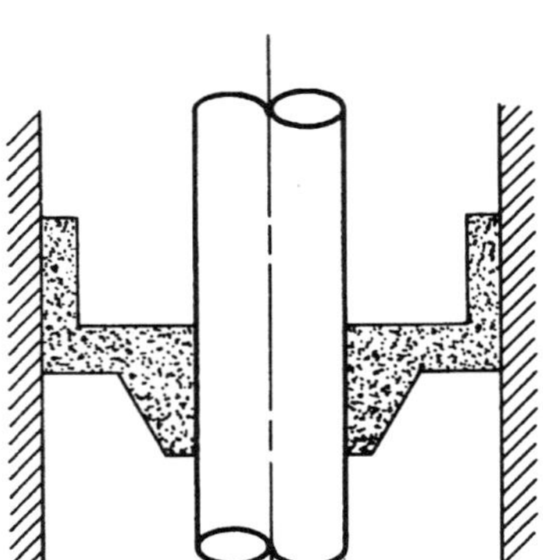

Fig. 7 Prototype part for illustration of preform design

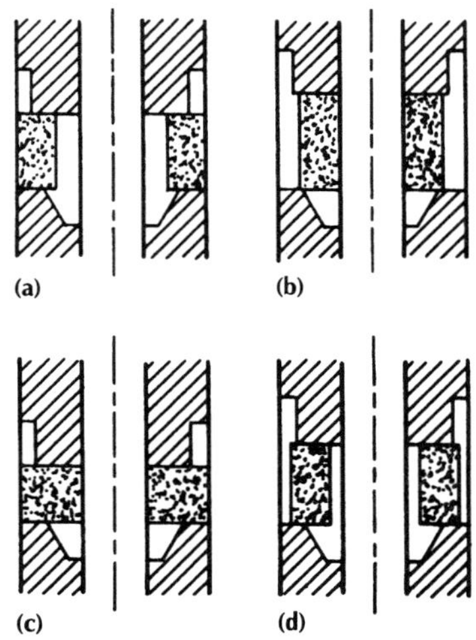

Fig. 8 Preform options for forging the part shown in Fig. 7. See text for discussion of (a) through (d).

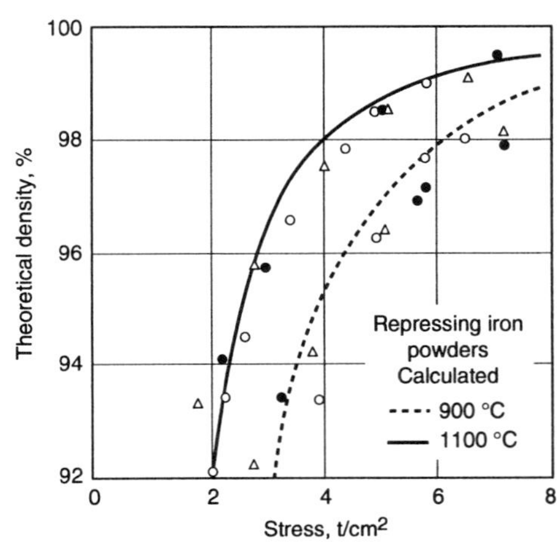

Fig. 9 Experimental and predicted increase in forming pressure with density during hot repressing of steel powder

Powder forging for mass-produced structural parts was developed to meet performance requirements beyond that available from conventionally pressed and sintered powder parts. A number of parts have been attempted by powder forging; however not all of these parts were successful, primarily for economic reasons. Currently, powder forged parts are produced mainly for automotive drive train applications. Powder forged connecting rods have been in use in U.S. automobiles since 1987, starting with the Ford Escort 1.9 liter engine. Powder forged connecting rods are now used in at least 13 engines among major U.S. automakers and are also used in some European and Japanese engines. The rods are typically forged from Fe-Cu-C powders to a minimum density of 7.84 g/cm^3. At this density they have an ultimate strength of 760 MPa (110 ksi), a yield strength of 550 MPa (80 ksi), and endurance limit of 275 MPa (40 ksi). Powder forged connecting rods eliminate the surface defects inherent in conventional forging, have a uniform microstructure, are dimensionally consistent, and have superior machinability. The uniformity of microstructure and material distribution also minimizes the amount of material removal required for balancing the connecting rods.

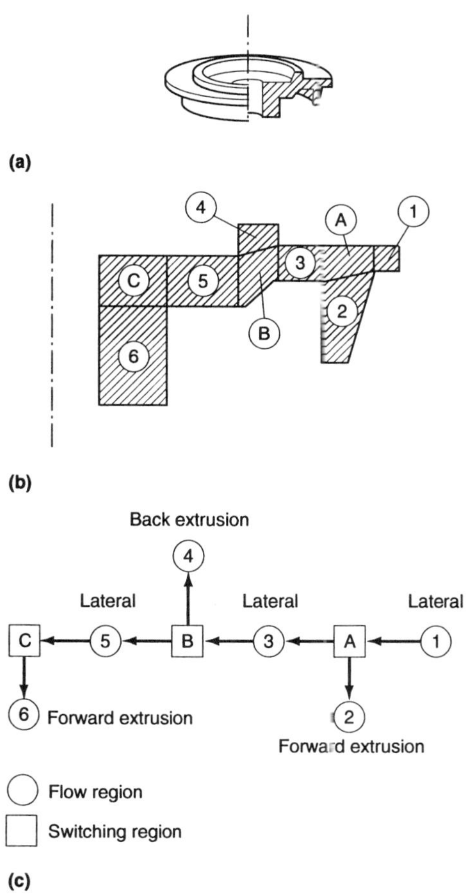

Fig. 10 An example of an axisymmetric part. (a) Isometric view. (b) Cross section showing regional subdivisions. (c) Link-node network representation of the part topology for expert system design decisions

Powder forging has also been used for specialty materials, such as tungsten, molybdenum, and beryllium. Since these materials cannot be cast easily, powder forging is the best option for processing fully dense net shapes. Powder forging concepts are also emerging for the fabrication of ceramic parts. With the recent development of superplastic ceramic materials, forming of complex shapes from simple ceramic preforms has been developed for high temperature bearing and seal applications (Ref 7).

Process Modeling

As with many other manufacturing processes, several computer models have been developed for analysis of powder forging, including finite element analysis and expert design systems. These computer tools aid design of the process, including preform design. Such models were difficult to develop because of the complications introduced by porosity in the material. The existence of porosity and accompanying density change during deformation preclude the use of conventional plasticity equations, which are based on zero volume change. In addition, heat transfer rates in the material depend on the local density level, which changes throughout the preform during the forging process. Recently, material models giving accurate representations of porous material behavior during plastic deformation have been developed. Coupled with modern numerical methods, these material models have led to reliable simulations of the deformation during powder forging (Ref 8). The fracture criterion given in Fig. 6 has also been embedded in the computer code so that fracture predictions can be made.

A reliable simulation code is particularly useful for preform design for powder forging of new parts. Starting with an initial guess of the preform shape, a finite element simulation of the forging of that shape will indicate areas of incomplete densification as well as locations at which fracture may occur. Then modified preform shapes can be attempted in an iterative approach to eventually derive a preform shape that leads to defect-free forged parts. This trial-and-error procedure, however, can be very tedious, especially for complex parts. For this reason, an expert design system was devised to provide an initial guess for the preform design, based on experience with previously forged parts. The part to be forged is first subdivided into regions of primary metal flow, such as back extrusion, forward extrusion, and lateral flow (Fig. 10). This subdivision into regions is done automatically by the program, using geometric reasoning applied to computer-aided design (CAD) data representing the part. Preform design rules accumulated from experience on other parts are then applied to each region. An additional set of rules is applied to evaluate material flow between regions. Next, a decision tree methodology is followed to optimize the preform shapes in each region, based on the objective of minimizing defects, wear, and tool loads. Smoothing functions then blend the preform shapes prescribed for each region. Finally, the design details are determined through limited applications of a finite element simulation code (Ref 9).

Hot Pressing

Material Preparation

Hot pressing is similar to powder forging in that powder material is compressed at high temperature between opposing punches, but hot pressing occurs at a much lower rate than forging. Also, lateral flow of material does not occur in hot pressing, whereas this is an important requirement in powder forging. Hot pressing is also similar to conventional powder compaction at room temperature since powder is placed in a die and compacted by uniaxial compression.

Unlike powder compaction, feeding loose powder into the die is not practical. Since hot pressing is carried out with dies heated to the pressing temperature, loose powder particles entering the die cavity heat up rapidly, causing particles to bond together or weld to the tool cavity and impede further flow of powder into the die. In addition, the large free surface area of loose powder increases the likelihood of oxidation of the powder at the high temperatures of hot pressing.

One option is to fill the cold die with loose powder at a separate station, and then move the filled die to the press where the die and powder are heated simultaneously before pressing. A second option is to preform the powder into a porous green compact for easy handling. This approach eliminates powder flow at high temperatures and minimizes oxidation problems, which outweighs the additional step of preform pressing. The compact, which may be preheated or at room temperature, is placed in the die, which may also be at room temperature or preheated to the hot pressing temperature. With the powder compact in the die, the powder and die must be heated to the hot pressing temperature. Heating can be done directly through induction or electrical resistance heating, or indirectly by heat transfer from the die, which is heated by other means.

Resistance heating of the powder is accomplished by passing a low voltage, high amperage current through the powder while pressure is applied. A variation of resistance heating involves discharging an electric spark through the powder. High temperatures generated at the points of contact between adjacent powder particles cause bonding. Heating of the die and then indirect heating of the powder compact is the most common method for heating. In this case the entire tool set may be heated by induction or a resistance wire heated muffle furnace.

As with other high temperature processes involving metal powders, a controlled environment must surround the powder material during hot pressing. This atmosphere can be a gas contained in the die cavity, or the gas can surround the entire die assembly inside a chamber. A vacuum can also be used as the hot pressing environment. Inert gases (such as argon) can be used when hot pressing powders such as beryllium or titanium that are particularly sensitive to contamination by oxygen or nitrogen.

Hot Press Equipment

A wide variety of hot pressing equipment configurations have been devised for both production and laboratory applications. These installations combine indirect or direct heating with various tooling configurations and means for containing the tooling in a controlled atmosphere. Several examples are given in Ref 10. Two examples are shown in Fig. 11 and 12. Figure 11 shows graphite tooling, heated by resistance, with a telescoping enclosure for containing an atmosphere. The entire fixture is fitted to a hydraulic press (Ref 11). Figure 12 is an example of a vacuum hot press used for beryllium consolidation. Graphite tooling is again used inside an induction coil for heating the tooling and powder. The assembly is mounted inside a double-walled chamber with vacuum seals for the rams and utility connections (Ref 12).

The material selected for hot pressing punches and dies is critical because these parts are subjected to high pressure and temperature for a sustained period of time. The punches must resist plastic deformation under the axial loads of pressing, while the die body must constrain the lateral pressure of pressing without yielding. As pointed out in the description of Fig. 2(b), as full density is approached, the lateral pressure nearly equals the applied axial pressure. As a result, the die must have a wall thickness great enough to withstand the circumferential stresses developed at peak axial loading.

Metals used for hot pressing tooling include molybdenum, a molybdenum alloy TZM (Mo-0.5Ti-0.1Zr), tungsten, superalloys, and alloy steels that are highly heat resistant. If steel or superalloy dies and punches are used, they must be kept at lower temperatures by water cooling through internal passages. Ceramic or graphite tooling is frequently used without cooling, but overloading must be prevented to avoid catastrophic failure of these brittle materials. Because thermal expansion occurs during heating, the dies and punches should be made of the same material. If the die expands at a greater rate than the punches, a gap will form between the die and punches into which pressed powder material will extrude, and lead to difficulty in ejecting the hot pressed material from the die. Conversely, if the punches expand at a greater rate than the die, the die will fit tightly around the punches, again preventing ejection of the hot pressed material.

The sustained processing time at high pressure and temperature during hot pressing also requires consideration of chemical reactivity between the powder material and tooling. Diffusion at the interface between the powder and tool surfaces may lead to welding of material to the die cavity. Graphite tools and molybdenum alloys require a controlled atmosphere to prevent oxidation. To avoid reactivity between the tooling and powder, as well as improve ejection of the pressed material from the die, die wall lubricants are frequently used. Colloidal graphite suspension in water or oil is applied to the die wall to reduce friction and wear. Ceramic or mica suspensions in hydrocarbons spray coated on the punch and die also inhibit diffusion bonding and welding.

Since hot pressing involves plastic deformation, creep, and diffusion, pressing times are extended to several minutes or even hours. For this reason hydraulic or pneumatic presses are used rather than mechanical presses. Generally a hydraulic press with a simple single action ram will suffice, as long as a means for ejecting the hot pressed material is provided. Hydraulic presses with side rams enable the use of split dies for easy extraction of hot pressed materials, particularly parts having multiple levels. In all cases, precise control of the axial load is necessary to ensure that a constant pressure or the required pressure-time profile is maintained during the pressing operation.

Applications

The material produced most widely by hot pressing is tungsten carbide. While small tungsten carbide parts, such as cutting tool inserts, are made by pressing and sintering, the density gradients resulting from such processes preclude their use for large parts. Tungsten carbide powders are hot pressed generally in graphite molds induction heated to 1400 °C (2550 °F) and pressed at pressures up to 17 MPa (2500 psi). Higher pressures may crack the graphite tooling, or cause the cobalt binder phase to extrude between the punches and die. If the pressure is too low, high levels of porosity remained and distortions will result during sintering to full density. Hot pressing is widely used for producing forming dies and wear resistant parts of simple geometries.

Hot pressing is also the preferred process for shapemaking of beryllium. Hot pressing of beryllium powders prevents grain coarsening, which is the major limitation in casting this material. Beryllium is hot pressed at 1050 °C (1920 °F) in graphite tooling in a vacuum. Hot pressed beryllium parts are used primarily in nuclear reactor modulators and in missile and aerospace applications.

Superalloys and titanium alloys have also been hot pressed to full density for simple geometry parts. However, serious problems occur due to reactions between these alloys and the tool materials. Such problems can be overcome by using hot isostatic pressing, where these interactions are more easily controlled.

Currently, hot pressing is a very important process in the development of new ceramics and metal matrix or ceramic matrix composites for a wide variety of applications. Materials include silicon carbide and silicon nitride in various metal

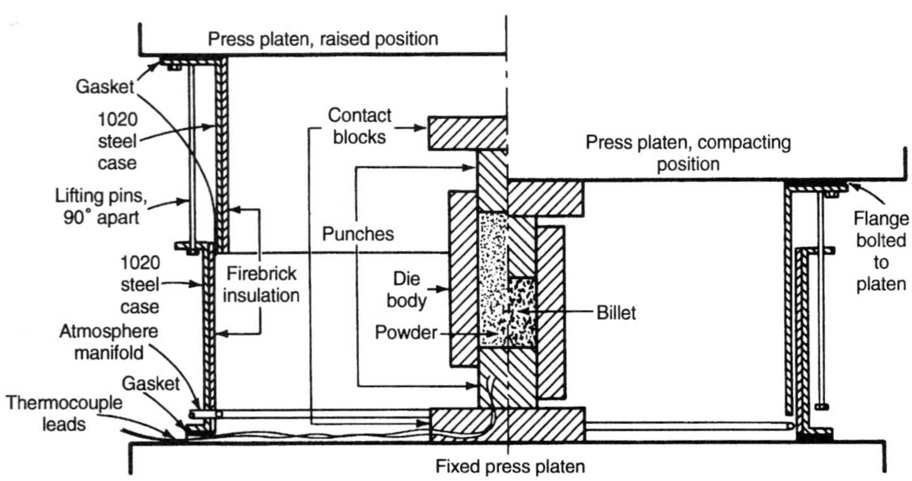

Fig. 11 Cross section of resistance heated graphite die assembly built into hydraulic press. Arrangement provides for telescoping controlled atmosphere enclosure and double-action hot pressing for steel, nickel alloy, and titanium alloy powders. Source: Ref 11

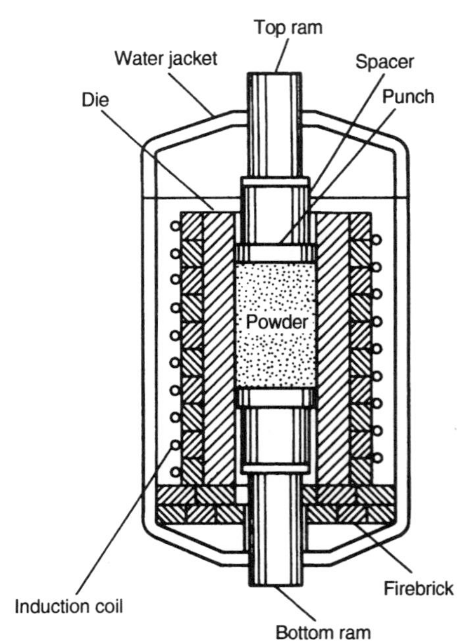

Fig. 12 Cross section of hydraulically activated, induction-heated graphite die assembly inside vacuum chamber. Arrangement permits double-action hot pressing of beryllium powder into blocks. Source: Ref 12

or intermetallic matrices. A relatively new material known as System 2 has been produced by the U.K. Atomic Energy Authority for protection of the next generation space shuttle, Skylon. The material consists of 40% silicon carbide continuous fibers in a refractory glass ceramic matrix that has been processed by hot pressing (Ref 13). The material is self healing at high temperatures and is ideal for hot structures where weight reduction is required. This material serves as an effective substitute for titanium- and nickel-base alloys for advanced aerospace applications. Alumina-aluminum silicon composites have been fabricated by in situ reaction of molten aluminum with aluminosilicate ceramics (Ref 14). This reactive metal penetration process is driven by a strongly negative Gibbs energy for reaction. In the aluminum-mullite system, aluminum reduces mullite to produce alpha alumina and elemental silicon. Ceramic-metal composites containing up to 30 vol% Al(Si) were prepared by reacting molten aluminum with dense aluminosilicate ceramic preforms, or by reactively hot pressing aluminum and mullite powder mixtures. Bulk metal matrix composites have also been produced by hot pressing blended prealloyed powders of aluminum alloy 6061 with 15 vol% of silicon carbide particles (Ref 15). In the T6 heat treated condition, yield strengths of 400 MPa (58,000 psi) were obtained with elongations of 2.5%. Sliding wear resistance of the metal matrix composite was also greatly superior to that of monolithic aluminum alloy 6061. A variety of ceramic components such as heat shields for space reentry vehicles are also produced by hot pressing in graphite molds heated by induction. Aluminum silicon carbide composites produced by hot pressing are used for kiln fixtures as well as high performance aerospace vehicle structures.

REFERENCES

1. H.A. Kuhn and B.L. Ferguson, *Powder Forging,* Metal Powder Industries Federation, 1988, p 74
2. Donald White, P/M in North America, *Int. J. Powder Metall.,* Vol 32 (No. 3), 1996, p 224
3. R.J. Dower and W.E. Campbell, The Toughness of Powder Metallurgy Forgings as a Function of Processing Route, *NEL Report* (Glasgow, Scotland), No. 632, March 1977
4. B.L. Ferguson, H.A. Kuhn, and A. Lawley, Fatigue of Iron-Base P/M Forgings, *Modern Developments in Powder Metallurgy,* Vol 9, Metal Powder Industries Federation, 1977
5. C.L. Downey and H.A. Kuhn, Application of a Forming Limit Concept to the Design of Powder Preforms for Forging, *Trans. ASME,* Vol 97, 1975, p 121
6. G. Bocksteigel and H. Olsen, Processing Parameters in the Hot Forming of Powder Preforms, Third European Powder Metallurgy Symposium, *Powder Metallurgy, Conf. Supplement, Part 1,* The Joint Powder Metallurgy Group, London, 1971, p 127–150
7. J. Wittenauer, Applications of Ceramic Superplasticity, *Mater. Sci. Forum,* Vol 243, 1997, p 653–662
8. A. Jinka and M. Bellet, Hot Forging of a P/M Connecting Rod: Three-Dimensional Computer Model, *Int. J. Powder Metall.,* Vol 32 (No. 3), 1996, p 255
9. H.A. Kuhn, J. Trasorras, and B.L. Ferguson, Automated Design of Net-Shape Forming Processes, *JOM,* April 1988, p 16
10. C.G. Goetzel, Hot Pressing, *Powder Metallurgy,* Vol 7, *Metals Handbook,* 9th ed., American Society for Metals, 1978, p 501–518
11. W.F. Courtis, "Spark Sintering Missile Component Feasibility," U.S. Naval Ordnance Contract NOO197-73-C-0081, Report LMSC-D-326418, 23 Aug 1973
12. W.G. Lidman and K.H. Moyer, Hot Pressing of Electrolytic Grade CR Beryllium, *Modern Developments in Powder Metallurgy,* Vol 3, Plenum Press, 1966, p 60–68
13. Super Ceramics for Next Generation Shuttle, *Am. Ceram. Soc. Bull.,* Vol 74 (No. 10), 1995, p 29
14. K.G. Ewsuk, S.J. Glass, R.E. Loehman, et al., Microstructure and Properties of Al_2O_3-Al (Si) and Al_2O_3-Al (Si) -Si Composites Formed by In Situ Reaction of Aluminum with Aluminosilicate Ceramics, *Metall. Trans. A,* Vol 27 (No. 8), 1996, p 2122–2129
15. T.N. Baker, A.J. Gorton, Y. Song, et al., Powder Processing of AA-6061 Aluminum Metal Matrix Composites using Hot Forging as Means of Consolidation, *Powder Metall.,* Vol 39 (No. 3), 1996, p 223–229

Pneumatic Isostatic Forging

Thomas W. Reddoch, AMETEK, Inc.

PNEUMATIC ISOSTATIC FORGING (PIF) is a technology for consolidating powder metal, or metal powder preforms to near or full density. The PIF process is a low-cost patent-pending technique for achieving high density and high performance in a range of materials. The short-time thermal exposure and lower processing temperature requirements relative to conventional processes provide a control of microstructure and related mechanical and magnetic properties.

The PIF process is a true forging process in which material is plastically deformed to achieve high density. The use of the term pneumatic is derived from the fact that argon or nitrogen gas is used as the pressure medium and is dispatched at pressures up to 415 MPa (60,000 psi), where the gas has fluid characteristics. The consolidation of the material is achieved by forging a workpiece with a gas hammer to high density. The workpiece is pressurized isostatically at a prescribed temperature in a uniform and rapid manner with a cycle time of 2 to 5 min. As such, PIF is a relatively high-strain-rate process that plastically deforms the workpiece. Selection of process parameters such as pressure, strain rate, temperature, and time at pressure is material dependent.

Pneumatic isostatic forging (PIF) and hot isostatic pressing (HIP) are similar to the extent that a gas-pressing medium applies isostatic pressure to the workpiece. Densification in HIP is achieved primarily through creep and diffusion with very limited use of plastic deformation. The PIF process uses plastic deformation and limited diffusion to achieve its goal. Furthermore, the strain rate for HIP is low and for PIF it is high.

Process Description

A flow diagram of the PIF process is shown in Fig. 1. Beginning on the left, P/M materials are prepared in one of two forms: a preform or a loose powder. A preform is consolidated to a near-net- or net-shape workpiece by one of several approaches, including conventional die pressing, cold isostatic pressing (CIP), or metal injection molding. The green pieces are often sintered before they are introduced to the PIF process. As a preform, the workpiece is prepared for PIF processing by a range of techniques including with and without coatings, limited encapsulation, and full encapsulation. The appropriate preparation is chosen to match the material and its initial compacted characteristics. All such choices are material dependent and proprietary in nature.

As an alternative to preform preparation, the free-standing powder is encapsulated in a thin-walled metal container and evacuated, typically in a simple geometric form as a cylinder or rectangular slab. The encapsulated material is then PIF densified. Various powders (including nickel compositions, copper compositions, and chromium, among others) have been processed by PIF.

After the material is prepared, it is preheated to a designated temperature. The temperature choice is selected based on the candidate material and is coordinated with a planned pressure cycle. As a typical value, the temperature choice will be 50 to 200 °C (120 to 390 °F) below a corresponding choice for a HIP transaction. Once the workpiece is in thermal equilibrium, it is transferred to the pressure system. Immediately following the transfer, the pressure system is energized to values up to 415 MPa (60,000 psi). A proprietary pressurization technique is used to increase the pressure and pressurization rate to the desired value. Once the desired pressure is achieved, it is held for periods up to 1 min. Depressurization strategy follows specific rules based on the materials being processed, but in most instances the pressure is relaxed within 1 to 5 min. Once consolidated, the workpiece can be cooled under pressure when seeking a specific desired end effect or the workpiece can be removed from the pressure chamber shortly after consolidation and cooled in a supplemental cooling station.

Densifying Materials by PIF. Because PIF is isostatic in nature and uses either argon or nitrogen gas as the pressure medium, it is often confused with hot isostatic pressing. The PIF process uses plastic deformation with high-strain-rate forces as its consolidation mechanism, whereas HIP uses very limited plastic deformation and primarily relies on diffusion and creep with low strain rate to achieve its end goals. Therefore, PIF is a true forging process utilizing a gas hammer to rapidly transfer energy to form the workpiece. Some materials cannot be forged to full density and may require the creep mechanism to achieve results, thereby requiring HIP.

Consolidation by PIF requires satisfying two critical conditions:

- The surface condition of the workpiece must be free of significant defects or holes and/or interconnected porosity where gas can be absorbed, causing a loss of pressure differential and therefore negating the effect of PIF. Experience has shown, however, that properties of high-pressure gas, particularly argon, enable full densification of preforms having densities as low as 87% of theoretical.

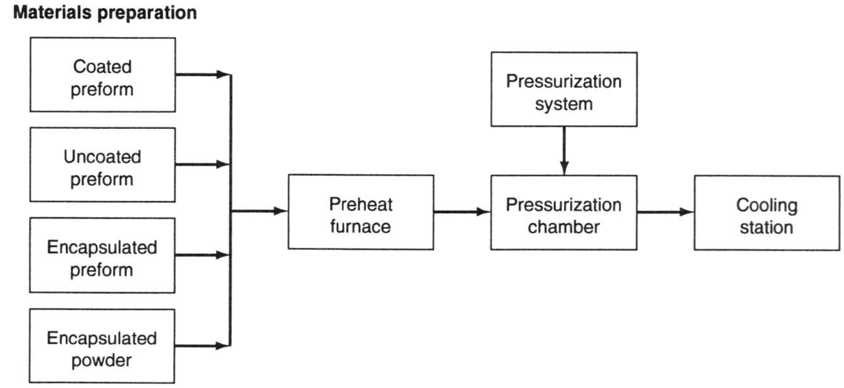

Fig. 1 Flow diagram of the PIF process

- In order to consolidate the workpiece to full density, its temperature must be of sufficient value to reduce the flow stress requirement of the material for forging such that the applied pressure or force "can collapse the material" in on itself or impart plastic flow. Experience indicates that some nonisostatic preform compaction techniques can produce uneven density profiles that can limit achieving full density by PIF.

Pressure/Strain Rate/Temperature/Time. The deployment of the PIF process is a careful orchestration of the process variables including pressure, strain rate, temperature, and time. Typical pressures used in the PIF process range from 310 to 415 MPa (45,000 to 60,000 psi). For each cycle, an end pressure is chosen where a pressure level is maintained for a period of 10 s to 2 min. Temperature plays the key role by reducing the flow stress requirement, thereby reducing the consolidating pressure and increasing the sensitivity of the deformation speed by means of the strain rate. Time is a factor in the preheat step where the material is heated to a homogeneous value before pressurizing. Typical temperatures include 525 °C (975 °F) for aluminum, 900 °C (1650 °F) for copper, and 1225 °C (2235 °F) for iron. A strain rate is chosen to develop the appropriate shear stress levels to consolidate the workpiece.

Dimensional Properties. A natural consequence of consolidation is the shrinkage of the workpiece. The changes in size and the character of the change are a direct result of the applied force and the prior conditions of the workpiece. Preexisting work hardening and density distributions within a workpiece significantly, if not uniquely, determine any efforts to consolidate a workpiece by PIF.

As a true isostatic compacting technology, PIF provides a uniform force in all directions to the workpiece. Therefore, the workpiece contracts in a volumetric manner. A typical consequence of nonisostatic compaction is the hourglass distortion of cylindrical forms. Fundamentally, PIF does not distort; however, properties installed in the workpiece from first-stage compaction and sintering influence the final dimensions and any distortions. It is well recognized that the constraining nature and uniaxial forces of die compaction produce density distributions in the workpiece. Sintering can amplify the effects of these density distributions. Applying supplemental treatments such as coining can further disturb the natural properties of the workpiece. None the less, PIF maintains dimensions within a few thousands of an inch from its starting condition.

Consolidation by PIF of die-compacted and sintered components tends to move the material in a proportional manner. As such, distortion is avoided when the same density distribution is maintained pre- and post-PIF. This occurs when the density distribution is uniform or when any area of the workpiece moves to the extent that it is either attaining just full density or less. Distortions during consolidation are a direct consequence of nonuniform properties. For example, isostatically or metal injected molded compactions have few, if any, density distributions; therefore, these specimens consolidate with little or no significant distortion.

The PIF process uses isostatic force to sustain dimensional properties; that is, no dies are used to constrain the workpiece. Tooling and fixturing can be used to influence the movement of the workpiece, thereby holding particular geometric positions similar to that provided by dies in first-stage compaction, coining, and powder forging. Care must be exercised to avoid bonding the workpiece to the fixture. Insulating materials and thermal properties of the fixture material are considered.

Estimates of Dimensional Shrinkage. The actual shrinkage of a workpiece by PIF is estimated using some simple relationships. Because PIF is volumetric movement in nature, a cubic variation in parameters prevails. The formula used to forecast dimensional change is:

$$d_f = d_i \sqrt[3]{\frac{D_i}{D_f}} \qquad \text{(Eq 1)}$$

where d_i and d_f are pre- and post-PIF dimensions (i.e., radius, length, width, or thickness), respectively, and D_i and D_f are pre- and post-PIF density, respectively. This empirical formula is very effective for estimating dimensional changes on geometries with a length-to-diameter ratio of 5 or less.

Some trial and error is required when applying the formula because the final density of the workpiece is a factor in the total dimensional change. Length-to-diameter ratio also influences specific dimensional changes. Materials consolidating to full density conform best to the formula estimates. Some materials and workpiece configurations do not attain full density, and therefore they must be calibrated to determine final density and the actual dimensional changes. Preforms with uniform density satisfy formula predictions very accurately.

Process Advantages. The pressure and temperature of the process are controlled and managed independently of one another, thereby providing an opportunity to control microstructure and properties of the material. As an additional consideration, lower processing temperatures and shorter exposure time to elevated temperature reduce grain growth in many materials. The short processing time at pressure expedites material throughput and maximizes the availability of the pressure vessel, therefore reducing the capital equipment requirements. Further, the pressure vessel size requirements are reduced. Material handling and processing requirements are simplified, thereby facilitating automated actions. Overall, the process is well suited to automated transactions where sustained throughput of workpieces is desired.

Process Limitations. The PIF process is used to enhance the properties of a workpiece through forging by improving the density and attaining various mechanical and magnetic properties. As a forging process, PIF relies on its pressing media to exert sufficient force to overcome any resistance movement by the material. In short, the force must overcome the flow stress requirements of the material in order to consolidate it. If the preform has density distributions, the PIF process will not significantly alter these distributions.

As a true forging process, plastic deformation by application of stress is the consolidation technique. If the material requires diffusion or creep to attain its consolidation goals, then PIF will not satisfy those needs.

Table 1 Pinion gears densified by PIF

Heat treat condition	Weight, g	Density, g/cm³ Average	Top	Middle	Bottom
As-sintered	133.25	7.340	7.343	7.270	7.346
Sinter/PIF	133.72	7.836	7.854	7.760	7.863
Sinter/PIF	133.92	7.837	7.836	7.812	7.862
Sinter/PIF	133.75	7.837	7.836	7.800	7.860
Presinter/PIF	132.19	7.802	7.849	7.712	7.851
Presinter/PIF	131.56	7.796	7.853	7.712	7.851

Materials Processed by PIF

A range of materials have been processed by PIF for a variety of applications. Early applications of the technique were for automotive applications for iron-base parts where high performance in either mechanical or magnetic properties was sought. Various aluminum applications for automotive use were also evaluated. In both applications, near-full density (98% or better) was achieved. Nickel and nickel aluminides have also been consolidated by PIF with good success (100% dense) by direct densification of powders in encapsulation containers. Various other nickel alloys have also been consolidated from powder preforms as well. Titanium and titanium/aluminum systems have also been consolidated very effectively by PIF. One such application has been sputtering targets. Other applications for sputtering targets involve direct consolidation of chromium powders. Work has been done on spinodal (copper/tin/nickel) alloys for thick-section material applications. The PIF process has also been used to consolidate various other materials including cobalt/chrome, stainless steel, aluminum silicon carbide, aluminum boron carbide, tungsten, and tungsten carbide, among others.

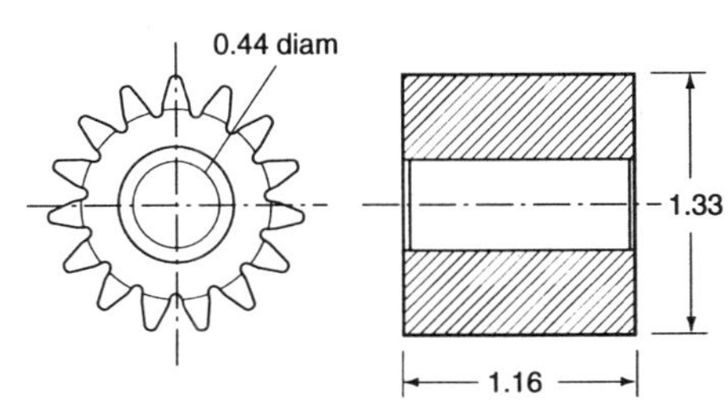

Fig. 2 Iron carrier pinion gear, dimensions in inches

Example 1: Iron Carrier Pinion Gear. This application focuses on the densification and microstructure properties of an iron carrier pinion gear (Fig. 2) manufactured from the iron composition FLN 4-4405. A product evaluation was done to determine the effectiveness of PIF at improving overall density and density at key areas of known lower value. A typical gear in the as-received condition has a density of 7.3 g/cm^3, which after sintering became 7.34 g/cm^3. Final density was evaluated under two sets of conditions:

- Press and delube (at a presinter of 815 °C, or 1500 °F)
- Press, delube (at a presinter of 815 °C), and sinter at 1150 °C (2100 °F)

Table 2 Pinion gear hardness (Rockwell B) in post-PIF condition

Heat treat condition	Section top/bottom	Hardness reading, HRB	Average
As-sintered	Top	83, 83, 84, 85	84
	Bottom	83, 84, 84, 85	84
Sinter/PIF	Top	100, 99, 100, 100	100
	Bottom	100, 100, 101, 102	101
Sinter/PIF	Top	100, 99, 100, 100	100
	Bottom	102, 101, 100, 101	101
Sinter/PIF	Top	101, 100, 99, 99	100
	Bottom	102, 101, 101, 101	101
Presinter/PIF	Top	89, 89, 88, 90	89
	Bottom	88, 89, 88, 86	88

The gear exhibits a density gradient with the center at the lowest value when compared with the top and bottom values.

The primary findings of the study are summarized in the tables and figures discussed below. The fully sintered gears provided the best overall density, at 7.84 g/cm^3 (see Table 1). Sectioning the gear into top, middle, and bottom shows that the density gradient present in the pre-PIF condition remained but density in each area significantly improved. The delubed (or presintered) gears showed lower average density than the fully sintered parts. The overall average density was 7.8 g/cm^3. The gradient remains although, as in the sintered gears, each area shows significant improvement.

Hardness is a significant characteristic of the quality of a gear. In Table 2, the hardness values are charted for the full sintered and delubed (or presintered) conditions after PIF. As a matter of comparison, a sinter/PIF gear shows improvement in hardness over sinter only. The presinter/PIF shows lower values for hardness simply because of the absence of sintering.

The metallographic observations also confirmed the findings in Table 1. Figure 3 shows the pronounced effect of combining sintering and PIF.

Example 2: Iron with 45P-1000B Alternator Rotor. As-received alternator rotors (Fig. 4) uniaxially pressed from iron powder containing 0.45 wt% P were densified by PIF to enhance magnetic and mechanical properties. The rotors were in various states including as-delubed, delubed and steam treated, delubed and electroless nickel coated (less than 1 mil), and debindered plus microwave sintered. Each specimen was presintered at 980 °C, 1150 °C, or 1205 °C (1800 °F, 2100 °F, or 2200 °F).

The PIF parameters evaluated were: pressures of up to 415 MPa (60,000 psi) and temperatures to 1200 °C (2190 °F) at hold times of 10 to 120 s. The most consistent results were obtained with pressures of 310 MPa (45,000 psi) at 1200 °C (1290 °F), with a 10 s hold time at temperature and pressure. Taguchi experiments were conducted to select the optimal parameters for compositions of 0.45, 0.60, and 0.80 wt% P, as well as to evaluate the roles of pressure, temperature, and hold times at the various parameters on the densification achieved. Magnetic and mechanical properties were evaluated as a function of density and processing parameters.

One key objective was the enhancement of magnetic properties during densification. Figure 5, in which B (flux density in kilogauss) is a function of H (field excitation in Oersted), compares specimens of single press, single sinter (SPSS), double press, double sinter (DPDS), and conventional wrought material (12L15) with

Table 3 Properties of aluminum-B_4C

Sample	Fabrication condition	Density	Interparticle bonding	Machining properties	Uniformity of B_4C dispersion
Billet 29	Cold isostatically pressed and low-temperature vacuum sintered	Average density 93%. Estimate density variation of 89 to 94% over entire piece	Limited by surface oxide films on the particles of the pressed billets	Limited by density and weak particle bonding properties	Very good
Billet 30	PIF at 310 MPa (45,000 psi) and forged without total encapsulation	99.5% average	Interparticle bonding was achieved with oxide coating breakup during forging	Very good machining and polishing properties	Very good after forging
Billet 31	As-forged at 310 MPa (45,000 psi) with billet encapsulated and evacuated in aluminum	99.5%+	Excellent interparticle bonding and oxide film breakup during forging	Very good machining and polishing properties	Very good after forging

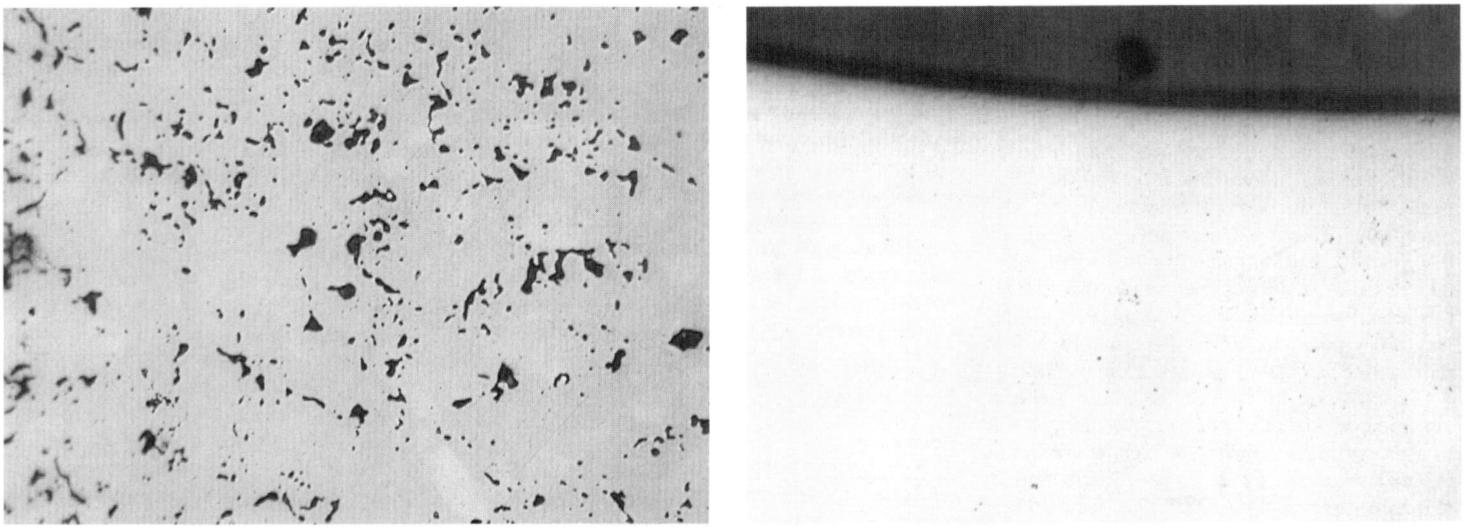

(a) (b)

Fig. 3 Unetched microstructure of iron carrier pinion gear. (a) As-sintered condition. (b) Top surface of the tooth after pneumatic isostatic forging. 48×

Fig. 4 Alternator rotor

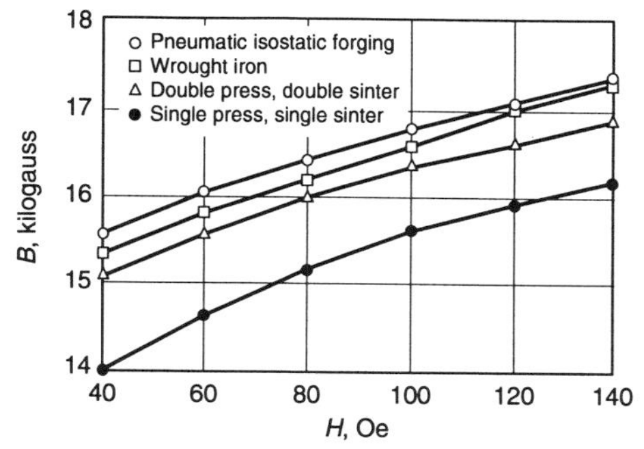

Fig. 5 Induction curve comparing the magnetic properties of sample specimens

those of SPSS/PIF processed samples. The average bulk density of the as-received rotor was 7.15 g/cm^3 with local density variations of 6.8 g/cm^3 to 7.25 g/cm^3. The bulk density after PIF was 7.81 g/cm^3. For a field excitation value of about 60 Oe, the PIF material has a flux density at about 9% larger than SPSS, 3% larger than DPDS, and a 1.5% larger value than wrought material. For larger values of field excitation (around 130 Oe), the PIF material and the wrought material approach one another in flux density while the SPSS and DPDS are sustained at their respective percentages.

Example 3: Aluminum-B_4C Billets. Pneumatic isostatic pressing has been used to achieve controlled microstructure, improved properties, and interparticle bonding in aluminum-, copper-, and titanium-matrix composites. The densification of aluminum composites by PIF through rapid plastic movement of mating particle surfaces allows the oxide films to break up on the aluminum during densification. This promotes interparticle bonding before the surface film can re-form, which would block interparticle bonding. The formation of the oxide film can occur in pressure-assisted diffusion-driven processes, such as HIP and sintering. These processes require vacuum processing to achieve density; however, this will not prevent oxide film formation during these processes. The mating of particle surfaces must be moved a sufficient amount during PIF to break up the oxide film and improve properties. A previously encapsulated, evacuated, and hot isostatically pressed aluminum-matrix composite would show only slight improvement in interparticle bonding if subjected to PIF.

Another advantage of PIF for aluminum-matrix composites is demonstrated in the analysis of predictable dimensional changes during PIF. This demonstrates the ability to fabricate net or near-net shapes from powder blends and retain the geometric spacing of the reinforcement material in the green-pressed shape. This ensures uniform fracture toughness properties in all directions, and a uniform stress state in the part. The uniform stress state in the part improves machining properties and minimizes distortion during heat treatment or secondary processing, such as joining by diffusion bonding or welding.

The as-received material had a bulk density at 93% of theoretical (pore-free) density from CIP and sintering. The density after PIF was 99.5% of theoretical. The PIF processing broke up oxide films and produced both interparticle bonding and high density levels while dispersing B_4C uniformly throughout the microstructure. All results are summarized in Table 3.

The mechanical properties with this uniform microstructure and B_4C dispersion retained during PIF can be further improved by heat treating. Using the PIF process can produce net and near-net shapes of aluminum alloy composites in a very effective manner. In addition, the as-received material can be densified without the encapsulation required by HIP, which permits near- and net-shape fabrication. Similar results with other additives such as nitrides and other carbides contained in an aluminum alloy matrix would be produced by PIF. The breakup of oxide films during PIF of P/M copper-matrix alloys has been accomplished previously. This was confirmed by improved thermal conductivity and electrical conductivity of PIF composites, compared to P/M composites fabricated by other techniques.

An evaluation of the densification and interparticle bonding evaluation of Al-B_4C is shown in Fig. 6(a, b, c). In the as-received condition (Fig. 6a), the most dense areas in the specimen are shown as light shades, as these areas polish and etch the best. The areas in Fig. 6(b) and 6(c) represent more typical microstructures, as polishing and staining have been accomplished with ease after PIF densification. Two forms of encapsulation were used: limited and complete (Fig. 6b and 6c, respectively).

Example 4: Stainless Steels Exhaust Flanges. Pneumatic isostatic forging has been shown to be very effective at densifying 400 series stainless steels. Exhaust flanges are typically made from the 400 series materials where they are die compacted and sintered. Sample specimens were pressed and sintered to about 91% density. In turn, the flanges underwent PIF at 310 MPa (45,000 psi) at 1225 °C (2235 °F) producing a 98.5% dense component.

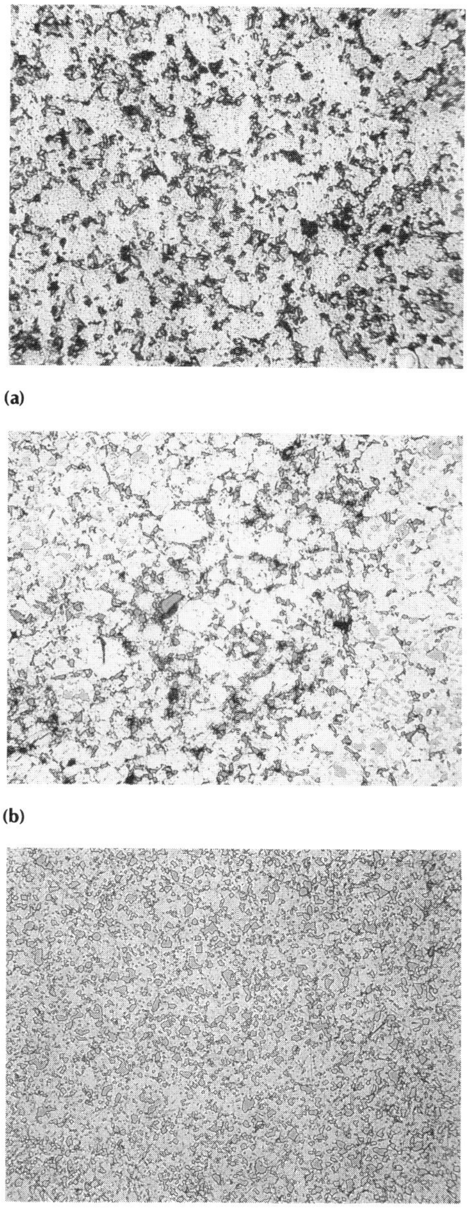

(a)

(b)

(c)

Fig. 6 Microstructure of AlB_4C sample. (a) As-received condition. (b) Sample after PIF with limited consolidation. (c) Sample after PIF with full encapsulation. 108×

SELECTED REFERENCES

- T.W. Reddoch, New Densification Technology Produces Very High Density, Near Net and Net Shape Metal and Metal Matrix Composite Parts Quickly and at Low Cost, *Mater. Technol.*, Vol 11 (No. 4), July–Aug 1996, p 134–137
- T.W. Reddoch and E.S. Hodge, "A New Densification Technology—Pneumatic Isostatic Forging (PIF)," presented at the 1996 World Congress on Powder Metallurgy & Particulate Materials (Washington, D.C., 16–21 June 1996)

Secondary Operations and Quality Control

Heat Treatment of Ferrous Powder Metallurgy Parts

Howard Ferguson, Metal Powder Products Company

As raw materials and processing methods continue to improve, more high-performance components requiring heat treatment are being converted to powder metallurgy due to its neat-shape capability and cost effectiveness. Powder metallurgy parts sintered to net shape after compaction typically contain a minimum of approximately 10% residual internal porosity in traditional press-and-sinter methods. With warm compaction, minimum residual porosity can be lowered to approximately 5%.

Unlike wrought steels, where hardening response is controlled primarily by chemical composition and grain size, P/M hardenability is significantly influenced by this interconnecting porosity, which makes the structure permeable to gases and fluids.

Because P/M processing can provide net-shape capability, the hardening process becomes much more critical when compared to near-net-shape wrought components where finish machining is usually applied after the harden-and-temper operations. With powder metallurgy, dimensional shrinkage normally occurs upon quenching and tempering. With simple shapes this shrinkage is predictable and can be accommodated in the compaction tool design.

However, as the P/M components become more complex in shape, such as the typical multilevel parts in production today, variations in porosity from one level to another can cause significant distortion in the part upon heat treating, which can affect the function or fit of the part in an assembly. When this occurs, a finish machining operation is usually required, which detracts from the cost benefit of powder metallurgy.

This permeability also influences the hardenability of the material. The interconnected porosity acts as an insulator and reduces the thermal conductivity of the material. When rapidly quenched from the austenite range, this permeability retards the cooling rate, resulting in mixed microstructure and inconsistent hardness measurements. These variations increase with part-shape complexity.

Effect of Porosity on Material Properties

Powder metallurgy materials are unique due to their permeable nature. This permeability reduces the theoretical density of the base metal, it changes the thermal conductivity as mentioned earlier, and it also affects other physical properties such as ferromagnetism and elastic modulus. Mechanical properties such as tensile strength, fatigue strength, and impact resistance are directly influenced by porosity.

Due to this influence, the hardenability of a P/M ferrous alloy is reduced in direct proportion with an increase in porosity. This is illustrated in Fig. 1, where the results of a series of Jominy tests on sintered carbon steel test bars are compared with a wrought steel of similar composition. The P/M specimens were prepared from atomized iron powder mixed with 0.90% graphite, compacted and sintered to various density levels ranging from 6.0 to 7.1 g/cm^3. These correspond to porosity levels ranging from 9 to 24%. The sintered bars along with a bar machined from wrought SAE 1080 steel having a composition similar to the P/M material, were austenitized at 870 °C (1600 °F) for 30 min in a neutral atmosphere. There were then water quenched in a Jominy fixture as described in ASTM A 255. Hardness readings were taken every 2.5 mm (0.10 in.) from the quenched end and plotted as shown in Fig. 1. This illustrates two important features of P/M heat treatment. Not only is hardenability directly affected by porosity through its influence on thermal conductivity, but there is also an important secondary effect because porosity also reduces the true hardness measurement of the material. A Rockwell or Brinell hardness measurement on a sintered material is referred to as apparent hardness due to the lower resistance to penetration of the indenter caused by the porosity.

A more practical effect of porosity is the limitation of the quenching medium to oil. Quenching in water or liquid salts can degrade the material because these entrained fluids can initiate internal corrosion. With the advent of gas pressure quench vacuum furnaces, P/M alloy steels with sufficient hardenability can be hardened without this problem of liquid entrapment.

Influence of Porosity on Case Depth

In P/M steel parts with interconnected porosity, case-hardening processes such as carburizing, carbonitriding, and nitrocarburizing become much more sensitive to process control due to the rapid penetration of the atmosphere gases. These high-carbon-potential atmospheres diffuse into the internal pore surfaces at the same rate as the external surfaces of the part, causing very rapid carburizing to a significant depth below the surface and a consequent loss of a sharp case/core transition zone.

The micrographs in Fig. 2 show microstructures of P/M test specimens that were carburized at

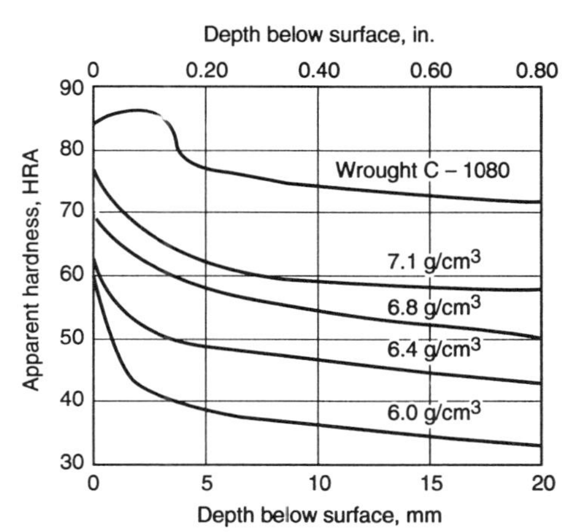

Fig. 1 Comparison of water quench Jominy tests (ASTM A 255) of sintered carbon steel test bars (density ranging from 6.0 to 7.1 g/cm^3) with wrought C-1080 steel with similar composition

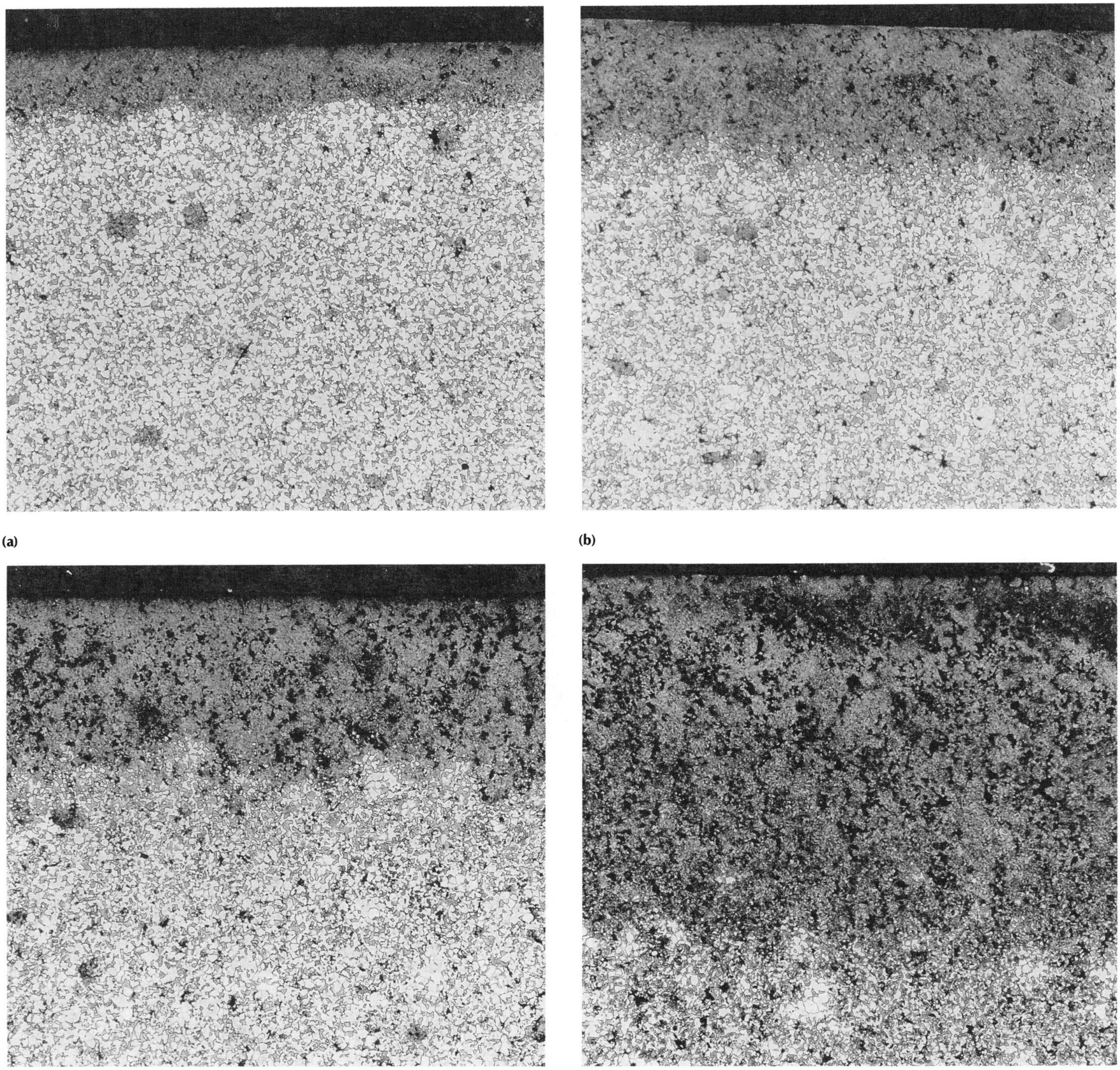

(a) (b) (c) (d)

Fig. 2 Microstructures of heat treated P/M test specimens carburized for 30 min at 865 °C (1590 °F) to density levels and case depths of: (a) 7.72 g/cm^3 and 0.23 mm (0.004 in.) (b) 7.45 g/cm^3, and 0.46 mm (0.018 in.) (c) 7.34 g/cm^{3}and 0.6 mm (0.024 in) (d) 7.08 g/cm^3 and 1.25 mm (0.050 in.)

865 °C (1590 °F) for 30 min. The test bars were processed to attain increasing density levels ranging from 7.0 to 7.7 g/cm^3 sintered density.

As shown, increasing density was accompanied by a proportional decrease in carbon penetration. At 7.2 g/cm^3, the interconnected pores begin to close off, thereby preventing the atmosphere gases from penetrating into the internal volume of the material. Plotting case depth versus density (Fig. 3) shows that for optimal control of case-hardening processes P/M parts should be specified at 7.2 g/cm^3 or higher.

The primary purpose of case hardening low-carbon steels is to provide a hard, wear-resistant surface while maintaining a tough, resilient core. This combination imparts optimal properties for gears and other highly stressed components. When carbon penetration is uncontrolled, such components can exhibit extreme embrittlement and dimensional distortion upon quenching.

Influence of Alloy Content on P/M Hardenability

Increasing the carbon content of a steel increases tensile strength and hardness in proportion with the added carbon. In P/M steels, this ratio is reached at the eutectoid composition of 0.8% C. Increasing carbon content above this level allows carbide networks to form at grain

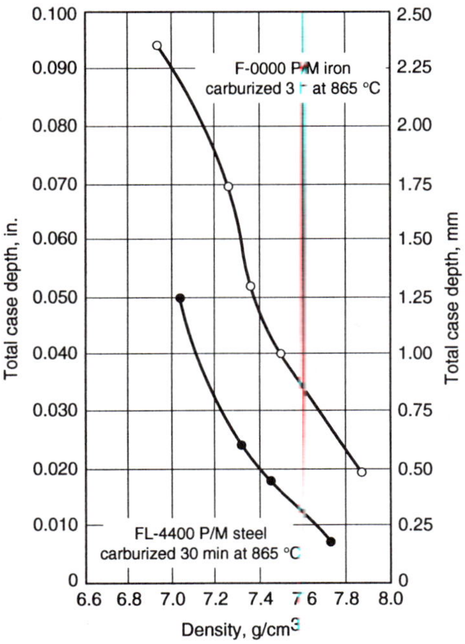

Fig. 3 Effect of increasing density on case depth for F-0000 P/M iron and FL-4400 P/M steel carburized at 865 °C (1590 °F) for 3 h and 30 min, respectively, with 0.95% carbon potential, 80 °C (175 °F) oil temperature, and 150 °C (300 °F) temper. Test bars for FL-4400 are shown in Fig. 2

boundaries and along pore channels, causing embrittlement and loss of strength.

When selecting a P/M alloy for subsequent heat treatment to improve strength, carbon content of the sintered material becomes the most critical variable. This is controlled by the graphite addition to the powder blend and the atmosphere control during the sintering operation. Hardenability is determined by part mass, density, and alloy content. The requirement for maximum strength improvement is a microstructure that is predominantly martensite at the surface, but contains sufficient transformation products in the core to provide a measure of toughness.

Figure 4 shows the influence of carbon content on the tensile strength of an iron-phosphorus (0.45 wt%) specimen sintered to 6.8 g/cm^3 density and subsequently reheated to 845 °C (1550 °F) for 15 min in a sealed quench furnace containing an endothermic atmosphere with a 0.85 C potential. The tensile specimens, which were machined from bars, were then oil quenched and tempered at 175 °C (350 °F) for 2 h.

As shown, sintered strength continues to increase with carbon content, but reaches its peak strength after hardening at approximately 0.40 to 0.50% C content. Also, elongation, which is reduced by the transformation stresses, is also decreased by increasing carbon levels in the material. To ensure strength improvement after hardening, the core microstructure should contain no more than 70% martensite, the remainder being a mixture of bainite and unresolved pearlite that can improve impact and fatigue resistance.

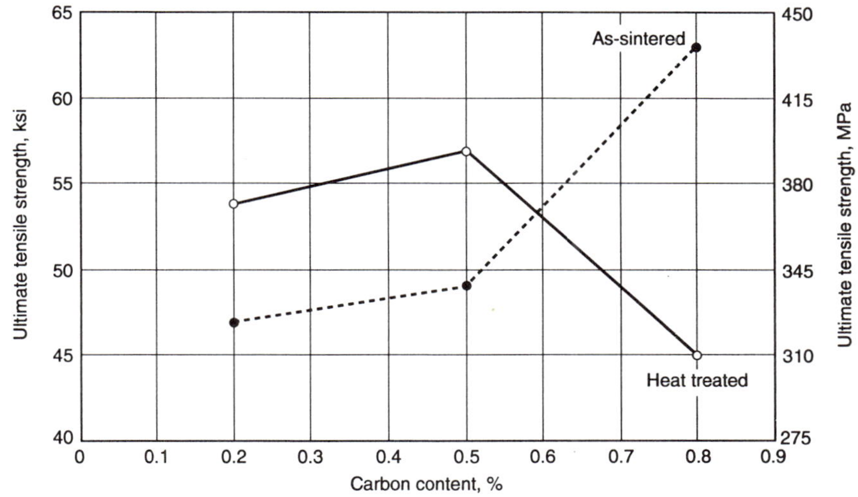

Fig. 4 Influence of carbon content on strength of iron-phosphorus alloy at 6.8 g/cm^3 sinter density

As alloying elements such as nickel, molybdenum, chromium, and copper are added, the optimal carbon content is lowered. In sintered steels, the most common alloying elements are copper and nickel.

Copper Content. Additions of copper increase both apparent hardness and tensile strength in the sintered condition. Upon hardening, it increases surface hardness, but loses toughness and strength due to its embrittling influence. This relationship between copper content and combined carbon is illustrated in Fig. 5. In the plain Fe-C compositions, peak strength occurs near the eutectoid composition in the as-sintered condition at approximately 0.65% C in the hardened-and-tempered condition. By adding copper to the sintered-steel composition, strength is significantly increased with increasing copper up to approximately 5% Cu. Upon heat treatment, however, optimal strength is found at lower carbon levels as the copper content is increased.

Nickel Content. Nickel increases tensile properties in sintered steels, but to a lesser degree than does copper. However, it provides significant strength improvement after heat treatment. This is due in part to the heterogeneous nature of nickel alloy steels made from blended elemental powders. Copper forms a liquid phase at the conventional sinter temperature of 1120 °C (2050 °F) and combines with the iron to form a solid solution that provides a considerable increase in strength and hardness. Nickel, having a much higher melting point, combines with iron by solid-state diffusion. This produces a duplex microstructure consisting of partially alloyed iron surrounded by nickel-rich islands. Upon quenching, the matrix transforms to martensite, but the nickel-rich areas remain austenitic, resulting in a strengthened matrix with some retention of toughness and ductility.

By alloying with both nickel and copper, the hardenability of a P/M steel can be substantially improved, as shown in Fig. 6. Here Jominy bars were pressed to a 6.7 g/cm^3 density with increasing amounts of copper and nickel additions. Carbon content was kept at 0.5% C. After sintering, all bars were austenitized at 850 °C (1560 °F) for 2 h before quenching. A dramatic increase in surface hardness was seen with a 2.5% Cu addition. Adding nickel produces a small increase in surface hardness, but a significant increase in depth of hardness. Many components that require optimal hardness and hardenability contain both copper and nickel additions.

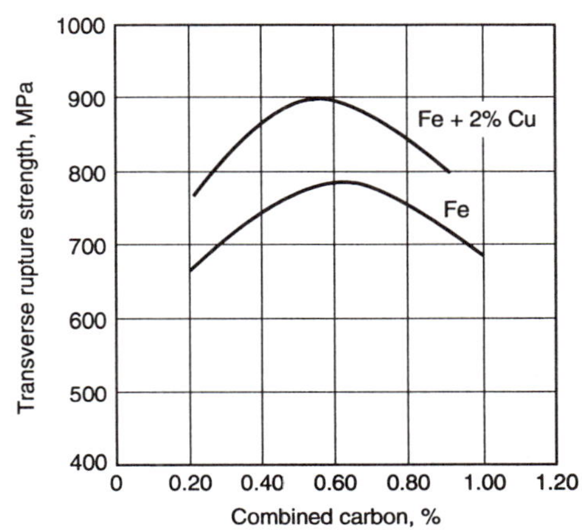

Fig. 5 Influence of copper and carbon content on strength of hardened P/M steel at 6.8% g/cm^3

Prealloy/Diffusion Alloyed Irons

Prealloyed irons are completely alloyed with hardenability agents such as manganese, nickel, molybdenum, and chromium to provide improved dynamic properties upon densification and heat treatment.

Diffusion alloyed irons are produced by core-ducing the iron oxide with the oxides of these elements to provide partial alloying. The advantage of partial alloying compared to prealloying is improved compressibility, allowing for higher green density at a given compaction pressure.

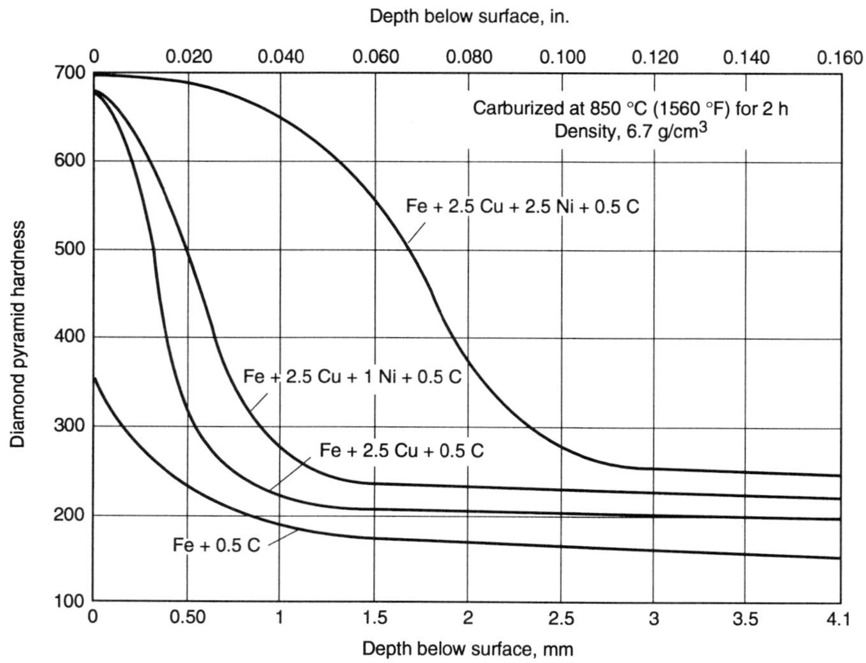

Fig. 6 Effect of copper and nickel addition on the hardenability of P/M steels. From Ref 8

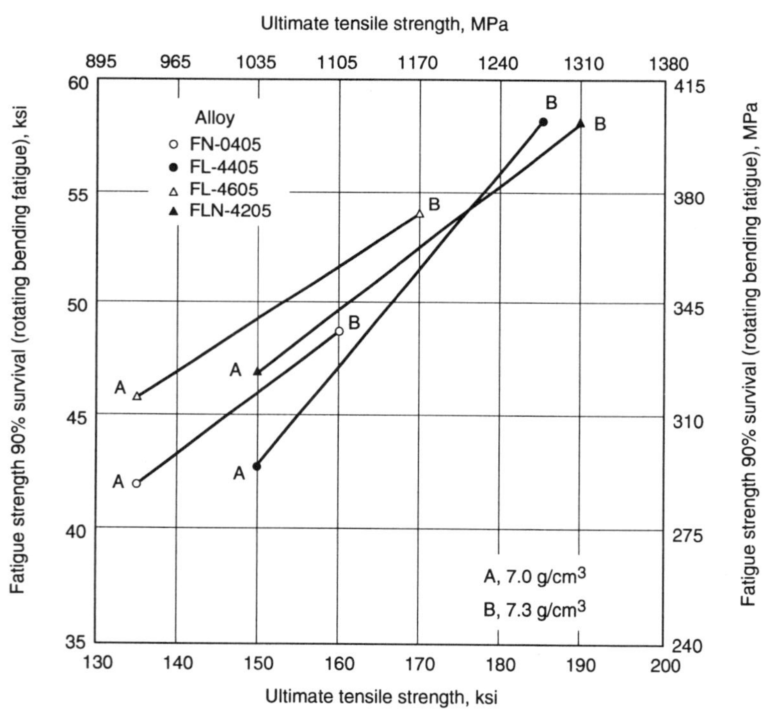

Fig. 7 Influence of alloy content on fatigue strength of heat treated low-alloy steels

In rotating bending fatigue (RBF) testing, as well as in component testing, it has been found that these alloys provide improved fatigue strength at higher densities than the preblended steel compositions. These alloys are frequently specified for highly stressed components where fatigue strength and impact resistance are required.

Figure 7 shows the improvement in fatigue strength compared to tensile strength at two density levels after hardening. As shown, the fatigue strength improvement is most pronounced with increased density and combination of a prealloy with elemental nickel to form a composite microstructure of tempered martensite and nickel-rich austenite.

The process route taken to make a component also is a significant factor in developing optimal properties. Table 1 shows the comparison of properties attained from an FL-4405 alloy by a single press sinter (SPS), a double press double sinter (DPDS), and a DPDS coupled with a high temperature sinter (HTS). As shown, HTS combined with high density provides a substantial increase in static and dynamic properties after heat treatment.

Preblended, prealloyed, and diffusion alloyed powders are being used in high-density components that are highly stressed during service. The component shape and process requirements usually dictate which type of powder is specified. If secondary machining is required, a less hardenable alloy may be necessary for improved machinability. All of these process requirements need to be considered to make the part cost effectively. The heat treatment must also be included in this analysis. These considerations need to be determined prior to design and manufacture of hard tooling. Dimensional change varies depending on the base powder type and composition as well as density and process conditions.

Figure 8 shows the dimensional changes occurring in three powder types at a given density with different graphite additions. Here, the transverse rupture bars of each material were sintered together at 1130 °C (2070 °F) for 30 min in $N_2 10\% H_2$ atmosphere then quenched from 870 °C (1600 °F) in a fast oil and temperature at 180 °C (350 °F) for 1 h. As shown, significant variation can occur, depending on type of powder and the extent of additives. It is very difficult to change powder compositions once tooling has been developed and maintain part tolerances.

Heat Treating Processes Applied To P/M Parts

The majority of P/M parts heat treated today are neutral hardened and tempered. This treatment was used to develop the properties shown in MPIF standard 35 (Ref 6). These steels are hardened primarily for wear resistance and improved core strength. They contain 0.5 to 0.8% C and usually have densities exceeding 6.8 g/cm^3. Most production P/M parts are normally hardened in batch-type sealed quench furnaces (rather than continuous belt furnaces) because they are net-shaped parts and should not touch or impinge upon one another during the heating cycle. This would tend to cause distortion and soft spots. The most widely used atmosphere for heat treating is endothermic gas with both methane and air additions to control carbon potential. Due to the rapid penetration of atmosphere into the interconnecting porosity, carbon diffusion is quite rapid and short cycle times at relatively low process temperatures are normally specified.

Medium- to high-density parts (6.8 to 7.2 g/cm^3) are better suited for batch-type sealed quench furnaces where the temperature profile can be easily controlled. With this equipment, parts can be loaded into the basket in the arrangement that best minimizes distortion. Also, control of carbon potential is more precise in batch equipment. The one drawback to a batch furnace is the

Table 1 Influence of density and sinter cycle on heat treat properties for 0.85 Mo-0, 13 Mn-0.40C alloy

Sinter cycle	Density g/cm^3	Sinter temperature °C	Sinter temperature °F	Atmosphere	Sintered ultimate tensile strength MPa	Sintered ultimate tensile strength ksi	Heat treated(a) ultimate tensile strength MPa	Heat treated(a) ultimate tensile strength ksi	Charpy impact J	Charpy impact ft · lbf	Apparent hardness, HRC	Microhardness, Surface	Microhardness, Core	Carbon content, % Surface	Carbon content, % Core
Press and sinter	7.02	1120	2050	N_2/H_2	515	75	1205	175	16	12	47	65	60	0.83%	0.57%
DPDS	7.42	1120	2050	N_2/H_2	635	92	1380	200	27	20	53	63	54	0.77%	0.44%
DPDS and HTS	7.46	1285	2350	N_2/H_2	690	100	1535	223	47	35	56 (surface)	63	51	0.73%	0.40%

(a) Heat treat: carburize at 900 °C (1650 °F) for 2 h, oil quench and temper 175 °C (350 °F) for 2 h

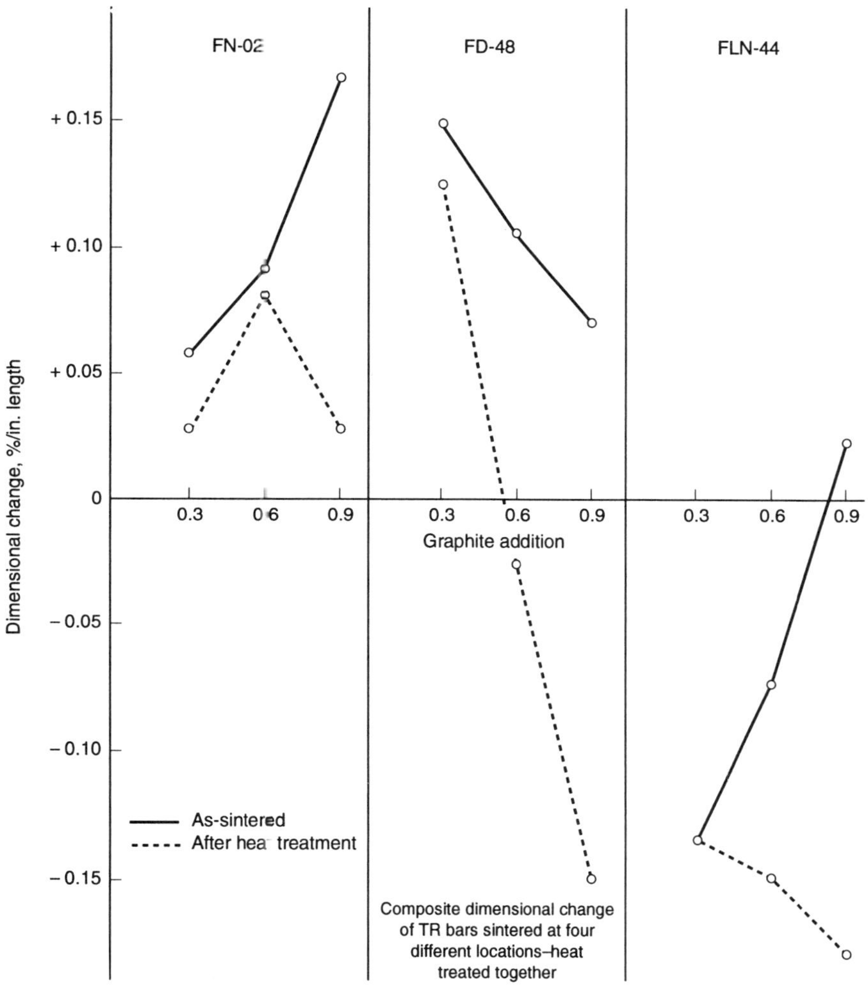

Fig. 8 Effect of carbon and heat treatment on size change. Density: 7.0 g/cm^3. TR, transverse rupture

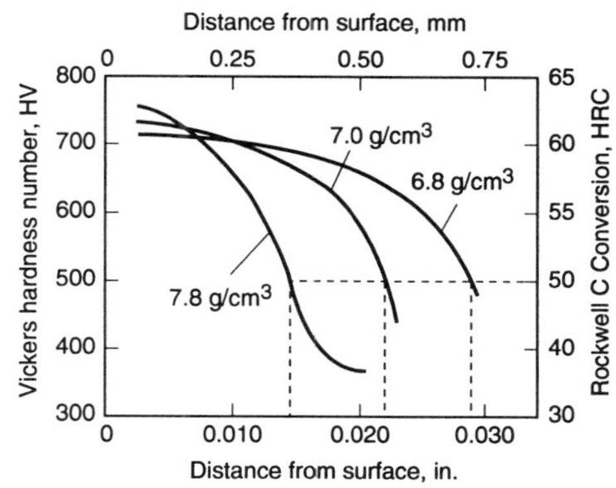

Fig. 9 Influence of porosity on carbon penetration and effective case depth during vacuum carburizing of an iron-carbon alloy at three density levels

longer transfer time to the quench. Many of the newer furnace designs have automated internal handling equipment to transfer the hot load to the quench. This can prolong the time to quench, causing a loss of temperature, and subsequent surface hardness.

When quenching for maximum hardenability, a fast oil is recommended. Quench oils with a General Motors quenchometer rating of 10 s or less is considered fast. Load size is also a critical factor in attaining uniform hardness. Most sealed quench furnaces are designed to quench 0.5 kg of steel per gallon of quench oil. For P/M parts, it is recommended that 11.5 to 15 L (3 to 4 gal) of quench oil be used to quench 0.5 kg of sintered steel. Overloading is probably the most common problem in attaining uniform heat treated properties.

Carburizing/Carbonitriding

These two case-hardening treatments are most commonly specified for applications that must withstand high-cycle loading and surface scuffing. Here, the need for high surface hardness combined with a relatively soft ductile core area to provide toughness. Most of the highly loaded gear applications require one or the other of these treatments.

In P/M applications, these treatments are used on high-density prealloy or partially alloyed steels where the combined carbon content is less than 0.5% C. Parts with porosity levels higher than 10% are not recommended for case hardening due to the rapid penetration of carbon and the difficult control of carbon potential in the furnace.

In wrought steel, case-hardening treatments are characterized by a uniform surface hardness and an effective case depth. This is usually defined as the depth at which a 50 HRC hardness is measured. Microhardness measurements can accurately show the hardness profile in a wrought steel section, but can be erratic when used on P/M steels where subsurface porosity can influence the hardness reading. It is recommended that at least three readings be taken at each depth level below the surface and averaged to determine effective case depth. Figure 9 shows the influence of porosity on carbon penetration and effective case depth during the vacuum carburizing of an iron-copper alloy at three density levels.

Carbonitriding differs from carburizing in that it provides both carbon and free nitrogen for diffusion. This is accomplished by the addition of anhydrous ammonia (≤ 2 vol%) to the furnace atmosphere. Nitrogen diffuses into the steel simultaneously with carbon, retarding the critical cooling rate on quenching, which leads to a more complete transformation to martensite. This produces a more consistent surface hardness profile, thereby improving the wear resistance of the P/M material.

Process temperatures for carbonitriding are usually lower than those required for carburiz-

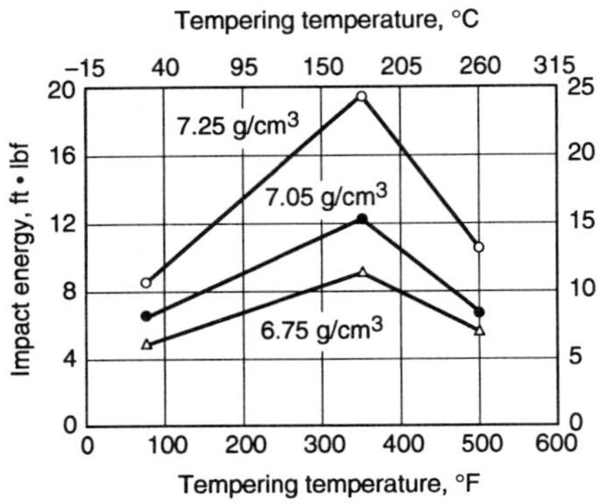

Fig. 10 Effect of density and tempering on impact energy

ing. This provides improved control of dimensions and less distortion of the part compared to carburizing. Care must be taken when adding ammonia; however, because excessive nitrogen diffusion into internal porosity can result in embrittlement and loss of strength.

Carbonitriding is a shallow-case treatment. Case depths greater than 0.5 mm (0.02 in.) deep are seldom specified. For this reason, cycle times are relatively short, usually 30 to 45 min depending on process temperature, case depth requirements, and density. Carburizing is the first part of the cycle followed by addition of ammonia. Process temperature is usually approximately 870 °C (1600 °F) depending on P/M composition and density.

As in neutral hardening, carbon control is a critical aspect of the treatment. Normally, carbon potentials of 1.0 to 1.2% are specified to maintain the carbon profile in the part. The usual (wrought) method is to keep carbon potential 0.85% to 0.95% C (to reduce the likelihood of producing mixed phases and retained austenite). In P/M parts, however, it may be difficult to maintain a uniform carbon content at the surface due to rapid penetration of gas through porosity. Decarburization can result if the effect of porosity is not adequately considered.

Tempering

Tempering is a postquench thermal treatment designed to reduce the stress imposed on a part by rapid quenching and martensite transformation. This stress, if not relieved, can increase the brittleness and notch sensitivity of the part, making it more prone to fracture. This stress is increased with increasing density level in the part. All parts with density levels above 6.7 g/cm^3 should be tempered after hardening. Recommended tempering temperatures for P/M parts range from 150 to 200 °C (300 to 400 °F). Above this range, entrained quench oil can ignite, creating a hazardous condition in the furnace. Tempering above 200 °C (390 °F) results in improved toughness and fatigue properties in the hardened part, at the expense of tensile strength and impact resistance; however, the tempering furnace would need to be specially adapted to handle the high volume of smoke generated by the volatilization of the quench oil. Tempering is also used to reduce the effects of retained austenite that can occur in alloy steels when rapidly quenched. This constituent has been shown to influence dimensional change and hardness of the material. Freezing by cryogenic treatment (below −100 °C, or −148 °F) for several hours transforms the retained austenite to martensite. This is usually followed by a second temper at 200 °C (390 °F) to stress relieve the newly formed martensite. Figure 10 shows the influence of tempering temperature on impact resistance of FL4205 alloy after hardening.

Induction Hardening

Spur gears, bevel gears, splined hubs, and cams are ideal components to utilized P/M production techniques. These parts usually require hard, wear-resistant surfaces in some areas, with retention of the ductility of the sintered matrix in the remainder of the part. Induction hardening is commonly specified for these applications. The process can be placed in an automated machining line to reduce handling and be a cost-effective hardening treatment when high volumes of parts are being produced. Because the inductance of P/M materials is typically reduced due to porosity, a higher power setting is normally required to reach a given depth of hardening compared to that used for a wrought material of similar composition. Furthermore, because the heat rapidly dissipates, a rapid transfer to the quench is mandatory.

As with wrought steels, the response to hardening by induction is dependent on combined carbon content, alloy content, and surface decarburization. This latter variable can be a major concern with P/M parts. Conventional belt-type sintering furnaces (depending on their construction) can allow decarburization to occur as the parts leave the hot zone and cool slowly through the 1100 to 800 °C (2010 to 1470 °F) temperature range. Newer designs are now able to compensate for this problem by installation of controlled cooling through the transformation range.

In most instances, induction-heated P/M parts are quenched in a water-based solution containing some type of rust preventative to forestall internal corrosion. In those applications where induction hardening is considered, densities above 90% should be specified. With a decrease in density, the resistivity of the steel increases and magnetic permeability decreases. For this reason, integral quench coils using a high-velocity spray quench are generally used to attain maximum surface hardness in the P/M part.

Nitrocarburizing

This process is rapidly gaining in popularity as an alternative treatment for P/M parts. Gaseous ferritic nitrocarburizing (FNC) is a surface treatment where both nitrogen and carbon are diffused into the surface of low-carbon steel parts under controlled conditions. To this extent the process resembles carbonitriding. It differs primarily in lower process temperature (570 to 630 °C, or 1060 to 1165 °F) and in atmosphere composition. As in carbonitriding (780 to 870 °C, or 1435 to 1600 °F), anhydrous ammonia is used as a source of nitrogen. Carbonitriding uses endothermic gas as a base with NH_3 addition for nitrogen. Nitrocarburizing, on the other hand, uses the NH_3 as the base gas and adds various hydrocarbons as the source for carbon. This process can be applied using various types of equipment, such as batch-type integral quench furnaces, pit-type furnaces, and plasma vacuum furnaces.

At the nitrocarburizing temperature range, no austenite transformation occurs, thereby significantly reducing the distortion that can occur on components that get oil quenched. Nitrogen and carbon are diffused into the surface of the steel part in sufficient concentration to form a thin layer of iron nitride on the surface, which appear as a white layer under metallographic examination. This white layer provides high hardness and lubricity for improved wear resistance. When applied properly, it can reduce the coefficient of friction of the steel surface to one-half its normal untreated value.

Because FNC is a diffusion process, its application to P/M parts is dependent on density. If the nitride compound is allowed to penetrate and form on an internal pore surface to any significant extent, a volume expansion as well as embrittlement can occur. For this reason, density of P/M parts should be 92% of the pore-free density. Because it requires warm compaction and/or additional re-pressing operations to reach this density level, a lower cost option would include steam treating prior to nitrocarburizing. This patented process can proceed in a conventional manner. The nitride layer, as mentioned before, improves the resistance of the part surface to adhesive wear. A subsequent oxidation process that can be applied during the nitrocarburizing cycle can also provide exceptional corrosion resistance as well as wear resistance.

A process that provides improved dimensional control is plasma nitrocarburizing. In this process, parts are processed in a vacuum chamber using a combination of direct and indirect plasma heating. This process is currently used in production of such parts as synchronizer hubs and cam lobes. The use of a plasma to deliver the nitrogen and carbon ions to the surface of the part allows for a more uniform control of surface concentrations and diffusion of the nitriding elements. It minimizes the nitridation of internal pore surfaces, thereby reducing the volume expansion that normally occurs during gaseous nitrocarburization.

This process can be applied to most iron blends and prealloys with uniform formation of the ε iron nitride hard phase at the surface of parts with sinter densities exceeding 6.9 g/cm^3. Below this density, porosity variations in the part can lead to nonuniform dimensional changes occurring on sintered iron transverse rupture bars

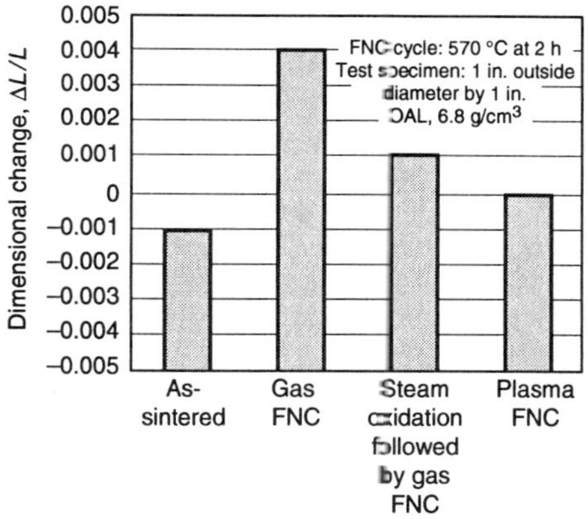

Fig. 11 Dimensional change in sintered iron after various nitriding treatments

Fig. 12 Microstructure of ferritic nitrocarburized FC-0208 (7.0 g/cm^3) hardened to file hard condition. 100×

that have been nitrocarburized by various methods (Fig. 11). Figure 12 shows the microstructure of a nitrocarburized P/M part. Figure 13 shows the microstructure of a P/M part that was cold forged to 95% pore-free density, nitrocarburized, and then underwent an oxidation treatment.

The FNC process is best used in applications where sliding wear and fretting are involved. Because the hard nitrided layer is relatively thin, the process should not be applied where high indentation or impact loading is involved.

The ε-nitride layer that is formed can attain a file hardness in excess of 60 HRC, depending on the alloy content of the steel. Indentation hardness testing is not recommended when evaluating this process.

Because no transformation occurs, the P/M parts can be air cooled without loss of surface hardness. Also, no oil absorption occurs, which leaves the porosity open and free of contamination. Nitrocarburizing also provides improved strength and reduced notch sensitivity in P/M parts. Figure 14 shows the fatigue improvement of two low-carbon P/M steels after nitrocarburizing.

Sinter Hardening

Powder metallurgy alloys that can transform to the hard martensite phase upon cooling from the sinter temperature have been commercialized. This technique, termed sinter hardening, is rapidly gaining popularity in applications requiring high strength and hardness immediately after sintering. This manufacturing route has proven to be cost effective for a number of parts because it eliminates the need for a postsinter heat treatment.

Prealloyed powders have been proven useful for this purpose. The ability of the alloy to promote the formation of martensite and/or bainite during cooling from the sinter temperature is the most important factor to optimize the microstructure and therefore the properties of the sintered parts.

Manganese is added to all commercial steels in the range of 0.25 to 1.0% to increase strength and hardenability of plain carbon steels. Chromium is also a popular hardening element, but it

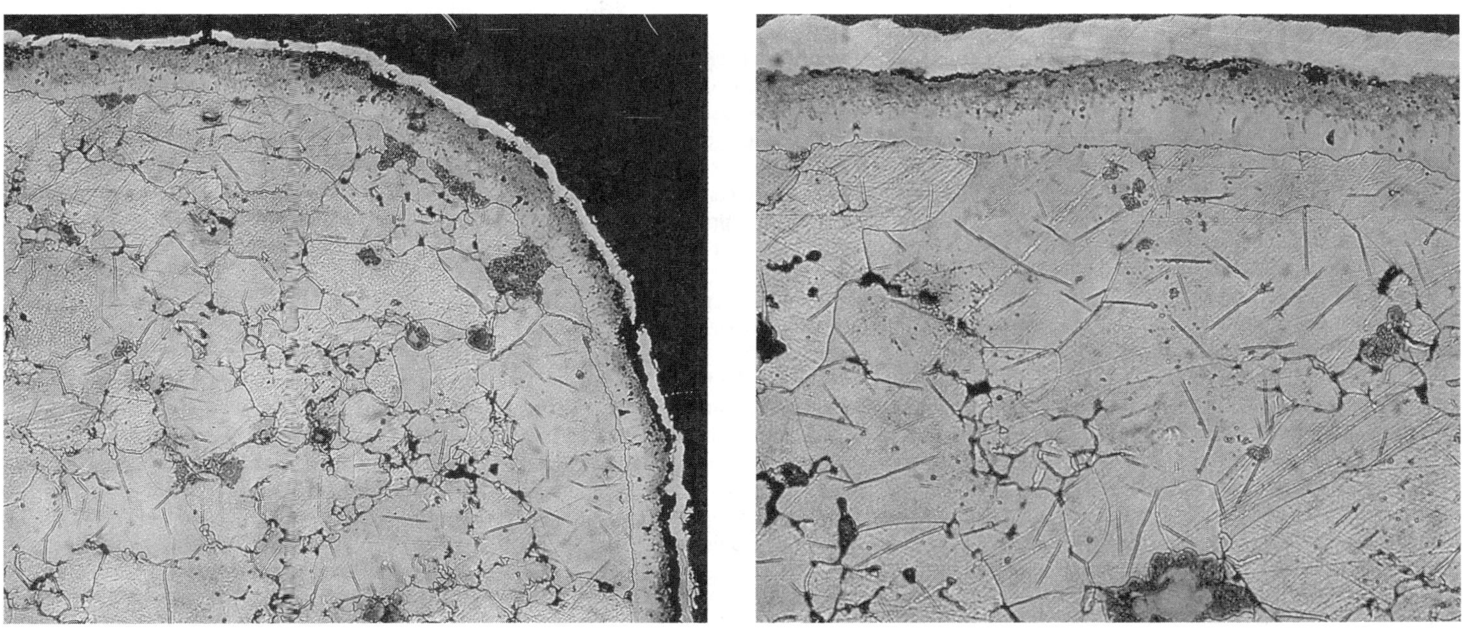

(a) (b)

Fig. 13 F-0000 P/M iron precision cold forged to 7.5 g/cm^3 and ferritic nitrocarburized to file hard condition. (a) Tooth corner. (b) Tooth, 400×

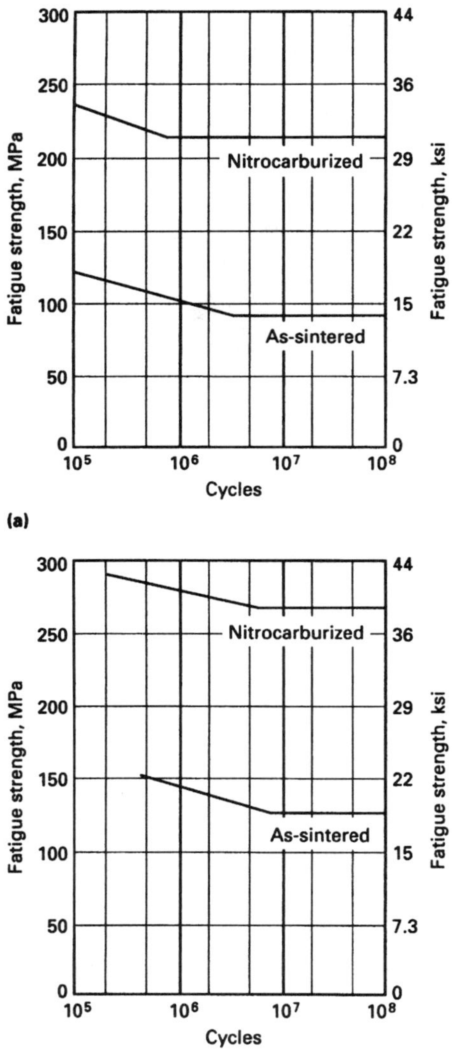

Fig. 14 Effect of nitrocarburization on the notched fatigue strength of two low-carbon P/M steels. (a) F-0000 carbon steel. (b) FC-0205 copper-carbon steel. P/M density: 7.1 g/cm^3. From Ref 1

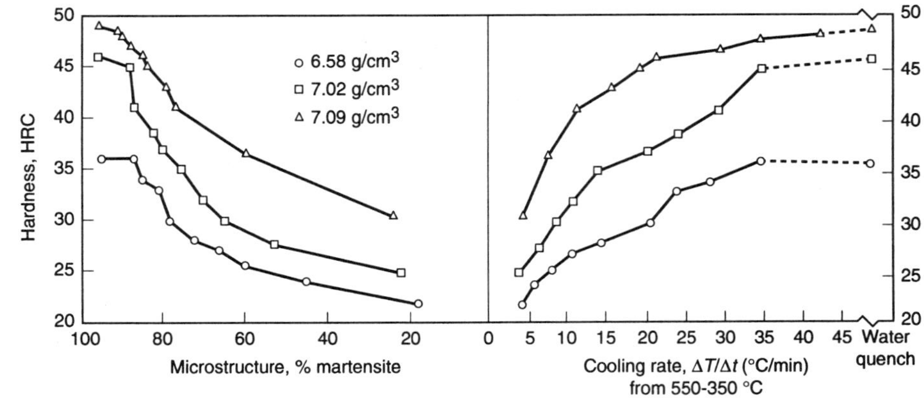

Fig. 15 Cooling rate-hardness-microstructure diagram to determine hardenability requirements. Source: Ref 17

is seldom used in prealloyed powders because of the difficulty in reducing the chrome oxide that can form at the surface of the particle and act as a barrier to interparticle diffusion. Molybdenum and nickel are most commonly used in low-alloy P/M steels because of their easily reducible oxides. These elements are very efficient in promoting increased strength and toughness. Sinter hardening requires controlled cooling after sintering in the austenite range (1120–1290 °C, or 2050–2350 °F). The factors that determine the ability of an alloy to sinter harden are cooling rate between 600 and 150 °C (1110 and 300 °F), which encompasses the martensite start (M_s) temperature of most P/M alloy steels, the alloy content, and the blend composition. The most commonly used alloy is type 4600 with additions of copper and carbon.

An optimal microstructure for a sinter-hardened material would show more than 90% martensite at the surface of the part with no less than 70% martensite in the core. This structure would provide the surface hardness and toughness of a quenched microstructure without the high stress concentration resulting from oil quenching.

Therefore, it is important to know the cooling rate and alloy content that provide the required hardenability for a part of a given mass and density. In a standard belt-type sinter furnace, an addition of 2% Cu and 0.9% C to the 4600 alloy may be necessary to provide sufficient hardenability for a given part density. With the new furnaces that are available with enhanced cooling capability, a lower alloy content or mix composition can be used to achieve the same level of hardenability.

A simple laboratory method has been developed to determine the effect of cooling rate, density, and alloy content on the hardness and microstructure of a sintered material. Cylindrical slugs of varying mass and density are sintered with thermocouples embedded at the surface and center of the slug. Another thermocouple is fixed on the belt to measure the ambient cooling rate of the furnace set at a known temperature and belt speed. Sectioning the slugs to characterize the microstructure and hardness profile at the location of the thermocouple provides the necessary information to construct a diagram as shown in Fig. 15, which shows the results for a type 4600 alloy. These microstructure-hardness-cooling rate diagrams can be constructed for various alloy blend compositions and density levels to find the most cost-effective material for a given application.

There are several advantages to sinter hardening besides cost effectiveness:

- Microstructures can be better tailored to provide optimal strength and hardness.
- The porosity remains free of oil, which is advantageous to parts operating in controlled environments.
- Distortion in complex-shaped parts is reduced compared to oil quenching.

This reduced distortion translates into improved dimensional stability and consequently higher yield and quality of production lots.

The major drawback of sinter hardening is that the alloys used require a high carbon content to achieve the necessary hardenability. As has been shown previously, high carbon levels reduce the mechanical properties of P/M steels after hardening. Table 2 shows a comparison of properties of a type 4600 alloy with 0.5% C that was quench and tempered with a 4600 alloy (with a 0.8% C level) that was sinter hardened. As with quenched alloys, sinter-hardened alloys require post tempering treatment to attain the optimal combination of hardness and toughness.

Table 2 Comparison of properties of quenched-and-tempered type 4600 alloy (0.5% C alloy) with a sintered hardened 4600 alloy (0.8% C)

Material code	Density, g/cm^3	Ultimate tensile strength, MPa	Ultimate tensile strength, ksi	Apparent hardness, HRC
FL-4605-120HT	6.95	895	130	34
FLC-4608-90HT	7.00	700	100	31

Steam Treating

Many P/M parts have traditionally been steam treated for improved wear resistance, corrosion resistance, and sealing capacity. In this process, P/M parts are heated in a specific manner under a steam atmosphere at temperatures between 510 and 570 °C (950 and 1060 °F) to form a layer of black iron oxide identified as magnetite, in the surface porosity, according to the chemical reaction:

$$3Fe + 4H_2O\,(\text{steam}) \leftrightarrow Fe_3O_4 + 4H_2\,(\text{gas})$$

Steam treating cannot truly be described as a heat treatment because no structural changes occur in the matrix. In the process, magnetite (Fe_3O_4) is formed at the interconnecting surface porosity, filling the porosity with a second phase. Magnetite has a hardness equivalent to 50 HRC.

The process itself is straightforward; the primary variables are temperature, time, and steam pressure. Caution must be exercised to prevent the formation of hydroxides and lower oxide such as ferrous oxide (FeO) and ferric oxide (Fe_2O_3), which is red rust (see Fig. 16). Also, adhesion of the surface oxide layer is an impor-

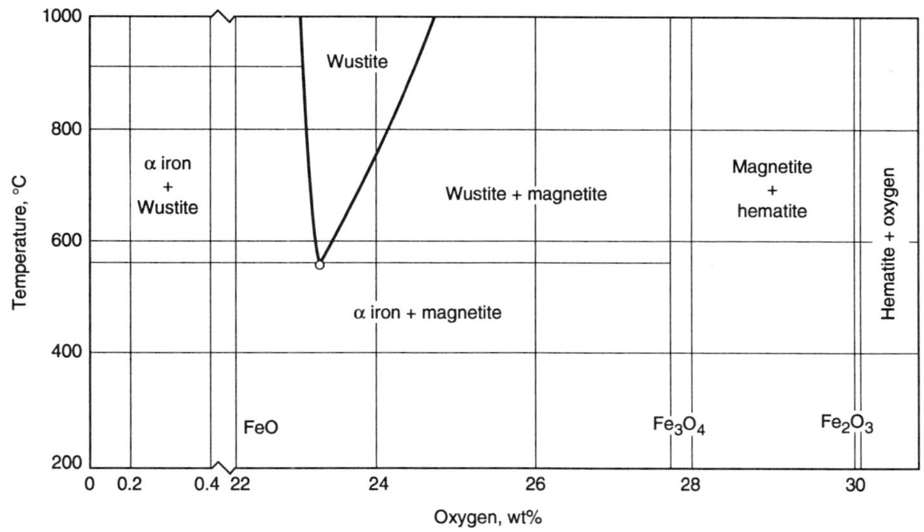

Fig. 16 Temperature-composition diagram for the iron-oxygen system. Source: Ref 11

Sintered density is an important consideration when applying steam treating for improved strength and hardness. Its ability to increase the wear resistance of the substrate material depends on the available porosity for oxidation. As density is increased, the amount of oxide formed is decreased, which minimizes the improvement in apparent hardness attributed to steam treating. This is shown in Fig. 17 for sintered iron.

The increase in density and apparent hardness produced by steam treating is illustrated in the micrograph of a sintered iron (Fig. 18). By filling the porosity with a hard second phase, the P/M steel offers a better support to the indentation hardness tester. Figure 19 illustrates that the transverse rupture strength is increased significantly by steam treatment for low-carbon P/M steels, but only modestly for high-carbon (0.8% C) P/M steels.

tant variable to control. Adhesion is strongly influenced by process time at temperature.

Spalling or flaking of the surface oxide layer can occur if the process temperature exceeds 570 °C (1060 °F) and process times exceed 4 h. The maximum thickness of the surface oxide layer should not exceed 7 μm (0.003 in.). Beyond this thickness, flaking can occur due to an increase in surface tensile stress.

The recommended procedure for steam treating is:

1. Preclean parts to remove any oil or lubricants that may have been absorbed into the porosity from prior machining, sizing, or finishing operations.
2. Load clean sintered parts in loosely packed baskets and place fixture into a furnace preheated to 315 °C (600 °F).
3. Heat parts in air until the center of the load has stabilized at the set temperature.
4. Introduce superheated steam at a line pressure of 35 to 105 kPa (5 to 15 psi) and allow furnace to purge for at least 15 min.
5. Increase furnace temperature to desired steam treatment temperature and hold for no longer than 4 h at heat.
6. On completion of the treatment, reduce furnace temperature to 315 °C (600 °F). When parts reach this temperature, the steam can be shut off and the parts unloaded.

Caution should be used when opening the furnace door after the steam cycle. As indicated above, hydrogen is produced during this process and can ignite. It is recommended that a nitrogen purge be applied prior to unloading. This process, when correctly applied, can impart improved surface properties and, depending on steel composition, increase compressive yield strength.

The above process parameters refer to the standard pit-type furnace. Currently, belt-type continuous furnaces have been developed, but have had limited success in developing consistent oxide content. Needless to say, the equipment must be properly sealed to prevent air contamination, or erratic oxide formation, and discoloration can occur.

In steam-treated P/M steels, the ductility is significantly reduced due to the internal stresses created by the formation of the iron oxide. Care must be taken when steam treating high-carbon P/M steels because these internal stresses can initiate microcracking and cause severe loss of ductility. Cases have been reported in which steam-treated parts accidentally dropped on the floor shattered in a brittle fashion. The best recommendation for preventing such an incident is to specify a maximum carbon content of 0.5% for P/M parts that are to be steam treated.

Guidelines for Heat Treating P/M Parts

Whether heat treating P/M parts is done in-house or by an outside commercial heat treater, some guidelines need to be followed to ensure that parts are properly treated. In the past, heat treatment of P/M parts was primarily specified to improve wear resistance. With the advent of high-compressibility powders and high-temperature sintering, heat treatments are now added to provide improved dynamic properties as well. Recommendations for heat treating porous P/M parts are:

- Always degrease parts prior to heat treating. In many manufacturing plants, P/M parts are dipped in a rust preventative after sintering. Also, many P/M parts are machined prior to

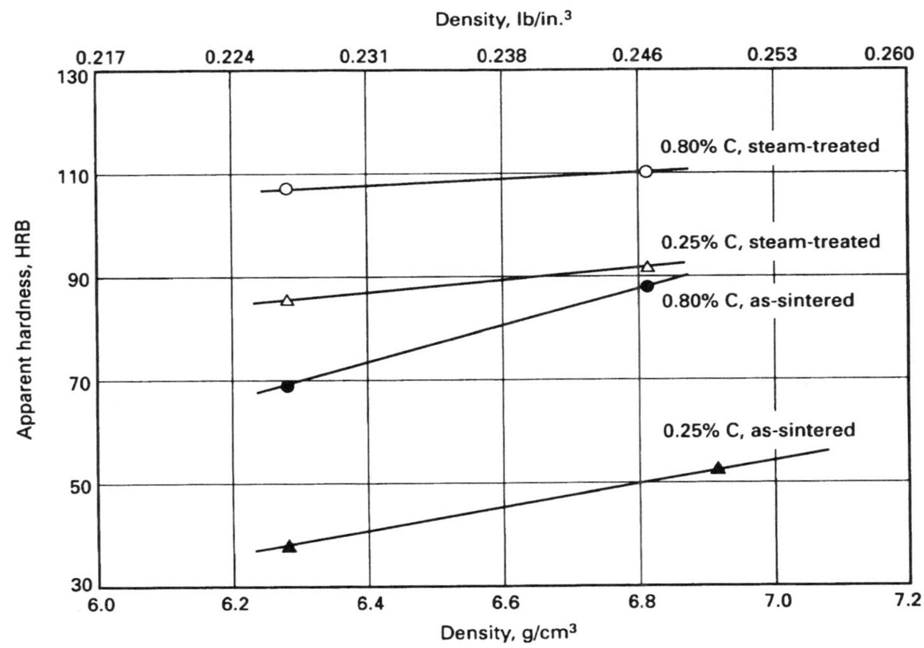

Fig. 17 Effect of steam treating on transverse rupture strength of sintered P/M carbon steels. From Ref 1

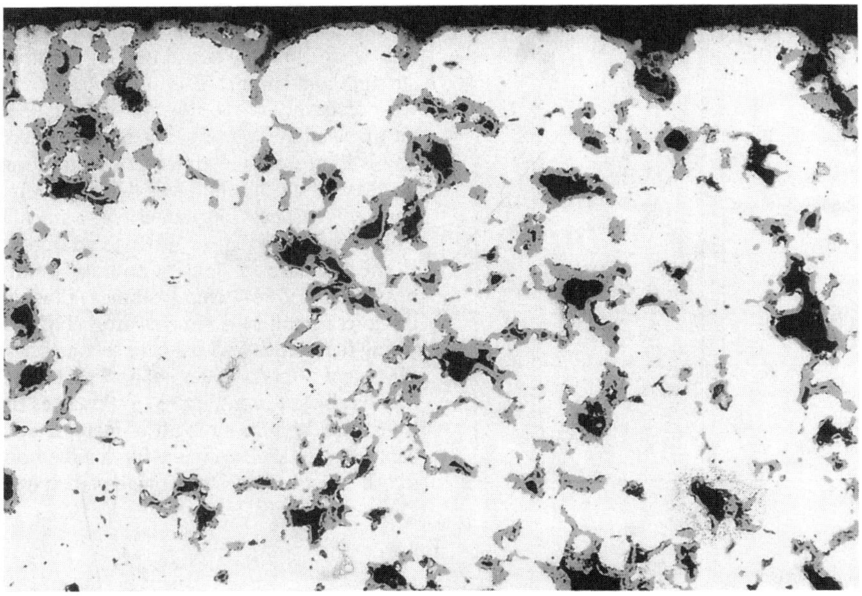

Fig. 18 Micrograph of steam-treated structure of alloy FC-0205 (6.75 g/cm^3). Unetched, 200×

heat treatment and can retain some of the lubricant. These oils can contaminate the heat treating atmosphere and cause discoloration.

- Do not use dense loads. Overlapping parts can cause soft spots and distortion. For best results, parts should be single spaced in layers.
- Use highly agitated quench oil with a quench-severity rating (*H*) of 0.7 to 1.0.
- When tempering, hold at 205 °C (400 °F) for 2 h before proceeding to higher temperatures. This minimizes the evolution of smoke and prevents ignition.

If it is planned to send parts to an outside commercial heat treater, a technical review of the part requirements and prior processing history is recommended. Many commercial shops have had bad experiences with powder metals because they did not fully understand the implications of porosity and tried to treat the parts as they would wrought steel. It is important that a good communication network is established with the heat treater so that they can be alerted to any structural changes in the P/M part caused by changes in powder lot, tooling, or sintering cycle.

A good procedure is to send sample lots to the heat treater for them to establish process capability. After the print requirements are met, the heat treater should run the parts in the same furnace under the exact same conditions as were established. Parts fabricators should give as much consideration to the heat treating process as would be given to primary processes. Reworking because of improper heat treatment not only adds more cost, but can also significantly affect the physical properties of the P/M part.

Mechanical Properties

The 1997 edition of MPIF standard 35 contains updated information on the heat treated properties of P/M materials. These properties are difficult to measure due to the extreme brittleness of the porous test specimens required for P/M materials. The MPIF standard only covers alloys containing 0.50% combined carbon, which is the maximum carbon level needed to produce the optimal heat treat properties.

Figure 20 charts the ultimate tensile values of various hardened and tempered P/M compositions (with a 7.0 g/cm^3 density, which represents the maximum strength that can be obtained with a single room temperature compaction process). As expected, strength increases as alloy content increases, with the prealloy irons providing the most consistent improvement in properties. Maximum strength, however, results from nickel additions to prealloyed iron, which produce a heterogeneous microstructure containing soft, nickel-rich areas. These soft areas dispersed throughout a hard martensite matrix provide both high tensile and fatigue strength (see Fig. 7).

Heat treatment, in many cases, is specified for wear resistance. For this, the highest achievable surface hardness is required. A different material selection may be advisable when maximum toughness is required for improved fatigue endurance. Figure 21 charts the apparent hardness of various alloys at 7.0 g/cm^3 sinter density that have been quenched and tempered. The FC0205 alloy provides the highest surface hardness (also at the lowest cost). This alloy is created by liquid phase sintering to form a copper-rich martensite upon quenching. This has been found to provide excellent wear and indentation resistance, although it does tend to reduce the ductility of the material.

When designing a P/M part of large mass and cross-sectional thickness that requires heat treatment, a knowledge of the hardenability of P/M alloys would be desirable. Jominy hardenability values have now been generated and published in the 1997 edition of MPIF standard 35 (Ref 6). As with wrought alloys, hardenability increases with both alloy and carbon content up to 0.80% C. However, these high-carbon alloys become very brittle upon hardening and, as stated before, a 0.50% C content is usually specified for optimal properties. Figure 22 charts the increase in hardenability with alloy content for materials containing 0.50% combined carbon and sintered to a 7.0 g/cm^3 density. As shown, the FLN4-4405 alloy, which contains 0.90% Mo in alloy and 4% Ni admixed can be hardened to a depth of 0.75 in. (19 mm). This is sufficient hardenability for the majority of the P/M parts being made currently.

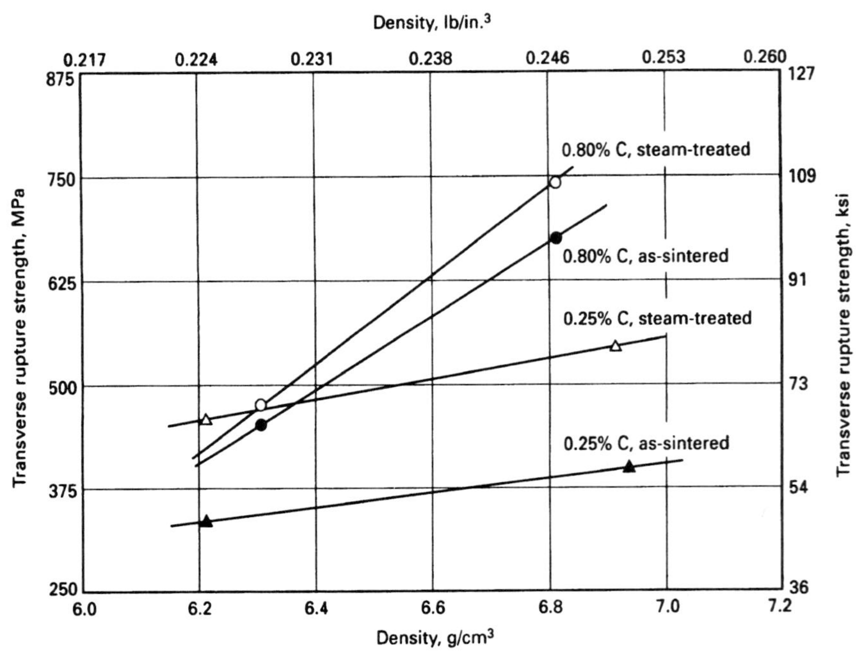

Fig. 19 Effect of steam treating on transverse rupture strength of sintered P/M carbon steels. From Ref 1

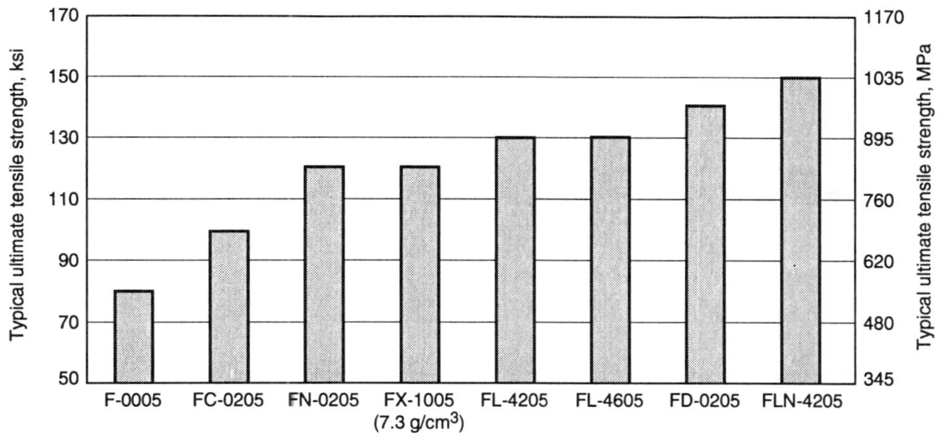

	F-0005	FC-0205	FN-0205	FX-1005 (7.3g/cm³)	FL-4205	FL-4605	FD-0205	FLN-4205
Copper content, %	…	2.0	…	15.00	…	…	1.50	…
Nickel content, %	…	…	2.0	…	0.45	1.80	1.80	2.00
Molybdenum content, %	…	…	…	…	0.65	0.65	0.50	0.65

Fig. 20 Strength comparison of heat treated P/M steels at 7.0 g/cm^3 density

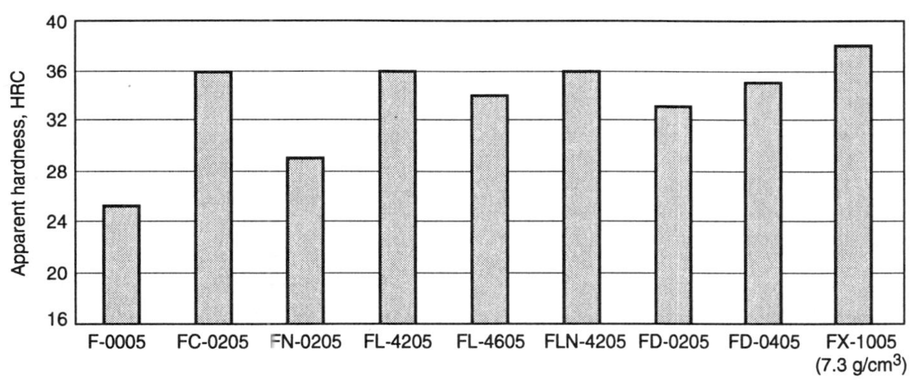

	F-0005	FC-0205	FN-0205	FL-4205	FL-4605	FLN-4205	FD-0205	FD-0405	FX-1005 (7.3 g/cm³)
Copper content, %	…	2.0	…	…	…	…	1.50	1.50	15.00
Nickel content, %	…	…	2.0	0.45	1.80	2.00	1.80	4.00	…
Molybdenum content, %	…	…	…	0.65	0.65	0.65	0.50	0.50	…

Fig. 21 Hardness comparison of heat treated P/M steels at 7.0 g/cm^3

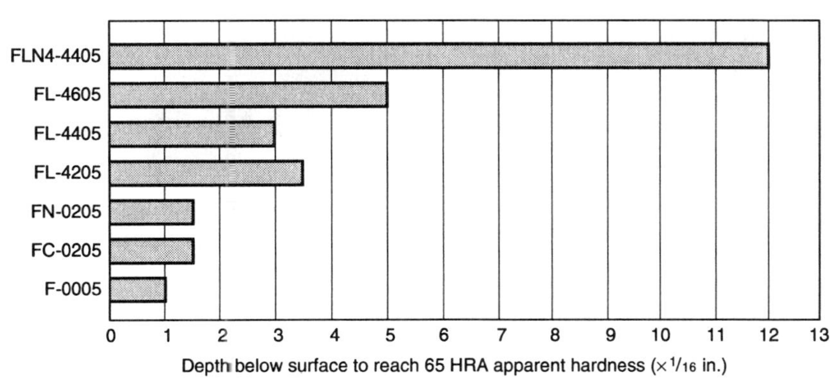

Fig. 22 Jominy hardenability of P/M steels at 7.0 g/cm^3

REFERENCES

1. H.A. Ferguson, Heat Treatment of Powder Metallurgy Steels, *Heat Treating,* Vol 4, *ASM Handbook,* ASM International, 1991, p 229–236
2. R. Burns, Production Presses and Tooling, *Powder Metallurgy,* Vol 7, *Metals Handbook, 9th Ed.,* American Society for Metals, 1984, p 229
3. J.M. Capus, Metal Powder Manufacturers Adapt to Changing Needs, *ASTM Stand. News,* Vol 18 (No. 3), 1900, p 54
4. *Powder Metallurgy Design Manual,* Metal Powder Industries Federation, 1989, p 65
5. S. Mocarski, D.W. Hall, J. Khanuja, and S.-I. Suh, Parts for Automotive Applications—Metal Powder Industries Federation, Part III, *Int. J. Powder Metall.,* Vol 25, 1989, p 103
6. *Materials Standards for P/M Structural Parts, Standard 35,* 1997 ed.
7. L.F. Pease III, Ferrous Powder Metallurgy Materials, Properties and Selection: Irons, Steels, and High-Performance Alloys, *ASM Handbook,* ASM International, 1990, p 801
8. H.A. Ferguson, Heat Treatment of P/M Parts, *Met. Prog.,* Vol 7 (No. 6), 1975, p 81–83
9. H.A. Ferguson, *Met. Prog.,* Vol 108 (No. 1), July 1975, p 66–69
10. A. de Rege, D. Pantano, and L.F. Pease III, Air Hardening of Low Alloy Powder Metallurgy Materials, *Advances in Powder Metallurgy (1989),* Vol 1, Metal Powder Industries Federation, 1989, p 223
11. L.F. Pease III, J.P. Colette, and D.A. Pease, Mechanical Properties of Steam-Blackened P/M Materials, *Modern Developments In P/M,* Vol 21, Metal Powder Industries Federation, 1988, p 275
12. R.J. Causton, J.A. Hamill, Jr., and S.O. Shah, Properties of Heat Treated P/M Alloy Steels, *Advances in Powder Metallurgy and Particulate Materials–1993,* Vol 4, Metal Powder Industries Federation, p 61
13. J.J. Fulmer and R.J. Causton, Tensile, Impact and Fatigue Performance of a New Water Atomized Low-Alloy Powder-Ancorsteel 85 HP, *Advances in Powder Metallurgy 1990,* Vol 2, Metal Powder Industries Federation, p 459
14. R.A. Ketterer, Production Aspects of Powder Metal Forging, *Advances in Powder Metallurgy 1989,* Vol 1, Metal Powder Industries Federation, p 321
15. *Standard Test Methods for Metal Powders Metallurgy Products,* 1993 ed., Metal Powder Industries Federation, 1993
16. H.A. Ferguson, "Steam Sealing for Nitrogen Treated Ferrous Part," U.S. Patent 4,738,730, 1988
17. H.A. Ferguson, G.L'Esperance, E. Duchesne, A. de Rege, "Effect of Mass/Cross Sectional Thickness of Sinter Hardening of Two Prealloyed Steels of Different Hardenability," *Advances in Powder Metallurgy and Particulate Materials,* Metal Powder Industries Federation, 1997
18. M. Kamada, Austempering of Sintered Low Alloy Steels, *Int. J. Powder Metall,* Vol 27, 1991, p 255

Welding and Joining Processes

Jack A. Hamill, Hoeganaes Corporation

JOINING P/M components increases the possibilities of manufacturing more complex configurations than conventional die-pressed geometries. Most P/M materials can be welded or joined successfully. However, not every potential P/M welding application is cost effective in terms of labor intensity, equipment, and overall manufacturing costs.

Several references detail the related joining processes and weldability of wrought materials (Ref 1, 2). However, very little comprehensive information is available regarding the application of these practices to P/M components. This article attempts to characterize the physical differences between P/M and wrought or cast materials as they apply to joining. In addition, acceptable procedures and techniques are discussed along with the weldability of various P/M materials.

Effects of Porosity on Properties. Perhaps the most important physical characteristic of a P/M component, which makes it uniquely different in regard to joining, is porosity (Ref 3). The pore volume or relative density has a pronounced influence on several important properties that affect the welding characteristics. These include:

- *Thermal conductivity* is dependent on the amount of porosity, and greater amounts of porosity change the heat transfer mechanism and ultimately influence the welding parameters.
- *Hardenability* is dependent on the rate of heat dissipation or thermal conductivity. Pores act as thermal insulators that slow the transfer of heat, making P/M components less hardenable than wrought parts of similar composition. However, because of potential densification associated with particle melting in the heat-affected zone (HAZ), the higher stress levels associated with greater hardenability can occur in P/M components and may increase the material's susceptibility to cracking.
- *Thermal expansion* is not directly associated with porosity, but this characteristic also plays an important role in determining susceptibility to cracking. Likely differences in composition between P/M, wrought, or cast materials result in dissimilar volume changes when metals are heated and subsequently cooled. Extreme differences in expansion or contraction rates increase the potential for cracking at the weld interface.

Porosity can also cause erratic fluctuations in welding performance because of entrapped oxides or impurities. Precautionary measures must be taken to eliminate any of the following residual elements that may remain in the pores as a consequence of related process conditions:

- Lubricants
- Quench oils
- Machining coolants
- Plating solutions
- Impregnating materials
- Cleaning or tumbling agents
- Free graphite or residual ash

Joining Methods

Joining is a comprehensive term used to denote all processes that affix one part to another. Most welding processes require the application of heat with some controlled melting of the base materials and/or filler metal (fusion), while others rely on surface diffusion and/or mechanical interlocking (solid state). Both solid-state and fusion welding processes have been used successfully in joining P/M components.

Lower-density (<6.5 g/cm^3) parts are most often joined using processes that do not involve molten weld metal and the associated solidification stresses. Because of fewer particle-to-particle bonds at these density levels, the relatively poor bond strength does not provide the ability to absorb high stresses generated by contracting weld metal. For example, Fig. 1 shows the result of a poor weldment design using P/M alloy F-0000 and wrought AISI 1010 stamping. The F-0000 P/M section has a density less than 6.2 g/cm^3, and an unusually large weld bead deposit results in a classic toe crack because of high weld metal solidification stresses.

The joining processes most often successful on lower-density parts are diffusion welding, sinter bonding, adhesive joining, and brazing. Parts at intermediate density levels (6.6 to 6.9 g/cm^3) can be joined using a wide assortment of processes. However, those that minimize the volume of fused metal, such as resistance projection, high-speed pulse welding (HSPW), or friction welding, enjoy the highest success rates. Parts exhibiting higher densities (>7.0 g/cm^3) typically have the same weldability as wrought materials and can be joined using fusion processes that include gas tungsten arc welding (GTAW), gas metal arc welding (GMAW), electron beam welding (EBW), and laser beam welding (LBW).

Resistance Projection Welding (RPW) is perhaps one of the most often used procedures for joining P/M parts. The joining surfaces are heated from the resistance of electrical currents, and the heated materials form a weld. Electrodes that carry the current also clamp the components under pressure to provide good electrical contact. The joint must be clean and the pores free of contaminants to obtain uniform weld soundness. Projection welding generally affords good dimensional control in the weld area because of the force applied by the contacts and the related

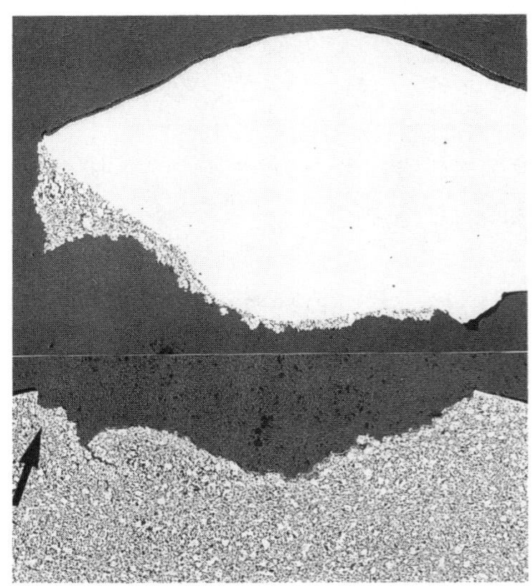

Fig. 1 Example of poor weldment design involving GMA welding of F-0000 P/M material (<6.2 g/cm^3) to AISI 1010 stamping. The unusually large weld bead deposit resulted in a classic toe crack because of high weld metal solidification stresses. Arrow indicates crack initiation site.

Fig. 2 Resistance projection welding application joining an intermediate density P/M spur gear to a low-carbon wrought steel stamping

fixturing of the components. An example of an RPW component is shown in Fig. 2.

In general, good weldment design requires particular attention be placed on the height and density of the projections. Other parameters including the current (kV · A) and pressure (psi) can be preset to achieve the best results and most often are very similar to those used for wrought materials. Approximate settings for current, weld time, and electrode force can be found in the *Resistance Welding Manual* (Ref 4).

An investigation involving projection welding of wrought materials to P/M parts at densities between 5.8 and 7.4 g/cm^3 found that lower density (<7.0 g/cm^3) components require greater projection heights (Ref 5). It was also found that steam-treated parts could not be satisfactorily welded at any density, and wrought chrome plated parts could not be welded to P/M components below 7.0 g/cm^3 density. All other materials tested, including plain iron, iron-carbon, Fe-Cu-C, and stainless steels, were welded successfully. In a similar investigation, RPW techniques were used to join high-carbon (0.4 to 0.8%) P/M parts, with density levels between 6.5 to 7.5 g/cm^3, to mild steel (Ref 6). The weldments were found to be of satisfactory quality and exhibited higher torsional strengths than standard keyed components. However, the highest torsional strengths were achieved after the components had been normalized.

High-speed pulse welding (HSPW/CD) is similar to RPW in many respects, but also exhibits several unique attributes. The major differences involve the capacitor stored energy supply with the resultant short-weld-pulse duration. Energy is concentrated toward the specific weld area using a quick pulse of 3 to 15 ms in the range of 150 j/mm^3 energy and subsequent pressure of 200 N/mm^3. However, specific weld parameters are dependent on system resistance, part shape, conductor connections, and material compositions.

The low-energy diffusion weld minimizes distortion, providing the ability to maintain reasonably tight tolerance control. Several studies (Ref 7, 8) indicate HSPW can successfully weld high-carbon materials and dissimilar material combinations to accommodate various performance characteristics. The ability to temper the weld joint using a second pulse, immediately after establishing the joint, helps reduce residual stresses while increasing strength and fatigue resistance.

Figure 3 shows a high-speed pulse weld of automotive engine bearing brace and windage baffle assembly. Brace rails are FN-0208 jointed to AISI 1010 steel stamping. Part design reduced vibrations induced from the bearing cap and improved engine performance.

Gas tungsten arc welding (GTA/TIG) uses a nonconsumable tungsten electrode to establish a direct current arc (for ferrous materials) that melts the material to be welded. An inert gas shield of argon, helium, or various commercial mixtures is used to protect the electrode and weld pool from oxygen, nitrogen, and hydrogen contamination and to stabilize the arc characteristics for improved welding performance. In some P/M applications, the process includes the use of a filler metal to counteract the shrinkage that results from particle melting and subsequent densification in the weld zone. The filler must be compatible with both materials to be joined in regard to thermal expansion characteristics, strength, and corrosion properties.

The GTA process provides suitable results in many situations because it allows a greater amount of control over the welding process. The travel speed, heat input, type of filler metal, and feed rate can be controlled to achieve the best results for each application. For this reason, GTA exhibits better weldability with higher-carbon materials because of the ability to control the heat input and subsequent weld hardness. Figure 4, for example, shows a FN-0205 cap nut welded to an alloy steel, hole-saw body. In this instance, GTAW was selected because of its ability to minimize distortion associated with high heat input.

One unique GTA application is the replacement of a casting by welding two P/M components together for use in a commercial truck differential (Ref 9). It was found that the welded P/M component exhibited higher and more consistent strength values than a bolted gray iron casting along with a 35% cost savings compared with the previous method of manufacture.

Gas metal arc welding (GMAW/MIG), unlike GTA, uses a continuous wire electrode that establishes the arc and furnishes filler metal into the weld pool. A shielding gas mixture, the composition of the mixture dependent on the mode

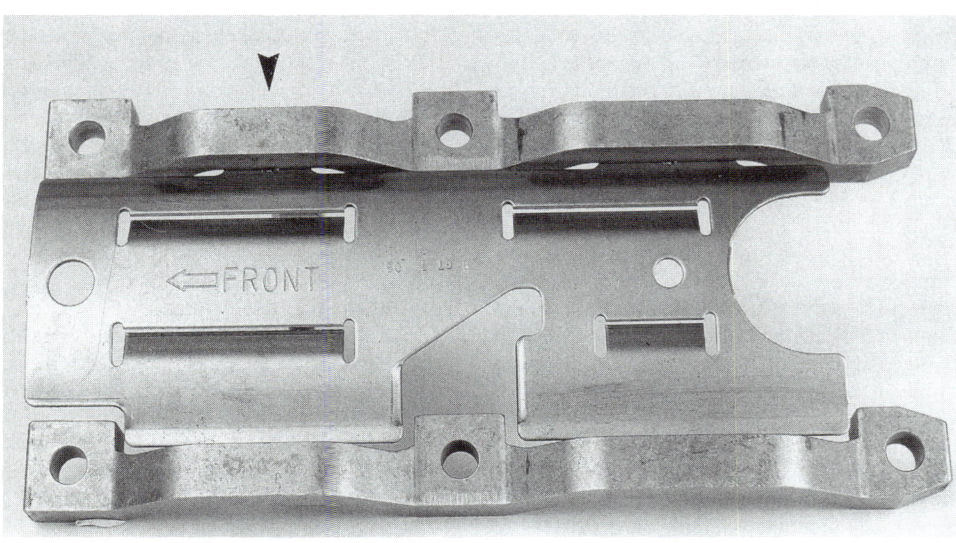

Fig. 3 High-speed pulse weld of automotive engine bearing brace and windage baffle assembly. Brace rails are FN-0208 joined to AISI 1010 steel stamping. High-speed pulse weld assembly replaced an aluminum forging and reduced overall cost by 40%. Courtesy of Process Equipment Company

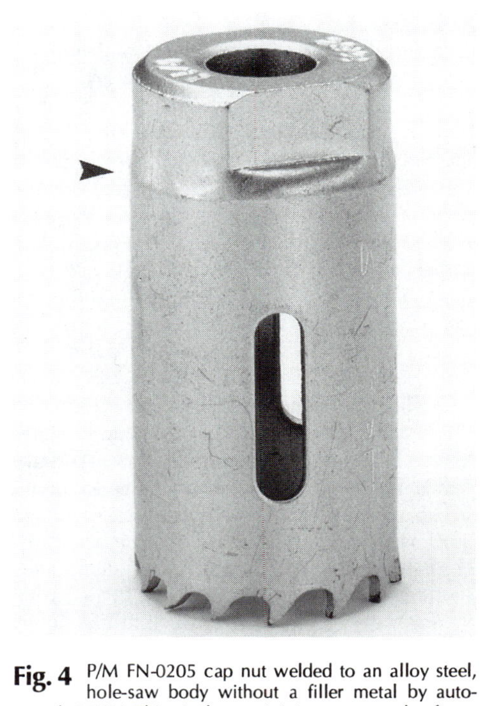

Fig. 4 P/M FN-0205 cap nut welded to an alloy steel, hole-saw body without a filler metal by automated GTAW. This single-step joining process also forms the seam of the cylindrical body.

of metal transfer and type of filler metal, is used to protect the molten weld pool from atmosphere contamination and to provide the correct arc characteristics. The short circuiting or dip metal transfer modes are particularly suited for welding P/M components because of the low energy input requirements that reduce the HAZ and minimize distortion. Typically, CO_2-based shielding gases are used with small diameter (0.89 to 1.14 mm) mild steel filler wire to weld sintered steel components. However, austenitic stainless steel fillers with inert gas shields can be used for higher carbon materials to reduce the potential for HAZ cracking. As in the GTA process, GMA affords the operator greater control over the entire welding process.

A study involving the GMA (MIG-CO_2) process concluded that various materials with admixed compositions of 0.3 to 0.9% C, 0 to 6.0% Ni, and 0 to 1.0% Mo could be welded successfully (Ref 10). The results indicate that sintered components appear to be less susceptible to hydrogen-induced cracking than wrought materials with similar hardenability. The same study also found copper-bearing materials (2.0 to 4.0%) could not be welded using the dip transfer-CO_2 process because of liquation cracking in the HAZ.

Friction welding (FRW), sometimes referred to as inertia welding, is a solid-state process that uses frictional heat generated when parts are rotated against one another. When the interface temperature of the part is heated into the forging range, an axial or upsetting force is applied to both sections. *Welding Handbook* (Ref 11) identifies the process variables as rotational speed, axial force, welding time, and upset force. Equipment power requirements range from 25 kV · A for smaller machines to 175 kV · A for larger units capable of joining 125 mm (5.0 in.) diam steel bars. Friction welding variables can be controlled by a computer program that makes the process less technique sensitive and highly automated. The time required to load and unload parts is often longer than the weld time, which can range from 0.5 to 15 s. The limiting factor in this process is the part configuration. Typically, components must accommodate the application of the rotational force at the joint interface. Materials with good dry bearing characteristics, for example, grades with free graphite in the pores, are not well suited to friction welding. The equipment, tooling, and automated material handling costs generally limit the use of this process to applications with long production runs.

Friction welding of iron-carbon and Fe-Cu-C sintered parts has been accomplished. An independent study (Ref 12) indicates this technique is capable of producing extremely favorable results when joining a full range of admixed compositions at various density levels to similar grades, dissimilar compositions, or wrought materials. Test results found that FRW exhibits lower hardness in the weld zone, in comparison with other processes, because of a slower cooling rate associated with this technique. The investigation determined that actual welding parameters are relatively flexible except for friction (axial) pressures and forging (upset) pressures. Higher friction pressures are likely to cold work a greater surface area when the forging pressure is applied. Excessive forging pressures can produce radial cracks that originate in the extruded flash and extend back into the base material. This situation can lower fatigue performance or ultimately result in joint failure. It has also been determined that welds often exhibit higher torsional load capacity in comparison with other joining techniques over a wide range of densities and material compositions.

Electron beam welding (EBW) is a sophisticated welding process that incorporates a beam of concentrated high-velocity electrons into a narrow, very intense source of heat. The weld generally is accomplished in a vacuum chamber at pressures ranging from medium to high vacuum, although sintered parts have been welded using a nonvacuum process. Electron beam welding has proved successful particularly on higher-density stainless steel P/M components. Iron-carbon sintered parts, on the other hand, have limited applications because of porosity and blowholes caused by escaping CO gas. An investigation found that as the welding speed slows, the width of the weld zone widens and the volume of escaping gases increases, causing more blowholes (Ref 13). It was also discovered that faster weld speeds produce a narrow HAZ with exceptionally high hardness values that would not be acceptable in most service environments. Considering the cost and disadvantages, it would be hard to justify this process for joining high-volume P/M structural components.

Laser beam welding (LBW) is somewhat similar to EBW, but it uses a focused laser to vaporize metal at the joint interface. Typically, CO_2 lasers are used with inert shielding gases such as helium. It is possible to add filler metal, but additions are not always necessary, particularly on higher-density components. The joint design has a major influence on the filler metal requirements. Welding parameters generally attempt to optimize the relation between the power setting and travel speed to improve the economics and provide a narrow weld seam that results in less internal stress. An investigation into the practical use of laser welding P/M components found it can be used successfully, particularly when joining higher-density parts with materials containing lower carbon levels (Ref 14).

This process can involve significant HAZ area shrinkage because of melting and resolidification in the narrow weld zone. In addition, the findings indicate the laser process is sensitive to the type of sintering atmosphere. The formation of blowholes in the weld zone has a relationship with the number of unreduced oxides in the component(s) to be joined. Specimens sintered in endothermic gas exhibited an unacceptable condition, whereas those sintered in hydrogen, dissociated ammonia, and vacuum all proved to be acceptable. The LBW process also requires that the medium hardenability material be preheated to 250 to 300 °C (480 to 570 °F) before welding to reduce martensite formation. The narrow HAZ with the larger adjoining material mass extracts heat away from the joint, which accelerates the cooling rate. Another observation is that oil-quenched materials do not prove successful because the oil residue in the pores vaporizes and forms blowholes. Good results, however, were obtained with 316L stainless steel P/M components. These weldments exhibited very low amounts of δ ferrite (<1.0%) in the weld zone, which probably relates to the narrow autogenous (without use of filler metal) fusion zone associated with the process (Ref 15).

Brazing is another very common process used to join P/M parts. This method allows P/M fabricators to use existing sintering furnaces and atmospheres to join P/M parts or P/M wrought or cast materials. Temperature requirements range from 1065 to 1150 °C (1950 and 2100 °F), depending on the brazing alloy composition, with atmospheres that include endothermic, dissociated ammonia, nitrogen-base, or vacuum. The capillary force of the pores requires special consideration when choosing a brazing alloy. When using a standard copper brazing material the porosity near the joint wicks the filler metal into the pores, leaving an insufficient amount to establish a satisfactory bond strength. For this reason, these parts must be copper infiltrated before brazing or pressed to a density greater than 7.2 g/cm^3 to limit interconnected porosity at the joint interface. An alternate method involves the use of a Ni-Cu-Mn alloy that increases liquidus temperature as iron from the mating components dissolves into the filler (Ref 16, 17). This fills the pores immediately adjacent to the brazement, leaving a sufficient amount of filler metal to establish suitable bond strength.

The preferred brazement design involves a lap rather than a butt-joint configuration. The optimal gap distance is between 0.05 and 0.13 mm. Consideration must be given to the thermal expansion characteristics of the material to be joined when attempting to maintain the proper gap distance at temperature. Ring preforms placed adjacent to the joint at the spot location of a brazing paste is preferred over sandwiching the filler between components. Fixturing may be necessary with complex configurations or when tighter dimensional control is required. Brazed parts can subsequently be quenched and tempered or steam treated without negatively impacting joint strength. The American Welding Society has established related guidelines under AWS C3.6-90, "Specification for Furnace Brazing" (Ref 18).

Diffusion/sinter bonding technique typically involves the use of materials that exhibit different sintering expansion characteristics due to composition or density level (Ref 19); for example, an outer ring having a propensity to show less growth during sintering than a press fit inner ring that has somewhat more growth in comparison to the mating part. The solid-state bond results from mechanical interlocking and alloy diffusion occurring at the interface of the mating parts. Similar results can also be achieved if two parts, placed in close contact, are copper infiltrated using a single-step process (Ref 20).

These methods have proved to be viable, cost-effective joining techniques that are currently used on production parts. However, diffusion/sinter bonding is limited to certain geometries and alloy compositions. The bond strengths are also somewhat lower than those achieved with other joining processes.

Adhesive Joining of P/M components is possible by gluing them together with epoxy resins, anaerobic or acrylic adhesives. The first two materials set up at room temperature, whereas the acrylic adhesive must cure at approximately 120 to 130 °C (250 to 265 °F). An important aspect of this procedure is that the surfaces be clean and the pores be free of any contaminants. This has effectively been accomplished using abrasives and ultrasonic cleaning with a final heating to drive off moisture. The use of alkali or acid-based cleaning agents is not recommended.

Two extensive studies identified several advantages of adhesive joining (Ref 21, 22):

- Stresses are distributed uniformly.
- Galvanic corrosion is minimized because the adhesive acts as a dielectric.
- Parts with smaller cross-sectional areas can be joined to larger sections.

Conversely, disadvantages include:

- Environmental temperatures cannot exceed 170 °C (340 °F) because substantial reductions in strength occur at elevated temperatures.
- Hostile environments, for example, high humidity and/or contact with chemically active materials, should be avoided.
- High peeling and/or shear stresses at the joint interface can result in failure.

The referenced investigations found that the best results were obtained by heat treating the adhesives using low applied coupling pressures with a minimum of 5 min at temperature. Steam treatment before adhesive joining along with increasing sintered density were found to enhance shear resistance.

P/M Welds

Successful P/M welding requires proper design and the proper choice of material(s) and joining process. This requires a thorough understanding of the performance characteristics for the application. These include, but are not limited to, the following:

- *Strength requirements:* applied load in tension, shear, torsion, and so forth
- *Dimensional restrictions:* potential problems with distortion and/or shrinkage
- *Environmental factors:* for example, galvanic corrosion from dissimilar metal combinations or effects of elevated temperature on the joint strength
- *Appearance:* surface condition
- *Economics:* cost effectiveness compared with other manufacturing methods

Once the application requirements have been established, consideration can be given to the major factors involved in the joining process.

Type(s) of Material. If strength requirements necessitate greater hardenability, control of hardness and related stresses in the weldment become more critical in welding. Also, the potential for any metallurgically incompatible elements must be eliminated. Compositions containing sulfur or high phosphorus levels are particularly suspect.

Joint design should ensure that the joint interface is not subjected to excessive loads or stress concentrations. In addition, the component geometry must lend itself to the type of joining process to be used, for example, the ability to press and control density in projections for RPW applications.

Process selection must be considered when evaluating the minimum strength requirement of the application. Typically, the fusion processes GTA, GMA, EBW, and LBW provide the highest joint strengths, while FRW and RPW are often near parity with these techniques. Brazing, diffusion, and adhesive bonding provide somewhat lower levels with respect to potential joint strength.

Fusion Weld Cracking. The most common problem with fusion welding involves cracking in or near the weld interface (Ref 23). Successful welding can be accomplished if proper consideration is given to why P/M weldments crack.

The formation of cracks in welded P/M components is most often associated with the stresses generated during solidification and cooling of weld metal (Ref 24, 25). The mass of material surrounding the weld puddle and HAZ resists the contraction forces as the metal cools, resulting in tensile stresses that initiate cracks. These stresses can be minimized by the use of several techniques:

- Preheating can drive off moisture (hydrogen) and lessen the thermal gradient across the weld zone.
- Postheating immediately after completing the weldment reduces stresses, particularly for high hardenability materials that form appreciable amounts of martensite in the weld metal and HAZ.
- Austenitic filler metals on steel or low-alloy components can be beneficial because of their superior toughness, strength, and lack of martensitic transformation.
- Processes that allow the operator to manipulate the amount of heat input can also assist in minimizing stresses (GTA, GMA).

Joint configuration plays a major role in stress formation. Mismatched joints, poor gap spacing, or narrow (low width-to-depth ratio) joints along with insufficient amount of filler metal to counteract densification can all have deleterious affects on the weldment. For example, Section A in Fig. 5 shows a throat crack resulting from increased volume of affected metal and related stress concentrations by using a large-diameter filler wire. In contrast, section B incorporates a

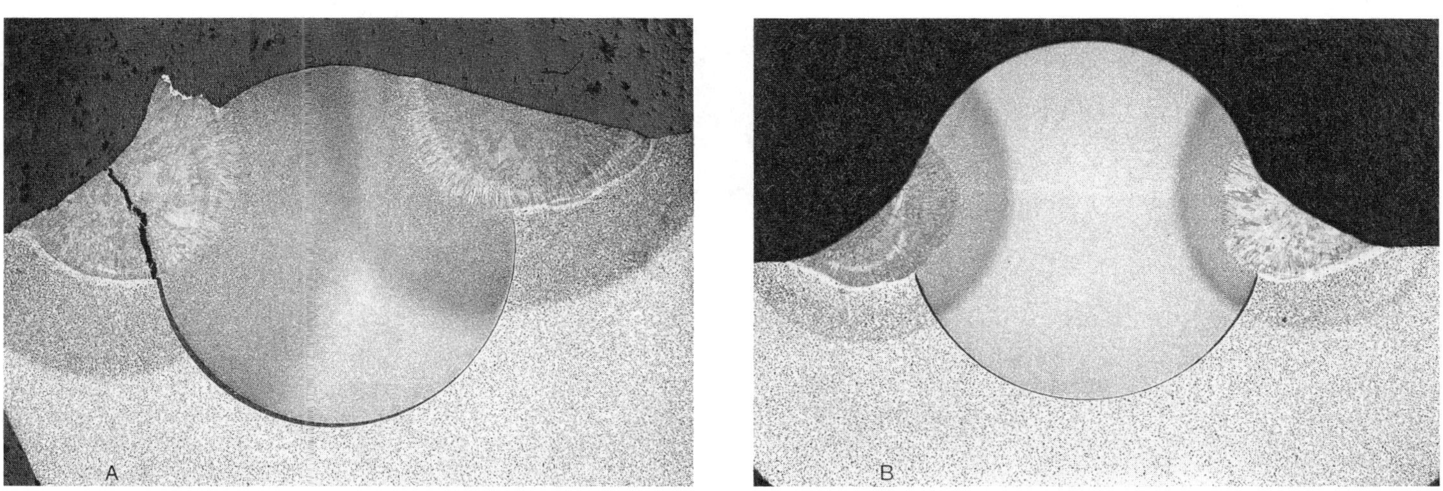

Fig. 5 P/M Alloy FL-4405 (6.95 g/cm^3) joined to low-carbon steel rod without different filler wire diameter. See text for details.

smaller-diameter filler wire and lower energy input, which provides a successful weldment.

In addition, fusion welding is not recommended for steam-treated or quenched-and-tempered P/M components. The oxides resulting from steam treatment act as contaminants in the weld zone, resulting in erratic performance and potential for cracking. Quenched-and-tempered parts, even with a sufficient temper to drive off entrapped quenchant, are not good candidates for welding (Ref 26). However, this is not to suggest that it cannot be accomplished. The most substantial drawback centers on the relatively high heat input associated with fusion processes that can change the structural constituents. This results in lower strength in the weld zone compared with the adjoining base material resulting in a stress gradient and potential fracture initiation site. It should also be noted that when welding infiltrated parts, free copper in the pores is likely to melt and subsequently migrate to the former austenitic grain boundaries.

Brazing Techniques

Joint Design. Lap joints are always preferred over butt designs. When using a lap joint, the amount of overlap is determined by a handy reference called the "rule of three," which suggests a lap distance of three times the thickness of the thinnest component. The fit between components should be between 0.05 and 0.13 mm to achieve the highest joint strength, which is typically 380 MPa (55 ksi) ultimate tensile strength. For some applications, the parts can be placed on top of or adjacent to one another without exercising any additional precautions. The exception involves brazing dissimilar metal parts having significant differences in density and/or composition. The density affects the thermal expansion coefficient whereas the composition influences dimensional change, both of which can influence the gap distance. The best procedure utilizes a method of gap width control, such as a fixed diameter wire or nipple projections placed between the mating surfaces. The brazing alloy should be placed immediately adjacent to the joint interface to assist in capillary action and flow into the joint. For large surface areas, shallow channels or gutters may be pressed into the joint face to assist in uniform flow (Ref 27).

Filler Metal. The most often used P/M brazing alloy is a patented (U.S. patent 4,029,476) Cu-Mn-Ni alloy powder (Ancorbraze 72) that can be admixed with flux and a lubricant for preforming into slugs/rings or dispensed as a paste. This material is suitable for use with iron, steel, low-alloy, or stainless P/M components. Typically, 3 to 4 g of filler metal is required for each square inch of surface area to be joined. The quantity is dependent on joint design, gap spacing, and component density. Closer gap spacing and higher component density favor lower amounts. Some applications requiring specific properties may require the use of a silver- or nickel-base brazing alloy for electrical, high-temperature, or corrosive environments.

Heating Cycle. Uniform heating of the components along with sufficient dwell time to equalize the temperature between the parts to be joined is important for proper flow and joint fill. A component having a lower mass or thinner section size will likely achieve the flow temperature before a larger mating part. This results in a greater capillary force toward the area having the highest temperature or heat energy that may cause poor joint fill. If this does occur, it is sometimes helpful to reposition the components on the furnace belt to assist in achieving uniform temperature between the parts. It may also help to increase the preheat temperature so parts can equalize before reaching the liquidus point (approximately 1065 °C, or 1950 °F, for Ancorbraze 72).

Atmospheres & Fluxes. Most of the commonly used sintering atmospheres can be used for P/M brazing. A flux is generally added to the braze alloy to assist in oxide reduction. In some circumstances, an additional borate- or fluoride-type flux may be swabbed onto the mating surfaces, particularly when brazing sintered parts, stainless steels grades, materials containing sulfur/manganese sulfide, and wetting large surface areas to be joined. It has also been determined that high CO_2 percentages in the sintering atmosphere can oxidize the fluxing agents and reduce the flowability of the brazing alloy.

By far, the most common difficulty experienced with furnace brazing P/M components is the lack of fill in the joint. This is generally associated with the following conditions:

- Improper cleaning or oxide reduction
- Excessive joint clearance
- Low sintering temperature or nonuniform part temperature
- Insufficient filler metal
- Entrapped lubricant, flux, or gases
- Movement of the mating parts before solidification

Diffusion/Sinter Bonding

This process requires an inner component that has a tendency to expand and an outer component with somewhat lower expansion rate or in some cases a propensity to shrink. The components are fitted together as green briquettes and sintered at conventional time and temperature. Elements that enhance diffusion—copper, carbon, phosphorus—also increase joint strength. Density gradients between components also promote diffusion. When choosing a combination for diffusion bonding, consideration must be given to the desired strength level for each component along with the respective dimensional change characteristics to ensure the parts can be made within the print tolerance and strength requirements.

P/M Materials for Joining

Most iron and steel powder compositions can be joined or welded without difficulty. However, some additions or material grades should be avoided if possible. In general, atomized iron grades have lower residual and tramp elements than sponge or other types of reduced iron powders. The cleanliness of these materials does not play a predominant role in weldment integrity, if held within acceptable limits. Nevertheless, the subtle influence of acid insolubles, oxides, and silicates over a period of time influences the ser-

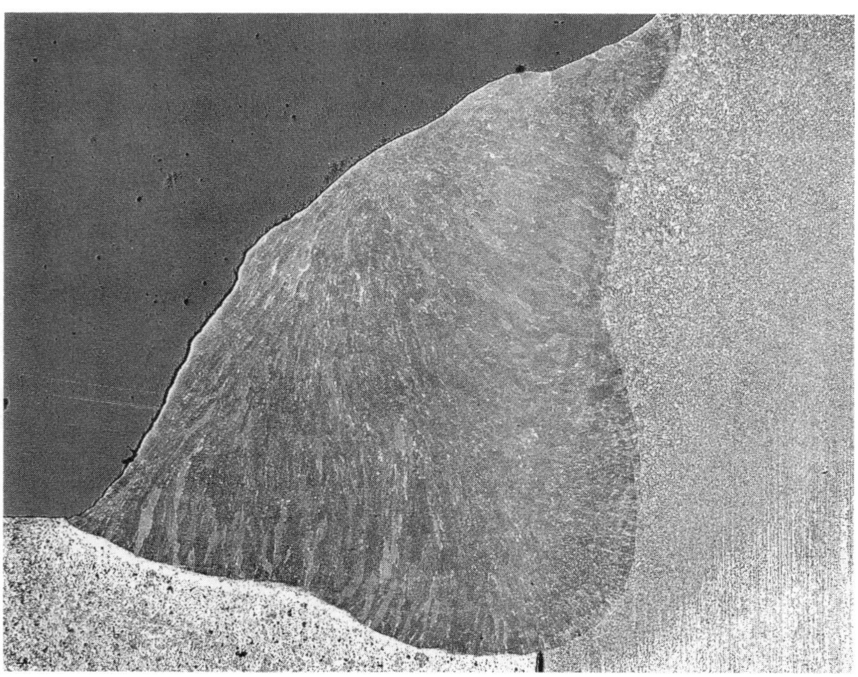

Fig. 6 High-strength GMA weldment between an FN-0205 steering gear (6.9 g/cm^3) and a AISI 1035 steel shaft using an E70S-type filler wire

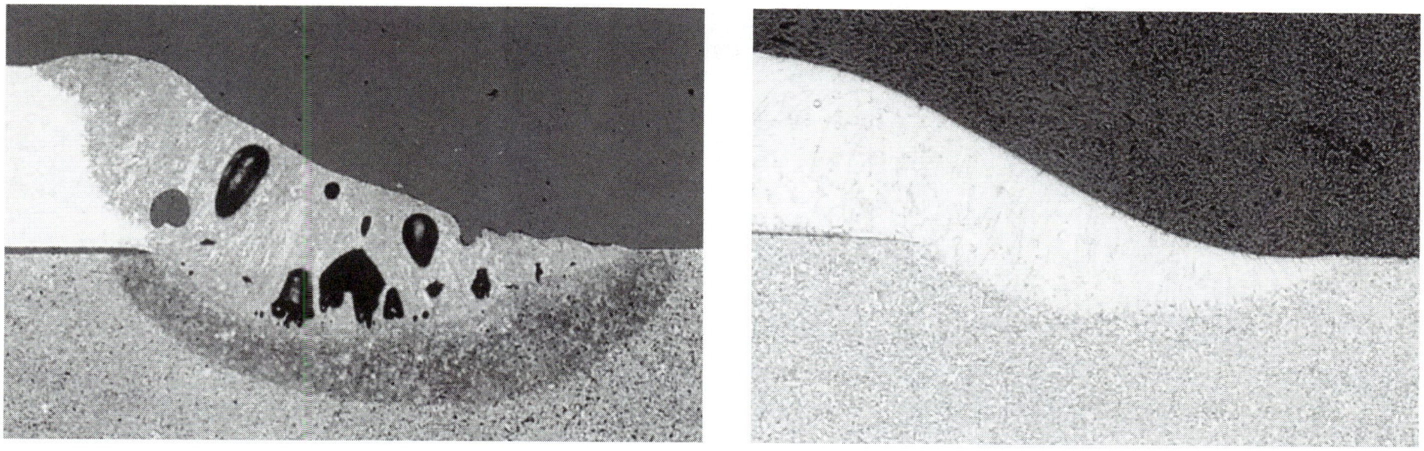

(a) (b)

Fig. 7 P/M stainless steel weld variation. (a) Unacceptable weld quality associated with high interstitial gas levels. (b) Weld microstructure representing the same weld parameters but a hydrogen-sintered, low-interstitial P/M stainless boss

Table 1 Influence of sintering conditions on properties

Property	Sinter environment				
	NH_3 at 1230 °C (2250 °F)	H_2 at 1260 °C (2300 °F)	Vacuum at 1150 °C (2100 °F)	Vacuum at 1205 °C (2200 °F)	Vacuum at 1260 °C (2300 °F)
Density, g/cm^3	6.57	6.70	6.68	6.95	7.02
Dimensions, %	...	−0.6	−0.8	−2.0	−3.5
Carbon, %	0.189	0.060	0.017	0.007	0.007
Nitrogen, %	0.339	0.042	0.018	0.003	0.001
Oxygen, %	0.376	0.144	0.225	0.209	0.176
Weldability	Poor	Good	Good	Good	Good

vice and fatigue performance. For this reason, the atomized grades are preferred for fusion, high-strength, and critical welding applications.

Carbon content has a pronounced influence on the overall weldability of a material. As a general rule, the carbon content should be held to as low a level as possible. However, carbon also greatly enhances the strength characteristics of a material. Joining processes and techniques can be developed to accommodate intermediate to high carbon levels that exhibit acceptable weld soundness and strength characteristics. For example, Fig. 6 illustrates a high-strength GMA weldment between an FN-0205 steering gear and an AISI 1035 steel shaft using an E70S-type filler wire.

Materials containing sulfur additions should be avoided for welding and brazing applications. The sulfur can migrate to the grain boundaries and may cause hot cracking when fusion welded. When brazing sulfur-bearing components, the manganese in the brazing alloy can form manganese sulfide in the presence of free sulfur, reducing the ability of the material to flow. If a machining enhancement is necessary, a more appropriate choice would include a manganese sulfide (MnS) addition.

Premixes with copper additions of 2.0% can be readily joined to other materials using most processes. The exception, however, involves compositions that include both sulfur and copper additions. Too high a copper content (4.0%) was found to lower the weldment strength to levels below the strength of the parent metal (Ref 28).

Phosphorus additions (Fe_3P), somewhat like sulfur, are not particularly attractive for fusion welding applications. The low melting Fe_3P addition may promote hot cracking in the weld zone. However, GTA and HPW processes have been used for a limited number of Ancorsteel 45P applications. Admixed additions of nickel to iron or steel powders generally enhance the toughness of the material and do not pose any particular difficulties involving weldability.

Stainless steel P/M components have been successfully welded using various joining processes (Ref 29). Gas metal arc welds of 316L P/M parts at various density levels, using 316L filler metal with an argon shield, provided good overall properties. The 303 free-machining grade and those identified as nitrogen strengthened are not good candidates for welding applications. The 410 martensitic grade can be welded, but precautionary measures must be observed with regard to its hardenability.

A growing number of automotive exhaust system flanges and bosses involve welded P/M stainless materials. Extensive testing programs involving static and dynamic characterization (Ref 30, 31) of flanges and welded assemblies indicate P/M stainless materials successfully meet all the appropriate performance criteria. However, during the initial prototype welding development, it was determined that sintering conditions have a pronounced influence on welding performance.

Production trials (Table 1) indicated a change from dissociated ammonia to pure hydrogen atmosphere substantially reduces the level of carbon, nitrogen, and oxygen when processing 409 stainless materials. Relatively high interstitial levels promoted the formation of fine martensite during welding, along with substantial outgassing that resulted in excessive porosity in the weldment. Weld trials using the hydrogen sintered parts proved successful (Fig. 7). The resulting weldments were free of porosity with completely ferritic microstructures.

Nickel-molybdenum admixed composition with a nominal 0.5% C, 5.0% Ni, 0.5% Mo having a sintered density of 7.0 g/cm3 was successfully welded using the GMA process and an austenitic filler metal without a preheat or postheat treatment (Ref 32). The friction/inertia process also provided a satisfactory, high-integrity weldment with this composition. The diffusion-bonded grades would respond in a similar manner if processed using the same techniques.

Joining Dissimilar Metals. Two considerations must be taken into account to achieve successful weldments. First, it is necessary to define the environmental conditions both metals must satisfy to ensure the best selection for the task. In addition, the physical compatibility of the metals must also be determined.

Compatibility involves several physical characteristics that affect the weldment. The properties to be considered are coefficients of expansion, electrochemical potential differences, melting temperatures, strength levels, and potential intermetallic compounds that can form between the two metals. When the coefficients of expansion are widely different, there will be internal stresses within the weldment that cannot be reduced by a postweld treatment. If these stresses are significant, the weld integrity will be jeopardized and the weldment prone to premature failure. Weldments involving dissimilar metals may also be subject to electrochemical corrosion. The differences in electrochemical potential indicate the susceptibility of the weldment to corrosion. If the alloys are at opposite spectra on the electrochemical scales, the nobility of one metal is far superior to the other and corrosion can be a potential problem. The solubility effects when joining certain metals can form brittle intermetallics when alloying occurs. The combinations most susceptible are steel and aluminum, magnesium or titanium; copper and aluminum; aluminum and magnesium; or similar combinations of titanium. When joining these alloys, caution should be exercised in choosing a welding procedure so as to prevent the formation of any brittle constituents.

REFERENCES

1. G.E. Linnert, *Welding Metallurgy,* Vol 1 and 2, American Welding Society, 1967
2. *Welding and Brazing,* Vol 6, *ASM Handbook,* ASM International, 1993
3. G.F. Bocchini, "The Influences of Porosity on the Characteristics of Sintered Materials," Technical Publication 860148, Society of Automotive Engineers, 1986
4. *Resistance Welding Manual,* Resistance Welding Manufacturers' Association
5. L.J. Johnson, G.J. Holst, and M.J. O'Hanlon, Projection Welding Powder Metal Components to Wrought Components, *1971 Fall P/M Conf. Proc.* (19–21 Oct, Detroit), S. Mocarski, Ed., Metal Powder Industries Federation, 1972, p 193–203
6. J.F. Hinrichs, P.W. Ramsey, and M.W. Zimmerman, "Joining Sintered Steel to Wrought Steel Using Various Welding Processes," Annual Meeting, American Welding Society, 1971, p 52
7. W. Johnson and M. VanHaaren, The ABC's of High-Speed Pulse Welding, *Weld. J.,* July 1995, p 47–49
8. W. Johnson et. al., "High Speed Pulse Welding of Powder Metallurgy Automotive Components," Technical Paper 950383, Society of Automotive Engineers, 1995
9. K. Couchman, M. Kesterholt, and R. White, Gas Tungsten Arc Welding of Powdered Metal Parts, *Process Seminar on Secondary Operations* (Int. Conf. PM '88, Orlando, FL), Metal Powder Industries Federation, June 1988, p 33–39
10. J.C. Thornley, CO_2 Welding Fe-Ni-Mo-C-Cu Sintered Steels, *Weld. Met. Fabr.,* Vol 12, Nov 1973, p 399–402
11. *Welding Handbook,* Vol 1, 8th ed., American Welding Society, 1987
12. J.E. Middle, Friction Welding of Sintered Fe:Cu:C Components, *Instit. Metallurgist,* Series 3, Vol 2 (No. 18), 1981, p 32–37
13. Y. Suezawa, H. Kuroda, and H. Kobayashi, An Investigation of Electron Beam Welding of Sintered Iron, *Trans. Jpn. Weld. Soc.,* Vol 10 (No. 2), 1979, p 3–9
14. E. Mosca, A. Marchetti, and U. Lampugani, Laser Welding of P/M Materials, *Int. Conf. PM 82* (Florence), Associazione Italiana di Metallurgia, 1982, p 193–200
15. D.J. Kotecki, Understanding Delta Ferrite, *Weld. Des. Fabr.,* Vol 62 (No. 12), 1990, p 33–36
16. W.K. Knopp, "Joining of P/M Structures," Technical Publication 740984, Society of Automotive Engineers, 1974
17. W.K. Knopp, Brazing Problems in Powder Metallurgy, *Int. J. Powder Metal. Powder Technol.,* Vol 11 (No. 1), 1975, p 63–65
18. "Specification for Furnace Brazing," AWC C3.6-90, American Welding Society
19. H. Akutsu and M. Iijima, Direct Diffusion Bonding Technique for Ferrous P/M Parts, *Modern Developments in Powder Metallurgy,* Vol 16, E.N. Aqua and C.I. Whitman, Ed., Metal Powder Industries Federation, 1985, p 195–208
20. K. Okimoto and T. Satoh, Joining of Iron Powder Compacts by an Infiltration Method, *Int. J. Powder Metall.,* Vol 23, 1987, p 163–169
21. A. DeIorio and R. Zocchi, The Feasibility of Joining Sintered Components by Using Structural Adhesives, *Proc. of PM/82* (Florence), Associazione Italiana di Metallurgia, 1982, p 201–210
22. A. DeIorio, M. Schiavone, and U. Siano, Influence of the Connection Technique on the Adhesive Tensile Strength in Sintered Component Joints, *Horizons of Powder Metallurgy,* Part II, W.A. Kaysser and W.J. Huppmann, Ed., European Powder Metallurgy Federation, 1986, p 697–701
23. R. Brosilow, Why Do Welds Crack, *Weld. Des. Fabr.,* Vol 63 (No. 2), Feb 1990, p 34–40
24. V. Podnozov, Cracking in Seam Welding of Porous Sintered Materials, *Svar. Proizvod.,* Vol 2, 1979, p 25–26
25. V.A. Novosel'tsev, L.G. Erchenko, and V.I. Mitin, Phase Transformation in the Welding of Iron-Graphite and Iron-Copper-Graphite Powder Materials, *Poroshk. Metall.,* Vol 26 (No. 10), 1987, p 48–52
26. R.W. Cummins, "The Effect of Welding Metallurgy on Design," Technical Publication 770529, Earthmoving Industry Conference, Society of Automotive Engineers, April 1977
27. M. Onoda, R. Kameda, and T. Koiso, "Application of Sinter-Brazing," Technical Publication 830395, Society of Automotive Engineers, 1983
28. P. Beiss, Finishing Processes in Powder Metallurgy, *Powder Metall.,* Vol 32 (No. 4), 1989, p 277–284
29. G.W. Halldin, S.N. Patel, and G.A. Duchon, Welding of 316L P/M Stainless Steel, *Progress Powder Metall.,* Vol 39, 1984, p 267–280
30. J.A. Hamill, F.R. Manley, and D.E. Nelson, "Fusion Welding P/M Components for Automotive Applications," Technical Paper 930490, Society of Automotive Engineers, 1993
31. G.R. Urffer and F. Garver, "Use of Stainless Steel P/M Components in Automotive Exhaust Systems: An Exhaust System Supplier's Point of View," Proceedings of PM^{2}TEC, Metal Powder Industries Federation, 1997
32. J.C. Thornley, Relative Weldabilities of Sintered and Wrought 5.0% Ni-Mo Steels, *Weld. Met. Fabr.,* Vol 11, Nov 1972, p 390–395

Techniques Improving Dimensional Tolerance

M.I. "Mike" Jaffe, Consultant

DIMENSIONAL TOLERANCE or size control in powder metallurgy parts manufacturing can be affected by virtually every stage of the process. The size change occurring during the sintering step is a result of the powder itself, the compacting, and the sintering conditions. Post-sintering operations such as sizing, machining, hardening, and finishing can all be affected by the condition of the parts as sintered. As would be expected, every stage can be affected by the previous stage.

Powder

Various methods of powder manufacture, even for the same type of material, influence the compacting, the sintered size, and the physical characteristics as well. Different manufacturing methods affect the particle size distribution, the particle shape, the particle hardness, and particle purity. All of these affect the apparent density (loose powder), the flow, the compressibility, the springback (incremental increase from die size) (Fig. 1) and the green (unsintered) strength of the compact. Variations in any of these parameters can have an effect on the size or tolerance control of the compact. The method of powder manufacture can affect size change in sintering, especially when the powders are blended with carbon, copper, nickel, or other alloys and combinations. The condition of the alloying powders such as particle size and oxide content also affect the size and physical characteristics of the sintered part.

The choice of the powders to be used may be based on many factors. These include price, availability, ease of compacting, green and sintered physical characteristics, and familiarity with the powder and its use. When good control of tolerances and physical properties is essential, the ability of the powder producer to make a material that is very consistent in all characteristics is a key factor.

The sintered size of a compact varies with the amount and type of the alloy, the density, and the sintering conditions. The size change plotted against the change in alloy or the change in density or the change in sintering conditions can be almost linear or a curve (Fig. 2–4). For optimal dimensional control, it is desirable to operate on the flattest part of a curve, if possible, as slight variations will have the least effect.

To obtain optimal part size control based on the powder, there must be very tight control on all stages of powder manufacture. It has been shown that even the water used for atomizing affects the sintered size change (Ref 1). High sodium water (softened) shows less growth when parts are sintered in dissociated ammonia (DA) compared with low sodium water and shows the opposite effect when sintered in a nitrogen atmosphere.

Most powder suppliers now use some form of statistical process control and may operate under an appropriate ISO standard to supply the user with the most consistent powder possible. Even when powder is shipped with extremely narrow control limits on all of the parameters that can affect the tolerance control of the part, there are variables beyond the manufacturing steps that can affect the part as pressed and sintered. Powder may be altered by shipping and transport conditions. Vibration and impact such as trucking long distances over poor roads can have an effect on the flow and apparent density. Powder in containers that have not been properly sealed or that have been left open for long periods may be changed by the atmosphere resulting in oxidation and/or moisture pickup.

If the part manufacturer blends or reblends the powder, variations can occur. Changes in blending time, the amount of load placed in the blender (percent of blender capacity), the type of blender, and the various lubricants can affect the resulting apparent density, flow rate, and green strength (Ref 2). Blend uniformity can be affected by the method of loading the materials into the blender, the accuracy of the weight of the blended material, and the blender time and action itself. Any of these variables can affect the compacted and/or the sintered part. The advantage of a single large blend by the powder producer is that all of such a large blend, 10,000 lb or more for example, should be quite consistent, whereas twenty 500 lb blends may show variation from blend to blend.

Compacting

At the press, the usual control measurements are at least for thickness, weight, and visual im-

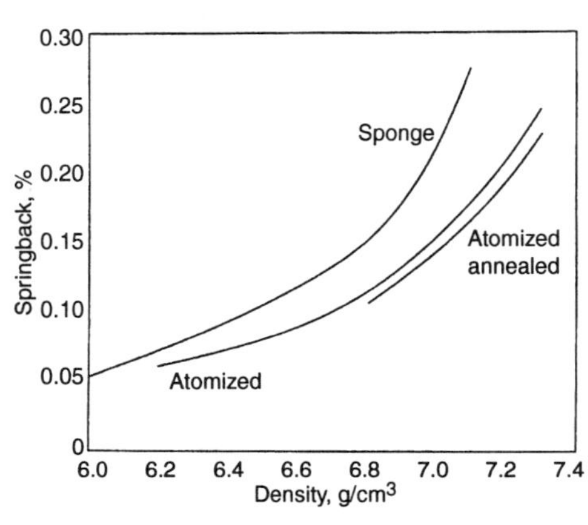

Fig. 1 Springback versus density. Source: Höganäs Metal Powder Data Sheets, Sweden

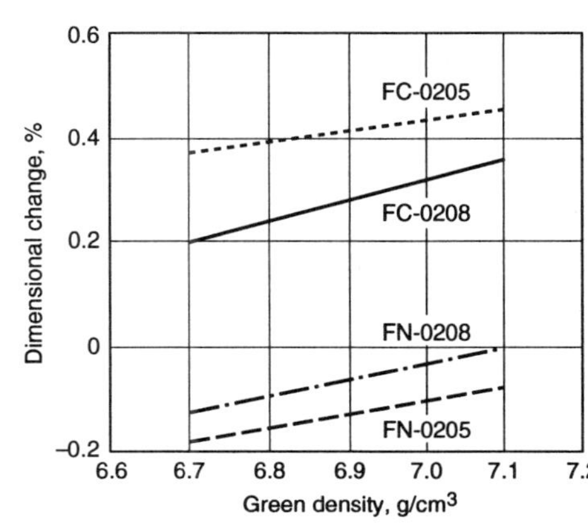

Fig. 2 Dimensional change for several alloys versus density. Source: Hoeganaes Corporation, Riverton, NJ

perfections. With some of the more advanced presses, compacting forces and/or part weight and/or tool positions may be monitored and be part of a closed-loop system to cause corrective action to be taken automatically.

In a multilevel part, several sections may be measured independently as well as the overall part. Independent thickness measurements are relatively simple, but individual sectional weight would require cutting or breaking the part accurately, which may be difficult. If a part such as a flange with a hub or a gear with a pinion (cluster gear) is controlled by two thicknesses (or by one-step thickness and an overall thickness) and the overall weight, it is likely that sectional weights and therefore their densities could vary although the overall weight is correct. For example, consider the stepped part in Fig. 5, given the following:

- Overall weight (WT 3) is 146.61 ± 1.47 g
- Overall density is 6.6 to 6.9 g/cm^3
- Nominal volume of flange is 19.769 cm^3
- Nominal overall volume is 21.560 cm^3

If the given weight of the flange (WT 1) is measured at 134.43 ± 1.34 g (determined by breaking off hub and smoothing the break flush to flange), then the hub density could theoretically range from 5.14 to 8.52 g/cm^3 (from much too soft to breaking tools), even if the part is made to flange weight, thickness tolerance, and overall weight requirements. This could lead to poor tolerance controls, variation in physicals properties, and possible damage to the tools.

Sectional densities of the compact can be controlled or compared by apparent hardness, although a very low or special load must be used to prevent the cracking of the part. Devices such as electronic probes can also be used to get relative readings or actual densities if properly calibrated for the conditions. Visual comparison of densities is possible with experience, but the surface finish of the tooling can have a large effect on the appearance of the compact. A rough visual indication of density can be obtained by quickly "swiping" a line across the part with a broad felt-tipped permanent marker. The ink penetrates faster in the lower-density sections than in the higher-density sections. Actual density measurements of the sections is the most straightforward procedure, but may be slower.

The uniformity of fill in even a simple flat part can be affected by a number of factors:

- The consistency of the powder itself is critical. Variations in flow rate or apparent density obviously lead to poor control of weight and/or thickness and require frequent adjustments.
- The speed of the press (which affects the time allowed for the powder to fill the die) and the dynamics of the powder flow can be critical. When the fill is narrow and deep as in a long, thin-walled cylinder, speed can have a great effect.
- Presses may have a system to cause the feeder to shake when it is over the die for fill. The amount and consistency of this shaking can affect the accuracy of the fill. If this shake is accomplished by a mechanical means such as a notched cam, loose linkage may allow for excess variation.
- If the press goes to fill before the feed shoe is over the die, the air in the die must be replaced by the powder in a smooth, consistent way, or air may be trapped, which adds another variable to filling and can affect the weight and therefore the density and resulting properties (Fig. 6).
- If the punches tend to stick or "hang up" the speed with which the press completes the fill, the fill time may vary, resulting in nonuniformity. The act of sticking may not be consistent on a part-to-part basis, resulting in substantial fluctuation in the weight and/or thickness or an occasional bad part.

In some press systems, the amount of powder in the hopper can have an effect on the weight of the powder fill. Some form of flow-control orifice between the feeder and the hopper can reduce this effect (Fig. 7). In most feeder systems, that side of the die that is nearest to the feeder may have a bit more fill as the feeder is over it for a longer period. If the fill is not uniform on a

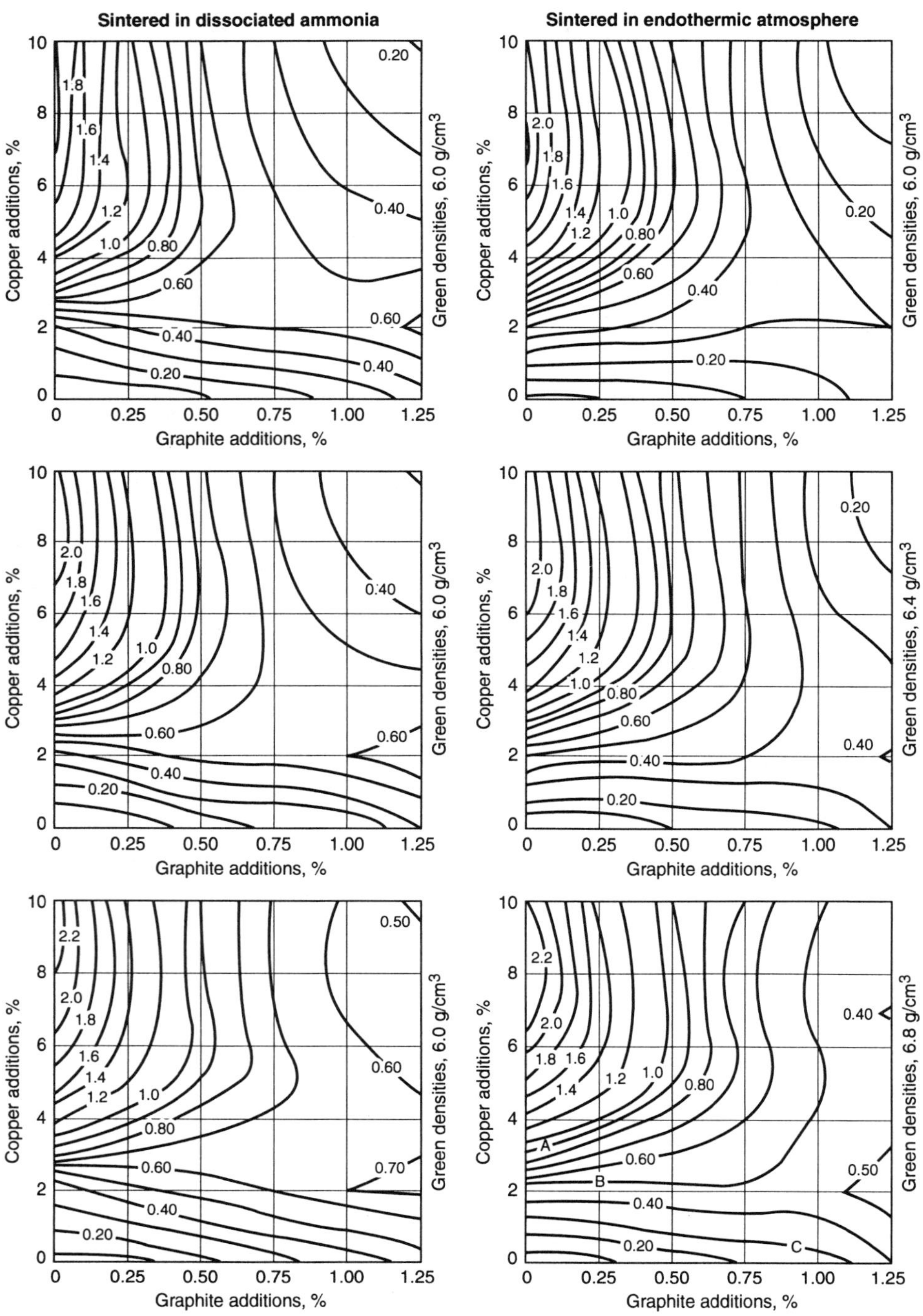

Fig. 3 Dimensional change versus density, copper, and graphite. Source: Hoeganaes Corporation, Riverton, NJ

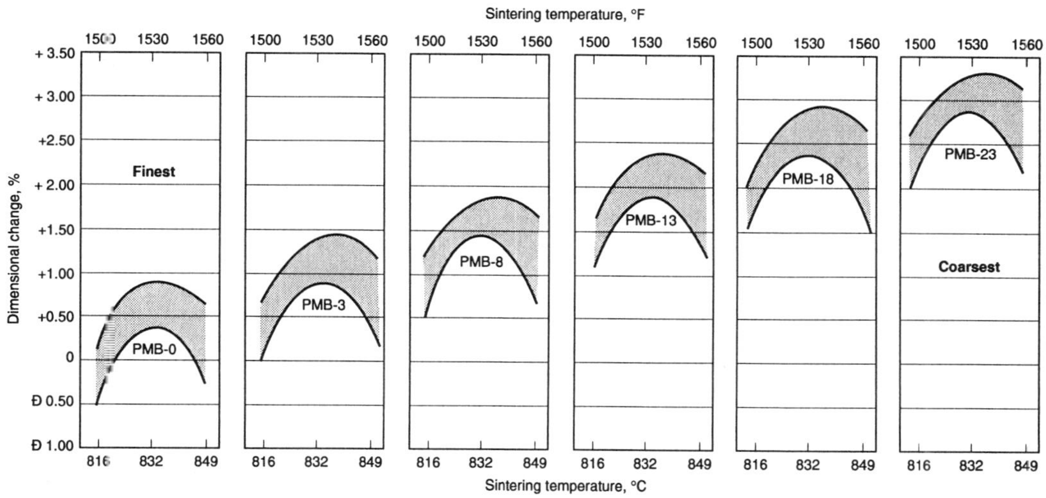

Fig. 4 Dimensional change versus sinter temperature for various bronze powders. The "PMB" designation indicate variation in particle size distribution. Source: AMG Americas

consistent basis, such as being too heavy on one side, the feeder can have some sort of mask placed in it to control the powder flow into the die. Such masks would usually be unique to a job and belong to the specific "tooling" for that job. There are now proprietary systems that are designed to ensure a very uniform fill (Ref 3).

Another item that can affect fill uniformity for ferrous parts or even nonferrous parts with some iron in them, would be any magnetization of the tools. This is particularly true for smaller parts with thin walls. As tools are often held in magnetic fixtures during toolmaking or checking operations, the possibility of retained magnetism is present unless good demagnetization techniques are used. Even pressing tools that have been carefully and effectively demagnetized can become magnetic if the mechanic tools and wrenches, such as "hex" wrenches, have become magnetic themselves. Plates and adapters in tool sets and even hardened press parts that have not been well demagnetized can have a dire effect on the operation. Magnetism of the tools can be easily checked with a "magnetometer" made for that purpose. Otherwise, a clean sheet of stiff white paper can be placed over the individual tool or over the top of the die if the tools are in the press; then iron powder is gently sprinkled on it. If the powder takes on the appearance of magnetic lines, there could be problems. A very gentle rap to the paper may help show these lines better.

Fill weight and compacting thickness can be affected by mechanical items in a press. For example, in a press that uses cams to determine fill levels or compacting height, an out-of-round or mechanically damaged cam follower can cause some type of cyclical variation. Powder buildup under punch supports or adapters can cause a change, usually a progressive change, not of a cyclical nature.

Part size change can be caused by the heat-up of the die and tools due to thermal expansion. A 100 mm die dimension could increase by almost 0.05 mm if a steel die is heated up by 24 °C. A

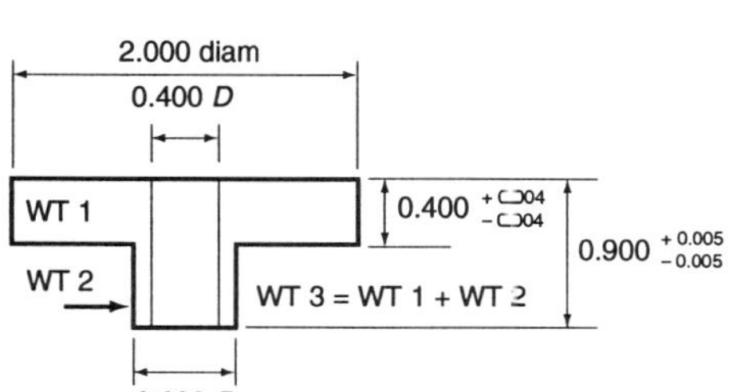

Fig. 5 Density control in a stepped part. The overall weight (WT 3) and dimensions are given, and the flange weight (WT 1) is measured by breaking off the hub and smoothing the break on the flange. With this method, however, the calculated values for hub weight (WT 2) are too inaccurate (see text for discussion).

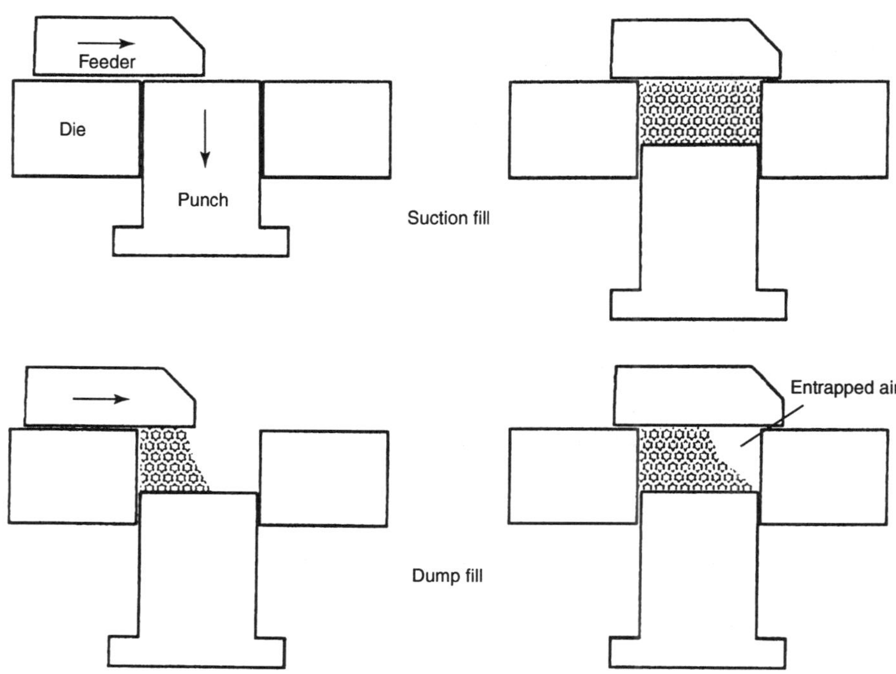

Fig. 6 Entrapped air during fill

carbide-lined die would expand somewhat less. Long or deep parts with a lot of die wall area such as a gear or sprocket tend to heat more than a plain round die because of its increased surface area with greater total friction. A well-polished carbide die minimizes friction and heat-up.

A cracked punch or punch holder or adapter can cause excessive variation and may get progressively worse until a complete failure occurs. If a press is run near its tonnage capacity, noticeable deflection or stretching can occur under load and this affects the thickness greatly if the weight varies. For example, in a 500 ton mechanical press, a plain cylindrical part 101.6 mm (4 in.) diam by 25.4 mm (1 in.) thick of atomized iron would weigh 1400 g when pressed to 6.8 g/cm^3. If the weight increased 1% to 1414 g the thickness could increase by about 0.076 mm (0.003 in.) in a typical mechanical press due to the stretching of the press. A part of half the diameter in the same press would increase only about 0.018 mm (0.0007 in.) for a 1% increase in weight. If a hydraulic press is used and is set to press to a fixed tonnage, the thickness will vary in proportion to the weight. Note that some hydraulic presses can press to an encoded dimension with high accuracy.

Presses are now available, in varying degrees of sophistication, that can monitor force and/or weight and react to these changes. This reaction can be based on a statistical trend and serve as a warning to the operator, or it can stop the press or cause the press to make an adjustment itself. Such systems can compensate for variations in powder characteristics and press effects. However, as with any system, better control of the powder or press actions gives better results. Some presses (usually hydraulic) can work on a computer numerical control (CNC) basis using precision encoders resolving to ±0.001 mm so that all levels or movements involved are at the correct location at the correct time and arrive there at the desired rate (Ref 4). For multilevel parts, this gives virtually absolute control. Computer numeral control presses are also able to compensate for press elongation. It should be noted that this type of equipment carries a high initial cost, but this may well be justified for many jobs.

There are press systems that can "learn" a force-distance curve for a particular job, and the press can react to changes from that curve in some appropriate way. Field tests using tonnage compensation and/or weight compensation have indicated reductions in standard deviations of up to 16%. Average long-term reduction of weight variation was 40% (Ref 5–7).

Most of the previous discussion refers to control of the thickness and/or weight of the pressed part, which is seen to depend on almost all of the features of the powder itself as well as on the uniformity of press conditions. The control of horizontal (die) dimensions brings in other factors. For a plain cylindrical part, the diameter in the as-pressed (green) condition is usually quite stable, but is affected somewhat by density as a more dense part tends to springback or pop out more as it is ejected from the die (Fig. 1). Various methods of powder production affect springback also. The harder, less pure powders spring back more.

As springback is generally a function of the size itself, a larger part changes more than a smaller part, both in horizontal and vertical dimensions. Other factors that affect the horizontal size, as compacted, are wear in the die or in cores. Wear is a long-term effect and depends on the abrasiveness of the powder material, the density, the relative movement in the die at compaction, the die material hardness and finish, and wear resistance. Under some conditions a die or other tool member that creates horizontal dimensions may crack, which will give a more sudden change than wear. If a die or punch that also acts like a die, or in some other way creates a horizontal dimension and is sensitive to cracks, this should be realized in the inspection procedures for the part and this dimension monitored as a "variable." In the example shown (Fig. 8), the lower punch acts not only as a punch for the gear but also as a die for the pinion. If the punch cracks (through the small hole or not), it will cause the width of the cracked pinion tooth to increase and may destroy the pinion accuracy. This type of crack would not affect the gear teeth directly, but could affect the centrality of the hole. It should be noted that the typical wear in a core rod is not easily measured by a micrometer, as the wear may occur as grooves and the effective decrease in the hole size could be more than the measured decrease in the rod size (Fig. 9).

Another horizontal variable could be the location of holes (Fig. 10) or the distance between holes. Although the diameter (horizontal dimension) of a hole is usually determined by the core rod—which is subject to wear—the location of the hole is affected by the clearance between the punch and the die and the clearance between the core rod and the punch. If several punches are nested together, the location of a central core rod can be a stack-up of clearances. Theoretically, the locational accuracy of the core can be between "0" and the grand total of all clearances

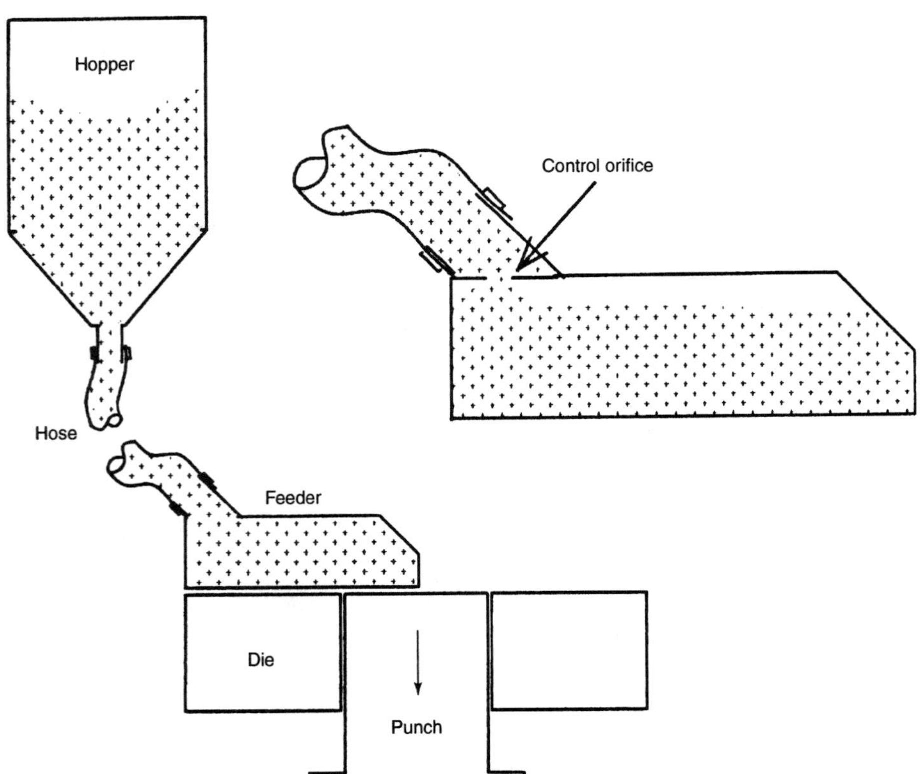

Fig. 7 Control orifice for more uniform fill

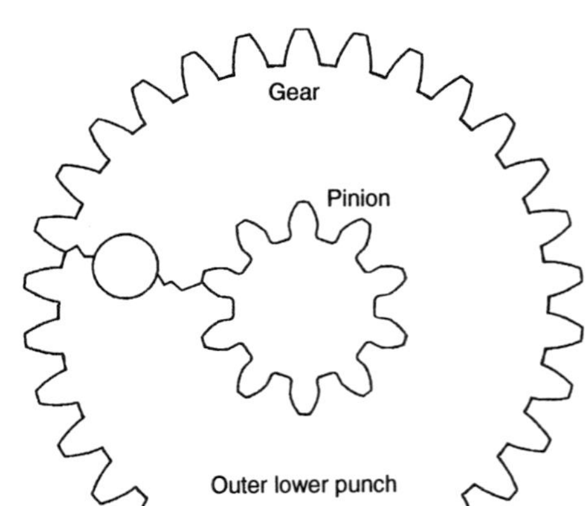

Fig. 8 Cracked punch affects size

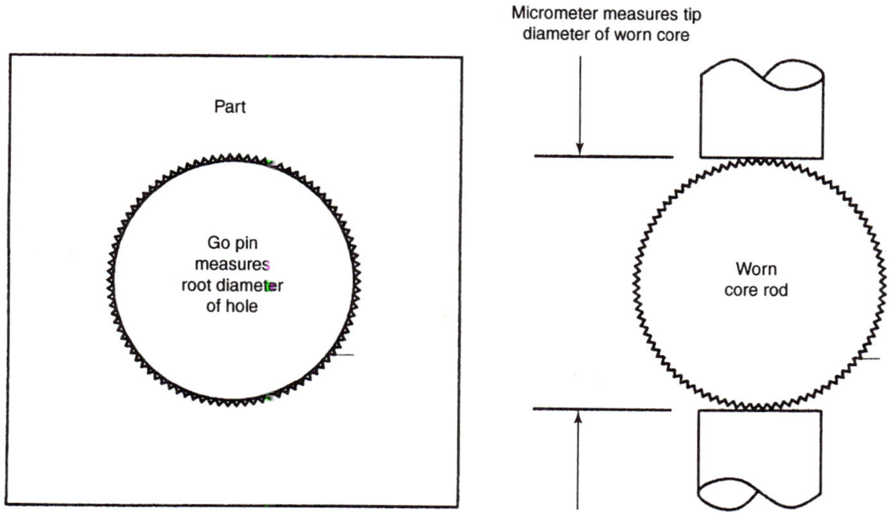

Fig. 9 Error in worn core measurement

(in the worst case). In practice, it will be somewhere in between. For this reason, the tolerance of hole locations or center distances can be more than the die-shape or the hole-shape tolerance. The size change that can occur in sintering must also be considered.

Bending or flexing of a core can affect the accuracy of its location and apparent size. A thin, relatively long core hole can be flexed by uneven powder density on different sides. For example, a hole central to a part of uniform thickness should have uniform pressure all around it and not flex, whereas a hole close to the edge may see higher density toward the inside and lower density to the outside due to differences in fill (see Fig. 11) and become "banana" shaped. A banana-shaped hole could seem undersize when measured with a "go" gage pin, although the true diameter of the hole could be correct at any one point in the hole (Fig. 12). This may often be corrected by fill compensation.

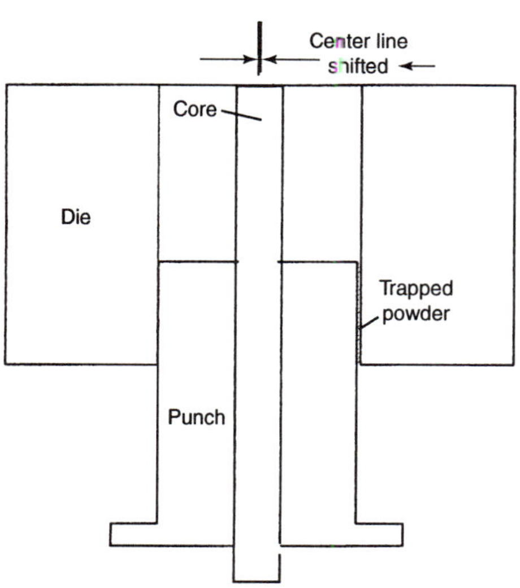

Fig. 10 Trapped powder affects hole location

Therefore the part, as it comes off the press, is seen to be influenced by:

- *Consistency of the powder itself:* flow, apparent density, screen sizes; purity, alloy, or trace elements; manufacturing processes; and springback
- *The feeding of the powder into the die:* powder "head"; speed, shake, timing, dwell; condition of the feeder, wiping action, powder left over the die level; and mechanical motions of the press tooling.
- *Changes in the tools themselves:* wear, cracking or other physical damage, cracking of punch bases and/or adapters, sticking, and variation in clearances due to powder entrapment (Fig. 10)
- *Changes in the press:* heat-up of the frame and tools and normal "common-cause" variations

To help diagnose problems of nonuniformity of a compact or sintered part, some reference to the position of the part in the tools is desirable. If the part is of an odd shape or not symmetrical, the position in the press may be obvious. If, however, the part is symmetrical—such as a disk, a cylinder, or a gear—there may be no way of verifying which side was up and the orientation of the part in the die. Forming the tools to put a small depression or other mark on the part indicates the position of the part as pressed. If care is taken in placing the mark, it may not have any harmful effects on the part. This helps to diagnose problems such as out-of-round parts, out-of-concentricity tools, nonuniform fill, chipping, cracking, and powder entrapping.

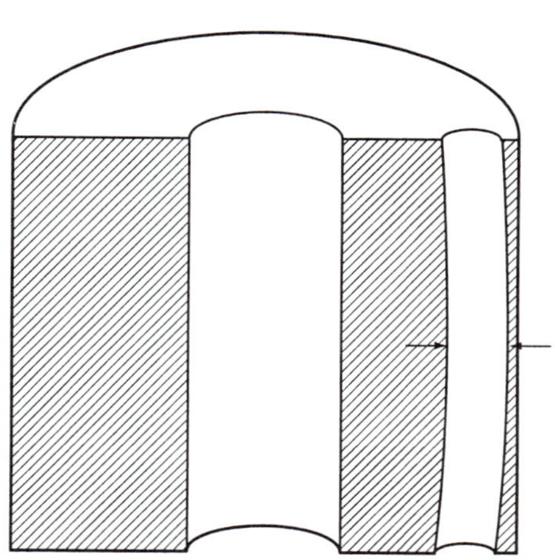

Fig. 11 Bending of a thin core near an edge

Handling

Once the part has been compacted, it must be removed from the die table. In some cases the part slides onto or is picked up and placed onto a belt or some conveyance that delivers it directly to the sintering furnace. This may not involve direct operator intervention, but inspection should be done on a periodic basis. If bad parts are being produced, they can be isolated from those parts going into the furnace to avoid contaminating the good with the bad. It is possible that the parts could have some form of automatic inspection for weight or size or both. Parts could be monitored by the press system itself, and bad parts could be rejected off to a side.

If the parts are left to slide into a container or onto a staging table or system, care must be taken to see that physical damage such as chips or cracks does not occur. If the part is of a "C" shape, care must be taken to ensure that there is no way that the opening in the "C" can be reduced or squeezed in. That can produce a variation in "die" shape. Large and heavy parts, particularly those with small projections that could be struck, are more susceptible to handling impact than small, lightweight parts. Care must be taken if the operator has to pick up a C-shaped

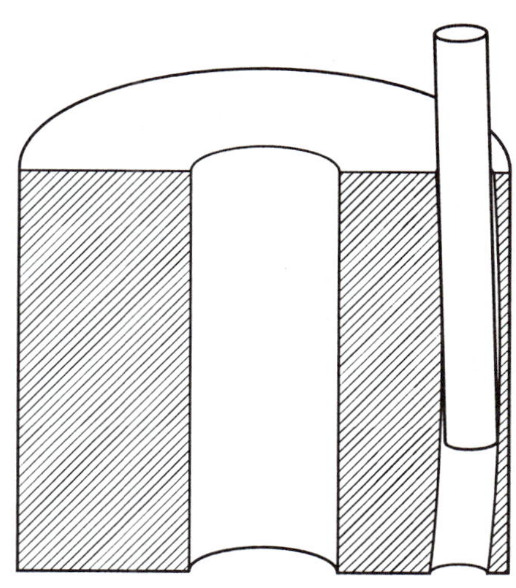

Fig. 12 Bent hole shows no-go. The pin entering the hole from the top will try to go in at a slight angle. Due to the "banana" shape, it could not go through. It would get stuck.

part, as squeezing it can cause dimensional changes or cracking (Fig. 13).

Sintering

Even if parts come off the press with extremely close dimensional and weight tolerances, the trip through the furnace can have a great effect on dimensional stability. The sintering operation is what develops the physical properties and size of the part, and many things happen in the furnace.

The final size and properties of the as-sintered part are influenced by the following:

- The material, its method of manufacture, and all basic attributes
- The alloys or additives and their uniformity
- The pressed density and the density distribution
- The pressed sizes
- The furnace conditions
- The time in delube or preheat, the time in high heat, and the cooling time and rate
- The temperatures reached at each stage
- The atmosphere, its composition, amount and flow rate, and flow pattern over the work
- The work load and its distribution in the furnace

For consistent work, the time-temperature profile that each part "sees" must be very uniform within the furnace and on an hour-to-hour and day-to-day basis. The part that comes off the press can be measured, weighed, evaluated, and examined as it is made. Parts being sintered can generally be measured and examined only after they leave the furnace, which means several hours of production before adjustments can be made based on the part itself.

Normal events that can occur to alter the sintering and therefore the tolerances could be:

- Variation in time due to change in belt speed. Most belt furnaces have a driving drum that has some form of rubber facing to grip the belt and usually a pinch roll to hold the belt tight to the drum. However, if the facing is worn or the gripping roller loses spring tension or the load varies too much, the effective speed can vary regardless of the speed of the drum.
- Speed control of the drive may have variables, and the setting of the speed may not be accurate. If the speed is measured by a ruler and a watch, errors of 5% or more are possible unless great care is taken.
- Inconsistent spacing of parts on furnace belts can alter the net effect of the heat-up and cooling rate.
- Variations in the temperature profile. In furnaces with a small number of control zones, variation in gas burners or change or deterioration of heating elements can cause a change in the temperature profile that may not be picked up by the temperature controller. More zones of control will minimize this, but not eliminate it. The newer electronic controls can be programmed to or actually learn how to obtain the most stable temperatures.

Parts that are relatively thin and flat, such as valve plates used in compressors, are often sintered placed on graphite or ceramic plates that are maintained very flat. Although this may reduce the yield of the furnace, it produces flatter parts than can be achieved by placing the part directly on the belt or on metal screens or trays that tend to go out of flat.

Re-Pressing (Sizing)

A part can be put back into a press and re-pressed for several end purposes:

- *To increase density.* There are practical limits to the density that can be obtained in the first pressing. The dry lubricant blended with the powder is low density, about 1 g/cm^3, and therefore occupies a lot of space. An iron compact with 1% zinc stearate as a lubricant has a theoretical solid density of 7.35 g/cm^3. Also, the iron particles receive some cold work during pressing. Sintering eliminates the lube and anneals the part to allow for substantial density increase upon re-pressing.
- *To calibrate size.* Ferrous and nonferrous parts can be sized or calibrated by re-pressing either in a compacting press or a press made specifically for sizing. Note, however, that if all of the controls on the powder, the compacting, and the sintering were sufficiently tight then there may be no need to size. Sizing, basically, compensates for the errors in these basic steps. The exception to this is the variation in sinter size caused by unavoidable density variations such as in bronze bearings. The part can be forced into a die that is somewhat smaller than the part and over a core that is somewhat larger than the hole. The part surfaces are therefore actually burnished to size. There may be little increase in density. A part can also be sized by making it to have some clearance when placed in a die that is close to or at the final size. The part reaches the desired size by filling out sideways as it is squeezed down.
- *To coin the surface(s).* This is similar to the conventional striking of a blank to make a coin or medallion. There is usually some change in configuration with coining such as the surface of a coin or creating text or a logo. Mechanical shapes such as small counterbores or depressions as well as small, raised shapes can also be created.

The actions described previously may well be combined. A part could be re-pressed to increase density and to improve size tolerance (calibrate) and also to change some shape feature. Bronze bearings can be sized by forcing the outside diameter (OD) and inside diameter (ID) into the tools, which burnish the surfaces so there is not much increase in density as that would reduce its ability to hold oil. If iron-base parts are sized with very little movement—such as being put into a die that is just about the size of the part—it may be difficult to hold tight tolerances. If the part density is increased sufficiently, it will be deformed beyond the elastic limit and tend to hold the size to a closer tolerance. For softer, lower-carbon parts or brass or copper parts, a reduction in thickness on the order of 10% has been used. For parts that are made of higher-carbon or alloy steels, the reduction in thickness may have to be less to avoid excess stress on the tools. Cams have been made with a "throw" (difference between the large and small radii) tolerance of no more than 0.0025 mm (0.0001 in.). These cams were made from plain iron and re-pressed. However, any attempt to add carbon or any other alloy to the iron would not allow that tolerance to be met on a consistent basis.

Holes in parts can be sized by re-pressing the entire part or by just burnishing the hole(s) with a core rod and with little or no decrease in the thickness. Roller burnishing is often used to bring a round hole into tolerance and generally works well as the inner surface of the hole can be densified locally by reducing the pore sizes without "smearing" or closing them off. This is of benefit if the hole is to function as a self-lubricated bearing.

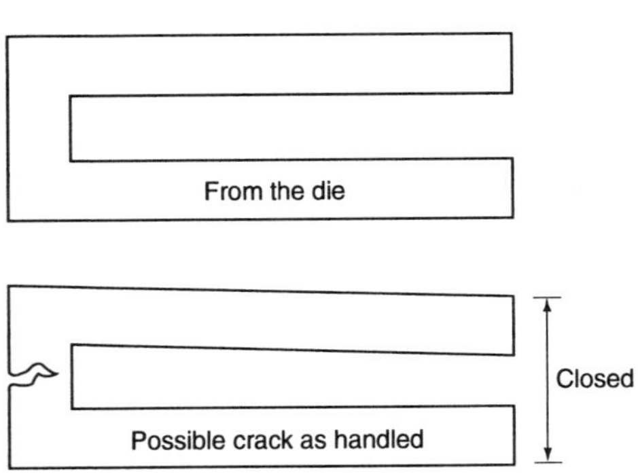

Fig. 13 Distortion of C-shaped part

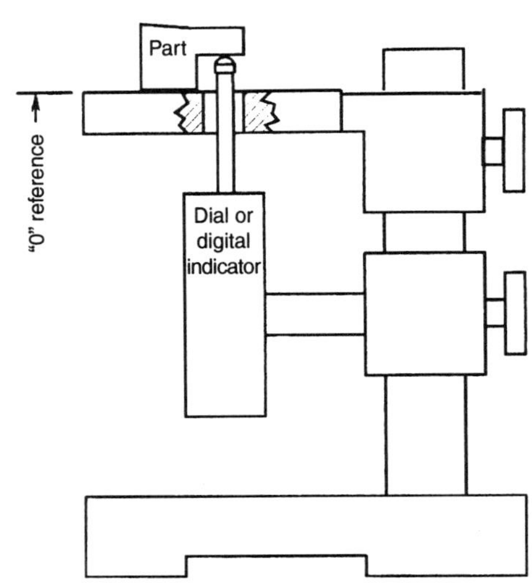

Fig. 14 Simple measurement of a step

When re-pressing a part, the lubrication is important as it is needed on the surfaces that contact the die or core rods. If a liquid or semiliquid lubricant is allowed to penetrate the pores, it is possible that hydrostatic pressure in the part, which can be a variable that can affect the ability to hold close tolerances, will be developed. Oils are usually applied by a quick dip or spray without allowing a soak time. Tumbling or coating the part with a dry lubricant such as molybdenum disulfide, zinc stearate, or other soap products eliminates the hydrostatic problem but may not do a good job of lubricating small holes and may affect the surface appearance of the part.

Re-pressing a part improves the tolerance on the die shape and the shape of holes. If the holes are not in the correct location or not concentric with the outside, sizing may not correct this, particularly if the material is relatively strong. Small diameter cores are subject to flexing. Sizing of a gear may improve the tooth shape and tooth-to-tooth errors, but may not improve runout or excessive total composite error based on runout. For parts that have a tight concentricity tolerance, such as a gear to produce AGMA class 8 or 9 or better, it may be necessary to bore or grind the ID. The concentricity of the pitch line to the OD of a P/M gear should be very good if the tools are made properly, so that the gear can usually be held on the OD to bore the ID with good results.

Machining

A P/M part can be machined to provide shapes that could not be created by P/M compacting or re-pressing, such as some undercuts, side holes, threading or tapping, and so forth. Machining, cutting, or grinding is also done to improve dimensional tolerances. As mentioned previously, a gear can have the hole bored or ground, which removes material to reestablish the centerline to improve concentricity tolerances as well as improving the size tolerance. Parts that require tight thickness tolerances and/or flatness can be surface or double-disk ground in a conventional manner. Likewise, parts that are of a suitable shape can be centerless ground to improve diameter or roundness. Although machining or grinding of P/M parts is similar to that of wrought or solid parts, there are some differences that should be noted:

- Due to porosity, P/M parts may absorb coolant or cutting or grinding fluids. These fluids should contain rust inhibitors.
- Brass or nickel-silver parts traditionally contained small percentages of lead, which improved machinability greatly. However, due to the restraints on the use of lead, other methods are being developed to improve machinability.
- Iron-base parts may have materials like manganese sulfide added to improve machinability.
- Resin impregnation of P/M parts can have a dramatic effect on increasing machinability.

Inspection Measurement

Proper and logical inspection procedures assist in maintaining optimal tolerance control. In powder manufacturing, the raw materials should be of a select grade. For atomized powders, the molten metal can be checked for proper chemistry before pouring and changes made to the melt as needed. Production lots of powders are usually tested and may be certified for screen analysis, apparent density, flow rate, compactability (density achieved by pressing at a specified tonnage), sinter size change under a specified condition, and chemical analysis for basic constituents. If the powder meets the agreed-on specifications the user should be assured of a reasonable consistency. It is obvious that tighter control of the powder results in better press-and-sinter tolerances, and this is an ongoing project with powder manufacturers. An audit by the parts producer of incoming powder certifications from time to time would help to ensure the required quality. Tests of flow rate, apparent density, and screen analysis are easy enough so that this should be done on a routine basis. Accurately controlled pressing of a tensile bar, a transverse bar, or a suitable ring that is then sintered in a laboratory furnace under standard time, temperature, and atmosphere conditions and then tested for strength, hardness, size change, and so forth is a very good way of ensuring powder control as well as a reference for sintering conditions.

Inspection procedures at the press should take into account the following variables.

Normal Variables. Weight and thickness are normal variables, and their running tolerances may be set by part requirements or by a process capability study that would yield statistical process control (SPC) limits for these parameters based on the material consistency and the equipment capability. This procedure indicates whether the equipment can, in fact, produce the required tolerances and yield a set of control limits that could indicate a process going out of control even if the attributes are well within the print specification range.

Weight can be measured manually by the operator or automatically by the pressing system. Electronic scales are the most practical, and some can be coupled to SPC monitoring systems. Automatic measuring by the press system can give a warning to the operator or go as far as making adjustments to the press fill. If the weight of a part is allowed to vary ±1% and the thickness is allowed to vary ±0.5%, the density could vary ±1.5% or a spread of 3%, which could create problems. See the section "Compacting" for a discussion of causes of weight variation.

Thickness can be measured by any of the standard measuring devices such as verniers or micrometers or dial gages (analog or digital), or laser devices. Although standard micrometers are very common, electronic digital types provide the best readability and minimize reading errors. If measurements are done manually, care must be taken to make the procedure as easy as possible and virtually "foolproof" for the operator. If there are several thicknesses to be measured that are more than a turn or two of the micrometer spindle, several micrometers or a dial indicator should be used. If a dimension, such as a step, that would be the difference of two normal micrometer or dial readings or require a depth micrometer is to be measured, a dial indicator referenced to one surface would give more accurate results with less chance of operator error (Fig. 14). For long runs of more complex parts, multipoint gaging setups can be created that measure many features at once and statistically process the data (Ref 8).

Slow-Change Variables. Slow changes in dimensions occur with wear on the tooling components. Where tolerances are nominal, a carbide die and carbide cores can produce millions of parts before wear is a problem. Tool steels containing large percentages of finely dispersed carbides (usually made by powder metallurgy) can do exceptionally well and be less prone to cracking. Most standard tool steels have a much shorter effective life. Therefore, the frequency of inspection for dimensions affected by tool wear depends on the tools themselves as well as on the powder used and its density. Although thickness and weight should be checked frequently, die and hole sizes need much less frequent inspection at the press. Steps in a part produced by either several different punches, or a step in the die, are subject to thickness variation and need frequent control. A step produced by a step in a punch face varies with punch wear and needs less frequent checking.

Sudden changes in dimension can be caused by the breaking or cracking of the punches or dies or of other tool support items. Experience on any particular job indicates those dimensions sensitive to this so that logical inspection can be set up. Often, when a part of a tool breaks or cracks, it can be heard or the press may indicate the problem by sound or by some change in those features normally checked.

By understanding the processes and how the variations or failures can occur, good tolerance control can be achieved efficiently, resulting in better P/M parts at lower cost.

REFERENCES

1. S. Masuhara and S. Kawai, Affect of Water Quality on Dimensional Change, *Advances in Powder Metallurgy,* Vol 2, Metal Powder Industries Federation, 1993, p 87
2. W. Fricke and D. Banyash, Blending Techniques to Enhance Apparent Density, *Advances in Powder Metallurgy,* Vol 1, Metal Powder Industries Federation, 1993, p 141
3. "Materials Innovation New Powder Delivery System," Materials Innovation Inc., W. Lebanon, NH
4. Dorst America, Inc., Bethlehem PA

5. "Adaptive Fill Control of a Compacting Press," Cincinnati, Inc., Cincinnati, OH
6. "Compacting Press Automatic Fill Compensation System," Cincinnati, Inc., Cincinnati, OH
7. " Tool Load Monitoring System for Multilevel Compacting Press," Cincinnati, Inc., Cincinnati, OH, see also *Advances in Powder Metallurgy,* Vol 2, 1993, p 11
8. R.L. Everett et al., "Electronic Gauging for Complex and Symmetrical PM Parts," Delco Remy America, Meridian, MS

SELECTED REFERENCES

- B. Bergquist and B. Hildenwall, Robust Sintering of Iron-Phosphorous P/M Steels, *Int. J. Powder Metall.,* Vol 33 (No. 5), 1997, p 33
- Y.T. Chen, R. Kiefer, M. Berlinger, B. Wasiczko, and H. Nayar, Atmosphere Effects on Dimensional Control of Fe-Cu-C Powder, *Advances in Powder Metallurgy,* Vol 2, 1993, p 103
- A. Griffo and R.M. German, Dimensional Control in the Sintering of Iron-Copper-Carbon via Particle Surface Area, *Int. J. Powder Metall.,* Vol 30 (No. 4), 1994, p 399
- K. Kulkarni, Dimensional Precision of MIM Parts under Production Conditions, *Int. J. Powder Metall.,* Vol 33 (No. 4), 1997, p 29
- P. Mathur and A. Lawley, Process Control Strategies in Powder Metallurgy, *Int. J. Powder Metall.,* Vol 28 (No. 2), 1992, p 121
- M.A. Phillips, A New Continuous Feedback Control System for Sintering Atmospheres, *Int. J. Powder Metall.,* Vol 29 (No. 1), 1993, p 15

Machinability of P/M Steels

R.J. Causton and T. Cimino, Hoeganaes Corporation

POWDER METALLURGY is considered a near-net shape manufacturing process, where designers seek to eliminate or significantly reduce secondary machining operations through die design, tool movements, and core rods. This ability to improve material utilization by reducing the need for machining operations has been a key to the success of P/M. However, improved machinability is becoming more important in extending the applications of P/M (Ref 1).

Machining is conducted for several reasons. Powder metallurgy methods have excellent shape-making abilities, but conventional high-volume techniques cannot make holes, or other re-entrants, normal to the pressing/compaction direction. Thus, cross holes required for lubrication must often be drilled after sintering. Accurate surfaces required for location or assembly also must sometimes be turned or milled. In addition, end users are increasing their use of subassemblies. Hence, P/M fabricators can supply a complete system, such as an oil pump rather than a gear, which can require machining operations previously undertaken by the end user.

When a parts fabricator supplies machined components, machining operations such as drilling, tapping, or boring can be the rate-limiting factor in the operation of an assembly line. Thus improved machinability of P/M materials to remove bottlenecks can then offer substantial increases in efficiency to both parts fabricator and end user. In cases where P/M steels compete with cast irons, machinability may be the leading factor in material selection. Thus, there is considerable interest and activity in measurement and improvement of machinability of P/M steels (Ref 2).

The relatively poor machinability of P/M steels compared to competing wrought products is usually considered to originate from a combination of factors, which include porosity, microcleanliness, microstructure, and knowledge. Some understanding of these factors is necessary to distinguish how the machinability of P/M steels is inherently different from wrought steels. These topics are discussed in this article, with a focus on the conditions of the base metal and additives to improve machinability. Coverage of the machining process is addressed in the article "Machining of Powder Metallurgy Materials" in this Volume.

The differences in machinability between wrought and P/M steels can be partially explained by the presence of porosity and the differences in microstructure required to obtain similar performance levels. The machinability of P/M steels can be improved significantly by systematic use of free-machining agents, cutting tools, and investigation optimizing machining conditions. In some applications, infiltration or polymer impregnation to close porosity will improve machinability significantly. In applications where machinability is key to the success of a P/M part, evaluation of machinability and possible improvements should begin at an early stage in part development and include part producer, tool suppliers, and powder producers.

This article shows that the machinability of sintered P/M steels can be improved by the use of free-machining agents that enhance machinability and the use of improved cutting tools. It appears that the use of green machining processes offers a means to improve machinability significantly. With correct choice of machining agent, cutting conditions, and cutting tools, it is possible to produce P/M steels with machinability equivalent to that of wrought steels; however, the optimization required can be time consuming. Where machinability is critical to the application and use of a P/M part, it should be considered at an early stage in the design process in cooperation with powder producers and cutting tool suppliers.

The Machining Process

The machining or metal cutting process is extremely complex. It has been widely studied for wrought (Ref 3–5) and P/M steels (Ref 6–9). The cutting process is generally illustrated by considering metal flow towards and over a simple cutting tool in an operation, such as turning or broaching (Fig. 1).

During cutting, tool movement into the workpiece generates a complex stress field around the tool tip. These stresses initiate cracks in the workpiece ahead of the tool tip. The cracks grow and join to separate the metal to be removed from the bulk of the workpiece. The layer of metal deforms, flows across the tool face, and is then removed from the cutting area. Ideally, the separated metal fractures so that it is removed easily from the cutting area as "chips." From this simplified description, it is apparent that the workpiece, cutting tool, and chips undergo high stresses and strain rates during the cutting process. These conditions, plus friction between the tool and chip, cause considerable local heating. In extreme cases, the heat can cause welding of chips to the tool face, local plastic deformation, or even melting of the tool or workpiece surface. Thus, both toolmakers and developers of engineering materials have attempted to improve the performance of materials during the cutting process without sacrificing other properties.

Effect of Porosity. The presence of porosity in P/M components significantly changes the cutting process (Ref 7, 8). The first and probably most significant effect is that cutting becomes discontinuous as the tool edge breaks out of the workpiece into pores. This action of successive small impacts on the cutting edge causes more rapid tool failure than continuous cutting operations (Fig. 2).

Porosity reduces the thermal conductivity of P/M steels, thus temperatures of the cutting zone and cutting edge can increase rapidly, which can

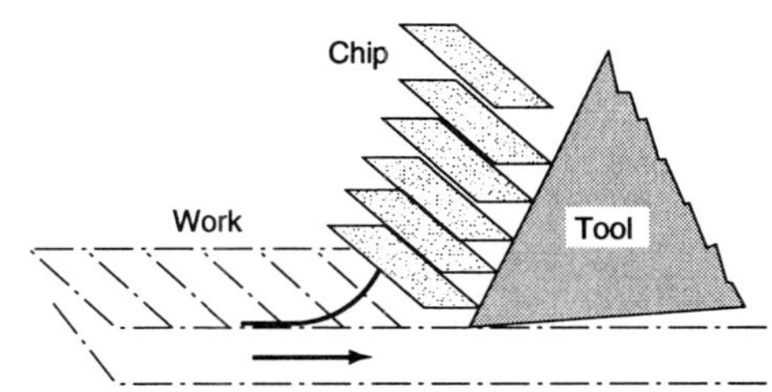

Fig. 1 Schematic of cutting process

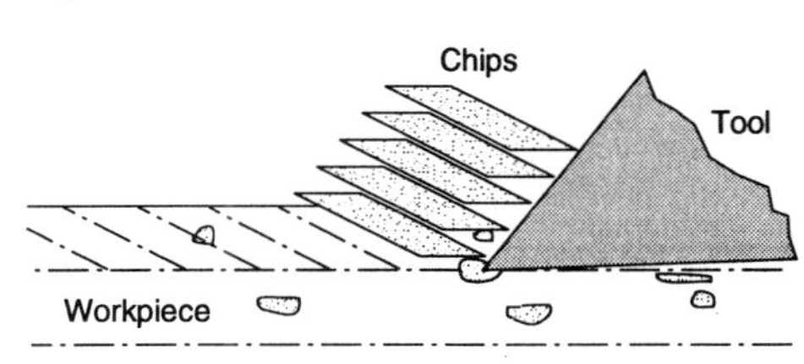

Fig. 2 Schematic of the effect of porosity on metal cutting

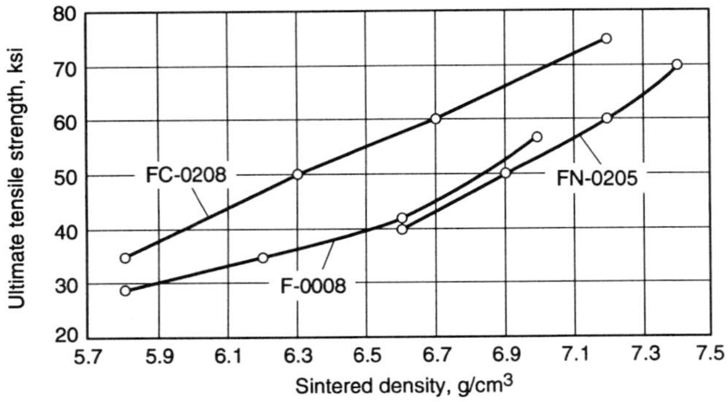

Fig. 3 Ultimate tensile strength of sintered P/M steels

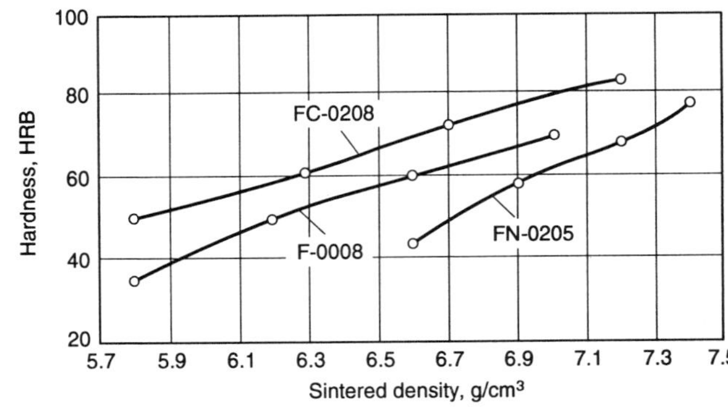

Fig. 4 Macrohardness of sintered P/M steels

accelerate tool wear and harden the workpiece. Local hardening of the workpiece can make subsequent finishing cuts more difficult. Interconnected porosity provides a path for cutting fluids to escape from the cutting area, which reduces their ability to cool and lubricate the cutting edge and can reduce their ability to wash chips from the cutting area. Such behavior can be critical to "spade" or "gun" drilling operations that depend upon controlled coolant flow. The porosity and remnants of prior particle boundaries inherent in P/M parts possess much greater surface area than wrought steels, which potentially increases physical and chemical reactions between the tool and workpiece that can accelerate wear.

Effect of Microcleanliness. The presence of undesirable nonmetallic inclusions is a cause of poor machinability (Ref 9, 10). Porosity also can contribute to poor microcleanliness and machining problems by allowing subsurface oxidation or carburization of pores during and particularly on cooling from sintering or heat treatment operations. In extreme cases, networks of oxide or carbide layers can significantly reduce machinability.

Influence of Microstructure and Properties on Machinability. Powder metallurgy steels employed in structural P/M parts possess somewhat different microstructure property relationships than wrought steels. The most critical effect that porosity has upon properties is to significantly reduce bulk properties, such as strength and macrohardness, from those intrinsic to the powder particles.

The strong effect of density or porosity upon bulk properties is illustrated in Fig. 3 and 4, which show ultimate tensile strength and hardness for three widely used P/M steels: F-0008, FN-0205, and FC-0208 (Ref 12). Neither the tensile strength nor apparent hardness of these steels is exceptional. They should not cause the severe machining problems associated with very soft annealed steels or high-strength steels.

The machinability of P/M steels is also influenced strongly by alloying and microstructures. The high carbon contents of sintered P/M steels result in microstructures with high volume fractions of pearlite in ferrite. In some cases, the microstructure can be pearlitic with grain boundary cementite. Cementite (Fe_3C) possesses a microhardness of approximately 1150 HK, and the cementite lamellae in the pearlite can conceivably cause significant abrasive wear of cutting surfaces. The ferrite regions, which can cause adhesive wear, possess much lower hardness of 100 to 150 HV.

Frequently, P/M steels are produced by mixing elemental additives, such as copper and nickel. Under practical sintering conditions, some of these elements may not dissolve completely. For example, nickel-rich martensite will possess different properties to the pearlitic areas if the nickel is not completely dissolved in a nickel steel such as FN-0405.

Metal Powder Industries Federation (MPIF) standard 35 also permits relatively wide ranges of carbon and alloy content in P/M steels, such as FN-0205 or FC-0208 (Ref 12). A change in chemistry from minimum to maximum can change their microstructure and properties. For FN-0205, increasing carbon from 0.4 to 0.6 wt% appears to increase the pearlite and martensite content. For FC-0208, increasing carbon from 0.6 to 0.9 wt% produces an almost completely pearlitic microstructure. Higher carbon additions produce a much finer, almost irresolvable pearlite. The changes observed in both alloy systems would be expected to reduce machinability.

One recent trend is to use prealloyed powders rather than pure iron (Ref 13). These P/M steels possess significantly higher mechanical properties due to their much higher hardenability. Consequently, they possess bainitic or partially martensitic microstructures in the sintered condition with tensile strengths of 415 to 690 MPa (60 to 100 ksi). Alloys of this type can also be sinter-hardened, which allows the development of high martensite contents during cooling from sintering. The properties of sinter-hardened steel approach or exceed those of heat treated FN-0205 P/M steel (Ref 14). It is to be anticipated that these high-strength steels will be difficult to machine with techniques developed for materials with a pearlitic microstructure.

Knowledge of Differences. The microstructure and properties of P/M steels differ significantly from those of competitive materials, such as cast irons or wrought steels. Therefore, the machinability of P/M steels will differ from that of iron and steel. However, the impact of the microstructural differences upon machinability is not well known or understood. Consequently, many P/M materials are machined with cutting conditions and tools that are optimized for the competing materials. When these do not produce good results, the conclusion is that P/M steel machinability is inferior to that of its competitors. The more correct conclusion is that P/M steels are different and frequently can be optimized by improvements to machining conditions and cutting tools. Thus, the knowledge required to machine P/M steels efficiently needs to be acquired and disseminated widely.

Machinability Measurement

Machinability is the ability or ease with which a material can be removed from the workpiece. It represents the interaction of the cutting tool, cutting conditions, and the workpiece. Measurement of machinability requires analysis of several interacting factors and is by nature complex and time consuming. Several different indices can be used to define machinability including tool forces, tool wear, tool life, and surface finish of the workpiece. For determination of optimal conditions, tests of specific machining operations, turning, milling, or drilling are required over a range of cutting conditions, speed, feed, and depth of cut (Ref 15). Thus, it is difficult to define a single index of machinability, although in some cases, such as drilling, materials such as brass, cast irons, and steels can be ranked by a single index (Ref 16).

The most thorough technique is to define the index (e.g., tool wear) that is considered critical, then measure the effect of systematic changes on machining variables, depth of cut, speed, and feedrate upon both the critical response and other factors as a function of machining (Ref 17). The results can be used to define tool life in minutes or number of pieces cut. The method is widely used for turning and milling operations, particularly during the development and optimization of cutting tools. The technique requires a large number of tests and is prone to considerable scatter. It is often difficult to decide which index is most critical (Ref 18). Several authors have used statistical techniques (Ref 19, 20) to

Table 1 Mean drill torque for 50 holes

Material	Torque, in · lbf		
	1380 rpm	2220 rpm	3480 rpm
Ancorsteel 1000	15.28	17.13	18.33
Ancorsteel 1000 MnS	13.96	12.63	14.62
Ancorsteel 1000 MoS_2	15.33	14.99	17.49

Table 2 Mean drill thrust for 50 holes

Material	Thrust, lbf		
	1380 rpm	2220 rpm	3480 rpm
Ancorsteel 1000	247.4	272.8	276.5
Ancorsteel 1000 MnS	242.2	220.4	225.5
Ancorsteel 1000 MoS_2	232.5	248.8	222.8

Table 3 Mean drill wear after 50 holes

Material	Wear, 0.001 in.			Holes drilled
	1380 rpm	2220 rpm	3480 rpm	at 3480
Ancorsteel 1000	13.78	15.33	15.00	18
Ancorsteel 1000 MnS	12.25	11.84	14.50	50
Ancorsteel 1000 MoS_2	9.25	10.70	15.00	11

reduce the number of experiments and clarify the interaction of machining variables and system responses.

Example: Machinability Test for Drilling FC-0205 Steel. Some of these points illustrated are for a machinability test using Ancorsteel 1000 with 1.75 wt% copper that evaluates the effects of drill speed and addition of free-machining agents (e.g., none, 0.35% MnS, 0.25% MoS_2) upon the machinability of FC-0205.

The test pieces were discs 4 in. in diameter by 1.25 in. thick, compacted to a green density of 6.8 g/cm^3, then sintered at 1121 °C (2050 °F) in 90% N_2/10%H_2 for 30 min. The drill test conditions are as follows:

- Drill: 0.250 in HSS, 118° point
- Lubricant: none
- Hole depth: 1 in.
- Speed: 1380, 2220, 3480 rpm
- Feedrate: 0.005 in./revolution

The 1 in. holes do not penetrate the workpiece completely; thus, there are no problems of breakout contributing to failure. The test is conducted using a computer numerically controlled (CNC) machining center that has been programmed to drill approximately 50 holes per disc. The drill torque and drilling forces are measured using a Kistler dynamometer (Kistler Instrument Group, Amherst, NY) mounted below the workpiece. The data logging system samples drill forces several times during the drilling of each hole. The drill is removed to measure drill wear at intervals of five holes. The test begins with an initial screening portion in which 50 holes are drilled at speeds of 1980, 2220, and 3480 rpm to identify major differences in the materials. The screening test shows differences among the test materials and does not indicate a single best material but rather several options depending upon conditions and criteria, such as torque, thrust, or drill wear employed.

Drill Torque. The data show that drill torque is somewhat characteristic of the material and increases steadily with the number of holes drilled (Table 1). The results show that 2220 rpm was an optimum drilling speed for the materials requiring lower torque. Both free-machining agents reduce drill torque compared to the base FC-0205 composition. The addition of 0.35% manganese sulfide produces lower drill torque than molybdenum disulfide.

Drill Thrust. There is little variation in drill thrust after 50 holes under the conditions employed (Table 2). Changing drill speed does not produce consistent changes in drill thrust. The data suggest that addition of manganese sulfide produced lower drill thrust at 2220 rpm, molybdenum disulfide produced lower drill thrust at 1380 and 3480 rpm.

Drill Wear. The measurements of drill wear (Table 3) show the greatest variation. The individual data show that drill wear increases with increasing number of holes until failure occurs. Drill wear tends to increase with increasing drill speed. Drilling at 3480 rpm causes rapid failure, so that only the material with manganese sulfide enables 50 holes to be drilled. However, at the lower speeds of 1380 and 2220 rpm, molybdenum disulfide produces the lowest wear after 50 holes.

Thus, definition of the "best" machining agent depends upon which index is critical to the machining operation or quality of the finished part. If the time available to drill the hole on a transfer line dictated drilling at the highest speed of 3480 rpm, the FC-0205 plus MnS is the only viable material. However, if the mechanical system permitted drilling at lower speed, molybdenum disulfide produced somewhat lower drill wear than manganese sulfide. If drill torque is used as an index of machinability, manganese sulfide is a better option than molybdenum disulfide.

Drill Life. The screening test showed that there is no clear "best" free-machining agent. It is thought that a longer test conducted at 2200 rpm would provide a clearer view of machinability. It is also of interest to compare the machinability of the P/M FC-0205 to materials with known free-machining response. It is considered that the FC-0205 plus MnS shows optimum performance at a drill speed of 2220 rpm. A test was conducted to compare the machinability of the FC-0205 plus MnS to those of known response including wrought 1215, a leaded free-machining steel (12L14), and an FC-0205 produced with a resulfurized powder. The wrought test pieces, using the same conditions, were cut from 4 in. diameter bars purchased from a steel stockholder.

Table 4 Results of long term test

Material	Holes to failure	Mean torque, in. · lbf	Mean thrust, lbf	Wear at 500 holes, 0.001 in.
FC-0205	95	16.56	276	29.1
FC-0205 MnS	>775	13.57	244	26.5
Resulfurized	>600	7.82	142.3	14.2
Wrought 12L14	>600	14.3	186.8	26.5
Wrought 1215	>600	15.58	177.7	19.1

The somewhat surprising results in Table 4 show that the machinability of the resulfurized powder is significantly better than that of the other materials systems, including the wrought steels. Drilling the resulfurized steel required lower drill torque and drill thrust and produced less tool wear at 500 holes than the other materials. The test confirmed that the machinability of the FC-0205 was not as good as the other materials. Drill failure occurred after only 95 holes, and the FC-0205 produced more wear just before failure at 95 holes.

The machinability of the FC-0205 plus MnS is significantly better than that of the FC-0205 and appears comparable to that of the wrought 1215 and 12L14. Drilling the FC-0205 plus MnS requires similar torque but higher thrust than the wrought steels. It appears to produce similar tool wear to that of the 12L14 but somewhat higher wear than the wrought 1215. The test shows under the drilling conditions chosen that the 1215 can possess slightly better machinability than the 12L14. The test also shows that by evaluating free-machining agents and machining conditions, the machinability of a P/M steel can be improved to that of a wrought steel.

Machinability Improvement

Machining P/M steels does present problems. Several different approaches to improve machinability are:

- Closure of porosity
- Green machining
- Presintering
- Microcleanliness improvement
- Free-machining additives
- Microstructure modification
- Tool materials

The effects of free-machining additives, microstructure modification, and tool materials are illustrated by controlled drilling tests conducted under laboratory conditions.

Closure of Porosity. Closing or sealing porosity improves the machinability of P/M steels significantly by changing the cutting process from intermittent to continuous. The reduction in vibration and chatter improves tool life and surface finish. Copper infiltration (Ref 8) and polymer impregnation (Ref 19) are efficient means to close porosity and can require an additional process step. Thus, they are most efficient when dictated by the end use, such as fluid power

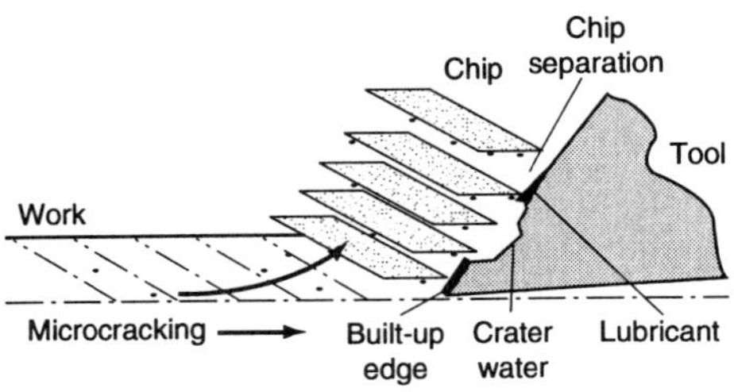

Fig. 5 Potential benefits of a machining agent

applications, that require a pore-free structure. However, the improvement in machinability can justify their use in severe machining operations or when a machining operation is the rate-limiting step in a process sequence.

Microcleanliness Improvement. The increase in the production and use of atomized rather than reduced iron powders has improved the microcleanliness of iron and low-alloy steel powders. Driven largely by the requirements of powder forging, the content of coarse nonmetallic inclusions in atomized powders has been reduced significantly (Ref 21). For an atomized FL-4600, the median frequency of inclusions greater than 100 μm in size (F_4) has been reduced from approximately 2.5 to 0.25 per 100 mm^2. The maximum frequency of inclusions greater than 100 μm was reduced from 9 to 1.3 inclusions per 100 mm^2. These improvements suggest that the incidence of edge damage due to the presence of coarse inclusions should be reduced significantly. Because powder forging practices are now employed to produce all atomized steel powders, P/M users of these powders have benefited.

Green Machining. One way to reduce the machining problems of P/M parts is to machine them in the green (i.e., as-pressed) condition prior to sintering. The lack of bonding between particles in green compacts results in low cutting forces. Such techniques are used in the processing of ceramic and hard metal powders. However, the green strength of metal powder compacts has been too low to withstand the cutting and clamping forces employed in machining operations.

The introduction of warm compaction technology (Ref 22) can change this perspective. The green strength of warm compacted parts is two to four times higher than that of conventional ferrous P/M parts (Table 5). This is sufficient to withstand both the cutting and clamping forces of modern machine tools. Research confirms (Ref 23) that green compacts produced with warm compaction can be machined with conventional cutting tools with low cutting forces. Drill testing (Table 6) shows that the cutting forces are relatively low. Both cutting forces and surface finish can be improved by changes to drill type and profile. These changes also alter the accuracy and surface finish of the drilled hole. Thus, the choice of tool will be a compromise between low cutting forces, surface finish, and tolerances.

Presintering the green compact at a lower temperature than the final sintering operation produces a compact of relatively low hardness and strength but with sufficient edge retention to be handled and machined. Thus, the machinability of a presintered compact can be substantially better than that of the sintered part. However, presintering introduces an additional step to the manufacturing process and increases cost. It can be justified where the properties of the as-sintered part make its machining difficult or impossible. For example, high-carbon sinter hardened steels can require grinding rather than machining. In this case, providing that the part application and tolerances permit, it can be desirable to machine the part in the presintered condition rather than perform a grinding operation. Similarly, if part design calls for a through hole normal to the compaction axis, drilling the presintered preform followed by final sintering can be the only way to produce the hole economically in a high performance P/M steel.

Free-machining agents are added to P/M steel to improve machinability (Ref 8, 24). These agents are thought to perform several functions during the cutting process (Ref 9), including initiation of microcracks at the chip/workpiece interface, chip formation, lubrication of the tool/chip interface, and prevention of adhesion between the tool and chips (Fig. 5).

Several materials including sulfur, molybdenum disulfide (Ref 24), manganese sulfide (Ref 11), and boron nitride (Ref 25) are used as free-machining agents for P/M steels. They are most frequently introduced as fine powder to powder premixes, but sulfur and manganese sulfide are also available as prealloyed powders (Ref 26–28).

Table 5 Green strength of ANCORDENSE premixes

Material	Green density, g/cm^3	Green strength, psi
M-1	7.29	4221
M-2	7.33	4800
M-3	7.37	7703
M-4	7.31	9454
M-5	7.15	6286

Table 6 Mean drilling forces for warm compacted test pieces

Drill type	Mean force, lbf
118° parabolic geometry	100.4
135° split point	45.3
135° split point-wide land parabolic flute	49.3
135° split point-wide land parabolic flute, coated	55.5

Material: Ancorsteel 85HP, 2% Ni, 0.4% graphite; green density, 7.33 g/cm^3. Machining: 0.375 in. HSS drill; speed, 3285 rpm; feed, 0.012in./revolution

Sulfur and molybdenum disulfide can have strong effects upon the dimensional change and strength of P/M steels (Fig. 6, 7). Their use should be considered at the part design stage rather than as a "retrofit" when machining problems become apparent. Manganese sulfide has smaller effects upon dimensional change and strength (Ref 13) and can be used to improve the machinability of existing premixes. The effects of several potential machining agents upon the machinability of P/M steels are described below.

Sulfides

Sulfides are probably the most frequently used free-machining agents in both wrought and P/M steels. Powder metallurgy offers more flexibility than wrought metallurgy. Sulfur can be prealloyed in the powder during the primary production process or admixed as sulfur or sulfides

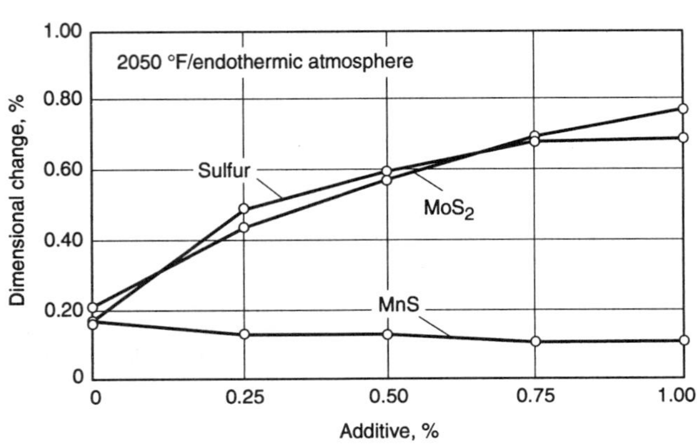

Fig. 6 Dimensional change of F-0008 atomized plus free-machining agents

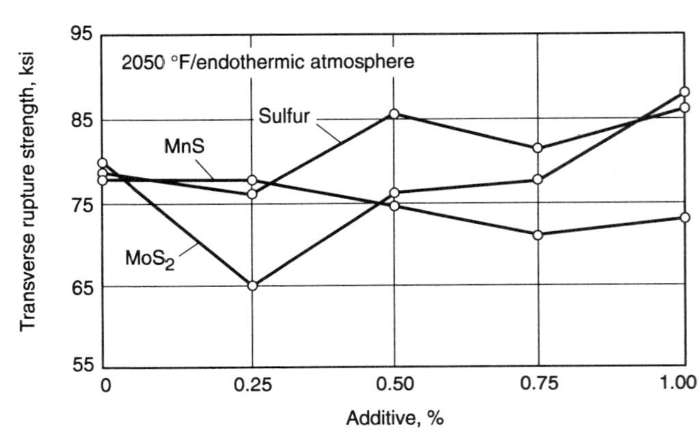

Fig. 7 Transverse rupture strength of F-0008 atomized plus free-machining steels

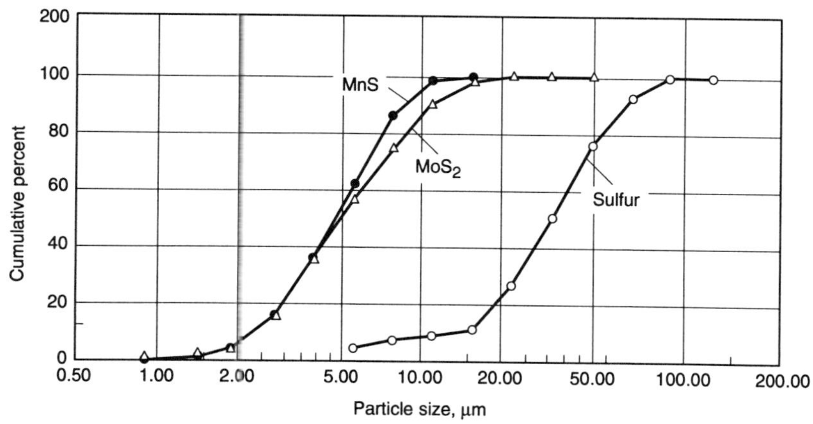

Fig. 8 Cumulative particle size distribution of free-machining agents

Table 7 Machinability of resulfurized iron powders

Premix	F-0008			FC-0208		
Base powder	MP 36S	MP 37	MP 35	MP 36S	MP 37	MP 35
Time for 25 holes, s	6.2	8.1	14.52	9.3	9.9	17.7
Holes to failure	203	82	25	40	32	21
Manganese, %	0.38	...	0.94	0.38	...	0.94
Sulfur, %	0.38	...	0.236	0.38	...	0.236

Test piece compacted to 6.6 g/cm^3. Sintered at 1121 °C (2050 °F). Cutting conditions: 0.25 in. HSS drill, 2300 rpm, 154 pound point loading. Source: Ref 26

during the preparation of a press-ready powder premix. Premixing offers more flexibility in the composition and amount of sulfide formed in the final compact. Prealloy produces a somewhat finer dispersion of sulfides within the powder particles.

Sulfur Prealloys. Several powder producers have supplied prealloyed or resulfurized powders, where sulfur is introduced during the primary powder production process. More recently, such powders have become a niche product to meet specific market needs, and admixed manganese sulfide has become a more widely used free-machining agent. It is possible that the powder manufacturing process offers a risk of cross contamination between resulfurized and nonresulfurized grades. Such contamination would introduce undesirable and easily detected sulfide inclusions into other high powder products rendering them unacceptable for high performance applications. Cross contamination can be most easily and efficiently minimized by introducing the sulfur or sulfide as late as possible in the premix stage.

Despite these problems, several powder producers (Ref 26, 27) offer prealloyed or resulfurized sulfur powders. These can offer significant improvements in machinability over nonresulfurized powders, at some loss in compressibility due to the solution-hardening effects of sulfur in iron.

By using a resulfurized powder for an F-0008 composition, Ancorsteel 1000M drill life was increased by about 50% compared to a similar composition with no free-machining additives:

- Ancorsteel 1000M: 128 holes to failure
- Ancorsteel 1000: 83 holes to failure
- Compaction: 6.8 g/cm^3
- Sintered: 1121 °C (2050 °F)
- Cutting conditions: 0.125 in. HSS drill, 3000 rpm, 0.003 in./revolution

One result (Table 7) shows that controlling both manganese and sulfur content of the prealloyed powder improves the machinability of iron powder indicated by the time required to drill holes and the number of holes drilled before failure. The use of resulfurized powders can be extended to reduced or sponge iron powders with similar beneficial effects upon drill life (Table 8).

Admixed Sulfides. Many different sulfides have been evaluated as free-machining additives for P/M steels. The additives most frequently used are manganese sulfide, sulfur, and molybdenum disulfide. Manganese sulfide made as an 0.35 to 0.6% addition to a premix is the most widely used. Additives usually take the form of fine high purity powders less than 50 µm in particle size as measured by laser particle size analysis (Fig. 8).

Effect of Sulfides upon Drill Life. The effect of sulfide free-machining agents upon drill life at an 0.5% addition is compared for three widely used premixes: F-0008, FC-0208, and FN-0205 in the as-sintered condition.

Machinability of F-0008. The machining agents improve drill life, indicated by holes completed before drill failure, when added to F-0008:

- Powder MH 1024: 45 holes to failure
- Powder MH 100: 22 holes to failure

Table 8 Effect of 0.5% machining agents on drill life

Condition	Drill life, holes to failure			
	No agent	Sulfur	Manganese sulfide	Molybdenum disulfide
Sponge	19	158	33	285
Atomized	83	418	890	718

Compacted to 6.8 g/cm^3 and sintered at 1121 °C (2050 °F)

Table 9 Effect of 0.5% machining agents on drill life

Condition	Drill life, holes to failure			
	No agent	Sulfur	Manganese sulfide	Molybdenum disulfide
Sponge	2	608+	81	108
Atomized	2	608+	72	249

Compacted to 6.8 g/cm^3 and sintered at 1121 °C (2050 °F)

- Compaction: 6.8 g/cm^3
- Sintered: 1121 °C (2050 °F)
- Cutting conditions: 0.125 in. HSS drill, 3000 rpm, 0.003 in./revolution

In the mixes made with atomized powder, all the machining additives increase drill life significantly. Manganese sulfide and molybdenum disulfide produce larger increases than sulfur. The improvements are not as large in the materials made with sponge powders where molybdenum disulfide produces the best drill life. Surprisingly, 0.5% manganese sulfide does not perform as well in the sponge test mixes in this sequence of tests.

Although drill life is the chosen indicator of performance, the machining agents produce other changes in cutting performance that could influence production use. During the drill test, it was apparent that different agents changed chip form; for example, molybdenum disulfide produces very small chips that are easily removed from the cutting area.

The machining agents appear to influence heat transfer to and plastic deformation of the workpiece during the test. As the drill accumulates heat during the test, some materials show local plastic deformation around the cutting area. In extreme cases, the cutting forces push material through the unsupported back of the workpiece, especially in mixes with sulfur and least apparent in molybdenum disulfide. Such extreme behavior is unacceptable in production. The changes in hole diameter that precede it are a significant factor in determining tool life.

Machinability of FC-0208. The free-machining additives all improve the machinability of the FC-0208 (Table 9). The 0.5% sulfur addition produces the largest improvement in both atomized and sponge compositions. The drills do not fail before all test material is consumed. Molybdenum disulfide is more effective than manganese sulfide under the chosen test conditions.

Effect of Sulfides. The drill test results show that the 0.5% sulfide addition improves ma-

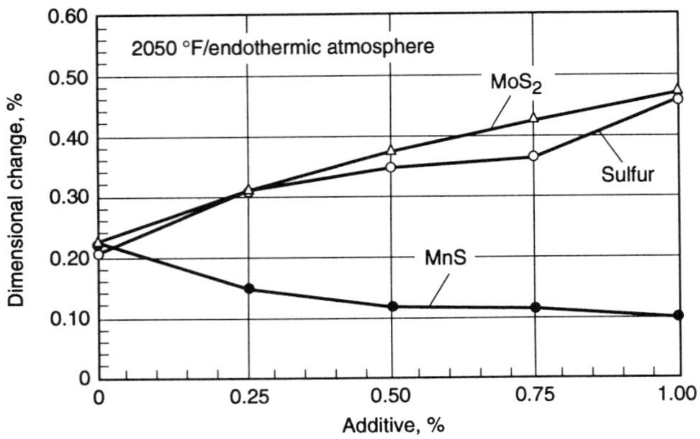

Fig. 9 Dimensional change of F-0008 sponge plus free-machining agents

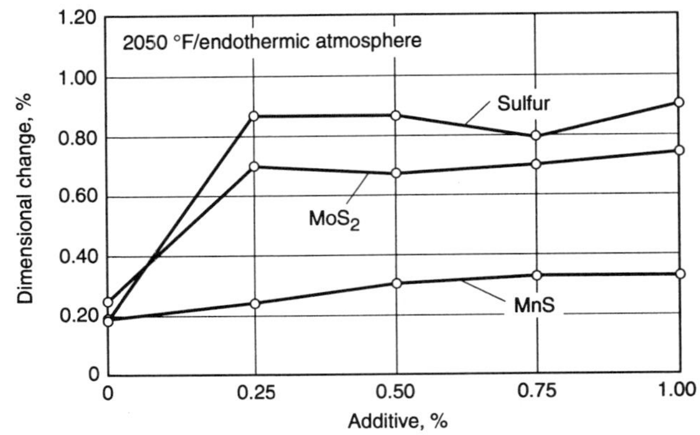

Fig. 10 Dimensional change of FC-0208 atomized plus free-machining agents

Table 10 Sulfur recovery for sintered F-0008 sponge

	Sulfur, %		Manganese sulfide, %		Molybdenum disulfide, %	
Addition, %	Added	Measured	Added	Measured	Added	Measured
0	0	0.01	0	0.01	0	0.01
0.25	0.25	0.17	0.09	0.11	0.10	0.04
0.50	0.50	0.41	0.18	0.18	0.20	0.12
0.75	0.75	0.60	0.28	0.26	0.30	0.18
1.00	1.00	0.76	0.37	0.35	0.40	0.25

Compaction in 1% zinc stearate to 6.1 g/cm^3. Sintering: to 1121 °C (2050 °F), endothermic, 30 min

chinability significantly. However, the addition of sulfides can change sintered properties, particularly dimensional change. The sulfides are not completely inert during the sintering process and can modify the sintering reactions.

Metallography

Free-machining agents have different effects upon the microstructures of the test premixes. These effects depend upon the nature of the base iron (sponge or atomized) and the premix composition (Ref 29) and are illustrated with reference to the F-0008 composition in Fig. 6 and 9. Sulfur appears to promote sintering of the test compositions and produces very round pores. A portion of the sulfur dissolves in the iron matrix and diffuses a short distance into the iron matrix. In sponge iron compositions, a portion of sulfur reacts with iron to form iron sulfides. Sulfur also appears to promote pore rounding during sintering.

Manganese sulfide (Ref 6) is considered to be stable in both iron-graphite and iron-copper-graphite premixes. Metallography indicates that the manganese sulfide occurs in pores or the fine pores remaining at prior particle boundaries. It appears to be almost inert during the sintering process with little evidence of diffusion into the iron matrix in the compositions examined.

The effects of molybdenum disulfide appear to be between those of sulfur and manganese sulfide and to depend upon the amount added to the test composition. It appears that a significant portion of the molybdenum disulfide reacts with the sintering atmosphere. At low MoS_2 additions, almost all of the addition transforms to molybdenum, which remains within pores and at particle boundaries. Some sulfur evaporates during the sintering process and some dissolves in the iron matrix and can form sulfides on cooling from sintering temperature. At higher sulfur additions, it appears that an equilibrium is reached between the sulfur present as molybdenum disulfide and that in the particles, so that more typical slight pore rounding occurs.

Stability of Sulfides

Sintering. Neither sulfur nor sulfides are completely stable during sintering. Comparison of the sulfur content of the F-0008 premix ingredients to that of the sintered test pieces shows that the sulfur recovery was lower than anticipated (Table 10).

The data show that manganese sulfide is stable under endothermic atmosphere sintering. The measured and predicted sulfur contents of the manganese sulfide mix agree well. As anticipated, a significant portion of the elemental sulfur is lost during sintering. The measured sulfur content in the molybdenum sulfide premix is also less than anticipated confirming the metallographic findings that a portion of the molybdenum disulfide decomposes to release sulfur to the iron matrix and sintering atmosphere. Similar trends were observed in the atomized F-0008 and FC-0208 compositions.

These results indicate that a portion of the sulfur can be released to the sintering atmosphere. It is possible that sulfur in the atmosphere could be absorbed by other compositions present in a production sintering furnace. Given the sensitivity of physical properties to sulfur content, it appears that considerable care should be exercised in furnace loading, atmosphere, and scheduling to avoid the possibility of contaminating a furnace or subsequent premix composition.

Stability during Handling. Manganese sulfide is hygroscopic. Several authors (Ref 29) have stressed that fine manganese powders should be kept in closed containers to prevent moisture absorption and oxidation of the manganese sulfide. If simple precautions are taken (e.g., closing containers and rapidly consuming all material in an open container), the sulfide is stable and provides very consistent results.

Stability during Machining. The potential for reactions between moisture and manganese sulfide can extend to the machining process. The majority of machining tests of P/M parts are conducted in the dry condition without a cutting fluid or lubricant. In contrast, many high volume machining operations employ a water-based cutting fluid for cooling and removing chips from the cutting area. In machining some lower density P/M components on high-speed transfer lines, reactions between water-based cutting fluids and manganese sulfides occur with detrimental results upon the machining operations (Ref 30).

Effects upon Sintered Properties

The experiments indicate that free-machining agents influence sintered microstructure. Thus, they will also influence sintered properties. The effects depend upon the machining agent, premix composition, and the sintering atmosphere. Generally, sulfur and molybdenum disulfide promote growth on sintering. In contrast, manganese sulfide has less effect and could slightly reduce growth. In general, free-machining agents improve the strength of F-0008 compositions but reduce that of FC-0208 significantly.

F-0008 Atomized Iron Powder. The effects of the machining agents upon the properties of F-0008 made with atomized iron powder are illustrated in Fig. 8 and 9. Increasing sulfur and molybdenum disulfide content tend to increase strength and dimensional change from die size.

Table 11 Effects of machining agents on hardness (HRB) of F-0008 atomized powder

Addition, %	Sulfur, %	Manganese sulfide, %	Molybdenum disulfide, %
0	54	53	51
0.25	52	53	52
0.50	56	53	58
0.75	53	53	65
1.00	56	55	66

Compacted to 6.8 g/cm^{3}and sintered at 1121 °C (2050 °F), endothermic atmosphere, 30 min

Table 12 TRS of F-0008 sponge powder plus free-machining agents

Agent addition, %	Transverse, rupture strength, ksi, with addition of: Sulfur	MnS	MoS
0	49.3	51.0	50.5
0.25	64.0	52.9	57.6
0.50	63.6	53.6	68.9
0.75	62.4	51.4	68.2
1.00	63.1	52.6	72.0

Compaction to 6.1 g/cm^3 and sintering at 1121 °C (2050 °F), endothermic atmosphere, 30 min

Table 13 Effect of machining agents on hardness (HRB) of F-0008 sponge

Addition, %	Sulfur, %	Manganese sulfide, %	Molybdenum disulfide, %
0	24	23	22
0.25	36	24	35
0.50	35	27	44
0.75	37	24	32
1.00	40	26	46

Compacted to 6.8 g/cm^3. Sintered at 1121 °C (2050 °F), endothermic atmosphere, 30 min

Table 14 Effect of machining agents on dimensional change of FC-0208 sponge

Additive, %	Dimensional change, % Sulfur	Manganese sulfide	Molybdenum disulfide
0	+0.27	+0.31	+0.31
0.25	+0.55	+0.35	+0.54
0.50	+0.50	+0.36	+0.44
0.75	+0.55	+0.38	+0.44
1.00	+0.56	+0.38	+0.41

Compaction of 1% zinc stearate to 6.1 g/cm^3. Sintering at 1121 °C (2050 °F) endothermic atmosphere, 30 min

Table 15 Effect of machining agents on TRS of FC-0208 sponge

Transverse rupture strength, ksi Sulfur	Manganese sulfide	Molybdenum disulfide
83.0	84.3	82.0
80.1	78.6	82.5
78.1	80.5	86.0
80.3	75.9	83.0
79.9	73.8	86.0

Compaction of 1% zinc stearate to 6.1 g/cm^3. Sintering at 1121 °C (2050 °F) endothermic atmosphere, 30 min

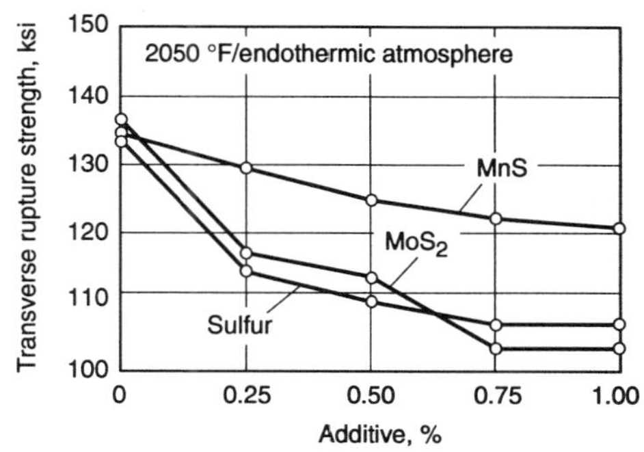

Fig. 11 Transverse rupture strength of FC-0208 atomized plus free-machining agents

Increasing manganese sulfide addition from 0 to 1% tends to decrease dimensional change and strength slightly. Sulfur and manganese sulfide have almost no effect upon macrohardness at a density of 6.8 g/cm^3 whereas increasing molybdenum disulfide content increases hardness. Both sulfur and molybdenum disulfide promote growth, in contrast, increasing manganese sulfide reduces growth slightly.

The effects of the machining agents upon transverse rupture stress (TRS) are slightly more complex than upon dimensional change. Increasing sulfur contents increases TRS slightly, whereas increasing manganese sulfide contents decreases TRS slightly. Adding molybdenum disulfide initially reduces TRS, but a 1% addition of MoS_2 increases the TRS of the F-0008 test premixes. Both sulfur and manganese sulfide increase the hardness of F-0008 test mix slightly (Table 11). Increasing molybdenum disulfide additions increase the hardness of the F-0008 atomized rapidly.

F-0008 Sponge Iron Powder. When added to F-0008 made with sponge iron, the free-machining agents had similar effects to those observed in the F-0008 composition made with atomized powder. Manganese sulfide was relatively neutral, while both sulfur and molybdenum disulfide increased the growth on sintering of the sponge-based F-0008 significantly. Adding manganese sulfide reduced growth from die slightly, the effect of the first 0.25% addition being most significant (Fig. 11).

Adding 0.25% of sulfur or molybdenum disulfide increases strength significantly (Table 12). Further increases in sulfur content do not change TRS significantly. Increasing molybdenum disulfide content to 1% increases TRS further. Manganese sulfide has no effect upon strength under the conditions tested. Adding the machining agents to F-0008 sponge causes similar changes in hardness to those observed in TRS. The first 0.25% addition of sulfur increases hardness significantly, but hardness increases slowly with further additions (Table 13). Molybdenum disulfide increases hardness significantly, but an increase beyond 0.5% causes little increase in hardness. The hardness of the F-0008 increases only slightly with manganese sulfide additions.

FC-0208 Atomized. For the FC-0208 made with atomized powder, all three agents increase growth significantly (Fig. 10). The effect of sulfur is slightly greater than manganese sulfide or molybdenum disulfide. The data indicate that additions above 0.5% cause little further increase in growth. All free-machining agents reduce the strength of the FC-0208 composition significantly (Fig. 11). The effect of manganese sulfide was less than that of sulfur and molybdenum disulfide.

FC-0208 Sponge. All machining agents increase the growth of the FC-0208 premixes made with sponge powder. They have somewhat less effect on TRS than observed in the FC-0208 premixes made with atomized powder (Table 14).

Sulfur additions cause the greatest increase in growth; molybdenum disulfide has a somewhat smaller effect, and manganese sulfide has the least effect. A 0.25% addition of sulfur or molybdenum disulfide causes a significant increase in growth that does not increase with further additions. The growth of FC-0208 sponge tends to increase slowly with increasing manganese sulfide content.

The effects of the free-machining agents upon the TRS of the FC-0208 sponge are relatively small (Table 15). Transverse rupture strength tends to increase slightly with increasing molybdenum disulfide. It decreases slightly with increasing sulfur content. Increasing manganese sulfide causes the most significant decrease in TRS at additions above 0.5%.

Nonsulfide Machining Agents. In wrought metallurgy, several other free-machining agents, such as lead and selenium, are employed. The P/M industry has made little use of these, possibly due to potential toxicity problems. Powder producers and parts makers (Ref 31–33) are taking advantage of the premixing operation to introduce free-machining agents that are incompatible with the processing of wrought steels. These agents are nonmetallic solids that are anticipated to assist in chip formation and lubrication of the chip cutting tool interface to reduce wear mechanisms. Unlike sulfide machining agents, these products are proprietary to specific producers and patent protected.

Enstatite is a soft mineral that shears easily under stresses but is relatively stable at sintering temperatures. K. Hayashi et al. (Ref 31) describe the use of enstatite in combination with manganese sulfide to enhance the machinability of sintered P/M steels.

Boron Nitride. Hexagonal boron nitride is a recognized solid lubricant with microstructure and frictional properties similar to those of graphite. The use of boron nitride as a free-machining agent in sintered P/M steels has been patented (Ref 32). Published data show that the addition of small quantities of boron nitride to sintered steels enhances machinability significantly in F-0008 and FC-0208 compositions (Table 16).

Graphite/Sulfur. Graphite is widely used as a solid lubricant in the production of porous sin-

Table 16 Drill life holes to failure

Material	ATOMET 29	ATOMET 29M
Fe-0.3%C	17	58
F-0005	56	144
F-0008	25	87
FC-0208	2	74

Compaction to 6.7 g/cm^3. Sinter at 1120 °C (2048 °F), 90% Ni/10% H atmosphere. Machining by 6.35 mm HSS drill, 4250 rpm. Source: Ref 32

Table 17 Effect of sulfur modification upon drill life for iron-2% copper

Graphite addition, %	Standard	Sulfur modified
0.8	<30	>600
1.0	<30	>600
1.2	<30	350

Compaction to 6.85 g/cm^3. Sinter at 1130 °C (2066 °F), nitrogen atmosphere. Machining in 1 mm HSS drill, speed 10,000 rpm, feed 0.012 mm/rev. Source: Ref 33

Table 18 Drill life for FC-0208

		Composition MnS (MnX), wt %						
Iron	Atmosphere	0	0 (0.35)	0 (0.50)	0.10 (0.25)	0.15 (0.35)	0.35 (0)	0.50 (0)
Ancorsteel 1000	Endothermic	2	220	186	91	134	107	312
	H_2/N_2	2	149	428	153	99	89	49
Ancor MH100	Endothermic	2	23	55	108	157	64	42
	H_2/N_2	4	30	270	100	584	122	48

Compaction to 6.8 g/cm^3. Machining in 0.125 in. drill, 3000 rpm, 0.006 in./rev feed

Table 19 Drill life for FN-0205

		Composition MnS (MnX), wt %						
Iron	Atmosphere	0	0 (0.35)	0 (0.50)	0.10 (0.25)	0.15 (0.35)	0.35 (0)	0.50 (0)
Ancorsteel 1000	Endothermic	9	250	275	1400	473	336	498
	H_2/N_2	8	380	816	1400	1251	331	352
Ancor MH100	Endothermic	9	62	120	111	416	54	152
	H_2/N_2	26	109	172	481	991	136	497

Compaction to 6.8 g/cm^3. Machining in 0.125 in. drill, 3000 rpm, 0.0006 in./revolution feed

Table 20 Drill life for Ancorsteel 85HP: 2% Ni, 0.5% graphite

	Composition MnS (MnX), wt %						
Iron	0	0 (0.35)	0 (0.50)	0.10 (0.25)	0.15 (0.35)	0.35 (0)	0.50(0)
Holes to Failure	1	1	2	2	3	4	2

Compaction to 7.0 g/cm^3. Sintering at 1121 °C (2050 °F), 75% H/25% Ni. Machining in 0.125 in. drill, 3000 rpm, 0.0006 in./revolution feed

Table 21 Effect of drill type upon drill life when cutting Ancorsteel 85HP: 2% Ni, 0.5% graphite

Steel type	Coating	Flute form	Point	Drill life
Base HSS	None	Standard	135°	2
M42	None	Standard	135°	2
Base HSS	TiN	Standard	135°	1
HSS	None	Parabolic	135°	1
Cobalt HSS	None	Parabolic	135°	3
M7HSS	TiN	Parabolic	135°	44
Cobalt HSS	TiN	...	Split	85

Compaction to 7.0 g/cm^3. Sinter at 1121 °C (2050 °F), 75% hydrogen/25% nitrogen. Machining at 0.125 in. drill, 3000 rpm, 0.006 in./revolution feed

tered bearings. However, its use in structural P/M parts is largely confined to that of an alloying agent intended to dissolve during sintering. Prealloyed sulfur has the ability to inhibit graphite solution so that some free graphite remains in the sintered microstructure without adversely reducing mechanical properties. Table 17 illustrates how this combination of prealloyed sulfur and graphite can produce a significant improvement in the machinability of sintered P/M steels (Ref 33).

MnX is a proprietary free-machining agent. Employed alone and in combination with manganese sulfide, it improves the machinability indicated by drill life of a wide range of P/M steels sintered in both endothermic and nitrogen hydrogen atmosphere (Ref 21, 34).

Atomized FC-0208. For FC-0208 made with atomized powder, a 0.50 wt% manganese sulfide addition produced best drill life when sintered in endothermic atmosphere (Table 18). However, when sintered in 75%H_2/25%N_2, 0.50 wt% MnX produces best performance. Earlier research (Ref 28) shows that 0.5 wt% additions of manganese sulfide can reduce strength and increase growth from die size. If this is unacceptable, then a 0.35 or 0.5 wt% addition of MnX increases drill life significantly. For FC-0208 premixes made with sponge powder, the combination of manganese sulfide and MnX produces longest drill life under both sintering atmospheres. The effects of the machining additives upon drill life are somewhat greater in H_2/N_2 atmospheres than endothermic.

FN-0205. Drill testing shows that the machining agents improve the machinability of FN-0205 significantly (Table 19). All free-machining agents increase the machinability of the FN-0205 test pieces under the conditions employed. The combination of manganese sulfide plus MnX produces longer drill life than either additive alone for both iron bases and both sintering atmospheres. In premixes made with atomized iron powder, it is possible that an 0.35 wt% total addition of the machining agents produces best results. For test premixes made with Ancor MH100, a 0.50 wt% total addition of MnS plus MnX produces the best tool life. These levels of machining agent have relatively minor effect upon the sintered properties of FN-0205.

Tool Materials

Both part and powder producers have made significant progress in developing higher strength P/M steels to compete with fully dense wrought steels and heat treated cast irons. The combinations of strength and ductility make these alloys more difficult to machine in the as-sintered condition requiring higher performance tool materials. However, tests show that these free-machining additives alone do not always improve machinability of higher strength materials, as seen in Table 20 for Ancorsteel 85HP. A series of cutting tests were conducted (Ref 21) to assess whether changes to tool material, tool geometry, or coating could improve cutting performance in this alloy. The drill test used the same conditions as previous tests; however, this test examined higher performance cobalt high-speed steels, the effects of titanium nitride coatings, and different tool geometries, such as parabolic flutes or "split points."

The results indicate that when combined, improvements to tool material, coating, and drill design improve drill life (Table 21). Individual improvements to high-speed steel, flute form, or coating had relatively little effect upon drill life under the test conditions employed. However, when combined, drill life increases significantly from an initial two holes for the standard drill to 85 holes for the optimum drill.

Attempts to improve drill life further by using solid carbide drills have been unsuccessful. The solid carbide drills appear to be too brittle for the test conditions and drilling system employed. The test results did show clearly that drilling of very high-strength P/M steels can be improved

Table 22 Comparison of mechanical properties and motor load

Cooling rate	Yield strength, ksi	Ultimate tensile strength, ksi	Hardness, HRB	Motor load, hp
8.75	36.7	36.7	48.9	4.39
16.7	50.5	50.5	57.2	4.55
39.8	58.3	58.3	73.0	4.76

Sintered density, 6.8 g/cm^3

Table 23 Ancorsteel 85 HP: 2 wt% Ni, 0.5 wt% graphite

Treatment	Temperature, °C (°F)	Time, h	Cool	Holes to failure	Hardness, HRB
Temper	149 (300)	1	Natural	4	78
	302 (575)	1	Natural	8	71
	454 (850)	1	Natural	14	82
	621 (1125)	1	Natural	6	74
Anneal	871 (1600)	1	Furnace	7	57
As-sintered		...	...	3	73

Compaction to 7.0 g/cm^3. Sintered to 1121 °C (2050 °F), 75% H/25% Ni. Machining at 0.125 in. drill, 3000 rpm, 0.006 in./revolution feed

using similar improvements in tool materials and design to those used for high-strength wrought steels.

Microstructure Modification

Wrought steels are frequently heat treated by producers to obtain optimal machinability. Annealing or normalizing treatments are used to produce a relatively coarse pearlitic or spheroidal microstructure that possesses good machinability. Higher carbon steels require longer heat treatment cycles intended to produce coarse carbides dispersed in ferrite. In contrast, low-carbon steels can be partially hardened to produce a microstructure that is less ductile and adhesive than fully annealed steels. Research has shown that reducing cooling rates from sintering can improve the machinability of FC-0208; however, the coarser microstructure reduces mechanical properties (Ref 35). A significant improvement in machinability indicated by drill motor load is only obtained at lower strength levels (Table 22).

A similar result is obtained in improving the machinability of a 0.85%Mo-2%Ni-0.5% graphite by tempering and annealing. The annealing treatments do not produce the anticipated improved machinability. The 0.85% molybdenum steel also proves to be very temper resistant (Table 23). It is noteworthy that the annealing treatment at 871 °C (1600 °F) changes the failure mode of the high-speed steel drill. The drill appears to adhere to the workpiece and snap rather than overheat.

The experiment indicates that the heat treatments employed did not change the microstructure and machinability of P/M steels sufficiently. The results do not justify the extra process step. Further study is necessary to define the optimal microstructure and heat treatment for machinability. It is possible that increasing tempering temperatures to 648 to 732 °C (1200 to 1350 °F) for subcritical annealing or controlled transformation annealing after sintering can improve machinability further. Because such treatments add an extra process step, they are justified only where parts require extensive machining prior to heat treatment.

REFERENCES

1. D. Madan, The Importance of Machinability in the Processing of P/M Parts, *Advances in Powder Metallurgy and Particulate Materials,* Vol 2, Metal Powder Industries Federation, 1995, p 8–55
2. H.I. Sanderow, J.R. Spirko, and R. Corrente, The Machinability of P/M Materials as Determined by Drilling Tests, *Advances in Powder Metallurgy and Particulate Materials,* Vol 2, Metal Powder Industries Federation, 1997, p 15–125
3. E.M. Trent, *Metal Cutting,* Butterworths, 1977
4. C.W. Kovach, Sulfide Inclusions and the Machinability of Steel, *Sulfide Inclusions in Steels,* American Society for Metals, 1975, p 459–479
5. V.A. Tipnis, Calcium Treatment for Improved Machinability, *Sulfide Inclusions in Steels,* American Society for Metals, 1975, p 480–495
6. R. Koos and G. Bockstiegel, The Influence of Heat Treatment, Inclusions, and Porosity on the Machinability of Powder Forged Steels, *Prog. Powder Metall.,* Vol 37, 1981, p 145–147
7. J.S. Agapiou and M.R. DeVries, Machinability of Powder Metallurgy Materials, *Int. J. Powder Metall.,* Vol 24 (No. 1), 1988, p 47–57
8. H. Chandler, *Machining of Powder Metallurgy Materials,* Vol 16, *Metals Handbook,* 9th ed., ASM International, 1989, p 879–892
9. S.A. Kvist, Turning and Drilling of Some Sintered Steels, *Powder Metall.,* Vol 12 (No. 24), 1969
10. M.P. Eisenmann and H.S. Nayar, Sintering of Machining Grade Ferrous Compacts for Improved Machinability, *Met. Powder Rep.,* 1985, p 96–98
11. D.S. Madan, An Update on the Use of Manganese Sulfide (MnS) Powder in Powder Metallurgy Applications, *Advances in Powder Metallurgy,* Vol 3, Metal Powder Industries Federation, 1991, p 101
12. MPIF Standard 35, *Materials Standards for P/M Structural Parts,* Metal Powder Industries Federation, 1997
13. J.A. Hamill, R.J. Causton, and S.O. Shah, High Performance Materials for P/M Utilizing High Temperature Sintering, *Advances in Powder Metallurgy and Particulate Materials,* Vol 5, Metal Powder Industries Federation, 1992, p 193–213
14. R.J. Causton, Sinter Hardening Low-Alloy Steels, *Advances in Powder Metallurgy and Particulate Materials,* Vol 5, Metal Powder Industries Federation, 1992, p 27–50
15. P.J. James, Factors Affecting Quality of Drilled Holes in Sintered Steels, *Powder Metall.,* Vol 37 (No. 2), p 133
16. *Machinery Handbook,* Industrial Press, 1984, p 1803
17. S. Berg, Machinability of Sintered Steels: Guidelines for Turning, Drilling, and Tapping, *Advances in Powder Metallurgy and Particulate Technology,* Vol 2, Metal Powder Industries Federation, 1997, p 15–145
18. G.T. Smith, Surface Integrity Aspects of Machinability Testing of Fe-C-Cu Powder Metallurgy Components, *Powder Metall.,* Vol 3 (No. 2), 1990, p 157
19. I. Sharif and K. Boswell, Prediction and Modelling of Surface Finish in Drilling of P/M Parts, *Advances in Powder Metallurgy and Particulate Technology,* Vol 2, Metal Powder Industries Federation, 1997, p 15–155
20. Y.T. Chen et al., Free-Machining P/M Alloy Optimization Using Statistical Analysis Techniques—The Effect of MnS Content and Particle Size, *Advances in Powder Metallurgy and Particulate Materials,* Vol 4, Metal Powder Industries Federation, 1992, p 269
21. R.J. Causton, Machinability of P/M Steels, *Advances in Powder Metallurgy and Particulate Materials,* Vol 2, Metal Powder Industries Federation, 1995, p 8–149
22. S. Luk, Metal Powder Compositions Containing a Binder Agent for Elevated Temperature Compaction, U.S. Patent 5,154,881, 13 Oct 1992
23. T.M. Cimino and S.H. Luk, Machinability Evaluation of Selected High Green Strength P/M Materials, *Advances in Powder Metallurgy and Particulate Materials,* Vol 2, Metal Powder Industries Federation, 1995, p 8–129
24. U. Engstrom, Machinability of Sintered Steels, *Prog. Powder Metall.,* Vol 38, 1982, p 417
25. M. Gagne, Sulfur Free Iron Powder Machinable Grade, *Advances in Powder Metallurgy,* Vol 3, Metal Powder Industries Federation, 1991, p 101
26. L.G. Roy et al., Prealloyed Powders for Improved Machinability in PM Parts, *Met. Powder Rep.,* Feb 1989

27. S. Hironori et al., 250 MSA Resulfurized High Green Strength Steel Powder, *Advances in Powder Metallurgy and Particulate Materials,* Vol 4, Metal Powder Industries Federation, 1997, p 15–27
28. R.J. Causton, T.M. Cimino, and H.M. Scanlon, Machinability Improvement of P/M Steels, *Advances in Powder Metallurgy and Particulate Materials*, Vol 7, Metal Powder Industries Federation, 1994, p 7–169
29. D. Madan and A. Fitzgibbon, Shelf Life of MnS Powder and MnS Containing Premixes, *Advances in Powder Metallurgy and Particulate Materials*, Vol 2, Metal Powder Industries Federation, 1995, p 8–177
30. O. Petterson, High Speed Turning and Boring of PM Carbon Steel, Paper 980629, Society of Automotive Engineers, 1998
31. K. Hayashi et al., Enhanced Machinability of Valve Guides Made from P/M Materials, *Advances in Powder Metallurgy and Particulate Materials,* Vol 4, Metal Powder Industries Federation, 1992, p 13–117
32. M. Gagne, Sulphur Free Iron Powder Machinable Grade, *Advances in Powder Metallurgy and Particulate Materials,* Vol 1, Metal Powder Industries Federation, 1989, p 365
33. S. Uenosono, S. Unami, and K. Ogura, A New Improvement in Machinability of P/M Steel Due to Retained Graphite Particles, *Advances in Powder Metallurgy and Particulate Materials,* Vol 2, Metal Powder Industries Federation, 1995, p 8–71
34. Additive Improves Machinability Threefold, *Met. Powder Rep.,* Sept 1993, p 35
35. J.J. Fulmer and J.M. Blanton, Effect of Microstructure on the Machinability of an MPIF FC-0208 Copper Steel, *Advances in Powder Metallurgy and Particulate Materials,* Vol 4, Metal Powder Industries Federation, 1992

Machining of Powder Metallurgy Materials

Sigurd Berg, Höganäs AB
Håkan Thoors, Swedish Institute for Metals Research
Bertil Steen, Swedish Institute for Production Engineering Research

POWDER METALLURGY is a near-net shape process capable of producing complex parts with little or no need for secondary operations, such as machining, joining, or assembly. However, the inability to produce certain geometrical figures (e.g., transverse holes, undercuts, and threads) frequently necessitates some machining, particularly drilling. Because of the growing use of P/M materials to increase productivity and reduce costs, the study of the machinability of these materials has become increasingly important, especially because powder metals are generally considered to have poor machinability in comparison with wrought or cast metals (Ref 1). In addition, there is a trend toward the design of complex P/M parts that call for machining, in contrast to the past practice to minimize machining (Ref 2).

The machinability of P/M components is dependent upon a number of factors, including workpiece and tool material properties, cutting conditions, machine and cutting tool parameters. Some P/M material and production process parameters (such as particle size, part geometry, porosity, and compaction and sintering methods) also influence machinability. In general terms, however, machining of P/M materials can be optimized or improved by:

- Taking advantage of the good machining properties of high-density parts (above 92% of theoretical density)
- Machining in the presintered condition
- Using additives that do not require sacrifices in powder processing or part properties
- Following the suggestions of published guidelines for material and machining methods
- Running machinability tests on P/M materials

These factors are briefly discussed in this article, along with more detailed discussions of the machining guidelines that improve the machinability of sintered steels. Additives and microstructural factors that enhance machining are discussed in more detail in the article "Machinability of Powder Metallurgy Steels" in this Volume.

General Guidelines

Representative machining methods are turning, milling, drilling, tapping, and grinding. The basic guideline is to start with cutting conditions for wrought and cast parts and tailor them for the application. The recommendation applies in particular to lower density parts. High-density parts (>92% of theoretical density) machine like wrought metals.

Smearing of self-lubricating porous parts can be a problem. Recommended practice involves the use of sharp tools and light cuts in single-point machining, such as turning or boring.

Coolants are preferred in most machining operations. Coolant pickup can be a problem. The rate of pickup is directly related to the amount of porosity. Ideally, all machining except grinding should precede deburring. Retained deburring abrasive can cause excessive tool wear. Ceramic and cubic boron nitride (CBN) inserts are usually run dry; performance is typically better than or at least equal to that obtained with coolants.

Material Selection. Powder metallurgy carbon steels are selected primarily for parts with moderate strength and hardness, combined with machinability. Iron-copper and copper steel materials are produced from admixtures of elemental iron powder and elemental copper powder with or without graphite powder (carbon). When secondary machining is required, combined carbon contents of less than 0.5% should be specified. Copper-infiltrated iron and steel materials offer improved machinability because of reductions in interrupted cuts, and machined parts have a smooth surface finish. Among stainless steels, SS-303 is preferred when parts require extensive secondary machining. Brass, bronze, and nickel silver parts usually have good machinability.

Additives. Free-machining benefits can be obtained by means of small additions to a standard powder composition. Additives for ferrous powders include lead, sulfur, copper, or graphite; for nonferrous powders, lead is used. The advantages of changing composition in this manner can be at least partially offset by side effects. Additions can cause problems, such as dimensional changes of parts during sintering and deterioration in the properties of parts.

Prealloyed manganese sulfide powders appear to avoid those shortcomings in ferrous alloys. Manganese content is intentionally high to ensure that all sulfur is present in the form of manganese sulfide inclusions. When these inclusions are extensively deformed in the shear plane and in the flow zone adjacent to the tool surface, they contribute to higher cutting speeds, longer tool life, good surface finish on parts, and lower tool forces. In addition, chips are more readily handled than those produced by conventional P/M materials. Oil or resin impregnation of porous P/M parts also improves machinability (see the article "Resin Impregnation of Powder Metallurgy Parts" in this Volume).

Design. Certain types of holes, undercuts, and threads are examples of features that cannot be accommodated by the P/M consolidation (pressing) process and therefore require machining. Holes in the direction of pressing, produced with core rods that extend up through the tools, are readily incorporated in parts, but side holes (those not parallel to the direction of pressing) cannot be made in the same way and are generally produced by secondary machining.

Undercuts on the horizontal plane (perpendicular to the die centerline) cannot be produced if they prevent the part from ejecting from the die (Fig. 1). Annular grooves around a part are produced by machining or by making the part in an assembly of two pieces. Likewise, a part with a reverse taper (larger on bottom than on top) cannot be ejected from a die. Because threads in holes and on outside diameters prevent a part from being ejected from a die, they cannot be made with conventional P/M methods; machining is required.

Machining Guidelines

The machining process is very complex, and tool performance is affected by the properties and condition of the workpiece and the cutting condition. For P/M materials, porosity is a major factor that reduces machinability. The cutting tool configuration (in terms of chip breaker profile, stability, and geometry of tool holder, insert style, etc.) also influences the wear processes that determine tool life.

In order to select the right tool and machining parameters, knowledge of the loads on the tool and the properties of the tool material together with an analysis of the wear mechanisms is necessary. The loads associated with the wear process (Ref 3) can be divided into four main groups:

- Mechanical load
- Thermal load
- Chemical load
- Abrasive load

To define cutting parameters, the loads on the tool must be controlled based on the active wear mechanisms. Typically, the loads on the edge of a cutting tool are different at different locations. Consequently, different wear mechanisms are activated and proceed at different rates at the various locations. The processes that influence tool life can be plotted schematically on a wear mechanism map (Fig. 2), which delineates the area's wear and "safe zones" for good tool performance.

The following sections describe machining conditions for common operations (e.g., turning, drilling, tapping, grinding, and milling). In addition, examples are given for machinability evaluations on the turning, drilling, and tapping of various sintered steels. These examples illustrate the influence of chemical composition, tool material, tool geometry, free machining additives, feed rate, cutting velocity, cutting conditions, and surface integrity while forming the basis of guidelines for the machining of sintered steels. Effects from microstructure, carbon content, density, machinability enhancing additives, etc. are examined. Guidelines for optimum machining parameters in turning, drilling, and tapping are stipulated for a wide spectrum of P/M steel conditions based on these examples.

Turning

Usually, parts with an average hardness of HRB 52 have machining properties similar to those of cast iron. At this hardness level, parts should be held rigid, and weak sections should be supported to prevent distortion. Compressed air is used to cool the tool and maintain swarf clearances. Jets are directed onto tool cutting edges and work surfaces. Liquid coolants cannot be used because parts must be kept dry and clean for subsequent sintering. Carbide tips of ISO designation K10 with a hardness of 92.6 HRA give good results and will accept some interruptions on the cutting surface.

Tools must be held rigid, and cutting edges must be sharp with rake angles of up to 3° positive on top and side and frontal clearances of 3 to 5°. Surface speeds of 105 to 120 m/min (350 to 400 sfm) and feeds of 0.050 to 0.10 mm/rev (0.002 to 0.004 in./rev) are satisfactory for form turning, but surface speeds can be increased to 180 to 210 m/min (600 to 700 sfm) in single-point turning. Feeds can be increased within the boundaries of economic tool life, the standard of accuracy, and surface finish requirements.

In machining fully sintered parts (average hardness, 90 HRB), K10 carbide tips give a satisfactory life with 0° top rake, 7° frontal clearance, and 5° side clearance. Single-point finish turning is used, with a stock removal of 0.125 to 0.20 mm (0.005 to 0.008 in.) of surface depth and 0.050 mm (0.002 in.) of feed per revolution. Surface speeds are 120 to 135 m/min (400 to 450 sfm), and tips require a radius of 0.20 to 0.25 mm (0.008 to 0.010 in.). Form turning is not advisable because of the work-hardening characteristics of the material in this state.

Abrasive flank wear is the dominating wear mechanism in turning. PVD-TiN coating of the hard metal (HM) inserts reduces the wear rate; CVD coatings (TiN, Al_2O_3) improves the performance even further. Oil impregnation improves the machinability in general, while cutting fluid is detrimental to the machinability.

The results for the following example show that the axial force is the dominating cutting force component after ~0.1 mm flank wear. The micro surface roughness (within the feed marks) is improved by increased density, and addition of MnS improves the macro surface integrity. Tool life is nearly independent on the feed rate in a range of 0.05 to 0.2 mm/rev. Alloy elements in general decrease the machinability. Correlation to hardness is not enough to explain the different machining performance of P/M material. Micro smearing from "soft" phases due to inhomogeneous microstructure is one explanation for P/M materials performance in relation to conventional steel. MnS addition for intermittent cutting has a strong effect on the machinability.

Example 1: Machinability Evaluation, Turning of Sintered Steel. Turning tests for a number of P/M materials were performed in a CNC turning lathe. The workpiece geometry was a thick-walled tube with an inner diameter of 35 mm, an outer diameter of 64 mm, and a height of

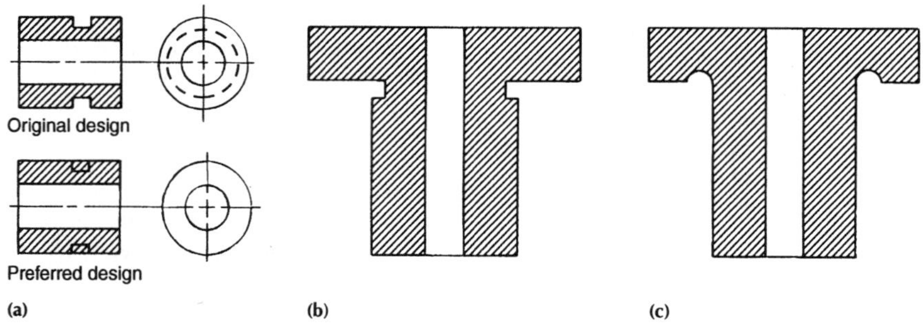

Fig. 1 P/M part design considerations. (a) Undercuts on horizontal plane cannot be produced in P/M process. Machining is required to obtain such features in parts. (b) Example of undercut in flange that is beyond capability of P/M process. (c) Alternative to part in (b) that can be made without secondary machining

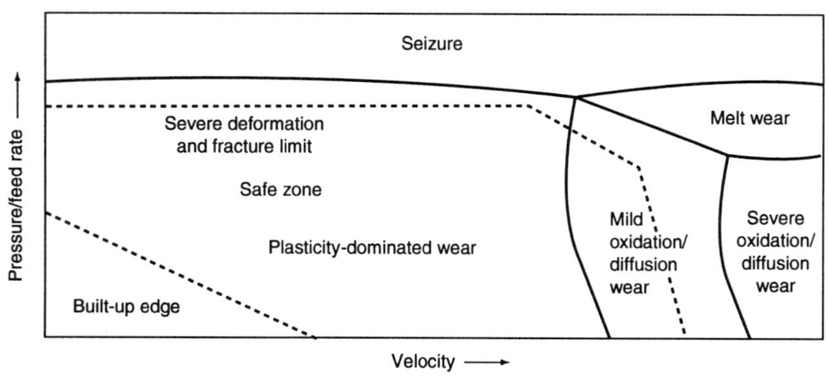

Fig. 2 Pressure/feed rate versus cutting speed in a wear mechanism map

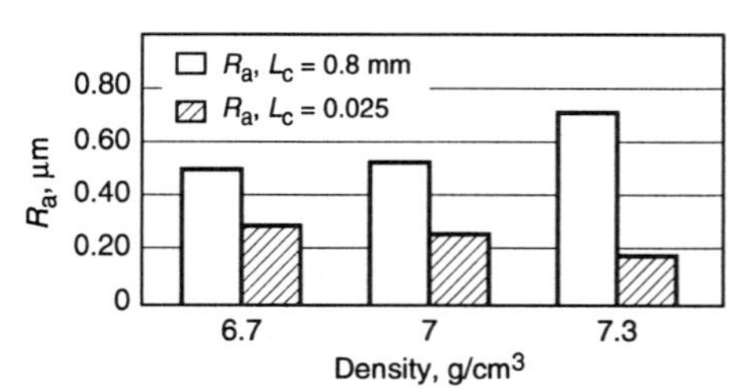

Fig. 3 Density influence on surface integrity for ASC100.29 2% Cu, 0.5% C; density variation; second edge wear, 0.2 mm

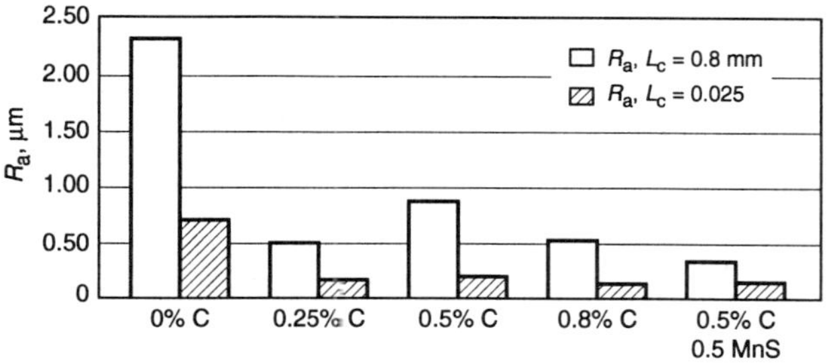

Fig. 4 Influence of carbon on the surface integrity for Distaloy AE carbon and MnS addition; second edge wear, 0.2 mm

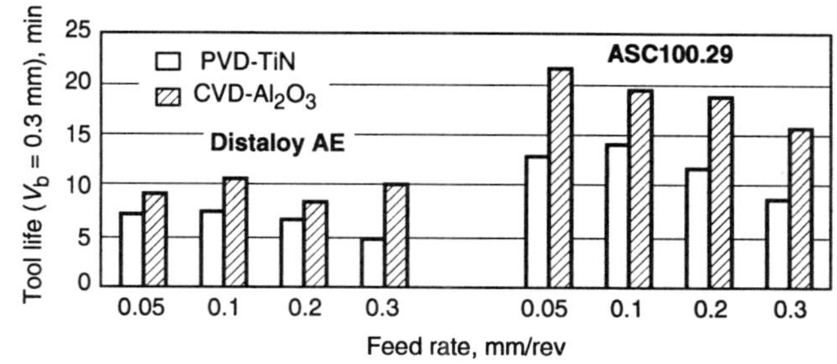

Fig. 5 Influence of feed rate and type of coating on tool life for Distaloy AE 0.5% C and ASC100.29 2% Cu, 0.5% C. Tool, CNMG 120408

62 mm. To study intermittent turning a synchronizing hub was used. All turning was performed as a facing operation. The materials were sintered at 1120 °C for 20 min. For materials where carbon is added, an endothermic atmosphere is used, while dissociated ammonia is used for the other materials. The solid wrought reference material was OVAKO 234S (DIN 16MnCr5) 0.5% C (HV~220) with the same workpiece geometry.

Initially, evaluation of tool material and cutting conditions was performed. Based on these results, a PVD-coated hard metal insert (ISO code CNMG120408-MF) was selected as standard. The depth of the cut was fixed to 0.5 mm. The force measurements were carried out using a Kistler three component piezo electric table and a digital oscilloscope (Kistler Instrument Group, Amherst, NY).

For P/M materials, the main tool wear mechanism during continuous turning is abrasive flank wear. A notch at the depth of cut that can limit the tool life is sometimes formed. The feed force and axial force are also more sensitive to changes in cutting data, wear, and the material type than the main cutting force. When the feed is increased, P/M materials show a smaller increase in these forces than the reference wrought material.

Measurements of the three cutting forces at different wear levels reveal that the axial force is the dominating force after ~0.15 mm flank wear. The investigations indicate that the wear rate accelerates after 0. 1 to 0.15 mm flank wear dependent on grade. The correlation between wear rate and axial force is significant. Key results are summarized below.

Surface Integrity. The surface roughness of the machined surfaces in Example 1 were evaluated in terms of R_a and R_z using a laser measuring station. The R_a value is the arithmetic average of all deviations of the roughness curve from the center line. The R_z value, mean roughness depth, is the mean of the maximum peak-valley distances in five successive lengths, L_c, of the roughness profile.

The density influence on R_a is shown in Fig. 3. Increased density will increase the macro R_a value while the micro R_a value is reduced. Looking at R_z it seems as if the value is constant in this density range.

Addition of carbon to atomized iron powder with diffusion bonded nickel (4%), copper (1.5%), and molybdenum (0.5%) improves R_a. In the range 0.25 to 0.8% C the micro surface roughness seems to be the same. Addition of MnS free machining additives reduces the macroscopic roughness R_a (Fig. 4). The reduction is even more pronounced when using R_z as a measure of the macroscopic surface.

Influence of Feed Rate. For conventional steel, the feed rate has a strong influence on the tool wear, and different wear mechanisms will be critical at different feeds. For P/M material there seems to be no change in the wear mechanism, and the tool life seems to be independent of the feed rate in a certain range. The tool grade and type of coating material determine the maximum feed rate.

For Distaloy AE 0.5% C, the maximum feed rate seems to be in the range of 0.2 mm/rev when using a PVD-TiN coated tool. By switching to a CVD multicoated (TiN, Al_2O_3) tool, the feed rate can be increased to 0.3 mm/rev (Fig. 5). For ASC100.29 2% Cu, 0.5% C, the maximum feed rate seems to be 0.2 mm/rev with a PVD-TiN or CVD-Al_2O_3 coated tool. In general CVD coating improves the performance in tool life by at least 30% compared to a PVD-TiN coated tool. The geometry of the tool has a large influence on the performance. Among the most important factors is a small tool edge radius, which decreases the axial force and prolongs the tool life.

Influence of Carbon Content. Carbon is the most common alloying element in ferrous powder metallurgy. The strengthening effect is due to the increased amount of pearlite in the microstructure. Above 0.8% precipitation of cementite at the grain boundaries will decrease the strength, which influences the machinability strongly as seen in Fig. 6.

For atomized iron powder with addition of 2% Cu, there is a clear difference between 0% C and 0.25% C. Smearing of workpiece material on the tool edge is believed to be the explanation for this. The surface for this grade is also considered to be rough, both in the macro and micro range, compared to 0.25% C addition. This effect decreases with cutting speed.

Influence of Cutting Fluid. The use of cutting fluid is common when machining conventional steel. The pores present in the microstructure of the P/M materials in combination with water from the cutting fluid can cause oxidation, which is detrimental to the mechanical properties. Oil impregnation is one alternative to enhance the machinability. In our turning studies, oil impregnation has a pronounced positive effect on P/M machinability, while the use of cutting fluid is detrimental.

The main reason for the observed deterioration of tool performance with cutting fluid can be traced back to the severe thermomechanical load cycle associated with the test mode. Introduction of cutting fluid can lead to the following:

- Severe fluctuation in the tool temperature due to the better heat transfer characteristics of the water-based cutting fluid leading to severe thermal cycling
- Probable reaction between rest products from the coolant and the atmosphere during the interruption period of the machining cycle

The inability of these effects to increase the tool wear when turning conventional steel with cutting fluid indicates that both the maximum tool temperature as well as the temperature difference within a machining cycle is higher when turning P/M materials. In Table 1 the flank wear (mm) after 40 and 90 passes is presented.

Intermittent Cutting. Initial tests to select the right tool for intermittent cutting revealed that a tougher tool material was needed than for continuous turning. The tool used for continuous cutting was subjected to chipping of the tool edge. The chipping was also reduced by the use of sharper tool geometry when machining P/M parts.

Addition of free machining additives such as MnS gave a clear improvement. The effect was emphasized at increasing cutting speed (Table

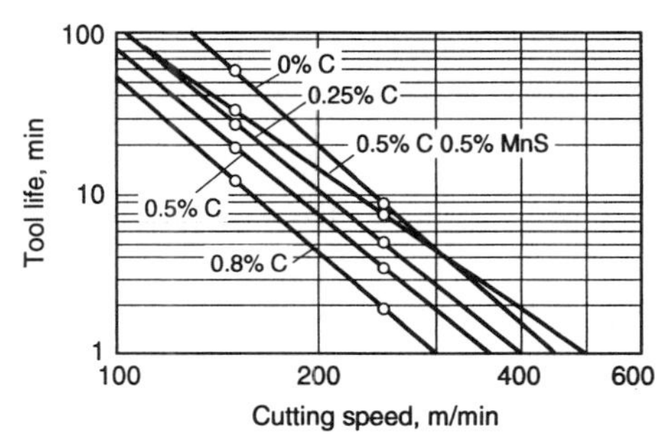

Fig. 6 Influence on carbon addition on tool life of Distaloy AE. Tool material: CNMG 120408, GC 1025. Cutting conditions: feed = 0.1 mm/rev; depth of cut = 0.5 mm; criteria, V_b = 0.3 mm; dry

2). For conventional steel, it is a common experience that the chip becomes thinner and the forces decrease as the cutting speed is raised. This drop in forces is caused by a decrease in contact area and in shear strength due to higher temperature. This seems also to be the case for P/M material. But it is still not clear if the effect arises from improvement in chip breaking or by reduction of the shear force.

Cubic Boron Nitride (CBN) Tools. In cases where the preservation of surface porosity is vital, CBN inserts are used, especially with low-porosity materials. When CBN tools are used, surface speeds can be increased from 600 to 1000 m/min (2000 to 3280 sfm) using the same rake and clearances as those for K10 tools. These tools will also accept some degree of interruption on the cutting surface. With K10, a coolant is necessary to keep the tool cool and to maintain swarf clearance. With CBN, the procedure used depends on the workpiece material.

Cutting speeds with CBN tools can vary widely from 250 to 2000 m/min (820 to 6500 sfm). Feed rates range from 0.050 to 0.075 mm/rev (0.002 to 0.003 in./rev) with depths of cut from 0.13 to 0.40 mm (0.005 to 0.015 in.) for steels with hardnesses in the range of 50 to 400 HB; usually, it is preferable to decrease the cutting speed to below 1000 m/min (3280 sfm) when the hardness is above 250 HB. These conditions are suitable for hot-pressed or cold-pressed ceramic tools. In cases in which uncoated carbide tools are used, cutting speeds should be less than 190 m/min (625 sfm) for machining most ferrous or nonferrous alloys.

Drilling

In drilling, speeds and feeds are 80 to 85% of those for wrought metals of the same composition. For long tool life, nitrided steel, high-speed steels containing cobalt, and carbide-tipped drills are recommended.

Low helix angle drills are not recommended for softer P/M materials because of their poor chip ejection characteristics. Drills with 40° helix angles had twice the tool life of those with 30° helix angles in work performed with soft P/M materials.

Large amounts of coolant are required in drilling medium- or low-density materials; coolant should operate effectively at the drilling point to reduce abrasive wear due to powder particles at the bottom of the hole. A single-nozzle coolant system does not work properly because small or powdered chips do not easily exit through the drill flutes. A ring design system, however, is effective in eliminating the chip-clogging problem. Oil hole drills are the most effective means for removing chips from the cutting zone.

Cutting speeds of up to 25 m/min (80 sfm) and feed rates up to 0.25 mm/rev (0.010 in./rev) are recommended for high-speed steel drills. Cutting speed and feed rate could be as high as 120 m/min (390 sfm) and 0.5 mm/rev (0.02 in./rev), respectively, when solid-carbide or carbide-tipped drills are used. Carbide indexable drills are efficient because margins are eliminated. Abrasive margin wear and the welding of powder chips are also eliminated.

Holes in planes beyond the capability of the P/M process are best made by drilling when parts are in the presintered (partially sintered) state. In this state, drilling properties are similar to those of cast iron. If it is necessary to qualify the position and size of holes after final sintering, a carbide reamer or carbide-tipped reamer should be used. Roller burnishing can be used to meet accuracy and surface finish requirements. Any size change in the operation is related to the preburnished surface finish and the size of the hole. A change of 0.019 to 0.025 mm (0.00075 to 0.001 in.) in diameter is representative for a hole about 25 mm (1 in.) in diameter.

A number of studies have been performed to determine ways of enhancing drill life during P/M machining (Ref 4–10). The following example is another. As shown in the following example, drill length is considered as a primary factor for improvements. Additives have strong influence on machinability for high performance materials. Tool life is nearly independent on the feed rate, which ranges from 0.05 to 0.16 mm/rev. Cutting fluid has no significant effect on the productivity. Distribution decreases for coated drills and for short drills.

Example 2: Machinability Evaluation, Drilling of Sintered Steel. Blind hole drilling test was performed on a wide spectrum of P/M qualities. Cylindrical blanks with a diameter of 80 mm and a height of 10 mm are used in the test. A survey test is performed in order to select the type of drill and cutting conditions. The main comparisons were made under dry condition using a HSS drill with a diameter of 4 mm and a point angle of 118°. Total breakdown of the drill is chosen as criterion, based on the fact that drilling is commonly used as a bulk removal operation. All materials are investigated in order to divide them into four groups. The classification criterion used is the time required to drill 100 holes. Influence of cutting fluid, density, alloying contents, feed rate, and drill type are evaluated for the selected group representatives.

All the tests were carried out in a numeric controlled machining center. Early in the testing procedure, the heat generated in the workpiece during the drilling cycle was recognized to have an effect on the result, and therefore, the time between inserts was increased and temperature measurements were carried out on the workpiece. Key results are noted below.

Tool Life. Evaluation for atomized iron with diffusion bonded nickel (4%), copper (1.5%), and molybdenum (0.5%), addition of 0.8% C regarding influence of length, type of drill, and coating of the drill is shown in Fig. 7. The length of the drill has a large influence on the performance. Evaluation of centered drilling shows improvements in distribution and in performance. It is believed that the first inlet is crucial for the performance. This has its origin from the inhomogeneous microstructure of P/M material.

Tool material and coating have a strong influence. Geometry changes exist among the three types of drills. Regarding feed rate, the performance is comparable with the result from the turning test presented for HM PVD-TiN tool, which ranges from 0.12 to 0.16 mm/rev. For Distaloy AE 0.5% C using HSS drill, the same tool life exists up to feed rate 0.16 mm/rev. This behavior is unique for P/M material.

Effect of Carbon Contents. Addition of carbon above 0.25% C for atomized iron powder or addition of 2% Cu will decrease the machinability. Micro smearing on the tool is believed to be the reason for the large decrease in machinability (Fig. 8).

Density, Cutting Fluid, and Additives. Influence from density is regarded as small in the range of 6.7 to 7.3 g/cm^3. Result from productivity evaluation of atomized iron powder with 2% Cu and 0.5% C at 100 holes reveals 5% decrease for the range. Use of cutting fluid has no significant effect on the productivity. Water as cutting fluid decreases the productivity. For high performance materials, additives have large influence on the machinability.

Tapping

Conventional tap drill charts should be followed to maintain 65 to 75% depth of thread. Two-flute taps are recommended for diameters up to 8 mm (5⁄16 in.). Three-flute taps should be used for diameters of 8 to 12.5 mm (5⁄16 to 1⁄2 in.). Spiral-point taps are desirable because they throw the chip out instead of driving it into the pores of the workpiece. Some experimenting in tapping P/M parts may be required to determine

Table 1 Influence of liquid coolant and oil impregnation on flank wear

		Flank wear, mm					
		Oil impregnated		Cutting fluid		Dry	
Material	Composition	40 passes	90 passes	40 passes	90 passes	40 passes	90 passes
Distaloy AE	0.5% C	0.046	0.098	0.1335	0.739	0.075	0.321
ASC100.29	2% Cu 0.5% C	0.034	0.04425	0.1215	0.39725	0.04375	0.055
OVAKO 2345	0.5% C	...	...	0.0495	0.0955	0.05775	0.07775

Table 2 Tool life for turning synchronizing hubs

Material	Composition	Tool life, min
Distaloy AE	0.5% C	3.36
	0.5% C, 0.5 MnS	31.62
Distaloy HP-1	0.5% C	2.79
ASC100.29	...	5.54
ASC100.29	2% Cu, 0.5% C	26.23

Cutting speed = 200 m/min; feed rate, 0.1 mm/rev; depth of cut = 0.5 mm. Criteria: V_b = 0.25 mm, dry. Tool: PVD-TiN coated which tap is best for a specific metal. As shown in the following example, chip clamping can degrade the performance, but it is improved by the use of cutting fluid. Evaluation of tapping with the use of cutting fluid for the first three holes and after tapping 50 holes is presented in Fig. 9. Tapping under dry condition reveals problems with chip clamping. Cutting fluid improves the performance. Carbon addition decreases the performance and additives like MnS decrease the torque. The distribution during the measuring length still indicates problems with chip clamping. Selection of tap geometry is considered to solve the problem.

Other Cutting Methods

Milling. Slot and side milling cutters are often used for machining P/M materials. Speeds of 70 to 100 m/min (230 to 330 sfm), feed rates of 0.005 to 0.1 mm (0.0002 to 0.004 in.) per tooth, and depths of cut of 0.13 to 0.4 mm (0.005 to 0.015 in.) are recommended in machining ferrous and nonferrous alloys with uncoated carbides. Higher speeds and feeds should be used in machining aluminum. Aluminum P/M alloys have better chip characteristics than their wrought counterparts. Chips are much smaller and are broken more easily, with little or no stringer buildup.

In face milling with uncoated carbides, cutting speeds of 90 to 120 m/min (295 to 395 sfm), feed rates of 0.05 to 0. 15 mm (0.002 to 0.006 in.) per tooth, and depths of cut of 0.12 to 0.4 mm (0.005 to 0.015 in.) are recommended for carbon and alloy steels and stainless steels; however, nonferrous materials can be cut at speeds up to 170 m/min (560 sfm) and feed rates as high as 0.1 mm (0.004 in.) per tooth.

Speeds of 25 to 50 m/min (80 to 165 sfm) are used in machining P/M iron, steel, stainless steel, copper, and brass with high-speed end mills. On the other hand, cutting speeds in the range of 100 to 200 m/min (330 to 655 sfm) are recommended in machining soft iron, steels, and aluminum with carbide tools. With harder steels, stainless steels, copper, and brass, speeds should be lowered to the range of 60 to 100 m/min (195 to 330 sfm).

Reaming. To control bore accuracy in P/M parts, reaming is sometimes used instead of pin sizing, ball sizing, or burnishing. Standard reamers are satisfactory; left-hand spiral reamers have also proved successful. The cutting edges should have the best possible finish to minimize edge buildup, which results in oversize holes. If the surface finish of the hole is not a factor, the drill should leave a reaming allowance, the amount depending on hole size. Guidelines can be used:

Hole diameter		Allowance	
mm	in.	mm	in.
≤6.5	≤0.25	0.050	0.002
6.5–12.5	0.25–0.50	0.050–0.10	0.002–0.004
12.5–25	0.50–1.0	0.10–0.15	0.004–0.006

If the surface finish is critical, reaming allowances should be doubled. When possible, reamers should be used in floating holders and run at 7.5 to 15 m/min (25 to 50 sfm). Recommended feeds are:

Hole diameter		Feed	
mm	in.	mm/rev	in./rev
≤6.5	≤0.25	0.15	0.006
8–12.5	0.30–0.50	0.18	0.007
14–19	0.55–0.75	0.25	0.01

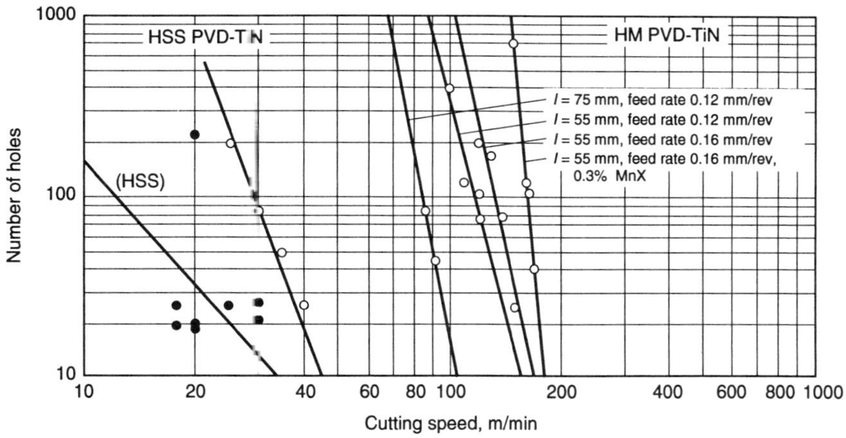

Fig. 7 Influence of feed rate, tool material, coating, and additives on machinability

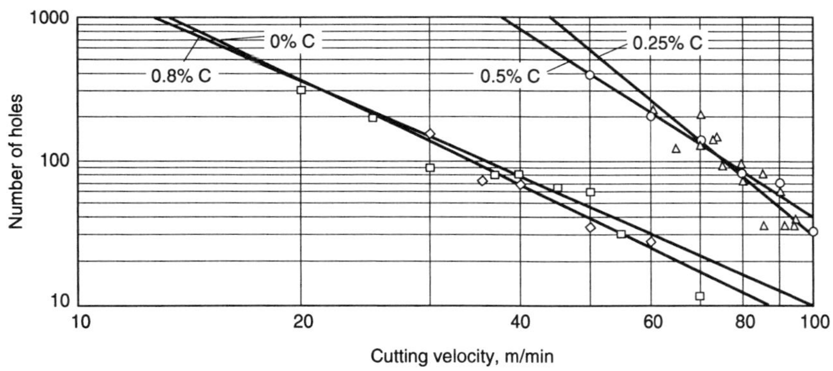

Fig. 8 Influence of carbon contents on machinability of ASC100.29 2% Cu. Drill, HSS; point angle, 118°; feed rate, 0.06 mm/rev; D = 4 mm; criteria, total failure

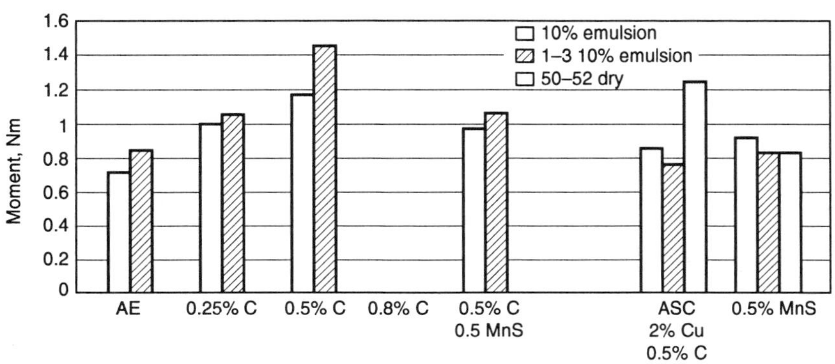

Fig. 9 Influence of carbon/MnS addition for the moment during tapping. M5 straight flute tap with 7.2 mm tap length. Bottom hole was used in the initial investigation.

Finishing

Burnishing. When the clearance between a shaft and a P/M bearing is ±0.012 mm (±0.0005 in.) or less, burnishing the bearing bores after they have been installed in the housing is preferred for correcting the bore size. No more than 0.002 mm/mm (0.002 in./in.) of diameter should be displaced, and the smallest amount of displacement that will produce the true diameter is desirable. The type of burnishing tool recommended for this operation is illustrated in Fig. 10.

Given a finished bore diameter of 38 mm, +0.005 mm/–0.0000 mm (1.500 in., +0.0002 in./–0.0000 in.) (B in Fig. 10), the diameter of the starting end of the burnishing tool then becomes 38.10 mm, –0.050 mm (1.5000 in., –0.0020 in.), or 38.05 mm (1.4980 in.), and bearings would be bored to 38.075 mm, +0.0125 mm/–0.0000 mm (1.4990 in., +0.0005/–0.0000 in.). Thus, there would be a minimum clearance of 0.025 mm (0.001 in.) at the entering end of the tool, and the first land would be a line-to-line fit. The tool then becomes progressively larger, and the bearing is expanded. If there were no springback, the operation would be stopped at the fourth or fifth tool land. However, the bearing would ordinarily be burnished to 0.010 mm (0.0004 in.) oversize to allow for springback.

Roller burnishing is a cold-working operation that compresses metal rather than removes it. The technique is suitable for sintered (not heat-treated) powder metal materials for which maintenance of open surface porosity is not critical. A significant improvement in surface finish can be obtained using a roller burnishing tool. In addition, the tool is adjustable to match individual product specifications as well as to compensate for wear on the rolls and mandrel. Both through holes and blind holes can be roller burnished. Hole size tolerance depends on the input tolerance of the hole; that is, a prepared tolerance of 0.050 mm (0.002 in.) can be reduced to 0.025 mm (0.001 in.), or a ±0.0025 mm (±0.0001 in.) tolerance can be held if the input tolerance is 0.010 mm (0.0004 in.). Surface finishes of 0.25 μm (10 μin.) are common after roller burnishing, A lightweight, low-viscosity lubricating oil is recommended for most P/M materials.

Honing and Lapping. Holes requiring extreme accuracy can be honed or lapped by normal techniques if retention of porosity is not required. However, size control of holes in P/M parts can usually be obtained more economically by reaming or burnishing.

High-density ferrous metal parts, especially when hardened, have been successfully honed and lapped using conventional procedures.

Diamond- and CBN-plated bore finishing tools are recommended for precise hole size control. These tools can be used on standard drilling or honing machines, as well as on multiple-spindle or numerically controlled machines. The use of an adjustable sleeve attached to a mating tapered mandrel increases tool life. The selection of diamond grit size determines the metal removal rate and the surface achieved. The amount of material to be removed from the hole diameter can be determined by: surface finish (start) – surface finish (after honing)/100,000 = required stock removal. If the existing finish is 1.25 μm (50 μin.) and the desired finish is 0.25 μm (10 μin.), then 0.010 mm (0.0004 in.) should be removed from the hole diameter.

Honing of infiltrated parts is seldom practical, because the stones become loaded. Neither lapping nor honing is recommended for porous parts, because either of these processes will cause the pores to become filled with abrasive particles. For special applications that require the use of lapping or honing, ultrasonic or solvent cleaning should be performed following grinding.

Grinding

Grinding of P/M parts can be very complex, especially when materials are low in density because in many cases preservation of surface integrity is essential. Usually, surface porosity decreases during grinding. A large amount of the generated powder chips is forced into pores, and many chips are welded due to the high temperature at the wheel/workpiece interface. When grinding is necessary to achieve dimensional functionality of a part, and surface porosity needs to be preserved, special processes such as ultrasonic or solvent cleaning are applied immediately after grinding. For rough applications, a downfeed of 0.025 to 0.075 mm (0.0010 to 0.003 in.) is recommended, while for finish passes, a maximum of 0.013 mm (0.0005 in.) should be used.

Stock removal rates should be either the same as or less than those used in finish turning of cast iron; wheels should be similar. It is important to keep a plentiful supply of coolant (containing an inhibitor) directed onto the wheel and the work to maintain a clean grinding wheel contact.

Grinding of P/M Tool Steels. The relative grindability of several conventional and P/M high-speed tool steels is illustrated in Fig. 11. The grinding ratio (volume of metal removal to the volume of wheel worn, as explained in the article "Principles of Grinding" in *Machining,* Volume 16, *ASM Handbook*) is clearly superior for the P/M tool steels. As expected, the grinding ratios generally decrease for both the conventional and the P/M tool steels as their alloy and carbon contents increase. The grinding conditions suggested for the CPM tool steels are similar to those recommended for conventional tool steels in the article "Machining of Tool Steels" in *Machining,* Volume 16, *ASM Handbook.* Some specific conditions for grinding CPM 10V are given in Table 3.

Tool Steels

Rapid solidification of the atomized powders used in the production of wrought P/M tool steels eliminates the segregation present in conventional tool steels and produces a very fine microstructure with a very uniform distribution of small carbides and nonmetallic inclusions. As a result, wrought P/M high-speed tool steels exhibit better machinability, dimensional control, and safety in heat treatment, grindability, and

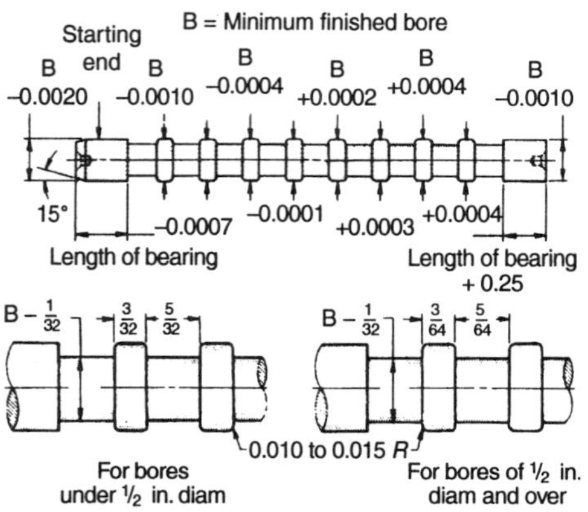

Fig. 10 Ball broach for burnishing bores in P/M parts. Dimensions given in inches

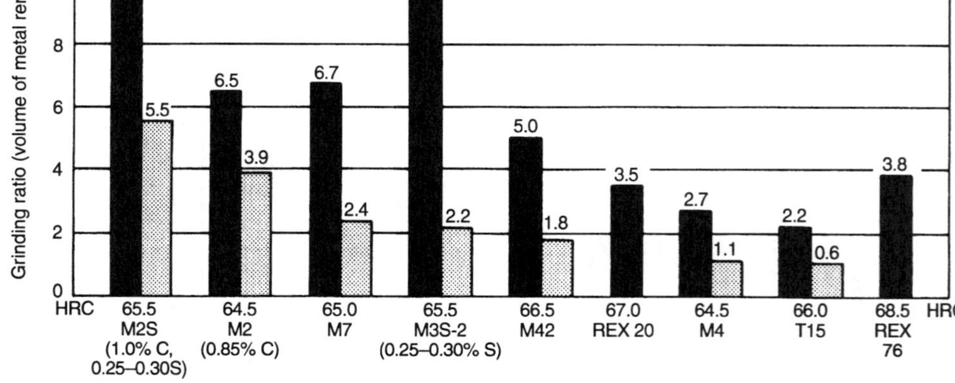

Fig. 11 Comparagraph showing the relative grindability of CPM and conventional high-speed tool steels. Source: Crucible Materials Corporation

Table 3 Grinding recommendations for CPM 10V cold-work tool steel

Toolroom grinding (sharpening)(a)

Abrasive: very sharp 38A or 32A
Grit sizes: 60 to 120 depending on removal and finish requirements
Grade: Grade I most effective, but grades as soft as G can work
Bond: Vitrified
Wheel example: Norton 32A60-I8VBE

Wet surface grinding(a)

Grit sizes: 100–150
Wheel example: CBN (Borazon) CB 120TBA

Internal grinding(a)

Grit sizes: 100–150
Wheel example: CBN (Borazon) CB 150VBA

Field reports concerning abrasives

Cubic boron nitride (Borazon) grinders must be rigid, in good condition, and able to mount wheels with very good accuracy.
Crystolon (silicon carbide) such as 39C60-I8VK are recommended.
Use very sharp, very friable aluminum abrasive that remains sharp during grinding, such as 38A.
We grinding recommended.

Note: Grinding wheel symbols and nomenclature are defined in the article "Grinding Equipment and Processes," *Machining,* Volume 16, *ASM Handbook.* (a) Based on in-house laboratory testing

Table 4 Typical machining conditions for P/M and conventional grades of AISI high-speed tool steels

Operation	Tool width or depth of cut		High-speed tooling				Carbide tooling			
			Speed		Feed		Speed		Feed	
	mm	in.	m/min	sfm	mm/rev	in./rev	m/min	sfm	mm/rev	in./rev
Single-point turning	3.8	0.150	18	60	0.38	0.015	91	300	0.38	0.015
	0.64	0.025	23	75	0.18	0.007	111	365	0.18	0.007
Drilling	6.4	¼	12	40	0.08	0.003	...	...	...	...
	13	½	12	40	0.13	0.005	...	...	...	...
	25	1	12	40	0.23	0.009	...	...	...	...
	50	2	12	40	0.33	0.013	...	...	...	...
Broaching	...	...	3	10	0.05	0.002	...	...	...	...
Face milling	3.2	0.125	20	65	0.20	0.008	78	255	0.30	0.012
	0.64	0.025	26	85	0.15	0.006	101	330	0.25	0.010
Cutoff	1.6	0.062	14	45	0.03	0.001	53	175	0.05	0.002
	3.2	0.125	14	45	0.03	0.001	53	175	0.08	0.003
	6.4	0.250	14	45	0.04	0.0015	53	175	0.11	0.0045
Cutting fluid	...	...	Sulfurized oil, light duty				Water-soluble oil			

edge toughness during cutting than conventional high-speed tool steels of the same composition. A variety of Anti-segregation process (ASP) and P/M tool steels are available.

As with conventional tool steels, P/M tool steels are generally machined in two stages: rough machining of the workpiece with the steel in the annealed condition, followed by finish machining (typically grinding) after heat treatment when the steel is in the hardened-and-tempered condition. Table 4 lists the typical cutting conditions for P/M and conventional AISI high-speed steels of similar composition.

An important advantage of the P/M process relates to the fact that the machinability and grindability of P/M tool steels can be improved by increasing their sulfur content to much higher than conventional levels without sacrificing toughness or cutting performance (see the article "Particle Metallurgy Tool Steels" in this Volume).

REFERENCES

1. K.H. Roll, Powder Metallurgy at the Turn of the New Century, *1987 Annual Powder Metallurgy Conf. Proc.,* Metal Powder Industries Federation, 1987
2. J.S. Agapiou and M.F. DeVries, Machinability of Powder Metallurgy Materials, *Int. J. Powder Metal., Powder Technol.,* Vol 34 (No. 1), 1988
3. A. Thelin, Verschleissmechanismen und Leistungen von Zerspanwerkzeugen, *VDI Berichte,* No. 762, 1989, p 111–126
4. J.S. Agapiou, G.W. Halldin, and M.F. DeVries, Drillability of 304 Stainless Steel P/M Material: Tool Wear and Life, *1987 Annual Powder Metallurgy Conf. Proc.,* Vol 43, Metal Powder Industries Federation, 1987, p 181
5. V.V. Podgorkov et al., Finish Machining of Sintered Iron and Copper Base Materials, *Sov., Powder Metall. Met. Ceram.,* Vol 13 (No. 8), 1974, p 674–677
6. S. Suzuki et al., Machinability of 4100 Series Sintered Steel Containing Sulfur, *1987 Annual Powder Metallurgy Conf. Proc.,* Vol 43, Metal Powder Industries Federation, 1987, p 511
7. U. Engstrom, Machinability of Sintered Steels, *Progress in Powder Metallurgy 1982,* Vol 38, *1982 National Powder Metallurgy Conf. Proc.,* Metal Powder Industries Federation, 1982, p 417
8. J.M. Capus and C. Fournel, Tool Wear Measurements in Machining of Sintered Ferrous Alloys, *Progress in Powder Metallurgy 1981,* Vol 37, *1981 National Powder Metallurgy Conf. Proc.,* Metal Powder Industries Federation, 1981, p 165
9. Y. Trudel, C. Ciloglu, and S. Tremblay, Selected Additives to Improve Machinability of Ferrous P/M Parts, *Modern Developments in Powder Metallurgy,* Metal Powder Industries Federation, Vol 15, 1984, p 775
10. A. deRege, G. L'Esperance, L.F. Pease, and L. Roy, Prealloyed MnS Powders for Improved Machinability, *Near Net Shape Manufacturing Conf.,* P.W. Lee and B.L. Ferguson, Ed., ASM International, 1988, p 57–68

Resin Impregnation of Powder Metal Parts

Charles M. Muisener, Research, Development & Engineering Group, Loctite Corporation

RESIN IMPREGNATION is a process that eliminates or reduces internal porosity of castings and P/M parts by saturating internal voids with liquid resins. The process has been practiced for many years on castings and P/M parts, and resin impregnation has, to a large extent, eliminated macroporosity (pore diameter > 125 μm) in castings. With further process improvements and low-viscosity resins capable of good penetration, impregnation is also capable of significantly reducing microporosity (pore diameter < 125 μm) (Fig 1).

The most common purpose of resin impregnation is to produce leak-tight parts. However, impregnation also:

- Enhances corrosion resistance
- Prepares porous surfaces for subsequent surface treatments or coatings
- Improves machinability

Depending on the product and application, sealant penetration can be minimal or deep, with interconnected porosity reduced up to 90%. When the application is preparation for surface finishing or plating of the part, penetration is usually minimal. It is much deeper for parts that need pressure retention and are machinable, including rugged hydraulic components for industrial pumps. Depth of penetration is controlled by the process time and exposure during impregnation, resin type, and the structure of the internal porosity.

Impregnation Technologies

The earliest method of impregnation involved lowering heated components into a bath of phenolic resin. This method has been effective in eliminating macroporosity in castings, but it has been abandoned with the development of vacuum or pressurized methods for the impregnation of micropores. Several different resin materials also have been used, including a variety of organic polymers such as phenolics and methacrylates and also inorganic materials such as aqueous sodium silicate. Four common types of impregnation resins are:

- Sodium silicate (also known as waterglass)
- Polyester resins
- Low-viscosity heat-curable resins
- Low-viscosity anaerobic resins

Of these, anaerobic resins are the most widely used for P/M parts, as described in the section "Impregnation Resins."

Impregnation Methods. The four common methods of resin impregnation are:

- Wet vacuum
- Wet-vacuum pressure
- Dry-vacuum pressure
- Pressure injection

In all methods, a cleaning pretreatment is essential for good penetration because contaminants in the porosity impede impregnation. Vapor degreasers are used, but costs and environmental considerations are factors. Aqueous-based washing systems can be used with either oven or vacuum drying. The use of vacuum drying with its short cycle times and the cooling of the compo-

Fig. 1 Microstructure of P/M iron with impregnated resin in microporosity. 200×

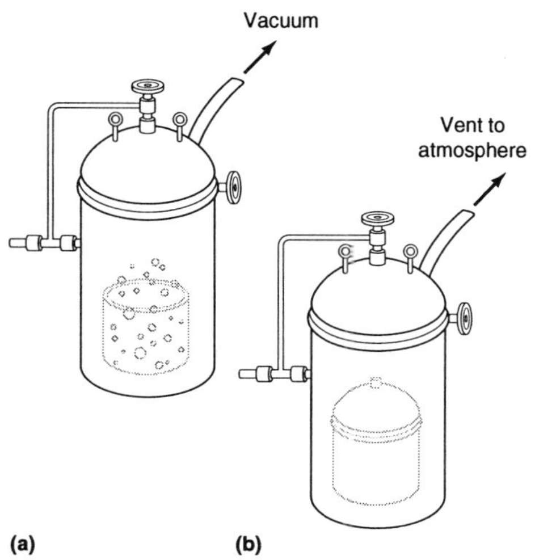

Fig. 2 Wet-vacuum impregnation. (a) Application of vacuum removes gas from internal pores. (b) Atmospheric air pressure assists penetration of resin into pores.

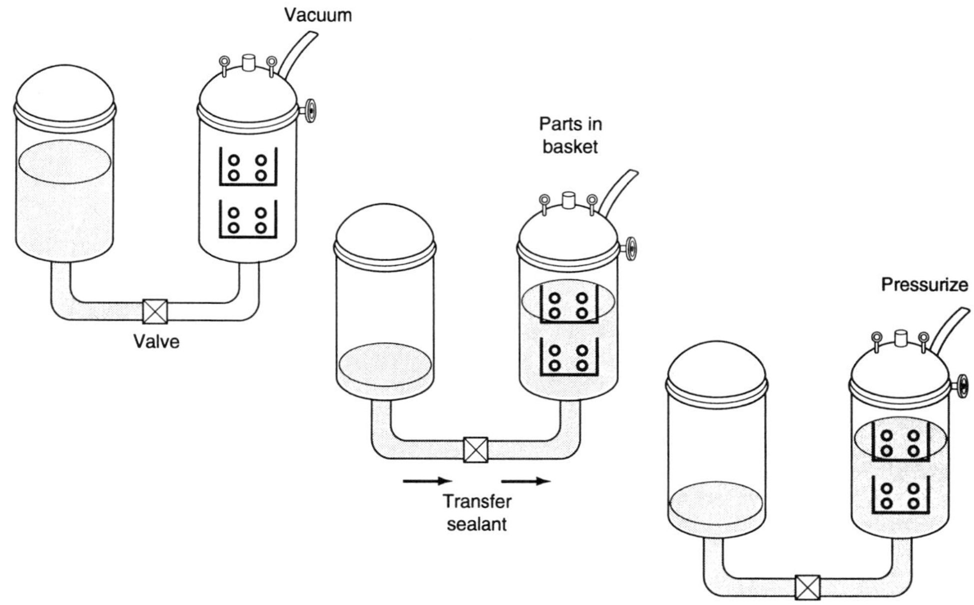

Fig. 4 Dry-vacuum-pressure impregnation

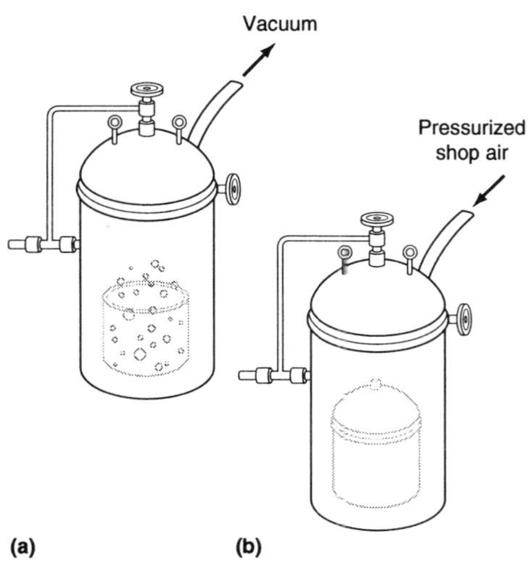

Fig. 3 Wet-vacuum-pressure impregnation. (a) Vacuum removes gas from pores. (b) Pressurized air increases the penetration rate of resin into internal pores.

nents enables direct integration with impregnation systems.

Wet vacuum is the fastest impregnation method, offering simplicity and rapid processing along with lower equipment costs. More than 95% of powder metal impregnation employs the wet-vacuum method.

The wet-vacuum process involves submerging parts in a tank of resin contained in a vacuum chamber. The vacuum cycle removes air from the tank and the pores within the part (Fig. 2a). Then the chamber is first returned to ambient pressure to allow resin to force into the pores (Fig. 2b).

Vacuums between 1 and 5 mbar held for 5 to 10 min can effectively extract air from the internal porosity. Often the components are held above the resin (or in a different vessel in other methods). High vacuum within the porosity improves the penetration of the resin and reduces exudation, caused by unextracted air expanding within the part during the curing phase. Exposing resins to vacuum at the same time degases and removes contaminants from the resin.

Penetration is a function of the resin properties, the structure of the porosity, and time. Using modern materials, even large dense parts can be fully impregnated during a cycle time of 10 to 15 min.

Impregnation is followed by rinsing for removal of surplus resin on the surface. Rinsing can also be followed by immersion of the part in a catalytic activator solution to cure the resin at the surface of each porosity. This creates a hardened plug at the outer portion of the pore, thereby trapping the remaining resin for anaerobic self-cure. Finally, the part is rinsed again to remove any activator from the surface. The total cycle time is approximately 20 min.

Wet-vacuum pressure is similar to the wet-vacuum method in that parts are submerged in sealant and a vacuum is applied to parts and sealant together (Fig. 3a). The difference is that resin penetration is assisted by pressurized air after release of the vacuum (Fig. 3b). Use of pressure improves penetration at the cost of adding air compressors to the system.

Dry-vacuum pressure is the slowest and most complicated vacuum impregnation method. It is traditionally used with high-viscosity sealants and high-density powdered metals. The impregnation cycle typically requires two tanks (Fig. 4), one to hold the sealant the other to process parts. Occasionally, a single tank with two separate process levels is employed. This application is sometimes specified with sealants where part porosity is very small and sealing requirements unusually stringent. This would apply to air-conditioner compressor components and some hydraulic pump parts. The process stages are:

- Draw vacuum in process tanks to remove air from pores
- Transfer sealant from storage tank to process tank and submerge parts
- Release vacuum and pressure process tank with compressed air pressure
- Remove parts, wash, and cure

In pressure injection the part is fixtured to seal passageways to internal cavities (Fig. 5a). Then it is pressurized with a sealant that flows through the porosity network until emerging on the exterior (Fig. 5b). This treatment is applied on automotive engine cylinder liners and other components in which pressure retention of fluids and gases is required.

Impregnation resins are mainly combinations of high-boiling-point methacrylate monomers formulated to meet the physical and chemical property demands of the impregnation process. The materials are low-viscosity fluids, which are cured within the porosity to form a polymer solid. The solid is achieved by formulating the monomer blend so that the resulting polymer has a cross-linked structure that is inherently inert and resistant to temperature, pressure, and process fluids.

The properties of the polymer are controlled by the blend of monomers used and the inclusion of minor ingredients such as corrosion inhibitor and antioxidants. Key monomer properties are:

- Low vapor pressure for vacuum methods
- Low viscosity for penetration (7 to 10 cP)
- Low surface tension for wetting and adhesion

Important properties of the polymer after curing are:

- Temperature resistance

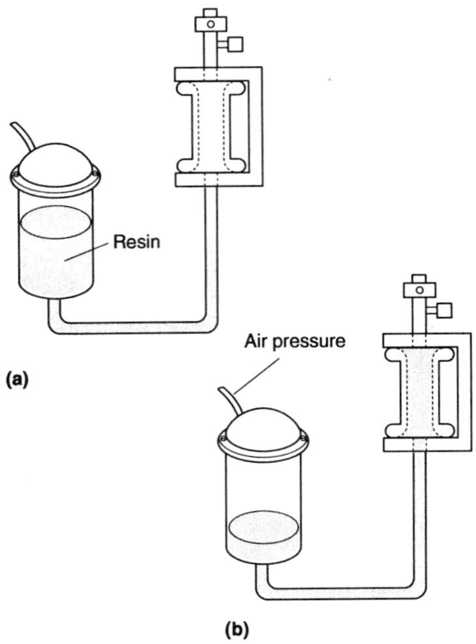

Fig. 5 Pressure injection impregnation. (a) Part is fixtured to direct resin flow through the part. (b) Pressurized sealant flows through porosity network.

- Chemical resistance
- Shrinkage on polymerization
- Adhesive strength
- Tensile strength
- Hardness

Temperature and chemical resistance are key factors. Typically, the temperature ranges of modern organic impregnation sealants are –55 to 200 °C (–65 to 400 °F). Pressure retention upper limits are governed by component designs. Many P/M components used in hydraulic applications are subject to constant conditions of more than 65 MPa (10,000 psi) with elevated operating temperatures.

Low shrinkage is required for good sealing performance in single impregnation, while good adhesive strength between sealant and metal contribute to performance under conditions such as temperature cycling and physical flexing.

As previously noted, there are four types of materials used in P/M impregnation. Sodium silicate is rarely employed, polyester resin uses toxic elements, and heat-cure resins are complex and difficult to handle. The most common resins are anaerobic sealants.

Sodium Silicate. Also known as waterglass, this inorganic material was the first widely used impregnation sealant. Although still found in some applications, particularly where extremely high temperatures are encountered, waterglass has been largely abandoned for most porosity-sealing procedures. Problems include slow processing, difficult cleanup of parts, and lengthened solidification of sealant, allowing parts to leak long after impregnation. Because of these difficulties, sodium silicate often requires multiple impregnation cycles.

Polyester Resin (Styrene). In use since the 1940s, this organic material provides reliable sealing, has good chemical resistance, and can operate well at temperatures up to 200 °C (400 °F). However, substantial disadvantages related to costs, productivity, and environmental health and safety, along with waste disposal issues, have virtually disqualified these resins from most industrial facilities.

Heat-Cure (Elevated-Temperature-Cure) Resins. This modern approach to porosity sealing uses various forms of organic methacrylate or polyester resins cured at temperatures below the boiling point of water. Sealants in this category generally are lower in viscosity than other styrene resins and often are applied with much simpler impregnation equipment. Basically, hot-water curing at 90 °C (195 °F) satisfies the requirements of manufacturers who need porosity sealing more effective than spray sealing without the maintenance requirements of an anaerobic system.

All heat cure systems require a sealant catalyst to render the sealant reactive. These additives typically are hazardous and must be measured and mixed with care in order to achieve desired curing. Many heat-cured materials also require a strong detergent solution for satisfactory washing of excess sealant from parts. Some bleedout of sealant almost always occurs as parts heat up and the impregnation material expands during cure, with a resultant reduction in sealing integrity and possible fouling of surfaces.

Also, an operator must be present to ensure that parts remain in the heated water long enough to reach proper temperature to completely cure the sealant. Failure to do so can result in parts leaving the sealing operation with inadequate curing. If this occurs the polymer will not fully cross link, and sealant performance is affected adversely.

Self-Curing Anaerobics. Anaerobic impregnation has consistently proven to be the most effective porosity sealing system. These organic materials self-cure at room temperature, eliminating the need for a hot-water treatment. This precludes sealant bleedout, fouling of surfaces, rework of parts, and the need for operators at cure stations. A highly cross-linked product, it will not liquefy and weep out of porosity. Many manufacturers rely on the superior sealing capability and consistent performance of anaerobics to ensure that all parts reaching the assembly line are leak free, with no testing before or after impregnation.

When impregnated into a part, the anaerobic no longer has a source of stabilizing air and, thus deprived of its oxygen supply, begins to solidify chemically. Prior to that, polymerization is prevented by light aeration at constant temperatures in the sealant storage vessel.

Contact with metal promotes curing, as does application of heat and chemical activators, but none of these are essential to the process. They regulate the speed of cure by controlling the rate of reaction, allowing the sealant system to be tuned for optimal results in each application. A unique activator rinse gives anaerobics much greater versatility in sealing a wide range of pore sizes, especially larger ones.

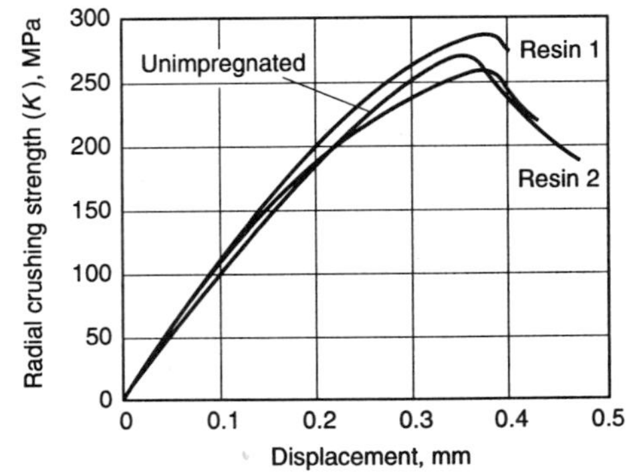

Fig. 6 Radial crushing strength of impregnated and unimpregnated P/M part. Source: Ref 1

Table 1 Radial crushing strength (K) comparison

Sample	Maximum load, N	Strength (K), MPa	Extension to maximum load, mm	Minimum deformation, mm
Unimpregnated	6504	258	0.382	0.432
Resin 1	7178	285	0.376	0.414
Resin 2	6768	269	0.352	0.472

Source: Ref 1

Anaerobic impregnation is generally recommended for volume manufacturing where the highest-quality control standards are required for sealing parts. The unique, self-curing capability of anaerobic sealants, along with the ability to regulate the rate of cure, has made this material the overwhelming choice for impregnating P/M parts. There are several anaerobic impregnation techniques such as wet vacuum, wet vacuum/ pressure, dry vacuum, pressure injection, and spray sealing.

Performance

Plastic impregnation generally has little or no effect on tensile strength and ductility, and often resin sealants for impregnated parts have very similar physical properties. Therefore, sealants are often considered roughly equivalent in terms of component strength and machinability.

However, there are important performance differences that have arisen with further developments and evaluation. For example, developments in improved sealing performance have coincided with adhesive strength improvement from 1 to 50 N/cm^2 (Ref 1). Tests of the radial crushing strengths for unimpregnated and impregnated components using different impregnants also show changes in both strength and ductility (Table 1, Fig. 6). Resin 1 has a 10% improvement in yield strength, while resin 2 shows improved ductility. The resin that improves the strength has good adhesive proper-

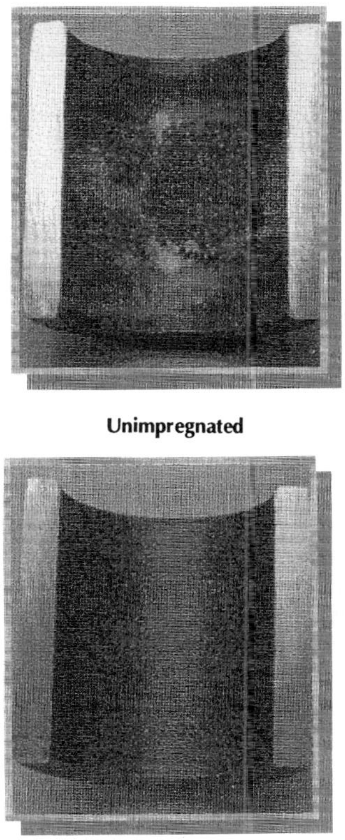

Fig. 7 Surface condition of P/M part after plating. (a)

ties, whereas the resin improving ductility forms the harder polymer. The harder polymer is more brittle, yet the components impregnated with resin 1 exhibit more pronounced brittle failure (Ref 1).

Corrosion Resistance. Corrosion and surface blemishes are a chronic problem in P/M parts. Pits, blisters, stains, and other imperfections break out because corrosives and industrial solvents are absorbed into pores. Even after surfaces have been treated with protective coatings and platings, the effects of internal corrosion may not appear until well after other such treatments as tumbling, spraying, painting, polishing, cleaning, and anodizing have been completed (Fig. 7).

Without a protective thermoset plastic impregnant, corrosive solvents soak into pores, eventually seeping upward and damaging the most hardened and lustrous surfaces. The fluids involved in P/M applications are numerous, including:

- Ethylene glycol
- Water
- Lubrication oil
- Jet fuels
- Carbon removal compounds
- Motor oils
- Hydrocarbon fluids
- Hydraulic fluids
- Machining fluids
- Acids
- Gasoline
- Alkaline cleaners

Machining Benefits. One of the more remarkable developments is the demonstrable improvements in machinability with impregnated metal parts versus those that remain untreated. The precise reason for this result is unclear, but the consensus points to a natural lubricity present in thermoset plastic resins used in most P/M impregnation processes. These cured polymers minimize tool chatter or vibration, heat buildup, and the interrupted cut associated with the machining of P/M parts. They also reduce chip thickness and adhesion, improve surface finishes, help achieve consistent finish dimensions, and improve dimensional control of parts.

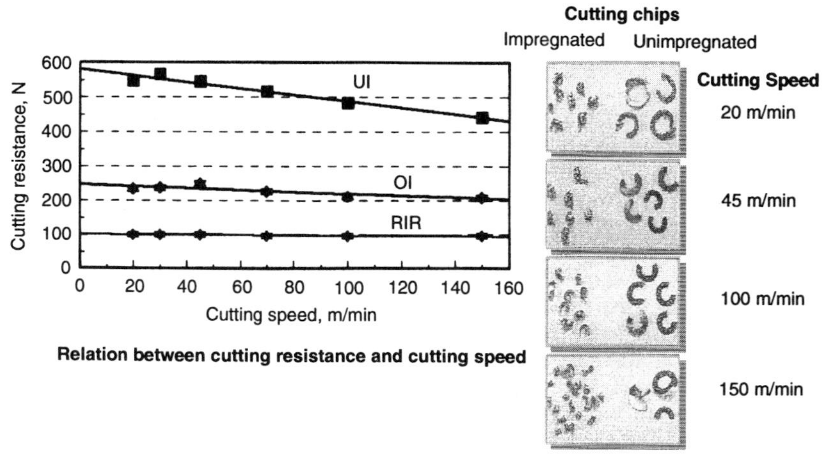

Fig. 8 Machining chart of P/M iron. UI, unimpregnated; OI, oil impregnated; RIR, resin-impregnated resinol

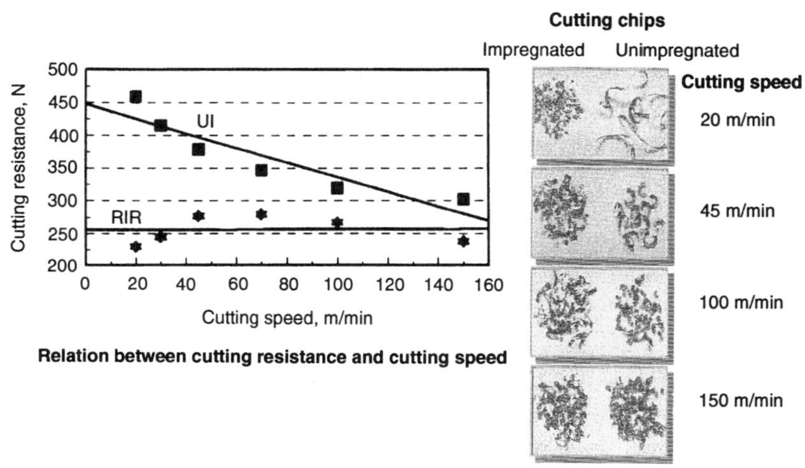

Fig. 9 Machining chart of P/M 304 stainless steel. UI, unimpregnated; RIR, resin-impregnated resinol

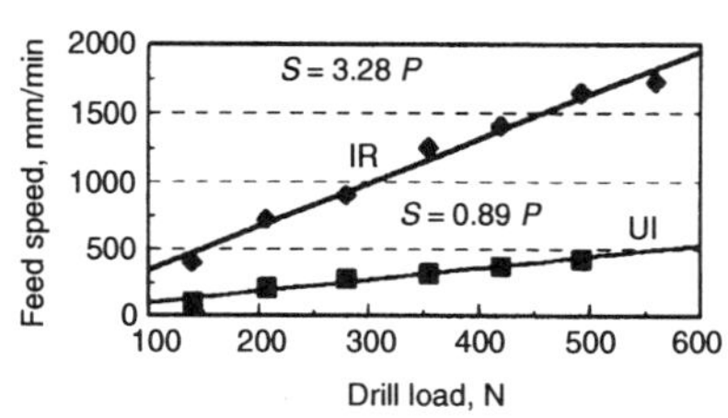

Fig. 10 Effect of resin impregnation on drilling. UI, unimpregnated; IR, impregnated resin. Efficiency of improved machinability: 370% = 3.28 + 0.89 × 100

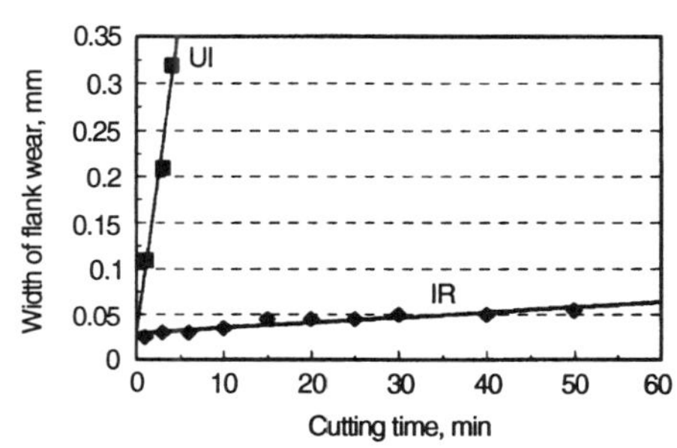

Test condition	Value
Density	7.0 g/cm^3
Cutting tool	Carbide, P20
Depth of cut, mm	1.0
Feed rate, mm/rev	0.1
Cutting speed, m/min	100

Fig. 11 Effect of resin impregnation on turning. Atomized iron powder SMF1015 impregnated with resinol. UI, unimpregnated; IR, impregnated resinol

By filling voids in the porous metal, the impregnating material promotes better chip formation and separation by the cutting tool. The plastic fill cushions the tool as it passes through metal, giving the tool edge an uninterrupted feel, thus extending its service life through a reduction in cutting force.

Many manufacturers impregnate parts solely to derive machining benefits. For example, a major equipment manufacturer verified benefits from impregnating parts with thermoset resins including higher-quality electroplating, pressure-tight sealing of hydraulic and pneumatic parts, and major improvements and cost savings in machinability. Tool life doubled in tapping and turning operations and tripled on parts with high hardness. Overall, perishable tooling costs dropped about 50% and, in some cases, the company is machining almost ten times as many pieces per tool.

Drilling is the most common machining operation used on P/M parts, and several studies (Ref 2–6) confirm lower drilling forces and longer tool life with impregnated parts. Some deterioration in surface roughness occurs (Ref 6), presumably due to melting of the impregnated resin. Nonetheless, several research programs definitively support resin impregnation as a method to significantly improve machinability. In one major program, exhaustive tests performed on various P/M alloys found that resins reduced drilling forces up to 75% in some cases (Fig. 8–11).

REFERENCES

1. M. Potts and G. Randell, Resin Impregnation of P/M Parts, *Advances in P/M Parts Production,* MPR Publishing, 1990, p 26.1–26.14
2. L.G. Roy, G. L'Espe'rance, P. Lambert, and L.F. Pease, Prealloyed Manganese Sulfide Powders for Improved Machinability in P/M Parts, *Progress in Powder Metallurgy,* Vol 43, Metal Powder Industries Federation, 1987, p 489–498
3. K.S. Chopra, Manganese Sulfide in Machining Grade Ferrous P/M Alloys, *Modern Developments in Powder Metallurgy,* Vol 21, 1988, Metal Powder Industries Federation, p 361–380
4. S. Nigarura, G. L'Espe'rance, L.G. Roy, A. DeRege, and L.F. Pease III, The Influence of Powder Processing on the Nature of Inclusions and Its Reaction to the Machinability of MnS Prealloyed P/M Parts, *Advances in Powder Metallurgy & Particulate Materials,* Vol 4, 1992, p 223–243
5. R.A. Chernenkoff, D.W. Hall, S. Mocarski, and M. Gagne, "Material Characterization of Powder-Forged Copper Steels," Technical Paper 901055, SAE International Congress and Exposition (Detroit, MI), Society of Automotive Engineers, March 1991
6. I. Shareef, Machinability of Resin Impregnated Sintered Steel, *Machining of Composites II,* ASM International, 1993, p 159–169

Planning and Quality Control of Powder Metallurgy Parts Production

Jack R. Bonsky, Cleveland State University, Advanced Manufacturing Center

QUALITY CONTROL involves several basic principles, where the planning of a quality control program requires an understanding of process variables and the reality of statistical variation in any process or measurement. Therefore, this article attempts to first summarize the basic concepts of statistical process control as a tool to measure, quantify, and analyze inherent variations. This understanding is considered essential in present-day industrial manufacturing.

In addition, the unique aspects of quality control planning in P/M parts production are also discussed in this article. Quality control for P/M parts production involves several process factors such as powder properties, press settings, tooling designs, and furnace conditions. These factors are discussed in terms of their impact on process variations and quality improvement. This article also discusses the methods that address P/M quality control in terms of:

- Defect control by prevention or detection
- Dimensional control for form and fit requirements

Defect and dimensional control are two major quality measures for P/M production, and each has a unique set of attributes for inspection and quality assessments.

Finally, it should also be noted that quality control also depends on an effective work environment that facilitates involvement and communication. This basic factor is often exemplified by the 14 Points for Management by W. Edward Deming, the eminent statistician who championed the role of statistical process control for industrial application. His 14 Points are:

- Create constancy of purpose for the improvement of product or service.
- Adopt the new philosophy of process control and variation reduction.
- Cease dependence on mass inspection for quality control.
- End the practice of awarding business on the basis of price.
- Improve constantly and forever the system of production and service in order to improve quality and productivity and thus continuously decrease costs.
- Institute thorough and better job-related trainings.
- Institutionalize leaderships.
- Drive out fears, so that everyone may work effectively for the company.
- Break down barriers between departments.
- Eliminate slogans, exhortations, and targets for the workforce that ask for zero defects and new levels of productivity.
- Eliminate work standards on the factory floor.
- Remove the barriers that rob employees at all levels in the company of their right to pride of workmanship.
- Institute a vigorous program of education and self-improvement.
- Put everybody in the organization to work to accomplish the transformation.

The 14 Points define essential institutional elements of quality and productivity improvement through statistical thinking and methods. It should also be noted that in the spirit of his 14 Points, Deming refines and improves these tenets.

Statistical Process Control Concepts

Statistical process control (SPC) is simply a method for monitoring the statistical variations of a process, where the objective is to control or reduce variations between upper and lower process limits. Control charts (Fig. 1) are used to monitor the statistical variations, but this is just one part of the overall SPC method. For the techniques of SPC (specifically Shewhart control charts such as the schematic in Fig. 1) to be successfully employed as an off-line problem identification and problem-solving tool, it is essential to keep in mind that it is a three-step process, as follows:

1. Use statistical signals to find improvement opportunities through the identification of process faults.

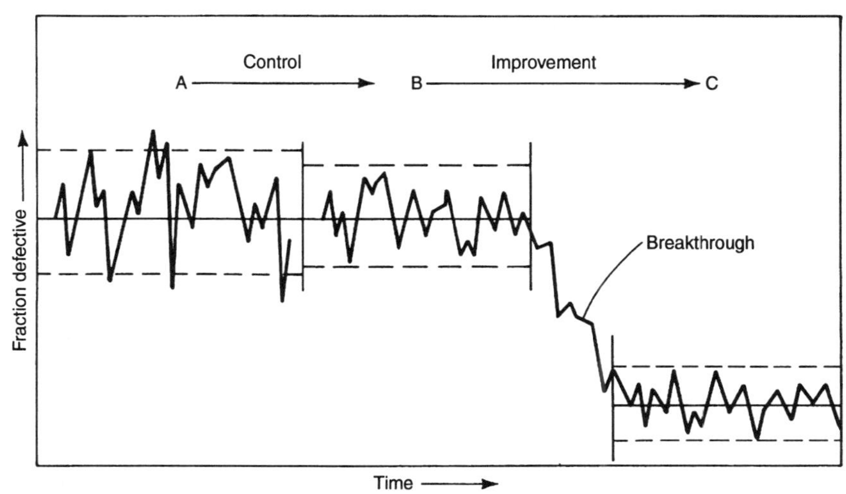

Fig. 1 Impact of having process initially in a state of statistical control versus improvement resulting from a breakthrough in performance

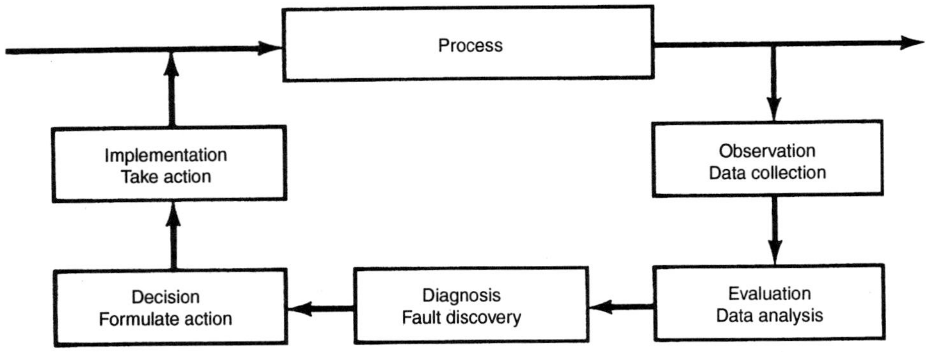

Fig. 2 Classical feedback control system view of SPC implementation

2. Use experience, technical expertise, and fault diagnosis methods to find the root cause of the fault that has been identified.
3. Develop an action plan to correct the fault in a manner that will enable any gains that are realized to be held.

This three-step process can be explained by using the classical feedback control system perspective, as shown in Fig. 2. There are five distinct stages in the generic control loop (Fig. 2), which facilitate the three-step process in the following way:

1. Use of statistical signals (observation and evaluation)
2. Fault diagnosis (diagnosis)
3. Action plan (decision and implementation)

However, bringing a process into a state of statistical control does not necessarily mean that a fundamental improvement has been achieved. Clearly, a bad situation has been rectified by bringing the process into control, and quality and productivity are enhanced. However, bringing a process into control simply means that the process is back to where it should have been to begin with. At this point, it is then possible to begin to assess the present ability of the process to realize the potential it was initially intended to have. It may be failing to realize this potential because the implementation of the process is flawed or because the design of the process itself is flawed. In either case, the root cause(s) of the chronic common cause problem must be identified and removed at the system level. This constitutes a breakthrough in performance; that is, an improvement in the process has taken place. The results of the essential steps leading to such a breakthrough are shown in Fig. 1.

This article focuses primarily on the key statistical concepts and definitions that are essential for appreciation of the SPC methods. There are several important concepts that should be understood, including the requirements of rational sampling and the definition of measurements by attributes (or defects). Once these basic definitions are established, the next step in the three-step process is an appreciation of the root cause or fault diagnosis for variations in P/M processing. The factors related to fault diagnosis for P/M quality control and planning are discussed in the section "P/M Process Planning" in this article.

Shewhart Control Chart

A succession of parts emanating from a process under statistical control will exhibit variability in their measurements because of a constant set of common causes. These variable measurements tend to collect into a predictable pattern of variation that can be easily described by a few simple statistical measures, namely, a mean (μ), a standard deviation (σ), and a frequency distribution (normal distribution, Poisson distribution, etc.). These measures stand as a model that predicts statistical behavior if the process is subject only to a constant set of common causes.

Many years ago, Dr. Walter Shewhart showed how data from a manufacturing process could be developed and interpreted through the use of very simple but useful statistical methods. Given a sufficient number of samples, the output of any process operating solely under a set of common causes can be shown to exhibit a normal distribution. Therefore, when taking a series of samples from a process that is in control, one may expect the observed values to fall within ±3 standard deviations of the process mean in 997 out of 1000 cases. By combining this logic with a simple system of charting measurement data versus time, an operator may easily identify the occurrences of special causes. If a point appears outside the 3 standard deviation limits, it is very likely that a special cause has had some effect on the process.

The charts generated by this system are known as Shewhart control charts. Each control chart actually consists of two charts for each characteristic being tracked. Figure 3 shows an example set of control charts. Each point on the charts represents a group of parts, called a sample subgroup, or just "sample" for short. Samples typically consists of three to six parts, though all samples for a given chart are the same size. The upper chart, called the $\overline{X}$ (X-bar) chart, shows the history of the process mean. The lower chart, called the range chart, or R chart for short, shows the history of variation within a sample. By using the $\overline{X}$ and R charts, significant shifts in both the process mean and spread can be detected as an indication of a process out of control.

The lines marked UCL and LCL represent the upper and lower control limits, respectively. These control limits are based on the mean value of the variable being charted ±3 standard deviations. In addition, the standard deviation used for calculating the control limits is based on variation *within* sample subgroups and not on variation within the entire population of parts sampled. Because each point appearing on an $\overline{X}$ chart is really the mean of several individual measurements, it would be improper to compare it to the 3σ limits for the entire population. For this reason, it is essential that an R accompany the $\overline{X}$ chart, so the R values for each sample subgroup can be compared. Once statistics of $\overline{X}$ and R are assembled for a number of sample subgroups, then charts for $\overline{\overline{X}}$ and $\overline{R}$ can also be assembled, where $\overline{\overline{X}}$ is the average of $\overline{X}$ from several sample groups, and $\overline{R}$ is the average range.

When calculating initial control limits, a large group of samples, consisting of at least 25 sample subgroups, is measured and the means and control limits calculated. If any points fall outside the control limits for the R chart, and they can be attributed to some special cause, then these points are eliminated from the chart and the control limits for both charts are recalculated. Next, this process is repeated with the $\overline{X}$ chart. Once the effects of all known special causes have been eliminated, the resulting control limits are considered to be the standard for future samples. Thus, the means and control limits are not recalculated with the addition of each new sample to the control charts. However, if the process experiences some improvement that causes a sustained shift on either the $\overline{X}$ or R chart, then the control limits should be recalculated based on the newer data.

In addition to the test for points appearing outside the control limits, there are a number of other items that can be applied to determine if a process is exhibiting normal random variation. Two common additional tests are the run test and the trend test. If there are eight or more consecutive points on one side of the mean for either chart, it is considered a run. If a run is detected, it is likely that there has been a shift in the process mean or variability (Ref 1). If six or more consecutive points exhibit a continuing upward or downward trend on either chart, it is considered a trend. If a trend is detected, it is likely that the process is drifting (Ref 2).

Calculations for Setting Up $\overline{X}$ and *R* Control Charts. Once the statistical basis for Shewhart control charts has been established, the first step in setting up $\overline{X}$ and R control charts is the selection of the samples. As previously noted, it is important that all samples be rational samples so that variation is attributable only to one constant system of common cause. Sampling from different machines, sampling over extended periods of time, and sampling from

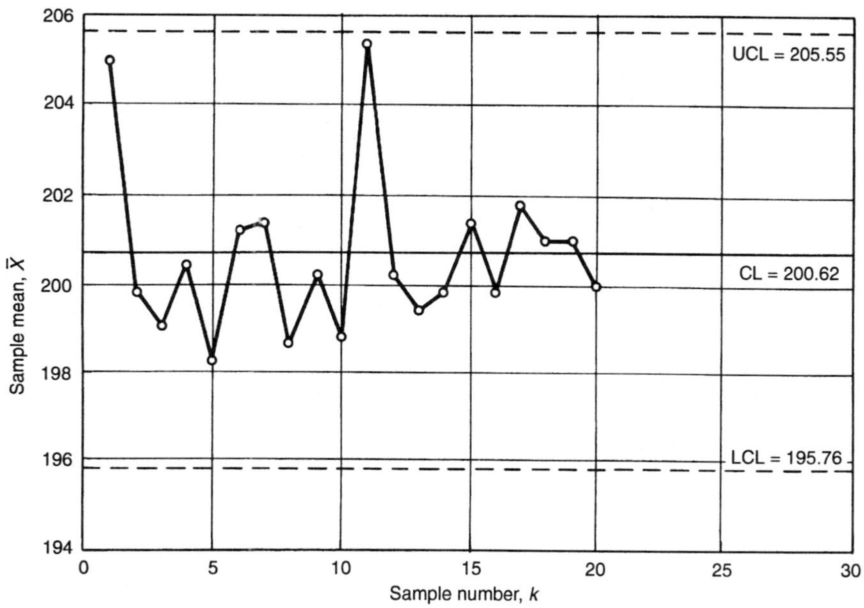

(a)

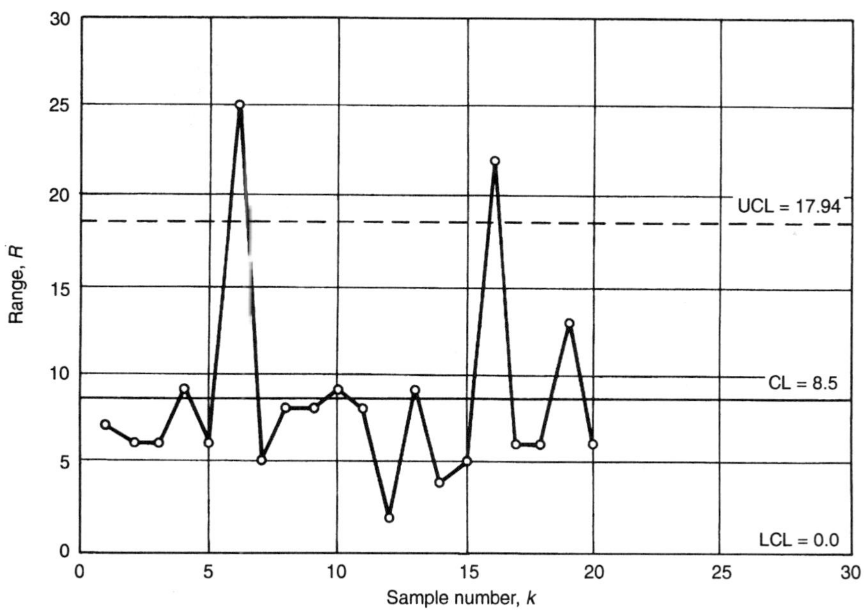

(b)

Fig. 3 Control charts for diameter measurements of a part. (a) $\overline{X}$-chart. (b) R-chart. Data are for $k = 20$, $n = 5$.

product combined from several sources are all nonrational sampling methods and must be avoided. Rational samples are discussed in more detail later in this section.

As a rule of thumb, 25 to 50 samples should be selected to provide a solid basis for the initiation of the control charts. This helps to ensure more precise estimation of the process mean and standard deviation. The sample/subgroup size should be relatively small (between $n = 3$ and $n = 6$).

With k rational samples of n each, the following steps can be used as a guide when constructing $\overline{X}$, R control charts with 3σ limits (the appropriate values for d_2, A_2, D_3, and D_4 in these calculations are obtained from Table 1):

1. Calculate the sample mean and sample range for each sample using $\overline{X} = \Sigma X/n$ and $R = X_{max} - X_{min}$
2. Calculate the grand mean of the n sample means and the average range using $\overline{\overline{X}} = \Sigma \overline{X}/k$ and $\overline{R} = \Sigma R/k$
3. Calculate the control limits for the R-chart. Although the true distribution of sample ranges is not normal and not symmetric, the symmetric limits are conventionally used for the R-charts. With assumed normal distribution for the individual measurements, the following formulas can be used for the calculation of the control limits: $UCL_R = D_4\overline{R}$ and $LCL_R = D_3\overline{R}$
4. Calculate the control limits for the $\overline{X}$-chart. Although the required standard deviation (the standard deviation of the sample mean, $\overline{X}$) for setting the limits is $\sigma_{\overline{X}}$, this value is conveniently estimated by $\overline{R}/(d_2\sqrt{n})$, where d_2 is a function of n. For $3\sigma_{\overline{X}}$ limits, one uses a factor called A_2, which is equal to $3/(d_2\sqrt{n})$ and can be found in Table 1. Thus, the control limits are calculated by: $UCL_{\overline{X}} = \overline{\overline{X}} + A_2\overline{R}$ and $LCL_{\overline{X}} = \overline{\overline{X}} - A_2\overline{R}$

Table 1 Factors for $\overline{X}$ and R control chart limits

Sample size, n	Factors for control limits: $\overline{X}$-chart A_2	R-chart D_3	R-chart D_4	Factor for calculating $\sigma_{\overline{X}}$ from range, R: d_2
2	1.880	0	3.267	1.128
3	1.023	0	2.573	1.693
4	0.729	0	2.282	2.059
5	0.577	0	2.114	2.326
6	0.483	0	2.004	2.534
7	0.419	0.076	1.924	2.704
8	0.373	0.136	1.864	2.847
9	0.337	0.184	1.816	2.970
10	0.308	0.223	1.777	3.078
11	0.285	0.256	1.744	3.173
12	0.266	0.283	1.717	3.258
13	0.249	0.307	1.693	3.336
14	0.235	0.328	1.672	3.407
15	3.472	0.223	0.347	1.653
16	0.212	0.363	1.637	3.532
17	0.203	0.378	1.622	3.588
18	0.194	0.391	1.608	3.640
19	0.187	0.403	1.597	3.699
20	0.180	0.415	1.585	3.735

Comparison of Tolerances and Control Limits. It is important to clearly differentiate between specification limits and control limits. The specification limits or tolerances of a part are:

- Characteristic of the part/item in question
- Based on functional considerations
- Related to/compared with an individual part measurement
- Used to establish the conformability of a part

The control limits on a control chart are:

- Characteristic of the process in question
- Based on process variability
- Dependent on sampling parameters, namely, sample size
- Used to identify presence/absence of special cause variation in the process

Control limits and tolerances must never be compared numerically and should not appear together on the same graph. Tolerances are limits on individual measurements and as such can be compared against the process as a whole as represented by many individual measurements collected in the form of a statistical distribution.

Process Capability Indices (C_p and C_{pk}). Once a process is in statistical control, it is common to measure process capability in units of standard deviation for the process. There are actually two measures, C_p and C_{pk}, both called the process capability index. The first measure, C_p, is based on the relationship between the upper and lower specification limits and the standard deviation process as follows:

$$C_p = \frac{USL_X - LSL_X}{6\sigma_X}$$

The minimum acceptable value for C_p is considered to be one.

However, because it is possible to achieve a C_p of greater than 1.0 while still producing a significant proportion of parts that do not meet specification limits, a revised version of the process capability index C_{pk}, has been widely adopted. This index, sometimes called the "actual process capability index," imposes a penalty for deviation of the process mean from the nominal value of the specification. The minimum value considered acceptable for C_{pk} is usually 1.33, although some industries are now beginning to demand 2.0 or better.

The C_{pk} index is defined as the difference between the process mean and its closest specification limit, divided by 3σ. To calculate this index, first the relationship between the process mean $\overline{X}$ and the specification limits in the units of standard deviations is determined:

$$Z_{USL} = \frac{USL - \overline{X}}{\sigma_X}$$

$$Z_{LSL} = \frac{LSL - \overline{X}}{\sigma_X}$$

Then the minimum of these two values is selected:

$$Z_{min} = \min[Z_{USL}, -Z_{LSL}]$$

The C_{pk} index is then defined by dividing this minimum value by 3:

$$C_{pk} = \frac{Z_{min}}{3}$$

Commonly, C_{pk} must be $\geqq 1.33$, but higher limits are also specified.

For example, Fig. 4 shows a process capability study of the overall length of a fully processed bevel gear (as heat-treated). The C_p of 1.75 shows that the inherent process variation is less than the maximum acceptable range for the product. The C_{pk} of 1.71 demonstrates that the process is properly targeted to produce an adequately centered finished product dimensional distribution.

Rational Sampling

Perhaps the most crucial issue to the successful use of the Shewhart control chart concept is the definition and collection of the samples or subgroups. This section discusses the concept of rational sampling, sample size, sampling frequency, and sample collection methods and reviews some classic misapplications of rational sampling.

Rational subgroups or samples are collections of individual measurements whose variation is attributable only to one unique constant system of common causes. In the development and continuing use of control charts, subgroups or samples should be chosen in a way that provides the maximum opportunity for the measurements within each subgroup to be alike and the maximum chance for the subgroups to differ from one another if special causes arise between subgroups.

Sample Size and Sampling Frequency Considerations. The size of the rational sample is governed by the following considerations:

- Subgroups should be subject to common cause variation. The sample size should be small to minimize the chance of mixing data within one sample from a controlled process and one that is out of control. This generally means that consecutive sample selection should be used rather than distributing the sample selection over a period of time. There area, however, certain situations where distributed sampling may be preferred.
- Subgroups should ensure the presence of a normal distribution for the sample means. In general, the larger the sample size, the better the $\overline{X}$ distribution is represented by the normal curve. In practice, sample sizes of three or more ensure a good approximation to normality.
- Subgroups should ensure good sensitivity to the detection of assignable causes. The larger the sample size, the more likely that a shift of a given magnitude will be detected.

When the above factors are taken into consideration, a sample/subgroup size of three to six is likely to emerge. Five is the most commonly used number because of the relative ease of further computation.

Sampling Frequency. The question of how frequently samples should be collected is one that requires careful thought. In many applications of $\overline{X}$ and R control charts, samples are selected too infrequently to be of much use in identifying and solving problems. Some considerations in sample frequency determination are the following:

- If the process under study has not been charted before and appears to exhibit somewhat erratic behavior, samples should be taken quite frequently to increase the opportunity to quickly identify improvement opportunities. As the process exhibits less and less erratic behavior, the sample interval can be lengthened.
- It is important to identify and consider the frequency with which occurrences are taking place in the process. This might include, for example, ambient condition fluctuations, raw material changes, and process adjustments such as tool changes or wheel dressings. If the opportunity for special causes to occur over a 15-min period is good, sampling twice a shift is likely to be of little value.
- Although it is dangerous to overemphasize the cost of sampling in the short term, clearly it cannot be neglected.

Common Pitfalls in Subgroup Selection. In many situations, it is inviting to combine the output of several parallel and assumed-to-be-identical machines into a single sample to be used in maintaining a single control chart for the process. Two variations of this approach can be particularly troublesome: stratification and mixing.

Stratification of the Sample. Here each machine contributes equally to the composition of the sample. For example, one measurement each from four parallel machines yields a sample/subgroup of $n = 4$. In this case, there will be a tremendous opportunity for special causes (true differences among the machine) to occur within subgroups.

When serious problems do arise, for example, for one or more of the machines, they will be very difficult to detect because of the use of stratified samples. This problem can be detected, however, because of the unusual nature of the $\overline{X}$-chart pattern (recall the previous pattern analysis) and can be rectified provided the concepts of rational sampling are understood.

The R-charts developed from such data will usually show very good control. The corresponding $\overline{X}$ control chart will show very wide limits relative to the plotted $\overline{X}$ values, and their control will therefore appear almost too good. The wide limits result from the fact that the variability within subgroups is likely to be subject to more than merely common causes.

Mixing Production from Several Machines. Often it is inviting to combine the output of several parallel machines/lines into a single stream of well-mixed product that is then sampled for the purposes of maintaining control charts.

If every sample has exactly one data point from each machine, the result would be the same as that of stratified sampling. If the sample size is smaller than the number of machines with different means or if most samples do not include data from all machines, the within-sample variability will be too low, and the between-sample differences in the means tend to be large. Thus, the $\overline{X}$-chart would given an appearance that the $\overline{X}$ values are too far away from the centerline.

P/M Process Planning

The first step in quality planning is a dialogue between engineering teams of the P/M part producer and the customer. At this stage, finished product (generally an assembled component) can be evaluated, and preliminary part prints may be exchanged. While the product is still "on

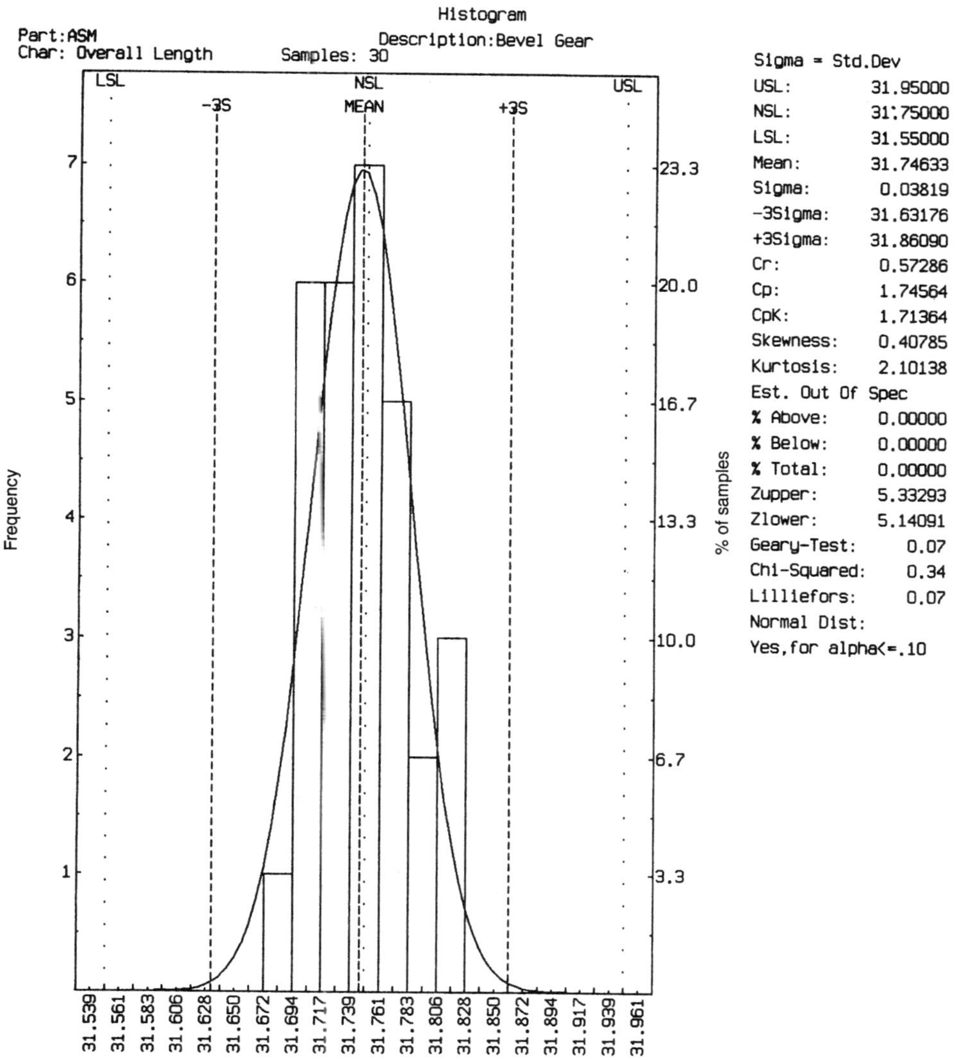

Fig. 4 Process capability for a heat treated bevel gear

paper" relatively inexpensive design changes should be considered to accommodate the P/M process and lower overall cost. Design factors in P/M manufacturing are addressed elsewhere in this Volume.

After final part prints have been agreed to, the P/M manufacturing process can be engineered. Several of the key factors include tooling dimensions, sintering parameters, and secondary operations such as heat treating and machining.

Dimensional and Tool Size Determination

When determining tooling sizes and process control limits, one must begin with a consideration of the final product, then work backward through the process. As parts move from compaction, to sintering, to secondary operations, the sizes usually change significantly. Additionally, the variation generally increases with each step in the process (Fig. 5). Notable exceptions are sized or machined parts, as these secondary processes are intended to improve tolerances and thus reduce variation.

The tool designer must draw upon material and processing knowledge to determine the size of the compaction tools. Commonly, the tooling designer begins with the finished part print and then "factors" the dimensions of all tooling members. For example, for a part with a 1.000 in. outside diameter and 0.800 in. inside diameter that shrinks through processing from compacting tooling size, an appropriate factor may be 1.005. The designer thus would size the die at 1.005 in. (1.000 in. × 1.005) and the core 0.804 in. (0.800 × 1.005) as shown schematically in Fig. 6.

The determination of the appropriate factor depends on the material, the processes used, and the parameters of the processes used. Powder manufacturers offer a great deal of baseline data for tooling size determination, but the best method is to draw upon previously tooled parts of similar processing and geometric configuration. When this previously generated data are not available, a pilot run using prototype tooling is an appropriate method of determining sizes. This reduces the likelihood of having to retool production tooling and can decrease the lead time for the first production run.

Complicated geometries can require the use of different factors for different portions of the tooling. For example, raised hubs formed by the top punch tend to have lower densities than the balance of the part. Lower densities have a tendency to shrink more, which means that the factor used should be higher than for the rest of the part geometry.

Sintering Parameters

Densification for a metallurgically sound part by sintering is determined by sintering temperature, time at temperature, and atmosphere composition. To ensure that metallurgical properties are not compromised, limits on the process inputs are normally established. General part categories, based on material and finished part application, are determined, and limits put on each of these categories. For example, the degree of sinter must be significantly better for structural parts such as gears compared to lightly stressed spacers.

When engineering the process for a new part, baseline, or trial, parameters are normally established. These settings are comfortably above the established minimums for the part category. This allows for minor adjustments during the initial sample process. Adjustments are sometimes needed to ensure that the required mechanical and physical properties are achieved.

Once the production parameters are established, it is best if they are kept nearly constant. Trying to use sintering to compensate for material or compacting variations leads to a high degree of run-to-run metallurgical variation and can contribute to part failures.

Secondary Operations

Secondary operations, discussed in more detail elsewhere in this Volume, must also be considered in process planning and control. Their relation with quality process control are briefly discussed below.

Heat Treat. As with sintering, baseline heat treating parameters are generally established based on the part material and the end use. Parameters are established during the initial sampling procedure, and adjustments from these settings are kept to a minimum to ensure consistent material properties.

Restrike. Whether the part is being re-pressed (densified), coined, or sized, the primary factor to producing good parts is the tool design. As with the design of compacting tools, the sizes may have to be factored to take into account subsequent processing that may change dimensions.

Plating. With powdered metal parts, virtually any of the commonly applied platings can be utilized. The primary difference between plating noncopper-infiltrated P/M versus wrought materials is that typically the parts first must be resin impregnated to fill the porosity before the application of the plating. Copper infiltration gener-

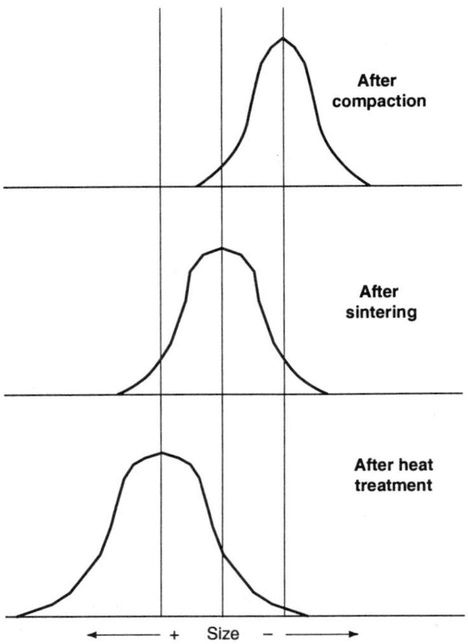

Fig. 5 Typical change in size distribution after sintering and heat treatment of a P/M compact. The distribution widens after additional processing steps.

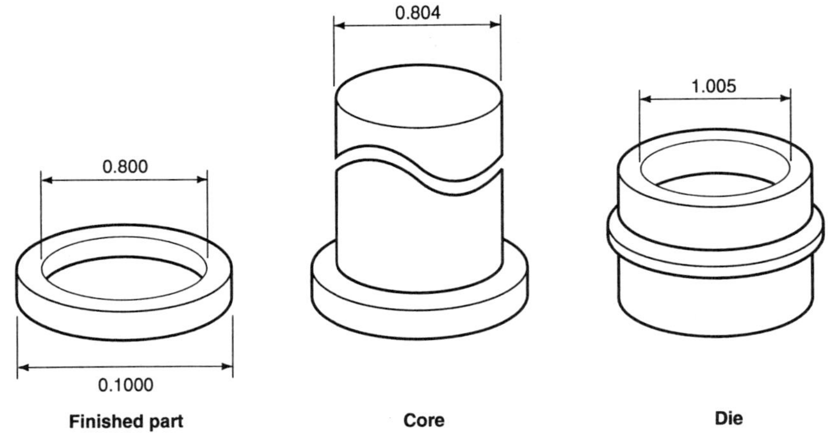

Fig. 6 Example of using tooling factor for die sizing. Factor used: 1.005; tooling not shown: upper punch, lower punch, various adapters, and so forth

ally fills enough of the porosity to eliminate the need for resin impregnation.

Machining. A great many powdered metal alloys are readily machinable. The primary difference between P/M versus wrought is that P/M machining is actually a series of interrupted cuts, due to its inherent porosity. Speeds, feeds, and coolant parameters are different, but overall throughput and process control philosophy is similar to the machining of non-P/M metals.

Quality Control and Inspection

Lot Traceability

The sophistication of a system for part traceability depends on customer requirements and the risk that the producer is willing to assume. Should a problem develop, discrete traceability to a specific production operation can help reduce the quantity that might be involved in a rejection or product recall.

As a minimum, most P/M suppliers maintain traceability to the material used. Each raw material (metal powder, lubricant, additive, etc.) is assigned a unique material lot number by its supplier. If the material is blended in-house, then it is common to assign a lot number based on the blended batch. This lot number is marked on all in-house processing containers and clearly designated on the finished goods containers as well. When the powder is received as a preblend, a new lot number may be designated, or the lot number created by the powder supplier may be used throughout processing.

Unlike the plastics molding and metal castings industries, easily changed lot designators cannot be inserted into the P/M compaction tools. High molding pressures make this sort of designator impractical. If as-molded designators are desired, they are high-strength portions of the tooling that require significant setup to change. Because of this expensive setup, generally as-molded designators are changed only for each run, or up to once per week.

Individual serialization of parts can be achieved with mechanical engravers. These engravers can be used after any of the operations, including on green compacts.

Powder Inspection

Consistency of powder characteristics is key to producing a quality finished part. Chemistry, cleanliness, particle shape, and size distribution are the primary drivers of powder performance.

A few, relatively simple checks are generally sufficient to ensure the consistency of incoming powder—the first check being flow rate.

Flow Rate Check. The determination of flow rate is generally measured according to MPIF Standard 03 (Ref 3). The procedure is summarized:

1. Obtain a sample of the powder. A good way to get a representative sample from a bulk pack or drum of powder is to use a Keystone Sampler (Fig. 7) according to MPIF Standard 01 (Ref 4). This manually operated device augers its way to the bottom of the container, then opens up to retrieve powder from throughout the container.
2. Load the flowmeter funnel with the powder while keeping a dry finger over the discharge orifice.
3. Start a stopwatch simultaneously with the removal of the finger from the orifice. Stop timing when the last of the powder leaves the flowmeter. The flow rate is recorded in elapsed time in seconds.

Flow rate is critical to press setup. Faster-flowing (shorter rate in seconds) powders can fill the die cavity faster and can allow for faster press speeds.

Apparent Density Check. Another simple method to ensure that the incoming powder is consistent lot to lot is to check apparent density. The most commonly followed method is MPIF Standard 04 (Ref 5), which is summarized:

1. A test specimen of metal powder is obtained. Again, the Keystone Sampler is a good tool for getting a representative sample from a container.
2. This entire specimen is loaded into the Hall flowmeter, and the powder should flow into the density cup.
3. Level the heaped powder in the density cup with a nonmagnetic spatula with the blade held perpendicular to the top of the cup. Do not jar the cup.
4. Tape the side of the cup slightly so that the powder settles, allowing the cup to be easily moved without spilling powder.
5. Transfer the powder to a balance and check its mass.
6. Apparent density is calculated as the ratio M/V, where M is the mass of powder from the density cup in grams, and V is the volume of the cup.

Apparent density determines the amount (weight) of the powder that fills the die. It is critical that setup personnel be aware of the apparent density value for a powder. It is especially critical that they know when a new container of powder has a large change in apparent density. The relative position of the die and lower punch(es) determines the amount of fill. If the press is set for a powder of apparent density, then loading a powder with higher apparent density into the press can have catastrophic results during pressing. The higher apparent density powder can increase the green density of the part to the point where the elastic limit of the tooling is exceeded, and the tools break.

Microscopic Evaluation. A simple check of the powder under a microscope is also a good safeguard against sending defective powder to the press. Powder discrepancies that can be de-

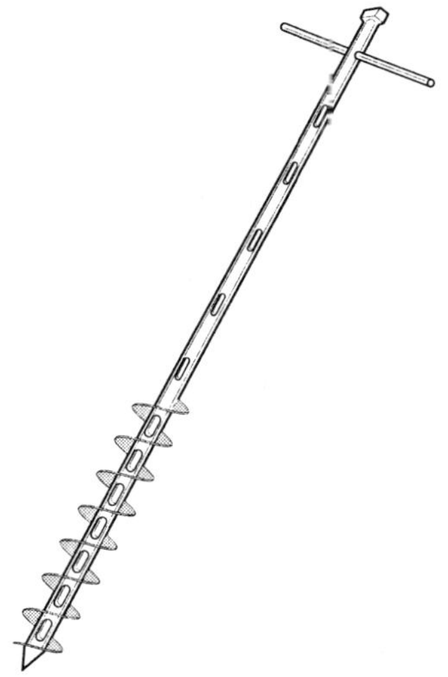

Fig. 7 Keystone sampler for measuring powder flow. Source: Ref 4, used with permission

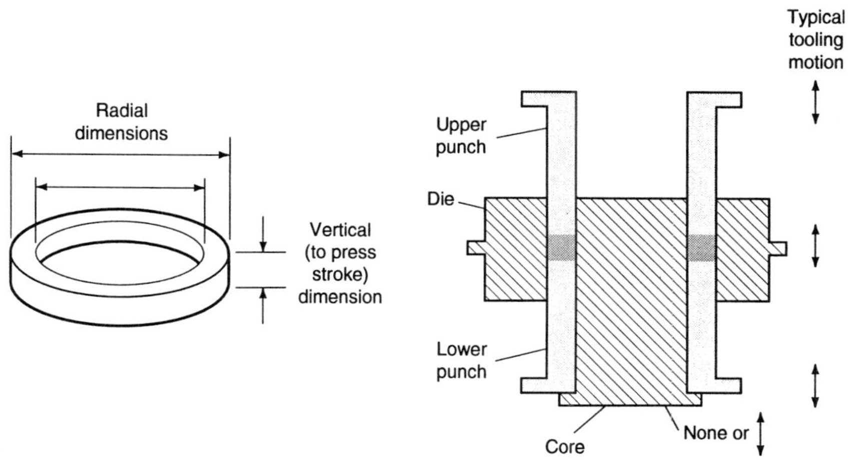

Fig. 8 Radial dimensions of part relative to typical die configuration and tool motion

tected vary from rust to gross anomalies such as large agglomerates of lubricant.

Process Control

A key aspect of process control philosophy is that shop floor operators be given as much "ownership" of their process as possible. Ideally, the operator gages the process, charts it, takes corrective action when it goes out of control, and makes the appropriate adjustments to the process when necessary.

Certain process control procedures, however, cannot be practically conducted by the production operators. Some measurements cannot be accurately determined on the production floor. Examples include green density checks and measurements requiring a coordinate measuring machine. Both of these checks are best left to the laboratory specialists. Additionally, some adjustments, most compaction press adjustments for example, are beyond the level of technical expertise of most operators. These adjustments require specially trained setup personnel.

Compacting Process Control. A variety of statistical process controls are used across the industry for the compaction. Charts successfully employed include X-bar, range, and median. This section briefly describes key variables that influence the implementation of statistical process control for powder compaction.

For conventional rigid die compaction, generally vertical-direction dimensions and weight are statistically charted. Vertical-direction dimensions are those that run in the same direction as the compacting motion (Fig. 8). These features vary significantly as a result of the various press inputs such as voltage, temperature, and mechanical movement. Radial-direction dimensions, those that run horizontally—or not in the direction of the press stroke—generally do not vary to a large degree throughout an individual production run. Therefore, monitoring of radial-direction dimensions is not adequate to show if the pressing action is in statistical control. For longer runs, typically in the 50,000-piece or longer range, a periodic spot check is a good idea to ensure that the tooling has not worn excessively. The exact frequency required for the radial dimension checks depend on the abrasiveness of the powder, part configuration, and the wear resistance of the tool steel.

Many newer presses are equipped with electronics that monitor and chart the various press characteristics. These characteristics include total press tonnage, loads at various press locations, air pressures, hydraulic pressures, and hydraulic temperatures. These outputs are sometimes used for true statistical control, but the majority of manufacturers use these outputs for simple, mechanically based, processing limits.

Green density is the density of the P/M component after compaction. A check of green density is a good verification method at the start of each production run and is checked periodically throughout the run for most parts. However, for single-level parts of less than 6.35 mm (¼ in.) length, an in-process density check is generally not required. Assuming that the control limits established will always produce correct density parts, so long as the process does not stray from these limits, the density will always be correct.

Generally, these density checks are performed once or twice per shift. The reason for the checks is that it is possible for the weight and lengths to be within control limits and the density in a portion of the part to be either over or under specification. For example, the process may shift, causing the flange density to increase and the body density to reduce. The overall density is correct, but portions of the part are high density and portions are low density. High degrees of density variations can cause cracks, break tools, and cause part service failures.

To ensure that the density is relatively homogenous, the green compact is sectioned and the density of each section is determined. The difference between the density sections is calculated and compared against established limits. The limits established may vary depending on a number of factors including tooling fragility and end-product use. For example, a density split limit of 0.2 g/cm^3 may be established on a two-level gear. If the flange is running 7.25 g/cm^3 and the body is running 7.00 g/cm^3 (for a 0.25 g/cm^3 split), the density should be adjusted at the compaction press. Parts run since the last density split check must be carefully evaluated to ensure that they are mechanically sound and will produce finished parts that are within print tolerances.

MPIF Standard 42 (Ref 6) is the industry standard. Application of this procedure for green density is summarized:

1. A test specimen is obtained.
2. Determine the mass of the specimen using a balance of adequate precision (Mass *A*).
3. Impregnate the specimen with oil.
4. Remove excessive oil.
5. Determine the mass of the impregnated test piece (Mass *B*).
6. Suspend the oil-impregnated specimen from the balance hook into water. Completely submerge the piece and be sure that all air bubbles are removed (Mass *C*).
7. Remove the specimen from the water and measure the mass of the balance hook as it is suspended in the same manner as it was when holding the test piece (Mass *E*).

Green density is calculated as:

$$\text{Green density} = \frac{A \cdot \rho_w}{B - (C - E)}$$

where ρ_w is a correction factor for water temperature.

Sintering Process Control. Unlike compaction, which usually has a cycle time of well under 1 min, the sintering process may take hours from beginning to end (furnace exit). On shorter runs, all of the green parts may be loaded into the furnace before a part is removed from the furnace exit. Because thousands of parts can be loaded before a part measurement can be made, a sample batch is usually run to ensure that the processing parameters are having their desired effect on the parts.

Throughout the production runs, parts are usually measured and charted to statistical control limits. Because the radial (from compaction stroke) dimensions are typically highly consistent in green compacts, these are normally good indicators of sintering performance. The variation noted in these measurements is largely from the sintering process; thus, they can be reliably used for assessing the state of sintering process control.

As with some of the newer compaction presses, many newer sintering furnaces have the capability to run true process control on the processing characteristics. With sintering, the outputs are temperature, speed, atmosphere flow rate, and so forth.

In addition to dimensional checks, part macrohardness is also generally monitored throughout processing. Control limits may be established based on the material type, the density being pressed, and the desired as-sintered properties.

Restrike Process Control. The movement of the material being repressed dictates the type of process control required. Control charts may be appropriate for vertical dimensions or radial dimensions, and sometimes both are monitored.

Heat Treat Process Control. Heat treating is similar to sintering in that radial dimensions are usually checked to ensure that the process is running well. For continuous-belt furnaces, in-process statistical charting is appropriate. The process can be adjusted throughout the run to maintain control.

With batch furnaces, all part-based verification for process control is after-the-fact inspection. The best way to ensure that the process is running properly is to have good, tight controls of the inputs (temperature, carbon content, oxygen, etc.). As with sintering, part macrohardness is usually checked. Control limits may be established based on the material type, density, sintering, and heat treating parameters.

Secondary Operations. For most other secondary operations, process control is conducted on P/M parts similarly to that of other metal fabrication techniques.

Inspection Considerations

Inspection techniques for finished parts should be based on concurrence between the customer and the supplier. There are many ways to gage most part print specifications, and the specific technique used can have a large effect on the results reported. A great many P/M part attributes may be measured with the same techniques used in other industries. There are, however, significant exceptions as briefly described below.

Hardness. In powder metallurgy there are generally two types of hardness specified—apparent hardness (macrohardness) and microhardness. The microhardness is the hardness of each particle of material, and the apparent hardness is the hardness of the surface—bridging across many particles and the porosity, too.

Apparent hardness is typically measured according to MPIF Standard 43 (Ref 7). The procedure is relatively straightforward and quick. The basics are:

1. Obtain a sample part of adequate thickness and parallel configuration (or, for cylindrical parts, a correction factor may be used).
2. The sample must be large enough so that the indenter marks from the hardness tester are at least three indenter diameters from any edge or previous impression.
3. Sand each face of the sample so that no burrs are present (burrs will cause erroneous readings), or be sure to use a holding fixture that avoids the burrs.
4. Take readings with a properly calibrated hardness tester.
5. Reject obvious outliers and report the average of at least five nonoutliers.

Typically, the outliers are on the low side. The cause of these occasional low readings is a chance happening that the hardness indenter falls right into a pore.

Microhardness is usually measured according to MPIF Standard 51 (Ref 8). The determination of microhardness is significantly more difficult than measuring apparent hardness and requires specialized equipment that many P/M users do not have on-site. The procedure involves:

1. Sectioning the part and making a polished mount for the evaluation.
2. Placing the mount in a special microhardness testing machine.
3. Under magnification, orienting the mount and making a diamond indenter mark precisely over a particle of the material.
4. Measuring the length of the penetration on the particle and converting this length to a hardness reading.

Microfinish. The porosity of P/M causes debate over the proper method of measuring microfinish. When using a standard cone stylus, unmachined P/M can give relatively high microfinish values. The individual particles are very smooth, but the probe path is interrupted by pores of varying sizes (Fig. 9). Because most parts are running against a mate that is much larger than the standard probe, many feel that the chisel probe provides a better gage as to the actual serviceability of the parts. The chisel probe bridges across the porosity and usually produces significantly lower microfinish values.

Physical Tests. Performing a physical test on finished parts is a great method of verifying that all processes ran properly, and the final parts will perform adequately. Crush, torque, impact, and tensile tests are commonly called out on P/M part prints. These tests greatly reduce the need for expensive, time-consuming, and sometimes ambiguous, microstructure analysis.

Defect Detection

The problem of forming defects in green parts during compaction and ejection has become more prevalent as parts producers have begun to use higher compaction pressures in an effort to achieve high-density, high-performance P/M steels. Several nondestructive inspection methods are practical for detecting defects as early as possible in the production sequence. The most promising nondestructive testing methods for P/M applications include electrical resistivity testing, eddy current and magnetic bridge testing, magnetic particle inspection, ultrasonic testing, x-ray radiography, gas permeability testing, and γ-ray density determination. The capabilities and limitations of each of the techniques are briefly summarized in Table 2.

The four most common types of defects in P/M parts are ejection cracks, density variations, microlaminations, and poor sintering. These defects are briefly described in this section, and SPC techniques related to defects are described in the next section of this article.

Ejection Cracks. When a part has been pressed, there is a large residual stress in the part due to the constraint of the die and punches, which is relieved as the part is ejected from the die. The strain associated with this stress relief depends on the compacting pressure, the green expansion of the material being compacted, and the rigidity of the die. Green expansion, also known as spring out, is the difference between the ejected-part size and the die size. A typical value of green expansion for a powder mix based on atomized iron powder pressed at relatively high pressure (600 to 700 MPa, or 45 to 50 tsi) is 0.20%. In a partially ejected compact, for example, the portion that is out of the die expands to relieve the residual stress, while the constrained portion remains die size and a shear stress is imposed on the compact. When the ability of the powder compact to accommodate the shear stress is exceeded, ejection cracks are formed.

The radial strain can be alleviated to a degree by increasing the die rigidity and designing some release into the die cavity. However, assuming that the ejection punch motions are properly coordinated, the successful ejection of multilevel parts depends to a large degree on the use of a high-quality powder that combines high green strength with low green expansion and low stripping pressure.

Density Variations. Even in the simplest tool geometry possible—a solid circular cylinder—conventional pressing of a part to an overall relative density of 80% will result in a distribution of density within the part ranging from 72 to 82% (Ref 12). The addition of simple features such as a central hole and gear teeth presents minor problems compared with the introduction of a step or second level in the part. Depending on the severity of the step, a separate, independently actuated punch can be required for each level of the part. During the very early stage of compaction, the powder redistributes itself by flowing between sections of the die cavity. However, when the pressure increases and the powder movement is restricted, shearing of the compact in planes parallel to the punch axis can only be avoided by proper coordination of punch motions. When such shear exists, a density gradient results.

The density gradient is not always severe enough for an associated crack to form upon ejection. However, a low-density area around an internal corner can be a fatal flaw because this corner is usually a point of stress concentration when the part is loaded in service.

Microlaminations. In photomicrographs of unetched part cross sections, microlaminations appear as layers of unsintered interparticle boundaries that are oriented in planes normal to the punch axis. They can be the result of fine microcracks associated with shear stresses upon ejection; such microcracks fail to heal during sintering. Because of their orientation parallel to the tensile axis of standard test bars, they have little influence on the measured tensile properties of the bars, but are presumed to be a cause of severe anisotropy of tensile properties.

Poor Sintering. When unsintered particle boundaries results from a cause other than shear stresses, they are usually present because of insufficient sintering time or sintering temperature, a nonreducing atmosphere, poor lubricant burn-off, inhibition of graphite dissolution, or a combination of these. Unlike microlaminations, defects associated with a poor degree of sintering are not oriented in planes.

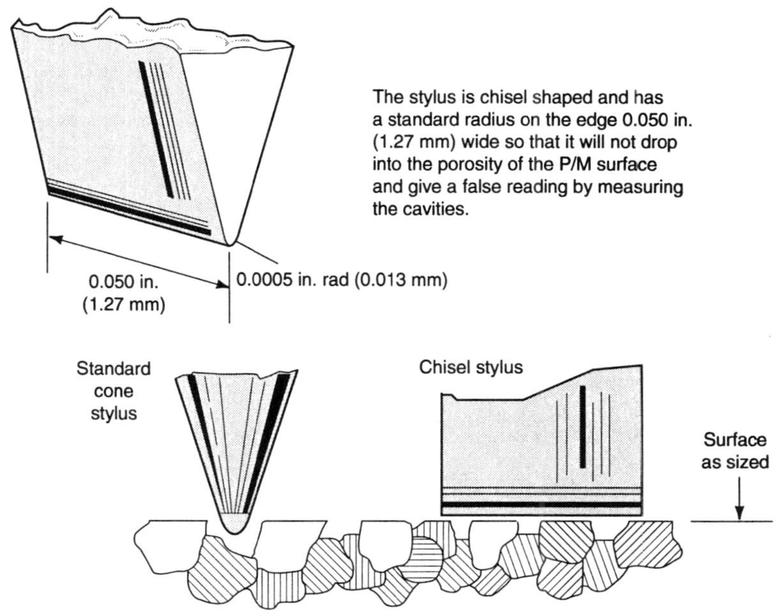

Fig. 9 Surface measurement with chisel stylus (a). (b) Effect of cone and chisel styli on an as-sized P/M surface. Source: Ref 9, used with permission.

Shewhart Control Charts for Defect Data

Many quality assessment criteria for manufactured goods are not of the variable measurement type. Rather, some quality characteristics are more logically defined in a presence-of or absence-of sense. In P/M, examples include chipped gear teeth, discolored plating, scratched surfaces, and powder accumulation on parts.

Such nonconformities or defects are often observed visually or according to some sensory criteria and cause a part to be defined simply as a defective part. In these cases, quality assessment is referred to as being made by attributes.

Many quality characteristics that could be made by measurements (variables) are often not done as such in the interest of economy. A go/no-go gage can be used to determine whether or not a variable characteristic falls within the part specification. Parts that fail such a test are simply labeled defective. Attribute measurements can be used to identify the presence of problems, which can then be attacked by the use of $\bar{X}$ and R control charts. The following definitions are required in working with attribute data:

- *Defect:* A fault that causes an article or an item to fail to meet specification requirements. Each instance of the lack of conformity of an article to specification is a defect or nonconformity.
- *Defective:* An item or article with one or more defects is a defective item.
- *Number of defects:* In a sample of n items, c is the number of defects in the sample. An item may be subject to many different types of defects, each of which may occur several times.
- *Number of defectives:* In a sample of n items, d is the number of defective items in the sample.
- *Fractional defective:* The fractional defective, p, of a sample is the ratio of the number of defectives in a sample of the total number of items in the sample. Therefore, $p = d/n$.

Operational Definitions

The most difficult aspect of quality characterization by attributes is the precise determination of what constitutes the presence of a particular defect. This is so because many attribute defects are visual in nature and therefore require a certain degree of judgment and because of the failure to discard the product control mentality. For example, a scratch that is barely observable by the naked eye may not be considered a defect, but one that is readily seen is. Furthermore, human variation is generally considerably larger in attribute characterization (for example, three different caliper readings of a workpiece dimension by three inspectors and visual inspection of a part by these same individuals yield anywhere from zero to ten defects). It is therefore important that precise and quantitative operational definitions be laid down for all to observe uniformly when attribute quality characterization is being used. The length or depth of a scratch, the diameter of a surface blemish, or the height of a burr.

The issue of the product control versus process control way of thinking about defects is a crucial one. From a product control point of view, scratches on a magnetic catch plate should be counted as defects only if they appear on visual surfaces, which would directly influence part function. From a process control point of view, however, scratches on a catch plate should be counted as defects regardless of where they appear because the mechanism creating these scratches does not differentiate between visual and concealed surfaces. By counting all scratches, the sensitivity of the statistical charting instrument used to identify the presence of defects and to lead to their diagnosis will be considerably increased.

A major problem with the product control way of thinking about part inspection is that when attribute quality characterization is being used, not all defects are observed and noted. The first occurrence of a defect that is detected immediately causes the part to be scrapped. Often, such data are recorded in scrap logs, which then present a biased view of what the problem may really be. One inspector may concentrate on scratch defects and will therefore tend to see these first. Another may think brightness is more critical, so his data tend to reflect this type of defect more frequently. The net result is that often such data may then mislead those who may be using it for process control purposes. Therefore, it is essential from a process control standpoint to carefully observe and note each occurrence of each type of defect.

p-Chart and c-Chart Analyses

Example 1: *p*-Chart Analysis for Fraction Defective during Tapping. Consider a tapping operation on a P/M bracket. Suppose the measures of quality conformance of interest are the

Table 2 Comparison of the applicability of various nondestructive evaluation methods of flaw detection in P/M parts

Method	Measured/detected	Applicability to P/M parts(a) Green	Applicability to P/M parts(a) Sintered	Advantages	Disadvantages
X-ray radiography	Density variations, cracks, inclusions	C	C	Can be automated	Relatively high initial cost; radiation hazard
Computed tomography	Density variations, cracks, inclusions	C	C	Can be automated; pinpoint defect location	Extremely high initial cost; highly trained operator required; radiation hazard
Gamma-ray density determination	Density variations	A	A	High resolution and accuracy; relatively fast	High initial cost; radiation hazard
Ultrasonic imaging: C-scan	Density variations, cracks	D	B	Sensitive to cracks; fast	Coupling agent required
Ultrasonic imaging: SLAM	Density variations, cracks	D	C	Fast; high resolution	High initial cost; coupling agent required
Resonance testing	Overall density, cracks	D	B	Low cost; fast	Does not give information on defect location
Acoustic emission	Cracking during pressing and ejection	C	D	Low cost	Exploratory
Thermal wave imaging	Subsurface cracks, density variations	D	C	No coupling agent required	Fat or convex surfaces only
Electrical resistivity	Subsurface cracks, density of variations, degree of sinter	A	A	Low cost, portable, high potential for use on green compacts	Sensitive to edge effects
Eddy current/magnetic bridge	Cracks, overall density, hardness, chemistry	C	A	Low cost, fast, can be automated; used on P/M valve seat inserts	Under development
Magnetic particle inspection	Surface and near-surface cracks	C	A	Simple to operate, low cost	Slow; operator sensitive
Liquid dye penetrant inspection	Surface cracks	C	D	Low cost	Very slow; cracks must intersect surface
Pore pressure rupture/gas permeability	Laminations, ejections, cracks, sintered density variations	A	A	Low cost, simple, fast	Gas-tight fixture required; cracks in green parts must intersect surface

(a) A, has been used in the production of commercial P/M parts; B, under development for use in P/M; C, could be developed for use in P/M, but no published trials yet; D, low probability of successful application to P/M. Source: Ref 10 and 11 (reprinted with permission)

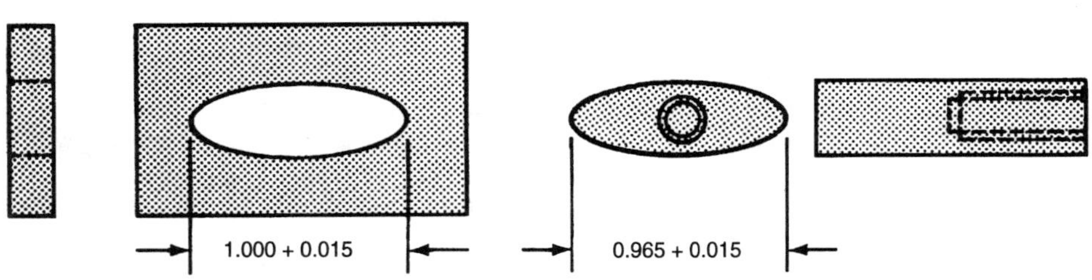

(a) (b)

Fig. 10 Pin and hole components statistically analyzed in Example 2. (a) Plate with hole. (b) Pin assembly. Dimensions given in inches

presence of threads (generally absent due to parts being inadvertently transferred into the finished parts bin without processing) and the size of the tapped threads as measured with a go/no-go plug (generally defective due to cutting tool wear).

To establish the control chart, rational samples of size $n = 50$ parts are drawn from production periodically (perhaps, each shift), and the sampled parts are inspected and classified as either defective (from either or both possible defects) or nondefective. The number of defectives, d, is recorded for each sample. The process characteristic of interest is the true process fraction defective p'. Each sample result is converted to a fraction defective:

$$p = \frac{d}{n} \quad \text{(Eq 1)}$$

The data (fraction defective p) are plotted for at least 25 successive samples of size $n = 50$. The individual values for the sample fraction defective, p, vary considerably, and it is difficult to determine from the plot at this point if the variation about the average fraction defective, $\bar{p}$, is solely due to the forces of common causes or special causes.

Control Limits for the p-Chart. It can be shown that for random sampling, under certain assumptions, the occurrence of the number of defectives, d, in the sample of size n is explained probabilistically by the binominal distribution. Because the sample fraction defective, p, is simply the number of defectives, d, divided by the sample size, n, the occurrence of values for p also follows the binominal distribution. Given k rational samples of size n, the true fraction defective, p', can be estimated by:

$$\bar{p} = \sum_{i=1}^{k} p_i \Big/ k \quad \text{(Eq 2)}$$

or

$$\bar{p} = \sum_{i=1}^{k} d_i \Big/ \sum_{i=1}^{k} n_i \quad \text{(Eq 3)}$$

Equation 3 is more general because it is valid whether or not the sample size is the same for all samples. Equation 2 should be used only if the sample size, n, is the same for all k samples.

Therefore, given $\bar{p}$, the control limits for the p-chart are then given by:

$$\text{UCL}_p = \bar{p} + 3\sqrt{\bar{p}(1-\bar{p})/n} \quad \text{(Eq 4a)}$$

$$\text{LCP}_p = \bar{p} - 3\sqrt{\bar{p}(1-\bar{p})/n} \quad \text{(Eq 4b)}$$

Thus, only $\bar{p}$ has to be calculated for at least 25 samples of size n to set up a p-chart. The binomial distribution is generally not symmetric in quality control applications and has a lower bound of $p = 0$. Sometimes the calculation for the lower control limit may yield a value of less than 0. In this case, a lower control limit of 0 is used.

***c*-Chart: Analysis for Number of Defects.** The p-chart deals with the notion of a defective part or item where defective means that the part has at least one nonconformity or disqualifying defect. It must be recognized, however, that the incidence of any one of several possible nonconformities would qualify a part for defective status. A part with ten defects, any one of which makes it defective, is on equal footing with a part with only one defect in terms of being defective.

Often it is of interest to note every occurrence of every type of defect on a part and to chart the number of defects per sample (c). A sample may only be one part, particularly if interest is focusing on final inspection of an assembled product, or inspection may focus on one type of defect or multiple defects.

The probability law that governs the incidence of defects is known as the Poisson law or Poisson probability distribution, where c is the number of defects per sample. It is important that the opportunity space for defects to occur be constant from sample to sample. The Poisson distribution defines the probability of observing c defects in a sample where c' is the average rate of occurrence of defects per sample

Construction of c-Charts from Sample Data. The number of defects, c, arises probabilistically according to the Poisson distribution. One important property of the Poisson distribution is that the mean and variance are the same value. Then given c', the true average number of defects per sample, the 3σ limits for the c-chart are given by:

$$CL_c = c' \pm 3\sqrt{c'}$$

Note that the standard deviation of the observed quantity c is the square root of c'. The Poisson distribution is a very simple probability model, being completely described by a single parameter c'.

When c' is unknown, it must be estimated from the data. For a collection of k samples, each with an observed number of defects c, the estimate of c' is:

$$\bar{c} = \sum_{i=1}^{k} c_i \Big/ k$$

Therefore, trial control limits for the c-chart can be established, with possible truncation of the lower control limit at zero, from:

$$UCL_c = \bar{c} + 3\sqrt{\bar{c}}$$

$$LCL_c = \bar{c} - 3\sqrt{\bar{c}}$$

Tolerance Control

The issue of part tolerancing and, in particular, the statistical assignment and assessment of tolerances are excellent examples of the need for design and manufacturing to understand what each other is doing and why. The best intentions of the design process can go unmet if the manufacturing process is not operated in a manner totally consistent with design intent. To more clearly appreciate the relationship that must exist between the design and manufacturing operations, some of the basic assumptions of the tolerancing activity and their relationship to the manufacturing process are examined. The following sections clearly point to the importance of statistical process control relative to the issue of process capability.

The key concepts in statistical tolerancing are:

- The use of a statistical distribution to represent the design characteristic and therefore the process output for the product/part in question relative to the design specifications
- The notion of random assembly, that is, random part selection from these part process distributions when more than one part is being considered in an assembly
- The additive law of variances as a means to determine the relationship between the variability in individual parts and that for the assembly

To assume that the parts can be represented by a statistical distribution of measurements (and for the assumption to hold in reality), the part processes must be in a state of statistical control. The following example illustrates the importance of statistical process control in achieving design intent in a tolerancing problem.

Example 2: Statistical Tolerance Model for Optimal Fit of a Pin Assembly into a Molded Hole of a P/M Plate. Figure 10 shows two simple parts: a plate with a hole and a pin that will ultimately be assembled to a third part but must pass through the hole in the plate. For the assembly, it is desired for function that the clearance between the plate hole and the pin be at least 0.015 in. but no more than 0.055 in.

To achieve the design requirement stated above, the nominal values and tolerance for the plate hole and pin were statistically derived and are shown in Fig. 10. To arrive at these tolerances, it was assumed that:

- The parts would be manufactured by processes that behave according to the normal distribution
- The process capabilities would be at least 6σ, the processes would be centered at the nominal values given in Fig. 10, and the processes would be maintained in a state of statistical control
- Random assembly would prevail

If these assumptions are met, the processes for the two parts, and therefore the clearance associated with assembled parts, would be as shown in Fig. 11, and the design intent would be met.

Suppose that despite the assumptions made and the tolerances derived, the processes manufacturing the pin and plate hole were not maintained in good statistical control. As a result, the parts actually more nearly follow a uniform/rectangular distribution within the specifications, as shown in Fig. 12. Such could have arisen as a result of sorting or rework of a more variable process(es), in which case the results are doubly distressing, that is, poorly fitting assemblies and increased cost to the system.

Figure 12 shows the distribution of the clearance if the hole and pin dimensions follow the uniform distribution within the specifications. The additive law of variances has been used to derive the variation in the clearance distributions but assuming the uniform distribution for the individual part processes. Some assemblies may not go together at all, some will fit quite tightly

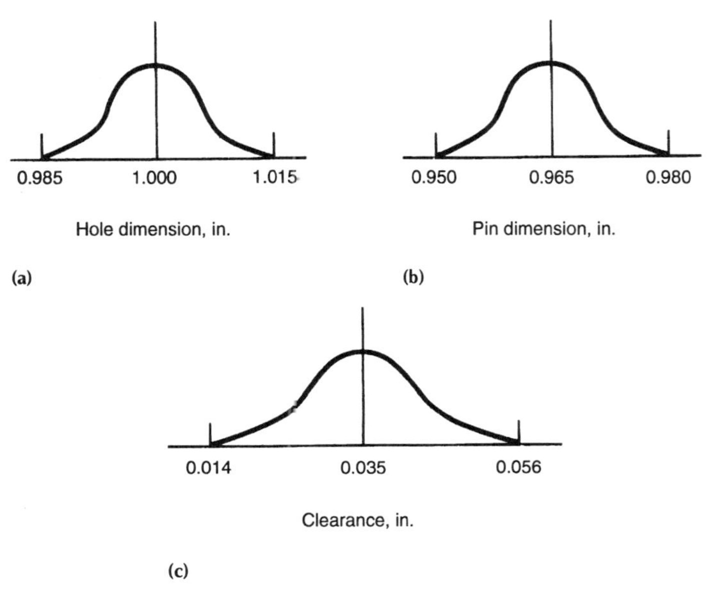

Fig. 11 Statistical basis for satisfying design intent for the hole/pin assembly clearance in Fig. 10. (a) Distribution of hole. (b) Distribution of pin. (c) Distribution of clearance

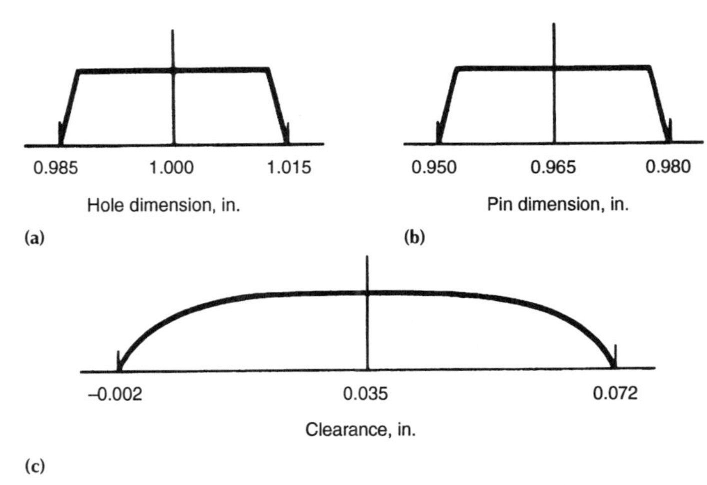

Fig. 12 Clearance implications of poor process control of plate hole and pin dimensions for components of Fig. 10. (a) Distribution of hole. (b) Distribution of pin. (c) Distribution of clearance

and may later bind if foreign matter gets into the gap, and others will fit together with a much larger clearance than desired.

The problem here is not a design problem. The plate hole and pin tolerances have been derived using sound statistical methods. However, if the processes are not in good statistical control and therefore not capable of meeting the assumptions made during design, poor-quality assemblies will follow. It should be noted that the altogether too common process appearance of a uniform distribution of measurements within the specifications can arise in several different ways:

- From processes that have good potential with regard to variation, but are not kept in good statistical control
- From unstable and/or large variation processes that require sorting/rework to meet the specifications
- From processes that are intentionally allowed to vary over the full range of the specifications to take advantage locally of wide specifications relative to the process variation

In all of the three cases mentioned above, additional costs will be incurred and product quality will be eroded. Clearly, SPC is crucial to the tolerancing issue in engineering design.

ACKNOWLEDGMENTS

Portions of this article are adapted from the following articles:

- Eric Rasmussen and David Zenger, Implementing Statistical Process Control for Powder Component Production, Characterization, Testing and Quality Control, *Advances in Powder Metallurgy and Particulate Materials—1994,* Vol 2, Metal Powder Industries Federation, p 252
- Richard DeVor and Tsong-how Chang, Statistical Quality Design and Control, *Nondestructive Evaluation and Quality Control,* Vol 17, *ASM Handbook,* 1989, p 719–753

REFERENCES

1. W.A. Levinson, Make the Most of Control Charts, *Chem. Eng. Prog.,* Vol 88 (No. 3), March 1992, p 86–91
2. R.E. DeVor, T.H. Chang, and J.W. Sutherland, *Statistical Quality Design and Control,* Macmillan, 1992
3. "Determination of Flow Rate of Free-Flowing Metal Powders Using the Hall Apparatus," Standard 03, Metal Powder Industries Federation
4. "Method for Sampling Finished Lots of Metal Powders," Standard 01, Metal Powder Industries Federation
5. "Method for Determination of Apparent Density of Free-Flowing Metal Powders Using the Hall Apparatus," Standard 04, Metal Powder Industries Federation
6. "Method for Determination of Density or Compacted or Sintered Metal Powder Products," Standard 42, Metal Powder Industries Federation
7. "Method for Determination of Apparent Hardness of Powder Metallurgy Products," Standard 43, Metal Powder Industries Federation
8. "Method for Determination of Microhardness of Powder Metallurgy Materials," Standard 51, Metal Powder Industries Federation
9. *P/M Design Guidebook,* Metal Powder Industries Federation, 1983, p 15
10. *Prevention and Detection of Cracks in Ferrous P/M Parts,* Metal Powder Industries Federation, 1988
11. R.C. O'Brien and W.B. James, Powder Metallurgy Parts, *Nondestructive Evaluation and Quality Control,* Vol 17, *ASM Handbook,* 1989, p 537
12. G. Taguchi, *On-Line Quality Control During Production,* Japanese Standards Association, 1981

Quality Control and Inspection of Secondary Operations

Pat Kenkel, John Engquist, and Mike Blanton, Burgess-Norton Manufacturing Company

QUALITY PLANNING TECHNIQUES for P/M secondary operations are no different than quality planning for any other manufacturing process—the thought process should be consistent throughout. Developing quality control and inspection for any P/M part begins with the quality planning phase of the product launch process. Quality planning begins well before the purchase order is issued. A thorough review of print requirements with the customer can minimize problems later in the process development. The dimensioning techniques used when designing a part will often determine types of gaging and quality control methods to be used. Geometric dimensioning and tolerancing (GDT) is widely used in design work to ensure the part meets the form, fit, and functional requirements. If properly used, GDT will minimize misinterpretations of the print and lead to the most economical gaging solutions for the part.

It is extremely useful to have mutual (customer/supplier) agreement on gaging concepts during the quality planning process. Quality requirements must also be discussed at this time to avoid capability and processing issues later on. Classification of characteristics (C of C) and the associated capability indices should be thoroughly understood to properly process the part. Secondary operations are often added to satisfy statistical process control requirements.

Another important step in the quality planning process is to plan and develop the process(es) *before* the first parts are produced—it is much less desirable to backtrack (reactive) in the quality planning activities than to act (proactive) on a well thought out plan with failure modes defined and methods provided to prevent and/or detect those potential failure modes. Two useful tools to help in the planning process are a failure mode and effects analysis (FMEA) and a control plan.

A FMEA process is a very good engineering tool that is used to detect problem areas in the process and to mistake-proof and/or assign the correct process controls. It should be noted that the FMEA process is used to flag potential causes of failure in the process: if the potential for failure is high, efforts should first be made to minimize the potential effects of the failure. Once the risk has been minimized, the appropriate process control can be assigned. Mistake-proofing should be used to reduce risk and ensure quality.

The control plan assigns control methods to the process in an effort to help minimize process and product variation. Features to be checked, method of inspection, control method, frequency of inspection, and responsibility are assigned for each operation. The control plan takes into consideration the FMEA results as well as any classification of characteristics that the customer has defined. Depending on the characteristic, the amount of control may vary, from simple first piece layouts (for tool generated dimensions) to 100% inspection on characteristics generated using incapable machinery.

Secondary Operations

The purpose of this article is not to discuss technology gains made in secondary operations for powder metallurgy, but to provide useful information in controlling and improving secondary processes using quality control and inspection methodologies. However, development of the following secondary operations is briefly discussed:

- Restrike—densification
- Restrike—sizing
- Heat treat
- Machining and grinding

Examples of FMEA and control plans are shown in Fig. 1 and 2, respectively, for the restrike—densification process to illustrate the use of such tools when assigning process control methods and inspection instructions to a process. Additional information on statistical process control concepts and terminology are also discussed in the article "Planning and Quality Control of Powder Metallurgy Parts Production" in this Volume.

Restrike—Densification. Restrike operations are frequently used to densify parts above the density attainable at compacting. In these instances, more emphasis is placed on attaining the desired material property through densification than on achieving tight dimensional limits (although dimensional control is also important). Adequate control must be placed on the process to ensure density requirements (and dimensions) are met. Because density checks are not readily performed in a production environment, dimensional controls may be used to control density if correlation can be proved to be sufficient. When densifying thin parts, greater dimensional and weight control may be required at compacting to avoid overdensification and damage to the restrike press and/or tooling. Press and tool load monitors are effective process control tools.

Restrike—Sizing. Sizing using a restrike operation is used to enhance dimensional characteristics on P/M parts. In this instance, the density of the part has already been met in the compacting and sintering operations. Sizing is used to meet tighter dimensional limits on diameters, lengths, tapers, and so forth. The quality control methods assigned for sizing operations depend on the characteristic sized and the capability of the machine in achieving the desired limits of size; highly capable processes require very little in-process or post-process inspection while marginally capable processes require significant and frequent inspection.

Heat Treat. The properties from heat treatment (such as hardness and tensile strength) are dependent on many process variables. Carbon potential, atmosphere, heat treat temperature, quench rate and temperature, and draw temperature affect the properties of parts. A FMEA of the heat treatment should be done to list potential failures caused by incorrect settings of the above inputs. To minimize potential process failures, process control of the heat treat operation should include monitoring the inputs as well as the outputs. Minimal variation of these parameters results in significantly reduced variation of product characteristics. Again, depending on the stability of the process, more or less dimensional and metallurgical inspection is required as post-process inspection.

Machining/Grinding. Although machining and grinding are different technologies, process control issues are similar for each. Variation in the process can and should be minimized in the selection of capable machinery, the design of the fixturing, and the selection of the tooling (inserts, cutters, grinding wheels, etc.). Process control and inspection techniques vary depending on the characteristics being generated, the capability of the process, and the quality requirements specified on the blueprint (classification of characteristics). Machine controls have allowed technologies such as automatic compensation and redundant tooling to significantly reduce process variation due to tool wear and other time-related variations. The selection of the machine to be used is a big factor when analyzing the FMEA and developing control plans.

Process Control Methods

Many methods are used to control part conformance in manufacturing processes. Control plans should be selected to minimize the financial impact to the process (either in the cost of the gage and/or the time required to gage the part), but also to allow for optimal control of the process (100% inspection controls the outgoing product, but typically is not a cost-effective inspection method). As stated above, the control plan must include the gaging method, the control method, the frequency, and the responsibility. A discussion of each area follows.

Gaging Method. The two types of commonly used gaging in the manufacturing environment are attribute gages and variable gages. Attribute gaging is used to check conformance or nonconformance for a specific dimension/tolerance. Variable gaging is used to record actual dimensional values for comparison to the tolerance specification. Both types of gages can be used to effectively control processes.

Typically, attribute gaging is used to control relational features (GDT features such as true positions, etc.) and dimensional features that are either incidental characteristics or produced by an extremely capable and stable process (i.e., C_{pk} values over 2.00) (see the article "Planning and Quality Control Powder Metallurgy Parts Production" in this Volume for definition of C_{pk} and other statistical control indices). Attribute gages range from inexpensive go/no-go plug gages to complex, part specific true position gages that can be very expensive. For well-centered processes, these gages can be useful in helping the operator control the process. It should also be noted that restricted tolerance attribute gaging, also called "narrow limit" gaging (50% or 75% of the tolerance band), can be used to monitor and compensate the process before nonconforming parts are produced.

One of the biggest advantages of powder metallurgy is the ability to press features that are complex. The secondary operations mentioned above refine pressed features and/or generate

Potential
Failure Mode and Effects Analysis
(Process FMEA)

Item: Coining - Repressing
Model Year/Vehicle(s): /
Core Team:

Process Responsibility: FGF
Key Date:

FMEA Number: 00040
Page 1 of 1
Prepared by:
FMEA Date (Orig.): 95.10.21 (Rev.): 95.10.21

Process Function / Requirements	Potential Failure Mode	Potential Effect(s) of Failure	Sev	Class	Potential Cause(s)/ Mechanism(s) of Failure	Occur	Current Process Controls	Detec	R.P.N.	Recommended Action(s)	Responsibility & Target Completion Date	Action Results: Actions Taken	Sev	Occ	Det	R.P.N.
Coin (Re-Pressing) Increase density beyond limits of conventional compacting capabilities.	High density.	Broken compacting tool, poor growth control, damaged coining tool.	5		Improper weight to length ratio.	2	High inspection frequency for weight and control chart for length at pressing operation and control chart for length at coin.	1	10							
	Low density.	Low apparent hardness, low crush strength, poor size control.	5		Improper weight to length ratio.	2	High inspection frequency for weight and control chart for length at pressing operation and control chart for length at coin.	1	10							
	Oversize / undersize length.	Non conformance at subsequent operation resulting in high or low density. Length control / compensation problems at I.D. grinding.	2		Improper machine adjustment.	2	High inspection frequency for weight and control chart for length at pressing operation and control chart for length at coin.	2	8							
	Tool control characteristic not acceptable.	Non conformance to blueprint at subsequent operation.	5		Tool not made to print.	1	First piece layout on new tools and setup inspection of major characteristics.	2	10							

Fig. 1 Brief example of a FMEA for restrike operation

new features that often have a geometric relationship to pressed features. With proper GDT, functional attribute gaging is a cost-effective way to control the relationship of features generated by secondary operations to those generated at the press.

Variable gaging allows the operator to record specific dimensional data that can then be analyzed and used to make decisions on process adjustments. Variable gages are used to control processes that are incapable and/or unstable over time (tool wear, etc.). Variable gaging is frequently used on significant characteristics that require a high degree of control. Variable gages can be standard components such as digital micrometers or complex gages that are part specific. In order for the variable gage to be effective for product acceptance decisions, the gage must have a minimum resolution of 10 divisions over the tolerance band being measured (0.010 in. tolerance band needs a gage capable of measuring in 0.001 in. increments).

Variable data is generally required by customers to evaluate the capability of a process. Gage resolution used to estimate process capability may be greater than the gage resolution required for product acceptance. It should also be noted that variable gaging is often used to establish process capability, but the process can then be monitored using attribute gages (if capability is high, for example). In these instances, gaging costs may still be high, but the cost impact to the process has been minimized by using attribute gages.

Control Method. The most common control method for manufacturing processes is statistical process control (SPC). Three common statistical tools are widely used in manufacturing to control processes. For ease of understanding their use and effectiveness, they are discussed below (in order of effectiveness, least to greatest).

Sampling tables were used in industry long before the current statistical approaches were made popular. Typical sampling tables, such as MIL-STD-105E (current standard ANSI/ASQC Z1.4—1993), were developed for use with attribute gaging and mandated some acceptable quality level (AQL) limit as a decision criteria for accepting or rejecting a lot of product. Statistical methods were used to determine what the sample sizes should be for a specific lot of material. Sampling tables can still be used as a quality control tool, but are usually used to control lots of product (post-process inspection plans), not the process itself.

A histogram/frequency distribution can be used in process control to visualize the pattern of variation being generated by an operation. This picture of the data is very useful in easily determining (1) the approximate central value or central tendency (average), (2) the spread of the values (variation), and (3) the relationship of the values to the process specifications (capability index). A histogram can turn machine-recorded values (Fig. 3) into a beneficial tool to monitor a process (Fig. 4).

Control Charts. Time-ordered data is an effective way to view process information. Control charts vary in the presentation of the data collected (individuals chart, *x*-bar and *R*, group charts, etc.), but are a useful tool to control processes over time. For example, if the same data used in the histogram in Fig. 4 is displayed as an individual chart (or average chart) (Fig. 5), the

MANUFACTURING CONTROL PLAN

Customer Part # : Part Name: Customer :
Supplier Part #: Date (Original): Date (Revised) : / /

OPER. SEQ.	PROCESS DESCRIPTION PROCESS PURPOSE	CONTROL ITEM & SPECIFICATION	DEPT. RESP.	FREQ. & SAMPLE SIZE	MEASURE-MENT METHOD	METHOD OF RECORDING - CONTROL METHOD	PROBLEM REACTION PLAN & RESP.
3	Press/Coin	Length	220	5 pcs/30 min's	0 - 1" Micrometer	X-bar & R Chart	Press Adj. Sort, Rework. Scrap Operator
		Parallelism	220	2 pcs. / 1 hour	0 - 1" Micrometrer	X-bar & R Chart	Sort, Rework. Scrap Operator
		Die-Controlled Features	220	At Set-up	Attribute Gauges	Audit	Sort, Rework. Scrap Operator
		Density	220	At Set-up	Water Displacem't	Audit	Press Adj. Operator
		No Cracks	225	5 pcs/Lot	Magnaflux	Frequency Distribution Chart	Sort, Rework, Scrap Operator
		Appearance	225	5 pcs/Lot	Visual Feel	Frequency Distribution Sheet	Sort, Rework, Scrap Operator

Fig. 2 Brief example of a control plan for restrike operation

ID size: 0.500 ± .005 in.									
.498	.501	.504	.502	.503	.504	.502	.505	.503	.500
.500	.499	.501	.502	.504	.499	.503	.502	.503	.502
.505	.502	.504	.504	.501	.503	.502	.500	.501	.501
.503	.501	.504	.501	.500	.502	.499	.502	.503	.503
.502	.501	.504	.502	.500	.502	.504	.501	.503	.503

Fig. 3 Data generated at a machine

Length	Frequency
0.495	
0.496	
0.497	
0.498	x
0.499	xxx
0.500	xxxxx
0.501	xxxxxxxxx
0.502	xxxxxxxxxxxx
0.503	xxxxxxxxxx
0.504	xxxxxxxx
0.505	xx

Fig. 4 Histogram of the same data as Fig. 3. A histogram gives a "picture" of the process that is easily analyzed by the operator.

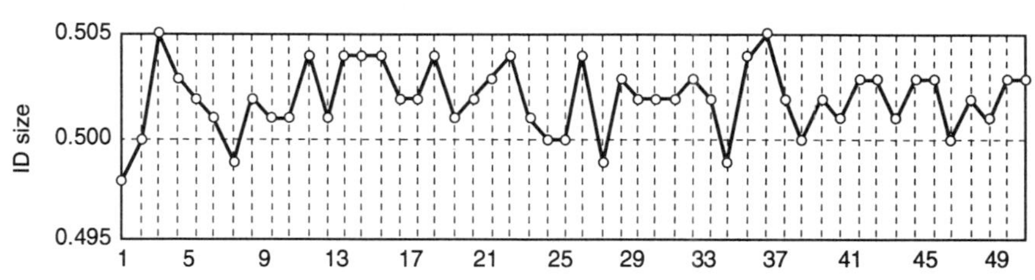

Fig. 5 x or *x*-bar chart for 49 diameter measurements

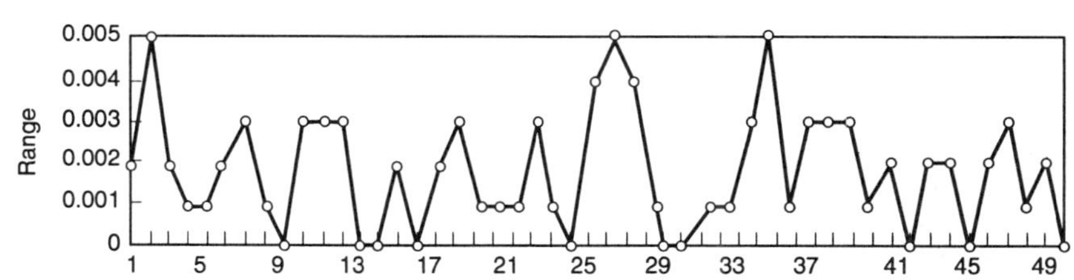

Fig. 6 Range chart of variations for measurements in Fig. 5

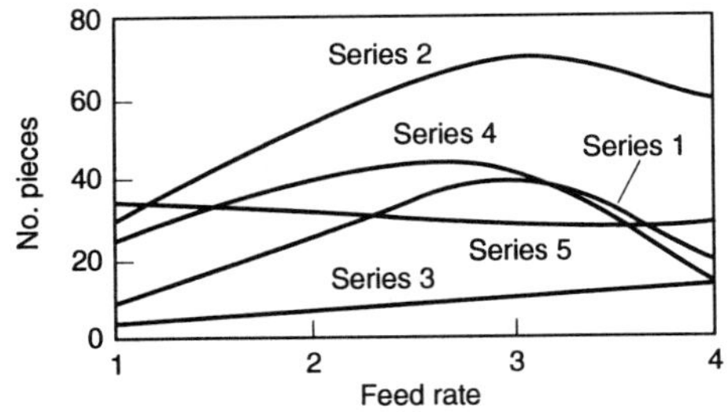

Fig. 7 Parts produced versus feed rate from lathe test

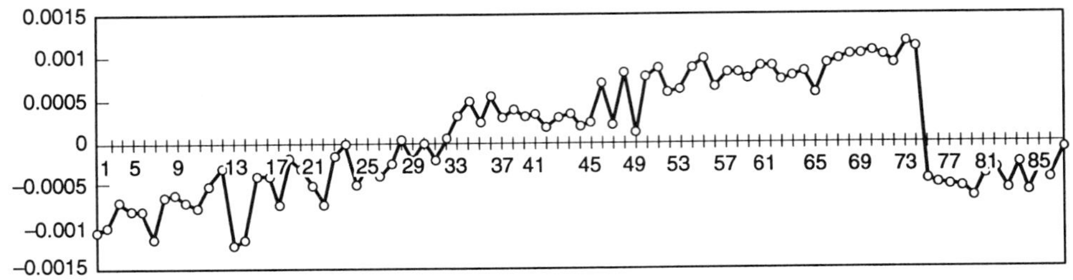

Fig. 8 Trend of dimensional change from tool wear

same information (average, variation, process performance) can be approximated. Because the control chart is time ordered, trends (decreasing or increasing values) can also be depicted. The range is determined as the difference in subsequent values and plotted on a range chart (Fig. 6). The range chart shows how stable the process is (the variation from one reading to the next).

Other control methods (nonstatistical) can be used to control manufacturing processes. If the features to be controlled are generated by the compacting or sizing tooling, an "audit" of the characteristic may be sufficient. The audit is usually done at setup (first piece) and/or at the end of the production run (last piece). This ensures that no dramatic change occurred during the process.

Another control method is 100% inspection. Inspection of all parts may be necessary if the process is incapable or out of control. If possible, 100% inspection should be avoided—it adds cost to the operation and is not an effective way to control the process (100% inspection is not 100% effective). The best control methods for capable and stable processes are statistically based. Mistake-proofing is a cost-effective approach to 100% inspection in some cases.

Frequency. The sampling frequency for a manufacturing process varies based on the classification of the characteristic, the capability of the process, and the stability of the process:

Process parameter	Inspection Frequency
Classification of characteristics	Increases as characteristics become more critical
Process capability	Decreases as capability increases
Process stability	Decreases as stability increases

A stable and capable process may only use tool control as a control method. The opposite extreme may require 100% inspection. Most process parameters can be controlled using statistical methods with either attribute or variable gaging and varying frequencies of inspection (statistical methods are discussed above).

Responsibility. Organizations assign quality and inspection responsibilities differently. The most effective inspections are done at the point of manufacture by the person(s) performing the manufacturing task. Real-time data collection and instant process feedback is the best control in any manufacturing environment. Obviously, this is not realistic in some instances (heat treating for example), so the responsibility in those situations must be assigned appropriately (to Inspection, for example). The main goal of collecting data should be to control the processes. It is useless to generate data that means nothing after the fact. Every effort should be made to utilize prevention activities, not detection activities, when developing control plans.

Continuous Improvement and Examples

The tools used in quality control and inspection can also be used to continually improve the processes mentioned above. The use of statistical analysis can be used to make small incremental improvements to processes. Design of experiments (DOEs) also utilize statistics to review data and make significant changes to improve existing processes.

Technology has dramatically improved information gathering, and consequently better, faster feedback can be made to the process to improve control. Improvements to equipment, machine tools, programmable logic controllers (PLCs) and computer numerical controllers (CNCs) have enhanced capabilities. Electronic data collection and analysis allows for communication between the machine and the data collector, thereby providing automatic compensation and reduced variation in the process. In-process gaging techniques can be used to measure the part as it is being processed (machined, ground, etc.). Below are examples where quality control and inspection techniques have been used to improve existing (or new) processes.

Case Study No. 1: Tool Life Study on CNC Lathe. A tool life study was conducted on a new CNC lathe. The lathe was purchased with an automatic tool compensation option. To implement automatic compensation, a study was done using statistical methods. A design of experiment was conducted to select the optimal tool insert and feed rate to be used (Fig. 7).

Once the best tool insert was selected, parts were 100% inspected and recorded using real-time data collection. Adjustments were made

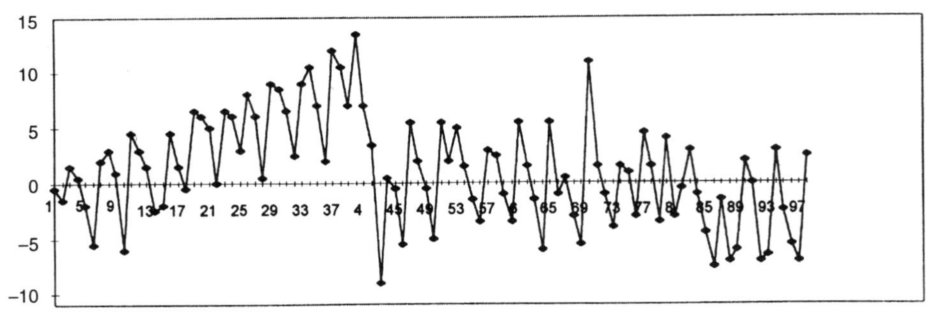

Fig. 9 Variation in runoff from startup test of a new CNC center

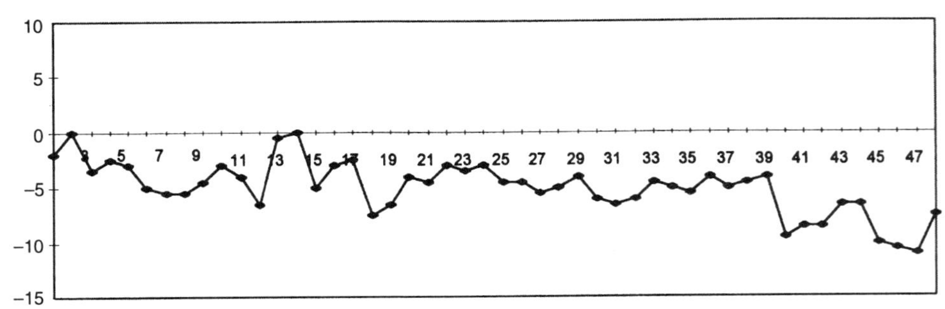

Fig. 10 Runoff variation after thermal adjustments and adjustments with tooling and machine controls

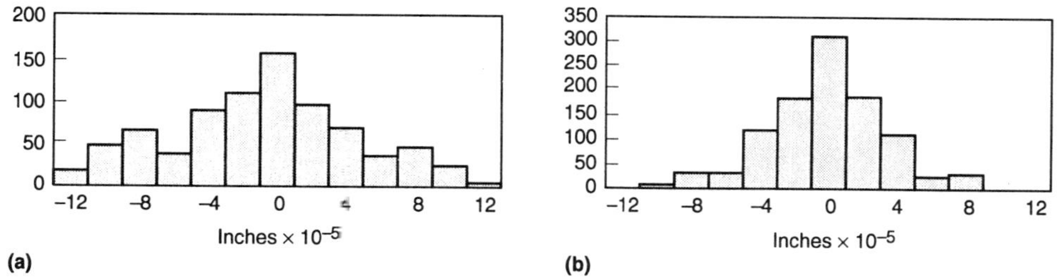

Fig. 11 Histogram of tolerance with OD grinder. a) Before process improvement. b) After process improvement.

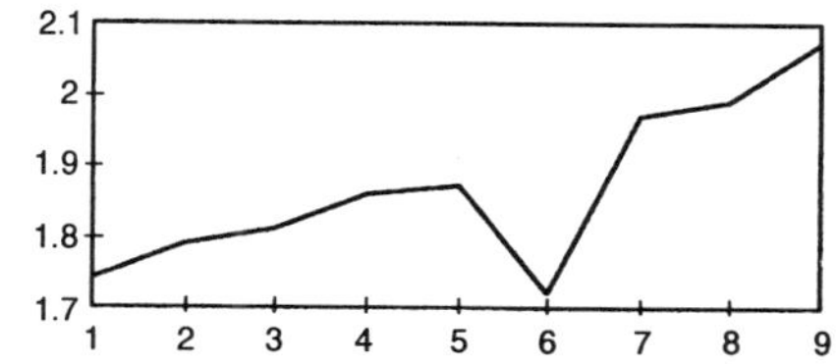

Fig. 12 Process quality index (C_{pk}) of improved grinding operation

only when the measured dimension reached the high limit of conformance. The manual adjustment targeted the low conformance limit of conformance. The machine continued to produce parts until the measured dimension was again at high limit where another manual adjustment was made. This mode of operation continued over several "manual" adjustments in order to determine the tool life of the insert and the size degradation of each cutting cycle. The study showed that tool life had a predictable trend (Fig. 8). With this information, the amount of compensation and the insert life span were entered into the lathe program. The lathe now runs with no operator input and less time-ordered variation than previously observed. Variation related to tool wear and operator "hunting" was virtually eliminated. Ongoing verification of the effectiveness of automatic compensation continues with real-time control charts.

Case Study No. 2: Capability Study—New Equipment Purchase. A runoff of a new CNC machining center was performed and all data recorded. The runoff criteria was met, but a pronounced pattern existed in the collected data (Fig. 9). The pattern consisted of four repeating readings that drifted over time. Automatic compensation was originally set at four pieces. It was determined that the amount of compensation was set too high, and an adjustment was made. The machine tool vendor was contacted about the repeating pattern, and upon investigation adjustments were made to the tooling and machine controls (thermal adjustments, etc.). The second run of parts revealed the results shown in Fig. 10. Again, a slight drift was observed over time. After one more adjustment to the automatic compensation, the process became capable and very stable.

Case Study No. 3: Capability Study—Existing Outside Diameter (OD) Grinder. An existing process on an OD grinder was set to run with a 0.0005 in. tolerance band on the OD. The process was stable, but marginally capable (Fig. 11a). The product was measured and charted on every machine cycle. The process drift was analyzed to determine wheel wear and offsets (and timing) were set. Using real-time data collection, a central tendency within 0.000005 in. was achieved (Fig. 11b). Ongoing process improvements with this process have led to dramatic quality improvements over time (Fig. 12).

Summary

The manufacture of P/M components represents focused technologies and creative innovations. When product tolerances and/or product features cannot be attained by conventional P/M processing (press-sinter), secondary operations are required. These operations may include repressing, heat treating, machining, grinding, and others. Quality requirements mandate control of operations throughout the manufacturing process. Although processes vary in technology, the quality control and inspection methodology is consistent.

A good quality planning process includes:

- Preliminary discussions between the customer and supplier (understanding of the product and the form, fit, and function of the component)
- Thorough understanding of the quality requirements (i.e., classification of characteristics, quality index requirements, etc.)
- The use of engineering tools such as FMEAs to plan manufacturing processes
- A control plan that includes gaging methods, control methods, inspection frequency, and responsibility

Consistent use of these tools and the methodology minimize the risks and failure modes in any manufacturing process. Process control methods are also used for continuous improvement.

SELECTED REFERENCES

- *Advanced Product Quality Planning (APQP) and Control Plan* reference manual, AIAG, 1995
- A.V. Feigenbaum, *Total Quality Control,* 3rd ed., McGraw-Hill, 1983
- E.R. Ott, *Process Quality Control,* McGraw-Hill, 1975

Testing and Evaluation of Powder Metallurgy Parts

FABRICATED P/M PARTS are evaluated and tested at several stages during the manufacturing for part acceptance and process control. This can include various types of tests, such as:

- Evaluation of dimensional changes
- Density measurements
- Hardness testing
- Mechanical testing
- Crack detection
- Metallography

This article briefly reviews the first five items, while metallography is discussed in the article "Metallography of Powder Metallurgy Materials" in this Volume.

Dimensional Evaluation

Dimensional accuracy of P/M sintered parts is determined with the same measurement techniques that are used for wrought materials. Other testing methods for P/M materials are specialized, however, such as determination of surface finish. For sintered parts, a chisel-pointed stylus is used to de-emphasize the effects of porosity. A conical stylus tends to engage porosity, thus giving an exaggerated measurement of roughness. Pores do not interfere mechanically with mating parts (Ref 1).

During the manufacture of sintered parts, dimensional change must be accommodated for during each processing step. Causes of these changes include:

- Elastic springback during ejection from tooling used for cold pressing
- Growth or shrinkage during delubrication, presintering, and sintering
- Elastic springback from tooling during cold repressing or sizing
- Thermal contraction from the tools used in hot forging or hot repressing
- Tool wear in cold or hot compacting
- Machining tolerances at secondary machining and associated tool wear
- Distortion during annealing
- Growth or shrinkage during carburizing, nitriding, or neutral hardening
- Shrinkage during tempering
- Growth during steam blackening

Parts manufacturers must be familiar with the amount of dimensional change to expect for the materials and equipment in use so that tooling can be produced that accommodates these changes and produces accurate parts. Understanding and controlling these factors is essential to commercial P/M parts manufacturing, as discussed in the following paragraphs.

Springback at Molding. Metal powders have varying yield points, and green compacts have varying elastic moduli. Thus, even if a tool set is perfectly rigid, the size of the ejected compact is larger than the tool cavity. This amount also varies, depending on molding pressure and powder characteristics. Furthermore, expansion is not uniform or isotropic, except for right circular cylinders. Dies exhibit some compliance, and split dies distort slightly in use, resulting in further changes in green dimensions. Thus, green dimensions depend on die design and construction.

Sintering Dimensional Change. Compacted elemental powders generally shrink during sintering. The compacts begin as unsintered objects that are larger than die size and eventually shrink below die size. This phenomenon is illustrated in Fig. 1 for pure iron (F-0000), based on the 89.61 mm (3.528 in.) dimension of a Metal Powder Industries Federation (MPIF) standard 10 tensile bar. Prealloyed powders generally shrink during sintering (see the curve for AISI 4680 steel in Fig. 2). Graphite additions to iron do not inhibit shrinkage from green dimensions, but they do cause additional elastic springback from the mold dimension. Such materials thus begin sintering ~0.5% larger than die size (see Fig. 2).

Elements such as carbonyl nickel accelerate the shrinkage of iron mixes (see FN-0208 in Fig. 2). In iron mixes, copper tends to cause growth. The higher the density of an iron particle (low surface area), the greater its dimensional growth with copper additions. Carbon dissolved in iron prior to copper diffusion inhibits growth (see Fig. 3 and 4). Growth of copper-tin premixes is shown in Fig. 5.

Growth of these bronze premixes is strongly density dependent, with densities near 7.0 g/cm^3 providing greater growth and densities near 6.0

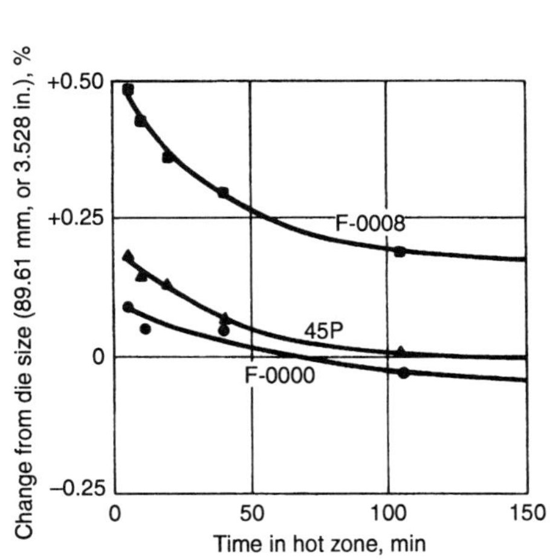

Fig. 1 Dimensional change on sintering of F-0000, F-0008, and 45P irons. Sintered at 1120 °C (2050 °F) in dissociated ammonia. Green density: 6.8 g/cm^3. 45P: atomized iron plus 0.45% phosphorus added as ferrophosphorus master alloy. F-0000: pure atomized iron. F-0008: atomized iron plus 0.9% graphite

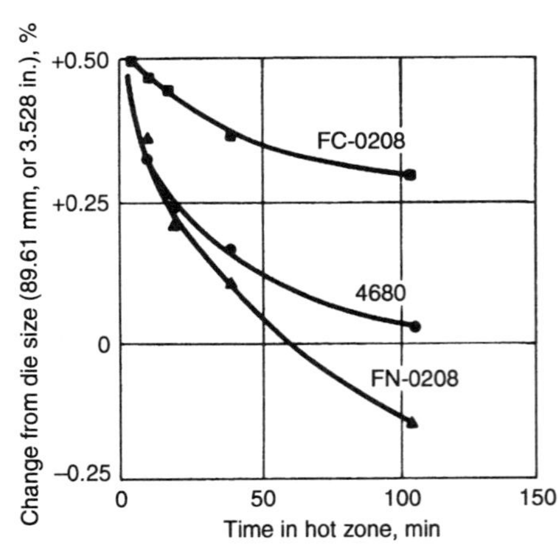

Fig. 2 Dimensional change on sintering of FC-0208, 4680, and FN-0308 irons. Sintered at 1120 °C (2050 °F) in dissociated ammonia. Green density: 6.8 g/cm^3. FC-0208: sponge iron plus 2% Cu plus 0.97% graphite. 4680: prealloyed AISI type 4600 steel plus 0.9% graphite. FN-0208: atomized iron plus 2% Ni plus 0.9% graphite

g/cm³ providing significantly less. Dimensional change data during sintering are complicated by the fact that dimensional change parallel to the pressing direction is not the same as dimensional change occurring perpendicular to the pressing direction.

Annealing Dimensional Change. Annealing dimensional change is similar to change occurring during extra sintering or stress relaxation, as compacts continue to shrink. Lower density compacts experience the most shrinkage. With pure copper that has been pressed, sintered, and repressed to high density, annealing in a hydrogen atmosphere that contains gases can cause growth or blistering. The hydrogen diffuses into the pores that are isolated from the surface. If hydrogen encounters residual oxygen, water vapor is formed, which causes expansion and growth. This phenomenon also can be caused by residual sulfur or the presence of lubricant. Consequently, the initial sintering of pure copper should remove as many impurities as possible while the pores are still open.

Dimensional Change during Heat Treating. Carbon and nitrogen absorbed during carburizing and nitriding of steel cause growth in areas in which they dissolve. During quenching, regions that form martensite experience growth. Thus, if a small part is through-hardened to all martensite, all regions (inside and outside diameter) experience outward expansion. Many sintered steels have modest hardenability, and only the outer 3.2 mm (0.125 in.) adjacent to surfaces forms martensite and expands.

On quenching a medium- or large-sized part (with a section >9.5 mm, or 0.4 in.), the interior remains as ferrite and fine pearlite, experiencing neither shrinkage nor growth. The outer surfaces expand outward, and the inner surfaces shrink inward. This phenomenon also is evident in case-hardened wrought parts. Thus, prediction of exact size change during heat treatment is difficult.

Steam Blackening. Sintered parts frequently are treated in steam at 540 to 595 °C (1000 to 1100 °F) for 1 to 4 h to fill the pores and coat the surface with a hard coating of black iron oxide. The coating causes a uniform growth of 0.0025 to 0.0050 mm (0.0001 to 0.0002 in.) similar to electroplating. The amount of blackening should be controlled, as measured by hardness and destructive break tests; excessive oxide coating thickness may lower impact properties. Thickness of the oxide layers also can be measured metallographically with a polishing procedure described in the section of this article on density measurement.

Evaluation of Dimensional Change in Incoming Powder. New lots of blended or raw powder are checked against internal standard lots to ensure consistent sintered dimensional change. Transverse-rupture bars 31.8 by 12.7 by 6.4 mm (1.25 by 0.50 by 0.25 in.) are molded at a fixed density or pressure from both the standard and test lot of powder. The two sets of bars are sintered simultaneously in a laboratory or production furnace. Dimensional change in the 31.8 mm (1.25 in.) length are checked against the requirements of American Society for Testing and Materials (ASTM) standard B 610.

Although dimensional change from sintering a bar made from the standard powder can differ from previous tests, comparable dimensional changes in the test bar made from incoming powder demonstrate the difference in the performance of the powders. Dimensional change in test and standard lots must agree to within a specified range (±0.1% of the bar length). These bars also can be used to evaluate sintered strength and hardness.

Dimensional Control. Table 1 illustrates typical dimensional tolerances of P/M materials. Separate tolerances apply to as-sintered, as-sized, and as-heat treated conditions. For concentricity between an inside diameter and an outside diameter, a total indicator reading of ≤0.075 mm (0.003 in.) is permitted. The distance between holes can be as great as 0.075 mm + 0.013 mm/mm (0.003 in. + 0.0005 in./in.). Gears can be molded to American Gear Manufacturers Association (AGMA) class 7, which is limited primarily by the concentricity of the bore to pitch line. If gears are held on the pitch line and bored more concentrically, AGMA class 10 or 11 is achieved.

Other processes, such as P/M hot forging, injection molding, and high-temperature sintering, produce wider tolerances than presented in Table 1. Powder metallurgy forged dimensional tolerances are given in Table 2. High-temperature sintering tolerances are given in Table 3. Injection-molded tolerances range from 0.075 to 0.10 mm/mm (0.003 to 0.004 in./in.), even though parts have experienced 12 to 15% linear shrinkage (Ref 5).

Table 1 Typical P/M tolerances (other than length)

	Condition					
	As-sintered		As-sized		As-heat treated	
Material	mm	in.	mm	in.	mm	in.
Brass	±0.089	±0.0035	±0.013	±0.0005	...	...
Bronze	±0.089	±0.0035	±0.013	±0.0005	...	...
Aluminum	±0.051	±0.002	±0.013	±0.0005	±0.013	±0.0005
Iron	±0.025	±0.001	±0.013	±0.0005	...	...
Copper alloy steel	±0.038	±0.0015	±0.025	±0.001	±0.038	±0.0015
Nickel alloy steel	±0.038	±0.0015	±0.025	±0.001	±0.038	±0.0015
Stainless steel	±0.025	±0.001	±0.013	±0.0005	...	...

Note: Up to 12.7 mm (0.500 in.). Length tolerance, ±0.102 mm (±0.004 in.), unless machined or ground. Source: Ref 1

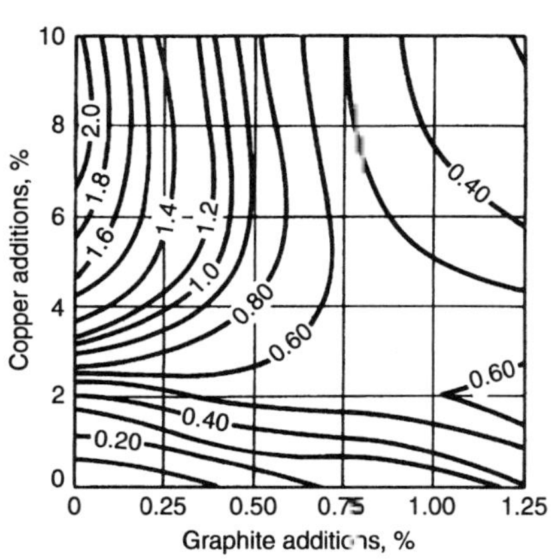

Fig. 3 Dimensional change on sintering of MH100 iron. Values inside the graph indicate the percentage change from die size. MH100 iron sintered 30 min at 1120 °C (2050 °F) in dissociated ammonia. Green density: 6.4 g/cm³. Source: Ref 2

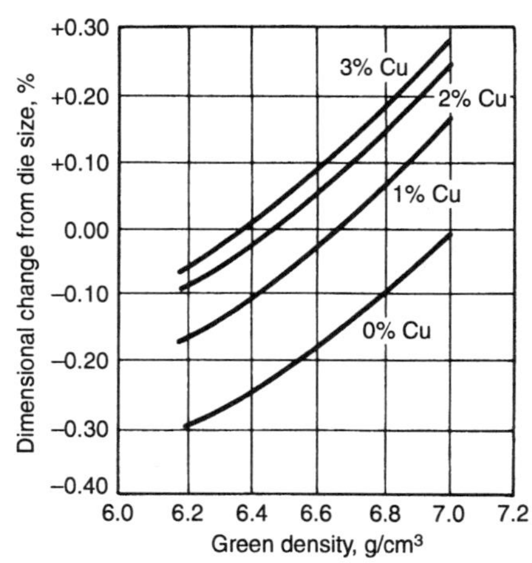

Fig. 4 Dimensional change on sintering of ATOMET 28 iron. ATOMET 28 powder plus 0.9% graphite plus 0.75% zinc stearate plus copper as shown. Green density ranges from 6.2 to 7.0 g/cm³. Sintered 30 min in dissociated ammonia at 1120 °C (2050 °F). Source: Ref 3

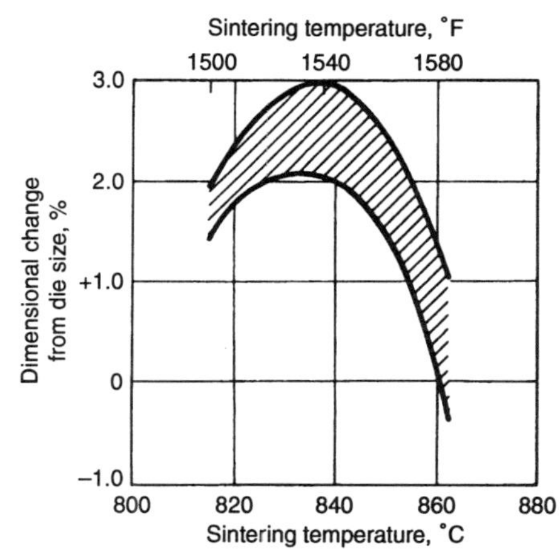

Fig. 5 Dimensional change on sintering of 90-10 bronze. PMB 18 premixed bronze sintered 15 min in hydrogen at various temperatures. Green density: 6.3 g/cm³. Source: Ref 4

Table 2 Tolerances on P/M forged parts

Parameter	Nominal dimension mm	Nominal dimension in.	Tolerance mm	Tolerance in.
Outside diameter	50.8	2.00	0.13	0.005
Outside diameter	50.8	2.00	0.25	0.010
Inside diameter	38.1	1.50	0.20	0.008
Thickness	25.4	1.00	0.38	0.015
Spline	25.4	1.00	0.23	0.009
Outside diameter	95.25	3.75	0.25	0.010
Inside diameter	63.5	2.50	0.25	0.010
Concentricity	95.25	3.75	0.10	0.004
Roundness	95.25	3.75	0.10	0.004
Thickness	15.8	0.625	0.25	0.010
Outside diameter	50.8–76.2	2.00–3.00	0.13	0.005
Outside diameter	25.4–50.8	1.00–2.00	0.10	0.004
Outside diameter	76.2	3.00	0.38	0.015
Outside diameter	50.8	2.00	0.13	0.005
Outside diameter	203	8.00	0.51	0.020
Thickness	25.4	1.00	0.25–0.634	0.010–0.025

Source: Ref 5

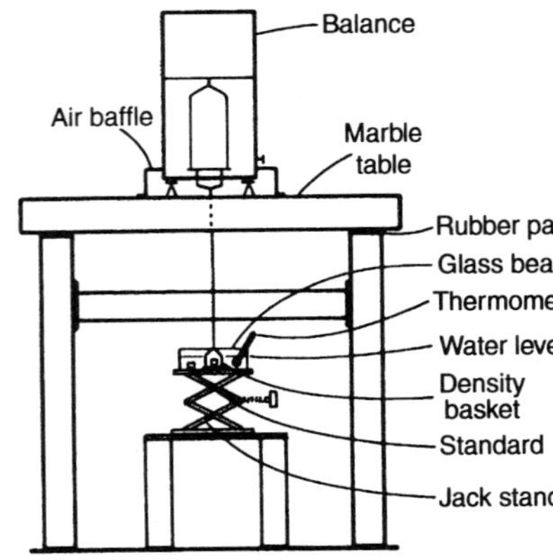

Fig. 6 Density measurement apparatus

Table 3 Dimensional tolerances of parts in the as-high-temperature sintered condition

Material	Nominal dimension mm	Nominal dimension in.	Tolerance mm	Tolerance in.
Composite	25.4	1.00	0.05	0.002(a)
3Si-Fe	19.0	0.75	±0.08	±0.003
4600	76.2	3.00	0.38	0.015
M-2	70.3	2.77	0.61	0.024
Low-alloy steel	22.2	0.88	0.08	0.003
Stellite	25.4	1.00	0.03	±0.001(b)

(a) Roundness. (b) Inside diameter sintered against a mandrel. Source: Ref 5

Measurement of Density

Density is the ratio of mass to volume. For a given material, degree of sintering, and heat treatment, density determines mechanical and physical properties. For example, higher density in sintered steels results in higher tensile strength, elongation, and impact resistance values. As-pressed, or green, density also influences growth or shrinkage that occurs during sintering. With nonuniform green density, parts grow or shrink nonuniformly, as in a thin-walled bronze bearing with a low-density region equidistant from the ends. This results in a significantly smaller diameter at midlength than at the ends and necessitates repressing or sizing for close dimensional control.

If cubes or right cylinders could be extracted from actual parts, linear dimensions could be measured and volume could be calculated easily. From the weight of a part, density can be easily calculated. This yields a value that, under ideal conditions, differs by 0.04 g/cm^3 (0.5%) from a reference (Ref 6). Unless the sintered part is directly molded to an easily measured shape, such as a transverse-rupture bar (31.8 by 12.7 by 6.4 mm, or 1.25 by 0.50 by 0.25 in.), this method of measuring linear dimensions is used infrequently.

Methods Based on Archimedes' Principle. Typical methods of measuring density depend on Archimedes' principle, in which hydrostatic forces in liquids exert buoyant forces proportional to the part volume. This measurement is standardized in ASTM B 328 (Ref 7), MPIF test method 42 (Ref 8), and International Standards Organization test method ISO 2738 (Ref 9). When an object is immersed in a liquid, the liquid exerts an upward buoyant force that is equal to the product of the object volume and the density of the liquid. The difference in weight between an object weighed in air and its weight when suspended in water is equal to the object volume in cubic centimeters times the density of water. Approximating the density of water as unity:

$$V = W_{air} - W_{water}$$

where V is the volume, cm^3; W_{air} is the weight in air, g; and W_{water} is the weight of object suspended in water less the weight of the suspending wire in water (tare), g. Density in g/cm^3 is then:

$$\text{Density} = W_{air} / (W_{air} - W_{water})$$

For unsintered materials molded with 0.75% lubricant, pores are well sealed, and water cannot penetrate. For such parts, the above calculation is suitable. It is also suitable for materials with pores that are sealed off from the surface (materials close to theoretical density). For most sintered materials that are 70 to 95% dense, water tends to infiltrate the pores during weighing in water. This minimizes the buoyancy effect of the water (that is, the liquid is acting on a smaller volume) and results in an erroneous calculation of low volume.

This low volume then causes an erroneously high density value. Infiltration of water into pores usually is accompanied by air bubbles escaping from the part. If the part is blotted to remove surface water and reweighed in air after weighing in water, any weight gain indicates that water has entered the pores. Although not a standard procedure, volume can be approximated as the weight in air after removing the part from the water, minus the weight in water.

To prevent infiltration of water, all three standard test methods require that the pores of the part be filled with oil. Oil impregnation is done after the part is weighed in air; this is carried out under vacuum or by immersion in hot oil. Oil prevents the water from entering the pores. The volume of the part is then determined as the part weight in air with oil in the pores, minus the weight of the oiled part suspended in water. Care should be taken to select an oil that is not soluble in water or not soluble in water plus wetting agent. Such oils also must exhibit superior demulsibility.

The precision of the ISO method is ±0.25%, regardless of sample density, and assumes a water density of 0.997 g/cm^3. Moyer (Ref 6) has reviewed the literature on precision methods of density determination (Ref 10–16) and has devised a method that provides accuracy to two or three decimal places, depending on sample porosity. The basic measuring apparatus is shown in Fig. 6. Requirements of precision density measurement include:

- Balance capable of measuring to the nearest 0.0001 g
- Vibration- and draft-free atmosphere
- Measurement of the density of the immersing liquid (water) by checking the density of a substance of accurately known density (four decimal places)
- Conversion of all densities back to 20 °C (68 °F) by compensating for thermal expansion of the sintered part
- Maintenance of liquid level at a constant height on the suspending wire
- Careful brushing of all bubbles from the test object

Using the above procedures, Moyer reports standard deviations of 0.0130 to 0.0005 g/cm^3 on 17 g parts with densities ranging from 5.12 to 7.85 g/cm^3, respectively.

To determine density variation from one point to another in a complex part, the available samples must be considerably <17 g. According to ASTM B 328, a minimum sample of 2 g is rec-

Table 4 Effect of sample size and weighing errors on density measurement

Balance accuracy (weighing error), g	Density range, g/cm³, for a sample weight (at 6.50 g/cm³) of: 0.1 g	1.0 g	10.0 g
0.01	0	±1.0459	±0.0922
0.001	±1.0459	±0.0922	±0.0091
0.0001	±0.0922	±0.0091	±0.0009

ommended, because a relatively high error rate results from measuring small samples. Table 4 shows density errors that can occur because of weighing errors, if all weighing errors are assumed to occur so as to maximize density errors. A 1 g sample on a balance accurate to 0.01 g (per ASTM B 328) could result in a density range from ±1.05 g/cm^3 to a mean of ~6.5 g/cm^3.

Metallographic estimates can be made of the area fraction of porosity, which is numerically equal to the volume porosity, and thus the density of sintered materials. The method is not standardized, and accuracy of results depends on the skill of the metallographer to define the correct area fraction of porosity. Frequently, the amount of porosity is exaggerated or minimized because of overpolishing or underpolishing (Ref 17). A suitable method for sintered iron-base materials is to:

- Remove oil and cutting fluids from the pores by Soxhlet extraction or heating in air or atmosphere to 315 °C (600 °F).
- Impregnate the pores with epoxy resin; generally does not fill all the pores, and polishing is required.
- Mount the sample in epoxy, bakelite, or other suitable medium.
- Wet grind through 600 mesh silicon carbide-coated paper.
- Open smeared pores by immersion etching in 2% nital for 1 min.
- For samples up to 12.7 by 12.7 mm (½ by ½ in.), polish with 1 μm alumina on Struers felt cloth or an equivalent long-napped cloth for 3 min on a 250 rpm, 20 cm (8 in.) diam wheel, using moderate to heavy hand pressure. This opens pores and exaggerates their presence.
- Restore pores to true area fraction by polishing for 3 min by hand, using a 1 μm diamond on Struers MOL cloth or equivalent, and moderate to heavy pressure. Use a 20 mm (0.8 in.) strip of diamond paste per specimen.
- Final polish for 30 to 60 s, using 0.05 μm alumina with Buehler Microcloth or equivalent long-napped cloth on a 20 cm (8 in.) diam polishing wheel at 125 rpm with light pressure; an automatic polisher is suitable.
- Wash, dry, and examine unetched at 200×.

A simple manual method to estimate the area fraction of porosity uses a 100 by 120 mm (4 by 5 in.) photograph of the sample taken at a magnification of about 200×. A grid of 13 by 17 thin black lines 6.4 mm (0.25 in.) apart should be constructed on clear vinyl to fit on the photograph. The grid contains 221 intersections. By placing the grid over the photograph, then counting the number of intersections that lie completely over the pores and adding one half of the number of intersections that are on the edge of the pores to obtain the effective number of intersections, the area fraction of porosity is calculated:

$$A_p = (221 - N_p) / 221$$

where A_p is the area fraction of porosity; and N_p is the effective number of grid intersections overlaying pores. For common iron-base materials, theoretical density is 7.87 g/cm^3; density (ρ) can be estimated as:

$$\rho = (1 - A_p) \times 7.87 \text{ g/cm}^3$$

This method can achieve a precision of about ±0.1 g/cm^3, depending on the laboratory. The advent of metallography with a television monitor and quantitative metallographic functions allows rapid measurement of area fractions of pores. This method is highly dependent on proper sample preparation.

Apparent Hardness and Microhardness

Porous materials exhibit wider variation in hardness testing than wrought counterparts. The entrance of the indenter into pores or groups of pores generally causes this effect. At least five consistent readings should be taken, in addition to any obviously high or low readings, which are discarded. The remaining five readings should be averaged.

Because most published data show typical hardness values, the buyer and seller must agree on specified or minimum values. The seller and user of P/M materials also should agree on which area or areas of a part are to be hardness tested. The average of five or more consistent readings must meet the standard hardness, not any single reading. Recommended scales for taking accurate measurements are summarized in Table 5.

Unlike fully dense materials, high hardness does not mean high tensile strength. As discussed previously, 15 to 30% fine pearlite in the iron-carbon sintered materials improves tensile strength: consequently, it is evident that high hardness values are not an indicator of high strength. Because P/M steels exhibit low hardenability, hardness specifications should be based on parts that have demonstrated successful service.

Although not yet widely used, the Rockwell B scale may be combined with a carbide ball for testing and controlling hardened parts. Scattering of data is minimized with the Rockwell B 1.58 mm ($\frac{1}{16}$ in.) diam ball, and it is useful up to 120 HRB (Ref 18, 19).

Microhardness of porous materials can best be measured with Knoop or diamond pyramid hardness indenters at loads of 100 g or greater. In atomized irons, particles exhibit minimal porosity; consequently, the Knoop indenter is suitable. It makes a very shallow indentation and is only infrequently disturbed by entering undisclosed pores. Care should be exercised in preparing the sample surface (see the guidelines given in the article "Metallography of Powder Metallurgy Materials" in this Volume. Use of the diamond pyramid indenter is particularly well suited to irons, which contain numerous fine internal pores. Because of its greater depth of penetration, the diamond pyramid indenter frequently encounters hidden pores. Microhardness testing and measurement of case depth are covered by Metal Powder Industries Federation standard MPIF 37.

Table 5 Common hardness scales

Material	Sintered hardness scale	Heat treated hardness scale
Iron	HRH, HRB	HRB, HRC
Iron-carbon	HRB	HRB, HRC
Iron-nickel-carbon	HRB	HRC
Prealloyed steel	HRB	HRC
Bronze	HRH	...
Brass	HRH	...

Fig. 7 Hardened and tempered Fe-0.8C alloy. Light areas are martensite (60 HRC); dark areas are fine pearlite (40 HRC). 4% picral etch

For examining tempered steels, a light etching in 2% nital for 6 s is recommended. Fine pearlite appears as dark areas, unresolved at 1000×. Martensite appears light colored, with little or no evidence of platelets or needles (Fig. 7). This degree of etching enhances clarity and contrast for measuring the diamond indentation. Subsequent darker etching clearly shows the martensite needles and any retained austenite or carbides. Picral (4%) is recommended for subsequent etching (15 s) by immersion.

For microhardness testing of sintered steels, care must be exercised in the definition of a case and the choice of microconstituent to be tested. For example, in a carburized and quenched nickel steel with 0.3% C in the core, martensite, fine pearlite, ferrite, and nickel-rich areas are visible. Nickel-rich areas should be ignored. The surface is nearly all martensite with some fine pearlite. If 50 HRC is defined as the case depth

Table 6 Comparison of the applicability of various nondestructive evaluation methods to flaw detection in P/M parts

Method	Measured/detected	Applicability to P/M parts(a)		Advantages	Disadvantages
		Green	Sintered		
X-ray radiography	Density variations, cracks, inclusions	C	C	Can be automated	Relatively high initial cost; radiation hazard
Computed tomography	Density variations, cracks, inclusions	C	C	Can be automated; pinpoint defect location	Extremely high initial cost; highly trained operator required; radiation hazard
Gamma-ray density determination	Density variations	A	A	High resolution and accuracy; relatively fast	High initial cost; radiation hazard
Ultrasonic imaging: C-scan	Density variations, cracks	D	B	Sensitive to cracks; fast	Coupling agent required
Ultrasonic imaging; SLAM	Density variations, cracks	D	C	Fast; high resolution	High initial cost; coupling agent required
Resonance testing	Overall density, cracks	D	B	Low cost; fast	Does not give information on defect location
Acoustic emission	Cracking during pressing and ejection	C	D	Low cost	Exploratory
Thermal wave imaging	Subsurface cracks, density variations	D	C	No coupling agent required	Flat or convex surfaces only
Electrical resistivity	Subsurface cracks, density variations, degree of sinter	A	A	Low cost, portable, high potential for use on green compacts	Sensitive to edge effects
Eddy current/magnetic bridge	Cracks, overall density, hardness, chemistry	C	A	Low cost, fast, can be automated; used on P/M valve seat inserts	Under development
Magnetic particle inspection	Surface and near-surface cracks	C	A	Simple to operate, low cost	Slow: operator sensitive
Liquid dye penetrant inspection	Surface cracks	C	D	Low cost	Very slow; cracks must intersect surface
Pore pressure rupture/gas permeability	Laminations, ejections, cracks, sintered density variations	A	A	Low cost, simple, fast	Gas-tight fixture required; cracks in green parts must intersect surface

(a) A, has been used in the production of commercial P/M parts; B, under development for use in P/M; C, could be developed for use in P/M, but no published trials yet; D, low probability of successful application in P/M

cut-off, the existence of a fine pearlite colony, which is below 50 HRC, would seem to indicate that the reading was taken beneath the case. Case may be defined as the point where there is <50% martensite by area, or where the hardness falls below the 50 HRC equivalent. By comparison, in testing a wrought material without porosity, a heavier load that makes a large indentation should be used, which averages the microstructures while pinpointing the 50 HRC equivalent. In a P/M steel, this would result in the pores being averaged in with the other phase.

The use of light loads (<100 g) causes errors when converting to the Rockwell C scale if the 500 g Rockwell conversion table is used (Ref 20). An alternative is to provide individualized calibrations by removing pieces from several Rockwell standard blocks, with hardnesses ranging from 20 to 60 HRC, and metallographically mounting the working faces. These standards are carefully polished, allowing a direct calibration to be made between the Rockwell C scale and the length of the indentation. When this calibration is graphed, an unknown sample can be taken, the length (in filar units) of the hardness indentation made at 100 g can be measured, and the filar units can be converted to Rockwell C values.

Mechanical Testing/Tensile Testing

MPIF standard 10 describes specimens for tensile testing. Specimens include flat unmachined test bars or machined rounds. Unmachined flats are more prone to slippage of grips or breakage in the gauge region.

For testing to be meaningful, it is important to verify that such bars are free of microlaminations (Ref 21), which requires careful metallographic evaluation. Highest quality bars are molded in well-bolstered die sets (890 kN, or 100 ton rating), with generous exit taper, high green strength powder, and top punch hold-down, if possible. The formation of microlaminations is accentuated by high molding pressures of 415 to 1380 MPa (30 to 100 tsi).

With all test bars, it is necessary to determine that substantial carburization or decarburization resulting from sintering or hardening does not exist. If bars are heat treated, the microstructure at the surface and in the interior should be described in the test report, because many P/M steels have low hardenability.

For heat treated materials, unmachined flat specimens tend to slip in the grips. A machined bar provides more accurate data. The machined bar shows an increase in apparent tensile strength of ≤50% compared to molded bars. Even with machined bars, some heat treated materials exhibit such low elongation that failure occurs prior to reaching 0.2% permanent deformation.

Transverse Rupture Strength. The transverse-rupture test breaks a 31.8 by 12.7 by 6.4 mm (1.25 by 0.5 by 0.25 in.) test bar as a simple beam. The test is theoretically valid only for perfectly brittle materials and measures the stress in the outer fiber at fracture. For many sintered steels, transverse-rupture stress is considered to be equal to twice the ultimate tensile strength. This is only an approximation. MPIF standard 35 presents transverse-rupture stress values that correspond with ultimate tensile stress values for common P/M materials.

The transverse-rupture test is useful for comparing and evaluating materials for strength, even if the bar bends before fracture. These bars are preferred, because they can be molded and sintered conveniently. Testing is faster than when using a tensile bar. When heat treated, the test bar does not experience distortion. This testing procedure is used mainly as a quality control tool to ensure the maintenance of minimum mechanical properties. During P/M part production, this method is used to evaluate properties of incoming powder, such as compressibility, sintered strength, and dimensional change.

Unnotched Charpy Impact Strength. A 10 by 10 by 76 mm (0.35 by 0.35 by 3.00 in.) molded bar is used for impact testing. The unnotched bar provides a more sensitive test, suitable for use on materials with an impact strength below 14 J (10 ft · lb). The bar is conventionally struck on the surface that contacted the die at molding.

Proof Testing. The most common method of demonstrating the strength of sintered parts is through mechanical tests that stress parts to failure. Qualification samples or first production lots are used to establish desired strength values; these data are incorporated in the part design specification.

For testing gears, several teeth are removed. The remaining teeth are loaded in a predetermined arrangement on a fixture. The load to fracture is recorded. To be meaningful, the destructive test must imitate service loading on the part.

Impact or drop weight tests also are used to evaluate materials. A drop weight test does not only use an acceptance or rejection evaluation. This testing procedure investigates the impacts above the acceptance level and below the rejection value. When a part does not break, each succeeding load is increased until breakage occurs. Thus, if load P does not break a part, 1.05 P is used on the succeeding impact.

Crack Detection

Crack detection is accomplished by various methods, such as mechanical proof testing, metallography, and filtered particle or magnetic particle inspection. Promising nondestructive testing methods for P/M applications also include electrical resistivity testing, eddy current,

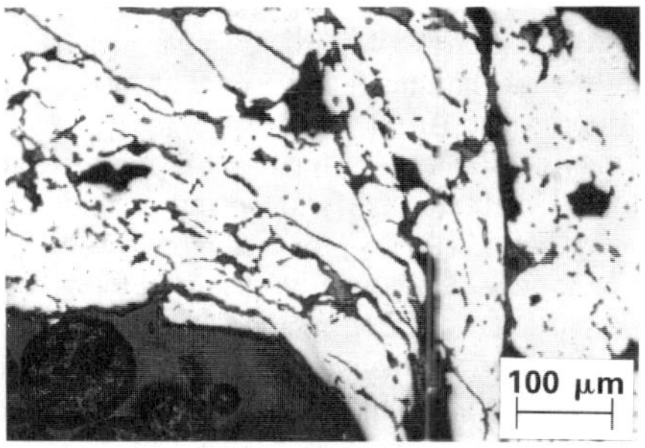

Fig. 8 Cracks and unbonded particles at the junction (radius) of a horizontal flange and vertical hub resulting from shear during compaction. Unetched

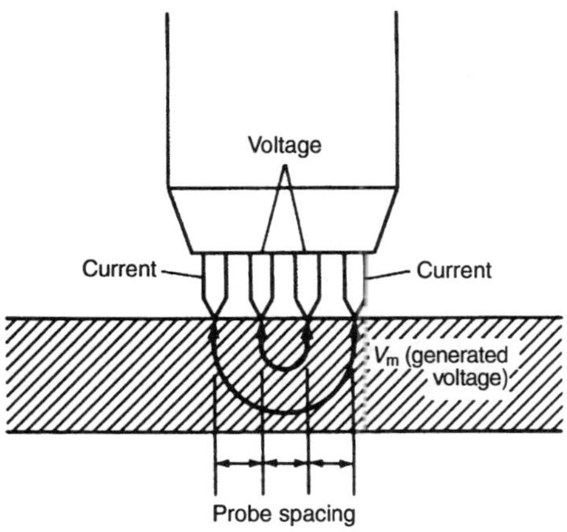

Fig. 9 Four-point probe used in the resistivity test. The outer probe pins are the current leads; the inner pins are the potential leads.

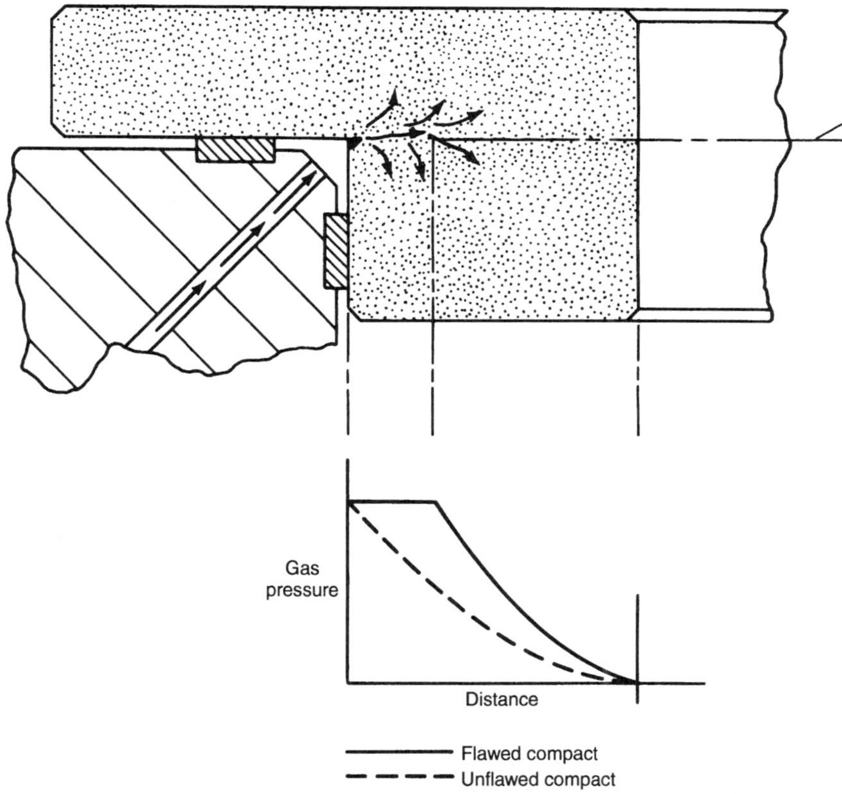

Fig. 10 Pore pressure rupture test for crack detection in green parts. Source: Ref 26

and magnetic bridge testing, magnetic particle inspection, ultrasonic testing, x-ray radiography, gas permeability testing, and γ-ray density determination. The capabilities and limitations of these techniques are summarized in Table 6.

The problem of forming defects in green parts during compaction and ejection has become more prevalent as parts producers have begun to use higher compaction pressures in an effort to achieve high-density, high-performance powder metallurgy (P/M) steels. Proper press setup for molding P/M parts also is critical to prevent cracking. In a flanged part that experiences a change in diameter, density in the hub and flange should be nearly equal. Unequal density leads to powder displacement from one part level to the next and to the formation of shear cracks. Such cracks often occur at 45° to the pressing direction and at surfaces at the junction (radius) between the hub and flange. At press setup, equal density should be obtained in the hub and flange. A high green strength powder and a press that maintains a small counter pressure on the top of the part during ejection from the tools (top punch hold-down) should be used.

Mechanical Proof Testing. A sampling of sintered parts can be broken to confirm the presence of a suspected cracking problem—for example, pushing flanges off hubs in such a manner that in addition to shear effect there is some stretch effect at the crack. The presence of a few unexplained low readings indicates that an initiating crack is present.

Metallography. Low-powered binocular microscopes can be used to detect cracks at changes in diameter. Metallography is a more time-consuming method. A sampling of parts are sectioned parallel to the pressing direction. When mounted and carefully polished to expose open pores and cracks, the presence of minute cracks is readily apparent (Fig. 8).

Liquid-Penetrant Crack Detection. During liquid-penetrant testing, the surface of the part is covered with a colored or fluorescent penetrating fluid that fills all cracks and some pores. Surface liquid is then washed off, leaving the penetrant in the cracks. A layer of developer powder, which acts as a blotter, is applied to the surface. The penetrant leaves the cracks and is absorbed by the developer layer near the crack. Crack indications are visible by color contrast or by fluorescence under ultraviolet light. Most sintered parts have porous surfaces that absorb and then release sufficient penetrant in all regions so that it is impossible to distinguish the crack from the porosity background.

The dye penetrant equipment found in P/M shops is generally used only for checking parts of the tooling and machinery for cracks. The dye does not preferentially reside at cracks in P/M parts, because the pore radius and the crack radius are equivalent. However, there might be an application for green parts because the surfaces of green parts are sealed against penetration by liquids through smearing of the metal powder against the die wall and through the formation of a thin coating of dry powder lubricant on the surface. Cracks intersecting the surface may form an opening in this layer that could be detected by the dye penetrant

Filtered Particle Crack Detection. One proprietary process of filtered particle crack detection (Partek) involves brief immersion of the test piece in a suspension of fluorescent particles. Particles are filtered and collect near the surface of cracks as the fluid enters. Cracks are clearly visible under black light. This one-step method is used to detect cracks in presintered porous tungsten carbide blanks (Ref 22).

To the extent that an unsintered part has open porosity, this method also can be used on green parts. Density cannot be too high, however, and excessive lubricant tends to clog the pores. Successful use of this method on presintered porous tungsten carbide blanks indicates that it may be suitable for sintered P/M parts with open pores into which fluid can enter. Small cracks fluoresce brightly, while large cracks are darker than the surrounding fluorescing surface.

Magnetic Particle Crack Detection. A magnetized part demonstrates abrupt changes in magnetic field when the field crosses a crack. When dry or wet magnetic particles are passed over the part, any magnetic leakage field at the part surface attracts and holds iron particles.

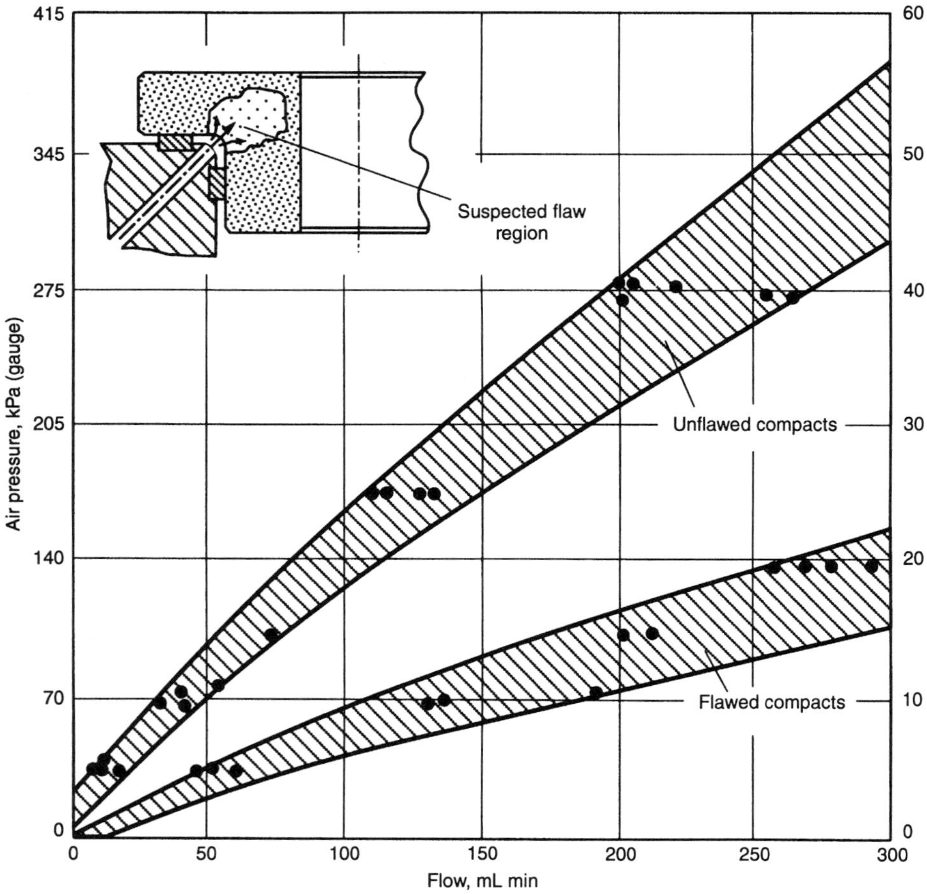

Fig. 11 Detection of flawed compact using the gas permeability technique. Source: Ref 26

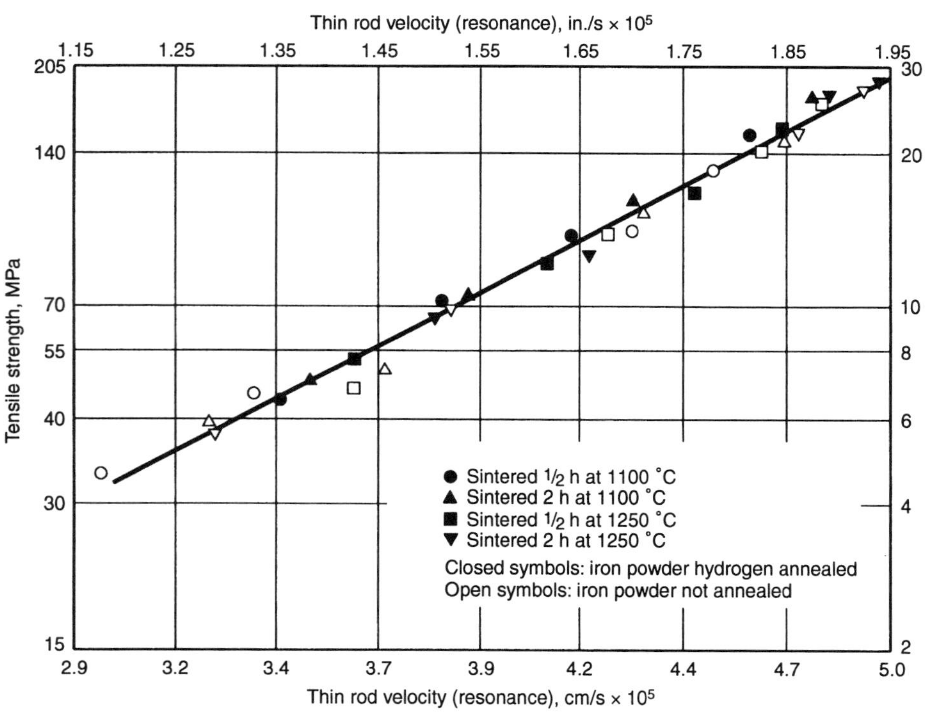

Fig. 12 Correlation of ultrasonic velocity with tensile strength of sintered steel. Source: Ref 29

These assemblies of particles over cracks are visible with the unaided eye or with black light on fluorescent particles.

This method also detects some near-surface cracks. Unsintered parts, however, are not adequately bonded to support a magnetic flux, and the method is consequently unsatisfactory. Magnetic particle detection methods have been successfully used for many years for inspecting medium-density sintered automotive parts, by both P/M parts producers and automotive manufacturers. It is possible to automate the inspection process by using digital image processing.

Direct Current Resistivity Testing. A voltage field within a conductive solid will create currents that are influenced by structural irregularities, including cracks and porosity. The arrangement shown in Fig. 9 is used to measure the voltage drop in a current field localized between two electrode probes. This method has been used to detect seeded defects in laboratory specimens. It has also been successfully applied to the production of sintered steel parts (Ref 23).

The direct current resistivity test can be used on any conductive material; it is not limited to ferromagnetic materials. Although further development is needed, resistivity measurements appear to be a promising technique for the nondestructive evaluation of both green and sintered P/M parts. In addition to detecting cracks in green parts, as well as part-to-part density variation, studies have shown that changes in resistivity due to poor carbon pickup during sintering were also detectable (Ref 23). Resistivity testing has also been used later in the processing sequence to screen heat-treated parts for incomplete transformation to martensite. Another study has yielded the relative density/conductivity relationship, suggesting that resistivity tests could be used as a rapid check for localized density variations (Ref 24). As with ultrasound, the elastic modulus and the toughness of porous steels can also be distinguished by resistivity checks (Ref 25).

Pore Pressure Rupture Testing of Green Compacts. A novel test is available for detecting ejection cracks in green compacts (Ref 26). A pressure seal is formed around a corner or area of a part where experience has shown that cracks are likely to occur. The area is then pressurized to ~3.5 MPa (500 psi) using a fixture such as that shown in Fig. 10. If a crack is present, the gas pressure in the crack will be sufficient to propagate the crack the rest of the way through the part. This would be classed as a proof test rather than a nondestructive test because the part is destroyed if defects are present.

The test can be used in a nondestructive manner on sintered parts. The gas permeability of the pressurized area is measured at reduced pressures, and the presence of cracks or low-density areas is indicated by high permeability, as shown in Fig. 11.

Ultrasonic

Many characteristics of solids can be determined from the behavior of sound waves propa-

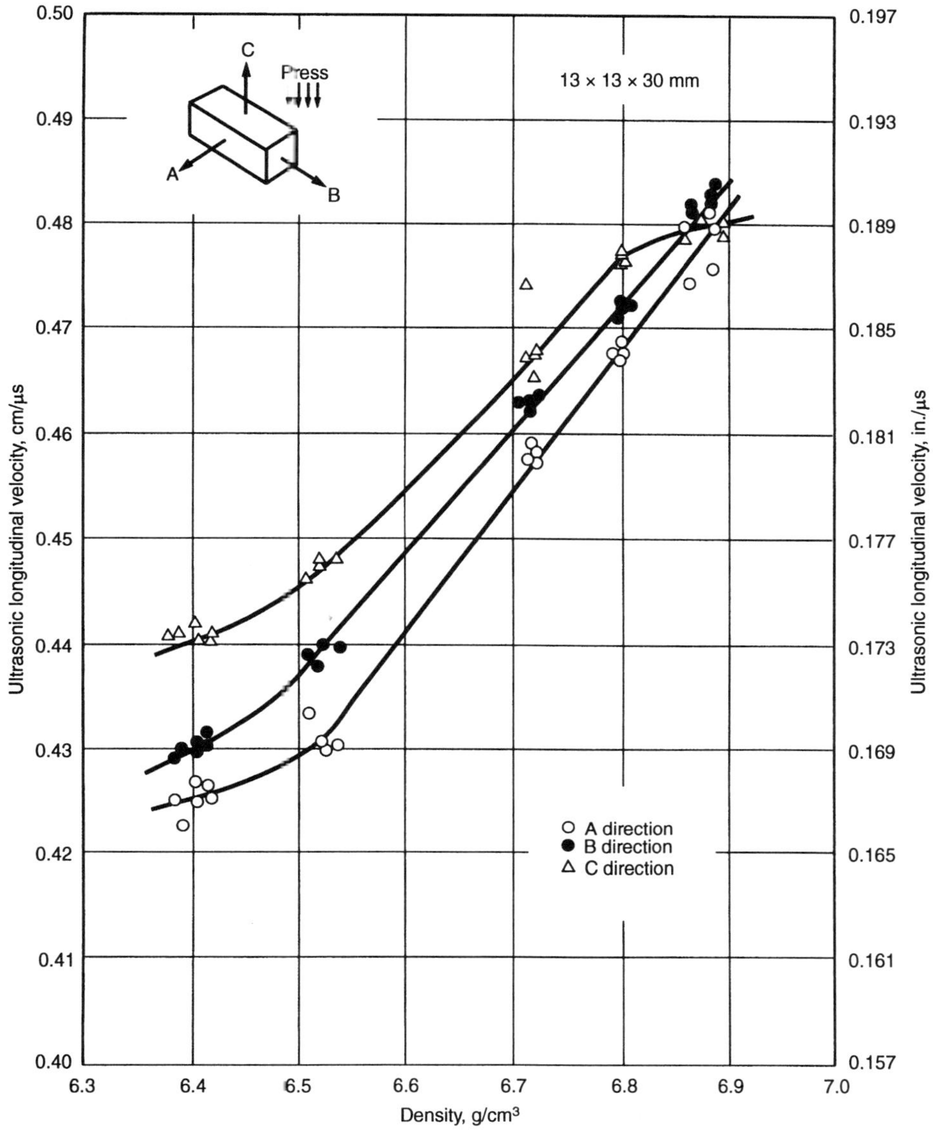

Fig. 13 Anisotropy of ultrasound velocity in sintered transverse rupture strength bars. Source: Ref 27

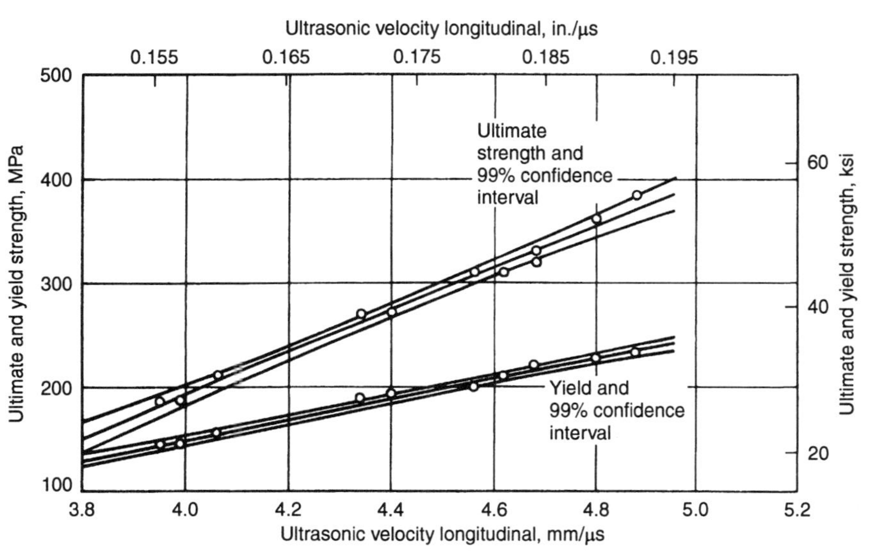

Fig. 14 Relationship of ultrasonic velocity and strength for Ancorsteel 1000B

gating in them. Ultrasonic signals impinged on a sample at one surface are transmitted at speeds and attenuated at rates determined by the density, modulus of elasticity, and continuity of the material.

Green Compacts. The characterization of green compacts by ultrasonic techniques appears to be hindered by problems of extreme attenuation of the incident signal. In one case, signals of 1 to 20 MHz were transmitted through an 8 mm (0.3 in.) thick compact of atomized iron with 0.2% graphite added. Attenuation did not allow back-wall echo measurement. Density was found to influence the transmitted intensity, with specimens at 95% relative density allowing some degree of transmission over the entire range of frequencies tested, while specimens at 87% relative density damped the incident signals entirely. It has also been shown that the velocity of ultrasound in green parts is highly anisotropic and that the experimental reproducibility is very poor (Ref 27).

Sintered Parts. In sintered parts, both the velocity of sound and their resonant frequencies have been related to density, yield strength, and tensile strength. Plain carbon steel P/M specimens were used in one series of tests, and the correlation was found to be close enough for the test to be used as a quick check for the degree of sintering in production P/M parts (Ref 28). Other work has demonstrated the relationship between sound velocity and tensile strength in porous parts (Fig. 12). The same types of relationships have also been documented in powder forgings (Ref 30).

Sintered parts have been found to transmit ultrasound according to the relationships shown in Fig. 13. The highest wave velocities occurred in the pressing direction. An additional distinction was found between the velocities in the longitudinal and lateral axes of an oblong specimen, and these results were shown to be reproducible between different powder lots and specimen groups. The anisotropy of velocity diminished at higher densities and disappeared above 6.85 g/cm^3.

The relationship between ultrasonic velocity and ultimate tensile strength or tensile yield strength in test bars is shown in Fig. 14 which illustrates this relationship for as-sintered 0.65% carbon steel. Table 7 gives sintering times and densities that resulted in various tensile properties and velocities. With similar data for a specific part, ultrasonic velocity can be easily measured, thereby determining the state of sintering or mechanical properties. It remains to be seen if this process is capable of detecting cracks and other flaws.

ACKNOWLEDGMENT

This article is adapted from L. Pease III, Inspection and Quality Control for P/M Materials, *Powder Metallurgy,* Vol 7, *ASM Handbook, 1985, p 480–491* and R.C. O'Brien and W.B. James, Powder Metallurgy Parts, *Nondestructive Evalu-*

Table 7 Characteristics of as-sintered FN-0208 alloy specimens prepared with Ancorsteel 1000B powder, 0.75% carbon, and 0.75% Acrawax C

Sintering time, min	Final density, g/cm^3	Ultrasonic velocity		Yield strength		Tensile strength	
		mm/µs	in./µs	MPa	ksi	MPa	ksi
5	6.26	3.95	0.1557	143	20.8	182	26.4
15	6.27	3.98	0.1568	142	20.6	182	26.4
30	6.27	4.06	0.1600	154	22.4	210	30.4
30	6.29	...	...	...	...	...	...
30	6.45(a)	4.42	0.1741	...	...	273	39.6
45	6.48	4.35	0.1711	183	26.6	266	38.6
5	6.76	4.56	0.1795	200	29.0	307	44.6
15	6.74	4.63	0.1821	210	30.4	307	44.6
30	6.74	4.69	0.1847	220	32.0	330	47.8
30	6.77	...	...	...	...	...	...
...	6.81(a)	4.88	0.1922	234	34.0	379	55.0
30	6.75	4.69	0.1846	218	31.6	317	46.0
45	6.89	4.83	0.1902	232	33.6	363	52.6

(a) Restruck

ation and Quality Control, Vol 17, *ASM Handbook,* 1989, p 536–545.

REFERENCES

1. P/M Design Guidebook, Metal Powder Industries Federation, 183, p 15
2. "Anchor MH100 Standard Molding Powder," Hoeganaes Corp., Riverton, NJ
3. "Aromet 28, Sintered Properties of P/M Copper Steels," Quebec Metal Powders Ltd., Sorel, Quebec, Canada
4. "Controlled Dimensional Change," SCM Metal Products, Cleveland
5. L. Pease III, "An Assessment of Powder Metallurgy Today, and Its Future Potential," Paper No. 831042, Passenger Car Meeting, Society of Automotive Engineers, Warrendale, PA, 1983
6. K. Moyer, Measuring Density of P/M Materials with Improved Precision, *Int. J. Powder Metall. Powder Technol.,* Vol 15 (No. 1), 1979, p 33
7. "Standard Test Method for Density and Interconnected Porosity of Sintered Metal Powder Structural Parts and Oil Impregnated Bearings," ASTM B 328, American Society for Testing and Materials, 1983
8. "Determination of Density of Compacted or Sintered Metal Powder Products," MPIF standard 42, Metal Powder Industries Federation, 1977
9. "Permeable Sintered Metal Materials—Determination of Density and Open Porosity," International Standards Organization Standard ISO 2738, available from American National Standards Institute, New York, 1982
10. N. Githens and J. Kane, "Study of Density Changes in Age-Hardening an Aluminum—5% Copper Alloy at 190 °C (374 °F)," senior thesis, Drexel University, Philadelphia, 1970
11. L. Tilton and J. Taylor, Accurate Representation of the Refractivity and Density of Distilled Water as a Function of Temperature. Research Paper RP971, *J. Res. Nat. Bureau Stand.,* Vol 18, Feb 1937, p 205–214
12. I. Hemins, Precision Density Measurements of Silicon, *J. Res. Nat. Bureau Stand—Anal. Phys. Chem.,* Vol 68A (No. 5), 1964, p 529–533
13. H. Bowman, W. Gallagher, and M. Schoonover, "The Development of a Working Density Standard," Instruments Society of America Pre-print 14.8–4.65, 20th Annual ISA Conference and Exhibit, Los Angeles, Oct 1965
14. H. Bowman and R. Schoonover, Procedure for High Precision Density Determinations by Hydrostatic Weighing, *J. Res. Nat. Bureau Stand.—Chem. Eng. Instrum.,* Vol 71C (No 3), 1967, p 179–198
15. H. Bowman, R. Schoonover, and C. Carroll, A Density Scale Based on Solid Objects, *J. Res. Nat. Bureau Stand.—Anal. Phys. Chem.,* Vol 78A (No. 1), 1974, p 13–40
16. H. Bowman, R. Schoonover, and C. Carroll, The Utilization of Solid Objects as Reference Standards in Density Measurements, *Metrologica,* Vol 10, 1974, p 117–121
17. L. Pease III, Metallography and Properties of Sintered Steels, in *Progress in Powder Metallurgy,* Vol 33, Metal Powder Industries Federation, 1977
18. W.Q. Judge, private communication, Engineered Sinterings and Plastics, Watertown, CT
19. G.E. Wood et al., Apparent Hardness Testing for Ferrous P/M Parts, *Prec. Metal Mag.,* Nov 1967, p 86
20. Nondestructive Inspection and Quality Control, Vol 11, *Metals Handbook,* 8th ed., American Society for Metals, 1976, p 17
21. L. Pease III, The Mechanical Properties of Sintered Steels and Derivation for MPIF Standard 35, in *Progress in Powder Metallurgy,* Vol 37, Metal Powder Industries Federation, 1981, p 113
22. Crack Test for P/M Compacts, *Prec. Metal Mold,* March 1967, p 66
23. R.A. Ketterer and N. McQiddy, "Resistivity Measurements on P/M Parts: Case Histories," Paper presented at the Prevention and Detection of Cracks in Ferrous P/M Parts Seminar, Metal Powder Industries Federation, 1988
24. E.R. Leheup and J.R. Moon, Electrical Conductivity Changes during Compaction of Pure Iron Powder, *Powder Metall.,* Vol 21 (No. 4), 1978, p 195
25. E.R. Leheup and J.R. Moon, Relationships between Density, Electrical Conductivity, Young's Modulus, and Toughness of Porous Iron Samples, *Powder Metall.,* Vol 21 (No. 4), 1978, p 1
26. I. Hawkes and C. Spehrley, Point Density Measurement and Flaw Detection in P/M Green Compacts, *Mod. Develop. Powder Metall.,* Vol 5, 1970, p 395
27. E.P. Papadakis and B.W. Petersen, Ultrasonic Velocity as a Predictor of Density in Sintered Powder Metal Parts, *Mater. Eval.,* April 1979, p 76
28. B. Patterson, C. Bates, and W. Knopp, Nondestructive Evaluation of P/M Materials, *Prog. Powder Metall.,* Vol 37, 1981, p 67
29. R.H. Brockelman, Dynamic Elastic Determination of the Properties of Sintered Powder Metals, *Perspect. Powder Metall.,* Vol 5, 1970, p 201
30. E.R. Leheup and J.R. Moon, Yield and Fracture Phenomena in Powder Forged Fe-0.2C and Their Prediction by NDT Methods, *Powder Metall.,* Vol 23 (No. 4), 1980, p 177

Metallography of Powder Metallurgy Materials

Leander F. Pease III and Douglas L. Pease, Powder-Tech Associates, Inc.

POWDER METALLURGY MATERIALS encompass enough differences to necessitate providing specific specimen preparation procedures in addition to those given in the Section "Metallographic Techniques" in *Metallography and Microstructures,* Volume 9 of the *ASM Handbook.* The major difference between parts made of metal powders and those made of wrought metal is the amount of porosity. Sintered materials generally exhibit 0 to 50% porosity, which affects mechanical properties and strongly interferes with metallographic preparation and interpretation of the structure. When examining micrographs, it is important to understand how the specimens were prepared. Careful metallographic preparation is significant in the analysis of sintered structures because the shape of the porosity is as important as the amount in judging sintered strength and degree of sintering.

In the metallographic preparation of most sintered specimens, the pores are smeared during sectioning, grinding, and rough polishing. This occurs to some degree even in materials whose pores have been filled with plastic resins. Proper polishing should open the smeared pores, then reveal their true shapes and area fractions. Routine metallography of the type used on a medium-carbon, ingot-base steel will not suffice. Because test method ASTM B 328 (Ref 1) cannot be used to measure density differences over short distances such as 0.25 to 5.25 mm, (0.01 to 0.25 in.), the true amount of porosity must be determined by image analysis to facilitate local density measurement. When the specimen is properly prepared, the area fraction of porosity will equal the volume fraction of porosity, and these must equal the porosity calculated from the actual measured density and pore-free density of a uniformly dense part:

$$V_p = (P_d - M_d)/P_d$$

where V_p is the volume fraction porosity, P_d is the pore-free density, and M_d is the measured density, for example, by ASTM B 328. Detailed information on density and porosity measurements may be found in the article "Surface Area, Density and Porosity of Powders" in this Volume.

Assuming that a part is uniformly 80% dense, in a properly prepared specimen, 20% of the area should appear as porosity. The surface of cold pressed and sintered parts will always be somewhat denser than the interior because of pressure losses caused by interparticle friction. However, parts that are P/M forged in tools at approximately 370 °C (700 °F) can have a chilled surface lower in density than the hotter, softer interior.

During sintering of cold-pressed compacts, the original particle boundaries disappear and result in a plane of fine pores that then grow into larger pores. In as-pressed parts, particle boundaries appear as thin gray lines in the metallographically mounted material. The progress of sintering can be judged by the disappearance of these boundaries. The original particle boundaries are similar to elongated, disk-shaped pores and have very sharp corners. These are extreme stress raisers. There is virtually no bonding across the original particle boundaries. Proper specimen preparation is required to distinguish residual original particle boundaries from the thin, gray boundaries that often appear between the edges of a particle and pores that were smeared during grinding and polishing. Therefore, an improperly prepared specimen with smeared porosity is often erroneously judged to be undersintered. If microhardness testing is to be performed, proper presentation of the porosity will result in fewer Knoop diamond indentations falling into hidden pores and thus in fewer wasted or incorrect readings. For additional information on microhardness testing of P/M materials, see the article "Testing and Evaluation of Powder Metallurgy Parts" in this Volume.

The open porosity in a mounted sintered part may trap water (moisture). During etching, this water may bleed out, resulting in staining. Water also corrodes some sintered materials. It may evaporate and then condense on the objective lens of the microscope, resulting in a foggy image. Etchants cause similar problems. Open porosity may trap abrasives and carry them onto subsequent cloths, which should hold only fine abrasives. The result is an increased tendency toward scratching of specimen surfaces. Filling the pores with epoxy resins alleviates these difficulties, but requires considerable technique.

Many of the interesting structures seen in P/M parts are caused by porosity and by the mixtures of elemental powders that constitute many alloys. These mixtures do not always result in homogeneous, well-diffused structures. Such heterogeneity is not necessarily detrimental and, in certain nickel steels and diffusion-alloyed steels, may be advantageous because the softer nickel-rich phases increase impact resistance and reduce the tendency toward ductility-limited tensile strength. It is therefore important to recognize when the observed heterogeneity is beneficial. The pores allow carburizing and nitriding gases to penetrate the interior of a sintered steel part, resulting in less-well-defined cases on carbon steel and the nitriding of 300 series stainless steels. The P/M steels are generally low in manganese. When the low alloys are prepared as elemental mixes with nickel and carbon, hardenability is lower than for fully dense, homogeneous low-alloy steels. Hardenability is not a problem in the fully dense prealloyed steels fabricated by forging or metal injection molding.

Sample Preparation

Sample Record. It is very useful to begin a metallographic investigation by making a photographic record of the sample prior to any sectioning. This may conveniently be done with a hand-held Polaroid (Polaroid Corp., Cambridge, MA) camera such as the Macro 5 SLR, which captures images from 0.2 to 3× or with a digital hand-held camera. In preparing a report on the sample, it is then a simple matter to mark the image with arrows to show the location of the sectioning. If the sample is then sectioned into several much smaller pieces, they may be similarly recorded for later identification.

Specimen Selection and Sectioning. It is usually necessary to examine a section that extends from the surface to the interior of the part and from top to bottom, in the pressing direction (a longitudinal section). The surface is susceptible to changes from (1) the sintering atmosphere, such as decarburization in a steel, (2) being sealed over with pure copper during infiltration, or sealed over by tumbling, or (3) shallow hardenability in elemental mixes of steels. Density can vary from point to point. During sintering, the protected bottom of the part "sees" a different atmosphere than the top, and the internal area may be more protected from the sintering atmosphere than an outer surface. Infiltrated parts have certain artifacts that appear where the infiltrant entered, for example, erosion or adhering copper alloy.

Improper tool design or press setup can cause cracks or laminations at different levels in a part, such as between a hub and flange; low-density regions normally occur just below flanges. If a part is overpressed in certain regions or if the tooling does not have the correct exit taper, microcracks can occur upon ejection. These are usually parallel to a punch face (normal to the direction of applied pressure) and are best seen in longitudinal sections that run parallel to the pressing direction. In a case-carburized gear, a section showing the plan view of the teeth would show the depth of case and whether or not the teeth were through-hardened. Therefore, likely potential defects in a P/M part influence selection of the planes of sectioning.

Because of the difficulty in opening all the smeared pores on a P/M part, the use of smaller sections is recommended. An easily polished specimen will measure less than 12 by 12 mm (½ by ½ in.). A soft abrasive wheel that breaks down easily will control overheating at cutoff and will not glaze easily. A composite wheel, 0.4 mm (0.015 in.) thick is especially recommended for specimens less than 25 mm (1 in.) in section and when very precise cuts are needed. Composite wheels consist of rubber-bonded aluminum oxide. Because of their 0.4 mm (0.15 in.) width, they cut with low force and quite quickly. Low-speed diamond wheels are very precise and damage free, but require long sectioning times. If the conventional abrasive wheel glazes or cuts slowly because the workpiece is too large, four small notches (3 cm^2, or 0.5 $in.^2$) can be removed using pliers at 90° positions around the periphery of the wheel. Because this weakens the wheel somewhat, it should always be operated in a well-guarded enclosure. A substantial flow of water that correctly strikes the interface between the wheel and the part is also required. Rust inhibitor should be added to recirculated water.

Sectioning an annularly shaped part parallel to its axis with the wheel advancing from the outer diameter toward the inner diameter may result in the wheel being captured when it penetrates the inner diameter. This often occurs in heat treated parts containing residual stresses and may cause wheel breakage. This can be avoided if the part is sectioned with the wheel moving along the axis of the part so that material on both sides of the inner diameter is cut at the same time. In general, it is preferable to secure in a vise the section of the specimen to be mounted. This section rarely contains burrs and may be mounted without further hand grinding.

Fluid Removal and Washing. Entrapped oil from heat treating or machining as well as water and rust inhibitors from the cutoff wheel must be removed from the pores before the specimen is mounted. Other sources of contamination include the sizing lubricants used at re-pressing and rust inhibitors added during tumble deburring of the parts. Failure to remove these contaminants will obscure the specimen surface when viewed under the microscope, bringing new meaning to the term "interference fringes." Entrapped oils also seem to interfere with polishing by slowing down the removal rate. If the specimen was not heat treated, contains no substantial amount of oil, and can tolerate heating to 260 to 370 °C (500 to 700 °F) for approximately 1 min, then the fluids can be removed on a hot plate under a hood. The specimens are easily heated until straw-colored or light blue. Water or small amounts of oil will evaporate or burn off quickly. At higher burnoff temperatures, a layer of oxide is created in the surface pores.

When the specimen cannot be heated as above, an extractor-condenser of the type shown in Fig. 1 or a Soxhlet apparatus may be used. The extractor-condenser consists of a flask, a siphon cup, and a condensing-coil unit that fits on the top of the flask. A solvent, such as toluene, is placed in the flask, and the specimens to be cleaned are placed in the siphon cup. Multiple specimens must first be coded for subsequent identification. For small specimens, a group photograph provides handy identification.

A cold-water line is connected to the condensing coil. The flask is heated to the boiling temperatures of the solvent. The solvent evaporates, and when the vapor contacts the cold condensing coil, the hot condensate drips into the siphon cup and onto the parts. When the siphon cup is filled to the level determined by the upper bend in the exit tube, it empties, returning solvent and dissolved oil to the boiling flask. Recycling allows a subsequent flow of clean solvent over the specimens. The oil and foreign matter remain in the flask and are periodically discarded along with the toluene.

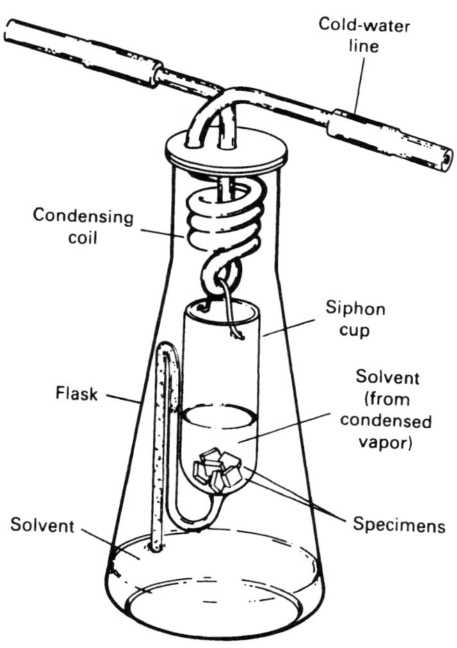

Fig. 1 Extractor-condenser used for washing P/M specimens to remove contaminants from pores

Six cycles of siphoning, requiring a total of approximately 1 h, will usually ensure removal of the oil. This method is also described in ASTM B 328 (Ref 1) and ISO 2738 (Ref 2). The latter test method includes the technique for removing oil when testing for total carbon. Because laboratory investigations often involve testing for carbon along with metallography, it may be efficient to use the more rigorous ISO method. Following extraction with the solvents, it is necessary to dry the parts for approximately 1 h at 120 °C (250 °F) to remove the solvent. When carbon content is to be measured, several repetitions of extraction and drying are carried out until the weight change between consecutive dryings is less than 0.01 wt%.

Mounting of Compacted Specimens. Epoxy resins, rather than wax and Bakelite (Union Carbide Corp., Danbury, CT) are especially preferred for filling the pores and mounting the specimens because the pores can be sealed as the multiple specimens are mounted. A convenient container for mounting is a length of 25 mm (1 in.) copper or aluminum tubing with an inner diameter of 32 mm (1¼ in.). This may be placed on a small, flat sheet of glass. To prevent resin from leaking under the edge of the tubing, a sheet of single-sided adhesive mounting paper may be placed under the tubing. The interior of the tube and the glass are coated with a mold-release agent. A convenient alternate two-piece cup could be machined from low-carbon steel (Fig. 2). The steel base can be ground flat in the lab as needed, and excess epoxy can be removed by sintering or heat treating the base. The bases of the plastic cups commercially available become concave after a few uses, but they are still very convenient to use for mounting specimens. The concave bases can be ground out by hand on a grinding lap. The cups, when new, result in mounts with clear sides, which is a convenience when looking into the mount to check location or orientation.

When more than one specimen is to be inserted in the same mount, each specimen must be identified for future reference. The use of an asymmetric arrangement of specimens or the mounting of a distinctive object such as a small, twisted piece of a paper clip, will permit easy, accurate reference. The use of the epoxy resin makes it easy to mount several small pieces in one mount. It is recommended that a sketch or map be drawn for each mount, for future reference. It is usually easier if the sketch shows the view of the polished surface, and the samples are

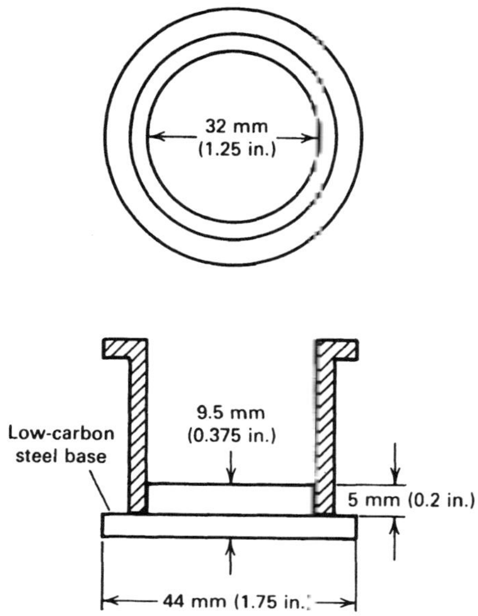

Fig. 2 Machined two-piece cup for mounting P/M specimens

carefully placed in the mounting cup in mirror image to the sketch.

The epoxy resin should be selected for low vapor pressure of resin and hardener, and any new resin should be tested in the vacuum chamber to note the pressure at which it bubbles. For example, some epoxies should not be used below 150 torr (20 kPa). This limits the amount of air that can be removed and the volume of pores that can be filled. Epoxy resins should be placed in disposable cups for stirring and mixing of resin and hardener. Volumetric measurement of the components with plastic syringes works reasonably well, but weighing the liquids on an electronic balance is easier. Epoxy resins should be selected for low viscosity. Rapid hardening is convenient, but must not be allowed to interfere with pore filling. Many resins heat cure in 1.5 to 2.0 h at 50 °C (120 °F), for example, Caldofix (Struers, Westland, OH). The Caldofix resin is crystal clear, which allows good viewing of the interior samples and ease of scribing the identification code on the back side.

Epoxy-Resin Impregnation. In one method, the resin is carefully poured over the specimens in their mounting cups to a depth of approximately 6 mm (¼ in.) above the sample with the cup filled to a total depth of 14 to 18 mm (⅝ to ¾ in.). The cups should be placed in a container that will catch any minor epoxy spills, and the container should be placed in a vacuum chamber. The air is then evacuated to the lowest pressure tolerated by the resin. The specimens should bubble for approximately 10 min. Air pressure is then restored and allowed to impregnate the specimens for 15 min. The specimens may then be cured overnight, or at an accelerated rate of 1.5 to 2.0 h at 50 °C (120 °F), or by following manufacturer's directions.

An alternate method for impregnation with epoxy resin involves suspending the specimen above the epoxy bath during air evacuation. This allows the air to exit the pores rapidly without causing air bubbling in the epoxy. It is similar to the technique used to fill the pores with oil in a sintered bearing. Another method is to hold the specimen above the oil magnetically or with a vacuum feedthrough manipulator. After 1 to 2 min at low pressure, no air will remain in the specimen, and it can be lowered into the epoxy bath, at which point the pressure is readmitted into the chamber. This method is limited by the vapor pressure of the epoxy resin and hardener. Commercial equipment with vacuum feedthrough evacuates air from the specimens and directs a stream of epoxy from outside the chamber into the specimen cup located in the vacuum. Again, the vapor pressure of the resins determines the lowest usable pressure.

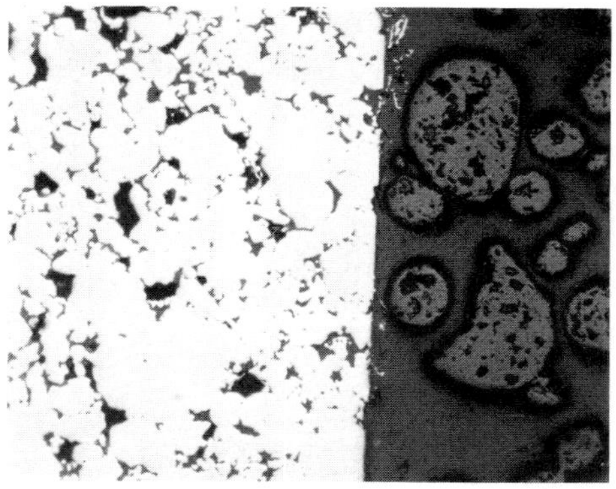

Fig. 3 Edge-retention technique in which dark Al_2O_3 granules (right) are added as a reinforcer to the epoxy resin. Not all of the pores are open, which indicates that Al_2O_3 additions necessitate extended polishing times. Fe-0.8C specimen (7.0 g/cm^3) pressed at 550 MPa (40 tsi) and sintered 30 min in dissociated ammonia at 1120 °C (2050 °F). 2% nital. 95×

Most mounts show a case effect where resin mainly enters the specimen through unmachined surfaces. The surface against the bottom of the specimen cup is almost always machined. Therefore, most mounts show evidence of epoxy resin in the surface pores, but the interior pores are rarely full. This effect is apparent to the unaided eye during initial polishing, because the outer edges appear more porous (like an orange peel) and the interior more mirrorlike.

Edge retention is achieved by adding light or dark −60 mesh alumina (Al_2O_3) granules to the epoxy resin (Fig. 3). Some of the ceramic may be mixed with resin and poured around the specimens to form a 1.6 mm ($\frac{1}{16}$ in.) thick layer. The rest of the mount is formed from clear resin poured on top of the mixture.

Alternatively, loose Al_2O_3 may be poured into the specimen cup to a depth of 1.6 mm ($\frac{1}{16}$ in.), surrounding the specimens, and the clear resin carefully poured on top. During vacuum evacuation, the resin flows in among the Al_2O_3 particles. This ceramic reinforces the epoxy and results in very little rounding of the specimen edge during polishing. Because the oxide greatly slows the rate of grinding and polishing, the times recommended below must be adjusted. A third technique involves forming a thin oxide-reinforced layer around the specimen itself by applying a thick, pasty mixture of resin and oxide to the specimen surface. In this way, the entire surface of the mount is not hardened, just a layer approximately 0.5 mm (0.020 in.) adjacent to the specimen. Therefore, the polishing time is not unduly increased.

Specimen Identification. The use of some epoxy resins, particularly when air cured, produces a surface that is soft and cloudy and inconvenient for scribing the specimen identification. The back of the specimen mount should be ground through 600 grit, then polished for 1 min using 1 μm Al_2O_3 on a medium-nap cloth. This yields a flat surface that can be scribed for clear identification. If it is necessary to look through the side of the mount to see the specimen inside the epoxy, the round-side surfaces must be polished as described above, or a new plastic mounting cup used.

Specimens that have been filled with wax can also be mounted in thermosetting Bakelite resins, Lucite (Clorox Co., Oakland, CA), or clear-liquid cold-mounting resins.

Mounting of uncompacted metal powders requires special procedures that include the use of epoxy resin and vacuum impregnation. A small amount of properly sampled powder per ASTM Standard B 215 (Ref 3) should be placed in the center of the specimen cup that will hold the epoxy resin. The pile should be approximately 13 mm (½ in) in diameter and approximately 3 mm (⅛ in.) deep at the center. The epoxy should be poured around the pile of powder without disturbing or segregating it. The cup should be filled with resin to a depth of 12 to 18 mm (½ to ¾ in.). The specimen is then evacuated at approximately 150 torr (20 kPa) for 10 min and returned to atmospheric pressure. Curing can take place at room temperature; higher temperatures hasten the process.

An alternate method is to mix the metal powder with a small amount of epoxy resin, then pour it into the bottom of the cup. This should be a thick, pasty mixture whose consistency will prevent the particles of metal from falling and segregating and will display several particles in a micrograph at 200×. If segregation is suspected, the specimen may be sectioned vertically after hardening. The sections can be remounted to note any segregation between top and bottom of the pile. This is important when attempting to measure the particle size distribution from a micrograph of the assembled powder particles.

Wax Impregnation, an Alternative to Epoxy Impregnation. After removal of fluid from the pores, subsequent abrasive, water, and etchants must be kept out of the pores. Wax impregnation may be used for specimens that are to be handheld or mounted in Bakelite. The specimens are soaked 2 to 4 h in a molten synthetic wax at 175 °C (350 °F). After cooling and removal of the surface wax, the specimens are ready to hand grind or to mount in Bakelite.

Ideally, wax impregnation should be carried out in a vacuum oven at the recommended temperature. The vacuum allows the air entrapped in

Fig. 4 Pressed and sintered Fe-0.8C alloy (6.8 g/cm^3), as-ground on 600 grit silicon carbide. Micrograph shows the closure of pores and flatness of specimen (the surface is shown at left). Arrows indicate closed pore edges. 95×

Fig. 5 Effect of polishing on pore opening in a pressed and sintered Fe-0.8C alloy: deliberately underpolished specimen. This region, which is adjacent to the specimen edge, shows all the pores open. Compare with Fig. 6 (center of specimen). 180×

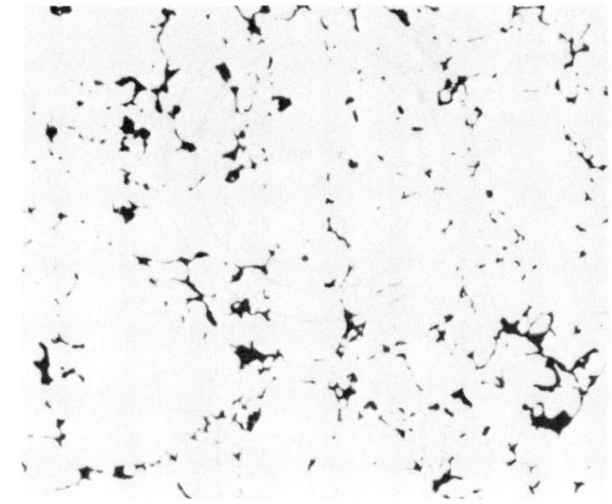

Fig. 6 Effect of polishing on pore opening in a presssed and sintered Fe-0.8C alloy: interior of same specimen as in Fig. 5. After 2 min of polishing, there are numerous smeared pores. Compare the amount of porosity with Fig. 5. This micrograph shows how the inner part of a specimen polishes more slowly than the edge. 180×

the pores to bubble up through the molten wax. The atmospheric pressure is returned to the system with the specimens immersed in the molten wax. The air pressure then should be allowed to act on the molten wax for 30 min to force it into the pores. Subsequently, the specimens may be cooled and the excess wax removed.

Automatic Grinding and Polishing

An automatic grinder can produce flat specimens using wet SiC papers of 220, 500, and 1200 mesh, 30 N (6.7 lbf) force per specimen, 300 mm wheel, 125 rpm, counterrevolving the specimen. Grind no more than 30 s on a paper, for six 30 mm (1¼ in.) mounts done at the same time. Before polishing, ultrasonically clean for about 5 min. The ultrasonic cleaner used for washing P/M specimens consists of a power supply and a small tank, which holds a water bath containing a small amount of surfactant such as liquid soap. The surfactant provides the energy coupling needed between the water bath and a beaker of solvent, such as alcohol, is suspended in the bath. The water bath and beaker together should fill the ultrasonic tank to within 25 mm (1 in.) of the top, (to provide protection to heating elements that the ultrasonic cleaner may contain). The specimen is placed inside the beaker and rests face down on cotton wadding or a paper towel. The power source produces high-frequency waves in the room-temperature water bath. The waves strike the beaker and force the alcohol into the pores of the specimen, removing foreign substances. Because the specimen is placed in the alcohol bath, the washing takes place in what soon becomes a contaminated solution. Debris accumulates in the cotton or on the paper towel, and the solution must be changed periodically. In addition to removing polishing grit when changing to successively finer media, it has also been reported that the ultrasonic action may help to open pores that were smeared at grinding (Ref 4).

Rough polishing consists of 10 min on short-nap cloth with a 3 μm diamond paste at 30 N (6.7 lbf) force on each specimen and 125 rpm on a 300 mm (11.8 in.) wheel. Final polishing for 1 to 2 min with 0.05 μm alumina on long-nap microcloth at 125 rpm, 1 N (4.4 lbf) force results in a mount free of scratches and flat to the edge. If the pores were well filled with epoxy resin, they will be seen in their true area fraction and shape. If the interior pores are not well filled, the above procedure must be supplemented by several repetitions of light etching (15 s immersion in 2% nital) and repolishing 1 to 2 min with the 0.05 μm alumina. To learn if the pores are still smeared, examine the mount at about 1000× and look for thin gray boundaries of the pore edges, for example, Fig 9.

Manual Grinding and Polishing

Grinding. Rough grinding of the mount must produce a planar surface for subsequent grinding and polishing. The preferred procedure involves using a water-cooled and lubricated wheel with 240, 400, and 600 mesh SiC papers. The paper is held to a rotating disk that makes use of the vacuum created by a thin layer of water under the sheet of abrasive paper. Alternatively, the paper can be held in place adhesively or by special flat magnets that grip a thin ferrous backing on the grinding paper. The same grinding may be carried out on wet papers placed on top of a sheet of glass. When using grinding wheels, the specimen is held in a fixed position on the wheel so that all scratches are in one direction, which requires even and moderate pressure. When changing papers, the specimen is rotated 90° to allow one to note the disappearance of the previous scratches. The use of one single sheet of SiC paper for more than two or three specimen mounts is not recommended because it leads to lack of flatness of the specimen surface. Grinding using 240 to 600 grit abrasives and moderate pressure at 125 to 250 rpm on 200 mm (8 in.) laps requires approximately 30 s for each paper. Using Al_2O_3 for edge retention (Fig. 3) gives a hard mount surface that requires longer grinding and polishing times.

It is important to grind a 30° bevel around the mount periphery. This allows the mount to pass smoothly over the subsequent polishing cloths and prevents the plowing aside of abrasives by a sharp edge. Failure to use a bevel slows polishing on long-nap cloths.

Polishing. Following grinding, the specimen is flat to the edges, and the pores are almost completely smeared (Fig. 4). Subsequent polishing generally rounds the specimen edges, because the mounting resin is much softer than the metal specimen. This rounding can be prevented by the use of embedded ceramic materials for edge retention, or such conventional techniques as plating of the specimen surface before mounting. Polishing must open all the pores, show true area fraction of porosity, remove scratches and disturbed metal, and minimize edge rounding. The presence of epoxy resin or wax in the pores facilitates opening the smeared pores, but does not eliminate the problem.

During polishing, the abrasives first open the pores closest to the specimen edges (Ref 5, 6), giving the impression that surfaces are less dense than interiors (Fig. 5 to 7). For ferrous materials, the fastest way to reveal the pores results in slight edge rounding but is adequate for routine work. These steps should be followed:

1. Etch 2 min in 2% nital by immersion. For materials other than low-alloy steels, use the customary agents. Etching before polishing initiates pore opening but does not exaggerate pore size.
2. Rough polish 2 min using 1 μm Al_2O_3 moderate hand pressure on a long-nap felt cloth (Ref 7). Use 250 rpm on a 200 mm (8 in.) diam wheel rotating the specimen counter to the wheel to prevent comet tails. The long-nap cloth and the fairly coarse Al_2O_3 rapidly open the pores (see Ref 8 for an example of another technique that requires 300 min).

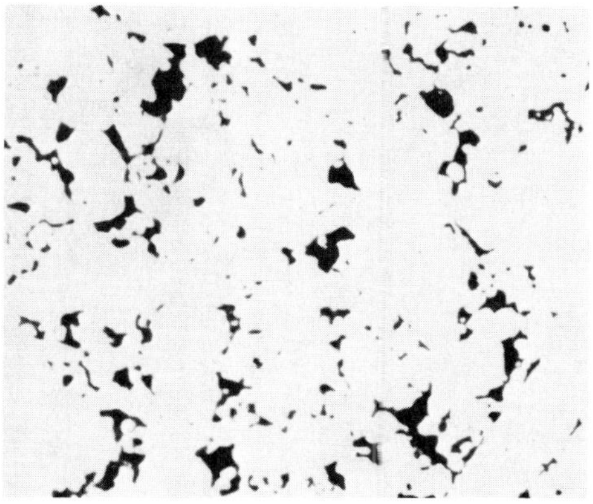

Fig. 7 Effect of polishing on pore opening in a pressed and sintered Fe-0.8C alloy: repolished version of Fig. 6 showing more pores in the center of the part (some remain smeared over). The densith appears highter than the true density of 6.8 g/cm^3. 180×

3. Repeat steps 1 and 2 once or twice. This procedure generally opens all the pores. To the unaided eye, the surface of the specimen should exhibit a uniform orange-peel appearance with no shiny specular (mirrorlike) regions. If necessary, repeat steps 1 and 2 until the surface is uniformly roughened. Even P/M forgings and metal-injection-molded parts at 98 to 99% of full density display porosity to the unaided eye.
4. This aggressive rough polishing exaggerates the pore area fraction by eroding and rounding pore edges. That is, the specimen will appear erroneously low in density. Final polishing must restore the true area fraction of porosity.
5. Polish 2 min using 1 μm diamond on a short-nap cloth at 250 rpm with moderate hand pressure. This sharpens pore edges and restores the pores to their true area fraction and eliminates most scratches, but leaves the edges of the specimen slightly rounded. A 19 mm (3/4 in.) long bead of diamond paste, weighing approximately 0.06 g (0.002 oz), is recommended for each 2 min of polishing. Use an alcohol-base solvent or thinner for the diamond paste so that it will wash off in water. Oily thinners penetrate the residual pores and bleed out of the specimen.
6. Final polish 30 s using a long-nap microcloth and 0.05 μm deagglomerated Al_2O_3. Use light hand pressure or an automatic polisher with 100 g weight (1 N, or 4.5 lbf) on the specimen at 125 rpm. This removes the fine scratches on most ferrous materials. The true area fraction of porosity of the surface is now restored. Reference 6 includes information on this method, and it also demonstrates that it is possible to open pores and show the correct porosity area fraction using a method that requires approximately 12 min and does not use diamond. It consists of 10 min of hand polishing using 1 μm Al_2O_3 on a synthetic suede, short-nap cloth at 250 rpm on a 200 mm (8 in.) lap and 2 min of light hand polishing using 0.05 μm Al_2O_3 at 125 rpm on the same type of cloth.

To produce a surface with no edge rounding, it is necessary to eliminate the 1 μm Al_2O_3 and long-nap cloth polishing. Instead, after the 2 min etch in 2% nital, step 5 should be repeated several times. More than five repetitions may be required to open all the pores on a large specimen, particularly if it is soft and undersintered. The wheel should be recharged with diamond at each repetition. However, the 2 min etching should not be repeated, because the 1 μm diamond paste does not rapidly remove etching effects.

Newly developed P/M materials require modified polishing procedures to show the correct area fraction of porosity. Such methods could be developed using standards of known density. Other modern methods of automatic polishing may open all pores, but should first be tested by preparing a specimen of known density, then measuring the area fraction of pores. The measured area fraction should agree with calculations made from the known density.

A vibratory machine in which the specimens circulate around a bowl with the abrasive slurry may also be used. The use of a short-nap chemotextile cloth (Texmet) and 0.3 or 1 μm Al_2O_3 will yield a specimen that is virtually free of edge rounding. However, for specimens that have been ground through 600 grit, this procedure requires approximately 3 h because of the slow material removal rate and the need to open all the pores.

The rate of material removal may be measured using a Knoop indenter mark as a reference (Fig. 8). First, a mark approximately 100 μm long is made in a known location on the specimen. A simple reference point in the interior of the specimen can be made using a Rockwell superficial indenter with the 15 kgf load. The Knoop mark is then placed approximately 0.4 mm (0.015 in.) away from the superficial indenter mark and at a known orientation to it. The Knoop mark is measured from a photograph or with the measuring stage of the microhardness tester. After polishing for a fixed time, such as 1 to 2 min, the Knoop mark is relocated and remeasured. The material removed normal to the specimen surface is the change in length of the Knoop diagonal divided by 30.51 (for a standard indenter). For a 25 by 25 mm (1 by 1 in.) specimen, polishing using a 250 rpm, 200 mm (8 in.) diam lap, 1 μm Al_2O_3 on a synthetic suede, short-nap cloth, and moderate hand pressure will remove 0.4 μm/min. A smaller specimen, such as 12 by 6 mm (0.5 by 0.25 in.) will polish at 1.45 μm/min. Additional information on material removal rates can be found in Ref 5.

Soft material, such as pure iron or copper, may still exhibit some fine scratches after the 0.05 μm Al_2O_3 polishing described above in step 6. One solution is to use a new long-nap cloth (microcloth) with adhesive backing attached to a flat glass plate or to a flat bench top. With the deagglomerated 0.05 μm Al_2O_3, charged onto the cloth at a ratio of 1 part (by volume) powder to 4 parts distilled water, the specimen should be polished in the abrasive slurry using approximately 50 light hand strokes straight back and forth. This will eliminate the fine scratches from the prior polishing; remaining scratches will be aligned parallel to the direction of polishing, and their source identified. This light final polishing does not cause comet tails or pore beveling. Colloidal silica, for example, Mastermet (Buehler, Ltd., Lake Bluff, IL), is excellent for a final polish.

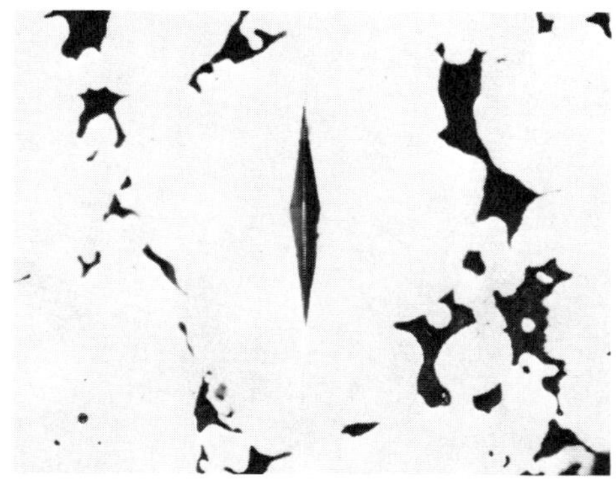

Fig. 8 Knoop indenter mark (100 gf) used as a reference to note the rate of material removal from the surface by measuring the change in length and depth of the indentation. Surrounding black pores in this unetched, pressed and sintered Fe-0.8C alloy (6.8 g/cm^3) are also revealed. 295×

Titanium alloys require a 4 min rough polishing using 1 μm Al_2O_3 on felt cloth at 250 rpm on a 200 mm (8 in.) diam lap with moderate hand pressure. This opens and slightly enlarges the pores. The autopolisher with 3 μm diamond on DAC cloth for 10 min at 30 N can be used. Final polish is colloidal silica on microcloth, at 1 N (4.5 lbf), 125 rpm, and for 1 min.

Metal Powder Particles

The powders, if mounted in epoxy resin as noted above, present a planar surface suitable for fine grinding. It is usually possible to begin using wet 600 grit SiC on revolving wheels or fixed sheets of paper. Approximately 30 s of grinding exposes enough particles to be viewed. The specimen is washed and then polished by hand for 2 min using 1 μm diamond on a short-nap cloth (MOL) at 250 rpm on a 200 mm (8 in.) diam lap. The use of an alcohol-base, water-soluble lubricant allows easy specimen cleaning in water.

The powders must not be overpolished, or the epoxy resin will be polished away between them and the particle edges will become rounded. Final polishing should be performed using deagglomerated 0.05 μm Al_2O_3 on a long-nap microcloth for 30 s with light hand pressure on a 200 mm (8 in.) diam, 125 rpm lap. Prolonged polishing or heavy hand pressure during final polishing rounds the particle edges.

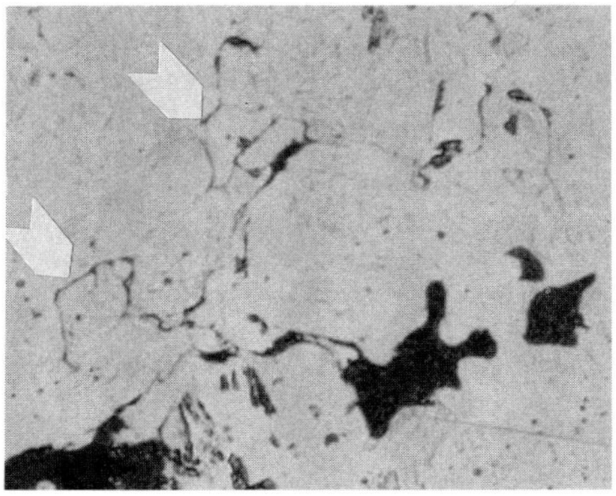

Fig. 9 Intermediate stage of polishing showing the edges of smeared pores, (see arrows), which have not yet been opened by the polishing. The material is pressed and sintered Fe-2%Cu-0.8%C, 6.8 g/cm^3. Magnifications of 500 to 1000× are required to see these features. These smeared pore edges must be removed by polishing and etching so that the original particle boundaries characteristic of undersintering can be seen, in the unetched condition, e.g., Fig. 27 to 29. Unetched. 960×

Fig. 10 Ferrous P/M specimen cut with the use of a coolant. No evidence of overheating. Nital. 12×

Fig. 11 Ferrous P/M specimen cut without the use of a coolant. Evidence of overheating (dark area at right edge of specimen). Nital. 12×

Because some particles have internal pores that may have been smeared, it is important to examine some of the particles, unetched at 500 or 1000×, for the thin, gray lines that are the edges of smeared pores (Fig. 9). If such undisclosed pores are noted, 15 s etching and 1 μm diamond polish must be repeated. Etching the mounted powders removes enough surface material to lower the particles below the surface of the epoxy resin, which then provides an opportunity to repolish them. The repolishing should always be performed using 1 μm diamond followed by 0.05 μm Al_2O_3 for 30 s on a long-nap cloth.

Aluminum powders and alloys must be polished using 1 μm diamond, rather than Al_2O_3, which reacts with aluminum powders. Final polishing is best done with colloidal silica on microcloth for 30 to 60 s. Hard powders, such as tungsten, may be examined as-polished using 1 μm diamond. Final polishing of very soft materials, such as pure copper, may be carried out using a fixed, long-nap cloth, as described previously for full-size specimens. The colloidal silica noted previously also works well on pure copper. Pure iron powder may require two or three 30 s fine polishings and light 15 s etching in 2% nital to remove fine scratches. The procedure also opens porosity in fine-porosity sponge iron.

Macroexamination

Macroexamination of sintered materials is not commonly practiced. In wrought or ingot-base materials, forging flow lines, oxide segregation, and stringers are studied extensively. These features are not usually found in P/M materials, but there are certain other uses for macroexamination.

In sectioning a heat treated P/M steel, care must be taken not to overheat the specimen and temper or reharden it locally. The etching performed during grinding and polishing to help open the pores will indicate any macroscopic striated darkening from overheating, as shown in Fig. 10 and 11. By revealing a lighter or darker surface than the interior, this same intermediate etching using 2% nital on steels will show if a specimen ground through 600 grit SiC paper is likely to have a decarburized or carburized surface layer. This effect is apparent to the unaided eye.

During manual polishing with 1 μm Al_2O_3, a porous appearing layer similar to orange peel develops near the outer edges of the specimen adjacent to the mounting medium. This layer spreads inward during subsequent polishing until the entire cross section appears uniformly porous to the unaided eye. If the surface is shiny or specular in certain regions, it almost surely will be found to contain pores that polishing has failed to open (Fig. 6 and 9). The penetration depth of the epoxy resin through the side surfaces of the specimen can be seen during polishing, because the regions where the epoxy is in the pores displays the orange peel appearance. The regions with unfilled pores are temporarily smeared over and shiny. After final polishing, the surface roughness from the pores is much diminished, but variations in density in a part are visible to the unaided eye.

Nital etching of elemental nickel steels reveals the nickel-rich areas as light-reflecting sparkles. Unsintered (green) or sintered, unprepared parts may be examined to 25× for cracks or the presence of added copper in iron. The fracture surface of a heat treated part shows varying degrees of discoloration in bands parallel to the outer surfaces. These dark bands are probably caused by oil impregnation during oil quenching, and the color is caused by in situ partial decomposition of the oil during quenching or tempering. The Metal Powder Industries Federation (MPIF) test method 37 for case depth (now obsolete) used the difference in the fracture appearance of the case region and the interior to measure case depth (Ref 9).

Scanning Electron Microscopy

The scanning electron microscope is a useful tool for examining metal powders, fracture surfaces, as-pressed and sintered surfaces formed by dies and punches, and, potentially, the as-polished sections used for optical microscopy. Scanning electron micrographs of iron, prealloyed steel, and stainless steel powders are shown in Fig. 12 to 24 in the series of representative micrographs in this article. In addition, Fig. 25 and 26 show the smooth outer surfaces of parts that were contacted by punches or dies during consolidation. Figure 26 illustrates the unsintered view of the side of a part that contacted the die. The powders have been pressed into close contact, and the boundaries between particles are readily visible. Sintering, which completes the bonding of adjacent powder particles, is traced in the accompanying optical micrographs (Fig. 27–31) that illustrate the disappearance of the particle boundaries and the rounding of pores in a diffusion-alloyed steel.

The development of bonding between metal particles may also be followed by examining their fracture surfaces. As bonds develop, the fracture shows cup-and-cone or dimpled-fracture regions where the bonds have been torn apart. The regions between the ductile cups and cones are the smooth surfaces of the original particles, which were not bonded to the adjacent particles. The progress of sintering may be followed through the increasing number and area fraction of the ductile torn regions. This is well illustrated in the scanning electron micrographs shown in Fig. 32 to 38. The energy spectra of the x-rays generated by the electron beam striking the atoms of the specimen can be analyzed to determine which elements are present at the fracture surface. The fracture surface may be scanned for the wavelength characteristic of a particular element and to record the intensity or concentration of the element as a function of location. This concentration appears on an x-ray

Table 1 Etchants used for examination of P/M materials

Etchant	Procedures and applications
2% nital: methyl alcohol plus 2% concentrated HNO_3	For as-sintered irons and steels (best for ferrite and low-carbon steels); immersed for 10–15 s; heat treated steels: 6–7 s
Concentrated picral: picric acid in methyl alcohol; some undissolved crystals remain in the container bottom	For higher carbon-containing materials to develop good contrast with carbides, pearlite, other eutectoid products, martensite, and retained austenite; etch by immersion, 15–20 s
Glyceregia: 10 mL HNO_3 conc, 15 mL HCl, 35 mL glycerol(a)	Show grain boundaries, twin boundaries, and carbides in austenitic and martensitic stainless steel; immerse for 1–2 min or swab lightly
4% $FeCl_3$ in H_2O	Develops red color in copper-rich regions in bronze; etch by swabbing, 10–20 s
2 g $K_2Cr_2O_7$, 4 mL NaCl, 8 mL H_2SO_4, 100 mL H_2O	Develops grain boundaries and small grain clusters in bronze; etch by swabbing, 10–20 s
Keller's reagent: 2.5 mL HNO_3, 1.5 mL HCl, 1.0 mL HF, 95 mL H_2O	For aluminum and aluminum alloys; immerse 8–15 s, wash in water; do not remove etchant products from surface
5% nital	For as-sintered tool steels; immerse 5 min
5 mL HNO_3 conc, 10 mL 48% HF, 85 mL H_2O	For titanium and titanium alloys; immerse 5 s
5 mL NH_4OH, 3 drops H_2O_2, 5 mL H_2O	For brasses; swab 20 s; make fresh solution every 20 min
450 mL H_2O, 25 mL H_2SO_4, 25 mL HNO_3, 25 g chromic oxide, 4 g NH_4Cl	For bronzes not etching clearly in the above $K_2Cr_2O_3$: Just before use, mix a few mL of this etchant with equal amount of 3% H_2O_2 and swab for 10 s

(a) Use hood for fumes and hand and eye protection when mixing this solution

map as a collection of dots. That display shows if the element is uniformly distributed or somewhat segregated, which is useful for monitoring the dissolution of copper or nickel in steel, or for checking for oxides of manganese or silicon.

Microexamination

Etchants commonly used for the metallographic examination of sintered metals are listed in Table 1. The initial use of the etchant, as noted previously, is to open the smeared porosity before and during the coarse polishing with 1 μm Al_2O_3. However, once fine polishing is completed, it is very important to examine sintered parts in the unetched condition. This displays the number and distribution of original particle boundaries present (Fig. 27–29). Etching reveals grain boundaries, which would be easily confused with the particle boundaries that also appear as thin gray or black lines. Cracks, density variation, oxide films or particles, and pore shape or rounding are easier to view in the absence of distracting etched features.

For iron and low-carbon sintered materials, 2% nital is preferred because it reveals the ferrite grain boundaries. It is applied by immersion for 10 to 15 s. The specimen is washed in running water, rinsed in methyl alcohol (optional), then dried in a cool air stream or wiped dry. (If any open porosity is present, the use of warm air may cause subsequent evaporation of entrapped water or alcohol onto the lens of the microscope.) For steels with medium-to-high carbon content or for heat treated structures, concentrated picral in methanol works well. It enhances the contrast of carbide platelets in the eutectoid in as-sintered structures.

Picral develops good contrast in the martensite needles and retained austenite of heat treated materials. The 2% nital also works well for these structures and does not stain hands and clothing. Heat treated specimens etched in nital should be immersed 6 to 8 s (swabbing causes streaks), rinsed in running water then alcohol, and then blown dry. This underetches the martensite, but enhances contrast between the fine pearlite (a dark, unresolved, nodular constituent) and the light-colored martensite. This is an optimal condition for checking microhardness, because the martensite is light, clearly seen, and forms a good background for measuring the length of the Knoop diamond indentation. The use of picral for 15 to 20 s or 2% nital for slightly longer enhances the contrast of martensite and retained austenite. For magnifications of 1000×, a lightly etched structure affords maximum clarity.

Sintered stainless steels are best etched in glyceregia using a protective hood and ventilation. Because this is a strong acid mixture, eye and hand protection are advised. The glyceregia is applied by immersion with the specimen surface upward; a fraction of a millimeter of the liquid is poured on top of the metal to be etched. When freshly prepared, the action is fairly slow and may require 1 to 2 min. The etchant may also be applied using a very light swabbing action. Glyceregia decomposes with time; it may turn orange and emit nitrous oxide. As the etchant darkens toward orange or brown, it becomes more reactive and less predictable. Shelf life may be extended by cool storage. Nonetheless, glyceregia rarely can be used for more than 2 h before disposal. Additional etching should be preceded by repolishing using 1 μm diamond and 0.05 μm Al_2O_3 to avoid unwanted etching artifacts.

In well-sintered stainless steel, moderate grain size, annealing twins, and grain boundaries free of precipitated carbides are visible. When carbides are present, the grain boundaries are not clean and straight, but ragged, broader, and fast etching. Prolonged etching causes the formation of pits, which may be confused with pores. The stainless steels are judged by freedom from precipitated carbides and original particle boundaries. Some of the new classes of high-temperature sintered stainless steels exhibit very well-rounded pores, which may even be isolated from each other. Such high degrees of sintering are rare in low-alloy steels.

Copper and bronze materials are etched in a 4% ferric chloride ($FeCl_3$) solution or the potassium dichromate ($K_2Cr_2O_7$) solution in Table 1. The $K_2Cr_2O_7$ etchant is preferred for bronze. During sintering of bronzes, the elemental tin melts, forming a series of increasingly higher temperature intermetallic compounds, then completely dissolves in the copper at approximately 785 to 845 °C (1450 to 1550 °F). The result of good sintering is a bronze with a moderate grain size and no free tin or blue-gray copper-tin intermetallic compounds. Intermetallic compounds are present during the early stage of sintering and in a final undersintered condition.

As sintering proceeds, the copper recrystallizes an its grains grow as the tin diffuses to form α-bronze. The presence of many reddish copper areas or large area fractions of fine-grained copper-rich or α-bronze indicates undersintering. The undersintered materials are the most difficult to etch and to show clear structures. The $K_2Cr_2O_7$ etchant should be applied (always by swabbing) for 15 s. The specimen is then placed under running water, and swabbing is continued for a few seconds. It may be necessary to polish and etch several times to obtain a clear structure free of distortion from grinding.

Some bronzes are made from mixtures of copper and copper-tin compounds. During sintering, the gray copper-tin compounds should dissolve in the copper and form α-bronze. If the sintering is not complete, the rounded gray copper-tin particle will be seen in the unetched microstructure. Interestingly, upon etching, they are preferentially attacked by the ($K_2Cr_2O_7$) reagent and are not visible in the etched structure. They are, therefore, easily overlooked. Other bronzes are not uniformly etched by the ($K_2Cr_2O_7$) reagent and the final etchant in Table 1 is recommended to reveal their grain structure.

Bronzes may contain up to 4% free graphite. The epoxy resin helps secure the graphite flakes for polishing and viewing. The use of 1 μm diamond and light final polishing with 0.05 μm Al_2O_3 on a long-nap cloth usually preserves the graphite.

Brasses and nickel silver are etched using a mixture of ammonium hydroxide (NH_4OH) diluted with 50% H_2O and a few drops of 10 to 30% hydrogen peroxide (H_2O_2) added at the time of etching. This is carried out by placing approximately 5 mL of the diluted ammonia solution in a watch glass and adding 3 to 4 drops H_2O_2. The mixture is swabbed on the brass with a cotton-tipped stick for 20 to 40 s (Ref 11). The nickel silver requires 1 to 2 min to etch. The etchant decomposes after 30 min, and a new mixture with the H_2O_2 should be prepared as required. The $K_2Cr_2O_7$ etchant noted above for bronze also works well on the brass and nickel silver.

Aluminum parts are etched using Keller's reagent, which is applied by swabbing with cotton for 15 to 30 s.

Titanium parts are etched by immersion for 15 s using 10% concentrated hydrofluoric acid (HF), 5 mL concentrated nitric acid (HNO_3), and 85 mL H_2O.

Tool steels are etched using 5% nital (5% HNO_3 concentrated in methyl alcohol). As-sintered materials should be swabbed for 5 min. As-annealed materials may etch within 20 s.

Microstructures of P/M Materials

Powder metallurgy materials include many alloy systems. A number of these systems are discussed in the sections that follow, with the main phases and compounds noted. Related information on the production, characterization and testing, consolidation, and applications and properties of these powder systems can be found in this Volume.

Iron-Base P/M Materials

Pure iron or very-low-carbon steel is a common structural or bearing material. The microstructure is predominantly ferrite, with modest amounts of pearlite in proportion to the minor amounts of carbon in solution. Several kinds of iron powder are commonly used, including sponge, atomized, electrolytic, and carbonyl powders (Fig. 12–22, 39–52). They have widely differing properties because of differences in surface area, residual alloying, internal porosity, and particle size. For additional information, see the article "Production of Iron Powder" in this Volume.

Sponge irons, which are produced by the reduction of iron oxide, are very irregularly shaped powders with high surface areas. They are more difficult to compress, but develop good green strength and excellent sintered strength at low densities. The pores inside the particles vary in size according to the temperature and time of reduction from the parent oxides. Low-temperature hydrogen reduction of mill scale leads to a fine sponge, and extended carbon reduction of ore concentrate results in a coarser sponge. These powders are characterized by moderate levels of unreduced oxides inside the particles. (Fig. 39–43, 45, 47–49, and 50).

Atomized iron powders press easily, but for densities up to 6.6 g/cm^3, the sponge irons result in higher sintered strength. Therefore, the atomized materials are usually found in materials pressed at 6.6 to 7.2 g/cm^3 (the pore-free density of iron is 7.87 g/cm^3). (Fig. 51).

Electrolytic Iron Powder. The high green strength, high compressibility, irregular particle shape, and high purity of electrolytic iron make it suitable for a number of applications, such as soft magnetic parts and enrichment of food products. Because of its high production costs, however, current usage is limited. ($2.00 to 4.25/lb versus $0.40/lb for atomized iron).

Carbonyl iron powders may have a particle size of 2 to 5 μm and are often used in metal-injection-molded P/M parts (see the article "Powder Injection Molding" in this Volume). Their high surface area and fine particle size allow the material to sinter to near-full density with large associated shrinkages. The resulting structure will be ferrite, with small rounded and isolated pores.

Iron-Graphite Mixtures. Sintering results in rapid diffusion of carbon into iron, with resulting steels containing up to 0.8% C. These steels show increasing amounts of pearlite with increasing carbon content calculated using the lever rule. The combined carbon for these materials is approximately 0.8% times the area fraction of pearlite. That area fraction does not include the area associated with porosity. For materials with hypereutectoid carbon contents (typically >0.8%), iron carbide networks appear in the grain boundaries, and the impact, tensile strength, and elongation are reduced. This carbide network is not to be confused with the divorced eutectoid carbide platelets that will appear occasionally in a grain boundary in the hypoeutectoid steels with less than 0.3% C. This effect is seen in ingot-based steels as well. Sintered iron bearings are fabricated with graphite in solution as well as present as free graphite flakes. The combined carbon is judged by the lever rule, which is important in quality control of iron-graphite bearing materials. Iron-carbon P/M structures are shown in Fig. 53 to 57.

Iron-Copper Alloys. Copper is frequently added to iron because it melts and rapidly dissolves, greatly increasing the strength of the iron. When copper melts, it is drawn by capillary action into the smallest available pores and capillaries. In an atomized iron powder, the copper flows between the particles that are pressed into close contact. It then dissolves in the iron at these points of contact. The copper activates the sintering of the particles that are in contact, resulting in rapid disappearance of particle boundaries and substantial neck growth.

The copper may separate the iron particles as it flows among them, causing growth of the part in 1 to 2 min. The subsequent dissolution of the copper and local lattice expansion at points of contact cause later growth. In an undersintered part with 2% added Cu, some of the residual copper may still be visible as a thin line between two iron particles. With sponge irons, the copper can flow into the fine pores inside the particles and thus not cause as much separation of particles. The high surface area also contributes to rapid sintering. These two effects are thought to explain why the sponge iron and copper mixtures do not expand as much on sintering as the mixes based on atomized iron.

In conventional sintering of iron-copper alloys (20 to 30 min at 1105 to 1120 °C, or 2025 to 2050 °F), at least 2% Cu disappear into solution in the iron. With 5% or more Cu, some free copper is always present as a copper-rich solid solution with the iron. Depending on the rate of cooling, copper-rich phases precipitate in the iron, and, vice versa. The copper-rich phases precipitate in the iron, darkening the ferrite; slow cooling increases darkening. This effect is limited to the outside of the iron particles, because the melted copper does not readily diffuse to the particle centers under conventional sintering conditions. Picral etching will help to stain the copper precipitate areas for easier identification. Iron-copper P/M structures are shown in Fig. 58 to 62.

Iron-Copper-Carbon Alloys. The most common of the moderate-strength, as-sintered alloys is Fe-Cu-C with 0.5 to 0.8% C and 2 to 5% Cu (Fig. 63 to 66). It combines the features described above for iron-carbon and iron-copper alloys. The carbon goes rapidly into solution in the iron (perhaps in 5 min at 1040 °C, or 1900 °F) and tends to prevent the sintering expansion prevalent in zero- or low-carbon iron-copper alloys. The combined carbon can be estimated by the lever rule, although the eutectoid may be as low as 0.75% C in this ternary system.

Copper-Infiltrated Steels. High-density iron-carbon alloys with 10 or 20% Cu are prepared by infiltrating the copper alloy into the porous steel matrix. Upon sintering and infiltrating, the copper alloy melts and flows into the iron-carbon matrix with which it is in contact. The copper tends to fill the highest-density, smallest-capillary regions of the matrix first. The coarse-pored lowest-density regions are filled last with whatever liquid copper remains. The structure often appears as islands of ferrite and pearlite with a continuous copper-alloy phase. The alloy of copper may include such elements as manganese and cobalt, which dissolve in iron and alter the alloy content of the steel matrix. Manganese increases the hardenability of the matrix. Elemental nickel contained in the matrix goes into solution in iron and copper, greatly increasing hardenability. Such materials may exhibit regions of martensite, even as furnace cooled. Copper-infiltrated steel structures are shown in Fig. 67 to 72.

Low-alloy steels of the 4600 series type are atomized as low-carbon materials with good compressibility. Because of their alloying elements, they display excellent hardenability and are usually used fully hardened. When viewed in the as-sintered condition, such materials exhibit ferrite and a eutectoid product that does not appear similar to the normal iron-carbon materials. The lamellae are more uniformly spread throughout the structure, and the tendency among the constituents to group into ferrite and pearlite is lessened, which complicates estimating the combined carbon content metallographically. However, this should be possible by devising metallographic photo standards of reference. The powder may contain up to 5% unalloyed iron as a contaminant. In the as-sintered structure, these free-iron particles do not tend to pick up carbon and thus stand out as ferrite (the carbon preferentially dissolves in the prealloyed steel powder particles). Upon quenching, the unalloyed particles are low in carbon and alloy content, do not harden, and are all ferrite or ferrite/pearlite mixtures. Low-alloy steel structures are shown in Fig. 73 to 80.

Iron-Phosphorus Alloys. The additions of iron phosphide (Fe_3P) to atomized iron results in the dissolution of phosphorus in amounts less than 1%. The phosphorus initiates a transient liquid-phase sintering reaction, then goes partly into solution in the iron, resulting in a material with excellent soft magnetic properties. Some of the phosphorus remains visible as a second phase with the ferrite. For magnetic properties, a low carbon content and freedom from pearlite are required. For optimal toughness and strength characteristics, a mixture of up to 1% P and up to 0.3% C is used. The phosphorus also causes pore rounding by virtue of the transient liquid phase, which gives the alloys their toughness and characteristic well-sintered appearance. Iron-phosphorus alloy structures are shown in Fig. 81 and 82.

Free-Machining Steels. The machinability of sintered irons and alloys is improved by adding sulfur. Historically, this has been accomplished by mixing fine sulfur powder (−325 mesh) into sponge iron. More recently, sulfur is dissolved in the liquid melt before atomizing (prealloyed sulfur) to form manganese sulfide (MnS) with carefully controlled amounts of manganese. Manganese sulfide particles have also been mixed with iron for a similar benefit. These additions result in particles of MnS in the pores as a gray phase, or a MnS phase inside the iron particles, if it was prealloyed. The use of high-hydrogen atmospheres at sintering desulfurizes a material to depths of 0.25 to 0.50 mm (0.01 to 0.02 in.), an effect whose analog in carbon is better known. Structures of P/M steels with additions of manganese and sulfur for enhanced machinability are shown in Fig. 83 to 85. Boron nitride powder is also mixed into iron in small amounts to improve machinability. Finally, the most effective enhancer of machinability is impregnation of the pores of the sintered material, using a plastic resin.

Nickel Steels. The most common high-strength heat treated materials are the nickel steels. In these mixtures, 2 or 4% elemental Ni is added to iron, along with 0.4 to 0.8% C and up to 2% Cu (optional). The usual nickel is very finely divided and is often prepared by carbonyl decomposition (production of nickel powder by carbonyl vapor metallurgy processing is discussed in the article "Production of Nickel-Base Powders" in this Volume). The copper is generally added for size control during sintering, because nickel induces shrinkage and copper causes expansion. The copper activates sintering, as noted previously in the section "Iron-Copper Alloys," and promotes the dissolution of nickel in the iron. Nickel-steel structures are shown in Fig. 86 to 98.

Nickel-rich regions comprise 20 to 50% of the area of these structures. The regions are extensive because the nickel content of their interiors has been diluted to 12% Ni by inward diffusion of iron. The nickel-rich regions tend to etch lightly. Their interiors often are unetched austenite, and their peripheries contain martensite or bainite with microhardnesses of 40 to 55 HRC, converted from 100 gf Knoop. The pearlite colonies are usually surrounded by a white band that appears similar to ferrite, but never contains eutectoid products. It is probably a higher-alloy diffusion zone. The austenitic cores of the nickel-rich regions increase toughness and strength in these alloys and tend to inhibit ductility. The undiffused nickel-rich regions figure significantly in the overall performances of the alloy. These islands with hard phases in the as-sintered condition contribute a degree of wear resistance that would not normally be expected.

It is difficult to assess the degree of sintering by studying the nickel-rich areas because copper additions greatly affect their extent and appearance. Sintering is best judged by the disappearance of original particle boundaries and by pore rounding. It is difficult to discern the combined carbon level in the nickel steels because of the presence of the nickel-rich regions, the white diffusion layer, porosity, and the probable lowering of the eutectoid carbon level by the nickel.

Diffusion-alloyed materials, such as Distaloy (Hoeganaes Corp., Riverton, NJ) are powders in which the alloying elements of molybdenum, nickel, and copper are added as finely divided elements or oxides to the iron powder. They are then coreduced with the iron powders at an annealing step, resulting in the firm attachment and partial diffusion of the elements to the iron. This partial alloying increases hardenability compared to elemental mixtures, yet these powders exhibit good compressibility. Bonding of the alloying elements also reduces the tendency toward powder segregation.

The sintered structures exhibit ferrite, pearlite, and nickel-rich regions such as those described previously for the elemental mixes, and the nickel-rich regions have all the benefits noted above. With added copper, additional partial hardening during sintering occurs. In Europe, this is used to advantage by producing medium-carbon alloys that are sold in the pressed, sintered, and sized conditions, but have good strength and impact resistance. This procedure avoids the distortions that can occur during normal heat treating. Diffusion-alloyed structures are shown in Fig. 99 to 101.

Sintered stainless steels are available in compositions that approximate AISI designations 303, 304, 316, 409, 410, 430, and 434. The austenitic materials display austenite grains and annealing twins. The most significant disadvantage may be decoration of the grain boundaries with chromium carbides, indicating loss of chromium from solution and reduction in corrosion resistance. The degree of pore rounding is the most important indication of strength and ductility. The materials are virtually always prepared from prealloyed powders; some variants contain added tin or copper for improved corrosion resistance. The 410 materials are often fabricated with 0.15% graphite mixed with prealloyed powders. This results in such high hardenability that the as-sintered structures are essentially all martensite and require tempering after sintering for optimal properties. Stainless steel P/M structures are shown in Fig. 102 to 113.

Powder metallurgy tool steels have long been used for tooling components such as punches and dies. These materials are produced by hot isostatic pressing of water-atomized, tool steel powders, resulting in a fully dense product with fine grain size and very fine, uniform carbide size. The product displays grindability that is superior to ingot-base tool steels. Such alloys as M2 and T15 are also available in molding grade powders. In addition to hot isostatic pressing, P/M tool steels can be fabricated by pressing to approximately 80% density, followed by vacuum sintering to full density. Tool steel powders of the M2 and T15 compositions can be cold pressed at 550 to 825 MPa (40 to 60 tsi), then liquid phase sintered to full density. For M2, sintering requires 1 h in vacuum at 1240 °C (2260 °F) at 100 to 1000 μm nitrogen or argon; T15 takes 1 h at 1260 °C (2300 °F) in the same vacuum. Temperature control within 5 °C (9 °F) may be required for product uniformity. The as-sintered T15 structures contain retained austenite, because of the high amount of carbon in solution, as well as primary M_6C and fine MC (vanadium carbide). The M2 structures contain mainly M_6C of varying small sizes against a matrix of retained austenite. The martensite start, M_s, temperature for these materials with the high carbon in solution is below room temperature. Upon annealing, the carbon precipitates out of solution onto the M_6C phase, reducing the carbon in the matrix. This structure may then be heat treated at 1150 to 1205 °C (2100 to 2200 °F), but heating and cooling times must be minimized to avoid putting too much carbon back into solution. Upon furnace cooling or air cooling, the matrix then forms martensite with the proper distribution of fine carbides (Ref 12). Powder metallurgy tool steel structures are shown in Fig. 114 to 116 (see also the article "Particle Metallurgy Tool Steels" in this Volume).

Nonferrous P/M Materials

As discussed in the section "Metal Powder Production and Characterization" in this Volume, a great many nonferrous metals are also produced in powder form, including:

- *Copper:* by reduction of oxides, atomization, electrolysis, and hydrometallurgical processing
- *Tin:* by atomization
- *Aluminum:* by atomization
- *Magnesium:* by mechanical comminution and atomization
- *Nickel:* by carbonyl vapormetallurgy, hydrometallurgy, and atomization
- *Cobalt:* by carbonyl vapor metallurgy, hydrometallurgy, reduction of oxides, and atomization
- *Silver:* by chemical precipitation, electrolysis, and reduction of oxides
- *Gold, platinum, and palladium:* by chemical precipitation
- *Tungsten and molybdenum:* by reduction of oxides

- *Metal carbides:* by carburization, Menstruum process, and exothermic thermite reactions
- *Tantalum:* by reduction of potassium tantalum fluoride and a sequence of electron beam melting, hydriding, comminution, and degassing (dehydriding)
- *Niobium:* by aluminothermic reduction of oxides
- *Titanium:* by reduction of oxides and atomization
- *Beryllium:* by reduction of vacuum-melted ingots by comminution
- *Composite powders:* by diffusion (alloy coating)

This section reviews copper-, titanium-, and aluminum-base P/M materials.

Copper-base alloys include pure copper for high-density electrical applications: 90Cu-10Sn bronzes for bearings and structural parts; brasses with 10, 20, and 30% Zn; and nickel silver (Cu-18Zn-18Ni). The brasses and nickel silvers are used for structural parts that require ductility, moderate strength, corrosion resistance, and decorative value. Copper exhibits a single-phase structure with some annealing twins. The most significant feature is the particle boundaries or their absence. There should be virtual freedom from particle boundaries from the surface to the center of the part. Bronzes should display all α-bronze with no gray copper-tin intermetallic compounds. Optimal mechanical properties and machinability dictate a minimum of reddish copper-rich areas and small grain clusters of bronze. Mixes containing admixed graphite show the mottled gray flakes in the pores of the part. Bearings exhibit varying degrees of sintering, depending on the final application. In general, however, a well-sintered bearing results in greater ease of oil impregnation. Bronze P/M structures are shown in Fig. 117 to 120.

Brasses and nickel silvers are generally single-phase structures. They should display good pore rounding and almost no original particle boundaries. Some of the materials may contain up to 2% Pb within the particles as an aid to machinability; this will appear as a fine, rounded gray phase (Fig. 121 and 122).

Titanium and titanium alloys such as Ti-6Al-4V are produced from metal powders in several ways. The powders may be prealloyed or may be an elemental mix of titanium and a master alloy of vanadium and aluminum. The latter can be pressed and vacuum sintered to an impermeable state, which may then be hot isostatically pressed to full density without a can. The prealloyed materials may be vacuum hot pressed or preformed, canned, and hot isostatically pressed to full density. Titanium alloys can also be consolidated by metal injection molding. Titanium alloy P/M structures are shown in Fig. 123 to 125.

Aluminum P/M alloys are pressed and sintered to 90 to 95% density. The common alloys are 201AB and 601AB. The alloys are prepared using low-alloy aluminum powder with additions of elemental or master alloy copper, magnesium, and silicon. During sintering, the additions cause a liquid phase to form that fluxes away the surface oxides and allows bonding between the aluminum particles. Sintering in nitrogen is performed at approximately 595 or 620 °C (1100 or 1150 °F) at a dew point of –50 °C (–60 °F) to prevent further oxidation of the aluminum. After sintering, the alloys are often solutionized and quenched, then repressed or coined before aging. The repressing densifies the material and establishes close dimensional tolerances. The materials may also be cold forged or rolled to varying reductions in thickness because of their favorable as-sintered ductility. Aluminum P/M structures are shown in Fig. 126 to 130.

Representative Micrographs

This section includes a discussion of unusual and/or defective structures. Also included are examples of heat treated materials and those subjected to other finishing operations, such as steam blackening. Alternate consolidation processes, such as P/M forging, hot isostatic pressing, metal injection molding, and liquid-phase sintering, are also illustrated.

Sintered parts may be undersintered, which is evidenced by the presence of excessive numbers of original particle boundaries. Undersintering is related to the normal pressed-and-sintered structural materials and their mechanical properties as shown in MPIF standard 35 or the various ASTM materials standardized by the B-9 Committee in ASTM Volume 02.05. In general, for ferrous materials, a field of view at 200× would not be expected to show more than approximately five small segments of original particle boundaries. The presence of larger numbers of particle boundaries would necessitate verifying the sintering conditions and the strength of the part. Figures 27 to 31 depict an increasing degree of sintering, as shown by the disappearance of particle boundaries.

High-temperature (1290 °C, or 2350 °F) sintered austenitic stainless steel does not exhibit particle boundaries, and the degree of rounding of the pores must be examined to compare sintering (Fig. 102 to 111). Injection-molded parts made of fine powders tend to sinter to a closed-pore state with no original particle boundaries (Fig. 131 and 132). Powder metallurgy forgings and hot isostatically pressed parts would not display such boundaries (Fig. 133 and 134).

In the etched condition, sintered steels may exhibit carburization or decarburization (Fig. 135 and 136). If parts of nonuniform section are pressed, density may vary, which may be noted and measured metallographically (Ref 15). If parts are overpressed, the particles will separate, showing microlaminations. Cracks may occur upon ejection, at the change in diameter between two sections of a part, such as between a hub and a flange (Fig. 137 and 138). Even in simple shapes, such as flat tensile bars, improper tool design can cause cracks, which then result in reduced mechanical properties (see Fig. 139).

Heat treated ferrous parts vary in structure from nearly all martensite at the surface to a mixture of martensite, ferrite, and 10 to 30% fine pearlite in the interior (Fig. 140 to 142). This fine pearlite improves tensile properties (Ref 16).

Microhardness testing must be limited to a particular phase when testing with the 100 gf Knoop indenter, for example, martensite. The heat treated structures may display retained austenite, carbides, and subsurface quench cracking (Fig. 143 to 145). Most P/M materials do not form a definite shallow case because of penetration of the carburizing gases. At densities above approximately 7.2 g/cm^3, a definite case tends to form if the core contains less than 0.2% C, as shown in Fig. 146.

Powder metallurgy parts can be finished by steam blackening. The degree of blackening, which should be controlled, generally lowers impact and tensile properties (Ref 17). The gray Fe_3O_4 layer (Fig. 147, 148) penetrates the pores and increases compressive strength and abrasive wear resistance. The thickness of the oxide layer can be measured metallographically, 10 μm (0.0004 in.).

Most P/M parts that are to be plated are first impregnated with a resin to prevent the corrosive plating solutions from entering and remaining in the pores. The resin is visible using optical metallography. The various plated layers are also visible, but polishing should be limited to 3 μm and 1 μm diamond on a short-nap (DAC) cloth to prevent rounding of the plated edge. To examine the original resin in the pores, the part should be mounted in an epoxy resin of contrasting color, for example, red. Also, impregnation with the red epoxy of a specimen that has no existing resin in the pores, allows the epoxy resin to be easily seen. The gray epoxy resins are difficult to image in the light microscope. Often, unfilled pores look dark or filled with debris. An epoxy-filled pore allows one to look through the epoxy at 1000× and see a specular reflection off the bottom.

Powder metallurgy parts can be joined to others by brazing, welding, or adhesive bonding; special precautions are necessary to prevent penetration of the brazing materials. Brazing of sintered steel is done using a special material with a melting point near the normal sintering temperature of iron. When it begins to penetrate iron pores, some iron dissolves in the liquid braze. This raises the solidus or melting point of the liquid, freezing occurs, and penetration proceeds no further into the porous material.

Manufacturers of sintered parts have occasion to examine raw materials (powders) metallographically. This is important because different production methods can result in powders with the same nominal chemistry, but disparate properties. Typical powder structures are shown in Fig. 12 to 26, 32 to 52, and 149 to 162. These micrographs are intended to simplify the task of examining a sintered part and attempting to deduce which powder was used to prepare it.

Figures 163 to 170 depict various normal and abnormal structures. Among these are undersintered structures (as shown in Fig. 163 and compared with an average sinter in Fig. 164), wear-resistant steels (Ancorwear 500) with high carbon contents (Fig. 165), gravity-sintered bronze filter powders (Fig. 166 and 167), sintered parts that were blistered during heating to the sintering temperature (Fig. 168), and parts that exhibit varying density from die-wall friction and the location in the die (Fig. 169 and 170).

Sealed Surfaces. Powder metallurgy parts can have their surface pores closed, either intentionally or inadvertently. Figure 171 shows the outside diameter of a tumbled P/M part in which the surface pores were closed to a depth of 50 μm (0.002 in.). Such closure can interfere with intended flow of fluids through the pores. Powder metallurgy parts, for example, gears and sprockets, are deliberately worked on the surface to densify the surfaces and increase the fatigue resistance. Machining P/M parts also is known to close off surface pores.

Determination of Crack Origin, Green or Sintered State. If a fracture occurs in the green state and the part is then sintered, the particles on the fracture surface will be rounded. If a fracture occurs after sintering, the surfaces of the particles show ductile cups, cones, and asperities, as in the scanning electron micrograph (SEM) view in Fig. 34, 36, and 37. If an SEM is not available, one can plate the fracture surface with 25 μm (0.001 in.) electroless nickel, Edgemet (Buehler Ltd., Lake Bluff, IL), in about 2 h. The plating is done on a specimen that was sectioned perpendicular to the fracture surface. The specimen is then mounted in epoxy resin, ground, and polished. The plating saves the details of the fracture surface, which is viewed at right angles to the fracture plane. In this way, any asperities and cones can be seen in profile view, at 1000×. It is best to create a deliberate fracture through a sintered region to serve as a comparison with the questioned fracture. Figure 172 shows the fracture that occurred in a green part, and it has rounded particle surfaces. Figure 173 shows a fracture that occurred in a sintered and heat treated part, and the asperities where sinter bonds were broken are shown by the arrows.

The plating above keeps epoxy resin from filling pores during sample mounting. This creates some difficulty at polishing, in trying to open all the pores. This is best accomplished at final polishing with 0.05 μm alumina on microcloth with a succession of 15 s nital etchings and 2 to 3 min polishing with 1 N force. Several iterations may be required. Smeared pores are most easily viewed at 1000× and are seen as thin gray lines in the unetched structure.

Sinter Hardened Steels. The FLC-4608-XXHT alloys contain enough copper, molybdenum, nickel, and carbon to harden to a largely martensitic structure during typical cooling from the sintering temperature. The structures show martensite with 0 to 40% colonies of dark fine pearlite, as shown in Figures 140 and 141. Such structures can be distinguished from austenitized and oil-quenched structures by the distribution of the fine pearlite. The sinter-hardened structures tend to have the fine pearlite distributed rather uniformly from surface to interior. Oil-quenched parts tend to be all martensite on the surface, and any fine pearlite is mainly formed in the interior. Sinter-hardened parts with more than 70% martensite should be tempered for 1 h at 175 °C (350 °F) in air to relieve brittleness and increase the tensile strength. Such tempering can double tooth strength, as compared with using the part in the as-sinter-hardened condition.

REFERENCES

1. "Standard Test Method for Density and Interconnected Porosity of Sintered Powder Metal Structural Parts and Oil-Impregnated Bearings," B 328, *Annual Book of ASTM Standards,* Vol 2.05, ASTM, 1998
2. "Permeable Sintered Metal Materials, Determination of Density, Open Porosity and Oil Content," ISO 2738, International Organization for Standardization, American National Standards Institute
3. "Standard Method of Sampling Finished Lots of Metal Powders," B 215, *Annual Book of ASTM Standards,* Vol 2.05, ASTM, 1998
4. G.A. Blann et al., Principles and Practices for Metallographic Preparation for Powder Metallurgy, *Microstr. Sci.,* Vol 22, 1995, p 135
5. L.F. Pease III, Inspection and Quality Control for P/M Materials, *Powder Metallurgy,* Vol 7, *Metals Handbook,* 9th ed., American Society for Metals, 1984, p 485–486
6. L.F. Pease III, Metallography and Properties of Sintered Steels, *Progress in Powder Metallurgy,* Vol 33, Metal Powder Industries Federation, 1977
7. O. Struglics, Hoeganaes Corp., Hoeganaes, Sweden, personal communication
8. S. Coleman and D. Tomkins, A Quantitative Assessment of the Mechanical Properties of Sintered Iron Micrographic Specimens, *Powder Metall.,* No. 2, 1976, p 53
9. "Effective Case Depth of Powder Metallurgy Products, Determination of," Standard 52, Metal Powder Industries Federation
10. S. Kaufmann, Ford Motor Company, Dearborn, MI, personal communication
11. P. Schmey, United States Bronze Powders, Inc., Flemington, NJ, personal communication
12. M. Svilar, SCM Metal Products, Cleveland, OH, personal communication
13. S. Abkowitz, Dynamet Technology, Burlington, MA, personal communication
14. G.F. Millsaps, Alcoa, Pittsburgh, PA, personal communication
15. L.F. Pease III, Inspection and Quality Control for P/M Materials, *Powder Metallurgy,* Vol 7, *Metals Handbook,* 9th ed., American Society for Metals, 1984, p 483
16. L.F. Pease III, The Mechanical Properties of Sintered Steels and Their Derivation for MPIF Standard 35, *Progress in Powder Metallurgy,* Vol 37, Metal Powder Industries Federation, 1981
17. L.F. Pease III, J. Collette, and D.A. Pease, Mechanical Properties of Steam Blackened P/M Materials, *Modern Developments in Powder Metallurgy,* Vol 18–21, Metal Powder Industries Federation, 1988, p 275
18. J. Hurst, C.I. Hayes, Inc., Cranston, RI, personal communication
19. F. Hanejko, Hoeganaes Corporation, Riverton, NJ, personal communication

SELECTED REFERENCES

- W.J. Hupmann and K. Dalal, *Metallographic Atlas of Powder Metallurgy*, Verlag Schmid, GmbH, 1986
- *Metallographic Handbook, Ferrous Powder Metallurgy,* Hoeganaes Corporation, 1995
- G.F. Vander Voort, *Metallography: Principles and Practice, McGraw Hill, 1984*

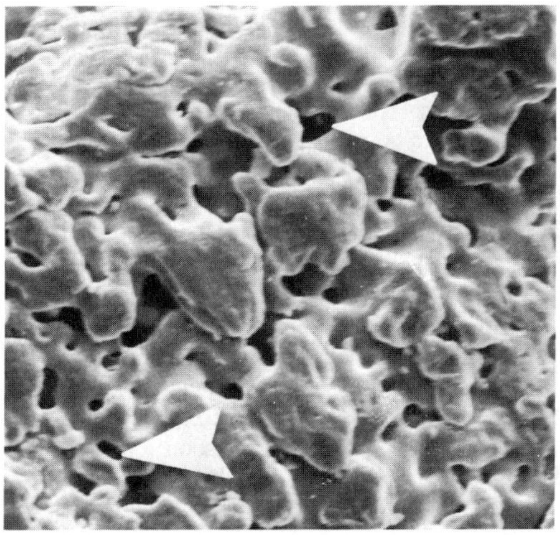

Fig. 12 Pyron 100, hydrogen-reduced sponge iron. A single particle, arrows indicate pores opening into the spongy interior. SEM. 1000×

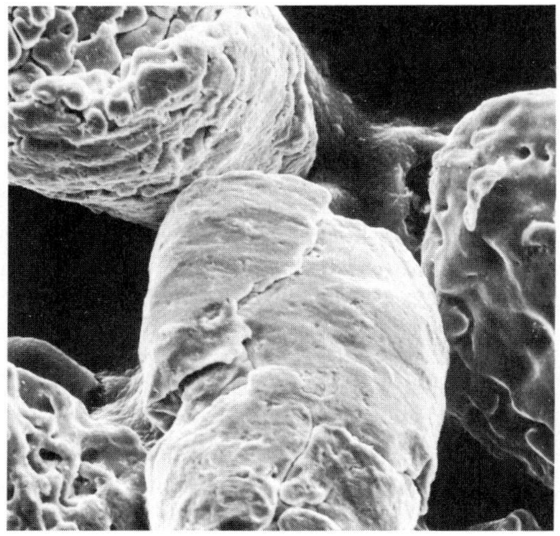

Fig. 13 Pyron D63, hydrogen-reduced sponge iron, exhibiting high apparent density. SEM. 750×

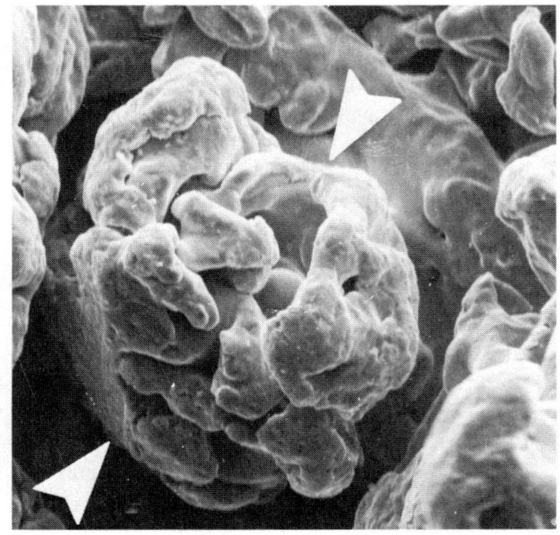

Fig. 14 MH-100, carbon-reduced iron ore. Arrows indicate one particle with coarse internal porosity. SEM. 750×

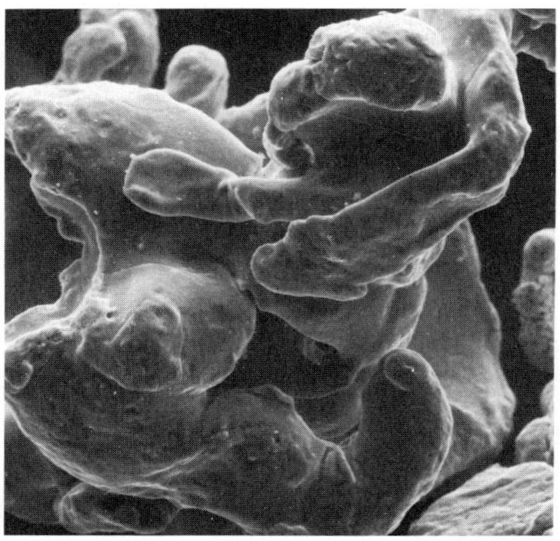

Fig. 15 Ancormet 101, carbon-reduced iron ore. One individual particle with coarse and extensive porosity is shown. SEM. 750×

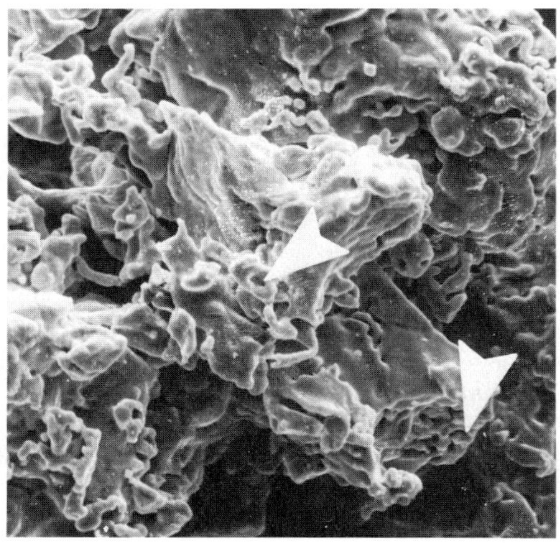

Fig. 16 Atomet 28 iron powder. Arrows indicate porosity in the spongy regions. SEM. 750×

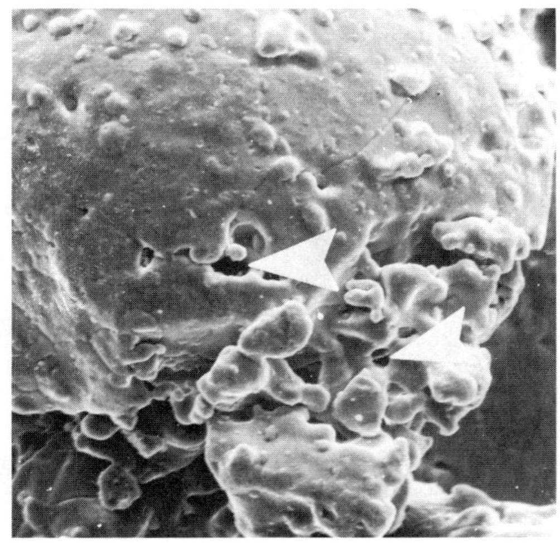

Fig. 17 MP35HD iron powder. Arrows indicate porosity in spongy regions. SEM. 750×

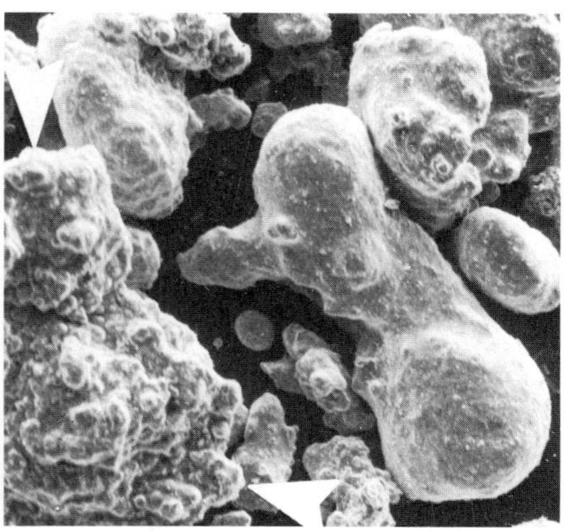

Fig. 18 Water-atomized iron. Arrows indicate this process can produce iron powder with a fair degree of irregularity or roughness on the surface. SEM. 190×

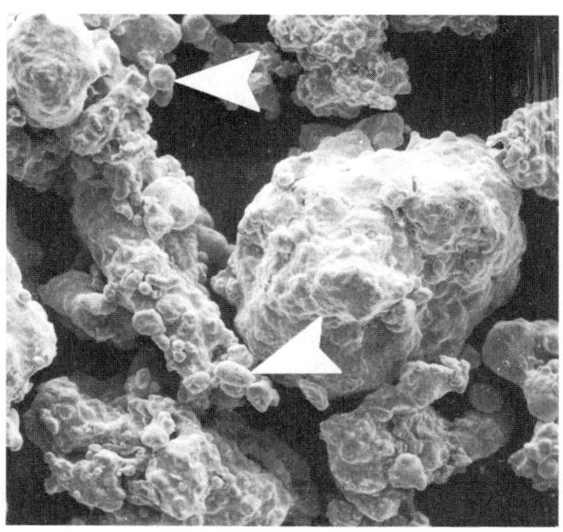

Fig. 19 Ancorsteel 1000, water-atomized and annealed iron powder. Arrows indicate small fines that were agglomerated onto the larger particles. SEM. 190×

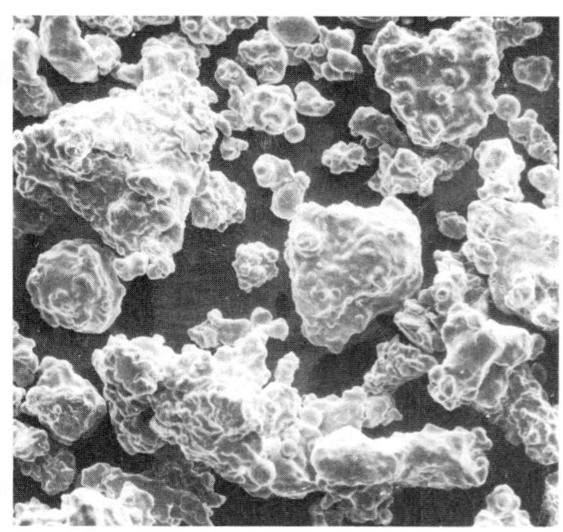

Fig. 20 Ancorsteel 1000B, water-atomized and double-annealed iron powder. SEM. 190×

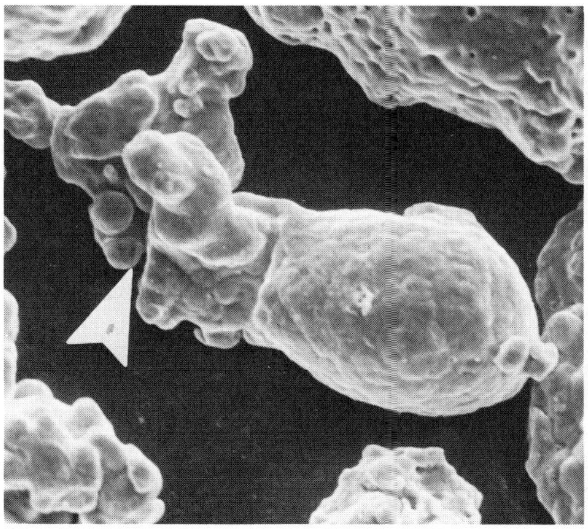

Fig. 21 Ancorsteel 4600V, water-atomized and annealed prealloyed steel powder. Note that some particles gain surface area and irregularity by agglomeration of fines (see arrow). SEM. 750×

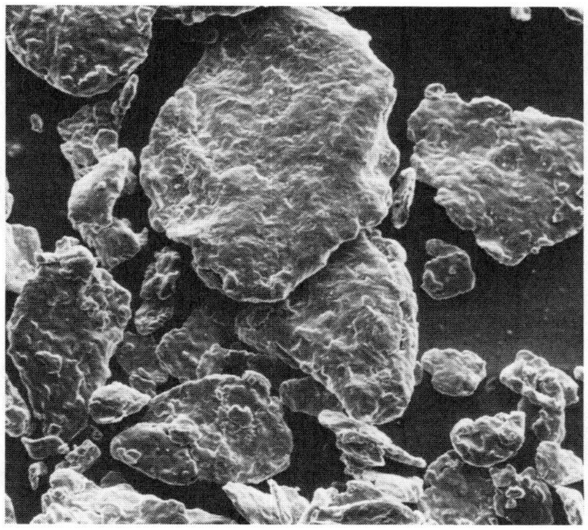

Fig. 22 SCM A283 electrolytic iron powder. Note the flaky shape characteristic of these powders. SEM. 190×

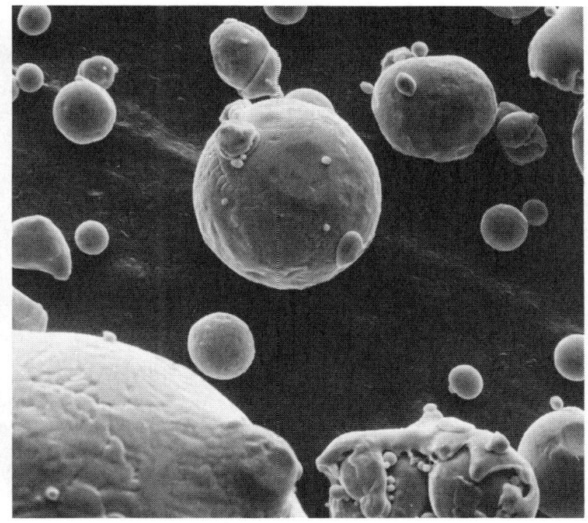

Fig. 23 Type 316, gas-atomized stainless steel powder. Note attached satellites. SEM. 750×

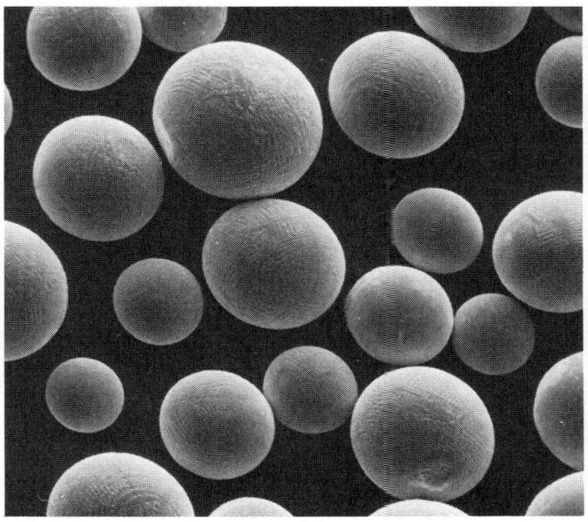

Fig. 24 Type 316L, rotating electrode processed stainless steel powder. Nearly perfect spheres with absence of satellite formation. SEM. 190×

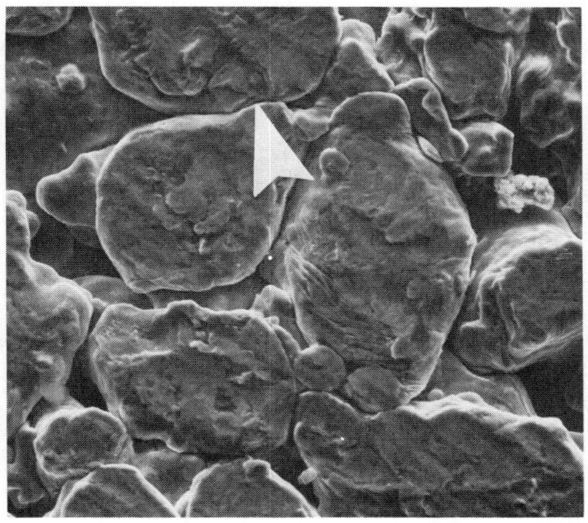

Fig. 25 Ancorsteel 1000 unsintered iron powder. Surface of part, which had been contacted by the upper punch at 275 MPa (20 tsi). Arrow shows the particle boundaries that will disappear during proper sintering. SEM. 750×

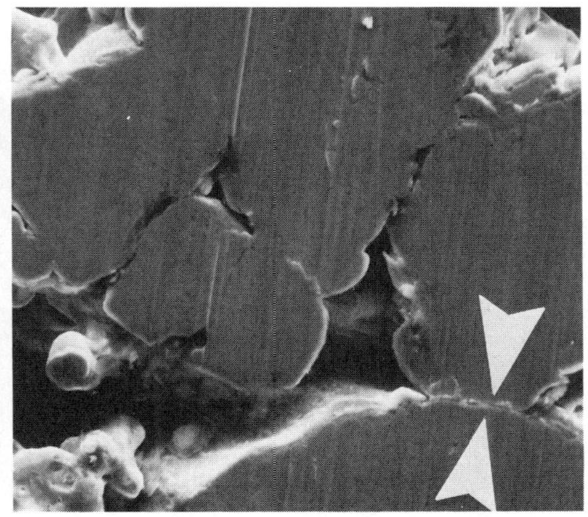

Fig. 26 Same as Fig. 25, but showing the view of the surface that was in contact with the die wall. Arrows show the boundary between particles that must be eliminated during sintering. SEM. 750×

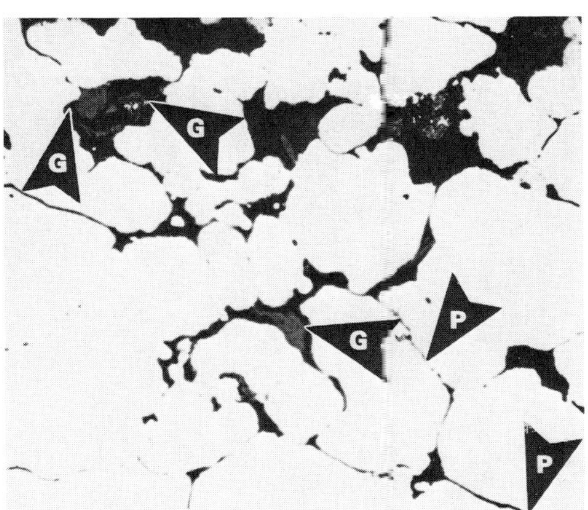

Fig. 27 Distaloy 4600 A (6.7 g/cm^3), pressed at 480 MPa, undersintered 5 min in dissociated ammonia in hot zone at 1120 °C (2050 °F). Arrows P: particle boundaries; arrows G: undiffused, gray flakes of graphite in pores. As-polished. 645×

Fig. 28 Same diffusion-alloyed steel as Fig. 27. Arrows P show the many original particle boundaries. Sintering longer will remove these low-strength boundaries. As-polished. 120×

Fig. 29 Same as Fig. 27, but sintered 15 min. Arrows P indicate persistence of original particle boundaries; arrows R, rounded pores (compare with Fig. 27). As-polished. 645×

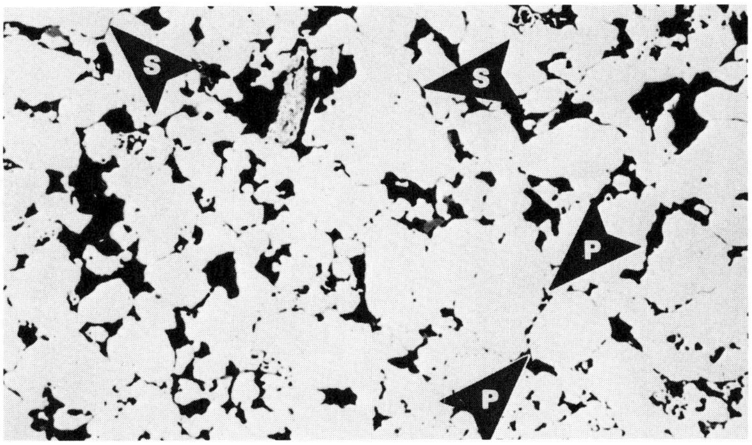

Fig. 30 Same as Fig. 29. Arrows S show segments of original particle boundaries that are shorter and less numerous than those in Fig. 28. Arrows P indicate a row of pores that show how original particles break down into planes of small voids, which coalesce or disappear from diffusion. As-polished. 180×

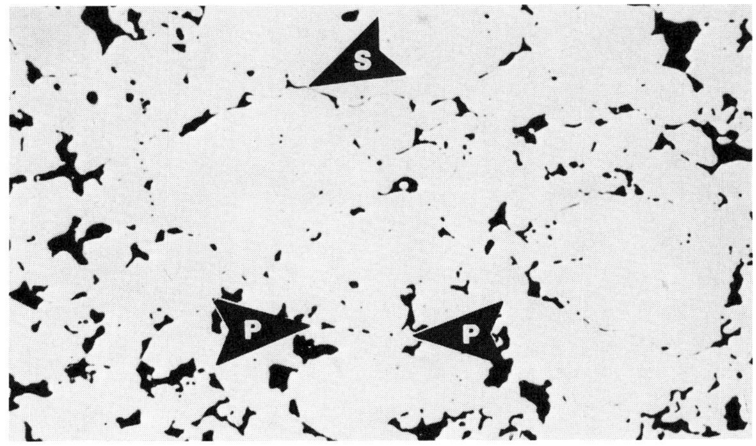

Fig. 31 Same as Fig. 27, but sintered approximately 37 min, which is longer than average. Structure still shows a few segments of original particle boundaries (arrow S). Arrows P indicate a row of pores at which a particle boundary is disappearing. As-polished. 180×

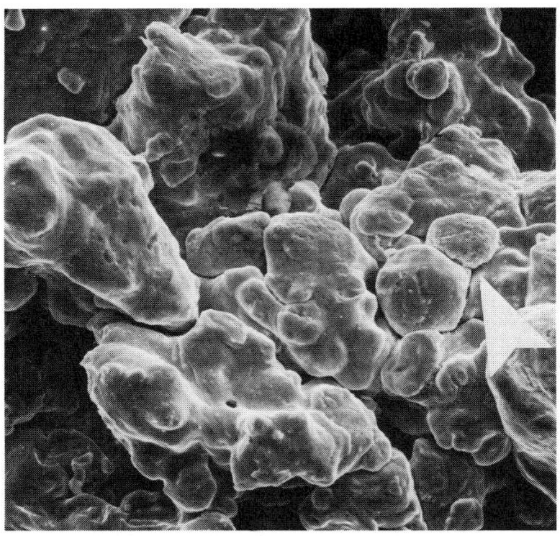

Fig. 32 Fracture surface of Ancorsteel 1000 iron powder (6.4 g/cm^3) pressed without lubricant at 275 MPa (20 tsi). Structure shows no evidence of cold welding or bonding of adjacent particles. Arrow indicates a triple particle boundary that will disappear during sintering. SEM. 750×. Source: Ref 10

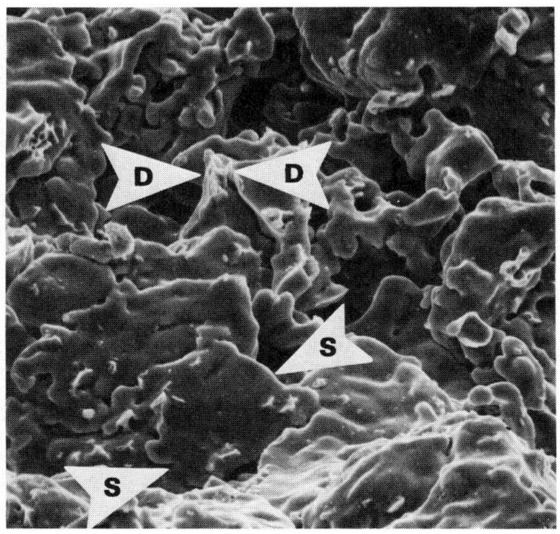

Fig. 33 Fracture surface of Atomet 28 iron powder pressed to 6.6 g/cm^3 and sintered 3 min in hot zone at 1120 °C (2050 °F) in dissociated ammonia. Arrow D shows where a bond has broken. Arrows S outline the smooth, rounded surface of a particle that did not bond to the adjacent particle above it. SEM. 750×. Source: Ref 10

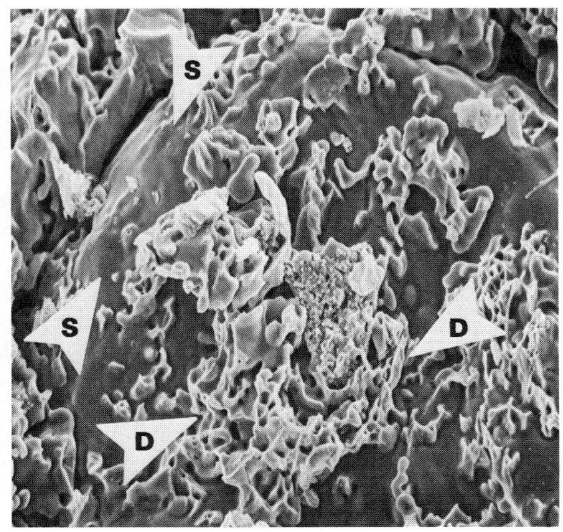

Fig. 34 Some as Fig. 33, but sintered 10 min in hot zone (approximately 1 to 3 min at 1120 °C, or 2050 °F). Arrows D show the ductile cup-and-cone fractures that occurred when this particle was torn from the adjacent one above it. Arrows S show the smooth surface of the particle that had not sintered to any adjacent particle. SEM. 750×. Source: Ref 10

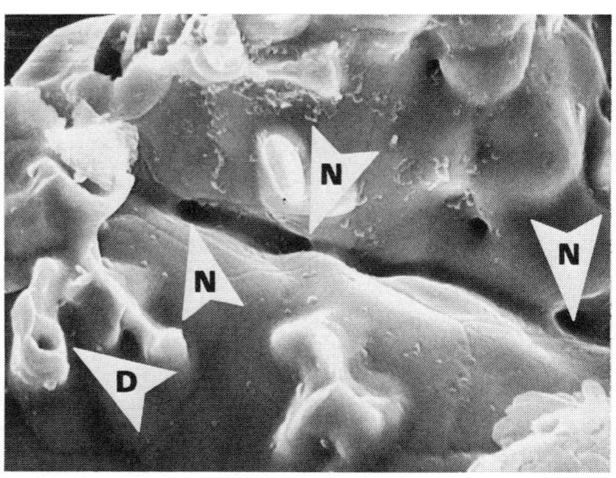

Fig. 35 Same as Fig. 34, but sintered 10 min in the hot zone at 1120 °C (2050 °F). Arrow D indicates a ductile cup-and-cone fracture where this particle was joined to the one above it. Arrows N show necks forming between two adjacent particles. These necks (solid regions) replace particle boundaries as sintering progresses. SEM. 2850×. Source: Ref 10

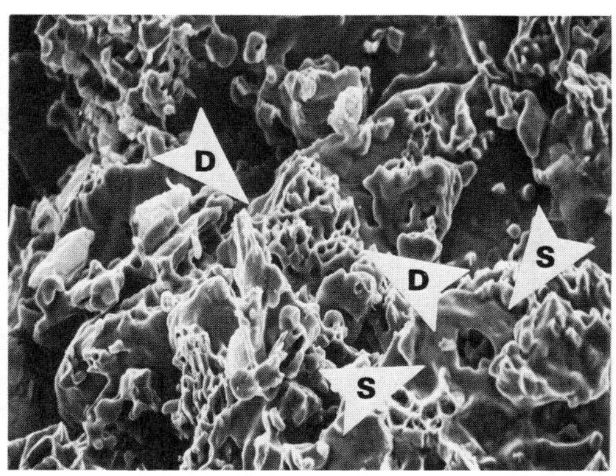

Fig. 36 Same as Fig. 33, but sintered 20 min in the hot zone. Arrows D show the development of ductile cup-and-cone fracture dimples formed when material was torn away from the adjacent particle. Arrows S indicate smooth surfaces where no adjacent particle bonding has occurred. SEM. 750×. Source: Ref 10

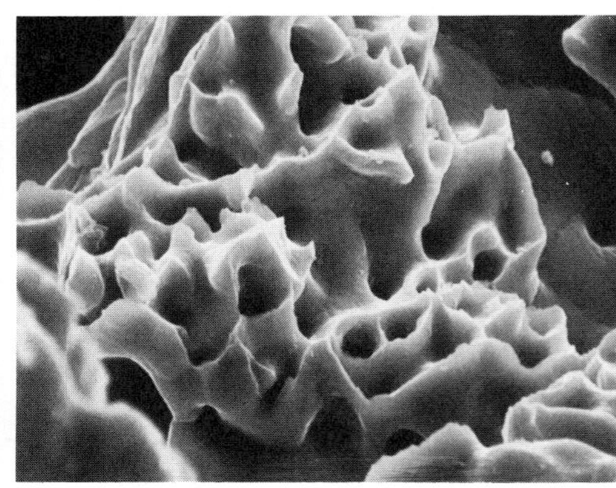

Fig. 37 Some material (Atomet 28) and processing as described in Fig. 36, but shown at higher magnification. Most of the field of view shows the ductile cup-and-cone fractures that occur when the material is torn apart. SEM. 2850×. Source: Ref 10

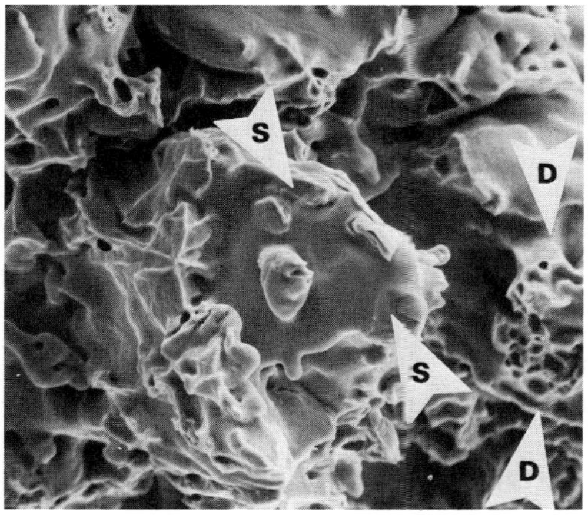

Fig. 38 Same as Fig. 33, but sintered [illegible]0 min in the hot zone. Approximately 50% of the area fraction is occupied by ruptured ductile bonds (arrows D). The remaining area consists of smooth surfaces of particles (arrows S) at which no bonding has occurred. SEM. 750×. Source: Ref 10

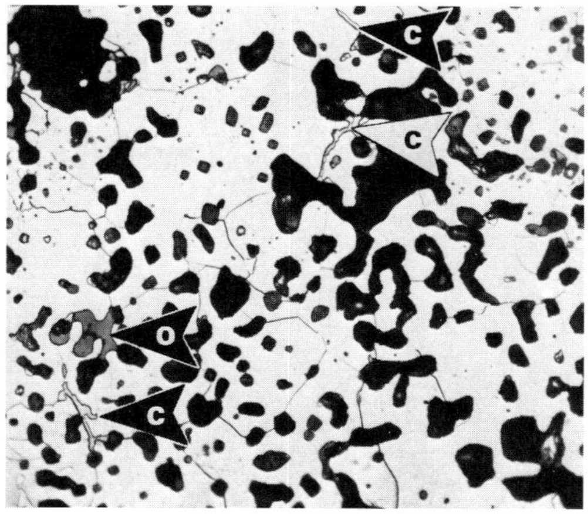

Fig. 39 Pyron D63 sponge iron (6.2 g/cm^3), pressed at 480 MPa (35 tsi) and sintered 30 min at 1120 °C (2050 °F) in dissociated ammonia. Mainly ferrite grain boundaries. Arrow O indicates a small, gray, unreduced oxide particle; arrows C, a few isolated carbide platelets. 2% nital. 645×

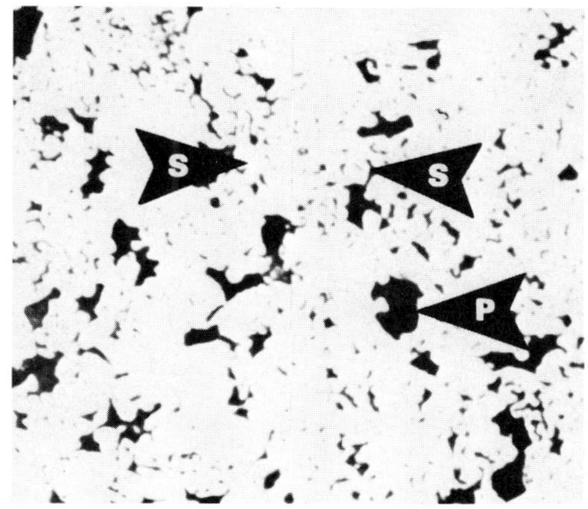

Fig. 40 Same as Fig. 39, but not etched. Arrows S surround a spongy particle having small, internal pores. Arrow P indicates a much larger pore between powder particles. Very few original particle boundaries are present. As-polished. 180×

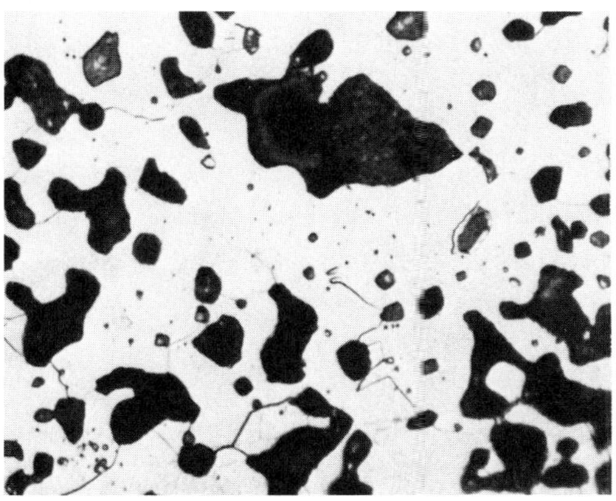

Fig. 41 Pyron 100 sponge iron (6.2 g/cm^3), pressed at 480 MPa (35 tsi) and sintered 30 min at 1120 °C (2050 °F) in dissociated ammonia. Average sinter, no residual particle boundaries. Dark areas are pores. 2% nital. 960×

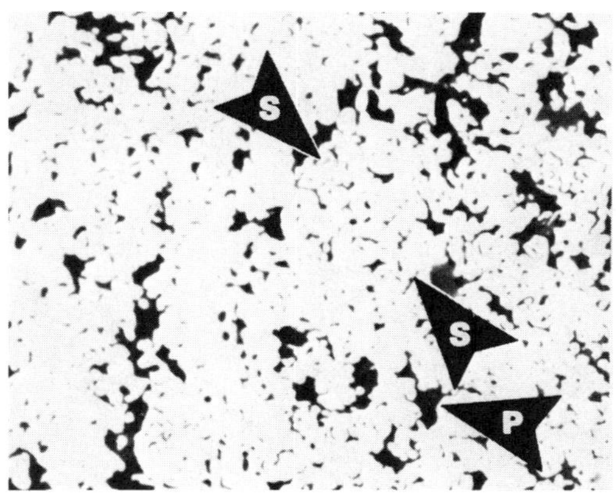

Fig. 42 Same as Fig. 41, but increased density (6.4 g/cm^3). Arrows S surround a spongy particle having small internal pores. Arrow P indicates a larger pore between particles. As-polished. 180×

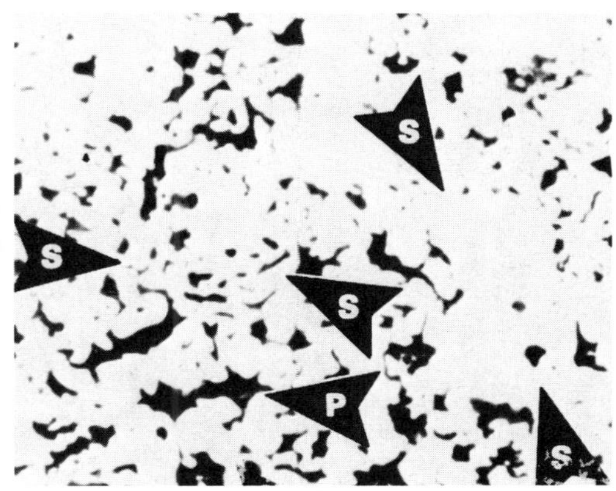

Fig. 43 MH-100 sponge iron (6.4 g/cm^3), pressed at 480 MPa (35 tsi) and sintered 30 min at 1120 °C (2050 °F) in dissociated ammonia. Arrows S indicate the various pore sizes in different particles. The larger pores (arrow P) are between the original particles. As-polished. 180×

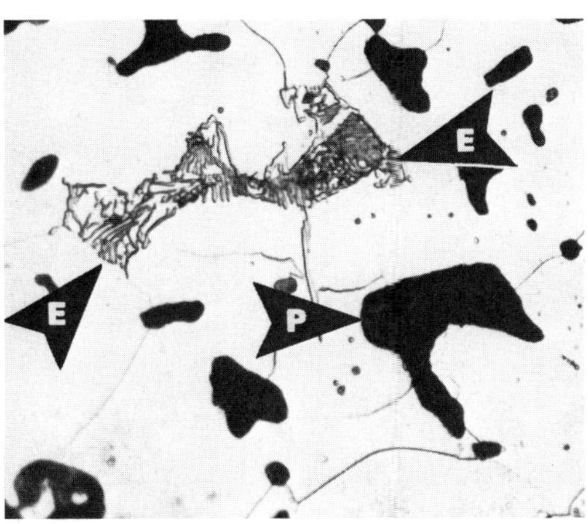

Fig. 44 Same as Fig. 43, but at higher magnification. Arrows E show eutectoid (pearlite) indicating <0.05% combined carbon. Arrow P indicates a pore. Structure is mainly ferrite. 2% nital. 960×

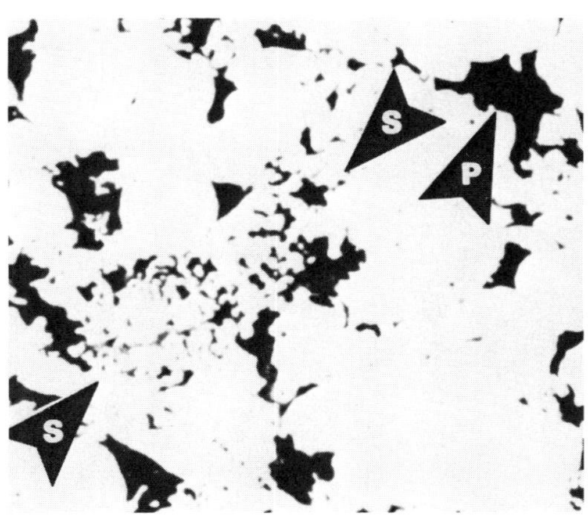

Fig. 45 MP35 iron powder (6.6 g/cm^3), pressed at 410 to 480 MPa (30 to 35 tsi) and sintered 30 min at 1120 °C (2050 °F) in dissociated ammonia. Arrows S surround a spongy region having pores inside the powder particles. Arrow P shows a pore between particles. As-polished. 180×

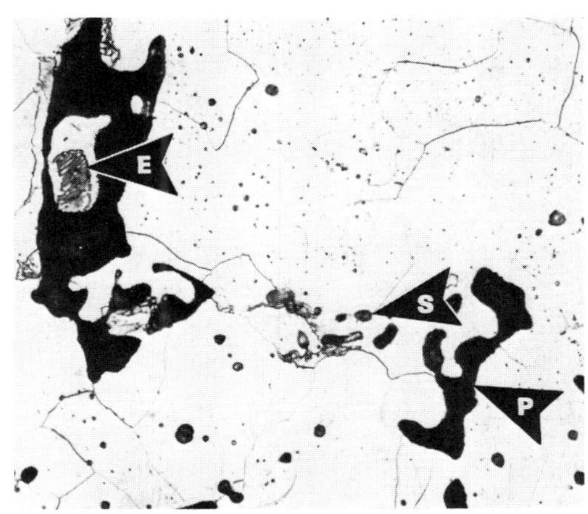

Fig. 46 Same as Fig. 45. Structure is mainly ferrite. Arrow E shows a colony of eutectoid (pearlite) indicating <0.05% combined carbon. Arrow S indicates pores with a spongy particle; arrow P, a pore between particles. 2% nital, 15 s. 645×

Fig. 47 Atomet 28 iron powder (6.7 g/cm^3), pressed at 410 to 480 MPa (30 to 35 tsi) and sintered 30 min at 1120 °C (2050 °F) in dissociated ammonia. Arrows S surround a spongy region. Arrow P indicates pores between particles. As-polished. 100×

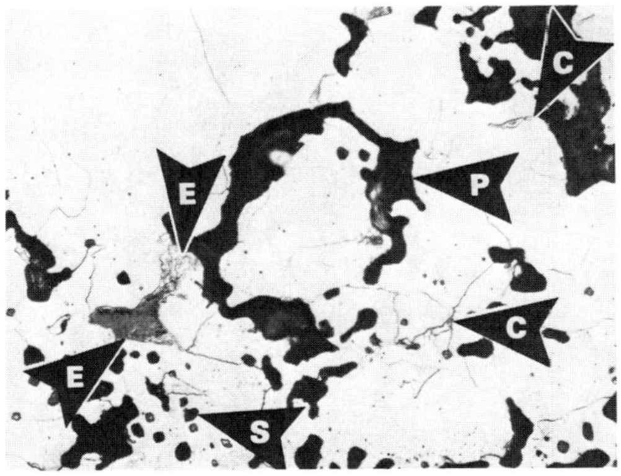

Fig. 48 Same as Fig. 47, but etched. Structure is mainly ferrite. Arrows E indicate eutectoid (pearlite); arrow C, isolated carbides or divorced eutectoid. Arrow P shows a pore between particles; arrow S, pores within a spongy particle. 2% nital. 545×

Fig. 49 Sponge iron blended with atomized low-carbon steel (6.4 g /cm^3), pressed at 410 to 480 MPa (30 to 35 tsi) and sintered 30 min at 1120 °C (2050 °F) in dissociated ammonia. Arrows A show an atomized particle with gas porosity. Arrows S indicate particles with interior porosity. Arrow P shows a pore between particles. As-polished. 100×

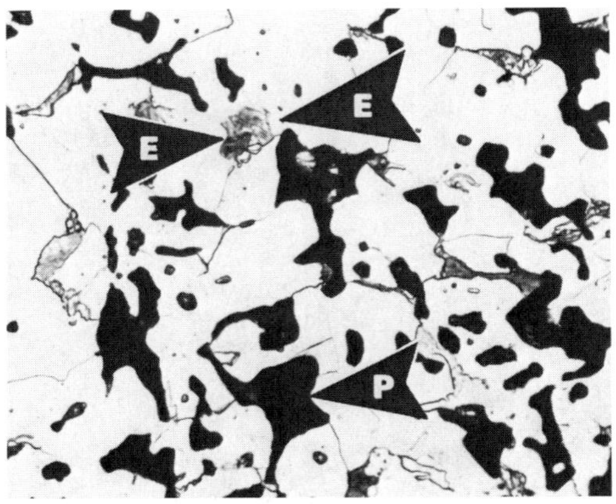

Fig. 50 Ancormet 100 sponge iron (6.4 g/cm^3) with same processing as Fig. 49. Structure is mainly ferrite. Arrows E indicate eutectoid (pearlite). The region shown in this micrograph is less dense than average for the specimen. Arrow P shows a pore between particles. 2% nital. 960×

Fig. 51 Ancorsteel 1000 atomized iron powder (6.7 g/cm^3), pressed at 410 to 480 MPa (30 to 35 tsi) and sintered 30 min at 1120 °C (2050 °F) in dissociated ammonia. Arrows S indicate residual particle boundary segments. Dark regions (arrows P) are pores. Arrows A surround a solid atomized particle. Arrow G shows a pore inside a particle formed during atomization. As-polished. 120×

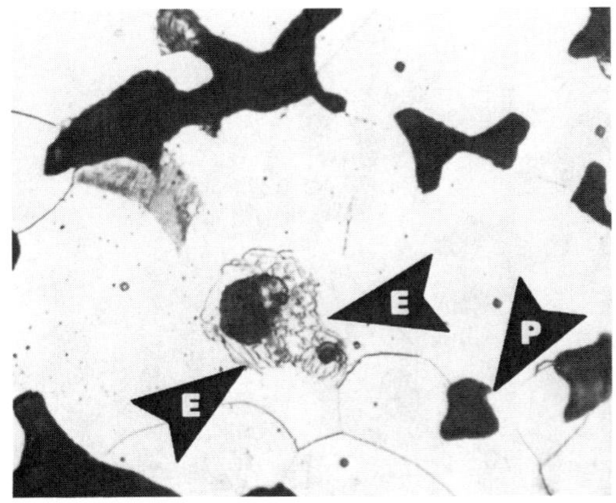

Fig. 52 Same as Fig. 51, but etched. Structure is mainly α-iron (ferrite). Dark regions (arrow P) are pores. Arrows E surround eutectoid (pearlite). Micrographs above 500× do not always show a representative area fraction of porosity and should not be used to estimate density. Micrographs of unetched structures at 180 to 200X are best suited for this purpose. 2% nital. 960×

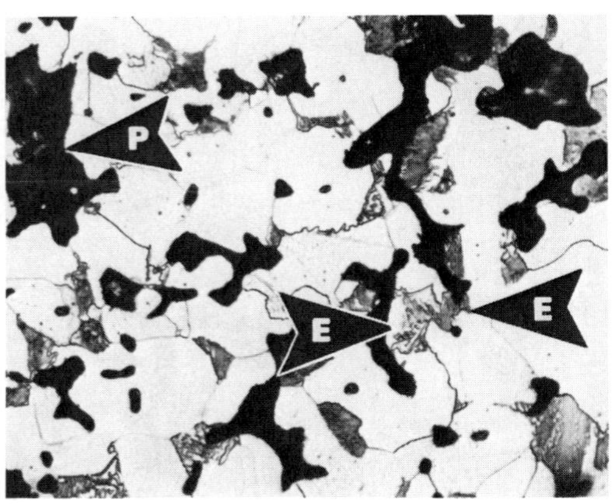

Fig. 53 Atomized iron powder with 0.3% graphite added to yield 0.1 to 0.2% combined carbon (6.7 g/cm^3). Pressed at 410 to 480 MPa (30 to 35 tsi) and sintered 30 min at 1120 °C (2050 °F) in dissociated ammonia. White regions are ferrite. Arrows E surround a colony of eutectoid (pearlite). Arrow P shows a pore. 2% nital. 545×

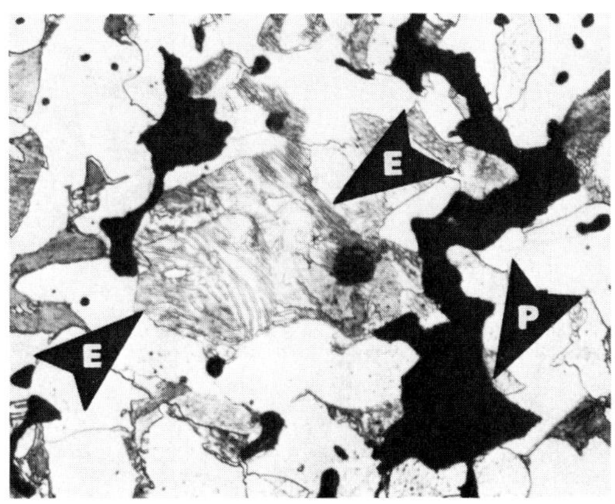

Fig. 54 Atomized iron with 0.4 to 0.5% C (6.7 g/cm^3). See Fig. 53 for processing. P/M steel with 138 MPa (20 ksi) minimum yield strength. Arrows E surround eutectoid (pearlite). White background consists of ferrite grains. Dark areas (arrow P) are pores. 2% nital. 545×

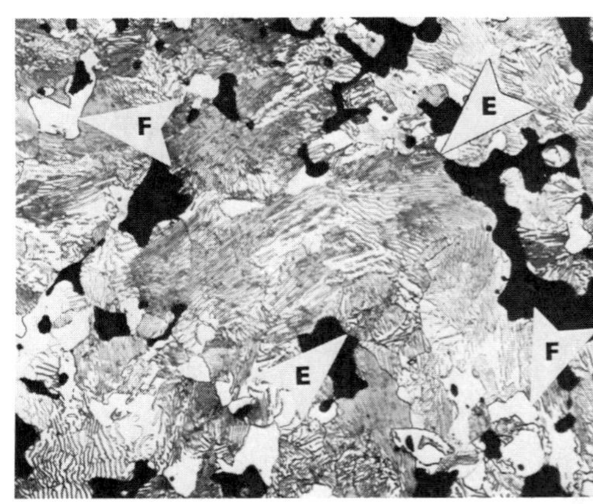

Fig. 55 Atomized iron with 0.8% C (6.7 g/cm^3). See Fig. 53 for processing. P/M steel with 207 MPa (30 ksi) minimum yield strength. Structure is mainly pearlite. Arrows E surround eutectoid (pearlite). Arrows F show a few grains of proeutectoid ferrite. 2% nital. 365×

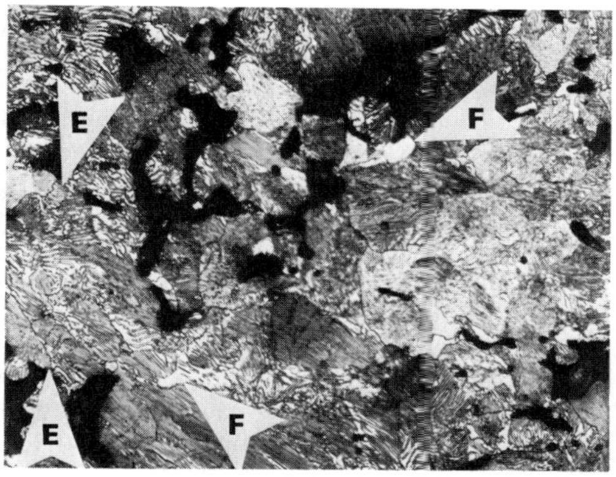

Fig. 56 Atomized iron with 1.0% combined carbon (6.7 g/cm^3). See Fig. 53 for processing. P/M steel with 207 MPa (30 ksi) minimum yield strength. Structure is mainly pearlite. Arrows E surround eutectoid (pearlite). Arrows F show a few grains of proeutectoid ferrite. 2% nital. 310×

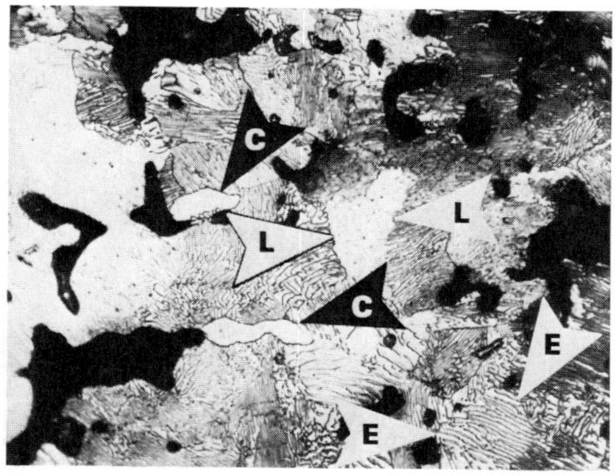

Fig. 57 Atomized iron with 1.3% graphite added to yield 1.1 to 1.2% combined carbon (6.7 g/cm^3). See Fig. 53 for processing. Structure is mainly eutectoid (pearlite), as shown by arrows E. Some pearlite etched very light (arrows L). Arrows C show areas of massive carbide from excessive graphite addition. 2% nital. 310×

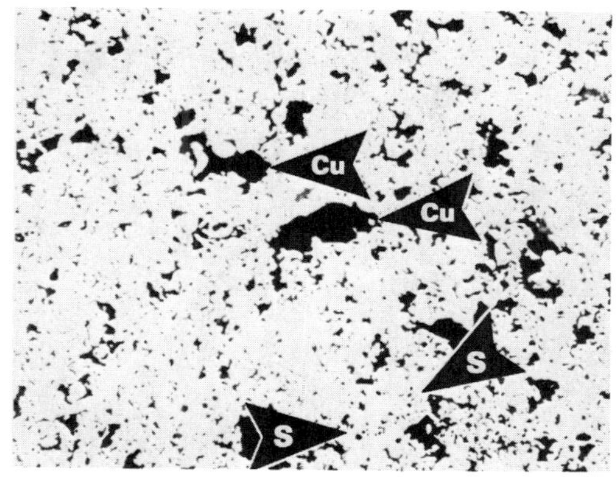

Fig. 58 Pyron 100 sponge iron with 2% Cu (6.3 g/cm^3). See Fig. 53 for processing. P/M copper steel with 124 MPa (18 ksi) minimum yield strength. Arrows Cu indicate pores that were originally occupied by copper particles before their melting and diffusion into the iron. Arrows S surround a spongy region showing pores inside the original iron powder particles. As-polished. 100×

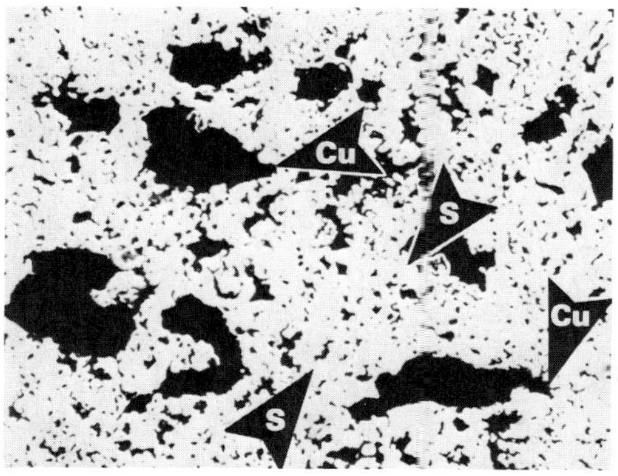

Fig. 59 Same as Fig. 58, but with 7% Cu added (6.7 g/cm^3 with 138 MPa, or 20 ksi, minimum yield strength). Arrows S surround the pores inside the sponge iron. Arrows Cu show pores that were once occupied by copper particles before they melted and filled other, smaller pores and diffused into the iron. As-polished. 310×

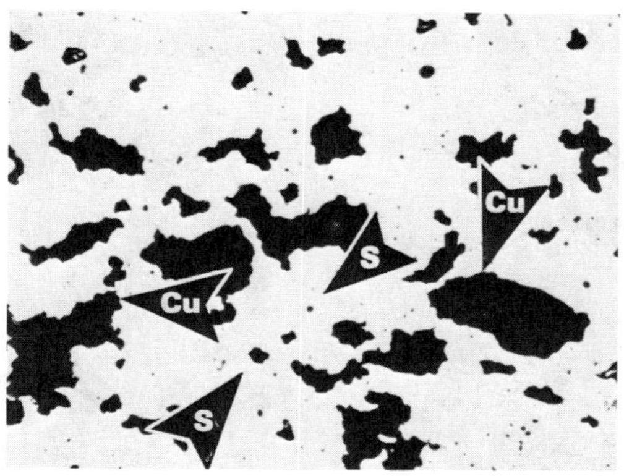

Fig. 60 Same as Fig. 58, but with 20% Cu added (6.3 g/cm^3). Arrows Cu show coarse pores formerly occupied by the copper particles. Arrows S indicate the few pores remaining that may have been present in the original iron particles. The coarse pore size facilitates oil flow in bearings. As-polished. 100×

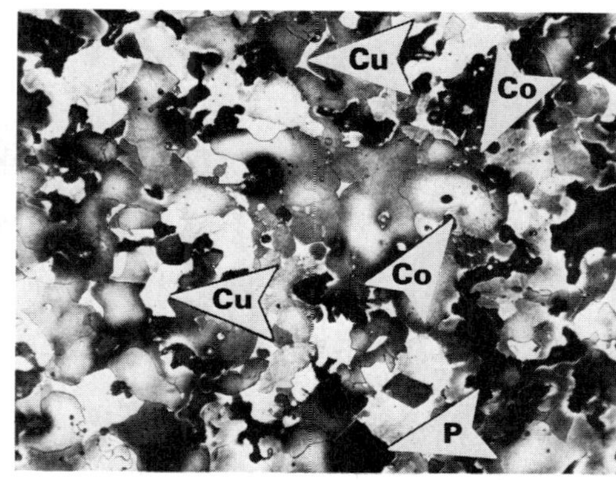

Fig. 61 Same as Fig. 58, but with 10% Cu added (6.3 g/cm^3) White areas (arrows Cu) are melted Cu-rich phase. Arrows Co show coring in iron particles. Darker areas on particle surfaces are where the copper has diffused, then precipitated as a Cu-rich phase upon cooling. The interiors are lighter, indicating no copper diffusion. Arrow P shows a pore previously occupied by a copper particle. 2% nital. 100×

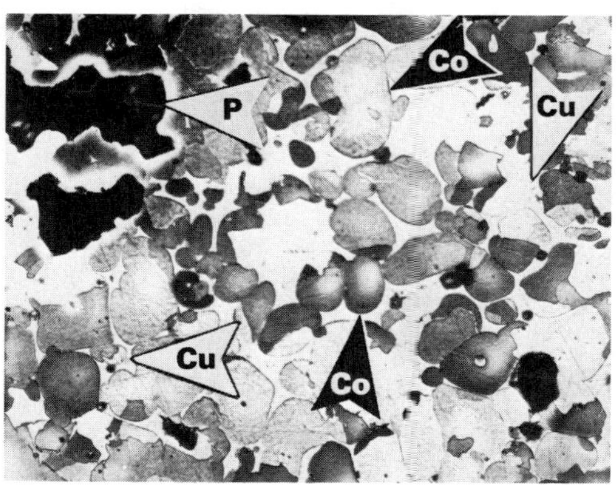

Fig. 62 Same as Fig. 61, but with 20% Cu added (6.7 g/cm^3). The copper is the white phase (arrows Cu). Note coring (arrows Co) as described in Fig. 61. These arrows indicate a string of four iron particles. Arrow P shows a large pore previously occupied by a copper particle. 2% nital. 100×

Fig. 63 Atomet 28 iron powder with 2% Cu and 0.5% C (6.7 g/cm^3 with 275 MPa, or 40 ksi, minimum yield strength). See Fig. 53 for processing. Arrows E show eutectoid (pearlite); arrow F, white ferrite grains. Arrows C indicate isolated or divorced eutectoid in the form of separate Fe_3C platelets. Arrows Co show coring from copper dissolving in the outer regions of iron particles. 2% nital. 310×

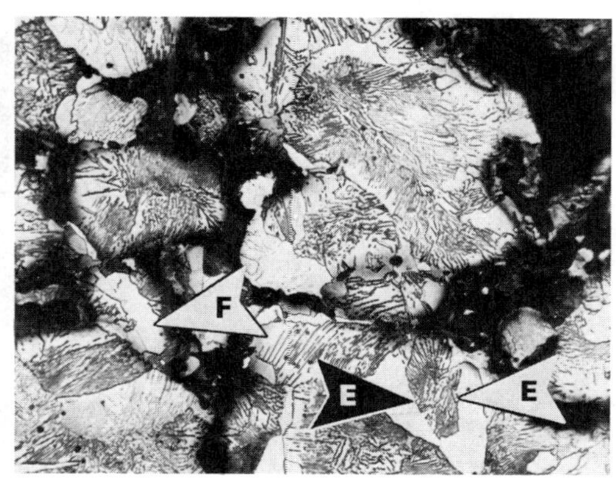

Fig. 64 Atomet 28 iron powder with 2% Cu and 0.8% C (6.7 g/cm^3). See Fig. 53 for processing. Mostly eutectoid (arrows E) with a small amount of white ferrite (arrow F). The shading results from copper coring in the iron. See Fig. 66 and 67 for effects of increased copper content on the resulting properties and structure. 2% nital. 310×

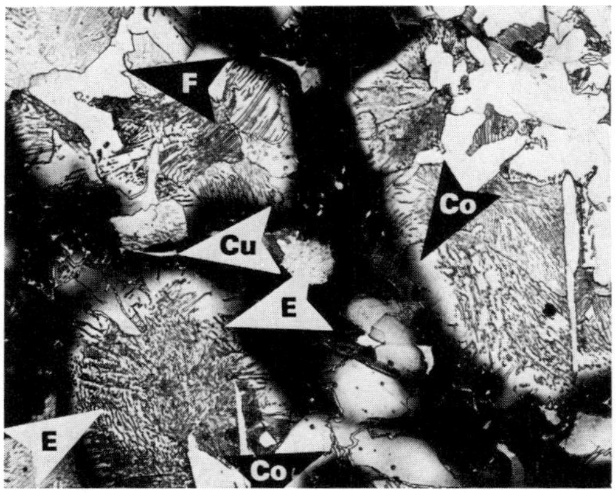

Fig. 65 Same as Fig. 63, but with 5% Cu and 0.5% C (6.7 g/cm^3 with 344 MPa, or 50 ksi, minimum yield strength). At 5% Cu, some residual copper-rich phase is shown as thin, light-colored slivers (arrow Cu). Arrows Co show bands of coring from copper diffusion into the iron. Arrows E indicate eutectoid (pearlite). Arrow F shows ferrite grain. 2% nital. 330×

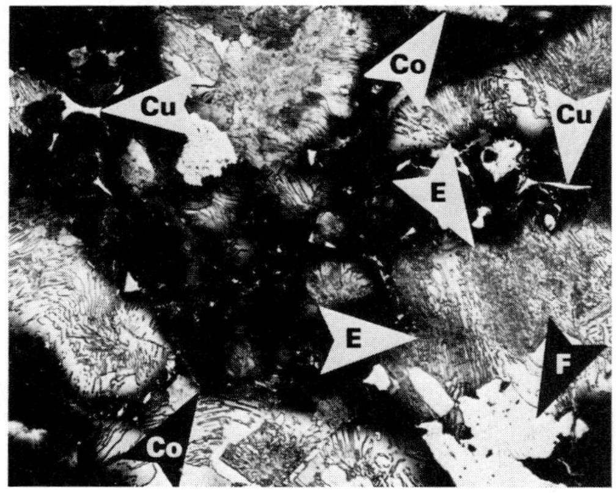

Fig. 66 Same as Fig. 63, but with 5% Cu and 0.8% C (6.8 g/cm^3 with 413 MPa, or 60 ksi, minimum yield strength). Arrows Cu show lenticular regions of melted copper-rich phase; arrows E, eutectoid (pearlite). Arrow F indicates a residual ferrite grain. Arrows Co show coring. 2% nital. 330×

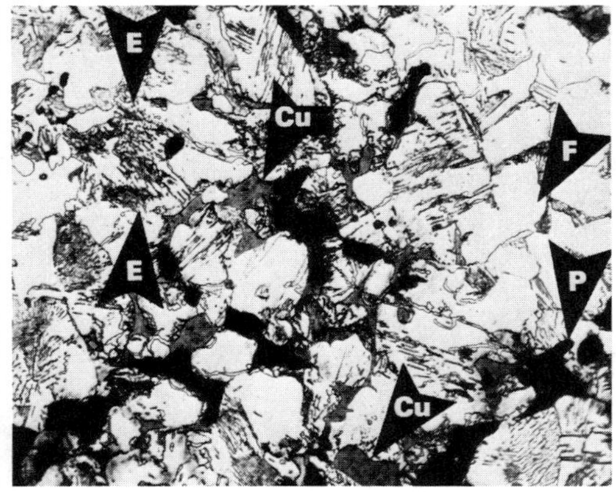

Fig. 67 Copper-infiltrated steel with 10% Cu and 0.5% C, pressed to 6.65 g/cm^3 and infiltrated with copper during 30 min sinter in hot zone at 1120 °C (2050 °F) in endothermic gas. Final density; 7.3 g/cm^3. Minimum yield strength: 276 MPa (40 ksi). Arrows Cu show copper phase; arrows E, eutectoid (pearlite); arrow F, white ferrite grain. Picral. 330×

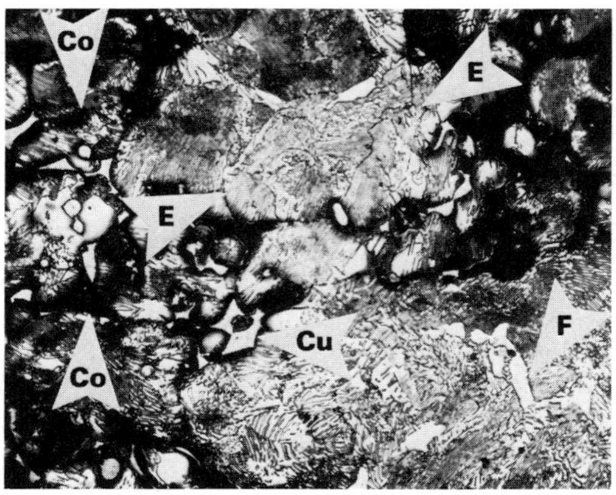

Fig. 68 Copper-infiltrated steel with 10% Cu and 0.8% C (7.3 g/cm^3 final density with 344 MPa, or 50 ksi, minimum yield strength). See Fig. 67 for processing. Arrows Cu show copper-rich region. Arrows E indicate eutectoid (pearlite); arrows Co show coring from partial diffusion of copper in the iron. Arrow Fe shows white ferrite grain. Picral. 330×

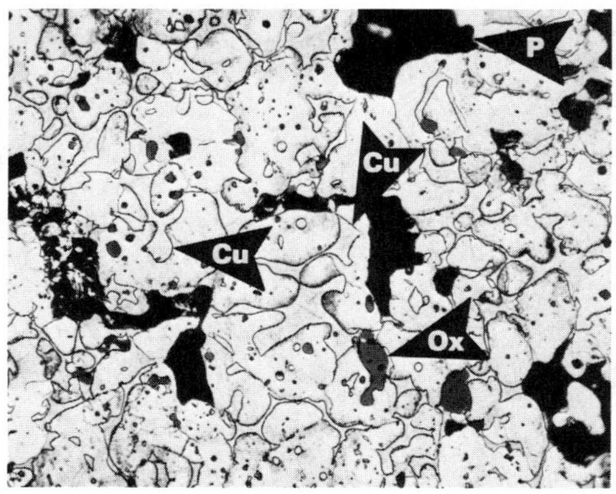

Fig. 69 Copper-infiltrated, hydrogen-reduced sponge iron with 20% Cu (7.3 g/cm^3 final density with 172 MPa, or 25 ksi, minimum yield strength). See Fig. 67 for processing. Arrows Cu show copper phase outlined against the white ferrite matrix. Arrow Ox indicates an unreduced oxide particle, and arrow P a residual pore. 2% nital. 330×

Fig. 70 Copper-infiltrated steel with 20% Cu and 0.8% C (7.3 g/cm^3 final density with 413 MPa, or 60 ksi, minimum yield strength). See Fig. 67 for processing. Arrow Cu shows the light-colored, melted copper-rich phase; arrows E, the eutectoid. Arrow P indicates residual pore. Picral, 30 s. 330×

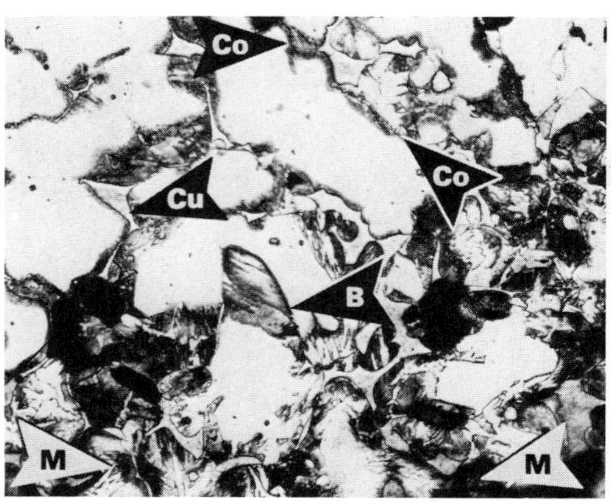

Fig. 71 Copper-infiltrated, H_2-reduced sponge iron with 10% Cu (7.3 g/cm^3 final density). Same processing as Fig. 67, but austenitized 60 min in endothermic gas at 845 °C (1550 °F), oil quenched, and tempered 1 h at 175 °C (350 °F). Arrow Cu shows melted copper-rich phase. Arrows Co indicate a cored region. Arrows M show mixed transformation products present 0.25 mm (0.010 in.) from the surface. Bainite (arrow B) may also be present. Picral. 330×

Fig. 72 Copper-infiltrated steel with 10% Cu and 0.8% C, infiltrated as per Fig. 67 and heat treated as per Fig. 71. Typical structure for interior of 5-mm (0.2-in.) diam tensile bar (minimum ultimate tensile strength of 758 MPa, or 110 ksi). Arrows M show martensite platelets; arrows Co, coring. Arrow C indicates copper-rich region. Arrows A show flecks of retained austenite. Picral. 585×

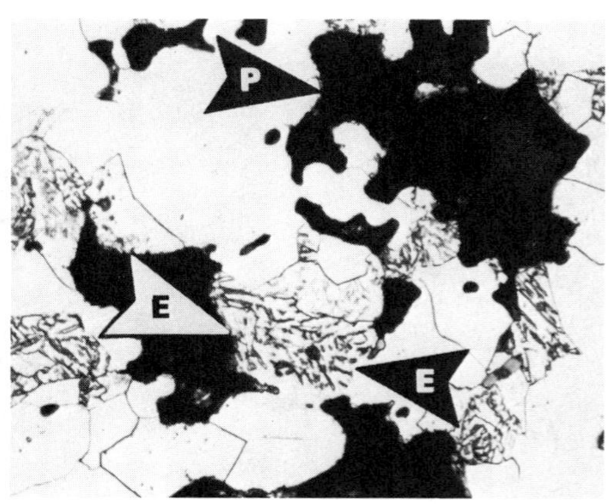

Fig. 73 Prealloyed low-alloy steel (Fe-0.42Ni-0.62Mo-0.2C, 6.7 g/cm^3), pressed at 410 to 480 MPa (30 to 35 tsi) and sintered 30 min at 1120 °C (2050 °F) in dissociated ammonia. Arrows E surround eutectoid region (background is white ferrite). Arrow P shows a pore. See also Fig. 74 and 75 for effects of increasing carbon content. 2% nital. 545×

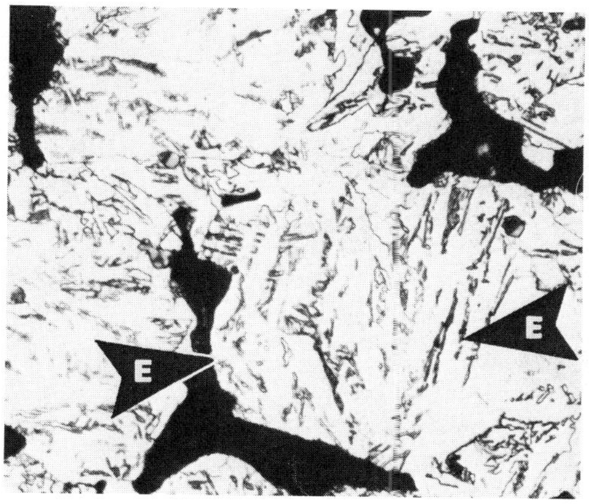

Fig. 74 Same as Fig. 73, but with 0.5% C. Note that the eutectoid (arrows E) now covers most of the field of view and that the distinction between eutectoid and white proeutectoid ferrite is not clear. 2% nital. 545×

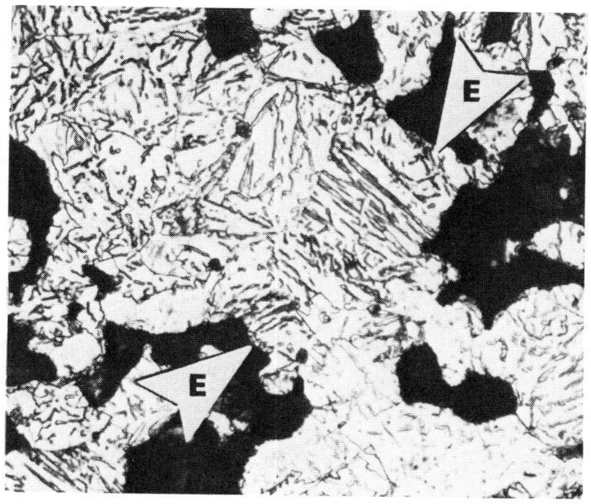

Fig. 75 Same as Fig. 73, but with 0.8% C. Arrows E show the typical eutectoid product. Dark areas are pores. Note that the density of carbide platelets in the eutectoid is higher than for Fig. 74. 2% nital. 545×

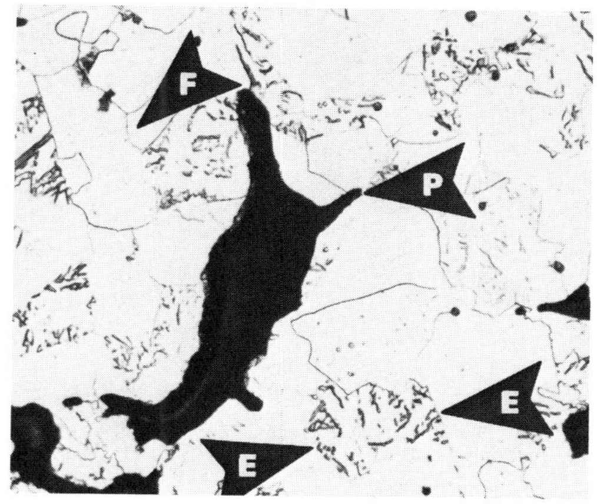

Fig. 76 Prealloyed, low-alloy steel (Fe-1.85Ni-0.60Mo-0.2C, 6.7 g/cm^3), processed as per Fig. 73. Matrix is white ferrite (arrow F). Eutectoid is found in colonies (arrows E). Arrow P shows a pore. Separation of eutectoid and ferrite is distinct, making it possible to estimate carbon content. 2% nital. 545×

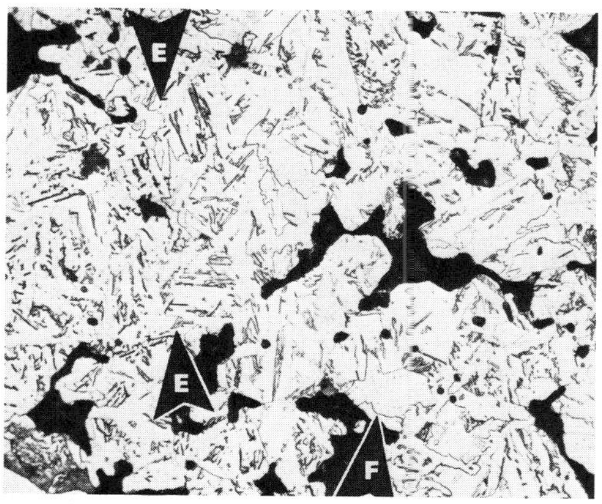

Fig. 77 Same as Fig. 76, but with 0.5% C. Arrows E outline the eutectoid product, which is so mixed with ferrite that it is difficult to distinguish proeutectoid ferrite. Arrow F shows one small grain that is probably white ferrite. 2% nital. 365×

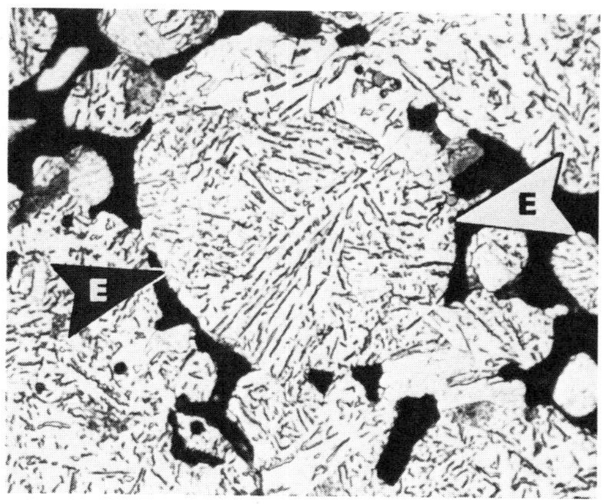

Fig. 78 Same as Fig. 76, but with 0.8% combined carbon. The increasing carbon content alters the appearance of the eutectoid (compare with Fig. 77). The carbine platelets in the eutectoid colonies (arrows E) are more numerous and more closely spaced than in Fig. 77. 2% nital. 545×

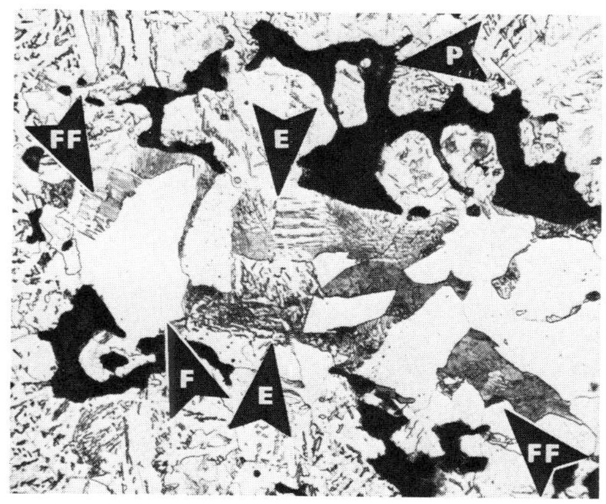

Fig. 79 Prealloyed steel (Fe-1.85Ni-0.6Mo-0.5C, 6.7 g/cm^3), with an impurity particle of unalloyed free ferrite (outlined by arrows FF), which has a carbon content of 0.3 to 0.4%. It is a mixture of proeutectoid ferrite (arrow F) and normal eutectoid (arrows E). Compare this eutectoid with the matrix eutectoid. Arrow P indicates a pore. 2% nital. 365×

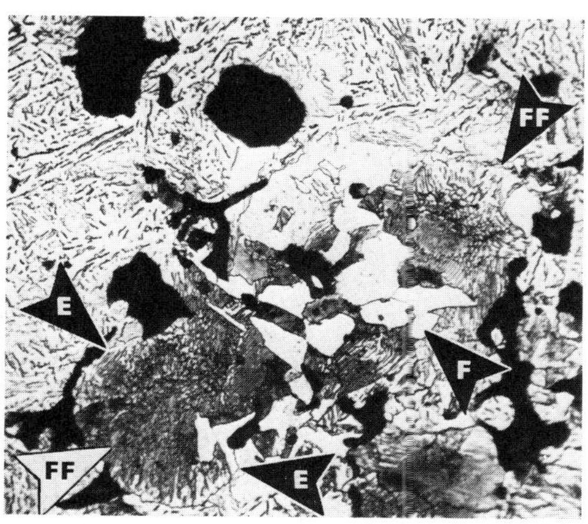

Fig. 80 Same as Fig. 79, with a similar unalloyed iron particle. Arrows FF outline the free ferrite, which has a carbon content of approximately 0.6%, judging by its pearlite (arrows E) and its white proeutectoid ferrite (arrow F). 2% nital. 365×

Fig. 81 Iron-phosphorus alloy (atomized iron mixed with Fe_3P to yield 0.45% P) for soft magnetic applications (6.7 g/cm^3). Pressed at 410 to 480 MPa (30 to 35 tsi) and sintered 30 min at 1120 °C (2050 °F) in dissociated ammonia. Phosphorus forms a liquid phase that causes pore rounding during sintering. As-polished. 180×

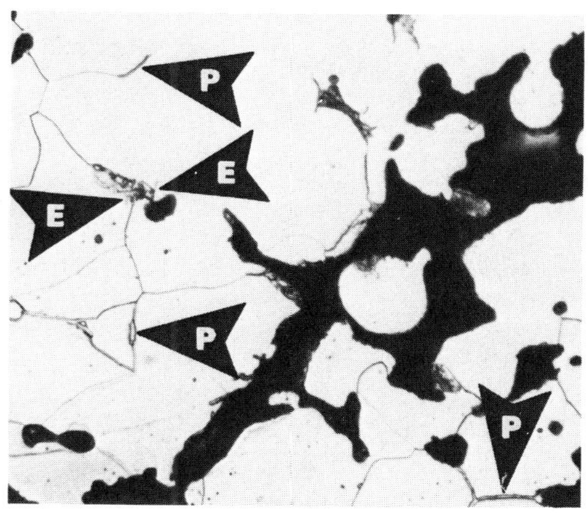

Fig. 82 Same as Fig. 81, but etched. Arrows E show fine two-phase eutectoid or eutectic of Fe-C-P. Arrows P indicate isolated phosphorus-rich phase in grain boundaries. Matrix is all white ferrite. 2% nital. 545×

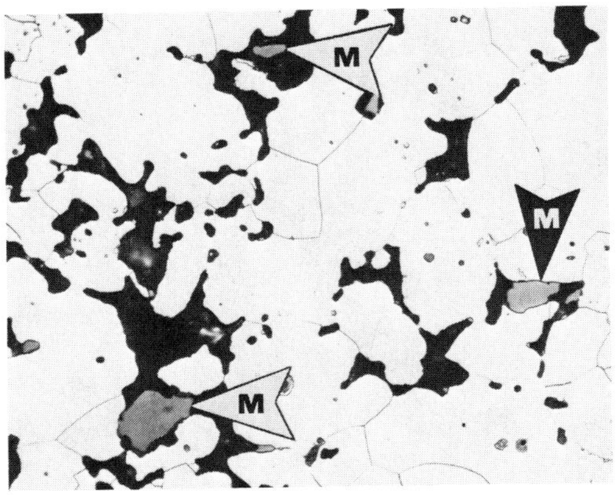

Fig. 83 Water-atomized iron powder with 0.5% MnS blended for increased machinability. See Fig. 81 for processing (6.7 g/cm^3 with 103 MPa, or 15 ksi, minimum yield strength). The MnS is the gray material (see arrows M) inside the darker pores. 2% nital. 330×

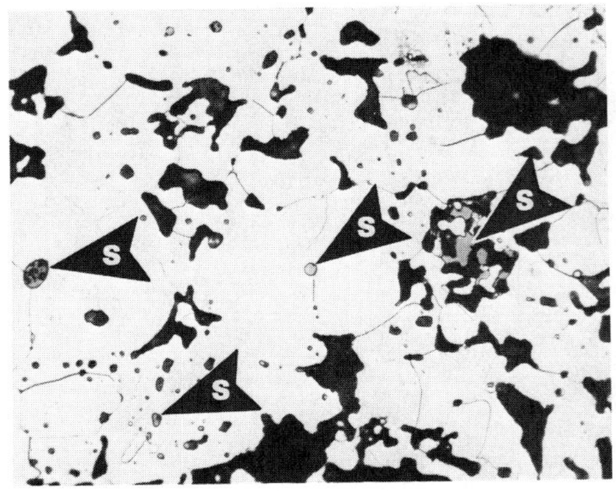

Fig. 84 Sponge iron powder with 0.27% S and 0.9% Mn added to the melt before particle formation. See Fig. 81 for processing (6.7 g/cm^3 with 103 MPa, or 15 ksi, minimum yield strength). Arrows S show gray MnS inclusions within the iron particles and also in the spongy areas. 2% nital.

Fig. 85 Carbon-reduced sponge iron powder with 0.5% S blended as elemental fines. See Fig. 81 for processing (6.4 g/cm^3 with 103 MPa, or 15 ksi, minimum yield strength). Arrows S show two-phase sulfide mixtures of MnS and FeS, which lie within the pores. 2% nital. 330×

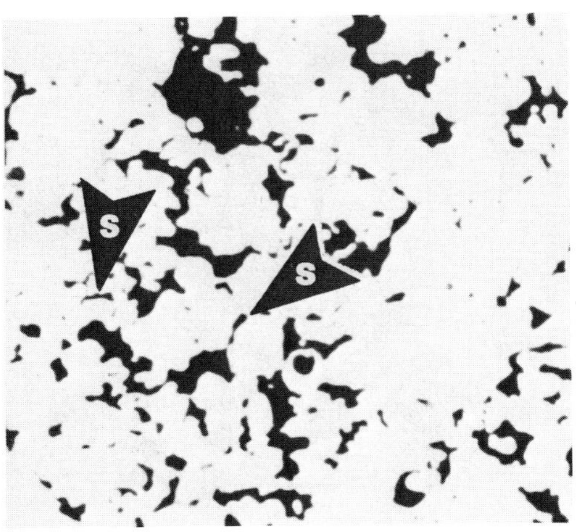

Fig. 86 Nickel steel (elemental mix of atomized iron, 2% Ni, and 0.8% combined carbon, 6.7 g/cm^3), pressed at 410 to 480 MPa (30 to 35 ksi) and sintered 30 min at 1120 °C (2050 °F) in dissociated ammonia. Arrows S indicate short segments of original particle boundaries. See also Fig. 87 to 90. As-polished. 180×

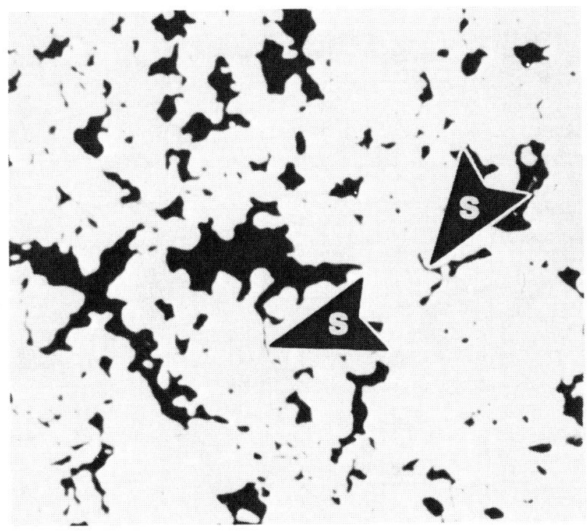

Fig. 87 Same nickel steel as shown in Fig. 86, but with 1% Cu added. A few short segments of original particle boundaries remain (arrows S). It is difficult to distinguish any microstructural effect of the copper. As-polished. 180×

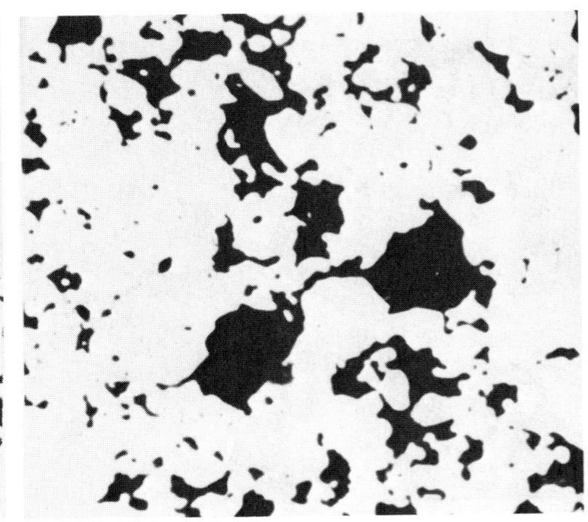

Fig. 88 Same nickel steel as shown in Fig. 86, but with 2% Cu added. The additional copper helped to eliminate original particle boundaries, round the pores, and eliminate small pores. As-polished. 180×

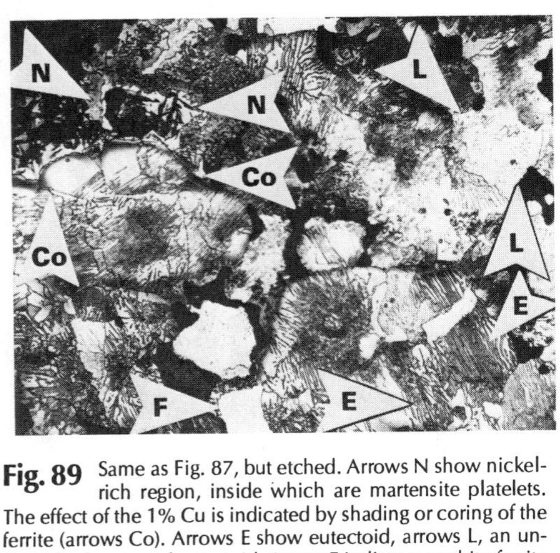

Fig. 89 Same as Fig. 87, but etched. Arrows N show nickel-rich region, inside which are martensite platelets. The effect of the 1% Cu is indicated by shading or coring of the ferrite (arrows Co). Arrows E show eutectoid, arrows L, an underetched region of eutectoid. Arrow F indicates a white ferrite grain. 2% nital. 310×

Fig. 90 Same as Fig. 88, but etched. The 2% Cu causes the nickel to dissolve, eliminating the nickel-rich regions. Arrows ? surround regions where the nickel probably was. These regions are darkened by the coring caused by copper dissolving there and contain carbide platelets. Another cored region is marked Co. Arrows E mark unresolved eutectoid, and arrow F indicates a white ferrite grain. 2% nital. 310×

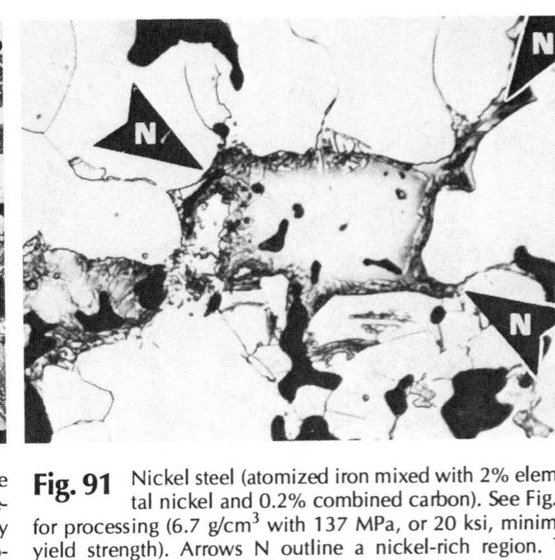

Fig. 91 Nickel steel (atomized iron mixed with 2% elemental nickel and 0.2% combined carbon). See Fig. 86 for processing (6.7 g/cm^3 with 137 MPa, or 20 ksi, minimum yield strength). Arrows N outline a nickel-rich region. The austenitic light-colored center is approximately 12% Ni. The outer edges are lower in nickel and higher in carbon. Dark regions are pores. 2% nital. 545×

Fig. 92 Same as Fig. 91, but with 0.5% C. Arrows N outline nickel-rich regions. The fine structure within these regions is martensite and bainite. Arrows E outline a eutectoid region. Arrow F indicates a white ferrite grain. Arrow D shows a white diffusion layer between the eutectoid and the nickel-rich areas. 2% nital. 310×

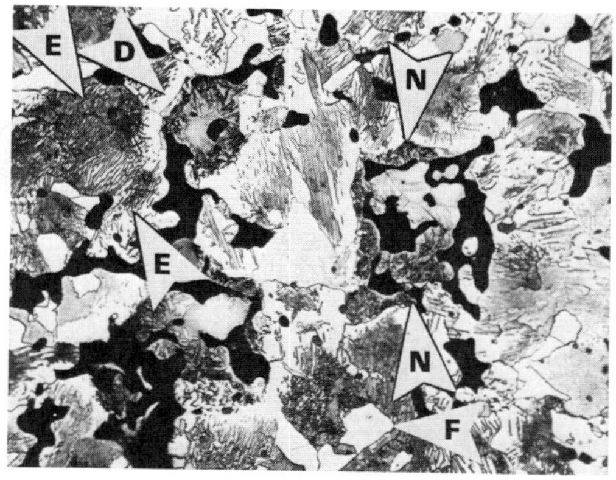

Fig. 93 Same as Fig. 91, but with 0.8% C. Structure is nearly all eutectoid (arrows E), with some nickel-rich regions (arrows N) and isolated ferrite (arrow F). See also Fig. 94 for higher magnification of this structure. 2% nital. 310×

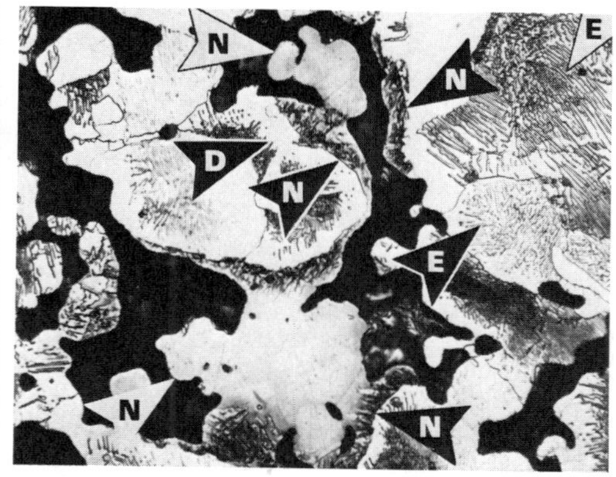

Fig. 94 Same as Fig. 93, but at higher magnification. Arrows N outline a nickel-rich region running through the structure along the pore surfaces, which probably forms a continuous three-dimensional network within the steel. Martensite needles are visible on the surface of the region. Arrows E show eutectoid; arrow D, white diffusion layer. 2% nital. 550×

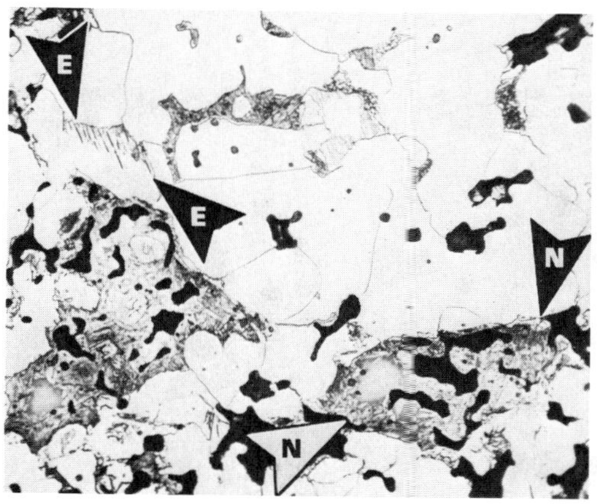

Fig. 95 Nickel steel (4% elemental carbonyl nickel admixed with atomized iron and 0.2% C, 6.7 g/cm^3). See Fig. 86 for processing. Arrows E show a small colony of eutectoid, but most of the carbon is tied up in the nickel-rich regions (arrows N). Background is white ferrite grains. 2% nital. 330×

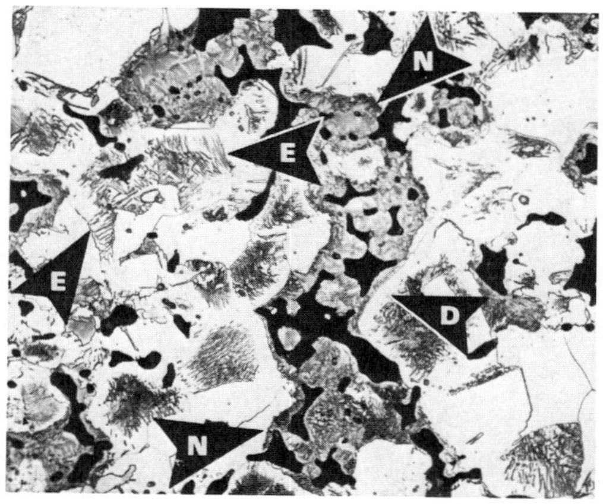

Fig. 96 Same as Fig. 95, but with 0.5% C. As-sintered part would have a minimum yield strength of 207 MPa (30 ksi) or 723 MPa (105 ksi) if hardened and tempered at 260 °C (500 °F). Arrows N show a nickel-rich region. Arrows E indicate eutectoid, and arrow D shows a white diffusion layer between nickel-rich regions and the eutectoid. 2% nital. 330×

Fig. 97 Same as Fig. 95, but with 0.8% C. Minimum yield strength: 276 MPa (40 ksi). Arrows N outline nickel-rich regions. Arrows E indicate eutectoid. Very little ferrite is present. 2% nital. 330×

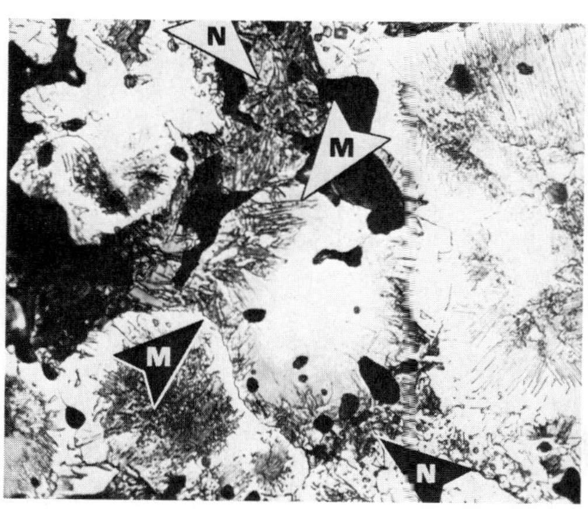

Fig. 98 Same as Fig. 97, but at higher magnification. A nickel-rich region is bounded by arrows N. Martensite platelets (40 to 50 HRC) are visible on the periphery of this region (arrows M). These convey extra wear resistance in the as-sintered condition. The remaining structure is eutectoid. 2% nital. 585×

Fig. 99 Diffusion-alloyed steel (Fe-1.75Ni-0.5Mo-1.5Cu-0.5C, 6.7 g/cm^3), pressed at 410 to 480 MPa (60 to 70 ksi) and sintered 30 min at 1120 °C (2050 °F) in dissociated ammonia. Arrows N show nickel-rich regions containing martensite or bainite platelets; arrows E, a eutectoid colony with a white diffusion layer on the periphery. Arrow F indicates ferrite. 2% nital. 330×

Fig. 100 Same as Fig. 99, but with 0.8% C. Arrows N show nickel-rich areas containing platelets of martensite or bainite. Arrow L indicates a light-etching eutectoid region, and arrow P a pore. Arrows E outline a colony of eutectoid containing thin carbide platelets. 2% nital. 330×

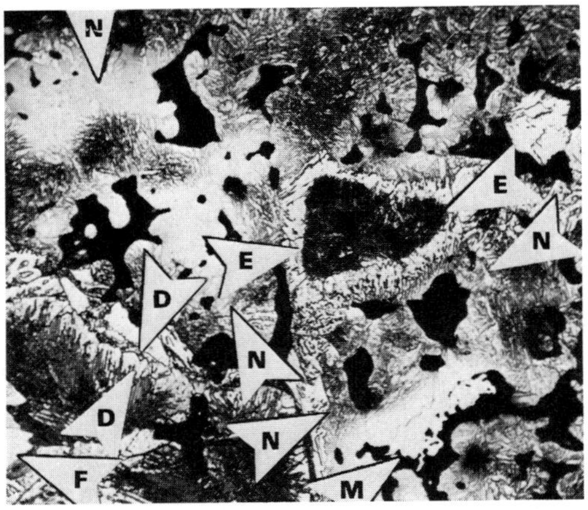

Fig. 101 Same as Fig. 99, but with 4% Ni. Structure shows extensive nickel-rich areas (arrows N), with light-colored austenite at their interiors and gray martensite at their peripheries. Unresolved eutectoid is shown by arrows E, and a diffusion layer by arrows D. Arrow F indicates white ferrite. Arrows M outline a residual, spongy, undiffused molybdenum particle. 2% nital. 365×

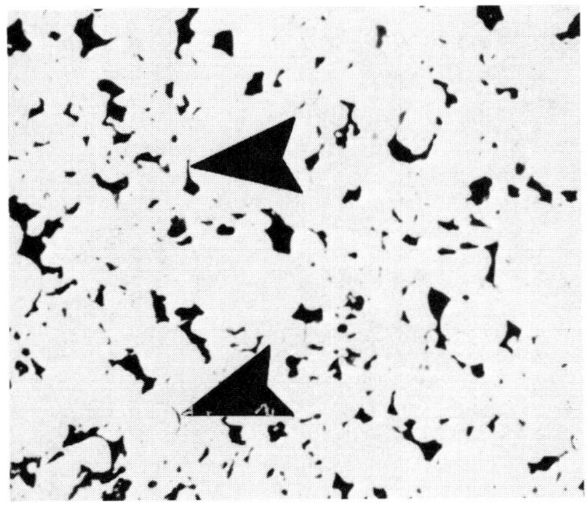

Fig. 102 Type 316 stainless steel (6.96 g/cm^3), pressed at 830 MPa (60 tsi) and sintered 30 min at 1150 °C (2100 °F) in dissociated ammonia. Structure is a tensile bar with tensile strength of 393 MPa (57 ksi) and elongation in 25 mm (1 in.) of 1%. Arrows show original particle boundaries. As-polished. 180×

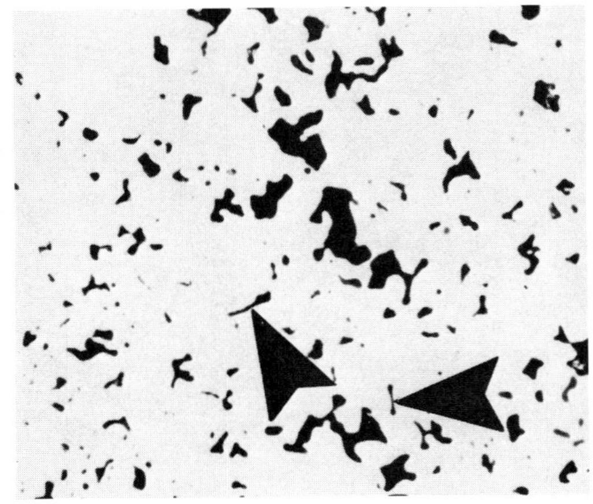

Fig. 103 Same as Fig. 102, but sintered 60 min at 1290 °C (2350 °F) in dissociated ammonia (7.15 g/cm^3). Many of the small pores have disappeared, and few particle boundary segments remain (see arrows). Specimen from tensile bar with a tensile strength of 432 MPa (62.7 ksi) and elongation in 25 mm (1 in.) of 11%. As-polished. 180×

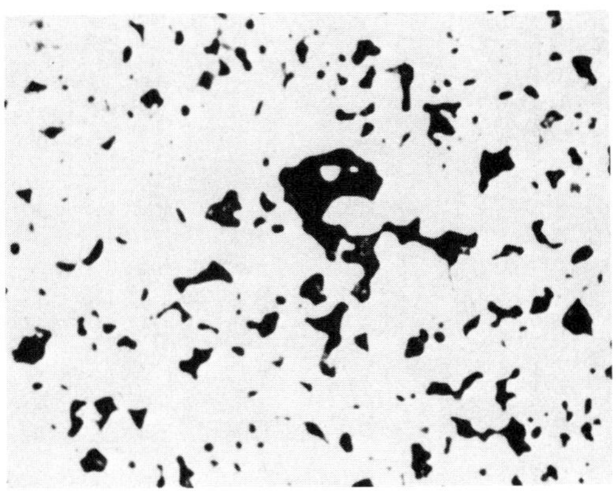

Fig. 104 Same as Fig. 102, but vacuum sintered at 1290 °C (2350 °F) for 60 min (7.26 g/cm^3). Pores are more rounded than the dissociated ammonia sinters in Fig. 102 and 103. Specimen from tensile bar with a tensile strength of 355 MPa (54.4 ksi) and elongation in 25 mm (1 in.) of 25%. As-polished. 180×

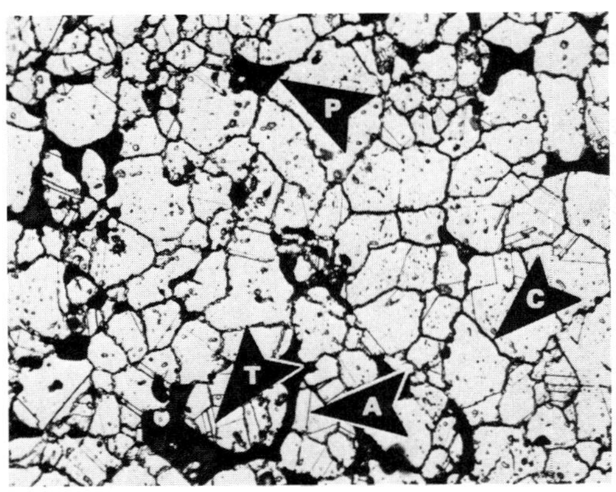

Fig. 105 Same as Fig. 102, but etched, showing the precipitated grain-boundary carbides (arrow C), which cause the irregular grain boundaries. Arrow A shows a normal austenite grain boundary, arrow T a twin boundary, and arrow P a pore. Glyceregia. 365×

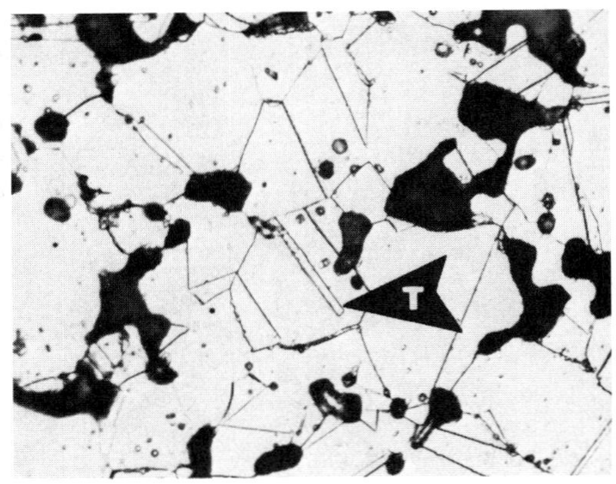

Fig. 106 Same as Fig. 103, but etched. Note that the austenite grain boundaries and twin boundaries (arrow T) are free of carbides due to the decarburizing effect of the high-temperature dissociated ammonia sinter. Glyceregia. 545×

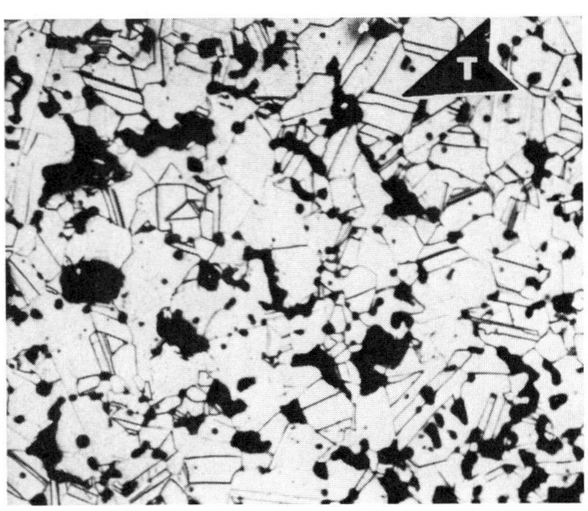

Fig. 107 Same as Fig. 104, but etched. Structure is free of grain-boundary carbides. Arrow T points to a typical twin boundary. Glyceregia. 180×

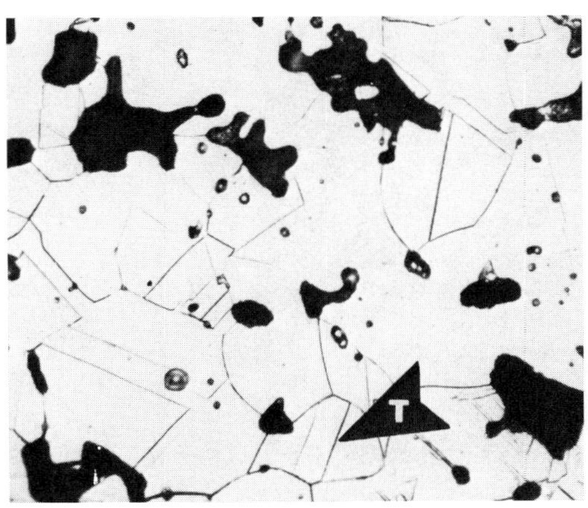

Fig. 108 Same as Fig. 107, but at higher magnification. Grain boundaries show no carbide precipitates. Arrow T shows an annealing twin. Glyceregia. 545×

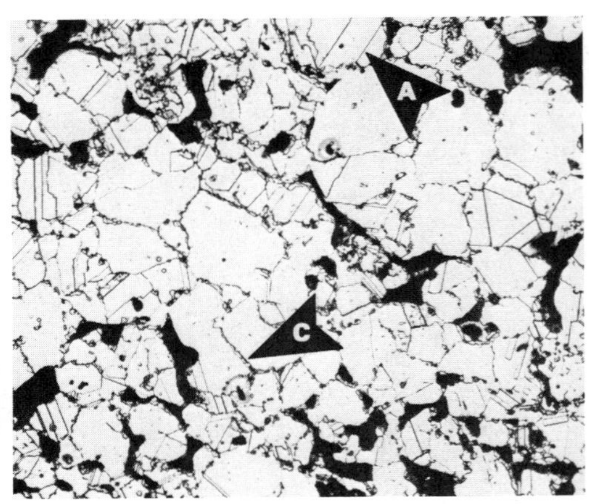

Fig. 109 Type 303 stainless steel (7.06 g/cm^3), pressed at 830 MPa (60 tsi) and sintered 60 min at 1120 °C (2050 °F) in dissociated ammonia. Residual chromium carbides are in the austenite grain boundaries (arrow C). Arrow A shows an austenite grain boundary free of carbides. Glyceregia. 360×

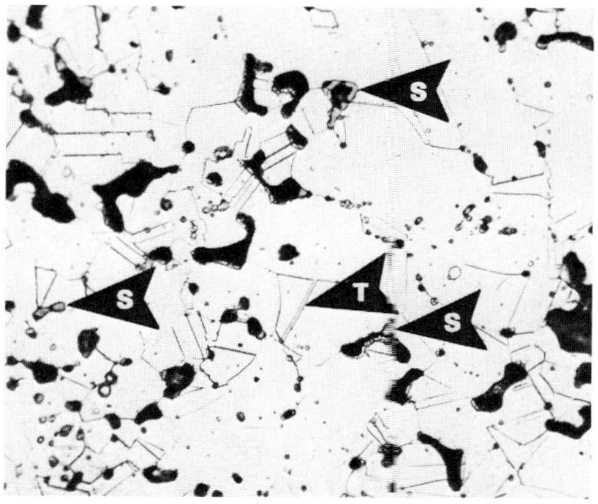

Fig. 110 Same as Fig. 109, but sintered 60 min at 1190 °C (2350 °F) in dissociated ammonia (7.23 g/cm^3). No carbides are present at grain boundaries. Arrows S indicate gray MnS inclusions added for machinability. Arrow T shows a typical annealing twin boundary. Glyceregia. 365×

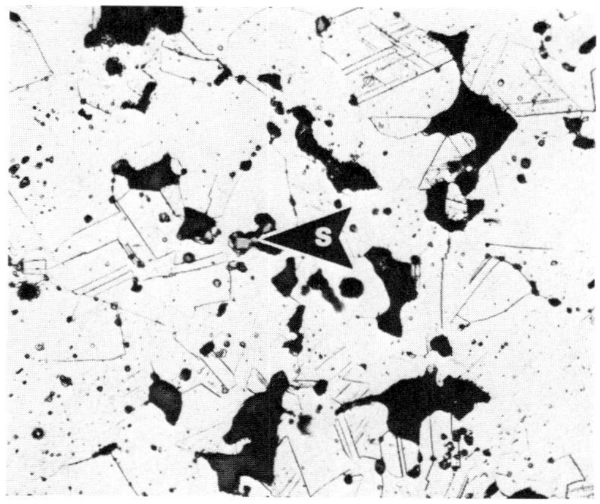

Fig. 111 Same as Fig. 109, but vacuum sintered at 1290 °C (2350 °F) for 60 min (7.29 g/cm^3). Arrow S shows a MnS particle. Glyceregia. 365×

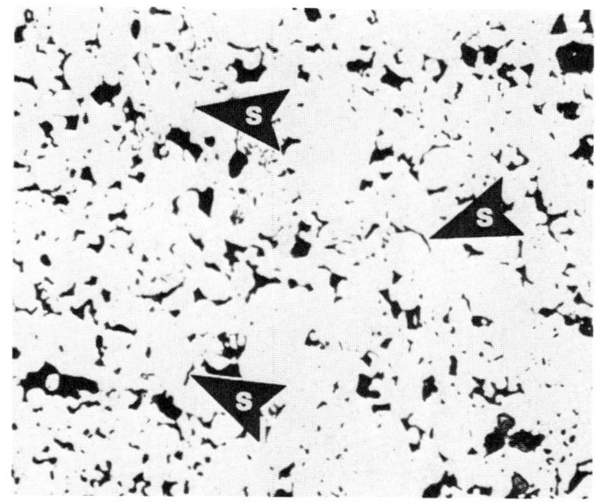

Fig. 112 Type 410 stainless steel with added 0.15% graphite (6.77 g/cm^3), pressed at 830 MPa (60 tsi), sintered 60 min at 1150 °C (2100 °F) in dissociated ammonia, and tempered 1 h at 175 °C (350 °F). Arrows S indicate original particle boundaries. Glyceregia. 120×

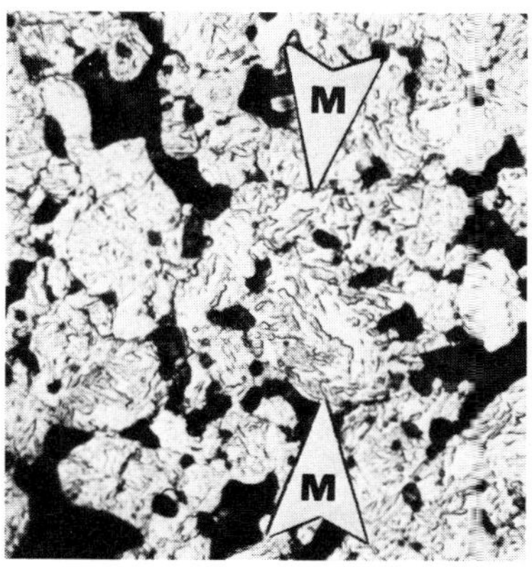

Fig. 113 Same type 410 stainless steel and pressing conditions as described in Fig. 112, but sintered 30 min at 1120 °C (2050 °F) and etched. Structure is all martensite (arrows M). Glyceregia. 545×

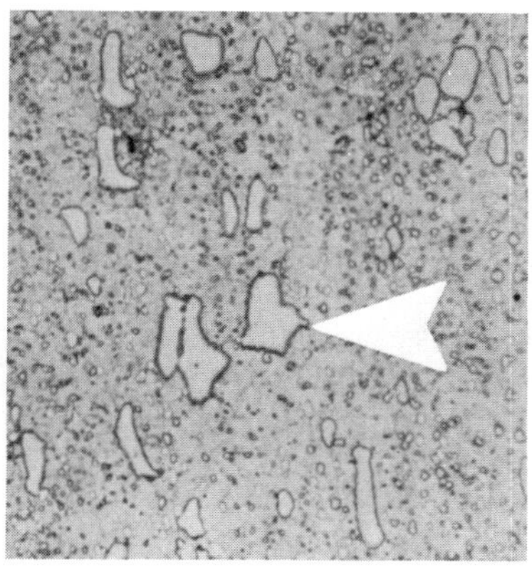

Fig. 114 Wrought T15 steel. Arrows show large primary M_6C carbide (compared with P/M tool steels in Fig. 115 and 116). Fine precipitate is vanadium carbide; matrix is martensite. 5% nital. 365×. Source: Ref 12

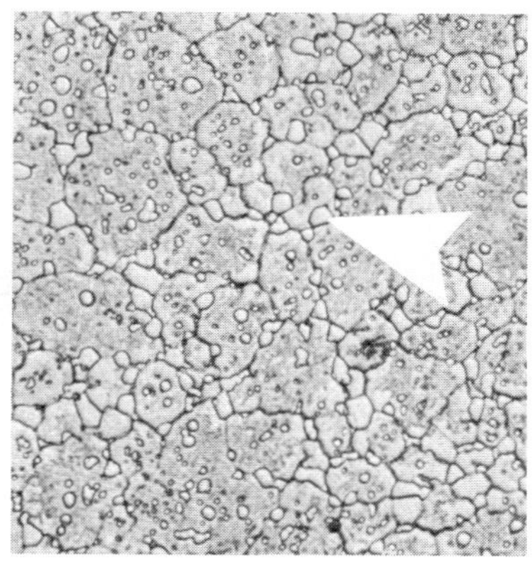

Fig. 115 P/M T15 tool steel, pressed at 550 MPa (40 tsi), and vacuum sintered 1 h at 1260 °C (2300 °F) to full density (8.25 g/cm^3). Arrow indicates primary M_6C carbide (compare with Fig. 114). 5% nital. 500×. Source: Ref 12

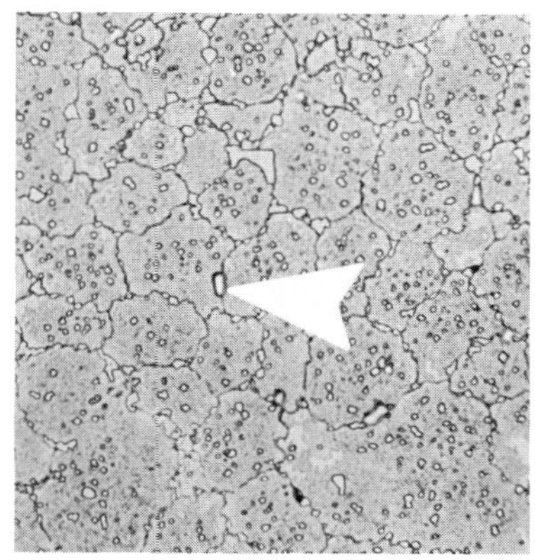

Fig. 116 P/M M2 tool steel (8.12 g/cm^3) pressed at 550 MPa (40 tsi) and vacuum sintered 1 h at 1240 °C (2260 °F). Arrow indicates small M_6C particle. Other small MC precipitates can be seen in the matrix of retained austenite, which also contains martensite. 5% nital. 400×. Source: Ref 12

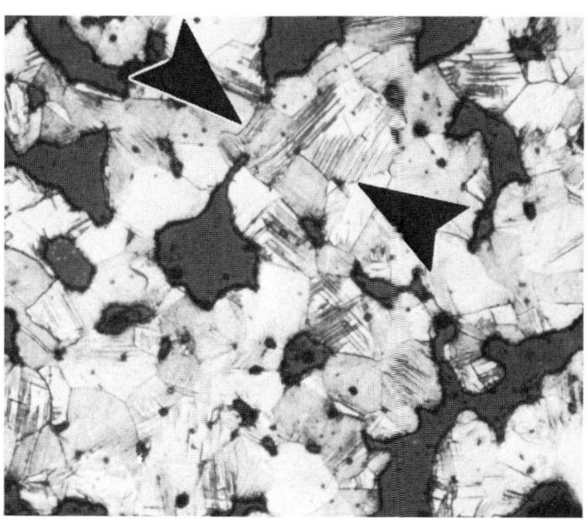

Fig. 117 Bronze (Cu-10Sn, 6.4 g/cm^3) pressed at 140 to 205 MPa (10 to 15 tsi), sintered (conditions not known), and sized for tolerances. Mostly all α-bronze grains. Gray areas are pores. Arrows surround slip line from cold working during sizing. $K_2Cr_2O_7$. 180×

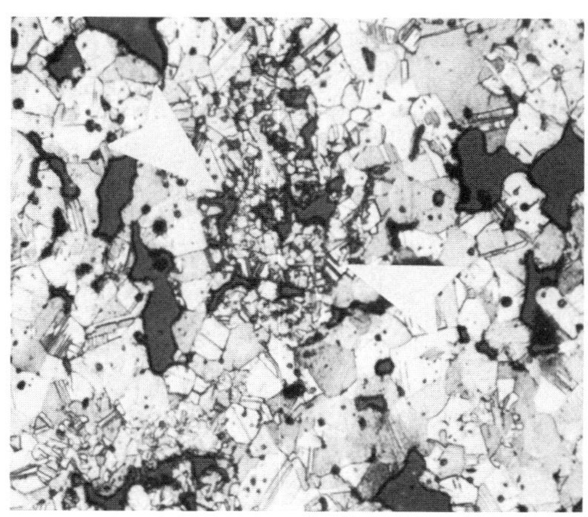

Fig. 118 Same material and processing as described in Fig. 117. Arrows indicate clusters of small grains that have not grown into larger α grains. Compare with Fig. 119, which shows the effect of a 4% graphite addition on the resulting structure. $K_2Cr_2O_7$. 180×

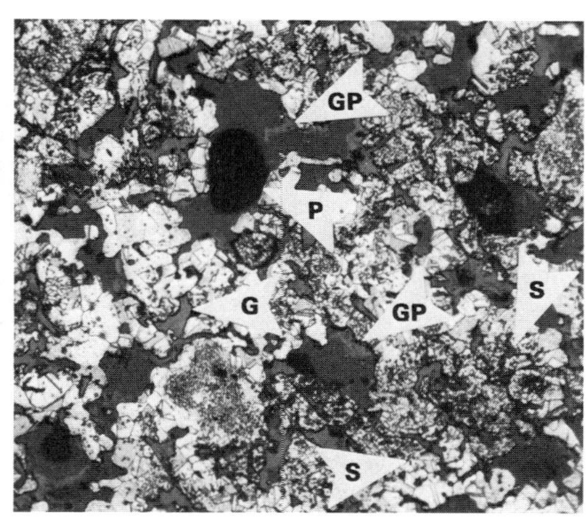

Fig. 119 Cu-10Sn bronze (6.4 g/cm^3) with 4% graphite. Pressed at 165 to 205 MPa (12 to 15 tsi), sintered 15 min at 845 °C (1550 °F), and sized. Arrows GP show graphite in pores surrounded by darker gray epoxy resin; arrow G, graphite flake in the matrix. Arrows S surround a small grain cluster; arrow P indicates a pore. $K_2Cr_2O_7$. 120×

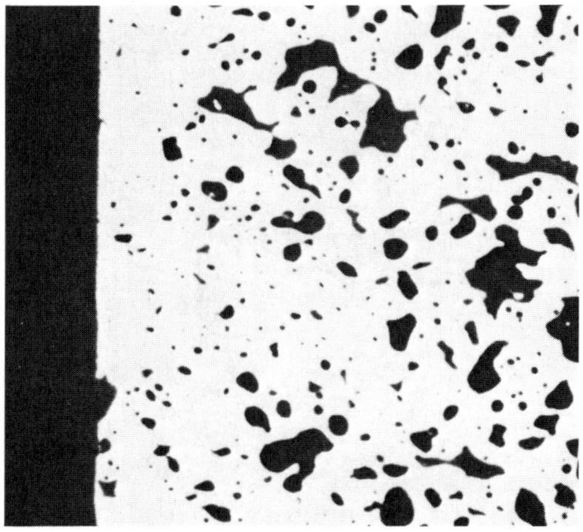

Fig. 120 Inner diameter (left) of a sized Cu-10Sn bronze bearing (see Fig. 119 for processing). Note that the pores on the ID have been closed by the sizing operation. As a result, oil would not have easy access from the interior to the ID. $K_2Cr_2O_7$. 180×

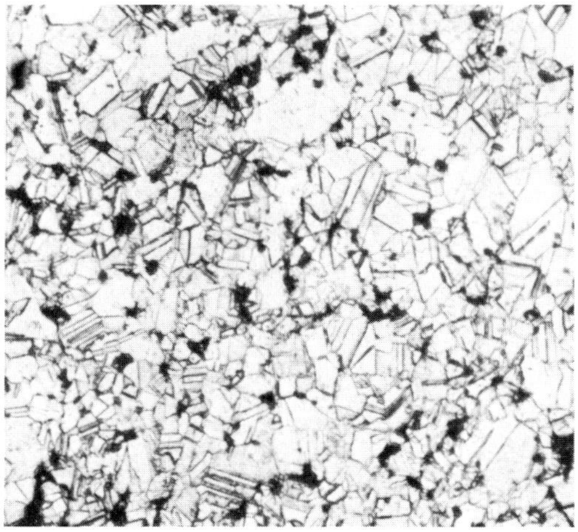

Fig. 121 Leaded brass (Cu-20Zn-2Pb, 8.1 g/cm^3), pressed and sintered 30 min at 870 °C (1600 °F), then repressed. Structure is all single-phase α-brass with a few annealing twins. Dark spots are well-rounded pores. 5 mL NH_4OH, 5 mL H_2O, and 3 drops 30% H_2O_2. 180×

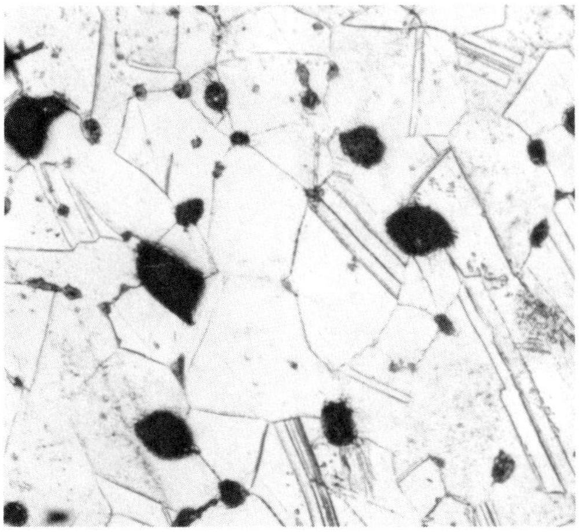

Fig. 122 Same as Fig. 121, but at higher sintering temperature and/or longer time, which results in much larger grains. The pores (dark areas) are coarser and more rounded than in Fig. 121. Etchant: same as Fig. 121. 180×

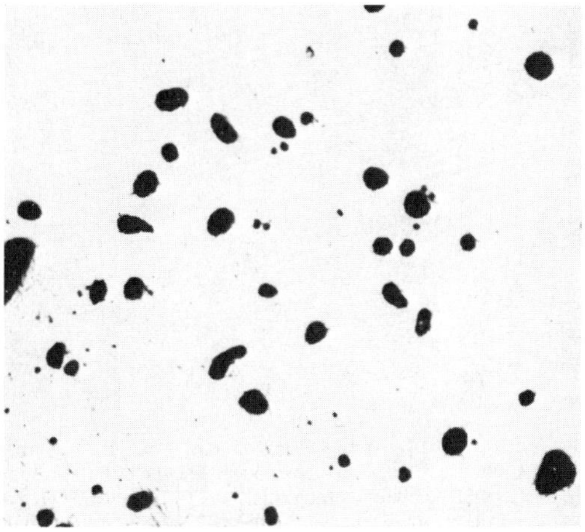

Fig. 123 Ti-6Al-4V, cold isostatically pressed and vacuum sintered to dissolve the master alloy additions of vanadium and aluminum (94% dense). Rounded pores indicate adequate sintering. As-polished. 180×. Source: Ref 13

Fig. 124 Same as Fig. 123, but etched. Dark areas are residual rounded pores. Matrix is transformed β-titanium. 85 mL H_2O, 5 mL HNO_3, and 10 mL HF. 180×. Source: Ref 13

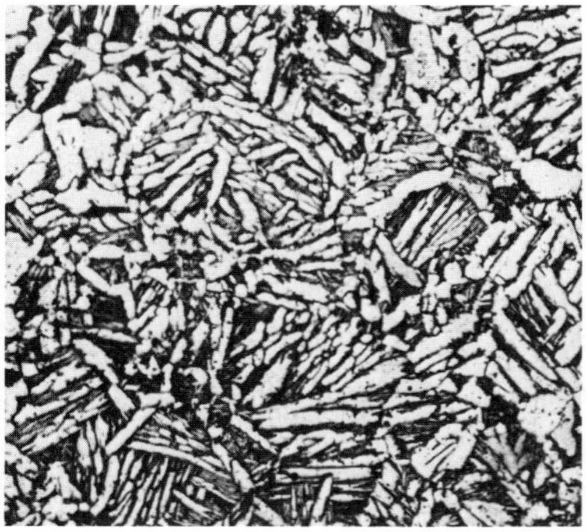

Fig. 125 Ti-6Al-4V, cold isostatically pressed, vacuum sintered, then hot isostatically pressed to 99.6% density. Structure shows all transformed β. 85 mL H_2O, 5 mL HNO_3, and 10 mL HF. 180×. Source: Ref 13

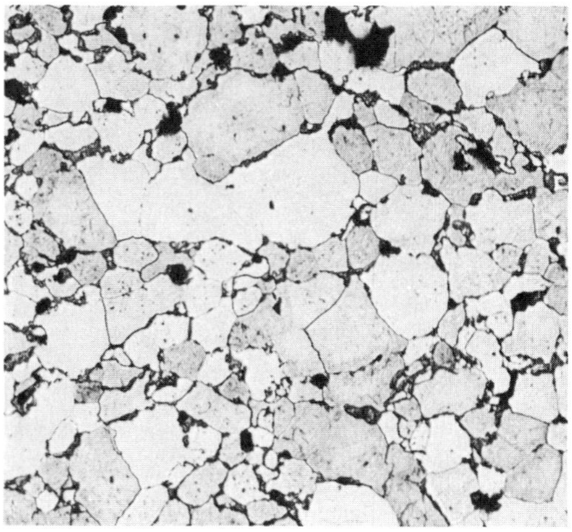

Fig. 126 Aluminum alloy powder (Al-4.4Cu-0.8Si-0.4Mg), pressed and sintered 30 min at 595 °C (1100 °F) in nitrogen. Typical as-sintered structure. Keller's reagent. 100×. Source: Ref 14

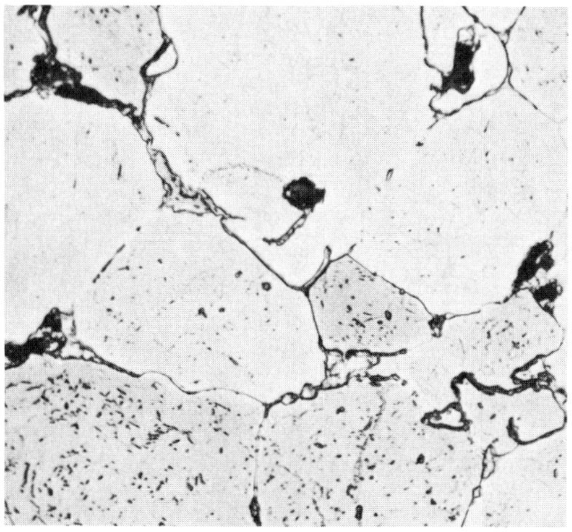

Fig. 127 Same alloy and processing as Fig. 126, but at higher magnification. Keller's reagent. 500×. Source: Ref 14

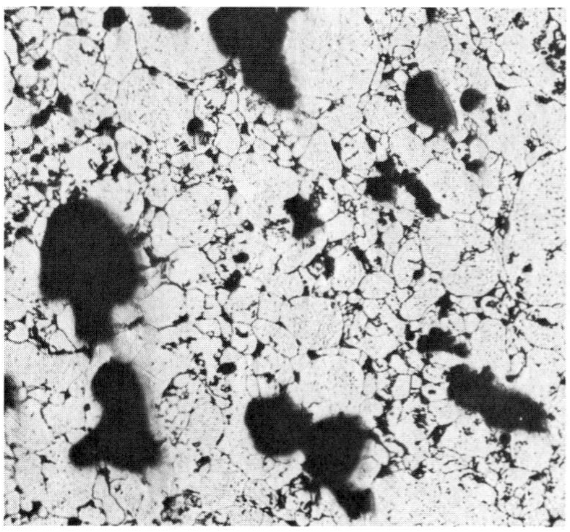

Fig. 128 Aluminum alloy powder (Al-1.0Mg-0.6Si-0.25Cu), pressed, sintered, and solution heat treated. Typical structure. Dark areas are pores. Keller's reagent. 100×. Source: Ref 14

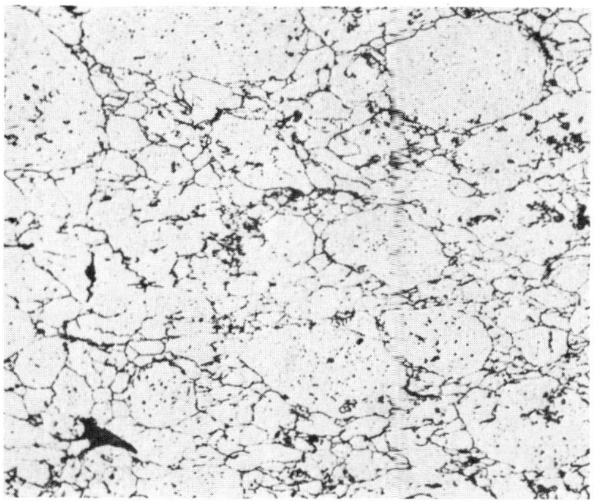

Fig. 129 Same alloy and processing as Fig. 128, but reduced 25% in thickness by forging, which closed up most of the porosity. Keller's reagent. 100×. Source: Ref 14

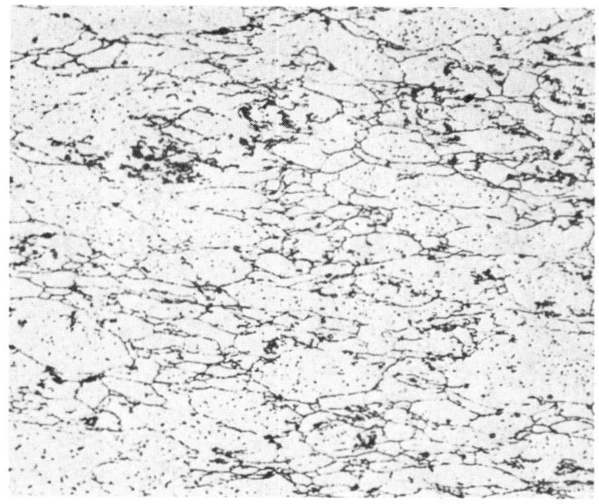

Fig. 130 Same as Fig. 128, but pressed, sintered at 620 °C (1150 °F), and reduced 50% in thickness by forging or rolling to eliminate porosity. Keller's reagent. 100×. Source: Ref 14

Fig. 131 Nickel steel (carbonyl Fe, 2% carbonyl Ni, and 0.3% C) injection molded and sintered (1120 °C, or 2050 °F) to 97% density. Structure shows fine isolated porosity (gray dots). At surface (bottom left) there is a 0.05 mm (0.002 in.) thick layer that exhibits no apparent porosity. As-polished. 180×

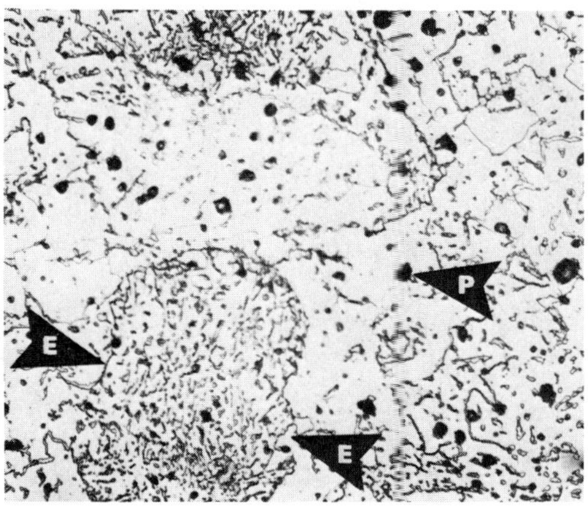

Fig. 132 Same as Fig. 131, but etched. The nickel has diffused into the iron, resulting in complete homogeneity. The eutectoid (arrows E) is similar to that found in 4600 prealloyed steel (Fig. 78). Arrow P indicates an isolated rounded pore. 2% nital. 365×

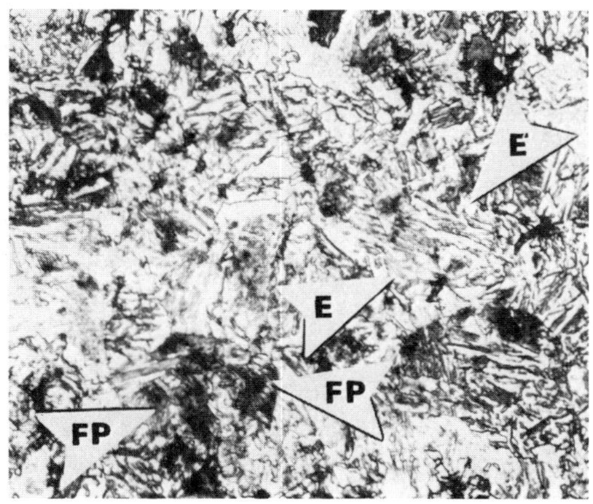

Fig. 133 Prealloyed steel powder (Fe-2Ni-0.5Mo-0.5C) pressed to 6.6 g/cm^3, sintered 30 min at 1120 °C (2050 °F), induction reheated, and hot forged (1095 °C, or 2000 °F) at 830 MPa (60 tsi) to over 99% of theoretical density. Arrows E show typical eutectoid; arrows FP, fine pearlite from rapid cooling of forged part. 2% nital. 545×

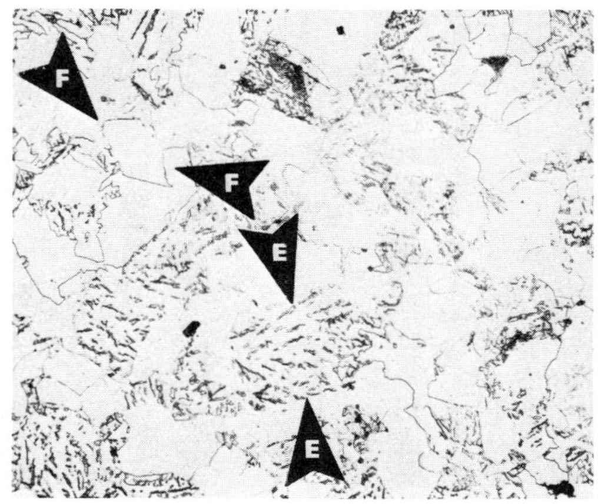

Fig. 134 Powder-forged gear (Fe-2.0Ni-0.5Mo-0.2C). See Fig. 133 for processing. Arrows E show eutectoid; arrows F, white ferrite grains and grain boundaries. Dark spots are pores. Part is 99% dense. 2% nital. 365×

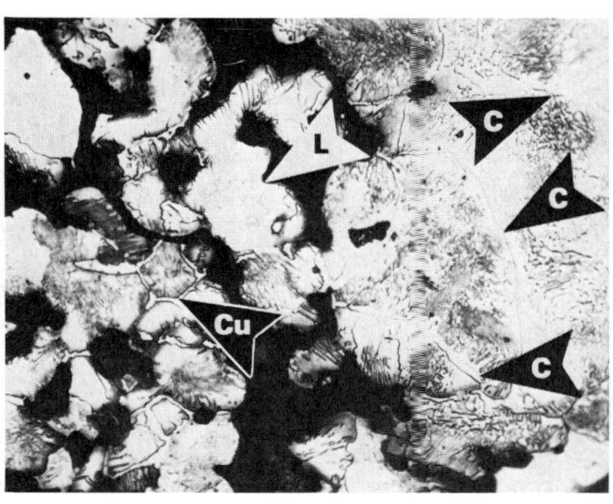

Fig. 135 Copper steel (Fe-2.0Cu-0.8C, 6.8 g/cm^3) that has formed blisters during delubrication and sintering. Arrows C indicate stringer (grain boundary) carbides resulting from carburization. Some light-colored copper is evident (arrow Cu). Arrow L shows a light-etching eutectoid region. 2% nital. 585×

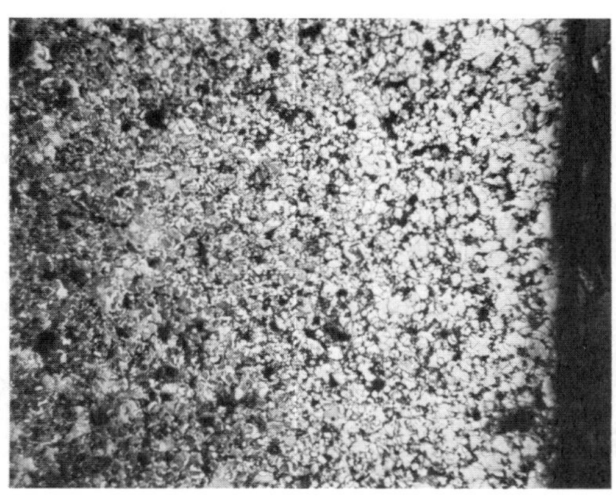

Fig. 136 Same as Fig. 135, but pressed at 480 MPa (35 tsi), sintered 30 min at 1120 °C (2050 °F) in dissociated ammonia, and decarburized. White ferrite grains near the surface (right) indicate loss of carbon. The interior (left) is a darker eutectoid, containing approximately 0.8% C; the surface is ~0.1% C. 2% nital. 30×

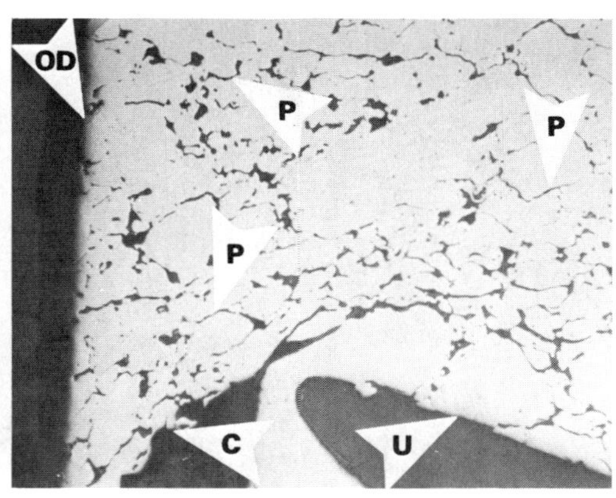

Fig. 137 Pressed and sintered sprocket gear (composition unknown). Arrow OD shows the outer edge of the tooth, and arrow U the underside. A chamfer has broken off the tooling, resulting in a reverse chamfer (arrow C) on the part. Arrow P indicates original particle boundary resulting from particle separation during part ejection. As-polished. 110×

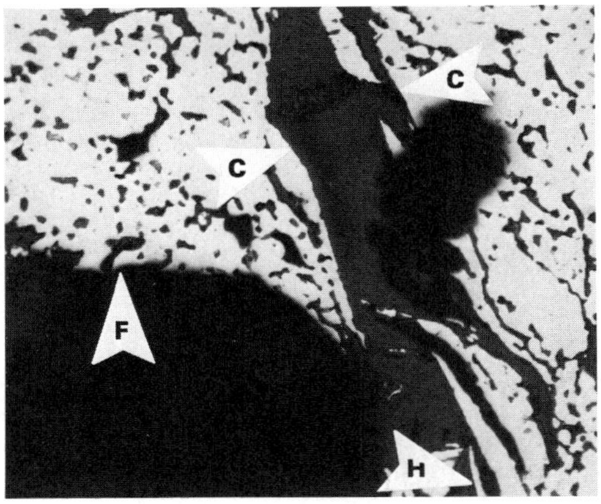

Fig. 138 View of the radius between a flange (arrow F) and a hub (arrow H) in a pressed and sintered part (composition unknown). The relative motion between the hub and flange was not adjusted correctly during pressing, resulting in a shear crack (arrows C). As-polished. 120×

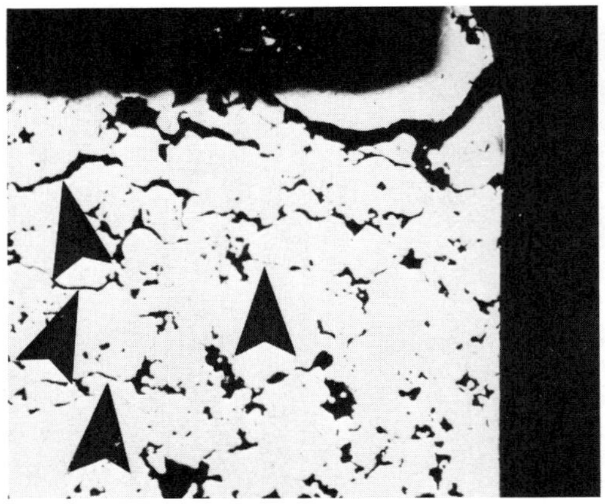

Fig. 139 A corner of an MPIF tensile bar (Fe-2Ni-0.8C, 6.8 g/cm^3) pressed at 550 MPa (40 tsi) and sintered 30 min at 1115 °C (2040 °F) in endothermic gas. Arrows indicate microlaminations that occurred during part ejection. These reduce the tensile strength of the bar by 30 to 40%. As-polished. 120×

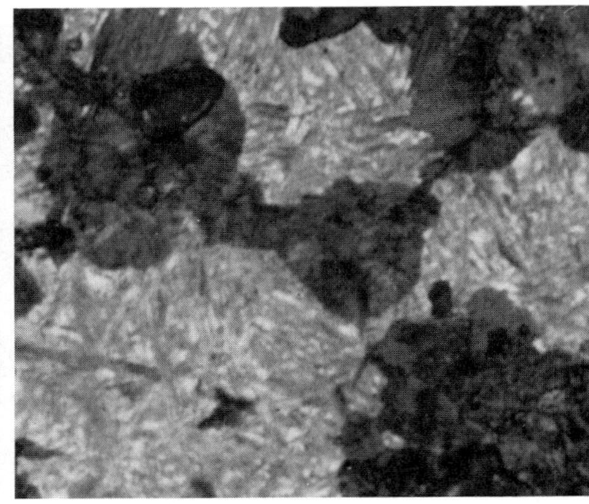

Fig. 140 Fe-08.C, pressed at 550 MPa (40 tsi) and sintered 30 min at 1120 °C (2050 °F) in dissociated ammonia (DA). Heat treated structure showing mixture of 50% fine pearlite (dark phase) and 50% martensite (lighter phase). 4% picral. 960×

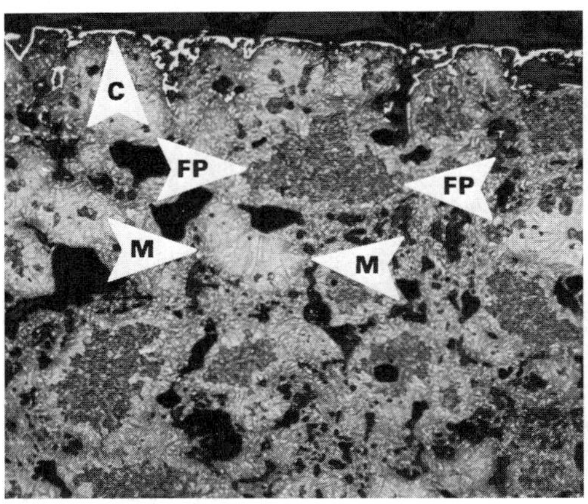

Fig. 141 Copper steel (Fe-2Cu-0.8C, 6.7 g/cm^3) pressed at 480 MPa (35 tsi), sintered 30 min at 1120 °C (2050 °F) in dissociated ammonia, austenitized 1 h at 870 °C (1600 °F), and oil quenched. Surface of part shows a thin, white layer of carbide (arrow C). The structure interior consists of 35% fine pearlite (arrows FP) and balance martensite (arrows M). Picral. 120×

Fig. 142 Fe-0.37Mn-1.80Ni-0.63Mo alloy powder with added 2% Cu and 0.8% C. Pressed at 480 MPa (35 tsi), sintered 30 min at 1120 °C (2050 °F) in a nitrogen-base atmosphere, and tempered 1 h at 175 °C (350 °F) to 6.85 g/cm^3. Structure shows fine pearlite (arrows P) and small amounts of white ferrite (arrow F) in a martensite matrix (arrows M). 2% nital. 365×

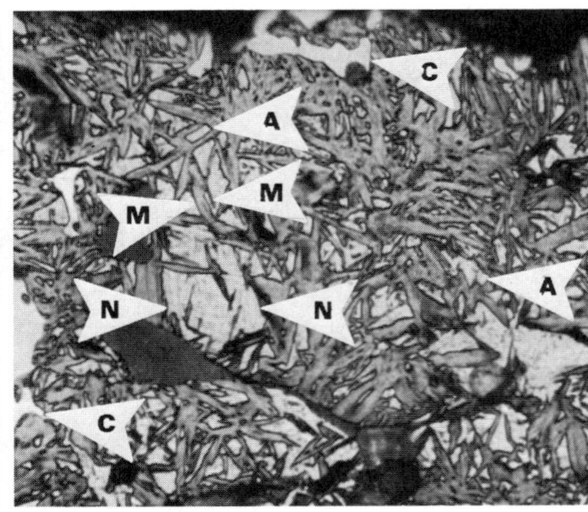

Fig. 143 Nickel steel (Fe-2Ni-0.8C, 6.7 g/cm^3). Pressed and sintered (normal processing) in endothermic gas, austenitized 1 h at 870 °C (1600 °F), oil quenched, and tempered 1 h at 260 °C (500 °F). Arrows C indicate carbide particles. Arrows M show a typical martensite platelet. Arrows N surround a nickel-rich region. Arrows A show light-colored regions of retained austenite. Picral. 645×

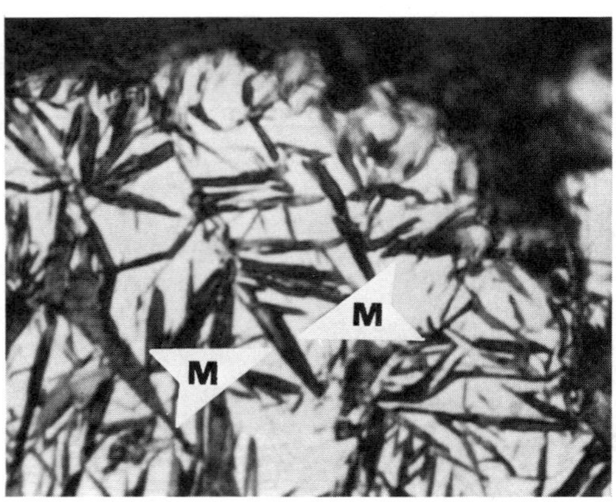

Fig. 144 Prealloyed steel (Fe-1.8Ni-0.5Mo-0.4C, 6.83 g/cm^3), pressed and sintered (normal processing) in dissociated ammonia, austenitized 1 h at 870 °C (1600 °F), oil quenched, and tempered 1 h at 175 °C (350 °F). Matrix (white areas) is retained austenite (excessive, 65 vol%). Arrows M indicate martensite platelets. Picral. 960×

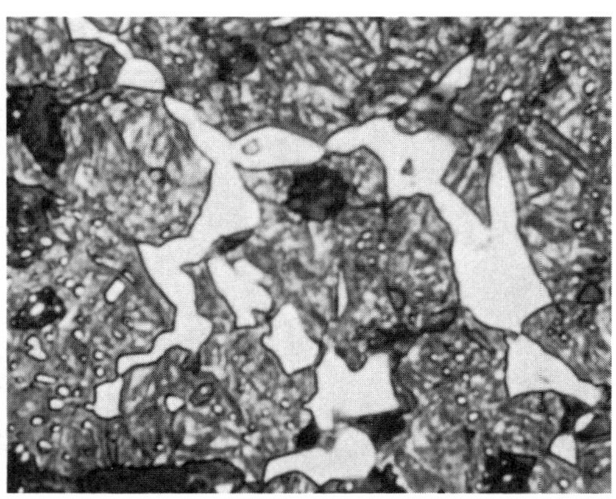

Fig. 145 Fe-1.5 graphite (6.7 g/cm^3), pressed and sintered under normal processing conditions and heat treated as described in Fig. 144. Structure shows massive carbides (sharply delineated light phase). Picral. 960×

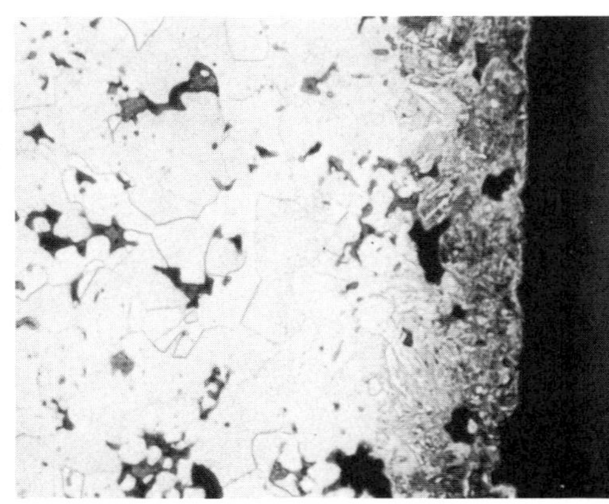

Fig. 146 Atomized iron powder, pressed to 6.8 g/cm^3, sintered in dissociated ammonia, austenitized 30 min in vacuum, gas carburized less than 5 min, and oil quenched. The 0.1 mm (0.004 in.) thick case is the dark martensite on right. Interior is all white ferrite. 2% nital. 180×. Source: Ref 18

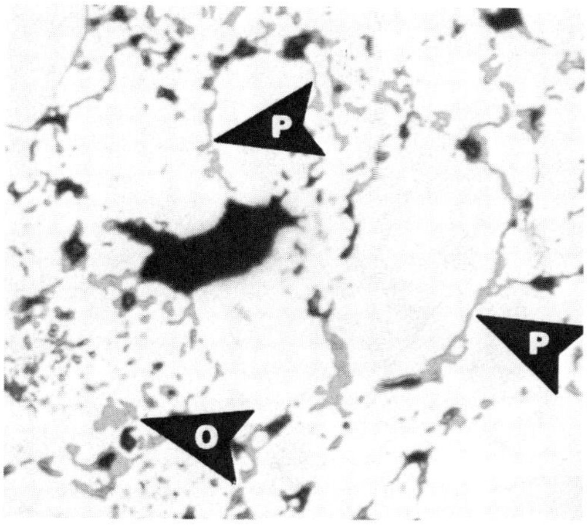

Fig. 147 Steam-blackened Fe-2Cu-0.8C part that was undersintered. Arrow O shows a spongy pore that has been filled with gray Fe_3O_4 during the steam process. Arrows P indicate original particle boundaries, which were filled by the steam process. As-polished. 180×

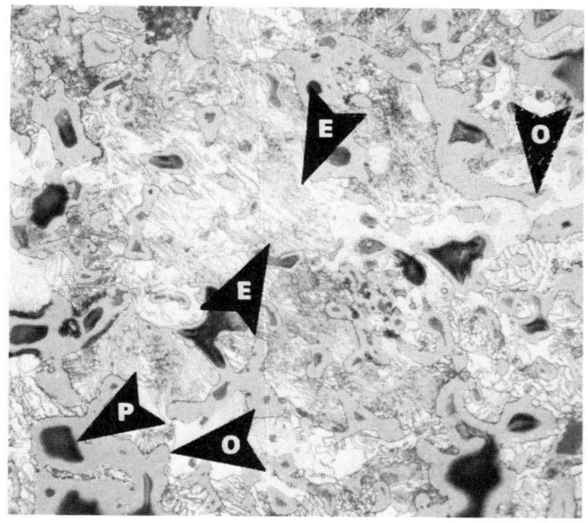

Fig. 148 Pressed and sintered Fe-0.8C steel (6.4 g/cm^3), which was steam blackened. The pores are nearly all filled with gray Fe_3O_4 (arrows O). Arrow P shows a pore not filled with oxide. Arrows E surround a eutectoid region. White areas are ferrite. 2% nital. 365×

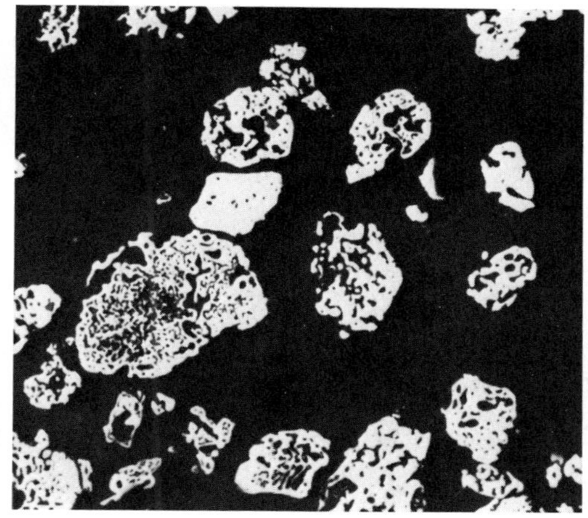

Fig. 149 Pyron D63, hydrogen-reduced mill scale, sponge iron powder (–100 mesh). Particles contain fine, internal pores and some coarse pores. As-polished. 180×

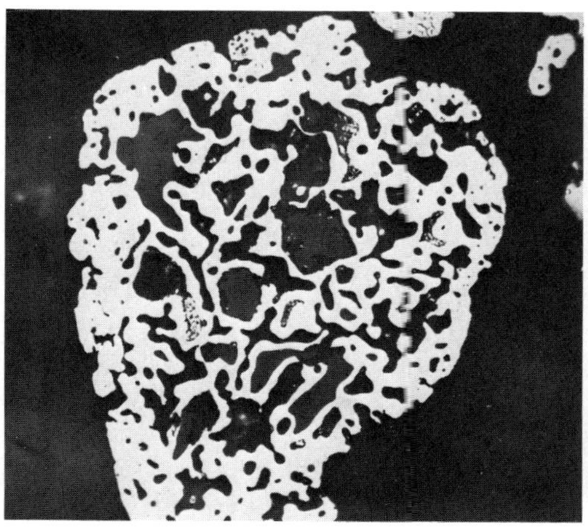

Fig. 150 Same as Fig. 149, but at higher magnification, showing details of the internal porosity and some unreduced oxides. As-polished. 615×

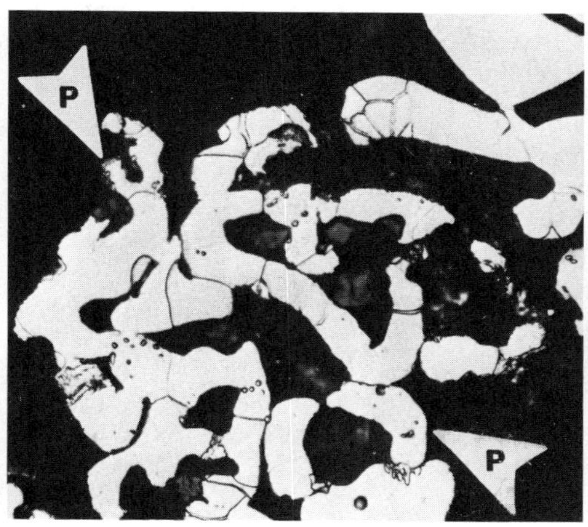

Fig. 151 MH-100 carbon-reduced sponge iron powder (–100 mesh). A few ferrite grain boundaries are shown, but no iron carbides. Arrows P surround a particle having extensive internal porosity. As-polished. 960×

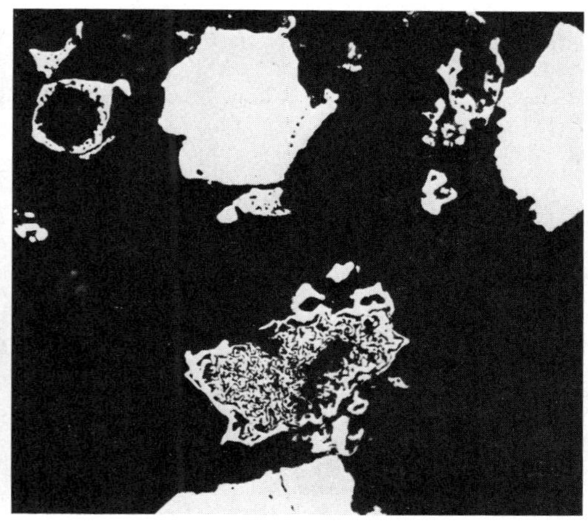

Fig. 152 Atomet 28 iron powder (granulated and oxidized, hydrogen reduced), showing particles of varying degrees of porosity. As-polished. 645×

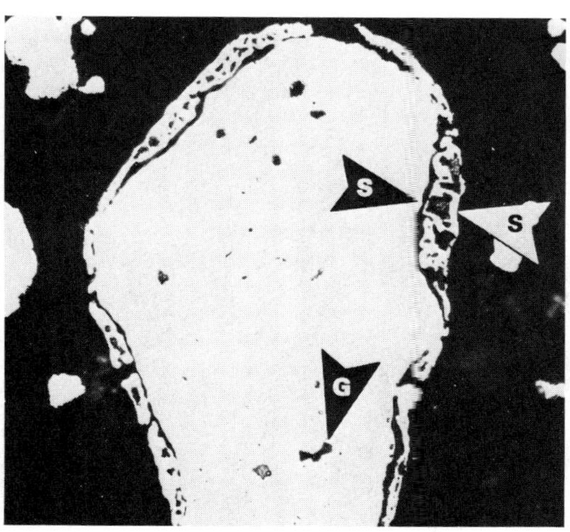

Fig. 153 Ancorsteel 1000 water-atomized, annealed iron. Solid particle with some gas porosity (arrow G) and some surface porosity (arrows S) caused the reduction of oxides picked up during atomization. As-polished. 645×

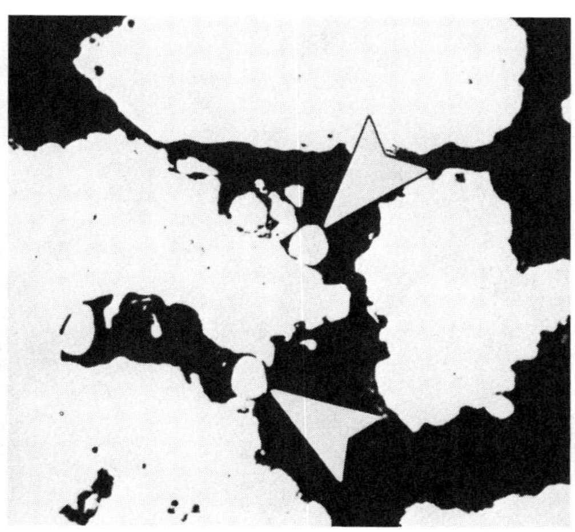

Fig. 154 Ancorsteel 1000 water-atomized iron (–80 mesh). Arrows indicate fine particles agglomerated to coarser ones during annealing. As-polished. 960×

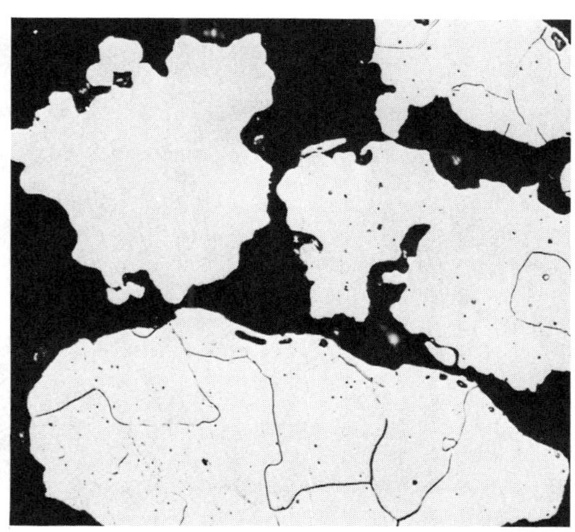

Fig. 155 Same as Fig. 154, but etched. Grain boundaries are ferrite, indicating low carbon content. Some particles do not etch. Note irregular surface of particle at upper left caused by agglomeration of smaller particles. 2% nital. 615×

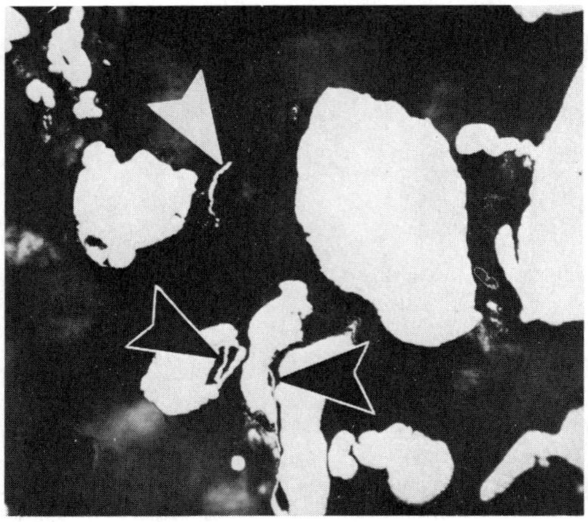

Fig. 156 316L stainless steel water-atomized powder (–100 mesh). Irregular particles with attached splatters of material (see arrows). As-polished. 645×

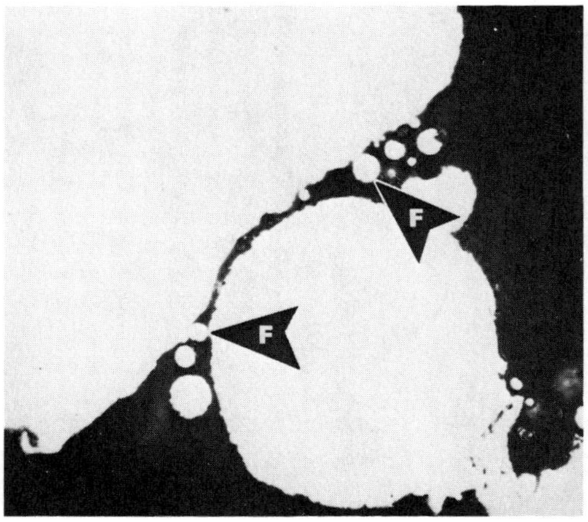

Fig. 157 316L stainless steel gas-atomized powder (–100 mesh). The wide range of particle sizes and the agglomeration of several small particles to a larger one (arrows F) during annealing is shown. As-polished. 645×

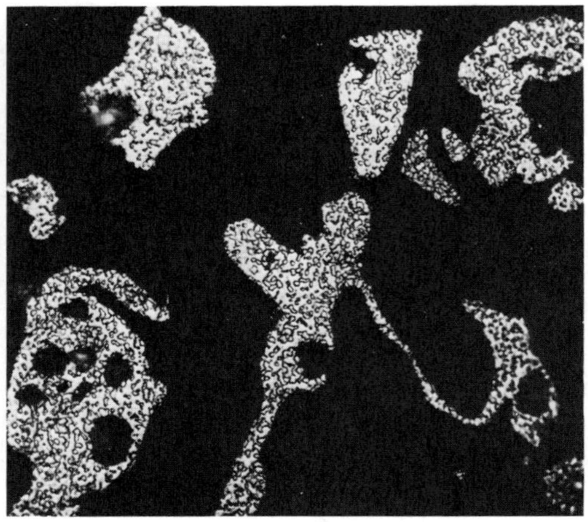

Fig. 158 M2 tool steel, water-atomized, vacuum-annealed powder. Irregular particles with a fine carbide phase. Equal parts 4% picral and 4% nital. 645×

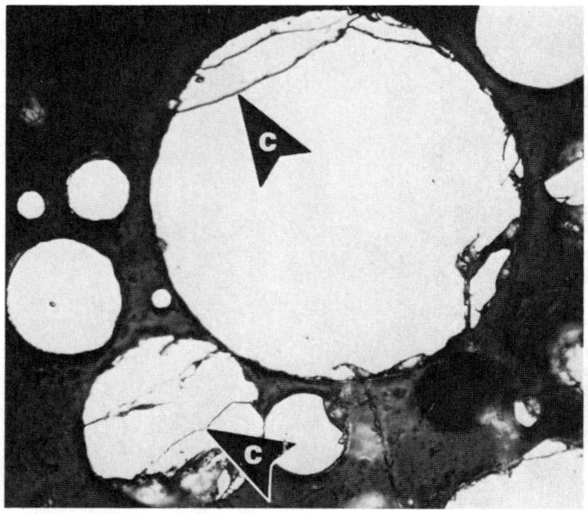

Fig. 159 Fe-31Si, water-atomized master-alloy powder (–100 mesh). Regular particle shape implies low green strength. Particles cracked (arrows C) during processing. As-polished. 645×

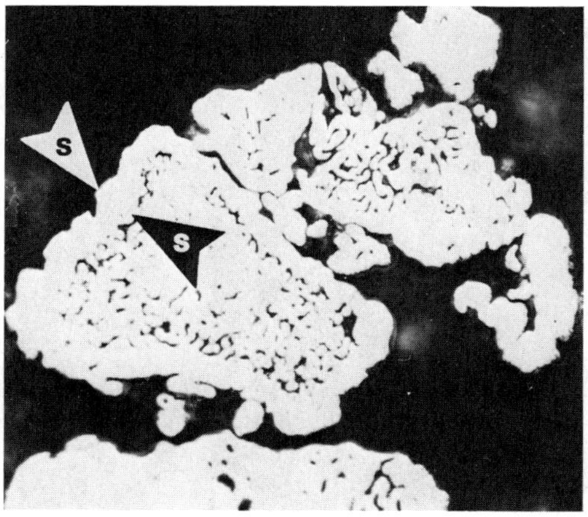

Fig. 160 100 RXM copper powder (–100 mesh) subjected to oxidation, grinding, and reduction cycle. Arrows S surround surface pores that have closed during reduction or annealing. As-polished. 645×

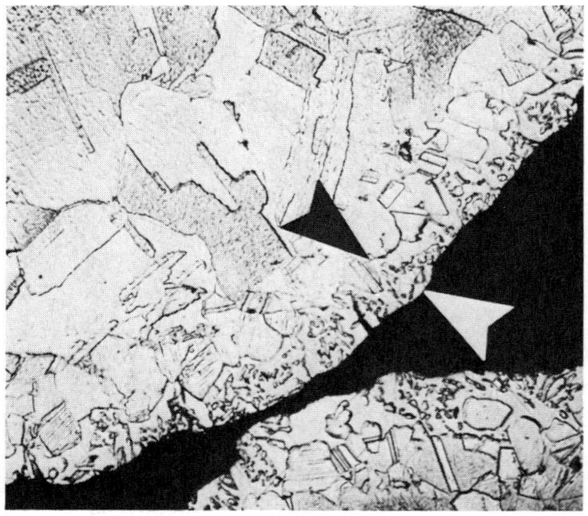

Fig. 161 92Cu-8Sn atomized bronze filter powder (–40+60 mesh). Surface of copper is cooled with tin, which is partly alloyed, forming intermetallics and some α-bronze. Arrows show depth of tin diffusion. As-polished. 645×

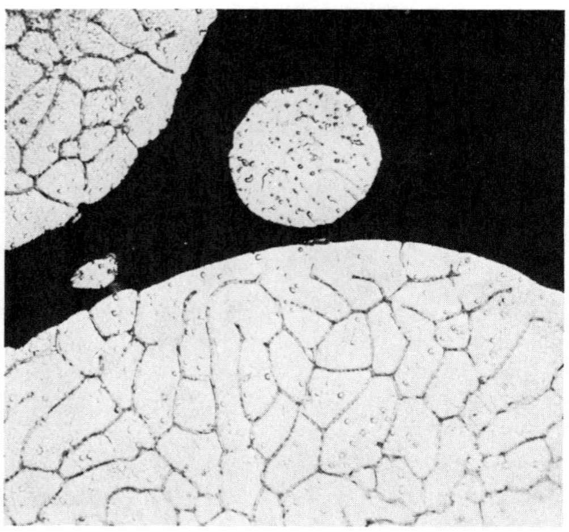

Fig. 162 601AC atomized aluminum powder. Very smooth regular particle surfaces with no evidence of the admixed elements. Keller's reagent. 660×

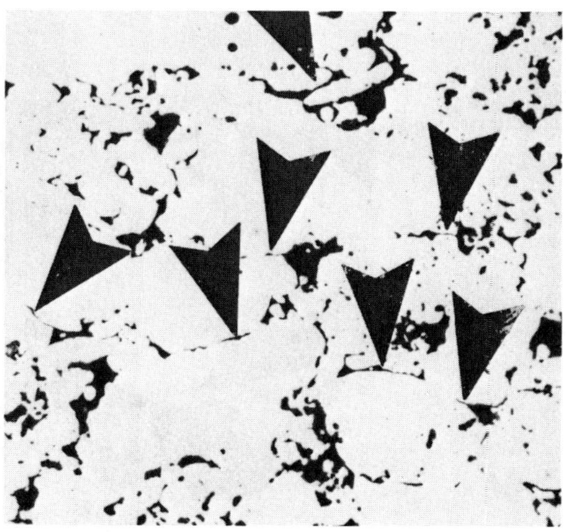

Fig. 163 Undersintered nickel steel (Fe-2Ni-0.8C, 7.0 g/cm^3), pressed at 550 MPa (40 tsi) and sintered 25 min at 1120 °C (2050 °F) in endothermic gas. Arrows indicate numerous original particle boundaries. As-polished. 180×

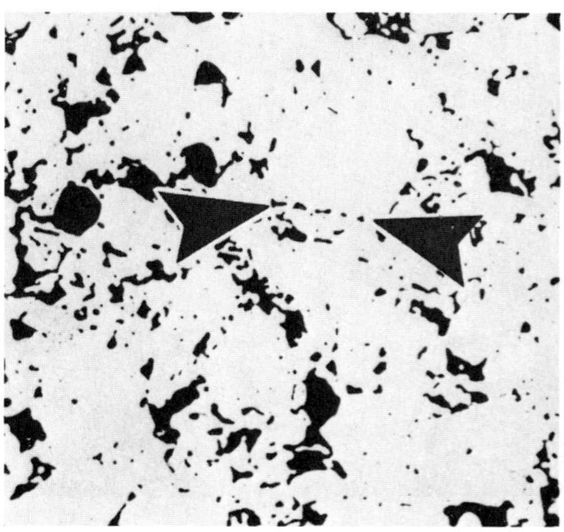

Fig. 164 Same as Fig. 163, but an average sinter (30 min at 1120 °C, 2050 °F, in endothermic gas). Arrows show a row of pores at which a particle boundary has broken up during sintering. As-polished. 180×

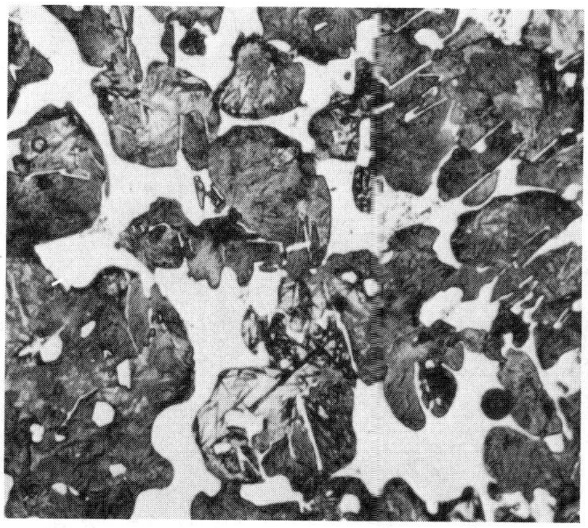

Fig. 165 Fe-5Cr-1Mo-2Cu-0.5P-2.5C pressed at 550 MPa; liquid phase sintered 30 min at 1110 °C in DA to 7.6 g/cm^3. White areas are primary M_3C for wear resistance. 2% nital. 545×. Source: Ref 19

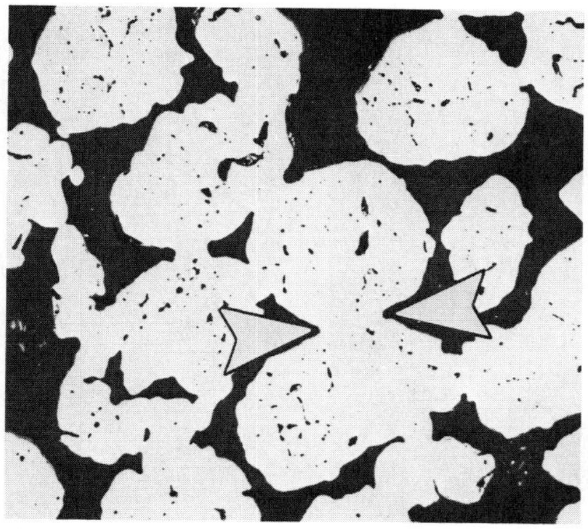

Fig. 166 Cu-8Sn filter powder, gravity sintered 30 min at 870 °C (1600 °F) in DA. Arrows N show bonding between particles. Dark areas are pores filled with epoxy. As-polished. 65×

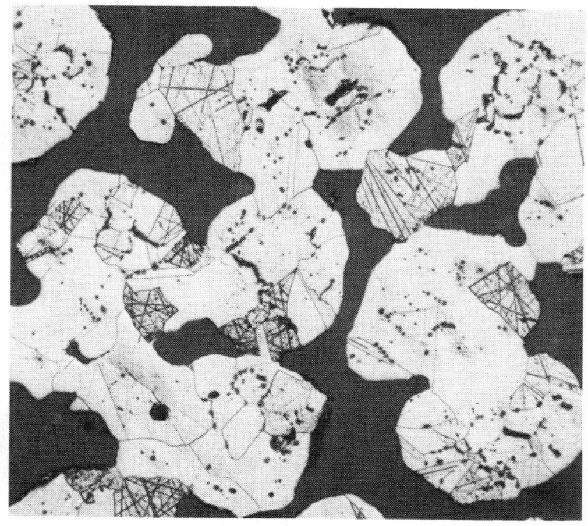

Fig. 167 Same as Fig. 166, but etched, showing well-sintered α-bronze grains. Cross-hatched grains result from residual scratches and etching. $K_2C_2O_7$. 65×

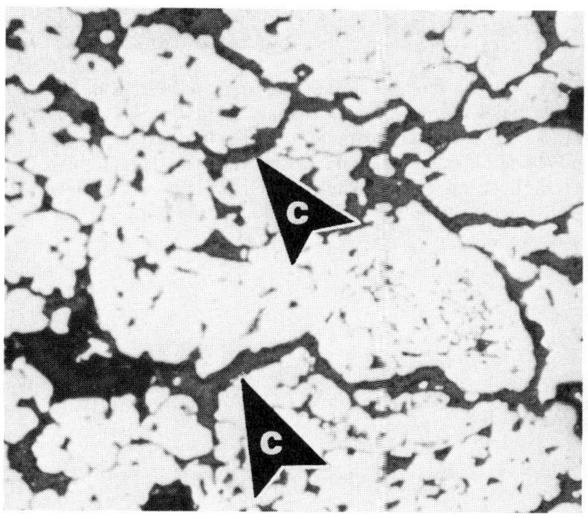

Fig. 168 Fe-2Cu-0.8C (6.8 g/cm^3) blistered during heat up to sintering because of soot that expanded the spaces between particles. Arrows C show resulting cracks. As-polished. 180×

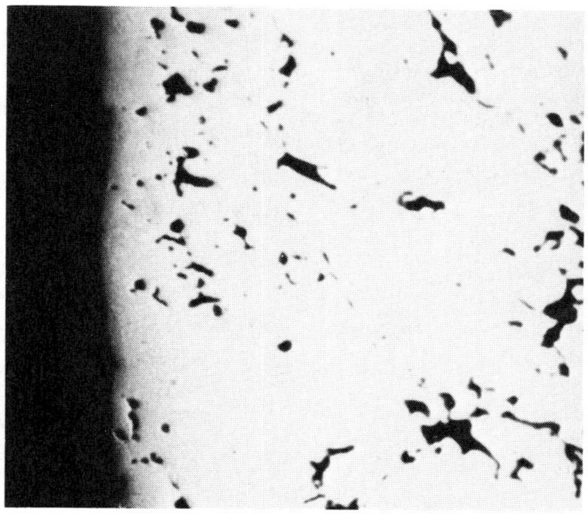

Fig. 169 High-density region (7.55 g/cm^3) molded above a die shelf. Density measured by the area fraction of pores. See Fig. 170. As-polished. 180×

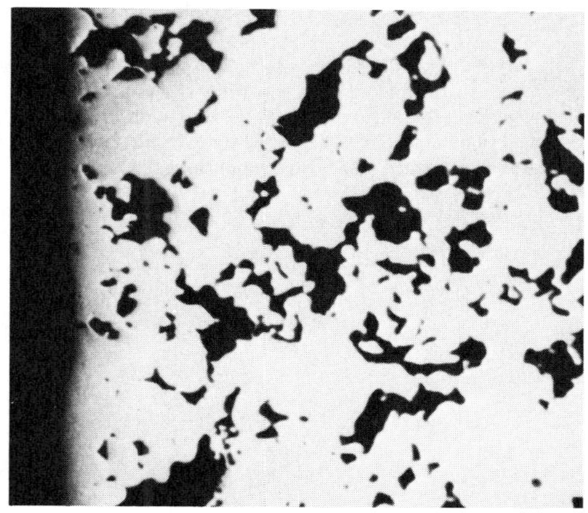

Fig. 170 Same nickel steel as Fig. 169, but a low-density region (5.98 g/cm^3) molded below the die shelf. As-polished. 180×

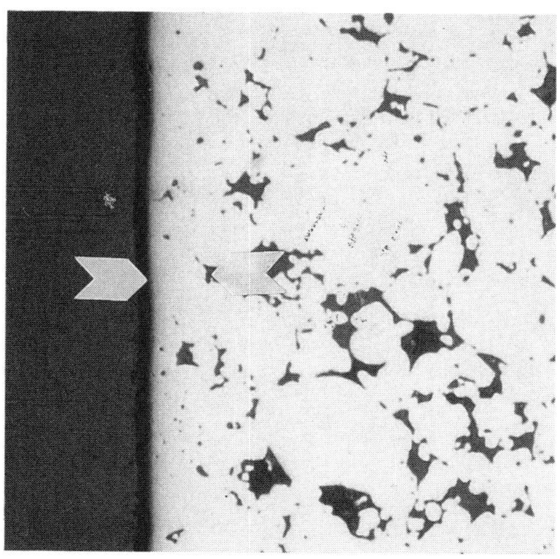

Fig. 171 Medium density sintered steel, typical sinter. The part was tumbled after sintering, which closed the surface pores to a depth of 0.002 in. as shown by the arrows. Unetched. 960×

Fig. 172 Fracture surface of a heat treated P/M steel where the fracture occurred in the unsintered state. Top arrow shows the compacted surface. Right arrow points to rounded particle surfaces with no indication of ductile cups and cones that would be typical of a fracture in the sintered state. White layer is electroless nickel plating used to preserve the surface details during polishing. Etched. Compare with Fig. 172. 960×

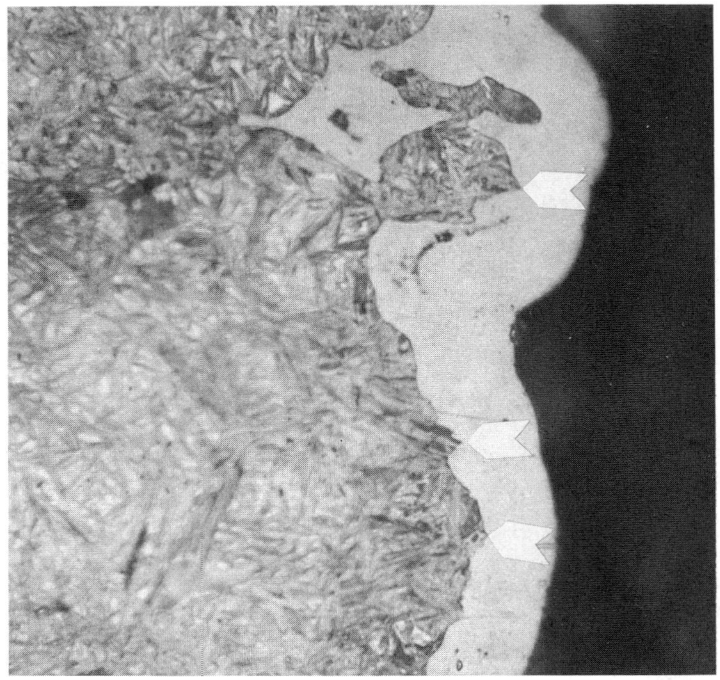

Fig. 173 Same material as Fig. 171, but a fracture surface deliberately created in the sintered and heat treated state. Arrow points to sharp asperities where the particles were torn away from the adjacent material on the opposite fracture surface. Such asperities are not created when a P/M part is fractured in the unsintered state. Even if they were so created, the asperities would become rounded during the usual sintering cycles. Etched. 960×

Materials Systems, Properties, and Applications

Ferrous Powder Metallurgy Materials

W.B. James, Hoeganaes Corporation
G.T. West, National Sintered Alloys Inc.

FERROUS POWDER METALLURGY processing is a net or near net-shaped production technology, which in many instances eliminates the need for secondary machining operations. Ferrous P/M offers a wide range of engineered materials for structural applications and satisfies close dimensional tolerance requirements for parts with complex geometries.

Figure 1 illustrates the continued growth of iron powder shipments in North America. From 1985 to 1995 the North American iron powder market grew at a compounded annual rate of 6.1%. In 1996, iron powder shipments were 318,227 metric tons (350,603 short tons). The ferrous P/M industry in North America has experienced sustained growth in the 1990s with growth projected to continue through the year 2005 (Ref 1). In comparison, for Japan, the total iron powder production in 1996, including exports, reached 157,356 metric tons (173,406 short tons), a decline of 4.6% (Fig. 2). At 103,258 metric tons (113,790 short tons), iron powder shipments during 1996 in Europe showed a marginal increase of 416 metric tons (459 short tons) compared with shipments in 1995. Figure 2 illustrates a comparison of international iron and steel powder shipments.

Automotive applications currently represent 69% of the ferrous structural parts market (Fig. 3) (Ref 1). The typical family vehicle contains 14 kg (31 lb) of P/M parts, and industry experts forecast that the P/M content in cars will reach 22.6 kg (50 lb) by the end of the decade. There are some models that already contain up to 18.1 kg (40 lb) of P/M parts. Other major markets are represented by household appliances; hardware; industrial motors/controls and hydraulics; and recreation, tools, and hobby applications.

When selecting a material for an intended application it is important that a "systems" approach, as illustrated in Fig. 4, is followed (Ref 2). Part designers need to involve part fabricators, tool designers, and material engineers from the outset of an application development program. Performance targets for the part need to be defined, as well as the application environment. The part fabricator can suggest design changes that will lead to a more robust P/M part, and the designer can assess the impact of the proposals on the anticipated performance of the part. It is important to determine from the outset whether the fabricator can make a net shaped part or whether secondary machining operations will be needed. This will have a major influence on the material selection for any proposed part processing sequence.

In addition, ferrous powder metallurgy also encompasses several production technologies other than traditional press and sinter operations. These include copper-infiltrated steels, powder forged steels, and injection molded steels, which are discussed in detail in separate articles in this Section. Various other options are also used to increase the performance of P/M parts. This may include additional processing (Fig. 5a) or alloying (Fig. 5b). This article briefly summarizes the general classification, mechanical properties, and applications of ferrous P/M materials for parts production. More detailed coverage on major alloys and processing methods are discussed elsewhere in this Volume and in Ref 4 and 5.

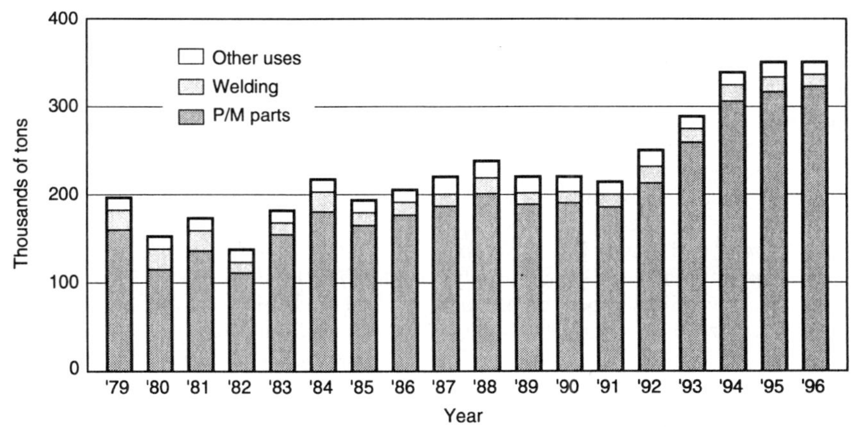

Fig. 1 Iron powder shipments in North America. Source: Ref 1

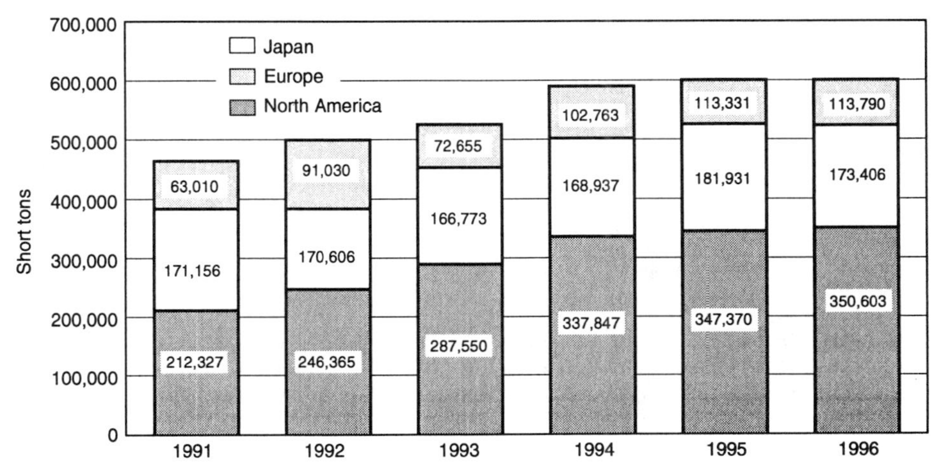

Fig. 2 International iron and steel powder shipments. Source: Ref 1

Alloying Methods

There are four principal alloying methods used in ferrous powder metallurgy:

- Admixed alloys
- Partial alloys (diffusion alloys)
- Prealloys
- Hybrid alloys

Each method will be described, and the benefits and disadvantages of each method will be reviewed along with the effect of processing on the microstructure and properties of the materials. In general, comments made with respect to sintering will relate to sintering at a temperature of 1120 °C (2050 °F) unless specified otherwise.

Admixed Powders. Alloying additions, in elemental form or as ferroalloys, such as Fe_3P, FeMn, or FeCr, can be mixed with an iron powder. This is generally the least expensive and most commonly used alloying method in ferrous powder metallurgy. When the mix is pressed, the additions are not alloyed with the iron powder; admixed materials therefore retain most of the compressibility of the base iron powder. The degree of alloying during sintering is limited by the diffusivity of the alloying elements in the iron at the sintering temperature. The resulting microstructures can be chemically and microstructurally heterogeneous.

Direct reduced sponge iron or water atomized iron powders can be used as the base powder for admixed materials. The density required to achieve the desired mechanical properties generally will dictate which type of powder is used. Sponge powders are selected typically for the density range of 6.4 g/cm^3 and lower, while atomized iron powders are used for parts with a density of 6.7 g/cm^3 and higher. Either type of powder can be used for the intermediate density range; the choice generally depends on the part geometry and the available compaction press tonnage.

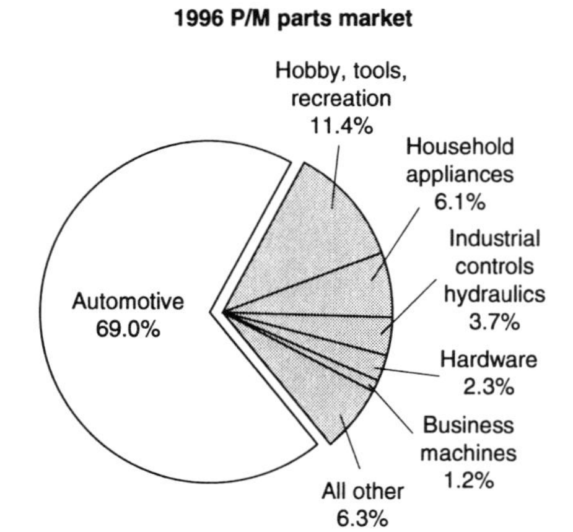

Fig. 3 Applications for ferrous P/M structural parts. Source: Ref 1

Partially Alloyed Powders. In partially alloyed powders, also known as diffusion alloyed powders, the alloy addition or additions are metallurgically bonded to either an elemental iron powder or to a prealloyed powder base (see the section "Prealloyed Powders" for an explanation of prealloyed powders). The diffusion alloying of the additions is controlled to provide adequate bonding of the additions to the base powder while limiting the extent of diffusion of the additives so they do not reduce the compressibility of the base powder appreciably.

While there is a wide range of diffusion alloyed powders available in Europe (Ref 6), there are only two grades in widespread commercial use in North America, South America, and Canada. Both are based on a highly compressible iron powder, contain nominally 1.5 wt% Cu and 0.5 wt% Mo, and differ with respect to their nickel contents. One powder contains 2 wt% Ni, while the other has a nickel content of 4 wt%. Parts made from partially alloyed powders have a complex as-sintered microstructure consisting

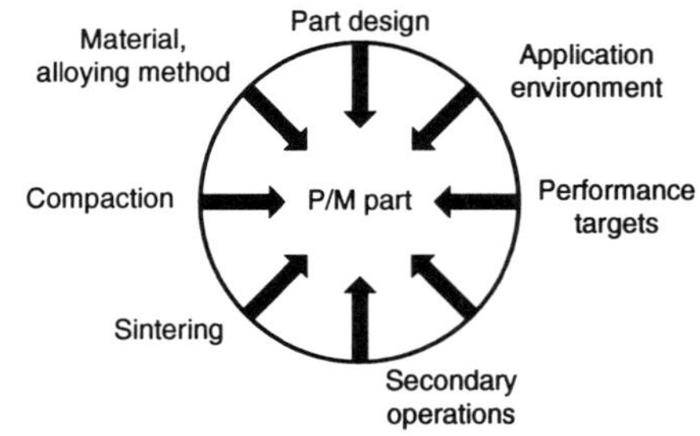

Fig. 4 "Systems" approach to P/M part development. Source: Ref 2

of areas of martensite, bainite, fine unresolved pearlite, nickel-rich ferrite containing blocky carbides, as well as dispersed nickel-rich areas of retained austenite (Fig. 6). This complex microstructure provides a combination of good tensile strength and tensile ductility as well as a material with excellent impact energy.

Prealloyed Powders. Prealloyed, low-alloy steel powders are particularly useful in the manufacture of P/M parts designed for high performance applications. Combinations of molybdenum, manganese, and nickel are added to molten iron during the steelmaking process to produce a chemically uniform powder. Practices are followed during melting and atomization to ensure the production of powders with nonmetallic inclusion contents suitable for the most stringent requirements. The reactivity of carbon, in the form of graphite, with the prealloyed powders is excellent, enabling close control of the combined carbon content during sintering.

These powders are used in a wide range of applications. Varying performance requirements are satisfied through a combination of alloy selection, carbon additions, and processing conditions. In general, the compressibility of prealloyed powders is lower than that of admixed or partially alloyed powders. However, prealloyed powders in which molybdenum is the principal

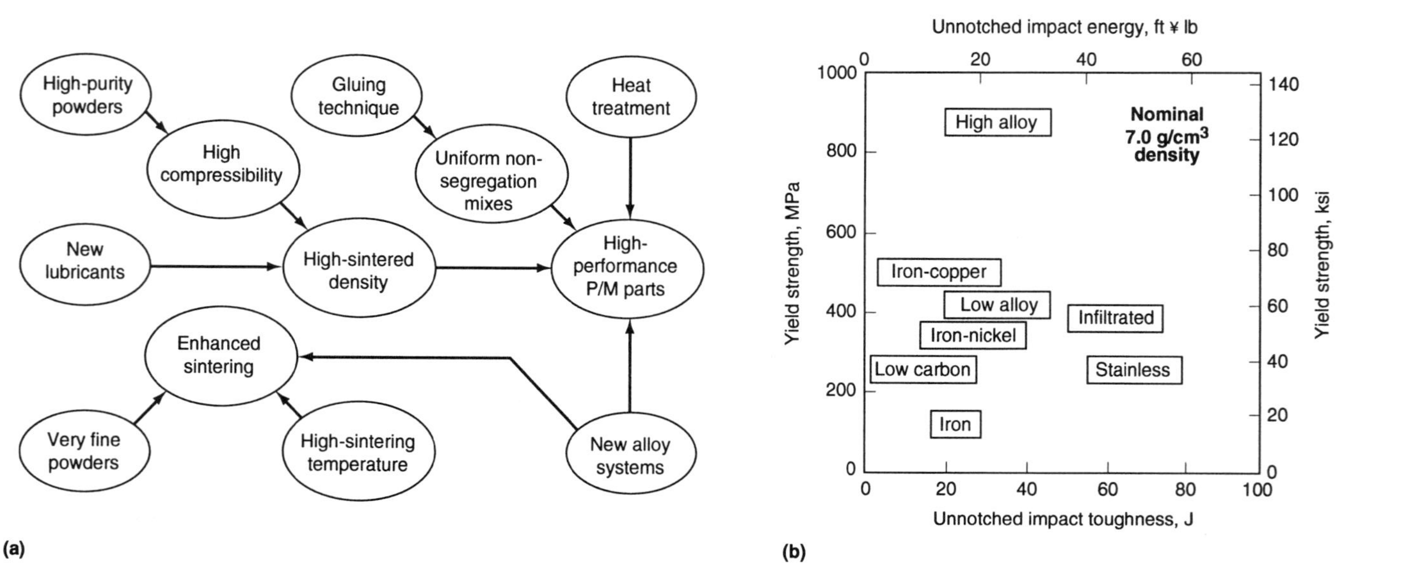

Fig. 5 General methods of achieving high-performance P/M steels. (a) Processing options. (b) Higher alloying to improve strength and toughness. Source: Ref 3 and 4

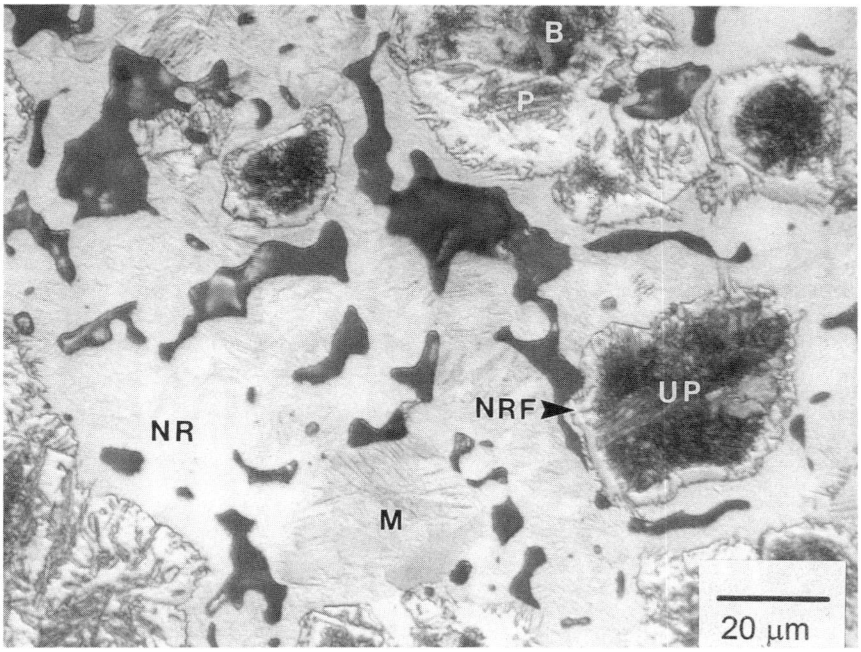

Fig. 6 As-sintered microstructure of a part made from a partially alloyed powder FD-0405: B, Bainite; P, pearlite; UP, unresolved pearlite; NR, nickel rich region; NRF, nickel rich ferrite; M, martensite

alloy addition have compressibilities that are comparable to those of iron powders (Ref 2).

The hardenability of the prealloyed materials is controlled by the amount and type of alloying employed and can be tailored to fit specific applications. Prealloyed materials have homogeneous microstructures with uniform apparent hardness. Unlike admixed, partially alloyed, and some hybrid alloys, alloying is not primarily dependent on diffusion processes. Just as with admixed, partially alloyed, and hybrid alloys (see the section "Hybrid Alloys"), higher sintering temperatures (>1120 °C, or 2050 °F) and longer sintering times (greater than 20 min at temperature) will, however, influence the number of pores that are present in the microstructure as well as their size and their shape. Pore coalescence occurs as sintering temperature and time are increased; large pores grow at the expense of the smaller pores. Increased time and temperature also lead to an increased amount of pore rounding.

Certain applications may require supplemental alloy additions via admixing to enhance the hardenability and mechanical properties of the material. Such materials are termed hybrid alloys.

Hybrid Alloy Powders. Hybrid alloys consist of either prealloyed or partially alloyed base powders to which elemental or ferroalloy additions are made to achieve the desired chemical composition. The practice of making nickel or copper additions to prealloyed powders has become widespread since the introduction of the highly compressible, prealloyed powders with molybdenum as the principal alloy addition (Ref 2, 7). These prealloyed powders are also particularly suitable for the addition of high-carbon ferroalloys to produce chrome-molybdenum, manganese-molybdenum, and chrome-molybdenum-manganese steels (Ref 8). Elemental nickel additions can be made to partially alloyed powders to provide even greater impact properties (Ref 9, 10). The compressibility and hardenability of hybrid alloy materials depends primarily on the compressibility and hardenability of the base powder employed. Hybrid materials generally have a heterogeneous microstructure with nonuniform apparent hardness. Higher sintering temperatures and longer sintering times generally lead to greater diffusion of the alloy additions and to more homogeneous microstructures.

Ferrous Powder Materials

Material code designations for ferrous P/M materials are found in MPIF Standard 35, *Materials Standards for P/M Structural Parts*. Similar designations have been adopted by ASTM and are summarized in specification B 783 (Ref 11). These standards are subject to periodic review, and users are cautioned to refer to the latest edition. Ferrous material designations have various prefixes:

- F, iron
- FC, iron-copper or copper steel
- FD, diffusion alloyed steel (partially alloyed)
- FL, prealloyed ferrous material except stainless steel
- FN, iron-nickel or nickel steel
- FX, copper infiltrated iron or steel
- SS, stainless steel (prealloyed)

Figure 7 provides examples of material designations. The designation FN-0205-35 (Fig. 7a) refers to a nickel steel with nominally 2 wt%Ni, 0.3 to 0.6 wt% combined carbon, and a guaranteed minimum 0.2% offset yield strength of 240

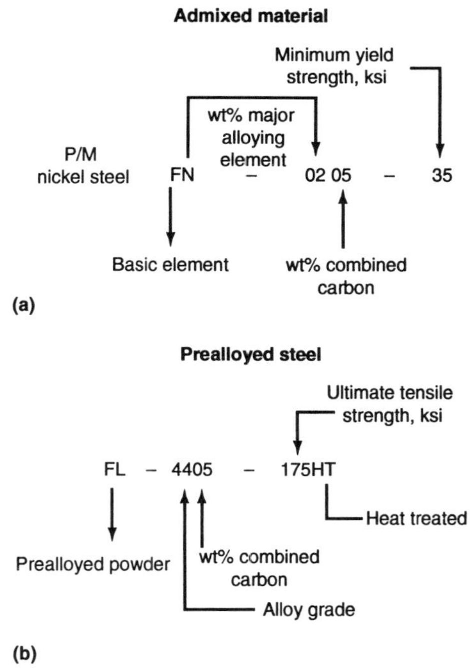

Fig. 7 Examples of ferrous P/M material designations. (a) FN-0205-35 and (b) FL-4405-175HT

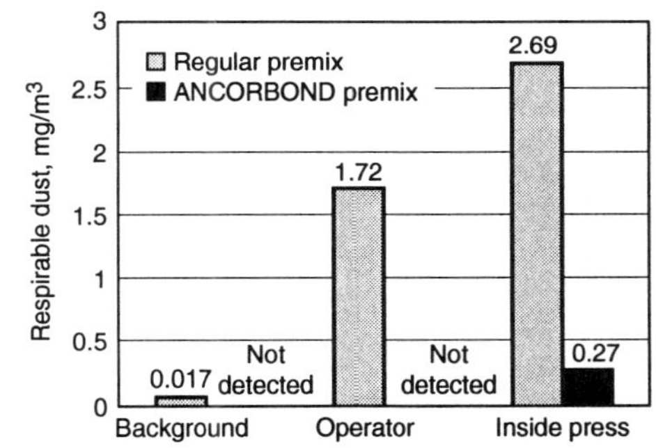

Fig. 8 Respirable dust in the work area of a P/M parts making plant. Source: Ref 2

MPa (35 ksi). The FN prefix designates an admixed nickel steel, and the 02 following the prefix indicates that there is 2 wt% of the major alloy addition (nickel). The 05 refers to the combined carbon content of the material. The code designations for various combined carbon contents are:

Combined carbon range, wt %	Code designation
0–0.3	00
0.3–0.6	05(a)
0.6–0.9	08

(a) The 05 designation for the FL materials (prealloyed) refers to 0.4 to 0.7 wt% combined carbon.

The designation FL-4405-175HT in Fig. 7(b) refers to a quench-hardened and tempered, prealloyed steel with a combined carbon content of 0.4 to 0.7 wt% and a guaranteed minimum ultimate tensile strength of 1210 MPa (175 ksi).

Table 1 Chemical composition of iron and carbon steels

Material designation	Chemical composition, wt% Fe	C	Element
F-0000	97.7	0.0	min.
	100.0	0.3	max.
F-0005	97.4	0.3	min.
	99.7	0.6	max.
F-0008	97.1	0.6	min.
	99.4	0.9	max.

Other elements: Total by difference equals 2.0 wt% maximum, which may include other minor elements added for specific purposes. Source: MPIF Standard 35

For as-sintered materials the material designation suffix refers to the guaranteed minimum 0.2% offset yield strength in thousands of pounds per square inch. For example, FC-0208-50 refers to an admixed copper steel with nominally 2 wt% Cu, a combined carbon content of 0.6 to 0.9 wt%, and a guaranteed minimum 0.2% offset yield strength of 340 MPa (50 ksi).

For heat treated materials the material designation suffix refers to the guaranteed minimum ultimate tensile strength in thousands of pounds per square inch. For example, FN-0205-130HT refers to an admixed nickel steel with nominally 2 wt% nickel, a combined carbon content of 0.3 to 0.6 wt%, and a guaranteed minimum ultimate tensile strength of 900 MPa (130 ksi).

Admixed Powders

Admixed materials are in widespread use throughout the ferrous P/M industry. They generally contain sufficient graphite to provide the desired carbon content after sintering because the typical base powders used are essentially carbon free. Base powders generally have low carbon contents because carbon forms an interstitial solid solution with iron and reduces the compressibility of the powder significantly. In

Table 2 Mechanical properties of iron and carbon steels

Minimum values: Material designation code	Minimum strength, ksi: Yield	Minimum strength, ksi: Ultimate	Typical values: Tensile properties: Ultimate strength, ksi	Tensile properties: Yield strength (0.2%), ksi	Elongation in 25 mm (1 in.), %	Elastic constants: Young's modulus, 10^6 psi	Elastic constants: Poisson's ratio	Unnotched Charpy impact energy, ft · lbf	Transverse rupture strength, ksi	Compressive yield strength (0.1%), ksi	Rockwell hardness: Macro (apparent)	Rockwell hardness: Micro (converted)	Fatigue limit (90% survival), ksi	Density, g/cm^3
F-0000-10	10	...	18	13	1.5	15.0	0.25	3.0	36	16	40 HRF	N/A	7	6.1
F-0000-15	15	...	25	18	2.5	17.5	0.25	6.0	50	18	60 HRF	N/A	10	6.7
F-0000-20	20	...	38	25	7.0	23.5	0.28	35.0	95	19	80 HRF	N/A	14	7.3
F-0005-15	15	...	24	18	<1.0	15.0	0.25	3.0	48	29	25 HRB	N/A	9	6.1
F-0005-20	20	...	32	23	1.0	16.5	0.25	4.0	64	31	40 HRB	N/A	12	6.6
F-0005-25	25	...	38	28	1.5	19.5	0.27	5.0	76	32	55 HRB	N/A	15	6.9
F-0005-50HT	...	50	60	(a)	<0.5	16.5	0.25	3.0	105	43	20 HRC	58 HRC	23	6.6
F-0005-60HT	...	60	70	(a)	<0.5	18.5	0.27	3.5	120	52	22 HRC	58 HRC	27	6.8
F-0005-70HT	...	70	80	(a)	<0.5	20.5	0.27	4.0	140	61	25 HRC	58 HRC	32	7.0
F-0008-20	20	...	29	25	<0.5	12.5	0.25	2.5	51	41	35 HRB	N/A	11	5.8
F-0008-25	25	...	35	30	<0.5	16.0	0.25	3.0	61	41	50 HRB	N/A	14	6.2
F-0008-30	30	...	42	35	<1.0	16.5	0.25	4.0	74	42	60 HRB	N/A	17	6.6
F-0008-35	35	...	57	40	1.0	20.5	0.27	5.0	100	42	70 HRB	N/A	25	7.0
F-0008-55HT	...	55	65	(a)	<0.5	16.5	0.25	3.0	100	42	22 HRC	60 HRC	26	6.3
F-0008-65HT	...	65	75	(a)	<0.5	16.5	0.25	4.0	115	58	28 HRC	60 HRC	30	6.6
F-0008-75HT	...	75	85	(a)	<0.5	19.5	0.27	4.5	130	75	32 HRC	60 HRC	34	6.9
F-0008-85HT	...	85	95	(a)	<0.5	21.5	0.27	5.0	145	86	35 HRC	60 HRC	38	7.1

HT, heat treated; N/A, not applicable. Suffix numbers represent minimum strength values in ksi (10^3 psi); yield in the as-sintered condition and ultimate in the heat treated condition. Mechanical property data derived from laboratory prepared test specimens sintered under commercial manufacturing conditions. Tempering temperature for heat treated materials: 177 °C (350 °F). (a) Yield and ultimate tensile strength are approximately the same for heat treated materials. Source: MPIF Standard 35

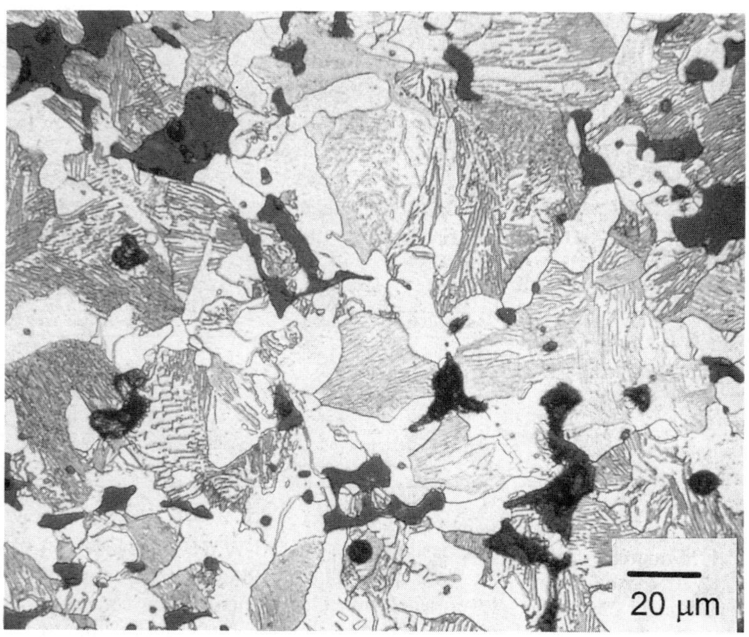

Fig. 9 Ferrite/pearlite microstructure of an as-sintered P/M carbon steel (F-0005)

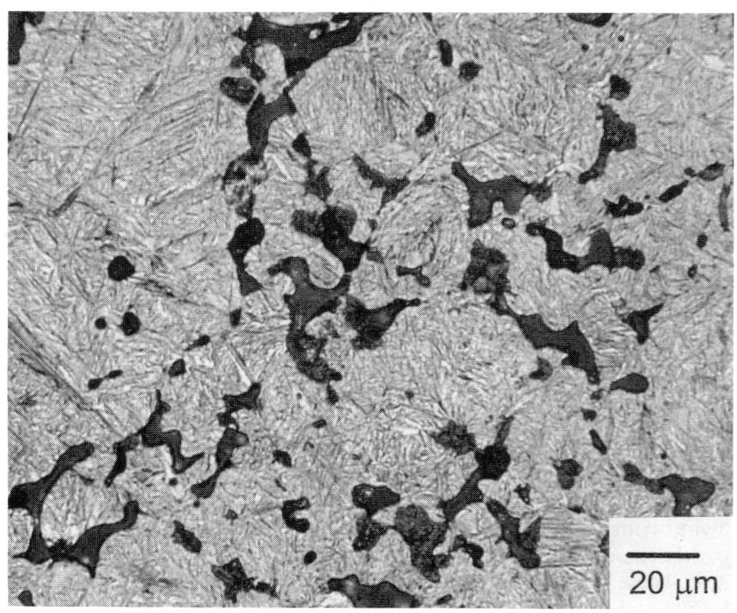

Fig. 10 Tempered martensitic microstructure of a quench-hardened and tempered P/M carbon steel (F-0008)

addition to graphite, the mixes contain a lubricant that aids particle rearrangement during the initial stages of compaction. The principal purpose of the added lubricant is, however, to reduce the friction between the powder particles and the compaction tooling members (die wall, core rods, and punches) and make ejection of the compacted part easier. Lubricant additions vary from 0.5 wt% to sometimes >1 wt%, depending on the part geometry, part density, and the surface area of the compact at the die-wall and core rod interfaces. Lubricant additions of 0.75 wt% are quite common. The lubricants used are generally metallic stearates (e.g., zinc stearate) or ethylene bisstearamides. They are white solids with pycnometric densities of ~1 g/cm^3. Because of their low density relative to that of iron, while they are added typically in quantities of about 0.75 wt%, they occupy ~5.5 vol% in the mixture. This has particular significance when parts are compacted at high pressures (Ref 12–14).

Because the various powders that comprise the admixed alloys have differing particle size and density, admixed alloys are particularly susceptible to dusting and segregation during handling and transfer to the die cavity. Numerous opportunities exist during the P/M process for dusting and segregation to occur (Ref 15). Compositional variations, which result from these demixing phenomena, cause inconsistencies in the green and sintered properties of P/M parts. The tendency for mixes to dust and segregate can be reduced significantly through the use of binder treated mixes (Ref 16–19). In addition to reducing dusting and segregation, binder treated mixes improve powder flow and die filling characteristics during part manufacturing. This results in more consistent part sectional densities and improved control of part mass. Improved flow and die fill uniformity result in reduced press cycle times and more consistency throughout the entire part manufacturing process.

Improved retention of alloy additions provides an economic advantage in terms of reduced amounts of green scrap, enhanced alloy efficiency, and a more pleasant and easier to maintain work environment. Plant cleanliness is dramatically improved, and the amount of respirable dust in the immediate vicinity of the compaction press is reduced by an order of magnitude (Fig. 8) (Ref 2). Binder treated premixes are produced using a proprietary mixing process that utilizes patented binders (Ref 16–19).

Iron and Carbon Steels. Table 1 summarizes the chemical compositions of the iron and carbon steels listed in MPIF Standard 35.

Unalloyed iron (F-0000) materials are used for lightly loaded structural applications and structural parts requiring self lubrication where strength is not critical. The combined carbon content of these materials can be from 0 to 0.3 wt%. Iron parts that are essentially carbon free are often used, particularly at higher part densities, for their soft magnetic properties (Ref 20–22).

P/M carbon steels with 0.3 to 0.6 wt% combined carbon (F-0005) possess moderate strength and apparent hardness. They are used where such properties combined with machinability are desired.

Higher carbon P/M steels (F-0008) have moderately higher strength compared with F-0005 materials, but are more difficult to machine. In the as-sintered condition carbon steels have a

Table 3 Chemical composition of iron-copper and copper steels

Material designation	Chemical composition, wt% Fe	Cu	C	Element
FC-0200	93.8	1.5	0.0	min.
	98.5	3.9	0.3	max.
FC-0205	93.5	1.5	0.3	min.
	98.2	3.9	0.6	max.
FC-0208	93.2	1.5	0.6	min.
	97.9	3.9	0.9	max.
FC-0505	91.4	4.0	0.3	min.
	95.7	6.0	0.6	max.
FC-0508	91.1	4.0	0.6	min.
	95.4	6.0	0.9	max.
FC-0808	88.1	7.0	0.6	min.
	92.4	9.0	0.9	max.
FC-1000	87.2	9.5	0.0	min.
	90.5	10.5	0.3	max.

Other elements: Total by difference equals 2.0 wt% maximum, which may include other minor elements added for specific purposes. Source: MPIF Standard 35

Table 4 Mechanical properties of various iron-copper and copper steels

Material designation code	Minimum strength, ksi: Yield	Minimum strength, ksi: Ultimate	Ultimate strength, ksi	Yield strength (0.2%), ksi	Elongation in 25 mm (1 in.), %	Young's modulus, 10^6 psi	Poisson's ratio	Unnotched Charpy impact energy, ft · lbf	Transverse rupture strength, ksi	Compressive yield strength (0.1%), ksi	Rockwell hardness Macro (apparent)	Rockwell hardness Micro (converted)	Fatigue limit (90% survival), ksi	Density, g/cm^3
FC-0200-15	15	...	25	20	1.0	14.0	0.27	4.5	45	18	11 HRB	N/A	10	6.0
FC-0200-18	18	...	28	23	1.5	16.5	0.25	5.0	51	21	18 HRB	N/A	11	6.3
FC-0200-21	21	...	31	26	1.5	16.5	0.25	5.5	56	23	26 HRB	N/A	12	6.6
FC-0200-24	24	...	34	29	2.0	19.5	0.27	6.0	63	26	36 HRB	N/A	13	6.9
FC-0205-30	30	...	35	35	<1.0	14.0	0.27	<2.0	60	50	37 HRB	N/A	13	6.0
FC-0205-35	35	...	40	40	<1.0	16.5	0.25	3.0	75	53	48 HRB	N/A	15	6.3
FC-0205-40	40	...	50	45	<1.0	17.5	0.25	5.0	95	57	60 HRB	N/A	21	6.7
FC-0205-45	45	...	60	50	<1.0	21.5	0.27	8.0	115	60	72 HRB	N/A	31	7.1
FC-0205-60 HT	...	60	70	(a)	<0.5	16.0	0.25	2.5	95	57	19 HRC	58 HRC	27	6.2
FC-0205-70 HT	...	70	80	(a)	<0.5	15.5	0.25	3.5	110	71	25 HRC	58 HRC	30	6.5
FC-0205-80 HT	...	80	90	(a)	<0.5	18.5	0.27	4.5	120	86	31 HRC	58 HRC	34	6.8
FC-0205-90 HT	...	90	100	(a)	<0.5	20.5	0.27	5.5	135	95	36 HRC	58 HRC	38	7.0
FC-0208-30	30	...	35	35	<1.0	12.5	0.25	<2.0	60	57	50 HRB	N/A	13	5.8
FC-0208-40	40	...	50	45	<1.0	16.5	0.25	2.0	90	62	61 HRB	N/A	17	6.3
FC-0208-50	50	...	60	55	<1.0	17.5	0.25	5.0	125	66	73 HRB	N/A	23	6.7
FC-0208-60	60	...	75	65	<1.0	22.5	0.28	7.0	155	71	84 HRB	N/A	33	7.2
FC-0208-50 HT	...	50	65	(a)	<0.5	15.0	0.25	2.5	95	58	20 HRC	60 HRC	25	6.1
FC-0208-65 HT	...	65	75	(a)	<0.5	17.5	0.27	3.5	110	72	27 HRC	60 HRC	30	6.4
FC-0208-80 HT	...	80	90	(a)	<0.5	18.5	0.27	4.5	130	91	35 HRC	60 HRC	35	6.8
FC-0208-95 HT	...	95	105	(a)	<0.5	21.5	0.27	5.5	150	105	43 HRC	60 HRC	40	7.1
FC-0505-30	30	...	44	36	<0.5	12.5	0.25	3.0	77	50	51 HRB	N/A	17	5.8
FC-0505-40	40	...	58	47	<0.5	16.5	0.25	4.5	102	54	62 HRB	N/A	22	6.3
FC-0505-50	50	...	71	56	<1.0	17.5	0.25	5.0	124	58	72 HRB	N/A	27	6.7
FC-0508-40	40	...	58	50	<0.5	13.0	0.25	3.0	100	58	60 HRB	N/A	22	5.9
FC-0508-50	50	...	68	60	<0.5	16.5	0.25	3.5	120	63	68 HRB	N/A	26	6.3
FC-0508-60	60	...	82	70	<1.0	18.5	0.27	4.5	145	68	80 HRB	N/A	31	6.8
FC-0808-45	45	...	55	50	<0.5	14.0	0.27	3.0	85	62	65 HRB	N/A	21	6.0
FC-1000-20	20	...	30	26	<1.0	14.0	0.27	3.5	53	33	15 HRB	N/A	11	6.0

HT, heat treated; N/A, not applicable. Suffix numbers represent minimum strength values in 10^3 psi; yield in the as-sintered condition and ultimate in the heat treated condition. Mechanical property data derived from laboratory prepared test specimens sintered under commercial manufacturing conditions. Tempering temperature for heat treated materials: 177 °C (350 °F). (a) Yield and ultimate tensile strength are approximately the same for heat treated materials. Source: MPIF Standard 35

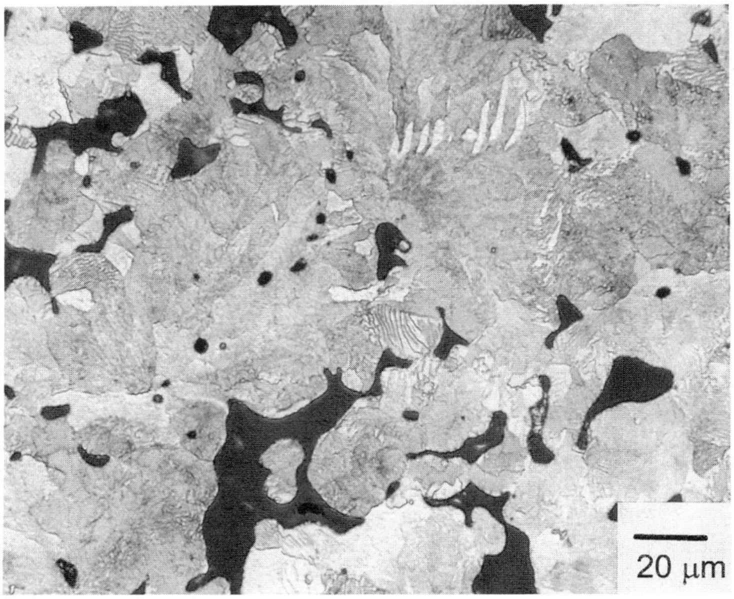

Fig. 11 Ferrite/pearlite microstructure of an as-sintered P/M copper steel (FC-0208)

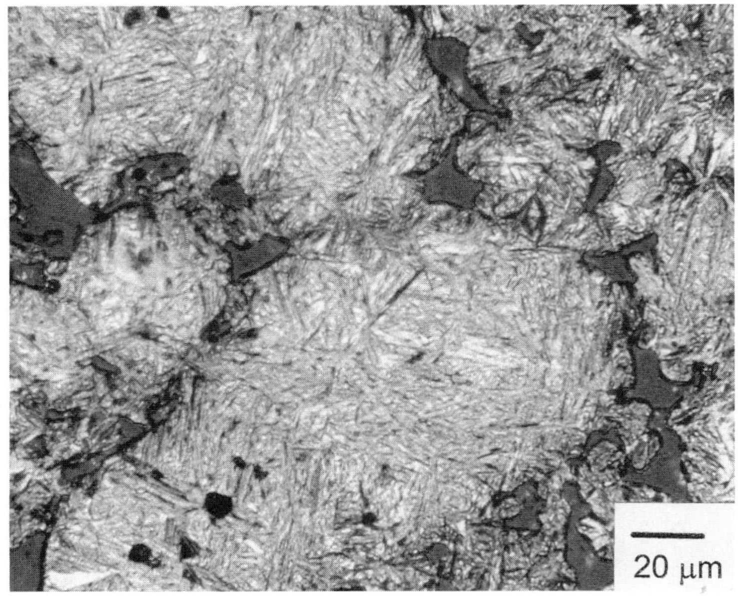

Fig. 12 Tempered martensitic microstructure of a quench-hardened and tempered P/M copper steel (FC-0208)

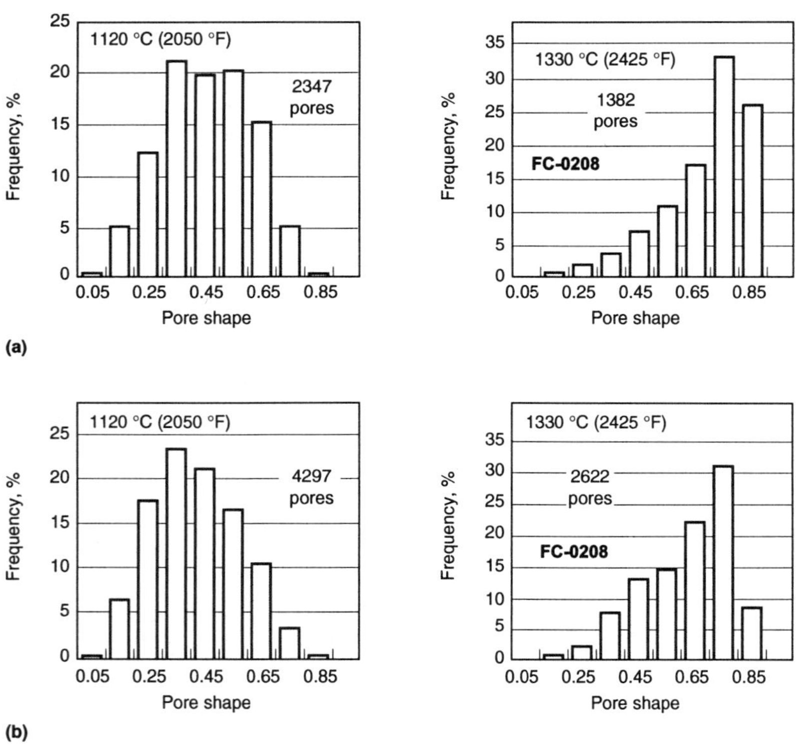

Fig. 13 Shape and size of porosity in P/M (a) and (b) copper (FC-0208) and (c) and (d) nickel (FN-0208) steels. Source: Ref 24

Table 5 Chemical composition of iron-nickel and nickel steels

Material designation	Fe	Ni	C	Cu	Element
FN-0200	92.2	1.0	0.0	0.0	min
	99.0	3.0	0.3	2.5	max
FN-0205	91.9	1.0	0.3	0.0	min
	98.7	3.0	0.6	2.5	max
FN-0208	91,6	1.0	0.6	0.0	min
	98.4	3.0	0.9	2.5	max
FN-0405	89.9	3.0	0.3	0.0	min
	96.7	5.5	0.6	2.0	max
FN-0408	89.6	3.0	0.6	0.0	min
	96.4	5.5	0.9	2.0	max

Chemical composition, wt%

Other elements: Total by difference equals 2.0 wt% maximum, which may include other minor elements added for special purposes. Source: MPIF Standard 35

ferrite/pearlite microstructure (Fig. 9). Both F-0005 and F-0008 steels can be heat treated to increase tensile strength, improve apparent hardness, and enhance wear resistance. Heat treated carbon steels have a martensitic microstructure (Fig. 10). Table 2 summarizes the mechanical properties of various carbon steels.

Powder metallurgy carbon steels often are steam treated to improve shelf life; close interconnected porosity; and increase compressive yield strength, apparent hardness, and wear resistance. Steam treatment, however, significantly degrades the tensile properties (Ref 23).

Iron-Copper and Copper Steels. Table 3 summarizes the chemical compositions of the iron-copper and copper steels listed in MPIF Standard 35.

Copper is added to increase strength, apparent hardness, and wear resistance. Copper steels (FC-0205 and FC-0208) are used in medium-strength structural applications. Copper steels with combined carbon contents of greater than 0.5 wt% are more difficult to machine than copper steels with lower combined carbon contents. In the as-sintered condition copper steels have a ferrite/pearlite microstructure. The FC-0208 material shown in Fig. 11 is almost of the eutectoid composition and is mostly pearlitic.

Copper steels can be heat treated to increase strength, apparent hardness, and wear resistance. Table 4 summarizes the mechanical properties of various copper steels.

Heat treated copper steels have martensitic microstructures (Fig. 12). Higher copper contents can be used for increased wear resistance when heat treatment is not practical.

Copper melts at 1082 °C (1980 °F), leaving a pore at the site of the prior copper powder particle. Admixed copper powders typically are –270 mesh (–53 μm). This is considerably coarser than the size of admixed nickel powders (8 μm), and the pore structure of copper steels is consid-

Fig. 14 Free copper in a P/M copper steel (FC-0508) that has not reached a temperature of 1082 °C (1980 °F) during sintering

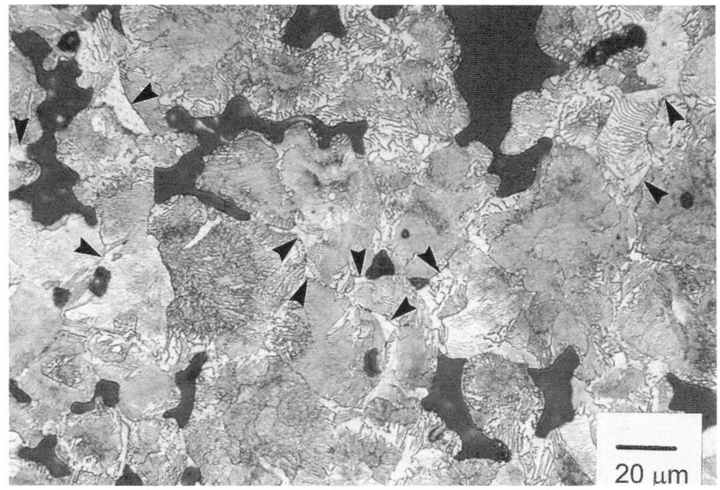

Fig. 15 Free copper that has precipitated from solution in a P/M copper steel (FC-0508)—see arrows

Fig. 16 Nickel-rich regions (NR) in a tempered martensitic structure of a quench-hardened and tempered P/M nickel steel (FN-0208)

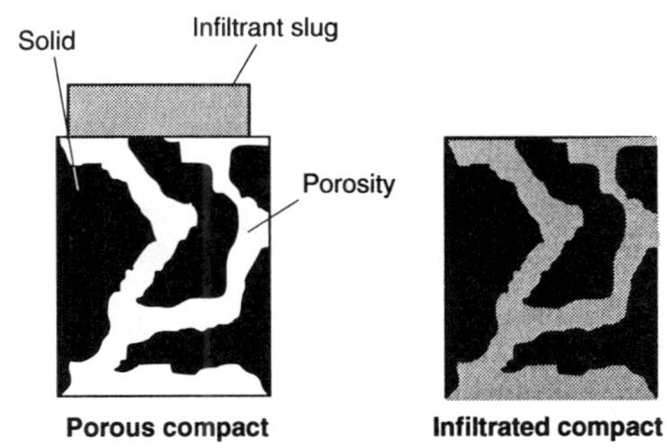

Fig. 17 Schematic of copper infiltration

erably different from that of admixed nickel steels (Fig. 13) (Ref 24). Sintered copper steels with 2 wt% Cu generally show little or no undissolved copper. At higher percentages, the copper will be seen as a separate phase. The form of the copper phase indicates whether the material has been heated to a temperature in excess of 1082 °C (1980 °F) (Fig. 14). When copper has precipitated from solid solution, there is a considerable difference in the shape and distribution of the copper. The copper is no longer in the form of discrete particles; typically, it follows grain boundary contours and fills small pores (Fig. 15).

Iron-Nickel and Nickel Steels. Table 5 lists the chemical compositions of the iron-nickel and nickel steels listed in MPIF Standard 35.

Typical admixed nickel additions in P/M nickel steels range from 2 to 4 wt%. Nickel does not diffuse very rapidly in iron at the typical sintering temperature of 1120 °C (2050 °F) and the sintered microstructure of these materials contains nickel-rich regions that improve toughness significantly, particularly in the heat treated condition (Fig. 16). P/M nickel steels are used mainly for heat treatable structural parts that need a combination of strength, wear resistance, and good impact toughness. Table 6 summarizes the mechanical properties of various nickel steels.

Copper Infiltrated Iron and Steel

Table 7 summarizes the chemical composition limits of the copper infiltrated irons and steels listed in MPIF Standard 35.

Infiltration is the process of filling the interconnected pores of a P/M part with an alloy of lower melting point than the base material. The lower melting point alloy is drawn into the pores by capillary force (Fig. 17). Infiltrants are generally alloys of copper with iron, zinc, and manganese (Ref 25). Infiltrant slugs can be placed on top of the P/M compact, beneath the compact, or both on top and beneath the compact. The mass of the infiltrant slug is calculated to fill the interconnected pore volume of the compact. The infiltration process can be either a single step or double step. In the single step process, the parts are infiltrated at the same time as they are sintered. It is extremely important for the green compact to be properly delubricated prior to reaching the high heat zone of the sintering furnace. Improper delubrication will lead to problems with infiltration. In double step infiltration, parts are sintered and cooled to room temperature under a protective atmosphere. Infiltrant slugs are then positioned, and the parts are subjected to a second sintering operation during which infiltration of the interconnected porosity occurs. Some parts have selected areas infiltrated to enhance properties in critical areas, a process known as selective infiltration.

Infiltration leads to higher and more uniform densities which in turn results in enhanced mechanical properties. Filling the interconnected porosity makes parts pressure tight for medium-pressure hydraulic applications. Infiltrated parts have improved machinability, better electrical and thermal conductivity, and can be plated. In-

Table 6 Mechanical properties of various iron-nickel and nickel steels

Minimum values			Typical values											
			Tensile properties			Elastic constants		Unnotched Charpy impact energy, ft · lbf	Transverse rupture strength, ksi	Compressive yield strength in 25 mm (0.1%), ksi	Rockwell hardness		Fatigue limit (90% survival), ksi	Density, g/cm^3
Material designation code	Minimum strength, ksi: Yield	Minimum strength, ksi: Ultimate	Ultimate strength, ksi	Yield strength (0.2%), ksi	Elongation (1 in.), %	Young's modulus, 10^6 psi	Poisson's ratio				Macro (apparent)	Micro (converted)		
FN-0200-15	15	...	25	17	3.0	16.5	0.25	10.0	50	16	55 HRF	N/A	10	6.6
FN-0200-20	20	...	35	25	5.0	20.5	0.27	20.0	80	18	75 HRF	N/A	13	7.0
FN-0200-25	25	...	40	30	10.0	23.5	0.28	50.0	105	20	80 HRF	N/A	15	7.3
FN-0205-20	20	...	40	25	1.5	16.5	0.25	6.0	65	33	44 HRB	N/A	14	6.6
FN-0205-25	25	...	50	30	2.5	19.5	0.27	12.0	100	38	69 HRB	N/A	18	6.9
FN-0205-30	30	...	60	35	4.0	22.5	0.28	21.0	125	42	69 HRB	N/A	22	7.2
FN-0205-35	35	...	70	40	5.5	24.5	0.28	34.0	150	45	78 HRB	N/A	26	7.4
FN-0205-80 HT	...	80	90	(a)	<0.5	16.5	0.25	3.5	120	77	23 HRC	55 HRC	26	6.6
FN-0205-105 HT	...	105	120	(a)	<0.5	19.5	0.27	4.5	160	90	29 HRC	55 HRC	35	6.9
FN-0205-130 HT	...	130	145	(a)	<0.5	21.5	0.27	6.0	190	99	33 HRC	55 HRC	42	7.1
FN-0205-155 HT	...	155	160	(a)	<0.5	22.5	0.28	7.0	215	103	36 HRC	55 HRC	47	7.2
FN-0205-180 HT	...	180	185	(a)	<0.5	24.5	0.28	9.5	250	112	40 HRC	55 HRC	54	7.4
FN-0208-30	30	...	45	35	1.5	17.5	0.25	5.5	85	41	63 HRB	N/A	16	6.7
FN-0208-35	35	...	55	40	1.5	19.5	0.27	8.0	105	47	71 HRB	N/A	20	6.9
FN-0208-40	40	...	70	45	2.0	21.5	0.27	11.0	130	53	77 HRB	N/A	25	7.1
FN-0208-45	45	...	80	50	2.5	23.5	0.28	16.0	155	59	83 HRB	N/A	28	7.3
FN-0208-50	50	...	90	55	3.0	24.5	0.28	21.0	170	62	88 HRB	N/A	32	7.4
FN-0208-80 HT	...	80	90	(a)	<0.5	17.5	0.25	4.0	120	99	26 HRC	57 HRC	29	6.7
FN-0208-105 HT	...	105	120	(a)	<0.5	19.5	0.27	4.5	150	124	31 HRC	57 HRC	38	6.9
FN-0208-130 HT	...	130	145	(a)	<0.5	20.5	0.27	5.5	185	136	35 HRC	57 HRC	46	7.0
FN-0208-155 HT	...	155	170	(a)	<0.5	22.5	0.28	7.0	220	162	39 HRC	57 HRC	54	7.2
FN-0208-180 HT	...	180	195	(a)	<0.5	24.5	0.28	8.0	250	188	42 HRC	57 HRC	62	7.4
FN-0405-25	25	...	40	30	<1.0	15.5	0.25	4.5	65	33	49 HRB	N/A	14	6.5
FN-0405-35	35	...	60	40	3.0	20.5	0.27	14.5	120	40	71 HRB	N/A	22	7.0
FN-0405-45	45	...	90	50	4.5	24.5	0.28	33.5	175	45	84 HRB	N/A	32	7.4
FN-0405-80 HT	...	80	85	(a)	<0.5	15.5	0.25	4.0	115	67	19 HRC	55 HRC	26	6.5
FN-0405-105 HT	...	105	110	(a)	<0.5	18.5	0.27	5.0	145	89	25 HRC	55 HRC	34	6.8
FN-0405-130 HT	...	130	135	(a)	<0.5	20.5	0.27	6.5	200	103	31 HRC	55 HRC	42	7.0
FN-0405-155 HT	...	155	160	(a)	<0.5	23.5	0.28	9.5	245	124	37 HRC	55 HRC	49	7.3
FN-0405-180 HT	...	180	185	(a)	<0.5	24.5	0.28	13.0	280	132	40 HRC	55 HRC	57	7.4
FN-0408-35	35	...	45	40	1.0	15.5	0.25	4.0	75	37	67 HRB	N/A	16	6.5
FN-0408-45	45	...	65	50	1.0	19.5	0.27	7.5	115	50	78 HRB	N/A	23	6.9
FN-0408-55	55	...	80	60	1.0	22.5	0.28	11.0	150	59	87 HRB	N/A	28	7.2

HT, heat treated; N/A, not applicable. Suffix numbers represent minimum strength values in 10^3 psi; yield in the as-sintered condition and ultimate in the heat treated condition. Mechanical property data derived from laboratory prepared test specimens sintered under commercial manufacturing conditions. Tempering temperature for heat treated materials: 260 °C (500 °F). (a) Yield and ultimate tensile strength are approximately the same for heat treated materials. Source: MPIF Standard 35

Table 7 Chemical composition of copper infiltrated irons and steels

Material designation	Chemical composition, wt%: Fe	Cu	C(a)	Element
FX-1000	82.8	8.0	0.0	min
	92.0	14.9	0.3	max
FX-1005	82.5	8.0	0.3	min
	91.7	14.9	0.6	max
FX-1008	82.2	8.0	0.6	min
	91.4	14.9	0.9	max
FX-2000	72.7	15.0	0.0	min
	85.0	25.0	0.3	max
FX-2005	72.4	15.0	0.3	min
	84.7	25.0	0.6	max
FX-2008	72.1	15.0	0.6	min
	84.4	25.0	0.9	max

Other elements: Total by difference equals 2.0 wt% maximum, which may include other minor elements added for specific purposes. (a) Carbon, on basis of iron only, may be a metallographic estimate. Source: MPIF Standard 35

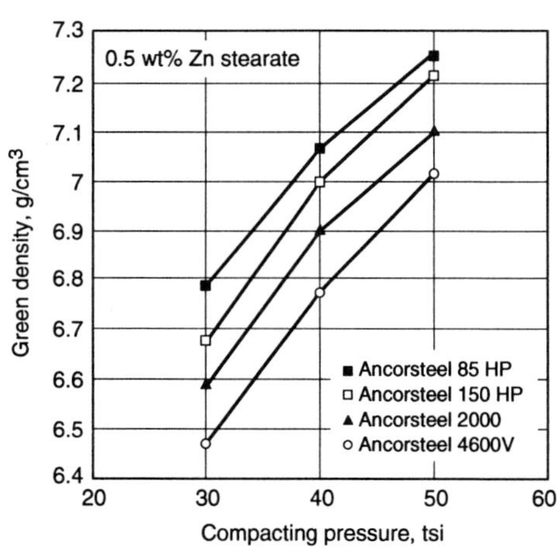

Fig. 18 Compressibility curves for a number of water-atomized, prealloyed low-alloy powders

filtration can also be used as a joining process to make multicomponent P/M assemblies during the sintering process.

Copper infiltrated parts can be used in the as-infiltrated condition, or they can be heat treated to improve strength, apparent hardness, and wear resistance. Table 8 summarizes the mechanical properties of various copper infiltrated irons and steels. More information is contained in the article "Copper Infiltrated Steel" in this Volume.

Phosphorus irons are admixed alloys made by adding fine (<20 μm and preferably close to 10 μm) ferrophosphorus particles to high compressibility iron powders (Ref 26–28). These alloys exhibit a good combination of tensile strength, ductility, and impact absorption energy. However, their primary use is for parts requiring superior soft electromagnetic performance. Compared with parts made from pure, high-compressibility iron powders, those made from phosphorus irons have higher resistivity, permeability, and induction and reduced coercive force. The presence of phosphorus also decreases the susceptibility of a P/M part to nitrogen aging (Ref 22). The typical soft magnetic properties and mechanical properties of phosphorus irons with two levels of phosphorus (0.45 and 0.80 wt%), sintered under various conditions, are summarized in Table 9 from ASTM A-839-96. Typical applications for parts made from the phosphorus irons include anti-lock brake sensors, magnetic solenoids, and plungers. More information on typical applications is found in the article "Magnetic Materials and Properties for Part Applications" in this Volume.

Partially Alloyed (Diffusion Alloyed) Powders

Table 10 summarizes the chemical compositions for various partially alloyed (diffusion al-

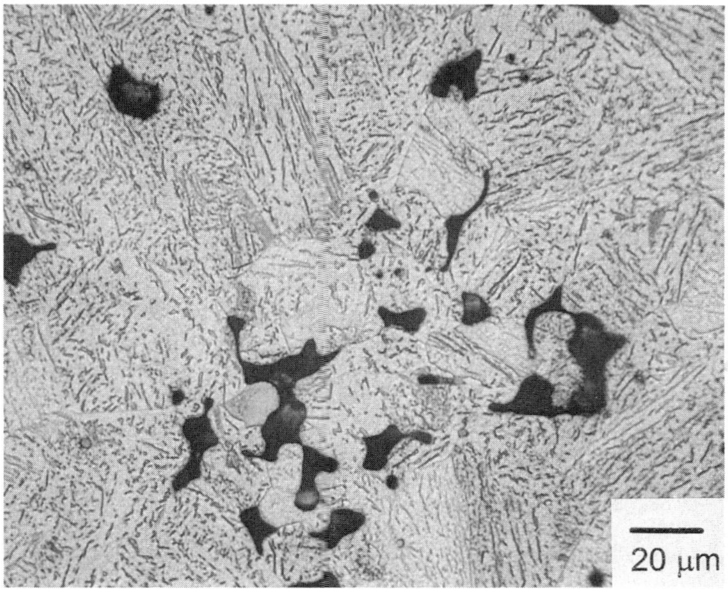

Fig. 19 As-sintered microstructure of a part made from a prealloyed steel material (FL-4405)

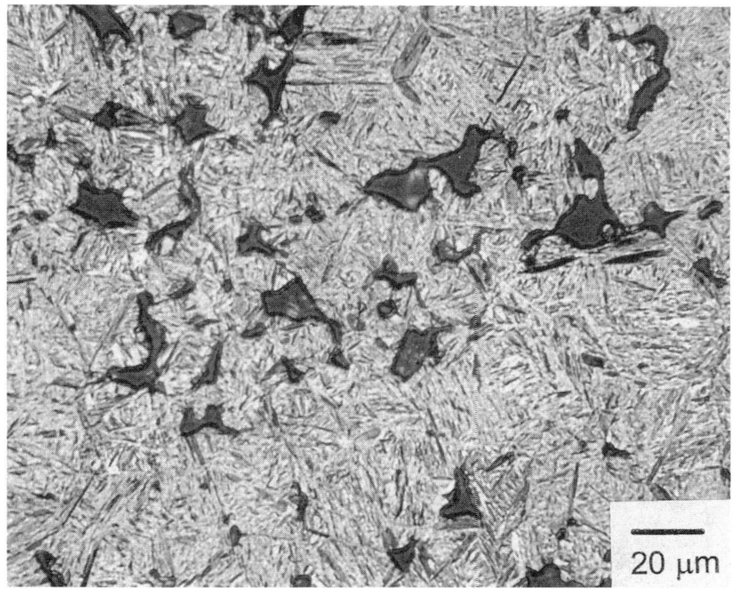

Fig. 20 Tempered martensitic microstructure of a quench-hardened and tempered part made from a prealloyed steel (FL-4605)

Table 8 Mechanical properties of various copper infiltrated irons and steels

Minimum values			Typical values											
			Tensile properties			Elastic constants		Unnotched Charpy impact energy, ft · lbf	Transverse rupture strength, ksi	Compressive yield strength (0.1%), ksi	Rockwell hardness		Fatigue limit (90% survival), ksi	Density, g/cm^3
Material designation code	Minimum strength, ksi: Yield	Minimum strength, ksi: Ultimate	Ultimate strength, ksi	Yield strength (0.2%), ksi	Elongation in 25 mm (1 in.), %	Young's modulus, 10^6 psi	Poisson's ratio				Macro (apparent)	Micro (converted)		
FX-1000-25	25	...	51	32	7.0	23.5	0.28	25.0	132	33	65 HRB	N/A	19	7.3
FX-1005-40	40	...	77	50	4.0	23.5	0.28	13.0	158	53	82 HRB	N/A	29	7.3
FX-1005-110 HT	...	110	120	(a)	<0.5	23.5	0.28	7.0	210	110	38 HRC	55 HRC	33	7.3
FX-1008-50	50	...	87	60	3.0	23.5	0.28	10.0	166	71	89 HRB	N/A	33	7.3
FX-1008-110 HT	...	110	120	(a)	<0.5	23.5	0.28	6.5	189	115	43 HRC	58 HRC	41	7.3
FX-2000-25	25	...	46	37	3.0	21.0	0.24	15.0	144	41	66 HRB	N/A	17	7.3
FX-2005-45	45	...	75	60	1.5	21.0	0.24	8.0	148	60	85 HRB	N/A	20	7.3
FX-2005-90 HT	...	90	100	(a)	<0.5	21.0	0.24	7.0	171	71	36 HRC	55 HRC	23	7.3
FX-2008-60	60	...	80	70	1.0	21.0	0.24	7.0	156	70	90 HRB	N/A	23	7.3
FX-2008-90 HT	...	90	100	(a)	<0.5	21.0	0.24	5.0	159	74	36 HRC	58 HRC	27	7.3

All data based on single-pass infiltration. HT, heat treated; N/A, not applicable. Suffix numbers represent minimum strength values in 10^3 psi; yield in the as-sintered condition and ultimate in the heat treated condition. Mechanical property data derived from laboratory prepared test specimens sintered under commercial manufacturing conditions. Tempering temperature for heat treated materials: 177 °C (350 °F). (a) Yield and ultimate tensile strength are approximately the same for heat treated materials. Source: MPIF Standard 35

loyed) steels listed in MPIF Standard 35. These materials are based on a highly compressible iron powder to which the alloy additions have been diffusion alloyed. While the dusting and segregation tendencies of the alloy additions are reduced through diffusion alloying, partially alloyed materials typically contain sufficient admixed graphite to provide the desired combined carbon content during sintering and lubricant to facilitate compaction. These additives are susceptible to dusting and segregation and, hence, mixes of partially alloyed powders are improved by binder treatment. The complex microstructures of these materials (Fig. 6) lead to a good combination of tensile strength, ductility, and toughness. Diffusion alloyed materials can be heat treated to improve strength, apparent hardness, and wear resistance. Table 11 summarizes the mechanical properties of various partially alloyed materials.

Prealloyed Powders

Table 12 summarizes the chemical compositions of various prealloyed, low-alloy steels listed in MPIF Standard 35, and the mechanical properties of materials made from prealloyed powders are summarized in Table 13. Prealloyed powders generally are used for parts that require heat treatment. The hardenability of the prealloyed materials is controlled by the amount and type of alloying employed. Nickel-molybdenum prealloyed powders have lower compressibility than prealloyed powders that use molybdenum alone as the principal alloying agent (Fig. 18). Parts made from prealloyed powders have a homogeneous microstructure and uniform apparent hardness. In the as-sintered condition, the carbides that form are not lamellar in nature (compare Fig. 19 with Fig. 9). In the as-sintered condition, prealloyed materials can be identified by this characteristic carbide morphology. In the heat treated condition, parts made from prealloyed powders exhibit a uniform tempered martensitic microstructure (Fig. 20).

Hybrid Alloy Powders

Hybrid alloy powders consist of prealloyed or partially alloyed (diffusion alloyed) powders to which elemental or ferroalloy additions have been made to produce the desired composition. In MPIF Standard 35 the term "hybrid alloy" is not currently used. However, some hybrid-alloy materials are covered in the sections covering low-alloy steels and sinter-hardened steels. The chemical composition of those hybrid alloys referenced in MPIF Standard 35 is summarized in Tables 12 and 14, and their mechanical properties are summarized in Tables 13 and 15.

The compressibility of hybrid alloys depends on the compressibility of the base powder used to constitute them. While many hybrid alloys are

Table 9 Typical soft magnetic properties and mechanical properties of phosphorus irons

	Type I powder (0.45% P)			Type II powder (0.8% P)		
Sintered at 1120 °C (2050 °F) in dissociated ammonia						
Sintered density, g/cm^3 (kg/m^3)	6.8 (6,800)	7.0 (7,000)	7.2 (7,200)	6.8 (6,800)	7.0 (7,000)	7.2 (7,200)
Maximum permeability (relative)	2,400	2,800	3,000	3,500	4,000	4,500
Maximum flux density, T (G)	1.05 (10,500)	1.20 (12,000)	1.25 (12,500)	1.10 (11,000)	1.20 (12,000)	1.30 (13,000)
Residual induction, T (G)	0.85 (8,500)	0.90 (9,000)	1.0 (10,000)	1.0 (10,000)	1.10 (11,000)	1.20 (12,000)
Coercive force, A/m (Oe)	140 (1.7)	140 (1.7)	140 (1.7)	120 (1.5)	120 (1.5)	120 (1.5)
0.2% offset yield strength, MPa (ksi)	210 (30)	220 (32)	240 (35)	280 (40)	310 (45)	340 (50)
Ultimate tensile strength, MPa (ksi)	280 (40)	310 (45)	340 (50)	330 (48)	340 (50)	380 (55)
Elongation in 25 mm (1 in.), %	5	7	9	2	3	3
Apparent hardness, HRB	40	45	55	55	65	70
Sintered at 1120 °C (2050 °F) in hydrogen						
Sintered density, g/cm^3 (kg/m^3)	7.0 (7,000)	7.2 (7,200)	7.4 (7,400)	7.0 (7,000)	7.2 (7,200)	7.4 (7,400)
Maximum permeability (relative)	3,500	4,000	4,500	4,000	4,500	5,000
Maximum flux density, T (G)	1.20 (12,000)	1.25 (12,500)	1.30 (13,000)	1.20 (12,000)	1.30 (13,000)	1.35 (13,500)
Residual induction, T (G)	1.0 (10,000)	1.10 (11,000)	1.25 (12,500)	1.05 (10,500)	1.15 (11,500)	1.30 (13,000)
Coercive force, A/m (Oe)	110 (1.4)	110 (1.4)	110 (1.4)	100 (1.3)	100 (1.3)	100 (1.3)
0.2% offset yield strength, MPa (ksi)	210 (30)	240 (35)	260 (38)	310 (45)	330 (48)	360 (53)
Ultimate tensile strength, MPa (ksi)	310 (45)	340 (50)	380 (55)	360 (53)	390 (57)	430 (62)
Elongation in 25 mm (1 in.), %	8	10	12	4	4	4
Apparent hardness, HRB	45	55	60	65	70	75
Sintered at 1260 °C (2300 °F) in hydrogen						
Sintered density, g/cm^3 (kg/m^3)	7.0 (7,000)	7.2 (7,200)	7.4 (7,400)	7.0 (7,000)	7.2 (7,200)	7.4 (7,400)
Maximum permeability (relative)	4,000	4,500	5,000	5,000	5,500	6,000
Maximum flux density, T (G)	1.25 (12,500)	1.30 (13,000)	1.35 (13,500)	1.30 (13,000)	1.35 (13,500)	1.40 (14,000)
Residual induction, T (G)	1.15 (11,500)	1.20 (12,000)	1.25 (12,500)	1.20 (12,000)	1.25 (12,500)	1.30 (13,000)
Coercive force, A/m (Oe)	95 (1.2)	95 (1.2)	95 (1.2)	80 (1.0)	80 (1.0)	80 (1.0)
0.2% offset yield strength, MPa (ksi)	220 (32)	230 (34)	260 (37)	310 (45)	330 (48)	360 (53)
Ultimate tensile strength, MPa (ksi)	310 (45)	340 (50)	380 (55)	340 (50)	380 (55)	400 (58)
Elongation in 25 mm (1 in.), %	10	12	15	2	2	2
Apparent hardness, HRB	45	55	60	65	70	75

Source: ASTM A 839, 1996

Table 10 Chemical composition of various partially alloyed (diffusion alloyed) steels

Material designation	Chemical composition, wt%					
	Fe	C	Ni	Cu	Mo	Element
FD-0205	93.15	0.3	1.55	1.3	0.4	min
	96.45	0.6	1.95	1.7	0.6	max
FD-0208	92.85	0.6	1.55	1.3	0.4	min
	96.15	0.9	1.95	1.7	0.6	max
FD-0405	90.70	0.3	3.60	1.3	0.4	min
	94.40	0.6	4.40	1.7	0.6	max
FD-0408	90.40	0.6	3.60	1.3	0.4	min
	94.10	0.9	4.40	1.7	0.6	max

Other elements: Total by difference equals 2.0 wt% maximum, which may include other minor elements added for specific purposes. Source: MPIF Standard 35

suitable for sintering at conventional temperatures (1120 °C, or 2050 °F), the mechanical properties of parts made from these materials often can be enhanced by high temperature sintering (>1150 °C, or 2100 °F) (Ref 7, 8, 29).

Mechanical Properties of Ferrous P/M Materials

The mechanical properties listed in MPIF Standard 35 relate to guaranteed minimum values and typical values.

Not all commercially available ferrous P/M materials are covered by current material standards. It takes time for the independent testing and review that is required before new materials can be added to the existing standards. When selecting materials for various applications, it is important therefore that P/M parts fabricators and design engineers discuss options with their material suppliers.

The mechanical properties of ferrous P/M materials are a function of part density and microstructure. The manner in which the P/M alloy is constituted has a direct influence on the hardenability of the material and hence, for a given process sequence, on the resulting microstructure. Figure 21 ranks the hardenability of various P/M steels at a density of 7.0 g/cm^3. In general, hardenability increases with density, as shown in Fig. 22 for two P/M steels. Prealloyed steels also attain greater hardenability than diffusion alloyed steels of comparable composition (Ref 30).

For a given microstructure the mechanical properties improve as density is increased. Tensile strength increases linearly with increase in density, while properties such as tensile ductility, impact energy, fatigue performance, and fracture toughness increase exponentially when the level of porosity reaches low levels. Various methods are used to achieve high densities for high-performance P/M parts, as described in the article "Mechanical Properties of High-Performance Powder Metallurgy Parts" in this Volume, for ferrous parts with densities >7.4 g/cm^3 for iron and low-alloy steel parts.

Figure 23 shows the effect of density on the tensile strength and tensile elongation of several P/M steels, from two different published sources (Ref 5 and 31). These typical values illustrate the linear dependence of strength on density and the exponential increase in elongation with density.

Strength and Toughness. If the yield strength of various P/M materials is plotted against their impact energy, as illustrated in Fig. 24(a) to (f), some interesting trends become apparent. Successive data points for each material represent a higher sintered density; the data from MPIF Standard 35 were used to construct the plots shown in Fig. 24.

In Fig. 24(a), the as-sintered copper steel (FC-0208) exhibits a higher tensile strength than the carbon steel (F-0008) over a similar range of impact energy. Increasing the combined carbon content of the copper steel increases the tensile strength and reduces the impact energy very slightly at the lower density level (FC-0208 versus FC-0205). The nickel steel (FN-0408) matches the tensile strength of the copper steel (FC-0208) but has better impact properties for a given level of tensile strength. The various nickel steels have the best impact energy values. For the nickel steels, increasing the combined carbon content improves the tensile strength at the expense of impact performance (FN-0208 versus FN-0205, and FN-0408 versus FN-0405). After quench hardening and tempering, carbon steel and copper steels have much lower tensile strength and impact properties than nickel steels (Fig. 24b).

Figure 24(c) and (d) compares the yield strength and impact energy of various nickel steels and partially alloyed (diffusion alloyed) materials. In the as-sintered condition, the diffusion alloyed materials have higher yield strengths than the nickel steels while maintaining comparable impact performance. In the quench hardened and tempered condition (Fig. 24d) tensile strength is increased at the expense of impact energy.

The modest yield strength and impact properties of the prealloyed materials (Fig. 24e) reflects why such materials are rarely used in the as-sintered condition. Quench hardening and tempering these materials increases their tensile strength at the expense of impact energy (Fig. 24f), although at higher density levels the FL-4405-HT and FLN-4205-HT materials maintain a good combination of tensile strength and impact energy.

Hardness. It should be noted that, for ferrous P/M materials, macroindentation hardness measurements result in a measure of apparent hardness. The hardness indentation reflects a composite of both matrix and porosity. Apparent hardness is thus a function of part density. A ferrous P/M material with an apparent hardness of 35 HRC may in fact have a matrix hardness of 60 HRC. The matrix hardness of materials with homogeneous microstructures, or the hardness of various constituents of heterogeneous microstructures, can be measured using microindentation hardness methods.

Table 11 Mechanical properties of various partially alloyed (diffusion alloyed) materials

Minimum values			Typical values											
			Tensile properties			Elastic constants		Unnotched Charpy impact energy, ft · lbf	Transverse rupture strength, ksi	Compressive yield strength (0.1%), ksi	Rockwell hardness		Fatigue limit (90% survival), ksi	Density, g/cm^3
Material designation code	Minimum strength, ksi Yield	Minimum strength, ksi Ultimate	Ultimate strength, ksi	Yield strength (0.2%), ksi	Elongation in 25 mm (1 in.), %	Young's modulus, 10^6 psi	Poisson's ratio				Macro (apparent)	Micro (converted)		
FD-0205-45	45	...	68	52	1.0	18.0	0.27	8.0	130	47	72 HRB	N/A	24	6.75
FD-0205-50	50	...	78	57	1.5	20.0	0.27	12.0	155	52	76 HRB	N/A	29	6.95
FD-0205-55	55	...	88	61	2.0	22.0	0.28	18.0	180	57	80 HRB	N/A	32	7.15
FD-0205-60	60	...	100	67	2.0	24.5	0.28	28.0	210	62	86 HRB	N/A	37	7.40
FD-0205-95 HT	...	95	105	(a)	<1.0	18.0	0.27	5.0	160	130	28 HRC	55 HRC	42	6.75
FD-0205-120HT	...	120	130	(a)	<1.0	20.0	0.27	7.0	190	155	33 HRC	55 HRC	52	6.95
FD-0205-140HT	...	140	150	(a)	<1.0	22.0	0.28	9.0	210	175	38 HRC	55 HRC	65	7.15
FD-0205-160HT	...	160	170	(a)	<1.0	24.5	0.28	11.0	240	200	45 HRC	55 HRC	75	7.40
FD-0208-50	50	...	69	58	<1.0	18.0	0.27	7.0	135	58	80 HRB	N/A	24	6.75
FD-0208-55	55	...	79	63	<1.0	19.5	0.27	9.0	155	62	83 HRB	N/A	33	6.90
FD-0208-60	60	...	92	68	1.0	21.5	0.27	12.0	180	67	87 HRB	N/A	38	7.10
FD-0208-65	65	...	103	73	1.0	23.0	0.28	17.0	195	72	90 HRB	N/A	46	7.25
FD-0405-55	55	...	86	62	1.0	18.0	0.27	11.0	160	56	80 HRB	N/A	25	6.75
FD-0405-60	60	...	103	66	1.0	21.0	0.27	20.0	195	63	85 HRB	N/A	28	7.05
FD-0405-65	65	...	123	70	2.5	24.0	0.28	27.0	230	73	91 HRB	N/A	40	7.35
FD-0405-100HT	...	100	110	(a)	<1.0	18.0	0.27	5.0	160	125	30 HRC	55 HRC	34	6.75
FD-0405-130HT	...	130	140	(a)	<1.0	21.0	0.27	7.0	200	150	35 HRC	55 HRC	50	7.05
FD-0405-155HT	...	155	165	(a)	<1.0	24.0	0.28	10.0	235	175	42 HRC	55 HRC	58	7.35
FD-0408-50	50	...	71	57	<1.0	17.5	0.25	9.0	130	62	85 HRB	N/A	22	6.70
FD-0408-55	55	...	90	62	1.0	20.0	0.27	13.0	165	68	89 HRB	N/A	27	6.95
FD-0408-60	60	...	110	67	1.5	22.5	0.28	18.0	200	73	93 HRB	N/A	38	7.20
FD-0408-65	65	...	125	71	2.0	24.5	0.28	22.0	230	80	95 HRB	N/A	48	7.40

HT, heat treated; N/A, not applicable. Suffix numbers represent minimum strength values in 10^3 psi; yield in the as-sintered condition and ultimate in the heat treated condition. Mechanical property data derived from laboratory prepared test specimens sintered under commercial manufacturing conditions. Tempering temperature for heat treated materials: 177 °C (350 °F). (a) Yield and ultimate tensile strength are approximately the same for heat treated materials. Source: MPIF Standard 35

Table 12 Chemical composition of various prealloyed and hybrid low-alloy steels

Material designation	Chemical composition, wt% Fe	C	Ni	Mo	Cu	Element
FL-4205	95.90	0.4	0.35	0.50	...	min
	98.75	0.7	0.55	0.85	...	max
FL-4405	96.35	0.4	...	0.75	...	min
	98.85	0.7	...	0.95	...	max
FL-4605	94.20	0.4	1.70	0.40	...	min
	97.50	0.7	2.00	1.10	...	max
FLN-4205	93.95	0.4	1.35(a)	0.49	...	min
	97.76	0.7	2.50(a)	0.85	...	max
FLN2-4405	93.35	0.4	1.00	0.65	...	min
	97.95	0.7	3.00	0.95	...	max
FLN4-4405	91.35	0.4	3.00	0.65	...	min
	95.95	0.7	5.00	0.95	...	max
FLN6-4405	89.35	0.4	5.00	0.65	...	min
	93.95	0.7	7.00	0.95	...	max
FLNC-4405	90.35	0.4	1.00	0.65	1.0	min
	96.95	0.7	3.00	0.95	3.0	max

Other elements: Total by difference equals 2.0 wt% maximum, which may include other minor elements added for specific purposes. (a) At least 1% of the nickel is admixed as elemental powder. Source: MPIF Standard 35

The Effect of Sintering Temperature on Mechanical Properties. A detailed review of the sintering of ferrous P/M materials is discussed in the article "Production Sintering Practices" in this Volume. During the sintering process any lubricant admixed to aid the compaction and ejection process is removed from P/M compacts in the preheat zone of the furnace. The time at temperature and the sintering temperature determine the extent to which diffusion occurs during the sintering process. Admixed, partially alloyed, and hybrid alloys depend on diffusion processes to alloy their various constituents during sintering. Sintering also generally serves to reduce any residual oxides and to effect solution of any admixed graphite. When compacts reach the end of the high-heat section of the sintering furnace, the hardenability of ferrous P/M material is determined by the extent of the diffusion that occurred during sintering. In the case of parts made from prealloyed powders, the hardenability is not primarily determined by the extent of diffusion processes but by the composition of the prealloyed powder and the sintered carbon content of the compact. For each alloying method the sintering time and sintering temperature will influence the number, the size, the size distribution, and the shape of the porosity present in the microstructure (Ref 24).

The mechanical property data for ferrous P/M materials that are summarized in MPIF Standard 35 relate to sintering at 1120 °C (2050 °F). Sintering at higher temperatures often enhances the mechanical and soft magnetic properties of materials (Ref 21, 22, 24, 29, 31, 32). For example, the tensile strength and elongation of samples compacted from the hybrid alloy FLN6-4405 reach 1240 MPa (180 ksi) and greater than 2% respectively when the samples are sintered at 1315 °C (2400 °F) for 45 min at temperature in a 75 vol% to 25 vol% dissociated ammonia atmosphere (Ref 29).

Figure 25 illustrates the effect of sintering temperature on the ultimate tensile strength and elongation of a variety of ferrous P/M materials (Ref 32). For each material the data points represent compacting pressures of 415, 550, and 690 MPa (30, 40, and 50 tsi), respectively. In general, tensile strength and elongation both increase with higher compacting pressure for sintering at both 1120 °C (2050 °F) and 1260 °C (2300 °F). The data for the higher sintering temperature are, in most instances, almost a direct extension of the data from the lower sintering temperature.

Sinter Hardening. The mechanical properties of ferrous powder metallurgy (P/M) materials are directly related to their density and microstructure. Many P/M parts are heat treated, in a secondary operation, to develop a tempered martensitic microstructure either in a surface layer, or throughout the part. The need for a secondary quenching operation can be avoided by "sinter hardening" the parts (Ref 9, 10, 33–39).

Ferrous P/M materials with sufficient hardenability will develop microstructures containing significant percentages of martensite in the as-sintered condition. Accelerated cooling techniques for sintering furnaces have been developed, which permits larger parts to be sinter hardened, or materials with lower hardenability to be used to produce sinter-hardened parts with smaller cross sections.

Conventionally sintered copper and nickel steels such as FC-0208 and FN-0205 usually yield microstructures consisting of ferrite and pearlite. More highly alloyed materials such as partially alloyed low-alloy steels (FD-0205, FD-0208, FD-0405, and FD-0408) can form some bainite and even some martensite in addition to ferrite and fine unresolved pearlite (Fig. 6). Sinter-hardened parts have been made from FLC-4608 using conventional sintering furnaces for many years. This is possible because of the hardenability of this material. If the sintering furnace is modified to permit the cooling rate to be increased, it is possible to sinter harden P/M

Table 13 Mechanical properties of various prealloyed and hybrid low-alloy materials

Minimum values			Typical values											
			Tensile properties			Elastic constants								
Material designation code	Minimum strength, ksi: Yield	Minimum strength, ksi: Ultimate	Ultimate strength, ksi	Yield strength (0.2%), ksi	Elongation in 25 mm (1 in.), %	Young's modulus, 10^6 psi	Poisson's ratio	Unnotched Charpy impact energy, ft · lbf	Transverse rupture strength, ksi	Compressive yield strength (0.1%), ksi	Rockwell hardness: Macro (apparent)	Rockwell hardness: Micro (converted)	Fatigue limit (90% survival), ksi	Density, g/cm³
FL-4205-35	35	...	52	42	1.0	18.5	0.27	6.0	100	42	60 HRB	N/A	20	6.80
FL-4205-40	40	...	58	47	1.0	20.0	0.27	9.0	115	47	66 HRB	N/A	27	6.95
FL-4205-45	45	...	66	52	1.5	21.5	0.27	12.0	125	52	70 HRB	N/A	32	7.10
FL-4205-50	50	...	73	58	2.0	23.5	0.28	17.0	150	56	75 HRB	N/A	40	7.30
FL-4205-80 HT	...	80	90	(a)	<0.5	16.5	0.25	5.0	135	80	28 HRC	60 HRC	30	6.60
FL-4205-100 HT	...	100	110	(a)	<0.5	18.5	0.27	7.0	160	110	32 HRC	60 HRC	37	6.80
FL-4205-120 HT	...	120	130	(a)	<0.5	20.5	0.27	8.0	185	140	36 HRC	60 HRC	44	7.00
FL-4205-140 HT	...	140	150	(a)	<0.5	22.5	0.28	12.0	215	170	39 HRC	60 HRC	50	7.20
FL-4405-35	35	...	52	42	1.0	17.5	0.25	6.0	100	39	60 HRB	N/A	20	6.70
FL-4405-40	40	...	58	47	1.0	19.5	0.27	11.0	125	45	67 HRB	N/A	27	6.90
FL-4405-45	45	...	66	52	1.5	21.5	0.27	16.0	140	52	73 HRB	N/A	32	7.10
FL-4405-50	50	...	73	58	2.0	23.5	0.28	22.0	165	56	80 HRB	N/A	40	7.30
FL-4405-100 HT	...	100	110	(a)	<1.0	17.5	0.25	5.5	160	135	24 HRC	60 HRC	34	6.70
FL-4405-125 HT	...	125	135	(a)	<1.0	19.5	0.27	7.0	200	155	29 HRC	60 HRC	42	6.90
FL-4405-150 HT	...	150	160	(a)	<1.0	21.5	0.27	9.0	230	175	34 HRC	60 HRC	48	7.10
FL-4405-175 HT	...	175	185	(a)	<1.0	23.5	0.28	14.0	280	195	38 HRC	60 HRC	58	7.30
FL-4605-35	35	...	52	42	1.0	18.0	0.27	6.0	100	42	60 HRB	N/A	20	6.75
FL-4605-40	40	...	58	47	1.0	20.0	0.27	11.0	120	45	65 HRB	N/A	27	6.95
FL-4605-45	45	...	66	52	1.5	22.0	0.28	16.0	140	52	71 HRB	N/A	32	7.15
FL-4605-50	50	...	73	58	2.0	24.0	0.28	22.0	165	57	77 HRB	N/A	40	7.35
FL-4605-80 HT	...	80	85	(a)	<0.5	16.0	0.25	4.5	130	91	24 HRC	60 HRC	29	6.55
FL-4605-100 HT	...	100	110	(a)	<0.5	18.0	0.27	6.0	165	114	29 HRC	60 HRC	37	6.75
FL-4605-120 HT	...	120	130	(a)	<0.5	20.0	0.27	8.0	195	139	34 HRC	60 HRC	46	6.95
FL-4605-140 HT	...	140	155	(a)	<0.5	22.5	0.28	12.0	230	170	39 HRC	60 HRC	53	7.20
FLN-4205-40	40	...	58	47	1.0	16.5	0.25	6.0	105	45	64 HRB	N/A	20	6.60
FLN-4205-45	45	...	66	52	1.0	18.5	0.27	8.0	125	50	70 HRB	N/A	27	6.80
FLN-4205-50	50	...	73	58	1.5	21.0	0.27	13.0	150	56	77 HRB	N/A	32	7.05
FLN-4205-55	55	...	87	63	2.0	23.5	0.28	22.0	175	60	83 HRB	N/A	40	7.30
FLN-4205-80 HT	...	80	90	(a)	<1.0	16.5	0.25	5.0	130	125	24 HRC	60 HRC	28	6.60
FLN-4205-105 HT	...	105	115	(a)	<1.0	18.5	0.27	7.0	170	145	30 HRC	60 HRC	36	6.80
FLN-4205-140 HT	...	140	150	(a)	<1.0	21.0	0.27	9.0	230	170	36 HRC	60 HRC	47	7.05
FLN-4205-175 HT	...	175	185	(a)	1.0	23.5	0.28	14.0	290	200	42 HRC	60 HRC	58	7.30

HT, heat treated; N/A, not applicable. Suffix numbers represent minimum strength values in 10^3 psi; yield in the as-sintered condition and ultimate in the heat treated condition. Mechanical property data derived from laboratory prepared test specimens sintered under commercial manufacturing conditions. Tempering temperature for heat treated materials: 177 °C (350 °F). (a) Yield and ultimate tensile strength are approximately the same for heat treated materials. Source: MPIF Standard 35

Table 14 Chemical composition of some hybrid low-alloy steels listed in MPIF Standard 35 as sinter-hardening materials

Material designation	Fe	C	Ni	Mo	Cu	Element
FLN2-4408	93.15	0.6	1.00	0.65	...	min
	97.75	0.9	3.00	0.95	...	max
FLN4-4408	91.15	0.6	3.00	0.65	...	min
	95.75	0.9	5.00	0.95	...	max
FLN6-4408	89.15	0.6	5.00	0.65	...	min
	93.75	0.9	7.00	0.95	...	max
FLN-4608	91.00	0.6	3.6(a)	0.39	...	min
	93.41	0.9	5.0(a)	1.10	...	max
FLC-4608	91.00	0.6	1.60	0.39	1.0	min
	96.41	0.9	2.00	1.10	3.0	max
FLC-4908	92.40	0.6	...	1.30	1.0	min
	95.10	0.9	...	1.70	3.0	max
FLNC-4408	90.15	0.6	1.00	0.65	1.0	min
	96.75	0.9	3.00	0.95	3.0	max

Other elements: Total by difference equals 2.0 wt% maximum, which may include other minor elements added for specific purposes. (a) At least 1 wt% of the nickel is admixed as elemental powder. Source: MPIF Standard 35

materials with less hardenability than FLC-4608. The current trend is toward lower cost, leaner alloys with good compressibility such as FLNC-4408. In addition to being essential for the sinter hardening of parts made from materials with less hardenability, furnaces with additional cooling capacity are beneficial for parts made from more highly alloyed materials as they permit sinter hardening of larger parts.

The tensile strength and apparent hardness of sinter-hardened materials are directly related to the percentage of martensite in the microstructure (Ref 33). In selecting materials, it is extremely important to understand all of the functional requirements of the intended application. Does the part need to meet a minimum hardness for an application in which wear is of primary concern? Does the part need to meet a specified tensile strength, elongation, fatigue strength, or impact energy value as well as a minimum hardness? Understanding whether the need is for hardness alone, or in combination with mechanical property requirements, is critical to selection of an appropriate material and process.

While it is possible to perform limited machining on sinter-hardened parts, they will, for the most part, be designed as net-shape parts. Typically, sinter-hardened parts need tempering to relieve stresses and improve toughness. The high apparent hardness of sinter-hardened parts makes sizing difficult if not impractical.

The Effect of Density on Mechanical Properties. The mechanical properties of ferrous P/M materials are directly related to their density. Mechanical properties improve as the amount of porosity is reduced. Tensile strength is almost a linear function of density while ductility, toughness, and fatigue strength increase significantly at low levels of porosity (Fig. 23). Increased density is generally achieved through the use of higher compacting pressures. However, there is a practical upper limit to the density that can be achieved through conventional compaction, and this is due to the presence of an admixed lubricant and the work hardening of the powder particles during compaction. Typically, to reach green densities of 7.1 g/cm³ and higher, ferrous P/M parts need to be either copper infiltrated, double pressed/double sintered, or powder forged. Recently, however, warm compaction processing has been developed to permit single compaction processing of parts to green densities of up to 7.3 g/cm³ (Ref 40–43) (see the article "Warm Compaction" in this Volume).

Applications of P/M Structural Parts

Automotive and off-highway vehicle applications dominate the ferrous P/M structural parts market (Fig. 3) (Ref 1). However, there are many other fields in which ferrous P/M parts are being

Table 15 Mechanical properties of sinter hardened FLC-4608

Minimum values			Typical values										
			Tensile properties			Elastic constants							
Material designation code	Minimum strength, ksi: Yield	Minimum strength, ksi: Ultimate	Ultimate strength, ksi	Yield strength (0.2%), ksi	Elongation in 25 mm (1 in.), %	Young's modulus, 10^6 psi	Poisson's ratio	Unnotched Charpy impact energy, ft · lbf	Transverse rupture strength, ksi	Rockwell hardness: Macro (apparent)	Rockwell hardness: Micro (converted)	Density, g/cm³
FLC-4608-50 HT	...	50	60	(a)	<1.0	16.5	0.25	5.0	120	20 HRC	55 HRC	6.60
FLC-4608-70 HT	...	70	80	(a)	<1.0	18.5	0.27	7.0	150	26 HRC	55 HRC	6.80
FLC-4608-90 HT	...	90	100	(a)	<1.0	20.5	0.27	9.0	190	31 HRC	55 HRC	7.00
FLC-4608-110 HT	...	110	120	(a)	<1.0	22.5	0.28	14.0	230	37 HRC	55 HRC	7.20

HT, heat treated. Suffix numbers represent minimum strength values in 10^3 psi; ultimate in the heat treated condition. Mechanical property data derived from laboratory prepared test specimens sintered under commercial manufacturing conditions. Tempering temperature for heat treated materials: 177 °C (350 °F). (a) Yield and ultimate tensile strength are approximately the same for heat treated materials. Source: MPIF Standard 35

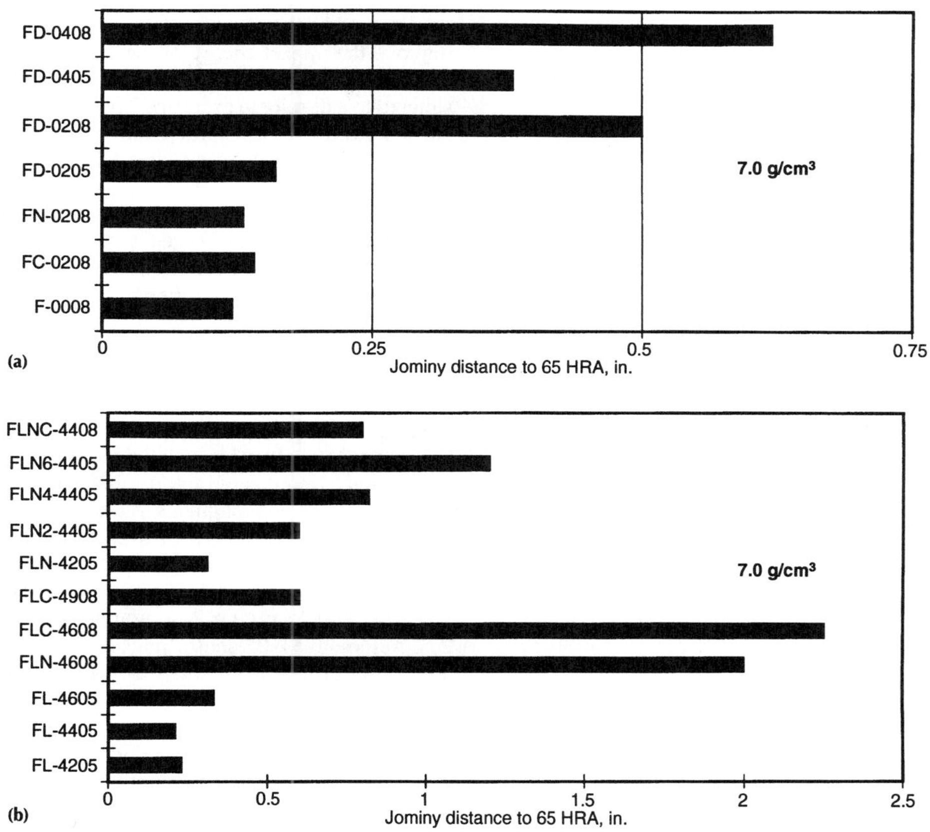

Fig. 21 Relative hardenability of various P/M alloy steels (70 g/cm³). (a) Admixed powders. (b) Prealloyed powders. Based on data from Ref 30

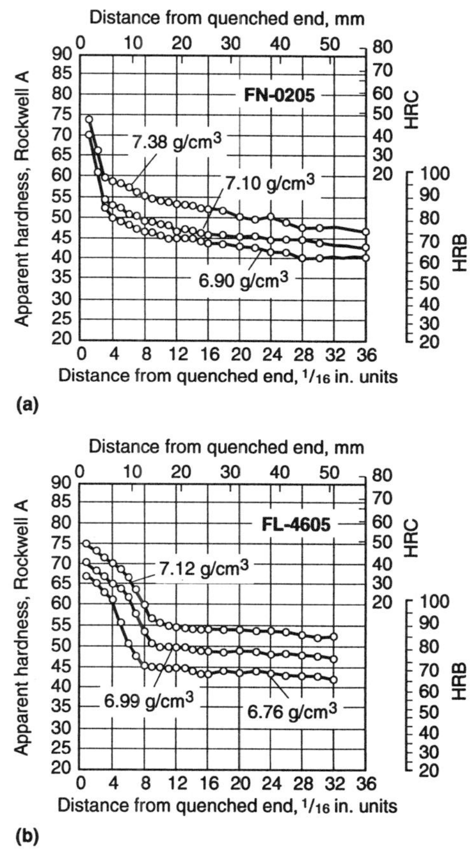

Fig. 22 Effect of density on hardenability of (a) FN-0205 P/M alloy and (b) FL-4605 P/M alloy. Source: Ref 30

used such as lawn and garden structural parts, parts for hand tools and hobby applications, parts for household appliances lock hardware parts, parts for industrial motors, controls, and hydraulic applications, and parts for business machine applications. Each of these business segments will be reviewed briefly with emphasis given to the largest application area, automotive parts. Various part applications are also illustrated in "Appendix 3: Examples of Powder Metallurgy Parts" in this Volume.

Automotive Applications

As previously noted, the typical family vehicle in North America contains 14 kg (31 lb) of P/M parts, and industry experts anticipate that this figure will reach in excess of 22 kg (~50 lb) by the end of the decade. Table 16 summarizes the application areas in which P/M parts are found (Ref 44, 45). Some major applications are noted in the following paragraphs.

Powder forged connecting rods currrently are used by Ford Motor Company, General Motors Corporation, and Chrysler Corporation. Among the first engines to use powder forged connecting rods were the Porsche 928 and the Toyota Camry engine. Powder forged rods have replaced both drop forged and cast rods in many applications because the use of powder forged rods significantly reduces the number of machining operations needed to produce a finished part ready for assembly (Ref 46, 47). Specifically, powder forged connecting rods can be fracture split rather than machined in order to separate the crank-end cap from the remainder of the rod. This process is needed to assemble the rod to the crankshaft, and fracture splitting allows easier processing and a better fit than machining. Another benefit of the powder forged rods is very consistent distribution of part mass, which eliminates grouping of rods into various weight categories (as required for cast and drop forged rods). Balance pads also can be eliminated or reduced on many applications.

Currently, most powder forged connecting rods are made from copper steels (P/F-10C45 to

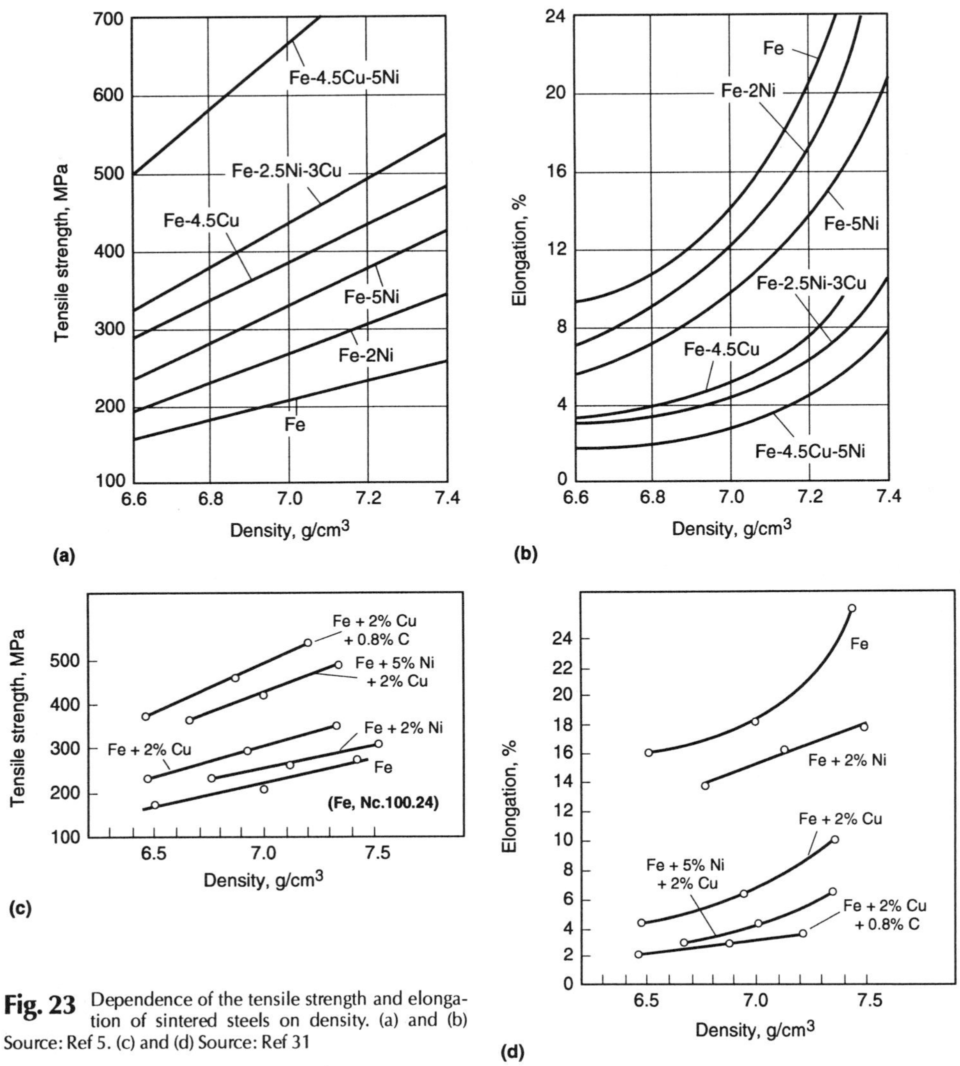

Fig. 23 Dependence of the tensile strength and elongation of sintered steels on density. (a) and (b) Source: Ref 5. (c) and (d) Source: Ref 31

P/F-10C60 or from materials such as P/F-11C45 to P/F-11C60 with manganese sulfide added to enhance machinability). These are admixed materials. The rods are used in the as-forged condition and have a ferrite/pearlite microstructure. The material used to make the powder forged rods has an ultimate tensile strength of ~830 MPa (120 ksi), a yield strength of 550 MPa (80 ksi), and a fatigue limit of 255 MPa (37 ksi). Some rods are produced from a prealloyed steel base powder and used in the as-forged condition. They have similar mechanical properties to the copper steel rods. Press and sintered (P/M) connecting rods have been developed with comparable performance to cast and drop forged rods (Ref 48–50). While engine tests have been satisfactory, no press-and-sintered rods are currently in commercial production. The advent of the warm compaction process with the possibility of machining the crank-end bolt holes in the "green" state offers promise for the production of P/M connecting rods.

P/M main bearing caps (Fig. 26) are currently used for some General Motors, Chrysler, and Honda engines in North America. Conversion and development programs are in progress at Ford Motor Company. The P/M bearing caps replace the traditional cast iron caps that are machined from a cast "loaf." The P/M caps are made from a modified FC-0205 material (an admixed material). The main bolt holes are designed into the P/M compact, while side holes are drilled and tapped in a secondary operation. Additional secondary machining operations include milling the surfaces of the feet of the caps and the snap width used to locate the caps. The P/M bearing caps have a sintered density of ~6.6 g/cm^3, and the material used to make the caps has an ultimate tensile strength of ~450 MPa (65 ksi), 3% elongation, and a fatigue limit of 160 MPa (23 ksi).

P/M carriers for planetary pinion gears (Fig. 27) are used in four-wheel drive transfer cases. The carriers shown in Fig. 27 are made from a 2% Cu steel admixed alloy (FC-0208) (Ref 52). The carriers have a sintered density of ~6.7 g/cm^3 and are made from two P/M parts that are sinter brazed together during the sintering operation. Selective copper infiltration is also carried out during sintering to enhance the strength, apparent hardness, and wear properties of the splines. The material used to make the carriers has an ultimate tensile strength of about 410 MPa (60 ksi) and an apparent hardness of 70 HRB. Powder metallurgy carriers are also used for transmission gears and, in conjunction with the helical pinion gears they contain, represent a major growth opportunity for ferrous powder metallurgy.

Steering column tilt levers (Fig. 28) are made using powder metallurgy. The levers lock the steering column tilt mechanism in position, and the parts require a good combination of strength, apparent hardness, impact toughness, and wear performance for this demanding application. The P/M levers are made from a hybrid alloy, a diffusion alloyed material (FD-0405) to which an additional 2 wt% Ni has been added to improve the impact energy of the material. The P/M tilt levers are double pressed and double sintered to a density in excess of 7.3 g/cm^3. They are subjected to a light surface carburizing treatment to enhance the wear characteristics of the grooved profile area.

Camshaft Lobes. An assembled camshaft (Ref 53) may have powder forged camshaft lobes made from a P/F-4680 material. The assembled camshaft using a hollow tube provides for improved distribution of lubricating oil, reduces the overall mass of the camshaft, and provides an engineered materials solution to the manufacture of the camshaft. Numerous varieties of assembled camshaft are in production.

Sensor Parts. Powder metallurgy is used for production of automotive antilock brake systems (ABS) sensors and engine sensors (Ref 54). Some alloys are made from iron or an iron-phosphorus alloy. The iron-phosphorus alloy is made by admixing ferrophosphorus powder (<20 μm particle size Fe_3P) to iron powder (Ref 22). Automotive ABS sensor rings (tone wheels) are also fabricated from P/M stainless steel (see the article "Powder Metallurgy Stainless Steels" in this Volume).

One-way clutch races (Fig. 29) (Ref 55) are powder forged. A dual material fill is used during compaction of the preforms used for powder forging. The inner portion of the race is made from P/F-4665, while the outer region is made from P/F-4615. Both materials are based on prealloyed powders to which an appropriate amount of graphite has been added. The races are quench hardened and tempered, and the use of dual materials results in a tempered martensitic microstructure with different hardness levels in the inner and outer regions of the heat treated forged parts. The races have a mass of 450 g and a forged density of 7.82 g/cm^3.

Other Applications

Lawn and Garden Applications. Application areas include power steering metering pumps, rockshaft servo cam and followers, engine oil pump gears, picker bar drive gears for cotton pickers, picker bar pivot studs, floating cams, dual sprocket assemblies, idler arms, feed rolls and shutoffs for grain drills, planetary gear

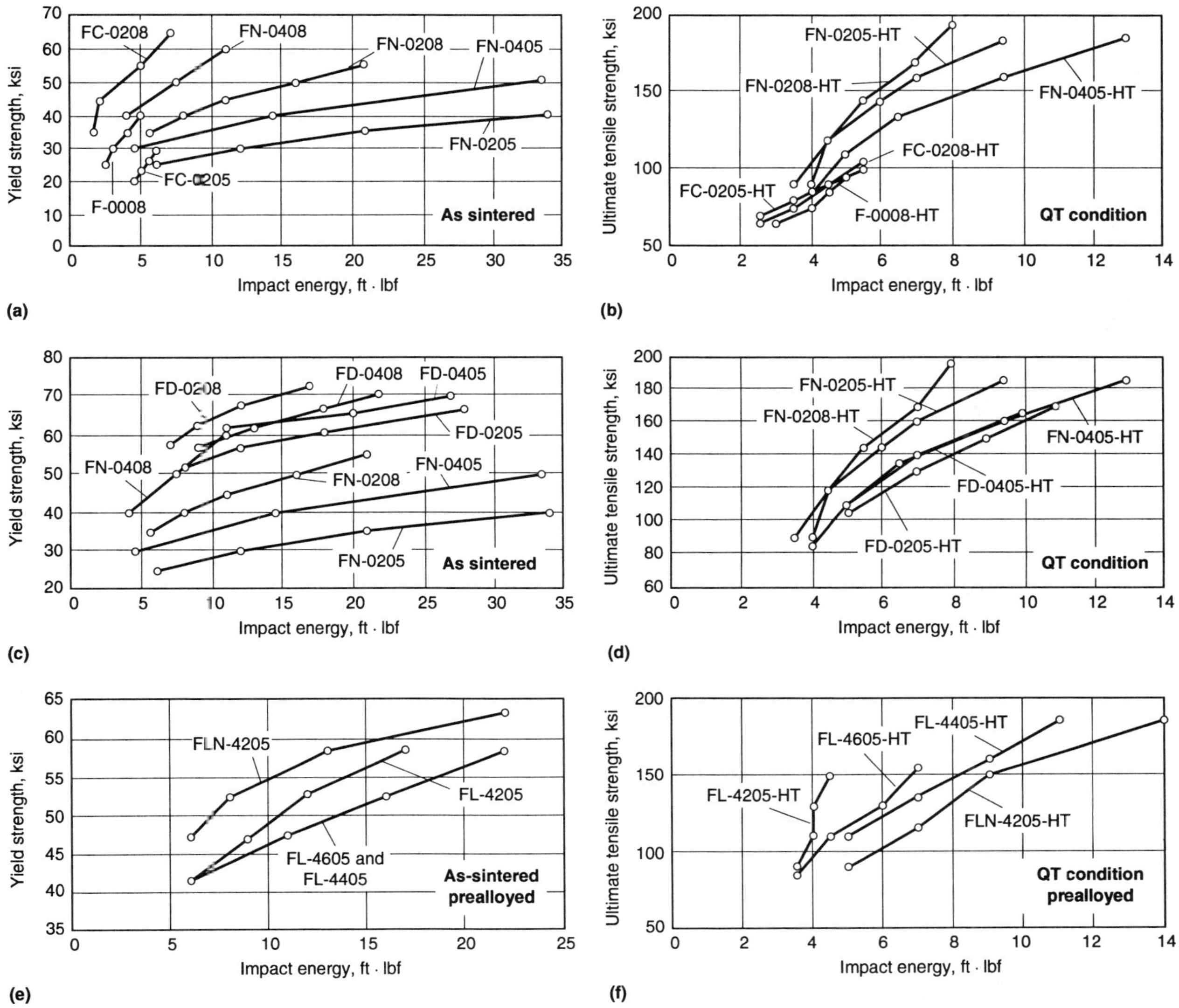

Fig. 24 Tensile strength versus impact energy for various ferrous P/M materials (MPIF Standard 35). (a) Compare carbon, copper, and nickel steels in the as-sintered condition. (b) Compare carbon, copper, and nickel steels in the quench-hardened and tempered condition. (c) Compare nickel steels and diffusion alloyed steels in the as-sintered condition. (d) Compare nickel steels and diffusion alloyed steels in the quench-hardened and tempered condition. (e) Compare prealloyed materials in the as-sintered condition. (f) Compare prealloyed materials in the quench-hardened and tempered condition

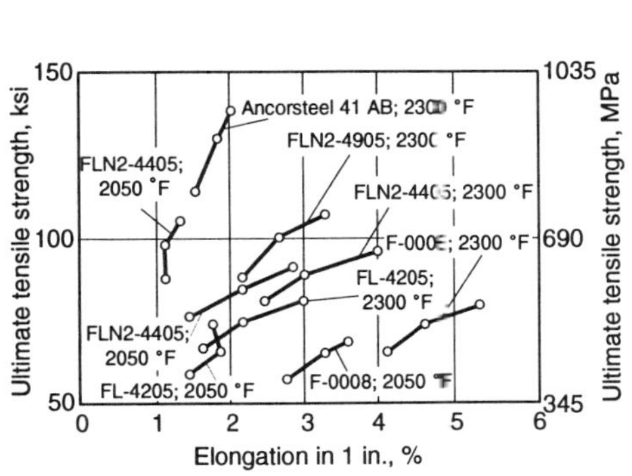

Fig. 25 Effect of sintering temperature on the ultimate tensile strength and elongation of various ferrous P/M materials. Source: Ref 33

assemblies, inner and outer 45° cams for combines plus auger bevel gears and hydraulic wheel motor manifolds, power take-off cams and followers for lawn and garden tractors, idler pivots, mower hubs, front-mounted mower neutral arms, fourth reduction gears, and rototiller gear assemblies. Several parts are illustrated in "Appendix 3: Examples of Powder Metallurgy Parts" in this Volume.

Recreation, Hand Tools, and Hobby Applications. Ferrous P/M parts are used in a wide variety of power hand tools. Figure 30 shows examples of some warm compacted parts, which include P/M power tool gears (Ref 56).

Household Appliances. Ferrous P/M parts used in appliances such as washers and dryers include eccentric gears and rack gears. Washing machine water pump impellers are made from stainless steel P/M powders. Powder metallurgy parts enjoy wide usage in domestic appliances, particularly washers and dryers, to reduce costs and increase service life of various components. These parts, which have solved strength, wear, and corrosion problems, fall into three general categories: stainless steel structural parts, iron-copper-graphite structural parts, and copper-tin bronzes used for bearings. Parts made from structural plastics compete with P/M parts in this market segment.

Hardware Applications. Many of the P/M parts used in lock hardware applications are made from nonferrous or stainless powders.

Business Machine Applications. Many of the P/M parts once used in business machines have been replaced by parts made from plastics, and many of the mechanical devices in business appliances have been replaced by solid state electronics. This segment currently represents just over 1% of the market for ferrous P/M structural parts.

REFERENCES

1. D.G. White, Challenges for the 21st Century, *Int. J. Powder Metall.*, Vol 33 (No. 5), 1997, p 45–54

2. W.B. James, Recent Developments in Ferrous Powder Metallurgy Alloys, *Int. J. Powder Metall.*, Vol 30 (No. 2), 1994, p 157–162
3. U. Engström and S. Allroth, A Newly Developed Sintered High Strength Material, *Met. Powder Rep.* Vol 41 (No. 11), 1986, p 815
4. R. German, *Powder Metallurgy of Iron and Steel,* John Wiley & Sons, 1998
5. A. Salak, *Ferrous Powder Metallurgy,* Cambridge International, 1995
6. P.F. Lindskog and G.F. Bocchini, Development of High Strength P/M Precision Components in Europe, *Int. J. Powder Metall.,* Vol 15 (No. 3), 1979, p 199–230
7. Hoeganaes Technical Data Sheets, Ancorsteel 150 HP, and Ancorsteel 85 HP, Hoeganaes Corporation, Riverton, NJ
8. W.B. James and R.J. Causton, Surface-Hardenable Heat Treated P/M Steels, *Advances in Powder Metallurgy & Particulate Materials,* Vol 5, J.M. Capus and R.M. German, Ed., Metal Powder Industries Federation, 1992, p 62–92
9. W.B. James, V.C. Potter, and T.F. Murphy, "Steering Column Tilt Lever—P/M Material Development," SAE Technical Paper 900381, SAE Congress and Exposition (Detroit, MI), Feb 1990
10. V.C. Potter, W.B. James, and T.F Murphy, Improved Dimensional Control and Elimination of Heat Treatment for Automotive Parts, *Advances in Powder Metallurgy,* Vol 3., L.F. Pease III and R.J. Sansoucy, Ed., Metal Powder Industries Federation, 1990, p 33–48

Fig. 26 Main bearing cap and end cap for GM 3.8 liter V-6 engine. Source: Ref 51

Table 16 Automotive applications for P/M parts

Application area	P/M part
Steering	Power steering pump cam ring
	Power steering pump rotor
	Steering column collar
	Locking wedge
	Passive restraint locking pawls
	Tilt lever
	Column lock bolt
	Power steering pump side plates
	Power steering pump vanes/slippers
	Steering column gears
	Steering end plate
	Steering rack guide
	Steering shift lever
Suspension	Ball joint bearing
	Shock absorber piston
	Shock absorber strut rod guides
	Clips
	Shock absorber cylinder ends
	Shock absorber compression valve
	McPherson strut rod guides
Transfer case	Clutch housing
	Driven sprocket
	Drive sprocket
	Inner mode hub
	Main shaft hub
	Planetary gear carrier
	Planetary carrier housing
Automatic transmission	Direct clutch hub
	Drive sprocket
	Driven sprocket
	Forward clutch hub
	Governor counterweight and sleeve
	Parking pawl
	Parking gear
	Planetary carrier gears
	Planetary carrier housing
	Planetary carrier spacer/tube
	Ring gear hub
	Support hubs
Automatic transmission (continued)	TCC sleeve
	TCC stator clutch cam
	Intermediate clutch hub
	One-way clutch inner cam
	One-way clutch inner race
	One-way clutch outer race
	One-way clutch outer cam
	Output shaft hub
	Pressure reaction plates
	Pump cam ring
	Pump gears/gerotor
	Pump rotors
	Pump stators
	Pump vanes
	TCC stator clutch race
	TCC turbine hub
Manual transmission	Detent/guide plate
	Shift fork
	Synchronizer blocking ring
	Synchronizer housing
	Synchronizer hub
	Shift levers
Engine	Balance shaft sprockets, gears
	Camshaft lobes
	Camshaft bearing caps
	Camshaft holddown bracket
	Camshaft retainer
	Connecting rods
	Crankshaft bearing caps
	EFI timing sensor ring
	Oil pump gears and rotors
	Rocker arm ball
	Rocker arm fulcrum
	Tensioner back plate
	Tensioner pulleys
	Timing sprocket, gears: intermediate
	Timing sprocket, pulleys, gears: camshaft
	Timing sprocket, pulleys, gears: crankshaft
	Valve guides
Engine (continued)	Valve seat inserts: intake and exhaust
	Water pump impeller
	Water pump pulley
	Water pump pulley hub or flange
Engine fuel system	Fuel injection sprockets
	Fuel injector components
	Fuel pump eccentric ring
	Fuel pump gears/gerotor
Engine-starter motor	Starter motor frame
	Starter motor pinions
	Starter motor plunger stop
	Starter motor pole piece
Platform body	Door hinges
	Electric remote mirror gears
	Headlight activator gears
	Instrument parts
	Windshield wiper drives
	Windshield wiper motor flux rings
Brake systems	ABS sensor ring
	Brake adjustor nuts
	Brake lock master cylinder
	Brake piston
Exhaust system	Exhaust flanges
	EGR base plate
	HEGOS bosses
HVAC	Air conditioner compressor clutch hub
	Air conditioner compressor swash plate
Interior fittings	Lock parts (striking plate/latch)
	Rearview mirror mounts
	Pinion gears (door windows)
	Seat safety belt locks
	Speedometer gear
	Airbag hardware
	Seat adjustment levers
	Seat adjustment racks
	Seat recliner parts
	Signal light lever
	Tachometer sensing ring

Source: Ref 44 and 45

Fig. 27 Powder metallurgy carriers for planetary pinion gears. Source: Ref 52

Fig. 28 Steering column tilt levers made from a hybrid P/M alloy (diffusion alloyed FD-0405 material with 2 wt% Ni). Source: Ref 9

11. ASTM B 783–93, Standard Specification for Materials for Ferrous Powder Metallurgy (P/M) Structural Parts, ASTM
12. D. Yarnton and T.J. Davies, The Effect of Lubrication on the Compaction and Sintering of Iron Powder Compacts, *Int. J. Powder Metall.,* Vol 8 (No. 2), 1972, p 51–67
13. G.F. Bocchini, "High Pressure Compaction, High Pressure Coining, and High Pressure Repressing of P/M Parts," paper presented during seminar at Prevention and Detection of Cracks in Ferrous P/M Parts, 1988 International Powder Metallurgy Conference (Orlando, FL), Metal Powder Industries Federation
14. W.B. James "Alloying Methods for High Performance Ferrous Powder Metallurgy Parts," Seminário Metalurgica do Pó (São Paulo, Brazil), Associação Brasileiro de Metais, Oct 1989, p 93–107
15. M.J. McDermott, P/M Parts Fabrication Experience with ANCORBOND (binder treated) Premixes, *Advances in Powder Metallurgy,* Vol 1, E.R. Andreotti and P.J. McGeehan, Ed., Metal Powder Industries Federation, 1990, p 209
16. U. Engström, Homogenous Iron Based Powder Mixtures Free of Segregation, U.S. Patent 4,483,906, 1984
17. F.J. Semel, Iron Based Powder Mixtures, U.S. Patent 4,834,800, 1989
18. S.H. Luk and J.A. Hamill, Jr., Dust and Segregation-Free Powders for Flexible P/M Processing, *Advances in Powder Metallurgy & Particulate Materials,* Vol 1, A. Lawley and A. Swanson, Ed., Metal Powder Industries Federation, 1993, p 153
19. F.J. Semel and S.H. Luk, Continuing Improvements in Binder Treatment Technology, *Advances in Powder Metallurgy & Particulate Materials,* Vol 4, Part 13, T.M. Cadle and K.S. Narasimhan, Ed., Metal Powder Industries Federation, 1996, p 353–362
20. C. Lall, Soft Magnetism, *Fundamentals for Powder Metallurgy and Metal Injection Molding,* Monographs in P/M Series No. 2, Metal Powder Industries Federation, 1992
21. C. Lall, The Effect of Sintering Temperature and Atmosphere on the Soft Magnetic Properties of P/M Materials, *Advances in Powder Metallurgy & Particulate Materials,* Vol 3, J.M. Capus and R.M. German, Metal Powder Industries Federation, 1992, p 129
22. F.G. Hanejko, H.G. Rutz, and C. Oliver, Effects of Processing and Materials on Soft Magnetic Performance of Powder Metallurgy Parts, *Advances in Powder Metallurgy & Particulate Materials,* Vol 6, J. M. Capus and R.M. German, Metal Powder Industries Federation, 1992, p 375–404
23. L.F. Pease III, J.P. Collete, and D.A. Pease, Mechanical Properties of Steam Blackened P/M Materials, *Modern Developments in Powder Metallurgy,* Vol 21, 1988, p 275–299
24. W.B. James, Fatigue Properties of Ferrous P/M Materials, Seminário Metalurgica do Pó (São Paulo, Brazil), Associação Brasileiro de Metais, Oct 1989, p 93–107
25. A.N. Ashurst, E. Klar, and H.R. McCurdy, Copper Infiltration of Steel—Part 1, Properties, *Met. Powder, Rep.,* June 1984, p 329; Copper Infiltration of Steel—Part 2, Applications, *Met. Powder Rep.,* Aug, 1984, p 438
26. P. Lindskog, The Effect of Phosphorus Additions on the Tensile, Fatigue, and Impact Strength of Sintered Steels Based on Sponge Iron Powder and High-Purity Atomized Iron

Fig. 29 Powder forged one-way clutch races. Source: Ref 55

Fig. 30 Examples of warm compacted parts (FD-0405). Densities from 7.2 to 7.5 g/cm^3. Source: Ref 56

Powder, *Powder Metall.,* Vol 16 (No. 32), 1973, p 374–386

27. P. Lindskog, J. Tengzelius, and S.Å. Kvist, Phosphorus as an Alloying Element in Ferrous P/M, *Modern Developments in Powder Metallurgy,* Vol 10, H.H. Hausner and P.W. Taubenblat, Metal Powder Industries Federation, 1977, p 97–128
28. J. Tengzelius, Avoiding Brittleness in Iron-Phosphorus Alloys, *Met. Powder Rep.,* Jan 1993, p 36
29. J.A. Hamill, Jr., R.J. Causton, and S.O. Shah, High Performance P/M Materials Utilizing High Temperature Sintering, *Advances in Powder Metallurgy & Particulate Materials,* Vol 5, J.M. Capus and R.M. German, Ed., Metal Powder Industries Federation, 1992, p 193
30. Clarence Skena et al., Hardenability Characteristics of P/M Alloy Steels, *Int. J. of Powder Metall.,* Vol 33 (No. 7), 1997, p 25–35
31. B. Kubicki, *Sintered Machine Elements,* Ellis Horword, 1995
32. H.G. Rutz, A.J. Rawlings, and T.M. Cimino, Advanced Properties of High Density Ferrous Powder Metallurgy Materials, *Advances in Powder Metallurgy & Particulate Materials,* Vol 3, Part 10, M. Phillips and J. Porter, Ed., Metal Powder Industries Federation, 1995, p 97
33. R.J. Causton , W.B. James, and J.J. Fulmer, Performance Characteristics of a New Sinter-Hardening Low Alloy Steel, *Advances in Powder Metallurgy,* Vol 5, L.F. Pease III and R.J. Sansoucy, Metal Powder Industries Federation, 1991, p 91
34. R.J. Causton and J.J. Fulmer, Sinter-Hardening Low Alloy Steels, *Advances in Powder Metallurgy & Particulate Materials,* Vol 3, J.M. Capus and R.M. German, Ed., Metal Powder Industries Federation, 1992, p 91–104
35. A.B. Davala, A.H. Graham, and R.J. Causton, Effect of Process Conditions upon Sinter-Hardening Response of FLC-4608 Materials, *Advances in Powder Metallurgy & Particulate Materials,* Vol 2, Part 14, R.A. McKotch and R. Webb, Ed., Metal Powder Industries Federation, 1997, p 81
36. H.G. Rutz, A.H. Graham, and A.B. Davala, Sinter-Hardening P/M Steels, *Advances in Powder Metallurgy & Particulate Materials,* Vol 1, Part 8, R.A. McKotch and R. Webb, Ed., Metal Powder Industries Federation, 1997, p 3
37. G. L'Espérance, E. Duchesne, and A. de Rege, Effect of Materials and Process Parameters on the Microstructure and Properties of Sinter Hardening Alloys, *Advances in Powder Metallurgy & Particulate Materials,* Vol 3, Part 11, T.M. Cadle and K.S. Narasimhan, Ed., Metal Powder Industries Federation, 1996, p 397
38. G. L'Espérance, S. Martel, A. de Rege, B. Kiefer, and Y.T. Chen, Evaluation of the Effect of Tempering on Microstructure and Properties of Three Widely Used Sinter Hardening Alloys, *Advances in Powder Metallurgy & Particulate Materials,* Vol 2, Part 8, M. Phillips and J. Porter, Ed., Metal Powder Industries Federation, 1995, p 39
39. F. Chagnon and Y. Trudel, Effect of Sintering Parameters on Mechanical Properties of Sinter Hardened Materials, *Advances in Powder Metallurgy & Particulate Materials,* Vol 2, Part 14, R.A. McKotch and R. Webb, Ed., Metal Powder Industries Federation, 1997, p 97
40. H.G. Rutz and S.H. Luk, Method of Making a Sintered Metal Component , U.S. Patent 5,154,881
41. H.G. Rutz and F.G. Hanejko, High Density Processing of High Performance Ferrous Materials, *Advances in Powder Metallurgy & Particulate Materials,* Vol 5, C. Lall and A. Neupaver, Ed., Metal Powder Industries Federation, 1994, p 117–133
42. S.H. Luk, H.G. Rutz, and M.A. Lutz, Properties of High Density Ferrous P/M Materials, *Advances in Powder Metallurgy & Particulate Materials,* Vol 5, J.M. Capus and R.M. German, Ed., Metal Powder Industries Federation, 1994, p 135–154
43. T.M. Cimino, A.J. Rawlings, and H.G. Rutz, Properties of Several ANCORDENSE Processed High Performance Materials, *Advances in Powder Metallurgy & Particulate Materials,* Vol 5, Part 13, T.M. Cadle and K.S. Narasimhan, Ed., Metal Powder Industries Federation, 1996, p 337–352
44. E.S. Boreczky, Hoeganaes Corporation, private communication
45. P.K. Johnson, 1992 Powder Metallurgy World Congress, *Int. J. Powder Metall.,* Vol 28 (No. 4), 1992, p 423–426
46. W.B. James, Powder Forging, *Reviews in Particulate Materials,* Vol 2, 1994, p 173
47. New Opportunities for P/M but Price is a Problem, *Met. Powder Rep.,* Vol 49 (No. 9), 1994, p 12–13
48. K. Richter, E. Hoffman, K. Lipp, and C.M Sonsino, Single-Sintered Con Rods—An Illusion?, *Met. Powder Rep.,* Vol 49 (No. 5), 1994, p 38–45
49. E. Ernst, U. Eilrich, and M. Weber, P/M—Connecting Rods: Porosity Versus Performance—An Inevitable Conflict?, SAE Technical Paper 960383, SAE Congress and Exposition (Detroit, MI), Feb 1996
50. P/M Parts Showcase, *Met. Powder Rep.,* Vol 49 (No. 5), 1994, p 6–9
51. P/M Parts Showcase, *Met. Powder Rep.,* Vol 49 (No. 7/8), 1994, p 8
52. P/M Parts Showcase, *Met. Powder Rep.,* Vol 50 (No. 1), p 5
53. R. Lugosi, M. Brauer, and J. Cook, "Assembled Camshaft for I.C. Engines with Forged Powder Metal Cams," SAE Technical Paper 870129, SAE Congress and Exposition (Detroit, MI), Feb 1987
54. S. Mocarski, D.W. Hall, J. Khanuja, and S.K. Suh, "Powder Metal Parts for Automotive Applications—Part III," SAE Technical Paper 890409, SAE Congress and Exposition (Detroit, MI), Feb 1989
55. P.K. Johnson, Winning Parts Show High Strength and Cost Savings, *Int. J. Powder Metall.,* Vol 22 (No. 4), 1986, p 267–272
56. Porite Product Brochure, "Warm Compaction," Porite Taiwan Co., 1996

Copper-Infiltrated Steels

Prasan K. Samal, OMG Americas
Erhard Klar, Consultant

THE CONVENTIONAL P/M PROCESS of cold compaction and sintering of metal powder parts generally results in materials that contain pores. Typically, such materials contain 12 to 18% pores, and most of these pores are interconnected. For some applications, such as the self-lubricating bronze bearings, this level of porosity is desirable, and for many other applications it is quite well acceptable. Presence of porosity, however, makes these materials substantially weaker than their wrought counterparts. The pores give rise to areas of stress concentration within the body, thus degrading the dynamic mechanical properties (e.g., fatigue and impact strength) and ductility of the material (Ref 1, 2).

Reduction or elimination of porosity in a P/M material usually requires the application of one of the relatively expensive nonconventional processes, such as hot isostatic pressing, P/M forging, metal injection molding, or hot pressing. These processes lead to fully or near fully dense (less than 2% porosity) materials that usually equal or exceed their wrought counterparts in mechanical properties and metallurgical structure. A recently developed process known as warm compaction can produce P/M steel parts having substantially reduced porosity (8 to 10%) while utilizing a modified, yet economical press-and-sinter technique (Ref 3). The more expensive double-press, double-sinter technique is also capable of producing these higher densities.

An alternative approach to producing substantially pore-free materials is based on molten metal infiltration. In this process, a porous metallic body, having a substantial degree of interconnected porosity (commonly known as the matrix), is infiltrated with another metal of lower melting point (commonly known as infiltrant) by means of a suitably designed heat treatment. The infiltrant is drawn into the interconnected pores of the matrix by capillary action, and upon its solidification a highly or fully dense composite structure is produced. The two most important classes of materials that are manufactured via infiltration methods are copper- and silver-infiltrated refractory metals and refractory carbides, and copper-infiltrated steels. This article focuses on copper-infiltrated steels. Infiltrated refractory metals are discussed in the article "Powder Metallurgy Electrical Contacts" and the article "Infiltration" in this Volume.

In its widely practiced form, sintering of the steel part and its infiltration are combined into one process, known as one-step infiltration. From a practical point of view, the process is designed to produce parts with incomplete infiltration, typically to approximately 92 to 94% of theoretical density. This lower density reduces the tendency of the infiltrant residue to adhere to the steel part, yet it produces mechanical strengths that are much higher than those of uninfiltrated P/M steels, hence making the process commercially attractive.

In a more recent version of copper infiltration of steels, residual porosity after infiltration is reduced to less than 1%. Also, because this modified process employs powder forging grade iron powder, the dynamic mechanical properties of these materials are comparable to those of the lower range of wrought steels (Ref 4). The manufacturing cost of these materials is less than that of double-pressed, double-sintered steels.

Basic Requirements

The infiltration technique is only applicable to material systems that meet certain key requirements. In the most simple case, infiltration may involve only two metals, but it is possible for combinations of more than two metals to equally satisfy the basic conditions. One of the key requirements is that the two metals or metal phases have sufficiently different melting points. This allows the matrix to be kept as a solid, while the infiltrant is heated above its melting point. Secondly, there should be no or limited solubility between the two metals or metal phases. This ensures free flow of infiltrant into the pores without causing significant changes in the structure of the matrix or the properties of the infiltrant. In cases where some mutual solubility exists, it is desirable to minimize the dissolution of the matrix in the infiltrant by modifying the composition of the infiltrant to match the equilibrium composition to that of the matrix alloy at the infiltration temperature. This minimizes the so-called erosion of matrix. In addition, it is generally desirable to limit the effects of mutual solubility by keeping the infiltration time period short and temperature of infiltration sufficiently low. For example, minimization of diffusion of copper into the grain boundaries of the steel matrix enhances the ductility and heat treatability of the copper-infiltrated steel. Additionally, it is essential that the two metals or metal phases do not form an intermetallic phase that may hinder the flow of the molten infiltrant into the pores. It is extremely important that the surface of the matrix be well wetted by the molten infiltrant during the infiltration process. This, in turn, requires that the process be well controlled to help maintain clean, oxide-free, and soot-free surfaces on the matrix. Wetting behavior of the molten infiltrant is also a function of its fluidity, and this is enhanced by good reducing conditions and by proper selection of infiltration temperature and infiltrant composition. Molten infiltrant is drawn into the interconnected pores of the matrix by capillary action (Ref 5). Capillary pressure (ΔP) varies with the inverse of pore diameter (d) as follows:

$$\Delta P = 2\gamma \cos \theta / d$$

where γ is the surface energy of the infiltrating liquid and θ is the contact angle between infiltration liquid and solid matrix. Hence for a given liquid, the capillary pressure can vary widely, depending on θ (the contact angle) and d (the pore diameter). With increased wetting the contact angle θ is lowered, increasing $\cos \theta$. Small-diameter pores have greater capillary force and hence are filled more completely than large-diameter pores.

Infiltration can be either combined with the sintering step (one step) or can be carried out separately (two steps), where sintering of the matrix is carried out separately prior to infiltration.

Conventionally (Partially) Infiltrated Steels

The conventional, that is, partial, infiltration of steel utilizes a matrix made by compaction of an iron powder graphite mix, in a manner similar to that used in the conventional P/M steel process. During the combined sintering/infiltration heat treatment carbon diffuses and alloys with iron to form the steel matrix prior to the melting and infiltration of the copper-base infiltrant. Because iron is partially soluble in copper at the infiltration temperature, commercial infiltrants almost always contain a small amount of iron (2 to 5%, typically), which minimizes dissolution of the matrix in the molten infiltrant.

Figure 1 shows that the ultimate tensile strength can be as high as two or three times that of an uninfiltrated material. However, due to the remaining porosity, the unnotched Charpy impact strength, although improved by several times over that of uninfiltrated steel, up to approximately 34 J (25 ft · lb), remains far below that of fully infiltrated steel. Because similar mechanical properties can be obtained less expensively with pressed-and-sintered nickel steels, commercial copper-infiltrated steels having densities of about 93 to 95% of theoretical usually combine some other features possible with infiltration (Table 1), such as the sealing of pores for pressure tightness, and secondary operations such as plating and brazing, joining of multiple component (green) assemblies, the improvement of machinability and electrical and thermal conductivity, and the reduction of noise in moving components. Table 2(a) lists the composition ranges of infiltrated steels, as specified by Metal Powder Industries Federation, and Table 2(b) provides an overview of typical mechanical properties of these materials.

Table 1 Features of infiltration

- Improved mechanical properties, both static and dynamic
- Improved machinability (interrupted cuts are reduced, smooth surface finish)
- Higher density
- Improved electrical and thermal conductivity
- Noise level reduction in moving components
- Pressure tightness (for use in medium-pressure hydraulic applications)
- Sealing of porosity for secondary operations (pickling, plating, brazing, painting)
- Multiple-component assemblies (assembly of green parts and bonding through common infiltration) of high-strength joints
- Selective properties through localized infiltration
- Good induction/flame-hardening characteristics (surface hardening in air without excessive internal oxidation of steel structure)
- Hard and wear-resistant surfaces with tough internal cores through clearly defined cases by carburizing or carbonitriding

Materials

Matrix. A mixture of iron powder, graphite, and a solid lubricant is used as the matrix material. Graphite provides the carbon content required to convert the iron powder into steel. There are two families of iron powders that are currently in use: oxide-reduced and atomized iron powder. The oxide-reduced powder particles tend to possess noninterconnected internal pores that cannot be infiltrated and often contain hard-to-reduce oxide inclusions. The atomized iron powders are cleaner (fewer inclusions) and possess none or few internal pores. As a result, atomized powders exhibit superior infiltrated properties compared to the reduced powders. The higher compressibilities of atomized powders also produce higher-density matrices that require less copper. Graphite powder added must be sufficiently fine and well dispersed so that it completely dissolves in the iron matrix during heating to the sintering temperature. Molten copper does not wet graphite, and hence if any graphite is present in the undissolved form at the time of infiltration, it can prevent infiltration. The amount of graphite added is dictated by the desired composition of the steel. It also affects the dimensional change of the infiltrated component. As shown in Fig. 2, low-carbon-containing steel matrices swell if they come in contact with copper during sintering. This is attributed to the greater wettability of molten copper with pure iron matrices and rapid diffusion of copper atoms along the grain boundaries (Ref 7). Most commercial practices employ graphite additions

Table 2(a) Composition ranges of infiltrated steels

Material designation	Chemical composition, % Fe	Cu	C(a)	Element
FX-1000	82.8	8.0	0.0	Minimum
	92.0	14.9	0.3	Maximum
FX-1005	82.5	8.0	0.3	Minimum
	91.7	14.9	0.6	Maximum
FX-1008	82.2	8.0	0.6	Minimum
	91.4	14.9	0.9	Maximum
FX-2000	72.7	15.0	0.0	Minimum
	85.0	25.0	0.3	Maximum
FX-2005	72.4	15.0	0.3	Minimum
	84.7	25.0	0.6	Maximum
FX-2008	72.1	15.0	0.6	Minimum
	84.4	25.0	0.9	Maximum

(a) Carbon on the basis of iron only, may be a metallographic estimate

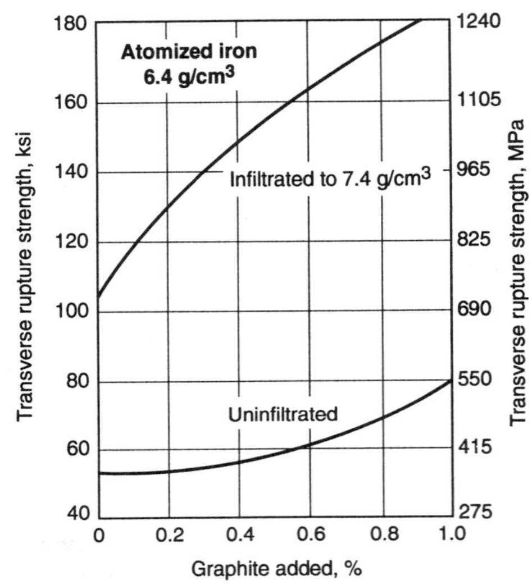

Fig. 1 Increase in transverse rupture strength of sintered steel due to infiltration as a function of amount of graphite added

Table 2(b) Mechanical properties of copper-infiltrated iron and steel

Material designation code(a)	Minimum strength(c) Yield MPa	Yield ksi	Ultimate MPa	Ultimate ksi	Typical properties: Tensile properties, Ultimate strength MPa	Ultimate strength ksi	0.2% yield strength MPa	0.2% yield strength ksi	Elongation, %	Elastic constants, Young's modulus GPa	Young's modulus 10^6 psi	Poisson's ratio	Unnotched Charpy impact energy J	ft·lbf	Transverse rupture strength MPa	ksi	0.1% compressive yield strength MPa	ksi	Rockwell hardness Macro (apparent)	Micro (converted)	Fatigue limit 90% survival MPa	ksi	Density, g/cm³
FX-1000-25	170	25	...	...	350	51	220	32	7.0	160	23.5	0.28	34	25.0	910	132	225	33	65HRB	N/A	130	19	7.3
FX-1005-40	280	40	...	...	530	77	340	50	4.0	160	23.5	0.28	18	13.0	1090	158	365	53	82HRB	N/A	200	29	7.3
FX-1005-110HT	...	...	760	110	830	120	(d)	(d)	<0.5	160	23.5	0.28	9	7.0	1445	210	760	110	38HRC	55 HRC	230	33	7.3
FX-1008-50	340	50	...	...	600	87	415	60	3.0	160	23.5	0.28	14	10.0	1140	166	490	71	89HRB	N/A	230	33	7.3
FX-1008-110HT	...	...	760	110	830	120	(d)	(d)	<0.5	160	23.5	0.28	9	6.5	1300	189	790	115	43HRC	58 HRC	280	41	7.3
FX-2000-25	170	25	...	...	320	46	260	37	3.0	145	21.0	0.24	20	15.0	990	144	280	41	66HRB	N/A	120	17	7.3
FX-2005-45	310	45	...	...	520	75	410	60	1.5	145	21.0	0.24	11	8.0	1020	148	415	60	85HRB	N/A	140	20	7.3
FX-2005-90HT	...	...	620	90	690	100	(d)	(d)	<0.5	145	21.0	0.24	9	7.0	1180	171	490	71	36HRC	55 HRC	160	23	7.3
FX-2008-60	410	60	...	...	550	80	480	70	1.0	145	21.0	0.24	9	7.0	1080	156	480	70	90HRB	N/A	160	23	7.3
FX-2008-90HT	...	...	620	90	690	100	(d)	(d)	<0.5	145	21.0	0.24	7	5.0	1100	159	510	74	36HRC	58 HRC	190	27	7.3

Note: All data based on single-pass infiltration. (a) Suffix numbers represent minimum strength values in ksi; yield in the as-sintered condition and ultimate in the heat treated condition. (b) Mechanical property data derived from laboratory prepared test specimens sintered under commercial manufacturing conditions. (c) Tempering temperature for heat treated (HT) materials: 177 °C (350 °F). (d) Yield and ultimate tensile strength are approximately the same for heat treated materials. Source: Ref 6 of 0.80 to 1.0%. If an application requires a smaller amount of graphite, possibly in the interest of a more ductile pearlitic structure, the dimensional change behavior of such a material needs to be taken into account in the designing of tooling. The matrix iron powder mix should contain sufficient amount of a solid lubricant, such as lithium stearate or Acrawax, to aid in the compaction process.

The pore volume and the pore structure of the matrix govern its capillarity and hence have a strong influence on the completeness of pore filling. Iron powders with different particle size and different particle size distributions can have very different performance even when processed at the same green density.

The infiltrant consists of mainly copper (more than 90% by weight) and small amounts of iron and other additives such as manganese. The iron content typically ranges from 2 to 5% and is intended to form a saturated solution in molten copper prior to infiltration (Ref 8). This minimizes dissolution of iron from the matrix in the molten copper, which erodes the surface of the matrix in the vicinity of the infiltrant slug. Iron may be present in the infiltrant either in alloyed or admixed form. The former is usually preferred because it is not prone to variations in composition due to powder segregation and also due to the fact that with an alloy there is no risk of any iron-free molten copper coming in contact with the matrix.

Commercial infiltrants contain various other additives, some proprietary. Their functions are mainly to enhance fluidity of molten copper and to aid in the formation of a nonadhering residue. Most commercial infiltrants are designed to form residues of varying sizes. The residue holds the molten copper during infiltration, allowing the matrix to gradually draw in the liquid. This prevents the molten copper from running off the edges of the part, into narrow crevices such as the teeth of a gear or lightning holes in gears. After infiltration, a small amount of the copper alloy is usually left in the residue due to the capillarity of the spongy residue mass. Also, if a part is supplied with any excess infiltrant, either due to part-to-part weight variation or density variation, the excess infiltrant is retained by the residue. The residue formed should be easily removable from the matrix and should drop off the infiltrated part when the part is tumbled.

There are a few commercial infiltrants that do not produce a residue. Their use requires precise control of infiltration conditions, particularly infiltration temperature, because of the change of solubility of iron in copper with temperature. Lack of control results in inferior infiltration and excessive erosion. The residue-forming property of manganese derives from its greater tendency for oxidation. However, if the dew point of the sintering atmosphere is very low (<13 °C, or 55 °F) some or all manganese may remain in the metallic form, in solution in molten copper, and find its way into the iron matrix. This can result in the sticking of the residue as well as excessive hardening of the steel matrix.

The infiltrant is conveniently used in a powder form, so as to permit a slug to be designed to match the geometry of the part and also to help promote the formation of a spongy residue.

The Process

The infiltrated density is determined by the green density of the matrix and the amount of the infiltrant used. The latter is often expressed as slug-to-matrix ratio. For each grade of infiltrant, the ideal slug-to-matrix ratio for a given final infiltrated density can be determined from its nomograph, which is usually supplied by the infiltrant manufacturer. Figure 3 is an example of such a nomograph (Ref 9). The infiltrant slug is most commonly compacted to a density of 6.5 to 6.8 g/cm^3. This range of density promotes the formation of an optimally porous spongy residue that also minimizes adhesion.

The location of the slug on the matrix is determined by a number of factors: criticalness of surface finish of the component, distance of travel of infiltrant, and density distribution of green matrix.

The matrix-slug green assemblies are heat treated in a continuous belt furnace for sintering and infiltration, either in a single operation or two. In a two-step operation, the green matrix is first sintered without the infiltrant and then passed through the furnace a second time with the infiltrant. The one-step process is the most commonly practiced process because it is more economical, and it does produce satisfactory results for a large variety of components.

The time-temperature requirements are quite critical in an infiltration process. Proper delubrication of the matrix is essential. Large parts require more care in delubrication. Typically, the green parts should remain for at least 10 min in the 425 to 650 °C (800 to 1200 °F) temperature range for effective delubrication.

Another key event that must occur prior to melting of the infiltrant is the dissolution of admixed graphite in the iron matrix. This occurs at temperatures above 980 °C (1800 °F). If the admixed graphite is not dissolved in the iron matrix prior to melting of the infiltrant, it will affect wetting of the pore surfaces by the molten infiltrant, thus resulting in incomplete infiltration. When the temperature of the assembly is at or above the melting point of the infiltrant, the actual infiltration process occurs. The process takes several minutes for completion, depending on the size of the part, fluidity of infiltrant, and the density and pore structure of the matrix. Most infiltrants are designed for use at a temperature of 1120 °C (2050 °F) and most often the time at this temperature is from 20 to 30 min. If a higher temperature is used, erosion of matrix near the location of slug placement occurs. If a lower temperature is used, infiltration may be incomplete and also strong adhesion of the slug may result.

The dew point of the sintering atmosphere plays a critical role in the process. Generally, a specific dew point range is recommended for each infiltrant. If an infiltrant is used in an atmosphere having a lower dew point than recommended, it may not form the desired residue and can cause sticking of residue. Also, low dew points can lead to soot formation, which in turn can prevent wetting and result in the formation of copper balls. Very high dew points lead to decarburization of the matrix and incomplete sintering of the matrix.

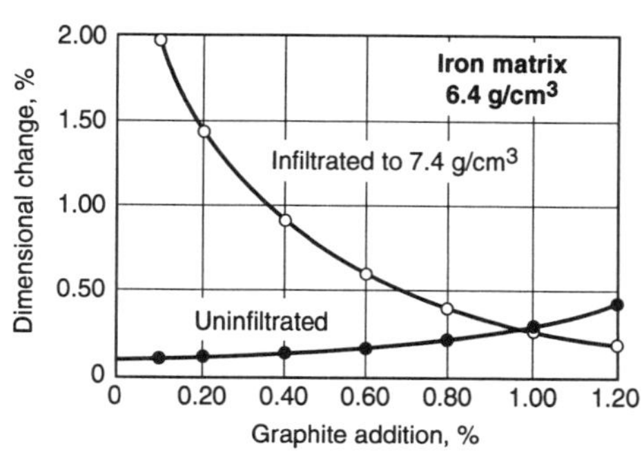

Fig. 2 Effect of amount of graphite added on the dimensional change of copper infiltrated and uninfiltrated sintered steel

Evaluation of Infiltrated Parts

Several criteria are used to judge the performance of an infiltration process. Depending on the application, some may be more critical than others. Some of the major criteria are described in the following paragraphs.

Completeness and Uniformity of Infiltration. In conventional infiltration, matrix pores are infiltrated to attain an infiltrated density of about 93 to 94% of theoretical. Areas of low green density, which usually contain large pores, remain uninfiltrated because of their weak capillarity. Parts possessing significant density gradients, therefore, possess wide hardness ranges af-

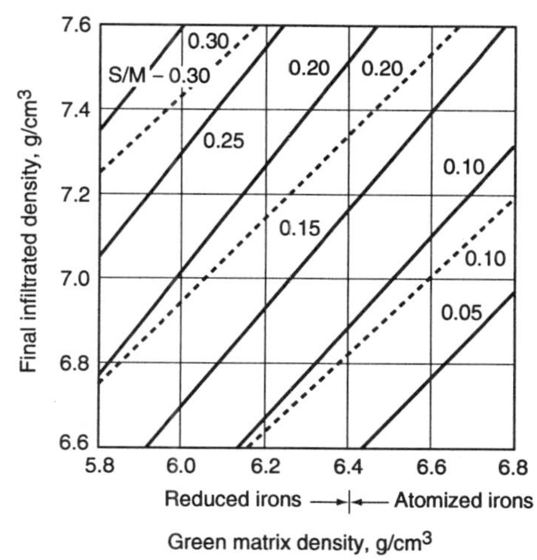

Fig. 3 Guidance chart to determine infiltrating slug weight required to achieve the desired infiltrated density. Solid lines, IP-174 and IP-190; dashed lines, IP-195-LD/HD and IP-204. To find the appropriate slug-to-matrix ratio (S/M), draw a line across the figure at the desired infiltrated density. Then draw a line at the matrix density. The proper S/M is where the lines cross. Multiply the slug-to-matrix ratio (S/M) by the matrix weight to estimate slug weight.

ter infiltration. Lower-than-expected hardness along with low efficiency can also be experienced if infiltration is carried out below the recommended temperature or for not long enough time or if poor wetting of the matrix by molten copper occurs due to oxidation, incomplete dissolution of graphite, or soot formation; or if fluidity of the infiltrant is low.

Adhesion of Residue. Strongly adherent residues may require machining, grinding, or chipping to remove them. Common causes of residue adhesion are excessive amount of infiltrant used, unsatisfactory oxidation of residue-forming constituents, shrinkage of residue around any kind of a projection on the surface of matrix, excessive amount of iron in the infiltrant in relation to infiltration temperature, incomplete infiltration, and incorrect matrix density.

Efficiency of infiltration is a measure of how well the infiltrant supplied was actually utilized to infiltrate the pores in the matrix. Hence, any amount of the infiltrant alloy that is left behind as copper balls or as a constituent of the residue directly accounts for a loss in efficiency. Based on its composition and infiltrating conditions, each infiltrant is expected to yield a specific efficiency. Infiltrants that contain large amounts of residue-forming elements typically are rated to have an efficiency of 90 to 92%, whereas a low residue infiltrant may have a rated efficiency of 99%. If a process yields a lower efficiency than what the infiltrant is rated for, then it is indicative of process deficiencies.

Process variables that commonly cause low efficiency are soot formation, leading to copper ball formation; too low a sintering temperature or too short an infiltration time, leading to incomplete infiltration; and use of an excessive slug-to-matrix ratio.

A simple way to determine efficiency is to compare the mass of the infiltrant slug used and mass of residue formed. This, however, requires that the residue be recovered wholly and weighed accurately, which is not always feasible. Hence, efficiency determinations are based on the mass of the green matrix, mass of green slug, and mass of the infiltrated matrix. In that case, calculation of efficiency must take into account the weight loss of the matrix due to lubricant burnoff, reduction of oxides in the matrix material, and other volatiles. This is conveniently achieved by sintering a test sample (called a blank) in the infiltration heat treatment cycle without an infiltrant slug placed on it. Infiltration efficiency can thus be calculated by the following equation:

$$\text{Efficiency, \%} = \frac{M_I - M_G\,(M_{BS}/M_{BG})}{I_G} \times 100$$

where M_G is the mass of the green matrix test sample; I_G is the mass of green infiltrant slug (placed on matrix test sample), M_I is the mass of infiltrated test sample, without residue; M_{BG} is the mass of matrix blank sample before sintering; M_{BS} is the mass of matrix blank sample after sintering.

Erosion refers to localized dissolution of the matrix material in the infiltrant that results in the formation of microscopic grooves and pits on the surface of the matrix. These grooves are often large enough to be seen with the naked eye. Erosion can lead to loss of mechanical strength, especially impact strength, besides being cosmetically undesirable. Erosion is commonly caused by inadequate concentration of iron in the molten infiltrant, too high an infiltration temperature, or the use of a relatively small contact area between the slug and matrix. Infiltrants that contain iron in the admixed form tend to produce more erosion compared to prealloyed infiltrants because the freshly melting copper dissolves some iron from the surfaces of matrix simultaneously with the dissolution of iron powder present in the slug.

Surface Appearance. In some applications, the surface appearance of the infiltrated part can

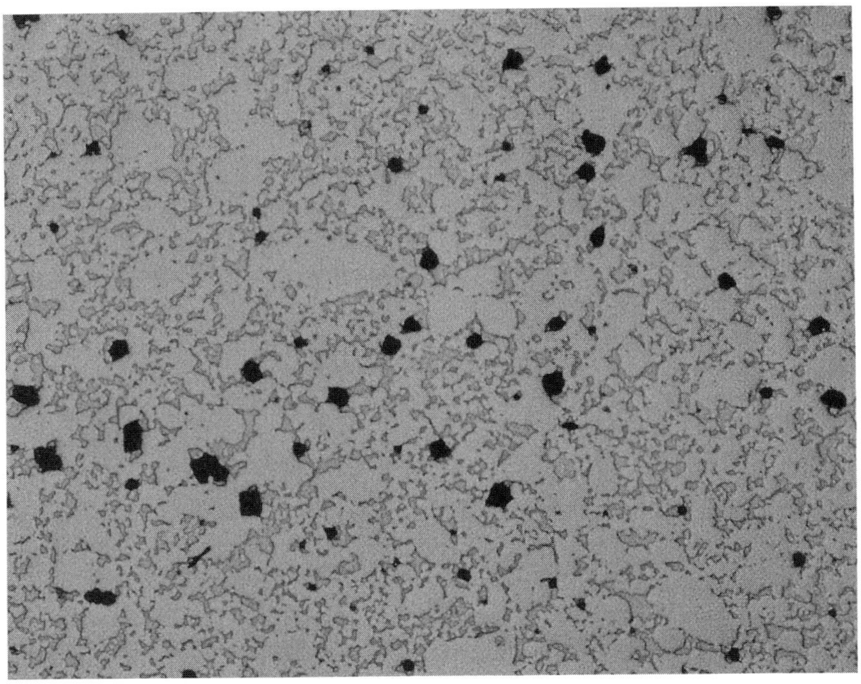

Fig. 4(a) Microstructure of a conventionally (partially) infiltrated steel showing the degree of pore filling. Unetched. 75×

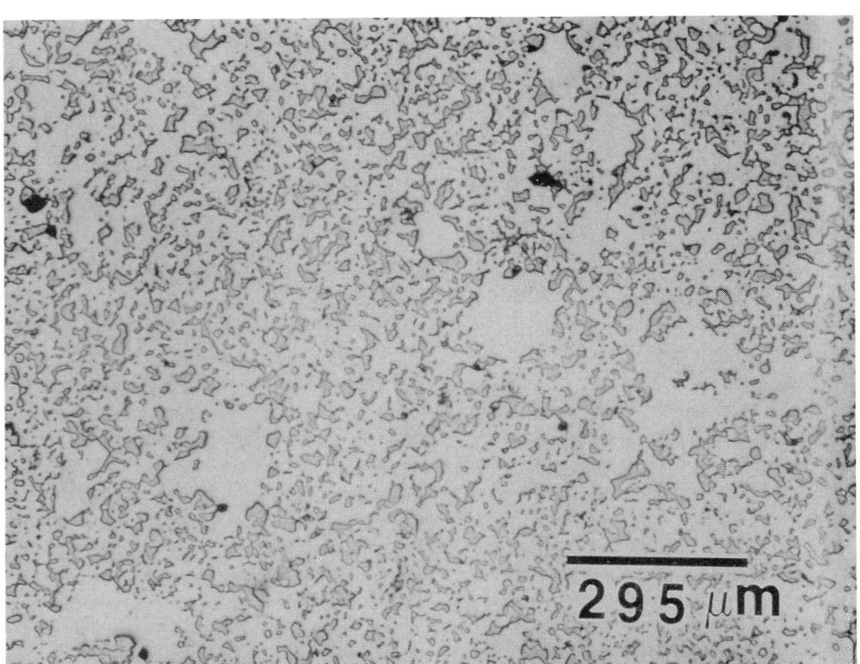

Fig. 4(b) Microstructure of a fully infiltrated steel showing completeness of pore filling. Unetched. 75×

be quite important. Typical unacceptable surface appearances are copper flashing, dirty or sooty surface, oxidized surface, and a steel gray color instead of the copper color.

Integrity of Residue. The residue formed may be a hard mass or a soft porous body. In some applications, a hard residue is undesirable because the residues can get entrapped in a crevice (such as between gear teeth or lightening hole), leading to problems in the final assembly of the part. Although a soft porous residue is generally preferred, if the residue tends to become too crumbly and fine it could lead to excessive dust generation in the manufacturing facility.

Selection of an infiltrant and infiltration process parameters (matrix materials, green densities, slug-to-matrix ratio, furnace atmosphere, dew point, temperature, and time, etc.) are based on specific performance requirements and overall process economics. The process parameters and infiltrant selected are often influenced by the geometry and size of the part because of their effect on density distribution and uniformity of part heat-up and on the availability of suitable location for slug placement.

Fully Infiltrated Steels

Properties similar to those of the lower range of wrought steels can be obtained if the infiltration process is conducted so that practically all pores are infiltrated. This technique bridges the gap in terms of both economy and properties between pressed-and-sintered and fully dense steels. Conventional state-of-the-art powders, compaction presses, and sintering equipment are used. Critical aspects include control of defects through:

- Selection of high-purity powders and erosion-free copper infiltrants
- Processing that controls maximum pore size
- Processing that ensures complete infiltration of the largest pores present
- Heat treatments which produce microstructures that give rise to ductile fracture (dimple rupture)

Residue adherence is minimized by selecting an infiltrant/matrix ratio that provides for full infiltration without an excess of copper.

Table 3 Properties of high-performance copper-infiltrated plain carbon steels

	At tempering temperature:		
	177 °C (350 °F)	621 °C (1150 °F)	704 °C (1300 °F)
Tensile strength, MPa (ksi)	1241 (180)	1034 (150)	758 (110)
Yield strength, MPa (ksi)	1206 (175)	965 (140)	655 (95)
Elongation (25 mm, or 1 in. length), %	<2	5	10
Unnotched Charpy impact energy, J (ft · lbf)	34 (25)	61 (45)	>325 (>240)
Fatigue limit(a), MPa (ksi)	448 (65)	345 (50)	276 (40)
Hardness	50 HRC	30 HRC	95 HRB

(a) $R = -1$; 133 Hz; machined and polished

Mechanical properties as well as fracture toughness and fatigue strength data of these fully infiltrated steels are shown in Table 3 (Ref 4). Figures 4(a) and (b) show the microstructures of a conventionally and a fully infiltrated steel, respectively.

REFERENCES

1. A. Ashurst, E. Klar, and H.R. McCurdy, Copper Infiltration of Steel, *Progress in Powder Metallurgy,* Vol 39, H.S. Nayar et al., Ed., Metal Powder Industries Federation, 1983, p 163–182
2. T. Kimura and H. Hamamoto, Strengthening of Iron Powder Compacts by Infiltration, *Modern Developments in Powder Metallurgy,* Metal Powder Industries Federation, 1997, p 135–148
3. Hoeganaes Offers Higher Density at Lower Cost, *Met. Powder Rep.,* July/Aug 1998, p 22–24
4. E. Klar, D.F. Berry, P.K. Samal, J.J. Lewandowski, and J.D. Rigney, Fracture Toughness and Fatigue Crack Growth Response of Copper Infiltrated Steels, *Int. J. Powder Metall.,* Vol 31 (No. 4), 1995, p 317–324
5. R.M. German, *Powder Metallurgy Science,* 2nd ed., Metal Powder Industries Federation, 1994, p 310
6. Standard 35, Metal Powder Industries Federation, 1997
7. F.V. Lenel, *Powder Metallurgy, Principles and Applications,* Metal Powder Industries Federation, 1980, p 315
8. J.E. Elliot, Characteristics of Copper Infiltrated Porous Iron, *Metallurgia,* Vol 52, 1955, p 226–234
9. "Copper Base Infiltrating Powders," *Tech. Bull.,* OMG Americas

Powder Metallurgy Stainless Steels

E. Klar and P.K. Samal, OMG Americas

COMMERCIAL PRODUCTION of sintered stainless steels in North America began in the late 1950s. After a typical steep initial expansion, growth from 1970 settled to an average compound growth rate of ~4.6%. In 1996, ~4500 short tons (4100 metric tons) of stainless steel powders were used. This represents only a small fraction of wrought stainless steels. As powder producers develop improved stainless steel powders (see the article "Production of Steel Powders" in this Volume), growth is expected to continue. Optimum sintering parameters are also being developed, as stainless steel parts producers continue to improve the control of their sintering operations (see the article "Production Sintering Practices" in this Volume). P/M companies and end users of P/M parts are working on the development of specifications and corrosion tests for sintered stainless steels. This article briefly reviews major applications of sintered stainless steel with a view toward the improvement of P/M stainless steel technology with respect to corrosion resistance. This is followed by processing characteristics and properties of sintered stainless steels. The last section of this article is a brief summary of the uses and properties of fully dense P/M stainless steels.

Applications

Stainless steel P/M parts are produced from powders in both the 300 (austenitic) series and the 400 (ferritic) series. The 300 series austenitic alloys are typically used in applications that require good corrosion resistance, while ferritic grades are used in applications that require magnetic properties or good thermal conductivity and/or durability in applications that involve thermal cycling (like automotive exhaust systems).

Automotive components are the most significant application sector, and three major automotive applications are briefly described in the following paragraphs. Substantial quantities of stainless steel powders also are used to make parts possessing controlled interconnected porosity features for filtration, metering of liquids and gas, and sound attenuation in telephones, microphones, and hearing aids. Substantial quantities of sintered stainless steel parts are also used in hardware, appliances, and electrical systems (Table 1). Figure 1 shows a few examples

Fig. 1 Examples of P/M stainless steel parts for various applications

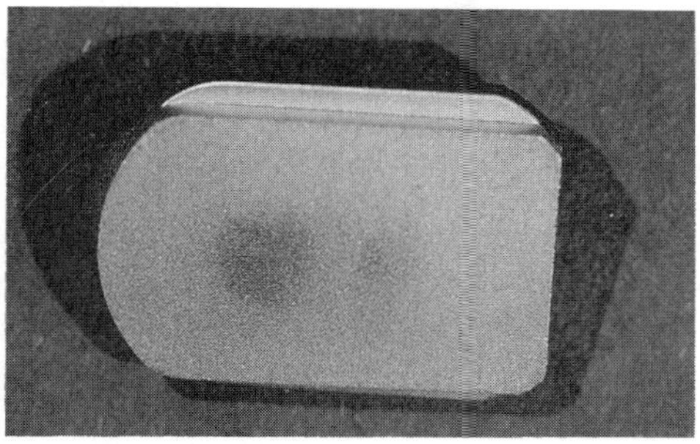

Fig. 2 Stainless steel bracket for automotive rear-view mirror

Table 1 Application of sintered stainless steels

Application area	Alloy
Automotive	
Exhaust system flanges	409L, 434L
Sensor bosses	409L, 434L
ABS sensor rings	410L, 434L
Rearview mirror mounts	434L, 430L, 316L
Brake components	434L
Seat belt locks	304L
Windshield wiper pinions	410L
Windshield wiper arms	316L
Manifold heat control valves	304L
Hardware	
Lock components	304L, 316L
Threaded fasteners	303L
Fasteners	303L, 304L, 316L
Quick-disconnect levers	303L, 316L
Spacers and washers	316L
Electrical and electronic	
Limit switches	410L
G-frame motor sleeves	303L
Rotatory switches	316L
Magnetic clutches	410L, 440A
Battery nuts	830
Electrical testing probe jaws	316L
Industrial	
Water and gas meter parts	316L
Filters, liquid, and gas	316L, 316L-SI
Recording fuel meters	303L
Fuel flow meter devices	410L
Pipe flange clamps	316L
Plumbing fixtures	303L
Sprinkler system nozzles	316L
Shower heads	316L
Window hardware	304L, 316L
Office equipment	
Nonmagnetic card stops	316L
Dictating machine switches	316L
Computer knobs	316L
Miscellaneous	
Coins, medallions	316L
Dental equipment	304L
Watch cases	316L
Fishing rod guides	304L, 316L
Photographic equipment	316L
Cam cleats	304L
Dishwasher components	304L
Can opener gears	410L

of sintered stainless steel parts from various industries.

Rearview Mirror Bracket. The first major automotive application of a sintered stainless steel was the rearview mirror bracket (Fig. 2). It was developed in the 1960s and today still ac-

Fig. 3 Sensor rings for anti-lock brakes

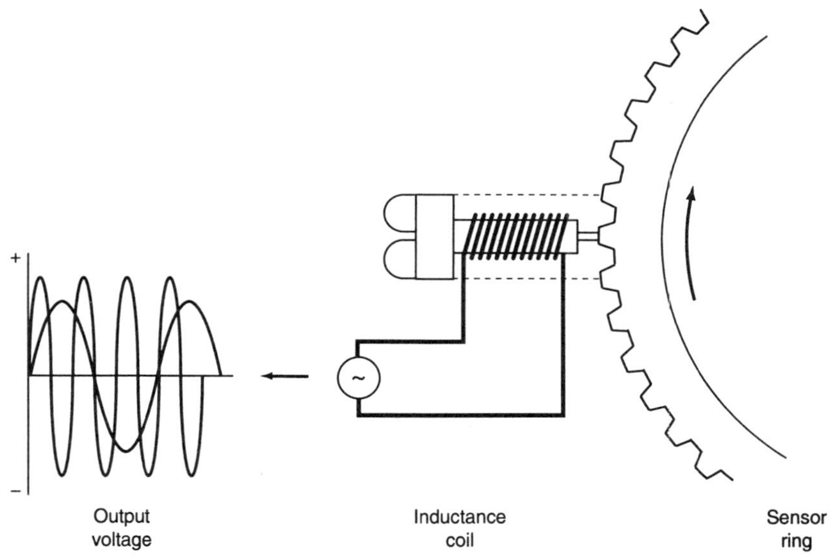

Fig. 4 Schematic of sensor system for anti-lock brakes

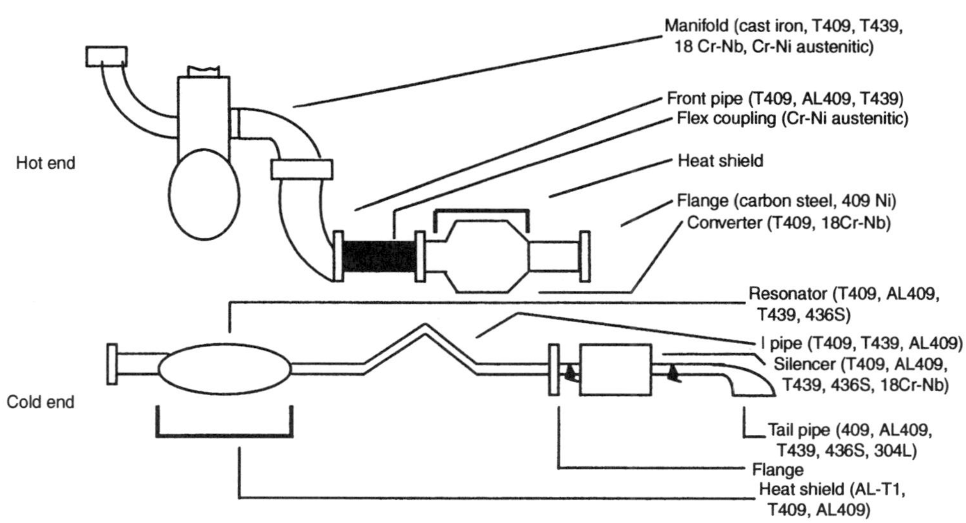

Fig. 5 Exhaust system components

Fig. 6 Powder metallurgy flanges and sensor bosses for automotive exhausts

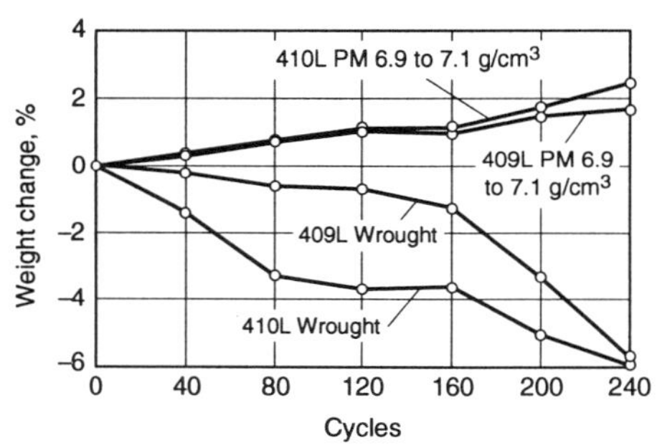

Fig. 7 Oxidation of wrought and P/M stainless steel flanges

counts for ~15% of all P/M stainless steel parts in North America. These brackets are made mainly from 430L and 434L by sintering in dissociated NH_3 or in H_2. Sintered density is ~7.1 g/cm^3. As might be expected, there are no demanding requirements regarding corrosion resistance. However, the thermal expansion coefficient of the bracket must be similar to that of glass as the metal bracket is glued to the windshield window.

Antilock brake system (ABS) sensor rings (Fig. 3) comprise the second major automotive application of sintered stainless steel parts. The application was developed in the 1980s and presently accounts for ~25% of the total volume of sintered stainless steel parts. As shown schemati-

Table 2 Characteristics of various grades of P/M stainless steels

Designation	Description	Characteristics
303L	Free machining austenitic grade	Designed for parts that require extensive secondary machining operations. It has high strength and hardness. This alloy has marginal corrosion resistance. Sulfur added for machinability
303LSC and Ultra 303L	Enhanced corrosion resistance version of 303L	These are copper and tin-modified versions of 303L alloy having all the characteristics of 303L, except for improved corrosion resistance. Hence, they combine machinability and moderate corrosion resistance.
304L	Basic austenitic grade	Most economical of austenitic grades. Used where material cost is large percentage of the total manufacturing cost. It has better corrosion resistance than 303L. Machinability is good. Copper and tin modified versions of 304L alloy (304LSC and Ultra 304L) are available for improved corrosion resistance.
316L	Standard austenitic grade	This alloy offers better corrosion resistance and machinability than 304L. With careful processing it can meet the corrosion resistance requirements of the more demanding applications. Copper and tin modified versions of 316L alloy offer even greater corrosion resistance than 316L alloy.
317L	Premium austenitic grade	It is a higher molybdenum content austenitic grade possessing excellent resistance to corrosion, especially to crevice corrosion (superior to 316LSC and Ultra 316L).
SS-100	Super premium austenitic grade	A highly alloyed austenitic grade superior to all other grades of P/M stainless steel in corrosion resistance. Its corrosion resistance equals to that of wrought 316L. In non-optimized sintering atmospheres it suffers a smaller loss of corrosion resistance, compared to other grades of P/M stainless steel.
409L	Weldable ferritic grade	A weldable grade of stainless steel containing niobium, which prevents sensitization. It is not recommended to make carbon additions to this grade. It is a magnetic alloy with good ductility and fair corrosion resistance.
410L	Standard ferritic/martensitic grade	This ferritic grade can be readily converted to a martensitic alloy by addition of small amounts of carbon prior to processing, which will also make it responsive to heat treatment. In the ferritic form the alloy is ductile and machinable, whereas in the martensitic form it is hard with reduced ductility. In the martensitic form it is used in wear resistant applications. Both forms of the alloy are magnetic. The martensitic form has the lowest corrosion resistance of all P/M stainless steel grades.
430L/434L	Premium ferritic grades	Used for applications requiring some corrosion resistance but where economics (or magnetic requirement) preclude use of austenitic grade. Within the specified levels of carbon and nitrogen of standard compositions (Table 3), these grades cannot usually be converted to a martensitic alloy. Color is compatible with chrome plate. Corrosion resistance is better than that of 410L. Machinability is slightly better than that of 410L.

Note: LSC and Ultra are proprietary grades of OMG Americas and Ametek, respectively

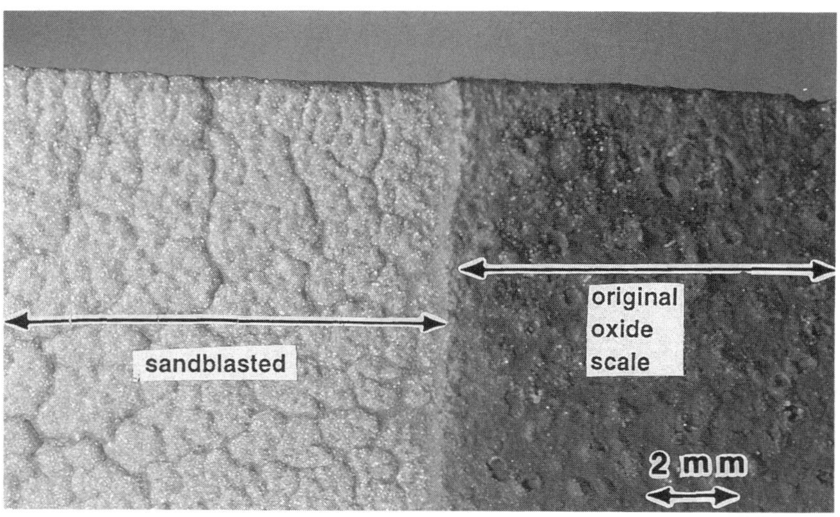

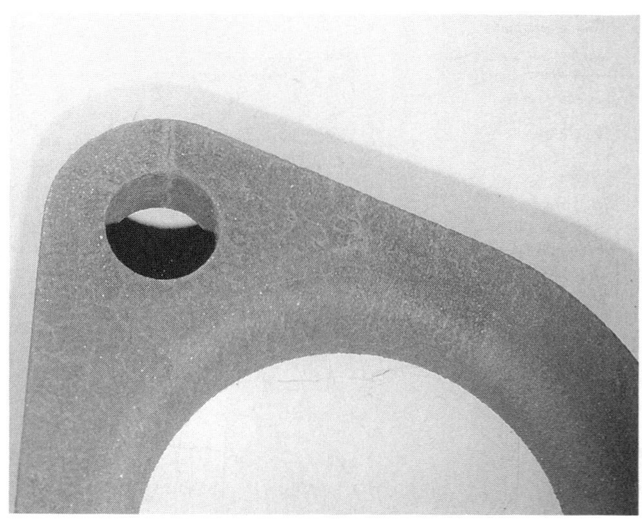

Fig. 8 Hairline cracks and fracture of P/M flange

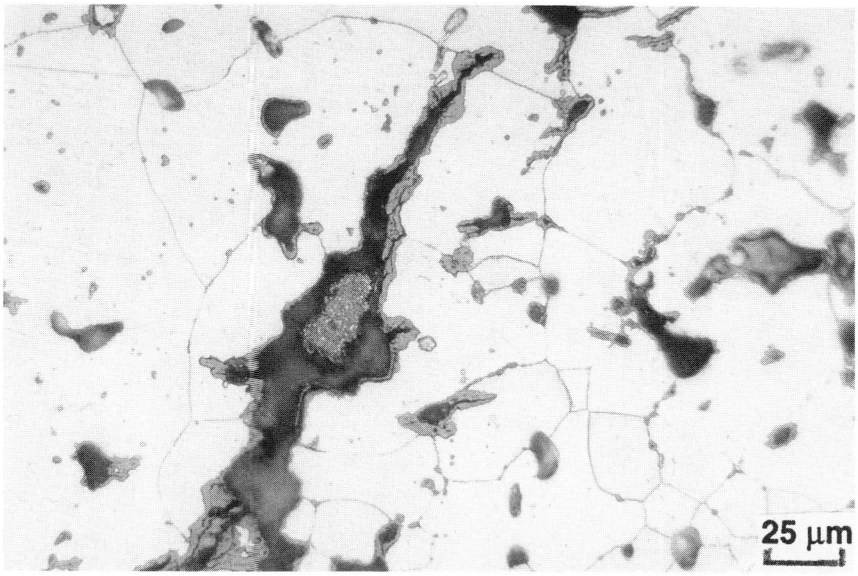

Fig. 9 Transgranular crack of a 410L P/M stainless steel flange after 100 thermal cycles

cally in Fig. 4, a sensor ring constitutes an integral part of an ABS. Rotating at the same speed as the car, a sensor ring generates voltages in a stationary induction coil. The frequency of the induced voltage depends on the rotational speed of the ring or the wheel. The voltage signal is used to control brake and ignition functions via a computer to prevent locking of the car wheels. Thus, in addition to adequate mechanical strength, ductility, and dimensional accuracy, the ring must possess adequate magnetic characteristics and sufficient corrosion resistance to survive their exposure to the elements of the road.

Various steel compositions and P/M techniques have been developed to achieve the previously mentioned objectives. Sensor rings made from mild steel require a protective coating to resist abrasion and corrosion (Ref 1). The majority of ABS sensor rings are now made from 410L, 434L, and modified 434L (18Cr, 2Mo) stainless steel by hydrogen or vacuum sintering. Nitrogen contents are preferably <50 ppm to ensure good magnetic response and adequate chloride corrosion resistance. Sintered densities range from 7.0 to 7.2 g/cm^3 (Ref 2). Stainless steel sensor rings processed in this manner can develop various amounts of rust as well as a small number of pits when subjected to a 100 h 5% NaCl salt spray test in accordance with ASTM B 117. However, extensive field testing in various locations across the United States and Canada has shown that these rings are still functioning properly after ten years and/or over 100,000 miles with only minor amounts of rust and no loss of mechanical strength and ductility (Ref 3). For aesthetic purposes, these rings can be nickel or chrome plated to give them a rust-free appearance.

Automotive Exhaust Systems. Because of stricter environmental concerns and consumer demands for extended service life, automotive exhaust systems are now increasingly made from stainless steel. For the connection of various components of an exhaust system (Fig. 5), as well as for measuring exhaust gas properties, flanges and sensor bosses (Fig. 6) are required, both of which can be manufactured advantageously by P/M technology. While some P/M stainless steel flanges are already in use in limited applications, the P/M industry is presently working on qualifying compositions and other properties to meet a number of specifications. Perhaps the most demanding of these specifications is a cyclic elevated temperature oxidation/corrosion test in which a part is heated in air to 650 °C (1200 °F), kept at this temperature for 90 min, then quenched in water for 1 min, and exposed for another 20 min in a humidity chamber (60 °C, or 140 °F; 85% relative humidity) for 200 cycles. Another critical test is a so-called "condensed corrosion test for welded parts" in which a flange that is welded to the exhaust pipe (typically of 409Cb composition) is soaked for 15 min in a solution of 5000 ppm sulfate plus 100 ppm chloride at a pH of 2.5. After drying in air for 75 min, the part is exposed for 20 h to 85% relative humidity at 60 °C (140 °F). This is followed by soaking for 90 min in air at 427 °C (800 °F). After 25 cycles, the P/M part is compared to a wrought welded 409Cb flange tested identically for signs of grain boundary corrosion due to sensitization.

Flanges made from various 300 and 400 series stainless steel compositions initially passed the "condensed corrosion" test (the degree of sensitization was small and similar to that of wrought 409Cb), but failed the "elevated temperature oxidation/corrosion" test. In spite of a slower rate of oxidation of the P/M flanges, in comparison to wrought flanges (Fig. 7), the P/M flanges, after 60 to 80 cycles, developed hairline cracks that grew and led to complete fracture (Fig. 8). The greater oxidation rate of the wrought parts was due to the oxide spalling off from the flanges, particularly upon quenching, resulting in a weight loss. The slower oxidation rate of a sintered flange was due to the oxide becoming anchored and locked into place in the surface

Table 3 Compositions of standard powder metallurgy stainless steels

Material	Chemical composition(a), wt%									
designation	Fe	Cr	Ni	Mn	Si	S	C	P	Mo	N	Nb
Austenitic grades											
SS-303N1, N2	bal	17.0–19.0	8.0–13.0	0–2.0	0–1.0	0.15–0.30	0–0.15	0–0.20	...	0.20–0.60	...
SS-303L	bal	17.0–19.0	8.0–13.0	0–2.0	0–1.0	0.15–0.30	0–0.03	0–0.20	...	0–0.03	...
SS-304N1, N2	bal	18.0–20.0	8.0–12.0	0–2.0	0–1.0	0–0.03	0–0.08	0–0.045	...	0.20–0.60	...
SS-304L	bal	18.0–20.0	8.0–12.0	0–2.0	0–1.0	0–0.03	0–0.03	0–0.045	...	0–0.03	...
SS-316N1, N2	bal	16.0–18.0	10.0–14.0	0–2.0	0–1.0	0–0.03	0–0.08	0–0.045	2.0–3.0	0.20–0.60	...
SS-316L	bal	16.0–18.0	10.0–14.0	0–2.0	0–1.0	0–0.03	0–0.03	0–0.045	2.0–3.0	0–0.03	...
Ferritic or martensitic grades											
SS-409L	bal	10.5–11.75	...	0–1.0	0–1.0	0–0.03	0–0.03	0–0.04	...	0–0.03	8XC–0.80
SS-410	bal	11.5–13.5	...	0–1.0	0–1.0	0–0.03	0–0.25	0–0.04	...	0.20–0.60	...
SS-410L	bal	11.5–13.5	...	0–1.0	0–1.0	0–0.03	0–0.03	0–0.04	...	0–0.03	...
SS-430N2	bal	16.0–18.0	...	0–1.0	0–1.0	0–0.03	0–0.08	0–0.04	...	0.20–0.60	...
SS-430L	bal	16.0–18.0	...	0–1.0	0–1.0	0–0.03	0–0.03	0–0.04	...	0–0.03	...
SS-434N2	bal	16.0–18.0	...	0–1.0	0–1.0	0–0.03	0–0.08	0–0.04	0.75–1.25	0.20–0.60	...
SS-434L	bal	16.0–18.0	...	0–1.0	0–1.0	0–0.03	0–0.03	0–0.04	0.75–1.25	0–0.03	...

(a) Maximum unless a range is specified. Other elements: total by difference equals 2.0% maximum, which may include other minor elements added for specific purposes

Table 4 Mechanical properties of standard powder metallurgy stainless steels

Minimum values				Typical values										
Material designation code(a)	Minimum strength, MPa		Minimum elongation, in 25.4 mm, %	Tensile properties			Elastic constants		Unnotched Charpy impact energy, J	Transverse rupture strength, MPa	Compressive yield strength (0.1%), MPa	Rockwell hardness		Density, g/cm³
	Yield	Ultimate		Ultimate strength, MPa	Yield strength (0.2%), MPa	Elongation in 25.4 mm, %	Young's modulus, GPa	Poisson's ratio				Macro (apparent)	Micro (converted)	
400 series (ferritic and martensitic) grades														
SS-410-90HT	...	620	0	720	...	<0.5	125	0.25	3	780	640	23 HRC	55 HRC	6.5
SS-410L-20	140	...	10.0	330	180	16.0	165	0.27	68	N/A	190	45 HRB	N/A	6.9
SS-430N2-28	190	...	3.0	410	240	5.0	170	0.27	34	N/A	230	70 HRB	N/A	7.1
SS-430L-24	170	...	14.0	340	210	20.0	170	0.27	108	N/A	230	45 HRB	N/A	7.1
SS-434N2-28	190	...	4.0	410	240	8.0	165	0.27	20	N/A	230	68 HRB	N/A	7.0
SS-434L-24	170	...	10.0	340	210	15.0	165	0.27	88	N/A	230	50 HRB	N/A	7.0
300 series austenitic grades														
SS-303N1-25	170	...	0	270	220	0.5	105	0.25	5	590	260	62 HRB	N/A	6.4
SS-303N2-35	240	...	3.0	380	290	5.0	115	0.25	26	680	320	63 HRB	N/A	6.5
SS-303N2-38	260	...	6.0	470	310	10.0	140	0.27	47	N/A	320	70 HRB	N/A	6.9
SS-303L-12	80	...	12.0	270	120	17.5	120	0.25	54	570	140	21 HRB	N/A	6.6
SS-303L-15	100	...	15.0	330	170	20.0	140	0.27	75	N/A	200	35 HRB	N/A	6.9
SS-304N1-30	210	...	0	300	260	0.5	105	0.25	5	770	260	61 HRB	N/A	6.4
SS-304N2-33	230	...	5.0	390	280	10.0	115	0.25	34	880	320	62 HRB	N/A	6.5
SS-304N2-38	260	...	8.0	480	310	13.0	140	0.27	75	N/A	320	68 HRB	N/A	6.9
SS-304L-13	90	...	15.0	300	120	23.0	120	0.25	61	N/A	150	30 HRB	N/A	6.6
SS-304L-18	120	...	18.0	390	180	26.0	140	0.27	108	N/A	190	45 HRB	N/A	6.9
SS-316N1-25	170	...	0	280	230	0.5	105	0.25	7	740	250	59 HRB	N/A	6.4
SS-316N2-33	230	...	5.0	410	270	10.0	115	0.25	38	860	300	62 HRB	N/A	6.5
SS-316N2-38	260	...	8.0	480	310	13.0	140	0.27	65	N/A	320	65 HRB	N/A	6.9
SS-316L-15	100	...	12.0	280	140	18.5	120	0.25	47	550	150	20 HRB	N/A	6.6
SS-316L-22	150	...	15.0	390	210	21.0	140	0.27	88	N/A	200	45 HRB	N/A	6.9

N/A, not applicable. (a) N1, nitrogen alloyed, good strength, low elongation, sintered at 1149 °C (2100 °F) in dissociated ammonia; N2, nitrogen alloyed, high strength, medium elongation, sintered at 1288 °C (2350 °F) in dissociated ammonia; L, low carbon, lower strength, highest elongation, sintered at 1288 °C (2350 °F) in partial vacuum, cooled to avoid nitrogen absorption. Suffix numbers represent minimum strength values in 10^3 psi. Yield strength in the as-sintered condition. Ultimate strength in the heated treated condition

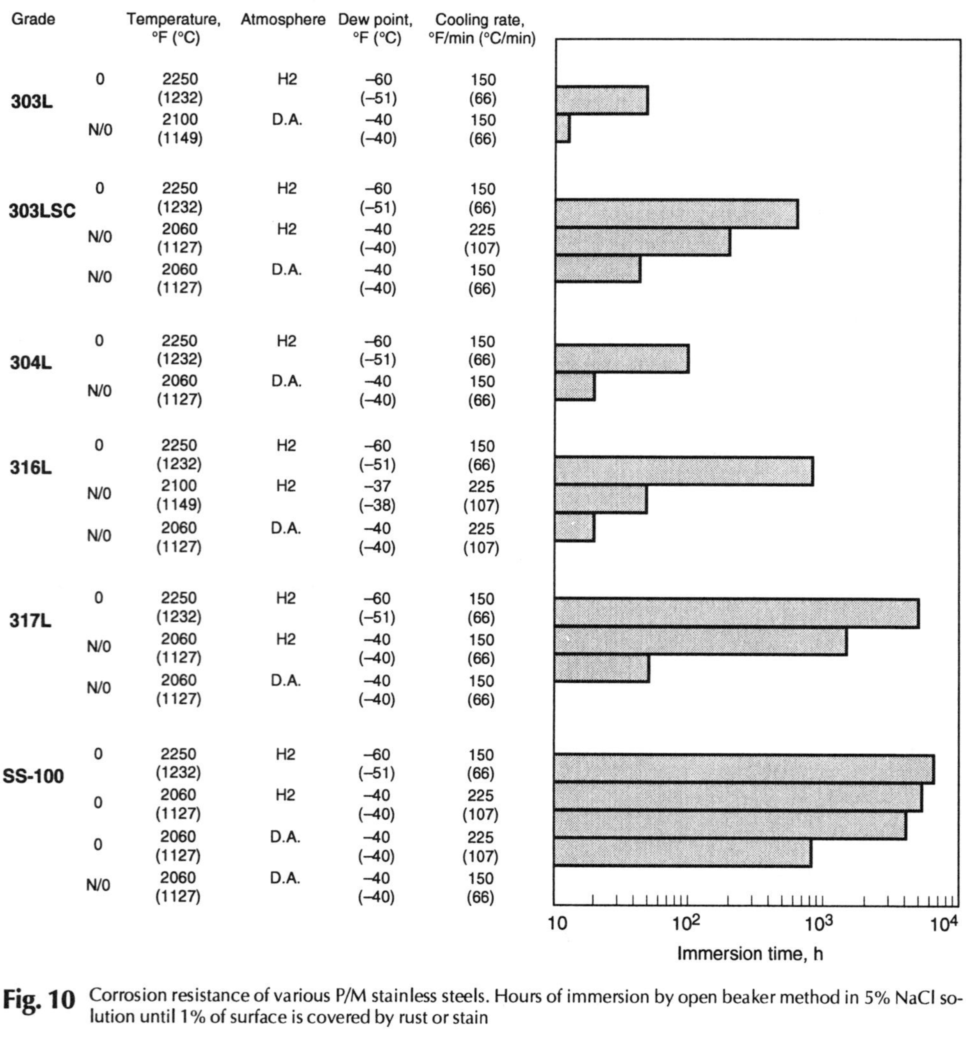

Fig. 10 Corrosion resistance of various P/M stainless steels. Hours of immersion by open beaker method in 5% NaCl solution until 1% of surface is covered by rust or stain

pore space, thus slowing the rate of oxidation due to the adhesion of the oxide layer. This mechanism caused the flange to increase weight. The failure of the P/M flanges was diagnosed as thermal fatigue, and due to the repetitive and cumulative effects of oxidation of surface-near pores and the stresses developed during quenching in water on account of different expansion coefficients of metal and metal oxide. Figure 9 shows a transgranular crack of a 410L flange after 100 cycles produced by this mechanism. This failure was overcome by increasing the density from a low range of 6.9 to 7.15 g/cm³ of the flange to a higher range of ~7.2 to 7.3 g/cm³, depending on composition. The higher density range further reduced the oxidation rate, kept the oxidation restricted to the outer surfaces of the flange, and also increased the strength of the flange.

Specifications and Properties

The selection of a P/M stainless steel alloy for a given application is based on a number of considerations. Corrosion resistance is a key property for the majority of uses, as is the case with wrought stainless steels. Additional reasons for selecting a P/M stainless steel are magnetic versus nonmagnetic characteristics, mechanical properties (room and elevated temperature), machinability, and hardenability. Ease of processing and raw material cost must also be taken into consideration. Table 2 summarizes the main characteristics of widely used P/M stainless steels. Composition ranges and properties of sintered stainless steels as standardized by the Metal Powder Industries Federation (MPIF) (Stan-

Table 5 Influence of sintering atmosphere on mechanical properties of type 316L pressed to 6.85 g/cm³

Property	Sintered in dissociated ammonia	Sintered in hydrogen
Ultimate tensile strength, MPa (psi)	365.4 (53,000)	288.2 (41,800)
Yield strength, MPa (psi)	274.4 (39,800)	183.4 (26,600)
Elongation in 25 mm (1 in.), %	7.0	10.9
Apparent hardness, HRB	67	47

Compacts sintered for 30 min at 1120 °C (2050 °F). Source: Ref 6

dard 35) in 1997, are shown in Tables 3 and 4, respectively. The information is practically identical to that of ASTM Standard B 783, published in 1993. The 1997 edition of MPIF Standard 35 features additional property values for higher sintered densities as well as property data for the ferritic alloys 430L and 434L.

Corrosion resistance of sintered stainless steel depends on a number of factors, particularly those related to sintering. Figure 10 illustrates how different sintering conditions can produce widely differing levels of corrosion resistance for the grades listed in Table 2. The importance of avoiding contamination and minimizing the presence, in the microstructure, of precipitates of chromium carbide, chromium nitride, silicon oxide, as well as controlling the formation of surface oxides and/or nitrides on cooling, are discussed in the article "Production Sintering Practices" in this Volume. Extensive coverage on P/M stainless steel corrosion is also in the article "Corrosion-Resistant Powder Metallurgy Alloys" in this Volume. As standards relating to the corrosion resistance of sintered stainless steel are not yet available, it is recommended that users of sintered stainless steel parts use published data as guidelines and conduct their own application-specific tests with respect to corrosion resistance.

Mechanical properties of sintered stainless steels are determined by the powder characteristics, the sintered density, and the sintering parameters. Table 4 lists typical and minimum mechanical properties specified in MPIF Standard 35. Medium-density stainless steels (6.2 to 6.6 g/cm³) usually are produced by pressing at 415 to 825 MPa (30 to 60 tsi) and sintering at 1120 to 1150 °C (2050 to 2100 °F) in dissociated ammonia, hydrogen, or vacuum. Superior elongation, impact resistance, and corrosion resistance are obtained by sintering at 1260 to 1315 °C (2300 to 2400 °F). Higher sintering temperatures and long sintering times produce more pore rounding and grain growth, both of which impart ductility and impact strength to the part. Lowering the carbon, nitrogen, and oxygen contents also improves ductility.

Mechanical properties of stainless steels depend on the atmosphere in which they are sintered. The most common commercial atmosphere for sintering stainless steel is dissociated ammonia (75 vol% hydrogen and 25 vol% nitrogen). Sintering in nitrogen-containing atmospheres results in the absorption of significant amounts of nitrogen, which produces increased strength and hardness, reduced ductility, and much less shrinkage in comparison to sintering

Table 6 Properties of martensitic stainless steels sintered at 1230 °C (2250 °F)

	410 material		420 material	
Processing				
Graphite, %	0	0.15	0.15	0.30
Atmosphere	D.A.	H_2	D.A.	H_2
Sintered plus tempered at 305 °C (600 °F)				
Ultimate tensile strength, psi	96,000	100,000	130,000	130,000
Yield strength, psi	84,000	82,000	120,000	120,000
Elongation, %	1.1	1.2	0.5	0.8
Apparent Rockwell hardness, HRC	30	27	30	28
Particle hardness, Vickers, HV (Rockwell, HRC)	606 (55)	656 (58)	678 (59)	715 (61)
Heat treated(a)				
Ultimate tensile strength, psi	...	120,000	130,000	140,000
Yield strength, psi	...	105,000	120,000	130,000
Elongation, %	...	1.5	0.5	0.8
Apparent Rockwell hardness, HRC	...	30	32	33
Particle hardness, Vickers, HV (Rockwell, HRC)	...	609 (55)	...	628 (56)

D.A., dissociated ammonia. (a) Reaustenitized at 1010 °C (1850 °F), quenched and tempered at 315 °C (600 °F). Source: Ref 7

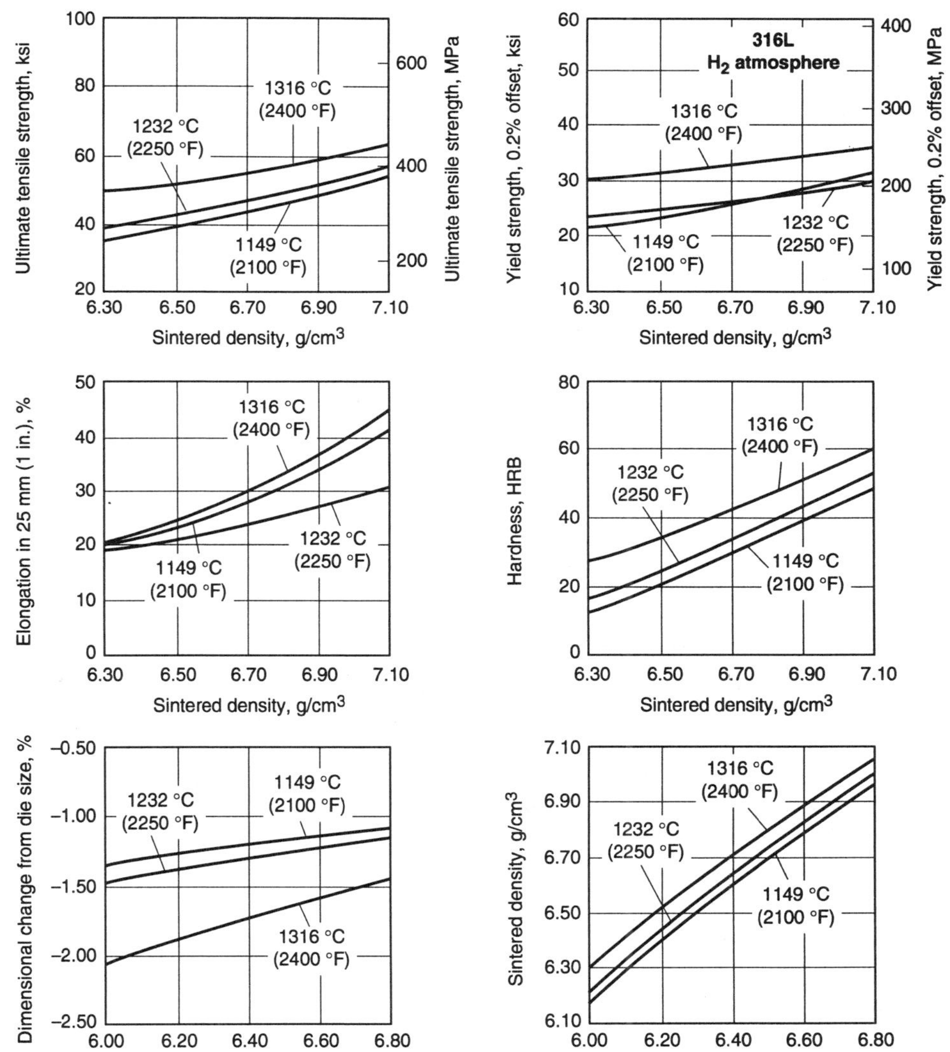

Fig. 11 Mechanical properties and densification behavior of sintered 316L as a function of sintering temperature and density. Sintered for 45 min in hydrogen atmosphere with dew point of approximately −45 °C (−50 °F)

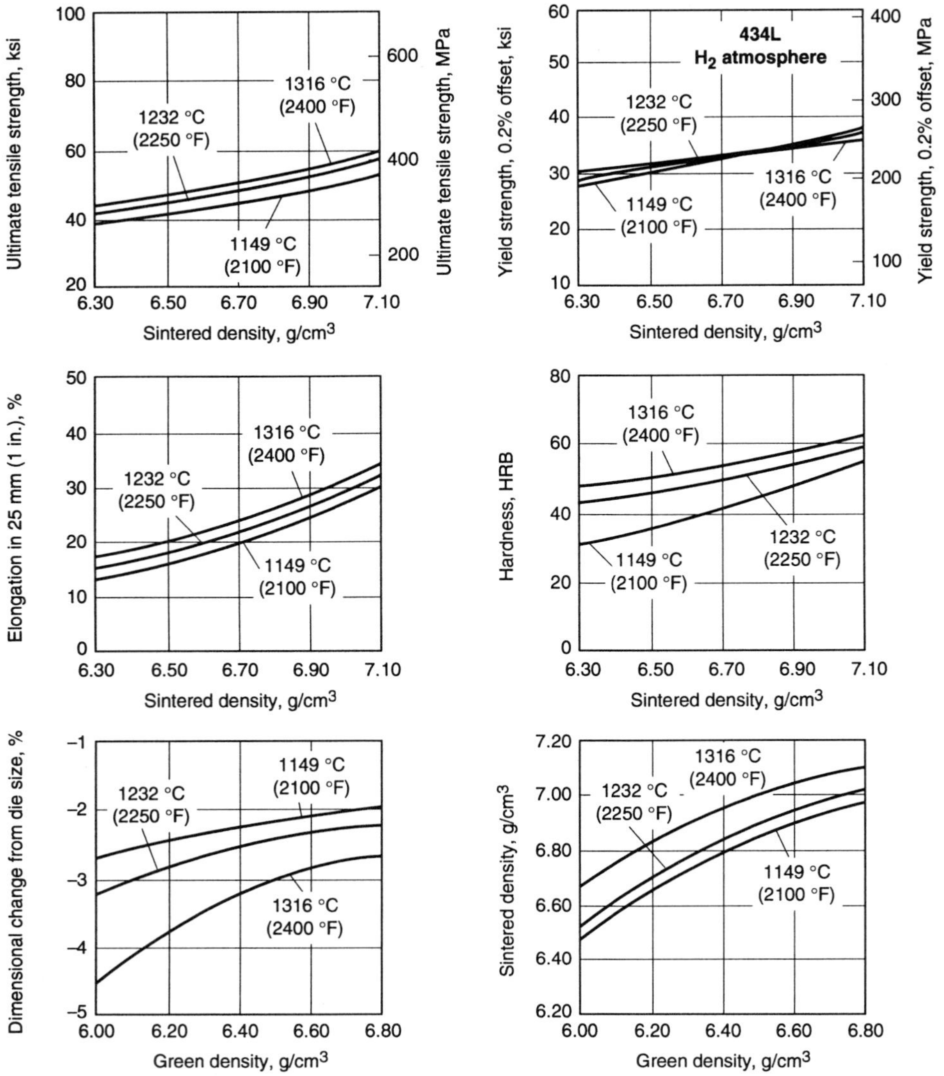

Fig. 12 Mechanical properties and densification behavior of sintered 434L as a function of sintering temperature and density. Sintered for 45 min in hydrogen atmosphere with −45 °C dew point

in hydrogen or under vacuum. This effect of sintering atmosphere is shown in Table 5 for austenitic 316 (Ref 6) and in Table 6 (Ref 7) for two martensitic steels.

Figures 11 to 14 summarize mechanical properties as well as dimensional changes of an austenitic (316L) and a ferritic (434L) stainless steel for various sintering temperatures and sintering atmospheres over a broad density range. The high sintering temperature (1232 and 1316 °C, or 2250 and 2400 °F) data were obtained from commercial sintering furnaces. The low temperature (1149 °C, or 2100 °F) data were obtained from a laboratory furnace. Sintering in hydrogen atmospheres with lower dewpoints increases densification and ductility.

The mechanical properties of vacuum-sintered stainless steels are similar to those of hydrogen-sintered stainless steels. If vacuum sintering is done with optimized additions of graphite (Ref 8, 9), low oxygen contents of 600 to 800 ppm will result with improved ductility and toughness. For minimizing interstitials (oxygen, carbon, nitrogen), high-temperature vacuum sintering (with a partial pressure of hydrogen) is superior to high-temperature hydrogen sintering.

Ferritic Stainless Steels and Magnetic Properties. The soft magnetic properties of powder metal (P/M) components are essentially dictated by the level of impurities, the resultant density, and the sintering practice. The optimum properties for soft magnetic performance are obtained by maximizing density and sintering temperatures and minimizing the contamination levels of carbon, nitrogen, and oxygen.

Tables 7 and 8 compare magnetic and mechanical properties of two stainless steels under various sintering conditions (Ref 10). Vacuum sintering produces the best product with regard to soft magnetic properties. This is attributed to the fact that both nitrogen and carbon were maintained below 50 ppm, while oxygen levels were below a few hundred ppm.

Sintering of low-chromium ferritic alloys in nitrogen-containing atmospheres promotes the formation of martensite upon cooling, resulting in low ductility, high hardness, and lower magnetic properties. High-temperature vacuum sintering (with a partial pressure of hydrogen) improves the mechanical and magnetic properties of ferritic stainless steels (Table 9) (Ref 11, 12).

Dimensional Shrinkage. Stainless steels exhibit greater shrinkage during sintering than do P/M irons. Figure 15 shows the dimensional change for type 304L sintered in dissociated ammonia, in which it can be seen that shrinkage increases with increasing temperature.

Sintering in either hydrogen or vacuum results in greater shrinkage. Sintering of low- and medium-density parts in hydrogen or under vacuum (particularly at the higher sintering temperatures and even more so for ferritic stainless steels) produces significant shrinkage, which in turn impairs the dimensional accuracy of a part. For this reason, such parts are often sized after sintering.

Table 7 Soft magnetic and mechanical properties of 410L stainless steel

Magnetic properties(a)											Sinter				Yield	Ultimate		
H_c, Oe	Maximum μ	B_m, 15 Oe, G	B_m, 25 Oe, G	B_m, 100 Oe, G	B_r, 15 Oe, G	B_r, 25 Oe, G	B_r, 100 Oe, G	Carbon, %	Oxygen, %	Nitrogen, %	density, g/cm^3	Sinter atmosphere	Time, min	Temperature, °F	stress, ksi	tensile strength, ksi	Elongation, %	Hardness, HRB
2	2200	10,400	10,900	13,000	9200	9400	10,500	0.0057	0.2100	0.0021	7.1	Vacuum	45	2300	30	51	9	53
6.8	...	1,500	2,800	8,100	1100	2100	6,300	0.0144	0.1800	0.1100	6.9	Hydrogen	45	2350	62	91.5	3	84
...	...	350	600	5,000	300	400	3,800	0.0667	0.2600	0.2800	6.5	75%H + 25%N	45	2280	...	...	...	96
48	50	400	700	4,500	...	...	3,500	0.0293	0.2703	0.4318	6.8	D.A.	45	2050	...	30	0	97
59	50	400	900	5,200	...	...	4,200	0.0424	0.2771	0.1360	6.9	D.A.	45	2200	...	52	0	100
42	90	1,150	2,350	7,450	...	...	5,800	0.0283	0.3000	0.1400	7.1	D.A.	45	2350	...	96	1	100
58	40	300	600	4,000	...	...	2,800	0.0462	0.2176	0.2475	6.8	Hydrogen	45	2050	...	17	0	95
63	60	400	800	5,600	...	...	4,200	0.0504	0.2546	0.2260	7.0	Hydrogen	45	2200	...	67	0	95
7.4	320	4,300	5,600	10,200	...	...	7,000	0.0094	0.2250	0.0808	7.1	Hydrogen	45	2350	42.5	68.5	4	74

D.A., dissociated ammonia. (a) B_m, maximum induction; B_r, residual induction; H_c, coercive force; m, permeability

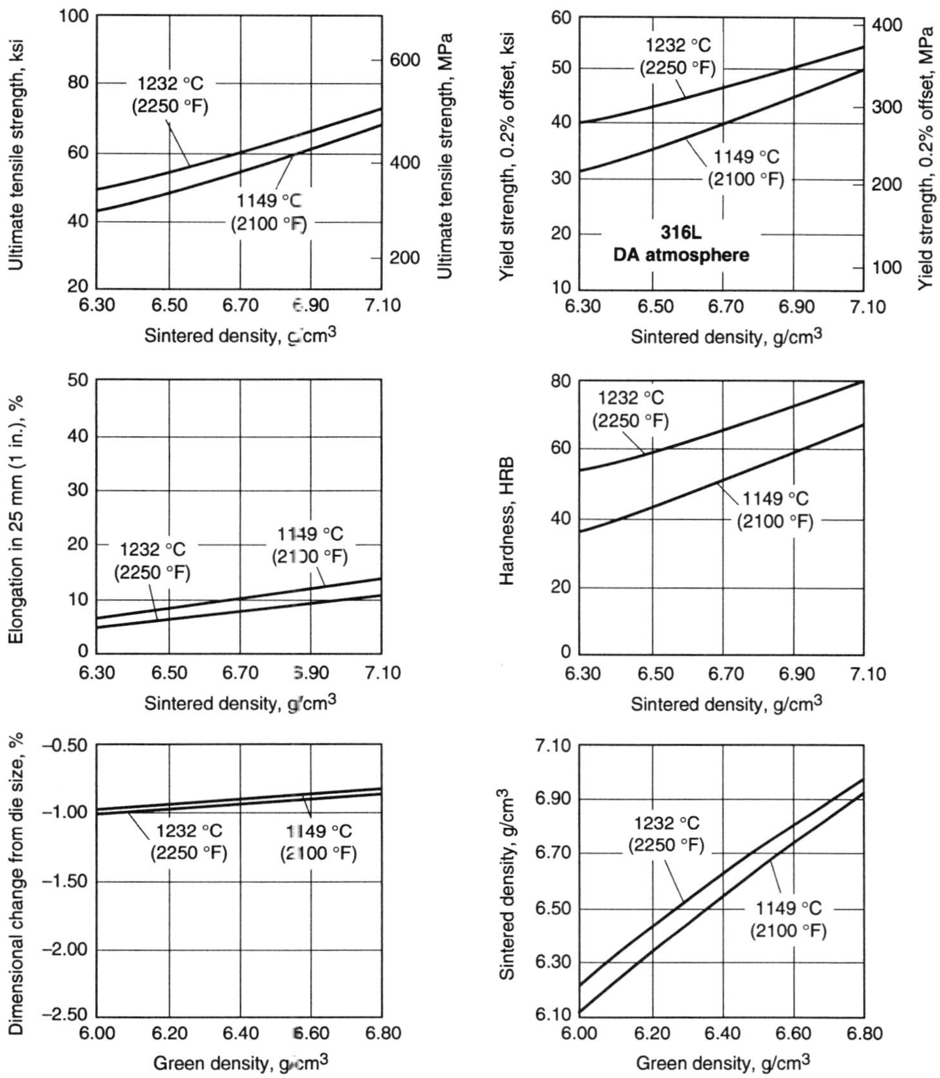

Fig. 13 Mechanical properties and densification behavior of sintered 316L as a function of sintering temperature and density. Sintered for 45 min in dissociated ammonia (DA) atmosphere with dew point of approximately –45 °C (–50 °F)

Machinability. With respect to machinability, sintered stainless steels are similar to wrought stainless steels. High-temperature sintering and sintering in hydrogen or vacuum result in improved machinability (Ref 13), whereas sintering in nitrogen-containing atmospheres impairs machinability on account of the greater hardness due to absorption of nitrogen. The addition of manganese sulfide to 304L has been used in the industry for improving the machinability. This approach, however, is at the expense of corrosion resistance. Tin-modified 303L has a good combination of machinability and corrosion resistance.

Elevated temperature properties of sintered stainless steels have recently become available in connection with the development of P/M flanges for automotive exhausts (Ref 14, 15). Figure 16 shows room and elevated temperature ultimate tensile strength, yield strength, and elongation results for three hydrogen-sintered ferritic stainless steels at 7.17 g/cm^3, compared with the same properties of the corresponding wrought steels. Figure 17 shows 676 °C (1250 °F) creep rupture strength data for two ferritic and two austenitic stainless steels at densities from 7.04 to 7.15 g/cm^3.

A systematic study of eight P/M stainless steels has also identified several key factors for successful application in automotive exhaust systems (Ref 14). A key factor is the importance of density in the specification of P/M stainless steel components. Low density alloys do not survive the cyclic elevated temperature oxidation/corrosion testing that simulates automotive exhausts. Testing indicates that a minimum density of 7.2 g/cm^3 for 400 series alloys is strongly recommended in thermal cyclic conditions for exhaust system components.

Alloy selection is also a key factor when specifying materials for cyclic applications involving thermal shock. In general, 400 series (ferritic) stainless steels at minimum 7.2 g/cm^3 density perform quite well in such cyclical applications. Austenitic 300 series alloys, on the other hand, are highly susceptible to cracking and failure in cyclical applications above 260 °C (500 °F).

Austenitic 300 series alloys maintain higher strength at elevated temperatures (Fig. 18) in comparison to 400 series alloys (Fig. 16a, b, c). However, ferritic grades have better thermal shock resistance than austenitic grades. This is due to the higher thermal conductivity of ferritic grades. The ductility of austenitic grades also decreases with temperature (Fig. 18c), while ductility of 400 series grades increases above ~500 °C (930 °F) (Fig. 16d).

Processing

All commercial compacting-grade stainless steel powders are produced by atomization. Water atomized powders (with irregular particle shapes) are typically used for conventional pressed-and-sintered parts, while gas atomized powders (with spherical shapes) are used for

Table 8 Soft magnetic and mechanical properties of 434L stainless steel

Magnetic properties(a)																			
H_c, Oe	Maximum μ	B_m, 15 Oe, G	B_m, 25 Oe, G	B_m, 100 Oe, G	B_r, 15 Oe, G	B_r, 25 Oe, G	B_r, 100 Oe, G	Carbon, %	Oxygen, %	Nitrogen, %	Sinter density, g/cm^3	Sinter atmosphere	Time, min	Temperature, °F	Yield stress, ksi	Ultimate tensile strength, ksi	Elongation, %	Hardness, HRB
2	1700	9500	10,100	12,000	8000	8400	9300	0.0068	0.2400	0.0037	7.0	Vacuum	45	2300	29	54	14	61
3.70	450	4800	6,100	9,700	3800	4400	6400	0.0369	0.2372	0.0600	6.9	Hydrogen	45	2350	...	...	...	...
7.60	100	1400	2,300	6,300	1100	1700	4600	0.0239	0.3000	0.2100	6.8	75%H + 25%N	45	2280	...	...	...	...
77	20	200	350	2,200	...	...	1600	0.0294	0.3118	0.6438	6.7	D.A.	45	2050	17	25	1	98
66	34	300	600	3,400	...	...	2500	0.0353	0.3071	0.1720	6.8	D.A.	45	2200	...	47.5	0	100
17	250	3700	4,950	8,700	...	...	5900	0.0090	0.3000	0.1200	7.0	D.A.	45	2350	47	69	3	80
9.2	90	1300	2,200	6,200	...	...	4500	0.0301	0.2774	0.2619	6.7	Hydrogen	45	2050	22	28	1	64
6.46	300	4000	5,200	8,500	...	...	5800	0.0108	0.2632	0.1566	6.9	Hydrogen	45	2200	24	29.5	1	63
2.76	450	5000	6,500	10,400	...	...	6700	0.0156	0.2489	0.1266	7.1	Hydrogen	45	2350	...	28	1	64

D.A., dissociated ammonia. (a) B_m, maximum induction; B_r, residual induction; H_c, coercive force; m, permeability. Source: Ref 14

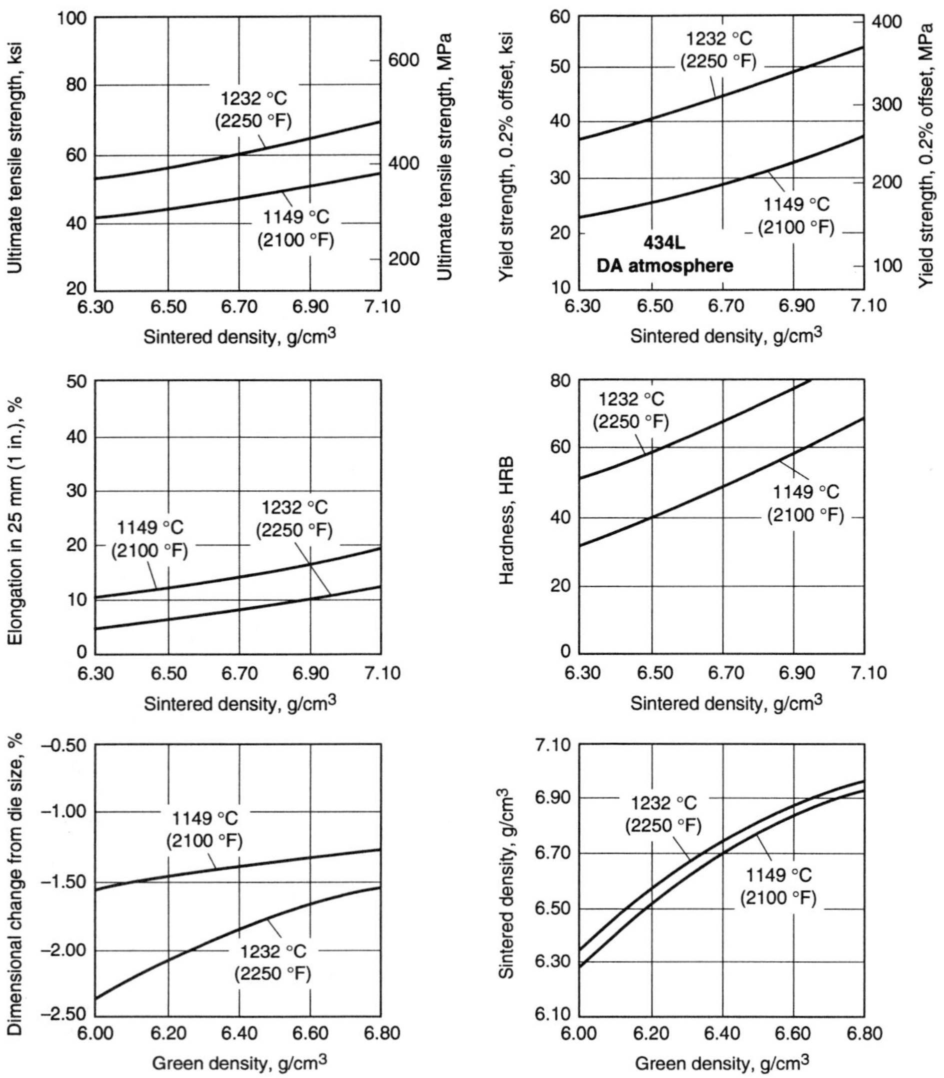

Fig. 14 Mechanical properties and densification behavior of dissociated ammonia (DA) sintered 434L as a function of sintering temperature and density. Sintered for 45 min, –45 °C dew point

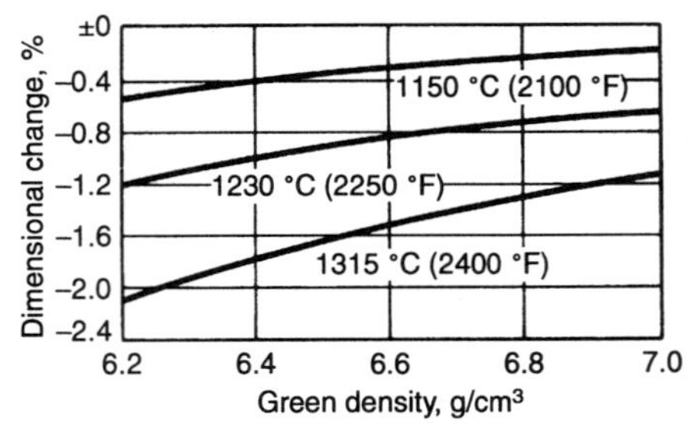

Fig. 15 Dimensional changes for type 304L stainless steel. Dimensional changes were determined on transverse-rupture bars sintered for 45 min in –40 °C (–40 °F) dew point dissociated ammonia and were calculated from die size.

Table 9 Magnetic and mechanical properties of two ferritic stainless steels vacuum sintered at 1260 °C (2300 °F) with a partial pressure of hydrogen

	410L	434L
Magnetic induction at an applied field of 15 Oe, G	10,500	9700
Residual induction, G	8,000	7700
Relative maximum permeability	1,900	1600
Coercive force, Oe	1.8	1.8
Sintered density, g/cm³	7.25	7.35
Ultimate tensile strength, ksi	53.2	58.3
Yield strength (0.2% offset), ksi	33.2	37.0
Elongation in 1 in., %	25	28
Apparent hardness, HRB	54	56
Residual elements, wt%		
C	0.001	0.009
S	0.004	0.009
N	0.007	0.015
O	0.063	0.012

Source: Ref 11

high-density consolidation. Table 10 lists chemical compositions, physical properties, and screen analysis of widely used powders.

Compaction. Design and processing of stainless steel P/M parts are subject to the same basic considerations as are other P/M materials. However, compared with low-alloy ferrous powders, stainless steel powders require higher compacting pressures and have lower green strength. Figure 19 shows typical compaction characteristics of austenitic and martensitic grades (Ref 6). Compaction pressures ranging from 550 to 830 MPa (40 to 60 tsi) are common in commercial practice. Carbide tooling is generally required.

The green strength of stainless steel compacts, which is about half that of P/M iron, is influenced by compaction pressure and the type of lubricant. Lubricants that provide high green strength, such as stearic acid (see Fig. 20), generally cause lower compactibility (Fig. 19). Therefore, lubricant selection is an important factor in determining successful application and fabrication of P/M stainless steels.

Stainless steel powders with lower carbon and nitrogen contents also improve compressibility and compactibility (green strength). In Japan, for example, over 1800 tonnes/year of stainless steel powder are used in P/M applications, of which 60% is 400 series powder without nickel. This volume of powder demand and the need for better mechanical properties has led to a 410L powder with lower carbon + nitrogen levels and better compactibility (Fig. 21). Powder annealing also improves green strength at similar carbon + nitrogen levels (Fig. 22) (Ref 16).

Sintering is the most important step in processing stainless steel P/M parts. Because carbon lowers corrosion resistance, the lubricant must be removed to prevent carbon diffusion into the part. As previously noted, medium-density stainless steels are sintered typically at ~1120 to 1150 °C (2050 to 2100 °F), except when improved mechanical properties and corrosion resistance are required. Higher sintering temperatures up to 1315 °C (2400 °F) are used for improved mechanical properties and corrosion resistance. More information on sintering of stainless steels is covered in the article "Production Sintering Practices" in this Volume.

Fully dense P/M stainless steels are made predominantly from spherically shaped, inert gas atomized stainless steel powders by extrusion, hot isostatic pressing, and metal injection molding (MIM). These techniques are often preferred for applications where high dynamic mechanical properties are important and/or where shape complexity (MIM) is a factor. A Swedish manufacturer of stainless steel uses substantial quantities of gas atomized stainless steel powder to make seamless tubing by extrusion. Stainless steel flanges, fittings, valve bodies, and pump components are made by hot isostatic pressing.

The mechanical properties, as well as corrosion resistance of parts produced by these techniques, are generally similar to those of their conventional wrought counterparts. In highly alloyed compositions, superior fatigue strength and superior corrosion resistance, particularly pitting resistance, have been reported and attributed to the more uniform microstructure, as well as to low sulfur content and spherical sulfide morphology (Ref 17) in comparison to ingot metallurgy (Ref 18).

High-density stainless steels can be produced by P/M forging, but high-temperature sintering

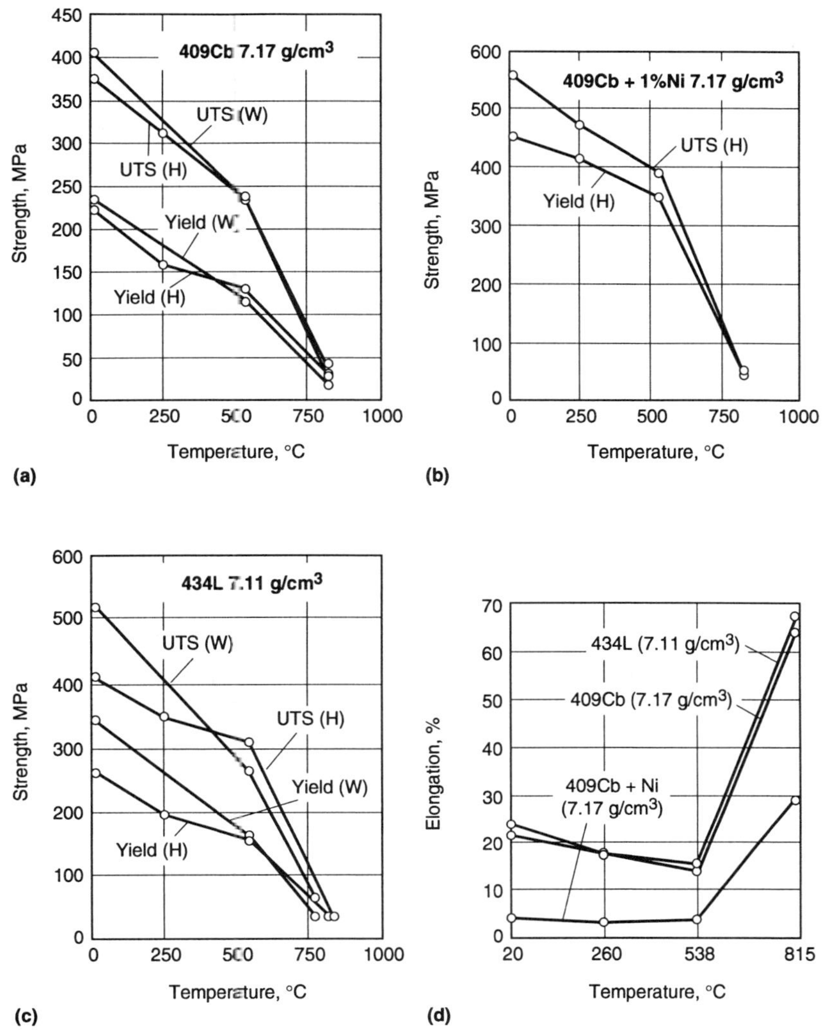

Fig. 16 Tensile properties of wrought (W) and hydrogen sintered (H) 409Cb and 434L stainless steels

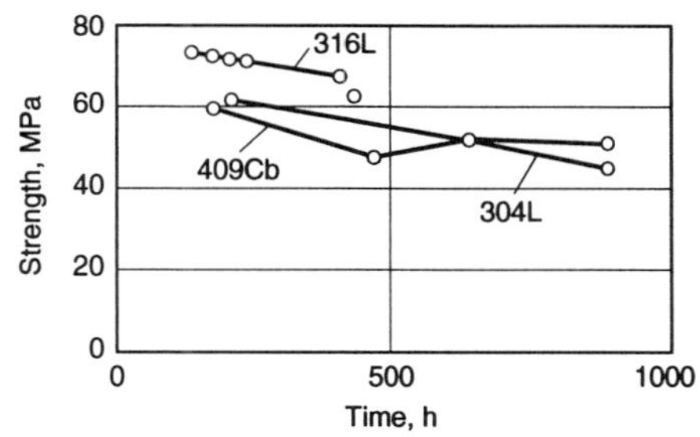

Fig. 17 Creep strength at 675 °C (1250 °F) of 304L, 316L, 434L, and 409Cb P/M stainless steels, hydrogen sintered

Table 10 Properties of stainless steel compacting-grade powders

	303L(a)	304L	316L	830(b)	410L	434L
Chemical analysis, %						
Chromium	17.5	18.5	16.5	20.5	12.0	17.0
Nickel	12.5	11.5	13.5	30.0	...	...
Molybdenum	...	...	2.1	2.5	...	1.0
Manganese	0.2	0.2	0.2	0.2	0.5	0.2
Silicon	0.7	0.8	0.7	1.0	0.8	0.8
Carbon	0.0[illegible]	0.02	0.02 (max)	0.02	0.02	0.02
Iron	bal	bal	bal	bal	bal	bal
Physical properties						
Apparent density, g/cm³	3.1	2.7	2.7	2.8	2.9	2.8
Flow rate, s/50 g	26	30	30	30	28	29
Screen analysis, %						
+100 mesh	1	1	1	1	1	1
−100 + 150 mesh	7	12	11	9	14	11
−150 + 200 mesh	13	20	18	14	20	17
−200 + 325 mesh	24	25	26	27	26	27
−325 mesh	55	42	44	49	39	44

Type 830 is used for the manufacture of P/M parts where superior resistance is of primary consideration. Parts made from type 830 exhibit improved resistance to oxidizing media and sulfuric acid. This grade is not recommended for conditions involving unstable chlorides. (a) 0.2% sulfur added for machinability. (b) 3.5% copper also present

is the preferred fabrication method. At high temperatures (1315 to1425 °C, or 2400 to 2600 °F), austenitic and ferritic stainless steels densify during sintering. With minor variations in composition, near theoretical density can be achieved in pressed and sintered parts. Table 11 (Ref 19) provides typical mechanical property data for nearly dense (>98% of theoretical density) stainless steels.

As noted, another form is tubing. Powder is packed in cans, sealed, cold isostatically pressed to reduce bulk, and extruded into tubing. The cold isostatically pressed cans can be hot isostatically pressed to full density, thus eliminating extrusion. Table 12 gives mechanical property data for extruded and hot isostatically pressed materials compared to wrought (ingot metallurgy) materials.

Metal Injection Molding. Stainless steels are the most common material used in MIM components, followed closely by iron-nickel steels. The MIM process is used to produce quantities of precision parts with complex shapes not conducive to production by the normal P/M method of pressing metal powder followed by sintering. The MIM process is also predictable because the green parts shrinks linearly upon binder removal and sintering. Shrinkage is ~20%, depending on the exact mixture used, and is precise from part to part. The standard part-to-part dimensional tolerance is ±0.003 in./in. The part surface is smooth and fine, thereby eliminating secondary machining operations in most cases.

Most MIM parts are required to have a density of 97 to 98% of theoretical to achieve the strength needed in their application. These high densities are made possible by using extremely fine spherical powders and by sintering at high temperature. Sintering can be incorporated directly onto a thermal, catalytic, or solvent debinding cycle. It is typically performed in a protective atmosphere or vacuum at a peak temperature that causes rapid elimination of the pores previously filled with binder. For steels and stainless steels, these temperatures are in the 1120 to 1350 °C (2050 to 2460 °F) range for times of 30 to 120 min, with linear shrinkages of 12 to 18% (Ref 20). The austenitic stainless steels such as Type 316, often specified for orthodontic braces and medical instruments, achieve highest density by sintering at 1350 °C (2460 °F).

As with traditional P/M, carbon control is critical in achieving good properties for MIM stainless steel parts. After molding, binders can add up to 5% C to the part, while the maximum allowed carbon in the final sintered part can be 0.03% or less for sintered stainless steel. Oxy-

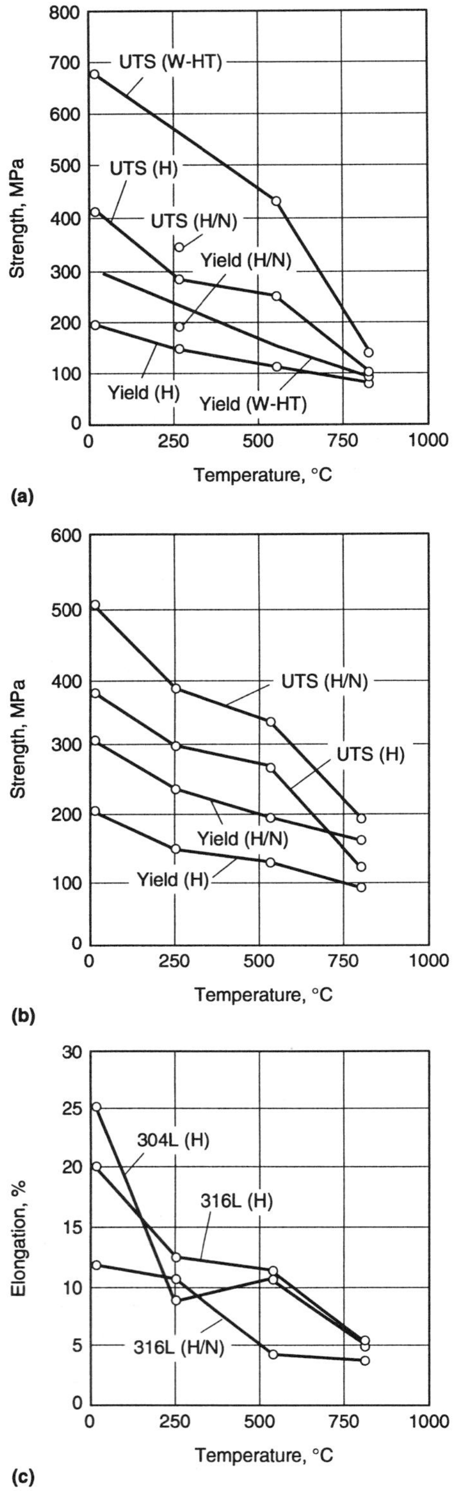

Fig. 18 Tensile properties of 304L and 316L austenitic stainless steels at high temperature. H, hydrogen sintered; H/N, hydrogen/nitrogen sinter (75% H_2/25% N); W-HT, wrought heat treated

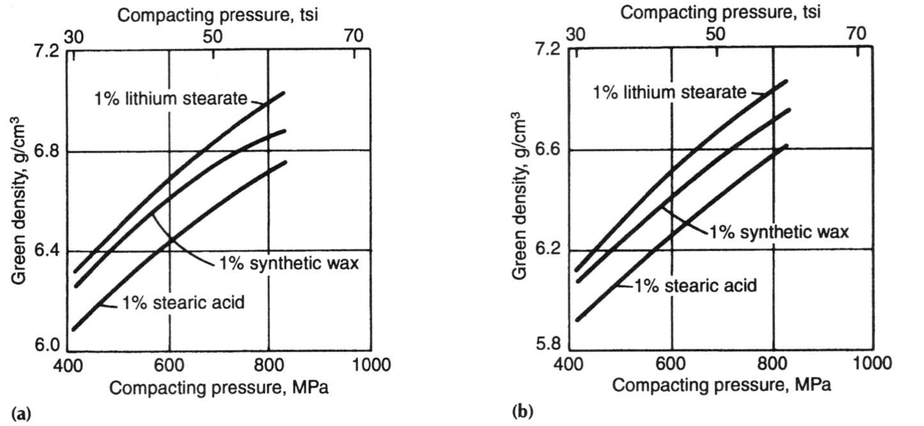

Fig. 19 Compactibility of stainless steel powders. (a) Type 316L austenitic stainless steel. (b) Type 410L martensitic stainless steel

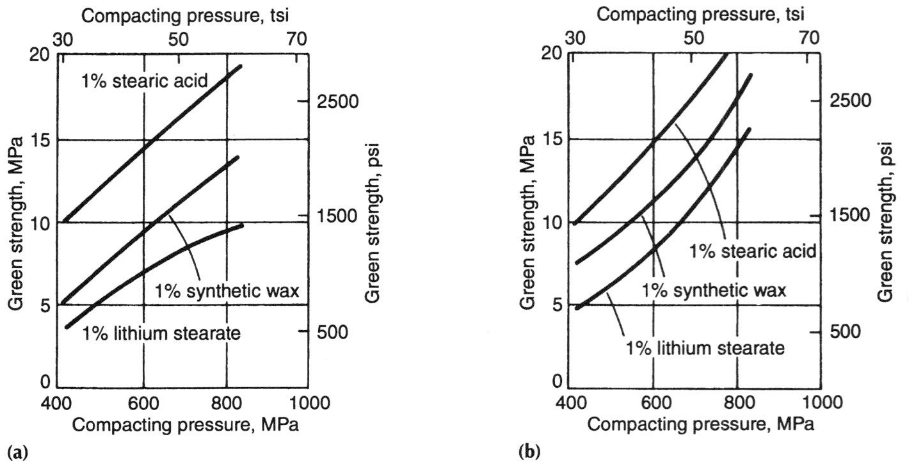

Fig. 20 Effect of lubricant and compacting pressure on green strength of stainless steel powder. (a) Type 316L austenitic stainless steel. (b) Type 410L martensitic stainless steel

gen control is also critical during debinding and subsequent sintering.

During binder extraction (at ~500 °C, or 930 °F), oxygen assists in carbon extraction through formation of CO and CO_2. However, the grain surfaces can become oxidized by the binder. These oxides are reduced again at sinter temperatures by carbon diffusing to the surface oxide. The oxide can be reduced by hydrogen at a temperature of 400 °C (750 °F). This temperature is lower than that needed for rapid reaction with carbon. Retention of nearly all the carbon can be achieved by introducing small quantities of hydrogen at 400 °C (750 °F).

REFERENCES

1. M. Hanada, N. Amano, Y. Takeda, Y. Saegusa, and T. Koiso, "Development of a Powder Metallurgy Sensor Ring for Use in an Antilock Brake System," SAE Technical Paper Series, Paper #890407, Int. Congress and Exhibition (Detroit, MI), 27 Feb to 3 March 1989
2. S. Shah, P.K. Samal, and E. Klar, "Properties of 410L P/M Stainless Steel Antilock Brake Sensor Rings," SAE Technical Paper Series, Paper #930449, Int. Congress and Exposition (Detroit, MI), 1–5 March 1993
3. S. Shah, J.R. McMillen, P.K. Samal, S.A. Nasser, and E. Klar, "On the Real Life Performance of Sintered Stainless Steel ABS Sensor Rings," SAE Technical Paper Series, Paper #970423, Int. Congress and Exposition (Detroit, MI), 24–27 Feb 1997
4. P.K. Johnson, Award-Winning Parts Show Broad Spectrum of Powder Metallurgy Applications, *Modern Developments in Powder Metallurgy,* Vol 20, 1988, p 3–12
5. P.K. Johnson, 1987 P/M Part-of-the-Year Design Competition Winners, *Prog. Powder Metall.,* Vol 43, 1987, p 915–924
6. R.W. Stevenson, P/M Stainless Steels, *Powder Metallurgy,* Vol 7, *Metals Handbook,* 9th ed., American Society for Metals, 1985, p 729
7. M. Svilar and H.D. Ambs, P/M Martensitic Stainless Steels: Processing and Properties, *Adv. Powder Metall.,* Vol 2, 1990, p 259–272
8. T. Tunberg, L. Nyborg, and C.X. Liu, Enhanced Vacuum Sintering of Water-Atomized Stainless Steel Powder by Carbon Addition, *Advances in Powder Metallurgy and Particulate Materials,* Vol 3, Metal Powder Industries Federation, 1992, p 383–396
9. R.M. Larsen and K.A. Thorsen, Removal of Oxygen and Carbon During Sintering of

Table 11 Typical mechanical properties of nearly dense P/M stainless steel

Based on high-temperature sintering.

Alloy	Condition	Ultimate tensile strength, MPa	Ultimate tensile strength, ksi	0.2% yield strength, MPa	0.2% yield strength, ksi	Elongation in 25 mm (1 in.), %	Hardness	Impact strength, J	Impact strength, ft · lb	Density, g/cm³	Theoretical density, g/cm³
Ultimet 04, 304	Sintered	593	86	248	36	36	80 HRB	10.8(a)	8(a)	7.8	7.9
Ultimet 16, 316	Sintered	687	99.6	308	44.7	26	94 HRB	8.1(a)	6(a)	7.7	7.8
	Solution treated and quenched	684	99.3	329	47.7	45	90 HRB	5.4(a)	40(a)	7.7	7.8
Ultimet 40C, 440C	Sintered	...	...	...	...	...	20–30 HRC	2.7(b)	2(b)	7.6	7.7
	Hardened and tempered	...	...	...	...	...	50–60 HRC	2.7(b)	2(b)	7.6	7.7

(a) Charpy V-notch. (b) Unnotched. Source: Ref 19

Table 12 Typical mechanical properties of fully dense stainless steel

Property	P/M material	Wrought material
Extruded 0.3 by 15.5 mm (0.1 by 0.61 in.) 317LM tube(a)		
Ultimate tensile strength, MPa (ksi)	693 (100)	693 (100)
0.2% yield strength, MPa (ksi)	324 (47)	353 (51)
Reduction in area, %	71	73
Elongation in 25 mm (1 in.), %	47	50
Hot isostatically pressed type 316		
Ultimate tensile strength, MPa (ksi)	579 (84)	...
0.2% yield strength, MPa (ksi)	288 (42)	...
Elongation in 25 mm (1 in.), %	58	...

(a) Gas-atomized powder, canned, cold isostatically pressed, and extruded. Source: Ref 19

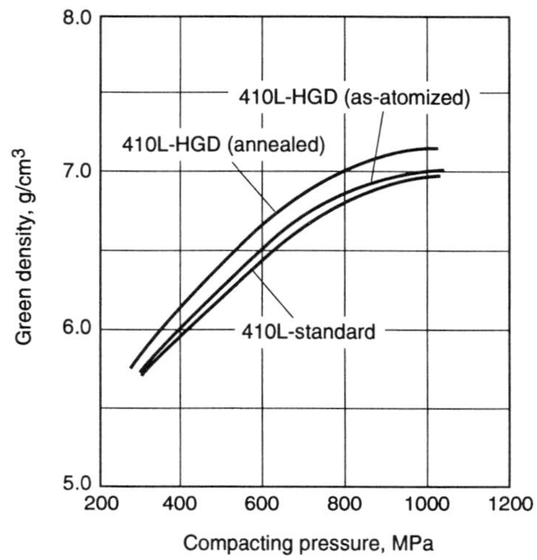

Fig. 21 The green densities of 410L mixed with 1% zinc stearate. HGD, high green density grade with lowered carbon and nitrogen levels

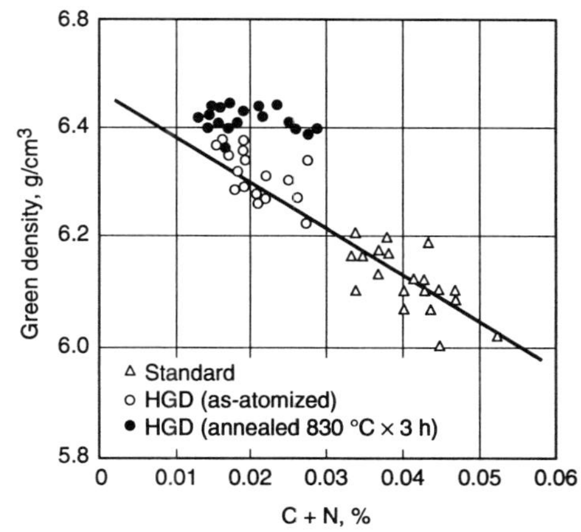

Fig. 22 Effect of annealing on the green density of 410L-HGD mixed with 1% zinc stearate, compacted at 490 MPa

Austenitic Stainless Steels, *Proc. of 1993 Powder Metallurgy World Congress* (Kyoto, Japan), Part 1, Japan Soc. of Powder and PM, 1993, p 377–380

10. C. Lall, Effect of Sintering Temperature and Atmosphere on Soft Magnetic Properties of P/M Materials, *Advances in Powder Metallurgy and Particulate Materials,* Vol 3, Metal Powder Industries Federation, 1992, p 129–156
11. K. Moyer, Selection of Powders and Processing for P/M Soft Magnetic Components, *Adv. Powder Metall.,* 1990, Vol 2, p 385–399
12. C. Lall, Fundamentals of High Temperature Sintering: Application to Stainless Steels and Soft Magnetic Alloys, *Int. J. Powder Metall.,* Vol 27 (No. 4), 1991, p 315–329
13. H.D. Ambs, Machinability Studies on Sintered Stainless Steel, *Adv. Powder Metall.,* Vol 3, 1991, p 89–100
14. T. Hubbard, K. Couchman, and C. Lail, "Performance of Stainless Steel P/M Materials in Elevated Temperature Applications," SAE Technical Paper Series, Paper No. 980326, Int. Congress and Exposition (Detroit, MI), 24–27 Feb 1997, Society of Automotive Engineers, 1998
15. T.R. Albee, P. DePoutiloff, G.L. Ramsey, and G.E. Regan, "Enhanced Powder Metal Materials for Exhaust System Applications," SAE Technical Paper Series, Paper No. 970281, Int. Congress and Exposition (Detroit, MI), 24–27 Feb 1997
16. Y. Okura and T. Kono, Developing Stainless Steel Powders to Meet Market Demands, *Met. Powder Rep.,* March 1993, p 47
17. G.S. Yurek, D. Eisen, and A.J. Garrat-Reed, *Metall. Trans. A,* Vol 13, 1982, p 473
18. C. Tornberg, "The Manufacture of Seamless Stainless Steel Tubes from Powder," Paper 8410-013, presented at the 1984 ASME Int. Conf. New Developments in Stainless Steel Technology (Detroit, MI), American Society of Mechanical Engineers, 1984, p 1–6
19. L. Pease III, Mechanical Properties of P/M Materials, *Powder Metallurgy,* Vol 7, *Metals Handbook,* 9th ed., American Society for Metals, 1985
20. R. German, *Powder Metallurgy of Iron and Steel,* John Wiley & Sons, 1998, p 107

Particle Metallurgy Tool Steels

R.B. Dixon, W. Stasko, and K.E. Pinnow, Crucible Research Center, Crucible Materials Corporation

PARTICLE METALLURGY processing of tool steels was first introduced commercially by the Crucible Materials Corporation in 1970, and it is currently the primary manufacturing method used for the development and production of high-performance high-alloy tool steels and tooling composites. The term *particle metallurgy* terminology applies exclusively to powder metallurgy (P/M) processes where gas-atomized and prealloyed powders are consolidated to full density by hot isostatic pressing (HIP). Particle metallurgy is used to distinguish such processing from traditional powder metallurgy techniques that typically employ mechanical pressing and high-temperature sintering to achieve varying degrees of densification in relatively small net-shape parts.

Particle metallurgy tool steels are produced and distributed internationally by a number of specialty steel manufacturers using various tradenames. Table 1 lists several of the major suppliers of these materials and the corresponding product tradename prefixes generally recognized in the marketplace. Some of the PM tool steel compositions are patented and are only available from a single supplier. Others not so protected are available from multiple producers or distributors often using registered tradenames. To avoid confusion concerning the use of multiple tradenames for similar PM tool steel compositions, an attempt is made in the balance of this article to use comparable AISI grade designations or nominal alloy descriptions wherever possible. Additional information on tool steel powders is also provided in the article "Production of Steel Powders" in this Volume.

Table 1 International suppliers of PM high-alloy tool steels

Company	Location	Tradename prefixes
Crucible Materials Corporation	United States	CPM
Carpenter Technology Corporation	United States	Micromelt PM
Bohler/Uddeholm Corporation	Austria	Vanadis/Isomatrix PM
Erasteel Kloster AB	Sweden	ASP
Bodycote Powdermet AB	Sweden	APM
Hitachi Metals	Japan	HAP
Daido Steel	Japan	DEX
Nachi Fujikoshi	Japan	FAX
Kobe Steel	Japan	KHA

Particle Metallurgy Process and Product Forms

As shown in Fig. 1, the basic PM process consists of induction melting of a prealloyed tool steel composition (with or without secondary refining), gas atomization to produce a rapidly solidified spherical powder, screening and encapsulation of the powder, and consolidation by HIP. There are a few select applications where it may be practical or technically necessary to use as-HIP product forms directly (e.g., large diameter hollows or composite bars). However, the more common practice is to HIP a large cylindrical "compact" that is then hot worked to round or rectangular billets for subsequent hot rolling and finishing to round bar, flat bar, or plate and sheet products; that is, the same product forms in which conventionally produced tool steels are stocked for resale to tooling or equipment component manufacturers. Starting compact sizes typically range from 36 to 51 cm (14 to 20 in.) in diameter by 152 to 254 cm (60 to 100 in.) in length, and they weigh approximately 909 to 2727 kg (2000 to 6000 lb). Consolidation by hot extrusion or other forging techniques is technically feasible, but generally not practical due to size limitations. It is also possible to produce PM tool steel near-net shapes using HIP technology, but it is usually more economical to machine directly from the solid. For special applications requiring unique material property combinations, or where it is not practical to fabricate the entire component from the PM tool steel, composites can be supplied by HIP-cladding a high-alloy PM tooling material to a technically different or less expensive substrate material.

Characteristics of Particle Metallurgy Microstructures

The distinguishing characteristic of wrought PM high-alloy tool steels compared to conventionally ingot cast and wrought materials of similar compositions is the uniform distribution and small size of the primary carbides that form in the PM tool steels during rapid solidification

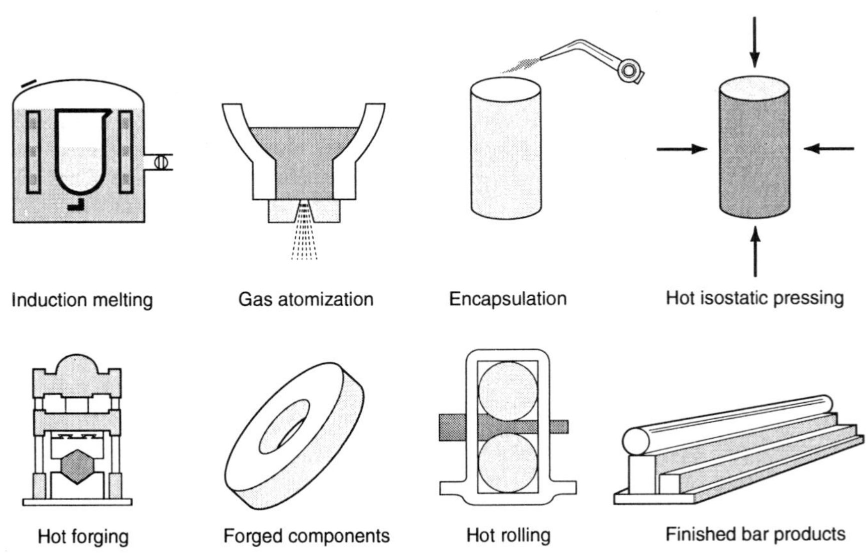

Fig. 1 Particle metallurgy processing schematic

of the powder particles during atomization from the melt. Figure 2 shows the primary carbide distributions observed in the heat treated microstructures of 51 mm (2 in.) diam wrought PM and conventional T15 high-speed steel (1.5C-12W-5V-5Co), where the primary carbides are a combination of vanadium-rich MC and tungsten-rich M_6C types. The PM microstructure shown would appear essentially the same for all product cross sections, while larger cross sections of high-alloy conventional tool steels exhibit inherently coarser carbides and a more pronounced carbide network due to slow ingot solidification and insufficient hot working to break up the as-cast structure.

For the same reasons that the primary carbides are uniformly distributed and very small in PM tool steels, so too are any sulfides that form during solidification. Examples of the sulfides observed in polished longitudinal cross sections of conventional and PM M35 high-speed steel containing about 0.10 and 0.22% S, respectively, are shown in Fig. 3. The ability to resulfurize to very high levels without detrimentally affecting hot workability or transverse toughness properties is a distinct advantage for PM processing, and it is done routinely for PM tool steel applications requiring good machinability.

Grindability and Toughness Improvements

The microstructural refinements achieved through PM processing of high-alloy tool steels provide potential benefits to both the tooling (or component) manufacturer and to the end user of the finished product. The two most important physical property and fabricating characteristics that have set high-alloy PM tool steels apart from conventional high-alloy tool steels are (a) good grindability in the heat treated condition and (b) excellent toughness properties in service. Although these characteristics apply to all PM tool steels, they are particularly evident in the wear-resistant high-vanadium tool steel compositions where conventional ingot metallurgy results in heavily segregated microstructures containing large, irregularly shaped, vanadium-rich MC primary carbides.

Figure 4 schematically illustrates why the PM tool steels exhibit improved grindability compared to conventionally produced tool steels with the same composition. In conventionally produced T15 high-speed steel, the large vanadium-rich MC carbides present in the heat treated microstructure are harder than the Al_2O_3 grains in the grinding wheel. These carbides, along with any other large carbides present, significantly impede the cutting action of the grinding wheel. The uniformly small carbide size inherent in PM T15 high-speed steel results in significantly less resistance to the cutting action of the much larger Al_2O_3 grains.

For tooling that is subjected to extremely abrasive wear conditions in service, it can be argued correctly that poor grindability characteristics would be beneficial for abrasive wear resistance. However, in most tooling applications the service conditions are typically a combination of sliding (adhesive) and mildly abrasive wear. Under these conditions, the PM tool steels have exhibited wear resistance at least comparable to conventional tool steels having the same composition and heat treated hardness. Figure 5 shows laboratory crossed-cylinder wear test results for M2, M4, and T15 high-speed steels, which contain increasing amounts of the MC carbide and therefore increasing wear resistance. As shown, the PM tool steels have at least comparable metal-to-metal wear resistance compared to the conventional steels, despite their excellent grindability.

Good toughness characteristics are critical to the successful application of high-alloy tool steels in many industrial operations. Low toughness or strength can cause catastrophic tool failures, or it can have a more subtle negative effect on cutting edge retention due to fine microchipping. To illustrate the potential benefit of PM processing for edge retention, Fig. 6 compares high-magnification scanning electron microscopy (SEM) photos of the edge "wear" experienced by conventional and PM high-speed steels in a machining operation. The relatively uniform

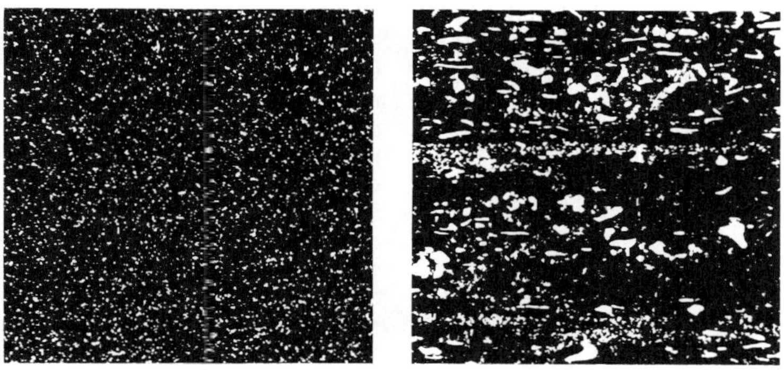

Fig. 2 Primary carbides in heat treated PM T15 (left) and conventional T15 (right) high-speed steels. 500×

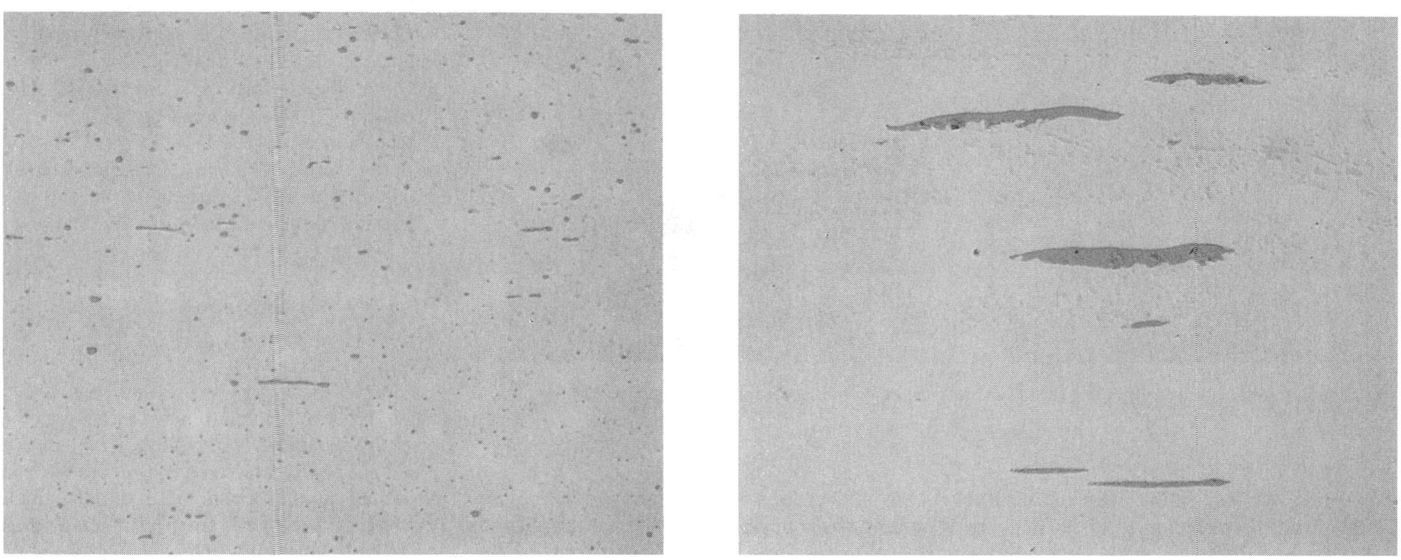

Fig. 3 Typical longitudinal sulfide distributions in PM (left, 0.22% S) and conventional (right, 0.10% S) high-speed steels. 500×

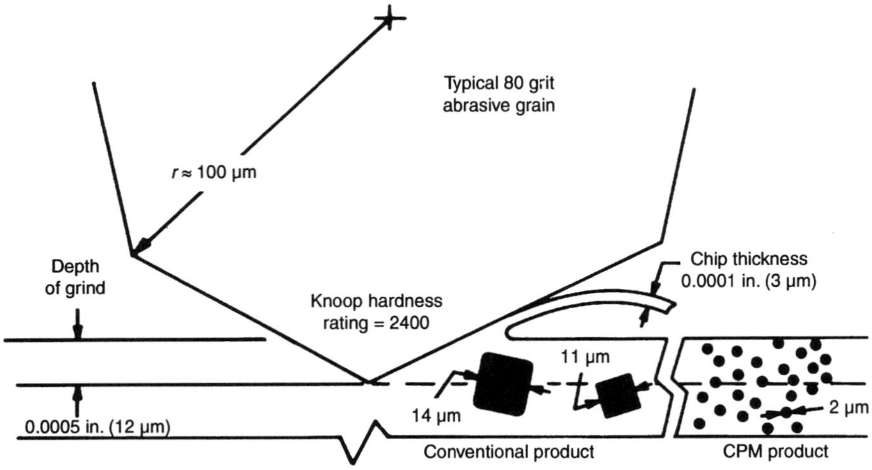

Fig. 4 Beneficial effect of fine carbide size on the grindability of PM tool steels

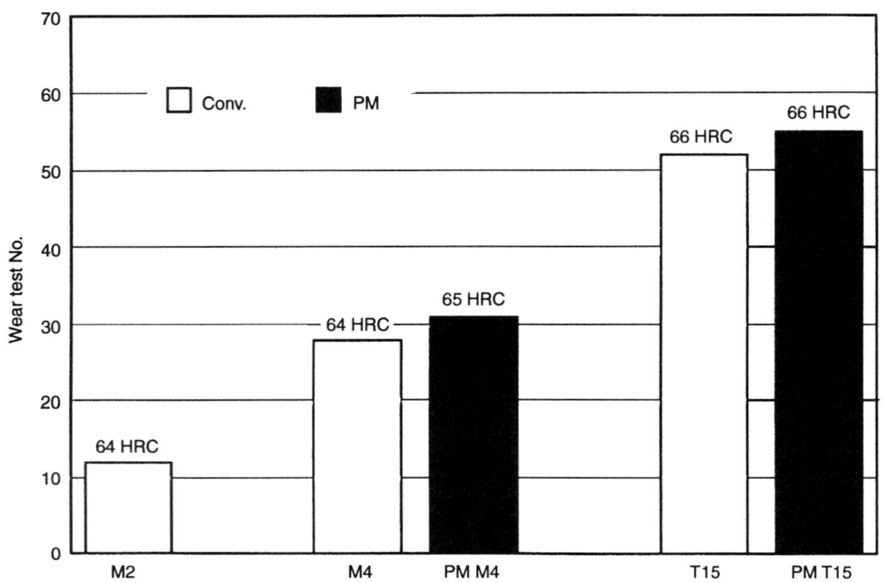

Fig. 5 Crossed-cylinder wear test results for M2, M4, and T15 high-speed steels

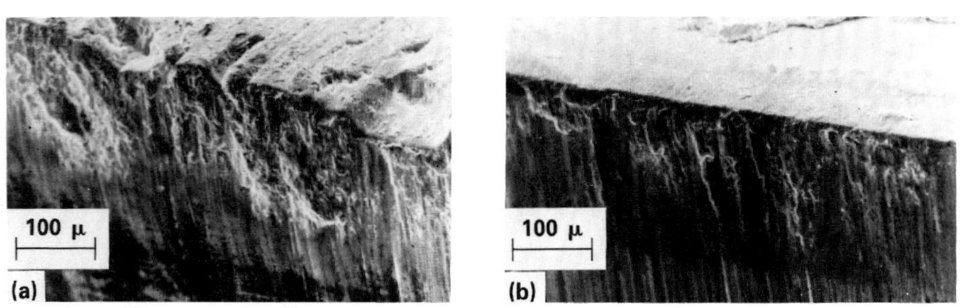

Fig. 6 SEM photos showing microchipping versus uniform wear at the cutting edges of (a) conventional M2 and (b) PM M3

wear observed on the cutting edge of the PM tool steel would be expected to result in more consistent tool performance and a better surface finish on the part. These edge retention benefits apply equally to high-speed machining and cold-work cutting operations and are particularly important when upgrading to higher-alloyed tool steel compositions is being considered to improve tool life. Table 2 lists longitudinal bend fracture strength and impact toughness test results for M2, M4, and T15 high-speed steels, which contain increasing amounts of vanadium-rich primary carbides for increased wear resistance. The data clearly show the toughness advantage of the PM steels for a given alloy composition and also indicate that the toughness penalty generally associated with upgrading to the higher-vanadium high-speed steel compositions can be avoided by using a PM tool steel. The better toughness of the PM tool steels is attributed to elimination of alloy segregation, carbide banding, coarse carbide size, and carbide angularity that are inherent in ingot-cast high-alloy tool steels.

Table 2 Longitudinal toughness comparison for M2, M4, and T15

Material	Hardness, HRC	Charpy C-notch impact, J (ft · lbf)	Bend fracture strength, MPa (ksi)
Conventional M2	65	18 (13)	3819 (554)
Conventional M4	64	14 (10)	3585 (520)
CPM M4	65	33 (24)	5377 (780)
Conventional T15	66	5 (4)	2151 (312)
CPM T15	67	19 (14)	466 (676)

Additional Processing Benefits

In addition to improving the grindability and toughness of high-alloy tool steels, PM processing results in other benefits that very often enable the tool manufacturer to economically supply higher-quality tools that are capable of more consistent tool performance. These benefits include:

- Better machinability in the annealed condition (enhanced with high sulfur additions)
- Consistent and more uniform heat treat response in larger cross sections
- Predictable size change (important for accurate unground tools as well as for high-temperature coating operations such as chemical vapor deposition and thermal diffusion)
- Greater capability to finish form grind or hard turn directly from solid
- More efficient and consistent electrodischarge machine (EDM) processing, with better surface finishes
- Better substrate quality for coating (plasma vapor deposition, chemical vapor deposition, thermal diffusion, etc.), nitriding, and plating

Particle Metallurgy Tool Steel Development

Particle metallurgy processing has also enabled the development of new and advanced tool steel compositions with unique properties that could not be manufactured economically by conventional ingot metallurgy. Examples include the following: super-high-speed steels with 30% or more total alloy content, cobalt-free super-high-speed steels, highly resulfurized tool steels for enhanced machinability, wear-resistant cold-work steels containing 3 to 18% V, wear/corrosion-resistant tool steels containing a combination of high-chromium and high-vanadium, and hot (warm) work tool steels for specialized applications. Several of these materials are discussed in later sections of this article.

Industrial Applications for Particle Metallurgy Tool Steels

Particle metallurgy tool steels are used in a wide range of applications within the traditional tool-and-die industry, as well as for equipment components and other nontooling wear parts. The general application categories for PM tool steels can be separated as follows:

- High-speed cutting and machining operations
- Cold working
- Processing and molding of engineering plastics
- Hot (warm) working and die casting
- Precision wear parts (nontooling)

The selection of a particular PM tool steel for a given application depends on a number of factors: the service life required, the operating conditions anticipated, compatibility with coating operations, material availability, cost, and so forth. The following sections describe the properties of the different classifications of PM tool steels and the types of applications where they are used.

Particle Metallurgy High-Speed Steels

Particle metallurgy tool steels used in the cutting tool industry are high-speed steel compositions capable of both achieving high room-temperature hardness for wear resistance (at least 64 HRC after heat treatment) and maintaining high hardness when exposed to the frictional heating encountered at the tool/workpiece interface during the cutting operation (commonly referred to as "red hardness").

The relative performance of the PM high-speed steels is strongly influenced by the alloy composition and heat treatment. The principal alloying elements found in varying amounts in high-speed steels are carbon, tungsten, molybdenum, vanadium, and cobalt. All high-speed steels also contain about 4% Cr to provide good hardenability in larger cross sections during heat treatment. Carbon is necessary for heat treat response, and it combines with the carbide-forming elements to form wear-resistant primary carbides. Tungsten and/or molybdenum are essential alloying elements for developing sufficient temper resistance to enable these materials to perform at "high speed" in cutting operations. Both elements form wear-resistant M_6C primary carbides that partially dissolve during heat treatment to provide a strong precipitation or "secondary" hardening response on tempering. In high molybdenum compositions (e.g., containing 9 to 10% Mo), it is also possible to form some M_2C carbides. Recognizing that tungsten and molybdenum can often be used interchangeably for alloying on an atomic basis and that their atomic weights differ by a factor of two, the combined effects of tungsten and molybdenum are often compared by calculating a "tungsten equivalency" (W_{eq}) equal to the tungsten content plus twice the molybdenum content (W + 2 Mo). Vanadium also contributes to secondary hardening during heat treatment, but its main function is to form primary MC carbides for wear resistance. Niobium would be expected to have an effect similar to that of vanadium, but it is not a commonly used alloying element for high-speed steels. Cobalt does not form carbides in tool steels, but significantly contributes to the hot hardness and tempering resistance of the matrix in compositions containing 5 to 8% Co and higher.

Table 3 lists a number of the commonly recognized PM high-speed steel compositions that are currently produced internationally. The table also includes the calculated W_{eq} and the attainable hardness capability for each alloy. Some of the compositions listed are PM modifications of traditional high-speed-steel grades (e.g., M3, M4, and T15) that are also produced conventionally. Notably absent from this table are the 1 to 2% V commodity high-speed steels such as M1, M2, M7, and M42. Although several of the latter materials have been periodically produced in PM form for specialized applications, the general trend in the cutting tool industry has been to take advantage of the grindability and toughness properties resulting from PM processing to upgrade to higher-performance higher-vanadium alloy compositions. Thus, the base alloy compositions listed are PM M3 and PM M4, which are essentially high-carbon and high-vanadium modifications of M2. With the exception of PM M35 (M2 + 5 Co), the balance of the compositions listed are generally classified as "super-high-speed" steels, which have high attainable hardness capability (66 HRC minimum) and excellent temper resistance for a combination of good wear resistance and high red hardness properties. The selection of a particular PM high-speed steel for a cutting or machining operation depends on the tool life requirements, the operating conditions (high wear, high frictional heating, or a combination of both), and whether or not the tool is to be coated (TiN, TiCN, etc.).

The wear resistance of a PM high-speed steel is determined by the heat treated hardness and by the amount and type (hardness) of primary carbide present in the heat treated microstructure. As discussed earlier, all of the PM

Table 3 Nominal compositions of PM high-speed steels

Tradename(s)	Designation				Composition, wt%					W_{eq}	Hardness, HRC
	AISI	UNS	JIS	Werk. No.	C	W	Mo	V	Co		
Wear-resistant high-speed steels containing 3 to 4% V											
ASP23, APM 23, CPM M3, Micromelt M3, FAX 31, DEX 20, KHA 32	M3	T11323	SKH53	1.3344	1.3	6.25	5	3	...	16.25	65–67
CPM M4, Micromelt M4, Isomatrix S690, HAP M4	M4	T11304	SKH54	...	1.4	5.75	5	4	...	15.75	65–67
Heat-resistant and super-high-speed steels containing 5 to 12% Co and 2 to 6.5% V											
CPM M35	M35	...	SKH 55	1.3243	1	6	5	2	5	16	65–67
CPM Rex 54	...	...	...	...	1.5	5.75	5	4	5	15.75	66–68
ASP30, APM30, CPM Rex 45, Micromelt HS 30, Isomatrix S790, FAX 38, DEX 40, HAP 40, KHA 30	...	...	...	...	1.3	6.25	5	3	8	16.25	66–68
CPM T15, Micromelt T15, FAX 55, DEX 61, HAP T15, KHA 50	T15	T12015	SKH10	1.3202	1.6	12	...	5	5	12	66–68
CPM Rex 76, Micromelt HS 76	M48	T11348	...	...	1.5	10	5.25	3	8.5	20.5	67–69
HAP 50, DEX 62	...	...	...	...	1.5	8	6	4	8	20	67–69
Isomatrix S390	...	...	...	...	1.6	11	2	5	8	15	66–68
ASP60, APM60, KHA 60	...	...	...	1.3241	2.3	6.5	7	6.5	10.5	20.5	67–69
DEX 80	...	...	...	...	2.1	14	6	5.5	12	26	68–70
HAP 70	...	...	...	...	2.2	12	9	5	12	30	69–71
Cobalt-free super-high-speed steels											
CPM Rex 20	M62	T11362	...	...	1.3	6.25	10.5	2	...	27.25	66–68
CPM Rex 25	M61	T11361	...	...	1.8	12.5	6.5	5	...	25.5	67–69

Note: All of the PM high-speed steels contain about 4% Cr for hardenability in large sections. Silicon, manganese, and sulfur contents are typically 0.50%, 0.30%, and 0.03% maximum, respectively. For select applications requiring improved machinability, sulfur contents are increased to 0.10 or 0.22% with corresponding increases in the manganese contents.

Table 4 Microhardness of primary carbides in PM tool steels

Carbide type	Knoop hardness
M(V,Nb)C	2200–2800
$M(W,Mo)_6C$	1550–1750
$M(Cr)_7C_3$	1300–1600

high-speed steels contain significant amounts of tungsten and/or molybdenum as well as vanadium, which combine with carbon to form tungsten or molybdenum-rich M_6C (occasionally some M_2C) and vanadium-rich MC primary carbides, respectively. The approximate microhardnesses of these complex carbides, as well as the chromium-rich M_7C_3-type primary carbide found in some cold-work die steels, which are discussed later, are given in Table 4. It follows that for a given heat treated hardness and a similar total primary carbide volume fraction, a high-speed steel with a higher percentage of the harder vanadium-rich (or niobium-rich) MC carbide exhibits better wear resistance. It is therefore very often possible to rank the PM high-speed steels in Table 3 in order of increasing wear resistance by simply looking at the relative vanadium contents. Thus, the PM high speed steels with 4% V generally outwear PM or conventional high-speed steels with 2 to 3% V, and steels with 5 to 6% V generally outwear the materials with 4% V. Exceptions to this vanadium content "rule of thumb" are special cases where the alloying results in a significant increase in the total volume fraction of primary carbides as well as the attainable hardness capability. An example of the latter is PM M48 (1.5C-10W-5Mo-3V-8.5Co), which contains about 33% total alloy content. This material is capable of attaining 67 to 69 HRC and also has a significantly greater volume fraction of primary carbides than the lower alloyed PM M3, PM M4, and PM M3 + 8Co grades. Other exceptions to this rule are the cobalt-free PM super high-speed steels, which by design also have significantly increased volume fractions of the M_6C primary carbide compared to the cobalt-bearing materials they were designed to replace. These materials are discussed in a later section.

The "red hardness" characteristic of a high-speed steel is dependent on the effect of alloying on the initial attainable hardness as well as on resistance to softening as the tempering or exposure temperature is increased. A measure of the relative red hardness capability is laboratory hot hardness, which is shown in Fig. 7 for a number of the PM high-speed steels. Note that the materials that exhibit higher initial attainable hardness also exhibit higher hot hardness as the testing temperature is increased. These same materials also exhibit greater resistance to tempering during heat treatment.

A unique class of materials are the cobalt-free PM super-high-speed steels that were first developed by the Crucible Materials Corporation in the late 1970s in direct response to a serious curtailment in the worldwide availability of cobalt raw materials. PM M61 (1.8C-12.5W-6.5Mo-5V) and PM M62 (1.3C-6.25W-10.5Mo-2V) were simultaneously designed as cobalt-free substitutes for T15 (1.5C-12W-5V-5Co) and M42 (1.1C-1.5W-9.5Mo-1V-8Co), which have historically been two of the more important cobalt-bearing cutting tool materials used in the aerospace industry and elsewhere. Replacing the solid-solution effects of cobalt on heat treating response and elevated-temperature properties was accomplished primarily by alloying with tungsten and/or molybdenum to form a significantly greater number of alloy carbides in the annealed microstructure, which partially dissolve during austenitizing for heat treatment to provide a strong secondary or precipitation hardening response on tempering. Thus, the PM M61

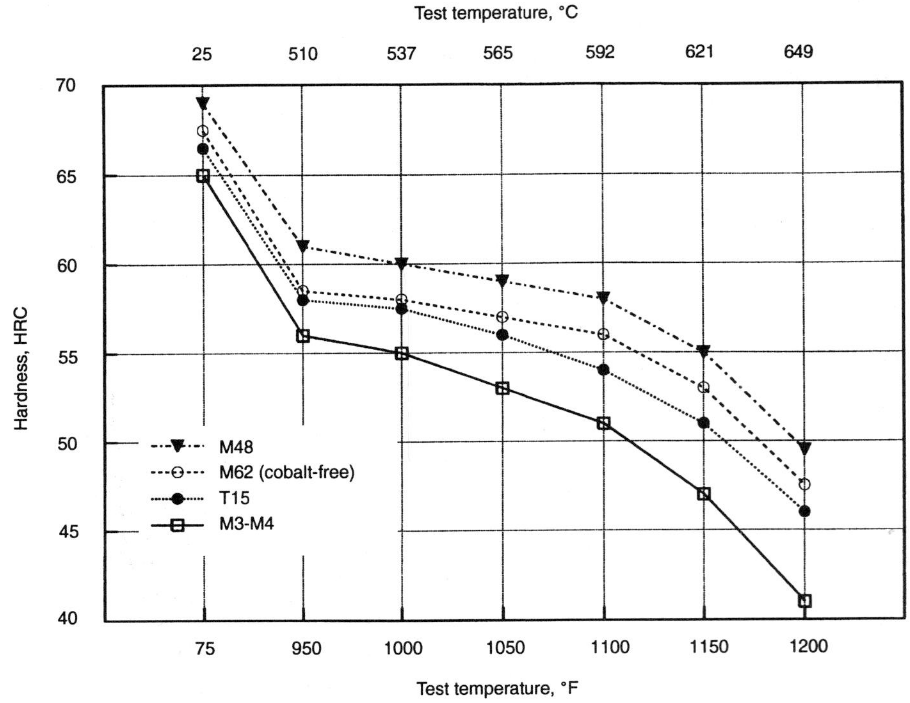

Fig. 7 Hot hardness of PM high-speed steels

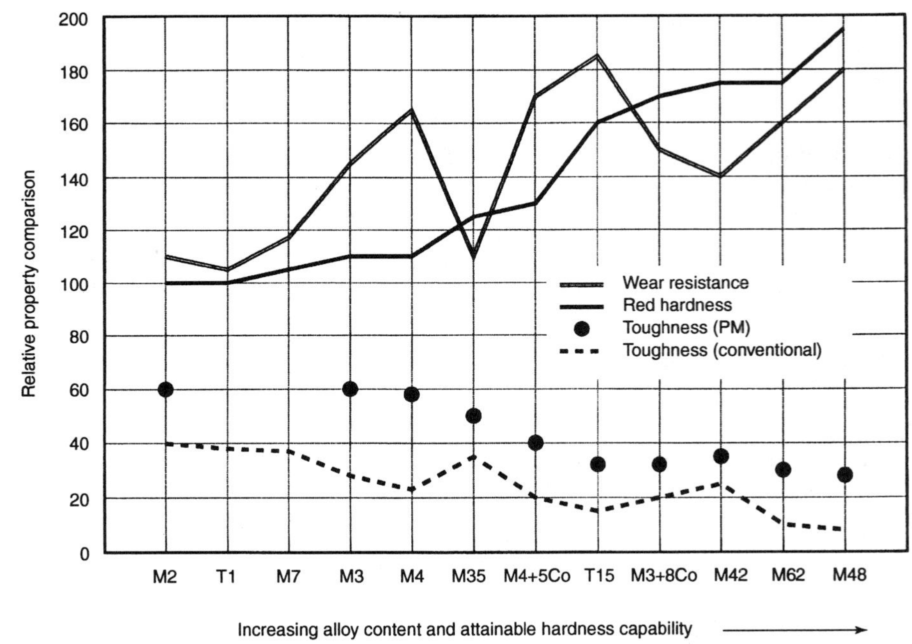

Fig. 8 Graphical comparison of high-speed steel properties

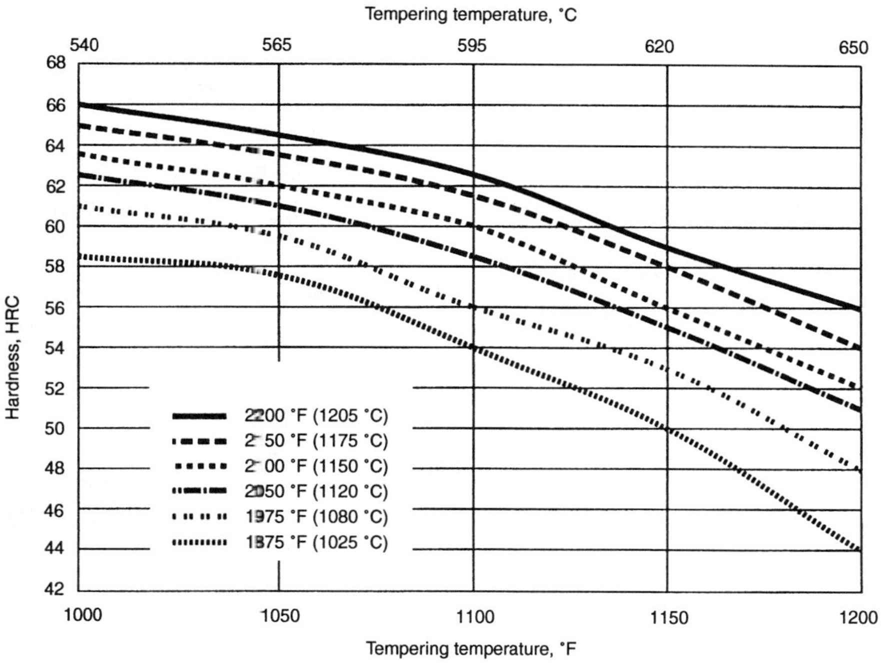

Tempering temperature		Heat treat response ± 1 HRC(a)					
°C	°F	1025 °C (1875 °F)	1080 °C (1975 °F)	1120 °C (2050 °F)	1150 °C (2100 °F)	1175 °C (2150 °F)	1205 °C (2200 °F)
As-quenched		59.5	62.5	64.5	65	65	63.5
540	1000	58.5	61	62.5	63.5	65	66
Optimum for maximum toughness and effective stress relieving							
550	**1025**	**58**	**60.5**	**62**	**63**	**64.5**	**65.5**
565	**1050**	**57.5**	**59.5**	**61**	**62**	**63.5**	**64.5**
595	1100	54	56	58.5	60	61.5	62.5
620	1150	50	53	55	56	58	59
650	1200	44	48	51	52	54	55
Minimum time at austenitizing temperature, min		45	30	20	15	10	5
Minimum number of tempers (2 h)		2	2	2	3	3	3

(a) Results may vary with hardening method and section size. Salt or oil quenching will give maximum response. Vacuum or atmosphere cooling may result in up to 1 to 2 HRC points lower.

Fig. 9 Heat treating guidelines for PM M4 high-speed steel

composition with a $W_{eq} = 25.5$ and no cobalt has heat treat response and temper-resistance characteristics comparable to T15 with 5% Co, and also better wear resistance due to a greater volume fraction of primary carbides in the heat treated microstructure. Similarly, the PM M62 composition with a $W_{eq} = 27.25$ and no cobalt has about the same attainable hardness, somewhat better temper resistance, and significantly better wear resistance compared to M42 with 8% Co. These highly alloyed cobalt-free compositions could not be economically produced or fabricated into tools without the benefits of PM processing.

The relative effects of attainable hardness, vanadium carbide content, and total alloying on the properties of the PM high-speed steels are summarized in the qualitative comparison chart shown in Fig. 8. As illustrated in the graph, red hardness increases as the total alloy content and corresponding attainable hardness capability increase. Wear resistance also generally increases with total alloy content and attainable hardness, but the most wear-resistant materials in each classification of high-speed steels are the higher vanadium compositions, for example, PM M4 and PM T15. The graph also shows that as the alloy content increases, the relative toughness decreases. As discussed earlier, however, the generally improved toughness of the PM materials compared to conventionally produced high-speed steels often enables upgrading without serious concerns about the actual toughness in service.

Another consideration when selecting a PM high-speed steel for a cutting tool application is whether or not the tool is to be coated (or recoated as part of tool refurbishing in service). Where a super-high-speed steel may have been required for wear resistance and to prevent localized softening and chip welding in the absence of a coating, these problems are often avoided by the application of a coating. Thus, it may be possible for the end user to consider downgrading to a lower alloy and therefore less expensive substrate material such as PM M3 or PM M4, which have an excellent combination of wear resistance and strength characteristics. However, if the successful performance of the coated tool is dependent on having a combination of high hot hardness and high wear resistance in the substrate (or in any uncoated areas), then alloy selection for the substrate material follows the same criteria used for uncoated tools with the higher alloyed PM compositions offering higher attainable hardness and hot hardness capability.

Basic heat treatments for PM high-speed steels involve preheating to about 790 to 845 °C (1450 to 1550 °F), austenitizing at high temperatures to dissolve enough of the carbides to achieve the desired heat treat response after tempering (without overheating), quenching at sufficiently fast rates to prevent reprecipitation of the dissolved carbides (but uniformly and not too fast to cause excessive distortion or cracking), and multiple tempering usually at a minimum of 540 °C (1000 °F) to eliminate retained austenite and provide adequate stress relieving. The traditional method of heat treatment for high-speed steels is to use salt baths or atmosphere furnaces with associated salt or oil quench baths. However, for environmental and safety concerns, many captive and commercial heat treating operations have converted to rapid-gas-quenching vacuum hardening furnaces employing a combination of high pressure and high volumetric mass flow to achieve quench rates approaching those previously achieved in salt.

An example of the type of hardening and tempering response information supplied by the steel producers is shown in Fig. 9 for PM M4. For tooling applications requiring maximum hardness, the austenitizing temperature typically selected for PM M4 is 1190 to 1205 °C (2175 to 2200 °F) to optimize the tempering response at 540 to 550 °C (1000 to 1025 °F) when attempting to achieve 64 to 66 HRC or 65 to 67 HRC. For tooling applications requiring additional toughness, and where a lower hardness is acceptable, it is common practice to "underharden" from lower austenitizing temperatures to take fewer of the carbides into solution and thereby maintain finer prior austenite grain size for improved toughness properties. Thus, for PM M4 cutting tools requiring a minimum application hardness of 63 to 64 HRC, a typical underhardening treatment might be to austenitize no higher than 1175 to 1190 °C (2150 to 2175 °F) and to multiple temper at 550 °C (1025 °F). For cold-work punch and die operations, where the required application hardness might be no higher than 60 to 62 HRC, PM M4 is typically austenitized from no higher than 1120 to 1150 °C (2050 to 2100 °F) and multiple tempered at a minimum of 550 °C (1025 °F) to maximize the toughness. It is particularly important to multiple temper as high as possible, or at a minimum of 550 °C (1025 °F), whenever extensive wire EDM cutting is anticipated to guarantee that the part is thoroughly stress relieved and thereby to prevent stress cracking. These heat treating guidelines apply to all high-alloy tool steels, whether they are produced using PM processing or by conventional ingot metallurgy. However, because there can be slight differences in the composition of similar steels supplied by different manufacturers, it is always advisable to refer to the specific heat treating procedures recommended by the material supplier.

Particle metallurgy cutting tool applications include the following: gear-cutting hobs, shaper cutters, and shaving cutters; broaches; form-relieved milling cutters; dovetail (flat) and circular form tools; and premium end mills. Figure 10 shows examples of cutting tools manufactured from the PM high-speed steels.

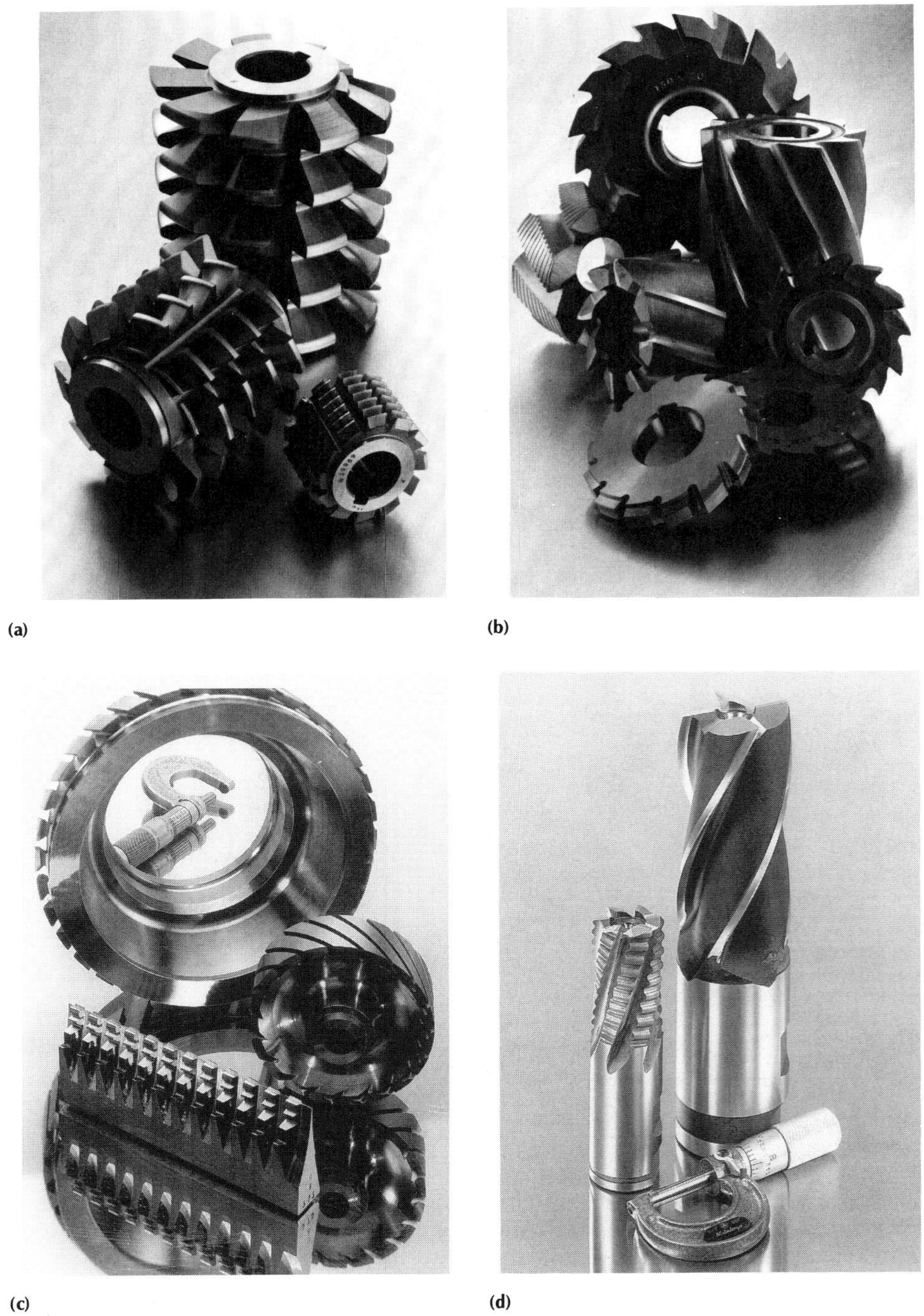

(a) (b) (c) (d)

Fig. 10 Examples of PM high-speed steel cutting tool applications. (a) Gear cutting hobs. (b) Milling cutters. (c) Broaches. (d) End mills

Cold-Work Particle Metallurgy Tool Steels

Many of the PM high-speed steels are also used in cold-work operations where a combination of good wear resistance and compressive strength (high hardness) is required. PM M3 and PM M4 are particularly well suited for cold-working applications because of their excellent combination of wear resistance, toughness, and grinding characteristics over a wide range of hardnesses.

Several of the international tool steel producers have also taken advantage of PM processing to develop new cold-work tooling materials alloyed primarily with high vanadium to maximize wear resistance. The first such material designed specifically for high-performance wear applications was Crucible particle metallurgy CPM 10V (hereafter referred to as PM 10V), which was introduced commercially in 1978. Also developed at about the same time was a lower-matrix carbon and slightly lower vanadium modification of PM 10V, called CPM 9V (hereafter referred to as PM 9V), which has lower attainable hardness but better toughness than PM 10V and can also be used in warm-working applications requiring resistance to heat checking. More recent PM tool steel development work has focused on the following: (a) even more wear-resistant ultrahigh vanadium compositions containing 15 to 18% V with up to 30% by volume of primary MC-type carbides, (b) low-to-intermediate carbide volume fraction materials moderately alloyed with vanadium and chromium to optimize the toughness properties while still maintaining good wear resistance, and (c) high-vanadium high-chromium compositions for wear applications that also require good corrosion resistance. The wear/corrosion-resistant PM tool steels are discussed in a later section.

Table 5 lists the nominal compositions for several of the commercially available PM cold-work tool steels as well as the PM M4 high-speed steel discussed previously. Also included are the compositions of conventionally produced A2, D2, and D7 cold-work tool steels. With the exception of PM 9V, all of the tool steels listed in the table are capable of attaining 58 to 62 HRC during heat treatment, which is the typical application hardness range for cold-work tooling. The heat treating procedures used for these materials are similar to those used for the PM high-speed steels. Depending on the alloy composition and desired hardness, the recommended austenitizing temperatures range from about 1024 to 1177 °C (1875 to 2150 °F). Tempering recommendations are also alloy dependent. In some cases, tempering at or below 524 °C (975 °F) is acceptable to maintain high hardness with adequate toughness. In other cases, however, a minimum tempering temperature of 538 °C (1000 °F) is recommended to eliminate retained austenite and to ensure thorough stress relieving for both optimal toughness and size stability. In all cases, multiple tempering is required.

For a given hardness, the relative wear resistance of both the PM and conventional cold-work tool steels is a function of the amount and type (hardness) of primary carbide present in the heat treated microstructure. Carbide size is only a factor under abrasive wear conditions and gen-

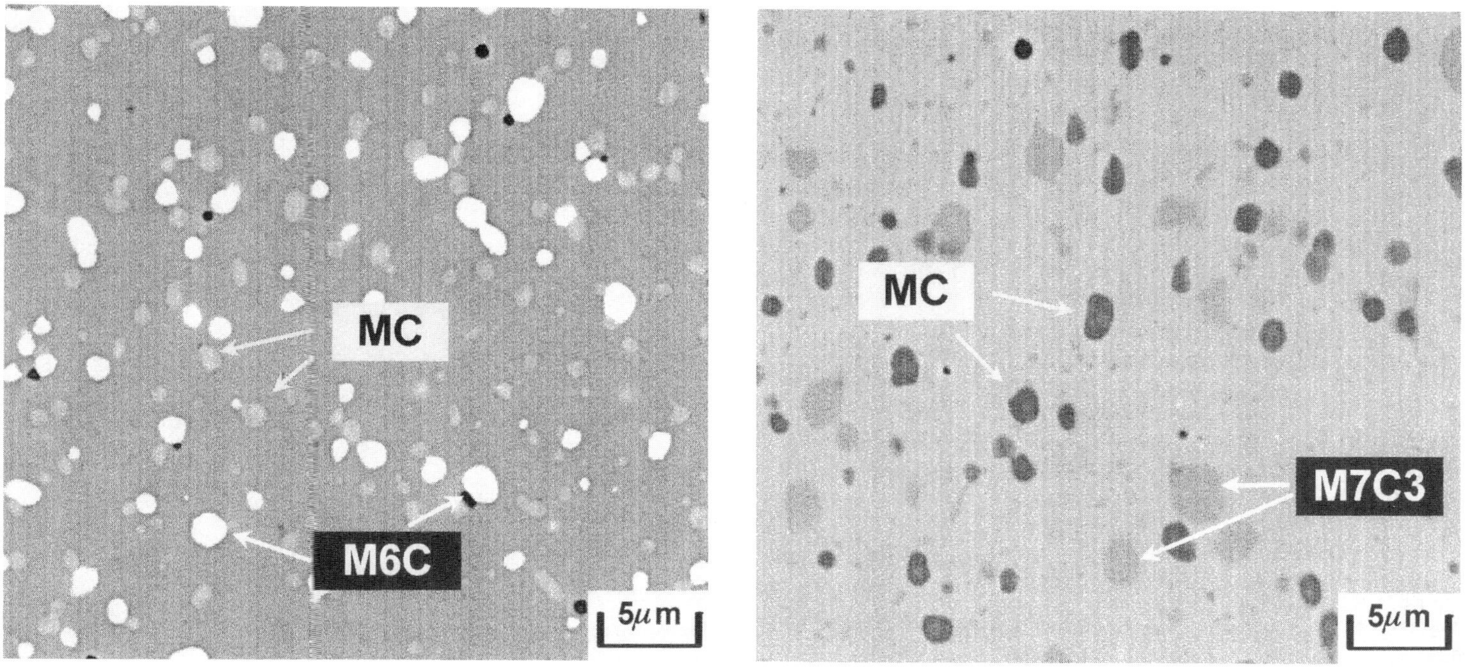

Fig. 11 Primary carbides in heat treated PM M4 (left) and PM 8Cr4V (right) wear-resistant tool steels containing approximately 12.5% total carbide volume

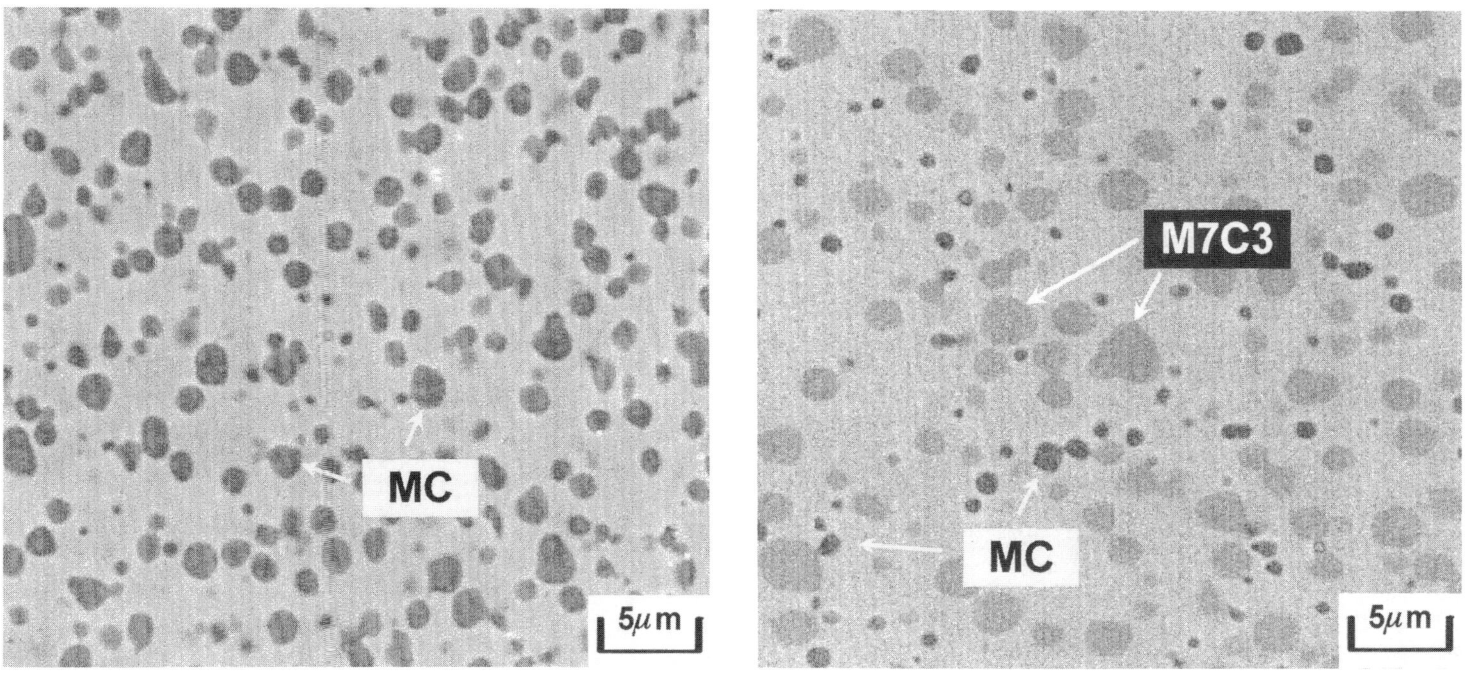

Fig. 12 Primary carbides in heat treated PM 15V (left) and PM 12Cr4V (right) wear-resistant tool steels containing approximately 23% total carbide volume

erally only applies to conventionally produced tool steels that have inherently coarser carbide size compared to the PM tool steels. Table 6 shows the results of SEM and image analysis of the primary carbides observed in heat treated samples of the alloys listed in Table 5. The total volume percent of primary carbides ranges from approximately 5% in PM 3V to 30% in PM 18V. The relative percentages of the primary carbide types present (MC, M_6C, M_7C_3) vary according to the alloying balance, with only PM 3V, PM 9V, PM 10V, PM 15V, and PM 18V having essentially all MC carbides. Although PM M4 and PM 8Cr4V have similar vanadium contents and total carbide volumes (about 12.5%), about two-thirds of the carbides in PM M4 are M_6C and about half the carbides in PM 8Cr4V are M_7C_3-type. These microstructural differences are illustrated in the photomicrographs in Fig. 11, where it can be seen that the various carbide types exhibit characteristically different electronic imaging in the SEM. Figure 12 shows a similar SEM metallographic comparison between PM 12Cr4V and PM 15V both of which contain about 23% total primary carbide volume. Whereas PM 15V contains all vanadium-rich MC-type carbides, PM 12Cr4V has predominantly M_7C_3-type due to the high chromium content. Table 6 also lists the approximate total primary carbide volumes for conventionally produced A2, D2, and D7. With less than 1% V, A2 and D2 tool steels contain predominantly M_7C_3-type primary carbides. Conventional D7 has essentially the same alloy composition as PM 12Cr4V and very similar carbide volumes as determined by SEM analysis.

Figures 13 and 14 show the results of dry crossed-cylinder (metal-to-metal) and pin-abrasion wear testing, respectively, plotted versus the primary carbide contents of the materials listed in Table 6. In the crossed-cylinder wear test, a tungsten carbide cylinder is pressed and rotated against a perpendicularly oriented and stationary test sample at a specified load. The volume loss of the test sample, which wears preferentially in this test, is determined at regular intervals and used to calculate a wear-resistance parameter based on the load and the sliding distance. In the pin-abrasion wear test, a small cylindrical test sample is rotated and pressed against an abrasive cloth at a fixed load. The abrasive cloth is attached to a movable table, which allows the test sample to be moved back and forth for a fixed distance in a nonoverlapping pattern across fresh abrasive. The weight loss of the sample (or lack thereof) is used as a measure of the wear resistance. The abrasive used for the test was 150 mesh garnet (Al_2O_3-$3FeO$-$3SiO_2$), which is considered only mildly to moderately abrasive compared to other abrasives, but which generally provides good differentiation in wear test results for hardened tool steels.

The metal-to-metal wear resistance of the PM cold-work tool steels increases dramatically as the volume percent of vanadium-rich primary MC carbides in the heat treated microstructure increases, as shown in Fig. 13. These results also indicate that simply increasing the total alloy content to achieve larger carbide volumes, or increasing the vanadium content alone, does not guarantee good metal-to-metal wear resistance in this test unless the composition is balanced to obtain significant percentages of the vanadium-rich MC carbide. This can be easily seen by comparing the wear test results for PM 3V, PM 9V, PM 10V, PM 15V, and PM 18V (which contain increasing percentages of only the MC carbide) with the wear test results for A2, D2, D7, and PM 12Cr4V (which contain ≤3 to 3.5% MC carbide). Conventional A2 and D2 tool steels contain little or no MC carbide and therefore exhibit very low metal-to-metal wear resistance. The wear test number does increase with an increase in total carbide volume for A2, D2, and D7 tool steels. However, D7 and PM 12Cr4V test only slightly better than PM 3V despite containing 4% V and having substantially greater total carbide volumes.

The influence of MC carbide content rather than total primary carbide volume in the crossed-cylinder wear test can be further illustrated by directly comparing the results for the PM tool steels with either similar vanadium contents or similar total carbide volumes, but very different MC carbide contents. For example, comparing PM 8Cr4V and PM 12Cr4V, which have about 4% V but different chromium contents, PM 8Cr4V exhibits slightly better wear resistance despite having only about half the total carbide volume. These results can be attributed to the 6.5% MC carbide content of PM 8Cr4V compared to about 3% MC carbide in PM 12Cr4V. Although the total carbide volumes of PM 12Cr4V and PM 15V are comparable (about 23%), the metal-to-metal wear resistance of PM 15V is substantially greater due to the fact that all of the primary carbides in PM 15V are vanadium-rich MC. A similar comparison can be made between PM M4 and PM 8Cr4V, which are both 4% V materials with similar total carbide volumes (about 12.5%). The lower wear test number obtained for PM 8Cr4V can be partially explained by the slight difference in test hardness, but the combination of MC and M_6C

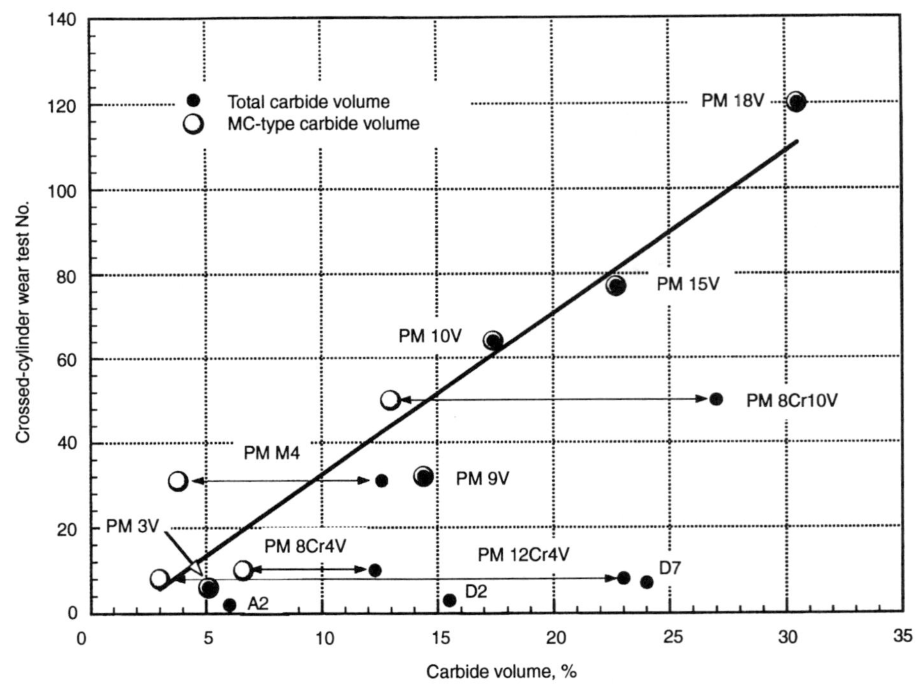

Fig. 13 Crossed-cylinder (metal-to-metal) wear test results for PM wear-resistant tool steels

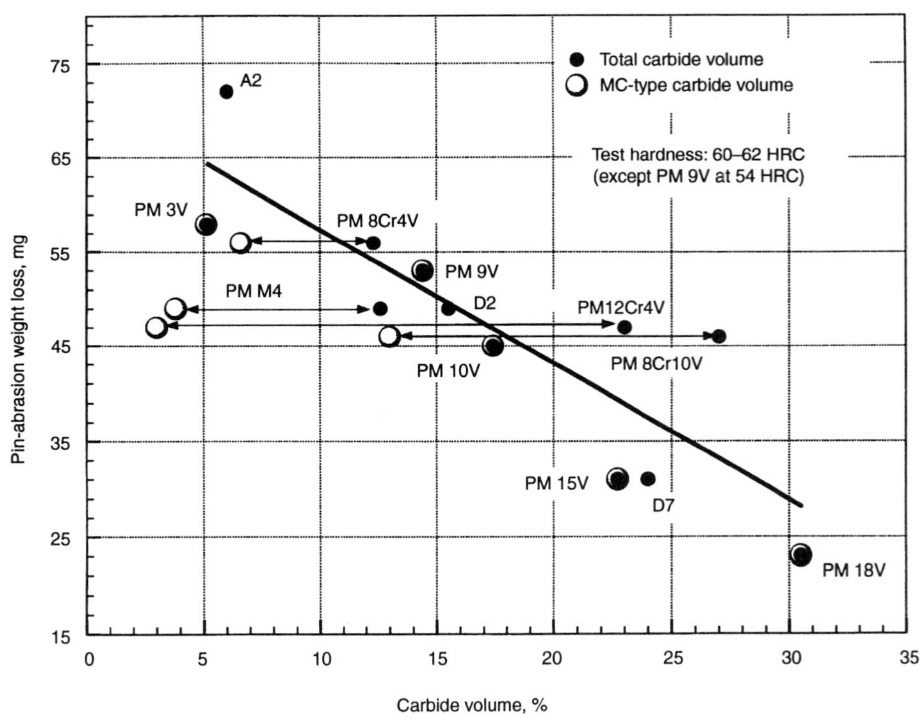

Fig. 14 Pin-abrasion wear test results for PM wear-resistant tool steels

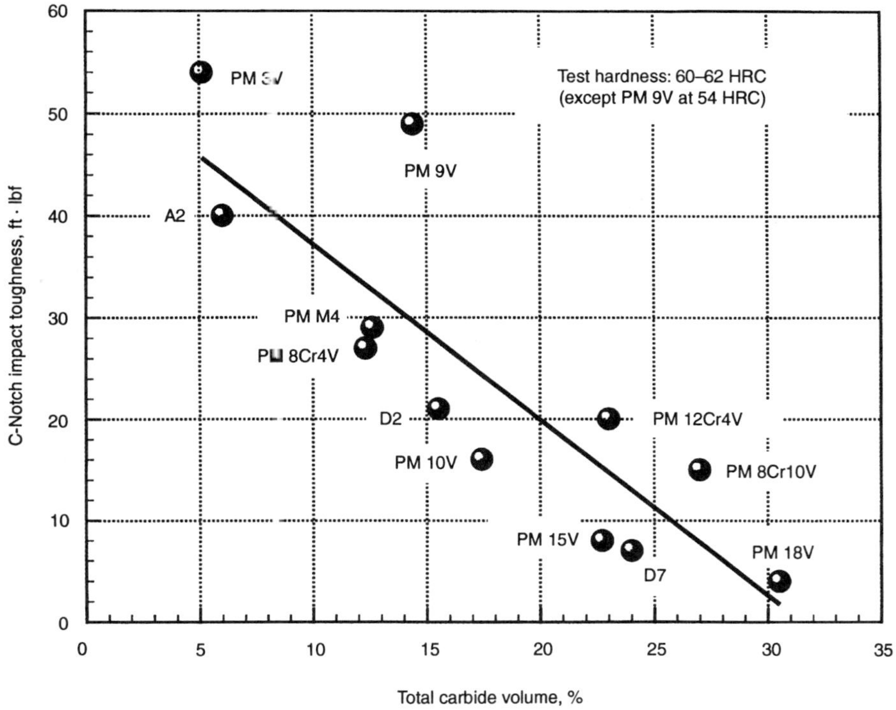

Fig. 15 Charpy C-notch impact toughness of PM wear-resistant tool steels

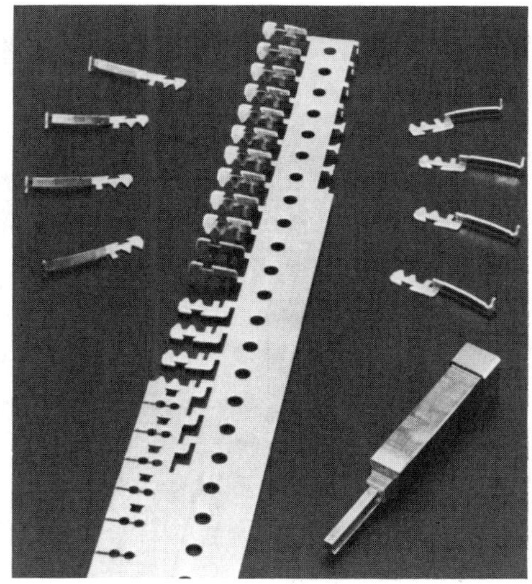

(a)

(b)

Fig. 16 Examples of PM tool steel punch and die applications. (a) Stamping and forming. (b) Powder metal compaction

Table 5 Nominal compositions of PM cold-work tool steels

Steel	AISI designation	Commercial equivalent	Composition, wt%: C	Cr	Mo	W	V
PM cold-work tool steels							
PM 3V	...	CPM 3V	0.80	7.50	1.00	...	2.75
PM M4	M4	CPM M4HC	1.40	4.00	5.25	5.75	4.00
PM 8Cr4V	...	Vanadis 4	1.50	8.00	1.00	...	4.00
PM 12Cr4V	D2	K190 PM	2.30	12.00	1.00	...	4.00
PM 9V	...	CPM 9V	1.80	5.25	1.30	...	9.00
PM 10V	A11	CPM 10V	2.45	5.25	1.30	...	9.75
PM8Cr10V	...	Vanadis 10	2.90	8.00	1.50	...	9.80
PM 15V	...	CPM 15V	3.50	5.25	1.30	...	14.50
PM 18V	...	CPM 18V	3.90	5.25	1.30	...	17.50
Conventionally produced (ingot-cast) cold-work tool steels							
A2	A2	...	1.00	5.25	1.15	...	0.30
D2	D2	...	1.55	11.50	0.80	...	0.90
D7	D7	...	2.35	12.00	1.00	...	4.00

carbides in PM M4 compared to the combination of MC and M_7C_3 carbides present in PM 8Cr4V is the primary reason for the difference in wear test results. A final comparison can be made between PM 10V and PM 8Cr10V, which have similar vanadium chemical analyses but very different total and MC carbide volumes due to differences in their carbon and chromium analyses. Some of the difference in total carbide volume can also be explained by the heat treatments employed to obtain 60 to 62 HRC; for example, a higher austenitizing temperature would have dissolved more of the M_7C_3 carbide in the PM 8Cr10V material. However, the wear test results correlate directly with the MC carbide content, with PM 10V exhibiting better metal-to-metal wear resistance due to its approximately 17.5% MC carbide content compared to about 13% MC in PM 8Cr10V. The fact that conventionally produced D7 and PM 12Cr4V, with similar alloy compositions and primary carbide contents, test comparably suggests that carbide size is not a major factor in this test.

There is a much better correlation between pin-abrasion weight loss and the total primary carbide volume for all of the materials tested, as shown in Fig. 14. The effects of primary carbide size and MC carbide content, however, are also evident. For example, conventionally produced D7, with an inherently large primary carbide size distribution compared to the PM materials, tests significantly better than PM 12Cr4V. In contrast, PM 15V with relatively small primary carbide size tests as well as D7. This can only be explained by the relative MC carbide content differences between these two materials. Similarly, both PM 12Cr4V and PM 8Cr10V test slightly worse than PM 10V, which has a significantly lower total carbide volume but substantially greater MC carbide content. The good abrasive wear test results for PM M4 relative to PM 8Cr4V can again be explained primarily by the combination of MC and M_6C carbides in PM M4. The fact that PM 9V tests very well at 54 HRC compared to the other PM tool steels at 60 to 62 HRC in both the crossed-cylinder and pin-abrasion wear tests is further testament to the strong influence of MC carbide content on wear resistance in these materials.

The toughness of the PM cold-work tool steels decreases with an increase in total primary carbide volume, relatively independent of the primary carbide type. This can be seen in Fig. 15,

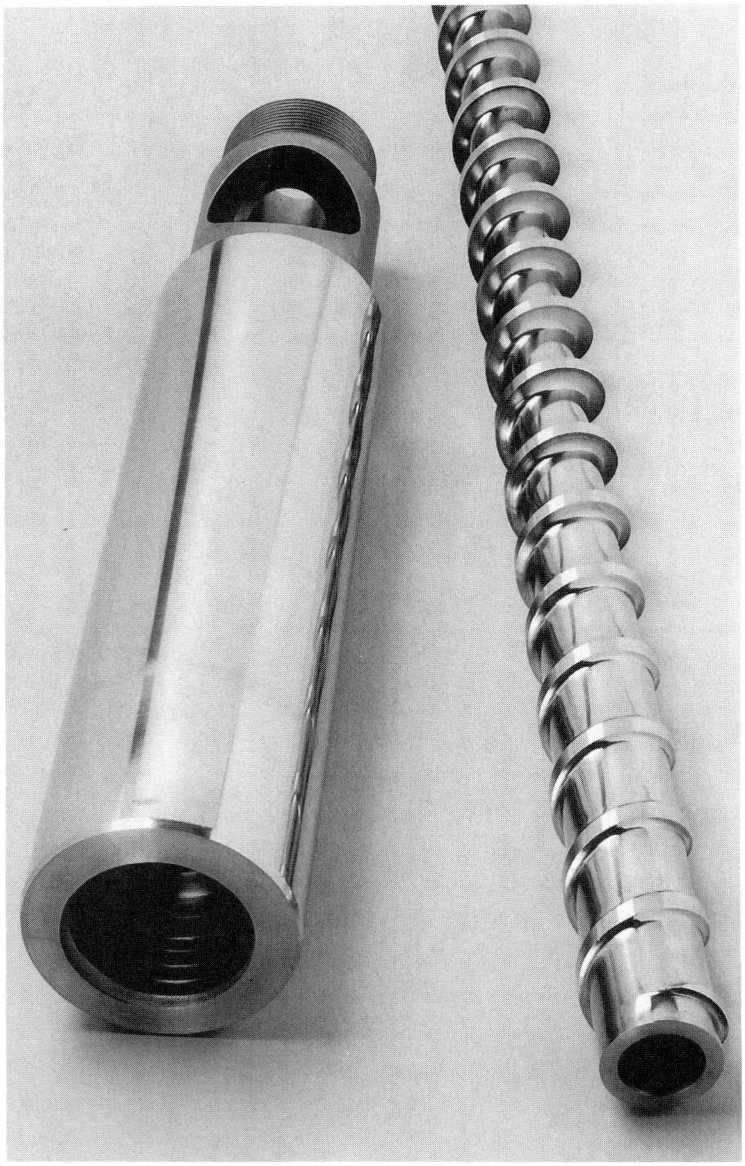

(a)

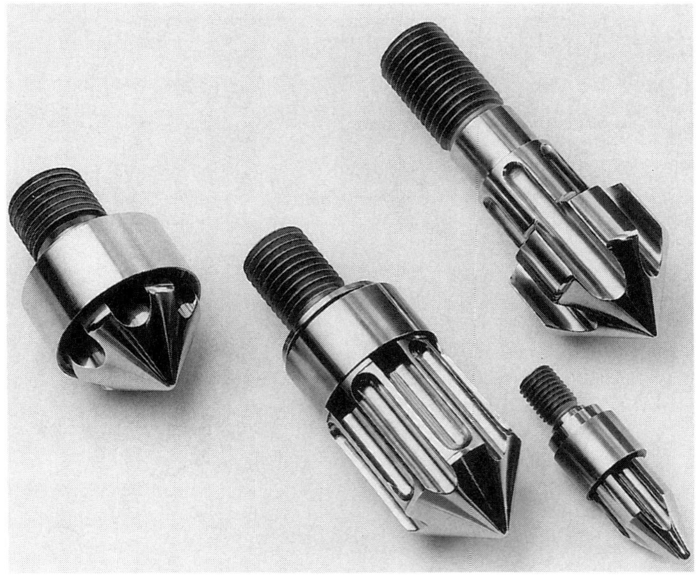

(b)

(c)

Fig. 17 Plastics processing equipment components. (a) Barrels and screws. (b) Nonreturn check ring valves. (c) Granulator and pelletizer knives

which shows the results of longitudinal Charpy C-notch impact testing versus total primary carbide volume for the PM cold-work tool steels as well as for PM M4, A2, D2, and D7. All of the materials were tested at 60 to 62 HRC, except for PM 9V which was tested at 54 HRC. With only 5% total carbide volume, PM 3V exhibits the highest impact toughness properties of the alloys tested, including A2, which has a comparable total primary carbide volume. PM M4 and PM 8Cr4V exhibit lower toughness values due to their increased carbide volumes, but are still considerably tougher than D2 and the higher-alloyed vanadium-rich PM grades. Despite their relatively high total carbide volumes, both PM 12Cr4V and PM 8Cr10V exhibit toughness comparable to the lower carbide volume PM 10V and D2 materials. The relatively low toughness values obtained for PM 15V and PM 18V may restrict their use in metalworking applications where shock loading is anticipated.

These high-performance PM tool steels are used for a wide range of cold-work punch-and-die applications such as long-run stamping and forming, fineblanking, and can manufacturing. The high-vanadium PM tool steels are particularly well suited for abrasive wear applications such as powder metal compaction, metal-injection molding, and equipment components for the processing of engineering plastics. The latter components include: screw elements, paddles, nozzles, and HIP-clad or solid barrel inserts for compounding; flat and circular pelletizer knives; screws, barrel inserts, nonreturn valves, mold inserts, and trim knives for injection molding; and finally, granulator knives for waste recycling. Examples of the types of tooling or wear parts used in these applications are shown in Fig. 16 and 17.

Wear/Corrosion-Resistant Particle Metallurgy Tool Steels

In many manufacturing operations, tooling and/or wear parts are subjected not only to wear caused by direct contact of the moving parts or by hard abrasive particles in the process media, but also to corrosive attack caused by moisture, acids, or other corrodents. This is particularly true of machinery components used in the high-volume compounding and injection molding of engineering plastics that contain large amounts

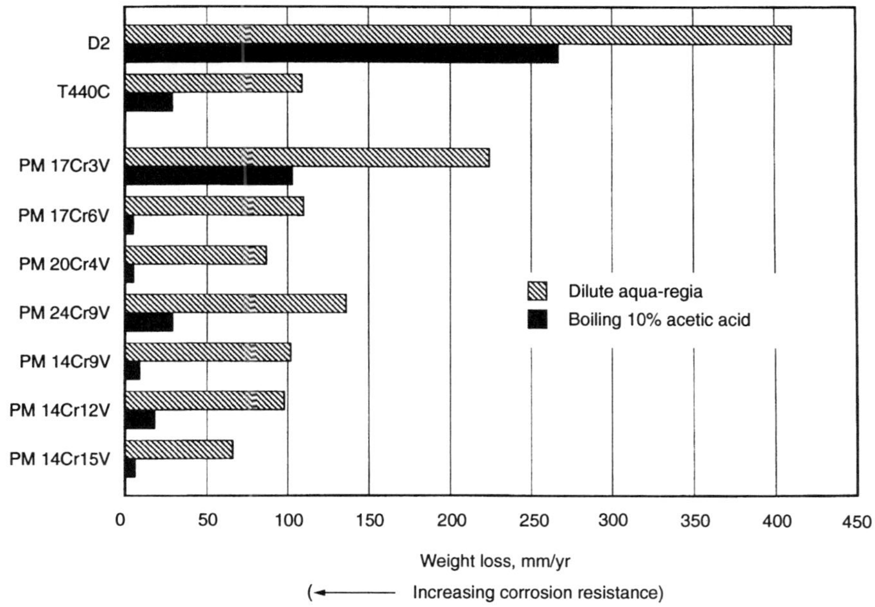

Fig. 18 Corrosion testing of PM wear/corrosion-resistant tool steels

of fillers that not only cause abrasive wear, but also frequently generate acidic conditions at the ambient or elevated temperatures required to process these materials. Similar operating conditions can also be encountered in the production of synthetics or in the processing and containerization of animal and consumer food products. Thus, the tool steels used in these operations must be both highly wear resistant and corrosion resistant, as well as sufficiently strong and impact resistant to withstand the operating stresses encountered.

To address these operating requirements, a number of high-performance wear/corrosion-resistant PM tool steels have been developed containing the following: (a) significant amounts of vanadium and chromium to form wear-resistant primary carbides, (b) a minimum of 11% matrix Cr after heat treatment to provide increased corrosion resistance compared to the PM tool steels containing 5 to 8% total Cr, and (c) an optimal level of carbon and/or nitrogen to form the wear-resistant carbides and to achieve a minimum of 56 to 58 HRC after heat treatment without forming excessive amounts of retained austenite or precipitating additional chromium-rich carbides, which would lower the desired matrix chromium. These materials usually also contain a minimum of 1% Mo to improve hardenability and resistance to pitting corrosion.

Table 7 lists the nominal compositions of several commercially available PM wear/corrosion-resistant tool steel grades along with the compositions of conventionally produced T440C stainless steel and D2 tool steel. As shown, the PM wear/corrosion-resistant materials have total chromium contents ranging from about 14 to 24%, vanadium contents ranging from about 3 to 15%, molybdenum contents ranging from about 1 to 3%, and carbon contents ranging from about 1.7 to 3.75%. The relative wear and corrosion resistance of these materials is strongly influenced by the partitioning of the alloying elements between the matrix and the primary carbides during solidification from the melt, by the volume fraction of primary carbides, and by heat treatment.

The heat treating procedures recommended for the PM wear/corrosion-resistant tool steels are similar to those recommended for the PM wear-resistant tool steels described in the previous section. All of the materials are designed to be air-hardenable, but may require oil quenching or high-pressure quenching in vacuum systems to minimize intergranular carbide precipitation or to avoid high-temperature transformation products in large cross sections. Austenitizing temperatures typically range from a minimum of about 1080 °C (1975 °F) to a maximum of about 1150 to 1175 °C (2100 to 2150 °F), which permits sufficient solutioning of the alloy carbides for heat treat response without resulting in excessive retained austenite levels during quenching. For optimal corrosion resistance, a maximum tempering range of about 260 to 400 °C (500 to 750 °F) is recommended. However, it may be necessary to employ multiple tempering at higher temperatures, for example, 540 to 550 °C (1000 to 1025 °F), in order to eliminate retained austenite for size stability or to guarantee more thorough stress relieving for wire EDM operations, and so forth. The use of higher tempering temperatures results in less corrosion resistance due to fine chromium carbide precipitation, but the corrosion resistance is still expected to be better than that of other tool steels containing less chromium in the matrix. The use of tempering temperatures in the range of about 425 to 525 °C (800 to 975 °F) should be avoided to prevent a loss in toughness.

When comparing these materials for potential corrosion resistance, it is important to recognize that higher total chromium content does not necessarily guarantee higher matrix chromium for better corrosion resistance, because a significant percentage of the chromium can be present as chromium-rich primary carbides that do not appreciably dissolve at the austenitizing temperatures used for heat treatment. Table 8 shows the results of SEM energy-dispersive x-ray spectrometry analysis of the matrix chromium for T440C and the high-chromium high-vanadium PM tool steels, where it can be seen that the matrix chromiums after heat treatment are comparable (about 12.5 to 13.5%) despite a wide range in total chromium contents (nominally 14 to 24%). Note that the PM 14Cr-9V material with only about 14% total Cr content has about the same matrix chromium as T440C as well as the other PM wear/corrosion-resistant tool steels, which have higher total chromium contents. This is accomplished by carefully "balancing" the carbon/alloy and chromium/vanadium ratios to preferentially form vanadium-rich MC carbides in the microstructure, thereby minimizing the formation of chromium-rich M_7C_3 carbides, and to simultaneously achieve a minimum matrix chromium of about 11 to 12%. In comparison, although D2 tool steel contains a total chromium content of about 11.5%, the matrix chromium of D2 is lower than that required for good corrosion resistance due to the formation of chromium-rich M_7C_3 primary carbides in the microstructure.

Figure 18 compares the corrosion resistance of the PM wear/corrosion-resistant tool steels to that of T440C and D2 tool steel in two accelerated corrosion tests: (a) the dilute aqua regia test in which samples are immersed for 3 h in a room-temperature aqueous solution containing 5 vol% nitric and 1 vol% hydrochloric acid, and (b) the boiling acetic acid test in which samples are immersed in a boiling aqueous solution of 10 vol% acetic acid for 24 h. The results show that D2 exhibits considerably less corrosion resistance than T440C in both corrosion tests. With the exception of PM 17Cr-3V, all of the PM wear/corrosion-resistant tool steels test about the same as T440C in the dilute aqua regia test. Similarly, all of the PM materials except PM 17Cr3V test about the same or slightly better than T440C in the boiling 10% acetic acid test. The somewhat poorer test results observed for the PM 17Cr3V material was determined to be due to an excess carbon balance in this material compared to the other PM materials tested. The fact that the three PM grades containing only about 14% Cr and 9 to 15% V exhibit corrosion resistance comparable to both T440C and the properly balanced 17 to 24% Cr PM grades further illustrates that relative corrosion resistance cannot be absolutely predicted by simply comparing total chromium contents.

As was the case for the PM wear-resistant tool steels, the wear resistance characteristics of the

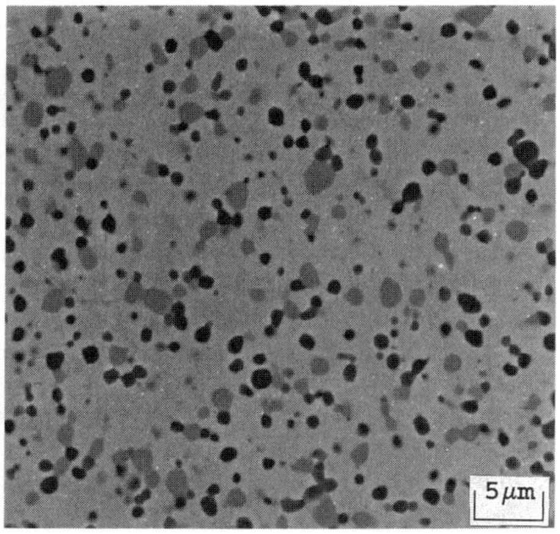

(a)

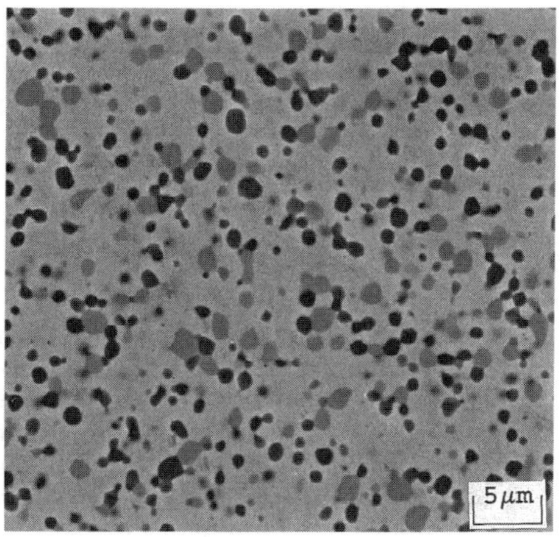

(b)

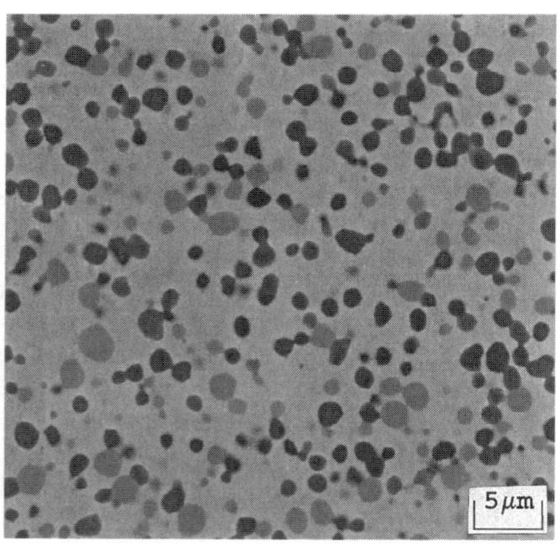

(c)

Fig. 19 Primary carbides in PM wear/corrosion-resistant tool steels containing 14% Cr and 9 to 14.5% V. (a) PM 14Cr9V (9.5% MC-13.5% M_7C_3). (b) PM 14Cr12V (12.5% MC-15.5% M_7C_3). (c) PM 14Cr15V (17% MC-14.5% M_7C_3)

Table 6 SEM image analysis of the primary carbides in PM cold-work tool steels

Steel	Heat treatment	Hardness, HRC	Carbide content, vol% MC	M_7C_3	M_6C	Total
PM cold-work tool steels						
PM 3V	1121 °C (2050 °F)/30 min + air cool + 524 °C (975 °F) temper (3 × 2 h)	60	4.7	0.4	...	5.1
PM M4	1163 °C (2125 °F)/4 min + oil quench + 565 °C (1050 °F) temper (3 × 2 h)	62	3.8	...	8.8	12.6
PM 8Cr4V	1021 °C (1870 °F)/30 min + air cool + 524 °C (975 °F) temper (2 × 2 h)	60	6.6	5.7	...	12.3
PM 12Cr4V	1121 °C (2050 °F)/30 min + oil quench + 260 °C (500 °F) temper (2 × 2 h)	59	3	20	...	23
PM 9V	1121 °C (2050 °F)/30 min + air cool + 565 °C (1050 °F) temper (2 × h)	54	14.4	...	...	14.4
PM 10V	1121 °C (2050 °F)/30 min + oil quench + 552 °C (1025 °F) temper (2 × 2 h)	61	17.4	...	...	17.4
PM 8Cr10V	1060 °C (1940 °F)/30 min + air cool + 552 °C (1025 °F) temper (2 × 2 h)	60	13	14	...	27
PM 15V	1177 °C (2150 °F)/10 min + oil quench + 552 °C (1025 °F) temper (3 × 2 h)	62	22.7	...	...	22.7
PM 18V	1121 °C (2050 °F)/30 min + oil quench + 552 °C (1025 °F) temper (2 × 2 h)	62	30.5	...	...	30.5
Conventionally ingot cast cold-work tool steels						
A1	Not reported	60	...	6	...	6
D2	Not reported	61	...	15.5	...	15.5
D7	Not reported	61	3	21	...	24

Table 7 Nominal compositions of wear/corrosion-resistant tool steels

Steel	Commercial equivalent	Composition, wt% C	Cr	V	Mo	Other
PM steels						
PM 14Cr-9V	CPM 420V (9V)	2.30	14	9	1	...
PM 14Cr-12V	CPM 420V (12V)	2.85	14	12	1	...
PM 14Cr-15V	CPM 420V (15 V)	3.25	14	14.5	1	...
PM 17Cr-6V	CPM 440VM	1.90	17	6	1	...
PM 17Cr-3V	Elmax PM	1.70	17	3	1	...
PM 20Cr-4V	Isomatrix M390	1.90	20	4	1	0.60%W
PM 24Cr-9V	Supracor	3.75	24	9	3	...
Conventional steel						
T440C	...	1.05	17	...	0.5	...
D2	...	1.55	11.5	0.8	...	...

Table 8 SEM analysis of heat treated wear/corrosion-resistant tool steels

	T440C	PM 17Cr3V	PM 20Cr4V	PM 17Cr6V	PM 24Cr9V	PM 14Cr9V	PM 14Cr12V	PM 14Cr15V
Approximate matrix chromium versus total chromium content, wt%								
Total	**16**	**17**	**20**	**17**	**24**	**14**	**14**	**14**
Matrix	13	11.5	13.5	13.5	13.5	12	12.5	12.5
Primary carbide content determined by image analysis, vol%								
M_7C_3	13.5	18.5	21	20	42	13.5	15.5	14.5
MC	...	2	3.5	4.5	4.5	9.5	12.5	17
Total	**13.5**	**20.5**	**24.5**	**24.5**	**46.5**	**23**	**28**	**31.5**

PM wear/corrosion-resistant tool steels are a function of the heat treated hardness, the total primary carbide volume, and the type (hardness) of carbide present. To maintain an optimal composition balance for good corrosion resistance, these materials are typically designed with the minimum carbon level necessary to achieve at least 56 to 58 HRC after heat treatment. Both the total primary carbide volume fraction and the types of carbide present are determined by the relative amounts of total chromium, vanadium, and/or other strong carbide-forming elements in the composition. Very high chromium levels promote the preferential formation of chromium-rich M_7C_3 primary carbides even when significant amounts of vanadium are also present. Conversely, high vanadium levels in the presence of low-to-moderate chromium levels promote the formation of the harder vanadium-rich MC carbides. Table 8 shows the results of SEM image analysis of the primary carbide types present in the PM wear/corrosion-resistant tool steels, where it can be seen that the alloys with nominally 17 to 24% total Cr and 3 to 9% V contain predominantly M_7C_3 carbides and less than about 4.5% by volume of the MC carbides. In contrast, the three PM grades containing only about 14% total Cr and 9 to 15% V by design contain significantly greater volume fractions of the MC carbide in addition to the M_7C_3 carbides

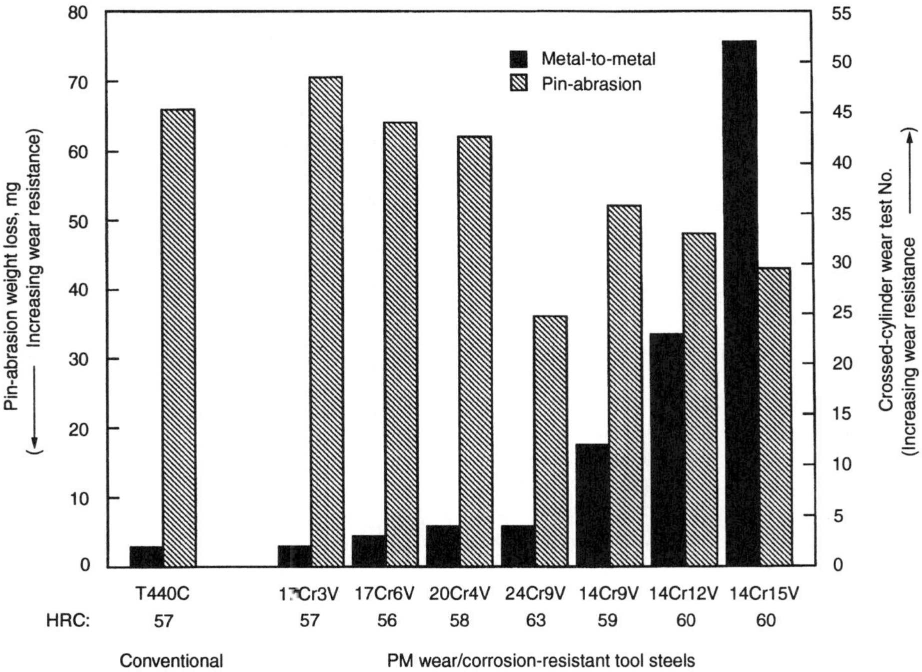

Fig. 20 Wear test results for PM wear/corrosion-resistant tool steels

that inevitably form at this chromium level. This effect of vanadium content on the total primary carbide volume and relative amount of MC carbide at the 14% Cr level is illustrated in Fig. 19, which shows SEM electronic images of the primary carbides observed in the heat treated microstructures. Carbide size can also affect wear resistance, particularly under abrasive wear conditions, but is generally not a factor for the PM tool steels as discussed in an earlier section.

Figure 20 shows the results of crossed-cylinder and pin-abrasion wear testing for several of the wear/corrosion-resistant PM tool steels as well as T440C. All of the materials except PM 24Cr9V were tested at either 56 to 58 or 58 to 60 HRC, which are the typical application hardnesses recommended by the material producers. PM 24Cr-9V was tested at its usual application hardness of 63 HRC. As is the case for the PM cold-work tool steels discussed in the previous section of this article, the wear resistance of the PM wear/corrosion-resistant tool steels generally improves with increasing total primary carbide volume and particularly with increasing MC carbide content. Taking into account the slight differences in test hardness, the PM 17Cr6V and PM 20Cr4V materials, with about 24.5% total carbide volume and about 3.5 to 4.5% MC, test about the same in both wear tests and somewhat better than either T440C or PM 17Cr3V, which have lower total carbide volumes and ≤2% MC. PM 24Cr9V, with about the same MC carbide content as PM 20Cr4V, tests about the same as PM 20Cr4V in the crossed-cylinder wear test but significantly better than all of the other materials in the pin-abrasion wear test. The latter result is due to the combination of the large total carbide volume and high test hardness of PM 24Cr9V. The significant influence of MC carbide content in both wear tests can be seen by comparing the wear test results for the high-vanadium 14% Cr materials as a group with those of the 17 to 20% Cr materials, which contain predominantly M_7C_3 carbides. Although T440C tests about the same as PM 17Cr-3V in the crossed-cylinder test and somewhat better in the pin-abrasion test (most likely due to the inherently larger primary carbide size of the conventionally produced material), all of the PM wear/corrosion-resistant tool steels are expected to be significantly more wear resistant than T440C in actual service.

The PM wear/corrosion-resistant tool steels are used in applications where a combination of wear resistance and good corrosion resistance are required. They are particularly well suited for the processing of corrosive engineering plastics and are used for many of the same equipment components where the PM wear-resistant materials are also used. Examples of barrels and screws, nonreturn check ring valves, and granulator and pelletizer knives that make use of these high-performance PM materials are shown in Fig. 17.

Hot Work Particle Metallurgy Tool Steels

The absence of alloy segregation, and the inherently fine carbide and sulfide size distributions typical for PM processing, make PM tool steels attractive for consideration in hot-work applications where standard or premium quality conventional H13 or other hot-work tool steel compositions are currently used. However, unlike the cutting tool and cold-work applications

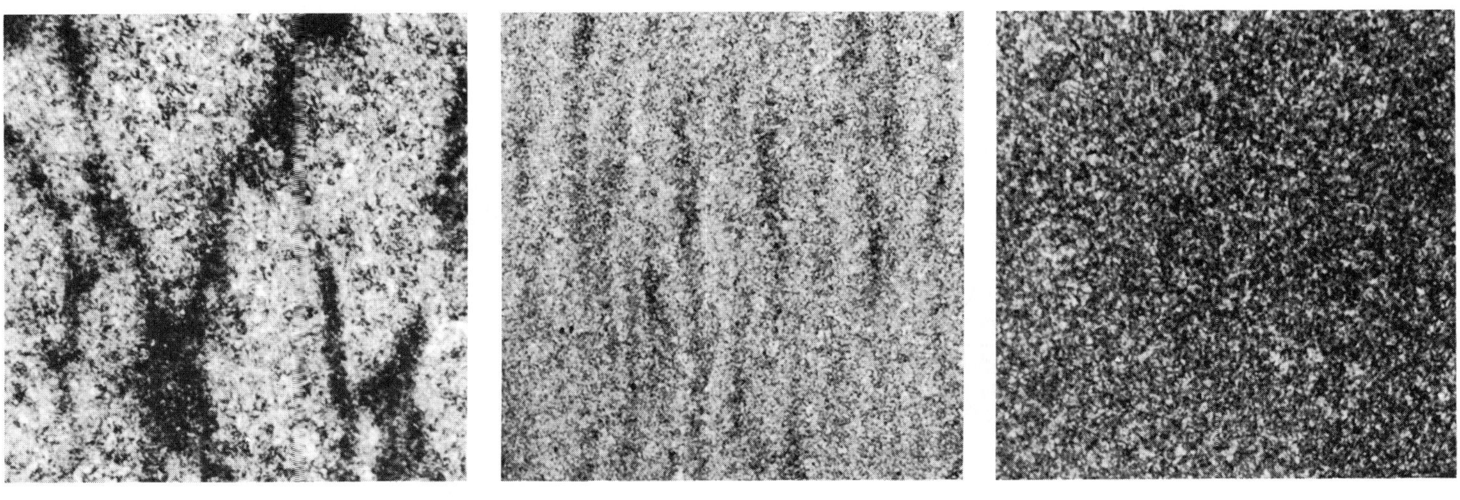

(a) (b) (c)

Fig. 21 Typical microstructures for (a) conventional H13, (b) premium quality H13, and (c) PM H13 tool steels. All at 50×

Table 9 Composition of premium quality and resulfurized H13 tool steels

Material	Composition, wt%						
	C	Mn	S	Si	Cr	Mo	V
NADCA PQ H13	0.40	0.35	0.005 max	1.00	5.25	1.35	1.00
Conventional H13-S	0.38	0.72	0.15	0.96	5.20	1.36	1.00
PM H13-S	0.36	0.34	0.17	1.00	5.50	1.60	0.85

where high hardness and high wear resistance are usually the more critical material property requirements, the most important requirements in hot-work applications are good thermal fatigue resistance and very high toughness. The latter two properties generally restrict the carbon content of the heat treated tool steel matrix to less than 0.40% and also limit the amount of total alloying that may be used to introduce primary carbides to the microstructure for additional wear resistance. Thus, PM alloy development in this application area has been limited.

One of the successful applications of PM processing technology has been the development of a resulfurized (free-machining), preheat treated PM H13 for use in die casting and plastic molding operations. These applications typically require a heat treated hardness of about 44 to 48 HRC for an optimal combination of strength, impact toughness, and heat checking resistance. The potential advantages for using PM H13-S compared with conventionally produced H13-S include the following:

- Fine microstructural control with no alloy segregation, which enables the safe and uniform preheat treatment of large bar cross sections
- Higher transverse impact properties and better thermal fatigue resistance due to the elimination of large sulfide stringers
- Significantly improved machinability and better resulting surface finishes in both the annealed and heat treated conditions

For the die maker, the prospect of being able to machine directly from the heat treated solid represents a significant cost and time savings due to elimination of premachining steps and the avoidance of distortion during heat treatment.

Figure 21 compares typical longitudinal microstructures observed in moderately sized block cross sections of standard quality conventional H13, premium quality H13, and PM H13. As can be seen, the conventionally produced material typically exhibits a banded microstructure due to normal alloy segregation in the cast ingot. Premium quality H13 exhibits a much more uniform but still slightly banded microstructure due to special processing during melting, ingot casting, and subsequent hot working. In comparison, the PM H13 microstructure is completely free of alloy segregation, very uniform, and fine grained. Table 9 shows the nominal composition of resulfurized PM H13 compared to the nominal compositions for premium-quality H13 (low S) produced to the North American Die Casting Association (NADCA) specification as well as conventionally produced H13-S. Table 10 shows the results of transverse impact testing for these same three materials. The detrimental effect of a high sulfur addition in conventional materials can be clearly seen, whereas the high-sulfur PM H13 material has impact properties comparable to the premium quality low-sulfur material. Figure 22 shows laboratory thermal fatigue testing results, where PM H13-S performed somewhat better-than-premium quality H13 and significantly better-than-conventional H13-S. Included in this study was a premium quality H13 test specimen that had been purposely given an EDM surface finish to illustrate the potentially detrimental effect of preexisting surface cracks on thermal fatigue resistance. These results indicate that tool life improvements could be realized in some die casting applications where the cavities can be machined directly from preheat treated PM H13-S, thereby eliminating the need for EDM.

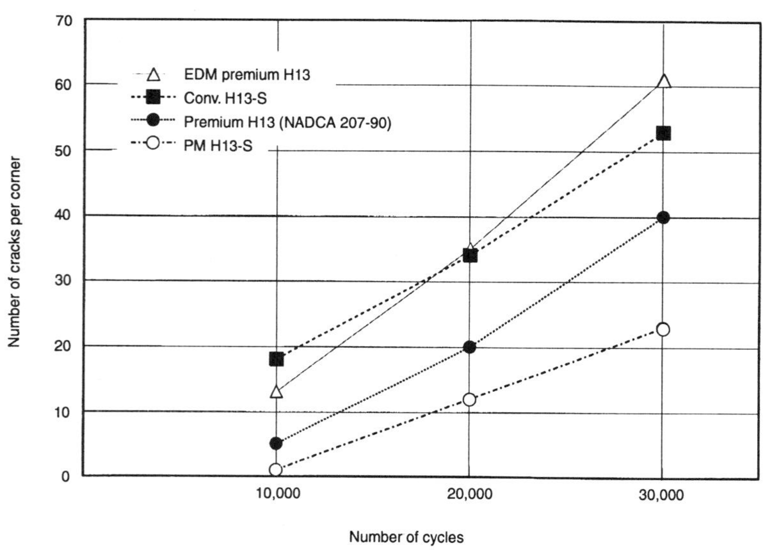

Fig. 22 Laboratory thermal fatigue test results for conventional (conv.) and PM H13 tool steels

Composite Bars, HIP Cladding, and Near-Net Shapes

In some tooling or wear part applications, it may not be practical or economical to produce the component from solid bar stock made entirely from a highly alloyed PM tool steel. In other applications, it may be desirable to produce a composite material to take advantage of the inherent property differences between the materials used to make the composite. In both cases, it is possible to HIP-clad PM tool steel powders to different solid substrate materials, HIP-clad other materials in powder or thin cross-section form to an existing PM tool steel bar product, or produce a composite bar by simultaneously HIP-consolidating two dissimilar prealloyed powders. Successful diffusion bonding during HIP at elevated temperatures and ultimately good bond strength in service generally require that the materials being diffusion bonded have mutual solid-state solubility without intermetallic formation, a wide temperature range for bonding, and compatible thermal expansions and elastic moduli. Consideration must also be given to the effects that the HIP cycle (both the elevated-temperature exposure and the relatively slow cooling rates employed) and/or any required post-HIP heat treatment may have on the basic properties of any nontool steel materials used to form the composite. Where the properties of the substrate or mating material are not critical, it is common to use carbon and low-alloy steels as the substrate materials. Where mechanical strength is a consideration, an air-hardenable tool steel such as H13 might be selected. Where corrosion resistance is required,

Table 10 Transverse Charpy V-notch impact testing of premium quality and resulfurized H13 tool steels

Material	Hardness, HRC	Charpy V-notch impact, J (ft · lbf)
NADCA PQ H13	46	13.5 (10)
Conventional H13-S	45	2.7 (2)
PM H13-S	45	11.5 (8.5)

Fig. 23 Bimetallic plastics extruder component machined from PM composite bar (PM 10V on outside-mild steel core)

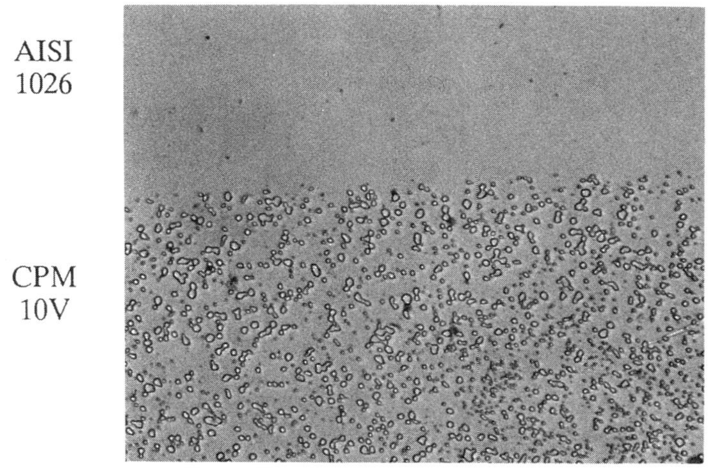

Fig. 24 Bond interface for PM 10V/mild steel composite. 500×

the substrate or mating material could be an austenitic or martensitic stainless steel.

Many of the advanced wear-resistant and wear/corrosion-resistant PM tool steels have been successfully produced in composite bar form or have been used in powder form for HIP cladding of prefabricated machine components. Figure 23 shows an example of one such application, where a low-alloy steel is HIP-clad to the inside diameter of a highly alloyed wear-resistant PM tool steel (in this case PM 10V) to produce composite bars that are subsequently fabricated into bimetallic extruder screw segments for the plastics industry. The low-alloy steel was specified for the ID to facilitate broaching of the splines and to provide some notch ductility during torque loading in operation. The excellent bond obtained between the PM 10V and the low-alloy steel is shown in Fig. 24.

Net-shape monolithic products can also be manufactured from the PM tool steels by either direct HIP consolidation of the powders or by warm/hot forging of billet stock. Most of the PM tool steels exhibit excellent hot-working characteristics and can be readily forged from billet into large cylindrical ring sections, simple solid shapes, or closed-die forgings. For large hollow bar applications, where material costs or the outside diameter (OD) or inside diameter (ID) size and length prohibit the use of forged rings or solid bar stock, as-HIP hollow bars can be produced. It is also possible to produce very intricate individual near-net-shape parts by direct HIP of specially designed containers filled with the desired PM tool steel powder. However, unless the desired final shape cannot be physically machined or otherwise fabricated, it is generally more economical to machine directly from the solid or hollow bar form. The composite part shown in Fig. 23 is a good example of where it is actually less expensive to machine the OD and ID profiles directly from the solid composite bar or hollow than it would be to manufacture the same shape as a near-net shape as-HIP part.

SELECTED REFERENCES

PM Tool Steels—General:

- K. Consemüller, B. Hribernik, and G. Schneider, Tool Steels Present and Future, *Progress in Tool Steels,* Proc. 4th Int. Conf. on Tooling, Ruhr University—Bochum, 11–13 Sept 1996, p 3–16
- W.B. Eisen, P/M Tool and High Speed Steel—A Comprehensive Review, *Advanced Particulate Materials and Processes,* Proc. 5th Int. Conf. on Advanced Particulate Materials and Processes, Metal Powder Industries Federation, 1997, p 55–63
- B. Hribernik, G. Hackl, and H. Lenger, Possibilities for the Application of Tool Steels Produced by Powder Metallurgy, *BHM,* Vol 134 (No. 11), 1989, p 334–341 (in German)
- K.E. Pinnow and W. Stasko, PM Tool Steels, *Metals Handbook,* Vol 1, 10th ed., ASM International, 1990, p 780–792

PM High-Speed Steels:

- W.T. Haswell, W. Stasko, and F.R. Dax, Cobalt-Free CPM High Speed Steels, *Processing and Properties of High Speed Tool Steels,* TMS/AIME, 1980, p 147–158
- R. Riedl et al., Developments in High Speed Tool Steels, *Steel Res.*, Vol 58 (No. 8), 1987, p 339–352
- L. Westin, Mechanical Properties of PM High Speed Steels Related to Heat Treatment and Hardness, *Metal Powder Rep.*, Vol 44 (No. 11), Nov 1989, p 768–773
- L. Westin and H. Wisell, Powder Metallurgy High Speed Steels, *Scand. J. Metall.,* Vol 25 (No. 1), Feb 1996, p 41–46

PM Wear/Corrosion-Resistant Tool Steels:

- P. Billgren and O. Sandberg, A New Generation of PM Steels for PM Compacting Tools and Other Tool Applications, *PM '94 Powder World Congress* (Paris), June 1994, p 1023–1026
- H. Makovec, G. Hackl, and B. Hribernik, Steels Produced by Powder-Metallurgy Methods for Cold Forming Technology, *New Materials, Processes, Experiences for Tooling*, Proc. Int. European Conf. on Tooling Materials (Interlaken), 7–9 Sept 1992, p 537–550
- L. Nordstrom and B. Johansson, Selection of Tool Steels for Cold-Work Applications, *New Materials and Processes for Tooling*, Proc. 2nd Int. Conf. on Tooling, Ruhr University—Bochum, 6–8 Sept 1989, p 193–208
- K.E. Pinnow, W. Stasko, J.J. Hauser, and R.B. Dixon, Properties of Advanced High Vanadium Wear/Corrosion-Resistant PM Tool Steels, *New Materials, Processes, Experiences for Tooling*, Proc. Int. European Conf. on Tooling Materials, (Interlaken), 7–9 Sept 1992, p 31–39
- W. Stasko, K.E. Pinnow, and W.B. Eisen, Development of Ultra-High Vanadium Wear Resistant Cold Work Tool Steels, *Advances in Powder Metallurgy and Particulate Materials,* Vol 5, Proc. 4th Int. Conf. on Advanced Particulate Materials and Processes, Metal Powder Industries Federation, 1996, p 17-179 to 17-188
- W. Stasko, K.E. Pinnow, and W.B. Eisen, Development of Improved PM Wear and Corrosion Resistant Tool Steels, *Progress in Tool Steels*, Proc. 4th Int. Conf. on Tooling, Ruhr University—Bochum, 11–13 Sept 1996, p 39–48
- W. Stasko, K.E. Pinnow, and R.B. Dixon, Particle Metallurgy Cold Work Tool Steels Containing 3-18% Vanadium, *Advanced Particulate Materials and Processes*, Proc. 5th Int.

Conf. on Advanced Particulate Materials and Processes, Metal Powder Industries Federation, 1997, p 401–409

- E. Tarney, Selecting High-Performance Tool Steels for Metalforming Tools, *Metal Forming*, Sept 1997, p 35–41

PM Hot-Work Tool Steels:

- H. Seilsdorfer, PM Hot Work Tool Steels, *Metall*, Vol 42 (No. 2), Feb 1988, p 146–152 (in German)
- E. Severson and C. Dorsch, Powder Metallurgy Produces a Prehardened Hot Work Tool Steel that is Superior to Premium Quality H13, Paper No. T97-083, *Transactions of the 19th Int. Die Casting Congress and Exposition,* North American Die Casting Association, 3–6 Nov 1997
- W. Stasko, C.J. Dorsch, and H.P. Nichols, "Advances in the Development of Particle Metallurgy Hot Work Tool Steels," Paper No. G-T89-051, *Transactions of the 15th Int. Die Casting Congress and Exhibition,* North American Die Casting Association, 1989

Direct HIP and HIP-Cladding of Tool Steels and Other Specialty Materials:

- E. Beyer, HIP—Tool Materials, *Powder Metall. Int.*, Vol 16 (No. 3), 1984, p 117–120
- P. Duerdoth and H. Seilsdorfer, Advanced Diffusion-Bonded HIP Composites, *Advances in Powder Metallurgy and Particulate Materials*, Vol 2, Metal Powder Industries Federation, 1996, p 5-25 to 5-34
- J.J. Eckenrod and K.E. Pinnow, P/M High Performance Stainless Steels for Near Net Shapes, *Advances in Powder Metallurgy and Particulate Materials*, Vol 4, Metal Powder Industries Federation, 1993, p 131–140
- P. Hellman, As-Hipped APM High Speed Steels, *Metal Powder Rep.*, June 1992, p 25–29
- P. Lülsdorf, Wear Protection With Hardenable Fe-Based PM-HIP Materials in Plastic Processing Machines, *Mater.wiss. Werkst.tech.*, Vol 26 (No. 10), 1995, p 524–535 (in German)

Powder Forged Steel

W. Brian James, Michael J. McDermott, and Robert A. Powell, Hoeganaes Corporation

POWDER FORGING is a process in which unsintered, presintered, or sintered powder metal preforms are hot formed in confined dies. The process is sometimes called P/M (powder metallurgy) forging, P/M hot forming, or is simply referred to by the acronym P/F. When the preform has been sintered, the process is often referred to as "sinter forging."

Powder forging is a natural extension of the conventional press and sinter (P/M) process, which has long been recognized as an effective technology for producing a great variety of parts to net or near-net shape. Figure 1 illustrates the powder forging process. In essence, a porous preform is densified by hot forging with a single blow. Forging is carried out in heated, totally enclosed dies, and virtually no flash is generated. This contrasts with the forging of wrought steels, in which multiple blows are often necessary to form a forging from bar stock and considerable material is wasted in the form of flash.

The shape, quantity, and distribution of porosity in P/M and P/F parts strongly influence their mechanical performance. The effect of density on the mechanical properties of as-sintered iron and powder forged low-alloy steel is illustrated in Fig. 2. Powder forging, therefore, is a deformation processing technology aimed at increasing the density of P/M parts and thus their performance characteristics.

There are two basic forms of powder forging:

- Hot upsetting, in which the preform experiences a significant amount of lateral material flow
- Hot repressing, in which material flow during densification is mainly in the direction of pressing, is sometimes referred to as hot restriking, or hot coining

These two deformation modes and the stress conditions they impose on pores are illustrated in Fig. 3.

In hot upset powder forging, the extensive unconstrained lateral flow of material results in a stress state around the pores that is a combination of normal and shear stresses. A spherical pore becomes flattened and elongated in the direction of lateral flow. The sliding motion due to shear stresses breaks up any residual interparticle oxide films and leads to strong metallurgical bonding across collapsed pore interfaces. This enhances dynamic properties such as fracture toughness and fatigue strength.

The stress state during hot repress powder forging consists of a small difference between vertical and horizontal stresses, which results in very little material movement in the horizontal direction and thus limited lateral flow. As densification proceeds, the stress state approaches a pure hydrostatic condition. A typical pore simply flattens, and the opposite sides of the pore are brought together under pressure. Hot repress forging requires higher forging pressures than does hot upset forging for comparable densification. The decreased interparticle movement compared with upsetting reduces the tendency to break up any residual interparticle oxide films and may result in lower ductility and toughness.

While powder forged parts are primarily used in automotive applications where they compete with cast and wrought products, parts have also been developed for military and off-road equipment.

The economics of powder forging have been reviewed by a number of authors (Ref 4–9). Some of the case histories included in the section "Applications of Powder-Forged Parts" in this article also compare the cost of powder forging with that of alternative forming technologies.

The discussion of powder forging in this article is limited to ferrous alloys. Information on the forging of aluminum, nickel-base, and titanium powders is available in other articles that follow in this Section.

Material Considerations

The initial production steps of powder forging (preforming and sintering) are identical to those of the conventional press and sinter P/M process. Certain defined physical characteristics and properties are required in the powders used in these processes. In general, powders are classified by particle shape, particle size, apparent density, flow, chemistry, green strength, and compressibility.

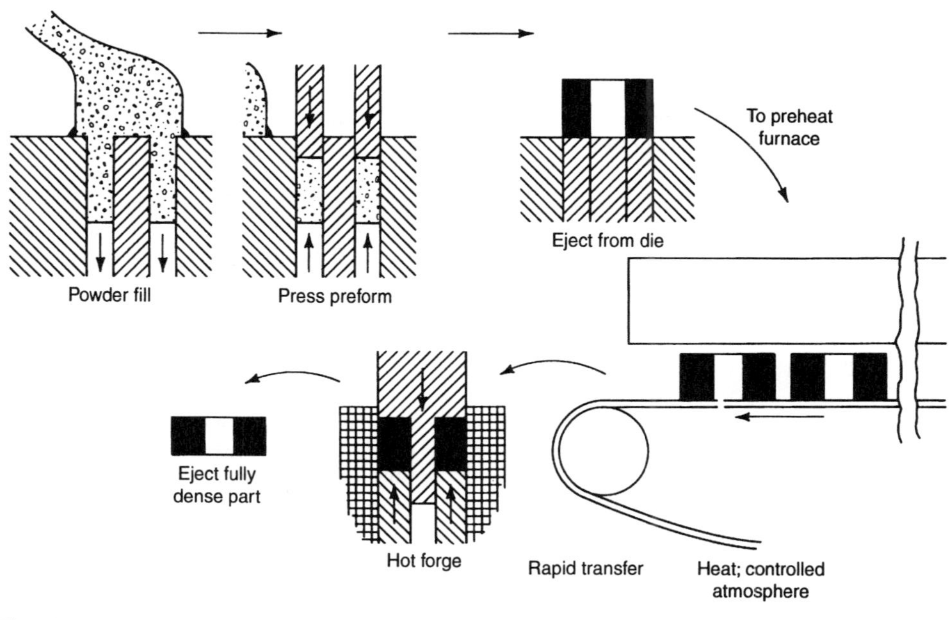

Fig. 1 The powder forging process

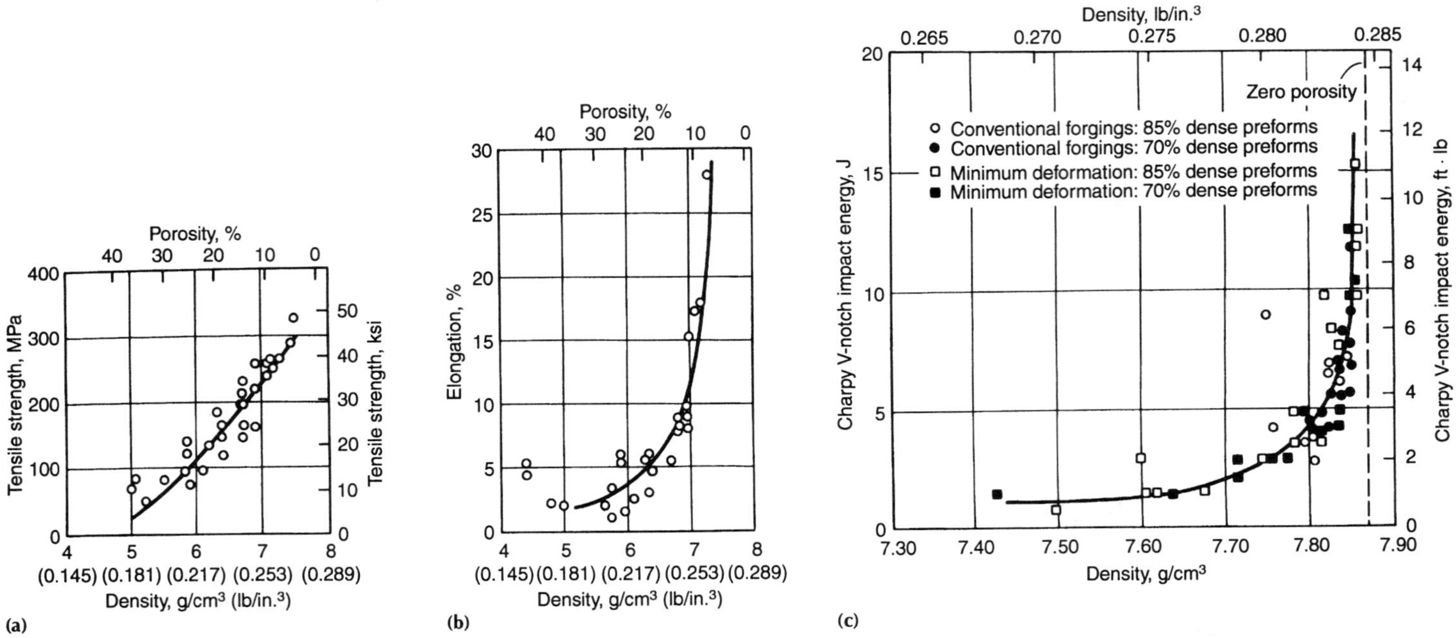

Fig. 2 Effect of density on mechanical properties. (a) and (b) As-sintered iron. Source: Ref 1. (c) Powder forged low-alloy steel. Source: Ref 2

Powder Characteristics. Shape, size distribution, apparent density, flow, and composition are important characteristics for both conventional P/M and powder forging processes. The shape of the particles is important in relation to the ability of the particles to interlock when compacted. Irregular particle shapes such as those produced by water atomization are typically used. In P/M parts, surface finish is related to the particle size distribution of the powder. In powder forging, however, the surface finish is directly related to the finish of the forging tools. This being the case, it might be considered possible to use coarser powders for powder forging (Ref 10). Unfortunately, the potential for deeper surface oxide penetration is greater when the proportion of coarser particles is increased. Typical pressing grades are –80 mesh with a median particle size of about 75 μm (0.003 in.). The apparent density and flow are important to maintain fast and accurate die filling. The chemistry affects the final alloy produced as well as the compressibility.

Green strength and compressibility are more critical in P/M than they are in P/F applications. Although there is a need to maintain edge integrity in P/F preforms, there are rarely thin, delicate sections that require high green strength. Because P/F preforms do not require high densities (typically 6.2 to 6.8 g/cm^3), the compressibility obtainable with prealloyed powders is sufficient. However, carbon is not prealloyed because it has an extremely detrimental effect on compressibility (Fig. 4).

Alloy Development. Several investigators have shown that forged conventional elemental powder mixes result in poor mechanical properties, such as fatigue resistance, impact resistance, and ductility (Ref 11, 12). This is almost entirely due to the chemical and metallurgical heterogeneity that exists in materials made by this method. To overcome this, very long diffusion times or higher processing temperatures are required to fully homogenize the material, particularly when elements such as nickel are used. Samples forged from prealloyed powder have also been shown to have better hardenability than samples forged from admixed powders (Ref 13). Fully prealloyed powders have therefore been produced by several manufacturers. Each particle in these powders is uniform in composition, thereby alleviating the necessity for extensive alloy diffusion.

Powder purity and the precise nature and form of impurities are also extremely important. In a conventional powder metal part, virtually all properties are considerably lower than those of equivalent wrought materials. The effect of inclusions is overshadowed by the effect of the porosity. For a powder forging at full density, as in a conventional forging, the dominant effect of residual porosity on properties is replaced by the form and nature of impurity inclusions.

The two principal requirements for powder forged materials are an ability to develop an appropriate hardenability to guarantee strength and to control fatigue performance by microstructural features such as inclusions.

Hardenability. Manganese, chromium, and molybdenum are very efficient promoters of hardenability, whereas nickel is not. In terms of their basic cost, nickel and molybdenum are relatively expensive alloying additions compared with chromium and manganese. On this basis, it would appear that chromium/manganese-base alloys would be the most cost-effective materials for powder forging. However, this is not necessarily the case, because these materials are highly susceptible to oxidation during the atomization process. In addition, during subsequent powder processing, high temperatures are required to reduce the oxides of chromium and manganese, and special care must be taken to prevent reoxidation during handling and forging. If the elements become oxidized, they do not contribute to hardenability. Nickel and molybdenum have the advantage that their oxides are reduced at conventional sintering temperatures. Alloy design is therefore a compromise and the majority of atomized prealloyed powders in commercial use are nickel/molybdenum based, with manganese present in limited quantities. The compositions of four commercial prealloyed, low-alloy steel powders are listed below.

Alloy	Typical composition(a), wt%		
	Ni	Mo	Mn
P/F-4600	1.8	0.55	0.20
P/F-4200	0.45	0.60	0.27
P/F-4400	...	0.85	0.15
P/F-4900	...	1.5	0.15

(a) Fe, balance

The higher cost of nickel and molybdenum along with the higher cost of powder compared with conventional wrought materials is often offset by the higher material utilization inherent in the powder forging process.

Material designations for powder forging are summarized in Table 1. These designations are taken from ASTM Specification B 848 "Standard Specification for Powder Forged (P/F) Ferrous Structural Parts."

More recently, P/F parts have been produced from iron powders (0.10 to 0.25 wt% Mn) with copper and/or graphite additions for parts that do not require the heat-treating response or high strength properties achieved through the use of

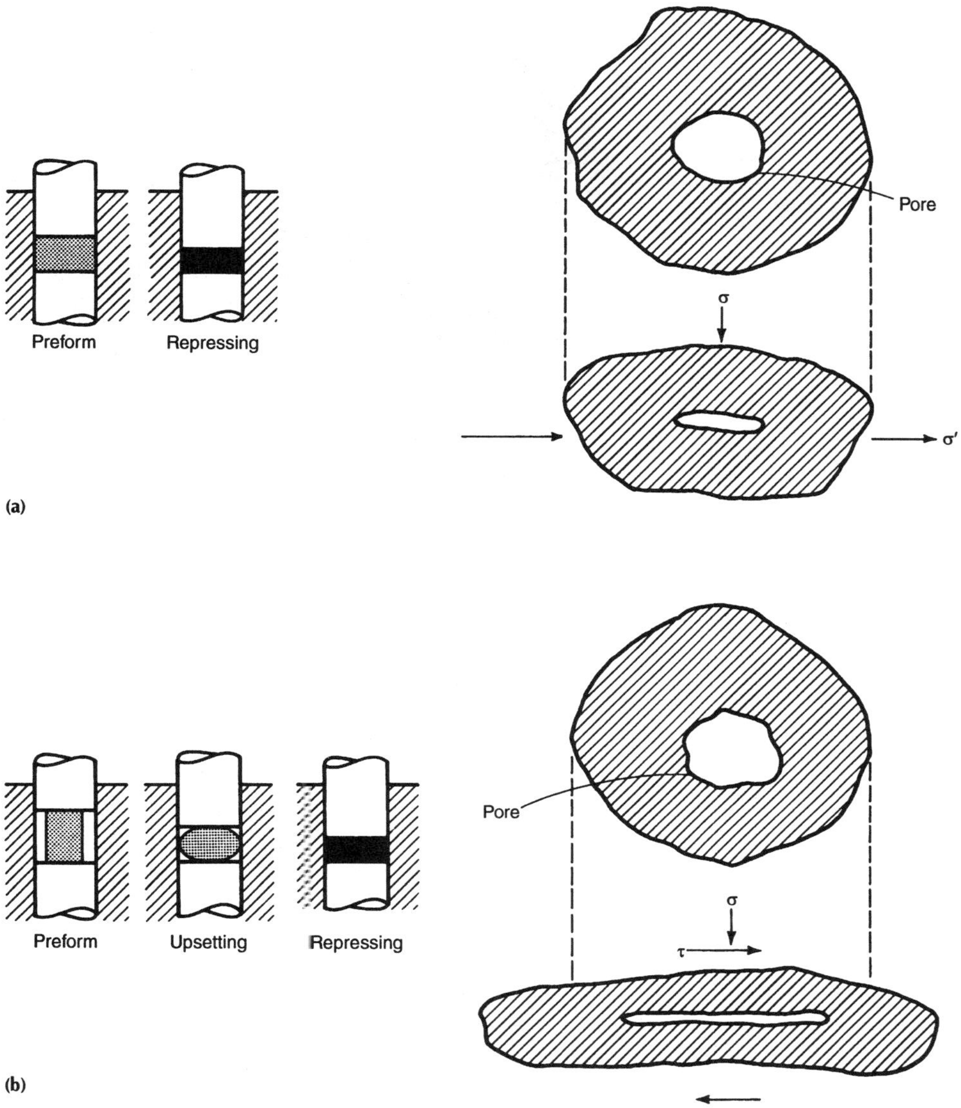

Fig. 3 Forging modes and stress conditions on pores for (a) repressing and (b) upsetting. Source: Ref 3

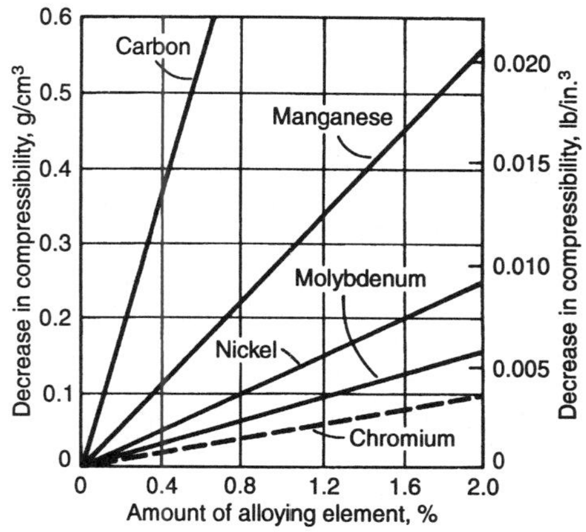

Fig. 4 Effect of alloying elements on the compressibility of iron powder. Source: Ref 10

Table 1 Material designations for powder forging

(See Table 12 for the chemical composition of the various materials)

Designation	Composition
P/F-10xx	Carbon steel (produced from atomized iron powder and graphite powder)
P/F-10Cxx	Copper-carbon steel (produced from atomized iron powder, copper, and graphite powders)
P/F-11xx	Carbon steel with manganese sulfide for enhanced machinability (produced from atomized iron powder, manganese sulfide, and graphite powders)
P/F-11Cxx	Carbon-carbon steel with manganese sulfide for enhanced machinability (produced from atomized iron powder, copper, manganese sulfide, and graphite powders)
P/F-42xx	Nickel-molybdenum steel (produced from prealloyed atomized iron-nickel-molybdenum powder and graphite powder)
P/F-46xx	Nickel-molybdenum steel (produced from prealloyed atomized iron-nickel-molybdenum powder and graphite powder)
P/F-44xx	Molybdenum steel (produced from prealloyed atomized iron-molybdenum powder and graphite powder)
P/F-49xx	Molybdenum steel (produced from prealloyed atomized iron-molybdenum powder and graphite powder)

the low-alloy steels. Detailed descriptions of alloy development for powder forging applications have been published previously (Ref 14, 15).

Inclusion Assessment. Because the properties of material powder forged to near full density are strongly influenced by the composition, size distribution, and location of nonmetallic inclusions (Ref 16–18), a method has been developed for assessing the inclusion content of powders intended for P/F applications (Ref 19–22). Samples of powders intended for forging applications are repress powder forged under closely controlled laboratory conditions. The resulting compacts are sectioned and prepared for metallographic examination. The inclusion assessment technique involves the use of automated image analysis equipment. The automated approach is preferred because it is not sensitive to operator subjectivity and can be used routinely to obtain a wider range of data on a reproducible basis.

In essence, an image analyzer consists of a good-quality metallurgical microscope, a video camera, a display console, a keyboard, a microprocessor, and a printer. The video image is assessed in terms of its gray-level characteristics, black and white being extremes on the available scale. The detection level can also be set to differentiate between oxides and sulfides.

Compared with wrought steels, only a limited amount of material flow is present in powder forged components. Inclusion stringers common to wrought steel are therefore not found in powder forged materials. Figure 5(a) illustrates an inclusion type encountered in powder forged low-alloy steels. The fragmented nature of these inclusions makes size determination by image analysis more complex than would be the case with the solid exogenous inclusion shown in Fig. 5(b). Basic image analysis techniques tend to count the inclusion shown in Fig. 5(a) as numerous small particles rather than as a single larger entity. Amendment of the detected video image is required to classify such inclusions; the method used is discussed in Ref 22 and forms the basis for ASTM Test Method B 796, "Standard Test Method for Nonmetallic Inclusion Level of Powder Forged Steel Parts."

Iron Powder Contamination. Water-atomized low-alloy steel powders are generally produced and processed in a plant that also manufactures pure iron powders. In the early days of alloy development when alloy powder production was limited, procedures were developed to minimize cross-contamination of powders. Considerable care is still taken to prevent cross-contamination, and iron powder contamination of low-alloy powders is typically <1%. Studies have shown that for "through hardening" applications, ≤3% iron powder contamination has little effect on the strength and ductility of powder forged material (Ref 23, 24).

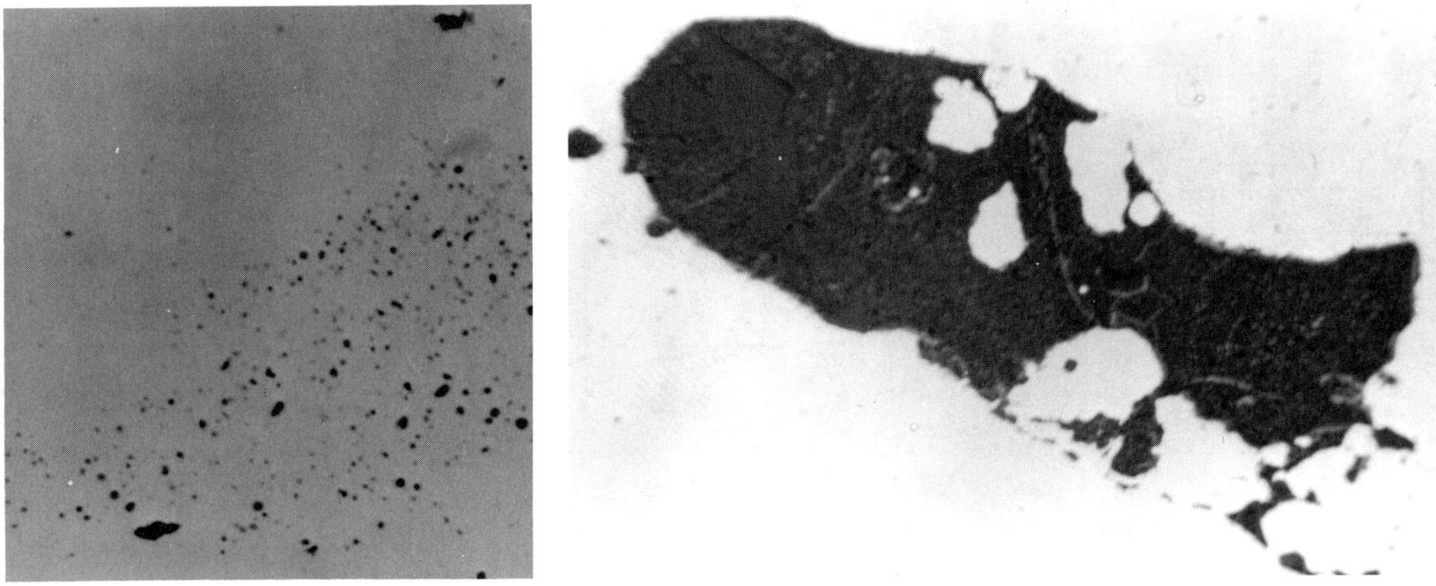

Fig. 5 Two types of inclusions. (a) "Spotty particle" oxide inclusion. 800×. (b) Exogenous slag inclusion. 590×. Source: Ref 21

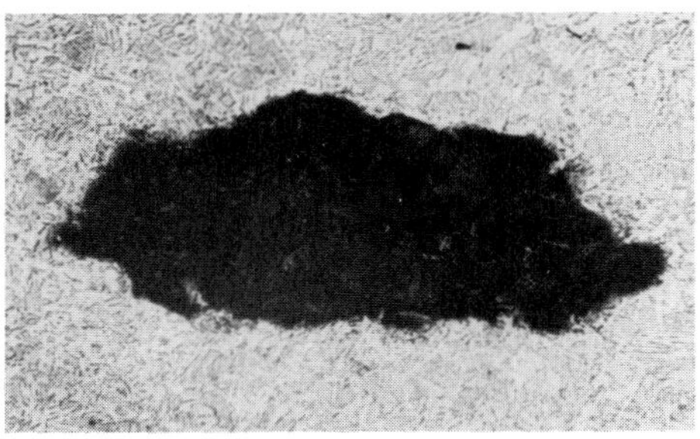

Fig. 6 Iron powder contamination of water-atomized low-alloy steel powder. Source: Ref 21

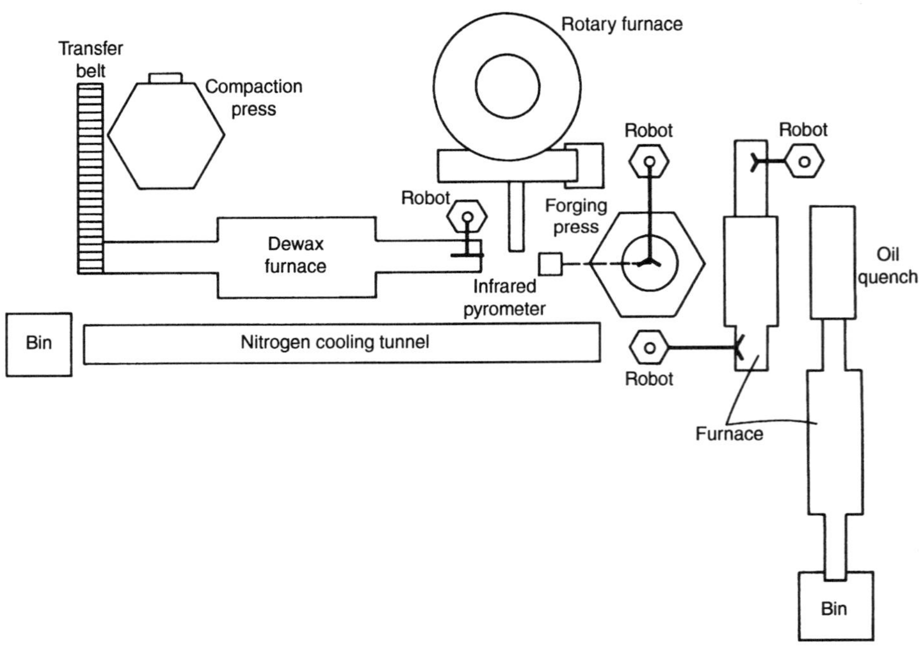

Fig. 7 A powder forging process line. Source: Ref 26

The compact used for inclusion assessment may also be used to measure the amount of iron powder particles present. The sample is lightly pre-etched with 2% nital. Primary etching is with an aqueous solution of sodium thiosulfate and potassium metabisulfite. This procedure darkens the iron particles and leaves the low-alloy matrix very light (Fig. 6) or, in the case of low-alloy contamination, darkens the matrix and leaves the low-alloy particles light.

The etched samples are viewed on a light microscope at a magnification of 100×. The total number of points of a 252 point grid that intersects iron particles for ten discrete fields is divided by the total number of points in the ten fields (2520) to determine the percentage of iron contamination. The approach forms the basis for ASTM B 795.

Process Considerations

Development of a viable powder forging system requires consideration of many process parameters. The mechanical, metallurgical, and economic outcomes depend to a large extent on operating conditions, such as temperature, pressure, flow/feed rates, atmospheres, and lubrication systems. Equally important consideration must be given to the types of processing equipment, such as presses, furnaces, dies, and robotics, and to secondary operations, in order to obtain the process conditions that are most efficient. This efficiency is maintained by optimizing the process line layout. Examples of effective equipment layouts for preforming, sintering, reheating, forging, and controlled cooling have been reviewed in the literature (Ref 4, 25). Figure 7 shows a few of the many possible operational layouts. Each of these process stages is reviewed in the following sections.

Preforming. Preforms are manufactured from admixtures of metal powders, lubricants, and graphite. Compaction is predominantly accomplished in conventional P/M presses that use closed dies. In order to avoid the necessity of thermally removing the lubricant, preforms can be compacted without admixed lubricants in an isostatic press. However, even though they pro-

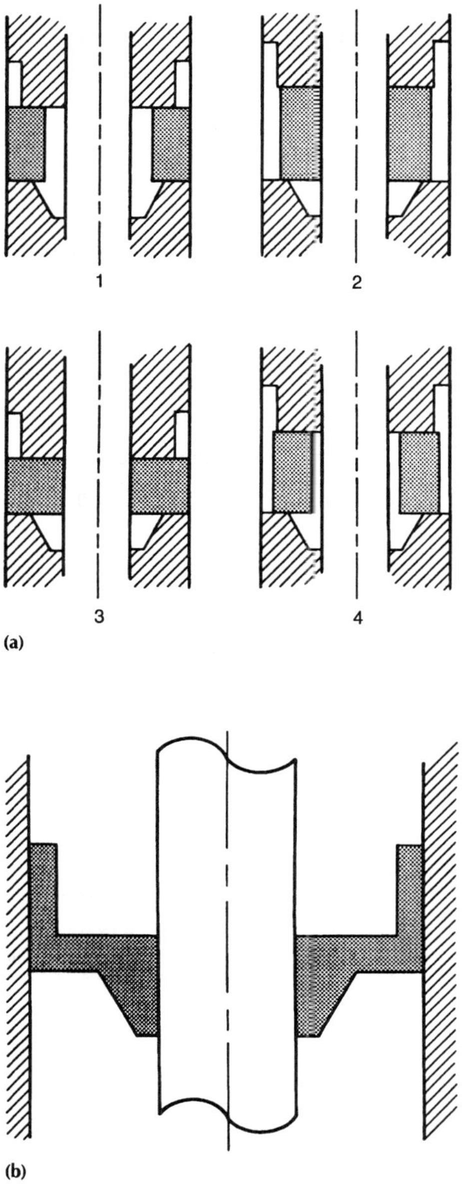

Fig. 8 (a) Possible configurations for the ring preform for forging the part shown in (b). See text for details. (b) Cross section of the part under consideration for powder forging. Source: Ref 31

duce uniform weight and density distributions, the pressure and rate limitations of high-production isostatic presses (414 MPa, or 60 ksi, pressure and 120 cycles per hour) have severely restricted their commercial use for compacting P/F preforms.

Control of weight distribution within preforms is essential to produce full density parts and thus maximize performance in the critical regions of the forged component. Excessive weight in any region of the preform may cause overload stresses that could lead to tool breakage.

Successful preform designs have been developed by an iterative trial and error procedure, using prior experience to determine the initial shape. More recently, computer-aided design (CAD) has been used for preform design (Ref 27–29). Preform design and workability analysis

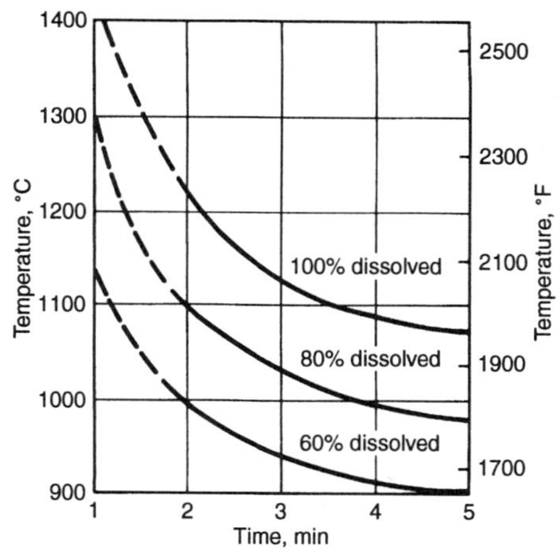

Fig. 9 Carbon dissolution as a function of time and temperature. Data are for an iron-graphite alloy at a density of 6.2 to 6.3 g/cm^3 (0.224 to 0.228 lb/in.^{3}). Source: Ref 34, Ref 8

is covered in detail by Ferguson and Kuhn (Ref 30).

Preform design is intimately related to the design and dimensions of the forging tooling, the type of forging press, and the forging process parameters. Among the variables to be considered for the preforming tools are:

- Temperature, that is, preform temperature, die temperature, and, when applicable, core rod temperature
- Ejection temperature of the forged part
- Lubrication conditions—influence on compaction/ejection forces and tooling temperatures
- Transfer time and handling of the preform from the preheat furnace to the forging die cavity

Correct preform design not only entails having the right amount of material in the various regions of the preform but also is concerned with material flow between the regions and prevention of potential fractures and defects.

An example of the effect of preform geometry on forging behavior can be taken from the work of C.L. Downey and H.A. Kuhn (Ref 31). Figure 8(a) shows four possible preforms that could be forged to produce the axisymmetric part having the cup and hub sections shown in Fig. 8(b).

Preform geometries 2 and 4 in Fig. 8(a) result in defective forgings due to cracking at the outer rim as metal flows around the upper punch radius. This occurs because the deforming preform is expanding in diameter as the metal flows around the corner, even though there is axial compression to help compensate for the circumferential tension. This type of cracking can be avoided by using a preform that fills the die with no clearance at the outside diameter, as in preforms 1 and 3.

Preform 3 can be rejected because it is similar to hub extrusion, and this may lead to cracking at the top surface of the hub. Allowing some clearance between the bore diameter of the preform and the mandrel eliminates this type of crack.

Table 2 Common powder forging problems and their probable causes

Forging problems	Probable causes
Surface oxidation	Extensive transfer time from furnace
Surface decarburization	Overly high forging temperature
	Entrapped liquid/graphite coating during reheat
	Excessive die lubrication (water)
	Oxidation during sintering or reheating
Surface porosity	Excessive contact time
	Low forging temperature
Tool wear	Low preform temperatures
	High or low tool temperature
	Excessive contact time
Poor tolerances	Temperature variations in tools and preforms
Excessive flash/tool jamming	Excessive preform temperature
	High preform weight/incorrect distribution
	Improper tool design
Excessive forging loads	Low preform temperature
	High preform weight
Low densities	Oxidation
	Low forging temperature/pressure
	Low preform weight
	Die chill
Cracks, laps	Improper tool or preform design
Improper die fill	Improper preform weight distribution/material flow

Preform 1 overcomes these problems. Use of this preform has resulted in defect-free parts, while the expected cracking occurred with use of the other preforms (Ref 31).

Sintering and Reheating. Preforms may be forged directly from the sintering furnace; sintered, reheated, and forged; or sintered after the forging process. The basic requirements for sintering in a ferrous powder forging system are lubricant removal, oxide reduction, carbon diffusion, development of particle contacts, and heat for hot densification. Oxide reduction and carbon diffusion are the most important aspects of the sintering operations. For most ferrous powder forging alloys, sintering takes place at ~1120 °C (2050 °F) in a protective reducing atmosphere with a carbon potential to prevent decarburization. The time required for sintering depends upon the number of sintering stages for delubrication, diffusion of carbon, reduction of oxides, and the type of sintering equipment used. Typical P/M sintering has been commonly performed at 1120 °C (2050 °F) for 20 to 30 min; these conditions may be required to help diffuse elements such as copper and nickel. In the prealloyed systems used for powder forging, only the diffusion of carbon is usually required. It has been shown that the time required to diffuse carbon and reduce the oxides is ~3 min at 1120 °C (2050 °F) (Ref 32–34). This is illustrated in Fig. 9. Increases in temperature will reduce the time required for sintering by improving oxide reduction and increasing carbon diffusion. Chromium-manganese steels have been limited in their use because of the higher temperatures required to reduce their oxides and the greater care needed to prevent reoxidation.

Any of the furnaces used for sintering P/M parts, such as vacuum, pusher, belt, rotary hearth, walking beam, roller hearth, and batch/box, may be used for sintering or reheating P/F preforms. Delubrication can be accomplished in any of these types of furnaces or in separate delubrication furnaces before entering the sintering furnace. Typically, belt, rotary hearth, and batch/box furnaces have been used for sintering and reheating preforms. However, the selection of sintering furnace largely depends upon the following conditions:

- Material being forged
- Size and weight of parts
- Forging process route (sinter/reheat versus sinter/forge)
- Forging temperature
- Atmosphere capabilities
- Delubrication capabilities
- Furnace loading capabilities/sintering rate
- Sintering time
- Robotics

The sintered preforms can be forged directly from the sintering furnace, stabilized at lower temperatures and forged, or cooled to room temperature, reheated, and forged. All cooling, temperature stabilization, and reheating must occur under protective atmosphere to prevent oxidation.

Induction furnaces are often used to reheat axisymmetric preforms to the forging temperature because of the short time required to heat the material. Difficulties may be encountered in obtaining uniform heating throughout asymmetric shapes because of the variation in section thickness.

Powder forging involves removing heated preforms from a furnace, usually by robotic manipulators, and locating them in the die cavity for forging at high pressures (690 to 965 MPa, or 100 to 140 ksi). Preforms can be graphite coated to prevent oxidation during reheating and transfer to the forging die. These dies are typically made from hot-work steels such as AISI H13 or H21. Lubrication of the die and punches is usually accomplished by spraying a water-graphite suspension into the cavity (Ref 35–37).

The forging presses commonly used in conventional forging (Ref 38–41), including hammers, high-energy-rate forming (HERF) machines, mechanical presses, hydraulic presses, and screw presses, have been evaluated for use in powder forging (Ref 8, 42). The essential characteristics that differentiate presses are contact time, stroke velocity, available energy and load, stiffness, and guide accuracy. Mechanical crank presses are the most widely used because of their short, fast strokes, short contact times, and guide accuracy. Hydraulic presses have also been used for applications in the 7.7 g/cm^3 density range, and screw presses are being used because of their lower cost and short contact times. More information on forging equipment is available in the articles "Hammers and Presses for Forging" and "Selection of Forging Equipment" in *ASM Handbook,* Volume 14, *Forming and Forging.*

Metal Flow in Powder Forging. Some of the problems encountered in powder forging and the probable causes are described in Table 2. These problems are related to the aforementioned sintering and reheating equipment and to the deformation processing described below.

Draft angles, which facilitate forging and ejection in conventional forging, are eliminated in powder forged parts. This means that greater ejection forces (~15 to 20% of press capacity as a minimum) are required for the powder forging of simple shapes. However, the elimination of draft angles permits P/F parts to be forged more closely to net shape. Figure 10 illustrates the ejection forces required for a P/F gear as a function of residual porosity (Fig. 10a) and preform temperature (Fig. 10b). To be suitable for powder forging, standard forging presses must be modified to have stronger ejection systems.

The deformation behavior of sintered, porous materials differs from that of wrought materials because porous materials densify during the forming operation. As a consequence, a porous preform will appear to have a higher rate of work hardening than its wrought counterpart. The work-hardening exponent, *m*, can be defined in terms of the true stress-true strain diagram:

$$\sigma = K\varepsilon^m \quad \text{(Eq 1)}$$

where σ is true stress, ε is true strain, and *K* is a proportionality constant. An empirical relationship between *m* and density ρ for a ferrous preform has been shown (Ref 43):

$$m = 0.31\rho^{-1.91} \quad \text{(Eq 2)}$$

where ρ is expressed as a fraction of the density of pore-free material. The value of *m* for pore-free pure iron is 0.31, and any excess over this value for porous iron is due to geometric work hardening.

A further consequence of the densification of porous preforms during deformation is reflected in Poisson's ratio for the porous material. Poisson's ratio is a measure of the lateral flow behavior of a material; for compression of a cylinder, it is expressed as diametral strain ε_d divided by height strain $-\varepsilon_z$. For a pore-free material, Poisson's ratio for plastic deformation ν is 0.5. This is a direct result of the fact that the volume of the material remains constant during deformation. For example, equating the volume of a cylinder before and after deformation (Ref 44):

$$H_0[(\pi D_0^2)/4] = H_f[(\pi D_f^2)/4] \quad \text{(Eq 3)}$$

where H_0 and H_f are initial and final cylinder heights, respectively, and D_0 and D_f are initial and final cylinder diameters, respectively.

Dividing by $H_f D_0^2$ yields:

$$H_0/H_f = (D_f/D_0)^2 \quad \text{(Eq 4)}$$

and taking logarithms

$$\ln(H_0/H_f) = \ln(D_f/D_0)^2 = 2\ln(D_f/D_0) \quad \text{(Eq 5)}$$

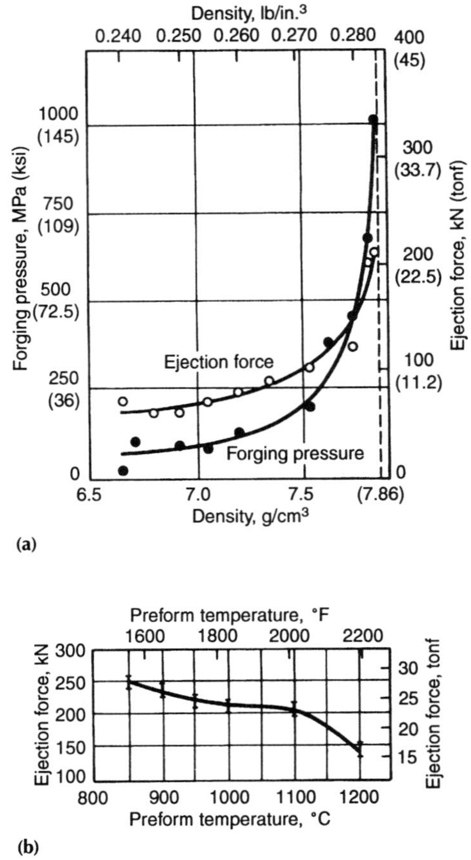

Fig. 10 (a) Forging pressure and ejection force as functions of density for the P/F-4600 powder forged gear shown in Fig. 12. Preform temperature: 1100 °C (2010 °F). (b) Ejection force after forging as a function of preform temperature for a powder forged gear. Forging pressure ranged from 650 to 1000 MPa (94 to 145 ksi).

or

$$-\varepsilon_z = 2\varepsilon_d \quad \text{(Eq 6)}$$

and, from the definition of Poisson's ratio:

$$v = \frac{-\varepsilon_d}{\varepsilon_z} = 0.5 \quad \text{(Eq 7)}$$

During compressive deformation of a sintered metal powder preform, some material flows into the pores, and there is a volume decrease. For a given reduction in height, the diameter of a sintered P/M cylinder will expand less than that of an identical cylinder of a pore-free material. Poisson's ratio for a P/M preform will therefore be less than 0.5 and will be a function of the pore volume fraction. H.A. Kuhn (Ref 44) has established an empirical relationship between Poisson's ratio and part density:

$$v = 0.5\rho^a \quad \text{(Eq 8)}$$

The best fit to experimental data is obtained with the exponent *a* = 1.92 for room-temperature deformation and *a* = 2.0 for hot deformation. The

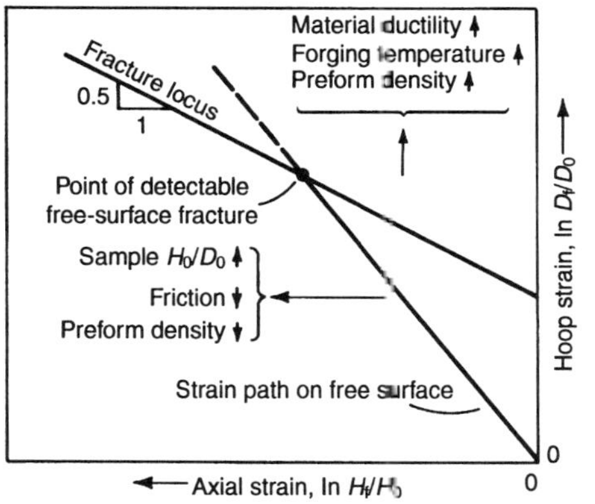

Fig. 11 Effects of forging variables on the workability of porous preforms in hot forging. Source: Ref 59

slight difference in this exponent may be due to work hardening (Ref 44).

In deformation processing of materials, plasticity theory is useful for calculating forming pressures and stress distributions. The above mentioned idiosyncrasies in the deformation behavior of sintered, porous materials have been taken into account in the development of a plasticity theory for porous materials. This has been of benefit in applying workability analysis to porous preforms (Ref 31, 43, 45–58).

A typical workability line is shown in Fig. 11, which also indicates the way processing variables affect the location of the line. The line has a slope of –0.5, and workability improves as the plane strain intercept (*y*-axis intercept) value increases. For a given material, workability can be improved through either temperature adjustment or a change in preform density. Figure 12(b) to (e) show the effects of temperature and pressure on the densification and forming of the powder forged gear shown in Fig. 12(a) (Ref 61). While Fig. 12(b) to (e) indicate that higher temperatures reduce the forging pressures required, Fig. 13 illustrates a region of forging pressure at lower temperatures that is comparable to that for higher-temperature forging. The ability to forge at lower temperatures may be beneficial in extending the life of the forging dies.

The data presented in Fig. 13 relate to a pure iron with no added graphite. The dramatic increase in the force required for densification ~900 °C (1650 °F) is due to the phase transformation from body-centered cubic (bcc) α iron to face-centered cubic (fcc) austenite. In this temperature range, the flow stress of austenite is higher than that of ferrite. However, although materials are fully austenitic at conventional conditions, in the data reported by Q. Jiazhong, O. Grinder, and Y. Nilsson (Ref 63), the mechanical properties of low temperature forged material are considerably inferior to those of material forged at higher temperatures (Table 3). The low-temperature forging resulted in incomplete densification, and this degraded the mechanical properties. G. Bockstiegel and H. Olsen observed a similar dependence of forged density on preform temperature (Ref 64). They pointed

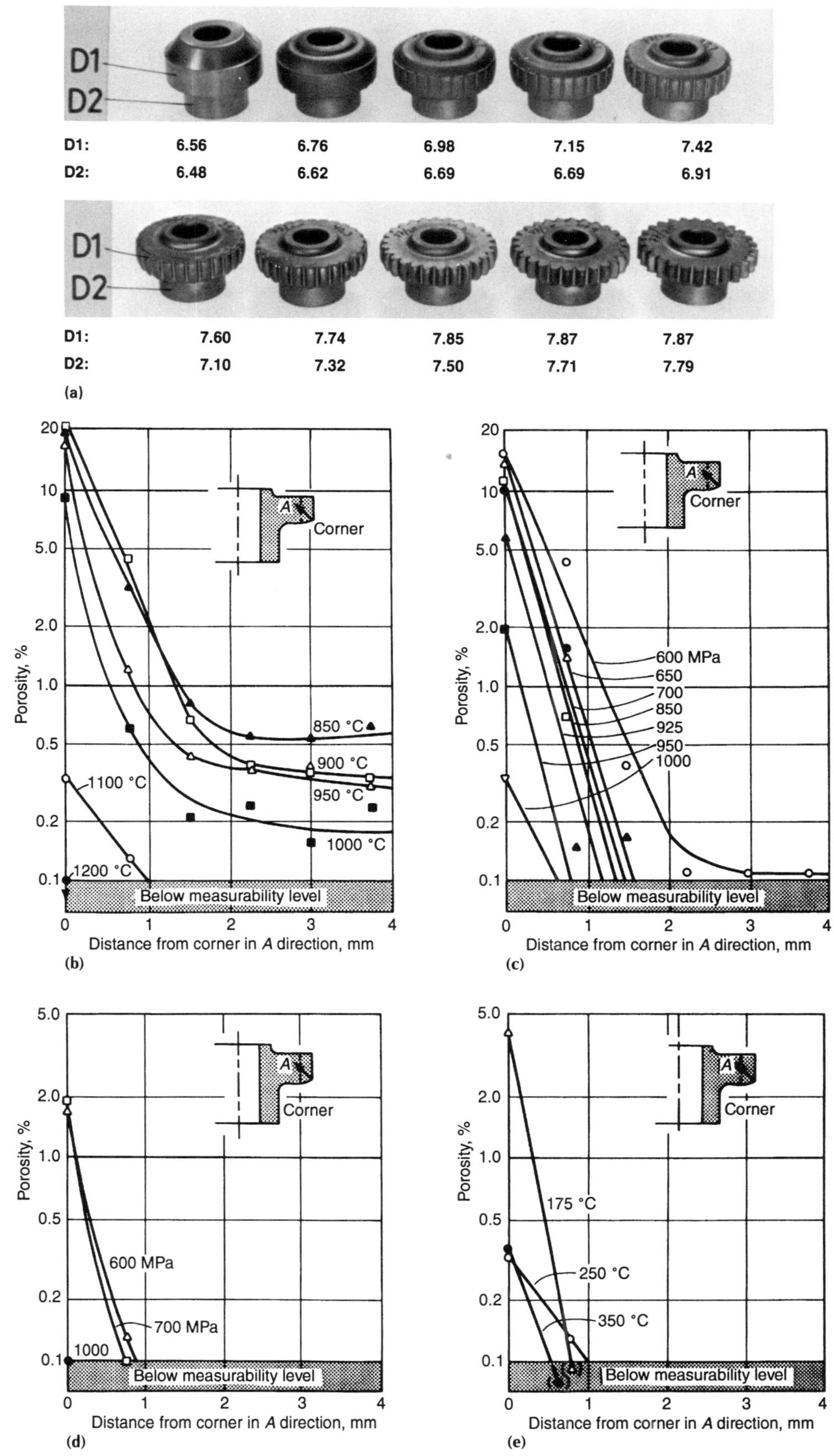

Fig. 12 Influence of process variables on residual porosity in critical corner areas of a powder forged gear tooth. (a) Powder forged gear; D1 and D2 are average densities in grams per cubic centimeter. (b) Preform temperature at a forging pressure of 1000 MPa (145 ksi). (c) and (d) Forging pressure at preform temperatures of 1100 °C (2010 °F) and 1200 °C (2190 °F), respectively. (e) Die temperature at a forging pressure of 1000 MPa (145 ksi) and a preform temperature of 1100 °C (2010 °F). Source: Ref 60, 61

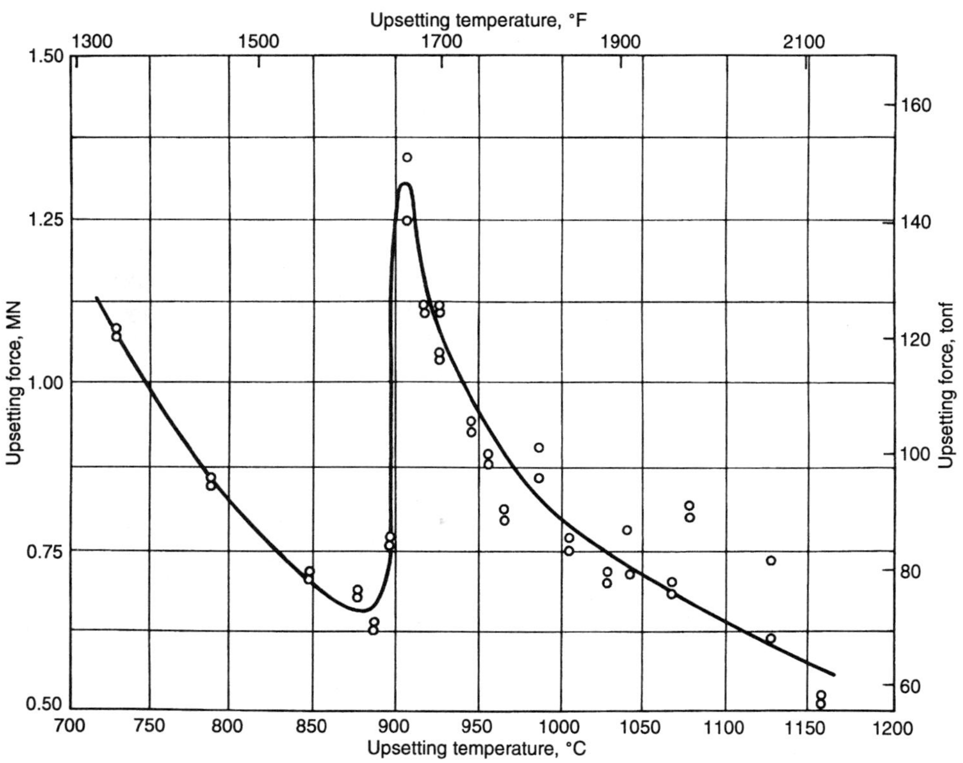

Fig. 13 Force required for a 50% reduction in height of water-atomized iron powder preforms as a function of deformation temperature. Source: Ref 62

out that the presence of free graphite might impede densification. During subsequent heat treatment, when the graphite goes into solution, it could leave fine porosity, which would degrade the mechanical properties of the material.

Metal flow can cause surface fractures. These are generally associated with contact forging temperatures (1000 to 1130 °C, or 1830 to 2065 °F), the flow stress of austenite at 1100 °C (2010 °F) is less than that of ferrite at 850 °C (1560 °F).

A similar low flow stress regime has been observed for prealloyed material (Fig. 14). However, depending on the amount of solution of graphite, the dip in the flow stress versus temperature curve becomes less pronounced and eventually is no longer observed. The presence of carbon in solution alters the phase distribution, and the observed flow stress depends on the relative proportions of ferrite and austenite in the microstructure.

In order to take any advantage of the low flow stress, the thermomechanical processing of preforms that contain added graphite must therefore be such that the graphite does not go into solution. Even under such conditions, in the data reported by Q. Jiazhong, O. Grinder, and Y. Nilsson (Ref 63), the mechanical properties of low temperature forged material are considerably inferior to those of material forged at higher temperatures (Table 3). The low-temperature forging resulted in incomplete densification, and this degraded the mechanical properties. G. Bockstiegel and H. Olsen observed a similar dependence of forged density on preform temperature (Ref 64). They pointed out that the presence of free graphite might impede densification. During subsequent heat treatment, when the graphite goes into solution, it could leave fine porosity, which would degrade the mechanical properties of the material.

Metal flow can cause surface fractures. These are generally associated with contact between the deforming preform and the forging tooling. Surface fracture problems can be avoided by changing the preform geometry or the lubrication conditions.

Frictional constraint at the interface between the preform and the forging die generates undesirable stress states in the preform that can lead to fracture. The types of fracture encountered in powder forging are free-surface fracture, die contact surface fracture, and internal fracture. Production of metallurgically sound forging requires the prediction and elimination of fracture. An excellent review of the subject is given in Ref 43, 46, 56, and 57.

Tool Design. In order to produce sound forged components, the forging tooling must be designed to take into account:

- Preform temperature
- Die temperature
- Forging pressure
- The elastic strain of the die
- The elastic/plastic strain of the forging
- The temperature of the part upon ejection
- The elastic strain of the forging upon ejection
- The contraction of the forging during cooling
- Tool wear

Specified part dimensional tolerances can only be met when the above parameters have been taken into account. However, there is still some flexibility in the control of forged part dimensions even after die dimensions have been selected. Higher preform ejection temperatures result in greater shrinkage during cooling. Increases in die temperature expand the die cavity and thus increase the size of the forged part. Therefore, if the forgings are undersize for a given set of forging conditions, a lower preform preheat temperature and/or a higher die preheat temperature can be used to produce larger parts. On the other hand, if the forged parts are oversize, the preform preheat temperature could be raised and/or the die temperature lowered to bring the parts to the desired size.

Secondary Operations. In general, the secondary operations applied to conventional com-

Table 3 Tensile and impact properties of P/F-4600 hot repressed at two temperatures

Repressing temperature		Repressing stress		Repressed density		0.2% offset yield strength		Ultimate tensile strength		Elongation,	Reduction in	Hardness(a),	Charpy V-notch impact energy	
°C	°F	MPa	ksi	g/cm³	lb/in.³	MPa	ksi	MPa	ksi	%	area, %	HV	J	ft·lbf
870	1600	406	59	7.65	0.276	1156	168	1634	237	2.6	2.8	519	2.9	2.13
870	1600	565	82	7.72	0.279	1243	180	1641	238	2.1	2.8	538	2.8	2.06
870	1600	741	107	7.78	0.281	1316	191	1702	247	2.4	2.4	564	3.1	2.29
870	1600	943	137	7.79	0.282	1349	196	1705	248	2.3	2.4	562	3.5	2.58
1120	2050	344	50	7.83	0.283	1364	198	1750	254	6.4	20.5	549	6.8	5.01
1120	2050	593	86	7.86	0.2840	1450	210	1777	258	6.7	17.3	566	6.2	4.57
1120	2050	856	124	7.87	0.2844	1592	231	1782	259	5.3	14.1	565	6.2	4.57
1120	2050	981	142	7.87	0.2844	1502	218	1788	260	5.5	12.3	572	6.0	4.42

(a) 30 kgf load

Table 4 Comparison of powder forging with competitive processes

Process	Range of weights, kg	Range of weights, lb	Height-to-diameter ratio	Shape	Material use, %	Surface roughness μm	Quantity required for economical production(a)	Cost per unit(b)
Powder forging	0.1–5	0.22–11	1	No large variations in cross section; openings limited	100	5–15	20,000	200
Precision forging	0.3–5	0.66–11	2	Any; openings limited	80–90	10–20	20,000	200
Cold forging	0.01–35	0.022–77	Not limited	Mostly rotational symmetry	95–100	1–10	5,000	150
Precision casting	0.1–10	0.22–22	Not limited	Any; no limits on openings	70–90	10–30	2,000	100
Sintering	0.01–5	0.022–11	1	No large variation in cross section; openings limited	100	1–30	5,000	100
Drop forging	0.05–1000	0.11–2200	Not limited	Any; openings limited	50–70	30–100	1,000	150

(a) For 0.5 kg (1.1 lb) parts. (b) Sintering = 100%

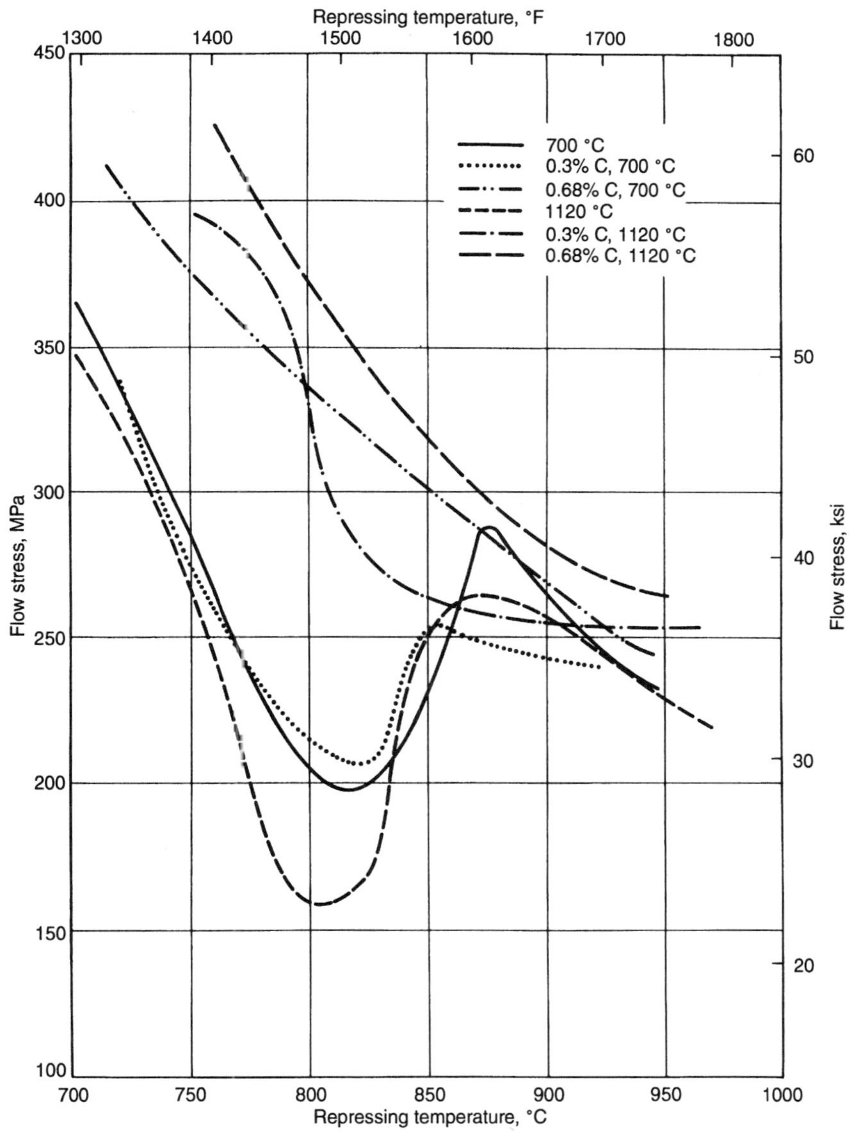

Fig. 14 Influence of hot repressing temperature on flow stress for P/F-4600 at various carbon contents and presintering temperatures. Data are for density of 7.4 g/cm^3 (0.267 lb/in.^{3}). Source: Ref 65, 66

ponents such as plating and peening, may be applied to powder forged components. The most commonly used secondary operations involve deburring, heat treating, and machining.

The powder forged components can require deburring or machining to remove limited amounts of flash formed between the punches and the die. This operation is considerably less extensive than that required for wrought forgings.

The heat treatment of P/M products is the same as that required for conventionally processed materials of similar composition. The most common heat-treating practices involve treatments such as carburizing, quench-and-temper cycles, or continuous-cooling transformations.

The amount of machining required for P/F components is generally less than the amount required for conventional forgings because of the improved dimensional tolerances, shown in Table 4. Standard machining operations can be used to achieve final dimensions and surface finish (Ref 67). One of the main economic benefits of powder forging is the reduced amount of machining required, as illustrated in Fig. 15.

In general, pore-free P/F materials machine as readily as conventional forgings processed to achieve identical composition, structure, and hardness. Difficulties are encountered, however, if P/F components are machined with the same cutting speeds, feed rates, and tool types as conventional components. These differences in machinability have been related to inclusion types and microporosity (Ref 16, 68). These studies conclude that P/F materials can exhibit equal or greater machinability than wrought steels. Improved machinability can be accomplished by the addition of solid lubricants such as manganese sulfide.

However, the presence of microporosity and low-density noncritical areas in P/F components leads to reduced machinability. The machinability behavior for these areas is similar to that of conventional P/M materials (Ref 69). The overall machinability of a powder forged component may be said to depend on the amount, type, size, shape, and dispersion of inclusions and/or porosity, as well as on the alloy and heat-treated structure.

Mechanical Properties

Wrought steel bar stock undergoes extensive deformation during cogging and rolling of the original ingot. This creates inclusion stringers and leads to planes of weakness, which affect the ductile failure of the material. The mechanical properties of wrought steels vary considerably according to the direction test pieces are cut from the wrought billet. Powder forged materials, on the other hand, undergo relatively little material deformation, and their mechanical properties have been shown to be relatively iso-

Table 5 Comparison of transverse and longitudinal mechanical properties of wrought steels

Material	Specimen orientation	Ultimate tensile strength		Yield strength, 0.2% offset		Impact energy		Fatigue endurance limit		Elongation, %	Reduction of area, %
		MPa	ksi	MPa	ksi	J	ft · lbf	MPa	ksi		
5046	Longitudinal	820	119	585	85	...	...	...	...	25.5	64
	Transverse	825	120	600	87	...	...	...	...	11.5	21
4340	Longitudinal	1095	159	1005	146	...	...	...	...	19.0	55
	Transverse	1095	159	1000	145	...	...	...	...	13.5	30
8620	Longitudinal	1060–1215	154–176	905–1070	131–155	...	...	...	...	12–15	53–57
	Transverse	1070–1240	155–180	905–1240	131–157	...	...	...	...	4–8	10–15
EN-16(a), lot Y	Longitudinal	920–980	133–142	...	...	100	74	310	45	17–19	60–62
	Transverse	910–950	132–138	...	...	10	7.4	250	36	5–12	8–24
EN-16(a), lot Z	Longitudinal	960–1000	139–145	...	...	100	74	400	58	17–18	58–62
	Transverse	950–970	138–141	...	...	10	7.4	290	42	7–10	6–15

(a) Composition of EN-16: Fe-1.7Mn-0.27Mo. Data on 5046 and 4340 are from Ref 71; data on 8620 are from Ref 72; data on EN-16 are from Ref 73.

Fig. 15 Comparison of material used for a conventionally forged reverse idler gear (top) and the equivalent powder forged part (bottom). Material yield in conventional forging is 31%; that for powder forging is 86%. 1 lb = 453.6 g. Source: Ref 61

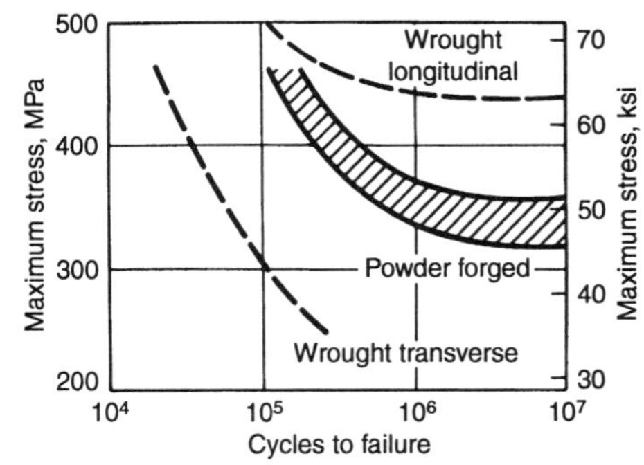

Fig. 16 Comparison of fatigue resistance of powder forged and wrought materials. Source: Ref 74

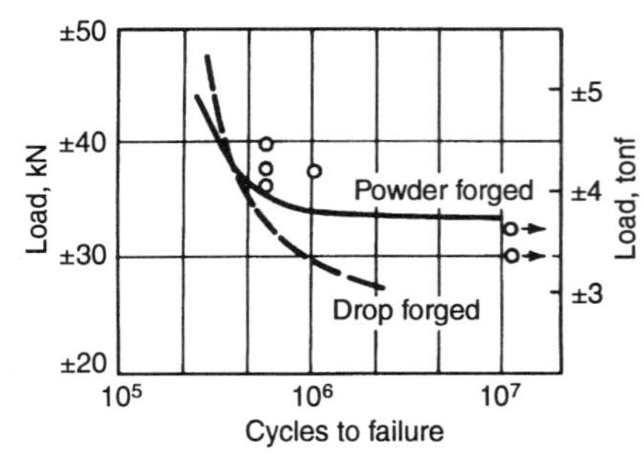

Fig. 17 Fatigue curves for powder forged and drop forged connecting rods. Source: Ref 75

tropic (Ref 70). The directionality of properties in wrought steel is illustrated in Table 5.

Mechanical properties of powder forged materials are usually intermediate to the transverse and longitudinal properties of wrought steels. The rotating-bending fatigue properties of powder forged material also have been shown to fall between the longitudinal and transverse properties of wrought steel of the same tensile strength (Ref 74). This is illustrated in Fig. 16.

While the performance of machined laboratory test pieces follows the intermediate trend described above, in the case of actual components, powder forged parts have been shown to have superior fatigue resistance (Fig. 17). This has generally been attributed not only to the relative mechanical property isotropy of powder forgings but also to their better surface finish and finer grain size.

The present section reviews the mechanical properties of powder forged materials. The data presented represent results obtained on machined standard laboratory test pieces. Data will be reported for four primary materials. The first two material systems are based on prealloyed powders (P/F-4600 and P/F-4200; see the section "Hardenability" in this article). The third material, based on an iron-copper-carbon alloy, was used by Toyota in 1981 to make P/F connecting rods; Ford Motor Company introduced powder forged rods with a similar chemistry in 1986 and powder forged copper-steel connecting rods are also currently used. Mechanical property data are therefore presented for copper and graphite powders mixed with an iron powder base to produce materials that generally contain 2 wt% Cu. Some powder forged components made from plain carbon steel are the fourth and final material for which mechanical property data are presented.

Forging Mode. It is well known that the forging mode has a major effect on the mechanical properties of components. With this in mind, the mechanical property data reported in this section were obtained on specimens that were either hot upset or hot repress forged. The forging modes used to produce billets for mechanical property testing are shown in Fig. 18.

Longitudinal test specimens 10 mm (0.4 in.) in diameter (for tensile and fatigue testing) and 10.8 × 10.8 mm (0.425 × 0.425 in.) square (for impact testing) were then cut from the forged billets. These specimen sizes represent comparable 10 mm (0.4 in.) diam ruling sections used for heat treatment and were the section sizes used unless otherwise noted.

Heat Treatments. There were three heat treatments used in developing the properties of the prealloyed powder forged materials: case carburizing, blank carburizing, and through-hardening (quenching and tempering).

Case carburizing was applied to materials with a nominal core carbon content of 0.20 to 0.25 wt%. Blank carburizing is intended to produce a microstructure similar to that found in the core of case carburized samples. The 0.20 to 0.25

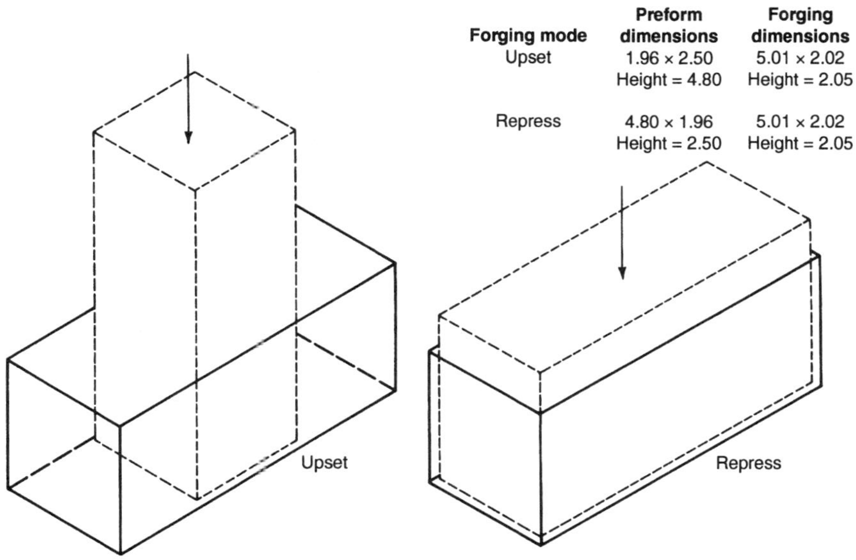

Fig. 18 Forging modes used in production of billets for mechanical testing. Dimensions, given in inches, are average values.

wt% C level results in a core hardness of 45 to 55 HRC.

Quenching and tempering was applied to achieve through-hardened microstructures over a range of forged carbon contents. A low-temperature temper or stress relief at 175 °C (350 °F) resulted in core hardnesses ~55 to 65 HRC for materials with carbon contents of ≥0.4 wt%. In addition, higher-temperature tempers were designed to achieve core hardnesses of 45 to 55 HRC and 25 to 30 HRC in these higher-carbon samples. Details of these heat treatments are given below.

Case Carburizing. Specimens were austenitized for 8 h at 955 °C (1750 °F) in an endothermic gas atmosphere with a dew point of −11 °C (+12 °F). They were then cooled to 830 °C (1525 °F) and stabilized at temperature in an endothermic gas atmosphere with a dew point of +2 °C (+35 °F).

The specimens were quenched in a fast quench rate oil with agitation at a temperature of 65 °C (150 °F). They were then stress relieved at 175 °C (350 °F) for 2 h. This heat treatment resulted in a case depth of ~1.52 mm (0.060 in.), with a 1.0 wt% carbon content in the case and a nominal core carbon of 0.25 wt%.

Blank Carburizing. The forged samples were austenitized for 2 h at 955 °C (1750 °F) in a dissociated ammonia and methane atmosphere. They were quenched with agitation in a fast quench rate oil at 65 °C (150 °F). The samples were reaustenitized at 845 °C (1550 °F) for 30 min in a dissociated ammonia and methane atmosphere, followed by oil quenching with agitation in oil held at 65 °C (150 °F). They were then stress relieved at 175 °C (350 °F) for 2 h in a nitrogen atmosphere.

Through-hardening. This quench and temper heat treatment consisted of austenitizing the specimens for 1 h at 955 °C (1750 °F) in a dissociated ammonia and methane atmosphere, followed by quenching with agitation in a fast quench rate oil at 65 °C (150 °F). The specimens were reaustenitized at 845 °C (1550 °F) for 30 min in a dissociated ammonia and methane atmosphere, followed by quenching with agitation in oil at 65 °C (150 °F). They were stress relieved for 1 h at 175 °C (350 °F) in a nitrogen atmosphere or tempered at the various temperatures listed in the tables. This procedure resulted in a uniform microstructure throughout the cross section.

Hardenability. Jominy hardenability curves are presented in Fig. 19 for the P/F-4600, P/F-4200, and iron-copper-carbon alloys. Testing was carried out according to ASTM A 255. Specimens were machined from upset forged billets that had been sintered at 1120 °C (2050 °F) in dissociated ammonia.

Tempering Response. Tempering curves (core hardness versus carbon content and tempering temperature) are presented in Fig. 20 for P/F-4200 and P/F-4600. The curves for P/F-4600 cover ruling sections of 10 mm (0.40 in.) to 25.4 mm (1.0 in.).

Tensile, Impact, and Fatigue Properties. Tensile properties were determined on test pieces with a gage length of 25.4 mm (1 in.) and a gage diameter of 6.35 mm (0.25 in.). Testing was carried out according to ASTM E 8 using a crosshead speed of 0.5 mm/min (0.02 in./min). Room-temperature impact testing was carried out on standard Charpy V-notch specimens according to ASTM E 23. Rotating-bending fatigue (RBF) testing was performed using single-load, cantilever, rotating fatigue testers. Dimensions of the RBF test specimen are shown in Fig. 21.

The tensile, impact, and fatigue data for the various materials are summarized in Tables 6 to 8 and Fig. 22 and 23.

The iron-copper-carbon alloys were either still-air cooled or forced-air cooled from the austenitizing temperature of 845 °C (1550 °F). Cooling rates for these treatments are shown in Fig. 24. The austenitizing temperature influences core hardness. These iron-copper-carbon alloys are often used with manganese sulfide additions for enhanced machinability. The tensile, impact, and fatigue properties for a sample with a 0.35 wt% manganese sulfide addition are compared with a material without sulfide additions in Table

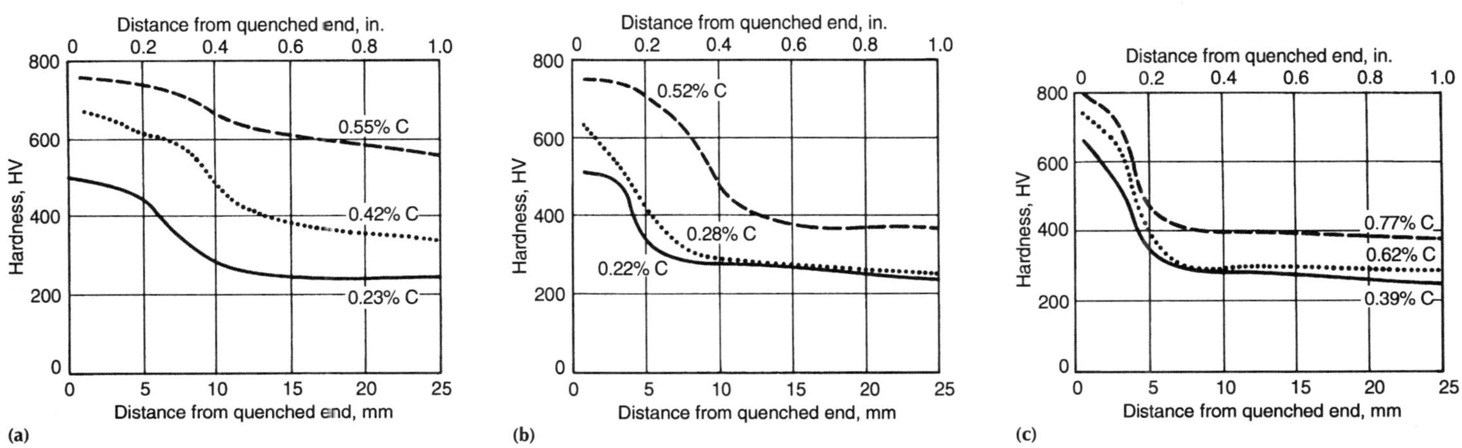

Fig. 19 Jominy hardenability curves for (a) P/F-4600, (b) P/F-2000, and (c) iron-copper-carbon materials at various forged-carbon levels. Vickers hardness was determined at a 30 kgf load.

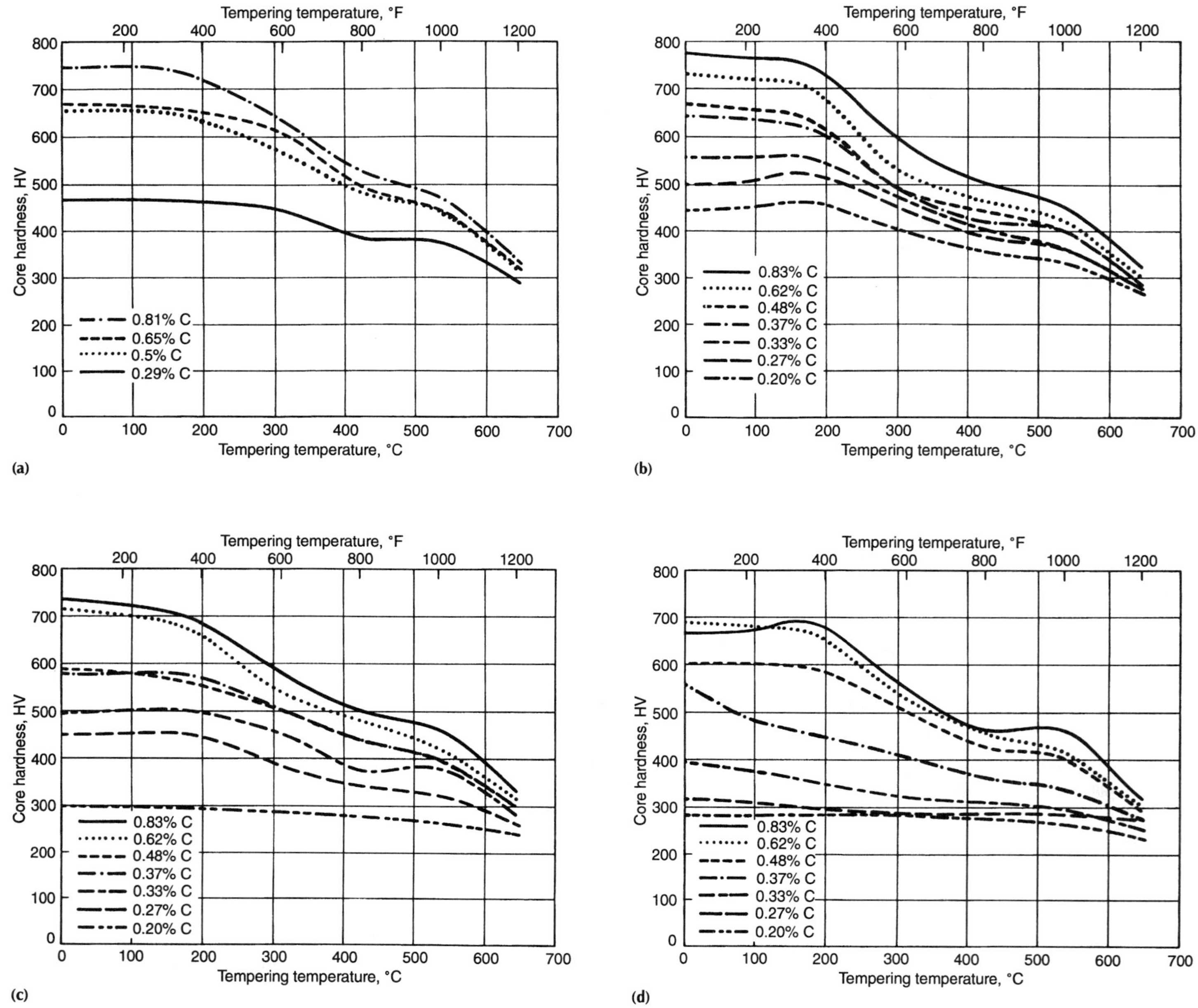

Fig. 20 Effect of tempering temperature and carbon content on the core hardness of (a) P/F-2000 for a ruling section of 10 mm (0.40 in.), and of P/F-4600 materials for ruling sections of (b) 10 mm (0.40 in.), (c) 19 mm (0.75 in.), and (d) 25.4 mm (1.0 in.)

9. The results obtained for a sulfurized powder sample are included for comparison. The tensile properties for iron-copper-carbon alloys with a range of forged carbon content are summarized in Fig. 23. Data from the samples with manganese sulfide and sulfurized powders are included for comparison. The manganese sulfide addition had little influence on tensile strength, whereas the sulfurization process degraded tensile properties.

Powder forged copper steels with tensile strengths ranging from 825 to 950 MPa and fatigue endurance limits from 186 to 292 MPa have been reported in the literature (Ref 77–82). Marra (Ref 79) conducted fatigue testing of powder forged connecting rods made from an admixed material with 2 wt% Cu, 0.60 wt% C, and 0.35 wt% MnS. The rods were developed for a General Motors V8 engine. The P/F rods showed improved fatigue life compared with that of wrought rods (AISI-SAE 1141) at an elevated test load.

Compressive Yield Strength. The 0.2% offset compressive yield strengths for P/F-4600 at various forged carbon levels and after different heat treatments are summarized in Table 10. A comparison of 0.2% offset tensile yield strength with the compressive yield strength for P/F-4600 with a range of carbon contents is given in Fig. 25 for samples stress relieved at 175 °C (350 °F).

Rolling-Contact Fatigue. Powder forged materials have been used in bearing applications. Rolling-contact fatigue testing is an accelerated bearing test used to rank materials with respect to potential performance in bearing applications. Rolling-contact fatigue testing of both case carburized and through-hardened P/F-4600 and P/F-2000 materials was carried out using ball/rod testers according to the procedure described in Ref 83. Weibull analysis data are summarized in Table 11.

Effect of Porosity on Mechanical Properties. The mechanical property data summarized in the previous sections are related to either hot repress or hot upset forged pore-free material. The general effect of density on mechanical properties was illustrated in Fig. 2, and the properties of material incompletely densified because of forging at 870 °C (1600 °F) were presented in Table 3. The tensile and impact properties of P/F-4600 with two levels of residual porosity are summarized in Fig. 26 and 27. In one instance, the material was at a density of 7.84 g/cm^3 and

Table 6 Mechanical property and fatigue data for P/F-4600 materials

Forging mode	Carbon, wt%	Oxygen, ppm	Ultimate tensile strength, MPa	Ultimate tensile strength, ksi	0.2% offset yield strength, MPa	0.2% offset yield strength, ksi	Elongation in 25 mm (1 in.), %	Reduction of area, %	Room-temperature Charpy V-notch impact energy, J	Room-temperature Charpy V-notch impact energy, ft·lbf	Core hardness, HV30	Fatigue endurance limit, MPa	Fatigue endurance limit, ksi	Ratio of fatigue endurance to tensile strength
Blank carburized														
Upset	0.24	230	1565	227	1425	207	13.6	42.3	16.3	12.0	487	565	82	0.36
Repress	0.24	210	1495	217	1325	192	11.0	34.3	12.9	9.5	479	550	80	0.37
Upset(a)	0.22	90	1455	211	1275	185	14.8	46.4	22.2	16.4	473	550	80	0.38
Repress(a)	0.25	100	1455	211	1280	186	12.5	42.3	16.8	12.4	468	510	74	0.36
Upset(b)	0.28	600	1585	230	1380	200	7.8	23.9	10.8	8.0	513	590	86	0.37
Repress(b)	0.24	620	1580	229	1305	189	6.8	16.9	6.8	5.0	464	455	66	0.29
Quenched and stress relieved														
Upset	0.38	270	1985	288	1505	218	11.5	33.5	11.5	8.5	554	...	...	...
Repress	0.39	335	1960	284	1480	215	8.5	21.0	8.7	6.4	...	...	...	...
Upset	0.57	275	2275	330	...	...	3.3	5.8	7.5	5.5	655	...	...	...
Repress	0.55	305	1945	282	...	...	0.9	2.9	8.1	6.0	...	...	...	...
Upset	0.79	290	940	136	...	...	0.0	0.0	1.4	1.0	712	...	...	...
Repress	0.74	280	1055	153	...	...	0.0	0.0	2.4	1.8	...	...	...	...
Upset	1.01	330	800	116	...	...	0.0	0.0	1.3	1.0	672	...	...	...
Repress	0.96	375	760	110	...	...	0.0	0.0	1.6	1.2	...	...	...	...
Quenched and tempered														
Upset(c)	0.38	230	1490	216	1340	194	10.0	40.0	28.4	21.0	473	...	...	...
Repress(c)	...	...	1525	221	1340	194	8.5	32.3	...	...	...	...	...	...
Upset(d)	0.60	220	1455	211	1170	170	9.5	32.0	13.6	10.0	472	...	...	...
Repress(d)	...	...	1550	225	1365	198	7.0	23.0	...	...	...	...	...	...
Upset(e)	0.82	235	1545	224	1380	200	8.0	16.0	8.8	6.5	496	...	...	...
Repress(e)	...	...	1560	226	1340	194	6.0	12.0	...	...	...	...	...	...
Upset(f)	1.04	315	1560	226	1280	186	6.0	11.8	9.8	7.2	476	...	...	...
Repress(f)	...	...	1480	215	1225	178	6.0	11.8	...	...	...	...	...	...
Upset(g)	0.39	260	825	120	745	108	21.0	57.0	62.4	46.0	269	...	...	...
Upset(g)	0.58	280	860	125	760	110	20.0	50.0	44.0	32.5	270	...	...	...
Upset(h)	0.80	360	850	123	600	87	19.5	46.0	24.4	18.0	253	...	...	...
Upset(j)	1.01	320	855	124	635	92	17.0	38.0	13.3	9.8	268	...	...	...

Sintered at 1120 °C (2050 °F) in dissociated ammonia unless otherwise noted. (a) Sintered at 1260 °C (2300 °F) in dissociated ammonia. (b) Sintered at 1120 °C (2050 °F) in endothermic gas atmosphere. (c) Tempered at 370 °C (700 °F). (d) Tempered at 440 °C (825 °F). (e) Tempered at 455 °C (850 °F). (f) Tempered at 480 °C (900 °F). (g) Tempered at 680 °C (1255 °F). (h) Tempered at 695 °C (1280 °F). (j) Tempered at 715 °C (1320 °F)

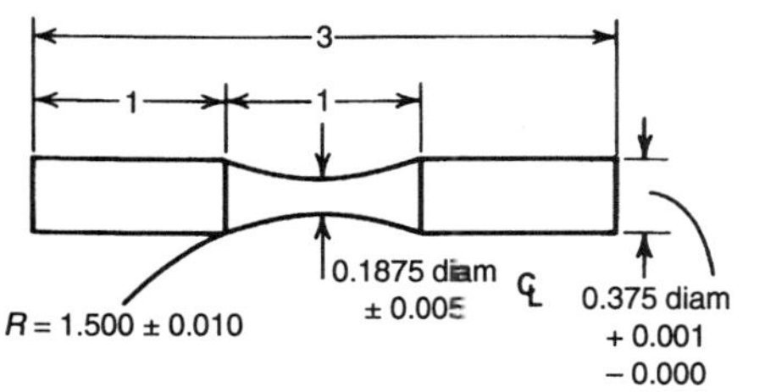

Fig. 21 Dimensions (in inches) of rotating-bending fatigue (RBF) test specimens

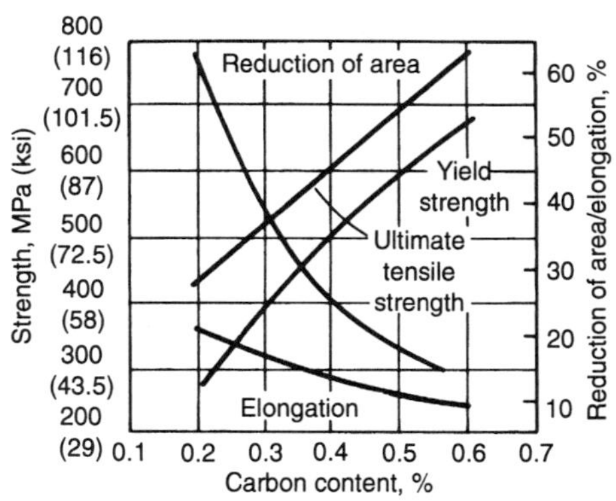

Fig. 22 Mechanical properties versus carbon content for iron-carbon alloys. Source: Ref 76

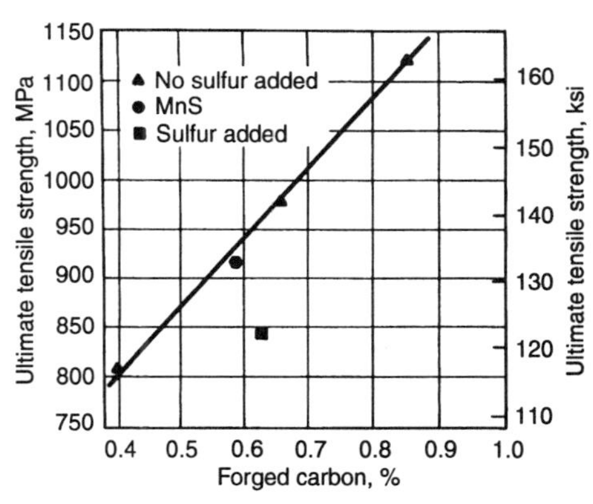

Fig. 23 Effect of sulfur and carbon on the ultimate tensile strength of iron-copper-carbon alloys. Samples were upset forged and forced-air cooled.

had a background of very fine porosity (Ref 84). The other series of samples had been purposely forged to a density of 7.7 g/cm^3 (Ref 85). The performance of these materials is compared with that for pore-free samples at two levels of core hardness: 25 to 30 HRC (Fig. 26) and 45 to 50 HRC (Fig. 27). At the lower hardness, porosity has no effect on tensile strength, but even fine microporosity significantly reduces tensile ductility and impact energy. Tensile ductility at the higher core hardness is slightly influenced by the fine microporosity, and is significantly reduced for the material with a density of 7.7 g/cm^3. The presence of porosity diminishes impact performance.

Recent studies published by Mashl and Smith have shown that, for quench-hardened and tempered impact forged P/F-4650 materials, residual porosity of as little as 0.1% can cause a significant decrease in tensile ductility (Ref 86). High tensile strength and ductility (1800 MPa and 40% reduction of area) have been achieved in P/F-4645 when the forged oxygen and sulfur contents are at very low levels (<100 ppm oxygen, and sulfur ~0.005 wt%) (Ref 87). Both tensile ductility and low cycle fatigue life are adversely affected by the presence of sulfide (MnS) and oxide inclusions (Ref 88). In a separate study (Ref 89), manganese sulfide added to P/F-4650 to serve as a machinability enhancer, was found to be detrimental to the ductility and low cycle fatige life of the material. Strength and ductility levels covered by the study ranged from 2070 MPa with 33% reduction in area for quenched and tempered material, to 690 MPa with 45% reduction in area for a furnace cooled sample.

Quality Assurance for P/F Parts

Many of the quality assurance tests applied to wrought parts are similar to those used for pow-

Table 7 Mechanical property data for P/F-4200 materials

Forging mode	Carbon, wt%	Oxygen, ppm	Ultimate tensile strength, MPa	Ultimate tensile strength, ksi	0.2% offset yield strength, MPa	0.2% offset yield strength, ksi	Elongation, in 25 mm (1 in.), %	Reduction of area, %	Core hardness(a), HV
Blank carburized									
Upset(b)	0.19	450	1205	175	...	...	10.0	37.4	390
Repress(b)	0.23	720	1110	161	...	...	6.3	17.0	380
Upset(c)	0.25	130	1585	230	...	...	13.0	47.5	489
Repress(c)	0.25	110	1460	212	...	...	11.3	36.1	466
Quenched and stress relieved									
Upset(b)	0.31	470	1790	260	...	...	9.0	27.3	532
Repress(b)	0.32	700	1745	253	...	...	4.0	9.0	538
Upset(b)	0.54	380	2050	297	...	...	1.3	...	694
Repress(b)	0.50	520	2160	313	...	...	2.0	...	653
Upset(c)	0.65	120	1605	233	...	...	...	...	710
Repress(c)	0.67	130	1040	151	...	...	...	...	709
Upset(b)	0.73	270	1110	161	...	...	...	...	767
Repress(b)	0.85	370	1345	195	...	...	...	...	727
Upset(b)	0.70	420	600	87	...	...	...	...	761
Repress(b)	0.67	320	540	78	...	...	...	...	778
Upset(c)	0.91	120	910	132	...	...	...	...	820
Repress(c)	0.86	120	840	122	...	...	...	...	825
Quenched and tempered									
Upset(d)	0.28	720	1050	153	895	130	10.6	42.8	336
Upset(e)	0.37	1200	1450	210	1385	201	10.2	33.0	447
Upset(e)	0.56	580	1680	244	7560	226	9.8	28.6	444
Upset(f)	0.70	760	1805	262	1565	227	5.0	11.8	531
Upset(g)	0.86	790	1425	207	1310	190	10.4	30.0	450
Upset(h)	0.26	920	835	121	705	102	22.6	57.6	269
Upset(j)	0.38	860	860	125	785	114	20.8	56.5	288
Upset(k)	0.55	840	917	133	820	119	17.8	49.5	305
Upset(l)	0.73	820	965	140	855	124	15.4	42.7	304
Upset(l)	0.87	920	995	144	850	123	15.6	33.9	318

(a) 30-kgf load. (b) Sintered in dissociated ammonia at 1120 °C (2050 °F). (c) Sintered in dissociated ammonia at 1260 °C (2300 °F). (d) Tempered at 175 °C (350 °F). (e) Tempered at 315 °C (600 °F). (f) Tempered at 345 °C (650 °F). (g) Tempered at 425 °C (800 °F). (h) Tempered at 620 °C (1150 °F). (j) Tempered at 650 °C (1120 °F). (k) Tempered at 660 °C (1225 °F). (l) Tempered at 675 °C (1250 °F)

der forged parts. ASTM specification B 848 covers material designations, chemical composition limits, and appropriate test methods for powder forged parts. Material designations are listed in Table 1 and chemical requirements are summarized in Table 12. Among the parameters specified are part dimensions, surface finish, magnetic particle inspection, composition, density, metallographic analysis, and nondestructive testing. These are discussed below.

Part Dimensions and Surface Finish. Typical tolerances for powder forged parts are summarized in Table 13. The as-forged surface finish of a powder forged part is directly related to the surface finish of the forging tool. Surface finish is generally better than 0.8 μm (32 μin.), which is better than that obtained on wrought forged parts. This good surface finish is beneficial to the fatigue performance of P/F parts.

Magnetic particle inspection is used to detect surface blemishes such as cracks and laps.

Composition. Parts are generally designed to a specified composition. The forged carbon and oxygen contents are of particular interest. The specified carbon level is required to achieve the desired heat treatment response, and forged oxygen levels have a significant influence on dynamic properties (Fig. 28).

Density. Sectional density measurements are taken to ensure that sufficient densification has been achieved in critical areas. Displacement density checks carried out per ASTM B 311 are generally supplemented by microstructural examination to assess the residual porosity level. Parts may be specified to have a higher density

Table 8 Mechanical property and fatigue data for iron-copper-carbon alloys (P/F-10C*xx*)

Forging mode	Carbon, wt%	Oxygen, ppm	Ultimate tensile strength, MPa	Ultimate tensile strength, ksi	0.2% offset yield strength, MPa	0.2% offset yield strength, ksi	Elongation in 25 mm (1 in.), %	Reduction of area, %	Room-temperature Charpy V-notch impact energy, J	Room-temperature Charpy V-notch impact energy, ft·lbf	Core hardness, HV30	Fatigue endurance limit, MPa	Fatigue endurance limit, ksi	Ratio of fatigue endurance to tensile strength
Upset(a)	0.39	250	670	97	475	69	15	37.8	4.1	3.0	228	...	...	...
Upset(b)	0.40	210	805	117	660	96	12.5	38.3	5.4	4.0	261	325	47	0.40
Repress(a)	0.39	200	690	100	490	71	15	35.4	2.7	2.0	227	...	...	...
Repress(b)	0.41	240	795	115	585	85	10	36.5	4.1	3.0	269	345	50	0.43
Upset(a)	0.67	170	840	122	750	109	10	22.9	2.7	2.0	267	...	...	...
Upset(b)	0.66	160	980	142	870	126	15	24.9	4.1	3.0	322	470	68	0.48
Repress(a)	0.64	190	825	120	765	111	10	24.8	3.4	2.5	266	...	...	...
Repress(b)	0.67	170	985	143	875	127	10	20.6	4.7	3.5	311	460	67	0.47
Upset(a)	0.81	240	1025	149	625	91	10	19.2	2.7	2.0	337	...	...	...
Upset(b)	0.85	280	1130	164	625	91	10	16.6	4.1	3.0	343	525	76	0.46
Repress(a)	0.81	200	1040	151	640	93	10	16.2	2.7	2.0	335	...	...	...
Repress(b)	0.82	220	1170	170	745	108	10	12.8	2.7	2.0	368	475	69	0.41

Sintered at 1120 °C (2050 °F) in dissociated ammonia, reheated to 980 °C (1800 °F) in dissociated ammonia, and forged. (a) Still-air cooled. (b) Forced-air cooled

Table 9 Mechanical property and fatigue data for iron-copper-carbon alloys with sulfur additions

Addition	Carbon, wt%	Oxygen, ppm	Sulfur, wt%	Ultimate tensile strength, MPa	Ultimate tensile strength, ksi	0.2% offset yield strength, MPa	0.2% offset yield strength, ksi	Elongation in 25 mm (1 in.), %	Reduction of area, %	Room-temperature Charpy V-notch impact energy, J	Room-temperature Charpy V-notch impact energy, ft·lbf	Core hardness, HV30	Fatigue endurance limit, MPa	Fatigue endurance limit, ksi	Ratio of fatigue endurance to tensile strength
Manganese sulfide	0.59	270	0.13	915	133	620	90	11	23.2	6.8	5.0	290	430	62	0.47
Sulfur	0.63	160	0.14	840	122	560	81	12	21.4	6.8	5.0	267	415	60	0.50
None	0.66	160	0.013	980	142	870	126	15	24.9	4.1	3.0	322	470	68	0.48

Sintered at 1120 °C (2050 °F) in dissociated ammonia, reheated to 980 °C (1800 °F) in dissociated ammonia, and forged

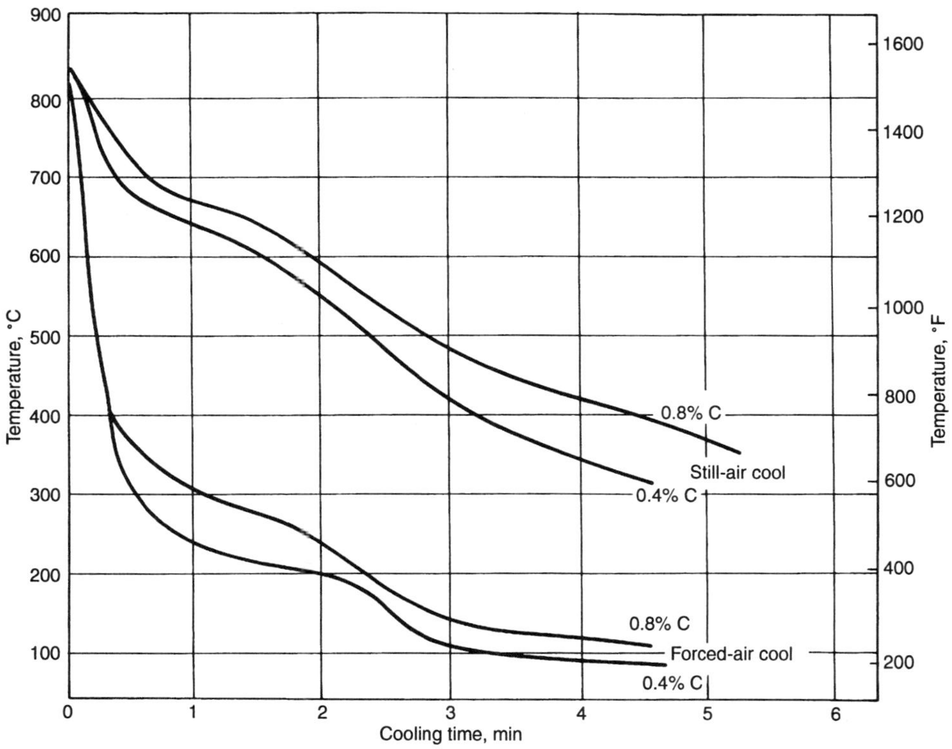

Fig. 24 Cooling rates used for iron-copper-carbon alloys

Table 10 Compressive yield strengths of P/F-4600 materials

Forged carbon content, wt%	Forged oxygen content, ppm	Heat treatment	Compressive yield strength (0.2% offset) MPa	ksi
0.22	460	Stress relieved at 175 °C (350 °F)	1240	180
0.22	350	Tempered at 370 °C (700 °F)	1155	168
0.22	440	Tempered at 680 °C (1255 °F)	575	84
0.29	380	Stress relieved at 175 °C (350 °F)	1440	209
0.35	430	Stress relieved at 175 °C (350 °F)	1670	242
0.43	410	Stress relieved at 175 °C (350 °F)	1690	245
0.41	410	Tempered at 370 °C (700 °F)	1360	197
0.41	460	Tempered at 680 °C (1255 °F)	680	99
0.46	480	Stress relieved at 175 °C (350 °F)	1780	259
0.44	380	Tempered at 370 °C (700 °F)	1275	185
0.44	400	Tempered at 680 °C (1255 °F)	685	100
0.57	330	Stress relieved at 175 °C (350 °F)	1980	287
0.66	400	Tempered at 440 °C (825 °F)	1325	192
0.60	330	Tempered at 680 °C (1255 °F)	700	101
0.75	300	Stress relieved at 175 °C (350 °F)	2000	290
0.80	480	Tempered at 455 °C (850 °F)	1355	196
0.77	410	Tempered at 695 °C (1280 °F)	700	101

Sintered at 1120 °C (2050 °F) in dissociated ammonia

in particular regions than is necessary in less critical sections of the same component.

For a given level of porosity, the measured density will depend on the chemistry, thermomechanical condition, and microstructure of the sample. The same sample will have a lower density in the quench-hardened and stress-relieved condition than it will after it has been fully annealed. This is extremely important to remember when establishing density levels in connection with part specifications. Empirical attempts to estimate the density of powder forged materials have been proposed by Cundill (Ref 92). More recently, however, methods for calculating the pore-free density of the alpha iron plus iron carbide and the virgin martensitic microstructures of plain-carbon and low-alloy steels have been developed by Smith (Ref 93). The calculations are based primarily on fundamental considerations of the crystal unit cell volume and the mass of each phase in the steel; these account for measured or estimated levels and types of nonmetallic inclusions. The calculation methods do not account for retained austenite or the presence of alloy carbides. The calculations also underestimate the pore-free densities of P/F materials containing admixed copper if significant compositional gradients exist in the material.

Smith and Mashl (Ref 94) compared the measured and calculated values for the density of powder forged steels. Their work showed that a fully virgin martensitic microstructure is difficult to obtain even in small sections that have been very rapidly quenched. They concluded that the alpha iron plus iron carbide ($\alpha + Fe_3C$) microstructure, which can be produced by slow cooling from austenite, or by extended spheroidization of a martensitic microstructure, should be used to assess the residual porosity level in P/F materials.

Metallographic Analysis. Powder forged parts are subjected to extensive metallographic evaluation. The primary parameters of interest include those discussed below.

The extent of surface decarburization permitted in a forged part will generally be specified. The depth of decarburization may be estimated by metallographic examination, but it is best quantified using microhardness measurements as described in ASTM E 1077.

Surface finger oxides are defined as oxides that follow prior particle boundaries into the forged part from the surface and cannot be removed by physical means such as rotary tumbling. An example of surface finger oxides is shown in Fig. 29. Metallographic techniques are used to determine the maximum depth of surface finger oxide penetration in accordance with ASTM method B 797.

Interparticle oxides follow prior particle boundaries. They can sometimes form a continuous three-dimensional network but more often will, in a two-dimensional plane of polish, appear to be discontinuous. An example is presented in Fig. 29. Metallographic techniques are used to detect their presence (ASTM B 797).

Most parts have what can be defined as functionally critical areas. The fabricator and end-user decide upon the maximum permissible depth of surface finger oxide penetration and whether oxide networks can be tolerated in critical regions. These decisions are then specified on the part drawing or in the purchase agreement as outlined in ASTM specification B 848.

The microstructure of a powder forged part depends on the thermal treatment applied after the forged part has been ejected from the die cavity. Most parts are carburized, quenched and stress relieved, or quenched and tempered. Other heat treatments used on wrought steels can also be applied to powder forged materials.

Alloyed or unalloyed iron contamination in powder forged parts can be quantified by ASTM method B 795 by using the etching procedure described in the section "Material Considerations" in this article.

The nonmetallic inclusion level in a powder forged part may also be quantified using the image analysis technique described in the section "Material Considerations." However, if the section of a component selected for inclusion assessment is not pore-free, image analysis procedures are not applicable (pores and oxide inclusions have similar gray level characteristics for feature detection). In fact, the presence of porosity makes it difficult for even visual quantitative determination of inclusion size. ASTM method B 796 is not applicable.

Nondestructive Testing. Although metallographic assessment of powder forged parts is common, it is also useful to have a nondestructive method for evaluating the microstructural integrity of components. It has been demonstrated that this can be achieved with a magnetic bridge comparator.

Magnetic bridge sorting can be used to compare the eddy currents developed within a forging placed in a coil that carries an alternating current with the eddy currents produced in a randomly selected reference sample from the same forging batch (Ref 21). Differences are indicated by the displacement of a light spot from its balanced position in the center of the measuring screen of the system. If the part being tested is similar to the reference sample, the light spot returns to the center of the screen. The screen can be arbitrarily divided into a number of zones, as illustrated in Fig. 30. Testing of randomly selected samples can then be used to establish a typical frequency distribution of components within a forged batch relative to the reference sample.

Once the frequency distribution has been established for a limited number of components within a forging batch, selected components that are representative of several zones on the screen are subjected to metallographic examination. Limited metallographic testing thus can be used to check the metallurgical integrity of parts from various zones.

Once acceptable zones have been defined, the entire forging batch can be assessed by means of the magnetic bridge. Components in unacceptable categories are automatically rejected. Experience with this technique minimizes the number of parts requiring sectioning for metallographic examination. Core hardness, surface decarburization, surface oxide penetration, and porosity can also be evaluated using this technique.

Magnetic bridge sorting, an adaptation of the technique used to test drop forged parts, enables potentially defective components to be elimi-

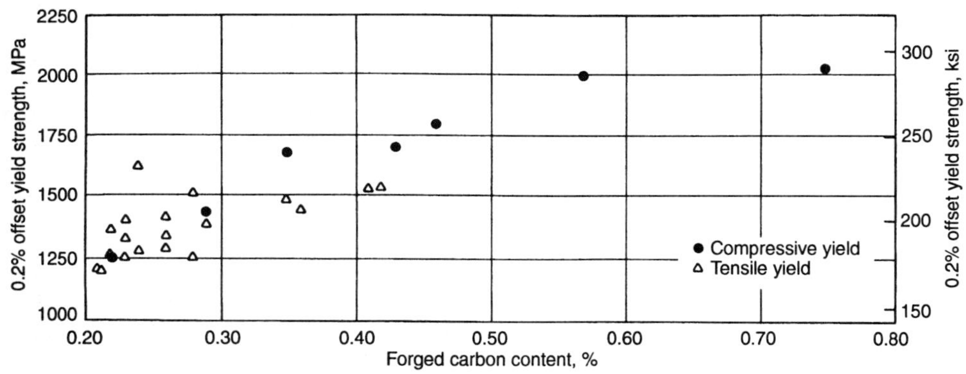

Fig. 25 Comparison of the tensile and compressive yield strengths of quenched and stress relieved P/F-4600 at various carbon levels

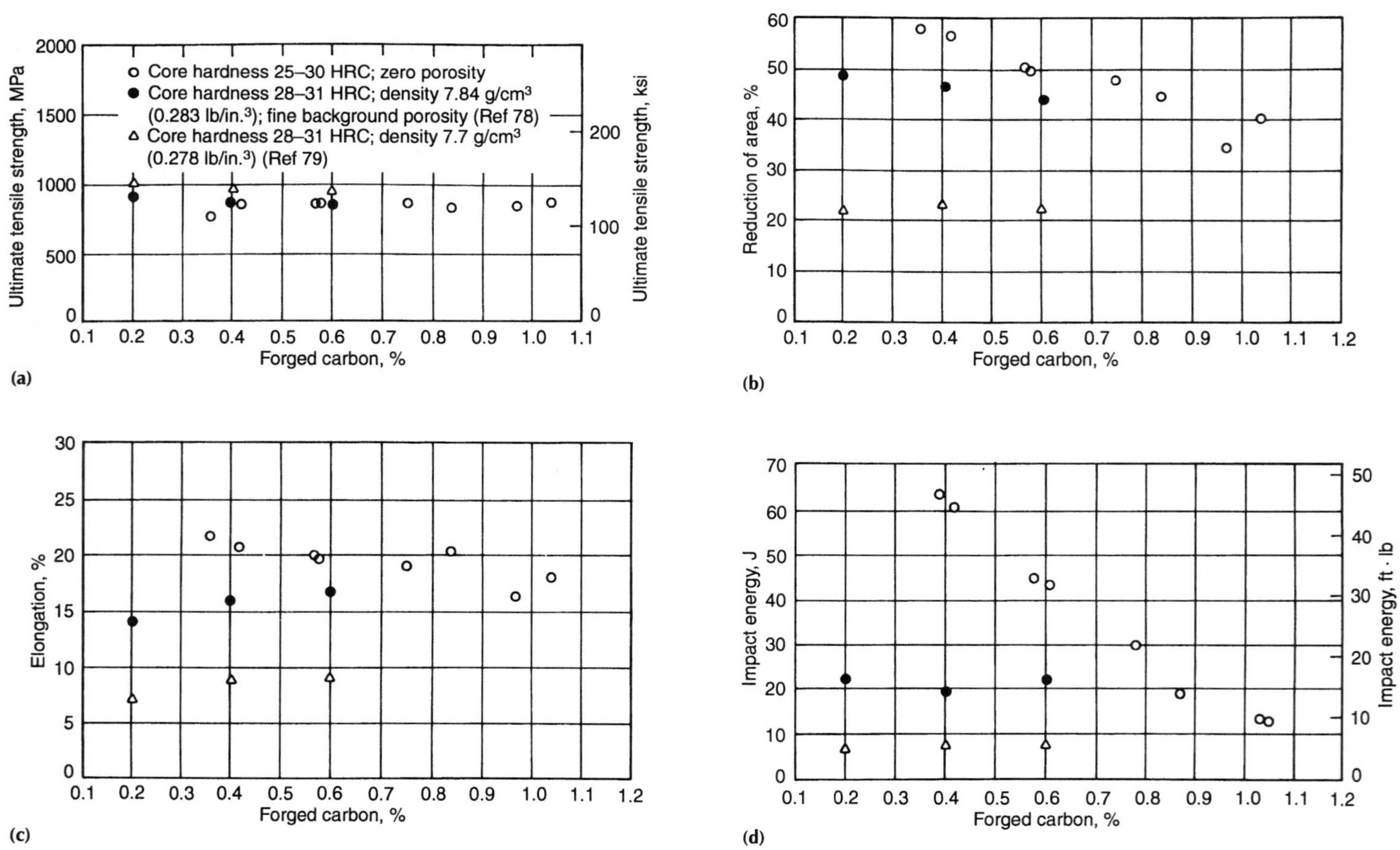

Fig. 26 Influence of density on the tensile and impact properties of P/F-4600 materials with core hardnesses of 25 to 30 HRC and 28 to 31 HRC. (a) Ultimate tensile strength. (b) Percent reduction of area. (c) Percent elongation. (d) Room-temperature impact energy. See also Fig. 27.

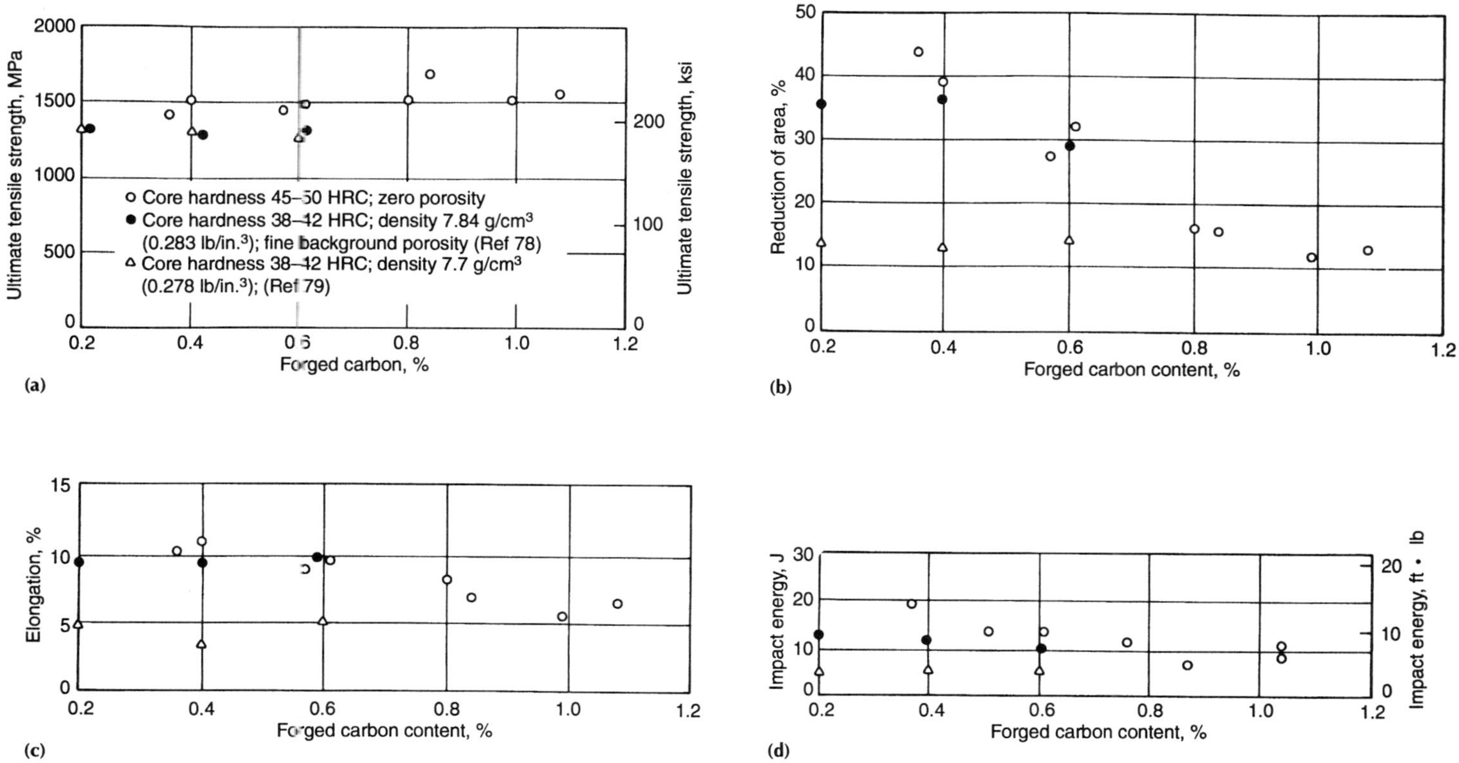

Fig. 27 Influence of density on the tensile and impact properties of P/F-4600 materials with core hardnesses of 38 to 42 HRC and 45 to 50 HRC. (a) Ultimate tensile strength. (b) Percent reduction of area. (c) Percent elongation. (d) Room-temperature impact energy. See also Fig. 26.

Table 11 Rolling-contact fatigue data for carburized and through-hardened P/F-4600 and P/F-4200

Sintering conditions	Forging mode	Carbon, wt%	Oxygen, ppm	Life to 10% failure rate, 10^6 cycles	Life to 50% failure rate, 10^6 cycles	Slope of Weibull plot	Surface hardness, HRC
Carburized P/F-4600							
1120 °C, DA(a)	Upset	...	...	4.31	12.59	1.78	...
1120 °C, DA	Repress	...	...	4.95	16.40	1.59	...
1260 °C, DA	Upset	...	...	4.27	16.70	1.38	...
1260 °C, DA	Repress	...	...	12.50	23.00	3.18	...
1120 °C, ENDO(b)	Upset	...	...	13.80	27.20	2.82	...
1120 °C, ENDO	Repress	...	...	6.37	22.24	1.52	...
Through-hardened P/F-4600							
1120 °C, DA	Upset	0.81	220	5.77	9.70	3.66	...
1120 °C, DA	Repress	0.81	210	6.35	11.16	3.35	...
1120 °C, DA	Upset	1.03	220	5.60	12.97	2.26	...
1120 °C, DA	Repress	0.98	330	3.89	11.31	1.78	...
1260 °C, DA	Upset	0.79	75	11.62	17.61	4.58	...
1260 °C, DA	Repress	0.78	85	9.00	18.38	2.66	...
1260 °C, DA	Upset	1.02	99	10.39	24.23	2.24	...
1260 °C, DA	Repress	0.99	110	3.96	17.53	1.27	...
Carburized P/F-2000							
1120 °C	Upset	...	...	1.13	6.06	1.13	64.0
1120 °C	Repress	...	...	1.34	5.30	1.38	63.0
1260 °C	Upset	...	...	2.79	8.28	1.74	63.5
1260 °C	Repress	...	...	1.11	6.52	1.07	63.0
Through-hardened P/F-2000							
1120 °C	Upset	0.67	450	1.75	5.93	1.56	60.5
1120 °C	Repress	0.70	460	1.97	6.28	1.64	61.0
1120 °C	Upset	0.84	345	0.59	3.14	1.14	62.0
1260 °C	Repress	0.86	425	2.22	7.49	1.56	61.0
1260 °C	Upset	0.64	190	4.32	10.40	2.16	...
1260 °C	Repress	0.66	160	3.45	9.55	1.86	60.0
1260 °C	Upset	0.84	200	4.04	11.53	1.81	61.0
1260 °C	Repress	0.84	195	2.54	11.16	1.28	61.0

1120 °C = 2050 °F. 1260 °C = 2300 °F. (a) DA, dissociated ammonia. (b) ENDO, endothermic atmosphere

nated from a batch of forgings. It also can be used to provide 100% inspection of the metallurgical integrity of a forging batch.

Applications of Powder Forged Parts

Previous sections in this article compared powder forging and drop forging and illustrated the range of mechanical property performance that can be achieved in powder forged material. The various approaches to the powder forging process were reviewed, as was the influence of process parameters on the metallurgical integrity of the forged parts. The present section concentrates on examples of powder forged components and highlights some of the reasons for selecting powder forged parts over those made by competing forming methods.

Example 1: Converter Clutch Cam. The automotive industry is the principal user of powder forged parts, and components for automatic transmissions represent the major area of application. One of the earliest powder forgings used in such an application is the converter clutch cam (Fig. 31). The primary reason powder forging was chosen over competitive processes was that it reduced manufacturing costs by 58%, compared with the conventional process of machining a forged gear blank. This cost saving resulted from substantially lower machining cost and lower total energy use.

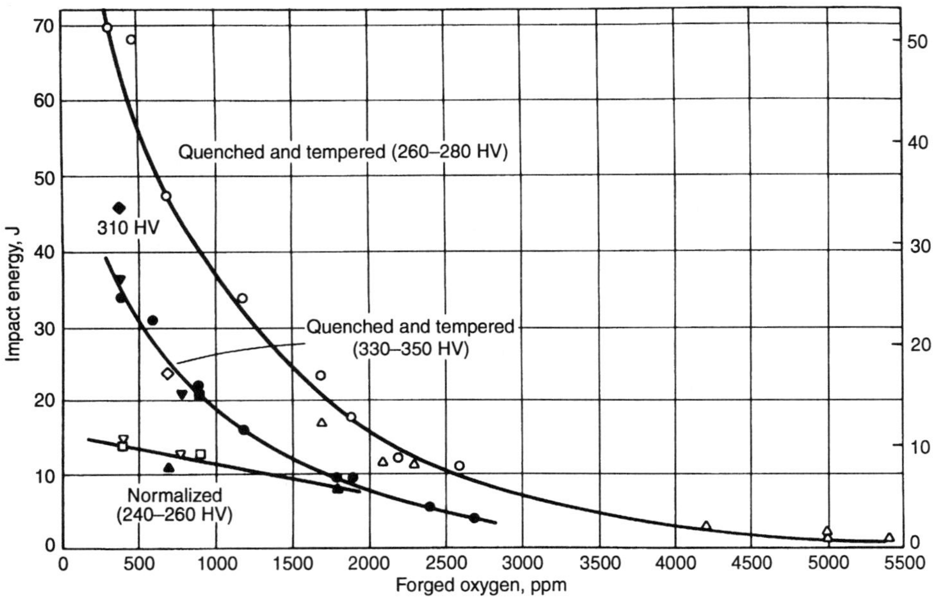

Fig. 28 Room-temperature impact energy as a function of forged oxygen content for various powder forged alloys. Heat treatments and hardnesses are indicated on the curves. Source: Ref 91

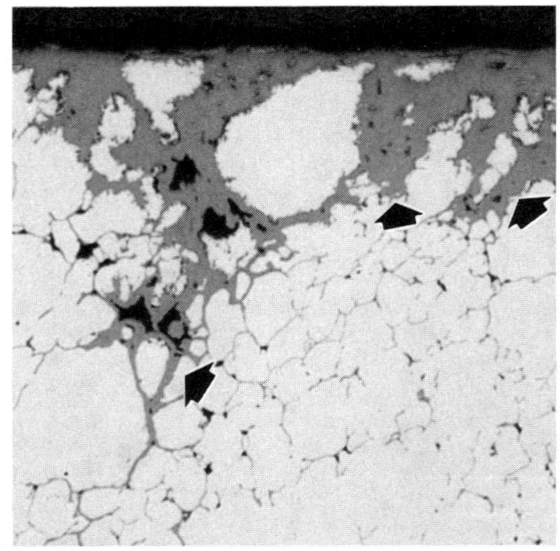

Fig. 29 Surface finger oxides (arrows at upper right) and interparticle oxide networks (arrow near lower left) in a powder forged material

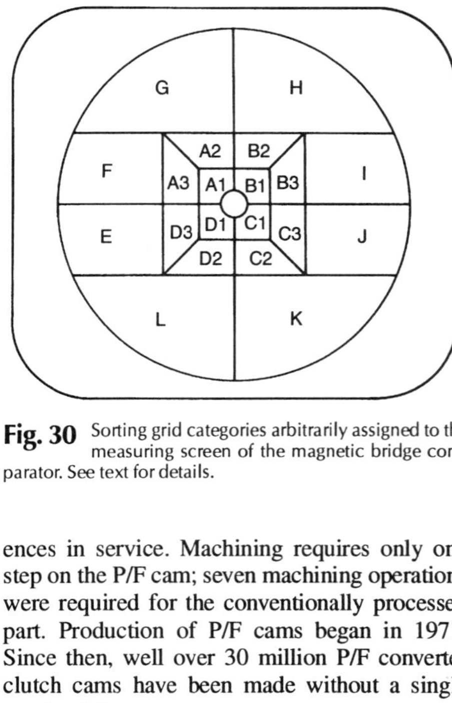

Fig. 30 Sorting grid categories arbitrarily assigned to the measuring screen of the magnetic bridge comparator. See text for details.

Table 12 Chemical composition requirements for powder forged parts (wt%)

Element	P/F-10*xx*	P/F-10C*xx*	P/F-11*xx*	P/F-11C*xx*	P/F-42*xx*	P/F-46*xx*	P/F-44*xx*	P/F-49*xx*
Ni (max)	0.10	0.10	0.10	0.10	0.40–0.50	1.75–2.00	0.10 max	0.10 max
Mo (max)	0.05	0.05	0.05	0.05	0.55–0.65	0.50–0.60	0.80–0.95	1.4–1.6
Mn	0.10–0.25	0.10–0.25	0.30–0.60(a)	0.30–0.60(a)	0.20–0.35	0.10–0.25	0.08–0.18	0.08–0.18
Cu	0.30 max	1.8–2.2	0.30 max	1.8–2.2	0.15	0.15	0.15	0.15
Cr (max)	0.10	0.10	0.10	0.10	0.10	0.10	0.10	0.10
S (max)	0.025	0.025	0.23(a)	0.23(a)	0.03	0.03	0.03	0.03
Si (max)	0.03	0.03	0.03	0.03	0.03	0.03	0.03	0.03
P (max)	0.03	0.03	0.03	0.03	0.03	0.03	0.03	0.03
C	(b)	(b)	(b)	(b)	(b)	(b)	(b)	(b)
O	(c)	(c)	(c)	(c)	(c)	(c)	(c)	(c)
Fe	bal	bal	bal	bal	bal	bal	bal	bal

(a) Covers manganese sulfide additions of from 0.3 to 0.5 wt%. The manganese content in solution is similar to P/F-10*xx* or P/F-10C*xx*, i.e., 0.10 to 0.25 wt%. (b) Carbon content specified by the purchaser. (c) When required, maximum oxygen content shall conform to the amount specified by the purchaser.

Table 13 Typical tolerances for powder forged parts

Dimension or characteristic	Description	Typical tolerance mm/mm	Typical tolerance in./in.	Minimum tolerance mm	Minimum tolerance in.
a	Linear dimension perpendicular to the press axis	0.0025	0.0025	0.08	0.003
b	Linear dimensions parallel to the press axis	±0.25	±0.10	0.20	0.008
c	Concentricity of holes to external dimensions	...	...	0.10	0.004
d	Surface finish	...	...	Normally better than 0.8 μm (32 μin.)	

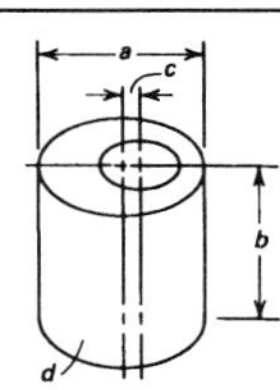

Source: Ref 90

Powder forged cams are made from a water-atomized steel powder (P/F-4200) containing 0.6 wt% Mo, 0.5 wt% Ni, 0.3 wt% Mn, and 0.3 wt% graphite. Preforms weighing 0.33 kg (0.73 lb) are compacted to a density of 6.8 g/cm^3. The preforms are sintered at 1120 °C (2050 °F) in an endothermic gas atmosphere with a +2 °C (+35 °F) dew point. The sintered preforms are graphite coated before being induction heated and forged to near full density (<0.2% porosity) using both axial and lateral flow. Alter forging, the face of the converter clutch cam is ground, carburized to a depth of 1.78 mm (0.070 in.), and surface hardened by means of induction. The part requires a high density to withstand the high Hertzian stress the inner cam surface experiences in service. Machining requires only one step on the P/F cam; seven machining operations were required for the conventionally processed part. Production of P/F cams began in 1971. Since then, well over 30 million P/F converter clutch cams have been made without a single service failure.

Example 2: Inner Cam/Race. A part that illustrates the complex shapes that can be formed on both the inner and outer surfaces of a powder forged component is the inner cam/race shown in Fig. 32 (Ref 95). The part is the central member in an automotive automatic transmission torque converter centrifugal lock-up clutch.

The inner cam/race is forged to a minimum density of 7.82 g/cm^3 from a P/F-4662 material. The part has a minimum quenched and stress-relieved hardness of 58 HRC and a tensile strength of 2070 MPa (300 ksi). The application imposes high stresses on the cams and splines.

Example 3: Internal Ring Gear. The powder forged internal ring gear shown in Fig. 33 is used in automatic transmissions for trucks with a maximum gross vehicle weight of 22,700 kg (50,000 lb) (Ref 96). The gear transmits 1355 N

Fig. 31 Powder forged converter clutch cam used in an automotive automatic transmission. Courtesy of Precision Forged Products Division, Borg Warner Corp.

Fig. 32 Powder forged inner cam/race for an automotive automatic transmission. Courtesy of Precision Forged Products Division, Borg Warner Corp.

Fig. 33 Powder forged internal ring gear used in automatic transmission for trucks of up to 22,700 kg (50,000 lb) gross vehicle weight. Courtesy of Precision Forged Products Division, Borg Warner Corp.

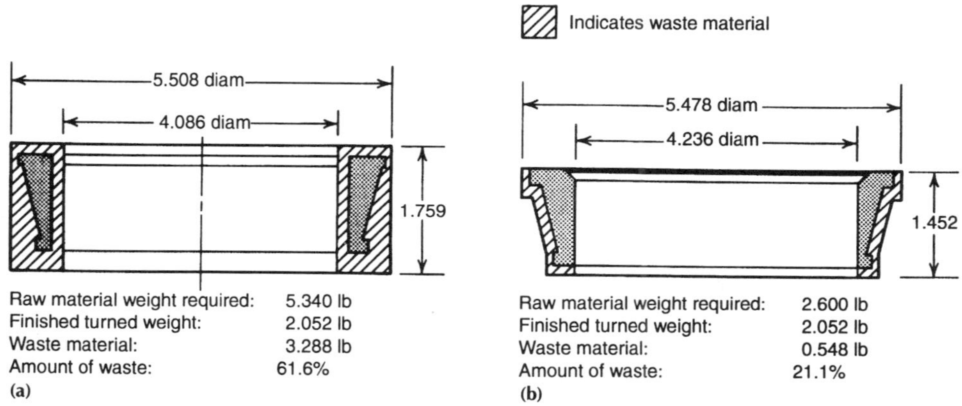

Fig. 34 Raw material utilization in the production of a tapered roller bearing race. (a) Produced from hot rolled tube stock. (b) Powder forged from preform. Source: Ref 100

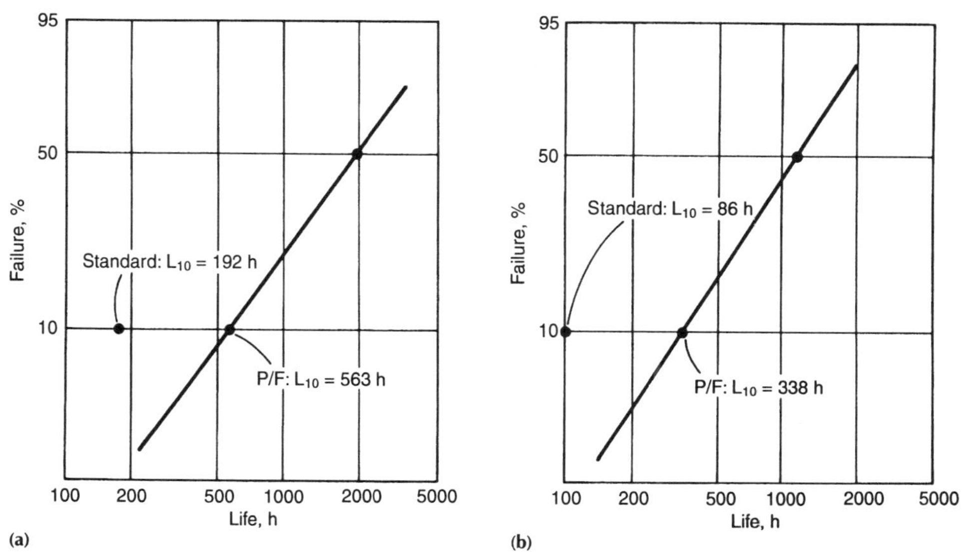

Fig. 35 Weibull plots of L_{10} life of P/F bearing races compared with L_{10} of wrought and machined races. (a) Cups. (b) Cones. Source: Ref 101

· m (1000 ft · lbf) of torque through the gear and spline teeth.

Originally, the gear was produced by forging an AISI 5140M tubing blank. The conventionally forged blank required rough machining, gear tooth shaping, spline machining, core heat treating, carburizing, and deburring. The only secondary operations required on the powder forged part are surface grinding, hard turning, shot blasting, and vibratory tumbling.

The P/F-4618 ring gear is produced to a minimum density of 7.82 g/cm^3. The part is selectively carburized using a proprietary process (Ref 97–99) and quench hardened. Minimum surface hardness is 57 HRC (2070 MPa, or 300 ksi, ultimate tensile strength), while the core hardness is 25 HRC (825 MPa, or 120 ksi, ulti-

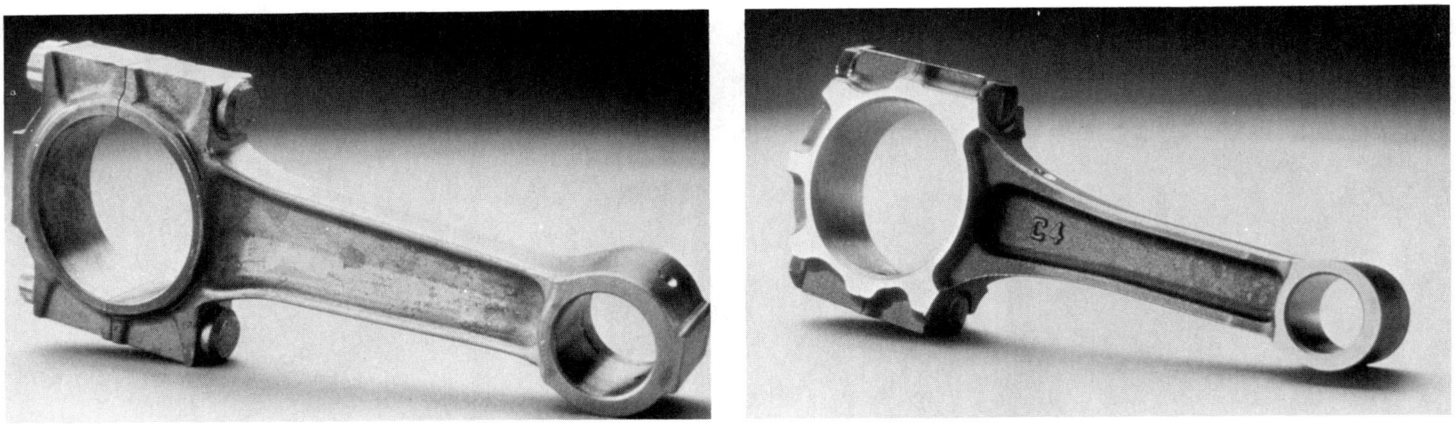

(a) (b)

Fig. 36 Powder forged connecting rods. (a) Rod for Porsche 928 V-8 engine. Note reduced size of balance pads. Courtesy of Powder Forging Division, GKN Forgings. (b) Rod for Toyota 1.9 L engine; balance pads are completely eliminated.

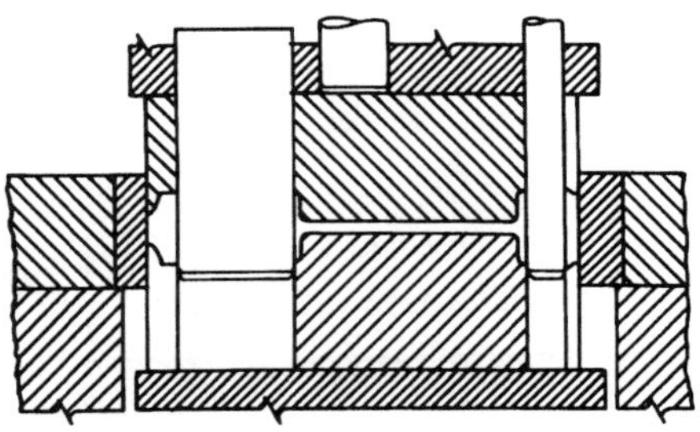

Fig. 37 Tooling used for powder forging of the Toyota connecting rod. Source: Ref 77

mate tensile strength). The internal gear teeth are produced to AGMA Class 7 tolerances.

Example 4: Powder Forged Tapered Bearing Race. The use of powder forging for production of tapered roller bearing races has resulted in considerable cost savings. The economy of the P/F process results from material savings, elimination of machining, energy savings from the elimination of subsequent carburizing, and raw material inventory reduction.

In some cases, up to 80% of the material is lost to machining when a bearing race is produced from bar stock. Material savings resulting from powder forging average 50% on bearing cup and cone production. In the example shown in Fig. 34, a material savings of 1.25 kg (2.74 lb) is realized using powder forging; nearly 62% of the feedstock is wasted when this component is machined from hot rolled tube stock.

In addition to the cost savings, the fatigue life of powder forged cups and cones was found to be greater than that of similar cups produced from wrought bearing steels (Fig. 35).

Example 5: Powder Forged Connecting Rods. Connecting rods were among the components selected for a number of powder forging development programs in the 1960s (Ref 5, 7, 18, 76, 102–105). However, it was not until 1976 that the first powder forged connecting rod was produced commercially. This was the connecting rod for the Porsche 928 V-8 engine (Fig. 36a).

The powder forged connecting rod for the Porsche 928 engine was made from a water-atomized low-alloy steel powder (0.3 to 0.4 wt% Mn, 0.1 to 0.25 wt% Cr, 0.2 to 0.3 wt% Ni, and 0.25 to 0.35 wt% Mo) to which graphite was added to give a forged carbon content of 0.35 to 0.45 wt%. The forgings were oil quenched and tempered to a core hardness of 28 HRC (ultimate tensile strength of 835 to 960 MPa, or 121 to 139 ksi), followed by shot peening to a surface finish of 11 to 13 on the Almen scale.

The preform was designed such that the powder forged component had <0.2% porosity in the critical web region. The powder forged connecting rod had considerably better fatigue properties than did conventional drop forged rods. Its weight control was good enough to allow a reduction in the size of the balance pads (Fig. 36a), resulting in ~10% weight saving (it weighed ~1 kg, or 2 lb). Powder forged connecting rods were subsequently used in both the Porsche 928 and 944 engines.

The first high-volume commercialization of powder forged connecting rods was in the 1.9 L Toyota Camry engine. In this design, the balance pads were completely eliminated (Fig. 36b). Despite the publication of the results of development trials in 1972 (Ref 104), it was not until the summer of 1981 that production rods were introduced (Ref 9, 77).

Toyota selected a copper steel (Fe-0.55 wt% C-2 wt% Cu) based on a water-atomized iron powder to replace conventional forgings, which had been made from a quenched and tempered 10L55 free-machining steel. The preform, which has a preshaped partial I-beam web section, has an average green density of 6.5 g/cm^3. The preform shape is such that forging is predominantly in the repressing mode. However, some lateral flow does take place where required in critical regions, such as the web.

Preforms are sintered for 20 min at 1150 °C (2100 °F) in an endothermic gas atmosphere in a specially designed rotary hearth furnace. During sintering, the preforms are supported on flat, ceramic plates. The preforms are allowed to stabilize at ~1010 °C (1850 °F) before closed-die forging.

Exposure of the preform to the atmosphere during transfer to the forging dies is limited to 4 to 5 s. The forging tooling is illustrated in Fig. 37. An ion nitriding treatment is applied to the punches and dies in the regions at which forging deformation occurs (Ref 9). The connecting rods are forged at the rate of 10 per min, and tool lives of over 100,000 pieces have been reported (Ref 106).

The forged rods are subjected to a thermal treatment after forging. This results in a ferrite/pearlite microstructure with a core hardness of 240 to 300 HV (30 kgf load). Subsequent operations include burr removal, shot peening, straightening, sizing, magnetic particle inspection, and finish machining.

Savings in material and energy are substantial for the powder forged rods (Ref 9). Billet weight for conventional forging is 1.2 kg (2.65 lb); the powder forging preform weighs 0.7 kg (1.54 lb) and requires little machining. In addition to the benefits in process economics, the variability in fatigue performance for the powder forged rods is reported to be half that of conventionally forged parts (Ref 77).

Good dimensional tolerance and precise control of part mass (±0.5%) require fewer machining operations for powder forged connecting rods compared with their drop forged or cast counterparts. In the case of P/F rods, a recent development involves the fracture splitting of the crank end cap from the rest of the connecting rod (Ref 107–112). Two approaches are used to accomplish this. In the technique patented by Sintermetallwerke Krebsöge (Fig. 38) (Ref 108), a V-shaped groove is shaped during compaction of the forging preforms and this groove is closed to form a short crack during the forging operation. This controlled defect initiates splitting of the end cap during the fracture splitting process. In an alternative procedure, used by Ford Motor Company, a notch is broached in the powder forged rod to facilitate splitting of the cap (Ref

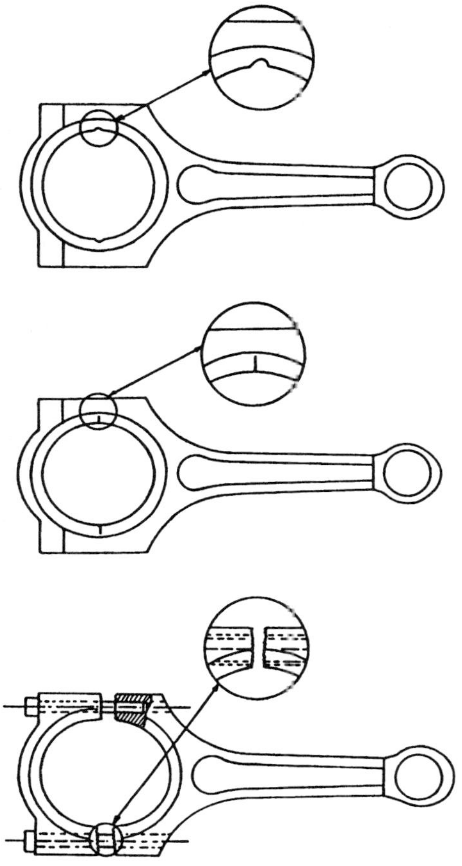

Fig. 38 Fracture splitting process used by GKN (Krebsöge). Source: Ref 108

110–112). A comparison of the conventional and fracture splitting methods is shown in Table 14.

GKN, Krebsöge of Radevermwald, Germany produces powder forged connecting rods used in the 2-cylinder engine for the new high performance BMW motorcycle-R1100 RS (Ref 107). The rods are made from the same prealloyed, low-alloy steel powder that was used to produce the Porsche connecting rods (Astaloy D). The material used to make the rods has a tensile strength of 1000 MPa, elongation of 11%, Young's modulus of 200 GPa, and a fatigue endurance limit (RBF) of 340 MPa. Krebsöge is also producing P/F connecting rods for a BMW V8 engine.

Chrysler Corp. is currently using powder forged connecting rods in new 2.0 and 2.4 L four-cylinder engines (Ref 113) while General Motors Corp. is using P/F rods in the 4.6 L V8 Northstar engine (Ref 114) and for 5.0 and 5.7 L V8 engines.

An interesting presentation reviewing the prospects for the production of P/M connecting rods was made during the international P/M conference in Paris (Ref 115). The authors concluded that pressed and sintered rods are a viable manufacturing option.

The results of a joint study by Ford and Krebsöge are summarized in Table 15 (Ref 107). For various connecting rod manufacturing processes, the study compared the mass of the connecting rods produced as well as the relative price of a finished rod. The study indicates that while aluminum or carbon fiber reinforced plastic connecting rods are attractive from the perspective of reduced mass they are currently too expensive compared with cast, drop forged, or powder forged rods. P/M rods appear to offer some promise. However, in order to match fatigue response, the P/M rods may not offer the same potential for reducing mass as that offered by the powder forged rods.

Table 14 Comparison of connecting rod production methods

Conventional method	Fracture splitting method
Alignment	Obsolete
Machining of outer contours	Obsolete
Deburring of outer contours	Obsolete
Preliminary grinding of side faces	Obsolete
Broaching of small and big end bearings	Obsolete
Machining of cap bolt bores	Machining of cap bolt bores
Drilling of oil bore	Drilling of oil bore
Machining of bolt head and nut seats	Obsolete
Separation of cap from rod	Fracture splitting
Grinding of rod/cap contact surfaces	Obsolete
Insertion of bolt sleeves	Obsolete
Assembly of cap and rod	Assembly of cap and rod
Finish-grinding of side faces	Finish-grinding of side faces
Final drilling of small and big end bearings	Final drilling of small and big end bearings
Milling of bearing shell positioning grooves	Milling of bearing shell positioning grooves
Honing of big end bearing	Honing of big end bearing
Insertion of small end bearing bush	Insertion of small end bearing bush
Insertion of big end bearing shells	Insertion of big end bearing shells
Machining of balance pads	Obsolete
Sorting according to weight	Obsolete

Table 15 Improvements in the relative cost and the mass of connecting rods by various processes

Process/ material	1987	1988	1989	1990	1991	1992
Casting						
Cost, %(a)	100	...	...	90	...	...
Weight, g	540	...	...	540	...	...
Drop-forging						
Cost, %(a)	120	120	...	120	...	...
Weight, g	580	550	...	500	...	...
Powder forging						
Cost, %(a)	130	120	100	...	80	...
Weight, g	515	480	460	...	440	...
Conventional sintering						
Cost, %(a)	130	...	100	...	80	...
Weight, g	650	...	600	...	580	...
Aluminum						
Cost, %(a)	500	...	...	...	...	300
Weight, g	300	...	...	...	...	280
Carbon fiber composite						
Cost, %(a)	800	...	...	...	...	400
Weight, g	280	...	...	...	...	260

(a) Cost of finished connecting rod (cast version = 100%)

REFERENCES

1. Ferrous Powder Metallurgy Materials, *Properties and Selection: Irons and Steels,* Vol 1, *Metals Handbook,* 9th ed., American Society for Metals, 1978, p 327
2. F.T. Lally, I.J. Toth, and J. DiBenedetto, "Forged Metal Powder Products," Final Technical Report SWERR-TR-72-5I, Army Contract DAAF01-70-C-0654, Nov 1971
3. P.W. Lee and H.A. Kuhn, P/M Forging, *Powder Metallurgy,* Vol 7, *Metals Handbook,* 9th ed., American Society for Metals, 1984, p 410
4. G. Bockstiegel, Powder Forging—Development of the Technology and Its Acceptance in North America, Japan, and West Europe, *Powder Metallurgy 1986—State of the Art,* Vol 2, Powder Metallurgy in Science and Practical Technology series, Verlag Schmid GmbH, 1986, p 239
5. P.K. Jones, The Technical and Economic Advantages of Powder Forged Products, *Powder Metall.,* Vol 13 (No. 26), 1970, p 114
6. G. Bockstiegel, Some Technical and Economic Aspects of P/M-Hot-Forming, *Mod. Dev. Powder Metall.,* Vol 7, 1974, p 91
7. J.W. Wisker and P.K. Jones, The Economics of Powder Forging Relative to Competing Processes—Present and Future, *Mod. Dev. Powder Metall.,* Vol 7, 1974, p 33
8. W.J. Huppmann and M. Hirschvogel, Powder Forging, Review 233, *Int. Met. Rev.,* (No. 5), 1978, p 209
9. C. Tsumuti and I. Nagare, Application of Powder Forging to Automotive Parts, *Met. Powder Rep.,* Vol 39 (No. 11), 1984, p 629
10. C. Durdaller, "Powders for Forging," Technical Bulletin D211, Hoeganaes Corp., Oct 1971
11. R.T. Cundill, E. Marsh, and K.A. Ridal, Mechanical Properties of Sinter/Forged Low-Alloy Steels, *Powder Metall.,* Vol 13 (No. 26), 1970, p 165
12. P.C. Eloff and S.M. Kaufman, Hardenability Considerations in the Sintering of Low Alloy Iron Powder Preforms, *Powder Metall. Int.,* Vol 3 (No. 2), 1971, p 71
13. K.H. Moyer, The Effect of Sintering Temperature (Homogenization) on the Hot Formed Properties of Prealloyed and Admixed Elemental Ni-Mo Steel Powders, *Prog. Powder Metall.,* Vol 30, 1974, p 193
14. G.T. Brown, Development of Alloy Systems for Powder Forging, *Met. Technol.,* Vol 3, May–June 1976, p 229
15. G.T. Brown, "The Past, Present and Future of Powder Forging With Particular Reference to Ferrous Materials," Technical Paper 800304, Society of Automotive Engineers, 1980
16. R. Koos and G. Bockstiegel, The Influence of Heat Treatment, Inclusions, and Porosity on the Machinability of Powder Forged Steel, *Prog. Powder Metall.,* Vol 37, 1981, p 145
17. B.L. Ferguson, H.A. Kuhn, and A. Lawley, Fatigue of Iron Base P/M Forgings, *Mod. Dev. Powder Metall.,* Vol 9, 1977, p 51
18. G.T. Brown and J.A. Steed, The Fatigue Performance of Some Connecting Rods Made by

Powder Forging, *Powder Metall.*, Vol 16 (No. 32), 1973, p 405

19. W.B. James, The Use of Image Analysis for Assessing the Inclusion Content of Low Alloy Steel Powders for Forging Applications, *Practical Applications of Quantitative Metallography,* STP 839, American Society for Testing and Materials, 1984, p 132
20. R. Causton, T.F. Murphy, C.-A. Blande, and H. Soderhjelm, Non-Metallic Inclusion Measurement of Powder Forged Steels Using an Automatic Image Analysis System, *Horizons of Powder Metallurgy,* Part II, Verlag Schmid, 1986, p 727
21. W.B. James, "Quality Assurance Procedures for Powder Forged Materials," Technical Paper 830364, Society of Automotive Engineers, 1983
22. W.B. James, Automated Counting of Inclusions in Powder Forged Steels, *Mod. Dev. Powder Metall.,* Vol 14, 1981, p 541
23. J.A. Steed, The Effects of Iron Powder Contamination on the Properties of Powder Forged Low Alloy Steel, *Powder Metall.,* Vol 18 (No. 35), 1975, p 201
24. N. Dautzenberg and H.T. Dorweiler, Effect of Contamination by Plain Iron Powder Particles on the Properties of Hot Forged Steels Made from Prealloyed Powders, *P/M '82 in Europe, Int. Powder Metallurgy Conf. Proc,* 1982, p 381
25. G. Bockstiegel, E. Dittrich, and H. Cremer, Experiences With an Automatic Powder Forging Line, *Proc. of the Fifth European Symp. on Powder Metallurgy,* Vol 1, 1978, p 32
26. W.B. James, New Shaping Methods for P/M Components, *Powder Metallurgy 1986—State of the Art,* Vol 2, Powder Metallurgy in Science and Practical Technology series, Verlag Schmid GmbH, 1986, p 71
27. H.A. Kuhn, S. Pillay, and H. Chung, Computer-Aided Preform Design for Powder Forging, *Mod. Dev. Powder Metall.,* Vol 12, 1981, p 643
28. S. Pillay and H.A. Kuhn, "Computerized Powder Metallurgy (P/M) Forging Techniques," Final Technical Report AD-A090 043, Rock Island Arsenal, Sept 1980

University of Pittsburgh, 1980

29. H.A. Kuhn and B.L. Ferguson, An Expert Systems Approach to Preform Design for Powder Forging, *Met. Powder Rep.,* Vol 40 (No. 2), 1985, p 93
30. H.A. Kuhn and B.L. Ferguson, *Powder Forging,* Metal Powder Industries Federation, 1990
31. C.L. Downey and H.A. Kuhn, Designing P/M Preforms for Forging Axisymmetric Parts, *Int. J. Powder Metall.,* Vol 11 (No. 4), 1975, p 255
32. P.J. Guichelaar and R.D. Pehlke, Gas Metal Reactions During Induction Sitering, *1971 Fall Powder Metallurgy Conf. Proc.,* Metal Powder Industries Federation, 1972, p 109
33. J.H. Hoffmann and C.L. Downey, A Comparison of the Energy Requirements for Conventional and Induction Sintering, *Mod. Dev. Powder Metall.,* Vol 9, 1977, p 301
34. R.F. Halter, Recent Advances in the Hot Forming of P/M Preforms, *Mod. Dev. Powder Metall.,* Vol 7, 1974, p 137
35. J.E. Comstock, How to Pick a Hot-Forging Lubricant, *Am. Mach.,* Oct 1981, p 141
36. T. Tabata, S. Masaki, and K. Hosokawa, A Compression Test to Determine the Coefficient of Friction in Forging P/M Preforms, *Int. J. Powder Metall.,* Vol 16 (No. 2), 1980, p 149
37. M. Stromgren and R. Koos, Hoganas' Contribution to Powder Forging Developments, *Met. Powder Rep.,* Vol 38 (No. 2), 1983, p 69
38. T. Altan, "Characteristics and Applications of Various Types of Forging Equipment," Technical Report MFR72-02, Society of Manufacturing Engineers
39. J.W. Spretnak, "Technical Notes on Forging," Forging Industry Educational and Research Foundation
40. *Forging Design Handbook,* American Society for Metals, 1972
41. J.E. Jenson, Ed., *Forging Industry Handbook,* Forging Industry Association, 1970
42. S. Mocarski and P.C. Eloff, Equipment Considerations for Forging Powder Preforms, *Int. J. Powder Metall.,* Vol 7 (No. 2), 1971, p 15
43. H.A. Kuhn, Deformation Processing of Sintered Powder Materials, *Powder Metallurgy Processing—New Techniques and Analyses,* Academic Press, 1978, p 99
44. H.A. Kuhn, M.M. Hagerty, H.L. Gaigher, and A. Lawley, Deformation Characteristics of Iron-Powder Compacts, *Mod. Dev. Powder Metall.,* Vol 4, 1971, p 463
45. H.A. Kuhn and C.L. Downey, Deformation Characteristics and Plasticity Theory of Sintered Powder Materials, *Int. J. Powder Metall.,* Vol 7 (No. 1), 1971, p 15
46. H.A. Kuhn, Fundamental Principles of Powder Preform Forging, *Powder Metallurgy for High Performance Applications, Proc. of the 18th Sagamore Army Materials Research Conf.,* Syracuse University Press, 1972, p 153
47. H.A. Kuhn and C.L. Downey, How Flow and Fracture Affect Design of Preforms for Powder Forging, *Powder Metall. Powder Technol.,* Vol 10 (No. 1), 1974, p 59
48. F.G. Hanejko, P/M Hot Forming, Fundamentals and Properties, *Prog. Powder Metall.,* Vol 33, 1977, p 5
49. H.F. Fischmeister, B. Aren, and K.E. Easterling, Deformation and Densification of Porous Preforms in Hot Forging, *Powder Metall.,* Vol 14 (No. 27), 1971, p 144
50. G. Bockstiegel and U. Bjork, The Influence of Preform Shape on Material Flow, Residual Porosity, and Occurrence of Flaws in Hot-Forged Powder Compacts, *Powder Metall.,* Vol 17 (No. 33), 1974, p 126
51. M. Watanabe, Y. Awano, A. Danno, S. Onoda, and T. Kimura, Deformation and Densification of P/M Forging Preforms, *Int. J. Powder Metall.,* Vol 14 (No. 3), 1978, p 183
52. G. Sjoberg, Material Flow and Cracking in Powder Forging, *Powder Metall. Int.,* Vol 7 (No. 1), 1975, p 30
53. H.L. Gaigher and A. Lawley, Structural Changes During the Densification of P/M Preforms, *Powder Metall. Powder Technol.,* Vol 10 (No. 1), 1974, p 21
54. P.W. Lee and H.A. Kuhn, Fracture in Cold Upset Forging—A Criterion and Model, *Metall. Trans.,* Vol 4, April 1973, p 969
55. C.L. Downey and H.A. Kuhn, Application of a Forming Limit Concept to the Design of Powder Preforms for Forging, *J. Eng. Mater. Technol., (Trans. ASME),* Series H, Vol 97 (No. 4), 1975, p 121
56. S.K. Suh, "Prevention of Defects in Powder Forging," Ph.D. thesis, Drexel University, 1976
57. S.K. Suh and H.A. Kuhn, Three Fracture Modes and Their Prevention in Forming P/M Preforms, *Mod. Dev. Powder Metall.,* Vol 9, 1977, p 407
58. C.L. Downey, "Powder Preform Forging—An Analytical and Experimental Approach to Process Design," Ph.D. thesis, Drexel University, 1972
59. B.L. Ferguson, "P/M Forging of Components for Army Applications," TriService Manufacturing Technology Advisory Group Program Status Review, 1979, p F1
60. G. Bockstiegel and M. Stromgren, "Hoganas Automatic PM-Forging System, Concept and Application," Technical Paper 790191, Society of Automotive Engineers, 1979
61. M. Stromgren and M. Lochon, Development and Fatigue Testing of a Powder Pinion Gear for a Passenger Car Gear Box, *Mod. Dev. Powder Metall.,* Vol 15, 1985, p 655
62. W.J. Huppmann, Forces during Forging of Iron Powder Preforms, *Int. J. Powder Metall.,* Vol 12 (No. 4), 1976, p 275
63. Q. Jiazhong, O. Grinder, and Y. Nilsson, Mechanical Properties of Low Temperature Powder Forged Steel, *Horizons of Powder Metallurgy,* Part II, Verlag Schmid, 1986, p 653
64. O. Bockstiegel and H. Olsen, Processing Parameters in the Hot Forming of Powder Preforms, *Powder Metallurgy, Third European Powder Metallurgy Symp.,* Conf. Supplement Part 1, 1971, p 127
65. Y. Nilsson, O. Grinder, C.Y. Jia, and Q. Jiazhong, "Hot Repressing of Sintered Steel Properties," STU 498, The Swedish Institute for Metals Research, 1985
66. O. Grinder, C.Y. Jia, and Y. Nilsson, Hot Upsetting and Hot Repressing of Sintered Steel Preforms, *Mod. Dev. Powder Metall.,* Vol 15, 1984, p 611
67. Surface Roughness Averages for Common Production Methods, *Met. Prog.,* July 1980, p 51
68. R. Koos, G. Bockstiegel, and C. Muhren, "Machining Studies of PM-Forged Materials," Technical Paper 790192, Society of Automotive Engineers, 1979
69. U. Engstrom, Machinability of Sintered Steels, *Powder Metall.,* Vol 26 (No. 3), 1983, p 137
70. F.G. Hanejko, Mechanical Property Anisotropy of P/M Hot Formed Materials, *Mod. Dev. Powder Metall.,* Vol 10, 1977, p 73

71. Closed-Die Steel Forgings, *Properties and Selection, Irons and Steels,* Vol 1, *Metals Handbook,* 9th ed., American Society for Metals, 1978, p 357
72. G.T. Brown, The Core Properties of a Range of Powder Forged Steels for Carburizing Applications, *Powder Metall.,* Vol 20 (No. 3), 1977, p 171
73. G.T. Brown and T.B. Smith, The Relevance of Traditional Materials Specifications to Powder Metal Products, *Mod. Dev. Powder Metall.,* Vol 7, 1974, p 9
74. G.T. Brown, Properties and Prospects of Powder Forged Low Alloy Steels Related to Component Production, *Powder Metallurgy: Promises and Problems,* Société Française de Métallurgie—Matériaux et Techniques, 1975, p 96
75. W.J. Huppmann and G.T. Brown, The Steel Powder Forging Process—A General Review, *Powder Metall.,* Vol 21 (No. 2), 1978, p 105
76. "GKN Powder Forging Materials Specification and Properties," Issue 2 GKN PowderMet, April 1978
77. K. Imahashi, C. Tsumuki, and I. Nagare, "Development of Powder Forged Connecting Rods," Technical Paper 841221, Society of Automotive Engineers, Oct 1984
78. Powder Metal Forgings, Isostatic Pressing Lead to Award-Winning Parts, *American Machinist and Automated Manufacturing,* July 1987, p 23
79. M. Marra and R. Williams, Mechanical Properties of a Powder Forged Material for Connecting Rods, *Advances in Powder Metallurgy 1989,* Vol 1, T.G. Gasbarre and W.F. Jandeska, Ed., Metal Powder Industries Federation, 1989, p 313
80. C. Tsumuki and I. Nagare, Application of Powder Forging to Automotive Parts, *Met. Powder Rep.,* Vol 39 (No. 11), 1984, p 629
81. A. Bhambri and L.F. Pease III, Mechanical Properties of P/F Connecting Rod Materials, *Modern Developments in Powder Metallurgy,* Vol 18, P. Ulf Gummeson and D.A. Gustafson, Ed., Metal Powder Industries Federation, 1988, p 155
82. R.A. Chernenkoff, D.W. Hall, and S. Mocarski, "Material Characterization of Powder-Forged Copper Steels," Technical Paper 910155, Society of Automotive Engineers, 1991
83. D. Glover, A Ball/Rod Rolling Contact Fatigue Tester, *Rolling Contact Fatigue Testing of Bearing Steels,* STP 771, J. Hoo, Ed., American Society for Testing and Materials, 1982, p 107
84. S. Buzolits, "Military Process Specification for Type 46XX Powder-Forged Weapon Components," Final Technical Report AD-E401-376, U.S. Army Armament Research and Development Center, 20 Aug 1985
85. S. Buzolits and T. Leister, "Military Specification for Type 10XX Powder-Forged Weapon Components," Final Technical Report AD-E401-412, U.S. Army Armament Research and Development Center, 14 Oct 1985
86. S.J. Mashl and D.W. Smith, Tensile Properties of High Strength Nearly Pore-Free Impact P/F-4650 Steel as Influenced by Forging Temperature and Macroscopic Strain, *Advances in Powder Metallurgy and Particulate Materials 1994,* Metal Powder Industries Federation, 1994
87. J.L. Seefelt, D.W. Smith, and A.A. Hendrickson, Tensile Properties of Impact Powder Forged 4645 Steel: Influence of Forging Conditions and Sulfur and Oxygen Contents," *Advances in Powder Metallurgy 1989,* Vol 1, T.G. Gasbarre and W.F. Jandeska, Ed., Metal Powder Industries Federation, 1989, p 411
88. J.L. Seefelt, D.W. Smith, and P.M. Machmeier, Further Examination of the Influence of Nonmetallic Inclusions on the Mechanical Properties of Impact Powder Forged 4645 Steel, *Advances in Powder Metallurgy 1990,* Vol 1, E.R. Andreotti and P.J. McGeehan, Metal Powder Industries Federation, 1990, p 323
89. S.J. Mashl and D.W. Smith, Properties of Powder Forged Steels Containing Admixed MnS, *Int. J. Powder Metall.,* Vol 28 (No. 3), 1992, p 279
90. Brochure, Powder Forging Division, GKN PowderMet, 1982
91. P. Lindskog and S. Grek, Reduction of Oxide Inclusions in Powder Preforms Prior to Hot Forming, *Mod. Dev. Powder Metall.,* Vol 7, 1974, p 285
92. R.T. Cundill, Relationships Between Oxides, Density and Porosity in Consolidated Steel Powders, *PM '82 in Europe,* Florence, 1982, p 145
93. D.W. Smith, Calculation of the Pore-Free Density of P/M Steels: Role of Microstructure and Composition, *Int. J. Powder Metall.,* Vol 28 (No. 3), 1992, p 259
94. D.W. Smith and S.J. Mashl, Pore-Free Density of Powder Forged Steel: Comparison of Measured and Calculated Values, *Int. J. Powder Metall.,* Vol 28 (No. 3), 1992, p 271
95. P.K. Johnson, Powder Metallurgy Design Competition Winners, *Int. J. Powder Metall.,* Vol 21 (No. 4), 1985, p 303
96. P.K. Johnson, Winning Parts Show High Strength and Cost Savings, *Int. J. Powder Metall.,* Vol 22 (No. 4), 1986, p 267
97. Method of Making Powdered Metal Parts, U.S. Patent 3,992,763
98. Method of Making Selectively Carburized Forged Powder Metal Parts, U.S. Patent 4,165,243
99. Method for Making Powder Metal Forging Preforms of High Strength Ferrous-Base Alloys, U.S. Patent 4,655,853
100. R.M. Szary and R. Pathak, Sinta-Forge an Efficient Production Process for High Fatigue Stress Components, *P/M Technical Conf. Proc.,* Hoeganaes Corp., Oct 1978
101. J.S. Adams and D. Glover, Improved Bearings at Lower Cost via Powder Metallurgy, *Met. Prog.,* Aug 1977, p 39
102. F.G. Hanejko and J. Muzik, Successful Applications and Processing Considerations for Powder Forming, *P/M Technical Conf. Proc.,* Hoeganaes Corp., Oct 1978
103. S. Corso and C. Downey, Preform Design for P/M Hot Formed Connecting Rods, *Powder Metall. Int.,* Vol 8 (No. 4), 1976, p 170
104. C. Tsumuki, J. Niimi, K. Hasimoto, T. Suzuki, T. Inukai, and O. Yoshihara, Connecting Rods by P/M Hot Forging, *Mod. Dev. Powder Metall.,* Vol 7, 1974, p 385
105. H.W. Antes, Processing and Properties of Powder Forgings, *Powder Metallurgy for High Performance Applications, Proc. of the 18th Sagamore Army Materials Research Conf.,* Syracuse University Press, 1972
106. Powder Forging Boosts P/M in Auto Industry, *Met. Powder Rep.* Vol 42, 1987, p 557
107. M. Weber, Comparison of Advanced Procedures and Economics for Production of Connecting Rods, *Powder Metall. Int.,* Vol 25 (No. 3), 1993, p 125
108. M. Weber, Method of Producing Powder Forged Components, U.S. Patent 4,923,674
109. M. Weber, "Cost Effective Finishing of Powder Forged Connecting Rods with the Fracture-Splitting Methods," Technical Paper 910157, Society of Automotive Engineers, 1991
110. P.Y. Hoag and D.A. Yeager, Making a Fractured Powder Metal Connecting Rod U.S. Patent No. 4,936,163
111. P.Y. Hoag and D.A. Yeager, Making a Fractured Powder Metal Connecting Rod, U.S. Patent 4,993,134
112. P.Y. Hoag and D.A. Yeager, Apparatus for Making a Powder Metal Connecting Rod, U.S. Patent 5,131,577
113. A. Wrigley, Powder Metal Auto Applications to Rise, *American Metal Market,* 12 Nov 1992
114. P.K. Johnson, Outstanding P/M Applications, *Int. J. Powder Metall.,* Vol 30 (No. 3), 1994, p 269
115. K. Richter, E. Hoffman, K. Lipp, and C.M. Sonsino, Single-Sintered Con Rods-An Illusion?, *Met. Powder Rep.,* Vol 49 (No. 5), 1994, p 38

SELECTED REFERENCES

- H.W. Antes, Cold Forging Iron and Steel Powder Preforms, *Mod. Dev. Powder Metall.,* Vol 4, 1971, p 415
- H.W. Antes, P/M Hot Formed Gears, *Met. Eng. Q.,* Nov 1974, p 8
- H.W. Antes and P.L. Stockl, The Effect of Deformation on Tensile and Impact Properties of Hot PM-Formed Nickel-Molybdenum Steels, *Powder Metall.,* Vol 17 (No. 33), 1974, p 178
- B.G.A. Aren, Optimizing the Preform Shape in Powder Forging a Linear Gear Profile, *Powder Metall. Int.,* Vol 7 (No. 1), 1975, p 12
- B.G.A. Aren, L. Olsson, and H.F. Fischmeister, The Influence of Presintering and Forging Temperature in Powder Forging, *Powder Metall. Int.,* Vol 4 (No. 3), 1972, p 1
- A.J. Ashley, P/M Forging Successes, *Met. Powder Rep.,* Vol 32 (No. 9), 1977, p 339

- F.A. Badia, F.W. Heck, and J.H. Tundermann, Effect of Compositional and Processing Variations on the Properties of Hot Formed Mixed Elemental P/M Ni Steels, *Mod. Dev. Powder Metall.,* Vol 7, 1974, p 255
- F.L. Bastian and J.A. Charles, Fracture Resistance of Some Powder Forged Steels, *Powder Metall.,* Vol 21 (No. 4), 1978, p 199
- G. Bockstiegel and C.-A. Blande, The Influence of Slag Inclusions and Pores on Impact Strength and Fatigue Strength of Powder Forged Iron and Steel, *Powder Metall. Int.,* Vol 8 (No. 4), 1976, p 155
- P. Bosse, R. Tremblay, and R. Angers, Hot-Pressing of Iron Powder and Preforms, *Int. J. Powder Metall.,* Vol 11 (No. 4), 1975, p 247
- W.J. Bratina, W.F. Fossen, D.R. Hollingberry, and R.M. Pilliar, Anisotropic Properties of Powder Forged Ferrous Systems, *Mod. Dev. Powder Metall.,* Vol 10, 1977, p 157
- G.T. Brown, Properties of Structural Powder Metal Parts—Over-Rated or Under-Estimated?, *Powder Metall.,* Vol 17 (No. 33), 1974, p 103
- G.T. Brown, Powder Forging: A Perspective, *Int. J. Powder Metall.,* Vol 21 (No. 3), 1985, p 193
- G.T. Brown, The History of Powder Forging, *Met. Powder Rep.,* Vol 41 (No. 1), 1986, p 54
- J.P. Cook, "Oxidation, Reduction, and Decarburization of Metal Powder Preforms," P/M Hot Forming Technical Data, Hoeganaes Corp., 1972
- J.P. Cook, "The Effect of Sintering Temperature and Flow on the Properties of Ni-Mo Steel Hot P/M Formed Material," Technical Paper 740982, Society of Automotive Engineers, 1974
- S. Corso and V. Giordano, Development of Differential Pinion Gear by PM Hot Forging Process, *Powder Metall.,* Vol 20 (No. 3), 1977, p 158
- A. Crowson, Surface Porosity in P/M Steel Forgings, *Prog. Powder Metall.,* Vol 34 and 35, 1978–1979, p 261
- G.W. Cull, Mechanical and Metallurgical Properties of Powder Forgings, *Powder Metall.,* Vol 13 (No. 26), 1970, p 156
- G.W. Cull, Some Practical Aspects of the Sinter-Forging Process, *Metall. Met. Forming,* April 1972, p 123
- R. Davies and M. Negm, The Effects of Some Process Variables on the As-Forged Properties of a Powder-Forged Ni-Mo Alloy Steel, *Powder Metall.,* Vol 20 (No. 1), 1977, p 39
- B. Dogan and T.J. Davies, The Effects of Inclusions on the Fracture Behavior of Powder Forged Steels, *Powder Metall. Int.,* Vol 15 (No. 1), 1983, p 11
- S.J. Donachie, Low Flow Stress Hot Forming of Ferrite, *Mod. Dev. Powder Metall.,* Vol 7, 1974, p 341
- S.J. Donachie and N.L. Church, Effect of Composition, Temperature, and Crystal Structure of the Flow Stress of P/M Forged Preforms, *Int. J. Powder Metall.,* Vol 10 (No. 1), 1974, p 33
- R.J. Dower and W.E. Campbell, The Toughness of P/M Forgings as a Function of Processing Route, *Mod. Dev. Powder Metall.,* Vol 10, 1977, p 53
- C.L. Downey and R.F. Halter, Design Criterion for P/M Hot Forming Dies, *Prog. Powder Metall.,* Vol 33, 1977, p 31
- P.C. Eloff and L.E. Wilcox, Fatigue Behavior of Hot Formed Powder Differential Pinions, *Mod. Dev. Powder Metall.,* Vol 7, 1974, p 213
- B.L. Ferguson, Ferrous Powder Metallurgy: Part II, Fully Dense Parts and Their Applications, *Powder Metallurgy—Applications, Advantages and Limitations,* E. Klar, Ed., American Society for Metals, 1983, p 86
- B.L. Ferguson, S.K. Suh, and A. Lawley, Impact Behavior of P/M Steel Forgings, *Int. J. Powder Metall.,* Vol 11 (No. 4), 1975, p 263
- H.F. Fischmeister, L. Olsson, and K.E. Easterling, Powder Metallurgy Review 6, Powder Forging, *Powder Metall. Int.,* Vol 6 (No. 1), 1974, p 30
- H.F. Fischmeister, G. Sjoberg, B.O. Elfstrom, K. Hamberg, and V. Mironov, Preform Ductility and Transient Cracking in Powder Forging, *Mod. Dev. Powder Metall.,* Vol 9, 1977, p 437
- J.R. Gleixner, Hot Forming P/M Parts, *Met. Prog.,* Dec 1983, p 33
- S.-E. Grek and R. Koos, Surface Oxides on Low Alloy Atomized Powders and Their Influence on Impact Properties of P/F Steels, *P/M '82 in Europe, Int. Powder Metallurgy Conf. Proc.,* 1982, p 393
- R.F. Halter, Pilot Production System for Hot Forging P/M Preforms, *Mod. Dev. Powder Metall.,* Vol 4, 1971, p 385
- F. Haneiko, "AISI 4000 Transverse and Longitudinal Impact Properties as a Function of Sintering Temperature and Deformation," Technical Paper 750951, Society of Automotive Engineers, 1975
- R.H. Hoefs, The Present Status of P/M Forging, Parts I and 2, *Precis. Met.,* May 1973, p 55, and June 1973, p 65
- G. Hoffmann and K. Dalal, Correlation Between Individual Mechanical Properties and Fracture Analysis of Hot Formed P/M Steels, *Mod. Dev. Powder Metall.,* Vol 10, 1977, p 171
- W.J. Huppmann, The Effect of Powder Characteristics on the Sinter-Forging Process, *Powder Metall.,* Vol 20 (No. 1), 1977, p 36
- W.J. Huppmann and L. Albano-Muller, Production of Powder Forged Parts of Complex Geometry, *Mod. Dev. Powder Metall.,* Vol 12, 1981, p 631
- Y. Ishimaru, T. Yamaguchi, Y. Saito, and Y. Nishino, Properties of Forged PM Ferrous Alloys, *Powder Metall. Int.,* Vol 37 (No. 5), 1982, p 252, and Vol 37 (No. 6), 1982, p 291
- M.P. Jarrett and P.K. Jones, "Automotive Forgings—Powder Leads to Higher Precision," Technical Paper 710119, Society of Automotive Engineers, 1971
- P.K. Jones and J.W. Wisker, "The Production of Precision Automotive Components by the Powder Forging Process—Present Situation and Future Prospects," Technical Paper 780361, Society of Automotive Engineers, 1978
- S.M. Kaufman, The Role of Pore Size in the Ultimate Densification Achievable During P/M Forging, *Int. J. Powder Metall.,* Vol 8 (No. 4), 1972, p 183
- S.M. Kaufman and S. Mocarski, The Effect of Small Amounts of Residual Porosity on the Mechanical Properties of P/M Forgings, *Int. J. Powder Metall.,* Vol 7 (No. 3), 1971, p 19
- M.J. Koczak, C.L. Downey, and H.A. Kuhn, Structure/Property Correlations of Aluminum and Nickel Steel Preform Forgings, *Powder Metall. Int.,* Vol 6, 1974, p 13
- T. Krantz, J.C. Farge, and P. Chollet, Hardenability and Mechanical Properties of Hot Forged Mn-Mo Steels Made from Prealloyed Powders, *Mod. Dev. Powder Metall.,* Vol 10, 1977, p 15
- K.M. Kulkami, P/M Forging Moves into Volume Production, *Mach. Des.,* 20 June 1985, p 74
- T.J. Ladanyi, G.A. Meyers, R.M. Pilliar, and G.C. Weatherly, Fracture Toughness of Powder Forged Cr-Mn Alloy Steels, *Metall. Trans. A,* Vol 6, 1975, p 2037
- L.-E. Larsson, M. Stromgren, and K. Svartstrom, Effects of Porosity and Matrix on the Hardness of Powder Forged Steel, *Mod. Dev. Powder Metall.,* Vol 15, 1985, p 585
- A. Lawley, Analysis of Mechanical Property—Structure Relations in Powder Forging, *Powder Metallurgy Processing—New Techniques and Analyses,* H.A. Kuhn and A. Lawley, Ed., Academic Press, 1978, p 139
- E.R. Leheup and J.R. Moon, Elastic Behavior of High Density Powder Forged Samples of Iron and Iron-Graphite, *Powder Metall.,* Vol 23 (No. 1), 1980, p 15
- E.R. Leheup and J.R. Moon, Yield and Fracture Phenomena in Powder-Forged Fe-0.2C and Their Prediction by NDT Methods, *Powder Metall.,* Vol 23 (No. 4), 1980, p 177
- G. Lusa, Differential Gear by P/M Hot Forging, *Mod. Dev. Powder Metall.,* Vol 4, 1971, p 425
- M.S. Maclean, W.E. Campbell, and R.J. Dower, An Insight into Mechanical Properties of Powder Metal Forging as a Function of Processing Route, *Powder Metall. Int.,* Vol 7, 1975, p 118
- S. Mocarski, Influence of Process Variables on Properties of Mod. 8600 and Manganese-Nickel-Molybdenum Low Alloy Hot Formed P/M Steels, *Mod. Dev. Powder Metall.,* Vol 7, 1974, p 303
- S. Mocarski and D. Hall, Properties of Hot Formed Mo-Ni-Mn P/M Steels with Admixed Copper, *Mod. Dev. Powder Metall.,* Vol 9, 1977, p 467
- K. Morimoto, K. Ogata, T. Yamamura, T. Yukawa, T. Saga, N. Yamada, and N. Sekiguchi, Transmission Spur Gear by Powder Forging, *Mod. Dev. Powder Metall.,* Vol 7, 1974, p 323

- K.H. Moyer, The Effect of Density on Impact Properties of Iron P/M Forgings, *Met. Eng. Q.*, Aug 1972, p 34
- K.H. Moyer, A Comparison of Deformed Iron-Carbon Alloy Powder Preforms with Commercial Iron-Carbon Alloys, *Mod. Dev. Powder Metall.*, Vol 7, 1974, p 235
- J. Muzik, "Steel Powders for the Powder Metallurgy (P/M) Forging Process," Technical Paper 720181, Society of Automotive Engineers, 1972
- L.F. Pease, "An Assessment of Powder Metallurgy Today and Its Future Potential," Technical Paper 831042, Society of Automotive Engineers, 1983
- T.W. Pietrocini, Hot Formed P/M Applications, *Mod. Dev. Powder Metall.*, Vol 7, 1974, p 395
- T.W. Pietrocini and D.A. Gustafton, Fatigue and Toughness of Hot-Formed Cr-Ni-Mo and Ni-Mo Prealloyed Steel Powders, *Int. J. Powder Metall.*, Vol 6 (No. 4), 1970, p 19
- R.M. Pilliar, W.J. Bratina, and J.T. McGrath, Fracture Toughness Evaluation of Powder Forged Parts, *Mod. Dev. Powder Metall.*, Vol 7, 1974, p 51
- K.A. Ridal and R.T. Cundill, Sinter Forging of Alloy Steel Components, *Metall. Met. Form.*, Aug 1971, p 204
- D.H. Ro, B.L. Ferguson, and S. Pillay, "Powder Metallurgy Forged Gear Development," Technical Report 13046, U.S. Army Tank-Automotive Command Research and Development Center, 1985
- S. Saritas and T.J. Davies, Reduction of Oxide Inclusions During Pre-Forging Heat Treatments, *P/M '82 in Europe, Int. Powder Metallurgy Conf. Proc.*, 1982, p 405
- S. Saritas and T.J. Davies, Fracture Behavior of Powder Forged Steels, *Mod. Dev. Powder Metall.*, Vol 15, 1985, p 599
- S. Saritas, W.B. James, and T.J. Davies, The Influence of Pre-Forging Treatments on the Mechanical Properties of Two Low Alloy Powder Forged Steels, *Powder Metall.*, Vol 24 (No. 3), 1981, p 131
- G. Sjoberg, Material Flow and Cracking in Powder Forging, *Powder Metall. Int.*, Vol 7, 1975, p 30
- H.M. Skelly, Properties of P/M Forgings Made by Six Methods, *Int. J. Powder Metall.*, Vol 14 (No. 1), 1978, p 33
- H.M. Skelly, Some Mechanical Properties of Powder-Forged Iron/Alloy Mixtures, *Powder Metall.*, Vol 22 (No. 2), 1979, p 41
- D.P. Townsend, "Surface Fatigue and Failure Characteristics of Hot Forged Powder Metal AISI 4620, AISI 4640, and Machined AISI 4340 Steel Spur Gears," NASA Technical Memorandum 87330, National Aeronautics and Space Administration, 1986
- M.V. Veidis, "Some Practical Aspects of Forging Sintered Metal Preforms," Technical Paper EM78–462, Society of Manufacturing Engineers, 1978
- M.V. Veidis, "The Forging of Sintered Preforms," Technical Paper MF79-129, Society of Manufacturing Engineers, 1979
- M. Weber, L. Albano-Muller, and W.J. Huppmann, "Truck Synchronizer Rings Produced by Powder Forging," Technical Paper 860153, Society of Automotive Engineers, 1986
- M. Weber and E. Brugel, Truck Synchronizer Rings Produced by Powder Forging, *Horizons of Powder Metallurgy,* Part I, Verlag Schmid GmbH, 1986, p 523
- G. Zapf, The Mechanical Properties of Hot-Recompacted Iron-Nickel Sintered Alloys, *Powder Metall.*, Vol 13 (No. 26), 1970, p 130

High-Temperature Sintering of Ferrous Powder Metallurgy Components

Howard I. Sanderow, Management & Engineering Technology

HIGH-TEMPERATURE SINTERING of ferrous components continues to be important in the P/M industry. Improvements in both production rates and properties are possible as sintering temperatures increases above 1120 °C (2050 °F). Rounding of the porosity results in increased strengths, especially impact strength. Higher diffusion rates also increase the strength and hardenability of admixed powder compositions. The greater oxide reduction that occurs at higher temperatures proves valuable in powder forging and other applications.

The specific benefits of high-temperature sintering as applied to ferrous P/M products can be categorized as follows:

- Improved mechanical properties
- Improved physical properties
- Development of liquid phase
- Ability to sinter active elements (Cr, Mn) in alloy steels

These four advantages of high-temperature sintering are discussed in this article for various ferrous P/M materials. A brief discussion of sintering stages and effects is also included.

Sintering Stages and Effects

During the sintering process a wide variety of physical, chemical and metallurgical phenomena occurs within the mass of metal powder particles. These effects are influenced by the sintering conditions (time, temperature, atmosphere), as well as the chemical composition of the powder mass. Because most of these phenomena are directly related to time and temperature, they have been listed in two categories. The first group primarily is associated with the early stages of sintering (conventional sintering practices), and the second group is related to the more advanced stages of sintering, typical of high-temperature methods.

Early stages of sintering (conventional) include the following mechanisms:

- *Homogenization:* The as-cast, dendritic structure of the atomized particles is removed and microsegration within the particles is eliminated; diffusion between powder particles begins to occur.
- *Alloying:* As the diffusion process continues, admixed additives begin to form alloyed structures with the base ferrous particles. For species such as carbon, this takes place early into the sintering process; for elements such as nickel or molybdenum, diffusion is much slower and takes longer times and higher temperatures to achieve a reasonable level of homogeneity.
- *Removal of gases/oxides:* Chemical reactions between the sintering atmosphere or admixed additives such as graphite and the surface oxides on the metal particles also begins early in the sintering cycle. This breakdown of oxides and removal of adsorbed gases cleanses the metal particle surfaces and promotes the diffusion process.
- *Particle bonding:* The formation of solid bridges or necks between individual or clusters of powder particles is the critical result of the early stages of sintering. These particle bonds give the powder mass integrity and mechanical strength.

Advanced stages of sintering (high temperature) influence several important characteristics, such as:

- *Densification:* As the sintering process continues at higher temperatures, the inherent porosity in the powder mass is reduced as pores are eliminated by bulk diffusion to grain boundaries. This reduction in the amount of porosity results in an increase in the density of the powder compact.
- *Porosity shape:* The remaining pores in the P/M structure lose their angular, irregular nature and become smooth, tending toward perfect spheres, as the sintering temperature increases.
- *Grain growth:* The individual powder particles lose their identity completely as grain boundaries move across prior particle boundaries. Larger grains replace the original fine particle structure.
- *Liquid phase:* Depending on the chemical constituents in the powder mass and the sintering temperature, a transient or permanent liquid phase may be formed. This liquid phase will accelerate particle rearrangement and diffusion, thereby aiding densification and pore elimination. For some additives, such as copper and phosphorus, liquid phase sintering will occur at conventional temperatures, while for silicon iron and the tool steels, high-temperature sintering is required. The phenomena attributed to the early stages of sintering will continue during high-temperature sintering. Thus a greater number and more complete particle bonding, as well as more homogeneous alloying, will occur during the advanced stages of sintering than found during the early stages.

The phenomenological changes that occur during sintering, are shown in Fig. 1 (a–d) for the microstructures of an austenitic stainless steel. Figure 1(a) shows the microstructure of the original compacted powder particles. The dendritic pattern in these particles is clearly seen. After sintering for 5 min at 1066 °C (1950 °F) (Fig. 1b), the cored dendritic structure is still visible, but the particles already show signs of homogenization. The structure after 30 min at 1121 °C (2050 °F) shown in Fig. 1(c), finds homogenization almost complete and particle-to particle bonding beginning to occur. Sintering for 2 h at 1316 °C (2400 °F) (Fig. 1d) produces a completely homogeneous structure, spheroidized porosity, and considerable grain growth. The original particle boundaries have completely disappeared with the formation of a single-phase austenitic structure, peppered with annealing twins.

Effect on Properties. Figure 2 illustrates the effect of sintering temperature on several physi-

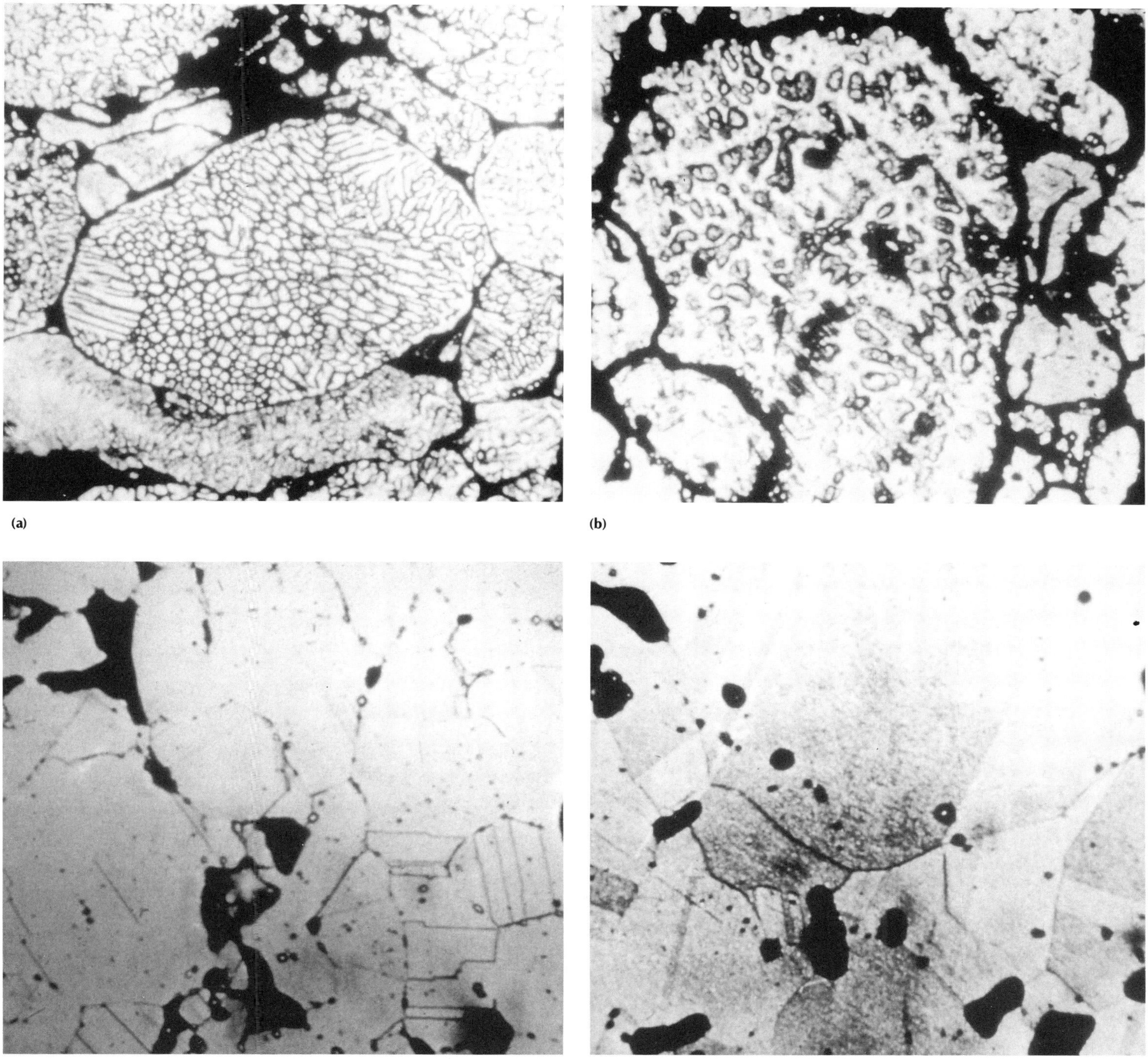

(a) (b) (c) (d)

Fig. 1 316 stainless steel. (a) As-compacted powder particles. (b) Sintered at 1066 °C (1950 °F) for 30 min. (c) Sintered at 1121 °C (2050 °F) for 30 min. (d) Sintered at 1316 °C (2400 °F) for 2 h. ~650×

cal and mechanical properties of the P/M compact (Ref 1). Although somewhat subjective in nature, the general trends are quite evident. As the degree of sintering improves, as indicated by a higher sintering temperature, the properties of the material also improve. The graph also demonstrates that the sensitivity of each of these properties to the sintering temperature can be quite different. The dynamic properties, such as fatigue and impact strength, are more sensitive to the sintering process than conductivity or mechanical strength.

Improved Mechanical Properties

As sintering temperatures are increased, the mechanical properties of ferrous P/M materials will improve due to more complete interparticle bonding, greater diffusion of alloying elements, and, to some extent, increased densification. Unfortunately, the deleterious effects of grain growth can negate the improvements from homogenization and densification if the grain size gets too large. In addition, the higher sintered density has been shown to be a more significant factor for improving dynamic properties than high-temperature sintering alone (Ref 2, 3). A comparison of conventional versus high-temperature sintering for four P/M steels will describe these effects further.

Nickel Steel (FN-0205/FN-0208). The mechanical properties of two standard nickel steel premixes, sintered at 1121 and 1260 °C (2050 and 2300 °F), are provided in Table 1 (Ref 4). Conventional sintering was performed in a nitrogen-base atmosphere belt furnace, while the high-temperature sintering used a nitrogen back-

Table 1 Effect of sintering conditions on the mechanical properties of FN 0205/FN0208

	FN0205		FN0208	
	Belt(a)	Vacuum(b)	Belt(a)	Vacuum(b)
Tensile strength, MPa (ksi)	380 (55)	550 (80)	450 (65)	760 (110)
Yield strength, MPa (ksi)	195 (28)	415 (60)	330 (48)	585 (85)
Elongation, %	4	7	2	4
Impact strength, J (ft · lbf)	19 (14)	38 (28)	11 (8)	33 (24)
Hardness, HRB	64	80	80	90
Sintered density, g/cm³	NR	7.32	NR	7.30

All samples pressed to a green density of 7.2 g/cm³. NR, not reported. (a) Sintered for 30 min at 1127 °C (2060 °F) in nitrogen/endothermic atmosphere. (b) Sintered for 2 h at 1260 °C (2300 °F) with nitrogen backfill. Source: Ref 4

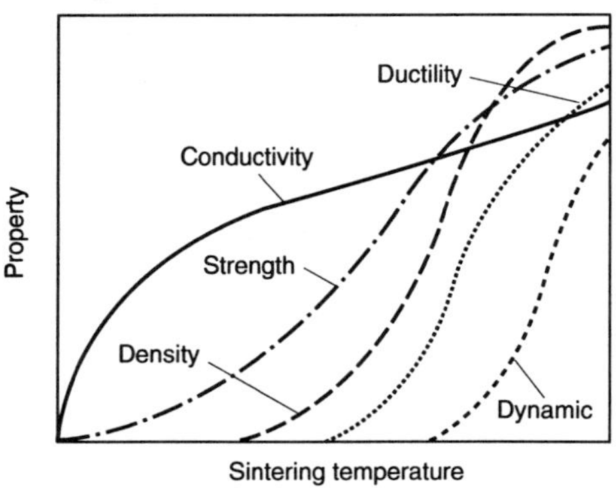

Fig. 2 The variation in compact properties with the degree of sintering as represented by the sintering temperature

Table 2 Effect of sintering conditions on the mechanical properties of copper steels

	FC0208(a)		Fe-2Cu-2.5Ni(b)		Fe-3Cu(c)	
	Belt	HTS	1120 °C	1280 °C	1120 °C	1250 °C
Tensile strength, MPa (ksi)	650 (94.1)	777 (112.7)	310 (45)	365 (53)	330 (47.7)	280 (40.6)
Yield strength, MPa (ksi)	505 (73)	605 (87.6)	195 (28)	275 (40)	385 (55.9)	315 (45.7)
Elongation, %	2	3	8	10	5	7
Impact energy, J(ft · lbf)	...	...	34 (25)	57 (42)	18.7 (13.8)	31.9 (23.5)
Endurance limit, MPa (ksi)	235 (34)	270 (39)	120 (17.4)	135 (19.4)	...	...
Sintered density, g/cm³	7.01	7.11	7.1	7.1	7.12	7.18

HTS, high-temperature sintering. (a) Belt: 1121 °C (2050 °F), 30 min in dissociated ammonia (Ref 5); HTS: 1329 °C (2425 °F), 60 min in dissociated ammonia + methane. (b) 1120 °C: 20 min in dissociated ammonia (Ref 6); 1280 °C: 40 min in 70% N_2, 30% H_2. (c) 2 h in hydrogen (Ref 7)

Table 3 Effect of sintering conditions on the mechanical properties of Distaloy

	4800A + 0.4 graphite(a)		AE + 0.5(b)	
	Belt	HTS	1120 °C	1250 °C
Tensile strength, MPa (ksi)	710 (103.2)	820 (119.1)	725 (104.9)	795 (115.2)
Yield strength, MPa (ksi)	440 (63.5)	535 (77.9)	415 (60.5)	485 (70.7)
Elongation, %	4	4	3	2.5
Impact energy, J(ft · lbf)	33 (24)	31 (23)	30 (22)	28 (21)
Endurance limit, MPa (ksi)	255 (37.2)	240 (34.5)	240 (35)	230 (33)
Sintered density, g/cm³	7.20	7.23	7.12	7.14

HTS, high-temperature sintering. (a) Belt: 1121 °C (2050 °F), 30 min in dissociated ammonia (Ref 2, 5); HTS: 1260 °C (2300 °F), 60 min in dissociated ammonia + methane. (b) 1120 °C: 30 min in endothermic atmosphere (Ref 3); 1250 °C: 30 min in 95% N_2, 5% H_2

Table 4 Effect of sintering conditions on the mechanical properties of 316 stainless steel

	1121 °C (2050 °F)	1204 °C (2200 °F)	1288 °C (2350 °F)	1204 °C (2050 °F)(N_2)
Tensile strength, MPa (ksi)	310 (45)	345 (50)	380 (55)	530 (77)
Yield strength, MPa (ksi)	205 (30)	185 (27)	170 (25)	360 (52)
Elongation, %	7	10	25	11

All sintered densities at 7.0 g/cm³. Sinter cycle: vacuum furnace, 2 h at temperature, argon backfill unless noted as nitrogen (N_2). Source: Ref 8

filled vacuum furnace. Tensile and yield strength showed a marked increase from the high-temperature sinter process, with the elongation and impact energy at least doubling for both materials. Because the sintered density of the belt-sintered samples was not reported but can be estimated at nearly equal the green density of 7.2g/cm³, improvements in mechanical properties are due mainly to more homogeneous distribution of the nickel and better particle bonding rather than just density increase alone.

Copper Steels. The data for three copper steels (Ref 5–7) listed in Table 2 does not show the significant improvements realized with the nickel steels. This improvement can be attributed to the liquid phase that has already formed when the copper melts at conventional sintering conditions; only modest improvements can be expected at the higher sintering temperatures. Note, however, the 70% increase in impact energy for both the Fe-2Cu-2.5Ni and Fe-3Cu materials when high temperature sintered. Fatigue values followed the small increase in tensile strength and remained close to 36% of the ultimate tensile strength, whether sintered conventionally or at higher temperatures.

Partially Prealloyed Steels (Distaloy). High-temperature sintering does not appear to offer much improvement for the Distaloy materials (Ref 2, 3, 5). The data in Table 3, taken from two Hoeganaes investigators, finds only a 10 to 20% increase in strength and no improvement in elongation, impact energy, or fatigue properties. This has been attributed to the greater chemical homogeneity of the nickel in the high-temperature sintered material, leaving fewer nickel-rich areas as sites for the higher strength bainite/martensite microstructural constituents. In this material system, the enhanced diffusion associated with high-temperature sintering helps remove the heterogeneity necessary for improved mechanical properties.

316 Stainless Steel. Prealloyed austenitic stainless steel P/M materials react favorably to high-temperature sintering conditions (Ref 8). The data in Table 4 illustrates the effect of three sintering temperatures on 316 stainless steel samples sintered in vacuum to a final density of 7.0 g/cm³. While the mechanical strength does not change significantly, the ductility more than tripled when the sintering temperature increased from 1121 to 1260 °C (2050 to 2300 °F). In addition, when nitrogen was used as the backfill gas, a small amount of nitrogen was dissolved in the austenitic structure resulting in a marked increase in strength and a lowering of ductility. Thus, high-temperature sintering can affect this alloy not only via the mechanism previously described by also by affecting the solubility of nitrogen in the austenitic steel matrix (Ref 9).

Improved Physical Properties

High-temperature sintering can improve the physical properties of ferrous P/M materials by means of the same phenomena that improve mechanical properties: more complete particle bonding, increased density, and a cleaner microstructure for similar sintering atmosphere conditions. Three examples will describe these effects.

Soft Magnetic Properties. The effect of high-temperature sintering on the structure sensitive magnetic properties of selected ferrous P/M materials has been described by several authors (Ref 10–12). Improved magnetization characteristics for high-temperature sintered material have been attributed to easier magnetic domain wall movement across particle boundaries and the decrease in interstitial impurities, such as carbon, nitrogen, and oxygen. Typical results for pure iron and Fe-0.8P have been listed in Table 5. The mass sensitive properties, B_{15} and Br, do not change at the higher sintering temperatures; in contrast, both coercive force and permeability show significant improvements for both materials even though sintered density remained constant.

Corrosion Resistance of 316 Stainless Steel. The corrosion resistance of P/M stainless steels has been recognized as inferior to wrought alloys of the same composition due to the presence of porosity (Ref 10). However, recent investigators have clearly established sintering conditions as a main contributor to the corrosion resistance of these alloys (Ref 9, 13).

Optimal sintering requires high sintering temperatures, low dew points, and control of interstitial elements in the alloy structure (nitrogen, oxygen, and carbon). Because many stainless

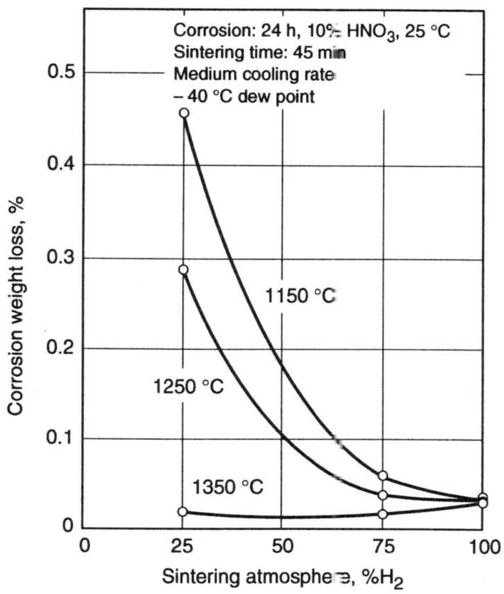

Fig. 3 Corrosion weight loss for 316 stainless steel as a function of sintering atmosphere for three sintering temperatures. Source: Ref 13

steels are sintered in nitrogen bearing atmospheres, rapid cooling from the sintering temperature is recommended to inhibit absorption of excess nitrogen that would lead to a depletion of chromium from the alloy matrix by the formation of chromium nitrides. This effect is illustrated in Figure 3 for 316 stainless steel (Ref 13), which shows much lower corrosion rates for higher sintering temperatures and lower nitrogen content sintering atmospheres. The higher temperature reduces the nitrogen solubility in the matrix, improves the deoxidizing effect of the hydrogen, and reduces the deleterious effect of tramp iron powder contamination by homogenizing the iron into the stainless steel matrix.

Table 5 Effect of sintering conditions on the magnetic properties of iron and iron-phosphorus

	Iron(a)		Iron-0.8% Phosphorus(b)	
	1121 °C (2050 °F)	1260 °C (2300 °F)	1121 °C (2050 °F)	1260 °C (2300 °F)
Magnetization at 15 Oe, kG	12.2	12.8	14.0	13.7
Remanent magnetization, kG	11.1	11.8	12.9	13.0
Coercive force, Oe	1.5	1.3	1.1	0.7
Maximum permeability	3300	4600	5900	9700
Sintered density, g/cm^3	7.2	7.2	7.35	7.37

(a) Sintered at 30 min in hydrogen (Ref 10). (b) Sintered at 1121 °C (2050 °F) in hydrogen, 1260 °C (2300 °F) in vacuum (Ref 11)

Leak Tightness. Because nearly all conventionally sintered ferrous P/M products contain interconnected porosity (unless copper infiltrated), these products must be sealed for those applications requiring leak tightness. Densification during high-temperature sintering can be used to achieve a final density, which eliminates the interconnected porosity and thus provides an impervious structure. A 316 stainless steel alloy, high-temperature vacuum sintered, was used for valve components without requiring any secondary sealing operations (Ref 8). This material successfully passed pressure tests at 250 psi and demonstrated a vacuum leak rate of less than 10^{-10} cm^3/s at a vacuum of 10^{-6} torr.

Development of a Liquid Phase

For several types of ferrous P/M alloys liquid phase sintering provides the means for more complete alloying, improved densification, and even attainment of full density. Some of the alloy systems reach liquid phase conditions at conventional sintering temperatures (e.g., iron-copper and iron-phosphorus). However, for the cases presented below, only high-temperature sintering procedures can produce the desired liquid phase conditions.

Iron-Silicon Alloys. The iron-silicon P/M alloys are typically prepared by blending a ferrosilicon master alloy with a pure iron powder. At conventional sintering temperatures little, if any, diffusion of the silicon into the iron matrix is observed, impeded to a large measure by the tenacious oxide film on the ferrosilicon particles. Raising the sintering temperature above 1200 °C will ensure formation of a transient liquid phase, which promotes rapid diffusion and homogeneity of the silicon into the base iron matrix. This effect can be seen in the microstructures shown in Fig. 4(a) and (b). The conventional belt sintered product retains the composite structure of soft iron particles and hard ferrosilicon particles (Fig 4a). The high-temperature sintered product (Fig. 4b) shows a homogeneous microstructure of iron with 3% silicon in solid solution throughout the matrix. Because this material is typically used for soft magnetic applications, it should be evident that without high-temperature sintering the expected magnetic properties cannot be developed.

Tool Steels. The production of fully dense high-speed tool steel P/M structures requires controlled high-temperature sintering at temperatures approaching the solidus temperature for the alloy. Studies by several investigators (Ref 14, 15) have noted the critical role of liquid phase formation for producing accelerated shrinkage and full dense structures. Overheating by as little as 5 °C will result in rapid grain

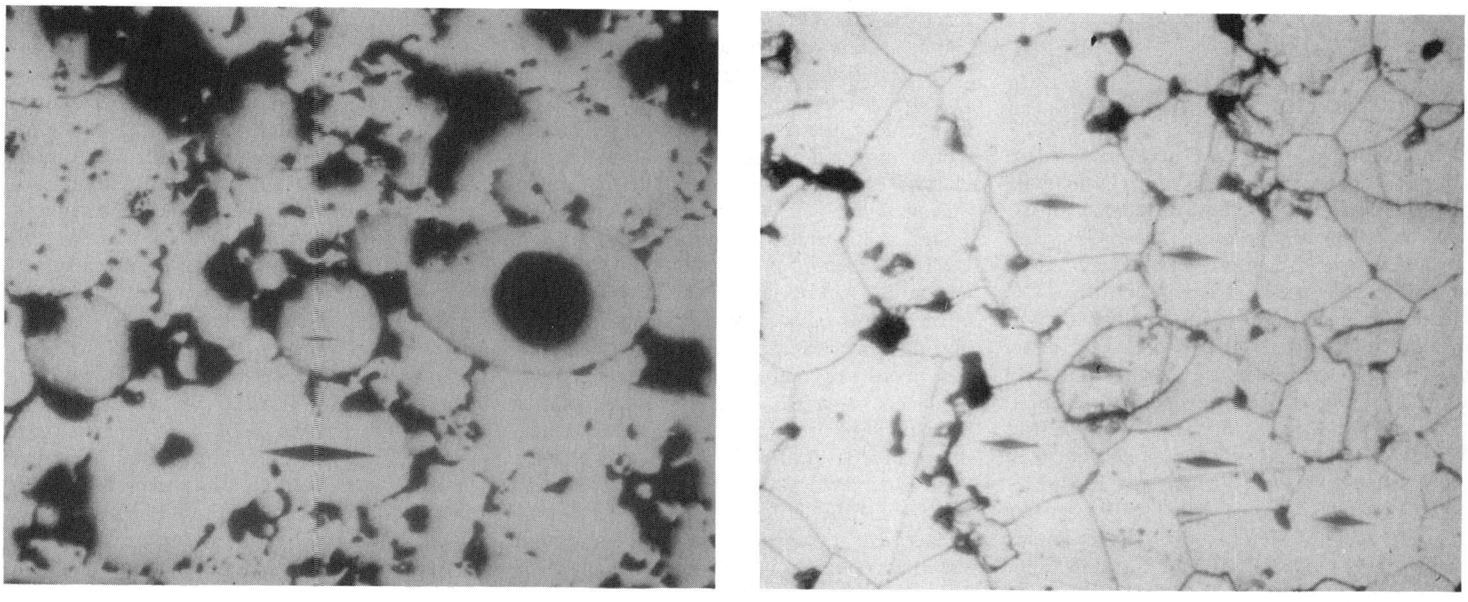

(a) (b)

Fig. 4 Microstructure of Fe-3%Si. (a) Sintered at 1121 °C (2050 °F) for 30 min. (b) Sintered at 1232°C (2250 °F) for 2 h. ~240×

growth and continuous carbide films between adjacent grains, primarily M_6C type carbides. When sintered below the solidus, the carbides dispersed in the structure impede grain growth and densification. On reaching this temperature, the carbides are dissolved, reducing the amount of inhibiting carbide dispersion, and thus grain growth and densification can occur. The desired structure of small grains and finely dispersed carbides can only be obtained if the amount of liquid phase formed during sintering is small and controlled (Ref 14).

The effect of sintering conditions on the densification of both M2 and T15 grade tool steels is listed in Table 6 (Ref 15). Quite clearly, precise control of sintering conditions, temperature as well as time, is critical for obtaining the desired full dense product with the proper microstructure. Due to differences in composition and carbides present, the T15 grade apparently has a wider sintering "window" than the M2 material, because an 8 °C rise in temperature produced an oversintered structure in M2, while the T15 required a 20 °C increment to produce this type of result.

Structural Alloy. R.M. German and his coworkers (Ref 16) studied the enhanced sintering of ferrous structural products, which led to the development of an alloy system that used both phosphorus and boron as liquid phase formers. A sintering temperature of 1200 °C was required to ensure formation of the iron-boron eutectic. The resultant iron-nickel-molybdenum base alloys developed excellent as-sintered strengths and good ductility with corresponding high sintered densities, as shown in the following table:

Alloy composition	Density, g/cm^3	Tensile strength, MPa (ksi)	Elongation, %
Fe-2.4Ni-3Mo-0.1P-0.25B-0.1C	7.30	720 (104.5)	6
Fe-4Ni-3Mo-0.1P-0.25B-0.1C	7.35	810 (117.5)	2

Sintering of Active Elements

The ability to sinter P/M steels using chromium and manganese as the principal alloying elements has been stymied by the need for exceptionally low dew points at conventional sintering temperatures. In order to reduce the oxides of these elements, a dew point of –50 °C would be needed, well below the range of commercial atmosphere sintering furnaces. However, if the sintering temperature can be raised to 1250 °C or higher, than a dew point of only –20 °C should be necessary. Work by Sumitomo (Ref 17) has shown effective sintering at 1250 °C in nitrogen-base atmospheres with a dew point of –10 °C for 4100 grade powder. Tensile strengths in excess of 1170 MPa (170 ksi) and an impact energy of 20 J (15 ft · lbf) were measured after heat treatment of this alloy. Therefore, these lower cost alloying elements (as compared with the more popular nickel, copper, and molybdenum) can be employed in prealloyed powders

Table 6 Effect of sintering conditions on the densification of high-speed tool steels

Alloy	Sintering conditions		Density, g/cm^3	Microstructure
M2	¼ h	1230 °C	7.45	Under sintered
	5 h	1233 °C	7.69	Under sintered
	¼ h	1240 °C	7.84	Close to optimum
	5 h	1242 °C	8.06	Close to optimum
	¼ h	1250 °C	8.01	Over sintered
	5 h	1250 °C	8.17	Over sintered
T15	¼ h	1247 °C	7.55	Under sintered
	5 h	1245 °C	7.78	Under sintered
	¼ h	1260 °C	8.10	Close to optimum
	5 h	1259 °C	8.18	Close to optimum
	¼ h	1278 °C	8.26	Over sintered
	5 h	1279 °C	8.28	Over sintered

(a) Theoretical density: 8.23 g/cm^3. (b) Theoretical density: 8.32 g/cm^3. Source: Ref 15

when coupled with high-temperature sintering techniques.

Process Control Requirements

With the benefits of high-temperature sintering clearly established for several ferrous alloy systems, the critical processing requirements must also be recognized and controlled in order to ensure this process can produce reliable and consistent P/M products. As with any multi-step manufacturing process, the change from conventional sintering to high-temperature sintering introduces new levels of variation in the final characteristics of the products, whether dimensional accuracy, mechanical properties, or physical conditions. Three sources of process variation requiring tighter control than under conventional sintering practices are the powder blend, green density, and sintering conditions.

Powder Blend. The raw material for any P/M product (i.e., the starting powder blend) is subject to variations from lot to lot. These variations can be seen in particle morphology, compressibility, apparent density, and segregation of admixed constituents. The two most critical to high-temperature sintering response are particle morphology and segregation. Changes in particle size distribution, especially the amount of fine particles, will accelerate the sintering response at higher temperatures more than conventional sintering practices. Higher fines content in the starting powder blend will result in greater densification, larger shrinkages in sintering, and somewhat improved mechanical properties. In order to control these effects, the powder producers must be aware that these powder blends will be used for high-temperature sintering and the level of variation that can be tolerated. The individual data sheets for each powder lot provided by the powder manufacturer must include data pertinent to the high-temperature sintering conditions being used, specifically sintered properties, in order for the parts producer to benefit from the data. In this way both the powder and parts producer will be assured that a specific powder lot can be used successfully for a high-temperature sintered product. Microsegregation of admixed additions is another key factor in powder control, especially for compositions that develop a liquid phase. Nonhomogeneous powder blends that cause excessive liquid formation in isolated areas can result in localized porosity, slumped or deformed shapes, and unusual or undesirable microstructures. Care must be taken by the powder supplier to insure con-trol of this type of problem; new bonding techniques for powder mixes can be helpful in reducing this type of segregation (Ref 18).

Green Density. Variation in green density, from part to part and within multi-level parts, is the responsibility of the parts fabricator. Allowances are made during compaction for the effect of power filling, tool motions, press configurations, and the actual powder being used to provide the necessary consistency in the final product. These small variations, which are acceptable at conventional sintering conditions, may no longer be satisfactory for parts sintered at higher temperatures. If excursions in green density are excessive, the result will be out-of-tolerance parts, distortion in part shape, and differences in physical/mechanical properties in different sections of the part.

An example of this effect can be seen by comparing Distaloy AG sintered at 1150 °C with Distaloy 4800A sintered at 1120 °C. Even though the nickel content is higher for the Distaloy AG (8% versus 4%), the dimensional change observed for the AG grade is 0.3% when compacting pressure varies from 30 to 60 tsi, while the 4800A material varies less than 0.1% over the same pressure range (Ref 19, 20). The higher temperature sintering for the Distaloy AG has contributed to this greater variation in size change. Because individual powder blends will react to high-temperature sintering conditions in a unique way, parametric studies by the parts fabricator are required to determine the allowable variation in green density that can be acceptable for the specific product design.

Sintering Conditions. The ability to control variations in sintering parameters (time, temperature, and atmosphere) become even more critical at high temperatures than conventional conditions. Because the physical phenomena discussed earlier are directly related to time and temperature (rate dependent), small variations relative to 1250 °C, for example, will have a greater effect on these rate controlled reactions than the same variation relative to 1120 °C. Such phenomena include diffusion, grain growth, and densification. If a liquid phase is formed, these controls are even more critical, as discussed earlier for high-speed tool steels. Both the parts fabricator and furnace manufacturer must be aware of these effects when specifying and designing the furnace itself, as well as the process control systems. Temperature uniformity within the sintering zone should be monitored frequently with a typical loading of parts in the furnace. It is not unusual to require a maximum temperature variation of ±5 °C within the hot zone. The time at temperature should also be monitored independent of the furnace control system to insure consistency in the conveyance mechanism. Variations in sintering time of six minutes or less are recommended (Ref 19). The effect of sintering atmosphere control, especially for carbon bearing alloys, is also critical at elevated temperatures. Decarburization reactions will be accelerated if dew point and carbon potential control are not properly maintained. Loss of carbon from the alloy will directly affect size control and final properties.

In order to appreciate the effect of high-temperature sintering on one parameter, size change, the following chart provides a comparison of the relative magnitude of shrinkage that can be expected for the variety of P/M alloys when high temperature sintered

Material system	Dimensional change, %
Low alloy steels	−0.01 to −0.02 (conventional)
Low alloy steels	−0.30 to −1.00
Stainless steels	−0.50 to −2.50
Tool steels	−5.00 to −8.00

Clearly, small process variations will create a greater effect on the final part size for the stainless steel and tool steel alloys than the low alloy steels.

REFERENCES

1. R.M. German, *Powder Metallurgy Science*, Metal Powder Industries Federation, 1984, p 175
2. R.C. O'Brien, Impact and Fatigue Characteristics of Select Ferrous P/M Materials, *Progress in Powder Metallurgy,* Vol 43, Metal Powder Industries Federation, 1987, p 749
3. B. Lindquist, Influence of Microstructure and Porosity on Fatigue Properties of Sintered Steels, *Modern Developments in Powder Metallurgy,* Vol 21, Metal Powder Industries Federation, 1988, p 67
4. H.I. Sanderow, H. Rodrigues, and J.D. Ruhkamp, New High Strength 4100 Alloy P/M Steels, *Progress in Powder Metallurgy,* Vol 41, Metal Powder Industries Federation, 1985, p 283
5. R.C. O'Brien "Fatigue Properties of P/M Materials," SAE Paper 880165, Society of Automotive Engineers
6. C.M. Sonsino et al., Influence of Homogeneity on the Fatigue Properties of Sintered Steel, *Int. J. Powder Metall.,* Vol 20 (No 1), Metal Powder Industries Federation, 1984, p 45
7. H. Danringer et al., High Pressure Compaction of Ferrous P/M Parts, *Met. Powder Rep.,* Nov. 1986, p 833
8. H.I. Sanderow, Vacuum Sintering of P/M Stainless Steel Components, *Modern Developments in Powder Metallurgy,* Vol 11, Metal Powder Industries Federation, 1976, p 229
9. M. Svilar and H. Ambs, 'NOC' Precipitation During the Sintering of Austenitic Stainless Steel and Their Effect on Corrosion, *Progress in Powder Metallurgy,* Vol 43, Metal Powder Industries Federation, 1987, p 163
10. F.V. Lenel, *Powder Metallurgy-Principles and Applications,* Metal Powder Industries Federation, 1980, p 535
11. K. Moyer and J Ryan, Iron-Phosphorus Alloys for Magnetic Application, *Progress in Powder Metallurgy,* Vol 43, Metal Powder Industries Federation, 1987, p 323
12. C. Lall and L. Baum, High Performance Soft Magnetic Components by P/M and MIM, *Modern Developments in Powder Metallurgy,* Vol 18, Metal Powder Industries Federation, 1988, p 363
13. G. Lei, R. German, and H. Nayar, Influence of Sintering Variables on the Corrosion Resistance of 316L Stainless Steel *Powder Metall. Int.,* Vol 15 (No. 2), 1983, p 70
14. M.T. Podob et al., The Mechanism of Sintering High Speed Steel to Full Density, *Modern Developments in Powder Metallurgy* Vol 13, Metal Powder Industries Federation, 1981, p 71
15. J.V. Bee et al., Phase Distributions during the Sintering of High Speed Steel Powders, *Progress in Powder Metallurgy,* Vol 41, Metal Powder Industries Federation, 1985, p 217
16. D.S. Madan, R. German, and C. Lall, High Strength Ferrous Alloys by Enhanced Sintering *Progress in Powder Metallurgy,* Vol 41, Metal Powder Industries Federation, 1985, p 307
17. M. Ichidate et al., Sintering of Oil Atomized Low Alloy Steel Powder, *Horizons of Powder Metallurgy,* Proc. of 1986 Int. P/M Conf. (Dusseldorf), Vol 1, Verlag Schmid GMBH, 1986, p 57
18. F.J. Semel, Properties of Parts Made from a Binder Treated 0.45% Phosphorus Containing Iron Blend, *Progress in Powder Metallurgy,* Vol 43, Metal Powder Industries Federation, 1987, p 723
19. J. Tengzelius, A New Powder for High Strength Applications, *Met. Powder Rep.,* Nov. 1988, p 757
20. Hoeganaes Distaloy Data Bulletin, June 1985

Conventional Aluminum Powder Metallurgy Alloys

ALUMINUM P/M PARTS are used in an increasing number of applications. The business machine market currently uses the greatest variety of aluminum P/M parts. Other markets that indicate growth potential include automotive components, aerospace components, power tools, appliances, and structural parts. Due to their mechanical and physical properties, aluminum P/M alloys provide engineers with flexibility in material selection and design. A variety of pressed and sintered aluminum P/M parts are shown in Fig. 1.

Conventionally pressed and sintered aluminum powder metal parts have been commercially available for many years. Sintered aluminum P/M parts are competitive with many aluminum castings, extrusions, and screw machine products that require expensive and time-consuming finishing operations. In addition, sintered aluminum P/M parts compete with other metal powder parts in applications where some of the attractive physical and mechanical properties of aluminum can be used. The following combination of properties make aluminum attractive for P/M parts:

- Light weight
- Corrosion resistance
- High strength
- Good ductility
- Nonmagnetic properties
- Conductivity
- Machinability
- Variety of finishes

In addition, P/M technology can be used to refine microstructures compared with those made by conventional ingot metallurgy (I/M), which often results in improved mechanical and corrosion properties. Microstructural refinement by P/M is made possible by two broad high-strength P/M technologies—rapid solidification (RS) and mechanical attrition (mechanical alloying/dispersion strengthening). The advantages of P/M stem from the ability of small particles to be processed. This enables:

- The realization of RS rates
- The uniform introduction of strengthening features, that is, barriers to dislocation motion, from the powder surfaces

The powder processes of rapid solidification and mechanical attrition lead to microstructural grain refinement and, in general, better mechanical properties of the alloy. In addition, RS can extend the alloying limits in aluminum by enhancing super-saturation and thereby enabling greater precipitation hardening without the harmful segregation effects from overalloyed I/M alloys. Moreover, elements that are essentially insoluble in the solid state, but have significant solubility in liquid aluminum, can be uniformly dispersed in the powder particles during RS. This can lead to the formation of novel strengthening phases that are not possible by conventional I/M, while also suppressing the formation of equilibrium phases that are deleterious to toughness and corrosion resistance. These topics and the development of dispersion strengthened aluminum-base composites by P/M methods are discussed in the article "Advanced Aluminum Powder Metallurgy Alloys and Composites" in this Volume.

This article focuses on more conventional aluminum P/M alloys and process methods. There are several steps in aluminum P/M technology that can be combined in various ways, but they will be conveniently described in three general steps:

- Powder production
- Powder processing (optional)
- Degassing and consolidation

Powder can be made by various RS processes including atomization, splat quenching to form particulates, and melt spinning to form ribbon. Alternatively, powder can be made by non-RS processes such as by chemical reactions including precipitation, or by machining bulk material. Powder-processing operations are optional and include mechanical attrition (for example, ball milling) to modify powder shape and size or to

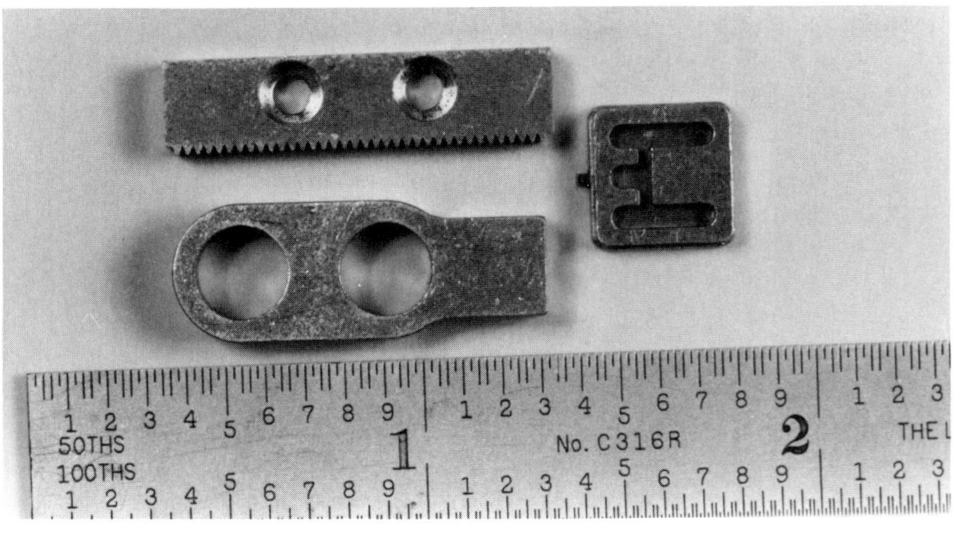

Fig. 1 Typical pressed and sintered aluminum P/M parts made from alloy 601AB. Top: gear rack used on a disc drive. Bottom: link flexure used on a print tip for a typewriter. Right: header/cavity block used on a high-voltage vacuum capacitor. Courtesy of D. Burton, Perry Tool & Research Company

introduce strengthening features, or comminution such as that used to cut melt-spun ribbon into powder flakes for subsequent handling.

Aluminum has a high affinity for moisture, and aluminum powders readily adsorb water. The elevated temperatures generally required to consolidate aluminum powder cause the water of hydration to react and form hydrogen, which can result in porosity in the final product, or under confined conditions, can cause an explosion. Consequently, aluminum powder must be degassed prior to consolidation. This is often performed immediately prior to consolidation at essentially the same temperature as that for consolidation to reduce fabrication costs. Consolidation may involve forming a billet that can be subsequently rolled, extruded, or forged conventionally, or the powder may be consolidated during hot working directly to finished-product form.

Powder Production

Atomization is the most widely used process to produce aluminum powder. Aluminum is melted, alloyed, and sprayed through a nozzle to form a stream of very fine particles that are rapidly cooled—most often by an expanding gas. Atomization of aluminum is discussed in more detail in the article "Production of Aluminum Powders" in this Volume.

Splat cooling is the process that enables cooling rates even greater than those obtained in atomization. Aluminum is melted and alloyed, and liquid droplets are sprayed or dropped against a chilled surface of high thermal conductivity—for example, a copper wheel that is water cooled internally. The resultant splat particulate is removed from the rotating wheel to allow subsequent droplets to contact the bare, chilled surface. Cooling rates of 10^5 K/s are typical, with rates up to 10^9 K/s reported.

Melt-spinning techniques are somewhat similar to splat cooling. The molten aluminum alloy rapidly impinges a cooled rotating wheel, producing rapidly solidified product that is often in ribbon form. The leading commercial melt-spinning process is the planar flow casting (PFC) process (Ref 1). The liquid stream contacts a rotating wheel at a carefully controlled distance to form a thin, rapidly solidified ribbon and also to reduce oxidation. The ribbon could be used for specialty applications in its PFC form but is most often comminuted into flake powder for subsequent degassing and consolidation.

Press and Sintered Aluminum Alloys

Commercially available aluminum powder alloy compositions (Table 1) consist of blends of atomized aluminum powders mixed with powders of various alloying elements such as zinc, copper, magnesium, and silicon. Press and sintered aluminum alloys are based on elemental blends. Prealloyed aluminum powders cannot be

Table 1 Compositions of typical aluminum P/M alloy powders

Grade	Composition, %				
	Cu	Mg	Si	Al	Lubricant
601AB	0.25	1.0	0.6	bal	1.5
201AB	4.4	0.5	0.8	bal	1.5
602AB	...	0.6	0.4	bal	1.5
202AB	4.0	...	...	bal	1.5
MD-22	2.0	1.0	0.3	bal	1.5
MD-24	4.4	0.5	0.9	bal	1.5
MD-69	0.25	1.0	0.6	bal	1.5
MD-76	1.6	2.5	...	bal	1.5

processed by conventional P/M press and sinter operations. The non-reducible aluminum oxide coating hinders the development of strong interparticle bonds.

The most common heat-treatable grades are comparable to the 2*xxx* and 6*xxx* series wrought aluminum alloys. Alloys 201AB and MD-24 are most similar to wrought alloy 2014. They develop high strength and offer moderate corrosion resistance. Alloys 601AB and MD-69 are similar to wrought alloy 6061. These alloys offer high strength, good ductility, corrosion resistance, and can be specified for anodized parts. Alloy 601AC is the same as 601AB, but does not contain an admixed lubricant. It is used for isostatic and die-wall-lubricated compaction. When high conductivity is required, alloy 602AB often is used. Conductivity of 602AB ranges from 24×10^6 to 28×10^6 S/m (42.0 to 49% IACS), depending on the type of heat treatment selected.

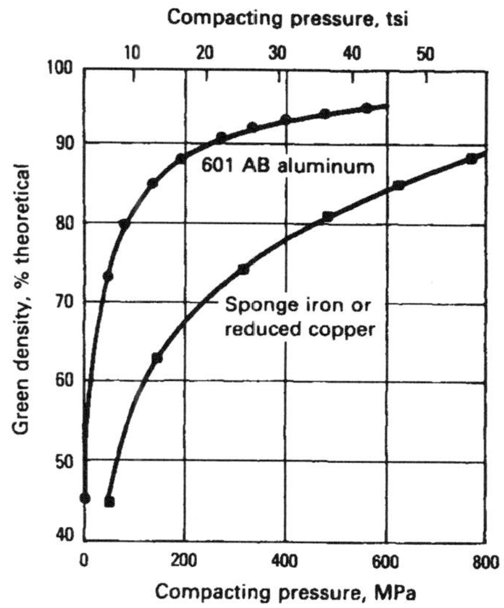

Fig. 2 Relationship of green density and compacting pressure

Aluminum P/M Part Processing

Basic design details for aluminum P/M parts involve the same manufacturing operations, equipment, and tooling that are used for iron, copper, and other metal-powder compositions.

Compacting. Aluminum P/M parts are compacted at low pressures and are adaptable to all types of compacting equipment. The pressure density curve, which compares the compacting characteristics of aluminum with other metal powders, indicates that aluminum is simpler to compact. Figure 2 shows the relative difference in compacting characteristics for aluminum and sponge iron or copper.

The lower compacting pressures required for aluminum permit wider use of existing presses. Depending on the press, a larger part often can be made by taking advantage of maximum press force. For example, a part with a 130 cm^2 (20 $in.^2$) surface area and 50 mm (2 in.) depth is formed readily on a 4450 kN (500 ton) press. The same part in iron would require a 5340 kN (600 ton) press. In addition, because aluminum responds better to compacting and moves more readily in the die, more complex shapes having more precise and finer detail can be produced.

Sintering. Aluminum P/M parts can be sintered in a controlled, inert atmosphere or in vacuum. Sintering temperatures are based on alloy composition and generally range from 595 to 625 °C (1100 to 1160 °F). Sintering time varies from 10 to 30 min. Nitrogen, dissociated ammonia, hydrogen, argon, and vacuum have been used for sintering aluminum; however, nitrogen is preferred because it results in high as-sintered mechanical properties (Table 2). It is also economical in bulk quantities, If a protective atmosphere is used, a dew point of –40 °C (–40 °F) or below is recommended. This is equivalent to a moisture content of 120 mL/m^3 (120 ppm) maximum.

Aluminum preforms can be sintered in batch furnaces or continuous radiant tube mesh or cast belt furnaces. Optimum dimensional control is best attained by maintaining furnace temperature at ±2.8 °C (±5 °F). Typical heating cycles for aluminum parts sintered in various furnaces are illustrated in Fig. 3.

Mechanical properties are directly affected by thermal treatment. All compositions respond to solution heat treating, quenching, and aging in the same manner as conventional heat-treatable alloys. More detailed information on sintering of aluminum can be found in the article "Production Sintering Practices" in this Volume.

Repressing. The density of sintered compacts may be increased by repressing. When repressing is performed primarily to improve the dimensional accuracy of a compact, it usually is termed "sizing"; when performed to improve configuration, it is termed "coining." Repressing

Table 2 Typical properties of nitrogen-sintered aluminum P/M alloys

Alloy	Compacting pressure		Green density		Green strength		Sintered density		Temper	Tensile strength(a)		Yield strength(a)		Elongation,	Hardness
	MPa	tsi	%	g/cm³	MPa	psi	%	g/cm³		MPa	ksi	MPa	ksi	%	
601AB	96	7	85	2.29	3.1	450	91.1	2.45	T1	110	16	48	7	6	55-60 HRH
									T4	141	20.5	96	14	5	80-85 HRH
									T6	183	26.5	176	25.5	1	70-75 HRE
	165	12	90	2.42	6.55	950	93.7	2.52	T1	139	20.1	88	12.7	5	60-65 HRH
									T4	172	24.9	114	16.6	5	80-85 HRH
									T6	232	33.6	224	32.5	2	75-80 HRE
	345	25	95	2.55	10.4	1500	96.0	2.58	T1	145	21	94	13.7	6	65-70 HRH
									T4	176	25.6	117	17	6	85-90 HRH
									T6	238	34.5	230	33.4	2	80-85 HRE
602AB	165	12	90	2.42	6.55	950	93.0	2.55	T1	121	17.5	59	8.5	9	55-60 HRH
									T4	121	17.5	62	9	7	65-70 HRH
									T6	179	26	169	24.5	2	55-60 HRE
	345	25	95	2.55	10.4	1500	96.0	2.58	T1	131	19	62	9	9	55-60 HRH
									T4	134	19.5	65	9.5	10	70-75 HRH
									T6	186	27	172	25	3	65-70 HRE
201AB	110	8	85	2.36	4.2	600	91.0	2.53	T1	169	24.5	145	24	2	60-65 HRE
									T4	210	30.5	179	26	3	70-75 HRE
									T6	248	36	248	36	0	80-85 HRE
	180	13	90	2.50	8.3	1200	92.9	2.58	T1	201	29.2	170	24.6	3	70-75 HRE
									T4	245	35.6	205	29.8	3.5	75-80 HRE
									T6	323	46.8	322	46.7	0.5	85-90 HRE
	413	30	95	2.64	13.8	2000	97.0	2.70	T1	209	30.3	181	26.2	3	70-75 HRE
									T4	262	38	214	31	5	80-85 HRE
									T6	332	48.1	327	47.5	2	90-95 HRE
202AB compacts	180	13	90	2.49	5.4	780	92.4	2.56	T1	160	23.2	75	10.9	10	55-60 HRH
									T4	194	28.2	119	17.2	8	70-75 HRH
									T6	227	33	147	21.3	7.3	45-50 HRE
Cold-formed parts (19% strain)	180	13	90	2.49	5.4	780	92.4	2.56	T2	238	33.9	216	31.4	2.3	80 HRE
									T4	236	34.3	148	21.5	8	70 HRE
									T6	274	39.8	173	25.1	8.7	85 HRE
									T8	280	40.6	250	36.2	3	87 HRE

(a) Tensile properties determined using powder metal flat tension bar (MPIF standard 10-63), sintered 15 min at 620 °C (1150 °F) in nitrogen

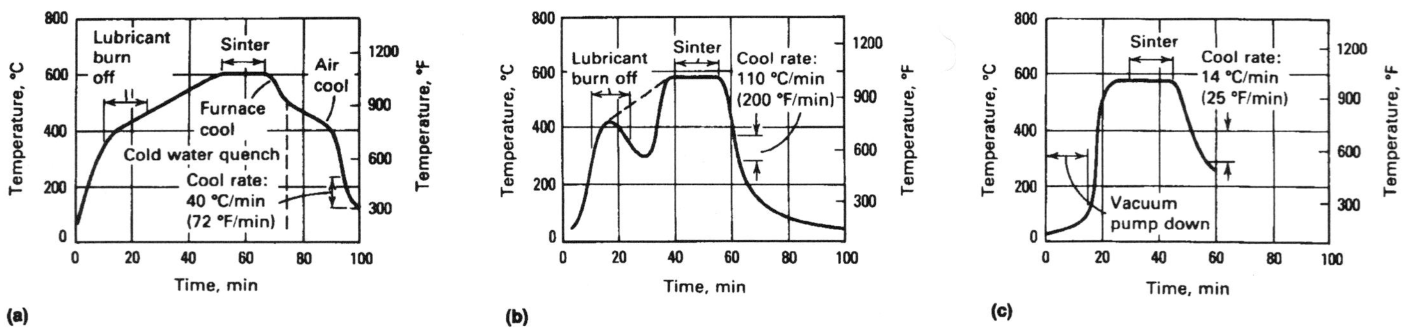

Fig. 3 Typical heating cycles for aluminum P/M parts sintered in (a) a batch furnace. (b) a continuous furnace. (c) a vacuum furnace

may be followed by resintering, which relieves stress due to cold work in repressing and may further consolidate the compact. By pressing and sintering only, parts of >80% theoretical density can be produced. By repressing, with or without resintering, parts of ≥90% theoretical density can be produced. The density attainable is limited by the size and shape of the compact.

Forging of aluminum is a well-established technology. Aluminum also lends itself to the forging of P/M preforms to produce structural parts.

In forging of aluminum preforms, the sintered aluminum part is coated with a graphite lubricant to permit proper metal flow during forging. The part is either hot or cold forged; hot forging at 300 to 450 °C (575 to 850 °F) is recommended for parts requiring critical die fill. Forging pressure usually does not exceed 345 MPa (50 ksi). Forging normally is performed in a confined die so that no flash is produced and only densification and lateral flow result from the forging step. Scrap loss is <10% compared to conventional forging, which approaches 50%. Forged aluminum P/M parts have densities of >99.5% of theoretical density. Strengths are higher than nonforged P/M parts, and in many ways, are similar to conventional forging. Fatigue endurance limit is doubled over that of nonforged P/M parts.

Alloys 601AB, 602AB, 201AB, and 202AB are designed for forgings. Alloy 202AB is especially well suited for cold forging. All of the aluminum powder alloys respond to strain hardening and precipitation hardening, providing a wide range of properties. For example, hot forging of alloy 601 AB-T4 at 425 °C (800 °F) followed by heat treatment gives ultimate tensile strengths of 221 to 262 MPa (32 to 38 ksi), and a yield strength of 138 MPa (20 ksi), with 6 to 16% elongation in 25 mm (1 in.).

Alloy 601AB. Heat treated to the T6 condition, 601AB has tensile properties of:

- UTS, 303 to 345 MPa (44 to 50 ksi)
- Yield strength, 303 to 317 MPa (44 to 46 ksi)
- Elongation, up to 8%

Forming pressure and percentage of reduction during forging influence final properties.

Alloy 201AB. In the T4 condition, alloy 201AB has tensile properties of:

- UTS, 358 to 400 MPa (52 to 58 ksi)

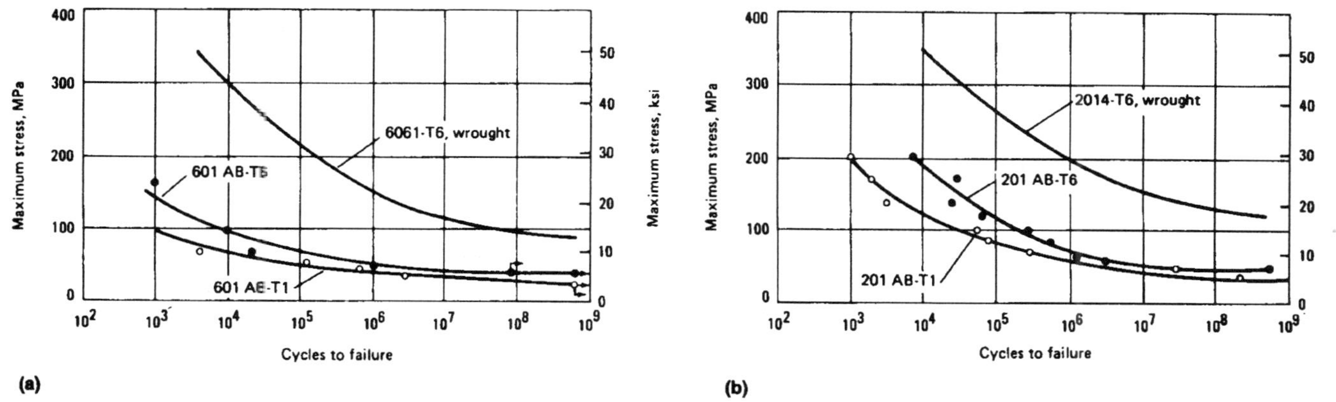

Fig. 4 Fatigue curves for (a) P/M 601AB (b) P/M 201AB

- Yield strength, 255 to 262 MPa (37 to 38 ksi)
- Elongation, 8 to 18%

When heat treated to the T5 condition, the tensile strength of 201AB increases from 393 to 434 MPa (57 to 63 ksi). Yield strength for this condition is 386 to 414 MPa (56 to 60 ksi), and elongation ranges from 0.5 to 8%.

Properties of cold-formed aluminum P/M alloys are increased by a combination of strain-hardened densification and improved interparticle bonding. Alloy 601AB achieves 257 MPa (37.3 ksi) tensile strength and 241 MPa (34.9 ksi) yield strength after forming to 28% upset. Properties for the T4 and T6 conditions do not change notably between 3 and 28% upset. Alloy 602AB has moderate properties with good elongation. Strain hardening (28% upset) results in 221 MPa (32 ksi) tensile and 203 MPa (29.4 ksi) yield strength. The T6 temper parts achieve 255 MPa (37 ksi) tensile strength and 227 MPa (33 ksi) yield strength. Highest cold-formed properties are achieved by 201AB. In the as-formed condition, yield strength increases from 209 MPa (30.3 ksi) for 92.5% density to 281 MPa (40.7 ksi) for 96.8% density.

Alloy 202AB is best suited for cold forming. Treating to the T2 condition, or as-cold formed, increases the yield strength significantly. In the T8 condition, 202AB develops 280 MPa (40.6 ksi) tensile strength and 250 MPa (36.2 ksi) yield strength, with 3% elongation at the 19% upset level.

Properties of Sintered Parts

Mechanical Properties. Sintered aluminum P/M parts can be produced with strength that equals or exceeds that of iron or copper P/M parts. Tensile strengths range from 110 to 345 MPa (16 to 50 ksi), depending on composition, density, sintering practice, heat treatment, and repressing procedures. Table 2 lists typical properties of four nitrogen-sintered P/M alloys. Properties of heat-treated, pressed, and sintered grades are provided in Table 3.

Impact tests are used to provide a measure of toughness of powder metal materials, which are somewhat less ductile than similar wrought compositions. Annealed specimens develop the highest impact strength, whereas fully heat-treated parts have the lowest impact values. Alloy 201AB generally exhibits higher impact resistance than alloy 601AB at the same percent density, and impact strength of 201AB increases with increasing density. A desirable combination of strength and impact resistance is attained in the T4 temper for both alloys. In the T4 temper, 95% density 201AB develops strength and impact properties exceeding those for as-sintered 99Fe-1C alloy, a P/M material frequently employed in applications requiring tensile strengths under 345 MPa (50 ksi).

Fatigue is an important design consideration for P/M parts subject to dynamic stresses. Fatigue strengths of pressed and sintered P/M parts may be expected to be about half those of the wrought alloys of corresponding compositions (see comparisons of two P/M alloys with two wrought alloys in Fig. 4). These fatigue-strength levels are suitable for many applications.

Electrical and Thermal Conductivity. Aluminum has higher electrical and thermal conductivities than most other metals. Table 4 compares the conductivities of sintered aluminum alloys with wrought aluminum, brass, bronze, and iron.

Machinability. Secondary finishing operations such as drilling, milling, turning, or grinding can be performed easily on aluminum P/M parts. Aluminum P/M alloys provide excellent chip characteristics; compared to wrought aluminum alloys, P/M chips are much smaller and are broken more easily with little or no stringer buildup. This results in improved tool service life and higher machinability ratings.

Powder Degassing and Consolidation

The water of hydration that forms on aluminum powder surfaces must be removed to prevent porosity in the consolidated product. Although solid-state degassing has been used to reduce the hydrogen content of aluminum P/M wrought products, it is far easier and more effective to remove the moisture from the powder. Degassing is often performed in conjunction with consolidation, and the most commonly used techniques are described below. The vari-

Table 3 Typical heat-treated properties of nitrogen-sintered aluminum P/M alloys

	Grades			
Heat-treated variables and properties	MD-22	MD-24	MD-69	MD-76
Solution treatment				
Temperature, °C (°F)	520 (970)	500 (930)	520 (970)	475 (890)
Time, min	30	60	30	60
Atmosphere	Air	Air	Air	Air
Quench medium	H_20	H_20	H_20	H_20
Aging				
Temperature, °C (°F)	150 (300)	150 (300)	150 (300)	125 (257)
Time, h	18	18	18	18
Atmosphere	Air	Air	Air	Air
Heat-treated (T6) properties(a)				
Transverse-rupture strength, MPa (ksi)	550 (80)	495 (72)	435 (63)	435 (63)
Yield strength, MPa (ksi)	200 (29)	195 (28)	195 (28)	275 (40)
Tensile strength, MPa (ksi)	260 (38)	240 (35)	205 (30)	310 (45)
Elongation, %	3	3	2	2
Rockwell hardness, HRE	74	72	71	80
Electrical conductivity, %IACS	36	32	39	25

(a) T6, solution heat treated, quenched, and artificially age hardened

Table 4 Electrical and thermal conductivity of sintered aluminum alloys, wrought aluminum, brass, bronze, and iron

Material	Temper	Electrical conductivity(a) at 20 °C (68 °F), % IACS	Thermal conductivity(b) at 20 °C (68 °F), cal/cm·s·°C
601AB	T4	38	0.36
	T6	41	0.38
	T61	44	0.41
201AB	T4	32	0.30
	T6	35	0.32
	T61	38	0.36
602AB	T4	44	0.41
	T6	47	0.44
	T61	49	0.45
6061 wrought aluminum	T4	40	0.37
	T6	43	0.40
Brass (35% Zn)	Hard	27	0.28
	Annealed	27	0.28
Bronze (5% Sn)	Hard	15	0.17
	Annealed	15	0.17
Iron (wrought plate)	Hot rolled	16	0.18

(a) Determined with FM-103 Magnatester. (b) Converted from electrical conductivity values

ous aluminum fabrication schemes are summarized in Fig.5.

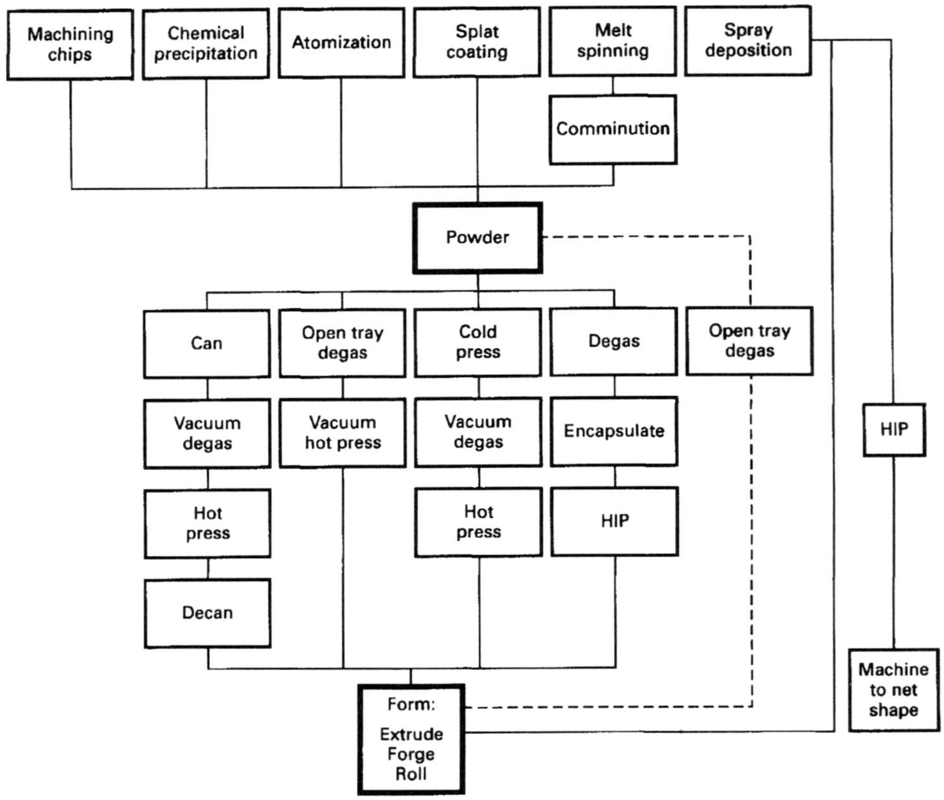

Fig. 5 Aluminum P/M fabrication schemes

Can Vacuum Degassing. This is perhaps the most widely used technique for aluminum degassing because it is relatively non-capital intensive. Powder is encapsulated in a can, usually aluminum alloys 3003 or 6061, as shown schematically in Fig. 6. A spacer is often useful to increase packing and to avoid safety problems when the can is welded shut. It has been found that packing densities are typically 60% of theoretical density when utilizing this method on mechanically alloyed powders. Care must be used to allow a clear path for evolved gases through the spacer to prevent pressure buildup and explosion.

To increase packing density, the powder is often cold isostatically pressed (CIP) in a reusable polymeric container before insertion into the can. Powder densities in the CIPed compact of 75 to 80% theoretical density are preferred because they have increased packing density with respect to loose-packed powder, yet allow sufficient interconnected porosity for gas removal. At packing densities of about 84% and higher, effective degassing is not possible for several atomized aluminum-alloy powders (Ref 2,3). Furthermore, one must control CIP parameters to avoid inhomogeneous load transfer through the powder, which can lead to excessive density in the outer regions of the cylindrical compact and much lower densities in the center. Such CIP parameters often must be developed for a specific powder and compact diameter,

The canned powder is sealed by welding a cap that contains an evacuation tube as shown in Fig. 6. After ensuring that the can contains no leaks, the powder is vacuum degassed while heating to elevated temperatures. The rate in gas evolution as a function of degassing temperature depends on powder size, distribution, and composition. The ultimate degassing temperature should be selected based on powder composition, considering tradeoffs between resulting hydrogen content and microstructural coarsening. For example, an RS-P/M precipitation-hardenable alloy that is to be welded would likely be degassed at a relatively high temperature to minimize hydrogen content (hotter is not always better). Coarsening would not significantly decrease the strength of the resulting product because solution treatment and aging would be subsequently performed and provide most of the strengthening. On the other hand, a mechanically attrited P/M alloy that relies on substructural strengthening and will serve in a mechanically fastened application might be degassed at a lower temperature to reduce the annealing out of dislocations and coarsening of substructure.

When a suitable vacuum is achieved (for example, <5 millitorr), the evacuation tube is sealed by crimping. The degassed powder compact can then be immediately consolidated to avoid the costs of additional heating. An extrusion press using a blind die (that is, no orifice) is often a cost-effective means of consolidation.

Dipurative Degassing. Roberts and coworkers at Kaiser Aluminum & Chemical Corporation have developed an improved degassing method called "dipurative" degassing (Ref 4). In this technique, the vacuum-degassed powder, which is often canned, is backfilled with a dipurative gas (that is, one that effectively removes water of hydration) such as extra-dry nitrogen,

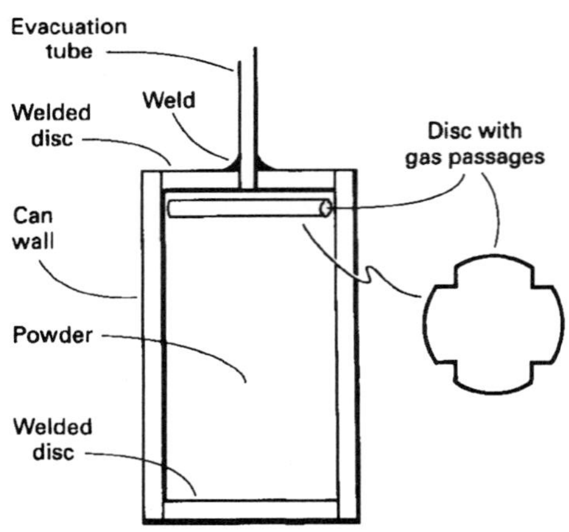

Fig. 6 Degassing can used for aluminum P/M process-

and then re-evacuated. Several backfills and evacuations can be performed resulting in lower hydrogen content. In addition, the degassing can often be performed at lower temperatures to reduce microstructural coarsening.

Vacuum Degassing in a Reusable Chamber. The cost of canning and decanning adversely affects the competitiveness of aluminum P/M alloys. This cost can be alleviated somewhat by using a reusable chamber for vacuum hot pressing. The powder or CIPed compact can be placed in the chamber and vacuum degassed immediately prior to compaction in the same chamber. Alternatively, the powder can be "open tray" degassed, that is, degassed in an unconfined fashion, prior to loading into the chamber. The processing time to achieve degassing in an open tray can be much less than that required in a chamber or can, thereby increasing productivity. Care must be exercised to load the powder into the chamber using suitable protection from ambient air and moisture. The compacted billet can then be formed by conventional hot-working operations.

Direct Powder Forming. One of the most cost-effective means of powder consolidation is direct powder forming. Degassed powder, or powder that has been manufactured with great care to avoid contact with ambient air, can be consolidated directly during the hot-forming operation. Direct powder extrusion and rolling have been successfully demonstrated numerous times over the past two decades (Ref 5, 6). It still remains an attractive means for decreasing the cost of aluminum P/M.

Hot Isostatic Pressing (HIP). In HIP, degassed and encapsulated powder is subjected to hydrostatic pressure in a HIP apparatus. Can vacuum degassing is often used as the precursor step to HIP. Furthermore, net-shape encapsulation of degassed powder can be used to produce certain near net-shape parts. Relatively high HIP pressures (~200 MPa, or 30 ksi) are often preferred. Unfortunately, the oxide layer on the powder particle surfaces is not sufficiently broken up for optimum mechanical properties. A subsequent hot-working operation that introduces shear-stress components is often necessary to improve ductility and toughness.

Rapid Omnidirectional Consolidation. Engineers at Kelsey-Hayes Company have developed a technique to use existing commercial forging equipment to consolidate powders in several alloy systems (Ref 7). Called rapid omnidirectional consolidation (ROC), it is a lower-cost alternative to HIP. In ROC processing, degassed powder is loaded into a thick-walled "fluid die" that is made of a material that plastically flows at the consolidation temperature and pressure, and which enables the transfer of hydrostatic stress to the powder. Early fluid dies were made of mild steels or a Cu-10Ni alloy, with subsequent dies made from ceramics, glass, or composites.

The preheated die that contains powder can be consolidated in <1 s in a forging press, thereby reducing thermal exposure that can coarsen RS microstructures. In addition, productivity is greatly increased by minimizing press time. Depending upon the type of fluid die material being used, the die can be machined off, chemically leached off, melted off, or designed to "pop off" the net-shape component while cooling from the consolidation temperature. For aluminum-alloy powders, unconfined degassing is critical to optimize the cost effectiveness of the ROC process. An electrodynamic degasser was developed for this purpose.

Just as in the case of HIP, the stress state during ROC is largely hydrostatic. Consequently, there may not be sufficient shear stresses to break the oxide layer and disperse the oxides, which can lead to prior powder-particle boundary (PPB) failure. Consequently, a subsequent hot-working step of the ROC billet is often necessary for demanding applications. More detailed information on this technique can be found in Ref 8.

Dynamic Compaction. Various ultrahigh strain-rate consolidation techniques, that is, dynamic compaction, have been developed and utilized for aluminum alloys (Ref 9). In dynamic compaction, a high-velocity projectile impacts the degassed powders that are consolidated by propagation of the resultant shock wave through the powder. The bonding between the powder particles is believed to occur by melting of a very thin layer on the powder surfaces, which is caused by the heat resulting from friction between the powder particles that occurs during impact. The melted region is highly localized and self-quenched by the powder interiors shortly after impact. Thus, dynamic compaction has the advantage of minimizing thermal exposure and microstructural coarsening, while breaking up the PPBs.

ACKNOWLEDGMENT

ASM International gratefully acknowledges J.R. Pickens, Martin Marietta Laboratories. Significant portions of this article were adapted from his article, "High Strength Aluminum P/M Alloys," *Properties and Selection: Nonferrous Alloys and Special-Purpose Materials,* Vol 2, *ASM Handbook,* 1990, p 201–215.

REFERENCES

1. M.C. Narisimhan, U.S. Patent 4,142,571, 1979
2. J.R. Pickens, J.S. Ahearn, R.O. England, and D.C. Cooke, High-Strength Weldable Aluminum Alloys Made from Rapidly Solidified Powder, *Proc. of the AIME-TMS Symp. on Powder Metallurgy* (Toronto), G.J. Hildeman and M.J. Koczak, Ed., 15–16 Oct 1985, p 105–133
3. J.R. Pickens, J.S. Ahearn, R.O. England, and D.C. Cooke, "Welding of Rapidly Solidified Powder Metallurgy Aluminum Alloys," End-of-year report on contract N00167-84-C-0099, Martin Marietta Laboratories and DARPA, Dec 1985
4. S.G. Roberts, U.S. Patent 4,104,061
5. G. Naeser, *Modern Developments in PM,* Vol 3, Hausner, Ed., 1965
6. D.H. Ro, "Direct Rolling of Aluminum Powder Metal Strip," Report for contract F33615-80-C-5161, 1 Sept 1980 to March 1981
7. J.R. Lizenby, "Rapid Omnidirectional Compaction," Kelsey-Hayes Company, Powder Technology Center, 1982
8. Rapid Omnidirectional Compaction, *Powder Metallurgy,* Vol 7, *Metals Handbook,* 9th ed., American Society for Metals, 1984
9. D. Raybould, Dynamic Powder Compaction, *Proc. of the Eighth Int. HERF Conference* (New York), American Society of Mechanical Engineers, I. Berman and J.W. Schroeder, Ed., 1984; also, in *Carbide Tool J.,* March/April 1984
10. D. Raybould and T.Z. Blazynski, Dynamic Compaction of Powders, in *Materials at High Strain Rates,* Elsevier, 1987, p 71–130

Advanced Aluminum Powder Metallurgy Alloys and Composites

Ram B. Bhagat, The Pennsylvania State University

POWDER METALLURGICAL PROCESSING provides much finer and homogeneous microstructure, better mechanical properties, and near-net shape parts producibility for aluminum alloys in comparison with ingot metallurgy (I/M). In addition to the conventional blending and consolidation of elemental or prealloyed powders into near-net shape parts, emerging processes such as mechanical alloying and rapid solidification (RS) create composite powders that upon subsequent consolidation provide significant improvements in room and elevated-temperature strength, fracture toughness, fatigue life, and corrosion resistance. The real advantage of P/M processing including spray forming is in the production of new alloys and composites with metallurgical structures and compositions that cannot be produced by I/M. Rapid solidification extends the solubility of alloying elements, particularly transition and rare earth elements, and refines the structure of intermetallic phases responsible for improved mechanical properties. Mechanical alloying is a dry, high-energy milling process producing dispersions of insoluble oxides and carbides that stabilize the microstructure leading to high strength at elevated temperatures in the consolidated materials. By blending the alloy powder with a strengthening phase, discontinuously reinforced aluminum-matrix composites containing insoluble dispersoids (oxides and carbides), particulates, whiskers, or fibers are produced for high-performance structural applications (Ref 1–6). Table 1 includes the chemical composition of aluminum P/M alloys and dispersion-strengthened composites.

Aluminum-matrix composites with reinforcement of particulates, whiskers, or chopped fibers have been widely studied worldwide. They are generally isotropic and less costly in comparison with continuous-fiber-reinforced aluminum-matrix composites. The discontinuously reinforced aluminum (DRA) composites are attractive for near-term commercial applications. Silicon carbide (SiC) or alumina-particle-reinforced aluminum composites have higher stiffness and, in general, high wear resistance in comparison with the unreinforced aluminum alloys. Silicon carbide whisker, however, is currently the most widely utilized reinforcement for the DRA composites for obtaining high resistance to creep and higher use temperatures. Early work on whisker reinforcement started in the 1960s. Brenner (Ref 7, 8) and Sutton (Ref 9, 10) used α-Al_2O_3 whiskers to fabricate metal-matrix composites (MMCs). These early composites were not attractive because of their relatively low strength and the high cost of the whisker. Later Divecha et al. (Ref 11) used SiC whiskers to fabricate aluminum-matrix composites with good mechanical properties. In 1973, Cutler (Ref 12) introduced his patented process of inexpensively producing SiC whiskers from pyrolized rice hulls. This renewed the interest in the development of whisker-reinforced MMCs (Ref 13–16). There have been concerns in handling fine particulates and whiskers because of health risks. For pathogenic reasons such as mesothelioma induction and fibrogenic and lung cancer potential, it has been found that it is the size and shape of any given particle that causes problems, as well as its chemistry (Ref 17, 18). Hence it is advisable to use particulate with diameters greater than 5 μm and to avoid contact with fibers or whiskers in the range 0.1 to 3 μm, with lengths greater than 5 μm. There has been only a limited interest in whiskers other than SiC for fabricating MMCs. Recently Imai, Nishida, and Tozawa (Ref 19) evaluated the mechanical properties of aluminum-matrix composites with Si_3N_4 whisker. The composites, fabricated by P/M hot pressing and subsequently extruded, showed good mechanical properties. Limited work has been reported on continuous-fiber-reinforced aluminum composites fabricated by P/M processing. Bhagat and coworkers (Ref 20–25) fabricated stainless steel wire (type 304) reinforced aluminum-matrix composites by hot pressing (800 K, 155 MPa, and 60 s). The composites showed attractive mechanical properties including high fracture toughness and fatigue life. These composites are suitable for structural applications requiring high fracture toughness and high resistance to impact damage.

Selection of suitable composition of the matrix material is important to meet mechanical and physical property requirements of the aluminum-matrix composites. Minor alloying additions in the wrought alloys are generally detrimental to the mechanical properties of the composites because of undesirable interfacial reaction (Ref 26–28) during the P/M consolidation. The P/M route for producing the discontinuously reinforced composites involves blending elemental or prealloyed powder with the reinforcement, followed by canning, vacuum degassing, and some form of consolidation, such as hot pressing or hot isostatic pressing (HIP), into a billet that is subsequently rolled, forged, or extruded into shapes. During the billet consolidation, however, the matrix alloy may experience incipient melting, creating both large insoluble and excess soluble intermetallics, which form upon resolidification of aluminum alloy matrix. These particles are deleterious to the fracture toughness and ductility of the DRA. This has led to the recent development of the leaner variants of the 2*xxx* and 6*xxx* series aluminum alloys (e.g., 2009 and 6090), which contain strengthening elements in concentrations no greater than their mutual solubilities, thereby leading to useful ductility and fracture toughness in the composites (Ref 29–31). It is important to note, therefore, that the compositional and microstructural requirements are different for dispersion-strengthened alloys and particle or whisker-reinforced aluminum composites.

In addition to fabricating shaped structures, deformation processing (extrusion, forging, or rolling) of the consolidated composites develops the best properties attainable by breaking up any preexisting oxide on the alloyed powder. Extrusion of whisker-reinforced composites leads to an approximate alignment of the whiskers in the extrusion direction. Production of useful shapes requires, at least to some extent, secondary processing such as machining and joining, which are not well established for the composites. Conventional machining of composites is very difficult because of the high hardness of the ceramic reinforcement. Expensive diamond or cubic boron nitride cutting tools are needed for effective machining. The high-cost conventional machining of MMCs can be eliminated by near-net shape manufacturing or by noncontact machining techniques such as laser processing (Ref 32) and abrasive waterjet cutting (Ref 33). Large-scale commercialization of the aluminum P/M alloys and composites is still small (compared to wrought and cast aluminum alloys), although new applications are emerging (see "Applications Outlook" in this article). Notwithstanding the high cost that remains a major concern for their use, there are other major technological challenges such as reproducibility/reliability, machining, joining, and recycling that need to be addressed.

Table 1 Chemical composition of aluminum P/M alloys including dispersion-strengthened composites

Alloy designation	Composition
Al-C	Al-3.0C
Dispal 2	Al-2.0C-1.0O
Al-Cu-C	Al-1.0Ci-1.5C
Al-Mg-C	Al-2.0Mg-1.5C
Al-Cr-X	Al-5.0Cr-2.0Zr-1.0Mn
Al-Ti-X	Al-3.0Ti-3.0Ce
Al-Be-X	Al-22.6Be-10.8Li
201AB	Al-4.4Cu-0.8Ci-0.5Mg-1.5 other
201AC	Al-4.4Cu-0.8Si-0.5Mg
202AB	Al-4.0Cu-1.5other
601AB	Al-0.25Cu-0.6Si-1.0Mg-1.5 other
601AC	Al-0.25Cu-0.6Si-1.0Mg
602AB	Al-0.6Mg-0.4Si-1.5 other
MD-76	Al-5.6Zn-2.5Mg-1.6Cu
7090	Al-8.0Zn-2.5Mg-1.0Cu-1.5Co-0.35O
7091	Al-6.5Zn-2.5Mg-1.5Cu-0.4Co
X7090	Al-0.12Si-0.15Fe-(0.6–1.3)Cu-(2–3)Mg-(7.3–7)Zn-(1.0–1.9)Co
X7091	Al-0.12Si-0.15Fe-(1.1–1.8)Cu-(1–3)Mg-(5.8–7.1)Zn-(0.2–0.6)Co
X7093	Al-9.0Zn-2.2Mg-1.5Cu-0.14Zr-0.10Ni
8009	Al-8.5Fe-2.4Si-1.3V
X8019	Al-8.3Fe-4.0Ce
Al-Fe-Ce	Al-8.0Fe-3.5Ce-0.35O
Al-Fe-Co	Al-(3–8)Fe-(2–7)Co-0.35O
Al-Fe-Ni	Al-8Fe-1.7Ni
Al-Fe-Mo	Al-8.0Fe-2.0Mo
Al-Fe-Ni-Co	Al-5Fe-3Ni-6Co
Al-Fe-Cr	Al-8.5Fe-1.5Cr-0.35O
XAP001	Al-0.2Fe-0.40C-0.08Si-6.0 oxide
XAP002	Al-0.28Fe-0.19C-0.10Si-8.0 oxide
XAP004	Al-0.29Fe-0.39C-0.10Si-14.0 oxide
IN9021	Al-4.0Cu-1.5Mg-1.2C-0.75O
IN9051	Al-4.0Mg-0.7C-1.4O
IN9052	Al-4.0Mg-0.75O-1.0C
905XL	Al-4.0Mg-1.3Li-0.4O-1.1C
Al-Li-Cu	Al-(2–3.5)Cu-0.5Mg-0.6Mn-(2.5–3.2)Li
Al-Mn-Co	Al-(3–8)Mn-(1.5–7)Ni
Al-Mn-Ni	Al-(3–8)Mn-(1.6–6.5)Co
Al-Ni-Co	Al-(2–5)Ni-(2–5)Co

Source: Ref 1–3

Conventional Consolidation

For wrought aluminum P/M alloys such as 601AB and 201AB (see Table 1), prealloyed aluminum powders are cold isostatically pressed into green compacts, degassed, and sintered in nitrogen atmosphere; sintering in dissociated ammonia or vacuum leads to relatively inferior mechanical properties. The cold compaction, degassing and sintering are done typically at 400 MPa, 450 °C, and 600 °C, respectively (Ref 1). The as-sintered products are usually not fully dense and have relatively low strengths. They require heat treatment and deformation processing (rolling, extrusion, or forging) for improved properties (see Table 2).

For full densification, powder particles are compacted isostatically and encapsulated in an aluminum canister. The compact is heated, and gases are evacuated from the canister. The heated canister/compact is pressed into a billet with 100% density. The canister is removed from the billet by machining. The billet is subsequently fabricated into parts by rolling, forging, or extrusion. The conventional P/M processing of the aluminum-matrix composites include the following steps. The reinforcement particulates/whiskers (see Table 3) are deagglomerated prior to mixing or blending in a suitable medium. Ultrasonic agitation of whiskers or fine particulates suspended in alcohol or wet milling in a polar solvent medium such as *n*-butanol is required to eliminate the problem of agglomeration. The deagglomerated mass is mixed with matrix alloy powder and subjected to a dry or wet blending process to obtain a uniform distribution of the reinforcement in the matrix. Improper blending leads to an inhomogeneous distribution of the reinforcement in the final product. The solvent is removed by drying the powder mixture in air. As an example, the powder mixture of 2124 Al and SiC particulates can be heated at 150 °C to remove *n*-butanol from the blend (Ref 34). The mixture is cold compacted in a die using a hydraulic press. Carbowax and stearic acid are suitable lubricants for the die wall to prevent wear. Level of densification is critical to provide some green strength for handling and yet leave behind an open interconnecting pore structure so as not to hinder degassing. Both the physically absorbed and chemically combined water on the matrix alloy powder particles are liberated in the form of water vapor during the degassing. If the degassing is incomplete, the postcompaction thermal exposure can give rise to hydrogen evolution due to reaction between the water vapor and aluminum. This causes blistering in the billet and reduces ductility and fracture toughness of the composite. Degassing temperature and time are critical parameters because lower times and temperatures lead to incomplete degassing, whereas higher times and temperatures lead to a chemical reaction between the matrix and the reinforcement that can result in a weak interface. Jain et al. (Ref 34) experimentally determined the optimal temperature and time of 723 K and 1 h, respectively, for degassing of the 2124/SiC_p green compact. Vacuum hot pressing or HIP is required for final consolidation of the degassed green compact. Full densification is obtained by optimizing the temperature, pressure, and the period of hot pressing. Further details on the conventional P/M processing of particulate or whisker-reinforced metal composites can be found elsewhere (Ref 34–44).

Table 2 Mechanical properties of nitrogen-sintered wrought aluminum alloys

Alloy	Yield strength, MPa	Ultimate tensile strength, MPa	Elongation, %
201AB-T6 (95% dense)	322	336	2
201AB-T6 (97% dense, rolled)	327	332	2
202AB-T6 (92.4% dense)	147	227	7.3
202AB-T8 (92.4% dense, cold formed, 19% strain)	250	280	3
601AB-T6(96% dense)	230	238	2
601AB-T6 (96% dense, rolled)	241	252	2
602AB-T6 (96% dense)	172	186	3

Source: Ref 1

Table 3 Typical properties of particulate, whisker, and discontinuous fiber reinforcements

Property	SiC_p	Al_2O_{3p}	TiB_{2p}	Si_3N_{4p}	Al_2O_3 (Saffil)	SiC_w	Si_3N_{4w}	$K_2O{\cdot}6TiO_{2w}$
Density, g/cm^3	3.21	3.97	4.5	3.18	3.3	3.19	3.18	3.3
Diameter, μm	...	...	...	...	3–4	0.1–1.0	...	...
Thermal expansion coefficient, 10^{-6}/K	4.3–5.6	7.2–8.6	8.1	3.0	~9	4.8	3.8	...
Tensile strength, MPa	100–800(a)	70–1000(a)	700–1000(a)	250–1000(a)	>2000	3,000–14,000	13,800	6,900
Young's modulus, GPa	200–480	380	514–574	304	300	400–700	379	274
Elongation, %	...	...	...	...	0.67	1.23	...	...
Composition	...	...	...	...	96%-Al_2O_3-4% SiO_2	>98% SiC	...	...

(a) Transverse rupture strength of bulk

Emerging Technologies

The primary advantage of P/M processing, as mentioned before, is due to the use of emerging technologies such as mechanical alloying, rapid solidification, and spray forming, which provide a greater flexibility in tailoring the composition and microstructure of dispersion-strengthened aluminum alloys and composites with or without particulate/whisker reinforcement.

Mechanical Alloying and Processing (MAP). Mechanical alloying is used for fabricating oxide-dispersion-strengthened alloys and discontinuously reinforced composites. Pure metal powder and alloying ingredients are mechanically alloyed using high-energy ball mills. During this process, a heavy working of powder particles results in the intimate alloying by a process of repeated welding, fracturing, and rewelding (Ref 45–56). In the case of aluminum alloys, carbon derived from process control agents and incorporated into the processed powder reacts with aluminum to form very fine carbides. These carbides and the fine oxide particles derived from the breakup of surface films on the initial powder particles create a dispersion that stabilizes the fine-grained microstructure. Therefore, a portion or all of the strengthening may be obtained from the ultrafine grain size stabilized by the oxide and carbide dispersions (see Ref 4). Mechanically alloyed (MA) product (i.e., the composite powder) is subsequently consolidated into suitable shapes. The mechanical alloying is also useful in developing materials with large contents of alloying additions for improving mechanical properties at elevated temperatures. This processing technique alleviates the problems of low solubility according to the phase diagram or possibility of forming low-melting equilibrium or nonequilibrium phases. Thus, an anomalous supersaturation can be achieved in systems such as aluminum-iron, aluminum-nickel, aluminum-chromium, and aluminum-titanium by mechanical alloying with or without particulate additions. Such materials have potential for improved strength at elevated temperatures. A novel use of the mechanical alloying is in the generation of amorphous alloys that, upon subsequent controlled crystallization, can generate nanometer-scale microstructures (Ref 57–59). The MA aluminum powder 9052 shows improved mechanical properties and corrosion resistance. MA Al-12.5Ti alloy has a stable, ultrafine grain structure containing fine Al_3Ti particles of 20 to 250 nm in size. This alloy has high stiffness, good ductility, and high strength at elevated temperatures. Bhaduri et al. (Ref 60) used mechanical alloying to prepare composite powders of 7010 Al from elemental powder with or without SiC particulates. The addition of SiC particulates appears to inhibit the progress of mechanical alloying. The composites had higher modulus but lower strength values than the matrix at room temperature. However, at temperatures >200 °C the composites exhibited higher tensile strength than the matrix. Other examples of MAP aluminum alloys and composites and their properties are discussed later.

Rapid Solidification and Processing (RSP). Rapid solidification is a "far-from-equilibrium" process that results in a significant undercooling of the molten metals or alloys leading to a refinement of microstructural features and constitutional effects such as extension of solubility limits, synthesis of novel crystalline phases, and formation of metallic glasses (Ref 3, 61–66). Fabrication of particle-reinforced MMCs by the RSP involves introduction of the particulates into the melt, proper mixing, melt spinning, and canning of the composite particles, followed by extrusion or forging or both into finished shapes. Rapid solidification processing has enabled the development of a new family of high-strength aluminum alloys for elevated-temperature applications. The RSP alloys such as those based on hypereutectic aluminum-iron compositions derive their high strengths from dispersion strengthening; the dispersoids resist dissolution and coarsening when the alloys are exposed to elevated temperatures. Aluminum 7091 and 7090 are Al-Zn-Mg-Cu alloys that are similar in composition to I/M 7175, but contain 0.4 and 1.45% Co, respectively. Cobalt forms Co_2Al_9 or $(Co,Fe)_2Al_9$ particles that are homogeneously dispersed. These dispersoids refine the grain size for improved high strength and ductility and enhance the resistance to stress-corrosion cracking (Ref 3). Kumpfert et al. (Ref 67) produced high-silicon-content (~20 wt%) aluminum alloys with high strength at room to elevated temperatures by combining RSP and the mechanical alloying. Courtreight et al. (Ref 68) used RSP for fabricating SiC-particle-reinforced aluminum composites with significantly improved mechanical properties.

A number of aluminum-base amorphous alloys have been fabricated by melt spinning (Ref 69–73). These aluminum glasses (with or without the presence of aluminum nanocrystals) show very high tensile strength and low density. The rapidly solidified and processed $Al_{70}Ni_{20}Zr_{10}$, $Al_{90}Fe_5Ce_5$, and $Al_{89}Ni_7Y_4$ metallic glasses with 25 vol% of nanocrystals can have yield stresses up to 1550 MPa, good ductility, and high elastic modulii (Ref 3, 74–76). Generally, these amorphous alloys have poor thermal stability. The problem of poor thermal stability can, however, be alleviated by compositional optimization as reported by Jin et al. (Ref 69). They found that iron in the range of 35 at.% in the rapidly solidified $Al_{86}Mm_4Ni_{1-x}Fe_x$ (at.%) alloys (Mm represents mischmetal, 49Ce-25La-18Nd-5.4Pr-2.6 other, wt%), provides optimal compositions having good thermal stability combined with high strength for practical structural applications.

Spray forming is a high-deposition-rate metal spraying process for the production of near-net shape parts at a relatively low cost. In this method, molten metal is rapidly atomized to form a fine spray of droplets that are deposited onto a stationary or moving collector. This forming process has proven to be a viable and cost-effective alternative to conventional metalworking technology for the production of metal preforms with properties superior to I/M. Spray forming has been demonstrated to create new materials such as new compositions of aluminum alloys (Ref 77–80), MMCs (Ref 81–87), and those suitable for semisolid forming (Ref 88). The spray-formed products are typically free from macrosegregation and prior-particle boundaries that are occasionally present in the conventional P/M products. The spray forming is useful in extending the maximum solute content of alloying elements and achieving a finer distribution of second-phase particles in an equiaxed fine-grain structure. Such homogeneous, low-segregation, fine-scale alloy microstructures are useful for subsequent rolling, forging, extrusion, and superplastic forming operations. The development of low-density aluminum-lithium alloys has a high potential for significant weight savings in aerospace structures. Alloys produced by conventional casting methods are currently limited to lithium levels of approximately 2.5 wt%. Above this level, it is difficult to avoid the formation of coarse second-phase particles and achieve low levels of hydrogen and alkali metal impurities in the cast ingots. Hydrogen and sodium are known to have an embrittling effect in aluminum-lithium alloys. A high hydrogen level also has detrimental effects on the weldability. These concerns can be eliminated by the spray-forming process that offers a number of significant advantages for producing aluminum-lithium alloys:

- Increasing solute content
- Refining of microstructure
- Elimination of oxide films
- Reduction of hydrogen and sodium levels

Examples of spray-formed aluminum-lithium alloys are UL30 and UL40 having increased levels of lithium and zirconium in comparison with one of the most commonly used aluminum-lithium alloys, 8090 (Al-2.5Li-1.0Cu-0.7Mg-0.12Zr). According to White et al. (Ref 89), UL30 (Al-3.0Li-1.0Cu-0.7Mg-0.3Zr) shows a 4.4% increase in specific modulus over 8090, and the strengthening is primarily by duplex particles of fine Al_3Zr (β′) and Al_3Li (δ′) and copper and magnesium in solid solution. These duplex particles form without the need of cold working, unlike 8090 in which stretching is required to enhance precipitation of the σ′ (Al_2LiMg) phase. Thus, tensile properties of unstretched UL30 can equal those of stretched 8090, with improved isotropy. By obviating the need for cold work, UL30 can be fabricated in near-net shapes by the spray forming process. Palmer et al. (Ref 78) have spray formed UL40 (Al-4Li-0.2Zr) with lithium content reaching the limit of solid solubility at the eutectic temperature. They found the as-sprayed grain structure equiaxed with grain sizes in the range 40 to 50 μm. They also found the presence of some relatively coarse δ phase (AlLi) at the grain boundaries and some needlelike δ phase in the grains. The amount of this fine δ phase was greatest at the spray-form center and decreased toward the surface, indicating that this phase precipitated as a result of slow cooling in the solid state. As expected, no coarse Al_3Zr particles were observed in the microstructure due to a relatively high solidification rate suppressing their formation. A small amount (<1%) of porosity was present in the as-sprayed materials, which disappeared during subsequent extrusion. The extrusion grain structure was very fine and highly elongated. Some of the intergranular δ phase was present in the microstructures. However, most of the intergranular δ phase and all of the intragranular needlelike δ phase were dissolved upon subsequent solution heat treatment.

Based on the transient heat transfer during droplet solidification via finite element analysis (Ref 90), Bhagat (Ref 91) developed a model for solidification and growth kinetics of the primary silicon in spray-formed (Osprey) Al-17Si alloy (Fig. 1). The predicted results were found to agree with the experimental results. Buhrmaster et al. (Ref 92) fabricated SiC-particle- or whisker-reinforced aluminum composites by spray casting. In this process, aluminum wire feedstock is melted and combined with SiC particulates or whiskers entrained in an inert gas. Upon striking a substrate or mold, the mixture of aluminum and SiC solidifies into a composite structure. The fabricated composites lacked uniform distribution of the reinforcement. Discontinuously reinforced MMCs were fabricated by a spray codeposition method wherein the reinforcement particulates or whiskers were introduced into the metal spray leading to their codeposition with the atomized metal onto the substrate (Ref 93, 94). Typical recovery efficiency of the spray codeposition method is 60 to 90%, depending on the product form such as hollow tube, forging stock, extrusion ingot, or plate. Precise control of gas pressures and particulate feed rates are required to ensure that a uniform distribution of particulate is produced within the matrix that may be 95 to 98% dense. The shape of the final product depends on the atomizing conditions and the shape and motion of the collector. Willis

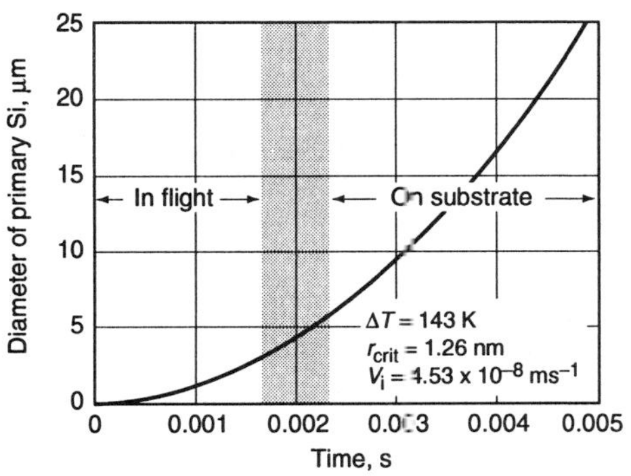

Fig. 1 Predicted size of primary silicon in an spray-formed (Osprey) Al-17Si alloy during in-flight cooling and subsequent cooling on the substrate. The prediction is based on a new and validated model for grain growth kinetics developed by the author. ΔT, r_{crit}, and V_i represent undercooling, critical radius of silicon, and instantaneous growth rate, respectively.

(Ref 93) reported uniform distribution of the particulates in a $2014/SiC/15_p$ composite (15 vol% SiC particulates) fabricated by the spray codeposition method. The average particle diameter was approximately 13 µm; the starting material was F600 SiC grit. Willis (Ref 93) also reported spray codeposition of aluminum-lithium 8090 reinforced with SiC particulates. The composites were subsequently extruded into shapes. The overall strength of the fabricated $8090/SiC_p$ composites in the peak aged condition did not show much improvement over the matrix material. However, the improvement in specific modulus for $8090/SiC_p$ (density = 2.62 g/cm^3) over the unreinforced aluminum alloy (density = 2.55 g/cm^3) is almost 50%. One of the primary reasons for the attractive properties of the spray-formed MMCs is the integrity of the particle/matrix interface. The interface is free from any precipitate or reaction product in Al/SiC composites according to Warner et al. (Ref 95) as reported in Ref 94. This is attributed to the relatively small (a few seconds) contact time between molten metal and SiC in the spray codeposition. Wu and Lavernia (Ref 96) fabricated $6061Al/SiC_p$ composites by spray-atomized codeposition technique. A maximum of 16% increase in stiffness of the composite over that of the unreinforced matrix was achieved. Other mechanical properties of the spray-formed composites are discussed later in this article.

In another variation of the spray-forming technology, Siemers et al. (Ref 97) used a low-pressure plasma deposition (LPPD) method for producing MMCs. In this process, a plasma (~10,000 to 20,000 K) is formed in the interior of a gun (commonly referred to as plasma gun) by the discharge of a high-intensity dc arc through a flowing gas mixture (argon or N_2 with additions of 2 to 15% H_2 or helium). Powder is injected into the plasma jet through one or more injector ports. The powder is then entrained in the high-velocity (Mach 2 to 3) plasma jet, melted, transported, and impacted on a substrate where the molten droplets solidify at cooling rates of 10^5 to 10^6 K/s. The LPPD process can be used for most metals, metal carbides, or metal oxides excluding those materials that decompose before melting. Siemers et al. (Ref 97) produced composites of metal-metal, metal-carbide, and metal-oxide combinations. They also produced graded layered structures of near-theoretical density to minimize the effects of interfacial stresses developed due to the thermal mismatch between the constituents. Both continuous and discontinuous composite laminates can be fabricated by the LPPD process. Continuous laminates, consisting of alternating layers of metal and reinforcement, are produced by first depositing metal, subsequently depositing the reinforcement, and repeating the sequence to build the desired final thickness. The proportions of the phases and the layer thickness can be varied over considerable ranges to produce tailored composites. For fabricating discontinuous laminates consisting of individual splattered particle lamellae (2 to 4 µm thick and 50 to 150 µm diameter), a powder blend of two or more different compositions can be injected into the plasma jet.

Other Technologies. In addition to the aforementioned emerging technologies, there are other P/M processing techniques such as XD process, solid free-form fabrication (SFF), and self-propagating high-temperature synthesis (SHS), which also show potential for developing high-performance aluminum alloys and composites including functionally graded materials. In the proprietary XD process (Martin Marietta, Baltimore, MD), elemental components of the desired reinforcing phase are mixed with or incorporated into the metallic matrix material. The mixture is heated to a high temperature ($>T_m$) where a self-propagating reaction takes place. The elemental constituents react exothermally, forming a *d*ispersion of submicroscopic reinforcing particles in the matrix. Because the particles of the reinforcing phase are formed by an exothermic reaction at an elevated temperature, they tend to be very stable through subsequent processing and use at high temperatures. However, if both elemental constituents of the particulate phase have high diffusivity in the matrix phase, then particle coarsening may occur at high temperatures. In addition, particle shape may change during heat treatment or use at elevated temperatures, if the surface energies are different. Christodoulou et al. (Ref 98) and Larsen et al. (Ref 99) used the XD process to produce a master-alloy material containing TiB_2 particles (TBA-70 sponge: 70 wt% TiB_2 and 30 wt% Al), which was subsequently used for fabricating the discontinuously reinforced aluminum composites.

Solid free-form fabrication facilitates integrated manufacturing, from computer-aided design (CAD) to finished parts, using only additive processing (Ref 100, 101). A solid or surface model is electronically sectioned into layers of predetermined thickness. The sections collectively define the shape of the part. Information about each section is then transmitted layer by layer to the SFF machine. Material is consolidated only at solid areas of the section. Subsequent layers are sequentially processed until the part is complete. It is this sequential, layered, or lithographic approach to parts manufacturing that defines the SFF technology. Selective laser sintering (SLS) and the selective laser reactive sintering (SLRS) are examples of the SFF technology for producing MMCs. Net-shape composite components can be directly produced from the elemental or prealloyed matrix powders mixed with suitable reinforcement by sintering, layer by layer, under a scanning laser beam; the reinforcement phase may be formed in situ.

Self-propagating high-temperature synthesis is a process in which the synthesis reaction proceeds through the reactant mixture in the form of a combustion wave (Ref 102–106). The progression of the wave is driven by the energetics of the reaction and the characteristics of the reactants. Some of the advantages of the SHS processing may include purity of the product, energy efficiency, formation of metastable phases, and simultaneous synthesis and densification. The SHS process thus has the potential to be a viable method for producing high-performance MMCs.

Functionally graded materials (FGMs) including MMCs create a continuous transition from an essentially ceramic material exposed to high temperatures to a partly or fully metallic material at the lower-temperature side of the structure (Ref 107–112). Such graded structures have been created in an AlN/Al composite using the conventional P/M technology (Ref 108, 113).

Table 4 Properties of P/M aluminum alloys and composites

Reinforcement contents are reported in vol%, if not otherwise specified. The composition of MMCs is described using a nomenclature system developed by the Aluminum Association.

Material	0.2% yield strength, MPa	Ultimate tensile strength, MPa	Elongation, %	Young's modulus, GPa	Reference
1100/SiC/10_p extruded	75	114	20	96	115
2009/SiC20_p-T8					
Longitudinal	462	593	5.2	109	116
Transverse	421	572	5.3	109	116
2014/SiC/20_p	466	493	2	100	2, 117
2014 Al	429	476	7.5	72	10
Alloy 201 (Al-4.4Cu-0.5Mg-0.8Si, wt%); aged 160 °C, 14 h	300	335	1.1	...	118
201/SiC(23 μm)/9_p	295	310	0.3	...	118
201/SiC(63 μm)/9_p	285	290	0.3	...	118
201/SiC(142 μm)/9_p	260	265	0.3	...	118
2080/SiC/15_p K_{Ic} = 33 MPa$\sqrt{m}$	365	483	7	100	2, 117
2080/SiC/20_p K_{Ic} = 22 MPa$\sqrt{m}$	393	517	6	110	117
2124/SiC/18_p solution treated					
At 485 °C	409	610	8.6(a)	...	114
At 545 °C	453	679	9.4(a)	...	114
At 585 °C	373	577	7.8(a)	...	114
2124/SiC/25_p-T6	496	738	5.0	117	31
6013/SiC/15_p-T6	434	517	6.3	101	116
6013/SiC/20_p-T6	448	538	5.6	110	116
6013/SiC/25_p-T6	469	565	4.3	121	116
6061 Al	330–347	365–386	13.6–14.4	70	29
6061/SiC/20_p	373	389–409	1.0–1.3	89	29
6061/Al_2O_3/20_p	317	354–361	3.6–4.1	95	29
6061/SiC/10_p/Al_2O_3/10_p	357	391	1.6	94	29
6061 Al	89	171	25.0	...	
At 473 K	85	127	29.3	...	32
At 723 K	20	23	46.2	...	32
6061/SiC/20_p (5μm)	123	224	7.3	...	32
At 473 K	119	163	16.7	...	32
At 723 K	23	25	20.1	...	32
6090/SiC/25_p-T6	393	483	5.5	117	31
6090/SiC/40_p-T6	427	538	2.0	138	31
7001/SiC/25_p-T6	655	724	1.5	117	31
7049/SiC/15_p	598	643	3	90	117
7075/SiC (5μm)/11_p K_Q = 15.7 MPa$\sqrt{m}$	570	630	6.6	...	119
7075/SiC (13 μm)/17_p K_Q = 13.6 MPa$\sqrt{m}$	595	645	3.5	...	119
7075/SiC (60 μm)/17_p K_Q = 18.5 MPa$\sqrt{m}$	501	504	0.6(a)	...	119
7075/SiC/15_p	556	601	4	95	117
7090 K_{Ic} = 31 MPa$\sqrt{m}$	579	621	9	74	2
7091 K_{Ic} = 46 MPa$\sqrt{m}$	545	593	11	72	2
X7093 K_{Ic} = 47 MPa$\sqrt{m}$	579	614	12	75	2
7475/SiC/25_p-T6	593	655	2.5	117	31
8009 K_{Ic} = 11 MPa$\sqrt{m}$	605	636	9	...	2
Monolithic 8009					
At 25 °C	407	448	17	88	30
At 149 °C	345	365	9	...	30
At 232 °C	296	310	11	...	30
At 371 °C	165	186	19	...	30
8009/SiC/11_p					
At 25 °C	496	544	5.5	102	
At 149 °C	400	427	3.5	...	
At 232 °C	310	338	4.2	...	
At 371 °C	193	220	7.0	...	
X8019	338	427	5	79	2
8090/SiC/15_p K_{Ic} = 14 MPa$\sqrt{m}$	499	547	3	101	117
8090/SiC/17_p-T6 designated as A817	...	542	4–6(a)	103	120
8090/SiC/18_p-T6	531	621	3.2	103	31
9052 K_{Ic} = 44 MPa $\sqrt{m}$	380	450	13	76	2
905 XL K_{Ic} = 30 MPa $\sqrt{m}$	450	520	9	...	2
Dispal 2	340	380	12	80	2
Al-Fe-Mo	393	512	10	82	2
Al-Cr-X	426	460	14	86	2
Al-Ti-X	255	294	22	...	2
Al-Be-X	483	510	20	96	2
Al-12Si (forged)	298	336	2.5	80	121
Al-25Si-AlN (reaction sintered)	...	350	...	80	122
Al-4Ni-2Fe-AlN (reaction sintered)	...	350	...	80	122
Al-7Si-20SiC-AlN (reaction sintered)	...	300	...	90	122

(a) Strain to failure

Mechanical Properties

The discontinuously reinforced aluminum composites, as mentioned earlier, are extruded, forged, or rolled to develop the best properties attainable by breaking up the prior-particle oxide skins. Usually a homogenization annealing is needed to dissolve the intermetallic inclusions in the matrix formed during vacuum hot pressing or HIP. Thomas and King (Ref 114) established an optimal solution treatment (545 °C, 6 h, 33 min) for an extruded P/M 2124/SiC/18_p composite for improved mechanical properties (see Table 4). In the DRA composites, the difference between the coefficients of thermal expansion (CTE) of the matrix aluminum alloys and the reinforcement is very significant. Compare the CTE of pure aluminum, which is 24×10^{-6}/K with those in Table 3 for particulate and whisker reinforcements. The CTE mismatch leads to a high dislocation density at interfaces, thereby creating complex residual stresses in the composite, putting matrix in tension and the reinforcement in compression, during the cooling from the processing temperature. The residual stresses adversely affect the mechanical behavior and dimensional stability of the composites. Li et al. (Ref 123) eliminated the residual stresses in P/M 6061/SiC/20_p by low-temperature treatments between −60 °C (ethanol + liquid nitrogen) and −196 °C (liquid nitrogen) and subsequent reheating to room temperature.

Fracture

The strengthening mechanisms and fracture of the dispersion-strengthened aluminum alloys are well understood (see Ref 4). However, the same is not true for the particulate- or whisker-reinforced aluminum composites exhibiting a complex frac-ture behavior. Silicon carbide particle-reinforced aluminum composites are of particular interest for stiffness critical applications. They possess high shear strength in comparison with matrix materials (see Table 5). The tensile properties of these composites, however, cannot be easily predicted. Mechanical properties included in Table 4 suggest that even similar materials have large differences in yield strength and ultimate tensile strength. The ductility, in general, is significantly lower than the matrix alloys. Wang et al. (Ref 118) developed a method to quantify particle cracking behavior as a function of strain in Al/SiC/9_p (23, 63, or 142 μm SiC particulates) composites fabricated by liquid-phase sintering. The particle crack area and the frequency of cracked particulates were found approxima-tely linear as a function of local strain along the direction of the tensile loading. The particle crack area at a given strain was not significantly affected by reinforcing particle size, but did increase with increasing matrix strength. The frequency of cracked particulates was affected by both reinforcing particle size and matrix strength. The particulates in the composites with high matrix strength are more highly stressed and, therefore, are more likely to

Table 5 Shear strength of SiC-particulate-reinforced aluminum composites

Composite	Shear strength, MPa
6061 Al (T-6)	207
25 vol% SiC/6061 Al (T-6)	277.9
30 vol% SiC/6061 Al (T-6)	289.6
7075 Al (T-6)	269
25 vol% SiC/7091 Al (T-6)	379.2
30 vol% SiC/7090 Al (T-6)	430.9
2124 Al (T-4, T-6)	283
2124 Al (T-81)	295
25 vol% SiC/2124 Al (T-4)	344.8

Source: Ref 37

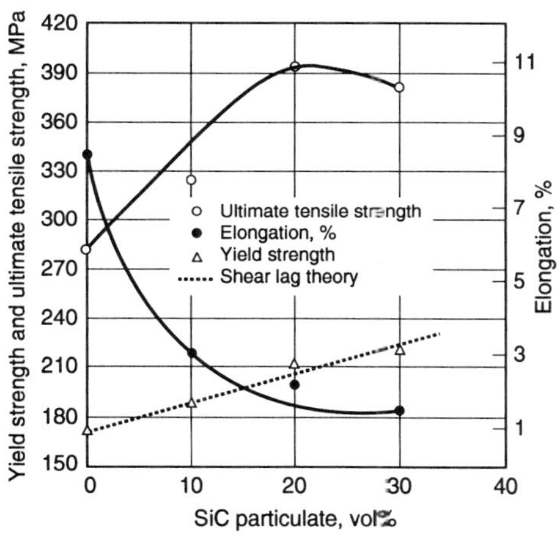

Fig. 2 Mechanical properties of SiC particle-reinforced 2124 Al composites (vacuum hot pressed and extruded). Source: Ref 34

fracture. Consider, for example, the mechanical properties of 2124/SiC$_p$ in Fig. 2 reported by Jain et al. (Ref 34) and those in Table 4 by Harrigan (Ref 31). Jain et al. (Ref 34) extruded 2124/SiC$_p$ at 773 K at a ram speed of 33 mm/s for an extrusion ratio of 16 to1 in the direct extrusion mode. The tensile strength initially increases with increase in the reinforcement volume fraction, peaks at 20 vol%, and then decreases (see Fig. 2). The maximum tensile strength obtained by Jain et al. is only about 50% of that reported by Harrigan for comparable small specimens. Similarly, the yield strength and the ductility are substantially lower for the composites fabricated by Jain et al. in comparison with that of Harrigan. Similar comparisons can be made for other systems. Table 4 includes the mechanical properties of heat treated 7075/SiC/11$_p$ and 7075/SiC/17$_p$ composites (Ref 119). In high-strength aluminum alloys such as 7000 series the reinforcement led to a decrease in yield stress and tensile strength, relative to the matrix alloy, together with reduced ductility. Doel et al. (Ref 119) discuss the fracture micromechanisms to explain the observed high toughness values for low ductility composites (see Table 4). It becomes clear that the choice of matrix compositions, reinforcement particle size, fabrication procedures, and even the test procedures play important roles in reconciling and reproducing desired level of mechanical properties.

Mechanical properties of the whisker-reinforced aluminum-matrix composites are included in Table 6. The whisker-SiC/Al composites (SiC$_w$/Al) have high specific strength and stiffness (Ref 41, 43, 128–131) as well as resistance to creep deformation (Ref 131). Komai et al. (Ref 127) found that apart from the elongation at failure, the mechanical properties of the 7075/SiC/20$_w$-T6 composite were superior to those of an unreinforced 7075-T6 alloy. There has been only a limited interest in whiskers other than SiC for fabricating MMCs. Recently Imai, Nishida, and Tozawa (Ref 19) evaluated the mechanical properties of extruded P/M aluminum-matrix composites with Si_3N_4 or $K_2O{\cdot}6TiO_2$ whiskers. The properties of these whiskers are included in Table 3. Elastic modulus of the composites is about 1.6 times that of the matrix. Tensile strength of the 6061/Si_3N_4/25$_w$ composite is 560 MPa, which is comparable to that of

Table 6 Properties of whisker-reinforced aluminum-matrix composites

The composition of MMCs is described using a nomenclature system developed by the Aluminum Association. Reinforcement contents are reported in vol%, if not otherwise specified.

Material	0.2% yield strength, MPa	Ultimate tensile strength, MPa	Elongation, %	Young's modulus, GPa	Reference
2009/SiC/15$_w$-T8					
Longitudinal	483	634	6.4	106	116
Transverse	400	552	8.4	98	116
2124/SiC/20$_w$ (wt%) K_{Ic} = 18.7 MPa$\sqrt{m}$	377	527	2.2(a)	110	124
2124-T8 K_{Ic} = 31 MPa$\sqrt{m}$	440	490	8(a)	...	125
6013/SiC/15$_w$-T6	469	655	3.2	119	116
6061-T6	225	290	17	70	126
6061/SiC/20$_w$	...	630	2.0	...	44
6061 SiC/20$_w$	440	585	4	120	126
6061 SiC/30$_w$	570	795	2	140	126
6061/Si_3N_4/25$_w$	...	560	...	...	19
7075/SiC/20$_w$	740	844	1.6	122	127

(a) Strain to failure

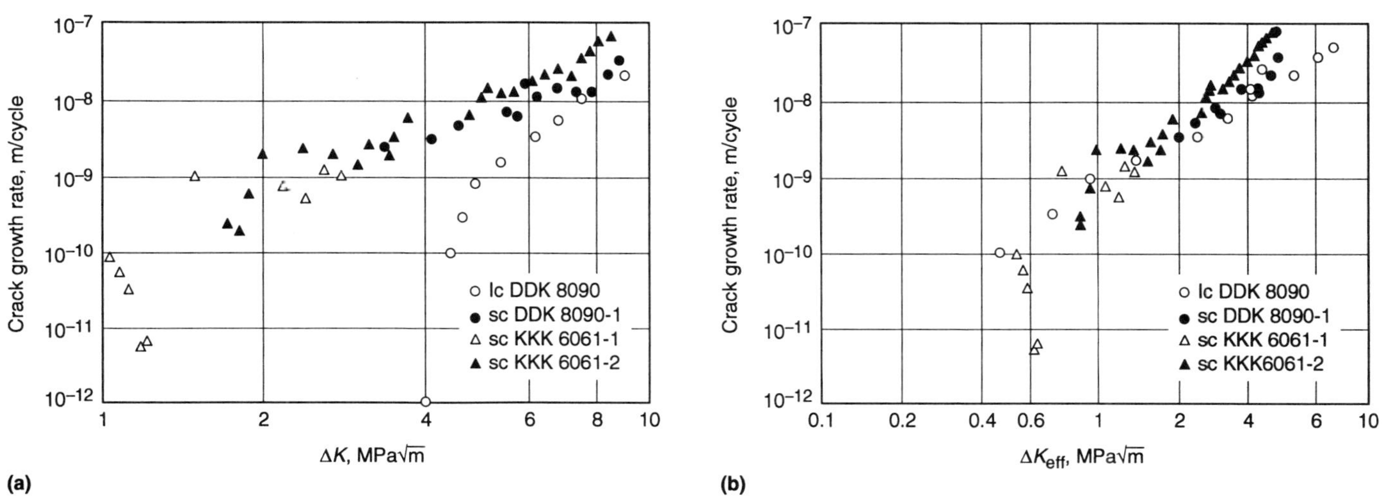

Fig. 3 The growth of large (lc) and small (sc) fatigue cracks compared on the basis of $\Delta K_{applied}$ (a) and ΔK_{eff} where closure effects have been removed (b). All data are from Ref 144 and 145 as reported in Ref 143.

the SiC whisker-reinforced aluminum composites (Table 6). The ultimate tensile strength of $K_2O{\cdot}6TiO_2$/Al is about 200 MPa at 573 K. Mechanical properties of the MAP and RSP aluminum alloys and composites are presented in Table 7. Table 8 includes the mechanical properties of the spray-formed aluminum alloys and composites.

Fatigue

A number of crack-tip shielding mechanisms including crack closure via asperity wedging in a coarse-particle-reinforced composite and crack bridging via uncracked ligaments in a composite reinforced with fine particulates have been identified during fatigue crack growth (Ref 141). The fatigue crack growth behavior in an Al/SiC/20_p composite having a strong interface has been examined by Shang and Liu (Ref 142). They suggest that the fatigue crack growth in the composite can be characterized as particle controlled. The interaction of the crack-tip stress field with the particulates results in microcracking, development of uncracked ligaments behind the crack tip, and local trapping of the crack front. Recently, Davidson (Ref 143) reviewed the mechanisms and fracture mechanics of fatigue crack initiation and the growth of small and large fatigue cracks and fracture toughness in aluminum alloys reinforced with SiC and alumina particulates. Fatigue cracks initiate within clumps of and at broken reinforcement particulates, and at intermetallic particles in the matrix. Small fatigue cracks grow faster than would be anticipated from the growth of large cracks and at lower stress-intensity range (ΔK). When ΔK for small and large cracks is adjusted for plasticity-induced fatigue crack closure to give ΔK_{eff}, the growth of all sizes of fatigue cracks are comparable (see Fig. 3). The source of fracture toughness in these composites is homogeneous crack-tip plasticity; fracture occurs at limiting values of strain or work dissipation within the crack-tip deformation zone. This is consistent with the results (see Fig. 4) reported by Doel et al. (Ref 119), who studied the mechanical properties of aluminum-base particulate MMCs fabricated by cospray deposition process (Ref 147). They found that as the particle size is increased from 13 to 60 μm, for a nominally constant volume fraction and for a given aging condition, the uniaxial tensile ductility falls, but the toughness increases.

Cernyar et al. (Ref 124) studied the fatigue of SiC whisker-reinforced aluminum composite under load control at 10 Hz, load ratios being 0.1 and 0.5. The composite exhibited a lower plane strain fracture toughness value than the matrix material, but a significantly higher resistance to fatigue crack growth in comparison with the wrought 2124 Al. For the composite, they obtained Paris law exponent n as 5.4, which is higher than that of the matrix (3 to 4), and a threshold stress intensity of 5.5 MPa$\sqrt{m}$ under a stress ratio R of 0.1. The complex fatigue damage mechanisms included whisker bridging, whisker pullout, crack deflection along the whisker/matrix interface, whisker tip voids, and whisker fracture. Komai et al. (Ref 127) investigated the fatigue behavior of 7075/SiC/20_w-T6. They found that a fatigue crack initiated at a whisker cluster or a crack at the specimen surface. They also found that the fatigue strength of the composite at a stress ratio of 0.1 was higher than that of an unreinforced 7075-T6 alloy. Narasimhan et al. (Ref 148) evaluated a P/M 6061/SiC/20_w composite for its thermomechanical fatigue performance at temperatures of 150 and 300 °C at a stress ratio of 0.1 under stress-controlled testing. They compared the high-temperature fatigue performance of the composite in terms of a stress-strain product plotted against cycles to fracture. The use of the stress-strain product term alleviates the problem of often reaching conflicting conclusions based on the conventional *S-N* type plots for strain-controlled and stress-controlled tests on composites.

Wear

The dispersion-strengthened aluminum alloys and composites generally have the potential for enhanced wear resistance over the matrix alloy (Ref 148–152). Murakami (Ref 3) has reported improved wear resistance of RSP Al-20Si-2Cu-

Table 7 Properties of MAP and RSP aluminum alloys and composites

The composition of MMCs is descried using a nomenclature system developed by the Aluminum Association. Reinforcement contents are reported in vol%, if not otherwise specified.

Material	0.2% yield strength, MPa	Ultimate tensile strength, MPa	Elongation, %	Young's modulus, GPa	Reference
MAP AA2014 (extruded)					
As extruded	...	450	...	...	132
-T6	...	563	...	...	132
MAP 7010/SiC/15_p (wt%)	...	182 at 573 K	...	...	133
MAP 7010					
At room temperature	412	498	1.85(a)	74.2	60
At 150 °C	...	377	...	...	60
At 150 °C	...	138	...	...	60
At 150 °C	...	80	...	...	60
MAP 7010Al	...	275 85 at 573 K	0.68(a)	79.7	134
At 150 °C	...	331	...	...	134
At 150 °C	...	176	...	...	134
At 150 °C	...	93	...	...	134
MAP 7010/SiC/20_p (wt%)					
At room temperature	394	414	0.59(a)	93.5	60
At 150 °C	...	313	...	...	60
At 150 °C	...	184	...	...	60
At 150 °C	...	131	...	...	60
RSP 7090	595	637	10	...	3
RSP X7090-T6E192 (extruded)	641	676	10	73.8	1
RSP X7091-T6E192 (extruded)	558	614	11	72.4	1
RSP AA8009	401	450	19	...	135
MAP					
AA8090 (extruded)	...	475.8	3.78	78.6	136
AA8090/SiC/8.2_p	...	488.5	2.45	88.3	136
AA8090/SiC/10_p	...	494.6	1.92	92.4	136
AA8090/SiC/15_p	...	519.3	1.25	104.5	136
MAP 9052Al K_{Ic} = 44 MPa$\sqrt{m}$	380	450	13	76	3
MAP Al-12.5Ti	...	...	...	104	3
At 220 °C	...	220	...	...	3
At 450 °C	...	150	...	...	3
RSP Al-10Fe-5Ce	357	453	8	...	136
RSP FVS1212 (Al-8.5Fe-1.3V-1.7Si)	522	572	23	...	135
RSP Al-8Fe-2.3Mo	460	510	7	...	135
RSP Al-25Si (extruded)	...	500	...	95	122
MAP Al-Li-Mg-O-C K_{Ic} = 45 MPa$\sqrt{m}$	450	510	10	...	137
Al-Li + 20wt% SiC_p	...	542	...	...	138
RS-MAP Al-12Si-5Mn-2Ti + 8 wt% SiC	450 360 at 423 K 240 at 573 K	...	...	...	67
RSP A356/SiC/15_p (10–15 μm)	296	358	6.1	93	68

(a) Strain to failure

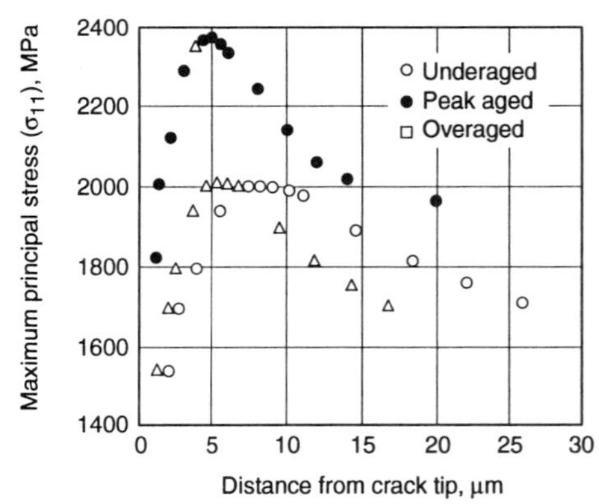

Fig. 4 Local maximum principal stress distributions in SiC-particle (5 μm) reinforced aluminum composites ahead of the crack tip, just prior to failure. From Ref 19 and 146 as reported in Ref 143

1Mg alloy. Jin et al. (Ref 151) evaluated the wear performance of RSP Al-8.5Fe-1.3V-1.7Si-Ti and Al-16Si-5Fe-3.5Cu-1.2Mg-SiC$_p$ against steel (65 HRC) over a range of sliding speed (0.16 to 1.26 m/s) under a load of 42 N at ambient and 150 °C under dry conditions. For both the materials, the wear rate first decreased with increase in sliding speed, reached a minimum and then increased at higher sliding speed. The transition sliding speed appeared to change the wear mechanism from abrasive wear (at low speed) to wear accompanying plastic deformation (at higher speed). Pin-on-disk wear tests on AA6061 composites containing Saffil, SiC$_p$, or mixtures of both ceramics against SiC grit and steel counterfaces, showed that composites containing only Saffil had inferior wear resistance to those containing the same volume fraction of SiC$_p$ (Ref 149). Bialo and Duszczyk (Ref 152) studied the wear of Al_2O_3/Al composite fabricated by liquid-phase sintering using a pin-on-disk apparatus for pressure (p) and sliding speed (v) range of 0.3 to 3 MPa and 0.1 to 5 m/s, respectively. They observed that the "pv" product is the most significant factor describing the wear process; stable friction and wear behavior of the material occurs only for pv below a threshold value of 3 MPa · m/s. They also observed that the reinforcement particle size has less significant effect on the wear rate than the pv product. Wang and Rack (Ref 153) have reported good wear resistance of SiC$_w$/Al. Dry sliding wear of DRA composites has been reviewed by Sannino and Rack (Ref 150). In general, DRAs offer substantially better wear resistance than the matrix alloy. However, under particular conditions and wear mechanisms, the wear performance of the overall DRA-metal couple is similar to or lower than the matrix alloy-metal couple. Furthermore, conflicting results regarding the effect of the different tribological parameters on the wear performance of DRAs are observed. This is primarily attributed to interactions of wear parameters; these interactions are currently not well understood.

High-Temperature Performance

The dispersion-strengthened aluminum alloys and composites are most attractive for their high-temperature strengths. Particulate or whisker reinforcement enhances their high-temperature capabilities further. Zedalis et al. (Ref 30), for example, developed high-temperature discontinuously reinforced aluminum composites for elevated-temperature applications by incorporating SiC particulate into a rapidly solidified, high-temperature Al-Fe-V-Si (alloy 8009) matrix. On a specific stiffness basis, this composite is competitive with Ti-6Al-4V and 17-4 PH stainless steel to temperatures approaching 753 K (see Fig. 5). In this section, the high-temperature performance of aluminum P/M alloys and composites are discussed in terms of their aging effects on tensile strength and toughness, strength at elevated temperatures, and creep.

Aging Effects

The aging characteristics of spray-atomized and codeposited 6061Al/SiC$_p$ composites are discussed by Wu and Lavernia (Ref 96). There was no evidence of any interfacial reaction products in the composites. They found an enhancement of the interfacial bond strength due to the interdiffusion of the alloying elements into the reinforcing particulates, upon heat treatment. Downes and King (Ref 154) evaluated the fracture behavior and fracture toughness of 8090/SiC/20$_p$ (wt%) as a function of matrix aging condition. The composite was produced by BP Metal Composites (Farnborough, UK) using a P/M route involving HIP, hot forging, and finally hot rolling to a plate thickness of 15 mm. They found the toughness values almost independent of reinforcement size (3, 6, and 23 μm). Aging at 443 K resulted in a monotonic decrease in toughness with increasing strength up to the peak condition, with no subsequent recovery in toughness on overaging.

Bhagat and coworkers (Ref 155–159) investigated the aging characteristics of SiC-whisker-reinforced 6061 Al composites (V_f = 0.20) received from Advanced Composite Materials Corporation (Greer, SC) in the form of 11.4 by 10.2 by 3.8 cm slabs. The composite was fabricated by a proprietary P/M method followed by extrusion in order to align the whiskers and increase tensile strength in the extrusion direction. The whiskers were found approximately aligned in the direction of extrusion as described in Ref 156. Both tensile dog-bone specimens and hardness samples were heat treated. These samples were placed into steel packets to protect against surface oxidation throughout the entire heat treatment process. The hot-worked condition of the as-received material was removed by first solutionizing the composite. A ramping period of 3 h (2.78 K/min) was used to obtain a stable solutionizing temperature of 803 ± 5 K. This temperature was held for 15 min, and then the samples were immediately water quenched. The

Table 8 Properties of spray-formed aluminum alloys and composites

The composition of MMCs is described using a nomenclature system developed by the Aluminum Association. Reinforcement contents are reported in vol%, if not otherwise specified.

Material	0.2% yield strength, MPa	Ultimate tensile straight, MPa	Elongation, %	Young's modulus, GPa	Reference
2014 Al	429	476	7.5	72	93
2014/SiC/10$_p$ spray codeposited, extruded, peak aged	457	508	1.8	81.2	94
2014/SiC/15$_p$ spray codeposited, extruded, peak aged	...	...	...	95	93
2014/SiC/20$_p$	466	493	2	100	2, 117
2080/SiC/20$_p$ K_{Ic} = 22 MPa$\sqrt{m}$	393	517	6	110	2, 117
2618Al spray codeposited, rolled, peak aged	345	400	7	72	94
2618/SiC$_p$ spray codeposited, rolled, peak aged	396	468	3.3	93.6	94
6061Al spray codeposited, extruded, peak aged	224	266	18.3	65.5	94
6061/SiC/13$_p$ spray codeposited, extruded, peak aged	317	356	4.9	89.5	94
6061/SiC/11.5$_p$-T6 spray atomized and codeposited	293	330	8.2	79.7	96
6061/SiC/14$_p$-T6 spray atomized and codeposited	294	330	9.0	...	139
6061/SiC/28$_p$-T6 spray atomized and codeposited	322	362	5.0	...	139
7049/SiC/15$_p$	598	643	3	90	2, 117
7075/SiC/15$_p$	556	601	4	95	2, 117
8090Al	480	550	5	79.5	93
8090/SiC$_p$ spray codeposited, extruded, peak aged	486	529	2.6	100	93
8090/SiC/15$_p$ K_{Ic} = 14 MPa$\sqrt{m}$	499	547	3	101	2, 117
Al-20Si-5Fe-2Ni					
Air atomized (Osprey), sintered at 550 °C, 1 h	...	346	0.78(a)	...	140
Nitrogen-atomized (Osprey), sintered at 550 °C, 1 h	...	360	0.82(a)	...	140

(a) Strain to failure

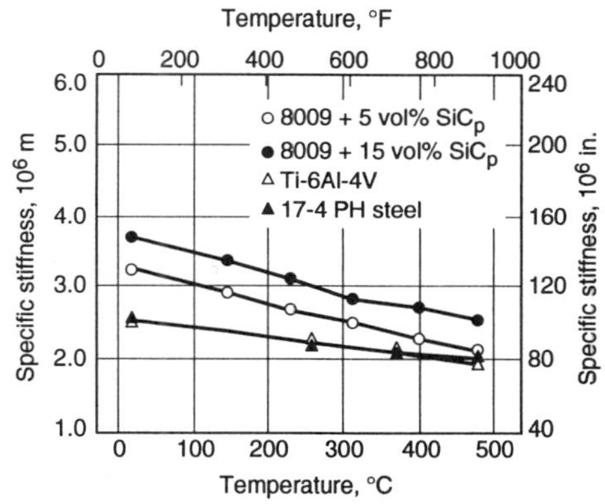

Fig. 5 Elevated-temperature specific stiffness of high-temperature discontinuously reinforced aluminum composites compared to that of titanium and stainless steel, reported in Ref 30

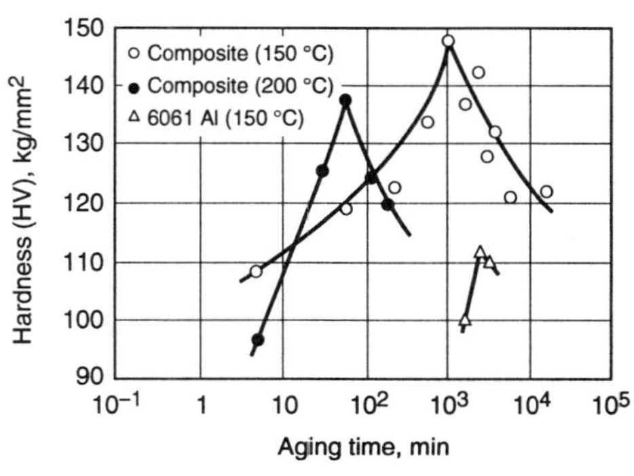

Fig. 6 Aging response of 6061/SiC/20$_w$ and wrought 6061 Al. The composite shows accelerated aging response.

solutionized samples were aged for periods of 5 min to 3 h at 473 K and for periods of 5 min to 500 h at 423 K. Following the aging, microhardness was measured at a load of 800 g using the Vickers hardness scale. Twenty values of microhardness were used to obtain the average hardness of the composite under various aging conditions.

Figure 6 includes the measured microhardness of SiC_w/6061 Al aged at 423 and 473 K. The peak microhardness value is 137.7 kg/mm^2 at 473 K, while at 423 K the peak microhardness value is 147.9 kg/mm^2. Peak aging times were 60 min and 18 h for aging temperatures of 473 K and 423 K, respectively. As a comparison, the peak microhardness for the resolutionized wrought 6061 Al is 111.7 kg/mm^2. Figure 6 demonstrates that the composite has significantly higher microhardness than the wrought 6061 Al. The high microhardness of the composite is attributed to the cumulative effect of the matrix material and whisker reinforcement. However, the trend of the aging response of the composite is primarily dependent on the aging response of the matrix. As the matrix hardens, the hardness of the composite increases until the peak is reached, followed by a drop in hardness due to the overaged matrix. The plots in Fig. 6 demonstrate the accelerated aging response of the composite as compared to the wrought 6061 Al. At the aging temperature of 423 K, peak hardness of the composite is achieved 25 h earlier, and it is 20% higher than the peak hardness of 6061 Al. For the composite, the lower aging temperature of 423 K causes the peak value of hardness to be 7% greater and 17 h later with respect to the higher aging temperature of 473 K. It is well known that the composite has a large dislocation density because of the thermal mismatch between matrix and fiber (Ref 160, 161). The high dislocation density speeds up the nucleation and growth of precipitates leading to the rapid precipitation hardening of the matrix in the composite (Ref 162). Such an accelerated aging for DRAs is also reported by other investigators (Ref 153, 163, 164).

The stiffness of the solutionized and aged composite remains constant around 105.68 GPa (as shown in Fig. 7), which is 28% lower than that of the as-received extruded composite having a stiffness of 147.55 GPa. Stiffness values of the aged composite are of practical significance for high-temperature applications, and they are about 53% higher than that of the wrought 6061-T6 Al. The tensile strength of SiC_w/6061 Al was measured as a function of aging time. Results in Fig. 8 show that the strength of the composite reaches a maximum value in the peak-aged condition. This shows that the dislocation entanglement and the finely dispersed precipitates lend maximum strength at the peak hardness. A 27% decrease in tensile strength is seen from the as-received condition. This is due to the solutionizing step that eliminated the prior hardening or strengthening due to the extrusion of the composite. Plots in Fig. 8 provide further evidence of the accelerated aging of the composite as discussed earlier. Maximum strength in the composite is reached 37 h earlier than 6061 Al with an increase of 40% in strength when measured from peak to peak. Maximum strength in the composite is maintained from 18 to 55 h; whereas 6061 Al peaks in strength at 55 h. Despite the overaging of the matrix (18 to 55 h), the strength of the composite remains almost constant. Aging the composite for 500 h reduces the ultimate tensile strength from 516 to 398 MPa, which still results in an increase of 27% over that of the identically aged wrought aluminum alloy. Ductility of the aged composite was found substantially higher than that of the as-received material which had a maximum elongation of 2%. Strain to failure was obtained by the strain versus load plot up to the fracture load. The

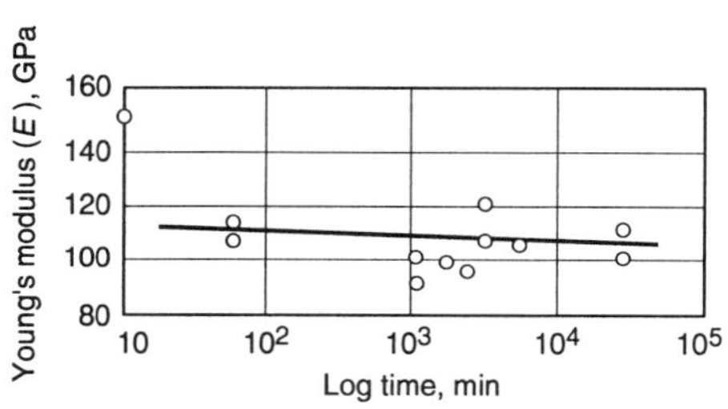

Fig. 7 Young's modulus versus aging time for solutionized 6061/SiC/20$_w$ aged at 423 K. After an initial drop, the Young's modulus is not much affected by the aging time.

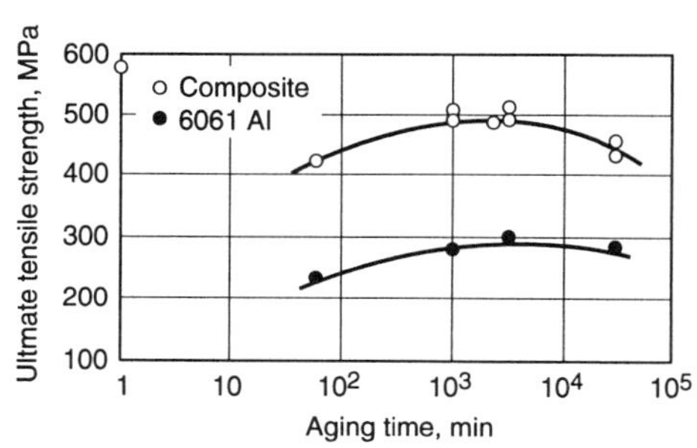

Fig. 8 Room-temperature tensile strength of solutionized 6061/SiC/20$_w$ and resolutionized 6061 Al aged at 423 K

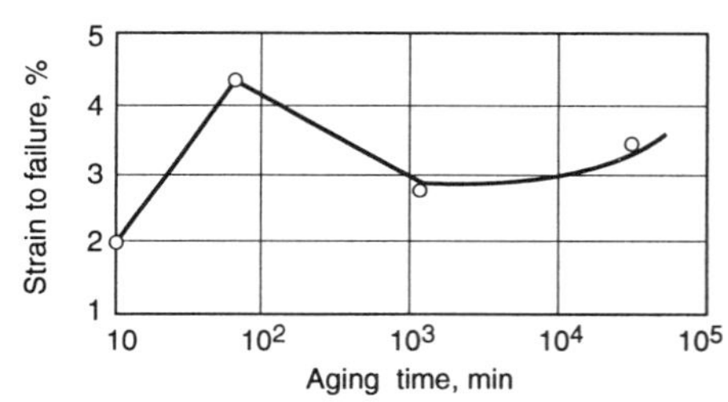

Fig. 9 Elongation to failure for solutionized 6061/SiC/20$_w$ aged at 423 K

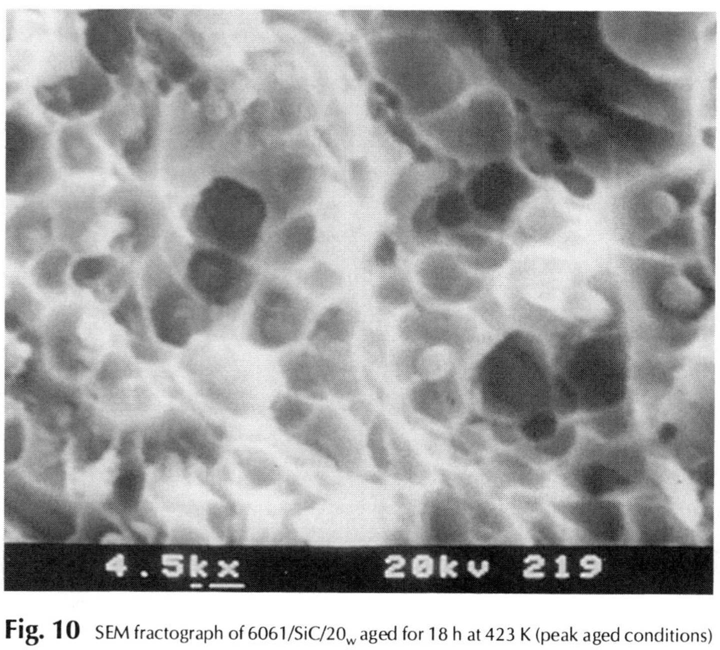

Fig. 10 SEM fractograph of 6061/SiC/20$_w$ aged for 18 h at 423 K (peak aged conditions)

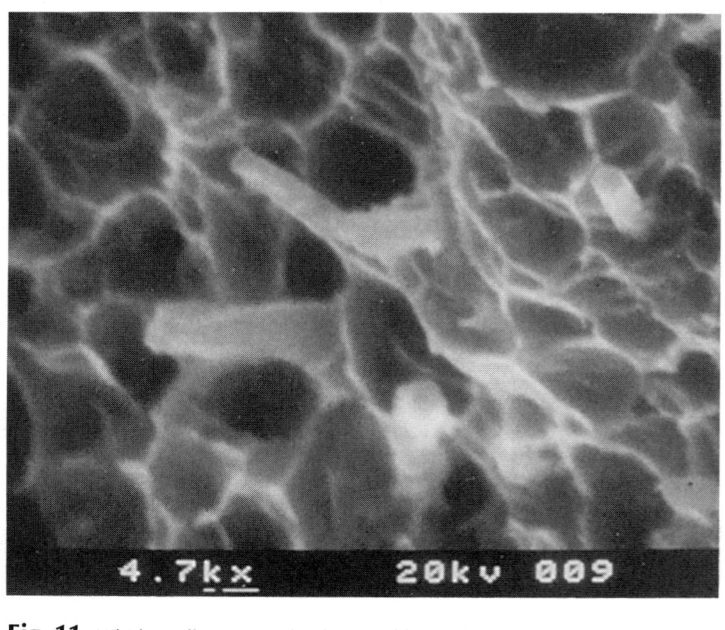

Fig. 11 Whisker pullout in 6061/SiC/20$_w$ aged for 500 h at 423 K

composite shows minimum strain to failure in the as-received condition (Fig. 9). The general trend of the elongation—rapid increase in ductility upon solutionization and aging, then decreasing, reaching a low value, and then gradually increasing with the aging time—is as expected. The work of fracture for the composite material is maximized in the overaged condition (500 h at 150 °C) having a value of 8.07 MJ/m^3 compared to the as-received condition, which has a work of fracture of 0.35 MJ/m^3. Ductile fracture as a result of microvoid coalescence can be seen in the tensile test fracture surface shown in Fig. 10. The fractograph represents a peak-aged composite showing that the whiskers serve as the initiation sites for microvoids. Nutt (Ref 165) has studied and modeled void nucleation at the whisker ends in SiC$_w$ aluminum composites. He found through transmission electron microscopy (TEM) analysis that the voids initially formed at the corners of the whiskers. Figure 11 shows a pulled-out whisker relatively free of interfacial reaction products.

Strength at Elevated Temperatures

Bhagat and coworkers (Ref 155–158) studied the elevated temperature mechanical properties of discontinuously reinforced aluminum-matrix composites. The unsolutionized SiC$_w$/6061 composite and wrought 6061-T6 Al specimens were subjected to elevated-temperature tensile testing over the range of 423 to 623 K. High-temperature strain gages were mounted to accurately determine Young's modulus and the ultimate tensile strength. Testing temperature was maintained within ±3 K of the desired set point. A uniform temperature distribution was achieved in the composite specimen by soaking at the testing temperature for 20 min prior to loading. Figures 12 and 13 depict the results of test temperatures on stiffness and ultimate tensile strength, respectively. Stiffness for the composite remains almost constant in the temperature range of 423 to 623 K, while strength decreases gradually. In the as-received condition, the composite exhibits a 111% improvement over 6061 Al in stiffness. At elevated temperatures, the stiffness of the composite exhibits a 56% improvement as compared to wrought 6061 Al. Comparing the ultimate tensile strength of the composite and 6061 Al, the whisker reinforcement at room temperature gives 114% improvement over 6061 Al, decreasing to only a 60% advantage at 623 K. The strength of the composite at elevated temperatures becomes matrix dominated as it is expected for discontinuously reinforced MMCs. An SEM fractograph showing ductile fracture because of microvoid coalescence is given in Fig. 14 for SiC$_w$/Al tested at 561 K. It can be seen in Fig. 14 that the whiskers form initiation sites for microvoids.

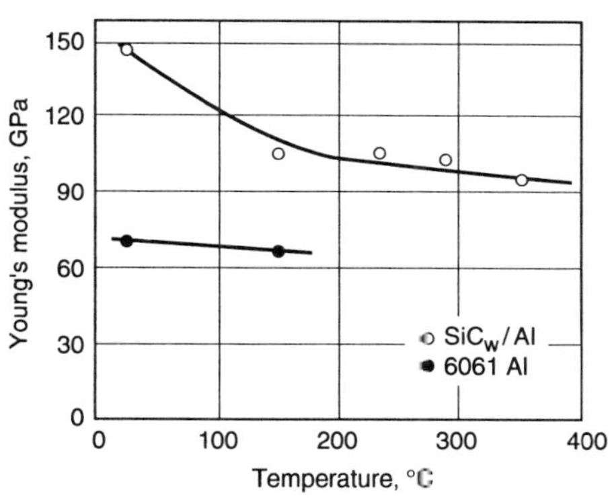

Fig. 12 High-temperature stiffness of 6061/SiC/20$_w$ and wrought 6061 Al

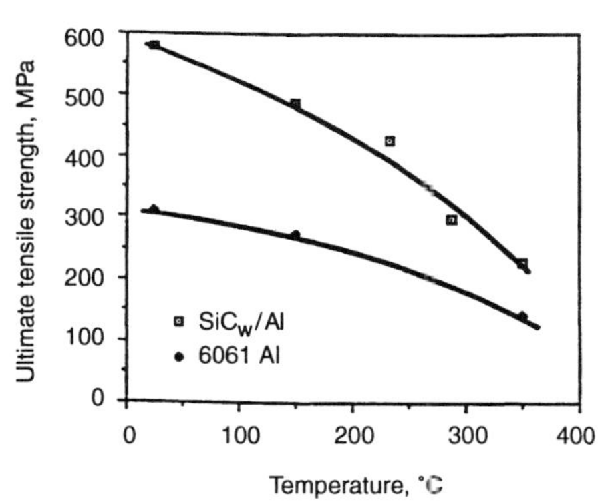

Fig. 13 High-temperature ultimate tensile strength of unsolutionized 6061/SiC/20$_w$ and wrought 6061-T6 Al

Creep

Several researchers have studied the creep behavior of aluminum-base alloys and composites (Ref 131, 157, 158, 166–183). The deformation mechanisms for creep of metals and alloys include:

- Dislocation glide due to slip
- Dislocation climb leading to subgrain formation
- Sliding of grain boundaries
- Diffusion of vacancies

These mechanisms may be related to creep deformations of the matrix in the composite, but the influence of the reinforcement is not well understood. The dispersion-strengthened alloys and reinforced composites, generally, show a change of stress exponent from a low to a relatively high value (see Table 9). The creep strength of these high-temperature aluminum alloys is significantly higher than the conventional wrought aluminum alloys. In addition, the creep resistance for the whisker-reinforced aluminum composites has been shown to increase with increasing volume fractions of the reinforcement (Ref 181). Their improved creep resistance can be attributed to the effective load transfer from the matrix to the whiskers and the inhibition of dislocation motion by the whiskers. Bhagat and coworkers (Ref 157, 158) performed the creep testing of 6061/SiC/20$_w$ composite on an 88.96 kN capacity Arcweld creep rupture machine. This machine is equipped with an automatic horizontal leveling arm and a three-zone Arcweld furnace. The temperature of the furnace was monitored with thermocouples located within the furnace and located next to the gage length. Displacements were measured by an LVDT attached to stainless steel grips. The creep tests were conducted over a temperature range of 505 to 623 K and for applied stresses of 60 to 200 MPa. The steady-state creep rates ($\dot{\varepsilon}_{ss}$) for the MMCs as a function of stress (σ) and temperature (T) are represented by the power-law creep equation, which has been historically used for the unreinforced matrix alloys:

$$\dot{\varepsilon}_{ss} = A\,\sigma^n \exp(-Q_a/RT) \quad \text{(Eq 1)}$$

The stress exponent (n), activation energy (Q_a) and the constant A of the creep power law (Eq 1) are important parameters in creep studies. The steady-state creep involves steady deformation at high temperatures and stresses. Bhagat et al. (Ref 158) experimentally determined that for an applied stress under 100 MPa, a well-behaved second-stage creep rate existed for the 6061/SiC/20$_w$ composite. The activation energy of the composite was determined by an isostress plot of the natural log of $\dot{\varepsilon}_{ss}$ and $1/T$ as shown in

Table 9 Steady-state creep results for P/M aluminum alloys and composites

Material	Volume fraction(a), %	Dispersoid size, nm	Grain size, μm	Stress exponent	Activation energy, kJ/mol	Reference
Al-Ti	8	54	0.5	7–8	240	168
Al-Fe-Ni	19–32	100	0.3	9.0–12.9	310–329	169
Al-Fe-Ni	19–32	100–160	0.2–0.3	9.6–12.9	...	170
Al-Fe-Ce	22	...	0.5	0.8–8.7	61–172	171
Al-Fe-Ce	25	...	0.5	1–8	84–142	172
Al-Fe-V-Si	36	30–80	<0.5	13–32	360	173
Al-Fe-V-Si-Er	27	87	0.4	14.7	342	174
Al-Cr-Zr	25	400	1.0	1.0	82	175
Al-Zr-V	5	30	0.5	14	...	176
Al-Zr-V	5–15	...	0.4–0.9	1	84	177
Al-Fe-W-Si	25–36	80	0.3	1.5–16	...	178
6061/SiC/20$_w$	...	...	...	...	390	131
6061/SiC/20$_w$ (solutionized)	...	...	...	20	390	179
6061/SiC/20$_w$ (unsolutionized)	...	...	...	16	76	158
2124/SiC/20$_w$	...	...	...	...	277 at 149–204 °C	180
					431 at 274–302 °C	180
Wrought 6061	...	...	...	3	...	179

(a) Volume fraction of dispersoids

Fig. 15. A comparison of the activation energy value obtained by Bhagat et al. (Ref 158) is made in Table 9 with the published results of other investigators. Differences between the values of the activation energy can be attributed to the solution treatment of the samples prior to the creep testing. Bhagat et al. (Ref 158) determined the stress exponent (n) of 16, representing the slope of the plot shown in Fig. 16. Figure 16 also shows a much smaller value of n in the low-stress regime, which can be ignored considering the high-strength capabilities of the composites. The high-stress regime n of 16 signifies that the composite is more stress sensitive compared to the wrought aluminum. This is true because the stress concentrations are developed at the whisker ends wherein the damage is initiated (see Fig. 17). The fractograph in Fig. 17 represents a creep-rupture surface of the composite tested at 561 K under a load of 100 MPa; the expected dimple morphology is noticeable. The steady-state creep rates for the whisker-reinforced composite, included in Table 10, clearly demonstrate that the composite has improved resistance to creep in comparison to the wrought 6061 Al.

Deformation Processing

Secondary processing such as rolling, forging, or extrusion in conjunction with suitable heat treatment is required to develop desired physical and mechanical properties of the aluminum alloys and the DRA composites in addition to producing useful shaped structures (Ref 35, 37, 115, 184). The formability of DRAs, however, is greatly restricted by the incompatible deformation characteristics of the soft matrix and the hard reinforcement. The optimal processing parameters for the matrix materials generally do not apply to the reinforced composites.

Rolling, Forging, and Extrusion

Near-net shape components of DRAs are produced by closed-die forging methods developed for high-strength aluminum alloys. Large components are forged directly from billet using a series of blocker and finish dies. Small parts can often be forged from extruded stock in fewer processing steps. Sheet and plate are hot rolled on conventional rolling mills from extruded bar or extruded-and-forged plate. Most P/M billets are extruded to rods or rectangular bars using lubricated conical or streamline dies. Rods can be reextruded into a variety of shapes using conventional shear-face dies. It has been reported that the forged aluminum parts have strength 40 to 60% higher than nonforged parts (Ref 1). Fatigue endurance limit is double that of nonforged P/M parts. Alloys 601AB, 602AB, 201AB, and 202AB are designed for forging. All of the aluminum P/M alloys respond to strain hardening and precipitation hardening, providing a wide range of properties. Properties of cold-formed P/M alloys are increased by a combination of strain-hardening densification and improved interparticle bonding.

Jokinen and Wiik (Ref 185) studied high temperature formability and mechanical properties of aluminum-matrix composites. They suggested that bulk forming operations applying large compressive forces (extrusion or die forging) are to be preferred when forming the composites. Bhat et al. (Ref 115) studied the constitutive flow behavior of $1100/SiC/10_p$ under hot-working conditions to generate a processing map. They suggested that the composites be extruded in the dynamic recrystallization regime at a temperature of 500 °C, wherein the dynamic recrystallization reconstitutes the microstructure and redistributes the prior-particle-boundary defects for improved formability. Cernyar et al. (Ref 124) rolled $2124/SiC/20_w$ into plate for improved properties. Various structural shapes, in addition to round and rectangular tubing, have been produced from composites containing up to 25 vol% SiC_w (Ref 116).

Fig. 14 SEM fractograph of unsolutionized $6061/SiC/20_w$ tensile tested at 561 K

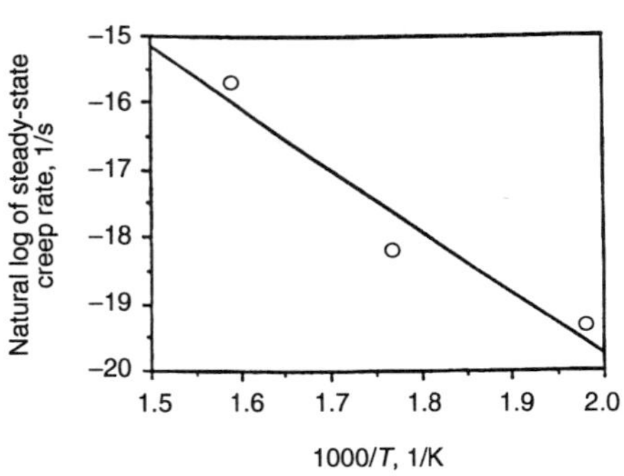

Fig. 15 Steady-state creep rate of SiC-whisker-reinforced 6061 Al composite at three temperatures. The slope of the plot is used for calculating the activation energy of the composite.

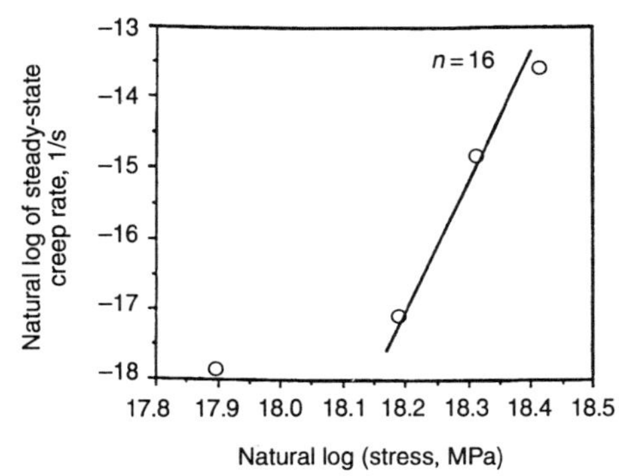

Fig. 16 Logarithmic steady-state strain rate versus logarithmic stress plot. The slope of the plot defines the stress exponent.

Superplastic Forming

Superplasticity is the ability of polycrystalline material to exhibit, in a generally isotropic manner, very high tensile elongation prior to failure. The high-strain-rate superplasticity (HSRS) is of great interest because it is expected to result in economically viable, near-net shape forming techniques for practical applications. Several P/M aluminum alloys and composites exhibit superplastic behavior at high strain rates over 10^{-2} s^{-1} (Ref 186–188). Figure 18 shows that the high-strain-rate superplasticity is associated with a very small grain size (<3 μm). It signifies that a fine grain size is a necessary condition for superplasticity, but it is not a sufficient condition for the observed HSRS phenomenon. Tensile elon-

Table 10 Steady-state creep rates for wrought 6061 Al and a P/M $6061/SiC/20_w$ composite

Temperature, °C	Stress, MPa	Steady-state creep rate, s^{-1}
Composite		
232	58.7	4.901×10^{-9}
293	59.1	1.287×10^{-8}
293	79.6	3.890×10^{-8}
298	89.6	3.758×10^{-7}
293	99.3	1.278×10^{-6}
356	60.0	1.546×10^{-7}
Wrought 6061 Al, 288	60.0	5.000×10^{-6}

gations of 1250 and 1400% have been achieved in the MAP IN9021 and 6061/SiC/20$_w$, respectively (see Table 11). Figure 19 shows that the mechanically alloyed IN9021 and IN9021/SiC/15$_p$ exhibit superplastic deformation. Mabuchi and Imai (Ref 189, 190) have reported fabrication of Si_3N_4-reinforced 2124 or 6061 Al composite having superplasticity. The 8090/SiC/17$_p$ composite in conjunction with superplastic forming has been used for fabricating net-shape components such as a module door for aircraft (Ref 120). Grain-boundary sliding is the principal deformation mechanism during the HSRS flow. The accommodation processes that must take place to allow continued grain boundary sliding include:

- Partial melting at grain boundaries
- Melting at interface
- Adiabatic heating as a result of large strain deformation at high strain rates
- Dynamic recrystallization
- Grain-boundary diffusion
- Lattice diffusion

These accommodation processes, however, are not well understood.

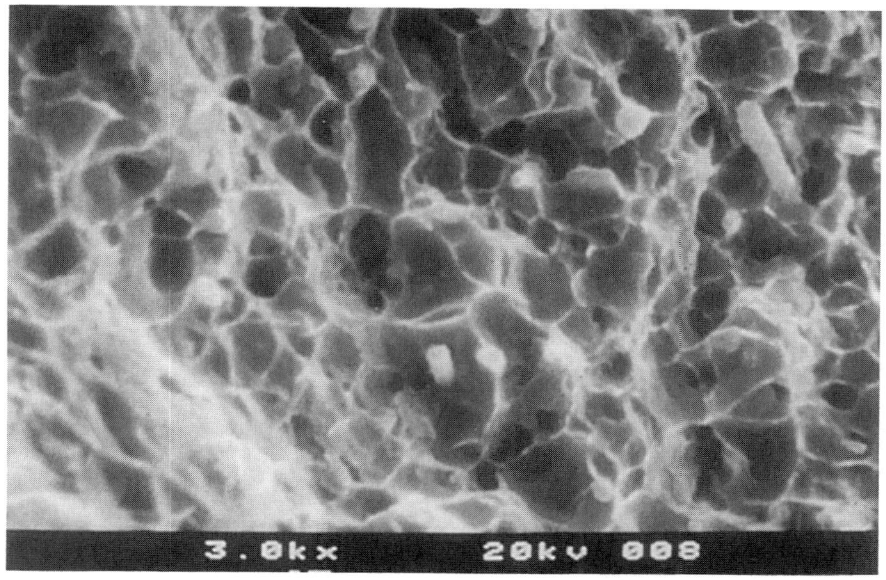

Fig. 17 Scanning electron fractograph of 6061/SiC/20$_w$ creep tested at 561 K under a load of 100 MPa.

Technological Challenges

In spite of significant advances in developing high-performance aluminum P/M alloys and composites, there remain major technological challenges that include reproducibility and reliability of performance, machining, joining, and recycling for making these materials attractive and affordable for large-scale commercial applications.

Reproducibility

Process models can be used to simulate the consolidation process, predict stress states and relative density distribution during and after processing, design special tools, and design processing schedules to ensure reproducibility, thus avoiding high cost of trial-and-error-based experiments (Ref 191, 192). For example, the mechanisms that contribute to densification of powder materials during HIP include plastic yielding, power-law creep, and lattice and boundary diffusion. Ashby and coworkers (Ref 193–196) have developed models for each densification mechanism, in each stage of densification, for the case of pure hydrostatic stress. These mechanisms are greatly influenced by the presence of ceramic particulates or whisker reinforcements. Even in the absence of any reinforcement, nonhydrostatic stress fields arise during consolidation due to the shielding effect of a canister, the geometry of a specimen, and the density and temperature gradients within a specimen. This led to a series of mechanism-based micromechanical modeling efforts (Ref 197–202). The presence of nonhydrostatic stress states necessitates the inclusion of deviatoric stresses in the constitutive laws governing consolidation leading to the development of PROSIM, a

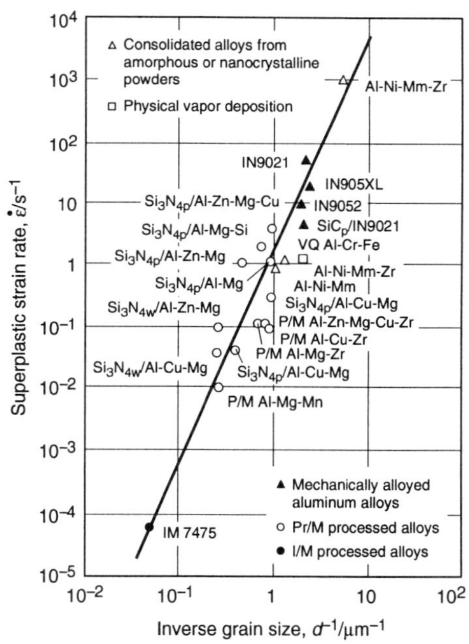

Fig. 18 The relationship between optimal superplastic strain rate and grain size for superplastic aluminum P/M alloys and composites; one I/M alloy is included for comparison. Source: Ref 187

general-purpose finite element program (Ref 191, 203, 204), and other computer programs (Ref 205) capable of analyzing deformation due to time-independent plasticity (plastic yielding) and time-dependent plasticity (power-law creep and diffusion creep). Bhagat and Rajesh (Ref 206) studied the consolidation of aluminum particles using Ashby's HIP map program. Based on the HIP map predictions, they found that the fine aluminum particles can be fully densified at an unrealistic low temperature by hot pressing (Fig. 20). This suggests that the constitutive relationships used in the HIP map program are not appropriate for fine particles (<1 μm).

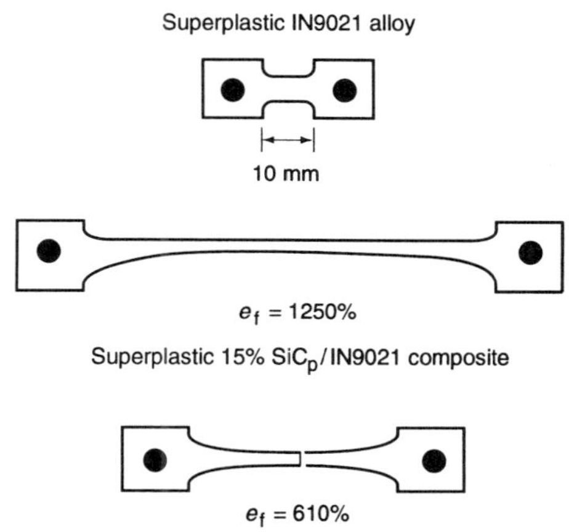

Fig. 19 Mechanically alloyed and processed samples deformed to fracture at their optimal superplastic conditions. Source: Ref 186

Machining and Joining

During conventional machining, the ceramic reinforcements dull the cutting tools, thereby reducing the machinability of the composites. The use of expensive diamond or cubic boron nitride cutting tools is not practical from cost consideration. Noncontact machining techniques such as laser processing (Ref 32) and abrasive waterjet cutting (Ref 33) appear attractive for machining these materials. Laser cutting of the SiC-whisker-reinforced aluminum composites is hampered by the formation of aluminum carbide as reported by Meinert and coworkers (Ref 207–212). They observed an aluminum carbide reaction layer on the cut surface of the composites. In addition, a significant amount of dross containing aluminum carbide remained attached to the bottom of the cut. This may be due to the

Table 11 Superplastic properties of P/M aluminum alloys and composites

Material	Grain size, μm	Solidus, °C	Test temperature, °C	Strain rate, s^{-1}	Stress, MPa	*m* value	Elongation, %
IN9021	...	495	475	1	30	...	300
IN9021	...	495	475	2	30	...	500
IN9021	0.5	495	550	50	22	0.5	1250
IN9021/SiC_p	...	495	550	5	5	0.5	600
IN9021/SiC_p	...	495	550	10	7	...	500
IN9052	0.5	580	590	10	18	0.6	330
IN905XL	...	...	550	20	5	...	200
IN905XL	0.4	...	575	20	12	0.6	190
2014/SiC/15_p	...	...	480	0.0004	...	0.4	395
2014/TiC	...	507	545	0.2	15	...	250
2024/SiC/10_p	...	...	515	0.0005	5	0.4	685
2024/SiC/20_w	...	...	100 ↔ 450(a)	0.0005	15	1.0	300
2124Al-0.6 wt% Zr	...	502	475	0.3	34.5	...	500
2124/SiC/20_w	...	502	525	0.3	10	0.33	300
2124/Si_3N_{4w}	...	502	525	0.2	10	...	250
6061/AlN	...	582	600	0.5	10	...	350
6061/SiC_w	...	582	550	0.2	6.5	...	300
6061/SiC/20_w	...	582	100 ↔ 450(a)	0.00001	7	1.0	1400
6061/Si_3N_4	...	582	545	0.5	20	...	450
7064/Si_3N_4	...	525	525	0.2	15	...	250
7075/SiC/27_w (wt%)	...	<538	500	0.2	40	...	...
7475Al-0.9 wt% Zr	...	<538	520	0.3	13	...	900
7475Al-0.7 wt% Zr	...	<538	520	0.05	5	...	900
7475/SiC/15_p	...	...	515	0.0002	...	0.38	442
7475/SiC/15_w	...	...	520	0.0002	7	>0.5	350
Al-3.2Li-1Mg-0.3Cu-0.18Zr	...	562	570	0.1	6	...	250
Al-Ni-Mm(b)	1.0	624	612	1	15	0.5	650
Al-Ni-Mm-Zr(b)	0.8	625	600	1	15	0.5	650
RS Al-Cr-Fe	0.5	623	625	1	20	0.5	505
Al-Cu-Zr	1.6	...	500	0.01	8	0.3	600
Al-Mg-Cr	3.4	...	575	0.01	3	0.5	510
Al-Mg-Zr	1.1	...	500	0.1	21	0.3	570
Al-Zn-Mg-Cu-Zr	1.2	...	515	0.07	8	0.3	1060
Al-Mg-Mn	3.5	572	575	0.003	0.7	0.5	660
Al-Cu-Mg/Si_3N_{4p} (1 μm)	2.0	580	500	0.1	5	0.3	640
Al-Mg/Si_3N_{4p} (1 μm)	1.0	593	555	1	6	0.3	700
Al-Mg-Si/Si_3N_{4p} (0.2 μm)	1.3	582	560	2	5	0.3	620
Al-Mg-Si/Si_3N_{4p} (0.5 μm)	1.9	585	560	1	6	0.3	350
Al-Mg-Si/Si_3N_{4p} (1 μm)	3.0	580	545	0.1	5	0.3	450
Al-Cu-Mg/Si_3N_{4w}	3.3	585	560	0.1	11	0.3	480

(a) Thermal cycling. (b) Mm, mischmetal. Source: Ref 186, 188

potential reaction between the molten aluminum and the SiC whiskers, forming a highly viscous mixture of aluminum carbide and silicon. The assist gas moves the mixture toward the bottom of the cut, but is unable to expel it completely, due to its high viscosity. This results in a thin layer of aluminum carbide on the cut surface and a buildup of aluminum carbide dross at the bottom of the cut. The amount of dross attached to the bottom of the cut can be quite significant; the average height of the dross attached to the bottom of a 2 mm thick sample of the 6061/SiC/20_w composite was 1.25 mm. Several different assist gases, power levels, and travel speeds were used in an attempt to minimize the dross formation. Variation of processing parameters had little effect on the cut quality. The formation of aluminum carbide appears to override the ability to obtain laser cuts with a clean surface.

The weldability of the aluminum P/M alloys and composites is limited by the residual hydrogen content and wide freezing range. The hydrogen is associated with the oxides on the surface of the metal powder. Upon melting, the adsorbed hydrogen coalesces and forms pores in the fusion zone and heat-affected zone (HAZ). Minimization of the fusion-zone porosity and retention of the base alloy microstructure in the weld zone are, therefore, the major factors precluding the application of many conventional fusion-welding techniques. These alloys require the use of autogenous electon beam and pulsed Nd:YAG (neodymium:yttrium-aluminum-garnet) laser welding, which show only occasional evidence of fusion-zone porosity (Ref 213). Ellis (Ref 214) summarizes advantages and disadvantages of various joining techniques for aluminum MMCs. Solid-state (both diffusion bonding and friction welding) methods have proved very successful in joining these materials, though geometry of components may preclude these techniques. Fusion-joining techniques are often hampered by interactions between the matrix and the reinforcement material. Laser processing, however, can offer many advantages over conventional techniques. Certain precautions are necessary prior to laser-beam welding of P/M MMCs (Ref 35, 215, 216). The composites should be thoroughly degassed by vacuum annealing to minimize porosity in the weld-fusion zone. In addition, weld energy input should be controlled for preventing formation of Al_4C_3 in the SiC/Al composites.

Similar to the problem encountered during laser cutting, laser welding of SiC-reinforced aluminum composites is hampered due to the following chemical reaction between liquid aluminum and solid SiC:

$$4Al(l) + 3SiC(s) \rightarrow Al_4C_3(s) + 3Si(s) \quad \text{(Eq 2)}$$

The formation of brittle aluminum carbide platelets reduces the strength of the weld. Further degradation of the weld takes place due to the dissolution of aluminum carbide in aqueous and moist environments. The formation of aluminum carbide leads to a substantial increase in the viscosity of the melt (Ref 217). The use of aluminum alloys containing relatively high amounts of silicon is one of the most commonly used methods for eliminating the formation of aluminum carbide during molten-metal fabrication techniques. Because the formation of aluminum carbide is inhibited by the high silicon content and a relatively low melt superheat, the fusion joining of these materials is possible if the weld pool temperature is kept below that at which aluminum carbide begins to form (Ref 218). In addition to the chemical reactions, several other factors may affect the quality of the weld. For example, the reinforcement materials generally do not melt or dissolve and may segregate in the weld during the laser processing. Another com-

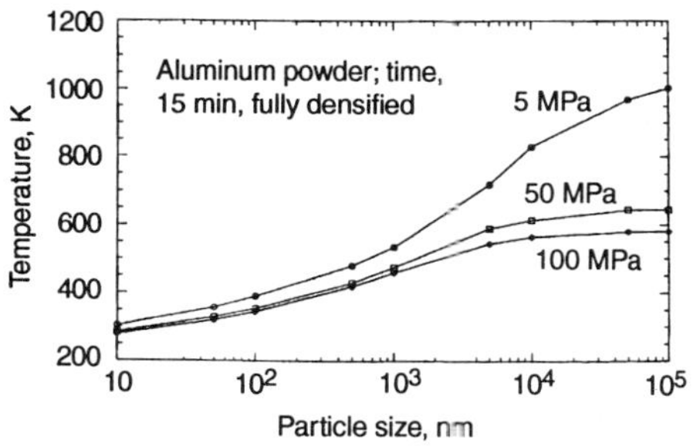

Fig. 20 Predicted temperatures according to Ashby's HIP map program for fully densifying aluminum powders by hot pressing. The plots represent 24 runs of HIP map considering eight different sizes of powder over three pressures. For each run of HIP map, the powder is assumed to have one fixed particle size.

plication is the increased viscosity of the molten pool due to the presence of the reinforcement. This can lead to a sluggish weld pool and incomplete joining.

Meinert and coworkers (Ref 207–212) used a commercially available P/M wrought 6061/SiC/20_w composite (Advanced Composite Materials Corporation, Greer, SC) to study the feasibility of laser processing. A Coherent General EFA 51, 1.5 kW, enhanced fast axial flow, CO_2 gas laser, with continuous wave (CW) and pulsed TEM_{00} mode capability was used during the investigation. A focal arrangement utilizing a 6.35 mm lens produced a focused spot size of 120 μm at an output wavelength of 10.6 μm, in the infrared range. Shielding or assist gas was provided coaxial to the beam, and a computer-controlled micropositioning system was employed to manipulate the test specimens during the laser processing. The assist gas discharges the molten metal from the cut, in the case of laser cutting. In contrast, the molten metal solidifies under an inert shielding gas, in the case of laser-beam welding. The shielding gas prevents the formation of unwanted reactions and gas absorption.

The amount of laser energy absorbed by the material is one factor that determines the effectiveness of the laser processing. In general, a high absorptivity improves the processing efficiency, thereby allowing deeper penetration and faster processing speeds. Aluminum has a relatively high reflectivity, which decreases its ability to effectively absorb, or couple with the laser beam. The reinforcement materials such as aluminum oxide and SiC absorb much more energy at 10.6 μm wavelength. Initial experiments revealed that energy absorbed by the composite materials was greater than the unreinforced aluminum. The improved energy absorption is attributed to the change in reflectivity of the bulk material by the addition of the reinforcement. There appears to be a correlation between the emissivity of the material and the effectiveness of laser processing. As mentioned before, the silicon addition has been successfully used to allow the fusion joining of SiC reinforced aluminum composites (Ref 218). Suppression of the formation of aluminum carbide, however, can be achieved by additions of titanium and zirconium,

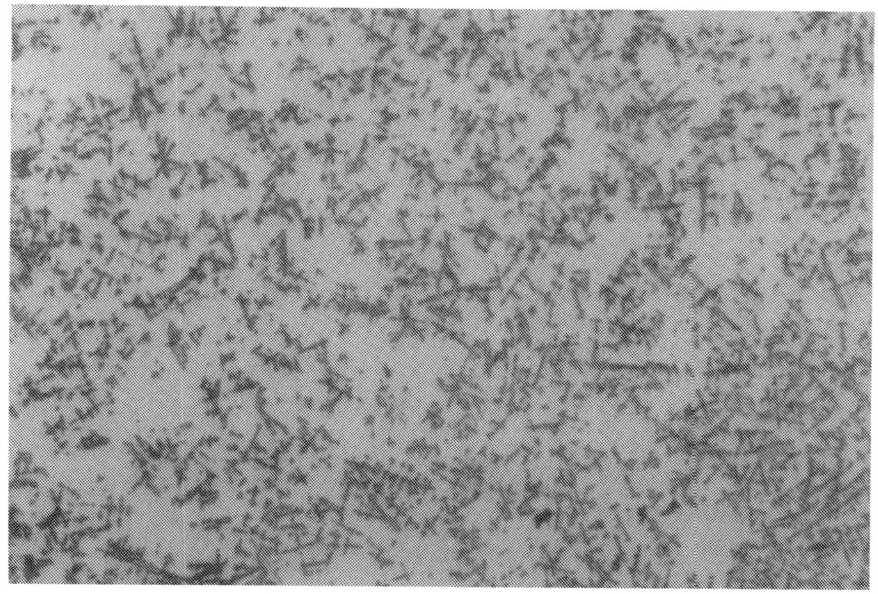

Fig. 21 Micrograph of a laser-welded 6061/SiC/20_w composite using titanium additions. Fine dendritic structures of TiC is evenly dispersed throughout the fusion zone.

which are strong carbide formers. Titanium carbide and zirconium carbide are also chemically stable compounds and are not subject to dissolution in water. Meinert and coworkers (Ref 207–212) laser welded 6061/SiC/20_w composite using commercially pure titanium additions that effectively suppressed the formation of aluminum carbide. Figure 21 shows very fine dendritic structures of TiC (1 to 5 μm) as confirmed by the electron microprobe analysis. The dendrites are evenly dispersed throughout the fusion zone. The analysis of the welds made with zirconium additions showed similar results.

Applications Outlook

In spite of a worldwide interest in developing the high-performance aluminum P/M alloys and composites, these materials have found only limited applications. With a few exceptions such as the spray-formed (Osprey) aluminum-silicon alloys for automotive parts, large-scale commercial use of the aluminum P/M alloys and composites currently does not exist. However, the outlook for potential applications of these materials in aerospace, automotive, business machines, power tools, appliances, and ordnance industries appears very bright. A few examples include:

- *Aerospace components*: Lightweight P/M aluminum alloys are attractive for major airframe primary-load-carrying structural members (Ref 219). The dispersion-strengthened aluminum-iron alloys are expected to replace conventional high-strength alloys and titanium alloys used in the manufacture of selected aerospace components such as fan and compressor cases, stiffeners, vanes and blades in gas turbine engines, fins, winglets, and rocket motor cases in missiles, and helicopter rotors. High-strength P/M 7090 is used for landing-gear assemblies (link, door actuators) (Ref 1). A high-stiffness SiC_p-reinforced 6092 Al composite has been fabricated into a ventral fin for F-16 jet fighter; the composite provides stiffness 50% higher than monolithic aluminum. The composite is expected to provide more than twice the expected life of the original fins (Ref 220). A module door for aircraft was fabricated by superplastic forming of 8090/SiC/17_p (Ref 120).
- *Automotive parts*: Due to their light weight, high stiffness, and wear resistance, the P/M aluminum-silicon alloys are attractive for automotive parts such as shock absorber, piston, drive shaft, brake rotor disk, engine cylinder block, autobody structures, chassis components, compressor vanes, connecting rod, and sensor housing. Ishijima et al. (Ref 121) developed forged P/M aluminum-silicon alloys for connecting rods for general-purpose engines. The fabricated connecting rods are found to be more than 30% lighter and thinner in comparison with conventional die-cast rods. Near-net shape compressor parts were fabricated by Takeda and Hayashi (Ref 122) using Al-AlN and Al-AlN-SiC composites by reaction sintering of rapidly solidified aluminum alloy/composite powder. The fabricated parts showed improved wear resistance.
- *Business machines:* Drive-belt pulleys, hubs, end caps, connecting collars, and gears
- *Appliances*: Sewing machine parts and thermostat control gears
- *Electrical and electronic applications*: Heatsinks, substrate/housing for microelectronics package, and spacers on structural electric transmission towers
- *Ordnance*: Rifle receiver forging, mortar and artillery projectile fuses, cartridge cases, artillery shells, and rocket warheads

REFERENCES

1. *Powder Metallurgy,* Vol 7, *Metals Handbook,* 9th ed., American Society for Metals, 1984
2. H. Buhl, Ed., *Advanced Aerospace Materials,* Springer-Verlag, 1992
3. Y. Murakami, Aluminum-Based Alloys, *Materials Science and Technology,* R.W. Cahn, P. Haasen, and E.J. Kramer, Ed., Vol 8, 1996, p 213–276
4. E. Nembach, *Particle Strengthening of Metals and Alloys,* John Wiley & Sons, 1997
5. J.-W. Kim and W.M. Griffith, Ed., *Dispersion Strengthened Aluminum Alloys,* TMS, 1988
6. E.W. Lee and N.J. Kim, Ed., *Light Weight Alloys for Aerospace Applications,* TMS, 1991
7. S.S. Brenner, *J. Appl. Phys.,* Vol 33, 1962, p 33
8. S.S. Brenner, *J. Met.,* Vol 14 (No. 11), 1962, p 808
9. W.H. Sutton and J. Chorn, *Met. Eng. Q.*, Vol 3 (No. 1), 1963, p 44
10. W.H. Sutton, Whisker Composite Materials—A Prospectus for the Aerospace Designer, *Astronaut. Aeronaut.,* Aug 1966, p 46
11. A.P. Divecha, P. Lare, and H. Hahn, "Silicon Carbide Whisker Metal Matrix Composites," AFML-Tr-69-7, May 1969
12. I.B. Cutler, "Production of Silicon Carbide from Rice Hulls," U.S. Patent No. 3,754,076, Aug 1973
13. D.L. McDaniels, *Metall. Trans. A,* Vol 16, 1985, p 1105
14. R.J. Arsenault, *Mater. Sci. Eng.,* Vol 64, 1984, p 171
15. C.R. Crow, R.A. Gray, and D.F. Hasson, *Proc. of ICCM5,* W.C. Harrigan, Jr., J. Strife, and A.K. Dhingra, Ed., AIME, 1985, p 843
16. A.P. Divecha, S.G. Fishman, and S.D. Karmarkar, *J. Met.,* Vol 9 (No. 12), 1981
17. G.H. Pigott and J. Ismael, *Proc. 5th Int. Symposium on Inhaled Particles,* W.H. Walton, Ed., Pergamon Press, 8–12 Sept 1980
18. M.F. Stanton and C. Wrench, *J. Nat. Cancer Inst.,* Vol 48 (No. 3), 1972, p 797
19. T. Imai, Y. Nishida, and Y. Tozawa, Mechanical Properties of Si_3N_4 and $K_2O{\cdot}6TiO_2$ Whisker Reinforced Aluminum P/M Composites, *Proc. Fourth Japan-U.S. Conf. on Composite Materials,* Technomic, 1988, p 109–117
20. R.B. Bhagat, The Effects of Hot Pressing Parameters on the Strength of Aluminum/Stainless Steel Composites, *Metall. Trans. A,* Vol 16, 1985, p 623–628
21. R.B. Bhagat, Growth Kinetics of Interface Intermetallic Compounds in Stainless Steel Fiber Reinforced Aluminum Matrix Composites, *Interfaces in Metal Matrix Composites,* A.K. Dhingra and S.G. Fishman, Ed., TMS, 1986, p 169–183
22. R.B. Bhagat, Low Cycle Fatigue Behavior of Aluminum/Stainless Steel Composites, *AIAA J.,* Vol 23 (No. 6), 1985, p 912–917
23. R.B. Bhagat and P. Ramakrishnan, Interfacial Interactions in Aluminum Matrix Stainless Steel Fiber Composites, *Progress in Science and Engineering of Composites,* Vol 2, T. Hayashi, K. Kawata, and S. Umekawa, Ed., ICCM-IV, Tokyo, 1982, p 1297–1305
24. R.B. Bhagat and P. Ramakrishnan, Stainless Steel Fiber Reinforced Aluminum P/M Composites, *Proc. 10th Plansee Seminar, Metallwerk, Plansee, Reutte* (Austria), Vol 1, 1981, p 385–393
25. R.B. Bhagat and P. Ramakrishnan, Fabrication of Aluminum Matrix Stainless Steel Fiber Reinforced Composites, *Proc. Fourth Biennial Seminar on Gas Turbine,* GTRE, Bangalore, India, 1979, p 425–430
26. H.J. Rack and J.W. Mullins, *High Performance Powder Aluminum Alloys II,* G. Hildeman and M. Koczak, Ed., TMS, 1986, p 155–171
27. F. Wawner, A.T. Chueng, and S. Bettadapur, "Metal Matrix, Carbon, and Ceramic Composites," Technical Conference Report No. 2357, J.D. Buckley, Ed., National Aeronautics and Space Administration, 1984, p 97–118
28. T.G. Nieh, R.A. Rainen, and D.J. Chellman, *Proc. Fifth Int. Conf. Composite Materials,* W.C. Harrigan, Jr., J. Strife, and A.K. Dhingra, Ed., TMS, 1985, p 825–842
29. P.S. Gilman, Discontinuously Reinforced Aluminum: Ready for the 1990s, *JOM,* Vol 43 (No. 8), 1991, p 7
30. M.S. Zedalis, J.D. Bryant, P.S. Gilman, and S.K. Das, High Temperature Discontinuously Reinforced Aluminum, *JOM,* Vol 43 (No. 8), Aug 1991, p 29–31
31. W.C. Harrigan, Jr., Scaling-Up Particulate Reinforced Aluminum Composites for Commercial Production, *JOM,* Vol 43 (No. 8), Aug 1991, p 32–35
32. G.P. Simpson and T. J. Culkin, Laser Cutting, *Machining,* Vol 16, *ASM Handbook,* 9th ed., ASM International, 1989, p 735–742
33. J.G. Sylvia, Abrasive Waterjet Cutting, *Machining,* Vol 16, *ASM Handbook,* 9th ed., ASM International, 1989, p 743–755
34. M.K. Jain, V.V. Bhanuprasad, S.V. Kamat, A.B. Pandey, V.K. Verma, B.V.R. Bhat, and Y.R. Mahajan, Processing, Microstructure and Properties of 2124 Al-SiCp Composites, *Int. J. Powder Metall.,* Vol 29 (No. 3), July 1993, p 267–275
35. H.J. Rack, Powder Techniques in Processing of Metal Matrix Composites, *Metal Matrix Composites: Processing and Interfaces,* R.K. Everett and R.J. Arsenault, Ed., Academic Press, 1991, p 83–101
36. H.J. Rack, P. Hood, P. Nishanen, and J.L. Cook, *Proc. Third Discontinuous Metal Matrix Workshop Group* (Santa Barbara, CA), Metal Matrix Composites Information Center, 1983
37. W.C. Harrigan, Jr., Metal Matrix Composites, *Metal Matrix Composites: Processing and Interfaces,* R.K. Everett and R.J. Arsenault, Ed., Academic Press, 1991, p 1–16
38. P.E. Hood and J.P. Pickens, "Silicon Carbide Whisker Composites," U.S. Patent No. 4463058, 31 July 1984
39. V.V. Bhanuprasad, B.V.R. Bhat, A.B. Pandey, K.S. Prasad, A.K. Kuruvilla, and Y.R. Mahajan, Effect of Processing Parameters on the Properties of Discontinuously Reinforced Al/SiC P/M Composites, *Int. J. Powder Metall.,* Vol 27 (No. 3), 1991, p 227
40. C.J. Skowronek, A. Pattnaik, and R.K. Everett, "Dispersion and Blending of SiC Whiskers in RSP Aluminum Powders," Naval Research Laboratory Memorandum Report 5750, Naval Research Laboratory, 1986
41. A.P. Divecha and S.G. Fishman, Mechanical Properties of Silicon Carbide Reinforced Aluminum, *Mechanical Behavior of Materials,* Vol 3, K.J. Miller and R.F. Smith, Ed., Pergamon Press, 1979, p 351–361
42. M.Y. Wu and O.D. Sherby, Superplasticity in a Silicon Carbide Whisker Reinforced Aluminum Alloy, *Scr. Metall.,* Vol 18, 1984, p 773–776
43. A.P. Divecha, S.G. Fishman, and S.D. Karmarkar, Silicon Carbide Reinforced Aluminum—A Formable Composite, *JOM,* Vol 33, Sept 1981, p 12–17
44. J.J. Shimizu, Kusui, A. Tanaka, and O. Iwao, Whisker Reinforced Composites Prepared from Wet Ballmilled Aluminum Powder, *Metal and Ceramic Matrix Composites: Processing, Modeling and Mechanical Behavior,* R.B. Bhagat, A.H. Clauer, P. Kumar, and A.M. Ritter, Ed., Minerals, Metals and Materials Society/AIME, 1990, p 31–38
45. J.S. Benjamin, *Metall. Trans.,* Vol 1, 1970, p 2943
46. G.H. Gessinger, *Powder Metallurgy of Superalloys,* Butterworths, 1984, p 213
47. P.S. Gilman and J.S. Benjamin, *Ann. Rev. Mater. Sci.,* Vol 13, 1983, p 279
48. R. Sundaresan and F.H. Froes, *JOM,* Vol 39 (No. 8), 1987, p 22
49. C.C. Koch, Processing of Metals and Alloys, *Materials Science and Technology—A Comprehensive Treatment,* Vol 15, R.W. Cahn, Ed., VCH, 1991, p 193
50. C. Suryanarayana and F.H. Froes, *Mater. Sci. Forum,* Vol 88–90, 1992, p 445
51. C.C. Koch, *Mater. Sci. Forum,* Vol 88–90, 1992, p 243
52. C. Suryanarayana, R. Sundaresan, and F.H. Froes, *Advances in Powder Metallurgy,* Vol 3, T.G. Gasbarre and W.F. Jandeska, Jr., Ed., Metal Powder Industries Federation, 1989, p 175
53. R. Sundaresan and F.H. Froes, *Met. Powder Rep.,* Vol 44, 1989, p 195
54. C. Suryanarayana, R. Sundaresan, and F.H. Froes, *Structural Applications of Mechanical Alloying,* F.H. Froes and J.J. deBarbadillo, Ed., ASM International, 1990, p 193
55. C. Suryanarayana and F.H. Froes, *Adv. Mater.,* Vol 5, 1993, p 96
56. A. Bose, *Advances in Particulate Materials,* Butterworth-Heinemann, 1995
57. C. Suryanarayana and F.H. Froes, *J. Mater. Res.,* Vol 5, 1990, p 1880

58. F.H. Froes, C. Suryanarayana, G.-H. Chen, A. Frefer, and G.R. Hyde, *JOM*, Vol 44 (No. 5), 1992, p 26
59. H.J. Fecht, E. Hellstern, Z. Fu, and W.L. Johnson, *Metall. Trans. A*, Vol 21, 1990, p 2333
60. A. Bhaduri, V. Gopinath, and P. Ramakrishnan, SiC Particulate Reinforced Aluminum Matrix Composites Prepared by Mechanical Alloying, *Advances in Powder Metallurgy & Particulate Materials,* Part 2, Metal Powder Industries Federation, 1996, p 3–12
61. S.K. Das, Rapidly Solidified P/M Aluminum and Magnesium Alloys—Recent Developments, *Rev. Part. Mater.,* Vol 1, 1993, p 1–40
62. C. Suryanarayana, F.H. Froes, and R.G. Rowe, *Int. Mater. Rev.,* Vol 36, 1991, p 85
63. H. Jones, *Rapid Solidification of Metals and Alloys,* Institute of Metallurgists, London, 1982
64. C. Suryanarayana and F.H. Froes, *JOM,* Vol 42 (No. 3), 1990, p 22
65. S.K. Das, R.L. Bye, and P.S. Gilman, *Mater. Sci. Eng. A*, Vol 134, 1991, p 1103
66. D.J. Skinner, R.L. Bye, D. Raybould, and A.M. Brown, *Scr. Metall.,* Vol 20, 1986, p 867
67. J. Kumpfert, G. Stanick, W. Kleinekathofer, and M. Thumann, *Structural Applications of Mechanical Alloying,* F.H. Froes and J.J. de Barbadillo, Ed., ASM International, 1990, p 163
68. E.L. Courtreight et al., *Adv. Mater. Process.,* Vol 138 (No. 5), 1990, p 71
69. H.W. Jin, C.R. Park, C.G. Park, and M.C. Kim, Mechanical Strength and Thermal Stability of Al-Mm-Ni-Fe Amorphous Alloys, *Light Weight Alloys for Aerospace Applications III,* E.W. Lee, N.J. Kim, K.V. Jata, and W.E. Frazier, Ed., TMS, 1995, p 303–310
70. A. Inoue, Y.H. Kim, and T. Masumoto, *Mater. Trans., JIM,* Vol 33, 1992, p 487
71. A. Inoue, Y. Horio, Y.H. Kim, and T. Masumoto, *Mater. Trans., JIM,* Vol 33, 1992, p 669
72. Y. He, G.M. Dougherty, G.J. Shiflet, and S.J. Poon, *Acta Metall. Mater.,* Vol 41, 1993, p 337
73. Y.J. Kim, A. Inoue, and T. Masumoto, *Mater. Trans., JIM,* Vol 32, 1991, p 559
74. A. Inoue and T. Masumoto, *Encyclopedia of Materials Science and Engineering,* Supplementary Vol 2, R.W. Cahn, Ed., Pergamon, 1990, p 660–667
75. Y. He, S.J. Poon, and G.J. Shiflet, *Science,* Vol 241, 1988, p 1640
76. Y.J. Kim, A. Inoue, and T. Masumoto, *Mater. Trans., JIM,* Vol 30, 1991, p 599
77. D.J. Chellman and T.D. Bayha, Property Behavior of High Temperature Spray Deposited Al-Cu-Mg-X Alloys, *Second Int. Conf. Spray Forming, ICSF2,* J.V. Wood, Ed., Woodhead Publishing, 1993, p 427
78. I.G. Palmer, D.J. Chellman, and J. White, Evaluation of a Spray Deposited Low Density Al-Li Alloy, *Second Int. Conf. Spray Forming, ICSF2,* J.V. Wood, Ed., Woodhead Publishing, 1993, p 385
79. H. Sano, N. Tokizane, Y. Ohkubo, and K. Sibue, Spray Formed Aluminum Alloy Components for Automotive Applications, *Second Int. Conf. Spray Forming, ICSF2,* J.V. Wood, Ed., Woodhead Publishing, 1993, p 363
80. T.D. Bayha and D.J. Chellman, Elevated Temperature Fracture Behavior of Spray Deposited 2618 Extrusions, *Advances in Powder Metallurgy & Particulate Materials,* Vol 5, Metal Powder Industries Federation, 1994, p 17
81. A.R.E. Singer, Metal Matrix Composites Made by Spray Forming, *Mater. Sci. Eng. A,* Vol 135, 1991, p 13–17
82. J.H. Tweed, J. Watling, J. Cook, J. Crook, A.D. Simpson, and I.W. Pearson, Characterization of the Uniformity and Filler Content of a Spray Formed 8090/SiC Composite, *Second Int. Conf. Spray Forming, ICSF2,* J.V. Wood, Ed., Woodhead Publishing, 1993, p 319
83. A. Lawley and D. Apelian, Spray Forming of Metal Matrix Composites, *Second Int. Conf. Spray Forming, ICSF2,* J.V. Wood, Ed., Woodhead Publishing, 1993, p 267
84. M. Gupta, J. Juarez-Islas, W.E. Frazier, F. Mohamed, and E.J. Lavernia, Microstructure, Excess Solid Solubility and Elevated-Temperature Mechanical Behavior of Spray-Atomized and Codeposited Al-Ti-SiCp, *Metall. Trans. B,* Vol 23, 1992, p 719–736
85. M. Gupta, F. Mohamed, and E.J. Lavernia, *J. Mater. Manuf. Proc.,* Vol 5, 1990, p 165–176
86. T.S. Srivatsan and E.J. Lavernia, Use of Spray Techniques to Synthesize Particulate-Reinforced Metal-Matrix Composites, *J. Mater. Sci.,* Vol 27 (No. 22), 1992, p 5965–5981
87. Y. Wu and E.J. Lavernia, Interaction Mechanisms between Ceramic Particles and Atomized Metallic Droplets, *Metall. Trans. A,* Vol 23 (No. 10), 1992, p 2923–2935
88. P.J. Ward, I.G. Elias, and H.V. Atkinson, Processing Parameters in the Spray-Forming and Subsequent Semi-Solid Forming of Hypereutectic Al/Si Based MMCs, *Second Int. Conf. Spray Forming, ICSF2,* J.V. Wood, Ed., Woodhead Publishing, 1993, p 303
89. J. White, K. Mingard, I.R. Hughes, and I.G. Palmer, Aluminum Alloys with Unique Property Combinations by Spray Casting, *Second Int. Conf. Spray Forming, ICSF2,* J.V. Wood, Ed., Woodhead Publishing, 1993, p 355
90. R.B. Bhagat and M.F. Amateau, Droplet Solidification and Microstructures Modeling for Aluminum-Lithium Alloys, *Advances in Powder Metallurgy & Particulate Materials,* Vol 7, Metal Powder Industries Federation, 1996, p 347–364
91. R.B. Bhagat, "Coarsening Kinetics of Solid Particles in the Spray-Formed (Osprey) Aluminum Alloy Preform," presented at 126th TMS Annual Meeting (Orlando), 9–13 Feb 1997
92. C.L. Buhrmaster, D.E. Clark, and H.B. Smartt, Spray Casting Aluminum and Al/SiC Composites, *JOM,* Vol 40 (No. 11), 1988, p 44–45
93. T.C. Willis, Spray Deposition Process for Metal Matrix Composites Manufacture, *Met. Mater.,* Aug 1988
94. T.C. Willis, J. White, R.M. Jordan, and I.R. Hughes, Microstructure and Properties of Al-SiC Composite Materials Produced by Spray Deposition, *Solidification Processing,* Institute of Metals, 1988, p 476–478
95. T.J. Warner et al., *ICCI II Proc.,* Case Western Reserve University (Cleveland), June 1988
96. Y. Wu and E.J. Lavernia, Spray-Atomized and Codeposited 6061Al/SiC$_p$ Composites, *JOM,* Vol 43 (No. 8), Aug 1991, p 16–23
97. P.A. Siemers, M.R. Jackson, R.L. Mehan, and J.R Rairden III, Production of Composite Structures by Low-Pressure Plasma Deposition, *Ceram. Eng. Sci. Proc.,* Vol 6 (No. 7–8), 1985, p 896–907
98. L. Christodoulou, P.A. Parrish, and C.R. Crowe, *High Temperature/High Performance Composites,* F.D. Lamkey, S.G. Fishman, A.G. Evans, and J.R. Strife, Ed., Vol 120, Materials Research Society, 1988, p 29
99. D.E. Larsen, M.L. Adams, S.L. Kampe, L. Chritodoulou, and J.D. Bryant, Influence of Matrix Phase Morphology on Fracture Toughness in a Discontinuously Reinforced XD Titanium Aluminide Composite, *Scr. Metall.,* Vol 24, 1990, p 851–856
100. H.L. Marcus and D.L. Bourell, Solid Preform Fabrication Finds New Application, *Adv. Mater. Process.,* Vol 144 (No. 3), Sept 1993, p 28–35
101. T. Studt, Rapid Prototyping Key to Fast Development, *R&D,* May 1994, p 55–56
102. Z.A. Munir, The Synthesis and Consolidation of Powders by Self-Propagating Combustion Methods, *Rev. Part. Mater.,* Vol 1, A. Bose, R.M. German, and A. Lawley, Ed., 1993, p 41–74
103. Z.A. Munir, *Am. Ceram. Soc. Bull.,* Vol 67, 1988, p 342
104. J.B. Holt and S.D. Dunmead, *Ann. Rev. Mater. Sci.,* Vol 21, 1991, p 305
105. Z.A. Munir and J.B. Holt, *J. Mater. Sci.,* Vol 22, 1987, p 710
106. S. Zhang and Z.A. Munir, *J. Mater. Sci.,* Vol 36, 1991, p 3380
107. A. Mortensen and M.J. Koczak, The Status of Metal Matrix Composite Research and Development in Japan, *JOM,* Vol 45 (No. 3), March 1993, p 10–18
108. M. Koizumi, Recent Progress of Functionally Gradient Materials in Japan, *Ceram. Eng. Sci. Proc.,* Vol 13, 1992, p 333–346
109. F.S. Petit, High Performance/High Temperature Materials in Japan, *Sci. Bull.,* Vol 17, 1992, p 119–130
110. R. Ford, Recent Developments in Functionally Gradient Materials, *Mater. Process. Rep.,* Vol 7, 1992, p 1–6
111. M. Sunakawa, "Current Status of R&D in Japan on Materials for Space Planes," Technical Paper 91-6096, 91-5096, American Institute of Aeronautics and Astronautics, 1991
112. G.P. Yiasemides and N.J.E. Adkins, The Application of the Graded Structures Concept to

Engineering Materials, *Advances in Powder Metallurgy & Particulate Materials,* Vol 9, Metal Powder Industries Federation, 1992, p 69–80

113. K. Atarashiya, K. Kurukawa, and T. Nagai, Functionally Gradient Material of the System Ni-MgO, Ni-NiO, Ni-Si_3N_4 or Al-AlN by Pressureless Sintering, *Ceram. Eng. Sci. Proc.,*Vol 13, 1992, p 400–407
114. M.P. Thomas and J.E. King, Improvement of the Mechanical Properties of 2124Al/SiC_p MMC Plate by Optimization of the Solution Treatment, *Compos. Sci. Technol.,* Vol 56, 1996, p 1141–1149
115. B.V.R. Bhat, Y.R. Mahajan, H.M.D. Roshan, and Y.V.R.K. Prasad, Processing Maps for Hot-Working of Powder Metallurgy 1100Al-10vol% SiC-Particulate Metal-Matrix Composite, *J. Mater. Sci.,* Vol 27, 1992, p 2141–2147
116. A.L. Geiger and J.A. Walker, The Processing and Properties of Discontinuously Reinforced Aluminum Composites, *JOM,* Vol 43 (No. 8), Aug 1991, p 8–15
117. G. Staniek and F. Lehnert, Powder Metallurgy of Aluminum Alloys, *Advanced Aerospace Materials,* H. Buhl, Ed., Springer-Verlag, 1992, p 47–58
118. B. Wang, G.M. Janowski, and B.R. Patterson, Particle Cracking in P/M Processed SiC-Reinforced Aluminum Matrix Composite Materials, *Advances in Powder Metallurgy & Particulate Materials,* Vol 6, Metal Powder Industries Federation, 1993, p 161–173
119. T.J.A. Doel, M.H. Loretto, and P. Bowen, Mechanical Properties of Aluminum Based Particulate Metal Matrix Composites, *Composites,* Vol 24 (No. 3), 1993, p 270–275
120. M. Hunt, Aerospace Composites, *Mater. Eng.,* Vol 108 (No. 6), June 1991, p 27–30
121. Z. Ishijima, H. Shikata, H. Urata, and S. Kawase, Development of P/M Forged Al-Si Alloy for Connecting Rod, *Advances in Powder Metallurgy & Particulate Materials,* Part 14, Metal Powder Industries Federation, 1996, p 3–13
122. Y. Takeda and T. Hayashi, Properties of Reaction-Sintered Al-AlN and Al-AlN-SiC Composite Alloys, *Advances in Powder Metallurgy & Particulate Materials,* Part 10, Metal Powder Industries Federation, 1996, p 41–49
123. H. Li, J.B. Li, L.Z. Sun, and Z.G. Wang, Modification of the Residual Stress State in a SiC_p/6061Al Composite by Low-Temperature Treatment, *Compos. Sci. Technol.,* Vol 57, 1997, p 165–172
124. J. Cernyar, F. Yang, and A. Saxena, Fracture and Fatigue Crack Growth Behavior of SiC_w/2124Al Composite, *J. Compos. Technol. Res.,* Vol 18 (No. 1), 1996, p 30–37
125. I.J. Polmear, *Light Alloys: Metallurgy of the Light Metals,* American Society for Metals, 1982, p 76–79
126. S. Krishnamurthy, Y.W. Kim, G. Das, and F.H. Froes, *Metal and Ceramic Matrix Composites,* R.B. Bhagat, A.H. Clauer, P. Kumar, and A.M. Ritter, Ed., TMS, 1990, p 145
127. K. Komai, K. Minoshima, and H. Ryoson, Tensile and Fatigue Fracture Behavior and Water-Environment Effects in a SiC-Whisker/7075 Aluminum Composite, *Compos. Sci. Technol.,* Vol 46, 1993, p 59–66
128. A. Sakamoto, H. Hasegawa, and Y. Minoda, Mechanical Properties of SiC Whisker Reinforced Aluminum Composites, *Fifth Int. Conf. Composite Materials, ICCM-V* (San Diego), W.C. Harrigan, Jr., J. Strife, and A.K. Dhingra, Ed., 1985, p 699–707
129. H. Kashiwaya, M. Morita, K. Hagata, T. Nishimura, T. Miyamoto, and Y. Ishiwata, Mechanical Characteristics of Bolts Made of SiC Whisker Reinforced 6061 Aluminum Alloy Composite, *Composites '86: Recent Advances in Japan and the United States,* K. Kawata, S. Umekawa, and A. Kobayashi, Ed., Proc. Japan-U.S. CCM III, Tokyo, 1986, p 529–536
130. D.F. Hasson, S.M. Hoover, and C.R. Crowe, Effect of Thermal Treatment on the Mechanical and Toughness Properties of Extruded SiC with Aluminum 6061 Metal Matrix Composite, *J. Mater. Sci.,* Vol 20, 1985, p 4147–4154
131. T.G. Nieh, K. Xia, and T.G. Langdon, Mechanical Properties of Discontinuous SiC Reinforced Aluminum Composites at Elevated Temperatures, *J. Eng. Mater. Technol.,* Vol 10, 1988, p 77–82
132. C.E. da Costa, J.M. Torralba, J.M. Ruiz-Prieto, J.M. Badia, V. Amigo, and W.C. Zapata, P/M 2014 Aluminum Alloy Obtained by Mechanical Alloying, *Advances in Powder Metallurgy & Particulate Materials,* Part 2, Metal Powder Industries Federation, 1996, p 23–31
133. P. Ramakrishnan, "Novel Processing of P/M Aluminum Matrix Composites," P/M World Congress (San Francisco), June 1992
134. M. Takeyama and C.T. Liu, Effects of Grain Size and Test Temperature on Ductility and Fracture Behavior of a B-Doped Ni_3Al Alloy, *Acta Metall.,* Vol 36 (No. 5), 1988, p 1241–1249
135. K. Sampath and W.A. Baeslack III, Joining Dispersion-Strengthened, Rapidly Solidified P/M Al Alloys, *JOM,* Vol 46 (No. 7), 1994, p 41–47
136. K.A. Khor, Z.H. Yuan, and F. Boey, Processing of Submicron SiC Reinforced Al-Li Composites by Mechanical Milling, *Processing and Fabrication of Advanced Materials IV,* T.S. Srivatsan and J.J. Moore, Ed., TMS, 1996, p 499–507
137. F.H. Froes, *Mater. Edge,* Vol 5, 1988, p 19
138. A. Bhaduri, V. Gopinathan, P. Ramakrishnan, and A.P. Miodownik, *Int. Conf. Advances in Materials and Process Processes,* ASM International, India Chapter, 16–19 Feb 1992
139. I.A. Ibrahim, F.A. Mohamed, and E.J. Lavernia, *Advanced Aluminum and Magnesium Alloys,* T. Khan and G. Effenberg, Ed., ASM International, 1990, p 745–754
140. J.L. Estrada and J. Duszczyk, Sintering of High Temperature Resistant P/M Aluminum Alloys, *Advances in Powder Metallurgy & Particulate Materials,* Part 11, Metal Powder Industries Federation, 1996, p 129–141
141. R.H. Dauskardt, R.O. Ritchie, and B.N. Cox, Fatigue of Advanced Materials: Part I, *Adv. Mater. Process.,* Vol 144 (No. 1), July 1993, p 26–31
142. J.K. Shang and G. Liu, *Role of Composite Interface in Fatigue Crack Growth, Control of Interfaces in Metal and Ceramic Composites,* R.Y. Lin and S.G. Fishman, Ed., Minerals, Metals and Materials Society/AIME, 1993, p 187–196
143. D.L. Davidson, Fatigue and Fracture Toughness of Aluminum Alloys Reinforced with SiC and Alumina Particles, *Composites,* Vol 24 (No. 3), 1993, p 248–255
144. S. Kumai, J.E. King, and J.F. Knott, Short and Long Fatigue Crack Growth in a SiC Reinforced Aluminum Alloy, *Fatigue Fract. Eng. Mater. Struct.,* Vol 13, 1990, p 511–524
145. T.J. Downes, D.M. Knowles, and J.E. King, The Effect of Particle Size on Fatigue Crack Growth in an Aluminum Based Metal Matrix Composite, *Fatigue of Advanced Materials,* R.O. Ritchie, R.H. Dauskardt, and B.N. Cox, Ed., p 395–407
146. R.M. McMeeking, Finite Deformation Analysis of Crack-Tip Opening in Elastic-Plastic Materials and Implications for Fracture, *J. Mech. Phys. Solids,* Vol 25, 1977, p 357–381
147. T.C. Willis, Spray Deposition Process for Metal Matrix Composites Manufacture, *Met. Mater.,* Vol 4, 1988, p 485–488
148. M.T. Narasimhan, R.B. Bhagat, and R.N. Pangborn, Thermomechanical Fatigue Performance of Silicon Carbide Whisker Reinforced Aluminum Composite, 1998
149. A.B. Gurcan and T.N. Baker, Wear Behavior of AA6061 Aluminum Alloy and Its Composites, *Wear,* Vol 188, 1995, p 185–191
150. A.P. Sannino and H.J. Rack, Dry Sliding Wear of Discontinuously Reinforced Aluminum Composites: Review and Discussion, *Wear,* Vol 189, 1995, p 1–19
151. C. Jin, K.-M. Cho, and I. Park, Wear Properties of High Temperature Al Alloys, *Light Weight Alloys for Aerospace Applications III,* E.W. Lee, N.J. Kim, K.V. Jata, and W.E. Frazier, Ed., TMS, 1995, p 43–55
152. D. Bialo and J. Duszczyk, Wear of Aluminum Matrix Composites Processed by Liquid Phase Sintering and Hot Extrusion, *Advances in Powder Metallurgy & Particulate Materials,* Part 16, Metal Powder Industries Federation, 1996, p 17–22
153. A. Wang and H.J. Rack, The Effect of Aging on the Abrasion Behavior of SiC_w/2124 Al Metal Matrix Composites, *Advanced Metal and Ceramic Matrix Composites: Processing, Modeling and Mechanical Behavior,* R.B. Bhagat, A.H. Clauer, P. Kumar, and A.M. Ritter, Ed., TMS, 1990, p 487–498
154. T.J. Downes and J.E. King, The Effect of Microstructure on the Fracture Toughness of a

Metal Matrix Composite, *Composites,* Vol 24 (No. 3), 1993, p 276–281

155. R.B. Bhagat and M.B. House, Elevated Temperature Mechanical Properties of Silicon Carbide Whisker Reinforced Aluminum Matrix Composites, *Mater. Sci. Eng. A,* Vol 144, 1991, p 319–326
156. R.B. Bhagat, M.B. House, and M.F. Amateau, Aging Characteristics and Mechanical Properties of Silicon Carbide Whisker Reinforced Aluminum Matrix Composites, *Proc. 1990 Powder Metallurgy Conf.* (Pittsburgh), May 1990
157. M.B. House, K.C. Meinert, and R.B. Bhagat, The Aging Response and Creep of DRA Composites, *JOM,* Aug 1991, p 24–28
158. R.B. Bhagat, M.F. Amateau, M. House, K.C. Meinert, and P. Nisson, Elevated Temperature Strength, Aging Response and Creep of Aluminum Matrix Composites, *J. Compos. Mater.,* Vol 26 (No. 11), 1992, p 1578–1593
159. R.B. Bhagat, M.F. Amateau, M.B. House, K.C. Meinert, and P. Nisson, Elevated Temperature Strength, Aging Response and Creep of Aluminum Matrix Composites, *Proc. Eighth Int. Conf. Composite Materials, ICCM VIII* (Honolulu), S.W. Tsai and G.S. Springer, Ed., SAMPE, Covina, CA, 15–19 July 1991, p 20C2–20C13
160. T. Christman and S. Suresh, Microstructural Development in and Aluminum Alloy-SiC Whisker Composite, *Acta Metall.,* Vol 36 (No. 7), 1988, p 1691–1704
161. M.J. Birt and W.S. Johnson, Characterization of the Tensile and Microstructural Properties of an Aluminum Metal Matrix Composite, *Fundamental Relationships between Microstructure and Mechanical Properties of Metal Matrix Composites,* P.K. Liaw and M.H. Gungor, Ed., TMS, 1990, p 71–88
162. T.G. Nieh and R.F. Karlak, Aging Characteristics of B4C-Reinforced 6061-Aluminum, *Scr. Metall.,* Vol 18, 1984, p 25–28
163. K.H. Oh, H.I. Lee, T.S. Kim, and T.H. Kim, Effect of SiC Whisker Reinforcement on the Precipitation Behaviors/Tensile Properties of Al-Cu-(Mg,Li) Alloys, *Fundamental Relationships between Microstructure and Mechanical Properties of Metal Matrix Composites,* P.K. Liaw and M.H. Gungor, Ed., TMS, 1990, p 115–126
164. S. Dionne, M.R. Krishnadev, and R. Bouchard, Effects of Heat Treatment on the Microstructure and Tensile Properties of A P/M SiC Particulate/7091 Al MMC, *Advanced Metal and Ceramic Matrix Composites: Processing, Modeling and Mechanical Behavior,* R.B. Bhagat, A.H. Clauer, P. Kumar, and A.M. Ritter, Ed., TMS, 1990, p 243–252
165. S.R. Nutt, Interfaces and Failure Mechanisms in Al-SiC Composites, *Interfaces in Metal-Matrix Composites,* A.K. Dhingra and S.G. Fishman, Ed., TMS, 1986, p 157–167
166. F.R.N. Nabarro and H.L. de Villiers, *The Physics of Creep,* Taylor & Francis, 1995
167. R.S. Mishra and A.K. Mukherjee, Creep Behavior of Rapidly Solidified and Processed Aluminum Alloys, *Light Weight Alloys for Aerospace Applications III,* E.W. Lee, N.J. Kim, K.V. Jata, and W.E. Frazier, Ed., TMS, 1995, p 319–332
168. R.S. Mishra, A.G. Paradkar, and K.N. Rao, Steady State Creep Behavior of a Rapidly Solidified and Further Processed Al-5 wt% Ti Alloy, *Acta Metall. Mater.,* Vol 41, 1993, p 2243–2251
169. M.K. Premkumar, M.J. Koczak, and A. Lawley, Elevated Temperature Mechanical Behavior of P/M Dispersion Strengthened Al-Fe-Ni Alloys, *High Strength Powder Metallurgy Aluminum Alloys II,* G. Hildeman and M.J. Koczak, Ed., TMS, 1986, p 265–284
170. M.K. Premkumar, A. Lawley, and M.J. Koczak, Mechanical Behavior of Powder Metallurgy Al-Fe-Ni Alloys, *Mater. Sci. Eng. A,* Vol 174, 1994, p 127–139
171. D. Legzdina and T.A. Parthasarathy, Deformation Mechanisms of a Rapidly Solidified Al-8.8Fe-3.7Ce Alloy, *Metall. Trans. A,* Vol 18, 1987, p 1713–1719
172. D. Legzdina and T.A. Parthasarathy, Effect of Cerium Content on the Deformation Behavior of Rapidly Solidified Al-Fe-Ce Alloys, *Metall. Trans. A,* Vol 21, 1990, p 2155–2158
173. G.M. Pharr et al., High Temperature Creep Deformation of a Rapidly Solidified Al-Fe-V-Si Alloy, *Dispersion Strengthened Aluminum Alloys,* Y.-W. Kim and W.M. Griffith, Ed., TMS, 1988, p 309–322
174. S.C. Khatri et al., Creep and Microstructural Stability of Dispersion Strengthened Al-Fe-V-Si-Er Alloy, *Mater. Sci. Eng. A,* Vol 167, 1993, p 11–21
175. R.S. Mishra, H. Jones, and G.W. Greenwood, Creep of a Rapidly Solidified and Further Processed Al-Cr-Zr Alloy, *Int. J. Rapid Solidif.,* Vol 5, 1990, p 149–162
176. Y.C. Chen, M.E. Fine, and J.R. Weertman, Microstructural Evolution and Mechanical Properties of Rapidly Solidified Al-Zr-V Alloys at High Temperatures, *Acta Metall. Mater.,* Vol 38, 1990, p 771–780
177. W.W. Park and H. Jones, The Creep Behavior of Rapidly Solidified Al-Zr-V Alloys at Low Stresses, *Mater. Sci. Eng. A,* Vol 134, 1991, p 1229–1233
178. J.E. Benci and W.E. Frazier, Evaluation of a New Aluminum Alloy for 700 °F Aerospace Applications, *Light Weight Alloys for Aerospace Applications,* E.W. Lee and N.J. Kim, Ed., TMS, 1991, p 231–245
179. T.G. Nieh, Creep Rupture of a Silicon Carbide Reinforced Aluminum Composite, *Metall. Trans. A,* Vol 15, Jan 1984, p 139–145
180. V.C. Nardone and J.R. Strife, Analysis of the Creep Behavior of Silicon Carbide Whisker Reinforced 2124 Al (T4), *Metall. Trans. A,* Vol 18, 1987, p 109–114
181. T. Morimoto, T. Yamaoka, H. Lilholt, and M. Taya, Second Stage Creep of SiC Whisker/6061 Aluminum Composite at 573 K, *J. Eng. Mater. Technol.,* Vol 110, April 1988, p 70–76
182. E.P. Barth, J.T. Morton, and J.K. Tien, Threshold for Creep Resistance in a Silicon Carbide Reinforced Aluminum Alloy, *Fundamental Relationships between Microstructures and Mechanical Properties of Metal Matrix Composites,* TMS, 1990
183. R.E. Smallman, *Modern Physical Metallurgy,* 4th ed., Butterworth, 1985
184. W.A. Spitzig, C.L. Trybus, and J.D. Verhoeven, Deformation-Processed Metal/Metal Composites, *Metal Matrix Composites: Processing and Interfaces,* R.K. Everett and R.J. Arsenault, Ed., Academic Press, 1991, p 151–179
185. A. Jokinen and B. Wiik, "Formability of Aluminum Alloy Matrix Composites," P/M World Congress (San Francisco), June 1992
186. K. Higashi, M. Mabuchi, and T.G. Langdon, High-Strain-Rate Superplasticity in Metallic Materials and the Potential for Ceramic Materials, *ISIJ Int.,* Vol 36 (No. 12), 1996, p 1423–1438
187. T.G. Nieh, J. Wadsworth, and O.D. Sherby, *Superplasticity in Metals and Ceramics,* Cambridge University Press, 1997
188. T.K. Ha and Y.W. Chang, High Temperature Deformation Behavior of 8090 Al-Li Alloy, *Light Weight Alloys for Aerospace Applications III,* E.W. Lee, N.J. Kim, K.V. Jata, and W.E. Frazier, Ed., TMS, 1995, p 281–289
189. M. Mabuchi and T. Imai, Superplasticity of Si_3N_4 Whisker Reinforced 6061 Aluminum at High Strain Rate, *J. Mater. Sci. Lett.,* Vol 9, 1990, p 761–762
190. T. Imai and M. Mabuchi, Superplasticity in Silicon Nitride Whisker Reinforced 2124 Aluminum Alloy Composite, *J. Jpn. Inst. Light Met.,* Vol 39, 1989, p 831–835
191. T.F. Zahrah and L. Christodoulou, Modeling and Design of P/M Consolidation Processes, *Advances in Powder Metallurgy & Particulate Materials,* Vol 3, Metal Powder Industries Federation, 1993, p 77–83
192. M.G. McKimpson and K.L. Paxton, Effect of Reinforcement Distribution on the Densification Behavior of Metal-Ceramic Composites at Isostatic Pressures Greater than 200 MPa, *Advances in Powder Metallurgy & Particulate Materials,* Vol 3, Metal Powder Industries Federation, 1993, p 215–228
193. A.S. Helle, K.E. Easterling, and M.F. Ashby, *Acta Metall.,* 1985
194. M.F. Ashby, *HIP 6.0 Background Reading,* Engineering Dept., Cambridge University, 1990
195. E. Arzt, *Acta Metall.,* Vol 30, 1982, p 1883
196. E. Arzt, M.F. Ashby, and K.E. Easterling, *Metall Trans. A,* Vol 14, 1983, p 211
197. L.T. Kuhn and R.M. McMeeking, *Int. J. Mech. Sci.,* Vol 34, 1991, p 563
198. N.A. Fleck, L.T. Kuhn, and R.M. McMeeking, *J. Mech. Phys. Solids,* Vol 40, 1991, p 1139
199. R.M. McMeeking and L.T. Kuhn, *Acta Metall. Mater.,* Vol 40, 1991, p 961

200. A.L. Gurson, *Int. J. Mech. Sci.*, Vol 14, 1977, p 215
201. P. Sofornis and R.M. McMeeking, *J. Appl. Mech.*, Vol 59, 1992, p 888
202. H. Riedel, *Ceramic Powder Science III*, G.L. Messing, Ed., 1990, p 619
203. W.P. Li, M.F. Ashby, and K.E. Easterling, *Acta Metall.*, Vol 35 (No. 12), 1987
204. T.F. Zahrah, C.J. Coe, and F.H. Charron, The Role of Process Models in Intelligent Hot Isostatic Pressing of Powder Materials, *Proc. Symposium on Applications of Mechanics and Materials Models to Design and Processing*, TMS, 4–5 March 1992
205. J. Xu and R. McMeeking, Modeling Powder Consolidation and the Formation of Composite Materials, *Advances in Powder Metallurgy & Particulate Materials*, Vol 3, Metal Powder Industries Federation, 1993, p 201–214
206. R.B. Bhagat and G. Rajesh, 1998 (unpublished work)
207. K.C. Meinert, Jr., R.B. Bhagat, and R. Martukanitz, "Laser Processing of Discontinuously Reinforced Aluminum Composites," Technical Report, File No. TR 94-001, Applied Research Laboratory, Pennsylvania State University, University Park, PA, 1994
208. R.B. Bhagat, R.P. Martukanitz, and K.C. Meinert, Jr., Laser Processing of Discontinuously Reinforced Aluminum Composites, *ARL Review*, 1993
209. K.C. Meinert, Jr., R.P. Martukanitz, and R.B. Bhagat, Laser Processing of Discontinuously Reinforced Aluminum Composites, *Proc. American Society for Composites Seventh Technical Conf. Composite Materials*, The Pennsylvania State University, University Park, PA, 13–15 Oct 1992, p 168–177
210. K.C. Meinert, Jr., R.P. Martukanitz, R.B. Bhagat, and J. Eckert, "Laser Welding of SiC Reinforced Aluminum Composites," presented at Aeromat '92 (Anaheim, CA), 18–21 May 1992
211. R.P. Martukanitz and R.B. Bhagat, Laser Processing of Discontinuously Reinforced Aluminum Matrix Composites, *The Metal Science of Joining*, M.J. Cieslak, J.H. Perepezko, S. Kang, and M.E. Glicksman, Ed., TMS, 1992, p 241–248
212. K.C. Meinert, Jr., R.P. Martukanitz, and R.B. Bhagat, Laser Cutting and Welding of Discontinuously Reinforced Aluminum Composites, *TMS Conf. Proceedings* (Denver), Spring 93
213. K. Sampath and W.A. Baeslack III, Weldability of a RS-PM Al-8Fe-2Mo Alloy, *Weld. J.*, Vol 72 (No. 8), 1993, p 416s–427s
214. M.B.D. Ellis, Joining of Aluminum Based Metal Matrix Composites, *Int. Mater. Rev.*, Vol 41 (No. 2), 1996, p 41–58
215. J. Ahern, C. Cook, and S.G. Fishman, *Met. Constr.*, Vol 14, 1982, p 192
216. T.S. Luhman, R.L. Williams, and K.B. Das, "Development of Joint and Joint Techniques for Metal Matrix Composites," AMMRC-TR 84-35, Army Laboratory, Boston, Aug 1984
217. D.M. Schuster, M.D. Skibo, and W.R. Hoover, Production and Semi-Fabrication of an Aluminum Composite Material: Extrusion and Casting Considerations, *Light Metal Age*, Feb 1989, p 15–19
218. *Arc Welding Guidelines*, Duralcan, 1991
219. J.T. Staley, J. Liu, and W.H. Hunt, Jr., Aluminum Alloys for Aerostructures, *Adv. Mater. Process.*, Oct 1997, p 17–20
220. M.V.D. Berg, Aluminum Composite Doubles Lifetime of F-16 Ventral Fin, *Adv. Mater. Process*, Oct 1996, p 7

Copper Powder Metallurgy Alloys and Composites

A. Nadkarni, OMG Americas, Inc.

COPPER-BASE POWDER METALLURGY (P/M) products rank third after iron and steel, and aluminum-base P/M products in terms of volume. The estimated shipments of copper and copper-base alloy powders in 1996 were about 20,500 Mg (23,000 short tons) in North America. The shipments in Europe were estimated to be about 14,000 Mg (15,600 short tons) while the shipments in Japan were estimated at about 6200 Mg (7000 short tons) (Ref 1).

The use of copper in the P/M industry dates back to the 1920s, when commercial porous bronze bearings were developed independently in the research laboratories of General Motors Corporation (Ref 2, 3) and Bound-Brook Oilless Bearing Company (Ref 4). These self-lubricating bearings still account for the major portion of P/M copper and copper alloy applications. Other important applications for copper and copper-base P/M materials summarized in Table 1 include friction materials, brushes, filters, structural parts, electrical parts, additives to iron powders (alloying as well as infiltration), catalysts, paints, and pigments.

This article briefly reviews copper-base P/M alloys and composites in terms of the powder types, their processing, and properties. Major applications (such as self-lubricating bearings and filters) are briefly described (Ref 5, 6). For more detailed information on application performance, see the articles "Porous Powder Metallurgy Materials" and "Powder Metallurgy Bearings" in this Volume.

In general, physical and mechanical properties of near-full (theoretical) density copper and copper alloy P/M structural parts are comparable to cast and wrought copper-base materials of similar composition. However, P/M copper parts vary in density from the low density typical of self-lubricating bearings or filters to the near-full density of electrical parts. The physical and mechanical properties depend greatly on the density as a percentage of theoretical density.

Copper Powders

Copper powders of 99+% purity are commercially available. The four major methods of producing such copper powders are:

- Atomization
- Oxide reduction
- Hydrometallurgy
- Electrolysis

Of the above methods, atomization and oxide reduction are presently practiced on a large scale globally. Hydrometallurgical and electrolytic copper powders have not been manufactured in the

Table 1 Applications of copper-base powders

Application	Powder type
Automotive and machinery	
Bearings and bushings	Bronze, copper-lead, and copper-lead-tin
Brake bands and linings	Copper, brass, copper-lead, copper-lead-tin
Bushings	Bronze
Control panels/instruments	Nickel-silver
Counterweights	Copper-tungsten
Chemical and industrial	
Chemical catalyst	Copper
Filters	Bronze
Flame arrestors	Bronze
Fungicide additive	Copper
Soil conditioning	Copper
Valves and pumps	Copper-nickel
Coatings and paints	
Antifouling paints	Copper
Conductive paints and plastics	Copper and brass
Decorative paints	Copper, brass, bronze
Lacquers	Brass, bronze
Mechanical (peen) plating	Copper, brass
Spray coating	Copper, brass
Vacuum metallizing	Copper
Construction hardware	
Conductive and nonsparking floors	Copper
Decorative plastics	Copper, bronze, and brass
Domestic water filters	Brass
Locks and keys	Brass, bronze
Nuts	Brass
Pipe joint compounds	Copper
Electrical and electronic	
Brushes	Copper
Brush holders	Nickel-silver
Contacts	ODS Copper
Heatsinks	Copper
Printed circuits	Copper
Semiconductor stud bases	Copper, dispersion-strengthened copper
Telephone components	Brass, bronze
Switch and contactor components	Copper, brass
Incandescent lamps	ODS Copper
X-ray, microwave tube parts	ODS Copper
Lubricants	
Antigalling pipe joint compounds	Copper
Copper lubricants	Copper
Plastic-filled metal	Copper, bronze
Lubricants (continued)	
Self-lubricating (oil-filled) parts	Bronze
Manufacturing and machining	
Abrasive wheel bonding	Copper
Brazing compounds	Copper, bronze, brass
Electrodischarge machining	Copper
Eletrochemical machining	Copper
Resistance welding electrodes	Copper, dispersion-strengthened copper
Ordnance	
Armor-piercing cores	Copper
Fuse parts	Brass
Projectile rotating bands	Copper, brass
Other equipment or uses	
Business machines	Brass
Coins, metals, medallions	Copper-nickel
Cordless devices	Copper
Fingernail polish	Copper
Lawn and garden equipment	Bronze
Photographic equipment	Bronze, brass, nickel-silver

United States since the early 1980s, but they are produced in Europe and Japan.

Table 2 shows a comparison of some of the typical fundamental powder characteristics of commercial copper powders made by the four production processes. Each process produces a unique particle shape and surface area.

Brief descriptions of each process and relevant powder properties for the four major processes are discussed below. For additional information on these areas, see the article "Production of Copper Powder" in this Volume.

Atomization. In this process, molten copper flows through a refractory nozzle, and the liquid stream is disintegrated into droplets by an impinging jet of water or gas. The droplets solidify into powder particles. The size and shape of these particles is governed by the atomizing medium, pressure, and flow rate. Gas atomization produces spherical particles, while the shape of water-atomized powder particles can be controlled from almost spherical to irregular by controlling the interaction between the water jet and the metal stream. Higher pressures and lower flow rates favor finer powders; average particle sizes less than 325 mesh (45 μm) are feasible.

Water is the preferred atomizing medium for producing copper powder. The atomized powder is often subjected to an elevated temperature reduction (to reduce any oxides formed during atomization) and agglomeration treatment to improve its compacting properties. Table 3 shows the typical properties of commercial water-atomized copper powders.

Oxide Reduction. This process involves oxidizing copper (particulate copper scrap, electrolytic copper, atomized copper, etc.) and milling the oxide into powder. The oxide powder is reduced at elevated temperature by solid or gaseous reducing agents, and the resulting sintered cake is milled into powder. The particle size is controlled through milling of the starting oxide and the reduced sinter cake. The milled copper powder particles are irregular and porous. A broad range of pore characteristics can be obtained by controlling the reduction conditions.

The particle size, shape, and pore characteristics determine the compacting properties of the powder and the part produced. Table 4 shows the typical properties of a commercial copper powder produced by the oxide-reduction process.

Hydrometallurgy. This process produces copper powder from copper ore concentrates or scrap copper. Copper is leached from these materials with sulfuric acid or ammoniacal solutions, and the resulting solution is separated from the residue by filtration. The solution is passed over iron scrap to precipitate copper as powder, generally called cement copper. The wet powder is dried in a reducing atmosphere, then milled, classified, and blended to obtain the desired particle size distribution.

The powder has fine particle size, with relatively low apparent density and high green strength. Low purity in general, and high iron content in particular, restrict the use of cement copper in P/M applications. Its irregular particle shape and high specific surface area, however, impart good green strength and make it useful in friction applications.

Electrolysis. In this process copper is electrodeposited to obtain a spongy powder deposit at the cathode rather than a smooth, adherent deposit. The electrolyte is typically an aqueous solution with a low copper ion concentration and a high sulfuric acid concentration. This, coupled with a high cathode current density, promotes formation of a loose powdery deposit. The anode and the cathode are pure electrolytic copper with the cathode being typically one tenth the size of the anode (Ref 7). After deposition, the powder is washed to remove all traces of the electrolyte and can be annealed in a reducing atmosphere. The reduced copper product is milled, screened, and blended to the desired particle size distribution.

Electrolytic copper powder is generally of high purity; however, impurities more noble than copper are codeposited. Powder morphology is typically dendritic. A wide range of powders with varying apparent densities and high green strengths can be obtained by electrolysis. For applications requiring a powder with a low apparent density and a high surface area, additions to the electrolytic bath can be made that decrease dendrite arm thickness.

Copper Alloy Powders

Alloy powders are available in various compositions, including brasses, nickel silvers, tin bronzes, aluminum bronzes, and beryllium bronzes. Alloy powders are produced by either (1) preblending copper powders with other elemental powders such as tin, zinc, or nickel or (2) prealloying during powder production.

Preblending. Preblended powders are mixtures of selected compositions, with or without lubricant, that form the desired alloy during sintering. The most common P/M copper alloy

Table 2 Characteristics of commercial copper powders

Type of powder	Composition, % Copper	Oxygen	Acid insolubles	Particle shape	Surface area
Electrolytic	99.1–99.8	0.1–0.8	0.03 max	Dendritic	Medium to high
Oxide reduced	99.3–99.6	0.2–0.6	0.03–0.1	Irregular; porous	Medium
Water atomized	99.3–99.7	0.1–0.3	0.01–0.03	Irregular to spherical; solid	Low
Hydrometallurgical	97–99.5	0.2–0.8	0.03–0.8	Irregular agglomerates	Very high

Source: Ref 6

Table 3 Properties of commercial grades of water-atomized copper powders

Copper, %	Chemical properties, % Hydrogen loss	Acid insolubles	Physical properties Hall flow rate, s/50 g	Apparent density, g/cm³	Tyler sieve analysis, % +100	−100+150	−150+200	−200+325	−325
99.65(a)	0.28	...	...	2.65	Trace	0.31	8.1	28.2	63.4
99.61(a)	0.24	...	...	2.45	0.2	27.3	48.5	21.6	2.4
99.43(a)	0.31	...	...	2.70	Trace	0.9	3.2	14.2	81.7
>99.1(b)	<0.35	<0.2	~50	2.4	<8	17–22	18–30	22–26	18–38

(a) Water atomized plus reduced. (b) Contains magnesium. Source: Ref 6

Table 4 Properties of commercial grades of copper powder produced by the oxide reduction process

Copper	Chemical properties, % Tin	Graphite	Lubricant	Hydrogen loss	Acid insolubles	Physical properties Apparent density, g/cm³	Hall flow rate, s/50 g	Tyler sieve analysis, % +100	+150	+200	+325	−325	Compacted properties Green density, g/cm³	Green strength, MPa (psi), at: 165 MPa (12 tsi)	6.30 g/cm³
99.53	...	...	...	0.23	0.04	2.99	23	0.3	11.1	26.7	24.1	37.8	6.04	6.15 (890)	...
99.64	...	...	...	0.24	0.03	2.78	24	...	0.6	8.7	34.1	56.6	5.95	7.85 (1140)(a)	...
99.62	...	...	...	0.26	0.03	2.71	27	...	0.3	5.7	32.2	61.8	5.95	9.3 (1350)(a)	...
99.36	...	...	...	0.39	0.12	1.56	...	0.1	1.0	4.9	12.8	81.2	5.79	21.4 (3100)(a)	...
99.25	...	...	...	0.30	0.02	2.63	30	0.08	7.0	13.3	16.0	63.7	...	...	8.3 (1200)(a)
90	10	...	0.75	...	...	3.23	30.6	0.0	1.4	9.0	32.6	57.0	6.32	...	3.80 (550)
88.5	10	0.5	0.80	...	...	3.25	12(b)	...	...	...	...	...	...	...	3.6 (525)

(a) Measured with die wall lubricant only. (b) Carney flow. Source: Ref 6

Table 5 Physical properties of typical brass, bronze, and nickel silver alloy compositions

Property	Brass(a)	Bronze(a)	Nickel silver(a)(b)
Sieve analysis, %			
+100 mesh	2.0 max	2.0 max	2.0 max
−100+200	15–35	15–35	15–35
−200+325	15–35	15–35	15–35
−325	60 max	60 max	60 max
Physical properties			
Apparent density	3.0–3.2	3.3–3.5	3.0–3.2
Flow rate, s/50 g	24–26	...	...
Mechanical properties			
Green density(c) at 415 MPa (30 tsi), g/cm^3	7.6	7.4	7.6
Green strength(c) at 415 MPa (30 tsi), MPa (psi)	10–12 (1500–1700)	10–12 (1500–1700)	9.6–11 (1400–1600)

(a) Nominal mesh sizes, brass, −60 mesh; bronze, −60 mesh; nickel silver, −100 mesh. (b) Contains no lead. (c) Compressibility and green strength data of powders lubricated with 0.5% lithium stearate. Source: Ref 6

made with preblended powders is tin bronze used in self-lubricating bearings. Typical bronze composition is 90Cu-10Sn, often containing up to 1.5% graphite. Some "dilute" bronze bearings contain various amounts of iron replacing some of the copper and tin. Copper-lead and steel-backed Cu-Pb-Sn materials, used to replace solid bronze bearings, also use preblended powders because lead is virtually insoluble in copper and cannot be prealloyed. Friction materials used in brakes and clutches contain disparate materials such as copper with several other components including lead, tin, iron, graphite, molybdenum disulfide, oxides, and so forth. These can only be made by using preblended powders.

Prealloying. Prealloyed powders are generally produced by melting the constituents to form a homogeneous alloy and atomizing the alloy melt by the methods similar to those used for the production of copper powder. They can also be produced by sintering preblended powders and grinding the materials to attain the desired powder characteristics. Some typical commercial prealloyed powders are discussed below.

Brass and Nickel Silver. Air atomization is generally used for making prealloyed powders of brass and nickel silver for use in high-density (>7.0 g/cm^3) components. The low surface tension of the molten alloys of these compositions renders the particle shape sufficiently irregular to make the powders compactible. Reduction of oxides is not necessary for the standard P/M grades of brass and nickel silver powders.

Commercial prealloyed brass and nickel silver powders are available in leaded and nonleaded compositions. Commercial brass alloys range from 90Cu-10Zn to 65Cu-35Zn. Leaded versions of 80Cu-20Zn and 70Cu-30Zn are most commonly used for the manufacture of sintered structural parts that may require secondary machining operations. The only commercially available nickel silver powder has a nominal composition of 65Cu-18Ni-17Zn, which is modified by addition of lead when improved machinability is required.

Bronze. Prealloyed bronze powders are not used widely for structural parts fabrication because their nodular particle form and high apparent density result in low green strength. However, blends of such powders with irregular copper powders and phosphorus-copper yield sintered parts with good mechanical properties.

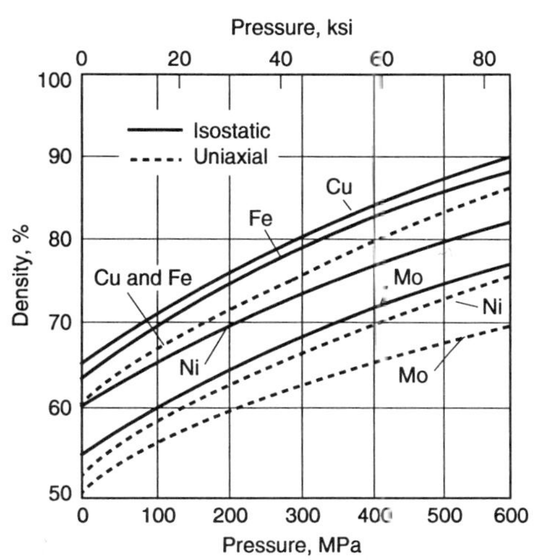

Fig. 1 Density as a function of pressure for isostatic and unidirectional pressing. Source: Ref 9

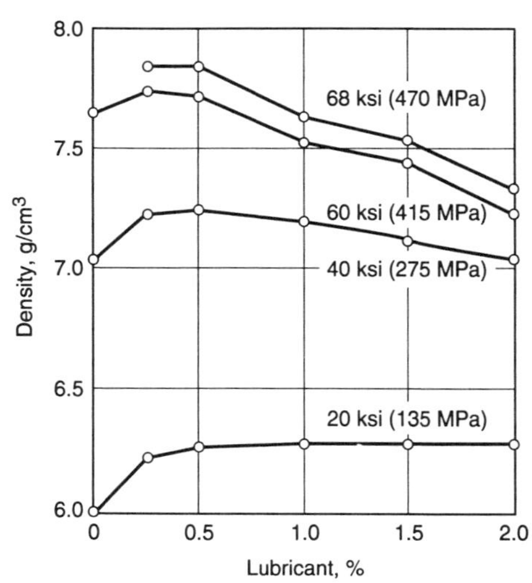

Fig. 2 Density of compact as a function of amount of lubricant (zinc stearate) in a copper-tin powder mix for various compacting pressures with ½ in. steel diam (1.7 μm finish) and ¼ in. compacts (double-end compression). Source: Ref 10

Table 5 shows typical properties of commercial grades of prealloyed brass, bronze, and nickel silver powders.

Powder Pressing

Copper and copper alloy powders are generally cold compacted in closed dies with top and bottom punches used to apply the desired pressure. During cold compaction the powder particles mechanically interlock with each other. Typical compacted densities are 80 to 90% of the theoretical density of the material. When higher densities are required, consolidation methods such as hot pressing, extrusion, hot isostatic pressing (HIP), and hot forging are used. The amount of densification depends on several processing variables such as compaction pressure, powder characteristics, types and amounts of lubricants, and so forth. These basic factors are briefly summarized below for copper-base powders, while more general coverage on consolidation is discussed in the Section "Shaping and Consolidation Technologies" in this Volume.

Compressibility is a measure of the extent to which a mass of powder can be densified by the application of pressure. The pressure required to achieve a given green density depends on the powder particle size distribution, inherent hardness of the material, lubricants, and compaction method (e.g., isostatic or uniaxial). At a given compacting pressure, the density of a green compact generally is:

- Lower for fine particles than for coarse particles (Table 6)
- Lower for hard powders than for soft powders (Fig. 1)
- Higher with lubricants, but there is an optimal lubricant level (Fig. 2)
- Higher for isostatic pressing than for uniaxial pressing (Fig. 1)

Copper powder is relatively soft compared to iron, nickel, and molybdenum powders and thus reaches a higher green density for a given compaction pressure (Fig. 1). Typical compacting pressures and compression ratios (green density of compact/apparent density of powder) for copper and copper alloy powders are summarized in Table 7. The ranges in Table 7 are only typical, as actual pressing requirements depend on the variables noted above.

Sintering

The compacted parts are sintered at elevated temperatures under protective conditions to avoid oxidation. During this process the powder particles are metallurgically bonded to each other. Sintering is generally understood as a three-phase process. Initially, neck growth between particles proceeds rapidly, but powder particles retain their identity. In the second phase, when most of the densification takes place, the pores become rounded and grain growth occurs, whereupon powder particles lose their separate identities and diffuse into each other. In the third phase, isolated round pores shrink, and densification proceeds at a slower rate.

The principal driving force for these changes is the decrease in free surface area (decrease in solid/gas interface) when particles grow together and voids shrink under the influence of surface tension or capillary forces. Material transport occurs during sintering by one or more mechanisms from among the following (Ref 12):

- Diffusional flow
- Evaporation/condensation
- Viscous flow
- Plastic flow

In most types of sintering, the most important mechanism is diffusion, which may occur through a lattice volume, along a free surface, or along a grain boundary. All three types of diffusion are governed by Fick's law, which is expressed as:

$$D = D_o \exp(-Q/RT)$$

where D is the diffusion coefficient, D_o is a material constant known as the jump frequency, Q is the activation energy for the type of diffusion (volume, surface, or grain boundary), R is the gas constant, and T is the absolute temperature. The activation energy for volume diffusion is the highest, the one for grain-boundary diffusion is lower, and the one for surface diffusion is still lower.

Volume and surface diffusion coefficients for copper, as a function of temperature, are shown in Fig. 3. Values for jump frequencies and activation energies for self-diffusion in copper and for the common alloying elements in copper are given in Table 8. Coefficients for grain-boundary diffusion lie between those for volume and surface diffusion. Generally, diffusion in metals proceeds more rapidly along grain boundaries (i.e., planar defects that are structurally only several atoms thick), and the jump frequency of atoms diffusing along the grain boundaries can be a million times greater than the jump frequency of regular lattice atoms at $2T_m/3$ (where T_m is the melting temperature). Because of the very high atomic mobility at grain boundaries relative to that in the bulk, grain-boundary diffusion plays a crucial role in the kinetics of microstructural changes during metallurgical processing such as sintering and solid-state reactions.

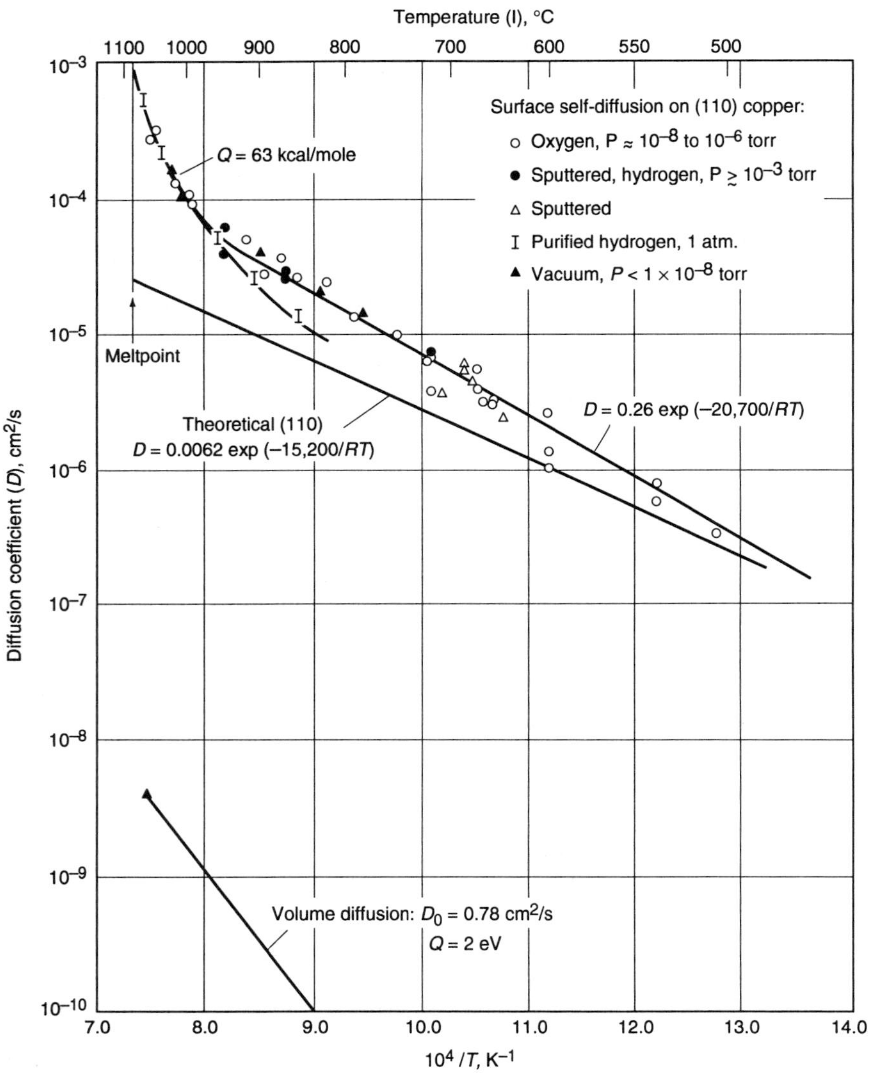

Fig. 3 Volume and surface diffusion coefficients of copper. Source: Ref 13

Activated Sintering. This process involves techniques aimed at accelerating the sintering rate by decreasing the activation energy for diffusion. It allows lower sintering temperatures, shorter sintering times, or better properties such as strength, electrical conductivity, and so forth. The rate of sintering can be modified by either physical or chemical treatments of the powder or compact or by adding reactive gases in the sintering atmosphere.

Activated sintering has been observed in copper powders having thin films of copper oxide on the particle surfaces (Ref 14). Oxide layers of 40 to 60 nm show improved properties in sintered compacts as shown in Table 9. The copper oxide is partly soluble in the copper matrix, and the oxygen from the oxide film diffuses into the copper, leaving behind highly active copper surfaces that sinter more efficiently. Thicker layers inhibit sintering as the inferior oxide to oxide contacts overcompensate for the activation effects. Small additions of copper oxide to the copper powders also show a similar activation effect.

Liquid-Phase Sintering. In this process, a compact made from a mixture of two or more powders is sintered at a temperature below the melting point of the high-melting-point constituent, but above that of the low-melting-point constituent. The sequence of events in heating a mixed phase begins with a mixture of elemental powders, such as copper-tin or copper-iron premixes. When the liquid forms, it flows to wet the solid particles and begins acting on the solid phase. Melt penetration between the solid particles can cause swelling of the compact. Along

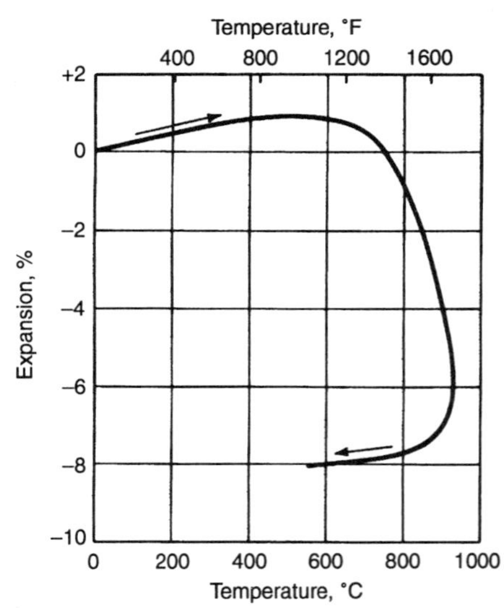

Fig. 4 Dimensional change during sintering of compacts from −74+43 mm copper powder. Pressed at 138MPa (20 ksi) and heated at a rate of 3.9 °C/min (7 °F/min) to 925 °C (1700 °F) and then cooled at the same rate. MPa = 6.8947 ksi

Table 6 Properties of pressed-and-sintered compacts of various copper powder sizes

Particle size	Compacting pressure, MPa	Compacting pressure, tsi	Green density, g/cm³	Sintered density, g/cm³	Sintered hardness, HRH
Copper, 2 μm	69	5	4.95	7.72	29
	138	10	5.54	7.80	38
	275	20	6.29	7.95	51
	414	30	6.94	8.18	61
Copper, 44–74 μm	69	5	5.38	6.58	1
	138	10	6.07	7.04	12
	275	20	6.88	7.67	31
	414	30	7.65	7.98	42
Copper, 2 μm, Tin, 44–74 μm	68	5	5.06	6.24	26
	138	10	5.75	6.24	...
	275	20	6.52	6.34	...
	414	30	7.01	6.40	33
Copper and tin, 44–74 μm	69	5	5.90	6.12	14
	138	10	6.45	6.18	...
	275	20	7.25	6.13	...
	414	30	7.65	6.04	17

Source: Ref 8

Table 7 Typical compacting pressure and compression ratios of various copper P/M parts

P/M parts	Compacting pressure, MPa	Compacting pressure, tsi	Compression ratio
Brass parts	414–689	30–50	2.4–2.6 to 1
Bronze bearings	193–275	14–20	2.5–2.7 to 1
Copper-graphite brushes	345–414	25–30	2.0–3.0 to 1
Pure copper parts	206–248	15–18	2.6–2.8 to 1

Source: Ref 11

with melt penetration, there is rearrangement of particles. The combination of wetting, liquid flow, and rearrangement contributes to a rapid change in the volume of the compact.

In the copper-tin system, the tin melts and alloys with copper to form a bronze, with accompanying expansion of the compact. In the iron-copper system, the copper melts, becoming saturated with iron, and the copper-iron alloy diffuses into the iron skeleton, causing expansion of the skeleton. Pores remain at the sites vacated by the copper.

In both the copper-tin and iron-copper systems, growth or shrinkage of compacts can be altered by the addition of carbon in the form of graphite. Thus, graphite can be used to control dimensional changes in these systems. In the Cu-Sn-C system, sintering is inhibited by mechanical separation of the constituents; as a result, expansion increases. In the Fe-Cu-C system, the amount of the liquid phase is increased by the formation of a ternary Fe-Cu-C eutectic that restricts expansion. See the article "Production Sintering Practices" in this Volume for detailed information regarding sintering of copper-base materials.

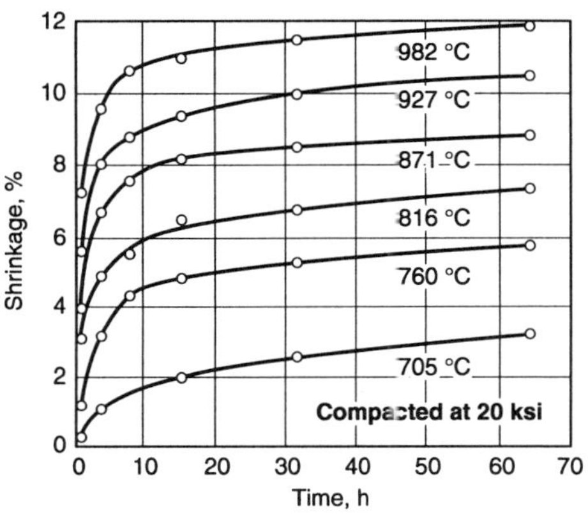

Fig. 5 Linear shrinkage of copper powder compacts from –75+44 μm sieve fraction. Source: Ref 17

Homogenization (Interdiffusion). When alloying involves blending of elemental powders, then the sintering step serves not only to increase the contact area between powder particles and promote densification, but also to partially or completely homogenize the alloy ingredients by interdiffusion between powder particles of the constituent elements. As in densification, homogenization is influenced by sintering time, sintering temperature, and particle size. Finer particles can greatly accelerate homogenization by reducing the diffusion distance.

Whether it is economical to produce an alloy by sintering compacts from elemental powders or from homogeneous alloy powders depends primarily on the coefficient for interdiffusion of the alloy ingredients. The difference in diffusion coefficients is a deciding parameter whether solid phase or transient liquid phase sintering can be used successfully. This influence of interdiffusion (homogenization) on densification depends on the difference between the diffusion coefficients of the single elements in a powder mixture. In a nickel-cobalt system ($D_{Co} \approx D_{Ni}$), for example, the interdiffusion exerts only a minor influence on densification. The shrinkage behavior of nickel-cobalt compacts lies between those of nickel and cobalt compacts. There is nearly linear correlation between shrinkage and concentration. Therefore, the densification process of the mixtures is dominated by the same

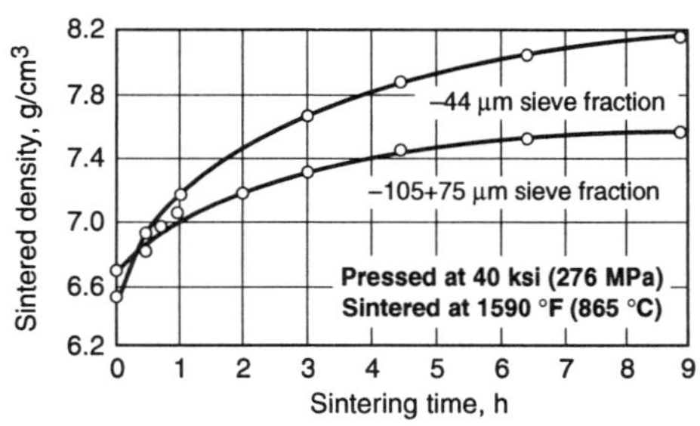

Fig. 6 Density of compacts from electrolytic copper powder. Source: Ref 18

defect-activated material transport mechanisms as in single-component systems.

However, in a copper-nickel system ($D_{Cu} > D_{Ni}$), the effect of interdiffusion is much stronger and depends on whether the component with the higher diffusivity (copper) is the minor or the major component. The faster diffusing species leaves porosity behind (Kirkendall porosity), causing expansion of the compact. An 80%Ni-20%Cu system shows more interdiffusion-induced porosity, while diffusion porosity is less pronounced in a 20%Ni-80%Cu system.

Densification. When compacts from a single metal powder are sintered, they undergo dimensional change as sintering proceeds. Figure 4 shows the dimensional changes that occur during sintering of copper powder compacts. In the initial stage, as the temperature increases, the compacts expand much like solid copper. The entrapped gases in the isolated pores and vaporized lubricant also contribute to this expansion, particularly in compacts pressed to high densities. When sintering begins, the compacts begin to shrink and reach a maximum at the peak sintering temperature. During cooling, the compacts contract like solid copper would. The total result of the three stages is generally a shrinkage and higher density.

Densification of P/M compacts during sintering depends on several variables, but principal factors are:

- Sintering temperature
- Sintering time
- Powder particle size
- Green density of the compact, which is primarily a function of compacting pressure

Typical sintering temperatures and times for various alloys are given in Table 10; copper alloys are generally sintered at much lower temperatures than iron- and nickel-base alloys. Typical shrinkage of a copper compact is shown in Fig. 5 for various sintering temperatures and times. The rate of shrinkage is initially high, but then decreases with increasing sintering time. Higher sintering temperatures promote a more rapid shrinkage than do longer sintering times.

Another factor affecting densification is the particle size of the powder. In Fig. 6, the sintered density of copper powder compacts pressed at 276 MPa (40 ksi) from two particle size fractions, –105 + 75 and –44 μm, and sintered at

Table 8 Diffusion constants in copper

Diffusing element	Volume diffusion		Surface diffusion	
	Dv_0, cm^2/s	Qv, kcal/mole	Ds_0, cm^2/s	Qs, kcal/mole
Copper (self-diffusion)	0.6–0.78	50.5–50.9	1000	39
Nickel in copper	6.5×10^{-5}	30	...	...
Tin in copper	4.1×10^{-3}	31	...	...
Zinc in copper	2.4×10^{-3}	30	...	...
Bronze	6×10^{-6}	25	800	49

865 °C (1590 °F) is plotted as a function of sintering time. Densification of compacts from the finer powder is faster than that of compacts from the coarser powder and, for sintering times above ½ h, the final density is considerably higher for the compacts from the finer powder.

A final factor affecting the densification of compacts from a single metal powder is the green density of the compact or the pressure at which the compact is pressed, which determines green density. In Fig. 7(a) and (b), dimensional change is plotted versus sintering temperature for various compacting pressures. Higher compacting pressures result in higher green and sintered densities, but the sintering shrinkage, or the change from green to sintered density, is smaller. As previously noted, expansion may also occur when soft metal powders are compacted at higher pressures as shown in Fig. 7(b) for compaction pressures of 550 and 620 MPa (80 and 90 ksi). This effect is associated with expansion of the entrapped gases in the isolated pores.

Pure Copper

Pure copper P/M parts are used mainly in electrical and electronic applications because of their high electrical conductivity. It is essential to use very pure copper powders (≥99.95% purity) or to bring about the precipitation of soluble impurities during sintering. As little as 0.023% Fe in solid solution in copper lowers its conductivity to 86% of that of pure copper. Small amounts of iron mechanically mixed with the copper powder lower the conductivity much less, unless the iron dissolves in the copper during sintering. If high-purity copper is used, or if soluble impurities are precipitated during sintering, it is possible to obtain the strength and conductivity values shown in Fig. 8.

Conductivity is directly related to porosity; the greater the void content (lower the density), the lower the conductivity. Electrical conductivity of pure copper parts pressed at moderate pressures of 205 to 250 MPa (15 to 18 tsi) and sintered at 800 to 900 °C (1500 to 1650 °F) varies from 80 to 90% IACS on a scale where conductivity of solid annealed copper is 100% IACS. The conductivity of solid copper can be reached or approached in P/M copper parts by sintering the pressed parts at higher temperatures, such as 930 to 1030 °C (1700 to 1900 °F), followed by repressing, coining, or forging.

Typical applications of pure copper parts in which high electrical conductivity is required include commutator rings, contacts, shading coils, nose cones, and electrical twist-type plugs. Copper powders also are used in copper-graphite compositions that have low contact resistance, high current-carrying capacity, and high thermal conductivity. Typical applications include brushes for motors and generators and moving parts for rheostats, switches, and current-carrying washers.

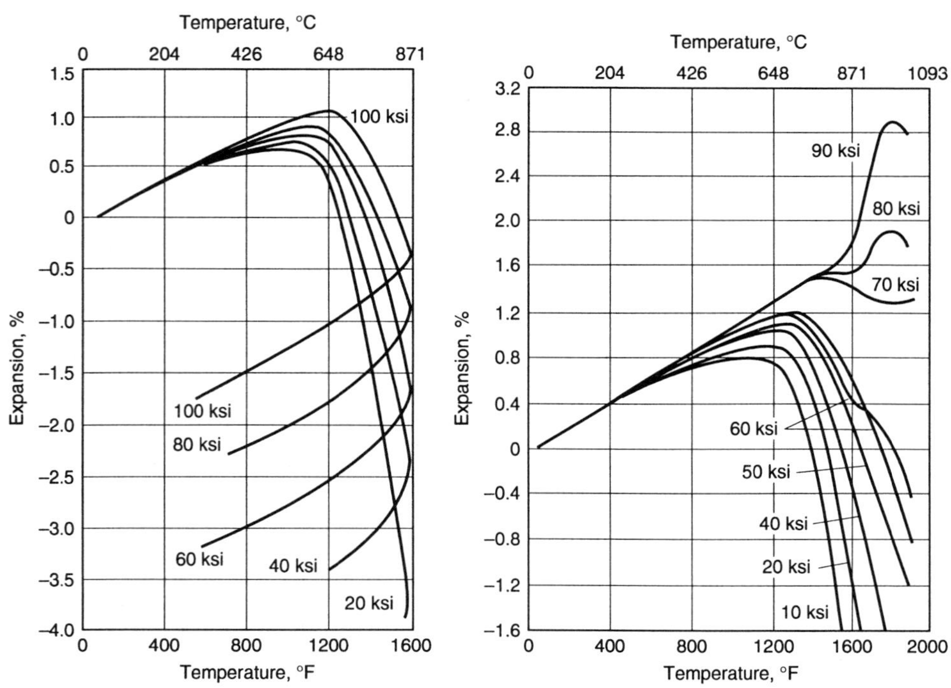

Fig. 7 Sintering curves for copper powder compacts at various compacting pressures. Source: Ref 19

Bronze

Powder metallurgy bronzes typically originate as premixes consisting of elemental copper and tin powders plus 0.5 to 0.75% dry organic lubricants such as stearic acid or zinc stearate. Some structural parts, however, requiring densities >7.0 g/cm^3 are fabricated from prealloyed powders. Prealloyed powders have higher yield strengths and work-hardening rates than premixed powders. Therefore, pressing loads required to achieve given green densities in prealloyed powders are higher than the pressures required for elemental powders. Differences in pressing characteristics of premixed and prealloyed powders are compared in Fig. 9.

Typical sintering furnace temperatures for bronze range from 815 to 870 °C (1500 to 1600 °F); total sintering time within the hot zone may range from 15 to 30 min, depending on the furnace temperature selected, required dimensional

Table 9 Effect of copper powder oxide thickness on strength of press-and-sintered compacts

Thickness of oxide film on powder, nm	Tensile strength of sintered compact(a)	
	MPa	ksi
Coarse copper powder		
0	29	4.2
40	51	7.4
80	38	5.5
120	21.5	3.1
160	15.7	2.2
200	7.8	1.1
Fine copper powder		
0	120	17.4
20	130	18.8
60	143	20.7
100	137	19.8
140	129	18.7
180	122	17.7
220	114	16.5

(a) Pressed at 4.4 tons/cm^2, sintered at 600 °C for ½ h in H_2. Source: Ref 15

Table 10 Typical sintering temperature and time of copper alloys and steels

Material	Temperature		Time, min
	°C	°F	
Bronze	760–870	1400–1600	10–20
Copper	840–900	1550–1650	12–45
Brass	840–900	1550–1650	10–45
Iron, iron-graphite, etc.	1010–1150	1850–2100	30–45
Nickel	1010–1150	1850–2100	30–45
Stainless steel	1095–1285	2000–2350	30–60

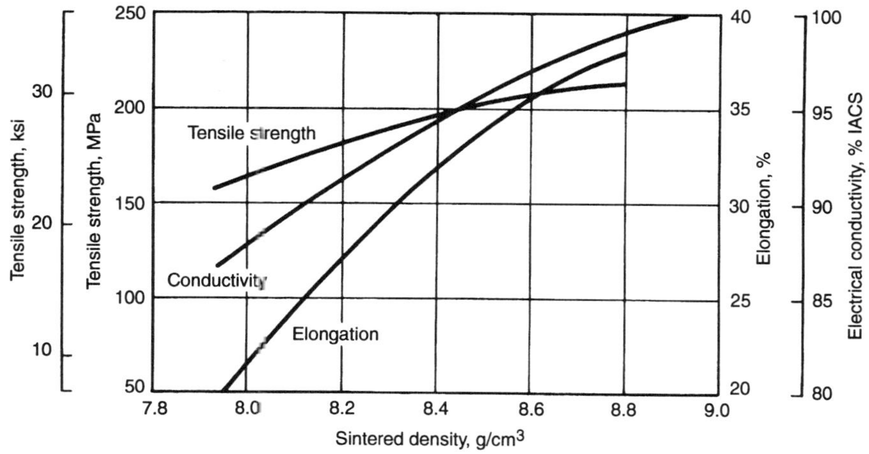

Fig. 8 Effect of density on electrical conductivity and tensile strength properties of P/M copper. Source: Ref 5

change, and most importantly, the presence of an optimal α bronze grain structure. Sintering atmospheres should be protective and reducing to facilitate sintering. Reduction of the copper oxides that may surround each copper powder particle and of tin oxide allow for increased diffusion rates. Typical strength/density data for 90Cu-10Sn sintered bronzes with and without graphite additions are shown in Fig. 10. Control of sintered dimensions in premix systems is achieved by manipulating sintering time and/or temperature.

Generally, copper-tin blends composed of relatively coarser powders sinter to higher growth values than a blend composed of finer powders. After powder blends have been tested and adjusted to provide an approximation of target dimensions, final adjustments are made during production sintering to obtain dimensional precision. Factors affecting the ultimate dimensional values include physical characteristics of the constituents and compacted density.

Bearings. Self-lubricating porous bronze bearings continue to consume the major portion of the copper powder produced each year. These bearings are made by pressing elemental powder blends of copper and tin, followed by sintering. The most widely used bearing material is 90Cu-10Sn bronze, often with the addition of up to 1.5% graphite. So-called dilute bronze bearings contain various amounts of iron. Dilution with iron reduces the cost of a bearing at the expense of some loss in performance.

Compaction pressures for the bronze powders range from about 140 to 415 MPa (20 to 60 ksi). Sintering is typically done in a continuous mesh belt furnace at temperatures between 815 to 870 °C (1500 to 1600 °F) for about 3 to 8 min at temperature. Typical furnace atmospheres are dissociated ammonia or endothermic gas. To obtain reproducible sintering results, it is important to carefully control time and temperature because of their influence upon the kinetics of the homogenization process, which in turn determines the dimensional changes taking place during sintering. Most bearings are sized for improved dimensional accuracy; typical sizing pressures may range from about 200 to 550 MPa (30 to 80 ksi). Bearings are sold either dry or saturated with oil. The pores are filled with oil by a vacuum impregnation process. Most common bearings range in density from 5.8 to 6.6 g/cm^3 dry or 6.0 to 6.8 g/cm^3 oil impregnated. This range corresponds to about 25 to 35% pore volume.

Figure 11 shows an assortment of bronze bearings. The most common shapes are simple or flanged bushings, but some have spherical external surfaces. Sizes range from about 0.8 to 75 mm (1/32 to 3 in.) in diameter. Typical applications of sintered bronze bearings are shown in Table 11.

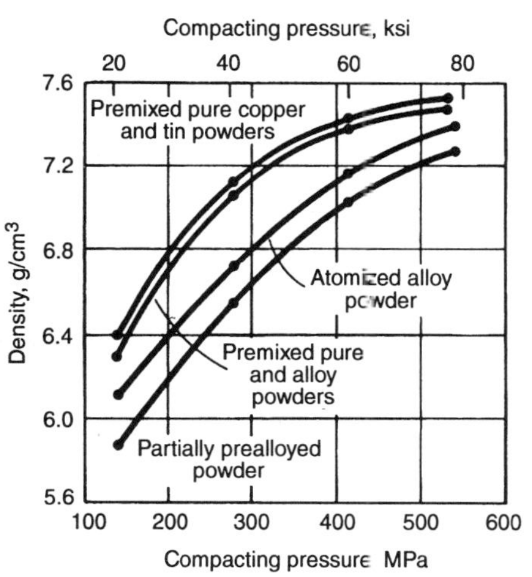

Fig. 9 Pressing characteristics of premixed and prealloyed 90Cu-10Sn powders. Source: Ref 20

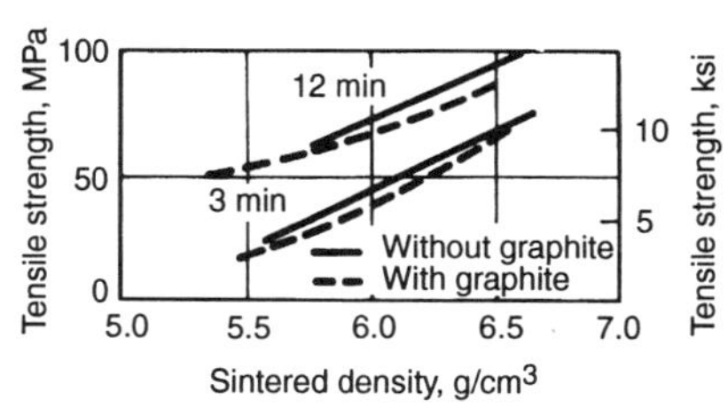

Fig. 10 Effect of density on the strength of copper-tin and copper-tin-graphite compacts. Source: Ref 21

Table 11 Applications of self-lubrication sintered bronze bearings (fractional horsepower electric motors)

Automotive components	**Farm and lawn equipment (continued)**
Starters	Cotton pickers
Light generators	Lawn mowers
Oil and water pumps	String cutters
Windshield wipers	Chain saws
Hood and window raisers	**Consumer electronics**
Heaters	Phonographs
Air conditioners	Record changers
Power antennae	Tape recorders
Power seat adjusters	**Business machines**
Home appliance	Typewriters
Dishwashers	Computers
Clothes dryers	Copiers
Washing machines	**Industrial equipment**
Sewing machines	Textile machines
Vacuum cleaners	Packaging machines
Refrigerators	Electric fans
Food mixers	Portable power tools
Farm and lawn equipment	Drills
Tractors	Saws
Combines	

Filters. Filters constitute one of the major applications for porous P/M parts. The ability to achieve close control of porosity and pore size is the main reason filters are made from metal powders. Most producers of nonferrous filters prefer atomized spherical powders with closely controlled particle size to allow production of filters within the desired pore size range. The effective pore size of filters generally ranges from 5 to 125 μm.

Tin bronze is the most widely used P/M filter material, but nickel silver, stainless steel, Cu-Sn-Ni alloys, and nickel-base alloys also are used. The

Fig. 11 Assorted P/M bronze bearings

major advantage of P/M bronze materials over other porous metals is cost. Porous P/M bronze filters can be obtained with tensile strengths ranging from 20 to 140 MPa (3 to 20 ksi) and appreciable ductility, up to 20% elongation. Powder metallurgy bronze also has the same corrosion resistance as cast bronze of the same composition and thus can be used in a wide range of environments.

Bronze filters usually are made by gravity sintering of spherical bronze powders, which are generally made by atomization of molten prealloyed bronze. These powders typically contain 90 to 92% Cu and 8 to 10% Sn. Filters made from atomized bronze have sintered densities ranging from 5.0 to 5.2 g/cm^3. To produce filters with the highest permeability for a given maximum pore size, powder particles of a uniform particle size must be used.

Although not widely used, coarser powders for bronze filters can be obtained by chopping copper wire and tumbling the choppings. Filters made from tin-coated cut copper wire with tin contents ranging from 2.5 to 8% are also used to a lesser extent. Filters made from these materials have sintered densities ranging from 4.6 to 5.0 g/cm^3.

During sintering the filters shrink slightly—as much as 8%. To avoid excessive shrinkage, filters from powders with fine particle size require lower sintering temperatures in the neighborhood of 815 °C (1500 °F). Because of the shrinkage during sintering, filters must be designed with a slight draft, so they can be removed from the mold.

Figure 12 shows an assortment of P/M bronze filters. Such filters are commonly used to filter gases, oils, refrigerants, and chemical solutions. They have been used in fluid systems of space vehicles to remove particles as small as 1 μm. Bronze diaphragms can be used to separate air from liquids or mixtures of liquids that are not emulsified. Only liquids capable of wetting the pore surface can pass through the porous metal part.

Bronze filter materials can be used as flame arrestors on electrical equipment operating in flammable atmospheres, where the high thermal conductivity of the bronze prevents ignition. They can also be used as vent pipes on tanks containing flammable liquids. In these applications, heat is conducted away rapidly so that the ignition temperature is not reached.

Structural Parts. Powder metallurgy bronze parts for structural applications frequently are selected because of corrosion and wear resistance of bronze. They are generally produced by methods similar to those used for self-lubricating bearings. Typical compositions of bronze structural parts (CT-1000) are included in Table 12, and the typical properties are shown in Table 13.

Figure 13 shows an assortment of P/M bronze structural parts. These parts are generally used in automobile clutches, copiers, outboard motors, and paint-spraying equipment.

Brass and Nickel Silver

In contrast to bronze structural parts, parts made from brass, leaded brass, and nickel silver are produced from prealloyed atomized powder. Compositions of some common brass and nickel silver alloys used in structural parts are shown in Table 12. The leaded compositions are used whenever secondary machining operations are required.

The alloy powders are usually blended with lubricants in amounts from 0.5 to 1.0 wt%. Lithium stearate is the preferred lubricant because of its cleansing and scavenging action during sintering. However, bilubricant systems are common, such as lithium stearate and zinc stearate, which minimize the surface staining attributed to excessive lithium stearate. Lubricated powders are typically compacted to 75% of theoretical density at 207 MPa (30 ksi) and to 85% of theoretical density at 415 MPa (60 ksi).

Sintering of brass and nickel silver compacts is normally performed in protective atmospheres. Dissociated ammonia, endothermic gas, and nitrogen-base atmospheres are most common. Temperatures range from 815 to 925 °C (1500 to 1700 °F) depending on alloy composition. To avoid distortion and/or blistering of the compacts, sintering temperatures should not exceed the solidus temperature of the alloy. Through multiple pressing and sintering operations, yield strength and hardness approaching those of the wrought alloy counterparts can be achieved. To minimize zinc losses during sintering, yet allow for adequate lubricant removal, protective-sintering-tray arrangements are used. Table 13 shows the typical properties of common brass and nickel silver P/M parts.

Next to bronze bearings, brasses and nickel silvers are the most widely used materials for structural P/M parts. Typical applications include hardware for latch bolts and cylinders for locks; shutter components for cameras; gears, cams, and actuator bars in timing assemblies and in small generator drive assemblies; and decorative trim and medallions. In many of these applications, corrosion resistance, wear resistance, and aesthetic appearance play important roles. Figure 14 shows some typical P/M brass components.

Copper-Nickel

Copper-nickel P/M alloys containing 75Cu-25Ni and 90Cu-10Ni have been developed for coinage and corrosion-resistance applications. The 75Cu-25Ni alloy powder pressed at 772 MPa (112 ksi) has a green density 89% of its theoretical density. After sintering at 1090 °C (2000 °F) in disassociated ammonia, elongation is 14%, and apparent hardness is 20 HRB. Re-pressing at 772 MPa (112 ksi) increases density to 95%. This alloy has the color of stainless steel and can be burnished to a high luster. The 90Cu-10Ni alloy has a final density of 99.4% under similar pressing-and-sintering conditions. It has a bright

Fig. 12 Assorted filters made from P/M bronze. Courtesy of Arrow Pneumatics, Inc.

Fig. 13 Assorted P/M bronze parts. Courtesy of Norddeutsche Affinerie

bronze color and also can be burnished to a high luster.

In one method of producing coins, medals, and medallions, a mixture containing 75% Cu and 25% Ni powders is blended with zinc stearate lubricant and compressed, sintered, coined, and resintered to produce blanks suitable for striking. These blanks are softer than rolled blanks because they are produced from high-purity materials. Therefore, they can be coined at relatively low pressures, and it is possible to achieve greater relief depth with reduced die wear.

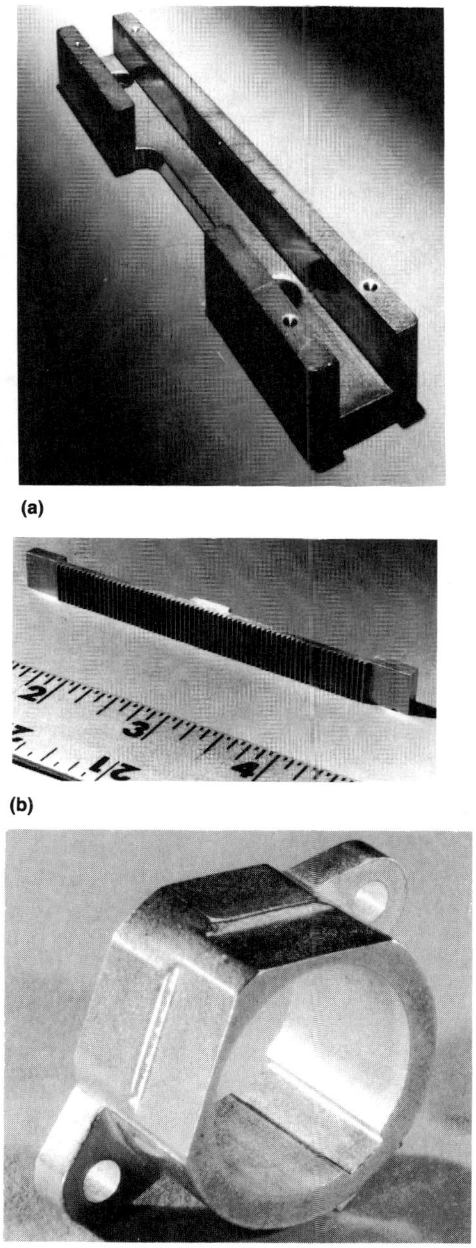

Fig. 14 P/M brass components. (a) Rack guide for rack-and-pinion steering column. (b) Leaded brass guide for stereomicroscope. (c) Leaded brass guide for microscope. Courtesy of Metal Powder Industries Federation

In another procedure, an organic binder is mixed with copper or copper-nickel powders and rolled into "green" sheets. Individual copper and copper-nickel sheets are pressed together to form a laminate, and blanks are punched from it. Blanks are heated in hydrogen to remove the organic binder and sinter the material. The density of the "green" blanks is low (45% of theoretical), but coining increases density to 97%. After pressing, the blanks are annealed to improve ductility and coinability.

Copper-Lead

Copper and lead, which have limited solubilities in each other, are difficult to alloy by conventional ingot metallurgy. Copper-lead powder mixtures have excellent cold pressing properties; they can be compacted at pressures as low as 76 MPa (11 ksi) to densities as high as 80% of theoretical density. After sintering, they can be re-pressed at pressures as low as 152 MPa (22 ksi) to produce essentially nonporous bearings.

Steel-backed copper or Cu-Pb-Sn P/M materials are sometimes used to replace solid bronze bearings. They are produced by spreading the powder in a predetermined thickness on a steel strip, sintering, rolling to theoretical density, resintering, and annealing. The end product has a residual porosity of about 0.25%. Blanks of suitable size are cut from the bimetallic strip, formed, and drilled with oil holes or machined to form suitable grooves. These materials include Cu-25Pb-0.5Sn, Cu-25Pb-3.5Sn, Cu-10Pb-10Sn, and Cu-50Pb-1.5Sn alloys.

Copper-Base Friction Materials

Sintered-metal-base friction materials are used in applications involving the transmission of motion through friction (clutches) and for deceleration and stopping (brakes). In these applications mechanical energy is converted into frictional heat, which is absorbed and dissipated by the friction material. Copper-base materials are preferred because of their high thermal conductivity; however, lower cost iron-base materials have been developed for moderate to severe duty dry applications.

Most friction materials contain copper powders blended with other metal powders, solid lubricants, oxides, and other compounds. These constituents are immiscible in each other and therefore can only be made by powder metallurgy. Compositions of some common copper-base friction materials are shown in Table 14.

Mixtures of the appropriate powders are carefully blended to minimize segregation of the constituents. Fine metal powders with high surface area are necessary to provide a strong and thermally conductive matrix. The blended powders are compacted at pressures ranging from 165 to 275 MPa (24 to 40 ksi).

Bell-type sintering furnaces are used where the friction facing is bonded to a supporting steel backing plate such as in clutch disks. The green disks are placed on the copper-plated steel plates and stacked. Pressure is applied on the vertical stack of disks. Sintering temperatures range from 550 to 950 °C (1020 to 1740 °F) in a protective atmosphere. Typical sintering times are 30 to 60 min. The sintered parts are typically machined for dimensional accuracy and surface parallelism.

The friction elements are usually brazed, welded, riveted, or mechanically fastened to the supporting steel members. They may also be pressure bonded directly to the assembly.

The operating conditions encountered by metal-base friction materials can be classified as dry/wet and mild/moderate/severe. Figure 15 shows some typical applications and the corresponding operating conditions. Some examples of copper-base friction elements are shown in Fig. 16.

Copper-Base Contact Materials

Electrical contacts are metal devices that make and break electrical circuits. Arcing, except in

Table 12 Compositions of copper-base P/M structural materials (brass, bronze, and nickel silver)

Material designation	Chemical composition, %					Element
	Cu	Zn	Pb	Sn	Ni	
CZ-1000	88.0	bal	...	...	...	min
	91.0	bal	...	...	...	max
CAP-1002	88.0	bal	1.0	...	...	min
	91.0	bal	2.0	...	...	max
CAP-2002	77.0	bal	1.0	...	...	min
	80.0	bal	2.0	...	...	max
CZ-3000	68.5	bal	...	...	...	min
	71.5	bal	...	...	...	max
CZP-3002	68.5	bal	1.0	...	...	min
	71.5	bal	2.0	...	...	max
CNZ-1818	62.5	bal	...	...	16.5	min
	65.5	bal	...	...	19.5	max
CNZP-1816	62.5	bal	1.0	...	16.5	min
	65.5	bal	2.0	...	19.5	max
CT-1000	87.5	bal	...	9.5	...	min
	90.5	bal	...	10.5	...	max

Note: Total by difference equals 2.0% max, which may include other minor elements added for specific purposes. Source: Ref 22 applications with low potential or current levels, is a major problem, particularly during opening of the contacts in a live circuit. The arc causes erosion of the contacts by blowing away the molten metal or by vaporizing the material. Welding can occur during closing of the contacts when small areas on the contact surfaces that are molten during arcing fuse together during closure (Ref 25).

Applications involving severe arcing and welding employ contacts made from refractory metals, such as tungsten and molybdenum, which have high melting and boiling points, and excellent resistance to arc erosion. Oxides are often used to prevent welding of the contacts. In both cases, a high conductivity metal such as silver or copper is used in conjunction with the refractory metal or oxide to provide the necessary current-carrying capability. Composites such as these are made by P/M techniques because the individual constituents are immiscible and cannot be made by conventional melt-cast process. Applications where arcing and welding are not severe utilize contacts made of pure metals or alloys. These are generally made by melting and casting followed by suitable metalworking processes.

Copper-base materials are used in electrical contacts because of their high electrical and thermal conductivities, low cost, and ease of fabrication. Their main drawbacks are poor resistance to oxidation and corrosion. Therefore, copper-base contacts are used in applications where the voltage drop resulting from the oxide film is acceptable or where it is possible to protect the contact, such as by immersion in oil or by enclosing the contact in a protective gas or vacuum.

Common copper alloys used in contacts include yellow brass (C27000), phosphor bronze (C51000), and copper beryllium alloys (C17200 and C17500). These are made by the melt-cast process and are limited to lower current applications where arcing and welding are not severe.

Composites of copper with refractory metals or their carbides are used in applications in which limited oxidation of the copper is acceptable or where oxidation is prevented by one of the methods mentioned above. Table 15 presents the compositions, properties, and typical applications for some P/M composite contact materials. The properties of the contacts depend on the manufacturing method used, and therefore the common methods used are also shown in the table. The specific method used depends on the composition of the composite. Generally, materials with 40% or less tungsten or its carbide are manufactured by the conventional pressing, sintering (generally below the melting point of copper), and re-pressing (PSR) technique. Materials containing more than 40% W are generally made by infiltrating (INF) the copper into either loose

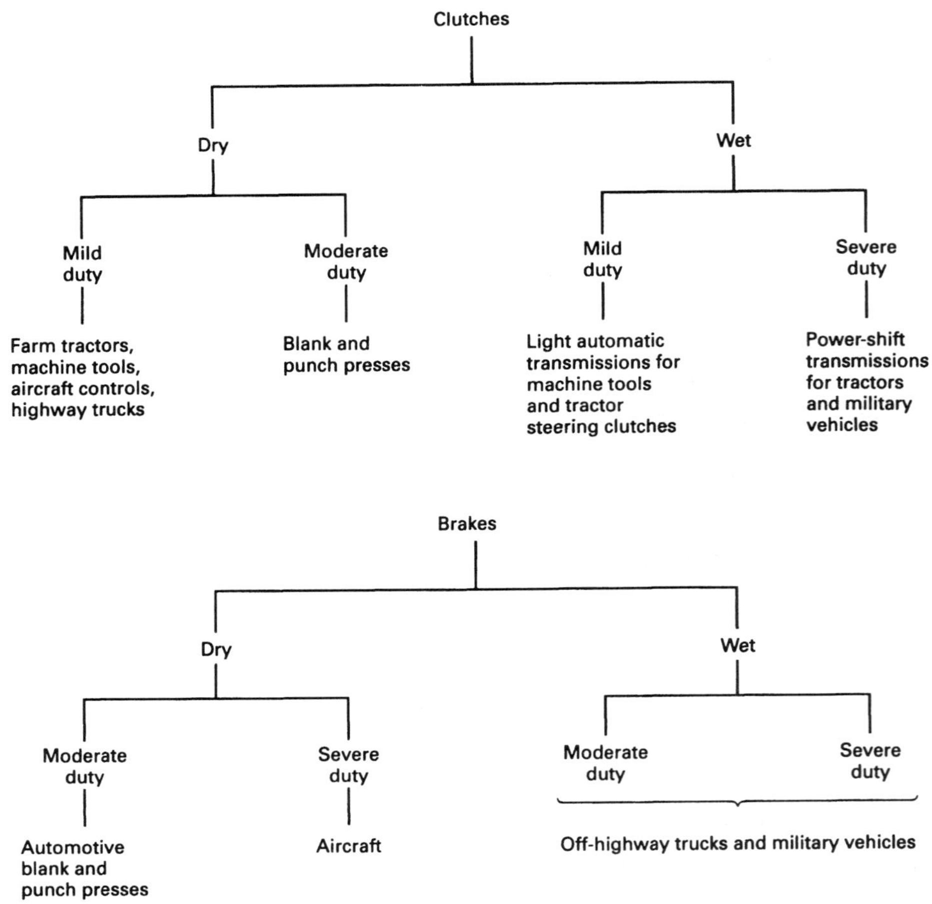

Fig. 15 Applications of sintered metal friction materials. Source: Ref 24

Table 13 Properties of copper-base P/M structural materials (brass, bronze, and nickel silver)

Mechanical property data derived from laboratory-prepared test specimens sintered under commercial manufacturing conditions

Material designation code(a)	Minimum yield strength, MPa	Minimum yield strength, ksi	Ultimate tensile strength, MPa	Ultimate tensile strength, ksi	0.2% yield strength, MPa	0.2% yield strength, ksi	Elongation in 25 mm (1 in.), %	Young's modulus, GPa	Young's modulus, 10^6 psi	Transverse rupture strength, MPa	Transverse rupture strength, ksi	Unnotched Charpy impact strength, J	Unnotched Charpy impact strength, ft · lbf	Density g/cm^3	Compressive yield strength, MPa	Compressive yield strength, ksi	Apparent hardness, HRH
			Typical values														
CZ-1000-9	62	9	124	18.0	65	9.5	9.0	52	7.5	270	39	(b)	(b)	7.60	(b)	(b)	65
CA-1000-10	70	10	138	20.0	76	11.0	10.5	69	10.0	315	46	(b)	(b)	7.90	(b)	(b)	72
CZ-1000-11	75	11	159	23.0	83	12.0	12.0	(b)	(b)	360	52	(b)	(b)	8.10	(b)	(b)	80
CZP-1002	(b)	(b)	(b)	(b)	(b)	(b)	(b)	(b)	(b)	(b)	(b)	(b)	(b)	(b)	(b)	(b)	(b)
CZP-2002-11	75	11	159	23.0	93	13.5	12.0	69	10.0	345	50	38	28.0	7.60	103	15.0	75
CZP-2002-12	83	12	207	30.0	110	16.0	14.5	83	12.0	480	70	76	56.0	8.00	110	16.0	84
CZ-3000-14	97	14	193	28.0	110	16.0	14.0	62	9.0	425	62	31	23.0	7.60	83	12.0	84
CZ-3000-16	110	16	234	34.0	131	19.0	17.0	69	10.0	590	86	51.5	38.0	8.00	90	13.0	92
CZP-3002-13	90	13	186	27.0	103	15.0	14.0	62	9.0	395	57	(b)	(b)	7.60	(b)	(b)	80
CZP-3002-14	97	14	217	31.5	115	16.5	16.0	69	10.0	490	71	(b)	(b)	8.00	(b)	(b)	88
CNZ-1818-17	117	17	234	34.0	140	20.0	11.0	75	11.0	500	73	32.5	24.0	7.90	172	25.0	90
CNZP-1816	(b)	(b)	(b)	(b)	(b)	(b)	(b)	(b)	(b)	(b)	(b)	(b)	(b)	(b)	(b)	(b)	(b)
CT-1000-13 (re-pressed)	90	13	152	22.0	110	16.0	4.0	38	5.5	310	45	5.4	4.0	7.20	186	27.0	82

(a) Suffix numbers represent minimum yield-strength values in ksi. (b) Additional data in preparation will appear in subsequent editions of MPIF standard 35. Source: Ref 22 tungsten powder or pressed-and-sintered tungsten compacts. Their counterparts using tungsten carbide are made by infiltrating the copper into loose powder because the tungsten carbide powder cannot be pressed into compacts.

Copper-Base Brush Materials

Brushes are components that transfer electrical current between the stationary and rotating elements in electric motors and generators. Most common brushes are made from composites of graphite and a conductive metal. The graphite provides the required lubrication, and the metal provides the current-carrying capability; copper and silver are preferred metals because of their high electrical conductivity.

Table 16 shows the compositions of some typical copper-graphite brush materials. The copper content varies from about 20 to 75%, the balance being graphite. Powder metallurgy is the only way to produce these materials because of the immiscibility of the two components. Copper powder used in the brushes could be made by oxide reduction, electrodeposition, atomization, or flaking.

The manufacture of brushes involves blending the copper and graphite powders. These are molded into brushes or large blocks, typically at pressures ranging from 100 to 200 MPa (15 to 30 ksi), providing green densities of 2 to 4 g/cm^3. The molded parts are sintered at 500 to 800 °C (950 to 1500 °F) in a protective atmosphere. Machining is performed if necessary to achieve the final dimensional tolerances.

All metal-graphite brush applications must operate at lower voltages than nonmetallic brush materials. Typical voltages range from near 0 to 30 V and can be emanating from a direct current power source such as a battery, a rectifier, or a sensing device such as a thermocouple.

Copper-graphite brushes are widely used in battery-powered tools that require high power outputs in small, lightweight packages. Typically, input voltage influences the metal content required. High voltages require a low metal content, whereas low voltages require a high metal content. Below 9 V, the metal content is usually higher than 80% while above 18 V, the metal content is generally below 50%.

Copper-graphite brushes are also used extensively in automotive applications including starter motors, blower motors, doorlocks, and windshield-wiper motors. The starter motors generally use high copper content grades to enable them to handle extremely high-current densities for short periods of time. Blower motors use lower copper content grades to extend service life to several thousand hours. The doorlocks and windshield-wiper motors use grades that are between the blower and starter motor grades.

Infiltrated Parts

Iron-base P/M parts can be infiltrated with copper or a copper alloy by placing a slug of the infiltrant on the part and then sintering above the melting point of the infiltrant. The molten infiltrant is completely absorbed in the pores by capillary action and a composite structure is created. The amount of infiltrant used is limited by the pore volume in the starting iron part and typically ranges between 15 and 25%.

Infiltration increases the density of the part, resulting in improved mechanical properties, corrosion resistance, electrical and thermal conductivities, machinability, and brazeability. Tensile strengths ranging from 480 to 620 MPa (70 to 90 ksi) can be achieved in iron-base parts infiltrated with 15 to 25% Cu.

Infiltration is used for iron-base structural parts that must have densities greater than 7.4 g/cm^3. Typical examples include gears, automatic transmission components, valve seat inserts, automobile door hinges, and so forth. More information is in the article "Infiltration" in this Volume.

ODS Copper Materials

Copper is widely used in industry because of its high electrical and thermal conductivities, but it has low strength, particularly when heated to high temperatures. It can be strengthened by using finely dispersed particles of stable oxides such as alumina, titania, beryllia, thoria, or yttria in the matrix. Because these oxides are immiscible in liquid copper, dispersion-strengthened copper cannot be made by conventional ingot metallurgy; P/M techniques must be used.

Manufacture. Oxide dispersion strengthened (ODS) copper can be made by simple mechani-

Table 14 Compositions of copper-base friction materials for wet and dry applications

Country	Composition, wt %							Use(a)
	Cu	Sn	Fe	Pb	Graphite	MoS_2	Other	
USSR	65–80	7–9	4–7	5–10	3–8	...	2–4 SiO_2	W, D
	70	9	4	6	4	...	3SiO_2, 3 asbestos	W
	60	10	4	5	4	...	9 asbestos, 8 bakelite powder	W
East Germany	81.5	4.5	...	5	4	...	5 mullite	W
	bal	...	...	5	12	...	8 MgO; 5 Ti	W, D
United States	60–75	4–10	5–10	...	3–10	3–12	2–7 SiO_2	D
	52.5	...	...	7.5	...	...	5 SiO_2; 15 Bi	W
	72	4.7	3.3	3.5	8.7	1.4	1.9 SiO_2; 0.2 Al_2O_3	W, D
	72	7	3	6	6	...	3 SiO_2; 4 MoO_3	D
	62	7	8	12	7	...	4 sand	D
	74	3.5	...	...	16	...	2 Sb; 4.5 SiO_2	D
United Kingdom	bal	3–10	5–10	1–10	0.8	≤4	1.5–4 SiO_2	W
West Germany	67.7	5.1	8	1.5	6.2	5	2.5 SiO_2; 3 Al_2O_3	D
	bal	4–15	5–30	...	20–30	...	3–10 Al_2O_3	W
Sweden	68.5	5.2	4.5	1.8	6.5	≤4	3.3 SiO_2; 3 Al_2O_3	W, D
	68.5	8	4.5	3	6	6	4 SiO_2	W, D
Italy	68	5.5	7	9	6	...	4.5 SiO_2	W, D
Austria	68	5	8	1.5	6.2	≤3	2.5 SiO_2; 3 Al_2O_3	W
	54.4	0.8	3.7	21.4	19	...	0.5 S; 0.04 Mn	D

(a) W, wet; D, dry. Source: Ref 23

(a)

(b)

(c)

Fig. 16 Copper-base P/M friction elements. (a) Grooved P/M friction elements for wet applications. (b) Copper-base P/M clutch plates (280 to 500 mm OD) used in power-shift transmissions for tractors. (c) Copper-base P/M friction pad cal mixing of the copper and oxide powders (Ref 27), by coprecipitation from salt solutions (Ref 28), by mechanical alloying (Ref 29), or by selective or internal oxidation (Ref 30). Dispersion quality and cost vary substantially among these methods; internal oxidation produces the finest and most uniform dispersion. Aluminum oxide is a common dispersoid used in the manufacture of ODS copper.

In internal oxidation, a copper-aluminum solid-solution alloy is internally oxidized at elevated temperature. This process converts the aluminum into aluminum oxide. For effective internal oxidation, oxygen must diffuse into the matrix (copper) several orders of magnitude faster than the solute element (aluminum). Because internal oxidation depends on diffusion of oxygen into the matrix, the reaction time is proportional to the square of the distance through which the oxygen must diffuse to complete the reaction. To hold reaction times within practical limits, diffusion distance must be small. In wrought form, internal oxidation can only be practical in thin wire or strip, which would severely limit the use of ODS materials. Powder metallurgy offers a unique solution to this problem, because powder particles can be internally oxidized rapidly and then consolidated into almost any shape.

The process involves melting a dilute solid-solution alloy of the aluminum in copper and atomizing the melt by use of a high-pressure gas such as nitrogen. The resulting powder is blended with an oxidant that consists primarily of fine copper oxide powder. The blend is heated to a high temperature; the copper oxide dissociates, and the oxygen produced diffuses into the particles of solid-solution copper-aluminum alloy. Because aluminum forms oxide more readily than copper, the aluminum in the alloy is preferentially oxidized to aluminum oxide. After complete oxidation of all the aluminum, excess oxygen in the powder is reduced by heating the powder in hydrogen or a dissociated ammonia atmosphere.

Full theoretical density is essential to realize the best potential properties of ODS copper. The powder is fabricated into fully dense shapes by various techniques. Mill forms, such as rod and bar, are made by canning the powder in a suitable metal container (generally copper) and hot extruding it to the desired size. Wire is made by cold drawing coils of rod. Strip is made either by rolling coils of extruded rectangular bar or by directly rolling powder with or without a metal container. Large shapes that cannot be made by hot extrusion are made by HIP of canned powder; alternatively, such shapes can be made by hot forging canned powder or partially dense compacted preforms.

Properties of the consolidated material depend on the amount of deformation introduced into the powder particles. Consequently, low-deformation processes such as HIP and, to a lesser extent, hot forging develop materials with lower strengths and ductilities than those produced by extrusion.

Finished parts can be made from consolidated shapes by machining, brazing, and soldering. Fusion welding is not recommended, because it causes the aluminum oxide to segregate from the liquid copper matrix, resulting in loss of dispersion strengthening. However, flash welding, in which the liquid metal is squeezed out of the weld joint, and electron beam welding, in which a small heat-affected zone is created, have been used successfully. Solid-state welding (with multiple cold upsets in a closed die) has also been used with success to join smaller coils into a large coil for wire drawing.

Properties. Oxide dispersion strengthened copper offers a unique combination of high strength and high electrical and thermal conductivities. More importantly, it retains a larger portion of these properties during and after exposure to elevated temperatures than any other copper alloy.

The properties of ODS copper can be modified to meet a wide range of design requirements by varying its aluminum oxide content and/or the amount of cold work. Figure 17 shows the ranges in tensile strength, elongation, hardness, and electrical conductivity obtained as a function of aluminum/aluminum oxide contents. These properties are typical for rod stock in the hot extruded condition. Cold work can be used to broaden the ranges in tensile strength, elongation, and hardness; its effect on electrical conductivity is minimal.

Three grades of ODS copper are commercially available. They are designated as C15715, C15725, and C15760 by the Copper Development Association. The nominal compositions of these three grades are:

Grade	Copper		Aluminum oxide	
	wt%	vol%	wt%	vol%
C15715	99.7	99.3	0.3	0.7
C15725	99.5	98.8	0.5	1.2
C15760	98.9	97.3	1.1	2.7

The free or reducible oxygen content of the three ODS coppers, generally about 0.02 to 0.05 wt%, is present in the form of dissolved oxygen

Table 15 Compositions, properties, and applications of copper-base electrical contacts

Nominal composition, %	Manufacturing method(a)	Density, g/cm^3		Electrical conductivity, % IACS	Hardness	Tensile strength		Modulus of rupture		Data source(b)	Application examples
		Calculated	Typical			MPa	ksi	MPa	ksi		
Tungsten carbide-copper											
50Cu	INF	11.39	11.00–11.27	42–47	90–100 HRF	...	...	1103	160	C,A	Arcing contacts in oil switches, wiping shoes in power transformers
44Cu	INF	11.77	11.64	43	99 HRF	...	...	1241	180	C	
30Cu	INF	12.78	12.65	30	38 HRC	...	...	...	...	...	
Tungsten-copper											
75Cu-25W	PSR	10.37	9.45–10.00	50–79	35–60 HRB	...	...	414	60	C,A	Current-carrying contacts
70Cu-30W	...	10.70	10.45	76	59–66 HRB	...	...	...	...	A	
65Cu-35W	...	11.06	11.40	72	63–69 HRB	...	...	...	...	A	Vacuum interrupter
60Cu-40W	...	11.45	11.76	68	69–75 HRB	...	...	...	...	A	Oil-circuit breakers, arcing tips
50Cu-50W	INF	12.30	11.90–11.96	45–63	60–81 HRB	...	...	...	...	A	
44Cu-56W	INF	12.87	12.76	55	79 HRB	434	63	827	120	C	
40Cu-60W	INF	13.29	12.80–12.95	42–57	75–86 HRB	...	...	...	...	A	Oil-circuit breakers, reclosing devices, arcing tips, tap change arcing tips, contactors
35Cu-65W	INF	13.85	13.35	54	83–93 HRB	...	...	...	...	A	
32Cu-68W	INF	14.20	13.95	50	90 HRB	...	...	896	130	C	
30Cu-70W	INF	14.45	13.85–14.18	36–51	86–96 HRB	...	...	1000	145	C,A	Circuit breaker runners, arcing tips, tap change arcing tips
26Cu-74W	INF	14.97	14.70	46	98 HRB	621	90	1034	150	...	
25Cu-75W	INF	15.11	14.50	33–48	90–100 HRB	...	...	...	...	A	Vacuum switches, arcing tips, oil-circuit breakers
20Cu-80W	INF	15.84	15.20	30–40	95–105 HRB	758	110	...	...	C	
15Cu-85W	PSR	16.45	16.0	20	190 HV(c), 260 HV(d)	...	...	...	...	M	
13.4Cu-86.6W	INF	16.71	16.71	33	20 HRC	621	90	1034	150	C	
10.4Cu-89.6W	INF	17.22	17.22	30	30 HRC	765	111	1138	165	C	

(a) PSR, press-sinter-re-press; INF, press-sinter-infiltrate; PS, press-sinter; PSE, press-sinter-extrude; IO, internal oxidation; PPSE, preoxidize-press-sinter-extrude; SF, oxidized from one direction. (b) A: Advance Metallurgy, Inc., McKeesport, PA. C: Contacts, Materials, Welds, Inc., Indianapolis, IN. E: Englehard Industries, Plainville, MA. G: Gibson Electric, Inc., Delmont, PA. S: Stackpole Carbon Co., St. Marys, PA. T: Texas Instruments, Inc., Attleboro, MA. M: Metz Degussa, South Plainville, NJ. W: Art Wire-Duduco, Cedar Knolls, NJ. (c) Annealed. (d) Cold worked. Source: Ref 25 and cuprous oxide. Alloys in this state are prone to hydrogen embrittlement at high temperatures. Low-oxygen (LOX) compositions are available in all of these grades, in which the reducible oxygen is converted to nonreducible oxide by adding up to 0.020 wt% B as the oxygen getter. These grades are then immune to hydrogen embrittlement and must be specified for applications in which the components are likely to be subjected to reducing atmospheres during manufacture or use.

Physical Properties. Because ODS copper contains small amounts of aluminum oxide as discrete particles in an essentially pure copper matrix, its physical properties closely resemble those of pure copper. Table 17 shows physical properties of the three commercial ODS coppers comparing them with oxygen-free (OF) copper. The melting point is essentially the same as for copper because the matrix melts and the aluminum oxide separates from the melt. Density, modulus of elasticity, and coefficient of thermal expansion are similar to those of pure copper.

High electrical and thermal conductivities are particularly interesting to design engineers in the electrical and electronics industries. At room temperature, these range from 78 to 92% of those for pure copper. Coupled with the high strengths of these materials, they enhance the current-carrying or heat-dissipating capabilities for a given section size and structural strength. Alternatively, they enable reduction of section sizes for component miniaturization without sacrificing structural strength or current and heat-carrying capabilities. At elevated temperatures, the decrease in electrical and thermal conductivities of ODS coppers closely parallel those of pure copper.

Room-Temperature Mechanical Properties. Table 18 shows the room-temperature mechanical properties of the three ODS coppers in available mill forms. These cover a wide range of sizes, typified by various amounts of cold work by drawing and rolling, for example. Oxide dispersion strengthened copper has strength comparable to many steels and conductivity comparable to copper.

Oxide dispersion strengthened copper has excellent resistance to softening even after exposure to temperatures close to the melting point of copper because the aluminum oxide particles are stable at these temperatures and retain their original size and spacing. These particles block dislocation and grain-boundary motion and thus prevent recrystallization, which is normally associated with softening. Figure 18 compares the softening behavior of C15715 and C15760 strip with OF copper (C10200) and a copper-zirconium (C15000) alloy. At common brazing and glass-to-metal sealing temperatures (above 600 °C) encountered in practice, ODS coppers retain much of their strength while OF copper and copper-zirconium lose most of their strength. Therefore, ODS copper is used in applications in which the component manufacture involves high-temperature operations such as brazing, glass-to-metal sealing, HIP diffusion bonding, and so forth.

Elevated-Temperature Mechanical Properties. ODS copper has excellent strength at elevated temperatures. Figure 19 shows the 100 h stress rupture strengths of C15760 and C15715 at temperatures up to 870 °C (1600 °F). Other high-conductivity copper-base materials are shown for comparison. Ranging from pure copper on the low end to precipitation-hardened alloys on the high end, there is a sharp drop in stress rupture strength in the 200 to 450 °C (400 to 850 °F) temperature range. Above 400 °C (750 °F), the ODS coppers are superior to any of the other alloys. Above 600 °C (1100 °F), the ODS coppers have rupture strengths comparable or superior to some stainless steels. Oxide dispersion strengthened copper has excellent thermal stability at high temperatures because the aluminum oxide particles retain their original particle size and spacing even after prolonged heating and do not allow recrystallization of the matrix. Cold work significantly enhances the stress rupture properties of ODS copper; the higher the temperature, the more noticeable the enhancement.

Applications. Oxide dispersion strengthened copper has gained wide market acceptance in several applications, and design engineers are continually developing new applications. The major applications are described below.

Resistance Welding Electrodes. Oxide dispersion strengthened copper electrodes are widely used for resistance welding in automotive, appliance, and other sheet metal industries. Sticking of the electrodes to the workpiece is a major problem when welding galvanized and other coated steels. This usually results in electrodes pulling off their holders and having to stop the assembly line to change the electrodes. Such interruptions are extremely costly. Oxide dispersion strengthened copper electrodes do not stick to galvanized and other coated steels and thus eliminate the costly downtime. Increasing use of coated steels in the automotive industry has prompted wide usage of ODS copper electrodes throughout the world. When used under optimal conditions, including heavy duty welding, these electrodes consistently outperform the conventional copper-chromium electrodes in weld life.

Metal-Inert-Gas Welding Contact Tips. Oxide dispersion strengthened copper is used in metal-inert-gas (MIG) welding contact tips. Its resistance to abrasive wear from the steel wire allows the tips to maintain the bore diameter and minimizes the wandering of the arc. This is very important in automated welding lines. The nonsticking property of ODS copper also minimizes the material buildup on the tip from the weld splatter.

Lead Wires. Oxide dispersion strengthened copper wire is used in leads for incandescent lamps. Its high-temperature strength retention capability enables glass-to-metal seals to be made without undue softening of the leads. This in turn eliminates the need for expensive molybdenum support wires without sacrificing lead stiffness. Superior strength of the leads allows reduction in lead diameter to conserve materials. Thinner lead wires also minimize heat loss from the filament, which enables the lamp to provide higher light output at lower wattage and results in a more energy efficient lamp.

Oxide dispersion strengthened copper wire also can be used in leads for discrete electronic components such as diodes. Advantages of ODS copper wire include its high-temperature strength

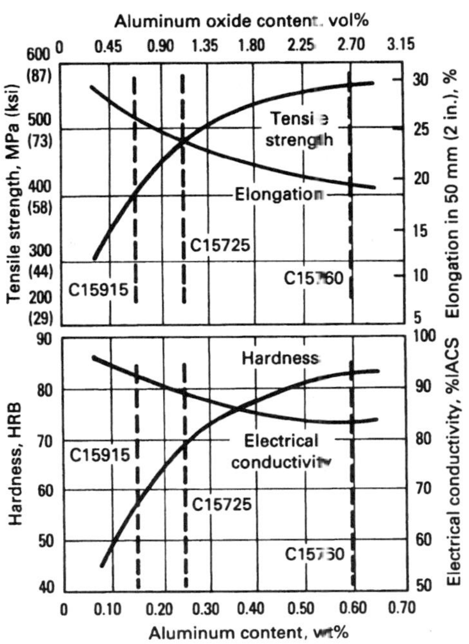

Fig. 17 Properties of three ODS coppers. Source: Ref 31

Table 16 Characteristics of typical graphite-metal brush materials

Grade No.	Nominal composition	Density, g/cm^3	Specific resistance, $\Omega \cdot m$	Specific resistance, $\Omega \cdot in.$	Maximum current density, A/m^2	Maximum current density, $A/in.^2$	Typical voltage, V	Scleroscope hardness
261C	21Cu-79C	2.2	0.024	0.0006	125,000	80	<72	28
261D	35Cu-65C	2.5	0.016	0.0004	125,000	80	<72	28
FQ	50Cu-50C	2.75	0.006	0.00015	130,000	85	<36	28
179P	65Cu-35C	3.5	0.0016	0.00004	190,000	125	<18	20
179V	75Cu-25C	4.0	0.0008	0.0002	235,000	150	<15	18
GHB	94metal-6C	6.0	0.0003	0.0000065	235,000	150	<6	6
GD	97metal-3C	6.5	0.0001	0.0000025	235,000	150	<6	5
22A-S	40Ag-60C	2.7	0.008	0.0002	150,000	100	<36	30
246	65Ag-35C	3.8	0.001	0.000025	190,000	125	<18	20
2-S	80Ag-20C	4.6	0.0008	0.00002	235,000	150	<9	23
1-S	93Ag-7C	7.0	0.0001	0.0000025	270,000	175	<6	10

Source: Ref 26

retention during brazing as well as hermetic sealing. The lead stiffness also enables multiple insertions in circuit boards.

X-Ray and Microwave Tube Components. Oxide dispersion strengthened copper rod and tube are used in x-ray and microwave tube components. An example is the stems for the rotating anodes in x-ray tubes where high strength retention after brazing and glass-to-metal sealing are important. The high thermal conductivity of ODS copper also allows more efficient heat removal, thus lowering the operating temperature and providing longer tube life and quieter tube operation.

Relay Blades and Contact Supports. These are current-carrying arms that move between the fixed contact points to "make" or "break" an electrical circuit. In a limit switch, the force that moves the contact arm (or contact support) to the "make" position is provided by mechanical means, such as a moving object. In a relay, the contact arm (relay blade) is moved by electromagnetic action. In most cases, the spring force of the arm itself moves the contact back to the "break" position.

Generally, the relay blades and contact supports have silver contacts brazed or riveted to them at the point where they touch the fixed contacts. The strength retention capability of ODS copper after exposure to elevated temperatures allows brazing of contacts to the blade without appreciable strength loss. Because of its higher electrical conductivity, ODS copper has replaced conventional copper alloys, such as phosphor bronze and beryllium copper, in some relays. Such relays can carry a higher current than was possible with the prior relays.

Sliding Electrical Contacts. Oxide dispersion strengthened copper bars are used in overhead sliding electrical contacts for high-speed electric trains. Its excellent resistance to abrasive wear against the electrical cable provides up to 10 times longer contact life and reduces the maintenance costs significantly. The higher the train speed, the greater the advantage ODS copper has over other copper-base materials.

Particle Accelerator Components. Particle accelerators are used for physics and materials research, medical diagnostics, and so forth. The high-energy particle beams are shaped and focused by using mirrors, lenses, and prisms in large doughnut-shaped hollow rings. The stray beams are absorbed by special x-ray absorbers. Oxide dispersion strengthened copper plates and bars are used in mirrors and x-ray absorbers because of their high thermal conductivity, high strength, creep resistance, and vacuum integrity.

Hybrid Circuit Package Components. Oxide dispersion strengthened copper strip is used as the base plate for hybrid circuit package cases.

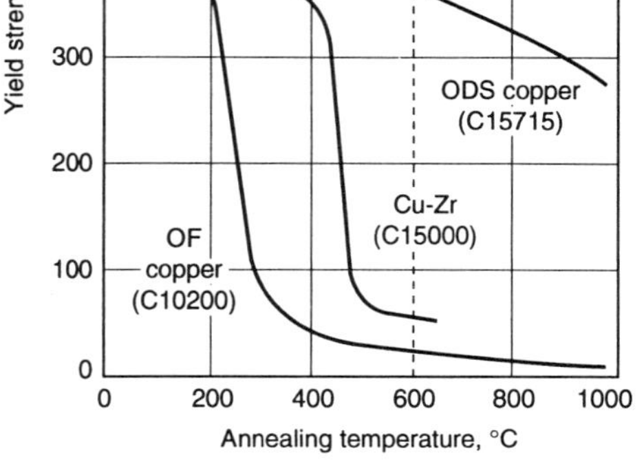

Fig. 18 Softening behavior of ODS copper compared to OF copper and copper-zirconium alloy. Source: Ref 34

Table 17 Physical properties of three ODS coppers and OF copper

Property	Material: CC15715	C15725	C15760	OF Copper
Melting point, °C (°F)	1083 (1981)	1083 (1981)	1083 (1981)	1083 (1981)
Density, g/cm³ (lb/in.)	8.90 (0.321)	8.86 (0.320)	8.81 (0.318)	8.94 (0.323)
Electrical resistivity at 20 °C (68 °F), Ω · mm²/m (Ω · circular mil/ft)	0.0186 (11.19)	0.0198 (11.91)	0.0221 (13.29)	0.017 (10.20)
Electrical conductivity at 20 °C (68 °F), M mho/m (%IACS)	54 (92)	50 (87)	45 (78)	58 (101)
Thermal conductivity at 20 °C (68 °F), W/m · K (Btu/ft · h · °F)	365 (211)	344 (199)	322 (186)	391 (226)
Linear coefficient of thermal expansion for 20 to 1000 °C (68 to 1830 °F), ppm/°C (ppm/°F)	16.6 (9.2)	16.6 (9.2)	16.6 (9.2)	17.7 (9.8)
Modulus of elasticity, GPa (10⁶ psi)	130 (19)	130 (19)	130 (19)	115 (17)

Source: Ref 32

Table 18 Typical room-temperature mechanical properties of three ODS coppers

Shapes	Thickness or diameter, mm	Thickness or diameter, in.	Temper or condition(a)	Tensile strength, MPa	Tensile strength, ksi	Yield strength, MPa	Yield strength, ksi	Elongation, %	Hardness, HRB
C15715									
Flat products	10	0.400	AC	413	60	331	48	20	62
	1.3	0.050	CW 88%	579	84	537	78	7	...
	0.6	0.025	CW 94%	620	90	579	84	7	...
	0.15	0.006	CW 98%	661	96	613	89	6	...
Plate	Up to 130	Up to 5.0	AC	365	53	255	37	26	62
	25	1.0	CW 60%	476	69	427	62	10	...
	16	0.625	CW 75%	483	70	455	66	10	...
Rod	29	1.125	AC	393	57	324	47	27	62
	19	0.75	CW 55%	427	62	407	59	18	68
	7	0.275	CW 94%	496	72	469	68	9	72
Wire	1.3	0.050	CW 99%	524	76	496	72	2	...
	1.3	0.050	HT 650 °C	400	58	351	51	10	...
	0.4	0.015	CW 99.9%	606	88	579	84	1	...
Rounds	Up to 760	Up to 30	AC	365	53	255	37	26	...
C15725									
Flat products	10	0.400	AC	434	63	345	50	21	72
	2.3	0.090	CW 78%	586	85	544	79	8	83
	0.15	0.006	CW 98%	675	98	613	89	6	...
Plate	Up to 130	Up to 5.0	AC	413	60	296	43	19	68
	25	1.0	CW 60%	496	72	441	64	9	...
	16	0.625	CW 75%	524	76	467	68	9	...
Rod	38	1.5	AC	441	64	358	52	24	73
	6.4	2.5	As Drawn	551	80	531	77	14	76
Rounds	Up to 760	Up to 30	AC	413	60	296	43	19	68
C15760									
Flat products	1.0	0.400	AC	517	75	413	60	13	81
	2.5	0.100	CW 75%	627	91	572	83	8	85
	0.15	0.006	CW 98%	737	107	655	95	6	...
Rod	14	0.54	AC	551	80	517	75	22	80
	13	0.5	CW 14%	572	83	544	79	16	83
	7	0.275	CW 74%	620	90	599	87	14	86
	7	0.275	HT 650 °C	579	84	544	79	18	80
	64	2.5	AC	496	72	475	69	4	80
Rounds	Up to 760	30	AC	469	68	331	48	4	76

(a) AC, as consolidated; CW, cold work % reduction in area; HT, heat treatment for 1 h. Source: Ref 33

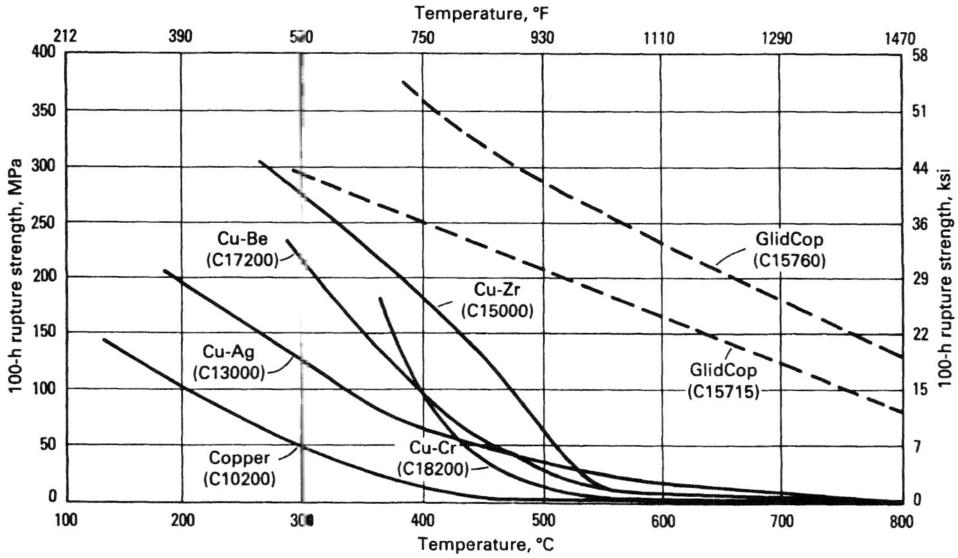

Fig. 19 Elevated-temperature stress-rupture properties of ODS copper compared to several high-conductivity copper alloys. Source: Ref 35

Manufacture of these cases involves brazing of the base plates to stainless steel and/or kovar components. The ability of ODS copper to retain its strength after such brazing operation makes it the material of choice when the package is used in high-stress environments. Oxide dispersion strengthened copper wire is also used in the leads for such packages because of its ability to retain its stiffness after the high-temperature ceramic sealing of the leads into the case.

Other Applications. Some other diverse applications of ODS copper include high field magnet coils, anode stems in chlorine cells, electrodischarge machining electrodes, high-speed motor and generator components, commutators, and so forth.

REFERENCES

1. Industry Statistics, Metal Powder Industries Federation, 1996
2. H.M. Williams, General Motors Corporation, U.S. Patent 1,556,658, 1925
3. H.M. Williams and A.L. Boegehold, General Motors Corporation, U.S. patents 1,642,347; 1,642,348; and 1,642,349, 1927
4. C. Claus, Bound-Brook Oilless Bearing Company, U.S. Patent 1,607,389, 1926
5. R.W. Stevenson, *Powder Metallurgy*, Vol 7, *Metals Handbook*, 9th ed., 1985, p 733–740
6. E. Klar and D.F. Berry, *Properties and Selection: Nonferrous Alloys and Special-Purpose Materials,* Vol 2, *ASM Handbook,* ASM International, 1990, p 392–402
7. F. Thummler and R. Oberacker, *Introduction to Powder Metallurgy,* The Institute of Materials, 1993, p 44–47
8. H.H. Hausner, *Mater. Meth.,* July 1946
9. R. Johansson, *Proc. 7 Plansee*
10. H.H. Hausner, *Handbook of Powder Metallurgy,* Chemical Publishing Co., Inc., 1973, p 142
11. J.L. Everhart, *Copper and Copper Alloy Powder Metallurgy Properties and Applications,* Copper Development Association
12. F.V. Lenel, *Powder Metallurgy Principles and Applications,* Metal Powder Industries Federation, 1980, p 247
13. *Diffusion,* American Society for Metals, 1973
14. F. Thummler and R. Oberacker, *Introduction to Powder Metallurgy,* The Institute of Materials, 1993, p 208
15. Sauerwald, Plansee seminar, 1952, p 201–202
16 Production Sintering Practices, *Powder Metallurgy*, Vol 7, *Metals Handbook*, 9th ed., 1985, p 310
17. F.V. Lenel, *Powder Metallurgy,* Metal Powder Industries Federation, 1980, p 214
18. F.V. Lenel, *Powder Metallurgy,* Metal Powder Industries Federation, 1980, p 219
19. H.H. Hausner, *Handbook of Powder Metallurgy,* Chemical Publishing Co., 1973, p 173
20. A. Price and J. Oakley, *Powder Metall.,* Vol 8, 1965, p 201
21. A.K.S. Rowley, E.C.C. Wasser, and M.J. Nash, *Powder Metall. Int.,* Vol 4 (No. 2), 1971, p 71
22. "P/M Materials Standards and Specifications," standard 35, Metal Powder Industries Federation, 1986–1987
23. W. Schatt, *Pulvermetallurgie Sinter und Verbundwerkstoffe,* VEB Deutscher Verlag fur Grundstoffindustrie, 1979, p 315
24. B.T. Collins, The U.S. Friction Materials Industry, *Perspectives in Powder Metallurgy,* Vol 4, Plenum Press, 1970, p 3–7
25. Y.S. Shen, P. Lattari, J. Gardner, and H. Wiegard, *Properties and Selection: Nonferrous Alloys and Special-Purpose Materials,* Vol 2, *ASM Handbook,* ASM International, 1990, p 841–868
26 C.P. Ferago, *Powder Metallurgy,* Vol 7, *Metals Handbook,* 9th ed., 1985, p 635
27. K.M. Zwilsky and N.J. Grant, *Trans. AIME,* Vol 221, April 1961, p 371–377
28. R.L. Crosby and D.H. Desy, U.S. Bureau of Mines Report 7266, June 1969, p 1–17
29. J.S. Benjamin, The International Nickel Company Inc., U.S. Patent 3,785,801, 1974
30. A.V. Nadkarni and E. Klar, SCM Corporation, U.S. Patent 3,779,714, 1973
31. A.V. Nadkarni, *Mechanical Properties of Metallic Components,* 1993, p 299
32. A.V. Nadkarni, *Mechanical Properties of Metallic Components,* 1993, p 297
33. Product literature, OAG Americas
34. A.V. Nadkarni, *Mechanical Properties of Metallic Components,* 1993, p 302
35. A.V. Nadkarni, High Conductivity of Copper and Aluminum Alloys, American Institute of Mining, Metallurgical, and Petroleum Engineers, 1984, p 95

Titanium Powder Metallurgy Alloys and Composites

Daniel Eylon, University of Dayton
F.H. (Sam) Froes, University of Idaho
Stanley Abkowitz, Dynamet Technology

THE DRIVE to develop net-shape technologies, such as casting and powder metallurgy (P/M), has been going on for many years. It has been spurred on by the desire to minimize alloy waste and to reduce or eliminate the cost of machining. This article focuses on the properties and applications of titanium P/M compacts.

Because of difficulties encountered with early melting practices, P/M was used in the beginning stages of titanium technology to produce alloy ingots (Ref 1). Titanium P/M has been developed as a net-shape technique only in the last 20 years (Ref 2–7). In general, P/M can be divided into two major categories:

- Elemental P/M, in which a blend of elemental powders, along with master alloy or other desired additions, is cold pressed into shape and subsequently sintered to higher density and uniform chemistry
- Prealloyed P/M, which is based on hot consolidation of powder produced from a prealloyed stock

In general, the blended elemental (BE) method produces parts at a low cost, but the parts often are less than fully dense; this technique is typically used for iron (Ref 8), copper (Ref 9), and heavy-metal (Ref 10) alloys. The prealloyed (PA) method is used for making fully dense high-performance components from aerospace alloys, such as nickel (Ref 11, 12), aluminum (Ref 13), and beryllium (Ref 14). Titanium P/M has incorporated both methods: BE is used to produce lower-cost parts that are less than fully-performance, full-density compacts (Ref 15). Recent developments in powder selection, compaction techniques, and postcompaction treatments have made it possible to obtain full density in titanium BE P/M. Properties exceeding those of ingot metallurgy (I/M) products have been achieved in BE and PA products.

This article highlights the properties and applications of both BE and PA titanium P/M compacts. It includes major recent developments that have led to improved performance in BE and PA products, but it does not cover the developments in titanium rapid solidification alloys that have not yet reached the commercialization stage. Detailed discussions of powder production methods and shape-making techniques are available elsewhere in this Volume. These processes are briefly described here in only enough detail to rationalize properties and applications.

Mechanical Properties of Titanium P/M Products

The mechanical properties of titanium P/M products depend on alloy composition and on the density and final microstructure of the compact. The compact density and microstructure depend on the nature of the powder, the specific consolidation technique employed, and postcompaction treatments such as secondary pressing or heat treatment. To date, most components produced by the various P/M methods have been made from Ti-6Al-4V, the most common aerospace titanium alloy. As a result, the majority of the P/M data available in the literature is for this alloy. However, these technologies are also very well suited for other alloys, such as the high-strength β alloys and the high-temperature near-α alloys. This article includes data on a wide variety of alloys to highlight the range of potential applications available for titanium P/M.

Blended Elemental Compacts

The blended elemental method, which is basically a pressing and sintering P/M technique, involves cold pressing or cold isostatic pressing (CIP) a blend of fine elemental titanium and master alloy powders that have been sintered. Titanium sponge fines (–100 mesh) are the most common elemental powder used in this process; these particles are obtained as by-products of the Hunter or Kroll reduction processes (Ref 4). The metallic titanium sponge produced by these processes is vacuum arc melted into ingots. The titanium sponge fines that are too small to be used in the melting process are available at a relatively low cost. This powder has an irregular shape (Fig. 1a), which makes it easy to cold press into green shapes. The powder is sintered at temperatures in the range of 1150 to 1315 °C (2100 to 2400 °F) in a vacuum to prevent gas contamination that can severely degrade compact properties. The high sintering temperature is needed to provide particle bonding and to homogenize the chemistry. It is well above the β transus (that is, the lowest equilibrium temperature at which the material is 100% β) of all common titanium alloys, and as a result, the compact microstructure in $\alpha + \beta$ alloys consists of colonies of similarly aligned coarse α plates (Fig. 1b). The plates are about 8 μm (320 μin.) wide and 25 μm (1000 μin.) long. The colonies are about 50 μm (0.002 in.) in diameter. The prior β grains are about 80 μm (0.0032 in.) in diameter. This microstructure is much finer than ingot material treated at the same temperature because of the inherent porosity of the powder compact (Ref 16). The porosity is the result of sodium chloride residues (Kroll process) in the sponge from the reduction process (Ref 4). The sponge fines contain from 0.12 to 0.15% Cl, and, as can be seen in Fig. 1(a), the resulting porosity cannot be entirely closed, even after secondary operations such as hot pressing or hot isostatic pressing (HIP) (Ref 17).

BE compacts have a green density of 85 to 90% after 415 MPa (60 ksi) cold pressing. After vacuum sintering, they have a density of 95 to 99%. Control of particle size and size distribution can produce compacts that are 99% dense. In ferrous, copper, and heavy-metal P/M alloys, 99% is considered to be full density; however, in titanium alloy compacts, such a level of residual porosity (Fig. 1a) will degrade both fatigue and fracture properties. A substantial effort has been made to entirely eliminate the porosity so that BE P/M parts can be used for fatigue-critical aerospace applications. Postsintering HIP densification can lead to 99.8% density and improved

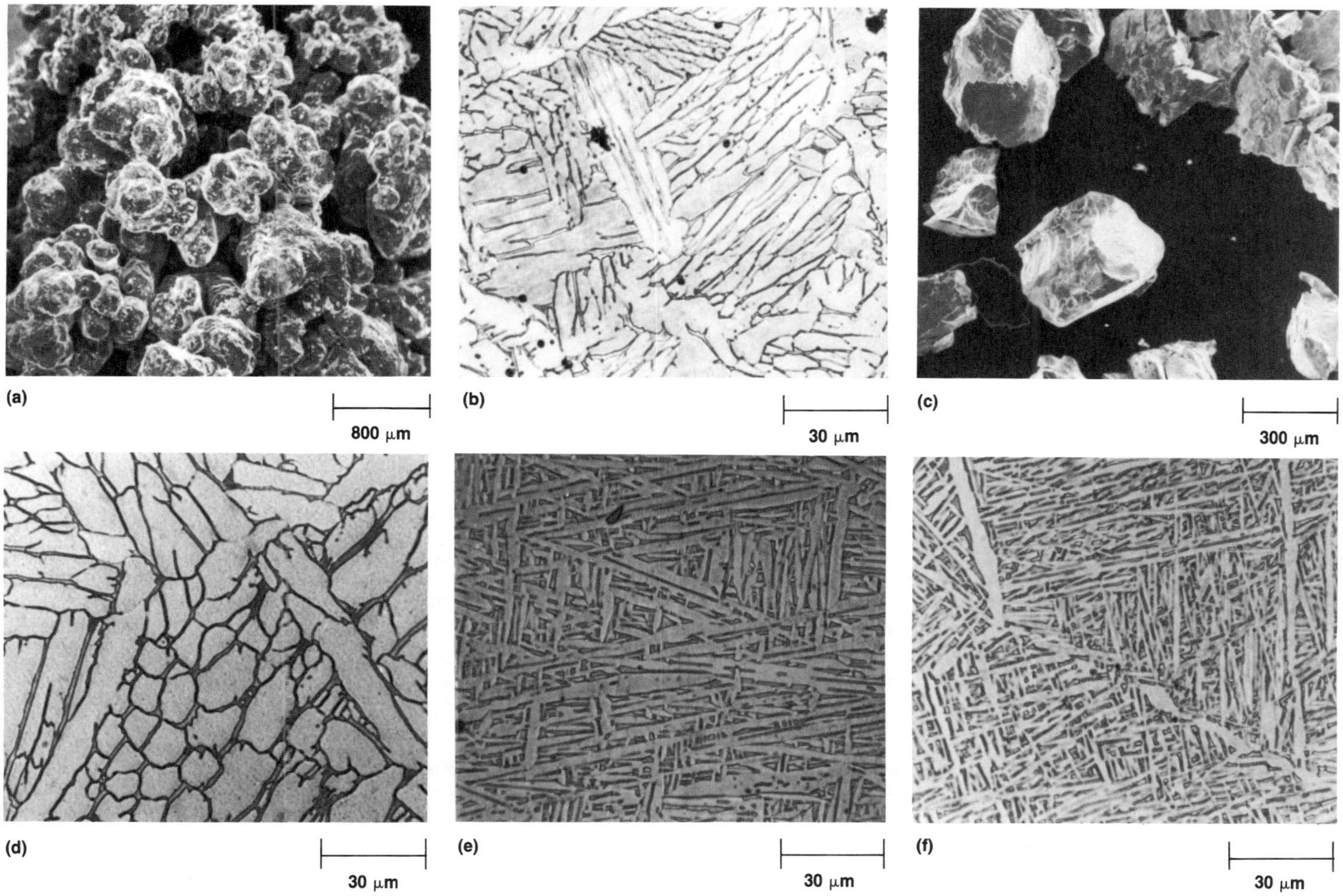

Fig. 1 Photomicrographs of titanium BE materials. (a) –100 mesh titanium sponge fines. (b) Microstructure of a pressed and sintered 99% dense Ti-6Al-4V compact. (c) Crushed hydrogenated-dehydrogenated titanium ingot or machine turnings. (d) Microstructure of a fully dense, pressed and sintered, and hot isostatically pressed Ti-6Al-4V compact. (e) Microstructure of a Ti-6Al-4V compact treated to produce a broken-up structure. (f) Microstructure of a Ti-6Al-4V compact treated with thermochemical processing

properties (Ref 4, 5). However, it is impossible to entirely eliminate the porosity with postsintering hot-pressing operations. During hot pressing, the chlorides present in the compact become volatile and create pockets of insoluble gas. Under the HIP pressure, these relatively large pressurized cavities (Fig. 2a) will break up into a multitude of submicron voids (Fig. 2b) (Ref 16–18) with sodium chloride in the center of the cavity (Fig. 2c). Both macrovoids and microvoids have an approximate hexagonal shape that is associated with basal-plane facets, which are the most energetically stable planes of the hexagonal close-packed structure (Ref 19). During cooling to room temperature, the gas in the voids transforms into cubic chloride crystals, as can be seen in the transmission electron microscopy image in Fig. 2(b).

To obtain pore-free 100% density material such as that produced by ingot metallurgy, the BE method must use chloride-free titanium powder (Ref 20). One source for such powder is commercially pure titanium ingot material or machine turnings embrittled by hydrogenation that are subsequently crushed and dehydrogenated. This powder is angular (Fig. 1c), and the sintered microstructure is much coarser (Fig. 1d) because of the lack of porosity during sintering.

Tensile Properties and Fracture Toughness. As with data for other titanium technologies, most of the published BE data are on Ti-6Al-4V. Table 1 is a comprehensive listing of tensile properties of Ti-6Al-4V BE compacts processed under a variety of conditions. Table 2 provides the more limited available information on the properties of additional titanium BE alloys.

Blended Elemental Ti-6Al-4V. As indicated in Table 1, most Ti-6Al-4V BE compact conditions exceed minimum MIL-T-9047 specifications. The process details for each condition are in the corresponding references listed in the table. The final shape of the compact can be achieved through a number of process sequences:

- Pressing and sintering
- Pressing and sintering plus HIP
- Pressing and sintering plus rolling
- CIP and sintering
- CIP, sintering, and HIP (collectively designated as cold and hot isostatic pressing, or CHIP)
- CIP, sintering, and rolling
- CIP, sintering, and forging

By controlling the process parameters, it is possible to obtain compacts with densities between 92 and 100% of the theoretical density. The yield and tensile strength of the compacts are proportional to the density (Fig. 3). The fracture toughness also increases with density (Fig. 4). Above 98% density, BE compacts have K_{Ic} values at the level of mill-annealed I/M materials. However, I/M materials with coarse lenticular microstructures similar to those of BE compacts (Fig. 1d) will have much higher K_{Ic} values (70 to 100 MPa$\sqrt{m}$, or 65 to 90 ksi$\sqrt{in.}$). The relatively lower K_{Ic} level of the BE compacts is probably the result of higher oxygen levels (Ref 26) and residual porosity.

Blended elemental compacts can be used as forging preforms (Ref 17). The strong effect of forging deformation on tensile strength and elongation is shown in Fig. 5.

Additional BE Alloys. The limited available mechanical test data for other BE alloys, such as Ti-6Al-2Sn-4Zr-6Mo, Ti-5Al-2Cr-1Fe, and Ti-4.5Al-5Mo-1.5Cr, are listed in Table 2. The most

Table 1 Tensile and fracture toughness properties of Ti-6Al-4V BE compacts processed under various conditions

Condition(a)	0.2% yield strength		Ultimate tensile strength		Elongation, %	Reduction in area, %	K_{Ic} or (K_Q)		Density, %	Chlorine, ppm	O_2, ppm	Ref
	MPa	ksi	MPa	ksi			MPa$\sqrt{m}$	ksi$\sqrt{in.}$				
Pressed and sintered (96% dense)	758	110	827	120	6	10	...	...	96	1200	...	21
Pressed and sintered (98% dense)	827	120	896	130	12	20	...	...	98	1200	...	21
Pressed and sintered (MR-9 process) (99.2% dense)	847	123	930	135	14	29	38	35	99.2	1200	...	21, 22
Pressed and sintered plus HIP	806	117	875	127	9	17	41	37	≥99	1500	2400	23, 24
CIP and sintered plus HIP	690	100	793	115	9	15	85	77	>99	...	...	97
	793	115	896	130	10	20	83	76	>99	...	...	...
CIP and sintered plus HIP(b)	896	130	965	140	12	22	...	...	99.8	...	...	...
Pressing and sintered plus α/β forged	841	122	923	134	8	9	...	...	≥99.4	1500	2400	24
Pressed and sintered plus α/β forged	951	138	1027	149	9	24	49	45	99	1200	...	26
Pressed and sintered (92% dense)	827	120	910	132	10	...	...	...	92	1500	2100	17
Plus α/β 30% isothermally forged	841	122	930	135	30	...	...	...	99.7	1500	2100	17
Plus α/β 70% isothermally forged	896	130	999	145	30	...	...	...	99.8	1500	2100	17
CIP and sintered plus HIP (low chlorine)	827	120	923	134	16	34	...	...	99.8	160	...	24
CIP and sintered plus HIP (ELCI)	882	128	985	143	11	36	...	...	100	<10	...	27
Plus BUS treated	951	138	1034	150	7	15	...	...	...	...	...	27
Plus TCP treated	1007	146	1062	154	14	20	...	...	...	...	...	30
Rolled plate, CIP and sintered plus HIP												
Mill annealed (L or TL)	903	131	958	139	10	26	(72)(c)	(65)(c)	≥99	200	1600	28, 29
Milled annealed (T or LT)	923	134	965	140	14	31	(71)(c)	(64)(c)	≥99	200	1600	28, 29
Recrystallization annealed (L or TL)	888	129	916	133	4	8	(75)(c)	(68)(c)	≥99	200	1600	28, 29
Recrystallization annealed (T or LT)	868	126	937	136	5	9	(67)(c)	(61)(c)	≥99	200	1600	28, 29
β annealed (L or TL)	841	122	937	136	10	26	(89)(c)	(81)(c)	≥99	200	1600	28, 29
β annealed (T or LT)	875	127	958	139	7	20	(92)(c)	(84)(c)	≥99	200	1600	28, 29
Minimum properties (MIL-T-9047)	827	120	896	130	10	25	...	...	...	...	...	4

(a) HIP, hot isostatic pressing; CIP, cold isostatic pressing; ELCI, extra-low chlorine powder; BUS, broken-up structure; TCP, thermochemical processing; L, longitudinal; TL, transverse longitudinal; T, transverse; LT, longitudinal transverse (TL and LT per ASTM E 399). (b) Forged 1010 °C + water quench (c) Precracked Charpy, K_v. Source: Ref 4, 17, 21–30, 97

Table 2 Tensile and fracture toughness properties of BE titanium alloy compacts processed under various conditions

Alloy and condition(a)	0.2% yield strength		Ultimate tensile strength		Elongation, %	Reduction in area, %	K_{Ic} or (K_Q)		Density, %	Chlorine, ppm	Ref
	MPa	ksi	MPa	ksi			MPa$\sqrt{m}$	ksi$\sqrt{in.}$			
Ti-5Al-2Cr-1Fe, pressed and sintered plus HIP	980	142	1041	151	20	39	...	...	≥99	310	31
Ti-4.5Al-5Mo-1.5Cr (Corona 5), pressed and sintered plus HIP	951	138	1000	145	17	39	(64)	(58)	≥99	310	31
Ti-6Al-2Sn-4Zr-6Mo, pressed and sintered, no STA or HIP	1068	155	1109	161	2	1	31	28	99	150	32
Ti-10V-2Fe-3Al											
Pressed and sintered, HIP (1650 °C, or 3000 °F), and STA (775–540 °C, or 1425–1005 °F)	1233	179	1268	184	9	...	30	27	99	1900	33
Pressed and sintered, HIP, and STA (750–550 °C, or 1380–1020 °F)	1102	160	1158	168	10	...	32	29	99	1900	33
Pressed and sintered, no STA or HIP	854	124	930	135	9	12	51	46	98	150	32
Ti-6Al-6V-2Sn, CHIP	931	140	1035	150	15	35	78	71	100	...	97

(a) HIP, hot isostatic pressing; STA, solution treatment and aging. (b) High modulus (Young's modulus of 20×10^6 psi, or 140 GPa). Source: Ref 31–34, 97

detailed work has been done on the Ti-10V-2Fe-3Al alloy, with some results reported at levels close to those for I/M materials (Ref 33). However, more data are needed for these alloys before reliable parameters for property levels and optimum processes can be established. In the case of Ti-10V-2Fe-3Al, a tensile strength of 1268 MPa (184 ksi) with 10% elongation can be achieved with BE methods.

Fatigue Strength and Crack Propagation. The fatigue life scatterband of chloride-containing Ti-6Al-4V BE compacts is compared in Fig. 6 to a mill-annealed I/M alloy. The effect of low chloride levels and postsintering treatments on fatigue strength is shown in Fig. 7. The effect of compact density on fatigue strength is shown in Fig. 8. The fatigue strength of BE compacts is inherently low when compared to that of mill-annealed I/M products because of the inherent chloride and related porosity of BE materials. Low fatigue strength limits the use of the lower-cost pressing and sintering or CIP and sintering processes to applications that are not fatigue critical, such as missile components. By increasing density through secondary pressing operations and through the use of chloride-free titanium powder, it is possible to further improve fatigue strength (Fig. 8). However, this method increases the cost of these products, thereby negating one of their primary advantages (Ref 15, 35).

Very limited data are available on the fatigue crack growth rate of Ti-6Al-4V BE compacts. Figure 9 shows that the fatigue crack growth rate of this material is between that of a β-annealed material and that of a mill-annealed I/M material (Ref 26). The BE material tested had a porosity of 1 to 2 vol%, which at this level seems not to adversely influence the fatigue crack growth rate.

Prealloyed Compacts

While BE compacts are produced and used in a wide range of densities, PA P/M parts are acceptable only at 100% density (Ref 2–5, 15, 35). The titanium PA powders are commercially available as spherical particles that have high tap density (65%) and good powder flow and mold fill characteristics. Two main production methods are used for making clean PA powder:

- Gas atomization (Ref 36)
- Plasma rotating-electrode process (PREP) (Ref 37), a modification of the older rotating-electrode process (REP) (Ref 38)

It is also possible to produce PA powders by comminution (Ref 39) and coreduction (Ref 40, 41) methods. However, because of insufficient mechanical property data, these techniques are not discussed in this article.

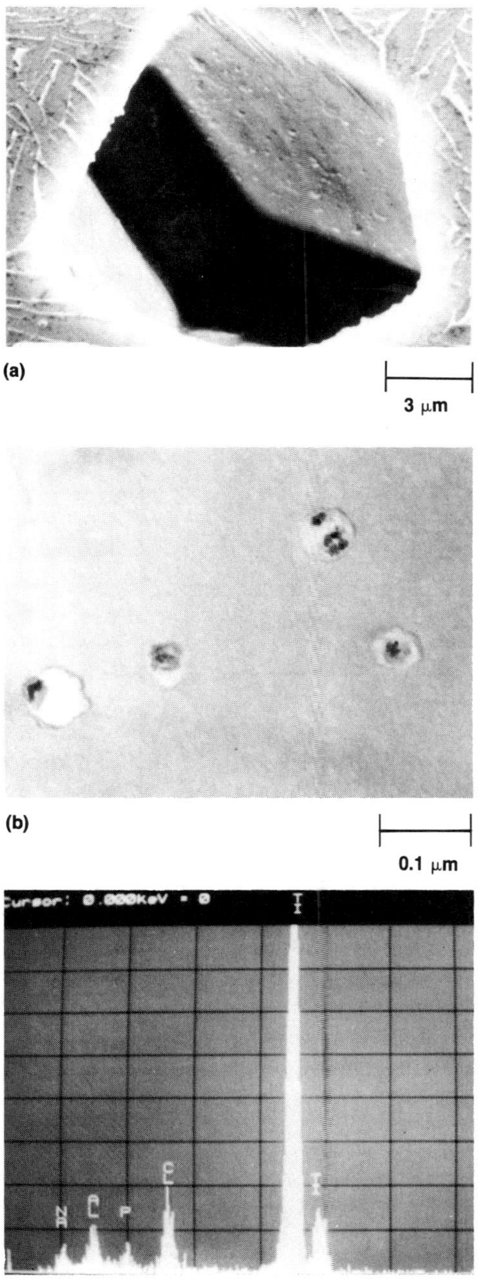

Fig. 2 Chlorine-induced porosity in a titanium BE compact. (a) Scanning electron microscopy photomicrograph of large-size residual porosity in a sectioned Ti-6Al-4V BE compact. (b) Transmission electron microscopy photomicrograph of a Ti-6Al-4V BE compact after postsintering HIP at 925 °C (1700 °F). (c) Chemical analysis showing sodium chloride contaminant at the center of a micropore

Hot isostatic pressing is the primary compaction method for PA powders (Ref 2–5, 15, 35), but vacuum hot pressing (VHP) (Ref 42, 43), extrusion (Ref 44), and rapid omnidirectional compaction (ROC) (Ref 45–48) have also been successfully used. The shape-making step is achieved by containing the powder in a shaped, evacuated, mild-steel can. The compaction is typically carried out at a temperature below the β transus to minimize reaction with the can. Processing in the α + β phase field results in a coarse low-aspect-ratio α structure (Fig. 10a). This material is most commonly compared to mill-annealed I/M material because of its microstructure and full density.

Powder cleanliness is one of the main factors governing the quality of PA compacts. Because of the full compact density, even a low level of contamination with foreign particles will lead to a substantial loss of inherent properties such as fatigue strength (Ref 49). As a result, only data obtained from clean powders are considered in this article. Also, because fully dense PA compacts are considered for more demanding applications than are the less-dense BE compacts, more mechanical test data have been developed within the aerospace industry on PA P/M than on BE P/M compacts. Only data considered to be typical are reviewed in this article. The majority of PA work has been done on Ti-6Al-4V.

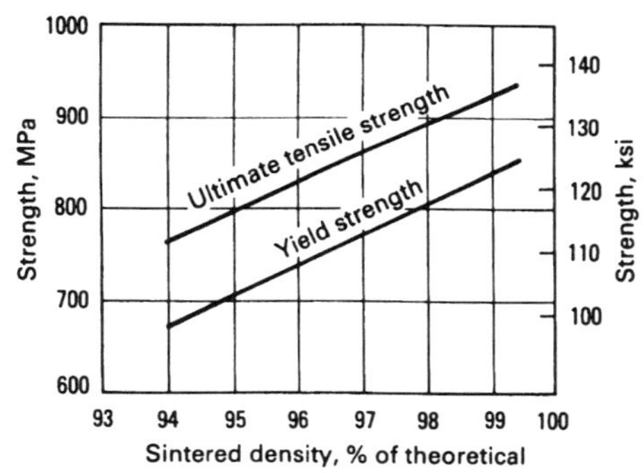

Fig. 3 Effect of sintered density on the yield and tensile strengths of press and sintered Ti-6Al-4V BE compacts. Source: Ref 22

Tensile Properties and Fracture Toughness. Table 3 is a comprehensive listing of the tensile properties of Ti-6Al-4V PA compacts processed under various conditions. Table 4 provides limited information on the properties of additional alloys. When the alloy compacts are produced using HIP (Ref 55), VHP (Ref 42), or ROC (Ref 48) at higher pressures but at lower temperatures, higher strength levels without losses in ductility are achieved. This result is due to the substantial microstructure refinement developed during high-pressure low-temperature powder processing. Similarly, post-compaction hot work, such as rolling (Ref 57) or forging (Ref 58), results in microstructural refinement that improves tensile strength and ductility. Property improvement after postcompaction treatments are discussed in the section "Postcompaction Treatments" in this article.

Additional PA Alloys. Table 4 shows the extent to which other PA titanium alloys have been studied. Almost all major I/M alloys have been evaluated in the PA P/M form. These include the high-strength metastable β alloys (Ref 33, 52, 65–68), the versatile α + β alloys (Ref 63, 64, 66), the high-temperature near-α alloys (Ref 59, 60), and the ordered titanium aluminide alloys (Ref 70–72). Of special interest is the alloy Ti-1.3Al-8V-5Fe (Ref 67–69). This alloy has a remarkable tensile strength of 1516 MPa (220 ksi) with 8% elongation. With conventional I/M methods, the high iron content of this alloy results in segregation problems. Powder metallurgy, on the other hand, produces fine-grain, homogeneous, and segregation-free products. Such alloys have the potential of expanding the market for titanium P/M technology.

Fatigue Strength and Crack Propagation. The smooth-bar fatigue life scatterband of

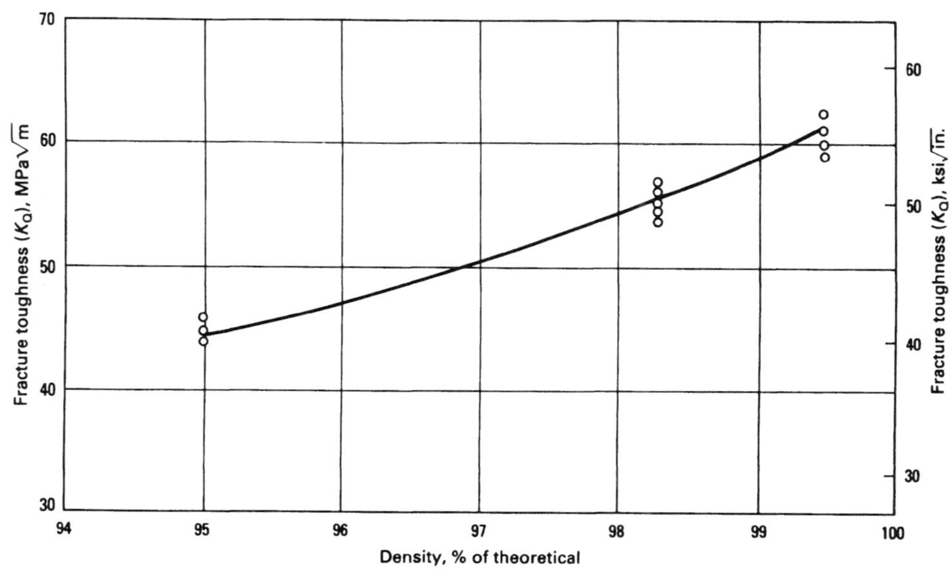

Fig. 4 Effect of density on the fracture toughness of press and sintered Ti-6Al-4V BE compacts. The values are not valid K_{Ic} and thus are labeled as K_Q. Source: Ref 21

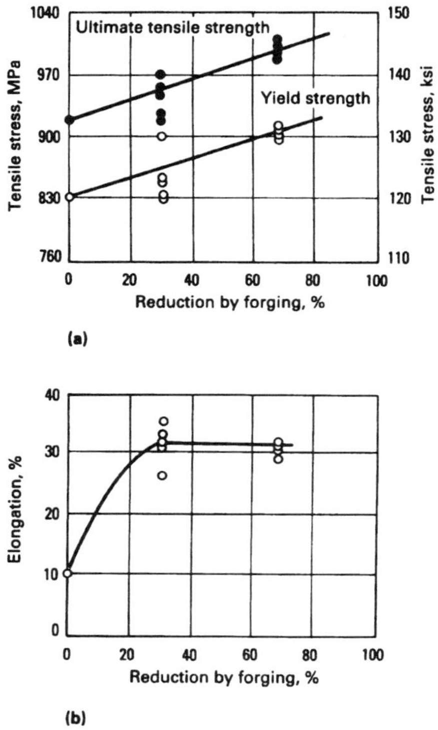

Fig. 5 Effect of forging deformation on Ti-6Al-4V BE compacts (Hunter reduction process sponge fines) isothermally forged at 925 °C (1700 °F). (a) Tensile strength and yield strength. (b) Tensile elongation. Source: Ref 17

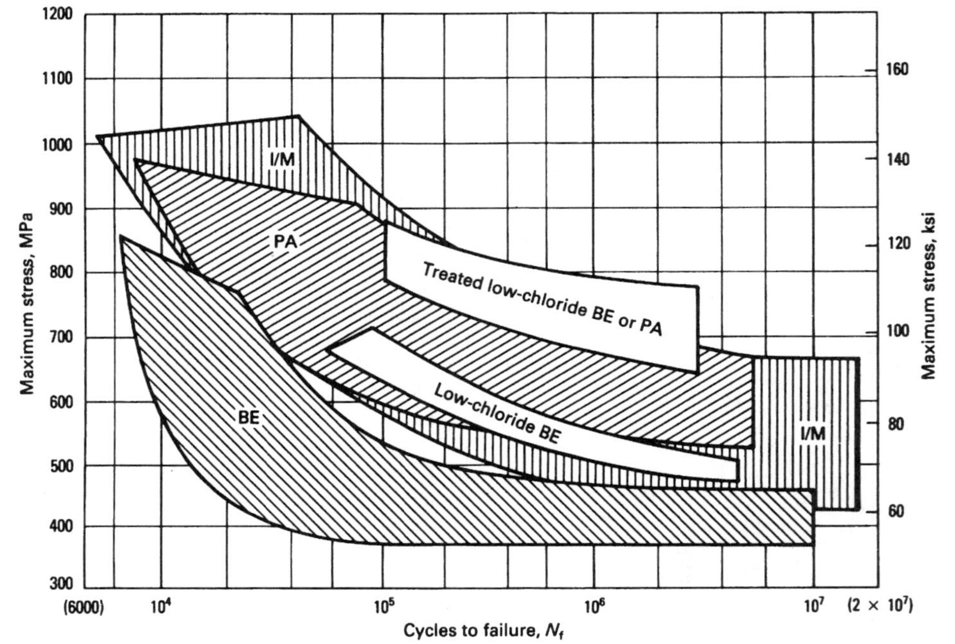

Fig. 6 Comparison of the room-temperature fatigue life scatterbands of BE and PA Ti-6Al-4V compacts to that of a mill-annealed I/M alloy

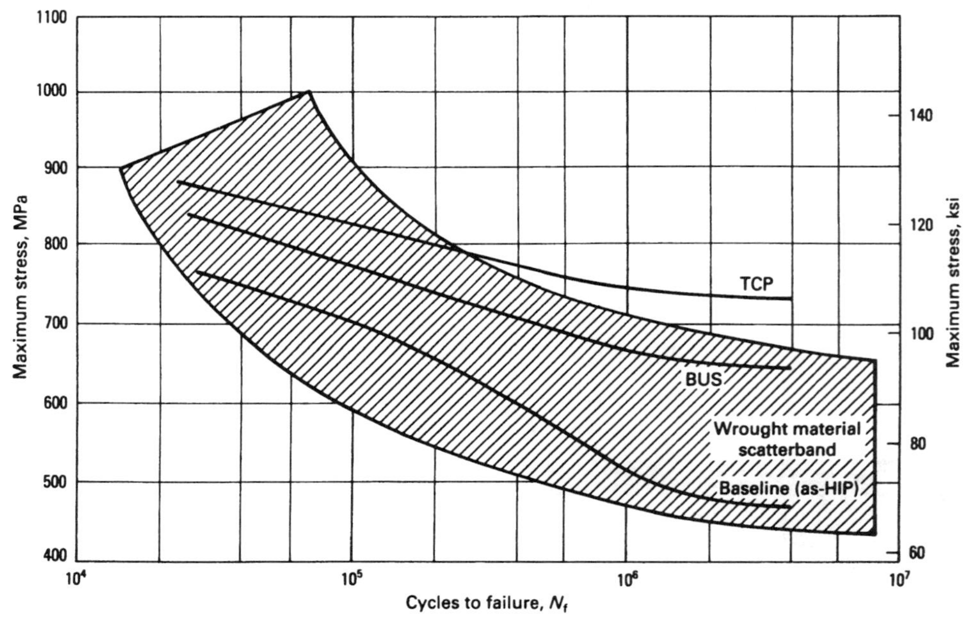

Fig. 7 Comparison of the fatigue strengths of fully dense extra-low chloride Ti-6Al-4V BE compacts with the scatterband for an I/M alloy. The BE compacts were tested in the as-HIP, broken-up structure (BUS), and thermochemically processed (TCP) conditions. Smooth axial fatigue data were obtained at room temperature. Stress ratio (*R*), 0.1; frequency (*f*), 5 Hz with triangular waveform. Source: Ref 27

Ti-6Al-4V PA compacts is compared to that of a mill-annealed I/M alloy in Fig. 6 (Ref 4). The P/M data were obtained by testing high-cleanliness REP and PREP compacts that had undergone hot isostatic pressing; some of the compacts received a postcompaction heat treatment. The data for the PA compacts are at equivalent levels to the best I/M results. Powder contamination must be avoided to maintain a high fatigue strength in these materials. The effect of 50, 150, and 350 μm (0.002, 0.006, and 0.014 in.) diameter contaminants on the fatigue strength of Ti-6Al-4V PA compacts is shown in Fig. 11 (Ref 49). Even 50 μm (0.002 in.) contaminant particles are sufficient to noticeably reduce fatigue strength. Figure 12 compares the fatigue characteristics of an actual P/M component to those of an I/M material. Figure 12(a) shows results for smooth-bar high-cycle fatigue; Fig. 12(b) covers notched-specimen high-cycle fatigue; and Fig. 12(c) shows data for strain-controlled low-cycle fatigue. In general, high-cycle fatigue results are in the range of 10^6 to 10^8 cycles to failure, and low-cycle fatigue results fall below 10^5 cycles to failure. For all three tests, the P/M alloy performance was comparable to or exceeded that of the I/M material. The component tested is an actual Ti-6Al-4V P/M part used in military airframes (Ref 61).

Figure 13 compares the fatigue crack growth rate of Ti-6Al-4V PA compacts to that of an I/M material with a similar composition and microstructure. Rates are at equivalent levels for both materials, even in PA material with a low level of contamination (Ref 74).

Postcompaction Treatments

Most P/M alloys that are subjected to postcompaction working or to lower-temperature consolidation display improved tensile and fatigue strengths as a result of microstructure refinement. However, in most cases, process economics do not allow subsequent working because it nullifies the objectives of a true net-shape technology. Therefore, only those postcompaction methods leading to microstructure refinement without the use of working will be considered in this article. Two approaches that meet this requirement are heat treatment and thermochemical processing (TCP).

Heat Treatment. In the case of BE Ti-6Al-4V, the only successfully used heat treatment has been the broken-up structure (BUS) treatment in which a β quench is followed by 850 °C (1560 °F) long-term annealing (Ref 27). After such treatment, the microstructure of the alloy is showing broken-up α phase in a matrix of β

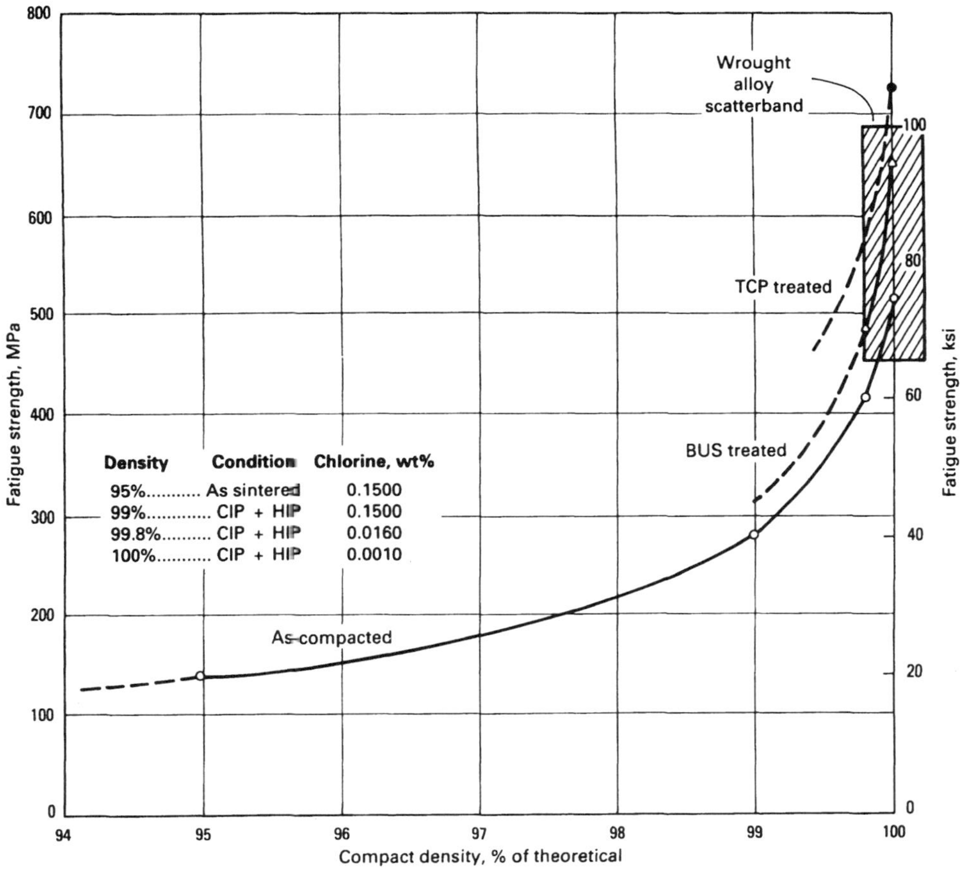

Fig. 8 Effect of compact density on fatigue strength of CIP and sintered Ti-6Al-4V BE compacts. Note that the higher densities are only possible in the low-chloride material. BUS, broken-up structure; TCP, thermochemical processing. Source: Ref 15

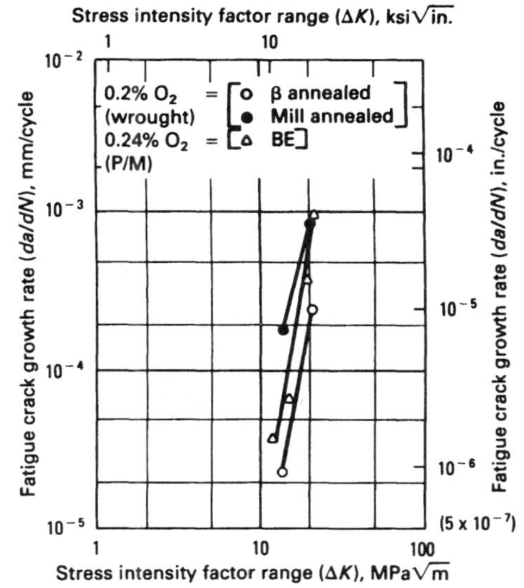

Fig. 9 Comparison of fatigue crack propagation rates of BE and I/M Ti-6Al-4V as a function of the stress intensity factor range at room temperature in air. Stress ratio (R), + 0.1 ($R = \sigma_{min}/\sigma_{max}$, where σ_{min} is the minimum stress and σ_{max} is the maximum stress); frequency (f), 5 Hz. Source: Ref 26

(Fig. 1e). This microstructure provides a significant improvement in both tensile and fatigue strengths (Fig. 7).

The BUS method is an improvement over standard heat treatments, which typically provide higher tensile properties but no increase in fatigue strength properties.

In the case of Ti-10V-2Fe-3Al BE compacts, β solution treatment and subsequent aging resulted in materials with good combinations of tensile strength and ductility (Ref 33). However, the K_{Ic} was found to be too low (Table 2), possibly because of the high chloride levels and the associated porosity. The Ti-6Al-4V PA compacts responded well to the BUS treatment (Fig. 10b) (Ref 51), as well as to solution treatment and aging (Ref 52). Figure 6 shows the improvement in fatigue strength of both BE and PA Ti-6Al-4V compacts as a result of a microstructure refinement brought about by heat treatment.

Thermochemical Processing. The TCP method (Ref 75) involves the use of hydrogen as a temporary alloying element to refine the microstructure of titanium alloys. This method is very suitable for net-shape products because no hot or cold work is needed to refine the microstructure. An example of the refinement obtained in Ti-6Al-4V P/M products can be seen by comparing Fig. 1(d) with Fig. 1(f) and Fig. 10(a) with Fig. 10(c). This microstructural refinement provides slightly higher strength levels than those typically obtained in I/M or conventional P/M materials (Table 3); more significantly, it substantially enhances the fatigue behavior of the P/M products (Fig. 6–8).

Applications of Titanium P/M Products

The two distinctively different titanium P/M technologies, the blended element and the prealloyed methods, not only produce compacts with different sets of properties, but also with two different price ranges. The relative low cost of titanium sponge fines and the volume production capability of the pressing and sintering technology allow the production of BE complex-shape

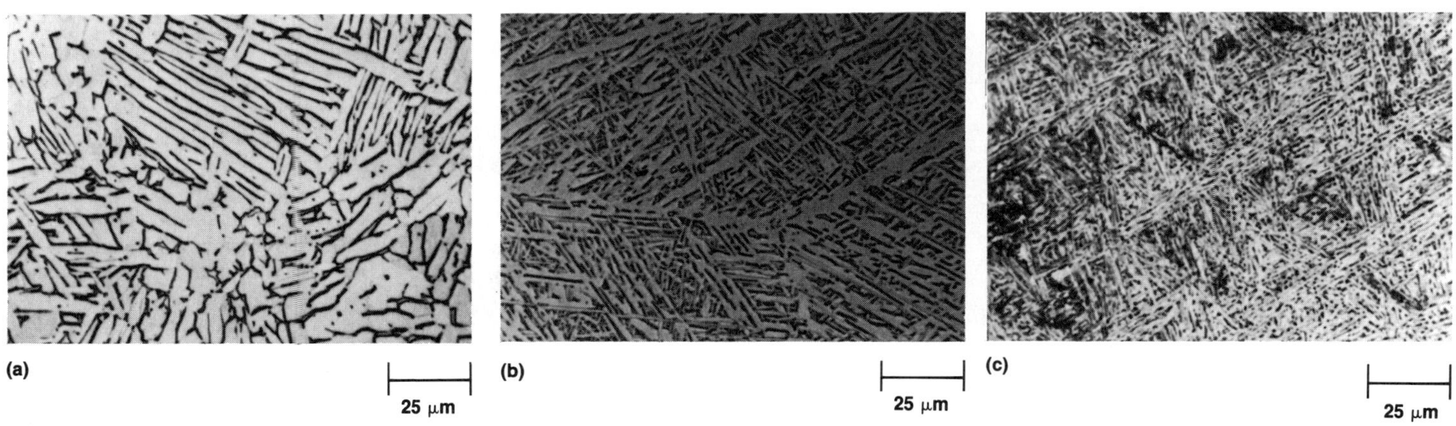

Fig. 10 Microstructure of Ti-6Al-4V PA compacts. (a) As-HIP. (b) Treated to produce a broken-up structure. (c) Thermochemically treated

Table 3 Tensile and fracture toughness properties of Ti-6Al-4V PA compacts processed under various conditions

Condition(a)	0.2% yield strength, MPa	0.2% yield strength, ksi	Ultimate tensile strength, MPa	Ultimate tensile strength, ksi	Elongation, %	Reduction in area, %	K_{Ic} or (K_Q), MPa$\sqrt{m}$	K_{Ic} or (K_Q), ksi$\sqrt{in.}$	Titanium PA powder preparation: Powder process	Compaction temperature, °C	Compaction temperature, °F	Other variables	Ref
HIP	861	125	937	136	17	42	(85)	(77)	PREP	925	1695	...	50
HIP (PSV) and β annealed	1020	148	1095	159	9	21	(67)	(61)	PSV	950	1740	975 °C (1785 °F) anneal	43, 50
HIP and BUS treated	965	140	1048	152	8	17	...	...	PREP	925	1695	...	51
HIP and TCP treated	931	135	1021	148	10	16	...	...	PREP	925	1695	...	30
HIP and annealed (700 °C, or 1290 °F) (REP)	820	119	889	129	14	41	(76)	(69)	REP	955	1750	...	52
HIP, annealed (700 °C, or 1290 °F), and STA (955–480 °C, or 1750–855 °F)	1034	150	1130	164	9	34	...	...	REP	955	1750	...	52
HIP and annealed (700 °C, or 1290 °F) (PREP)	882	128	944	137	15	40	(73)	(67)	PREP	955	1750	...	53
ELI; HIP (as-compacted)	855	124	931	135	15	41	(99)	(90)	REP	955	1750	1300 ppm O_2	54
ELI; HIP and β annealed	896	130	951	138	10	24	93	85	REP	955	1750	1020 °C (1870 °F) anneal	54
HPLT and HIP (as-compacted)	1082	157	1130	164	8	19	...	...	PREP	650	1200	315 MPa (46 ksi)	55
HPLT, HIP, and RA (815 °C, or 1500 °F)	937	136	1013	147	22	38	...	...	PREP	650	1200	315 MPa (46 ksi)	55
HIP and rolled (955 °C, or 1750 °F) (T)	958	139	992	144	12	35	...	...	REP	925	1695	75% rolling reduction	56
HIP, rolled (955 °C, or 1750 °F), and β annealed													
L or LT	820	119	896	130	13	31	73	66	REP	925	1695	75% rolling reduction	56
T or TL	813	118	896	130	11	23	61	55	REP	925	1695	75% rolling reduction	56
HIP, rolled (950 °C, or 1740 °F), and STA (960–700 °C, or 1760–1290 °F)	924	134	1041	151	15	35	...	...	REP	950	1740	60% rolling reduction	57
HIP, forged (950 °C, or 1740 °F), and STA (960–700 °C, or 1760–1290 °F)	1000	145	1062	154	14	35	...	...	REP	915	1680	56% forging reduction	58
VHP (830 °C, or 1525 °F) (as-compacted)	945	137	993	144	19	38	...	...	REP	830	1525	...	42
VHP (760 °C, or 1400 °F) (as-compacted)	972	141	1014	147	16	38	...	...	REP	760	1400	...	42
ROC (900 °C, or 1650 °F) (as-compacted)	882	128	904	131	14	50	...	...	PREP	900	1650	As-ROC	46
ROC (900 °C, or 1650 °F) and RA (925 °C, or 1695 °F)	827	120	882	128	16	46	...	...	PREP	900	1650	925 °C (1695 °F) RA	46
ROC (650 °C, or 1200 °F) (as-compacted)	1131	164	1179	171	10	23	...	...	PREP	600	1110	As-ROC	48
ROC (600 °C, or 1100 °F) and RA (815 °C, or 1500 °F)	965	140	1020	148	15	43	...	...	PREP	600	1110	815 °C (1500 °F) RA	48
Minimum properties (MIL-T-9047)	827	120	896	130	10	25	...	...	...	...	...	...	4

(a) HIP, hot isostatic pressing; PSV, *pulverization sous vide* (powder under vacuum), French-made powder (Ref 42); BUS, broken-up structure; TCP, thermochemical processing; REP, rotating-electrode process; STA, solution treated and aged; PREP, plasma rotating-electrode process; ELI, extra-low interstitial; HPLT, high-pressure low-temperature compaction; RA, recrystallization annealed; T, transverse; L, longitudinal; LT, longitudinal-transverse; TL, transverse-longitudinal; VHP, vacuum hot pressing; ROC, rapid omnidirectional compaction. Source: Ref 30, 42, 43, 46, 48, 50–58

aerospace alloy parts at a cost of under \$100/kg (\$45/lb). Prealloyed powders, on the other hand, require an expensive melt stock, ultraclean handling, and expensive compaction tools. As a result, the higher-performance fully dense PA parts are currently priced above \$200/kg (\$90/lb), although it is projected that volume production will bring this price down substantially. The differences in density, property, and price target these two technologies to different application markets. Because of their lower cost, more BE components than PA parts are currently in use. The introduction of gas-atomized powder (Ref 36) is expected to lower powder costs and make PA products more cost competitive.

Blended Elemental Products. On the low end of the density scale (20 to 80%), commercially pure (CP) titanium filters are produced for electrochemical and other corrosion-resistant applications. Higher-density pressed and sintered CP titanium parts are made commercially for the chemical-processing industry.

For more demanding applications, Ti-6Al-4V BE components with densities from 98% to close to 100% are produced by CIP and sintering and by pressing and sintering methods. Very complex shapes, such as impellers or pivot fittings, can be produced by CIP using elastomeric molds (Ref 7). Part size is currently limited to a maximum length of 610 mm (24 in.) by the availability of CIP equipment. The possible dimensional tolerances for small parts are ±0.5 mm (±0.02 in.). The pressing and sintering method is more volume oriented.

Prealloyed Products

The relatively high product cost of PA parts has thus far limited the consideration of potential applications of PA technology for the most part to the manufacture of critical aerospace components. As previously discussed, property levels have been attained in properly processed PA compacts that match those of high-quality I/M parts. The decision-making process for using PA P/M is based primarily on economic considerations.

Future Trends in Titanium P/M Technology

Clearly, sufficient data are now available to allow both prealloyed and blended elemental Ti-6Al-4V P/M compacts to be used with confidence. However, cost remains a major concern. Use of the PA P/M approach is difficult to justify for parts with approximately the same, or perhaps even slightly higher, mechanical property levels. Only a significantly lower cost for PA compacts would enable them to replace reliable I/M materials. The production of a low-cost powder either by a scaled-up gas atomization process or by a direct chemical method would be a significant breakthrough. The BE P/M technique can be cost effective for less critical parts, and its increased use in this area is likely. However, as with the PA material, the production of fully dense chloride-free BE products is likely to be stymied by cost unless a breakthrough occurs.

The trend with other conventional titanium P/M alloys is likely to follow that previously described for the Ti-6Al-4V alloy. An exception is the high-strength Ti-1.3Al-8V-5Fe (Ti-185) alloy (Ref 67–69), which cannot be satisfactorily made by the I/M approach because of the segregation of the iron. This alloy should be strictly classified with rapid solidification alloys (discussed below) that require rapid transformation from the liquid to solid states.

The ordered intermetallic titanium aluminides are much more difficult to process and machine than the conventional titanium alloys. In fact, their production characteristics approach those of the superalloys. Thus, the cost benefits to be gained by a net-shape P/M approach are great (particularly for the equiatomic titanium aluminide, TiAl), and these benefits could accelerate the acceptance of titanium P/M methods, particularly with the development of a lower-cost powder.

Research is being conducted to expand the boundaries of conventional titanium P/M technology. Efforts are in progress to evaluate the possibility of using advanced techniques, such as rapid solidification (Ref 76, 77), mechanical al-

Table 4 Tensile and fracture toughness properties of PA titanium alloy compacts processed under various conditions

Alloy and condition(a)	0.2% yield strength, MPa	0.2% yield strength, ksi	Ultimate tensile strength, MPa	Ultimate tensile strength, ksi	Elongation, %	Reduction in area, %	K_{Ic} or (K_Q), MPa$\sqrt{m}$	K_{Ic} or (K_Q), ksi$\sqrt{in.}$	Titanium PA powder preparation: Powder process(b)	Compaction temperature, °C	Compaction temperature, °F	Other variables	Ref
Ti-5.5Al-3.5Sn-3Zr-0.25Mo-1Nb-0.25Si (IMI 829)													
HIP and STA (1060–620 °C, or 1940–1150 °F)	951	138	1089	158	18	22	...	...	PREP	1040	1905	...	59
ROC and STA (1060–620 °C, or 1940–1150 °F)	909	132	1034	150	18	20	...	...	PREP	...	...	α + β ROC	59
Ti-6Al-5Zr-0.5Mo-0.25Si (IMI 685)													
HIP and STA (1050–550 °C, or 1920–1020 °F)	970	141	1020	148	11	19	...	...	PREP	950	1740	...	60
Ti-6Al-2Sn-4Zr-2Mo													
HIP and STA (1050–550 °C, or 1920–1020 °F)	924	134	1034	150	17	36	...	...	PREP	910	1670	...	61
Ti-6Al-2Sn-4Zr-6Mo													
HIP, forged (920 °C, or 1690 °F), and annealed (705 °C, or 1300 °F)	1165	169	1296	188	11	37	...	...	REP	900	1650	920 °C (1690 °F), 70% forging reduction	62
Ti-6Al-6V-2Sn													
HIP and annealed (760 °C, or 1400 °F)	1008	146	1055	153	18	37	59	54	PREP	900	1650	...	63, 64
Ti-5Al-2Sn-2Zn-4Cr-4Mo (Ti-17)													
HIP and STA (800–635 °C, or 1470–1175 °F)	1123	163	1192	173	8	11	...	...	REP	915	1680	...	52
Ti-4.5Al-5Mo-1.5Cr (Corona 5)													
HIP and aged (705 °C, or 1300 °F)	944	137	999	145	13	...	(75)	(68)	REP	845	1555	(c)	65
HIP and aged (760 °C, or 1400 °F)	916	133	971	141	14	...	(79)	(72)	REP	845	1555	(c)	65
Ti-10V-2Fe-3Al													
HIP and STA (745–490 °C, or 1375–915 °F)	1213	176	1310	190	9	13	...	...	PREP	775	1425	...	33
HIP, forged, and STA (750–495 °C, or 1380–925 °F)	1286	186	1386	201	7	20	28	25	PREP	775	1425	750 °C (1380 °F), 70% forging reduction	33
HIP, forged, and STA (750–550 °C, or 1380–1020 °F)	1065	155	1138	165	14	41	55	50	PREP	775	1425	750 °C (1380 °F), 70% forging reduction	33
ROC (as-compacted)	965	140	1007	146	16	54	...	...	PREP	650	1200	...	48
ROC and STA (760–510 °C, or 1400–950 °F)	1296	188	1400	203	6	26	...	...	PREP	650	1200	...	...
Ti-11.5Mo-6Zr-4.5Sn (Beta III)													
β HIP and STA (745–510 °C, or 1375–950 °F)	1288	187	1378	200	8	18	...	...	PREP	760	1400	...	66
Ti-1.3Al-8V-5Fe													
β extruded and STA (705 °C, or 1300 °F)	1392	202	1482	215	8	7	...	...	PREP	760	1400	...	67
β extruded and STA (770 °C, or 1420 °F)	1461	212	1516	220	8	20	...	...	GA	760	1400	...	68
β HIP and STA (675 °C, or 1245 °F)	1315	191	1414	205	5	10	...	...	GA	725	1335	...	69
Ti-24Al-11Nb													
HIP (1065 °C, or 1950 °F) and STA (1175 °C, or 2145 °F)	510	74	606	88	2	2	...	...	PREP	1065	1950	...	70
HIP (925 °C, or 1700 °F) and STA (1175 °C, or 2145 °F)	696	101	765	111	2	2	...	...	PREP	925	1695	...	71
Ti-25Al-10Nb-3Mo-1V													
ROC (as-compacted)	710	103	854	124	5	6	...	...	PREP	1050	1920	...	72

(a) HIP, hot isostatic pressing; STA, solution treated and aged; ROC, rapid omnidirectional compaction. (b) PREP, plasma rotating-electrode process; REP, rotating-electrode process; GA, gas atomization. (c) Weld study sample. Source: Ref 33, 48, 59–72

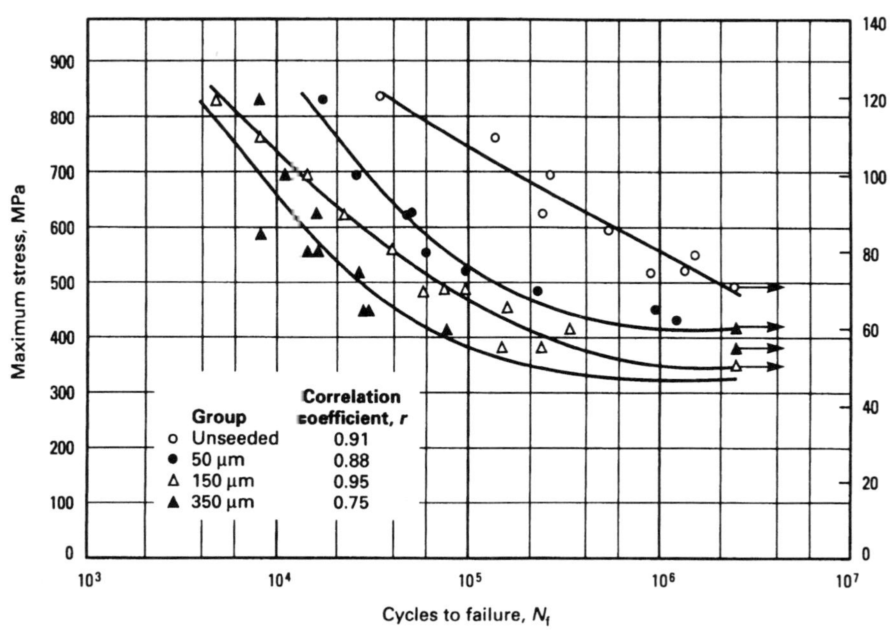

Fig. 11 Effect of contaminant particles on the room-temperature fatigue strength of Ti-6Al-4V PA compacts. Unseeded compacts are compared with SiO_2-seeded PREP compacts. Stress ratio (R), +0.1; frequency (f), triangular waveform load/time cycle at 5 Hz. Source: Ref 49

loying (Ref 78), and nanostructures (Ref 79), to enhance the behavior of titanium-base materials.

Work on rapid solidification has focused on increasing the temperature capability of both terminal alloys and intermetallic compositions by dispersion strengthening. However, while some improvements have been made, they are not considered to be significant enough to warrant the extra cost and concern over product quality assurance associated with a P/M method. The rapid solidification technique has two major drawbacks. The first is that it has been unable to produce more than about 6 vol% of second-phase particles. Second, rapid solidification results in a smaller than desirable β grain size and a lack of elongated α phase formed on cooling into the α-β phase field after a β anneal. Unfortunately, a β anneal, which corrects the second drawback, results in unacceptable coarsening of the dispersoids.

Mechanical alloying of titanium alloys is at a very early stage, but it does exhibit the potential to increase the volume percentage of dispersoids for elevated-temperature applications. In addition, there are indications that the normally immiscible titanium and magnesium can be com-

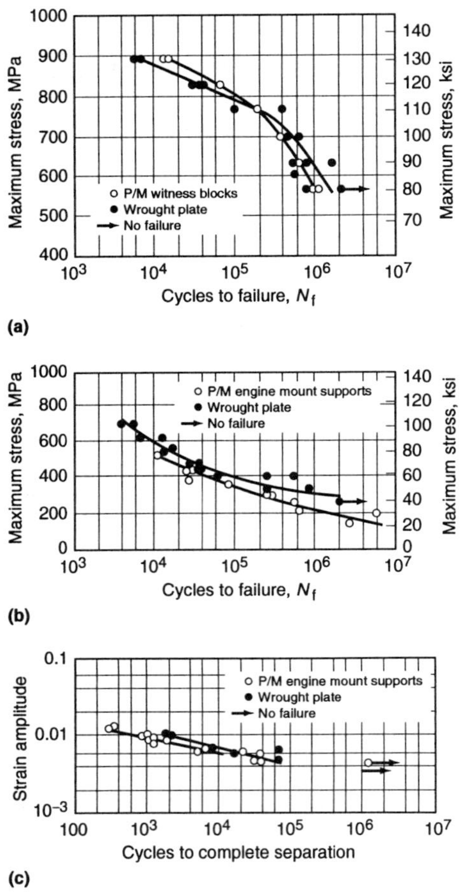

Fig. 12 Comparison of the fatigue strengths at room temperature in air of PREP HIP Ti-6Al-4V P/M components to those of I/M products. (a) Load-controlled smooth-specimen high-cycle fatigue for large bars (13 mm, or ½ in., in diameter). Stress ratio (*R*), 0.1. (b) Load-controlled notched-specimen high-cycle fatigue. Stress concentration factor (K_t), 3; stress ratio (*R*), 0.1. (c) Strain-controlled low-cycle fatigue for small specimens (6.4 mm, or ¼ in., in diameter). Stress ratio (*R*), –1. Source: Ref 73

bined by mechanical alloying to produce a low-density titanium alloy.

Very preliminary results on the mechanical alloying of titanium-magnesium and titanium-eutectoid formers such as nickel and copper suggest that a very fine nanoscale microstructure (~10^{-9} m scale) can be obtained; such a microstructure could have novel physical and mechanical properties. Further information on powder production and P/M applications of titanium alloys is in the article "Production of Titanium Powder" in this Volume.

Particulate Reinforced Titanium

S. Abkowitz, Dynamet Technology Inc.

Since the late 1980's, the unique benefits of the solid-phase processing of P/M titanium alloys have been used in the development of discontinuously reinforced titanium matrix composites. For example, titanium carbide and titanium diboride powders have been added to both commercially pure titanium and the common titanium alloys utilizing the BE technique (Ref 34, 80–83). The resultant family of particulate-reinforced metal matrix composites (PR-MMCs) provide improved wear-resistance, elastic modulus, creep, fatigue, and corrosion properties with less than a 3% change in alloy density. Elongation and, to a lesser extent, fracture toughness decrease in the tradeoff. The following examples typify several titanium PR-MMCs, both in current production and under development.

Among the first titanium-base PR-MMC materials introduced to the commercial market was CermeTi-C (Dynamet Technology Inc., Burlington, MA). These composites are created by the BE technique with the addition of titanium carbide (TiC) to prominent titanium alloys, such as Ti-6Al-4V and Ti-6Al-6V-2Sn. The titanium alloy matrix is formed during sintering by diffusion-driven solid state alloying, typically at temperatures less than 250 °C (450 °F) above the β transus, or 1230 °C (2245 °F) for Ti-6Al-4V. As opposed to melt processing, the thermal decomposition of the TiC reinforcing particulate (melting point 3065 °C) can be limited to a practical extent during the sintering cycle. The resultant PR-MMC displays improved strength and elastic modulus, reflecting the load-sharing contribution of the tightly-bonded TiC particulate (density of 4.92 g/cm^3, elastic modulus of 448 GPa, and hardness of 3200 HV (Ref 84).

From 5 to 20 wt% TiC is added to commonly used titanium alloys. Typical properties are summarized in Table 5. As an alternative to HIP or in addition to HIP operation, hot working (including forging, rolling, and extrusion) further increases the density and mechanical properties of the titanium alloy PR-MMC. The mechanically homogenized particulate distribution and the refined acicular α-β microstructure are retained by air cooling from the forging or extrusion temperature with a commensurate increase in tensile strength and ductility, as shown in Table 5. High-temperature rigidity and yield strength are shown in Fig. 14 and 15.

The addition of TiB_2 particulate by the BE technique is also done (Ref 85, 86). During normal sintering at 1200 °C (2190 °F), the TiB_2 particulate, in contact with the elemental titanium powder, reacts extensively or completely to form TiB. Upon cooling, the resultant TiB phase platelets extend beyond the boundaries of the prior titanium particles and bridge multiple α-β colonies. Figure 14 shows the effect of elevated temperature on the modulus of TiB_2-Ti-6Al-4V composite.

In an effort to increase high-cycle fatigue properties over those of the BE matrix alloy, two BE titanium alloy matrix materials, Ti-6Al-2Sn-4Zr-2Mo and Ti-6Al-1.7Fe-0.1Si, have been evaluated with a 5 wt% TiB_2 particulate reinforcement (Ref 87, 88). Starting with low residual chloride, hydride-dehydride Ti powder, and

Table 5 Comparison of mechanical properties of BE P/M titanium alloys reinforced with TiC or TiB_2 particulate

BE titanium alloy matrix	Yield strength, MPa	Yield strength, ksi	Ultimate tensile strength, MPa	Ultimate tensile strength, ksi	Elongation, %	Reduction in area, %	Modulus, GPa	K_Q or K_{IC}, MPa$\sqrt{m}$	Oxygen, ppm
Ti-6Al-4V (min)(a)(b)	828	120	895	130	10	25	114	...	2000 max
Ti-6Al-4V ELI (min)(c)	759	110	828	120	10	25	114	...	1300 max
Ti-6Al-4V BE CHIP(d)	690	100	793	115	9	15	114	85	1500
Ti-6Al-4V BE CHIP(d)	793	115	896	130	10	20	114	83	2500
Ti-6Al-6V-2Sn (min)	965	140	1034	150	14	23	112	75	2000
Ti-6Al-6V-2Sn CHIP(d)	931	140	1034	150	15	35	112	78	2800
Cerme Ti-C-10/64 CHIP(d)	945	137	1000	145	2	2	134	38	3000
Cerme Ti-C-10/64 (e, f)	1069	155	1138	165	2	2	134	45	3000
Cerme Ti-C-10/662 (e, f)	1034	150	1055	153	1	1	134	40	3000

(a) AMS-4930RevD. (b) ASTM B 348 grade 5. (c) ASTM B 348 grade 23. (d) Cold and hot isostatic pressing, CHIP. (e) Extruded 960 °C + air cooled. (f) Forged 1030 °C + air cooled

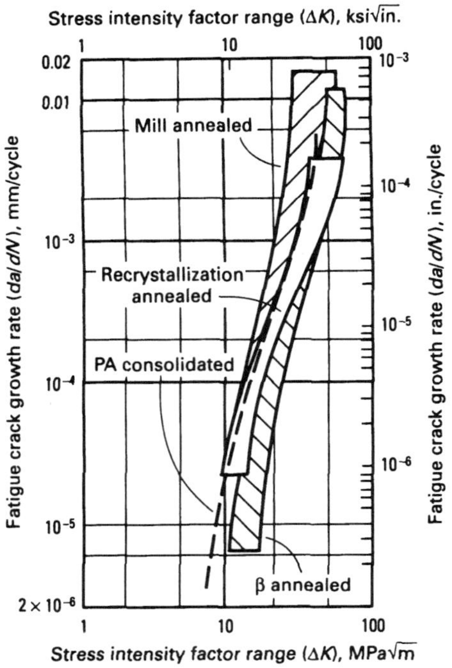

Fig. 13 Comparison of the fatigue crack growth rate at room temperature in air of Ti-6Al-4V PA compacts with that of an I/M alloy material. Stress ratio (*R*), 0.1; frequency (*f*), 5 to 30 Hz (5 Hz for a PA compact). Source: Ref 74

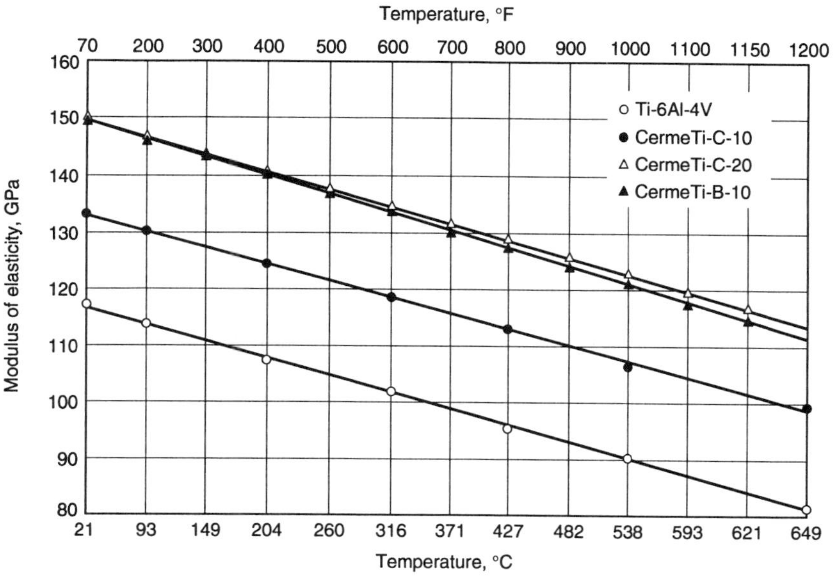

Fig. 14 Modulus of elasticity for Ti-6Al-4V and particulate reinforced composites of Ti-6Al-4V

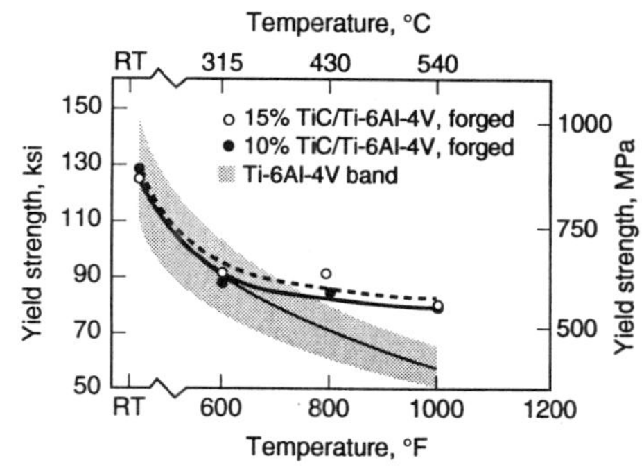

Fig. 15 Yield strength of Ti-C reinforced Ti-6Al-4V after forging. RT, room temperature

master alloy powders, the blend was mixed with 5 wt% TiB_2 and processed as follows:

- CIP at 400 MPa
- Sintered at 1200 °C (2190 °F) for 3 h
- HIP at 930 °C and 200 MPa

The in situ formation of TiB platelets, with an average measured length of 30 μm and width of 2 μm, regardless of matrix, appeared to decrease the aspect ratio of the α platelets as well. After water-quenching from 1050 °C and a 930 °C anneal, the 10% TiB reinforced Ti-6Al-2Sn-4Zr-2Mo matrix achieved high cycle (10^7) fatigue strength of 590 MPa, a 35% increase in tensile strength, and a modulus of 140 GPa (Table 4).

At the Toyota Central Research and Development Laboratories, Inc., Saito et al. have continued efforts to reduce the cost of BE TiB_2-based PR-MMCs of titanium alloys (Ref 89–93). When 5 wt% TiB is added to BE-derived, metastable β matrix, composed of Ti-4.3Fe-7.0Mo-1.4Al-1.4V, a derivative of Timetal LCB (TIMET, Denver, CO), the resultant 10%TiB reinforcement forms a stable, crystallographically-oriented, coherent interface boundary, increasing the strength, stiffness, hardness, fatigue properties, and heat resistance of the matrix (Table 4). Deformation flow stress of this material at 700 °C approximates that of medium carbon steel, aiding producibility. The authors claim that production cost estimates for automotive engine parts (such as connecting rods or gears), produced from this PR-MMC material, almost compete with steel, but require a stable supply of low-cost titanium powder.

Wear resistance, specific (density-normalized) stiffness, and specific strength are primary considerations in the selection of titanium PR-MMC material for typical applications. The combination of near-net sizing, the absence of an alpha case in the CHIP condition, and machinability similar to the matrix alloy enhance the producibility of BE titanium PR-MMCs. Prospective performance-driven applications of these BE titanium PR-MMC materials include automotive valves, connecting rods, sporting goods, aerospace and defense missile structures, and industrial tooling.

Future Trends. The practical utilization of titanium PR-MMC materials is growing on three fronts: design, matrix and particulate composition, and producibility. The potential to design and build BE structures with controlled gradients in physical (thermal expansion, density) and mechanical (wear, toughness, fatigue) properties requires close technical interaction between product designer and producer. Because the BE CHIP process allows multiple layers of different titanium alloys and composites to be built up in specific areas before cold isostatic pressing, the concept holds promise as a means of optimizing structural and environmental performance. As examples of increasing design awareness, particulate reinforced composites are under evaluation for clad missile fins and engine valves.

The optimization of BE alloy and particulate compositions will continue under increasing market pressure to reduce production cost without diminishing specific mechanical properties. The ongoing differentiation between automotive and aerospace PR-MMCs continues as the trade-offs between mechanical property requirements and production economics becomes more focused. Efforts to reduce the cost of the BE titanium matrix constituents, to limit impurities (oxygen and chlorine), and to refine P/M process parameters for consistent production densities above 99% are technical goals of both markets.

Producibility and shape-making technology considerations are driving research in superplastic forming and investment casting of titanium-base PR-MMCs. Elongations of more than 200% have been achieved in TiC reinforced materials in ongoing research in transformation mismatch superplasticity (Ref 94–96). Recent investment casting trials of TiC composite of Ti-6Al-4V showed improved fluidity, castability, and weldability versus Ti-6Al-4V in thin-walled components.

Titanium Aluminides

D. Eylon, University of Dayton

Interest in titanium aluminide alloys based on the $\alpha_2 Ti_3Al$ and γTiAl phases began in the 1970s, and powder metallurgy was one of the first alloy processing solutions due to the difficulty in processing alloy cast ingots. Plasma rotating electrode process (PREP) powder and gas atomized (GA) powders were good solutions to achieve uniform chemistry and microstructure materials (Ref 98, 99) for both α_2 (Ref 100, 101) and γ based (Ref 102, 103) alloys. An earlier work on rapidly solidified titanium alloy particulate material (Ref 104) showed that lower temperature HIP compaction can be achieved by increasing the pressure from 100 to 300 MPa, thus retaining the ultrafine particulate structure, which leads to higher strength and ductility. Similar practice resulted in α_2(Ti-25Al-10Nb-3V-1Mo) and γ(Ti-48Al-2Cr-2Nb) powder compacts with tensile properties better than wrought (Ref 100, 103) due to the retention of fine equiaxed grain structure. The fine equiaxed structure PM compacts were also suitable preforms for forging (Ref 105–108) and sheet rolling (Ref 109, 110, 111). The γ alloy P/M rolled sheets are of a very high quality and are now marketed by Plansee in Austria (Ref 110).

REFERENCES

1. H.B. Bomberger, F.H. Froes, and P.H. Morton, Titanium—A Historical Perspective, *Titanium Technology: Present Status and Future Trends*, F.H. Froes, D. Eylon, and H.B. Bomberger, Ed., Titanium Development Association, 1985, p 3–17

2. F.H. Froes, D. Eylon, G.E. Eichelman, and H.M. Burte, Developments in Titanium Powder Metallurgy, *J. Met.*, Vol 32 (No. 2), Feb 1980, p 47–54
3. F.H. Froes and D. Eylon, Titanium Powder Metallurgy—A Review, *Titanium Net-Shape Technologies*, F.H. Froes and D. Eylon, Ed., The Metallurgical Society of AIME, 1984, p 1–20
4. F.H. Froes and D. Eylon, Powder Metallurgy of Titanium Alloys—A Review, *Titanium, Science and Technology*, Vol 1, G. Lutjering, U. Zwicker, and W. Bunk, Ed., DGM, 1985, p 267–286; *Powder Metall. Int.*, Vol 17 (No. 4), 1985, p 163–167, continued in Vol 17 (No. 5), 1985, p 235–238; *Titanium Technology: Present Status and Future Trends*, F.H. Froes, D. Eylon, and H.B. Bomberger, Ed., Titanium Development Association, 1985, p 49–59
5. F.H. Froes and D. Eylon, Powder Metallurgy of Titanium Alloys, *Int. Mater. Rev.*, in press
6. F.H. Froes and D. Eylon, Production of Titanium Powder, *Metals Handbook*, Vol 7, *Powder Metallurgy*, 9th ed., American Society for Metals, 1984, p 164–168
7. F.H. Froes, D. Eylon, and G. Friedman, Titanium P/M Technology, *Metals Handbook*, Vol 7, *Powder Metallurgy*, 9th ed., American Society for Metals, 1984, p 748–755
8. Automotive Applications, *Metals Handbook*, Vol 7, *Powder Metallurgy*, 9th ed., American Society for Metals, 1984, p 617–621
9. R.W. Stevenson, P/M Copper-Based Alloys, *Metals Handbook*, Vol 7, *Powder Metallurgy*, 9th ed., American Society for Metals, 1984, p 733–740
10. T.W. Penrice, Kinetic Energy Penetrators, *Metals Handbook*, Vol 7, *Powder Metallurgy*, 9th ed., American Society for Metals, 1984, p 688–691
11. G.H. Gessinger, *Powder Metallurgy of Superalloys*, Butterworths, 1984
12. B.L. Ferguson, Aerospace Applications, *Metals Handbook*, Vol 7, *Powder Metallurgy*, 9th ed., American Society for Metals, 1984, p 646–651
13. R.W. Stevenson, Aluminum P/M Technology, *Metals Handbook*, Vol 7, *Powder Metallurgy*, 9th ed., American Society for Metals, 1984, p 741–748
14. Beryllium P/M Technology, in *Metals Handbook*, Vol 7, *Powder Metallurgy*, 9th ed., American Society for Metals, 1984, p 755–762
15. F.H. Froes and D. Eylon, Titanium Powder Metallurgy—A Review, in *PM Aerospace Materials*, Vol 1, MPR Publishing, 1984, p 39–1 to 39–19
16. F.H. Froes, C.M. Cooke, D. Eylon, and K.C. Russell, Grain Growth Blended Elemental Ti-6Al-4V Powder Compacts, *Sixth World Conference on Titanium*, Part III, P. Lacombe, R. Tricot, and G. Beranger, Ed., Les Editions de Physique, 1989, p 1161–1166
17. I. Weiss, D. Eylon, M.W. Toaz, and F.H. Froes, Effect of Isothermal Forging on Microstructure and Fatigue Behavior of Blended Elemental Ti-6Al-4V Powder Compacts, *Metall. Trans. A*, Vol 17 (No. 3), 1986, p 549–559
18. H.I. Aaronson, D. Eylon, and F.H. Froes, Observations of Superledges Formed on Sideplates during Precipitation of Alpha from Beta Ti-6%Al-4%V, *Scr. Metall.*, Vol 21 (No. 11), 1987, p 1421–1425
19. G. Welsch, Y.-T. Lee, P.C. Eloff, D. Eylon, and F.H. Froes, Deformation Behavior of Blended Elemental Ti-6Al-4V Compacts, *Metall. Trans. A*, Vol 14 (No. 4), 1983, p 761–769
20. D. Eylon, R.G. Vogt, and F.H. Froes, Property Improvement of Low Chlorine Titanium Alloy Blended Elemental Powder Compacts by Microstructure Modification, *Progress in Powder Metallurgy*, Vol 42, compiled by E.A. Carlson and G. Gaines, Metal Powder Industries Federation, 1986, p 625–634
21. P.J. Andersen, V.M. Svoyatytsky, F.H. Froes, Y. Mahajan, and D. Eylon, Fracture Behavior of Blended Elemental P/M Titanium Alloy, *Modern Developments Powder Metallurgy*, Vol 13, H.H. Hausner, H.W. Antes, and G.D. Smith, Ed., Metal Powder Industries Federation, 1981, p 537–549
22. J. Park, M.W. Toaz, D.H. Ro, and E.N. Aqua, Blended Elemental Powder Metallurgy of Titanium Alloys, *Titanium Net Shape Technologies*, F.H. Froes and D. Eylon, Ed., The Metallurgical Society of AIME, 1984, p 95–105
23. S. Abkowitz, Isostatic Pressing of Complex Shapes from Titanium and Titanium Alloys, *Powder Metallurgy of Titanium Alloys,* F.H. Froes and J.E. Smugeresky, Ed., The Metallurgical Society of AIME, 1980, p 291–302
24. S. Abkowitz, G.J. Kardys, S. Fujishiro, F.H. Froes, and D. Eylon, Titanium Alloy Shapes from Elemental Blend Powder and Tensile and Fatigue Properties of Low Chloride Compositions, *Titanium Net Shape Technologies,* F.H. Froes and D. Eylon, Ed., The Metallurgical Society of AIME, 1984, p 107–120
25. R.R. Boyer, J.E. Magnuson, and J.W. Tripp, Characterization of Pressed and Sintered Ti-6Al-4V Powders, *Powder Metallurgy of Titanium Alloys,* F.H. Froes and J.E. Smugeresky, Ed., The Metallurgical Society of AIME, 1980, p 203–216
26. Y. Mahajan, D. Eylon, R. Bacon, and F.H. Froes, Microstructure Property Correlation in Cold Pressed and Sintered Elemental Ti-6Al-4V Powder Compacts, *Powder Metallurgy of Titanium Alloys,* F.H. Froes and J.E. Smugeresky, Ed., The Metallurgical Society of AIME, 1980, p 189–202
27. D. Eylon, R.G. Vogt, and F.H. Froes, Property Improvement of Low Chlorine Titanium Alloy Blended Elemental Powder Compacts by Microstructure Modification, *Progress in Powder Metallurgy,* Vol 42, compiled by E.A. Carlson and G. Gaines, Metal Powder Industries Federation, 1986, p 625–634
28. P.R. Smith, C.M. Cooke, A. Patel, and F.H. Froes, in *Progress in Powder Metallurgy,* Vol 38, J.G. Bewley and S.W. McGee, Ed., Metal Powder Industries Federation, 1983, p 339–359
29. P.R. Smith, F.H. Froes, and C.M. Cooke, in *Materials and Processes—Continuing Innovations,* Vol 28, Society for the Advancement of Material and Process Engineering, 1983, p 406–421
30. C.F. Yolton, D. Eylon, and F.H. Froes, Microstructure Modification of Titanium Alloy Products by Temporary Alloying with Hydrogen, *Sixth World Conf. on Titanium,* Part III, P. Lacombe, R. Tricot, and G. Beranger, Ed., Les Editions de Physique, 1989, p 1641–1646
31. M. Hagiwara, Y. Kaieda, and Y. Kawabe, Improvement of Mechanical Properties of Blended Elemental α-β Ti Alloys by Microstructural Modification, *Titanium 1986, Products and Applications,* Vol II, Titanium Development Association, 1987, p 850–858
32. J.E. Smugeresky and N.R. Moody, Properties of High Strength, Blended Elemental Powder Metallurgy Titanium Alloys, *Titanium Net-Shape Technologies,* F.H. Froes and D. Eylon, Ed., The Metallurgical Society of AIME, 1984, p 131–143
33. R.R. Boyer, D. Eylon, C.F. Yolton, and F.H. Froes, Powder Metallurgy of Ti-10V-2Fe-3Al, *Titanium Net-Shape Technologies,* F.H. Froes and D. Eylon, Ed., The Metallurgical Society of AIME, 1984, p 63–78
34. S. Abkowitz and P. Weithrauch, Trimming the Cost of MMC, *Adv. Mater. Proc.,* Vol 136 (No. 1), July 1989, p 31–34
35. F.H. Froes, H.B. Bomberger, D. Eylon, and R.G. Rowe, Potential of Titanium Powder Metallurgy, *Competitive Advances in Metals and Processes,* Vol 1, R.J. Cunningham and M. Schwartz, Ed., Society for the Advancement of Material and Process Engineering, 1987, p 240–254
36. C.F. Yolton, Gas Atomized Titanium and Titanium Aluminide Alloys, in *Powder Metallurgy in Aerospace and Defense Technologies,* Metal Powder Industries Federation, 1989
37. E.J. Kosinski, The Mechanical Properties of Titanium P/M Parts Produced From Superclean Powders, *Progress in Powder Metallurgy,* Vol 38, J.G. Bewley and S.W. McGee, Ed., Metal Powder Industries Federation, 1983, p 491–592
38. P.R. Roberts and P. Loewenstein, Titanium Alloy Powders Made by the Rotating Electrode Process, *Powder Metallurgy of Titanium Alloys,* F.H. Froes and J.E. Smugeresky, Ed., The Metallurgical Society of AIME, 1980, p 21–35
39. J.P. Laughlin and G.J. Dooley III, The Hydride Process for Producing Titanium Alloy Powders, *Powder Metallurgy of Titanium Alloys,* F.H. Froes and J.E. Smugeresky, Ed., The Metallurgical Society of AIME, 1980, p 37–46
40. J.A. Megy, U.S. Patent 4,127,409, Nov 1978
41. G. Buttner, H.-G. Domazer, and H. Eggert, U.S. Patent 4,373, 947, Feb 1983
42. W.H. Kao, D. Eylon, C.F. Yolton, and F.H. Froes, Effect of Temporary Alloying by Hydrogen (Hydrovac) on the Vacuum Hot Pressing and Microstructure of Titanium Alloy Powder Compacts, *Progress in Powder Metallurgy,* Vol 37, J.M. Capus and D.L. Dyke, Ed.,

Metal Powder Industries Federation, 1982, p 289–301
43. J. Devillard and J.-P. Herteman, Evaluation of Ti-6Al-4V Powder Compacts Fabricated by the PSV Process, *Powder Metallurgy of Titanium Alloys,* F.H. Froes and J.E. Smugeresky, Ed., The Metallurgical Society of AIME, 1980, p 59–70
44. I.A. Martorell, Y.R. Mahajan, and D. Eylon, "Property Modification of Ti-10V-2Fe-3Al by Low Temperature Processing," Unpublished report, 1987
45. C.A. Kelto, Rapid Omnidirectional Compaction, *Metals Handbook,* Vol 7, 9th ed., *Powder Metallurgy,* American Society for Metals, 1984, p 542–546
46. Y.R. Mahajan, D. Eylon, C.A. Kelto, T. Egerer, and F.H. Froes, Modification of Titanium Powder Metallurgy Alloy Microstructures by Strain Energizing and Rapid Omnidirectional Compaction, *Titanium, Science and Technology,* Vol 1, G. Lutjering, U. Zwicker, and W. Bunk, Ed., DGM, 1985, p 339—346; *Powder Metall. Int.,* Vol 17 (No. 2), 1985, p 75–78; *Titanium Net-Shape Technologies,* F.H. Froes and D. Eylon, Ed., The Metallurgical Society of AIME, 1984, p 39–51
47. Y.R. Mahajan, D. Eylon, C.A. Kelto, and F.H. Froes, Evaluation of Ti-10V-2Fe-3Al Powder Compacts Produced by the ROC Method, *Progress in Powder Metallurgy,* Vol 41, H.I. Sanderow, W.L. Giebelhausen, and K.M. Kulkarni, Ed., Metal Powder Industries Federation, 1986, p 163–171; *Met. Powder Rep.* Vol 41 (No. 10), Oct 1986, p 749–752
48. D. Eylon, C.A. Kelto, A.F. Hayes, and F.H. Froes, Low Temperature Compaction of Titanium Alloys by Rapid Omnidirectional Compaction (ROC), *Progress in Powder Metallurgy,* Vol 43, compiled by C.L. Freeby and H. Hjort, Metal Powder Industries Federation, 1987, p 33–47
49. S.W. Schwenker, D. Eylon, and F.H. Froes, Influence of Foreign Particles on Fatigue Behavior of Ti-6Al-4V Prealloyed Powder Compacts, *Metall. Trans. A,* Vol 17 A (No. 2), 1986, p 271–280
50. J.-P. Herteman, D. Eylon, and F.H. Froes, Mechanical Properties of Advanced Titanium Powder Metallurgy Compacts, *Titanium, Science and Technology,* Vol 1, G. Lutjering, U. Zwicker, and W. Bunk, Ed., DGM, 1985, p 303–310; *Powder Metall. Int.,* Vol 17 (No. 3), 1985, p 116–118
51. L. Levin, R.G. Vogt, D. Eylon, and F.H. Froes, Fatigue Resistance Improvement of Ti-6Al-4V by Thermo-Chemical Treatment, *Titanium, Science and Technology,* Vol 4, G. Lutjering, U. Zwicker, and W. Bunk, Ed., DGM, 1985, p 2107–2114
52. R.E. Peebles and C.A. Kelto, Investigation of Methods for the Production of High Quality, Low Cost Titanium Alloy Powders, *Powder Metallurgy of Titanium Alloys,* F.H. Froes and J.E. Smugeresky, Ed., The Metallurgical Society of AIME, 1980, p 47–58
53. R.E. Peebles and L.D. Parsons, Study of Production Methods of Aerospace Quality Titanium Alloy Powder, *Titanium Net-Shape Technologies,* F.H. Froes and D. Eylon, Ed., The Metallurgical Society of AIME, 1984, p 21–28
54. G.R. Chanani, W.T. Highberger, C.A. Kelto, and V.C. Petersen, Application of Titanium Powder Metallurgy for Manufacture of a Large and Complex Naval Aircraft Component, *Powder Metallurgy of Titanium Alloys,* F.H. Froes and J.E. Smugeresky, Ed., The Metallurgical Society of AIME, 1980, p 279–290
55. D. Eylon and F.H. Froes, HIP Compaction of Titanium Alloy Powders at High Pressure and Low Temperature (HPLT), *Met. Powder Rep.,* Vol 41 (No. 4), April 1986, p 287—293; *Titanium, Rapid Solidification Technology,* F.H. Froes and D. Eylon, Ed., The Metallurgical Society, 1986, p 273–289
56. R.F. Geisendorfer, Powder Metallurgy Titanium 6Al-4V Plate, *Powder Metallurgy of Titanium Alloys,* F.H. Froes and J.E. Smugeresky, Ed., The Metallurgical Society of AIME, 1980, p 151–162
57. R.F. Vaughan and P.A. Blenkinsop, A Metallurgical Assessment of Ti-6Al-4V Powder, *Powder Metallurgy of Titanium Alloys,* F.H. Froes and J.E. Smugeresky, Ed., The Metallurgical Society of AIME, 1980, p 83–92
58. D. Eylon, F.H. Froes, D.G. Heggie, P.A. Blenkinsop, and R.W. Gardiner, Influence of Thermomechanical Processing on Low Cycle Fatigue of Ti-6Al-4V Powder Compacts, *Metall. Trans. A,* Vol 14, 1983, p 2497–2505
59. N.R. Osborne, D. Eylon, and F.H. Froes, Compaction and Net-Shape Forming of Ti-829 Alloy by P/M ROC Processing, in *Advances in Powder Metallurgy,* compiled by T.G. Gasbarre and W.F. Jandeska, Metal Powder Industries Federation, 1989
60. B. Borchert, H. Schmid, and J. Wortmann, Microstructure and Strength of P/M Ti-685, *Titanium, Science and Technology,* Vol 1, G. Lutjering, U. Zwicker, and W. Bunk, Ed., DGM, 1985, p 295–302
61. V.K. Chandhok, J.H. Moll, C.F. Yolton, and G.R. McIndoe, Advances in P/M Titanium Shape Technology Using the Ceramic Mold Process, *Overcoming Material Boundaries,* Vol 17, Society for the Advancement of Material and Process Engineering, 1985, p 495–506
62. I. Weiss, F.H. Froes, D. Eylon, and C.C. Chen, Control of Microstructure and Properties of Ti-6Al-2Sn-4Zr-6Mo Powder Forgings, *Titanium Net-Shape Technologies,* F.H. Froes and D. Eylon, Ed., The Metallurgical Society of AIME, 1984, p 79–94
63. R.H. Witt and I.G. Weaver, Titanium P/M Components for Airframes, *Titanium Net-Shape Technologies,* F.H. Froes and D. Eylon, Ed., The Metallurgical Society of AIME, 1984, p 29—38
64. R.H. Witt and W.T. Highberger, Hot Isostatic Pressing of Near-Net Titanium Structural Parts, *Powder Metallurgy of Titanium Alloys,* F.H. Froes and J.E. Smugeresky, Ed., The Metallurgical Society of AIME, 1980, p 255–265
65. D.W. Becker, W.A. Baeslack III, and F.H. Froes, Welding of Corona 5 P/M Product, *Powder Metallurgy of Titanium Alloys,* F.H. Froes and J.E. Smugeresky, Ed., The Metallurgical Society of AIME, 1980, p 217–228
66. C.F. Yolton, P/M Beta Titanium Alloys for Landing Gear Applications, *Progress in Powder Metallurgy,* Vol 42, compiled by E.A. Carlson and G. Gaines, Metal Powder Industries Federation, 1986, p 635–653
67. R.G. Vogt, D. Eylon, and F.H. Froes, Production of High Strength Beta Titanium Alloy through Powder Metallurgy, *Titanium, Rapid Solidification Technology,* F.H. Froes and D. Eylon, Ed., The Metallurgical Society, 1986, p 195–199
68. R.R. Boyer, E.R. Barta, C.F. Yolton, and D. Eylon, P/M of High Strength Titanium Alloys, *Powder Metallurgy in Aerospace and Defense Technologies,* Metal Powder Industries Federation, 1989
69. C.F. Yolton and J.H. Moll, Evaluation of a High Strength Rapidly Solidified Titanium Alloy, *Progress in Powder Metallurgy,* Vol 43, compiled by C.L. Freeby and H. Hjort, Metal Powder Industries Federation, 1987, p 49–63
70. C.F. Yolton, T. Lizzi, V.K. Chandhok, and J.H. Moll, Powder Metallurgy of Titanium Aluminide Components, *Progress in Powder Metallurgy,* Vol 42, compiled by E.A. Carlson and G. Gaines, Metal Powder Industries Federation, 1986, p 479–488
71. V.S. Moxson and G.I. Friedman, Powder Metallurgy of Titanium Aluminides, *Progress in Powder Metallurgy,* Vol 42, compiled by E.A. Carlson and G. Gaines, Metal Powder Industries Federation, 1986, p 489–500
72. N.R. Osborne, W.J. Porter, and D. Eylon, Unpublished report, 1989
73. A.S. Sheinker, G.R. Chanani, and J.B. Bohlen, Evaluation and Application of Prealloyed Titanium P/M Parts for Airframe Structures, *Int. J. Powder Metall.,* Vol 23 (No. 3), 1987, p 171–176
74. S.W. Schwenker, A.W. Sommer, D. Eylon, and F.H. Froes, Fatigue Crack Growth Rate of Ti-6Al-4V Prealloyed Powder Compacts, *Metall. Trans. A,* Vol 14 (No. 7), July 1983, p 1524–1528
75. F.H. Froes and D. Eylon, Thermochemical Processing (TCP) of Titanium Alloys by Temporary Alloying with Hydrogen, *Hydrogen Effects on Material Behavior,* A.W. Thompson and N.R. Moody, Ed., The Metallurgical Society, 1990
76. F.H. Froes and R.G. Rowe, Rapidly Solidified Titanium, *Rapidly Solidified Alloys and Their Mechanical and Magnetic Properties,* Vol 58, B.C. Giessen, D.E. Polk, and A.I. Taub, Ed., Materials Research Society, 1986, p 309–334
77. R.G. Rowe and F.H. Froes, Titanium Rapid Solidification—Alloys and Processes, *Processing of Structural Metals by Rapid Solidification,* F.H. Froes and S.J. Savage, Ed., ASM International, 1987, p 163–173

78. R. Sundaresan and F.H. Froes, Mechanical Alloying, *J. Met.,* Vol 39 (No. 8), Aug 1987, p 22–27
79. F.H. Froes and C. Suryanarayana, Nanocrystalline Metals for Structural Applications, *J. Met.,* June 1989, p 12–17
80. S. Abkowitz, H.L. Heussi, and H.P. Ludwig, Titanium Carbide/Titanium Alloy Composite and Process for Powder Metal Cladding,U.S. Patent 4,731,115, 15 March 1988
81. S. Abkowitz, P.F. Weihrauch, and S.M. Abkowitz, Particulate-Reinforced Titanium Alloy Composites Economically Formed by Combined Cold and Hot Isostatic Pressing, *Ind. Heat.,* Vol LX (No.9), Sept, 1993, p 32–37
82. S.M. Abkowitz, P. Weihrauch, S. Abkowitz, and H. Heussi, The Commercial Application of Low-Cost Titanium Composites, *JOM,* Vol 47 (No. 8), Aug. 1995, p 40–41
83. S.M. Abkowitz et al., P/M Titanium Matrix Composites: From War Games to Fun & Games, *Eighth World Conf. on Titanium* (Birmingham, England), 23–26 Oct 1995
84. *Properties and Selection: Stainless Steels, Tool Materials, and Special-Purpose Metals,* Vol 3, *Metals Handbook,* 9th ed., American Society for Metals, 1980, p 543
85. S. Abkowitz, H.L. Heussi, H.P. Ludwig, D.M. Rowell, and S.A. Kraus, Titanium Diboride/Titanium Alloy Metal Matrix Microcomposite Material and Process for Powder Metal Cladding, U.S. Patent 4,906,430, 6 March 1990
86. S. Abkowitz, H.L. Heussi, H.P. Ludwig, D.M. Rowell, and S.A. Kraus, Titanium Diboride/Titanium Alloy Metal Matrix Microcomposite Material and Process for Powder Metal Cladding, U.S. Patent 4,968,348 6 Nov 1990
87. M. Hagiwara, S. Emura, J. Takahashi, Y. Kawabe, and N. Arimoito, *Titanium '92, Science and Technology,* Vol 3, F.H. Froes and I.L. Caplan, ed., TMS, 1992 p 2487–2494
88. M. Hagiwara, S. Emura, Y. Kawabe, and S.J. Kim, Low Cost Synthesis of P/M Ti Alloys and Ti-Based Particulate Composites with Improved High Cycle Fatigue Strength, *Proc. 1998 TMS Annual Meeting* (San Antonio, TX), TMS, in publication
89. T. Saito, T. Furata, T. Yamaguchi, and K. Ogino, *Proc. 1993 Powder Metallurgy World Congress—PM '93,* Y. Bando and K. Kosuge, Ed., JPMA-JSPM, Vol 1, 1993, p 642
90. T. Saito, T. Furata, and T. Yamaguchi, Development of Low Cost Titanium Matrix Composite Metallurgy and Technology of Practical Titanium Alloys, S. Fujishiro, D. Eylon, and T. Kishi, TMS, 1994
91. T. Saito, T. Furuta, and T. Yamaguchi, Fatigue Properties of TiB Particle Reinforced P/M Titanium Matrix Composite, *Recent Advances in Titanium Metal Matrix Composites,* F.H. Froes and J. Storer, Ed., TMS, 1995, p.133
92. T. Furata and T. Saito, Fatigue Properties of TiB Particle Reinforced P/M Titanium Matrix Composite, *Proc. of 4th Conf. on P/M Aerospace, Defense, and Demanding Applications,* F.H Froes, Ed., Metal Powder Industries Federation, 1995, p 173–180
93. T. Saito, H. Takamiya, and T. Furuta, Thermechanical Properties of P/M β Titanium Metal Matrix Composite, *Mater. Sci. Eng. A,* Vol 243 (No. 2), 1998, p 273–278
94. C.M. Bedell, P. Zwigl, and D.C. Dunand, in *Superplasticity and Superplastic Forming,* A.K. Ghosh and T.R. Bieler, Ed., TMS, 1995, p 125–133
95. D.C. Dunand and C.M. Bedell, Method of Enhancing Superplasticity in Composites, U.S. Patent 5,413,649, 1995
96. D.C. Dunand and C.M. Bedell, *Acta Mater.,* Vol. 44, 1996 p 1063–1076
97. Dynamet Technology Inc., Unpublished research, 1995
98. K.R. Teal, D. Eylon, and F.H. Froes, Compaction of Titanium Aluminide Powders by Hot Isostatic Pressing (HIP), *Sixth World Conference on Titanium,* Part II, P. Lacombe, R. Tricot, and G. Beranger, Ed., Lés Editions de Physiques, 1989, p 1121–1125
99. W.J. Porter, N.R. Osborne, D. Eylon, and J.P. Clifford, Rapid Omnidirectional Compaction (ROC) of Titanium Aluminide Prealloyed Powders, *Advances in Powder Metallugy, 1990,* Vol 2, E.R. Andreotti and P.J. McGeehan, Ed., Metal Powder Industries Federation, 1990, p 243–257
100. N.R. Osborne, W.J. Porter, and D. Eylon, The Mechanical Properties of Powder Metallurgy Ti-829 and Ti-25Al-10Nb-3V-1Mo Produced by Rapid Omnidirectional Compaction, *SAMPE Q.,* Vol 22 (No. 4), 1991, p 21–28
101. W.J. Porter, N.R. Osborne, and D. Eylon, High Integrity Alpha Two Titanium Aluminide Compacts Produced by Prealloyed Powder Metallurgy, *MPR,* Vol 46 (No. 10), 1991, p 48–55
102. D. Eylon, C.M. Cooke, and Y.-W. Kim, Microstructure Control of Gamma Titanium Aluminides Through Powder Metallurgy, *Microstructure Science,* Vol 19, M.R. Lauthan and D.A. Wheeler, Ed., ASM International, 1992, p 711–730
103. C.F. Yolton and D. Eylon, Effect of Processing and Heat Treatment on Microstructure and Mechanical Properties of Gamma Titanium Aluminide Powder Compacts, *Titanium '92 Science and Technology,* Vol II, F.H. Froes and I.L. Caplan, Ed., TMS, 1993, p 1083–1090
104. D. Eylon and F.H. Froes, Compaction of Rapidly Solidified Titanium Alloy Powders, *Hot Isostatic Pressing—Theories and Applications,* T. Garvare, Ed., Centek Publishers, Sweden, 1988, p 197–204
105. D. Eylon, C.F. Yolton, and D.U. Furrer, Improved Forgeability and Properties of TiAl Alloys by Prealloyed Powder Metallurgy, *Proceedings of PM '93,* Japan, 1993
106. D. Eylon, C.M. Cooke, C.F. Yolton, W.T. Nachtrab, and D.U. Furrer, Microstructural Refinement and Property Improvement of TiAl Alloys Through Powder Metallurgy and Powder Forging, *First International Conference on Processing Materials for Properties,* H. Henein and T. Oki, Ed., TMS, 1993, p 241–244
107. S.L. Semiatin, G.R. Cornish, and D. Eylon, Hot Compression Behavior and Microstructure Evolution of Ti-48Al-2Sr-2Nb Prealloyed Powder Compacts, *Mater. Sci. Eng. A,* Vol 185, 1994, p 45–53
108. C.F. Yolton, D. Eylon, and D.U. Furrer, Microstructural Refinement of Gamma Titanium Aluminide Through Powder Metallurgy for Improved Forgeability and Mechanical Properties, *Proceedings of the 50th Anniversary SAMPE Symposium,* Society for the Advancement of Material and Process Engineering
109. D. Eylon, C.F. Yolton, H. Clemens, P. Schretter, and P.E. Jones, Forging, Rolling, and Net-Shape Processing of Gas-Atomized Ti-48Al-2Cr-2NB Powder Compacts, *PM '94, Powder Metallurgy World Congress,* Vol II, June 1994, p 1271–1275
110. H. Clemens, W. Glatz, P. Schretter, C.F. Yolton, P.E. Jones, and D. Eylon, Sheet Rolling of Ti-48Al-2Cr-2Nb Prealloyed Powder Compacts, *Gamma Titanium Aluminides,* Y.-W. Kim, R. Wagner, and M. Yamaguchi, Ed., TMS, 1995, p 555–562
111. J.P. Beckman, H. Clemens, P. Schretter, W. Glatz, P.E. Jones, D. Eylon, and C.F. Yolton, Effect of Processing on Properties of Rolled Gamma Titanium Aluminide Sheet Made from Powder Metallurgy Plate, *Titanium 95 Science and Technology,* Proceedings of the 8th World Conference on Titanium, The Institute of Materials, 1995, p 217–224

Powder Metallurgy Superalloys

John H. Moll and Brian J. McTiernan, Crucible Research, Crucible Materials Corporation

THE TERM SUPERALLOY is used to describe those materials with high strength at high temperatures that are used primarily in turbine engines for aircraft and power generation. Superalloys are predominantly nickel-base alloys that are strengthened by solid-solution elements including molybdenum, tungsten, cobalt, and precipitation of a Ni_3Al-type compound designated γ'. There are several iron-, iron-nickel, and cobalt-base compositions that also meet the general description of superalloys.

The development of superalloys has been driven by the need for increasingly higher strength at higher temperatures for advanced aircraft engines. As the strength of these materials increased, hot workability decreased. Additional problems were encountered as the result of increased segregation associated with more complex alloying and the need for larger ingots for larger turbine disks. A solution to these problems was to minimize segregation through rapid solidification of the metal by atomization to powder and consolidation by methods that do not melt the powder particles but still attain full density. A further benefit of the P/M process is that the consolidated products were superplastic and amenable to isothermal forging. Thus, force requirements were greatly reduced and near-net forgings could be made. Near-net shapes are also made by hot isostatic pressing (HIP). Whereas P/M alloys cost more than conventional wrought alloys, they have allowed designers to design to higher creep strength and tensile capability while maintaining the expected cyclic life of components. The value added to the system in terms of the performance benefit gained by the system operating temperature and weight reduction more than balances the increased cost of P/M application.

Initial P/M superalloy compositions were modifications of existing alloys such as IN-100, Astroloy, and René 95. These alloys have been used in aircraft engines for more than 20 years and are still the most widely used P/M superalloys. With segregation essentially eliminated, alloy designers were able to develop a number of new P/M alloys, such as AF115 and AF2-1DA, with exceptionally high strength. Unfortunately the defect tolerance of these alloys was generally

Table 1 Engine systems utilizing forged P/M superalloys

Engine	Number produced through 1996
GE T-700	9674
GE F-404/414	3077
GE-F110	2259
GE 90	38
PW F100	6496
PW 2000	951
PW 4000	1819

Table 2 Engines and airframe systems utilizing as-HIP P/M superalloys

GE Aircraft Engines		CFM International		AlliedSignal	
Turbine	Airframe	Turbine	Airframe	Turbine	Airframe
CF6-80C2/E1	B747-400, 767AWACS	CFM56-3B1/B2	B 737-300	GTC-131-3[A]	B2
F108-CF-100	KC-135-R	CFM56-3B2/3C	B 737-400	GTC-131-9[D]	MD-90
F110-GE-100	F-16 C/D/G	CFM56-3B4	B 737-500	GTC-131-9[B]	B737-X
F110-GE-129	F-16-D/G	CFM56-5A1	A-320-100/-200	GTC-131-9[A]	Airbus A319, A320, A321
F110-GE-400	F-14B	CFM56-5A/5B	A-319	GTCP-331-200	B757/B767 B747-200B
F118-GE-100	B2	CFM56-5B1/B2	A-321-100	GTCP-331-250	A300, A310, C17A
F404-GE-400/402	F/A-18 C/D	CFM56-5C2/C4	A-340-200/300	GTCP-331-350	Airbus A330, A340
F404-GE-400/402	F-117A	CFM56-7	B737-700	GTCP-331-500	B777
T-700-GE-401	Bell AH 1W	CFM56-8	B 737-600/800	RE-220	Gulfstream V Global Express CRJ-700
T-700-GE-401	Sikorsky SH-60B/F				
T-700-GE-700	Sikorsky UH-60A/L				

Table 3 Composition of P/M superalloys

	Composition, wt %													
Alloy	C	Cr	Mo	W	Ta	Ti	Nb	Co	Al	Hf	Zr	B	Ni	Other
René 95	0.07	13.0	3.5	3.5	...	2.5	3.5	8.0	3.5	...	0.05	0.01	bal	...
IN-100	0.07	12.5	3.2	...	...	4.3	...	18.5	5.0	...	0.04	0.02	bal	0.75V
LC Astroloy	0.04	15.0	5.0	...	...	3.5	...	17.0	4.0	...	0.04	0.025	bal	...
N18	0.02	11.5	6.5	...	...	4.3	...	15.5	4.3	0.5	...	0.015	bal	...
René 88DT	0.03	16.0	4.0	4.0	...	3.7	0.7	13.0	2.1	...	0.03	0.015	bal	...
Udimet 720	0.025	16.0	3.0	1.25	...	5.0	...	14.7	2.0	...	0.03	0.020	bal	...
IN-706	0.02	16.0	...	...	...	1.7	3.0	...	0.15	...	...	...	bal	38Fe
IN-718	0.02	18.0	3.0	...	...	0.9	5.0	...	0.45	...	...	0.004	bal	18Fe
AF 115	0.05	10.5	2.8	6.0	...	3.9	1.7	15.0	3.8	2.0	...	...	bal	...
AF2-1DA-6	0.04	12.0	2.75	6.5	1.5	2.8	...	10.0	4.6	...	0.10	0.015	bal	...
PA101	0.10	12.5	...	4.0	4.0	4.0	...	9.0	3.5	1.0	...	...	bal	...
MERL 76	0.02	12.4	3.2	...	...	4.3	1.4	18.5	5.0	0.4	0.06	0.02	bal	...
TMP-3	0.07	10.8	3.1	3.4	...	2.8	3.9	6.9	3.9	...	0.05	0.01	bal	...
SR3	0.03	13.0	5.1	...	...	4.9	1.6	12.0	2.6	0.2	0.03	0.015	bal	...
KM4	0.03	12.0	4.0	...	...	4.0	2.0	18.0	4.0	...	0.03	0.03	bal	...

not acceptable. Defect tolerance is the ability of a material to resist the growth of a crack initiated by a defect in the microstructure. Current P/M superalloy development centers on somewhat lower strength but highly defect-tolerant compositions such as N18, René 88DT, and Udimet 720.

Current Use of P/M Superalloys

The first power plant to employ widespread use of extruded and isothermally forged nickel-base superalloy turbine was the Pratt & Whitney F100. It began operational service on the F-15 Eagle fighter in 1974. Currently, most rotating aircraft turbine components are produced via extrusion and isothermal forging. A typical aircraft engine will utilize anywhere from 680 to 2040 kg (1500 to 4500 lb) of nickel-base superalloy powder isothermal forgings. Table 1 lists engines utilizing P/M superalloy forgings (Ref 1).

In addition to the widespread use of extruded and isothermally forged components, more than 100,000 as-hot isostatically pressed P/M superalloy components are flying daily in a variety of military and commercial engines worldwide. Table 2 lists engine power plants and airframes utilizing as-hot isostatically pressed P/M superalloy components.

One of the most interesting and current developments utilizing nickel-base superalloys is the AlliedSignal dual-alloy turbine wheel concept for aircraft auxiliary power units (Ref 2–4). This design, which began development in 1979, HIP bonds a cast MAR-M-247 blade ring to an as-HIP low-carbon Astroloy hub. This provides an assembly with a fine-grained, fatigue-resistant hub and an investment-cast, creep-resistant blade ring. Since the introduction into commercial service, more than 1000 assemblies on the GTCP 331 APU engine alone have accumulated more than 12 million cycles without failure. Based upon this experience, nine other turbine applications have either been introduced or are in the process of being designed and certified by AlliedSignal. Nearly 10,000 as-HIP turbine disks with integral or inserted blades are also in use in other APU applications, and this usage continues to increase.

Powder Metallurgy Superalloy Processing

Conventional processing of P/M superalloys involves production of spherical prealloyed powder, screening to a suitable maximum particle size, blending the powder to homogenize the powder size distribution, loading the powder into containers, vacuum outgassing and sealing the containers, and then consolidating the powder to full density. For most applications, the consolidated powder is consolidated by extrusion and subsequently forged to enhance me-

Fig. 1 As-HIP René 95 turbine disks. As-HIP shape (upper left), sonic shape (upper right), finished machined disks (bottom). Courtesy Crucible Compaction Metals

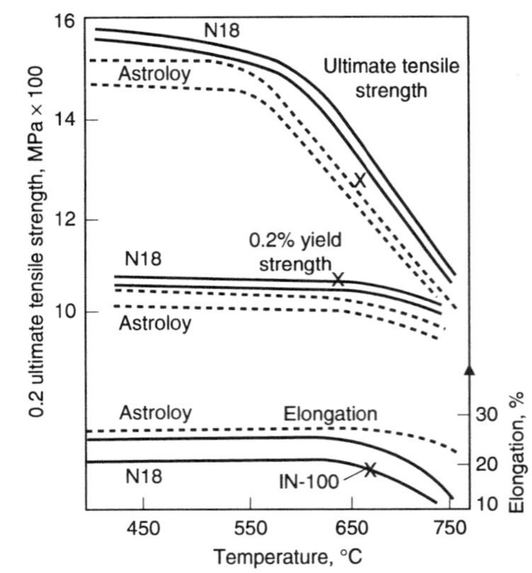

Fig. 2 Tensile properties of extruded plus isothermally forged P/M N18 compared to P/M Astroloy and P/M IN-100 (N18 heat treatment: 1165 °C (2130 °F)/4 h/cooled at 100 °C/min + 700 °C (1292 °F)/24 h/air cool + 800 °C (1472 °F/4 h/air cool). Source: Ref 13

Table 4 Tensile properties of as-HIP P/M René 95

Temperature		0.2% yield strength		Tensile strength		Elongation,	Reduction
°C	°F	MPa	ksi	MPa	ksi	%	of area, %
Subsolvus treated(a)							
21	70	1260	183	1653	240	18	18
204	400	1233	179	1591	231	12	12
427	800	1205	175	1571	228	11	11
537	1000	1192	173	1619	235	13	14
649	1200	1171	170	1515	220	14	16
Supersolvus treated(b)							
21	70	1171	170	1592	231	14	16
204	400	1137	165	1481	215	9	12
427	800	1109	161	1557	226	12	14
537	1000	1075	156	1551	225	14	16
649	1200	1027	149	1433	208	14	15
760	1400	951	138	1158	168	10	12

(a) Heat treatment: 1135 °C (2075 °F) + 760 °C (1400 °F)/8 h. (b) Heat treatment: 1175 °C (2150 °F) + 760 °C (1400 °F)/8 h. Source: Ref 7

Table 5 Creep rupture properties of as-HIP P/M René 95

Temperature		Stress		Time to 0.2%	Time to
°C	°F	MPa	ksi	creep, h	rupture, h
Subsolvus treated(a)					
649	1200	847	123	50	340
740	1300	551	80	20	340
Supersolvus treated(b)					
649	1200	847	123	80	1500
704	1300	551	80	280	3000

(a) Heat treatment: 1135 °C (2075 °F) + 760 °C (1400 °F)/8 h. (b) Heat treatment: 1175 °C (2150 °F) + 760 °C (1400 °F)/8 h. Source: Ref 7 chanical properties, particularly high-temperature strength and ductility. For some applications, powder consolidated by HIP is utilized without further thermomechanical processing. In either case, the resulting product is fully dense and has properties that are generally superior to cast and wrought material.

Powder Production and Processing. Virtually all powder used to produce P/M superalloys is prealloyed, meaning that the powder is made from the molten state and each powder particle is essentially a mini-ingot with the same composition as the molten alloy. The principal commercial powder-making processes are gas atomization and vacuum atomization. In gas atomization, high-purity raw materials are vacuum induction melted and subsequently atomized using gas. Typically, the equipment consists of a top-melt chamber where high-purity raw materials are vacuum induction melted. The metal is delivered to a nozzle to form a small metal stream that is atomized by high-pressure gas using either a continuous annular-type nozzle or discrete individual nozzles. The resulting powder, which is highly spherical, is collected in a bottom chamber and/or cyclone collector. Argon gas is normally used to atomize superalloys. Nitrogen gas atomization can also be used to reduce cost and to eliminate entrapped argon. The use of nitrogen usually requires slight modifications in composition such as lower carbon content, to accommodate the nitrogen increase associated with the atomization process.

Vacuum atomization is a commercial practice based on the principle that, when a molten metal is supersaturated with gas under pressure and is then suddenly exposed to a lower pressure, the gas expands, comes out of solution, and causes the liquid metal to be atomized. The equipment consists of two vertical chambers connected by a transfer tube. The metal is vacuum induction melted in the lower chamber. When the lower chamber is pressurized with gas, the molten metal is forced upward through the transfer tube. As the metal rises into the upper chamber, it is atomized by escaping gases. The resulting powder falls to the bottom of the upper chamber and then through a powder drain into a collection container. See the article "Atomization" in this Volume.

Superalloy powder can also be produced using the plasma rotating electrode process (PREP) and the centrifugal atomization process. In the PREP process, a plasma arc is used to melt the end of a rapidly rotating prealloyed bar. The high-speed rotation of the bar causes liquid droplets to spin from the end of the bar and solidify in flight through a helium atmosphere. In centrifugal atomization, a molten stream of prealloyed metal is poured onto a spinning disk and atomized by centrifugal force. The emitted liquid particles are further broken up and subsequently cooled by high-pressure helium gas.

Following powder production, the powders are screened to remove oversized particles and then blended to obtain a uniform size distribution. All powder handling is performed to minimize the possibility of introducing foreign material into the powder. This involves the use of specially designed stainless steel containers, valves, and inert handling or air handling of powder in clean rooms. Powder is normally screened to minimize the inclusion size in the final part. Depending on the application, powder sizes ranging from –60 mesh (–250 μm) to –325 mesh (–45 μm) are typically used. Prior to being loaded into containers for consolidation, superalloy powders are evaluated for cleanliness by techniques such as

Table 6 Tensile properties of extruded and isothermally forged P/M René 95

Temperature		2% yield strength		Tensile strength			Reduction
°C	°F	MPa	ksi	MPa	ksi	Elongation, %	of area, %
21	70	1254	182	1675	243	20	23
204	400	1233	179	1578	229	22	25
427	800	1206	175	1627	236	21	23
538	1000	1192	173	1599	232	17	18
649	1200	1172	170	1454	211	11	13
760	1400	1068	155	1137	165	6	8

Note: Heat treatment: 1120 °C (2050 °F) + 760 °C (1400 °F)/8 h. Source: Ref 7

Table 7 Creep rupture properties of extruded and isothermally forged P/M René 95

Temperature		Stress		Time to 0.2%	Time to
°C	°F	MPa	ksi	creep, h	Rupture, h
649	1200	847	123	10	600
704	1300	551	80	90	270

Note: Heat treatment: 1120 °C (2050 °F) + 760 °C (1400 °F)/8 h. Source: Ref 7

Table 8 Typical low-cycle fatigue data for extruded plus isothermally forged P/M

Temperature		Cycles to failure
°C	°F	(N_f)
400	750	130,000
540	1000	160,000

Note: Heat treatment: 1107 °C (2025 °F)/1 h/oil quench + 760 °C (1400 °F)/8 h/air cool. Tested at 0.7% strain, 30 CPM. Source: Ref 8

Table 9 Mechanical properties of P/M IN-100 disk made by extrusion and isothermal forging

Room-temperature tensile properties	
Tensile strength, MPa (ksi)	1591 (231)
Yield strength, MPa (ksi)	1102 (160)
Elongation, %	26
Reduction in area, %	26
Tensile properties at 700 °C (1300 °F)	
Tensile strength, MPa (ksi)	1267 (184)
Yield strength, MPa (ksi)	1068 (155)
Elongation, %	18
Reduction in area, %	20
Stress-rupture properties at 732 °C (1350 °F) and 654 MPa (95 ksi)	
Life, h	35
Elongation, %	13
Creep properties at 700 °C (1300 °F) and 551 MPa (80 ksi)	
Time to 0.1%, h	102
Time to 0.2%, h	134

Note: Heat treatment: 1121 °C (2050 °F/2 h/oil quench + 871 °C (1600 °F)/40 min/air cool + 982 °C (1800 °F)/45 min/air cool + 760 °C (1400 °F)/4 h/air cool. Source: Ref 10, 11

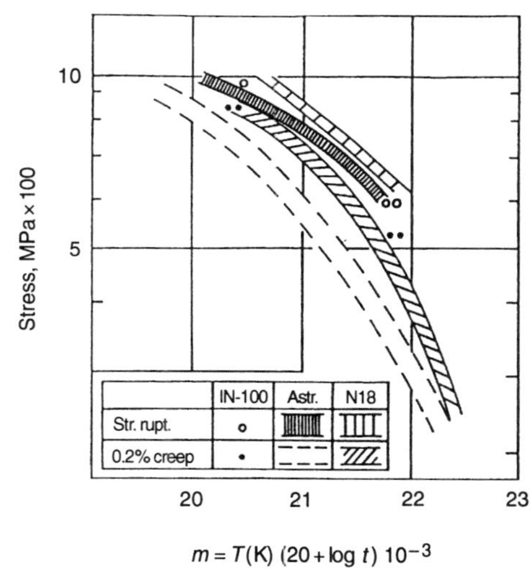

Fig. 3 Larson-Miller creep rupture curves for extruded plus isothermally forged P/M N18 compared to P/M Astroloy and P/M IN-100 (N18 heat treatment: 1165 °C (2130 °F)/4 h/cooled at 100 °C/min + 700 °C (1292 °F)/24 h/air cool + 800 °C (1472 °F)/4 h/air cool). Astr., Astroloy; Str., rupt., stress rupture. Source: Ref 13

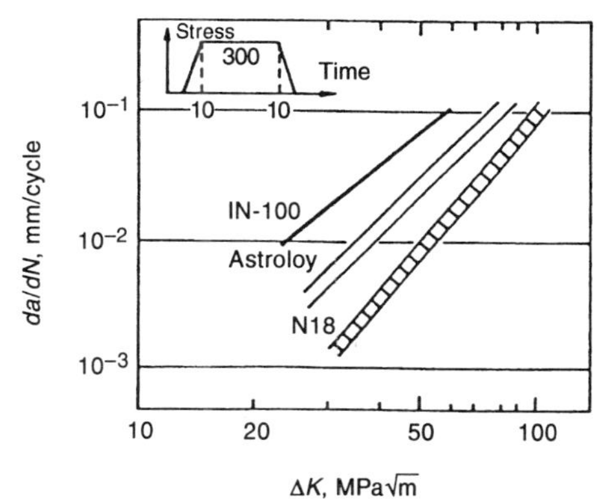

Fig. 4 Crack growth rates at 650 °C (1202 °F) for extruded plus isothermally forged P/M N18 compared to P/M Astroloy and P/M IN-100 (N18 heat treatment: 1165 °C (2130 °F)/4 h/cooled at 100 °C/min + 700 °C (1292 °F)/24 h/air cool + 800 °C (1472 °F)/4 h/air cool). Source: Ref 13

water elutriation, which separates nonmetallic inclusions from the powder for counting, sizing, and identification. A consolidated sample of the powder can also be evaluated by conducting large bar (e.g., 1.27 mm, or 0.500 in., diameter × 5.08 mm, or 2.0 in., gage section) fatigue tests. Cleanliness is evaluated on the basis of fatigue life and fracture origin.

Containers for consolidation are made from stainless steel or mild steel. Exhaustive procedures are use to ensure that the containers are clean before powder loading. A "final filter" is commonly used to rescreen the powder as it enters the container as a final in-process control to ensure that no oversized particles are in the consolidated part. At some point in the processing, the powder is subjected to a vacuum and heated to remove air and adsorbed moisture. This can be accomplished during loading or after loading the powder into a container. In the former case, powder is loaded from an evacuated container into an evacuated consolidation container and then the container is sealed. In the latter case, powder is loaded in air into a consolidation container which is subsequently cold and hot outgassed and then sealed. Various combinations or modifications of these two outgassing techniques are used by different manufacturers.

Powder Consolidation. Superalloy powders are consolidated to full density using a combination of high temperature and high pressure. The primary consolidation methods for superalloys are HIP, extrusion, or a combination of the two. For HIP, the powder containers can be simple geometric shapes or complex near-net shapes. The product can be a billet suitable for subsequent extrusion and/or forging. The HIP product can also be a near-net forging preform or a near-net machining preform to be used without additional hot working. The HIP parameters generally used are 1040 to 1205 °C (1900 to 2200 °F) at 105 to 205 MPa (15 to 30 ksi) for 3 to 5 h. An example of a HIP near-net sonic shape and the resulting turbine disk is shown in Fig. 1.

Extrusion of superalloy powders can be done either as the extrusion of loose powder or extrusion of a preconsolidated billet. The extrusion of loose powder requires a specialized container with a nose plug designed to protect the evacuation stem on the container from rupturing prior to entering the extrusion die. Extrusion ratios used for direct extrusion of powder are at least 7 to 1. With extrusion of preconsolidated billet, the powder may be preconsolidated by HIP, a separate forging press, or in the extrusion press

Table 10 Tensile properties of as-HIP P/M LC Astroloy

Test temperature		Yield strength		Tensile strength		Elongation,	Reduction of area,
°C	°F	MPa	ksi	MPa	ksi	%	%
22	72	930	135	1345	195	22	24
540	1000	861	125	1240	180	22	24
650	1200	827	120	1137	165	22	32
760	1400	792	115	1068	155	22	34

Note: HIP: 1205 °C (2200 °F)/100 MPa (15 ksi)/4 h. Heat treatment: 1115 °C (2040 °F)/2–4 h/AC + 870 °C (1000 °F)/8 h AC + 980 °C (1800 °F)/4 h/AC + 650 °C (1200 °F) 24 h/AC + 760 °C (1400 °F)/8 h/AC. Source: Ref 12

Table 11 Stress rupture properties of as-HIP P/M LC Astroloy

Test temperature		100 h rupture stress	
°C	°F	MPa	ksi
650	1200	889	129
700	1300	703	102
760	1400	496	72

Note: HIP: 1205 °C (2200 °F)/100 MPa (15 ksi)/4 h. Heat treatment: 1115 °C (2040 °F)/2–4 h/AC + 870 °C (1000 °F)/8 h AC + 980 °C (1800 °F)/4 h/AC + 650 °C (1200 °F) 24 h/AC + 760 °C (1400 °F)/8 h/AC. Source: Ref 12

Table 12 Tensile properties of extruded + isothermally forged P/M René 88DT

Temperature		0.2% yield strength		Tensile strength		Elongation, %	Reduction of area, %
°C	°F	MPa	ksi	MPa	ksi		
22	72	1143	166	1232	179	20	25
200	400	1074	156	1143	166	15	19
425	800	1067	155	1053	153	18	20
540	1000	1046	152	1088	158	17	20
650	1200	1019	148	1377	200	18	21
760	1400	971	141	1170	170	13	17

Note: Heat treatment: 1150 °C (2100 °F) + 760 °C (1400 °F)/8 h. Source: Ref 7

Table 13 Creep rupture properties of extruded plus isothermally forged René 88DT

Temperature		Stress		Time to 0.2% creep, h	Time to rupture, h
°C	°F	MPa	ksi		
650	1200	847	123	1200	2000
700	1300	550	80	820	7000

Note: Heat treatment: 1150 °C (2100 °F) + 760 °C (1400 °F)/8 h. Source: Ref 7

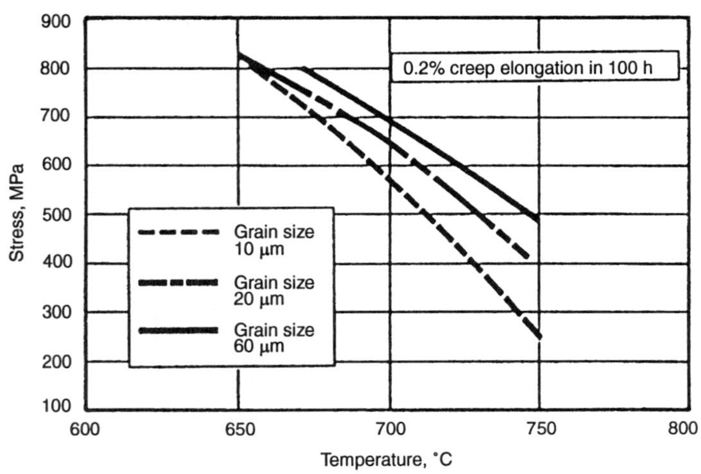

Fig. 5 Effect of grain size on the creep capability of extruded plus isothermally forged P/M N18. Source: Ref 15

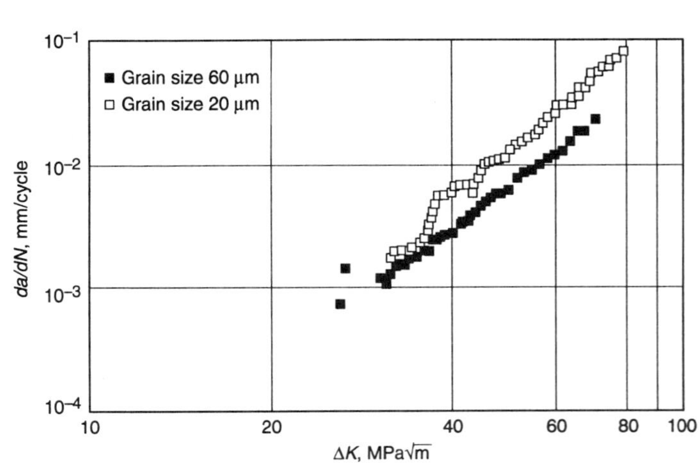

Fig. 6 Effect of grain size on the 750 °C (1382 °F) fatigue crack growth rate of extruded plus isothermally forged P/M N18. Source: Ref 15

against a blank die. Extrusion ratios for solid billet are normally at least 3 to 1. Typical extrusion temperatures for P/M superalloys range from 1040 to 1175 °C (1900 to 2150 °F).

The postconsolidation processing of P/M superalloys is similar to that used for ingot products, but the P/M material has a number of advantages. The uniform grain size and lack of segregation in P/M material generally improves forgeability, machinability, and ultrasonic inspectability. Most P/M superalloys exhibit superplasticity and are normally isothermally forged at low strain rates to reduce force requirements and to produce close-tolerance forgings. Conventional high-strain forging is difficult to accomplish due to the high strength and cracking tendencies of these materials. However, recent work has shown that the high-strain-rate formability of Udimet 720 can be markedly improved by HIP at a temperature slightly below the solidus of the alloy (Ref 5). The improvement is attributed to the elimination of grain boundaries, which are coincident with prior-particle boundaries. As a result, conventional forging and ring rolling of P/M billet is deemed practical. The ultrasonic inspectability (background noise level) of P/M alloys, due to their homogeneity, is superior to most conventional cast plus wrought superalloys. As a result, smaller flaws can be detected in the P/M alloys.

Heat treatments applied to P/M materials are also similar to those used for ingot product. In general, this involves a solution treatment followed by multistep aging treatments to stabilize carbides and precipitate and stabilize γ′ (generally, a precipitate of the form Ni_3Al). Maximum short-time strength is developed with a "subsolvus" solution treatment at a temperature slightly below the γ′ solvus temperature. As discussed later, recent studies have shown, however, that fatigue crack growth resistance and creep strength of P/M superalloys can be significantly improved with a "supersolvus" solution treatment slightly above the γ′ solvus temperature. The result is improved defect tolerance in these high-strength materials. The supersolvus treatment is being applied to a number of P/M alloys.

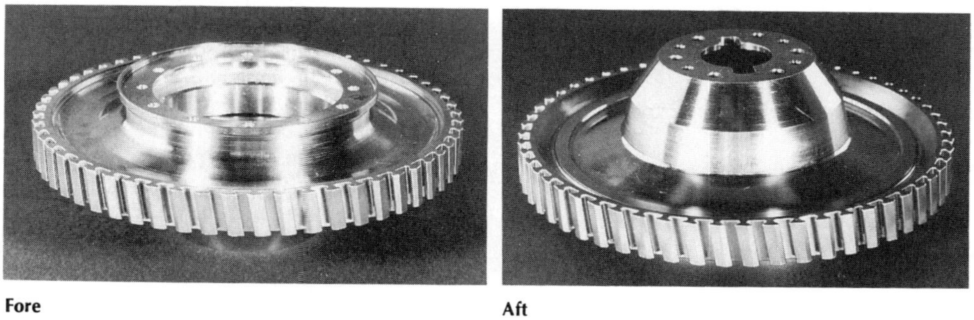

Fig. 7 Fully machined P/M Udimet 720 turbine disk produced by extrusion plus isothermal forging. Courtesy of Allison Engine Company

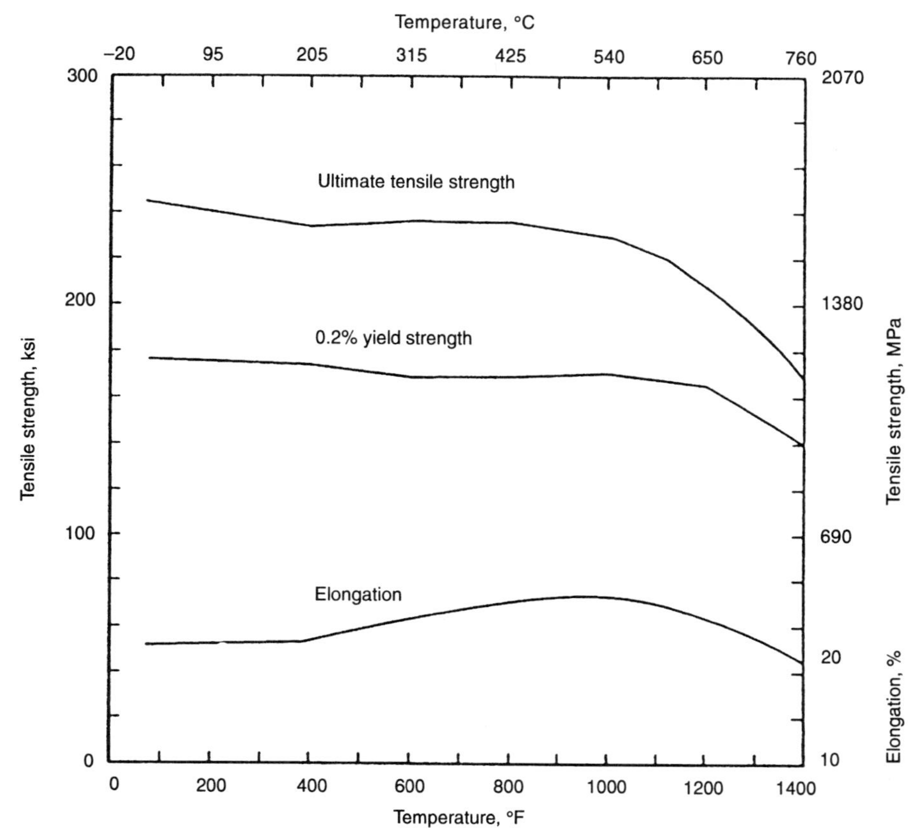

Fig. 8 Tensile properties of P/M Udimet 720 turbine disk produced by extrusion plus isothermal forging (heat treatment: 1090 °C (2000 °F) + two-step age). Source: Ref 18

Composition and Mechanical Properties of P/M Superalloys

Table 3 gives the compositions of P/M superalloys. Alloys that are in commercial production include René 95, IN-100, LC Astroloy, N18, and René 88DT. A number of alloys are in earlier stages of P/M development. These include Udimet 720, Inconel 706, Inconel 718, AF115, and KM4. Other alloys, such as PA101, MERL 76, and AF2-1DA, were originally considered for P/M applications but are not currently being used commercially. The characteristics and properties of individual alloys are discussed in the following paragraphs.

René 95 is one of the highest-strength alloys available for service in the range of 425 to 650 °C (800 to 1200 °F). The alloy also provides excellent high-temperature creep resistance and low-cycle fatigue life. René 95 was originally

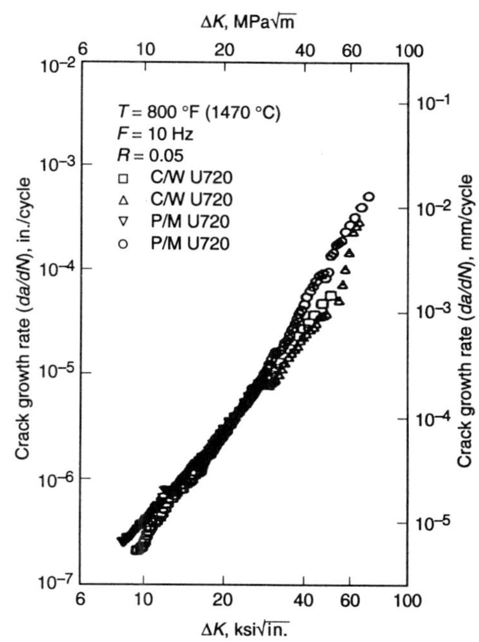

Fig. 9 Fatigue crack growth rates for P/M Udimet 720 turbine disk produced by extrusion plus isothermal forging (heat treatment: 1090 °C (2000 °F) + two-step age) compared to cast and wrought (C/W) Udimet 720. Source: Ref 18

developed as a cast and wrought material, but was soon changed to a P/M material due to forging difficulties and inconsistent mechanical properties. Initially, the P/M version was developed for use in the as-hot isostatically pressed (i.e., no hot working) plus heat treated condition (Ref 6). For maximum short-time strength, HIP René 95 is solution treated slightly below the γ' solvus temperature (subsolvus treatment) and rapidly cooled using a molten salt quench followed by two-step aging. This treatment results in a fine equiaxed structure (ASTM 11) with a fine dispersion of γ'. Solution treatment above the γ' solvus (supersolvus treatment) provides increased creep resistance and stress rupture strength through a slightly increased grain size (ASTM 7–8). Tensile and creep rupture properties for hot isostatically pressed René 95 are given in Tables 4 and 5.

Although as-hot isostatically pressed René 95 is in current use, most René 95 production involves extrusion of preconsolidated powder followed by isothermal forging and heat treatment. Tensile and creep rupture properties for thermomechanically processed P/M René 95 are given in Tables 6 and 7. Typical low-cycle fatigue properties are given in Table 8.

IN-100. Powder metallurgy IN-100 is a modified version of what was originally a casting alloy (Ref 9). It is now widely used as a compressor and turbine disk material by Pratt & Whitney Aircraft. Powder metallurgy IN-100 hardware is produced by extrusion of preconsolidated powder followed by isothermal forging and heat treatment. Typical properties are given in Table 9 (Ref 10, 11).

LC Astroloy is a low-carbon version of cast plus wrought Astroloy. The alloy offers a unique combination of excellent ductility and high-temperature strength. As-hot isostatically pressed Astroloy is used by AlliedSignal in dual-alloy turbine disks as well as one-piece hubs with inserted blades for aircraft auxiliary power units (Ref 2, 4). Processing of the alloy usually involves powder consolidation by HIP or direct extrusion plus heat treatment. Typical properties of HIP plus heat treated LC Astroloy are given in Tables 10 and 11.

N18 is a defect tolerant modification of LC Astroloy developed by SNECMA for high-pressure turbine engine disks (Ref 13, 14). The alloy exhibits high strength as well as good creep resistance and excellent creep fatigue crack growth behavior up to 650 °C.

Alloy N18 is useful in both bore and rim locations at temperatures up to ~650 °C. The alloy is normally produced by extrusion followed by isothermal forging and a subsolvus (1165 °C) heat treatment. Under these processing conditions, the resulting grain size is 12 µm (ASTM 10). Typical tensile, creep rupture, and crack growth rate properties are compared with those of LC Astroloy and IN 100 in Fig. 2 to 4. The properties of the alloy are markedly affected by grain size (Ref 15, 16). Increasing the grain size to 60 µm (ASTM 6) using a supersolvus treatment (1200 °C) increases creep (Fig. 5) and fatigue crack growth resistance (Fig. 6) without major change in low-cycle fatigue resistance. Larger grains can decrease low-cycle fatigue resistance; thus grain size control during processing is most important.

René 88DT (damage tolerant) is a nickel-base P/M superalloy that was developed by General Electric to meet the higher-temperature requirements of advanced engines (Ref 17). The specific goals relative to René 95 were a 50% reduction in cyclic fatigue crack growth rate, a 13 °C

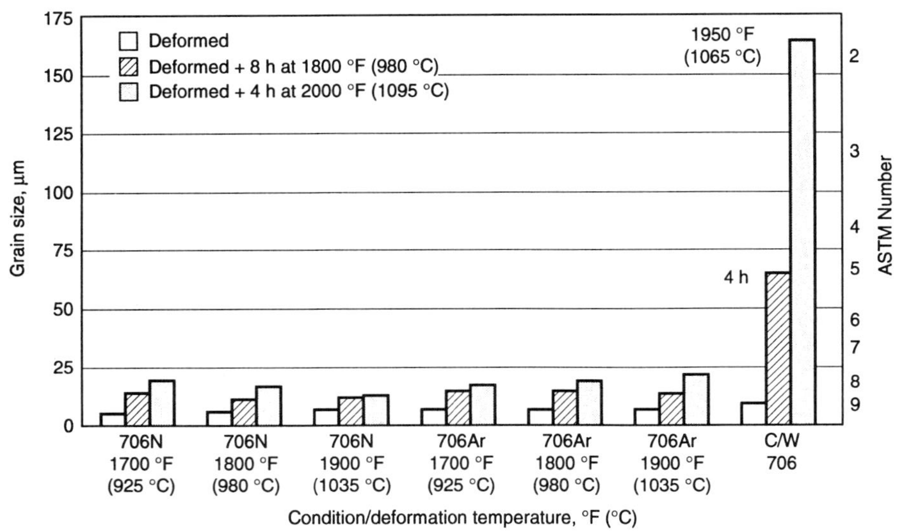

Fig. 10 Grain size of HIP plus forged P/M 706 and cast and wrought (C/W) 706 after heat treating. Source: Ref 21

Table 14 Room-temperature properties of as-HIP P/M 706

Grade(a)	Mesh size	HIP temperature °C	HIP temperature °F	0.2% yield strength MPa	0.2% yield strength ksi	Ultimate tensile strength MPa	Ultimate tensile strength ksi	Tensile elongation, %	Reduction of area, %	Impact energy J	Impact energy ft · lb
N706	–60	1130	2065	1025	149	1325	192	23	30	35	26
N706	–60	1065	1950	1035	150	1325	192	19	25	27	20
A706	–140	1130	2065	1035	150	1330	193	20	27	26	19
A706	–140	1065	1950	1040	151	1345	195	21	29	27	20

Note: Heat treatment: 980 °C (1800 °F)/ 1 h + 732 °C (1350 °F)/10 h + 620 °C (1150 °F)/8 h. (a) N706, nitrogen atomized; A706, argon atomized. Source: Ref 21

Table 15 Room-temperature properties of as-HIP plus forged P/M 706

Grade(a)	Mesh size	HIP temperature °C	HIP temperature °F	0.2% yield strength MPa	0.2% yield strength ksi	Ultimate tensile strength MPa	Ultimate tensile strength ksi	Tensile elongation, %	Reduction of area, %	Impact energy J	Impact energy ft · lb
N706	–60	1130	2065	860	125	1205	177	23	44	50	37
N706	–60	1065	1950	855	125	1206	177	23	44	49	36
A706	–140	1130	2065	855	124	1205	177	22	40	41	30
A706	–140	1065	1950	845	123	1190	175	23	39	38	28

Note: Heat treatment: 980 °C (1800 °F)/ 1 h + 732 °C (1350 °F)/10 h + 620 °C (1150 °F)/8 h. (a) N706, nitrogen atomized; A706, argon atomized. Source: Ref 21

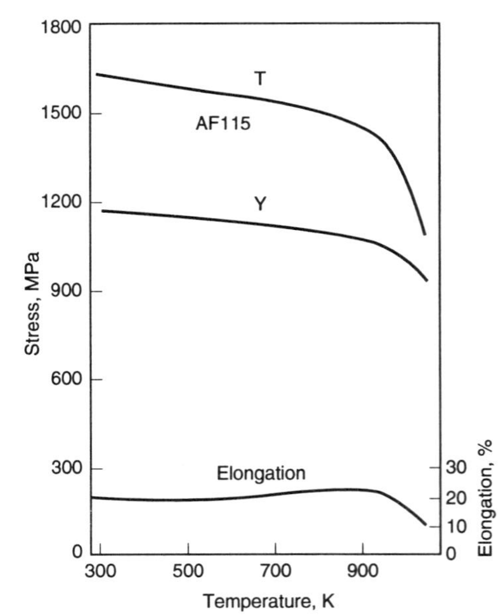

Fig. 11 Tensile properties of as-HIP P/M AF115. HIP: 1180 °C (2155 °F)/3 h/102 MPa (15 ksi). Heat treatment: 1175 °C (2150 °F) + 760 °C (1400 °F) air cool. Source: Ref 25

(23 °F) minimum improvement in creep and stress rupture strength, and a maximum 10% reduction in tensile strength. The alloy is currently used in production engine disk hardware (e.g., high- and low-pressure turbine disks and forward outer seals). Typical properties of René 88DT are given in Tables 12 and 13.

Udimet 720 was originally developed as a wrought turbine blade alloy for industrial turbines. In the cast and wrought form it is being used as a turbine disk alloy. More recently, the alloy has been evaluated as a P/M material (Ref 18, 19). Reportedly, P/M Udimet 720 has excellent fatigue crack growth resistance and is being strongly considered as a disk material in small to medium gas turbine engines as well as aircraft auxiliary power units. Udimet 720 can be produced by HIP or by extrusion plus isothermal forging. Figure 7 shows a fully machined P/M Udimet 720 turbine disk produced by extrusion plus isothermal forging. A comparison of the costs for P/M and cast and wrought Udimet 720 disks showed more than 20% cost reduction with the P/M process (Ref 18). Tensile properties for P/M Udimet 720 are shown in Fig. 8, and the results of fatigue crack growth rate tests of P/M and cast and wrought material are shown in Fig. 9. Low-cycle fatigue tests ($R = 0.0$, $K_t = 1.0$, $F =$ 20 cpm at 425, 540, and 650 °C (800, 1000, and 1200 °F) have shown that the P/M Udimet 720 has mean lives that are higher than cast and wrought Udimet 720.

IN-706 is an iron-nickel-base superalloy that is used for very large forged turbine disks in land-based turbine engines for power generation. Cast and wrought processing for these applications includes vacuum induction melting plus electroslag remelting plus vacuum arc remelting to minimize melt-related defects and segregation. Powder metallurgy is being developed for IN-706 in an effort to further minimize segrega-

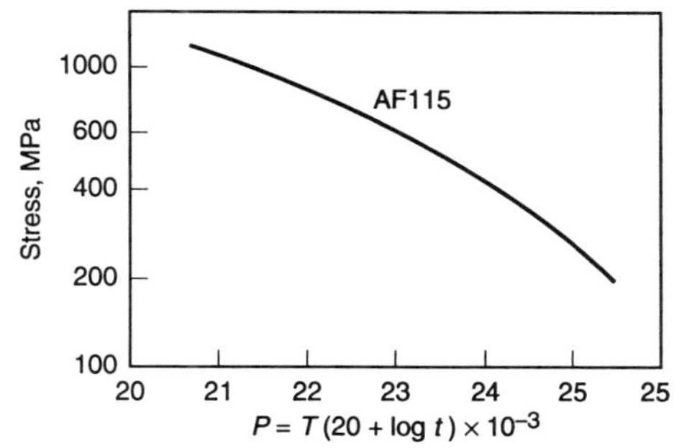

Fig. 12 Stress rupture strength of as-HIP P/M AF115. HIP: 1180 °C (2155 °F)/3 h/102 MPa (15 ksi). Heat treatment: 1175 °C (2150 °F) + 760 °C (1400 °F) air cool. *T*, in K; *t*, in h. Source: Ref 25

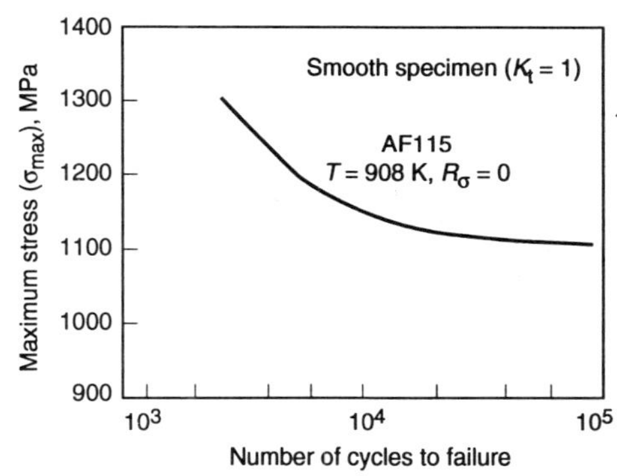

Fig. 13 Low-cycle fatigue life of as-HIP P/M AF115 at 635 °C (1175 °F). HIP: 1180 °C (2155 °F)/3 h/102 MPa (15 ksi). Heat treatment: 1175 °C (2150 °F) + 760 °C (1400 °F) air cool. Source: Ref 25

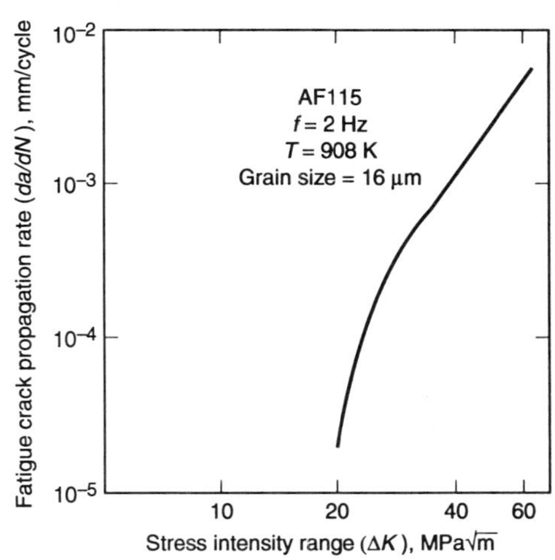

Fig. 14 Fatigue crack growth rate of as-HIP P/M AF115 at 635 °C (1175 °F). HIP: 1180 °C (2155 °F)/3 h/102 MPa (15 ksi). Heat treatment: 1175 °C (2150 °F) + 760 °C (1400 °F) air cool. Source: Ref 25

Table 16 Room-temperature fracture toughness and 750 °C (1380 °F) and 900 °C (1650 °F) low-cycle fatigue properties of HIP plus forged P/M 706

Alloy(a)	Mesh size	HIP temperature, °C	HIP temperature, °F	J_{Ic}, $10^{-2} \cdot$ Nm/m^2	J_{Ic}, in · lb/in.^{2}	K_{Ic}, MPa$\sqrt{m}$	K_{Ic}, ksi$\sqrt{in.}$	Low-cycle fatigue(b): Cycles to failure at 750 °C (1380 °F)	Low-cycle fatigue(b): Cycles to failure at 900 °C (1650 °F)
N706	−60	1130	2065	...	...	...	...	34,312	32,820
N706	−60	1065	1950	9.3	533	145	133	26,515	52,423
A706	−140	1130	2065	7.5	426	130	119	41,846	64,055
A706	−140	1065	1950	7.4	420	129	118	27,623	45,916

Note: Heat treatment: 980 °C (1800 °F)/1 h + 732 °C (1350 °F)/10 h + 620 °C (1150 °F)/8 h. (a) N706, nitrogen atomized; A706, argon atomized. (b) Tested at 0.7% strain. Source: Ref 21

Table 17 Tensile properties of P/M AF2-1DA-6 produced by extrusion plus isothermal forging

Temperature, °C	Temperature, °F	0.2% yield strength, MPa	0.2% yield strength, ksi	Tensile strength, MPa	Tensile strength, ksi	Elongation, %	Reduction of area, %
21	70	1138	165	1586	230	12	12
760	1400	1000	145	1227	178	8	10
870	1600	793	115	1206	175	10	10

Note: Heat treatment: 1204 °C (2200 °F)/2 h/air cool + 1066 °C (1950 °F)/2 h air cool + 760 °C (1400 °F)/16 h/air cool. Source: Ref 28

Table 18 Stress rupture properties of P/M AF2-1DA-6 produced by extrusion plus isothermal forging

Temperature, °C	Temperature, °F	Stress, MPa	Stress, ksi	Rupture life, h	Elongation, %	Reduction of area, %
760	1400	662	96	50	6	7
816	1500	483	70	55	7	7

Note: Heat treatment: 1204 °C (2200 °F)/2 h/air cool + 1066 °C (1950 °F)/2 h air cool + 760 °C (1400 °F)/16 h/air cool. Source: Ref 28

Table 19 Tensile properties of as-HIP P/M PA101 produced as part of a dual-property wheel

Temperature, °C	Temperature, °F	0.2% yield strength, MPa	0.2% yield strength, ksi	Tensile strength, MPa	Tensile strength, ksi	Elongation, %	Reduction of area, %
20	70	945	137	1472	213	15	14
650	1200	896	130	1315	191	10	13
760	1400	875	127	1088	159	11	15

Note: Heat treatment: 1120 °C (2050 °F)/2 h/fan cool + 840 °C (1550 °F)/4 h/air cool + 760 °C (1400 °F)/16 h/air cool. Source: Ref 32 tion and to lower the cost of these large forgings (Ref 20, 21).

In the P/M work, the effects of varying process parameters (argon and nitrogen atomization, powder size, HIP temperature, forging conditions, and heat treatment) have been studied. Tables 14 and 15 give the results of room-temperature tensile and Charpy V-notch tests of P/M IN-706 in the HIP plus heat treated and HIP forged plus heat treated conditions, respectively. Typical data for cast plus wrought IN-706 are included for comparison. Table 16 gives fracture toughness and low-cycle fatigue data for forged P/M IN-706. Compared to cast and wrought material, the P/M material exhibits good fracture toughness and excellent fatigue resistance. The work conducted to date indicates that P/M IN-706 ages more rapidly than cast and wrought IN-706. As a result, some heat treatment modifications may be required for the P/M version of IN-706.

In addition to excellent mechanical properties, P/M IN-706 is highly resistant to grain growth at elevated temperatures. This is shown in Fig. 10, which compares the grain growth characteristics of IN-706 with cast and wrought IN-706. The grain growth resistance of P/M IN-706 may permit the use of higher forging temperatures and lower forging forces than that used for cast and wrought material. Powder metallurgy would also permit larger finish cross sections and more reliable ultrasonic inspection.

IN-718 is similar to IN-706, but has higher elevated-temperature strength. Powder metallurgy IN-718 is being considered as a disk material in the next generation of land-based power generation turbine engines (Ref 22). Research is currently in progress, but no mechanical property data have been reported in the literature. The potential advantages of P/M IN-718 are similar to those described above for P/M IN-706.

AF115 was developed in the 1970s as a very-high-strength P/M disk alloy for 1400-F service (Ref 23), but the alloy is not currently specified for any engine system. However, a lower-carbon version of the alloy (0.05% versus the original 0.15%) continues to be considered for disk application in the as-HIP and HIP plus forged conditions (Ref 24–26). Typical properties for the alloy in the as-HIP condition are given in Fig. 11 to 14. Properties in HIP plus forged condition are discussed below in the section "Dual-Alloy Turbine Disks/Wheels" in this article.

AF2-1DA-6 is one of the strongest nickel-base superalloys available for use in the temperature range of 650 to 980 °C (1200 to 1800 °F). Typical tensile rupture properties are given in Tables 17 and 18. Powder metallurgy AF2-1DA-6 has been evaluated as a turbine disk material (Ref 29). However, the alloy is not currently specified for any turbine engine application.

PA101 is a hafnium-containing modification of the cast alloy IN-792 (Ref 30). The alloy is not currently specified for any engine system, but it has been successfully evaluated as one component of a dual-alloy turbine wheel for small turbine engines (Ref 31, 32). Tensile and stress rupture properties of as-HIP PA101 in a dual-property wheel are given in Tables 19 and 20.

MERL 76 is a P/M alloy that was developed for as-HIP fabrication of turbine engine compo-

Table 20 Stress rupture properties of as-HIP P/M PA101 produced as part of a dual-alloy wheel

Temperature		Stress		Rupture	
°C	°F	MPa	ksi	life, h	Elongation, %
650	1200	862	125	98	5
760	1400	586	85	70	10

Note: Heat treatment: 1120 °C (2050 °F)/2 h/gas fan cool + 840 °C (1550 °F/4 h/air cool + 760 °C (1400 °F)/16 h/air cool. Source: Ref 32

Table 21 Tensile properties of Inconel 690 bar produced from powder by DLF

Material condition	0.2% yield strength		Tensile strength		Elongation, %
	MPa	ksi	MPa	ksi	
DLF—as deposited	448	65	668	97	49
DLF + 930 °C (1700 °F)/ 1 h	489	71	689	100	46
DLF + 980 °C (1800 °F)/ 1 h	475	69	682	99	46
DLF + 1040 °C (1900 °F)/ 1 h	448	65	668	97	47
DLF + 1090 °C (2000 °F)/ 1 h	385	56	648	94	52
Conventionally processed hot-rolled rod	372	54	737	107	50

Source: Ref 40

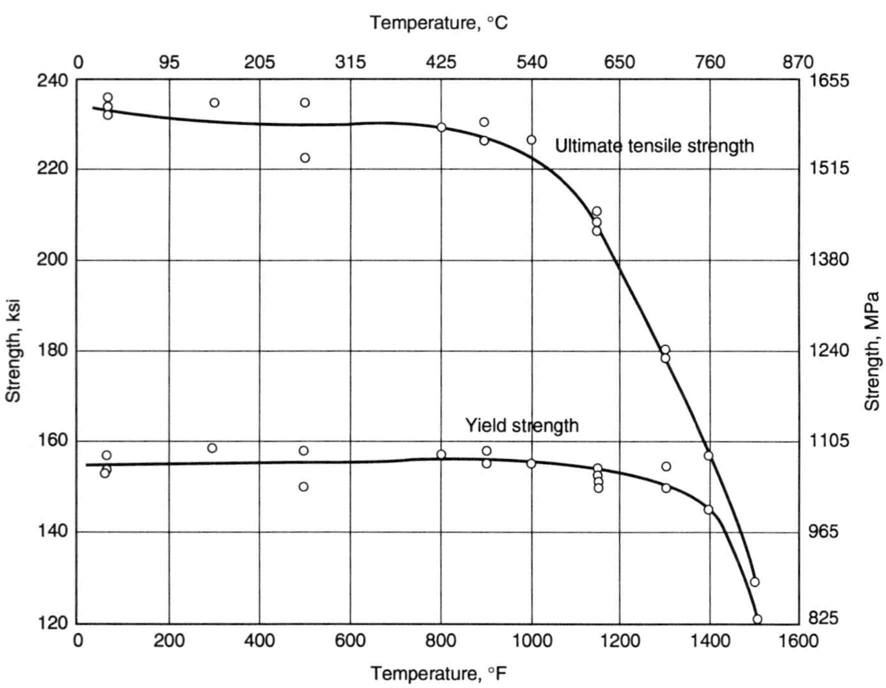

Fig. 15 Tensile and yield strength for as-HIP MERL 76. Heat treatment: 1105 °C (2125 °F)/2 h/oil quench + 870 °C (1600 °F)/40 min/air cool + 982 °C (1800 °F)/45 min/air cool + 650 °C (1200 °F)/24 h/air cool + 760 °C (1400 °F)/16 h/air cool. Source: Ref 33

Table 22 Mechanical properties of PIM Inconel 718

Property	PIM Inconel 718	AMS 5596
Room-temperature tensile		
Tensile strength, MPa (ksi)	1350 (196)	1241 (180)
Yield strength, MPa (ksi)	1139 (165)	1034 (150)
Elongation, %	14	12
540 °C (1000 °F) tensile		
Tensile strength, MPa (ksi)	1119 (162)	...
Yield strength, MPa (ksi)	975 (141)	...
Elongation, %	14	...
650 °C (1200 °F) tensile		
Tensile strength, MPa (ksi)	1049 (152)	999 (145)
Yield strength, MPa (ksi)	904 (131)	827 (120)
Elongation, %	10	5
650 °C (1200 °F) stress rupture		
Rupture life, h	50	...
Rupture elongation, %	4	...
High-cycle fatigue (R = −1)		
Runout stress(a) at 430 °C (800 °F), MPa (ksi)	379 (55)	333 (48)
Runout stress(a) at 540 °C (1000 °F), MPa (ksi)	379 (55)	54 (369)
Runout stress(a) at 650 °C (1200 °F), MPa (ksi)	448 (65)	47 (326)

Note: Heat treatment: 950 °C (1750 °F)/1 h/air cool + 718 °C (1315 °F)/8 h/furnace cool at 100 °F per h to 620 °C (1150 °F)/6 h/air cool. (a) Runout = 10^7 cycles. Source: Ref 48

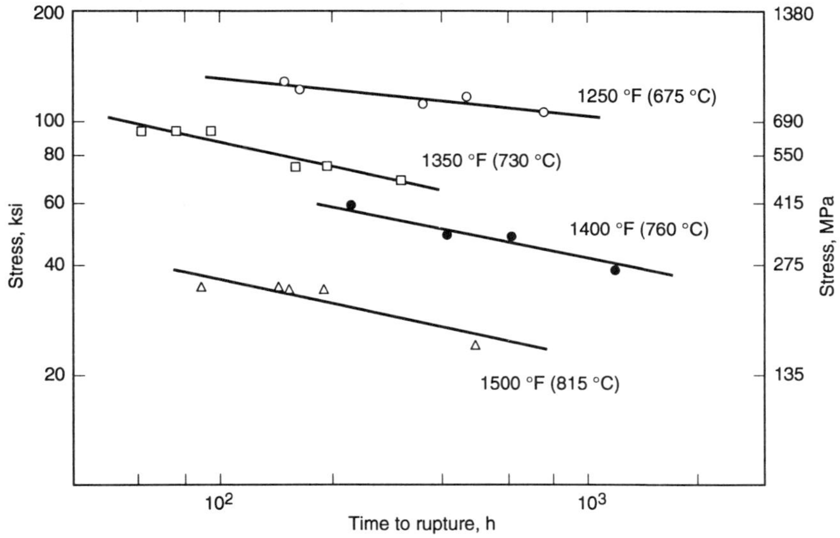

Fig. 16 Stress rupture properties of as-HIP MERL 76. Heat treatment: 1105 °C (2125 °F)/2 h/oil quench + 870 °C (1600 °F)/40 min/air cool + 982 °C (1800 °F)/45 min/air cool + 650 °C (1200 °F)/24 h/air cool + 760 °C (1400 °F)/16 h/air cool. Source: Ref 33

Fig. 17 Section of a dual-alloy property turbine wheel produced by HIP bonding a cast IN-713C blade ring to a P/M LC Astroloy hub. Courtesy of AlliedSignal Engine Division

nents (Ref 33). It was designed to have properties in the HIP condition similar to those of isothermally forged IN-100. There are no current applications for MERL 76. Figures 15 and 16 give tensile and stress rupture properties, respectively, for the alloy in the as-HIP plus heat treated condition.

Specialized P/M Superalloy Processes

Dual-Alloy Turbine Disk/Wheels. There are significantly different operating conditions for various locations in gas turbine wheels and disks (Ref 34). The rims of these parts operate in the 650 to 750 °C (1202 to 1382 °F) temperature range where creep and rupture strength are limiting properties. The bores of these parts operate at temperatures below 550 °C (990 °F). In this temperature region, high tensile strength and low-cycle fatigue resistance are required. Conventional processing of disks and wheels uses a single alloy to meet these varied requirements. As a result, some compromise in properties in either the rim or the bore often has to be tolerated, thereby limiting the service conditions of these parts. Several P/M approaches have been developed to resolve this problem.

AlliedSignal has developed one such process and qualified it for use in auxiliary power unit engines (Ref 2). The process involves a preconsolidated P/M LC Astroloy hub that is HIP bonded to a cast IN713LC blade ring to produce an axial turbine wheel (Fig. 17). The development allowed the high-cycle fatigue requirements of the engine to be met. A similar process has been reported for manufacturing a small axial wheel by HIP bonding a P/M PA101 bore to a cast MAR-M246 blade ring (Ref 32).

Kobe Steel has developed a process that uses two P/M superalloys: AF 115 for the rim and TMP-3 for the bore (Ref 35). In this process, the bore alloy is HIP consolidated. The rim is then consolidated and bonded to the bore in a second HIP cycle. The resulting assembly is isothermally forged to a turbine disk configuration. Subsequent heat treatment develops the required coarse-grained microstructure in the rim alloy and retains the fine-grained structure in the bore alloy. Figures 18 to 20 give the tensile, stress rupture, and low-cycle fatigue properties of the resulting forging. The tensile properties of the bore and rim section are very similar. As desired, the rim section (AF 115) had the highest strength, whereas the bore section TMP-3 exhibited the highest resistance to low-cycle fatigue.

General Electric has developed a dual-alloy disk process that utilizes two advanced high-strength P/M superalloys designated KM4 and SR3 as bore and rim alloys, respectively (Ref 36). The process involves forge bonding to join the two components. The advantages of the process are the high degree of mechanical work

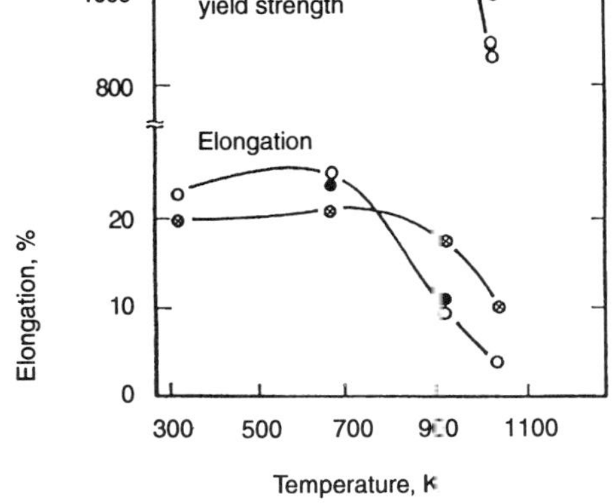

Fig. 18 Tensile properties of a TMP-3 (bore)/AF115 (rim) dual-alloy turbine disk. Source: Ref 35

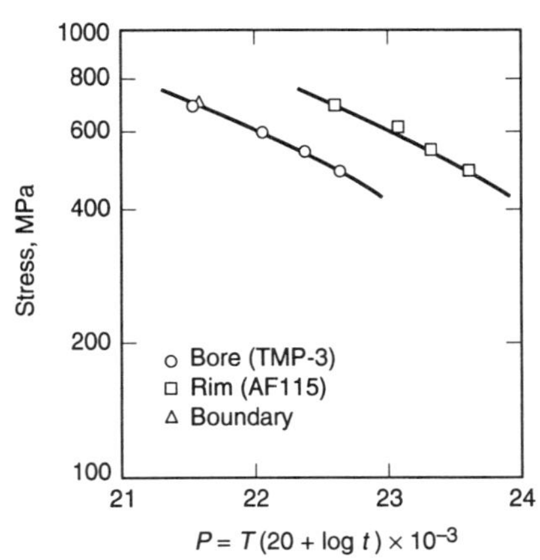

Fig. 19 Stress rupture strength of a TMP-3 (bore)/AF115 (rim) dual-alloy turbine disk. T, in K; t, in h. Source: Ref 35

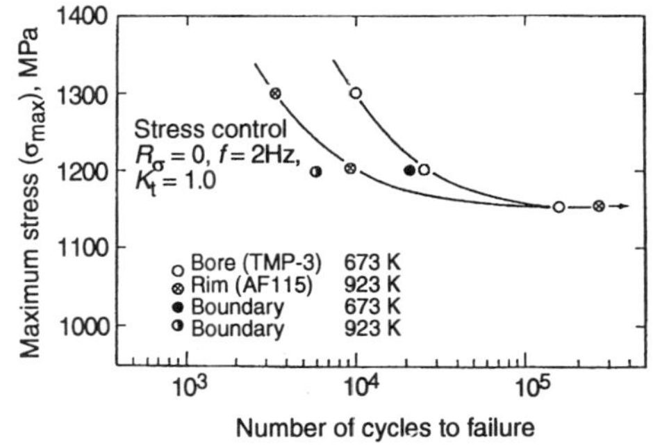

Fig. 20 Low-cycle fatigue life of a TMP-3 (bore)/AF115 (rim) dual-alloy turbine disk. Source: Ref 35

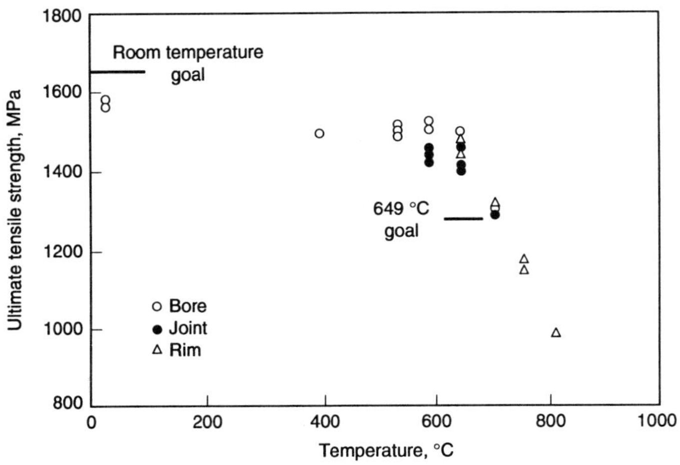

Fig. 21 Tensile strength of a KM4 (bore)/SR3 (rim) dual-alloy turbine disk produced by forge bonding. Source: Ref 36

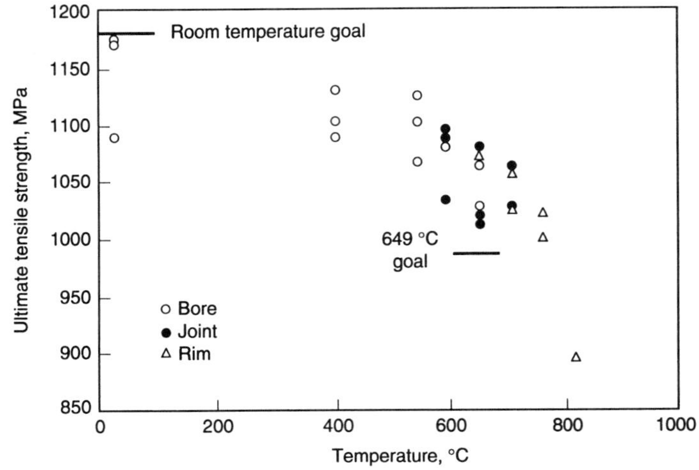

Fig. 22 Yield strength of a KM4 (bore)/SR3 (rim) dual-alloy turbine disk produced by forge bonding. Source: Ref 36

at the joint and the expulsion of bondline material, which are key factors in developing a high-integrity joint. Figures 21 to 23 give the results of tensile and creep tests of the dual-alloy wheel. The results indicate no loss in strength in the joint region.

Other dual-alloy processing scenarios are possible. For example, various combinations of alloys and various solid-state joining techniques such as friction inertia welding, hot uniaxial pressing, HIP, and extrusion (Ref 37).

Laser Assisted Rapid Prototyping and Manufacturing. Several processes are being developed that produce a part directly from a three-dimensional computer-aided design (CAD) model. The attraction for these processes is that they require no hard tooling and, as a result, considerable time and cost can be saved in prototyping or manufacturing a metal article. In these processes, a laser is used to sinter or melt input powder one layer at a time until the entire part is built up layer upon layer. Two basic techniques are used: selective laser sintering and direct metal deposition.

In selective laser sintering (Ref 38), a thin layer of powder is laid out on a substrate. A computer-guided laser then traces the first planer layer of the part and sinters only that powder that lies in that plane of the part. The part is then indexed downward the equivalent of one powder layer. A new layer of powder is then placed on the part and the laser traces the second plane of the part and sinters only the powder that lies in that plane of the part. The process is repeated until the part is complete. Figure 24 shows the major components of selective laser sintering equipment.

Heating the metal to sintering temperatures can adversely affect the mechanical properties of the material, and residual stresses developing upon cooling can induce distortion. An alternate approach has been developed to minimize these difficulties using polymer-coated metal powder as the raw material. The laser is then used only to soften the polymer and form necks with other coated particles that then hold the metal particles in the desired shape. The "green" parts are porous and must be further treated to remove the polymer and sinter the metal. Full density can then be attained by HIP (Ref 39).

Direct metal deposition is similar to selective laser sintering with the exception that the powder is introduced into the laser beam, melted, and deposited. The CAD-based model is used to index the laser and/or the part as the metal is laid down one layer at a time. Variations of the process are referred to as directed light fabrication (DLF) (Ref 27, 40), laser engineered net shaping

Table 23 Composition of superalloys produced by mechanical alloying

	Composition, wt%											
Alloy	Cr	W	Mo	Al	Ti	Ta	C	B	Zr	Fe	Y_2O_3	Ni
MA 754	20	...	...	0.3	0.5	...	0.05	...	...	1.0	0.6	bal
MA 6000	15	4.0	2.0	4.5	2.5	2.0	0.05	0.01	0.15	...	1.1	bal
MA 785	30	...	...	0.3	0.5	...	0.05	...	...	...	0.6	bal

Source: Ref 63–65

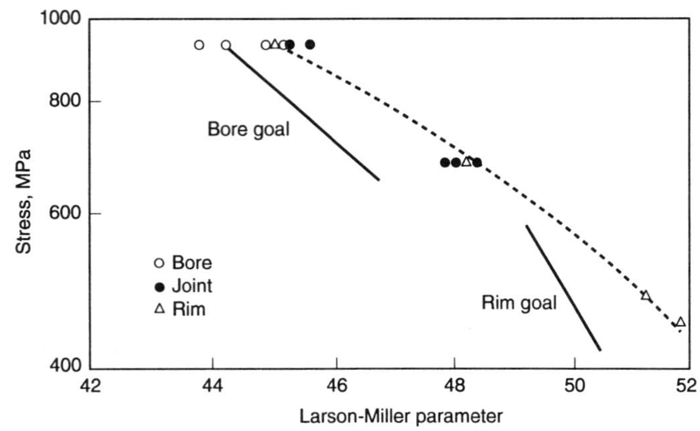

Fig. 23 Time to 0.2% creep data for a KM4 (bore)/SR3 (rim) dual-alloy turbine disk produced by forge bonding. Source: Ref 36

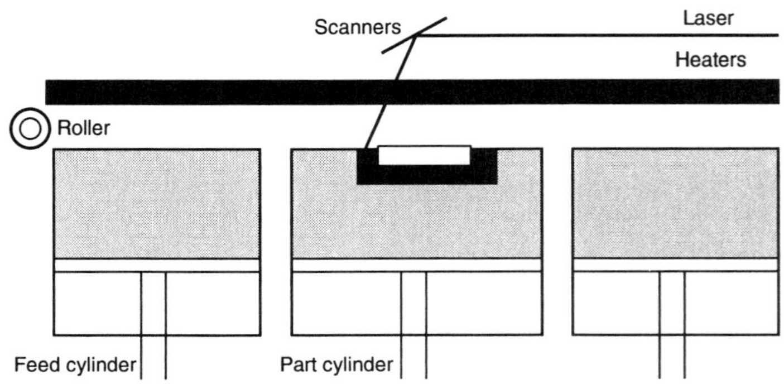

Fig. 24 Major hardware components of selected laser sintering equipment. Source: Ref 38

(LENS) (Ref 41), laser-aided direct metal deposition (LADMD) (Ref 42), or laser aided direct deposition of metals (Ref 43). Figure 25 shows a schematic of the DLF process. Figure 26 shows an Inconel 690 hexagon shape made by the DLF process.

Currently, there is considerable activity aimed at the full development of laser-assisted rapid prototyping and manufacturing. Much of the work has been conducted using stainless steels (Ref 27) and die steels (Ref 44). However, some trials have been conducted on nickel-base alloys such as Inconel 690, Inconel 625, IN-100, and IN-718. Table 21 gives the properties of Inconel 690 bar produced by the DLF process. Some cracking difficulties have been encountered with the γ'-strengthened alloys indicating that some specialized techniques may be required for precipitation-hardened alloys. For more information, see the article "Powder Metallurgy Methods for Rapid Prototyping" in this Volume.

Powder injection molding (PIM) is a relatively low-cost process for producing large quantities of small net-shape complex parts. The process involves mixing powder with a suitable binder, pelletizing the mixture, injection molding into a die, debinding, sintering (plus HIP if required), and heat treatment (Ref 45). The process is used to produce a wide variety of parts from steels, stainless steels, nickel alloys, and cobalt alloys, but has been only recently considered for use with superalloys for aerospace applications. This work has centered primarily on IN-718 (Ref 46–49). As shown in Table 22, the tensile, stress rupture, creep and high-cycle fatigue properties of PIM IN-718 can meet the minimum requirements of aerospace materials specification AMS 5596. Based on the results of this work, PIM IN-718 is believed to be a viable process for demanding applications with an expected cost reduction of 50% (Ref 48). In another study (Ref 50), PIM was evaluated using MERL 76 and Udimet 700. After HIP, densities of nearly 100% were attained and tensile strengths were at acceptable levels. The oxygen and carbon contents were relatively high, but the distribution of nonmetallic inclusions was fine and uniform.

Spray forming involves atomizing a stream of molten metal into droplets and collecting the droplets, an approximately 50/50 mixture of liquid and solid, on a substrate before they fully solidify. The process is capable of producing various shapes such as billet, tubes, disks, and sheet (Ref 51–58). Spray-formed preforms can have a density up to about 99.8% of theoretical, but the material is normally HIP and/or hot worked to fully densify and improve properties. Compared to conventional P/M, the advantage of the process is the potential of lower cost because powder handling, canning, and the initial consolidation step are eliminated. Disadvantages of the process are a coarser structure and the inability to control inclusion size through particle sizing. The process was originally developed by Osprey Metals and is currently being commercially used to produce billet and tubular shapes of steels and corrosion-resistant alloys.

The application of spray forming to superalloys is currently being developed. The process is being considered for producing engine hardware such as disks and ring-shaped components. For superalloys, the process utilizes vacuum induction melting or electroslag melting to minimize inclusions.

In a process developed by Howmet, designated Spraycast-X, metal is vacuum induction melted, argon atomized to a spray, and deposited onto a preheated rotating mild-steel mandrel. The resulting preform is then fully densified by HIP. Further processing may involve ring rolling or forging. In conjunction with Pratt & Whitney, the process is being evaluated for use in manufacturing engine hardware. A spray plus HIP plus ring-rolled IN-718 PW4000 high-pressure turbine case has been successfully tested (Ref 58).

A process developed by General Electric (Ref 52) uses electroslag remelting of a vacuum induction electrode and a ceramic free-metal delivery system to minimize inclusion content (Fig. 27). Argon gas can be used for atomization, but

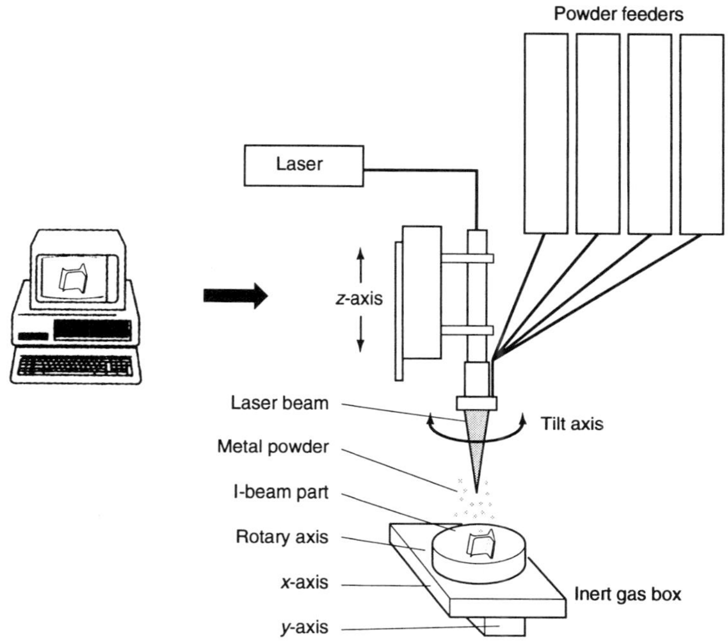

Fig. 25 Schematic of DLF system. Source: Ref 40

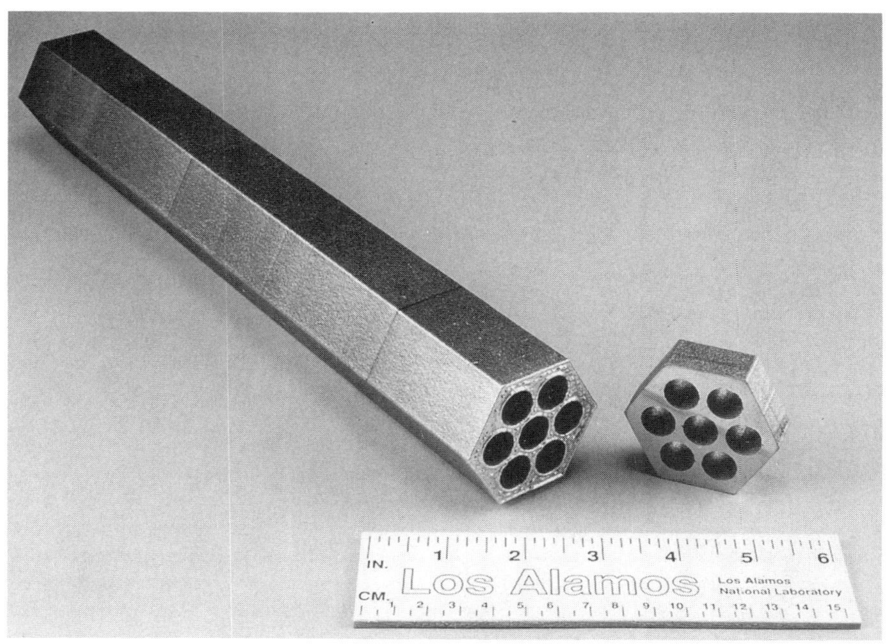

Fig. 26 Inconel 690 hexagon shape with hole array made by DLF. Courtesy of Los Alamos National Laboratories

gas entrapped in the structure can lead to pore formation with subsequent high-temperature exposure such as that encountered in forging, heat treatment, or welding. For this reason, nitrogen gas atomization is being explored (Ref 56, 57). Nitrogen reacts with most superalloys to form stable nitrides. As a result, pore formation does not occur during high-temperature exposure.

A number of different alloys have been spray formed and evaluated. These include René 95 (Ref 56), 718 (Ref 53, 57, 58), René 41 (Ref 53, 55), MERL 76 (Ref 56), Astroloy (Ref 56), Waspaloy (Ref 53), AF2-IDA (Ref 56), and AF115 (Ref 56). In most instances, the properties of spray-formed plus HIP and/or hot working are equal to or improved over those of cast and wrought material. For example, Fig. 28(a) to (c) compare the properties of spray-formed IN-718 with conventional wrought material. Figure 29 compares low-cycle fatigue data of spray formed plus electron beam welded René 41 with that of conventional wrought material. The low-cycle fatigue strength of the base metal was excellent, but that of the weld material was low apparently due to argon pore formation during welding.

Mechanical alloying is a dry, high-energy ball milling process that can be used to produce composite metal alloy powders with a uniform dispersion of refractory metal particles in a complex alloy matrix. The process occurs by repeated welding, fracturing, and rewelding of powder metal particles. The resulting mechanically alloyed powder is subsequently consolidated and then thermomechanically treated to optimize grain structure and properties (Ref 59–62). Several alloys have been developed using a γ' strengthened matrix with an yttrium oxide dispersion. The resulting alloys have an exceptional strength at very high temperatures. Compositions of superalloys produced by mechanical alloying are given in Table 23. Typical properties are given in Table 24 and 25 (Ref 63–65).

Technical Issues

The major technical issue for P/M superalloys, and cast and wrought superalloys as well, has been and continues to be crack initiation and growth. The useful life of heavily stressed superalloy disks in critical rotating components in aircraft engines can be reduced by the presence of even small inclusions that can act as crack initiators under low-cycle fatigue loading (Ref 66). It is important to note, however, that the low-cycle fatigue behavior of superalloys is being continu-

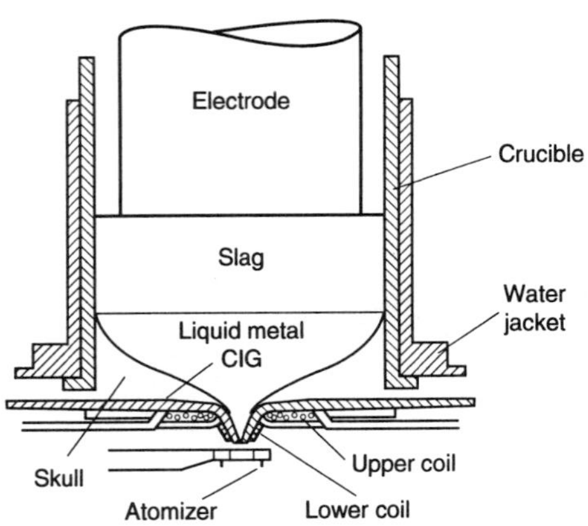

Fig. 27 Schematic of General Electric's clean metal spray forming concept. Source: Ref 52

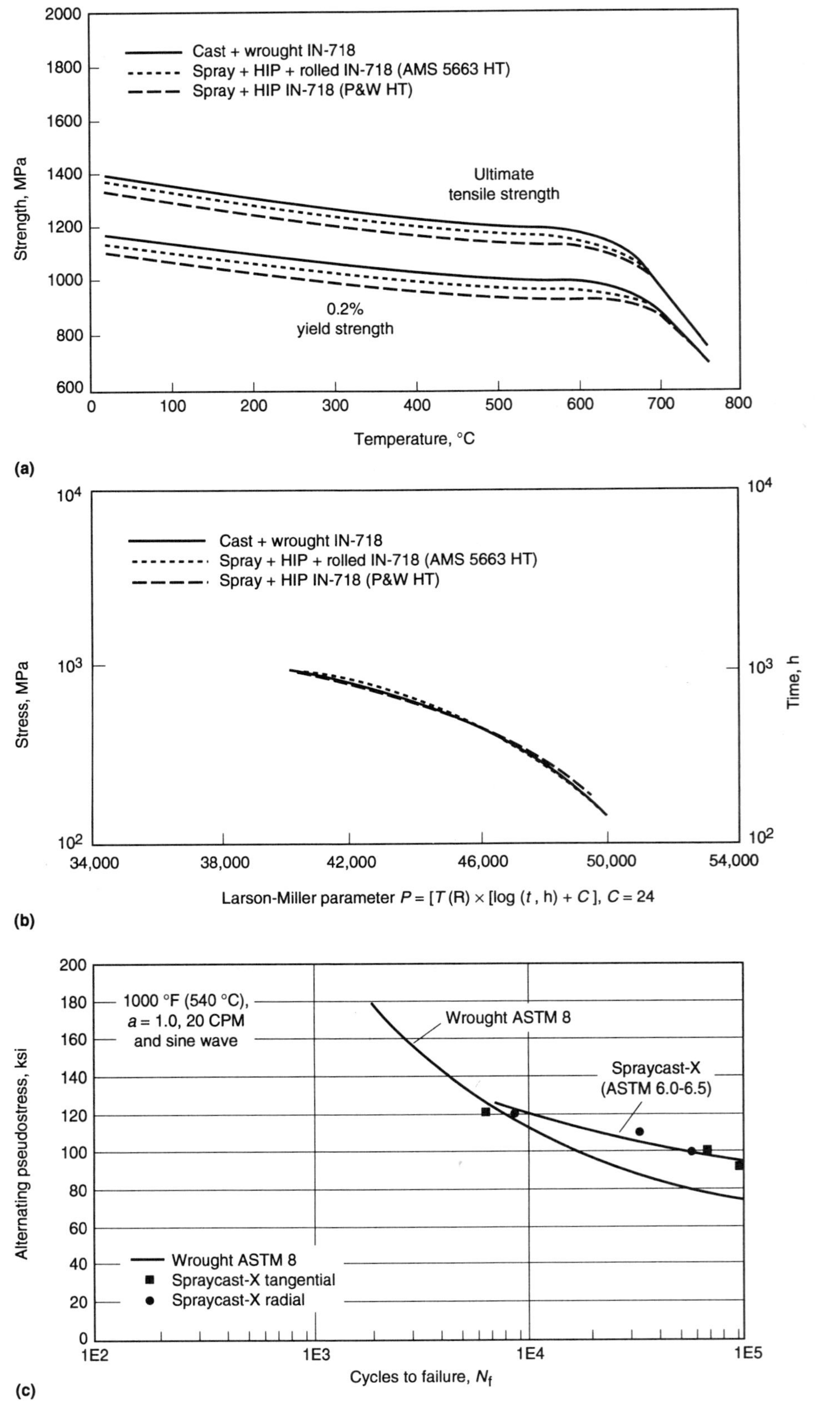

Fig. 28 Comparison of Spraycast-X and conventional cast and wrought IN-718. (a) Tensile properties. (b) Stress rupture properties. (c) Low-cycle fatigue properties. Source: Ref 58

ally improved through inclusion control, alloy development, and heat treatment modifications. Furthermore, designers are now using probabilistic lifing strategies to more reliably predict the useful life of engine hardware. The P/M process is uniquely suited to utilize all four of these methodologies.

Inclusion Control. Figure 30 shows an example of the improvement in low-cycle fatigue behavior of a production P/M superalloy in recent years. This improvement is primarily the result of three major factors. First, the size of potential inclusions has been reduced by reducing the powder particle size. Early P/M superalloy materials were made using powder with a maximum particle size of 250 μm. Most P/M superalloys are now made using a maximum particle size of 105 μm or smaller. From an economic standpoint, this has been made possible by major improvements in fine-powder yields. A second factor is a decrease in the frequency of inclusions. Figure 31 shows an example of how the occurrence of surface-initiated failures in low-cycle fatigue tests has decreased dramatically in recent years, indicating a major reduction in inclusion frequency.

The third factor in inclusion control is minimization or elimination of certain types of inclusions. In early P/M superalloys there were four types of potential defects: (1) metallic inclusions, (2) reactive nonmetallic inclusions that result in heavy prior-particle boundary outlining, (3) nonreactive nonmetallic inclusions, and (4) pores and voids. The first two of these tended to be particularly troublesome because they can grow in size with thermal exposure. However, a recent evaluation of low-cycle fatigue fracture initiation sites in P/M Udimet 720 has shown that the predominant initiation features were either grain-boundary facets or microporosity formed by entrapped argon (Ref 18). Less than 30% of the specimens failed at nonmetallic inclusions. These were identified as either discrete particles of zirconia or agglomerated particles of alumina, silica, or titanium oxide. The indication is that metallic and reactive nonmetallic inclusions have been dramatically reduced in P/M superalloys. Other work has shown that no prior-particle-boundary outlining type defects have been observed at low-cycle fatigue fracture initiation sites in HIP René 95 in the last six years (Ref 8).

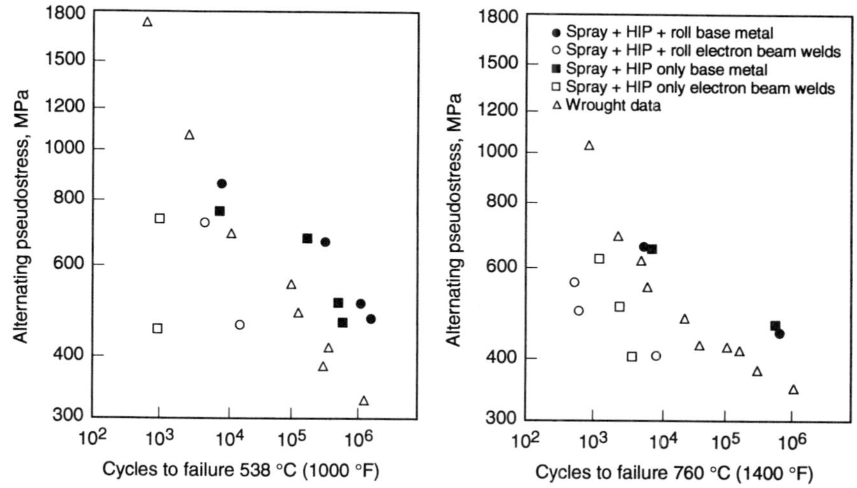

Fig. 29 Low-cycle fatigue data for spray formed René 41 (alternating pseudostress = strain range × Young's modulus/2). Source: Ref 55

It should be noted that a general condition of prior-particle-boundary outlining can occur under conditions that are not related to inclusions. Certain alloys develop particle outlining due to their composition and physical metallurgy. For example, Astroloy can develop γ′ decoration at prior-particle boundaries upon slow cooling from above the γ′ solvus. The problem can be corrected by resolutioning and rapid cooling. Astroloy, IN-100, and Inconel 718 are prone to form oxycarbonitrides at prior-particle boundaries during consolidation (Ref 67). Usually the condition is minimized by reducing the carbon content of the P/M version of the alloy. The condition can also result from exposure of powder to air at an elevated temperature. The resulting oxygen at the surface apparently facilitates precipitation of oxycarbonitrides at the powder surface. Nitrogen atomization can also result in particle outlining for the same reason.

Prior-particle-boundary outlining caused by the above described compositional and metallurgical factors need not be of great concern. The condition can be avoided by alloy selection, alloy modification, or proper selection of processing parameters. In the event that the condition occurs as the result of a processing error, it is throughout the consolidated material and is readily detected by routine metallography. Furthermore, the condition may not necessarily be detrimental, particularly if the precipitate at the powder boundary forms as discrete particles and not as a continuous film. As previously described, recent work has shown that nitrogen-atomized IN-706, with prior-particle-boundary outlining, has properties that are at least equal to those of argon-atomized IN-706 (Ref 20, 21).

Alloy Development and Heat Treatment. In recent years, considerable effort has been made to improve the defect tolerance of P/M superalloys through alloy development and heat treatment modification. For example, N18 is a modification of Astroloy that exhibits improved creep behavior and fatigue crack growth resistance (Ref 13, 14). Similarly, René 88DT was developed as a more damage-tolerant material than René 95 (Ref 17). As previously discussed, the defect tolerance of nickel-base alloys can also be improved by solution treatment slightly above the γ′-solvus temperature. This "supersolvus" treatment can result in significant improvement in creep strength and fatigue crack growth resistance (Ref 15, 16). These new developments may not have been possible without the compositional uniformity afforded by the P/M process.

Probabilistic Lifing Strategies. The major advantages of P/M superalloys are that they con-

Table 24 Longitudinal tensile properties of superalloys produced by mechanical alloying

	Temperature		Tensile strength		Yield strength			Reduction of
Alloy	°C	°F	MPa	ksi	MPa	ksi	Elongation, %	area, %
MA754	21	70	965	140	586	85	21	33
	649	1200	600	87	476	69	25	44
	871	1600	248	36	214	41	31	58
	1093	2000	148	22	134	19	12	24
MA6000	21	70	1294	188	1284	186	4	3
	760	1400	1156	168	781	113	6	12
	982	1800	407	59	344	50	12	35
	1093	2000	222	32	192	28	9	31
MA758	21	70	949	138	560	81	27	...
	700	1292	546	79	371	54	29	...
	800	1472	339	49	214	31	47	...
	1000	1832	173	25	151	22	29	...

Source: Ref 63–65

Table 25 Longitudinal stress rupture properties of superalloys made by mechanical alloying

	Temperature		100 h rupture stress	
Alloy	°C	°F	MPa	ksi
MA 754	650	1200	282	41
	870	1600	172	25
	1090	2000	103	15
MA 6000	760	1400	482	70
	1090	2000	151	22

Source: Ref 63, 64

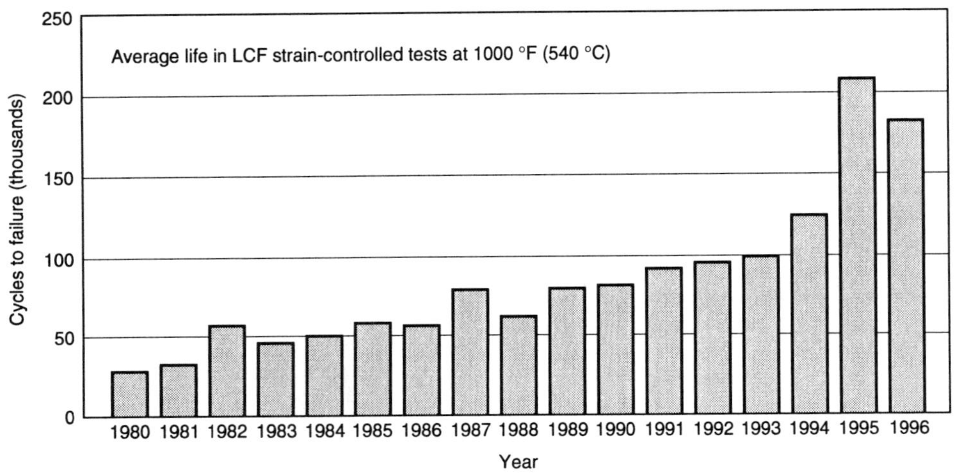

Fig. 30 Average low-cycle fatigue life for a P/M superalloy during 1980 through 1996

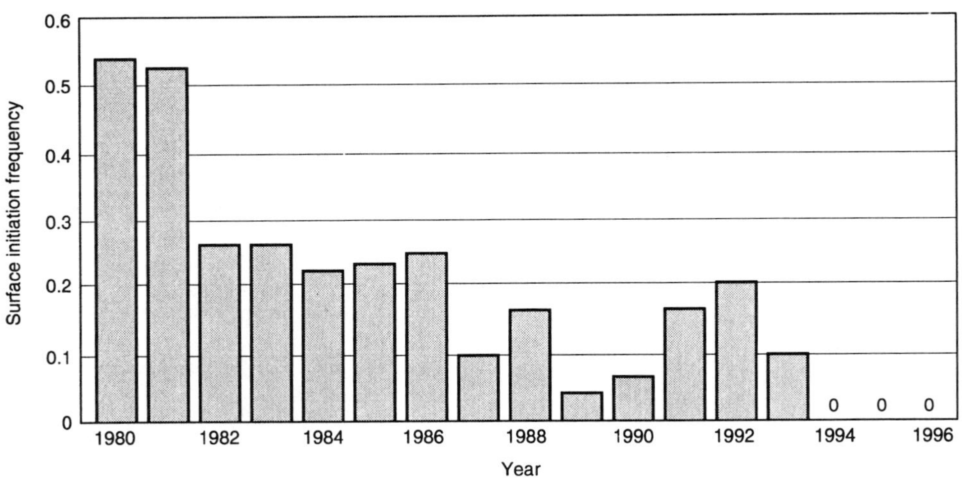

Fig. 31 Frequency of surface-initiated failure in 540 °C (1000 °F) strain-controlled low-cycle fatigue tests of a P/M superalloy during 1980 through 1996

tain less segregation and a finer grain size than conventionally produced alloys. The major benefit of these characteristics is that they deliver a higher fatigue life at elevated temperatures and high stresses. Although it is difficult to directly quantify, it is probable that P/M superalloys contain a lower frequency of large inclusions when compared with cast and wrought product. Designers must use sophisticated means to predict the inclusion size and frequency that will occur in production. Then, some degree of confidence must be calculated that any large inclusion that may occur in production can be detected by available ultrasonic or other nondestructive evaluation techniques after part fabrication. The improved ultrasonic inspectability of P/M alloys is a major benefit to the life prediction analyses of these alloys.

Probabilistic lifing strategies are being used to more accurately estimate part life. These techniques account for the probability of an inclusion being located at, or near, the part surface, and the probability of the inclusion being larger than a critical flaw size. This method of fatigue life calculation initially lowered original estimates of P/M part life when based primarily on laboratory test results. However, as process improvements have been realized that reduce inclusion size and frequency, probabilistic lifing strategies have enabled designers to rely on increased and improved fatigue life predictions for P/M superalloys (Ref 68–71).

REFERENCES

1. Gas Turbine Forecast, *Forecast International 1997,* AlliedSignal, Newton, CT, Aug 1997
2. G.S. Hoppin, III and W.P. Danesi, Manufacturing Processes for Long-Life Gas Turbines, *J. Metal.,* July 1986, p 20–23
3. Gas Turbine Forecast, *Forecast International 1997,* AlliedSignal Model 331, Newton, CT, 1997, p 2
4. AlliedSignal Engine Division, private communication, Sept 1997
5. G.E. Maurer, W. Castledine, F.A. Schweizer, and S. Mancuso, *Development of HIP Consolidated P/M Superalloys for Conventional Forging to Gas Turbine Engine Components,* Superalloys 1996, R.D. Kissinger et al., Ed., The Minerals, Metals and Materials Society, 1996, p 645–652
6. J.L. Bartos and P.S. Mathur, Development of Hot Isostatically Pressed (As-HIP) Powder Metallurgy René 95 Turbine Hardware, *Superalloys: Metallurgy and Manufacture, Proc. of the Third Int. Symp.,* B.H. Kear et al., Ed., Claitor's Publishing Division, 1976, p 495–508
7. General Electric Aircraft Engines
8. Crucible Compaction Metals, Oakdale, PA
9. M.M. Allen, RL. Athey, and J.B. Moore, Nickel-Base Superalloy Powder Metallurgy—State of the Art, *Progress in Powder Metallurgy,* Vol 31, Metal Powder Industries Federation, 1975
10. J.E. Coyne, W.H. Couts, C.C. Chen, and R.P. Roehm, Superalloy Turbine Components—Which is the Superior Manufacturing Process: As-HIP, HIP + Isoforge or Gatorizing of Extrusion, *Powder Metallurgy Superalloys—Aerospace Materials for the 1980's,* Vol 1, MPR Publishing, 1980
11. Pratt & Whitney Aircraft, private communication, 1997
12. "Crucible Nickel Base Superalloys; Low Carbon Astroloy," Crucible Compaction Metals, Oakdale, PA
13. C. Ducrocq, A. Lasalmonie, and Y. Honnorat, N 18, A New Damage Tolerant PM Superalloy for High Temperature, *Superalloys 1988,* The Metallurgical Society, 1988
14. J.H. Davidson, G. Raisson, and O. Faral, The Industrial Development of a New PM Superalloy for Critical High Temperature Aeronautical Gas Turbine Components, *Int. Conf. on PM Aerospace Materials 1991,* MPR Publishing, 1992
15. J.C. Lautridou and J.Y. Guedou, Heat Treatment Upgrading on PM Superalloy N18 for High Temperature Applications, *Materials for Advanced Powder Engineering,* Part II, D. Coutsouradis et al., Ed., *SNECMA,* 1994, p 951–960
16. M. Soucail, M. Marty, and H. Octor, Development of Coarse Grain Structures in a Powder Metallurgy Nickel Base Superalloy N18, *Scr. Mater.,* Vol 34 (No. 4), 1996, p 519–525
17. D.D. Krueger, R.D. Kissinger, and R.G. Menzies, Development and Introduction of a Damage Tolerant High Temperature Nickel-Base Disk Alloy, René 88 DT, *Superalloys 1992,* S.D. Antolovich et al., Ed., The Minerals, Metals and Materials Society, 1992
18. K.A. Green, J.A. Lemsky, and R.M. Gasior, Development of Isothermally Forged P/M Udimet 720 for Turbine Disk Applications, *Superalloys 1996,* R.D. Kissinger et al., Ed., The Minerals, Metals and Materials Society, 1996, p 697–703
19. H. Hattory, M. Takekawa, D. Furrer, and R.J. Noel, Evaluation of P/M U720 for Gas Turbine Application, *Superalloys 1996,* R.D. Kissinger et al., Ed., The Minerals, Metals and Materials Society, 1996, p 705–711

20. U. Habel, J.H. Moll, F.J. Rizzo, and J.J. Conway, Microstructure and Properties of HIP P/M 706, *Advanced Particulate Materials and Processes,* F.H. Froes et al., Ed., Metal Powder Industries Federation, 1997, p 447–455
21. U. Habel, F.J. Rizzo, J.J. Conway, R. Pishco, V.M. Sample, and G.W. Kuhlman, First and Second Tier Properties of HIP and Forged P/M 706, *Superalloys 718, 625, 706 and Various Derivatives,* E.A. Loria, Ed., TMS, 1997, p 247–256
22. A.S. Watwe, J.M. Hyzak, and D.M. Weaver, Effect of Processing Parameters on the Kinetics of Grain Coarsening in P/M 718, *Superalloys 718, 625, 706 and Various Derivatives,* E.A. Loria, Ed., TMS, 1997, p 237–246
23. J.L. Bartos, "Development of a Very High Strength Disk Alloy for 1400F Service," Air Force Materials Laboratory, Wright-Patterson Air Force Base, Dec 1974
24. H. Takigawa, N. Kawai, K. Iwai, S. Furuta, and N. Nagata, Process Development for Low-Cost, High-Strength PM Ni-Base Superalloy Turbine Disk, *Met. Powder Rep.,* Vol 44 (No. 9), Sept 1989
25. K. Iwai, S. Furuta, and T. Yokomaku, Mechanical Properties of Ni-Base Superalloy Disks Produced by Powder Metallurgy, *Met. Powder Rep.,* Vol 43 (No. 10), Oct 1988, p 664–666
26. K. Iwai, S. Furuta, T. Yokomaku, and H. Murai, Mechanical Properties of Ni-Base Superalloy Disks Produced by Powder Metallurgy, *R&D Kobe Steel Eng. Rep.,* Vol 37 (No. 3), 1987, p 11–14
27. G.K. Lewis, D.J. Thoma, R.B. Nemec, and J.O. Milewski, Directed Light Fabrication of Near-Net Shape Metal Components, *Advances in Powder Metallurgy & Particulate Materials—1996,* Vol 4, Parts 13–15, compiled by T.M. Cadle and K.S. Narasimhan, Metal Powder Industries Federation, p 15-65 to 15-76
28. "P/M CAP AF 2-IDA-6," Cytemp Specialty Steel Div., Preliminary Data Sheet, 11 May 1972
29. D.F. Gray, Mechanical Properties of Thick Section AF2-1DA-6 Powder Metal Turbine Rotors, *Rapidly Solidified Materials,* American Society for Metals, 1985, p 387–395
30. B. Ewing, F. Rizzo, and C. ZurLippe, Powder Metallurgy Products for Advanced Gas Turbine Applications, *Superalloys Processing, Proc. Second Int. Conf.,* Metals & Ceramic Info. Center, 1972
31. PM Aerospace Materials, *Met. Powder Rep.,* Vol 38 (No. 10), Oct 1983
32. B.A. Ewing, A Solid-to-Solid HIP-Bond Processing Concept for the Manufacture of Dual-Property Turbine Wheels for Small Gas Turbines, *Superalloys 1980,* J.K. Tien et al., Ed., American Society for Metals, 1980, p 169–178
33. D.J. Evans and R.D. Eng, Development of a High Strength Hot-Isostatically-Pressed Disk Alloy, MERL 76, *Modern Developments in Powder Metallurgy,* Vol 14, Metal Powder Industries Federation, 1980, p 51–63
34. J.M. Hyzak and S.H. Reichman, *Advances in High Temperature Structural Materials and Protective Coatings,* National Research Council of Canada, 1994, p 126–146
35. K. Iwai, S. Furuta, H. Takigawa, O. Tsuda, and N. Kanamaru, Dual-Structure PM Ni-Base Superalloy Turbine Disk; *PM Aerosp. Mater.,* 1991, MPR Publishing, 1992, p 3–1
36. D.P. Mourer, E. Raymond, S. Ganesh, and J. Hyzak, Dual Alloy Disk Development, *Superalloys 1996,* R.D. Kissinger et al., Ed., The Minerals, Metals and Materials Society, 1996, p 637–643
37. Y. Bienvenu, M.L. Dupont, G. Lemaître, and F. Schwartz, A Study of the Powder Metallurgy Processing of Hybrid Nickel Based Superalloy Components, *Powder Metall.,* 1994, p 2053–2056
38. U. Lakshminarayan and K.P. McAlea, Advances in Manufacturing Metal Objects by Selective Laser Sintering (SLS™), *Advances in Powder Metallurgy & Particulate Materials—1996,* Vol 4, compiled by T.M. Cadle and K.S. Narasimhan, Metal Powder Industries Federation, p 15-129 to 15-138
39. B. Badrinarayan and J.W. Barlow, Effect of Processing Parameters in SLS of Metal-Polymer Powders, *Solid Freeform Fabrication Symposium Proc.,* University of Texas, Austin, 1995, p 55–63
40. G.K. Lewis, J.O. Milewski, D.J. Thoma, and R.B. Nemec, Properties of Near-Net Shape Metallic Components Made by the Directed Light Fabrication Process, 8th Solid Freeform Fabrication Symposium, University of Texas, Austin, 11–13 Aug 1997
41. D.M. Keicher, J.A. Romero, C.L. Atwood, J.E. Smugeresky, M.L. Griffith, F.P. Jeantette, L.D. Harwell, and D.L. Greene, Free Form Fabrication Using the Laser Engineered Net Shaping (LENS™) Process, *Advances in Powder Metallurgy & Particulate Materials—1996,* Vol 4, Parts 13–15, compiled by T.M. Cadle and K.S. Narasimhan, Metal Powder Industries Federation, p 15-119 to 15-127
42. D.M. Keicher and J.E. Smugeresky, The Laser Forming of Metallic Components Using Particulate Materials, *JOM,* May 1997, p 51–54
43. J. Mazumder, J. Koch, K. Nagarthnam, and J. Choi, Rapid Manufacturing by Laser Aided Direct Deposition of Metals, *Advances in Powder Metallurgy & Particulate Materials—1996,* Vol 4, Parts 13–15, compiled by T.M. Cadle and K.S. Narasimhan, Metal Powder Industries Federation, p 15-107 to 15-118
44. J. Mazumder, J. Choi, K. Nagarthnam, J. Koch, and D. Hetzner, The Direct Metal Deposition of H13 Tool Steel for 3-D Components, *JOM,* May 1997, p 55–60
45. R.M. German, *Powder Injection Molding,* Metal Powder Industries Federation, 1990
46. J.J. Valencia, J. Spirko, and R. Schmees, Sintering Effect on the Microstructure and Mechanical Properties of Alloy 718 Processed by Powder Injection Molding, *Superalloys 718, 625, 706, and Various Derivatives,* E.A. Loria, Ed., The Minerals, Metals and Materials Society, 1997, p 753–762
47. A. Bose, J.J. Valencia, J. Spirko, and R. Schmees, "Powder Injection Molding of Inconel 718 Alloy," National Center for Excellence in Metalworking Technology, Operated by Concurrent Technologies Corporation under contract No. N00140–92-C-BC49 to the U.S. Navy
48. R. Schmees, J. Spirko, and Dr. J. Valencia, "Powder Injection Molding (PIM) of Inconel 718 Aerospace Components," Pratt & Whitney, West Palm Beach, FL, and Concurrent Technologies Corporation, Johnstown, PA
49. K.F. Hens, J.A. Grohowski, R.M. German, J.J. Valencia, and T. McCabe, Processing of Superalloys via Powder Injection Molding, *Advances in Powder Metallurgy and Particulate Materials,* Vol 4, compiled by C. Lall and A.J. Neupaver, MPIF/APMI Int., 1994, p 137–148
50. E. Lang and M. Poniatowski, Production of Metallic Turbine Parts by the Powder Metallurgical Injection Molding Technique, *Mater. Technol. Testing,* Vol 18 (No. 10), Oct 1987, p 337–344
51. Spray Casting: A Review of Technological and Scientific Aspects, Book Series on Power Metallurgy, Vol 3, *Current Status of P/M Technology,* I. Jenkins and J.V. Wood, Ed., Institute of Metals, 1990
52. M. Hull, Spray Forming Poised to Enter Mainstream, *Powder Metall.,* Vol 40 (No. 1), 1997, p 23–26
53. R.P. Dalal and P.D. Prichard, Thermomechanical Processing of Spraycast-X Superalloys, *Spray Forming 2,* Woodhead Publishing, 1993, p 141–153
54. W. Reichelt et al., *Spray Deposition—An Innovative Method for Innovatory Products 49th International Congress on the Technology of Metals and Materials* (Brazil), Vol 5, Associacao Brasileira deMetalurgia e Materiais, 1995, p 161–170
55. E.S. Huron, Properties of Sprayformed Superalloy Rings, *Spray Forming 2,* Woohead Publishing, 1993, p 155–164
56. K.M. Chang and H.C. Fiedler, Spray-Formed High-Strength Superalloys, *Sixth Int. Symposium on Superalloys,* D.N. Duhl et al., Ed., TMS, Sept 1988
57. M.G. Benz, T.F. Sawyer, F.W. Clark, and P.L. Dupree, *Properties of Superalloys Spray Formed at Process Flow Rates of Less than 20* cm^3/S, GE Research & Development Center, Physical Metallurgy Laboratory 90CRD145, Aug 1990
58. N. Paton, T. Cabral, K. Bowen, and T. Tom, SPRAYCAST IN718 Processing Benefits, *Superalloys 718, 625, 706, and Various Derivatives,* E.A. Loria, Ed., TMS, 1997, p 1–16
59. S.K. Kang and R.C. Benn, Characterization of INCONEL Alloy MA 6000 Powder, *Metall. Trans. A,* Vol 18 (No. 5), May 1987, p 747–758
60. J.S. Benjamin, Dispersion Strengthened Superalloys by Mechanical Alloying, *Metall. Trans.,* Vol 1, Oct 1970, p 2943–2951

61. W. Betteridge and S.W.K. Shaw, Overview Development of Superalloys, *Mater. Sci. Technol.,* Vol 3, Sept 1987, p 682–694
62. G.M. McColvin and M.J. Shaw, Inco Alloys International Ltd., Volume Manufacturing and Applications of Mechanically Alloyed Materials, *Mater. Sci. Forum,* Vol 88–90, 1992, p 235–242
63. Inconel Alloy MA 754, *Alloy Dig.,* 1990, ASM International, Rev. March 1990, May 1977
64. Inconel Alloy MA 6000, *Alloy Dig.,* 1980, ASM International, July 1983
65. Inconel Alloy MA 758, *Alloy Dig.,* 1990, ASM International, May 1996
66. J.C. Lautridou, J.Y. Guedou, and Y. Honnorat, Effect of Inclusions on LCF Life of PM Superalloys for Turboengine Discs, *High Temperature Materials for Power Engineering,* Kbuwer, 1990, p 1163–1172
67. J.H. Moll, V.C. Petersen, and E.J. Dulis, Powder Metallurgy Parts for Aerospace Applications, *Powder Metallurgy, Applications, Advantages and Limitations,* E. Klar, Ed., American Society for Metals, 1983, p 247–298
68. J. Grison and L. Rely, Fatigue Failure Probability in a Powder Metallurgy Ni-Base Superalloy, *Eng. Fract. Mech.,* Vol 57 (No. 1), 1997, p 54
69. P.G. Roth, "Probabilistic Rotor Design System (PRDS) Phase 2," WL-TR-97-2046, Aero Propulsion and Powder Directorate, Wright Laboratory, Wright-Patterson Air Force Base, 1997, p 9
70. J.C. Latridou, A. deBussac, and F. Soniak, A Probabilistic Model For Fatigue Life Prediction of PM Ni-Base Superalloys Containing Inclusions, *The Int. Conf on Fatigue and Fatigue Thresholds,* Vol 3, U.K. Engineering Material Advisory Services, 1993
71. E.S. Huron and P.G. Roth, The Influence of Inclusions on Low Cycle Fatigue Life in P/M Nickel-Base Disk Superalloy, *Superalloys 1996,* R.D. Kissinger et al., Ed., The Minerals, Metals and Materials Society, 1996, p 359–368

Refractory Metals

THE REFRACTORY METALS include niobium (also known as columbium), tantalum, molybdenum, tungsten, and rhenium. With the exception of two of the platinum-group metals, osmium and iridium, they have the highest melting temperatures and lowest vapor pressures of all metals. The refractory metals are readily degraded by oxidizing environments at moderately low temperatures, a property that has restricted the applicability of the metals in low-temperature or nonoxidizing high-temperature environments. Protective coating systems have been developed, mostly for niobium alloys, to permit their use in high-temperature oxidizing aerospace applications.

Refractory metals at one time were limited to use in lamp filaments, electron tube grids, heating elements, and electrical contacts; however, they have since found widespread application in the aerospace, electronics, nuclear and high-energy physics, and chemical process industries. Typical applications are summarized in Table 1.

The refractory metals are distinguished by several common characteristics, including high density, high melting point, and superior resistance to wear and acid corrosion. Tungsten, for example, has a density over twice that of iron and a melting point of 3410 °C (6170 °F), the highest of any element. These metals have body-centered cubic crystal structures (with the excep-

Table 1 Commercial applications of refractory metals and alloys by industry

Application	Material	Application	Material
Aerospace and nuclear industries		**Electronics industry (continued)**	
Counterweights (aircraft, inertial guidance systems)	Tungsten alloys	Heat sinks	Molybdenum, tungsten
Solid-propellant rockets		Backing wafers, semiconductors	Molybdenum, tungsten
2650–2750 °C (4800–4980 °F) flame temperature	Molybdenum, tungsten	Filaments, ion gages, photoflash	Rhenium, W-Re
3425–3550 °C (6195–6420 °F) flame temperature	Silver and copper-infiltrated tungsten	**Process industries**	
Lifting and guidance structures for glide reentry vehicles	Cb-752(a), FS-85(a), C-129Y(a)	Heating and cooling coils	Tantalum, Ta-Nb
Leading edges and nose caps for hypersonic flight vehicles	Cb-752(a), FS-85(a), Ta-10W(a)	Shell and tube heat exchangers	Tantalum
Thrust chambers	C-103(a)	Condensers	Tantalum
Radiation nozzle extensions	C-103(a), FS-85	Tantalum-clad steel vessels	Tantalum
Jet engine components		Distillation towers	Tantalum
Augmenter liners	C-103(a)	Valves for hot sulfuric acid service	Molybdenum, tantalum, Ta-Nb
Center body	Cb-752(a)	Expansion joints (bellows)	Tantalum
Flame holders	Cb-752(a)	Glass-processing equipment	Tantalum
Rocket nozzles	FS-85(a), Ta-10W(Ta-Hf clad)	Crucibles, all sizes up to 1 m (3 ft) diameter × 1.3 m (4 ft) high	Tungsten, tantalum, Ta-40Nb
Thermal shields	C-129Y(a)	Spinnerettes, textile industry	Tantalum, niobium
Porous ionizer plates	Tungsten	Thermocouple protection tubes	Tantalum-coated copper or steel
Heat shields and cesium vapor inlet tubes (ion engine)	Tantalum	Rupture discs	Tantalum
Fasteners	Nb-Ti	Thermowells	Tantalum-coated copper, tantalum
Honeycomb structures	Molybdenum, Cb-752, Ta-10W	Spargers, funnels, jet ejectors	Tantalum
Hot gas tubing	Ta-10W(a)	Bayonet heaters	Tantalum
Hot gas bellows	C-103(a)	Pumps for hydrogen chloride service at 200 kPa (30 psi) and 150 °C (300 °F)	Tantalum (exposed parts)
Solid propellant expansion nozzle	C-103(a)	Cathodic protection electrodes	Niobium
Nuclear and high-energy physics		**Special equipment**	
Linear accelerators, microwave cavities	Niobium	Furnace parts	
Superconductors	Nb-Ti, Nb_3Sn	Heating elements, shields, boats, trays, platens, fixtures	Tungsten, molybdenum, tantalum
Liquid metal containers and piping	Nb-1Zr	Susceptors (induction furnace)	Tantalum
Electronics industry		Extrusion dies	Tungsten, molybdenum
Capacitors	Tantalum powder, foil, wire	Piercing points, hot punches	Tungsten, molybdenum
Capacitor cases	Tantalum strip	Cups	Tungsten, molybdenum, tantalum
Rectifiers, railway signals	Tantalum	Fasteners (nuts, screws, studs, rivets)	Tungsten, molybdenum, C-3009, C-129Y
Battery chargers	Tantalum	Die casting molds, cores	Molybdenum, tungsten
Transducers	Molybdenum, tungsten	Vacuum-metallizing coils, boats	Tungsten, molybdenum
Electron tube parts		Springs	Tungsten, molybdenum, tantalum
Heaters	Tungsten, W-Re	Boring bars	Tungsten, molybdenum
Supports	Molybdenum	Surgical implants	Tantalum
Cathodes	Tantalum	Instruments	Tantalum
Anodes	Molybdenum, tungsten	Electroplating equipment	Tantalum
Superconducting wire	Nb-Ti, Nb_3Sn	Thermocouples, spot weld electrodes	W, W-Re alloys
X-ray targets	Tungsten, molybdenum, rhenium, composite W-Mo	Cathodes, plasma generator	W-1Ni
Electrodes (mercury switches)	Molybdenum, tungsten	Rapid-fire gun barrels	C-3009(a)
Thin-film substrates	Molybdenum, tungsten	Sodium vapor lamp electrodes	Nb-1Zr
Electrical contacts	Tungsten, rhenium, W-Ag, W-Cu		

(a) Parts are silicide-coated in use.

Table 2 Nominal compositions of commercially important refractory metal alloys

Alloy designation	Composition, %														
	Nb	Ta	Mo	W	Re	Zr	Hf	Ti	Y	C	ThO_2	Si	K	Al	O
Niobium alloys															
Nb-1Zr	bal	...	...	...	...	1	...	...	...	...	...	...	...	...	...
FS-85	bal	27.5	...	11	...	1	...	...	...	...	...	...	...	...	...
Cb-752	bal	...	...	10	...	2.5	...	...	...	...	...	...	...	...	...
C-103	bal	...	...	...	...	...	10	1	...	...	...	...	...	...	...
C-129Y	bal	...	...	10	...	...	10	...	0.15	...	...	...	...	...	...
C-3009(a)	bal	...	...	10	...	...	30	...	...	...	...	...	...	...	...
Cb-Ti superconductor	...	...	...	...	...	...	...	46.5	...	...	...	...	...	...	...
Tantalum alloys															
63 metal	0.15	bal	...	2.5	...	...	...	...	...	...	...	...	...	...	...
Ta-10W	...	bal	...	10	...	...	...	...	...	...	...	...	...	...	...
T-111	...	bal	...	8	...	...	2	...	...	...	...	...	...	...	...
T-222	...	bal	...	10	...	...	10	...	...	0.01	...	...	...	...	...
Ta-40Nb	40	bal	...	...	...	...	...	...	...	...	...	...	...	...	...
61 metal (P/M)	...	...	bal	...	7.5	...	...	...	...	...	...	...	...	...	...
Molybdenum alloys															
Mo-0.5Ti	...	...	bal	0.02	...	...	...	0.5	...	...	...	...	...	...	...
TZM	...	...	bal	0.02	...	0.1	...	0.5	...	...	...	...	...	...	...
Tungsten alloys															
W-ThO_2 alloys															
W-1 ThO_2	...	...	...	bal	...	...	...	...	...	...	1	...	...	...	...
W-2 ThO_2	...	...	...	bal	...	...	...	...	...	...	2	...	...	...	...
W-Mo alloys(b)															
W-2 Mo	...	...	2	bal	...	...	...	...	...	...	...	...	...	...	...
W-15 Mo	...	...	15	bal	...	...	...	...	...	...	...	...	...	...	...
W-Re alloys(c)															
W-1.5 Re	...	...	...	bal	1.5	...	...	...	...	...	...	...	...	...	...
W-3 Re	...	...	...	bal	3	...	...	...	...	...	...	...	...	...	...
W-25 Re	...	...	...	bal	25	...	...	...	...	...	...	...	...	...	...
Doped W	...	...	...	bal	...	...	...	...	...	...	...	50 ppm	90 ppm	15 ppm	35 ppm

(a) C-3009 is in the literature as a family of alloys ranging from 9–15% W and ≤5% Ti. The hafnium content is constant at 30%. (b) Various molybdenum contents; two most common alloys listed. (c) Various rhenium contents to ≤26%; three most common alloys listed

tion of rhenium, whose crystal form is hexagonal). All are subject, however, to severe oxidation above 500 °C (930 °F) and must be protected for service by coatings or nonoxidizing atmospheres.

The refractory metals are extracted from ore concentrates, processed into intermediate chemicals, and then reduced to metal. The refractory metals, except for niobium, are produced exclusively as metal powders, which are consolidated by sintering and/or melting. The process for niobium differs only in that the metal is most commonly reduced by aluminothermic reduction of oxide. In this process, oxide impurities slag from the molten niobium. Powder production is covered in the article "Production of Refractory Metal Powders" in this Volume.

Selection and Fabrication

Table 2 lists nominal compositions of commercially prominent refractory metal alloys. Selection of a specific alloy from the refractory metal group often is based on fabricability rather than on strength or corrosion resistance. Niobium, tantalum, and their alloys are the most easily fabricated refractory metals. They can be formed, machined, and joined by conventional methods. They are ductile in the pure state and have high interstitial solubilities for carbon, nitrogen, oxygen, and hydrogen. Because of the high solubilities in niobium and tantalum, these embrittling contaminants normally do not present problems in fabrication. However, tantalum and niobium dissolve sufficient amounts of oxygen at elevated temperatures to destroy ductility at normal operating temperatures. Therefore, elevated-temperature fabrication of these metals is used only when necessary. Protective coatings or atmospheres are mandatory unless some contamination can be tolerated. The allowable level of contamination, in turn, determines the maximum permissible exposure time in air at elevated temperature.

Molybdenum, molybdenum alloys, tungsten, and tungsten alloys require special fabrication techniques. Fabrication involving mechanical working should be performed below the recrystallization temperature. These materials have limited solubilities for carbon, nitrogen, oxygen, and hydrogen. Because the residual levels of these elements required to prevent embrittlement are impractically low, the microstructure must be controlled to ensure a sufficiently low ductile-to-brittle transition temperature (DBTT).

Hot forging or extrusion is used for breaking down ingots into rounds or rectangular sheet bar. These bars, as well as sintered products, are processed into sheet, plate, foil, tubing, and bar. Table 3 gives typical mill-processing temperatures for the refractory metals.

Most refractory metals and alloys also are available as wire. Tungsten wire, for example, which comes in diameters as small as 0.0102 mm (0.0004 in.), is used as fiber reinforcement in composite materials in which the matrix is any one of various ductile alloys. Tantalum wire is used extensively in capacitor manufacturing and in surgical applications.

Recently, chemical equipment has been fabricated from steel plate explosively clad with tantalum. Forming and welding methods have been developed for fabrication of the clad plate into reactor vessels, tanks, and other types of chemical equipment. Explosive bonding produces a metallurgical bond at the tantalum/steel interface where bond efficiency is over 98%.

Cleaning is critical throughout the fabrication process, especially for niobium and tantalum. Cleaning should both precede and follow welding, heat treating, or any thermal process. Cleaning for these metals is accomplished in a hot alkaline solution (minimum 10 min exposure), followed by a chemical cleaning in a mixture of hydrofluoric, nitric, and sulfuric acids. A coupon of the same alloy must accompany each lot of hardware to record material removal. The coupon should have a reference point for measurement. Each cleaning operation should remove approximately 0.0025 mm (0.0001 in.) per side. The activity of the acid should be checked before the hardware is placed in the acid. If the oxide is thicker than 0.0025 mm (0.0001 in.) per

Table 3 Mill-processing temperatures for refractory metals

Metal or alloy	Forging: Temperature(a) °C	Forging: Temperature(a) °F	Forging: Typical total reduction, %	Extrusion: Temperature(a) °C	Extrusion: Temperature(a) °F	Extrusion: Typical reduction ratio	Rolling: Temperature(a) °C	Rolling: Temperature(a) °F	Rolling: Typical total reduction between anneals, %
Niobium and niobium alloys									
Niobium	980–650	1800–1200	50–80	1095–650	2000–1200	10:1	315–205	600–400	50 breakdown
							20	70	90 finish
Nb-1Zr	1205–980	2200–1800	50–80	1205–980	2200–1800	10:1	315–205	600–400	50 breakdown
							20	70	80 finish
FS-85	1315–980	2400–1800	50	1315–980	2400–1800	4:1	370–205	700–400	40 breakdown
							20	70	50–65 finish
Cb-752	1205–980	2200–1800	30	1315–980	2400–1800	4:1	370–260	700–500	50 breakdown
							20	70	65–75 finish
C-103	1315–980	2400–1800	50	1315–980	2400–1800	8:1	205	400	50 breakdown
							20	70	60–70 finish
C-129Y	1315–980	2400–1800	50	1315–980	2400–1800	4:1	425	800	50 breakdown
							20	70	60–70 finish
Tantalum and tantalum alloys									
Tantalum	<500	<930	50–80	1095	2000	10:1	370–260	700–500	80 breakdown
	20	70	Finish	...	...	...	20	70	90 finish
Ta-10W	1260–980	2300–1800	50	1650–1425	3000–2600	10:1	370–260	700–500	80 breakdown
	1095–815	2000–1500	Finish	...	...	...	20	70	90 finish
T-222	1260–1205	2300–2200	50	2040–1650	3700–3000	10:1	370–260	700–500	75 breakdown
							20	70	50–75 finish
Molybdenum and molybdenum alloys									
Molybdenum	1315–1150	2400–2100	50	1760–1370	3200–2500	8:1	1205	2200	50 breakdown
	925–815	1700–1500	Finish	...	...	...	870	1600	90–75 finish
Mo-0.5Ti	1425–1260	2600–2300	50	1815–1480	3300–2700	8:1	1205	2200	50 breakdown
	1315–1150	2400–2100	Finish	...	...	...	870	1600	75 finish
TZM	1480–1315	2700–2400	50	1815–1540	3300–2800	8:1	1350–1205	2460–2200	50 breakdown
	1370–1205	2500–2200	Finish	...	...	...	1000–980	1830–1800	60
							315	600	10 finish
Tungsten									
Tungsten	1815–1595	3300–2900	20	1925–1650	3500–3000	9:1	1450–1400	2640–2550	50 breakdown
	1315–1010	2400–1850	Finish	...	...	...	1370–980	2500–1800	90 finish

(a) Where a range is given, the higher temperature is the typical starting temperature and the lower temperature is the minimum working temperature for that process.

Table 4 Recommended postweld annealing treatments for selected refractory alloys

Alloy	Annealing temperature(a): Gas tungsten arc welds °C	°F	Annealing temperature(a): Electron beam welds °C	°F
C-103	1315	2400	1315	2400
Cb-752	1205	2200	1315	2400
FS-85, C-129Y	1315	2400	1205	2200
T-111, T-222	1315	2400	1315	2400
Ta-2.5W	1260	2300	1260	2300
Ta-10W	Not annealed		Not annealed	

(a) 1 h at temperature

side, it is likely that the entire component has been contaminated.

After cleaning, the part should be handled with clean, lint-free white gloves, and the edges to be welded should be wrapped in a clean, lint-free material such as plastic. Welding should commence as soon as possible after cleaning; the time between cleaning and welding should never exceed 4 h. All weld tooling should be thoroughly cleaned with methyl ethyl ketone (MEK) or an equivalent residue-free compound; an argon or helium cover gas should be kept flowing at 0.6 to 1.1 m^3/h (20 to 40 ft^3/h) during welding. If copper tooling is used, it must be chromium plated to avoid copper contamination.

Welding. All refractory metals can be joined by electron beam welding, gas tungsten arc welding, or resistance welding. Two major problems are encountered in joining: chemical changes due chiefly to atmospheric contamination and microstructural changes resulting from thermal cycling. The latter changes include grain growth and different stages of precipitation hardening (solution, precipitation, and overaging). Preheating and postheating generally are required to minimize deleterious effects arising from precipitation hardening as well as from the residual stresses normally induced by welding.

Although recrystallization and grain growth are unavoidable in weldments of wrought tungsten and molybdenum, proper choice of welding process and procedure can localize these effects. Electron beam (EB) welding has proved effective in achieving full weld penetration with an extremely narrow heat-affected zone. As larger EB chambers become available, size limitations for electron beam welding have become less restrictive. Chambers capable of handling hardware up to 1.5 m (5 ft) in the longest dimension are in commercial use.

Rhenium welds made by inert gas or EB methods are extremely ductile and can be formed further at room temperature. Care must be taken during welding to protect the rhenium against oxidation, however. All other refractory metals suffer losses in ductility and increases in DBTT when welded, but niobium and tantalum alloys are less affected than are molybdenum and tungsten alloys. Tantalum and niobium alloys generally retain greater than 75% joint efficiency after gas tungsten arc welding. Preheating is not required, but postweld annealing can restore large amounts of ductility and toughness to commercial alloys. Table 4 summarizes recommended postweld annealing treatments for selected refractory metal alloys; Table 5 lists recommended welding conditions.

For niobium and tantalum alloys, joint design is particularly important. The surface to be melted must be twice the thickness of the thickest component; that is, if welding a 1.5 mm (0.060 in.) part to a 1.0 mm (0.040 in.) part, weld height (thickness) must be 3.0 mm (0.120 in.). Weld tooling should be of hard chromium-plated copper. No copper can contact the refractory metal. Before use, the hard chromium plate should be wire brushed to check adhesion. If the chromium flakes, the tooling must be stripped and replated.

For gas tungsten arc welding, the weld zone should be well flushed with inert cover gas before striking the arc, and the fusion zone should be allowed to cool below 205 °C (400 °F) with gas coverage. Prior to welding, all burrs should be removed by draw filing with a file used only

Table 5 Typical conditions for welding 0.9 mm (0.035 in.) refractory metal sheet

	Gas tungsten-arc welds							Electron beam welds							
	Speed		Clamp spacing		Current(a),	Arc gap		Speed		Clamp spacing		Deflection(b)		Voltage,	Current,
Alloy	mm/min	in./min	mm	in.	A	mm	in.	mm/min	in./min	mm	in.	mm	in.	kV	mA
C-103	760	30	9.5	3/8	80	1.5	0.06	1270	50	4.8	3/16	1.3	0.050	150	3.2
C-129Y	760	30	9.5	3/8	110	1.5	0.06	1270	50	13	1/2	1.3	0.050	150	4.1
Cb-752	760	30	9.5	3/8	87	1.5	0.06	380	15	4.8	3/16	1.3	0.050	150	3.3
FS-85	380	15	9.5	3/8	90	1.5	0.06	1270	50	4.8	3/16	1.3	0.050	150	4.4
T-111	380	15	9.5	3/8	115	1.5	0.06	380	15	13	1/2	1.3	0.050	150	3.8
T-222	760	30	6.4	1/4	190	1.5	0.06	380	15	13	1/2	1.3	0.050	150	4.5
Ta-10W	190	7.5	6.4	1/4	118	1.5	0.06	380	15	13	1/2	1.3	0.050	150	3.8

(a) Direct current, straight polarity. (b) Beam deflection at 60 cycles parallel to weld direction

Table 6 Typical brazing filler metals and service temperatures

Filler metal	Maximum service temperature °C	°F
For niobium alloys		
Si-Cr-Ni	980	1800
4Be-48Zr-48Ti	925	1700
Zr-6Be-19Nb	925	1700
Ti-0.5Si	1370	2500
Zr-0.1Be-16Ti-25V	1205	2200
V-35Nb	1205	2200
Ti-50Zr	1650	3000
Titanium	1760	3200
Ti-33Cr	1370	2500
Ti-3Al-11Cr-13V	1650	3000
For tantalum alloys		
Hf-7Mo	2095	3800
V-20Nb-20Ta	1870	3400
Ti-15Ta-25V	1650	3000
Ta-10Hf-(15–70)Nb	2205	4000
Nb-(30–50)Hf	2205	4000
Ta-10Hf	2205	4000
Nb-1.3B	1925	3500
Copper	980	1800
For molybdenum alloys		
Ti-3Be-25Cr	1595	2900
Pt-Mo	1650	3000
Zr-Ti	1230	2250
Cu-Au	815	1500
Ni-Cu	1205	2200
V-35Nb	1205	2200
Ti-30V	1370	2500
Ti-13Ni-25Cr	1760	3200
Co-10Ni-15W-20Cr	1315	2400
For tungsten		
Ag-Mn	870	1600
V-Nb-Ta	1925	3500
V-Ti-Ta	1925	3500
W-25Os	2205	4000
W-3Re-50Mo	2205	4000
Mo-5Os	1925	3500
Niobium	1650	3000
Tantalum	2205	4000
For rhenium		
Ag-Cu	730	1350
Vanadium	1900	3450

for one particular alloy. All hand welding should be accomplished in chambers that are evacuated prior to backfilling with argon and/or helium.

Although pure niobium shows no evidence of an aging reaction, Nb-1Zr undergoes abrupt losses in strength and ductility when treated at 815 to 980 °C (1500 to 1800 °F) for up to 500 h. Welds are subject to such embrittlement but can be restored to a ductile condition by postweld vacuum annealing at 1040 to 1205 °C (1900 to 2200 °F) for 3 h. This treatment produces overaging, preventing embrittlement on subsequent heating at a lower temperature.

In contrast to welds in niobium and tantalum, which retain good ductility, welds in molybdenum and tungsten are brittle (<50% joint efficiency), and thus these metals are difficult to join. Before welding, molybdenum and tungsten must be preheated above their ductile-to-brittle transition temperatures to prevent fracture. Sections in thicknesses of 0.64 mm (0.025 in.) and less demand special attention in this respect and, at best, present serious cracking problems. Welds in these metals are always brittle, and joint efficiency depends on the reinforcing effect of the weld bead. Resistance welding is feasible, but some problems with electrode sticking can arise. Resistance Welding Manufacturers' Association (RWMA) class I copper electrodes show the least susceptibility to sticking. Projection welding can result in relatively high mechanical properties.

Tungsten is the most difficult refractory metal to join for satisfactory high-temperature service. Welding, especially the EB process, offers the best compromise for joining tungsten for service at high temperatures. Mechanical joints are unsatisfactory unless molybdenum fasteners are used. Diffusion bonding is impractical because of severe tooling problems. Brazing for relatively low-temperature applications is done using precious metals (silver, palladium, and platinum alloys) and transition metals (nickel and manganese alloys) as filler metals.

Table 6 lists typical brazing filler metals and their maximum service temperatures for all refractory metal systems. Molybdenum brazing has received much attention. Brazed molybdenum honeycomb configurations are used for structural and heat shield applications at temperatures from 1370 to 1650 °C (2500 to 3000 °F). Low-temperature brazing processes have been developed for titanium-zirconium-molybdenum (TEM) alloy. The high remelt temperatures of the filler metals listed in Table 6 permit relatively high service temperatures.

Niobium and its alloys may be silicide-coated with a chromium- and titanium-containing material before being subjected to temperature-oxidizing environments; the preferred braze alloy is Ti-33Cr because it is compatible with the coating. Foil, 0.13 mm (0.005 in.) thick, is fit metal-to-metal to ensure good filleting of the joint. The foil should be held in place with resistance spot welds. Cleanliness of the mating surfaces will ensure good flow of the alloy. A clean vacuum furnace is heated to 1315 °C (2400 °F) and held for 5 min, then increased to 1480 °C (2700 °F) and held for 8 min. The parts must be furnace cooled to 205 °C (400 °F) before exposure to air. After brazing, the hardware should be pickled and diffusion treated at 1315 °C (2400 °F) in a vacuum for a period of 16 h. When possible, the parts should be wrapped in tantalum foil to minimize contamination.

Diffusion bonding also is used to join refractory metals, primarily niobium and tantalum. The same rationale is required as for brazing. Vanadium foil, 0.05 to 0.08 mm (0.002 to 0.003 in.) thick, is placed in the joint and weighted using molybdenum or tungsten tooling. Diffusion bonding with vanadium is considered superior to brazing because a bimetallic system is not necessary, and the joint is microstructurally clean because vanadium forms a continuous solid solution with niobium and tantalum.

Explosive Bonding. Recently, chemical equipment has been fabricated from steel plate explosively clad with tantalum. Forming and welding methods have been developed for fabrication of the clad plate into reactor vessels, tanks, and other types of chemical equipment. Explosive bonding produces a metallurgical bond at the tantalum/steel interface where bond efficiency is over 98%.

Properties and Applications

Typical physical properties for pure refractory metals are summarized in Table 7. Figures 1 and 2 compare the temperature-dependent ultimate tensile strengths and elastic moduli of the refractory metals. The values for hexagonal close-packed (hcp) rhenium are quite different from those of the other metals, which are body-centered cubic (bcc).

Tungsten- and molybdenum-base alloys must be hot worked, whereas tantalum and niobium are easily cold worked. Rhenium is used primarily as an alloy additive. Most molybdenum and niobium is used as alloy additions in steels, irons, or superalloys. Other applications are noted in the following sections.

Table 7 Mechanical and physical properties of pure refractory metals

Property	Niobium	Tantalum	Molybdenum	Tungsten	Rhenium
Structure and atomic properties					
Atomic number	41	73	42	74	75
Atomic weight	92.9064	180.95	95.94	183.85	186.31
Density at 20 °C (70 °F), g/cm^3 (lb/in.^{3})	8.57 (0.310)	16.6 (0.600)	10.22 (0.369)	19.25 (0.695)	21.04 (0.760)
Crystal structure	bcc	bcc	bcc	bcc	hcp
Lattice constants, nm					
a	0.3294	0.3303	0.3147	0.3165	0.27609
c	...	...	...	...	0.45829
Slip plane at room temperature	110	110	112	...	0001–1010
Thermal properties					
Melting temperature, °C (°F)	2468 (4474)	2996 (5425)	2610 (4730)	3410 (6170)	3180 (5755)
Boiling temperature, °C (°F)	4927 (8901)	5427 (9801)	5560 (10040)	5700 (10290)	5760 (10400)
Vapor pressure at 2500 K, mPa (torr)	5.3 (4 × 10^{-5})	0.11 (8 × 10^{-7})	80 (6 × 10^{-4})	0.0093 (7 × 10^{-8})	0.17 (1.3 × 10^{-6})
Coefficient of expansion, near RT(a), μm/m · K (μin./in. · °F)	7.3 (4.1)	6.5 (3.6)	4.9 (2.7)	4.6 (2.6)	6.7 (3.7)
Specific heat at 20 °C (70 °F), kJ/kg · K (Btu/lb · °F)	0.268 (0.0643)	0.139 (0.0333)	0.276 (0.0662)	0.138 (0.0331)	0.138 (0.0331)
Latent heat of fusion, kJ/kg (Btu/lb)	290 (125)	145–174 (62–75)	270 (115)	220 (95)	177 (76)
Latent heat of vaporization, kJ/kg (Btu/lb)	7490 (3202)	4160–4270 (1790–1840)	5123 (2160)	4680 (2010)	3415 (1470)
Thermal conductivity, W/m · K (Btu/ft · h · °F)					
At 20 °C (70 °F)	52.7 (30.4)	54.4 (31.4)	142 (81.9)	155 (89.4)	71 (41)
At 500 °C (930 °F)	63.2 (36.5)	66.6 (38.4)	123 (71.0)	130 (75)	...
Electrical properties					
Electrical conductivity at 18 °C (64 °F), %IACS(b)	13.2	13.0	33.0	30.0	8.1
Electrical resistivity, at 20 °C (70 °F), nΩ · m	160	135	52	53	193
Electrochemical equivalent, mg/C	0.1926	0.375	0.166	0.318	0.276
Hall coefficient, nV · m/A · T	0.09	0.095	...	...	...
Magnetic properties					
Magnetic susceptibility (volume) at 25 °C (75 °F), mks system	28 × 10^{-6}	10.4 × 10^{-6}	1.17 × 10^{-8}	4.1 × 10^{-8}	0.37 × 10^{-6}
Optical properties					
Total emissivity at 1500 °C (2730 °F), %	0.19	0.21	0.19	0.23	...
Spectral emittance at λ = 650 nm, %	0.37	0.49	0.37	0.43	...
Additional properties					
Poisson's ratio at 25 °C (75 °F)	0.38	0.35	0.32	0.28	0.49
Elastic modulus, GPa	103	185	324	400	469
Ductile-to-brittle transition temperature (DBTT), K	<147(c)	<25(c)	...	250(d)	...

(a) RT, room temperature. (b) IACS, International Annealed Copper Standard. (c) Viscous with iron purity. (d) Value for as-drawn material; DBTT for annealed tungsten is 325 K

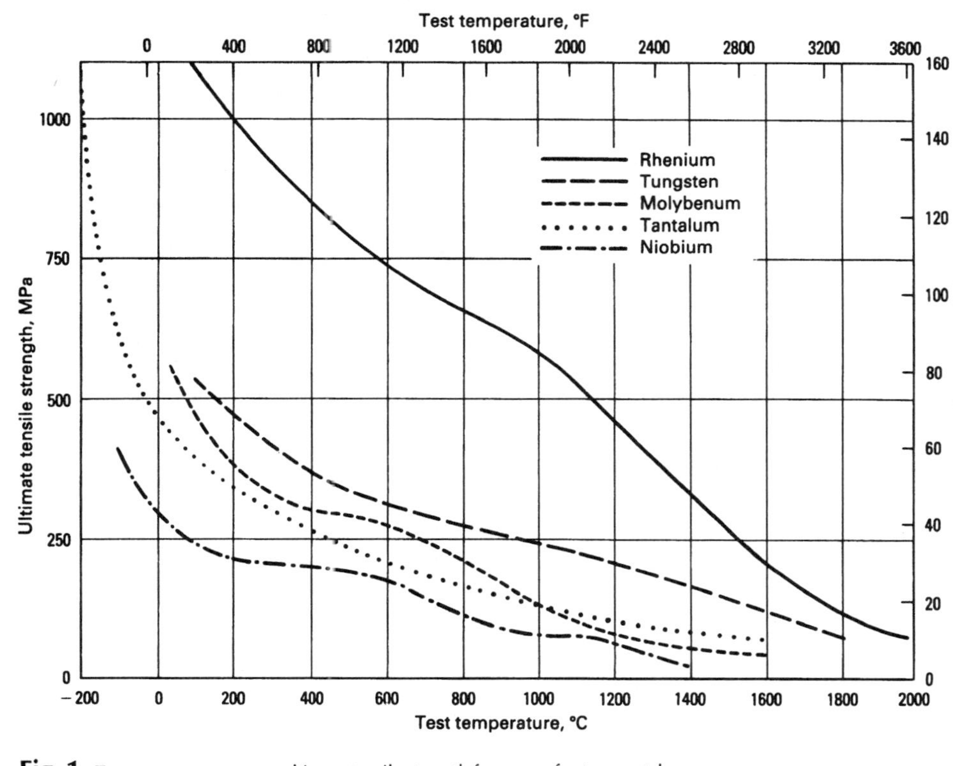

Fig. 1 Test temperature versus ultimate tensile strength for pure refractory metals

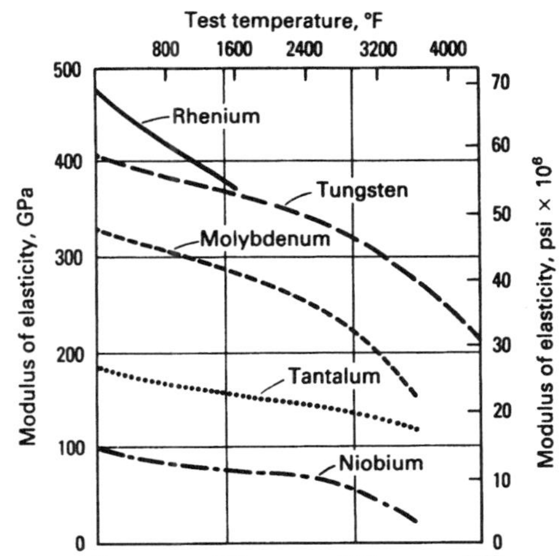

Fig. 2 Test temperature versus modulus of elasticity for pure refractory metals

Tungsten

Despite difficulties in fabrication, tungsten is used in more applications than any other refractory metal besides tantalum. Tungsten is consumed in four forms:

- Tungsten carbide
- Alloying additions
- Pure tungsten
- Tungsten-base chemicals

Tungsten carbide accounts for about 65% of tungsten consumption. It is combined with cobalt as a binder to form the so-called cemented carbides, which are used in cutting and wear applications (see the article "Cermets and Cemented Carbides" in this Volume). Metallic tungsten and tungsten alloy mill products account for about 16% of consumption. Tungsten and tungsten alloys dominate the market in applications for which a high-density material is required, such as kinetic energy penetrators, counterweights, fly-wheels, and governors. Other applications include radiation shields and x-ray targets.

In wire form, tungsten is used extensively for lighting, electronic devices, and thermocouples. Tungsten chemicals make up approximately 3% of the total consumption and are used for organic dyes, pigment phosphors, catalysts, cathode-ray tubes, and x-ray screens.

The high melting point of tungsten makes it an obvious choice for structural applications exposed to very high temperatures. Tungsten is used at lower temperatures for applications that can use its high elastic modulus, density, or shielding characteristics as an advantage.

Tungsten and tungsten alloys can be pressed and sintered into bars and subsequently fabricated into wrought bar, sheet, or wire. Many tungsten products are intricate and require machining or molding and sintering to near-net shape and cannot be fabricated from standard mill products. Further information on fabrication and forming is discussed in the article "Refractory Metals and Alloys" in *Properties and Selection: Nonferrous Alloys and Special Purpose Materials,* Volume 2, *ASM Handbook.*

Tungsten mill products can be divided into three distinct groups on the basis of recrystallization behavior. The first group consists of EB-melted, zone-refined, or arc-melted unalloyed tungsten; other very pure forms of unalloyed tungsten; or tungsten alloyed with rhenium or molybdenum. These materials exhibit equiaxed grain structures upon primary recrystallization. The recrystallization temperature and grain size both decrease with increasing deformation.

The second group, consisting of commercial grade or undoped P/M tungsten, demonstrates the sensitivity of tungsten to purity. Like the first group, these materials exhibit equiaxed grain structures, but their recrystallization temperatures are higher than those of the first-group materials. Also, these materials do not necessarily exhibit decreases in recrystallization temperature and grain size with increasing deformation. In EB-melted tungsten wire, the recrystallization temperature can be 900 °C (1650 °F) or lower, whereas in commercially pure (undoped) tungsten it can be as high as 1205 to 1400 °C (2200 to 2550 °F).

The third group of materials consists of AKS-doped tungsten (that is, tungsten doped with aluminum-potassium-silicon), doped tungsten alloyed with rhenium, and undoped tungsten alloyed with more than 1% ThO_2. These materials are characterized by higher recrystallization temperatures (>1800 °C, or 3270 °F) and unique recrystallized grain structures. The structure of heavily drawn wire or rolled sheet consists of very long interlocking grains.

Tungsten rod or wire has many uses, including light bulb filaments, electron emitters, welding electrodes, electrical contacts, and reinforcing filaments for metal-matrix composites. The rod is fabricated by isostatic pressing of powder and sintering, usually by resistance heating, in a hydrogen atmosphere. The rod is reduced to wire or other forms by hot swaging and is cut to length with diamond saws.

Non-sag wire, prepared from tungsten powder containing about 0.1% potassium aluminum silicate, is particularly useful for hot filaments. During sintering or annealing, the aluminum and silicon volatize, producing bubbles several angstroms in size. On subsequent working, these bubbles align and promote the formation of interlocking, elongated recrystallized grains.

Additions of thoria or zirconia stabilize the grains and strengthen tungsten rod and wire. Because these compositions can contain up to 1 to 2% oxide, they are, strictly speaking, tungsten alloys. The thoria or zirconia is added as an aqueous nitrate salt solution to tungstic oxide (blue oxide) before reduction. The material is then dried and reduced in hydrogen.

Because thoriated tungsten is stronger than pure tungsten, higher sintering temperatures and longer sintering times are necessary. Swaging and drawing is similar to that performed on pure tungsten except that the rate of work hardening is greater, requiring more intermediate anneals.

Non-swag wire is better when distortion is a problem; however, thoria or zirconia dispersion-strengthened tungsten has superior hot strength and electron emission characteristics. Consequently, thoriated tungsten is widely used for inert gas welding electrodes.

Although silicide and noble metal coatings are effective oxidation-resistant coatings, protective atmospheres such as helium or argon are used to protect tungsten heating and incandescent elements and welding electrodes. Superior heat resistance also makes tungsten an ideal candidate for trays and radiation shields in vacuum or controlled-atmosphere furnaces.

Superior electrical conductivity, combined with resistance to wear and spark erosion, makes tungsten, tungsten-copper, and tungsten-silver useful for electrical contacts. Tungsten is also used in contacts for automotive ignition systems, broadcasting equipment, light switches, and voltage-control thyratron devices.

Tungsten heavy metals contain 90 to 97% W, and the remainder is either copper and nickel or iron and nickel plus additional elements. Densities are high, ranging from 16.8 to 18.5 g/cm^3. These compositions are easier to machine than pure tungsten. Tensile strength is several times higher than carbon steel and twice that of brass or cast iron, ranging from 760 to 970 MPa (110 to 140 ksi). Young's modulus for tungsten heavy metals is from 270 to 340 GPa (40 to 50×10^3 ksi).

Tungsten heavy metals have low thermal expansion coefficients (about half that of iron). They retain dimensional stability over a wide temperature range and exhibit excellent corrosion resistance. These materials can be fabricated economically by P/M techniques, including liquid phase sintering and infiltration. In the latter process, a porous tungsten powder compact is placed in contact with a second, lower melting point metal, which is then heated to its melting point. The liquid metal infiltrates the pores of the tungsten compact. Heavy metal parts are readily machined to close tolerances. More information on processing and properties is contained in the article "Tungsten Heavy Alloys" in this Volume.

Tantalum

Currently, the largest use of tantalum is in electrolytic capacitors. Tantalum P/M anodes are used in solid and wet electrolytic capacitors, and precision tantalum foil is used, although to a lesser extent, in foil capacitors. Tantalum is also used in chemical process equipment, such as heat exchangers, condensers, thermowells, and lined vessels; most notably, it is used for the condensing, reboiling, preheating, and cooling of nitric acid, hydrochloric acid, sulfuric acid, and combinations of these acids with many other chemicals. Because of its high melting point, tantalum is used for heating elements, heat shields, and other components in high-temperature vacuum furnaces, and it has found some use for trueing of grinding wheels. Tantalum and its alloys have been used in specialized aerospace and nuclear applications and have found increasing use in military components. Because of its corrosion resistance to body fluids, it is used in prosthetic devices and in surgical staples. Tantalum is used as an alloying element in superalloys. Tantalum carbide is an important constituent in complex cemented carbides used in cutting tools.

Tantalum Powder. Several methods for reducing tantalum compounds to tantalum metal have been developed, but sodium reduction of K_2TaF_7 to produce tantalum metal powder is the most commonly used today. The product of the sodium reduction can then be further refined by melting. The powder can also be pressed and sintered into bar or sold as capacitor-grade powder. By varying the parameters of sodium reduction (for example, time, temperature, sodium feed rate, and diluent), powders of various particle sizes can be manufactured.

A wide range of sodium-reduced capacitor powders are currently available, with unit capacitances ranging from 5000 $\mu F \cdot V/g$ to greater than 25,000 $\mu F \cdot V/g$. Capacitor powders are also manufactured from hydrided, crushed, and degassed EB-melted ingot. These melt-grade

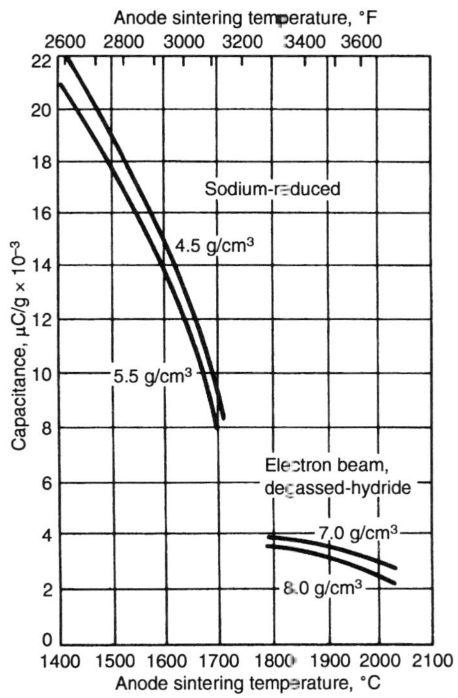

Fig. 3 Capacitance of sodium-reduced and electron beam melted, degassed-hydride tantalum powder. For 30 min anode sintering temperature at anode green densities commonly used for each powder

powders have higher purity than the sodium-reduced types and have better dielectric properties. However, unit capacitance is usually lower for EB-type powder.

Additions of small amounts (10 to 100 ppm) of sintering inhibitors, such as phosphorus, are frequently employed. These agents allow finer powders to be used than would be possible in the absence of an inhibitor. Alternatively, higher sintering temperatures can be used without loss of surface area, thereby aiding vaporization of undesirable impurities. Dielectric properties are thereby improved.

Sodium-reduced tantalum powders are used for 80 to 90% of capacitor applications. Electron beam melted, degassed-hydride powders account for the remainder. Sodium-reduced material is used in high-capacitance devices at lower voltage ratings (<50 V) because these powders can be produced in finer particle sizes, and the material has improved pressing characteristics. The latter permits anodes to be pressed at lower green density.

Consequently, sodium-reduced powders are more efficient, requiring less powder to achieve a given capacitance rating in a finished device. However, electron beam melting provides degassed-hydride powders with an additional purification step. Because of this higher purity, the anodic oxide film can be electrolytically formed, or stressed, to higher voltages without breakdown. Therefore, electron beam melted powders frequently are required for high-reliability capacitors rated for 50 to 125 V use. Figure 3 compares capacitance at several anode sintering temperatures for sodium-reduced and electron beam melted, degassed powders.

Although the dielectric properties of niobium oxide are inferior to those of tantalum oxide, niobium has been evaluated as a means of producing low-cost capacitors. Considerable niobium powder development work is still required before niobium capacitors will be offered commercially.

Pressing and Consolidation. Tantalum and its alloys are produced in semifinished metallic form by further processing of the sodium-reduced powders. The powder is isostatically pressed into bars, which can then be electron beam melted or sintered at high temperature under vacuum. Tantalum ingots up to 305 mm (12 in.) in diameter can be produced by electron beam melting.

Pressing and Sintering. Manufacturing of the tantalum pellet begins by cold pressing the powder into a low-density pellet. The capacitance rating of the device determines the powder weight required. On the average, the weight per pellet is about 0.2 g (0.007 oz). The pellet is then sintered in high vacuum (10^{-4} torr). The sintering temperature is set low enough to retain as much of the porous structure as possible, while sintering the "necks" of adjacent particles to form a continuous path of conducting metal.

Further surface purification of the metal also occurs during sintering by volatilization of the monoxides of tramp cations, the presence of which can otherwise cause defects in the anodic oxide film. Sintering temperatures usually range from 1450 to 2000 °C (2640 to 3630 °F). Selection of temperatures is determined by the type of powder and device required; there is a trade-off between loss of surface area and improved purity as temperature increases.

The EB-melting process utilizes evaporation, volatilization of suboxides, and carbon deoxidation as purifying reactions. All of these reactions are more favorable in high vacuum (<130 MPa, or 10^{-3} torr). Because of its high melting point and very low vapor pressure, tantalum can be produced with a purity exceeding 99.95%. In addition, certain metallurgical properties can be imparted to the ingot by the addition of a vacuum arc remelt of the EB-melted ingot. One such property is grain size refinement.

Powder Metallurgy Mill Products. Unalloyed tantalum and tantalum alloy ingots can be broken down by either forging or extrusion. Arc cast ingots should be extruded only after upsetting and side forging. Powder metallurgy sintered bars can be rolled directly without any prior breakdown.

Powder metallurgy products differ from those produced from cast materials primarily because surface oxides are present in the starting powders. These oxides are present in the microstructure of the P/M material. Recrystallization is inhibited and finer grain metal is produced. The product is usually harder and has higher yield and tensile strength but somewhat less ductility.

Powder metallurgy tantalum has superior deep-drawing properties, but it should not be

Table 8 Typical properties of tantalum and tantalum-base alloys

Grade(a)	Hardness, HV	Density, g/cm³	Density, lb/in.³	Melting point, °C	Melting point, °F	Temperature, °C	Temperature, °F	Tensile strength, MPa	Tensile strength, ksi	Yield strength, MPa	Yield strength, ksi	Elongation, %	Modulus of elasticity, GPa	Modulus of elasticity, 10^6 psi
Commercial pure tantalum, EB melted	110	16.9	0.609	3000	5430	20	70	205	30	165	24	40	185	27
						200	390	190	27.5	69	10	30	...	...
						750	1380	140	20	41	6	45	160	23
						1000	1830	90	13	34	5	33	...	...
Commercial pure tantalum, P/M	120	16.6	0.600	3000	5430	20	70	310	45	220	32	30	185	27
63 metal, EB melted	130	16.7	0.602	3005(b)	5440(b)	20	70	345	50	230	33	40	195	28
						200	390	315	46	195	28	33	...	...
						750	1380	180	26	83	12	22	...	...
						1000	1830	125	18	69	10	20	...	...
Ta-10W, EB melted	245	16.8	0.608	3030	5490	20	70	550	80	460	67	25	205	30
						200	390	515	75	400	58	...	...	...
						750	1380	380	55	275	40	...	150	22
						1000	1830	305	44	205	30	...	...	...
Ta-7.5W, P/M														
Wire	325	16.8	0.606	3025(b)	5477(b)	20	70	1035	150	1005	146	6	200	29
Sheet	400	16.8	0.606	3025(b)	5477(b)	20	70	1165	169	875	127	7	200	29
Ta-40Nb, EB melted	...	12.1	0.437	2705	4900	...	...	275	40	193	28	25	...	...

(a) EB, electronic beam; P/M, powder metallurgy. (b) Estimated

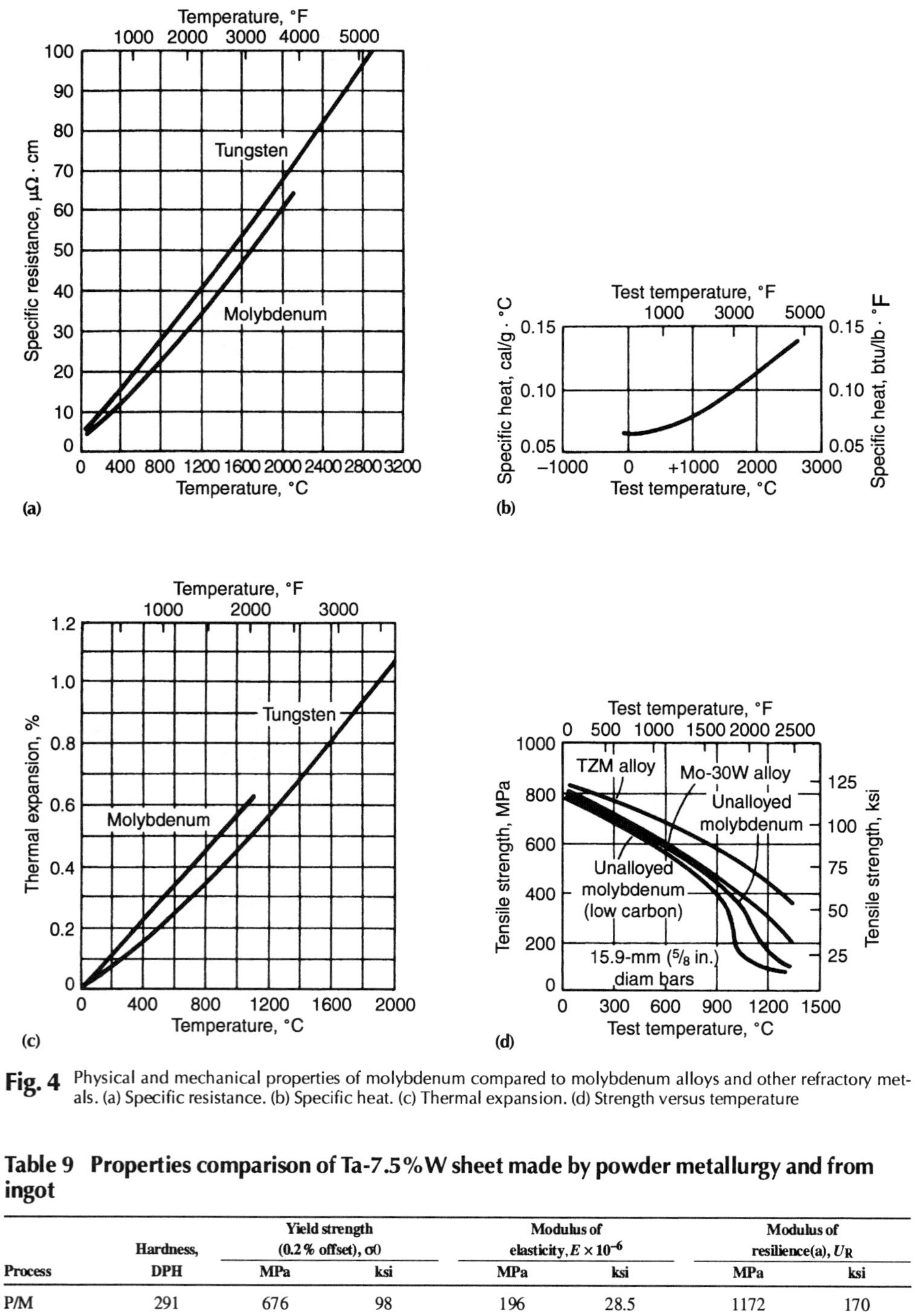

Fig. 4 Physical and mechanical properties of molybdenum compared to molybdenum alloys and other refractory metals. (a) Specific resistance. (b) Specific heat. (c) Thermal expansion. (d) Strength versus temperature

Table 9 Properties comparison of Ta-7.5%W sheet made by powder metallurgy and from ingot

Process	Hardness, DPH	Yield strength (0.2% offset), σ_0		Modulus of elasticity, $E \times 10^{-6}$		Modulus of resilience(a), U_R	
		MPa	ksi	MPa	ksi	MPa	ksi
P/M	291	676	98	196	28.5	1172	170
Ingot	150	483	70	196	28.5	593	86

Measured for 2.1 mm (0.083 in.) thick, annealed and recrystallized sheet. (a) $U_R = \sigma_0^2 / 2E$ (see Ref 1)

welded because of the porosity that forms in the heat-affected zone. However, EB-melted tantalum can be used for various welded products, including welded and drawn tubing.

Powder metallurgy mill products begin by cold isostatic pressing the powder into bars. The bars are then direct resistance sintered by clamping the ends firmly between water-cooled copper terminals in a vacuum furnace at 1.33×10^{-2} Pa (10^{-4} torr). Heating occurs by the passage of an electric current.

For tantalum, temperatures above 2300 °C (4170 °F) are required to achieve full density and adequate purification (removal of interstitial impurity elements). Sintered bars are then rolled or drawn at room temperature, with appropriate intermediate vacuum anneals.

Tantalum products can subsequently be manufactured by standard cold-working techniques, such as rolling, drawing, tube reducing, and swaging. Typical reductions between anneals are 75 to 80%, but reductions in excess of 95% are not uncommon. Rolled sheet having a controlled, predominant {111} crystallographic texture is being produced because of its superior drawing and forming characteristics. The texture control is achieved by a specific thermomechanical process history.

Microalloying. Oxide microalloying additions, such as thoria or yttria, are sometimes intentionally added to control grain size and strength after recrystallization. Small quantities of silicon are also added, particularly to bars that are subsequently drawn to capacitor wire.

Silicon aids in the retention of ductility of the wire by preventing embrittlement of the tantalum lead-wire embedded in the capacitor pellet. During anode vacuum sintering, oxygen diffuses from the surrounding powder into the embedded wire, which has lower oxygen, and collects at grain boundaries, thus causing embrittlement. Silicon acts as a getter for the excess oxygen, There is evidence that the microstructure is modified with precipitated oxides in the form of globules rather than platelets. A portion of the silicon vaporizes harmlessly as volatile silicon monoxide.

Tantalum Alloys. Selected mechanical and physical properties for commercially prepared pure EB-melted and P/M tantalum and tantalum alloys are given in Table 8.

Tungsten and hafnium are the alloying additions commonly added to improve the strength properties of tantalum. Although tantalum powder can be an intermediate, most tantalum alloys are melted. As examples, Ta-2.5%W is used for the fabrication of corrosion-resistant heat exchangers, valves, and other chemical equipment. Ta-10%W, although less ductile, is a harder, stronger, and more wear-resistant alloy and is particularly suited for the fabrication of furnace hardware, missile parts, nozzles, and fasteners.

A Ta-7.5%W alloy, useful for springs, bellows, and leaf springs in severe corrosion environments, such as dry chlorine, bromine, or hydrochloric acid, is made by powder metallurgy. For springs, the modulus of resilience, U_R, is an important parameter. Resilience is defined as the "ability of a material to absorb energy when deformed elastically and retain it when unloaded" (Ref 1). Because the yield strength of the P/M product is higher than ingot material of the same tungsten content, its modulus of resilience is also considerably higher, as shown in Table 9.

Molybdenum

More molybdenum is consumed annually than any other refractory metal. Molybdenum ingots, produced by melting of P/M electrodes, are extruded, rolled into sheet and rod, and subsequently drawn to other mill product shapes, such as wire and tubing. These materials can then be stamped into simple shapes. Molybdenum is also machined with ordinary tools and can be gas tungsten arc and electron beam welded, or brazed.

Molybdenum has outstanding electrical and heat-conducting capabilities and relatively high tensile strength (Fig. 4a and b). Thermal conductivity is approximately 50% higher than that of steel, iron, or nickel alloys. It consequently finds wide usage as heatsinks. Its electrical conductivity is the highest of all refractory metals, about one third that of copper, but higher than nickel, platinum, or mercury. The coefficient of thermal expansion of molybdenum plots almost linearly with temperature over a wide range (Fig. 4c). This characteristic, in combination with its heat-conducting capabilities, accounts for its use in bimetal thermocouples. Methods of doping molybdenum powder with potassium aluminosilicate to obtain a non-sag microstructure comparable to that of tungsten also have been developed.

Applications. The major use for molybdenum is as an alloying agent for alloy and tool steels, stainless steels, and nickel- or cobalt-base superalloys to increase hot strength, toughness, and corrosion resistance.

In the electrical and electronic industries, molybdenum is used in cathodes, cathode supports for radar devices, current leads for thoria cathodes, magnetron end hats, and mandrels for winding tungsten filaments. Molybdenum is also used as a filler metal for brazing tungsten. Molybdenum resistance heating elements are used in electric furnaces that operate at temperatures up to 2205 °C (4000 °F).

Molybdenum is important in the missile industry, where it is used for high-temperature structural parts, such as nozzles, leading edges of control surfaces, support vanes, struts, reentry cones, heat-radiation shields, heat sinks, turbine wheels, and pumps. Molybdenum alloys are particularly well-suited for use in airframes because of their high stiffness, high recrystallization temperature, retention of mechanical properties after thermal cycling, and good creep strength. Alloy Mo-0.5Ti has been used in many aerospace applications, but TZM is preferred where higher hot strength is needed.

In the metalworking industry, molybdenum is used for die casting cores; for hot work tools such as piercer points and extrusion and isothermal forging dies; for boring bars, tool shanks, and chill plates; and for tips on resistance welding electrodes. It is also used for cladding, for equipment for trueing grinding wheels, for molds, and for thermocouples.

Molybdenum has also been useful in the nuclear, chemical, glass, and metallizing industries. Service temperatures for molybdenum alloys in structural applications are limited to a maximum of about 1650 °C (3000 °F). Pure molybdenum has good resistance to hydrochloric acid and is used for acid service in chemical process industries.

The mechanical properties of molybdenum and molybdenum alloys greatly depend on the amount of working performed below the recrystallization temperature and on the ductile-to-brittle transition temperature. The minimum recrystallization temperature for molybdenum is 900 °C (1650 °F). For optimum ductility, parts should be given at least 50% reduction in area. The metal exhibits superior strength and hardness at elevated temperatures (Fig. 4d). However, when hot strength is required, a molybdenum alloy rather than the element is the material of choice. Titanium and zirconium additions to molybdenum produce alloys with hot strength and recrystallization temperatures above those of unalloyed molybdenum.

Table 10 Room- and elevated-temperature tensile properties of 380 μm (15 mil) molybdenum wire

Material designation	Composition	Temperature °C	Temperature °F	Ultimate tensile strength MPa	Ultimate tensile strength ksi	Elongation, %
Unalloyed molybdenum	Mo	20	70	1350	196	4.1
		1000	1830	305	44	2.4
		1100	2010	140	20	10.3
		1200	2190	115	17	12.5
MT-104	Mo-0.5Ti-0.08Zr-0.01C	20	70	1565	227	3.1
		1000	1830	1020	148	2.7
		1100	2010	795	115	3.2
		1200	2190	675	98	2.8
Mo + 45 W	Mo-45W	20	70	1980	287	3.6
		1000	1830	1095	159	2.3
		1100	2010	950	138	2.3
		1200	2190	745	108	2.2
HCM	Mo-1.1Hf-0.07C	20	70	1795	260	2.9
		1000	1830	1270	184	3.4
		1100	2010	1185	172	3.3
		1200	2190	1035	150	3.0
HWM-25	Mo-25W-1.0Hf-0.035C	20	70	1935	281	3.2
		1000	1830	1350	196	3.3
		1100	2010	1225	178	3.1
		1200	2190	1075	156	4.6
HWM-45	Mo-45W-0.9Hf-0.03C	20	70	2135	310	3.6
		1000	1830	1460	212	3.3
		1100	2010	1295	188	2.6
		1200	2190	1170	170	2.4

Table 10 summarizes the chemistry and basic mechanical property data for 0.38 mm (0.015 in.) wire fabricated from unalloyed molybdenum and several of the newer molybdenum alloys. Additional property data for molybdenum alloys appear in the compilations that follow.

Molybdenum Alloy TZM. The molybdenum alloy of greatest technological importance is the high-strength, high-temperature alloy TZM. The material is manufactured either by P/M or arc-cast processes. The composition of TZM is:

Element	%
Carbon	0.01–0.04
Titanium	0.40–0.55
Zirconium	0.06–0.12
Oxygen	<0.0025
Hydrogen	<0.0005
Nitrogen	<0.002
Iron	<0.010
Nickel	<0.002
Silicon	<0.008
Molybdenum	bal

Figure 5 shows typical strength-to-density ratios for TZM compared to other refractory materials. At about 1100 °C (2010 °F), its strength is about twice that of unalloyed molybdenum. The alloy is ideal for structural applications at 1000 °C (1830 °F) under conditions where unalloyed molybdenum normally is used.

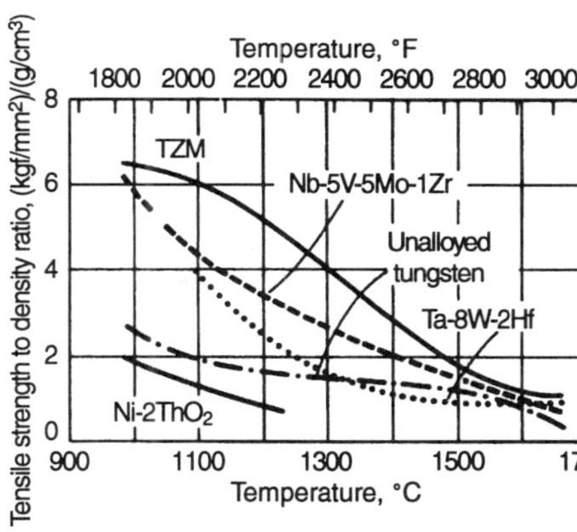

Fig. 5 Strength-to-density ratio for TZM compared to refractory alloys

TZM has a higher recrystallization temperature and higher strength and hardness at room and at elevated temperatures than unalloyed molybdenum. It also exhibits adequate ductility. Its superior mechanical properties are due to the dispersion of complex carbides in the molybdenum matrix. TZM is well suited to hot work applications because of its combination of high hot hardness, high thermal conductivity, low thermal expansion, and high resistance to heat checking compared to hot work steels. Major uses include:

- Die inserts for casting aluminum, magnesium, zinc, and iron
- Rocket nozzles
- Die bodies and punches for hot stamping
- Tools for metalworking (due to the high abrasion and chatter resistance of TZM)

Table 11 Typical composition of niobium and C-103 niobium alloy powder made by the hydride-dehydride process

Element	Analysis, ppm	
	Niobium	C-103
Niobium	≥99.7(a)	687.2(a)
Oxygen	1820	1980
Tantalum	800	2800
Hafnium	<20	9.8(a)
Zirconium	<20	1800
Titanium	20	0.91(a)
Carbon	500	194
Iron	100	200
Aluminum	<20	<20
Nitrogen	197	62
Silicon	30	<20
Copper	<40	<40
Cobalt	<10	<10
Boron	<1	<10
Hydrogen	150	50
Nickel	<20	<20
Molybdenum	<20	100
Tungsten	<50	1100
Other elements(b)	<20	<20

(a) Analysis in %. (b) Other elements include cadmium, chromium, magnesium, manganese, lead, tin, vanadium, and zinc. Source: Fansteel, Inc.

- Heat shields for furnaces, structural parts, and heating elements

In an attempt to improve the high-temperature strength of P/M TZM alloys, alloys have been developed in which titanium and zirconium carbide is replaced by hafnium carbide. Such an alloy can contain 1.0% Hf and 0.06% C. It has a tensile strength of 590 MPa (85 ksi) at 1315 °C (2400 °F), compared to 480 MPa (70 ksi) for TZM. Creep rate at 1205 °C (2200 °F) at a stress of 330 MPa (48 ksi) is 0.038%/h, compared to 0.05%/h for TZM.

Other Molybdenum Alloys. Alloys of molybdenum and rhenium are more ductile than pure molybdenum. An alloy with 35% Re can be rolled at room temperature to more than 95% reduction in thickness before cracking. For economic reasons, molybdenum-rhenium alloys are not widely used commercially. Alloys of molybdenum with 5 and 41% Re are used for thermocouple wires.

Niobium Alloys

Commercial application of niobium is dominated by its use as an alloying element in steels. Almost 75% of all niobium metal is used as minor alloying additions in low-alloy steel. Another 20 to 25% is used as alloy additions in nickel-base superalloys and heat-resisting steels. Only 1 to 2% of all niobium is used in the form of niobium-base alloys and pure niobium metal including superconducting niobium-titanium alloy, which accounts for over one-half of all niobium alloys produced (Ref 2).

Originally, niobium metal was produced by powder metallurgy methods which involved high temperature vacuum sintering and carbon reduction. In the early 1960s, however, aluminothermic reduction and electron beam purification

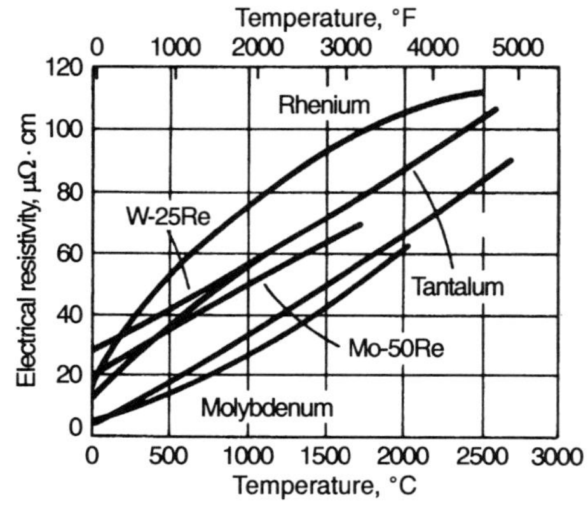

Fig. 6 Electrical resistivity versus temperature for rhenium, its alloys, and other refractory metals

became the standard practices (Ref 2). At present, the major process for the recovery of niobium is the aluminothermic reduction of pyrochlore concentrates to ferroniobium. Niobium metal is purified by a chlorination process wherein volatile $NbCl_5$ is distilled and then hydrolyzed to the oxide. The metal is then recovered by a second aluminothermic reduction:

$3Nb_2O_5 + 10Al$ {{———>} up 60 back 320 roman size 7 accelerator } $6Nb + 5Al_2O_3$

During the exothermic reaction, oxide impurities slag from the molten niobium. Carbothermic reduction has also been practiced.

Another niobium recovery process involves the collection of niobium oxide as a by-product in the processing of tantalum ores. The ore is digested in mixed acids containing hydrofluoric acid and sulfuric acid, and solvent extraction is employed for purification. After aluminothermic reduction of the recovered oxide, the metal is further purified and consolidated by EB melting.

Powder Production. Powders are produced from ingot by hydriding, crushing, and dehydriding; in addition, some recent efforts have been directed toward producing complex metastable alloy powders, such as niobium-aluminum and niobium-silicon alloys, by liquid metal atomization and rapid quenching. The particle structure of degassed hydride niobium powder is completely analogous to that of a tantalum powder produced in a similar process for capacitors. Typical compositions of niobium and C-103 alloy powder made by this process are compared in Table 11. Niobium powders produced by the hydride-crush-degas process are not normally used for capacitors, and thus milling to a very fine particle size (that is, a high surface area) is not required. Normally, powders are crushed to pass an 80-mesh screen, and a mean particle size of 10 to 15 μm (400 to 600 μin.) is typical. Although thermal agglomeration is feasible and is commonly employed to form clusters for very fine tantalum powders, it is rarely necessary for niobium.

Niobium powder is frequently used as the starting material to blend with alloying agent

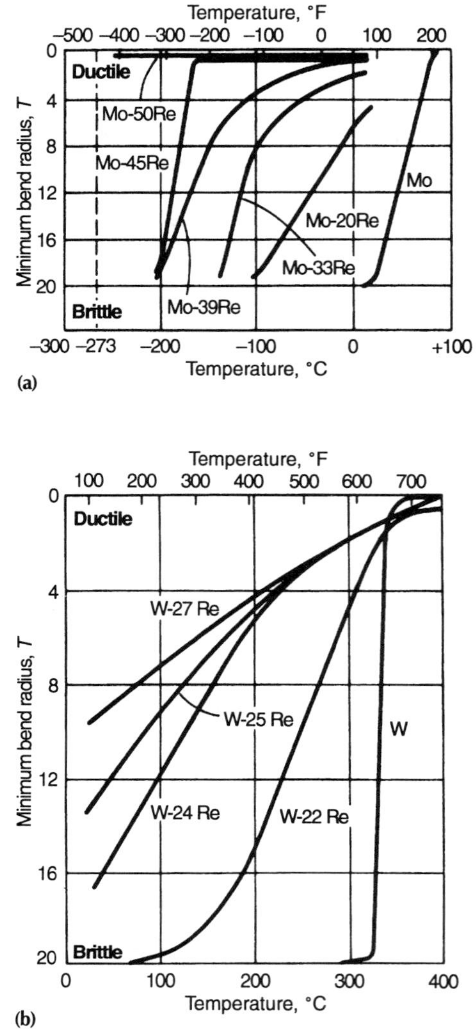

Fig. 7 Effect of alloying additions of rhenium on ductility. (a) Rhenium-molybdenum alloys. (b) Rhenium-tungsten alloys

powders. The blend is pressed to bars and melted, thereby promoting alloy homogeneity. Niobium alloy scrap, which is reduced to powder by the hydride-dehydride process, can also be incorporated into the alloy blends. One P/M dispersion-strengthened alloy containing 0.5% TiO_2 is claimed to have improved yield strength and high-temperature grain stability.

Product Forms. Most commercial niobium alloys are wrought forms. Primary (bulk) working is generally done by high-temperature extrusion or forging. Secondary working produces a variety of mill products such as sheet, foil, rod, wire, and tubing.

Powder metallurgy methods have been evaluated for common niobium alloys, such as C-103 and Nb-30Hf-9W. It has been determined that the high production cost of atomized niobium powder prevents economic production of net-shape parts. However, hydride-dehydride niobium powders are used to produce commercial quantities of hot working preforms of alloy Nb-30Hf-9W (Ref 2). More information on niobium applications is discussed in Ref 2 and "Refractory Metals and Alloys" in *Properties and Selection: Nonferrous Alloys and Special Purpose Materials,* Volume 2, *ASM Handbook.*

Rhenium

Among the elements, rhenium has the highest melting point, except for tungsten and carbon. Its density is exceeded only by osmium, iridium, and platinum. A ductile-to-brittle transition temperature does not exist in pure rhenium. Rhenium is the only refractory metal that does not form carbides.

Rhenium has a high electrical resistivity over a wide temperature range. Figure 6 compares electrical resistivity values for tungsten and rhenium. Rhenium typically exhibits higher resistivity values than tungsten. This characteristic, combined with a low vapor pressure, makes it ideally suited for filament applications; additionally, it maintains ductility and is not affected by the oxidation/reduction cycle experienced in these applications, as is tungsten.

One of the largest applications for rhenium is for mass spectrometer filaments. These are available in commercial (99.99%) and zone-refined (99.995%) purities.

Rhenium is not attacked by molten copper, silver, tin, or zinc. It dissolves readily in molten iron and nickel, but is stable in the presence of aluminum. Rhenium has a significant hardening effect on platinum.

At elevated temperature, rhenium resists attack in hydrogen and inert atmospheres. It is resistant to hydrochloric acid and seawater corrosion and to the mechanical effects of electrical erosion.

Rhenium is widely used in combination with platinum in catalysts, particularly for selective hydrogenation and crude oil reforming. Rhenium catalysts display high resistance to poisons, such as nitrogen, sulfur, and phosphorus.

Rhenium Alloys. Rhenium is a beneficial alloying addition with other refractory metals. Rhenium greatly enhances the ductility and tensile strength of refractory metals and their alloys (Fig. 7). This tendency is maintained on heating above the recrystallization temperature.

Rhenium alloys are used in nuclear reactors, semiconductors, electronic tube components, thermocouples, gyroscopes, miniature rockets, electrical contacts, thermionic converters, and other commercial and aerospace applications. Tungsten-rhenium alloys, applied by vapor deposition, are used to coat the surface of molybdenum targets in x-ray tube manufacture. Other rhenium alloys (with tungsten or molybdenum) are used for filaments, grid heaters, cathode cups, and ignitor wires in photoflash bulbs.

ACKNOWLEDGMENT

This article was adapted from "Refractory Metals and Alloys" in *Properties and Selection: Nonferrous Alloys and Special-Purpose Materials,* Volume 2, *ASM Handbook* by John B. Lambert (Chair), John R. Rausch, Sam Gerardi, Charles Pokross, Walter A. Johnson, Toni Grobstein, Robert Titran, and Joseph R. Stephens.

REFERENCES

1. G.E. Dieter, Jr., *Mechanical Metallurgy,* McGraw-Hill, 1961
2. C.C. Wojcik, Processing, Properties, and Applications of High-Temperature Niobium Alloys, *High Temperature Silicides and Refractory Alloys,* Symposium Proceedings, Vol 322, Materials Research Society, 1994, p 519–530

Tungsten Heavy Alloys

S.G. Caldwell, Teledyne Metalworking Products

TUNGSTEN HEAVY ALLOYS (WHAs) are a category of tungsten-base materials that typically contain 90 to 98 wt% W. Most commercial WHAs are two-phase structures, the principal phase being nearly pure tungsten in association with a binder phase containing the transition metals plus dissolved tungsten (Fig. 1). As a consequence, WHAs derive their fundamental properties from those of the principal tungsten phase, which provides for both high density and high elastic stiffness. It is these two properties that give rise to most applications for this family of materials.

The current uses of WHAs are quite varied, spanning a wide range of consumer, industrial, and government applications that include:

- Damping weights for computer disk drive heads
- Balancing weights for ailerons in commercial aircraft, helicopter rotors, and for guided missiles
- Kinetic energy penetrators for defeating heavy armor
- Fragmentation warheads
- Radiation shielding, radioisotope containers, and collimation apertures for cancer therapy devices
- High performance lead-free shot for waterfowl hunting
- Gyroscope components
- Weight distribution adjustment in sailboats and race cars
- Heavy casings for downbore logging of oil wells
- Weights in premium golf clubs for improving consistency in response
- Low chatter, high stiffness boring bars, and toolholders for metalworking
- Electrical contacts

When compared to other high density metals, the unique combination of characteristics WHAs offer the materials engineer can easily be seen in Table 1. While lead is generally considered a very dense material, WHAs can easily exceed its density by 60% or more. Lead has been the metal of choice for low-cost density-related applications in past years. Currently, toxicity concerns drive a slow but steady process of lead replacement in many instances. Lead, however, is excluded from many density applications on the basis of its extremely low strength at room temperature, readily recrystallizing under applied stress. Room temperature creep eliminates any use in which it must be tightly fastened or where it is needed in large, thin cross-section geometries.

Many applications that require high gravimetric density for balance weights, inertial masses, or kinetic energy penetrators or high radiographic density for radiation shielding and collimation necessitate rather large bulk shapes. Such a requirement eliminates all but a few candidates on the basis of prohibitive cost, typically reducing the choice of very dense alloys down to either tungsten- or uranium-base materials. Uranium alloys, like lead, are eliminated from an increasing number of potential applications based on toxicity considerations, with uranium-base materials requiring a license except for very small quantities. While the precious metals listed possess attractive densities and offer essentially no toxicity, their cost is prohibitive for all but a few density applications.

WHAs are the preferred form of tungsten for most mass/density related requirements in that large components can be easily fabricated from compacted elemental metal powders by liquid-phase sintering (LPS) in H_2 at moderate temperatures, yielding complete densification at temperatures far below the melting point of tungsten. Figure 1 shows a typical LPS microstructure for WHA. In contrast, pure tungsten must be solid-state sintered at temperatures on the order of 2000 to 3000 °C and achieve densities only slightly greater than 90% without subsequent thermomechanical processing. A number of restrictions on component size and geometric complexity, absent for LPS WHAs, correspondingly apply. Secondly, WHAs are readily machinable by common metalworking tools and techniques, whereas pure sintered tungsten has limited machinability and may require shaping via grinding. Only a very few ap-

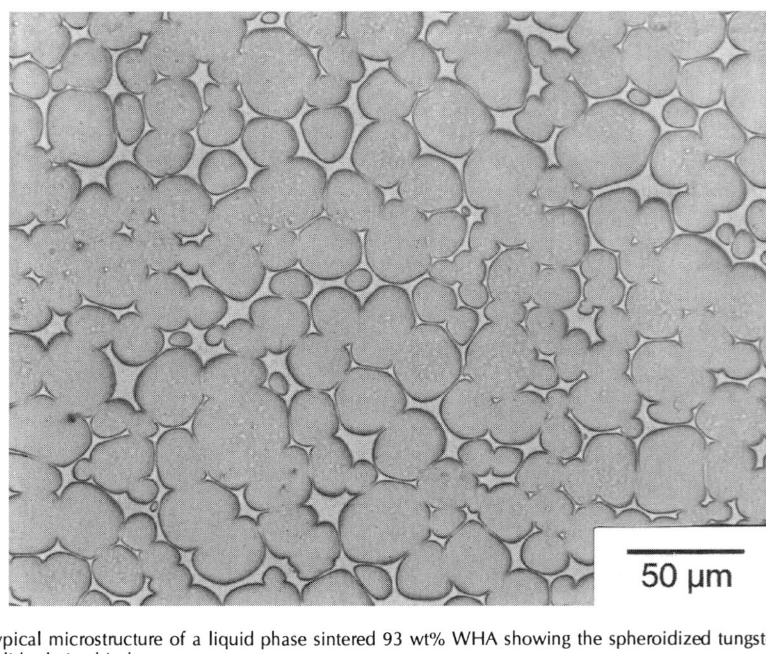

Fig. 1 Typical microstructure of a liquid phase sintered 93 wt% WHA showing the spheroidized tungsten phase and solid solution binder

Table 1 Comparison of high-density metals and alloys

Material	Density, g/cm^3	Melting temperature, °C	Young's modulus, GPa	Cost	Toxicity
Iron (for reference)	7.86	1536	207	Low	Low
Lead	11.4	328	Recrystallization at RT	Low	High
Mercury	13.5	−39	Liquid at RT	Moderate	High
Tantalum	16.6	2996	186	High	Low
WHAs (typical)	17.0–18.5	(1450)	310–380	Moderate	Moderate
DU-0.75Ti	18.6	1135	160	Moderate	High
Uranium	18.9	1132	160	Moderate	High
Tungsten	19.3	3410	410	Moderate	Low
Gold	19.3	1064	80	Very high	Very low
Platinum	21.4	1772	147	Very high	Very low
Iridium	22.5	2443	524	Very high	Very low

RT, room temperature

Table 2 Property comparison of unworked alloys

Alloy	Ultimate tensile strength, MPa	Elongation, %	Hardness, HRC	Charpy toughness, J
91W-6.3Ni-2.7Fe	940	35	29	300
91W-6Ni-3Co	960	40	31	470+

plications that require high-temperature service would favor the pure metal.

Composition

WHAs typically consist of 90 to 98 wt% W in combination with some mix of nickel, iron, copper, and/or cobalt. The bulk of WHA production falls into the 90 to 95%W range. The choice of alloy composition is driven by several considerations. The primary factor is the density required by the given application. Further considerations include corrosion resistance, magnetic character, mechanical properties, and postsinter heat treatment options. The first WHA developed was a W-Ni-Cu alloy. Alloys of this ternary system are still occasionally used today, primarily for applications in which ferromagnetic character and electrical properties must be minimized. W-Ni-Cu alloys otherwise offer inferior corrosion resistance and lower mechanical properties than the present industry standard W-Ni-Fe alloys.

The majority of current uses for WHAs are best satisfied with the W-Ni-Fe system. Alloys such as 93W-4.9Ni-2.1Fe and 95W-4Ni-1Fe represent common compositions. Mechanical properties of the unworked alloy vary with the nickel-to-iron ratio, as seen in Fig. 2. Yet alloy selection must take into account the necessity for postsinter heat treatment of extreme nickel-to-iron ratio alloys. Optimal mechanical properties of unworked material are obtained in the vacuum heat treated and quenched state. Due to the body-centered cubic (bcc) structure of the principal tungsten phase, WHAs are sensitive to hydrogen embrittlement. There are two common postsinter heat treatments for WHAs. Vacuum heat treatment is useful for reduction of hydrogen embrittlement and resolutionization/quench for reduction of segregation-induced embrittlement. Yet for cost-sensitive applications, one or both of these processing steps may not be desired. Further, very large components with thick cross sections present a formidable challenge for effective postsinter heat treatment.

Assuming first no postsinter thermal processing, the binder composition is limited both on the nickel-rich and iron-rich sides by the formation of embrittling intermetallics, restricting the range of common use to $1.5 \leq Ni/Fe \leq 4$. The exact compositional interval is a function of cooling rate from sintering. Alloys with $Ni/Fe \leq 1.2$ typically possess very low mechanical properties due to the formation of a μ-phase (Fe_7W_6). Once formed, it cannot be resolutionized at practical processing temperatures, being stable to above 1600 °C. W-Ni-Fe alloys with $Ni/Fe > 5$ offer superior mechanical properties, but form nickel-rich intermetallics, predominantly Ni_4W.

If postsinter vacuum heat treatment is not to be performed, alloys with nickel-to-iron ratios of 1.5 to 2 typically give as good properties as higher ratios. For components that can be vacuum heat treated, nickel-to-iron ratios in the range 2.3 to 4 should be chosen, as better mechanical properties are generated. For applications such as kinetic energy penetrators that demand the utmost in mechanical performance, W-Ni-Fe alloys should be resolutionized and quenched. This processing step reduces both the segregation of interstitial species and the presence of any nickel-rich intermetallics and thereby opens the possibility for use of alloys with nickel-to-iron ratios in the range of 6 to 15, which provide noticeably improved mechanical properties.

The addition of cobalt to a W-Ni-Fe alloy is a common approach for slight enhancement of both strength and ductility. The presence of cobalt within the alloy provides solid-solution strengthening of the binder and slightly enhanced tungsten-matrix interfacial strength. Cobalt additions of 5 to 15% of the nominal binder weight fraction are most common. Higher levels of addition risk the formation of a mixed μ-phase ($[Fe,Ni]_7W_6$), which would then require high-temperature resolution/quench of the material.

For extremely demanding applications, even higher mechanical properties are obtainable from the W-Ni-Co system with nickel-to-cobalt ratios ranging from 2 to 9. Such alloys require

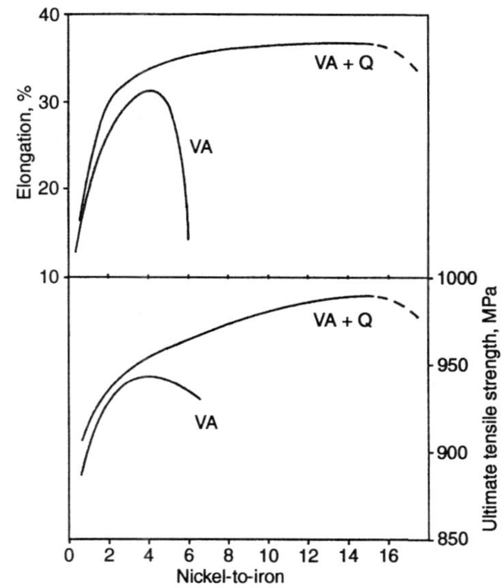

Fig. 2 Mechanical property variation as a function of Ni/Fe ratio and heat treatment condition for LPS 93W-Ni-Fe alloy. VA, vacuum anneal; VA + Q, vacuum anneal and quench

resolution/quench, however, due to extensive intermetallic (Co_3W and others) formation on cool down from sintering. These alloys not only provide greater quasi-static tensile strength and ductility, but substantial improvement in fracture toughness as well (Table 2). The use of W-Ni-Co alloys is generally limited to defense applications requiring state-of-the-art mechanical properties.

A number of specialty WHAs are known as well. An example is the W-Mo-Ni-Fe quaternary alloy, which utilizes molybdenum to restrict tungsten dissolution and spheroid growth, resulting in higher strengths (but reduced ductility) in the as-sintered state. There are also a number of alloy systems in various stages of development for kinetic energy penetrators that are intended to provide a WHA that will undergo high deformation rate failure by shear localization in a manner similar to quenched and aged U-0.75Ti for more efficient armor defeat. These alloys to date have not exhibited a property set of interest for industrial applications, however.

Consolidation

WHAs are typically formulated from elemental powders with mean particle sizes in the range 1.5 to 7 μm. Blend homogeneity is essential for microstructural uniformity and predictable sintering to full density. Blended powder is sometimes mixed with an organic binder such as 1 to 2 wt% paraffin and granulated. While this procedure is required for even die-cavity filling for some component shapes (especially flat forms), it should be employed only when necessary as the introduction of organics necessitates a debinding operation in the process sequence. It is essential that any organic binder be capable of clean thermal decomposition and that the process be carried to completion as residual carbon in the sintering environment can lead to interstitial embrittlement and in the worst case, WC formation. Unlike WC grade powder, WHA blends tend to be readily pressable without a binder using standard techniques such as uniaxial compaction and isostatic pressing at pressures of ~200 MPa. Even without organic binders, the green strength of WHA pressings is generally quite acceptable.

WHAs are most commonly processed to full density by solid-state sintering (SSS) or LPS in H_2. The former process is ideal for parts requiring low thermal distortion ("slumping" due to capillary action and/or gravity) and need only minimal mechanical properties. This is often required for precision net-shape parts. Where practical, sintering at a slightly higher temperature allows the formation of an entirely different microstructure capable of significantly higher mechanical properties due to the occurrence of a liquid phase. In the common W-Ni-Fe compositions, liquid-phase conditions are achieved slightly above 1440 °C. Liquid-phase sintering for 90 to 97% W alloys is generally carried out in the range 1470 to 1580 °C, whereas SSS is performed at ~1430 to 1440 °C. Linear shrinkages of ~20% are typical.

Vacuum sintering of WHAs in batch-type furnaces is used only rarely due to the inferior microstructure and mechanical properties obtained. Additionally, if liquid-phase conditions are reached, the high vapor pressure of the transition metals present may cause rapid deterioration of tungsten or molybdenum furnace components in the furnace hot zone. Vacuum sinter furnaces utilizing graphite elements and heat shielding are unsuitable due to excessive carbon transport to the WHA parts, resulting in embrittlement or superficial WC formation.

It is essential that the rate of temperature rise experienced by a pressing be gradual, so as to avoid thermal shock or premature surface pore closure. Below ~1000 °C, there is significant reduction of metal oxides on transition metal particles and general desorption of O_2 and other species. At ~1000 °C and slightly higher, tungsten oxides are readily reduced by the hot H_2 and significant interdiffusion of metallic species occurs, initially by surface diffusion. As the furnace temperature exceeds ~1200 °C, surface closure of interconnected porosity is underway. It is essential that the heating rate is sufficiently slow so as to permit internal deoxidation of the pressing prior to this point. While hydrogen at these temperatures possesses high mobility through the bcc tungsten structure, the H_2O formed by reaction relies on interconnected porosity for escape. If surface porosity is closed prematurely, this molecular specie will be trapped, giving rise to residual porosity. In more extreme cases of high oxygen powders, larger voids may form, a process that can be greatly minimized by the use of higher dew point H_2.

Commercial production of WHA components most commonly utilizes continuous pusher-type furnaces in that such designs offer high throughput rates and good temperature uniformity. Pressings are transported through the furnace on ceramic slabs or partially buried in molybdenum boats containing alumina sand. Stoker furnaces in general produce parts with slightly higher mechanical properties than what is obtainable for batch furnace sintering due to better internal material cleanup by the action of the hot flowing H_2, effectively sweeping evolved species from the vicinity of the parts.

While hot isostatic pressing (HIP) and rapid consolidation techniques are also available densification options, their use is very rare because the short processing times generally result in microstructures that yield inferior mechanical properties, often at a substantially higher cost as well. Problems with porosity, common in many P/M materials, is virtually nonexistent in WHAs provided LPS is correctly executed using high quality raw materials.

Liquid-phase sintering provides a superior means of densifying pressings in which a microstructure consisting of spheroids (~40 to 60 μm in size) of nearly pure tungsten in an austenitic binder phase of transition metals plus dissolved tungsten forms via Ostwald ripening. Such a structure provides better mechanical properties than SSS structures due to less angular grains, lower tungsten-tungsten contiguity for a given tungsten content, and a greater percentage of metastable tungsten in the austenitic solid-solution binder. The amount of tungsten retained for common cooling rates from sintering is substantially greater than equilibrium values and varies with the binder composition as seen in Table 3. The slower the cooling rate from sintering temperature, the lower the percentage of retained metastable tungsten in the binder phase.

Table 3 Metastable tungsten retained in solution

Alloy	% W in solid solution
91W-6.3Ni-2.7Fe (SSS)	~18
91W-6.3Ni-2.7Fe (LPS)	~23
91W-6Ni-3Co (LPS)	~40

Postsinter Heat Treatment

A postsinter heat treatment is generally required for any WHA component that has significant mechanical requirements. The most common is a H_2 outgassing treatment employed to virtually eliminate hydrogen embrittlement. During sintering, especially if a liquid phase was allowed to form, WHA parts absorb sufficient hydrogen so as to significantly limit ductility. By subjecting the sintered components to an isothermal treatment in a low p_H environment, absorbed hydrogen is given sufficient thermal activation to diffuse to a free surface, recombine into dimer form (H_2), and migrate from the furnace environment. This heat treatment is generally performed at 1000 to 1100 °C for several hours, depending on section thickness. While this operation could be performed in flowing N_2, vacuum heat treatment is usually the most convenient means of protecting the parts from oxidation during processing and is therefore commonly referred to as vacuum annealing. This procedure typically increases the ultimate tensile strength of an alloy by 5% and the elongation up to 40% as compared to the as-sintered property set. While greatly reducing hydrogen embrittlement, the additional time at moderate temperature increases both the segregation of interstitial species to grain boundaries and the amount of intermetallic precipitation in susceptible alloy systems.

While vacuum annealing provides a method of improving low strain rate properties over those of the as-sintered state, dynamic properties such as fracture toughness may in fact be lowered. Resolutionization and quench can be used for dynamic property recovery. This second postsinter step addresses two separate but related embrittlement mechanisms. First, this procedure minimizes segregation-induced embrittlement caused by sulfur, phosphorus, and carbon by providing thermal activation for redistribution of these species into the bulk and preventing reconcentration in the boundary region by imposition of rapid cooling to a temperature at which diffusion is ineffective. This mechanism is operative for all WHAs. Resolution is performed in the vicinity of 1100 °C for 1 to 2 h, followed immediately by a water quench. Secondly, in alloy systems containing intermetallic precipitates, resolution and quench also provides a means of property recovery. The treatment temperature must be increased to ~1200 °C for effective resolution of nickel-base compounds. Quenched WHAs typically exhibit another 5% gain in ultimate tensile strength, 10 to 30% gain in elongation, and ~300% gain in fracture toughness (as evaluated with unnotched Charpy impact) over vacuum annealed material. Resolution and quench make possible the use of high ductility/toughness alloys such as high nickel-to-iron ratio W-Ni-Fe alloys and all W-Ni-Co alloys. All WHAs benefit from such treatment, though not all part geometries may be suitable for such postsinter processing, however. All applications requiring high dynamic properties should specify quenched WHA. While uncommon in commercial practice, it is possible in modern cold-wall sintering furnaces to combine H_2 sintering, vacuum annealing, and quenching into a single energy and time efficient cycle.

Mechanical Processing

For nearly all density- and mass-driven applications, the mechanical properties of unworked WHAs are more than adequate. Yet for many other applications, the strength and hardness of unworked material may be insufficient. Through a combination of mechanical and thermal processing, a wide range of property sets are available. WHAs can be processed for maximum ductility with tensile elongations exceeding 40% or for strength levels exceeding 1650 MPa with a hardness of 50 HRC. As can be seen in Fig. 3, there is a continuous set of mechanical proper-

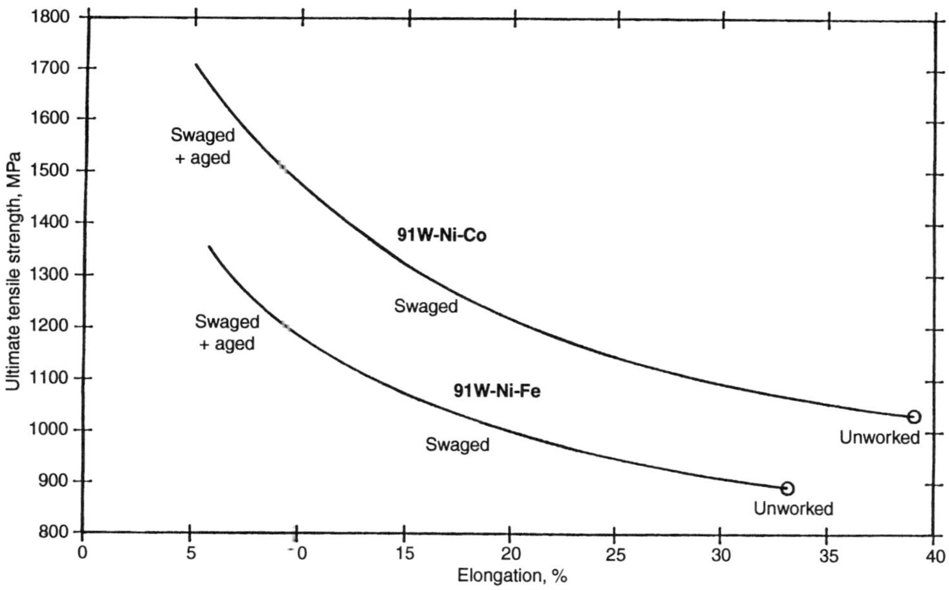

Fig. 3 Strength and ductility can be varied over a substantial range by postsinter processing. The higher the initial mechanical properties, the greater the practical range of processing options.

ties available via the trade-off in ductility and toughness for strength and hardness. The greater the initial unworked property set, the greater the range of processing options. The most commonly used means of increasing the strength and hardness of WHAs is by rotary swaging, in which the diameter of a bar is reduced by rapid short-stroke hammering, generally in a four-die configuration. This cold (or occasionally warm) working results in an increase in bar length and imparts an axial orientation to the microstructure. The extent of swaging possible for a given alloy is a function of its ductility. Table 4 provides a guideline for maximum percentages of reduction in area (RA) via swaging for selected 90 to 93% W alloys, valid for initial bar diameters of 5 cm and smaller.

Exceeding the listed %RA values often results in microcracking of the material. Such induced microstructural damage results in diminished returns for greater degrees of working. A more successful approach to achieving higher levels of strength and hardness involves taking material worked from 15% RA to the values shown in Table 4 and subjecting the material to an isothermal treatment—typically 1 to 2 h at 350 to 550 °C. This gives rise to significant strain aging of both the binder phase and the tungsten phase. Attempts at aging in the range 700 to 1000 °C may cause embrittlement in some alloys due to intermetallic precipitation. Aging at ~1200 °C recovers essentially all of the induced strain hardening and essentially returns the material to its original condition, though some microstructural orientation remains. The original mechanical property set will not be recovered, however. The binder no longer contains the same extent of tungsten supersaturation. As tungsten content of the WHA and/or starting bar diameter increases, the limiting value of %RA must be decreased. There are a number of subtle variations that can only be described generically. These include die geometry and feed rate into the swage as they determine the profile of induced deformation and the extent of work. The effectiveness of such processing is seen in Table 5.

Although swaging is the most commonly used deformation processing technique for WHAs, alternative methods of working include extrusion (conventional or hydrostatic), forging, rolling, or upsetting for strength and hardness enhancement. Hydrostatic extrusion is capable of successfully achieving very high reduction ratios, producing a fibrous microstructure. Upsetting, a commonly used practice for introducing work into short rods (such as preforms), differs from other techniques in terms of direction of material flow. This difference must be noted for applications in which the Bauschinger effect may prove important.

Microstructure Considerations

In LPS WHAs, the microstructure of most commercial alloys is dual phase and takes the form of spheroids of nearly pure tungsten in an austenitic (face-centered cubic) matrix of the selected transition metals plus dissolved tungsten. In a mature LPS structure of W-Ni-Fe, spheroids are typically 40 to 60 μm in size with a degree of connectedness, termed contiguity. For a given composition, mature spheroid size is a function of sintering temperature, higher temperature favoring larger size. Below approximately 91% W, the binder phase is a continuous structure within the solid. Such alloys display noticeably higher ductility and toughness and also are capable of more rapid work hardening. Tungsten spheroids typically contain less than 0.2 wt% Ni and 0.1 wt% Fe in common alloys. In such alloys, three types of interfaces exist: tungsten-tungsten boundaries, binder grain boundaries, and tungsten-matrix interfaces. Of these three structures, the tungsten-tungsten contact regions are the weakest and therefore serve as a principal limiting factor in the overall strength of the alloy. Thus, any microstructural characteristics that affect these features strongly influence mechanical behavior. Segregation of impurities to this interface serves to weaken its strength. Likewise, an increase in contiguity, resulting in a greater fraction of tungsten-tungsten boundary area, also lowers mechanical properties. It has been demonstrated that the critical flaw size is comparable to the mean spheroid diameter for common alloys.

Contiguity is determined by the weight percent of tungsten, the sintering temperature, and the thermochemistry of the binder phase as to tungsten solubility and wetting behavior. High tungsten content alloys exhibit high contiguity due to the shape accommodation that occurred during LPS as tungsten spheroids grew to impingement (Fig. 4). As the microstructure becomes more brittle, the alloy can undergo less strain hardening prior to fracture. Thus, contiguity and interfacial embrittlement serve to limit both the ductility and strength of WHAs.

Some applications for WHAs demand densities below 17 g/cm^3. Such uses, which may often require high bend ductility as well, are best met with SSS material with tungsten contents in the range of 60 to 90%. Solid state sintering provides minimal sintering distortion and near full densification with a more refined microstructure than that available from LPS. A typical SSS microstructure is shown in Fig. 5. The tungsten phase exhibits a mean size of ~10 to 15 μm in a continuous austenitic binder solid solution. The high volume fraction of ductile binder and the low tungsten-tungsten contiguity of this type result in high deformability. Such material can be easily rolled into thin sheet providing high mechanical properties.

Table 4 Swaging limitations for various WHAs

Alloy family	Maximum %RA swage
W-Ni-Fe (2.3 ≤ Ni/Fe ≤ 4)	25
W-Ni-Fe-Co (typical)	30
W-Ni-Co	40

Table 5 Typical properties for 93W-4.9Ni-2.1Fe

Condition	Ultimate tensile strength, MPa	Elongation, %	Hardness, HRC	Charpy toughness, J
As-sintered	860	15	29	...
Vacuum annealed	910	25	29	...
Quenched	930	32	29	230
Swaged 15% RA	1115	16	39	150
Swaged 25% RA	1200	14	40	110
Swaged 25% + aged 500 °C/1 h	1400	7	44	80

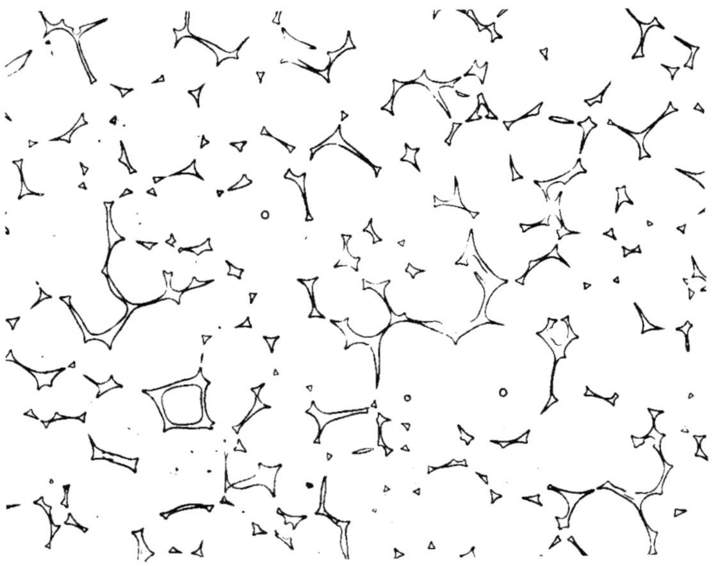

Fig. 4 High tungsten content WHAs display substantial amounts of tungsten-tungsten contiguity and shape accommodation during Ostwald ripening, as seen in this example of 98% W

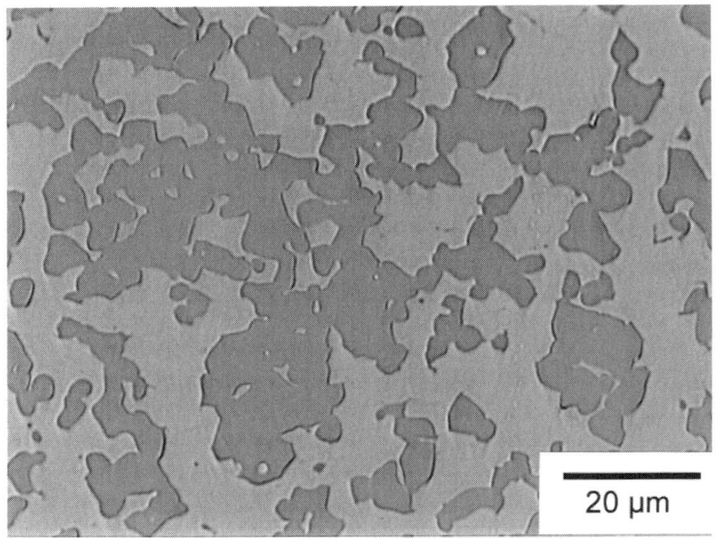

Fig. 5 Typical microstructure of a SSS 70W-Ni-Fe alloy exhibiting fine, polyhedral tungsten grains in a continuous matrix phase

Whether SSS or LPS, the sintered structure of WHAs can be described as a metal-matrix composite, as each phase presents a distinctly different set of mechanical properties. This structure gives rise to notch sensitivity, and therefore all components requiring high mechanical integrity should avoid grooves, threads, or corners with radii smaller than ~0.5 mm. Otherwise, stress concentration will result in reduced mechanical strength. This becomes important also in mechanical property measurement, as specimen preparation, finishing, and marking techniques appropriate for steel specimens may not be suitable for WHA samples.

Notch sensitivity is also the reason why WHA toughness is most commonly evaluated using unnotched Charpy specimens per ASTM E 23, as a much higher level of discrimination is possible. Low-toughness WHA typically tests at ~8 J notched and ~55 J unnotched. The same alloy in a higher toughness condition would test at ~10 J notched and ~215 J unnotched. In the notched testing mode, statistical noise would mask any difference, whereas unnotched test results are clearly separated. The use of subscale Charpy bars are sometimes used, but results cannot be compared to full-scale values by ratio of areas as the span of the Charpy anvil remains fixed and hence provides a larger relative span for subscale specimens, leading to artificially high values of absorbed energy to failure.

WHAs also display strain-rate sensitivity. When tested at medium to high strain rates, WHAs provide a significantly higher dynamic yield strength than suggested by quasi-static values. A 90% WHA exhibits a strain-rate sensitivity exponent, *m*, of ~0.012.

During tensile testing, unworked WHAs generate a load-displacement curve with a high-slope linear elastic region, a rather gentle transition into plasticity and work hardening, and minimal rise in engineering strength to failure. The 0.2% offset yield stress is approximately 67% of the ultimate tensile strength. A Poisson ratio of ~0.29 is commonly encountered. Work hardening follows power-law behavior with an exponent of $n \approx 0.51$ for 90% WHA. The rate of work hardening increases slightly with tungsten content. Load-displacement curves of heavily worked WHA exhibit sharp knees with a virtual equivalence of ultimate tensile strength and 0.2% offset yield stress.

Toughness of WHA is readily revealed by both the absorbed energy to failure numbers from Charpy testing as well as the resultant fracture morphology when tested in unnotched mode. Optimal mechanical properties are associated with alloy conditions in which the fracture surface will mainly consist of cleaved tungsten spheroids surrounded by binder that has undergone significant elongation, failing in distinct knife edges between the tungsten grains. Embrittled material typically exhibits distinct tungsten-tungsten spheroid contact zones and matrix decohesion. Highly embrittled conditions, such as from intermetallics forming a peritectic reaction layer on the tungsten phase, typically exhibit an almost glassy fracture of the binder phase (Fig. 6).

In addition to the mechanical difference between the two phases, there are significant electrochemical differences as well. While WHAs are generally considered corrosion resistant, corrosion can nevertheless occur due to local galvanic activity between the tungsten phase, and the electrochemically dissimilar austenitic binder phase if a suitable electrolyte is present. Highly alkaline solutions readily attack the tungsten phase whereas the binder is more readily attacked by acidic solutions. Corrosion can occur under ambient conditions in the presence of very high humidity, especially condensing moisture, and in the presence of ionic salt solutions.

Table 6 Variation of density and Young's modulus with tungsten content

Wt% W	Density, g/cm^3	Young's modulus, GPa
90	17.2	345
91	17.4	350
93	17.8	360
95	18.2	370
97	18.6	385
98	18.8	390
100	19.3	410

Physical Properties

In that the principal phase in WHAs is nearly pure tungsten, it follows that the physical properties of WHAs should be very close to that of elemental tungsten. A comparison of density and elastic stiffness for various W-(3Ni/Fe) alloys and pure tungsten is provided in Table 6. The (longitudinal) sonic velocity of 90 to 95% WHAs is approximately 5150 m/s, slightly lower than that of pure tungsten. In as-sintered material with a bulk hardness of 29 HRC, the microhardness of the tungsten phase is ~500 HV (25 gf)—approximately that of pure tungsten. The matrix, however, possesses a hardness of only ~300 HV (25 gf).

Thermal Properties

While primarily comprised of the highest melting point metal, WHAs are limited in service temperature by nature of the transition metal binder phase and the formation of intermetallics. There are other factors as well that serve to limit useful service temperature. Under ambient conditions, oxidation alone may impose a temperature ceiling of ~300 to 400 °C. If a WHA bar is in a mechanically worked condition, a temperature as low as ~300 °C will result in noticeable strain aging in as short a time as a few hours, resulting in altered properties. Even if a protective atmosphere is employed, service temperatures ~600 °C and higher are to be avoided as these composite materials exhibit "hot-short" behavior. The ultimate limit on service temperature is imposed by the threshold of formation of a liquid phase, which can occur in some W-Ni-Fe alloys as low as ~1450 °C.

While the coefficient of thermal expansion of pure tungsten is very low at 4.4 ppm/°C (for 20 to 200 °C), the coefficient of thermal expansion for common WHAs is only slightly higher at ~5.0 ppm/°C. Thermal conductivity for a typical commercial WHA is approximately 120 W/m · K, with a corresponding electrical conductivity of ~14% IACS. Specific heat decreases slightly with increasing tungsten content, ranging from approximately 0.037 cal/g · C for 90 W to 0.032 cal/g · C for 97% W. While thermal properties are principally determined by the tungsten phase, service temperature limitations are determined by the binder.

Applications

Virtually all of the applications for WHAs depend upon its high density. WHAs are used in an increasingly wide array of industrial, defense, and consumer products. In addition to impressive gravimetric density, WHAs offer a very useful radiographic density that generally exceeds that of lead, but is slightly lower than that of depleted uranium (DU), depending on the type of radiation and its energy. For 1 MeV γ-radiation, linear absorption by 95W-Ni-Fe is approximately 50% higher than by lead. Additionally, WHAs prove superior for most shielding and collimation applications over lead in that they provide excellent mechanical strength, are readily machined, resist routine handling damage, and pose no toxicity considerations. Consequently, WHAs are widely used as injection syringe shields, bulk radioisotope storage containers, collimator components in cancer therapy devices, and precision shielding for radiation detector assemblies. Some typical configurations are pictured in Fig. 7.

A rapidly growing area of use is consumer sporting goods. While WHAs have been employed as high-density dart bodies for decades, large volume consumption is now being realized as these materials are being used for golf club weights and waterfowl shot. In the course of development for premium irons and drivers, club heads are being designed with WHA inserts that provide intelligently concentrated mass so as to increase the "sweet spot" and hence club performance. The remainder of the club head is fashioned from a low density titanium alloy, so as to keep overall club head weight comparable to former all-steel designs. Some putter designs utilize WHA masses on the extreme ends of the head, thereby increasing the rotational moment of inertia and adding greater directional stability for slightly off-center strikes of the ball. Also of large volume usage is the application of lead-free shot for hunters of wetland birds. A WHA of sintered tungsten-iron provides an excellent density match to lead alloy shot. The tungsten-iron shot does not deform as it passes through the choke of the gun, permitting improved down-range patterning. Currently, shotshells are designed with a special provision for protecting the bore of the gun, thereby overcoming a serious problem present with early tungsten-base shotshells.

For several decades WHA has been used in various designs of kinetic energy penetrators for the defeat of armor on account of its high density and favorable mechanical properties. Designs have ranged from small- and medium-caliber (5.56 to 35 mm bore) fin-stabilized short-rod penetrators to high length-to-diameter (L/D ~ 20+) fin-stabilized medium- and large-caliber (35 to 120+ mm bore) rods. These massive projectiles are typically launched at velocities of 1200 to 1650 m/s, with larger designs delivering many megajoules of kinetic energy to the target. Currently, other ordnance concepts require lengths exceeding 1 m and masses greater than 500 kg. Smaller munitions devices in current use employ prescored cases of WHA loaded with high explosive. Upon detonation, the case disintegrates into a controlled dispersion of very high velocity fragments of prescribed size and flight characteristics for both antimateriel and antipersonnel effects. WHA has largely supplanted WC-Co as the material of choice for non-DU penetrators.

While political pressure is mounting against the continued used of DU for ordnance applications, its use persists in large caliber due to the 5 to 10% penetration performance advantage commonly observed over WHA. This slight difference is viewed as critical in that it determines the maximum effective engagement distance to enemy tanks. While much development work has been devoted to developing a WHA with a shearable matrix that could simulate the penetration and erosion behavior of DU-0.75Ti rods, satisfactory results to date have not been obtained. All current penetrator designs employ W-Ni-Fe(-Co) alloys with the exception of the highest performance large-caliber rods that are W-Ni-Co. While higher mechanical properties do not necessarily imply better terminal ballistic performance, high properties are nevertheless desirable for WHA penetrator materials. The higher the property set, the less support required for long rods during launch. The reduced parasitic weight of the launch package means a greater percentage of the energy of a given gun system deliverable to the target.

In addition to making the strongest possible alloys, controlled embrittlement is useful for addressing another class of ordnance requirements. By carefully controlling the degree of embrittlement of a WHA, a penetrator can be fabricated that dependably withstands chambering and launch stresses, but fragments into a controlled dispersion of fragments on impact with metal or composite sheets, depositing a large amount of

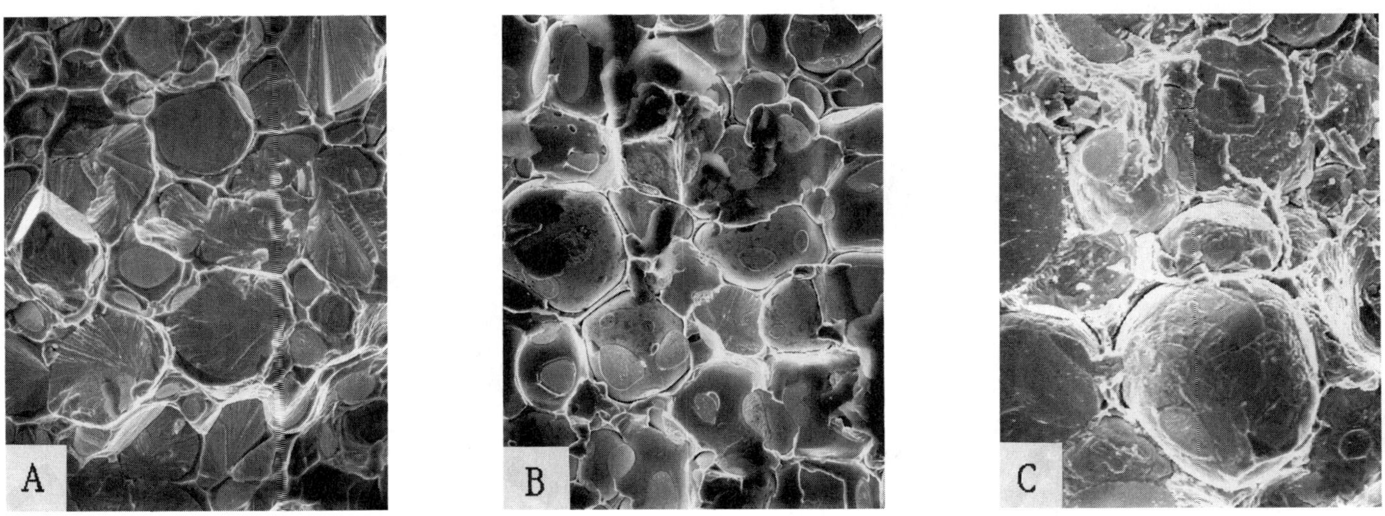

Fig. 6 Fracture morphology resulting from mechanical testing clearly reveals substantial differences in alloy properties: (A) high toughness material displaying predominant tungsten cleavage and ductile matrix failure, (B) alloy with tungsten-matrix interfacial embrittlement but good matrix ductility, and (C) extremely brittle failure due to presence of intermetallic precipitates

energy in a localized volume. Termed frangible rounds, these penetrators are useful in damaging large areas of airframes in anti-aircraft warfare. Such rounds suffer minimal velocity fall-off to target and provide a cost-effective means for defeat of spaced light armor and aerostructures.

In answer to an opposite concern, WHA sheet by virtue of its high areal density provides a very effective ballistic shielding material against high-velocity particles even when reasonably thin. WHA has an advantage over DU in that it is not pyrophoric, a consideration that is always present when using DU, whether in machining of parts or for the end use.

WHA is widely used for high stiffness boring bars, long extension toolholders, and chatter-free tooling. When compared to more standard iron-base toolholder materials, WHA offers a 70% gain in stiffness and WC-10Co a 295% gain, as seen in Table 7. While carbide provides superior stiffness, WHA is much easier to machine to form and offers better vibration damping. The elastic mismatch of the two phases, a problem in other contexts, proves advantageous for increased damping. Additionally, the higher density of the WHA provides better inertial damping than that available from the less dense carbide.

One of the most common applications for WHAs is logically balancing weights and inertial masses. The strength and hardness levels of WHAs permit secure mechanical fastening of components, a prime consideration for such applications as aileron balance weights in commercial aircraft. Sizes and configurations of current WHA balancing weights span the spectrum from fractional gram masses for balancing hard disk drive heads to large keel masses for ships weighing up to a metric ton or more. Race cars, commercial aircraft, and guided missiles all benefit from the combination of high density, good mechanical properties, and low toxicity provided by these materials. WHAs are used for gyroscope rotors and flywheels, as the density of these materials allows much greater mechanical energy storage than from materials such as steel. For applications in which WHA weights are exposed to weather, cleaning fluids, or other harsh environments, corrosion can be effectively prevented by epoxy paint or plating with nickel or chromium, all of which are supplanting cadmium plating as more environmentally acceptable alternatives.

Machinability

A prime advantage of WHAs over pure tungsten is their ability to be machined to complex geometries using common metalworking tools and techniques. WHAs in general machine in a manner similar somewhat to gray cast iron. WHAs tend to form short chips on machining. The exceptions are highly ductile, lower tungsten content alloys that may tend to form continuous chips and require more attention to chip breaking. The machining characteristics of such WHAs resemble those of superalloys. As all WHAs possess high elastic stiffness, cutting forces are higher than for most metalcutting operations, thus requiring carbide tooling, rigid support, and adequate spindle torque.

Fig. 7 WHAs can be easily machined into a variety of high density, high tolerance components such as radiation collimators and precision balance weights

Turning operations are best performed with ISO K10 type inserts. Sharp insert edges are vital for minimizing cutting force. Very rough cuts should be made with negative rake angle inserts though tooling forces will be high. Roughing passes typically utilize depths of cut of 0.7 to 3 mm, feed rates of 0.13 to 0.25 mm/rev, with surface speeds of 60 to 90 m/min. For turning and milling operations, the use of a coolant/lubricant is optional. Finishing cuts are typically made using 0.13 to 0.38 mm depths of cut and feed rates of 0.08 to 0.25 mm/rev. Large nose radius inserts, high spindle speed, and light feed rates contribute to the finest surface finish. Tightest tolerance finishing passes are made with positive rake inserts. WHAs can easily be machined dry. If a cutting fluid is to be used, it should be a water-based, nonalkaline formulation.

Multi-insert cutter heads should be employed for milling WHAs. ISO K10 inserts with a positive rake generally prove most useful. Roughing can be performed at surface speeds of 60 to 120 m/min and feed rates of 0.13 to 0.25 mm per insert. Finishing passes are often conducted at 90 to 150 m/min and reduced feeds of 0.08 to 0.25 mm per insert.

Grinding is best performed with vitrified-bond alumina or silicon carbide wheels of medium hardness. The ductile binder phase present in WHAs prevents the use of diamond wheels due to loading.

Tapping is the most challenging of machining operations for WHAs. Tapping of threads finer than 6-32 may prove impractical. The use of coated cobalt steel taps with two (at most three) heavy flutes and positive rake for increased clearance is highly recommended. Sulfonated oils are effective tapping lubricants. Approaches to minimizing tap/work contact area include the use of large pilot hole sizes and 60% thread area engagement should be considered. Holes 1/4-20 and finer may require dropping to 50% engagement due to the high stiffness of the alloy and the correspondingly high torsional stress induced in the flutes and shank. Holes should be tapped to completion without back threading. For larger holes, single-point threading may prove the best option. Designs should always utilize the coarsest thread possible.

Joining

Mechanical fastening techniques should be employed where possible so as to minimize the uncertainties associated with thermal bonding techniques, most of which relate to the formation

Table 7 Comparison of various toolholder materials

Material	Modulus, GPa	Damping	Density, g/cm^3
4*xxx* steel	210	Low	8
92WHA	360	Moderate	17.5
WC-10Co	620	Low	14.5

of embrittling intermetallics. Additionally, thermal bonding of worked WHA is not feasible due to property alteration that will result as a loss of strain hardening or at lower temperatures, overaging.

Due to differences in both chemistry and melting point of the tungsten phase and the binder, WHAs are not readily weldable. Brazing of WHA to dissimilar metals is possible using copper or commercial brazing alloys based on copper, nickel, or silver. In virtually every case, the mechanical properties of the joint are lower due to the formation of intermetallics. Diffusion bonding is also readily performed, but is subject to the same limitations as brazed joints.

When it becomes necessary to join sintered WHA bodies together to form larger components, this is readily accomplished by sinter bonding, in which the pieces are placed in contact and heated sufficiently to permit liquid-phase formation. If done properly, this technique yields an imperceptible joint with no local degradation in properties.

Environmental Considerations

As environmental and safety considerations become evermore important drivers in the design of new products and selection of appropriate materials, the fact that WHAs have only a slight percentage of their total weight in slightly toxic metals such as nickel and cobalt provides an incentive for their selection over other common high density materials such as lead and DU. Whereas the latter two materials pose definite toxicity concerns, less than 7% of the total weight of a typical WHA is comprised of metals such as nickel and cobalt.

Life cycle considerations are also favorable for WHAs as these materials are readily recyclable by both chemical and oxidation/reduction means. Both WHA machining scrap and spent components serve as raw materials for future alloy production instead of posing a disposal problem. WHAs are therefore an ideal materials choice for applications requiring a combination of high density, good machinability, moderate cost, minimal toxicity, high stiffness, and reasonable thermal stability.

SELECTED REFERENCES

- A. Bose, D. Sims, and R.M. German, Test Temperature and Strain Rate Effects on the Properties of a Tungsten Heavy Alloy, *Metall. Trans. A*, Vol 19, 1988, p 487–494
- S.G. Caldwell, Variation of Ni/Fe Ratio in W-Ni-Fe Heavy Alloys: A Current Perspective, *Tungsten & Tungsten Alloys—1992*, Metal Powder Industries Federation, 1993, p 89–96
- G.B. Dudder and W.E. Gurwell, Hydrogen Embrittlement Effects on Tensile Properties of Tungsten Heavy Alloys, *Tungsten and Tungsten Alloys—Recent Advances*, A. Crowson and E.S. Chen, Ed., The Minerals, Metals, and Materials Society, 1991, p 161–167
- H. Hofmann, "Konstitution des W-Ni-Fe Systems und Mechanische Eigenschaften W-Reicher W-Ni-Fe-Legierungen," Ph.D. dissertation, Technischen Universitat Berlin, 1983
- R.E. Oakes, Jr. and M.W. Moyer, "A Review of Techniques Used to Characterize Cemented Tungsten Alloys for Penetrator Applications," Report Y-2083, Oak Ridge National Laboratory Y-12 Facility, 1977
- B.H. Rabin and R.M. German, Microstructure Effects on Tensile Properties of Tungsten-Nickel-Iron Composites, *Metall. Trans. A*, Vol 19, 1988, p 1523–1532
- J. R. Spencer and J.A. Mullendore, "Relationship between Composition, Structure, Properties, Thermo-Mechanical Processing, and Ballistic Performance of Tungsten Heavy Alloys," Report MTL TR 91-44, U.S. Army Materials Technology Laboratory, Watertown, 1991
- S.W.H. Yih and C.T. Wang, *Tungsten*, Plenum Press, 1979

Cermets and Cemented Carbides

METAL AND CERAMIC composites include two types of materials known as cermets and cemented carbides. The most outstanding example of the desirable properties obtained from combining metal and ceramic materials involves the hard-metal types made from cemented carbides. Cemented carbides have enjoyed a steady expansion over the past six decades. Over that time, the development of metal-ceramic tool materials has moved away from the early tungsten-base carbides to carbide- and nitride-base compositions of increasing complexity (Table 1). This includes several types of cermet materials.

Cermets

Cermet is an acronym that is used world wide to designate "a heterogeneous combination of metal(s) or alloy(s) with one or more ceramic phases in which the latter constitutes approximately 15 to 85% by volume and in which there is relatively little solubility between metallic and ceramic phases at the preparation temperature" (Ref 1, 2). A good definition of the term ceramic can be found in the *Ceramic Glossary* (Ref 3): "Any of a class of inorganic, nonmetallic products which are subject to a high temperature during manufacture or use. Typically, but not exclusively, a ceramic is a metallic oxide, boride, carbide, or a mixture or compound of such materials; that is, they include anions that play important roles in atomic structures and properties." With particular reference to cermets, this definition of the ceramic component could be broadened to include nitrides, carbonitrides, and silicides.

Cermets originally were used for cutting tool applications. Some 45 years ago (Ref 2, 4), they began to be considered for use in more taxing applications, such as propulsion systems. The expectations were that the refractory behavior, strength, and corrosion resistance of the ceramic phase could be mated advantageously on a proportional basis with the high ductility and thermal conductivity of the metallic phase, and that some superior new materials would become available for a multitude of high-temperature applications.

A significant application of cermets involves cutting tool materials that utilize titanium carbide or titanium carbonitrides as the hard refractory phase. Frequently, molybdenum carbide (Mo_2C) and other carbides are also built into these cermet formulations. The cratering and flank wear resistance properties of the titanium carbide and titanium carbonitride cermet tool materials are better than those of the conventional cemented-carbide (that is, cobalt-bond tungsten carbide) tool. In comparison to ceramic cutting tools, these cermets permit heavier cuts, which, at high speed, results in greater amount of metal removal at a comparable level of tool life. Cermets clearly possess characteristics of a cutting tool material that is capable of filling the gap between conventional cemented carbides and ceramics (see *Machining*, Volume 16 of the *ASM Handbook*).

Table 1 History of cermet product development and marketing

Year	Composition	Trademark	Manufacturer
1930–1931	WC-Co	G1	Krupp-Widia
1930	TiC-Mo_2C-(Ni, Mo, Cr)	Titanit S	Metallwerk Plansee
1930	TaC-Ni	Ramet	Fansteel Corporation
1933	TiC-TaC-Ni	...	Siemens AG
1938–1945	TiC-VC-(Fe, Ni, Co)	...	Metallwerk Plansee
1949–1955	TiC-(NbC)-(Ni, Co, Cr, Mo, Al)	WZ	Metallwerk Plansee
	TiC-(Nb, Ta, Ti)C-(Ni, Mo, Co)	Kentanium	Kennametal
1952–1954	TiC-(steel, Mo)	Ferro-TiC	Sintercast (Chromalloy)
1960	TiC-(Ni, Mo)	...	Ford Motor Company
1970	Ti(C, N)-(Ni, Mo)	Experimental alloys	Technical University Vienna
1974	(Ti, Mo) (C, N)-(Ni, Mo)	Spinodal Alloy	Teledyne Firth Sterling
1975	TiC-TiN-WC-Mo_2C-VC-(Ni, Co)	KC-3	Kyocera
1977–1980	TiC-Mo_2C-(Ni, Mo, Al)	...	Ford Motor Company, Mitsubishi
1980–1983	(Ti, Mo, W) (C, N)-(Ni, Mo, Al)	...	Mitsubishi
1988	(Ti, Ta, Nb, V, Mo, W) (C, N)-(Ni, Co)-Ti_2AlN	TTI, TTI 15	Krupp-Widia

Source: Ref 4 and Kennametal, Inc.

Classification of Cermets

Cermets can be classified according to their hard refractory component. In this system, the principal categories of cermets are determined by the presence of six components: carbides, carbonitrides, nitrides, oxides, borides, and miscellaneous carbonaceous substances.

The metallic binder phase can consist of a variety of elements, alone or in combination, such as nickel, cobalt, iron, chromium, molybdenum, and tungsten; it can also contain other metals, such as stainless steel, superalloys, titanium, zirconium, or some of the lower-melting copper or aluminum alloys. The volume fraction of the binder phase depends entirely upon the intended properties and end use of the material. It can range anywhere from 15 to 85%, but for cutting tool applications it is generally kept at the lower half of the scale (for example, 10 to 15 wt%) (Ref 5).

The metallic bond for each cermet is selected in order to produce the desired structure and properties for the specific application. The iron group metals and their alloys dominate as in the cemented tungsten carbide class of hard metals; nickel and, to a lesser extent, cobalt and iron possess a desirable combination of relatively high hardness and good ductility. However, the binder for a cermet can also be chosen from the group of more reactive metals, such as titanium or zirconium, or it can be selected from series of refractory metals that includes chromium, niobium, molybdenum, and tungsten. Lower-melting metals and alloys, primarily those based on copper and aluminum, round out the list of binders at the bottom of the temperature scale. Aluminum, however, is more commonly associated with metal-matrix composites.

Carbide-base cermets are by far the largest category of cermets, even if the term is used in its narrower sense and excludes the broad field of cemented-carbide cutting tools and wear parts based on tungsten carbide (WC). Since the inception of cermet technology, the dominant concept has been that of a material based on TiC as the primary hard and refractory constituent, with the bonding provided by any of a variety of lower-melting ductile metals or alloys (much the same as those used for cemented tungsten carbides). The TiC cermets have found use in tool and wear resistance applications; in selected high-stress, high-temperature systems; and in corrosive environments. Cermets based on SiC and B_4C, which generally are classified as metal-matrix composites, have gained considerable industrial significance in wear and corrosion resistance, or antifriction, applications; they are also used in nuclear reactor applications. Cermets with a chromium carbide (Cr_3C_2) base have been used for a variety of corrosion resistance applications and as gage blocks; however, they have apparently lost much of their industrial usage.

Carbonitride-base cermets can be produced with or without additions of various other carbides (of which Mo_2C is the most important); they are bonded with the common cemented-carbide binders. At present, these materials are the primary cermets for tool applications. Their enhanced strength, which makes them suitable for high-speed cutting tools, is based on a greatly improved bond between the hard carbide grains and the binder metal. The improved bond is a consequence of a miscibility gap in the quaternary TiC, TiN, MoC, and MoN system that results in a so-called spinodal decomposition into two isostructural phases (Ref 6) with inherently better wettability to the binder (Ref 5).

Nitride-base cermets constitute a special class of tool materials. Titanium nitride (TiN) and especially cubic boron nitride (CBN) produce excellent cutting materials if they are combined with a hard binder metal. Titanium nitride and zirconium nitride (ZrN) bonded with their respective metallic elements have been developed for special heat- and corrosion-resistant purposes.

Oxide-base cermets constitute a category that includes UO_2 or thorium dioxide (ThO_2), which are used for a major fissions component in nuclear reactor fuel elements; Al_2O_3 or other highly refractory oxides, used for components in liquid-metal manipulation (for example, pouring spouts) and general furnace parts; and SiO_2, used for a minor constituent in friction elements. Combinations of Al_2O_3 with TiC are suitable for hot-machining tools.

Boride-base cermets have a boride of one of the transition metals as the dominant phase.

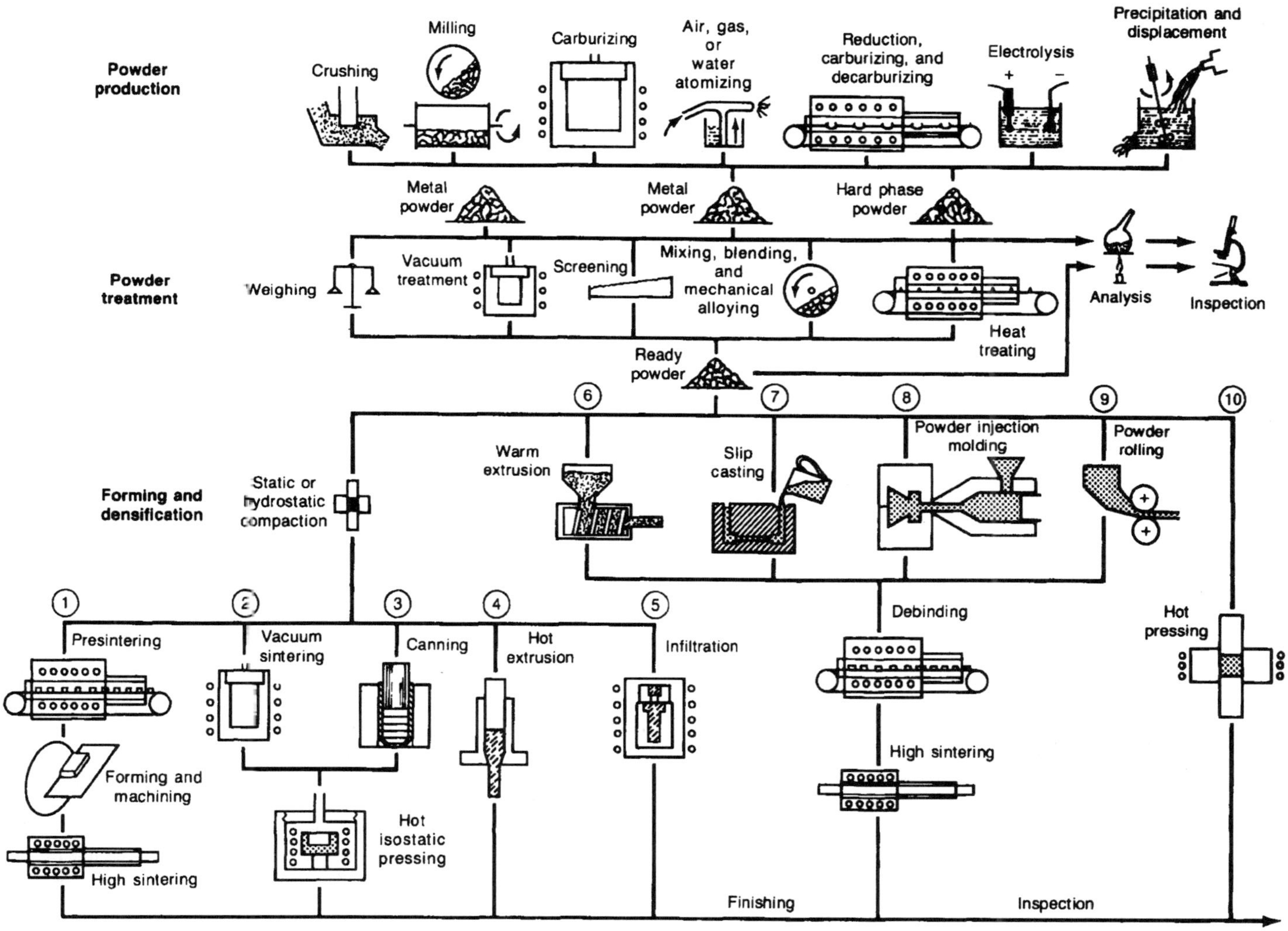

Production method	Products	Production method	Products
1. Presintering	Cemented-carbide parts and cermets	6. Warm extrusion	Cemented-carbide rods or other slender cermet parts
2. Vacuum sintering	Steel-bonded carbides (standard pieces) and cermets	7. Slip casting	Cermets with high proportions of hard phase
3 Canning	Steel-bonded carbides (special pieces)	8. Powder injection molding	A wide variety of cermet compositions
4. Hot extrusion	Aluminum cermets with moderate amounts of hard-phase additions	9. Powder rolling	Aluminum, copper, and other nonferrous metals with moderate additions of hard-phase components
5. Infiltration	TiC parts with nickel- or cobalt-base infiltrants and other cermets with about 55–85 vol% hard phase	10. Hot pressing	A wide variety of cermet compositions

Fig. 1 Powder metallurgy production methods for cermet and cemented-carbide products

These cermets provide excellent high-temperature corrosion resistance to attack by active metals, such as aluminum, in the molten or vapor state. A combination of ZrB_2 and SiC is resistant to erosion from the propulsion gases of chemical rockets.

Carbon-containing cermets are materials that contain graphite in varying proportions. They are used for electrical brushes and contacts or as minor constituents to provide some lubrication in friction elements. Also included in this category are diamond particles within metal matrices that are used in special tools.

Cermet Fabrication

The methods used for powder preparation, forming, firing or sintering, and post-treatments of cermets generally are similar to conventional ceramic and powder metallurgy (P/M) processing techniques. Figure 1 is a flow chart of the various P/M techniques applicable to cermets. Table 2 summarizes the relative characteristics of the major forming methods that are practiced in producing carbide-base and most other types of cermets. The principal processes are cold forming and sintering, pressure sintering, and infiltration.

The cold-pressing process includes static uniaxial and isostatic multiaxial compaction. The powder mixtures are compacted at pressures of 35 to 100 MPa (5 to 14.5 ksi). The predominant method involves pressing dry wax-lubricated powder in hardened steel dies with double-action opposing punches. For long rods or tubes of uniform cross section, these dies are used for the extrusion of a paste in which the powder particles are embedded in suitable organic binder or wax. To form complex or large shapes, the dry powder is placed in a pliable mold and compacted from all sides by hydrostatic pressure inside a sealed, reinforced steel cylinder.

Powder Preparation. The first step in the cermet production process is the mixing and milling of the ingredient powders. The mixtures, consisting of the hard-phase substance in powder form and the pure metal or metals in the proper proportions required for the composition of the binder alloy, are milled in ball mills. The balls are made of tungsten carbide or, more frequently, of a highly sintered cermet. The mill can be lined with the same type of cermet material to reduce the possibility of mixture contamination. In addition to conventional ball mills, high-energy vibratory ball mills and attrition mills are used. With the later type of mills, substantial savings in milling time, energy, and floor space can be achieved. During the milling process, the hard phase particles are comminuted and thoroughly coated with binder metal. Organic liquids such as hexane are used in the process to minimize the rise in temperature and prevent oxidation. After the powders have been milled to a particle size of 325 mesh or finer, the mixtures are dried before further processing and use. A lubricant is then added, and those powder mixtures destined for compaction in automatic presses are agglomerated so that they can flow freely from the hopper to the compacting die.

Static Cold Pressing. Cold-pressing methods for cermet powder mixtures generally follow the well-known powder-compacting techniques used in conventional powder metallurgy. Small cermet parts needed in reasonably large quantities are compacted in special hard-metal dies by automatic presses with double-action opposing punches (Fig. 2). Whether solid or segmented, the dies are shrunk into a strong, tough, heat-treated steel retainer.

The automatic compaction cycle consists of filling powder in the die, compacting the powder, ejecting the compact, and removing it. Two methods are used for ejecting the compact from the die. In the first method, the lower punch moves upward and pushes the bottom of the pressed piece to the level of the die table. In the second method, the die is forced down (withdrawn) over the lower punch until the bottom of the compact is on level with the top of the die (Fig. 3). The second method is gaining favor with many specialists because it allows building shorter, less expensive tools, and because it provides better support for the fragile compact during ejection.

Another interesting variation, particularly for the compaction of fragile cermets, is provided by the so-called anvil press (Fig. 2b). In this method, the powder is pressed against the anvil by the upward motion of a single lower punch. The anvil is then laterally removed to allow ejection of the compact by the continued upward motion of the punch. Because this cold-pressing

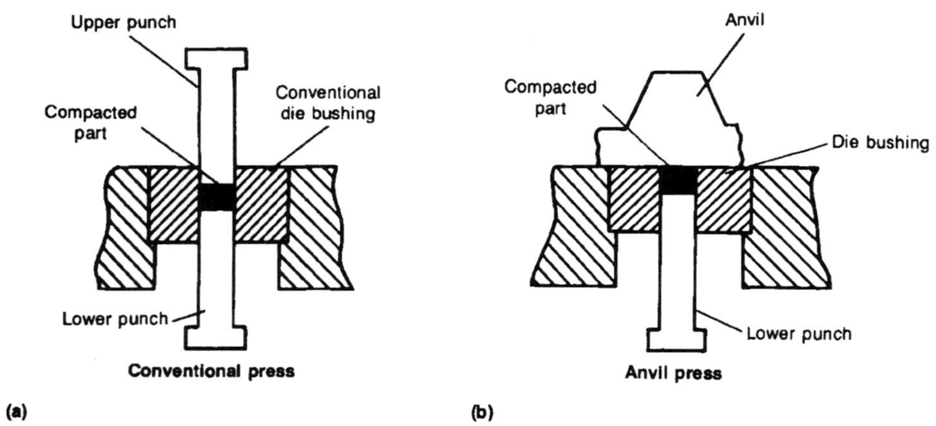

Fig. 2 Static cold pressing with (a) a conventional press and (b) an anvil press. The anvil press has no upper punches; therefore, the misalignment, breakage, and wear problems associated with those punches are eliminated. Courtesy of PTX-Pentronix, Inc.

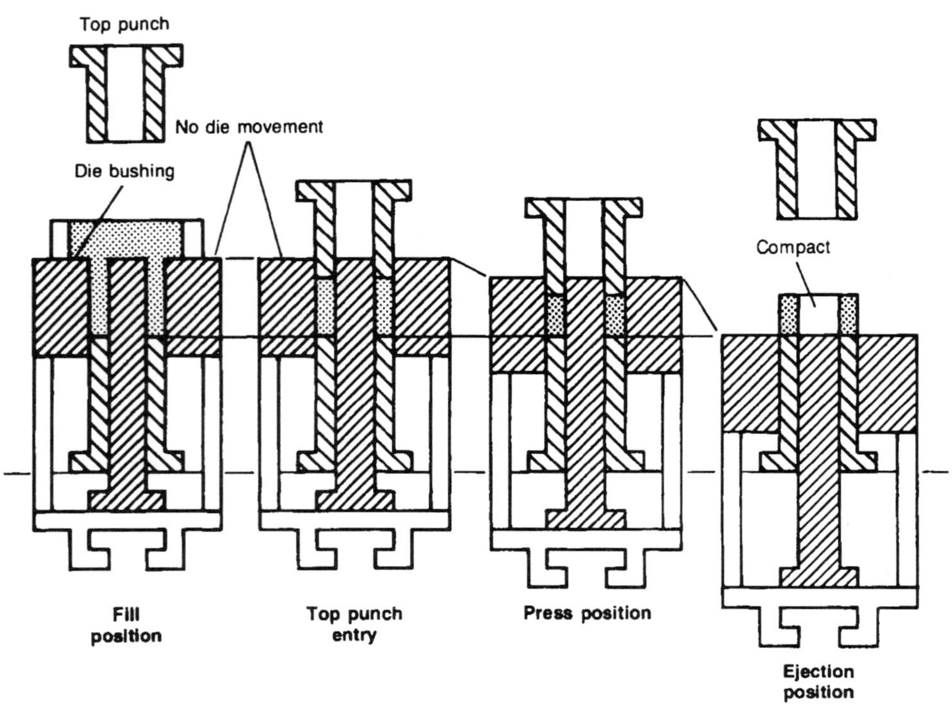

Fig. 3 Withdrawal press cycle with controlled die motion (top and bottom pressure). Courtesy of Dorst America

technique is a single-action pressing method, an anvil press is acceptable only for relatively thin single-level pieces. Yet, where applicable, the anvil press saves tool costs; in addition, it reduces the ejection path to a minimum, which is an important factor for fragile cermets.

Medium-to-large rectangular pieces are compacted on hydraulic presses in multiple-section dies held together by powerful steel frames. These frames are capable of counteracting the large internal forces that are exerted on the powder mass and transmitted radially to the die sections. Often, at the completion of the cycle, the dies are opened for the careful removal of the fragile cermet compact.

For round pieces it is preferable to use double-action hydraulic presses with a built-in ejection action. These presses can use either one of the two ejection methods:

- If the die is mounted on a hydraulically activated floating platen, a single-action press will adequately provide the effect of a double-action press; in this case, ejection is accomplished by the withdrawal method.
- When a full-power double-action hydraulic press is available (obviously a heavier and more expensive machine than the single-action press), the die can be mounted on the stationary platen. The compact is ejected by raising the lower punch to the level of the die table at the completion of the pressing cycle.

Cold compaction of simple cermet shapes gives generally good results when using adequately lubricated powder mixtures, well-designed tooling, and sturdy presses. Even though the compacts are fragile, they should nevertheless be firm, have adequate green strength and well-formed edges, and be free of laminations or other internal defects. Depending on the composition of the cermet, compaction problems can arise that require modifications in the equipment or the process. Generally, the higher the proportion of the hard phase, the greater the difficulties that arise in compaction; these difficulties may not be evident until after sintering. Also, more problems can be expected with iron-, nickel-, or cobalt-base cermets than with the softer, more malleable aluminum-base compositions. Larger pieces, as measured both in diameter and height, intensify the problems associated with composition. Air entrapment, bridging, laminations, intermittent voids, and variable density throughout the compact are just a few of the problems encountered in the cold compaction of cermets (Ref 7).

Some of these problems can be overcome by adding more lubricant to the mixture, by increasing the taper of the compacting die, or by slowing down the compaction process. Die wall lubrication between pressings often helps to eliminate compaction or ejection problems. Adequate preloading of the die and die lapping in the pressing direction are other precautions that help to overcome problems.

Even for the simplest forms, such as cylindrical or rectangular single-level blanks, not all problems encountered can be solved by the conventional static cold-pressing technique. This is one reason why P/M specialists depend heavily on other, more sophisticated forming processes for the fabrication of a wide variety of products.

Cold Hydrostatic Pressing. High-quality cermet compacts require uniform densification throughout. This can be ideally accomplished by cold hydrostatic pressing. In this method, pressure is applied simultaneously and uniformly from all sides toward the center of gravity of the powder mass while all friction of powders against the die wall is completely eliminated. In order to compact simple or even relatively complex shapes by this method, dry powders are filled into a pliable mold. The powders are settled, and air is removed on a vibrating table; the mold is then sealed and placed into a reinforced steel cylindrical vessel filled with a fluid. After the vessel is closed, hydraulic pressure is built up, thereby compressing the powder contained in the mold. The two hydrostatic pressing methods frequently used for pressing cermets are the wet-bag method and the dry-bag method.

The wet-bag method involves placing one or more powder-filled molds inside a hydrostatic pressing vessel (Fig. 4). The powder-filled mold is pliable and is placed within a perforated container for support. Inside the vessel, the powder-containing mold is completely surrounded by hydraulic fluid. Depending on the size of the vessel and the individual mold, often a number of molds can be placed in the vessels and compacted simultaneously. The entire process—loading the vessel with one or several molds, building up and holding the pressure, releasing the pressure, and reopening and unloading the vessel—is relatively slow. Moreover, filling the mold, assembling the mold, loading the molds into the pressure vessel, and unloading and removing the pressed piece after completion are slow, manual processes that require meticulous attention to detail.

Dry-bag pressing uses a flexible mold that is permanently sealed in the pressure vessel (Fig. 5). After the mold cavity has been filled with a controlled quantity of powder, the cover plate is closed, and hydraulic pressure is applied. After the pressure is released, the molded piece is removed, and a new cycle commences. Dry-bag pressing is a much faster production process than the wet-bag method and lends itself to automation. Much of this technology has been developed for producing near-net shape ceramic pieces, for example, automotive spark plug bodies. Each dry-mold setup requires special engineering and development (Ref 8).

Advantages and Disadvantages. Hydrostatic pressing offers the following advantages (Ref 7):

- Pressed cermet pieces have a uniform density regardless of size and shape.
- Wet-bag method is well suited for large pieces and often is the only practical method for pressing such pieces.
- Slender pieces with high ratios of length to cross section are feasible.
- Mold cost is low compared to that of rigid compacting dies. Low production quantities can thus be economically produced, especially by the wet-bag method.

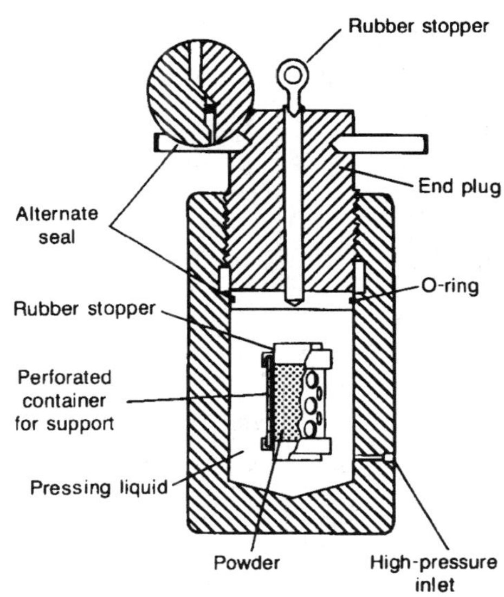

Fig. 4 Schematic of a cold hydrostatic vessel with a wet-bag powder mold. Source: Ref 7

Table 2 Cermet forming techniques

Technique	Size capability	Shape capability	Mold or die requirements	Production	
				Rate	Labor
Static cold pressing	Limited by press capacity	Prismatic shapes without undercuts	Hardened steel or carbide dies	High	Low to moderate
Hydrostatic cold pressing	Limited by capacity of pressure vessel	Simple or complex shapes	Rubber molds	Low	High
Powder rolling	Limited width, long length	Flat and thin	...	High	Low
Warm extrusion	Limited by equipment size	Long pieces with uniform cross section	Hardened steel or carbide dies	High	Low
Powder injection molding	Small pieces	Complex shapes with undercuts	Hardened steel or carbide dies	High	Low
Static hot pressing	Limited by press capacity	Prismatic shapes without undercuts	Graphite or ceramic molds	Low	High
Hot isostatic pressing	Limited by capacity of pressure vessel	Simple or complex shapes	...	Medium	Low to high
Hot extrusion	Depending on press capacity	Long pieces with uniform cross sections	Alloy steel dies	High	Low
Infiltration	Depending on equipment	Intricate shapes feasible	Graphite and ceramic molds	Low	High

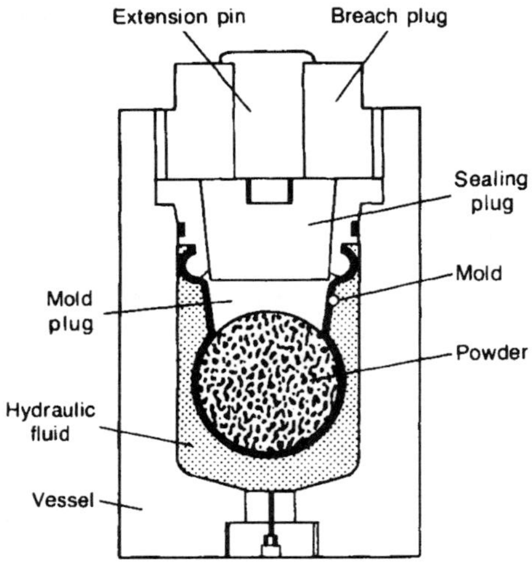

Fig. 5 Schematic of dry-bag hydrostatic pressing equipment. Courtesy of Olin Energy Systems

- Undercuts and varying cross sections are feasible with either the dry-bag or wet-bag method.
- Little or no lubricant is required.
- The process is well suited for research and development work.

The disadvantages of hydrostatic pressing are:

- Dimensional control of compacts if limited. Mold design must accommodate the radial and axial shrinkage caused by hydrostatic pressing as well as the shrinkage that occurs during subsequent sintering.
- Surfaces of compacts are less smooth than those of die-pressed pieces.
- A high liquid-phase sintering step or encapsulation is necessary before hydrostatically pressed cermet pieces can be densified by hot isostatic pressing.
- Equipment cost is high, and equipment utilization can be low.
- Labor cost is relatively high.

For difficult-to-press cermet compositions with high loading of the hard phase and/or relatively hard metal and alloy binders, cold hydrostatic pressing often is a convenient production method; sometimes it is the only reliable method for working with certain compositions.

Warm Extrusion of Cerment Powder Mixtures. The process of warm extruding cemented ultrafine carbide powder with an admixture of plasticizers has been known for many years. It is successfully used for cermets as well as for forming simple prismatic shapes that have a high ratio of length to cross sections. Cylindrical and triangular shapes and other cross sections can be readily extruded; even tubes are feasible (Ref 9).

Depending on the plasticizer used (for example, polystyrene with an admixture of diphenyl and diphenyl-ether), extrusion requires a temperature somewhere between 160 and 175 °C (320 and 350 °F). Slow and complete debinding under vacuum prior to high sintering is critical in order to avoid distortion, cracking, or microporosity. Screw extruders similar to those used in the plastics industry are adapted to this process (Fig. 6). For the production of a high-quality product, hot isostatic pressing is recommended.

Power rolling (roll compacting) is a well-known forming process in conventional powder metallurgy that may find application in cermet production. In this process, cermet powder mixtures are fed from a hopper into the gap of a rolling mill and emerge as a continuous strip or sheet. While the horizontal arrangement of the rolls is most convenient for feeding the powder from the hopper into the roll gap (Fig. 7), the vertical arrangement is preferable for feeding the emerging fragile strip horizontally into a series of subsequent operations. The vertical arrangement requires more carefully engineered devices for feeding the powder uniformly into the roll gap (Ref 7).

In contrast to the starting material in slab rolling, the loose powder used in powder rolling has no strength before entering the roll gap and must flow freely or be forced into the gap. During the roll compacting process, the density and physical properties of the powder mixture change. For cermet powder compositions, the feasibility of roll forming a strip of sufficient density and strength depends upon a number of factors, including, but not limited to, the roll diameter and speed, the degree of loading of the cermet mixture with hard-phase substances, the ductility of the metallic phase, and the amount of plasticizer added to the mixture. The presence of the hard component adds to the friction of the powders against the roll and to the internal friction of the powder mixture during the compacting step. This is a favorable characteristic of cermet powders for roll forming; however, it is offset to some extent by the inherently low green strength of the resultant sheet or strip.

The sheet thickness that can be compacted with a given diameter roll is quite limited. A ratio of roll diameter to strip thickness between 600 to 1 and 100 to 1 seems to be the range for various metal powders (Ref 7). It is reasonable to assume that the middle-to-lower range applies to cermets. Special devices are required to prevent the powders from flowing laterally out of the roll gap. A uniform flow of powders over the entire width of the roll is essential for obtaining uniform density in the roll-formed strip. Edge cracking can occur, particularly with heavier strips. An optimum strip thickness has to be established experimentally. Thicker strips are too stiff to be coiled, and thinner ones are too fragile.

Fig. 6 Warm extrusion of cermet powder mixtures. Courtesy of Dorst America

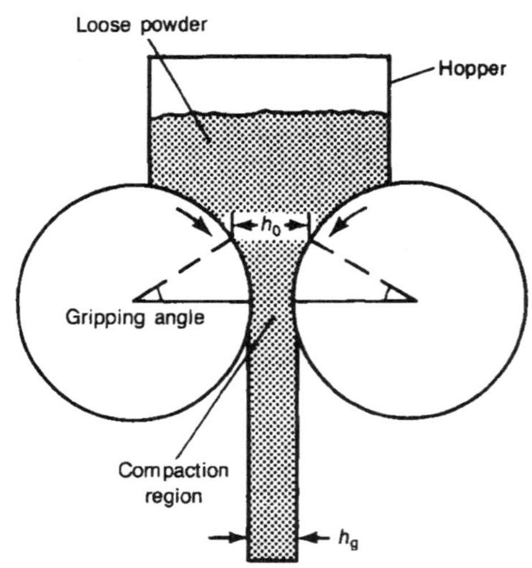

Fig. 7 Schematic of powder rolling with saturated feed and horizontal roll arrangement. Compression ratio, h_0/h_g. Source: Ref 7

Rolling speed is another variable that can only be optimized through experimentation with a given cermet powder composition. Pure metal powders without hard phase have been roll compacted at speeds of 30 m/h (100 ft/h). It remains to be seen whether an output anywhere near this order of magnitude can be obtained with cermets.

A complete powder-rolling line for continuous operation includes debinding and sintering furnaces, rerolling stands, and, if necessary, one or more reannealing furnaces. Up-coiling equipment is needed at the end of the line. This equipment constitutes a major capital investment that is warranted only by a large and continuous demand for the product. Although labor costs for such an operation are low, it may be some time before this line production method finds applications in cermets. A simpler arrangement (Fig. 8) is feasible if, after a debinding step (not shown) and continuous atmosphere sintering, a product emerges of sufficient strength and ductility to permit upcoiling. Roll-compacting arrangements have been proposed for producing a sandwich-type strip consisting of two layers of different compositions.

Slip casting, a method for forming metal powders into a desired shape, follows a technique that has been used for ceramics for a long time. This method uses an aqueous suspension of cermet powders (the slip) that is poured into a porous plastic mold. The liquid is absorbed by the mold, and the powder is deposited on the mold wall. In the case of hollow shapes, the excess slip is drained off after the deposit reaches the required wall thickness; for solid parts the slip must remain and slowly dry.

The water-base slip has low viscosity in order to facilitate pouring, yet it should be stable during standing in order to avoid demixing. Demixing can be a serious problem with cermet powders, particularly those having a substantial difference of specific weight between the hard phase and the binder metal; it can lead to differences in composition and properties from one end of a cermet part to another. Variations in composition can lead to cracks during drying or subsequent sintering. In order to control the viscosity of the slip at the optimum level, it is generally necessary to use a deflocculant and to control the pH. After slow drying, the slip cast part needs a debinding step followed by high sintering. The resultant part has a higher density than the tap density of its original powder mixture. The fine powders that frequently are used to facilitate slip casting can lead to superior properties in the sintered part (Ref 7).

Slip casting and mold making together are more art than technology. They require knowledge of parameters such as slip viscosity and suspension stability, wetting agents and deflocculants, and slip-mold interaction and mold release. Other important parameters are wall-building rate and casting crack formation. Slip casting requires only a small investment; however, it is labor intensive and is not well suited for mass production. At the present state of technology, cermet parts of a certain complexity are more likely to be suitable for injection metal molding than for slip casting. The former process is more capital intensive, but it is better suited for a medium-to-large production volume.

The P/M injection molding (MIM) process has evoked a great deal of interest since it was first developed in the early 1970s. Commercialization has been slow, mostly because of the long cycle that is required from concept to the point of shipping acceptable parts to a customer. Intensive research and application engineering continues in many laboratories, and more rapid growth is expected in the future (Ref 10, 11).

On a laboratory basis, cermet parts have been made by this process, and its commercialization is underway, particularly in the field of cemented carbides. However, the bulk of current MIM experience is in the area of structural ferrous and nonferrous parts.

The powder injection molding process for cermets (Fig. 9) involves mixing and blending the ingredient metal and hard-phase powders with a suitable polymer binder and then granulating the mixture. The granulated product is heated and injection molded under pressure. The polymer imparts viscous flow characteristics to the mixture to aid in forming, mold filling, and uniform packing. After demolding, the binder is removed, and the remaining cermet structure is densified by sintering and, perhaps, by hot isostatic pressing (Ref 10).

Binder compositions and debinding techniques are the main differences among the various MIM processes. There is no universal binder. A primary requirement of the binder is that it allow flow and packing into the mold cavity. It must wet the powder, and it should be designed to minimize debinding time and defects. A multiple-component binder that is not chemically intersoluble allows for progressive extraction in debinding. As one compound is removed and the pores partially opened, the remaining binder holds the particles in place and maintains the shape of the compact. The remainder then vaporizes through the open pores without generating an internal vapor pressure that might cause compact failure. Waxes with additives are most frequently used as binders. The phases of the molding operation are:

- Clamping and filling of the mold

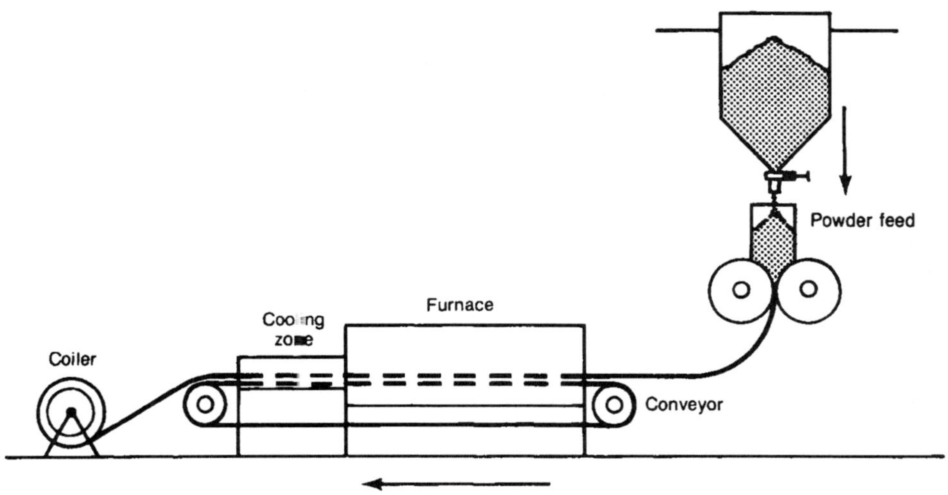

Fig. 8 Powder rolling process with strip reeled into individual rolls after first sintering treatment. Source: Ref 7

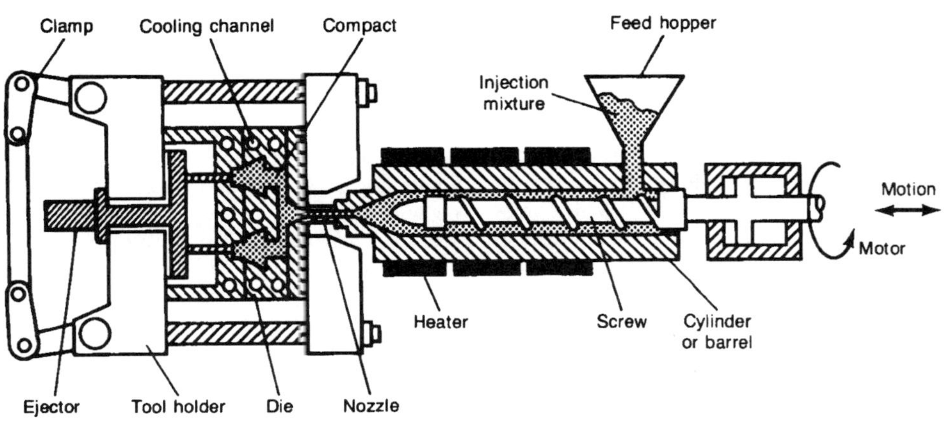

Fig. 9 Mold and injection mechanism for the MIM process. Source: Ref 10

- Maintaining pressure while the compact becomes solid
- Retraction of filling mechanism
- Opening of the mold and ejection of the compact

Mold filling depends on the viscous flow of the feedstock into the mold cavity. The viscosity depends on temperature, shear rate, binder chemistry, powder interfacial chemistry, and loading (Ref 10).

Thermal debinding is the most frequently used technique, but capillary wicking and solvent extraction can also be considered as an alternative method. Complete debinding is required before commencing the sintering cycle. Most cermets require a liquid-phase sintering cycle to achieve complete densification of the compact. A modern furnace that combines debinding, high-vacuum sintering, and a final pressure-sintering cycle can accomplish all of these steps economically.

Applications and Advantages of the MIM Process for Cermets. In recent years, major progress has been made in using the MIM process for the production of heat engine components, military hardware, computers, and aerospace and automotive components. The powder injection molding process offers new opportunities in advanced materials manufacturing (Ref 11), and it offers potential advantages for use in cermet manufacturing technology. General aspects of applying the MIM process in cermet manufacturing include:

- In principle, the MIM process is applicable to cermets without modification of the injection molding machines or the typical mold designs (Fig. 9).
- The production of small- to medium-size complex shapes by the MIM process is feasible, provided that the geometry of the shapes allows for demolding. When this requirement is met, multiple levels, reentrant angles, and undercuts can be accommodated.
- On small parts, tolerances of ±3 μm/mm (0.003 in./in.) after sintering can be obtained on conventional P/M parts with the MIM process. Larger tolerances would be needed on cermets to allow for shrinkage when liquid-phase sintering is needed for complete densification.
- Small runs (of as few as 2000 parts) are feasible for conventional P/M parts. Because of the higher price level of cermets and the relatively high cost of competitive forming techniques, runs of similar or even smaller size could be economically attractive for cermet parts produced by the MIM process.
- With proper debinding techniques and a liquid-phase sintering cycle (perhaps followed by hot isostatic pressing), high-quality parts with good physical properties could be produced by the MIM process.
- Excessive mold wear caused by the hard phase during the injection molding of a cermet composition, particularly one with high loading, does not seem to be a serious problem in the MIM process. Future experience will demonstrate if such a mold wear problem exists and to what extent it affects the economics of using the MIM process for cermets.

Cermet Sintering and Consolidation

Sintering. Not all cermets require liquid-phase sintering, but the majority use this process to convert the green compacts into solid, strong, and dense products. Sintering temperatures depend entirely on the ceramic-metal system involved and on the choice between solid- and liquid-phase sintering. Typical temperatures range from 850 to 1050 °C (1560 to 1920 °F) for products that contain a bronze, silver, or copper metal matrix; 1300 to 1500 °C (2370 to 2730 °F) for cemented carbides and borides; and 1700 to 2200 °C (3100 to 4000 °F), or even higher, for certain ceramic oxide-base cermets.

For applications requiring fine machining and grinding, as in many cemented-carbide parts and tools, presintering is performed at 1000 to 1100 °C (1830 to 2010 °F) to bond the metallic contact points and give enough green strength to the body so that it can withstand rough machining. Allowance is made for the substantial shrinkage that occurs during subsequent sintering.

Depending on the green density, cermet compacts can shrink during liquid-phase sintering by as much as 18 to 26% linear (45 to 60% by volume). In systems with good sinterability, virtually all porosity is eliminated (Ref 12).

During all sintering processes, particularly during the liquid-phase process, many complicated metallurgical phenomena take place that depend on temperature, furnace atmosphere (hydrogen, inert gas, or vacuum), and the dynamics of the particular ceramic-metal system. For example, metals change into alloys; the hard phase partially dissolves in the liquid phase and changes the composition of the latter phase; portions of the liquid phase can diffuse into the hard phase; and reprecipitation of some elements dissolved in the liquid phase can take place during the cooling portion of the cycle. Also, if carbon is present in the furnace atmosphere (perhaps from the furnace furniture), it will reach with oxygen or other elements. The phenomena that occur during the sintering of WC-cobalt systems have been investigated very thoroughly over a long period of time. Ample literature on the basic system and many of its alloy variations is available (Ref 12).

Mechanism of Liquid-Phase Sintering. While not strictly a cermet, the liquid-phase sintering of heavy alloys consisting of tungsten-nickel-copper has interesting ramifications that are applicable to the cermet field (Ref 7). This liquid-phase sintering process includes these principal features:

- The hard phase is partially soluble in the liquid phase during sintering. At temperature, the liquid phase is limited so that the compacts keep their shape.
- The sintering temperature must be high enough so that an appreciable amount of liquid phase is present. When these conditions are met, densification takes place.
- The finer the particle size of the hard phase is, the more rapid and complete the densification of the compact is.
- Final density is independent of the compacting pressure. To reach theoretical density, compacts pressed at low pressure will shrink correspondingly more than those pressed at high pressure.
- The microstructure will show grain growth when compared to the particle size of the original hard-phase powder. This grain growth can be appreciable and is dependent on the sintering time and temperature.

When the original hard-phase particles are angular (for example, titanium carbide), they can become rounded during the sintering process. However, this is not always the case because some angular hard substances (for example, tungsten carbide) seem to possess shape memory. During the reprecipitation of dissolved elements from the liquid phase during cooling, angular contours reappear on the hard-phase particles of these substances.

Furnaces. Continuous high-temperature sintering furnaces have been used in the cemented-carbide and refractory metal industries for many years. They are equipped with a hydrogen or protective atmosphere with a low dew point to reduce residual oxygen in the compacts and to prevent oxidation. Continuous pusher-type furnaces equipped with silicon carbide or molybdenum heating elements have been particularly successful and are used for sintering a large volume of small parts. Batch-type vacuum furnaces have become very popular in the last 30 to 40 years. When there is a choice between pusher-type continuous furnaces and batch-type furnaces, the equipment and operating costs favor the former.

In a typical operation, the parts are laid out without packing on graphite plates that are stacked with spacers within the furnace. When direct contact between the graphite and the compacts is undesirable, the plates are lined with an inert ceramic. It is essential to use vacuum equipment when highly reactive powder mixtures are sintered. The optimum level of vacuum to be reached during liquid-phase sintering varies greatly depending on the hard-phase binder system being treated. Several advanced furnace designs provide for initial operation using a hydrogen atmosphere, with a switch to vacuum at a later stage in the sintering cycle. Others provide a pulsating cycle of hydrogen pressure alternating with vacuum.

Static Hot Pressing. Hot pressing is a cermet production method in which the pressure and temperature are applied simultaneously. The powder mixtures are either compacted directly in the hot press mold or prepressed cold in dies and then transferred to the hot press tools of the pressure-sintering furnace. Pressures are considerably lower than for the cold press method. They can range from deadweight loads up to 3 MPa

(500 psi) for pressure sintering (of friction elements, for example), or from 10 to 35 MPa (1500 to 5000 psi) for hot pressing; the lower end of the hot-pressing range applies to liquid-phase systems.

Sintering temperatures are reached by induction or resistance heating of the mold, or by direct induction or resistance heating of the powder compact. In the former case, the mold material consists of graphite. This is the more practical process because usually no controlled atmosphere supply is required. The latter method requires ceramic molds, which are sensitive to thermal shock, break easily on product removal, and are costly to produce accurately to the dimensions of the mold opening. The advantage of direct compact heating—that the tooling and surrounding area can remain cool—can be offset by a temperature gradient and the resulting microstructure segregation effects in the product. For most systems with readily oxidizing metal matrices, a controlled atmosphere is required.

The densification effect of conventional static hot pressing is more pronounced than the effect that can be achieved by cold pressing and subsequent sintering. Heating the cermet powder mixture increases its plasticity and produces larger areas of inter-particle contact. The surface shearing action that occurs during the process mechanically disrupts surface oxide films and generates clean bonding surfaces (Ref 13). Shape limitations are similar to those for static cold pressure. Prismatic single-level pieces that have no undercuts or re-entrant angles are preferred; however, shallow details in the punch faces are acceptable. Very large pieces are well suited for this process.

A typical graphite induction-heated vacuum furnace (Fig. 10) with graphite tooling is capable of double-action hot compacting or repressing of a 125 mm (5 in.) diam billet at pressures up to 90 Mg (100 tons) and temperatures up to 2300 °C (4200 °F) in a high-vacuum or controlled atmosphere. The fully pressed compact is ejected from the die while still hot to reduce cooling time, minimize sticking, and prolong mold life. Items produced in this type of furnace include WC-Co draw dies and friction elements, Al-B_4C billets, stable oxide components, and some boride-base cermets.

Among the various cermet production processes, static hot pressing in such a furnace is the only reliable single-step method for producing a fully dense, high-quality, near-net shape compact from a cermet mixture. However, graphite die life is limited, and cermet compositions that do not react with graphite are preferred. In order to avoid a cermet-graphite reaction, ceramic molds can be used for hot pressing, although they are more fragile and costly than cermet molds. The complete hot press setup (including a vacuum system or atmosphere generator power system, hydraulic system, controls, and instrumentation) is expensive. In addition, static hot pressing is labor intensive because products are pressed one at a time. Therefore, hot isostatic pressing may be more appropriate for many sensitive high-temperature consolidation tasks involving small- and medium-size pieces. For large and very large pieces, vacuum hot pressing is often used because large equipment is readily available (Ref 14). For example, Fig. 10 shows a 225 Mg (250 ton) vacuum hot press with double-action bottom rams and a 1070 mm (42 in.) square platen; the press has a maximum operating temperature of 1315 °C (2400 °F).

Fig. 10 Production-scale 225 Mg (250 tons) vacuum hot press. Courtesy of Vacuum Industries, Inc.

Hot isostatic pressing (HIP) has become increasingly popular as a means for producing carbide-base and other cermets of very high and uniform density. Internal flaws and micro- or macroporosity are virtually eliminated in the resultant product. Isostatic pressing is a batch process accomplished in water-cooled pressure vessels capable of withstanding internal pressures of up to 210 MPa (30 ksi). Heating up to a temperature of 1600 °C (3000 °F) is achieved with a high-frequency or resistance furnace mounted inside the pressure vessel. The pressure medium is an inert gas, usually argon. The pressure medium can also be helium (Ref 8), which at the pressure employed (100 to 150 MPa, or 15 to 20 ksi) has a density close to that of water.

Hot isostatic pressing was originally developed for use in gas pressure-assisted diffusion bonding processes such as the encapsulation of nuclear fuel elements (uranium oxide, for example) in a zircaloy sheath. This was soon followed by applications such as powder consolidation and densification of difficult-to-sinter substances and cermet composition (Ref 13). Hot isostatic pressing is most successfully applied in the cemented-carbide industry and in the manufacture of steel-bonded titanium carbide. Notwithstanding the ease with which these cermets sinter to high density, they often have slight localized porosity in the range up to 50 μm; on occasion they have voids in the range from 0.25 to 2.5 mm (0.01 to 0.1 in.) caused by random or accidental contamination.

Hot isostatic pressing is an improvement over static hot pressing in that it eliminates the need for costly and highly perishable molds. However, before compacts are submitted to isostatic pressing, they need to have a sufficiently dense structure (at least at the skin) to inhibit gas penetration. Compacts with lower density and interconnecting pores require a gas-tight encapsulation of some sort before being treated. Three methods are in common use to accomplish this encapsulation (Ref 13). In the first method, compacts are formed and sintered to 95% or more of theoretical density, and the resultant continuous, dense surface structure acts as an impenetrable envelope to the high-pressure gas. Alternatively, the density is raised to a sufficiently high level that no interconnecting porosity remains within the compact. This method is used primarily for small- and medium-size pieces.

The second method involves a steel can that is prepared in accordance with the desired form of the compact. The can is filled with a powder mixture that is densely packed by vibration or pressing. (A cold-compacted or hydrostatically pressed piece could be encapsulated in the steel can instead of the powder mixture.) After loading, the can is closed tightly by welding and then evacuated. This method is most commonly used for medium- and large-size pieces, in situations where the additional cost of a perishable can is economically bearable.

The third method is the Ugine-Sejournet process, in which vitrified glass is used to encapsulate the compact. This method may be more economical than the steel can method.

None of the methods for preventing the high-pressure gas from penetrating the compact are inexpensive. Fortunately, the HIP process is flexible enough to allow for the simultaneous hot isostatic pressing of a number of freestanding or encapsulated compacts. The process cost can be apportioned according to the volume occupied by each piece in the available furnace space.

The cemented-carbide and P/M products that undergo HIP processing are of substantially higher quality than those produced by any other process. The higher quality is a result of the near theoretical densities produced by HIP processing. Pieces with near theoretical densities have high strength levels and reliable physical properties. Hot isostatic pressing is a capital-intensive batch process; when the encapsulation method is used, it is labor intensive as well.

Hot extrusion of cermet billets is unique among the various cermet processes because it is essentially a solid-state process. All of the other previously discussed densification processes involve a liquid phase. During hot extrusion of powdered material, large hydrostatic compression forces occur. A unidirectional force component first compresses the powder material to full density and then forces the material through the die. Depending on the configuration of the front surface of the extrusion die, a large shear component can absorb as much as one-half of the total energy needed for extrusion. The total amount of one-step deformation (ratio of the cross section of the billet to the cross section of the die opening) is much larger than in any other cermet hot-working process.

Hot extrusion is an attractive forming and densification process for cermets. Unfortunately, its application in this broad field has serious limitations caused by the loading of the nonmetallic material, the choice of the metallic component, and the degree of interaction between the two substances. Precise limitations on the volumetric amount of nonmetallic substance have not been established, but when this substance exceeds about 18 to 25 vol%, the composite material exhibits hot shortness to such an extent that the problems of edge cracking, internal lamination, and distortion of the extruded product become intolerable. When the nonmetallic phase occurs in the form of very fine powders, the hot shortness problems are aggravated; they are also aggravated when the nonmetallic substance interacts with the metallic component at the extrusion temperature to form new phases or eutectoids. How to deal with these complex problems for each particular pairing of cermet components is beyond the scope of this review. A careful study of the constitution diagram of the binder metal and the hard phase before undertaking any serious work is recommended.

Like metal alloy billets, cermet billets are much easier to extrude when aluminum or aluminum alloys are used for the metallic components. Billet densification and extrusion occur at lower temperatures and pressures. The expensive canning and decanning process can generally be avoided, as can the use of a protective atmosphere. Moreover, straightening and finishing of the extruded product often can be performed at room temperatures. Defects due to hot shortness in extruded aluminum-base products occur only at higher levels of loading with hard-phase substances.

Compared with cermets containing aluminum or aluminum alloys, cermets with iron, nickel, cobalt, or alloys of these metals as the metallic phase are more technically demanding and more costly to extrude. For example, billet heating and hot compacting require a controlled atmosphere, billet canning is practically unavoidable, and a glass process is required for lubrication and reduction of die wear. Postextrusion finishing, such as the removal of can material, straightening, finish rolling, and so on, also requires more costly processes. The extruded cermet product can be expected to be harder, stiffer, less malleable, and more brittle than the metal binder component. Because of the higher extrusion temperature, metallurgical interaction between the matrix metal or alloy and the hard-phase cermet component is far more likely to occur, particularly if the hard phase cermet component is a nonmetallic compound of the carbide or boride group. Oxide-base hard-phase components are less likely to interact with the metallic component.

The undesirable interaction between the hard substance and the metallic component of the cermet is easier to control with the solid-state extrusion process than with any densification process involving a liquid phase. Also, in spite of the aforementioned limitations and problems, hot extrusion continues to attract the attention of product development engineers as a possible production method for certain cermets, particularly those with an aluminum base. It has the potential to be a relatively low labor cost, yet capital-intensive, mass-production process for high-technology rod or strip material.

Combination Sintering-Compacting. Combination debinding, sintering, and pressure consolidation furnaces have been developed in an attempt to simplify the manufacturing process for cermets and similar products. As stated before, a debinding step is essential before sintering green products that contain admixed lubricants, organic binders, or plasticizers. These additives are needed in varying proportions for static cold pressing, warm extrusion, powder roll compacting, slip casting, and injection metal molding of cermets. After debinding and during liquid-phase sintering, the green compacts shrink to nearly complete density. When densification progresses to the point that pressurized gas can no longer penetrate into the compact, hot isostatic compacting occurs. Gas compacting the already-sintered dense cermet at high gas pressures and at a temperature near that of liquid-phase formation improves the product quality by eliminating all residual porosity, internal flaws, and defects.

Recent experience with WC-Co compacts has shown that using a lower isostatic pressure of only 2.7 MPa (390 psig) can produce compacts with strength and densification nearly equal to those of compacts produced by the high-pressure HIP process (Table 3). Based on these findings, multimode single-chamber pressure furnaces have been developed that are capable of operating in vacuum, with partial pressure, and with positive gas pressure up to 10 MPa (1500 psig); the furnaces operate at temperatures between 1450 and 2200 °C (2640 and 3390 °F) (Ref 15).

This new furnace concept of combining three operations in one cycle offers several advantages over separate sintering and hot isostatic pressing operations:

- Debinding, sintering, and densification under pressure take place in one cycle and in a single vessel.
- During the various stages of the process, the compacts do not come in contact with air.
- Controlling the cycle with an electronic microprocessor ensures automatic operation and

Table 3 Comparison of transverse rupture strength for various cemented carbides after hot isostatic pressing and pressure sintering

Processing method	Density, g/cm^3	Density, lb/in.^{3}	Hardness, HRA	Transverse rupture strength, MPa	Transverse rupture strength, ksi
6% cobalt					
Vacuum sintering	14.87	0.537	91.6	2180	316
Vacuum sintering and HIP	14.90	0.538	91.9	2645	384
Pressure sintering	14.89	0.538	92.0	2480	360
9% cobalt					
Vacuum sintering	14.59	0.527	91.2	2170	315
Vacuum sintering and HIP	14.58	0.527	91.2	2380	345
Pressure sintering	14.63	0.529	91.2	2843	412
12% cobalt					
Vacuum sintering	14.09	0.509	89.9	2140	310
Vacuum sintering and HIP	14.07	0.508	90.0	2515	365
Pressure sintering	14.11	0.510	90.8	2565	372

Source: Ref 15

a high degree of program reproducibility (Fig. 11).

- Transfer of parts from one process step to another is avoided, saving labor cost and process time.
- Combining processes saves energy.

Infiltration is a process that is similar to liquid-phase sintering, except that the solid phase is first formed into a porous skeleton body, and the liquid-metal phase is introduced during sintering from the outside and allowed to penetrate the pore system. Excessive shrinkage associated with in situ liquid-phase sintering is avoided, and dimensional stability of the product is obtained, except for about 1% growth that is due to a thin surface film formed by the liquid metal.

This technique is used for systems of two or more components that have widely differing melting temperatures. Aside from hot pressing, infiltration is the only powder processing method that can obtain essentially full density of a near-net shape. All of the other densification processes involve substantial shrinkage and thus destroy shape and dimensional accuracy. By machining, hydrostatic pressing, or powder injection molding the skeletal preforms prior to infiltration, complexities in part design, such as undercuts, re-entrant angles, and multiple levels, can be realized to an extent not possible in parts of comparable high density that are made by extrusion or hot pressing. The other unique feature of infiltration is that—under suitable conditions of low contact angles and limited solubilities between the high- and low-melting phases— systems of completely intertwined continuous networks can be obtained. This is of considerable importance for making products that must combine high thermal or electrical conductivity with acceptable levels of strength and abrasion or erosion resistance.

The procedure used for TiC cermets involves two steps (Ref 17). First, an approximately 60% dense carbide skeleton body of near-net shape is formed by mixing the TiC powder with a small percentage of nickel binder and wax, cold pressing the mixture at about 35 MPa (5000 psi) into a slab, vacuum sintering the slab at about 1300 °C (2370 °F), and then machining the contour (for example, a turbine blade). The second step consists of inserting the skeleton shape into a mold assembly that contains the metal in a ceramic tundish on top and provides for the gravity feeding of the liquid to the skeleton at the preferred contact faces (Ref 18).

The mold assembly is made of graphite, and its cavity is lined with a refractory ceramic in powder form that interfaces with the TiC skeleton. The ceramic liner is chosen so that it does not react with the titanium carbide up to infiltration temperature and also so that it shrinks at a controlled rate, permitting the formation of a uniform gap all around. The mold assembly is heated in a vacuum furnace to about 1400 to 1500 °C (2550 to 2730 °F), well above the melting temperature of infiltrating alloys, such as 80Ni-20Cr and 70Co-24Cr-6Mo. During infiltration, the liquid metal first fills the gap between the liner and the skeleton exterior by capillary forces and then penetrates the interior of the porous TiC part. After furnace cooling, the fully infiltrated product can be readily extracted by fragmenting the sintered ceramic liner without degrading the graphite mold assembly, which can be reused.

Graded cermet parts can be produced by varying the density of the TiC skeleton through the use of a special die filling and multiple-step pressing. For example, a turbine blade can be made that has a high concentration of titanium carbide and, therefore, high strength at the center of the foil and in the transition to the root. The turbine blade also has a metal-rich jacket around the foil and especially at the mechanical shock-sensitive blade edges, as well as around the serrated root needed for blade attachment to the turbine disk.

The infiltration process has been successfully applied to other cermet systems; it has been especially successful when used with interfacial reactions. An example of such an application is the production of complex ceramics that are reinforced by microscopic-size platelets of another compound and bonded by a third species (Ref 19). The preformed ceramic is a metalloidal carbide, the binder is a relatively high-melting reactive metal, and the platelets are the reaction product of the carbide and binder. The unique composite microstructure of such a platelet-reinforced ceramic is obtained by gravity infiltration of the molten metal into a porous preform or bed of the carbide. In the case of a B_4C ceramic and a zirconium metal infiltrant, controlled oxidation at the contact faces produces a new phase, ZrB_2, that precipitates abundantly in the form of platelets that reinforce the ceramic. A similar result is achieved with SiC preforms or fillers that are infiltrated with molten aluminum metal under oxidizing conditions; the reaction product in this case is Al_2O_3 (Ref 19).

The container or mold used in the infiltration process is made from graphite, and it is shaped in accordance with the configurations of the desired product; the process is conducted in an argon atmosphere. The infiltration and reaction temperature depends on the melting point and liquidity of the metal. It can be as high as 2000 °C (3630 °F) in the system involving zirconium. The time to complete penetration and reaction is in the 1 to 2 h range, and the end product typically contains 5 to 15% residual binder metal. The platelet-reinforced infiltrated ceramic system of the type $ZrB_2/ZrC_x/Zr$ exhibits a good combination of high strength, high fracture toughness, and high thermal conductivity. This combination makes it an interesting candidate material for rocket engine components and wear parts (Ref 20). Other systems that have been successfully produced or are potentially workable by the infiltration process include TiB_2 ceramics combined with nickel as second phase (Ref 21), TiC with steel (Ref 22), WC with cobalt (Ref 23), AlN with aluminum (Ref 24), and Al_2O_3 with aluminum (Ref 25).

Infiltration processing of boron carbide and boride-reactive metal cermets has also been used successfully in the development of high-strength, hard, and lightweight products that offer an interesting combination of toughness with high thermal and electrical conductivity (Ref 26). The process involves the infiltration of molten reactive metals, particularly aluminum, into chemically treated boron carbide, or it can use metal-boride starting constituents, such as powders or low aspect ratio fibers, that have been consolidated into a porous ceramic precursor sponge. This process is an alternative to the infiltration of the molten aluminum into thermally modified precursor sponges. Conventional or colloidal chemistry is used in the chemical reaction-controlled casting and infiltration procedures. The potential also exists for consolidating the precursor sponges by injection molding in either a single-step or a two-step process. The key to the process lies in controlling the surface chemistry of the starting constituents. In the two-step process, the first step is the production of highly

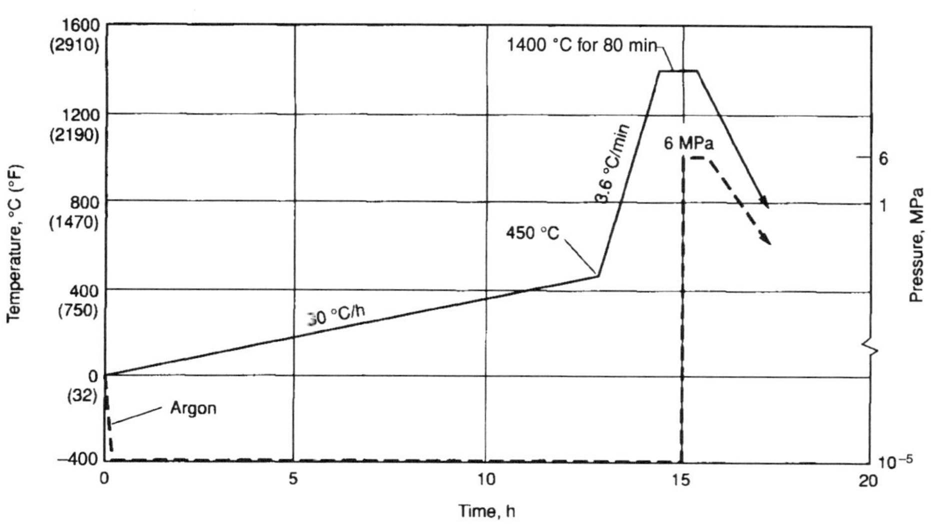

Fig. 11 Schematic cycle diagram for low-pressure dewaxing and overpressure sintering. Source: Ref 16

configured geometries that can be molded by using chemically pretreated binders. In the second step, the binder is volatilized from the precursor, leaving it as a skeleton ready for infiltration.

Cemented Carbides

Cemented carbides belong to a class of hard, wear-resistant, refractory materials in which the hard carbide particles are bound together, or cemented, by a soft and ductile metal binder. These materials were first developed in Germany in the early 1920s in response to demands for a die material having sufficient wear resistance for drawing tungsten incandescent filament wires to replace the expensive diamond dies then in use. The first cemented carbide to be produced was tungsten carbide (WC) with a cobalt binder. Although the term cemented carbide is widely used in the United States, these materials are better known internationally as hard metals.

Tungsten carbide was first synthesized by the French chemist Henri Moissan in the 1890s. There are two types of tungsten carbide: WC, which directly decomposes at 2800 °C (5070 °F), and W_2C, which melts at 2750 °C (4980 °F). Early attempts to produce drawing dies from a eutectic alloy WC and W_2C were unsuccessful, because the material had many flaws and fractured easily. The use of powder metallurgy techniques by Schroeter in 1923 paved the way for obtaining a fully consolidated product. Schroeter blended fine WC powders with a small amount of iron, nickel, or cobalt powders and pressed the powders into compacts, which were then sintered at approximately 1300 °C (2400 °F). Cobalt was soon found to be the best bonding material. Over the years, the basic WC-Co material has been modified to produce a variety of cemented carbides, which are used in a wide range of applications, including metal cutting, mining, construction, rock drilling, metal forming, structural components, and wear parts. Approximately 50% of all carbide production is used for metal cutting applications.

This section discusses the manufacture and composition of cemented carbides and their microstructure, classifications, physical and mechanical properties, and applications. The current status of cemented carbides in nonmetal cutting applications will also be covered.

Manufacture

Cemented carbides are manufactured by a powder metallurgy process consisting of a sequence of steps in which each step must be carefully controlled to obtain a final product with the desired properties, microstructure, and performance. The steps include:

- Processing of the ore and the preparation of the tungsten carbide powder
- Preparation of the other carbide powders
- Production of the grade powders
- Compacting or powder consolidation
- Sintering
- Postsinter forming

The sintered product can be directly used or can be ground, polished, and coated to suit a given application.

Preparation of Tungsten Carbide Powder. There are two methods by which tungsten carbide powders are produced from the tungsten-bearing ores. Traditionally, tungsten ore is chemically processed to ammonium paratungstate and tungsten oxides. These compounds are then hydrogen-reduced to tungsten metal powder. The fine tungsten powders are blended with carbon and heated in a hydrogen atmosphere between 1400 and 1500 °C (2500 and 2700 °F) to produce tungsten carbide particles with sizes varying from 0.5 to 30 μm. Each particle is composed of numerous tungsten carbide crystals. Small amounts of vanadium, chromium, or tantalum are sometimes added to tungsten and carbon powders before carburization to produce very fine (<1 μm) WC powders.

In a more recently developed and patented process, tungsten carbide is produced in the form of single crystals through the direct reduction of tungsten ore (sheelite) (Ref 27). The ore is mixed with iron oxide, aluminum, carbon, and calcium carbide. A high-temperature exothermic reaction ($2Al + 3FeO \leftrightarrow Al_2O_3 + 3Fe$) at about 2500 °C (4500 °F) produces a molten mass that, when cooled, consists of tungsten carbide crystals dispersed in iron, and a slag containing impurities. The crystalline WC (Fig. 2) is then chemically separated from the iron matrix.

Tungsten-titanium-tantalum (niobium) carbides are used in steel-cutting grades to resist cratering or chemical wear and are produced from metal oxides of titanium, tantalum, and niobium. These oxides are mixed with metallic tungsten powder and carbon. The mixture is heated under a hydrogen atmosphere or vacuum to reduce the oxides and form solid-solution carbides such as WC-TiC, WC-TiC-TaC, or WC-TiC-(Ta, Nb)C. The menstruum method can be used to produce WC-TiC solid solution. In this method, the individual carbides are dissolved in liquid nickel. Solid-solution carbides are then precipitated during cooling.

Production of Grade Powders. Cemented carbide grade powders may consist of WC mixed with a finely divided metallic binder (cobalt, nickel, or iron) or with additions of other cubic carbides, such as TiC, TaC, and NbC, depending on the required properties and application of the tool. Intensive milling is necessary to break up the initial carbide crystallites and to blend the various components such that every carbide particle is coated with binder material. This is accomplished in ball mills, vibratory mills, or attritors that use carbide balls. The mills are usually lined with carbide sleeves, although mills lined with low-carbon steel or stainless steel are also used.

Milling is performed under an organic liquid such as heptane or acetone to minimize heating of the powder and to prevent its oxidation. The liquid is distilled off after the milling operation. A solid lubricant such as paraffin wax is added to the powder blend in the final stages of the milling process or later in a blender. The lubricant provides a protective coating to the carbide particles and prevents or greatly reduces the oxidation of the powder. The lubricant also imparts strength to the pressed or consolidated powder mix.

After milling, the organic liquid is removed by drying. In a spray-drying process commonly used in the cemented carbide industry, a hot inert gas such as nitrogen impinges on a stream of carbide particles. This produces free-flowing spherical powder aggregates.

Powder Consolidation. A wide variety of techniques is used to compact the cemented carbide grade powders to the desired shape. Carbide tools for mining and construction applications are pill pressed (pressure applied in one direction) in semiautomatic or automatic presses. Metal cutting inserts are also pill pressed, but may require additional shaping after sintering. Cold isostatic pressing, in which the powder is subjected to equal pressure from all directions, followed by green forming, is also a common practice for wear and metal forming tools. Rods and wires are formed by the extrusion process.

Unlike most other metal powders, cemented carbide powders do not deform during the compacting process. Generally, they cannot be compressed to much above 65% of the theoretical upper limit for density. Despite this low green

Table 4 Properties of refractory metal carbides

Carbide	Hardness, HV (50 kg)	Crystal structure	Melting point °C	Melting point °F	Theoretical density, g/cm³	Modulus of elasticity GPa	Modulus of elasticity 10^6 psi	Coefficient of thermal expansion, μm/m · K
TiC	3000	Cubic	3100	5600	4.94	451	65.4	7.7
VC	2900	Cubic	2700	4900	5.71	422	61.2	7.2
HfC	2600	Cubic	3900	7050	12.76	352	51.1	6.6
ZrC	2700	Cubic	3400	6150	6.56	348	50.5	6.7
NbC	2000	Cubic	3600	6500	7.80	338	49.0	6.7
Cr_3C_2	1400	Orthorhombic	1800(a)	3250	6.66	373	54.1	10.3
WC	(0001) 2200 (10$\bar{1}$0) 1300	Hexagonal	~2800(a)	5050	15.7	696	101	(0001) 5.2 (10$\bar{1}$0) 7.3
Mo_2C	1500	Hexagonal	2500	4550	9.18	533	77.3	7.8
TaC	1800	Cubic	3800	6850	14.50	285	41.3	6.3

(a) Not congruently melting, dissociation temperature. Source: Ref 8 density, carbide manufacturers have developed the technology for achieving good dimensional tolerances in the sintered product.

Sintering and Postsintering Operations. The first step in the sintering process is the removal of the lubricant (dewaxing) from the powder compact. The pressed compacts are normally set on graphite trays coated with a graphite paint. The compacts are first heated to about 500 °C (900 °F) in a hydrogen atmosphere or vacuum using either semicontinuous or batch-type graphite furnaces.

After lubricant removal, the compacts are heated in a vacuum (0.1 Pa, or 10^{-3} torr) to a final sintering temperature ranging from 1350 to 1600 °C (2460 to 2900 °F), depending on the amount of the cobalt binder and the desired microstructure. The dewaxing and sintering operations can also be performed in a single vacuum cycle using furnaces equipped to condense the lubricant and remove it from the heating chamber.

During the final sintering operation, the cobalt melts and draws the carbide particles together. Shrinkage of the compact ranges from 17 to 25% on a linear scale, producing a virtually pore-free, fully dense product.

In the 1970s, the cemented carbide industry took advantage of hot isostatic pressing (HIP) technology, in which vacuum-sintered material is heated again under a gaseous (argon or helium) pressure of 100 to 150 MPa (15 to 20 ksi). The temperatures of this additional process are 25 to 50 °C (45 to 90 °F) below the sintering temperature. The high temperatures and pressures employed in the HIP furnace remove any residual internal porosity, pits, or flaws and produce a nearly perfect cemented carbide.

The latest advancement in sintering technology is the sinter-HIP process, which was developed in the early 1980s (Ref 28). In this process, low-pressure hot isostatic pressing (up to about 7 MPa, or 1 ksi) is combined with vacuum sintering, and the pressure is applied at the sintering temperature when the metallic binder is still molten. With this process, void-free products can be produced at costs only slightly higher than those of vacuum sintering. More information on sintering cemented carbides is in the article "Production Sintering Practices" in this Volume.

Postsinter Forming. A large number of cemented carbide products are shaped after sintering because of surface finish, tolerance, and geometry requirements. This forming operation is both time consuming and expensive. The sintered material is formed with metal-bonded diamond or silicon carbide wheels, turned with a single-point diamond tool, or lapped with diamond-containing slurries.

Machining Applications

The performance of cemented carbide as a cutting tool lies between that of tool steel and cermets. Compared to tool steels, cemented carbides are harder and more wear resistant, but also exhibit lower fracture resistance and thermal conductivities than tool steel. Cermets, on the other hand, are more wear resistant than cemented carbides, but may not be as tough. Any comparison of cermets and cemented carbides, however, depends on the percent of binder material and the type and size of carbide grains. Cermets and cemented carbide cutting tools are described in more detail in Volume 16, *Machining*, of the *ASM Handbook*.

The performance of carbide cutting tools is strongly dependent on composition and microstructure, and the properties of cemented carbide tools depend not only on the type and amount of carbide but also on the carbide grain size and the amount of binder metal. The basic physical and mechanical properties of refractory metal carbides used in the production of cemented carbide tools are given in Table 4. Tungsten carbide and molybdenum carbide have hexagonal crystal structures, whereas the carbides of titanium, tantalum, niobium, vanadium, hafnium, and zirconium are cubic. They undergo no structural changes up to their melting points.

Tungsten Carbide-Cobalt Alloys. The first commercially available cemented carbides consisted of tungsten carbide particles bonded with cobalt. These are commonly referred to as straight grades. These alloys exhibit excellent resistance to simple abrasive wear and thus have many applications in metal cutting. Table 5 lists the representative properties of several straight WC-Co alloys.

The commercially significant alloys contain cobalt in the range of 3 to 25 wt%. For machining purposes, alloys with 3 to 12% Co and carbide grain sizes from 0.5 to more than 5 μm are commonly used.

The ideal microstructure of WC-Co alloys should exhibit only two phases: angular WC grains and cobalt binder phase. Representative microstructures of several straight WC-Co alloys are shown in Fig. 12. The carbon content must be controlled within narrow limits. Too high a carbon content results in the presence of free and finely divided graphite (Fig. 13), which in small amounts has no adverse effects in machining applications. Deficiency in carbon, however, results in the formation of a series of double carbides (for example, Co_3W_3C or Co_6W_6C), commonly known as η phase, which causes severe embrittlement. Because the formation of η phase involves the dissolution of the original carbides into the cobalt binder, η phase appears as an irregularly shaped phase in the microstructure (Fig. 14).

Submicron Tungsten Carbide-Cobalt Alloys. In recent years, WC-Co alloys with submicron carbide grain sizes have been developed for applications requiring more toughness or edge strength. Typical applications include indexable inserts and a wide variety of solid carbide drilling and milling tools. Grain refinement in these alloys is obtained by small additions (0.25 to 3.0 wt%) of tantalum carbide, niobium carbide, vanadium carbide, or chromium carbide. Additions can be made before carburization of the tungsten or later in the powder blend. Vanadium carbide is the most effective grain growth inhibitor. Chromium carbide, in addition to being an efficient grain growth inhibitor, imparts excellent mechanical properties. Tantalum carbide is not as effective as vanadium carbide or chromium carbide in grain refinement.

Alloys Containing Tungsten Carbide, Titanium Carbide, and Cobalt. The tungsten carbide-cobalt alloys, developed in the early 1920s, were successful in the machining of cast iron and nonferrous alloys at much higher speeds than were

Table 5 Properties of representative cobalt-bonded cemented carbides

Nominal composition	Grain size	Hardness, HRA	Density, g/cm³	Density, oz/in.³	Transverse strength, MPa	Transverse strength, ksi	Compressive strength, MPa	Compressive strength, ksi	Modulus of elasticity, GPa	Modulus of elasticity, 10^6 psi	Relative abrasion resistance(a)	Coefficient of thermal expansion, μm/m · K, at 200 °C (390 °F)	Coefficient of thermal expansion, μm/m · K, at 1000 °C (1830 °F)	Thermal conductivity, W/m · K
97WC-3Co	Medium	92.5–93.2	15.3	8.85	1590	230	5860	850	641	93	100	4.0	...	121
94WC-6Co	Fine	92.5–93.1	15.0	8.67	1790	260	5930	860	614	89	100	4.3	5.9	...
	Medium	91.7–92.2	15.0	8.67	2000	290	5450	790	648	94	58	4.3	5.4	100
	Coarse	90.5–91.5	15.0	8.67	2210	320	5170	750	641	93	25	4.3	5.6	121
90WC-10Co	Fine	90.7–91.3	14.6	8.44	3100	450	5170	750	620	90	22	...	...	...
	Coarse	87.4–88.2	14.5	8.38	2760	400	4000	580	552	80	7	5.2	...	112
84WC-16Co	Fine	89	13.9	8.04	3380	490	4070	590	524	76	5	...	...	...
	Coarse	86.0–87.5	13.9	8.04	2900	420	3860	560	524	76	5	5.8	7.0	88
75WC-25Co	Medium	83–85	13.0	7.52	2550	370	3100	450	483	70	3	6.3	...	71
71WC-12.5TiC-12TaC-4.5Co	Medium	92.1–92.8	12.0	6.94	1380	200	5790	840	565	82	11	5.2	6.5	35
72WC-8TiC-11.5TaC-8.5Co	Medium	90.7–91.5	12.6	7.29	1720	250	5170	750	558	81	13	5.8	6.8	50

(a) Based on a value of 100 for the most abrasion-resistant material possible with high-speed steel tools, but were subject to chemical attack or diffusion wear when cutting steel. As a result, the tools failed rapidly at speeds not much higher than those used with high-speed steel. This led to the development of WC-TiC-Co alloys.

Tungsten carbide diffuses readily into the steel chip surface, but the solid solution of tungsten carbide and titanium carbide resists this type of chemical attack. Unfortunately, titanium carbide and WC-TiC solid solutions are more brittle and less abrasion resistant than tungsten carbide. The amount of titanium carbide added to tungsten carbide-cobalt alloys is therefore kept to a minimum, typically no greater than 15 wt%. The carbon content is less critical in WC-TiC-Co alloys than in WC-Co alloys, and the η phase does not appear in the microstructure unless carbon is grossly inadequate. In addition, free graphite rarely occurs in these alloys.

Steel-Cutting Grades of Cemented Carbide Alloys. The WC-TiC-Co alloys have given way to alloys of tungsten carbide, cobalt, titanium carbide, tantalum carbide, and niobium carbide. The tungsten carbide-cobalt alloys containing TiC, TaC, and NbC are called complex grades, multigrades, or steel-cutting grades. Adding TaC to WC-TiC-Co alloys partially overcomes the deleterious effects of TiC on the strength of WC-Co alloys. Tantalum carbide also resists cratering and improves thermal shock resistance. The latter property is particularly useful in applications involving interrupted cuts. Tantalum carbide is often added as (Ta, Nb)C because the chemical similarity between TaC and NbC makes their separation expensive. Fortunately, NbC has an effect similar to TaC in most cases. The relative concentrations of tantalum carbide and niobium carbide in these alloys are dependent on the raw material used, the desired composition, the properties, and the microstructure.

Unlike the WC-Co alloys, the microstructure of WC-TiC-(Ta,Nb)C-Co alloys shows three phases: angular WC grains, rounded WC-TiC-(Ta,Nb)C solid-solution grains, and cobalt binder. The solid-solution carbide phase often exhibits a cored structure, indicating incomplete diffusion during the sintering process. Representative microstructures of several WC-TiC-(Ta,Nb)C alloys are shown in Fig. 15. The size and distribution of the phases vary widely, depending on the amounts and grain sizes of the raw materials employed and on the method of manufacture. Similarly, the properties of these complex alloys also vary widely, as indicated in Table 5 for a few representative steel-cutting grades.

Classification

There is no universally accepted system for classifying cemented carbides. The systems most often employed by producers and users are discussed below. Each system has inherent strengths and weaknesses in describing specific materials, and for this reason close cooperation between user and producer is the best means of selecting the proper grade for a given application.

C-Grade System. The U.S carbide industry uses an application-oriented system of classification to assist in the selection of proper grades of cemented carbides. This C-grade system does not require the use of trade names for identifying specific carbide grades (Table 6). Although this

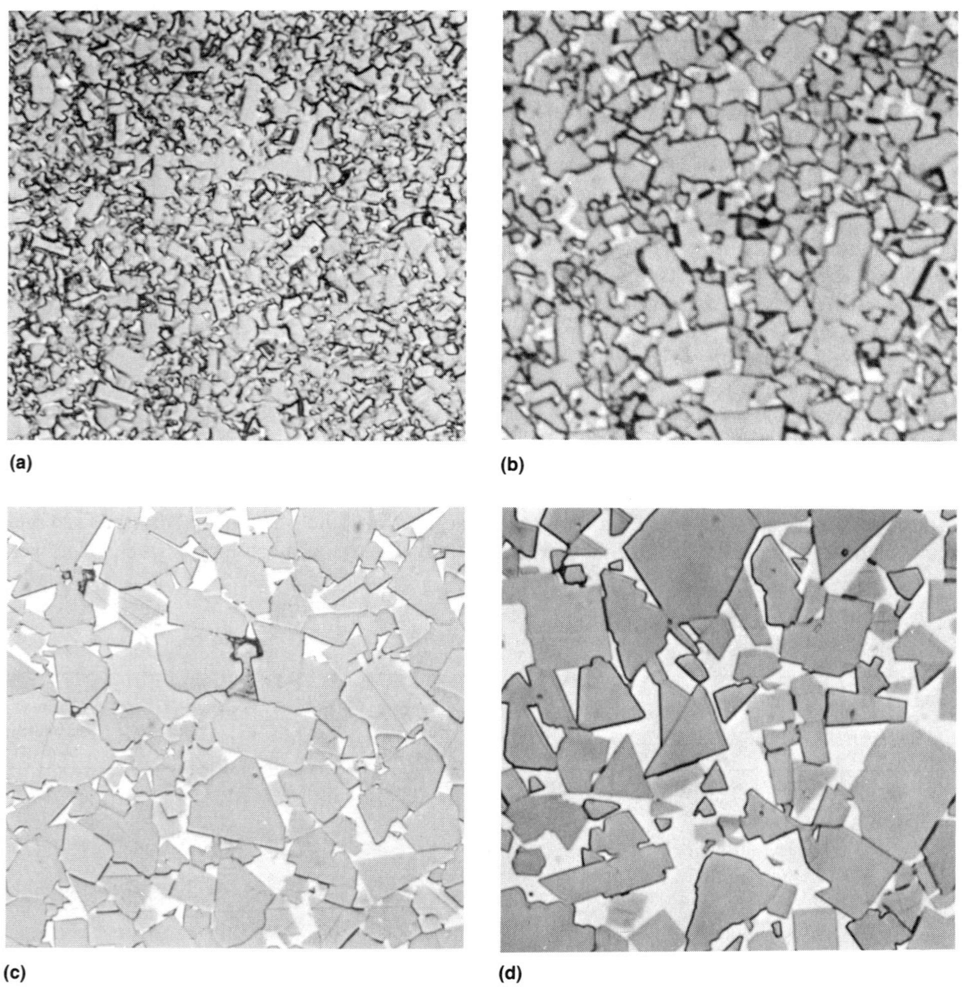

Fig. 12 Microstructures of straight WC-Co alloys. (a) 97WC-3Co alloy, medium grain size. (b) 94WC-6Co alloy, medium grain. (c) 94WC-6Co alloy, coarse grain. (d) 85WC-15Co alloy, coarse grain. All etched with Murakami's reagent for 2 min. 1500×

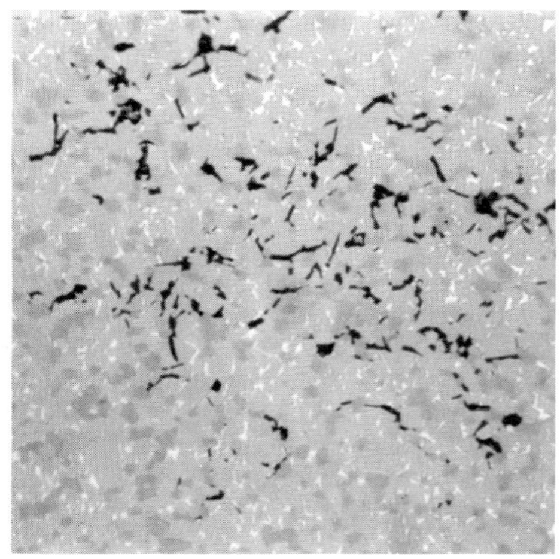

Fig. 13 Free graphite in a tungsten carbide alloy. Black areas contain graphite and are an example of C-type porosity. Polished 86WC-8 (Ta,Ti,Nb)C-6Co alloy. 1500×

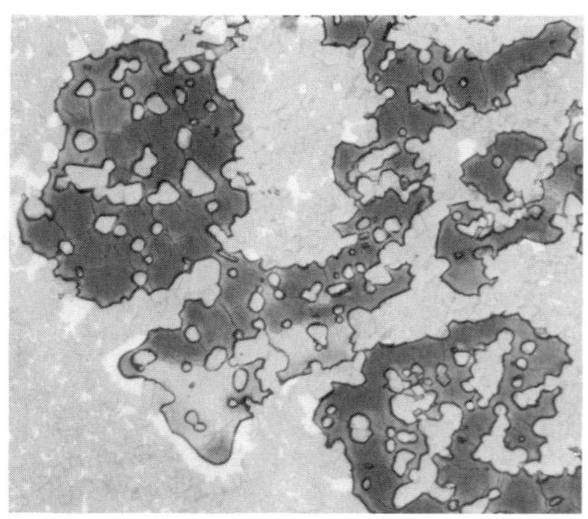

Fig. 14 η phase microstructure. Micrograph shows a $(Co_3W_3)C$-η phase in detail. η phase appears as various shades of gray with clearly defined grain boundaries. Light gray WC particles surrounded by η phase are rounded because of the solubility of tungsten carbide in the binder. 85WC-8(Ta,Ti,Nb)C-7Co alloy etched with Murakami's reagent for 3 s. 900×

classification simplifies tool application, it does not reflect the material properties that significantly influence selection of the proper carbide grade. Additionally, the definitions of work materials involved in this classification scheme are imprecise. There is also no universal agreement on the meanings of the terms used to describe the various application categories. Despite these limitations, the C-grade classification has been successfully used by the manufacturing industry since 1942.

ISO Classification. In 1964 the International Organization of Standardization (ISO) issued ISO Recommendation R513 "Application of Carbides for Machining by Chip Removal." The basis for the ISO classification of carbides is summarized in Table 7.

In the ISO system, all machining grades are divided into three color-coded groups:

Table 6 C-grade classification of cemented carbides

C-grade	Application category
Machining of cast iron, nonferrous, and nonmetallic materials	
C-1	Roughing
C-2	General-purpose machining
C-3	Finishing
C-4	Precision finishing
Machining of carbon and alloy steels	
C-5	Roughing
C-6	General-purpose machining
C-7	Finishing
C-8	Precision finishing
Nonmachining applications	
C-9	Wear surface, no shock
C-10	Wear surface, light shock
C-11	Wear surface, heavy shock
C-12	Impact, light
C-13	Impact, medium
C-14	Impact, heavy

- Highly alloyed tungsten carbide grades (letter P, blue color) for machining steel
- Alloyed tungsten carbide grades (letter M, yellow color, generally with less TiC than the corresponding P series) for multipurpose use, such as steels, nickel-base superalloys, and ductile cast irons
- Straight tungsten carbide grades (letter K, red color) for cutting gray cast iron, nonferrous metals, and nonmetallic materials

Each grade within a group is assigned a number to represent its position from maximum hardness to maximum toughness. P-grades are rated from 01 to 50, M-grades from 10 to 40, and K-grades from 01 to 40. Typical applications are described for grades at more or less regular numerical intervals. Although the coated grades had not been developed at the time the ISO classification system was prepared, one should be able to classify them as easily as the uncoated grades.

Selection

As cutting tool materials, tungsten carbide materials may be separated into two types. First and most common is the simple two-phase type consisting of tungsten carbide with a cobalt binder (C-1 through C-4). If metal removal involves generation of a short, weak, discontinuous chip, as in the machining of cast iron, this type of material is most effective. These carbides exhibit extreme hardness and excellent resistance to abrasive wear. However, if the workpiece material yields a strong, continuously curling chip, such as with steel, the alloyed grades C-5 through C-8 must be used.

Tungsten carbide and cobalt usually are alloyed with additional carbide(s), usually titanium carbide and /or tantalum carbide. These alloying additions delay crater development and increase tool life.

Straight tungsten carbide grades are most resistant to abrasive wear and have the greatest strength for a given hardness value. Two possible variables affect mechanical properties: grain size of the tungsten carbide particles and the percentage of cobalt. The following factors should be considered in grade selection:

- Finer grain size results in higher hardness.
- Lower cobalt content results in higher hardness.
- Higher hardness results in greater abrasive wear resistance.
- Lower cobalt content results in lower strength.
- Finer grain size results in lower strength.

Straight tungsten carbide grades are classified in the C-1 through C-4 range. They are used to machine most cast irons, 200 and 300 series stainless steels, most high-temperature alloys, nonferrous alloys, and nonmetals.

Alloyed tungsten carbide grades are used where the predominant wear factors are adhesion and diffusion due to high cutting forces and high heat. The following factors should be considered in grade selection.

Higher titanium carbide content results in:

- Greater resistance to cratering
- Higher hot hardness
- Higher wear resistance
- Lower strength

Higher tantalum carbide content results in:

- Greater resistance to thermal deformation and cratering
- Higher hot hardness
- Lower wear resistance
- Lower strength

Alloyed tungsten carbide grades are classified in the C-5 through C-8 range. These grades are used to machine most carbon and alloy steels, 400 and 500 series stainless steels, and malleable and ductile irons.

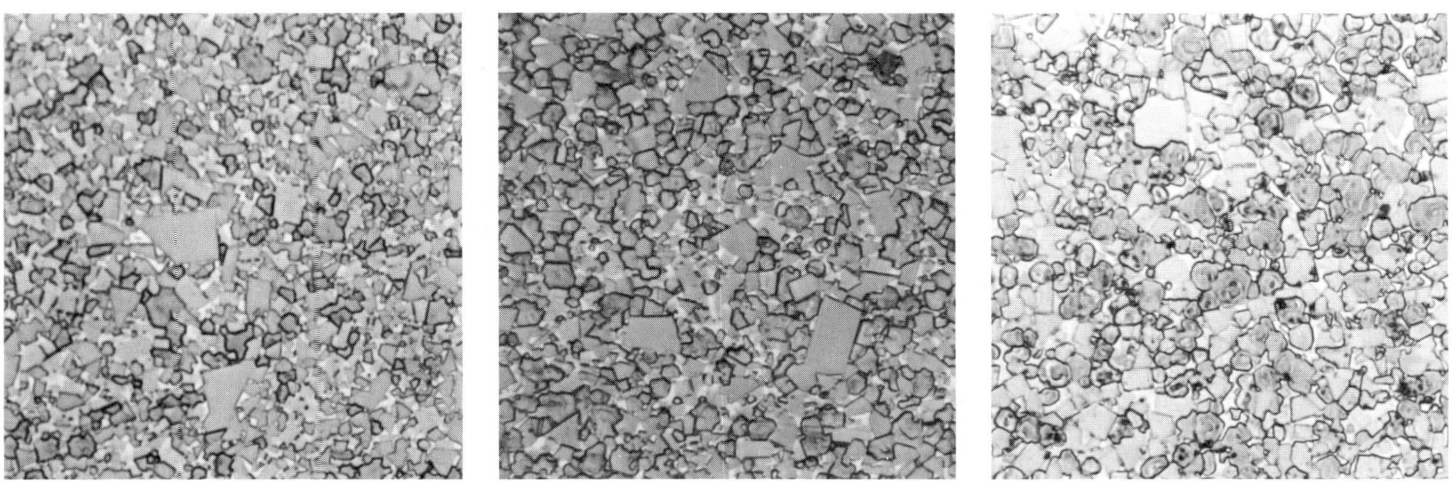

(a) (b) (c)

Fig. 15 Representative microstructures of steel-cutting grades of cemented tungsten carbide. (a) 85WC-9(Ta,Ti,Nb)C-6Co alloy, medium grain size. (b) 78WC-15(Ta,Ti,Nb)C-7Co alloy, medium grain. (c) 73WC-19(Ta,Ti,Nb)C-8Co alloy, medium grain. The gray, angular particles are WC, and the dark gray, rounded particles are solid-solution carbides. The white areas are cobalt binder. All etched with Murakami's reagent for 2 min. 1500×

Wear Applications

Many applications exist for cemented carbides in noncutting areas. These may be generally categorized as wear parts. Properties that are paramount for wear part applications include abrasion resistance, impact strength, and corrosion resistance. Due to the physical property requirements, suitable materials for wear part applications include tungsten carbide and cobalt.

Tantalum carbide may be added in some applications for antigalling and corrosion resistance. Nickel may be substituted for cobalt in corrosion-resistant applications. Titanium carbide, widely used in metal cutting grades, is never used in wear applications, because it lowers strength and abrasion resistance.

Cemented carbides used for wear applications may be divided into the following categories:

Light impact, high abrasion resistance applications:

- Blast nozzles
- Spray nozzles
- Reamers
- Gun drills
- Wear pads
- Powder compaction dies
- Glass cutters
- Plastic extrusion dies
- Guide rings
- Circuit board drills
- Paper slitters

Medium impact, medium abrasion resistance applications:

- Tape slitters
- Metal draw dies
- Aluminum extrusion dies
- Shear knives
- Guide rolls
- Snowplow blades

High impact, low abrasion resistance applications:

- Cold header dies
- Stamping dies
- Bar mill rolls

Corrosion resistance applications:

- Seal rings
- Nozzles
- Ball valves

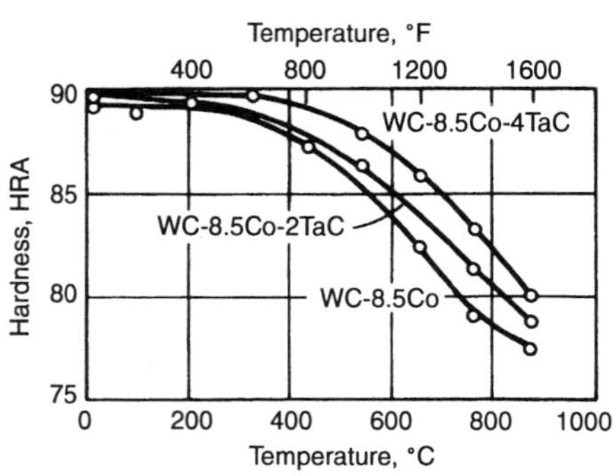

Fig. 16 Effect of tantalum carbide additions on hot hardness of tungsten carbide/cobalt cemented carbides

- Metering chokes
- Aluminum can tooling

Specialty applications:

- Nuclear components
- Nonmagnetic guidance gyros
- Nonmagnetic dies for magnets

Light Impact, High Abrasion Resistance Applications. A large number of applications capitalize on the abrasion resistance of cemented carbides. These include sliding friction, light uninterrupted cutting, or particle abrasion applications, which involve minimal mechanical shock. Generally only straight tungsten carbide/cobalt grades are used. High abrasion resistance grades typically have low cobalt contents (4 to 6%) and fine tungsten carbide particle sizes (1 to 2 μm). High abrasion resistance grades, which are classified in the C-9 through C-12 range, have the lowest impact strengths.

Submicron grades with 6 to 10% cobalt also are used for these applications. Abrasion-resistant grades have the lowest thermal shock resistance. As such, care should be taken when brazing cemented carbides to other materials such as steel, which have significant differences in thermal expansion. The difference in thermal expansion rates can result in fracture during brazing.

Medium impact, medium abrasion resistance applications require a combination of abrasion resistance and impact strength. Generally only straight tungsten carbide/cobalt grades (C-13 classification) are used for these applications. Grade composition ranges from 9 to 13% Co, and tungsten carbide particle size ranges from 2 to 3.5 μm. Submicron grades with from 10 to 15% Co also are suitable for this usage. Tantalum carbide may be added to prevent galling or to improve elevated-temperature properties (Fig. 16). Due to the higher cobalt content and coarser tungsten carbide particle size, these grades have improved thermal shock properties.

High Impact, Low Abrasion Resistance Applications. In many high-impact applications, cemented carbides compete with tool steel. Suitable

Table 7 ISO R513 classification of carbides according to use for machining

Designation (a)	Groups of application	
	Material to be machined	Use and working conditions
P 01	Steel, steel castings	Finish turning and boring; high cutting speeds, small chip section, accuracy of dimensions and fine finish, vibration-free operation
P 10	Steel, steel castings	Turning, copying, threading, and milling; high cutting speeds, small or medium chip sections
P 20	Steel, steel castings, malleable cast iron with long chips	Turning, copying, milling, medium cutting speeds and chip sections; planing with small chip sections
P 30	Steel, steel castings, malleable cast iron with long chips	Turning, milling, planing, medium or low cutting speeds, medium or large chip sections, and machining in unfavorable conditions(b)
P 40	Steel, steel castings with sand inclusion and cavities	Turning, planing, slotting, low cutting speeds, large chip sections with the possibility of large cutting angles for machining in unfavorable conditions(b)
P 50	Steel, steel castings of medium or low tensile strength, with sand inclusion and cavities	For operations demanding very tough carbide: turning, planing, slotting, low cutting speeds, large chip sections, with the possibility of large cutting angles for machining in unfavorable conditions(b) and work on automatic machines
M 10	Steel, steel castings, manganese steel, gray cast iron, alloy cast iron	Turning, medium or high cutting speeds; small or medium chip sections
M 20	Steel, steel castings, austenitic or manganese steel, gray cast iron	Turning, milling; medium cutting speeds and chip sections
M 30	Steel, steel castings, austenitic steel, gray cast iron, high-temperature resistant alloys	Turning, milling, planing; medium cutting speeds, medium or large chip sections
M 40	Mild free-cutting steel, low-tensile steel, nonferrous metals, and light alloys	Turning, parting off, particularly on automatic machines
K 01	Very hard gray cast iron, chilled castings of over 85 scleroscope hardness, high-silicon aluminum alloys, hardened steel, highly abrasive plastics, hard cardboard, ceramics	Turning, finish turning, boring, milling, scraping
K 10	Gray cast iron over 220 HB, malleable cast iron with short chips, hardened steel, silicon aluminum alloys, copper alloys, plastics, glass, hard rubber, hard cardboard, porcelain, stone	Turning, milling, drilling, boring, broaching, scraping
K 20	Gray cast iron up to 220 HB, nonferrous metals: copper, brass, aluminum	Turning, milling, planing, boring, broaching, demanding very tough carbide
K 30	Low-hardness gray cast iron, low-tensile steel, compressed wood	Turning, milling, planing, slotting, for machining in unfavorable conditions(b) and with the possibility of large cutting angles
K 40	Softwood or hardwood, nonferrous metals	Turning, milling, planing, slotting, for machining in unfavorable conditions(b) and with the possibility of large cutting angles

(a) In each letter category, low designation numbers are for high speeds and light feeds; higher numbers are for slower speeds and/or heavier feeds. Also, increasing designation numbers imply increasing toughness and decreasing wear resistance of the cemented carbide materials. (b) Unfavorable conditions include shapes that are awkward to machine; material having a casting or forging skin; material having variable hardness; and machining that involves variable depth of cut, interrupted cut, or moderate to severe vibrations.

cemented carbide grades (C-14 classification) have cobalt contents ranging from 15%, with 3 to 4 μm tungsten carbide particle sizes for lamination dies, to 25% with 4 to 5 μm tungsten carbide particle sizes for cold header dies. Generally, submicron grades are not used in impact applications.

The 25% Co binder content is the highest available because of the difficulty maintaining shape during liquid-phase sintering. Tantalum carbide is used in some applications for galling resistance. High cobalt grades (18% and above) may be machined with aluminum oxide and silicon carbide wheels and cemented carbide tooling. Below 17 to 18% Co, diamond tools are required.

Corrosion Resistance Applications. Cemented carbides frequently are used for their combined corrosion and abrasion resistance. Typically, the binder metal corrodes. Finer tungsten carbide particle size and lower cobalt content improve corrosion resistance. Grades with binder metal contents as low as 1.5% are used.

Nickel binder grades with fine tungsten carbide grain sizes are finding increasing usage. Important corrosion resistance applications include the use of cemented carbides for valve stems and seats, chemical seal rings, and for aluminum can tooling for use with synthetic lubricants. Grades with small chromium additions are used for sour gas applications (natural gas with high hydrogen sulfide). Ultimately, the optimum cemented carbide composition for a corrosion application depends on the intended service environment.

Specialty Applications. Many wear part applications require the special properties of cemented carbide in addition to abrasion resistance and impact strength. These include nuclear and nonmagnetic applications. In parts for the nuclear industry (seal rings, wear pads, and hardfacing), nickel binder grades are suitable, because nickel has a much shorter radioactive half-life than cobalt.

For nonmagnetic applications (guidance gyros, dies for ceramic magnets, and tape player heads), nickel binder grades with high dissolved tungsten or chromium contents are used. Cobalt binder grades are ferromagnetic. Another promi-

Table 8 Test methods for determining the properties of cemented carbides

Property	Test method: ASTM/ANSI	CCPA(a)	ISO
Abrasive wear resistance	B 611	P112	...
Apparent grain size	B 390	M203	...
Apparent porosity	B 276	M201	4505
Coercive force	...	...	3326
Compressive strength	E 9	P104	4506
Density	B 311	P101	3369
Fracture toughness	...	...	...
Hardness, HRA	B 294	P103	3738
Hardness, HV	E 92	...	3878
Linear thermal expansion	B 95	P108	...
Magnetic permeability	A 342	P109	...
Microstructure	B 657	M202	4499
Poisson's ratio	E 132	P105	...
Transverse rupture strength	B 406	P102	3327
Young's modulus	E 111	P106	3312

(a) Cemented Carbides Producers Association

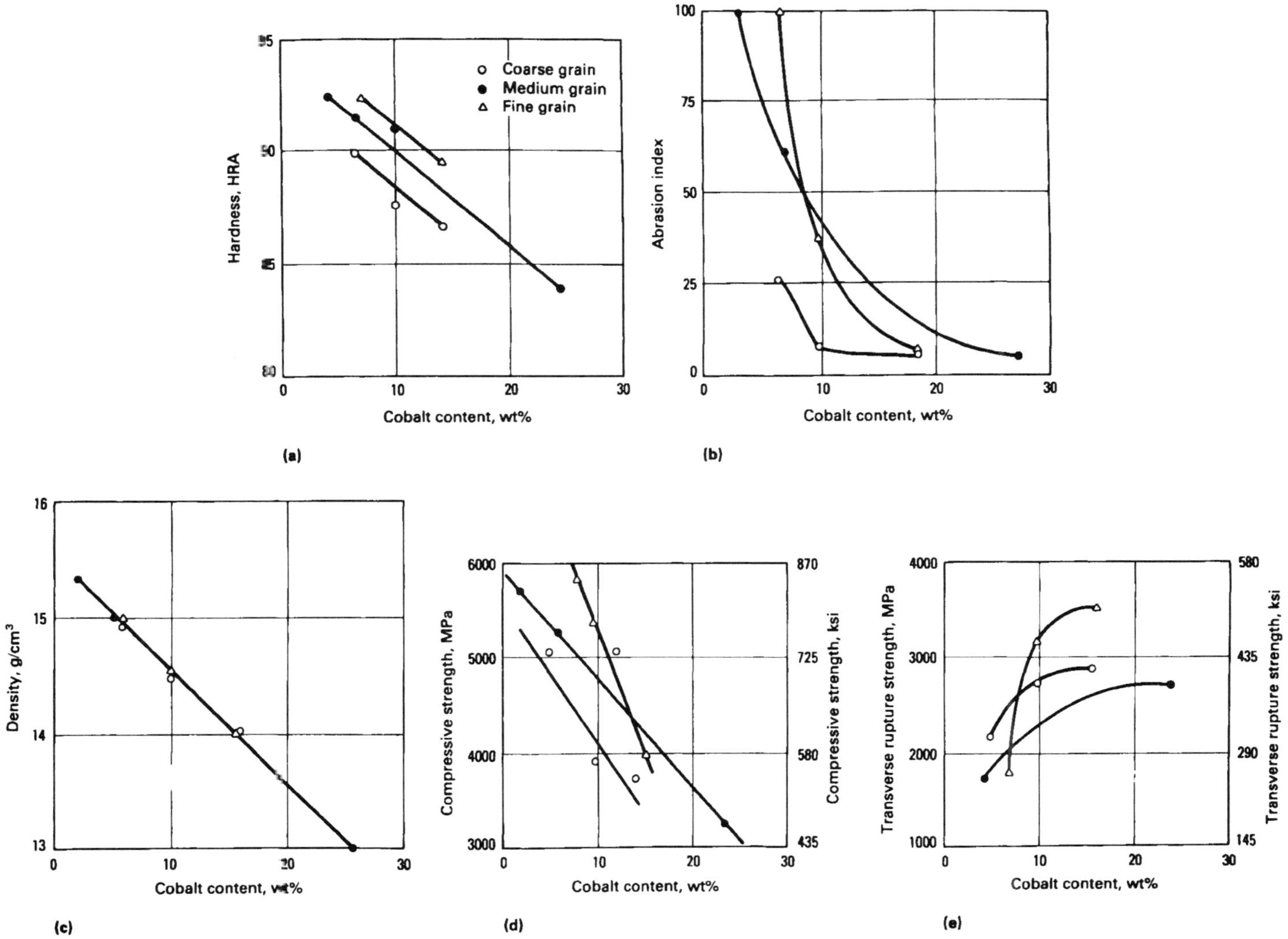

Fig. 17 Variation in properties with cobalt content and grain size for straight WC-Co alloys. (a) Variation in hardness. (b) Variation in abrasion resistance. (c) Variation in density. (d) Variation in compressive strength. (e) Variation in transverse rupture strength

nent application for cemented carbides is gage blocks. A chromium carbide/nickel grade typically is used. This material has a low coefficient of thermal expansion, is corrosion resistant, nonmagnetic, and abrasion resistant, and exhibits high reflectivity when lapped.

Properties

A number of industry, national, and ISO standards have been developed for determining the selected properties of cemented carbides (Table 8).

Hardness determines the resistance of a material to abrasion and wear. It is affected not only by composition but also by the level of porosity and microstructure. For straight WC-Co alloys of comparable WC grain size, hardness and abrasion resistance decrease with increasing cobalt content (Fig. 17a and b). However, because both composition and microstructure affect hardness, cobalt content and grain size must be considered. At a given cobalt level, hardness improves with decreasing WC grain size (Fig. 17a).

In cemented carbides, hardness is measured by the Rockwell A-scale diamond cone indentation test (HRA) or by the Vickers diamond pyramid indentation test (HV). Both tests are performed on a finely ground, lapped, or polished planar surface placed at right angles to the indentor axis. The Rockwell A test employs a load of 60 kg, whereas a range of loads can be used in the Vickers test. For cemented carbides used in machining applications, hardness values range from 88 to 94 HRA and from 1100 to 2000 HV.

Although the Rockwell scale has been used for decades as a measure of hardness, a true indication of the resistance to plastic deformation in metal cutting operations can be obtained only by measuring hardness at elevated temperatures. Measurements of hardness over a wide range of temperatures are therefore valuable for tool selection.

Hardness testers with high-temperature capability (up to 1200 °C, or 2200 °F) are commercially available and are being increasingly used by the cemented carbide industry. Figure 18 shows hot hardness data for a number of cemented carbides. The hardness of these materials decreases monotonically with increasing temperatures.

Compressive Properties. One of the unique properties of cemented carbides is their high compressive strength. Uniaxial compression tests can be performed on straight cylindrical samples or on cylinders having reduced diameters in the middle to localize fracture. The compressive strengths of cemented carbides are greater than those of most other materials. Typical values of compressive strength range from 3.5 to 7.0 GPa (0.5 to 1.0 × 10^6 psi).

The ductility of cemented carbides is generally low at room temperature, so there is little difference between their yield strength and fracture strength. At higher temperatures, however, these materials exhibit a small but finite amount of ductility. Measurement of yield strength is therefore more appropriate at elevated temperatures. High-temperature compressive yield strength is typically measured at 0.2% offset strain. Compression tests are performed in a high-temperature furnace (typically with resistance heating) under a vacuum or in an inert atmosphere. Figure 19 shows yield strength data for selected straight and alloyed WC-Co grades. Like hardness, the compressive yield strengths of cemented carbides decrease monotonically with increasing temperature; the rate of decrease depends on the composition and the microstructure. As in metallic materials, fine-grain alloys tend to lose their yield strengths more rapidly with increasing temperature than coarse-grain grades, although at room temperature the former can exhibit high yield strengths.

Transverse Rupture Strength. The most common method of determining the fracture strength of cemented carbides is the transverse rupture test. In this test, a rectangular test bar is placed across two sintered carbide support cylinders, and a gradually increasing load is applied by a third carbide cylinder at the midpoint between the supports. Transverse rupture strength is determined from the dimensions of the test bar, the distance between the supports, and the fracture load. A disadvantage of this test is the large scatter in the experimental data resulting from surface defects introduced into the test specimens during processing. Nevertheless, it is an excellent quality control test, and it is particularly useful for large carbide components. Figure 17(e) shows the variation in transverse rupture strength with cobalt content.

Fracture toughness is less sensitive than transverse rupture strength to such extrinsic factors as specimen size, geometry, and surface finish. Fracture toughness is measured by the critical stress intensity factor (K_{Ic}), which indicates the resistance of a material to fracture in the presence of a sharp crack and thus provides a better measure of the intrinsic strength of the cemented carbide than transverse rupture strength. A variety of specimen geometries have been used in this test, including the single-edge notched beam, the double cantilever beam, the compact tension specimen, and the double torsion specimen. The carbide industry in the United States generally uses commercial equipment for fracture toughness evaluation. The fracture toughness of cemented carbides increases with cobalt content and with WC grain size (Fig.

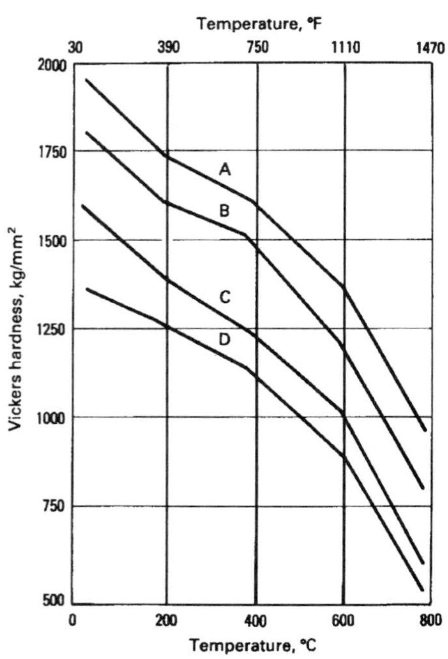

Fig. 18 Variation in microhardness with temperature. Microhardness is based on a 1 kg load, and all alloys are of medium WC grain size. A, 97WC-3Co alloy; B, 94WC-6Co; C, 80WC-12(Ti,Ta,Nb)C-8Co; D, 86WC-2TaC-12Co

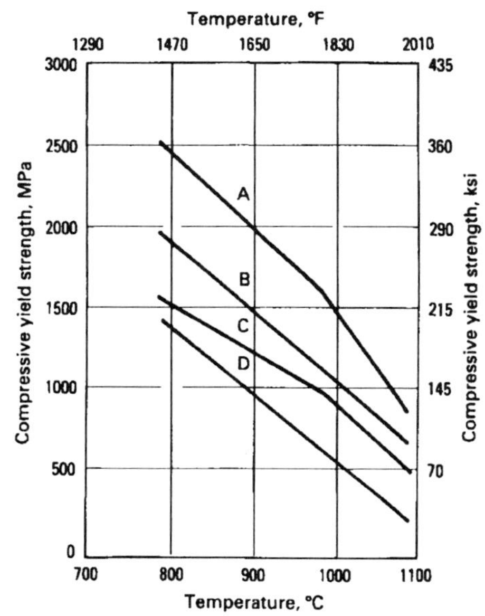

Fig. 19 Variation in compressive yield strength with temperature. Measured at 0.2% offset strain; all alloys characterized by medium grain size. A, 73WC-22(Ti,Ta,Nb)C-5Co alloy; B, 80WC-12(Ti,Ta,Nb)C-8Co; C, 86WC-8(Ti,Ta,Nb)C-6Co; D, 86WC-2TaC-12Co

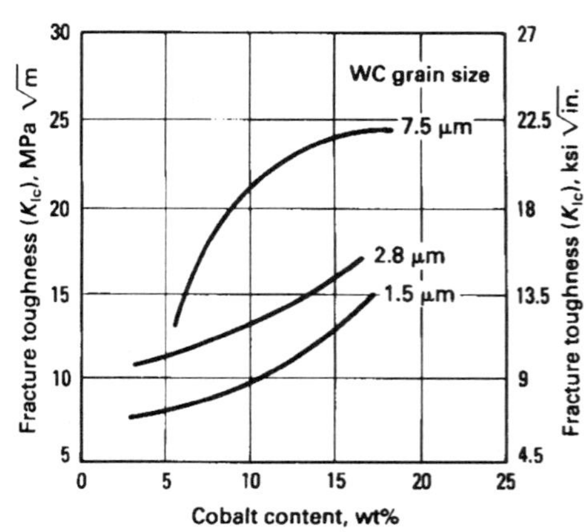

Fig. 20 Variation in toughness (K_{Ic}) with cobalt content for WC-Co alloys with different WC grain sizes. Source: Ref 29

20). On the other hand, cubic carbide additions lessen the fracture toughness of WC-Co alloys.

As with the other mechanical properties, attention is focused on the development of test techniques to evaluate fracture toughness at elevated temperatures. Figure 21 shows K_{Ic} data for a number of cemented carbides from room temperature to about 1000 °C (1800 °F). Depending on the composition of the cemented carbide, the K_{Ic} parameter is insensitive to temperature, up to about 600 °C (1100 °F), but increases rapidly at higher temperatures. This behavior is reminiscent of the ductile-to-brittle transition observed in quenched-and-tempered steels.

The density, or specific gravity, of cemented carbides is very sensitive to composition and porosity in the sample and is widely used as a quality control test. Density values of cemented carbides range from 15 g/cm^3 for low-cobalt straight WC-Co alloys to about 10 or 12 g/cm^3 for highly alloyed carbide grades (Fig. 17c).

Magnetic Properties. Tungsten carbide-cobalt alloys lend themselves to the analysis of magnetic properties because cobalt is ferromagnetic. The properties measured are magnetic saturation and coercive force. Both free cobalt and solid solutions of cobalt and tungsten contribute to magnetization. The magnetic saturation of pure cobalt is 201×10^{-6} T m^3/kg. With additions of tungsten to cobalt, the magnetic saturation decreases steadily from 201×10^{-5} to about 151×10^{-6} T m^3/kg. In this range, the cemented carbide is characterized by two or three phases (WC and Co, or WC, a solid-solution carbide, and Co). Values below 151×10^{-6} T m^3/kg indicate the presence of η phase. The solubility of tungsten in cobalt is inversely proportional to carbon content. Lower-carbon alloys have more tungsten dissolved in cobalt and are characterized by lower magnetization. Magnetic saturation thus provides an accurate measure of the changes in carbon content in the cemented carbide alloy and is widely used as a quality control test.

The coercive force varies considerably with increasing sintering temperature and indicates the structural changes that take place during sintering. The coercive force of WC-Co alloys reaches a maximum at the optimum sintering temperature and decreases at higher temperatures because of grain growth. Therefore, measurement of the coercive force permits control of the sintering process. The factors influencing the coercive force are complex, varied, and interactive. For a given cobalt and carbon content, the coercive force provides a measure of the degree of distribution of the carbide phase in the microstructure. Commercial units are available for the rapid measurement of magnetic saturation and coercive force in cemented carbides.

Porosity. The properties of a cemented carbide are dependent on its density, which in turn is critically dependent on composition and porosity. Porosity is evaluated on the as-polished material. The American Society for Testing and Materials (ASTM) has established a standard procedure (B 276) that rates three types of porosity:

Table 9 Thermal shock resistance parameters for cemented carbides

Composition, wt %	WC grain size	Transverse rupture strength, σ MPa	ksi	Thermal conductivity (k), W/m · K	Young's modulus, E GPa	10^6 psi	Coefficient of thermal expansion (α), μm/m · K	Thermal shock resistance (σk /Eα), kW/m
97WC-3Co	Medium	1590	230	121	641	93	5.0	60
94WC-6Co	Medium	2000	290	100	648	94	5.4	57
90WC-10Co	Fine	3100	450	80	620	90	6.0	67
71WC-12.5TiC-12TaC-4.5Co	Medium	1380	200	35	565	82	6.5	13
72WC-8TiC-11.5TaC-8.5Co	Medium	1720	250	50	558	81	6.8	23

- Type A, covering pore diameters less than 10 μm
- Type B, covering pore diameters between 10 and 25 μm
- Type C (Fig. 13), covering porosity developed by the presence of free carbon

Type A porosity is rated at a magnification of 200×, while types B and C porosity are rated at 100×. The degree of porosity is given by four numbers ranging in value from 02 to 08. The number provides a measure of pore volumes as a percentage of total volume of the sample.

Thermal Shock Resistance. Resistance to thermal shock is an important property that determines tool performance. No laboratory test has yet been developed that can consistently predict the resistance to thermal shock of a tool. However, empirical parameters have been suggested that can be used to evaluate tool materials for their probable resistance to thermal shock (Ref 30). A commonly used parameter is $\sigma k/E\alpha$, where σ is the transverse rupture strength, k is the thermal conductivity, E is Young's modulus, and α is the coefficient of thermal expansion. Table 9 lists representative values of this parameter for a number of WC-Co alloys. In general, the higher the value of $\sigma k/E\alpha$, the better the thermal shock resistance.

Abrasive Wear Resistance. Most producers of cemented carbides use a wet-sand abrasion test to measure abrasion resistance. In this test, a sample is held against a rotating wheel for a fixed number of revolutions while the sample and wheel are immersed in a water slurry containing aluminum oxide particles. Comparative rankings are reported, usually on the basis of a wear rating based on the reciprocal of volume loss (Fig. 17b).

Abrasion resistance is also measured by Cemented Carbide Producers Association procedure CCPA-112 with an abrasion resistance index as shown in Fig. 22. The lower the binder content and finer the tungsten carbide grain size, the harder and more abrasion resistant is the final product. Small additions of tantalum carbide (less than 1%) do not affect abrasion resistance, because tantalum carbide is a softer carbide than tungsten carbide. Large additions of tantalum carbide, however, lower abrasion resistance. Nickel binder grades have lower abrasion resistance than comparable cobalt binder grades.

Although standard test procedures are available, carbide producers have not agreed on a single test method, and so the values of abrasion resistance cited in the literature vary widely. Because of this variance, it is almost impossible to make valid comparisons among test results reported by different producers. It is also fallacious to use abrasion resistance as a measure of the wear resistance of cemented carbide materials when they are used for cutting steel or other materials; abrasion resistance in a standard test does not correspond directly to wear resistance in machining operations.

Generally, the abrasion resistance of cemented carbides decreases as cobalt content or grain size is increased (Fig. 17b). Abrasion resistance is

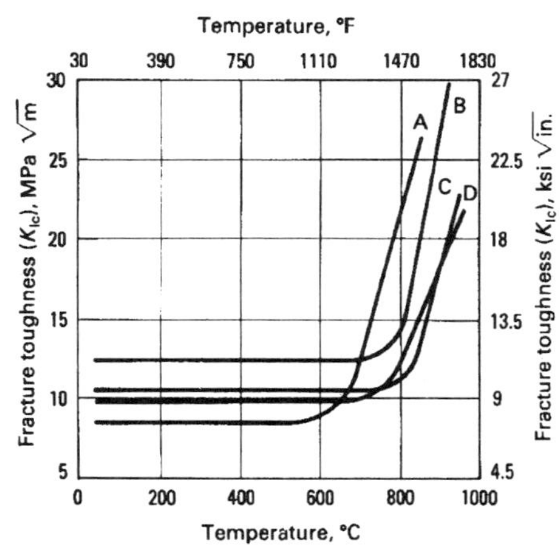

Fig. 21 Variation in fracture toughness (K_{Ic}) with temperature for a number of WC-Co base alloys. A, 86WC-2TaC-12Co; B, 85WC-9(Ti,Ta,Nb)C-6Co; C, 80WC-12(Ti,Ta,Nb)C-8Co; D, 96WC-4Co

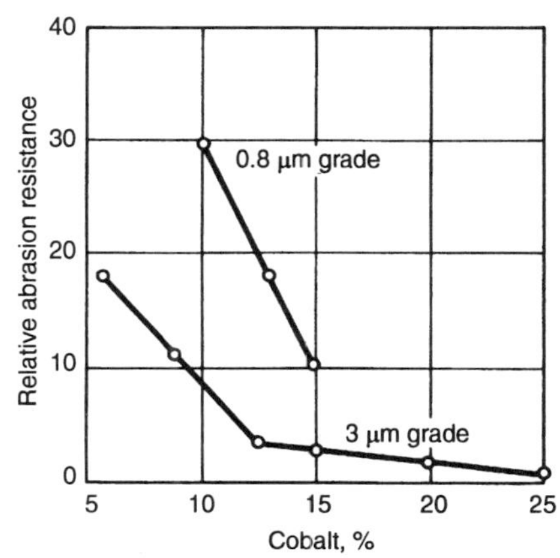

Fig. 22 Effect of tungsten carbide grain size and cobalt content on the relative abrasion resistance of cemented carbides

Table 10 Corrosion rates of cemented carbide test samples immersed for 48 h at 22 °C (72 °F)

	Tungsten carbide grain size,	Corrosion rate in:					
		10% nitric acid		10% hydrochloric acid		10% sulfuric acid	
Material	μm	g/m²/h	mg/in.²/h	g/m²/h	mg/in.²/h	g/m²/h	mg/in.²/h
94WC-6Co	1.5	2.45	1.58	0.09	0.06	0.19	0.12
85WC-15Co	3	18.85 (a)	12.16 (a)	0.14	0.09	1.4	0.91
85WC-15Co	0.8	19.82 (a)	12.79 (a)	0.12	0.08	0.68	0.44
94WC-6Ni	1.5	0.88	0.57	0.03	0.02	0.05	0.03

(a) Tungsten carbide surface layer spalled

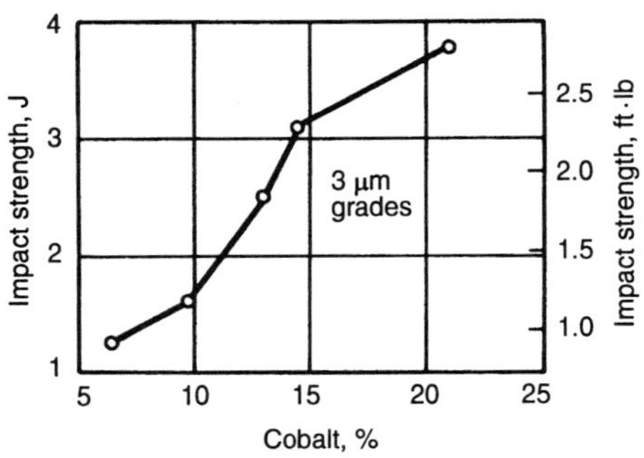

Fig. 23 Effect of cobalt content on the impact strength of cemented carbides. Average tungsten carbide grain size: 3 μm

also lower for complex carbides than for straight WC grades having the same cobalt content.

Impact strength, or shock resistance, is a function of cobalt binder content and tungsten carbide grain size. Higher cobalt levels and larger tungsten carbide grain sizes increase impact strength (Fig. 23). Tantalum carbide additions do not affect impact strength, however. Conditions that increase impact strength also increase thermal shock resistance

Corrosion Resistance. In corrosive attack of cemented carbides, the tungsten carbide component usually is not affected. The binder metal corrodes, exposing the angular tungsten carbide grains. A roughened surface results, which may cause tool wear or work material galling. To increase corrosion resistance, less binder metal should be exposed to attack. This can be accomplished by using a finer tungsten carbide grain size and lower binder metal content. Finer tungsten carbide grain size also yields more tungsten carbide surface area and results in a higher dissolved tungsten content in the binder metal, with improved corrosion resistance. Frequently, nickel binder grades exhibit higher corrosion resistance than comparable cobalt binder grades (Table 10). Small additions of chromium to both nickel and cobalt increase corrosion resistance, but cause embrittlement of the cemented carbide.

ACKNOWLEDGMENTS

- The section on cermets was adapted from an article in Volume 2 of the *ASM Handbook*, 1990, p 978–1007, by John L. Ellis, Consultant, and Claus G. Goetzel, Consultant and Lecturer.
- The section on cemented carbides was adapted from an article in Volume 2 of the *ASM Handbook*, 1990, p 951–977, by A.T. Santhanam, P. Tierney, and J.L. Hunt of Kennametal Inc.

REFERENCES

1. ASTM Committee C-21, "Report of Task Group B on Cermets," American Society for Testing and Materials, 1955
2. J.R. Tinklepaugh and W.B. Crandall, Chap. 1 in *Cermets*, Reinhold, 1960
3. E.C. Van Schoick, Ed., *Ceramic Glossary*, The Ceramic Society, 1963
4. R. Kieffer and F. Benesovsky, *Hartmetalle*, Springer-Verlag, 1965, p 437–489
5. P. Ettmayer and W. Lengauer, The Story of Cermets, *Powder Metall. Int.*, Vol 21 (No. 2), 1989, p 37–38
6. E. Rudy, Boundary Phase Stability and Critical Phenomena in Higher Order Solid Solution Systems, *J. Less-Common Met.*, Vol 33, 1973, p 43–70
7. F.V. Lenel, *Powder Metallurgy, Principles and Applications*, Metal Powder Industry Federation, 1980
8. P. Popper, *Isostatic Pressing*, British Ceramic Research Association, Heyden & Sons Ltd., 1976
9. R. Kieffer and P. Schwarzkopf, *Hartstoffe and Hartmetalle*, Springer-Verlag, 1953
10. R.M. German, Molding Metal Injection, *Powder Injection Molding*, Metal Powder Industries Federation, 1989
11. L.F. Pease III, Present Status in PM Injection Molding (MIM): An Overview, *Progress in Powder Metallurgy*, Vol 43, Metal Powder Industries Federation, 1987
12. K.J.A. Brookes, *World Directory and Handbook of Hardmetals*, 2nd ed., Engineer's Digest Publications, 1979
13. E. Lardner, Metallurgical Applications of Isostatic Hot Pressing, Chap. 10 in *High Pressure Technology*, Marcel Dekker, 1977
14. Vacuum Hot Press Furnaces for Powder Compaction, *Met. Powder Rep.*, Vol 37 (No. 11), 1982
15. S.W. Kennedy, "Development in Combination Debinder/Pressure Consolidation Furnace," Technical Note, Vacuum Industries Inc., 1989
16. R.E. Bauer, Sinter-HIP Furnaces Sintering and Compacting in a Combined Cycle, *Modern Developments in Powder Metallurgy*, Metal Powder Industries Federation, 1988
17. C.G. Goetzel, Infiltration Process, *Cermets*, Reinhold, 1960, p 73–81
18. H.W. Lavendel and C.G. Goetzel, Recent Advances in Infiltrated Titanium Carbides, *High Temperature Materials*, R.F. Heheman and G.M. Ault, Ed., John Wiley & Sons, 1959, p 140–154
19. W.B. Johnson, T.D. Claar, and G.H. Schiroky, Preparation and Processing of Platelet Reinforced Ceramics by the Directed Reaction of Zirconium with Boron Carbide, *Ceram. Eng. Sci. Proc.*, Vol 10 (No. 7/8), 1989
20. T.D. Claar, W.B. Johnson, C.A. Anderson, and G.H. Schiroky, Microstructure and Properties of Platelet Reinforced Ceramics Formed by the Directed Reaction of Zirconium with Boron Carbide, *Ceram. Eng. Sci. Proc.*, Vol 10 (No. 7/8), 1989
21. V.J. Tennery, C.B. Finch, C.S. Yust, and G.W. Clark, Structure-Property Correlations for TiB2-Based Ceramics Densified Using Active Liquid Metals, *Proc. of the Int. Conf. on the Science of Hard Materials*, Plenum, 1983
22. C.G. Goetzel and L.P. Skolnick, Some Properties of a Recently Developed Hard Metal Produced by Infiltration, *Sintered High-Temperature and Corrosion-Resistant Materials*, F. Benesovsky, Ed., Pergamon Press, 1956, p 92–98
23. R. Kieffer and F. Benesovsky, The Production and Properties of Novel Sintered Alloys (Infiltrated Alloys), *Berg Hüttenmänn. Monatsh.*, Vol 94 (No. 8/9), 1949, p 284–294
24. D.K. Creber, S.D. Poste, M.K. Aghajanian, and T.D. Claar, AlN Composite Growth by Nitridation of Aluminum Alloys, *Ceram. Eng. Sci. Proc.*, Vol 9 (No. 7/8), 1988, p 975
25. M.S. Newkirk, H.D. Lesher, D.R. White, C.R. Kennedy, A.W. Urquhart, and T.D. Claar, Preparation of Lanxide Ceramic Matrix Composites: Matrix Formation by the Directed Oxidation of Molten Metals, *Ceram. Eng. Sci. Proc.*, Vol 8 (No. 7/8), 1987, p 879–882
26. D.C. Halverson, A.J. Pyzik, I.A. Aksay, and W.E. Snowden, Processing of Boron Carbide-Aluminum Composites, *Advanced Ceramic Materials*, Preprint UCRL-93862, Lawrence Livermore National Laboratory, 1986
27. P.M. McKenna, U.S. Patent 3,379,503, 1968
28. R.C. Lueth, Advances in Hardmetal Production, *Proc. of the Metal Powder Report Conf.* (Lucerne, Switzerland), Vol 2, MPR Publishing Services Ltd., 1983
29. J.L. Ellis, Forming of Dispersion Type Aluminum Base Powder Metallurgy Nuclear Products, *Progress in Powder Metallurgy*, Vol 18, Metal Powder Industries Federation, 1962
30. W.D. Kingery, H.K. Bowen, and D.R. Uhlmann, *Introduction to Ceramics*, 2nd ed., John Wiley & Sons, 1960, p 828

Powder Metallurgy Beryllium

James M. Marder, Brush Wellman

BERYLLIUM, due to its unique mechanical and physical properties, is used in various nuclear, x-ray, aerospace, and consumer fields. Interest in beryllium parts was stimulated with the advent of the nuclear energy program. Current and past uses include diverse applications such as:

- Moderator material in nuclear reactors
- Neutron source in nuclear weapons
- Lightweight and dimensionally stable material in inertial guidance systems
- Use in reentry systems with the ability of beryllium to efficiently absorb heat
- Heatsink material for brakes on C5A, F-14, B-1B, F-18, and S3A aircraft
- Structural material in aerospace applications that require superior strength, thermal diffusivity, and rigidity at elevated temperatures in conjunction with light weight

Other applications include precision instruments; x-ray windows; satellite structures including antennas, control rods, small rocket nozzles, mechanical fasteners, and high-speed rotating shafts. Consumer products include laser mirrors for bar code readers. The wide acceptance of beryllium by the optical industry for use as support structures and mirror substrates is directly related to the density, stiffness, and reflectivity of beryllium.

During the last 20 years, efforts have increased to overcome the problem of brittleness associated with beryllium. New grades of beryllium with strength and ductility significantly superior to those available in the 1960s have been developed. Breakthroughs in P/M technology have been achieved with the use of balanced iron and aluminum levels, the introduction of impact attritioning as a means of beryllium powder production, and the use of cold and hot isostatic pressing techniques for consolidation. Additional information on beryllium processing can be found in the article "Production of Beryllium Power" in this Volume.

Physical Properties

Beryllium, with a density of 1.8477 g/cm^3, is a light metal with a combination of physical properties that is not found in any other metal or alloy. Beryllium has a high melting point (1283 °C, or 2341 °F), high modulus of elasticity (up to 300 GPa, or 44×10^6 psi), high specific heat (1.88 kJ/kg · K, or 0.45 Btu/lb · °F at room temperature), and a low coefficient of thermal expansion (average of 10.66×10^{-6}/°C in the temperature range of 2 to 38 °C, or 36 to 100 °F). The strength-to-weight ratio for wrought beryllium is higher than titanium. The following properties make beryllium one of the more desirable structural materials in the aerospace field:

Density, g/cm^3	1.8477
Melting point, °C (°F)	1283 (2341)
Modulus of elasticity, GPa (10^6 psi)	275–300 (40–44)
Specific heat, kJ/kg · K (Btu/lb · °F)	1.88 (0.46)
Thermal conductivity, W/m · K (Btu · ft/ft^2 · h · °F)	190 (104)
Linear coefficient of thermal expansion, 10^{-6}/°C	
At 25–100 °C	11.6
At 25–300 °C	14.5
At 25–600 °C	16.5
At 25–1000 °C	18.4

The following physical characteristics also contribute to the versatility of beryllium and its wide usage in a variety of applications:

- A protective oxide surface film, similar to that of aluminum, is formed in air.
- Oxidation at elevated temperature is not excessive below about 700 °C (1290 °F).
- Its electrical conductivity is 40% IACS.
- It is nonmagnetic.
- It reflects light (optical reflectivity, 50%; ultraviolet reflectivity, 55%; infrared (10.6 μm) reflectivity, 98%).
- It transmits sound (velocity of sound in beryllium is 12,600 m/s, 41,300 ft/s, 2½ times that of steel).
- It is transparent to x-rays.

The low thermal neutron absorption cross section of beryllium (0.009 b) and high neutron scattering cross section (6.9 b) make it useful as a moderator and a reflector for nuclear reactors. Beryllium is particularly well suited for reactor applications in which weight and space economy is important. Neutron economy and flux density are increased by the use of beryllium in nuclear reactors.

Mechanical Properties

The mechanical properties of beryllium are consistent with its application as an aerospace structural material. Tensile properties for various grades of beryllium are presented in Table 1. It is apparent that the yield and ultimate strengths compare quite favorably with those of aluminum alloys and are comparable to structural titanium alloys on a strength-to-weight basis.

Historically, the use of beryllium in structural applications was somewhat limited by low ductility values. Within the past twenty years, however, ductility levels have significantly increased, so that elongation values of 4 and 5%, a typical value for many titanium and aluminum alloys,

Table 1 Room-temperature tensile properties of commercial beryllium grades

Grade	Pressing method	Orientation	Ultimate tensile strength, MPa	Ultimate tensile strength, ksi	0.2% Offset yield strength, MPa	0.2% Offset yield strength, ksi	Elongation, %
S-200F	VHP	L	382	55.4	263	38.1	3.4
		T	407	59.0	262	38.0	6.1
S-200FH	HIP	L	441	64.0	345	50.0	4.9
		T	414	60.0	345	50.0	4.9
S-65	VHP	L	412	59.7	270	39.2	3.0
		T	454	65.8	273	39.6	5.4
S-65H	HIP	L	471	68.3	322	46.7	4.9
I-220	VHP	L	439	63.7	337	48.9	2.6
I-220H	HIP	L	523	75.9	448	65.0	3.7

VHP, vacuum hot pressed; HIP, hot isostatic pressing; L, longitudinal; T, transverse are no longer considered highly unusual. A great deal of process development was required in order to achieve this level of ductility. Improvements in purity were superseded by the realization that balancing the levels of impurity elements such as iron and aluminum was a more effective method of eliminating embrittling phases. In addition, powder making techniques such as impact grinding and inert gas atomization were shown to be effective in reducing the anisotropy of the final consolidated product. Use of hot isostatic pressing (HIP) as a powder consolidation process has likewise improved the isotropy and grain size control required to impart significant ductility. Methods to improve beryllium ductility at room temperature have included control of preferred orientation, improved purity, reduction of inclusions (beryllium oxide content, for example), altering inclusion distribution, and reducing grain size.

Preferred Orientation. Hot upsetting or rolling beryllium tends to align the basal planes parallel to the plane of rolling. This alignment of basal planes improves the ductility in the plane of sheet or block, because longitudinal stresses are mostly sustained by prism planes. Stresses resolved across the basal planes also are lower for a given applied load.

Grain Size. Strength of beryllium is related directly to grain size (Fig. 1) which is primarily a function of the particle size, oxide content, and consolidation temperature of the powder used. Historically, attempts to produce fine-grained, high-ductility beryllium by reducing the size of the source powder have met with only moderate success. Although the grain size is reduced, conventional powder manufacturing techniques such as disk mill attritioning and ball milling increase oxide content rapidly as particle size of the powder decreases. This has an embrittling effect, which often nullifies the effect of reduced grain size.

The impact attrition system, which is illustrated in Fig. 2, currently is used for production of beryllium powder. In this system, coarse powder from the feed hopper is fed into the gas stream, and the gas-borne powder is carried through the nozzle, where the velocity of the gas and powder mixture increases. The powder impacts the beryllium target and is comminuted. Gas-borne powder is transported to the primary classifier, where coarse particles drop out, and fines are carried to the secondary classifier. In this stage, the coarse particles (the product) drop out, and ultrafines (powders with high oxides and intermetallic contaminant contents) are discarded. High-pressure gas is turned off when the feed hopper is empty. When the system returns to ambient pressure, the valve at the bottom of the primary classifier is opened to deposit the coarse particles into the feed hopper for the next cycle.

Impact attritioning has enabled consistent control of powder composition by reducing impurity contamination and oxidation of powder particles. The process also results in improved powder configuration and morphology, resulting in improved isotropy and cleaner microstructure of the consolidated product.

Microalloying

Impurity elements that affect the behavior of beryllium include aluminum, iron, silicon, magnesium, copper, cobalt, nickel, oxygen, and carbon. Aluminum and iron can cause deleterious effects in beryllium. When present at the grain boundaries, aluminum causes reduced ductility in components for high-temperature applications.

Iron forms an intermetallic compound, $FeBe_{11}$, which results in precipitation hardening. Yield point phenomena observed in beryllium have been attributed to the pinning of dislocations by $FeBe_{11}$ precipitates. If the pinning of dislocations persists until the ultimate strength is reached, the metal fractures with little plastic deformation. However, moderate yield points that peak before the fracture stress may have little effect on the mechanical behavior of beryllium and may be of considerable advantage where elastic limit is a design factor. Aluminum and iron, when present together in beryllium, can form an $FeAlBe_4$ compound. The proper balancing of iron and aluminum content can eliminate the embrittling effects of aluminum or $FeBe_{11}$ precipitation without resorting to ultra-high-purity materials.

Silicon acts as a sintering aid in the consolidation of beryllium powder. In practice, the hot pressing of beryllium powder with low silicon content (<100 ppm) to full density is difficult because of the pressure tolerable with graphite dies. The high-pressure capability of hot isostatic pressing often is used for the consolidation of low-silicon, high-purity electrolytic powders. Alternatively, the material is doped with silicon either by elemental additions at the vacuum melting stage or by adding trichlorosilane at the vacuum hot pressing stage.

When present at grain boundaries, silicon, along with other low-melting constituents such as aluminum and magnesium, also affects the morphology, size, and distribution of beryllium

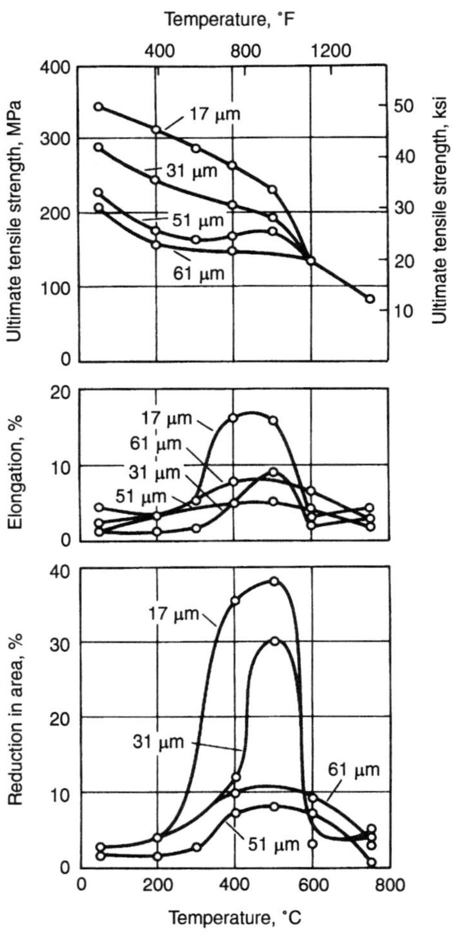

Fig. 1 Effect of grain size on the elevated-temperature tensile properties of vacuum hot pressed beryllium

Table 2 Nominal compositions of nuclear and structural grades of beryllium powders

Chemical composition, %	Nuclear grade S-65C	Structural grade S-200F
Beryllium assay(a)	99.0	98.0–98.5
Beryllium oxide(b)	1.0	1.5
Aluminum(b)	0.60	0.10
Boron(b)	0.00020	...
Cadmium(b)	0.0002	...
Calcium(b)	0.01	...
Carbon(b)	0.10	0.15
Chromium(b)	0.01	...
Cobalt(b)	0.0005	...
Copper(b)	0.015	...
Iron(b)	0.0800	0.13
Lead(b)	0.002	...
Lithium(b)	0.0003	...
Magnesium(b)	0.0600	0.08
Manganese(b)	0.012	...
Molybdenum(b)	0.002	...
Nickel(b)	0.02	...
Nitrogen(b)	0.02	...
Silicon(b)	0.06	0.06
Silver(b)	0.0010	...
Other metal impurities(b)	...	0.04

(a) Minimum. (b) Maximum

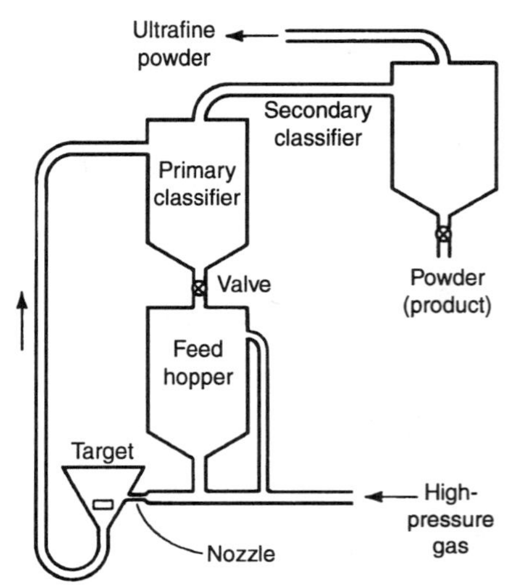

Fig. 2 Impact attrition mill

oxide. In general, these elements cause agglomeration and growth of beryllium oxide particles and also affect recrystallization, grain growth, and creep strength of beryllium. Copper, cobalt, and nickel exhibit extensive solubility in beryllium.

Among the nonmetallic impurities, only carbon and oxygen are present to any appreciable extent. Both have minimal solubility in beryllium. Carbon occurs as the cubic refractory compound beryllium carbide, which usually is found within the grains. This carbide accelerates the corrosion of beryllium, as it slowly hydrolyzes in the presence of moisture.

Oxygen forms stable beryllium oxide, which is primarily a grain-boundary constituent. However, it may also be found in the matrix. Typical concentrations of beryllium oxide in beryllium range from 0.7 to about 2% in commercial products.

Powder Production and Consolidation

The production of beryllium powder begins with extraction of beryllium from ore by chemical, thermal, and mechanical processing. Extraction produces beryllium "pebble," a relatively impure product containing magnesium fluoride and other impurities. These pebbles are vacuum remelted and cast into ingots, which exhibit a coarse grain size, low ductility, and low resistance to fracture. The vacuum cast ingots are cut up into chips by lathe turning. Chips are subsequently comminuted by impact attritioning (Fig. 2) to a particle size suitable for P/M fabrication. Table 2 gives the nominal compositions for nuclear and structural grades of beryllium powders. Production of other grades of beryllium is possible by varying the powder particle size, chemical composition, and processing pressure and temperature. Beryllium is one of the few metals that is fabricated exclusively by P/M techniques. Beryllium powders are consolidated to near-theoretical density by:

- High-vacuum/high-temperature sintering
- Vacuum hot pressing
- Cold isostatic pressing followed by vacuum hot pressing
- Cold isostatic pressing followed by pressureless sintering
- Cold isostatic pressing followed by hot isostatic pressing
- Direct hot isostatic pressing

The most common methods of consolidation, vacuum hot pressing and hot isostatic pressing, are described below.

Vacuum hot pressing was the most frequently used technique for producing commercial beryllium billet. Graphite dies are commonly used because they have a low coefficient of thermal expansion, which simplifies removal of the beryllium block after cooling. Graphite is available in sufficiently large sizes to produce vacuum hot pressings at more than 560 mm (22 in.) in diameter. Pressings that are too large in diameter for graphite dies are produced in nickel-alloy dies. IN-100 dies allow pressings up to 1830 mm (72 in.) in diameter.

Beryllium powder is vibratory loaded into the dies to achieve a density of about 55%. The die assembly is then placed in a vacuum hot pressing chamber. Vacuum, heat, and pressure are applied concurrently. Typical pressing temperatures range from 1000 to 1100 °C (1830 to 2010 °F), and pressure ranges from 3.5 to 14 MPa (500 to 2000 psi) with vacuum conditions below 500 torr, depending on powder characteristics and size of the compact. High-purity and fine particle size powder require much higher pressures.

Hot isostatic pressing offers advantages over vacuum hot pressing in that higher pressures are available and the limitations imposed by graphite dies are eliminated. Hot isostatic pressing has superseded vacuum hot pressing as the highest production volume process. The 3.5 to 14 MPa (500 to 2000 psi) pressure normally considered the maximum for graphite dies is marginal for some grades of beryllium and inadequate for high-purity grades. Hot isostatic pressing is generally done at 103 MPa (15 ksi) at temperatures varying from 760 to 1100 °C (1400 to 2010 °F).

High-purity beryllium powders lack sufficient concentration of impurity elements such as aluminum and silicon that act as sintering aids in the vacuum hot pressing of normal-purity powders. Although normal-purity beryllium powder can be vacuum hot pressed to nearly theoretical density by pressing at about 7 MPa (1000 psi) at 1050 °C (1920 °F), high-purity powder can be consolidated to only 98% density at the same temperature even when employing pressures as high as 24 MPa (3500 psi) or more.

These pressures are close to the upper limit at which consolidation can be carried out using graphite die tooling. Hot isostatic consolidation at pressures of about 103 MPa (15 ksi) and a temperature of 1100 °C (2010 °F) offers the only effective means of reliably consolidating high-purity powders to full theoretical density. High-temperature isostatically pressed beryllium has somewhat lower strength and better elongation. Optimal properties of strength and ductility are obtained by isopressing at lower temperatures.

Improved isotropy of both microstructure and mechanical properties is a major advantage of isostatic pressing. Because pressure is applied from all sides, shrinkage is uniform in all directions, and directionality originating from the pressing operation is prevented. However, there is some preferential alignment of powder particles in the starting powder bed, which results in some anisotropy in the hot isostatically pressed part. When hollow parts are hot isostatically pressed against a mandrel (usually made of copper), there is more shrinkage in the thickness direction, because friction between the mandrel and the part impedes free movement in the axial direction. Thus, the basal planes become aligned in the circumferential direction.

Isostatic consolidation of high-purity powder products improved physical and mechanical properties, including fine grain size, well-distributed oxide content, clean microstructure, good tensile and elongation properties, a smooth stress-strain curve that is free of yield point, a high degree of isotropy, good compression creep

Table 3 Typical tensile properties of beryllium

Material	Test orientation	Ultimate tensile strength		Yield strength(a)		Elongation(b), %
		MPa	ksi	MPa	ksi	
Block						
Normal-purity (hot-pressed)	L	370	53	266	38	2.3
structural grade(c)	T	390	56	273	39	3.6
Thermal or brake grade	L	294	42	196	28	2.7
	T	322	46	196	28	4.6
High-purity (isostatically	L	455	65	287	41	3.9
pressed)	T	455	65	287	41	4.4
High-oxide instrument	L	476	68	406	58	1.5
grade (hot pressed)	T	511	73	413	59	2.7
Fine grain size (isostatically	L	580	84	407	59	3.7
pressed)	T	587	85	407	59	4.2
Sheet (1–6.4 mm, or 0.040–0.250 in., thick)						
Normal-purity powder	...	531	77	372	54	16
Normal-purity ingot	...	352	51	172	25	7
Extrusions						
Normal-purity powder	...	655–690	95–100	345–518	50–75	8–13
High-purity powder	...	828–655	95–120	345–518	50–75	8–13
Forgings (no longer produced)						
Normal purity	...	483–600	70–87	435–600	63–87	0–4.5
Wire (0.05–0.64 mm, or 0.002–0.025 in., diameter) (no longer produced)						
High-purity ingot	...	966	140	793	115	3

L, longitudinal; T, transverse. (a) 0.2% offset. (b) Elongation in 50 mm (2 in.), except 250 mm (10 in.) for wire. (c) Structural grade contains about 1.8% BeO and thermal or brake grade about 0.9%. Source: Ref 1

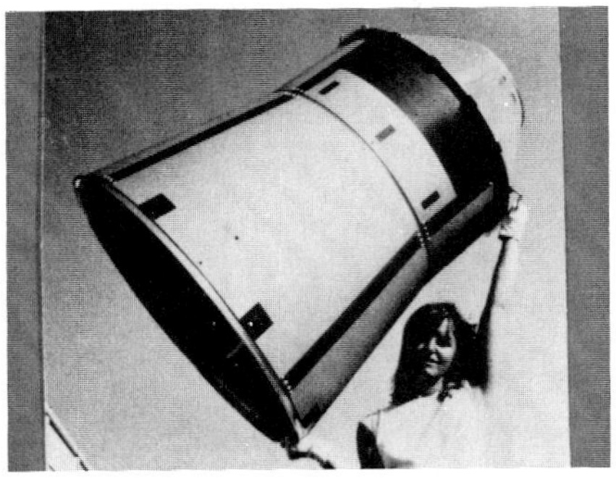

Fig. 3 Beryllium thrust tube assembly for the Japanese CS-2 Satellite. Courtesy of Brush Wellman

Fig. 4 Mast-mounted sight system (positioned above rotor) using a beryllium optical bench

properties, good high-temperature properties, and no grain growth below 1260 °C (2300 °F).

Wrought Product Forms

After powder is consolidated, beryllium is readily processed by conventional metalworking methods. Wrought forms of beryllium with optimal properties are produced from vacuum hot pressed P/M billets by conventional hot-working techniques. Wrought products exhibit improved mechanical properties in the predominant direction of metal flow. Mechanical properties of wrought products depend greatly on crystallographic orientation. Biaxial or triaxial deformation procedures frequently are used to produce parts with balanced directional properties.

Sheet, plate, and foil are the most commonly produced wrought forms of beryllium. These flat-rolled products are produced by encasing beryllium rolling blocks in steel jackets, followed by cross rolling at a moderate temperature. Biaxial deformation is accomplished by alternating the direction of the rolling pack through the rolling mill. This procedure produces flat-rolled products with uniform biaxial properties. See Table 3 for typical tensile properties of beryllium sheet. The standard condition that sheet and plate are supplied in is hot rolled, stress relieved, ground, and pickled.

Beryllium foil is available in thicknesses from 0.025 to 0.5 mm (0.001 to 0.019 in.) and panel sizes up to 125 by 300 mm (5 by 12 in.). Beryllium foil is also available in disk form in diameters up to 200 mm (8 in.) for x-ray window applications.

Beryllium extrusions are available with consistent mechanical properties. Sections are supplied to dimensions that are within commercial tolerances. Mechanical properties of extrusions are superior to the mechanical properties of hot pressed block in the direction of metal flow. Directional properties are produced in varying degree, as a function of crystallographic orientation.

Input billets for extrusions usually are machined from hot pressed block. Vacuum cast ingots and unsintered powder can be extruded for custom applications. Extrusion billets are jacketed in low-carbon steel cans that have shaped nose plugs and are extruded through a steel die between 900 to 1065 °C (1650 to 1950 °F). Warm extrusions are produced at about 425 °C (800 °F) when the volume of the product is large enough to justify the special tooling. Steel jackets are chemically removed from the extrusions.

Rod, tubing, and structural shapes are also available. Rod is produced in sizes from 9.5 to 135 mm (0.375 to 5.250 in.) in diameter. Tubing is available from 6.4 mm (¼ in.) outside diameter by 10 mm (0.040 in.) wall thickness up to 150 mm (6 in.) outside diameter by 3.20 mm (0.125 in.) wall thickness. A wide variety of structural and special shapes are also available. Current applications include structural shapes for aerospace applications, finned tubing for nuclear reactors, bar stock for forging and finish machining, fastener stock, rotating shafts, fuel element cladding, and draw stock for wire. Table 3 lists typical tensile properties of beryllium extrusions.

Applications of Beryllium

Aerospace Applications. The low density and high modulus of elasticity of beryllium make it an ideal choice as a structural member in aerospace components. The development of higher ductility grades of beryllium in various wrought forms, including sheet metal, has increased the potential application of beryllium in the aerospace field. Figure 3 illustrates a beryllium thrust tube (cone) assembly for the Japanese CS-2 satellite. Tests to determine the structural load-carrying capacity indicated that this structure has carried uniform loads at and/or above yield strength limits. Posttest analysis and inspection of the beryllium assembly indicated that it was flight qualified; loads were increased to over 160% of the theoretical limit. Another aerospace application is the Hughes MODIS satellite, which uses a beryllium structure as well as a beryllium optical system.

The low density, high elastic modulus, and dimensional stability of beryllium are also exploited in the optical bench for the McDonnell Douglas mast-mounted helicopter sight system. Laser, television, and infrared sensors are mounted to the bench, which must maintain precise alignment for proper operation of the target detector and designator system. Figure 4 shows the location of the mast-mounted sight on a Bell-OH-58D helicopter.

X-ray and Nuclear Applications. Because it has excellent transparency to soft x-rays and other radiation such as gamma rays, beryllium is used in x-ray tube windows. Because of its low atomic weight, beryllium passes x-rays 17 times better than an equivalent thickness of aluminum. Beryllium x-ray windows allow the use of long-wave x-rays that have greater density. Most of these applications require disks that are thinner than 0.25 mm (0.01 in.) and are cut from sheet or machined from vacuum hot pressed blocks.

The stiffness of beryllium and its transparency to x-rays are combined in medical radiation de-

Table 4 Physical properties of AlBeMet 162

Elastic modulus at 25 °C (77 °F), GPa (10^6 psi)	200 (29)
Elastic modulus at 200 ° C (390 °F), GPa (10^6 psi)	195 (28)
Poisson's ratio	0.167
Density, g/cm^3	2.01
Linear coefficient of thermal expansion at 25–100 °C, 10^{-6}/°C	16.35
Thermal conductivity at 25 °C (77 °F), W/m · K (Btu · in. · ft^2 · h · °F)	212 (14.7)
Specific heat, J/g · °C (Btu/lb · °F)	1.500 (0.36)
Electrical resistivity, μΩ · cm	3.43
Solidus temperature, °C (°F)	644 (1190)

Table 5 Specification mechanical properties for AlBeMet 162 aluminum-beryllium alloy

Specification and product form	0.2% Offset yield strength		Ultimate tensile strength		Elongation,
	MPa	ksi	MPa	ksi	%
AMS 7913, sheet	276	40	386	56	5
AMS 7911, HIPed block	193	28	262	38	2
AMS 7912, extrusions (longitudinal)	276	40	358	52	7

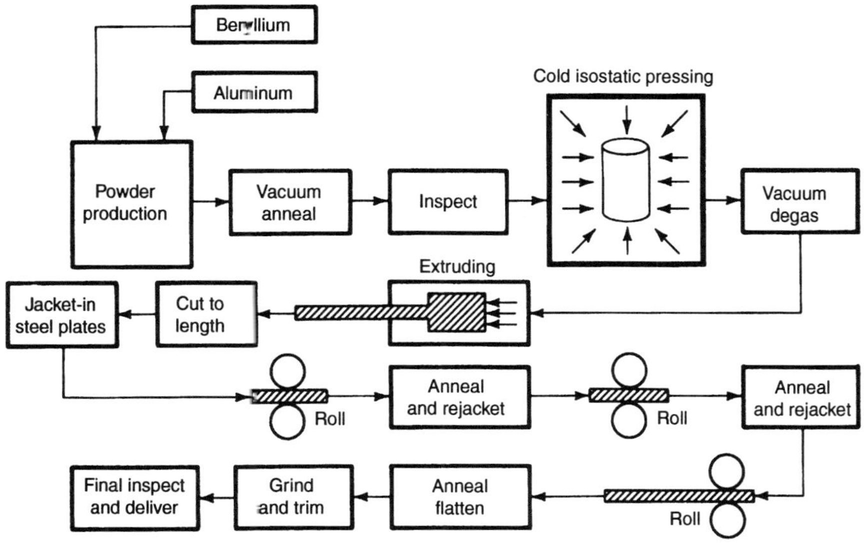

Fig. 5 AlBeMet aluminum-beryllium alloy sheet and plate manufacturing sequence

Fig. 6 Aluminum-beryllium alloy AlBeMet 160 disk drive actuator arm. Courtesy of Brush Wellman

tection devices such as computerized axial tomographic scanners and other related equipment with pressurized detection chambers. A stiff beryllium window resists the internal gas pressure with minimum deflection. The capability of rolling beryllium to a thin foil permits the transparency of beryllium to electromagnetic radiation to be used to the greatest advantage.

The growth of the nuclear energy field has provided new applications for beryllium components. Beryllium continues to be used as a reflector material in research and fast reactors. Beryllium has also been used to some degree in power reactors, for moderators, reflectors, and fuel cladding material.

Instrument Applications. Beryllium is used in precision-machined components in inertial guidance systems. Since its initial use in the mid-1950s, beryllium has come to be accepted as the baseline structural material for precision navigational instruments. It is now used in precision navigational aides for aircraft, spacecraft, missiles, ships, and submarines. The high modulus, high precision elastic limit, and dimensional stability under high G-forces make beryllium the material of choice for this application.

High-Temperature Applications. Beryllium has the highest specific heat of all structural materials, good thermal conductivity, relatively high melting point, and resistance to oxidation up to about 700 °C (1290 °F). It is a highly efficient heatsink, heat shield, and aircraft and automobile brake material. Beryllium is also used for making rocket engine nozzles and switchgear components. Beryllium has been used successfully for load-carrying disc brakes in military cargo and fighter aircraft and for making thrust chambers of rocket engines.

Nuclear Fusion Application. Beryllium has a unique combination of properties that dramatically improves the control of nuclear fusion energy reactors such as the Joint European Torus (JET). The combination of the excellent oxygen gettering performance of beryllium with its low atomic number eliminates plasma density disruptions, increases the time scale of magnetohydrodynamic effects, and reduces runaway photoneutron production. The JET has used S-65C beryllium extensively in belt limiters, target plate tiles, and the radiofrequency antenna.

Aluminum-Beryllium Alloys. AlBeMet 162 is an alloy of aluminum with 68 wt% beryllium. It was first used for the ventral fin of the SR-71 aircraft, when it was known as Lockalloy. Renewed interest in this alloy as part of the National Aerospace Plane effort led to its widening use in aerospace and commercial applications. The elastic and thermal-management properties of AlBeMet are significantly better than those of aluminum, although they do not equal those of beryllium. Physical properties of AlBeMet 162 are presented in Table 4, and mechanical properties of the three commercial available forms—extrusions, sheet, and hot isostatically pressed block—are presented in Table 5. The production process for sheet and plate and extrusions is shown in Fig. 5.

The properties of AlBeMet make it a meaningful candidate for many applications in which aluminum or titanium could also be used. In addition, fabrication such as welding and joining of AlBeMet can be easier than with beryllium. This makes it an attractive alternative when the properties of beryllium are well in excess of required values. Current aerospace uses include electronic modules for the F-22 tactical fighter, and antenna and structures of a variety of commercial communication satellites such as the OrbComm system. In commercial products, AlBeMet served as the voice coil actuator in high-performance hard disk drives (Fig. 6) in the early 1990s because of its light weight and high stiffness. In addition, the brake calipers of Formula I automobiles have been manufactured in AlBeMet and used by many racing teams.

REFERENCE

1. D.R. Floyd and J.N. Lowe, *Beryllium Science and Technology,* Plenum Press, 1979, p 321

SELECTED REFERENCES

- G.A. Baglyuk and G.E. Mazharova, Powder Hot Pressing Materials on the Basis of Nonferrous Metals, *Sov. Eng. Res.*, Vol 9 (No. 7), July 1989, p 64–67
- L.M. Beltz and D. Chellman, The Effects of Beryllium Content on Properties of Powder Metallurgy Aluminum—Beryllium Alloys, *Processing, Properties and Applications*, Vol 4, *Advances in Powder Metallurgy & Particulate Materials—1993*, Metal Powder Industries Federation, 1993, p 191–200
- R.B. Bhagat, Emerging P/M Metal Matrix Composites, *Particulate Materials and Processes*, Vol 9, *Advances in Powder Metallurgy & Particulate Materials*, Metal Powder Industries Federation, 1992, p 139–146
- L. Buisson, P. Bracconi, and C. Bonnet, Physical-Chemical Processes Involved in the Powder Metallurgy of Beryllium, *Powder Metallurgy World Congress (PM '94)*, Vol III, Editions de Physique, 1994, p 2233–2236
- R.G. Castro, P.W. Stanek, K.E. Elliott, and L.A. Jacobson, Plasma Spraying of Beryllium and Beryllium-Aluminum-Silver Alloys, *Processing and Fabrication of Advanced Materials III*, Minerals, Metals, and Materials Society/AIME, 1994, p 487–499
- P. Clement, *How the Beryllium Industry Is Building New Markets by Applying Isostatic*, MPR Publishing Services, 1991, p 11
- P. Clement and C. Boyer, HIP Processing of Powders, *Powder Processing Education: For The Year 2000,* Minerals, Metals, and Materials Society/AIME, 1994, p 239
- T.P. Clement, T.B. Parsonage, and M.B. Kuxhaus, Near-Net-Shape Processing Cuts Cost of Beryllium Optics, *Adv. Mater. Process.*, March 1992, p 37–40
- F. Grensing, D. Hashiguchi, and J. Marder, Fracture Toughness of Vacuum Hot Pressed

Beryllium Powder, *Advances in Powder Metallurgy 1990*, Vol 2, Metal Powder Industries Federation, 1990, p 27–36

- F.C. Grensing and D. Hashiguchi, Properties of Wrought Aluminum—Beryllium Alloys, *Processing, Properties and Applications*, Vol 4, *Advances in Powder Metallurgy & Particulate Materials—1993*, Metal Powder Industries Federation, 1993, p 179–189
- D.H. Hashiguchi, T.P. Clement, and J.M. Marder, Properties of Beryllium Consolidated by Several Near-Net Shape Processes, *J. Mater. Shaping Technol.*, Vol 7 (No. 1), 1989, p 23–31
- D.H. Hashiguchi and J.M. Marder, Property Evaluation of Spherical Beryllium Powder Billets, *Advances in Powder Metallurgy 1990*, Vol 2, Metal Powder Industries Federation, 1990, p 37–49
- B.G. Naik, V.K. Singhal, and A. Geddam, Electroless Nickel Coating of Powder Metallurgy Components: Problems in Coating of Beryllium, *PMAI Newsl.*, Vol 14 (No. 2), March 1988, p 24–28
- P.I. Stoev, I.I. Papirov, G.F. Tikhinskii, and A.A. Vasil'ev, Diagrams of Beryllium Isostatic Pressing, *Fiz. Met. Metalloved.*, Vol 78 (No. 1), July 1994, p 9–19
- P.I. Stoev, I.I. Papirov, G.F. Tikhinskii, and A.A. Vasil'ev, Diagrams of Isostatic Pressing of Beryllium, *Phys. Met. Metallogr. (USSR)*, Vol 78 (No. 1), July 1994, p 4–10
- B.S. Sun, X.H. Song, and L.H. Liu, Beryllium Mirrors Made by HIP-NNS Process, *Powder Metall. Technol. (China)*, Vol 11 (No. 1), Feb 1993, p 53–58
- B.S. Sun, Developments in Beryllium P/M Technology, *Powder Metall. Technol. (China)*, Vol 12 (No. 2), May 1994, p 126–130

Mechanical Properties of High-Performance Powder Metallurgy Parts

John C. Kosco, Keystone Powdered Metal Company

POWDER METALLURGY was first developed for processing refractory and other special materials and has now expanded into a competitive forming method for a wide variety of parts in automobiles, aircraft, lawn and garden equipment, sporting goods, and appliances. In terms of parts production, P/M bearings were the first part application, in the late 1920s, in response to a need for low-cost bearings for the electric motor business. In this case, it was found that the porosity inherent in early P/M materials could be filled with oil so a self-lubricating bearing material resulted.

The P/M bearing remains a major product to this day, but the scope of powder metallurgy has expanded from that beginning to encompass a wide variety of applications from low-density filters to fully dense structural components with properties equal to and, in some cases superior to, wrought alloys. For high-performance P/M structural materials, the single most significant characteristic of P/M materials is density and its effect on mechanical properties as shown in Fig. 1 and 2 for various iron compacts. There is a slow increase in properties as density increases from 6.8 to 7.4 g/cm^3. At that point, properties (especially Izod impact, Fig. 2) increase exponentially as density approaches the theoretical limit of steel (about 7.87 g/cm^3).

The traditional and least expensive method for producing P/M parts is a single press and sinter. By cold pressing and sintering only, ferrous P/M parts are produced to density levels of about 6.4 to 7.1 g/cm^3, about 80 to 90% of theoretical density, which for iron or low-carbon steel is 7.87 g/cm^3. Additional processing, such as infiltration or double press/double sinter, can be used to increase densities at somewhat higher cost relative to single press-and-sinter operations (Fig. 3). However, higher density is crucial in obtaining sufficient mechanical properties for high-performance parts. Therefore, this article briefly reviews P/M processes for higher-density parts followed by a compilation of mechanical

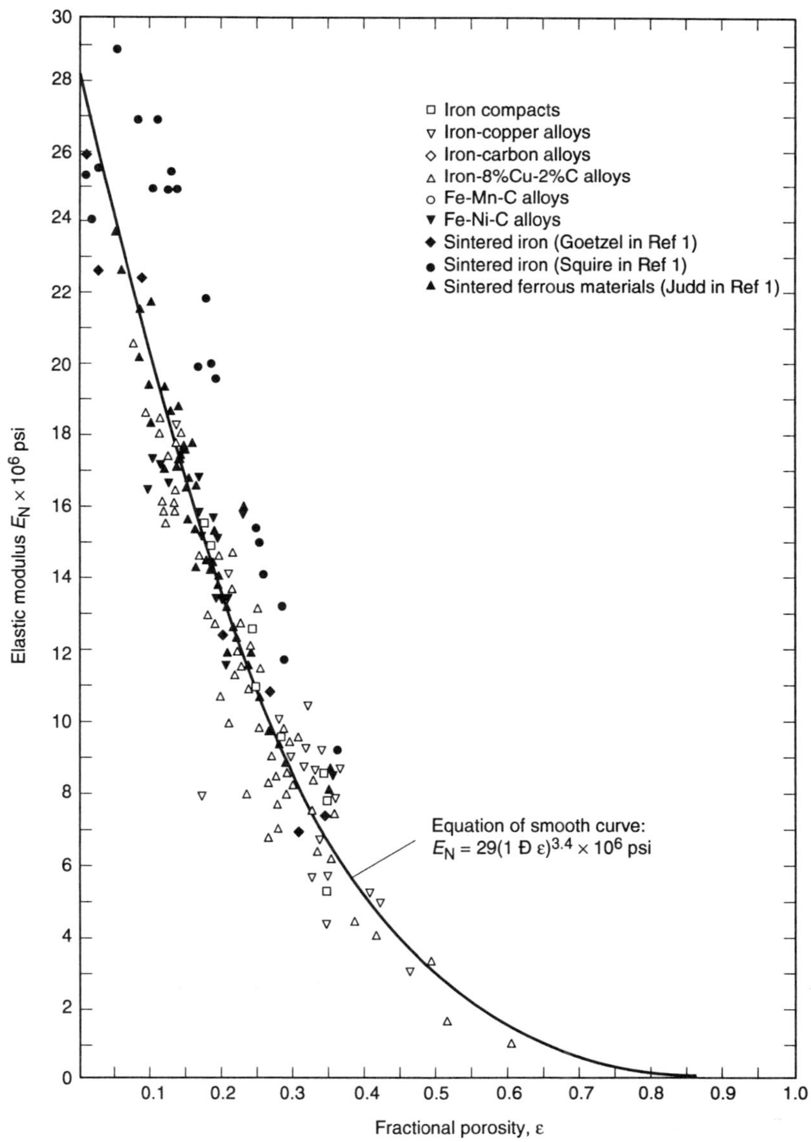

Fig. 1 Elastic modulus of ferrous sintered materials as a function of density. Source: Ref 1

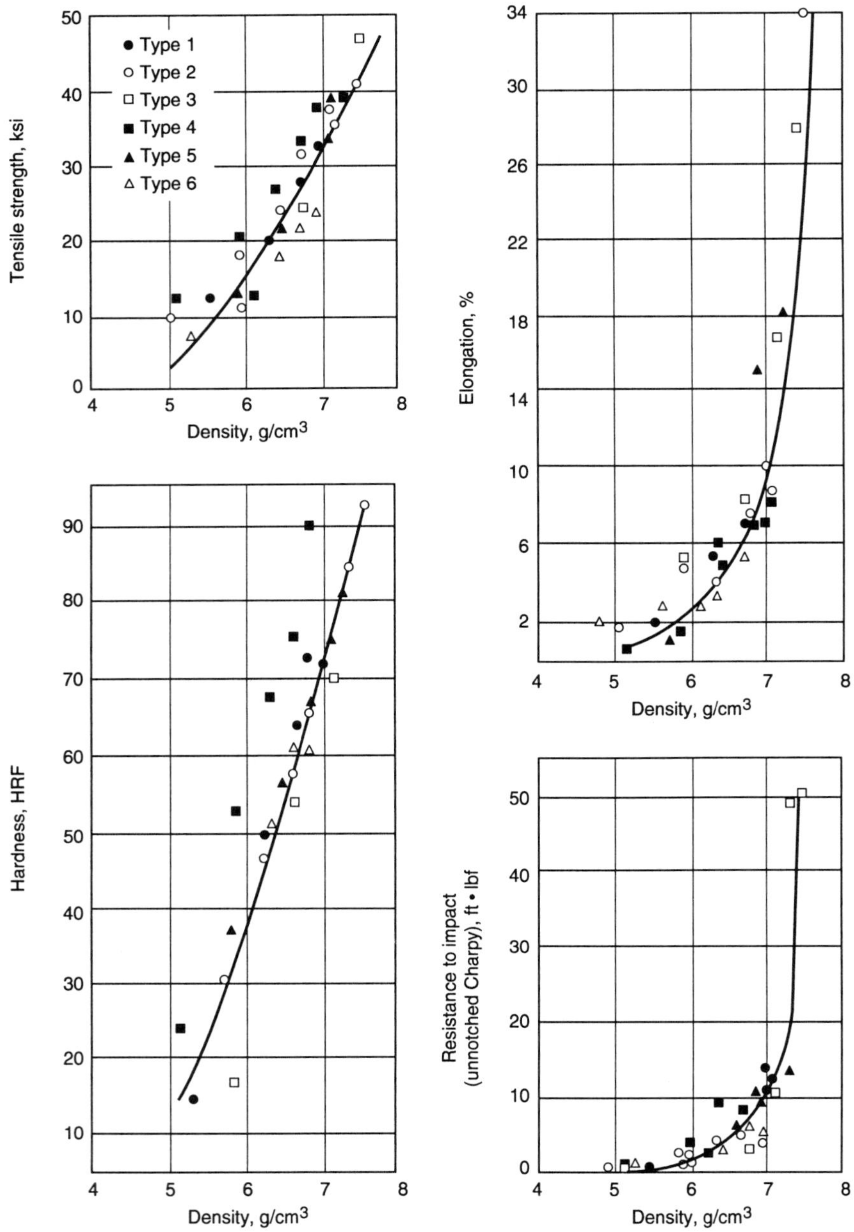

Fig. 2 Mechanical properties of iron compacts as a function of sintered density. Source: Ref 2. (See also Fig. 1 in the article "Hot Isostatic Pressing of Metal Powders.")

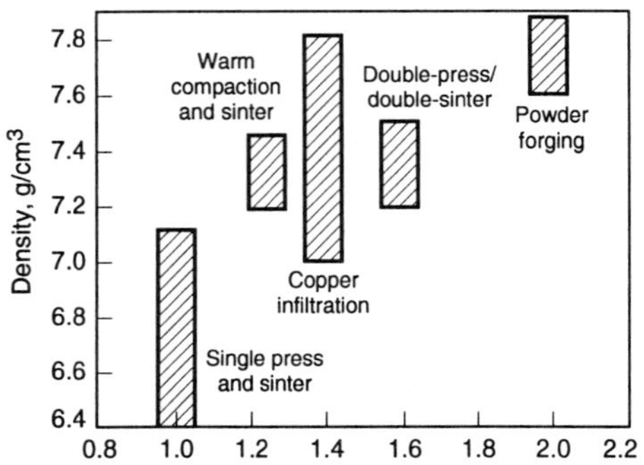

Fig. 3 Relative cost versus ferrous part density of several P/M processes. Note: Cost also depends on die/part size (e.g., see Fig. 2 in the article "Hot Isostatic Pressing of Metal Powders.")

property data as a function of density and processing. The mechanical property data are intended to show the influence of composition, processing, and heat treatment for ferrous parts. Other P/M materials and additional data on ferrous P/M materials are covered elsewhere in this Volume.

High-Density Processing

Powder metallurgy parts with theoretical densities less than 75% are considered to be low density, those above 90% are high density, and those in between are classified as medium density. Generally, structural parts have densities ranging from 80 to above 95%. The types of P/M processes and corresponding densities for ferrous structural parts are:

Process	Method	Density, g/cm^3
Single press sinter	Mold at 415–690 MPa (30–50 tsi) and sinter 1120–1260 °C (2050–2300 °F)	6.8–7.0
Sinter harden	Same as single-press sinter but using an alloy that will harden while cooling in the sintering furnace	6.8–7.1
Copper infiltration	Infiltration of porosity with copper	7.0–7.3
Warm mold	Uses a polymeric binder and molding at about 150 °C (300 °F)	7.2–7.4
DP/DS	Double press/double sinter	7.2–7.5
HFA	High-fatigue alloy via mold-sinter-size-heat treat	7.5–7.65
HF	Hot formed, forged at 760–980 °C (1400–1800 °F)	7.7–7.85

This section briefly summarizes only those processes capable of reaching at least 95% theoretical density, which would be a density in excess of 7.4 g/cm^3 for iron and low-alloy steel compacts. This excludes copper-infiltrated parts, even though Fig. 3 indicates densities in excess of 7.4 g/cm^3 for copper-infiltrated parts. Densities of copper-infiltrated steels can exceed 7.4 g/cm^3 (see the article "Infiltration" in this Volume). However, standard grades of copper-infiltrated steels have typical densities around 7.3 g/cm^3 (see "Appendix 1" in this Volume). From a practical point of view all commercially infiltrated steel parts have a density of less than 7.5 g/cm^3. Higher densities can be achieved, but special processing requirements have to be met, otherwise severe sticking of residue is experienced. A process has been developed for such high density infiltration, but it is not commercial (see the article "Copper-Infiltrated Steel" in this Volume).

Double Press/Double Sinter

For high-density P/M parts, secondary operations such as re-pressing are commonly used. In the re-pressing process, pressed parts are presintered 760 to 900 °C (1400 to 1650 °F). The presintered parts are then re-pressed to a higher density and sintered at a conventional temperature. Typical density levels of presintered parts are 6.8 to 7.0 g/cm^3, and this level is increased to 7.3 to 7.5 g/cm^3 after re-pressing. Consequently, mechanical properties of re-pressed P/M parts are increased, allowing them to be used for intermediate-stress applications.

A wide range of properties can be achieved with the double-press/double-sinter (DP/DS) process by varying the molding and sizing pressures. In this process, an iron/carbon-alloy element powder mix is compacted, then sintered at about 870 °C (1600 °F) to anneal the compact without putting carbon into solution. It is then remolded (sized) to increase density and improve dimensional control and sintered at conventional sintering conditions, 1120 to 1260 °C (2050 to 2300 °F) in a protective atmosphere. Table 1 illustrates a typical molding and sizing matrix and the properties obtained with the spe-

Table 1 Molding and sizing matrix for elemental nickel steel (2Ni-0.5C) powder FN-0205

Mold pressure			Sized at:			
MPa	tsi	Property (a)	0 MPa (0 tsi)	415 MPa (30 tsi)	620 MPa (45 tsi)	825 MPa (60 tsi)
415	30	Sintered density, g/cm³	6.67	7.13	7.32	7.43
415	30	Hardness, HRC	28	40	45	48
415	30	Modulus of rupture, MPa (ksi)	1040 (151)	1580 (229)	1820 (264)	1895 (275)
415	30	Izod impact toughness, J (ft · lbf)	3.8 (2.8)	6.1 (4.5)	8.1 (6.0)	8.8 (6.5)
415	30	Size change in sintering, in./in.	0.001	0.002	0.002	0.003
415	30	Ultimate tensile strength, MPa (ksi)	570 (83)	870 (126)	...	...

(a) Sintered at 1120 °C (2050 °F) and heat treated

Table 2 Equidensity properties of FN-0205

Molding/sizing pressures			Sinter density (a),	Rockwell	Modulus of rupture		Izod impact toughness		Ultimate tensile strength	
MPa	tsi	Condition	g/cm³	hardness	MPa	ksi	J	ft · lbf	MPa	ksi
415/690	30/50	Sintered	7.[illegible]0	77 HRB	1105	160	...	...	...	...
550/620	40/45	Sintered	7.[illegible]0	77 HRB	1205	175	...	...	...	...
690/550	50/40	Sintered	7.[illegible]0	77 HRB	1190	173	...	...	...	...
415/345	30/25	Sintered	7.[illegible]0	68 HRB	90	130	...	...	...	...
690/0	50/0	Sintered	7.[illegible]0	68 HRB	815	118	...	...	...	...
415/345	30/25	QT400 (b)	7.[illegible]0	38 HRC	1500	220	5.4	4.0	825	120
690/0	50/0	QT400 (b)	7.[illegible]0	38 HRC	1480	215	5.2	3.8	780	113

(a) Sintered at 1120 °C (2050 °F). (b) Quenched and tempered with temper at 400 °F

Table 3 Effect of high-temperature sintering on impact toughness of FN-0205

Molding/sizing pressures		Sintered density,	Sinter temperature		Sinter	Izod impact toughness	
MPa	tsi	g/cm³	°C	°F	atmosphere	J	ft · lbf
690/0	50/0	7.05	1140	2080	N_2	5.2	3.8
690/415	50/30	7.2[illegible]	1140	2080	N_2	9.5	7.0
690/825	50/60	7.4[illegible]	1140	2080	N_2	10.8	8.0
690/0	50/0	7.1[illegible]	1260	2300	Vacuum	11.3	8.3

cific combinations involved. A complete listing of DP/DS grades can be found in MPIF standard 35 from the Metal Powder Industries Federation.

Figure 4 is an example of data on hardness, modulus of rupture (transverse bending strength), and ultimate tensile strength (UTS) as influenced by the mold/size history of the part. Referring to the UTS plot, it is apparent that one could obtain a UTS value of 448 MPa (65 ksi) either by molding at 690 MPa (50 tsi) and sintering or by molding at 415 MPa (30 tsi), presintering, sizing at 415 MPa (30 tsi), and sintering. Normally, one would favor the single mold-sinter cycle for economics, but factors such as the complexity of the part and cost of the die might provide reason to go the other route. Table 2 summarizes typical properties for FN-0205 (2Ni-0.5C) steel processed to equal densities via different combinations of mold/size pressures. Regardless of the manner in which the density was achieved, properties are essentially the same. Figure 5 summarizes hardness, modulus of rupture and impact properties for heat treated 2Ni steel. Again, a range of properties is possible depending on processing conditions.

As with wrought materials, alloying affects P/M properties. Figure 6(a) plots the sintered density for two common P/M alloys, FN-0205 and FL-4205. The former is made by blending elemental atomized iron with elemental nickel and graphite powders. FL-4205 is similar in chemistry, although it does contain 0.5% Mo, but the major difference is that 4205 is prealloyed prior to atomization. The final alloy is obtained by blending in graphite. The presence of the nickel and molybdenum in solution hardens the base powder and makes it harder than pure iron. As a result, it is more difficult to densify the prealloyed powder. Figure 6(a) gives data for these two alloys, both molded at 550 MPa (40 tsi), presintered and sized. Note that higher sizing pressures are required to obtain equal sintered density with FL-4205. In Fig. 6(b-d), properties for FN-0205 and FL-4205 are plotted as a function of density. It is obvious that the alloy addition has a beneficial effect at equal density. However, this is not necessarily true for equal processing conditions.

In recent years, there has been a shift toward high-temperature sintering capacity in the P/M industry. Belt furnaces are limited to about 1150 °C (2100 °F) max, but new pusher and walking beam and vacuum furnaces now permit sintering to 1315 °C (2400 °F). The higher temperatures lead to increased diffusion and rounding of porosity with property improvements as indicated in Table 3. More information on high-temperature sintering is provided in the article "High-Temperature Sintering" in this Volume.

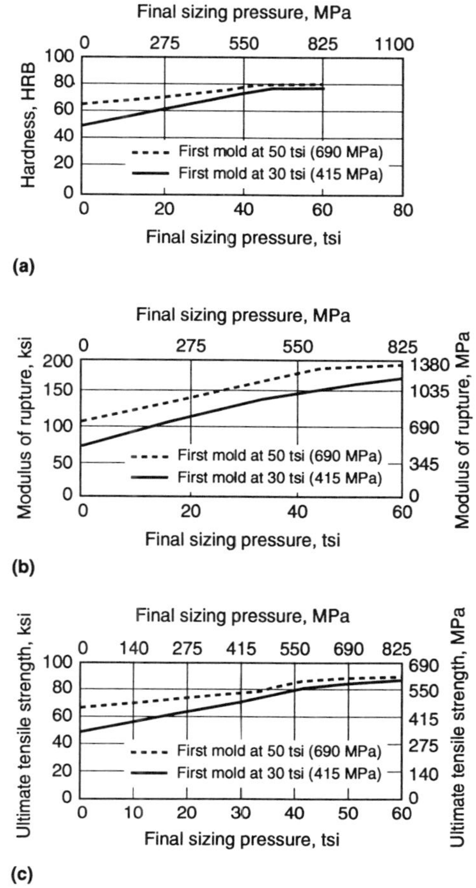

Fig. 4 Effect of sizing pressure on mechanical properties of DP/DS nickel steel (2Ni-0.5C) FN-0205 with a final sinter at 1120 °C (2050 °F). (a) Hardness. (b) Modulus of rupture. (c) Ultimate tensile strength

Warm Compaction

Warm compaction is a relatively new P/M production method that utilizes heated tooling and powder during the compaction step. The powder and tooling are typically heated between 130 and 150 °C (260 and 300 °F). In order for the powder premix to perform at these temperatures, proprietary lubricants have been developed that provide better compressibility and lower die ejection forces than conventional lubricants. Lubricants can also incorporate polymeric binders to limit segregation and provide enhanced flow characteristics of the powder premix (see the article "Warm Compaction" in this Volume).

Warm compaction has several advantages including better compressibility and higher green strengths than conventional cold pressing and sintering. The increased green strength reduces green chipping and cracking due to part handling prior to sintering and makes possible the crack-free compaction of complex multiple-level parts. Additionally, the higher green strength provides an opportunity to machine the P/M part in the green state. This capability is critical in the use of high-performance alloy systems that achieve high hardness in the as-sintered state.

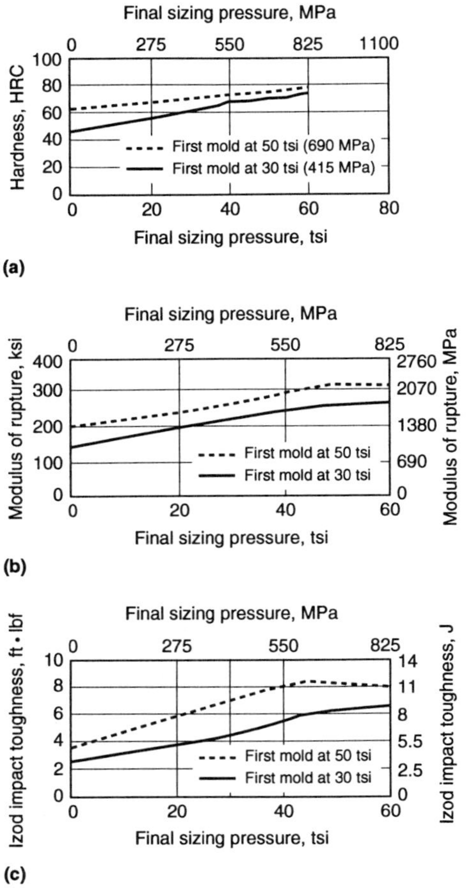

Fig. 5 Effect of sizing pressure on mechanical properties of DP/DS quenched-and-tempered nickel steel (FN-0205/QT400) with a final sinter at 1120 °C (2050 °F). (a) Hardness. (b) Modulus of rupture. (c) Izod impact toughness

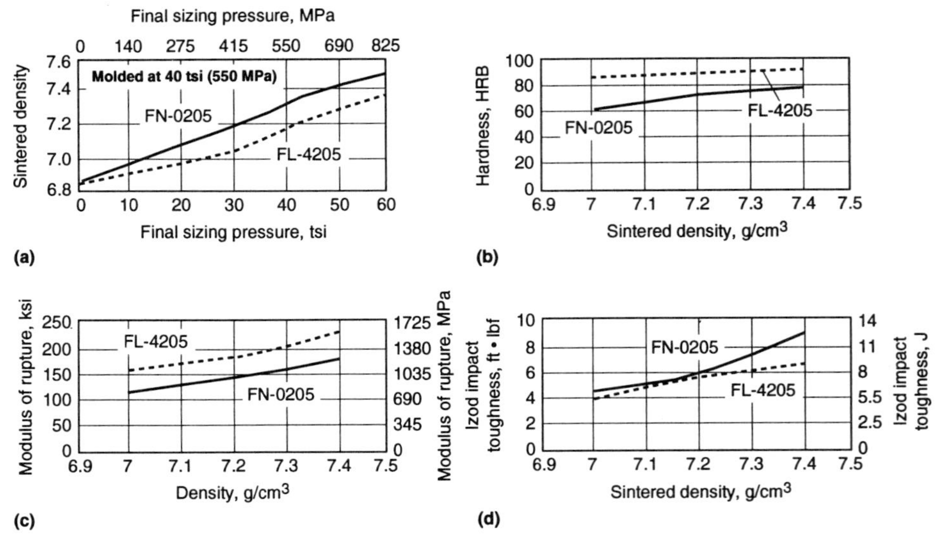

Fig. 6 Comparison of elemental nickel steel (2Ni-0.5C) FN-0205 with prealloyed nickel-molybdenum steel FL-4205. (a) Sintered density versus sizing pressure. (b) Hardness versus density. (c) Modulus of rupture. (d) Izod impact toughness. Final sintering at 1120 °C (2050 °F)

However, warm compaction also enables P/M fabricators to single press and single sinter (SP/SS) ferrous P/M parts to densities slightly above 7.4 g/cm^3. Final densities depend to a large degree on the compaction pressures and chemistry. Single-press/single-sinter warm compaction is also competitive with conventional DP/DS densities (Fig. 7).

Warm compaction can also be combined with DP/DS methods to achieve densities well above 7.4 g/cm^3 for ferrous P/M parts. In DP/DS warm compaction, the first press is at typical warm compaction temperatures followed by a presinter similar to conventional DP/DS. The second compaction can be at ambient temperature, followed by the final sinter. A comparison of properties and densities are summarized in Fig. 7 and Table 4.

Surface Densification

Surface densification is a promising method for developing high density in P/M parts. To a great extent, the structure and properties of the zone within about 0.25 to 1 mm (0.010 to 0.040 in.) of the surface determine the performance of components. For example, the wear properties of a sliding member depend primarily on the surface and much less so on the core. This principle is applied with much success in coated cutting tools and flame sprayed parts. In heavily loaded gears and cams, the highest Hertzian stresses are usually manifest within 0.05 to 0.25 mm (0.002 to 0.010 in.) of the surface so, again, properties of the surface dominate performance. Thus it may not be necessary to densify the entire part; instead, selective densification of specific areas may suffice.

A number of techniques have been developed to increase the density of near-surface areas to wroughtlike levels. These include hot working, rolling, swaging, shot peening, and other such processes. Each of these provides heavy mechanical working to a localized zone with the net effect that pores are eliminated. In addition, the surfaces of those voids are forced into close contact and bonded together.

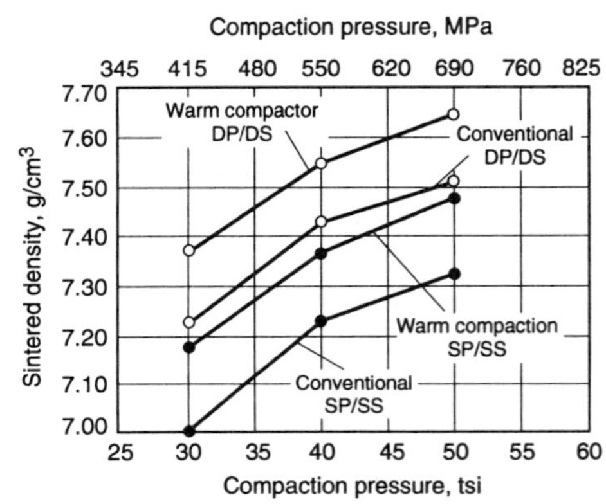

Fig. 7 Sintered density versus compaction pressure for a prealloyed nickel-molybdenum steel powder (Ancorsteel 85HP) by warm compaction and conventional cold press and sinter. Final sinter at 1260 °C (2300 °F) for 30 min, 90% Na/10% H_2 atmosphere

Fully Dense Ferrous Parts

Ferrous parts of near and full theoretical density can be produced by a variety of thermomechanical methods. These methods involve various amounts of heat and mechanical work to aid densification of the powder mass. For production of low-alloy steel parts, powder forging and injection molding have received the most attention. Both of these powder process technologies are formulated specifically to yield a product as close to full density as possible. This contrasts significantly with the conventionally processed products where attainment of full density is not the primary goal.

Powder Forging

Powder forging involves fabrication of a preform by conventional press-and-sinter processing, followed by hot forging of the porous preform into a final shape with substantial densification. Forging is generally performed in one blow in confined dies to eliminate flash formation and achieve net shapes. Thus, parts containing through holes and complex configurations can be manufactured with little or no material loss due to trimming, hole punching, machining, or grinding.

Powder forging is used to obtain higher density in P/M parts for high-stress applications, where pores must be minimized or completely eliminated by hot upsetting and/or hot re-pressing. In hot upsetting, the preform experiences a significant amount of lateral material flow, which results in a combination of normal and shear stresses around the pores. Thus, spherical pores become flattened and elongated in the direction of lateral flow. The sliding motion due to shear stresses breaks up any residual interparticle oxide films and leads to strong metallurgical

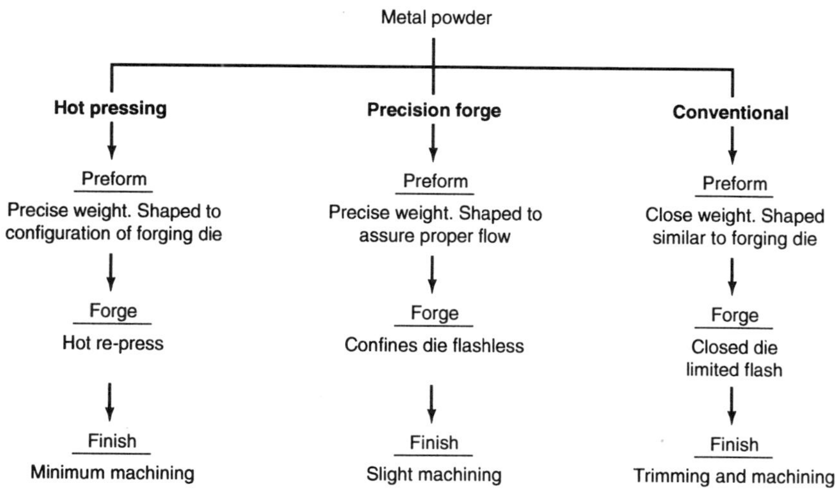

Fig. 8 Typical sequences of P/M forging processes. Hot pressing is closely related to the present P/M practice. Precision forging combines the advantages of P/M and forging. The last process is an extension of the present forging practice.

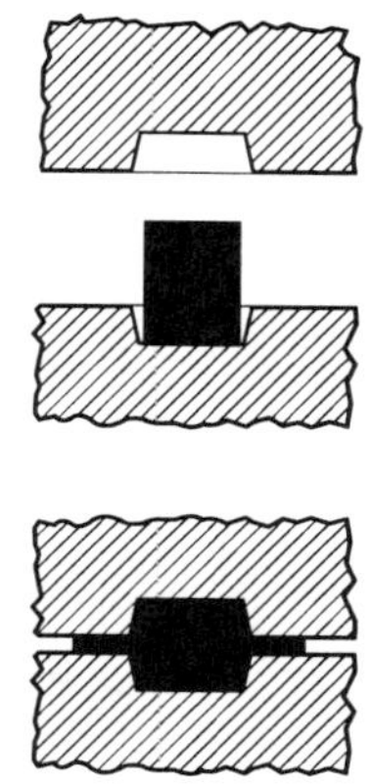

Fig. 9 Forging with flash such as in conventional forging

Table 4 Effect of SP/SS and DP/DS processing on the density and transverse rupture strength

Base material	Processing technique	Compaction pressure MPa	Compaction pressure tsi	Sintered density, g/cm³	Transverse rupture strength MPa	Transverse rupture strength ksi	Hardness, HRB
Ancorsteel 85HP(a)	Warm compaction, SP/SS	415	30	7.18	1100	159.5	83
		550	40	7.36	1325	192.1	88
		700	50	7.47	1340	194.6	91
	Warm compaction, DP/DS	415	30	7.36	1340	194.6	89
		550	40	7.54	1495	216.7	92
		700	50	7.63	1530	221.7	94
	Cold compaction, SP/SS	415	30	7.01	995	144.3	79
		550	40	7.22	1145	166.0	85
		700	50	7.32	1280	185.7	88
	Cold compaction, DP/DS	415	30	7.22	1240	180.2	87
		550	40	7.42	1460	212.0	92
		700	50	7.50	1530	221.9	94
Distaloy 4800A(b)	Warm compaction, SP/SS	415	30	7.04	1205	174.8	90
		550	40	7.27	1375	199.6	94
		700	50	7.35	1450	210.1	97
	Warm compaction, DP/DS	415	30	7.20	1505	218.5	94
		550	40	7.40	1690	245.4	97
		700	50	7.49	1695	246.0	100
	Cold compaction, SP/SS	415	30	6.90	975	141.4	81
		550	40	7.08	1265	183.5	87
		700	50	7.20	1340	194.1	91
	Cold compaction, DP/DS	415	30	7.12	1375	199.4	90
		550	40	7.34	1645	238.4	94
		700	50	7.46	1800	261.1	96

(a) Ancorsteel 85HP is a prealloyed steel powder containing 2.0% Ni, 0.85% Mo, 0.4% graphite, and 0.6% lubricant. Compacts were sintered at 1260 °C (2300 °F). (b) Distaloy 4800A is a diffusion-alloyed steel powder containing 4.0% Ni, 1.5% Cu, 0.50% Mo, 0.5% graphite, and 0.6% lubricant. Compacts were sintered at 1120 °C (2050 °F). Source: Hoeganaes Corporation

bonding across collapsed pore interfaces. This enhances dynamic properties such as fracture toughness and fatigue strength.

In hot re-pressing, densification is mainly in the direction of pressing. This results in very little material movement in the horizontal direction and thus limited lateral flow. As densification proceeds, the stress state approaches a pure hydrostatic condition. A typical pore simply flattens, and the opposite sides of the pore are brought together under pressure. Hot re-press forging requires higher forging pressures than does hot upset forging for comparable densification. The decreased interparticle movement compared with upsetting reduces the tendency to break up any residual interparticle oxide films and may result in lower ductility and toughness.

Powder Forging Methods. The development of a viable powder forging system requires the consideration of many process parameters. The mechanical, metallurgical, and economic outcomes depend to a large extent on operating conditions, such as temperature, pressure, flow/feed rates, atmospheres, and lubrication systems. Equally important consideration must be given to the types of processing equipment, such as presses, furnaces, dies, and robotics, and to secondary operations, in order to obtain the process conditions that are most efficient. These considerations are detailed in the article "Powder Forging and Hot Pressing" in this Volume.

In general, however, there are three methods of forging from metal powder compacts (Fig. 8):

- *Hot pressing:* Preform for precise weight, hot re-press and machine
- *Precision forging:* Preform for weight and shape, closed-die forge flashless and finish machine
- *Conventional forging:* Preform for weight and shape, forge with flash, trim and machine

These general methods are briefly described below.

Hot repressing requires a preform shape close to that of the final part except for its height in the forging direction. However, the weight of the preform needs to be tightly controlled. In this process, the friction between die and preform during hot forging is high, and therefore, the pressure needed to get complete densification would also be high, which would involve rapid wear of forging tools. Hot re-pressing is used in applications where densities on the order of 98% of theoretical density are satisfactory.

Precision forging is the most widely used method in industry. Precision forging is a flashless process, and the shape of the preform is simpler than that of the final part. The desired final shape is produced to closely controlled dimensions in the hot forging step. The lateral flow during deformation improves densification. Because this process is precise, tight control of preform weight and shape is required.

Conventional forging is a process to be considered more from the point of view of the conventional forge shop. This process involves flash, and the part is not produced to net shape (Fig. 9). In this case, relative to the other two processes, weight and shape control is not as critical.

Mechanical Properties of Ferrous Powder Forged Materials. Unlike the extensive deformation in wrought metalworking, powder forged

Table 5 Typical mechanical properties of P/M forged low-alloy steels

All materials are in the hardened-and-tempered condition unless otherwise indicated.

Material	Processing	Ultimate tensile strength, MPa	Ultimate tensile strength, ksi	0.2% yield strength, MPa	0.2% yield strength, ksi	Elongation in 25 mm (1 in.), %	Reduction in area, %	Charpy V-notch impact energy, J	Charpy V-notch impact energy, ft · lbf	Hardness	Fracture toughness (K_{Ic}), MPa $\sqrt{m}$	Fracture toughness (K_{Ic}), ksi $\sqrt{in.}$	Density, % of theoretical	Ref
Fe-2MCM-0.67C(a)(b)	...	960	139.3	590	86	...	12	...	...	98 HRB	...	...	...	3
Fe-2MCM-0.67C(a)	...	1900	275.6	1500	218	...	4.5	...	...	49 HRC	...	...	...	...
4120	Sintered at 1315 °C (2400 °F), re-pressed	701	101.7	616	89.4	14	46	38	28	20–25 HRC	...	...	100	4
1520	Sintered at 1315 °C (2400 °F), re-pressed	936	135.7	...	...	9	13	39	29	20–25 HRC	...	...	100	...
4130	Gas atomized, –65 mesh	1586	230	1303	189	5	3	10	7.5	46 HRC	49	45	100	5
4640	Gas atomized, –65 mesh	...	...	...	...	...	...	7	5	55 HRC	36	33	100	5
	Water atomized	...	...	...	...	...	...	7	5	42 HRC	37	34	100	5
	Sintered at 1200 °C (2190 °F)	1040	150.8	1000	145	20	40	36	26	310–350 HV	...	...	99	6
Fe-2Ni-0.35C	Mixed elemental powders	938	136	600	87	13	44	...	13	31 HRC	...	...	99	7
Fe-0.55Ni-0.32Mo-0.47Mn-0.23Cr-0.30C	Sintered at 1200 °C (2190 °F)	1020	147.9	970	141	17	37	46	34	...	...	...	...	6
Fe-3Cu-0.5C-0.3S	...	873	127	...	...	6.5	...	...	...	274 HV	...	...	99	8
Fe-9Cu-0.34Mn-0.43Ni-0.65Mo-0.31C	...	1675	245	1410	205	13	31	19	14	49 HRC	...	...	99	9
Fe-0.35Mn-0.57Mo-1.95Ni-0.5C	...	1200	174	1120	162	10	19	30	22	475 HV	...	...	99	10
4630 modified	Sintered at 1205 °C (2200 °F)	148	215	1331	193	6	10	8	6	42 HRC	...	...	98	11

(a) MCM is a master alloy containing 20% Mn, 20% Cr, 20% Mo, and 7% C. (b) As-sintered condition

Table 6 Mechanical property and fatigue data for P/F-4600 materials

Sintered at 1120 °C (2050 °F) in dissociated ammonia unless otherwise noted

Forging mode	Carbon, %	Oxygen, ppm	Ultimate tensile strength, MPa	Ultimate tensile strength, ksi	0.2% offset yield strength, MPa	0.2% offset yield strength, ksi	Elongation in 25 mm (1 in.), %	Reduction of area, %	Room-temperature Charpy V-notch impact energy, J	Room-temperature Charpy V-notch impact energy, ft · lbf	Core hardness, HV30	Fatigue endurance limit, MPa	Fatigue endurance limit, ksi	Ratio of fatigue endurance to tensile strength
Blank carburized														
Upset	0.24	230	1565	227	1425	207	13.6	42.3	16.3	12.0	487	565	82	0.36
Re-press	0.24	210	1495	217	1325	192	11.0	34.3	12.9	9.5	479	550	80	0.37
Upset(a)	0.22	90	1455	211	1275	185	14.8	46.4	22.2	16.4	473	550	80	0.38
Re-press(a)	0.25	100	1455	211	1280	186	12.5	42.3	16.8	12.4	468	510	74	0.36
Upset(b)	0.28	600	1585	230	1380	200	7.8	23.9	10.8	8.0	513	590	86	0.37
Re-press(b)	0.24	620	1580	229	1305	189	6.8	16.9	6.8	5.0	464	455	66	0.29
Quenched and stress relieved														
Upset	0.38	270	1985	288	1505	218	11.5	33.5	11.5	8.5	554	...	...	...
Re-press	0.39	335	1960	284	1480	215	8.5	21.0	8.7	6.4	...	...	...	...
Upset	0.57	275	2275	330	...	...	3.3	5.8	7.5	5.5	655	...	...	...
Re-press	0.55	305	1945	282	...	...	0.9	2.9	8.1	6.0	...	...	...	...
Upset	0.79	290	940	136	...	...	...	...	1.4	1.0	712	...	...	...
Re-press	0.74	280	1055	153	...	...	...	...	2.4	1.8	...	...	...	...
Upset	1.01	330	800	116	...	...	...	...	1.3	1.0	672	...	...	...
Re-press	0.96	375	760	110	...	...	...	...	1.6	1.2	...	...	...	...
Quenched and tempered														
Upset(c)	0.38	230	1490	216	1340	194	10.0	40.0	28.4	21.0	473	...	...	...
Re-press(c)	...	...	1525	221	1340	194	8.5	32.3	...	...	...	...	...	...
Upset(d)	0.60	220	1455	211	1170	170	9.5	32.0	13.6	10.0	472	...	...	...
Re-press(d)	...	...	1550	225	1365	198	7.0	23.0	...	...	...	...	...	...
Upset(e)	0.82	235	1545	224	1380	200	8.0	16.0	8.8	6.5	496	...	...	...
Re-press(e)	...	...	1560	226	1340	194	6.0	12.0	...	...	...	...	...	...
Upset(f)	1.04	315	1560	226	1280	186	6.0	11.8	9.8	7.2	476	...	...	...
Re-press(f)	...	...	1480	215	1225	178	6.0	11.8	...	...	...	...	...	...
Upset(g)	0.39	260	825	120	745	108	21.0	57.0	62.4	46.0	269	...	...	...
Upset(g)	0.58	280	860	125	760	110	20.0	50.0	44.0	32.5	270	...	...	...
Upset(h)	0.80	360	850	123	600	87	19.5	46.0	24.4	18.0	253	...	...	...
Upset(i)	1.01	320	855	124	635	92	17.0	38.0	13.3	9.8	268	...	...	...

(a) Sintered at 1260 °C (2300 °F) in dissociated ammonia. (b) Sintered at 1120 °C (2050 °F) in endothermic gas atmosphere. (c) Tempered at 370 °C (700 °F). (d) Tempered at 440 °C (825 °F). (e) Tempered at 455 °C (850 °F). (f) Tempered at 480 °C (900 °F). (g) Tempered at 680 °C (1255 °F). (h) Tempered at 695 °C (1280 °F). (i) Tempered at 715 °C (1320 °F)

materials undergo relatively little material deformation, and their mechanical properties have been shown to be relatively isotropic. The mechanical properties of powder forged materials are usually intermediate to the transverse and longitudinal properties of wrought steels. The rotating-bending fatigue properties of powder forged materials have also been shown to fall between the longitudinal and transverse properties of wrought steel of the same tensile strength.

While the performance of machined laboratory test pieces follows the intermediate trend described above, in the case of actual components powder forged parts have been shown to have superior fatigue resistance. This has gener-

Table 7 Mechanical property data for P/F-4200 materials

Forging mode	Carbon, %	Oxygen, ppm	Ultimate tensile strength, MPa	Ultimate tensile strength, ksi	0.2% offset yield strength, MPa	0.2% offset yield strength, ksi	Elongation in 25 mm (1 in.), %	Reduction of area, %	Core hardness, HV30
Blank carburized									
Upset(a)	0.19	450	1205	175	...	...	10.0	37.4	390
Re-press(a)	0.23	720	1110	161	...	...	6.3	17.0	380
Upset(b)	0.25	130	1585	230	...	...	13.0	47.5	489
Re-press(b)	0.25	110	1460	212	...	...	11.3	36.1	466
Quenched and stress relieved									
Upset(a)	0.31	470	1790	260	...	...	9.0	27.3	532
Re-press(a)	0.32	700	1745	253	...	...	4.0	9.0	538
Upset(a)	0.54	380	2050	297	...	...	1.3	...	694
Re-press(a)	0.50	520	2160	313	...	...	2.0	...	653
Upset(b)	0.65	120	1605	233	...	...	...	...	710
Re-press(b)	0.67	130	1040	151	...	...	...	...	709
Upset(a)	0.73	270	1110	161	...	...	...	...	767
Re-press(a)	0.85	370	1345	195	...	...	...	...	727
Upset(a)	0.70	420	600	87	...	...	...	...	761
Re-press(a)	0.67	320	540	78	...	...	...	...	778
Upset(b)	0.91	120	910	132	...	...	...	...	820
Re-press(b)	0.86	120	840	122	...	...	...	...	825
Quenched and tempered									
Upset(c)	0.28	720	1050	153	895	130	10.6	42.8	336
Upset(b)	0.37	1200	1450	210	1385	201	10.2	33.0	447
Upset(d)	0.56	580	1680	244	7560	226	9.8	28.6	444
Upset(e)	0.70	760	1805	262	1565	227	5.0	11.8	531
Upset(f)	0.86	790	1425	207	1310	190	10.4	30.0	450
Upset(g)	0.26	920	835	121	705	102	22.6	57.6	269
Upset(h)	0.38	860	860	125	785	114	20.8	56.5	288
Upset(i)	0.55	840	917	133	820	119	17.8	49.5	305
Upset(j)	0.73	820	965	140	855	124	15.4	42.7	304
Upset(j)	0.87	920	995	144	850	123	15.6	33.9	318

(a) Sintered in dissociated ammonia at 1120 °C (2050 °F). (b) Sintered in dissociated ammonia at 1260 °C (2300 °F). (c) Tempered at 175 °C (350 °F). (d) Tempered at 315 °C (600 °F). (e) Tempered at 345 °C (650 °F). (f) Tempered at 425 °C (800 °F). (g) Tempered at 620 °C (1150 °F). (h) Tempered at 650 °C (1200 °F). (i) Tempered at 660 °C (1225 °F). (j) Tempered at 675 °C (1250 °F)

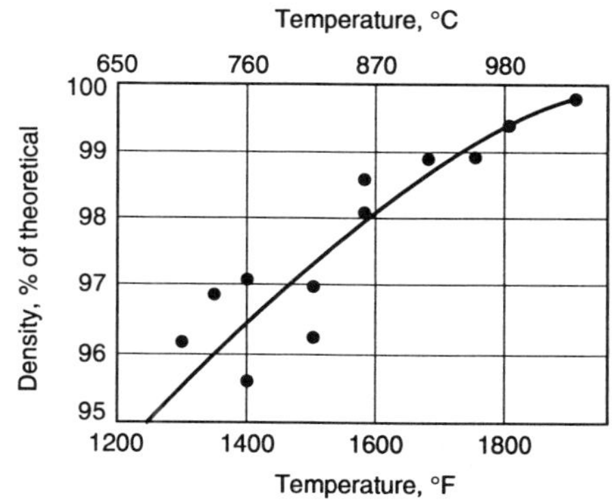

Fig. 10 Hot compressibility of low-alloy steel powder compacts under a forging pressure of 440 MPa (32 tsi). Source: Ref 13

ally been attributed not only to the relative mechanical property isotropy of powder forgings, but also to their better surface finish and finer grain size.

Typical mechanical properties of various powder forged steels are listed in Table 5. Additional mechanical data obtained from standard machined testpieces are shown in Tables 6 to 9 for four primary powder forged materials. The first two material systems are based on prealloyed powders (P/F-4600 and P/F-4200). The other alloys are based on Fe-Cu-C, which was used by Toyota in 1981 to make powder forged connecting rods; Ford Motor Company introduced powder forged rods with a similar chemistry in 1986.

Effects on Mechanical Properties (adapted from Ref 12). In powder forging, porosity occurs near part surfaces where die chilling occurs and also where cracking occurs during the early stages of forging due to the presence of tensile stresses. Preform redesign and providing suitable friction conditions aid in eliminating or reducing residual porosity.

The effects of small amounts of residual porosity on the mechanical properties of P/M forgings were first investigated by Kaufman and Mocarski in the 1970s (Ref 13). Compacted preforms of two different size iron powders and 0.57% graphite were forged at a constant pressure of 440 MPa (32 tsi). Density variations were produced by forging between 700 to 1040 °C (1300 and 1900 °F). Figure 10 shows almost a linear relationship, with density increasing with an increase in forging temperature. Figure 11 shows that the yield strength decreases rapidly as porosity concentration increases for normalized steel and for samples held at a temperature below the eutectoid and spheroidized condition.

Particle size distribution in the initial powder does not have much effect if samples are forged at the same temperature. The effects of porosity and second-phase dispersions were additive, but independent of one another. Tensile strength was found to monotonically decrease at about the same rate as yield strength with pore concentration (Fig. 11).

Table 8 Mechanical property and fatigue data for iron-copper-carbon alloys

Sintered at 1120 °C (2050 °F) in dissociated ammonia, reheated to 980 °C (1800 °F), in dissociated ammonia, and forged

Forging mode	Carbon, %	Oxygen, ppm	Ultimate tensile strength, MPa	Ultimate tensile strength, ksi	0.2% offset yield strength, MPa	0.2% offset yield strength, ksi	Elongation in 25 mm (1 in.), %	Reduction of area, %	Room-temperature Charpy V-notch impact energy, J	Room-temperature Charpy V-notch impact energy, ft · lbf	Core hardness, HV30	Fatigue endurance limit, MPa	Fatigue endurance limit, ksi	Ratio of fatigue endurance to tensile strength
Upset(a)	0.39	250	670	97	475	69	15	37.8	4.1	3.0	228	...	...	...
Upset(b)	0.40	210	805	117	660	96	12.5	38.3	5.4	4.0	261	325	47	0.40
Re-press(a)	0.39	200	690	100	490	71	15	35.4	2.7	2.0	227	...	...	...
Re-press(b)	0.41	240	795	115	585	85	10	36.5	4.1	3.0	269	345	50	0.43
Upset(a)	0.67	170	840	122	750	109	10	22.9	2.7	2.0	267	...	...	...
Upset(b)	0.66	160	980	142	870	126	15	24.9	4.1	3.0	322	470	68	0.48
Re-press(a)	0.64	190	825	120	765	111	10	24.8	3.4	2.5	266	...	...	...
Re-press(b)	0.67	170	985	143	875	127	10	20.6	4.7	3.5	311	460	67	0.47
Upset(a)	0.81	240	1025	149	625	91	10	19.2	2.7	2.0	337	...	...	...
Upset(b)	0.85	280	1130	164	625	91	10	16.6	4.1	3.0	343	525	76	0.46
Re-press(a)	0.81	200	1040	151	640	93	10	16.2	2.7	2.0	335	...	...	...
Re-press(b)	0.82	220	1170	170	745	108	10	12.8	2.7	2.0	368	475	69	0.41

(a) Still-air cooled. (b) Forced-air cooled

Table 9 Mechanical property and fatigue data for iron-copper-carbon alloys with sulfur additions

Sintered at 1120 °C (2050 °F) in dissociated ammonia, reheated to 980 °C (1800 °F) in dissociated ammonia, and forged

Addition	Carbon, %	Oxygen, ppm	Sulfur, %	Ultimate tensile strength, MPa	Ultimate tensile strength, ksi	0.2% offset yield strength, MPa	0.2% offset yield strength, ksi	Elongation in 25 mm (1 in.), %	Reduction of area, %	Room-temperature Charpy V-notch impact energy, J	Room-temperature Charpy V-notch impact energy, ft · lbf	Core hardness, HV30	Fatigue endurance limit, MPa	Fatigue endurance limit, ksi	Ratio of fatigue endurance to tensile strength
Manganese sulfide	0.59	270	0.13	915	133	620	90	11	23.2	6.8	5.0	290	430	62	0.47
Sulfur	0.63	160	0.14	840	122	560	81	12	21.4	6.8	5.0	267	415	60	0.50
None	0.66	160	0.013	980	142	870	126	15	24.9	4.1	3.0	322	470	68	0.48

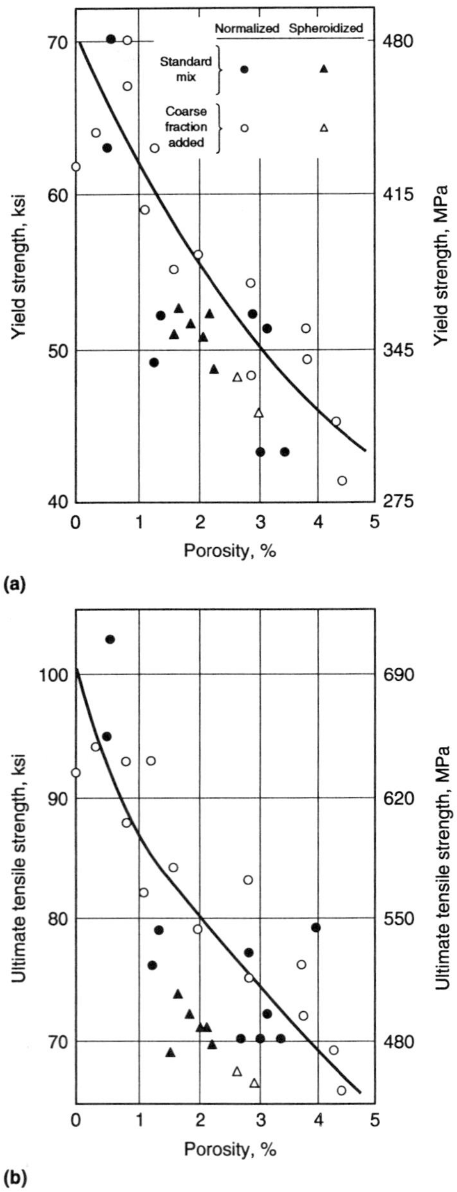

Fig. 11 Effect of porosity on (a) yield strength and (b) ultimate tensile strength. Source: Ref 13

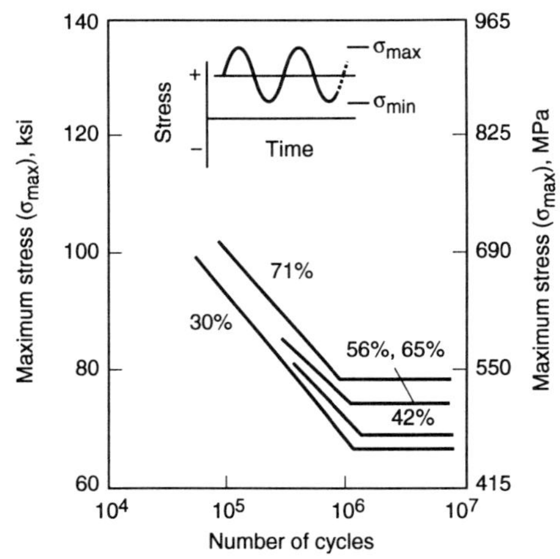

Fig. 12 Axial fatigue *S-N* curves for full-density powder forgings of 4620 as a function of height strain, as-forged condition. Source: Ref 17

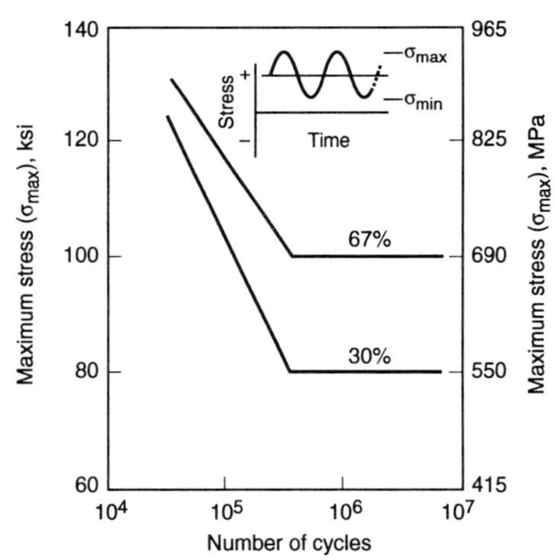

Fig. 13 Axial fatigue *S-N* curves for full-density powder forgings of 4620 as a function of height strain, heat treated condition. Source: Ref 17

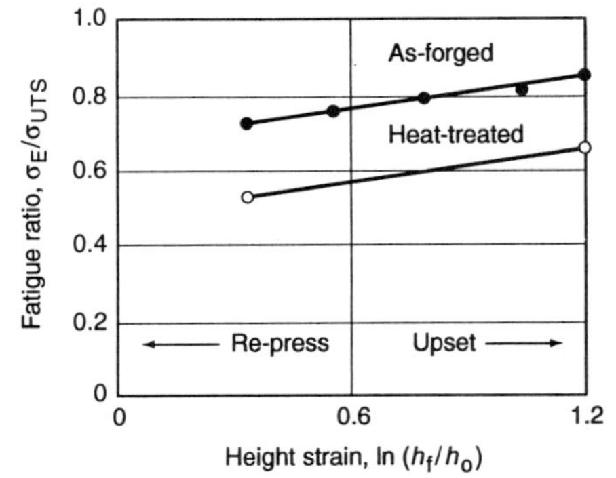

Fig. 14 Endurance ratio of 4620 powder forgings in axial fatigue as a function of height strain. Source: Ref 17

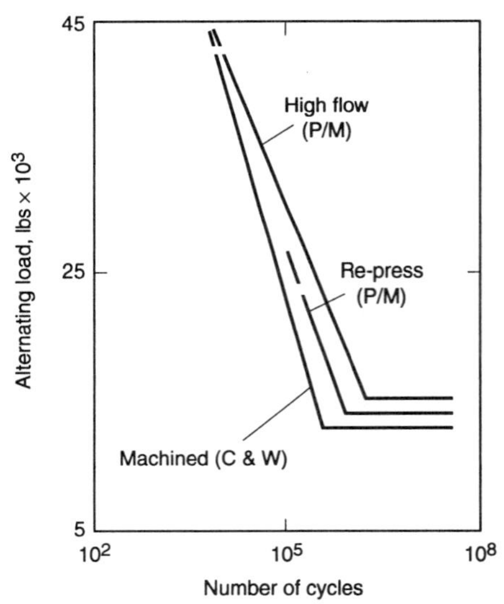

Fig. 15 *S-N* curves for simulated axial fatigue on prototype powder forged parts. C and W indicate cast and wrought, respectively. Source: Ref 17

Toughness. In steels, the lower initial preform density (75 to 83% of theoretical) results in optimal toughness in fully dense forgings. This is attributed to the fact that the lower density preforms have a significantly larger amount of interconnected porosity; during sintering, the oxide reducing gas can penetrate the lower density preforms to a greater extent, thus lowering the final oxide content prior to forging and thereby raising the final toughness after forging. Observations of Ferguson et al. (Ref 14) have shown that toughness (Charpy energy) is enhanced by lateral flow produced by upsetting in fully dense ferrous powder forgings.

Fatigue strength and resistance to crack propagation under conditions of cyclic loading also are enhanced by lateral flow at full density. An extensive experimental study and analysis were conducted by Ferguson et al. (Ref 15, 16) in terms of microstructure and stress state using specimens hot forged in plane-strain conditions. Figures 12 and 13 represent the increase in fatigue strength with increased strain (lateral flow) in as-forged and heat treated conditions from axial fatigue tests (Ref 17). Figure 14 represents the increase in endurance ratio in the case of upset versus re-pressed specimens in the forged and heat treated conditions. Figure 15 compares the *S-N* curves for a high-flow P/M prototype part with corresponding parts made by re-pressing and parts machined from a cast and wrought stock. Axial and rotating bend fatigue data are discussed in Ref 17.

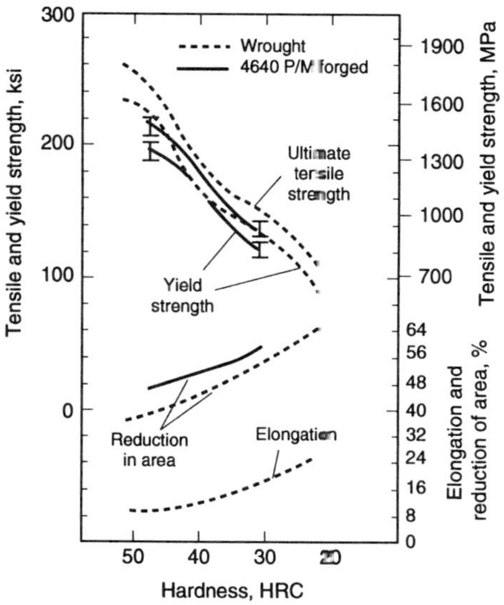

Fig. 16 Tensile properties of P/M forged 4640 and typical properties of wrought 4640. Source: Ref 18

Table 10 Mechanical properties of MIM low-alloy steels

Material designation/condition	Minimum values: Ultimate tensile strength, MPa	ksi	0.2% yield strength, MPa	ksi	Elongation in 25 mm (1 in.), %	Typical values: Ultimate tensile strength, MPa	ksi	0.2% yield strength, MPa	ksi	Elongation in 25 mm (1 in.), %	Density, g/cm³	Apparent hardness(a)
MIM-4600 as-sintered	255	37	110	16	20.0	290	42	125	18	40	7.6	45 HRB
MIM-4650 as-sintered	380	55	170	25	11.0	440	64	205	30	15	7.5	62 HRB
MIM-4650 quenched and tempered	1480	215	1310	190	<1.0	1655	240	1480	215	2	7.5	48 HRC (55 HRC)
MIM-2700 as-sintered	380	55	205	30	20.0	415	60	255	37	26	7.6	69 HRB

(a) Where applicable, the matrix (converted) hardness is also given in parentheses. Source: MPIF Standard 35, *Materials Standards for Metal Injection Molded Parts,* 1993–1994 edition

Table 11 Typical mechanical properties of nearly dense P/M stainless steel

Based on high-temperature sintering

Alloy	Condition	Ultimate tensile strength, MPa	ksi	0.2% yield strength, MPa	ksi	Elongation in 25 mm (1 in.), %	Hardness	Impact strength, J	ft·lbf	Density, g/cm³	Theoretical density, g/cm³
Ultimet 04, 304	Sintered	593	86	248	36	36	80 HRB	10.8(a)	8(a)	7.8	7.9
Ultimet 16, 316	Sintered	687	99.6	308	44.7	26	94 HRB	8.1(a)	6(a)	7.7	7.8
	Solution treated and quenched	684	99.3	329	47.7	45	90 HRB	5.4(a)	40(a)	7.7	7.8
Ultimet 40C, 440C	Sintered	...	...	...	...	...	20–30 HRC	2.7(b)	2(b)	7.6	7.7
	Hardened and tempered	...	...	...	...	...	50–60 HRC	2.7(b)	2(b)	7.6	7.7

(a) Charpy V-notch. (b) Unnotched. Source: Ref 20

Heat treatment of powder forged steels is similar to that of wrought steel. For ferrous forgings, conventional quench-and-temper cycles can be used to heat treat parts. Figure 16 shows that tensile and yield strengths are comparable for 4640 bar stock and 4640 powder forged material heat treated to the same hardness. Hardenability is slightly lower in powder forged steels, owing to the finer grain sizes of powder forged parts as compared to wrought forms. Typical hardenability curves for some common powder forged steels are in Fig. 17.

Injection Molding

Finely divided (1 to 10 μm) powders can be mixed with organic binder and injection molded like plastics. After binder removal, metal preforms can be sintered to near-full density. Mass transport mechanisms during sintering are enhanced by the high surface area of the powder and the elevated temperatures of 1150 to 1315 °C (2100 to 2400 °F). Densification is accompanied by more than 10% linear shrinkage. High final densities, often 95 to 99%, result in improved dynamic properties compared to conventionally pressed and sintered materials.

Current metal injection molded (MIM) ferrous grades include both prealloyed powders and admixtures of iron powder and alloying elements such as nickel, molybdenum, and carbon. Chemical compositions of MPIF ferrous grades include:

MPIF designation	Ni	Mo	C	Other	Fe
MIM-4600	1.5–2.5	0.0–0.5	0.0–0.1	2.0	bal
MIM-4650	1.5–2.5	0.0–0.5	0.4–0.6	2.0	bal
MIM-2700	6.5–8.5	0.0–0.5	0.0–0.1	2.0	bal

Depending on the fineness of the starting powder and the sintering temperature, densities of 93 to 98% of full density are obtained. Minimum and typical mechanical property values for 95 and 96% dense MIM low-alloy steels are given in Table 10. As these data indicate, ductility of

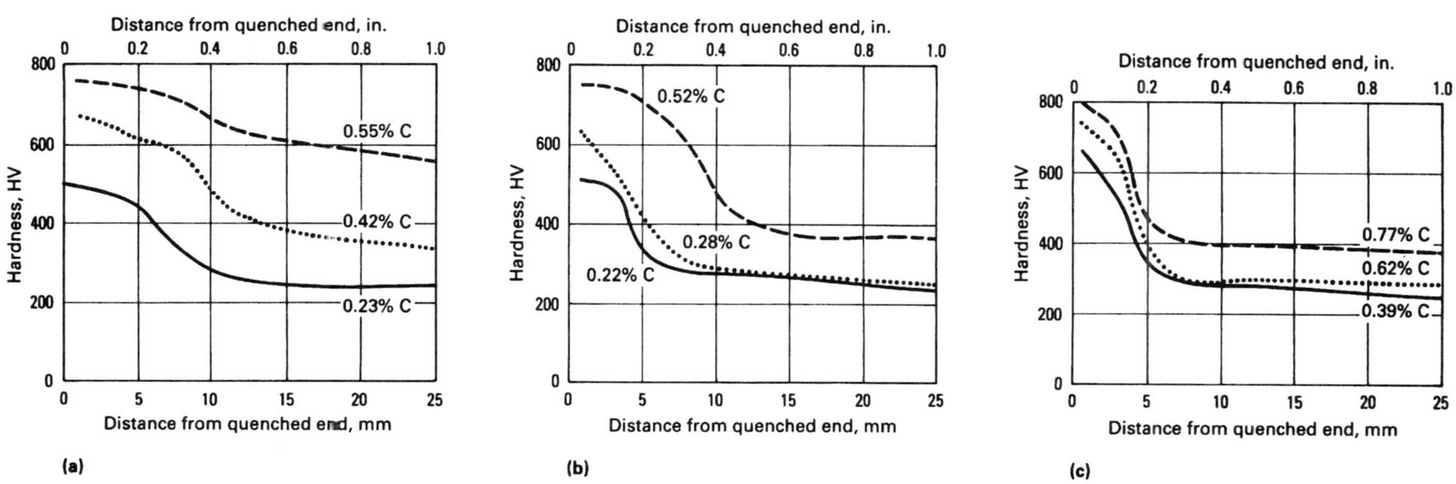

Fig. 17 Jominy hardenability curves for (a) P/F-4600, (b) P/F-4200, and (c) Fe-Cu-C materials at various forged carbon levels. Vickers hardness was determined at a 30 kgf load. Testing was carried out according to ASTM A 255. Specimens were machined from upset forged billets that had been sintered at 1120 °C (2050 °F) in dissociated ammonia.

Table 12 Typical mechanical properties of fully dense stainless steel

Property	P/M material	Wrought material
Extruded 0.3 by 15.5 mm (0.1 by 0.61 in.) 317LM tube(a)		
Ultimate tensile strength, MPa (ksi)	693 (100)	693 (100)
0.2% yield strength, MPa (ksi)	324 (47)	353 (51)
Reduction in area, %	71	73
Elongation in 25 mm (1 in.), %	47	50
Hot isostatically pressed type 316		
Ultimate tensile strength, MPa (ksi)	579 (84)	...
0.2% yield strength, MPa (ksi)	288 (42)	...
Elongation in 25 mm (1 in.), %	58	...

(a) Gas-atomized powder, canned, cold isostatically pressed, and extruded. Source: Ref 21

MIM-processed parts is very high, with elongation values of 30% or higher.

Fully Dense Stainless Steels

High-density stainless steels can be produced by P/M forging, but high-temperature sintering is the preferred fabrication method. At high temperatures (1315 to 1425 °C, or 2400 to 2600 °F), austenitic and ferritic stainless steels densify during sintering. With minor variations in composition, near-theoretical density can be achieved in pressed-and-sintered parts. Table 11 provides typical mechanical property data for nearly dense (>98% of theoretical density) stainless steels.

An alternative process for forming stainless steel into mill shapes, such as tubing, is to begin with clean, gas-atomized powder. Powder is packed in cans, sealed, cold isostatically pressed to reduce bulk, and extruded into tubing. The cold isostatically pressed cans may be hot isostatically pressed to full density, thus eliminating extrusion. Table 12 gives mechanical property data for extruded and hot isostatically pressed materials compared to wrought (ingot metallurgy) materials.

REFERENCES

1. G.D. McAdam, *J. Iron Steel. Inst.*, Vol 168, 1951, p 346
2. A. Squire, *Trans. AIME,* 1947, p 171, 485
3. W.J. Huppmann and L. Albano-Muller, Production of Powder Forged Parts of Complex Geometry, *Modern Developments in Powder Metallurgy,* Vol 12, Metal Powder Industries Federation, 1981, p 631
4. F. Hanejko, Mechanical Properties of Powder Forged 4100 and 1500 Type Alloy Steels, *Modern Developments in Powder Metallurgy,* Vol 12, Metal Powder Industries Federation, 1981, p 689
5. R.M. Pilliar et al., Fracture Toughness Evaluation of Powder Forged Parts, *Modern Developments in Powder Metallurgy,* Vol 7, Metal Powder Industries Federation, 1974, p 51
6. P. Lindskog, Reduction of Oxide Inclusions in Powder Preforms Prior to Hot Forming, *Modern Developments in Powder Metallurgy,* Vol 7, Metal Powder Industries Federation, 1974, p 285
7. F. Badia, F. Heck, and J. Tundermann, Effect of Composition and Processing Variations on Properties of Hot Formed Mixed Elemental P/M Nickel Steels, *Modern Developments in Powder Metallurgy,* Vol 7, Metal Powder Industries Federation, 1974, p 255
8. C. Tsumuki et al., Connecting Rods by P/M Hot Forging, *Modern Developments in Powder Metallurgy,* Vol 7, Metal Powder Industries Federation, 1974, p 385
9. S. Mocarski and D.W. Hall, Properties of Hot Formed Mo-Ni-Mn P/M Steels with Admixed Copper, *Modern Developments in Powder Metallurgy,* Vol 9, Metal Powder Industries Federation, 1977, p 467
10. S. Saritas, W.B. James, and T.J. Davies, Influence of Preforging Treatments on the Mechanical Properties of Two Low Alloy Powder Forged Steels, *Powder Metall.,* Vol 3, 1981, p 131
11. T.W. Pietrocini and D.A. Gustafson, Fatigue and Toughness of Hot Formed Cr-Ni-Mo and Ni-Mo Prealloyed Powders, *Modern Developments in Powder Metallurgy,* Vol 4, Metal Powder Industries Federation, 1971, p 431
12. R. Duggirala and R. Shiupuri, Effects of Processing Parameters in P/M Steel Forging on Part Properties: A Review, Part II: Forging of Sintered Compacts, *J. Mater. Eng. Perform.,* Vol 1 (No. 4), 1992, p 505–506
13. S.M. Kaufman and S. Mocarski, Effect of Small Amounts of Residual Porosity on the Mechanical Properties of P/M Forgings, *Int. J. Powder Metall.,* Vol 7, 1971, p 19
14. B.L. Ferguson, S.K. Suh, and A. Lawley, *Int. J. Powder Metall. Powder Technol.,* Vol 11, 1978, p 263–275
15. B.L. Ferguson, "Toughness and Fatigue of Iron-Base P/M Forgings," Ph.D. dissertation, Drexel University, 1976
16. B.L. Ferguson, H.A. Kuhn, and A. Lawley, *Modern Developments in Powder Metallurgy,* Vol 9, Metal Powder Industries Federation, 1977
17. H.A. Kuhn and A. Lawley, *Powder Metal Processing—New Techniques and Analyses,* Academic Press, 1978, p 160–165
18. F.T. Lally, I.J. Toth, and J. Dibenedetto, Forged Metal Powder Products, *Progress in Powder Metallurgy,* Vol 9, Metal Powder Industries Federation, 1972, p 276–302
19. MPIF Standard 35, *Materials Standards for Metal Injection Molded Parts,* 1993-1994 edition, Metal Powder Industries Federation
20. High Technology Materials, Amstead Research Laboratories, Bensenville, IL, private communication, 1983
21. C. Aslund, "Fully Dense Stainless Steel Products Compete Successfully with Forged Products," presented at Metal Powder Industries Federation National Powder Metallurgy Conference (New Orleans), Metal Powder Industries Federation, 1983

Fatigue and Fracture Control for Powder Metallurgy Components*

Randall M. German and Richard A. Queeney, The Pennsylvania State University

POWDER METALLURGY (P/M) is one of the most diverse approaches to metalworking. The main attraction of P/M technology is the ability to fabricate high-quality, complex parts to close tolerances in an economical manner. In essence, P/M converts a metal powder from a semifluid state into a strong, precise, high-performance shape. Key steps include the shaping or compaction of the powder and the subsequent thermal bonding of the particles by sintering. These two steps can be combined into a single operation, for example in hot powder forging or hot isostatic pressing. All P/M processes are fairly automated with relatively low energy consumption, high material utilization, and low capital costs. These characteristics align P/M with current concerns over manufacturing productivity. Consequently, the field is experiencing growth and progressively replacing traditional metalforming operations over a wide range of applications and materials. As illustrations, P/M is used in the fabrication of lamp filaments (tungsten), dental restorations (precious metals), self-lubricating bearings (bronze), automotive transmission gears (steels), armor piercing projectiles (tungsten alloys), welding electrodes (copper), nuclear power fuel elements (uranium dioxide), orthopedic implants (cobalt and titanium alloys), high-temperature filters (stainless steels), aircraft brake pads (iron-copper-tin-carbon), rechargeable batteries (nickel), and jet engine components (superalloys).

There are three basic approaches to powder metallurgy processing (Ref 1). The most common method, termed pressing and sintering, is to fill a die cavity with loose powder and apply a uniaxial compaction pressure to the powder. This pressure deforms and densifies the powder to approximately 85 to 90% of theoretical density. Subsequently, the pressed powder is heated to a temperature where atomic diffusion gives rise to interparticle bonding, but with little densification. Accordingly, the final product is porous. Such a technique is in widespread use for forming moderately complex shapes for mechanical systems using ferrous powders. The open continuous pore structures that exist in these P/M products dominate fracture and fatigue behavior. Additionally, many filters, electrodes, capacitors, batteries, and other porous structures are formed in a similar manner using low compaction pressures. Such high-porosity structures should not be employed in a fatigue-sensitive environment.

Alternatively, small particles are shaped into useful components at low pressures with the assistance of an organic binder, such as wax, by injection molding, tape casting, or extrusion. The particle-packing density is relatively low, typically only 60% of theoretical. After shaping, the binder is removed by either heat or solvent extraction, and the powder is densified by sintering at a high temperature. These approaches give a final density that is usually between 94 and 100% of theoretical. They are slower and more costly than die compaction, but they deliver greater shape complexity and improved mechanical response measures. Thus, techniques such as powder injection molding are in widespread use for computer, biomedical, and firearm applications, especially using stainless steels. Because the pores are small, closed, and spherical, they have less detrimental effect on fracture and fatigue properties.

Finally, a powder can be subjected to a combination of heat and stress simultaneously. This allows full densification and is widely employed in the fabrication of structural metals, composites, and high-temperature alloys. Variations include forging, hot pressing, hot isostatic pressing, extrusion, and roll forming. Alloys fabricated this way are usually based on aluminum, titanium, steel, nickel, or refractory metal systems, but include composites and intermetallics. Because there is no residual porosity, fracture and fatigue properties are totally dependent on the microstructure, especially any inhomogeneities or contaminants. When properly performed, these processes result in full-density P/M products that have mechanical responses superior to those of their wrought equivalents, largely because of the microstructure homogeneity.

P/M Materials

Many metals are available via P/M techniques. Aluminum and its alloys are highly compressible as powders; green densities of 90% of theoretical are common. They can be sintered or hot consolidated using extrusion, forging, hot pressing, and hot isostatic pressing. As summarized in Table 1, typical strengths for the press and sinter approach are in the 200 MPa (29 ksi) range with 2% elongation to fracture. Higher strengths are available by dispersion strengthening and deformation processing, including hot isostatic pressing and extrusion. In the best cases, fatigue endurance limits (or fatigue strength at about 10^7 cycles) approach about 200 MPa (29 ksi). Some of the high-performance rapidly solidified P/M

Table 1 Properties attainable in aluminum P/M alloys

Composition(a), wt %	Fabrication(b)	Density, g/cm^3	Yield strength, MPa	Tensile strength, MPa	Elongation, %
4Mg-0.80Si-1.1C	MA + forged	...	550	570	2
4Cu-1.5Mg-0.80Si-1.1C	MA + forged	...	580	600	11
0.4Si-0.6Mg	Cold forged	2.66	90	180	11
4.4Cu-0.8Si-0.5Mg	P+S	2.64	200	250	3
0.4Cu-1.0Mg-0.6Si	P+S	2.45	176	183	1
0.4Cu-1.0Mg-0.6Si	P+S	2.58	230	238	2
4Ti	MA + HIP	2.74	325	380	11
8Fe-2Mo	HIP	2.89	470	490	7

(a) Balance Al. (b) MA, mechanically alloyed; HIP, hot isostatically pressed; P+S, pressed and sintered

*Adapted from *Fatigue and Fracture*, Vol 19, *ASM Handbook*

Table 2 Common ferrous P/M alloy classes

Designation	Composition
Pure iron (steel)	max 1% C
Copper steel	1–22% Cu, max 1% C
Iron-nickel	1–3% Ni, max 2.5% Cu, max 0.3% C
Nickel steel	1–8% Ni, max 2.5% Cu, max 1% C
Low-alloy steel	0.3–2% Ni, 0.5–1% Mo, 0.4–0.8% C
Infiltrated steel	8–25% Cu, max 1% C
Phosphorus steel	0.4–0.8% P, low C
Sinter-hardened steel	1–3% Cr, 1–2% Mn, 2% Ni, 0.4–0.8% C

Table 3 Sample mechanical properties for Fe-2Cu-0.8C P/M alloys

Pressed and sintered, 1120 °C, ½ h, N_2-H_2 atmosphere

Density, g/cm^3	6.65	6.85	7.15
Porosity, %	14.2	11.8	7.9
Hardness, HRB	70	75	85
Yield strength, MPa	365	400	415
Tensile strength, MPa	425	495	620
Elongation, %	1.3	1.8	2.5
Transverse rupture strength, MPa	890	1025	1325
Fatigue strength, MPa	168	198	266

Table 4 Density and heat treatment effects on the properties of an Fe-10Cu-0.3C P/M alloy

Density, g/cm^3	6.4	6.4	7.1	7.1
Thermal condition	As-sintered	Heat treated	As-sintered	Heat treated
Hardness (scale)	50 (HRB)	25 (HRC)	80 (HRB)	40 (HRC)
Yield strength, MPa	280	...	395	655
Tensile strength, MPa	310	380	550	690
Elongation, %	0.5	0.5	1.5	0.5
Fatigue strength, MPa	115	145	210	260
Impact energy, J	4	...	11	...
Elastic modulus, GPa	90	90	130	130

Table 5 Mechanical property comparison for Ti-6Al-4V processed by various P/M techniques

Process	Porosity, %	Yield strength, MPa	Tensile strength, MPa	Elongation, %	Reduction in area, %
Blended elemental P+S	2	786	875	8	14
Blended elemental HIP	<1	805	875	9	17
Prealloy HIP	0	880	975	14	26

P+S, pressed and sintered; HIP, hot isostatically pressed

Table 6 Examples of the mechanical properties of tungsten heavy alloys sintered at 1500 °C to 100% density

Composition, wt %	Density, g/cm^3	Hardness, HRA	Yield strength, MPa	Tensile strength, MPa	Elongation, %
97W-2Ni-1Fe	18.6	65	610	900	19
93W-5Ni-2Fe	17.7	64	590	930	30
90W-7Ni-3Fe	17.1	63	530	920	30
86W-4Mo-7Ni-3Fe	16.6	64	625	980	24
82W-8Mo-8Ni-2Fe	16.2	66	690	980	24
74W-16Mo-8Ni-2Fe	15.3	69	850	1150	10

products provide excellent strength retention to high temperatures.

Copper, brass, and bronze are sintered from particles, where the typical applications are not fatigue sensitive. Cemented carbides, such as WC-Co and TiC-Ni, are sintered using a liquid phase to deliver a full-density structure, often by the application of high pressure at the end of the sintering cycle. The elimination of residual pores has considerable impact on fracture resistance, giving a fracture strength in the 1700 to 3000 MPa (245 to 435 ksi) range. Unfortunately, the basic materials are brittle, so fracture toughness is usually in the range of 10 to 20 MPa$\sqrt{m}$, depending on cobalt (or other matrix phase) content and carbide grain size.

Stainless steel P/M products are usually selected for their corrosion resistance. However, they are capable of highly variable final response measures, depending on the composition, density, and microstructure. For precipitation-hardenable alloys such as 17-4 PH, yield strengths of 1100 MPa (160 ksi) with 12% elongation are possible with fatigue endurance limits in the 500 MPa range (72 ksi). Alternatively, for austenitic stainless steels such as 316L, a sintered yield strength of 250 MPa (36 ksi) and considerable ductility (30% or more) are common.

By far the largest segment of the P/M applications rely on iron-base alloys. Generally, a powder is pressed in uniaxial tooling to near-final dimensions, but not to full density. Dimensional control during sintering is very important, and usually size can be held to within ±0.025 mm of specification, with concentricity to 0.1 mm, squareness to 0.05 mm, and density to 0.1 g/cm^3. Strength typically exhibits a small scatter of ±35 MPa (±5 ksi) and elongation exhibits a scatter of ±2%. In most sintered structural steel components, over 90% of the composition is iron. Table 2 gives examples of common P/M alloy compositions. In all cases, the mechanical properties increase with the final density.

Iron-copper-carbon compositions are the most common in production, because copper forms a liquid phase during sintering that greatly aids particle bonding. This system illustrates the properties possible with P/M. Copper and graphite (carbon) are mixed with iron, and during sintering the copper forms a liquid phase. Wrought materials of equivalent compositions are not possible due to extensive segregation in the molten state. The mechanical properties are degraded by whatever pores remain after sintering. Tables 3 and 4 provide examples of the property degradation by listing the hardness, strength, ductility, and impact energy versus density for two Fe-Cu-C alloys. Table 3 shows a density effect for an Fe-2Cu-0.8C alloy, while Table 4 includes both density and heat treatment effects for an Fe-10Cu-0.3C alloy. In these tables the fatigue life was measured at 10^7 fully reversed cycles ($R = -1$). Note that, for example, hardness and strength actually change less with increases in density than does the fatigue endurance strength. This reflects the greater sensitivity of the dynamic properties to pore structure as compared with the quasistatic tensile properties.

Nickel is another common addition to ferrous P/M alloys for improved strength. In low concentrations, phosphorus is used due to its potent hardening of iron and formation of a liquid phase at temperatures above 1050 °C (1920 °F). The liquid aids sintering, pore spheroidization, and alloy hardening, but usually these additives are selected for magnetic properties, not mechanical properties, with a popular composition containing 0.45% P. Most of these alloys are formed by mixing powders that are alloyed as part of the sintering cycle, because of the higher compressibility of the elemental powders as compared with that of prealloyed powders.

There are several other widely employed P/M materials. Tool steels are usually fabricated to full density by liquid-phase sintering or hot isostatic pressing. Cobalt-base alloys, titanium alloys, and superalloys are fabricated to full density by hot isostatic pressing. Table 5 compares the mechanical properties of Ti-6Al-4V alloys fabricated by three processing routes. A very useful group of P/M alloys are the tungsten heavy alloys. These are based on W-Ni-Fe mixtures that are densified by liquid-phase sintering. Table 6 gives the typical mechanical properties of sintered tungsten heavy alloys. These are full-density products, but despite the high-sintered density they lack good fatigue properties due to the two-phase microstructure. Like tungsten, most of the other refractory metals (molybdenum, tantalum, titanium, chromium, niobium, and rhenium) are fabricated from powders.

Low-cost composites are fabricated using P/M techniques. Particle reinforcement requires mixing of the constituents and consolidation to full density. A popular combination is Al-SiC. Sample mechanical properties of this composite as fabricated by vacuum hot pressing are given in Table 7, where the SiC content ranges from 0 to 40 vol%. Again, the fatigue properties of these

Table 7 Mechanical properties of Al-SiC P/M composites

6061 alloy matrix, densified by vacuum hot pressing

SiC, vol%	Elastic modulus, GPa	Density, g/cm^3	Yield strength, MPa	Elongation, %
0	69	2.71	[illegible]30	20
15	97	2.77	[illegible]35	6
20	103	2.80	[illegible]50	5
25	114	2.83	[illegible]75	4
30	121	2.85	[illegible]10	0
40	138	2.91	[illegible]79	<1

composites are not very attractive because of the differential elastic modulus between the reinforcement and matrix.

Porosity Effects

An inherent physical characteristic of P/M materials is the presence of pores. The role of porosity in determining fatigue endurance in powdered metals is akin to that of porosity that is induced through metal solidification in casting or welding. However, the porosity that is characteristic of sintered powdered metals, and of those materials subsequently deformation processed, may differ in character and influence from solidification porosity. In all cases, porosity catches the attention of the design engineer because it immediately conjures up images of classical stress concentrators. In addition, porosity featuring sharp re-entrant corners, a possibility for marginally equilibrium-sintered powder particle boundaries, may be more accurately viewed as crack precursors. Pore geometry can be altered by modifications to the sintering cycle, such as a longer hold time or higher temperature, wherein the smoother pores improve strength, fatigue life, and fracture resistance.

Ductility is sensitive to the pore structure, but impact, fracture, and fatigue behavior have the greatest sensitivities. In general, dynamic strength responses are the most sensitive. Even in those materials possessing full density, inferior properties can occur due to microstructural defects. Recent applications have pushed P/M into very demanding applications, a good example being the automotive connecting rod. Such rods are formed by hot forging a porous P/M preform using an alloy of Fe-2Cu-0.8C. They weigh nearly 650 g. The ultimate tensile strength is 825 MPa (120 ksi), with a yield strength of 550 MPa (80 ksi) and fatigue endurance limit of 255 MPa (37 ksi). With expansion of P/M fabrication technology into dynamically loaded components, there arise performance limitations associated with fracture and fatigue. For static tensile properties, there is a good basis for predicting the effect of residual porosity on strength (Ref 2–5). Recent research has had an emphasis on fatigue and fracture behavior to fill the database void.

Porosity is especially important to the high-cycle fatigue life (Ref 6–9). Pores play a role in both crack initiation and propagation, typically increasing the threshold stress intensity for crack initiation but lowering the resistance to crack propagation (Ref 10). For an alloy of Fe-2Ni-0.8C the fatigue endurance limit at 10^7 reverse cycles is between 200 and 250 MPa at a density of 7.1 g/cm^3 (approximately 10% porosity). That is approximately 35% of the tensile strength, and many porous injection-molded materials exhibit similar ratios of fatigue strength to tensile strength. Closed pores are less detrimental than those that are interconnected and open to the external surface. Surface pores act as the preferred site of fatigue crack initiation by acting as stress concentrators (Ref 11). The simplest view of a pore would be that of a spherical hole embedded in a continuous matrix whose response parameters are identical with that of a fully dense form of exactly the same metallurgical state. Since the maximum, most positive, principal stress is the controlling load parameter for high-cycle fatigue endurance, the embedded pore in a tensile field is a reasonably approximate model for a stressed porous metal without interacting stress fields (one of less than 10% porosity). The local pore-dominated stress field can be expected to control local events such as crack initiation and threshold stress intensity values (ΔK_{th}). For an embedded pore, the local stress fields are related to the far-field (applied) design stresses (Ref 12):

$$\sigma_{local} = \sigma_{design}\left[\frac{27-15\nu}{2\,(7-5\nu)}\right] \quad \text{(Eq 1)}$$

Note that, for the average steel, $\nu = 0.30$, and the local stresses are magnified by a factor of about 2. The concentrated stress field does not persist far beyond the embedded pore, being reduced to 105% of the design value at a distance of $2a$ (where a is the pore radius) from the pore center, thus possibly exerting little influence on far field parameters.

If the pore features sharp re-entrant corners, the elastic concentrated stresses are more accurately predicted from an elliptical hole example. However, these same pores are unlikely to exhibit smooth ellipsoidal morphologies, and their

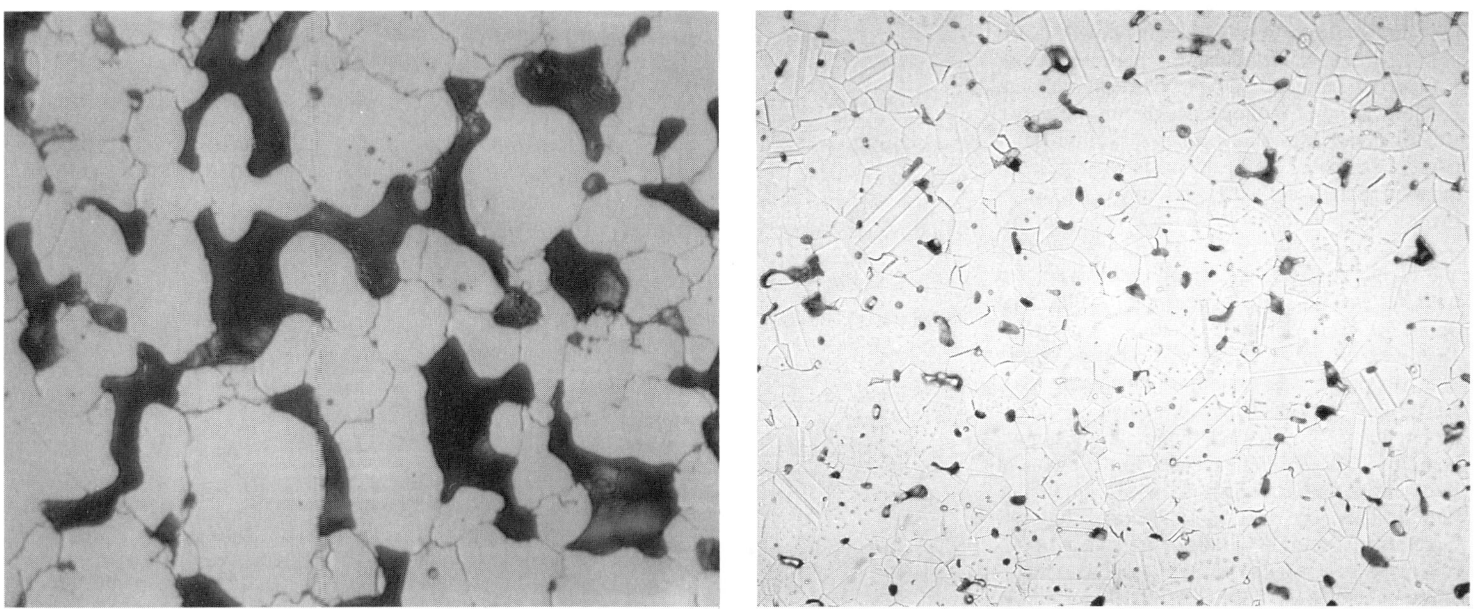

(a) (b)

Fig. 1 Two stainless steels fabricated by P/M, as demonstrations of the microstructure variations possible by tailoring the powder, compaction, and sintering variables. (a) A high-porosity microstructure useful for filtration, formed by press and sinter. 1000×. (b) A closed-porosity, high-density microstructure useful for mechanical components, formed by injection molding and high-temperature sintering. 200×

concentration effects can be more usefully predicted by the proportionality (Ref 13):

$$\sigma_{local} \approx 2\sigma_{design}\sqrt{\frac{a}{\rho}} \quad \text{(Eq 2)}$$

Here, a is the major pore length normal to the maximum principal stress direction, and ρ is the radius of curvature at the sharp corner. Suffice to say that the calculated stress concentration value can be large for a nonspheroidized pore.

Due to the stress-raising properties of pores, and the fact that most fatigue failures originate at free surfaces, treatments aimed at surface densification and serendipitous surface strengthening (e.g., coining, shot peening, ausrolling) raise the fatigue endurance limit (Ref 14–16). As a consequence, the current models for fatigue response in porous sintered materials have a major dependence on the pore microstructure: the models address the total porosity, alloying homogeneity in near-pore regions, pore size, pore shape, and interpore separation distance (Ref 6, 17–24). Round pores provide improved resistance to crack propagation. Pores act as linkage sites through which cracks can propagate. Microstructure-based fatigue models for ferrous alloys have had to address seemingly contradictory porosity effects: round pores retard stable fatigue crack propagation but increase crack extension growth rates by contributing linkage sites. While extant theories successfully explain some porosity effects on crack propagation, no total predictive model has been created that embraces microstructure effects (Ref 18, 20, 22). The presence of pores in the reversed plastic zone that is the site of propagating crack damage does not lend itself to facile analysis.

Pore structure changes are obtainable through processing and material variations: powder variables (particle size distribution, particle shape); compaction variables (type of lubricant, amount of lubricant, tool motions, maximum pressure); sintering variables (hold time, maximum temperature, atmosphere); and postsintering treatments. Figure 1 compares the pore structure in two sintered stainless steels to emphasize this point. Smaller particles result in faster sintering and higher strengths and toughness. Associated with the smaller particle sizes are smaller final pores. At lower densities (around 6.6 g/cm^3 or 18% porosity in steels), a high-relative content of small particles is beneficial to fatigue resistance, while at higher densities (over 7.1 g/cm^3 or less than 10% porosity), larger particles prove beneficial. The difference relates to the ligament size between pores, which is the determinant of fatigue. There are three pore microstructure parameters relevant to the fatigue resistance of porous P/M materials: pore size, pore curvature, and pore spacing. These largely reflect the role of stress concentration with respect to the advancing fatigue crack, as noted above (Ref 6). Thus, lower porosity contents, smoother (rounder) pores, and wider interpore separations increase the fatigue endurance strength.

Table 8 Representative P/M materials, processing cycles, and fatigue endurance limit

Composition, wt%	Processing	Density, g/cm^3	Testing	Endurance limit, MPa
Al-5Cu-0.5Mg-0.8Si	P+S, 600 °C, 1 h	2.6	...	53
Fe	P+S, 1120 °C, ½ h	6.0	Rotating $R=-1$	39
Fe	P+S, 1120 °C, ½ h	6.7	Bending $R=0$	67
Fe	P+S, 1120 °C, ⅔ h	6.9	Rotating $R=-1$	102
Fe	P+S, 1150 °C, 1 h	7.2	Axial $R=-1$	65
Fe	P+S, 1120 °C, ½ h	7.3	Rotating $R=-1$	145
Fe	P+S, 1250 °C, 2 h	7.6	Rotating $R=-1$	181
Fe	HIP, 1100 °C, 200 MPa	7.86	Axial $R=-1$	230
Fe-17Cr-4Cu-4Ni (17-4 PH)	PIM, 1350 °C, 2 h, HT	7.5	Rotating $R=-1$	517
Fe-1.5Cu-0.6C	P+S, 1120 °C, ½ h, HT	7.0	Bending $R=0$	390
Fe-2Cu-0.5C	P+S, 1120 °C, 30 min	7.1	Rotating $R=-1$	125
Fe-2Cu-0.8C	P+S, 1120 °C, ½ h	6.7	Rotating $R=-1$	165
Fe-2Cu-0.8C	P+S, 1120 °C, ½ h	7.0	Rotating $R=-1$	234
Fe-2Cu-0.8C	P+S, 1120 °C, ½ h	7.15	Rotating $R=-1$	241
Fe-2Cu-0.8C	P+S, 1330 °C, 1 h	7.1	Rotating $R=-1$	270
Fe-2Cu-2Ni-0.8C	P+S, 1120 °C, ½ h, HT	7.0	Rotating $R=-1$	240
Fe-2Cu-2Ni-0.8C	P+S, 1120 °C, ½ h, HT, SP	7.0	Rotating $R=-1$	282
Fe-2Ni-0.5C	PIM, 1250 °C, 4 h, HT	7.7	Rotating $R=-1$	239
Fe-2Ni-0.8C	P+S, 1120 °C, ½ h, HT	7.12	Rotating $R=-1$	159
Fe-2Ni-0.8C	P+S, 1175 °C, ½ h	6.9	Rotating $R=-1$	192
Fe-2Ni-0.5Mo-0.5C	P+S, 1120 °C, ½ h, HT	7.0	Rotating $R=-1$	350
Fe-2Ni-0.5Mo-0.5C	DP + DS, 1260 °C, 30 min, HT	7.4	Rotating $R=-1$	425
Fe-2Ni-0.5Mo-0.5C	P+S, 1120 °C, 30 min, HT	6.8	Rotating $R=-1$	345
Fe-2Ni-0.5Mo-0.4C	PF, 1150 °C	7.9	Rotating $R=-1$	780
Fe-2Ni-1Mo-0.9C	P+S, 1275 °C, 1 h, HT	...	Rotating $R=-1$	390
Fe-2Ni-1.5Cu-0.5Mo-0.5C	HIP, 1160 °C, 3 h, 105 MPa, HT	7.86	Rotating $R=-1$	480
Fe-4Ni-1.5Cu-0.5Mo-0.6C	P+S, 1120 °C, ½ h	7.1	Rotating $R=-1$	129
			Bending $R=0$	148
Fe-4Ni-1.5Cu-0.5Mo-0.6C	P+S, 1120 °C, 2 h	7.1	Rotating $R=-1$	135
Fe-4Ni-1.5Cu-0.5Mo-0.6C	P+S, 1250 °C, ½ h	7.1	Rotating $R=-1$	147
Fe-4Ni-1.5Cu-0.5Mo-0.6C	P+S, 1250 °C, 2 h	7.1	Rotating $R=-1$	195
			Bending $R=0$	266
Fe-7Ni	PIM, 1250 °C, 1 h	7.71	Rotating $R=-1$	236
Tool steel (Fe-8Co-6.3W-5Mo-3V-4Cr-1.3C)	HIP, 1150 °C, hot roll, HT	8.0	Axial $R=-1$	950
Ni	P+S, 1000 °C, 1 h	7.8	Rotating $R=-1$	70
Ni	DP + DS, 1300 °C, 3 h	8.5	Rotating $R=-1$	121
Ni_3Si	Hot extrude, HT	...	Axial $R=-1$	579
Ti-6Al-4V	HIP, 925 °C, 3 h, 200 MPa	4.46	Axial $R=-1$	475

P+S, pressed and sintered; PIM, powder injection molded and sintered; HIP, hot isostatically pressed; HT, heat treated; SP, shot peened; DP + DS, double pressed and double sintered; PF, powder forged

Table 9 Mechanical properties of pressed and sintered Fe-1.8Ni-0.5Mo-0.5C P/M compacts

Double pressed and double sintered, 1120 °C, ½ h, tested as-sintered

Porosity, %	Density, g/cm^3	Elastic modulus, GPa	Yield strength, MPa	Ultimate strength, MPa	Elongation, %	Notched impact energy, J	Fracture toughness, $MPa\sqrt{m}$
16	6.6	110	280	350	2	3	19
10	7.1	145	370	460	3	4	28
5	7.4	180	425	610	5	4	38
0	7.9	190	590	800	19	12	65

Table 10 Representative P/M materials, processing cycles, and fracture toughness

Composition, wt%	Processing	Density, g/cm^3	Fracture toughness, $MPa\sqrt{m}$
Fe-4.4Cr-9.2Co-7.2V-3.7Mo-9.2W-2.7C	P+S, 1150 °C, 1 h	8.1	13
Fe-1.5Cu-2Ni-0.8C	P+S, 1120 °C, ½ h	6.8	40
Fe-1.8Ni-0.5Mo-1.5C	P+S, 1120 °C, ½ h	6.6	15
Fe-1.8Ni-0.5Mo-1.5C	P+S, 1150 °C, ½ h	6.8	26
Fe-1.8Ni-0.5Mo-1.5C	P+S, 1120 °C, ½ h	7.1	24
Fe-1.8Ni-0.5Mo-1.5C	DP + DS, 1100 °C, ½ h	7.5	21–38
Fe-1.8Ni-0.5Mo-1.5C	HF, 1100 °C	7.85	64
Fe-0.8P-0.3C	P+S, 1120 °C, ½ h	7.0	22
Fe-0.8P-0.3C	DP + DS, 1120 °C, ½ h	7.8	20
Ti-6Al-4V	HIP, 925 °C, 3 h, 200 MPa	4.46	65

P+S, pressed and sintered; HIP, hot isostatically pressed; DP + DS, double pressed and double sintered; HF, hot forged

Fatigue cracks have been successfully analyzed with regard to their propagation response, and the cyclic growth of a crack can be predicted by the modified Paris growth law:

$$\frac{da}{dN} = A\,(\Delta K_{eff})^n \quad \text{(Eq 3)}$$

where the material parameters A and n must be experimentally determined for any given material microstate, including different distributions of porosity. The stress analytical variable ΔK_{eff}, the effective stress intensity range, factors out that portion of the total stress range that relieves the stresses holding the crack flanks closed, the opening stress range $\Delta\sigma_{op}$, leaving only the stress range component that displaces the crack faces relative to each other. In the presence of appreciable levels of mean stress, the relation between load design parameters and fatigue crack propagation rates is given by:

$$\frac{da}{dN} = \frac{C\Delta K^n}{(1-R)K_c - \Delta K} \quad \text{(Eq 4)}$$

Here, the total stress-intensity factor range ΔK is the load variable, but the fatigue ratio $R = \sigma_{min}/\sigma_{max}$ and the fracture toughness K_c, or K_{Ic} for a low-ductility sintered member, enter into the fatigue crack propagation response, as do the material constants C and n.

Regardless of which of the three fatigue crack propagation relations are relevant in a particular service context, their collective utility lies in predicting member lifetimes, or precise segments of that lifetime. Thus, in the case of the Paris law:

$$N = \int_0^{N_f} dN = \int_{a_0}^{a_c} \frac{da}{A(\Delta K)^n} \quad \text{(Eq 5)}$$

The calculated endurance cycles to failure, N, can be from an initial flaw size a_0 that may be the minimum detectable to final fracture at $a = a_c$, where:

$$K_{Ic} = \sigma_{max}\sqrt{\pi a_c} \quad \text{(Eq 6)}$$

Again, the fracture toughness of the material plays a role in determining structural endurance. The fracture resistance K_{Ic}, or K_R in the case of more ductile sintered materials (Ref 25), is known to be porosity sensitive (Ref 26). As a first estimate, the sensitivity is about a 100 $MPa\sqrt{m}$ gain in toughness per percentage point of porosity reduction in quenched and tempered steels. Copper-infiltrated steels have toughnesses that run from 40 to 50% those of wrought steels (Ref 27, 28). However, their static strengths are equivalent, reflecting the inability of the steel skeleton to absorb the same level of strain energy release as a fully dense body of the same material.

The material parameters in the modified Paris equation (Eq 3) are sensitive to the porosity state, a not unexpected result since the coefficient and exponent are related to the plastic zone size r_p ahead of the advancing crack, the only region of irreversible deformation. The coefficient A in Eq 3 increases with increasing porosity fraction (Ref 29), resulting in faster growth rates for comparable ΔK_{eff} values. With higher coefficient measures but constant exponent n values, the threshold stress intensity range ΔK_{th}, below which no crack growth is thought to occur, also decreases with porosity increase. However, ΔK_{th} values for fully dense materials are sufficiently low that it is not at all clear in what way they could be successfully employed in design practice if service stresses are to be set at appreciable fractions of the yield or tensile strength. The reversed plastic zone size is given by (Ref 30):

$$r_p = \frac{K_I^2}{6\pi\sigma_y^2} \quad \text{(Eq 7)}$$

When the plastic zone size is of the same size as the average pore diameter, it is effectively enlarged by the high strain field in the vicinity of the pore. The net result of the enlarged effective plastic zone is reflected in higher values of the exponent n (Ref 11), and the enlarged zone is even more pronounced in the response of short cracks driven by the locally raised stress/strain fields associated with design stress concentrators (Ref 15). Copper-infiltrated steels (Ref 27, 28) are as fatigue and fracture resistant as fully dense wrought medium-strength steels. Although the performance standards are not up to those of fully dense martensitic steels, copper infiltration represents a considerable cost savings over forging to full density, while maintaining the cost advantage inherent in press and sinter P/M.

Typical fatigue endurance limits (or fatigue strengths at about 10^7 cycles) are collected in Table 8 for several P/M materials. This compilation includes several ferrous alloys, reflecting the high interest in P/M fatigue for automotive applications. There are variations in density, alloying, and sintering cycles to show the relative effects on sintered properties. Systematic testing of various alloys has shown that slight changes in the particle size distribution or alloying homogeneity can affect these properties. For example, in the Fe-2Ni-0.8C alloy system, shifts in just the iron powder source lead to ±33 MPa (±4.8 ksi) variations in the fatigue endurance strength. Accordingly, the values in Table 8 are for relative ranking purposes only and cannot be used as an accurate basis for design of fatigue-sensitive components.

Fracture studies of P/M alloys are usually restricted to impact testing, and often this is performed in the unnotched condition because of the low toughness of porous materials. Fracture toughness measurements on P/M materials are relatively rare. Tables 9 and 10 summarize some prior findings. Table 9 demonstrates the density effect on tensile properties and K_{Ic}, for an Fe-Ni-Mo-C steel. Note that as the porosity decreases, the strength essentially doubles, ductility increases substantially (a nearly tenfold gain), the impact energy goes up by a factor of 4, and fracture toughness increases threefold. Table 10 collects several examples of the mechanical properties of ferrous alloys and one titanium alloy as representative values obtainable via P/M. Clearly, porosity negatively affects the fracture toughness in most materials. The fracture toughness of P/M steels is essentially a linear function of density (Ref 18), with sensitivities of about

100 MPa$\sqrt{m}$ gain in toughness per percentage point of porosity reduction. In low-density P/M materials, the fracture crack propagation is rapid because the pore structure amplifies the stress and provides an easy path. At low porosity levels, an advancing fracture crack can be blunted by the pores, effectively forming microcracks that improve toughness (Ref 31).

Other Factors Determining Fatigue and Fracture Resistance

It is well established that porosity is the major detriment to fatigue life for P/M materials. Beyond porosity, the sintered microstructure is a factor. Even in full-density materials fabricated by hot isostatic pressing, microstructure has a role. Weak links in the microstructure become evident during fracture. For porous structures, these weak links prove to be microstructural inhomogeneities, typically resulting from incomplete diffusional homogenization. Often powders (such as iron, nickel, and graphite) are mixed in the compaction stage. During heating the intent is for the mixed powders to homogenize to form a uniform microstructure, but this often is inhibited by too short a hold time at the peak temperature. Consequently, the alloying elements are poorly distributed and give point-to-point composition and microstructure changes, which are especially evident in postsintering heat treatment response. Accordingly, during fatigue or fracture, the weak links become the preferred failure paths. Most notable are the negative effects from inadequate homogenization of carbon to ensure uniform strength (Ref 18).

Little is known about the sensitivity of fatigue and fracture to loading conditions for P/M material. Table 11 compares the 2×10^6 fatigue endurance strengths for Fe-1.5Cu-0.6C at 7.1 g/cm^3 density using bending and axial fatigue tests. The table also includes a comparison with two loading stress ratios (half-cycle and fully reversed, $R = -1$) and two notch conditions (unnotched and notched) (Ref 18). In these cases the unnotched loading shows little sensitivity to axial versus bending fatigue, but a large sensitivity is evident in the presence of notches. The notch sensitivity factor is reported to range between 0.32 and 0.43 for many of the common pressed and sintered P/M alloys (Ref 32).

Full-density materials also suffer from residual microstructure artifacts that degrade the microstructure. In hot isostatically compacted powders, the achievement of 100% density is still insufficient to guarantee competitive fracture toughness, fatigue life, or even impact toughness. Thermally induced porosity is a subtle problem in many full-density P/M products. After consolidation the material is pore free, but it may contain small quantities of adsorbed gas. Once the product is put into high-temperature heat treatment or service, this residual gas precipitates to form pores if there is no compressive stress. In hot isostatically pressed titanium alloys, gas precipitation reportedly gives a 10 to 20% decrement in fatigue endurance strength (Ref 33).

Another difficulty rests in slight contaminants located on the interfaces that were previously particle surfaces, a feature termed prior particle boundary decorations. Figure 2 shows such decorations in a fully densified P/M steel. Improper powder handling or cleaning prior to consolidation are the primary detriments. These contaminants remain on the powder interfaces, even though the structure is fully densified. Consequently, a small contamination film runs throughout the structure, providing an easy fracture path that is often traced to a trivial impurity level. The fracture path is along the prior particle boundaries and has a characteristic morphology, as shown in Fig. 3. In hot isostatically pressed Ti-6Al-4V there is substantial fatigue life improvement due to removal of the contaminant, with a change from 450 MPa endurance strength to 600 MPa due to powder cleaning prior to consolidation (Ref 34).

One option for limiting the detrimental effects from prior particle boundary decorations is to forge the structure after hot isostatic pressing, a process often used in producing aerospace structures to ensure ultimate reliability. The forging operation upsets the microstructure and breaks apart the continuous films of contamination. The alternative is to resort to clean handling and processing, where the powder is produced by rapid solidification and kept under inert condi-

Table 11 Fatigue properties of Fe-1.5Cu-0.6C

Notch factor, K	Stress ratio, R	Endurance strength, MPa at 2×10^6 cycles
Axial		
1.0	−1	165
	0	130
2.8	−1	84
	0	64
Bending		
1.0	−1	160
	0	127
2.8	−1	137
	0	102

Note: Sintered to 7.2 g/cm^3 at 1120 °C for 30 min. Elastic modulus, 153 GPa; yield strength, 418 MPa; tensile strength, 483 MPa. Source: Ref 15

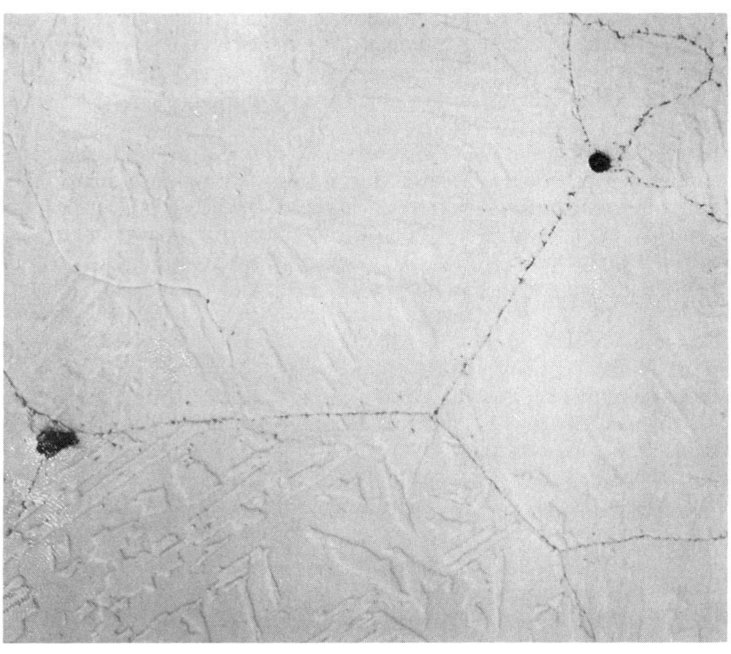

Fig. 2 Prior particle boundary precipitates formed on a hot isostatically pressed steel as the result of contamination during powder fabrication. 500×

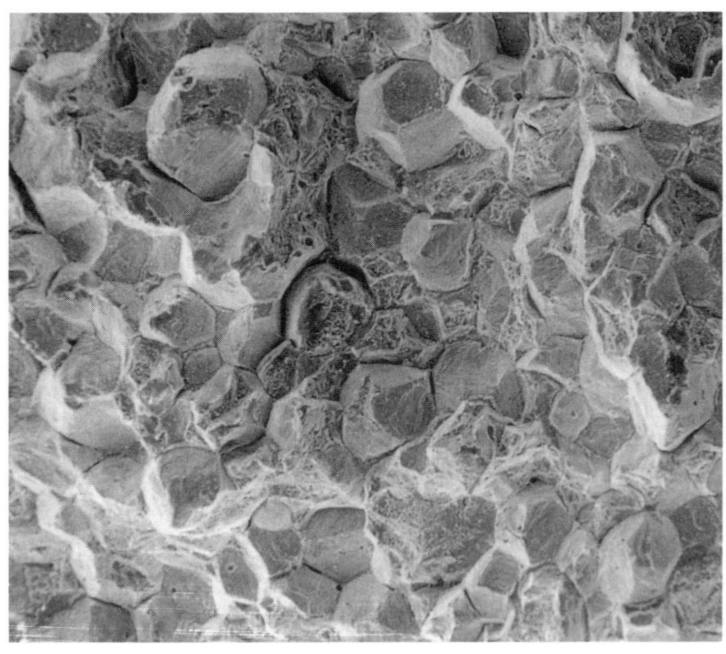

Fig. 3 A fracture surface showing preferential failure along prior particle boundaries. 150×

tions during handling. These steps, which minimize segregation and contamination, are employed in the production of aerospace components, microelectronic structures, and high-performance filters.

Steps to Improve Fatigue and Fracture Resistance

Surface pores are particularly detrimental to sintered materials with respect to fatigue life (Ref 35). Accordingly, carbonitriding and other surface strengthening and sealing treatments are most useful. Common treatments include shot peening, case hardening, repressing and resintering, coining, sizing, surface ausrolling, and postsintering heat treatments. For example, in pressed and sintered ferrous alloys, the endurance limit can be increased on small cross sections (6 by 6 mm) by at least 20% through shot peening. Carbonitriding is even more effective and can double the fatigue endurance limit. A typical carbonitride cycle involves heating to 940 °C in a mixture of ammonia and carbon dioxide for 4 h to form a 0.5 mm deep carbon-rich layer. Surface grinding is another approach to improved fatigue strength. The larger the cross section of the material, the less benefit possible from surface treatments, because bulk material states will dominate mechanical response.

The double press and double sinter approach was largely the only viable option for improving density and strength in traditional press and sinter P/M. This is more costly and involves extra tooling. A newly employed technique for improved fatigue life and fracture toughness in pressed and sintered ferrous alloys is to sinter at higher temperatures. The typical sintering temperature for steels is about 1120 °C, largely because of conveyor belt limitations in the furnaces. New materials of construction (ceramic belts) and new conveyor mechanisms (pusher plate and walking beam designs) allow higher-temperature processing regimes. Additionally, vacuum sintering usually is not limited in temperature, so it is viable for high-performance components. There is more sintering densification at the higher temperatures, so the density gain alone improves properties. Induced changes in the pore shape and size also improve fracture and fatigue properties. Several examples of the property gains are evident in Table 8. In a comparison of density gains versus sintering temperature effects, it is usually concluded that a change from 1120 to 1280 °C is equivalent to a density gain from 7.1 to 7.4 g/cm^3 in terms of both fracture toughness and fatigue endurance strength.

For small components the surface treatments are most useful, because the compressive forces extend through a major portion of the microstructure. However, for the porous materials with large cross sections, the need is to sinter at higher temperatures to improve fatigue and fracture. Further, designs that minimize density gradients will assist in minimizing fatigue failure. The high-density regions have a higher fatigue strength, and the difference in strength with density often results in failure at the interface between high- and low-density regions. For fatigue-sensitive components, the tolerable range of densities is less than 0.05 g/cm^3 within the structure. As with all fatigue-sensitive components, consideration must be given to surface finishing and processing optimization. The keys to improved performance are reduction in the total porosity, elimination of segregation and contamination, and manipulation of the pore microstructure.

There are some unique design opportunities where the microstructure of P/M materials offers a fatigue advantage in spite of the porosity. Because many alloys are formed from mixed powders, there is often an enrichment of the alloying addition near pores. The resulting higher strength aids local strength and fatigue life. Surprisingly, when properly exploited this effect gives higher fatigue and fracture strength to porous structures formed from mixed elemental powders, compared to those formed from prealloyed powders (Ref 36). Unfortunately, pores reduce the elastic modulus, strength, ductility, hardness, and other mechanical properties, so a high-density structure usually proves most successful in fatigue-sensitive applications. Indeed, the high sensitivity to porosity mandates that porosity be tightly controlled (density held within 0.05 g/cm^3) in regions of high stress concentrations. At a given strength level the P/M steels exhibit less notch sensitivity than wrought steels, because pores inherently act to blunt cracks and redistribute the load, especially under complex loading. Open pores, which dominate the sintered microstructure for densities below 92% of theoretical, retard fatigue crack propagation. Surface densification of P/M steels from shot peening, sizing, coining, surface rolling, or carbonitriding all prove beneficial in improving fatigue strength because of pore closure and surface compressive stresses.

Safety Factors for P/M Materials

Several factors contribute to scatter in the fracture and fatigue properties of P/M materials. There is the obvious error in testing, where the range of highest fatigue strength to lowest fatigue strength for a single test condition may be 4%. Further, there is typically a notch sensitivity and test error. Consequently, a safety factor of 1.4 is often cited as appropriate for sintered P/M alloys (Ref 18). This means that a peak cyclic stress of about 70% of the fatigue endurance strength is the maximum recommended in service.

REFERENCES

1. R.M. German, *Powder Metallurgy Science*, 2nd ed., Metal Powder Industries Federation, 1994
2. B. Karlsson and I. Bertilsson, Mechanical Properties of Sintered Steels, *Scand. J. Metall.*, Vol 11, 1982, p 267–275
3. G.F. Bocchini, The Influences of Porosity on the Characteristics of Sintered Materials, *Rev. Powder Met. Phys. Ceram.*, Vol 2, 1985, p 313–359
4. R. Haynes, The Mechanical Behavior of Sintered Metals, *Rev. Deform. Behav. Mater.*, Vol 3, 1981, p 1–101
5. S.H. Danninger, G. Jangg, B. Weiss, and R. Stickler, Microstructure and Mechanical Properties of Sintered Iron, Part 1: Basic Considerations and Review of Literature, *Powder Met. Inter.*, Vol 25, 1993, p 111–117
6. B. Weiss, R. Stickler, and H. Sychra, High-Cycle Fatigue Behaviour of Iron-Base PM Materials, *Metal Powder Report*, Vol 45, 1990, p 187–192
7. R. Haynes, Fatigue Behaviour of Sintered Metals and Alloys, *Powder Met.*, Vol 13, 1970, p 465–510
8. S. Oki, T. Akiyama, and K. Shoji, Fatigue Fracture Behavior of Sintered Carbon Steels, *J. Japan Soc. Powder Met.*, Vol 30, 1983, p 229–234
9. W.B. James and R.C. O'Brien, High Performance Ferrous P/M Materials: The Effect of Alloying Method on Dynamic Properties, *Progress in Powder Metallurgy*, Vol 42, Metal Powder Industries Federation, 1986, p 353–372
10. H. Danninger, G. Jangg, B. Weiss, and R. Stickler, The Influence of Porosity on Static and Dynamic Properties of P/M Iron, *PM into the 1990's*, Vol 1, *Proceedings of the World Conference on Powder Metallurgy*, Institute of Materials, London, 1990, p 433–439
11. J. Holmes and R.A. Queeney, Fatigue Crack Initiation in a Porous Steel, *Powder Met.*, Vol 28, 1985, p 231–235
12. S. Timoshenko and J.N. Goodier, *Theory of Elasticity*, McGraw-Hill, 1951, p 359–362
13. F.A. McClintock and A.S. Argon, *Mechanical Behavior of Materials*, Addison Wesley, 1966, p 412
14. C.M. Sonsino, F. Muller, V. Arnhold, and G. Schlieper, Influence of Mechanical Surface Treatments on the Fatigue Properties of Sintered Steels under Constant and Variable Stress Loading, *Modern Developments in Powder Metallurgy*, Vol 21, Metal Powder Industries Federation, 1988, p 55–66
15. C.M. Sonsino, G. Schlieper, and W.J. Huppmann, How to Improve the Fatigue Properties of Sintered Steels by Combined Mechanical and Thermal Treatments, *Modern Developments in Powder Metallurgy*, Vol 16, Metal Powder Industries Federation, 1985, p 33–48
16. J.H. Lange, M.F. Amateau, N. Sonti, and R.A. Queeney, Rolling Contact Fatigue in Ausrolled 1%C 9310 Steel, *Inter. J. Fatigue*, Vol 16, 1994, p 281–286
17. H. Kuroki and Y. Tokunaga, Effect of Density and Pore Shape on Impact Properties of Sintered Iron, *Inter. J. Powder Met. Powder Tech.*, Vol 21, 1985, p 131–137

18. F.J. Esper and C.M. Sonsino, *Fatigue Design for PM Components*, European Powder Metallurgy Association, Shrewsbury, UK, 1994
19. K.D. Christian and R.M. German, Relation between Pore Structure and Fatigue Behavior in Sintered Iron-Copper-Carbon, *Inter. J. Powder Met.*, Vol 31, 1995, p 51–61
20. I. Bertilsson, B. Karlsson, and J. Wasen, Fatigue Properties of Sintered Steels, *Modern Developments in Powder Metallurgy*, Vol 16, Metal Powder Industries Federation, 1985, p 19–32
21. R.C. O'Brien, Impact and Fatigue Characterization of Selected Ferrous P/M Materials, *Progress in Powder Metallurgy*, Vol 43, Metal Powder Industries Federation, 1987, p 749–775
22. P.S. Dasgupta and R.A. Queeney, Fatigue Crack Growth Rates in a Porous Metal, *Inter. J. Fatigue*, Vol 3, 1980, p 113–117
23. T. Prucher, Fatigue Life as a Function of the Mean Free Path between Inclusions, *Modern Developments in Powder Metallurgy*, Vol 18, Metal Powder Industries Federation, 1988, p 143–154
24. K.D. Christian, R.M. German, and A.S. Paulson, Statistical Analysis of Density and Particle Size Influences on Microstructural and Fatigue Properties of a Ferrous Alloy, *Modern Developments in Powder Metallurgy*, Vol 21, Metal Powder Industries Federation, 1988, p 23–39
25. I.J. Mellanby and J.R. Moon, The Fatigue Properties of Heat-Treatable Low Alloy Powder Metallurgy Steels, *Modern Developments in Powder Metallurgy*, Vol 18, Metal Powder Industries Federation, 1988, p 183–195
26. J.T. Barnby, D.C. Ghosh, and K. Dinsdale, Fracture Resistance of a Range of Steels, *Powder Met.*, Vol 16, 1973, pp 55–71
27. E. Klar, D.F. Berry, P.K. Samal, J.J. Lewandowski, and J.D. Rigney, Fracture Toughness and Fatigue Crack Growth Response of Copper Infiltrated Steels, *Inter. J. Powder Met.*, Vol 31, 1995, p 316–324
28. R.A. Queeney, Fatigue and Fracture Response of Metal-Infiltrated Sintered Powder Metals, *Proceedings of ICM3*, Vol 3, Pergamon Press, Oxford, 1979, p 373–381
29. D.A. Gerard and D.A. Koss, The Influence of Porosity on Short Fatigue Crack Growth at Large Strain Amplitudes, *Inter. J. Fatigue*, Vol 13, 1991, p 345–352
30. P.C. Paris, *Fatigue—An Interdisciplinary Approach*, Syracuse University Press, 1964, p 107–117
31. W. Pompe, G. Leitner, K. Wetzig, G. Zies, and W. Grabner, Crack Propagation and Processes Near Crack Tip of Metallic Sintered Materials, *Powder Met.*, Vol 27, 1984, p 45–51
32. A.F. Kravic, The Fatigue Properties of Sintered Iron and Steel, *Inter. J. Powder Met.*, Vol 3 (No. 2), 1967, p 7–13
33. R.L. Dreshfield and R.V. Miner, Effects of Thermally Induced Porosity on an as-HIP Powder Metallurgy Superalloy, *Powder Met. Inter.*, Vol 12, 1980, p 83–87
34. F.H. Froes and C. Suryanarayana, Powder Processing of Titanium Alloys, *Reviews in Particulate Materials*, Vol 1, A. Bose, R.M. German, and A. Lawley, Ed., Metal Powder Industries Federation, 1993, p 223–276
35. J.M. Wheatley and G.C. Smith, The Fatigue Strength of Sintered Iron, *Powder Met.*, Vol 6, 1963, p 141–153
36. U. Engstrom, C. Lindberg, and I. Tengzelius, Powders and Processes for High Performance PM Steels, *Powder Met.*, Vol 35, 1992, p 67–72

Wear Resistance of Powder Metallurgy Alloys

Zhigang Fang, Smith Tool, Smith International, Inc.

WEAR is the progressive loss of material due to relative motion of two or more surfaces. Wear of materials belongs to the science and technology field of tribology, which includes the study of friction, wear, and lubrication. For many years, study of tribology has focused on mechanical system optimization with respect to friction characterization, component design, and lubrication. An equally important subject, which may be less amenable to quantitative analysis, is the dependence of friction and wear behavior on intrinsic properties and microstructure of materials. While wear resistance of a material is a function of its hardness and toughness, large differences in wear resistance can occur at the same hardness. In other words, for a set of intrinsic mechanical properties, different microstructures can yield different friction and wear behavior. Powder metallurgy technology stands unique in its flexibility to create desired microstructures and compositions, which would be difficult or impossible by conventional processes, to meet demands of friction and wear properties.

Engineering materials subjected to wear applications include different metal alloys, ceramics, and polymers. Powder metallurgy is a class of processing technologies. More and more engineering materials and mechanical components are being fabricated by P/M processes. Any material processed by P/M that suffers wear loss is the subject of this article. However, it is a formidable task to cover all materials that can be manufactured using P/M techniques and subjected to wear during application. Instead, this article focuses on effects on wear of property and microstructure factors that are unique to P/M materials, such as metal alloys and metal matrix composites.

There are two approaches to reduce wear: reduce friction between rubbing objects or increase wear resistance of the materials involved. In sliding wear applications, introduction of friction reduction elements into material composition could be the most effective approach in certain specific applications. Reduction of friction can be accomplished by lubricant or compositional changes of the material; however, increasing wear resistance is the subject of this article.

There are numerous books including handbooks on wear behavior of materials. Suggested reading materials listed in Ref 1–5 discuss general theory on wear behavior of materials. Metallic components fabricated using P/M technology follow the general rules and trends of those theories. What P/M technology offers is special composition and microstructure that cannot be fabricated by other conventional processing methods. Factors that effect wear properties and are unique to P/M materials are discussed in this article; however, a brief discussion on wear classifications will help in understanding the effects of those factors.

Wear Classifications

Wear can be classified by four main mechanisms: adhesive, abrasion, surface fatigue, and tribochemical reactions. Wear can also be classified by different types of relative movements between mating surfaces, such as sliding, rolling, oscillation, impact, and erosive wear.

Adhesive wear is defined as the formation and breaking of interfacial adhesive bonds (e.g., cold-welded junctions). The tendency to incur adhesive wear depends on physical and chemical properties of the materials in contact. There are five different mechanisms proposed for adhesive wear: (1) mechanical interlocking, (2) diffusion theory, (3) electronic theory, (4) adsorption theory, and (5) chemical adsorption theory. Adhesive wear is considered the primary wear mechanism for sliding wear, but it is not equivalent to sliding wear because other mechanisms contribute to wear loss during relative sliding motion between two objects. Common terms such as material transfer, galling, scuffing, and scoring are examples of adhesive wear, although scuffing and scoring can mean scratching by abrasive particles.

Abrasive wear is the dislodging of material caused by hard particles between or embedded in the surfaces in relative motion, or by hard protuberances on one or both of the relatively moving surfaces. Mechanisms responsible for abrasive wear include microploughing, microcutting, and microcracking on the surfaces of metals. Abrasion is further divided into two-body or three-body abrasion. Abrasion can also be described by high-stress or low-stress abrasion. In general, during high-stress abrasion, abrasive particles are crushed, while during low-stress abrasion, the particles remain unbroken. Erosion is one kind of abrasion which refers to wear caused by impact of particles. The hard particles striking the surface are carried by a gas stream or entrained in a flowing liquid. The terms "gouging" and "scratching" are often used to describe wear scars left by abrasive action.

Surface fatigue results from repeated alternating loading of solid surfaces. Most often it occurs during rolling and impact contacts, which causes cyclic surface stressing. Mechanisms responsible for surface fatigue wear are undersurface crack formation and flaking material. The commonly observed pits are typical of surface fatigue.

Tribochemical reactions proceed by continual removal and new formation of reaction layers on the contacting surfaces, such as oxide layers formed in the presence of oxygen. The reaction from rubbing surfaces can be caused by gas or liquid in the environment.

Factors Affecting Wear Resistance

Root causes of any wear behavior are microstructure and chemical composition, as for any mechanical property. However, wear behavior is best understood from three different perspectives: fundamental mechanical properties of the material, microstructure features, and specific wear environments and mechanisms. In many cases, wear resistance of a material is largely the function of its intrinsic mechanical properties (i.e., elastic modulus, strength, hardness, and fracture toughness). Depending on wear action, different mechanical properties dominate the rate of wear loss, and different microstructure is desired. However, microstructure can also affect wear resistance independently without changing chemical composition or altering basic relationships between mechanical properties and microstructure. For example, large differences in wear resistance can occur at the same hardness. In other words, changing microstructure features can alter the rate of wear loss of a material with or without concurrent change in common mechanical properties. The relationships among wear resistance and mechanical properties and microstructure are also affected by specific wear environments (i.e., specific wear mechanisms). For example, wear resistance against sliding abrasive particles has different requirements for mechanical properties and microstructure than against impact fatigue or some other wear mechanism. Wear resistance depends on interrelated mechanical, microstructural and environmental factors (Fig. 1).

Mechanical Properties. Mechanical properties affecting wear resistance of a material include hardness, fracture toughness, elastic modulus, tensile/compressive strength, and impact fatigue strength. In many cases, wear resistance is primarily a function of hardness not of toughness, especially for metal alloys which have sufficient fracture toughness. The higher the hardness, the higher the resistance to abrasion by hard particles, surface plastic deformation, debonding, and microploughing. For wearing systems where plastic deformation is the main mechanism and impact is not prevalent, hardness controls wear rate of the material. In cases where fracture toughness of the material is relatively low, the wear application involves impact and repeated loading condition changes, and wear mechanism involves microfracturing, the role of fracture toughness in determining wear rate becomes more important. As a general principle, both wear mechanisms—plastic flow and fracture, can result in excavation and detachment of material. Raising the material hardness will reduce plastic flow but can increase the danger of fracture. Hence, it is more often that some compromise between high hardness and high fracture toughness gives the best wear resistance.

Microstructure. Materials can be categorized as those having homogeneous microstructures and those having multiphase inhomogeneous materials. Homogeneous P/M materials are similar to what can be achieved by traditional processing techniques, such as casting, forging, and heat treatment. Homogeneous microstructure indicates that the material is composed of one single phase or multiple phases that act like a single phase structure with respect to wear. Wear resistance is primarily proportional to its hardness and fracture toughness. Alloys or composites composed of at least two different phases have microstructures formed either by blending different powders together or one or more phase forms in situ. The boundary between a homogeneous microstructure and multiphase structure can be gray. For example, cemented tungsten carbide (WC-Co) is a two-phase composite, except for those with extremely fine grains and small mean free path (MFP). Powder metallurgy tool steel is mostly a homogeneous material, although its wear resistance depends on carbide particles in its microstructure. The various carbide particles are in situ precipitates. The two-phase or multiphase microstructure makes P/M technique unique with regard to tailoring microstructure to phases that cannot be created by casting, forging, or heat treatment in order to achieve desired wear resistance. Microstructure factors that affect wear include volume fraction, particle size, particle size distribution, MFP, and the ratio between MFP and abrasive particle sizes.

Volume fraction of hard phase particles in a multiphase microstructure is the most important factor affecting mechanical properties such as strength and fracture toughness. Wear resistance is influenced by volume fraction's influence on strength and toughness. No other direct effects of volume fraction on wear resistance have been reported. In general, wear resistance is proportional to volume fraction of hard phase particles. However, wear resistance levels off when hard phase particles reach certain volume fraction in some alloy systems and applications.

Particle size or grain size and particle size distribution of hard phase particles affect both microstructure and fundamental mechanical properties. At any given volume fraction, finer particle size results in smaller mean free path (MFP), i.e., smaller spacing between hard phase particles. When MFP is sufficiently small, the finer the particle size, the more the composite resembles homogeneous material (i.e., the higher the hardness, and the higher the wear resistance). A larger spacing is likely to cause preferential wear of the softer matrix which leads to uprooting of hard particles. In such cases, larger particle size is beneficial for wear resistance. Size distribution also affects wear resistance. For a given average particle size, wider particle size distribution is likely to give larger MFP. However, the effect of different powder size blend is strongly related to wear applications. In some cases, multiple wear mechanisms are in effect. Therefore, different powder particle sizes may be needed for different roles.

Mean free path (MFP) is a function of particle size and volume fraction of hard particles given by:

$$\lambda = 4 \cdot (1 - V_V)/S_V$$

The effects of MFP on wear resistance reflect the combined influence of hard phase volume fractions and particle sizes. MFP affects basic mechanical properties including hardness, strength, and fracture toughness. As a general rule of thumb, hardness and strength increase with the decrease in MFP. Wear resistance is proportional to hardness in most cases; therefore, wear resistance is inversely proportional to MFP. The increase of wear resistance as MFP decreases is attributed to the resistance to plastic deformation, which accounts for various wear mechanisms, especially during abrasive and adhesive wear situations. However, in cases where microchipping and fracturing is responsible for material loss, an optimum MFP has to be achieved to maximize wear resistance. Microchipping and cracking resistance is proportional to MFP. The balance between microchipping and local plastic deformation resistance yields optimum wear resistance. Another important factor during abrasive wear is the ratio between MFP and abrasive wear particles. When abrasive particle size is smaller than mean free path between hard phase particles, preferential wear of metal phase between them is dominant. When abrasive particle size is larger than MFP, overall hardness of the microstructure becomes a dominant factor.

Type of Hard Phase Particles. Many hard particles are known to be used in wear applications, including carbides, borides, nitrides, oxides, and other ceramics. Among carbides, tungsten carbide (WC), chromium carbides, and titanium carbide (TiC) are most widely used. Vanadium

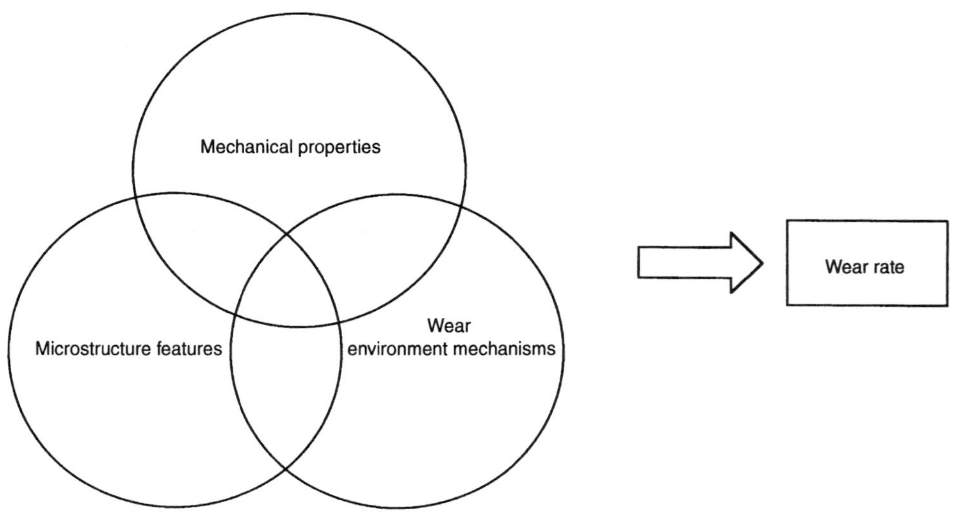

Fig. 1 Factors affecting wear resistance of materials

carbide (VC), molybdenum carbides, and tantalum carbides are often used as additives. Titanium diboride is known to be extremely wear resistant. Usage of TiB_2 powder has been on the rise recently. Relative hardness of hard phase particles to that of abrasive particles is an important factor attributed to the interaction between the microstructure and wear environment.

Wear Environment/Mechanisms. Factors to consider include mechanical loading conditions, wearing mechanisms, harshness of abrasive particles, and hardness and size/shape of abrasive particles. In cases of abrasive wear, relative size between the hard phase particles, d_c, and abrasive particle d_a, is an important factor. In general, if the $d_c < d_a$, rate of wear loss is greater than it would be if $d_c > d_a$. Wear mechanisms depend on the wear environment and the material. In cases of sliding wear, relative hardness between the two mating surfaces, the chemical affinity between the two mating surfaces, and abrasive particles contained in the materials are all important factors that affect the final outcome. Depending on the wearing system, different mechanical properties (e.g., resistance against plastic deformation, crack formation or crack propagation) have to be considered. Depending on which wear mechanism is dominant, different material properties and microstructure may be desired. Other environment factors, such as corrosive fluid and potential chemical interactions, need also be considered.

Wear-Resistant P/M Alloys

Wear resistant P/M alloys include cemented tungsten carbide, P/M steels, and P/M metal-matrix composites.

Cemented Tungsten Carbide

Cemented tungsten carbide and cermet are the most widely used P/M materials for wear applications. Cemented tungsten carbide and cermet are used in all types of wear applications including abrasive wear, sliding wear, and erosive wear situations. Cemented tungsten is a composite of tungsten carbide, WC, and cobalt metal, Co. In most commercially available grades of WC-Co, cobalt contents range from 6.0 to 16.0%. There are applications using as low as 3% cobalt or as high as 25% cobalt, although those are not common. Tungsten carbide grain size range from submicron to 10 μm. Commercially available ultrafine WC powder has nanometer size grains such as NanoCarb™ (Nanodyne, Inc., New Jersey).

Most cemented tungsten carbide manufacturers have their own designations for different grades that are used by suppliers and customers. For metal removal cutting tool applications, there are internationally accepted grade designation systems that are used as category reference only. For example, the P-M-K system found in Ref 5, and the "C" system in which C-1 through C-4 are used for abrasion resistant grades and C-5 through C-8 are used for crater and deformation resistant grades. C-1 through C-4 are straight WC-Co grades, and C-5 through C-8 contain cubic carbides, such as TiC and TaC for cutting tool applications.

A schematic flowchart of processes for cemented tungsten carbide is illustrated in Fig. 2. One of the critical factors and an occasional problem in industry today is the carbon control during liquid phase sintering. In recent years, low pressure (around 1000 psi) hot isostatic pressing has been used in combination with the standard vacuum liquid phase sintering to improve quality and properties of cemented tungsten carbide products.

Fundamental mechanical properties of cemented tungsten carbide include hardness, fracture toughness, transverse rupture strength (TRS), compressive strength, and wear resistance. Hardness and TRS are widely used in day-to-day engineering activities for selection and quality control, while fracture toughness is mostly used in developing new grades. Hardness is proportional to compressive strength and inversely proportional to fracture toughness. Transverse rupture strength depends on both strength and toughness of the material. When there is significant volume percent of defects such as porosity in microstructure, TRS becomes very defect-sensitive. Figure 3 shows the interrelationships between hardness, fracture toughness, and TRS. Transverse rupture strength is affected by both crack initiation and crack propagation processes; therefore, it exhibits an apex as hardness increases. Wear resistance is more closely related to hardness. Figure 4 demonstrates that the higher the hardness, the higher the abrasive wear resistance. It also demonstrates that the higher the fracture toughness, the lower the wear resistance. The trade-off between wear resistance and toughness has to be made when selecting cemented tungsten grades for specific engineering applications.

Microstructure of most cemented tungsten carbide belongs to the category of dual-phase composites, in which tungsten carbide grain is the hard phase, and cobalt metal in between WC

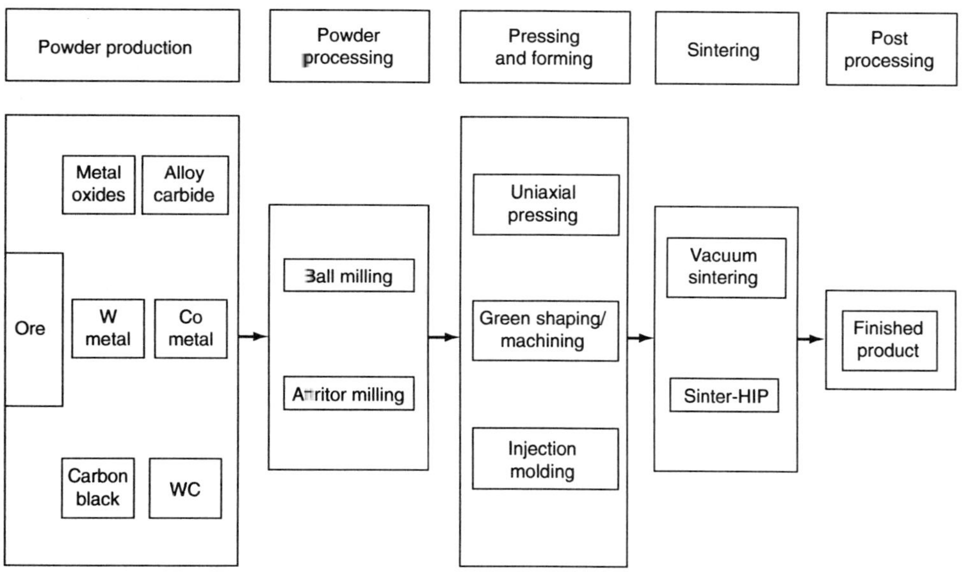

Fig. 2 Schematic flow chart of manufacturing processes for cemented tungsten carbides

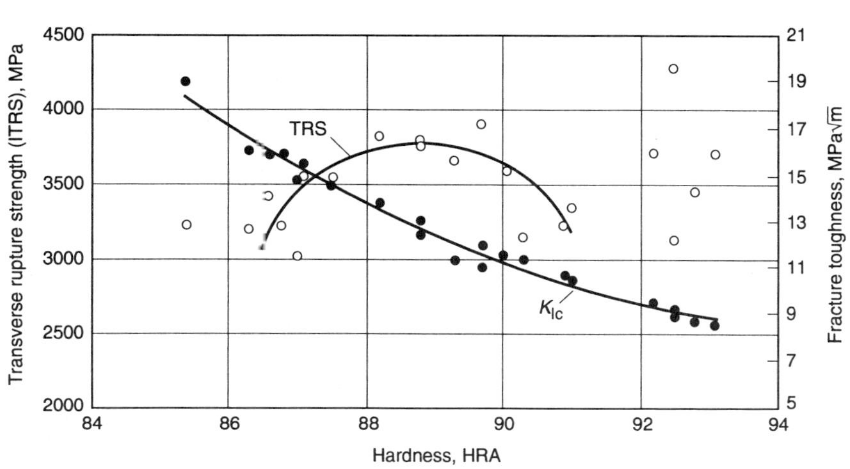

Fig. 3 Mechanical properties of WC-Co alloys

grains are the ductile phase. In submicron super fine grain, the MPF is so small that the composite material behaves more like homogeneous materials. Because cobalt content normally ranges from 6 to 20% (8 to 30% vol), it can also be categorized into ductile phase reinforced brittle materials. Hardness of the materials decreases and toughness increases as cobalt content increases. Grain size of WC also affect properties. As grain size increases, hardness decreases and toughness increases. Influence of microstructure on properties is reflected in the dependence of hardness, toughness, and wear resistance on MFP. The larger the MFP, the lower the hardness and the higher the toughness (Fig. 5). In considering wear, different mechanical properties (e.g., resistance against plastic deformation, crack formation or crack propagation) have to be considered.

In the sequence from crack formation to crack propagation and finally fracture, crack propagation plays the determining part in most metallic alloys. In contrast, crack formation seems to be the most important part in ceramics. For cemented tungsten carbide, both plastic deformation and microfracturing mechanisms are important. However, the dependence of wear resistance on hardness suggests that its resistance to local deformation is dominant at the onset of a wear process. Due to limited ductility, initial plastic yield on the wear surface evolves into microfracturing quickly. The surface plastic yield is apparently controlled by the cobalt content. Preferential wear of cobalt also occurs as cobalt content increases. Another important mechanism is the pullout of carbide grains, WC, as a result of preferential wear of cobalt phase. The larger the grain size, the less the tendency of carbide grain pullout. Therefore, within a certain range of hardness and toughness, larger grain size grades can be more wear resistant than finer grain size grades at an equivalent hardness level.

Major categories of applications of cemented tungsten carbide include metal removal cutting tools, rock and earth drilling tools, drawing and punching dies and punches, wear components, and other engineering specialty applications.

For metal cutting tools, wear progresses very quickly under high temperature and heavy load. Tool life is limited to a very small amount of wear loss, usually indicated by the change in radius of the cutting edge. Crater wear on the top surface of a cutting insert is the result of a chemical reaction between metal chips and tungsten carbide under high heat. Titanium carbide and tantalum carbide additions are prevalent in most cutting tool grades to combat crater wear. Modern cutting tools are often coated by chemical vapor deposition (CVD) with TiN, TiCN, TiC, and in combination with Al_2O_3. Physical vapor deposition (PVD) is utilized to coat drills, end mills, etc. with TiN, TiCN, and CrN.

Cutting tools are categorized based on their specific cutting application, such as general purpose grades, rough turning grades, or finish turning grades. Trade-offs are often made in designing as well as selecting grades between chipping resistance and wear resistance. For example, KC-850 (Kennametal, Inc., PA) is a general purpose turning grade used in industry which combines good wear resistance with good chipping resistance. GC4015 (Sandvik Coromant Co., Fair Lawn, NJ) is a finishing turning grade with high hardness and crater resistance but moderate chipping resistance. It is not suitable for interrupted turning. Twist drills often use fine grain (submicron) high hardness grade. RTW 2606 (Roger Tool Works, Inc., Rogers, AR) is a grade for printed circuit board drills with a hardness in the range of 93.5 to 95 Ra.

P/M cutting tool materials also include the cermet. Cermet is commonly used to describe (TiC, TiCN)-(Ni, Ni/Mo) alloys. Cermet has better high temperature hardness than cemented tungsten carbide. They are mostly used for finishing turning cutting applications. A comprehensive review on cutting tool materials and their wear resistance can be found in the article "Friction and Wear of Cutting Tools and Cutting Tool Materials" in Vol 18 of the *ASM Handbook.*

Earth drilling tools include petroleum drilling bits, mining bits, coal mining tools, and construction tools that use cemented tungsten carbide as cutting inserts. Earth boring bits operate under very heavy load and in extremely abrasive environments. Cutting structure of such tools must have very high compressive strength, very high abrasive wear resistance, and sufficient fracture toughness. Cemented tungsten carbide offers the best choice. In typical grades used for earth boring applications, cobalt content ranges from 6 to 16%, and grain size range from 1 to 10 μm. Hardness of these grades ranges from 85.0 to 91.0 HRA. Corresponding fracture toughness is from 10 to 18 MPa$\sqrt{m}$. Selection of a specific grade depends on the application. For example, 6% cobalt grades with relatively low toughness are often used in applications where mechanical impact is minimum and sliding abrasive wear resistance is the main concern. In applications where mechanical impact is unavoidable, grades with higher fracture toughness (by increase cobalt content or grain size) are used.

Abrasive wear is considered the primary mechanism for earth boring applications. However, depending on different types of rock drilling tools (rotary, percussive, or rotary and per-

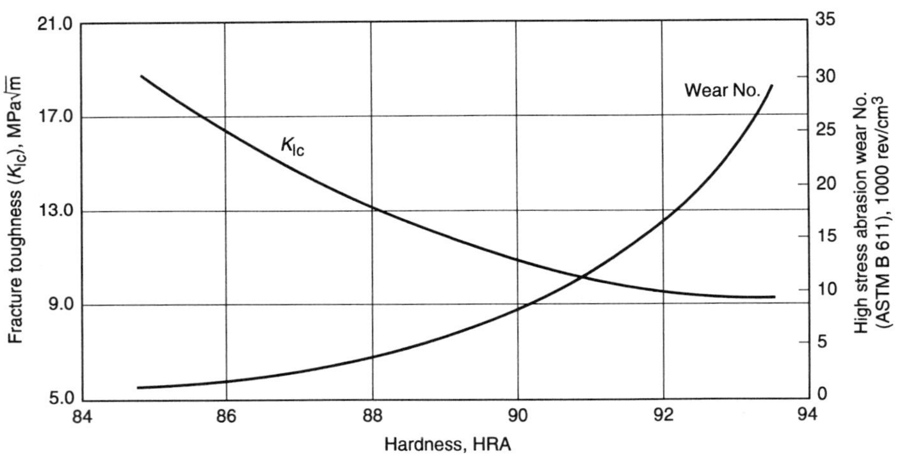

Fig. 4 Dependence of wear resistance of WC-Co alloys on their hardness and fracture toughness

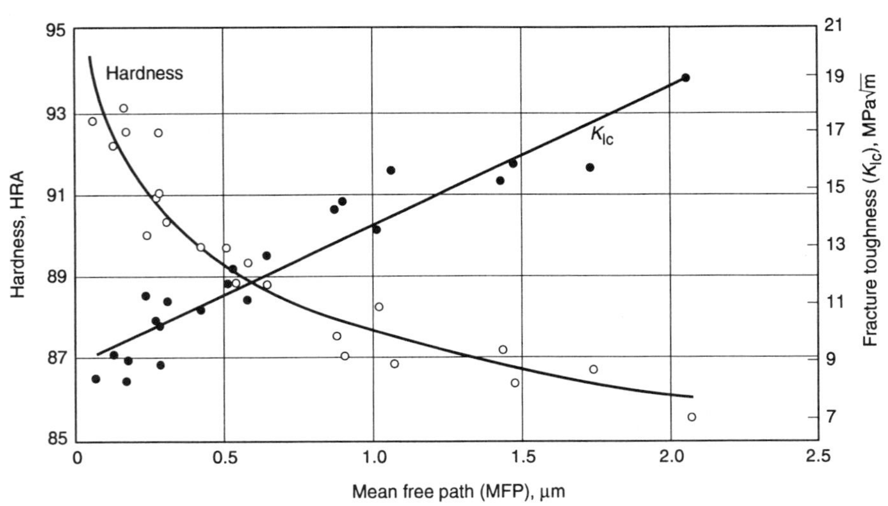

Fig. 5 Dependence of hardness and fracture toughness on mean free path of tungsten grains in WC-Co alloys

cussive), impact and/or impact fatigue can play a major role in causing material loss during drilling. Contribution of impact fatigue and abrasion to wear can be separated to a certain extent. Figure 6 shows the dependence of impact wear and abrasion wear on cobalt content, respectively. The impact wear resistance shows a maximum of 6 to 8% Co, which is similar to the relationship between compressive strength and transverse rupture strength versus cobalt content. For abrasion resistance during rock drilling, the general principle for the effects of grain size, cobalt content, and MFP applies. Very hard cemented carbides containing a high volume fraction of carbides exhibit lower abrasive wear loss on fine grain structures. Softer, more ductile cemented carbides, containing less volume fraction of carbides, exhibit lower wear loss in the case of coarse grain structures. In short, the abrasive wear resistance is generally related to hardness and cobalt content, while the impact wear appears to be closely related to bulk compressive strength and transverse rupture strength. Another important factor affecting tool life during earth boring is the size of abrasive particles and its relative ratio to grain size or mean free path of the microstructure. Figure 7 shows that there is optimum λ/D_{eff} where wear loss is minimum, where D_{eff} is the effective diameter of abrasive particles. The optimum values of λ/D_{eff} for fine grain and coarse materials are different indicating differences in wear mechanisms.

Metal forming tools and dies, structural components, and a wide variety of wear parts constitute roughly one-third of total tonnage of cemented tungsten carbide shipped. Metal forming tools and dies include hardmetal dies for drawing wire, rod, bar, tube, and special sections. For wire drawing dies, crucial properties are abrasive wear resistance and compressive strength. High surface finish requirements often dictate selection of fine grain and low cobalt grades. However, for larger cross section drawing, such as rods and special sections, toughness is important. The biggest market for pressing dies is the carbide industry itself, because nearly all cutting inserts and cemented carbide components are formed by the uniaxial die pressing method. Other metal forming tools include pressing dies used for synthetic diamond and cubic boron nitride manufacturing, and rolls for the rolling of steel rod and sections, wire, tape, and foils. Foil rolls require perfect mirror surface finish. Fine grain and porosity free microstructure, which yields improved rupture strength and fatigue resistance of rolls, is extremely important.

Structural components made of cemented carbide include mechanical seals, compressor plungers, lathe jaws, boring bars, grinding spindles, pulverizing pins, and bearing journals. Wear parts include nozzles, guides, plungers, balls, slitting wheels and cutters for paper industry, and many other applications.

Wear for most metal forming tools, structural, and wear components are classified as "sliding" wear based on their mechanical function. Almost all wear mechanisms could occur during sliding wear applications. However, depending on specific application, only one or two mechanisms may be dominant. For most cemented carbide sliding wear applications, abrasive wear and adhesive wear are the primary mechanisms. For metal rolls, obviously, rolling fatigue should be considered in analyzing wear mechanisms.

P/M Steels

There are various P/M tool steels engineered with respect to wear applications. The steel compositions vary in carbon and alloying elements such as manganese and phosphorous contents. By far the most recognized P/M steels for wear applications, however, are P/M tool steels. Compared to conventional tool steels, P/M grades offer distinctive advantages in terms of finer grain size and homogeneous microstructure with less segregation. The powder metallurgy approach also lends to near-net shape manufacturing important for components made of tool steel from the perspectives of machinability and cost. A list of commercially available P/M tool steel grades with their typical composition and characteristics are found in Table 1 and in the article "Particle Metallurgy (PM) Tool Steels" in this Volume. A number of P/M high speed steels have been developed that cannot be made by conventional methods because of their high carbon, nitrogen, or alloy contents. Examples include CPM Rex 20, CPM Rex 25, CPM Rex 76, and ASP 60. The P/M hot isostatic pressing (HIP) process is the only way to produce such highly alloyed high speed steel (HSS) as Rex 76 and ASP 60.

Figure 8 schematically illustrates P/M processing routes for tool steels. The most critical processing steps are manufacturing of prealloyed powder by atomization and consolidation by HIP. There are two competing processes in making powder: water atomization and gas atomization. Typical advantages of gas atomization over water atomization include cleanliness and particle shape control. Tool steels produced by HIP are normally conducted in argon at 7 to 200 MPa pressure and 1000 to 1200 °C. Powder metallurgy tool steels and other ferrous P/M alloys can also be manufactured by nonpressurized sintering techniques, such as solid state, transient, or super-solidus liquid phase sintering. Standard sintering techniques have the distinct advantage of low cost and therefore wide application areas; but, in general, substantial compromise is made in terms of mechanical properties due to presence of significant levels of porosity.

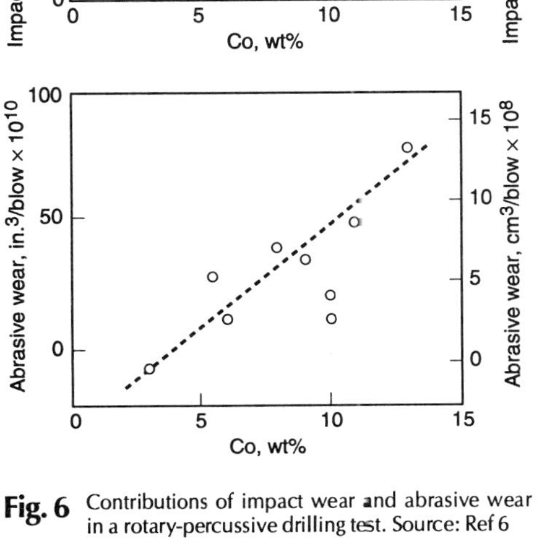

Fig. 6 Contributions of impact wear and abrasive wear in a rotary-percussive drilling test. Source: Ref 6

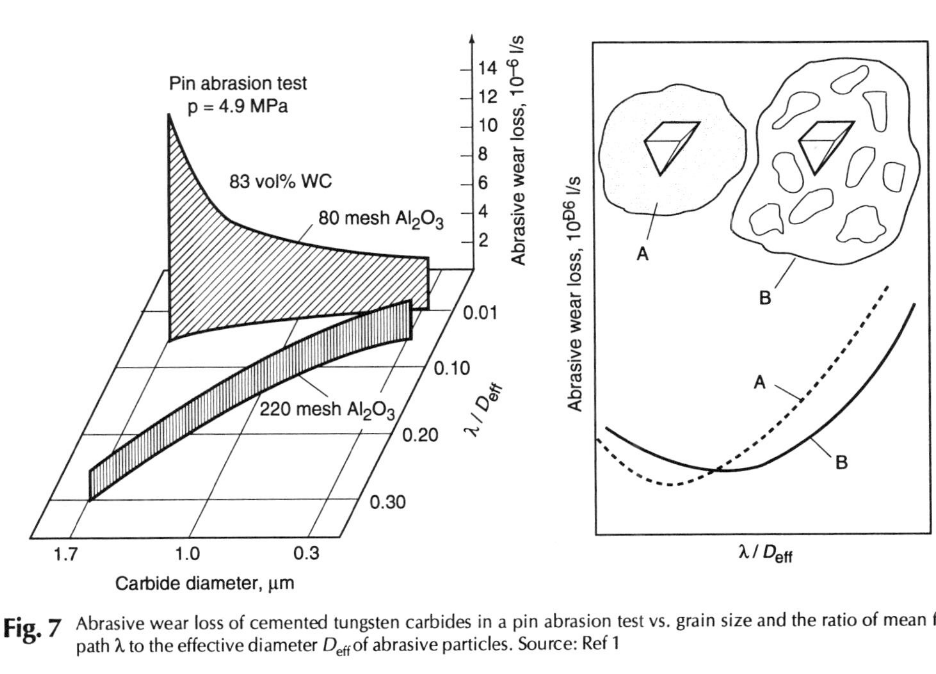

Fig. 7 Abrasive wear loss of cemented tungsten carbides in a pin abrasion test vs. grain size and the ratio of mean free path λ to the effective diameter D_{eff} of abrasive particles. Source: Ref 1

Table 1 Commercial P/M tool steel compositions

Trade name	AISI designation	Constituent elements, %: C	Cr	W	Mo	V	Co	S	Other	Hardness, HRC
High-speed tool steels(a)										
ASP 23	M3	1.28	4.20	6.40	5.00	3.10	...	...	...	65–67
ASP 30	...	1.28	4.20	6.40	5.00	3.10	8.5	...	...	66–68
ASP 60	...	2.30	4.00	6.50	7.00	6.50	10.50	...	...	67–69
CPM Rex M2HCHS	M2	1.00	4.15	6.40	5.00	2.00	...	0.27	...	64–66
CPM Rex M3HCHS	M3	1.30	4.00	6.25	5.00	3.00	...	0.27	...	65–67
CPM Rex M4	M4	1.35	4.25	5.75	4.50	4.00	...	0.06	...	64–66
CPM Rex M4HS	M4	1.35	4.25	5.75	4.50	4.00	...	0.22	...	64–66
CPM Rex M35HCHS	M35	1.00	4.15	6.00	5.00	2.00	5.0	0.27	...	65–67
CPM Rex M42	M42	1.10	3.75	1.50	9.50	1.15	8.0	...	...	66–68
CPM Rex 45	...	1.30	4.00	6.25	5.00	3.00	8.25	0.03	...	66–68
CPM Rex 45HS	...	1.30	4.00	6.25	5.00	3.00	8.25	0.22	...	66–68
CPM Rex 20	M62	1.30	3.75	6.25	10.50	2.00	...	...	...	66–68
CPM Rex 25	M61	1.80	4.00	12.50	6.50	5.00	...	...	...	67–69
CPM Rex T15	T15	1.55	4.00	12.25	...	5.00	5.0	0.06	...	65–67
CPM Rex T15HS	T15	1.55	4.00	12.25	...	5.00	5.0	0.22	...	65–67
CPM Rex 76	M48	1.50	3.75	10.0	5.25	3.10	9.00	0.06	...	67–69
CPM Rex 76HS	M48	1.50	3.75	10.0	5.25	3.10	9.00	0.22	...	67–69
HAP 10	...	1.35	5.0	3.0	6.0	3.8	...	...	...	64–66
HAP 40	...	1.30	4.0	6.0	5.0	3.0	8.0	...	...	...
HAP 50	...	1.50	4.0	8.0	6.0	4.0	8.0	...	...	...
HAP 60	...	2.00	4.0	10.0	4.0	7.0	12.0	...	...	...
HAP 70	...	2.00	4.0	12.0	10.0	4.5	12.0	...	...	...
KHA 33N	...	0.95	4.0	6.0	6.0	3.5	...	...	0.60N	65–66
Cold-work tool steels										
CPM 9V	...	1.78	5.25	...	1.30	9.00	...	0.03	...	53–55
CPM 10V	All	2.45	5.25	...	1.30	9.75	...	0.07	...	60–62
CPM 440V	...	2.15	17.50	...	0.50	5.75	...	...	...	57–59
Vanadis 4	...	1.50	8.00	...	1.50	4.00	...	...	...	59–63
Hot-work tool steels										
CPM H13	H13	0.40	5.00	...	1.30	1.05	...	...	...	42–48
CPM H19	H19	0.40	4.25	4.25	0.40	2.10	4.25	...	...	44–52
CPM H19V	...	0.80	4.25	4.25	0.40	4.00	4.25	...	...	44–56

(a) HCHS, high carbon, high sulfur; HS, high sulfur

Lower strength and lower hardness dictate that wear resistance is lower than that of HIP alloys.

Powder metallurgy tool steels utilize the same basic heat treatment as their conventional counterparts, but they tend to respond more rapidly and with better predictability to heat treatment because of their more uniform microstructure and finer carbide size. Powder metallurgy tool steels also have superior machinability to conventional tool steels due to generally higher sulfur content in P/M alloys.

Wear resistance of P/M tool steels is largely associated with its hardness. Figure 9 illustrates the relationship between impact toughness and hardness. Balance in toughness and wear resistance must be considered in selecting a specific tool steel grade for applications. Figure 10 is a representative view of the wear resistance of a range of sintered alloy steels against average hardness. Variations in wear resistance versus hardness are due to specific alloys with respect to carbide contents, carbide particle sizes, size distributions, and shapes.

P/M tool steel is essentially a homogeneous material. Volume fraction of carbides in a tool steel is <50% (10 to 20% in most cases) and particle size of carbide particles is usually very small, ranging from submicron to a few microns. Although wear resistance is proportional to volume fraction of carbides, studies show that the rate of increase in wear resistance diminishes >30% by volume for specific ferrous-base alloy systems and applications.

On the other hand, the tool steels including high speed steels can be considered as composite materials with a martensitic steel matrix and a harder, particulate reinforcement consisting of primary carbides. Wear resistance of P/M tool steels is therefore similar to that of cemented tungsten carbide. Microstructure factors controlling wear resistance include carbide volume fraction, carbide type, carbide particle size and size distribution, carbide particle shape, and steel matrix microstructure.

Carbide types in tool steels include MC, M_2C, and M_6C types, where M could be V, W, Mo, or Cr. The presence of M_2C or MC type carbides is considered to impart a greater wear resistance to these materials than M_6C type carbides. Vanadium carbide plays an important role in P/M tool steel. PM 10V and PM 15V are primary examples of P/M tools steels with high vanadium content of 9.75 and 14.5 vol% respectively, which is impossible for conventional tool steels. The microstructure of PM 15V contains ~22 vol% primary vanadium carbides with the majority having a size of 3 μm or finer. Table 2 compares high vanadium P/M tools to others. It is apparent that wear resistance increases rapidly as vanadium content increases at the expense of impact toughness. In general, carbide particle sizes range from submicron to a few microns (<10 μm), depending on processing history including heat treatment. Particle size and volume distributions are important in the wear behavior, a high proportion of M_2C/MC carbides of large size (6 μm) give a greater wear resistance.

Compared to cemented tungsten carbide, tool steels are perceived as tougher and less wear resistant. Figure 11 illustrates relative wear resistance of P/M tool steels compared to cemented tungsten carbide with 8% cobalt (Ref 10). However, although the relation is generally true, attention must be given to the specific environment in which wear occurs. For example, the American Society for Testing and Materials (ASTM) has two widely used abrasive wear tests. ASTM G 65 is considered a low stress abrasion test, during which wear coupons are

Table 2 Wear resistance and toughness of high vanadium content steels and other wear resistant steels

Grade	Hardness, HRC	Pin abrasion resistance weight loss, mg	Cross cylinder wear resistance, $\times10^{10}$ lb/in.^{2}	Toughness Charpy C-notch impact energy, ft/lb
PM 15V	64	24	124	8
PM 10V	64	32	90	14
PM M4	61	49	31	31.5
PM D7	59	46	11	22
PM 8Cr4V	60	56	30	27

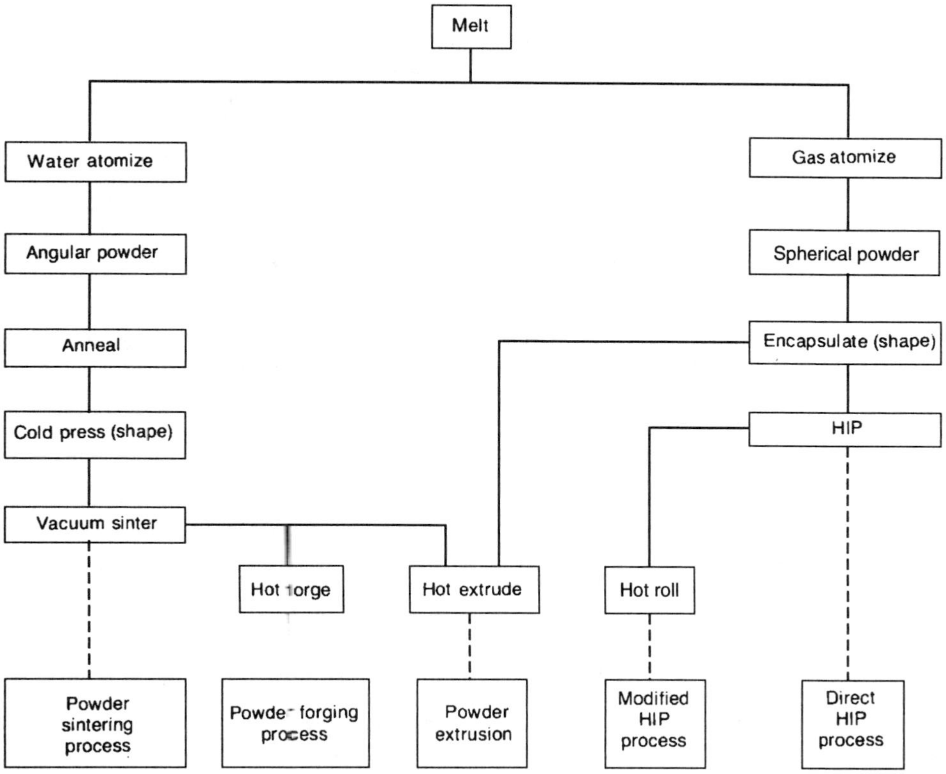

Fig. 8 Typical processing flow chart of P/M tool steels. Source: Ref 7

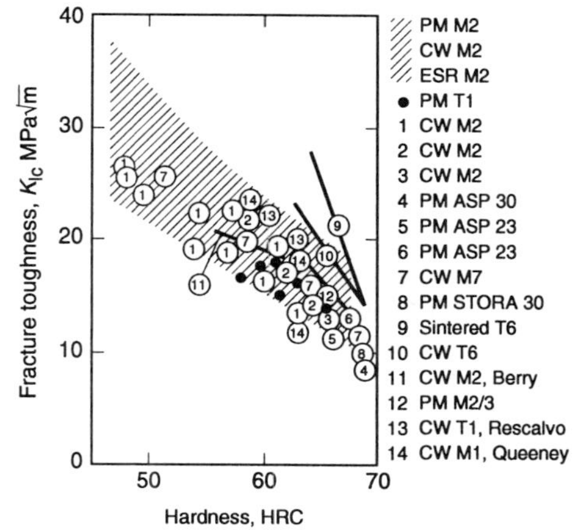

Fig. 9 Compilation of fracture toughness of tool steels.

abraded by loose sand particles against a lightly loaded rubber wheel under dry conditions. ASTM B 611 is considered a high stress abrasion test, during which wear coupons are abraded by alumina against a heavily loaded steel wheel under wet medium. Table 3 compares wear resistance of a few typical tool steels to WC-Co alloys with average grain size ~4 μm using the two different abrasion tests. It shows that cemented tungsten carbides have higher wear resistance under both low stress and high stress conditions. Tool steels have significantly lower wear resistance under the low stress dry sand abrasion condition. However, the differences between tool steels and WC-Co alloys are much smaller under high stress abrasion conditions. It indicates that toughness plays a larger role in high stress abrasive resistance than in low stress abrasion resistance. This is the result of complex wear mechanisms and their interrelationship with hardness or fracture toughness. It is also the result of complex interaction between wear environment and microstructure features, such as carbide particle size, volume fraction, and MFP. Therefore, in selecting materials for wear applications, not only wear tests data are considered, other overall effects of basic mechanical properties and actual wear environment have to be considered too. Table 4 qualitatively compares properties of P/M tool steel to cemented tungsten carbide.

Thermal fatigue cracking resistance and corrosion resistance are sometimes the primary cause for material loss and fracture. Final selection of a material for an application depends on the mechanisms responsible for material loss versus consideration of combined effects of all properties.

Applications of P/M tool steels and high speed steels include metal removal cutting tools and tooling for metal forming and shaping operations. Cutting tools include milling inserts, hole machining reamers, taps, drills, and broaching tools. Typical P/M tool steel grades for cutting tools include ASP 30, ASP 60, and the CPM Rex M3 and M4. At similar hardness levels, P/M grades offer superior impact toughness to similar grades processed by conventional means. Table 5 compares CPM Rex 20 and CPM Rex M42 to conventional M42. As the result of better basic mechanical properties and more uniform finer microstructure, wear life of P/M tool steel tools are longer than their counterparts. Figure 12 is one example of the tool life of milling inserts of ASP 30 and ASP 60 compared to conventional M42.

Cold- and hot-work rolls are another primary application of P/M tool steels. High vanadium

Table 3 Wear resistance comparison between P/M tool steels and typical WC-Co grades

Materials	Wear resistance under low stress abrasion (ASTM-G65), 1000 rev/cc	Wear resistance under high stress abrasion (ASTM B 611), 1000 rev/cc
WC-10%Co, grain size 5 μm	0.435	3.3
WC-16%Co, grain size 6 μm	0.238	1.8
A2	0.087	1.95
D2	0.186	1.8
CPM M4	0.590	1.9
S7	0.036	1.95

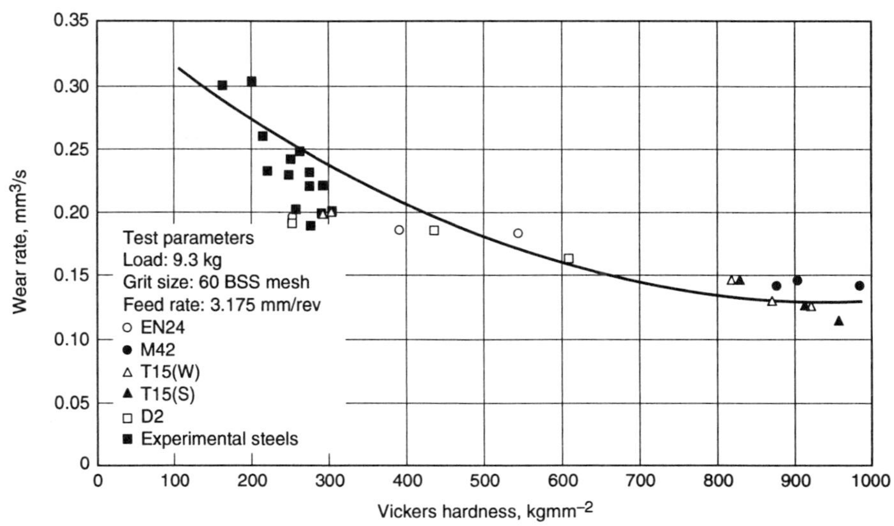

Fig. 10 Wear rates of a range of sintered alloy steels against average hardness. Source: Ref 9

P/M tool steel such as CPM 9V and 10V are used for high wear and cold work applications. Similar to cutting tools, the more uniform microstructure of P/M cold-work steel yields better toughness. Higher vanadium content makes it possible that higher wear resistance is achieved at an improved toughness simultaneously. CPM 10V has proved to be more wear resistant than any commercially available high-alloy tool steel. CVP 10V has far superior wear resistance and equivalent impact toughness when compared to conventional D2, M2, and CPM M4 (Fig. 13). For hot-work applications, superior thermal fatigue resistance is very important. The absence of segregation in P/M tool steels is an advantage over conventional alloys, because premature failure due to thermal fatigue is often attributed to segregation and heterogeneous microstructure. H13 and H19 are typical tool steel grades for hot-work applications. The P/M versions, P/M H13 and H19 have more uniform properties and equivalent or better toughness. Again, high vanadium content can improve wear resistance and toughness.

P/M Metal Matrix Composites

The concept of metal-matrix composite (MMC) refers to a large class of materials with at least one type of reinforcement situated in a metal matrix. In general, the scale of the reinforcement is large compared to that of the microstructure. The reinforcement can be in the form of continuous fiber, short fiber, or particulate. Although it is recognized that MMCs offer considerable potential for enhanced wear resistance, understanding of their wear characteristics is still far from complete. This is due in part to the inherent complexity of many wear processes, but the problem is compounded by an interplay with microstructure variables in MMC, such as reinforcement content, size, orientation, interface strength, etc.

Many MMCs are fabricated by powder metallurgy, in other words, the composites are formed by consolidating constituent powders together. Particulate MMC is especially suitable for P/M processing. In a narrower sense, MMC refers to modern low-density high-strength alloy systems such as aluminum- or titanium-base composites (e.g., Al-SiCp, Al-Al_2O_3, Ti-SiCp, and Ti-6Al-4V-SiCp). However, it is clear from the previous discussion that many P/M materials for wear resistant applications bear the same basic microstructure characteristics of MMCs. Therefore, many P/M materials can be considered MMC, and many particulate MMCs can be considered P/M materials. Hence, general principles governing wear behavior of two or multiphase P/M materials apply to many MMCs.

Key factors affecting wear resistance of MMC include volume fraction of reinforcement particles, particle size of the reinforcement, and the diameter of reinforcement relative to the size of the abrading particles. Figure 14 illustrates that specific wear rates of aluminum-base composites decrease with the increase of volume fraction of different hard particle reinforcements. This behavior is true under either abrasive or adhesive wear situations. For a given volume fraction of particles, composites that contain harder particles exhibit a lower wear rate. Similar results can be expected for tool steel-base composites such as steel-base TiC particle composites. Figure 15 demonstrate the effects of the size of the hard particles. Larger ceramic particles and higher contents give greater wear resistance. Larger abrasive particles cause significantly more wear to composites because of the increasing chances of fracturing of reinforcement particles under the heavy load exerted by large abrading particles. In the case of erosive wear, situations are more complex. Factors such as the impact angle play a very important role as well as volume fraction. For example, it was found that the minimum erosive wear rate appeared at 34 vol% binder in WC-Co cermets. MMC can also contain soft reinforcement. Al-Sn alloys, for example, have long been used as bearing alloys. In general, when improvements to the wear resistance are obtained in this way, it is normally accompanied by a reduction in the coefficient of friction.

Hardfacing and Thermal Spray

Applications. Metal alloy powders are used for hardfacing and thermal spray coatings for wear-resistant applications. Hardfacing is the application of hard, wear-resistant material to the surface of a component by welding, thermal spraying, or a similar process for the main purpose of reducing wear. Metal alloy powders used for hardfacing include WC-Co powders, other carbide-metal alloy powders, cobalt-base wear-

Table 4 Qualitative comparison of mechanical properties of P/M tool steel and cemented tungsten carbide

Property	P/M tool steel	P/M tool steel	WC-Co	WC-Co
Hardness	50–58 HRC	59–68 HRC	>88 HRA	<87 HRA
TRS	Medium	High	Medium/high	Medium
K_{Ic}	High	Medium	Low	Medium
Impact resistance	High	Medium-low	Low	Medium
Wear resistance	Low	Medium	High	Medium
Thermal shock resistance	Low	Low	High	Medium

Table 5 Charpy C-notch impact and bend strength of CPM alloys comparing to conventional M42 alloy

Alloy grade	Hardness, HRC	Charpy C-notch impact energy, J	Bend fracture strength, MPa
CPM Rex 20	67.5	16	4006
CPM Rex M42	67.5	16	4006
Conventional M42	67.5	7	2565

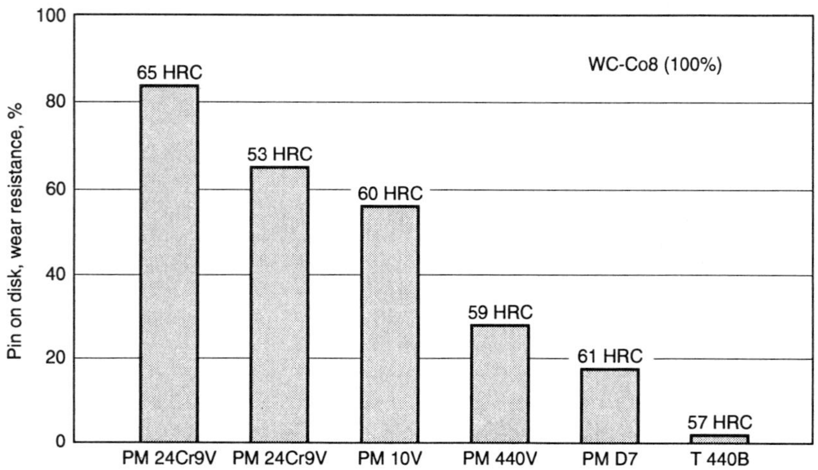

Fig. 11 Relative comparison of pin-on-disc wear resistance of PM 24Cr9V, PM 440V, PM 10V, PM D7, and T440B stainless to WC-8Co. Source: Ref 10

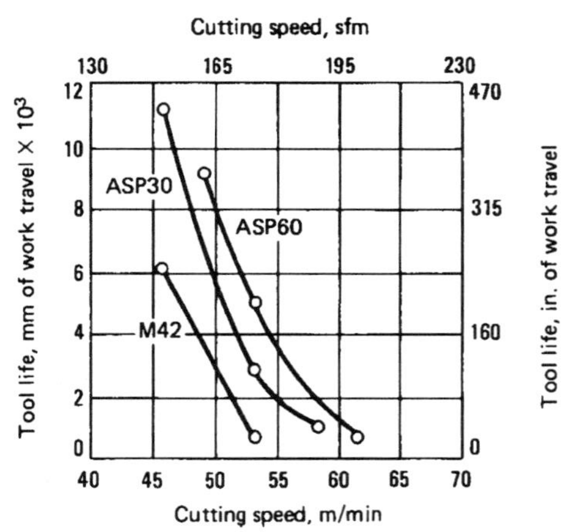

Fig. 12 Comparison of P/M tool steel ASP 30 vs. conventional M42 during end mill tests on Ti-6Al-4V. Source: Ref 7

resistant alloy powders such as Stellite (Deloro Stellite, Inc., Belleville, ON) series of alloys, and some nickel- or iron-base wear-resistant alloy powders.

Thermal spray coatings are themselves a major engineering field. Numerous different techniques for applying hardmetal powders onto a substrate can be categorized as thermal spray coating. Thermal spray techniques include but are not limited to traditional oxyacetylene torch spray, plasma thermal spray, detonation gun (D-gun and super D-Gun) thermal spray, and more recent technologies such as high velocity oxy-fuel (HVOF) thermal spray. Figure 16 is a schematic illustration of the plasma spray coating process. Common features of thermal spray coatings, differing from other hardfacing processes (such as welding processes), are thickness of thermal spray coating is usually limited to 0.025 to 0.5 mm and bonding between coating and substrate is primarily mechanical. Key advantages of thermal coatings over welding are geometry, flexibility, and process temperature of substrate components can be maintained so low that the coating can be applied to finished products.

Hardfacing materials are often applied by various welding techniques. The welding techniques for hardfacing include oxyacetylene torch, gas metal arc, gas tungsten arc, plasma arc, and plasma transfer arc welding. Plasma transferred arc process is illustrated in Fig. 17. Characteristics of weld-on hardfacing are that coatings are thicker than thermal spray coatings, and coatings are metallurgically bonded to the substrate. Welded-on wear-resistant hardfacing has wide ranging applications from earth moving equipment to precision bearing and seal surfaces.

Another category of hardfacing techniques is cladding. Hardfacing materials can be cladded onto substrates by furnace fusing prearranged layers of loosely bonded hardmetal onto substrate (Fig. 18). Two layers of metal and metal carbide tapes are stacked over the substrate. The tape is made from powder bonded by polytetrafluoroethylene (PTFE) or other polymers. One of the two layers is a carbide powder tape and the other is high temperature brazing alloy powder tape, such as a nickel-base brazing alloys. Coating is formed by heating the assembly to brazing temperature and infiltrating the brazing alloy into the carbide layer. The final coating of nickel alloy bonded carbide particles is bonded to the substrate during the process. Cladding can also be done by HIP. Advantages of furnace cladding include better dimensional and thickness control. Disadvantages are that the

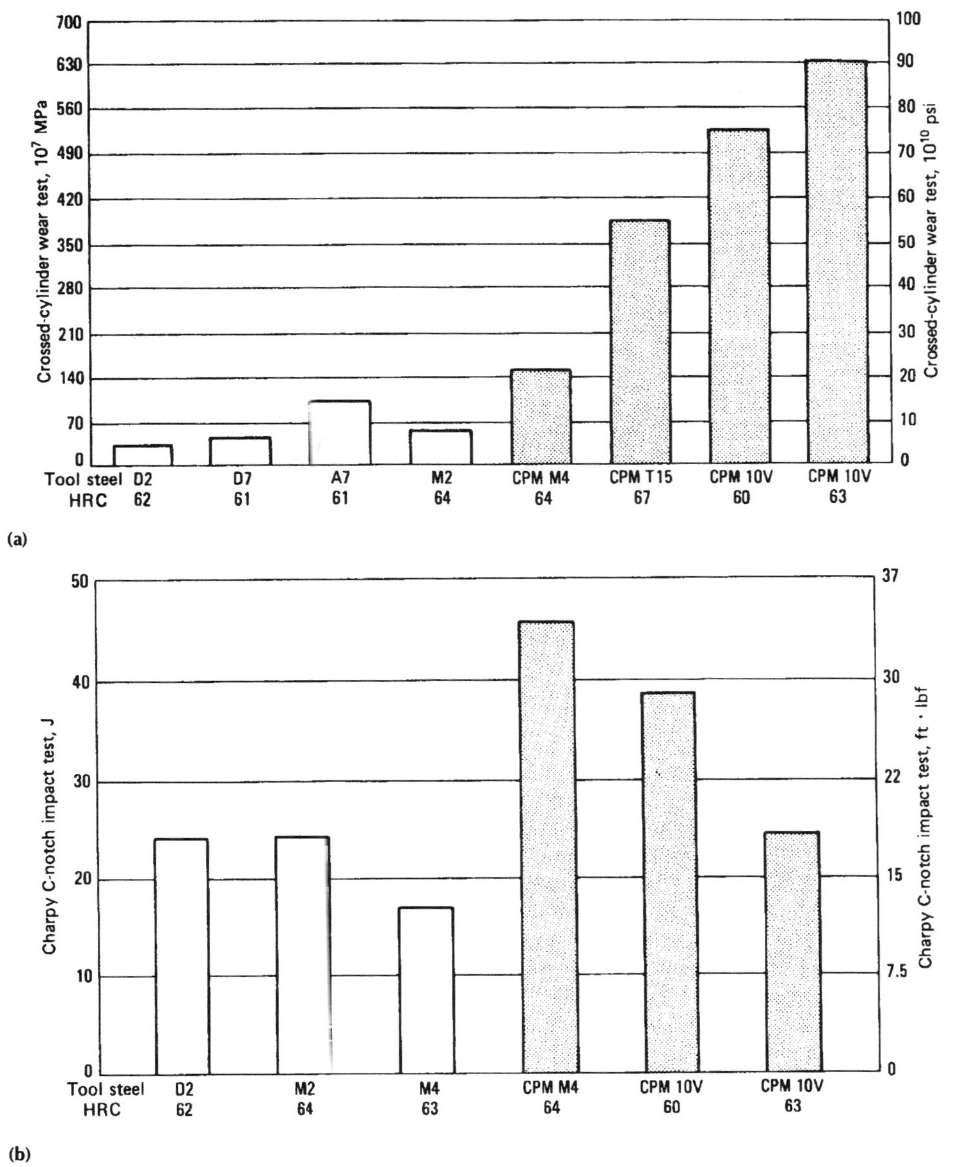

Fig. 13 (a) Wear resistance and (b) impact toughness of CPM10V compared to conventional, P/M, other tool steels at indicated hardness. Source Ref 7

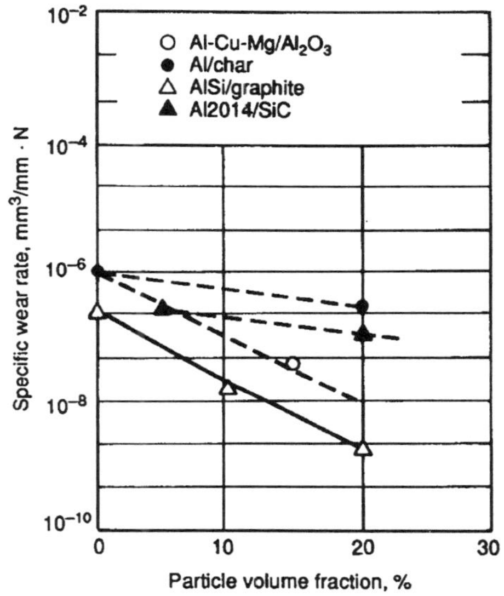

Fig. 14 Wear rate of aluminum matrix composites vs. volume fraction of reinforcement particles during a sliding test against steel. Source: Ref 11, 12

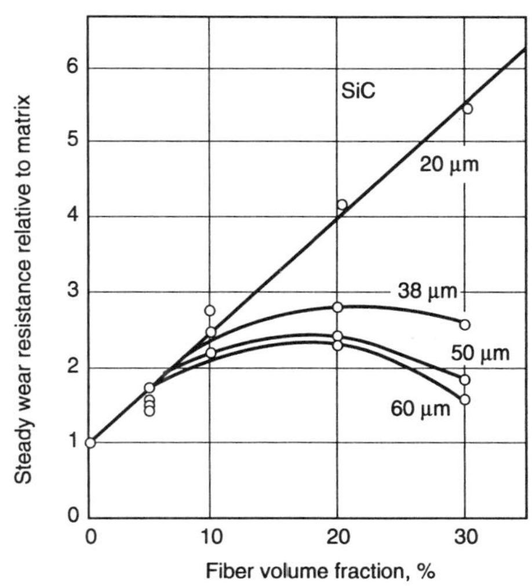

Fig. 15 Dependence of wear rate of 6061 Al-Safill (δ-alumina) composites on volume fraction of reinforcement and particles size of abrasives. Source: Ref 11

entire components are exposed to high temperature (typically 1000 to 1250 °C) for an extended period of time.

Metal carbides are the most widely used materials for wear-resistant coatings and hardfacings. Among various types of carbides, tungsten carbide is the most popular one. Chromium carbide is also very common. Other carbides, such as vanadium carbide and molybdenum carbides, are often the choice for additives.

Sintered cemented tungsten carbide (WC-Co) powders are widely used for thermal spray coatings and torch welding hardfacing. WC-Co powder can be a product of crushing bulk WC-Co alloys or specifically made spherical shaped pellets. Tungsten carbide powders for thermal spray coatings usually contain from 6 to 18% cobalt. Table 6 lists typical compositions of carbide powders for thermal spray coatings. Cr_3C_2 powder and its alloys are also popular for thermal applications. Particle sizes of WC-Co and other alloy powders for thermal spray coatings range from 5 to 150 μm. Powders for torch welding hardfacing ranges from 150 μm to 1.2 mm.

Macrocrystalline tungsten carbide powders are a special kind of tungsten carbide powder manufactured by high temperature thermit process during which ore concentrate is converted directly to WC. Macrocrystalline tungsten carbide maintains carbon content of 6.13% by weight, the correct stoichiometric amount. Macrocrystalline tungsten carbide is grown in crystals ranging from 1 μm to 5 mm. Coarse mesh sizes of macrocrystalline carbide are extensively used in abrasion and erosion protection applications. The finer size is employed as wear rate modifiers. Coarse grain tungsten carbide powder can also be obtained by conventional dry-carburizing tungsten metal method. Examples include commercial grades of MAS 2000, MAS 3000, and MAS 5000 by H.C. Starck (Newton, MA). Particle sizes of these powders range from 20 to 50 μm.

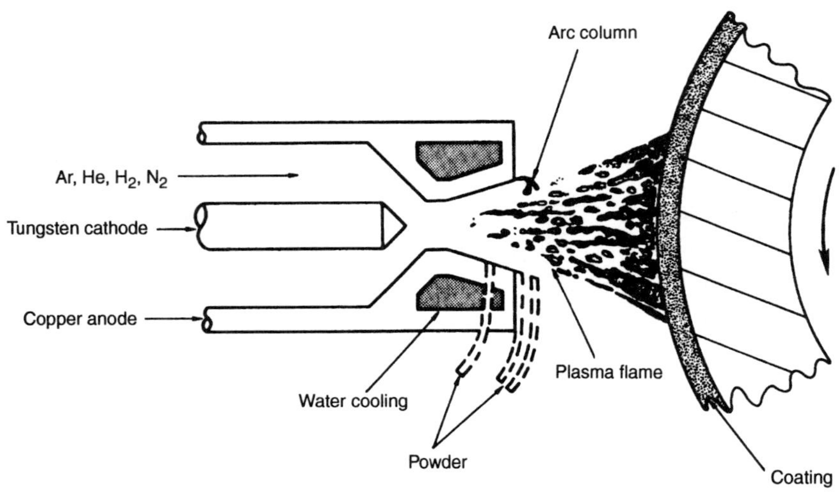

Fig. 16 Schematic of plasma spray hardfacing process

Cast tungsten carbide is another category of carbide that has wide applications in hardfacing. Cast carbide refers to eutectic of WC and W_2C that can range in carbon content from 3.5 to 4.5% by weight. It is manufactured by melting mixtures of tungsten metal, tungsten carbide, and carbon ingredient at temperatures above 3000 °C. The melt is cast into billets, which are then crashed into specified size ranges of powder. Cast carbide can be crashed to as coarse as 20 mesh and as fine as –325 mesh powder. Cast carbide, mostly used in torch welding hardfacing and infiltrated drag bit matrix body, is often used in combination with sintered tungsten carbide and other carbides to enhance wear resistance. Addition of cast carbide tends to degrade toughness of coating.

Spherical cast tungsten carbide is a relatively new product in the marketplace with increasing applications. Spherical cast carbide has all the basic

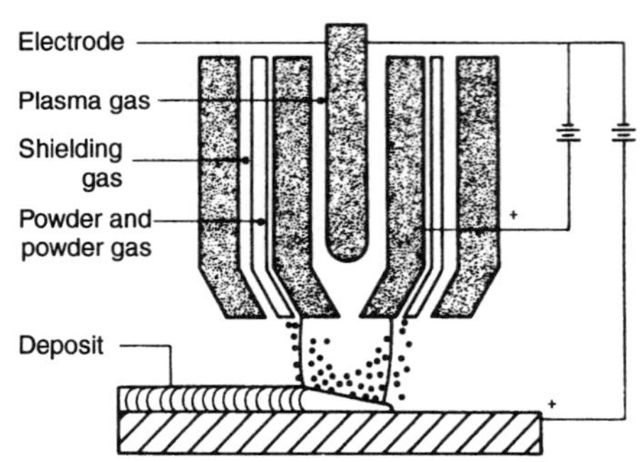

Fig. 17 Schematic of the plasma transferred arc hardfacing process

Table 6 Typical thermal spray alloy powders for wear applications

Powders	Composition						
	WC	Cr_3C_2	Co	Ni	Fe	Cr	Other
WC-12Co	bal	...	12	...	...	...	...
WC-10Co-4Cr	bal	...	10	...	...	4	...
WC-17Co	bal	...	17	...	...	...	...
WC-10Ni	bal	...	...	10	...	...	...
Cr_3C_2	...	bal	...	...	...	...	...
Cr_3C_2-25NiCr	...	bal	...	...	...	...	24–26NiCr
Cr_3C_2-FeCrAlY	...	bal	...	...	...	...	14–16FeCrAlY
Ni-718	...	...	...	bal	...	17–21	Mo2.8–3.3, Nb4.8–5.5, Fe 14–21, Ti 0.75–1.15
Stellite 6	...	...	bal	...	...	28	4W-1Si-1C
Stellite 1	...	...	bal	...	...	31	12.5W-1Si-2.5C
Stellite 12	...	...	bal	...	...	29	8.5W-1.5Si-1.5C
410 stainless steel	...	...	...	...	bal	12.5	...

Table 7 Typical cobalt-, nickel-, and iron-base hardfacing alloys

Alloys category	Content	Hardness	Abrasion resistance	Impact resistance	Adhesion resistance	Applications
Iron based low-alloy	0.3–1.0C, 1.7–7.0W, 3.5–5.0Cr, 4.0–9.0Mo, bal Fe	55–60 HRC	More wear resistant than low carbon steels	Moderate	Good	Cutting tools, sheer blades, reamers, dies, ingot tongs
Iron-chromium	3.0–5.0C, 4.0–8.0Mn, 26–32Cr, 2.0Mo, bal Fe	51–62 HRC	Excellent for low stress abrasion	Good for light impact	Excellent	Agricultural machinery, coke shutes, brick-making equipment
Cobalt-chromium	0.7–3.0C, 2Mn, 3–14W, 3Ni, 19–33Cr, 5Fe, bal Co	28–65 HRC	Excellent in low stress abrasion. 5–20% of wear rate of carbon steel	Fair. Deteriorates as carbon increases	Excellent for low stress. Poor for high stress	High temperature, high corrosion applications, pump shafts, engine valves
Nickel-chromium	0.3–1.0C, 1.0–1.5Co, 8–18Cr, 1.25–5.5Fe, 2–4.5B, 1.25–5.5Si, bal Ni	24–62 HRC	Excellent in low stress. Increases with carbon content	Good for light impact. Decreases as carbon increases	Excellent for low stress	Seal rings, cement pump screws, valves, cams
Tungsten carbide	60% carbide granules, 40% steel tube	90–95 HRA (granules), 30–60 HRC matrix	Superior to all others	Good under light impact	Poor	Cutting teeth and edge holding surfaces of rock drill bits, quarrying, digging, earth moving equipment

characteristics of crashed cast carbide. Spherical cast carbide has extremely high wear resistance and higher resistance to chipping and cracking than crushed cast carbide due to its spherical shape. Spherical cast carbide is manufactured by patented solidification technique (Ref 14), during which molten eutectic carbide droplets are rapidly cooled. Spherical cast carbide can be used alone or in combination with other carbides.

Hardfacing Alloy Powders

Cobalt-base hardfacing alloy powders are best known as Stellite powders. Typical compositions and wear resistance of Stellite powders are found in Table 7. Cobalt alloys for hardfacing have advantages for use in combined wear-resistant, corrosion-resistant, and high temperature applications. Cost of cobalt-base alloys tends to be higher than nickel- and iron-base alloy powders. Cobalt alloys are typically used for overlay applications, deposited readily by plasma transfer arc process.

Nickel-base hardfacing alloy powders are usually Ni-Cr-B-Si system based. Additions of boron and silicon to nickel suppress the melting point so that less heat is required to deposit the nickel-base hardfacing alloy onto substrates. However, excessive additions of boron and silicon lower plasticity. Most of these alloys contain nickel boride (Ni_3B), chromium borides (CrB, Cr_5B, and Cr_2B), and other complex borides as hard phases. Typical alloy compositions and wear resistance are listed in Table 7. Deposit hardness in these alloys is as high as 60 HRC, depending on the chromium, boron, and silicon contents. Nickel-base hardfacing alloys are particularly popular in oil pumping and in glass, ceramic, cement, and plastic industries.

Iron-base hardfacing alloys can be classified into Pearlitic steels, Austenitic steels, Martensitic steels, and high-alloy irons. The high alloy irons usually contain high carbon and high chromium content. More iron-base hardfacing alloys are used in various industries than cobalt- and/or nickel-base hardfacing alloys, primarily due to the lowest cost and highest availability. Typical composition and wear resistance comparison is shown in Table 7.

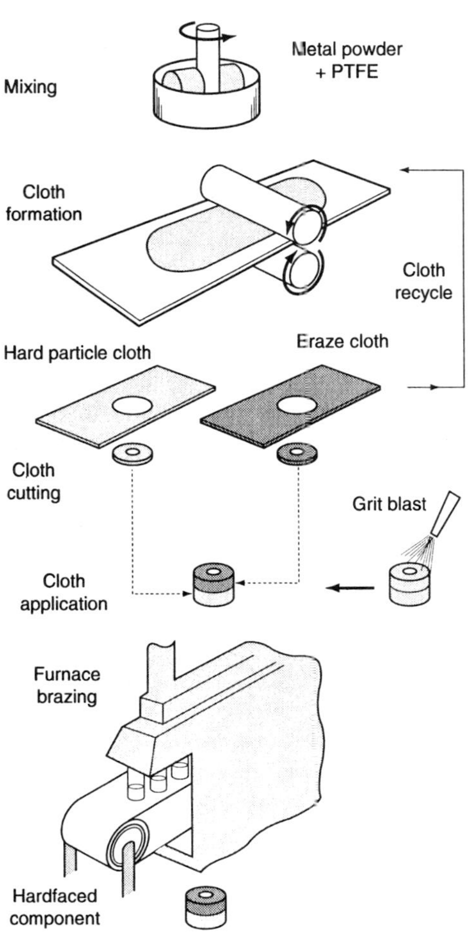

Fig. 18 Flow chart of a tape cladding process. Source: Ref 13

Full Density Cobalt Alloys

Alloys based on cobalt chromium-tungsten-carbon, such as the Stellite family of alloys, are extremely versatile, wear-resistant materials (Ref 15, 16). Full density P/M components also can be manufactured by hot isostatic pressing or extruding operations (Ref 17, 18) for near-net shape processing and refined microstructures.

Material Descriptions and Applications. Table 8 gives the nominal chemical analysis for the most popular Stellite P/M materials. The microstructure of Stellite alloys contains com-

Table 8 Nominal chemical analysis of most widely used Stellite P/M materials

Alloy	Composition, %									
	Cr	W	Ni	Mo(a)	Si(a)	Mn(a)	Fe(a)	C	B(a)	Other(a)
Stellite 3	31	12.5	3(a)	...	1	1	3	2.4	1	1
Stellite 6	29	4.5	3(a)	1.5	1.5	1	3	1.2	1	2
Stellite 12	30	8.5	3(a)	...	1	1	3	1.5	1	3
Stellite 19	31	10.5	3(a)	...	1	1	3	1.9	1	2
Stellite 31	25.5	7.5	10.5	...	1	1	2	0.5	...	2
Stellite 190	26	14	3(a)	1	1	1	5	3.1	1	2
Stellite 98M2	30	18.5	3.5	0.8	1	1	5	2	1	2, 4.2 V
Stellite Star J	32.5	17.5	3(a)	...	1	1	3	2.5	1	2

(a) Maximum; all materials cobalt base

Table 9 Physical properties of Stellite materials

Property	Stellite trade designation						
	3	6	12	19	31	98M2	Star J
Density, g/cm^3(a)	8.40	8.20	8.28	8.35	8.45	8.45	8.58
Melting range, °C (°F)	1213–1285 (2215–2345)	1260–1357 (2300–2475)	1255–1341 (2290–2445)	1239–1299 (2260–2370)	1340–1396 (2445–2545)	1224–1275 (2235–2325)	1215–1299 (2220–2370)
Coefficient of thermal expansion, 10^{-6}/K							
At 0–100 °C (32–212 °F)	12.3	13.9	11.9	12.8	14.0	10.8	12.0
At 0–400 °C (32–750 °F)	12.9	14.7	13.4	13.8	14.5	11.7	12.3
At 0–700 °C (32–1290 °F)	13.4	15.8	14.4	14.8	16.0	13.0	13.2
At 0–1000 °C (32–1830 °F)	16	17.4	15.8	16.9	17.0	14.1	15.1
Magnetic permeability at 200 Oersteds and 22 °C (72 °F)	<1.2	<1.2	<1.2	<1.2	...	<1.2	<1.2

(a) P/M materials are typically 97 to 100% dense

Table 10 Mechanical properties of Stellite materials

Property	Stellite P/M trade designation							
	3	6	12	19	31	190	98M2	Star J
Transverse rupture strength, MPa (ksi)								
At 22 °C (72 °F)	966 (140)	1725 (250)	...	1898 (275)	...	932 (135)	1035 (150)	863 (125)
Ultimate tensile strength, MPa (ksi)								
At 22 °C (72 °F)	863 (125)	897 (130)	...	1035 (150)	828 (120)	621 (90)	794 (115)	523 (75)
At 1000 °C (1830 °F)	725 (105)	828 (120)	...	...	676 (98)	518 (75)	725 (105)	539 (78)
At 1200 °C (2190 °F)	690 (100)	766 (110)	...	...	614 (89)	518 (75)	690 (100)	569 (82.5)
At 1400 °C (2550 °F)	621 (90)	518 (75)	...	...	545 (79)	518 (75)	656 (95)	573 (83)
Elongation, %								
At 22 °C (72 °F)	<1	<1	...	<1	4	<1	<1	<1
At 1000 °C (1830 °F)	<1	1	...	...	14	<1	<1	<1
At 1200 °C (2190 °F)	<1	1	...	...	13	<1	<1	<1
At 1400 °C (2550 °F)	1	10	...	...	16	<1	<1	<1
Hardness, HRC								
At 22 °C (72 °F)	54	40	44	49	32	58	58	56
At 1000 °C (1830 °F)	46	37	...	...	...	54	...	52
At 1200 °C (2190 °F)	39	30	...	...	...	46	...	43
At 1400 °C (2550 °F)	28	15	...	...	...	34	...	31

Table 11 Guide for machining Stellite alloys

	Speed		Roughing				Finishing			
			Feed		Depth of cut		Feed		Depth of cut	
Operation	mm/min	ft/min	mm/rev	mils/rev	mm	mils	mm/rev	mils/rev	mm	mils
Turning(a)	9–15	30–50	0.203–0.305	8–12	0.127–5.08	5–205	0.152–0.254	6–10	≤0.635	≤25
Facing(a)	9–15	30–50	0.203–0.305	8–12	0.127–2.54	5–100	0.152–0.254	6–10	≤0.635	≤25
Boring(a)	6–15	20–50	0.127–0.203(b)	5–8	0.635–2.54(b)	25–100	0.127–0.203	5–8	≤0.635	≤25
Drilling(c)	3–8	10–26	...	...	...	...	0.025–0.127(d)	1–5	...	...
Reaming(e)	3–9	10–30	0.025–0.127(f)	1–5	0.762–1.52	30–60	0.025–0.076(f)	1–3	0.254–0.635	10–25
Threading	Use single-point C-3 tungsten carbide tools at 3–8 m/min (10–26 ft/min)									

(a) C-3 tungsten carbide tools should be used. Coolant is water-base fluid diluted 15 parts water to one part fluid. Tools for facing and boring are basically the same as turning tools, except for greater clearances where needed. (b) Depends on bar and size of tool. (c) C-2 tungsten carbide twist drills should be used, with drill web kept as thin as possible. Coolant is same as in (a). (d) Depends on size of drills. (e) C-3 tungsten carbide tools should be used. Reamers should have a 45° side cutting edge angle. Coolant is same as in (a). (f) Per tooth

plex combinations of M_7C_3, M_6C, and $M_{23}C_6$ carbides embedded in a cobalt-chromium-tungsten superalloy matrix.

For highly abrasive conditions, alloys such as Stellite 3, 98M2, and Star J are preferable because they contain a large volume fraction of carbides. The lower carbon grades are more ductile and are suitable for applications involving shock loading or impact.

Typical applications include precision balls, bearing race blanks, valve seat inserts, saw tips, cutters, spacer bushings, and wear pads. Stellite 3 is also used in medical instruments. Stellite 12 P/M has proved to be exceptional for cutting timber.

Material Properties. Tables 9 and 10 show the physical and mechanical properties of Stellite materials. The physical properties were measured on castings but are thought to be representative of P/M parts. It should be noted that Stellite materials are only weakly magnetic under most conditions.

The excellent hot hardness of the materials is particularly important to prevent the dulling of cutting tools where frictional heating is typically extreme at the sharp edge. It may be noted that some grades, such as Stellite 31, are relatively soft and ductile. This grade is used when high-temperature strength, shock loading, fatigue, and fretting are concerns.

Heat Treatment. Stellite P/M materials are not generally considered heat treatable nor is heat treatment required. Stress relief treatment involves a slow heat-up and soaking at 900 °C (1650 °F) for at least 2 h, followed by furnace cooling. Stellite 6 can be solution annealed by soaking at 1200 °C (2190 °F) for 2 h followed by rapid cooling. This maximizes the corrosion resistance and ductility of Stellite 6. Stellite 31 can be age hardened using a 2 day aging treatment at 800 °C (1470 °F).

Joining. Stellite materials can be joined to other materials by brazing or welding. Flame brazing is typically done with silver solder and an oxyacetylene flame. Vacuum furnace brazing uses a nickel-base brazing alloy such as AMS 4777.

Gas tungsten arc and gas metal arc resistance welding are commonly used for Stellite P/M materials. For the fusion processes, a preheat of 810 °C (1490 °F) is required for the harder grades, such as Stellite 3, 19, 98M2, and Star J. The softer grades, such as Stellite 6 and 31, should be preheated to 540 °C (1005 °F). Furnace cooling is required for the harder grades, whereas cooling in still air is satisfactory for the softer grades. Stellite 25 or Nistelle W filler metals are recommended for joining Stellite material to mild steel or stainless steel.

Table 12 General recommendations for grinding wheels

Purpose	Recommended wheels	Coolant
Peripheral wheels, Stellite alloy deposits	A60JV	Emulsifying oil
Form grinding (finer grits recommended for sharper forms)	A100GV, A120KV, A150GV	Emulsifying oil
Ring wheels		
Mixed Stellite alloy and steel	A46DZB	3.8 L (1 gal) good grade of water-soluble oil plus 190 L (50 gal) water
Stellite alloy only	A50BB	...
Segmental wheels		
Mixed Stellite alloy and steel	A46DZB	0.45 kg (1 lb) sal soda to 190 L (50 gal) water
Stellite alloy only	A50BB	...
Tool and cutter grinding		
Cup wheels for peripheral wheels (see above)	A46JV	3.8 L (1 gal) good grade of water-soluble oil plus 190 L (50 gal) water
Internal grinding bores greater than ~100 mm (4 in.) diam	A54JV, A54MV	3.8 L (1 gal) good grade of water-soluble oil plus 190 L (50 gal) water
Centerless grinding less than ~115 mm (4.6 in.) diam	A60L-V	3.8 L (1 gal) good grade of water-soluble oil plus 190 L (50 gal) water
Off-hand grinding	A46MV	Dry
Cutting off Stellite alloy	A46QB	1 part water-soluble oil to 60 parts water
Surface grinding, general Stellite alloy grinding		
For wheels ≤305 mm (12 in.)	A60JV	1 part water-soluble oil to 60 parts water
For wheels >305 mm (12 in.)	A46JV	...

Note: Wheels carrying the same marking from different manufacturers do not necessarily have the same cutting action. Wheel descriptions based on ANSI B74.13-1970 and BS 4481

Machining. With the proper machine setup, Stellite alloys can be rough machined. Alloys having hardnesses of <55 HRC can usually be machined, whereas alloys having hardnesses >60 HRC are normally ground. For better surface finishes, any of the Stellite wear-resistant alloys should be ground.

These alloys are usually machined with tungsten carbide tools. Tools for turning should have a 5° end relief, 10° side relief, and a side cutting edge angle of 45°. Tools for facing and boring are essentially the same, except for greater clearance where needed. For best results in drilling, the drill web should be kept as thin as possible, using carbide-tipped drills. In reaming, a 45° side cutting edge angle should be used. The tapping of holes is not recommended for any of the harder alloys, but threads can be produced by electrical discharge machining techniques.

Table 11 is a guide to machining these alloys; it has been written to cover the machining of all the alloys and does not necessarily represent the optimum conditions for each. In general, the harder alloys should be machined, starting at the lower end of the speed and feed ranges shown in Table 11. Softer alloys or more rigid setups will allow higher speeds and feeds.

Grinding. Whenever close tolerances are required, grinding is recommended for finishing Stellite wear-resistant alloys. Recommended wheels and coolants are listed in Table 12. Grinding speeds should be kept between 14 and 31 m/s (45 and 100 ft/s). Alloys being ground dry should not be quenched because this may cause surface checking. Care should be taken in grinding Stellite 98M2, Star J, and Stellite 3 P/M because they are more sensitive to thermal shock than are the other alloys, and flood cooling is recommended.

ACKNOWLEDGMENT

The section "Full Density Cobalt Alloys" was adapted from the article "P/M Cobalt-Base Wear-Resistant Materials" by R. Dawson and

E.M. Foley, Deloro Stellite, Inc., in *Properties and Selection: Irons, Steels, and High-Performance Alloys,* Vol 1, *ASM Handbook,* 1990, p 977–980.

REFERENCES

1. K.-H. Zum Gahr, *Microstructure and Wear of Materials,* Elsevier, 1987
2. I.M. Hutchings, *Tribology—Friction and Wear of Engineering Materials,* Edward Arnold, 1992
3. K.C. Ludema, *Friction, Wear, Lubrication,* CRC Press, 1996
4. *Friction, Lubrication, and Wear Technology,* Vol 18, *ASM Handbook,* ASM International, 1992
5. K.J.A. Brooks, *World Directory and Handbook of Hardmetals and Hard Materials,* 5th ed., International Carbide Data, 1992
6. J. Larsen Basse, *Powder Metall.,* Vol 16 (No. 31), 1973, p 1–32
7. *Properties and Selection: Irons, Steels, and High-Performance Alloys,* Vol 1, *Metals Handbook,* 10th ed., ASM International, 1990
8. H.F. Fischmeister and L.R. Olsson, Cutting Tool Materials, ASM International, 1981, p 111–131
9. D.S. Coleman, J. Bates, Q.A. Shaikh, and P.R. Brewin, *Proc. 12th Int. Plansee Seminar '89,* Metallwerk Plansee GmbH (Reutte, Austria), Vol 2, 1989
10. K.E. Pinnow, W. Stasko, J.J. Hauser, and R.B. Dixon, Proc. Advances in Powder Metallurgy & Particulate Materials (San Francisco), Vol 6, Metal Powder Industries Federation, June 1992
11. C.T. Clyde, *Introduction to Metal Matrix Composites,* Cambridge Press, 1991
12. *Friction, Lubrication, and Wear Technology,* Vol 18, *ASM Handbook,* ASM International, 1992, p 804
13. Conforma Clad, Inc., New Albany, IN, technical literature
14. E. Findeisen et al., Process of Manufacturing Cast Tungsten Carbide Spheres, U.S. Patent 5,089,182, 18 Feb 1992
15. *Properties and Selection: Nonferrous Alloys and Special-Purpose Materials,* Vol 2, *ASM Handbook,* ASM International, 1990, p 428–445
16. K.C. Anthony, Wear Resistant Cobalt-Base Alloys, *J. Met.,* Feb 1983, p 52–60
17. K.M. Kulkarni, P.J. Walsh, and E.M. Foley, Full Density P/M Parts of Stellite Alloys, *Met. Powder Rep.,* Dec 1983, p 681–684
18. E.M. Foley, Powder Metallurgy Processing Produces Wear and Corrosion Resistant Cobalt-Base Alloys with Improved Properties, *Proc. of the P/M Defense Technology Seminar,* Metal Powder Industries Federation and American Powder Metallurgy Institute, Dec 1986

Corrosion-Resistant Powder Metallurgy Alloys

Barbara Shaw, Penn State University

POWDER METALLURGY COMPONENTS produced from corrosion-resistant alloys constitute a relatively small, but rapidly growing, segment of the P/M market. Applications for corrosion-resistant P/M alloys, especially stainless steel alloys, span a variety of industries including aerospace, automotive, chemical processing, medical, and recreational. Stainless steels, nickel-base superalloys, and titanium and titanium-base alloys are the primary corrosion-resistant P/M alloys in use today. Among these, P/M stainless steels currently find the widest range of applications.

The corrosion resistance of stainless steels (SS), nickel-base superalloys, and titanium and titanium-base alloys can be attributed to the natural, protective oxide layer (passive film) that forms on the surfaces of these metals. This protective oxide film is subject to localized breakdown, yielding pitting, crevice corrosion, and/or intergranular corrosion of the underlying substrate. Sintered stainless steels, nickel-base superalloys, and titanium and titanium-base alloys also derive their corrosion resistance from protective surface oxide films. However, in the case of the sintered metals, inherent pores/crevices are part of the material. Depending on their dimensions, these pores can act as initiation sites for breakdown of the protective oxide leading to the establishment of crevice corrosion.

Pitting resistance, especially in wrought stainless alloys, is often increased by the addition of passivity-enhancing elements to the base alloy. For wrought stainless steels, passivity is typically enhanced through the addition of molybdenum, nickel, or more chromium. These same alloying additions also enhance the pitting resistance of P/M stainless steels. While passivity-enhancing alloying additions also increase crevice-corrosion resistance, they are often not completely effective, and significant alteration of crevice dimensions (increasing its width and/or decreasing its depth), or total elimination of the crevice, is needed to deter crevice corrosion. Another form of localized attack that corrosion-resistant alloys are susceptible to is intergranular corrosion. This form of degradation is commonly encountered in wrought stainless steels and is the result of microstructural differences in composition in the grain-boundary region of the alloy. These unwanted compositional differences result in what is called a sensitized microstructure, and they are usually the result of improper heat treating or welding of the alloy. Powder metallurgy alloys can acquire sensitized microstructures during welding or, for unwelded components, as a result of the sintering/cooling cycle used to produce the alloy.

Susceptibility to pitting, crevice corrosion, and intergranular corrosion is typically evaluated by exposure testing or through the use of one or more electrochemical test methods. Exposure tests provide information on the number of pit or crevice sites present on the specimen surface and the maximum pit or crevice depth or weight loss resulting from attack of sensitized regions of a microstructure. Electrochemical test methods indicate how the passive characteristics of one alloy compare to those of another and provide insight into the mechanisms responsible for corrosion. Parameters that are frequently used to assess passive behavior include the pitting potential, the protection potential, and the passive current density.

Much of the corrosion literature for P/M alloys, including corrosion-resistant alloys, consists of simple immersion and salt-spray cabinet test results. Data from these tests can provide useful information for some simple systems and will be included in this article; however, the results of these tests are often not discriminating enough to separate similar types of alloys, and they do not provide information concerning the nature or mechanisms of corrosion in P/M alloys. In fact, the mechanism of degradation in an artificial environment, such as the salt-fog chamber, is often quite different than the operative mechanism in a natural environment. This difference in mechanism could lead to design decisions based on totally inappropriate data. Electrochemical test methods provide another means for evaluating corrosion resistance, and they can also be used to elucidate important mechanistic information. The corrosion of P/M alloys is often interpreted in simplistic terms; however, in reality the corrosion of sintered stainless steels, nickel-base superalloys, and titanium and titanium-base alloys is more likely the result of somewhat complicated electrochemical interactions. An understanding of these interactions and the mechanisms of corrosion is essential to the continued development of corrosion-resistant P/M alloys and exploitation of their inherent processing advantages. This understanding can only come from carefully planned and executed experiments.

In the following sections, the corrosion characteristics of P/M stainless steels, nickel-base superalloys, and titanium and titanium-base alloys are presented and discussed. Whenever possible, the corrosion data for P/M alloys are compared to that of their wrought counterparts, and a wide array of test results is presented. Sections of this article have been derived from Erhard Klar's article "Corrosion of Powder Metallurgy Materials," which appears in Volume 13 of the *ASM Handbook.* Readers are strongly encouraged to read his article for further information on the corrosion of P/M stainless steel alloys and superalloys.

Evaluating the Corrosion Resistance of P/M Alloys

A wide array of experimental techniques is currently being used to assess the corrosion resistance of P/M alloys. These tests range from simple laboratory exposure studies in open containers to more sophisticated electrochemical approaches aimed at gaining insight into the underlying mechanism(s) of P/M alloy corrosion. This section introduces several of the more commonly employed tests and provides specific examples of the use of these methods in evaluating the corrosion resistance of P/M stainless steels. More detailed information on corrosion testing, in general, is available in Volume 13 of the *ASM Handbook* and in Volume 3.02 of the ASTM standards.

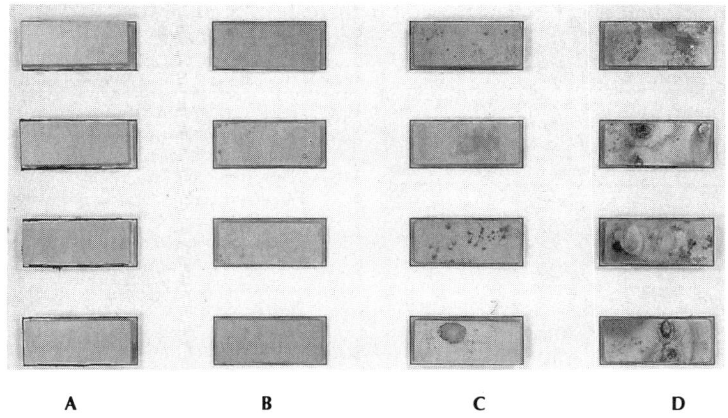

Fig. 1 Photographic chart of sintered stainless steel transverse rupture specimens tested in 5% aqueous NaCl by immersion. Definition of ratings: A, sample free from any corrosion; B, up to 1% of surface covered by stain or rust; C, 1 to 25% of surface covered by rust; D, more than 25% of surface covered by rust. Source: Ref 1. Courtesy of Prasan Samal, OMG Americas

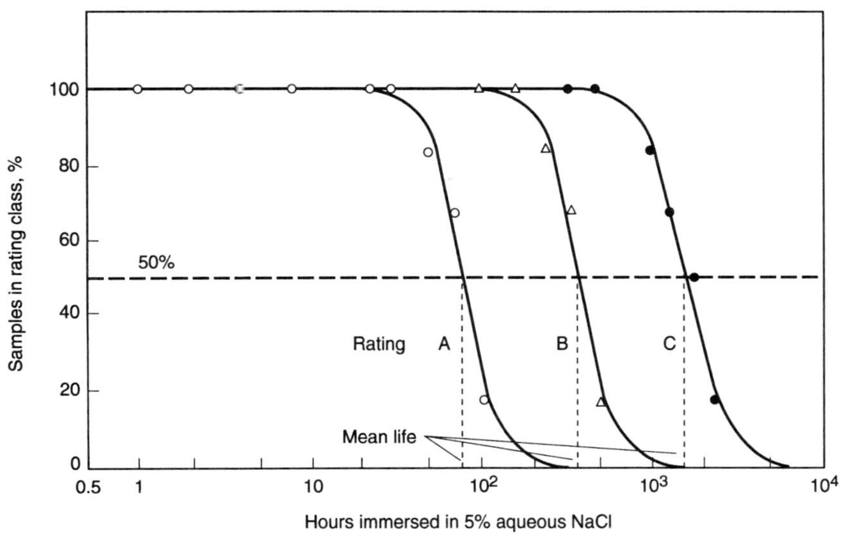

Fig. 2 Plot of percentages of replicate specimens with a given rating versus immersion time. Source: Ref 1

Results from Exposure Testing

Exposure tests are commonly employed for ranking the performance of a group of related alloys and assessing quality control. Ideally, these tests should be conducted in actual service environments. Unfortunately, very few references in the literature cite service or environmental exposure test results for P/M materials; instead, much of the P/M literature cites results from immersion and salt-spray cabinet testing.

Immersion Test Results. Most immersion test results found in the P/M literature are for exposures in either acid or chloride environments. The concentration of the aggressive species (i.e., chloride or hydrogen ions) in these immersion tests is typically greater than that expected during actual service. The increased concentration of the aggressive species serves to accelerate the test; however, the degree of acceleration is unknown and no real consensus exists on the best concentration to use. Several investigators have used a 5% NaCl solution for the chloride exposures. Immersion test results have been reported in a number of acid solutions including 1% HCl, 30% H_2SO_4, 10% HNO_3, 20% HNO_3, at both ambient and elevated temperatures. Unfortunately, no single acid electrolyte is used consistently.

Erroneous results can occur with some specimen preparation practices that are used on wrought specimens. Surface polishing prior to testing, for example, can produce erroneous results because polishing tends to distort or close off surface pores that might normally be present on the surface of the actual component. Cleaning and degreasing porous P/M materials prior to testing is another practice that needs to be considered carefully prior to testing. Degreasing, a practice that is routinely used for wrought specimens, can trap liquids in the pores of P/M specimens and care must be taken to remove this liquid prior to the initiation of a test. As with all corrosion exposure tests, it is advisable to test specimens under their anticipated service conditions.

Evaluation of P/M specimens after exposure testing is also a subject of some controversy; however, several studies have followed a visual rating system proposed by Klar (see "Corrosion of Powder Metallurgy Materials," in Volume 13 of the *ASM Handbook).* This rating system involves visually assessing the amount of surface staining or corrosion on a test coupon and categorizing its degradation in one of four ways: A, no attack of the surface, categorized as 0% attack; B, attack of 1% or less of the surface; C, attack of 1 to 25% of the surface, and D, attack of more than 25% of the surface. A photographic chart showing a series of stainless steel specimens after exposure to a 5% solution and characterization using the A through D rating system (Ref 1) is shown in Fig. 1. Another way of using the data acquired from such an inspection is illustrated in Fig. 2, which shows a plot of the percentage of replicate specimens holding a particular rating versus the log of the time of exposure (Ref 1). This plot reveals that the mean life of a 316L P/M specimen immersed in 5% NaCl solution, accepting no surface rust or staining, is approximately 80 h (the intersection of the 50% line with the 1% surface rust/stain curve). If 1% surface rust/stain is acceptable, then the mean life is extended to longer than 300 h. If 1 to 25% surface rust/stain is acceptable, the mean life is extended to more than 1000 h. Corrosion-resistance data for a few different stainless steels, in terms of the hours needed to attain up to 1% surface rust or stain ("B" rating on Klar's scale), is presented in Table 1 (Ref 2). Not surprisingly, the more highly alloyed stainless steels exhibited the best performance. Because strict adherence with the first category in Klar's rating scheme (no rust or rust stain) is difficult to achieve for P/M specimens, this category is often broadened to allow for the first spot of rust or stain on the specimen surface. Some researchers find it more convenient to use 0.1% rust or rust stain for this first category instead of no attack (Ref 3–5).

Measurement of weight loss following exposure or immersion in a corrosive media is a standard method used for assessing corrosion resistance. While weight-loss measurements are relatively straightforward and easy to apply to wrought metals, they can be more difficult to apply to P/M materials because of their inherent porosity. Porosity traps the test solution and the corrosion products within the pores, making it difficult to accurately record small weight losses, even when elaborate cleaning and drying procedures are employed. Most researchers have been successful in obtaining meaningful weight–loss values for porous P/M alloys after immersion testing in acid solutions, but weight losses in less aggressive solutions (such as 0.1 *M* NaCl or 5% NaCl) are not available. Figure 3 (Ref 6) shows weight loss data for P/M 316L stainless steel as a function of sintering atmosphere, after immersion in a 10% HNO_3 solution for 24 h. Following testing, the specimens in this study were rinsed three times in distilled water and dried in a box oven at a temperature less than

Table 1 Sintering conditions, chemical analysis, and corrosion resistances of hydrogen-sintered stainless steel. Specimens held 45 min at 1127 °C (2060 °F)

Stainless steel grade	Sample No.	Sintering conditions: Dew point, °C	°F	Cooling rate, °C/min	Properties of sintered parts: Density, g/cm³	Composition, wt%: N	O	C	Corrosion resistance (5% aqueous NaCl by immersion): B-rating(a), h	% of maximum
316L	1	−32	−25	540	6.7–6.8	0.04	0.21	0.033	2	~24
	2								50	
	3								60	
	4								80	
316L	5	−32	−25	54	6.7–6.8	0.03	0.21	0.032	1	~12
	6								1	
	7								30	
	8								60	
316LSC	9	−40	−40	187	6.7	0.06	0.21	0.045	1800	~60
	10								1800	
	11								1800	
316LSC	12	−44	−45	187	6.8	0.04	0.19	0.043	1200	~100
	13								2900	
	14								2900	
	15								2900	
317L	16	−44	−45	187	6.75	0.04	0.24	0.052	600	~17
	17								200	
	18								1600	
	19								1600	
SS100	20	−44	−45	187	6.7	0.05	0.15	0.049	6000	~100
	21								6000	
	22								6000	
	23								>6000	
SS100	24	−40	−40	187	6.7	0.07	0.14	0.11	7800	~100
	25								7800	
	26								7800	

(a) Average of hrs specimens are in test solution until 1% of surface is covered by stain or rust. Source: Ref 2

Table 2 Procedure used to clean P/M specimens after acid exposure testing

Step No.	Procedure
1	Heat P/M samples to 70 °C.
2	Submit samples to a vacuum of (−100 kPa) to evaporate solution in the pores.
3	Repeat steps 1 and 2 two times.
4	Fill pores of the P/M samples with distilled water to wash out the pores.
5	Repeat steps 1 and 2 three times.
6	Impregnate samples with ethyl alcohol to absorb water.
7	Repeat steps 1 and 2 once.
8	Store in a dessicator.

Source: Ref 7

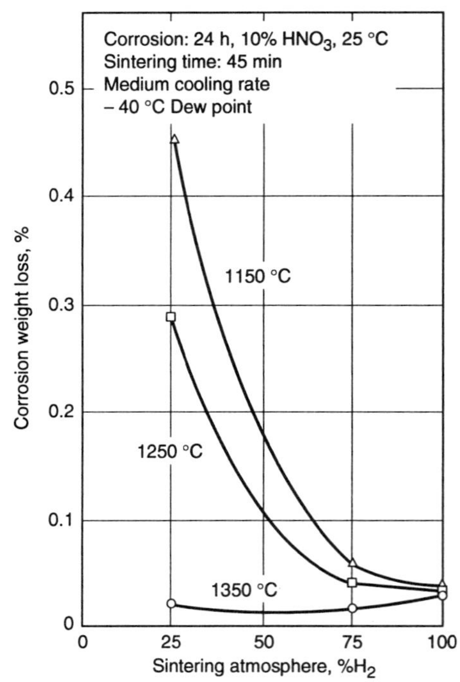

Fig. 3 Corrosion weight loss for 316L in 10% HNO_3 shown as a function of the sintering atmosphere for three sintering temperatures. Source: Ref 6

Table 3 Properties and corrosion rates of sintered stainless steels AISI 304L and AISI 316L

Corrosion rates are given in g/dm² day after exposure to 0.5 *M* H_2SO_4 at a temperature of 25 °C. Wrought specimens are denoted with a W and sintered specimens are marked A, B, and C.

Code	Sintering atmosphere	Sintering temperature, °C	Volume mass, (kg/m³) × 10³	Rockwell hardness, HRB	Porosity total, %	Pore size average, μm	Corrosion rates(a), Immersion time: 7 days	14 days	21 days
304L									
W	...	...	...	...	...	...	2.5×10^{-2}	1.2×10^{-2}	3.4×10^{-2}
A	Vacuum	1200	6.52	44	12.0	1.99	2.1×10^{-1}	1.2×10^{-1}	4.0×10^{-1}
B	$N_2 + H_2$	1120	6.39	72	11.7	1.25	7.2×10^{-1}	3.7×10^{-1}	7.3×10^{-1}
C	$N_2 + H_2$	1190	6.43	78	10.9	0.79	1.4	9.3×10^{-1}	1.3
316L									
W	...	...	...	...	...	...	1.8×10^{-3}	9.2×10^{-4}	6.1×10^{-4}
A	Vacuum	1200	6.43	38	10.4	0.98	1.2×10^{-1}	6.9×10^{-2}	2.1×10^{-1}
B	$N_2 + H_2$	1120	6.34	63	12.7	0.98	1.8	1.5	1.6
C	$N_2 + H_2$	1190	6.36	69	11.4	0.79	2.3	2.0	2.3

(a) (g/dm² × day) in 0.5 *M* H_2SO_4, T = 25 °C, of compact (W) and sintered samples. Source: Ref 9

100 °C. The drying was continued until the point where no further water evaporation was detected in the weight-loss measurements. A more elaborate cleaning and drying procedure, employed after similar testing, is described in Ref 7 and is outlined in Table 2. Researchers occasionally report weight gains following corrosion testing; however, the utility of such values in assessing corrosion resistance is unknown and is, perhaps, not of much value.

Weight-loss data are commonly used to determine the corrosion rate of a material. The conversion of weight-loss data into corrosion penetration rates is accomplished via the relationship given by (provided that the density of the material, the exposed surface area, and the time of exposure are known):

$$\text{mm/yr} = 87.6\ W/DAT \qquad \text{(Eq 1)}$$

where D is the density in g/cm³, W is the weight loss in mg, A is the surface area in cm², and T is the time in hours.

Corrosion-penetration rates of less than 0.02 mm/yr are considered outstanding, while those of 0.1 to 0.5 mm/yr, 0.5 to 1 mm/yr, and 1-5 mm/yr are considered good, fair, and poor, respectively (Ref 8). Weight-loss data are sometimes reported in the units of mg/dm²day (mdd) or g/dm²day (gdd). Conversion between these two corrosion rate measures is accomplished via the following equations:

$$\text{mm/yr} = 56.6\ \text{mdd/specific gravity or } 0.0566\ \text{gdd/specific gravity} \qquad \text{(Eq 2)}$$

An example of corrosion rate data for 304L and 316L wrought and P/M sintered stainless steel after immersion in 0.5 *M* H_2SO_4, as a function of immersion time and sintering conditions, is presented in Table 3 (Ref 9). The wrought alloys clearly exhibited the lowest corrosion rates, followed by the vacuum-sintered alloys.

Elevated temperature or boiling acid solutions are often used to assess susceptibility of wrought stainless steels to intergranular corrosion, and procedures for this type of testing are covered in ASTM methods A 262 and A 763. Because the solutions called for in these ASTM methods are quite aggressive to porous P/M specimens, a variety of modifications have been employed by

Table 4 Average intergranular corrosion rates via ASTM A 763 practice Z

Specimen	Density, (g/cm^3)	Sintering temperature °C	Sintering temperature °F	Corrosion rate(a), mm/month
P/M434L	7.00	1200	2200	4.68
P/M434L	7.20	1290	2350	3.77
Wrought 434L	...	...	...	2.47
P/M 409Cb	6.90	1200	2200	63.01(b)
P/M 409Cb	7.07	1290	2350	23.06
Wrought 409	7.73	...	...	19.83
Welded wrought 409	...	...	...	12.37
Welded P/M 434L	...	1200	2200	2.73
Welded P/M 434L	...	1290	2350	3.66

(a) Average of two tests. (b) After testing, pieces of the specimen were so badly dissolved that they could not be collected. Source: Ref 10

Table 5 Average intergranular corrosion rates for P/M and wrought 434 and 409 according to modified ASTM A 763

Specimen	Sintering temperature °C	Sintering temperature °F	Corrosion rate(a), mm/month
P/M 409Cb	1290	2350	1.22
Wrought 409	...	...	0.226
P/M 434L	1290	2350	0.231
Wrought 434L	...	...	0.002(b)

(a) Average of three tests. (b) The weight losses used to calculate corrosion rates were within the expected error range of the scale. Source: Ref 10

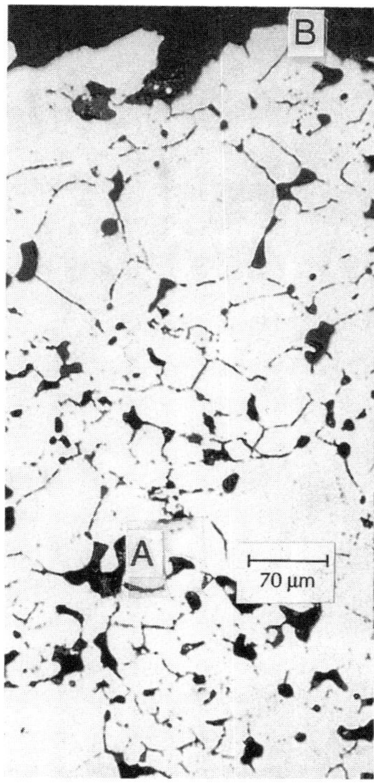

Fig. 4(a) Metallographic cross section of a 1290 °C (2350 °F) sintered P/M 434L specimen subjected to ASTM A 763 that reveals regions of: A, subsurface intergranular attack; and B, surface uniform attack

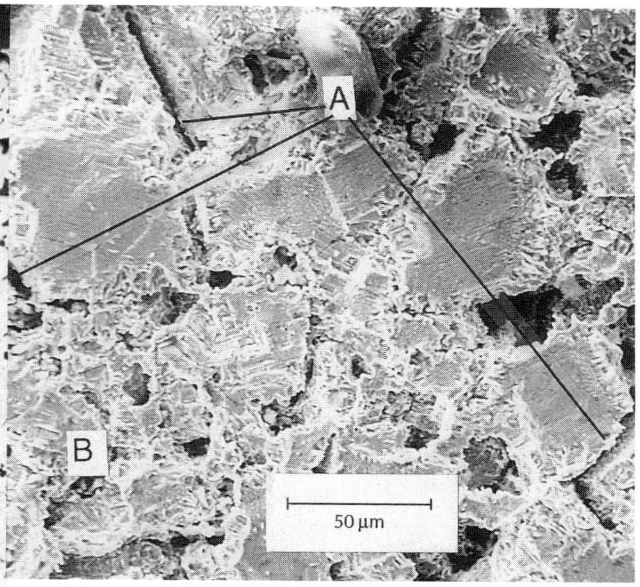

Fig. 4(b) Scanning electron micrograph of the surface of a 1290 °C (2350 °F) sintered P/M 434L specimen subjected to ASTM A 763 that reveals regions of: A, intergranular attack; and B, uniform attack

Table 6 Total iron concentrations after immersion testing in 400 mL of 5 wt% NaCl solution

Sintering conditions and resulting properties of stainless steel samples							Average total iron concentration in solution(c), ppm		
Designation	Alloy	Sintering conditions(a)	Sintered density(b), g/cm^3	Surface, wt% C	Surface, wt% N	Surface, wt% O	1 day	4 days	8 days
304N2	304 SS	1316 °C DA(d)	6.32	0.002	0.550	0.170	1.18	5.28	↑
304N2 9	304 SS	1316 °C DA(d)	6.90	0.016	0.560	0.190	0.61	4.01	6.33
304L	304 SS	1288 °C Vac	6.31	0.002	0.010	0.220	0.57	4.19	7.44
304L 9	304 SS	1288 °C Vac	6.89	0.004	0.010	0.200	0.20	1.57	3.92
304N2 LSC	304 LSC	1316 °C DA(d)	6.49	0.006	0.220	0.180	↓	↓	↓
304N2 LSC 9	304 LSC	1316 °C DA(d)	6.91	0.010	0.220	0.150	↓	↓	0.12
304L LSC 9	304 LSC	1288 °C Vac	6.92	0.006	0.078	0.450	↓	0.12	0.50
304N2 PLUS	304 PLUS	1316 ° DA(d)	6.40	0.008	0.140	0.170	↓	↓	↓
304L PLUS	304 PLUS	1288 °C Vac	6.41	0.004	0.044	0.280	0.15	0.19	0.20
316N2	316 SS	1316 °C DA(d)	6.54	<0.001	0.570	0.150	1.13	4.06	7.17
316N2 9	316 SS	1316 °C DA(d)	6.83	0.010	0.430	0.190	1.37	3.57	5.53
316L	316 SS	1288 °C Vac	6.46	0.007	0.007	0.250	0.18	0.42	0.48
316L 9	316 SS	1288 °C Vac	6.86	0.006	0.014	0.230	↓	0.50	1.62
316N2 LSC	316 LSC	1316 °C DA(d)	6.45	0.008	0.180	0.100	↓	↓	↓
316N2 LSC 9	316 LSC	1316 °C DA(d)	6.87	0.010	0.210	0.210	↓	↓	0.03
316L LSC	316 LSC	1288 °C Vac	6.49	0.002	0.052	0.120	↓	0.33	0.76
316L LSC 9	316 LSC	1288 ° C Vac	6.63	0.005	0.091	0.250	↓	0.06	0.52
316N2 PLUS	316 PLUS	1316 °C DA(d)	6.42	0.004	0.110	0.120	↓	↓	↓
316L PLUS	316 PLUS	1288 °C DA(d)	6.45	0.007	0.045	0.150	↓	↓	↓
316X	316 SS	1288 °C Vac	6.65	0.005	0.001	0.295	↓	↓	0.10
316X LSC	316 LSC	1288 °C Vac	6.69	0.006	0.002	0.174	↓	↓	0.03
316X PLUS	316 PLUS	1288 °C Vac	6.47	0.007	0.008	0.215	↓	↓	↓
100N2	100 SS	1316 °C DA(d)	6.29	0.006	0.820	0.039	1.88	4.39	5.70
100N2 9	100 SS	1316 °C DA(d)	6.82	0.010	0.610	0.054	0.78	3.45	4.16
100L	100 SS	1288 °C Vac	6.25	<0.001	0.011	0.140	↓	↓	↓
100L 9	100 SS	1288 °C Vac	6.71	<0.001	0.011	0.140	↓	↓	↓
100X	100 SS	1288 °C Vac(e)	6.72	0.007	0.001	0.144	↓	↓	↓

(a) DA, dissociated ammonia; Vac, vacuum. (b) Density was measured with oil impregnation per ASTM B328. (c) ↑, above the detection range; ↓, below the detection range. (d) Actual furnace atmosphere was 91.8 vol% D.A. + 8.2 vol% N_2. (e) Followed by a hold at 1150 °C for 10 min. Source: Ref 12

investigators to reduce the aggressiveness of the specified solutions. These modifications include reducing the time of the test, reducing the concentration of the acid, and/or reducing the temperature of the solution. Unfortunately, no consensus on the appropriate approach for intergranular corrosion testing of P/M alloys has been reached. Weight-loss values for both P/M and wrought 434L and 409 stainless steels after testing according to ASTM method A 763 practice Z are presented in Table 4 (Ref 10). Practice Z of ASTM method A 763 calls for exposure of the test specimens in a boiling solution of 6 wt% $CuSO_4$ + 16 wt% H_2SO_4 for a period of 24 h. In this test, the specimens are embedded in copper shot that is poured into the bottom of the test flask. The standard practice Z calls for 180° bending of the tested specimen, because weight loss for wrought specimens is often not large enough to determine corrosion rates. In the case of P/M parts, however, weight loss was found to be effective in measuring corrosion rates and, while 180° bending of the P/M specimens was possible prior to testing, bend testing resulted in fracture after only a few degrees of bending. Examination of the specimens after intergranular corrosion testing, via ASTM A 763 practice Z, revealed uniform corrosion of the P/M alloys in addition to the anticipated grain-boundary attack, as Fig. 4(a) and (b) reveal for a P/M 434 alloy. After modifying this test method by reducing the concentration of sulfuric acid, as suggested in Ref 10, weight losses were reduced to the values reported in Table 5. Examination of the specimens after testing revealed that corrosion was predominantly along the grain boundaries, as the scanning electron micrograph in Fig. 5(b) reveals.

Another means of quantifying the degree of corrosion of a metal is to measure the amount of corroded material that is in solution following

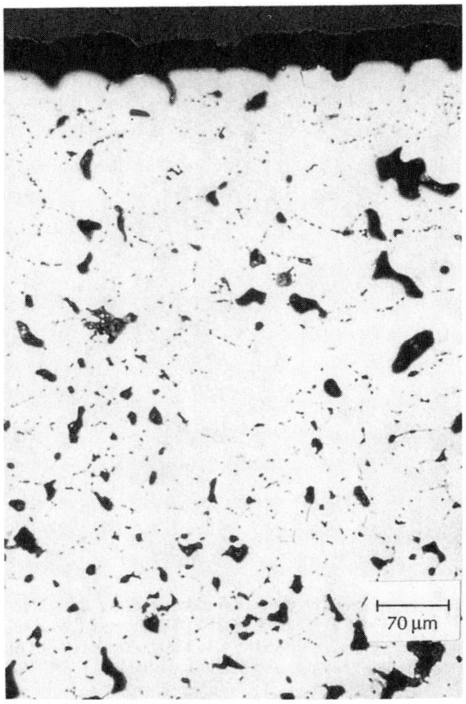

Fig. 5(a) Metallographic cross section of a 1290 °C (2350 °F) sintered P/M 434L specimen subjected to the modified intergranular test that reveals subsurface intergranular attack and no uniform surface attack

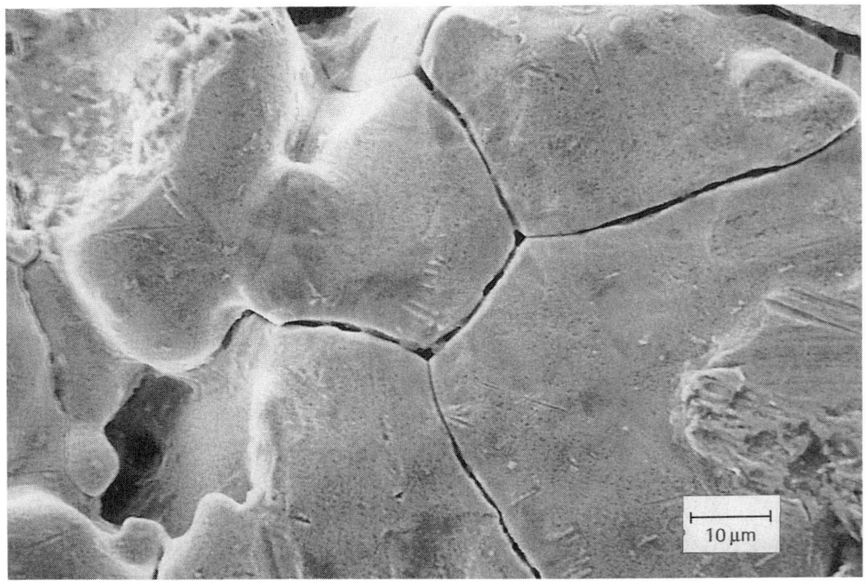

Fig. 5(b) Scanning electron micrograph of the surface of a 1290 °C (2350 °F) sintered P/M 434L specimen subjected to ASTM A 763 that reveals intergranular attack and no attack of grain interiors

immersion for a given period of time. A colorimetric method for evaluating the corrosion of P/M stainless steels, which utilizes a commercially available test kit, has been described in Ref 11. In this method, the iron concentration in solution is measured as a function of time using a commercial test kit containing self-filling ampoules, activator solutions, and a photometer. First, a sample of the test solution is placed in a sample cup and an activator, a solution containing a mixture of thioglycolic acid and ammonia, is added to the solution. The activator dissolves most forms of particulate iron and reduces ferrous (Fe^{3+}) ions within the solution to ferric (Fe^{2+}) ions. The partially evacuated ampoule containing an acidic, buffered solution of 1,10-phenanthroline is inserted into the test solution, and its tip is snapped allowing both solutions to mix and fill the ampoule. After mixing, the ampoule is then inserted into a calibrated photometer and the transmission of light is measured by percentage. Iron content in parts per million (ppm), within the range of 0.01 to 7.6 ppm is then calculated based on a conversion chart provided with the test kit. Examples of the average total iron concentration measured as a function of exposure time for a variety of sintered stainless alloys exposed to a 5 wt% NaCl solution are presented in Table 6 (Ref 11). In each case, specimens of identical dimensions were exposed to 400 mL of a 5 wt% NaCl solution in closed containers, and the values reported were the average of five readings. A nearly linear iron dissolution rate was observed for most of the alloys during the eight-day test, as Fig. 6 illustrates (Ref 11). Good correlation was noted between the results of the colorimetric test and salt-spray and 5% NaCl immersion tests (Ref 12), as Fig. 7 and 8 reveal.

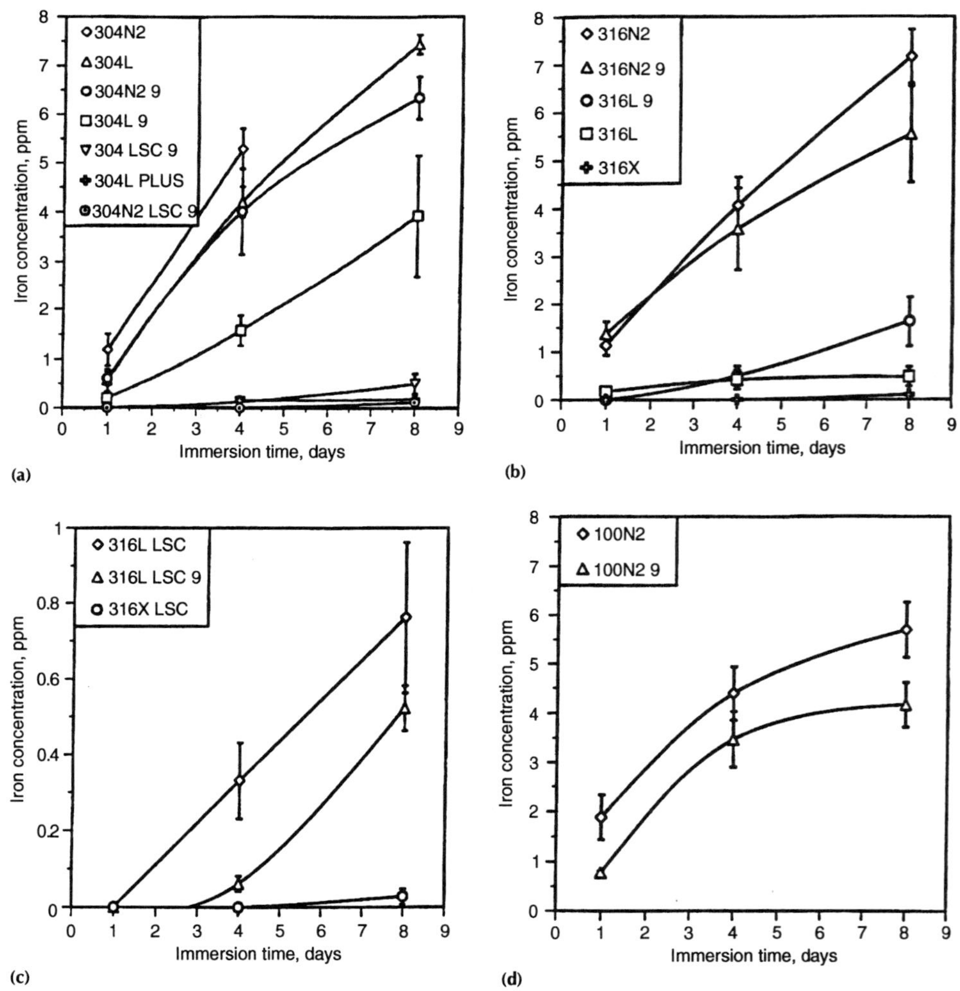

Fig. 6 Total dissolved iron in solution for alloys showing detectable concentrations after 1, 4, and 8 days. (a) 304 alloys. (b) 316 regular alloys. (c) 316 special alloys. (d) SS 100 alloys. Source: Ref 11

Immersion testing can also be used in conjunction with other experimental methods, such as mechanical tests, to assess corrosion resis-

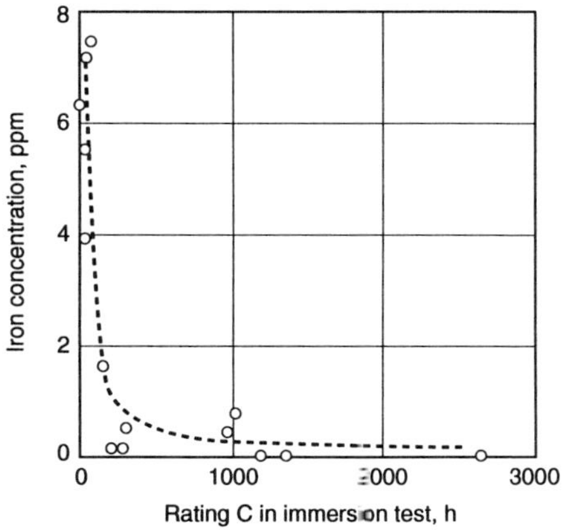

Fig. 7 Correlation between results of the colorimetric test for various alloys after eight days and the corresponding times to achieve a C rating in a 5 wt% NACE NaCl immersion test. Source: Ref 11

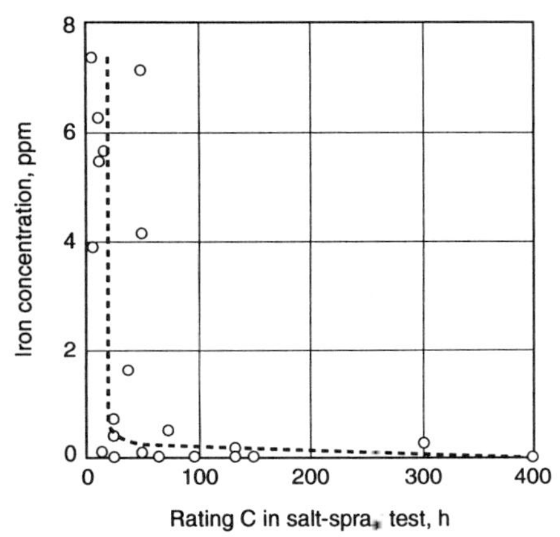

Fig. 8 Correlation between results of the colorimetric test for various alloys after eight days and the corresponding times to achieve a C rating in the salt-spray test. Source: Ref 11

Table 7 Corrosion tests on 316L sintered stainless steel specimens exposed to 5% NaCl solution at 20%C

Specimens were sintered in various atmospheres at 1150 °C for 1 h. Passivation 1 was carried out for 30 min at 80 °C in a 16 vol% HNO_3 nitric acid solution. Passivation 2 was carried out for 60 min at 70 °C in a 40 vol% HNO_3 solution.

Sintering atmosphere	Properties after sintering but before corrosion testing: Transverse rupture strength, MPa	ksi	Angle of bend, degrees	Treatment after sintering	Period, days	Properties after corrosion testing: Transverse rupture strength, MPa	ksi	Angle of bend, degrees	Appearance
Vacuum	460	67.0	>111(a)	None	4	425	61.6	>107(a)	No attack
					15	450	65.1	>109(a)	No attack
				Passivation 1	4	430	62.3	100	Slight rusting
					15	440	63.8	108	Slight rusting
				Passivation 2	4	432	62.8	91	Slight rusting
					15	437	63.4	95	Slight rusting
Hydrogen	515	74.6	112	None	4	475	68.9	44	Rusted
					15	460	66.8	55	Rusted
				Passivation 1	4	480	69.8	80	Rusted
					15	470	68.2	62	Rusted
				Passivation 2	4	380	55.3	45	Rusted
					15	469	68.1	60	Rusted
Dissociated ammonia	730	106.1	17	None	4	657	95.3	14	Rusted
					15	700	101.4	12	Rusted
				Passivation 1	4	655	95.1	17	Rusted
					15	650	94.2	10	Rusted
				Passivation 2	4	729	105.8	16	Rusted
					15	665	96.8	15	Rusted

(a) Specimen did not break. Source: Ref 13

Table 8 Ferroxyl test results for P/M and wrought stainless steels

AISI No.	Type, treatment	Visual rating(a) in ferroxyl test with chlorine added (% Cl^-): 0.05%	0.10%	0.50%	Pitting potential(b), mV SCE
316L	P/M, 1120 °C 20 min, dissociated ammonia (−27 °C)	3	3	...	75
316L	P/M, 1250 °C 30 min, H_2 (−35 °C)	...	1	3	250
316L	P/M, 1120 °C 30 min, H_2 (−70 °C)	...	0	3	375
316L	P/M, 1120 °C 120 min, H_2 (−70 °C)	...	0	3	375
316L	P/M, 1250 °C 120 min, H_2 (−70 °C)	...	0	1	600
431	Wrought	3	3	...	270
303	Wrought	3	3	...	325
304	Wrought	1	3	...	470
316	Wrought	...	1	2	590

(a) Rating scale: 0, no attack and no blue spots; 1, light attack and very weak blue spots without growth; 2, moderate attack with blue spots with slow growth, no needle growth or large accumulation of blue dye; 3, severe attack, blue spots with growth, needle growth or accumulation of corrosion product on surface. (b) 0.1% Cl^-, pH 5, 30 °C, 5 mV/min. Source: Ref 14

tance. Table 7 illustrates the results from such a test conducted on sintered 316L stainless steel specimens (Ref 13). Prior to mechanical testing via a three-point loading transverse rupture test, the specimens in this study were immersed in a 5% NaCl solution at 20 °C. Three sintering atmospheres were evaluated, and some of the specimens were passivated in acid solutions prior to chloride exposure. Under the sintering conditions used (laboratory alumina tube furnaces, no purification of the gaseous atmospheres, and dew points in the range of −40 to −60 °C upon entering the furnace), only the vacuum sintered specimens showed resistance to corrosion. These data also reveal that the two passivating treatments were detrimental to corrosion resistance. Strengths of specimens sintered in vacuum and in hydrogen were similar. A similar investigation using a 20% NaCl solution, in lieu of the 5% solution, found that the corrosion resistance of 316L stainless steel was the same (Ref 13).

Ferric Chloride and Ferroxyl Exposures. The ferric chloride test method, ASTM G 48, for characterizing the resistance of wrought stainless steels and related alloys to pitting and crevice corrosion is sometimes used for evaluating P/M materials. While this method may be useful in assessing the susceptibility of wrought materials to crevice corrosion, its use for evaluating crevice-corrosion susceptibility in sintered stainless steel alloys is questionable because these materials already contain inherent crevices. Applying a crevice former to a specimen that already contains a crevice may facilitate the initiation of crevice corrosion. Because crevice width and length are critical factors in establishing crevice corrosion, the application of one crevice on top of another seriously alters the intent and validity of the original test. However, if the ferric chloride solution is simply being used as an aggressive oxidizing solution and no crevice formers are used, such as the test described in method A, then the data for P/M specimens may be of value.

It should be noted that some researchers have found the exposures in the 6% $FeCl_3$ solution suggested in ASTM G 48 to be too aggressive for the characterization of sintered stainless steels. An alternative solution that has been used for exposure studies is a hexacyanoferrate solution (ferroxyl) that may or may not contain added chloride (Ref 14). A benefit of this solution, in addition to being an oxidizing agent, is that it can be used as an indicator of contamination by iron because the solution turns blue in the presence of ferrous ions. A typical test involves exposure to the hexacyanoferrate solution for 24 h and a visual inspection of the specimens using the following rating scale (Ref 14):

Category	Amount of attack	Visual inspection results
0	No attack	No blue spots
1	Light attack	Very weak blue spots without growth
2	Moderate attack	Blue spots with slow growth—no needle growth or large accumulation of blue dye
3	Severe attack	Blue spots with growth—needle growth or accumulation of the corrosion product on the surface

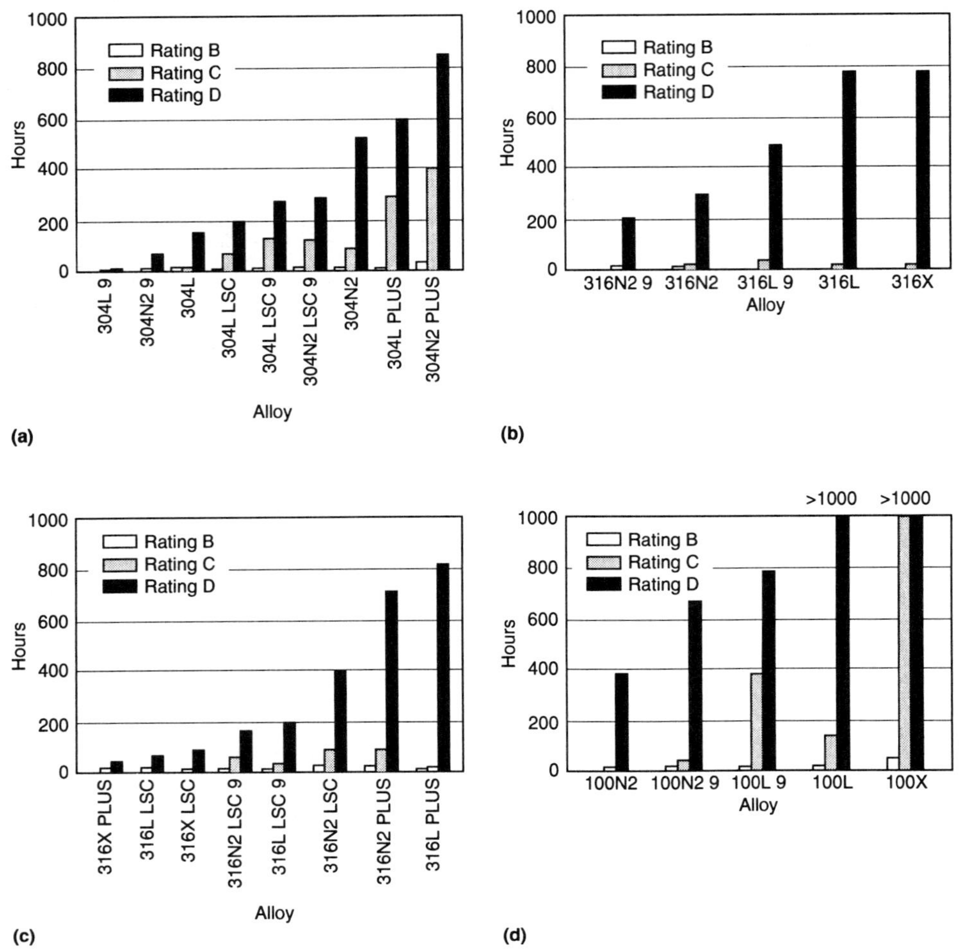

Fig. 9 Salt-spray test results. (a) 304 alloys. (b) 316 regular alloys. (c) 316 special alloys. (d) SS 100 alloys. B rating, attack of 1% or less of the surface; C rating, attack of 1 to 25% of the surface; D rating, attack of more than 25% of the surface. Source: Ref 11

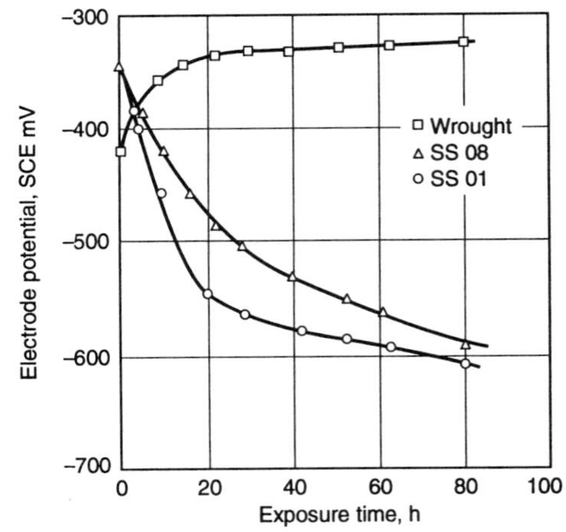

Fig. 10 Variation in open-circuit electrode potential with time for wrought and P/M 316L stainless steel in a 3% NaCl solution. Source: Ref 15

Table 9 Salt-spray cabinet test results for P/M stainless steels

P/M material designation				Corrosion tests(a)	
AISI	Ames	MPIF	Density, g/cm^3	A, hours to 0.1% stain	B, hours to 1% stain
303	SFN-Cr18-N11-64	SS-303N-125	6.4	20	1–5 (200–500)
	SFAN-Cr18-N11-66	SS-303N2-35	6.6	20	N/A
	SFA-Cr18-N11-70	Approx SS-303L-12	7.0	24	N/A
304	SFN-Cr19-N10-64	SS-304N1-30	6.4	20	50–100 (500–1200)
	SFAN-Cr19-N10-66	SS-304N2-33	6.6	20	N/A
	SFA-Cr19-N10-70	Approx SS-304L-13	7.0	24	N/A
316	SFN-Cr17-N12-M1-64	SS-31N1-25	6.4	20	200–500 (500–1500)
	SFAN-Cr17-N12-M1-66	SS-316N2-33	6.6	20	N/A
	SFA-Cr17-N12-M1-70	Approx SS-316L-15	7.0	24	N/A
410	SFAN-Cr12-67	Approx SS-410-90HT	6.7	5	N/A
	SFA-Cr12-70	...	7.0	5	N/A
420	SFAN02-Cr12-67	Approx SS-410-90HT	6.7	5	N/A
434	SFA-Cr17-M1-70	...	7.0	24	N/A
632	SFA-Cr16-NS-M1-70	...	6.9	22	N/A

(a) A, 5% aqueous NaCl salt spray, per ASTM B 117; B, 5% aqueous NaCl immersion, per ASTM G31. N/A, data not available. Source: Ref 3

An example of the results obtained on sintered 316L stainless steel (produced under different sintering conditions) and a few wrought stainless steels (which were tested with an artificial crevice formed by an O-ring) are provided in Table 8 (Ref 14).

Salt Spray Cabinet Test Results. Despite the controversy that exists concerning the applicability of salt-spray cabinet exposure results compared to natural environments, the test remains a standard in many communities. In relation to other types of corrosion data for P/M materials, a substantial amount of salt-spray cabinet data exists. In fact, when conducted according to ASTM method B 117, the test is probably the most commonly used experiment for assessing corrosion in P/M metals and alloys. Examples of the times needed to notice 0.1% surface rust for a variety of stainless steels after salt-spray cabinet testing are presented in Table 9 (Ref 3). An inspection at 100 h revealed no attack of wrought 316L specimens. Similar data, obtained using the Klar rating system, for sintered 304L, regular 316L, special 316L alloys, and SS 100 alloys after ASTM B 117 testing for times in excess of 1000 h are presented in Fig. 9 (Ref 11).

Electrochemical Testing

As stated earlier, electrochemical methods of corrosion testing yield information concerning corrosion mechanisms as well as information on corrosion rates. Because of their inherent porosity, some special issues arise in the electrochemical testing of sintered P/M materials. These issues include alteration of surface porosity resulting from polishing, insufficient removal of degreasing solvents from pores prior to testing, unknown or unreported surface areas, and increased time needed to establish steady-state conditions. Irrespective of the particular electrochemical test being conducted, care should be exercised in addressing each of the aforementioned issues.

Open Circuit Potential versus Time. Prior to initiating an electrochemical test it is customary to monitor the open circuit or corrosion potential for a period of time. Monitoring of the potential should be conducted for a long enough period of time to ensure that steady-state conditions exist between the metal and its environment. The length of time needed to establish a steady-state open circuit potential can vary from a few minutes to several weeks, depending on the particular metal/environment combination. Important information concerning the nature of corrosion can sometimes be determined from open-circuit potential versus time behavior.

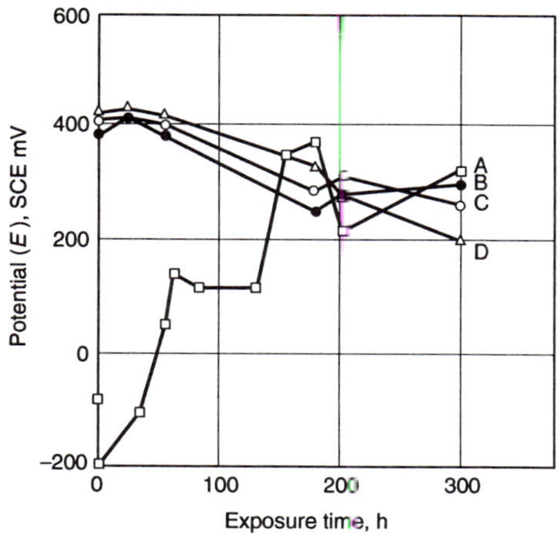

Fig. 11 Open-circuit potential of various passive hot-pressed and sintered samples in 1 N H_2SO_4 solution at ambient temperature: A, 316 stainless steel; B, 316 stainless steel containing 0.5 wt% Ni; C, 316 stainless steel containing 2 wt% Pt; D, 316 stainless steel containing 2 wt% Pd. Source: Ref 17

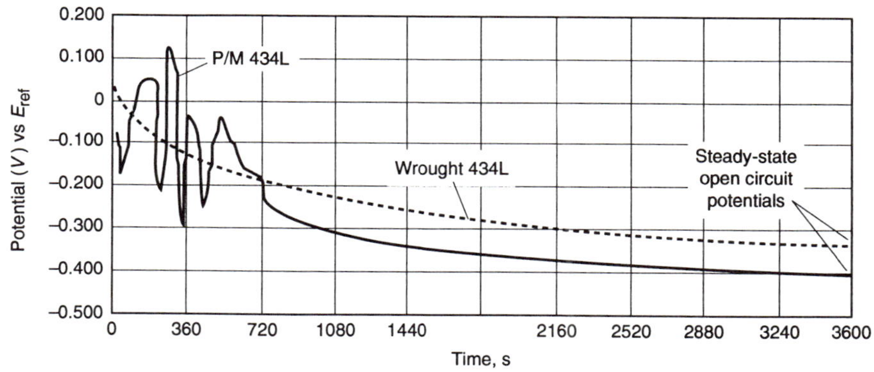

Fig. 12 Open-circuit potential versus time data for wrought 434L and P/M 434L sintered at 1290 °C (2350 °F). Source: Ref 18

Comparison of open circuit potential versus time behavior for wrought and P/M alloys of the same nominal composition typically reveals lower potentials for the P/M alloys—indicating a more active electrochemical state, as Fig. 10 reveals for porous 316L sintered stainless steel (Ref 15, 16). Open-circuit potential behavior can also provide an indication of the role that alloying additions play in altering corrosion resistance. The data in Fig. 11 show that the addition of either nickel, platinum, or palladium to sintered stainless steel increases the open-circuit potentials of the alloy, suggesting that these alloying additions enhance passivity, at least temporarily (Ref 17). In some cases, the initial open-circuit potential versus time behavior for P/M materials reveals significant instabilities, supposedly resulting from the infiltration of solution into the porous alloy, as shown in Fig. 12 (Ref 18). Instabilities such as these, despite their origin, are the reason that the open-circuit potentials are monitored prior to initiating electrochemical experiments that rely on steady-state behavior, such as polarization tests.

Polarization experiments are the most commonly used electrochemical methods for assessing corrosion behavior. Of specific interest in the study of corrosion-resistant P/M materials are the following polarization techniques: anodic polarization, cyclic polarization, polarization resistance, and electrochemical potentiokinetic reactivation. Prior to a discussion of electrochemical corrosion testing of P/M materials, it is necessary to briefly review some basic information concerning polarization curves and to define a few terms related to polarization behavior and passivity. The reader is encouraged to consult Ref 19 for more detailed information.

Figure 13 shows anodic polarization behavior for an active/passive metal (Ref 20). With regard to corrosion-resistant P/M alloys, anodic polarization curves provide useful information concerning the potential range over which a material is passive, the amount of protection afforded by

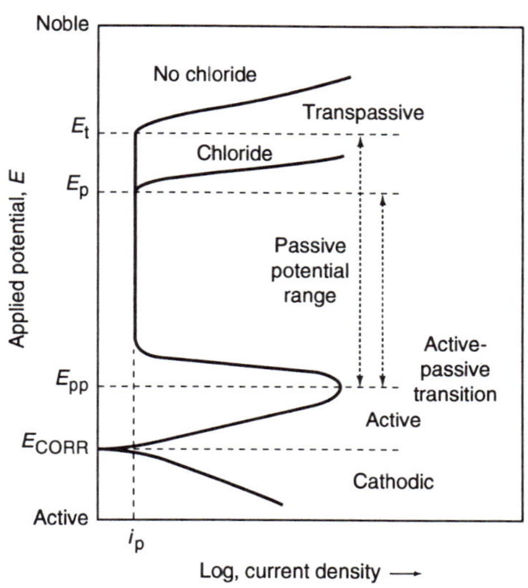

Fig. 13 Polarization curve for a stainless steel in a sulfuric acid solution. E_t, transpassive potential; E_{corr}, corrosion potential Source: Ref 20

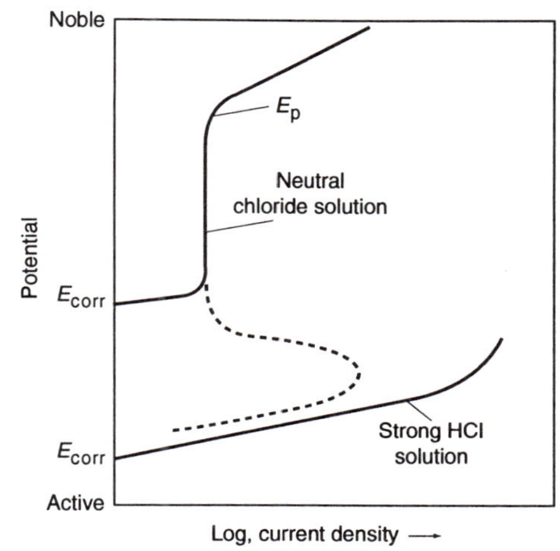

Fig. 14 Anodic polarization curves for stainless steel in a neutral chloride solution showing self-passivating behavior (top curve, solid line) and active dissolution in a strong hydrochloric acid solution (bottom curve). Source: Ref 20

the film, the susceptibility to pitting, and the ease with which passivity is obtained. Starting at the open-circuit potential and moving in the anodic direction, many metals undergo an initial period of active dissolution, prior to passive film formation, as the potential is scanned in the positive direction. The current density just prior to the onset of passive film formation is termed the "critical current density" and is often denoted by i_{crit}. The potential marking the onset of passivity is termed the primary passivation potential and it is denoted by E_{pp}. Passivity is observed in the region of the anodic polarization curve where the current density drops significantly. It is desirable for passivity to extend over a broad potential range and for the current density in this range, termed the passive current density (i_p), to be very low (typically, a few μA/cm^2 or less). If aggressive species, such as chloride ions, are not present, then the end of the passive region of the curve is determined by the onset of transpassive dissolution of the film or the initiation of another anodic reaction, such as oxygen evolution, on the specimen surface. The passive film is subject to pitting corrosion when chloride or another aggressive species is present. The onset of pitting is marked by a sudden increase in the current density—the potential where this increase occurs is called the pitting potential, E_p, or the breakdown potential, E_b. The higher the pitting potential, the less susceptible the material is to pitting. It should be noted that the pitting potential as measured in a potentiodynamic polarization experiment is a function of scan rate. As a result, care must be exercised in picking a scan rate to ensure that a slow enough scan rate is being used to maintain steady-state conditions at the metal electrolyte interface and to allow for the induction time needed for the onset of pitting. In a polarization curve, such as the one shown in Fig. 13, the potential provides the driving force for a reaction to occur and the current density (the current/surface area of the specimen) is a measure of the corrosion rate (Ref 20). For some metal/solution combinations, the protective passive film is present on the metal surface initially and polarization is not required for its formation. The term "self-passivating" is used to describe such metals, and their polarization behavior is illustrated in Fig. 14 (Ref 20).

Anodic polarization curves for wrought and P/M 434L stainless steel in 0.1 M NaCl are presented in Fig. 15 (Ref 18). In this particular ex-

Table 10 Data obtained from anodic polarization testing of sintered AISI 316L, raw 316L powder, and wrought 316 as a function of sintering conditions

Electrochemical properties were measured in 0.1% Cl^-, pH 5, 30 °C, with a scan rate of 5 mV/min.

	H_2 (−30 °C)				H_2 (−70 °C)				Vacuum					
	1120 °C		1250 °C		1120 °C		1250 °C		1120 °C		1250 °C		Raw 316L	Wrought
Sintering(a)	30 min	120 min	30 min	120 min	30 min	120 min	30 min	120 min	30 min	120 min	30 min	120 min	powder	316L
Density, g/cm^3	6.62	6.68	6.71	6.84	6.62	6.68	6.71	6.84	6.67	6.73	6.76	6.86	...	8.00
Nitrogen, ppm	400	320	220	60	470	190	110	70	410	220	90	20	700	
Oxygen, ppm	2400	2400	2200	1500	2300	2000	1900	1700	2200	2200	2100	1800	1900	
Carbon, ppm	230	220	190	130	240	250	170	110	60	60	20	10	180	300
I_{peak}, μA/cm^2	150	90	87	83	10	10	7	9	4	7	8	9	...	0
I_p, μa/cm^2	29	21	28	19	14	10	12	11	9	13	12	7	...	0.5
E_{pit}, mV SCE	250	243	243	333	345	370	330	395	368	410	363	405	...	665(b)
E_{stp}, mV SCE	269	213	188	163	238	275	188	163	263	238	175	150	...	538(c)
NSS 1, h	36	60	48	24	1392	1278	1260	1512	1056	1008	420	240	...	1512
NSS 2, h	13	24	13	2	1512	1140	1260	60	1512	1008	324	24	...	...

E_{stp} stepwise polarization. NSS; time to corrosion in neutral salt-spray test; 1, no pretreatment; 2, specimens filled with test solution. (a) Measured with a crevice-free electrode. (b) Measured with a creviced electrode. Source: Ref 1

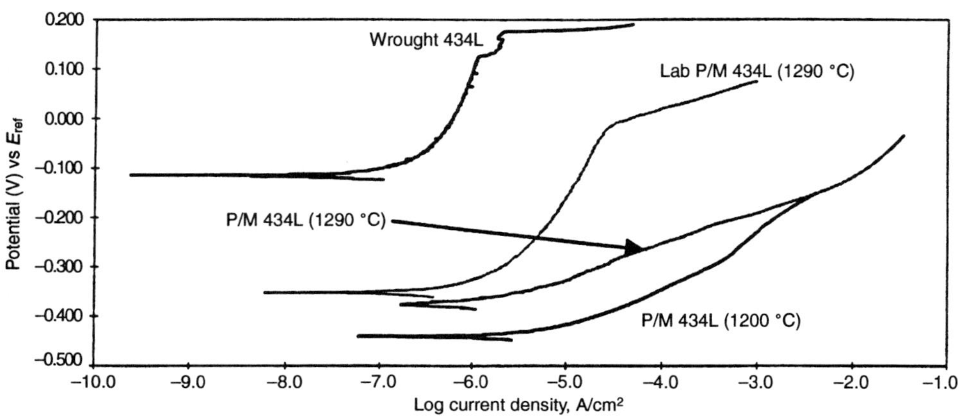

Fig. 15 Anodic potentiodynamic polarization scans for wrought 434L, laboratory-sintered P/M 434L sintered at 1290 °C (2350 °F), P/M 434L sintered in an industrial furnace at 1200 °C (2200 °F), and P/M 434L sintered in an industrial furnace at 1290 °C (2350 °F). Source: Ref 18

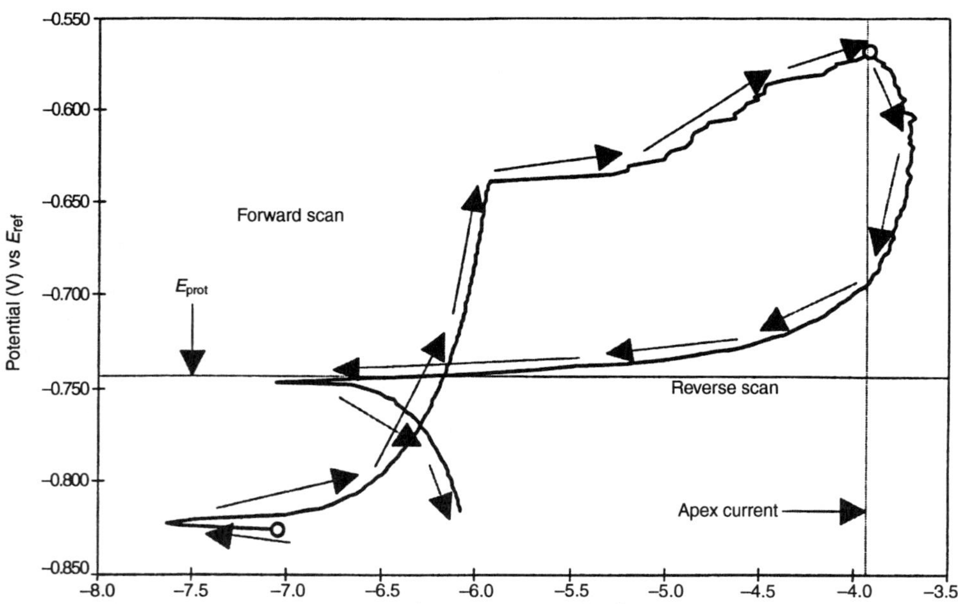

Fig. 16 Sample cyclic polarization curve with labeled values and regions. E_{prot}, protection potential. Source: Ref 21

ample, the current density was based on the apparent surface area of the specimen rather than the actual wetted surface area of the specimen. Measurement of the actual wetted surface area would be needed for precise determination of current-related polarization characteristics, such as the passive current density and critical current density. In addition, the actual wetted surface area is needed for an accurate determination of the corrosion rate. Unfortunately, much of the P/M electrochemical corrosion literature fails to measure and/or report actual surface areas. As a result, care must be taken when basing judgments on variations in current density (i_p, i_{crit}, or i_{corr}) of an order of magnitude or less, unless the actual surface area of the specimen was measured. Anodic polarization characteristics (i_p, i_{crit}, E_p, E_{corr}) are often summarized in tabular form, as shown in Table 10 for wrought and sintered 316L stainless steel (Ref 1).

Because E_p is subject to scan rate effects and is a stochastic parameter, other more conservative parameters are often used to characterize the safe upper potential bounds for use of a passive metal. The most frequently used parameter in this category is the protection potential. A cyclic polarization experiment in which the potential scan direction is reversed after reaching a predetermined critical current or potential is used to determine the protection potential, as shown in Fig. 16 (Ref 21). Pourbaix originally defined the protection potential as the potential at which the current on the reverse scan returns to a low value. This low value is traditionally reported as the potential at which the reverse scan intersects the forward scan. Cyclic polarization experiments are frequently conducted on P/M materials, and an example of the difference in protection potential observed between wrought and porous 316L stainless steel is shown in Fig. 17 (Ref 15).

A few types of electrochemical measurements can be used to determine corrosion rates, as discussed in Ref 19. One of these methods, polarization resistance, involves polarizing the specimens a few millivolts above and below the open-circuit potential and measuring the current

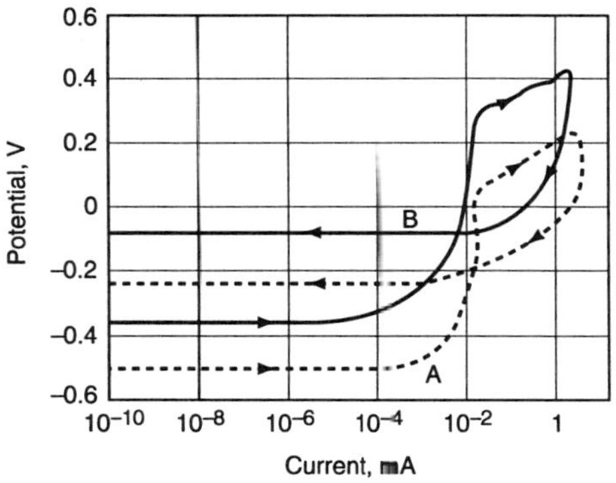

Fig. 17 Anodic potentiodynamic polarization curves for 316L stainless steel. A, P/M specimen; B, wrought specimen. Source: Ref 15

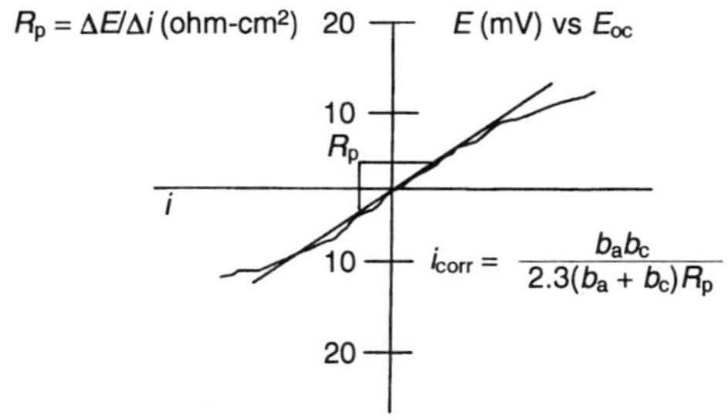

Fig. 18 Sample polarization resistance data showing determination of i_{corr} from R_p. The slope at E_{oc} is called the polarization resistance, R_p. The Tafel slopes, b_a and b_c, must be obtained from anodic and cathodic polarization experiments or estimated.

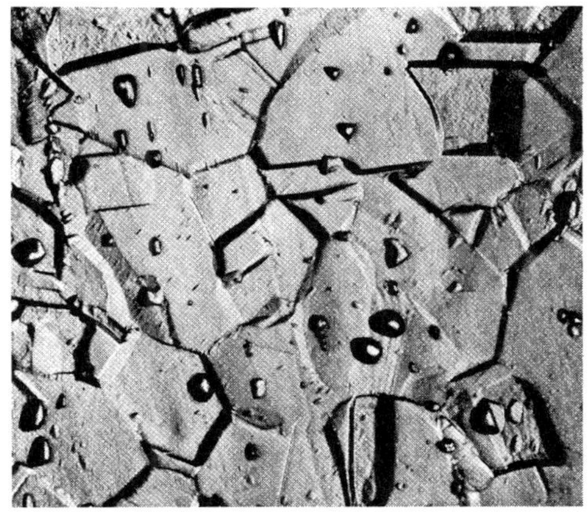

(a)

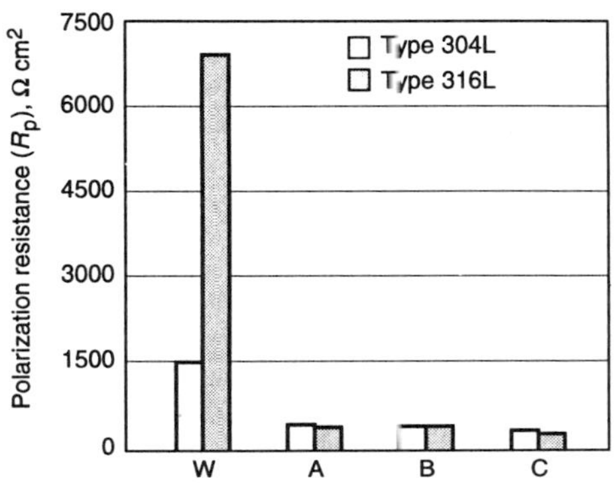

Fig. 19 Polarization resistance, R_p, values for wrought (W) and sintered (A, B, C). A, B, and C correspond to the same sintering conditions as shown in Table 3. Type 304L and 316L samples in 0.5 M H_2SO_4 solution at T = 25 °C. A, B, and C correspond to the same sintering conditions as shown in Table 3 (Ref 9).

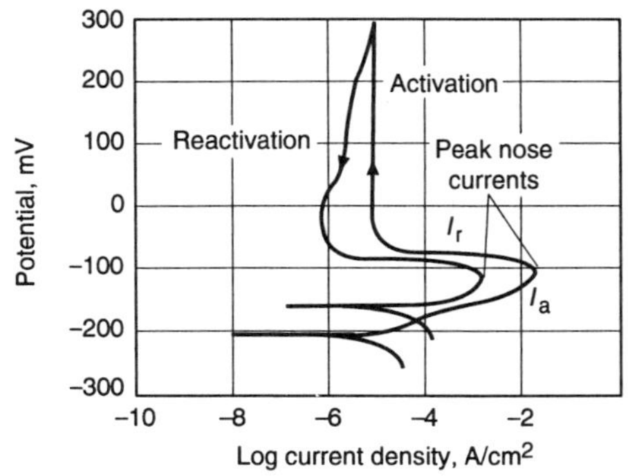

Fig. 20 I-V curve typical of a double-loop EPR experiment

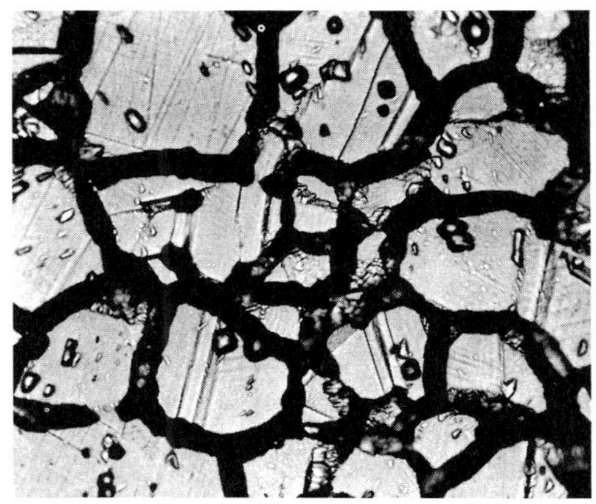

(b)

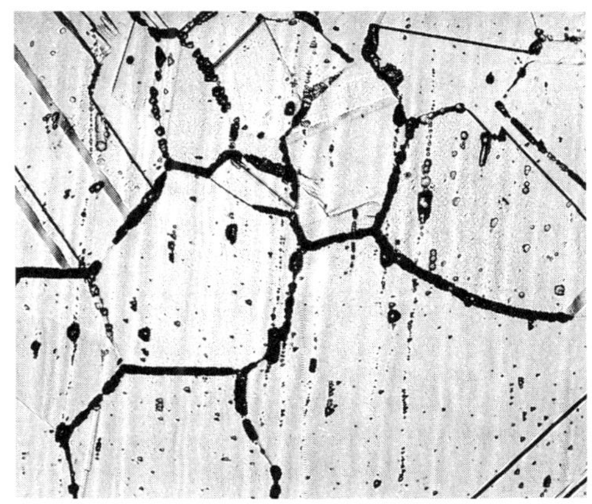

(c)

Fig. 21 (a) Oxalic acid etch (500×). Step structure. Etched 1 A/cm² for 1.5 min. (b) Oxalic acid etch (500×). Ditch structure. Etched 1 A/cm² for 1.5 min. (c) Oxalic acid etch (250×). Dual structure. Etched 1 A/cm² for 1.5 min. Source: Ref 22 reprinted with permission

response. The data are plotted using linear scales and the slope of the potential versus current density data at the open-circuit potential ($\Delta E/\Delta i$) is taken as illustrated in Fig. 18. This slope is called the polarization resistance, R_p, and it is inversely proportional to the corrosion current density:

$$i_{corr} = (1/2.3R_p)[(b_a + b_c)/(b_a + b_c)] \quad \text{(Eq 3)}$$

where b_a is the anodic Tafel slope and b_c is the cathodic Tafel slope. This can be used for the corrosion rate:

$$\text{Corrosion rate} = (3.26 \times 10^{-3})(i_{corr})(EW)/d \quad \text{(Eq 4)}$$

where EW is the equivalent weight of the metal or alloy in grams, d is the density of the metal or alloy in g/cm, and i_{corr} is the corrosion current density in mA/cm². A comparison of polarization resistance values for P/M and wrought 316 and 304 stainless steels is shown in Fig. 19 (Ref 9).

Sensitization in stainless steels can be evaluated by two electrochemical potentiokinetic reactivation methods (EPR): the single-loop method and the double-loop method. A more detailed description of intergranular corrosion testing is offered in Ref 22. The single-loop EPR method is a rather time-consuming test because, in addition to the electrochemical experiment, the grain-boundary area of the specimen must be determined. The double-loop EPR method was developed as a more rapid and reliable means of assessing susceptibility to intergranular corrosion. In the double-loop method, the potential is scanned (at a rate of 6 V/h) in the anodic direction from the open-circuit potential to +300 mV versus a saturated calomel electrode (mV SCE). At +300 mV, the direction of the scan is reversed and the potential is decreased (at the same scan rate) until the corrosion potential is reached. Susceptibility to sensitization is determined from the ratio of the maximum current on the reverse scan (I_r) to that measured on the forward scan (I_a), as illustrated in Fig. 20. The I_r / I_a ratios have been correlated with the step, dual, and ditch structures observed during oxalic acid etch screening of wrought stainless steels to identify sensitization. These three structures are presented in Fig. 21. Step structures (Fig. 21a) are free of grain-boundary attack resulting from sensitization. If the grain boundaries contain enough carbides to encircle the grain completely, then grains are undermined upon exposure to the etch and a ditch structure is observed (Fig. 21b). If the ditches do not completely encircle the grains (Fig. 21c), then a dual classification is used to describe the structure. Current ratios in the double-loop EPR test in the range of 0.0001 to 0.001 correspond to step structures, ratios of 0.001 to 0.05 correspond to dual structures, and ratios of 0.05 to 0.3 correspond to ditch structures. Double-loop EPR curves for P/M 316L stainless steels with and without sensitized microstructures are shown in Fig. 22 (Ref 14). In this example, the sensitized micro-

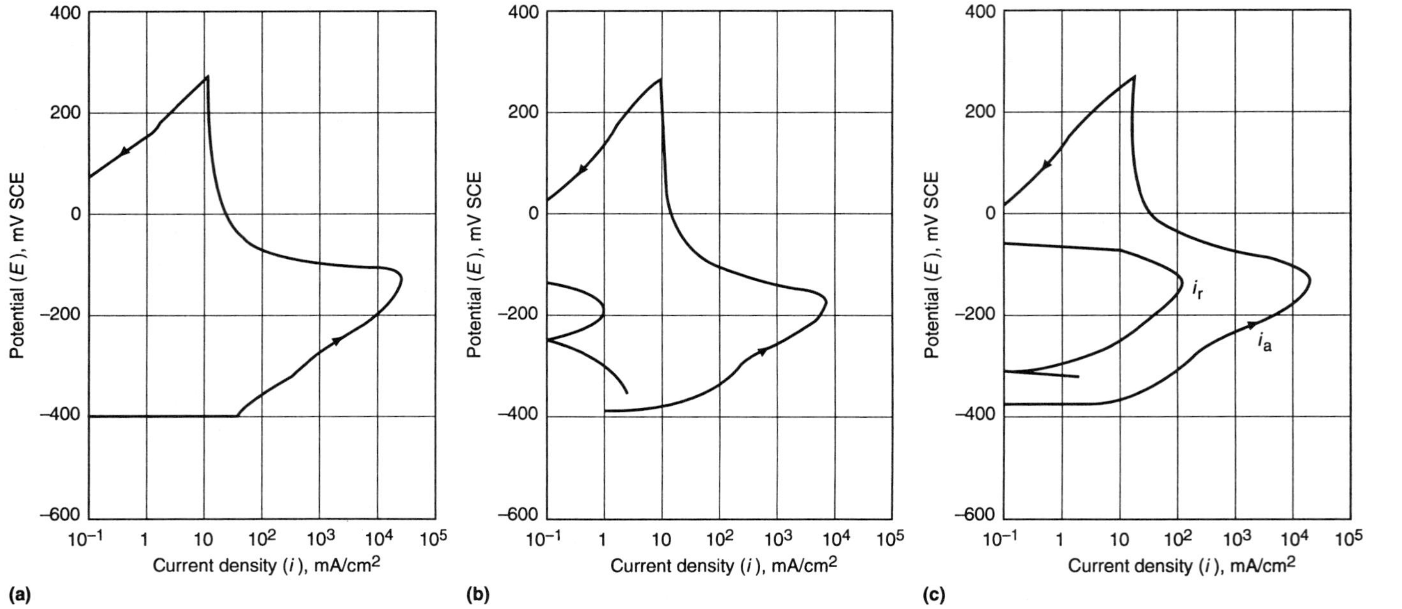

Fig. 22 Polarization curves for 316L P/M steels obtained by the EPR double-loop technique in 0.5 *M* H_2SO_4 + 0.1 *M* KSCN (30 °C). (a) Steel without sensitization. (b) Sensitized steel with 1850 ppm N. (c) Liquid phase sintered steel with addition of boron. Source: Ref 14

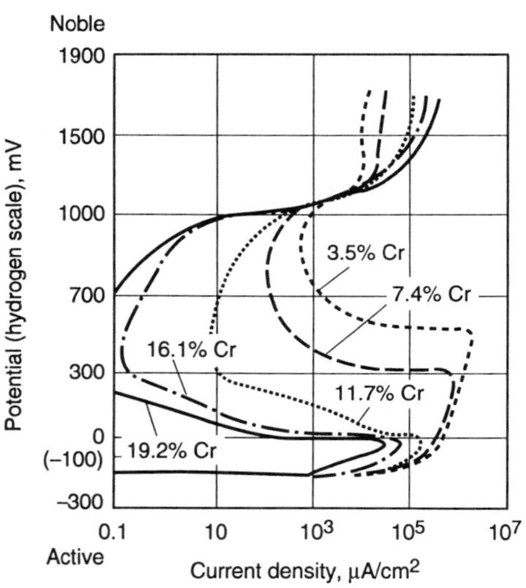

Fig. 23 Effect of chromium content of FeNiCr alloys on their anodic polarization behavior in 2 *N* H_2SO_4 at 90 °C. The nickel content was in the range of 8.3 to 9.8%. Source: Ref 20

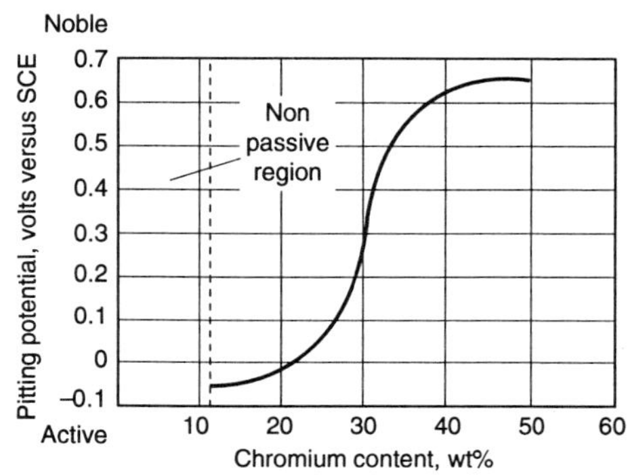

Fig. 24 Effect of chromium content on pitting potential of FeCr alloys in deaerated 0.1 *N* NaCl at 25 °C. Source: Ref 20

structures in curves in Fig. 22(b) and (c) resulted from high nitrogen contents and liquid phase sintering using a boron addition, topics that are discussed in more detail in subsequent sections of this article.

P/M Stainless Steels

It was stated earlier that the corrosion resistance of the metals and alloys that are considered in this article is a result of the protective passive film that forms on their surfaces. This is certainly true for stainless steels, whose passive characteristics result from the alloying of iron with chromium. For wrought alloys, the influence of chromium concentration on the passivity of iron in a 2 *N* H_2SO_4 solution at 90 °C is shown in Fig. 23 (Ref 20). As the chromium content is increased, the primary passivation potential, E_{pp}, the critical current density, i_{crit}, and the passive current density, i_p, are reduced—extending the potential region over which the alloy is passive, reducing dissolution of the alloy, and enabling the alloy to become self-passivating, respectively. When chloride is present, chromium increases the pitting potential, as revealed in Fig. 24 (Ref 20).

Figure 23 clearly shows that a significant improvement in passivity occurs as the chromium content approaches 12%. While there is some debate over the minimum chromium content needed to call an iron-base alloy a stainless steel, most consider that at least 12% is needed to enhance passivity and, thus, corrosion resistance significantly. Passivity results from the presence of a stable, protective oxide barrier on the alloy surface. For stainless steels, this protective barrier is largely composed of chromium oxide and significantly limits dissolution of the underlying metal. Corrosion resistance in stainless steels is compromised when the chromium concentration of the alloy, or any part of its microstructure, falls below a threshold concentration of 10 to 12%.

The primary difference in the corrosion resistance of wrought and P/M stainless steels is due to the inherent porosity of the P/M materials. Of special concern with regard to the corrosion resistance of sintered stainless steels is their susceptibility to localized breakdown of the passive film in the form of crevice corrosion, pitting, intergranular corrosion, and stress-corrosion cracking. Development of corrosion-resistant sintered stainless steels depends on increasing one's understanding of how passivity is attained and sustained in porous metals.

Despite the fact that P/M stainless steel production accounted for only 2% of the overall U.S. stainless steel market in 1993, research on the corrosion resistance of these alloys has been increasing, and it is obvious that an enormous opportunity exists for their future development and utilization. To date, the compositions of commercial P/M stainless steels have been largely based on those of their wrought counterparts. A listing of the currently produced commercial, water-atomized P/M stainless steel alloys is presented in Table 11 (Ref 23). As stated earlier, applications for these P/M alloys are wide ranging with austenitic grades accounting for approximately two-thirds of the total usage. Table 12 summarizes a number of the current applications for the P/M stainless alloys identified in Table 11 (Ref 23). It should be noted that automotive applications, especially exhaust systems, are responsible for much of the recent, increased interest in and demand for corrosion-resistant P/M stainless steel alloys.

Table 11 Compositions of commercial P/M stainless steels

Alloy	Composition, %											
	Cr	Ni	Si	Mo	Cu	Su	Mn	C	S	P	Fe	O (ppm)
Austenitic grades												
303	17–18	12–13	0.6–0.8	...	...	...	0.3(a)	0.03(a)	0.1–0.3	0.03(a)	bal	...
304L	18–19	10–12	0.7–0.9	...	...	...	0.3(a)	0.03(a)	0.03(a)	0.03(a)	bal	1000–2000
304LSC	18–20	10–12	0.8–1.0	...	2(b)	1(b)	0.3(a)	0.03(a)	0.03(a)	0.03(a)	bal	...
316L	16.5–17.5	13–14	0.7–0.9	2–2.5	...	...	0.3(a)	0.03(a)	0.03(a)	0.03(a)	bal	1000–2000
Martensitic grade												
410L	12–13	...	0.7–0.9	...	...	...	0.1–0.5	0.05(a)	0.03(a)	0.03(a)	bal	1500–2500
Ferritic grades												
430L	16–17	...	0.7–0.9	...	...	...	0.3(a)	0.03(a)	0.03(a)	0.03(a)	bal	...
434L	16–18	...	0.7–0.9	0.5–1.5	...	...	0.3(a)	0.03(a)	0.03(a)	0.03(a)	bal	...

(a) Maximum. (b) Typical. Source: Ref 23

Table 12 Applications for P/M stainless steels

Part	Alloy
Aerospace	
Seatback tray slides	316L
Galley latches	316L
Jet fuel refueling impellers	316L
Foam generators	316L
Agriculture	
Fungicide spray equipment	316L
Appliances	
Automatic dishwasher components	304L
Automatic washer components	304L
Garbage disposal components	410L
Pot handles	316L
Coffee filters	316L-Si
Electric knives	316L
Blenders	303L
Can opener gears	410L
Automotive	
Rearview mirror mounts	316L, 434L
Brake components	434L
Seat belt locks	304L–434L
Windshield wiper pinions	410L
Windshield wiper arms	316L, 434L
Manifold heat control valves	304L, 434L
Exhaust system flanges	304L, 434L
Coupling for a water pump	316L
Solenoid spacer for fuel injector	316L
Sealing washer for water pump	420L
ABS rings	434L
Building and construction	
Plumbing fixtures	316L
Spacers and washers	316L
Sprinkler system nozzles	316L
Shower heads	316L
Window hardware	304L, 316L
Thermostats	410L
General construction	303L
Electrical and electronic	
Limit switches	410L
G-frame motor sleeves	303L
Rotary switches	316L
Magnetic clutches	410L, 440A
Battery nuts	830
Electrical testing probe jaws	316L
Hardware	
Lock components	304L, 316L
Threaded fasteners	303L
Fasteners	316L
Quick-disconnect levers	303L, 316L
Industrial	
Water and gas meter parts	316L
Filters, liquid and gas	316L-Si
Recording fuel meters	303L
Fuel flow meter devices	410L
Pipe flange clamps	316L
High polymer filtering	316L-Si
Jewelry	
Coins, medals, medallions	316L
Watch cases	316L
Watch band parts	316L
Marine	
Propeller thrust hubs	316L
Cam cleats	304L
Medical	
Centrifugal drive couplings	316L
Dental equipment	304L
Hearing aids	316L
Anesthetic vaporizers	316L
Office equipment	
Nonmagnetic card stops	316L
Dictating machine switches	316L
Computer knobs	316L
Recreation and leisure	
Fishing rod guides	304L, 316L
Fishing rod gear ratchets	316L
Photographic equipment	316L
Soft drink vending machines	830, 316L
Travel trailer water pumps	316L
Computers	
Support frame for computer CPU	303L
Bearing holder for hard disk	304L
Pulley for computer application	316L
Bearing housing for hard disk	410L
Chemical	
Filters	304L-Si, 316L
High-corrosion resistance filters	830
Cartridge assemblies	316L-Si

Source: Ref 23

Early P/M stainless steel developmental efforts were aimed at improving powder atomization and compacting properties, while interest in enhanced corrosion resistance has been more recent. In fact, the understanding of the corrosion of P/M stainless steels and the concomitant development of corrosion-resistant P/M alloys (specifically, P/M alloys that have been especially developed for their corrosion resistance) is still rudimentary. Several factors are responsible for the differences between the corrosion resistance of P/M stainless steels and their wrought counterparts. Next to the properties of the powder itself (which needs to have carbon, oxygen, and nitrogen levels less than or equal to values 0.02%, 0.2%, and 0.03%, respectively), the chief factor governing the production of corrosion-resistant P/M alloys is the sintering process. Several sintering variables are of significant importance in the production of corrosion-resistant P/M alloys. Inadequate dissipation of lubricant results in the incorporation of unwanted carbides, and low sintering temperatures encourage chromium nitride formation; both conditions can deplete the alloy of the chromium needed to maintain passivity and aid in the establishment of microgalvanic cells in the microstructure. Low sintering temperatures encourage chromium nitride formation and deplete the alloy of the chromium needed to maintain passivity and establish microgalvanic cells within the microstructure. Insufficient cooling rates contribute to chromium carbide and chromium nitride formation. High dew points encourage oxide formation that has been associated with diminished corrosion resistance. Carbon and iron contamination from furnaces leads to carbide formation and iron incorporation in the alloy. Sintering atmospheres can either promote or degrade corrosion resistance; vacuum and hydrogen atmospheres typically yield the best results and dissociated ammonia usually produces the worst results. Unfortunately, a significant portion of the literature on the corrosion of P/M alloys, stainless steels included, lacks specific details concerning processing conditions, making performance comparisons among different alloy groups difficult or impossible.

Austenitic stainless steels owe their corrosion resistance to chromium, nickel, and other passivity enhancing alloying additions such as molybdenum. As stated earlier, chromium is the primary element responsible for the corrosion resistance of stainless steels. Nickel stabilizes the austenitic structure and promotes repassivation, especially in reducing environments. Molybdenum stabilizes the passive film, especially in combination with chromium, and improves pitting and crevice-corrosion resistance. Nitrogen enhances the strength and corrosion resistance of austenitic wrought stainless steels, but is linked to sensitization in stainless steels. These elements, and a few others, influence polarization behavior and enhance passivity in wrought alloys in the manner depicted in Fig. 25 (Ref 20).

Intergranular corrosion of 300 series stainless steels has long been associated with sensitization resulting from welding or improper heat treating. If the alloys are held in the temperature range of 550 to 850 °C, carbides can form at the grain boundaries depleting the alloy adjacent to the grain boundary of chromium, as the chro-

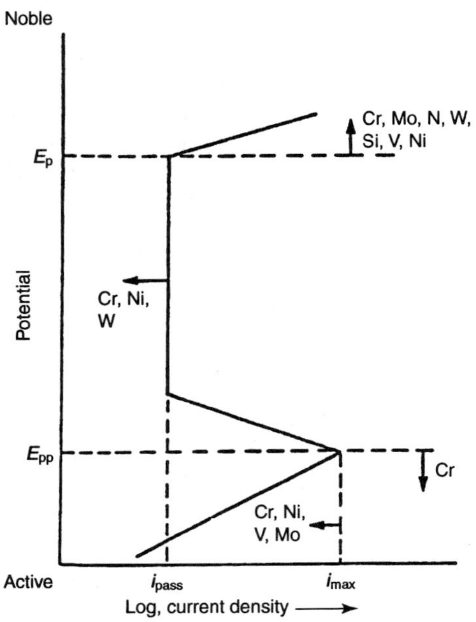

Fig. 25 Summary of the effect of alloying elements in stainless steels on the anodic polarization behavior. Source: Ref 20

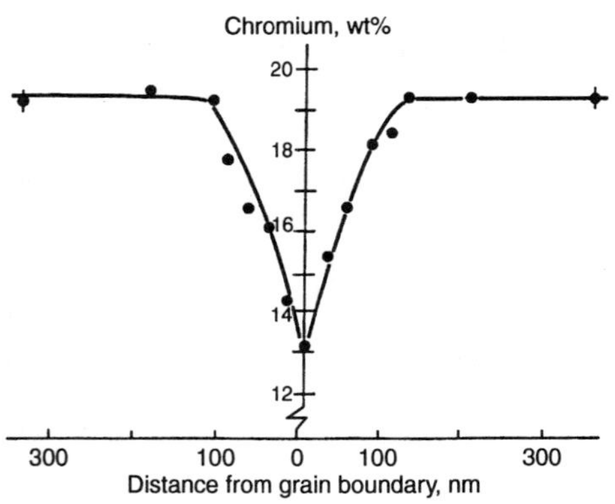

Fig. 26 Chromium concentration profile across a grain boundary between $M_{23}C_6$ carbides in type 304 (0.039% C) stainless steel heat treated for 10 h at 700 °C (Ref 24)

mium profile for a sensitized 304 stainless steel shown in Fig. 26 reveals (Ref 24). The region adjacent to the grain boundary with the lower chromium content has a lower corrosion resistance than either the precipitates at the grain boundary or the rest of the grain. These areas of dissimilar chromium concentration can result in the establishment of localized galvanic cells within the microstructure that lead to corrosion of the chromium-depleted regions adjacent to grain boundaries, as Fig. 27 illustrates (Ref 23). In P/M alloys, sensitization can occur during sintering as well as during welding operations. Stainless steel alloys with carbon and nitrogen contents that approach 0.03% have been shown to form chromium precipitates at the grain boundaries (Ref 24). In addition, sintering or cooling (especially from 870 to 425 °C, or 1600 to 800 °F) in an atmosphere that contains nitrogen can lead to the formation of chromium nitrides. When these chromium carbides and chromium nitrides form, they deplete the adjacent microstructure of chromium. Hence, a galvanic interaction ensues between the more corrosion-resistant, high-chromium concentration regions of the microstructure and the less-corrosion-resistant, chromium-depleted regions.

As revealed in Table 12 (Ref 23), the 300-series alloys account for the largest percentage of applications for stainless P/M alloys. Among wrought stainless steels, 304L is the most commonly used alloy because of its excellent corrosion resistance and lower cost compared to other austenitic stainless steels. To date, 316L is the most commonly used P/M stainless steel because of its improved corrosion resistance in comparison to sintered 304L stainless steel.

Table 13 Corrosion resistances of sintered and wrought austenitic stainless steels

Sintered stainless steels	Corrosion test(a)	Corrosion resistance rating(b), h	Sintered density, g/cm³	Comments: Sintering atmosphere(c)	Sintering temperature °C	Sintering temperature °F	Sintering time, min	Type of furnace(d)
303L	I	5	6.7–6.8	DA	1150	2100	60	L
303LSC(e)	I	500	6.7–6.8	DA	1150	2100	60	L
304L	I	100	6.7–6.8	DA	1150	2100	60	L
316L	I	500	6.7–6.8	DA	1150	2100	60	L
	I	...	6.7–6.9	Vac	1205	2200	60	L
	NSS	...	6.7	H_2	1150	2100	30	Ind
	NSS	...	6.3	H_2	1150	2100	30	Ind
	NSS	...	6.6–6.7	Vac	1120	2050	30	L
303LSC(e)	I	500	6.7–6.8	DA	1150	2100	60	L
	I	1700	6.7–6.9	Vac	1205	2200	60	L
317	I	4400	6.7–6.9	Vac	1205	2200	60	L
	NSS	...	6.7	H_2	1150	2100	30	Ind

(a) *I*, by immersion in 5% NaCl; NSS, neutral salt-spray test (ASTM B 117; ISO 4540-1980(E)). (b) Time in hours until 1% of surface of specimen is covered with stain or rust. (c) H_2, hydrogen; DA, dissociated ammonia; Vac, vacuum. (d) L, laboratory; Ind, industrial. (e) Proprietary grades of SCM Metal Products. Source: Ref 1

Table 14 Corrosion properties of sintered steels produced from prealloyed powders (–100 mesh) with different alloy compositions

Sintering, °C/min	Type	PRE	E_{pit}(a), mV SCE	E_{stp}(b), mV SCE	NSS1	NSS2
1120/30/H_2	316L	25	500	225	600	8
	317L	30	725	725	>1500	9
	18-18-6(c)	37	275	275	48	4
	SS-100(d)	37	575	400	>1500	9
	17-25-8(e)	42	550	425	>1500	9
1250/120/H_2	317L	25	500	150	96	7
	317L	30	500	350	14	6
	18-18-6	37	450	275	50	5
	SS-100	37	>800	>800	>1500	10
	17-25-8	42	675	450	355	8

PRE, Pitting resistance equivalent (%Cr + 3.3% Mo + 16% N) in wt%; NSS1, time to first sign of corrosion, salt-spray test in 5% NaCl; NSS2, rust rating after 1500 h of testing where 10 = no corrosion and 0 = surface half covered with corrosion products. (a) 0.1%Cl⁻, pH 5, 30 °C, 5 mV/min. (b) 0.1% Cl⁻, pH 5, 30 °C, 25 mV/8h. (c) 18.3% Cr, 18.3% Ni, 5.6% Mo, 1.7% Cu, 1.3% Sn, 0.78% Si, 0.23% Mn, bal Fe. (d) 20% Cr, 17.0% Ni, 5.0% Mo, 0.75% Si, >0.15% Mn, bal Fe. (e) 16.3% Cr, 24.3% Ni, 7.7% Mo, 0.81% Si, 0.25% Si, bal Fe. Source: Ref 14

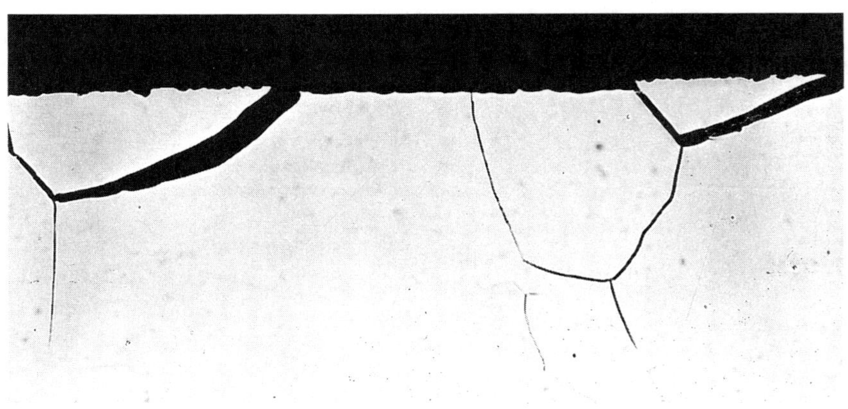

Fig. 27 Intergranular attack in a sensitized austenitic alloy produced by exposure to a boiling sulfuric acid-ferric sulfate solution. Prolonged exposure causes grains to detach from surface. 100×. Source: Ref 24, reprinted with permission

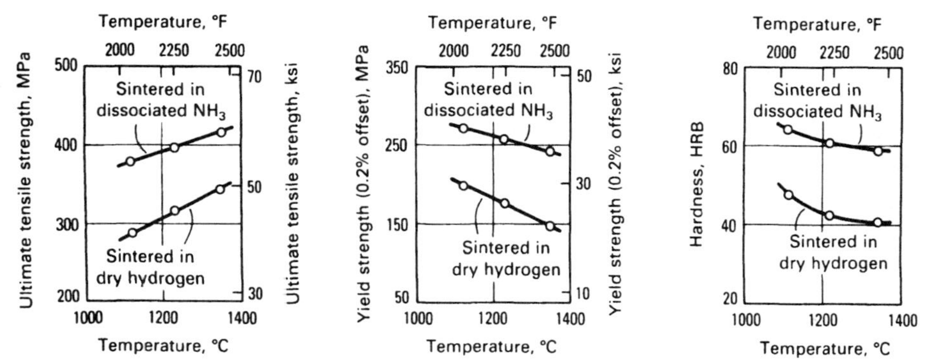

Fig. 28 Effect of sintering temperature on tensile and yield strengths and apparent hardness of type 316L stainless steel. Parts (density: 6.85 g/cm^3) were sintered for 30 min in various atmospheres. Source: Ref 23

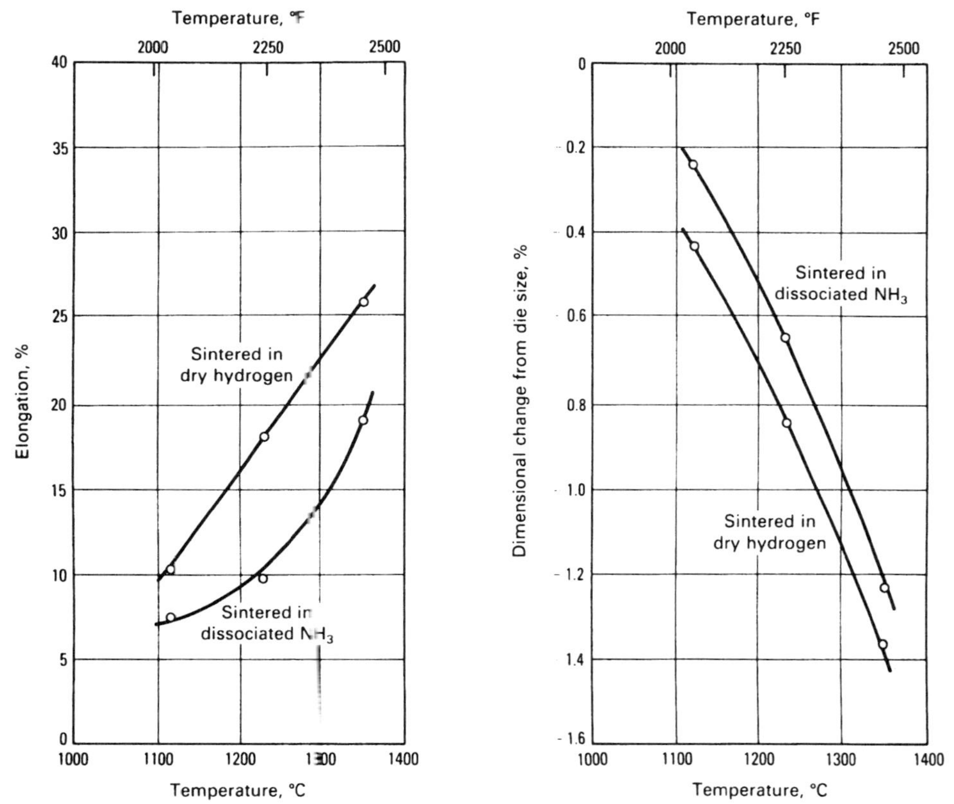

Fig. 29 Effect of sintering temperature on elongation and dimensional change during sintering of type 316L stainless steel. Parts (density: 6.85 g/cm^3) were sintered for 30 min in various atmospheres. Source: Ref 23

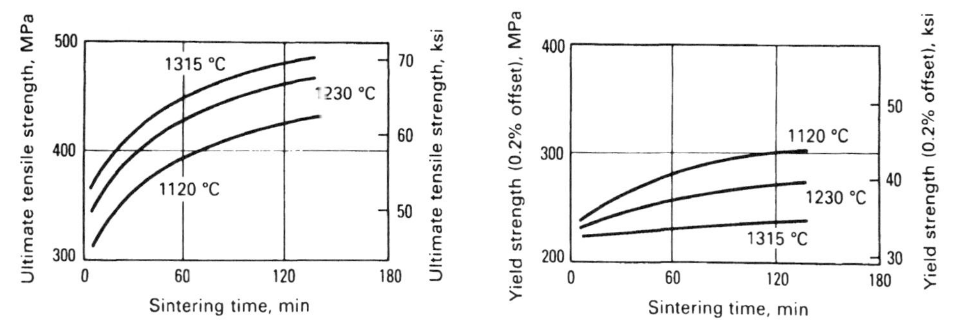

Fig. 30 Effect of sintering time on tensile and yield strengths of type 316L stainless steel. Parts were pressed to 6.85 g/cm^3 and sintered at various temperatures in dissociated NH_3. Source: Ref 23

Low-carbon-grade alloys are essential for P/M stainless components because of sensitization that can accompany the sintering process.

Corrosion data for 300 series P/M stainless steels is largely centered on the susceptibility of these alloys to pitting/crevice corrosion, intergranular corrosion, and processing-induced degradation. This last item is the subject of the next section of this article. It must be stated that while there is an expanding base of literature on the subject of corrosion of sintered stainless, a number of studies fail to mention important details related to either processing or testing of the alloys. Processing details such as dew-point temperatures, sintering and delubing times, and information related to cooling of the alloys are often missing. All of these items are critical to the production of corrosion-resistant stainless steels, and the lack of such information limits the utility of a number of the studies that have been conducted. Testing details that are sometimes neglected include surface area measurements (needed for determining corrosion rates), test specimen preparation and test cleaning procedures, and the inclusion of wrought counterparts or control materials in test plans. This information is especially important because sintered alloys contain porosity. Corrosion resistances of a variety of austenitic stainless steel P/M alloys are shown in Tables 13 and 14 (Ref 1, 14).

Mechanical properties of 316L stainless steel as a function of some of the more important processing parameters are presented in Fig. 28 through 30 (Ref 23). As these figures reveal, nitrogen-containing atmospheres result in the absorption of considerable amounts of nitrogen, which increases strength, decreases ductility, and, as is seen in the following paragraphs, influences P/M alloy corrosion resistance. Mechanical properties for a few P/M stainless steel alloys sintered in dissociated ammonia are presented in Table 13 (Ref 1).

Ferritic and Martensitic Stainless Steels. The 400 series alloys are typically less heavily alloyed than the austenitic grades and, as a result, they usually exhibit inferior corrosion resistance. In addition to the lower pitting and crevice-corrosion resistance resulting from lower concentrations of passivity enhancing elements, ferritic stainless steels are also more susceptible to sensitization and intergranular corrosion. Ferritic stainless steels exhibit a greater affinity for sensitization than austenitic stainless steels because the solubility limit of carbon in the austenite phase is greater than in the ferrite phase. Hence, the precipitation of carbides is more prevalent in ferritic microstructures (Ref 18).

Influence of Processing Parameters on the Corrosion Resistance of P/M Stainless Steels

Influence of Iron or Steel Contamination on Corrosion Resistance. It should come as no surprise that the corrosion resistance of P/M stainless steels is seriously degraded if iron or steel particles become incorporated into the alloy. The potential difference between iron or steel and stainless steel is typically on the order of several hundred millivolts and easily results in

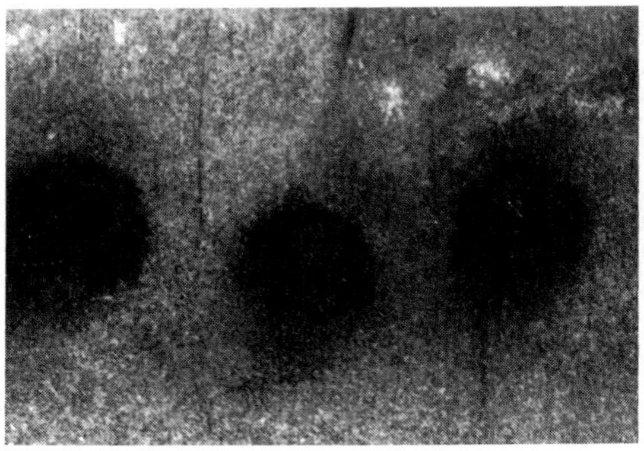

Fig. 31 Small circles of rust around iron particles embedded in the surface of sintered type 316L stainless steel after testing in 5% aqueous NaCl. 35×

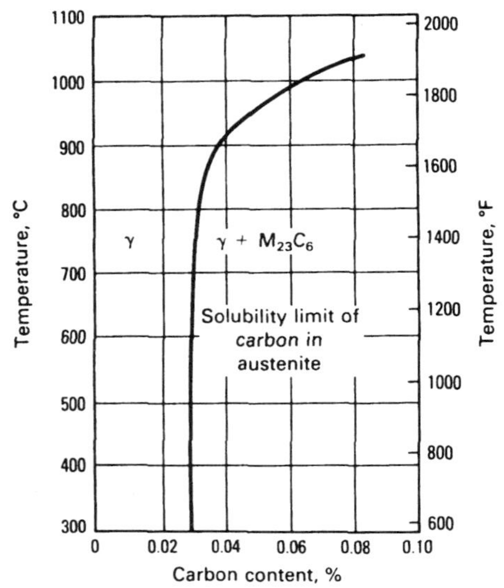

Fig. 32 Solid solubility of carbon in an austenitic stainless steel. Source: Ref 24

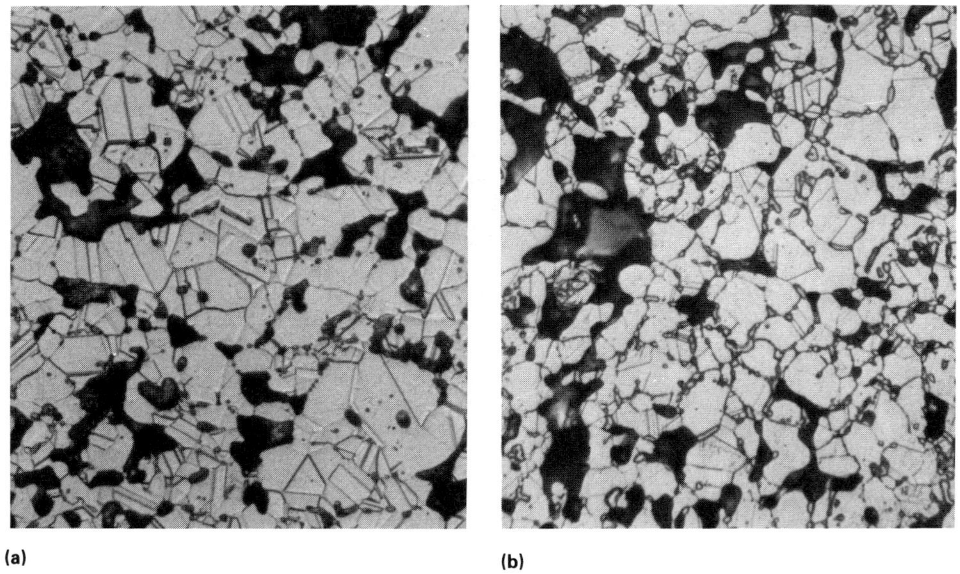

Fig. 33 Microstructures of type 316L stainless steel sintered in hydrogen at 1150 °C (2100 °F). (a) Low carbon content. (b) Excessive carbon content. Both 400×

the establishment of galvanic or dissimilar metal corrosion within the contaminated component. There are numerous possible contamination sources: contamination of the initial powder at the supplier; inadvertent introduction during mixing/blending, feeding, or pressing operations; incorporation of airborne particles during processing or storage; and inadequate furnace cleaning. Cleanliness is of the utmost importance and separate or dedicated equipment is often used for the production of stainless components. Figure 31 shows an example of the appearance of iron-contaminated sintered 316L stainless steel after exposure to a 5% NaCl solution. Rusting became apparent within minutes of exposure to the chloride-containing solution.

Corrosion resulting from iron or steel contamination is perhaps the worst and, ironically, most avoidable corrosion problem encountered with P/M stainless steels. A concentrated copper sulfate solution can be used to easily detect iron, or an iron alloy, present in a stainless steel powder or on the surface of a sintered part. Dissolved copper from a copper sulfate solution readily plates out on the anodic (lower potential) iron sites, making them easy to see at low magnification.

Influence of Lubricant and Carbon. As Fig. 32 suggests (Ref 24), carbide formation (especially, chromium carbide formation) with concomitant sensitization is an issue when the carbon content of an austenitic stainless steel exceeds 0.03%. In order to resist intergranular corrosion, water-atomized stainless steel powders have carbon contents greater than 0.03%. Unfortunately, other sources of carbon are associated with processing sintered stainless steels. These sources include the carbon resulting from inadequate organic lubricant dissipation and carbon contamination (soot) from insufficiently cleaned furnaces. Microstructures from two sintered 316L stainless steels, one below and one above the critical 0.03% concentration, are shown in Fig. 33. Thin, undecorated grain boundaries are observed in the low-carbon stainless steel, whereas, heavily decorated grain boundaries are observed for the high-carbon stainless steel. In insufficiently cleaned furnaces, loose, adherent soot can fall onto the surface of stainless steel parts or moisture from the sintering atmosphere can react with soot and form carbon monoxide and carburize the stainless steel. If care is not taken to limit the uptake of carbon, sensitization of the microstructure can occur and severely compromise the overall corrosion resistance of the alloy. The influence of carbon content on the pitting potential for a number of different sintered stainless steels is shown in Fig. 34 (Ref 14). Sensitization can be minimized with proper lubricant dissipation, a clean furnace, and low initial carbon concentration in the powder. It should be noted that when optimal sintering conditions are used, differences in corrosion resistance have not been noted as a function of lubricant type, as Table 15 reveals (Ref 25).

Carbon contents in excess of 0.03% can be of benefit when stainless steels are vacuum sintered. In vacuum sintering, the excessive carbon is used for the reduction of some oxides on the water-atomized stainless steel, improving strength, ductility, and corrosion resistance. Microstructures of vacuum-sintered 430L stainless steel with and without the addition of 0.2% graphite are shown in Fig. 35. The graphite-containing stainless steel exhibited clean grain boundaries, while the alloy without graphite had grain boundaries containing carbides.

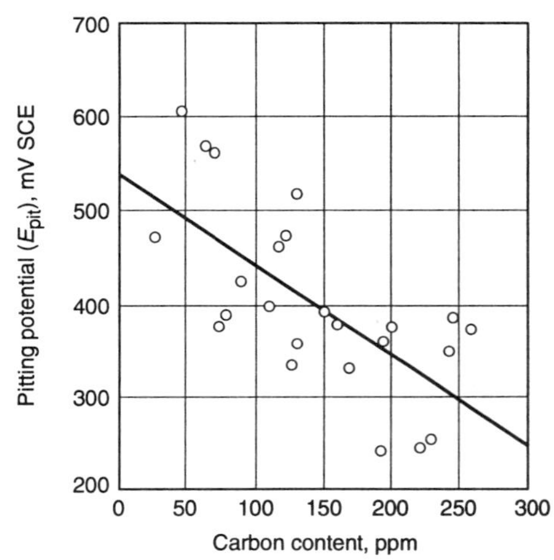

Fig. 34 Influence of carbon concentration on the pitting potential for a number of different materials. Source: Ref 14

Influence of Nitrogen and Sintering Atmosphere. Dissociated ammonia is a commonly used sintering atmosphere because it costs less than other sintering atmospheres. However, sintering in dissociated ammonia usually leads to the pickup of nitrogen by the stainless steel—a factor that can enhance susceptibility to corrosion in a manner analogous to that observed with chromium carbide formation. For wrought stainless steels, enhanced passivity is observed with increased nitrogen content. However, if chro-

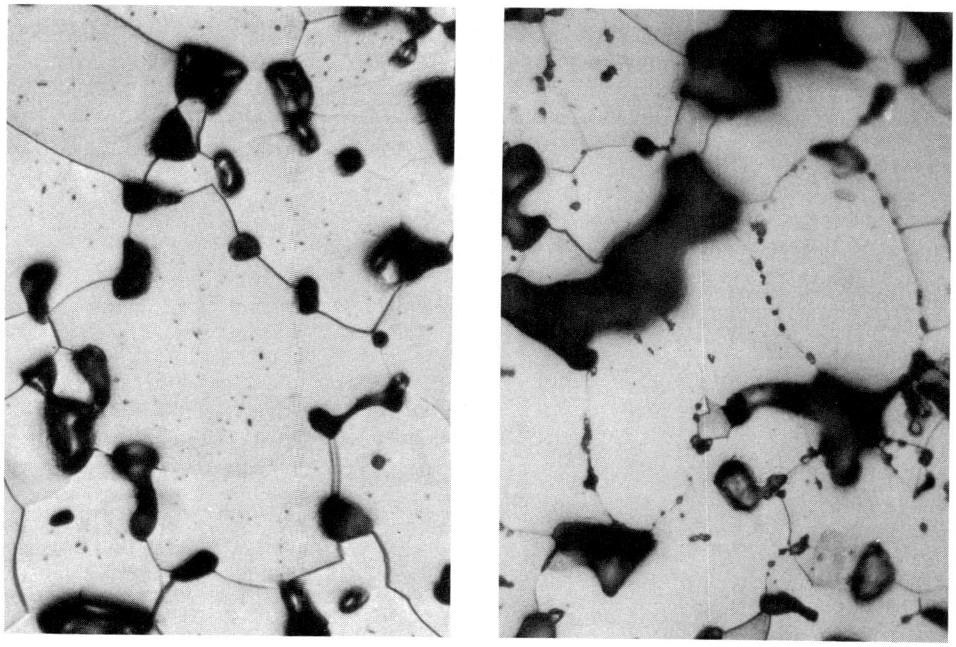
(a) (b)

Fig. 35 Cross sections of vacuum-sintered (30 min at 1330 °C, or 2430 °F) type 430L stainless steel. (a) No oxides are present in grain boundaries after addition of 0.2% graphite. (b) Small, gray, rounded oxide particles in grain boundaries (no graphite added)

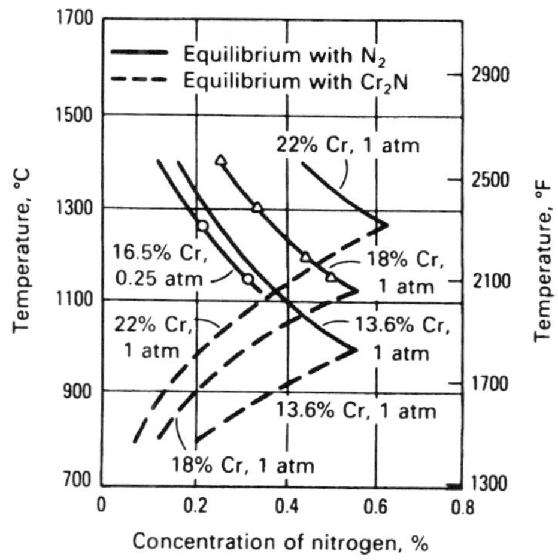

Fig. 36 Solubility of nitrogen in austenitic stainless steel in equilibrium with gaseous nitrogen or Cr_2N. Source: Ref 23

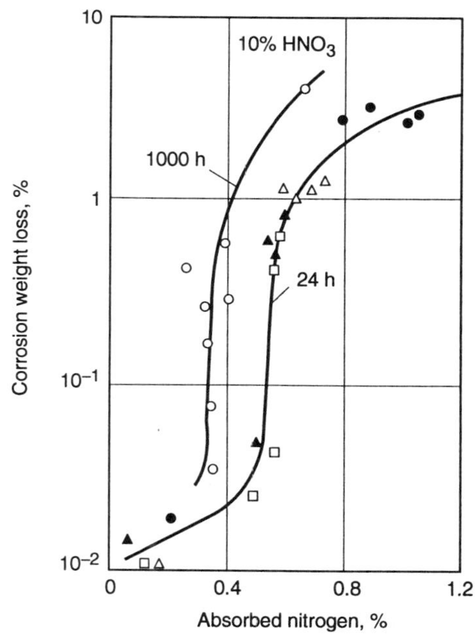

Fig. 37 Effect of the absorbed nitrogen content during sintering on the corrosion resistance of 316L stainless steel sintered at 1150 °C in several atmospheres; corrosion rate is given in terms of weight loss resulting from immersion in 10% HNO_3. Source: Ref 3

Table 15 Effect of binder/lubricant on the corrosion resistance of sintered 316L stainless steel in deaerated 1000 ppm Cl buffered with acetate at 30 °C (pH = 5)

The dew point of the gas atmospheres in the furnace was approximately –30 °C.

Binder	Sintering, °C/min	Atmosphere	I_{pass}, A/cm²	E_{pit}, mV SCE	Salt spray, h
Acraw	1120/20	DA	45	65	20
Metlb			18	230	24
Acraw	1160/45	H_2	3	455	30
Metlb			2	400	620
Acraw	1250/30	H_2	16	230	24
Metlb			16	230	24
Acraw	1120/30	Vacuum	4	390	500
Metlb			4	380	500
Acraw	1200/50	Vacuum	8	475	560
Metlb			5	425	560
Acraw	1295/30	Vacuum	2	405	240
Metlb			2	500	330

Acraw, Acrawax; DA, dissociated ammonia; Metlb, Metallub. Source: Ref 25

Table 16 Corrosion properties of 316L stainless steel sintered in hydrogen or nitrogen/hydrogen mixtures

Sintering, °C/min	Nitrogen content in atm, %	Nitrogen, ppm	i_{peak}(a), μA/cm²	I_{pass}(a), μA/cm²	E_{pit}(a), mV SCE	E_{stp}(b), mV SCE	NSS1, h	NSS2	I_r/I_a(c) × 1000
1120/30	0	360	1[illegible]	11	375	100	>1500	8	0.0
	5	1710	1[illegible]	11	475	350	>1500	9	0.7
	10	2100	1[illegible]	11	525	350	>1500	9	7.2
	25	5670	33[illegible]	34	325	225	24	4	28.9
1250/120	0	20	[illegible]	10	600	375	990	8	0.0
	5	1350	1[illegible]	9	550	300	864	6	0.0
	10	1850	10[illegible]	12	600	375	240	3	0.1
	25	7180	400	160	–25	–25	24	2	390

(a) 0.1% Cl⁻, pH 5, 30 °C, 5 mV/min. (b) 0.1% Cl⁻, pH 5, 30 °C, 25 mV/8h. (c) EPR test in 0.5 M H_2SO_4 + 0.01 M KSCN, 30 °C. Source: Ref 14

mium nitrides precipitate, the sensitized stainless steel is susceptible to intergranular corrosion. Equilibrium solubilities for nitrogen in austenitic stainless steels with different chromium contents are presented in Fig. 36 (Ref 23). The concentration of dissolved nitrogen depends on the amount of nitrogen in the atmosphere, the sintering temperature, and the cooling rate of the sintered alloy. An example of the influence of nitrogen concentration in the sintering atmosphere on the corrosion resistance of 316L stainless steel is shown in Table 16 (Ref 14). These results reveal that up to the point where supersaturation associated with sensitization occurs, no significant difference in corrosion behavior was noted. In addition, these results show that the EPR test is very

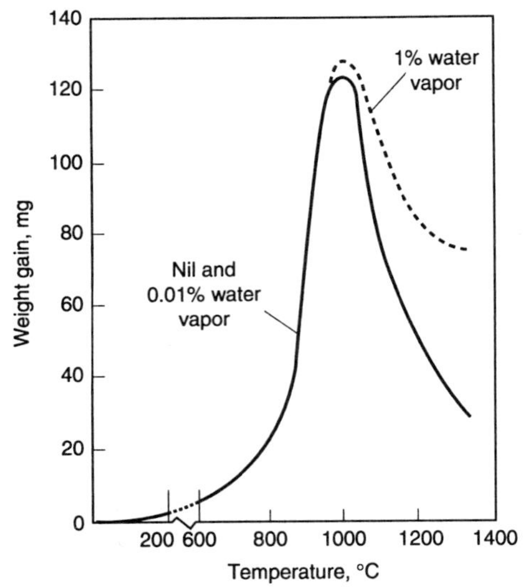

Fig. 38 Increase in weight of specimens heated in nitrogen containing various amounts of water vapor. Source: Ref 13

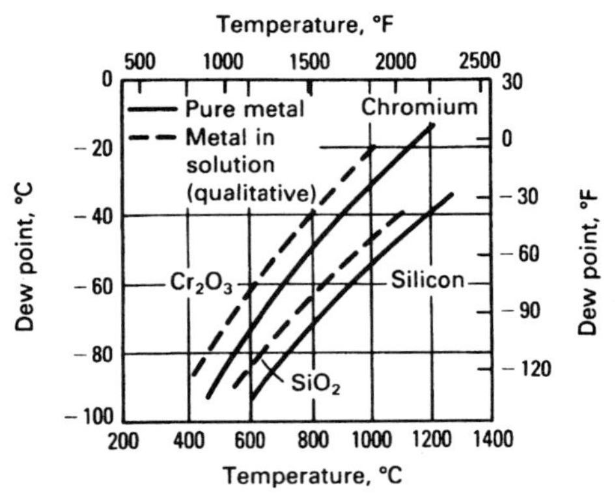

Fig. 39 Redox curves for chromium and silicon alone and in solution. Source: Ref 23

Table 17 Influence of heating and cooling rates on the corrosion resistance of 316L stainless steel specimens sintered at 1150 °C in dissociated ammonia

Heating rate, °C/min	Cooling rate, °C/min	Weight increase, mg/g	Result of corrosion test in 5% NaCl solution
22	22	3.3	Corroded in 1 day
67	67	2.5	Corroded in 1 day
22	100	3	Slight attack in 4 days
5	200	2.9	No attack in 5 days(a)
22	200	3	No attack in 5 days
200	200	2.3	No attack in 5 days

(a) Test continued to 12 days with no attack. Source: Ref 13

sensitive to the identification of nitride formation. Reference 26 reported that chromium nitrides were not present in 316L when its nitrogen content was lower than 0.4 wt%. This point is supported by the weight loss data after exposure to 10% HNO_3 (as a function of absorbed nitrogen content) for sintered 316L stainless steel shown in Fig. 37 (Ref 3). The data presented in this figure were obtained using several sintering atmospheres. An example of the influence of temperature on dissolved nitrogen concentration is shown in Fig. 38 (Ref 13). The data in this figure were obtained by continuously measuring weight gain during heating in dry nitrogen (nitrogen containing 0.01% water and 1% water). In the 700 to 1000 °C temperature range a large absorption of nitrogen occurs. A maximum of 9 mg per gram of stainless steel was absorbed—24 mg/g would be required to convert all of the chromium in the alloy to Cr_2N (Ref 13). An example of the influence of cooling rate, and hence the dissolved nitrogen concentration, on the corrosion rate of 316L stainless steel is shown in Table 17 (Ref 13). While the rate of heating had no influence on corrosion of the alloy, cooling rates in excess of 100 °C/s inhibited or eliminated corrosion. Chromium nitride sensitization with concomitant loss of corrosion resistance is not limited to 316L stainless steel. Other stainless steels (both 300 and 400 series alloys) are subject to loss of corrosion resistance when nitrogen-containing sintering atmospheres are used. Table 2 showed weight-loss data for sintered 304L and 316L stainless steels as a function of sintering atmosphere and revealed that nitride formation lowered corrosion resistance (weight loss after 5% NaCl exposure).

Table 18 Corrosion properties of 316L stainless steel sintered in hydrogen with a dew point of −35 or −70 °C at different combinations of time and temperature

Sintering, °C/min	i_{peak}(a), µA/cm² at dew point: −35 °C	i_{peak}(a), µA/cm² at dew point: −70 °C	i_{pass}(a), µA/cm², at dew point: −35 °C	i_{pass}(a), µA/cm², at dew point: −70 °C	E_{pit}(a), mV SCE at dew point: −35 °C	E_{pit}(a), mV SCE at dew point: −70 °C	NSS1, at dew point −35 °C	NSS1, at dew point −70 °C	NSS2, at dewpoint −35 °C	NSS2, at dewpoint −70 °C
1120/30	150	10	29	11	250	375	36	>1500	5	9
1250/30	105	7	20	12	325	325	288	1260	4	8
1120/120	120	10	25	10	325	375	48	1272	5	7
1250/120	83	4	19	9	325	500	24	96	1	7

(a) 0.1% Cl⁻, pH 5, 30 °C, 5 mV/min. Source: Ref 14

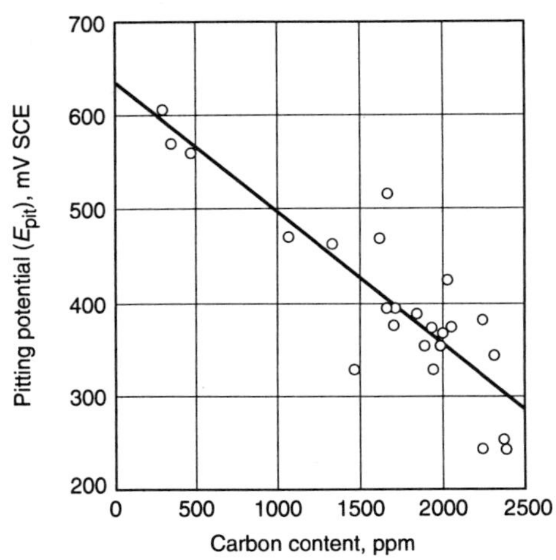

Fig. 40 Influence of the oxygen content on the pitting potential for a number of different materials of sintered AISI 316L. Source: Ref 25

Influence of Oxygen and Water Vapor/Dew Point on Corrosion Resistance. The influence of oxygen on the corrosion resistance of sintered stainless steels can, perhaps, best be understood by visualizing the structure of the as-sintered material. The as-received powders contain oxygen, much of which resides on the surface of the powder. Reduction of these oxides in industrial furnaces is not always complete, and the grain boundary oxides within the as-sintered structure provide paths for easier corrosion of the alloy. The role these grain-boundary oxides play in the corrosion of sintered metals is likely similar to the role that such oxides play in the degradation of thermal sprayed coatings (Ref 27, 28). Upon cooling, high oxygen affinity elements oxidize when they reach the metal-oxide equilibrium temperature, and the water content (dew point) of the atmosphere determines the stability of the oxides according to Fig. 39 (Ref 23). This figure reveals that the oxides are more easily formed at lower temperatures. The negative effect of high dew point on the corrosion resistance of sintered 316L stainless steel is further shown by the data in Table 18.

Table 19 Effect of water-vapor content on the corrosion resistance of stainless steel specimens sintered at 1150 °C in hydrogen

Water-vapor content, ppm	Result of corrosion test in 5% NaCl solution
30	No attack in 8 days
45	No attack in 8 days(a)
90	Rusted after 3 days
110	Stained in 3 h
150	Rusted in 1½ h

(a) Test continued to 14 days with no attack. Source: Ref 13

A number of investigators have observed decreased corrosion resistance with increasing oxygen content (Ref 14, 18, 25, 29). Figure 40 reveals that the pitting potential of P/M 316L stainless steel is found to decrease with increasing oxygen content (Ref 25). Immersion data in a 5% NaCl solution shows an identical trend, as seen in Fig. 41 (Ref 23). High dew points (greater than −34 °C) resulted in high oxygen concentrations within sintered alloys leading to a reduction in mechanical properties and corrosion resistance. This is not surprising in light of the type of microstructure that is attained in high dew point sintering atmospheres and illustrated in Fig. 43 (Ref 22). The P/M 316L stainless steel specimen shown in Fig. 42 exhibited a lack of interparticle bonding, resulting from the high

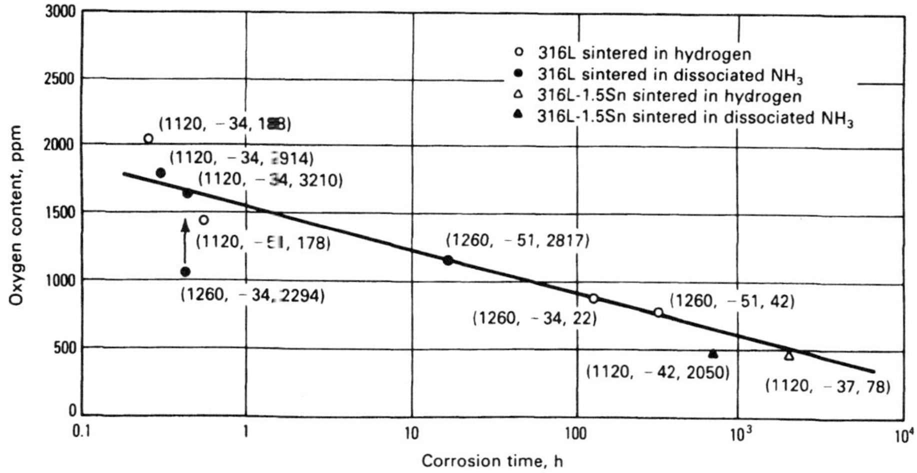

Fig. 41 Effect of oxygen content on corrosion resistance of sintered type 316L and tin-modified type 316L (sintered density: 6.65 g/cm^3; cooling rate: 75 °C/min, or 135 °F/min). Parenthetical values are sintering temperature (°C), dew point (°C), and nitrogen content (ppm), respectively. Time indicates when 50% of specimens showed first sign of corrosion in 5% aqueous NaCl. Source: Ref 23, 30

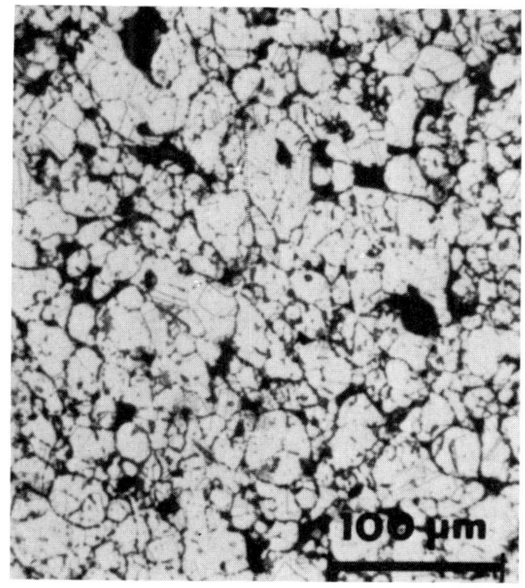

Fig. 42 Microstructure of type 316L stainless steel sintered in a high-dew-point atmosphere. Oxygen content: 5100 ppm; sintered density: 7.5 g/cm^3. Etched with Marble's reagent. 200×

Table 20 Effect of sintering temperature and time on the corrosion resistance of sintered 316L stainless steel

The materials were sintered in hydrogen with a dew point of −70 °C.

Sintering, °C/min	Open pores, %	Nitrogen, %	Oxygen, %	Carbon, %	I_{pass}, A/cm^2	E_{pit}, mV SCE	Salt spray, h
1120/30	17.4	380	2230	250	3.4	383	>1500
1250/30	16.3	110	1980	130	3.1	357	>1500
1120/240	15.8	70	1640	130	3.0	508	>1500
1250/240	13.8	20	450	70	1.8	561	260

Source: Ref 25

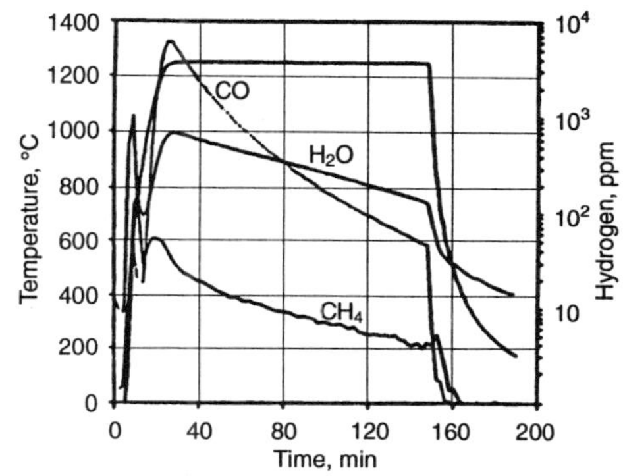

Fig. 43 Gas composition and progress of reactions for a sintering experiment performed with pure hydrogen. Source: Ref 14

Table 21 Effect of cooling rate on the corrosion resistance of 316L stainless steel specimens sintered at 1150 °C in a hydrogen atmosphere containing 100 ppm water vapor

Heating rate, °C/min	Cooling rate, °C/min	Result of corrosion test in 5% NaCl solution
5	22	Corroded in 2 days
200	67	Attack started after 1 day; severe attack after 8 days
200	200	No attack after 3 days; slight staining after 5 days

Source: Ref 13

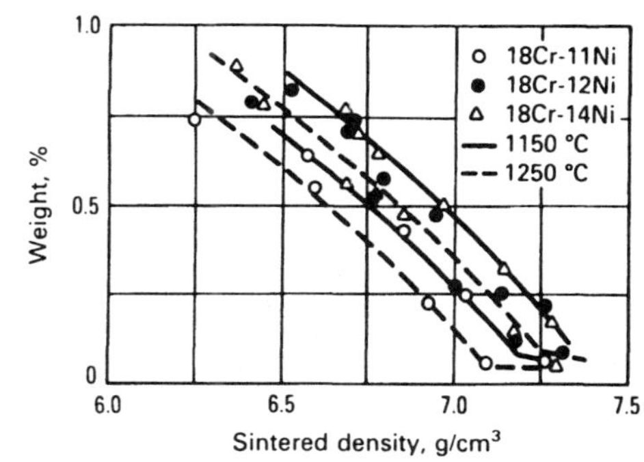

Fig. 44 Relationship between sintered density and weight loss of three austenitic stainless steels in 40% HNO_3 solution. Source: Ref 31

grain boundary oxides, that lead to poor mechanical and corrosion properties. Immersion data for sintered 316L stainless steel in a 5% NaCl solution, as a function of water vapor content of the hydrogen sintering atmosphere, is shown in Table 19 (Ref 30). When the water vapor content was 45 ppm or lower, no corrosion was noted after eight days of exposure.

Influence of Sintering Temperature, Sintering Time, and Cooling Rate on Corrosion Resistance. Improved electrochemical corrosion resistance has been noted for sintered 316L stainless steel with increased sintering time, as Table 20 reveals (Ref 25). These improvements in corrosion resistance were attributed to the reduced nitrogen, oxygen, and carbon levels observed for the specimens after the longer sintering time. Carbon, oxygen, and CO, H_2O, and CH_4 concentrations in a hydrogen sintering gas as a function of time at temperature are presented in Fig. 43. Earlier, the significant influence of cooling rate on corrosion resistance is shown in Table 17. When there is sufficient water vapor to cause corrosion with slow cooling, it appears that a fast cooling rate (200 °C/min) retards corrosion, as shown in Table 21 (Ref 13).

Influence of Porosity/Alloy Density. Sintered stainless steels are used in low-density forms (e.g., in filters) and in a wide variety of forms requiring higher-density alloys. The literature on P/M stainless steels in acid solutions reveals that corrosion resistance improves with increasing density, as Fig. 44 shows (Ref 31). In saline solutions the situation is not as clear—some researchers have reported that increasing density is beneficial, while others have reported it to be detrimental. These discrepancies are believed to be a result of differences, from study to study, in pore morphology and alloy density. This point is illustrated in Fig. 45, which shows the corrosion resistance of sintered 316L stainless steels as a function of density (Ref 1, 32). At low sintered densities, the network of pores, including boundary oxides and particle boundaries, is rather open and discourages the formation of the occluded cell environment associated with crevice-corrosion initiation in stainless steels. At relatively high sintered densities, this network is tighter and encourages both the establishment of an aggressive environment within the crevice and a high potential drop down the

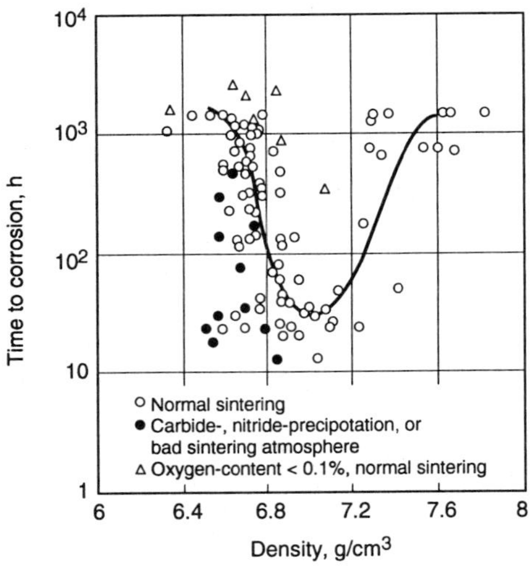

Fig. 45 Effect of sintered density upon corrosion resistance of sintered 316 type alloys. Source: Ref 1, 32

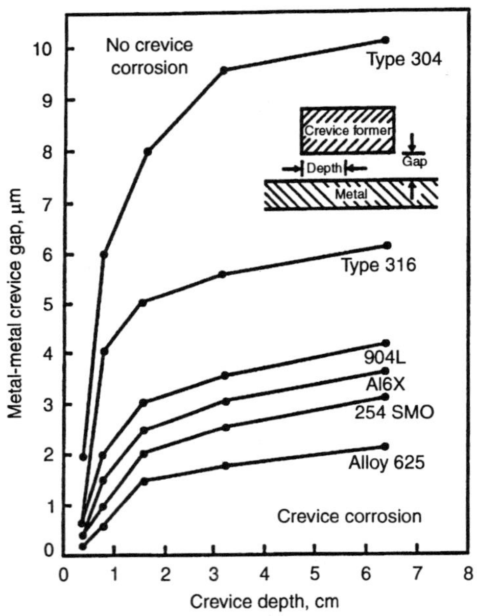

Fig. 46 Effect of crevice gap and depth on the initiation of crevice corrosion in various stainless steels and alloy 625. The gaps and depths below and to the right of the curve for each material define crevice geometries where initiation of crevice corrosion is predicted by the mathematical model of T.S. Lee and R.M. Kain, NACE Corrosion 83 Conference proceeding paper 69, 1983. Source: Ref 24

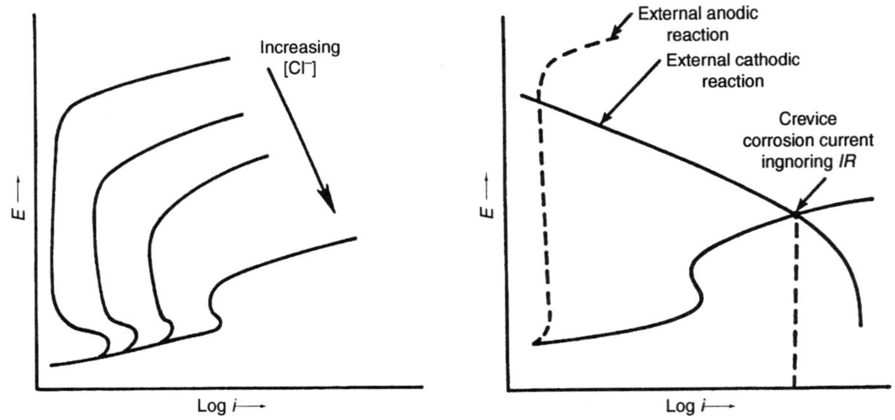

Fig. 47 Crevice-corrosion initiation resulting from lowering of E_p with increasing chloride concentration. Source: Ref 33

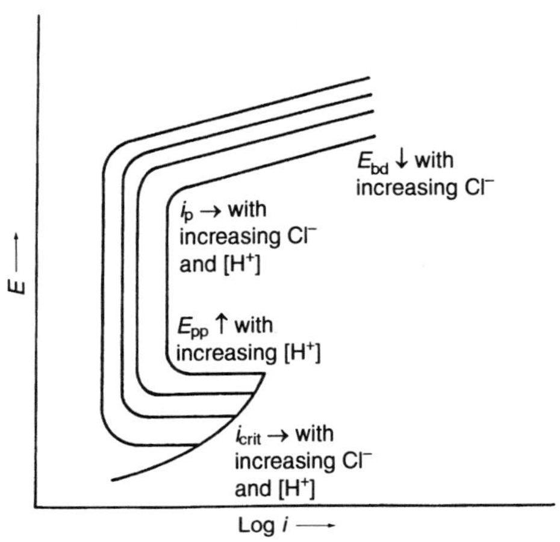

Fig. 48 Changes in anodic polarization behavior that occur as the environment in the crevice becomes increasingly aggressive. Source: Ref 33

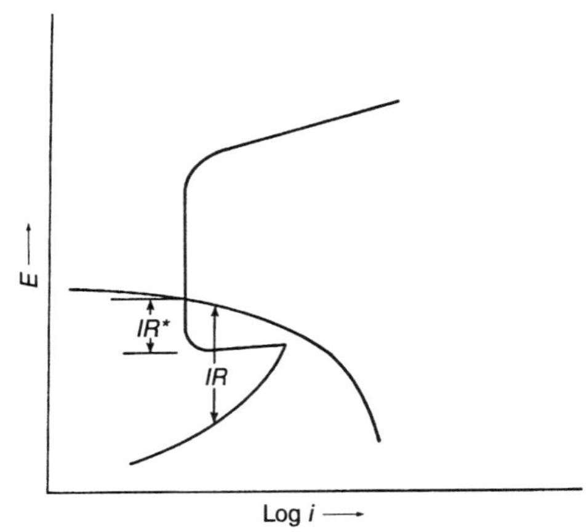

Fig. 49 Crevice-corrosion initiation by an *IR*-induced mechanism. When the *IR* drop down the crevice exceeds *IR**, crevice corrosion initiates. Source: Ref 33

crevice. At very high sintered densities, crevice-corrosion susceptibility decreases as the porous network is closed off with increasing alloy density. Data showing the percentage of open pores as a function of sintered density and the resulting time to first rust during salt spray exposure are presented in Table 20 (Ref 25).

In the corrosion literature for wrought alloys, it is well recognized that the aspect ratio of a crevice (the ratio of its width to its length) is a critical parameter in the establishment of crevice corrosion. Figure 46 illustrates that for narrow crevice gaps, crevice corrosion initiates at shallow depths, whereas, for wider crevice gaps, attack initiates deeper within the crevice (Ref 24). As a result, narrow and/or long crevices are likely initiation sites for crevice corrosion. In sintered stainless steels the inherent porous nature of the material provides a narrow tortuous electrolyte path that both encourages crevice-corrosion initiation and sustains its propagation.

Several changes occur in the occluded cell environment of a crevice during the initiation stages of crevice corrosion: oxygen depletion within the crevice that establishes a separation of anodic and cathodic sites (where the cathode is largely outside the crevice and the anode is inside the crevice), a lowering of the pH of the solution within the crevice by hydrolysis reactions, and migration of chloride into the crevice to maintain charge neutrality. For some of the less heavily alloyed stainless steels, the reduction in E_p resulting from the increased chloride concentration is enough to initiate crevice corrosion, as the schematic polarization curve and mixed potential analysis in Fig. 47 reveal (Ref 33). The crossover point in the mixed potential analysis indicates dissolution of the metal within the crevice via a pitting type of attack. Changes in the anodic polarization behavior for the stainless steel within the crevice, resulting from acidification and increase in chloride ion concentration of the crevice solution, are illustrated in Fig. 48 (Ref 33). As the aggressive nature of the crevice solution increases, i_{crit}, E_{pp}, and i_p increase, while E_p decreases. A mixed potential analysis of crevice-corrosion initiation (using the cathodic polarization behavior for the stainless steel in the environment outside of the crevice and the anodic polarization behavior for the stainless steel in the environment inside the crevice) is depicted in Fig. 49 (Ref 33). This analysis reveals that when the potential drop associated with the tortuous electrolyte path of a crevice exceeds a certain value, *IR** in this illustration, crevice corrosion is initiated by active dissolution of the metal in the crevice. While the mechanisms just described were proposed for wrought materials containing intentional or unintentional crevice formers, such as O-rings or gaskets, they are equally applicable to P/M materials containing inherent porosity. In fact, the extremely tortuous path provided by the pores, oxides, and particle boundaries in a sintered stainless steel provide what could be viewed as the ultimate geometry for establishing and, perhaps, even studying crevice corrosion. Clearly,

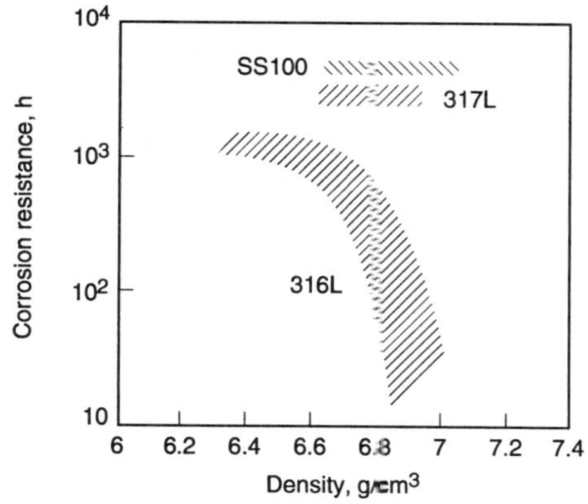

Fig. 50 Effect of density on corrosion resistance of sintered austenitic stainless steel. Source: Ref 32

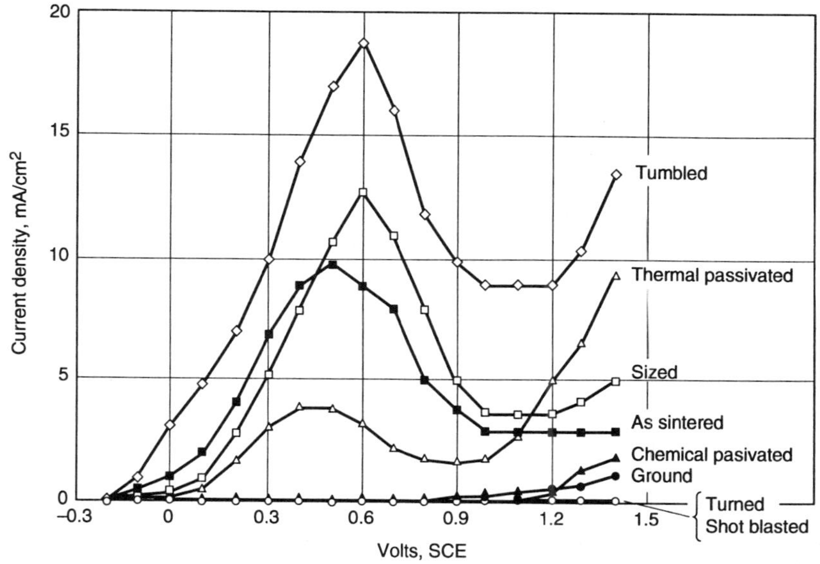

Fig. 51 Anodic potentiodynamic polarization curves for 316L stainless steel in 0.1 N NaCl/0.4 N $NaClO_4$ as a function of surface finishing. Source: Ref 3

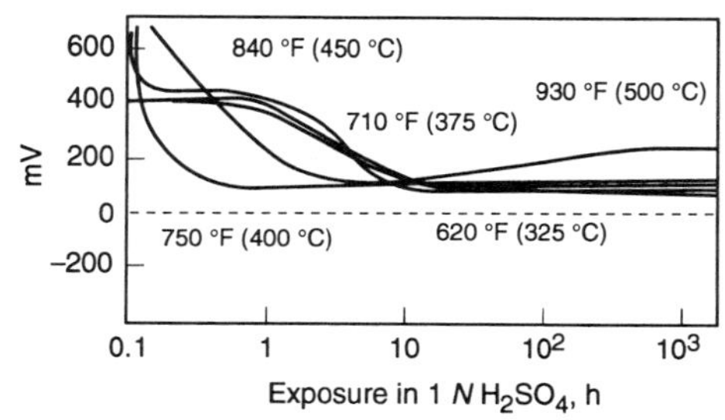

Fig. 52 Rest (open-circuit) potential measurements for sintered 316L thermally prepassivated at temperatures between 325 and 500 °C. Source: Ref 35

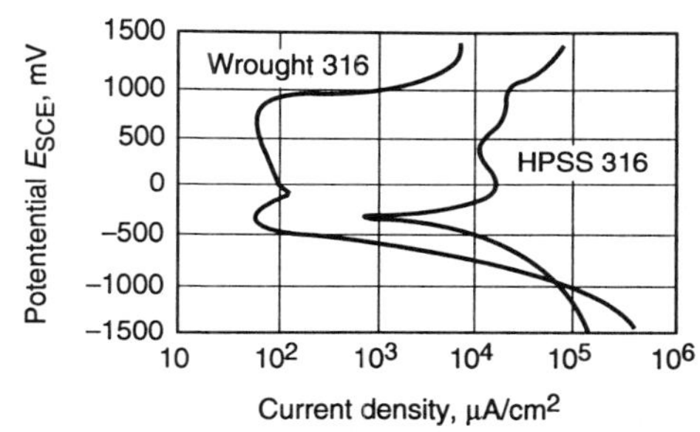

Fig. 53 Potentiodynamic polarization curves of samples tested in H_2SO_4, scanning rate 2 mV/S (1 N H_2SO_4). Source: Ref 29

by gaining a better understanding of the influence that processing parameters play in establishing pore morphologies susceptible to crevice corrosion, it will be possible to alter pore morphology to discourage crevice-corrosion initiation.

Another means of altering the susceptibility of sintered stainless steels to crevice attack is to alter alloy composition. By altering alloy composition the hydrolysis reactions responsible for lowering the pH within the crevice can be influenced, and these changes can be used to discourage crevice-corrosion initiation, as Fig. 50 reveals (Ref 32). The enhanced crevice-corrosion resistance of the 317L and SS100 alloys is attributed to their higher molybdenum concentrations. It should be noted that the addition of molybdenum does not make these alloys immune to crevice corrosion because it is possible to initiate crevice corrosion in neutral pH crevice environments without the presence of chloride when the aspect ratio of the crevice is severe enough; it simply makes initiation more difficult.

Approaches Used to Improve the Corrosion Resistance of Sintered Stainless Steels. A variety of means have been employed to improve the corrosion resistance of sintered stainless steels—some of which simply alter the number of open pores, others are aimed at both reducing the number of open pores and enhancing passivity of the alloy. These approaches include the following finishing processes: tumbling, grinding and shot blasting, passivating treatments, liquid phase sintering, double pressing and sintering, and the addition of alloying elements, such as copper, tin, and noble metals (Ref 34).

One group of researchers evaluated the influence of several finishing processes on the anodic polarization behavior of P/M 316L stainless steel in a 0.1 N NaCl/0.4 N $NaClO_4$ solution (ASTM B 627). As shown in Fig. 51, the investigation revealed the following results: tumbling likely smears the pores and is ineffective at improving corrosion resistance; coining/sizing introduces residual stresses in the surface of the alloy that may increase corrosion; grinding, turning, and shot blasting can seal surface pores and improve corrosion resistance; and thermal and chemical passivation processes can alter the thickness and/or composition of the passive film, thus enhancing corrosion resistance (Ref 3). In another investigation, thermal passivation in the temperature range of 325 to 500 °C for 30 min also showed promise for enhancing passivity, and thus corrosion resistance, of sintered 316L stainless steel exposed to a 1 N H_2SO_4 solution, as Fig. 52 reveals (Ref 35).

Closing of porosity through operations such as double pressing and double sintering (DPDS), while expensive, has been found to significantly reduce the amount of open porosity and yield anodic polarization behavior in aggressive acid solutions similar to that of wrought materials. Figure 53 shows anodic polarization data for wrought and hot pressed and hot sintered 316 stainless steel in 1 N H_2SO_4 (Ref 29). When a DPDS operation is used, a P/M 316 stainless steel alloy of nominal composition exhibits a degree of passivity almost identical to that of wrought 316 (Ref 36). Another, more thorough, means for reducing or eliminating open porosity in P/M materials is liquid phase sintering. Sintering additives such as boron, NiB, BN, and CrB have been found to be effective in producing dense microstructures with enhanced salt-spray corrosion resistance, as Table 22 reveals (Ref 14). Unfortunately, sensitized microstructures, as evidenced by the high I_r / I_a ratio listed in Table 22 are a by-product of liquid phase sintering (Ref 14). Careful development of liquid phase sintering additives for stainless steels will be needed in order to produce dense alloys that are not susceptible to intergranular corrosion. Injection molding also shows promise for producing dense P/M alloys with enhanced corrosion resistance. Figure 54 compares weight-loss values for injection-molded and wrought 14-4PH stainless steel as a function of exposure time in either full strength chlorine bleach or a 10% $FeCl_3$ solution (Ref 37). The injection-molded P/M compacts had densities of 96 to 97% of the wrought density and exhibited weight losses comparable to those of the wrought 17-4PH stainless steel.

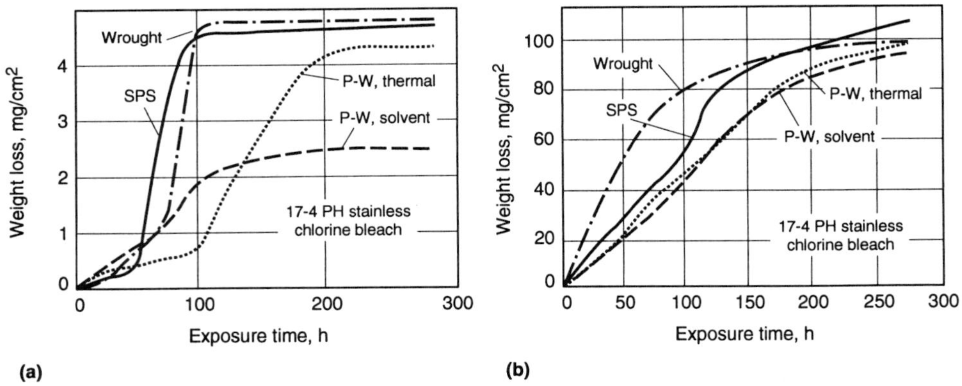

Fig. 54 Comparison of the corrosion weight loss versus exposure time in (a) ferric chloride and (b) chlorine bleach tests at room temperature. P-W, polymer wax binder, 69% paraffin wax, 20% polypropylene, 10% carnuba wax, 1% stearic acid; SPS, 73% acetanilide, 18% polystyrene, and 9% stearic acid. Source: Ref 37

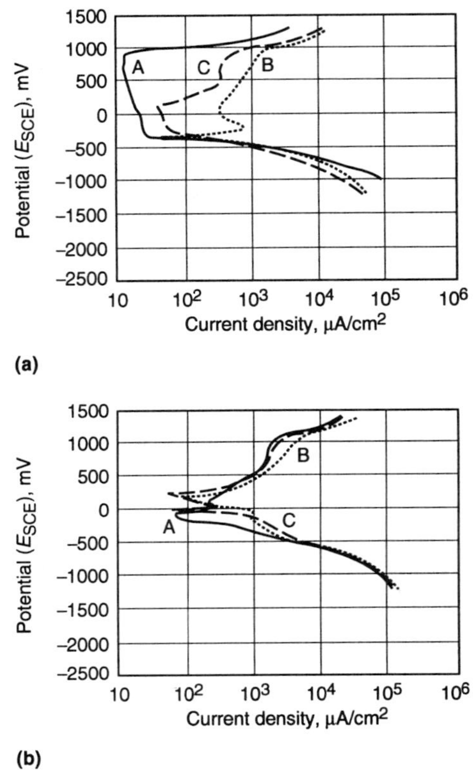

Fig. 55 Potentiodynamic polarization curves of as-received samples tested in 1 *N* H_2SO_4, scanning rate 1 mV/s. (a) A, wrought plate stainless steel of type 316; B, hot-pressed sintered stainless steel; C, hot-pressed-and-sintered stainless steel containing 0.25% Cu. (b) A, Hot-pressed-and-sintered stainless steel containing 1 wt% Cu; B, hot-pressed-and-sintered stainless steel containing 3 wt%; C, hot-pressed-and-sintered stainless steel containing 5 wt% Cu. Source: Ref 38

Table 22 Corrosion properties of liquid phase sintered 316L steels with the addition of boron-base sintering additives

All stainless steels were sintered at 1250 °C for 60 to 120 min in pure hydrogen.

Additive	Density, g/cm^3	Open pores, %	E_{stp}(a), mV SCE	NSS1, h	NSS2	I_r/I_a $\times 10^3$
None	6.86	8.2	150	96	7	0.0
0.2% B (–38 μm)	7.83	0.1	500	>1500	10	4.4
1% BN (–63 μm)	7.61	0.2	400	762	9	2.5
1% NiB (–38 μm)	7.67	0.1	525	>1500	10	3.8
1% CrB (–38 μm)	7.64	0.1	550	>1500	10	2.5

(a) 0.1% Cl^-, pH 5, 30 °C, 25 mV/8 h. Source: Ref 14

A number of alloying additions or infiltrants have been explored as means for enhancing passivity in P/M stainless steels. Among the most popular are copper and tin. Infiltration of P/M 316L and 304L alloys with either copper or bronze have been observed to increase corrosion resistance in boiling and room-temperature acid solutions (see Fig. 55). As this figure reveals, similar results were noted for the bronze and copper in the room-temperature acid, while the bronze infiltrant gave the best results in the boiling acid. Enhanced corrosion resistance was attributed to the elimination of open-connected porosity and the effect of the alloying. It should be noted that in both cases the infiltrants diffused either partially (bronze at all infiltration percentages and copper at the 6% infiltration level) or totally (copper at 4 and 6% infiltration levels) into the matrix. Table 1 illustrates the beneficial influence of copper alloying additions on the corrosion resistance of 316L stainless steel. Additional data supporting the beneficial influence of copper alloying additions on hot-pressed-and-sintered 316 stainless steel are provided in Fig. 38 (Ref 35). The improved passivity of the copper-containing alloys is attributed to its depolarizing influence on the oxygen reduction reaction. Depolarization of the cathodic, oxygen-reduction reaction enables the alloys to passivate more easily. Figure 56 shows the benefits of tin additions to the polarization behavior of 316 stainless steel (Ref 39). The addition of tin decreases the passive and critical current densities of the hot-pressed specimens by two orders of magnitude. Improvements in corrosion resistance, such as the ones just described, led to the development of copper- and tin-modified stainless steel powder chemistries, which are currently being marketed under the trademarks of Ultra 303L, Ultra 304, and Ultra 316. The enhanced corrosion resistance of these alloys, when compared to those of nonmodified 303L, 304L, and 316L, is evident in Fig. 55 and 56. When both copper and tin are prealloyed into the powder, the alloys are marketed with an "LSC" designation. The improved corrosion resistance for 316LSC in comparison to that of nonmodified 316L is presented in Fig. 57 (Ref 23).

As this section has revealed, a number of processing variables exert a strong influence on the corrosion resistance of sintered stainless steels. A listing of the most important variables and their impact on corrosion performance is presented in Table 23 (Ref 23).

P/M Superalloys

Development of P/M superalloys began in the 1960s with the search by the aerospace industry (and later the electric power industry) for stronger high-temperature alloys in order to operate engines at higher temperatures and thus improve fuel efficiency.

Initially, lower production costs were a major objective in exploring the P/M approach. Later, specific advantages linked to the P/M approach, such as the use of more complex and greater volume fractions of dispersoids, reduced segregation, improved workability, and reducing the cost of consolidating superalloy powders, particularly of oxide-dispersion-strengthened (ODS) superalloys, through the development of suitable forging techniques (Ref 40). Efforts are also underway to exploit the advantages of microcrystallinity and extended solid solutions of rapid solidification technology.

Uses and State of Commercialization. Table 24 summarizes the uses of P/M superalloys in terms of components, engine use, and reasons for using P/M technology. Other uses of superalloys include nuclear reactors, heat exchangers, furnaces, sour gas well equipment and other high-temperature applications.

Manufacturing of P/M Superalloys. An important prerequisite for making P/M superalloys that possess reliable dynamic properties is the use of clean powders. Years of intensive work were spent in identifying and controlling the problems related to unclean powders. Today, argon and vacuum (also known as soluble gas process) atomization, as well as atomization by the rotating electrode process, are known to be suitable for producing powders with the required low oxygen content and low degree of contami-

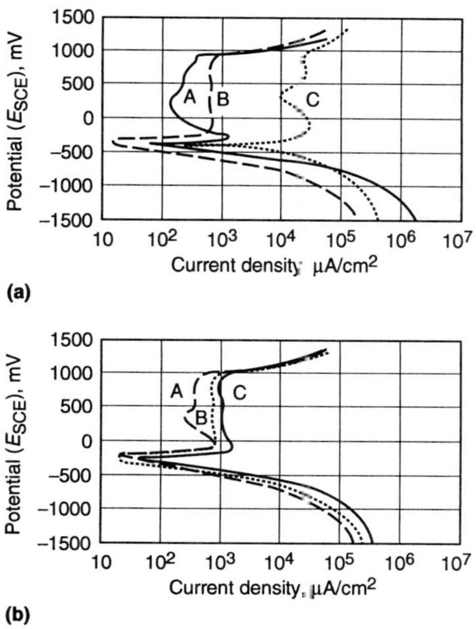

Fig. 56 Potentiodynamic polarization curves of samples tested in 5 N H_2SO_4, scanning rate 2 mV/s. (a) A, Wrought plate stainless steel of type 316; B, 2 wt% Sn; C, HPHS 316. (b) Hot-pressed-and-sintered sample containing A, 3 wt% Sn; B, 2 wt% Sn; C, 0.5 wt% Sn. Source: Ref 39

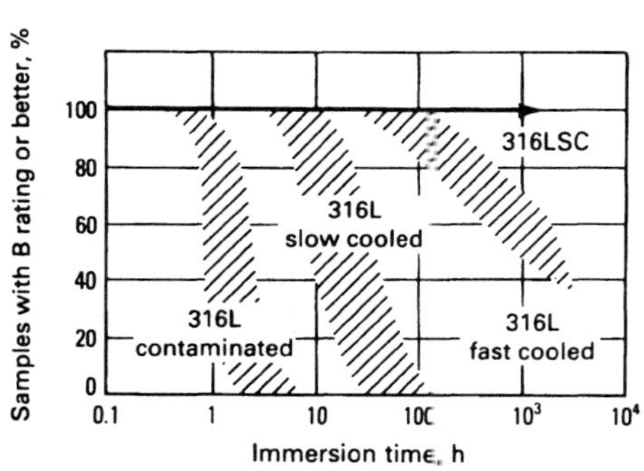

Fig. 57 Typical corrosion behavior of regular and copper-tin modified (type 316LSC) sintered type 316L stainless steel sintered in dissociated NH_3 under various conditions of cooling and contamination. B rating indicates that <1% of the specimen surface is covered by stain. Testing of the modified stainless steel was terminated after 1500 h. Source: Ref 23

nation. The so-called prior particle boundary (PPB) problem, that is, the presence of carbides segregated at PPBs, was solved through the development of low-carbon alloys. Special equipment is used for removing ceramic particles and particles containing entrapped argon. Some of these problems are minimized or avoided in ODS alloys made by mechanical alloying. In mechanical alloying, elemental and master alloy powders as well as refractory compounds are mechanically alloyed by high-energy milling (Ref 42, 43).

Two established powder consolidation techniques for P/M superalloys are hot isostatic pressing (HIP) and isothermal forging. Both P/M methods permit the manufacture of so-called near-net shape parts with attendant improved material use and reduced machining

Table 23 Effect of iron, carbon, nitrogen, and oxygen on corrosion resistance of sintered austenitic stainless steels

Variable	Origin of problem	Effect on corrosion resistance	Suggested solutions
Iron	Contamination of prealloyed powder with iron or iron-base powder at powder or parts producer's facility	Lowering of corrosion resistance by more than 99% due to galvanic corrosion	Utmost cleanliness at both powder and parts producer's manufacturing facilities, preferably separate and dedicated equipment and facilities
Carbon	Inadequate lubricant removal; carburizing sintering atmosphere soot in sintering furnace; high-carbon powder	Inferior resistance to intergranular corrosion	Use L-grade designation of stainless steel powder. Ensure adequate lubricant removal (before sintering). Use clean soot-free sintering furnace and carbon-free sintering atmospheres; carbon content of sintered part should be ≤0.03%.
Nitrogen	Sintering in dissociated NH_3 or other nitrogen-containing atmosphere combined with slow cooling	Inferior resistance to intergranular corrosion	Reduce percentage of nitrogen in sintering atmosphere. Use fast cooling of parts preferably >150 to 200 °C/min (>270 to 360 °F/min) through critical temperature range (700 to 1000 °C, or 1290 to 1830 °F). Use higher sintering temperature. Use intermediate dew points (–37 to –45 °C, or –35 to –50 °F) in cooling zone of furnace. Use tin-modified stainless steel powders.
Oxygen	Excessive oxygen in powder; excessive dew point of sintering atmosphere; slow cooling after sintering	Inferior resistance to general corrosion	Use low oxygen content powder, preferably <2000 ppm. Control dew point within sintering furnace to ensure reducing conditions. Fast cooling, preferably >200 °C/min (360 °F/min) For nitrogen-containing atmospheres, use dew point of –37 to –45 °C (–35 to –50 °F) in cooling zone. For sintering in H_2, ensure that water vapor content of atmosphere is below 50 ppm.
Density of sintered part	High sintered density	Inferior resistance to crevice corrosion	Use lower density to increase pore size and circulation of corrodent. In acidic environments, corrosion resistance improves with increasing density due to a decrease of specific surface area.

Source: Ref 23

Table 24 Aerospace applications of P/M superalloys

P/M superalloy	Component	Engine	Aircraft/ manufacturer	Reason for using P/M technology: Cost reduction	Improved properties
IN-100	Turbine disks, seals, spacers	F-100	Pratt & Whitney	X	X
René 95	Turbine disks, cooling plate	T-700	Helicopter/G.E	...	...
	Turbine disks, compressor shaft	F-404	F-18 Fighter	X	...
	Vane	F-404	G.E.	...	...
	High-pressure turbine blade retainer, disks, forward outer seals	F-101	...	X	...
Astroloy	High-pressure turbine disks	JTSD-17R Turbofan	...	X	...
Merl 76	Turbine disks	Turbofan	...	X	X
Inconel MA-754	Turbine nozzle vane	F-404	F-18 Fighter	...	X
	High- and low-pressure turbine vanes	Selected engines	...	...	X
Stellite 31	Turbine blade dampers	TF 30-P100	USAF F-111F	X	...
Inconel MA-6000E	Turbine blades	TFE 731	...	...	X

Source: Ref 41

Table 25 Nominal compositions of several P/M superalloys

Alloy	Composition, %															
	C	Cr	Mo	W	Ta	Ti	Nb	Co	Al	Hf	Zr	B	Ni	Fe	V	Y_2O_3
IN-100	0.07	12.5	3.2	...	...	4.3	...	18.5	5.0	...	0.04	0.02	rem	...	0.75	...
René 95	0.07	13.0	3.5	3.5	...	2.5	3.5	8.0	3.5	...	0.05	0.01	rem	...	...	...
MERL 76	0.02	12.4	3.2	...	...	4.3	1.4	18.5	5.0	0.4	0.06	0.02	rem	...	...	...
AF 115	0.05	10.5	2.8	6.0	...	3.9	1.7	15.0	3.8	2.0	...	...	rem	...	...	...
PA101	0.1	12.5	...	4.0	4.0	4.0	...	9.0	3.5	1.0	...	...	rem	...	...	...
Low-carbon Astroloy	0.04	15.0	5.0	...	...	3.5	...	17.0	4.0	...	0.4	0.025	rem	...	...	...
MA 754	0.05	20.0	...	...	...	0.5	...	...	0.3	...	...	...	rem	...	...	0.6
MA 956	...	20.0	...	...	...	0.5	...	...	4.5	...	...	...	...	rem	...	0.5
MA 6000	0.05	15.0	2.0	4.0	2.0	2.5	...	...	4.5	...	0.15	0.01	rem	...	...	1.1
Stellite 31	0.5	25.5	...	7.5	...	...	...	rem	...	...	...	...	10.5	2.0	...	...

Table 26 Relative increase in fatigue crack growth rates after 15 min at 650 °C (1200 °F)

$\Delta K = 30\ MPa\sqrt{m}$ ($27\ ksi\sqrt{in.}$)

Alloy	Relative increase in crack growth rates	Grain size, μm	Chromium content, %
René 95 after HIP + forge	242	50–70	12.8
IN-100	43.4	4–6	12.0
HIP-consolidated MERL 76	41.5	15–20	12.0
NASA II B-7	335	4–6	8.9
HIP-consolidated Astroloy	3.3	50–70	15.1
Waspaloy	4.0	40–150	19.3
Astroloy after HIP + forge	3.5	50–100	14.7

Source: Ref 46

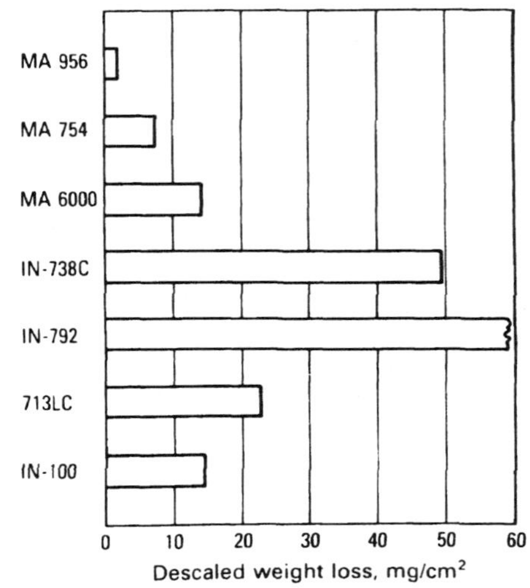

Fig. 59 Comparison of the oxidation resistance of ODS alloys MA 956, MA 754, and MA 6000 with that of other superalloys. Testing conditions: 504 h at 1100 °C (2010 °F) in air containing 5% H_2O. Temperature was cycled between test temperature and room temperature every 24 h. Source: Ref 46

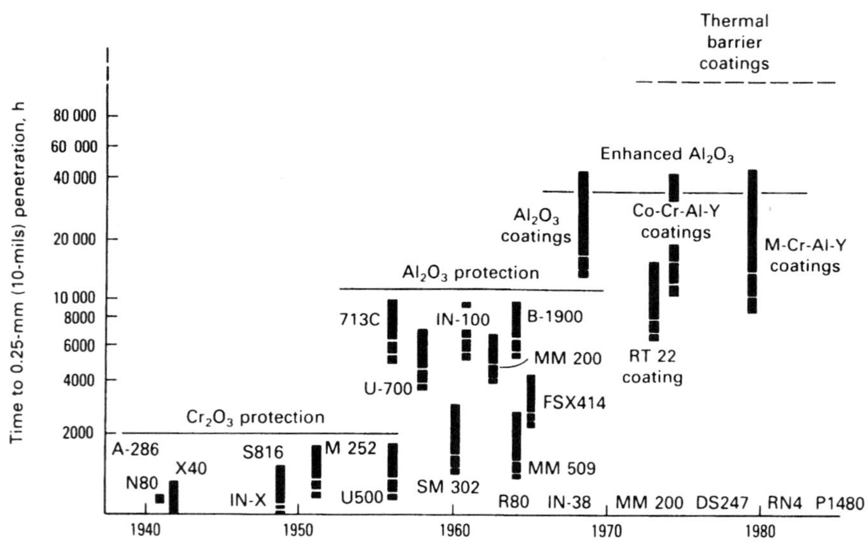

Fig. 58 Advancing steps in the protection of superalloys against oxidation of high temperatures showing life (in h) to 0.25 mm (10 mils) penetration of 980 °C (1800 °F). Source: Ref 45

costs. Powder metallurgy forging exploits the improved forgeability deriving from the higher incipient melting temperature and reduced grain size of P/M material. Hot compaction by extrusion leads to very fine grain size, improved hot ductility, and superplasticity.

Depending on the application of a superalloy part, the powder consolidation process can be controlled in order to yield either a fine or a coarse grain size. Fine grain size is preferred for intermediate temperatures (up to about 700 °C, or 1290 °F) because of its higher strength and ductility at these temperatures. For high-temperature blade and vane applications, however, a large grain size (ASTM 1 to 2) provides superior creep strength due to reduced grain-boundary sliding. Grain coarsening of ODS alloys is achieved through special heat treatments after consolidation (Ref 40).

Compositions and Properties. Table 25 shows the compositions of the best known P/M superalloys. Many have the same compositions as cast alloys but are manufactured similarly to wrought alloys. The important P/M superalloys—IN-100, René 95, and Astroloy—were adapted in the P/M process by reducing their carbon content and by adding stable carbide formers to eliminate the problem of PPB carbides. To facilitate HIP, alloy compositions were modified to increase the temperature gap between the γ' solvus (above which HIP has to be carried out for increasing grain size) and the solidus temperature.

Oxidation. Nickel-, cobalt-, and iron-base superalloys use the selective oxidation of aluminum or chromium to develop oxidation resistance (Ref 44). These alloys are therefore often referred to as Al_2O_3 or Cr_2O_3 formers depending on the composition of the oxide scale that provides protection. Alloy composition, surface conditions, gas environment, and cracking of the oxide scale affect the selective-oxidation process (Ref 44). Figure 58 shows the development of superalloys in terms of the progress achieved against high-temperature oxidation.

Cyclic oxidation causes protective scales of Al_2O_3 and Cr_2O_3 to crack and spall. Regeneration of the scales will eventually result in the complete depletion of chromium and aluminum. The length of time for which superalloys are Al_2O_3 or Cr_2O_3 formers under given conditions is very important because of the subsequent appearance of less protective oxides. The impor-

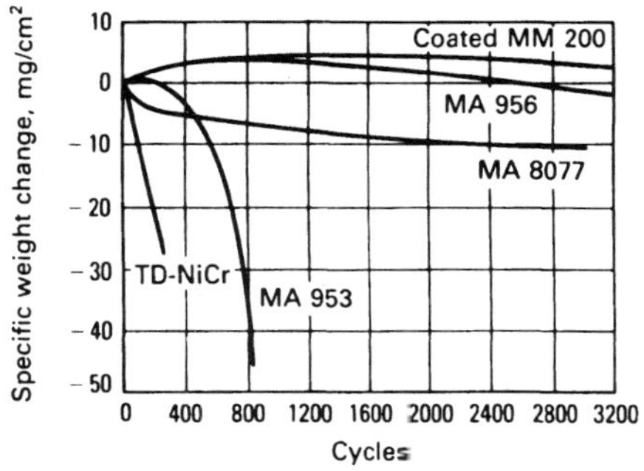

Fig. 60 Cyclic oxidation of ODS alloys MA 956, MA 8077, MA 953, and TD-NiCr compared to that of coated alloy MM 200. Testing conditions: held at 1100 °C (2010 °F) for 1 h and cooled by a 3 min air blast. Source: Ref 46

tance of chromium for imparting oxidation resistance is demonstrated in Table 26, which lists fatigue crack growth rates for different alloys.

Figures 59 and 60 show comparisons of the oxidation resistances of superalloys with and without oxide dispersions. Many studies have confirmed the beneficial effect of dispersed oxides on oxidation resistance. The lower oxidation rates of ODS alloys have been attributed to the reduced time required to form a continuous Cr_2O_3 scale due to the presence of dispersed oxides, which act as nuclei for oxidation (Ref 47). Based on marker studies with platinum, one investigation attributed the beneficial effect of oxide dispersions to the predominant, inward diffusion of oxygen ion (O^{2-}) and a slowdown of the chromic ion (Cr^{3+}) diffusion (Ref 48). The latter may be caused by the blocking of the dispersions in the Cr_2O_3 scale. With the dispersed oxides becoming dissolved in the scale, it also appears possible that trivalent ions, such as yttrium (Y^{3+}) and lanthanum (L^{3+}), will reduce the number of vacant cation sites, thus lowering the diffusivity of Cr^{3+}.

Dispersed oxides may also improve scale adhesion because of the thinner scale or because of increased porosity or smaller grain size in the oxide scale (Ref 46). It was reported that at 1300 °C (2370 °'F) the outer regions of the Al_2O_3 film of MA 956 (Fe-20Cr-4.5Al-0.5Ti-0.5Y_2O_3) became enriched with titanium, giving rise to a continuous layer of titanium-rich oxide (Ref 49). Pegging of the oxide by titanium carbide particles and the irregular metal/oxide interface is said to contribute to the good spalling resistance of the alloy.

In oxidation tests in air and in an inert atmosphere at 1260 °C (2300 °F) for MA 754 (Ni-20Cr-0.5Ti-0.5Y_2O_3-0.3Al-0.05C), Ni-20Cr (cast/wrought), and an ODS nickel-chromium alloy, subsurface porosity was attributed to the oxidation of chromium and aluminum (Kirkendall porosity), and thermally induced porosity was excluded as a cause (Ref 50). This type of porosity decreases with improving oxidation resistance of the alloy.

Results of cyclic oxidation tests at 1100 °C (2010 °F) for MA 956, TD-NiCr, and Hastelloy X are given in Table 27. The superior resistance of MA 956 is attributed to a very stable Al_2O_3 film and parabolic oxidation for more than 500 h. Tables 28 and 29 list sulfidation and carburization resistance data for MA 956. As in the case of oxidation, the alloy shows marked superiority to the other alloys tested. Tables 30 and 31 provide similar data for alloy MA 6000E (Ni-15Cr-4.5Al-4W-2Mo-2Ta-2.5Ti-1.1Y_2O_3) (Ref 51). The functions of the various alloying elements are as follows:

- Aluminum, titanium, and tantalum for γ' hardening
- Y_2O_3 for high-temperature strength and stability
- Aluminum and chromium for oxidation resistance
- Titanium, tantalum, chromium, and tungsten for sulfidation resistance
- Tungsten and molybdenum for solid-solution strengthening

Hot Corrosion. The requirement of hot corrosion resistance of superalloys derives from the use of sodium- and sulfur-containing fuels and the presence of salt in the air necessary for combustion. Under such conditions, combustion gases often leave deposits of sulfates or chlorides with metallic constituents of sodium, calcium, magnesium, or potassium on the surfaces of superalloys. The resulting corrosion problems are particularly severe if these condensed phases are liquid. Hot corrosion may occur in gas turbines, boiler tubes, and incinerators. Typically (and similar to what happens in pure oxidation), hot corrosion of superalloys occurs in two stages: a slow rate initiation stage, followed by a propagation stage of rapid degradation. The difference, compared to oxidation, is that the conditions causing hot corrosion simply shorten the time in which superalloys form protective Al_2O_3 or Cr_2O_3 scales by selective oxidation (Ref 44). Factors affecting the length of the initiation stage (at the end of the initiation stage, the superalloy must be removed from service because of the start of excessive corrosion) include alloy composition, alloy fabrication conditions, gas composition and velocity, deposit composition and its physical state, amount of deposit, temperature, temperature cycles, erosion, and specimen geometry.

When a protective scale dissolves into a liquid deposit, so-called fluxing reactions can occur with the appearance of other basic or acidic nonprotective reaction products. Propagation may also be caused by components from the deposit that can accumulate in the deposit or the alloy and thus cause a nonprotective scale to form. Chlorine and sulfur produce such effects, and hot corrosion caused by the latter is known as

Table 27 Cyclic oxidation resistance of superalloys at 1100 °C (2010 °F)

504 h test in atmosphere of air containing 5% H_2O; temperature cycled between 1100 °C (2010 °F) and room temperature every 24 h

Alloy	Weight change, mg/cm² Undescaled	Weight change, mg/cm² Descaled	Metal loss, μm	Maximum attack, μm
MA 956	0.99	−1.57	2	15
TD-NiCr	−4.66	−12.52	20	33
Hastelloy X	−11.81	−20.62	50	256

Source: Ref 51

Table 28 Sulfidation resistance of superalloys at 925 °C (1700 °F)

312 h test in burner rig with air-to-fuel ratio of 30 to 1; fuel contained 0.3% S and 5 ppm seawater. Temperature cycle: 58 min at temperature, followed by 2 min cool to room temperature

Alloy	Weight change, mg/cm² Undescaled	Weight change, mg/cm² Descaled	Metal loss, μm	Maximum attack, μm
MA 956	1.04	−0.17	5	18
TD-NiCr	−1.69	−11.57	25	129
Hastelloy X	−2.25	−6.83	33	132

Source: Ref 51

Table 29 Carburization resistance of superalloys at 1095 °C (2000 °F)

100 h test in atmosphere of hydrogen containing 2% methane

Alloy	Weight change, mg/cm² Undescaled	Weight change, mg/cm² Descaled	Metal loss, μm	Maximum attack, μm
MA 956	0.07	−0.42	10	10
Incoloy 800	33.74	29.89	132	7615
Alloy 814	0.82	−0.73	13	363

Source: Ref 51

Table 30 Sulfidation resistance of superalloys at 925 °C (1700 °F)

Tested in burner rig with air-to-fuel ratio of 30 to 1; fuel contained 0.3% S and 5 ppm seawater. Temperature cycle 58 min at temperature, followed by 2 min cool to room temperature

Alloy	Exposure time, h	Descaled weight loss, mg/cm²	Maximum attack, μm
MA 6000E	312	−11.11	24
IN-100	48	−367.36	169
Alloy 713LC	168	−488.63	328
IN-738C	312	−9.73	28

Source: Ref 51

Table 31 Cyclic oxidation resistance of superalloys at 1100 °C (2010 °F)

504 h test in air containing 5% H_2O; temperature cycled from 1100 °C (2010 °F) to room temperature every 24 h

Alloy	Descaled weight change, mg/cm²
MA 6000E	−14.12
IN-100	−7.27
Alloy 713LC	−22.08
IN-738C	−49.51

Source: Ref 51

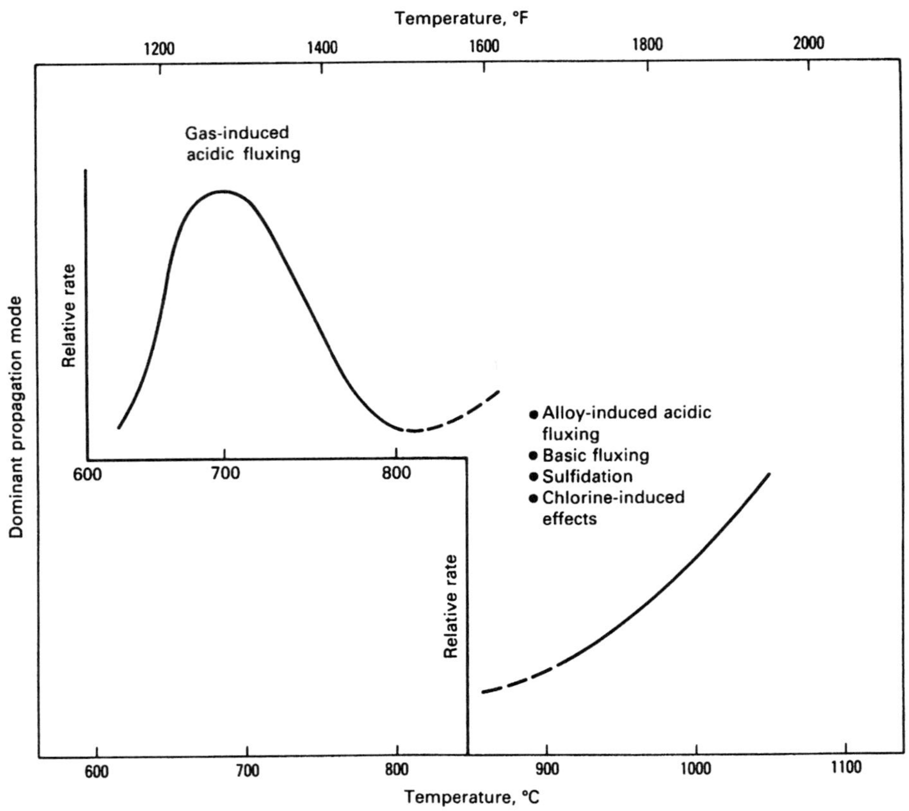

Fig. 61 Schematic showing the temperature regimes over which different propagation modes are most prevalent.

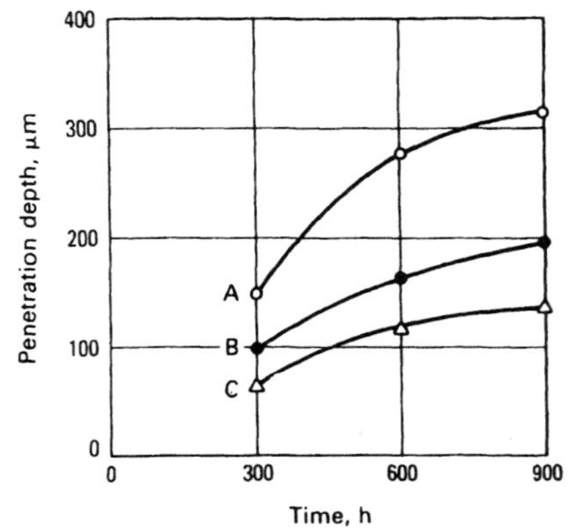

Fig. 62 Comparison resistance of alloy IN-738LC in hot (850 °C, or 1560 °F) gases. A, IN-738LC; B, IN-738LC with Y_2O_3 dispersion, annealed at 1270 °C (2320 °F); C, IN-738LC with Y_2O_3 dispersion, annealed at 1100 °C (2010 °F). Source: Ref 40

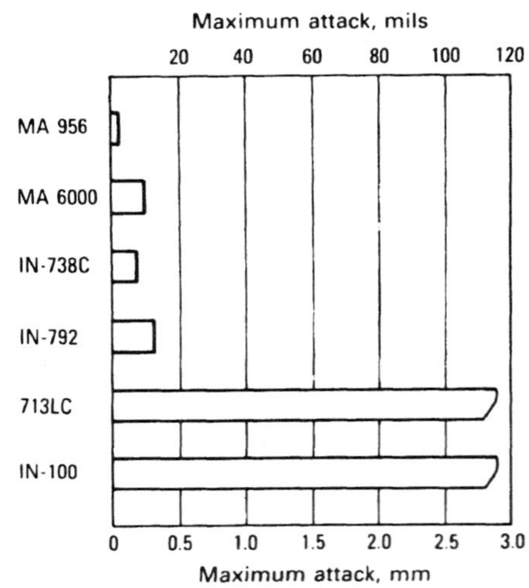

Fig. 63 Comparison of the corrosion resistance of MA 6000 and MA 956 with that of other superalloys. Tested in a burner rig for 312 h using a 30-to-1 air-to-fuel ratio. Fuel contained 0.3% S and 5 ppm seawater, and specimens were held at temperature for 58 min of each hour, then cooled 2 min with an air blast. Source: Ref 46

Table 32 Dependence on carbide microstructure of creep crack growth rates of alloy IN-750 in four environments

Grain-boundary carbide microstructure	Crack growth rate (da/dt), mm/min(a)			
	Helium	Air	Helium + 4% methane	Helium + 3% SO_2
Blocky	7×10^{-4}	1.95×10^{-2}	7×10^{-4}	1.5×10^{-1}
Cellular	1.15×10^{-3}	2.6×10^{-3}	3.1×10^{-3}	1.3×10^{-1}
None	7×10^{-4}	6.3×10^{-2}	7×10^{-4}	1.6×10^{-1}

(a) $K = 35$ MPa$\sqrt{m}$ (32 ksi$\sqrt{in.}$). Source: Ref 46, 55

sulfidation. Figure 61 shows the temperature ranges over which the various hot corrosion propagation modes are important.

Some superalloys corrode in several modes. For example. hot corrosion of IN-738 proceeds by alloy-induced acidic fluxing, but is preceded by other propagation modes, including a basic fluxing mode. The higher chromium content alloys IN-738 and IN-939 were developed to improve the hot corrosion resistance of land-based gas turbines. Carbide stabilization through tungsten and tantalum and delay of $M_{23}C_6$ formation in service were expected to allow the large chromium content to impart improved hot corrosion resistance. Increasing the chromium and decreasing the Al_2O_3, however, lowered γ' solution temperatures and strength, which necessitated the use of coatings. The use of coatings led to the current use of enhanced aluminum, that is, carefully balanced coating alloys (based on nickel, iron, or cobalt with chromium, aluminum, and other active elements). Generally, all superalloy load-bearing parts used at very high temperatures under dynamic conditions are coated (Ref 45). Nevertheless, coatings generally last longer on more corrosion-resistant base materials.

In a model study, IN-738 was used to demonstrate the effect of grain size and Y_2O_3 dispersions on hot corrosion behavior (Ref 52). Under gas turbine simulated hot gas corrosion test conditions at 850 and 950 °C (1560 and 1740 °F) (Fig. 62), the presence of a Y_2O_3 dispersion lowered the corrosion rate. At 950 °C (1740 °F), a finer grain size further reduced the corrosion rate, which was thought to be mainly due to a higher diffusion rate of chromium and aluminum. The effect of the dispersion was predominant at 850 °C (1560 °F). Reduced sulfate formation at 850 °C (1560 °F) was attributed to the likely formation of yttrium oxysulfide.

In a study of the oxidation and hot corrosion resistance of P/M LC Astroloy and IN-100, isostatically pressed samples were found to be moderately attacked in a sulfate-chloride environment and heavily corroded by pure sodium sulfate (Na_2SO_4) (Ref 53). Heat treatment and the use of coarse powder (62 to 150 µm for Astroloy and 88 to 200 µm for IN-100) lowered the susceptibility to catastrophic corrosion. Additions of yttrium to IN-100 improved the corrosion resistance in pure sulfate, but were detrimental when NaCl was present. Therefore, yttrium additions to IN-100 cannot be recommended for marine turbines. It was concluded that in many cases impregnation coatings must be considered for components made of IN-100 alloys.

As a part of an evaluation of improved alloys for use in oil and gas drilling at depths of 6100 m (20,000 ft), HIP nickel-base alloy Inconel 625

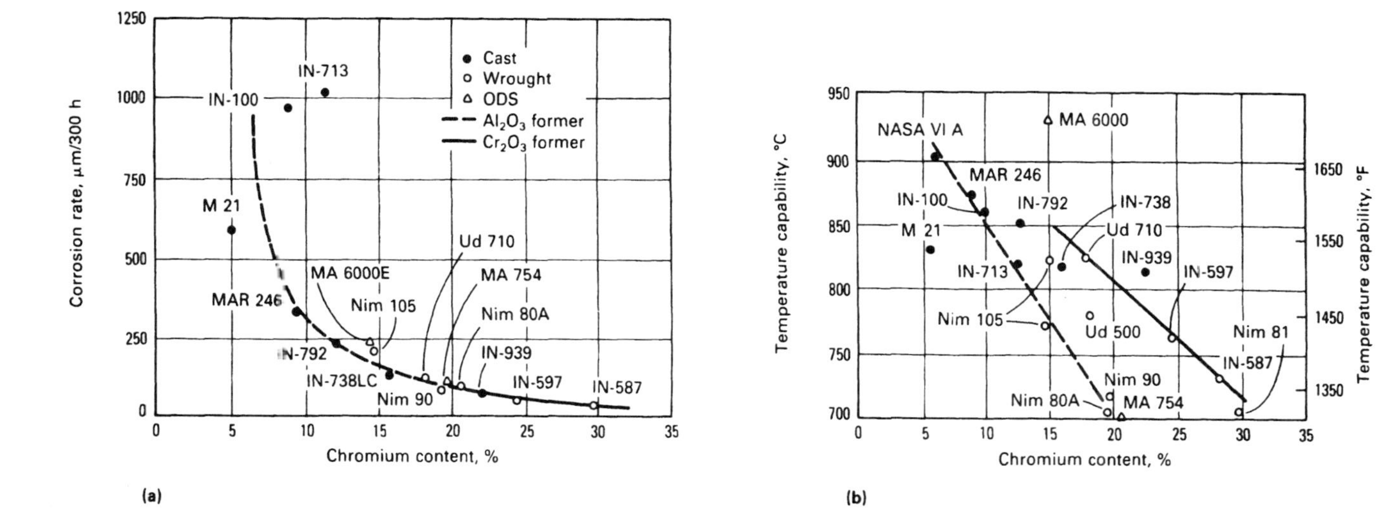

Fig. 64 Corrosion rate (a) and temperature capability (b) of MA 6000, MA 754, and non-ODS superalloys as a function of chromium content. A stress of MPa (29 ksi) was applied to the specimens during the 10,000 h test. Source: Ref 46

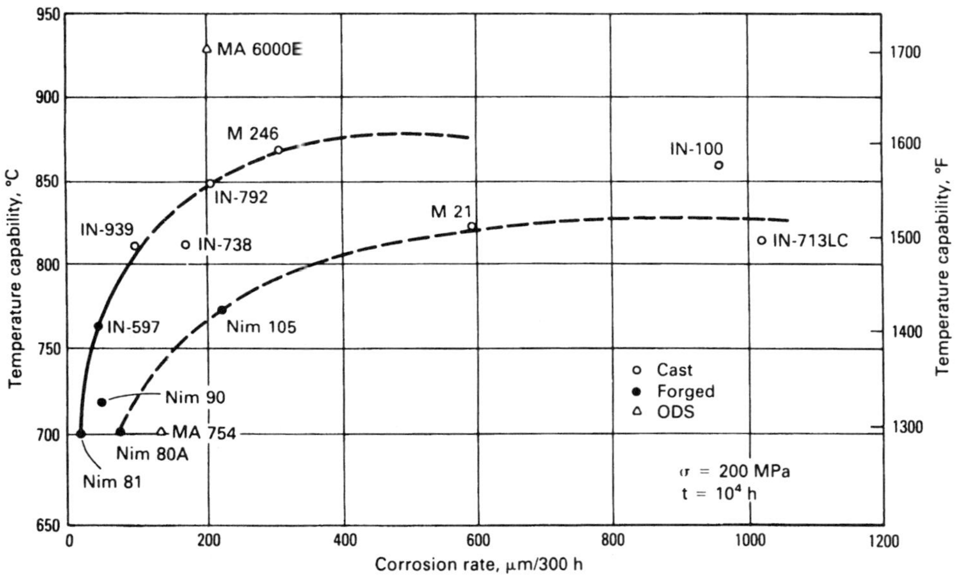

Fig. 65 Temperature capability as a function of corrosion rate for various superalloys. Same data as in Fig. 69. Source: Ref 46

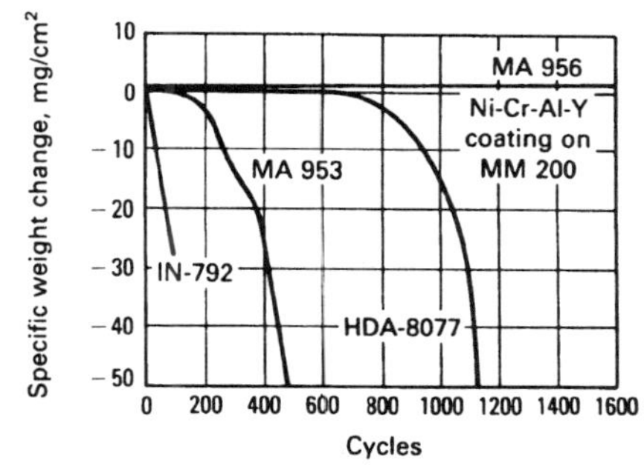

Fig. 66 Hot corrosion of alloys MA 953, HDA 8077, and MA 956 compared to that of some non-ODS alloys. Test conditions: 900 °C (1650 °F), 1 h, followed by a 3 min air blast, 5 ppm sea salt. Source: Ref 46

was studied in a simulated deep, hot, sour cell environment (Ref 54). The alloy demonstrated resistance to pitting and crevice corrosion, sulfide stress cracking, chloride stress-corrosion cracking (SCC), and elevated-temperature anodic stress cracking. Hot isostatically pressed Inconel 625 exhibited essentially the same corrosion resistance as wrought Inconel 625.

Fatigue and Creep Crack Growth. Fatigue crack growth rates of nickel-base superalloys measured at frequencies above 0.1 Hz, at intermediate temperatures, and at an intermediate stress intensity range, ΔK, were found to be several times higher than those measured in inert atmospheres (Ref 55). The buildup of corrosion products with decreasing ΔK, however, was thought to enhance crack closure, thus reducing the effective stress intensity range and leading to fatigue thresholds higher than those in inert environments.

Table 32 shows creep crack growth rates of IN-750 with various grain-boundary carbide microstructures (Ref 46, 55). Aggressive environments (helium + 3% SO_2 and air) produce order of magnitude increases over the rates in inert gas. In general, the reaction of both oxide dispersoid-free P/M superalloys and cast and wrought superalloys to aggressive environments is similar. This suggests that crack growth is governed mainly by microstructure and alloy chemistry.

Oxide-Dispersion-Strengthened Alloys. Mechanical alloying has removed many constraints on the development of new superalloys. Many new ODS alloys were designed specifically for corrosion resistance because alloying requirements for precipitation strengthening can be greatly reduced. Superior creep, corrosion, and erosion resistance at high temperatures have been claimed to enable the use of lower-grade fuels (Ref 56). Figures 63 to 66 show the corrosion resistances of ODS alloys MA 956, MA 6000, and MA 754 compared to several conventional superalloys.

Coatings for ODS Alloys. As mentioned above, for extended high-temperature service, superalloys require additional protection through coatings. The use of aluminide coatings appears to be unsatisfactory due to the development of subsurface Kirkendall porosity and early spalling of the protective scale. Kirkendall porosity decreases with increasing aluminum content of the substrate alloy as well as with decreasing grain size (Ref 56). Only limited information exists on the properties of Cr-Al-Y coatings (Ref 46) and on diffusion barrier coatings (Ref 57–60).

ACKNOWLEDGMENTS

As noted earlier, much of the data of Dr. Erhard Klar were used in this review and the author would like to acknowledge him and thank him for this contribution. The author would also like to thank Mr. Michael Baran, currently at Gamry Instruments, for the contribution of his work on ferritic stainless steels.

REFERENCES

1. E. Klar and P.K. Samal, *Powder Metals, Corrosion Tests and Standards Manual,* R. Baboian, Ed., ASTM, 1995, p 551–557
2. E. Klar and P.K. Samal, Effect of Density and Sintering Variables on the Corrosion Resistance of Austenitic Stainless Steels, *Advances in Powder Metallurgy and Particulate Materials,* Vol 3, M. Phillips and J. Porter, Ed., Metal Powder Industries Federation, American Powder Metallurgy Institute, 1995, p 3–17
3. C. Molins, J.A. Bas, J. Planas, and S.A. Ames, P/M Stainless Steel: Types and Their Characteristics and Applications, *Advances in Powder Metallurgy and Particulate Materials,* Vol 5, J.M. Capus, Ed., Plenum Press, 1992, p 345–357
4. J.A. Bas, J. Peñafiel, A. Bolarin, and M.P. Latre, Determination Methods of the PM Stainless Steels Corrosion Resistance, *Advances in Powder Metallurgy and Particulate Materials*, Plenum Press, p 47-59
5. D. Itzhak and E. Aghion, An Anodic Behavior Study of an Analogical Sintered System of Austenitic Stainless Steel in H_2SO_4 Solution, *Corros. Sci.,* Vol 24 (No. 2), 1984, p 145–152
6. G. Lei, R.M. German, and H.S. Nayar, Corrosion Control in Sintered Austenitic Stainless Steels, reprinted from *Progress in Powder Metallurgy,* Vol 39, Metal Powder Industries Federation, 1984, p 55–74
7. A. Tremblay and R. Angers, *Corrosion Resistance of 316L P/M Stainless Steel, Advances in Powder Metallurgy and Particulate Materials—1994,* Vol 7, C. Lall and A.J. Neupaver, Ed., Metal Powder Industries Federation, American Powder Metallurgy Institute, p 225–237
8. D.A. Jones, *Principles and Prevention of Corrosion,* 2nd ed., Prentice Hall, 1996
9. G. Scavino, E. Angelini, M. Rosso, and F. Rosalbino, Comparison of Corrosion Resistance Properties of PM Steels, *Computer Methods and Experimental Measurements for-Surface Treatment Effects,* Computational Mechanics Publications, 1993, p 337–346
10. M. Baran, A.E. Segall, B.A. Shaw, et al., "Evaluation of P/M Ferritic Stainless Steel Alloys for Automotive Exhaust Applications," presented at the P/M Tec '97 Conf. (Chicago, IL), Metal Powder Industries Federation, June 1997
11. D.W. Yuan, J.R. Spirko, and H.I. Sanderow, Colorimetric Corrosion Testing of P/M Stainless Steel, *Int. J. Powder Metall.,* Vol 33 (No. 2), 1997
12. D.W. Yuan, T. Prucher, and H.I. Sanderow, "An Evaluation of the Relative Corrosion Resistance of P/M Stainless Steel Alloys," Technical Paper No. 950391, Society of Automotive Engineers, 1995
13. R.L. Sands, G.F. Bidmead, and D.A. Oliver, The Corrosion Resistance of Sintered Austenitic Stainless Steel, *Modern Developments in Powder Metallurgy,* Vol 2, H.H. Hausner, Ed., Plenum Press, 1966, p 73–83
14. E. Maahn, S.K. Jensen, R.M. Larsen, and T. Mathiesen, Factors Affecting the Corrosion Resistance of Sintered Stainless Steel, *Advances in Powder Metallurgy and Particulate Materials,* Vol 7, C. Lall, Ed., Plenum Press, 1994, p 253–271
15. T. Raghu, S.N. Malhotra, and P. Ramakrishnan, Corrosion Behavior of Porous Sintered Type 316L Austenitic Stainless Steel in 3% NaCl Solution, *Corrosion,* Vol 45 (No. 9), 1989, p 698–704
16. D.H. Ro and E. Klar, Corrosion Behavior of P/M Austenitic Stainless Steels, *Modern Developments in Powder Metallurgy,* Vol 2, H.H. Hausner, Ed., Plenum Press, 1980, p 247–287
17. P. Peled and D. Itzhak, The Surface Composition of Sintered Stainless Steel Containing Novle Alloying Elements Exposed to a H_2SO_4 Environment, *Corros. Sci.,* Vol 32 (No. 1), 1991, p 83–90
18. M. Baran, "An Evaluation of a P/M Ferritic Stainless Steel Automotive Exhaust Flange," Master's dissertation, The Pennsylvania State University, 1997
19. Laboratory Testing, *Corrosion,* Vol 13, *ASM Handbook,* ASM International, 1987, p 212–228
20. A.J. Sedriks, Effects of Alloy Composition and Microstructure on the Passivity of Stainless Steels, *Corrosion,* Vol 42 (No. 7), 1986, p 376–388
21. L. Campbell, "Corrosion Behavior of an Alumina-Reinforced Aluminum Metal Matrix Composite," Baccalaureate thesis, The Pennsylvania State University, 1996
22. M.A. Streicher, *Corrosion Tests and Standards Manual,* R. Baboian, Ed., ASTM, 1995, p 197–217
23. E. Klar, Corrosion of Powder Metallurgy Materials, *Corrosion,* Vol 13, *Metals Handbook,* ASM International, p 824–845
24. A.J. Sedriks, *Corrosion of Stainless Steels,* 2nd ed., John Wiley & Sons, 1996
25. E. Maahn and T. Mathiesen, Corrosion Properties of Sintered Stainless Steel, *UK Corrosion Proc.* (Manchester, UK), Vol 2, Institute of Corrosion, 1991, p 1–15
26. K. Frisk, A. Johansson, and C. Lindberg, Nitrogen Pick Up During Sintering Stainless Steel, *Advances in Powder Metallurgy and Particulate Materials,* Vol 3, J.M. Capus and R.M. German, Ed., Metal Powder Industries Federation, 1992, p 167
27. B.A. Shaw and G.S. Morton, "Thermal Spray Coatings—Marine Performance and Mechanisms," ASM International Thermal Spray Technology Proc. of the National Thermal Spray Conf. (Cincinnati, Ohio), 24–27 Oct 1988, ASM International
28. B.A. Shaw, "Characterization of the Corrosion Behavior of Aluminum, Zinc and Zinc-Aluminum Thermal Spray Coatings," Master's dissertation, Johns Hopkins University, 1985
29. D. Itzhak and E. Aghion, Corrosion Behavior of Hot-Pressed Austenitic Stainless Steel in H_2SO_4 Solutions at Room Temperature, *Corros. Sci.,* Vol 23 (No. 10), 1983, p 1085–1094
30. E. Klar, M. Svilara, C. Lall, and H. Tews, Corrosion Resistance of Austenitic Stainless Steels Sintered in Commercial Furnaces, *Advances in Powder Metallurgy and Particulate Materials—1992,* Vol 5, J.M. Campus and R.M. German, Ed., Metal Powder Industries Federation, American Powder Metallurgy Institute, p 411–426
31. F.M.F. Jones, The Effect of Process Variables on the Properties of Type 316L Powder Compacts, *Progress in Powder Metallurgy,* Vol 30, Metal Powder Industries Federation, 1970, p 25–50
32. E. Klar and P.K. Samal, Optimization of Vacuum Sintering Parameters for Improved Corrosion Resistance of P/M Stainless Steels, *Advances in Powder Metallurgy and Particulate Materials,* Vol 7, C. Lall, Ed., Plenum Press, 1994, p 239–251
33. B.A. Shaw, "Crevice Corrosion of a Ni-Cr-Mo-Fe Alloy in Natural and Chlorinated Seawater," Ph.D. dissertation, Johns Hopkins University, 1988
34. P. Peled and D. Itzhak, The Effect of Noble Alloying Elements Ag, Pt, and Au on the Corrosion Behavior of Sintered Stainless Steel in an H_2SO_4 Environment, *Corros. Sci.,* Vol 28 (No. 10), 1988, p 1019–1028
35. M.H. Tikkanen, Corrosion Resistance of Sintered P/M Stainless Steels and Possibilities for Increasing It, *Scand. J. Metall.,* Vol 11, 1982, p 211–215
36. P. Peled and D. Itzhak, The Corrosion Behavior of Double Pressed, Double Sintered Stainless Steel Containing Noble Alloying Elements, *Corros. Sci.,* Vol 30 (No. 1), 1990, p 59–65
37. R.M. German and D. Kubish, Evaluation of Injection Molded 17–4 PH Stainless Steel Using Water Atomized Powder, *Int. J. Powder Metall.,* Vol 29 (No. 1), 1993, p 47–62
38. D. Itzhak and P. Peled, The Effect of Cu Addition on the Corrosion Behavior of Sintered Stainless Steel in H_2SO_4 Environment, *Corros. Sci.,* Vol 26 (No. 1), 1986, p 49–54
39. D. Itzhak and S. Harush, The Effect of Sn Addition on the Corrosion Behavior of Sintered Stainless Steel in H_2SO_4, *Corros. Sci.,* Vol 25 (No. 10), 1985, p 883–888
40. G.H. Gessinger, Recent Developments in Powder Metallurgy of Superalloys, *Powder Metall. Int.,* Vol 13 (No 2), 1981, p 93–101
41. R.F. Singer, Recent Developments and Trends in High Strength P/M Materials, *Powder Metall. Int.,* Vol 17 (No. 6), 1985, p 284–288
42. L.R. Curwick, The Mechanical Alloying Process: Powder to Mill Product, *Frontiers of High-Temperature Materials,* J.S. Benjamin, Ed., Proc. International Conference on Oxide Dispersion Strengthened Superalloys by Mechanical Alloying, Inco Alloy Products, 1981, p 3–10

43. J.S. Benjamin and T.E. Volin, The Mechanism of Mechanical Alloying, *Metall. Trans.*, Vol 5, 1974, p 1929–1934
44. F.S. Pettit and G.H. Meier Oxidation and Hot Corrosion of Superalloys, *Superalloys, 1984,* M. Gell et al., Ed., Proc. the Fifth International Symposium on Superalloys, The Metallurgical Society, 1984, p 651–687
45. C.T. Sims, A History of Superalloy Metallurgy for Superalloy Metallurgists, *Superalloys, 1984,* M. Gell et al., Ed., Proc. the Fifth International Symposium on Superalloys, The Metallurgical Society, 1984, p 309–419
46. G.H. Gessinger, *Powder Metallurgy of Superalloys,* Butterworths, 1984
47. J. Stringer, B.A. Wilcox, and P.I. Jaffee, *Oxid. Met.,* Vol 5, 1972, p 11
48. C.S. Giggins and F.S. Pettit, The Oxidation of TD Ni Cr (Ni-20 Cr-2 Vol. pct ThO_2) Between 900 and 1200 C, *Metall. Trans.,* Vol 2, 1971, p 1071–1078
49. F. Perry, Oxide-Dispersion-Strengthened P/M Alloys Produced for Severe Service Applications, *Ind. Heat.,* Vol 49 (No. 5), May 1982, p 22–25
50. J.H. Weber and P.S. Gilman, Environmentally Induced Porosity in Ni-Cr Oxide Dispersion Strengthened Alloys, *Scr. Metall.,* Vol 18, 1984, p 479–482
51. J.H. Weber, *High Temperature Oxide Dispersion Strengthened Alloys,* Proc. 25th National SAMPE Symposium and Exhibition Society for the Advancement of Material and Process Engineering, 1980, p 752–763
52. G.H. Gessinger, *High Temperature Alloys for Gas Turbines,* D. Coutsouradis et al., Ed., Applied Science, 1978, p 817
53. P.L. Antona, A. Bennani, P. Cavalloti, and O. Ducati, Heat Treatments and Oxidation Behavior of Some P/M Ni-Base Superalloys, *European Symposium on Powder Metallurgy,* Vol 1, Jernkontoret Activity Group B, 1978, p 137–142
54. W.K. Uhl, M.R. Pendley, and S. McEvoy, "Evaluation of HIP Nickel-Base Alloys for Extreme Sour Service," Paper 219, presented at Corrosion/84, New Orleans, LA, National Association of Corrosion Engineers, April 1984
55. S. Floreen, Effects of Environment on Intermediate Temperature Crack Growth in Superalloys, *Proc. AIME Symposium* (Louisville, KY), American Society of Mining, Metallurgical, and Petroleum Engineers, 1981
56. G.A.J. Hack, Inconel Alloy MA 6000—A New Material for High Temperature Turbine Blades, *Met. Powder Rep.,* Vol 36 (No. 9), Sept 1981, p 425–429
57. D.H. Boone, D.A. Crane, and D.P. Whittle, *Thin Solid Films,* Vol 84, 1981, p 39
58. F.R. Wermuth and A.R. Stetson, Report NASA CR-120852, National Aeronautics and Space Administration, 1971
59. M.A. Gedwill, T.K. Glasgow, and LS. Levine, "A New Diffusion Inhibited Oxidation Resistant Coating for Superalloys," NASA TM 82687, National Aeronautics and Space Administration, 1981
60. T.K. Glasgow and G.J. Santoro, Oxidation and Hot Corrosion of Coated and Bare Oxide Dispersion Strengthened Superalloy MA 755E, *Oxid. Met.,* Vol 15 (No. 314), April 1986, p 251–276

SELECTED REFERENCES

- L. Fedrizzi, A. Molinari, F. Deflorian, A. Tiziani, and P.L. Bonora, Corrosion Study of Industrially Sintered Copper Alloyed 316L Austenitic Stainless Steel, *Br. Corros. J.,* Vol 26 (No. 1), 1991, p 46–50
- M. Hanada, Y. Takeda, N. Amano, et al., Development of a PM Sensor Ring for Use in an Antilock Brake System, *Met. Powder Rep.,* Vol 44 (No. 10), 1989, p 695–698
- C. Lall and M. Svilar, The Corrosion Resistance of P/M Stainless Steels and Selected Alloys in Methanol-Based Fuels, *P/M Steels,* J.M. Capus, Ed., Vol 5, Plenum Press, 1992, p 427–435
- G. Lei and R.M. German, Corrosion of Sintered Stainless Steels in a Sodium Chloride Solution, *Modern Developments in Powder Metallurgy,* E.N. Aqua, Ed., Vol 16, Plenum Press, p 261–275
- G. Lei, R.M. German, and H.S. Nayar, Influence of Sintering Variables on the Corrosion Resistance of 316L Stainless Steel, *Powder Metall. Int.,* Vol 15 (No. 2), 1983, p 70–76
- S.N. Malhotra and P. Ramakrishnan, Corrosion Behavior of PM Stainless Steel Filters, *Met. Powder Rep.,* 1991, p 48–51
- T. Mathiesen and E. Maahn, Evaluation of Sensitization Phenomena in Sintered Stainless Steels, *Powder Metallurgy World Congress (PM),* Paris, Vol 3, Les Editions de Physique, 1994, p 2089–2092
- H.S. Nayar, R.M. German, and W.R. Johnson, The Effect of Sintering on the Corrosion Resistance of 316L Stainless Steel, *Progress in Powder Metallurgy,* Vol 37; Metal Powder Industries Federation, p 1–7
- P. Peled, S. Harush, and D. Itzhak, The Effect of Ni Addition on the Corrosion Behavior of Sintered Stainless Steel in H_2SO_4, *Corros. Sci.,* Vol 28 (No. 4), 1988, p 327–332
- J.R. Pickens, Techniques for Assessing the Corrosion Resistance of Aluminum Powder Metallurgy Alloys, *Rapidly Solidified Powder Aluminum Alloys,* ASTM, 1986 p 381–409
- J.H. Reinshagen and T.J. Bockius, Stainless Steel Based P/M Alloys with Improved Corrosion Resistance, *Advances in Powder Metallurgy and Particular Materials,* Metal Powder Industries Federation, Vol 3, 1995, p 19–30
- J.H. Reinshagen and G.D. Flick, Improved Corrosion Resistant Stainless Steel Based P/M Alloys, *Advances in Powder Metallurgy and Particulate Materials—1996,* Metal Powder Industries Federation, Vol 5, 1995, p 61–71
- J.H. Reinshagen and A.J. Neupaver, Fundamentals of P/M Stainless Steels, *Advances in Powder Metallurgy,* Vol 2, Plenum Press, 1989, p 283–295
- F. Velasco, J.R. Ibars, J.M. Ruiz-Roman, et al., Improving the Corrosion Resistance of Power Metallurgy Austenitic Stainless Steels through Infiltration, *Corrosion,* Vol 52 (No. 1), 1996, p 47–52

Magnetic Materials and Properties for Part Applications

Kenneth H. Moyer, Magna-Tech P/M Labs

MAGNETIC POWDER METALLURGY PARTS are produced predominantly from "soft" (or temporary) magnetic materials, which generate a magnetic force only in the presence of an externally applied magnetic field. Commercial soft magnetic P/M alloys are commonly produced from high-purity iron or various ferrous alloy types such as Fe-2Ni, Fe-3Si, Fe-0.45P, Fe-0.6P, and 50Ni-50Fe. The use of P/M techniques to make magnetically soft components are effective for applications in which complicated magnetic parts would otherwise require considerable machining. In some applications, P/M fabrication eliminates all machining operations and saves a substantial amount of the total cost, compared to conventional manufacturing.

Powder metallurgy and metal injection molding (MIM) also have other advantages for fabrication of magnetic materials. One major advantage includes economical blending of small-lot specialty alloys. Powder metallurgy also has the advantage of processing materials that would be difficult for fabrication as wrought metal forms. For example, iron-phosphorus alloys cannot be made in wrought form because of "hot shortness," while Fe-P alloys can be processed effectively by P/M methods for soft magnetic applications. Powder metallurgy also is the method used for the relatively new permanent magnetic materials based on combinations of neodymium, iron, and boron (introduced in 1983). The high reactivity of neodymium or other rare earths and their alloys requires the suppression of contamination during alloy preparation and processing. In particular, oxidation by O_2 or H_2O or both must be kept to a minimum. Any oxidation of the alloy occurring during processing depletes the alloy of the rare-earth components. This usually results in the production of an alloy having unfavorable magnetic qualities. To obtain powder compacts with maximum magnetization, the powder is magnetically aligned and pressed.

Basic Magnetism

In simple terms, magnetism is a field force that can be produced either by permanent ("hard") magnetic materials or by an electric current generated in a coil of wire that surrounds a section of iron. The magnetic field is defined quantitatively in terms of the magnetic field strength (or the magnetizing force), designated as H, with the unit of oersteds (Oe) or amps/meter (A/m).

The magnetic response of a material is characterized by the induced magnetic field that is caused by an applied magnetic field, as shown in Fig. 1. The current in the primary winding produces an external magnetic field, which then induces a magnetic field in the toroid and the secondary circuit. This method characterizes the magnetic response in terms of magnetic induction (B) versus the applied field (H) as shown in Fig. 1(b). These basic characteristics of magnetic response are briefly summarized in this introductory section.

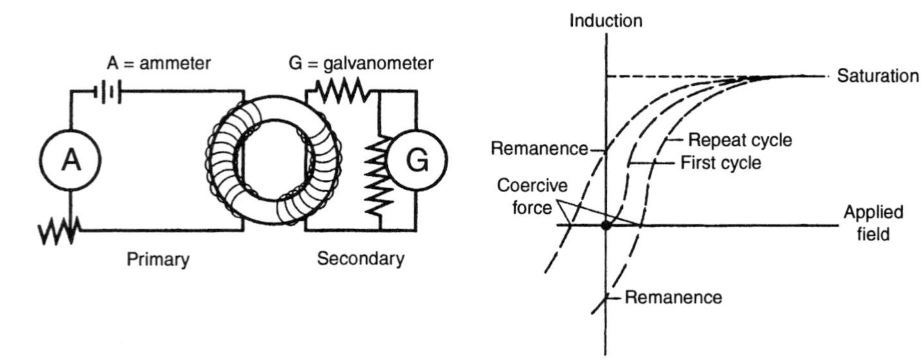

Fig. 1 Standard toroid test of magnetic response. (a) Test setup. (b) Hysteresis response. Source: Ref 1

Intensity of Magnetization and Magnetic Induction. The intensity of magnetization (I) is defined in terms of the number of unit north poles and south poles, m, within a given cross-sectional area. A rectangular bar of length l and cross-sectional area a with m unit north and south poles at each end would have an intensity of magnetization defined as $m/a = I$. The intensity of magnetization also can be likened to the flow of electrons through a wire where the concentration and direction of the electrons are known at any point.

In a magnetic field, flux lines pass from a magnetized material into the air at the north pole, re-enter at the south pole, and then pass through the material to the north pole to form a closed loop. The total number of lines crossing a given area at right angles is defined as the flux in the area. The flux per unit area is defined as the flux density or the magnetic induction, B. Both the magnetizing force or the applied field, H, and the intensity of magnetization, I, contribute to the lines of induction, but in magnetic materials, the intensity of magnetization, I, is generally much larger. The magnetic induction is therefore defined by the formula: $B = H + 4\pi I$. The cgs unit of magnetic induction is gauss (G), the SI unit is tesla (T). A tesla is equivalent to 10^4 G.

Magnetism and Permeability. When a piece of iron is brought near a permanent magnet or a magnetic field of an electrical current, the magnetization induced in the iron by the applied field is described by a magnetization curve obtained by plotting either the intensity of magnetization or the magnetic induction, B, as a function of the applied field, H. A typical magnetization curve is shown in Fig. 2(a). However, if the direction of magnetization is reversed, the magnetization curve does not retrace itself but has a property known as hysteresis. Owing to this hysteresis, a loop rather than a curve is generated (Fig. 2b). However, unless losses are realized, the magnetization curve should be reproduced repeatedly.

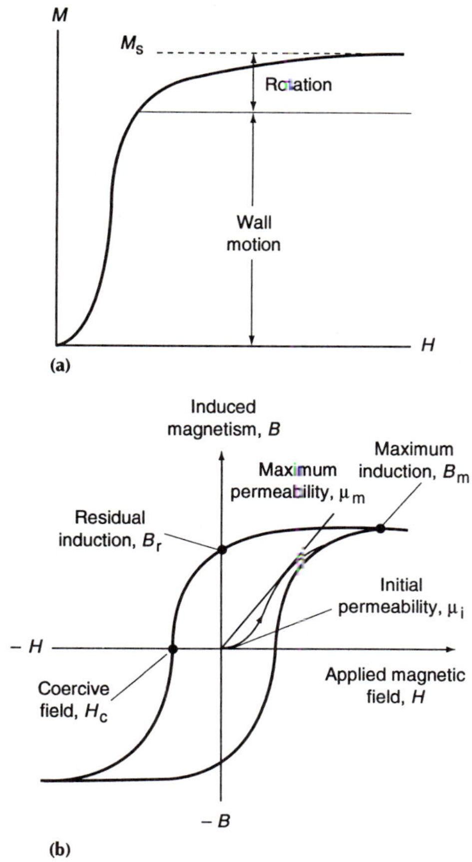

Fig. 2 Schematic of (a) magnetization curve (b) hysteresis loop. Source: Ref 1, 2

Table 1 Room-temperature magnetic properties of selected materials

Material/condition	Density, g/cm³	Maximum induction, B_{max} T	Maximum induction, B_{max} kG	Residual induction, B_r T	Residual induction, B_r kG	Maximum permeability, μ_{max}	Coercive force, H_c A/m	Coercive force, H_c Oe
Wrought metals								
Unalloyed cobalt	8.8	1.87(a)	18.7(a)	0.49	4.9	245	708	8.9
Unalloyed iron	7.87	2.15(a)	21.5(a)	1.18	11.8	5,000	80	1.0
Unalloyed nickel	8.9	0.616(a)	6.16(a)	0.3	3.0	1,240	167	2.1
Wrought alloys								
Fe-3%Si	7.65	2.0	20	...	...	8,000	56	0.7
Fe-50%Ni	8.2	1.6	16	0.8	8.0	70,000	4.0	0.05
430F stainless	7.62	1.2	12	0.6	6.0	2,000	160	2.0
Soft magnetic P/M Iron (ASTM A 811)								
H = 15 Oe	6.6	0.9	9	0.8	8	1,800	175	2.2
H = 15 Oe	6.9	1.0	10	0.9	9	2,100	...	...
H = 15 Oe	7.2	1.2	12	1.1	11	2,500	160	2.0
Pure P/M iron(b)								
At H = 15 Oe	7.0	1.09	10.9	0.97	9.7	2,400	135	1.7
At H = 25 Oe	7.0	1.15	11.5	1.0	10.0	2,400	135	1.7
At H = 100 Oe	7.0	1.35	13.5	1.1	11.0	2,400	135	1.7
P/M Fe-0.45%P(b)								
At H = 15 Oe	7.0	1.19	11.9	1.12	11.2	3,300	127	1.6
At H = 25 Oe	7.0	1.24	12.4	1.14	11.4	3,300	127	1.6
At H = 100 Oe	7.0	1.43	14.3	1.24	12.4	3,300	127	1.6
P/M Fe-50%Ni(b)								
At H = 15 Oe	7.2	0.99	9.7	0.81	8.1	12,300	22	0.28
At H = 25 Oe	7.2	1.01	10.1	0.84	8.4	12,300	22	0.28
At H = 100 Oe	7.2	1.14	11.4	0.88	8.8	12,300	22	0.28

(a) Magnetic saturation (B_{sat}) values. (b) Samples sintered at 1250 °C (2280 °F), 45 min in dissociated ammonia. Source: *ASM Handbook,* Vol 2, 1990, Ref 1, 3

The behavior of any magnetic material can be defined by the hysteresis loop and the permeability. The ratio B/H is defined as the permeability, μ. This value indicates the relative increase in flux or magnetic induction caused by the presence of a magnetic material. Of the many useful permeabilities, the initial permeability, μ_0, and the maximum permeability, μ_{max}, are of most importance for P/M applications. The initial permeability, μ_0, is simply the slope of the initial portion of the magnetization curve, say to 0.01 T (100 G). The maximum permeability, μ_{max}, is the slope to the tangent of the magnetization curve (Fig. 2). After μ_{max} is achieved, the permeability falls off rapidly and finally reaches a value of 1. At this point, the magnet is said to be saturated. In other words, even though the magnetization curve will continue to rise, the contribution to magnetization will derive only from the applied field. The magnetic induction at a permeability of 1 is defined as the saturation induction (B_{sat} or B_m in Fig. 2b). Each magnetic material at a temperature of 20 °C will have a specific constant value for the magnetic saturation. In the case of iron, the value is 2.15 T (21,500 G).

Hysteresis Loop. When the applied magnetic field (H) is cycled over time, the magnetic induction lags the applied field and defines a hysteresis loop (Fig. 2b). There are two magnetic properties of interest that occur as a result of the hysteresis. These properties include the remanence or residual induction, B_r, and the coercive field, H_c. The remanent magnetization is the specific value of the magnetic induction when the applied field has been reduced to 0. In other words, the remanent magnetization is the magnetic induction at an applied field of 0, or the intercept of the B-axis. The coercive field, H_c, is the applied field required to reduce the magnetic induction to a value of 0. This value is the intercept of the H-axis. Both of these properties, including magnetic induction, are important to the performance of any magnetic material. These properties are identified in Fig. 2b as specific points on the hysteresis curve.

Demagnetizing Fields, Air Gaps. Most magnetic devices are not solid but contain discontinuities, such as air gaps. When a discontinuity or an air gap is present, it creates a magnetic field in a direction opposite to the applied field. The net effect is to reduce or subtract from the magnetic induction that results from the application of the applied field. Thus, the true applied field causing the magnetic induction is somewhat lower than the applied field, owing to any demagnetizing forces that are present, such as air gaps. The true applied field can therefore be expressed as follows: $H = H_a - \Delta H$. The demagnetizing field is therefore approximately proportional to the intensity of magnetization. In other words: $\Delta H = NI$, where N is defined as the demagnetizing factor, and where:

$$\Delta H = \frac{N(B - H)}{4\pi}$$

The effect of air gaps is to shear the magnetization curve. In other words, air gaps counteract the positive effects of purifying materials and growing grains (for example, by the use of high-temperature sintering during P/M processing).

Ferromagnetism and Factors Affecting Magnetic Quality. The magnetic properties of magnetic materials depend on the chemical composition, cold deformation, and heat treatment. There are essentially three elements and their respective alloys that are truly magnetic. These ferromagnetic materials are iron, nickel, and cobalt. Iron and cobalt have high magnetic saturations (Table 1). Because of the expense, scarcity, and difficulty of sintering, cobalt and its alloys are not generally selected, except for special applications. Iron, on the other hand, is inexpensive and quite common. Nickel has a magnetic saturation of only about 0.6 T (6 kG). Owing to its much lower magnetic saturation with respect to iron and its higher cost, nickel and its alloys are used mainly when quick response is required.

In crystalline materials, the basis for ferromagnetism lies in the alignment of magnetic moments from noncompensated electron spins in the 3d shell of the transition series elements, such as iron, nickel, and cobalt (Table 2). In ferromagnetic materials that are below their Curie temperature, the magnetic moments of adjacent atoms are "coupled" parallel to each other.

Table 2 Magnetic moments of neutral atoms of 3*d* transition elements

Unpaired 3*d* electrons	Atom	Number of electrons	Electronic configuration 3*d* orbitals					*a*/*D* ratio(a)
3	V	23	↑	↑	↑			2
5	Cr	24	↑	↑	↑	↑	↑	1
5	Mn	25	↑	↑	↑	↑	↑	2
4	Fe	26	↑↓	↑	↑	↑	↑	2
3	Co	27	↑↓	↑↓	↑	↑	↑	2
2	Ni	28	↑↓	↑↓	↑↓	↑	↑	2
0	Cu	29	↑↓	↑↓	↑↓	↑↓	↑↓	1

(a) The ratio of atomic spacing, *a*, and diameter, *D*, of the 3*d* orbitals determines the occurrence of positive exchange energy for ferromagnetism. Source: Ref 4

For a small volume of material, all of the individual magnetic moments are aligned in one direction. This small volume is magnetized to saturation and is known as a magnetic domain. An adjacent volume of material, also magnetized to saturation, can have the summation of its magnetic moments point in another direction. Where two such volumes meet with differing alignments, a domain boundary wall must exist. The total magnetization of a sample of material is the net vector summation of all the individual component domain magnetization vectors.

The parallel "coupling" of magnetic dipoles in iron, cobalt, and nickel occurs from an "isotropic exchange energy," which is the mechanism responsible for ferromagnetism that was a complete mystery until the advent of quantum mechanics by Heisenberg. In very simple terms, parallel "coupling" of magnetic dipoles in iron, cobalt, and nickel is due to a positive exchange energy that occurs when the ratio of atomic spacing (a) to the diameter (D) of the third orbital is between 1.4 and 4.7 (Ref 4). Thus iron, cobalt, and nickel are ferromagnetic, but manganese and chromium are not (Table 2).

Factors Affecting Permeability and Hysteresis Losses. Permeability, coercive field, and hysteresis loss are affected by impurities within the alloy, by cold deformation such as sizing, and by any heat treatment that is provided. Impurities that are most harmful to these alloys include carbon, nitrogen, oxygen, and sulfur. All P/M parts also contain a lubricant to facilitate compaction. If this lubricant is not properly removed during the presinter portion of the sintering cycle, the structure-sensitive properties (permeability, coercive field, and hysteresis loss) may be significantly degraded. Sintering furnace condition and choice of protective atmosphere are also important. Hydrogen or vacuum is preferred. However, hydrogen/nitrogen mixtures, exothermic gas, endothermic gas, and dissociated ammonia can be selected as the protective atmosphere. Any of the latter can cause nitriding or carburizing, which will degrade the structure-sensitive magnetic properties. Because most powders that are used are water atomized and therefore contain surface oxides, these surface oxides must be reduced during the sintering cycle or else degradation of structure-sensitive magnetic properties will occur.

Knowledge of phase diagrams is important to the development of alloys for magnetic applications. Alloys most commonly selected are single-phase alloys. For instance, more than 0.008 wt% C or more than 0.006 wt% N causes carbide or nitride formation in α-iron. The more carbon contamination or the more nitriding that occurs, the greater the degradation of the structure-sensitive magnetic properties. In phosphorus irons, additions of >1.5 wt% P result in formation of phosphorus iron intermetallics. In silicon irons, >3 wt% Si makes the alloys difficult to compact; however, as much as 6 wt% Si has been added to P/M silicon iron alloys. Soft magnets should be processed from single-phase compositions of high purity.

If second phases form, the structure-sensitive magnetic properties can be affected over time from aging. At the elevated temperatures required to sinter the parts, iron can dissolve considerable carbon or nitrogen into the structure, which consists of austenite (γ-phase). After sintering and cooling, the structure-sensitive properties can appear to be unaffected. However, over time in service, the properties degrade as a result of the nitrogen or carbon precipitating from solution. Some carbon or nitrogen can also be trapped in the iron, or a reversible action can occur. The net result is a degradation of the structure-sensitive magnetic properties over a period of time (aging). If the alloy is cooled rapidly after sintering, the carbon or nitrogen is trapped in solution in the iron, which in turn strains the iron lattice, causing degradation of the structure-sensitive magnetic properties. This latter effect is small, but it occurs nonetheless.

Magnetic P/M Materials

Most P/M materials are designed for soft magnetic properties with little or no retentivity—that is, if they are magnetized in a magnetic field and then are removed from that field, they lose most, if not all, of the magnetism they exhibited while in the field. Magnetically soft materials that are produced in large quantities include high-purity iron, low-carbon steels, silicon steels, iron-nickel alloys, iron-cobalt alloys, and soft-magnetic ferrites.

In contrast, permanent magnets are normally used in a single magnetic state and have stable magnetic fields. Permanent magnet materials include a variety of metals, intermetallics, and ceramics. Commonly included are certain steels, Alnico, CuNiFe, Fe-Co alloys containing V or Mo, Pt-Co, hard ferrites, and cobalt-rare earth alloys. Only a few P/M alloys are useful as hard magnetic materials. In this regard, the major alloy is Fe-Nd-B rapidly solidified powders.

This section briefly describes the common soft magnetic P/M alloys including high-purity iron Fe-2Ni, Fe-3Si, Fe-0.45P, Fe-0.6P, and Fe-50Ni. Table 3 compares the typical magnetic properties of P/M compositions. Values in Table 3 are only typical as considerable variability occurs in these properties, especially the magnetic force needed to demagnetize an alloy. Part of this variation is due to differences in sintered density, but impurity control is also a major factor. Like strength, the magnetic characteristics are sensitive to contaminants.

Pure Iron. Iron powder can be produced in many forms. All are magnetic, some more so than others. In early applications, sponge iron was used to produce magnets. The magnetic properties of these early magnets were not as good as those of current magnets because the sponge iron magnets contained as much as 2 wt% insolubles, generally in the form of oxides. Besides being impure, the sponge iron powder was also harder and therefore could not be compacted to densities above 6.2 g/cm^3. Therefore, magnetic induction was limited, owing to the low sintered density. The structure-sensitive properties—permeability and coercive field—were limited by the impurities. If improved magnetic properties were required, either carbonyl iron or electrolytic iron was selected, at a great increase in cost of the powder.

Today, inexpensive water-atomized iron powder is readily available. This powder has <1 wt% impurities present, normally as surface oxides. Because the powder is pure, the structure-sensitive magnetic properties are greatly enhanced. In fact, these properties are governed by the sintering conditions that are provided. Given a quality sinter, there is no reason why these structure-sensitive magnetic properties will not equal those of the highest-purity wrought iron magnets.

In addition to water-atomized iron being pure, it is also soft because there are no hard inclusions present, and it is annealed under optimal conditions. As a result, water-atomized iron powder can be compacted to green densities approximately 6.8 to 7.2 g/cm^3. With a conventional sinter of 1120 °C (2050 °F) for 30 min, magnetic properties listed as typical in ASTM specification A 811 can be attained (Ref 3). These magnetic properties are listed in Table 1. Also, listed in Table 1 are magnetic properties of

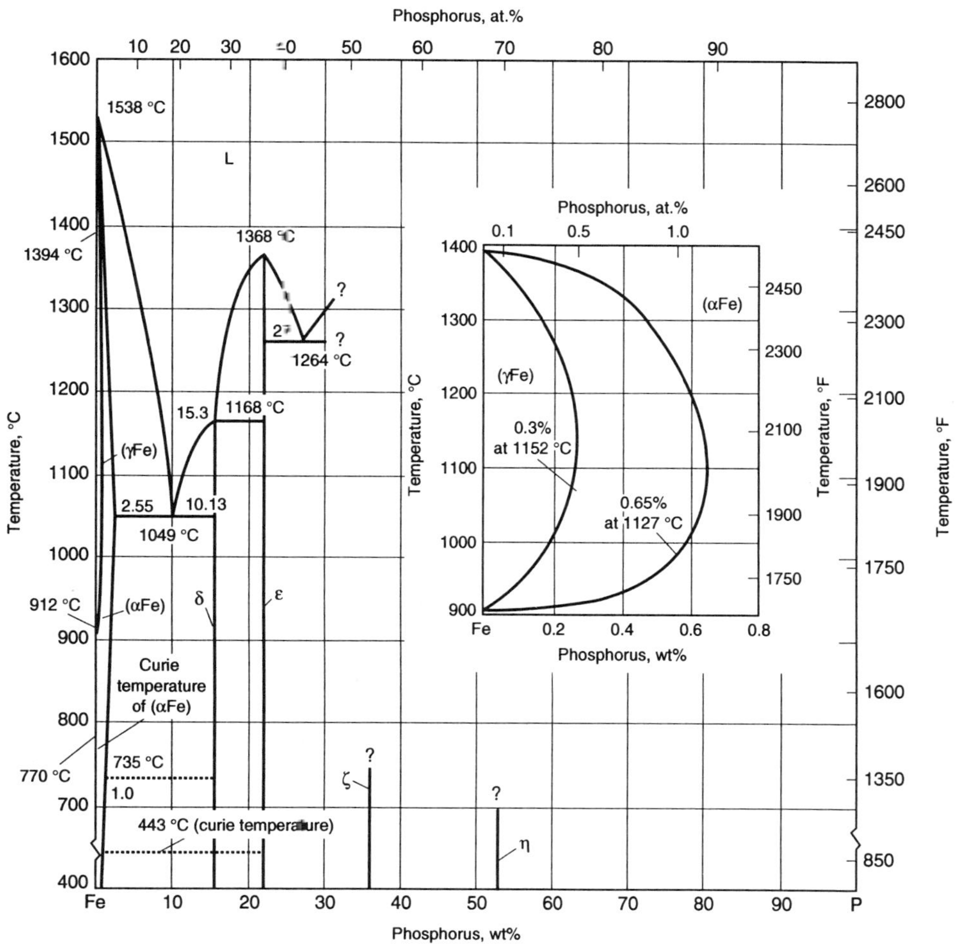

Fig. 3 Iron-phosphorus phase diagram. Source: Ref 5

Table 3 Typical magnetic properties of P/M alloys

Alloy	Resistivity, μΩ · cm	Maximum magnetization (B_{max}), T	Remanence magnetization (B_r), T	Coercive force, Oe	Maximum permeability
410L	90	1.2	4.1	3.5	960
430L	93	1.1	4.8	2.7	630
434L	104	1.1	5.7	2.3	1,300
Fe	20	1.6	1.2	1.4	4,000
Fe-49Co-2V	40	2.0	5.6	3.0	3,000
Fe-50Co	60	1.7	0.56	2.0	2,000
Fe-5Mo	...	1.6	1.3	1.8	2,900
Fe-2Ni	15	1.5	1.2	1.0	...
Fe-50Ni	40	1.3	0.9	0.3	25,000
Fe-0.45P	21	1.4	1.2	1.1	4,000
Fe-0.8P	23	1.8	1.3	0.4	6,100
Fe-3Si	45	1.4	1.3	0.9	4,500
Fe-6.5Si	81	1.3	1.2	0.3	4,000

Except for 400 series stainless steels at approximately 6.9 g/cm^3, all others are high-density products over 7.4 g/cm^3. Source: Ref 1

other selected materials. The properties that are presented are not limiting, but they are those values that are achieved in present commercial parts. However, if the P/M parts are forged to remove the pores, the saturation magnetism of iron of 2.15 T (21,500 G) will be realized. Similarly, if parts are sintered at high temperatures >1260 °C (>2300 °F), and they are held at temperature for long periods of time, maximum permeabilities of 350,000 and coercive fields as low as 0.8 A/m (0.01 Oe) are possible. The control is dependent on both grain growth and impurities within the material.

Iron is selected to produce parts because of the dimensional control that is realized when iron is sintered at 1120 °C (2050 °F) in standard belt furnaces for 30 min. These parts do not have the best magnetic properties but are adequate as inexpensive flux paths for magnetic devices. In reality, the P/M iron parts serve as inexpensive replacements for carbon steel parts. Properties are limited at these sintering temperatures because grain growth and sintered densities are limited by the limited diffusion at these sintering temperatures.

Phosphorus Irons. Although the attributes of alloying iron with phosphorus formerly were well known, wrought products, when hot rolled, became embrittled when phosphorus was alloyed with the iron in significant amounts. However, because phosphorus is normally admixed with water-atomized iron powder as ~10 μm Fe_2P or Fe_3P powder, P/M technology can take advantage of phosphorus additions to iron. When compacted, the compressibility of the soft iron is realized without loss of green density because the hard, small 10 μm Fe_2P or Fe_3P particles rearrange themselves among the larger iron particles as compaction occurs. When sintered, the phosphorus iron intermetallic compounds melt at a temperature of 1050 °C (1920 °F) and diffuse into the iron, forming a solid solution of phosphorus in iron (Fig. 3). The liquid phase enhances diffusion rates and assists in the rearrangement of the pores and particle boundaries, thereby further densifying the iron. If a normal belt furnace sinter of 1120 °C (2050 °F) for 30 min is used, then sintered densities as high as 7.2 g/cm^3 can be realized (Table 3). If the sintering temperature is increased to 1260 °C (2300 °F), and the parts are sintered for 1 h in a partial pressure of hydrogen, a sintered density as high as 7.6 g/cm^3 can be realized. Because magnetic induction is a linear function of the sintered density, the magnetic induction realized from the phosphorus iron alloys is higher than that realized from pure iron parts. However, the drawback to the use of phosphorus iron parts is that, owing to the liquid phase sintering that occurs, dimensional control is more difficult. When sintered at 1120 °C (2050 °F) in belt furnaces, dimensional control is not too difficult because the normal phosphorus addition of 0.45 wt% provides what is commonly known as a zero-growth part. However, if the phosphorus content is increased to 0.8 wt% or if high-temperature sintering is specified to improve density and structure-sensitive magnetic properties, shrinkage becomes greater and dimensions are more difficult to control. The problems with dimensional control and added cost of high-temperature sintering generally cause a compromise to be made between achievable magnetic properties and commercial manufacturing practice.

From the iron-phosphorus binary phase diagram in Fig. 3, it can be seen that the maximum solubility of phosphorus in iron is ~1.5 wt%. Theoretically, phosphorus can be alloyed to this wt%; however in P/M technology, phosphorus additions to 2 wt% have been investigated (Ref 6). Increased alloying of phosphorus in iron results in several advantages to improve magnetic properties.

Resistivity is one physical property that is improved by alloying phosphorus with iron. If 0.45 wt% P is alloyed, resistivity is approximately

Table 4 Typical direct-current magnetic properties of phosphorus iron alloys

	Compacting pressure					
	0.45% P compacted at pressures of:			0.80% P compacted at pressures of:		
Property	410 MPa (30 tsi)	550 MPa (40 tsi)	685 MPa (50 tsi)	410 MPa (30 tsi)	550 MPa (40 tsi)	685 MPa (50 tsi)
Sintered density, g/cm^3	6.80	7.10	7.23	6.88	7.12	7.28
Induction, for a field of 1200 A/m (15 Oe), T (kg)	1.06 (10.6)	1.19 (11.9)	1.27 (12.7)	1.12 (11.2)	1.23 (12.3)	1.31 (13.1)
Residual induction, T (kG)	0.87 (8.7)	0.99 (9.9)	1.08 (10.8)	1.01 (10.1)	1.13 (11.3)	1.20 (12.0)
Maximum relative permeability	2400	2800	3100	3680	4240	4640
Coercive force, A/m (Oe)	135 (1.7)	135 (1.7)	127 (1.6)	112 (1.4)	112 (1.4)	112 (1.4)

Source: Ref 3

Table 5 Magnetic data on ferritic stainless steels

Steel type	Density, kg/m^3	Maximum induction (B_m), T	Residual induction (B_r), T	Relative permeability, μ_r	Coercive field (H_c), A/m
Wrought					
430F, B_m at H = 796 A/m, B_s = 1.42 T	7620	1.2	0.6	2000	159
Powder metal					
410L, H = 1990 A/m:					
1260 °C, vacuum 45 min	7100	1.09	0.94	2200	169
1288 °C, H_2	7100	5.6	...	320	590
430L					
H = 1194 A/m, 1121 °C, H_2, 30 min	6450	0.73	0.47	1000	182
	6670	0.79	0.51	1043	185
H = 1194 A/m, 1121 °C DA, 30 min	5810	0.034	0.003	11	231
	6130	0.041	0.005	13	294
	6420	0.045	0.005	14	279
H = 1194 A/m, 1260 °C, H_2	7250	1.05	0.80	1900	159
H = 1194 A/m (vacuum + backfill of listed gas):					
1121 °C, H_2	6690	0.81	0.74	1200	207
1232 °C, H_2	7110	0.98	0.89	1800	167
1121 °C, H_2 + N_2	6490	0.016	0.009	...	740
1232 °C, H_2 + N_2	6990	0.038	0.024	...	669
434L					
H = 1990 A/m:					
1260 °C, vacuum 45 min	7000	1.01	0.84	1700	159
1288 °C, H_2	7100	6.5	...	450	220
H = 1194 A/m:					
1288 °C, DA, 30 min	6430	0.728	0.463	1092	160
1288 °C, H_2, 30 min	6650	0.791	0.483	1165	151
H = 1194 A/m: 1121 °C, DA, 30 min	5830	0.053	0.009	416.8	414
	6030	0.063	0.013	719.7	477
	6290	0.079	0.019	424.9	533
1288 °C	7100	0.89	0.44	1275	119
H = 1194 A/m, 1232 °C, H_2	7350	0.97	0.77	1600	143

DA, dissociated ammonia. Source: Ref 6

doubled. If 0.8 wt% P is alloyed, the resistivity is approximately 24 m$\Omega \cdot$ cm. Higher resistivity is required to minimize eddy current losses at low levels so that heat generation will not destroy the magnetic device. The low resistivities of the phosphorus irons do not enable these alloys to be used for most alternating-current applications. However, if 0.8 wt% P is used, then the phosphorus irons can be considered for some pulsed direct-current applications.

It is therefore readily understandable that phosphorus irons have become the workhorse for specification of material for magnetic parts. Depending on impurity content, sintering conditions, and heat treatment, these alloys can compete with most other P/M magnetic materials for demanding devices that require improved performance for direct current, and for many devices that require a pulsed direct current for application. Owing to low resistivity and the inability to produce thin sheet for laminations, the alloys are not normally considered for alternating-current applications. Typical magnetic properties of the two phosphorus iron alloys considered for current commercial applications are shown in Table 4.

It is possible to enhance structure-sensitive magnetic properties (permeability and coercive field). If a 0.8 wt% alloy is sintered at 1260 °C (2300 °F) for 1 h in a partial pressure of hydrogen, maximum permeabilities of 15,000 and coercive fields as low as 55 A/m (0.7 Oe), are possible (see Moyer, 1990 in "Selected References"). These values are comparable to those of commercial 3 wt% silicon iron. In reality, even more improved structure-sensitive properties are possible if impurity products, such as carbon, nitrogen, oxygen, and sulfur are maintained at low levels, and heat treatment is provided to grow grains. However, practical difficulties (cost, time, dimensional control) prevent industrial application.

Ferritic Stainless Steels. The automotive market consumes approximately three-fourths of the P/M parts that are fabricated today. This statement is also true for direct-current magnetic parts. As newer applications emerge, magnetic part applications can involve higher operating temperatures and require improved corrosion resistance. The optimal material selection for these applications are the ferritic stainless steels. Although corrosion resistance is enhanced by this selection, the magnetic properties are sacrificed at the expense of this requirement.

Four ferritic stainless steels (409L, 410L, 434L, and 430L), are commercially available for magnetic applications. The 410L powder nominally contains 12 wt% Cr as an alloying addition. The 434L powder contains nominally 17 wt% Cr and 2 wt% Mo. The latter alloy is more corrosion resistant but more expensive. In addition, because magnetic induction depends on the quantity of iron present in the alloy, the magnetic induction of the ferritic stainless steels is lower than that of either iron or the phosphorus irons.

Most stainless steel powders are generally water atomized, although gas-atomized powders are used for parts made by the MIM process. Because of the solid-solution strengthening of chromium in the iron matrix, the powder particles are harder than iron powder, which limits compressibility. Therefore, sintered densities >7.2 g/cm^3 are not common. The lower densities also limit the magnetic induction of these alloys.

Present state-of-the-art production generally requires a high-temperature sinter for these alloys to attain their best level of performance magnetically. For highest magnetic properties, the sintering temperature is normally 1260 °C (2300 °F) or greater for at least 1 h in either vacuum or hydrogen (Table 5). This sintering temperature is required because chromium oxides present on the surfaces of these water-atomized powders do not readily reduce at temperatures lower than 1260 °C (2300 °F).

High-temperature sintering presents some problems in addition to increased cost. If ferritic stainless steel parts are high-temperature sintered, shrinkage of at least 1% is realized. These high shrinkage rates again make dimensional control difficult. Some parts fabricators therefore elect to sinter at either 1205 °C (2200 °F) or at 1230 °C (2250 °F) to reduce the problem of dimensional control. More recently, parts fabricators supplying parts to the automotive industry have even used belt furnaces to sinter ferritic stainless steels at 1150 °C (2100 °F). In all cases, lower temperatures result in lower sintered densities and higher oxygen contamination of the parts. This being the case, magnetic properties and corrosion resistance are sacrificed.

In addition to compromises with respect to sintering temperature, many parts fabricators

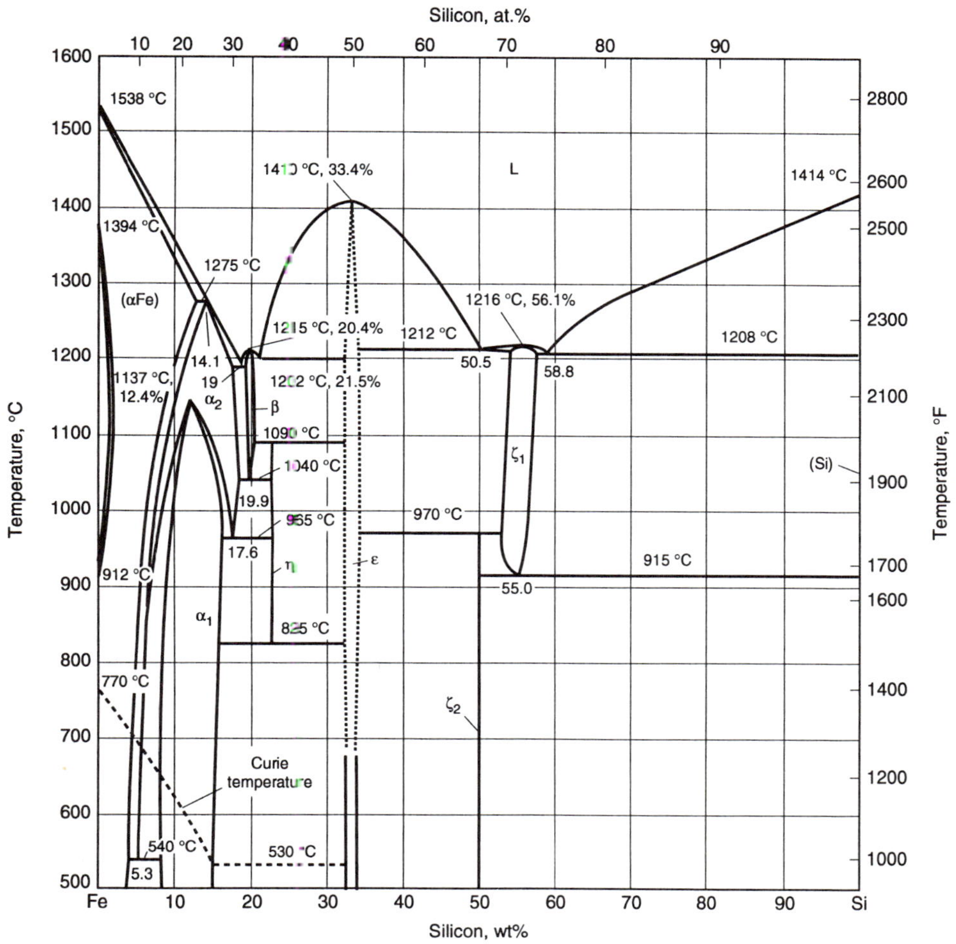

Fig. 4 Iron-silicon phase diagram. Source: Ref 5

Table 6 Magnetic data on 50 nickel/50 iron

Treatment	Density, kg/m^3	Maximum induction (B_m), T	Residual induction (B_r), T	Relative permeability, μr	Coercive field (H_c), A/m
Wrought					
ASM Handbook, Vol 2	8200	1.60	0.80	70,000	3.98
Powder metal					
H = 1990 A/m, 1260 °C, DA, 45 min(a)	6800	0.93	0.71	...	20.7
	7100	1.09	0.80	...	19.9
	7500	1.27	0.94	21,000	19.1
H = 1990 A/m, 1260 °C, vacuum	7400	1.08	0.86	10,600	23.9
At temperature of:					
1121 °C	7100	1.15	0.85	11,000	31.4
1288 °C	7300	1.29	0.90	16,000	23.9
H = 1194 A/m:					
1246 °C, vacuum 2 h	7300	1.12	0.70	...	23.9
1246 °C, vacuum 2 h	7500	1.23	0.75	...	23.9

(a) Resistivity 0.78, 0.69, 0.60 μΩ·m at 6800, 7100, and 7500 kg/m^3, respectively. DA, dissociated ammonia. Source: Ref 7

elect to sinter in dissociated ammonia, nitrogen/hydrogen gas mixtures, nitrogen, exothermic, or endothermic gas. These atmospheres are rich in nitrogen, carbon, or both. Chromium in the ferritic stainless steel readily reacts with the components of these atmospheres to form carbides or nitrides in grain boundaries, thereby depleting the chromium protective film. These carbides or nitrides lower structure-sensitive magnetic properties and corrosion resistance and also decrease ductility.

50 Nickel/50 Iron. Nickel irons are usually the alloys selected when exceptional structure-sensitive properties are required. The alloy composition generally selected is 50 wt% Ni combined in solid solution with 50 wt% Fe. Although this alloy is most predominantly selected, nickel/iron combinations of all ranges have been selected for magnetic applications. Nickel irons are the most expensive of all the commonly used magnetic alloys today. The nickel irons are selected when fast response is required in magnetic properties. Fe-50Ni has a high maximum permeability (Table 3).

Because structure-sensitive magnetic properties are generally required, these alloys are normally high-temperature sintered at 1260 °C (2300 °F) or higher, in either vacuum or hydrogen, for at least 1 h (Table 6). These more demanding sintering conditions are necessary to promote diffusion of the substitional elements (iron, nickel) and to aid grain growth and purification (removal of carbon, nitrogen, oxygen, and sulfur). If the above conditions are met, relative maximum permeability as high as 10,000, and a coercive field as low as 15 A/m (0.2 Oe) are possible in the as-sintered condition.

These alloys, however, are sensitive to heat treatment. Generally, if annealing is required, the wrought practice of heating to a temperature of 1120 to 1175 °C (2050 to 2150 °F) in dry hydrogen for 4 h is employed. Using this practice, the relative maximum permeability can be increased to 21,000; however, the coercive field is not changed. It must be stressed at this point that further work on heat treatment is needed to improve the structure-sensitive magnetic properties of the P/M alloy and make them comparable to the wrought product. Typical magnetic properties of the 50 nickel/50 iron alloy are shown in Table 6.

Silicon Irons. Wrought silicon irons have been selected for magnetic applications since the 1940s. These alloys normally contain 1.5 to 3 wt% Si. They are used primarily for motors, generators, and relays. Alloys containing from 3 to 5 wt% have found service in high-efficiency motors and in transformers. Barrett, Brown, and Hadfield in 1900 published a paper which stated that 2 to 2.5 wt% Si decreased the coercive field of the alloy to half that of iron. Silicon steels were first produced in Germany for commercial applications in 1903. Commercial silicon iron was also produced in the United States in the same year. Its features included increased permeability, lower hysteresis loss, lower eddy current losses owing to higher resistivity, and no aging.

The solubility of silicon in ferrite extends to ≅16 mol% (Fig. 4). Although the solubility is high, providing an advantage of high resistivity, most wrought alloys are limited to 3 wt% Si because of the difficulty of deforming alloys containing greater additions of silicon.

Although P/M magnetic parts had been fabricated commercially from 3 wt% silicon iron since the 1950s, only a few parts fabricators considered producing these parts. First, the silicon irons are made by admixing hard intermetallic compounds with a soft base iron matrix. In the case of the silicon irons, sufficient master alloy is added to yield 3 wt% Si upon sintering. Master alloys containing 17% Si can cause significant die wear.

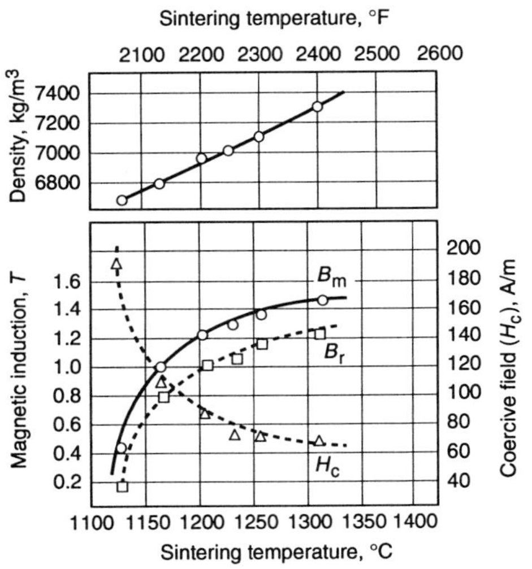

Fig. 5 Magnetic properties of Fe-3%Si as a function of sintering temperature in dissociated ammonia. Source: Ref 7

Table 7 Magnetic data on metal injection molded alloys

Alloy	Density, kg/m³	Maximum induction (B_m), T	Residual induction (B_r), T	Coercive field (H_c), A/m
Fe-2% Ni, 1316 °C	7670, 97%	1.51	1.29	82
Fe-50% Ni, 1316 °C	7660, 94%	1.27	0.42	16
Fe-3% Si, 1316 °C	7550, 98%	1.50	1.21	45
Then, H_2 annealed	7550, 98%	1.47	1.14	47
Fe-3% Si, 1232 °C	7540, 98%	1.44	0.62	60
Fe-6% Si, 1316 °C	7540, 99%	1.32	1.12	46
Then, H_2 annealed	7410, 98%	1.37	1.22	46
Fe, 1371 °C, DA	7600, 97%	1.53	1.37	294
Fe, 1371 °C, DA	7550, 96%	1.55	1.34	183
Fe-3% Si, 1371 °C, DA	7550, 98%	1.45	1.07	57
Fe-3% Si, 1371 °C, DA	7550, 98%	1.50	1.21	51
Fe-6% Si, 1371 °C, DA	7420, 99%	1.37	1.21	40
Fe-50% Ni, 1371 °C, DA	7660, 93%	1.27	0.42	16
430L stainless steel, 1371 °C, vacuum	7400, 95%	1.15	0.54	202
Fe-3% Si	...	1.48	0.88	62
		1.46	1.10	56
Fe-50% Ni	...	1.22	0.53	16
		1.04	0.64	19
430L stainless steel	...	0.85	0.22	119

DA, dissociated ammonia. Source: Ref 7

In addition to an increased die wear problem, problems in sintering these alloys are also encountered. Silicon is even more readily oxidized than chromium. Therefore, high-temperature sintering is a must. If sintering temperatures below 1205 °C (2200 °F) are employed, the oxide film encasing the silicon master alloy is not reduced, and the master alloy particles do not diffuse into the ferrite, but they remain as small spheres within the basic ferrite matrix (Ref 8). A liquid phase develops (Fig. 4) below 1260 °C (2300 °F); therefore if a hydrogen atmosphere is employed as the protective atmosphere, the oxide film is reduced, and a proper sinter can be provided between 1260 and 1315 °C (2300 and 2400 °F). Because there is a large concentration of the master alloy present in the parts, sintering is rapid, and there is a danger of melting occurring if the sintering temperature selected approaches 1315 °C (2400 °F). Owing to the large amount of liquid phase that occurs, dimensional control is also more difficult and sizing may be required.

Provided that the parts fabricator is aware of the practical difficulties of processing silicon irons, enhanced magnetic properties can be realized from these alloys at reasonable cost. In addition to the above-mentioned disadvantages, the magnetic saturation of the silicon irons is lower than that of iron. Iron has a saturation magnetization of 2.15 T (21,500 G) compared to 1.95 T (19,500 G) for Fe-3 wt% Si. Therefore, iron or phosphorus iron is a better choice for applications such as flux paths. However, the resistivity of a 3 wt% silicon iron is ≅65 μΩcm, which is five times that of pure iron. High resistivity lowers hysteresis and eddy current losses. Therefore, if a pulsed direct current or a 60 Hz application is specified, the 3 wt% silicon iron is a logical choice for this application, provided that the part is not too thick. The only commercial P/M material with better structure-sensitive magnetic properties is 50 nickel/50 iron, but this is much more expensive. Typical magnetic properties of the silicon-iron alloy are shown in Fig. 5. Work continues to develop alloys with increasing silicon content for commercial applications.

Metal Injection Molding

Metal injection molding (MIM) is a process that has recently been developed to produce parts with complex geometry. Gas-atomized powders are normally selected to fabricate these parts because they are purer and finer. Iron and nickel powders produced by the carbonyl process are also widely selected. Gas-atomized and carbonyl powders are considerably more expensive than water-atomized powder. To date, the expense has restricted parts made by the injection molding process to those of a size less than that of a golf ball. However, as developments in this technology progress, there should be less restriction on the size of parts that can be produced using MIM. One part already being developed using MIM technology is a stainless steel magnetic fuel injector.

There is no restriction of the materials that can be selected to be metal injection molded, provided they are magnetic (alloys containing as base iron, nickel, or cobalt). Therefore, parts for magnetic applications have been made from all of the alloy systems discussed above. In addition, there is no restriction that limits development of new alloy systems for magnetic applications.

Normally, magnetic parts are produced from fine (10 μm) pure gas-atomized powders. These fine powders are mixed with a system of polymers to form a mixture with the texture of toothpaste when heated. The feedstock is then heated and injected under pressure into a die cavity that is of the geometry of the desired part. The molded part is heated to remove the polymers and then sintered. Some processing techniques may use a solvent to extract the binder prior to sintering. Nonetheless, MIM is not an easy process, and more detailed information is contained in the article "Powder Injection Molding" in this Volume.

For magnetic parts, much is to be gained by selecting MIM as the processing technology, provided that the increased cost can be justified. Because fine, pure powders and high-temperature sintering are normally considered in conjunction with MIM processing, densities approaching theoretical density are possible. Therefore, magnetic inductions close to saturation magnetic induction are possible. This means that magnetic inductions equivalent to those of wrought alloys are possible for alloys of similar composition. In addition, new alloys can be developed to take advantage of additional alloying elements that cannot be considered using wrought fabrication technology.

Although metal injection molded parts shrink much more than conventional P/M parts, dimensional control becomes less of a problem because the shrinkage is more uniform. The uniform shrinkage results because the part has uniform stress after injection and, if binder removal is done correctly, rearrangement of the particles occurs. Coupled with enhanced diffusion across pure reduced particle surfaces, this rearrangement results in a high-density part of more uniform dimensions.

The purity of the fine gas-atomized powders that are employed is also helpful in maintaining improved structure-sensitive magnetic properties. If the binder is removed properly and if the parts are sintered correctly, the relative maximum permeability and coercive field should equal those of the best processed conventional P/M parts. An added benefit is the magnetic induction should be equivalent to that of wrought parts, because the part density is so high.

Table 8 Effect of sintering atmosphere on magnetic properties

	0.45 mol% phosphorus iron 2050 °F	
Property	$75H_2/25N_2$	H_2
Carbon, %	<0.01	<0.01
Nitrogen, %	0.012	<0.0010
Oxygen, %	0.070	0.054
Sulfur, %	0.009	0.004
Density, g/cm^3	7.31	7.37
Induction at 15 Oe, kG	13.1	13.7
Residual induction, kG	11.4	12.6
Maximum permeability	2900	5100
Coercive force, Oe	1.8	1.2

A sampling of magnetic properties that are possible from MIM technology is shown in Table 7.

Optimizing Magnetic Properties

Powders. Most magnetic parts are fabricated from water-atomized metal powders. The major exception to this is metal injection molded parts, where gas-atomized powders are selected, as a rule. Because purity is important, carbon, nitrogen, and sulfur are maintained at very low levels if structure-sensitive magnetic properties are to be optimized. The major contaminant in the powders is oxygen in the form of surface oxides. In the case of ferritic stainless steel powders, as much as 2000 ppm of oxygen is concentrated on the surface of the water-atomized powders. Base iron powder normally contains 1000 ppm or less of oxygen. Gas-atomized powders contain even less oxygen. Nevertheless, these surface oxides must be readily reduced to provide a good sinter of the part.

The water-atomized powders are generally classified as –100 mesh powders and have a normal distribution. The gas-atomized powders, or the admixed additions, are normally 10 μm in size. All additions must also be as free as possible of carbon, nitrogen, oxygen, and sulfur. Ferritic stainless steels and the 50 nickel/50 iron powders are prealloyed. In the case of the phosphorus irons and the silicon steels, alloying is accomplished by admixing the alloying components. For the phosphorus irons, generally a 10 μm Fe_3P powder is admixed with the base iron powder, although a Fe_2P powder can also be selected as the alloying addition. Some practitioners have found, however, that Fe_2P powder is more abrasive and contributes to increased die wear. For the silicon irons, a master alloy containing ≅17 wt% Si is generally selected as the alloying addition.

Compaction. Generally, parts are compacted to the highest green density possible because the magnetic induction is dependent on the sintered density. This is a linear function; the higher the density, the higher the magnetic induction. For flux paths, the higher the magnetic induction, the greater the amount of flux that can be carried in a given cross section. The limitation that normally controls the green density that can be achieved is the compaction pressure. Normally, most parts fabricators do not wish to use compaction pressures ≥620 MPa (>45 tsi) because tool breakage becomes a problem. High compacting pressures (1000 MPa) do not improve the magnetic properties of Fe-Si alloys (Ref 9). For certain applications with Fe-Si alloys, it is recommended to combine low compacting pressures and long sintering time (24 h) or medium compacting pressures (600 MPa) and short sintering time (2 h) (Ref 10). The density of complex or long parts also can vary appreciably within the cross section of the part. It is therefore important that the parts fabricator and the user agree on the critical density of the section of the part that is to be fabricated. Generally, green densities of 6.8 to 7.2 g/cm^3 are desired.

Sintering. Of all the variables to consider in selection of the sintering conditions, the selection and control of the sintering atmosphere is the most critical. For example, dissociated ammonia is often used instead of hydrogen for cost reasons. This may introduce a contamination problem from the nitrogen in dissociated ammonia. When parts are sintered in dissociated ammonia, structure-sensitive magnetic properties can be seriously degraded (Table 8). In this example, rings fabricated from 0.45 wt% phosphorus iron were sintered in hydrogen, and companion rings were sintered in dissociated ammonia, at 1120 °C (2050 °F). As can be seen, contamination levels of both sets of parts are equivalent, except for nitrogen and oxygen contamination. Note that the sintered density of the parts sintered in hydrogen was higher. As a result, the magnetic induction of the parts sintered in hydrogen was also higher. The major changes in properties are seen in the structure-sensitive magnetic properties, relative maximum permeability, and coercive field. The relative maximum permeability of the parts sintered in hydrogen was doubled, and the coercive field was significantly lower. Minimization of carbon, nitrogen, oxygen, and sulfur is a must if structure-sensitive magnetic properties are to be achieved. Even if the proper sintering temperature is selected, magnetic properties will be severely degraded if the sintering atmosphere selection is incorrect.

Many kinds of sintering atmospheres are available. In most cases, cost is the prime consideration in the selection of the sintering atmosphere. Sintering atmospheres that can be considered include exothermic gas, endothermic gas, nitrogen, mixtures of nitrogen and hydrogen, dissociated ammonia, hydrogen, vacuum, and inert gas, such as argon. Of these atmospheres, exothermic and endothermic gases contain both carbon and nitrogen. If magnetic parts are sintered in these gases, carbon and/or nitrogen contamination are possible. If either occurs, magnetic properties are severely degraded.

Nitrogen, mixtures of nitrogen, and dissociated ammonia all contain some concentration of nitrogen. Nitrogen contamination can be most severe in magnetic parts because the nitrogen can be in solution in the ferrite or precipitated as nitrides. If in solution, the user may mistakenly believe that the magnetic properties are acceptable, only to find that with time and temperature of operation, the nitrogen has precipitated from solution as nitrides, and magnetic performance is severely degraded.

If structure-sensitive magnetic properties are a requirement, the choice of sintering atmosphere becomes hydrogen, vacuum, or an inert gas like argon. Of these, hydrogen is particularly recommended because hydrogen reduces carbides, nitrides, and oxides. Therefore, if highest purity of the part is desired, hydrogen is the choice. Vacuum or an inert gas is acceptable; however, the atmosphere is neutral and therefore does not assist in reduction of carbides, nitrides, or oxides. It is hydrogen that should be selected if structure-sensitive magnetic properties are required.

Iron and phosphorus-iron parts are normally sintered in belt furnaces at 1120 °C (2050 °F). For optimal magnetic properties, however, a high-temperature sinter at about 1260 °C (2300 °F) should be used. When high-temperature sintering is selected, shrinkage becomes a problem, and often the sintered parts must be sized to bring them to required dimensions. If sizing is required, the part is mechanically cold worked. An anneal should then be provided, if magnetic properties are to be optimized. Normally, parts sintered in belt furnaces will have sintered densities that range from 6.8 to 7.2 g/cm^3. Parts that are high-temperature sintered can have sintered densities as high as 7.6 g/cm^3. Many parts fabricators also use what is referred to as repressing to densify parts that require high magnetic induction. If this process is selected, an anneal must be provided or else magnetic properties are degraded.

Other factors to be considered in selecting a parts fabricator are more subtle. Contamination again is the problem. Many parts fabricators elect to sinter all parts in a common furnace. Because magnetic parts are more demanding in performance, parts fabricators that produce magnetic parts usually dedicate a furnace or furnaces to sintering only magnetic and stainless steel parts. Dedication of the furnace restricts the magnetic parts from becoming contaminated with carbon from other structural steel parts that may have preceded the magnetic parts. In addition to segregation of furnaces, some parts fabricators are also segregating presses to compact stainless steel parts. If parts containing iron powder precede the compaction of the stainless steel parts, there is the possibility of stray iron particles being embedded in the stainless steel parts. If this embedding of iron particles occurs, the stainless steel parts will not appear to be corrosion resistant because rust spots will develop in the parts contaminated with iron.

High-temperature sintering must be used if optimal structure-sensitive magnetic properties are required. High-temperature sintering permits grain growth and also helps to reduce impurities in the parts. Both must be achieved if structure-sensitive magnetic properties are a requirement.

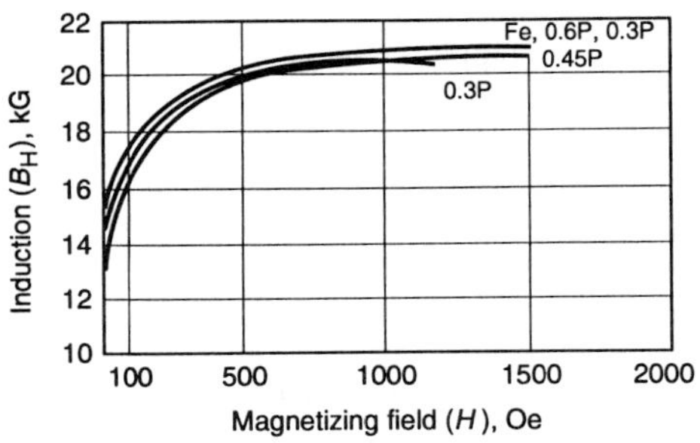

Fig. 6 Magnetization curves for hot repressed iron-phosphorus alloys. Source: Ref 12

High-temperature sintering of ferritic stainless steels, 50 nickel/50 iron, and silicon irons is always necessary if magnetic properties are to be optimized. Ferritic stainless steels contain chromium as the major alloying element. The powders are almost always water atomized and therefore contain 2000 ppm O_2 as surface oxide. This oxide is a protective Cr_2O_3 film. Unless a temperature in excess of 1205 °C (2200 °F) is achieved, there is little likelihood that the oxide film will be reduced and that adequate sintering will occur. In the case of the silicon irons, an SiO_2 oxide is on the surface of the master alloy particles. This oxide is even more difficult to reduce than the Cr_2O_3. It is therefore imperative that the silicon irons be sintered at a temperature of at least 1260 °C (2300 °F). The 50 nickel/50 iron alloy also requires a sintering temperature of at least 1260 °C (2300 °F). The alloy is strongly structure sensitive, and unless grain growth occurs and the alloy is cooled properly, structure-sensitive properties will be degraded.

Effects on Magnetic Properties

Porosity and all kinds of impurities in the form of inclusions generate the same magnetic interactions as small-size air gaps inside ferromagnetic materials. Their effects include demagnetization and shifting of magnetic domain borders during magnetization. Residual stresses may also degrade the magnetization characteristic of the sintered part. These general factors are summarized here.

Alloying. As previously noted, cobalt, iron, and nickel are the base elements for any alloy in magnetic applications. Other elements can be included to provide special properties, but the addition of other elements reduces the saturation magnetization. The addition of 0.45 or 0.8 wt% P does little to reduce the saturation magnetization of iron. In general, phosphorus lowers the magnetic induction by about ~0.05 T (500 G) (Fig. 6). In contrast, an addition of 3 wt% Si lowers the saturation magnetization to 1.95 T (19,500 G).

Effect of Porosity. Most magnetic parts that are manufactured contain some degree of porosity. If no pores were present in the material and a proper sinter were provided, the magnetic induction and the remanent magnetization should be equivalent to a wrought product of comparable composition. Pores are merely an absence of material that is available to be magnetized. Therefore, the magnetic induction and the remanent magnetization decrease linearly as the amount of porosity increases. This behavior is shown in Fig. 7(a) for iron sintered at 1260 °C (2300 °F) in hydrogen atmosphere for 1 h. The effect of density on magnetic properties of sintered iron is shown in Fig. 7(b) and 7(c).

Purity. The major impurities in magnetic materials that degrade magnetic properties include carbon, nitrogen, oxygen, and sulfur. Small amounts of any of these impurities can severely degrade structure-sensitive magnetic properties. Most water-atomized powders, as well as sintering furnaces, are relatively free of sulfur contamination. Therefore, sulfur is not discussed in detail as a contaminant.

Oxygen is normally overlooked as a contaminant because oxygen is not a contaminant in wrought steels. However, powders are mainly water atomized and have large surface area. Oxygen then becomes a major problem in P/M alloy magnets. In iron and phosphorus iron, oxygen can be reduced to relatively low levels. To best accomplish this, hydrogen should be selected as the furnace atmosphere, and high-temperature sintering should be considered. The curve shown in Fig. 8 shows the effect of sintering temperature on oxygen reduction for parts that were sintered in hydrogen.

Oxygen contamination in silicon irons and stainless steels is most troublesome. High-temperature sintering is mandatory if any reduction is to be achieved. Hydrogen as a furnace atmosphere is also a must, or else no reduction of oxygen will result. Unless oxygen is reduced, the customer has wasted his money paying for these two special-purpose alloys. In ferritic stainless steel, structure-sensitive magnetic properties equivalent to wrought ferritic stainless steels are possible through high-temperature sintering in hydrogen. However, corrosion resistance may not be adequate for some applications. For silicon irons, the SiO_2 film surrounding the master alloy particles are not reduced, and the silicon does not enter into solid solution in the ferrite.

Carbon is generally introduced into the part by improper burnoff of lubricant, furnace contamination, or choice of an exothermic or endo-

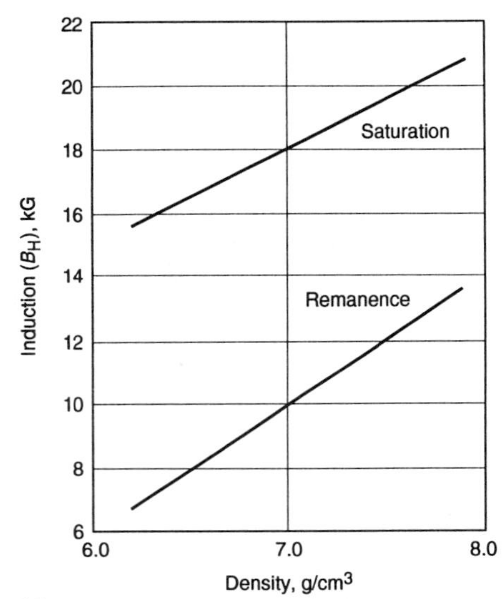

(a)

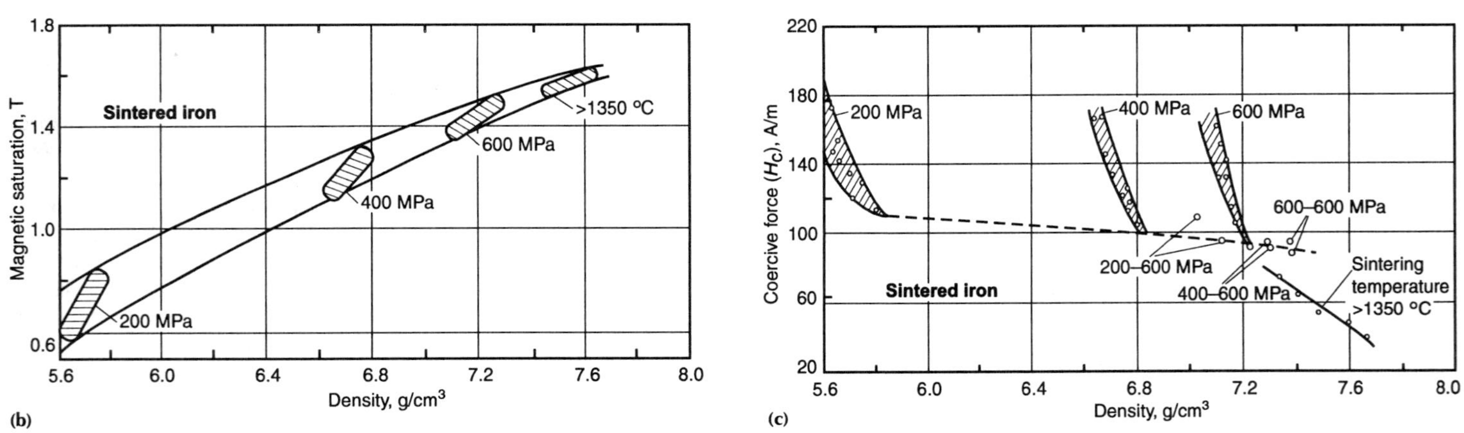

(b) (c)

Fig. 7 Density effects on magnetic properties. (a) Effect on magnetic induction and remanent magnetization (1260 °C sinter/lh/H_2). (b) Magnetic saturation. (c) Coercive force of sintered iron as a function of density and sintering density (hatched areas indicate identical compaction but varying sintering times from 0.5 to 16 h and temperature from 1100 to 1300 °C. Source: Ref 10, 11

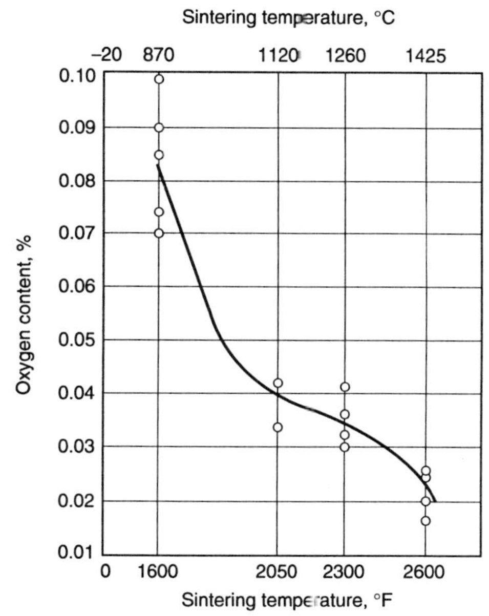

Fig. 8 Effect of sintering temperature on oxygen content. Source: Ref 12

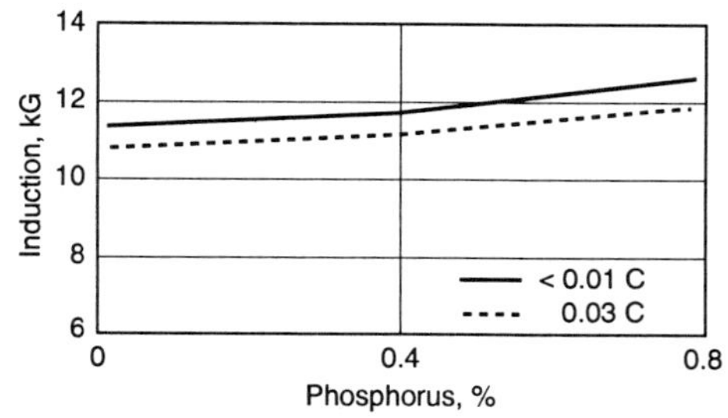

Fig. 9 Effect of carbon pickup on induction. Source: Ref 11

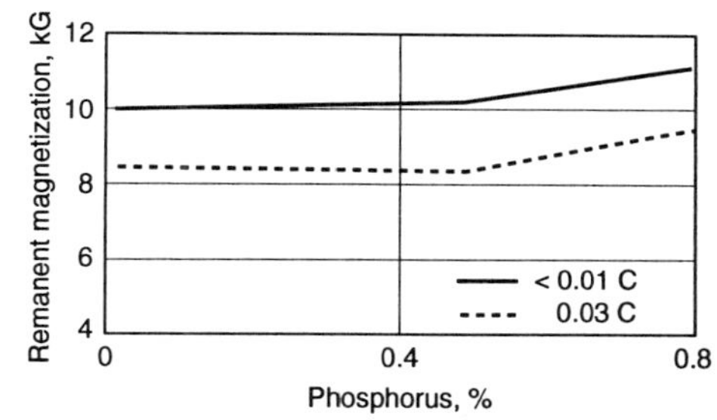

Fig. 10 Effect of carbon pickup on remanent magnetization. Source: Ref 11

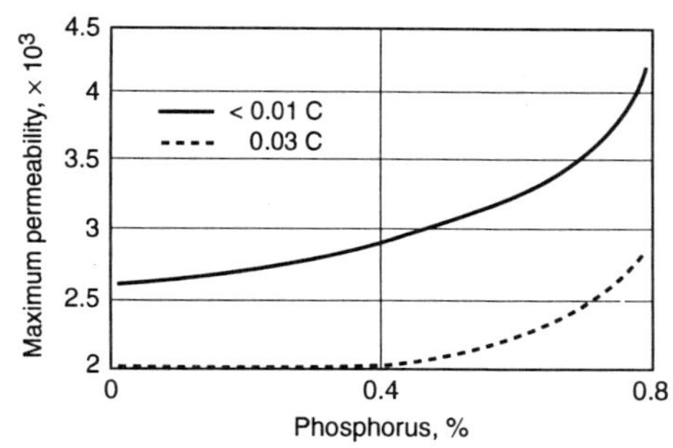

Fig. 11 Effect of carbon pickup on permeability. Source: Ref 11

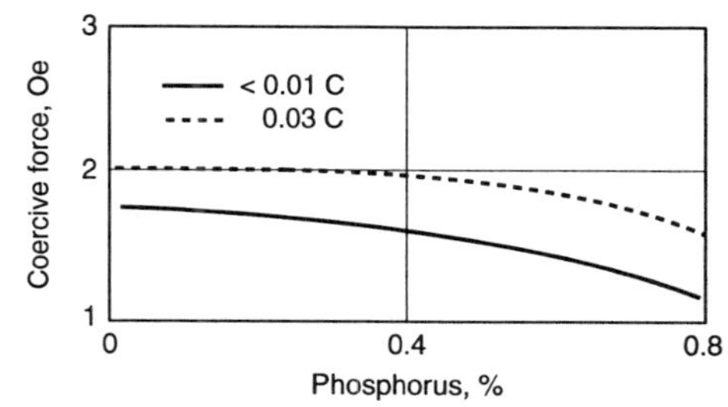

Fig. 12 Effect of carbon pickup on coercive force. Source: Ref 11

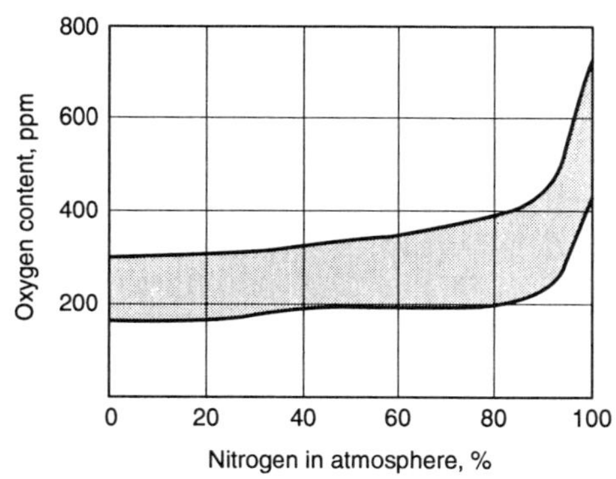

Fig. 13 Effect of nitrogen base atmospheres on reduction of oxygen. Source: Ref 11

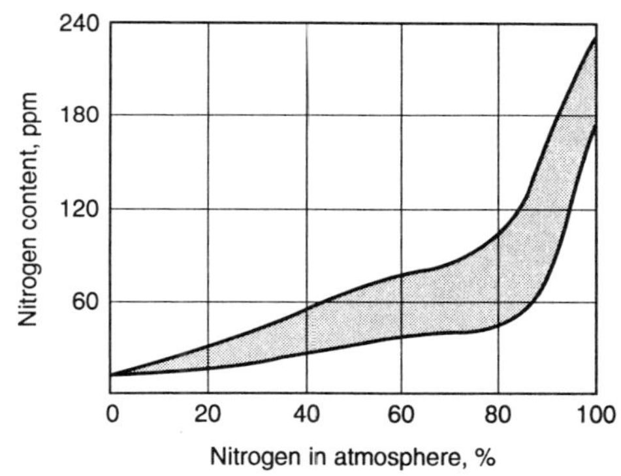

Fig. 14 Nitrogen content of compacts sintered in nitrogen-base atmospheres. Source: Ref 11

thermic atmosphere. Most fabricators of magnetic parts use dedicated furnaces for sintering magnetic parts and select protective atmospheres that do not include carbon. However, most part fabricators burn off lubricant in the preheat section of the sintering furnace. If a belt furnace is employed, the belt speed is fixed by the length of the hot zone of the furnace. If large parts are sintered, the speed may be too fast to permit all of the lubricant to void from the part, resulting in carbon entrapment within the part. This entrapment of carbon is generally the reason for carbon contamination in currently used parts. However, most parts fabricators that produce parts for magnetic applications are well aware of this and generally avoid contaminating parts with carbon.

Carbon contamination can occur when furnaces are not dedicated to sintering of parts that do not contain carbon. In addition, green parts contain lubricant. Unless this lubricant is completely voided from the parts, carbon contamination results.

Figures 9 through 12 demonstrate the effect that a small quantity of carbon can have on magnetic properties. The example is of a furnace that had previously been used to sinter parts containing carbon prior to sintering magnetic parts. The example can also serve for situations where small amounts of lubricant are trapped in the parts, owing to improper burnoff. The data are inclusive for both iron and phosphorus irons. Figure 10 shows that this small amount of contamination does not have a serious effect on the magnetic induction because the amount of iron available for magnetization controls. However, Fig. 10 to 12 show the effect on structure-sensitive magnetic properties. As little as 0.03 wt% C seriously degrades these properties.

Nitrogen is another contaminant that damages structure-sensitive magnetic properties. As with carbon, the solubility of nitrogen is of the order of 0.006 wt%. Any amount greater than this causes nitrides to precipitate from solution. However, at common sintering temperatures, the amount of carbon and nitrogen that can be in solution in iron is greatly increased. On cooling, this amount of carbon or nitrogen cannot be maintained in solid solution, and therefore with time carbides or nitrides precipitate as a second phase. When this happens, structure-sensitive magnetic properties are severely degraded. Generally, selection of the protective sintering atmosphere is the main cause for nitrogen contamination. If structure-sensitive magnetic properties are required, nitrogen, nitrogen hydrogen mixtures, and dissociated ammonia should not be considered as the furnace protective atmosphere.

Nitrogen from these furnace atmospheres is the main cause of nitrogen contamination in iron, phosphorus irons, and ferritic stainless steels. It is to be remembered that nitrogen is not reducing like hydrogen. Therefore when nitrogen is selected, not only does nitrogen contamination result, but surface oxides are not reduced either (Fig. 13). The effect on iron-base alloys of nitrogen contamination resulting from nitrogen/hydrogen mixtures is shown in Fig. 14.

Aging Effects. Both carbon and nitrogen have an effect on magnetic properties after aging. Sometimes a part can behave like a high-purity part after sintering because, at elevated temperatures such as those selected to sinter iron or phosphorus iron parts, large concentrations of carbon or nitrogen can enter into solid solution in the austenite and form carbides or nitrides as the material cools. At room temperature, only 0.008 wt% C and only 0.006 wt% N can remain in solution in the ferrite. A slow cooling rate permits precipitation of the carbides or nitrides; magnetic properties are degraded, and the parts may be rejected. With a fast cooling rate, carbon or nitrogen can be retained in solution. In this case, the magnetic properties are not degraded,

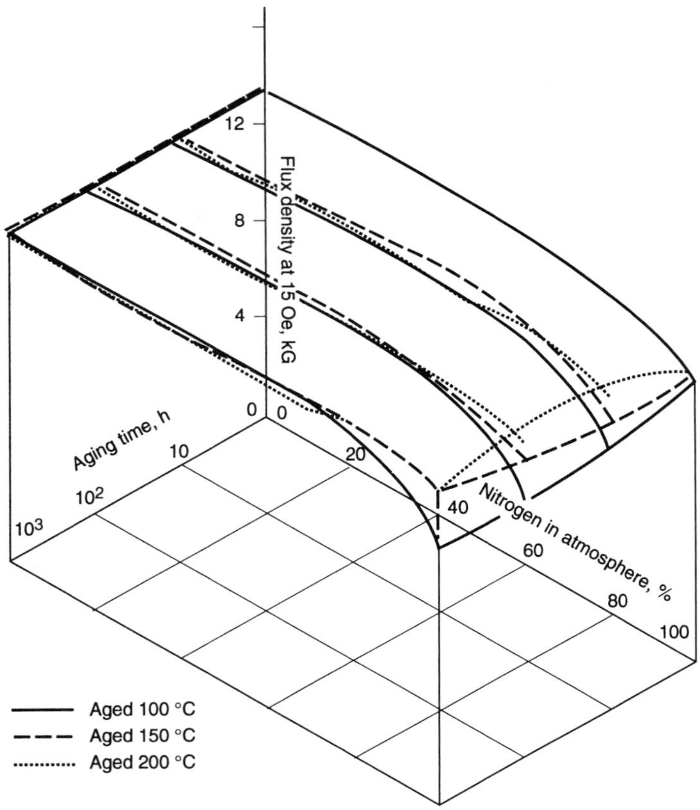

Fig. 15 Effect of aging on induction of iron compacts. Source: Ref 11

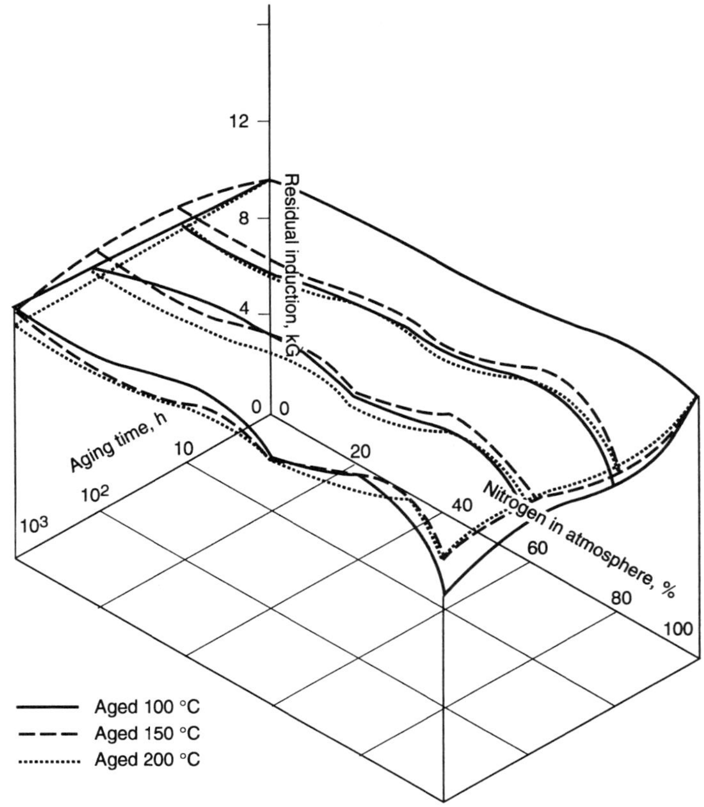

Fig. 16 Effect of aging on the residual induction of iron compacts. Source: Ref 11

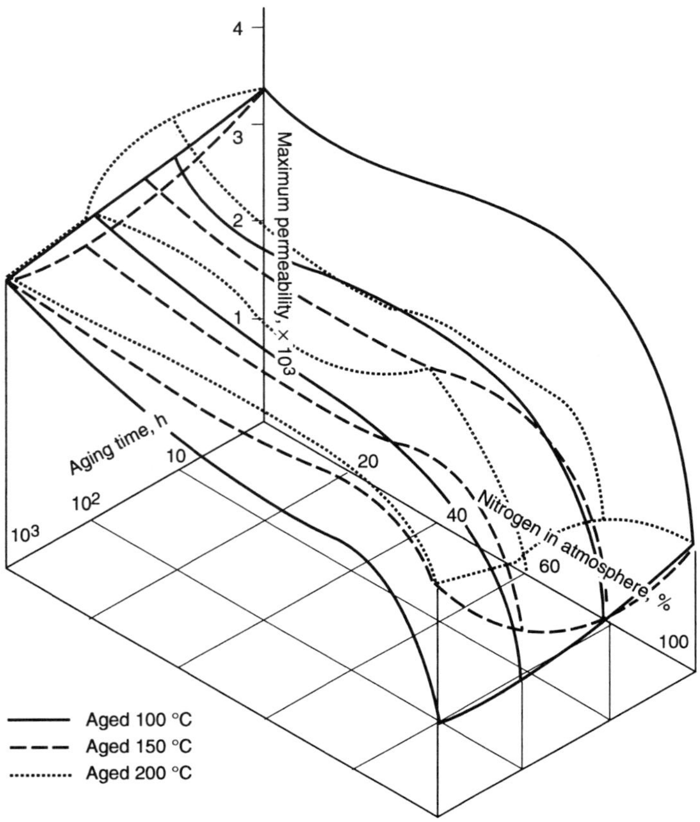

Fig. 17 Effect of aging on the maximum permeability of iron compacts. Source: Ref 11

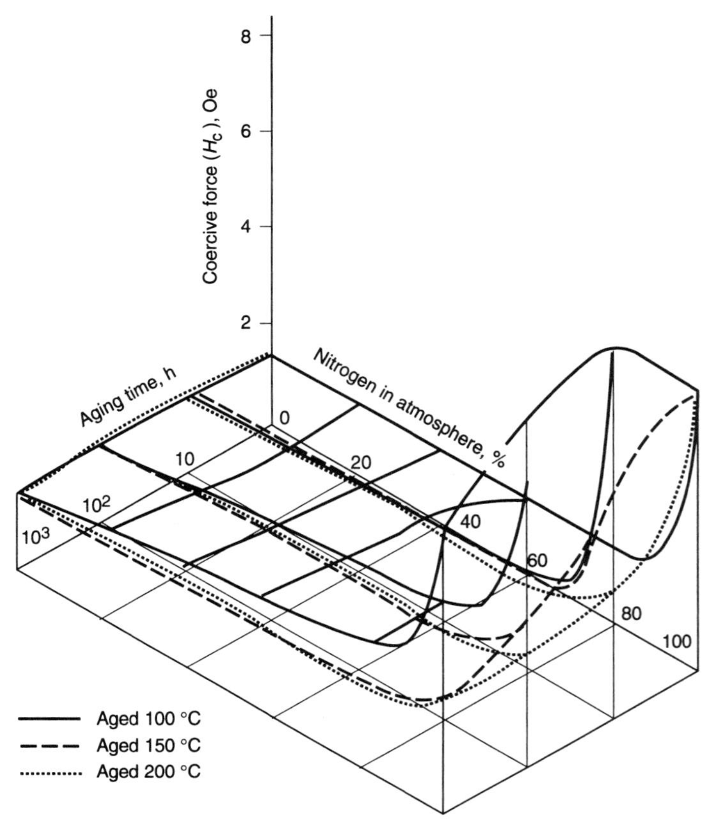

Fig. 18 Effect of aging on the coercive force of iron compacts. Source: Ref 11

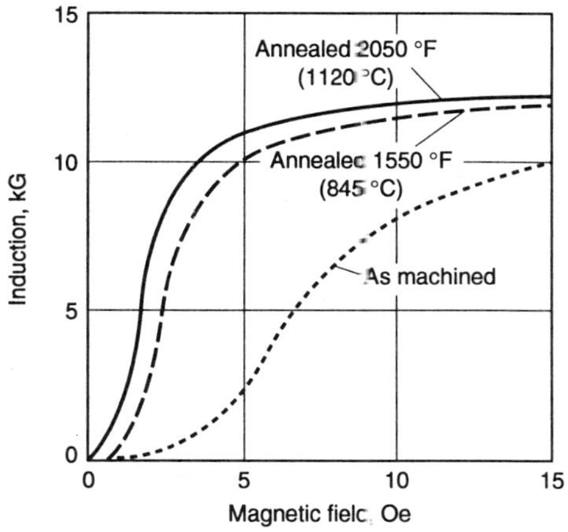

Property	As machined	845 °C (1550 °F) annealed	1120 °C (2050 °F) annealed
Carbon	0.029	<0.01	<0.01
Nitrogen	0.017	0.0079	<0.0010
Oxygen	0.18	0.090	0.039
Sulfur	0.014	0.013	0.007
Magnetic saturation, kG	18.9	...	...
Induction at 15 Oe, kG	10.0	11.9	12.1
Remanent magnetization, kG · Oe	8.5	10.8	11.2
Maximum permeability, kG · Oe	0.8	2.5	3.6
Coercive force, Oe	5.0	2.3	1.5

Fig. 19 Magnetic properties of iron cores after annealing treatment. Source: Ref 11

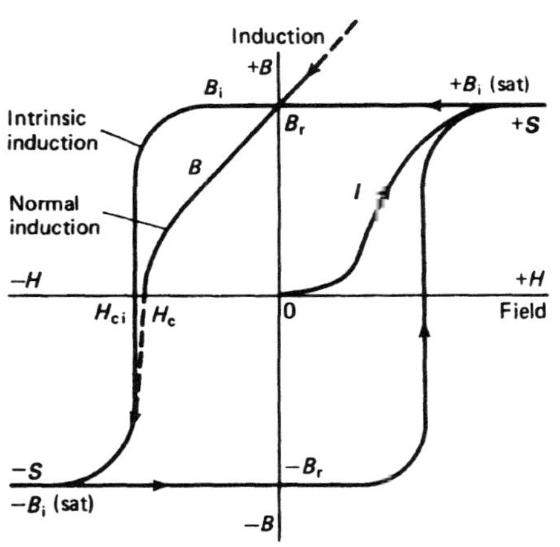

Fig. 20 Major hysteresis loop for a permanent magnet material

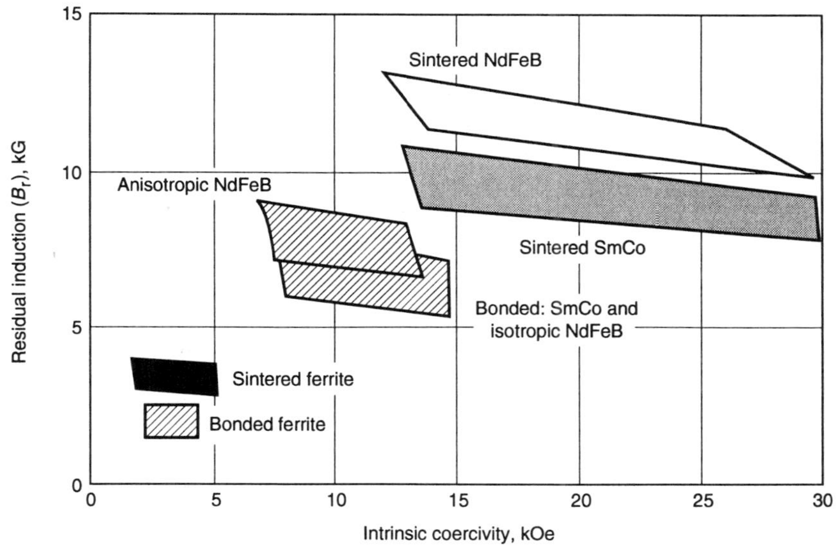

Fig. 21 Typical magnetic properties of bonded and sintered ferrites. Source: Ref 12

and the user can mistakenly believe that the part has acceptable magnetic properties. With time and temperature in service, the carbides and nitrides precipitate from solution, thereby degrading magnetic properties.

Two kinds of tests are employed to detect whether aging will occur in magnetic parts. An accelerated test heats the parts for 100 h at 150 °C, with a retest for magnetic properties. The alternative test heats the parts for 600 h at 100 °C and retest. Figures 15 through 18 show the effects on magnetic properties of aging times at 100, 150, and 200 °C for materials sintered in nitrogen/hydrogen mixtures. Contamination does not appreciably affect the magnetic induction and residual induction (Fig. 15, 16). However, structure-sensitive properties (relative maximum permeability, coercive field, and residual induction) are significantly affected if parts are contaminated with carbon or nitrogen (Fig. 17, 18).

Effect of Residual Stresses. Secondary operations include any machining operation, sizing, and repressing. If any of these operations are required, mechanical stresses are induced into the alloy. Mechanical stresses degrade structure-sensitive magnetic properties. Therefore, an anneal is required to eliminate any mechanical stresses that may have been induced during secondary processing. Most parts fabricators have furnaces that are operating at the selected sintering temperatures. The parts fabricators simply reprocess the stressed parts a second time through the sintering furnace. For optimal magnetic properties, however, the protective atmosphere must be hydrogen.

Initially, when a part is evaluated for production, it is not produced to shape because dies are expensive. Rather, a slug similar in size to the part is compacted, sintered, and machined into the dimensions of the part. Machining induces mechanical stresses that degrade magnetic properties (Fig. 19). Unless the machined parts are annealed to remove the mechanical stresses, the measured magnetic properties will not be representative of the sintered part that will ultimately be produced. The misleading data can often result in extra expense owing to specification of excessive copper windings, weight, or permanent magnet requirements needed to activate the magnet.

Effect of Surface Treatment. Most ferrous magnetic parts are normally coated for corrosion resistance prior to service. Normally electroless nickel is used because nickel does not degrade magnetic properties. Prior to plating, the porous parts are infiltrated with a polymer to seal the porosity. Other elements can be used for plating as well, such as zinc. Because the plate is thin, magnetic properties are not degraded significantly. However, some parts fabricators prefer to steam treat, carburize, or nitride magnetic parts. In these instances, the surface of the part becomes contaminated and magnetic performance suffers.

Permanent Magnets

Contrary to soft magnetic materials, permanent magnets cannot be easily demagnetized once they are magnetized. They provide a steady magnetic field and are unaffected by stray fields.

As with soft magnets, many alloys can be selected to provide the required steady magnetic field. These alloys include metals, intermetallics, and ceramic materials. Commonly included are high-carbon steels, Alnicos, CuNiFe, iron cobalt alloys containing vanadium or molybdenum, platinum cobalt, hard ferrites, cobalt rare earths, and Fe-B-Nd alloys.

Permanent magnets are characterized by a high coercive force (H_c) and residual induction (B_r). These alloys are processed to provide the maximum energy product, which is the numerical value of $B_x \cdot H_x$ in the second quadrant (top left) of the hysteresis curve (Fig. 20). The energy product is measured by fitting the largest rectangle to the demagnetization curve, found in the second quadrant of the hysteresis loop (Fig. 20). It is normally expressed as the product $(B_dH_d)_{max}$ and determines the quality of the permanent magnet. Higher contents of iron, cobalt, or nickel result in magnetization. Highest saturation is found in the iron cobalt alloys. The ability to resist demagnetization is of utmost importance. In this case, unlike soft magnets, precipitates, mechanical strain, material imperfections, and fine particles are all desirable in the microstructure.

Hard Ferrites. There are two kinds of ferrites: hard and soft. This section discusses only hard ferrites. Hard ferrites usually have a general formula of $MO{\cdot}6Fe_2O_3$ and have a hexagonal

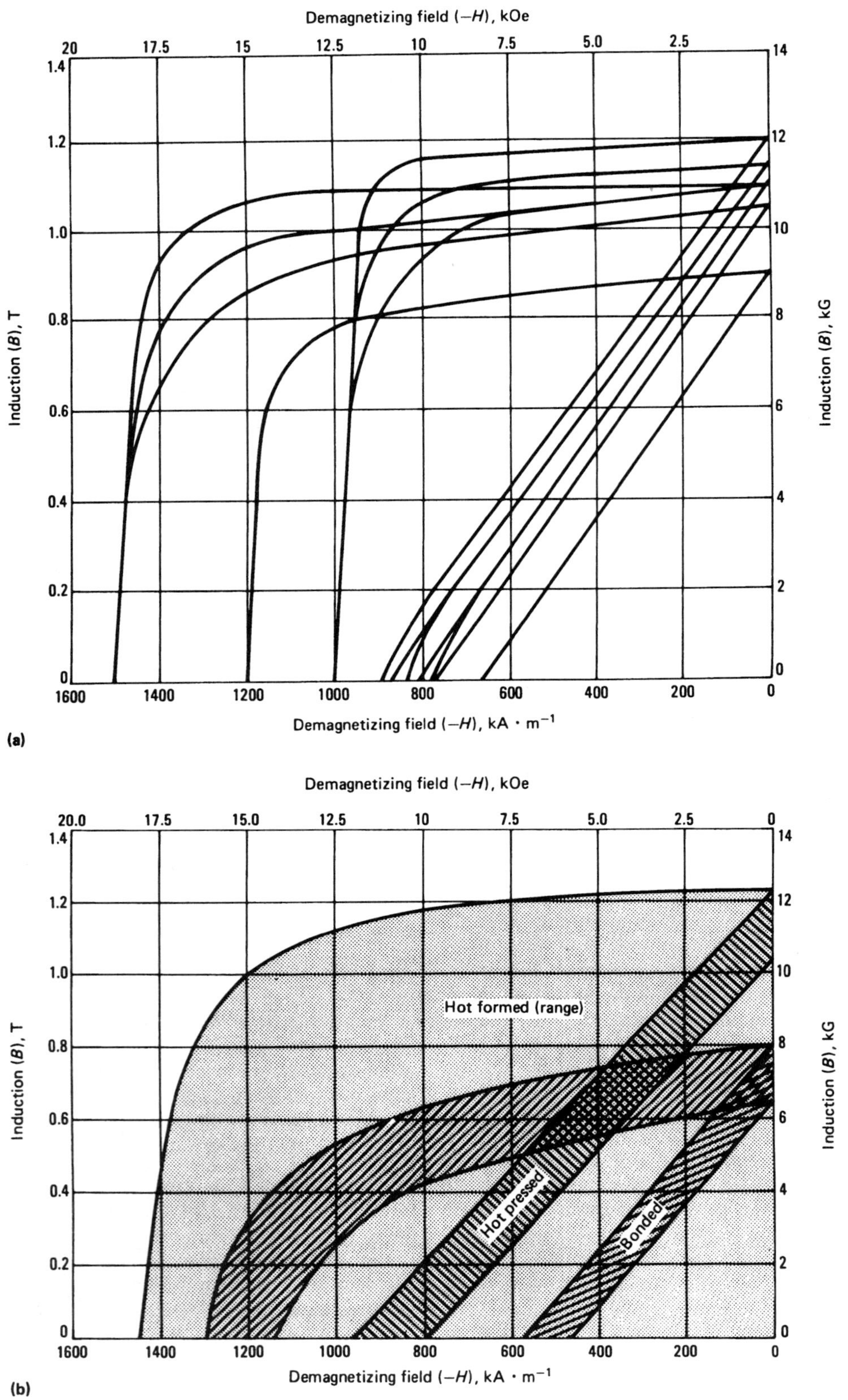

Fig. 22 Demagnetization curves for Nd-Fe-B magnet alloys. (a) Sintered. (b) Prepared from rapidly solidified ribbon. Source: "Permanent Magnet Materials," *ASM Handbook,* Vol 2, 1990, p 790

structure. Two hard ferrites, barium ferrite ($BaO{\cdot}6Fe_2O_3$) and strontium ferrite ($SrO{\cdot}6Fe_2O_3$) are available. The latter is generally favored for current applications.

The ferrites are produced by blending Fe_2O_3 with $SrCO_3$ and other additives. The blend is then calcined to form a hexaferrite. The hexaferrite is wet milled and then either wet or dry pressed in a magnetic field. The green part is then dried, fired in air, ground, and inspected. Even today almost 90% of permanent magnets are made from ferrites, because of their low cost.

Many ferrite applications require only polymer-bonded magnets. In this case, the magnetic powder is blended with various polymers. The mixture is then rolled into a sheet, cured, and slit or punched into a desired shape. If the polymer selected is a nitrile rubber, flexible magnets are produced.

Permanent hard ferrite magnets are selected mainly because of their low cost. Although they have only a moderate flux field, they have a high resistance to demagnetization. They are brittle and have moderate temperature stability. Hard ferrite magnets are used in loudspeakers, generators, relays, motors, telephone ringers, receivers, door closers, seals, and latches.

Alnicos. The Alnicos consist of a family of alloys containing aluminum, nickel, and cobalt. Next to the ferrites, these alloys are the most widely used today, again, owing to low cost. The alloys have a high-energy product, of the order of 39.8 to 71.6 kJ/m^3 (5 to 9 MG · Oe), a high remanent magnetization of 0.7 to 1.35 T (7 to 13.5 kG), and a moderate coercivity of 40 to 160 kA/m (500 to 2010 Oe). The alloys are iron base with additions of aluminum, nickel, cobalt, and about 3 wt% Cu. Titanium is also alloyed with the high coercivity alloys.

In the case of the sintered alloys, the alloying elements are blended, compacted, delubed, sintered, heat treated, thermal aged, and surface finished. They can be compacted to very close tolerances and sintered to high strength. The alloys are extremely temperature stable, and magnetic orientation can be changed through a reheat treatment process.

Rare-Earth Alloys. Rare-earth magnets have maximum energy products up to 240 kJ/m^3 (30 MG · Oe) and coercivities to 3200 kA/m (40 kOe). $SmCo_5$ single-phase magnets are widely used. The high coercivity of the rare earths is based on nucleation and pinning of domain walls at surfaces and at grain boundaries. Therefore, powders milled to particle sizes ranging from 1 to 10 μm are used for the alloy composition. The blended fine particles are compacted into a green part in a magnetic field, sintered, heat treated, machined to size, coated, and magnetized. Typical energy products range from 130 to 160 kJ/m^3 (16 to 20 MG · Oe) with proper processing.

In precipitation-hardened $Sm(Co, Cu)_{7.5}$, copper is substituted for some of the cobalt. This produces a fine, coherent precipitate (~10 nm) when aged at 750 to 930 °F (400 to 500 °C). The coherent precipitates pin domain walls at the fine-precipitated particles. Again, a magnetic field can be employed in conjunction with the compacting cycle to align the magnetic field with the applied field. Additions of small amounts of iron and zirconium can further increase the coercivity.

Rare-earth magnets are used in medical devices, such as thin motors in implant pumps, valves, and in aiding eyelid motion. Rare-earth

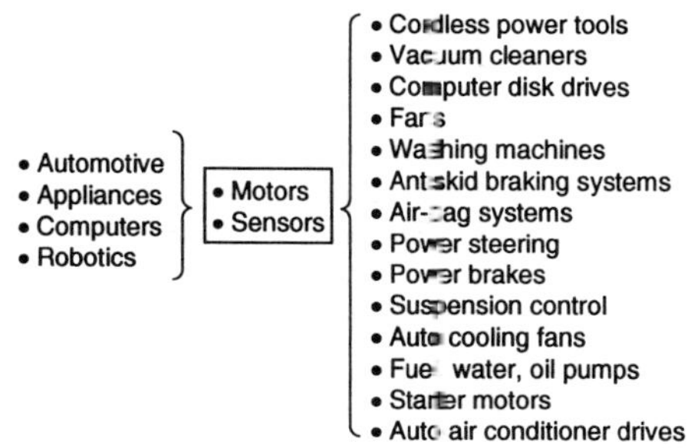

Fig. 23 Potential rare-earth bonded applications. Source: Ref 12

magnets are also used in electronic watches and traveling wave tubes. Direct-current and synchronous motors and generators include rare-earth magnets for size reduction. The reasons for usage include high flux density, high coercive field, high energy for size, and temperature stability.

Neodymium Iron Boron. Neodymium is also a rare-earth element. However, it is more abundant and less costly than other rare earths. An alloy close to the composition of $Nd_{15}Fe_{77}B_8$ is melted and then rapidly solidified by casting the molten stream on a cold rotating disk. This rapid solidification produces a thin amorphous ribbon. The ribbon, in turn, is chopped and crushed into 200 μm particles. The finer particles, being less stable, can oxidize and create a fire hazard. The particles can be bonded with polyamides and injection molded to form magnets. The bonded magnets can have a maximum energy product of typically 95.5 kJ/m^3 (12 MG · Oe), a residual induction of typically 0.76 T (7.6 kG), and a coercive field of typically 501 kA/m (6.3 kOe).

The crushed powder can also be formed into dense magnets by hot pressing. These magnets may have a maximum energy product as high as 60 MG · Oe. Disadvantages are that the alloy has a maximum operating temperature of only 170 °C, and the magnet must be coated for corrosion resistance.

Typical magnetic properties of the bonded and of the sintered magnets are shown in Fig. 21. Demagnetization curves are shown in Fig. 22. Uses for the bonded magnets are shown in Fig. 23. Automotive applications are listed in Table 9. Household applications of magnets include numerous devices for products such as air conditioners, electric shavers, garage door openers, lawn mowers, refrigerators, video cassette recorders, and washers and dryers.

Ferromagnetic Cores

Ferromagnetic cores are used for alternating-current applications where the frequency is high, generally within the kilohertz or the megahertz range. Iron, soft ferrites, or molybdenum Permalloy powder (B&D Industrial & Mining Services, Jasper, AL) have generally been used for these applications. Applications include power supply transformers, audio transformers, alternating current and radio-frequency filter inductors, broadband RF transformers, narrow-band RF transformers, and damping networks. Cores are generally produced in the form of rods, toroids, pot cores, and other forms.

Table 9 Automotive applications for magnets

Top usage: Currently, 20–30 per vehicle; future, 100 per vehicle

Door locks	Seat recline
Windows	Seat supports
Mirrors	Heater blower
Starter	Water pump
Radiator fan	Oil pump
Idle speed	Air conditioning
Cruise control	Electric braking
Air pump	Differential control
Brake assist	Wipers
Steering	Spoilers/aerodynamics
Transmission control	Headlight wipers
Antilock	Headlight direction
Suspension	Entertainment
Fuel pump	Sunroof
Seat base	Traction control

The purpose of the core is to generate as high a magnetic flux as possible. Thus the induction B should be as high as possible, and the permeability should be high as well. Besides magnetic qualities, however, eddy current losses are also a critical concern at high frequencies. Therefore, particles should be as fine as possible, and insulation of particles is critical.

At high frequencies, eddy currents losses are a problem by the relation:

$$E\ (\text{loss}) = \frac{k f^2 B^2 t^2}{\rho}$$

where f is frequency; B, flux density; t, thickness; ρ, resistivity; and k, a constant. Because thickness is also a squared term within the equation expressing the eddy current losses, it is imperative that fine particles be used for these applications and that each particle be adequately insulated. The other variable that can be controlled is the resistivity of the material. Because the resistivity is a reciprocal of eddy current losses, the resistivity should be as high as possible.

The form specified by the designer dictates the power level of the circuit in which the core is used, physical dimensions of the assembled transformer or inductor, and fabrication time. Generally, core assemblies use insulated bobbins to contain the windings for high-volume production. Pot cores, U cores, and I-E cores are generally used for these applications. Choice between core material is usually iron powder or ferrite. Because ferrites are easier to saturate, they are generally selected for direct-current to direct-current converters and magnetic amplifiers. The permeability of ferrites is also higher than that of iron, which permits an upper range of 90. With ferrite, it is possible to obtain a permeability as high as 5000. However, core temperature stability is also a requirement, and the greater the permeability, the more unstable the core will be at the higher frequencies. Stability is extremely important in circuits, such as narrow-band filters, narrow-band tuned transformers, and oscillators. Iron powder is a better choice for these applications because it offers a higher efficiency (Q factor), a low permeability, and therefore better stability over a wide range of flux levels and temperatures.

Table 10 Magnetic performance data of plastic-coated iron powder

Material	Initial permeability	Maximum permeability	Coercive force, Oe	Induction at 40 Oe, G
SC120	120	425	4.7	11,200
SC100	100	400	4.8	10,900
TC80	80	210	4.7	7,700

Iron powder cores consist of fine powders that are insulated from each other by a binder compound, such as a phosphate, which insulates the particles from each other. A lubricant can also be added to facilitate die compaction. The mix of coated particles is compacted into a desired shape at relatively high pressure to compact the core to the greatest density without disrupting the insulating coating. The core can then be baked to provide additional strength. The useful frequency range versus the Q of an inductor wound on the core is determined by the powder characteristics, size, and density. Extremely fine particles that distribute uniformly within the insulating coating permit application for very high frequency (VHF) applications and higher.

Recently, powder producers have modified the process by coating iron particles with a thin layer of an insulating thermoplastic. Coatings of the polymer are applied using fluid-bed technology. Hot powder compaction is then utilized to produce compacts with higher densities than cores made using prior technology. Strength for handling is also greatly enhanced by the new technology.

Typical magnetic properties of the warm compacted insulated powder are included in Table 10. Permeability, maximum induction, and resistivity vary as the amount of the insulating coating. At high frequencies, the polymer-insulated warm-compacted powder appears to have Q values and permeabilities that will provide competition for electronic circuits. Compared to laminations, however, the eddy current losses and the low induction for equivalent applied fields and frequencies limits the performance of these insulated powders in applications below 400 Hz.

REFERENCES

1. R. German, *Powder Metallurgy of Iron and Steel,* John Wiley & Sons, 1998
2. C. Lall, Fundamentals of Magnetism, Chapter I, *Soft Magnetism,* Metal Powder Industries Federation, 1992, p 1–27
3. *Magnetic Properties,* Vol 03.04, *Annual Book of ASTM Standards,* ASTM
4. W.F. Smith, Magnetic Materials, Chapter 11, *Principles of Material Science and Engineering,* McGraw-Hill, 1986, p 604

5. D.T. Hawkins and R. Hultgren, Constitution of Binary Alloys, *Metallography, Structures and Phase Diagrams,* Vol 8, *Metals Handbook,* 8th ed., American Society for Metals, 1973, p 251–376
6. B. Weglinski and J. Kaczmar, Effect of Fe3P Addition on Magnetic Properties and Structure of Sintered Iron, *Powder Metall.,* Vol 23 (No. 4), 1980, p 210–216
7. C. Lall, Soft Magnetic Properties of Selected Alloys, Chapter III, *Soft Magnetism,* Metal Powder Industries Federation, 1992, p 49–87
8. L.W. Baum, Jr., "Theoretical and Practical Considerations for P/M Production of Magnetic Parts," presented at P/M Technical Conference (Philadelphia), Hoeganaes Corporation, Oct 1978
9. J. Kaczmar and B. Weglinski, *Int. J. Powder Metall. Powder Technol.,* Vol 17 (No. 1), 1984, p 25
10. A. Salak, *Ferrous Powder Metallurgy,* Cambridge International, 1995, p 188, 283
11. E. Klar, *Powder Metallurgy Applications, Advantages and Limitations,* American Society for Metals, 1983
12. J. Ormerod and S. Constantinides, *P/M Magnetic Materials and Applications,* short course, 22–23 June 1993, Metal Powder Industries Federation, 1993

SELECTED REFERENCES

- *Advances in Powder Metallurgy and Particulate Materials,* Metal Powder Industries Federation, 1989–1997
- R.M. Bozorth, *Ferromagnetism,* Institute of Electrical and Electronics Engineers, 1993
- B.D. Cullity, *Introduction to Magnetic Materials,* Addison-Wesley, 1972
- M.F. DeMaw, *Ferromagnetic-Core Design and Application Handbook,* Prentice-Hall, 1981
- K.H. Moyer, Selection of Powders and Processing for P/M Soft Magnetic Components, *Advances in Powder Metallurgy,* Vol 2, Metal Powder Industries Federation, 1990, p 385–399
- P/M Special Materials and Applications, Structure Related Applications, *Modern Developments in Powder Metallurgy,* Vol 11, 1976, p 355–384
- P/M Technical Conference (Philadelphia), Oct 1978, Hoeganaes Corporation
- *Prog. Powder Metall.,* selected articles, 1986 and 1987
- *Properties and Selection: Stainless Steels, Tool Materials and Special-Purpose Metals,* Vol 3, *Metals Handbook,* 9th ed., American Society for Metals, 1980, p 597–639
- B. Weglinski, Sinters for Magnetic Applications, *Rev. Powder Metall. Phys. Ceram.,* Vol 5 (No. 1), 1992
- B. Weglinski, Soft Magnetic Powder Composites—Dieolectromagnetics and Magnetodielectrics, *Rev. Powder Metall. Phys. Ceram.,* Freund Publishing Limited, 1990

Powder Metallurgy Electrical Contact Materials

ELECTRICAL CONTACTS are metal devices that make and break electrical circuits. Contacts are made of either elemental metals, composites, or alloys that are made by the melt-cast method or manufactured by P/M processes. Powder metallurgy facilitates combinations of metals that ordinarily cannot be achieved by alloying.

A majority of contact applications in the electrical industry utilize silver-type contacts, which include the pure metal, alloys, and powder metal combinations. Other contact materials include copper, copper-base materials, platinum group metals, tungsten, molybdenum, and mercury.

This article describes the processing, properties, and performance of electrical contacts based on P/M or hybrid composite technologies with refractory metals and compounds. Copper-base P/M contact materials are described in the article "Copper Powder Metallurgy Alloys and Composites" in this Volume.

Tungsten and Molybdenum

Most tungsten and molybdenum contacts are made in the form of composites with silver or copper as the other principal component. Tungsten, which was one of the earliest metals other than copper and silver adopted for electrical contact applications, has the highest boiling point (5930 °C, or 10,700 °F) and melting point (3110 °C, or 5625 °F) of all metals; it also has very high hardness at both room and elevated temperatures. Therefore, as a contact material, it offers excellent resistance to mechanical wear and electrical erosion. Its main disadvantages are low corrosion resistance and high electrical resistance. After a short period of operation, an oxidized film will build up on tungsten contacts, resulting in very high contact resistance. Considerable force is required to break through the film, but high pressure and considerable impact cause little damage to the underlying metal because of its high hardness. Tungsten contacts are used in switching devices with closing forces of more than 20 N (4.5 lbf) and in circuits with high voltages and currents not more than 5 A, such as automotive ignitions, vibrators, horns, voltage regulators, magnetos, and electric razors. In low-voltage dc devices, tungsten is always used as the negative contact and is paired with a positive contact made of precious metal.

Tungsten rods or strips that are consolidated by swaging or rolling from sintered powder compacts have very poor ductility. They cannot be cold worked, in contrast to other contact materials. Tungsten disks are usually cut from rods or punched from strips and then brazed directly to functional parts such as breaker arms, brackets, or springs.

Properties such as grain size, grain configuration, and the degree of fibrous structure, which affect contact behavior, are controlled by using special swaging methods and annealing cycles. Tungsten disks usually are supplied with a ground finish, but they can also be electrochemically polished to obtain high-luster surfaces.

The high boiling and melting points of molybdenum—5560 °C (10,040 °F) and 2610 °C (4730 °F), respectively—are second only to those of tungsten and rhenium. Molybdenum is not used as widely as tungsten because it oxidizes more readily and erodes faster on arcing than tungsten. Nevertheless, because the density of molybdenum (10.2 Mg/m^3, or 0.369 $lb/in.^3$) is about half that of tungsten (19.3 Mg/m^3, or 0.697 $lb/in.^3$), use of molybdenum is advantageous where mass is important. Its cost by volume is also lower.

In addition to its use in make-break contacts, molybdenum is widely used for mercury switches because it is not attacked, but only wetted, by mercury.

Like tungsten, molybdenum strips and sheets are made by swaging or rolling sintered powder compacts. Disks made from rods or sheets are brazed to blanks of other structural components. Table 1 lists the properties of tungsten and molybdenum, and Fig. 1 and 2 show the effect of temperature or diameter on various properties.

Composite Materials

Composite electrical contacts are made from three categories of materials:

- Those that contain refractory constituents such as tungsten or molybdenum carbide
- Those that contain semirefractory constituents such as cadmium oxide, magnesium oxide, and tin oxide
- Those that contain elements (such as silver and nickel) that do not conventionally alloy, but which are formed by P/M processes to produce contact materials with unique properties

The various types of composite contact materials (Table 2) generally have a base material of silver, copper, or refractory metals and their carbides. The refractory-base and silver-base contacts are used in switching devices operated in air. Copper-base composite contacts are used in vacuum and oil-switching devices.

Table 1 Typical properties of tungsten and molybdenum

Some of the physical properties of tungsten and molybdenum vary considerably with cross-sectional area and grain structure.

Tungsten	
Hardness, HRA (HV)	70 (385)
Modulus of elasticity, GPa (10^6 psi)	
At 20 °C (68 °F)	405 (59)
At 1000 °C (1830 °F)	325 (47)
Density, g/cm^3	19.3
Melting point, °C (°F)	3410 (6170)
Boiling point, °C (°F)	5900 (10650)
Specific heat, J/kg (Btu/lb · °F), at 20 °C (68 °F)	140 (0.033)
Thermal conductivity, W/m · K (Btu/ft · h · °F), at 20 °C (68 °F)	130 (75)
Coefficient of linear thermal expansion, μm/m · K, at 20 °C (68 °F)	4.43
Specific resistance, nΩ · m at 20 °C (68 °F)	5.5
Electrical conductivity, % IACS, at 20 °C (68 °F)	31
Molybdenum	
Hardness, HRA (HV)	58 (210)
Modulus of elasticity, GPa (10^6 psi)	
At 20 °C (68 °F)	325 (47)
At 1000 °C (1830 °F)	270 (39)
Density, g/cm^3	10.22
Melting point, °C (°F)	2622 (4750)
Boiling point, °C (°F)	4800 (8672)
Specific heat, J/kg (Btu/lb · °F), at 20 °C (68 °F)	270 (0.065)
Thermal conductivity, W/m · K (Btu/ft · h · °F), at 20 °C (68 °F)	155 (89)
Coefficient of linear thermal expansion, μm/m · K, at 20 °C (68 °F)	5.53
Specific resistance, nΩ · m at 20 °C (68 °F)	5.2
Electrical conductivity, % IACS, at 20 °C (68 °F)	33

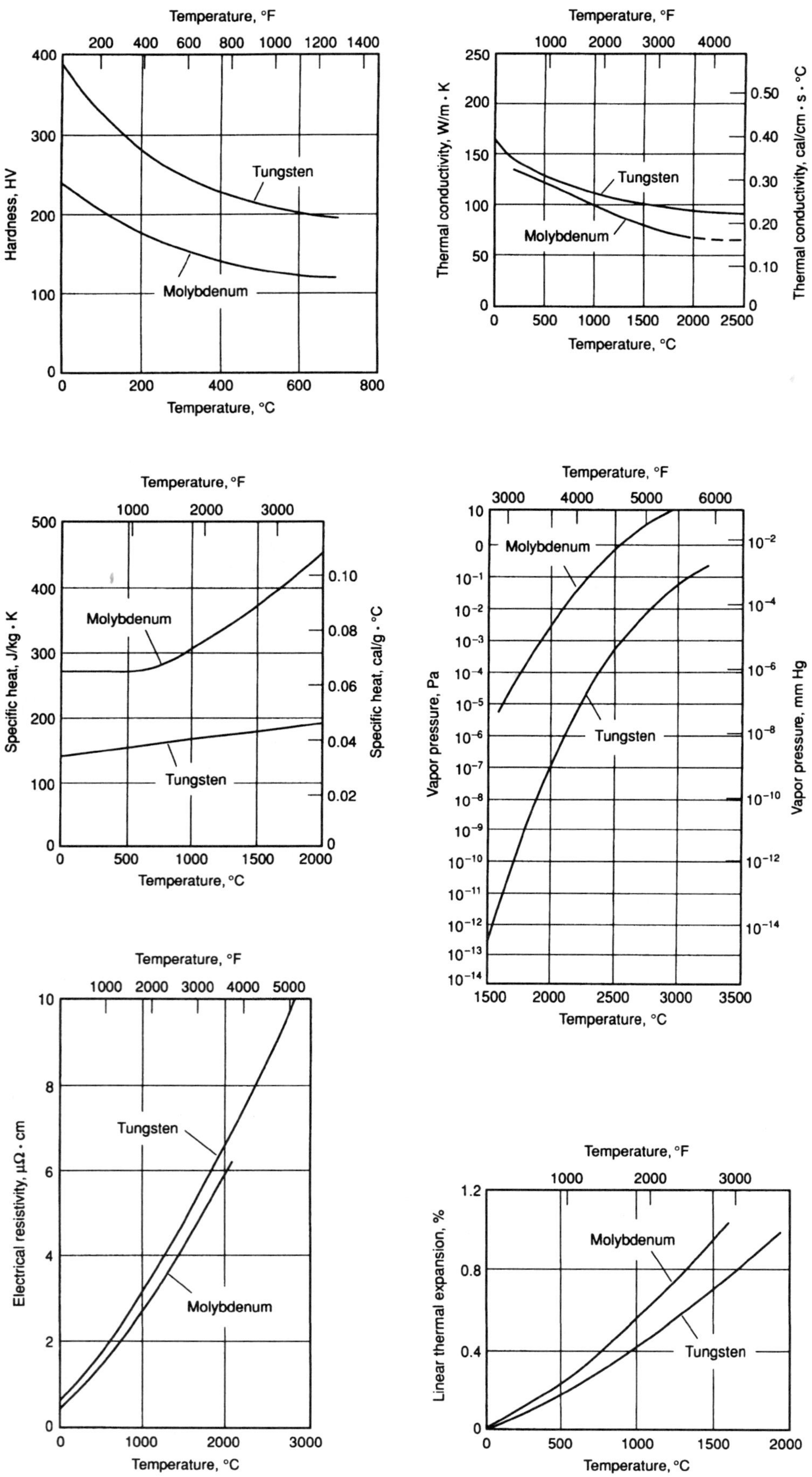

Fig. 1 Variation of properties with temperature for tungsten and molybdenum

Table 2 presents the compositions and properties of various composite contact materials. Because manufacturing methods affect the properties of materials with the same composition, the manufacturing methods are also given in Table 2. The most common methods of producing composite electrical contact materials are described in the section "Composite Manufacturing Methods" in this article.

Data published by contact manufacturers usually include density, hardness, and electrical conductivity (Table 2). These data provide designers of electrical devices with the basic properties of a composite contact. Other properties, such as contact resistance, may depend on operational parameters such as force (Fig. 3).

Characteristics that relate directly to failure modes such as arc erosion or material transfer are usually described in a qualitative manner. Very few quantitative data pertaining to these characteristics have been published because these properties depend on several test parameters. For instance, the arc erosion rate is affected by various mechanical factors:

- Opening force and opening speed
- Closing force and closing speed
- Bouncing of the movable contact
- Wiping distance
- Gap between opposing contacts

or electrical factors:

- Current—both amperage and whether ac or dc
- Voltage
- Power factor (inductive/capacitive)

Because each variable can greatly affect the arc erosion rate of a composite, it is virtually impossible to define a universal test to evaluate erosion rate.

Published data on erosion rate and welding frequency usually are collected under very specific conditions. They are valid only for qualitative description in a specific set of circumstances and cannot be extrapolated to suit other applications. The only means of learning how a composite will perform in a specific application is to test it extensively in the device in which it will be used. Examples of test data are given in Fig. 3 to 6.

Refractory Metal and Carbide-Base Composites

Refractory metals and their carbides are distinguished by high melting and boiling points and high hardness, but poor electrical and thermal conductivities and poor oxidation resistance. In pure elemental form, refractory metals perform well only under low-current conditions.

Forming a composite can compensate for these drawbacks. For example, the development of composite contact materials involving silver or copper with tungsten or molybdenum or their carbides has resulted in materials that can withstand higher currents and more arcing than other contact materials, without experiencing sticking or rapid erosion. The refractory metal content

Table 2 Properties of composites for electrical make-break contacts

Nominal composition, %	Manufacturing method(a)	Density, g/cm^3 Calculated	Density, g/cm^3 Typical	Electrical conductivity, % IACS	Hardness	Tensile strength MPa	Tensile strength ksi	Modulus of rupture MPa	Modulus of rupture ksi	Data source(b)	Application examples
Molybdenum-silver											
90Ag-10Mo	PSR	10.47	10.38	65–68	35–40 HRB	...	...	...	...	A	Air conditioner controls
80Ag-20Mo	PSR	10.44	10.36	59–62	38–42 HRB	...	...	...	...	A	Light- and medium-duty applications, automotive circuit breakers
75Ag-25Mo	PSR	10.42	10.33	58–61	44–47 HRB	...	...	...	...	A	
70Ag-30Mo	PSR	10.41	10.31	56–60	46–48 HRB	...	...	...	...	A	
65Ag-35Mo	PSR	10.39	10.30	55–64	49–55 HRB	...	...	...	...	A	Automatic circuit protectors, starting switches
60Ag-40Mo	PSR	10.38	10.28	55–62	55–62 HRB	...	...	...	...	A	
50Ag-50Mo	INF	10.35	10.10–10.24	45–52	70–80 HRB	...	...	758	110	C, A	Air- and oil-circuit breakers, arcing tips, traffic signal relays, home circuit breakers
	PSR	10.35	10.14	50	65 HRB	...	...	552	80	C, A	
45Ag-55Mo	INF	10.33	10.10–10.32	44–58	75–82 HRB	...	...	...	...	A	
40Ag-60Mo	INF	10.32	10.10–10.22	42–49	80–90 HRB	...	...	...	...	C, A	Aircraft switches, breaker arcing tips, electric raisers, air- and oil-circuit breakers
	PSR	10.32	10.12	45	50–68 HRB(c)	...	...	676	98	C, A	
35Ag-65Mo	INF	10.30	10.00–10.08	40–45	82–92 HRB	...	...	...	...	A	
30Ag-70Mo	INF	10.29	10.00–10.31	35–45	85–95 HRB	414	60	931	135	C, A	Air-circuit breakers, low-erosion arcing tips
25Ag-75Mo	INF	10.27	10.27	31–34	93–97 HRB	414	60	958	139	C, A	
20Ag-80Mo	INF	10.26	10.23–10.26	28–32	96–98 HRB	407	59	965	140	C, A	Arcing contacts, heavy-duty electrical applications
15Ag-85Mo	INF	10.24	10.18	28–31	97–102 HRB	...	...	...	...	G	
10Ag-90Mo	INF	10.23	10.13	27–30	97–102 HRB	...	...	...	...	G	Semiconducting material
Ag-CdO											
97.5Ag-2.5CdO	PSR	10.42	10.21	85	22 HRF(c)	110(c)	16(c)	...	...	C	Aircraft circuit breakers, aircraft relays, automotive relays, truck controls, snap switches, contractors, motor controllers, circuit breakers, governor relays
	PSE	10.42	10.42	95	37 HRF(c)	131(c)	19(c)	...	...	E, C	
					60 HRF(d)	172(d)	25(d)	...	...		
95Ag-5CdO	PSR	10.35	9.50–10.14	80–90	32 HRF(c)	110(c)	16(c)	...	...	C, A	
	PSE	10.35	10.35	92	40 HRF(c)	131(c)	19(c)	...	...	E, C	
					70 HRF(d)	172(d)	25(d)	...	...		
	IO	10.35	10.35	80	40 HRF((c)	186(c)	27(c)	...	...	E, C	
					75 HRF(d)	241(d)	35(d)	...	...		
	PPSE	10.35	10.35	85	70 HRF(c)	207(c)	30(c)	...	...	E, C	
					90 HRF(d)	248(d)	36(d)	...	...		
90Ag-10CdO	PSR	10.21	9.30–9.80	72–85	42 HRF(c)	103(c)	15(c)	...	...	C, A	
	PSE	10.21	10.21	84–87	46 HRF(c)	172(c)	25(c)	...	...	W, E, A, C, T	
					80 HRF(d)	228(d)	33(d)	...	...		
	IO	10.21	10.21	75	45 HRF(c)	186(c)	27(c)	...	...	W, E, C, M	
					81 HRF(d)	262(d)	38(d)	...	...		
	SF	...	...	...	...	...	...	...	...	E, W	
	PPSE	10.21	10.21	82	71 HRF(c)	269(c)	39(c)	...	...	E, C	
88Ag-12CdO	PSE	10.3	10.2	81.0	90 HRF(d)	317(d)	46(d)	...	...		
					60 HV(c)	...	...	...	...	E, M, W	
					95 HV(d)	...	...	...	...		
87Ag-13CdO	...	...	9.20	43	56 HRF(c)	...	...	...	...	A	
86.7Ag-13.3CdO	IO	10.11	10.11	68	48 HRF(c)	200(c)	29(c)	...	...	E, C	
					90 HRF(d)	324(d)	47(d)	...	...		
86.5Ag-13.5CdO	PPSE	10.11	10.11	75	70 HRF(d)	276(c)	...	...	...	E, C	
85Ag-15CdO	PSR	10.06	8.60–9.58	55–75	35 HRF(c)	83(c)	12(c)	...	...	E, C, A, M	
	PSE	10.06	9.90–10.06	55–75	57 HRF(c)	193(c)	28(c)	...	...	E, T, C, A, M, W	
					80 HRF(d)	241(d)	35(d)	...	...		
	IO	10.06	10.06	65	50 HRF(c)	207(c)	30(c)	...	...	W, C, E	Pressure and temperature controls
					85 HRF(d)	269(d)	39(d)	...	...		
	SF	...	...	...	...	...	...	...	...	E, W	
	PPSE	10.06	10.06	72	70 HRF(c)	276(c)	40(c)	...	...	C	
					90 HRF(d)	331(d)	48(d)	...	...		
83Ag-17CdO	IO	10.01	10.01	62	52 HRF(c)	214(c)	31(c)	...	...	C, E	Aircraft circuit breakers, aircraft relays, truck controls, contactors circuit breakers, governor relays
					88 HRF(d)	276(d)	40(d)	...	...		
	PPSE	10.01	10.01	70	70 HRF(c)	276(c)	40(c)	...	...	C, E	
					90 HRF(d)	352(d)	51(d)	...	...		
80Ag-20CdO	PPSE	9.93	9.93	68	70 HRF(c)	276(c)	40(c)	...	...	C, E	
					90 HRF(d)	354(d)	50(d)	...	...		
75Ag-25CdO	PPSE	9.79	9.79	60	...	...	...	...	...	C, E	
Silver-graphite											
99.75Ag-0.25C	PSR	10.41	9.70–10.40	95–103	33–45 HRF(c)	172(c)	27(c)	...	...	C, A, S	Automotive regulators, low voltage make-break contacts, sliding contacts
					70–73 HRF(d)	255(d)	37(d)	...	...		
99.5Ag-0.5C	PSR	10.31	9.60–10.30	92–102	26–44 HRF(c)	169(c)	24.5(c)	...	...	C, A, S	
					69–72 HRF(d)	252(d)	36.5(d)	...	...		
99.25Ag-0.75C	PSR	10.22	10.21	90–100	39 HRF(c)	165(c)	24(c)	...	...	C	
					70 HRF(d)	247(d)	35.8(d)	...	...		
99Ag-1C	PSR	10.13	9.40–10.12	87–99	24–36 HRF(c)	162(c)	23.5(c)	...	...	C, A, S	
					68–69 HRF(d)	241(d)	35(d)	...	...		

(continued)

(a) PSR, press-sinter-re-press; INF, press-sinter-infiltrate; PS, press-sinter; PSE, press-sinter-extrude; IO, internal oxidation; PPSE, preoxidize-press-sinter-extrude; SF, oxidized from one direction. (b) A: Advance Metallurgy, Inc., McKeesport, PA. C: Contacts, Materials, Welds, Inc., Indianapolis, IN. E: Engelhard Industries, Plainville, MA. G: Gibson Electric Inc., Delmont, PA. S: Stackpole Carbon Co., St. Marys, PA. T: Texas Instruments Inc., Attleboro, MA. M: Metz Degussa, South Plainville, NJ. W: Art Wire-Duduco, Cedar Knolls, NJ. (c) Annealed. (d) Cold worked

Table 2 (continued)

Nominal composition, %	Manufacturing method(a)	Density, g/cm^3 Calculated	Density, g/cm^3 Typical	Electrical conductivity, % IACS	Hardness	Tensile strength MPa	Tensile strength ksi	Modulus of rupture MPa	Modulus of rupture ksi	Data source(b)	Application examples
Silver-graphite											
98.5Ag-1.5C	PSR	9.96	10.04	97	33 HRF(c)	152(c)	22(c)	...	...	C, A	
					66 HRF(d)	231(d)	33.5(d)	...	...		
98Ag-2C	PSR	9.79	9.15–9.57	82–90	22 HRF(c)	...	...	...	...	C, A, S	
					65 HRF(d)	...	...	...	...		
97Ag-3C	PSR	9.46	8.80	55–62	20 HRF(c)	...	...	...	...	A, S	
					69 HRF(d)	...	...	...	...		
97Ag-3C	PSE	9.10	8.90	86	42 HV	...	...	...	...	W	Mate with other contact materials in circuit breakers
96Ag-4C	PSE	9.15	8.8	79	41 HV	...	...	...	...	W	
95Ag-5C	PSR	8.88	8.30–8.68	55–62	25 HRF(d)	...	...	...	...	C, A, S	
	PSE	8.88	8.84	75	40 HRF(d)	...	...	...	...	W, C	
93Ag-7C	PSR	8.37	7.80	50–57	15 HRF(c)	...	...	...	...	C, A, S	
					45 HRF(d)	...	...	...	...		
90Ag-10C	PSR	7.69	6.30–7.20	43–53	13 HRF(c)	...	...	...	...	W, C, A, S	
					30 HRF(d)	...	...	...	...		
Silver-iron											
90Ag-10Fe	PSR	10.16	9.60–10.25	87–92	48 HRF(c)	214(c)	31(c)	...	...	C, A	Wall switches, thermostat controls
					81 HRF(d)	272(d)	39.5(d)	...	...		
Silver-nickel											
99.7Ag-0.3Ni	...	10.49	...	100	53 HR15T(c)	...	...	...	...	T	...
					79 HR15T(d)	...	...	...	...		
95Ag-5Ni	PSR	10.41	9.80–10.41	80–95	32 HRF(c)	165(c)	24(c)	...	...	C, A, S	Appliance switches
					84 HRF(d)	...	...	...	...		
90Ag-10Ni	PSR	10.31	9.70–10.32	75–90	35 HRF(c)	172(c)	25(c)	...	...	W, C, S, A, E	Low rating line starters
					89 HRF(d)	...	...	...	...		
85Ag-15Ni	PSR	10.22	9.50–10.02	66–80	40 HRF(c)	186(c)	27(c)	...	...	W, C, A, S,	Circuit breakers
					93 HRF(d)	...	...	...	...	A, E	
80Ag-20Ni	PSR	10.13	9.30–9.50	63–75	52–59 HRF(c)	...	...	...	...	W, E, A, S	
					80 HRF(d)	...	...	...	...		
75Ag-25Ni	PSR	10.05	9.20	59	61 HRF(c)	...	...	...	...	S	
70Ag-30Ni	PSR	9.96	9.40–9.53	55–56	42 HRF(c)	...	...	...	...	W, C, S, A	
					87 HR(d)	...	...	...	...		
65Ag-35Ni	PSR	9.88	9.00	49	26 HR30T(c)	...	...	...	...	S	Circuit breakers, disconnect switches
60Ag-40Ni	PSR	9.80	8.90–9.60	44–47	40 HR30T(c)	241(c)	35(c)	...	...	W, C, S, A	
					92 HR30T(d)	414(d)	60(d)	...	...		
	PSE	9.80	9.60	60	46 HR30T(c)	...	...	...	...	S	
55Ag-45Ni	PSR	9.71	8.80	41	25 HR30T(c)	...	...	...	...	S	
50Ag-50Ni	PSR	9.63	9.00	38	50 HR30T(c)	...	...	...	...	S	
45Ag-55Ni	PSR	9.56	8.50	35	30 HR30T(c)	...	...	...	...	S	
40Ag-60Ni	PSR	9.48	8.80	32	35 HR30T(c)	...	...	...	...	S	Circuit breakers
					97 HR(d)	...	...	...	...		
	PSE	9.48	9.30	40	68 HR30T(c)	...	...	...	...	S	
35Ag-65Ni	PSR	9.40	8.60	30	40 HR30T(c)	...	...	...	...	S	
30Ag-70Ni	PSR	9.32	8.50	27	40 HR30T(c)	...	...	...	...	S	Transformer protectors, contactors, relays
25Ag-75Ni	PSR	9.25	8.20	24	40 HR30T(c)	...	...	...	...	S	
20Ag-80Ni	PSR	9.17	8.00	21	35 HR30T(c)	...	...	...	...	S	
Ag-SnO$_2$											
92Ag-8SnO$_2$	PSE	10.08	10.00	88	58 HV(c)	205–230	30–33.5	...	...	E, M, W	Light switches, relays, motor vehicle switches
					92 HV(d)	...	...	...	...		
90Ag-10SnO$_2$	PSE	9.98	9.97	82	64 HV(d)	215(c)	31(c)	...	...	E, M	Light switches, relays, motor vehicle switches
					98 HV(d)	...	...	...	...		
88Ag-12SnO$_2$	PSE	9.70	9.68	72	72 HV(c)	...	...	...	...	E, M, W	Low-voltage motor contactors and switches rated to 10 A. Low-voltage circuit breakers rated to 100 A
					105 HV(d)	...	...	...	...		
Ag-ZnO											
92Ag-8ZnO	PSE	9.81	9.80	77	60–65 HV(c)	...	...	...	...	E, M, W	Low-voltage circuit breakers rated to 200 A
Tungsten carbide-silver											
65Ag-35WC	INF	11.86	11.53–11.85	55–60	50–65 HRB	272	39.5	483	70	C, A	Aircraft contactors, lighting relays, low-voltage switches, circuit breakers
	PSR	11.86	11.10–11.80	50–60	50–62 HRB	...	...	...	...	C, A	
60Ag-40WC	PSR	12.09	11.40–11.92	46–55	60–70 HRB	...	...	...	...	A	
58Ag-42WC	PSR	12.17	11.86–11.97	50–55	75–85 HRB	...	...	...	...	C	
50Ag-50WC	INF	12.56	12.12–12.50	43–52	75–85 HRB	276	40	793	115	C, A	
40Ag-60WC	INF	13.07	12.70–12.92	40–47	90–100 HRB	379	55	827	120	C, A	Heavy-duty circuit breakers
38Ag-62WC	INF	13.18	12.92–13.29	35–38	90–100 HRB	552	80	...	...	C	
35Ag-65WC	INF	13.35	12.90–13.18	30–37	95–105 HRB	...	...	...	...	A	Semiconducting material
20Ag-80WC	PSR	14.23	13.2	19	400 HV(c)	...	...	...	...	M	Suppression of tungstate formation in low-voltage and high-voltage circuit breakers
					470 HV(d)	...	...	...	...		

(continued)

(a) PSR, press-sinter-re-press; INF, press-sinter-infiltrate; PS, press-sinter; PSE, press-sinter-extrude; IO, internal oxidation; PPSE, preoxidize-press-sinter-extrude; SF, oxidized from one direction. (b) A: Advance Metallurgy, Inc., McKeesport, PA. C: Contacts, Materials, Welds, Inc., Indianapolis, IN. E: Engelhard Industries, Plainville, MA. G: Gibson Electric Inc., Delmont, PA. S: Stackpole Carbon Co., St. Marys, PA. T: Texas Instruments Inc., Attleboro, MA. M: Metz Degussa, South Plainville, NJ. W: Art Wire-Duduco, Cedar Knolls, NJ. (c) Annealed. (d) Cold worked

Table 2 (continued)

Nominal composition, %	Manufacturing method(a)	Density, g/cm³		Electrical conductivity, % IACS	Hardness	Tensile strength		Modulus of rupture		Data source(b)	Application examples
		Calculated	Typical			MPa	ksi	MPa	ksi		
Tungsten-silver											
90Ag-10W	PSR	11.00	10.30–11.20	90–95	20–33 HRB	...	...	...	...	C, A	Controls, automatic circuit protectors, wall switches
85Ag-15W	PSR	11.27	10.60–11.30	85–90	25–38 HRB	...	...	...	...	A	Current-carrying contacts in circuit breakers, light-duty contactors
80Ag-20W	PSR	11.55	10.90–11.70	80–85	30–43 HRB	...	...	...	...	A	
70Ag-30W	PSR	12.16	12.00	72–80	40–47 HRB	...	...	...	...	A	
65Ag-35W	PSR	12.48	12.1	68	80 HV(c)	...	...	...		M	
					90 HV(d)	...	...	...	...		
60Ag-40W	PSR	12.84	12.10–12.60	60–65	50–60 HRB	...	...	...	...	A	
35Ag-65W	INF	14.92	14.20–14.77	45–53	80–93 HRB	...	...	827	120	C, M	Automotive starting switches, circuit breakers
	PS	14.92	13.90–14.20	47–50	85–87 HRB	...	...	...	...	C	
	PSR	14.92	14.65–14.74	47–50	55–65 HRB	...	...	572	83	C	
30Ag-70W	INF	15.42	15.02	40–50	85–93 HRB	...	...	...	...	A	Motor starters, aircraft equipment, circuit breakers, contactors, computers, arcing tips
27.5Ag-72.5W	INF	...	15.56	49	90 HRB	483	70	896	130	C	
	PSR	...	15.44	...	58–68 HRB(c)	...	...	586	85	C	
25Ag-75W	INF	15.96	15.25–15.40	40–50	85–95 HRB	...	...	...	...	A	
20Ag-80W	INF	16.53	16.18	35–40	91–100 HRB	...	...	...	...	A	
15Ag-85W	INF	17.14	16.60–17.05	32–41	90–100 HRB	448	65	758	110	A, C	Motor governors
10Ag-90W	PSR	17.81	17.25	29–35	95–105 HRB	379	55	758	110	A	Semiconducting material
Tungsten carbide-copper											
50Cu	INF	11.39	11.00–11.27	42–47	90–100 HRF	...	...	1103	160	C, A	Arcing contacts in oil switches, wiping shoes in power transformers
44Cu	INF	11.77	11.64	43	99 HRF	...	...	1241	180	C	
30Cu	INF	12.78	12.65	30	38 HRC	...	...	...	...	...	
Tungsten-copper											
75Cu-25W	PSR	10.37	9.45–10.00	50–79	35–60 HRB	...	...	414	60	C, A	Current-carrying contacts
70Cu-30W	...	10.70	10.45	76	59–66 HRB	...	...	...	...	A	
65Cu-35W	...	11.06	11.40	72	63–69 HRB	...	...	...	...	A	Vacuum interrupter
60Cu-40W	...	11.45	11.76	68	69–75 HRB	...	...	...	...	A	Oil-circuit breakers, arcing tips
50Cu-50W	INF	12.30	11.90–11.96	45–63	60–81 HRB	...	...	...	...	A	
44Cu-56W	INF	12.87	12.76	55	79 HRB	434	63	827	120	C	
40Cu-60W	INF	13.29	12.80–12.95	42–57	75–86 HRB	...	...	...	...	A	Oil-circuit breakers, reclosing devices, arcing tips, tap change arcing tips, contactors
35Cu-65W	INF	13.85	13.35	54	83–93 HRB	...	...	...	...	A	
32Cu-68W	INF	14.20	13.95	50	90 HRB	...	...	896	130	C	
30Cu-70W	INF	14.45	13.85–14.18	36–51	86–96 HRB	...	...	1000	145	C, A	Circuit breaker runners,arcing tips, tap change arcing tips
26Cu-74W	INF	14.97	14.70	46	98 HRB	621	90	1034	150	...	
25Cu-75W	INF	15.11	14.50	33–48	90–100 HRB	...	...	...	...	A	Vacuum switches, arcing tips, oil-circuit breakers
20Cu-80W	INF	15.84	15.20	30–40	95–105 HRB	758	110	...	...	C	
15Cu-85W	PSR	16.45	16.0	20	190 HV(c)	...	...	...	...	M	
					260 HV(d)	...	...	...	...		
13.4Cu-86.6W	INF	16.71	16.71	33	20 HRC	621	90	1034	150	C	
10.4Cu-89.6W	INF	17.22	17.22	30	30 HRC	765	111	1138	165	C	
Tungsten-graphite-silver											
48Ag-51.75W-0.25C	PSR	13.21	13.38	65	55 HRB	...	...	552	80	C	Circuit breakers, arcing tips
46Ag-53W-1C	PSR	13.58	12.85	55	85 HRB	...	...	...	...	C	
45Ag-50W-5C	PSR	11.00	10.60	37–43	45–55 HRB	...	...	621	90	A	
Complex composite contacts											
88Ag-10Ni-2C	PSR	9.63	9.37	70	26 HRF(c)	...	...	...	...	C, A	Sliding contacts
					64 HRF(d)	...	...	...	...		
25Ag-50Fe-25Cu	PSR	8.67	8.52	21	84 HRF(c)	...	...	...	...	C	Circuit breakers
					94 HRF(d)	...	...	...	...		

(a) PSR, press-sinter-re-press; INF, press-sinter-infiltrate; PS, press-sinter; PSE, press-sinter-extrude; IO, internal oxidation; PPSE, preoxidize-press-sinter-extrude; SF, oxidized from one direction. (b) A: Advance Metallurgy, Inc., McKeesport, PA. C: Contacts, Materials, Welds, Inc., Indianapolis, IN. E: Englehard Industries, Plainville, MA. G: Gibson Electric Inc., Delmont, PA. S: Stackpole Carbon Co., St. Marys, PA. T: Texas Instruments Inc., Attleboro, MA. M: Metz Degussa, South Plainville, NJ. W: Art Wire-Duduco, Cedar Knolls, NJ. (c) Annealed. (d) Cold worked

can vary from 10 to 90%, although 40 to 80% usually is used in air- and oil-immersed circuit breaker devices. Refractory metals offer good mechanical wear resistance and resistance to arcing. The silver and copper provide the good electrical and thermal conductivities.

Because silver and copper do not alloy with tungsten, molybdenum, or their carbides, P/M processes are required in fabrication. Depending on the composition, refractory metals containing silver or copper contact materials are made either by pressing and sintering or by the press-sinter-infiltrate method. When infiltration is used, either all refractory metal powder is compacted to shape, or a small amount of silver or copper powder is blended with the refractory metal, compacted, and sintered in a reducing atmosphere. The sintered compact is then returned to the furnace; silver or copper is added to act as the infiltrant.

Most infiltrated composite contacts use silver as the infiltrant because of its excellent thermal and electrical conductivities, as well as its superb oxidation resistance. Copper infiltrant, which costs less but has very poor corrosion resistance, is used for composites that operate in noncorrosive environments such as oil, vacuum, or inert atmospheres. At temperatures above the melting point of the infiltrant, the liquid metal penetrates and fills the interconnecting voids of the pressed-and-sintered compact. Densities of 96 to 99% of theoretical can be achieved by this process. Infiltrated contact materials find use as current-carrying contacts in air- and oil-immersed circuit breakers, heavy-duty relays, automotive starters, and switches. Lower properties can be obtained by pressing and sintering.

In a material made by infiltration, the function of the infiltrant (silver or copper) is twofold. First, because silver or copper does not alloy with tungsten, molybdenum, or carbides, the conductivity of the composite depends strictly on the volume percentage of infiltrant. Second, during arcing, the high temperature melts the infiltrant; consequently, the heat of fusion ab-

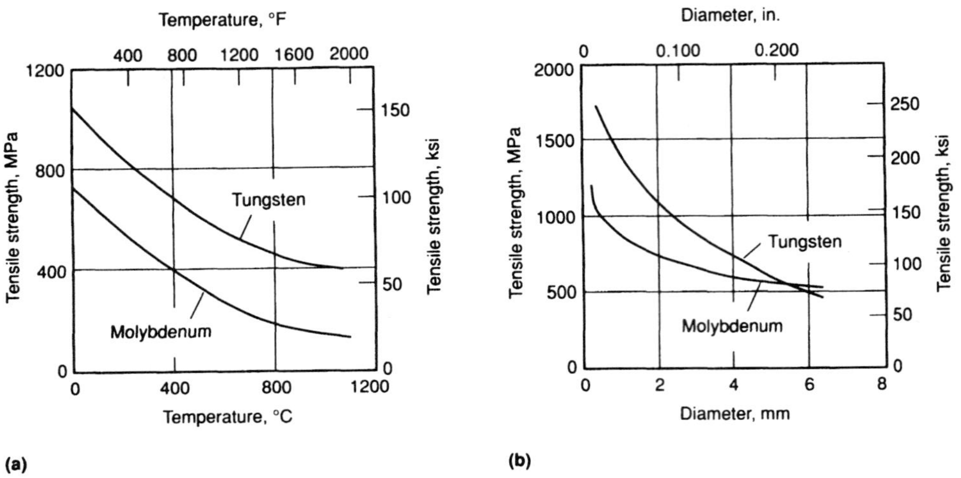

Fig. 2 Tensile strength of tungsten and molybdenum. (a) Variation with temperature. (b) Variation with diameter for tungsten and molybdenum rod

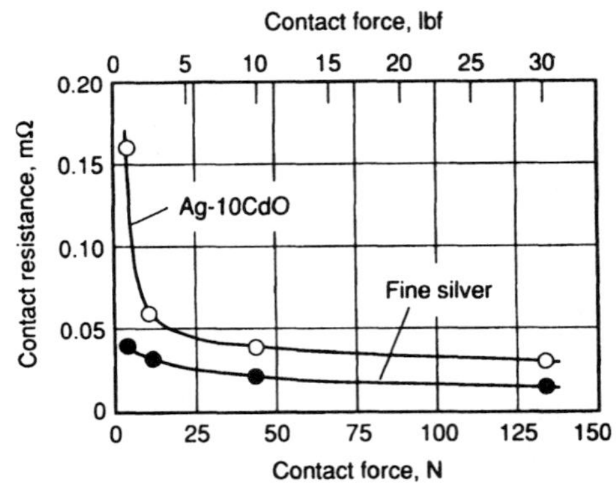

Fig. 3 Contact resistance versus force for fine silver and Ag-CdO contacts. Unarced contacts were 12.7 mm (½ in.) in diameter with a 38 mm (1½ in.) spherical radius. Resistance measurements were made with ac current at 50 A and 60 Hz.

sorbs (quenches) a portion of the heat generated by the arc. Theoretically, the skeleton, which is made of a high-melting element, will not begin to melt until all the low-melting component evaporates. The refractory skeleton also prevents molten infiltrant from flowing by capillary action. Because of this, erosion loss of the contact is low. Properties (such as the erosion data in Fig. 4 and 6) of the contact vary with the composition of the composite. A composite with high skeletal composition has high hardness and better wear resistance, but lower current-carrying capacity. On the other hand, a high-silver composite possesses high electrical and thermal conductivities and undergoes lower temperature rise, but is softer.

Compositions of Refractory Component. There is a lower limit for the composition of the skeleton material. Generally, when the amount of refractory or carbide is less than about 30 vol%, it is difficult to form a sound and uniform skeleton to accommodate the amount of silver. For practical purposes, the skeleton material should amount to a minimum of 50 wt% for tungsten and molybdenum and 35 wt% for tungsten carbide. Any composite containing lesser amounts than these limits should be made by the press-sinter-re-press method and should be considered a silver-base composite in which the function of the refractory material is to reinforce the silver matrix. For compounds with 60% or less tungsten, the classical method of mixing the powders, pressing, sintering (generally below the copper melting point), and re-pressing might also be used. Materials with 60 to 80% W are generally produced by infiltration, either of loose tungsten powder or of a pressed-and-sintered tungsten compact.

Tungsten, tungsten carbide, and molybdenum powders are the most commonly used materials for making skeletons for infiltrated contacts. Composites with tungsten skeletons have the best arc-interrupting and arc-resisting characteristics and the best arc-erosion resistance. Their antiwelding properties are moderate (Fig. 5). High-energy devices usually use silver-infiltrated composites having a tungsten skeleton.

Composites with tungsten carbide skeletons have better resistance to welding, better anticorrosion properties, and more stable contact resistance compared with other infiltrated composites. Devices that handle switching arcs usually use composites based on tungsten carbide skeletons.

For a combination of properties, or sometimes for a special requirement, a skeleton made of a mixture of tungsten and tungsten carbide is used. The blended powder contains either the mixture of tungsten and tungsten carbide or a mixture of tungsten and graphite. In the latter case, the graphite and part of the tungsten react to form tungsten carbide during sintering.

Composites with molybdenum skeletons have relatively low contact resistance and behave well in circuit-interrupting devices. For the same current-carrying capacity, a molybdenum-base composite costs less than the other two, but the antiwelding and anticorrosion properties of molybdenum-base composites are inferior to those with tungsten or tungsten carbide skeletons. Figure 6 compares the erosion characteristics of a molybdenum composite with tungsten and tungsten carbide composites.

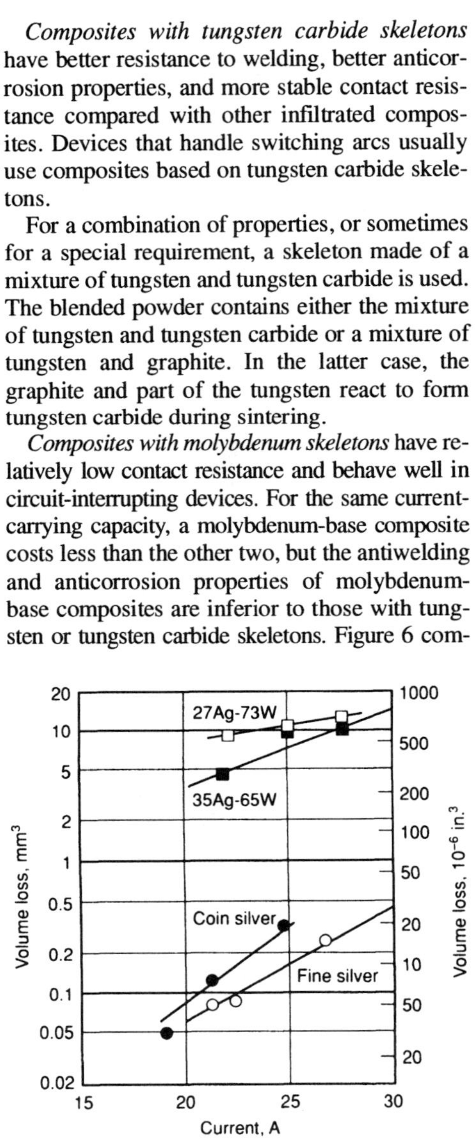

Fig. 4 Contact erosion characteristics of silver and silver-tungsten contacts. Test conditions were 115 V, 60 Hz, and 1.0 power factor for 100,000 operations at 60 operations/min. Closing and opening speeds were 38 mm/s (1½ in./s). Closing force was 980 mN (0.22 lbf) and opening force, 735 mN (0.165 lbf).

Silver-Base Composites

The main advantage of a silver composite over a silver alloy is that the bulk conductivity of a silver composite depends generally on the percentage of silver by volume. An alloying element in solution greatly decreases the conductivity of silver. For instance, the volume of silver in Ag-15CdO composite is less than that in Ag-15Cd alloy, yet the electrical conductivity of the former (65% IACS) is much greater than that of the latter (35% IACS).

In silver composites, the second phase forms discrete particles that are dispersed in the silver matrix. The dispersed phase improves the matrix in two ways. First, it increases the hardness of the composite material in a manner similar to dispersion hardening. Second, in the region where two mating contacts touch upon closure, the second phase particles reduce the surface area of silver-to-silver contact. This greatly reduces the tendency to stick or weld. In cases where the contacts do weld, the second-phase oxide particles (which are weaker and more brittle than silver) behave as slag inclusions and reduce the strength of the weld, allowing the

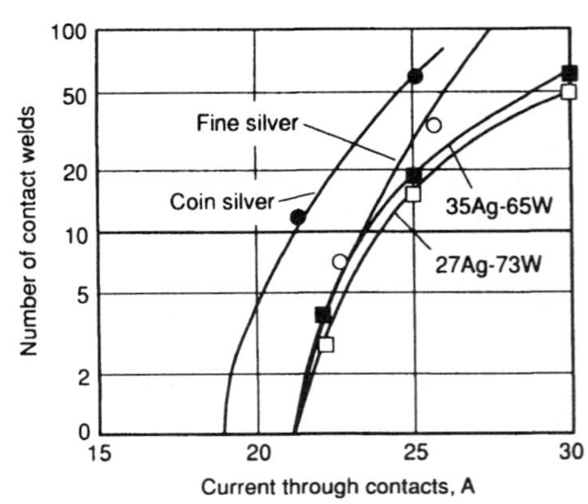

Fig. 5 Contact welding characteristics of silver and silver-tungsten contacts. Operation characteristics are the same as for Fig. 4.

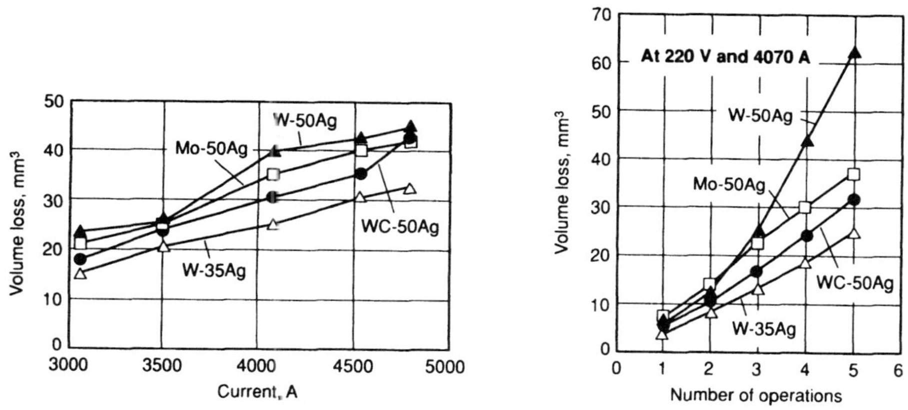

Fig. 6 Results of short-circuit tests on silver-tungsten, silver-molybdenum, and silver tungsten carbide

device contact-separating force to pull the contacts apart.

Silver-base composites can be divided into two types: type 1 uses a pure element or carbide as the dispersed phase; type 2 uses oxides as the dispersed phase. In both types, the hardness increases and the conductivities decrease as the volume fraction of dispersed phase increases, and vice versa.

Silver-Base Composites with a Pure Element or Carbide. In type 1, the dispersed phase functions as a hardener and improves the mechanical properties of the silver matrix. The dispersed phase also promotes improved electrical performance such as antiwelding properties. Elements used include tungsten, tungsten carbide, molybdenum, nickel, iron, graphite, and mixtures of these materials.

Silver-Tungsten and Silver-Molybdenum Composites. Silver composites (made by the press-sinter-re-press method using tungsten, tungsten carbide, and molybdenum as the dispersed phases) show electrical conductivities similar to those of infiltrated composites of the same components. However, their mechanical properties are inferior because the dispersed phases do not form a refractory skeleton.

Silver-Nickel Composites. One of the elements typically combined with silver by P/M processes is nickel. Nickel is more effective as a hardening agent than copper; consequently, silver nickel is considerably harder than coin silver. At the same time, nickel does not increase contact resistance appreciably, particularly in combinations that include 15 wt% Ni or less. Silver nickel is combined in proportions ranging to about 40 wt%.

Composites with nickel as the dispersed phase resist mechanical deformation or peening under impact and possess good antiwelding properties. Silver-nickel composite contacts can be used as both members of a contact pair. Sometimes, a silver-nickel composite is used as the moving contact operating against a stationary contact of a different composite such as silver-graphite.

The combinations most widely used are 60Ag-40Ni and 85Ag-15Ni. These materials are very ductile and can be formed in all of the shapes in which silver contacts are used, including very thin sheets for facing large contact areas. This material is ideal for use under heavy sliding pressures. It does not gall like fine silver and coin silver, but instead takes on a smooth polish. It is therefore suitable for sliding contact purposes, as well as for make-break contacts. Silver nickel can handle much higher currents than fine silver before it begins to weld. It has a tendency to weld when operated against itself. Therefore, it is frequently used against silver graphite.

The 60Ag-40Ni composite is the hardest material in the silver-nickel series. It is the most suitable for sliding contact in which pressure is high. This alloy also has the lowest rate of wear under sliding action. It is less ductile than silver-nickel materials containing less nickel, but it is still sufficiently ductile for all conventional manufacturing processes.

The 85Ag-15Ni composite is the most widely used material in the silver-nickel series. Because of its ideal mechanical properties, 85Ag-15Ni is an ideal material for motor-starting contactors and is superior in this type of application to fine silver, coin silver, and copper. It is also suitable as a general-purpose contact for various types of relays and switches.

The contact resistance of clean 85Ag-15Ni contacts that have not operated under load tend to be slightly lower for fine silver. However, in make-break circuits, silver tends to gradually increase contact resistance. This increase is not necessarily permanent, as contact resistance varies with the effects of arcing on the contacts. Generally, average resistance is higher than the initial resistance before the contacts operate. The contact resistance of 85Ag-15Ni is similar, except that it usually varies within a narrower range. Exhibiting nearly constant contact resistance is more important than possessing low contact resistance.

85Ag-15Ni exhibits a lower contact resistance and is also harder than coin silver.

Another advantage of 85Ag-15Ni is its low flammability; that is, it makes a smaller arc than other materials. In testing of more than 40 contact materials, 85Ag-15Ni exhibited the lowest arc energy. Low arc energy is important in that the ability to break a circuit with as little flame as possible is desirable. This characteristic was primarily responsible for the adoption of 85Ag-15Ni for relays in aircraft electrical systems.

Silver-Graphite Composites. Graphite is also combined with silver by P/M techniques. Graphite in silver-base composites serves as a good lubricant, reducing the damage caused by frictional forces. Silver-graphite composites are used chiefly as sliding or brush contacts. These materials have high resistance to welding and are also used as make-break contacts. In circuit breakers, they are usually paired with silver-nickel composites.

The most frequently used composition is 95Ag-5C, although graphite compositions ranging from 0.25 to 90% with the remainder silver have been used. This material was developed as a circuit breaker contact material. The addition of graphite prevents welding. Frequently, 95Ag-5C is used in combination with silver-nickel or silver-tungsten contacts. It is also used in combination with pure nickel contacts and with fine silver contacts. Silver graphite is soft compared to other types of contact materials, and electrical and mechanical erosion is more rapid.

95Ag-5C has been widely used as a material for contacts in molded-case circuit breakers, sliding contacts, and contact brushes. This material is only moderately ductile and can be rolled into sheets and punched into contacts of various shapes. However, it cannot be headed to make solid rivets or bent to any great extent without cracking. It can be coined to a moderate extent. 95Ag-5C contacts can be individually molded. Depending on size, shape, and quantity, contacts of this material are either punched from rolled slabs, extruded, or individually molded from powders. Copper is combined with graphite as a substitute for silver in certain applications.

A modified form of silver graphite is silver-nickel-graphite. Typical compositions are 88Ag-10Ni-2C and 77Ag-20Ni-3C. These materials are substantially harder than 95Ag-5C and exhibit superior wear resistance, but offer less protection against welding. Like 95Ag-5C, they can be manufactured from slabs or by molding individually.

Composites of silver iron exhibit good antiwelding and good wear characteristics when used in creep-type thermostat devices. These materials have poor corrosion resistance.

Silver-Base Composites with Dispersed Oxides. Type 2 silver-base composites use semirefractory oxides as the dispersed phase. These silver-base composites are produced by a variety of methods such as internal oxidation, preoxidation, and conventional P/M processes (see the section "Composite Manufacturing Methods" in this article).

The semirefractory component of type 2 silver-base composites includes metal oxides such as CdO, SnO_2, or ZnO. In general, the semirefractory constituents promote nonsticking qualities or provide increased resistance to wear.

The Ag-CdO group of electrical contact materials is the most widely used of all the silver semirefractory contact materials. The addition of

5 to 15% CdO to silver imparts excellent nonsticking and arc quenching qualities.

Because of its resistance against arc erosion and its low contact resistance, which does not increase even after switching, Ag-CdO has proved to be a universally good contact material for many switching devices. Ag-CdO contact materials are well suited for contactors and motor starters, but are also used in circuit breakers, relays, and switches with medium to low currents.

Ag-CdO material has antiwelding and antierosion properties united with constant resistance, examples of its main advantage of well-combined properties. Another favorable quality is that it has good workability. It can be fabricated by either the internal oxidation (least costly), preoxidation, or P/M methods. The Ag-CdO material can also be cold reduced or rolled quite easily. For instance, Ag-15CdO material can endure more than 70% cold reduction.

Ag-SnO₂, which is used widely in Europe as a contact material, is a class of composite materials that has the potential to replace Ag-CdO composites in many electrical contact applications. However, general comparisons of Ag-SnO_2 contacts with Ag-CdO contacts are difficult because results may depend on the specific conditions of testing. Previous concerns on the toxicity of CdO, which was one of the motivations for using Ag-SnO_2 contacts, have also been relaxed in Japan and Europe. The toxicity of CdO must be distinguished from the highly toxic nature of cadmium.

Like Ag-CdO contacts, Ag-SnO_2 contacts can be produced by internal oxidation or P/M techniques. One drawback of the Ag-SnO_2 composite is that a third element (such as indium) must be added to achieve internal oxidation when the silver alloys contain more than 4% Sn. The oxidized material also does not allow a high level of cold reduction because of its brittleness. Therefore, a press-sinter-re-press method or extruded method is the most feasible way to fabricate Ag-SnO_2 although extruded products are more brittle than extruded Ag-CdO powder of similar compositions. For example, extended Ag-10SnO_2 can be subject to a maximum of 30% cold reduction compared to more than 60% for Ag-12CdO.

Another drawback is the higher-temperature rise of Ag-SnO_2 contacts (as compared to Ag-CdO) after arcing. This troublesome characteristic has, however, been eliminated with Ag-SnO_2 materials made by P/M methods.

Table 2 lists three grades of commercially available Ag-SnO_2 composite contact materials. Ag-SnO_2 contact materials cannot be easily brazed or welded. To be able to braze Ag-SnO_2 contacts, they are made with at least two layers, the contact layer and brazable or weldable fine silver layer. The brazing alloy can be applied separately in the shape of paste, wire, or foil, or it is already clad onto the semifinished product.

Ag-ZnO is another composite material that has been tested and marketed for contact applications. Ag-ZnO, like Ag-SnO_2 composite, cannot take high cold reduction because of the brittleness of the oxidized material. When internal oxidization is used, the maximum zinc content cannot exceed 6% for good oxidation. Typical applications of a commercially available Ag-ZnO composite are listed in Table 2.

Multiple-Component Composites. There is no ideal material to meet all conditions for contact applications. If required by manufacturers of switching devices, contact manufacturers can offer composite materials consisting of as many as four or five components. Most of these composites serve only special purposes. They are not universally accepted and generally cost more. Two common three-component composites are listed in Table 2.

Composite Manufacturing Methods

The methods used to manufacture composite contact materials can be classified into three major categories:

- Standard P/M processes, for producing composites from materials that cannot be conventionally alloyed
- Internal oxidation processes, for producing silver-base composites with dispersed oxides
- Hybrid consolidation, which is a combination of the internal oxidation and P/M consolidation processes

Powder Metallurgy Methods

Infiltration is used exclusively for making refractory metal and carbide-base composite contact materials. Metal powder or carbide powder is first blended to the desired composition with or without a small amount of binder to impart green strength, then is pressed and sintered into a skeleton of the required shape. Silver or copper is then infiltrated into the pores of the skeleton. This method produces the most densified composites, generally 97% or more of theoretical density. Complete densification is not possible because of the presence of some closed pores in the sintered skeleton. After infiltration, the contact is sometimes chemically or electrochemically etched so that only pure silver appears on the surface. The contact thus treated has better corrosion resistance and performs better in the early stages of use.

Press-Sinter. For small refractory metal contacts (not exceeding about 25 mm, or 1 in., in diameter), a high-density material can be obtained by pressing a blended powder of exact final composition into shape and then sintering it at the melting temperature of the low-melting-point component (liquid-phase sintering). In some cases, an activating agent such as nickel, cobalt, or iron is added to improve the sintering effect on the refractory metal particles. For this process, powders of much finer particle size are required so that more bonding surface exists. However, the skeleton formed by this process is weaker than that formed by the infiltration process. Formation of the skeleton usually shrinks the apparent volume of the refractory portion of the composition, thus bleeding out the molten component onto the surface of the finished contact.

Press-Sinter-Repress. The press-sinter-repress process is used for all categories of contact materials, especially those in the silver-base category. Blended powders of the correct composition are compacted to the required shape and then sintered. Afterward, the material is further densified by a second pressing (re-pressing). Sometimes the properties can be modified by a second sintering or annealing. The versatility of this process makes it applicable for contacts of any configuration and of any material. However, it is difficult to obtain material with as high a density as is obtained with other processes. Material thus produced also may have weak bonding between particles.

Press-Sinter-Extrude. Blended powder of final composition is pressed into an ingot and sintered. The ingot is then extruded into wires, slabs, or other desired shapes. The extruded material may be subsequently worked by rolling, swaging, or drawing. Material made by this method is usually fully dense.

The press-sinter-extrude process is used mostly for silver-base composites. Other processes used for manufacturing silver-base composite contacts are direct extrusion or direct rolling of loose powder. Although they appear to be uncommon, they are economically feasible if the equipment is properly designed and built.

Internal Oxidation

Silver-base composites with dispersed metal oxides can be produced by internal oxidation. In this process, a silver alloy (such as a silver-cadmium alloy) is first cast into ingots, which are rolled into strips or fabricated further into the finished product form. The silver alloy material is then heated in air or oxygen, so that the oxygen diffuses into the alloy and forms metal oxide particles (such as CdO in the case of a silver-cadmium alloy) dispersed in the silver matrix.

Internal oxidation is used in the production of a substantial portion of Ag-CdO composites. The initial silver-cadmium alloy can be internally oxidized either in strip or finished product form. The silver-cadmium alloy is heated between 800 to 900 °C (1470 to 1650 °F) in a furnace with air, oxygen-enriched air, or pressurized oxygen. Under this condition, the oxygen species diffuse into the silver-cadmium alloy and oxidize the cadmium species. Upon the completion of the oxidation, the cross section of the material will display a microstructure of CdO particles embedded in a silver matrix. Contact parts are punched from the strip and then coined into required shapes.

The size of the CdO particles and the uniformity of their dispersion are dependent on the temperature and the partial pressure of oxygen. Reduced temperature decreases coalescence of cadmium prior to being oxidized and thereby causes a finer dispersion of CdO. Increasing the partial pressure of oxygen in the furnace increases the diffusion rate of oxygen into the silver. This also causes a finer CdO dispersion by

Table 3 Comparison of Ag-CdO material made by different methods

Properties	Press-sinter-re-press	Press-sinter-extrude	Internal oxidation	Preoxidize-press-sinter-extrude
Performance characteristics				
Resistance to arc erosion	3	2	1	1
Resistance to sticking and welding	1	1	2	2
Low contact resistance and temperature rise	1	1	1	1
Arc interruption	3	2	1	1
Resistance to corrosion	1	1	1	1
Material characteristics				
High mechanical properties	3	2	2	1
Resistance to annealing	3	2	2	1
Electrical and thermal conductivity	2	1	1	1
Flexibility of composition	2	2	2	1
Uniform cadmium oxide distribution	1	1	3	1

Note: 1 indicates that under most conditions this is the preferred material; 2 indicates that under most conditions the material is preferable to 3, but not as good as 1; 3 indicates that the material may be acceptable, but under typical operating conditions it is not as good as 1 or 2.

reducing the time available for the cadmium to coalesce.

During the internal oxidation of a silver-cadmium alloy, the cadmium species become depleted in zones when the oxygen front moves into the silver-cadmium alloy. The cadmium atoms before the oxygen front immediately diffuse into the zone against the oxygen front. As the oxidation front moves from the surfaces of the strip toward the center, the concentration of the cadmium species becomes increasingly dilute as compared to the original composition. Hence, after the oxidation is completed, the cross section will display a significant oxide-deficient or oxide-depleted zone in the center of the contact body.

For some applications the presence of the depletion zone is detrimental, requiring its removal or displacement from the center. There are two common methods to achieve this result. In the first, an oxidation barrier, such as ceramic glaze, is applied to one surface so that the oxidation can proceed from only one side. The second method is to laminate two silver-cadmium sheets of the same size and to form a package by welding along the four edges. After oxidation, the sheets are separated. The oxide-deficient zone will appear on one side (the inner side of the package) of each sheet.

Package rolling is another technique for reducing the size of the depletion region. In this method, very thin silver-cadmium sheets are first oxidized. Then a number of sheets (for example, 16 sheets) are stacked together and hot-bond rolled into one slab. The cross section of the final product displays very thin depleted zones equal to the number of sheets.

Hybrid Consolidation

Various hybrid techniques use a combination of internal oxidation and P/M methods. These methods are used to produce a finer average oxide size and/or a more uniform distribution of cadmium oxides in the matrix of a Ag-CdO composite. Hybrid consolidation methods include preoxidized-press-sinter-extrude and coprecipitation. Table 3 compares Ag-CdO composites manufactured by different methods

The preoxidized-press-sinter-extrude process combines the oxidation process and the press-sinter-extrude process. Commercially, it is called "preoxidized process." The purpose of this method is to redistribute the oxide-deficient center of Ag-CdO composites.

The preoxidized process is used exclusively for making Ag-CdO material. Alloys are reduced to small particles in the shape of flakes, slugs, or shredded foil. These particles are oxidized and then consolidated with the press-sinter-extrude process. Material made by this method is more uniform than the same material made by conventional internal oxidation. Mechanical properties are superior to those of the same material made by the press-sinter-re-press method.

The Ag-CdO particulates are made by one of four methods and then are pressed into ingots, sintered, and extruded according to standard metallurgical method. There are four processes to prepare the particulates:

- *Granulated wire:* Silver-cadmium alloy is first made into wire and oxidized. The oxidized wire is then chopped into granules with a length of about 3 mm (1/8 in.).
- *Low-pressure water atomization:* The molten silver-cadmium alloy is atomized by water at a pressure of 100 to 200 kPa (15 to 30 psi). The approximately quarter-inch particulates are in the form of thin twisted flakes. Then the flakes are oxidized for consolidation.
- *High-pressure water atomization:* The molten silver-cadmium alloy is atomized with high water pressure, usually higher than 2750 kPa (400 psi). The powder sizes range between 40 mesh (420 μm) and 270 mesh (53 μm). Then the alloy powders are oxidized before consolidation.

Coprecipitation. Conventional blending or mechanical mixing of silver and CdO powders begins by dissolving the proper amounts of silver and cadmium metals in nitric acid. Compounds of silver and cadmium coprecipitate from the solution when the pH value of the solution is changed by adding either hydroxide or carbonate solutions. During subsequent calcination at about 500 °C (930 °F), the compound mixture decomposes to form a mixture of silver and CdO. Alkali metal content can be controlled in the ppm range by adequate washing. Controlled amounts of sodium, potassium, and lithium may enhance electrical life. Excessive amounts of these elements can lead to rapid erosion, restrike, and generally poor electrical life. Depending on device design, the range may be from 10 to 300 ppm. Contacts are consolidated from this mixture by conventional P/M methods. The microstructure of contacts made by this method displays a finer particle size and a more uniformly dispersed CdO phase than material made by conventional blending.

Availability

Silver, gold, platinum, palladium, and most of their ductile alloys are available as stamped contacts. Except for material from which contact disks are produced, a variety of stock sizes are not maintained because, in general, no two applications are identical. For disks, material in strip form, varying in thickness from 0.25 mm to 1 mm (0.010 to 0.040 in.) is available.

Except for tungsten and molybdenum, contact materials are ductile enough so that they can be produced in all contact forms. Tungsten and molybdenum and some of the P/M materials have lower ductility and are available in fewer forms. The commercially available forms of common electrical contact materials are listed in Table 4.

Silver, gold, platinum, palladium, and nearly all the alloys of these metals, as well as tungsten, molybdenum, and the various sintered products of silver and the refractory metals, can be used to produce steel-back contacts. Steel-back contacts have been made in the form of screws, rivets, or buttons for projection welding.

Powder metallurgy materials are available with final densities up to 99% of theoretical and with high-conductivity surfaces as inserts or overlays. Both P/M and wrought materials may be attached to appropriate carriers by brazing, welding, or diffusion bonding, even though they contain CdO. For percussion welding of Ag-CdO contacts, backing is not required. For resistance welding, a fine silver backing is needed.

Powder metallurgy contacts are available with a silver matrix for air and oil applications, and with a copper matrix for oil applications only. The second phase may be tungsten, molybdenum, graphite, tungsten carbide, CdO, or ZnO.

Materials with a high content of refractory metal (50% or more) are usually made by infiltration. In this process, the refractory metal powder is pressed into the desired size and shape with a controlled porosity. The compact is sintered at high temperature in a reducing atmosphere, and then molten silver or copper is infiltrated into the porous sintered compact.

Compositions of a lower refractory content are made by blending powders, pressing them to a desired size and shape, sintering the pressed compact at a high temperature, and re-pressing

Table 4 Commercially available forms of electrical contact materials

Alloy	Product form: Solid rivet	Wire	Strip	Tape	Disks	Attached(a)	Composite weld disks(b)	Clad(c)	Rings	Brushes	Manufacture method: Melting	P/M
100 Ag	X	X	X	X	X	X	X	X	X	...	X	...
100 Pd	X	X	X	X	X	X	X	X	X	...	X	...
100 Au	X	X	X	X	X	...	...	X	...	...	X	...
100 Ru	...	...	X	...	X	X	...	...	...	...	X	X
100 Ir	...	X	X	...	X	X	...	...	...	...	X	X
100 Pt	X	X	X	...	X	X	X	X	...	X	X	X
100 Os	...	...	...	...	X	X	...	...	...	...	...	X
100 Rh	...	X	X	...	X	...	...	...	Plate	...	X	...
92.5Ag-7.5Cu	X	X	X	...	X	X	X	X	...	...	X	...
90Ag-10Cu	X	X	X	X	X	X	X	X	X	...	X	...
75Ag-24.5Cu-0.5Ni	X	X	X	X	X	X	X	X	X	...	X	...
72Ag-28Cu	...	X	X	...	...	...	...	X	...	X	X	...
99Ag-1Pd	X	X	X	X	X	X	X	X	...	X	X	...
97Ag-3Pd	X	X	X	X	X	X	X	X	...	X	X	...
90Ag-10Pd	X	X	X	X	X	X	X	X	...	...	X	...
90Ag-10Au	X	X	X	...	X	X	...	...	...	...	X	...
97Ag-3Pt	X	X	X	...	X	X	...	...	X	...	X	...
85Ag-15Cd	X	X	X	...	X	X	X	X	...	...	X	...
95Ag-5CdO	X	X	X	X	X	X	X	X	...	...	X	X
90Ag-10CdO	X	X	X	X	X	X	X	X	...	...	X	X
85Ag-15CdO	X	X	X	X	X	X	X	X	...	...	X	X
90Ag-10Fe	X	...	X	...	X	X	X	...	...	...	...	X
90Ag-10W	X	...	...	...	X	X	X	...	...	...	...	X
50Ag-50WC	...	...	...	...	X	X	X	...	...	...	...	X
65Ag-35WC	...	...	...	...	X	X	X	...	...	...	...	X
75Ag-25Zn	...	...	X	...	X	...	...	...	...	...	X	...
85Ag-15Ni	...	...	...	...	X	X	X	...	X	...	...	X
70Ag-30Ni	...	...	...	...	X	X	X	...	...	...	...	X
70Ag-30Mo	...	...	...	...	X	X	X	...	...	...	...	X
97Ag-3 graphite	...	...	...	...	X	X	X	...	...	...	...	X
95Ag-5 graphite	...	...	...	...	X	X	X	...	...	X	...	X
60Pd-40Ag	X	X	X	...	X	X	X	X	...	...	X	...
60Pd-40Cu	X	X	X	...	X	X	X	...	...	X	X	...
95Pd-5Ru	X	X	X	...	X	X	X	...	...	...	X	...
75Au-25Ag	X	X	X	...	X	X	X	X	...	...	X	...
90Au-10Cu(Coin)	X	X	X	...	X	X	X	X	X	...	X	...
95Pt-5Ru	...	...	X	...	X	X	X	...	...	...	X	...
90Pt-10Ru	...	...	X	...	X	X	X	...	...	...	X	...
90Pt-10Rh	...	...	X	...	X	X	X	...	...	...	X	...
90Pt-10Ir	...	X	X	...	X	X	X	...	...	...	X	...
85Pt-15Ir	...	X	X	...	X	X	X	...	...	...	X	...
80Pt-20Ir	...	X	X	...	X	X	X	...	...	...	X	...
75Pt-25Ir	...	X	X	...	X	X	X	...	...	...	X	...
65Pt-35Os	...	...	...	...	X	X	X	...	...	...	...	X
69Au-25Ag-6Pt	X	X	X	...	X	X	X	...	X	X	X	...
60Ru-35Ir-5Pt	...	...	...	...	X	X	X	...	...	...	X	...

(a) Contact disks attached to screws, rivets, blades and bars. (b) Composite welding-type buttons produced for resistance welding attachment. Backings of nickel, Monel, and steel. (c) Clad materials including overlay, throughlay, edgelay, strip of precious metal on (or in) base metal

to size the parts and to increase the density of the compact.

Although parts usually are molded to final shape, they are sometimes finished by machining or grinding to obtain special shapes or unusually close tolerances. When only a few parts are needed, they may be machined from bars to save the cost of expensive dies for pressing a P/M compact.

Sintered materials are available in sizes ranging from rectangles or disks about 0.8 mm (1/32 in.) thick by 3 mm (1/8 in.) square or 3 mm (1/8 in.) in diameter to bars 200 mm (8 in.) long. Most are available in widths up to 75 mm (3 in.) and thicknesses up to 12.5 mm (1/2 in.). For some materials, these dimensions may be exceeded.

In small sizes (up to about 12.5 mm, or 1/2 in., square), the higher refractory compositions are frequently made with serrated surfaces coated with excess silver or silver-base brazing alloy so that they can be attached easily to a backing by welding or by resistance brazing in a welding machine. Larger sizes are generally attached to backing by silver alloy brazing. The materials are often pretinned with silver-base brazing alloy in a controlled atmosphere for good wetting of the contact material.

Many of the low-refractory contact materials are fabricated as disks, rectangles, and special-contour facings that also are attached to backing by silver alloy brazing. Most materials with 90% or more silver are ductile enough to allow fabrication as rivets and to allow assembly by staking or spinning. Among silver-graphite materials, those containing more than 0.5% graphite cannot be satisfactorily cold headed into rivets.

Ag-CdO mixtures are available as round, rectangular, or special-shape facings, and also as rivets.

ACKNOWLEDGMENT

This article was adapted from Y.-S. Shen, P. Lattari, J. Gardner, and H. Wiegard, "Elecrical Contact Materials," *Properties and Selection: Nonferrous Alloys and Special-Purpose Materials,* Vol 2, *ASM Handbook,* 1990, p 840–868.

Porous Powder Metallurgy Technology

Mark Eisenmann, Mott Corporation

THE TECHNOLOGY of fabricating lower density, porous P/M materials can fulfill a wide variety of applications. A rigid, permeable structure can be created using P/M technology by forming a network of sintered powder particles and interconnected pore channels. Using similar manufacturing equipment and technology as structural P/M components, porous P/M materials are normally sintered to densities between 25 and 85% of theoretical mean density (TMD). Structural P/M components are typically 85 to 99.9% of TMD. These unique engineered materials provide specialized products for applications such as filtration, fluid flow control, self-lubricating bearings, spargers, and battery electrodes.

Porous P/M materials are specified when special characteristics are required such as good mechanical properties, rigidity, corrosion resistance, uniform porosity, and controlled permeability. For example, porous bronze, stainless steel, or nickel alloys are often selected for service in elevated-temperature and pressure environments. Porous materials such as papers, synthetic fibers, and plastics are commonly selected for service in less severe conditions due to their lower cost (Ref 1). However, in applications where disposal costs are significant due to the nature of the service environment, renewable porous P/M materials offer excellent cleanability for economical reuse. Other porous metal materials made from sintered fibers, foam, perforated sheet, and screen may be selected as alternatives to porous P/M materials, depending on the application (Ref 2). Metal fiber/screen composites, foam/powder composites, screen/powder composites, and fiber/powder composites are also sintered to form porous media (Ref 3–5). Porous ceramics are utilized in applications that require higher-temperature service or more extreme corrosion resistance than porous metals. However, porous ceramics are limited in their application because they have substantially lower mechanical properties than do metals and because of fabrication difficulties that result when joining to metal hardware (Ref 6).

Porous P/M materials accounted for about 2% of the overall P/M market based on market share by application. Approximately 70% of the overall P/M market is comprised of structural components for the automotive industry (Ref 7). Porous P/M materials are also estimated to account for less than 2% of the overall filtration and fluid control markets that are currently dominated by lower cost, disposable porous media. Despite the small market size currently served by porous P/M materials, this material science is an emerging technology driven by increased environmental concerns, improved product quality demands, and more complex processing of industrial products. The unique materials, characteristics, fabrication methods, and applications of porous P/M technology are summarized in this article.

Materials Selection

Materials can be selected from a wide variety of P/M materials, depending on the combination of application requirements and economics. The porosity is determined by the powder particle shape, the powder size distribution, the powder surface texture, and other powder characteristics that are dependent on the material processing method. Because the powder characteristics are one of the major factors in determining the porous material properties of the finished component, reproducible powder characteristics and manufacturing methods are critical to production of a consistent product.

The four most common porous P/M materials are bronze, stainless steel, nickel, and nickel-base alloys. Other materials such as titanium, aluminum, copper, platinum, gold, silver, iron, iron aluminide (Fe_3Al), niobium, tantalum, and zirconium are fabricated into porous materials from powder (Ref 8). The size distribution of the particles can be a direct result of the powder manufacturing process or can be altered by sieving or air classification techniques. The characteristics, sintering conditions, and commercial availability of a powder determine whether the desired material can be manufactured into a porous structure.

Bronze powders consisting of 89 to 90% Cu, 10 to 11% Sn, and 0.1 to 0.5% P are the most common material for porous bronze P/M components such as filters and self-lubricating bearings. Small amounts of phosphorous can be added to improve mechanical strength (Ref 9). Bearings can be produced from preblended elemental powder mixes or prealloyed powders. These 89/11 or 90/10 bronze powders are available from three main processes. Gas or air atomization of prealloyed bronze yields fine, spherical particles that can be processed to low densities by gravity sintering, as shown in Fig. 1. Spherical bronze particles can also be obtained by tin coating of gas-atomized copper powder. Rounded particles are made from cutting tin-coated copper wire with a composition of 93 to 97% Cu and 3 to 7% Sn (Ref 10) as shown in Fig. 2. Bronze powder particle sizes range from 10 to 850 μm in diameter depending on the powder type.

Stainless steel and nickel-base alloys are often specified for increased service life when more corrosion resistance, temperature resistance, or strength is required. Due to economic considerations, the most common stainless steel (SS) materials chosen are the extra-low-carbon, austenitic grade, type 316L, or from the nonstandard, 2 to 3% Si modified grade, known as 316B. The additional silicon serves as a sintering aid by lowering the sintering temperature and controlling the porosity and mechanical properties. Other austenitic, 300 series grades, such as 304L and 347, along with the ferritic grades such as 430 are used in a few specialized applications where the required corrosion resistance or mechanical properties are superior to 316L SS. Nickel-base alloys such as Hastelloy C22, Inconel 600, Monel 400, and Hastelloy X are utilized for extreme corrosion or thermal resistance.

Stainless steel and nickel-base alloy powders are produced by water or gas atomization (refer to Sections on production of stainless steel and high-alloy powders and production of nickel-base powders in this Volume). Water atomization produces powders with rounded, irregular shapes that can be processed by compaction and sintering, as shown in Fig. 3 and 4. These particles interlock when compacted to form a part with sufficient green strength for handling. The water-atomization process yields a wide variation in particle size distribution that must be separated into much narrower size ranges for subsequent processing into porous components. As the powder size decreases, the particle shape

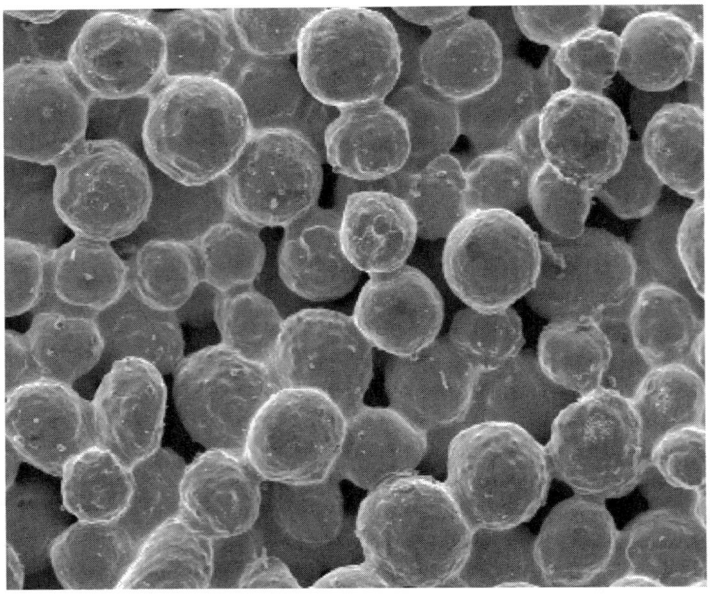

Fig. 1 Bronze particles made by tin coating atomized copper powder with a size range of 45 to 100 μm that are gravity sintered to 64% dense in order to yield a 10 μm filter grade disk. 100×

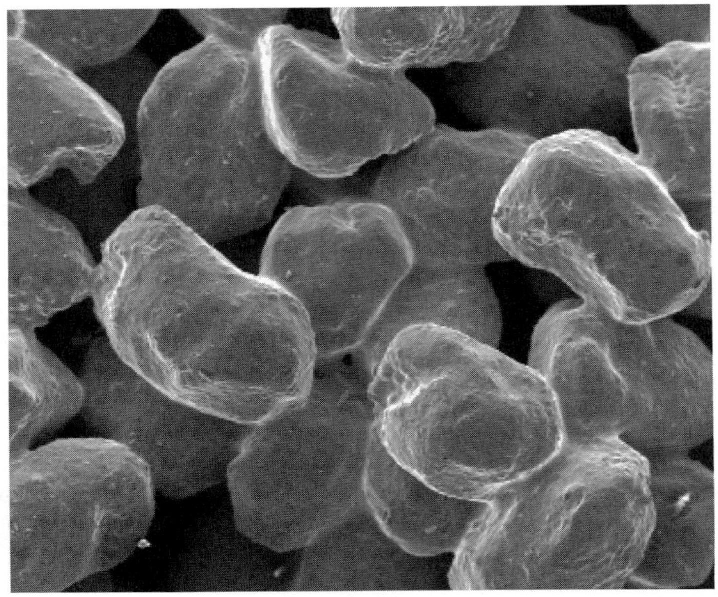

Fig. 2 Bronze particles made from tin-coated, cut copper wire with a size range of 250 to 425 μm that are gravity sintered to 55% dense in order to yield a 40 μm filter grade disk. 100×

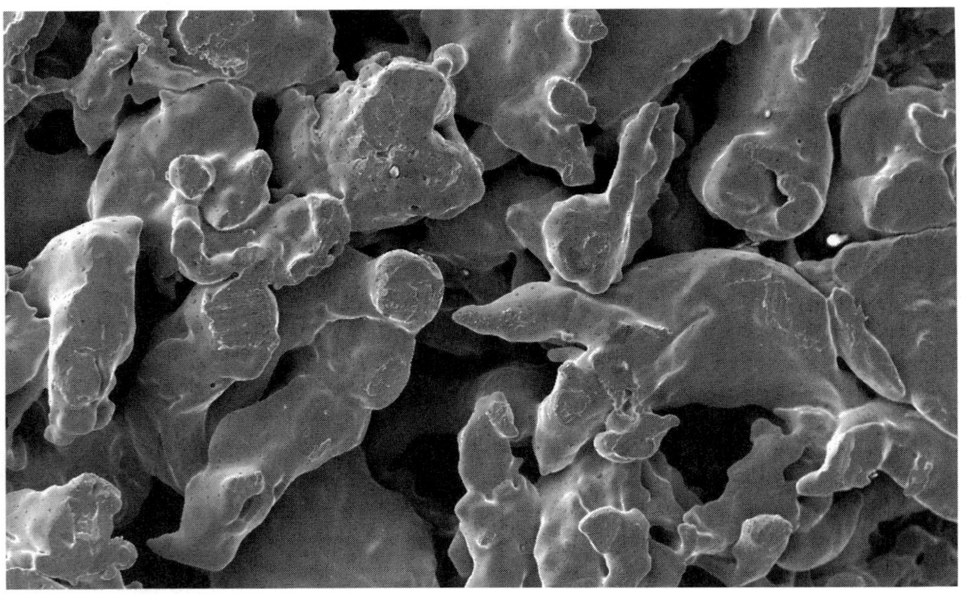

Fig. 3 Water-atomized, 316L SS powder particles with a size range of 300 to 600 μm that are compacted and sintered to 45% dense in order to yield a 40 μm filter grade sheet. 100×

becomes more rounded and higher compaction forces are required to produce components with adequate green strength. Some nickel-base alloys such as Hastelloy C22, tend to atomize into more rounded particles than similar size distributions of 316L SS and thus require increased compaction force to obtain similar green strength. Material characteristics such as hardness can also influence the compressibility of the powder. Water atomization typically produces particle sizes that range from as fine as 100 nm to as coarse as 1000 μm. The particles less than 10 μm in diameter are normally more spherical in shape and tend to have limited cold compaction green strength unless combined with an additive.

Stainless steel and nickel alloy powders are also produced by the gas-atomization process, which typically yields particles with spherical shapes. This powder cannot be readily compacted without additives. Components can be fabricated by vibratory filling of molds and gravity sintering. The mechanical properties of gravity sintered, spherical powders are normally much lower than other fabrication methods that use compaction. Therefore, gravity sintering is only used in special applications that require higher porosity or unique shapes. In addition, irregular-shaped powders, and powders with binders, can be processed using gravity sintering to obtain low-density products.

Carbonyl or water-atomized nickel powders are rapidly growing in acceptance for use as porous P/M products. The thermal decomposition of nickel carbonyl produces very fine (0.5 to 10 μm) particles attached in a filamentary chain-type structure. Very narrow particle size distributions are possible using this process, which can result in extremely uniform pore size materials after processing, as shown in Fig. 5. Although these powders do not flow well, components with higher porosity and finer pore sizes are made using very low compaction pressures less than 6.9 MPa (1000 psi) (Ref 11). For nickel components with larger pore sizes, water-atomized, irregular-shaped particles with diameters greater than 20 μm are normally selected instead of carbonyl nickel because the latter process produces more rounded particles as the particle size increases (Ref 12).

The atomization process can be varied to produce particles with other shapes designed to optimize porosity and permeability. Potato-shaped powders consist of elongated, rounded particles with smooth surfaces. Coarser grades of potato-shaped powders can be compacted directly to form shaped parts while finer grades are normally combined with a binder to achieve adequate green strength. Other processes have compacted and sintered mixtures of irregular-shaped particles with up to 25% spherical particles. The smoother surface and packing characteristics of potato-shaped powders and spherical powders can result in significant permeability increases, especially in the coarser filter grades (Ref 13).

The characteristics of the four most commonly used porous metal powder materials are briefly summarized to illustrate the different powder

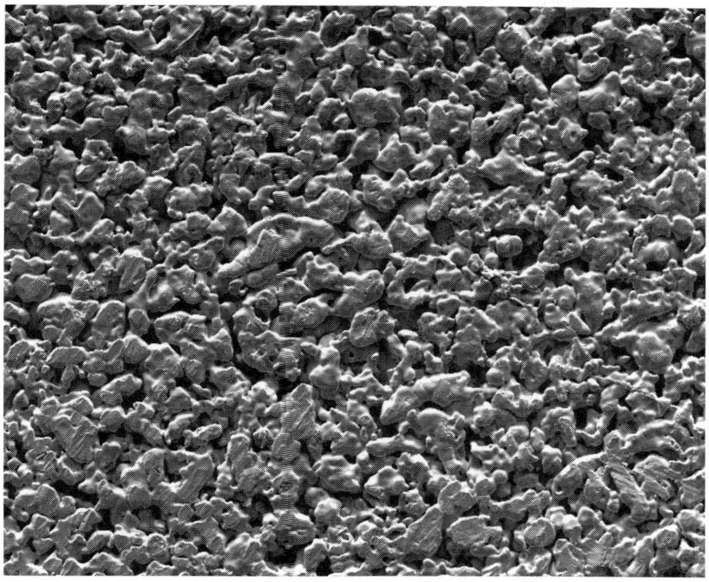

Fig. 4 Water-atomized, 316L SS powder particles with a size range of 45 μm and less that are compacted and sintered to 75% dense in order to yield a 0.5 μm filter grade sheet. 100×

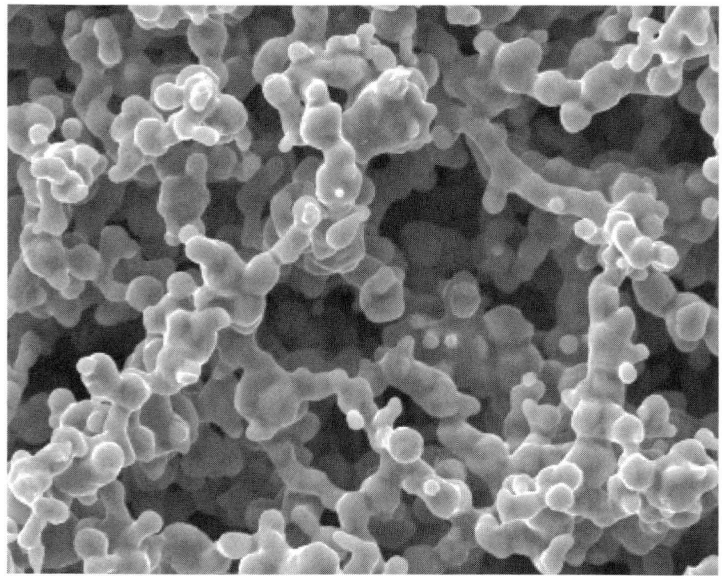

Fig. 5 Carbonyl nickel powder particles with a size range of 2 to 4 μm that are compacted and sintered to 35% dense in order to yield a 0.1 μm filter grade cup. 2000×

particle characteristics available to create porous structures. Characteristics of P/M materials are more completely described in previous sections of this Volume and other references (Ref 14, 15).

Processing Methods

After the material is selected, the P/M processing method is critical for determining the final mechanical properties and porous characteristics of the porous component. Selection of the fabrication method is dependent on the powder characteristics and the type of porosity required by the application. The preparation of the powders, the use of additives, the compaction methods, and the sintering conditions must be carefully controlled to produce uniform and repeatable porous characteristics. Secondary operations are normally required to enhance metallurgical properties and to allow easier adaptation to the intended application.

Powder preparation begins with the separation of the powder particles in the desired size distribution. Depending on the amount of secondary processing, commercial powders are available with either wide or narrow particle size distributions. For example, a wide particle size distribution such as −150 μm (−100 mesh screen) can be generated depending on the powder manufacturing process. However, secondary sieving or screening can classify a much narrower particle size distribution such as between 75 and 150 μm (−100/+200 mesh). Screening must be carefully monitored because there are significant variations in the screen openings or because the screen openings are blocked with fine particles. Therefore, significant amounts of particles that fall outside of the desired particle size range are not removed by screening. The porous part manufacturers process the commercially available powders into even narrower size distributions by removing oversize and undersize particles or by adding another screen to produce a tighter cut, such as between 106 and 150 μm (−100/+140 mesh). Vibratory or ultrasonic screening methods are normally used for particles greater than 20 μm (600 mesh screen) and other separation methods such as air classification are used for particles less than 20 μm. Porous P/M components manufactured from narrow particle size distributions have excellent reproducibility and uniformity of the final porosity.

Once the desired particle size distribution is obtained, the powder must be properly blended prior to use in order to avoid segregation and to maximize uniformity. The apparent and tap densities of the powder cuts are normally measured to meet tight specifications because these characteristics control further processing steps such as powder flow and die filling. Excessive blending can change the powder characteristics by generating additional fine powder particles or by changing the particle surface. Additives such as lubricants, pore formers, or binders are precisely mixed with the powder if required by the processing method (Ref 16, 17). Significant amounts of powder with controlled particle size distributions are used as loose pack beds for applications such as polymer filtration. However, most applications require further processing to develop a more rigid porous structure. The powder cleanliness and chemical analysis must also be carefully monitored to maintain the chemistry requirements of the material and to avoid contamination with other materials being processed. Because these materials are often used in aggressive environments, contamination from iron, carbon, aluminum, sulfur, and other elements could cause significant reduction in mechanical properties or corrosion resistance. Dedicated powder processing equipment and manufacturing facilities are the best defense against contamination. Other trace impurities such as oxygen, alumina, silica, and nitrogen levels from the powder manufacturing process can also affect the final mechanical properties.

Compaction methods and sintering are normally required to produce porous materials from stainless steel, nickel-base alloys, and nickel materials in order to achieve the best combination of mechanical strength, metallurgical properties, and porosity. Lower-density, higher-permeability parts from these alloys can also result from using a gravity-sintering process without compaction of the powder when the application does not require higher strength levels of compacted and sintered parts. Die compaction, isostatic pressing, and roll compaction increase the green strength of the part by cold welding of the particle to particle contact areas due to plastic deformation. As compaction force is increased, the density and green strength of the porous part increases, which results in finer porosity and lower permeability than methods that do not use compaction prior to sintering.

However, when compared to compacted structural P/M parts, the lower density and green strength limit the shape complexity of a porous component. For porous materials, uniform density is critical to the function of the parts because the largest porosity and greatest permeability are in the lowest-density region. Density variations of more than ±2.5% within the same component cause significant differences in performance.

Die compaction and sintering are the most common methods of improving mechanical properties for porous parts with length-to-diameter ratios of less than 5 to 1. Disks, cups, bushings, and other shapes can be readily compacted by filling the powder into dies and using hydraulic or mechanical P/M presses. Typically, porous materials are manufactured to achieve the lowest-density part in order to maximize porosity and permeability.

Minimum green handling strength can be obtained by using lower compaction forces in the range of 700 to 2100 kg/cm^2 (5 to 15 tons/in.^{2}). By comparison, higher-density structural P/M components require compaction pressures of 4200 to 8400 kg/cm^2 (30 to 60 tons/in.^{2}) (Ref 18). Porous parts with 20 to 50% green densities usually have minimal green strength and require careful handling; they often have to be manually picked off the press and placed in a sintering tray. Higher-green-density parts in the 50 to 80% range have sufficient green strength to allow the press feed shoe to eject the part off the die table and into a container. For low-density parts made from coarse powders, press rates can be less than 5 parts/min and part handling is critical to final yield rates. With proper sintering, die compaction can hold dimensional tolerances within ±1%.

Die compaction with lower tooling forces can be accomplished with die-wall lubrication or adding less than 0.5% lubricant to the powder. In the case of coarser water-atomized stainless steel powders, only 0.1 to 0.3% lubricant addition to the powder is generally used. Because smaller particle sizes of this same material are generally more rounded and have better particle packing (smaller pores), slightly higher pressing forces and lubricant additions are required for adequate green strength. Waxes and stearates are commonly added to the powder to reduce die-wall friction and tool wear. However, because increasing the lubricant percentage also reduces green strength, minimizing lubricant additions is often a better alternative than obtaining maximum tool life service. Tooling must be kept in a highly polished condition with tight tolerances to avoid smearing of the part surface during ejection. If smearing or galling of the part surface occurs, significant reductions in the final surface porosity size and in the overall part permeability occur. Improper monitoring of lubricant additions and tooling conditions can lead to problems with the uniformity and the reproducibility of the porosity.

Isostatic compaction is a common method of compaction for producing components such as tubes that have a length-to-diameter ratio greater than 3 to 1. Isostatic compaction can produce porous components with more uniform density than can die compaction (Ref 19). The wet bag cold isostatic pressing (CIP) process uses a hydraulic fluid to apply pressure ranging from 3.5 to 550 MPa (500 to 80,000 psi) to the tooling, which is sealed after powder filling. As the hydraulic fluid is gradually pressurized, the soft polyurethane outer liner compresses the powder against the metal core rod to obtain sufficient green strength. This "outside-in" pressing process results in a green part with a smooth inner surface and a rougher outer surface. If the pressing direction is reversed by using an expandable polyurethane inner bladder to compress the powder against a metal outer sleeve, a tube with a smooth outer surface can be obtained. Other near-net shapes such as cones, funnels, and tubes with splines can be manufactured using the CIP process (Ref 20). For example, a 25 mm (1 in.) outside diameter (OD) by 19 mm (3/4 in.) inside diameter (ID) by 76 mm (3 in.) long tube would have uniform density along the entire length, whereas a similar-sized, die-compacted bushing would have high-density ends and a lower-density center. Hot isostatic pressing (HIP) combines the advantages of CIP and sintering into one process and is utilized in applications where complex shapes are needed or when a material has low green strength for handling.

Roll compaction of porous nickel sheet materials have been commonly used in the production of battery materials (Ref 21, 22). Nickel and other powders are rolled to a thickness ranging from 0.13 to 3.8 mm (0.005 to 0.150 in.) and widths up to 1 m (39.37 in.), depending on the green strength and requirements of the application. Controlling the powder feed is critical to the process to obtain the best uniformity of the final product. Attempting to roll sheet with larger cross-sectional thickness can lead to delamination or low-density center regions that cause poor mechanical properties. Thin cross sections or powders with low green strength limit the handling of the porous sheet prior to sintering. The rolled sheet can be directly fed into a sintering furnace to minimize handling problems in certain high-volume applications. Roll compaction is used to produce an economical, uniform density material with good mechanical properties and tight dimensional tolerances (±2% on thickness). Width and length tolerances are determined by secondary machining, cutting, or shearing operations. Shapes or patterns can often be designed into the rolls, or parts can be stamped from the finished porous sheet.

Other forming methods used to manufacture porous metal components include centrifugal slurry casting (Ref 23, 24), blow molding (Ref 25), metal injection molding (Ref 25), extrusion (Ref 26), and gravity filling of shaped molds. Sheet materials can also be made using other processes such as a liquid lay-down method, which uses a powder and binder slurry (Ref 27). Complicated shapes and lower-density components are formed using additives or pore formers such as water and/or thermoplastic binders that allow sufficient handling strength. Removal of the additives prior to sintering is normally accomplished in the preheat zone of the furnace or in a separate controlled bake-out process. Metal spraying (Ref 28) can create a controlled porous structure with or without additives by spraying molten metal onto a base material to combine compaction and sintering into one process.

Sintering of porous metal is a critical balance between maximizing material properties and maximizing the open porosity and permeability. However, because permeability and material properties such as strength and ductility are generally inversely related, the desired balance of these characteristics normally occurs in a very small processing window (Ref 29, 30).

Sintering requires the proper compromise of temperature, time at temperature, and atmosphere to arrive at the desired porosity characteristics. Porous components that are not adequately sintered exhibit poor mechanical properties due to lower density and insufficient interparticle neck growth. Porous components that are exposed to excessive sintering conditions have lower permeability and higher densities than desired.

The preheat and cooling portions of the sintering cycle must also be closely controlled to achieve the proper metallurgical properties. Controlling the preheat conditions ensures adequate burnoff of additives and lubricants as well as minimizing the distortion of the parts. The cooling conditions must be designed to provide maximum corrosion resistance and to avoid oxidation.

Sintering temperature must be selected by considering the material, the powder shape, and the powder particle size distribution. Sintering is normally accomplished at 70 to 90% of the material melting temperature. Finer powder particles require a lower sintering temperature because the surface energy driving force to initiate bond growth is much higher than for a coarser particle. Sintering at too high a temperature also causes the formation of very large pores and nonuniform porosity just prior to melting. Controlling the furnace temperature within ±1% of the optimal sintering temperature achieves the best porosity uniformity and reproducible properties. Furnace design, furnace loading, and furnace hot-zone uniformity are critical to the production of porous metal components. Even slight lot-to-lot variations due to differences in the powder shape or the powder particle size distribution can often require adjustment of the sintering temperature to maintain the balance of properties and permeability.

Sintering time must be monitored to allow for a minimum exposure time at the desired sintering temperature. Sintering for at least 30 to 60 min at the maximum sintering temperature is recommended for most materials for sufficient bond formation and growth. Inadequate sintering time can lead to large variations in part shrinkage and final density, causing porosity and permeability variations. The sintering time at temperature must allow for the temperature of the furnace load to stabilize at the desired temperature, especially when batch size and furnace recovery can widely vary. Furnace profiles and recovery rates for various load sizes for each furnace must be considered when determining appropriate sintering time. Shrinkage and density measurements can be used to control sintering time and temperature if the furnace atmosphere is consistent. Excessive sintering unnecessarily reduces permeability due to pore size reduction and pore closure without significantly improving mechanical properties.

Sintering atmosphere selection is critical for determining the metallurgical properties of the porous metal product. Because porous materials have much higher surface area than a similar-size structural part, the atmosphere has more contact with surfaces throughout the part rather than just near the surface. Porous parts also contain relatively large amounts of trapped air in the pores that must be removed by purging or good atmosphere circulation in the furnace. Often, higher flow rates of an atmosphere or longer vacuum pump down times are required to displace the air prior to sintering. The sintering cycle can be subdivided into three main processes: preheating, sintering, and cooling. Each process of the sintering cycle has unique atmosphere requirements that provide the optimal properties of the porous metal product just as with structural P/M components (Ref 31, 32).

The preheating cycle must remove any lubricants or binders from the green compact and normally requires an oxidizing atmosphere. A separate furnace is often used to burn off these additives in order to avoid contamination of the sintering furnace. Preheating in the same furnace with a reducing atmosphere is often sufficient to remove low levels of additives because the open porous network allows more complete removal of the thermal decomposition products than does a structural part. Specifically, mechanical strength and corrosion resistance properties are highly dependent on the interaction with the sintering atmosphere. Porous materials are commonly sintered either in reducing atmospheres such as nitrogen-hydrogen mixtures, hydrogen, dissociated ammonia, or vacuum. Inert atmospheres such as dry argon can be used to sinter some of the specialty porous materials such as aluminum, refractory metals, and precious metals. Reactive materials such as titanium and zirconium require good vacuum sintering with a high-purity, inert backfill gas. Nitrogen-containing atmospheres can have a nitriding effect on some materials such as porous stainless steels. Series 300 stainless steels can benefit from increased tensile strength with slightly lower ductility in a reactive nitrogen-sintering atmosphere such as dissociated ammonia. However, reduced corrosion resistance may result unless the cooling rate is carefully controlled to minimize the formation of chromium nitrides (Ref 33). A hydrogen atmosphere with a low dew point or vacuum sintering with a hydrogen backfill during cooling produces the best combination of corrosion resistance and mechanical properties for porous stainless steels. The cooling process cycle usually requires a reducing or oxygen-free, inert gas atmosphere for best metallurgical properties. When heavy oxidation or nitriding is prevented during cooling, a protective passive surface layer will form and good mechanical properties will result. Carbon-containing atmospheres are not normally utilized in processing porous materials because higher carbon levels are usually detrimental to corrosion resistance.

Gravity sintering refers to one of the simplest methods of sintering to improve the mechanical properties of a porous component. Diffusion or liquid-phase bonds form between the particles when loose (noncompacted) powders are heated to a temperature near the solidus temperature (Ref 34). Localized melting of the particle surfaces form necks between the powder particle contact surfaces, and the amount of shrinkage during sintering controls the characteristics of the interconnected porosity. As an example, spherical porous bronze powders are poured and vibrated into a graphite or an oxidized stainless steel mold without compaction. The bronze or tin-coated copper powder is then sintered at 800 to 1000 °C (1472 to 1832 °F) for 20 min in a reducing atmosphere to form a liquid phase between the particles (Ref 9). Upon cooling, a strong metallurgical bond is formed at the contact points between the powder particles. Disks, cups, bushings, or other simple shapes can be produced using this method. Typically, parts have a 1° draft angle to facilitate part release from the mold and have dimensional tolerances of ±3%.

Secondary operations are widely used to modify the part shape, the porosity, and the surface characteristics of a porous part. Porous parts can often be formed to net shapes for use without secondary operations. However, the lower green strength and density of porous parts can limit the complexity of the green part and subsequent handling prior to sintering. The part shape complexity can be enhanced by sizing, machining, and forming. Secondary sizing processes can improve the roundness, flatness, and other dimensional tolerances that cannot be held during sintering. Porous P/M materials must not be overstressed during sizing because cracking or closure of the pores will occur. Sizing operations increase the density of the part and reduce the surface porosity in the tool contact areas.

Secondary machining operations are used to produce features such as threads or tighter dimensional tolerances that cannot be readily controlled during compaction. Conventional machining of porous materials closes off the majority of the surface porosity by metal smearing. Machining can be utilized to blind off the porosity and fluid flow in selected areas, depending on the application. Controlled chemical etching of a machined surface can restore the surface porosity and still retain the tighter dimensional tolerances produced by machining. Care must be taken to avoid contamination of the porous material during machining or chemical reactivating. Machining methods without coolants and special cleaning processes are often used to remove the contamination after processing. Other machining methods such as electrical discharge machining (EDM) are also used to avoid closing off the surface porosity. However, removal of any machining fluids again requires special cleaning methods. Conventional machining, EDM, and laser methods are used to cut porous parts from larger parts or sheets.

Forming of porous materials by cold working using dies or calendering rolls is used to change the part shape. Flat sheet can be formed into tubes and seam welded. Cold working during forming increases the density, hardness, and tensile strength of the porous material, with a corresponding decrease in ductility. Forming is also used to make special shapes and assemblies. Light sizing and swaging operations alter the surface layer to produce parts with tighter dimensional tolerances than as-pressed components. The surface porosity is flattened during forming, and the permeability and the surface pore size are decreased. However, porous P/M materials are difficult to straighten or flatten because of their elastic properties, which result in significant "spring back" behavior just prior to crack initiation.

Joining methods such as brazing, welding, gluing, sinterbonding, and mechanical interlocking are used to create porous metal assemblies. The conditions for joining lower-density materials to other porous materials or to solid materials are very different from joining solid materials. For example, brazing materials and processing conditions must be selected to avoid wicking of the braze material away from the joint area. Often a braze material is selected to form a higher melting point eutectic upon alloying with the porous base material in order to solidify the braze material in the joint area.

Welding of porous materials to other porous materials or to solid materials becomes much more complex than welding solid materials. The addition of filler metal, the reduction of weld heat input and the use of special welding gases are required to overcome welding problems associated with higher porosity and gases trapped in the pores. Good weld joint design allow proper heat distribution to prevent leaks and large holes. Weld purge gases and material cleanliness are important to avoid contamination and to provide good corrosion resistance. Tungsten inert gas, electron beam, and laser welding processes are used to obtain full penetration of weld joints with relatively narrow heat-affected zones. Postweld heat treatment processes such as annealing or stress relieving are recommended for good metallurgical properties.

Heat treatment or thermal cycles in reducing atmospheres can also be used for sinterbonding porous parts by processing at 10 to 65 °C (50 to 150 °F) below the sintering temperature in order to maintain the porosity and permeability. Sinterbonding is a thermal processing method used to form a metallurgical diffusion bond between two parts such as between a porous part and assembly hardware. The hardware is generally the same material as the porous component and allows the sinterbonded assembly to be more easily adapted to the application than if only a porous part was used. The diffusion bonds formed are significantly lower than welded or brazed joints because the sinterbonds are only formed at the local contact points. However, the sinterbonds are significantly stronger than just an interference fit alone and can ensure that the porous component cannot be easily removed. Sinterbonding can be accomplished during a sintering cycle if proper allowances for shrinkage are considered. The sintered porous part must have adequate strength to allow for a press-fit assembly without developing cracks. The press-fit interference for parts less than 25.4 mm (1 in.) OD is between 0.025 and 0.1 mm (0.001 and 0.004 in.). The hardware surface should have at least a 0.4 μm (16 μin.) surface finish so that the assembly interface does not have any gaps that may form openings larger than the pores in the porous metal component. The assemblies are normally checked for joint integrity after sinter-bonding.

Thermal processing can also be used for removal of oxidation, burnoff of certain contaminants introduced by secondary operations, and for passivation of the metal surfaces. Oxidation color from the heat-affected zone of welds or from the heat of a cutting operation must often be removed by heat treatment in a reducing atmosphere for corrosion resistance or aesthetic reasons. Organic contaminants that are captured in the pores from prior processing steps or improper handling can be vaporized and removed from the pores by thermal processing. Heat treatment in a reducing atmosphere can also restore a uniform oxide layer upon cooling that forms a passive surface on the porous metal in order to increase corrosion resistance.

Porous sintered metals can be coated with a material such as silicone to alter the surface characteristics. Silicone treatment of the pores changes the hydrophilic nature of the surfaces to a hydrophobic surface that still maintains some permeability. Metal oxide coatings have been successfully formed on the porous surfaces to increase corrosion resistance or to decrease the pore size by applying a thin layer of submicron particles of a metal oxide such as silica, titanium dioxide, or alumina (Ref 35, 36). These fine metal oxide particles can be bonded to a coarser porous metal substrate to produce a dual-porosity structure with good permeability and excellent corrosion resistance.

Porous Material Characteristics

Design engineering information is readily available for many of the porous material characteristics such as density, pore size and distribution, bubble point, permeability, and mechanical properties (Ref 8, 10, 17). However, many of the unique characteristics of porous P/M materials such as elevated service temperature and corrosion resistance properties in specialized environments require additional design engineering considerations. The density, pore size, pore size distribution, and permeability determine the overall performance of a porous material in an application. The interrelationship of these four characteristics often allows the design engineer to specify only one or two of these factors. For example, testing has shown that the permeability of a porous part can be accurately predicted if the density and the initial powder particle characteristics are known for a particular processing method (Ref 37). The initial powder particle size controls the pore size and distribution when sintered to a specified density. The permeability is related to the pore size and pore distribution. Other porous material characteristics can be estimated from wrought (or full-density P/M) products because higher-density products have more extensive test data available in the literature. Limited availability of design engineering data for porous materials often results in the need for actual testing in the intended application prior to use.

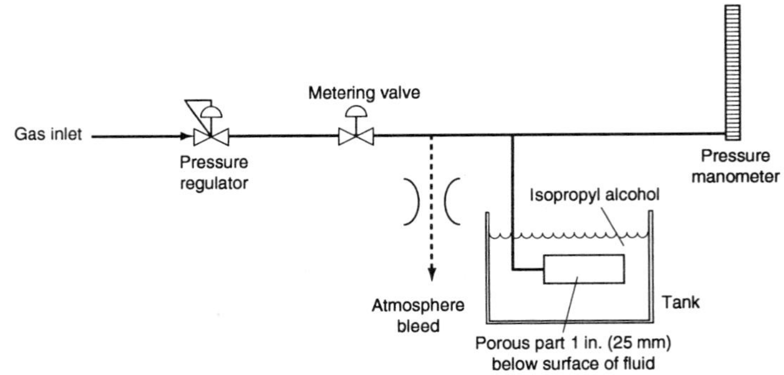

Fig. 6 Bubble-point testing apparatus

Material density is an important characteristic for predicting mechanical properties and permeability. As the density decreases (and porosity increases), the mechanical properties are reduced and the permeability is increased. Porosity consists of interconnected channels and isolated pores. Isolated pores that are either closed or dead-end channels do not contribute to the overall permeability, and the percentage of isolated pores must be considered if density comparisons are used to specify a component. Nearly all of the porosity in a component with an overall density less than 75% of TMD is interconnected to the surface pores and is usable for filtration and fluid-flow paths (Ref 38). As the overall density increases from 75 to 92% of TMD, the percentage of isolated pores increases rapidly and the functional interconnected porosity is depleted.

Density measurement methods for porous materials are described by International Organization for Standardization (ISO) Document 2738 (Ref 39), which includes the use of Archimedes principle. The porosity must be fully sealed prior to immersion in water using this method. Overall density can be estimated by weighing and volume calculation for simple part geometry. Density can also be measured by optical analysis of carefully prepared surfaces. The cutting, polishing, and etching techniques during metallographic preparation must be monitored to avoid smearing or excessive removal of the surface porosity. Density uniformity throughout a porous product must be considered when measuring overall density because the forming process may introduce large variations. For example, a coaxial pressed cup could have a density variation of more than 5 to 10% when comparing the lower-density side walls versus the higher-density end cap.

Pore characteristics such as size, shape, and distribution can be determined using many different measurement techniques. A comparison of the various pore characterization methods reveals that while the test method is usually consistent to show relative differences between various porous material grades, the test methods do not always agree (Ref 40). Because it is normally possible to correlate the various test methods, technical information regarding pore characteristics must be carefully compared to determine if the pore characteristics are equivalent.

One of the most common pore size characterization methods is a "bubble-point" test, which measures the pressure required to release the first bubble from the surface of a porous material submerged in a liquid and pressurized from one side as shown in Fig. 6. The liquid selected for this testing must have good wetting characteristics with relatively low viscosity and surface tension, such as isopropyl alcohol. Several standard test methods, such as ASTM E 128 (Ref 41) and ISO 4003 (Ref 42), have been developed to measure the pressure at which the first bubbles form and break away from one area. The maximum interconnected pore diameter can be estimated using Poiseuille's capillary law and an additional pore shape correction factor for noncylindrical pores:

$$d = 4KR \cos B/P$$

where d is the pore diameter, K is the shape correction factor, R is the surface tension of the wetting liquid, B is the contact angle of liquid and porous material, and P is the pressure at first bubble point (Ref 43). The shape correction factor, K, is largely dependent on the shape of the pores and the porous material. The shape correction factor can vary from about 0.2 for more spherical pores (e.g., bronze) to 0.4 for more irregular-shaped pores (e.g., 316L SS). Therefore, the bubble-point test is not a highly accurate measurement of pore size, but serves only as a reproducible, nondestructive test for ranking porous material grades by the largest pore size. The maximum pore size of the porous media calculated from the bubble-point test is not a direct indication of filtration performance, but the results can be correlated to more complex filter-retention testing. The filter industry often refers to the diameter of the largest pore determined by the bubble-point method as a micron rating. This terminology should not be confused with the diameter of a particle retained by a filter or the absolute filtration rating. The bubble-point test is not a good indicator of the gas or liquid filtration rating because filtration is dependent on many other variables, such as the uniformity and distribution of the pores, the media thickness, and the type of solids being filtered. The bubble-point method is also an excellent integrity test of a material that has been subjected to secondary operations such as welding or forming in order to determine if any larger pores were created.

Mercury porosimetry and a mean flow pore size distribution measurement technique developed by Wenman and Miller use the Washburn Equation to calculate the pore diameter as the pressure is gradually increased (Ref 44, 45). Mercury porosimetry calculates the interconnected pore size and pore distribution of material by measuring the pressure required to force mercury into the pores and the volume of mercury in the pores. The other method determines the pore size distribution using a fluid that fully prewets the porous sample and then measures the pressure differential as the air flow is gradually increased. As the pressure increases, the fluid is forced from the larger pores first, and the air flow permeability begins to increase until the sample is dried. A comparison of the wet and dry permeability curves allows the calculation of the pore size and distribution. ASTM F 316 (Ref 46) details the test method for determining the maximum pore size by bubble-point testing and the pore size distribution by mean flow pore testing. The capillary action of the porosity is dependent on pore size and is an important property for applications that require fluid wicking and storage.

Permeability is a measure of the flow resistance through a porous material and is one of the most important characteristics for design engineers specifying porous materials. The fluid flow through a porous material is normally expressed as a volumetric flow rate at a specified pressure differential for a given surface area (e.g., standard liters per minute per square centimeter at $dP = 1000$ kg/cm^2 of nitrogen). The nature of the fluid flow can be characterized as diffusional, laminar, or turbulent depending on flow velocity. For very low flow rates and small pores or low absolute pressures, diffusional flow is the predominant mechanism and is not dependent on the fluid viscosity because the mean free path of the molecular collisions is larger than the pores. Fluid flow that is unidirectional along parallel flow paths with uniform velocity is considered laminar or viscous flow. Turbulent flow occurs when flow is high enough to cause irregular, random flow paths. Other types of flow mechanisms termed slip and inertial flow occur in the transition regions among diffusion, viscous, and turbulent flow (Ref 47) For example, inertial flow mechanisms must be considered to

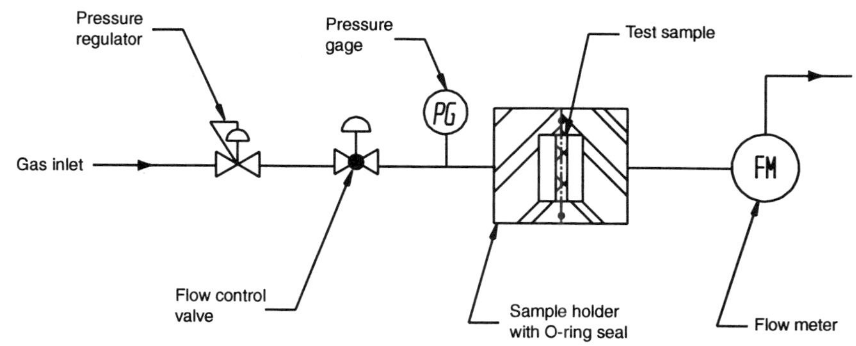

Fig. 7 Schematic of gas permeability testing apparatus

account for the energy losses and increased flow resistance of the gas at higher flow rates as it changes direction in the tortuous pore network flow path. For laminar flow of liquids, the volumetric flow rate, Q, can be predicted by Darcy's Law:

$$Q = dPAC_v/tV$$

where dP is the differential pressure, A is the surface area, C_v is the coefficient of viscous permeability, t is the material thickness, and V is the dynamic viscosity of the fluid. In the case of compressible gases, Darcy's law is modified to allow for the gas volume change with pressure as shown below:

$$Q = AC_v(P_1^2 - P_2^2)/2\,P_2tV$$

where P_1 is the upstream pressure and P_2 is the downstream pressure (Ref 48). For higher pressures and gas velocities, the flow rate through a porous material can be predicted by a modified version of Darcy's law, which includes correction factors for inertial and slip flow (Ref 48). Although the permeability can often be accurately modeled, the complex gas flow physics of the porous structures limits the accuracy of the theoretical calculation of fluid flow, especially in the extreme low- and high-flow regions. As the gas density increases due to an increase in pressure or a decrease in temperature, the permeability decreases. As the viscosity of a gas increases due to an increase in pressure or temperature, the permeability also decreases. Pressure increases result in only a slight increase in the gas viscosity, but significantly increase the gas density. Temperature increases typically result in an increase in the gas viscosity and a reduction in the gas density. When comparing different porous materials, the accuracy of theoretical calculations is often reduced due to the interactions of these variables along with the various flow mechanisms operating in the microchannels of the porous structure.

As shown above, the flow rate is also dependent on the surface area and the thickness of a porous structure. As surface area is increased, the differential pressure required to produce the same flow rate is decreased. Also, as the thickness of a material is decreased, the pressure differential is reduced to obtain the same flow rate. When comparing two porous materials, higher permeability indicates that a lower pressure differential was required to produce the same flow. Therefore, less surface area of a material with a higher permeability is required to produce the same fluid flow rate as a porous material with lower permeability. The economics become more favorable as the porous material surface area is reduced to meet the desired flow rate. The pressure differential and permeability are important design characteristics for determining overall flow rates for applications such as filter systems.

Permeability testing is one of the best methods for specifying the characteristics of a porous material by measuring the fluid flow rate and pressure drop. Permeability if a function of the gas or liquid testing fluid and requires close control of the fluid test conditions. Gas permeability is dependent on many variables such as density, viscosity, temperature, and pressure conditions. Many references have investigated gas-flow dynamics and have proposed models for predicting flow rates as conditions are varied (Ref 30, 48–50). The permeability of a fluid through a rigid porous material can be determined by standard test methods as described by ISO 4022 (Ref 51) and ASTM 128 (Ref 41). A typical gas permeability test schematic diagram is shown in Fig. 7. The flow path position of the volumetric flow measuring device is critical in order to avoid inaccurate measurements if an excessive back pressure or vacuum is created. The flow rate is measured by controlling the pressure differential and correcting the measured flow to standard temperature and pressure conditions using the ideal gas law to correct for gas volume changes. The atmospheric pressure and gas temperature can be measured in order to correct the measured gas flow rate back to standard temperature (21.1 °C, 70 °F, or 294.1 K) and pressure (760 mm Hg, or 29.92 in. Hg). Accurate flow rate comparisons can then be made between porous parts tested with nonstandard conditions. A general formula that accounts for volume changes and corrects the measured gas flow rate back to standard conditions is given below:

$$Q_s = Q_m(P_m / P_s)(T_s / T_m)$$

where Q_s is the standard flow rate, Q_m is the measured flow rate, P_m is the measured pressure

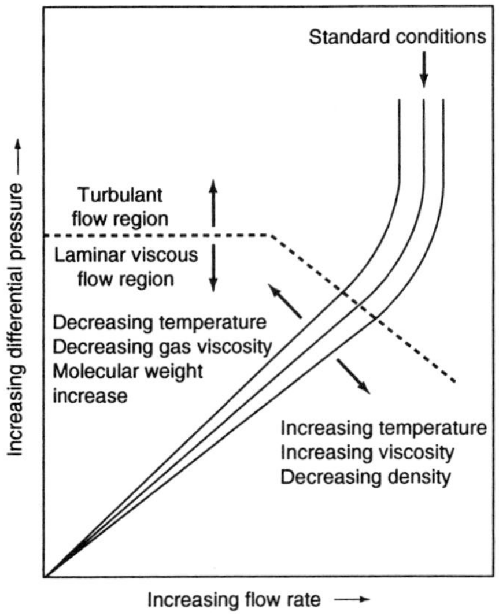

Fig. 8 Typical gas flow versus pressure drop through porous sintered metal restrictors

Table 1 Typical mechanical properties of bronze and 316L SS filters

Material(a)	Filter grade(b), µm	Density, %	Minimum ultimate tensile strength, MPa	Minimum ultimate tensile strength, ksi	Elongation, %	Shear strength(c), MPa	Shear strength(c), ksi
Bronze	10	75	48	7	8	130	18.8
	20	68	41	6	6	110	15.9
	40	62	35	5	4	100	14.5
	90	57	28	4	3	75	10.9
	150	54	21	3	2	40	5.8
	250	52	14	2	2	30	4.3
316L SS	0.2	80	180	26	6	250	36.2
	0.5	75	159	23	6	230	33.3
	2	65	117	17	5	190	27.5
	5	62	90	13	5	160	23.2
	10	60	69	10	4	130	18.8
	20	55	48	7	3	100	14.5
	40	50	31	4.5	2	80	11.6
	100	40	27	3.9	2	70	10.1

(a) For comparison purposes, wrought bronze (90% Cu/10% Sn) has a theoretical mean density of 8.8 g/cm^3 with a minimum UTS of 300 MPa (43 ksi) and 20% elongation. 316L SS bar stock (cold finished and annealed per ASTM A 276) has a theoretical mean density of 8.0 g/cm^3 with a minimum UTS of 482 MPa (70 ksi) and 30% elongation. Data are for commercial bronze filter disks and 316L SS sheet. (b) Filter grades in micrometers as estimated by bubble-point test method. (c) Shear strength is the punching force divided by the sheared edge area per DIN standard V 30910 (Ref 55).

in mm Hg, P_s is the standard pressure, T_m is the measured temperature, and T_s is the standard temperature. Corrections to standard conditions are significant because a small pressure change of 25.4 mm Hg (1 in. Hg) or a small temperature change of 2 °C (4 °F) can result in approximately, a 1% difference in the flow rate.

Additional correction factors can further improve the accuracy of permeability testing by accounting for changes in the gas viscosity and the interactions of the gas with the pore structure as temperature and pressure are varied. As temperature increases, the gas viscosity increases and the gas density decreases, which in turn increases the measured gas flow rate relative to standard conditions as shown in Fig. 8. Also, as the gas contracts to enter or expands to exit the pore structure, energy losses occur to reduce the flow (Ref 49). When measuring the flow of one gas and calculating the flow rate equivalent of a different gas, the flow rate is adjusted for differences in gas viscosity or specific gravity. Gas flow conversions are important when testing of the actual gas is not practical due to safety or handling reasons. For example, the flow rate of nitrogen is often measured and used to estimate the actual flow rate of a toxic or corrosive gas at the same temperature and pressure conditions.

The volumetric flow rate can also be calculated by dividing the mass flow rate by the fluid density. The mass flow rate of gases can be measured with an accuracy of about ±2% by devices that correlate the gas flow rate to the cooling rate of the gas as it flows past a hot wire. These devices are based on calculating the gas flow rate by measuring the amount of energy required to keep the wire at a reference temperature and knowing the thermal conductivity of the gas. Mass flow rates do not need to be corrected back to standard conditions. Volumetric flow rates that are corrected back to standard conditions can achieve an accuracy of better than ±1% by collecting the fluid in a known volume as a function of time. Volumetric flow rate devices such as positive displacement piston meters are highly accurate primary standards. Mass flow rate devices are slightly less accurate and are secondary standard devices.

Liquid permeability is also sensitive to variations in temperature, pressure, viscosity, and density. Liquids flowing through porous materials are highly dependent on the viscosity and the ability of the fluid to wet the pore structure. Liquid flow rates are often difficult to reproduce if the fluid does not have consistent wetting properties. For example, isopropyl alcohol is preferred for liquid flow testing of porous metal materials, whereas water does not wet well. Water can also have large differences in the mineral content depending on the source, and these quality differences can make flow rate comparisons difficult, especially for tighter porous structures. The permeability of gases and liquids through a rigid porous material can be determined by specialized testing as described by ISO 4022 (Ref 51) and ASTM E 128 (Ref 41).

Mechanical properties such as ductility, tensile strength, shear strength, collapse strength, burst strength, and fatigue life of porous materials are highly dependent on the porosity and the processing method (Ref 52, 53). Table 1 illustrates typical values for tensile strength, elongation, and shear strength of 90% Cu/10% tin bronze discs and 316L stainless steel sheet as a function of density and filter grade as rated by the bubble-point method (Ref 8, 10, 54). Mechanical properties increase significantly as the pore size and the percentage of porosity decreases (Ref 54, 56–58). Alternatively, permeability decreases as the pore size and the percentage of porosity decreases. Therefore, an optimal balance of mechanical properties and permeability must be achieved to meet the application. Processing methods and materials can usually be selected to create a porous P/M material that meets the minimum mechanical properties and provides the maximum permeability. In some more severe applications involving high-pressure differentials, corrosive environments, and/or high-temperature service, mechanical properties are maximized at the expense of permeability in order to maximize service life. Enhanced mechanical properties also extend the number of cleaning and reuse cycles in some applications.

Other material properties such as thermal conductivity, thermal expansion, fatigue, electrical conductivity, and magnetic properties are also highly dependent on porosity and generally decrease as porosity increases (Ref 59, 60). For example, thermal expansion can be directly related to the porosity by the equation (Ref 59):

$$C_t = C_o D_f^{1/3}$$

where C_t is the effective thermal expansion of the porous material, C_o is the bulk thermal expansion of the wrought material, and D_f is the fractional density. Other theoretical relationships have been proposed to cover the mechanical properties over a broad range of porous materials and porosity levels, but actual testing is recommended to determine the properties because slight variations in porosity and processing methods can produce significant changes (Ref 61, 62).

Elevated-temperature and corrosion properties are often important criteria for selection of porous P/M materials versus other porous materials such as plastics and papers. The maximum service temperatures for several P/M materials are listed in Table 2. The application environment is also critical to obtaining maximum service life because the atmosphere may be oxidizing, reducing, or inert in nature. Mechanical properties at an elevated temperature may be significantly reduced by accelerated corrosion reactions in atmospheres that contain ammonia, carbon, chlorine, fluorine, hydrogen, moisture, sulfur, or other reactive materials. For example, catastrophic oxidation of 316L SS can occur at a

Table 2 Elevated temperature service for selected porous filter materials

Material	Maximizing temperature oxidizing atmosphere		Maximum temperature reducing atmosphere	
	°C	°F	°C	°F
Bronze	150	304	400	750
316L SS	400	750	538	1000
Inconel 600	593	1100	815	1500
Hastelloy X	788	1450	927	1700

lower temperature than expected due to the formation of low-melting-point eutectic such as vanadium pentaoxide if vanadium is used as a reaction catalyst. The presence of lead and molybdenum can also result in catastrophic oxidation (Ref 63). The corrosion resistance of 316L SS is also reduced in ammonia service at 540 °C (1004 °F) due to the formation of a brittle nitride layer and in the presence of sulfur due the formation of iron and nickel sulfides. Temperature fluctuations and localized hot spots can cause accelerated oxidation due to spalling or cracking of the passive layer.

Corrosion resistance of porous P/M materials is a very complex subject due to the large number of variables that change the material behavior (Ref 64). Material variables include composition, porosity characteristics (density, pore size, pore distribution, and pore surface condition), and the effect of interstitial elements (carbon, nitrogen, and oxygen). Processing variables include delube conditions, sintering conditions (time, temperature, atmosphere composition, dew point, and cooling rate), and secondary operations (cold work, machining, welding, and heat treatment (Ref 65). Environmental variables include fluid composition, velocity, aeration, temperature, process cycle, and exposure conditions. Corrosion failure of porous materials is normally due to localized attack of the sinter-bonds due to intergranular corrosion, crevice corrosion, stress-corrosion cracking, and pitting mechanisms (Ref 66, 67). There is limited design engineering data that predict the corrosion rates of porous materials in various environments. In general, porous materials have significantly higher corrosion rates due to their larger surface area than the comparable wrought material. In order to predict the service life of porous materials without actual testing, corrosion handbooks (e.g., Ref 68) for solid materials offer a reference point. As a general rule, avoid selecting a porous material for an environment that rates the corrosion resistance of the solid material as excellent. Even a very low corrosion rate for a solid material at the specified conditions can often still be too high for a porous material that obtains its mechanical properties from the very small bonds at the particle contact points. The corrosion resistance of a solid is often rated using a low fluid velocity in contact with the surface layer. Increased fluid velocity in the pore structure and aeration of the fluid may decrease the corrosion resistance of a porous material if the passive layer on each powder particle is disturbed. Also, pitting- and crevice-corrosion mechanisms can occur in porous materials that have dead-end pore channels. Pitting corrosion can also occur if a corrosive liquid remains in the pores to create a localized concentration cell. Inadequate cleaning or prolonged storage without drying after removal from service can cause pitting failure.

There are many methods to enhance the corrosion resistance of porous materials. Proper alloy selection results in the best protection against corrosion although the economic considerations of the more expensive alloys may be prohibitive. One advantage of the P/M process is that special alloys or custom blends can be made. For example, more nickel, chromium, and manganese can be added to improve corrosion resistance of a 300 series stainless steel rather than being limited to standard alloys. Bronze filters can be plated with nickel or tin to improve corrosion resistance. While it is difficult to predict actual service life in a particular application, common corrosion tests can give some indication of corrosion reactions. Evaluating corrosion resistance using standard immersion (weight-loss) testing and salt-spray tests can often predict the behavior of porous materials (Ref 69). One of the best indicators of corrosion resistance uses porous tensile bar samples suspended in the actual solution. While the fluid-flow velocity effects are not considered in this test, mechanical testing and optical analysis can detect the effects of corrosion. Electrochemical testing can also be used to indicate the potential of a sample to exhibit passive behavior in a solution (Ref 66, 69). For stainless steels, indicator tests to determine the presence of unalloyed iron or carbide precipitation at grain boundaries determines proper processing. Sensitization of 316L SS due to slow cooling in nitrogen-containing atmospheres during sintering or carbide precipitation during welding can lead to intergranular corrosion (Ref 70). Welding should be followed by an annealing or stress relieving heat treatment to increase corrosion resistance and mechanical properties. Proper handling to minimize contamination and corrosion initiation sites (e.g., fingerprints, scratches, dents, foreign material) can also improve corrosion resistance.

Filtration properties are dependent on the porous material characteristics, the surface area available for filtration, and the process conditions of the application. Filtration and separation are intricate technologies that can be classified into six general areas: gas-solid, liquid-solid, gas-liquid, gas-gas, solid-solid, and liquid-liquid separation. Porous P/M parts are normally used for solid particle separation from a process stream, but have significantly different filtration properties in gas or liquid service (Ref 71). The filter efficiency or filter rating in a gas stream is higher than in liquids. There are additional gas-liquid separations possible with porous P/M parts such as bronze filters, which separate water from air due to surface tension differences. Filters are specified and compared by their desired performance characteristics, which include high permeability, low-pressure drop, retention efficiency of specified particle sizes, particle-loading capacity and resistance to blinding. The desire to have high permeability must be balanced with the minimum mechanical properties required for filtration. As the porosity is reduced, the density of a filter increases, resulting in higher strength and lower permeability. Predicting filter performance in an application is difficult because there are many process variables such as the nature of the particles (size, shape, compressibility, and composition), particle concentration, fluid flow rate, viscosity, vibration, service temperature, and pressure.

Surface- and depth-type filter mechanisms perform the separation of solids from a process stream with porous P/M media. Surface filtration is characterized by the formation of a layer of particles (cake filtration) on the upstream surface of porous media due to sieving, direct interception, and bridging of the particles. As the filter cake forms, the layer of particles begins to filter the process stream, and the porous media act more as a support. Wire mesh screens are examples of surface-type filters that are generally used to retain coarser particles (>140 μm). Finer filtration retention can be accomplished with screens, but the higher pressures required to drive the fluid often pushes the finer particles into the screen openings to cause blinding. Thin layers of fiber metal bonded to the surface of a screen can provide even finer levels of filtration with improved permeability (Ref 72). Surface filters can be easily cleaned by reversing the flow direction if properly sized. Depth filtration is a more complex form of filtration than surface filtration because it is characterized by the retention of particles within the interconnected pore structure. Particles are retained by a variety of mechanisms including direct interception, inertial impaction, diffusion, adsorption, electrostatic attraction, and gravitational settling (Ref 73). In gas-solid separation, depth filtration depends primarily on capturing particles by interception, diffusion, electrostatic deposition, and impaction mechanisms (Ref 74). These mechanisms are less important in liquid-solid separation because the electrostatic repulsion and the hydrodynamic fluid flow forces allow certain particle sizes to navigate the pore channels and not be captured. Hydrophobic and electrostatic adsorption can also be important in liquid filters, depending on the nature of the electrical charge interactions of the filter media, the contaminants, and the liquid. Sieving mechanisms also occur in depth filters depending on the size and distribution of the contaminants.

Porous P/M materials offer a combination of surface and depth filtration for filtration requirements less than 140 μm. Porous P/M materials are fabricated in relatively thin cross sections as low as 0.12 mm (0.005 in.) for use as surface filters. Typically, thicker wall cross sections are fabricated to obtain a strong, rigid media that can withstand the higher operating pressures associated with depth-type filters. The permeability decreases, and the corresponding drop in pressure across the filter increases as the thickness of

Table 3 Major application areas of porous P/M materials

Filtration/separation technology		
Polymer processing	Chemical processing	Petrochemical
Caustic/sizing recovery	Catalyst recovery	Fluid cracking catalysts
Nylon and polyester fibers	Process gases and liquids	Slurry oils
Gel shearing	Fluid-bed reactor products	Flue gases
Power generation	Mineral processing	Pharmaceutical
Condensate water polishing	Coal, silica, metal oxides	Process steam
Fossil-fuel waste streams	Calciner and incinerator off-gas	Oxygen filters
Radioactive material refining	Catalyst manufacturing	Autoclave vent filters
Semiconductor	Fluid power	Chemicals
Particle removal process gas	Protect hydraulic valves	Acids, solvents, and inks
Bulk gas delivery systems	Pneumatic equipment	Adhesives and greases
Purifier media retainers	Water removal in air lines	Precipitates, salts, and carbon
Analysis instruments	Food and beverage	Other filter areas
Gas/liquid chromatography	Removal of yeast from beer	Oil burners
Gas sampling	Juice/sugar clarification	Aircraft and marine fuel
Sensor protection	Cooking oil, corn syrup	Paper and pulp
Flow-control devices and flow restrictors		
Flow restrictors	Calibrated leaks	Flow-rate timing devices
Vents and breathers	Gas and liquid metering	Fuses, Vacuum delay valves
Surface area devices		
Batteries	Fuel cells	Ionizers
Electrodes	Biomedical implant/prostheses	Cathode dispensers
Reservoirs and capillary attraction		
Self-lubricating bearings	Die-wall lubrication	Heat sinks
Heat exchangers	Liquid wicks, evaporation	Applicators
Transpiration cooling	Printer ink reservoirs	Atomizers
Dampening		
Sound attenuation	Telephone transmitters	Pressure snubbers
Exhaust silencers/mufflers	Flame arrestors/quenching	Vibration reduction
Hearing aids	Pressure equalization	Microphone frequency
Boundary layer control		
Air bearings	Turning bars for film/webs	Polymer fiber cooling
Flotation devices	Fluidizer plates for fluid beds	Friction reduction
Material handling/transport	Vacuum plates	Mold release
Gas-liquid contacting		
Aerators	Spargers	Bubblers
Agitation of liquids	Chlorine and oxygen bleaching	Enhances chemical reactions
Oxygenation, fish farming	Oxygen and volatile stripping	Hydrogenation
Bioremediation	Water/oil separation	Fermentation
Steam injectors/heat transfer	pH control	Carbonators

the media is increased. Filtration ratings can be determined from standardized testing under tightly controlled conditions in order to compare various filter grades and materials. Standardized filter efficiency tests using monosized glass beads, test dusts, or polydisperse salt particles are used to more accurately determine filter ratings (Ref 58, 75–77). Other common industrial tests such as the Beta Rating Test use a known contaminant in a multipass filter challenge method (Ref 78). Filter challenge testing eliminates some of the confusion when filters are described by absolute, nominal, or mean filter ratings based on a correlation of the pore size determined by the bubble-point test. These descriptions can only offer a reference point for the actual filter efficiency at a given particle size because a "10 micron" rated filter may retain anywhere from 50 to 99.9999999% of 10 μm particles, depending on the application and test method. In addition, this filter rating does not indicate the filtration efficiency of the most penetrating particle (Ref 74). Particles that are smaller or larger than 10 μm can be more difficult to retain, depending on the test conditions and filter media. Testing the filter in the application is recommended to determine the actual performance and efficiency by measuring the downstream contaminants.

Cleaning and subsequent reuse of porous P/M materials is a major advantage over disposable media that generally have low strength, limited temperature capability, and corrosion resistance and cannot be cleaned practically. The economic justification for using porous metals often depends on extending the service life with numerous cleaning cycles. Regeneration of porous metal media with blocked porosity can be accomplished with a variety of cleaning techniques and solutions depending on the nature of the foulant. Cleaning methods include thermal removal, ultrasonic immersion, reverse flow methods, high-velocity waterjets, and chemical processing with soaking and rinsing baths. Cleaning solutions can range from mild detergents to harsher caustic solutions or diluted acids, depending on the type of blockage. The cleaning solution and technique must be selected to avoid damage or corrosion of the base metal. Combinations of methods and solutions can be selected for optimal cleaning. Some of the reverse flow, gas, and liquid blowback methods can be performed in situ while other methods such as furnace cleaning, salt baths, and ultrasonic baths require removal of the porous part. Cleaning service organizations determine the best method to clean the part without damage based on testing and experience. After cleaning the part, the cleaning solutions must be thoroughly flushed and dried. Visual inspection, bubble-point testing, and permeability testing are performed to determine the effectiveness of the cleaning method to return the porous part to the original manufacturer specifications. Measuring the weight before and after cleaning can also be used to evaluate cleaning performance as long as corrosion is not an issue. Gathering data on each serialized part prior to service provides a baseline for measuring cleaning effectiveness. Mechanical cleaning methods such as wire brushing, scraping, sand blasting, or glass bead blasting must be avoided to prevent pore smearing and further entrapment of the contaminants. The porous part cannot always be fully cleaned without damaging the base material, but the part can be sufficiently cleaned to be returned to service.

Applications

A summary of the major applications of porous P/M materials are listed below. Filtration, flow control, battery electrodes, and self-lubricating bearings are the largest commercial applications. Many specialized applications have been developed to take advantage of the unique characteristics of porous materials, as shown in Table 3.

Filtration and separation technology applications (Ref 2, 8, 77–80) are the largest market for porous P/M materials. Many industrial applications require the fine filtration capability, superior mechanical properties, and corrosion resistance of porous metals, especially at higher temperature and pressures. Particle retention, media uniformity, absence of particle shedding, and cleanability are critical to the filter system operation. Porous metals can also be used to support finer filter membranes or used as filter septum when they are coated with ion-exchange resins.

Laminar flow control devices utilizing porous P/M flow restrictors (Ref 2, 8, 80–82) are more accurate and reliable than other volumetric-flow-limiting products such as orifices and micrometering valves that operate at higher fluid velocities. Orifice technology normally relies on controlling gas flow in the choked flow regime by requiring that the inlet pressure be at least double the outlet pressure. Porous P/M materials are more resistant to contamination and plugging because there are hundreds of small flow paths available instead of a single orifice. Flow restrictors are used to provide a constant flow for a given set of conditions (pressure, temperature, and fluid) and are mechanically strong devices that do not have mechanical or electrical components that can wear or require calibration.

A few applications of porous P/M materials rely on the extremely high internal surface area and porosity available (Ref 2, 22, 83). For example, the conductivity of the electrodes for alkaline batteries and for fuel cells can be significantly increased by large contact area with the reaction fluid. Biomedical implants use the large surface area to allow tissue growth into the porous structure to allow good joint attachment.

Reservoirs and capillary-attraction devices are other major application areas (Ref 2, 8, 81, 83). Self-lubricating bearings that store lubricants in the pores and protect the surface from wear are one of the oldest commercial applications of porous P/M materials. Heat removal devices for microelectronics packaging and for cooling devices used in satellites are some of the newest application areas.

Sound, pressure, and vibration dampening components are also common industrial applications (Ref 2, 80–82). Porosity acts as an acoustical impedance and dampens certain sound frequencies while others pass through. A pressure surge in a fluid delivery line is also dampened by reducing the fluid velocity as it passes through the small interconnected channels. Many pressure gages have porous metal pressure snubbers to dampen the vibration or pressure change rate and to provide smoother operation of the indicating needle.

Boundary layer control devices such as air flotation bars and vacuum hold-down plates are highly specialized applications (Ref 2, 8, 81, 82, 84). A boundary layer of air for noncontact turning of thin films, tapes, or webs can be formed on the outside diameter of a fine-porosity tube when the inside cavity is pressurized.

In gas-liquid contacting applications (Ref 84, 85), flowing gas through a fine-grade, porous P/M material with high surface area produces small bubbles that result in high-efficiency transfer contact of the gas to the liquid. Finer pore size materials are used to prevent the liquid from penetrating back into the pores at lower internal pressures. Conventional drilled pipe spargers produce larger bubbles that do not dissolve and react as well as the finer bubbles generated by porous metals.

Overall, the application of porous P/M technology is rapidly emerging as a cost-effective, renewable, industrial resource that offers diverse solutions to engineering challenges.

REFERENCES

1. C. Dickenson, *Filters and Filtration Handbook,* 3rd ed., Elsevier Science, 1994, p 66–82
2. V. Tracey, Porous Materials: Current and Future Trends, *Int. J. Powder Metall. Powder Technol.,* Vol 12 (No. 1), 1976, p 25–35
3. V. Tracey and N. Williams, The Production and Properties of Porous Nickel for Alkaline Battery and Fuel Cell Electrodes, *Electrochem. Technol.,* Vol 3 (No. 1–2), 1965, p 17–25
4. M. Eisenmann, A. Fischer, H. Leismann, and R. Sicken, P/M Composite Structures for Porous Applications, *Proc. 1988 Int. P/M Conf.,* Metal Powder Industries Federation, 1988, p 637–652
5. P. Koehler, Porous Metal Article and Method of Making, U.S. Patent No. 4,613,369, 1986
6. C. Dickenson, *Filters and Filtration Handbook,* 3rd ed., Elsevier Science, 1994, p 184–188
7. D. White, Challenges for the 21st Century, *Int. J. Powder Metall.,* Vol 33 (No. 5), 1997, p 45–54
8. Porous Metal Products for OEM Applications, *Mott Technical Handbook,* Mott Corporation, 1996, Sections 1000–9000
9. H. Neubing, Properties and Sintering Behavior of Spherical Tin Bronze Powders for the Manufacture of Filters, *Int. J. Powder Metall.,* Vol 18 (No. 4), 1986, p 4
10. Bronze Filter Powders, ACu Powder International Technical Brochure, Acu Powder International, 1997
11. N. Bagshaw, M. Barnes, and J. Evans, The Properties of Porous Nickel Produced by Pressing and Sintering, *Powder Metall.,* Vol 10 (No. 19), 1967, p 13–31
12. R. German, *Powder Metallurgy Science,* 2nd ed., Metal Powder Industries Federation, 1994, p 96
13. G. Hoffman and D. Kapoor, Properties of Stainless Steel P/M Filters, *Int. J. Powder Metall.,* Vol 12 (No. 4), 1976, p 371–386
14. R. German, *Powder Metallurgy Science,* 2nd ed., Metal Powder Industries Federation, 1994, p 28–81
15. M. Phillips and J. Porter, Comp., *Advances in Powder Metallurgy and Particulate Materials,* Part 1, Metal Powder Industries Federation, 1995
16. C. Helliker and T. O'Sullivan, Process for Fabricating Porous Nickel Bodies, U.S. Patent No. 4,255,346, 1980
17. Porous Metal Council, *Porous Metal Design Guidebook,* Metal Powder Industries Federation, 1980
18. R. German, *Particle Packing Characteristics,* Metal Powder Industries Federation, 1989, p 219–252
19. L. Albano-Muller, Filter Elements of Highly Porous Sintered Metals, *Powder Metall. Int.,* Vol 14, 1982, p 73–79
20. F. Lenel, *Powder Metallurgy—Principles and Applications,* Metal Powder Industries Federation, 1980, p 143–154
21. V. Tracey, The Roll-Compaction of Metal Powders, *Powder Metall.,* Vol 12 (No. 24), 1969, p 598–612
22. N. Williams and V. Tracey, Porous Nickel for Alkaline Battery and Fuel Cell Electrodes: Production by Roll-Compaction, *Int. J. Powder Metall.,* Vol 4 (No. 2), 1968, p 47–62
23. P. Koehler, Seamless Porous Metal Article and Method of Making, U.S. Patent No. 4,822,692, 1989
24. P. Koehler, Seamless Porous Metal Article and Method of Making, U.S. Patent No. 4,828,930, 1989
25. P. Neumann, V. Arnhold, K. Heiburg, and R. Rohlig, Porous Metal Products with Special Properties, *Powder Metal Materials Colloquium,* Soc. Franc. de Metal. et de Mater., April 1992, p 4-1 to 4-6
26. L. Mott, Process for Making Porous Metallic Bodies, U.S. Patent No. 2,792,302, 1957
27. P. Koehler, S. Geibel, and M. Whitlock, Liquid Laydown Process and Metal Filter, U.S. Patent No. 5,149,360, 1992
28. Porous Metal Council, From Powder to Porous Metal Parts, *Design Engineering,* Metal Powder Industries Federation, 1981, p 56–89
29. V. Tracey, Sintering of Porous Nickel—Theoretical and Practical Considerations, *Modern Developments in Powder Metallurgy,* Vol 12, Metal Powder Industries Federation, 1980, p 423–438
30. V. Morgan, Sintered Metal Filters, *Hydraulic Pneumatic Power,* Trade and Technical Press, 1974, p 323–324
31. H. Nayar, Production Sintering Atmospheres, *Powder Metallurgy,* Vol 7, *Metals Handbook,* American Society for Metals, 1984, p 340–350
32. D. Garg, K. Berger, D. Bowe, and J. Marsden, Effect of Atmosphere Composition on Sintering of Bronze, *Gas Interactions in Non Ferrous Metal Processing,* Minerals, Metals and Materials Society, 1996, p 17–26
33. H. Nayar, R. German, and W. Johnson, Effect of Sintering on the Corrosion Resistance of 316L Stainless Steel, *Proc. 1981 P/M Conf.,* Metal Powder Industries Federation, 1981, p 653–667
34. V. Tracey, Sintering of Porous Nickel, *Powder Metall.,* Vol 26 (No. 2), 1983, p 89–92
35. Graver Separations Brochure Number 5-106, Graver Chemical Company, 1996
36. A. Mulder et al., SiO_2-Membranes on Porous Sintered Metal Substrates, Starting from Silicone Solutions, *Key Eng. Mater.,* Vol 61–62, 1991, p 411–414
37. D. Smith, J. Smugeresky, and B. Meyers, "The Dependence of Permeability and Filtration on Pore Morphology in Consolidated Particulate Media," Report SAND87-8227, Sandia National Laboratories, 1987, p 1–74
38. D. Smith, E. Brown, J. Smugeresky, and T. McCabe, Characterization of Controlled Density P/M Structures for Filtration Applications, *Proc. 1985 P/M Conf.,* Metal Powder Industries Federation, 1985, p 653–667
39. "Permeable Sintered Metal Materials—Determination of Density, Oil Content, and Open Porosity," Standard 2738, International Standards Organization, 1987
40. R. Iacocca and R. German, A Comparison of Powder Particle Size Measuring Instruments, *Int. J. Powder Metall.,* Vol 33 (No. 8), 1997, p 35–48
41. "Standard Test Methods for Maximum Pore Diameter and Permeability of Rigid Porous Filters for Laboratory Use," E 128, ASTM, 1994
42. "Permeable Sintered Metal Materials—Determination of Bubble Test Pore Size," Standard 4003, International Standards Organization, 1977

43. V. Morgan, Filter Elements by Powder Metallurgy, *Symposium on Powder Metallurgy,* The Iron and Steel Institute, 1956, p 81–89
44. R. Lines, A Pore Man's Guide, *Filtration News,* Eagle Publications, Vol 10 (No. 2), 1992, p 40–44
45. R. Wenman and B. Miller, *Particle Size Analysis,* John Wiley & Sons, 1985, p 583–589
46. "Standard Test Methods for Pore Size Characteristics of Membrane Filters by Bubble Point and Mean Flow Pore Test," F 316, ASTM, 1986
47. R. German, *Particle Packing Characteristics,* Metal Powder Industries Federation, 1989, p 353–390
48. R. German, Gas Flow Physics in Porous Metals, *Int. J. Powder Metall.,* Vol 15 (No. 1), 1979, p 23–30
49. "Flow of Fluids through Valves, Fittings and Pipe," Technical paper 410, Crane Company, 1979
50. P. Johnston, "Pore Sizes in Filter Media—A Review of Meanings and Size Distributions," American Filtration Society Meeting on the Pore, 1991, p 1–12
51. "Permeable Sintered Metal Materials—Determination of Fluid Permeability," Standard 4022, International Standards Organization, 1987
52. R. German, *Powder Metallurgy Science,* 2nd ed., Metal Powder Industries Federation, 1994, p 380–386
53. F. Lenel, Chapter 15, *Powder Metallurgy—Principles and Applications,* Metal Powder Industries Federation, 1980
54. "High Porosity Sintered Materials," Krebsoge Technical Brochure, Pressmetall Krebsoge, 1989, p 1–25
55. "Sintered Metal Materials, Part 2: Materials for Filters," Standard V 30910, Normenausschuss Pulvermetallurgie, Deutsches Institue fur Normung e.V., 1986
56. A. Palfalvi et al., Influence of Porosity on the Mechanical Characteristics of Sintered Porous Stainless Sheets, *Powder Metall. Int.,* Vol 20 (No. 4), 1988, p 16–19
57. U. Lindstedt, B. Karlsson, and R. Masini, Influence of Porosity on Deformation and Fatigue Behavior of P/M Austenitic Stainless Steel, *Int. J. Powder Metall.,* Vol 33 (No. 8), 1997, p 49–61
58. V. Tracey, Effect of Sintering Conditions on Structure and Strength of Porous Nickel, *Powder Metall.,* No. 2, 1979, p 45–48
59. R. German, *Powder Metallurgy Science,* 2nd ed., Metal Powder Industries Federation, 1994, p 391
60. V. Antsiferov et al., Electric Resistance and Thermal Conductivity of Highly Porous Permeable Cellular Materials, *Poroshk. Metall.,* 1989, p 668–673
61. "Determination of Properties of Sintered Bronze P/M Filter Powders," Standard 39, Metal Powder Industries Federation, 1983
62. "Tension Test Specimens for Pressed and Sintered Metal Powders," Standard 10, Metal Powder Industries Federation, 1963
63. J. Davis, Ed., *ASM Specialty Handbook: Stainless Steels,* ASM International, 1994, p 211–212
64. D. Ro and E. Klar, Corrosion Behavior of P/M Austenitic Stainless Steels, *Modern Developments in Powder Metallurgy,* Metal Powder Industries Federation, 1981, p 247–267
65. G. Lei, R. German, and H. Nayar, Influence of Sintering Variables on the Corrosion Resistance of 316L Stainless Steel, *Powder Metall. Int.,* Vol 15, 1983, p 70–76
66. R. Lula, Stainless Steel, *An Introduction to Stainless Steel,* American Society for Metals, 1986, p 129–151
67. S. Malhotra and P. Ramakrishnan, Corrosion Behavior of P/M Stainless Steel Filters, *Met. Powder Rep.,* 1991, p 48–51
68. K. Pruett, *Chemical Resistance Guide for Metals and Alloys,* Compass Publications, 1995
69. H. Sanderow, Corrosion Resistance and Test Methods, *Stainless Steel Short Course,* Metal Powder Industries Federation, 1996
70. T. Gooch, Corrosion Behavior of Welded Stainless Steel, *Welding Research Supplement to the Welding Journal,* American Welding Society, 1996, p 135s–154s
71. "Pall Porous Metals Filters Guide," Product Bulletin PSS 700a, Pall Corporation, 1978
72. L. Ortino, Porous Metal Materials in Industrial Filtration, *Advanced Filtration and Separation Technology,* Issue on Fine Particle Filtration and Separation, Vol 4, 1991, p 30–39
73. J.-K. Lee, K. Rubow, and B.Y.H. Liu, Particle Retention by Microporous Membrane Filters in Liquid Filtration: A Predictive Model and Experimental Study, *Proc. Institute of Environmental Sciences,* Institute of Environmental Sciences, 1992, p 297–308
74. K. Rubow and C. Davis, Particle Penetration Characteristics of Porous Metal Filter Media for High Purity Gas Filtration, *Proc. 37th Annual Technical Meeting of the Institute of Environmental Sciences,* Institute of Environmental Sciences, 1991, p 834–840
75. K. Rubow, D. Prause, and M. Eisenmann, A Low Pressure Drop Sintered Metal Filter for Ultra-High Purity Gas Systems, *Proc. 43rd Annual Technical Meeting of the Institute of Environmental Sciences,* Institute of Environmental Sciences, 1997
76. H. Sandstedt and J. Weisenburger, Cartridge Filter Performance and Micron Rating, *Filtration Separation,* Vol 22 (No. 2), 1985, p 101–105
77. C. Moreland and B. Williams, "Selecting Polymer Filtration Media," presented at the Fiber Producer Conf., PTI Technologies, 1980
78. J. Eleftherakis and R. Webb, Industrial Filtration Test Methods, *Filtrat. News,* Sept–Oct 1995, p 46–49
79. "The Pall Gas Solid Separation System for the Chemical Process, Refining and Mineral Industries," Product Bulletin GSS-1, Pall Corporation, 1987, p 3–5
80. M. Busche, Porous Metals Filters Liquids, Cut Noise, Dampen Vibration, *Materials in Design Engineering,* Vol 65 (No. 2), 1966, p 80–83
81. J. Snyder, P/M Porous Parts, *Powder Metallurgy,* Vol 7, *ASM Handbook,* American Society for Metals, 1984, p 696–700
82. W. Mossner, "Applications and Properties of Controlled Porosity P/M Parts," SSI Sintered Specialties, 1986
83. F. Lenel, *Powder Metallurgy—Principles and Applications,* Metal Powder Industries Federation, 1980, p 359–380
84. W. Johnson and M. Shenuski, Controlling Fluid Flow with Porous Metals, *Mach. Design.,* Jan 1987, p 3
85. Porous Metal Products for OEM Applications—A Guide to Advanced Gas Sparging and Gas-Liquid Contacting, *Mott Technical Handbook,* Mott Corporation, 1996, Section 11000

Metallic Foams

J. Banhart, J. Baumeister, and M. Weber, Fraunhofer-Institute for Applied Materials Research, Bremen

CELLULAR MATERIALS find widespread use in everyday life for cushioning, insulating, damping, constructing, and filtering purposes and in many other applications. Highly porous materials are known to have a high stiffness combined with a very low specific weight. For this reason, cellular materials frequently occur in nature as constructional materials.

Even pure metals and metallic alloys can be produced as cellular solids or metal foams. In the past, metal foams were prepared by adding a foaming agent to a molten metal after properly adjusting the viscosity of the melt (Ref 1, 2). The foaming agent is usually a powdered metal hydride, for example, TiH_2, which releases hydrogen gas when heated to temperatures above approximately 400 °C. As soon as the foaming agent comes into contact with the molten metal, it decomposes such that there is little time to achieve a homogeneous distribution of the gas-releasing powder. Because this process is difficult to control, more widespread application could not be achieved with this technology.

A new P/M process for production of metal foams has been developed at the Fraunhofer-Institute for Applied Materials Research (Bremen, Germany) (Ref 3, 4). This enlarges the application range of cellular materials with the advantageous mechanical and thermal properties of metal foams and fewer ecological problems in comparison to polymer foams. The process has been developed for aluminum foams and is currently being extended to other metals and alloys such as tin- and zinc-base foams (Fig. 1). These nonferrous foams are the subject of this article.

For the production of steel foams, the type of foaming agent must be changed to prevent excessive oxidation. The foaming process for steel also has to take place in an inert atmosphere or in a vacuum. Using steel foams, the applicable temperature range could be extended. As an example, the exhaust manifold of car engines could be manufactured from this material. Due to the strongly reduced thermal conductivity of the manifold, it would require less time to reach the normal operating temperature of the exhaust catalyst, leading to a reduction in emissions.

Production Methods

The P/M process for production of nonferrous foams uses conventional foaming agents with compacted metal powders. Commercially available powders (such as aluminum, zinc, tin, or lead) are mixed with a foaming agent by conventional means, such as a tumbler mixer. In this simple manner, a very homogeneous distribution of the gas-releasing powder is obtained without the necessity of agitating a pool of molten metal. Subsequent to mixing, the powder blend is compacted to give a dense, virtually nonporous, solid semifinished product. Several compaction methods can be employed that range from uniaxial pressing to powder extrusion and even to roll compaction. The result of the densification step is a foamable material that, upon heating to temperatures within the range of the melting point, expands into a highly porous cellular solid with a closed-pore structure. This means that each particle of the foaming agent is embedded in a gas-tight metallic matrix so that, when decomposition of the foaming agent begins, the released gas cannot escape via some interconnected residual porosity.

Numerous shapes can be produced by this method (Ref 3, 5). The semifinished form can be processed by conventional techniques such as rolling, swaging, or extrusion to provide rods, sheets, profiles, and so forth, if desired. Merely heating this material to the melting point initiates the foaming process so that it is also possible to obtain complex-shaped foamed articles. Arbitrarily shaped hollow molds are filled with the foamable material and subsequently heated to effect foaming. It was found that thin-walled molds are particularly advantageous because they allow for a close temperature control of the foaming process.

Using appropriate heating methods, it is possible to selectively expand metallic foams. This means that certain regions of the material can be kept at a high density that facilitates joining to other materials.

Sandwich-type structures also can be produced via several methods. The simplest one is to glue sheets of conventional materials to a sheet of foamed metal. However, due to the low thermal stability of the resins, a metallic bonding

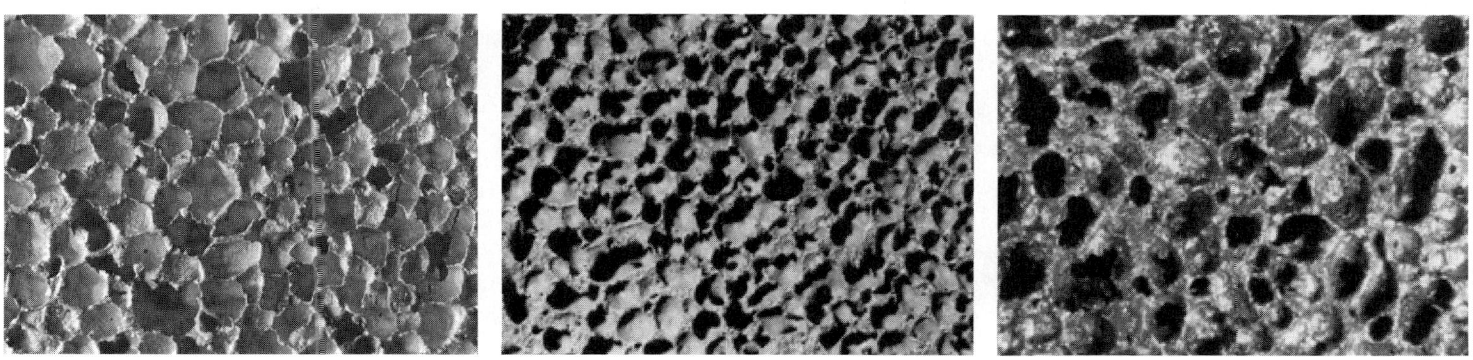

(a) (b) (c)

Fig. 1 Optical micrographs of (a) aluminum (4 to 1), zinc (6 to 1), and lead (4 to 1) foams

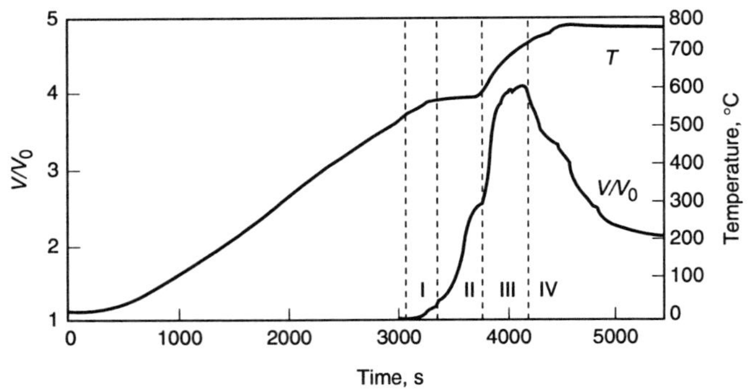

Fig. 2 Expansion characteristics of Al-12Si (heating rate: 10 K/min).

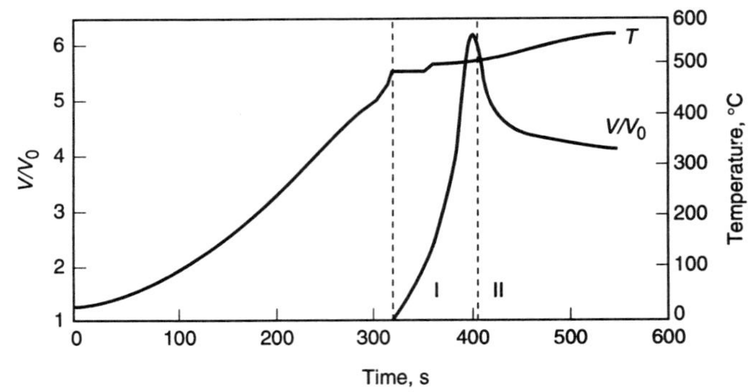

Fig. 3 Expansion characteristics of Zn-4Cu (heating rate: 100 K/min)

of the sheets might be preferred. For example, this can be achieved by using a roll-cladding process.

Foaming agents are chosen to have a decomposition temperature that matches the melting point of the metal or alloy. Various foaming agents have been shown to yield good results for the metals in consideration (aluminum, zinc, lead). Among the foaming agents used are metal hydrides such as titanium, zirconium, and magnesium hydride, which have decomposition temperatures between 280 and 600 °C. Figure 1 shows typical cross sections of aluminum, zinc, and lead foams obtained.

Foaming Kinetics. Metal foam parts are usually produced by heating up a piece of foamable precursor material in a hollow mold and allowing the expanding melt to fill the volume. Foaming times and temperatures can be varied, and their influence on the foam properties studied. This procedure, however, does not yield any direct information about the foaming kinetics, that is, the time dependence of the volume of the expanding metallic melt and the corresponding temperatures. Therefore, an experiment was designed (Ref 6) that allows for measurement of such data. The device consists of a metal tube with a lid on one side and a movable piston on the other side. The position of the piston is measured with a magnetic detector and read out into a computer. Moreover, a thermocouple measures the temperature inside the tube. For an expansion experiment, a piece of foamable material is placed into the tube, which is then closed by the piston. The entire device is heated up at a selected heating rate. The expanding melt causes a displacement of the piston that, together with the temperature reading, generates a pair of functions $V(t)$ (volume) and $T(t)$ (temperature) that characterizes the expansion kinetics.

The expansion behavior of metal foams has been investigated for a variety of metals and alloys. Three examples are shown here: the foaming of an Al-12Si alloy, the expansion of Zn-4Cu, and the expansion of pure lead.

The expansion of Al-12Si is depicted in Fig. 2. As can be seen, the volume expansion takes place in four stages: first, before the melting temperature of the alloy (577 °C) is reached, the expansion is small, resembling the volume increase also found for samples without foaming agent. The actual expansion starts when the metal starts to melt and softens and the released gas can inflate the pores. In this stage II the volume increases to 2.5 times the initial volume. The expansion even accelerates as the liquidus temperature is exceeded, and the now rising temperature releases more and more gas (stage III). After the maximum expansion has been achieved, the foam collapses (stage IV) due to the now-decreasing pressure in the cells as the foaming agent is exhausting.

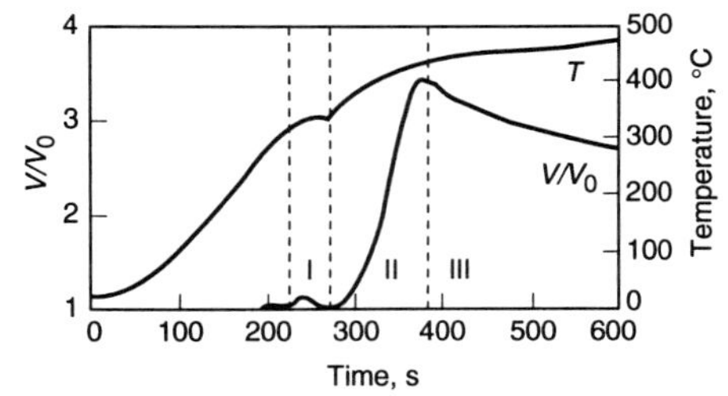

Fig. 4 Expansion characteristics of lead (heating rate: 100 K/min)

The expansion of the zinc-copper foam is shown in Fig. 3. Here, the expansion takes place in just two stages: expansion (I) and collapse (II). There is hardly any sign of an expansion below the melting temperature, and the main expansion is not separated into two stages. The reason for this can be seen in the much higher heating rate that was applied in this case. One sees that the expansion is time controlled and that foaming is ruled by a complex interplay between melting and decomposition processes. Note that the temperature shows a kink at the melting temperature in Fig. 3. This can be attributed to an alloying process between zinc and copper, which were contained in the powder mixture as elementary powders. For lead foams, the expansion takes place in three stages (Fig. 4): first there is a slight expansion after the melting point has been reached (I), then the expansion accelerates as the temperature increases after all the metal has become molten (II) and, finally, the foam collapses (III). The reason for the slight decrease of volume between stages I and II is not yet clear and is being investigated.

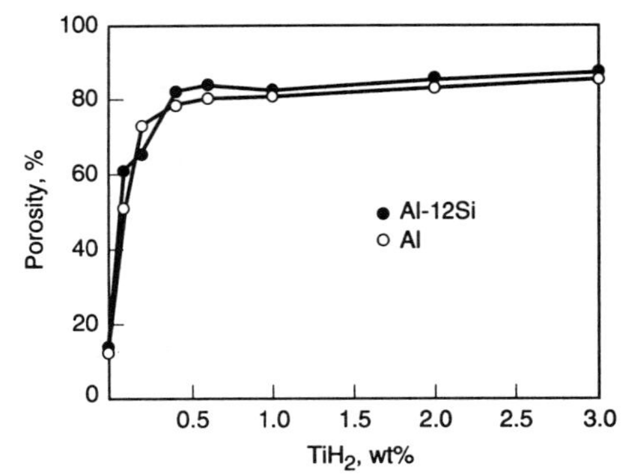

Fig. 5 Maximum porosity achieved when foaming aluminum alloys with various contents of foaming agent. Solid circles, Al-12Si; open circles, aluminum

Influence of Foaming Agent Content. It is important to know how much of the foaming agent is necessary to obtain the desired density of the foamed metal and how low the achievable densities are. In order to get this information, foamable precursor material was prepared with various contents of foaming agent for two different alloys. The content of the foaming agent was varied between 0 and 3%, the alloys chosen were pure aluminum (99.5% Al) and eutectic (Al-12Si). Each piece of precursor material was expanded to its maximum volume at the same temperature, and each measurement was repeated three times. The averaged results are shown in Fig. 5. The samples without any foaming agent exhibit the lowest expansion rate, which is about 15%. This expansion comes from the P/M preparation of the precursor material, which produces a high content of oxides, hydroxides, and gas in the powder compact that are then released upon heating. Already a small content of 0.1% TiH_2 causes an expansion of more than 50% in volume, producing a structure that justifies the name "foam." The maximum expansion is obtained for foaming agent contents of more than 0.7% and is independent of the foaming agent content up to 3%. Therefore, in practical applications foaming agent contents up to 1% are chosen. The maximum porosities that can be achieved are 80% for aluminum and 85% for Al-12Si. Higher porosities can be produced

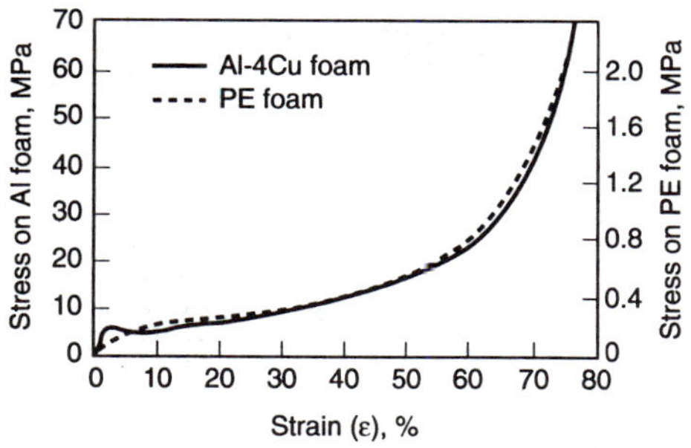

Fig. 6 Stress-strain curve of an Al-4Cu foam (solid line, left scale) and of a polyethylene (PE) foam (broken line, right scale)

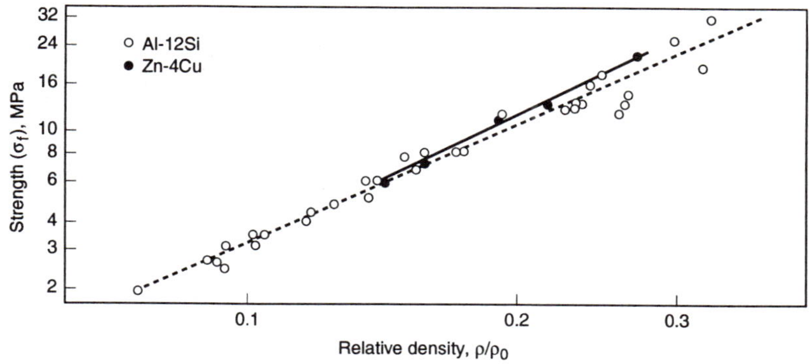

Fig. 7 Compression strength of a series of Al-12Si (open circles) and Zn-4Cu (solid circles) alloys. Test specimens were 30 by 30 by 40 mm. Testing was performed at 5 mm/min. As the transition from the initial linear increase of stress to the plateau regime is not defined unambiguously, the stress at 10% (Zn) or 20% (Al) deformation was taken as the compression strength (σ_f).

by changing the alloy composition or by adding reinforcement particles, but not by increasing the content of the foaming agent.

Properties of Metal Foams

Like any other cellular solid, metal foams are characterized by a very low specific weight. Using the P/M technological approach described above, density values ranging between 0.5 and 1 g/cm^3 are usually obtained with aluminum foams, although values down to 0.2 g/cm^3 and up to 2 g/cm^3 can even be achieved. Due to its closed porosity, aluminum foam floats upon water.

Mechanical properties of foams are usually determined by compression testing. To show the superior strength of metal foams, a polyethylene foam (density, 0.12 g/cm^3) is compared to an Al foam (Al-4Cu alloy: density 0.45 g/cm^3) in Fig. 6. In both materials, the initial porosity is about 83 to 87%. For better comparison, the strength scale of the PE foam was enlarged by a factor of 30. The curves show a behavior that is typical for highly porous cellular solids: an initial, approximately linear regime is followed by an extended plastic collapse plateau, truncated by a densification response at high strains during which the stress again increases steeply. Due to this special form of the compressive stress-strain curve, foamed materials have a high capacity to absorb great amounts of energy at a relatively low strength level.

To characterize the energy absorption behavior of foamed materials, it is convenient to determine that efficiency η. For a given strain, this parameter is defined as the ratio of the actual absorbed energy to the energy that would be absorbed by an ideal absorber. Because an ideal cushioning material would exhibit a "rectangular" stress-strain curve, a real absorber can only approximate this behavior, giving rise to values below 1, or 100%, for the efficiency.

Compression Strength. Foamed materials show a universal compression behavior that makes them valuable for energy absorption applications: when loaded they first deform elastically until at a certain stress plastic deformation of the cell walls starts. This deformation typically takes place at nearly constant stress until the entire structure is densified. The stress that marks the beginning of the foam collapse is called the compression strength and can be extracted from the stress-strain diagram. Stress-strain curves were measured on specimens of metal foams with various densities (Ref 7).

The compression strength of a series of Al-12Si and Zn-4Cu alloys of various densities are shown as a function of foam density in Fig. 7. Simple theory based on cubic cells predicts a dependence of compression strength from density for an open-pore foam of the form (Ref 8):

$$\sigma_f = 0.3 \cdot \sigma_{ys}\left(\frac{\rho_f}{\rho_s}\right)^n$$

where ρ_f and ρ_s are the densities of the foamed and the massive material, respectively, and σ_{ys} is the yield strength of the matrix material. For closed pores, the formula is more complicated, but experience shows that the mechanical properties are determined by the thick cell edges and not by the thin membranes so that the equation for the open porosity works well. In order to allow for a direct reading of the exponent *n*, the data were plotted logarithmically in Fig. 7; the value of *n* can then be obtained by fitting the experimental data to a straight line, thus yielding:

Alloy	Exponent, *n*	Yield strength (σ_{ys}), MPa
Al-12Si	1.78	660
Zn-4Cu	2.00	990

The value for the exponent is in fairly good agreement with the value derived from the cubic model (1.5), whereas the value for σ_{ys} is much too high in both cases. One should keep in mind that the determination of the parameters *n* and σ_{ys} is not very accurate because of the considerable scatter especially for Al-12Si. As the pore structure of the Zn-4Cu foams is rather homogeneous, the relation between the foam density and its compression strength can be established in a more reliable way for these alloys. The results indicate that the cubic model gives a good description of the relative variation of compression strengths as the density is varied, but is not able to predict the absolute value of such strengths. This finding is compatible with earlier measurements on aluminum-base foams, where exponents between 1.5 and 3 were found, but the parameter σ_{ys} was always too high.

Elastic Modulus. The determination of the elastic modulus of foamed metals should not be carried out the convenient way, that is, by evaluation of the slope in the elastic portion of the stress-strain curve. Even in this early stage of loading there are some plastic contributions that result from setting of the specimen, heterogeneities of the pore structure, and other defects. It is therefore preferable to apply elastic loading of the samples, for example, by means of vibrational testing (as described below) or ultrasonic methods.

The elastic modulus of foamed Al-12Si was investigated using vibrational bending tests. The resonance frequency (first order) of rectangular specimen 250 by 10 by 5 mm^3 in size was determined, and the resulting storage modulus calculated. Because the values are independent of frequency, it can be concluded that the results represent the static modulus.

Considering the simplified nature of the cubic model, it is surprising that the predictions from this model describe the elastic modulus to a quite reasonable degree. The equation for the elastic modulus states that (Ref 8):

$$E_f + E_s\left[\left(\varphi\frac{\rho_f}{\rho_s}\right) + (1-\varphi)\left(\frac{\rho_f}{\rho_s}\right)\right]$$

where E_f is the modulus of foamed material, E_s is the modulus of the massive material, and ϕ describes the contribution of material in the cell edges, and where (1 – ϕ) is the contribution from cell faces. A straight line with a slope of 2 would be expected in a double-log plot of the foam modulus versus density. From Fig. 8, it can be seen that this type of correlation has actually been observed (solid line in Fig. 8). An even better approximation of measured values is obtained by taking ϕ = 0.88 (dashed line). This

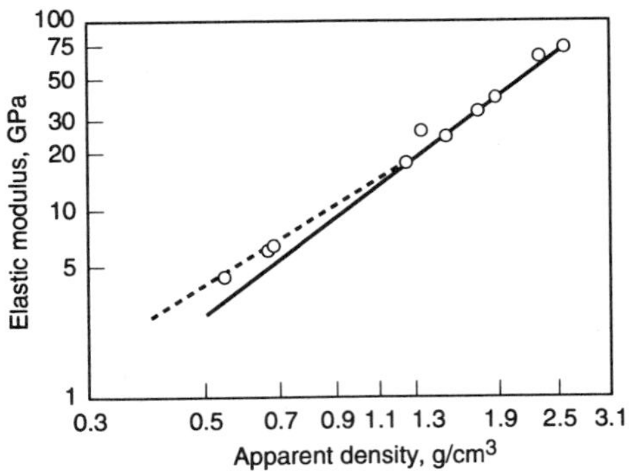

Fig. 8 Elastic modulus of Al-12Si foams as a function of the apparent density

means that most of the material in the aluminum foams investigated is concentrated in the cell edges rather than in the cell faces.

Mechanical Damping. Measuring loss factors of metallic materials is a nontrivial task due to the relatively low level of damping and variables affecting the loss factor. Such variables are the friction between sample and bearings and air friction. Damping also depends on strain amplitude. Reproducible results without these contributions can be obtained by using flat strips of metal foam, which have an attached head. The whole structure is made of one block of foam. The head is used to clamp the sample with a considerable force and to reduce the slip of the sample relative to the mount to a minimum.

The sample was forced into vibrations by magnetic excitation. The resonance amplitude was monitored with a capacitive detector. By sweeping the applied frequency, resonance curves could be recorded. The resonance frequency v and the width of the resonance Δv were determined, thus yielding a loss factor:

$$\eta = \Delta\frac{v}{v}$$

Various samples were made of Al-12Si foam by foaming blocks of this material and cutting out the samples. The density covered a wide range between 0.67 and 2.65 g/cm^3 (unfoamed precursor material). In this density range, η varied from 20×10^{-4} (at 0.67 g/cm^3) to 12×10^{-4} (at 2.65 g/cm^3).

In order to study the frequency dependence of the damping, the samples were shortened step-by-step, thus making the resonance frequency higher. The loss factor was measured for each length in the second mode of flexural vibration at the frequency determined by the cantilever length. If was found that the loss factor is not a smooth function of frequency for any of the samples, but fluctuates strongly as the frequency changes. The probable explanation is that the loss factor is nearly frequency independent for aluminum foams in reality and that frequency variations are not due to microstructural, frequency-dependent effects but are rather caused by some inhomogeneous material distribution

Fig. 9 Sandwich part containing a porous aluminum foam core and two aluminum face sheets (length, 1 m; total width, approximately 60 cm; thickness, 8 to 10 mm). Courtesy of Karmann, Germany

Fig. 10 Sandwich part with steel face sheets and a porous aluminum foam core. Courtesy of Studiengesellschaft Stahlanwendungen, Germany

that is characteristic for the particular sample. Therefore, the damping capacity of a highly porous metallic foam is not a simple materials property, but also characteristic of the part itself.

Damping is strongest in the foams with the lowest densities. It is well known that any kind of porosity enhances damping due to the stress concentration and what is called mode conversion around pores. This mechanism applies to microporosity as well as to the large pores found in metal foams. Especially in thin membranes an external force, even if small, can lead to high, multiaxial stresses that induce atomic rearrangements and viscouslike sliding of atoms relative to each other.

Another mechanism is of macroscopic origin: the cell walls of metallic foams sometimes contain cracks and holes that may either originate from the gas released by the foaming agent during the foaming process or from the contraction of the foamed body while cooling to room temperature after foaming. These defects may give rise to an extra frictional contribution caused by the relative movement of the crack or hole walls with respect to each other.

Compact precursor materials that are not foamed have a much higher loss factor than conventional cast aluminum materials. This is due to the P/M production of this material, which yields an aluminum matrix with embedded silicon particles, remnants of the oxide films that covered each aluminum powder particle and also foaming agent particles. This causes extra friction at the interfaces of these particles. It is also well established that P/M materials have higher damping capacities than cast ones.

Applications

One application of metal foams is for energy absorption. Aluminum foams can also be used for lightweight constructions. The replacement of honeycomb structures with sheets of foamed aluminum may lead to reduced costs and more isotropic properties. For example, the German automotive supplier Karmann is considering replacement of the bulkhead and the rear seat wall of convertible cars by sandwiches containing an aluminum foam core and aluminum face sheets (Fig. 9). The parts are up to ten times stiffer in comparison to the conventional steel parts used in the past and weigh 30 to 50% less.

Steel is also used for sandwich face sheets. In Fig. 10, steel sheets were roll cladded to a sheet of foamable aluminum and then deep drawn to yeild a three-dimensional structure. Subsequent foaming created the porous core.

Because the elastic modulus can be varied within a wide range via the choice of the foam density, it is possible to match the resonance frequency of foamed parts. In this way, detrimental vibrations can be suppressed. Complex-shaped parts of foamed metals can also be used to encapsulate components that are hot or produce noise. In this case, the high thermal stability of foamed metals can be exploited.

Closed-cell foams are especially well suited to use as floating structures because of their high damage tolerance. These structures retain their buoyancy even when locally damaged. In particular, metal foams are able to withstand higher pressures or higher temperature than plastic foams.

REFERENCES

1. J.A. Ridgeway, "Cellarized Metal and Method of Producing the Same," U.S. Patent 3,297,431, 1967
2. S. Akiyama et al., "Foamed Metal and Method for Producing the Same," European Patent Application EP 0 210 803 Al, 1986
3. J. Baumeister, "Method for Producing Porous Metal Bodies," German Patent DE 40 18 360, 1990
4. J. Baumeister and H. Schrader, "Methods for Manufacturing Formable Metal Bodies," German Patent DE 41 01 630, 1991
5. J. Baumeister, J. Banhart, and M. Weber, *Int. Conf. on Materials by Powder Technology,* F. Aldinger, Ed., DGM Informationsgesellschaft Oberursel (Germany), 1993, p 501
6. M. Weber, Ph.D. thesis, Technical University Clausthal, 1995
7. M. Weber, J. Baumeister, J. Banhart, and H.D. Kunze, *Proc. Powder Metallurgy World Congress PM94* (Paris), Les Editions de Physique, 1994, p 585
8. L.J. Gibson and M. Ashby, *Cellular Solids,* Oxford, 1988

SELECTED REFERENCE

- Baumeister et al., U.S. Patent 5,151,246, 29 Sept 1992

Friction Powder Metallurgy Materials

FRICTION MATERIALS are the components of a mechanism that converts mechanical energy into heat upon sliding contact. The conversion product, heat, is absorbed or dissipated by the friction material. The coefficient of friction, an index of shearing force of the contacting parts, determines the degree of performance of the friction material. The required level of the coefficient of friction depends on the operating conditions and the end use of the product.

Sintered metal friction materials have been used as brake disks, especially for heavy-duty application. Because of their good breaking performance and low wear rate under high temperatures and heavy-duty conditions, sintered friction materials have become more important. Sintered friction materials typically comprise:

- Sinterable metal powders (e.g., copper and iron)
- Friction modifiers, such as abrasive particles (e.g., alumina, silica, and mullite) and lubricants (e.g., graphite and molybdenum disulfide)

Metallic friction materials are used in heavy-duty applications, such as in aircraft brake linings and as clutch facings on tractors, heavy trucks, earth-moving equipment, and heavy presses. There are two principal types of applications or operating conditions for metallic friction materials: "wet" and "dry." Under wet conditions, the friction components, such as clutches in powershift and automatic transmissions, are immersed in oil. Dry operating conditions involve direct contact of friction components without oil, such as in aircraft brakes and standard clutches.

A low apparent density (AD) is a key characteristic of metal powders for friction applications. For example, powders with high surface area and internal porosity enhance brake formulations by:

- *Lowering costs:* Lowest apparent density reduces brake weight and lowers total material usage.
- *Improving brake surface:* High internal porosity ensures a more homogeneous brake surface, and high surface area improves particle bonding with the matrix.
- *Increasing molded strength:* Higher green strength is achieved when compacting powders with lower apparent densities (Fig. 1).

Highly porous and irregular iron powders with very low apparent densities can be produced by the hydrogen reduction (Pyron) process. A comparison of iron powder apparent density follows:

- Hydrogen-reduced iron powder, 1.20 to 2.5 g/cm^3
- CO-reduced powder, 2.30 to 2.60 g/cm^3
- Atomized powder, 2.95 to 3.25 g/cm^3

Manufacturing

Metallic friction materials are produced by compacting and sintering mixes of metal powders and friction-producing ceramic materials such as silicon dioxide or aluminum oxide. Friction materials consist of a dispersion of a friction-producing ingredient in a metallic ma rix. Originally, a copper-tin alloy was used as the metallic matrix, in which copper powder, tin powder, the friction-producing ingredient in powder form, and other ingredients that modify the frictional behavior were mixed, compacted, and sintered. This matrix material is still used, but other compositions such as copper-zinc matrix materials have been developed, mainly for oil-immersed (wet) applications. The copper-zinc materials have the ability to maintain a strong, yet porous, matrix that retains oil. The more porous copper-zinc materials have higher friction coefficients and higher energy absorption capacity than copper-tin materials.

Raw Material Blending. Sintered metal friction materials involve a wide variety of compositions. The choice of the composition depends on the nature of the application, such as the use and energy-power requirements. In general, metallic friction materials can be classified as either copper or iron base. Table 1 provides the range of compositions of metallic friction materials. Proportions of the components greatly affect physical properties of the materials. By varying the percentages of the individual components, different coefficients of friction may be achieved. Before mixing, powders generally are brush screened to break up agglomerated particles. Mixing usually is done in cone-type blenders. During mixing, small amounts of additives are introduced, mainly to prevent possible segregation of components by specific gravity. These additions are light-fraction oils that are volatilized easily during sintering.

Compacting. Most compacting of powders is done in hydraulic presses. Compacting pressure varies from 165 to 276 MPa (12 to 20 tsi) and is determined by the type of powder and the compacted density required. Before starting a production run, a series of density checks should be made. Low densities often cause handling problems, especially with thin cross sections. This can be overcome by increasing pressure. Surface parallelism is a major consideration and, in gen-

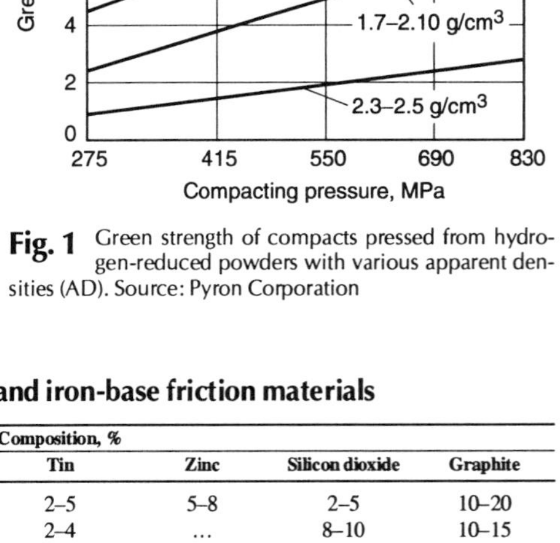

Fig. 1 Green strength of compacts pressed from hydrogen-reduced powders with various apparent densities (AD). Source: Pyron Corporation

Table 1 Nominal compositions of copper-base and iron-base friction materials

Premix	Composition, % Copper	Iron	Lead	Tin	Zinc	Silicon dioxide	Graphite
Copper base	65–75	...	2–5	2–5	5–8	2–5	10–20
Iron base	10–15	50–60	2–4	2–4	...	8–10	10–15

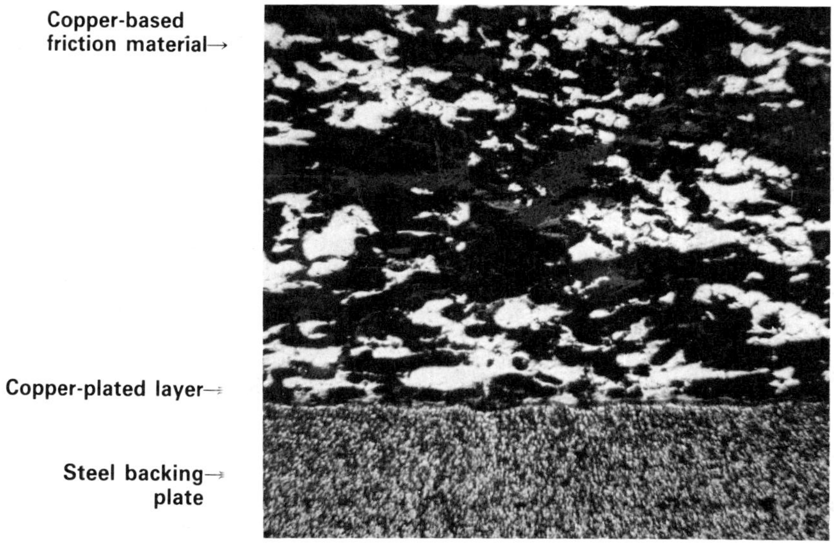

Fig. 2 Structure of a sintered copper-base friction material

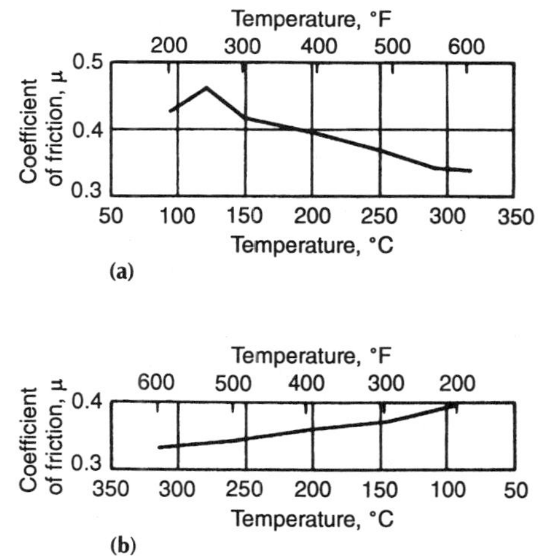

Temperature °C	Temperature °F	Coefficient of friction
Fig. 5(a)		
93	200	0.425
120	250	0.460
150	300	0.415
180	350	0.405
205	400	0.395
230	450	0.380
260	500	0.360
290	550	0.350
315	600	0.345
Fig. 5(b)		
315	600	0.330
260	500	0.340
205	400	0.360
150	300	0.370
93	200	0.400

Source: Raymark Corporation

Fig. 5 Effect of temperature on the coefficient of friction for a proprietary iron-base compound for brake usage

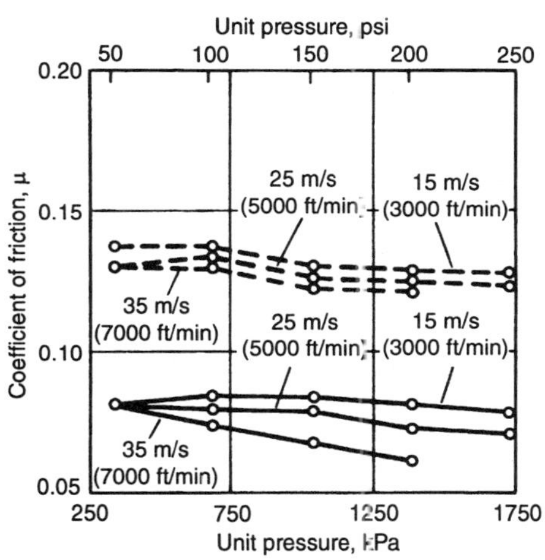

Fig. 3 Effect of unit pressure and rubbing speed on the coefficient of friction in a wet system. Solid line indicates dynamic coefficient of friction. Dashed line indicates static coefficient of friction.

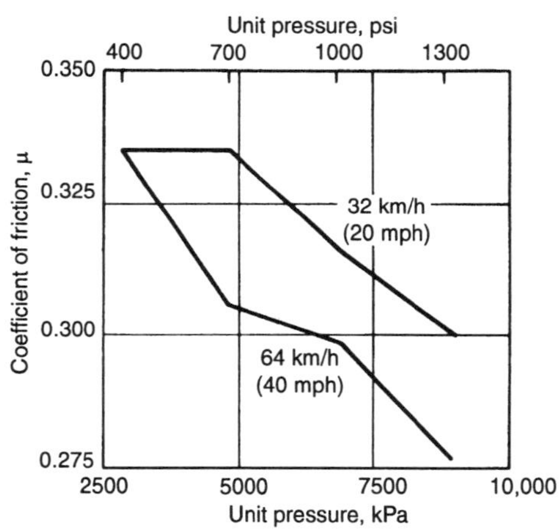

Unit pressure kPa	Unit pressure psi	Coefficient of friction
32 km/h (20 mph) line		
2760	400	0.336
4825	700	0.336
6900	1000	0.317
8965	1300	0.291
64 km/h (40 mph) line		
2760	400	0.336
4825	700	0.306
6900	1000	0.298
8965	1300	0.277

Source: Raymark Corporation

Fig. 4 Effect of pressure and speed on the coefficient of friction in an iron-base dry braking surface

eral, parts are kept in the 0.05 to 0. 10 mm (0.002 to 0.004 in.) tolerance range.

Sintering of Clutch Plate Facings. Green compacts, placed on supporting steel backing plates, are stacked in sintering furnaces. Backing plates or cores generally are cleaned and copper plated to achieve good bonding with the friction material. The type of sintering furnace used depends on the shape of the parts being produced. Bell-type furnaces are generally used for clutch discs because of the compression requirement. However, for disc brake pads and other odd shapes, the use of bell-type furnaces is not essential because compression is not required during sintering.

In bell furnaces, pressure is applied on the vertical stack of discs to reach the desired sintered density level and to prevent warpage or distortion. However, excess pressure causes high sintered densities and loss of low-melting-point metals from the friction material. Sintering is carried out at temperatures of 550 to 1000 °C (1020 to 1830 °F) in a protective atmosphere to prevent oxidation. Figure 2 shows the structure of a copper-base friction material, copper-plated layer, and steel backing plate after sintering at 650 °C (1200 °F) for 2 h.

Final Operations. Dimensional accuracy and additional design considerations are of major importance for applications. Parts are machined after sintering to meet dimensional specifications.

Grooving is required in the production of friction facing. Various types of grooves may be used, and each performs in a different manner. Two of the most frequently used types of grooves are combined in one friction facing (Ref 1). The purpose for grooving varies from wet friction to dry friction materials, but grooves are very important in wet clutch systems. In wet applications, the heat energy is partly removed by the oil. Grooves allow cooling oil to flow across the surface of the friction faces, yet allow fast oil runoff during engagement to minimize oil thickness on the surfaces and maximize the coefficient of friction.

In dry systems, grooves serve somewhat different purposes than in wet systems. These include (a) prevention of the crushing effect of thermal expansion during high application temperatures, as in the disc brake pads of heavy-duty vehicles; (b) removal of operation debris;

and (c) transfer of water and other liquids if the surface becomes wet.

Friction Applications

The coefficient of friction (μ) is the most important property in selection of a facing material. Friction may be expressed as static or dynamic values. The static coefficient of friction is the friction value of two surfaces at zero speed, while the dynamic coefficient of friction is measured at speeds greater than zero. The coefficient of friction is a function of conditions such as rubbing speed, pressure, and temperature. The relationship between rubbing speed and the coefficient of friction is shown in Fig. 3. With increasing rubbing speeds of the two surfaces, the coefficient of friction tends to drop. Applied pressure has a similar effect on the stability of the coefficient of friction. In wet friction applications, any increase in the temperature of the two rubbing surfaces above 150 °C (300 °F) will cause a substantial drop in the coefficient of friction. This temperature is commonly referred to as "breakdown temperature."

In wet applications, wear rates are usually low; therefore, thinner facings may be produced. Wear rate depends on factors such as temperature, number of engagements, mating or coupling plate surface finish, coupling plate, and facing material.

Dry friction applications are simple and lightweight compared to wet applications. Their major disadvantage is their erratic behavior in wet environments (Ref 2). For instance, on passenger cars, considerable fade is encountered when water enters the brake assembly. Figure 4 shows the effect of pressure and speed on the dynamic coefficient of friction of an automobile disc brake. Similar to wet friction materials, the drop in the dynamic coefficient of friction with pressure is noticeable.

Dry friction materials vary in composition from wet friction materials. In wet friction compounds, the matrix of the sintered material is copper, but in dry friction compounds, the percentage of copper is reduced and iron is increased. The effect of temperature on dry friction material is shown in Fig. 5(a) and (b). Between 95 and 315 °C (200 and 600 °F), the drop in the dynamic coefficient of friction is noticeable. After a certain level, the rate of decrease slows and eventually stabilizes.

In some materials, a slight increase in the coefficient of friction with increasing temperature is found after stabilization. This is mainly because of a change in surface morphology and the high-temperature effect of graphite in the facing material. In many cases, the drop in the friction level is recoverable, as shown in Fig. 5(b). As the temperature drops, the coefficient of friction increases. At 95 °C (200 °F), frictional recovery is almost complete. However, the level of recovery depends on the material used.

Clutch and brake applications include:

Dry friction materials

- Earth moving equipment
- Agricultural equipment
- Cranes and hoists
- Lift trucks
- Highway trucks (clutches)
- Aircraft (brakes)

Wet friction materials

- Earth moving equipment
- Agricultural equipment
- Military
- Lift trucks (clutches)

ACKNOWLEDGMENT

This article was adapted from an article by S. Ozsever of the Raymark Corporation titled "P/M Friction Materials" in *Powder Metallurgy,* Volume 7 of the *ASM Handbook.*

REFERENCES

1. R.L. Fish, "Wet Friction Applications: Some Design Considerations," technical report, Raymark Corp., Stratford, CT
2. W. Jenson, Friction Materials, *Machine Design*, Jan 1972, p 108–113

SELECTED REFERENCES

- Friction and Antifriction Materials, *Perspectives in Powder Metallurgy*, Vol 4, Plenum Press, New York, 1970
- R.H.T. Dixon and A. Clayton, *Powder Metallurgy for Engineers*, Machinery Publishing Co., Ltd., London, 1971
- J.R. Zimmerman, Clutches and Brakes, *Standard Handbook of Machine Design*, McGraw-Hill, 1996

Powder Metallurgy Bearings

Norbert A. Arnold, Keystone Carbon Co., Victor C. Straub, Keystone Powdered Metal Company,
and Michael Schloder, Specialty Pressed Components, Inc.

SELF-LUBRICATING BEARINGS are one of the oldest industrial applications of porous P/M parts, dating back to the mid-1920s (Ref 1–3). They remain the highest volume part produced by the P/M industry. The major advantage of porous bearings is that porosity in the bearing acts as an oil reservoir. The pores are filled with a lubricant that comprises about 25 vol% of the material.

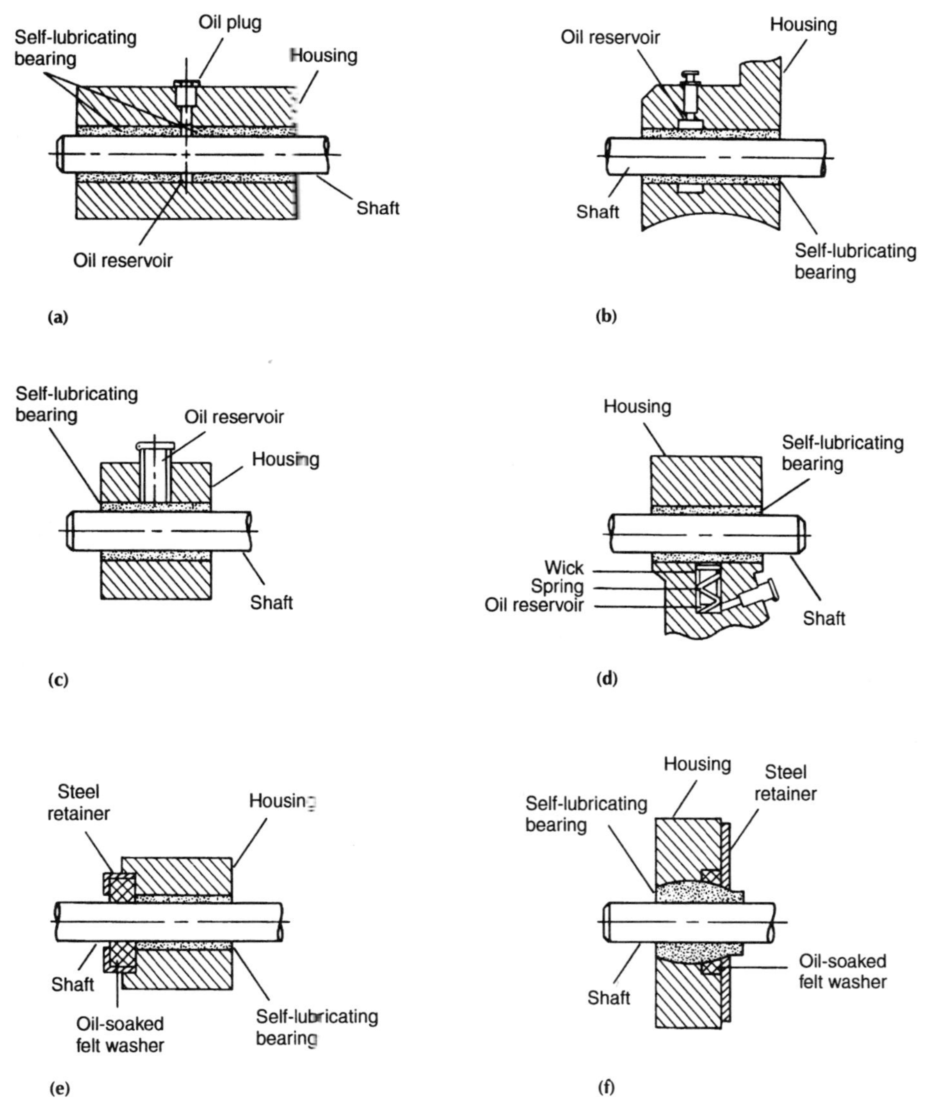

Fig. 1 Supplementary lubrication of porous bearings. (a) Oil reservoir created in the space between two bearing ends. (b) Oil reservoir around bearing. (c) Oil reservoir above bearing. (d) Oil reservoir below bearing. (e) Oil-soaked felt washer to provide additional lubrication. (f) Oil-soaked felt washer with self-aligning bearing

When the journal in an oil-impregnated self-lubricating bearing starts to turn, friction develops, the temperature rises, and oil is drawn out of the press because of the greater coefficient of expansion of the oil compared with the metal and because of the hydrodynamic pressure differential in the oil film between the journal and the bearing. When rotation stops and the bearing cools, the oil is reabsorbed by capillary action.

For many self-lubricating bearings, lubricant contained in the pores of the bearing remains for the entire service life of the bearing. On some heavy-duty bearing applications, an oil reservoir that feeds additional oil through the bearing wall may be provided on the outside diameter of the bearing. Figure 1 shows typical examples of arrangements for supplementary lubrication of porous bearings. The most common shapes of self-lubricating bearings are shown in Fig. 2.

Every component that requires rotary motion relies on a bearing of some type. P/M self-lubricating bearings can be found in almost every part requiring rotary motion. Examples of the wide usage of P/M bearings include:

- *Automotive:* Heater motors, window lift motors, air conditioners, windshield wiper motors, power antenna motors, trunk-closing motors, seat adjuster motors, and tape deck motors
- *Portable power tools:* Drills, reciprocating saws, jigsaws, and sanders
- *Home appliances:* Washers, dryers, refrigerators, blenders, mixers, food processors, fans, and clocks
- *Consumer electronics:* Phonographs, high-fidelity equipment, stereo equipment, tape players, and video recorders
- *Business machines:* Typewriters, computers, and copiers
- *Farm and lawn equipment:* Tractors, combines, cotton pickers, lawn mowers, string cutters, and chain saws
- *Marine equipment:* Outboard motors

Fig. 2 Typical bronze self-lubricating bearings

Table 1 Chemical composition of self-lubricating sintered bronze bearings

Element	Grade 1	Grade 2	Grade 3
Copper	87.2–90.5	85.7–90.2	83–88.5
Tin	9.5–10.5	9.5–10.5	9.0–10.0
Graphite	0–0.3	0.5–1.8	2.5–5.0
Iron, max	1.0	1.0	1.0
Total other elements by difference, max	1.0	1.0	1.0

Source: Ref 4

Table 2 Density requirements of sintered bronze bearings, grades 1 and 2

Type	Density, g/cm^3
1	5.8–6.2(a)
2	6.4–6.8
3	6.8–7.2
4	7.2–7.6

(a) Maximum density limit of 6.2 g/cm^3 has been established on type 1 to ensure meeting an oil content of 27% min. Satisfactory bearings can also be produced between type 1 and 2. These bearings have slightly higher strength constants and slightly lower oil content. Source: Ref 4

Table 3 Density requirements of sintered bronze bearings, grade 3

Type	Density, g/cm^3
1	5.8–6.2
2	6.2–6.6

Source: Ref 4

Bearing Compositions

Porous self-lubricating bearings are divided into four groups:

- Sintered bronze bearings
- Iron-base sintered bearings
- Iron-bronze (diluted bronze) sintered bearings
- Iron-graphite sintered bearings

The original and most widely used P/M bearing material is 90%Cu-10%Sn bronze with or without the addition of graphite (1% fine natural graphite is often added to enhance processing as well as improve bearing properties and bearing operation). The 90%Cu-10%Sn bronze material is superior in bearing performance to the iron-base and iron-bronze compositions, which are lower in cost and used in less severe applications.

Sintered bronze bearings, which are covered in American Society for Testing and Materials (ASTM) standard B 438, are made primarily from elemental copper, tin, and graphite powders. These bearings are available in three grades, which consist of the basic bronze material with different percentages of graphite as shown in Table 1. Each grade is available in four types (except grade 3, which is available in two types), determined by density range. The density ranges are specified in Tables 2 and 3.

A maximum density limit of 6.2 g/cm^3 has been established on type 1 to ensure a minimum oil content of 27%. Satisfactory bearings can also be produced between type 1 and 2—these bearings have slightly higher strengths with slightly lower oil content. Oil contents of the sintered bronze grades are shown in Table 4.

Sintered bronze bearings are also specified in Metal Powder Industries Federation (MPIF) standard 35. Refer to Table 5 for the MPIF specification for sintered bronze bearings.

Iron-base sintered bearings, which are covered in ASTM B 439, consist of four iron-base compositional grades (Table 6). Grades 1 and 2 are iron-carbon materials, while grades 3 and 4 are iron-copper materials. Density and oil content of these grades are specified in Table 7.

Iron-copper-carbon bearings are also specified in MPIF standard 35. Refer to Table 8 for the MPIF specification for sintered Fe-Cu-C bearings.

Iron-bronze sintered bearings (diluted bronze) are covered in ASTM B 612. They are Fe-Cu-Sn graphite oil-impregnated bearings of one composition, commonly referred to as diluted bronze. This bearing grade is an excellent bearing material for standard applications where shaft speed is below 10,000 rpm and the bearing is subject to nominal loading conditions. It has the unique feature of being bronze in appearance but lower in material cost. The microstructure of a diluted bronze P/M alloy is shown in Fig. 3. The density range of this bearing grade is 6.0 to 6.4 g/cm^3 and the oil content is greater than 17%. Chemical composition of this bearing material is shown in Table 9.

The diluted bronze sintering bearings are also specified in MPIF standard 35. Refer to Table 10 for the MPIF specifications for diluted bronze sintered bearings.

Table 4 Oil content of sintered bronze bearings

	Minimum oil content, vol%		
Type	Grade 1	Grade 2	Grade 3
1	27	25	11(a)
2	19	17	...(b)
3	12	9	...
4	9	7	...

(a) At 3% graphite, type 1 will contain 14% min oil content. (b) At 3% graphite, type 2 will contain 8% min oil content. At 5% graphite, type 2 will contain only a minimal amount of oil. Source: Ref 4

Table 5 MPIF specification for sintered bronze bearings

		Minimum values(a)						
	Material	Strength constant	Oil content	Density (O_{wet})(c), g/cm^3		Chemical composition, wt%		
Material	description code	(K), ksi	(P_0)(b), vol%	min	max	Element	min	max
Bronze (low graphite)	CT-1000-K19	19	24(d)	6.0	6.4	Copper	87.2	90.5
						Tin	9.5	10.5
						Graphite	0	0.3
						Other(e)	...	2.0
	CT-1000-K26	26	19	6.4	6.8	Copper	87.2	90.5
						Tin	9.5	10.5
						Graphite	0	0.3
						Other(e)	...	2.0
	CT-1000-K37	37	12	6.8	7.2	Copper	87.2	90.5
						Tin	9.5	10.5
						Graphite	0	0.3
						Other(e)	...	2.0
	CT-1000-K40	40	9	7.2	7.6	Copper	87.2	90.5
						Tin	9.5	10.5
						Graphite	0	0.3
						Other(e)	0	2.0
Bronze (medium graphite)	CTG-1001-K17	17	22(f)	6.0	6.4	Copper	85.7	90.0
						Tin	9.5	10.5
						Graphite	0.5	1.8
						Other(e)	...	2.0
	CTG-1001-K23	23	17	6.4	6.8	Copper	85.7	90.0
						Tin	9.5	10.5
						Graphite	0.5	1.8
						Other(e)	...	2.0
	CTG-1001-K33	33	9	6.8	7.2	Copper	85.7	90.0
						Tin	9.5	10.5
						Graphite	0.5	1.8
						Other(e)	...	2.0
	CTG-1001-K34	34	7	7.2	7.6	Copper	85.7	90.0
						Tin	9.5	10.5
						Graphite	0.5	1.8
						Other(e)	0	2.0
Bronze (high graphite)	CTG-1004-K10	10	11	5.8	6.2	Copper	82.5	87.5
						Tin	9.5	10.5
						Graphite	3.0	5.0
						Other(e)	...	2.0
	CTG-1004-K15	15	(g)	6.2	6.6	Copper	82.5	87.5
						Tin	9.5	10.5
						Graphite	3.0	5.0
						Other(e)	...	2.0

(a) These data are based on material in the finished condition. (b) Minimum oil content will decrease with increasing density. Those shown are valid at the upper limit of the density given. (c) Oil-impregnated. Assumes an oil density of 0.875 g/cm^3. (d) For oil content of 27% min, density range will be 5.8–6.2 g/cm^3, and K constant will be 105 MPa (15 ksi). (e) Iron 1wt% max. (f) For oil content of 25% min, density range will be 5.8–6.2 g/cm^3 and K constant will be 90 MPa (13 ksi). (g) At maximum graphite (5%) and density (6.6 g/cm^3), this material will contain only a trace of oil. At 3% graphite and 6.2–6.6 g/cm^3 density, it may contain 3–10 vol% oil. Source: Ref 5

Table 6 Chemical composition of iron-copper carbon sintered bearings

Element	Grade 1	Grade 2	Grade 3	Grade 4
Copper	0–1.5	0–1.5	9.0–11.0	18.0–22.0
Iron	bal	bal	bal	bal
Total other elements by difference, max	2.0	2.0	2.0	2.0
Combined carbon(a) (on basis of iron only)	0.3 max	0.3–0.6	0.3 max	0.3 max

(a) The combined carbon may be a metallographic estimate of the carbon in the iron. Source: Ref 6

Table 7 Density and oil content of iron-copper sintered bearings

	Density, g/cm^3		
Grade	min	max	Oil content, volume
1 and 2	5.6	6.0	21
3 and 4	5.8	6.2	19

Source: Ref 6

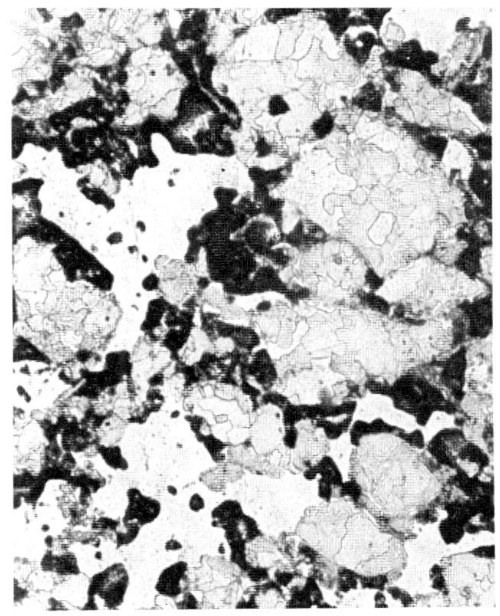

Fig. 3 Microstructure of diluted bronze P/M bearing material. 150×

Iron-graphite sintered bearings, which are covered in ASTM B 782, consist of two iron-graphite base grades. These two material grades are the newest bearing materials that are finding widespread use in applications in the industry. Grade 1 has a very low coefficient of friction coupled with quiet operation. This grade has been very popular in numerous applications where quiet operation, low coefficient of friction, and normal operating loads are encountered. It also has the added feature of lower cost than the bronze or diluted bronze materials with equivalent performance under normal operating conditions. Grade 2 has found a unique use in high-speed applications, but it is important to note that hardened shafts are required for good operational success. Refer to Table 11 for chemical composition and Table 12 for density and oil content specifications.

The iron-graphite grades are also specified in MPIF standard 35. Refer to Table 13 for the MPIF specifications for the iron-graphite bearings.

Blending and Mixing

Mixing or blending of fine elemental powders normally is done in a double-cone blender. For bronze bearings, mixtures of electrolytic, reduced, or atomized copper powders, atomized tin powders, and natural graphite powders are used. The particle size distribution of the copper powder is 100% −100 mesh and between 40 to 70% −325 mesh. The tin powder is 95% −325 mesh. Both reduced and atomized iron powders are used for iron-base bearings. The powders are combined with a fine lubricant such as stearic acid, zinc stearate, or Acrawax. The lubricant, which burns off during sintering, acts as a die release.

Blending of elemental and/or prealloyed powders is done for a predetermined time to produce a homogeneous mixture of all components and to supply a mix that has a specific density, which is necessary for fixed fill die cavities. Thorough mixing is required to achieve uniform strength, density, and dimensional control during sintering. The idea is to obtain uniform melting and diffusion of low melting constituents such as tin in bronzes or copper in iron-base alloys.

Compacting. The mix or preblend of raw materials is then transferred to a hopper, which is attached to a mechanical or hydraulic press. Loose powder flows by gravity (some vibration may be required) to the die cavity, where it is compacted to the desired shape and density. Compacting pressure for typical bearing compositions varies from 138 to 413 MPa (10 to 30 tsi). Depending on final dimensional accuracy, the green compact, or bearing, is molded 1 to 5% oversize to allow for material movement during sizing.

Tools used to press the loose powder are precision tooling. The die and core pin normally are made of tungsten carbide, while the upper and lower punches are made of high-grade tool steel.

Table 8 MPIF specification for Fe-Cu-C bearings

Material description code	Minimum values (a) Strength constant (*K*), ksi	Oil content (P_0), vol%	Density (D_{wet})(b) min	max	Chemical composition, wt% Element	min	max
FC-0205-K20	20	22	5.6	6.0	Iron	93.5	98.2
					Copper	1.5	3.9
					Carbon(c)	0.3	0.6
					Other	...	2.0
FC-0305-K35	35	17	6.0	6.4	Iron	93.5	98.2
					Copper	1.5	3.9
					Carbon(c)	0.3	0.6
					Other	...	2.0
FC-0208-K25	25	22	5.6	6.0	Iron	93.2	97.9
					Copper	1.5	3.9
					Carbon(c)	0.6	0.9
					Other	...	2.0
FC-0208-K40	40	17	6.0	6.4	Iron	93.2	97.9
					Copper	1.5	3.9
					Carbon(c)	0.6	0.9
					Other	...	2.0
FC-0508-K35	35	22	5.6	6.0	Iron	91.1	95.4
					Copper	4.0	6.0
					Carbon(c)	0.6	0.9
					Other	...	2.0
FC-0508-K46	46	17	6.0	6.4	Iron	91.1	95.4
					Copper	4.0	6.0
					Carbon(c)	0.6	0.9
					Other	...	2.0
FC-2008-K44	44	22	5.6	6.0	Iron	75.1	81.4
					Copper	18.0	22.0
					Carbon(c)	0.6	0.9
					Other	...	2.0
FC-2008-K46	46	17	6.0	6.4	Iron	75.1	81.4
					Copper	18.0	22.0
					Carbon(c)	0.6	0.9
					Other	...	2.0

(a) These data are based on material in the finished condition. (b) Oil-impregnated. Assumes an oil density of 0.875 g/cm^3. (c) Metallurgically combined carbon based on iron content. Source: Ref 5

Table 9 Chemical composition of diluted bronze sintered bearings

Element	Composition, %
Iron	54.2–62.0
Copper	34.0–38.0
Tin	3.5–4.5
Total carbon	0.5–1.3
Combined carbon(a)	0.5 max
Total other elements	2.0 max

(a) On basis of iron. Ref 7

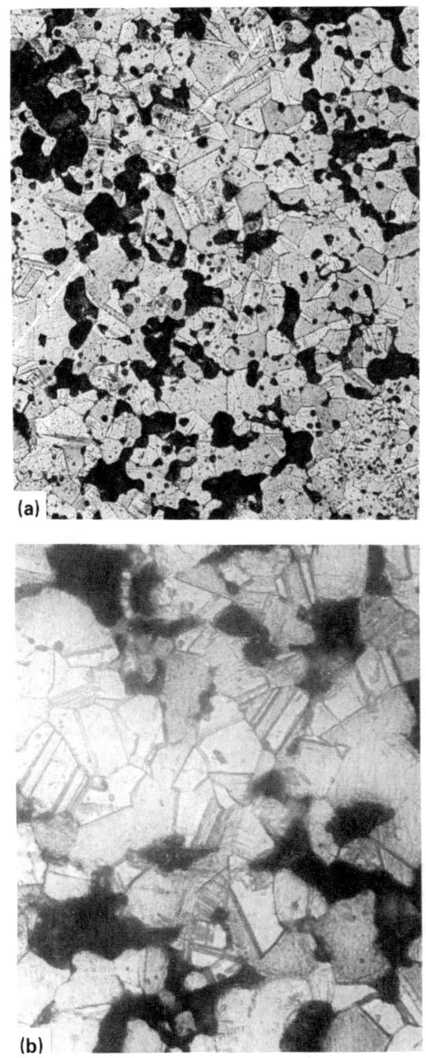

Fig. 4 Alpha bronze microstructure in 90%Cu-10%Sn P/M bearing alloy. (a) 150×. (b) 300×

Sintering transforms the green bearing into a metallurgically bonded alloy. The green bearing, an elemental mixture of copper and tin, is transformed into a true α bronze alloy at a temperature that is below the melting point of copper but above the melting point of tin. Under these liquid-phase conditions, sintering occurs without an appreciable change in the size of the bearing.

Sintering temperature for bronze may vary from 830 to 900 °C (1525 to 1650 °F), depending on the time at temperature (normally 3 to 8 min) and the desired sintered size. Suitable bronze bearings have microstructures that are essentially α bronze, as shown in Fig. 4. Diluted bronze bearings require sintering temperature of 870 to 980 °C (1600 to 1800 °F), whereas iron-copper bearings should be sintered above the melting point of copper—normally in the range of 1105 to 1125 °C (2020 to 2060 °F). See the article "Production Sintering Practices" in this Volume for additional information.

Sizing. Most precision bearings require a sizing operation after sintering. Sizing is done in highly polished, dimensionally accurate tools. Sizing pressures range from 206 to 551 MPa (15 to 40 tsi), depending on the amount of sizing required and the composition of the sintered part. Sizing is done primarily to improve dimensions; however, it cannot be done effectively without increasing density.

Typical sized inside diameter tolerances range from 0.0076 mm (0.0003 in.) for spherical bearings up to 16 mm (0.63 in.), with an overall length of up to 19 mm (0.8 in.). Straight-wall bearing inside diameter tolerances normally are 0.03 mm (0.001 in.) per 25 mm (1 in.) of diameter for lengths up to 38 mm (1.5 in.).

Impregnation. Powder metal bearings appear solid, but are actually 20 to 25% porous. This characteristic—porosity—makes parts made by P/M processing ideal bearing materials. Porosity consists of thousands of small capillaries that are interconnected throughout the structure of the bearing and act as a built-in reservoir for the lubricant. These pores can be filled with a lubricant that is sufficient for the life of the unit for most applications. However, for longer bearing life, oil-saturated felt or wicking can surround the bearing.

The pores are impregnated with oil normally under vacuum. Bearings are placed in the tank or autoclave, the lid is closed, and the vacuum pump is turned on. Pressure is reduced to at least 709 torr (28 in. Hg) before oil is permitted to flow over the bearings. Oil is introduced into the tank without shutting off the vacuum pump, and the tank, now filled with bearings and oil, is evacuated for another few minutes or until the pressure is again reduced to 709 torr (28 in. Hg).

After obtaining the desired vacuum, the vacuum pump is shut off, a valve is opened, and the oil is forced out of the impregnating tank to the storage reservoir. A properly impregnated bearing contains sufficient oil to fill 90% of the available interconnecting porosity or approximately 20 vol% oil (standard density 6.4 to 6.8 g/cm^3 bearing) and 2½ to 3 wt% oil.

The principal considerations in selecting a lubricant are the temperature of operation, bearing load or pressure between moving parts, operating speed, clearance between mating parts, and

Table 10 MPIF specifications for diluted bronze sintered bearings

		Minimum values(a)							
		Strength constant (*K*), ksi		Oil content	Density (D_{wet})(b), g/cm^3		Chemical composition, wt%		
Material	Material description code	min	max	(P_0), vol%	min	max	Element	min	max
Iron/bronze 60/40, diluted bronze	FCTG-3604-K16	16	...	36	22	6.0	Iron	54.2	62.0
							Copper	34.0	38.0
							Tin	3.5	4.5
							Total carbon(c)	0.5	1.3
							Other	...	2.0
	FCTG-3604-K22	22	50	17	6.0	6.4	Iron	54.2	62.0
							Copper	34.0	38.0
							Tin	3.5	4.5
							Total carbon(c)	0.5	1.3
							Other	...	2.0
Iron/bronze 40/60, diluted bronze	CFTG-3806-K14	14	35	22	5.6	6.0	Iron	50.2	58.0
							Copper	36.0	40.0
							Tin	5.5	6.5
							Total carbon(c)	0.5	1.3
							Other	...	2.0
	CFTG-3806-K22	22	50	17	6.0	6.4	Iron	50.2	58.0
							Copper	36.0	40.0
							Tin	5.5	6.5
							Total carbon(c)	0.5	1.3
							Other	...	2.0

(a) These data are based on material in the finished condition. (b) Oil-impregnated. Assumes an oil density of 0.875 g/cm^3. (c) Metallurgical combined carbon, 0.5 wt% max. Source: Ref 5

Table 11 Chemical composition of iron-graphite sintered bearings

	Composition, %	
Element	Grade 1	Grade 2
Iron, min	94.5	94.5
Graphite	2.0–3.0	1.5–2.5
Total other elements by difference, max	2.0	2.0
Combined carbon	0–0.5	0.5–1.0

Source: Ref 8

Table 12 Density and oil content of iron-graphite sintered bearings

	Density, g/cm^3		Minimum oil
Grade	min	max	content, vol%
1	5.6	6.0	18
2	5.6	6.0	18

Source: Ref 8

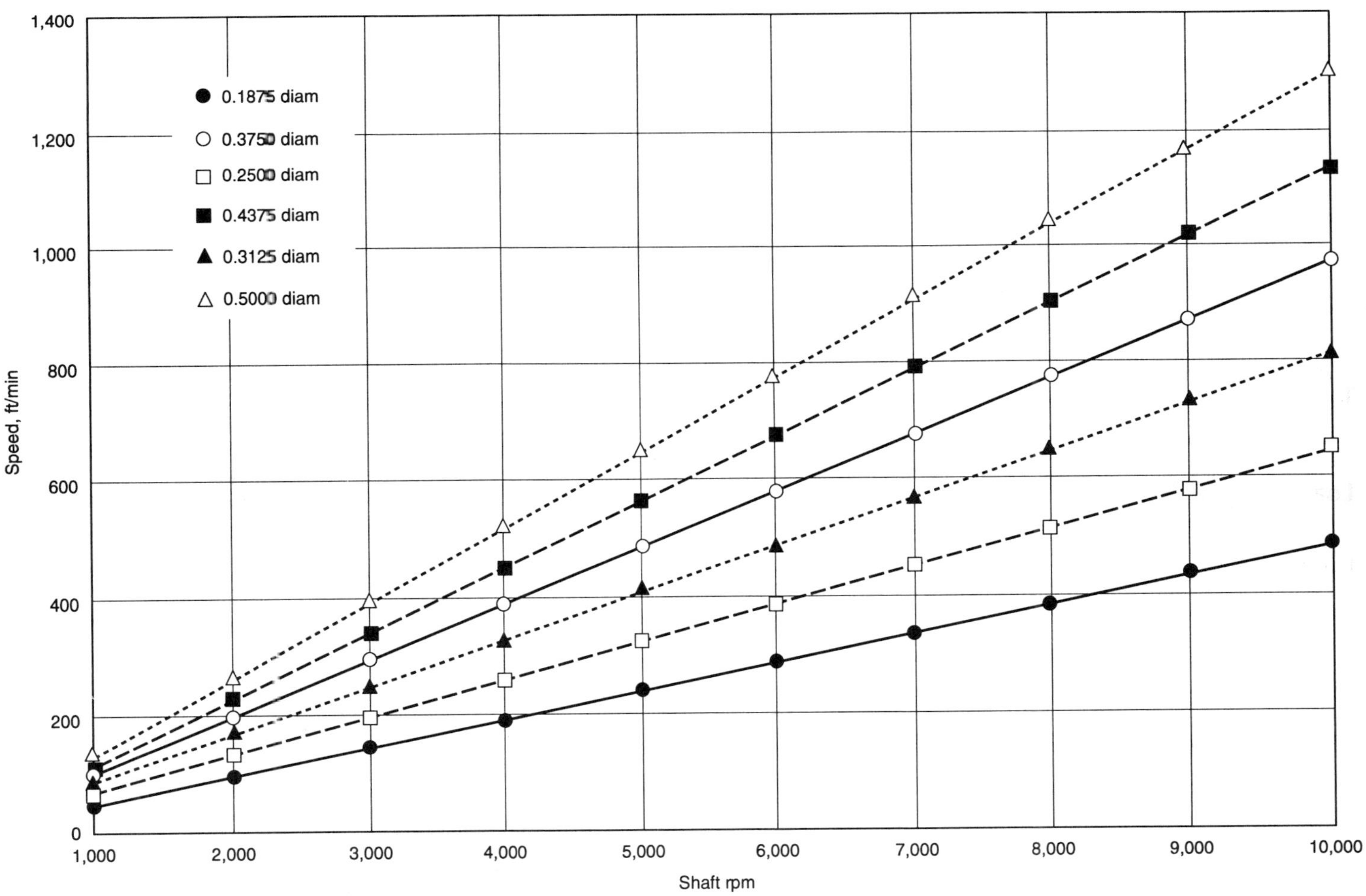

Fig. 5 Bearing speed information for shaft revolutions per minute versus speed at varying diameters

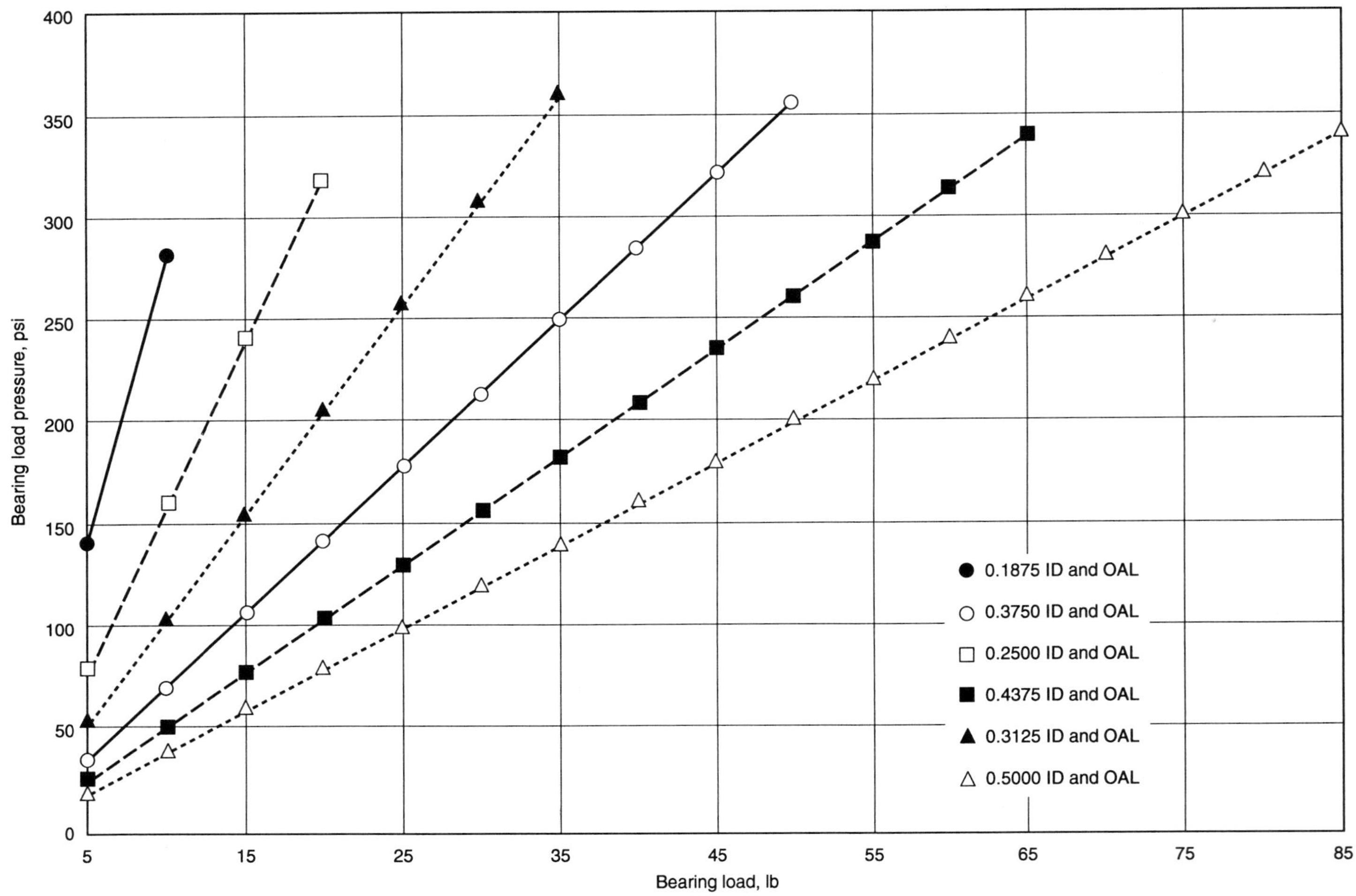

Fig. 6 Bearing load pounds to psi at varying diameters and overall length (OAL)

atmospheric conditions such as air temperature, dust, and moisture. In general, the most versatile type of lubricant for applications requiring self-lubricating bearings is a high-grade turbine oil. Several standard grades of oils for self-lubricating bearings, along with their properties, are listed in Table 14.

Load-Carrying Capacities

The *PV* formula should be used when determining the load-carrying capacity of self-lubricating bearing materials. The formula used to determine loads and speeds is:

$$PV = \frac{W}{Ld} \times \frac{\pi dn}{12} = \frac{\pi Wn}{12L}$$

where *P* is the load pressure on the projected bearing area (bearing inside diameter × length), *V* is the surface velocity of the shaft in feet per minute, and *W* is the bearing load in pounds. *L* is the bearing length in inches, *d* is the inside diameter of the bearing (also in inches), and *n* is the shaft speed in revolutions per minute. P/M self-lubricating bearing materials have a permissible *PV* factor of 50,000. Factors of *PV* above 50,000 are possible with special considerations. It should be noted that thrust bearings are the exception, where a maximum *PV* factor of 10,000 is recommended. Supplemental lubrication increases the permissible *PV*. Adverse conditions, such as the presence of dirt or abrasive particles, misalignment of a shaft, or an operating temperature in excess of 93 °C (200 °F) reduce the permissible *PV* factor.

Table 13 MPIF specification of iron-graphite sintered bearings

Material description code	Minimum values(a) Strength constant (*K*), ksi min	max	Oil content (P_0), vol %	Density (D_{wet})(b), g/cm^3 min	max	Chemical composition, wt % Element	min	max
FG-0303-K10	10	30	18	5.6	6.0	Iron	94.5	98.0
						Graphite(c)	2.0	3.0
						Carbon(d)	0	0.5
						Other	...	2.0
FG-0303-K12	12	35	12	6.0	6.4	Iron	94.5	98.0
						Graphite(c)	2.0	3.0
						Carbon(d)	...	0.5
						Other	...	2.0
FG-0308-K12	20	45	18	5.6	6.0	Iron	94.5	98.0
						Graphite(c)	1.5	2.5
						Carbon(d)	0.5	1.0
						Other	...	2.0
FG-0308-K22	22	55	12	6.0	6.4	Iron	94.5	98.0
						Graphite(c)	1.5	2.5
						Carbon(d)	0.5	1.0
						Other	...	2.0

(a) These data are based on material in the finished condition. (b) Oil-impregnated. Assumes an oil density of 0.875 g/cm^3. (c) Graphitic carbon, also known as free graphite. (d) Metallurgically combined carbon. Source: Ref 5

Table 14 Standard lubricants for porous bearings

SAE No.	Viscosity(a) at 38 °C (100 °F), s	Viscosity(a) at 99 °C (210 °F), s	Viscosity index	Pour point °C	Pour point °F	Flash point °C	Flash point °F	Recommended use
30	400	60	105	−9.5	15	240	465	General purpose
10	146–150	44	113	−12	10	205	400	High speeds, low loads
5	72	37	150	−70	−90	210	410	Low temperature or wide temperature variation
90	1076	89	90	−18	0	215	415	High loads, low speed, extreme-pressure lubricant (iron-base bearings only)

(a) Saybolt universal seconds. Source: Ref 7

Table 15 Permissible loads for sintered bronze self-lubricating bearings

Shaft velocity(a)		Permissible loads, grades 1 and 2							
		Type I		Type II		Type III		Type IV	
m/s	ft/min	MPa	psi	MPa	psi	MPa	psi	MPa	psi
Slow, intermittent	Slow, intermittent	22	3200	28	4000	28	4000	28	4000
0.13	25	14	2000	14	2000	14	2000	14	2000
0.26–0.51	50–100	3.4	500	3.4	500	3.9	550	3.9	550
Over 0.51–0.77	Over 100–150	2.2	365	2.2	325	2.5	365	2.5	365
Over 0.77–1.02	Over 150–200	1.7	280	1.7	250	1.9	280	1.9	280

Note: With a shaft velocity of less than 0.255 m/s (50 ft/min) and a permissible load greater than 6.89 MPa (1000 psi), an extreme-pressure lubricant should be used. With good heat dissipation and heat-removal techniques, higher *PV* ratings can be obtained. (a) For shaft velocities over 1.02 m/s (200 ft/min), permissible loads may be calculated: $P = 50{,}000/V$ ($P = 1.75/V$). Source: Ref 4

Table 16 Permissible loads for iron-based sintered self-lubricating bearings

Shaft velocity(a)		Permissible loads			
		Grades 1 and 2		Grades 3 and 4	
m/s	ft/min	MPa	psi	MPa	psi
Slow, intermittent	Slow, intermittent	25	3600	55	8000
0.13	25	12	1800	20	3000
0.26–0.51	50–100	3.1	450	4.8	700
Over 0.51–0.77	Over 100–150	2.1	300	2.8	400
Over 0.77–1.02	Over 150–200	1.6	225	2.1	300

(a) For shaft velocities over 1.02 m/s (200 ft/min), permissible load may be calculated using $P = 50{,}000/V$. Source: Ref 6

Table 17 Permissible loads for iron-bronze sintered self-lubricating bearings

Shaft velocity(a)		Permissible load	
m/s	ft/min	MPa	psi
Slow, intermittent	Slow, intermittent	28	4000
0.13	25	14	2000
0.26–0.51	50–100	2.8	400
Over 0.51–0.77	Over 100–150	2.1	300
Over 0.77–1.02	Over 150–200	1.4	200

(a) For shaft velocities over 1.02 m/s (200 ft/min), permissible load may be calculated using $P = 40{,}000/V$. Source: Ref 7

When determining bearing size, the shaft diameter is established by the requirements of the application. For the best performance, the bearing outside diameter should be at least 25% larger than the inside diameter (ID) (optimal oil reservoir and heat dissipation). Figures 5 and 6 can be used to determine bearing load. From Fig. 5, determine the shaft speed in feet per minute. From Fig. 6 where the bearing ID and length are the same, determine the load in pounds per square inch. If the load times the speed exceeds 50,000 *PV*, the bearing length must be increased.

Example. A bearing with 0.5 in. ID is running at 1800 rpm and supporting a load of 75 lb. From Fig. 5, the speed would be 230 ft/min. From Fig. 6, the load on a 0.5 ID × 0.5 OAL (overall length) bearing would be 300 psi. Then 300 × 230 = 69,000 *PV*. Increasing the bearing length to 0.70 in., an estimation based on the amount the 50,000 *PV* was exceeded, 40% with a safety margin, would reduce the load to:

75 lb/0.5 × 0.7 = 215 psi × 230 ft/min = **49,450 *PV***

Permissible loads for P/M bearings are given in Tables 15 to 17.

REFERENCES

1. H.M. Williams, U.S. Patent 1,556,658, 1925
2. H.M. Williams and A.L. Boegehold, U.S. patents 1,642,347; 1,642,348; and 1,642,349, 1927
3. C. Claus, U.S. Patent 1,607,389, 1926
4. ASTM B 438, ASTM
5. MPIF 35, Metal Powder Industries Federation
6. ASTM B 439, ASTM
7. ASTM B 612, ASTM
8. ASTM B 782, ASTM

Powder Metallurgy Gears

George T. Shturtz, Carbon City Products

POWDER METALLURGY (P/M) is a flexible metalworking process for the production of gears. The P/M process is capable of producing close tolerance gears with strengths to 1240 MPa (180 ksi) at economical prices in volume quantities. Spur, helical, bevel, face, spur-helical, and helical-helical gears are produced by P/M techniques. The process is particularly attractive when the gear contains depressions, through holes, levels, or projections.

The first application of P/M gears was for oil pumps in 1937/1938 (Ref 1). Gear pumps constitute a common type of application. Depending on application stress levels, different materials and density levels are used. For example, iron-carbon or Fe-Cu-C materials with densities of 6.0 to 6.8 g/cm^3 are used in gear pumps for low-pressure applications such as engine lubrication and automatic transmissions. For higher stress applications up to 20 MPa (3 ksi), alloy steels at the minimum density of 7.1 g/cm^3 are used. Powder metallurgy material requirements for gear pump applications at various stress levels are given in Table 1.

With the considerable progress in P/M technology over the years, the quality and application of P/M gears has increased. Progress in production, compaction, and sintering has resulted in new techniques such as infiltration, powder forging, surface rolling, and several finishing treatments. These developments, as discussed elsewhere in this Volume, have enabled the P/M parts to compete successfully with wrought metal counterparts.

Figure 1 summarizes the various P/M manufacturing methods used in gear production. Selection of a process method depends on factors such as dimensional accuracy and gear properties. In terms of dimensional accuracy, AGMA (American Gear Manufacturers Association) quality classes 5 to 7 can be obtained through conventional press-and-sinter operations. When AGMA quality levels above 7 are required, secondary operations such as coining, burnishing, and grinding typically are employed (see the section "Gear Tolerances" in this article for more information).

Performance of any P/M part is influenced primarily by density. Theoretical density of steel is around 7.85 g/cm^3, and Table 2 summarizes density levels obtained from various P/M processes for steel. All of these P/M methods are employed for gear manufacture, including various surface treatment such as carbonitriding, shot peening, and steam oxide treatments (Ref 2). More detailed information is given in specific articles in the Section "Shaping and Consolidation Techniques" in this Volume for the various processes listed in Table 2.

A wide variety of base metals are available in powder form, such as brass, bronze, iron, low-alloy steels, and stainless steel. The selection is governed by required mechanical properties, physical properties, service conditions, and cost. Table 3 lists commonly used P/M materials and broad applications of standard grades. Customized mixing and blending of elemental powders also provides a variety of possibilities for the development of alloy compositions formulated for specific mechanical and physical properties.

Table 1 Material requirements for P/M pump gear applications

Application	Material code
Output pressure less than 0.69 MPa (100 psi) and light service	FC-0208 type; as-sintered; 5.8–6.2 g/cm^3
Output pressure less than 6.9 MPa (1000 psi) but subject to high vibration and heavy service; hardened by heat treating	FC-0208 or FN-0106; 6.4–6.8 g/cm^3
Medium high pressure 6.9–8.6 MPa (1000–1250 psi) for short intervals; hardened by heat treating	FC-0208 type; 6.2–6.5 g/cm^3
General-purpose hydraulic pumps up to 10.3 MPa (1500 psi) continuous service	FN-0106 type; heat treated; 6.8–7.1 g/cm^3
General-purpose hydraulic pumps up to 17.2 MPa (2500 psi) continuous service	AISI 4630 type; heat treated; 7.2–7.6 g/cm^3

Source: Ref 1

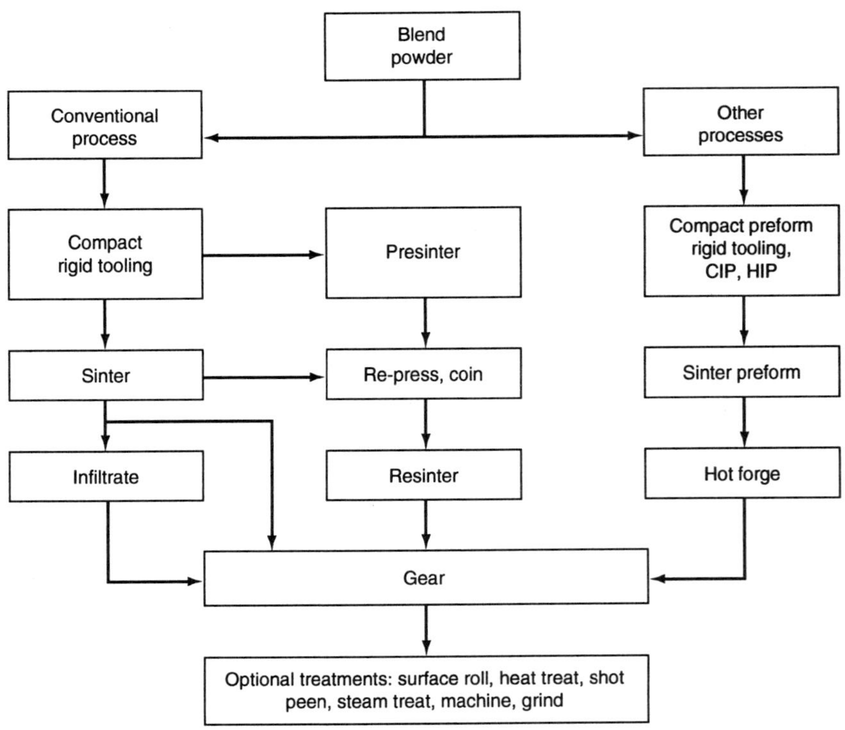

Fig. 1 P/M gear production process. CIP, cold isostatic pressing; HIP, hot isostatic pressing

Capabilities and Limitations

Powder metal gears are used in various applications for appliances, office machines, machinery, machine tools, and automotive products. The P/M process offers various advantages over traditional gear manufacturing for various applications.

In traditional methods, gears are manufactured from blanks obtained by machining of castings, forgings, and rolled bar stock. These blanks are machined by various methods such as milling, hobbing, and shaping. For precision and high-speed applications, the gear teeth are finished by secondary operations such as shaving and grinding. Gear production by these methods has the following disadvantages:

- Processing times are long, particularly with difficult-to-machine alloys.
- Material utilization is very low.
- Machined gear teeth have an unfavorable grain flow pattern with flow lines intersecting the gear tooth profile.

In comparison to the traditional gear manufacture process, the P/M process offers several advantages, particularly the elimination of machining and scrap losses for the manufacture of various gear types such as helical, bevel (both straight and spiral), rack, face, internal and external spur gears, and also compound gears. Internal configurations (splines, keys, keyways) also can be formed simultaneously with the gear profile during P/M manufacture, eliminating subsequent machining operations. Thus, P/M gear manufacturing allows efficient utilization of materials for moderate- to high-volume production. The minimum quantity required to make the P/M process economically competitive with machining depends on the size and complexity of the part and its accuracy and other property requirements. While powder metallurgy is best suited and most widely known for high-volume production quantities, there are numerous applications where small quantities (a thousand parts) can still offer cost benefits over traditional manufacturing methods.

General process advantages of P/M gear manufacture include:

- Economy in mass production
- Repeatability and uniformity of part features and dimensions
- Production of multilevel gears
- Close control of density or, conversely, porosity to suit a particular application
- Reduction or elimination of secondary operations
- Improved surface finishes on gear teeth by reducing or eliminating the machining marks (or "scoring") that can be imparted by conventional methods
- Self-lubricating ability from impregnation of oils and lubricants
- Noise reduction from the sound dampening qualities of the pore structure of a P/M component, high surface finish and tooth form consistency also can have an effect on overall reduction of noise in comparison to cut gear sets of comparable quality
- Weight reduction by incorporating weight saving or lightening holes (or similar shapes) in the initial tool design

Limitations of the P/M process for gear manufacturing include (Ref 2):

- Not economical in low- and medium-quantity production
- Not normally suitable for production of worn gears, herringbone gears, and helical gears with helix angle exceeding 35° because of various pressing and tooling considerations
- In view of the press capacities required for compaction, the sizes of gears are limited

Table 2 Density range for P/M steel consolidation methods

P/M method	Density range, g/cm^3	Notes
Pressed-and-sintered steel	6.9–7.1	Conventional powder metallurgy
Double processing (press, presinter, restrike, full sinter, heat treat)	7.2–7.4	Restrike (or repress) tooling has tolerances closer to finish dimensions than the tooling for the first strike
Warm compaction (single pressed)	7.2–7.5	Specially formulated powders to obtain higher green strength at lower compaction pressure
Warm compaction (double pressed)	7.4–7.7	...
Cold densification (press, sinter, cold form, heat treat)	7.6–7.8	Cold forming improves fatigue and wear resistance
Powder forging (press, sinter, forge, heat treat)	7.6–7.8+	Limited to spur, bevel, and face gears although trials with helical gears have been done (Ref 2)
Powder injection molding	7.8+	...
Copper infiltrated steel	Near full density	Copper infiltration also improves machinability
Roll densification	7.6–7.8	Increases localized densification for wear and fatigue resistance

Fig. 2 Examples of powder metallurgy gear forms. Courtesy of Carbon City Products

Table 3 Commonly used metal powder grades

Material	Chemistry	Applications
F-0000	Pure Fe	Low strength at high density, soft magnetic properties
F-0005	Fe + 0.5% C	Moderate strength, medium-carbon steel
FC-0208	Fe + 2% Cu + 0.8% C	Higher-strength structural components
FN-0205	Fe + 2% Ni + 0.5% C	High strength (heat treated), good impact resistance
FX-2008	Fe + 20% Cu + 0.8% C	Copper-infiltrated steel, high strength, machinable
SS-316L	316 stainless steel	Good toughness, corrosion resistance
SS-410	410 stainless steel	Good hardenability, abrasion resistance
CZ-2002	Brass	Good toughness, elongation, corrosion resistance
CT-1000	Bronze	Structural and bearing applications
CNZ-1818	Nickel silver	Decorative, tough, ductile, corrosion resistance

- The gear thickness or face width is restricted by the requirement that the depth of the die cavity and the stroke of the press must be at least 2 to 2½ times the gear thickness

Gear Forms

The size of P/M gears range from as small as a few millimeters to more than 300 mm in diameter. For all practical purposes the face width of P/M gears is limited to less than 75 mm (3 in.). The commonly used compaction presses have insufficient press stroke to consolidate the powder to greater than this length (face width).

The following types of gears can be produced by the powder metallurgy process:

- Spur gears
- Bevel gears
- Face gears
- Hypoid gears
- Helical gears of helix angle not exceeding 35°

Combination gears or multiple gears in which two or more gear forms are combined into one component are a specialty of the P/M process.

Some examples of P/M gear design are shown in Fig. 2 to 5. True involute forms (Fig. 3a) that are expensive to make by other methods can be easily made by the P/M process. The P/M process also is particularly advantageous for gear designs that have irregular curves, radial projections, keyways, or recesses, because these features are easily obtainable without requiring secondary machining operations. Full tooth face width for compound gears and sufficient fillet radii at the root diameter for tip clearance can be attained without allowances for cutter clearance.

Spur gears are the most common type of P/M gear in use today. The P/M process is capable of producing spur gears of many tooth forms and modifications. If an electrodischarge (EDM) electrode can be ground or cut, the mating form can be duplicated in P/M dies.

Compared to wrought metal gears, the face width or thickness of a spur gear is restricted in powder metallurgy in order that the density and mechanical properties are maintained fairly uniform throughout the section. Fine pitch (module < 0.75 mm) spur gear teeth present powder filling problems, which are more serious if the face width is large.

Helical gears are also commonly made by the P/M process. However, in contrast to metal-cutting processes, the helix angle of P/M gears is limited to approximately 35°. As with spur gears, many tooth forms and modifications are possible.

Helical gears have not been produced by powder forging, but recent production tests have been performed on a unique tooling method for powder forging of helical gears. One production trial involved the powder forging of a helical gear with a 31° helix angle and a normal pressure angle of 15° for an automotive transmission part (Ref 4). Experimental work later carried out by the German Research Society showed the superiority of the finished product in terms of toughness, tooth root fatigue, and contact fatigue (Ref 5). During powder forging of a helical gear, the die is the only helical part of the tool.

Bevel and face gears are readily made by the P/M process, but differ somewhat from spur and helical gears in mechanical property flexibility. Due to the nature of the compaction process, sections such as face gear teeth cannot be made to high densities without resorting to copper infiltration or repressing. Most bevel and face gear forms can be made providing they are not undercut, but the maximum mechanical properties attainable will be somewhat less than those that can be achieved for spur and helical gears.

All face tooth forms must be such that the green compact does not lock into the punch. For this reason undercut gear forms cannot be produced. Under the action of the punch, the crown area of face gear teeth also normally receives less compaction than the root zone. This results in an undesirable density variation, which can be offset by copper infiltration or repressing.

Another method to offset density variations is by putting a nonfunctional gear form with its teeth in the offset position so that a tooth gap comes below a tooth crown; this allows full punch pressure to be exerted on the tooth crowns of the functional gear, enabling a better density distribution (Ref 3).

Combination Gears. Powder metallurgy gear manufacture allows the production of combination gears or multilevel (two-to-four level) gears resulting in a single part with two or more gear forms. All the gear forms can be positioned close to one another, thus making the part very compact. This helps in saving space and handling and assembly costs. Combination gears can be produced by either of the two methods:

- Compacting the part in the combined form
- Compacting separately, assembling them in green state and sintering

Gear Tolerances

Tolerances of P/M gears are determined by the compaction tooling, compaction process, sintering process, and the P/M alloy. Tooth form and dimensions remain relatively stable throughout the process. The density variation in the compact should not exceed 0.2 g/cm^3, otherwise, dimensional control will be difficult during sintering.

Powder metallurgy gears are produced typically in AGMA 5 to 7 quality levels. Higher AGMA classification is possible, depending on the size of the compacted gear and the additional secondary operations. Higher AGMA classification increases the cost of manufacturing due to the secondary (machining) or restrike operations that are added to meet the tighter tolerances. In general, traditional press-and-sinter operations are economical for achieving AGMA quality levels 5 to 7. It is not advisable to produce higher quality levels through only compaction and sintering, as closer tolerances and tool wear will impact costs. Where higher accuracy is required, secondary operations such as sizing, shaving, burnishing, and grinding are required. Sizing (repressing) is the most common method because it improves mechanical properties through cold working.

Because dimensional variation is a function of part geometry, material, density, and subsequent thermal operations, dimensional tolerances may vary by application.

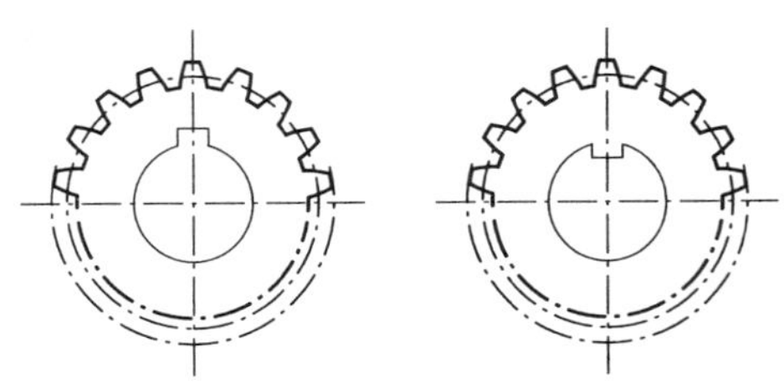

Fig. 4 Keyways and integral keys. Source: Ref 3

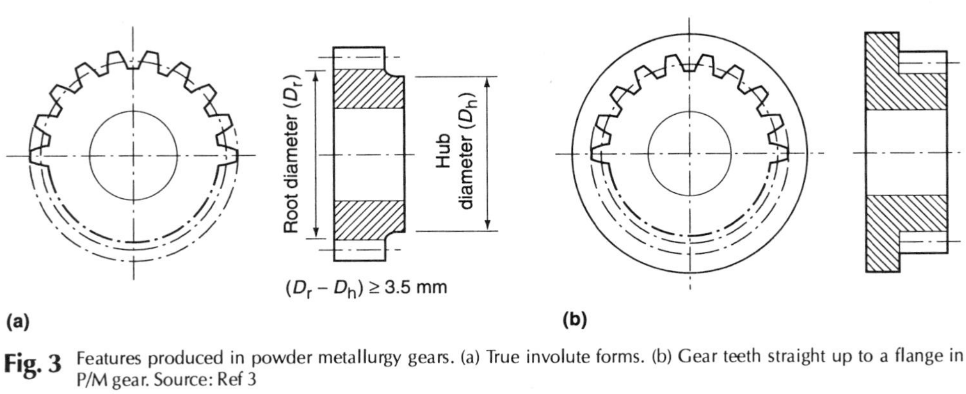

Fig. 3 Features produced in powder metallurgy gears. (a) True involute forms. (b) Gear teeth straight up to a flange in P/M gear. Source: Ref 3

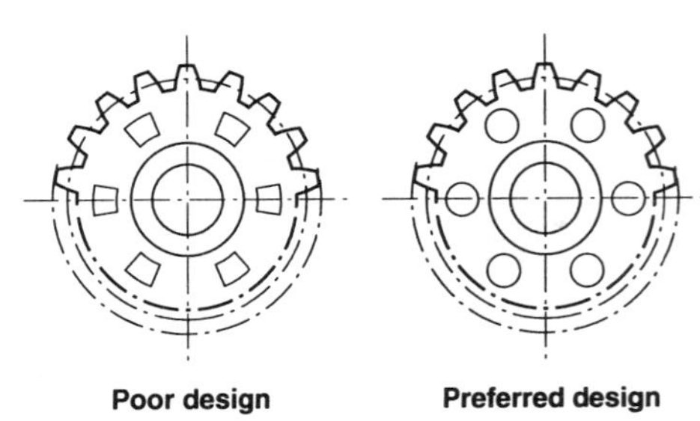

Fig. 5 Weight reducing holes on P/M gear. Source: Ref 3

Powder metallurgy gear tolerances are determined by the compaction tooling, compaction process, sintering techniques, and the alloy. The tolerances can be divided into functional tolerances that directly affect gear function and nonfunctional elements that pertain to portions of the gear that are more properly termed general machine design considerations. Nonfunctional elements include such items as hub design dimensions, fastening designs such as keyways, set screw holes, inside diameters for press fits, and other special body configurations.

Functional gear tolerances are controlled by the compaction tooling, the processing methods, and gear material. Because most P/M gears are pressed in carbide dies, the tooth form and dimensions remain relatively stable and constant throughout the process. Concentricity of the gear outside diameter and pitch diameter to the bore are also controlled by the tooling. If the pressing core rod or punches are eccentric to the die, the parts will be eccentric. Once the part is pressed, the material characteristics and sintering process determine the dimensional tolerance capability of the outside diameter, pitch diameter, test radius, and bore diameter.

Because of the number of variables within the P/M process it is not possible to state absolute dimensional tolerance capabilities. However minimum tolerances that one should expect for ferrous-base P/M gears without secondary operations are listed in Table 4 for spur and helical gears and in Table 5 for bevel gears. Typical tolerances for single press and sinter are listed in Table 6.

Gear Design and Tooling

Properly designed gears and compaction tooling are the key factors in the economical production of P/M gears with consistent quality. Blended metal powder is usually compacted at room temperature, at pressures between 275 to 690 MPa (40 to 100 ksi) of projected surface area. The tooling for a single-level gear includes a die, an upper punch, and a lower punch. If the particular component to be formed requires a bore or other inside-diameter configuration, a core rod is also part of the tooling. Multilevel compound gears or other complex structural parts may have two or more upper and/or lower punches (Fig. 6).

To form the gear, the die is filled with powder at a ratio of approximately two times the parts thickness; for example, a spur gear that has a thickness of 19 mm (0.75 in.) will be compacted form a column of 38 mm (1.50 in.) thick. During the compaction cycle of the more commonly used type of press, the upper punch enters the die while the lower punch, which in the fill position seals off the bottom of the die, remains stationary. As the upper punch travels downward, the compressive forces cause the die assembly to move downward in relationship to the lower punch, resulting in the same effect as if the lower punch were moving upward during the compaction stroke. The rate of movement and pressure of the upper punch and the motion of the die are relatively equal to ensure uniform density within the compacted preform. After completion of the compaction cycle, the upper punch retracts from the die, and the lower punch initiates an upward motion, ejecting the preform from the dies.

The shape of the compacted preform is determined by the shape of the tooling and the axial motion of the compaction press. Two main factors influence part design: the flow behavior and characteristics of the metal powders and the movement of the tools within the pressing cycle. Metal powders do not flow hydraulically, and the allowance for friction between the powder particles themselves and with the moving tool members must be factored in the final P/M part design. The pressing action from both top and bottom largely governs the shape, length, and dimensional details of the preform. The clearance tolerance of the moving tool members, relative to the inside diameter or bore, contributes to a larger degree of eccentricity than with machined gears. Details, or features that increase the number of tool members compound the eccentricity. Concentricity can be improved with the addition of secondary operations, such as grinding the bore in relationship to the pitch diameter.

The compacted preform is sintered typically at temperatures of approximately 1120 °C (2050 °F) to obtain tensile properties around 125 to 620 MPa (18 to 90 ksi), depending on the material composition. The addition of heat treatment can increase ultimate tensile properties of a compacted-and-sintered steel component to 1240 MPa (180 ksi).

Design guidelines for P/M gears include the following considerations.

Ejection from the Die. Some features must be eliminated from the final part (or added through secondary machining operations) when they inhibit ejection of the preform from the die. For example, undercuts, reverse angles, details at right angles to the direction of pressing (e.g., holes, grooves), threads, diamond knurls and reentrant angles would interfere with smooth ejection.

Die Filling During Compaction. Gear design should allow for the movement of the metal powders throughout the tool members during the compaction cycle. Metal powders do not flow hydraulically. Therefore, extremely thin-walled sections, very narrow grooves, and deep counterbores should be avoided because the metal powders may not completely fill these details within the tool members.

Tool life can be improved by avoiding narrow deep grooves, very sharp edges, complete spherical profiles, and knife-thin tool thicknesses. Often design simplification of these features allows for a more robust or "practical" set of com-

Table 4 Minimum tolerance capabilities for spur and helical P/M gears

Tooth-to-tooth composite tolerance	AGMA class 6
Total composite tolerance	AGMA class 6
Test radius	0.05 mm + 0.002 × diameter
Over-pin measurement	0.10 mm + 0.002 × diameter
Lead error	0.001 mm/mm
Perpendicularity—face to bore	0.05 mm + 0.001 × diameter
Parallelism	0.025 mm + 0.001 × diameter
Outside diameter	0.10 mm + 0.002 × diameter
Profile tolerance	0.008 mm

Tolerances with single press and sinter without secondary operations. Source: Ref 3

Table 5 Minimum tolerance capabilities for bevel P/M gears

Tooth-to-tooth composite tolerance	AGMA class 6
Total composite tolerance	AGMA class 6
Outside diameter	0.10 mm + 0.002 × diameter
Perpendicularity–face to bore	0.05 mm + 0.001 × diameter

Tolerances with single press and sinter without secondary operations. Source: Ref 3

Table 6 Typical tolerances of single pressed and sintered gears

See Tables 4 and 5 for specific gear forms.

Feature	Tolerance (without secondary operations) mm	in.
Length	±0.125	±0.005
Flatness on end	0.05	0.002
Concentricity	±0.075	±0.003(a)
Inside diameter	±0.050	±0.002(a)
Profile tolerance	±0.0075	±0.0003
Tooth-to-tooth error	±0.025	±0.001
Circular tooth error	±0.05	±0.002
Parallelism on ends	±0.04	±0.0015
Lead error	±0.025	±0.001
Total composite error	±0.08	±0.0032(b)
Outside diameter tolerance with pitch diameter up to:		
25 mm (1 in.)	0.075	0.003(c)
50 mm (2 in.)	0.1	0.004(c)
75 mm (3 in.)	0.125	0.005(c)
100 mm (4 in.)	0.15	0.006(c)

(a) Tolerance depends on feature size. (b) Tolerance depends on gear size and number of levels. (c) Total tolerance on outside diameter and measurement over wires

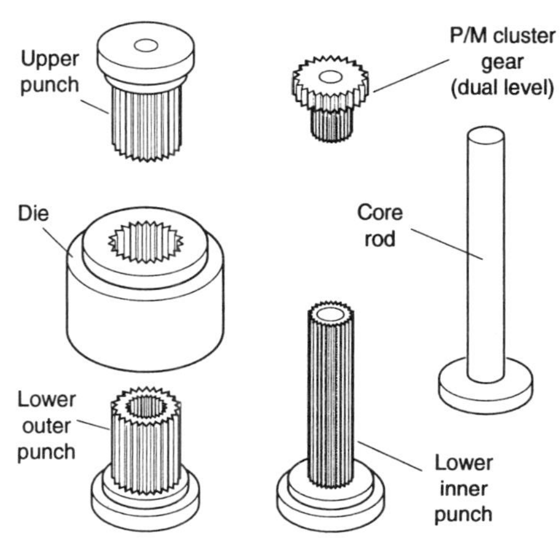

Fig. 6 Compaction tooling for compound gear. Source: Ref 6

paction tools without adding machining operations.

Clearance. Hubs or bosses for gears, sprockets, or cams can be readily produced, but the designs should ensure the maximum permissible material between the outside diameter of the hub and the root diameter of gear or sprocket features (Fig. 7) (Ref 6). Another consideration for gear design is to maintain sufficient clearance between the inside diameter and the root diameter, which can range from 0.9 mm (0.035 in.) on small pinions to 7.5 mm (0.30 in.) for more demanding applications. Design consideration should be given to maximize the clearance between the gear root diameter and compound gears, pinions, or features (i.e., cams, etc.) that are incorporated parallel to the gear teeth.

Thickness Changes. The configuration of the gear should limit major changes in section thickness (Fig. 8) for compound gear geometries such as gear-pinion and spur-face combinations, which are well within the capability of the P/M process. Uniform density and high strength is best achieved by limiting the number of section thickness changes (levels) that are designed into the preform. The number of levels a part may have is determined by the specific type of compaction press and/or the design of the compaction tooling.

Right-Angle Intersections. The configuration of the gear should minimize right-angle intersections. Radii should be incorporated at right-angle intersection of section thickness changes. These radii improve the integrity of the preform. Gear engagement should be designed to occur above the radius. Often raised surface features can be added to the compacted preform to assist in this engagement.

Edge Detail. The configuration of the gear should incorporate edge detail (i.e., chamfers on top and bottom of the gear teeth and on outside and inside diameters) (Fig. 9). This edge detail of chamfer has two main purposes. First, it increases the density of the teeth. Higher tooth density results in improved mechanical properties, particularly strength. It also reduces the adverse effects of burrs. The chamfer detail will in most cases keep the burr within the overall thickness (width) of the gear. These burrs are a result of the fit clearance of the tool members and, if necessary, can be removed by a vibratory finishing (tumbling) operation.

Gear Performance

In general terms, gear teeth loading is akin to normal loading of two parallel cylinders with a combined sliding and rolling. In rolling contact, two elements roll in contact with each other without slippage. Both elements rotate at the same velocity. Gear teeth have sliding contact over most of their surfaces, with rolling contact occurring only at the pitch line. The amount of sliding varies with the distance from the pitch line. When the rolling and sliding are in the same direction, less damage occurs than with negative slide (Ref 7).

Gears can be damaged by several mechanisms such as bending fatigue or by surface fatigue mechanisms such as pitting or spalling. The origin of pitting occurs where the combination of pressure and negative sliding are the highest. On spur gears, this is usually at the lowest point of single-tooth contact on the dedendum of the drive gear (Ref 8).

As with any other material the choice of a P/M alloy is governed by the magnitude and nature of the transmitted load, the speed, the life requirements, the environment, the type of lubrication, and the gear and assembly precision. Allowable loads on contact surfaces of gears are limited by the occurrence of pitting, in most cases, provided they are not run at very high speeds to cause scoring (Fig. 10). Even at light loads, gear teeth experience relatively high concentrated stresses because contact areas are very small. In general, Hertz equations are used to design for contact stresses, and these equations consider compressive loads on stationary contact surfaces.

Spalling damage of wrought and P/M alloys has similar appearance but different crack propagation mechanisms. In wrought alloys, most of the time is spent in the crack initiation phase (under high-cycle conditions) with relatively quick crack propagation once the crack is big enough. In P/M alloys, the porosity blunts crack growth to the surface. Until the crack can propagate to the surface, a band of subsurface cracking continues. The depth of the band is probably related to the depth of the maximum shear stress.

Selecting Gear Materials. Required strength is the first factor when selecting a gear material. In spite of the approximations involved, one can generally arrive at the correct material by using the methods outlined in the AGMA standards for rating the strength of spur, helical, and straight bevel gear teeth (Ref 9–11). A good approximation of compressive load carrying ability can be found by using the methods outlined in the AGMA standards for surface durability of spur, helical, and bevel gear teeth (Ref 12–14). In addition, the allowable contact stress S_{ac}, the elastic modulus, and Poisson's ratio of the P/M material must be determined.

Allowable Contact Stress. Tensile strength, elastic modulus, and Poisson's ratio for most ferrous P/M materials is given in MPIF standard 35. In the absence of data for allowable contact stress of ferrous P/M materials, an approximation of S_{ac} is 70 MPa (10 ksi) below the ultimate tensile strength (S_{ac} = UTS – 10 ksi).

Allowable bending stress (S_{at}) for a P/M material can be approximated in the absence of data at 0.3 of the ultimate tensile strength of ferrous-base P/M materials (S_{at} = 0.3 UTS). The only caution one should be aware of is that tensile data may be typical properties. The designer should consult the supplier or specifications for minimum property data.

Impact and Fatigue Strength. Most P/M gears are limited in performance by their impact and

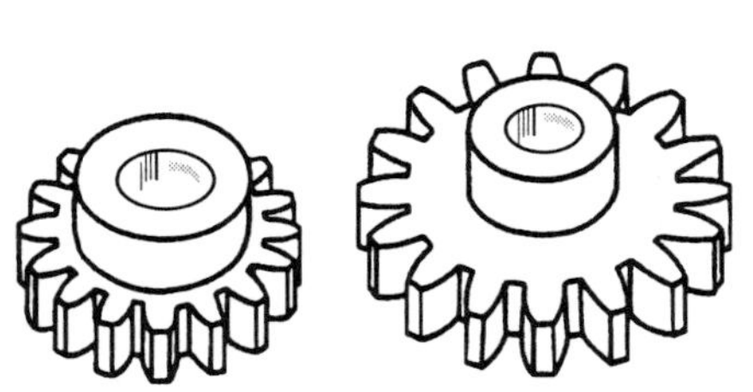

Fig. 7 Hub feature. Source: Ref 6

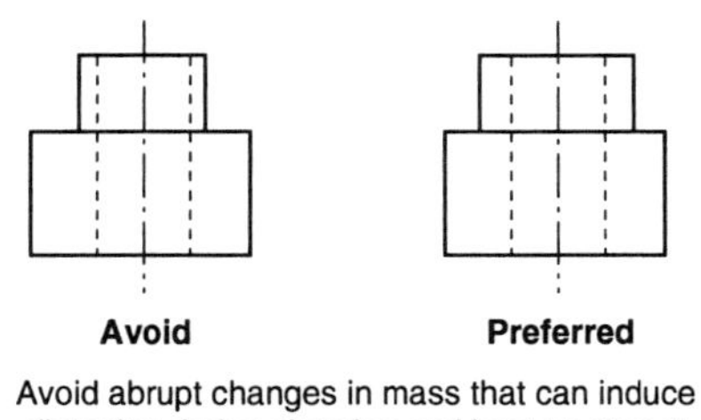

Fig. 8 Preferred design for section thickness changes. Source: Ref 6

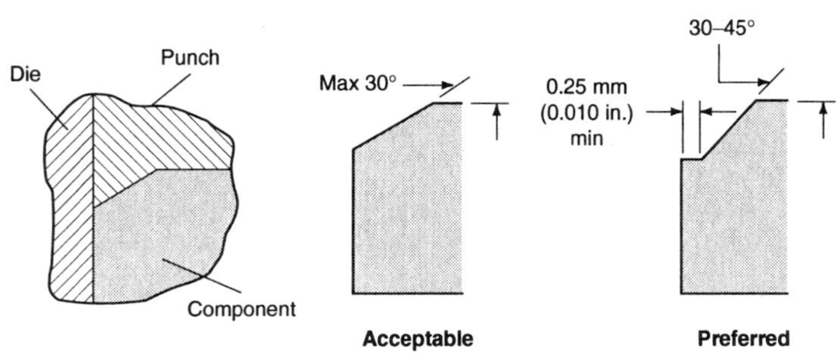

Fig. 9 Edge radii design features. Source: Ref 6

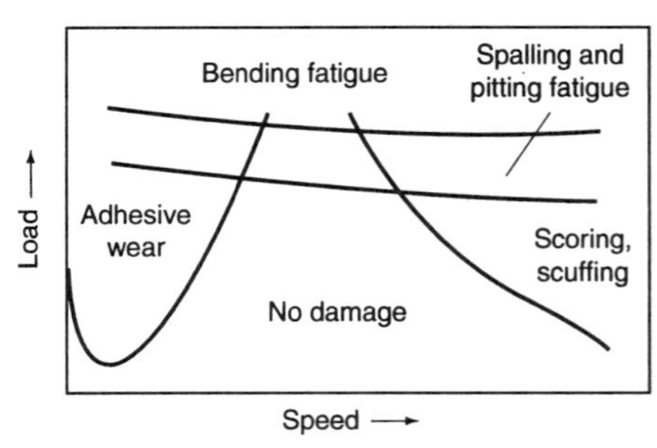

Fig. 10 Schematic of gear damage mechanisms. Source: Ref 15

fatigue strength as opposed to their compressive yield strength and surface durability. Surface treatments such as shot peening, material cleanliness, and other factors influence contact fatigue. Additional information on these topics is contained in the article "Contact Fatigue" in *ASM Handbook, Fatigue and Fracture,* Volume 19. Detailed information on P/M gear fatigue testing and evaluation is discussed further in Ref 15 and 16. Additional information on fatigue is also contained in the article "Fatigue and Fracture Control for Powder Metallurgy Components" in this Volume.

Quality Control and Inspection

Inspection is one of the most important stages in gear production. Inspection and testing is done on dimensional specifications, mechanical properties, and surface durability (hardness). The information required to describe a gear for the P/M process and subsequent inspection methods are essentially the same as for a machine-cut gear. The part drawing should include:

- Number of teeth
- Diametral pitch (normal diametral pitch for a helical gear)
- Pressure angle (normal pressure angle for a helical gear)
- Tooth form for spur, helical, and straight bevel gears
- Helix angle and rotation (right or left hand) for a helical gear
- Pitch, face, root, and back angles for straight bevel gears

To assist in the qualification (i.e , inspection) the drawing should also include:

- Outside diameter (as a reference dimension for straight bevel gears)
- Root diameter
- Pitch apex to back face

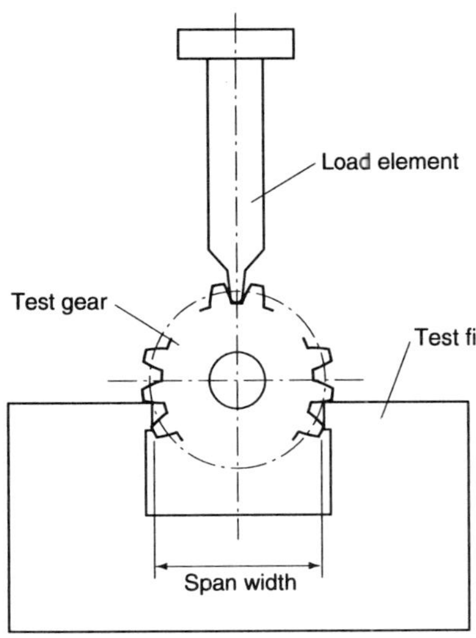

Fig. 11 Spur gear tooth strength test. Source: Ref 17

- Total composite tolerance
- Tooth-to-tooth tolerance (not appropriate for straight bevel gears)
- Test radius

Measurement over wires (MOW) and measurement between wires (MBW) (for internal gears) are convenient measurable criteria that should also be included as drawing specifications. Master gear information (specification and/or data) should be included on the drawing. Tip radius detail for spur gears is also necessary. It is advisable that right-angle (rolling) fixtures be developed to check mesh and backlash. Often several ID (inside diameter) pins are required to confirm mesh, after/during critical processing operations (minimally, compaction, sintering, and heat treatment). Typically, these inspection fixtures are specified and purchase at the time the tooling is manufactured.

In production, it is common practice to check the face width, outside diameter, over-pin measurement, total composite error, tooth-to-tooth error, perpendicularity of face-to-bore, parallelism, and inside diameter of the gears. Also the tooth strength and macroscopic hardness are usually checked. The sampling plan and frequency of measurement varies and depends on the criticality of the application. A joint committee with representatives from the American Gear Manufacturers Association (AGMA) and the Metal Powder Industries Federation (MPIF) have formulated a "Specification for Powder Metallurgy Gears," which delineates the minimum data necessary to adequately specify P/M gears. This specification is available from AGMA or MPIF.

General gear terminology includes:

- *Pitch circle:* the circumference of a gear measured at the point of contact with the mating gear
- *Pitch diameter:* the measurement across the gear within the circumference (pitch circle)
- *Diametral pitch (DP):* the ratio of the number of teeth in a gear to each inch of its pitch diameter. For example, a 30-tooth gear with a 125 mm (5 in.) pitch diameter would have 6 teeth/in. of pitch diameter, therefore would be a 6 DP gear.
- *Active profile:* the part of the gear tooth profile that actually comes in contact with the profile of the mating gear while in mesh
- *Addendum:* the height of the tooth above the pitch circle
- *Backlash:* the amount of movement or "play" between mating teeth of two gears in mesh
- *Center distance:* distance between the axes of rotation of two mating gears
- *Circular thickness:* the length of arc between the two sides of a gear tooth at the pitch circle
- *Dedendum:* the depth of the tooth below the pitch circle
- *Face width:* the length of the teeth in an axial plane
- *Fillet radius:* the radius of the fillet curve at the base of the gear tooth
- *Pressure angle:* the angle between the line of action and a line tangent to the pitch circle at the pitch point
- *Tooth thickness:* the thickness of the tooth measured at the pitch circle

Gear inspection terms include:

- *Runout (radial):* the total variation of the radial distance of the gear teeth from the center or bore
- *Runout (axial):* the total variation of gear teeth along its axis, measured from a reference plane perpendicular to its axis. Also referred to as "wobble" or face runout
- *Lead:* the axial advance of a gear tooth in one revolution (360°)
- *Lead variation:* the condition in a gear where some teeth vary in lead, plus or minus from the average lead
- *Total composite error:* the total variation of the center distance in one complete revolution
- *Tooth-to-tooth variation:* the variation of the center distance in a section of its circumference (the section of the circumference is an angle of 360/N° where N is the number of gear teeth

Mechanical Testing. Mechanical properties are also measured, and in many cases bench life and impact tests or field tests are made to check the gear functionality. Because tensile strength of a gear cannot be measured directly, it is appropriate to establish a minimum tooth breakage strength that can be used as a production acceptance criteria from either sample gears made from a die or gears cut from a P/M slug.

One quick and accurate means of measuring tooth strength of a spur gear is shown in Fig. 11. By mounting the gear between fixed supports and applying a vertical load, one can measure a tooth breakage strength. By testing a number of gears that perform successfully in the particular application, it is possible to calculate statistical limits for the tooth breakage strength that can be used as a future material strength acceptance. Helical and bevel gears are usually evaluated in torque-type tests and the same type of statistical analysis applied. The use of statistics is particularly expedient in gear strength testing because proper sampling allows process control to a normal statistical distribution (i.e., a $\pm 3\sigma$) minimum acceptance criteria ensures 99.85% confidence of adequate strength.

Surface durability is a specified and measured minimum Rockwell hardness because this macroscopic hardness relates to the compressive yield strength of the materials. The minimum Rockwell hardness depends on the material and can be established either from samples or from the supplier's historical information.

In cases where scoring may be a problem, in particular with heat treated gears, it is appropriate to specify and measure a microscopic hardness. This is accomplished with a diamond indenter, and the microscopic or individual particle hardness of the material is measured. The particle hardness specification is generally determined from historical data for the producer's material.

In contrast to most metals, P/M materials have both a macro- and microhardness as a result of porosity within the structure. Because the P/M structure consists of individual hard particles bonded together with interdispersed voids, the conventional Rockwell indention registers a composite hardness. Because sliding and surface contact involves microscopic contact (individual asperities) the particle or microhardness of the P/M material influences the score resistance of the P/M material more so than the macro Rockwell hardness. As a generality, when a P/M gear is run against a wrought gear, the P/M particle hardness should be no harder than the hardness of the wrought gear to prevent wear of the mating gear.

REFERENCES

1. P.W. Lee, Press and Sintered Parts and Their Applications, *Powder Metallurgy Applications Advantages and Limitations*, Erhard Klar, Ed., American Society for Metals, 1983, p 82
2. S.D.K. Saheb and K. Gopinath, Powder Metallurgy Gears—A Brief Review, *Powder Metall. Sci. Technol.,* Vol 1 (No. 4), July 1990, p 43–66
3. S.D.K. Saheb and K. Gopinath, Tooling for Powder Metallurgy Gears, *Powder Metall. Sci. Technol.,* Vol 2 (No. 3), April 1991, p 25–42
4. *Met. Powder Rep.*, Vol 43, 1988, p 536
5. W. Konig, G. Rober, K. Vossen, and M. Stromgren, Powder Forging of Helical Gears for Car Manual Gear Boxes—Concept and Properties, *Met. Powder Rep.,* Vol 45 (No. 4), 1990, p 269–273
6. *Powder Metallurgy Design Manual,* Metal Powder Industries Federation, 1994, p 193
7. G. Parrish and G.S. Harper, *Production Gas Carburizing,* Pergamon, 1985, p 20
8. *Gear Design Manufacturing and Inspection Manual,* SAE No. 841083, Society of Automotive Engineers, 1990, p 46
9. "AGMA Standard for Rating the Strength of Spur Gear Teeth," AGMA 220.02, American Gear Manufacturers' Association, 1966
10. "AGMA Standard for Rating the Strength of Helical and Herringbone Gear Teeth," AGMA 221.02, American Gear Manufacturers' Association, 1965
11. "AGMA Standard for Rating the Strength of Straight Bevel and Zero Bevel Gear Teeth, AGMA 222.02, American Gear Manufacturers' Association, 1964
12. "AGMA Standard for Surface Durability of Spur Gear Teeth," AGMA 210.02, American Gear Manufacturers' Association, 1965
13. "AGMA Standard for Surface Durability of Helical and Herringbone Gear Teeth," AGMA 211.02, American Gear Manufacturers' Association, 1969
14. "AGMA Standard for Surface Durability Formulas for Straight Bevel and Zero Bevel Gear Teeth," AGMA 212.02, American Gear Manufacturers' Association, 1964
15. T. Prucher and H. Sanderow, Surface Fatigue of P/M Alloys, *Characterization, Testing, and Quality Control,* Vol 2, *Advances in Powder Metallurgy and Particulate Materials,* Metal Powder Industries Federation, 1994, p 99–111
16. R. Gnanamoorthy and K. Gopinath, Surface Durability Studies on As-Sintered Low Alloy Steel Gears, *Advances in Powder Metallurgy & Particulate Materials—1992,* Vol 6, Metal Powder Industries Federation, 1992, p 237–250
17. W. Smith, Ferrous Based Powder Metallurgy Gears, *Gear Manufacture and Performance,* American Society for Metals, 1974, p 257–269

SELECTED REFERENCES

- AGMA & MPIF Develop Standards, Information Sheet for Powder Metal Gears, *Gear Technol., J. Gear Manuf.,* Sept/Oct 1996
- R. Moderow, Gear Inspection and Measurement, *Gear Technol., J. Gear Manuf.,* July/Aug 1992
- MPIF Standard 35, *Material Standards for P/M Structural Parts,* Metal Powder Industries Federation, 1997
- *Powder Metallurgy Design Manual,* 2nd ed., Metal Powder Industries Federation, 1995
- *Powder Metallurgy Design Solutions,* Metal Powder Industries Federation, 1993
- H. Sanderow, Powder Metal Gear Design and Inspection, *Gear Technol., J. Gear Manuf.,* Sept/Oct 1996
- G. Shturtz, The Beginner's Guide to Powder Metal Gears, *Gear Technol., J. Gear Manuf.,* Sept/Oct 1995

Metal and Alloy Powders for Welding, Hardfacing, Brazing, and Soldering

Revised by Deepak S. Madan, F. W. Winter Inc. & Co.

METALS AND ALLOY POWDERS are used in welding, hardfacing, brazing, and soldering applications. These applications include powders for hardface coatings, powders for use in the manufacture of welding stick electrodes and flux-cored wires, and additives in brazing pastes or creams. This article briefly reviews these applications and the specific powder properties and characteristics they require.

Powders for Manufacture of Stick Electrode and Flux-Cored Wires

Metal powders used in electrode coatings include a wide range of ferroalloys and pure metals. Some widely used metal powders for this application include chromium, iron, manganese, and nickel. A large variety of ferroalloys are used in the manufacture of electrodes: ferro-aluminum, ferro-boron, high-carbon ferro-chromium, low-carbon ferro-chromium, ferro-manganese, ferro-molybdenum, ferro-niobium, ferro-silicon, ferro-silicon-zirconium, ferro-titanium, ferro-titanium-boron, ferro-vanadium, and ferro-tungsten. In addition, carbide powders like boron carbide and chromium carbide are also used for this application. Further, a number of mineral and flux powders are used in the manufacture of stick electrodes and flux-cored wires; these powders include calcium carbonate, fluorspar, alumina, feldspar, iron oxides, titanium dioxide, and rutile.

Of the metal and alloy powders listed above, iron, chromium, manganese, ferro-chromium, and ferro-manganese powders are all used in fairly large quantities. The iron powder used for welding applications is typically water atomized. The chromium, manganese, ferro-alloys, and carbide powders are manufactured by mechanical size reduction.

Powder flow, particle shape, apparent density, and size distribution are very important characteristics for powders used for stick electrodes and flux-cored wires. In general, excessive fines in these powders are not desirable. In most cases, the chemistry of the material is extremely crucial and has a significant impact on the final weld-metal properties.

Stick Electrodes versus Flux-Cored Wires

For the manufacture of stick electrodes, typically various metal, alloy, and flux powders are combined with a binder to create a coating mixture. The coating is extruded onto steel rods of appropriate composition to create the correct type of electrode. The coating is baked and dried prior to packaging.

For the manufacture of flux-cored wires, however, various metal, alloy, and flux powders are combined to create a filler material. A steel strip of an appropriate composition is drawn into a U-shaped wire. The filler material is next deposited into this U-shaped wire, and the wire is drawn into a cylindrical shape and folded, and its diameter is decreased by a series of drawing operations to create a flux-cored welding wire.

The coating of the stick electrodes and the filler material of the flux-cored wires influence various welding parameters and also the final chemistry of the weld-metal.

Functions of Electrode Coatings or Filler Material

There are five general functions controlled or influenced by electrode coatings or filler material: welding parameters, slag property control, weld deposit quality, metal transfer, and compatibility with extrusion or drawing equipment. Success in fulfilling these functions depends on the proper balance of the constituents of the electrode coating.

Control of Welding Parameters. The electrode coating or filler determines the type of welding current and the arc stability. For example, core wire used for an E6010 direct current electrode can be the same as that used in an E7024 alternating current/direct current electrode, but the coatings are different. Arc voltage and optimum current are functions of the type and formulations of the coating or filler material; therefore, electrodes of the same classification produced by different manufacturers may require minor current adjustments for optimum welding in field applications. To prevent side arcing, the coating should be a good insulator.

Slag Property Control. Control of the many properties of a molten slag is very difficult. Wetting characteristics of the slag help determine the contour of the deposit and the degree of slag coverage. Expansion of the slag markedly affects the slag removal. Fluidity or viscosity of the slag alters the out-of-position characteristics of the finished electrode. Slag freezing point affects bead contour and is a controlling factor in determining the acceptance of the electrode for vertical and overhead positions. All these interrelated properties are controlled by the electrode-coating composition. Only extensive laboratory testing can produce a coating or filler material with the optimum balance of all these properties.

Weld Deposit Quality. The electrode coating or filler material controls the metallurgical structure and quality of the weld deposit. Service requirements demand that every weld deposit must be free from porosity and slag inclusions. The coating must contain a sufficient quantity of deoxidizing material to degas the molten metal and to ensure that all resulting oxide and silicate inclusions float off in the slag. The final weld-metal chemistry and properties are a direct function of the coating or filler material formulation.

Compatibility with Extrusion and Drawing Equipment. Modern high-speed extrusion equipment places rigid demands on electrode coating ingredients. A proper balance of particle size among the various ingredients is essential for ease of extrusion, ability to withstand mechanical damage, and a pore structure that permits immediate and continuous drying of the extruded coating. In the case of flux-cored wires, again, the size distribution of the various ingre-

Table 1 Functions and composition ranges of coating constituents on low-carbon steel arc welding electrodes

Coating constituent	Function of constituent: Primary	Function of constituent: Secondary	E6010, E6011	E6012, E6013	E6020, E6022	E6027	E7014	E7016	E7018, E7048	E7024	E7028
			Composition range in electrode coating of class, %								
Cellulose	Shielding gas	Slag modifier	25–40	2–12	1–5	0–5	2–6	...	...	1–5	...
Calcium carbonate	Shielding gas	Slag modifier	...	0–5	0–5	0–5	0–5	15–30	15–30	0–5	0–5
Fluorspar	Slag modifier	Fluxing agent	...	...	...	...	...	15–30	15–30	...	5–10
Dolomite	Fluxing agent	...	...	...	...	...	...	...	...	...	5–10
Titanium dioxide (rutile)	Slag modifier	Arc stabilizer	10–20	30–50	0–5	0–5	20–35	15–30	...	20–35	10–20
Potassium titanate	Arc stabilizer	Slag former	(a)	(a)	...	...	...	...	0–5	...	0–5
Feldspar	Slag modifier	Stabilizer	...	0–20	5–20	0–5	0–5	0–5	0–5	...	0–5
Mica	Extrusion aid	Stabilizer	...	0–15	0–10	...	0–5	...	0–5	0–5	...
Clay	Extrusion aid	Slag former	...	0–10	0–5	0–5	0–5	...	...	...	...
Silica	Slag modifier	...	...	...	5–20	...	...	...	...	...	...
Manganese oxide	Slag modifier	Alloying	...	...	0–20	0–15	...	...	...	...	...
Iron oxide	Slag modifier	...	...	...	15–45	5–20	...	...	...	...	...
Iron powder	Deposition rate	Contact welding	...	...	...	40–55	25–40	...	25–40	40–55	40–55
Ferrosilicon	Deoxidizer	Alloying	...	...	0–5	0–10	0–5	5–10	5–10	0–5	2–6
Ferromanganese	Alloying	Deoxidizer	5–10	5–10	5–20	5–15	5–10	2–6	2–6	5–10	2–6
Sodium silicate	Binder	Fluxing agent	20–30	5–10	5–15	5–10	0–10	0–5	0–5	0–10	0–5
Potassium silicate	Arc stabilizer	Binder	(a)	5–15(a)	0–5	0–5	5–10	5–10	5–10	0–10	0–5

(a) Used in E6011 and E6013 electrodes to permit welding with alternating current

Table 2 Typical properties of reduced and water-atomized iron welding powders

Property	Powder 1(a)	2(a)	3(a)	4(b)	5(b)
Chemical composition, %					
Iron, total	98.20	98.0	98.0	98.0	98.0
Hydrogen loss	0.40	0.45	0.45	0.50	0.50
Carbon	0.05	0.05	0.05	0.05	0.05
Sulfur	0.01	0.015	0.015	0.025	0.025
Phosphorus	0.01	0.01	0.01	0.01	0.01
Sieve analysis, U.S. standard, %					
+40	...	Trace	Trace	2	2
–40+80	1	54	55	61	62
–80+200	48	44	43	32	33
–200	...	2	2	5	3
–200+325	24	...	...	...	...
–325	27	...	...	...	...
Apparent density, g/cm^3	2.5	2.4	2.8	3.2	3.65

(a) Reduced iron powder. (b) Water-atomized iron powder

dients of the filler material are critical in determining the final diameter of the wire.

Electrode Coating and Filler Material Constituents

The constituents of coating and filler materials are divided into four categories: (a) binders and extrusion aids; (b) arc stabilizers; (c) fluxes and slag modifiers; and (d) alloying materials. Some of the materials might fall into two or more of the above categories if their overall effects are considered. For example, sodium and potassium silicate can be equally effective as binders, but potassium silicate is preferred where additional arc stability is desired. Titanium dioxide and titania-bearing materials frequently are used to improve arc stability. However, such additions can markedly alter the properties of the resultant slag.

Binders and extrusion aids commonly used by the electrode industry include liquid sodium and potassium silicates, celluloses, clays, bentonites, talc, and mica. The alkaline silicates and celluloses provide dry strength, green strength, or assist in controlling the drying characteristics. Other inorganic materials such as talc, mica, clay, or bentonite can be used to increase plasticity, that is, slippage under extrusion pressure.

Arc stabilizers are predominantly titanium- or potassium-bearing materials. Titanium dioxide, rutile, potassium titanate, ilmenite, and potassium carbonate or hydroxide are the most common. Lithium-bearing materials have arc-stabilizing properties, but their cost is somewhat higher.

Fluxes and Slag Modifiers. Almost all of the more common minerals and fluxes used by the ceramics industry have been employed as additions to electrode coating or filler material compositions. In commercial practice, calcium carbonate, fluorspar, silica flour, alumina, feldspar, manganese oxides, iron oxides, wallastonite, and zirconium-bearing materials appear to be the most widely used.

Alloying Materials. A wide variety of metal, ferroalloys, and carbide powders are commonly used in electrode coatings and filler materials. In addition to providing the necessary alloying contents, materials such as ferro-manganese and ferro-silicon also serve as deoxidizers.

As an example, Table 1 gives compositions of electrode coatings for classes of electrodes used in shielded metal arc welding of low-carbon steel and lists the primary and secondary functions of the coating constituents. The thickness of the coatings varies from 10 to 55% of the total diameter of the coated electrode, depending on the type of coating.

Types of Iron Powder for Welding

Iron powder is typically one of the major components for electrode coatings and filler materials. In general, iron powders used for welding applications are of coarser size compared to those used for powder metallurgy applications. The iron powders used for welding applications are produced by a variety of methods. A brief summary of these production methods follows. Detailed information of these processes can be found elsewhere in this Volume.

Reduced iron is the oldest iron powder production method in which pure magnetite (Fe_3O_4) is reduced by a carbonaceous material. Table 2 lists compositions and properties of reduced iron welding-grade powders.

For certain welding applications, especially for small-diameter electrodes, fine grades (–100 mesh) of sponge iron powder having an apparent density of 2.50 g/cm^3 are used. This powder has a large specific surface due to the microporous particles and irregular grain shape. This contributes to increased electrical conductivity, good arc stability, and smooth arc action of the welding electrode.

Coarser sponge iron powder (–40+200 mesh) having an apparent density of 2.40 g/cm^3 are used in the manufacture of certain low-hydrogen electrodes. It is also used in rutile and acid electrodes with low to medium deposition efficiency.

Table 3 Electrode classification of iron powder, titania, and low-hydrogen electrodes

Iron powder content, %		AWS classification	Type of covering	Capable of producing satisfactory welds in positions shown(a)	Type of current(b)
E60 series: minimum tensile strength of deposited metal in as-welded condition, 410 MPa (60 ksi) or higher					
10 max		E6010	High cellulose, sodium	F, V, OH, H	DCEP
10 max		E6011	High cellulose, potassium	F, V, OH, H	ac or DCEP
10 max		E6012	High titania, sodium	F, V, OH, H	ac or DCEN
10 max		E6013	High titania, potassium	F, V, OH, H	ac, DCEN, or DCEP
25		E6020	High iron oxide	H-fillets	ac or DCEN
				F	ac, DCEN, or DCEP
30–45		E6027	Iron powder, iron oxide	H-fillets	ac or DCEN
				F	ac, DCEN, or DCEP
E70 series: minimum tensile strength of deposited metal in as-welded condition, 480 MPa (70 ksi) or higher					
U.S.	European				
18	25	E7014	Iron powder, titania	F, V, OH, H	ac, DCEN, or DCEP
25–35	44	E7018	Iron powder, low hydrogen	F, V, OH, H	ac DCEP
45–50	65	E7024	Iron powder, titania	H-fillets, F	ac, DCEN, or DCEP
50	65	E7028	Iron powder, low hydrogen	H-fillets, F	ac DCEP

(a) Abbreviations indicate these welding positions: F, flat; H, horizontal; H-fillets, horizontal fillets; V, vertical; OH, overhead. (b) DCEN, direct current electrode negative (straight polarity); DCEP, direct current electrode positive (reverse polarity)

Table 4 Classification of low-hydrogen, high-strength electrodes

Iron powder content, %	AWS classification	Type of coverage	Capable of producing satisfactory welds in positions shown(a)	Type of current(b)
E70 series: minimum tensile strength of deposited metal in stress-relieved condition, 480 MPa (70 ksi)				
25–35	E7018-X	Iron powder, low hydrogen	F, V, OH, H	ac or DCEP
35	E7020-X	High iron oxide	H-fillets	ac or DCEN
			F	ac, DCEN, or DCEP
20–45	E7027-X	Iron powder, iron oxide	H-fillets	ac or DCEN
			F	ac, DCEN, or DCEP
E80 series: minimum tensile strength of deposited metal, 550 MPa (80 ksi)				
30	E8018-X	Iron powder, low hydrogen	F, V, OH, H	ac or DCEP
E90 series: minimum tensile strength of deposited metal, 620 MPa (90 ksi)				
30	E9018-X	Iron powder, low hydrogen	F, V, OH, H	ac or DCEP
E100 series: minimum tensile strength of deposited metal, 690 MPa (100 ksi)				
25	E10018-X	Iron powder, low hydrogen	F, V, OH, H	ac or DCEP
E110 series: minimum tensile strength of deposited metal, 760 MPa (110 ksi)				
20	E11018-X	Iron powder, low hydrogen	F, V, OH, H	ac or DCEP
E120 series: minimum tensile strength of deposited metal, 830 MPa (120 ksi)				
15–20	E12018-X	Iron powder, low hydrogen	F, V, OH, H	ac or DCEP

(a) Abbreviations indicate these welding positions: F, flat; H, horizontal; H-fillets, horizontal fillets; V, vertical; OH, overhead. (b) DCEN, direct current electrode negative (straight polarity); DCEP, direct current electrode positive (reverse polarity)

Atomized Iron. Water-atomized iron powder is also used in electrode coatings. Properties and compositions of these powders are given in Table 2. The low carbon and iron oxide content produces a smooth arc and a slag that is easy to remove. This type of powder, which has an apparent density of 3.20 g/cm^3, is designed for production of low-hydrogen and titania-type electrodes using 50% iron powder in the coating.

High-density (≥3.50 g/cm^3) atomized iron powder has relatively smooth particles that help reduce manufacturing costs and facilitate the use of high-depositions-efficiency electrodes. Low-hydrogen electrodes (Type E7028) can be made with coatings containing 65% iron powder. Such electrodes have deposition efficiencies to 225%. This type of iron powder also can be used for E6027, E6024, E7024, and gravity feed electrodes having deposition efficiencies from 190 to 210%. Classifications of coated electrodes are listed in Tables 3 and 4.

In the manufacture of electrodes, the higher-density, relatively smooth particles require less silicate, allow higher extrusion pressure, and have faster extrusion rates, better brushing, and less swelling in baking. The resulting high-strength coating resists chipping and cracking.

Use of high-density iron powder electrodes saves time and cost through high deposition rates, longer lead lengths, better penetration, no undercutting, and slag that is self-removing. Powder characteristics promote smooth arc action and better recovery of metallics, providing uniform physical properties.

Other Metal and Alloy Powders for Welding

Typically, metal and alloy powders other than iron powder, which are used for welding applications, are made via mechanical size-reduction.

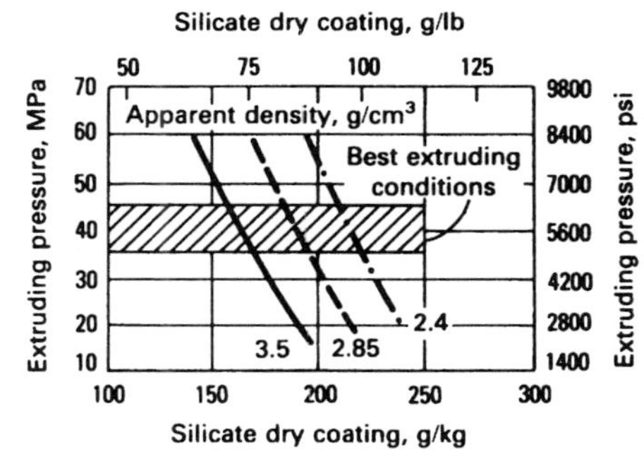

Fig. 1 Effect of silicate content on extrusion pressure

These include chromium, manganese, all ferroalloys, and the carbide powders. In these cases, the starting material is in a lump state, typically 4 in. $\times D$ in size. The starting raw material is crushed using a series of crushers to a size of less than $\frac{1}{4}$ in. Next the material is milled and screened to generate the required size distribution. Powder sizes used in the welding industry include 50 × 325 mesh, 80 × 325 mesh, <100 mesh, <120 mesh. The selection of a particular size is dictated by the welding consumable being manufactured.

Influence of Powder Characteristics on the Manufacture of Welding Electrodes

Currently, welding electrodes are produced by a continuous high-speed production system. Consistent physical properties of electrode coating and filler material components help to attain shorter throughput times. Extrusion and drawing problems tend to multiply as coating thickness and the amount of metal powders added to the coating are increased.

As an example, extrusion pressure is influenced greatly by the physical properties of iron powder and the amount of silicate added to the electrode coatings. Figure 1 shows results from extrusion tests on a rutile high-yield electrode. By using 65% of iron powder having a 3.50 g/cm^3 apparent density at a coating-to-core-wire ratio of 2 to 1 and 170 g silicate/kg (75 g silicate/lb) of dry coating material, a welding efficiency yield of 185% was obtained from this electrode. A pressure range of 35 to 45 MPa (5000 to 6500 psi) provided the best conditions. The ideal extrusion pressure of 39 MPa (5600 psi) was arrived at by testing the following additions of silicate:

Iron powder apparent density, g/cm^3	Amount of silicate added, g
3.50	150
2.85	185
2.40	215

The slopes of the curves in Fig. 1 show that comparatively small changes in quantity of added silicate or in the physical properties of the

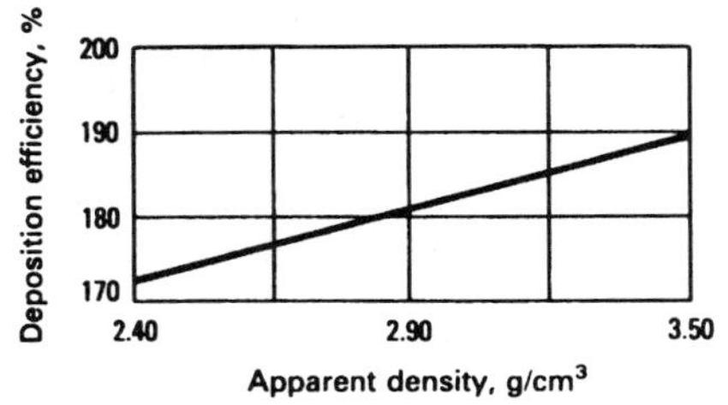

Fig. 2 Deposition efficiency (yield) of a coated electrode as a function of apparent density of iron powder addition. Electrodes extruded at 39 MPa (5600 psi)

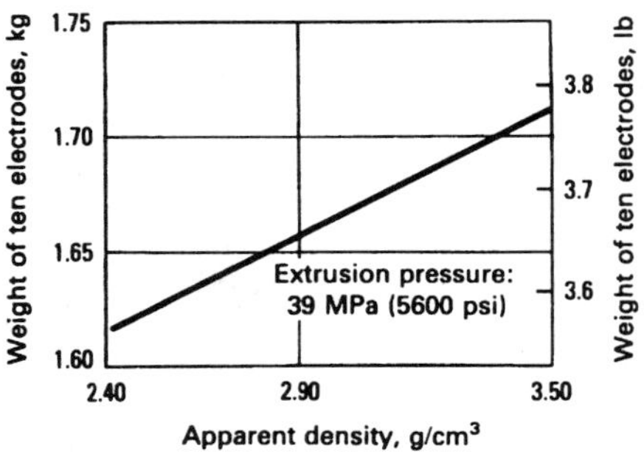

Fig. 3 Weight of a coated electrode (after baking) as a function of the apparent density of the iron powder addition

iron powder make the extrusion pressure fall beyond the best extrusion conditions. Pressures that are too high would cause longitudinal cracks immediately after extrusion. Transversal cracks might appear after baking. Conversely, an extrusion pressure that is too low produces a coating that is too soft. In such a case, the electrode can easily be deformed during handling and transportation.

For those high-yield electrodes investigated, it was found that when silicate addition was unchanged, a powder with smooth surface structure gave a comparatively lower extrusion pressure. When adding large portions of iron powder, rough, irregularly shaped grains might critically augment the internal friction within the coating material, thereby increasing the extrusion pressure. Use of a smoother, heavier powder makes possible the use of similar additions of silicate or other plasticizing ingredients. This is desirable technically and economically, especially in the production of this type of high-yield electrode.

Basic electrodes are somewhat of an exception to this observation. Cracks in the coating can appear after the electrode is baked; they might also be due to the coating composition or the relatively high temperature applied. The coating can be reinforced through large silicate additions, counteracting the tendency to form contraction cracks during baking. It has also been observed that light, fine iron powder allows an increased addition of silicate without rendering the coating too soft after extrusion. With basic electrodes, improved arc stability is desirable. One way to attain this is by increasing the addition of potassium silicate; however, such an in-

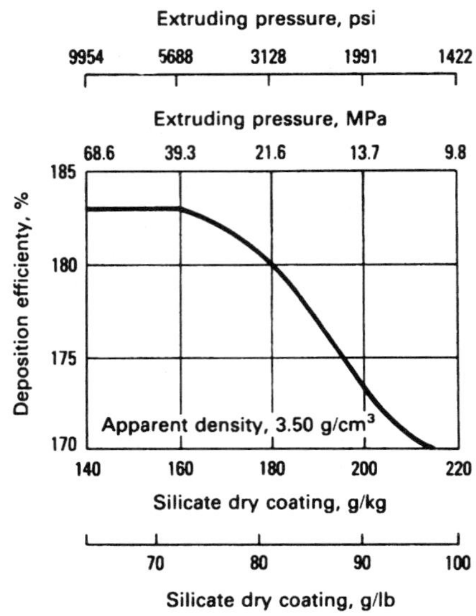

Fig. 4 Effect of silicate content and extrusion pressure on deposition efficiency of coated electodes. Electrodes extruded with high-density (3.5 g/cm³) iron powder

crease can make the coating more hygroscopic, which is a disadvantage. Therefore, when choosing raw materials for welding electrodes, the advantages and disadvantages must be evaluated.

Performance of Powder-Coated Electrodes

The deposition efficiency (yield) of an electrode is calculated as follows:

$$(A \div B) \times 100 = R$$

In this equation, A is the deposited weld metal, B is the weight of melted core wire, and R is the deposition efficiency (yield).

Both the deposition efficiency (yield) and performance of welding electrodes can be influenced by the amount of silicate added and the apparent density of the iron powder used in the coating. The tests on which Fig. 2 to 5 are based were carried out with a high-yield rutile electrode. Normally, such an electrode will have a yield of 180 to 185%, provided that 61% of high-density iron powder and approximately 150 g silicate/kg (70 g silicate/lb) of dry coating are added. This electrode is characterized by very good welding performance when operating on horizontal fillet welds.

Figure 2 illustrates yield as a function of the apparent density of the iron powder used for electrodes extruded at 39 MPa (5600 psi). Approximately 15% lower yield was obtained from the electrode made with iron powder that had an apparent density of 2.4 g/cm^3, as compared to an electrode made with iron powder that had an apparent density of 3.5 g/cm^3.

Figure 3 illustrates data that explain this observation. The actual weight of the electrode after baking varies with the apparent density of the

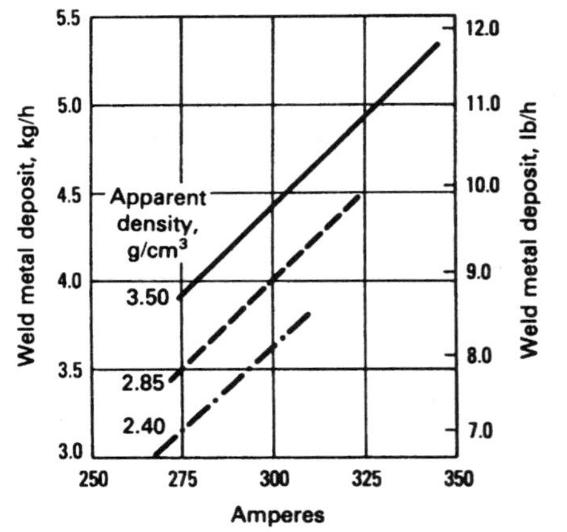

Fig. 5 Amount of weld metal deposited (melting rate) as a function of welding amperage for exrruded coated electrodes. Electrodes contain iron powder of varying densities, as indicated in the figure.

powder. The nature of the curve is the same as that in Fig. 2, indicating the oxidation losses are independent of powder particle shape and apparent density. However, this only applies when the amounts of carbon and iron oxide in the iron powders are fairly equal. When using iron powder with very large amounts of both carbon and iron oxide, the deposition efficiency (yield) can drop considerably because of increased spatter losses.

Figure 4 shows the deposition efficiency (yield) of electrodes extruded with high-density iron powder (3.5 g/cm^3) as varying amounts of silicate were added. The deposition efficiency decreases sharply when the amount of silicate is increased. The correct amount of silicate addition with respect to extruding pressure would be 160 g/kg (72 g/lb) of dry mix. With such an addition, the electrode will give a yield of 175%.

Figure 5 shows the melting rate calculated as weight (kilograms or pounds) of weld deposit per hour as a function of the welding amperage applied on various electrodes. The electrodes were extruded by using iron powder with varying apparent densities. Faster welding speeds can be achieved when using high-density iron powders. The amperage considered maximally applicable in normal welding practice is higher for electrodes made with high-density iron powders; thus, the differences between various iron powders are more pronounced under practical conditions, such as welding of horizontal V-fillet welds or horizontal fillet welds.

The following conclusions can be drawn from the examples in Fig. 2 to 5:

- The amount of silicate added for a specific type of electrode and a specific extrusion pressure depends on the apparent density, particle shape, and sieve analysis of the powder.
- Deposition efficiency and melting rate rise when the silicate content is reduced and the apparent densities of the iron powder are increased; therefore, only high-density iron powder welding grades are recommended for

high-yield electrodes that have deposition efficiencies greater than 160%.

- Narrow specification ranges for physical properties, particle shape, apparent density, and sieve analysis of the iron powder are necessary for reproducible production of welding electrodes.

Metal Powders for Hardfacing

Hardfacing is the application of hard, wear-resistant material to the surface of a component by welding, thermal spraying, or a similar process, mainly to reduce wear. The control of wear is growing in importance because wear adversely affects the performance of equipment and components. Wear can cause inefficient performance of equipment or costly downtime. Hard-facing is used on components of original equipment, as well as for the repair and rebuilding of worn components. Hard-facing filler (consumable) material is available in different forms, such as powder, solid welding rods or wires, and tubular rods or wires.

This section discusses hardfacing processes and applications that directly use powders. The use of metal powders in the fabrication of consumable electrodes for hardfacing has already been discussed in the section above. In addition, applications in which corrosion or oxidation resistance is of prime importance, in addition to surface modification techniques such as flame hardening, nitriding, and ion implantation, are not covered in this article.

Hardfacing of a selected alloy onto a base metal (substrate) provides a protective surface layer that resists wear or loss of material by adhesion, galling, abrasion, impingement erosion, cavitation erosion, or fretting.

Adhesive wear is caused by the sliding action between two metallic components. At some transition load, metallic bonds form between the surface asperities of mating materials and wear becomes severe. The term "galling" is used to describe conditions in which adhesive wear is severe enough to cause seizure of the moving components.

Abrasive wear is caused by the cutting action of sliding abrasives between mating surfaces that suffer scratches or gouges. Impingement erosion refers to the type of material loss that is caused by the cutting action of moving particles. Cavitation erosion is caused by shock waves that result from collapse of air bubbles on the metal surface in turbulent fluid flow. Fretting is a combination of mild adhesive (oxidative) and abrasive wear caused by very small amplitude vibrations at mechanical connections.

In severe abrasive wear service, such as rock crushing and pulverizing, several pounds of material can be rapidly worn away. In control valves, however, a few thousandths of an inch of wear is intolerable. Wear and corrosion are encountered in knives used in the food processing industry and in valves and pumps handling corrosive fluids. Diesel engine valve seats are subjected to severe conditions involving erosion fatigue and hot corrosion. Commercially available hardfacing materials help solve these industrial problems.

The choice of filler material requires careful consideration of service conditions, available equipment, hardfacing processes, equipment and process cost, hardfacing operation, and equipment downtime. Technical assistance from the powder supplier is strongly recommended, as is the testing of samples under operating conditions.

Many of these material systems are used as powders because of their specific advantages. A wide range of compositions is possible through powder processing; some of these cannot be made by conventional techniques. Powder use allows uniformity of composition and microstructural features. Powder processes minimize heat input to the base metal and hence dilution of the deposit. Furthermore, in many applications, powder techniques provide more effective coating of intricate surfaces.

General Powder Characteristics

Gas and water atomization are the most widely used methods for producing hardfacing powders. However, for a growing number of specialized applications, powders are made by milling and crushing operations. Some examples of powders made by crushing and milling include tungsten carbide powders, ferro-chromium powders, chromium carbide powders, some grades of nickel-base powders, and other friable materials. These methods generate broad ranges of particle size and size distribution; therefore, screening (sieving) and air separation are used to create specific size distributions for special grades of powder.

Figure 6 shows typical particle size ranges of different grades of powder used for various hardfacing processes. Approximately 85 wt% of any grade of powder falls within the ranges shown. For the same process, however, equipment made by different manufacturers can require different particle size ranges.

As an example, the plasma spray operation requires fine powders in the 10–45 μm size range because of the high jet speed and short dwell time of the particles in the heating zone (even though the zone temperatures are high). However, special dispensing devices are required to prevent powder flow problems. In some applications, like plasma transferred arc (size range, 45–150 μm), use of excessive fines is damaging to the deposit because fines tend to have higher surface oxides and burn-off tendencies, causing smoking. On the other hand, certain hardfacing techniques, such as bulk welding processes, require much coarse powder in the 100–600 μm range.

Type of application also governs powder size requirements. For instance, thin overlays typically use fine powders. Therefore, the user must consider the process, equipment, and final application before selecting powder size. Powder shape and cleanliness are also important considerations in any hardfacing operation. For most applications, the deposit must be nearly free of porosity or entrapped fluxes. The use of spherical powders is beneficial to the service and integrity of the coatings. Spherical powder flows better than angular powder because of minimal interparticle friction; this is important when using gravity powder-feed devices.

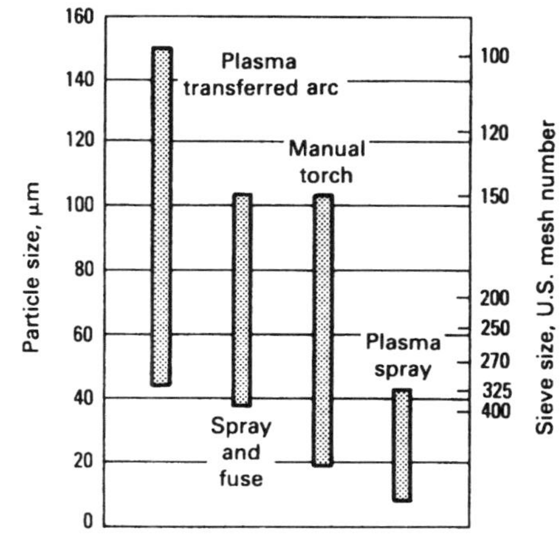

Fig. 6 Powder size ranges for different hardfacing processes. At least 85% of the powder by weight is in the size range shown.

Larger atomized particles sometimes have smaller particles, or satellites, attached. If excessive, these satellites can cause flow problems; if loosely attached, they can break up during spraying, resulting in oxidation and smoking. Gas-atomized powders generally are spherical in shape and prone to satellite formation. Water-atomized powders can be irregular or spherical in shape, depending on the alloy used and the quenching rate. Surface oxidation is a potential problem with water-atomized powder (particularly in cobalt-base alloys). Self-fluxing nickel-base powders are routinely made by this process, because oxygen pickup is minimal and the process is more economical.

Apparent density and flow rate also depend on the characteristics mentioned above. Irregular particle shape, numerous satellites, or a narrow size distribution lowers apparent density and slows the flow rate of the powder. Equipment with gravity powder-feed devices requires powders that flow faster. In any specific application, compatibility between the equipment and operating conditions, in addition to reproducibility of these powder properties, is a consideration.

Deposit hardness depends on the powder composition, as well as on the conditions used for depositing. Hardness range frequently is reported in relation to powder characteristics. The bulk hardness of a deposit, as measured on the Rockwell C scale, for example, depends on the deposit matrix hardness; it depends as well on the type and quantity of hard constituents, such as carbides and borides, or phases, such as martensite or Laves phase intermetallics.

Table 5 Selection guidelines for hardfacing alloy systems

Surfacing material type	Composition, %	Hardness		Maximum service temperature		Resistance to wear(a)	
		Hot	Cold	°C	°F	Abrasion	Impact
Iron-base, low alloy	0.3–1.0 C, 1.0–7.0 W, 3.0–5.0 Cr, 4.0–9.0 Mo, bal Fe	47 HRC at 595 °C (1100 °F) 30 HRC at 650 °C (1200 °F)	55–60 HRC (30 HRC, annealed)	595	1110	Superior to low-carbon steel	Moderate; increases with temperature
Iron-manganese	0.5–0.9 C, 11.0–16.0 Mn, 2.75–6.0 Ni, or 0.5–1.5 Mo, bal Fe	...	170–230 HB (workhardens to 450–550 HB)	260–315	500–600	Moderate	Excellent
Iron-chromium	3.0–5.0 C, 4.0–8.0 Mn, 26.0–32.0 Cr, 2.0 Mo, bal Fe	47 HRC at 480 °C (900 °F) 5 HRC at 650 °C (1200 °F)	51–62 HRC	425–480(b)	800–900(b)	Excellent for low-stress abrasion	Good for light impact
Cobalt-chromium	0.7–3.0 C, 2 Mn, 3–14 W, 3 Ni, 25–33 Co, 5 Fe, bal Cr	20–39 HRC at 650 °C (1200 °F)	23–58 HRC	980	1800	Excellent in low-stress; wears 5–20% as fast as carbon steel	Fair; deteriorates as carbon increases
Nickel-chromium	0.3–1.0 C, 1.0–1.5 Co, 8.0–18.0 Cr, 1.25–5.5 Fe, 2.0–4.5 B, 1.25–5.5 Si, bal Ni	19–42 HRC at 540 °C (1000 °F)	24–62 HRC	955	1750	Excellent in low-stress; increases with carbon content	Good for light impact; decreases as carbon increases
Copper-base	0.25–5.0 Fe, 0.01–15.0 Al, 0.02 Zn, 0.04–4.0 Si, 9.0 Sn, bal Cu	...	70–320 HB	205	400	Poor	CuAl: good, decreases with increased Al content; CuSi: good; CuSn: poor; CuZn: very poor
Tungsten-carbide	60% carbide granules, 40% steel tube	41–47 HRC at 650 °C (1200 °F)	90–95 HRA (granules) 30–60 HRC (bonding matrix)	540	1000	Excellent	Good under light impact

(continued)

(a) Ratings are relative between material types. (b) Softens but retains other properties up to 908 °C (1800 °F)

Hardness is also influenced by the degree of dilution. For a specific base metal and powder alloy combination, hardness normally decreases with increased dilution of the deposit by the base metal. Dilution of 5% means that the deposit contains 5% base metal and 95% hardfacing powder alloy. Dilution increases as the base metal receives more heat input and more of base metal melts. The plasma transferred arc process provides more dilution than the manual torch process, whereas plasma spray provides essentially no dilution. More dilution can make the deposit tougher and more resistant to cracking.

In practice, hardfacing processes and techniques should be selected to control dilution to less that 20%. In a multilayer deposit, the fist layer experiences the largest degree of dilution and is the softest. Generally, deposit hardness can be used to verify powder quality only if the testing method is rigidly standardized; even then, hardness may differ from the hardness of the deposit on the components to be hardfaced.

Powder Composition

Chemical composition of hardfacing powders is the most important influence on the properties of the deposit. The main criteria in selecting an alloy for a given hardfacing application are its response to the type of wear encountered, the base metal to be hardfaced, and material cost. Other influential properties are impact strength, corrosion, oxidation, and thermal behavior.

Generally, impact resistance of hardfacing alloys decreases as the carbide content increases. Consequently, in applications in which a combination of impact and abrasion resistance is desired, a compromise between the two must be achieved. In applications in which impact resistance is important, austenitic manganese steels are used to build up worn parts.

Frequently, wear is accompanied by aqueous corrosion from acids and alkalis, as in the chemical processing and petroleum industries or in flue gas scrubbers. Few iron-base hardfacing alloys possess the necessary corrosion resistance in such aqueous mediums. As a result, nickel- or cobalt-base surfacing alloys generally are recommended when a combination of corrosion resistance and wear resistance is required. For example, knives used to cut tomatoes in a food processing plant will last many times longer if their edges are made of a cobalt-base alloy rather than tool steel.

Oxidation and hot corrosion resistance of iron-base alloys is also generally poor. Typically, boride-containing nickel-base alloys do not contain sufficient chromium in the matrix to resist oxidation. Hence, Laves phase or carbide-containing nickel- or cobalt-base alloys typically are recommended for applications in which wear resistance combined with oxidation or hot corrosion resistance is required.

The ability of an alloy to retain strength at elevated temperatures is important for wear applications such as hot forging dies or valves for service at 870 °C (1600 °F), as well as service in coal gasification or liquefaction applications. Iron-base alloys with martensitic structures lose hardness at elevated temperatures. Generally, the high-temperature strength retention of a hardfacing alloy increases with its tungsten or molybdenum content. In applications requiring elevated-temperature strength and wear resistance, cobalt-base alloys or Laves phase alloys are recommended.

Wear-resistant materials typically are classified by type and alloy content, which largely determine their properties. These materials have structures consisting of hard phases (carbides, borides, or Laves phases) in a matrix that is hardened to a level that depends on the desired alloy properties. Table 5 provides a comprehensive guide to selection of surfacing alloy systems. Powder alloys can be custom made for various applications by using the same composition, but varying the sieve size or distribution. Because American Welding Society (AWS) designations do not exist for many of the hardfacing alloys discussed in this article, tradenames will be used. It should be noted that many of these compositions are available from various manufacturers under different names.

Cobalt-Base Powders

The two types of commercially available cobalt-base hardfacing alloys are carbide-containing alloys and alloys containing Laves phase, which are described later in this article. Carbide-containing cobalt-base alloys have been widely used since the early 1900s, when a cobalt-base alloy with the nominal composition of Co-28Cr-4W-1.1C was first developed. Typically, this alloy, referred to in the hardfacing industry as Alloy No. 6, is a eutectic alloy containing 16 to 17 vol% M_7C_3-type carbides in a cobalt-base matrix alloyed with chromium and tungsten.

The lightest colored phase in the microstructure is the matrix, which has a face-centered cubic (fcc) crystal structure. The dark phase is M_7C_3; the exact composition is $(Cr_{0.85}, Co_{0.14}, W_{0.01})_7C_3$, which has a hexagonal crystal structure. The carbide phase in Alloy No. 6 forms between matrix dendrites. These carbides are

Table 5 (continued)

Surfacing material type	Resistance to wear(a)		Mechanical properties	Machinability	Microstructure	Heat treatment	Applications
	Corrosion	Adhesion					
Iron-base, low alloy	Fair to poor	Good	Good compressive strength; takes a high polish for low coefficient of friction	Good; anneal, machine, then retreat to harden	Martensite, austenite, carbides	Surfacing: preheat base metal to 150 °C (300 °F)	Where hardness and toughness are required to 595 °C (1100 °F); cutting tools, shear blades, reamers, dies, ingot tongs
Iron-manganese	Low (similar to carbon steel)	Excellent	Low yield point deformation	Poor; grind if required	Austenite saturated with carbon; iron-manganese carbides at grain boundaries in thin, flat, brittle plates	Heat to reaustenitize; water quench	High-impact service; rail tracks, frogs, switches; use on carbon or manganese steel base
Iron-chromium	Resists oxidation except in moist air; poor in liquid service	Excellent	Low coefficient of friction; high strength	Difficult to grind	Hard carbides in austenite matrix	None	Agricultural machinery, coke shutes, brick-making equipment
Cobalt-chromium	Excellent for salt, steam, and some acids up to 980 °C (1800 °F)	Excellent for low-stress; poor for high-stress; increases with carbon content	Excellent metal-to-metal; takes a high polish, low coefficient of friction	Poor	Chromium, tungsten, cobalt carbides	None	High-temperature, high-corrosion applications; pump shafts, engine valves
Nickel-chromium	Excellent for salt, steam, some acids and alkalis	Excellent for low-stress	Excellent metal-to metal wear	Poor; finish by grinding	Chromium carbides, chromium borides, in eutectic matrix	Preheat to 315–425 °C (600–800 °F) to prevent cracking; slow cool	Seal rings, cement pump screws, valves, cams
Copper-base	Oxide coating, good protection from atmosphere; good for many acids, bases, and saltwater	Good	Elastic limit 35–1170 MPa (5–170 ksi)	Good	See composition	None	Gears, sheaves, cams, bearings, wear plates, dies for forming stainless steel, carbon steel, aluminum
Tungsten-carbide	Similar to steel; low resistance in oxidizing mediums	Poor	Yield same as high carbon steel	Unmachinable; finish with diamond grinding wheel	W and WC granules in a ferritic matrix	Carburizing and hardening may harden base metal	Cutting teeth and edge holding surfaces of rock drill bits; quarrying, digging, earthmoving equipment

(a) Ratings are relative between material types. (b) Softens but retains other properties up to 908 °C (1800 °F)

Table 6 Chemical compositions and hardnesses of typical cobalt- and nickel-base hardfacing alloys

Alloy	Composition, wt%										Hardness	
	B	C	Co	Cr	Fe	Mo	Mn	Ni	Si	W	HRC	Diamond pyramid
Cobalt-base alloys												
Alloy No. 1	...	2.5	bal	30	3 max	1 max	1 max	3 max	1	12.5	51–58	550–685
Alloy No. 6	...	1.1	bal	28	3 max	1 max	1 max	3 max	1	4	39–43	380–425
Alloy No. 12	...	1.4	bal	29	3 max	1 max	1 max	3 max	1.4	8	47–51	480–550
Alloy No. 21	...	0.25	bal	27	2 max	5.5	1 max	2.8	2 max	...	28–32	285–320
Alloy No. SF1(a)	2.2	1.3	bal	19	3	...	1 max	13	3	13	54–58	600–685
Alloy No. SF6(a)	1.7	0.7	bal	19	3	...	1 max	13.5	2.3	7.5	43–46	425–460
Alloy No. SF12(a)	1.8	1	bal	19	3	...	1 max	13	2.8	9	48–50	490–540
Alloy No. SF20(a)	3	1.3	bal	19	2 max	...	1 max	13	2.8	15	60–65	740–900
Nickel-base alloys												
Alloy No. 22	1.5	0.2 max	...	...	1 max	...	...	bal	2.8	...	19–24	200–250
Alloy No. 35	1.6	0.3	...	3.7	1.2	...	...	bal	3.1	...	32–37	320–365
Alloy No. 40	1.7	0.35	...	7.5	1.5	...	...	bal	3.5	...	35–42	350–415
Alloy No. 50	2.4	0.45	...	11	3	...	...	bal	4	...	49–52	525–575
Alloy No. 60	3.5	0.8	...	15.5	4	...	...	bal	4.3	...	59–62	710–790

(a) Spray-and-fuse powders. Source: Cabot Corp.

often referred to as eutectic carbides. Generally, resistance to abrasive wear is imparted by the presence of carbides, whereas resistance to corrosion and/or elevated-temperature hardness retention is imparted by the matrix. Galling and metal-to-metal wear properties are similarly determined by the matrix alloy. The hardness of Alloy 6, which is designated by AWS as RCoCrA, is 35 to 43 HRC.

Nominal compositions and average deposit hardness data for a number of alloys in the Co-Cr-W-C system are given in Table 6. These alloys retain high hardness at elevated temperatures primarily from carbide and solid-solution strengthening rather than precipitation hardening. They contain up to 30% Cr, up to 15% W, and up to 3.0% C. Some of these alloys also contain combinations of nickel, silicon, molybdenum, boron, and manganese.

Alloys No. 1 (AWS RCoCrC) and 12 (AWS RCoCrB) are higher hardness modifications of Alloy No. 6 because of additional carbon and tungsten contents. For example, Alloy No. 1 contains primary M_7C_3, eutectic M_7C_3 plus matrix, and secondary M_6C. Alloy No. 12 is similar to Alloy No. 6, except it has more M_6C and, therefore, contains interdendritic eutectic carbide (tungsten and chromium rich). As shown in Table 7, the weight percent of carbides decreases, as does abrasion/erosion resistance, when comparing these alloys. Specific abrasive and adhesive wear, hot hardness, and corrosion rates for some of these alloy designations are given in Tables 8 and 9.

Cobalt-base alloys, when used in overlay applications, are prone to cracking during cooling.

Table 7 Carbide percentage and hardness of cobalt-base alloys

Alloy	Carbides, wt%	Hardness, HRC
Alloy No. 1	28–29	48–55
Alloy No. 12	16–17	43–50
Alloy No. 6	12–13	35–43
Alloy No. 21	<5	28–32

As the weight percent of carbides decreases, so does abrasion/erosion resistance. Source: Cabot Corp.

Table 8 Wear data for cobalt- and nickel-base alloys

Alloy	Average hot hardness (DHP), kg/mm^2			Adhesive wear volume loss(a), mm^3		Average abrasive wear(b)		Galling test(d) (self mated) at 4080 kg (9000 lb), degree of damage, mm^3
	At 540 °C (1000 °F)	At 650 °C (1200 °F)	At 760 °C (1400 °F)	At 68.2 kg (150 lb) load	At 95.5 kg (210 lb) load	Volume loss, mm^3	Wear coefficient(c)	
Cobalt-base alloys								
Alloy No. 1	465	390	230	0.6	0.7	46	15×10^{-4}	0.63
Alloy No. 6	275	260	185	2.6	9.5	66	13×10^{-4}	1.88
Alloy No. 12	325	285	245	2.4	7.6	53	11×10^{-4}	...
Alloy No. 21	145	135	115	5.2	10.3	70	8×10^{-4}	No damage
Nickel-base alloys								
Alloy No. 60	440	250	115	0.2	0.2	11	3×10^{-4}	13.7
Alloy No. 50	...	...	...	0.3	0.3	12	3×10^{-4}	...
Alloy No. 40	270	195	80	0.19	0.22	21	5×10^{-4}	...

(a) Average of two or more tests in air against a case hardened SAE 4520 steel ring (63 HRC). (b) Tested for 2000 rpm at a load of 13.6 kg (30 lb) using a 230 mm (9 in.) diam rubber wheel and dry sand. (c) Wear coefficient (K) is calculated from $V = KPL/h$, where V = wear volume in mm^3, P = load in kg; L = sliding distance in mm; and h = diamond pyramid hardness. (d) Pin on block test; reciprocating motion through 120°. Surface is analyzed via profilometry. Source: Cabot Corp.

Typically, this problem can be overcome by preheating the base metal prior to application. These alloys are deposited readily by plasma transferred arc, which is a low-dilution, high-deposition-rate process. Cobalt-base hardfacing powders are unsurpassed for use in combined wear-resistant, corrosion-resistant, and high-temperature applications. Cost of these materials is affected by the prevailing price of cobalt and tungsten and tends to be higher than nickel- and iron-base alloy powders. Several alloys with lower cobalt content, or no cobalt, have been developed as low-cost replacements for cobalt-base grades. However, these substitutes have not found broad commercial acceptance.

Nickel-Base Powders

Commercially available nickel-base hardfacing alloys can be divided into three groups: boride-containing alloys, carbide-containing alloys, and Laves phase-containing alloys, which are described later in this article. Based on the Ni-Cr-B-Si system, these alloys are finding increased usage for wear-resistant surfacing applications, particularly in oil pumping and in glass, ceramic, cement, plastic, and steel production. The oil pumping and glass industries are the major users of nickel-base powders.

Additions of boron and silicon to nickel lower the melting point so that an oxyacetylene torch can be easily used. Nickel and boron form a Ni-Ni_3B eutectic at 1092 °C (1998 °F). Silicon additions further lower the eutectic. However, excessive additions of boron and silicon lower plasticity. Because silicon forms a substitutional solid solution with nickel, and because the solubility of boron in nickel is small (0.1 wt%), the abrasion resistance of these alloys is a function of the amount of hard borides present. Most of these alloys contain nickel boride (Ni_3B), chromium borides (CrB, Cr_5B, and Cr_2B), and other complex borides as hard phases.

Typical alloy compositions are listed in Table 6. Deposit hardness in these alloys is as high as 60 HRC, depending on the chromium, boron, and silicon contents. Usually, the boron content varies from 1.5 to 3.5%, while chromium varies from 0 to 17%. Alloys containing large amounts of boron, such as Ni-14Cr-4Si-3.4B-0.75C, are extremely abrasion resistant, but have poor impact toughness.

The Ni-Cr-B-Si alloys are self fluxing. The low-melting eutectic combines with the surface

Table 9 Corrosion rates for typical cobalt- and nickel-base alloys

Alloy	Corrosion rates for: 30% formic acid at 66 °C (150 °F)	30% boiling acetic acid	5% sulfuric acid at 66 °C (150 °F)	65% nitric acid at 66 °C (150 °F)
Cobalt-base alloys				
Alloy No. 1	E	G	S	G
Alloy No. 6	E	E	E	U
Alloy No. 12	E	G	E	E
Alloy No. 21	E	E	E	E
Nickel-base alloys				
Alloy No. 60	G	U	U	U
Alloy No. 50	S	U	U	U
Alloy No. 40	S	U	U	U

Five 24 h test periods; determined in laboratory tests; samples should be tested under actual plant conditions. E (excellent) 0.13 mm/yr (5 mils/yr); G (good) 0.13 to 0.51 mm/yr (5 to 20 mils/yr); S (satisfactory) greater than 0.51 to 1.27 mm/yr (20 to 50 mils/yr); U (unsatisfactory) greater than 1.27 mm/yr (50 mils/yr). Source: Cabot Corp.

oxides on powders to form borosilicates, which promote wetting of the substrate. Some manufacturers add copper and molybdenum to improve corrosion and pitting resistance. However, hot tearing has been reported due to copper additions. These alloy powders have good abrasive and metal-to-metal (adhesive) wear resistance, although hot hardness and corrosion resistance are slightly inferior to those of cobalt-base alloys (Tables 8, 9). The most common use of Ni-Cr-B-Si powders is in spray-and-fuse and manual torch applications in the glass industry. They are also used in the plasma transferred arc process.

The use of carbide-containing nickel-base alloys has been limited. The most popular and widely used alloys in this group are included in the Ni-Cr-Mo-C system. A typical alloy in this system is Hastelloy Alloy C, which has a nominal composition of Ni-17Cr-17Mo-0.12C and a hardness level of 95 HRB. This alloy, which has good inherent corrosion resistance, is normally deposited by the plasma spray process.

Carbide-containing nickel-base alloys of the Ni-Cr-Mo-Co-Fe-W-C system are gaining popularity as low-cost alternatives to cobalt-base alloys. A typical alloy in this system is Haynes Alloy 716, which has a nominal composition of Ni-11Co-26Cr-29Fe-3.5W-3Mo-1.1C-0.5B and a hardness of 32 HRC. These alloys, depending on precise composition, contain M_7C_3- or M_6C-type carbides. They are normally deposited by the plasma transferred arc process.

Table 10 Approximate hardness of carbides and select materials

Material	Hardness DPH	HK	Mohs
Diamond	...	8000	10
SiC	3200	2750	9.2
W_2C	3000	2550	+9
VC	2800		+9
TiC	2800	2750	+9
Cr_3C_2	2700		
Alumina	...	2100	9
WC	2400	1980	+9
Cr_7C_3	2100	...	...
$Cr_{23}C_6$	1650	...	...
Mo_2C	1570	...	8
Zircon	...	1340	...
Fe_3C	1300	Cementite	...
Quartz	1000	800	7
Lime	...	560	...
Glass	...	500–600	...

Carbides

Carbide hardfacing powders are used for severe abrasion and cutting applications. Tungsten carbide is the most widely used carbide in these types of powders. Comminuted tungsten carbide usually has a mixture of WC and W_2C carbides. (Tungsten carbide/cobalt powder contains WC and M_6C carbides.) Fine carbides are used when smooth surfaces and improved impact resistance are desired.

Larger particles yield a rougher surface and improved abrasion resistance, but lower toughness, which often results in fracture and subsequent loss of carbides. Some commercial grades also contain percentages of cobalt and nickel to improve aqueous corrosion or high-temperature oxidation resistance of the deposit. However, oxidation of tungsten carbide particles limits the upper service temperature to 600 °C (1100 °F). These powders are used primarily in plasma spray and manual torch processes, with small amounts used in plasma transferred arc and spray-and-fuse powders. Table 10 lists the hardnesses of various carbides and select materials.

Iron-Base Powders

Iron-base hardfacing alloys are more widely used than cobalt- and/or nickel-base hardfacing alloys and constitute the largest volume used of hardfacing alloys. Although iron-base powders are most frequently used as a filler material in consumable electrode hardfacing processes, they are also used in hardfacing techniques that use powder as the consumable. Iron-base hardfacing alloys offer low cost and a broad range of desirable properties.

Most equipment that undergoes severe wear, such as crushing and grinding equipment and earth-moving equipment, is usually very large, rugged, and often subject to contamination. Parts subjected to wear usually require downtime for repair. For this reason, they are usually hardfaced with the lowest cost and most readily available materials. As a result, literally hundreds of iron-base hardfacing alloys are in use today.

Due to the large number of alloys involved, iron-base hardfacing alloys are best classified by their suitability for different types of wear and their general microstructures rather than by chemical composition. Most iron-base hardfacing alloys can be divided into pearlitic steels, austenitic steels, martensitic steels, or high-alloy irons. Properties of several typical iron-base hardfacing alloys of major types are listed in Table 11 as a general guide for selection of hardfacing materials.

Table 11 Composition and hardness of select iron-base hardfacing alloys

Nominal composition	Nominal hardness, DPH	Nominal hardness, HRC	Unlubricated sliding wear volume(a), mm^3	Abrasive wear volume(b), mm^3	Density, g/cm^3
Pearlitic steels					
Fe-2Cr-1Mn-0.2C	318	32	0.5	55	7.75
Fe-1.7Cr-1.8Mn-0.1C	372	38	0.6	67	7.47
Austenitic steels					
Fe-14Mn-2Ni-2.5Cr-0.6C	188 HRB	88 HRB	0.4	86	7.75
Fe-15Cr-15Mn-1.5Ni-0.2C	230	18	0.3	113	7.75
Martensitic steels					
Fe-5.4Cr-3Mn-0.4C	544	52	0.4	54	7.47
Fe-12Cr-2Mn-0.3C	577	54	0.3	60	7.47
High-alloy irons					
Fe-16Cr-4C	595	55	0.3	13	7.47
Fe-30Cr-4.6C	560	53	0.2	15	7.19
Fe-36Cr-5.7C	633	57	0.1	12	7.47

(a) Wear measured from tests conducted on Dow-Corning LFW-1 against 4620 steel ring at 80 rpm for 2000 revolutions varying the applied loads. (b) Wear measured from dry sand rubber wheel abrasion tests. Tested for 2000 revolutions at a load of 14 kg (30 lb) using a 23 cm (9 in.) diam rubber wheel and AFS test sand

Pearlitic steels are essentially low-carbon steels with minor adjustments in composition to achieve weldability. These alloys contain low carbon (0.25%) and low amounts of other alloying elements, resulting in a pearlitic structure. Pearlitic steels are useful as buildup overlays, used primarily to rebuild machinery parts back to size. Examples include shafts, rollers, and other parts in heavy machinery subjected to rolling, sliding, or impact loading. Typically this group of alloys has high impact resistance and low hardness (25 to 35 HRC), as well as excellent weldability.

Austenitic Steels. Austenite in this group of alloys is usually stabilized by manganese additions. Austenitic iron-base hardfacing alloys essentially are modeled after Hadfield steels. Most commercially available alloys in this category can be broadly subdivided into low-chromium and high-chromium alloys.

Low-chromium alloys usually contain up to 4% Cr, 12 to 15% Mn, and some nickel or molybdenum. Low-chromium austenitic steels generally are used to rebuild machinery parts subjected to high impact, such as impact crushers or shovel lips. Low-chromium austenitic steels are not recommended for joining manganese steel parts due to the possibility of cracking. Welding to plain carbon steels should be carefully executed. Martensite sometimes forms in zones that are low in manganese content, thereby embrittling the interdiffusion zone.

High-chromium austenitic steels, which normally contain 12 to 17% Cr in addition to about 15% Mn, were developed to preclude diffusion zone embrittlement. Furthermore, the as-deposited hardness of high-chromium steels is higher (16 to 20 HRC) than that of low-chromium austenitic steels (86 to 88 HRB). High-chromium austenitic steels typically are used for rebuilding manganese steel and carbon steel parts subjected to high metal-to-metal pounding, such as railroad frogs and steel mill wobblers, as well as for joining manganese steels.

Martensitic Steels. Alloys in this category are designed to form martensite on normal air cooling of the weld deposit. As a result, these steels are often termed "self-hardening" or "air hardening," and they resemble tool steels with hardness in the range of 40 to 45 HRC. The carbon content of the martensitic steels ranges up to 0.5%. Other elements such as molybdenum, tungsten, nickel (up to 3%), and chromium (up to 15%) are added to increase hardenability and strength and to promote martensite formation. Manganese and silicone usually are added to aid weldability. The major hardfacing applications for martensitic steels include unlubricated metal-to-metal rolling or sliding parts such as undercarriage parts of tractors. The impact resistance of martensitic steels is inferior to that of pearlitic or austenitic alloys, but there is a compensating increase in hardness and resistance to abrasive wear.

High-Alloy Irons. This group of alloys is referred to as irons because of their similarities to cast irons. These alloys contain large amounts of chromium and/or molybdenum carbides in what is essentially a martensitic matrix. The carbon content ranges from 2 to 6%, providing large amounts of carbides. Accordingly, resistance to abrasion is improved considerably over the low-carbon iron-base alloys. Impact resistance and toughness of these alloys, however, are correspondingly lower. In certain alloys and for certain applications, carbon contents of the powders can be as high as 9% with chromium contents around 68%.

High-alloy iron hardfacing deposits usually are limited to one or two passes in thickness due to an extremely high susceptibility to form tension cracks. In some alloys containing 4 to 8% Mn or up to 5% Ni, the austenitic phase is stabilized in the matrix, which reduces the tendency to crack on cooling. Most high-alloy irons have extremely high compressive strength, but very low tensile strength. The hardness range for high-alloy irons is typically 52 to 62 HRC.

Special Compositions

Each manufacturer produces many different powder compositions, some of which are unique. Several examples of powders with unique properties are discussed below.

Laves Phase Alloys. Laves phase is a type of topologically close-packed intermetallic compound. Historically, the presence of Laves phase has been avoided in most alloys due to its detrimental effect on mechanical properties. However, in the early 1960s, the usefulness of Laves phase in resisting metal-to-metal wear was discovered; subsequently, alloys containing Laves phase have become commercially available.

The deposits have smooth surfaces that can be finished to a 2 μm (78 μin.) surface finish. Alloys are particularly well suited to applications in which adequate lubrication is a problem. Laves phase materials maintain mechanical properties over a wide range of temperatures. Their main uses are for plasma spray hardfacing operations such as plasma spray coating of gas turbine blades.

There are two different Laves phase-containing cobalt-base alloys commercially available for hardfacing applications: Co-28Mo-8Cr-2Si and Co-28Mo-17Cr-3Si. These alloys contain at least 50 vol% of Laves phase bound in a cobalt-base matrix that is alloyed with chromium and molybdenum.

Laves phase has a hexagonal structure similar to M_7C_3 carbides, but a hardness value between 1000 and 1200 DPH, which is less than that of carbides. Consequently, the Laves phase-containing alloys are less abrasive to mating materials than carbide-containing alloys in metal-to-metal wear situations.

Only one Laves phase-containing nickel-base alloy is commercially available: Ni-32Mo-13Cr-3Si. This alloy, like most nickel-base alloys, is difficult to weld using the oxyacetylene process, but can be readily welded using gas tungsten arc welding or the plasma transferred arc process. It can also be applied using the plasma or detonation gun spray technique. Although it has excellent metal-to-metal wear resistance and moderate abrasive wear resistance, it exhibits poor impact resistance.

Nickel-Aluminum Powders. Protective coatings based on nickel-aluminum composite powders are deposited by plasma spray techniques. Such composite powders are inherently heterogeneous, in that each particle consists of a nickel core and an aluminum coating. The main advantage of using nickel-aluminum powders is their superior bond strength, which results from exothermic reactions that occur within or on the surface of the powders as they pass through the plasma spray flame. Some diffusion also results in the substrate (in the absence of heat), further improving the overall coating bond strength. These composite powders are also referred to as self-bonding, one-step, or exothermic hardfacing powders.

Ceramics are being used increasingly as hardfacing powders. In addition to resistance to wear, they can have other desirable properties such as insulation and chemical resistance. The processes used for depositing ceramics include plasma spray and detonation gun techniques. The material can be a singe oxide, mixed oxides, or other type of ceramic. Examples of some materials in this category are alumina (Al_2O_3) and chromium oxide (Cr_2O_3). Ceramic coatings have high hardnesses, ranging up to 1200 DPH, and very little porosity. In addition, these coatings can be ground and lapped to obtain a very smooth surface finish.

High-Temperature Powders. A relatively new type of hardfacing powder is that based on the (Fe, Ni, Co)-Cr-Al-Y system. These yttrium-containing powders are used for selective coating of gas turbine blades by the plasma spray process. The coating is highly resistant to high-temperature oxidation and corrosion and is claimed to be superior to coatings made by the pack cementation, diffusion coating technique.

Hardfacing Processes

For satisfactory performance of the coating, selection of the proper hardfacing process is equally as important as selection of the proper hardfacing material. Process selection requires careful consideration of many factors, such as required coating thickness, metallurgical integrity of the coating, size and shape of the component to be hardfaced, compatibility with the base metal, welder skill, cost of filler material, and equipment downtime.

The term "hardfacing" is sometimes used to describe true welding processes, in which both the filler material and a small amount of the base metal are molten and bonded metallurgically. Such processes are preferred for applications that require dense and relatively thick coatings with excellent bonding between the hardfacing material and the base metal. Thermal spray processes are used for applications that require a thin, hard coating applied with minimum work distortion and where a purely mechanical bond between the hardfacing material and the base metal is acceptable. Unless otherwise stated, the term "hardfacing" is used in this section to include both the welding and thermal spray types of processes.

As discussed earlier, the properties of the hardfacing material are affected by dilution, which in turn varies with the hardfacing process. The success of a hardfacing operation depends on factors such as surface preparation of the base metal, preheating conditions, and specific process parameters.

Spray-and-Fuse Process

The spray-and-fuse process is a two-step process in which powdered coating material is deposited by conventional thermal spraying, usually using either a combustion gun (or torch) or a plasma spray gun, and subsequently fused using either a heating torch or furnace. Coatings usually are made of nickel or cobalt self-fluxing alloys to which hard particles, such as tungsten carbide, can be added for increased wear resistance. Deposits ranging from 0.51 to 2.5 mm (0.020 to 0.100 in.) thick can be made by building up several layers at a rate of 0.13 to 0.25 mm (0.005 to 0.010 in.) per pass.

Deposit thickness is controlled by the traverse speed of the torch, speed of rotation (when done between centers on cylindrical parts), powder flow, and the number of layers applied. Deposit rates of 3.5 to 5.5 kg/h (8 to 12 lb/h) can be achieved. Spray-and-fuse hardfacing is particularly effective on small cylindrical parts, as well as on irregular shapes. Cylindrical parts can be sprayed and fused between centers on a converted lathe. Work travels at about 0.8 m/s (150 sfm).

Several types of combustion powder spraying systems are used for the initial deposition step, differing primarily in the method used to feed the powder into the flame. Compressed air can be used as a propellant for the powder, which is injected directly into the flame through a central orifice in the nozzle. Another system feeds the powder into aspirating gas positioned ahead of the nozzle prior to ignition. As shown in Fig. 7, gravity feed can be used to feed the powder directly into the flame. Powder is then conveyed to the workpiece by the force of the combustion gases.

Acetylene combined with oxygen typically is used as the fuel gas, although other fuel gases such as propane and MAPP gas are sometimes used. Powder can also be encapsulated in plastic tubing for use with wire-type spray equipment. Alternatively, a plasma spray torch rather than a combustion gun can be used. In all setups, the powdered alloy becomes molten or semimolten as it passes through or into the combustion flame or plasma effluent, thus facilitating the required initial bond prior to fusing.

Fusion is usually accomplished in a separate operation using an oxyacetylene torch with a multitip flame designed to spread the heat over a wide area. Alternatively, fusion can be accomplished by induction heating, or by heat treatment in a vacuum or reducing atmosphere furnace. As-sprayed coatings are stressed; consequently, fusing should immediately follow spraying to eliminate the possibility of cracking of the coating.

In the spray-and-fuse process, the component, usually cleaned by grit blasting, is heated uniformly to approximately 345 °C (650 °F). The torch is then placed over the area where fusion is to initiate, and the temperature of this area is raised to about 760 °C (1400 °F). Torch movement is then stopped, and the heat is concentrated at the starting point.

The temperature is raised at this position until the surface exhibits a shiny molten layer (glazing). The torch is then moved along the coating to fuse it uniformly, as indicated by glazing: heating should not be excessive, or loss of shape or running may result. Fusing should not be interrupted, and the entire coating should be fused at once. The as-deposited powder coating is bonded mechanically, and upon fusing, a high-quality metallurgical bond also results between the deposit and the base metal.

The spray-and-fuse process can be used to coat a wide range of materials. However, some materials, such as tool steels, austenitic stainless steels, and steels with carbon contents higher than 0.25%, require preheating and careful cooling after fusing to prevent cracking the deposit. Use of type 303 stainless steel or other free-machining metals containing sulfur, manganese, phosphorus, lead, or selenium can result in porous overlays due to the emanation of gas during fusion; these materials usually are not considered for this process. Titanium- or aluminum-bearing alloys (type 321 stainless steel) are also not considered because the highly stable oxide layers on these base metals interfere with proper bonding of the overlay to the base metal during the fusion operation.

Self-fluxing nickel-base powders are the most common hardfacing materials used with the spray-and-fuse process. As shown in Table 6, these powders are available for a wide range of deposit hardness. The lower hardness grades find extensive use in the glass mold industry. The higher hardness grades are used extensively on pump components, particularly in the oil industry.

The quality and properties of the deposit depend not only on the hardfacing powder composition, but also on the particle size distribution

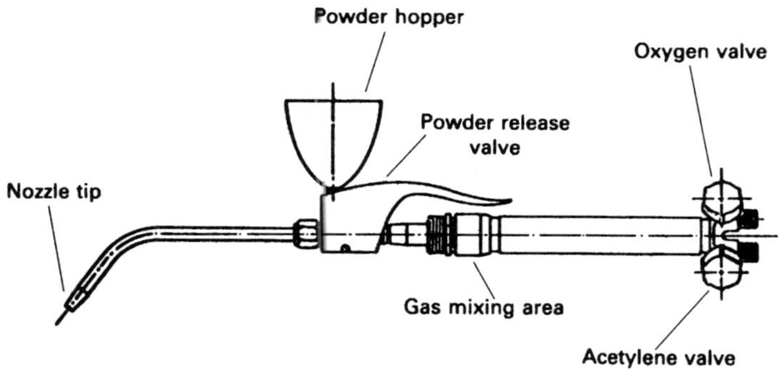

Fig. 7 Spray-and-fuse torch in which the powder is conveyed by gravity into the flame and conveyed to the workpiece by the force of the combustion gases

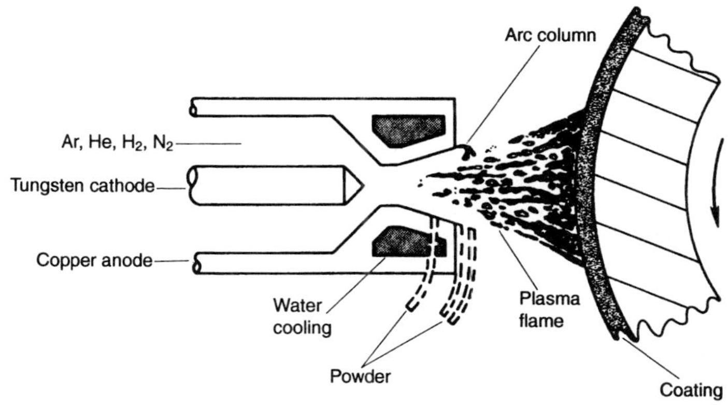

Fig. 8 Schematic of plasma

and the care exercised in making and fusing the deposit. Porosity up to 5% is common in the majority of hardfacing applications. The spray-and-fuse process is also used with cobalt-base powders (Table 6), and with carbide-containing composite powders. Composite powders are more difficult to use, and their deposits exhibit more porosity.

Spray-and-fuse powders generally are applied to cylindrical parts such as pump shafts, packing gland sleeves, and pistons. Other typical applications include ammonia applicators, cams, fan blades, glass molds, pump sleeves and pump plungers, valve seats and stems, bushings, sprockets, wire drawing capstan rings, overlays on dies and molds, and casting repairs.

Plasma Spray Process

In the plasma spray process, a plasma gas stream heats and propels powder particles into the workpiece surface. Because plasma temperatures are extremely high, the plasma spray process can be used to apply refractory coatings that cannot be applied by the spray-and-fuse process.

Although plasma spray coatings have inherent porosity and are principally mechanically bonded to the substrate, they generally have higher density and better adhesion than is achieved with single-step flame spraying. Powders can be applied manually or with mechanized equipment at rates ranging from less than 0.45 kg/h (1 lb/h) to over 14 kg/h (30 lb/h), depending on power of the equipment and type of powder used. In most applications, coating thickness is in the range of 0.13 to 0.4 mm (0.005 to 0.015 in.); depending on the application, much thicker coatings, such as thermal barrier coatings, can be applied. Usually, the workpiece is maintained below 150 °C (300 °F), resulting in little or no distortion of the component and no dilution of the coating by the substrate. Very smooth deposits can be obtained, and finishing may not be required.

Figure 8 schematically depicts a typical plasma spray gun, showing alternative powder inlet positions. A gas, usually argon or nitrogen (or a mixture of these) with hydrogen or helium, flows through a water-cooled copper anode, which serves as a constricting nozzle. An arc is maintained internally from an axial rear tungsten electrode.

Powder introduced into the plasma stream is heated to a molten or semimolten state and propelled at high velocity onto the substrate. Plasma spray deposits with some high-velocity (<305 m/s, or 1000 ft/s) torches can have bond strengths in excess of 70 MPa (10 ksi). The powder is fed at a precisely regulated rate into an inert carrier gas stream and is usually introduced into the plasma stream either in the diverging portion of the nozzle or just beyond the exit.

Usually, the carrier gas is the same composition as the primary plasma gas (argon and/or nitrogen). The choice of powder-feed rate is important. If the powder is fed faster than it can be properly heated, deposition efficiency decreases rapidly, and the coating contains trapped, unmelted powder. In contrast, if the feed rate is too low, operation costs significantly increase.

Materials for the plasma spray process must be in powder form and must be sized appropriately to achieve melting. For the highest coating density, powders are sized between about 44 μm (325 mesh) and 10 μm. The lower limit ensures free flow. Coarser powder can be sized up to 100 μm, with typical distribution falling between 88 μm (170 mesh) and 44 μm (325 mesh). Powder shape ranges from spherical to acicular, though reduced surface area and consistent flow make spherical shapes the best.

Any material that melts without subliming and is available as a properly sized powder has potential as a plasma spray process coating. Plasma spray coatings can be elemental (aluminum, molybdenum, nickel, or chromium), alloyed (nickel-, iron-, or cobalt-base), compounds (Al_2O_3, Cr_2O_3, or Cr_3C_2), composites (nickel-aluminum, cobalt-bonded tungsten carbide, TiO_2, or clad Al_2O_3), or mechanical blends.

In one process variation, the plasma gun and workpiece are enclosed in a vacuum chamber, and the entire operation is carried out in an inert atmosphere at low pressure (50 torr). Advantages of this low-pressure plasma technique include improved bonding and density of the deposit, improved control over coating thickness (even with an irregular work surface), and higher deposit efficiency. However, facility cost is higher, and workpiece size is limited by the vacuum chamber size. This modification is suitable for coating turbine airfoils, blade tips, and shroud segments.

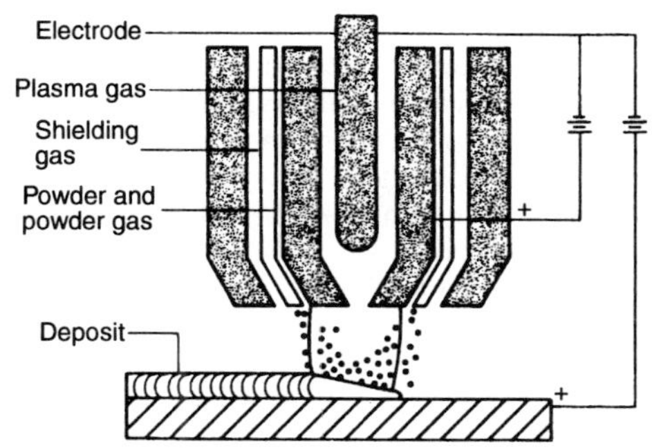

Fig. 9 Schematic of the plasma transferred arc hardfacing process

Very fine powder (–325 mesh) generally is used for low-pressure plasma spraying to achieve maximum density. If high-purity coatings are also desired, then the powder must be manufactured under inert cover and atomized in an inert gas to preserve purity.

Suitable applications for plasma spraying include those requiring a thin, smooth, wear-resistant coating, with low heat buildup in the workpiece. This process is used for hardfacing seal rings of aircraft jet turbine engines and for other turbine components. The coating provides a thermal barrier, resists high-temperature erosion, and forms a tight seal between rotating parts. Other applications include bottling machine parts, mixer and feeder parts, shafts and sleeves, wear rings, arc disks, and thread guides.

Plasma Transferred Arc

Plasma transferred arc surfacing, as shown schematically in Fig. 9, is a welding process in which the powder is introduced into a combined arc/plasma stream to form a molten pool on the workpiece. The resulting deposit is homogeneous and dense, and provides excellent metallurgical bonding with the base metal (workpiece). This process uses a constricted arc. The alloy is carried from a powder feeder to the plasma torch in a stream of argon gas, which forms the plasma and is directed away from the torch into the arc effluent, where it is melted and fusion bonded to the base metal. A direct current power source connecting the tungsten electrode and the workpiece provides energy for the transferred arc. A second direct current power source connecting the tungsten electrode and the arc-constricting orifice supports a nontransferred arc. This nontransferred arc supplements the heat of the transferred arc and serves as a pilot arc to initiate the transferred arc. Argon is passed through a gas diffuser to provide a blanket of shielding gas in and around the arc zone.

Deposits ranging from 0.64 to 3.2 mm (0.025 to 0.125 in.) thick can be produced rapidly in a single pass. Penetration into the base metal can be controlled closely to very low levels. However, a dilution of 5 to 20% in the deposit is typical. Powder used for plasma transferred arc surfacing typically ranges from about 150 to 45 μm in size. Coarser powders can be used, because the transferred arc can easily melt the powder. Use of particles finer than 45 μm must be limited to prevent smoking and oxide inclusions in the deposit.

The plasma transferred arc process is well suited to automation operation and is therefore a frequent choice for high-volume applications such as valves. The exhaust valves of internal combustion engines are subjected not only to mechanical degradation (due to impact and sliding stresses) but also to hot corrosion from the by-products of combustion. In diesel engines, hot corrosion is particularly detrimental, especially if low-grade (residual) fuels are used. To combat hot corrosion and mechanical wear of valve seat faces that must maintain integrity for efficient engine operations, these components are commonly coated with cobalt-base alloys.

Manual Powder Torch Welding

Also termed "powder welding," manual powder torch welding is a true welding process in which powder application and fusion occur in a single operation using a special oxyacetylene torch. Powder is fed from a small hopper mounted on a gas welding torch into the fuel gas supply and conveyed through the flame to the workpiece surface. By following general oxyacetylene surfacing practice, smooth, thin, dense deposits are obtained. Deposit thickness is controlled by powder flow rate and movement of the torch. A dilution of 1 to 5% in the deposit occurs from this process.

Deposits ranging from 0.76 to 3.2 mm (0.030 to 0.125 in.) in thickness can be applied at a rate of 1.3 mm (0.050 in.) per pass. Sweating of the base metal during surfacing generates a fusion-type bond with the hardfacing alloy. Ease of application and material recovery are enhanced because of close control of the powder flow into the welding flame.

Oxyacetylene torch flame adjustment is essential to produce good powder deposition. The amount of excess acetylene required for hardfacing varies depending on alloy composition. The amount of excess acetylene is conveniently measured in terms of the flame geometry as the ratio between the length of the oxyacetylene feather and the inner cone. The length of the inner cone, indicated by the value *X*, is the distance from the torch tip to the extreme end of the inner cone.

The length of the oxyacetylene feather is measured from the torch tip to the extreme end of the intermediate zone, generally expressed in multiples of the length of the inner cone. A neutral flame is thus classified as 1×. Cobalt-base powders are usually applied using 2× to 3× flame. The oxyacetylene feather is two to three times the length of the inner cone.

Tungsten carbide-containing composite powders should be applied using a 1.5× flame. Nickel-base powders are welded with a neutral or slightly oxidizing flame. Because flame characteristics change somewhat when powder is being sprayed, the powder feed handle should be depressed and final flame adjustment made while the powder is being sprayed. Alloys commonly welded with this process contain boron and silicon, which slag off any oxide skin to provide a clean weld pool.

The manual torch process is well suited for small repairs or for hardfacing of small areas requiring a wide range of hardness (about 20 to 60 HRC). Typical applications of manual torch welding include glass dies and molds, brick molding equipment, components for cement industry, cast iron parts for foundries, feed screws, hammer mill hammers, and pump parts.

Detonation Gun

The detonation gun spray process is markedly different from other flame spraying processes and was initially developed for the deposition of hard, wear-resistant materials such as oxides and carbides. The extremely high particle velocities achieved in the detonation gun result in coatings with higher density, greater internal strength, and superior bond strength than can be achieved with conventional plasma spraying or single-step flame spraying.

This process, which is depicted schematically in Fig. 10, uses a controlled series of explosions of a mixture of oxygen and acetylene to blast the powder onto the surface of the workpiece. The powder to be sprayed is introduced into the long chamber in the mixture of oxygen and acetylene.

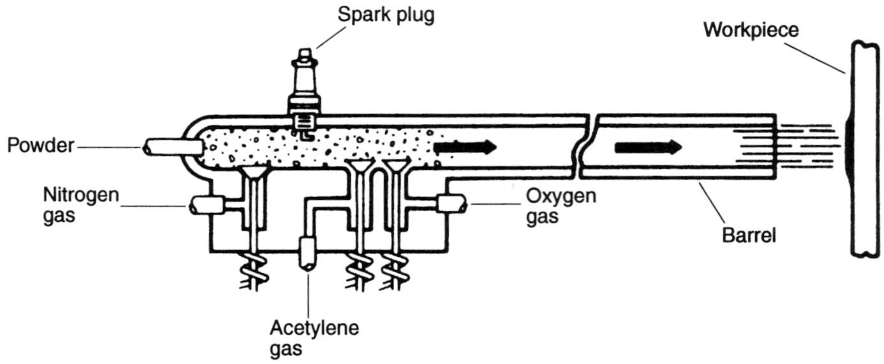

Fig. 10 Schematic of detonation gun hardfacing process

When the gas mixture is ignited, the detonation products rapidly heat and accelerate the powder particles. Temperatures of 3300 °C (6000 °F) are reached inside the gun.

The process is capable of spraying a wide range of both metallic and nonmetallic powders with thicknesses of up to 0.5 mm (0.020 in.). The coating produced is similar to, but denser than, a plasma spray coating. It has an excellent bond strength and can be finished to an excellent surface finish. The workpiece can be processed below 150 °C (300 °F) in a carbon dioxide spray cooling system. This process has found wide acceptance in aerospace and oil exploration industries.

Jet Kote Surfacing System

This is a high-pressure combustion thermal spray process, in which the spray gun burns a fuel (either MAPP gas or propane) with a high volume of oxygen within a combustion chamber. Using a nitrogen carrier gas, the powder is injected into the combustion products and is rapidly heated and accelerated to impact at high speed against the work surface to be coated. Unlike the detonation gun process, the Jet Kote system (Cabot Corp.) is based on continuous combustion within the gun.

Systems with high operating pressure provide a smooth, dense coating that is comparable to coatings produced by the detonation gun process. The process can also use both metallic and ceramic powders comparable to those used in plasma spraying. One of the main advantages of the Jet Kote system is low capital equipment cost compared to plasma spraying. Typical applications include those in which high-quality, thermal spray coatings are required, such as in the oil exploration industry.

Bulk-Welding Process

The bulk-welding process, also referred to as joint welding, is an "add-on" to conventional submerged arc welding. It utilizes the existing heat produced by a conventional submerged arc weld to melt powders ahead of the melt pool to increase overall deposition rates. The process uses a powder composition that matches the feed wire composition.

The quantity of powder used is governed by the application. The powder metal to wire ratio

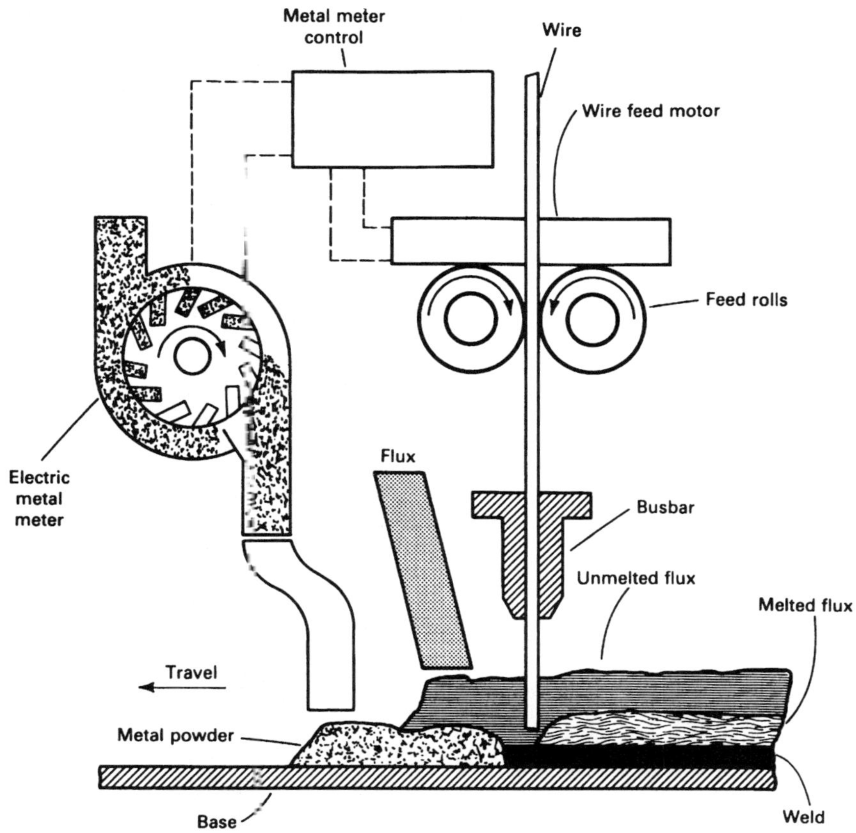

Fig. 11 Schematic of the bulk-welding process using metal powder joint fill

can be as high as 2 to 1; however, for most applications, this ratio is 1 to 1. A schematic of the process is shown in Fig. 11. Advantages associated with the bulk-welding process are:

- Deposition rate with bulk-welding can be increased two or three times over single-wire submerged arc welding without increasing heat input. The deposition rate depends on the amount of powder metal used in relation to the wire.
- Flux consumption per pound of deposited metal is reduced considerably.
- Less distortion is evident due to lower heat input.
- Accurate powder metering prevents spreading and other related problems, such as unmelted powder and rough edges, and improves overall deposition efficiency.
- A reduction in weld passes is possible. One-pass welding can be done on most applications.

The process is used for shipbuilding, heavy machinery, bridges, vessels, and building fabrication in which partial- or full-penetration weld and fillet welds are required. Most powders used are iron base. However some cobalt-base powders are also used.

Conforma Clad Process

This process uses a flexible cloth made of polytetrafluoroethylene and a hardfacing powder of desired composition. The main benefit derived from this technique is that the cloth can be cut into any required size and placed on the surface to be coated, regardless of the profile of the workpiece. During subsequent furnace brazing, polytetrafluoroethylene is vaporized, and the hardfacing powder forms the deposit layer. The powder particles can be irregular or spherical in shape and can vary in size from 1 to 150 μm. Typical powder compositions are similar to the nickel-base alloys 40, 50, and 60 and cobalt-base alloy SF6 in Table 6. In addition, composites of these nickel-base alloys with tungsten carbide or chromium carbide are also produced.

Deposits produced by this process are highly dense and wear resistant. The primary advantage is the ability of Conforma Clad process to build up thick coatings (up to 2.5 mm or 0.1 in. thick) on selected areas of parts. Typical applications of the process are for agricultural equipment and valves.

Brazing and Soldering Powders

Metal Powders are used extensively as filler metals in the brazing and soldering industries. These powders offer a convenient method of applying filler metal to parts, although alternative filler metal forms (wire, rod, clad sheet, foil, preforms, and plating) also are used. For brazing, filler metals have a liquidus temperature above 450 °C (840 °F) and below the solidus of the base metal. For soldering, filler metals (solders) have a liquidus temperature below 450 °C (840 °F) and below the solidus of the base metal. In both brazing and soldering, the filler metal, when heated to a suitable temperature with a suitable flux or protective atmosphere, is distributed between closely fitted surfaces of the joint by capillary action.

Powder Types

Atomized Powders. Most powders currently are produced by atomization because it offers a convenient and cost-effective means of producing large quantities of powder. Atomization consists of melting the metal and alloy in an induction melting furnace and then pouring it into a crucible, which has a hole in its bottom and is fitted with a gas nozzle at the bottom of the hole. In most atomizing processes, a high-pressure stream of fluid (usually water or gas) impinges on the molten metal stream, breaking it into droplets.

Materials such as solders can be readily air atomized, while most silver-base, nickel-base, and other alloys are atomized with a high-pressure argon or nitrogen gas. Powders are subsequently screened to the desired particle size distribution. Most brazing filler metals are sieved to one of four standard sieve analyses as follows:

Sieve size	Size analysis, %
100 mesh	
−60	100 min
−100	95 min
140 C mesh	
+100	Trace
+140	10 max
−325	20 max
140 F mesh	
+100	Trace
+140	10 max
−325	55 max
325 mesh	
+200	Trace
+325	10 max
−325	90 min

Source: AWS specification A5.8

Brazing filler-metal atomizer powders are available in the following types:

- Nickel alloys
- Cobalt alloys
- Silver alloys
- Gold alloys
- Copper-phosphorus alloys
- Copper
- Aluminum-silicon alloys

These powders are generally screened to American Society for Testing and Materials standard ASTM B 214 ("Standard Method for Sieve Analysis of Granular Metal Powders"), using sieves with size to ASTM E 11 ("Standard Specification for Wire-Cloth Sieves for Testing Purposes"). Table 12 lists the compositions of various brazing powders, as well as their respective solidus, liquidus, and brazing temperatures.

Fine prealloyed solder powders (–325 mesh) are used to make soldering pastes and creams, in which the solder alloy is suspended in the fluxing material. The ratio of solder to flux is 80 to 90 wt%. A standard specification for solder metal can be found in ASTM B 32. Table 13 lists the compositions of some typical solder alloy powders used in soldering pastes and creams, as well as their respective solidus, liquidus, and soldering temperatures.

Ball Milled Powders. Historically, nickel filler metals have been produced by ball milling. In this process, alloy shot is ball milled into fine powders. Currently, atomization techniques are used to produce these powders.

Mechanically Filed Powders. Early filler metal powders made of silver and gold filler metals were mechanically filed using standard commercial filing machines. These powders, however, were not as readily usable in paste form since the long, thin filings do not flow through a thin nozzle; consequently, mechanical filing, while still available, is decreasing in usage and has been converted mostly to atomized powder manufacture.

Oxide Powders. Most past-type copper filler metals are made of red copper oxide that has been milled to a fine powder. This powder is then mixed with a vehicle that provides a convenient paste for the filler metal application. Because copper oxide dissociates readily, it makes a good filler metal for furnace brazing of carbon steels in strongly reducing atmospheres such as combusted city gas (nitrogen/carbon monoxide/hydrogen) or prepared mixed atmospheres (nitrogen-hydrogen).

Preforms from Metal Powders. Preforms are used because they are readily applied with automated equipment. Preforms are easily produced from wire and sheet; consequently powders are generally not used to make preforms other than stamping from transfer tape materials, which are plastic-bonded sheets with an adhesive on one side. Solder powders generally are not used for making preforms.

Flux-Powder Mixtures. Brazing fluxes used in flux-powder applications are generally proprietary. These fluxes usually are composed of various mixtures of some of the following: borates of sodium, potassium, and lithium; fused borax; fluoroborates of potassium or sodium; fluorides of sodium, potassium, and lithium; chlorides of sodium, potassium, and lithium; boric acid (frequently calcined); alkalis such as potassium hydroxide and sodium hydrides; wetting agents; and water.

Depending on the desired reactivity of the brazing flux and its melting characteristics, these materials are used in various quantities. Some fluxes are suitable for direct manufacture of flux-powder pastes, while other flux powder pastes are made using a long-chain viscous organic vehicle that can exhibit thixotropic properties.

Solder fluxes are categorized as corrosive, intermediate, and noncorrosive fluxes. These fluxes are mixed with fine solder powder to form solder creams and pastes. Corrosive fluxes contain one or more of the following components:

- Zinc chloride
- Ammonium chloride
- Stannous chloride
- Sodium or potassium chloride
- Lithium chloride
- Aluminum chloride
- Sodium or potassium fluoride
- Boron trifluoride
- Hydrochloric acid
- Hydrofluoric acid
- Orthophosphoric acid
- Fluoboric acid

Intermediate-type powder/paste fluxes contain one or more of the organic acids (lactic acid, citric acid, stearic acid, and oxalic acid, for example) or organic hydrohalides (glutamic acid hydrochloride, aniline hydrochloride, and hydrazine hydrochloride, for example). Amines and other compounds such as urea, glycerol and diethylene diamine, or triamine also are used. For flux-powder pastes, one or more of these fluxing agents is added to the powder and is mixed with a long-chain viscous organic vehicle that can exhibit thixotropic properties.

Noncorrosive flux-powder pastes contain rosin as the main constituent, which is active at soldering temperatures but inactive at room temperature. Occasionally, additives are used to increase the reactivity of the rosin.

Paste Mixtures

Brazing paste mixtures usually are prepared for furnace brazing in a protective atmosphere. Consequently, they do not require flux additions. For special-purpose applications, a small amount of flux occasionally can be added.

Copper oxide powder pastes are extremely fine; consequently, vehicles such as ethylene glycol and mineral oil are used to form the paste mixture. On heating in the presence of copper or copper oxide, these vehicles chemically crack, leaving a residue of carbon. At the brazing temperature, carbon reacts with the water in the higher dew point (–1 to 26 °C, or 30 to 80 °F) furnace brazing atmosphere to produce carbon monoxide gas, eliminating the carbon from the surface of the part. If the furnace atmosphere is too dry, the carbon remains on the surface of the part.

Paste mixtures of nickel, silver, and gold filler metals use a long-chain organic vehicle that can exhibit thixotropic properties. These vehicles or binders are vaporized on heating and are forced out of the furnace by the gas atmosphere. If sufficient moisture exists in the furnace atmosphere (approximately –1 to 26 °C, or 30 to 80 °F, dew point) the vehicle reacts with the moisture to produce hydrogen and carbon monoxide. In a high-vacuum furnace (approximately 10^{-3} torr and below), the vehicle vaporizes when heated and is forced out into the cold portions of the furnace and usually is pumped out of the system.

Application of Filler Metal

Filler metal can be easily applied with the use of automated brazing machines by using powder in a paste mixture of either flux and/or binder, as required by the particular type of equipment. If the parts are to be brazed in an automated machine with air-gas torch heating, the use of flux is required to provide a suitable braze and to protect parts from oxidation, while allowing brazing or soldering filler metal to wet and flow over the part. When powder is to be used in automated equipment or in furnaces with controlled atmospheres, flux is not required. A binder is used to hold the powder to the work; when heated, it is gassed off and pumped out of the system.

Brazing and soldering powders, pastes, and creams are applied in different manners. One method consists of the use of an inexpensive plastic bottle that allows manual application of the filler metal during brazing or soldering furnace operations. The operator puts the parts on the furnace belt, then drops a quantity of filler metal in the proper location at the joint. More sophisticated types of application equipment apply the filler metal pneumatically or volumetrically to the parts to be brazed or soldered on an automated brazing or soldering machine. Powder and flux can be added either separately or as a single material on the brazed joint as the parts move through the machine.

Solders generally are supplied in paste or cream form, which can be applied by screening, brushing, extruding, or rolling onto the substrate surfaces. Solder pastes and creams, which are used widely in microelectronic component soldering, are often applied using a syringe. Other methods of application of brazing filler metal include spraying, dipping, applying with a syringe, or applying dry powder and wetting with a binder.

Mechanical Properties

The mechanical properties of joints made with powdered filler metals have properties comparable to those of joints made from solid rod, wire, and foil, provided that the powders are clean and free of oxides, and that the protective flux or atmosphere is maintained as necessary to prevent oxidation of the specific metal or alloy. Solder pastes and creams containing fluxes protect these powders from oxidation.

Brazing filler metals that contain chromium, silicon, boron, and other highly oxidizing ele-

Table 12 Compositions and properties of typical brazing alloy powders

Filler metal	Composition(a), %													Other elements total	Solidus temperature		Liquidus temperature		Brazing temperature range	
	Cr	B	Si	Fe	C	P	S	Al	Ti	Mn	Cu	Zr	Ni		°C	°F	°C	°F	°C	°F
Nickel alloys																				
BNi-1	13.0–15.0	2.75–3.50	4.0–5.0	4.0–5.0	0.6–0.9	0.02	0.02	0.05	0.05	...	...	0.05	bal	0.50	977	1790	1038	1900	1066–1204	1950–2200
BNi-1a	13.0–15.0	2.75–3.50	4.0–5.0	4.0–5.0	0.06	0.02	0.02	0.05	0.05	...	...	0.05	bal	0.50	977	1790	1077	1970	1077–1204	1970–2200
BNi-2	6.0–8.0	2.75–3.50	4.0–5.0	2.5–3.5	0.06	0.02	0.02	0.05	0.05	...	...	0.05	bal	0.50	971	1780	999	1830	1010–1177	1850–2150
BNi-3	...	2.75–3.50	4.0–5.0	0.5	0.06	0.02	0.02	0.05	0.05	...	...	0.05	bal	0.50	982	1800	1038	1900	1010–1177	1850–2150
BNi-4	...	1.5–2.2	3.0–4.0	1.5	0.06	0.02	0.02	0.05	0.05	...	...	0.05	bal	0.50	982	1800	1066	1950	1010–1177	1850–2150
BNi-5	18.5–19.5	0.03	9.75–10.50	...	0.10	0.02	0.02	0.05	0.05	...	...	0.05	bal	0.50	1079	1975	1135	2075	1149–1204	2100–2200
BNi-6	...	...	...	...	0.10	10.0–12.0	0.02	0.05	0.05	...	...	0.05	bal	0.50	877	1610	877	1610	927–1093	1700–2000
BNi-7	13.0–15.0	0.01	0.10	0.2	0.08	9.7–10.5	0.02	0.05	0.05	0.04	...	0.05	bal	0.50	888	1630	888	1630	927–1093	1700–2000
BNi-8	...	...	6.0–8.0	...	0.10	0.02	0.02	0.05	0.05	21.5–24.5	4.0–5.0	0.05	bal	0.50	982	1800	1010	1850	1010–1093	1850–2000

Filler metal	Composition(a), %													Other elements total	Solidus temperature		Liquidus temperature		Brazing temperature range	
	Cr	Ni	Si	W	Fe	B	C	P	S	Al	Ti	Zr	Co		°C	°F	°C	°F	°C	°F
Cobalt alloys																				
BCo-1	18.0–20.0	16.0–18.0	7.5–8.5	3.5–4.5	1.0	0.7–0.9	0.35–0.45	0.02	0.02	0.05	0.05	0.05	bal	0.50	1121	2050	1149	2100	1149–1232	2100–2250

Filler metal	Composition(a), %								Other elements total	Solidus temperature		Liquidus temperature		Brazing temperature range	
	Ag	Cu	Zn	Cd	Ni	Sn	Li	Mn		°C	°F	°C	°F	°C	°F
Silver alloys															
BAg-1	44.0–46.0	14.0–16.0	14.0–18.0	23.0–25.0	...	...	...	...	0.15	607	1125	618	1145	618–760	1145–1400
BAg-1a	49.0–51.0	14.5–16.5	14.5–18.5	17.0–19.0	...	...	...	...	0.15	627	1160	635	1175	635–760	1175–1400
BAg-2	34.0–36.0	25.0–27.0	19.0–23.0	17.0–19.0	...	...	...	...	0.15	607	1125	702	1295	702–843	1295–1550
BAg-2a	29.0–31.0	26.0–28.0	21.0–25.0	19.0–21.0	...	...	...	...	0.15	607	1125	710	1310	710–843	1310–1550
BAg-3	49.0–51.0	14.5–16.5	13.5–17.5	15.0–17.0	2.5–3.5	...	...	...	0.15	632	1170	688	1270	688–815	1270–1500
BAg-4	39.0–41.0	29.0–31.0	26.0–30.0	...	1.5–2.5	...	...	...	0.15	671	1240	779	1435	779–899	1435–1650
BAg-5	44.0–46.0	29.0–31.0	23.0–27.0	...	...	...	...	...	0.15	677	1250	743	1370	743–843	1370–1550
BAg-6	49.0–51.0	33.0–35.0	14.0–18.0	...	...	...	...	...	0.15	688	1270	774	1425	774–871	1425–1600
BAg-7	55.0–57.0	21.0–23.0	15.0–19.0	...	...	4.5–5.5	...	...	0.15	618	1145	652	1205	652–760	1205–1400
BAg-8	71.0–73.0	bal	...	...	...	...	...	...	0.15	779	1435	779	1435	779–899	1435–1650
BAg-9	64.0–66.0	19.0–21.0	13.0–17.0	...	...	...	...	...	0.15	671	1240	718	1325	718–843	1325–1550
BAg-10	69.0–71.0	19.0–21.0	8.0–12.0	...	...	...	...	...	0.15	691	1275	738	1360	738–843	1360–1550
BAg-13	53.0–55.0	bal	4.0–6.0	...	0.5–1.5	...	...	...	0.15	718	1325	857	1575	857–968	1575–1775
BAg-13a	55.0–57.0	bal	...	...	1.5–2.5	...	...	...	0.15	771	1420	893	1640	871–982	1600–1800
BAg-18	59.0–61.0	bal	...	...	...	9.5–10.5	...	...	0.15	602	1115	718	1325	718–843	1325–1550
BAg-20	29.0–31.0	37.0–39.0	30.0–34.0	...	...	...	...	...	0.15	677	1250	766	1410	766–871	1410–1600
BAg-21	62.0–64.0	27.5–29.5	...	...	2.0–3.0	5.0–7.0	...	...	0.15	691	1275	802	1475	802–899	1475–1650
BAg-22	48.0–50.0	15.0–17.0	21.0–25.0	...	4.0–5.0	...	...	7.0–8.0	0.15	680	1260	699	1290	699–830	1290–1525
BAg-24	49.0–51.0	19.0–21.0	26.0–30.0	...	1.5–2.5	...	...	...	0.15	660	1220	707	1305	750–843	1305–1550
BAg-25	19.0–21.0	39.0–41.0	33.0–37.0	...	...	...	...	4.5–5.5	0.15	738	1360	790	1455	790–845	1455–1555
BAg-26	24.0–26.0	37.0–39.0	31.0–35.0	...	1.5–2.5	...	...	1.5–2.5	0.15	707	1305	802	1475	800–870	1475–1600
BAg-27	24.0–26.0	34.0–36.0	24.5–28.5	12.5–14.5	...	...	...	...	0.15	607	1125	746	1375	745–860	1375–1575
BAg-28	39.0–41.0	29.0–31.0	26.0–30.0	...	...	1.5–2.5	...	...	0.15	649	1200	710	1310	710–843	1310–1550

Filler metal	Composition(a), %				Other elements total	Solidus temperature		Liquidus temperature		Brazing temperature range	
	Au	Cu	Pd	Ni		°C	°F	°C	°F	°C	°F
Gold alloys											
BAu-1	37.0–38.0	bal	...	...	0.15	991	1815	1016	1860	1016–1093	1860–2000
BAu-2	79.5–80.5	bal	...	...	0.15	891	1635	891	1635	891–1010	1635–1850
BAu-3	34.5–35.5	bal	...	2.5–3.5	0.15	974	1785	1029	1885	1029–1091	1885–1995
BAu-4	81.5–82.5	...	...	bal	0.15	949	1740	949	1740	949–1004	1740–1840
BAu-5	29.5–30.5	...	33.5–34.5	35.5–36.5	0.15	1135	2075	1166	2130	1166–1232	2130–2250
BAu-6	69.5–70.5	...	7.5–8.5	21.5–22.5	0.15	1007	1845	1046	1915	1046–1121	1915–2050

Filler metal	Composition(a) %			Other elements total	Solidus temperature		Liquidus temperature		Brazing temperature range	
	P	Ag	Cu		°C	°F	°C	°F	°C	°F
Copper-phosphorus alloys										
BCuP-2	7.0–7.5	...	bal	0.15	710	1310	793	1460	732–843	1350–1550
BCuP-3	5.8–6.2	4.8–5.2	bal	0.15	643	1190	813	1495	718–816	1325–1500
BCuP-4	7.0–7.5	5.8–5.2	bal	0.15	643	1190	718	1325	691–788	1275–1450
BCuP-5	4.8–5.2	14.5–15.5	bal	0.15	643	1190	802	1475	704–816	1300–1500
BCuP-6	6.8–7.2	1.8–2.2	bal	0.15	643	1190	788	1450	732–816	1350–1500
BCuP-7	6.5–7.0	4.8–5.2	bal	0.15	643	1190	771	1420	704–816	1300–1500

(continued)

Table 12 (continued)

Filler metal	Composition(a), % Cu	Zn	Sn	Fe	Mn	Ni	P	Pb	Al	Si	Other elements total	Solidus temperature °C	°F	Liquidus temperature °C	°F	Brazing temperature range °C	°F
Copper																	
BCu-1a	99.0 min	...	...	...	...	...	...	...	...	...	0.30	1083	1981	1083	1981	1093–1149	2000–2100
BCu-2	86.5 min	...	...	...	...	...	...	...	...	...	0.50	1083	1981	1083	1981	1093–1149	2000–2100

Filler metal	Composition(a), % Si	Cu	Fe	Zn	Mg	Mn	Al	Other elements each	total	Solidus temperature °C	°F	Liquidus temperature °C	°F	Brazing temperature range °C	°F
Aluminum-silicon alloys															
BAlSi-4	11.0–13.0	0.30	0.8	0.20	0.10	0.15	bal	0.05	0.15	577	1070	582	1080	582–604	1080–1120

(a) Single values are maximum percentages, unless otherwise indicated. Source: AWS A5.8, "Specification for Brazing Filler Metals"

Table 13 Compositions and properties of typical soldering alloy powders

ASTM designation	Composition, % Tin	Silver	Antimony	Bismuth(a)	Copper(a)	Iron(a)	Aluminum(a)	Arsenic(a)	Lead	Solidus temperature °C	°F	Liquidus temperature °C	°F	Soldering range °C	°F
10 B	8–12	...	0.2–0.5	0.25	0.08	0.02	0.005	0.02	bal	268	514	299	570	329–369	624–696
20 B	18–22	...	0.2–0.5	0.25	0.08	0.02	0.005	0.02	bal	183	361	277	531	307–347	585–657
30 B	28–32	...	0.2–0.5	0.25	0.08	0.02	0.005	0.02	bal	183	361	255	491	285–325	545–617
40 B	38–42	...	0.2–0.5	0.25	0.08	0.02	0.005	0.02	bal	183	361	238	460	268–308	514–586
50 B	48–52	...	0.2–0.5	0.25	0.08	0.02	0.005	0.03	bal	183	361	216	421	213–282	415–540
60 B	58–62	...	0.2–0.5	0.25	0.08	0.02	0.005	0.03	bal	183	361	190	374	220–282	428–540
63 B	61–65	...	0.2–0.5	0.25	0.08	0.02	0.005	0.03	bal	183	361	183	361	212–253	414–487
62 P	61.5–67.5	1.75–2.25	0.3–0.5	0.25	0.08	0.02	0.005	0.005	bal	179	355	179	355	209–249	409–481
96.5 TS	95–96	3.3–3.7	0.2–0.5	0.25	0.08	0.02	0.005	0.005	...	221	430	221	430	251–291	484–556

Other solder alloys are also available in paste form. Typical compositions, impurity levels, and designations are given in ASTM B 32. (a) Maximum level

ments experience some oxidation when heated to the brazing temperature. When the furnace atmosphere is not properly controlled, the large surface area of the brazing powder particles may pick up enough oxygen to encapsulate each particle or may remove most of the boron, causing poor brazing. This constitutes inadequate quality control and is easily avoidable.

Powder particle size can affect the flow of the brazing filler metal when it contains elements with stable oxides such as chromium, silicon, boron, aluminum, or titanium. The high surface area of –325 mesh powders, which contain elements with oxides that are difficult to dissociate, can pick up sufficient oxygen from the atmosphere to reduce the flow of the filler metal. The degree of oxygen pickup depends on the specific element present and the partial pressure of oxygen present in the brazing atmosphere during the heating portion of the cycle.

For example, if Ni-19Cr-10Si filler metal (BNi-5) is heated in a marginal atmosphere on type 304 stainless steel base metal, the base metal can emerge bright and clean. As the brazing temperature is approached, however, the chromium and silicon in the powder filler metal will oxidize to some extent. The degree of oxidation depends on the partial pressure of oxygen in the atmosphere surrounding the powder. In a high-quality atmosphere, the degree of oxidation is so minimal that there is no deleterious effect on the melting and flow characteristics of the brazing filler metal.

As the partial pressure of oxygen in a brazing atmosphere increases, more surface oxidation of the powder occurs; smaller particles may be completely oxidized and thus will not melt and flow when the brazing temperature is reached. Larger particles may not be completely oxidized and thus will melt and flow; however, oxides and residual unmelted powders remain at the point of application. Unmelted powder results from the loss of silicon through oxidation.

Because high-surface-area brazing filler metal powders (–325 mesh) are more susceptible to oxidation, the majority of brazing operations utilize 140 F mesh powders. It should be noted, however, that if the brazing atmosphere is strictly controlled to maintain a low partial pressure of oxygen, –325 mesh powders that contain one or more of the elements having stable oxides can be used. For a lower quality atmosphere (higher partial pressure of oxygen), a coarser powder such as 140 C mesh is used.

As a result, brazing filler metal powder that is of a given mesh size and that contains one or more readily oxidizable elements is a good indicator of atmosphere quality. Every time the melting and flow characteristics of one of these filler metals is tested, the quality of the brazing atmosphere is being tested to an even greater extent. For this reason, extreme care must be exercised when interpreting the results of melting and flow tests. With proper control of the furnace brazing cycle (including atmosphere), each type of filler metal, regardless of its form (wire, sheet, or powder) will have the same mechanical properties in the joint.

Flame Cutting

Metal powder cutting is a technique that supplements an oxyfuel torch with a stream of iron or blended iron-aluminum powder to facilitate flame cutting of difficult-to-cut materials. The powdered material propagates and accelerates the oxidation reaction, as well as the melting and spalling action of the materials to be cut. Metal powder cutting was developed after World War II to provide an effective method of cutting stainless steels. These highly alloyed materials resist conventional oxyfuel cutting because a refractory oxide layer forms when stainless steel is heated in the presence of oxygen.

In all oxyfuel cutting processes, cutting is accomplished through a thermochemical reaction in which preheated metal is cut, or removed, by rapid oxidation in a stream of pure oxygen. During metal powder cutting, finely divided iron powder is carried by dry compressed air or nitrogen into the cutting oxygen stream along the line of the cut. Combustion of this powder increases the temperature of the heated zone; consequently, the refractory oxides are melted and fluxed by the action of the powder particles.

The extreme heat created by the burning of the powder enables cutting without preheating of the

Table 14 Typical properties of iron powder used for metal powder cutting

Screen analysis		
U.S. mesh	Opening, mm	%
On 70	212	Trace
−70+100	150	5
−100+140	106	28
−140+200	75	23
−200+325	45	24
−325	...	20
Chemical analysis, %		
Iron		98
Carbon		0.20 max
Manganese		0.10 max
Hydrogen loss		1.0 max
Silica acid insolubles		0.30 max
Physical properties		
Apparent density		2.9 g/cm^3
Flow rate		26 s/50 g

metal. As the cut progresses, the heat and fluxing action of the burning iron powder enable the cutting oxygen stream to oxidize the base metal continuously.

Powder Selection and Production

Many metallic and nonmetallic powders have been used for cutting operations. The ideal powdered solid fuel, whether metallic or nonmetallic:

- Exhibits nonhygroscopic action (does not readily absorb moisture)
- Liberates considerable heat during combustion
- Generates combustion products that flux refractory oxides
- Generates nonirritating and nontoxic combustion products
- Provides good process economy

Nonmetallic powders, such as sodium bicarbonate, remove refractory oxides by forming liquid slags that have a fluxing effect. Such powders, however, give off little or no heat. Consequently, cutting speeds are significantly slower than for metallic powders. The most suitable metallic powders for cutting applications are iron and iron-aluminum blends. For metal powder cutting of stainless steel, iron powder can be used. For cutting oxidation-resistant materials other than stainless steels, blends containing iron and atomized −100+325 mesh aluminum powder are preferable.

Iron powder used for metal cutting applications is produced by water atomization, or by a combination of granulation of molten iron by high-pressure water jets and chemical reduction. According to the Metal Powder Industries Federation, 2% of all iron powder produced in North America is used for cutting applications. Table 14 provides data on iron cutting powder produced by the Quebec Metal Powder (QMP) process. Cutting-grade iron powder of similar density and composition also is produced by the Domfer Process.

Table 15 Typical properties of atomized aluminum powder for metal powder cutting and lancing

Screen analysis	
U.S. mesh	%
On 100	Trace
−100 + 200	52–67
−200 + 325	31–42
−325	0–4
Chemical analysis(a), %	
Aluminum	99.7
Iron	0.18–0.25 max
Silicon	0.12–0.15 max
Other metallics, each	0.01–0.03 max
Other metallics, total	0.15 max
Physical properties	
Apparent density	1.1 g/cm^3
Tap density	1.4 g/cm^3
Surface area	0.10–0.20 m^2/g

(a) Chemical analysis excludes 0.4% aluminum oxide, which exists on the surface of the particles.

Aluminum powder is sometimes added to iron powder to form blends used for cutting and lancing concrete, firebrick, difficult-to-cut nonferrous metals, and refractory-laden scrap, such as slag deposits on ladle pouring spouts. Table 15 provides data on cutting-grade aluminum powders that are produced by air atomization of molten aluminum. Detailed information on the production and properties of atomized aluminum powder can be found in the article "Production of Aluminum and Aluminum Alloy Powder," in this Volume.

Powder blends consisting of iron and 10 to 40 wt% Al are used to cut a variety of materials. Although preblended compositions consisting of iron and 15 wt% Al are available, iron and aluminum powders can be custom blended for specific applications, as shown in Table 16.

To blend powders properly, components should be poured simultaneously into a dry, nitrogen-filled container and tumbled for about 5 min. Additional blending occurs as the iron-aluminum powder is screened in the powder dispenser. Slight nonuniformities in the mixture have little effect on cutting performance.

Equipment

Metal powder cutting processes require a means of introducing and burning powder in the cutting oxygen stream after the powder is conveyed to the cutting torch. To accomplish this, a typical powder cutting equipment setup includes the following components:

- Dispenser to supply powder during cutting
- Source of dry compressed air or nitrogen to convey powder from the dispenser to the cutting oxygen stream
- Powder nozzle or attachment to introduce the powder into the cutting oxygen stream
- Cutting torch (blowpipe) to control the gases required to burn the oxygen/powder mixture

Table 16 Iron-aluminum blends used for cutting and lancing difficult-to-cut materials

Nominal composition	Applications
Fe-10-15Al	Refractory scrap, thin nonferrous sections up to 25 mm (1 in.) thick
Fe-25Al	Aluminum, brass, bronze
Fe-30Al	Nickel, Monel, Inconel, Hastelloy, and concrete
Fe-40Al	Copper, brass, and bronze heavy sections 15 cm (6 in.) thick and up

Powder dispensers can be of two general types, pneumatic and vibratory. Generally, all dispensers consist of a compressed air/nitrogen filter and regulator, a powder hopper, and a means of expelling the powder from the dispenser. In a pneumatic dispenser, powder flow to the ejector unit is the result of gravitational force and pressure differential from compressed air (or nitrogen) passing through the ejector, which draws powder from the ejector baffle plate into the gas stream. Vibratory dispensers allow a quantity of powder to be dispensed from a hopper that is governed by a vibrator. Powder flow can be controlled by adjusting the amplitude of vibration. Vibratory dispensers are generally used where uniform and accurate powder flow is required. Typical applications include precision cutting of materials, such as stainless steel, as well as production of high-quality, sharp top edges on cuts in carbon steels.

Powder-Cutting Apparatus. Attachments for metal powder cutting are designed for use with conventional oxyfuel cutting torches. Attachments are available in three types: single-tube, multijet, and dual-tube. In addition to powder cutting attachments, specially designed powder cutting torches are available.

Applications

Metal powder cutting is frequently used in cutting, scarfing, and lancing of oxidation-resistant materials. This type of metal cutting was originally developed for cutting stainless steels. The alloying elements (chromium and nickel) that impart desirable properties to stainless steels have made these steels difficult to process by conventional oxyfuel cutting. Chromium is the most detrimental; when oxyfuel cutting is applied to stainless steels, chromium, which has a high affinity for oxygen at elevated temperatures, immediately forms the highly refractory chromium oxide on the faces of the kerf and prevents further oxidation. These refractory oxides are melted and fluxed by the combustion of the powder particles during metal powder cutting.

Nonferrous metals with the following thicknesses are suitable candidates for powder cutting:

Material	Thickness cm	Thickness in.
Aluminum	25	10
Nickel	15	6
Brass and bronze	25	10
Copper	15	6
Hastelloy	16.5	6.5
Inconel and Monel	25	10

Powder Scarfing. In steel mills, powder metal cutting is used to scarf metals with alloy contents that are too high for oxyfuel scarfing. Typical applications include powder scarfing of large, bloomed ingots and small, unrolled ingots, slabs, and billets. Conditioning of these product forms by powder scarfing is less expensive and less time consuming than mechanical methods such as grinding.

Powder lancing is a piercing process in which the energy of the powder reaction is applied to oxygen lancing, permitting rapid, effective piercing of many materials that are difficult to pierce with a standard oxygen lance. These include iron and steel containing inclusions, reinforced concrete, firebrick, cinder block, aluminum billets, and sand and metal incrustation inside large castings.

Typical powder lancing applications include:

- Removal of blast furnace bosh plates
- Removal of large masses of iron (salamanders) that are deposited at the base of the blast furnace
- Cleaning of furnace lining
- Tapping furnace to remove slag
- Cleaning of soaking pits
- Removal of ladle skulls
- Piercing holes in reinforced concrete walls and floors

The efficiency of the powder lance in piercing reinforced concrete is preferred to the time-consuming, costly alternative of drilling with pneumatic tools.

Specialty Applications of Metal Powders

Stephen Claeys, Pyron Corporation
Steven Lampman, ASM International

METAL POWDERS are used in many diverse applications besides the traditional production of pressed and sintered parts. The primary market for metal powder is the production of powder metallurgy (P/M) parts, which are dominated primarily by iron and copper powders. Iron powders constitute the largest share, where over 80% of iron powders (on a tonnage basis) are used for P/M parts. Copper and copper-base powders follow iron in volume of powder used in P/M fabrication. Approximately 45% of all copper powder is used in the manufacture of self-lubricating bearings. It also is used as elemental additions in ferrous P/M parts. Another 40% is used for making P/M parts for builders' hardware, such as locksets, door latches, and other components that require a decorative finish as well as corrosion resistance.

Although iron and copper powders represent the largest share of P/M powder applications, they do not represent the entire range of P/M usage. Table 1 gives major applications of metal powders. This includes many nontraditional applications that do not involve parts production. For example, approximately 10% of iron powders (on a tonnage basis) are used for welding electrodes. Another 2% of the market is iron powders used for cutting, scarfing, and lancing operations (where the iron powder is injected into oxyacetylene flames to raise the temperature of the cutting zone). The remainder (8%) is used in various applications, including pharmaceuticals, food enrichment, copier powders, and other miscellaneous applications.

Nonferrous powders are also used in diverse applications. For example, copper powders are used as pigments for marine paints and as additives for metallic flake paints. Aluminum powder is also used extensively for explosives and pyrotechnics. This article briefly reviews some of these diverse and emerging applications of ferrous and nonferrous powders.

Primary emphasis is on various chemical applications of metal powders. The extremely high surface area of powders in general, and sponge powders in particular. has made them very valuable for uses where the rate of a chemical reaction is important. Not only can reaction rates be maximized by the use of powders, but they can be tailored very precisely, by varying particle size, size distribution, and morphology. For instance, fine iron powder is used in many applications where other metal ions need to be removed from a solution, especially copper. Important characteristics here are the fineness of the iron powder (<325 mesh), high surface area (sponge powder), and low cost.

Mirror backings are another application area of iron powder. Mirror backings are produced by laying down a coating of silver nitrate solution on glass, followed by a reduction of the nitrate by a sugar containing solution. The very thin layer of silver that is deposited is then backed by copper precipitated out of a cospray of a copper sulfate solution and fine iron powder suspended in water. This process was developed over 100 years ago and is still used for producing virtually all mirrors.

Copper also is receiving increasing attention from an environmental standpoint due to its toxicity. It can be precipitated out of many chemical process effluent streams by the addition of fine iron powder. This not only protects the environment but recovers a valuable resource. Conversely, the toxicity of copper can be put to good use. Copper is a potent biocide and is added to boat-bottom paints to prevent encrustation by various forms of marine life. Copper oxide has been used in the past, but is only suitable for epoxy-based, color-limited paints. Recent latex-based paints have been developed that use metallic copper and extend the color range available.

Copier Powders

Copier powders perform three essential functions in electrostatic image development: (a) they charge the finer toner particles to the proper level and appropriate polarity so that they can be attracted to the latent electrostatic image; (b) they physically transport toner from the developer reservoir, or sump, to the photoreceptor surface, where they are presented for development and, ultimately, are deposited in charged image areas; and (c) they act as a cleansing medium to remove toner particles that have been spuriously deposited in nonimage areas. In the cleansing function, copier powders also serve to scavenge background toner deposits, depending on machine design configuration. To accomplish these functions, electrostatic carrier powders must have the proper chemical, electrical, physical, and magnetic properties.

Core Powder

The majority of copier powders are coated with a thin resin to improve copy quality and life (see the section "Coatings and Surface Treatments"). The core powder is formed from sand, glass, iron, steel, nickel, aluminum, or ferrite materials. Ranging from approximately 40 to 600 μm in diameter, these granules can be irregular or spherical (often called "shot") in shape. These developer beads carry much smaller toner particles (from 5 to 30 μm in diameter) on their surfaces. They bring the toner powder close enough to the charged, latent image for development to occur. Electrostatic attraction causes the fine toner particles to be stripped from the carrier bead and to be deposited in image areas on the photoreceptor drum or copy sheet.

Iron and Steel. Irregularly shaped iron powders were first used in the early electrofax machines of the 1960s. These magnetic brush devices required a conductive, ferromagnetic granule to carry toner. Cascade developers that used spherical steel powders became prominent in the late 1960s. Magnetic brush devices use a variety of metal powder shapes, including sponge, fragment, and flake iron powders.

Ferrite. The introduction of spherical ferrite copier powders was a major advancement in the field of carrier bead technology. Spherical ferrites and spheroidized magnetite powders offer certain advantages over iron carrier core materials. Due to low remanence, these powders have improved flow properties, which enable rapid release from the magnetic brush roller when the field is relaxed. They are more resistant to changes in resistivity, even when uncoated, because the material is already in its fully oxidized

Table 1 Major applications of metal powders

Application	Powder used
Abrasive flashing	
Shot cleaning	Iron, steel, stainless steel
Abrasive wheels	Iron, steel, copper, tin, nickel
Alloy production	
Steels, electroslag steel, free-machining steel, cast iron, ductile iron	Nickel, lead, ferro-tungsten, ferro-silicon, ferro-molybdenum, ferro-manganese, iron, aluminum, tungsten, molybdenum, selenium, tellurium
Agriculture	
Animal feed	Iron
Animal medication	Cobalt
Chelate fertilizers	Iron
Farm machinery	Iron, steel, copper, bronze
Food enrichment	Iron, copper, manganese
Fungicides	Copper
Lawn and garden equipment	Iron, steel, iron alloy, copper, bronze
Seed cleaning	Iron
Seed coating	Aluminum
Soil conditioning	Iron, copper
Aerospace	
Brake linings	Copper, lead, tin, high-nickel alloys, graphite, iron
Cores (electronic components)	Iron, MolyPermalloy
Counterweights	Tungsten, copper, nickel, iron
Filters for fuel, hydraulic fluids, air	Bronze, stainless steel, nickel
Hardware	Aluminum, beryllium, titanium, iron
Heat shields	Beryllium, tungsten
Heat shield coating	Aluminum
Jet engine components	Superalloys
Magnetic shields	Aluminum, nickel, cobalt
Measuring aircraft drift	Aluminum
Rocket fuels	Aluminum
Repairing of components	Nickel chromium, nickel aluminum, stainless steel
Fretting/wear spraying	Molybdenum, copper nickel alloys, cobalt alloys, chrome carbide
Automotive(a)	
Air conditioners	Iron, steel
Alternator regulator, contacts, pole pieces	Platinum alloy, iron
Body solder	Steel, aluminum, lead, lead alloys
Bushings, bearings	Copper, lead, tin, iron, aluminum, graphite, bronze
Contacts	Tungsten, silver
Filters	Stainless steel, bronze
Fuel pump parts	Iron, copper, tool steel
Polychrome body finishes	Aluminum
Shock absorbers	Iron, aluminum
Spark plug (body)	Iron
Spark plug (corrosion protection)	Zinc
Transmission parts	Iron, copper, steel
Truck signal flares	Aluminum
Brake bands, linings	Copper, iron, lead, tin, graphite, brass, zinc
Tire studs	Tungsten carbide
Valve inserts	Tool steels, stainless steel
Building and construction	
Aerated concrete	Aluminum, iron
Asphalt roof coating	Aluminum
Caulking compound	Aluminum
Conductive and non-spark flooring	Copper
Decorative plastics and linoleum for floors, walls, and counter tops	Iron, brass, copper, aluminum, stainless steel
Lancing	Iron, aluminum
Protective coatings for canvas awnings and decks	Aluminum, zinc
Pipe joint compounds	Zinc, lead, copper
Carbides	Tungsten, titanium, tantalum, niobium
Catalysts	
Gasoline synthesis	Platinum, nickel
Hydrogenation	Platinum, platinum alloys, tungsten, molybdenum, rhenium
Promote combustion in space heaters	Platinum
Alkyls	Aluminum
Purifying hydrogen	Ruthenium, palladium
Reacting hydrogen and oxygen	Palladium
Raney catalyst	Nickel, aluminum
Rocket fuels	Iron
General chemical catalysts	Iron, copper, copper oxides
Cemented carbides	Cobalt, molybdenum, titanium, tungsten, nickel, tantalum, niobium
Chemical	Aluminum, lead, tin, copper, iron, nickel, manganese
Coatings	
Anti-fouling paints	Copper, copper oxide
Conductive paints and plastics, Silver, copper, nickel	
Corrosion-resisting paints	Stainless steel, aluminum, zinc, lead
Decorative paints	Aluminum, brass, bronze, zinc, stainless steel, lead, copper
Fabric coatings	Aluminum
Heat-reflecting paint	Aluminum
Hard facing	Cobalt alloys, nickel alloys,tungsten, molybdenum, stainless steel, tungsten carbide
Lacquers	Silver, brass, bronze, aluminum
Mechanical (peen) plating	Zinc, lead, tin, brass, copper, cadmium
Slurry coating	Nickel
Roof coatings	Aluminum
Spray coating	Iron, aluminum, zinc, tin, nickel alloys, copper, bronze, stainless steel, tungsten carbide, titanium carbide, silver, cadmium
Vacuum metallizing	Aluminum, copper, zinc
Coinage	Nickel, copper-nickel, titanium
Diffusion coating	Aluminum-titanium alloys, chromium, zinc, hafnium
Electrical and electronic	
Acoustical elements to improve microphone response	Stainless steel
Batteries	Nickel, zinc, silver, iron, lead, graphite
Breathers	Stainless steel
Brushes	Copper, silver, graphite
Cathode-ray tubes	Graphite, nickel
Contacts	Copper, silver, tungsten, molybdenum, iron, tin, platinum alloys, graphite, carbon
Crystal supports	Silver, copper, tungsten
Delay lines, computers	Iron, MolyPermalloy
Fuel cells	Nickel, zinc
Motor pole pieces	Iron, silicon-iron
Printed circuits	Copper, silver, palladium, gold, platinum
Relays	Iron, nickel, molybdenum
Semiconductors	Lead
Solenoids	Iron
Telephone components	Iron, brass, bronze, palladium, stainless steel
Terminals	Nickel
Thin-film resistors	Palladium, silver
Electrical discharge machining	
Electrodes, preforms	Tungsten, copper, silver
Electrochemical machining	
Electrodes	Copper, silver, tungsten
Hardware	
Lock components	Brass, bronze, iron, stainless steel
Heat treatment	
Furnace heating elements	Platinum alloys, tungsten, molybdenum
Furnace shielding	Aluminum, tungsten
Infiltration	Copper, brass, silver, manganese
Protection tubes	Platinum alloys
Thermocouples	Platinum alloys, tungsten, molybdenum, rhenium
Industrial (general)	
Air tool exhaust muffler	Stainless steel, bronze
Bearings and bushings	Copper, tin, lead, bronze
Bonded asbestos brake linings	Zinc, graphite, brass
Cutting tools	Tungsten, cobalt
Filters (liquids, gases)	Bronze, nickel, stainless steel, superalloys, titanium
Flame cutting and scarfing	Iron, aluminum
Fluids for magnetic clutches	Stainless steel (magnetic)
Friction material	Copper, tin, zinc, iron, lead, graphite
Gas engine exhaust purification	Platinum
Grinding wheels	Brass, bronze
Sound deadening plastic	Lead, graphite
Industrial explosives	
Mining	Aluminum
Joining	
Brazing	Copper, nickel, silver, cobalt, brass, gold, aluminum, nickel-cobalt-iron-chromium alloy, nickel-silicon-boron alloys
Coated and tubular electrodes for arc welding	Iron, nickel, chromium
Resistance welding electrodes	Tungsten, copper, silver, molybdenum, manganese
Soldering	Lead, tin, solder alloys
Thermit welding	Aluminum
Lubricants	

(continued)

Table 1 (continued)

Application	Powder used
Anti-galling pipe joint compound	Copper, lead, zinc, graphite
Copper lubricants	Copper
Greases	Lead, graphite
High-temperature lubricants	Aluminum, graphite
Plastic-filled metals	Iron, copper
Magnetic	
Clutches	Stainless steel (magnetic), iron
Inks for paper and tape	Iron
Permanent magnets	Iron, nickel, cobalt, aluminum, barium, ferrite, iron-aluminum alloys, molybdenum
Pole pieces, relay parts	Iron, iron-cobalt, silicon-iron
Soft magnetic parts	Iron, nickel-iron, silicon-iron
Traveling wave tube magnets	Platinum, cobalt
Medical and dental	
Dental amalgam	Silver, gold, alloys
Prosthetics	Superalloys
Insulin production	Zinc
Operating room air filters	Stainless steel, Monel, cobalt alloys
Pharmaceutical final filters	Stainless steel, Monel
Porcelain overlay dental restorations	Palladium
Prevention of infection of open wounds	Aluminum
Surgical gauzes and pins	Tantalum, zirconium
Surgical implants	Nickel and cobalt alloys
Orthopedic external fixation system	Titanium
Radiation shielding	Tungsten
Metal recovery	
Copper cementation	Iron
Gold cementation	Zinc
Metals from solution	Aluminum
Nondestructive testing	
Magnetic particle inspection	Iron
Nuclear engineering	
Control rods and reflectors	Zirconium, beryllium, hafnium, uranium
Filters for liquid sodium-potassium, gas, heavy water	Stainless steel, nickel alloys
Fuel elements	Iron, stainless steel
High-density concrete	Iron
High-density rubber	Lead, brass
Shielding	Aluminum
Shielding (gamma ray)	Tungsten, nickel, copper, lead
Shielding (neutron)	Boron, nickel, iron, lead
Computers	Iron
Office equipment	
Copying machine parts	Iron, stainless steel, bronze, aluminum
Facsimile equipment	Aluminum, steel, bronze, stainless steel
Recording tapes	Iron
Toner for electrostatic copying machine	Iron
Business machines	Steel, iron, stainless steel, brass, aluminum
Ordnance (military)	
Ammunition	Graphite
Armor-piercing cores	Tungsten, copper, nickel, cobalt
Anti-personnel bombs	Iron
First fire mixes and fuzes	Aluminum, graphite
Frangible bullets	Iron, lead
Fuze parts	Brass, stainless steel, iron, steel
Incendiary bombs, flares, tracers, torpedoes, etc.	Iron, aluminum, copper, cerium, magnesium, zinc, silicon
Missile filters	Nickel-base alloys
Projectile rotating bands	Copper, iron, brass
Proximity fuze cup	Nickel
Rocket launcher parts	Stainless steel, aluminum
Solid missile fuel	Aluminum, magnesium
Personal products	
Cigarette lighter flint	Cerium alloy
Cordless electric toothbrush and razor	Copper, nickel, brass, iron, bronze
Cosmetics	Zinc, aluminum
Enriched foodstuffs and cereals	Iron
Personal products (continued)	
Fingernail lacquer	Aluminum, copper
Floating soap	Aluminum
Mirror silvering	Iron
Pen points	Platinum, ruthenium, tungsten, stainless steel
Photography flash bulbs	Zirconium, cerium
Poker chips	Brass, bronze, copper-nickel
Vitamins	Iron
Petroleum and petrochemical	
Hydrocracking	Platinum alloys
Hydrogenation of olefines	Platinum
Petroleum fuels reforming	Platinum
Purifying hydrogen	Ruthenium
Oil drilling	Tungsten carbide, tungsten alloys
Plastic reinforcing	
Body solders	Steel, aluminum, lead
Cements for repairing castings and metal parts	Iron, stainless steel, aluminum, silver
Hull-smoothing cements	Iron, aluminum
Tools and dies	Iron, aluminum
Printing inks	
Metallic inks for offset, letterpress, rotogravure, etc.	Copper, brass, aluminum
Process industries	
Aluminum alkyls	Aluminum
Catalyst for promoting synthesis of water	Palladium
Coatings for papers	Aluminum, copper, tin
Control of beer fermentation	Stainless steel
Decolorizing alum solutions	Aluminum
Filters in chemical reactors	Stainless steel
Hydrogenation, etc.	Platinum metals, aluminum, nickel
Optical glass processing	Platinum alloy
Rayon production	Platinum alloy
Synthesis of caffeine	Iron
Synthesis of hydrocarbons	Iron
Pyrotechnics	Aluminum, graphite, iron, magnesium
Radio and television	
Cores	Iron, MolyPermalloy
Getters (TV tubes)	Barium, nickel
Permanent magnets	Iron, nickel, cobalt, aluminum, barium, ferrites, rare earths
Printed circuits	Copper, silver
Receiving tubes	Tungsten, molybdenum, nickel
Recreation	
Golf clubs	Tungsten, iron, brass
Sporting darts	Tungsten
Hunting knives	Brass, stainless steel, iron, nickel, silver
Shotguns	Iron, stainless steel
Fishing rod reels	Iron, brass, stainless steel
Outboard motors	Brass, steel
Sailboat hardware	Stainless steel, bronze
Self-lubricating parts	
Oil-filled	Copper, tin, lead, iron, aluminum, graphite, bronze
Plastic-filled	Iron, copper, silver
Sheet and strip	
Alloy	Cobalt, iron, nickel, aluminum, tungsten, molybdenum
Glass sealing	Iron, cobalt, nickel alloy
Nickel, cobalt	Nickel, cobalt
Steel	Iron
Ships	
Anti-fouling paints	Copper, copper oxide
Hull-smoothing cements	Iron
Ship-bottom paints	Aluminum, zinc
Soundproofing	
Acoustical plastics	Lead, graphite
Waterproofing	
Concrete	Iron, aluminum
Roof coatings	Iron, aluminum

(a) About 50% of all P/M parts are used in automobiles, many in engine and transmission applications.

state. An increase in resistivity causes a decrease in the solid fill. Furthermore, spherical ferrite copier powders are easier to reclaim from an exhausted developer mix, either by solvent washing or by high-temperature firing. As metal oxides, however, they have higher resistivities than metallic copier powders.

Sand. Uniformly graded sand was the first carrier core material used in office copiers based on cascade development. In these machines, the raw sand, or "flint shot," is coated with a dyed resin lacquer composition. In the United States, many of these older copiers and duplicators have been refitted with magnetic brush-type developer assemblies. Because the newer models use iron beads, the requirement for flint shot is relatively small. However, sand cascade machines are still in operation in many overseas markets.

Glass. Glass shot is a substitute for sand in cascade developer systems. This shot consists of extremely fine spherical beads. Although improved shape offers certain advantages in mixing and flow characteristics, higher cost has limited its use.

Aluminum. Another copier powder used in cascade development is aluminum shot. Initially, its low bulk density had strong appeal for equipment designers because it required minimal torque for mixing and blending. Also attractive were its low cost and conductive properties that yielded improved print quality over insulating-type cascade carrier core materials. Aluminum is not widely used as a core material, however, because it is nonmagnetic and has been superseded by the more advanced magnetic brush-type xerographic systems.

Nickel. Spherical or nodular-shaped nickel powder enjoys limited application as a carrier core material. Although relatively expensive, this material is not prone to oxidation and tends to maintain a uniform resistivity throughout the life of the developer mix.

Coatings and Surface Treatments

Carrier coatings serve as a release layer to prevent toner from impacting (filming) on the bead surface. Additionally, they provide a smooth, uniform carrier surface for toner attraction. Carrier coatings inhibit oxidation of steel powders and reduce the moisture sensitivity of the developer mix.

Carrier coatings prolong the life of the developer mix and protect the photoreceptor from abrasive carrier core materials. They control charge polarity through the use of certain dyes or other charge control agents that can be incorporated in the coating layer. Also, carrier coatings enhance the triboelectric effect.

Controlled electrical discharge is the basis of the xerographic process. Proper carrier coating is an effective way of exercising this control. Coatings vary from 0.2 to 4.0% of the core material weight, depending on the porosity of the carrier bead. The more porous carriers, such as reduced (sponge) iron, require a heavier coat weight.

The use of external additives that are dry blended with the developer mix is another method of controlling electrical discharge. These additives usually are incorporated in small amounts, from 0.25 to 2.0 wt%. They improve powder flow, modify triboelectric charge generation, increase carrier life, and improve photoreceptor performance. Furthermore, the proper additive can prevent toner accumulation on the carrier surface, the photoreceptor drum, or both. Amorphous silica is the most frequently used external additive because of its excellent performance as a promoter of powder flow or as a flow lubricant. Zinc stearate is used frequently in the toner industry as an anticaking and drum cleaning agent.

Carrier coatings and treatments include a broad variety of methods and materials, ranging from the use of raw or solvent-washed developer beads to the use of a controlled oxidation process to treat the surface of iron and steel powders. "Blued" iron and steel powder is frequently used as a carrier in the electrostatic copier industry.

Surface resistivity is lowest with uncoated or oxidized carriers. As the surface coating becomes thicker, the resistivity increases, thus making the particle less capable of effecting complete discharge of the developed image through the grounded magnetic brush. Lacquer-coated carriers provide the best compromise between print quality (low resistivity) and developer life (good release properties). A variety of polymeric materials have been used effectively for carrier core coatings. Styrene-acrylic copolymers, vinyl chloride-acetate copolymers, cellulose acetate butyrate, and nitrocellulose compositions are perhaps the most popular.

Powder Coating Processes. Four processes are used frequently for copier powder coating: surface oxidation, lacquer coating, fluorinated polymer coating, and powder coating. Of these, the first three currently are the most popular. Some original equipment manufacturers use a combination of oxidation and lacquer coating techniques.

Surface Oxidation. Iron and steel copier powders can be "blued" by a heat treatment to produce a surface layer of iron oxide. The process is carried out in an oven at approximately 285 to 290 °C (550 °F) for 15 to 20 min. Control of both air flow and moisture content is critical for ensuring color uniformity.

Equally important is powder circulation, which provides uniform exposure to the oxidizing environment for the entire surface of each bead. This can be accomplished by treating the metal powder in shallow, open trays with periodic agitation. Vibrating trays can be used to keep the steel powder in constant motion. Alternately, tube furnaces have been employed in which the carrier is contained inside a slowly rotating cylinder.

Lacquer Coating. Resin solutions, usually in the range of 10 to 20% solids, are applied to raw or oxidized carrier core materials in one of four equipment types: vibrating tub, cement mixer, vacuum tumble dryer, or fluidized bed coater. In each process, the lacquer is prepared separately by dissolving the appropriate resin, charge-orienting dyes, and other additives in a volatile solvent, typically toluene or methyl ethyl ketone, at elevated temperature.

The raw copier powder is heated separately to approximately 80 to 95 °C (175 to 200 °F) prior to introducing the hot lacquer solution. The lacquer is added rapidly to the agitated powder in a vibrating tub or cement mixer. Good ventilation is essential for safe operation. As the lacquer dries, the solid-liquid blending operation passes through a critical phase in which a tendency for agglomeration exists. The dry powder must be screened to remove doublet, triplet, and other multiparticle clumps.

Particle agglomeration can be controlled by using either a vacuum tumble dryer or a fluidized bed coater. Because they are closed systems, these devices afford a means for recovering the lacquer solvent. They are more economical to operate than either the vibrating tub or the cement mixer, both of which discharge to the atmosphere. The lacquer solution is normally injected slowly, and the highly agitated powder charge provides a better mechanism for separating individual granules during the coating and drying cycles.

Fluorinated Polymer Coating. The fluidized bed coater is used almost exclusively to apply polytetrafluoroethylene coatings to carrier core materials. Polytetrafluoroethylene resins require a high-temperature curing cycle, usually ranging from 285 to 345 °C (550 to 650 °F). These temperatures are beyond the capability of most tumble dryers.

Electrostatic powder coating is the same as the powder painting process used to coat appliances, lighting fixtures, and various metal parts. It is a solventless coating method in which the coating resin, in the form of a finely divided powder, is sprayed onto or blended with the agitated carrier beads at room temperature. An applied electrical field or triboelectric force causes the fine resin particles to cling to the surface of each bead. During subsequent heating, the resin particles melt and coalesce to produce a uniform coating.

Flake Pigments

Metallic flake pigments are additives that are mixed directly into paints or other coating vehicles. Aluminum, copper, and gold bronze constitute the vast majority of metallic flake products sold in the United States. However, other metallic flake pigments are produced.

Stainless steel flake pigments have been available since the late 1940s. Originally, stainless steel flake was available in a variety of alloys, but currently only the 316L alloy is available. Stainless steel flake is used in high-durability coatings that are exposed to highly corrosive atmospheres. Weathering improves the appearance of stainless steel by actually polishing the coating, thus making it smoother and brighter.

Zinc flake has been used in corrosion-resistant paints because of the chemical properties of zinc; however, zinc dust is inexpensive and can perform the same function.

Nickel and silver are also available in flake form. Because of their superior conductivities, they are employed mainly in electrical applications.

Aluminum Flake Pigments. Aluminum pigments for coatings generally are classified into four types—leafing paste, leafing powder, nonleafing paste, and nonleafing powder, These types are further classified by fineness. The American Society for Testing and Materials (ASTM) has issued specifications for aluminum flake pigments (ASTM D 962, Part 28). Testing methods are described in ASTM D 480. Specifications are given for screen analysis of the coarse portion of the product, percentage of nonvolatiles, and leafing values, however, specific surface area of the flake and optical properties of each type are not covered. Aluminum flake powders used in noncoating applications, such as pyrotechnics and commercial explosives, also are not covered by these specifications.

Finely divided aluminum is extremely reactive with acids, bases, chlorinated solvents, and water. One of the by-products of the reaction is hydrogen gas, which can produce bulging and possible rupture of closed production containers. As an air-borne dust cloud, finely divided aluminum is potentially explosive. A dust cloud of aluminum powder will flash if an ignition source is present, and an explosion will occur if the concentration is within explosive limits.

Aluminum flake pigments should not be exposed to high-shear, high-speed agitation during mixing because this tears the delicate flakes, strips the protective lubricant coating from the particle surfaces, and creates reaction sites. This type of agitation reduces the amount of leafing and reduces the brilliance of the coating of leafing aluminum pigments. High temperatures during mixing affect the thin stearic acid layer on leafing pigments, resulting in loss of leafing properties and luster.

Leafing is the ability of the flake to rise to the surface of a coating and remain there in spite of a higher specific gravity than the coating vehicle. Flakes have a thin layer of a lubricant, such as stearic acid, and when they rise to the coating surface through currents caused by the evaporation of the solvent, they are held there by the surface tension of the vehicle. A minimum of 25 dynes/cm surface tension is required for successful leafing. Higher surface tension solvents such as high aromatic solvents are suitable for leafing.

Lead and zinc dryers deleaf the flakes and have been used to change leafing grades to nonleafing paste. Good leafing vehicles typically are simple compositions, such as a varnish. Nitrocellulose is complex and renders a leafing grade of aluminum nonleafing.

Roof coatings comprise the largest market for leafing aluminum paste. Aluminum roof coatings are either fibrated or nonfibrated asphalt. In nonfibrated systems, a coarse grade of aluminum paste (ASTM D 962 type 2 class C) is used. Coarse leafing grades have a rapid rate of leafing, resulting in a continuous metallic film on the surface of nonfibrated asphalt. The use of fine grades of flaked powder tends to produce a spotty appearance. Typical aluminum paste pigmentation quantities for nonfibrated asphalt are 240 to 360 g/L (2 to 3 lb/gal).

In fibrated systems that contain approximately 5% fiber, such as asbestos, a fine grade of aluminum paste is required for a bright metallic appearance. Fibrated coatings are applied in thick layers so that speed of leafing is not as important as the surface area of the pigment flakes. However, different asphaltic cutbacks can be used in roof coatings. In fibrated coatings, typical pigmentation amounts are in the range of 120 to 360 g/L (1 to 3 lb/gal), with the majority around 180 g/L (1.5 lb/gal).

Aluminum is well suited for roofing applications because it provides a moisture barrier and has high reflectivity. Black asphalt roof coatings absorb and transmit heat into the building, resulting in increased air conditioning costs in the summer. Aluminum reflects much of the heat, resulting in lower interior temperatures, as well as extending the life of the roof.

Aluminum paints exhibit special properties, such as moisture resistance, good durability, and high reflectivity, which account for their wide industrial usage. Moisture resistance enhances protection of structural steel, while high reflectivity increases the usefulness of aluminum paint coatings on highway bridges and guard rails. Usually, aluminum paint is used as a topcoat because it does not exhibit the corrosion resistance of pigments such as zinc dust.

However, urethane primers containing nonleafing aluminum are gaining acceptance. Urethane paints increase coating life and chemical resistance. Structural steel coatings usually contain a standard lining paste (ASTM D 962 type 2 class B). Aluminum pigmentation rates vary between 180 and 240 g/L (1.5 and 2.0 lb/gal).

Aluminum pastes are used in heat-resistant coatings on surfaces such as smoke stacks. The use of silicone resins has resulted in coatings with good durability at temperatures up to 400 °C (750 °F). Aluminum pastes should be a standard leafing lining paste (ASTM D 962 type 2 class B) or a fine lining paste (ASTM D 962 type 2 class A). For these paste coatings, aluminum pigmentation is typically between 120 and 180 g/L (1 and 1.5 lb/gal).

Water-dispersible aluminum flake pastes are available for use in water-reducible paints. Flake grades have added surfactants that allow the mineral spirits in the paste to mix with water.

Automotive coatings comprise the largest market for nonleafing pastes. Properties required for this application include acid stain resistance, sparkle, seeding, hiding power, and gloss. Standard grades of nonleafing aluminum are produced from commercially pure aluminum (a minimum of 99% Al). When aluminum particles in a coating are exposed to the atmosphere, as in the one-coat finishes normally used in automotive coatings, they may react and stain the coating. High-purity aluminum (99.98% Al) is less reactive than commercially pure grades and offers some protection against staining. Grades of nonleafing aluminum pigment that use high-purity aluminum are referred to as "acid stain resistant."

When flake particles align parallel to the surface of the paint film, the appearance is typical of a standard leafing pigment. When the flake orients at various angles to the surface, light reflects in a scattering pattern, which is described as "sparkle." The sparkle effect causes extremely fine flakes to appear much larger, giving a more metallic appearance without seeding problems. Seeding problems occur when flakes project above the surface of the paint film.

High gloss in automotive coatings is achieved by using an aluminum flake pigment with a short flake diameter. To obtain the combination of a highly metallic appearance and high gloss, more than one aluminum pigment frequently must be used in a coating.

Printing inks use aluminum pigment to simulate the richness of silver at a much lower cost. Aluminum pigmented inks can be used in offset, letterpress, flexogravure, and rotogravure printing. Rotogravure inks pigmented with 10 to 20% Al are printed at extremely fast speeds and use fast-drying solvents. The brilliance of the aluminum depends on the size of the flake. The larger the flake size, the higher the brilliance of the coating. However, maximum usable flake dimensions depend on the depth and width of the etching on the printing cylinder surface.

Ink is picked up in the etching and transferred to the paper. Coarse particles overfill the etching and are removed by a doctor blade before reaching the paper. Leafing is not important in this application because the ink is applied and dried quickly. However, most tinted metallic inks used in rotogravure printing require leafing grades of aluminum flake powder.

Flexogravure inks are similar to those used in rotogravure. A standard lining grade is recommended. If maximum hiding effect is required, fine lining grades should be used.

Letterpress inks pigmented with 20 to 30% Al require a fine lining grade of leafing paste or flake. There is more transferal of the ink in this application, and optimum transfer properties are obtained with pigments that have short flake diameters. Letterpress inks dry more slowly than rotogravure and flexogravure inks, so leafing properties are important. All pastes contain mineral spirits or a combination of mineral spirits and a high aromatic solvent. Offset printing requires fine lining grades and is similar to letterpress in that a short flake diameter and high leafing are important.

Plastics. Aluminum is used in plastics to provide a silver color and opacity. Aluminum flake powders are normally blended to a paste consistency with a compatible plasticizer and sold to a compounder. This product is mixed with a plastic and extruded, usually in pellet form. The pelletized colorant is then mixed at low concentrations with a higher melting plastic, followed by heating and extruding into final form. The fineness of powder used depends on the effect desired; coarse grades provide high brilliance, and fine grades provide higher opacity at the sacrifice of higher brilliance.

Copper and Gold Bronze Pigments. Gold bronze pigments are brass alloy flake powders ranging in composition from 70Cu-30Zn to 90Cu-10Zn. Copper and gold bronze flake pigments are made by either stamping or dry ball milling. The recent trend has been toward ball milling.

The color shades of gold bronze are determined by the alloy composition used and by

Table 2 Properties of reactive materials

Material	Density, g/cm^3	Melting point (approximate) °C	Melting point (approximate) °F	Boiling point °C	Boiling point °F	Heat of fusion(a) kJ/mol	Heat of fusion(a) cal/g mol	Heat capacity J/g · °C	Heat capacity cal/g · °C	Combustion temperature(b) °C	Combustion temperature(b) °F
Aluminum	2.70	660	1220	2467	4472	0.790	2550	0.90	0.215	3500	6300
Magnesium	1.74	649	1200	1090	1994	8.954	2160	1.02	0.244	3100	5600
Titanium	4.54	1660	3020	3287	5948	6.90	5000	0.27	0.064	3000	5400
Zirconium	6.51	1852	3365	4377	7910	15.450	5500	0.52	0.124	4500(c)	8100(c)

(a) Source: Ref 1, p B-231 to B-235. (b) The adiabatic (no gain or loss of heat) combustion temperature in oxygen at 100 kPa (1 atm) pressure. (c) Metals such as zirconium produce temperatures that approach the maximum possible in a chemical reaction.

Table 3 Heats of formation of inorganic oxides

Reaction	Heat of formation, ΔH_0(a) g cal/mol	kJ/mol	g cal/g
$2\,Al(solid) + 3/2\,O_2(gas) = Al_2O_3$	−404 080	−1691.5	−3962
$Mg(solid) + 1/2\,O_2(gas) = MgO$	−144 090	−603.2	−3602
$Ti(\alpha) + O_2(gas) = TiO_2$	−228 360	−955.9	−2858
$Zr(\alpha) + O_2(gas) = ZrO_2$	−262 980	−1100.8	−2134
$C(graphite) + O_2 = CO_2$	−93 690	−392.2	−2129

(a) Negative values mean heat is given off. Source: Ref 1

subsequent heat treating. Heating brass flakes results in partial oxidation of the flake surface, imparting a deepening of color. The higher the level of zinc, the greener the shade; the lower the zinc level, the redder the shade. The red shade is generally termed "pale gold," and the green shade is called "rich gold." Partially oxidized or heat treated powders are referred to as "deep golds" or "colored bronze."

Unlike aluminum, gold bronze powders are not color stable and may tarnish on exposure to the surrounding environment. Tarnish-resistant grades of gold bronze are produced, but heat treating usually results in lower brilliance and opacity,

Decorative Applications. Gold bronze is used primarily for decorative purposes, such as cigarette packages, cigar boxes, wrapping paper, greeting cards, and picture frames. Coarse grades (having up to 60% +325 mesh) are used in lithograph dusting, in which an adhesive is applied to the surface to be printed, and dry powder is dropped onto the wet adhesive. Excess powder is then lightly brushed off, and the powder is retained only where the adhesive was applied.

Paints. Ready mixed paints pigmented at approximately 360 g/L (3 lb/gal) use standard fineness pigments, which yield a smooth, brilliant surface. Leafing is required for this application, and the standard fineness grades (having 1 to 5% +325 mesh) exhibit good leafing properties. Aluminum ready mix paints require a nonreactive coating vehicle, such as varnish.

Aerosol-propelled paints demand a fine lining powder (having less than 0.1% +325 mesh) with good leafing properties. Coarser grades should be avoided because they contain large particles that would clog the spray nozzle. The sprayed paint film should have a continuous film of metal flake on its surface (leafing). Brilliance is determined by flake size, leafing value, and amount of polishing performed on the flake during manufacturing.

Gold bronze and copper powders react with vinyl resins, which causes gelling. Many types of stabilizers have been used, such as organotin, organozinc, barium, cadmium, and zinc complexes. A stabilizer should be chosen according to application.

Gold bronze and copper react with nitrocellulose lacquers, causing greening and gelling of the lacquer. Citric acid has been useful in improving stability, but must be evaluated for each application. Most gold bronze nitrocellulose lacquer mixes are used immediately after production because of inherent product stability problems.

Plastics utilize gold bronze and copper pigments for aesthetic reasons. Unlike aluminum, these pigments usually undergo one heating cycle. Gold bronze and copper powders do not pose the same explosion hazard as aluminum powders and can be mixed dry with the plastic, rather than molded into a pellet. Also, gold bronze oxidizes with heat, so heating must be kept to a minimum, or flake discoloration will occur. The fineness of the powder used depends on the desired effect. Coarse grades produce brilliance, and fine grades produce high opacity.

Printing ink pigmentation is similar to that of aluminum inks, but the pigmentation rate is much higher. Typical rotogravure and flexogravure inks contain between 30 and 40% gold bronze, due to the difference in density between brass and aluminum. Letterpress and offset ink pigmentation is approximately 50%. Almost all of the gold bronze powders used in printing inks are 99.9% −325 mesh.

One method of evaluating flake pigments is by a water coverage technique, in which the surface area of flake coatings is approximated. A fixed weight of flake is spread on water and mixed until the flake layer is approximately one flake thick. The area of the flake is then measured and calculated into square centimeters per gram. Powders can also be evaluated by measuring the particle size by optical or mechanical methods.

Both particle size and water coverage are important controlling factors. In rotogravure and flexogravure printing, a large size particle results in high brilliance, while a high water coverage powder results in high opacity. In letterpress and offset inks, small flakes have the best transfer properties, producing higher opacity and a more brilliant finish.

Fuels

Metal powders are used as fuels in solid propellants, pyrotechnic devices, explosives, and similar applications. These applications require properties of high density, high flame temperature, and high energy output per unit of expended fuel. The most widely used type of metal powder for propellants, pyrotechnics, and explosives is aluminum powder. Other metals used, particularly pyrotechnics, include magnesium, zirconium, titanium, tungsten, manganese, beryllium, and cerium. Table 2 lists properties of aluminum, magnesium, zirconium, and titanium.

In fuel applications, the metals undergo highly exothermic (heat-producing) chemical reactions to yield stable oxides. Table 3 gives data for the heats of formation of some oxides. Comparison of the values for the metals given with those for carbon shows that all four produce more heat per unit than carbon. Carbon is the basis of many common fuels because it is easily ignited, safe, reasonably stable, abundant, inexpensive, and its products of combustion are gaseous.

For many metal-oxygen systems, the combustion temperature is limited to the boiling point of the oxide. Because aluminum and magnesium form oxides with high heats of formation and very low dissociation pressures, they are extremely stable. Dissociation pressure for a reaction such as :

$$Al_2O_3\,(solid) \rightarrow 2Al\,(solid) + {}^3\!/_2\,O_2\,(gas)$$

is the partial pressure of oxygen in equilibrium with the metal and oxide phases at a given temperature. This equilibrium oxygen pressure is referred to as the dissociation pressure of the oxide and is an indication of its thermal stability.

Solid Propellants. To obtain higher energy in solid propellants, metal powders such as aluminum, boron, or beryllium may be added. Beryllium, however, is not currently used, because of toxicity problems. Solid propellants used in rocket propulsion systems are composite structures that consist of a solid oxidizer, solid fuel, and polymeric fuel binder, mixed and formed into a monolithic grain or unit. These grains, which range from several millimeters to several meters in length, are usually cylindrical in shape and have various configurations to facilitate the type and rate of burning desired. Recent advances include the development of hybrid fuel-oxidizer systems that spontaneously ignite in the presence of metal powders (Ref 2).

Ammonium perchlorate is the major oxidizer used in solid propellants, but ammonium nitrate also has limited special applications. Examples of binders are polyesters, polyethers, and polybutadienes terminated with chemically reactive groups such as carboxyl or hydroxyl. Two advantages of using aluminum as a fuel in solid propellants are its high combustion energy and its ability to reduce water and carbon dioxide to lower molecular weight (MW) gases during burning. Low molecular weight gases are desirable because the energy output of a propellant is inversely proportional to the square root of the average molecular weight of product gases. Thus, a mixture of diatomic hydrogen (MW = 2), carbon monoxide (MW = 28), and diatonic nitrogen (MW = 28) is more energetic than a mixture of water (MW = 18), carbon dioxide (MW = 44), and diatomic nitrogen (MW = 28). The disadvantages are vaporization and ignition difficulties, slow burning rate, formation of reaction products of higher density than the metal, and, often, lower combustion efficiencies.

The fuel on the Space Shuttle solid rocket boosters is aluminum powder mixed with ammonium perchlorate in a rubber binder. Important characteristics here are the particle size and especially the distribution of the particles within the binder mix. Aluminum as a rocket fuel has a very high specific impulse, but suffers from some obvious disadvantages in a solid rocket application, not the least of which is the lack of controllability, or the ability to be throttled up or down. Engines have been proposed that use aluminum powder in a slurry, which when mixed with a suitable oxidizer, allows throttle control.

Iron powder has several novel applications as a fuel source. In a reversal of the process that uses hydrogen to reduce iron oxide to iron, iron powder itself has been proposed as a energy storage medium. The familiar reduction reaction, which produces water vapor, occurs readily at temperatures near 1000 °C. However, at temperatures near 300 °C, the reverse reaction is favored, where water vapor will oxidize iron, with the liberation of hydrogen. This concept is used in the steam treating (blackening) of P/M parts. It has been proposed that iron powder could be used to "fuel" vehicles by passing water vapor over a bed of heated powder and burning the hydrogen produced. The spent (oxidized) powder would be returned to a centralized facility for reduction.

Automobile air bag inflators use sodium azide and iron oxide powder as the propellant. It is important in this application that the gas produced be inert because it is exhausted into the interior of a vehicle during use. The reaction is:

$$2NaN_3 + 0.33\, Fe_2O_3 \rightarrow 0.67\, Fe + Na_2O + 3N_2$$

Early inflators were either too explosive in their release of gas, or burned too slowly. By tailoring the particle size distribution of the ingredient powders, the burn rate has been controlled.

Hand warmers for use in cold weather use iron powder as fuel. Because iron oxidizes in a slow and controllable fashion, these devices can be manufactured quite easily and inexpensively with a guaranteed measure of safety. In an extension of this application, foods packaged in self-heating containers are now available. These were first developed for the military, but have found a wide demand in the consumer marketplace.

Specifications for Aluminum. Atomized aluminum is the most widely used metal powder in the propellant, explosive, pyrotechnic, and incendiary fields. The most important factors affecting properties of aluminum are method of manufacture, chemical purity or grade, particle size or class distribution, surface area, and apparent density. Both spheroidal and flake powders are used. Aluminum flake powders are produced by milling regular atomized powders under controlled conditions. Some flake powders are coated with stearic acid and show good stability against gassing in an aqueous environment.

Dedusted grades have been specially treated to reduce the dust hazards associated with flake powders. The treatment consists of adding a nominal 0.1% Teflon, which after processing, weakly binds the aluminum flake powder and prevents dust formation. Further processing in slurries breaks the Teflon bonds, restoring the powder to its effective state. This treatment does not disturb the integrity of the stearic acid coating on the flakes, and the powders maintain good sensitizing properties.

Magnesium powders gained importance during World War II, when large quantities were used for pyrotechnics and explosives. They currently are used to (a) manufacture Grignard reagents, which are organometallic halides such as ethyl magnesium chloride (C_2H_5MgCl) that are used in organic synthesis to produce pharmaceuticals, perfumes, and other fine chemicals; (b) effect chemical reductions, as in the manufacture of beryllium and uranium; (c) act as a light source in flares and photoflash bombs; (d) modify metallurgical properties of other metals, such as in the production of ductile iron and the removal of sulfur from hot iron products of blast furnaces; and (e) function as additives in electric welding electrode flux. At one time, magnesium powders were extruded into sections for structural applications. However, they are no longer used to produce such parts.

Almost all powder is produced from pure magnesium, although some magnesium-aluminum alloy powder containing up to 50% aluminum is used for photoflash bombs and for desulfurizing hot metal. Powder is produced by a variety of methods of comminution of solid or liquid metal.

Table 4 Types and forms of magnesium powder

Type	Manganese(a), %	Form
Type I, grade A	99.80 min	Shavings, turnings, flakes, or combinations of these
Type I, grade B	(b)	Same as above
Type II	99.80 min	Oblong chip-like particles with rounded edges
Type III	99.80 min	Granular or spheroidal particles
Type IV	95.50 min	Shavings, turnings, flakes, or combinations of these

(a) Present in manufacturing stock. (b) To meet specification

Magnesium powders produced by chipping and milling are used for military flares, fireworks, and production of uranium. In addition, these powders are used in the chemical and pharmaceutical industries and in the iron- and steelmaking industries, where they are used as desulfurizing agents and inoculants for the nodulation of gray iron. Another method produces powder of less than 325 mesh for flares by grinding.

Magnesium has been used to a small extent as a fuel in solid propellants. Compared with aluminum, it has a lower specific impulse, and there are hazards in handling the fine powder. Substitution of particles of a magnesium-aluminum blend reduce the tendency to agglomerate, but the specific impulse decreases with increasing magnesium content of the particles.

Important properties of magnesium powder include form, composition, granulation, and apparent density. Information on the forms and chemical requirements is given in Tables 4 and 5, respectively.

Manganese. Manganese powder is employed in delay cartridges (grade I) and pyrotechnic mixtures (grades II and III). Delay cartridges consist of a primer (an explosive charge that can be set off by percussion), an ignition charge (the delay column or delay train) that burns at a predetermined rate, a transfer charge that transfers the energy of the delay column to the main charge, and the main charge itself. Table 6 gives requirements of manganese powders.

Tungsten. The major use of tungsten powder is as a fuel in long-burning delay compositions. These compositions, which are discussed later in this article, have burning times greater than 8 s/cm (20 s/in.). Table 7 shows the particle size distribution.

Zirconium. This powder is used in pyrotechnic and special ignition compositions. Its combustion behavior differs from that of aluminum in several respects. In the combustion of zirconium, the oxide—ZrO_2—a stable combustion product, dissolves in molten zirconium metal, which can react with nitrogen to form unstable intermediates. Zirconium is more easily ignited than aluminum, possibly because the oxide is a less protective coating than Al_2O_3, While alumi-

Table 5 Chemical requirements for magnesium powders

Chemical requirement	Composition, % — Type I, Grade A	Type I, Grade B	Type II	Type III	Type IV
Volatile at 105 °C (221 °F), max	0.10	0.10	0.10	0.10	0.10
Oil and grease, max	0.02	0.02	0.02	0.02	0.02
Carbides, max	0.004	0.004	0.004	0.004	0.004
Material insoluble in sulfuric acid, max	0.15	...	0.15	0.15	0.15
Metallic iron, max	0.05	...	0.05	0.05	0.05
Ferric oxide, max	0.25	...	0.25	0.25	0.25
Aluminum(a), max	...	10.0	...	...	2.50
Free metallic magnesium, min	96.0	...	96.0	98.0	95.50
Aluminum plus free metallic magnesium, min	...	96.0	...	...	...

(a) Calculated as percentage of pure aluminum, not as aluminum oxide (Al_2O_3) or other compound

Table 6 Requirements for manganese powder

Chemical or size requirement	Composition, min % — Grade I	Grade II	Grade III
Total manganese	99.5	98.0	95.0
Free metallic manganese	99.0	...	...
Granulation			
−200 sieve	...	98.0	99.0
−230 sieve	100	...	...
−325 sieve	98	...	...
Particles <10 μm diam	50 (max)	...	...

Table 7 Particle size distribution for tungsten powder

Particle size distribution, μm	Weight, % — As supplied, min	As supplied, max	Lab milled, min	Lab milled, max
Type I, agglomerated				
0–5	6	18	...	...
5–10	22	38	...	...
10–15	20	35	...	...
15–20	12	25	...	...
20–25	0	25	...	...
Type II, deagglomerated				
0–1	0	2	0	5
1–2	1	5	1	11
2–3	5	14	10	20
3–4	10	20	15	27
4–5	11	23	15	27
5–6	11	21	10	20
6–7	8	17	4	12
7–25	bal	bal	bal	bal

num melts before it ignites, zirconium particles ignite before they melt. Zirconium is highly reactive with a variety of oxidizing agents. For example, the reaction

$$Zr + \text{halogen} \rightarrow ZrX_4$$

is self-sustaining and exothermic and can be initiated at relatively low temperatures.

Table 8 gives the chemical requirements for zirconium. Particle size and particle size distribution for the two types, which are extremely important, are shown in Table 9. The average particle size requirements of type II powders are 2.5 ± 1.0 μm for class 1, 2.0 ± 0.3 μm for class 2, and 3.0 ± 1.0 μm for class 3. The ignition gain percentage, which is the percent gain in weight when a sample of zirconium is ignited in a crucible over the full heat of a Meker burner for 15 min, is another critical parameter of zirconium powders. The ignition gain requirements for type II zirconium powders are 30.2 to 33.0% for classes 1, 2, and 3 and 30.0 to 33.0% for class 4.

Because of the safety problems encountered in storing and handling zirconium, special labeling and marking of containers are essential. These special labeling and marking requirements for zirconium are:

- Zirconium metal powder (minimum 25% water)
- Danger! Flammable solid. May explode if water content is 10% or below
- Keep wet in storage—dry powder may be ignited by friction, static electricity, or heat.
- Do not attempt to loosen or remove material from container with any tool.
- Keep away from heat, sparks, and open flame.
- Keep from freezing.
- In case of spillage, keep wet and remove carefully.
- In case of fire, smother with foam-type extinguisher or sand.
- Do not use carbon tetrachloride, carbon dioxide extinguishers, or water.
- Wear goggles or face shield and fire retardant clothing when handling.

Titanium. Thermal ignition of titanium-base pyrotechnics can be controlled by encapsulating the metal powders in titanium oxide. As the powder is heated, the oxide dissolves to form a reactive surface (Ref 3).

Fillers

Metal powders are used as fillers in various materials, such as polymers, epoxies, or other binder systems. Additives for polymers are common, but the use of iron powder in hydraulic cement dates to antiquity. It is still in use today to create hydraulic and waterproof cements, which expand on curing. An interesting extension of this application is the use of 70% iron powder mixed with 30% portland cement to create an amazingly strong filler putty. This inexpensive material can be used to repair machine bases, repair holes in concrete, and set posts and supports. The concrete used for some nuclear reactor containment vessels also has additions of iron powder to impart high density, high strength, and neutron absorbing capabilities.

Metal powders have numerous applications as fillers in polymers and adhesives. Prealloyed bronze powder is added to Teflon (i.e., DuPont de Nemours & Co., Inc., Wilmington, DE) and other plastics to provide strength and wear resistance. Many high-strength epoxy adhesives also contain fine metal particles. As with any particle strengthening system, these act to increase the crack path length and energy absorbed during crack propagation, but in addition, also improve high-temperature properties due to increased thermal conductivity and creep resistance. High-temperature silicone adhesives and sealants also contain copper powder or fibers for enhanced thermal conductivity. Silicones are also manufactured using copper powder as a catalyst.

Most inexpensive bronze "castings," especially for statues and figurines, are not cast from melted bronze, but are made in rubber molds filled with copper, bronze, or brass powders in a plastic binder.

Radar absorbing "stealth" coatings can be made with iron and other metal powder additives in a plastic binder. The radar waves become "trapped" and dissipated by repeated internal reflections and absorbtions within the coating.

Metal Powder Fillers for Polymers. Metal powders are added to polymers primarily to change the appearance and/or the physical properties of the polymer. Increased demand in the automotive industry and related fields for high-performance parts that are lighter in weight than conventional metals has led to the use of polymer-based composites. A large number of these composites contain metal powders that match the metallic appearance of the components they replace.

In other applications where the physical properties of the polymers are modified for a specific end use, metal powders are used to change the insulative nature of the covalently bonded polymer or to increase electron density. Physical properties and functions that are affected in specialty polymers include:

- Electrical conductivity
- Thermal conductivity
- Electromagnetic shielding
- Anti-static surfaces

Powder metal-filled plastics are manufactured in various shapes and sizes, depending on end use. Sheet, films, rods, fibers, molded parts, and liquids are available. Metal forming operations such as compression molding, extrusion, injection molding, cold drawing, deep drawing, thermoforming, and stamping are used to produce polymers. Where a special network of metal powder is desired within the polymer, compression molding is preferred because it does not disturb the relative position of the particles. Typical polymers used in these composites in-

Table 8 Chemical requirements for zirconium powder

Chemical requirement	Composition, %				
	Type I	Type II			
		Class 1	Class 2	Class 3	Class 4
Total zirconium, min	96.0	94.0	95.0	95.0	94.0
Calcium, max	0.05	0.05	0.05	0.05	0.10
Iron, max	0.30	0.20	0.03	0.05	0.20
Aluminum, max	0.10	0.30	0.15	0.10	0.30
Hydrogen, max	0.20	0.20	0.20	...	0.17
Chloride as Cl, max	0.03	...	...	...	...
Silicon, max	0.10	...	...	...	...
Tin, max	0.75	...	...	...	...

clude polyvinyl chloride, polymethyl methacrylate, polyethylene, polycarbonate, polyurethane, polyvinyl acetate, teflon, nylon, epoxy, and bakelite. Filler powder metal varies depending on the application, but may be made of pure metals such as copper, nickel, silver, aluminum, and iron, or alloys such as stainless steel.

Molded articles are usually made by mixing the metal and polymer powders thoroughly (in a V-mixer, for example) for sufficient time to allow the buildup of static charge on the surface of the polymer powder, which aids in attracting the smaller metal particles. To minimize shear, compression molding of the homogeneous mixture follows by applying sufficient heat and pressure to fuse the polymer particles together in the desired shape. The polymer temperature is controlled to be above the glass transition temperature, but not high enough above the melting point to cause degradation of the polymer. With ultrahigh molecular weight polyethylene (average molecular weight of 2 to 8 million), the molding temperature must be at least 50 to 70 °C (120 to 160 °F) above its melting point to guarantee maximum particle-particle fusion and minimum porosity. Generally, high molecular weights result in high polymer viscosities that require high molding temperatures. Typical molding conditions for polyvinyl chloride are 150 °C (300 °F) and 70 to 100 MPa (10 to 15 ksi). Control of pressure, temperature, and time is important to minimize porosity in the molded part. This process is particularly suited for molding thermoset polymers, such as bakelite, where the final curing occurs during the heating cycle.

Extrusion and injection molding of the metal and polymer mixture also are widely used because these processes provide continuous production and superior economics compared to compression molding. The polymer, however, experiences high shear rates in these processes, which cause metal particle segregation and orientation along the flow lines. Shear usually results in the appearance of crack-like flaw lines in the filled injection molded parts when two separate flow patterns of molten polymer merge. These flaw lines are areas with minimum density and are almost devoid of filler particles. In injection molded parts, a skin of unfilled polymer exists where the liquid polymer adheres to the cold mold surface during solidification. This surface layer may have to be removed if a conductive outer surface is desired.

Metal powders also can be dispersed in liquid polymers or solutions, such as in metallic paints containing aluminum flakes. Additionally, cast resins such as epoxy, polyesters, polyurethanes, and silicone rubber may contain metal particles in the liquid phase. Dispersion is accomplished by stirring before the viscous slurry is poured into the mold. Some segregation is expected, although viscosity of the resin can be adjusted to minimize segregation. The resin is then cured by the addition of a suitable catalyst at either ambient or elevated temperature. An added refinement of this technique calls for the orientation of nickel particles in the liquid phase by an external magnetic field, which has been reported to enhance the electrical conductivity of the composite.

Food Enrichment

Recently, the use of iron for food enrichment has been estimated at approximately 40% ferrous sulfate (dried), 50% elemental iron powder, and 10% other iron salts. Although water-soluble ferrous sulfate, which is abundant and inexpensive, may seem to be the favored iron source, ferrous sulfate and other salts are too reactive, leading to food enrichment problems such as off flavors, odors, color changes, and chemical change or reaction during the baking or cooking process. Some of these difficulties reportedly are overcome by encapsulation of the ferrous sulfate with food triglycerides, such as hydrogenated vegetable oils, but the practice is not widely used, possibly due to the added cost. Normally, the use of ferrous sulfate is limited to food products that have a short expected shelf life.

By contrast, elemental iron powder has enjoyed a 50% usage in food enrichment, mostly because of its stability in all food environments. By far, the greatest usage is in enriched flour and bread, in which the amount of iron added is approximately 37 and 29 mg/kg (17 and 13 mg/lb), respectively. Although the quantity seems small, these and other food product applications amount to a 910 metric tons/year (1000 tons/year) elemental iron powder requirement in the United States alone. The total usage is increasing rapidly as overseas countries adopt U.S. standards. There's even a new application for enrichment of cattle feed that could be a larger market than human food enrichment.

Table 9 Particle size requirements for zirconium powder

Particle size, µm	Sieve No.	Composition, wt%		
		Class 1	Class 2	Class 3
Type I				
840	−20	99 min	...	...
250	−60	5 max	...	...
177	−80	...	100	...
149	−100	...	98 min	...
88	−170	...	...	100
74	−200	...	50 max	98 min
44	−325	...	25 max	...
<20	(a)	...	...	...
<10	(a)	...	2.0 max	10.0 max
Type II				
125	−120	100	100	99.5 min
74	−200	99 min	99 min	...
20	−26(b)	96 min	96 min	...
<9	(a)	85 min	85 min	...
3	(a)	70–90 min	70–90 min	...
0.75	(a)	12–30	12–30	...

(a) Subsieve particle size. (b) Buckbee Mears sieve

Negative properties of elemental iron powder, although not considered serious, include its dark gray color, which has a slight graying effect on the color of flour and bread, and its higher density compared to the milled cereal grain, which may lead to segregation if incorporation is not performed carefully. Finally, the reported relative bioavailability (RBV) of the different forms of commercial elemental iron powder varies anywhere from 10 to 90% that of ferrous sulfate. The usefulness of an iron source to combat or prevent anemia is expressed in terms of RBV, in which the adsorption of the iron source is compared to that of ferrous sulfate, which has an RBV arbitrarily assigned at a value of 100. A higher RBV number indicates a better iron source. In general, higher surface area powders (i.e., sponge) have higher RBV's.

Environmental Remediation

Iron as a Supplement for Phytoplankton Growth. One of the most dramatic uses of iron powder relates to the problem of global warming (Ref 4). The crystal-clear water in much of the tropical oceans is a result of the fact that the water is poor in nutrients. In areas of the world that receive the most sunshine, phytoplankton growth is inhibited primarily by the lack of iron. This was clearly and dramatically demonstrated in a recent experiment where 500 kg of dissolved iron spread over 100 square km of Pacific ocean soaked up 350,000 kg of CO_2 by dramatically enhancing plankton growth (Ref 4). It is hoped that the effect on global CO_2 levels will be to lower them permanently by the burial on the ocean floor of the carbon containing skeletons of these plankton. Although dissolved iron was used in the experiment, iron powder has a role in the preparation of the solutions, and it has been proposed that iron powder could be employed directly if used in a low density form (i.e., sponge powder), perhaps with a biodegradable coating that would allow the particles to float near the surface of the ocean and dissolve slowly.

Toxin Removal. Hexavalent chromium, a very toxic form of the element, can be removed from soil by digging a deep trench and filling it with iron powder. The contaminated water can be pumped, or allowed to flow naturally, through this wall. Movement through the iron bed reacts the Cr(VI) to Cr(III)Fe(III) oxyhydroxide, which is more easily removed (Ref 5). Hydrocarbons in groundwater also can be effectively oxidized by the highly reactive oxygen atom catalytically liberated from hydrogen peroxide by iron powder (Ref 5).

Material Substitution

Brake Liners. When it was discovered that asbestos fibers can cause lung cancer, it became important to find a replacement material for friction applications, especially for auto brakes because of the enormous volume of use. The "metallic brakes" found on virtually all front wheel drive cars are now made with metal powders, primarily iron, but also copper. Sponge iron powders, especially hydrogen reduced, low-density versions, have been found to be not only adequate substitutes, but are far superior to asbestos in many respects. The excellent friction characteristics of sponge powders derives from the many edges the particles present when cut in a cross section. The irregular sponge particles also have a very high green strength and lock into the binder system of the pad to a greater extent than spherical particles.

Even though the use of metals has allowed the elimination of asbestos in auto brakes, the use of copper in this application may have created its own problems. Copper levels in the San Francisco Bay have been rising in recent years, and the cause is thought to be the copper from auto brakes. As the brakes wear, the dust formed is washed into the bay by rainwater through storm drains. Copper is a potent biocide, but can be effectively removed by cementation with iron powder. There have been proposals to treat the runoff by this process.

Munitions. Concerns about lead contamination have come to the forefront lately, and this will probably become the next rallying point for environmentally driven materials substitution. Lead-containing munitions and bullets have several serious detrimental environmental aspects. Lead bullets have been banned at numerous indoor practice ranges because of lead fumes. As lead replacements, tungsten-copper and tungsten-iron composites are experiencing very rapid growth in the area of munitions, especially in the sport, hunting, and target practice sector. In sporting applications, tungsten composites can give the same performance and ballistic characteristics as lead, something the early steel replacement materials did not provide. In military applications, high tungsten composites can deliver much higher impact and penetration energies than other materials.

In target practice and training applications, powder composite bullets also can be made with various levels of frangibility in order to control the breakup of the bullet. Tungsten-copper and tungsten-iron composites have been designed to be environmentally friendly, while at the same time providing a measure of safety. The bullets are designed to have the ballistic characteristics of lead and break apart just behind the target on a backing plate, with the residue being recyclable. Some military training ammunition, used in higher volume than live ammunition, is made to break apart shortly after leaving the weapon. City police agencies also use frangible ammunition to avoid ricocheting bullets in crowded city environments.

Magnetic and Electrical Applications

Weed seeds can be separated from crop seed by dampening both with water and sprinkling with iron powder. The iron sticks to the weed seeds, which are characteristically more fibrous and wettable, followed by magnetic separation (Ref 6).

Fluid seals are created using iron powder suspended in oil, surrounded by magnetic fields that contain and control the slurry. These seals are gas tight and leak very little oil.

Magnetorheological fluids are made with suspensions of iron or ferritic stainless steel in oil. They are used in the construction of very precise and adjustable torque and braking units controlled by electromagnets.

Magnetorheological solids have been made from iron particles in an elastomeric polymer matrix. These solids change shape with the application of magnetic fields.

The National Aeronautics and Space Administration (NASA) has developed a pistonless compressor that moves fluids by the sequential magnetization of separated beds of ferritic stainless steel powder (Ref 7).

In another NASA invention, silver-filled epoxy can be used in several surge-arrestor applications. By varying the fill amount, such materials can be made to be insulators at low voltages, but switch to low resistance conductors at predetermined higher voltages, thus shunting to ground any unwanted voltage spikes (Ref 8).

ACKNOWLEDGMENTS

Portions of this article were adapted from articles in *Powder Metallurgy,* Vol 7, *Metals Handbook,* 1984, p 580–616 by Barry Van Hoose, Alcan Ingot and Powders; Mae I. Fauth, Naval Surface Weapons Center; and Ihab Kamel, Drexel University.

REFERENCES

1. *Handbook of Chemistry and Physics,* CRC Press, 1985
2. S.R. Jain, Spontaneously Igniting Hybrid Fuel-Oxidiser Systems, *Defence Science Journal,* Vol 45, 1995, p 5–16
3. L.W. Collins, Thermal Ignition of Titanium Based Pyrotechnics, *Combustion and Flame,* Vol 41, 1981, p 325–330
4. R. Monastersky, Iron Versus the Greenhouse, *Sci. News,* Vol 148, 30 Sept, 1995, p 220
5. A. Shanley, Ed., Beyond Pump and Treat, *Chemical Eng.,* March 1996, p 33
6. N.R. Brandenburg, Magnetic Separation of Seeds, Technical Bulletin 137, Agricultural Experiment Station, Oregon State University, May 1977
7. NASA Technical Briefs, June 1995, p 76
8. NASA Technical Briefs, June 1996, p 45

Appendices

Appendix 1: Mechanical Properties of Ferrous Powder Materials

Table 1 Typical mechanical properties of ferrous powder materials

Minimum strength values (in ksi) are specified by the two- or three-digit suffix of the material designation code in the first column of the table. Typical values are given in the remaining columns.

Material designation code	Ultimate strength, MPa	Ultimate strength, ksi	0.2% offset yield strength, MPa	0.2% offset yield strength, ksi	Elongation in 25 mm (1 in.), %	Elastic modulus, GPa	Elastic modulus, 10^6 psi	Transverse rupture strength, MPa	Transverse rupture strength, ksi	Impact energy(a), J	Impact energy(a), ft · lbf	Apparent hardness(b)	Fatigue strength(c), MPa	Fatigue strength(c), ksi	Density, g/cm^3
Iron and carbon steel															
F-000-10(d)	120	18	90	13	1.5	105	15.0	250	36	4	3.0	40 HRF	46	7	6.1
F-0000-15(d)	170	25	120	18	2.5	120	17.5	340	50	8	6.0	60 HRF	65	10	6.7
F-0000-20(d)	260	38	170	25	7.0	160	23.5	660	95	47	35.0	80 HRF	99	14	7.3
F-0005-15(d)	170	24	120	18	<1.0	105	15.0	330	48	4	3.0	25 HRB	60	9	6.1
F-0005-20(d)	220	32	160	23	1.0	115	16.5	440	64	5	4.0	40 HRB	80	12	6.6
F-0005-25(d)	260	38	190	28	1.5	135	19.5	520	76	7	5.0	55 HRB	100	15	6.9
F-0005-50HT(e)	410	60	(f)	(f)	<0.5	115	16.5	720	105	4	3.0	20 HRC (58 HRC)	160	23	6.6
F-0005-60HT(e)	480	70	(f)	(f)	<0.5	130	18.5	830	120	5	3.5	22 HRC (58 HRC)	190	27	6.8
F-0005-70HT(e)	550	80	(f)	(f)	<0.5	140	20.5	970	140	5	4.0	25 HRC (58 HRC)	220	32	7.0
F-0008-20(d)	200	29	170	25	<0.5	85	12.5	350	51	3	2.5	35 HRB	80	11	5.8
F-0008-25(d)	240	35	210	30	<0.5	110	16.0	420	61	4	3.0	50 HRB	100	14	6.2
F-0008-30(d)	290	42	240	35	<1.0	115	16.5	510	74	5	4.0	60 HRB	120	17	6.6
F-0008-35(d)	390	57	260	40	1.0	140	20.5	690	100	7	5.0	70 HRB	170	25	7.0
F-0008-55HT(e)	450	65	(f)	(f)	<0.5	115	16.5	690	100	4	3.0	22 HRC (60 HRC)	180	26	6.3
F-0008-65HT(e)	520	75	(f)	(f)	<0.5	115	16.5	790	115	5	4.0	28 HRC (60 HRC)	210	30	6.6
F-0008-75HT(e)	590	85	(f)	(f)	<0.5	135	19.5	900	130	6	4.5	32 HRC (60 HRC)	240	34	6.9
F-0008-85HT(e)	660	95	(f)	(f)	<0.5	150	21.5	1000	145	7	5.0	35 HRC (60 HRC)	280	38	7.1
Iron-copper and copper steel															
FC-0200-15(d)	170	25	140	20	1.0	95	14.0	310	45	6	4.5	11 HRB	70	10	6.0
FC-0200-18(d)	190	28	160	23	1.5	115	16.5	350	51	7	5.0	18 HRB	72	11	6.3
FC-0200-21(d)	210	31	180	26	1.5	115	16.5	390	56	7	5.5	26 HRB	80	12	6.6
FC-0200-24(d)	230	34	200	29	2.0	135	19.5	430	63	8	6.0	36 HRB	87	13	6.9
FC-0205-30(d)	240	35	240	35	<1.0	95	14.0	410	60	<3	<2.0	37 HRB	90	13	6.0
FC-0205-35(d)	280	40	280	40	<1.0	115	16.5	520	75	4	3.0	48 HRB	100	15	6.3
FC-0205-40(d)	340	50	310	45	<1.0	120	17.5	660	95	7	5.0	60 HRB	140	21	6.7
FC-0205-45(d)	410	60	340	50	<1.0	150	21.5	790	115	10	8.0	72 HRB	210	31	7.1
FC-0205-60HT(e)	480	70	(f)	(f)	<0.5	110	16.0	660	95	3	2.5	19 HRC (58 HRC)	190	27	6.2
FC-0205-70HT(e)	550	80	(f)	(f)	<0.5	105	15.5	760	110	5	3.5	25 HRC (58 HRC)	210	30	6.5
FC-0205-80HT(e)	620	90	(f)	(f)	<0.5	130	18.5	830	120	6	4.5	31 HRC (58 HRC)	230	34	6.8
FC-0205-90HT(e)	690	100	(f)	(f)	<0.5	140	20.5	930	135	7	5.5	36 HRC (58 HRC)	260	38	7.0
FC-0208-30(d)	240	35	240	35	<1.0	85	12.5	410	60	<3	<2.0	50 HRB	90	13	5.8
FC-0208-40(d)	340	50	310	45	<1.0	115	16.5	620	90	3	2.0	61 HRB	120	17	6.3
FC-0208-50(d)	410	60	380	55	<1.0	120	17.5	860	125	7	5.0	73 HRB	160	23	6.7
FC-0208-60(d)	520	75	450	65	<1.0	155	22.5	1070	155	9	7.0	84 HRB	230	33	7.2
FC-0208-50HT(e)	450	65	(f)	(f)	<0.5	105	15.0	660	95	3	2.5	20 HRC (60 HRC)	170	25	6.1
FC-0208-65HT(e)	520	75	(f)	(f)	<0.5	120	17.5	760	110	5	3.5	27 HRC (60 HRC)	210	30	6.4
FC-0208-80HT(e)	620	90	(f)	(f)	<0.5	130	18.5	900	130	6	4.5	35 HRC (60 HRC)	240	35	6.8
FC-0208-95HT(e)	720	105	(f)	(f)	<0.5	150	21.5	1030	150	7	5.5	43 HRC (60 HRC)	280	40	7.1
FC-0505-30(d)	300	44	250	36	<0.5	85	12.5	530	77	4	3.0	51 HRB	114	17	5.8
FC-0505-40(d)	400	58	320	47	<0.5	115	16.5	700	102	6	4.5	62 HRB	152	22	6.3
FC-0505-50(d)	490	71	390	56	<1.0	120	17.5	850	124	7	5.0	72 HRB	186	27	6.7

(continued)

(a) Unnotched Charpy test. (b) Apparent hardness measures a combination of matrix hardness plus the effect of porosity. Where applicable, the matrix (converted) hardness is also given in parentheses. (c) Fatigue limit for 10^7 cycles from reverse-bending tests (R.R. Moore testing machines). (d) The suffix number represents the minimum yield strength (in ksi) for the material in the as-sintered condition. (e) The suffix number for the heat treated (HT) or sinter-hardened conditions represents the minimum yield strength (in ksi). Tempering temperature for most heat treated materials is 175 °C (350 °F); nickel steels are tempered at 260 °C (500 °F). (f) Yield strength and ultimate tensile strength are approximately the same for heat treated materials. Source: MPIF Standard 35, 1997 edition

Table 1 (continued)

Material designation code	Ultimate strength MPa	Ultimate strength ksi	0.2% offset yield strength MPa	0.2% offset yield strength ksi	Elongation in 25 mm (1 in.), %	Elastic modulus GPa	Elastic modulus 10^6 psi	Transverse rupture strength MPa	Transverse rupture strength ksi	Impact energy(a) J	Impact energy(a) ft · lbf	Apparent hardness(b)	Fatigue strength(c) MPa	Fatigue strength(c) ksi	Density, g/cm^3
Iron-copper and copper steel (continued)															
FC-0508-40(d)	400	58	340	50	<0.5	90	13.0	690	100	4	3.0	60 HRB	152	22	5.9
FC-0508-50(d)	470	68	410	60	<0.5	115	16.5	830	120	5	3.5	68 HRB	179	26	6.3
FC-0508-60(d)	570	82	480	70	<1.0	130	18.5	1000	145	6	4.5	80 HRB	217	31	6.8
FC-0808-45(d)	380	55	340	50	<0.5	95	14.0	590	85	4	3.0	65 HRB	144	21	6.0
FC-1000-20(d)	210	30	180	26	<1.0	95	14.0	370	53	5	3.5	15 HRB	80	11	6.0
Iron-nickel and nickel steel															
FN-0200-15(d)	170	25	120	17	3.0	115	16.5	340	50	14	10.0	55 HRF	70	10	6.6
FN-0200-20(d)	240	35	170	25	5.0	140	20.5	550	80	27	20.0	75 HRF	91	13	7.0
FN-0200-25(d)	280	40	210	30	10.0	160	23.5	720	105	68	50.0	80 HRF	103	15	7.3
FN-0205-20(d)	280	40	170	25	1.5	115	16.5	450	65	8	6.0	44 HRB	100	14	6.6
FN-0205-25(d)	340	50	210	30	2.5	135	19.5	690	100	16	12.0	59 HRB	120	18	6.9
FN-0205-30(d)	410	60	240	35	4.0	155	22.5	860	125	28	21.0	69 HRB	150	22	7.2
FN-0205-35(d)	480	70	280	40	5.5	170	24.5	1030	150	46	34.0	78 HRB	180	26	7.4
FN-0205-80HT(e)	620	90	(f)	(f)	<0.5	115	16.5	830	120	5	3.5	23 HRC (55 HRC)	180	26	6.6
FN-0205-105HT(e)	830	120	(f)	(f)	<0.5	135	19.5	1110	160	6	4.5	29 HRC (55 HRC)	240	35	6.9
FN-0205-130HT(e)	1000	145	(f)	(f)	<0.5	150	21.5	1310	190	8	6.0	33 HRC (55 HRC)	290	42	7.1
FN-0205-155HT(e)	1100	160	(f)	(f)	<0.5	155	22.5	1480	215	9	7.0	36 HRC (55 HRC)	320	47	7.2
FN-0205-180HT(e)	1280	185	(f)	(f)	<0.5	170	24.5	1720	250	13	9.5	40 HRC (55 HRC)	370	54	7.4
FN-0208-30(d)	310	45	240	35	1.5	120	17.5	590	85	7	5.5	63 HRB	110	16	6.7
FN-0208-35(d)	380	55	280	40	1.5	135	19.5	720	105	11	8.0	71 HRB	140	20	6.9
FN-0208-40(d)	480	70	310	45	2.0	150	21.5	900	130	15	11.0	77 HRB	170	25	7.1
FN-0208-45(d)	550	80	340	50	2.5	160	23.5	1070	155	22	16.0	83 HRB	190	28	7.3
FN-0208-50(d)	620	90	380	55	3.0	170	24.5	1170	170	28	21.0	88 HRB	220	32	7.4
FN-0208-80HT(e)	620	90	(f)	(f)	<0.5	120	17.5	830	120	5	4.0	26 HRC (57 HRC)	200	29	6.7
FN-0208-105HT(e)	830	120	(f)	(f)	<0.5	135	19.5	1030	150	6	4.5	31 HRC (57 HRC)	260	38	6.9
FN-0208-130HT(e)	1000	145	(f)	(f)	<0.5	140	20.5	1280	185	7	5.5	35 HRC (57 HRC)	320	46	7.0
FN-0208-155HT(e)	1170	170	(f)	(f)	<0.5	155	22.5	1520	220	9	7.0	39 HRC (57 HRC)	370	54	7.2
FN-0208-180HT(e)	1340	195	(f)	(f)	<0.5	170	24.5	1720	250	11	8.0	42 HRC (57 HRC)	430	62	7.4
FN-0405-25(d)	280	40	210	30	<1.0	105	15.5	450	65	6	4.5	49 HRB	100	14	6.5
FN-0405-35(d)	410	60	280	40	3.0	140	20.5	830	120	20	14.5	71 HRB	150	22	7.0
FN-0405-45(d)	620	90	340	50	4.5	170	24.5	1210	175	45	33.5	84 HRB	220	32	7.4
FN-0405-80HT(e)	590	85	(f)	(f)	<0.5	105	15.5	790	115	5	4.0	19 HRC (55 HRC)	180	26	6.5
FN-0405-105HT(e)	760	110	(f)	(f)	<0.5	130	18.5	1000	145	7	5.0	25 HRC (55 HRC)	230	34	6.8
FN-0405-130HT(e)	930	135	(f)	(f)	<0.5	140	20.5	1380	200	9	6.5	31 HRC (55 HRC)	290	42	7.0
FN-0405-155HT(e)	1100	160	(f)	(f)	<0.5	160	23.5	1690	245	13	9.5	37 HRC (55 HRC)	340	49	7.3
FN-0405-180HT(e)	1280	185	(f)	(f)	<0.5	170	24.5	1930	280	18	13.0	40 HRC (55 HRC)	390	57	7.4
FN-0408-35(d)	310	45	280	40	1.0	105	15.5	520	75	5	4.0	67 HRB	110	16	6.5
FN-0408-45(d)	450	65	340	50	1.0	135	19.5	790	115	10	7.5	78 HRB	160	23	6.9
FN-0408-55(d)	550	80	410	60	1.0	155	22.5	1030	150	15	11.0	87 HRB	190	28	7.2
Low-alloy steel															
FL-4205-35(d)	360	52	290	42	1.0	130	18.5	690	100	8	6.0	60 HRB	140	20	6.80
FL-4205-40(d)	400	58	320	47	1.0	140	20.0	790	115	12	9.0	66 HRB	190	27	6.95
FL-4205-45(d)	460	66	360	52	1.5	150	21.5	860	125	16	12.0	70 HRB	220	32	7.10
FL-4205-50(d)	500	73	400	58	2.0	160	23.5	1030	150	23	17.0	75 HRB	280	40	7.30
FL-4205-80HT(e)	620	90	(f)	(f)	<0.5	115	16.5	930	135	7	5.0	28 HRC (60 HRC)	210	30	6.60
FL-4205-100HT(e)	760	110	(f)	(f)	<0.5	130	18.5	1100	160	9	7.0	32 HRC (60 HRC)	260	37	6.80
FL-4205-120HT(e)	900	130	(f)	(f)	<0.5	140	20.5	1280	185	11	8.0	36 HRC (60 HRC)	300	44	7.00
FL-4205-140HT(e)	1030	150	(f)	(f)	<0.5	155	22.5	1480	215	16	12.0	39 HRC (60 HRC)	340	50	7.20
FL-4405-35(d)	360	52	290	42	1.0	120	17.5	690	100	8	6.0	60 HRB	140	20	6.70
FL-4405-40(d)	400	58	320	47	1.0	135	19.5	860	125	15	11.0	67 HRB	190	27	6.90
FL-4405-45(d)	460	66	360	52	1.5	150	21.5	970	140	22	16.0	73 HRB	220	32	7.10
FL-4405-50(d)	500	73	400	58	2.0	160	23.5	1140	165	30	22.0	80 HRB	280	40	7.30
FL-4405-100HT(e)	760	110	(f)	(f)	<1.0	120	17.5	1100	160	7	5.5	24 HRC (60 HRC)	230	34	6.70
FL-4405-125HT(e)	930	135	(f)	(f)	<1.0	135	19.5	1380	200	9	7.0	29 HRC (60 HRC)	290	42	6.90
FL-4405-150HT(e)	1100	160	(f)	(f)	<1.0	150	21.5	1590	230	12	9.0	34 HRC (60 HRC)	330	48	7.10
FL-4405-175HT(e)	1280	185	(f)	(f)	<1.0	160	23.5	1930	280	19	14.0	38 HRC (60 HRC)	400	58	7.30
FL-4605-35(d)	360	52	290	42	1.0	125	18.0	690	100	8	6.0	60 HRB	140	20	6.75
FL-4605-40(d)	400	58	320	47	1.0	140	20.0	830	120	15	11.0	65 HRB	190	27	6.95
FL-4605-45(d)	460	66	360	52	1.5	150	22.0	970	140	22	16.0	71 HRB	220	32	7.15
FL-4605-50(d)	500	73	400	58	2.0	165	24.0	1140	165	30	22.0	77 HRB	280	40	7.35
FL-4605-80HT(e)	590	85	(f)	(f)	<0.5	110	16.0	900	130	6	4.5	24 HRC (60 HRC)	200	29	6.55
FL-4605-100HT(e)	760	110	(f)	(f)	<0.5	125	18.0	1140	165	8	6.0	29 HRC (60 HRC)	260	37	6.75
FL-4605-120HT(e)	900	130	(f)	(f)	<0.5	140	20.0	1340	195	11	8.0	34 HRC (60 HRC)	320	46	6.95
FL-4605-140HT(e)	1070	155	(f)	(f)	<0.5	155	22.5	1590	230	16	12.0	39 HRC (60 HRC)	370	53	7.20
FLN-4205-40(d)	400	58	320	47	1.0	115	16.5	720	105	8	6.0	64 HRB	140	20	6.60
FLN-4205-45(d)	460	66	360	52	1.0	130	18.5	860	125	11	8.0	70 HRB	190	27	6.80
FLN-4205-50(d)	500	73	400	58	1.5	145	21.0	1030	150	18	13.0	77 HRB	220	32	7.05
FLN-4205-55(d)	600	87	430	63	2.0	160	23.5	1210	175	30	22.0	83 HRB	280	40	7.30
FLN-4205-80HT(e)															

(continued)

(a) Unnotched Charpy test. (b) Apparent hardness measures a combination of matrix hardness plus the effect of porosity. Where applicable, the matrix (converted) hardness is also given in parentheses. (c) Fatigue limit for 10^7 cycles from reverse-bending tests (R.R. Moore testing machines). (d) The suffix number represents the minimum yield strength (in ksi) for the material in the as-sintered condition. (e) The suffix number for the heat treated (HT) or sinter-hardened conditions represents the minimum yield strength (in ksi). Tempering temperature for most heat treated materials is 175 °C (350 °F); nickel steels are tempered at 260 °C (500 °F). (f) Yield strength and ultimate tensile strength are approximately the same for heat treated materials. Source: MPIF Standard 35, 1997 edition

Table 1 (continued)

Material designation code	Ultimate strength, MPa	Ultimate strength, ksi	0.2% offset yield strength, MPa	0.2% offset yield strength, ksi	Elongation in 25 mm (1 in.), %	Elastic modulus, GPa	Elastic modulus, 10^6 psi	Transverse rupture strength, MPa	Transverse rupture strength, ksi	Impact energy(a), J	Impact energy(a), ft · lbf	Apparent hardness(b)	Fatigue strength(c), MPa	Fatigue strength(c), ksi	Density, g/cm^3
Low-alloy steel (continued)															
FLN-4205-105HT(e)	790	115	(f)	(f)	<1.0	130	18.5	1170	170	9	7.0	30 HRC (60 HRC)	250	36	6.80
FLN-4205-140HT(e)	1030	150	(f)	(f)	<1.0	145	21.0	1590	230	12	9.0	36 HRC (60 HRC)	320	47	7.05
FLN-4205-175HT(e)	1280	185	(f)	(f)	1.0	160	23.5	2000	290	19	14.0	42 HRC (60 HRC)	400	58	7.30
Sinter-hardened steel															
FLC-4608-50HT(e)	410	60	(f)	(f)	<1.0	115	16.5	830	120	7	5.0	20 HRC (55 HRC)	...	...	6.60
FLC-4608-70HT(e)	550	80	(f)	(f)	<1.0	130	18.5	1030	150	9	7.0	26 HRC (55 HRC)	...	...	6.80
FLC-4608-90HT(e)	690	100	(f)	(f)	<1.0	140	20.5	1310	190	12	9.0	31 HRC (55 HRC)	...	...	7.00
FfLC-4608-110HT(e)	830	120	(f)	(f)	<1.0	155	22.5	1590	230	19	14.0	37 HRC (55 HRC)	...	...	7.20
Diffusion-alloyed steel															
FD-0205-45(d)	470	68	360	52	1.0	125	18.0	900	130	11	8.0	72 HRB	170	24	6.75
FD-0205-50(d)	540	78	390	57	1.5	140	20.0	1070	155	16	12.0	76 HRB	200	29	6.95
FD-0205-55(d)	610	88	420	61	2.0	150	22.0	1240	180	24	18.0	80 HRB	220	32	7.15
FD-0205-60(d)	690	100	460	67	2.0	170	24.5	1450	210	38	28.0	86 HRB	260	37	7.40
FD-0205-95HT(e)	720	105	(f)	(f)	<1.0	125	18.0	1100	160	7	5.0	28 HRC (55 HRC)	290	42	6.75
FD-0205-120HT(e)	900	130	(f)	(f)	<1.0	140	20.0	1310	190	9	7.0	33 HRC (55 HRC)	360	52	6.95
FD-0205-140HT(e)	1030	150	(f)	(f)	<1.0	150	22.0	1450	210	12	9.0	38 HRC (55 HRC)	450	65	7.15
FD-0205-160HT(e)	1170	170	(f)	(f)	<1.0	170	24.5	1650	240	15	11.0	45 HRC (55 HRC)	520	75	7.40
FD-0208-50(d)	480	69	400	58	<1.0	125	18.0	930	135	9	7.0	80 HRB	170	24	6.75
FD-0208-55(d)	540	79	430	63	<1.0	135	19.5	1070	155	12	9.0	83 HRB	230	33	6.90
FD-0208-60(d)	630	92	470	68	1.0	150	21.5	1240	180	16	12.0	87 HRB	260	38	7.10
FD-0208-65(d)	710	103	500	73	1.0	160	23.0	1340	195	23	17.0	90 HRB	320	46	7.25
FD-0405-55(d)	590	86	430	62	1.0	125	18.0	1100	160	15	11.0	80 HRB	170	25	6.75
FD-0405-60(d)	710	103	460	66	1.0	145	21.0	1340	195	27	20.0	85 HRB	200	28	7.05
FD-0405-65(d)	850	123	480	70	2.5	165	24.0	1590	230	37	27.0	91 HRB	280	40	7.35
FD-0405-100HT(e)	760	110	(f)	(f)	<1.0	125	18.0	1100	160	7	5.0	30 HRC (55 HRC)	230	34	6.75
FD-0405-130HT(e)	970	140	(f)	(f)	<1.0	145	21.0	1380	200	9	7.0	35 HRC (55 HRC)	340	50	7.05
FD-0405-155HT(e)															
FD-0408-50(d)	490	71	390	57	<1.0	120	17.5	900	130	12	9.0	85 HRB	150	22	6.70
FD-0408-55(d)	620	90	430	62	1.0	140	20.0	1140	165	18	13.0	89 HRB	190	27	6.95
FD-0408-60(d)	760	110	460	67	1.5	155	22.5	1380	200	24	18.0	93 HRB	260	38	7.20
FD-0408-65(d)	860	125	490	71	2.0	170	24.5	1590	230	30	22.0	95 HRB	330	48	7.40
Copper-infiltrated steel															
FX-1000-25(d)	350	51	220	32	7.0	160	23.5	910	132	34	25.0	65 HRB	133	19	7.3
FX-1005-40(d)	530	77	340	50	4.0	160	23.5	1090	158	18	13.0	82 HRB	200	29	7.3
FX-1005-110HT(e)	830	120	(f)	(f)	<0.5	160	23.5	1450	210	9	7.0	38 HRC (55 HRC)	230	33	7.3
FX-1008-50(d)	600	87	410	60	3.0	160	23.5	1140	166	14	10.0	89 HRB	230	33	7.3
FX-1008-110HT(e)	830	120	(f)	(f)	<0.5	160	23.5	1300	189	9	6.5	43 HRC (58 HRC)	280	41	7.3
FX-2000-25(d)	320	46	260	37	3.0	145	21.0	990	144	20	15.0	66 HRB	122	17	7.3
FX-2005-45(d)	520	75	410	60	1.5	145	21.0	1020	148	11	8.0	85 HRB	140	20	7.3
FX-2005-90HT(e)	690	100	(f)	(f)	<0.5	145	21.0	1180	171	9	7.0	36 HRC (55 HRC)	160	23	7.3
FX-2008-60(d)	550	80	480	70	1.0	145	21.0	1080	156	9	7.0	90 HRB	160	23	7.3
FX-2008-90HT(e)	690	100	(f)	(f)	<0.5	145	21.0	1100	159	7	5.0	36 HRC (58 HRC)	190	27	7.3

(a) Unnotched Charpy test. (b) Apparent hardness measures a combination of matrix hardness plus the effect of porosity. Where applicable, the matrix (converted) hardness is also given in parentheses. (c) Fatigue limit for 10^7 cycles from reverse-bending tests (R.R. Moore testing machines). (d) The suffix number represents the minimum yield strength (in ksi) for the material in the as-sintered condition. (e) The suffix number for the heat treated (HT) or sinter-hardened conditions represents the minimum yield strength (in ksi). Tempering temperature for most heat treated materials is 175 °C (350 °F); nickel steels are tempered at 260 °C (500 °F). (f) Yield strength and ultimate tensile strength are approximately the same for heat treated materials. Source: MPIF Standard 35, 1997 edition

Appendix 2: Powder Metallurgy Standards

Table 1 Cross-index of powder metallurgy standards

Standard title	Designation(a)			
	ASTM	MPIF	ISO	Other
Metal powder specifications				
Aluminum Pigment, Powder and Paste, for Paint	D 962	...	...	TT-P-320B
Aluminum Powder, Flaked, Grained and Atomized	...	...	...	MIL-A-512A
Antimony, Powdered, Technical	...	...	...	MIL-A-10841B
Brass Powder (for sintering)	...	...	...	MIL-B-11552B
Bronze Powder, Gold	D 267	...	...	TT-P-340
Copper Powder (for use in ammunition)	...	...	...	JAN-C-768
Copper Powder for Use in Antifouling Paints	D 964	...	...	...
Iron Powder (for pyrotechnics)	...	...	...	MIL-I-12058A
Magnesium-Aluminum Alloy, Powdered	...	...	...	JAN-M-454
Magnesium Powder (for use in ammunition)	...	...	...	JAN-M-382A
Manganese, Powdered (for use in ammunition)	...	...	...	JAN-M-476A
Nickel Powder (for use in ammunition)	...	...	...	JAN-N-412A
Powders, Metal, Atomized (Aluminum, Magnesium, Magnesium-Aluminum)	...	...	...	MIL-P-14067B
Tin, Pulverized	...	...	...	JAN-T-458
Titanium, Powdered	...	...	...	MIL-T-13405
Tungsten, Powdered	...	...	...	MIL-T-13827
Zinc Dust (for use in pyrotechnics)	...	...	...	JAN-Z-365
Zinc Dust (metallic zinc powder); Dry (paint pigment)	D 520	...	...	TT-P-460 No. 1
Zirconium-Nickel Alloy Powder (for use in delay compositions)	...	...	...	MIL-Z-11410A
Zirconium, Powdered (for use in ammunition)	...	...	...	JAN-Z-399A
Methods of powder characterization and testing				
Acid-Insoluble Content of Copper and Iron Powders	E 194	6	4496	...
Bronze Filter Powder	...	39	...	...
Chemical Analysis of Zinc Dust (Metallic Zinc Powder)	D 521	...	...	...
Compressibility of Metal Powders in Uniaxial Compaction	B 331	45	3927	...
Compression Testing of Metallic Materials at Room Temperature	E 9	...	...	...
Density, Apparent of Free-Flowing Metal Powders	B 212	4	3923/1	...
Density, Apparent of Metal Powders Using the Arnold Meter	703	48	...	...
Density, Apparent, of Non-Free-Flowing Metal Powders	B 417	28	3923/1	...
Density, Apparent, of Powders of Refractory Metals and Compounds by Scott Volumeter	B 329	...	3923/2	...
Flow Rate of Metal Powders	B 213	3	4490	...
Green Strength for Compacted Metal Powder Specimens	B 312	15	3995	...
Hydrogen Loss of Copper, Tungsten, and Iron Powders	E 159	2	4491	...
Iron Content of Iron Powder	...	7	...	...
Measuring Dimensional Change of Metal Powder Specimens Due to Sintering	B 610	44	4492	...
Particle Distribution of Refractory Metal-Type Powders by Turbidimetry	B 430	...	...	...
Particle Size Distribution of Metal Powders and Related Compounds by Light Scattering	B 822	...	...	...
Particle Size Distribution of Refractory Metals and Their Compounds by X-Ray Monitoring of Gravity Sedimentation	B 761	...	...	...
Particle Size, Average, of Metal Powders by Fisher Subsieve Sizer	...	32	...	MIL-STD-1233
Particle Size, Average, of Powders of Refractory Metals and Their Compounds by the Fisher Subsieve Sizer	B 330, C 721	...	...	...
Safety Requirements for the Construction, Safeguarding, Care and Use of P/M Presses	...	47	...	...
Sampling and Testing Aluminum Powder and Paste	D 480	...	...	...
Sampling Finished Lots of Metal Powders	B 215	1	3954	...
Sieve Analysis of Granular Metal Powders	B 214	5	4497.2	MIL-STD-1233
Subsieve Analysis of Granular Metal Powders by Air Classification	B 293	12	...	...
Tap Density of Metal Powders	B 527	46	3953	MIL-STD-1233

(continued)

(a) Standards referenced include those of the American Society for Testing and Materials (ASTM), Metal Powder Industries Federation (MPIF), Society of Automotive Engineers (SAE), and federal agencies, including the military. The standards may be obtained from the following sources: ASTM—ASTM, 100 Barr Harbor Drive, West Conshohocken, PA 19428–2959; MPIF—Metal Powder Industries Federation, 105 College Road East, Princeton, NJ 08540; SAE—Society of Automotive Engineers, 400 Commonwealth Drive, Warrendale, PA 15096; JAN, MIL, TT—Superintendent of Documents, U.S. Government Printing Office, Washington, D.C. 20402; CCPA—Cemented Carbide Producers Assn., 2130 Keith Building, Cleveland, OH 44115. Source: *1998 Annual Book of ASTM Standards,* Volume 2.05, ASTM, 1998

Table 1 (continued)

Standard title	Designation(a)			
	ASTM	MPIF	ISO	Other
Methods of powder characterization and testing (continued)				
Tension Test Specimens for Pressed and Sintered Metal Powders	E 8	10	2740	...
Testing Copper-Base Infiltrating Powders	...	49	...	...
Particle Size Analysis of Particulate Substances in the Range of 0.2 to 75 Micrometers by Optical Microscopy	E 20	...	...	...
De-Agglomeration of Refractory Metal Powders and Their Compounds Prior to Particle Size Analysis	B 859	...	...	...
Reporting Particle Size Characteristics of Pigments	D 1366	...	...	...
Metal powder bearings				
Iron-Base Sintered Bearings (Oil-Impregnated)	B 439	35	...	SAE J471d, MIL-B-5687D, MS21782
Iron Bronze Sintered Bearings (Oil-Impregnated)	B 612	35	...	...
Iron-graphite sintered bearings	B 782	35	...	...
Materials Standards and Specifications for Powder Metallurgy Mechanical Components and Oil Impregnated Bearings plus Code Designations and Coding Systems	B 438, B 439	35	...	SAE J471d, MIL-B-5687D
Sintered Bronze Bearings (Oil-Impregnated)	B 438	35	...	MIL-B-5687D, MS17795, MS17796, MS21783
Sintered Bronze Bearings (Oil-Impregnated) (Metric)	B 438M	...	...	...
Density and Interconnected Porosity of Sintered Powder Metal Structural Parts and Oil Impregnated Bearings	B 328	42	2738, 3369	...
Determination of Oil Content	B 328	...	2737	...
Determination of Radial Crushing Strength	...	...	2739	...
Magnetic powder cores and ceramic magnets				
Permeability of Feebly Magnetic Materials	A 342	...	...	CCPA P-109
Acceptance Standards for Electronic Cores	...	17	...	...
Coding Methods for Color Identification of Magnetic Powder Cores	...	36	...	...
Determination of Iron Powder Toroidal Core Inductance	...	33	...	...
Electronic Iron Powder Core Mechanical Test Specifications	...	22	...	...
Iron Powder Core Dimensional Specifications	...	11	...	...
Nominal Dimensions of Molybdenum-Permalloy Powder Cores	...	38	...	...
Metal powder structural parts				
Copper-Infiltrated Sintered Carbon Steel Structural Parts	B 303	35	...	SAE No. 870, 872
Materials for Ferrous Powder Metallurgy (P/M) Structural Parts	B 783	35	...	SAE J471d
Materials for Nonferrous Powder Metallurgy (P/M) Structural Parts	B 823	35	...	...
Powder Forged (P/F) Ferrous Structural Parts	B 848	...	...	...
Powder Metallurgy (P/M) Boron Stainless Steel Structural Components	B 853	...	...	...
Powder Metallurgy (P/M) Titanium Alloy Structural Components	B 817	...	...	...
Sintered Aluminum Structural Parts	B 595	...	...	...
Sintered Austenitic Stainless Steel Structural Parts	B 525	35	...	...
Sintered Brass Structural Parts	B 282	35	...	MIL-B-12128C
Sintered Bronze Structural Parts	B 255	35	...	...
Sintered Carbon Steel Structural Parts	B 310	35	...	SAE No. 850, 851, 853
Sintered Copper Steel Structural Parts	B 426	35	...	SAE No. 864
Sintered Copper Structural Parts for Electrical Conductivity Applications	B 715	...	...	...
Sintered Iron-Copper Structural Parts	B 222	35	...	SAE No. 862, 863
Sintered Metal Powder Parts (bearings and structural parts)	...	...	...	SAE J471d
Sintered Nickel Silver Structural Parts	B 458	35	...	...
Sintered Nickel Steel Parts	B 484	...	...	SAE 471d
Tungsten Base, High Density Metal (Sintered or Hot Pressed)	...	...	...	MIL-T-21014C
Testing of metal powder structural parts				
Determining the Percentage of Alloyed or Unalloyed Iron Contamination Present in Powder Forged (P/F) Steel Parts	B 795	...	...	...
Microhardness and Case Depth of Powder Metallurgy (P/M) Parts	B 721	51, 52	4507	...
Nonmetallic Inclusion Level of Powder Forged (P/F) Steel Parts	B 796	...	...	...
Surface Finger Oxide Penetration Depth and Presence of Interparticle Oxide Networks in Powder Forged (P/F) Steel Parts	B 797	...	...	...
Coefficient of Friction and Wear of Sintered Metal Friction Materials Under Dry Clutch Conditions	B 526	...	...	...
Density and Interconnected Porosity	B 328	42	2738, 3369	...
Density of Sintered Metal Friction Materials	B 376	...	...	...
Dynamic Coefficient of Friction and Wear of Sintered Metal Friction Materials Under Dry Conditions	B 460	...	...	...
Frictional Characteristics of Sintered Metal Friction Materials Run In Lubricants	B 461	...	...	...
Hardness of Sintered Metal Friction Materials	B 347	...	...	...
Tension Test Specimens for Pressed and Sintered Metal Powders	E 8	10	2740	...
Transverse Rupture Strength of Sintered Metal Friction Materials	B 378	...	...	...
Determination of Apparent Hardness of Sintered Metal Powder Products	...	43	4498	...
Determination of Impact Strength of Sintered Metal Powder Specimens	...	40	5754	...
Determination of Transverse Rupture or Bending Strength of Sintered Metal Powder Test Specimens	B 528	41	3325	...
Determining the Case Hardness of Powder Metallurgy Parts	...	37	4507	
Young's Modulus	...	...	3312	...
Cemented carbides				

(continued)

(a) Standards referenced include those of the American Society for Testing and Materials (ASTM), Metal Powder Industries Federation (MPIF), Society of Automotive Engineers (SAE), and federal agencies, including the military. The standards may be obtained from the following sources: ASTM—ASTM, 100 Barr Harbor Drive, West Conshohocken, PA 19428–2959; MPIF—Metal Powder Industries Federation, 105 College Road East, Princeton, NJ 08540; SAE—Society of Automotive Engineers, 400 Commonwealth Drive, Warrendale, PA 15096; JAN, MIL, TT—Superintendent of Documents, U.S. Government Printing Office, Washington, D.C. 20402; CCPA—Cemented Carbide Producers Assn., 2130 Keith Building, Cleveland, OH 44115. Source: *1998 Annual Book of ASTM Standards,* Volume 2.05, ASTM, 1998

Table 1 (continued)

Standard title	Designation(a)			
	ASTM	MPIF	ISO	Other
Tungsten Carbide, Crystalline	...	...	...	MIL-T-13366
Abrasive Wear Resistance of Cemented Carbides	B 611	...	...	...
Compression Testing for Cemented Carbides	...	...	4506	CCPA P-104
Density of Cemented Carbides	B 311	...	...	CCPA P-101
Determination of Poisson's Ratio at Room Temperature for Cemented Carbides	...	...	...	CCPA P-105
Determination of Young's Modulus at Room Temperature for Cemented Carbides	...	...	...	CCPA P-106
Diametral Compression Testing of Cemented Carbides	...	...	...	CCPA P-115
Electrical Resistivity of Cemented Tungsten Carbides	B 421	...	...	CCPA M-203
Grain Size, Apparent, and Distribution of Cemented Tungsten Carbides	B 390	...	...	CCPA P-107
Hardness Testing of Cemented Carbides	B 294	...	3738	CCPA P-103
Linear Thermal Expansion for Cemented Carbides	...	...	...	CCPA P-108
Metallographic Determination of Microstructure in Cemented Carbides	B 657	...	...	...
Metallographic Sample Preparation of Cemented Carbides	B 665	...	...	...
Porosity, Apparent, in Cemented Carbides	B 276	...	...	CCPA M-201
Tension Testing of Cemented Carbides	...	...	...	CCPA P-113
Transverse Rupture Strength of Cemented Carbides	B 406	...	3327	CCPA P-102
Short Rod Fracture Toughness	B 771	...	...	...
Miscellaneous				
Brushes, Electrical Contacts, and Carbon Stock, Electrical Contact Brush	...	...	...	MIL-B-003743D
Glossary for Metal Powder Compacting Presses and Tooling	...	31	...	...
Metallized Surfaces on Ceramic, Specification for	F 44	...	...	...
Precision Electroformed Sieves, Specification for	E 161	...	...	...
Terms Used in Powder Metallurgy	B 243	9	3252	...
Wire-Cloth Sieves for Testing Purposes, Specification for	E 11	...	...	RR-S-366

(a) Standards referenced include those of the American Society for Testing and Materials (ASTM), Metal Powder Industries Federation (MPIF), Society of Automotive Engineers (SAE), and federal agencies, including the military. The standards may be obtained from the following sources: ASTM—ASTM, 100 Barr Harbor Drive, West Conshohocken, PA 19428–2959; MPIF—Metal Powder Industries Federation, 105 College Road East, Princeton, NJ 08540; SAE—Society of Automotive Engineers, 400 Commonwealth Drive, Warrendale, PA 15096; JAN, MIL, TT—Superintendent of Documents, U.S. Government Printing Office, Washington, D.C. 20402; CCPA—Cemented Carbide Producers Assn., 2130 Keith Building, Cleveland, OH 44115. Source: 1998 Annual Book of ASTM Standards, Volume 2.05, ASTM, 1998

Appendix 3: Examples of Powder Metallurgy Parts

POWDER METALLURGY PARTS are used in a wide variety of applications for automobiles, machinery, appliances, aircraft, consumer items, electrical instruments, and other manufactured products. The parts produced by powder metallurgy (P/M) methods can also be unique and innovative. As a small sample of various applications, this appendix illustrates some typical examples of P/M parts and recent award winning parts from the annual parts competition series of the Metal Powder Industries Federation (MPIF).

General Examples

Powder metallurgy parts are used for many types of common assembly items, such as bearings, gears, sprockets, cams, manifolds, bushings, and connector parts. Typical applications are illustrated in the following examples.

Example 1: Drive Gear for Mailhandling Machine. A flat-belt drive gear weighing 490 g (17.3 oz), which is compacted to a density of 7.45 g/cm^3, is shown in Fig. 1. This type of gear is used in high-speed mailhandling and inserting machines. This part, consisting of electrolytic iron preblended with 0.35% C, replaced a two-piece gear assembly made of American Iron and Steel Institute (AISI) 1030 steel, thus resulting in a total cost savings of 66%.

The powder was first compacted on a 5340 kN (600 ton) press to a density of 6.8 to 6.9 g/cm^3 at a compaction pressure of 414 MPa (30 tsi). It was then presintered and repressed to a density of 7.45 g/cm^3 at 965 MPa (70 tsi). After repressing, the part was sintered at 1120 °C (2050 °F) for 30 min.

As shown in Fig. 1, considerable machining (turning) of the hub was required to generate the final shape. In use, this gear rotates and a magnetic clutch rides on the machined hub. When the operator trips the cycle control on the machine, the clutch closes around the small hub diameter, and the load is transferred to the P/M gear. Pressing of a large hub ensures a uniform density in the cross-sectional area, but requires additional machining of the P/M blank.

After machining, the hub area is carbonitrided to a case depth of 0.127 mm (0.005 in.). Case hardness is 700 DPH (minimum), while core hardness is 550 DPH. To maintain a 0.025 mm (0.001 in.) diameter tolerance between the hub and the magnetic clutch, the hub diameter is ground to final dimensions. Typical properties for this P/M part are:

Tensile strength, MPa (ksi)	414–586 (60–85)
Yield strength, MPa (ksi)	365–414 (53–60)
Elongation, %	1–2
Transverse-rupture strength, MPa (ksi)	930 (135)

Example 2: Sprocket Assemblies on Agricultural Equipment. The sprocket assemblies in Fig. 2 are parts in seed and herbicide metering equipment. The dual sprocket shown in Fig. 2(a) is made of two P/M sprockets that are staked together. This provides the flexibility to produce different gear ratios from pairs of single sprockets, rather than manufacturing a complete dual sprocket for each ratio required. The P/M staked

Fig. 1 Iron-carbon P/M drive gear. (a) Finished machined, hardened, ground part. The hub diameter is machined oversized (+0.254 mm or +0.010 in.) to allow for final grinding. (b) P/M blank as-pressed

	11-tooth sprocket	19-tooth sprocket
MPIF material	F-0005-S	F-0005-S
Carbon, %	0.26–0.60	0.26–0.60
Iron, %	97.40 min	97.40 min
Density, g/cm^3	6.9–7.3	6.9–7.3
Finished weight, kg (lb)	0.227 (0.50)	0.680 (1.50)
Additional requirements	Stake sprockets together	

(a)

	Sprocket	Retainer
MPIF material	FC-0208-P	Sheet steel
Carbon, %	0.60–1.00	...
Copper, %	1.00–2.50	...
Iron, %	94.5 min	...
Density, g/cm^3	6.1–6.5	...
Finished weight, kg (lb)	0.091 (0.20)	...

Note: Assembled in green condition and brazed together during sintering

(b)

Fig. 2 Powder metallurgy sprocket assemblies in an agricultural planter. (a) Dual sprocket assembly. (b) Sprocket and retainer assembly

	Arm	Hub
MPIF material	F-0008-P	FC-0408-P
Carbon, %	0.61–1.00	0.60–1.00
Copper, %	...	2.00–6.00
Iron, %	97.00 min	91.00 min
Density, g/cm^3	6.1–6.5	6.1–6.4
Heat treatment	Steam blue after sintering	
Finished weight, kg (lb)	0.132 (0.30)	0.132 (0.30)

Fig. 3 Idler arm

assembly replaced a machined and welded one at a significant cost savings. The dual sprocket fits on a shaft that is fixed to the seed-metering device. A drive chain powers one of the dual sprockets to rotate the sprocket assembly and in turn drive the seed-metering device.

The other sprocket on the assembly drives a sprocket assembly on the herbicide/insecticide meter of the planter through use of an additional chain. This sprocket assembly (Fig. 2) is produced by compacting the P/M sprocket and then brazing the sheet metal retainer to it during sintering.

Example 3: Idler Arms. Figure 3 illustrates a planter component that is made from two P/M parts assembled in the green state and sintered together into a single piece. This assembly is used to apply tension through a nylon roller to the chain that drives the dual sprocket assembly discussed in Example 2.

Example 4: Eccentric Gear. Figure 4 shows an eccentric gear for a washing machine that is made of a blend of 2% copper, 0.5% graphite, and the balance iron, which is compacted, sintered, sized, and machined. An intermediate level of graphite was chosen to produce an as-sintered part that could be sized readily to ensure an accurate gear tooth profile. Lower carbon content also improves machinability. Density of this part is 6.5 g/cm^3.

Fig. 4 Eccentric gear for a washing machine made of iron alloy powder containing 2% copper and 0.5% graphite. Center hole inside diameter: approximately 19 mm (0.76 in.); eccentric outside diameter: approximately 57 mm (2.28 in.); total thickness: approximately 28.50 mm (1.14 in.); eccentric thickness: approximately 15.90 mm (0.63 in.); pitch diameter: approximately 115 mm (4.59 in.)

This gear is made from atomized powder to obtain optimum compressibility. Because this gear is a two-level part, green strength is critical to eliminate cracking at juncture. Press motion and tool design also are critical in preventing hub cracking.

Sintering is done in an endothermic atmosphere for 30 min at 1105 °C (2025 °F). Dew point is monitored to prevent carburization of the surface layers so as not to affect sizing or machining. After sintering, the part is deburred in a steel medium to facilitate sizing. Prior to sizing, the part is dipped in soluble sizing wax, which also acts as a long-term storage and rust protection medium.

Sizing enhances the tooth form and pitch diameter tolerances. After sizing, the center hole is bored for concentricity, and the outside of the hub is machined to correct any ovality. This part replaced a class 30 gray iron casting that was machined from a rough blank.

Example 5: Fourth Reduction Gears. Lawn and garden tractors require gears that can withstand extremely high wear and heavy loading. For example, the fourth reduction gear shown in Fig. 5 is a drive gear in a six-speed transmission. A crush load of 4000 kg (8800 lb) per tooth is specified. High tooth density is required of the P/M part. To provide this feature, repressing the part to a density of 7.3 to 7.5 g/cm^3 would seem to be the solution. However, the part is too large to be compacted on a 4.45 MN (500 ton) press. To overcome this limitation, the part is compacted to three densities: 6.4 to 6.6 g/cm^3 in the hub, 6.6 to 6.8 g/cm^3 in the inner flange, and 6.9 to 7.0 g/cm^3 in the tooth area.

After presintering, the gear is repressed only in the tooth area to a density of 7.3 to 7.5 g/cm^3. Carbon content is kept low to improve compressibility during repressing, after which the teeth are carburized.

MPIF material	FN-0205
Carbon, %	0.5
Nickel, %	2
Molybdenum, %	0.5
Density, g/cm^3	
Tooth	7.3–7.5
Inner flange	6.6–6.8
Hub	6.4–6.6
Tooth apparent hardness, HRC	50–56
Tooth crush strength, kN (ton)	40 (4.5)
Finished weight, kg (lb)	1.042 (2.300)

Fig. 5 Fourth reduction gear

Example 6: Bronze Clutch Lever Bearing. A relatively large number of nonferrous P/M parts are used in automobiles, but they represent nominal weight and generally are similar to parts used in nonautomotive applications. These include bushings for radio controls, windshield wiper motor bushings and gears, distributor bushings, and other annular parts that are similar to bushings in other applications. Figure 6 is a copper alloy P/M clutch lever bearing, which is considered a large P/M part. A considerable volume of copper metal powder has been used in the past as a journal bearing matrix material.

Fig. 6 Clutch lever bearing. Composed of 90Cu-10Sn bronze. Weight: 20 g (0.05 lb). Secondary operation: oil impregnation

Example 7: Hollow Titanium Bolts. Figure 7 shows hollow titanium bolts produced by P/M methods. Originally, this bolt was to be manufactured from a high-strength plastic for an aerospace application. The high-temperature environment, however, made use of plastic bolt materials impractical. Because design weight was based on use of plastic, the replacement material had to maintain the light weight, with no sacrifice of strength at elevated temperatures. These requirements were met by designing and fabricating a hollow bolt with a hex shape in the head and a round bore through the remainder of the bolt. This part was made by cold isostatic

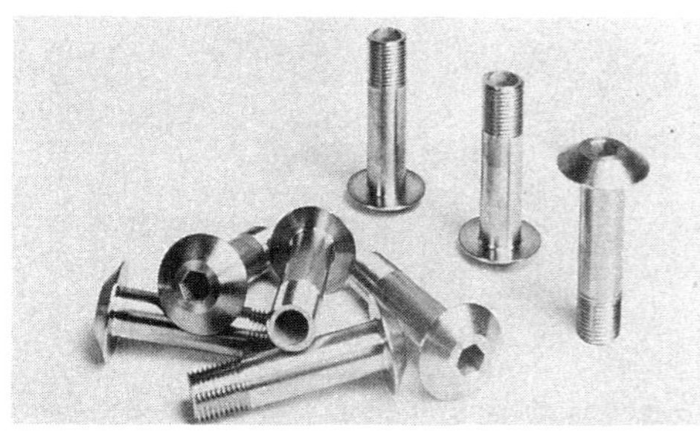

Fig. 7 Hollow bolts. Courtesy of Imperial Clevite Technology Center

MPIF material	FX-2008-T
Carbon, %	0.60–1.00
Copper, %	15.00–25.00
Iron, %	69.00–84.39
Density, g/cm^3	7.1–7.6
Copper infiltrate	2.7 kg (6.0 lb) wafer
Finished weight, kg (lb)	12.973 (28.60)
Leak testing	Pressurize-alternate ports with 415 kPa (60 psi) nitrogen, leakage must not exceed 75 L/min (19.8 gal/min)

Fig. 8 Hydraulic wheel motor manifold

pressing and sintering. Finishing included a finish cut on the outer diameter and the addition of threads.

Example 8: Manifolds for Hydraulic Cam Lobe Motors. Four-wheel drive agricultural combines have hydraulic motors on two wheels. Centrally located within the motor is a large P/M part that acts as a manifold (Fig. 8). This manifold weighs approximately 15 kg (32 lbs) and features 30 close-tolerance radial oil passages.

The manifold was designed for P/M processing because the machining required with other processes would be cost prohibitive. Using P/M technology, this is not an easy part to manufacture. The tooling is complex and features a dual upper punch motion, which forms the 30 radial oil passages. Two lower punches and a stepped core rod impart a taper to the manifold.

The manifold is compacted on a 8.9 MN (1000 ton) press and is copper infiltrated during sintering to provide a final density of 7.1 to 7.6 g/cm^3. Copper infiltration ensures pressure tightness and adds lubricity in operation.

Some machining is required before the manifold can be used. However, this is limited to turning the inside and outside diameters and drilling several holes. Because of the complexity of this part, cost savings are high, even at low volumes (5000 to 10,000 units).

Example 9: Power Steering Metering Pumps. In fully hydraulic power steering, there is no mechanical connection between the steering wheel and the front wheels. The operator steers by controlling a pressurized column of oil with a power steering metering pump. This pump activates the steering valve, which in turn directs the steering motor.

The power steering metering pump in Fig. 9 consists of a cover, base, body, and two gears, all of which are made by P/M processes. Gray cast iron was considered for these parts, but it was rejected because of the capital investment required for equipment to do additional machining. Minimal machining of the P/M parts is required and is limited to lapping and machining of a cross hole. The cover, base, and body must be leakproof to perform in this pressurized hydraulic environment. This is achieved through epoxy impregnation of these parts. Pressurized nitrogen leak testing is specified for the cover, base, and body. Crush testing is specified for the gears.

MPIF Parts Recognition

The Metal Powder Industries Federation holds an annual P/M Part-of-the-Year Design Competition to select outstanding powder metallurgy products in five categories: ferrous, nonferrous, stainless steel, injection molding, and overseas. The following sections briefly describe parts from recent competitions.

1995 Parts Competition

Steel differential case in Fig. 10 is a complex, five-level nickel-steel part that serves as two halves of an inter-axle differential case used in a heavy-duty tandem axle assembly on class 8 trucks. The part has a Rockwell B hardness of 85, minimum yield strength of 410 MPa (60 ksi), ultimate tensile strength of 710 MPa (100 ksi), and a fatigue limit of 270 MPa (40 ksi). The P/M case replaces a casting that required extensive machining.

The 900 g (2 lb) halves are arc welded together to form the case, which distributes input power between the front and rear axles in the tandem. Designed to reduce both the cost and size of the assembly, the smaller case results in lower trailer bed height and increased cargo capacity. It is considered a critical application because the case must withstand rotary and thrust forces. The part was designed and fabricated by Eaton Corp. Truck Components of North America and Metal Powder Components Inc.

A multilevel aluminum capstan (Fig. 10) is part of an automatic ticket stamping machine at bus stations and subways. Use of P/M technology saved 70% compared with the cost of a machined part. Critical tolerances include concentricity and the D-shaped internal diameter. Tensile strength is 235 MPa (34 ksi), and yield strength is 230 MPa (33 ksi). Secondary operations include machining, drilling, tapping, and hard-coat anodizing. The part was made by Pennsylvania Pressed Metals Inc.

A two-piece spacer and guide assembly (Fig. 10) made of stainless steel alloy 410L is used in an automobile fuel injection system. The P/M spacer reduces cost by 25% compared with the original machined-and-welded design.

The part has a Rockwell B hardness of 43, elongation of 10%, ultimate tensile strength of 325 MPa (50 ksi), and yield strength of 200 MPa (30 ksi). During molding, a stainless steel alloy 302 guide ring is introduced into the molding tool and compacted into the part. Secondary operations include coining the fuel holes and tumbling. The spacer was made by Windfall Products Inc.

Injection-Molded Fixed Blade. A complex high-speed steel fixed blade (Fig. 10) used in a hair clipper was produced by injection molding. The fixed blade operates next to a movable blade that is made by ceramic injection molding. The P/M blade was formerly stamped out of steel, but metal injection molding reduced production time and resulted in a blade with 70 HRC hardness and improved wear resistance. The blade was made by Hitachi Metal AFT Ltd.

A copper-steel rotatory compressor frame (Fig. 10) is used in a room air conditioner. The complex part has a surface hardness of 75 HRB, a tensile strength of 300 MPa (45 ksi) and a yield strength of 160 MPa (25 ksi). The P/M frame offers a 25% cost savings compared with other

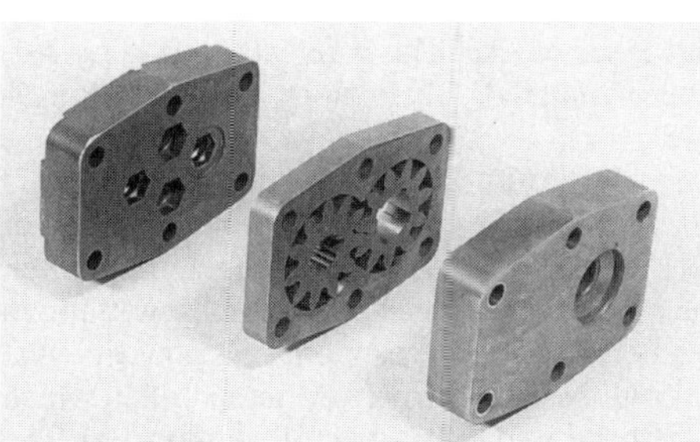

Fig. 9 P/M parts for power steering metering pump

	Cover	Base	Body	Gear	Gear
MPIF material	FC-0408-P	FC-0408-P	...	FC-0408-P	FC-0408-P
Carbon, %	0.60–1.00	0.60–1.00	0.8	0.60–1.00	0.60–1.00
Copper, %	2.00–6.00	2.00–6.00	2.0	2.00–6.00	2.00–6.00
Nickel, %	...	...	0.40–0.50	...	...
Manganese, %	...	...	0.25–0.35	...	...
Molybdenum, %	...	...	0.55–0.65	...	...
Iron, %	91.00(a)	91.00(a)	...	91.00(a)	91.00(a)
Density(a), g/cm^3	6.4	6.4	6.4	6.4	6.4
Apparent hardness(a)	55 HRB	55 HRB	...	25 HRC	25 HRC
Particle hardness(a), HRC	...	...	55	55	55
Compressive strength, MPa (ksi)	...	...	515 (75)	515 (75)	515 (75)
Crush strength, kN (ton)	...	...	...	26.7 (3.0)	26.7 (3.0)
Impregnation	Epoxy	Epoxy	Epoxy	...	...
Finished weight, kg (lb)	1.400 (3.10)	1.460 (3.20)	0.460 (1.00)	0.125 (0.28)	0.159 (0.35)
Leak testing	Must withstand 690 kPa (100 psi) nitrogen for 5 min without leaking				

(a) Minimum

manufacturing methods. The part is made by Hitachi Powdered Metals Co. Ltd.

Power ratchet parts (Fig. 11) are made of infiltrated steel, nickel steel, and iron. Able to withstand up to 120 J (90 ft-lb) of torque, the ratchet contains eight powder-metal parts. The P/M components also provide the wear characteristics required for high-speed use over long periods of time.

Other fabrication methods were considered, including investment casting, stamping, forging, and machining. However, they would have cost more and were incapable of providing sufficient flatness, consistent dimensional control, fine detail in the product logo, and a quality finish. The part was made by Asco Sintering Co.

A phosphorus magnetic iron integral rotor (Fig. 11) is used in a high-speed mechanical printer and offers a 90% cost saving compared with investment casting and machining. The material has minimum yield strength of 210 MPa (30 ksi) and a Rockwell B hardness of 68. Secondary operations include drilling and tapping the side hole, coining, resin impregnation, and a dri-lube coating. The 59-tooth form is accurate to within half a degree. The part was made by Ceromet Inc.

Copper Strip. Parts are stamped from full-density spinodal copper-nickel-tin strip that replaced a beryllium-copper alloy. The strip is stamped into nonsparking, nonmagnetic, corrosion-resistant safety tools, such as knives, scrapers (Fig. 11), and spatulas. The tools are used in potentially hazardous situations where the nonsparking characteristic is vital, such as petroleum, natural gas, and chemical process industries. The P/M stamping replaced a cast and rolled alloy. The material has ultimate strength of 1035 MPa (150 ksi), yield strength of 965 MPa (140 ksi), and Rockwell C hardness of 30. The copper-nickel-tin strip is made by Ametek Inc.

An aluminum-alloy timing gear (Fig. 11) is used in an outboard motor. This marks the first time that an aluminum P/M alloy has been used in a marine timing gear application. The gear has a 5% elongation, ultimate tensile strength of 140 MPa (20 ksi), yield strength of 90 MPa (15 ksi), and a Rockwell H hardness range of 60 to 65. Secondary operations include sizing and anodizing. The part was made by Webster-Hoff Corporation.

Fig. 10 1995 P/M Part-of-the-Year Design Competition winners. Background: Ferrous inter-axle differential case. Foreground: Left, aluminum capstan. Center, stainless steel spacer and guide assembly; injection molded hair clipper blade. Right, rotatory compressor frame. Courtesy, MPIF

A stainless steel rotor and stator for a pepper mill (Fig. 11) requires close inside and outside diameter tolerances. The helical design is another key dimensional feature. Combining coarse and fine cutting teeth, which have a Rockwell C hardness of 30, is also critical to the product's function. Originally designed for powder metallurgy, the parts offer significant cost savings over machining. The parts were made by Pacific Sintered Metals Co.

316 stainless steel valve handle assembly (Fig. 11) is made for a chemical process application. The powder metal assembly replaced a handle formed by plastic injection molding.

The parts must pass a 5000 h salt water test and meet a critical wall thickness specification. The P/M components have an average ultimate tensile strength of 260 MPa (40 ksi) and a yield strength of 190 MPa (30 ksi). Powder metallurgy offered more than 50% cost savings over other manufacturing techniques. The assembly was manufactured by Intech Metals Inc.

A trigger guard (Fig. 11) made from an AISI 4650 low alloy steel for a sporting rifle is a metal injection molded part. The net-shape part has ultimate tensile strength of 1650 MPa (240 ksi) and yield strength of 1480 MPa (215 ksi), and it saves more than 50% over the cost of conventional machining. The component is especially critical in terms of the multiaccess holes, logo, and self-supported arch. The trigger guard was made by Advanced Forming Technology.

A magnetic latching post (Fig. 11) is made from a 17-4 PH stainless steel for a high-density disk memory storage system. Special features of this injection molded part include a complex shape with variation in the cross-sectional area, a tapered hole that mates into a straight hole, and an unsupported bridge area with a thin slot. The part requires only a light coining operation on the unsupported area to provide the air gap that establishes the magnetic field for the latching mechanism. It has a typical Rockwell C hardness of 27, yield strength of 660 MPa (100 ksi), and ultimate tensile strength of 875 MPa (130 ksi). The part was made by Parmatech Corp.

Miniature parts (Fig. 11) used in a 1.8 in. hard disk drive are injection molded. They are complex blocks that perform multiple magnetic functions. The 1.47 g block is made of a nickel-steel alloy and is plated with nickel when used with other blocks. The parts were made by Advanced Materials Technology Pte. Ltd., Singapore.

An aluminum-beryllium substrate (Fig. 11) is used as a heat exchanger in the radar system of the U.S. Air Force F-22 twin-engine jet fighter. The part is a vacuum-brazed liquid flow-through assembly made of gas-atomized powders that are cold isostatically pressed into a billet, which is extruded into a slab and rolled into sheet. The material offers dimensional stability, low coefficient of thermal expansion, high strength and modulus, and low density. Its low coefficient of thermal expansion reduces the fatigue load on the solder joints, which increases the life of the entire avionics assembly. The part was made by Brush Wellman Inc.

1996 Parts Competition

Magnetic pole piece (Fig. 12) is a bimetallic duplex component that is an essential part of a magetic steering assist unit for cars. The complex part is formed via a very innovative simultaneous multipowder compaction process. The inner and outer poles are made from iron- phosphorus powder, and the bridge is made of stainless steel (MPIF SS-304NI-30). A specially designed multiactuated feeder system fills the dies with the two different powders.

The part must withstand a test load of 6.4 kN (1440 lb), which is applied to the hub section while supporting the stainless steel rim. More than 400,000 of these parts are produced annually. The part is made by Zenith Sintered Products Inc.

An unleaded brass upper trip (Fig. 12) is used in an agricultural sprinkler system. The complex, six-level net-shape part replaced a die casting and functions along with a P/M lower trip. Use of powder metallurgy provides a 60% cost savings and an ultimate tensile strength of 190 MPa (30 ksi), a yield strength of 110 MPa (16 ksi), a transverse rupture strength of 39 MPa (60 ksi), and an elongation of 14%. The part is made by Ceromet Inc.

A support base assembly (Fig. 12) is used for a missile guidance system. The complex SS-3161-15 three-level part provided substantial cost savings over a machined part with very tight tolerances for weight and flatness. The flatness and profile of the six support pads are held within 0.0075 mm (0.0003 in.).

The part is used in conjunction with a P/M magnetic iron pole and armature. Assembly requires extremely tight tolerances, and the thin sections demand close density control and very fragile tooling. Secondary operations include CNC milling, lapping, tapping, and plastic impregnating. The part is made by Capstan Pacific.

Vending machine parts (Fig. 12) are injection molded and are used in an electric switchbox and vending machine coin lock. The intricate stainless steel components, which have many slits and holes, replaced zinc die castings because of superior corrosion resistance and strength. The parts have a tensile strength of 520 MPa (75 ksi), a yield strength of 220 MPa (32 ksi), and a 50% elongation. The parts were made by Hitachi Metals AFT Ltd.

Brazed Planetary Carrier. A three-piece planetary carrier (Fig. 12) is used in a four-wheel drive transfer unit. The complex P/M carrier weighs 2.4 kg (5.4 lb), and it is brazed during sintering using a nickel-copper-manganese brazing alloy. The carrier has a tensile strength of 390 MPa (57 ksi), a brazed joint strength range of 390 to 490 MPa (57 to 72 ksi), and a hardness of 60 HRB. Only minor machining is required. Compared with machined cast iron, the P/M carrier offers a cost and weight savings of approximately 20%. The internal gear of this carrier works as a synchronizer with the driveshaft gear in the engine. The part is made by Sumitomo Electric Industries Inc.

Transmission Sprockets. High performance auto transmission sprockets (Fig. 13) are selectively densified ferrous parts. After compacting and sintering, the teeth and journal are selectively densified by a proprietary cold working process, and then vacuum carburized. Eight parts are used in the transmission.

Vacuum carburizing produces a precisely controlled case on the critical journal tooth flank surfaces with minimum distortion. This provides high strength and exceptional durability for the bearing and tooth surfaces.

The parts have an ultimate tensile strength of 860 MPa (125 ksi), a yield strength of 825 MPa (120 ksi), and a tooth and journal surface hardness of more than 60 HRC. The endurance of the powder metallurgy sprockets exceeds that of induction-hardened malleable iron. The part is made by Stackpole Limited Automotive Gear Division.

A camshaft sprocket assembly (Fig. 13) made by ICM/Krebsoge, consists of a P/M steel sprocket and a timing sensor ring made from low-carbon A620 steel, which is attached to the P/M sprocket by "rollerizing" the P/M hub over the stamping flange. Inside diameter tolerances are held to 2.025 mm (0.080 in.). The sensor ring provides the magnetic pickup to tell the onboard computer the position of the camshaft in relation to the position of the crankshaft. The nickel-steel P/M sprocket has a tensile strength of 380 MPa (55 ksi), a minimum yield strength of 240 MPa (35 ksi), and a hardness range of 60 to 75 HRB.

Nickel-silver polarizing keys (Fig. 13) used in electrical rack and panel connectors on aircraft flight data recorders are thin-walled fragile parts with an ultimate tensile strength of 230 MPa (35 ksi), a yield strength of 140 MPa, (20 ksi), and a hardness of 85 HRH, and a 14% elongation.

The parts withstand a 500 connect/disconnect cycle test without measurable wear, and are also tested for electrical conductivity. The net-shape

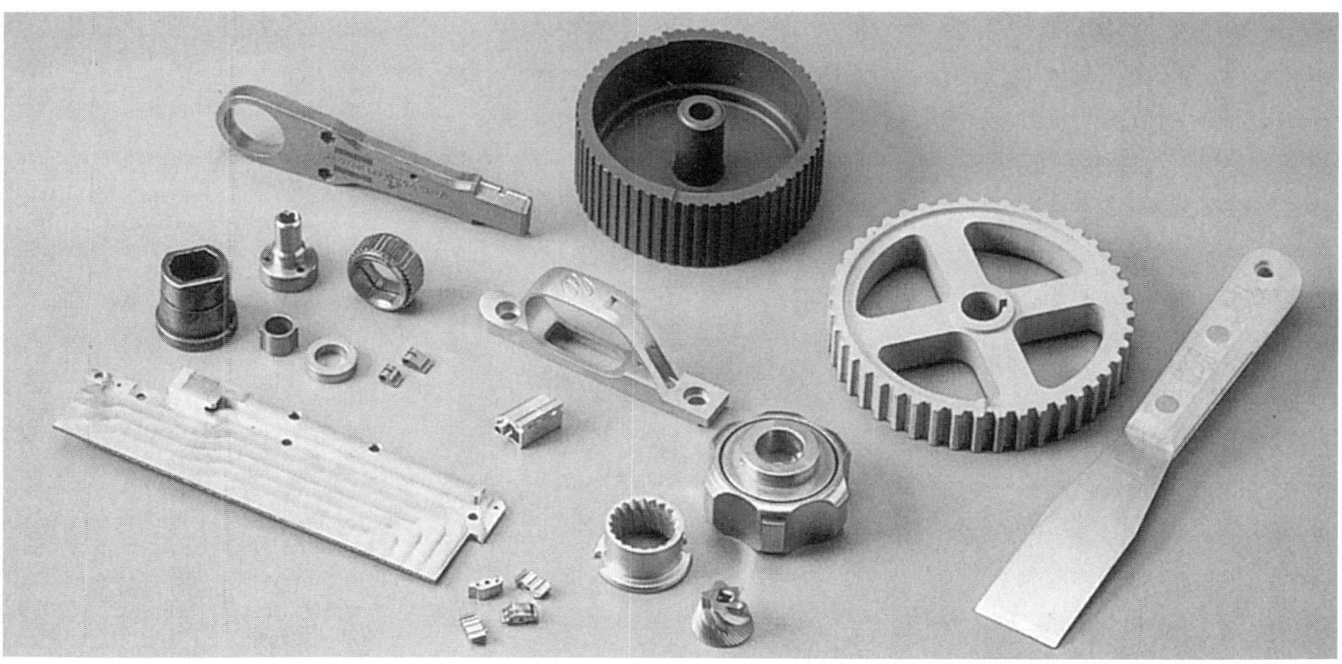

Fig. 11 1995 Awards of Distinction and Award of Achievement winners. Background, left to right: eight power ratchet steel parts and integral rotor. Nonferrous timing gear and scraper. Center, Injection molded trigger guard and latching post. Center right, Stainless steel stator & rotor, and valve handle assembly. Left foreground, aluminum-beryllium heat exchanger. Center foreground, four nickel-steel disk drive parts. Courtesy, MPIF

part requires no secondary operations and demonstrates a high degree of reliability. The keys can be found in almost every flight data recorder in commercial and military aircraft produced in the United States. The parts are made by Asco Sintering Corp.

Stainless steel palate expander (Fig. 13) is a complex stainless steel (MPIF SS-316NI) part for a device that expands the palates of young children with small mouths. The assembly consists of four P/M parts. Left and right P/M housings are spot welded together. Two guide pins are press-fitted and staked into left and right P/M plates along with the assembly of the adjusting screws. The orthodontic appliance may be expanded with the twist of an Allen wrench inserted into the front, a much safer procedure than that used with previous designs.

The parts have an ultimate tensile strength of 350 MPa (50 ksi) and give an estimated cost savings of 80% over other forming methods, such as machining. Secondary operations are limited to tapping and burnishing. The part is made by Precision Powdered Metal.

A foldable paper-hole punch (Fig. 13) is made by metal injection molding (MIM) of 17-4PH stainless steel. The assembly consists of two complex shapes, which have a tensile strength in the range 1100 to 1195 MPa (160 to 175 ksi), a yield strength in the range 980 to 1100 MPa (140 to 160 ksi), and hardness of 38 HRC.

The MIM powder metallurgy parts replaced machined and stamped components that required welding. MIM provided a cost savings of approximately 50%. Secondary operations are limited to reaming, grinding, and tumbling. The part is made by Metalor 2000, Israel.

A copper-steel flange pulley (Fig. 13) made by Hitachi Powdered Metals Co. is a multifunctional pulley with a sensor cam/flange that was designed especially for P/M. It has a minimum tensile strength of 390 MPa (55 ksi), a minimum yield strength of 270 MPa (40 ksi), and hardness of 55 HRB.

A metal-fiber filter (Fig. 13) is used in automotive airbag inflators. The metal-fiber filter diffuses and cools the gas that inflates the airbag. The filters take the brunt of the explosive force of the pyrotechnic fuel, which creates enough volume of extremely hot gas (>800 °C, or 1470 °F) to fill the airbag. Besides cooling and diffusing the gas flow, the filter prevents hot particles as small as 1 μm (0.00004 in.) in diameter, and other undesirable respirables, from entering the airbag.

The sintered media can be sheared, rolled, pleated, and/or welded without cracking, splitting, or destroying its porous properties. Extremely fine metallic fibers fabricated through a bundle drawing process provide an integral, controlled-porosity medium.

The stainless steel fibers have diameters as fine as 4 μm (0.00016 in.) in a sintered web matrix combined with a stainless wire mesh. The sintered media strips are provided in a preformed, crimped condition. This facilitates handling by the customer, who flanges and roll-forms the strips to various diameters, and then welds the overlap seam. The part is made by Memtec America Corp.

Fig. 12 1996 Powder Metallurgy Part-of-the-Year Competition winners: Left to right, front row: Vending machine parts and guidance system support base. Second row: sprinkler upper trip and bimetallic magnetic pole piece. Back: Planetary carrier. Courtesy, MPIF

1997 Parts Competition

Warm-Compacted Turbine Hub. The first automotive powertrain application fabricated through warm compaction technology is a steel torque converter turbine hub (Fig. 14). The hub is formed by means of the EI-Temp warm compaction system from powder in an 825 ton (7.3 MN) compacting press.

The part has a minimum density of 7.2 g/cm^3 (0.26 lb/in.^{3}), a tensile strength of 807 MPa (117 ksi), a yield strength of 428 MPa (62 ksi), a fatigue limit of 242 MPa (35 ksi), and a 17 HRC apparent hardness.

Subjected to severe durability testing, it must withstand an internal spline torque of 1210 N · m (890 ft · lb) for one million cycles. Powder metallurgy replaced a machined and heat-treated forged part. It provided a cost savings of more than 30% and an assembly savings of more than 50%. The part is made by Chicago Powdered Metal Products.

Brass latchbolt (Fig. 14) is made to a density of 7.9 g/cm^3 (0.28 lb/in.^{3}) and has an ultimate tensile strength of 210 MPa (30 ksi). Its yield strength is 110 MPa (16 ksi), elongation is 15%, and hardness is 84 HRH.

The latchbolt is assembled with mating parts by crimping the tail section, requiring strength and ductility in the thin-walled areas. Secondary operations are limited to drilling a 2.4 mm (0.94 in.) hole, tumbling, and resin impregnation. A contour tolerance of 0.15 mm (0.006 in.) must be held. The art must pass a stringent pendulum ram test where ram blows are applied to the lock. The part is made by ASCO Sintering Company.

Automotive exhaust manifold flanges (Fig. 14) are made of stainless steel alloy 434. The manifold flange connects the manifold to the exhaust pipe that leads to the catalytic converter. The exhaust converter outlet flange connects the exhaust pipe leading from the catalytic converter to a flange welded to the manifold. The three-level manifold flange (6.9 g/cm^3, or 0.24 lb/in.^{3})

required special tooling to form the "tulip" shape.

Powder metallurgy replaced a two-piece stamped and welded assembly, which leaked exhaust fumes. The P/M part dramatically reduces exhaust leakage at the manifold/exhaust system junction. Alloy 434 P/M stainless steel has tensile properties superior to wrought 409 stainless steel at elevated temperatures, as well as improved corrosion and oxidation properties. Formed to a minimum density of 7.0 g/cm^3 (0.25 lb/in.^{3}), the exhaust converter outlet flange is supplied with two cold-headed wrought bolts, which are press-fitted into it and shipped as an assembly. Deburring is the only secondary operation performed on both parts. The components are made by SSI Technologies Inc.

Complex MIM gun parts (Fig. 14) are used for seven parts that go into .22 and .38 caliber revolvers. The frame, weighing 240 g (8.46 oz), is made from nickel-steel and has a minimum hardness of 25 to 30 HRC. The other parts are made from 4140 steel. All parts have a density of 7.6 g/cm^3 (0.27 lb/in.^{3}). The barrel must withstand heavy shock and wear. Metal injection molding technology replaced investment casting and forging to achieve better finishes and closer tolerances at more than 40% cost savings. Secondary operations include heat treating and tapping, threading and rifling the barrel, and applying a black oxide or chrome finish. Testing of the parts includes firing 1000 overload rounds. The parts are made by Taurus International Manufacturing Inc.

Warm compacted ratchet-wheel set for a new angle grinder (Fig. 14) is made to a density of more than 7.0 g/cm^3 (0.25 lb/in.^{3}), has an ultimate tensile strength of 760 MPa (110 ksi), impact strength of 22 J (16 ft · lb) and a fatigue strength of 300 MPa (40 ksi). The warm compacted thrust ring and flange are formed to a density exceeding 7.2 g/cm^3 (0.26 lb/in.^{3}) and have an ultimate tensile strength of 1100 MPa (160 ksi), an impact strength of 24 J (18 ft · lb), and a fatigue strength of 414 MPa (60 ksi). Powder metallurgy provided almost 50% cost savings and very high precision. Secondary operations include heat treating and minor machining on the wheel; the thrust ring is machined, and the flange is heat treated. The part is made by Porite Taiwan Co. Ltd.

Clutch Plates. P/M steel notch plate and pocket plate (Fig. 15) is for a new one-way clutch that revolutionizes powertrain/drivetrain designs. The P/M plates offer significant cost savings over alternative manufacturing processes. The new clutch is more economical than conventional "sprag and roller ramp" designs, and is less sensitive to vibration, unbalanced loading, and centrifugal force.

The two-level notch plate has a minimum density of 7.0 g/cm^3 (0.25 lb/in.^{3}), a minimum ultimate tensile strength of 720 MPa (105 ksi), and a minimum apparent hardness in the range 20 to 25 HRC. The three-level pocket plate has a minimum density of 6.9 g/cm^3 (0.24 lb/in.^{3}) and a minimum ultimate tensile strength of 900 MPa (130 ksi). Both parts are heat treated, machined, and tumbled.

A Stroker load cycle and overrun test are performed on the assembly. The impact load specification is 20 cycles at 1580 J (1200 ft · lb), and the Stroker requirement is 200,000 cycles at 790 J (600 ft · lb) and 14,000 cycles at 1580 J (1200 ft · lb). The P/M clutch also withstood a maximum ultimate torque of 3950 J (3000 ft · lb), more than six times the torque that the more conventional roller ramp clutch could withstand. The P/M notch and pocket plates are assembled with four struts, four springs, and a threaded weir to form the one-way clutch. The design is patented by Epilogics Inc., and the part is made by Metal Powder Components.

Brass optical sensing part (Fig. 15) made by FMS Corp. holds a photoelectric emitter and receiver in place. The part has thin walls and four pyramid-shaped "locking ribs" located on the center bridge sections. These protrusions allow the part to be staked into a plastic molded frame.

The part is repressed and sized after sintering, then resin-impregnated, deburred, and tin plated. It has a minimum yield strength of 80 MPa (12 ksi) and a density of 8.0 g/cm^3 (0.30 lb/in.^{3}). The 0.635 mm (0.025 in.) wall thickness required high-precision tooling for compacting and sizing. A process capability index, C_{pk}, of 1.33 is required on critical dimensions, which include 1.3 mm ±0.13 mm (0.050 ±0.005 in.) on bridge thickness and 2 mm +0.10 mm/–0.05 mm (0.080 + 0.004/–0.002 in.) on the distance from peak-to-peak on the gripping ribs.

Mortise lock parts (Fig. 15) are 304 stainless steel with a minimum density of 6.4 g/cm^3 (0.23 lb/in.^{3}). The stainless steel parts have a 61 HRB hardness, a yield strength of 207 MPa (30 ksi), and an ultimate tensile strength of 296 MPa (43 ksi). Brass forging, stainless casting, and zinc die casting were also considered, but could not approach the economies of powder metallurgy. The part also provides high hardness, wear resistance, and net shape. The new mortise lock design can change function and handling in the field without disassembly and extra parts, which means the lock can be installed on a door regardless of the way it is swung. The lock also contains five copper-infiltrated P/M steel parts. The parts are made by Ceromet.

Three intricate 17-4PH stainless steel parts (Fig. 15) are metal injection molded parts for a new endoscopic surgical stapler. The thin-walled parts are produced very close to their final shape.

The heat-treated parts have an ultimate tensile strength of 1180 MPa (170 ksi), a yield strength of 1100 MPa (160 ksi), a 36 HRC hardness, a 3% elongation, and a 7.5 g/cm^3 (0.26 lb/in.^{3}) density. The receiver has a 0.7 mm (0.028 in.) wall section, which forms a slot held to between 8.8 and 8.9 mm (0.34 to 035 in.) over the entire length. The clevis holds a roundness of 11 .7 mm

Fig. 13 1996 Awards of Distinction winners. Left to right, front row: palate expander assembly, polarizing keys, hole punch. Second row, flange pulley and filter. Third row: transmission sprockets. Back: camshaft sprocket. Courtesy of MPIF

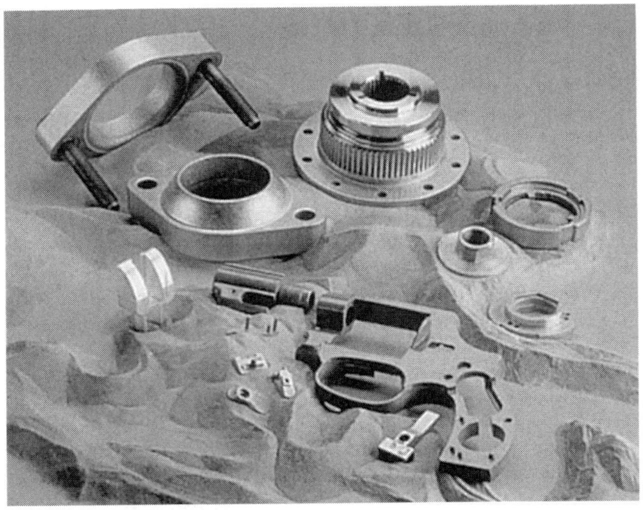

Fig. 14 1997 P/M Part-of-the-Year Competition winners. Left to right, foreground: Latchbolt and seven gun parts. Left to right, background: Exhaust manifold flanges, turbine hub, and a power tool ratchet wheel set, which includes the wheel, flange, and thrust ring

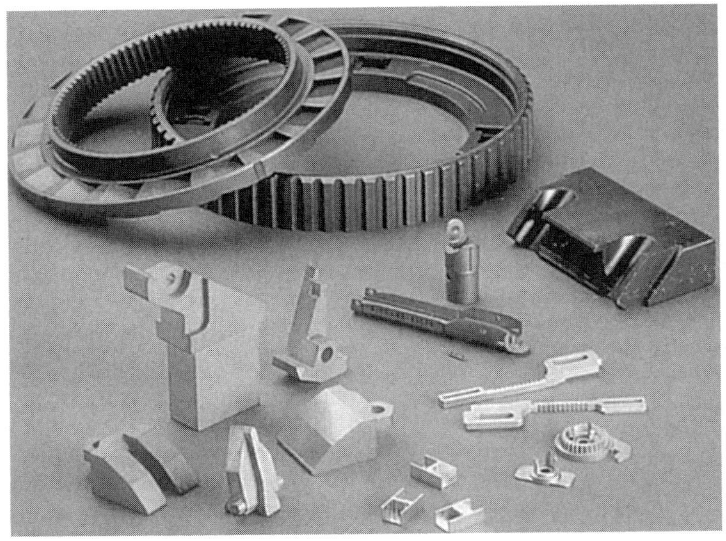

Fig. 15 1997 Award of distinction winners. Left to right, foreground: Five stainless steel mortise lock parts; three opto holders; and metal injection molded latch door, gear, and racks. Center: Metal injection molded link, receiver, and clevis; and padlock body. Background: Notch plate and pocket plate

±0.05 mm (0.460 in. ±0.002 in.) on a 0.864 mm (0.034 in.) wall section. The link holes are held to 1.19 mm ±0.019 mm (0.047 ±0.00075 in.). The endoscopic linear cutter combines a 45° bilateral articulation with a 360° rotation, providing excellent maneuverability and tissue access for endoscopic procedures, such as appendectomies and lung resections. The parts are made by PC Advanced Forming Technology.

MIM Parts for Locking Mechanism. A gear, rack, and latch door (Fig. 15) are used as a locking door unit gear mechanism for closing small battery doors in a portable terminal. The gear and latch door are made from a P/M nickel steel to a density of 7.75 g/cm^3 (0.27 lb/in.^{3}). They have a tensile strength range of 414 to 448 MPa (60 to 65 ksi), a yield strength of 276 to 310 MPa (40 to 45 ksi), and a 65 to 75 HRB hardness range.

The rack is made from 316L stainless steel to a density of 7.70 g/cm^3 (0.27 lb/in.^{3}) and has a tensile strength range of 470 to 515 MPa (68 to 75 ksi), a yield strength range of 150 to 172 MPa (22 to 25 ksi), and a 65 HRB hardness. Originally, the parts were produced from plastic and aluminum, but they failed. Metal injection molding provided a more reliable design and a cost savings of up to 50% compared to machining. The parts are made by Metalor 2000.

A net-shape steel padlock body (Fig. 15) is used in a gear-shift lever locking mechanism. Produced by Gino Olivares s.r.l., Gessate-Milan, Italy, to a density range of 6.4 to 6.6 g/cm^3 (0.23 to 0.24 lb/in.^{3}), the part required complex tooling. It has a 65 HRB hardness along with good ductility. Close tolerances are required in the hole and height of the part. The padlock body is the housing for a cylinder and two latches that activate the locking mechanism. It is wrapped by hardened steel, which protects the inner parts. Secondary operations are limited to burnishing and oil impregnation.

ACKNOWLEDGMENT

The examples in this article were adapted from *Metals Handbook*, Vol 7, *Powder Metallurgy* and from *Advanced Materials & Processes* descriptions of the MPIF parts competition.

Metric Conversion Guide

This Section is intended as a guide for expressing weights and measures in the Système International d'Unités (SI). The purpose of SI units, developed and maintained by the General Conference of Weights and Measures, is to provide a basis for worldwide standardization of units and measure. For more information on metric conversions, the reader should consult the following references:

- "Standard for Metric Practice," E 380, *Annual Book of ASTM Standards,* American Society for Testing and Materials, 100 Barr Harbor Drive, West Conshohocken, PA 19428-2959
- "Metric Practice," ANSI/IEEE 268-1982, American National Standards Institute, 1430 Broadway, New York, NY 10018
- *The International System of Units,* SP 330, 1986, National Institute of Standards and Technology. Order from Superintendent of Documents, U.S. Government Printing Office, Washington, DC 20402-9325
- *Metric Editorial Guide,* 4th ed. (revised), 1985, American National Metric Council, 1010 Vermount Avenue NW, Suite 1000, Washington, DC 20005-4960
- *ASME Orientation and Guide for Use of SI (Metric) Units,* ASME Guide SI 1, 9th ed., 1982, The American Society of Mechanical Engineers, 345 East 47th Street, New York, NY 10017

Base, supplementary, and derived SI units

Measure	Unit	Symbol
Base units		
Amount of substance	mole	mol
Electric current	ampere	A
Length	meter	m
Luminous intensity	candela	cd
Mass	kilogram	kg
Thermodynamic temperature	kelvin	K
Time	second	s
Supplementary units		
Plane angle	radian	rad
Solid angle	steradian	sr
Derived units		
Absorbed dose	gray	Gy
Acceleration	meter per second squared	m/s^2
Activity (of radionuclides)	becquerel	Bq
Angular acceleration	radian per second squared	rad/s^2
Angular velocity	radian per second	rad/s
Area	square meter	m^2
Capacitance	farad	F
Concentration (of amount of substance)	mole per cubic meter	mol/m^3
Conductance	siemens	S
Current density	ampere per square meter	A/m^2
Density, mass	kilogram per cubic meter	kg/m^3
Electric charge density	coulomb per cubic meter	C/m^3
Electric field strength	volt per meter	V/m
Electric flux density	coulomb per square meter	C/m^2
Electric potential, potential difference, electromotive force	volt	V
Electric resistance	ohm	Ω
Energy, work, quantity of heat	joule	J
Energy density	joule per cubic meter	J/m^3
Entropy	joule per kelvin	J/K
Force	newton	N
Frequency	hertz	Hz
Heat capacity	joule per kelvin	J/K
Heat flux density	watt per square meter	W/m^2
Illuminance	lux	lx
Inductance	henry	H
Irradiance	watt per square meter	W/m^2
Luminance	candela per square meter	cd/m^2
Luminous flux	lumen	lm
Magnetic field strength	ampere per meter	A/m
Magnetic flux	weber	Wb
Magnetic flux density	tesla	T
Molar energy	joule per mole	J/mol
Molar entropy	joule per mole kelvin	J/mol · K
Molar heat capacity	joule per mole kelvin	J/mol · K
Moment of force	newton meter	N · m
Permeability	henry per meter	H/m
Permittivity	farad per meter	F/m
Power, radiant flux	watt	W
Pressure, stress	pascal	Pa
Quantity of electricity, electric charge	coulomb	C
Radiance	watt per square meter steradian	$W/m^2 \cdot sr$
Radiant intensity	watt per steradian	W/sr
Specific heat capacity	joule per kilogram kelvin	J/kg · K
Specific energy	joule per kilogram	J/kg
Specific entropy	joule per kilogram kelvin	J/kg · K
Specific volume	cubic meter per kilogram	m^3/kg
Surface tension	newton per meter	N/m
Thermal conductivity	watt per meter kelvin	W/m · K
Velocity	meter per second	m/s
Viscosity, dynamic	pascal second	Pa · s
Viscosity, kinematic	square meter per second	m^2/s
Volume	cubic meter	m^3
Wavenumber	1 per meter	1/m

Conversion factors

To convert from	to	multiply by
Angle		
degree	rad	1.745 329 E−02
Area		
in.^{2}	mm^2	6.451 600 E+02
in.^{2}	cm^2	6.451 600 E+00
in.^{2}	m^2	6.451 600 E−04
ft^2	m^2	9.290 304 E−02
Bending moment or torque		
lbf · in.	N · m	1.129 848 E−01
lbf · ft	N · m	1.355 818 E+00
kgf · m	N · m	9.806 650 E+00
ozf · in.	N · m	7.061 552 E−03
Bending moment or torque per unit length		
lbf · in./in.	N · m/m	4.448 222 E+00
lbf · ft/in.	N · m/m	5.337 866 E+01
Current density		
A/in.^{2}	A/cm^2	1.550 003 E−01
A/in.^{2}	A/mm^2	1.550 003 E−03
A/ft^2	A/m^2	1.076 400 E+01
Electricity and magnetism		
gauss	T	1.000 000 E−04
maxwell	μWb	1.000 000 E−02
mho	S	1.000 000 E+00
Oersted	A/m	7.957 700 E+01
Ω · cm	Ω · m	1.000 000 E−02
Ω circular-mil/ft	μΩ · m	1.662 426 E−03
Energy (impact, other)		
ft · lbf	J	1.355 818 E+00
Btu (thermochemical)	J	1.054 350 E+03
cal (thermochemical)	J	4.184 000 E+00
kW · h	J	3.600 000 E+06
W · h	J	3.600 000 E+03
Flow rate		
ft^3/h	L/min	4.719 475 E−01
ft^3/min	L/min	2.831 000 E+01
gal/h	L/min	6.309 020 E−02
gal/min	L/min	3.785 412 E+00
Force		
lbf	N	4.448 222 E+00
kip (1000 lbf)	N	4.448 222 E+03
tonf	kN	8.896 443 E+00
kgf	N	9.806 650 E+00
Force per unit length		
lbf/ft	N/m	1.459 390 E+01
lbf/in.	N/m	1.751 268 E+02
Fracture toughness		
ksi $\sqrt{\text{in.}}$	MPa $\sqrt{\text{m}}$	1.098 800 E+00
Heat content		
Btu/lb	kJ/kg	2.326 000 E+00
cal/g	kJ/kg	4.186 800 E+00
Heat input		
J/in.	J/m	3.937 008 E+01
kJ/in.	kJ/m	3.937 008 E+01
Length		
Å	nm	1.000 000 E−01
μin.	μm	2.540 000 E−02
mil	μm	2.540 000 E+01
in.	mm	2.540 000 E+01
in.	cm	2.540 000 E+00
ft	m	3.048 000 E−01
yd	m	9.144 000 E−01
mile	km	1.609 300 E+00
Mass		
oz	kg	2.834 952 E−02
lb	kg	4.535 924 E−01
ton (short, 2000 lb)	kg	9.071 847 E+02
ton (short, 2000 lb)	kg × 10^3(a)	9.071 847 E−01
ton (long, 2240 lb)	kg	1.016 047 E+03
Mass per unit area		
oz/in.^{2}	kg/m^2	4.395 000 E+01
oz/ft^2	kg/m^2	3.051 517 E−01
oz/yd^2	kg/m^2	3.390 575 E−02
lb/ft^2	kg/m^2	4.882 428 E+00
Mass per unit length		
lb/ft	kg/m	1.488 164 E+00
lb/in.	kg/m	1.785 797 E+01
Mass per unit time		
lb/h	kg/s	1.259 979 E−04
lb/min	kg/s	7.559 873 E−03
lb/s	kg/s	4.535 924 E−01
Mass per unit volume (includes density)		
g/cm^3	kg/m^3	1.000 000 E+03
lb/ft^3	g/cm^3	1.601 846 E−02
lb/ft^3	kg/m^3	1.601 846 E+01
lb/in.^{3}	g/cm^3	2.767 990 E+01
lb/in.^{3}	kg/m^3	2.767 990 E+04
Power		
Btu/s	kW	1.055 056 E+00
Btu/min	kW	1.758 426 E−02
Btu/h	W	2.928 751 E−01
erg/s	W	1.000 000 E−07
ft · lbf/s	W	1.355 818 E+00
ft · lbf/min	W	2.259 697 E−02
ft · lbf/h	W	3.766 161 E−04
hp (550 ft · lbf/s)	kW	7.456 999 E−01
hp (electric)	kW	7.460 000 E−01
Power density		
W/in.^{2}	W/m^2	1.550 003 E+03
Press capacity		
See Force		
Pressure (fluid)		
atm (standard)	Pa	1.013 250 E+05
bar	Pa	1.000 000 E+05
in. Hg (32 °F)	Pa	3.386 380 E+03
in. Hg (60 °F)	Pa	3.376 850 E+03
lbf/in.^{2} (psi)	Pa	6.894 757 E+03
torr (mm Hg, 0 °C)	Pa	1.333 220 E+02
Specific heat		
Btu/lb · °F	J/kg · K	4.186 800 E+03
cal/g · °C	J/kg · K	4.186 800 E+03
Stress (force per unit area)		
tonf/in.^{2} (tsi)	MPa	1.378 951 E+01
kgf/mm^2	MPa	9.806 650 E+00
ksi	MPa	6.894 757 E+00
lbf/in.^{2} (psi)	MPa	6.894 757 E−03
MN/m^2	MPa	1.000 000 E+00
Temperature		
°F	°C	5/9 · (°F − 32)
°R	°K	5/9
Temperature interval		
°F	°C	5/9
Thermal conductivity		
Btu · in./s · ft^2 · °F	W/m · K	5.192 204 E+02
Btu/ft · h · °F	W/m · K	1.730 735 E+00
Btu · in./h · ft^2 · °F	W/m · K	1.442 279 E−01
cal/cm · s · °C	W/m · K	4.184 000 E+02
Thermal expansion		
in./in. · °C	m/m · K	1.000 000 E+00
in./in. · °F	m/m · K	1.800 000 E+00
Velocity		
ft/h	m/s	8.466 667 E−05
ft/min	m/s	5.080 000 E−03
ft/s	m/s	3.048 000 E−01
in./s	m/s	2.540 000 E−02
km/h	m/s	2.777 778 E−01
mph	km/h	1.609 344 E+00
Velocity of rotation		
rev/min (rpm)	rad/s	1.047 164 E−01
rev/s	rad/s	6.283 185 E+00
Viscosity		
poise	Pa · s	1.000 000 E−01
stokes	m^2/s	1.000 000 E−04
ft^2/s	m^2/s	9.290 304 E−02
in.^{2}/s	mm^2/s	6.451 600 E+02
Volume		
in.^{3}	m^3	1.638 706 E−05
ft^3	m^3	2.831 685 E−02
fluid oz	m^3	2.957 353 E−05
gal (U.S. liquid)	m^3	3.785 412 E−03
Volume per unit time		
ft^3/min	m^3/s	4.719 474 E−04
ft^3/s	m^3/s	2.831 685 E−02
in.^{3}/min	m^3/s	2.731 177 E−07
Wavelength		
Å	nm	1.000 000 E−01

(a) kg × 10^3 = 1 metric ton or 1 megagram (Mg)

Abbreviations and Symbols

A ampere
Å angstrom
ABC activity-based costing
ac alternating current
AI artificial intelligence
AMS Aerospace Material Specification
ANSI American National Standards Institute
ASTM American Society for Testing and Materials
at.% atomic percent
atm atmosphere (pressure)
b Burgers vector
bal balance
bcc body-centered cubic
bct body-centered tetragonal
Btu British thermal unit
C Coulomb; heat capacity
C-C carbon-carbon
CAD computer-aided design
CAE computer-aided engineering
CAM computer-aided manufacturing
cd candela
CIP cold isostatic pressing
cm centimeter
CMC ceramic-matrix composite
CNC computer numerical control
cpm cycles per minute
cps cycles per second
CPS creative problem solving
CSA Canadian Standards Association
cSt centiStokes
CTE coefficient of thermal expansion
CTOD crack tip opening displacement
CVD chemical vapor deposition
CVI chemical vapor infiltration (impregnation)
CVN Charpy V-notch (impact test or specimen)
dB decibel
DBTT ductile-brittle transition temperature
dc direct current
dhcp double hexagonal close-packed
diam diameter
DIN Deutsche Institut für Normung
DTA differential thermal analysis
DWTT drop weight transition temperature
e charge of an electron; natural log base, 2.71828
E modulus of elasticity; Young's modulus; potential
E_s secant modulus
EDM electrical discharge machining
emf electromotive force
EMI electromagnetic interference
Eq equation
ESC environmental stress cracking
esu electrostatic units
et al. and others
eV electron volt
f fiber
f focal length; frequency
f_n normal load
F Faraday constant; force
fcc face-centered cubic
fct face-centered tetragonal
FEA finite-element analysis
FEM finite-element modeling
Fig. figure
FM figure of merit
FMEA failure modes and effects analysis
ft foot
FTA fault tree analysis
g gram
G gauss
G mean grain size; shear modulus
G_{Ic} interlaminar fracture toughness (mode I, peel; mode II, shear; mode III, scissor shear)
gal gallon
GPa gigapascal
gpd grams per denier
gr grain
h hour
h thickness
H henry
H change in height; degree of homogenization; enthalpy; hardness; height; magnetic field
HAZ heat-affected zone
HB Brinell hardness
hcp hexagonal close-packed
HDT heat-deflection temperature
HERF high-energy-rate forging
HIP hot isostatic pressing
HK Knoop hardness
hp horsepower
HR Rockwell hardness (requires a scale designation, such as HRC for Rockwell C hardness)
HV Vickers hardness
Hz hertz
i current (measure of number of electrons)
I current; emergent intensity
IC integrated circuit
ID inside diameter
in. inch
IPD integrated product development
IPTS International Practical Temperature Scale
IR infrared (radiation)
ISO International Organization for Standardization
ITS International Temperature Scale
IV intrinsic viscosity
J joule
JIS Japanese Industrial Standard
k karat
k Boltzmann constant; notch sensitivity factor; thermal conductivity; wave number
K Kelvin
K bulk modulus of elasticity; coefficient of thermal conductivity; empirical constant; interface reaction; mean integrated thermal conductivity
K_I stress-intensity factor
K_{Ic} mode I critical stress-intensity factor; plane-strain fracture toughness
K_{Id} dynamic fracture toughness
K_{Iscc} threshold stress intensity for stress-corrosion cracking
K_c plane-stress fracture toughness
K_f stress-concentration factor
K_t stress-concentration factor
K_{th} threshold crack tip stress-intensity factor
kg kilogram
km kilometer
kN kilonewton
kPa kilopascal
ksi kips (1000 lb) per square inch
kV kilovolt
kW kilowatt
l mean free path; length
L liter; longitudinal direction; liquid
L length
lb pound
lbf pound-force
LCA life cycle analysis (or assessment)
L/D length-to-diameter ratio
LED light-emitting diode
LEFM linear-elastic fracture mechanics
ln natural logarithm (base e)
log common logarithm (base 10)
m matrix; meter
m ion mass; Weibull modulus
M metal atom
M molecular weight
mA milliampere
MeV megaelectronvolt
mg milligram

Mg megagram
MIM metal injection molding
min minimum; minute
MJ megajoule
mL milliliter
MLT marketing lead time
mm millimeter
MMC metal-matrix composite
mod modified
mol% mole percent
MOR modulus of resilience; modulus of rupture
mPa millipascal
MPa megapascal
mpg miles per gallon
mph miles per hour
ms millisecond
MS megasiemens
mT millitesla
MTBF mean time between failures
MTTF mean time to failure
mV millivolt
MV megavolt
MW molecular weight
n growth exponent
n integral number
N Newton
N fatigue life (number of cycles)
NASA National Aeronautics and Space Administration
NBS National Bureau of Standards
NC numerical control
NDE nondestructive evaluation
NDI nondestructive inspection
NDT nondestructive testing
NIST National Institute of Standards and Technology
nm nanometer
No. number
NTC negative temperature coefficient
OD outside diameter
Oe oersted
OEM original equipment manufacturer
OHA operating hazards analysis
ORNL Oak Ridge National Laboratory
OSHA Occupational Safety and Health Administration
oz ounce
p page
p pressure
P applied load; power; pressure
Pa pascal
PCE pyrometric cone equivalent
PDS product design specification
PDT product design team
P/F powder forging
pH negative logarithm of hydrogen-ion activity
P/M powder metallurgy
ppb parts per billion
ppba parts per billion atomic
ppm parts per million
ppt parts per trillion
psi pounds per square inch
psia pounds per square inch absolute
psid pounds per square inch differential
psig pounds per square inch gage
PTC positive temperature coefficient
PVD physical vapor deposition
QC quality control
QFD quality function deployment
QLF quality loss function
r radius vector in a plane normal to the axis
r particle radius; radius of curvature; rate of reaction; reflectivity; spherical particle radius
R roentgen
R average particle radius; gas constant; radius; ratio of the minimum stress to the maximum stress; resistance; reliability
R_a surface roughness in terms of arithmetic average
RA reduction of area
rad absorbed radiation dose; radian
RCF rolling contact fatigue
RE rare earth
Ref reference
rem remainder
rf, RF radio frequency
RH relative humidity
rms, RMS root mean square
ROM rough order of magnitude; rule of mixtures
rpm revolutions per minute
RT room temperature
RTI relative thermal index
s second
S siemens
S distance traveled
SAE Society of Automotive Engineers
scfm standard cubic foot per minute
SEM scanning electron microscopy
sfm surface feet per minute
SG standard grade
SI Système International d'Unités
sineh sine hyperbolic
S/N signal-to-noise (ratio)
S-N stress-number of cycles
SPC statistical process control
SPF superplastic forming
sp gr specific gravity
SRIM structural reaction injection molding
std standard
STEM scanning transmission electron microscopy
STM scanning tunneling microscopy
Sv sievert
t thickness; time
T Tesla; transverse direction
T absolute temperature; temperature; tenacity; total dispersion; transmittance
T_g glass transition temperature
T_m melt temperature
tan equal to ratio of the loss modulus to the storage modulus
TBC thermal barrier coating
TEM transmission electron microscopy
TGA thermogravimetric analysis
TMA thermomechanical analysis
TP thermoplastic
TQM total quality management
TS thermoset
tsi tons per square inch
TTT time-temperature-transformation
UL Underwriters' Laboratories
UNS Unified Numbering System
UTS ultimate tensile strength
UV ultraviolet
v workpiece velocity
V volt
V_f volume fraction of fiber
V_m volume fraction of matrix
V_v volume fraction of void content
VI viscosity index
vol volume
vol% volume percent
w whisker
W watt
W total radiation; width
wt% weight percent
x axial distance
X neck diameter
Y scale of microstructural segregation
YAG yttrium aluminum garnet
yr year
z ion charge
Z atomic number; impedance
° angular measure; degree
°C degree Celsius (centigrade)
°F degree Fahrenheit
0° fiber direction
90° perpendicular to fiber direction
α coefficient of thermal expansion
γ shear strain; surface energy; surface tension
γ_b interfacial energy
γ_s surface energy
δ an increment; a range; change in quantity; grain boundary width; thickness of liquid boundary
Δ an increment; a range; change in quantity
ΔT temperature difference or change
ε strain
$\dot{\varepsilon}$ strain rate
η loss coefficient; viscosity
θ angle; geometrical constant
θ_i angle of incidence
θ_r angle of refraction
κ dielectric constant
λ thermal conductivity
μ friction coefficient; linear attenuation coefficient; magnetic permeability; the mean (or average) of a distribution
μin. microinch
μm micrometer (micron)
μs microsecond
ν Poisson's ratio; velocity
π pi (3.141592)
ρ density; resistivity
σ Stefan-Boltzmann constant; strength; stress; tensile stress; tensile strength
σ_a applied stress
σ_f applied axial load
Σ summation of
τ shear stress
ϕ angle of internal friction; dihedral angle; porosity; power
Φ reaction rate (kinetic function); volume concentration of impurity ions
ψ damping; geometrical constant
ω frequency
Ω atomic volume; electrical conductivity; molecular volume of solute; ohm

∂ partial derivative
$\partial\eta/\partial t$ kinetic function
⇄ direction of reaction
÷ divided by
= equals
≈ approximately equals
≠ not equal to
≡ identical with
> greater than
>> much greater than
≥ greater than or equal to
∞ infinity
∝ is proportional to; varies as
∫ integral of
< less than
<< much less than
≤ less than or equal to
± maximum deviation
− minus; negative ion charge
× diameters (magnification); multiplied by
· multiplied by
/ per
% percent
+ plus; positive ion charge
√ square root of
~ approximately; similar to

Greek Alphabet

Α, α alpha
Β, β beta
Γ, γ gamma
Δ, δ delta
Ε, ε epsilon
Ζ, ζ zeta
Η, η eta
Θ, θ theta
Ι, ι iota
Κ, κ kappa
Λ, λ lambda
Μ, μ mu
Ν, ν nu
Ξ, ξ xi
Ο, ο omicron
Π, π pi
Ρ, ρ rho
Σ, σ sigma
Τ, τ tau
Υ, υ upsilon
Φ, φ phi
Χ, χ chi
Ψ, ψ psi
Ω, ω omega

Index

A

B

C

D

E

F

G

H

J

K

L

M

N

O

P

S

T

U

V

W

X

Y

Z